Catalogue of Useful Plants of Colombia

Catalogue of Useful Plants of Colombia

Edited by
Raquel Negrão
Alexandre Monro
Carolina Castellanos-Castro
Mauricio Diazgranados

Kew Publishing
Royal Botanic Gardens, Kew

First published in 2022 by Royal Botanic Gardens, Kew, Richmond, Surrey, TW9 3AB, UK, in association with the Instituto de Investigación de Recursos Biológicos Alexander von Humboldt, Calle 28A #16-29, Bogotá, Colombia.
www.kew.org

ISBN 978-1-84246-774-9
eISBN 978-1-84246-775-6

Distributed on behalf of the Royal Botanic Gardens, Kew in North America by the University of Chicago Press, 1427 East 60th Street, Chicago, IL 606037, USA.

British Library Cataloguing in Publication Data
A catalogue record for this book is available from the British Library.

Design: Nicola Thompson, Culver Design
Production Manager: Jo Pillai
Proofreading: Sharon Whitehead

Printed in Great Britain by Halstan & Co Ltd

MIX
Paper | Supporting responsible forestry
FSC® C016278

For information or to purchase all Kew titles please visit shop.kew.org/kewbooksonline or email publishing@kew.org
For Instituto Humboldt publications visit http://repository.humboldt.org.co/

Kew's mission is to understand and protect plants and fungi, for the wellbeing of people and the future of all life on Earth.

Kew receives approximately one third of its funding from Government through the Department for Environment, Food and Rural Affairs (Defra).
All other funding needed to support Kew's vital work comes from members, foundations, donors and commercial activities, including book sales.

Corresponding author: Mauricio Diazgranados, Herbarium, Royal Botanic Gardens Kew, Richmond TW9 3AE, United Kingdom
E-mail: M.Diazgranados@kew.org

Front cover photo credit: Néstor Plata © RBG Kew; **back cover background photo credit:** Mauricio Diazgranados © RBG Kew; **back cover plants photo credits:** *Lupinus bogotensis* (Mauricio Diazgranados © RBG Kew), *Espeletia jimenez-quesadae* (Mauricio Diazgranados © RBG Kew), *Typha domingensis* (Ori Fragman-Sapir ©), *Passiflora cumbalensis* (David Granados © RBG Kew), *Leopoldinia pulchra* (Mauricio Diazgranados © RBG Kew), *Cardiospermum halicacabum* (Wolfgang Stuppy © RBG Kew), *Anacardium occidentale* (Wolfgang Stuppy © RBG Kew), *Draba litamo* (Mauricio Diazgranados © RBG Kew), *Datura metel* (Laura Kor © RBG Kew), *Cattleya trianae* (Mauricio Diazgranados © RBG Kew)

Funded by:

Newton-Caldas Fund

MINISTERIO DE CIENCIA, TECNOLOGÍA E INNOVACIÓN

Implemented by:

BRITISH COUNCIL

Contents

	Preface	7
	Executive Summary	11
	Foreword	15
	Acknowledgements	17
	List of Contributors	19
Chapter 1	The Useful Plants and Fungi of Colombia (UPFC) project: delivering botanical knowledge to support conservation and sustainable development	21
Chapter 2	Assessing extinction risk coverage and identifying geographical hotspots to guide conservation action for Colombia's useful plants	35
Chapter 3	Endemic Medicinal Plants of Colombia	49
Chapter 4	The hidden food basket of Latin America: an overview of Colombian edible plant diversity and its distribution	63
Chapter 5	Insecticide Plants of Colombia	79
Chapter 6	Useful Plants of the Colombian Amazon	93
Chapter 7	Colombian plants in Kew's Economic Botany Collection	103
Chapter 8	Bioeconomy: a sustainable and responsible option for making the most of Colombian plant resources	113
Chapter 9	Sustainable value chains and development pathways for natural ingredients in Colombia: the case of naidí (*Euterpe oleracea* Mart.)	121
Chapter 10	A taxonomic summary of useful plants in Colombia	135
Chapter 11	Notes on the geographic distribution of the useful plants of Colombia	149
Chapter 12	Annotated checklist of useful plants of Colombia	165
	Checklist	171
	Laminae	475
	Index of Families	529
	Index of Genera	531
	Index of Synonyms	541
	Index of Vernacular/Common Names	989

Espeletia nemekenei Cuatrec (Asteraceae =Compositae) used as Medicinal in Colombia.
Mauricio Diazgranados

Preface

Raquel Negrão, Alexandre Monro, Carolina Castellanos-Castro, Mauricio Diazgranados

"Existence begins as sunlight
Much of life on earth is a solar being
From the plant that drinks its golden rays
To make the sugars that nourish all animal life
We are all powered by energy from the sun"[1]
(…)
"In this room, the ingredients of life are made possible
In a ceaseless cycle of transformation" [1]
(…)
"In a glade and in a thicket
In a desert and woodland
They are part of a legacy that imbricates all life on earth:
Their exhale with your inhale
Their inhale with your exhale"[2]
(…)
"We are all environments
Walking oceans
Teeming with life
United by an unending rhythm
That blurs the boundaries between all breathing bodies
And submerges us in torrent of being"[3]

Excerpts from a poetic narrative by Daisy Lafarge presented at Observations on Being (developed by Marshmallow Laser Feast for Coventry City of Culture 2021): 1 "The World's First Breath"; 2 "From Breathing Cell to Breathing Planet"; 3 "The Tides within us". Daisy Lafarge is a poet from the UK and studied at the universities of Edinburgh and Glasgow. Her collection of poetry *Life Without Air* (Granta Books) was published in 2020 and shortlisted for the T.S. Eliot Prize 2020.

Plants and fungi fulfil our everyday needs, and it depends on us to use them sustainably. The consequence of nature's entangled evolution is that our future, from our first to our last breath, depends on all life forms on the planet and in this network, plants and fungi are found at the base. They offer us everything we need to survive and to enjoy life with comfort and joy: from the basic needs of oxygen, food, fuels, clothes, housing, water cycling, and erosion control, to the thermal comfort of shade, medicines for the body and mind, gratifying properties such as fragrances, the aesthetic and mental health values of flowers and landscapes, the entertainment or delight provided during cultural-social celebrations, spiritual tools for use in rituals (such as *aguardiente*, *tobacco*, *ayahuasca*, *mambe*, *rapé*) and daily pleasures such as a coffee or a chocolate bar.

There are also several examples of how plants have played a remarkable part in our cultural and social history, influencing largest-scale human population shifts, and allowing written as well as visual and oral communication. The primitive hardware for much of our intellectual development was made of natural resources: clay tablets, animal skin, and plant fibres especially have been used for thousands of years. Paper or paper-like materials made from plants were developed mainly from papyrus, rice, insect-based silk, hemp, bamboo, Kapa bark cloth (paper mulberry), and cotton-based materials, providing a portable and durable means to "spread the word" and allowing shifts in societal organisation. After centuries of traditional travellers disclosing paper producing cultures (from Egypt, China, passing by the Middle East and entering Europe from the Italian city of paper, Fabriano), vegetable-based writing materials (e.g., cotton sheets) started to be transformed on a much larger scale. Eventually, printing techniques were upgraded for the large-scale production of books, magazines, and newspapers, facilitating the diversification of communication, languages, knowledge, and art. From the same production centre and using the techniques developed to transform cotton sheets, the Italians also developed a more sophisticated, "lighter!" monetary system. This was the first and still the main producer of paper money (it is worth visiting the paper museum [Museo della Carta e della Filigrana] in Fabriano). You may argue that we are already in the age of digital money and *"intelligent"* virtual systems. Still, we would argue that we are not even partially independent of plant-based resources as primary resources for a continuous evolutionary process. Amid the biodiversity crisis, more than ever, our survival requires that we sustain the natural world and plants more than our short-term wealth (especially the wealth of those who are already economically comfortable). In other words, our origin, persistence, and future on this planet rely fully on plants and fungi, which must be protected.

Conservation and the sustainable use of biodiversity necessarily incorporates rural livelihoods. Since 2015, the United Nations has been implementing the 2030 Agenda for Sustainable Development, which comprises 17 Sustainable Development Goals and 169 targets that complete the Millennium Development Goals. The 2030 Agenda aims to integrate and balance the three dimensions of sustainable development (economic, social, environmental) and to reinforce human rights. The inclusion of livelihoods in conservation is a particularly important development in the post-2020 Global Biodiversity Framework, as it defined new global targets and actions to support a sustainable, healthy future for people and nature". Also, expressive activist movements responding to climate change, extinction, and ecocide, have increased the accumulation and diffusion of scientific knowledge and public awareness. The post-2020 Global Biodiversity Framework also emphasises the key role of indigenous and local communities using traditional knowledge to protect nature and develop sustainable strategies.

This book is our contribution to dealing with one of the challenges of the Anthropocene Age, how to use plants and plant-dominated landscapes sustainably. RBG Kew's Science Strategy 2021–2025 "sets out a series of ambitious commitments to urgently help stop biodiversity loss and accelerate understanding of the potential of plants and fungi to help address challenges such as food insecurity, climate change and deforestation". Recent publications (Bridle, 2020; Marteau *et al.*, 2021) have shown that a quarter of greenhouse gas emissions come from food and transport. Behaviour changes by individuals, commercial entities and policy makers that incentivise people to eat more locally grown produce and have more meat-free days could help us to fight climate change. Our *Catalogue of Useful Plants of Colombia* demonstrates the great diversity of plants available as local foods, medicines, fuel sources and, materials, valuable resources that could help us to achieve net zero in all domains. It will hopefully contribute to Colombia's ambitious plan to develop a green economy working with traditional communities and local initiatives in an inclusive approach that, especially considers the voices of traditional communities, and their needs and rights.

Our more practical goal with this publication is to inform you about the richest natural resources available, thereby potentially helping us to achieve a better society in which natural resources are valued and used and avoiding sustainably further habitat and biodiversity loss. We want to offer you (scientists, students, and academics in general) a comprehensive guide to the sample of this biodiverse beautiful planet that is found in Colombia. With almost 29,000 plant species and thousands more yet to be described, Colombia is one of the most biodiverse countries in the world.

The *Catalogue of Useful Plants of Colombia* showcases a successful partnership between Colombia and the UK (see Chapter 1), which has enabled us to document the strengths (on uses, biological diversity, geographical distributions, and business interests) as well as a lack of knowledge about less well known uses and species conservation (chapters 2, 3, 4, 5, 6, 10 and 11). Chapters on the history of Kew's Economy Botany Collections (chapter 7) and the development of bioeconomy in Colombia (chapter 8) look to the past to understand the present and point us towards better future, as does the successful example of 'Colombia's Bio' – green growth development model in the form of naidí (*Euterpe oleracea*) (chapter 9). Finally, we explain how the checklist was assembled and what kind of resources are presented for all of the 7,472 species of useful plants in Colombia (chapter 12). Following the checklist, there is a section showing 500 images as examples of species and uses.

USING THIS CATALOGUE AND THE COL*PLANTA* ONLINE PORTAL

A large amount of the information on Colombia's plant diversity present in this Catalogue is aimed mainly at an academic audience. To access additional information, users can follow the hyperlink on the species name, read the species QR code, or type the scientific, common name, genus, family, or descriptor of a plant of interest or its uses in the search bar on the home page of the Col*Plant*A portal (http://www.colplanta.org) (see Chapter 12). The search results will show the users relevant plant species profiles, which they can further explore. Each plant profile encompasses the classification of the plant, the status of its name (whether accepted or considered a synonym), and information on habit, description, ecology, distribution, uses, common names, and conservation status (see Chapter 12). Images of wild plants and herbarium specimens are also featured where available, alongside additional information about the species. The Col*Plant*A portal has advanced search and filtering options that allow users to build customised queries based on their needs.

Both the *Catalogue of Useful Plants of Colombia* and the Col*Plant*A portal are open-access resources built by experts, using accredited information on the useful plants of Colombia. Both resources are designed to disseminate information on useful plants to the community, supporting research, learning, development, and innovation for sustainable growth, economic development, and well-being. The basic data can be used as a source of information about Colombian plants. Also, both resources represent essential baseline information that will underpin conservation actions, environmental management, legislation, and policy.

Although so many botanical and conservation tasks appear to be carried out by scientists, ecologists, government bodies, and non-governmental organisations, there is a multitude of key people behind the scenes, involved in documenting, communicating, and conserving plant diversity, and playing single or multiple roles: from the teachers who show children, for the first time, bean seeds germinating on cotton, to horticulturalists, gardeners, reserve managers, field biologists and helpers, data compilers, curators and many others.

There are still many unknown challenges related to climate change. Very painfully, the Covid-19 pandemic has shown that it is better for human societies to be more collaborative instead of competitive, sharing benefits and commitments, accessing the diverse knowledge available, and taking emergency actions to protect all living beings when needed.

References

Bridle SL (2020). *Food and climate change without the hot air: change your diet: the easiest way to help save the planet*. UIT Cambridge Ltd, 256 p.

Marteau TM, Chater N & Garnett EE (2021). Changing behaviour for net zero 2050. *BMJ*, *375*, n2293. https://doi.org/10.1136/bmj.n2293

Orthrosanthus chimboracensis (Kunth) Baker (Iridaceae). In Colombia, this species presents Medicinal, Material and Environmental uses.
Mauricio Diazgranados

Executive Summary

Colombia is the second most biodiverse country in the world. Its biological resources are key assets in the country's efforts to transition into a more sustainable, competitive, and inclusive economy. However, to increase the contribution of the bioeconomy to national development, a crucial step is to consolidate the current state of knowledge of the rich diversity of useful plants in the country.

Arising from the collective efforts of the Alexander von Humboldt Biological Resources Research Institute (Colombia) and the Royal Botanic Gardens, Kew—under the Useful Plants and Fungi of Colombia Project (UPFC)—this Catalogue is presented as a source of baseline information about the great diversity of useful plant species in Colombia. Before this project, Colombia did not have a updated national list of useful plants, with only a few lists of species by use category (i.e., medicines, edible plants, and fibres), often with conflicting taxonomy and incomplete geographic coverage, being available. The Catalogue presents a comprehensive panorama, including the results of studies developed across the country by several scientific groups from Colombia and the UK.

The substantial progress, represented by this catalogue, resulted from a successful partnership between Colombia and the UK, showcased in the introductory chapter (Chapter 1). The authors highlight the achievements of the Useful Plants and Fungi of Colombia (UPFC) project, which has provided a framework to develop and promote a market for useful indigenous species and their high-value products, whilst protecting the surrounding natural resources. With a multinational team, the project compiled and generated knowledge on over 36,000 species of plants and fungi, and developed pathways for tackling socio-environmental challenges, contributing to Colombia's green transformation. To date, it has produced over 100 outputs, including books, booklets, scientific journal publications, technical reports, websites, online portals, and educational tools.

Drawing on the Checklist of Useful Plants of Colombia (CUPC), Chapter 2 proposes conservation actions based on (1) taxonomic gaps in extinction risk assessments and (2) geographic hotspots for native useful plants of Colombia (UPC). From the analysis, 544 species of conservation concern—threatened with extinction and/or endemic to Colombia—were identified. However, the analysis also underlined that 45% of the native useful plants of Colombia lack extinction risk assessments. The proportion of unassessed species was significantly higher for 13 of the 256 useful plant families, results that can be used to prioritise extinction risk assessments. There are also variations between categories of uses, and 78 units of analysis (covering 89,901 km^2) were identified as hotspots for native useful plants of Colombia. In this chapter, the authors discuss how the hotspots can contribute to the identification of Important Plant Areas in Colombia, as well as the importance of stakeholder engagement to develop and implement conservation actions and sustainable uses of plants.

One of the main findings of this checklist is that medicinal species represent the most diverse group. Chapter 3 presents a long-term study (from 2014 to 2021) on plant species that are used as medicines in Colombia and their relationship with traditional, scientific, and technological knowledge. Before the checklist was compiled, the authors had documented a total of 3,005 medicinal plants used in Colombia, of which 1,719 are native (including 204 endemics), 558 exotic, and 728 species of uncertain geographical distribution. The lack of knowledge on the conservation status and sustainable uses of most of the Colombian endemic medicinal species was verified: only a few restricted species had sufficient information available to allow their categorisation.

Chapter 4 addresses the second largest use category, the edible plant species, and their conservation status. The chapter concludes that the high diversity of Colombian edible plant species is not currently matched by adequate *in-situ* and *ex-situ* conservation actions. Chapter 4 also identifies significant knowledge gaps regarding the conservation status of some of the most important edible plant genera. The biogeographic analysis highlights the 'Andes' and 'Páramo' bioregions as diversity hotspots for edible plants, emphasising the urgent need for their preservation. The authors concluded that further research is critical to understand the complex interactions between people, food, and the natural environment, whilst at the same time, conservation efforts need to consider the intrinsic value of edible plant diversity and the full potential for socio-economic development at the local and national levels.

A special category of use is presented in Chapter 5, which reports an insightful long-term study on insecticide activity. To date, no publication has detailed this category of use in Colombia and besides identifying and characterising the plant species used as insecticides, this chapter also assesses the state of traditional and scientific knowledge on these species. The authors demonstrate that 632 species used as 'insecticides' were identified in Colombia. For none of these species, however, is there sufficient scientific knowledge to underpin their application as natural insecticides.

Chapter 6 presents a synthesis of the current state of knowledge about the Useful Plants of the Colombian Amazon, based on information gathered from field expeditions and records added to the Colombian Amazon Herbarium (COAH) database managed by the Amazonian Scientific Research Institute SINCHI. Useful plants were classified into 12 categories of uses, by origin (native or endemic), and according to their extinction risk assessments. In total, 2,286 useful plants were recorded in the Colombian Amazon. Similar to the results found from the CUPC, Chapter 6 found that the medicinal (883) and food (690) categories presented the highest numbers of species documented for the Colombian Amazon. Challenges to achieving the sustainable use of species in this region are discussed from the economic botany perspective.

The history of Kew's Economy Botany Collections is presented in Chapter 7, looking to the past to understand the present. This collection was first founded as the Museum of Economic Botany in 1847. Among a total collection of 100,000 plant raw materials and plant products, 320 specimens originate in Colombia. Recent research has enabled botanical names to be updated and provenance data retrieved through archival research. The Colombian specimens mirror the changing scientific relationship between Colombia and Kew from the mid-nineteenth century to today.

For a megadiverse country such as Colombia, sustainable bioeconomy requires a completely different circular way of producing natural products, with intensive use of science to add value to biological resources and biological processes, as is presented in Chapter 8. Basic concepts of circular bioeconomy, value chains and biotechnology are described, and some examples of bioeconomy-focused biotechnological developments from the Research Centre for Agro-industrialisation of Tropical Aromatic and Medicinal Plant Species CENIVAM are presented. The authors concluded that integrating bioprospecting results with those from analytical chemistry and molecular biology (bioactivity assays) led to the detection of promising vegetal species to be used as ingredients for new consumer products. Field projects have established the conditions for biomass production under sustainable agricultural practices, as well as for their rural processing to afford essential oils, extracts, biocontrol solutions, composting material, and biofuel. Multiple challenges still need to be addressed, such as establishing commercial chains with reinvestment into the countryside, the achievement of origin recognition, road improvement, internet coverage, water supplies, and stable legislation.

A successful example from Colombia's Bio program, using the *green growth model* for the development of a real, sustainable value chain for natural ingredients in Colombia, is the case of naidí (*Euterpe oleracea*), which is presented in Chapter 9. From semi-structured interviews involving community-based organisations, private companies, NGOs, international organisations and universities, this case study assesses the full range of activities and relationships involved in creating value. The main results highlight that the factors which characterise the value chain consist of trusted relationships between small-scale producers, a low distribution of profits among local collectors, limited profitability supported by external subsidies, and low regional and national demand. The main results suggest that this value chain could be improved in the short- and medium-term by increasing the capacity of cold chains and storage centres, fostering economic incentives for implementing agroforestry systems, developing strategies for promoting traditional diets by using its natural ingredients, and improving technology and business capabilities to enhance its commercialisation.

Chapters 10 and 11 provide descriptive summaries of the taxonomic and geographical distribution of species, based on the uses in the database to produce the checklist of Useful Plants of Colombia (UPC) detailed in chapter 12. Chapter 10 provides notes on the taxonomic coverage of this checklist, highlighting the plant groups richest in useful plant species, categories of use, and phylogenetic spread. Potential taxonomic gaps are also identified. The checklist of useful plants comprises 7,472 species, 2,140 genera, and 258 families, representing all major plant groups. Most of the species are native (78.6%), and 6.4% are endemic to Colombia. At least 14% of the useful species are commercially cultivated, and 8.6% are introduced and naturalised. The families that are richest in useful plant species are Fabaceae, Asteraceae, and Poaceae. The genera that are richest in useful plant species are *Solanum*, *Inga*, and *Passiflora*. The categories of use with the most associated plant species are medicines (5,108 spp.), human food (3,806spp.), and materials (2,363spp.).

Chapter 11 characterises the geographic distribution of the species listed in the checklist by describing distribution patterns at varying biological and geographical scales. In total, the plant species with documented uses are represented by 435,230 publicly available georeferenced records. Of the 13 bioregions assessed, the humid forests of the Andes made up the bioregion with by far the largest numbers of records, species, genera, and families. The humid forests of the Amazonia and Caribe followed in terms of species richness. At the regional level, Antioquia has the largest number of useful species, followed by Cundinamarca and Valle del Cauca, whilst at the municipality level, Solano (Caquetá), Manizales (Caldas), and Leticia (Amazonas) report the highest species richness. However, records are unevenly distributed across both taxonomic groups and geographic areas. Within all use categories, the species richness decreased with elevation. Potential drivers, implications and knowledge gaps are presented, and the current understanding of the distribution of useful plants can be used as a baseline to target future research efforts, supporting plant conservation planning and efforts across Colombia.

Finally, a comprehensive checklist of the useful plants of Colombia is presented in Chapter 12. The chapter describes how the checklist was built and what kind of resources are presented for all useful species of Colombia. The list includes supraspecific taxa, accepted species and authors, common names, species origin, geographic information (habitats, regions, departments, and elevation range), conservation status (assessments of the accepted name), and level 1 of the category of use.

Tracking species as their names change through time and maintaining the link to their uses, distributions, conservation status and common names, is an endless task. This is the reason for having the printed version of the catalogue linked to a dynamic online portal. For full details on the species, readers should scan the QR codes and visit the species profiles in https://www.colplanta.org, which include morphological descriptions, geographic distribution and maps, synonymy, notes on uses, links to herbarium specimens, images of live plants, illustrations, economic botany items and anatomical laminae, bibliography, and additional sources.

The checklist will enable you to access its supplementary material, to explore open questions and opportunities to develop innovative ideas on alternative processes for consumption and production, novel technologies, and species conservation, to promote a social and environmental positive impact, and to inspire new generations.

We hope that this Catalogue and Col*Plant*A (as its dynamic resource) will be used by the audience as a useful referential for decision-making, production, and demand building. We expect these resources to meet a broad audience (consumers, interested public, politicians, producers, regulators, scientists, stakeholders, and students), allowing Colombia as a Nation to achieve its Mission on Green Growth, and to use and conserve its natural resources in an integrated, equitable and sustainable way.

Flower of *Passiflora ambigua* Hemsl. (Passifloraceae) used as human food in Colombia.
Mauricio Diazgranados

Foreword

Back in 2016, during the State Visit that President Juan Manuel Santos paid to the UK, there was an important event at the Natural History Museum, where HRH The Prince of Wales, President Santos and other high-level UK Government stakeholders discussed how to transform the scientific collaboration between the UK and Colombia. That episode marked the beginning of the UK Government's Newton-Caldas Fund support to an inspirational programme called *Colombia BIO,* aimed at understanding Colombia's vast biodiversity to ignite an unprecedented systematic international scientific collaboration, the epicentre of which was in rural areas of Colombia formerly affected by years of armed conflict.

At that event, I met two members of the Kew Science team, Professor Monique Simmonds OBE and Dr Mauricio Diazgranados Cadelo, who have unintentionally contributed to both my professional and personal lives, transforming part of my own relationship with nature. In that meeting, the three of us drafted the first ideas to conceptualise what months later became Col*Plant*A, a project led by Royal Botanic Gardens, Kew, within the British Council's Professional Development & Engagement strand of their Newton-Caldas Fund work. This project was carried out in partnership with what we know today as the Ministry of Science and Technology in Colombia, Minciencias.

Col*Plant*A, an online tool to provide open information about Colombian plants and their uses, was developed to complement other work that the Royal Botanic Gardens, Kew was already doing with Colombian scientists and research institutions with the support of the UK Government. Once Col*Plant*A was established, the scientific team at Kew knew there was still work to do and worked hard to win a new Newton-Caldas Fund grant, administered by the British Council, which was announced during the visit of President Iván Duque to the UK as a guest of the Government, back in 2019. That, in my view, was the origin of the Useful Plants and Fungi of Colombia project.

Part of my role at the British Embassy in Colombia is to connect the UK scientific ecosystem with its peer equivalent in Colombia. The Useful Plants and Fungi of Colombia project is a perfect example of how peers exchange knowledge and produce equal partnerships, within an epistemic exercise to put science at the front of a race to act against a changing climate that demands immediate solutions from all of us.

Those solutions—we learn from the work of institutions such as RBG Kew, Instituto Alexander von Humboldt and a wide range of Colombia and UK academic institutions—have a lot to do with plants and the ways in which we use them. The more we learn about these uses, actual or prospective, the better equipped we are to play our individual role in solving a collective challenge such as climate change.

I feel immensely proud to see research products such as the one you are reading now, and I am hugely grateful for the opportunity I have had to contribute to making this possible, almost literally from the inception stage. In a year when life as we used to know it has been radically changing, I truly believe in science not only as the way to find the answers to complex global development challenges but also as a superpower to transform lives in entire communities.

This is perhaps one of the most important benefits of a research product like this. Offering Colombian communities the opportunity to learn about—and sustainably live from—their surrounding nature is, in itself, a paramount achievement that history will put into perspective in the years to come. I celebrate this unique piece of scientific collaboration and I have no doubt about the potential of this work to contribute not only to Colombia but also to the rest of global society. This book is testament to a deep bilateral scientific relationship I have had the immense honour to energise through my work, so I really thank those who made it possible and do hope you enjoy it and find it useful.

Luis Calzadilla Waldmann

Head of Science and Innovation
British Embassy Colombia

Inflorescence of *Puya trianae* Baker (Bromeliaceae), a species from the páramos with Environmental and Animal Food uses.
Mauricio Diazgranados

Acknowledgements

This work was supported by a Professional Development & Engagement grant under the Newton-Caldas Fund partnership. The grant is funded by the UK Department for Business, Energy and Industrial Strategy (BEIS) and the Colombian Ministry of Science, Technology and Innovation (MinCiencias) and delivered by the British Council. For further information, please visit www.newtonfund.ac.uk.

The *Catalogue of Useful Plants of Colombia* could have not been realised without the remarkable efforts of many individuals and the support of several institutions. It represents one of the main products of the project Useful Plants and Fungi of Colombia (UPFC), led by the Royal Botanical Gardens, Kew (RBG Kew), with the collaboration of the Alexander von Humboldt Biological Resources Research Institute (Colombia).

This book is substantially based on the data from the Col*Plant*A portal, which is developed and maintained by RBG Kew, with the collaboration of various Colombian partners. The editors and authors are very grateful for the support of the funders regarding the production costs of this publication, RBG Kew for providing the research structure and internal services, the Humboldt Institute for the contribution of data and literature, and Kew Publishing for the final copy-editing and layout.

The editors and authors wish to acknowledge their appreciation of all staff, past and present, of the institutions involved in this catalogue, for their dedicated research that resulted in the contents of this work, and of the numerous specialists who contributed to the elaboration and revision of the updated lists, and with valuable information on local and regional uses of the species.

Special acknowledgements are due to the team working on taxonomic reconciliation and data management, especially Bob Allkin, Tiziana Cossu, Julia Carretero, Laura Kor, and Benedetta Gori, and to Joaquim d'Souza and Malcolm Stone for providing the PURLs that are used to make the QR Codes. Many thanks to those who were working dedicatedly to provide invaluable support on the compilation of the preliminary lists, references, specimens, common names, and all information related to species and their respective uses and conservation status, in both the UK and Colombia: Laura Green, Ellie Graves, Priscila Reis, Germán Torres-Morales, Alejandra Aguilar-Giraldo, Daniel Jiménez-Pastrana, Lina Isabel Guevara-Ruiz, Henry Agudelo and Fabio Ávila. The authors also wish to thank those who contributed chapters and images, and those who acted as reviewers and made valuable suggestions: Amalia Díaz, Benedetta Gori, Cristina López-Gallego, Dairon Cárdenas, David Hammond, Federico Padilla, Felipe García, Henry Yesid Bernal, Jennyfer Andrea Aldana, Justin Moat, Mark Nesbitt, Nelson Salinas, Phillipa Ryan, Rafael F. Almeida, Steven Bachman, Tiziana Ulian, and William Milliken.

We are very grateful to all those who contributed species and field images (Benedetta Gori, Carolina Castellanos-Castro, Dairon Cárdenas-López, Elena Stashenko, Elizabeth Hodson, Laura Green, Laura Kor, Mateo Fernández Lucero, Mauricio Diazgranados, Mónica Andrea Flórez Pulido, Priscila dos Reis and Ellie Graves), as well as to the Spatial Analysis team from RBG Kew who produced the species distribution maps available in the Col*Plant*A species profiles, especially Carolina Tovar, Ian Ondo, and Justin Moat. We are also grateful to Daisy Lafarge for allowing us to use excerpts from a poem she presented at the 'Observations on Being' exhibition developed as part of the Coventry City of Culture 2021.

Thank you to Kew's IT team for helping with IT and computational issues, specially Reinis Rozkalns, Erkki Mellin and Chris Soh.

We would like to express our deepest thanks to our Project Manager, David Hammond, the Project Officer Kaitalin White and the Communication Specialists Camila Gutiérrez and María del Pilar Mira Pontón, for their enormous support of the Useful Plants and Fungi of Colombia project, which was extended during the entire production process for the publication of the Catalogue.

Sadly, one of our contributing authors and colleagues, Dairon Cárdenas-López, passed away during the production of this catalogue. We dedicate this volume to him, his family and his life as a field botanist, scientific researcher, and highly regarded Director of the herbarium at the Amazonian Scientific Research Institute (SINCHI). We celebrate his valuable contribution and dedication to advancing our understanding of the flora of the Colombian Amazon, including the collection of more than 45,000 botanical specimens.

Finally, we would like to thank the successive directors and researchers of the RBG Kew, the Humboldt Institute and the Ministry of Environment and Sustainable Development of Colombia, as well as the British Embassy in Colombia and the Colombian Embassy in the UK, for their continuous support. Many other people and institutions gave their support in one way or another to the preparation of this catalogue.

Schoenocephalium teretifolium (Flor de Inírida de Verano), in white sand savannas of Inírida.
Mateo Fernandez Lucero

List of Contributors

CONTRIBUTORS	INSTITUTION
Alba Nohemí Téllez A.	Pontificia Universidad Javeriana
Alejandra Aguilar-Giraldo	Instituto de Investigación de Recursos Biológicos Alexander von Humboldt
Alex Monro	Royal Botanic Gardens, Kew
Andrés A. Barona-Colmenares	Instituto Amazónico de Investigaciones Científicas SINCHI
Benedetta Gori	Royal Botanic Gardens, Kew
Carlos Alberto Cortés	Instituto de Investigación de Recursos Biológicos Alexander von Humboldt
Carolina Castellanos-Castro	Instituto de Investigación de Recursos Biológicos Alexander von Humboldt
Catalina Mesa Sánchez	Pontificia Universidad Javeriana
Cristina Lopez-Gallego	Universidad de Antioquia
Dairon Cárdenas-López †	Instituto Amazónico de Investigaciones Científicas SINCHI
Daniel Jiménez-Pastrana	Instituto de Investigación de Recursos Biológicos Alexander von Humboldt
David Hammond	Royal Botanic Gardens, Kew
Elena Stashenko	Centre for Research on Tropical Aromatic and Medicinal Plant Species (CENIVAM) – Industrial University of Santander
Elizabeth Hodson de Jaramillo	Pontificia Universidad Javeriana
Fabio Ávila	Jardín Botánico de Bogotá José Celestino Mutis
Felipe García	Instituto de Investigación de Recursos Biológicos Alexander von Humboldt
Germán Eduardo Torres-Morales	Instituto de Investigación de Recursos Biológicos Alexander von Humboldt
Henry Agudelo	Universidad Nacional de Colombia
Henry Yesid Bernal M	Pontificia Universidad Javeriana
Hernando Echeverri-Sanchez	Royal Botanic Gardens, Kew; University College London
Jennyfer Andrea Aldana Mejía	Universidade de São Paulo
Joaquim de Souza	Royal Botanic Gardens, Kew
Julia Carretero	Royal Botanic Gardens, Kew
Kaitalin White	Royal Botanic Gardens, Kew
Laura Kor	Royal Botanic Gardens, Kew
Lina Isabel Guevara-Ruiz	Instituto de Investigación de Recursos Biológicos Alexander von Humboldt
Luis Calzadilla Waldmann	British Embassy Colombia
Mabel Tatiana Rojas	Instituto de Investigación de Recursos Biológicos Alexander von Humboldt
María del Pilar Mira	Royal Botanic Gardens, Kew
Mark Nesbitt	Royal Botanic Gardens, Kew
Mauricio Diazgranados	Royal Botanic Gardens, Kew
Natalí Sánchez-Garzón	Royal Botanic Gardens, Kew
Nicolás Castaño-Arboleda	Instituto Amazónico de Investigaciones Científicas SINCHI
Nórida Marín-Canchala	Instituto Amazónico de Investigaciones Científicas SINCHI
Raquel Negrão	Royal Botanic Gardens, Kew
Tiziana Cossu	Royal Botanic Gardens, Kew
Tiziana Ulian	Royal Botanic Gardens, Kew

Brosimum alicastrum Sw. seeds.
Néstor Plata

Chapter 1

The Useful Plants and Fungi of Colombia (UPFC) project: delivering botanical knowledge to support conservation and sustainable development

Mauricio Diazgranados[1]*, David Hammond[1], Tatiana Rojas[2], Kaitalin White[1], María del Pilar Mira[1], Carolina Castellanos-Castro[2], Camila Gutiérrez[1] & Tiziana Ulian[1]

[1] Royal Botanic Gardens, Kew
[2] Instituto de Investigación de Recursos Biológicos Alexander von Humboldt

***Corresponding author:** m.diazgranados@kew.org

Keywords: bioeconomy, fungi uses, green growth, plant uses, sustainable development

ABSTRACT

Colombia ranks second in the world in biodiversity and is recognised as one of the most ethnically diverse countries. Despite its biocultural richness, the country is marked with vast social inequality and rural poverty. Following decades of internal armed conflict, the country's 2016 Peace Agreement has provided new opportunities for its socio-economic growth, representing either a threat to Colombian biodiversity or an opportunity for sustainable development based on its treasured natural resources. The Useful Plants and Fungi of Colombia (UPFC) project aimed to develop pathways to enhance nature's contribution to people in Colombia by increasing, consolidating and making accessible the knowledge on its useful plants and fungi for the benefit of local communities. The project has provided a framework to develop and promote a market for useful indigenous species and their high-value products whilst protecting the surrounding natural resources. It has produced over 140 dissemination outputs, including books, booklets, scientific journal publications, technical reports, websites, online portals, and educational tools. Also, it has delivered capacity-building events, reaching a broad audience. With the participation of a multinational team, the project compiled and generated knowledge on over 36,000 plants and fungi and developed pathways for tackling socio-environmental challenges and contributing to Colombia's green transformation.

RESUMEN

Colombia ocupa el segundo lugar en el mundo en biodiversidad y es reconocido como uno de los países con mayor diversidad étnica. A pesar de su riqueza biocultural, el país está marcado por una gran desigualdad social y pobreza rural. Tras décadas de conflicto armado interno, el Acuerdo de Paz de 2016 del país ha brindado nuevas oportunidades para su crecimiento socioeconómico que pueden representar tanto una amenaza para la biodiversidad colombiana como una oportunidad para un desarrollo sostenible basado en sus preciados recursos naturales. El proyecto Plantas y Hongos Útiles de Colombia (UPFC) tuvo como objetivo desarrollar vías para mejorar la contribución de la naturaleza a las personas en Colombia, mediante el aumento, la consolidación y la accesibilidad del conocimiento sobre sus plantas y hongos útiles para el beneficio de las comunidades locales. El proyecto ha proporcionado un marco para desarrollar y promover un mercado para especies autóctonas útiles y sus productos de alto valor, al mismo tiempo que se protegen los recursos naturales circundantes. Así, este proyecto ha producido más de 140 productos de difusión, incluidos libros, folletos, publicaciones en revistas científicas, informes técnicos, sitios web, portales en línea y herramientas educativas. También se han realizado eventos de creación de capacidad, llegando a una amplia audiencia. Con la participación de un equipo multinacional, se ha recopilado y generado conocimiento sobre más de 36.000 plantas y hongos, y se desarrollaron vías para abordar los desafíos socioambientales y contribuir con la transformación verde de Colombia.

COLOMBIA'S RICH BIODIVERSITY IN DANGER

Our life and well-being rely on the environmental goods and services provided by plants and fungi. Not only do they provide us with products, such as food, medicines, natural fibres, fuel, building materials, and cosmetics, but they also act to purify the air, enrich the soil, protect against erosion, regulate water flow and quality, and provide habitats for animals, while acting as a major store of global carbon among other functions (Millennium Ecosystem Assessment, 2005). However, conversion of natural habitats—primarily for farming—is fuelling land degradation, undermining the well-being of two-fifths of humanity, and raising the risk of migration and conflict (IPBES, 2018). Therefore, the preservation and sustainable use of Colombia's biodiversity are essential to human well-being and future economic prosperity.

TABLE 1. Plant and fungal diversity of useful vs non-useful taxa in Colombia, based on data collected by this project.

Numbers	Plants	Useful plants	Fungi	Useful fungi	Plants & fungi	Useful plants & fungi
Species	28,947	7,472	7,140	374	36,087	7,790
Genera	3,583	2,140	1,763	192	5,346	2,318
Families	395	258	448	99	843	356

TABLE 2. Origin and conservation status of plant and fungal diversity in Colombia, based on data collected in this project.

Trait	Plants	Useful plants	Fungi	Useful fungi	Plants & fungi	Useful plants & fungi
Cultivated	1,077	1,077	N/A	N/A	N/A	N/A
Naturalised	677	638	N/A	N/A	N/A	N/A
Native	27,134	5,830	1,264	25	28,398	5,855
Endemic	9,473	478	206	8	9,679	486
Threatened	1,061	249	10	3	1,071	252

Colombia is one of the most biodiverse countries on Earth, comprising at least 28,947 plant species (24% endemic; >1,000 threatened), belonging to 3,583 genera and 395 families (Table 1). The country ranks second in terms of diversity of plants worldwide and is a centre of origin of tomatoes, peppers, potatoes, chillies, and many other crops and their wild relatives (Diazgranados *et al.*, 2020, Khoury *et al.*, 2016).

Its fungal diversity is also impressive, with at least 7,140 species belonging to 1,763 genera and 448 families (Table 2). Given that the observed ratio between fungi and plant species in well-documented areas is 9.8:1, Colombia could harbour up to 300,000 fungi species, representing 9% of the global diversity (Gaya *et al.*, 2021).

Colombia is also described as one of the most ethnically diverse countries globally, consisting of at least 85 ethnic groups, with 68 recognised native languages. The knowledge associated with useful plants and fungi can be as rich as the cultural variety. The country has been considered a "cradle for modern ethnobotany", with a plethora of recent studies on this topic (Bernal *et al.*, 2011). However, ethnobotanical knowledge is still vastly under-documented in this region (Cámara-Leret *et al.*, 2014).

Regrettably, the country lost just under 6 million hectares of forest between 1990 and 2015, at an average rate of 237,000 hectares per year (FAO, 2015). This deforestation is mainly due to agricultural expansion, urban development and illegal mining, driven by the race for economic growth following the country's peace process, and inadequate political decision-making that is often not supported by science (Eufemia *et al.*, 2019, Sabater *et al.*, 2017, Salazar *et al.*, 2021). Because this impressive cultural and biological biodiversity is under threat, researchers and conservationists are in a race to protect it (Andrade, 2011, WWF-Colombia, 2017). As deforestation proceeds, the disappearance of traditional knowledge on conservation, use, and management of plants and fungi accelerates with biodiversity loss.

A recent review of studies across four Andean countries (Colombia, Ecuador, Peru, and Bolivia) evaluated the sustainability of wild plant use in the region and found that unsustainable harvesting and the loss of useful wild plant species repeatedly arose (Kor *et al.*, 2021). Colombia has the highest percentage of studies reporting unsustainable outcomes (86%). The authors identified five main factors contributing to the unsustainable use of plants: 1) biology; 2) land tenure and access rights; 3) loss of knowledge, resources, and capacity; 4) economic and market pressures due to overexploitation; and 5) institutional structures, policy, and legislation.

With its unique and extraordinary biocultural diversity and richness, Colombia can become an example for achieving the Sustainable Development Goals (SDG) (United Nations, 2015) by reducing inequality and poverty through the sustainable use of its natural resources, reducing the degradation of its plants and fungi. Although native non-crop plants and fungi have great potential to improve livelihoods and support economic development, knowledge on the most useful native species of Colombia remains highly dispersed, largely inaccessible, and susceptible to disappearing over time.

CAN THE SUSTAINABLE USE OF BIODIVERSITY BECOME PART OF THE SOLUTION?

Despite the unique biological and cultural richness of Colombia, vast social inequality and marked poverty in rural areas have triggered more than six decades of internal conflict. Plant and fungal diversity and its associated cultural knowledge could tackle socio-environmental challenges by boosting rural employment and incomes, improving people's livelihoods, reducing inequalities, and supporting peace. Improving access to critical biological information and developing markets for natural ingredients, while meeting sustainability benchmarks, can boost the stake of marginalised communities in the national economy by adding value to their traditional knowledge. Simultaneously, this process would empower meaningful participation, ultimately leading to more stable and equitable socio-economic outcomes. This approach tackles several of the SDGs, most notably those related to the alleviation of poverty, improvement of health, empowerment of the disenfranchised, sustainability of economic growth and promotion of peace (United Nations, 2015).

In recent years, Colombia has established several policies promoting the sustainable use of biodiversity to reduce the social gap and to consolidate peace whilst following a green growth approach to economic development (Baptiste *et al.*, 2017). The country has stated its intention to develop as a bioeconomy, which is "an economy that efficiently and sustainably manages biodiversity and biomass to generate new value-added products, processes, and services based on knowledge and innovation" (CONPES, 2018). The commitment of the Colombian government to green growth has been demonstrated through the Colombia Bio programme and its principal aim to foster the development of the bioeconomy through science, technology, and innovation.

As an offspring of this over-arching national effort, the Useful Plants and Fungi of Colombia project (UPFC) was supported by a Professional Development & Engagement grant under the Newton-Caldas Fund partnership. The £2.5 million grant was funded by the UK Department for Business, Energy, and Industrial Strategy (BEIS) and the Colombian Ministry of Science, Technology, and Education (MinCiencias) and delivered by the British Council over 28 months from November 2019 to February 2021.

THE USEFUL PLANTS AND FUNGI OF COLOMBIA

The overall goal of the UPFC was to develop pathways to enhance nature's contribution to people in Colombia by increasing, consolidating, and making accessible the knowledge on its useful plants and fungi for the benefit of local communities. In addition, the project promoted a market for useful indigenous species and their high-value products to motivate the sustainable use of biodiversity whilst protecting the surrounding natural resources.

This project documented and broadly disseminated knowledge on the useful plants and fungi of Colombia across a wide range of audiences. Outputs targeted policy and decision-makers in government, local NGOs, scientists, educators, and private and public companies, aiming to boost bioeconomic development. The project's long-term vision was to support Colombia's economic future by promoting the sustainable expansion of its bioeconomy, based on its unique plant and fungal diversity, whilst conserving its natural resources and associated traditional knowledge.

The UPFC project was structured as four work packages (WP), each with a specific objective:

1. document knowledge on useful plants and fungi of Colombia (WP1)
2. synthesise and disseminate knowledge on useful plants and fungi of Colombia using various platforms for different audiences (WP2)
3. develop a framework for creating a sustainable value chain network (VCN) from plant and fungal diversity (WP3)
4. apply a VCN framework to improve local communities' livelihoods in three pilot areas (WP4).

DOCUMENT KNOWLEDGE ON USEFUL PLANTS AND FUNGI OF COLOMBIA (WP1)

Native non-crop plants and fungi have great potential to improve livelihoods and economic development in the country. Still, high-quality information on their uses and properties remains scarce or inaccessible. This WP aimed to gather and compile this information to boost Colombia's bioeconomy and benefit local communities.

Significant effort to document the diversity of Colombian flora and fungi has been carried out in recent years (Andrade, 2011, Bernal *et al.*, 2012, Bernal *et al.*, 2019, Diazgranados *et al.*, 2020, Gaya *et al.*, 2021). Still, hundreds of sources ofpublished information had not been compiled and made readily available. The project brought together a binational team of researchers who collected and compiled data from available datasets and publications; digitised data that were not readily available; scanned specimens, illustrations, and records from scientific collections (from herbaria, fungaria, and libraries); generated an image bank of useful plants and fungi of Colombia; developed species distribution maps and models; and created species profiles available online through the project's portals.

The gathered information was analysed to identify taxonomic and geographic gaps and assessed for quality and potential contribution to conservation practices, among other aspects. In all, six research questions on useful plants and fungi were addressed, each with a scientific output (https://in-colombia.org/):

- How can we assess the comprehensiveness (in quantity and quality) of the information available for each species?
- Are there taxonomic gaps in the available information?
- Are there geographic gaps?
- What are the most likely factors influencing observed taxonomic and geographic patterns?
- Which geographic areas are important for conservation?
- Can traditional knowledge erosion be mitigated through Citizen Science?

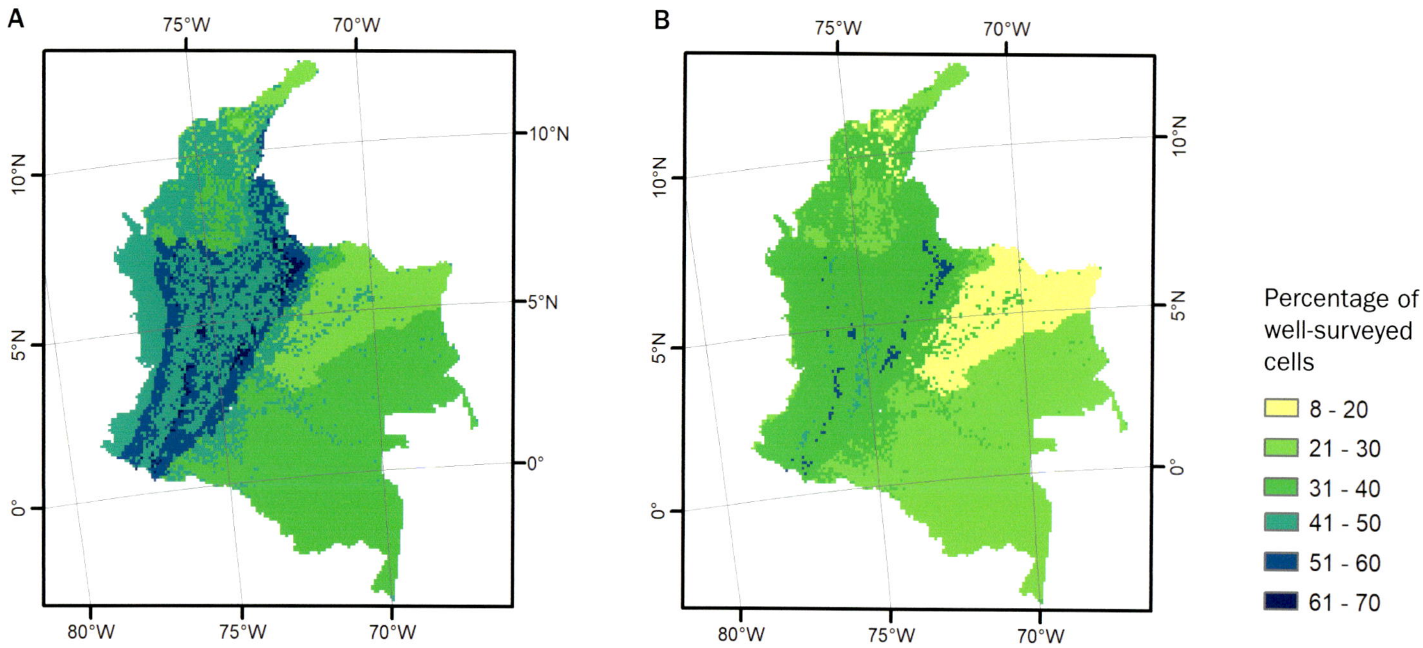

FIGURE 1. Survey completeness for all vascular **A** and useful **B** plants by bioregion (Bystriakova *et al.*, 2021). Percentage of 10×10 km well-surveyed grid cells, defined as those with ≥ 25 observations, calculated by bioregion.

Results from initial activities helped us to identify areas of the country that were poorly surveyed for useful plants and plants in general (Bystriakova *et al.*, 2021), informing subsequent fieldwork and ethnobotanical priorities for WP1 (Figure 1).

In addition to the research questions, three tools were designed to manage and collect data on useful plants and fungi, specifically: 1) the *shinyCCleaner* tool, to clean georeferenced records in the R environment (Ondo *et al.*, pers. comm.); 2) an online survey built-in ESRI Survey123, to collect ethnobotanical information from mobile devices; and 3) a Data Management tool, with modules for data entry and exportation, taxonomic reconciliation, classification of uses, etc. (available online at https://upaf-uat.kew.org/).

SYNTHESISING AND DISSEMINATING KNOWLEDGE ON USEFUL PLANTS AND FUNGI OF COLOMBIA (WP2)

Access to information about plants and fungi and their uses underpins our collective effort to further understand Colombia's diversity, allowing us to better conserve species, habitats, and ecosystems and use them sustainably to the benefit of society. Knowledge must be available for different audiences to empower Colombia to protect and sustainably utilise its valuable plants and fungi. This WP aimed to build a comprehensive synthesis and disseminate this through various media, including online portals, scientific journal publications, books, booklets, reports, magazine and news articles, social media streams, and educational tools.

This wide range of dissemination outputs and formats aimed strategically to reach a broad audience of policymakers, researchers, and practitioners, including those in governmental, monitoring, and developmental aid agencies, regulatory bodies, conservation organisations, biobusinesses, schools, and universities. Potential users included: innovators, developers, bioeconomists, herbal and food entrepreneurs (including plant and fungi importers, suppliers and retailers), biotechnologists, bioenergy experts, agronomists, forestry engineers, crop specialists, horticulturalists, arborists, weed scientists, forest managers, wildlife managers, health professionals, biopharmaceutical scientists, medicine regulators, conservationists, naturalists, photographers, environmental consultants, library managers, students, professors, researchers (biologists, botanists, ecologists, chemists, agronomists, biomedical researchers, ethnobiologists, etc.), scientific editors, people interested in culture and gardening, and the general public.

The project built two websites (https://www.kew.org/upfc, in English; and https://in-colombia.org/; in Spanish) to inform the public and to disseminate project outputs. In addition, it built on the previously created Col*Plant*A portal (http://colplanta.org/), further synthesising knowledge of Colombian plants and improving the user experience, content, and infrastructure (Diazgranados *et al.*, 2020). It also created a companion portal for Colombian fungi, Col*Fungi* (http://colfungi.org/) (Figure 2). Each portal was accompanied by an explanatory booklet (Diazgranados *et al.*, 2020, Gaya *et al.*, 2021) designed for a general audience.

Comprehensive annotated checklists for both plants and fungi, along with a range of thematic chapters, are presented here and in the Catalogue of Fungi of Colombia (de Almeida *et al.*, 2022) as a reference for scientists and long-term preservation of knowledge.

A survey analysis was conducted to reveal how much Colombian society knows about their native diversity of plants and fungi and their uses as part of this WP. An online national survey was developed to achieve this, which collected information on how Colombians generally use plants and fungi. Semi-structured interviews of 20 experts from different sectors of society, including academia, civil society, NGOs, and the private sector, were also undertaken. These experts provided information regarding the state of knowledge of useful plants and fungi. They advised on the limitations and opportunities of various studies and the inclusion of plants and fungi in value chains.

In consideration of socio-cultural components, a Spanish-language storytelling series, "*Somos Historias*", was developed to identify and promote practices and relationships that underlie key the value chains of key plants. The co-creation of short videos that portrayed the practices, knowledge, and relationships of local producers with these plants and their environment promoted value chains and increased project visibility to a wider audience, while also gathering qualitative information on the local use of the species. Finally, the UPFC project published over 50 blogs and notes in the press to raise awareness on the importance of useful plants and fungi for the well-being and prosperity of Colombians and humanity.

BUILDING ON COL*PLANTA*

A significant output under WP2 was work carried out to improve "Col*Plant*A – Colombian resources for Plants made Accessible" (http://colplanta.org/), an authoritative, expert-driven, open access, an online portal for botanical

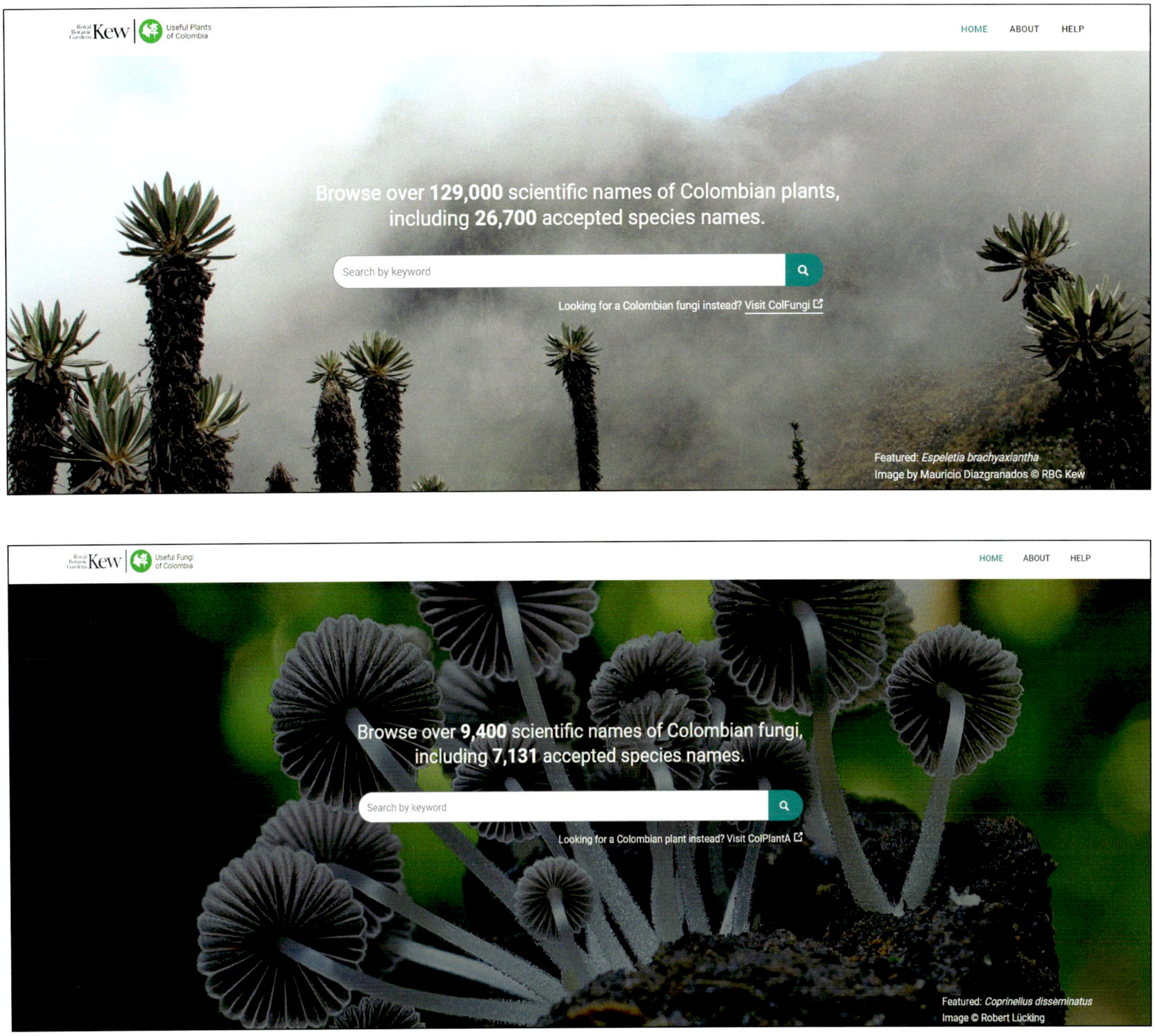

FIGURE 2. Col*Plant*A (https://colplanta.org/) and Col*Fungi* (https://colfungi.org/) site homepages.

information on Colombian plants that was officially launched in August 2019 during the 10th Colombian Botany Meetings as part of a previous project (Diazgranados *et al.*, 2019, Diazgranados *et al.*, 2020).

Col*Plant*A is a content aggregator site – it collects information from Kew's databases and various external sources and displays it in one place while preserving the appropriate links and citations. Therefore, it is a comprehensive resource that for the first time, brings together high-quality taxonomic information with a wide range of data and information sources. In doing so, Col*Plant*A unlocks valuable information, enriches the data and its potential uses, and maximises the visibility of local resources. Only publicly available content is used and, as part of the content-gathering process, steps are taken to ensure that publishers have respected the IBCD Nagoya Protocol (to ensure the fair and equitable sharing of benefits arising from the utilisation of genetic resources) as well as Colombian laws relating to traditional knowledge.

Col*Plant*A holds an online profile for each plant species known from Colombia, aggregating information from other databases and portals, and displaying previously inaccessible information. It links to type specimens from RBG Kew, species descriptions, maps on geographic distribution, published plant uses, sustainable practices, conservation status, photographs, common names, and information on ecology and natural history.

Currently, Col*Plant*A holds more than 27,400 accepted species profiles (including 19,644 with images), 145,000 scientific names, and circa 20,000 distribution maps for Colombian plants (Col*Plant*A (http://colplanta.org/), consulted on 18 February 2022). Since its launch, the portal has hosted approximately 31,500 unique users in 97 countries, averaging 1,400 sessions by 1,260 users each month. The site has been crawled and indexed by Google, with roughly 86,900 web pages indexed (SEOQuake, consulted on 18 February 2022). The portal is linked to the Plants of the World Online (POWO) database – an RBG Kew Strategic Output – and any change to POWO is updated automatically in Col*Plant*A.

Col*Plant*A supports the identification and understanding of the benefits provided by plant diversity, utilising new technologies, concepts, and social frameworks. Also, it helps to underpin the wider goals of Colombia's green growth by providing evidence of links between plants, uses and sustainable development goals.

DIGITAL OUTREACH OPPORTUNITIES

The project reached a range of audiences and created new alliances and opportunities for collaboration through an array of varying virtual events and social media activities.

Despite the recent challenges presented by the coronavirus pandemic, the project found new methods to engage with its target audiences through virtual events, such as Hay Festival Cartagena Virtual, Hay Festival Digital Queretaro 2020, Hay Festival Cartagena 2022

FIGURE 3. The VCN website ("Red de Ingredientes Naturales – Colombia"; https://redin-colombia.org/), displaying the registration form for connecting to the market chain. The site also includes a blog, a serving BOT, and a practical guide to understanding the legal processes in Colombia for developing and commercialising natural ingredients or products of any kind from non-crop plants and fungi.

the Colombian Chef Masterclass series, and virtual workshops, reaching over one million viewers. In addition, it delivered more than 23 virtual conferences, five seminar presentations, three teaching modules, 30 socialisation workshops, and fifteen radio interviews. It had more than 90,000 engagements in its social accounts (Facebook and Twitter), and the project published ten blogs and more than 90 popular articles and media notes, all of which helped to broaden the project's recognition.

BUILDING A VALUE CHANGE NETWORK TO SUPPORT COLOMBIA'S BIOECONOMY (WP3)

Innovative projects to deliver natural ingredients or other bioproducts to the market often fail because a viable market has not been established. The project developed a value chain network (VCN) to provide pathways to develop such markets. This VCN was centred on a digital platform that facilitates interaction between local, national, and international actors in the production, refinement, distribution and exportation of plant- and fungi-based products. The VCN aims to enable the connections between plant and fungal diversity, knowledge of its uses and local livelihoods which depend on it. It is the first VCN focused on useful plants and fungi for the country and the continent.

The VCN connects scientists, technologists, innovators, producers, and consumers. It enables the flow of materials downstream (e.g., the supply of natural ingredients produced by local communities to consumers) and information upstream (e.g., consumer demand to producers) whilst linking products to species names on the basis of a taxonomic backbone. This functionality allows two-way linking between the VCN platform and the scientific portals (Col*Plant*A and Col*Fungi*), enabling users to browse specific uses, read interesting plants and fungi species profiles, and contact product providers directly. A website for the VCN (https://www.redin-colombia.org/) was created in Spanish to maximise outreach among the Colombian audience (primary users) (Figure 3).

In addition, the project delivered the first comprehensive practical guide (174 pages) to understanding the Colombian legal processes for developing and commercialising natural ingredient products from non-crop plants and fungi (Rojas *et al.*, 2021, available at https://redin-colombia.org/guide/). This freely available guide is in Spanish and is organised around four central roles within the value chain: producer, refiner, national wholesaler/retailer, and exporter. Content focuses on permissions, regulatory compliance, and certification concerning three types of natural ingredients: harvested non-timber forest products, nursery-grown products, and genetic material products. It articulates processes related to intellectual property rights and benefit-sharing agreements, a threatened species, among others related to national and international regulations, compliance, and market access. Eleven semi-structured interviews were conducted with staff from the Ministry of Environment and Sustainable Development, ProColombia, natural product companies, and NGOs.

Two online tools were created to accompany the guide:

1. a serving BOT, which is an automated tool designed to guide users to obtain information from within the guide by answering a series of targeted questions regarding their needs and the type of commercial role they play within the value chain (e.g., producer, refiner, distributor, retailer, exporter). The serving BOT is accessible via the VCN website (https://www.redin-colombia.org/)
2. an online quiz in Spanish, drawing upon the information contained within the guide, that tests users' knowledge regarding documentation, certification and responsible agencies related to the production, distribution, and exportation of bioproducts from Colombia. Access to the live quiz platform in Kahoot can be found through the link: https://kahoot.it/challenge/00115304 or in the practical guide.

To establish a baseline for the current status of the Colombian bioeconomy, researchers produced a 124-page report (Rojas *et al.*, 2020), also freely accessible at https://in-colombia.org/. Highlights included:

1. a prioritisation exercise for potential species in Colombia's bioeconomy
2. a broad directory of companies engaged in the use of prioritised species of native plants and fungi (25 in total)
3. a Science, Technology, and Innovation capability analysis around the Bioeconomy and Natural Ingredients sector
4. a multi-criteria value chain infrastructure for biodiversity-based products
5. a case study of the value chain analysis for the first prioritised species, naidí (*Euterpe oleracea* Mart.), in the Pacific Coast region of Colombia
6. a search and analysis of patents and publications related to *naidí* in Colombia and the rest of the world

Finally, to inform the VCN, a standard for controlled vocabulary on natural ingredients and bioproducts was developed, which included a harmonised table and accompanying explanatory report. The standard will allow their classification, indexing, cataloguing, description, and analysis. It was based on the harmonisation of three classification structures linking natural ingredient products: Kew's Economic Botany Data Collection Standard (EBDCS), the UN Central Product Classification scheme (CPC) and the *Departamento Administrativo Nacional de Estadistica*'s Clasificacion Central de Productos (DANE CPC). These three structures were reconciled on the basis of each hierarchy's concept, definition, and rules to identify a common, core standard linking botanical (e.g., species), national and international bioproduct classification structures.

FIGURE 4. Map of urban centres in which the main collaborators are based, and pilot areas selected for the implementation of the VCN framework: Bahía solano (Chocó), Becerril (Cesar), and Otanche (Boyacá). Photograph credits: Mateo Hernández (Becerril), Mauricio Diazgranados (Otanche), and Daniel Uribe (Bahía Solano).

IMPLEMENTING A VALUE CHANGE NETWORK IN THREE PILOT AREAS OF COLOMBIA (WP4)

Three biodiverse pilot areas of Colombia that had been affected recently by the armed conflict were selected to test the potential of the VCN framework to improve impoverished communities' livelihoods (food security, health, and prosperity): Bahía Solano (Chocó), Becerril (Cesar) and Otanche (Boyacá) (Figure 4).

The selection of the areas considered several criteria within five categories: environmental, social, governance, business, and economic factors (Diazgranados *et al.*, 2021).

The implementation included:

1. assessment of the useful local plants and fungi resources
2. evaluation of the market status based on local native diversity
3. assessment and modelling of current and potential outcomes of the VCN implementation
4. identification of the most promising species and types of natural ingredients
5. the gleaning of lessons from the implementation and community feedback

The project expected that implementing the VCN would improve family livelihoods over the middle- to long-term through its role in facilitating diversified, market-based approaches to income generation.

The social impact assessment carried out during the implementation of the VCN will allow the measurement of middle- and long-term changes in cultural, economic, and social outcomes. The outcome monitoring activities implemented at the time of writing include participant surveys (collecting data on gender, age, locality, ethnic group, family composition, etc.) and analysis of statistics relating to the potential implementation. The project published a booklet to connect experiences in local communities within the three pilot areas with a broader, more general audience. It included highlights, notes on the prioritisation of species, stories from local initiatives and experiences (Quiñones-Hoyos *et al.*, 2021).

THE UPFC TEAM

The UPFC project was led by RBG Kew in collaboration with the Instituto de Investigación de Recursos Biológicos Alexander von Humboldt in Colombia. It directly involved 85 people (50 from Kew and 35 from the Humboldt Institute) and received contributions from 34 other researchers representing 26 institutions across Colombia and Europe. Diverse and gender-balanced participation was accomplished in all activities to ensure a more inclusive environment and in line with SDG5 (Empower women). We have been collaborating to include minorities such as indigenous people, Afro-Colombians and the LGBT+ community, always following RBG Kew's and Humboldt Institute's policies on diversity and inclusion. Of 115 project participants, 64 (56%) were female and 51 (44%) were male. The project deemed this gender balance to have been appropriate given the traditional tendency of Botany to be a male-dominated discipline.

REACHING COMMUNITIES ACROSS SOCIETY

The UPFC project built and maintained connections with other government and non-government initiatives. Through various fora and media, objectives and outcomes were communicated to the Presidency of the Republic, the National Committee of Sustainability, the Green Business Bureau of the Ministry of Environment, the Programa de Transformación Productiva (PTP) of the Ministry of Commerce, ProColombia, the Instituto de Investigaciones Ambientales del Pacífico John von Neumann (IIAP) and the Green Growth Mission (GGM) created by the Colombian National Planning Department in 2017.

The GGM, together with the Biointropic Corporation and Suricata Consulting, identified seven factors to boost the country's bioeconomy: 1) research and technological development; 2) regulatory framework; 3) market dynamics; 4) human capital; 5) financing and investment; 6) environmental or ecosystem services; and 7) infrastructure. By contributing directly to factors 1–4, the project supported GGM's mission and the Green Growth Policy, i.e., "Promote by 2030 the increase in productivity and economic competitiveness of the country, while ensuring the sustainable use of natural capital and social inclusion, in a compatible way with the climate".

It engaged the Colombian National Board of Natural Ingredients (Programa de Interés estratégico-PINE), an initiative led by the Program Colombia + Competitiva of SwissContact as part of a 5-year plan, to boost natural ingredients for cosmetics. The Humboldt Institute leader of the board's six sub-activities regarding the governance and competitiveness of native plant species' value chains were key in maintaining this interaction.

The project also connected with the private sector, such as Crepes & Waffles, the Chamber of Food Industry, the National Association of Industrials of Colombia (ANDI), Corpocampo, Naidí Pacífico SAS, Fondo Acción, Planeta CHB SA, Partnerships for Forest, MUCHOCOL, UNIANDES (Faculty of Economics), Apsacesar, Envol-vert, Selvacéutica S.A.S., E3 Asesorías, Corporación Boyapaz, Alianza Quinchas, among many others.

The botanical work, particularly around the initial consolidation of the Col*Plant*A portal, involved botanists from several Colombian organisations, including the Asociación Colombiana de Herbarios (ACH), the Grupo Etnobotánico Latinoamericano (GELA) and several universities such as Universidad de Antioquia, Universidad de la Amazonia, Universidad de los Llanos, Universidad de Nariño, Universidad de Pamplona, Universidad del Cauca, Universidad Distrital "Francisco José de Caldas", Universidad Nacional de Colombia, Universidad Pedagógica y Tecnológica de Colombia, and Universidad Tecnológica del Magdalena.

FIGURE 5. UPFC fieldwork activities in the pilot areas. Top left: collecting useful plants and fungi with the help of local *"sabedores"* (wisdom people). Top right and bottom: species prioritisation workshops with local communities. Photograph credits: Mateo Fernández (top left), Sabina Bernal (top right and bottom right), and Edgar Padilla (bottom left).

Fieldwork was carried out through a participatory approach involving local organisations, social leaders, and communities from the veredas of Caño Rodrigo and Río Maracas in Becerril (Cesar), and Altazor, Betania, Camilo, Cunchalita, Curubita, El Carmen and Nazareth in Otanche (Boyacá) (Figure 5).

In conclusion, the UPFC project played a pivotal role by reaching communities across society, promoting the conservation and sustainable use of the country's plants and fungi, combining traditional and scientific knowledge to boost the national bioeconomy, and raising awareness about the importance of these resources for the country. On a broader scale, it represents a sustainable development model of green growth in megadiverse nations worldwide.

UNEXPECTED IMPACTS AND CHALLENGES DUE TO COVID-19

The pandemic of 2020–21 had significant impacts on where and how the UPFC project team carried out their work due to the restriction of travel and facility access, recruitment delays, and the shortage of human resources. Mitigation plans proved effective in buffering some of the effects, especially regarding WP1, WP2 and WP3 workflows and timelines, which used increased teleconferencing and online stakeholder consultations with several cross-cutting contingency actions. Nevertheless, capacity building (e.g., the internship and research visitor programmes) and fieldwork activities had to be scaled down considerably, and future funds could help to strengthen the implementation of the VCN at the national scale.

PATHWAY TO IMPACT ON ECONOMIC DEVELOPMENT AND SOCIAL WELFARE

This project has developed novel pathways to enhance nature's contribution to people in Colombia. It has made available 'state-of-the-art' knowledge on the nation's useful plants and fungi, while unveiling how this information can be used to promote and develop markets for these species and their high-value products.

Beyond the delivery of specific outputs, a broader set of project outcomes were achieved. These include:

- a significant increase in online, open-access knowledge on useful plants and fungi of Colombia
- increased capacity of Colombian scientists to research plant and fungal diversity and uses
- increased public perception and awareness about the importance of plant and fungal diversity and their uses to support green growth in Colombia
- improved opportunities for alternative local income streams based on value-added bioproducts through VCN development, at least in the three pilot areas
- supported expansion of a natural ingredients market at the national level
- opportunities opened for further improvement to people's livelihoods in the three pilot areas
- informed development of a bioeconomy, bringing together science, policy, and industry.

The predicted spillover effects included:

- increased interest by Colombian institutions in expanding plant and especially fungal research
- enhanced recognition of market possibilities for natural ingredients based on native species of plants and fungi at the Colombian national level
- incentive to develop specific areas of commercial businesses based on natural ingredients from native Colombian plants and fungi, e.g., by the British and Colombian Chamber of Commerce, ProColombia, etc.
- expansion of Colombian bioeconomy by strengthening industrial endeavours to add value to useful plants and fungi products.

In addition to the contributions to the United Nations SDGs, the project outputs and activities also contributed to several objectives of the Convention of Biological Diversity (CBD) relating to the targets defined from the Global Strategy for Plant Conservation (GSPC) 2010–2020, particularly those linked to ecosystem goods and services (GSPC1), recognition, respect, and maintenance of indigenous and local community knowledge (GSPC 5), and improved awareness of the urgency of plant conservation and public participation (GSPC 6).

As most communities in rural Colombia are male-dominated, with men focused on farming, mining and construction, the project emphasised pathways for women to play lead roles in the obtention and commercialisation of natural ingredients, potentially bringing additional income to the household. Thus, the implementation of the VCN could improve gender equality and opportunities, supporting SDG5 (Empower women).

The project pathway to impact on economic development and social welfare was based on a stepwise approach:

- Short-term (1–3 years): the project documented and broadly disseminated knowledge on useful plants and fungi of Colombia, reaching both general and academic audiences, as well as policy and decision-makers. It also contributed to capacity building. Last, by implementing the VCN in three pilot areas, the project aimed to support improvements to communities' livelihoods by establishing a baseline that can be monitored and provide initial training to communities.
- Mid-term (3–10 years): the project outputs will increase awareness in Colombian society about the importance of the country's diversity of useful plants and fungi, and will inform policy and decision-makers in government and non-government organisations, thereby boosting sustainable biobusinesses and supporting national and international-scale connections. The VCN may improve the livelihoods of communities at the national scale, providing new opportunities to reduce poverty and gender inequality, and to improve health and well-being.
- Long-term (10–15 years): Colombia will rise as a successful example of a developing country that is expanding its bioeconomy based on its unique plant and fungal diversity while preserving its natural environment.

FUTURE PERSPECTIVES AND FINAL REMARKS

The project has contributed significantly to the Colombian Green Growth Policy's scope to "Promote by 2030 an increase in productivity and economic competitiveness of the country, while ensuring the sustainable use of natural capital and social inclusion, in a compatible way with the climate" (CONPES, 2018). While the transformation of a country is a very complex, long-term process, usually involving all levels of society, the UPFC project is confident that its outputs and outcomes have supported Colombia's transformation towards a greener, more sustainable economy. This has been achieved by enhancing critical baseline knowledge on plants and fungi and re-imagining how access and practical use of this information could be improved through open-access digital platforms and tools.

The impact of the UPFC project will continue, as the information generated here has been made broadly available to Colombian companies that have an interest in the sustainable development of biodiversity-based products (bio-based), typically the agricultural, pharmaceutical, food, human health, and personal care sectors. Colombia's green transformation has begun!

Acknowledgements

The UPFC project thanks all the researchers and contributors who made this project possible: Adriana Corrales, Aida Marcella Vasco-Palacios, Alba Nohemi Téllez A., Alejandra Aguilar, Alejandra Osejo, Alex Monro, Aline Dufat, Amalia Díaz, Ana Cristina Bolaños-Rojas, Ana Esperanza Franco-Molano, Ana Garrido, Andrea Baquero, Angélica Herrera, Angie Karina Rengifo Fernandez, Anna Haigh, Benedetta Gori, Bibiana Moncado, Bob Allkin, Camilo Garzón, Carlos Cortés, Carolina Quiñones, Carolina Soto, Carolina Tovar, Catalina Mesa Sánchez, Clara Morales, Cristian Lasso-Benavides, Dairon López, Daniel Jiménez, Daniel López, David Bishop, David Díaz-Escandón, Diana Marcella Moreno, Diego Simijaca-Salcedo, Edier Soto-Medina, Eduardo Toledo Romero, Edwin Rios, Ehidy Rocio Peña-Cañón, Elizabeth Hodson, Ellie Graves, Emmerson Pastás, Ester Gaya, Fabiola Eugenia González-Cueller, Felipe García, Germán Torres, Guilana Furci, Harrie J.M. Sipman, Henry Agudelo, Henry Yesid Bernal M., Hugh Pritchard, Ian Ondo, James Morley, Jenny Williams, Jennyfer Andrea Aldana Mejía, Joaquim d'Souza, Joseph Ruff, Julia Carretero, Justin Moat, Laura Green, Laura Kor, Lee Davies, Lina Guevara, Louise Colville, Luis Fernando Coca, Lynn Parker, Malcolm Stone, Marcela Cely, Margarita María Jaramilo-Ciro, Mariana Pizano Noguera, Mario Murcia, Mark Nesbitt, Martha Lucía Ortiz-Moreno, Mateo Fernández, Mauricio Ramírez-Castillón, Mauricio Salazar-Yepes, Meike Piepenbring, Melanie-Jayne Howes, Mimi Tanimoto, Nadia Bystriakova, Natalia Vargas-Estupiñán, Nataly Gómez-Montoya, Nestor García, Nestor R. Salinas, Nicolas Black, Pablo Hendigo, Paola Acosta, Paola Andrea Torres-Andrade, Paul Kersey, Paul Wilkin, Priscilla Reis, Rafael Felipe de Almeida, Raquel Negrao Baldoni, Rene Camacho, Robert Lücking, Rose Taylor, Sabina Bernal, Santi Gómez, Sarah Phillips, Silvia Restrepo, Tatiana Sanjuan, Tim Utteridge, Tiziana Cossu, Vicky Murphy, Viviana Motato-Vásquez, Wiebke Hillebrecht, William Baker, William Milliken, and Yeina Milena Niño-Fernández. The project also thanks the local communities and authorities in the pilot areas, as well as all the stakeholders involved.

References

Andrade CMG (2011) Estado del conocimiento de la biodiversidad en Colombia y sus amenazas: consideraciones para fortalecer la interacción Ciencia-Política. *Rev. Acad. Colomb. Cienc. Exactas Fis. Nat.*, 35 (137): 491-507.

Baptiste B, Pinedo-Vasquez M, Gutierrez-Velez VH, Andrade GI, Vieira P, Estupiñán-Suárez LM, Londoño MC, Laurance W, Lee TM (2017) Greening peace in Colombia. *Nat. Ecol. Evol.*, 1: 102. doi: 10.1038/s41559-017-0102

Bernal HY, Quevedo Sánche GF, García Martínez H (2011) *Pautas para el conocimiento, conservación y uso sostenible de las plantas medicinales nativas en Colombia: Estrategia nacional para la conservación de plantas.* Bogotá, Instituto de Investigación de Recursos Biológicos Alexander von Humboldt. 232p.

Bernal R, Galeano G, Sarmiento H, Gutiérrez M (2012) Nombres Comunes de las Plantas de Colombia. Available at: http://www.biovirtual.unal.edu.co/nombrescomunes/

Bernal R, Gradstein S, Celis M (eds) (2019) *Catálogo de plantas y líquenes de Colombia*. Instituto de Ciencias Naturales, Universidad Nacional de Colombia. Available at: http://catalogoplantasdecolombia.unal.edu.co/

Bystriakova N, Tovar C, Monro A, Moat J, Hendrigo P, Carretero J, Torres-Morales G, Diazgranados M (2021) Colombia's bioregions as a source of useful plants. *PLoS ONE* 16(8): e0256457. doi: 10.1371/journal.pone.0256457

Cámara-Leret R, Paniagua-Zambrana N, Balslev H, Macía MJ (2014) Ethnobotanical knowledge is vastly under-documented in northwestern South America. *PLoS ONE,* 9(1): e85794. doi: 10.1371/journal.pone.0085794

CONPES (2018) Política de Crecimiento Verde. Consejo Nacional de Política Económica y Social (Documento CONPES 3934). Available at: https://colaboracion.dnp.gov.co/CDT/Conpes/Económicos/3934.pdf.

de Almeida RF, Luecking R, Vasco-Palacios AM, Gaya E & Diazgranados M (2022) *Catalogue of Fungi of Colombia*. Royal Botanic Gardens, Kew. Kew Publishing, London. 512p.

Diazgranados M, Ávila F, Álvarez M, Ambrose E, Baines D, Barker A, Bishop D, Carretero J, Clamp N, Cook F, Crowe J, Dennehy Z, Di Sacco A, Echeverri H, Fabriani F, Forest F, Granados D, Graves E, Green L, Iacona J, López I, Marrugo S, Mattana E, Monro A, Moreno D, Murray L, Nesbitt M, Nic Lughadha E, Pérez-Martínez L, Phillips S, Pinzón L, Piquer-Doblas M, Reis P, Roberts A, Rodríguez Á, Romero R, Sánchez N, Smith R, Suárez C, Torres G, Turner R, Way M, Ulian T (2019) *ColPlantA: Colombian resources for Plants made Accessible. ColPlantA: Recursos Colombianos de Plantas hechos Accesibles.* London, Royal Botanic Gardens, Kew28p. doi: 10.34885/81

Diazgranados M, Allkin B, Ávila F, Baker W, Bishop D, Black N, Bystriakova N, Carretero J, Castellanos-Castro C, Cely M, Colville L, Cossu T, Davies L, Díaz A, Dsouza J, García F, Gaya E, Graves E, Green L, Haigh A, Hammond DS, Hendrigo P, Herrera A, Hillebrecht W, Howes M-J, Iggulden D, Kersey P, Kor L, Mattana E, Milliken W, Mira M, Moat J, Monro A, Morley J, Murphy V, Nesbitt M, Ondo I, Parker L, Phillips S, Piquer M, Pritchard HW, Reis P, Rojas T, Ruff J, Stone M, Tanimoto M, Torres G, Tovar C, Turner R, Taylor R, Utteridge T, Vargas N, Weech M-H, White K, Wilkin P, Williams J, Ulian T (2020) *ColPlantA: Colombian Resources for Plants made Accessible. 2nd edition,* London, Royal Botanic Gardens, Kew. 33p.

Diazgranados M, Hammond D, Rojas T, Castellanos-Castro C, Díaz A, Hernández M, Quinoñez C, Kor L, Ulian T (2021) Selection of pilot areas in Colombia for the implementation of a value chain network on useful plants and fungi. *UPFC Report Series.* London, Royal Botanic Gardens, Kew & Bogotá, Instituto de Investigaciones en Recursos Biológicos Alexander von Humboldt.

Eufemia L, Bonatti M, Castro-Nunez A, Lana M, Morales H, Sieber S (2019) Colombia's inadequate environmental goals. *Science,* 364: 444-445.

FAO (2015) *Global Forest Resources Assessment 2015. Desk reference.* Food and Agriculture Organization of the United Nations (FAO). Available at: http://www.fao.org/3/i4808e/i4808e.pdf

Gaya E, Vasco-Palacios AM, Vargas-Estupiñán N, Lücking R, Carretero J, Sanjuan T, Moncada BBA, Bolaños-Rojas AC, Castellanos-Castro C, Coca LF, Corrales A, Cossu T, Davis L, Dsouza J, Dufat A, Franco-Molano AE, García F, Gómez-Montoya N, González-Cuellar FE, Hammond D, Herrera A, Jaramillo-Ciro MM, Lasso-Benavides C, Mira MP, Morley J, Motato-Vásquez V, Niño-Fernández Y, Ortiz-Moreno ML, Peña-Cañón ER, Ramírez-Castrillón M, Rojas T, Ruff J, Simijaca D, Sipman HJM, Soto-Medina E, Torres G, Torres-Andrade PA, Ulian T, White K, Diazgranados M (2021) *ColFungi: Colombian Resources for Fungi Made Accessible.* London, Royal Botanic Gardens, Kew. 37p.

IPBES 2018: IPBES (2018) *Summary for Policymakers of the IPBES Assessment Report on Land Degradation and Restoration.* Intergovernmental Science-Policy Platform on Biodiversity and Ecosystem Services (IPBES). Available at: https://www.ipbes.dk/udgivelser-og-viden/ipbes-udgivelser/

Khoury CK, Achicanoy HA, Bjorkman AD, Navarro-Racines C, Guarino L, Flores-Palacios X, Engels JMM, Wiersema JH, Dempewolf H, Sotelo S, Ramírez-Villegas J, Castañeda-Álvarez NP, Fowler C, Jarvis A, Rieseberg LH, Struik PC (2016) Origins of food crops connect countries worldwide. *P Roy Soc B-Biol Sci,* 283: 20160792.

Kor L, Homewood K, Dawson TP, Diazgranados M (2021) Sustainability of wild plant use in the Andean Community of South America. *Ambio*, 50: 1681-1697. doi: 10.1007/s13280-021-01529-7

Millennium Ecosystem Assessment (2005) Ecosystems and human well-being. Washington, DC. 156p. https://www.millenniumassessment.org/documents/document.356.aspx.pdf

Quiñones-Hoyos C, Rengifo-Fernández A, Bernal-Galeano S, Peña R, Fernández M, Rojas T, Diazgranados M (2021) *Una mirada a las plantas y los hongos útiles en tres áreas biodiversas de Colombia*. London, Royal Botanic Gardens, Kew & Bogotá, Instituto de Investigaciones en Recursos Biológicos Alexander von Humboldt.

Rojas MT, Cortés C, Noguera M, Acosta P, Diazgranados M (2021) *Guía práctica para potencializar el uso sostenible de los ingredientes naturales en Colombia*. London, Royal Botanic Gardens, Kew & Bogotá, Instituto de Investigaciones en Recursos Biológicos Alexander von Humboldt. 170p.

Rojas T, Cortés C, Noguera M, Ulian T, Diazgranados M (2020) *Evaluación del estado de los desarrollos bioeconomicos colombianos en plantas y hongos*. London, Royal Botanic Gardens, Kew & Bogotá, Instituto de Investigación de Recursos Biológicos Alexander von Humboldt.123p.

Sabater S, González-Trujillo JD, Elosegi A, Rondón JCD (2017) Colombian ecosystems at the crossroad after the new peace deal. *Biodivers. Conserv.*, 26: 3505–3507 doi: 10.1007/s10531-017-1415-8

Salazar A, Salazar JF, Sánchez-Pacheco SJ, Sanchez A, Lasso E, Villegas JC, Arias PA, Poveda G, Rendón ÁM, Uribe MR, Pérez JC, Dukes JS (2021) Undermining Colombia's peace and environment. *Science,* 373: 289-290.

United Nations (2015). *Transforming our World: the 2030 Agenda for Sustainable Development,* Document A/70/L.1. UNGA, New York.

WWF-Colombia (2017) *Colombia Viva: un país megadiverso de cara al futuro. Informe 2017.* Cali, WWF-Colombia. 166p.

Field expedition with community participants in the Municipality of Otanche.
Laura Kor

Chapter 2

Assessing extinction risk coverage and identifying geographical hotspots to guide conservation action for Colombia's useful plants

Laura Kor[1,2], Cristina Lopez-Gallego[3], Carolina Castellanos-Castro[4] & Mauricio Diazgranados[1]*

[1] Research Department, Royal Botanic Gardens, Kew, Richmond, TW9 3AE, UK
[2] Department of Geography, Bush House NE, King's College London, London, WC2B 4BG, UK
[3] Instituto de Biología, Universidad de Antioquia, Medellín, Colombia
[4] Instituto de Investigación en Recursos Biológicos Alexander von Humboldt, Bogotá, Colombia

***Corresponding author:** m.diazgranados@kew.org

Keywords: Conservation planning; Conservation prioritisation; Important Plant Areas; Red List; *in-situ* conservation

ABSTRACT

Colombia is one of the most biologically and culturally diverse countries in the world, supporting over 7,000 plant species with documented human uses. However, large gaps remain in understanding the country's biodiversity and its role in human welfare. Simultaneously, global conservation strategies increasingly highlight the importance of the protection and sustainable use of socially, economically, and culturally valuable species. Drawing on the Checklist of Useful Plants of Colombia (CUPC), this chapter proposes conservation actions based on: (1) taxonomic gaps in extinction risk assessments and (2) geographic hotspots for native useful plants of Colombia (UPC). We identified 544 species of conservation concern – threatened with extinction and/or endemic to Colombia. However, 45% of native UPC lack extinction risk assessments. The proportion of unassessed species was significantly higher in 13 of the 256 useful plant families. We propose that these families should be prioritised for extinction risk assessments. There were also variations between use categories, with 73% of fuel species assessed compared to 42% of species with environmental uses. Existing records of native UPC were distributed across all major ecosystems in Colombia. Applying three definitions of hotspots and using a regionalised approach that combines habitat and municipality boundaries, we identified 78 units of analysis (89,901 km^2) as hotspots for native UPC. *In-situ* conservation gaps exist, with 48 hotspots having less than 25% of their land within Colombia's current protected area or Indigenous and Community Conserved Area systems. We discuss how these hotspots can contribute to the identification of Important Plant Areas in Colombia. While prioritisation is required to target limited conservation planning and resources efficiently, stakeholder engagement will be crucial to develop and implement conservation action and sustainable use for the benefit of people and nature.

RESUMEN

Colombia es uno de los países del mundo con mayor diversidad biológica y cultural, al albergar más de 7.000 especies de plantas con usos documentados. Sin embargo, siguen existiendo grandes vacíos en la comprensión de la diversidad biológica del país y su papel en el bienestar humano. Al mismo tiempo, las estrategias de conservación global de la biodiversidad destacan cada vez más la importancia de la protección y el uso sostenible de especies de valor social, económico y cultural. A partir de la Lista de Verificación de Plantas Útiles de Colombia (CUPC), este capítulo propone acciones de conservación basadas en (1) vacíos taxonómicos en las evaluaciones de riesgo de extinción y (2) puntos geográficos críticos para las plantas nativas útiles de Colombia (UPC, por su sigla en inglés). Identificamos 544 especies de interés para la conservación – amenazadas de extinción o endémicas de Colombia. Sin embargo, el 45% de las UPC nativas carecen de evaluaciones de riesgo de extinción, con una proporción de especies no evaluadas significativamente mayor en 13 de las 256 familias de plantas útiles. De esta manera se propone priorizar estas familias para futuras evaluaciones de riesgo de extinción. También se encontraron variaciones en este aspecto entre las categorías de uso, con el 73% de las especies usadas para combustible evaluadas comparado con el 42% de las especies con usos ambientales. Los registros existentes de las UPC se distribuyeron a largo de los principales ecosistemas de Colombia. Aplicando tres definiciones de puntos críticos y utilizando un enfoque regionalizado que combina hábitats y límites de municipios, identificamos 78 unidades de análisis (89.901 km^2) como puntos críticos para las UPC nativas. Se identificaron vacíos en términos de áreas gestionadas para la conservación *in-situ*, con 48 regiones críticas que tienen menos del 25% de su área terrestre dentro de un área protegida actual de Colombia o dentro de territorios colectivos indígenas o de afrodescendientes. Discutimos cómo estos puntos críticos pueden contribuir a la identificación de Áreas de Importancia para la Conservación de las Plantas (IPAs por sus siglas en inglés) en Colombia. Si bien se requiere priorización para orientar de manera eficiente la ejecución de los limitados recursos para conservación, la participación de las partes interesadas será crucial para desarrollar e implementar acciones de conservación y uso sostenible para las personas y la naturaleza.

INTRODUCTION

Plants are the primary producers in almost all ecosystems, providing the structure and resources needed for life on earth. Additionally, more than 40,000 plant species have a documented human use, from food and fuels to plants with ritual or religious significance (Diazgranados *et al.*, 2020). Despite their importance, two in five plant species globally are threatened with extinction, driven by habitat conversion, overexploitation and climate change (Nic Lughadha *et al.*, 2020). Effective plant conservation is therefore urgently needed for the wellbeing of both people and biodiversity.

Colombia is one of the most biodiverse countries on earth (Corzo *et al.*, 2010). The region is also rich in cultural diversity, with thousands of plant uses documented across Colombia and the wider Andean region (Bernal *et al.*, 2011, Paniagua-Zambrana & Bussmann., 2020, Kor *et al.*, 2021). Following decades of internal conflict, the 2016 Peace Agreement in Colombia has provided an opportunity to improve biodiversity knowledge and simultaneously has generated socio-economic changes, with impacts on natural resources (Bridge Colombia, 2017). This combination of factors makes Colombia a particularly pertinent country for socio-ecological approaches to plant conservation. The identification of conservation priorities is a crucial step in this process to enable targeted conservation planning and resource allocation.

This chapter summarises current plant conservation efforts before discussing and evaluating useful plant conservation in Colombia. We aim to propose conservation priorities based on useful plant species by highlighting gaps in extinction risk assessments and identifying geographic hotspots. We draw on the Checklist of Useful Plants of Colombia (CUPC; Chapter 12) to ask: (1) which useful plant species are known to be at risk of extinction? (2) what are the gaps in extinction risk assessments? (3) where are the hotspots for useful plant species? and (4) how well are they represented in national protected areas and Indigenous and Community Conserved Areas (ICCA)?

International framework for plant conservation

The Convention on Biological Diversity (CBD) is a major international agreement that guides plant conservation efforts. Plants have their own taxonomically focused strategy that was adopted in 2002, the Global Strategy for Plant Conservation (GSPC), which aims to "halt the continuing loss of plant diversity" (CBD, 2012). This strategy was updated in 2010 in line with the Aichi Biodiversity Targets, and the Post-2020 Global Biodiversity Framework is currently being developed (CBD, 2021). While all plant conservation efforts can benefit useful plants, more targets in the Draft Post-2020 GSPC directly highlight socio-economically important species and sustainable use than ever before

TABLE 1. Preliminary 2030 targets of the Global Strategy for Plant Conservation draft that have direct relevance to useful plant species (values denoted as "*a significant percentage*" are still to be confirmed at the time of writing).

Target	Detailed aims
4	By 2030, there has been at least a [50%] reduction in the number of plant species threatened by international trade and by unsustainable levels of harvesting.
9	9a: By 2030, *a significant percentage* (to be defined) of areas under agriculture, aquaculture and forestry are managed sustainably, ensuring the conservation of associated wild and crop plant diversity 9b: By 2030, *a significant percentage* of crop varieties, landraces, forest genetic resources, crop wild relatives (CWR) and other domesticated socio-economically and culturally valuable plant species are conserved *ex-situ*, and viable populations are effectively managed *in-situ* to prevent genetic erosion and safeguard their genetic diversity.
12	By 2030, *a significant percentage of* countries are benefitting from the exchange of plant materials and associated traditional knowledge to support plant conservation, ecological restoration, and sustainable use.
13	By 2030, at the latest, plant diversity values have been integrated into rural and urban development and poverty reduction strategies and planning processes and have been implemented in natural capital and other national accounting mechanisms and reporting systems worldwide.
19	19d: By 2030, the value of plant diversity and responsibility for its protection is universally recognised by the world's people, including the ecosystem services they provide and the steps that can be taken to conserve and use plants sustainably.
20	By 2030, with the full and effective participation of indigenous and local communities, at all relevant levels, the traditional knowledge, innovations, and practices of indigenous and local communities relevant for the conservation and sustainable use of plant diversity are respected, safeguarded, and preserved to support customary and cultural use of these resources.

(Table 1). These conventions and strategies demonstrate international recognition of the importance of the direct use of plants and the need to recognise and conserve local and indigenous knowledge for biodiversity conservation and sustainable development.

An additional international treaty that contributes to the conservation of plants is the Convention on International Trade in Endangered Species of Wild Fauna and Flora (CITES). CITES was established to regulate the international trade in specimens of wild animals and plants and currently includes over 32,800 plant species, most of which are of socio-economic importance.

Important Plant Areas

With global conservation efforts commonly focused on area-based protection, Targets 2 and 19b of the Post-2020 GSPC are also relevant to useful plant conservation. They aim for the identification of "the most important areas for plant diversity [to be] identified" and "adequately protected for plant conservation". Traditional approaches to protected areas often involved fortress-conservation strategies, which restricted access to ecosystem services and ignored the role of local and indigenous communities (Palomo *et al.*, 2014). The Important Plant Area (IPA) programme, established to support GSPC targets, does not confer national legal protection and can instead be used as a tool to support national and regional strategic planning and to drive community-led sustainable management (Darbyshire *et al.*, 2017). This tool is crucial as we increasingly recognise that for scientific findings to translate into sustainable use, there needs to be an understanding and integration of the social, ethical, economic, political, and religious factors which influence conservation outcomes (Cunningham, 2001, Kor *et al.*, 2021).

IPA identification is based on three globally consistent criteria, first published by Anderson (2002). The criteria were updated in 2017 to include socially, economically, or culturally valuable species – or useful plant species (Darbyshire *et al.*, 2017). However, the application of IPA criteria to useful plants is limited, with a notable exception being IPAs for medicinal plants in the Himalayas (Hamilton *et al.*, 2007).

PLANT CONSERVATION IN COLOMBIA

As a party to the CBD, Colombia has developed policies and strategies for biodiversity conservation, including a National Biodiversity Strategy and a NSPC – National Strategy for Plant Conservation (or ENCP, *Estrategia Nacional para la Conservación de Plantas* in Spanish). The latter has driven the country's plant conservation policy and action, including the publication of a Catalogue of Plants and Lichens (Bernal *et al.*, 2020) and a series of Red Books of Plants in Colombia (Garcia *et al.*, 2007). Risk assessments for species are ongoing and follow the methodologies of the International Union for Conservation of Nature's (IUCN) Red List of Threatened species (hereafter, IUCN Red List). The implementation of Colombia's NSPC was reviewed by the Humboldt Institute and the National Network of Botanical Gardens in 2010 (García Martínez *et al.*, 2010), with the publication of an Action Plan in 2017 (Castellanos-Castro *et al.*, 2017). Aligned to the NSPC, a diagnosis and guidelines for the conservation of medicinal plants have been developed (Bernal *et al.*, 2011), and conservation action plans for taxonomic groups of charismatic and socio-economic importance, such as orchids, palms, cycads, and some timber trees, have been published.

Other conservation measures in Colombia include a National System of Protected Areas (SINAP), with public national protected areas consisting of 122 sites within nine categories (UNEP-WCMC, 2020). Colombia also has a large national system of indigenous and afro-descendant reserves with legally recognised collective land tenure and co-management strategies (MADS, 2017), plus much smaller public and private protected areas within regional and local systems. Some international conservation areas have been designated, including Important Bird Areas (IBAs) and RAMSAR sites. However, IPAs (or AIPs, Áreas de Importancia para la Conservación de Plantas in Spanish) are yet to be identified at the national level.

A methodology for IPA identification in Colombia has been proposed (Diazgranados & Castellanos-Castro 2021). This methodology draws on global IPA criteria, with adjustments to account for the country's rich diversity and the lack of reliable species distribution data. Identifying IPAs in Colombia may help promote the use of plant and fungi information in biodiversity conservation planning, which has traditionally focused on vertebrate species (Corlett, 2016). Additionally, useful plants can drive IPA identification based on the presence of threatened and/or range-restricted useful plant species under criterion A or sites holding concentrations of useful species under criterion B (Darbyshire *et al.*, 2017). Such an approach would support the NSPC Action Plan, which recognises that continued efforts are required to meet targets on sustainable use.

Conservation priority setting

The global and national level structures described above can help to direct overarching conservation approaches. However, prioritisation is also important to ensure that resources and funds for conservation action are targeted effectively to prevent biodiversity loss (Wilson *et al.* 2009, Arponen, 2012). This prioritisation can be undertaken at different biological and geographical scales and can apply a range of methods.

At the species level, conservation prioritisation has considered a range of factors such as evolutionary uniqueness and economic and cultural importance (Brehm *et al.*, 2010, Arponen, 2012). Extinction risk assessments have also become one of the most often applied criteria for prioritisation, based on the urgency of action to avoid the loss of a species. The IUCN Red List of Threatened Species (IUCN, 2021) is the most comprehensive and recognised source on extinction risk, with replicated protocols to produce red lists at the national level. However, its limitations include taxonomic and geographic bias in the species assessed, with many gaps persisting in plant assessments (Nic Lughadha *et al.*, 2020).

Spatial approaches are also widely used in conservation priority setting to identify protected or other conservation areas, as well as biodiversity hotspots (Wilson *et al.*, 2009). The principles of Systematic Conservation Planning (SCP) are often applied (Margules & Sarkar 2007). However, despite the gaps in extinction risk assessments, spatial prioritisation is often based on the presence of threatened species listed as Vulnerable (VU), Endangered (EN) or Critically Endangered (CR) on the IUCN and other red lists.

MATERIALS AND METHODS

Useful plants of conservation concern

We compiled a preliminary list of useful plant species of conservation concern in Colombia based on their current extinction risk and geographical distribution. Uses were based on the Checklist of Useful Plants of Colombia (CUPC), produced as part of the Useful Plants and Fungi of Colombia project (UPFC, 2020), with nomenclatural reconciliation of plant names undertaken to match extinction risk assessments with the CUPC (Chapter 12). The CUPC was compiled from relevant databases, reports, and scientific literature, identifying 7,472 species within 2,140 genera and 258 families (Figure 1). Uses were grouped within the ten categories applied in the World Checklist of Useful Plant Species (Diazgranados *et al.*, 2020), based on the 'Level 1 States' of the *Economic Botany Data Collection Standard* (EBDCS) (Cook, 1995).

A stepwise process was used to identify species of conservation concern (Figure 1). First, only species that are native to Colombia were considered. Second, native species were only included if known to be: (i) internationally threatened; (ii) nationally threatened; or (iii) endemic. Native and endemic species were identified using the Colombian Catalogue of Plants and Lichens (CCPL) (Bernal *et al.*, 2020) and Plants of the World Online (POWO, 2019). International and national conservation status was based on the IUCN Red List (IUCN, 2021) and on a list of national extinction risk assessments carried out by the IUCN Colombian Plant Specialist Group, respectively. We considered CR, EN or VU species as threatened. Additionally, as this study focuses on socially, economically, or culturally valuable species, IUCN Extinct (EX) or Extinct in the wild (EW) species were also included.

Extinction risk assessment gaps

Extinction risk assessments have not been undertaken for many plant species, representing gaps in our knowledge of their risk of extinction (Nic Lughadha *et al.*, 2020). We calculated the proportion of Colombian useful plant species assessed at the international (IUCN Red List) or national level within each plant family and use category. We then used the Chi-square test and post-hoc analyses applying the standardised residual method (Beasley and Schumacker, 1995) to identify statistical variations in the proportion of species assessed in each family, thereby enabling families with the largest species gaps in extinction risk assessments to be identified. Only families with 15 or more reported useful plant species were included in the analyses. Equivalent statistical analyses between use categories were not undertaken as species can fall within multiple categories.

Distribution data for useful plants

Existing public distribution data were used to identify hotspots for all known native useful plants of Colombia. All 5,830 native useful plants were queried against the Global Biodiversity Information Facility (GBIF) names backbone, and occurrence records linked to matched names and synonyms were downloaded from GBIF on 15 July 2021 (GBIF Occurrence Download, 2021). Data were cleaned to remove records with coordinates which were: outside Colombia; duplicated; assigned to country centroids or institutions; non-terrestrial; rounded to less than one decimal place; with equal or zero longitude and latitude; and within 1 km of the capital. This data cleaning resulted in 366,936 records for 5,241 native useful species (90.1% of total species).

Identifying hotspots for useful plants

To identify hotspots for useful plants, we defined our units of analysis (UA) based on ecological and political criteria. These were determined by splitting the 30 synthesised ecosystems defined by the *Mapa de Ecosistemas Continentales Marinos y Costeros de Colombia* (MADS *et al.*, 2017) at the borders of Colombia's 1,122 municipalities, resulting in 8,930 polygons. Occurrence records for native useful plants of Colombia (UPC) were joined to this multipolygon layer to enable analyses. This *a priori* regionalisation approach follows the proposed IPA methodology for Colombia (Diazgranados & Castellanos-Castro 2021). While the alternative grid cell approach has been applied to IPA identification in other countries and has certain advantages (Hamidah *et al.*, 2020), it makes it difficult to ensure comprehensiveness of ecosystem coverage – a key concept in Systematic Conservation Planning (Margules & Sarkar, 2007).

We applied three methods for hotspot identification, with UAs falling under one or more of these definitions highlighted as hotspots for UPC. First, 'national 5%' hotspots are the UAs with the highest species richness and account for 5% of the total land area of Colombia. This method involved calculating the number of native UPCs recorded in each UA, sorting the UAs in decreasing order of richness, then selecting the UAs with the highest species richness until 5% of Colombia's terrestrial area (c. 56,825 km^2) was reached. This threshold is based on consistent findings that the richest 1–5% of the land area can represent a substantial proportion of species (Orme *et al.*, 2005) and reflects the approach taken by Chi *et al.* (2017) to identify conservation priorities for medicinal plants in China.

The next two methods took an ecosystem-level approach, further enabling a more comprehensive representation of the diversity of habitats in Colombia. 'Ecosystem richest' hotspots are defined as the UAs that support the highest species richness of each ecosystem type. This method is based on criterion B1 for areas of exceptional botanical richness in the methodology for IPA identification in Colombia, with one UA per ecosystem type identified as a hotspot (Diazgranados & Castellanos-Castro 2021). Finally, 'ecosystem 5%' hotspots are the UAs with the highest species richness and account for 5% of the total land area of each ecosystem type. This method applied the 'national 5%' methodology at the ecosystem level, thereby combining the ecosystem-specific approach recommended by the IPA methodology with a threshold previously applied for hotspot identification at the national level in other countries (Chi *et al.*, 2017). For the ecosystem-dependent analyses, all aquatic ecosystem types were combined, and ecosystems classified as 'other areas' and 'no information' were not included, resulting in a total of 19 ecosystem types (Figure 3).

Spatial conservation gaps for useful plants

To detect *in-situu* conservation gaps, all identified hotspots for UPC were overlaid with the distribution of protected areas and ICCA in Colombia. Hotspots were categorised depending on percentage overlap with these area-based designations: (1) high conservation gap (≤25% overlap); medium conservation gap (26–75% overlap); and low conservation gap (>75% overlap). Protected area data were downloaded from the World Database on Protected Areas (UNEP-WCMC, 2020) and include areas designated at the national, regional, and local levels. ICCA areas were downloaded from Agencia Nacional de Tierras (2017).

Data reconciliation, preparation, and analyses were primarily conducted in R v.3.4.3 (R Core Team, 2020), using the packages *dplyr* v.1.0.6 (Wickham *et al.*, 2021), *rgbif* v3.5.2 (Chamberlain *et al.*, 2021), *rgdal* v1.5–23 (Bivand *et al.*, 2021), and *CoordinateCleaner* (Zizka *et al.*, 2019). Geographic analyses were conducted using a combination of R and ArcGIS Pro 2.8.0 (Esri Inc., 2021).

RESULTS

Useful plants of conservation concern and risk assessment gaps

Native useful plant species in Colombia occurred across 232 families, with the most species-rich families being Fabaceae (530), Asteraceae (300), and Rubiaceae (257). We identified 544 useful plant species of conservation concern—as defined in Section 3.2.1—within 111 families. Of these, 294 species were identified as internationally or nationally threatened, while 250 unassessed species were included as endemics (Figure 1). As only 55% of UPC have

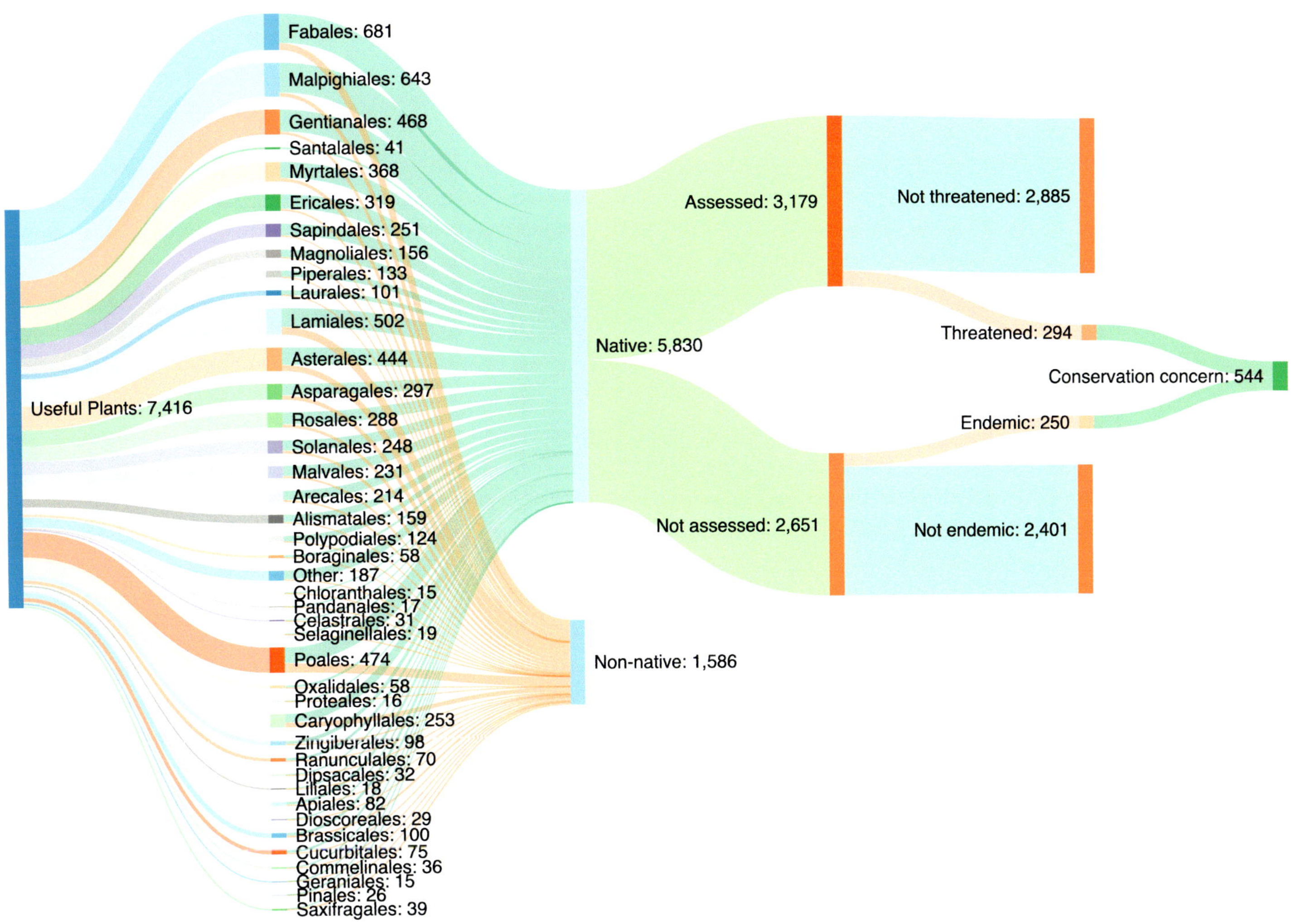

FIGURE 1. Breakdown of orders of useful plants of Colombia and the stepwise process used to determine the species of conservation concern.

FIGURE 2. Proportion of useful plant species threatened at national and/or global levels or lacking extinction risk assessments across **A** families and **B** use categories. Numbers show total species per group, with some species in multiple-use groups. In **A**, only families that significantly contributed to the chi-square statistic (χ^2), which tested differences in the proportion of species assessed, are shown, with remaining families grouped in 'other'; * indicates families that are significantly under represented in Red Lists.

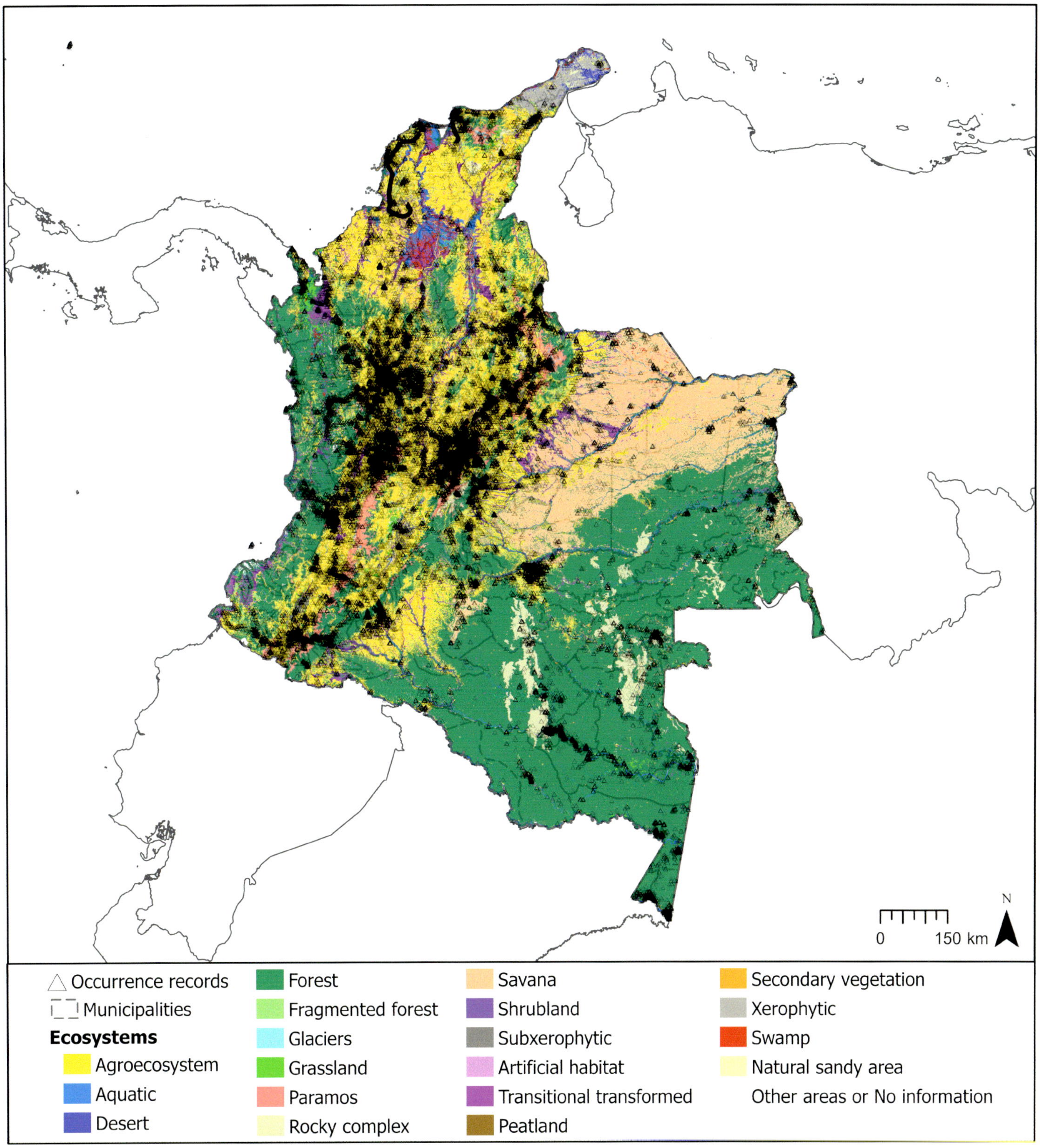

FIGURE 3. Ecosystem map of Colombia (MADS *et al.*, 2017) and distribution of native useful plants of Colombia, based on occurrence points downloaded from GBIF (GBIF Occurrence Download, 2021).

TABLE 2. Hotspots for useful plant species of Colombia by designation type and percentage overlap with existing protected areas (PA) or Indigenous and Community Conserved Area (ICCA).

Hotspot type	Number of UAs	Total area (km²)	Native UPC recorded	PA overlap (%)	ICCA overlap (%)	PA or IR overlap (%)
National 5%	3	52,384	1,644	44.3	26.3	70.5
Ecosystem 5%	76	77,593	4,039	44.4	16.5	56.6
Ecosystem richest	19	60,074	2,894	49.1	20.1	69.1
All hotspots	78	89,901	4,134	36.5	24.2	60.4

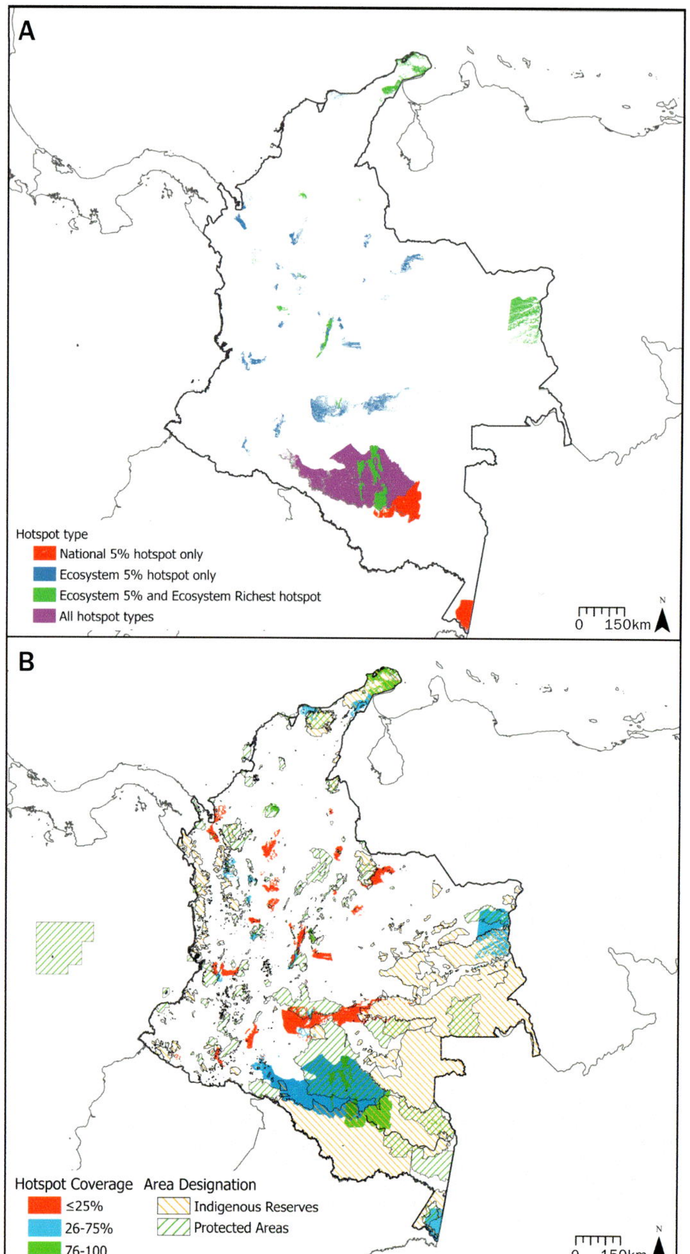

FIGURE 4. A Units of analysis identified as hotspots for useful plant species in Colombia, and the hotspot definitions they fall under. **B** Percentage overlap of hotspots with protected areas or Indigenous and Community Conserved Areas (ICCAs).

been assessed, the list of taxa under conservation concern is preliminary and could be extended in the future. A total of 151 families supported fewer than 15 native UPC and were not included in the following analyses.

The proportion of useful plant species with extinction risk assessments at the national or international level was unevenly distributed across families ($\chi^2 = 1213$, df = 94, $p < 0.01$). On average, 48.7% of species were assessed per family, with 28 families significantly contributing to the unevenness of coverage in risk assessments (χ^2 post hoc analysis based on standardised residuals $p < 0.01$). Thirteen of those had a significantly smaller proportion of species with extinction risk assessments than other families, while the rest had a significantly higher proportion of species assessed (Figure 2A). The 13 families highlighted in Figure 2 with a lower proportion of assessments can be used as a basis to prioritise UPC for further extinction risk analysis.

Variation was also seen between the Level 1 use categories. Species with environmental and medicinal uses were under represented in extinction risk assessments, with 42% and 47% assessed, respectively. This scenario was in contrast to species in the fuels, invertebrate food, and materials categories, which had levels of assessment that were higher than the average (73%, 62% and 61% assessed, respectively). As many species occur in multiple-use categories, variables are not mutually exclusive, with the Chi-square test of significance not applied.

Geographic hotspots and conservation gaps

Records of native useful plants of Colombia were distributed across 967 municipalities (Figure 3), with Solano (Caquetá department) found to support the greatest diversity. Hotspots were distributed across all ecosystem types, with agroecosystems (4,523 species) and forest habitats supporting the most species (4,405).

In total, 78 UAs were identified as hotspots for UPC (Figure 4a), classified under one or more of the three definitions applied, and covering a total area of 89,901 km² across 57 municipalities. Hotspots had a mean size of

1,152.6 km^2, ranging from 2.5 km^2 (swamp habitat in Santa Fé de Bogotá municipality, Cundinamarca) to 40,076.5 km^2 (forest habitat in Solano municipality). The identified hotspots are known to support 4,134 native UPC (70.9% of total) while covering 7.9% of the country's terrestrial land area. Sixty-eight of the hotspots include records for UPC that are of conservation concern (threatened and/or endemic; Figure 1), including a total of 353 species (64.9% of all UPC of conservation concern; Figure 1). This includes 67 species listed as VU on the IUCN Red List, 41 listed as EN, 11 species assessed as CR, four species categorised as EW and EX (two species in each category), and 226 endemic species to Colombia. However, only 36.5% of the total hotspot area is currently included in the protected areas system of Colombia, with 24.2% represented in ICCAs (Table 2).

Hotspots were ranked for their conservation priority based on the percentage gap in coverage by existing protected areas or ICCAs (Figure 4B). Forty-seven of the hotspots (16,775 km^2) overlap with existing designated areas by less than 25% and are therefore considered as high conservation priority. These hotspots include ten different ecosystem types distributed across 37 municipalities. Thirteen hotspots (17,870 km^2) overlap with designated areas by more than 75%, while the remaining 18 hotspots (55,256 km^2) are considered as medium priority.

DISCUSSION

Despite plant conservation efforts, studies indicate that targets for useful wild plant species have not been met, and overexploitation continues to be a major driver of extinction (Khoury *et al.*, 2019). Based on the Checklist of Useful Plants of Colombia, we present priorities to guide two key actions to contribute to conservation in the country: extinction risk assessments at the species level and area-based management at the ecosystem level. To inform this process, we compiled a list of useful plant species of known conservation concern in Colombia, highlighted families and use groups that are under-represented in extinction risk assessments and identified geographic hotspots and area-based conservation gaps for useful plant species in Colombia. These are important first steps in the long-term conservation and sustainable use of plants in one of the most biodiverse countries in the world (UN-WCMC, 2014).

Useful plants of conservation concern and their risk assessments

We identified 544 UPC that are known to be endemic and/or threatened at national or global levels. However, with 45% of native useful plant species not represented in either the Colombian or IUCN Red Lists (Figure 1), many more UPC may be at risk of extinction. This level of extinction risk assessment is greater than the average coverage of global assessments across plant species (roughly 30%), reflecting findings that useful plants are over represented in red listing (Nic Lughadha *et al.*, 2020). However, it is important to fill the remaining gaps, as extinction risk assessments often form the basis of conservation policy, planning, and action. Gaps in assessments may therefore lead to key species being ignored in conservation efforts, including spatial conservation prioritisation.

Undertaking risk assessments is resource-intensive, requiring a high level of specialist knowledge and robust data. To target limited conservation resources efficiently, our analyses highlighted 13 families in which significantly fewer useful plant species have been assessed (Figure 2). This count includes some of the most species-rich groups both in Colombia and globally, such as Asteraceae and Poaceae. Meanwhile, at the level of use categories, species with medicinal and environmental uses were found to be the most under represented in extinction risk assessments – two of the top three most species-rich categories in the country. We propose that these families and use groups should be considered as priorities for ongoing extinction risk assessments in Colombia with a view to informing subsequent conservation prioritisation.

While red listing is an important first step for prioritisation, extensive further work across disciplines is required to achieve sustainable use and management at species- and location-specific levels. This includes developing knowledge of plant biology and harvesting practices; understanding relevant economics and market pressures; recognising and supporting land rights holders and enabling management which facilitates responsible use; building capacity and providing resources for sustainable harvesting practices; and developing institutional structures and policies that support such practices (Kor *et al.*, 2021).

Geographic hotspots and conservation gaps for useful plants

In total, 78 hotspot UAs for useful plants of Colombia were highlighted, falling under any one of three hotspot definitions applied. These areas should be prioritised for conservation measures that incorporate sustainable plant use and provide a first step towards identifying a national IPA network for socially, economically, and culturally valuable species.

By focusing on useful species, this study supports national priorities to integrate biodiversity conservation into Colombia's post-conflict development and the NSPC Action Plan, whose main objectives include sustainable use and protecting useful plant species (Castellanos-Castro *et al.*, 2017). Additionally, it supports global conservation efforts (CBD, 2021). A recent assessment of the GSPC's progress highlighted that while there have been many successes in global plant conservation, the targets where the least progress have been made include *in-situ* conservation and sustainable use (Sharrock, 2020).

***In-situ* conservation gaps**

Our results showed that UPC are not well represented in Colombia's current protected area system, with just 36.5% of the total hotspot area covered by local, regional, or national PAs in the country (UNEP-WCMC, 2020). Adding ICCAs to this analysis lessoned the conservation gap, but 48 of the 78 hotspot UAs remained as high priorities, with less than 25% of their land area occurring within any designated areas. While indigenous reserves and afro-descendant territories are primarily designated for recognising and protecting ethnic and biocultural diversity, they present great relevance to biodiversity conservation, with many occurring in ecosystems of rich biodiversity and 62% of land use associated with forest protection (Arango, 2017).

Conservation areas focusing on useful plants remain scarce globally. However, their potential has been shown in Colombia, with the *Santuario de Flora Plantas Medicinales Orito Ingi-Ande* national park designated as the first and only protected area for medicinal plants in the country (IUCN, 2008). It is located in territories used by the Kofanes community in southwestern Colombia, who first proposed the concept. Species characterisation was undertaken by them in collaboration with a wide range of stakeholders over several years. The hotspots presented in this chapter are therefore only the first step in the process of conservation designation, and designation is by no means a guarantee of sustainable plant use, with a multitude of social, political, and economic, as well as further ecological considerations, required. Additionally, data challenges arose which reflect those commonly met in conservation planning.

Data and methodological limitations

Our analyses were based on existing species occurrence records available through large, global datasets (GBIF Occurrence Download, 2021). However, records for many of the priority UPC were scant or missing altogether. Additionally, the geographical distribution of occurrence records was unequal, favouring ecosystems such as agroecosystems and forests, which are more easily accessible or of greatest economic and research interest. The high occurrence of useful plant species in agroecosystems (Section 3.3.2) may also result from historic records that are no longer present following land-use conversion. As a result, while many UPC are not represented in the hotspots identified, other useful species reported as present may no longer be in the locations highlighted, with targeted ground-truthing required in some areas before designating conservation protection. This scenario is exacerbated by the uncertainty associated with occurrence information, which tends to lack taxonomic verification. Therefore, the data used may also include misidentified species, with case studies finding that rates of species misidentification in plant occurrence information can range from <1% to 17% (Meyer *et al.*, 2016). These biases, gaps and uncertainties are common challenges that have affected conservation planning globally, including IPA networks (Blasi *et al.*, 2011, Hamidah *et al.*, 2020).

Further conservation action and prioritisation based on the high-level gaps identified in this chapter should include wider sources of data, such as herbarium records and geographic data from published literature. In addition, to improve data availability more generally, it is important to continue supporting national and regional herbaria in increasing the digitisation and public accessibility of their records, which can include important specimens and provide local data for conservation planning and a wide range of other research (Sweeney *et al.*, 2018). These actions are a resource-intensive procedure and require constant curation to ensure that high-quality data are provided digitally. Supporting participatory research can also help document the vast amount of traditional knowledge held by local experts that should form the basis for conservation decision-making.

In addition to the data challenges, the methodologies applied for hotspot identification represent just some of many potential approaches, each with associated benefits and limitations. While we applied three methods, all were based on measures of overall richness of native useful plant species in Colombia. These methods did not account for species' severity of extinction risk within the UAs, with 32% of UPC of conservation concern (Figure 1) not represented within the hotspots. This approach was partly chosen due to current gaps in extinction risk assessments. However, future analyses could instead focus on hotspots for species of conservation concern, in line with Criterion A of the IPA methodology, which focuses on the presence of threatened and/or range-restricted plant species (Darbyshire *et al.*, 2017). In addition, the UAs used were of varying size, which can create a bias towards larger areas due to the fundamental species-area relationship. Alternative approaches applying grid cells as UAs (e.g., Hamidah *et al.*, 2020) or accounting for ratios such as species density per area and species numbers per record could inform further analyses and provide useful comparisons to the approach taken in this chapter.

From conservation priorities to action

This chapter presents the first steps in identifying gaps and suggesting priorities for the conservation of useful plant species in Colombia. However, undertaking extinction risk assessments and identifying geographic hotspots for useful plants are not sufficient to achieve their conservation. Other important steps must be taken to transform the hotspots identified into locally relevant, sustainably managed areas and to use information from extinction risk assessments to inform effective conservation prioritisation and action. Crucially, a wide range of socio-economic considerations must be determined in collaboration with a range of stakeholders, including levels of government, land right-holders, and community members, to ensure that overexploitation of useful species does not occur (Margules and Sarkar, 2007).

Social-Ecological Systems (SES) science is increasingly used as a framework to support transdisciplinary approaches in environmental management and biodiversity conservation (Virapongse *et al.*, 2016). Such an approach is particularly important when considering useful plants, whose sustainable use and conservation require the involvement of resource users and is dependent on market forces and national policies (Kor *et al.*, 2021). Simultaneously, linking ethnobotany and conservation enables the recognition and incorporation of local, traditional, and indigenous knowledge in biodiversity management and can support the conservation of 'knowledge networks' and cultural heritage.

In-depth botanical and ecological knowledge data are also required to strengthen the definition of IPA boundaries and to enable the development of effective management plans for the sustainable use of species. Despite including both ecosystem and municipality boundaries in our definition of UAs, some identified hotspots were extremely large (Section 3.3.2), with further geographic refinement required to determine feasible units for conservation management. These hotspots should be supported by targeted field data collection and consultation with experts. Meanwhile, *ex-situ* conservation is also vital to the long-term survival of threatened species, providing insurance against extinction and the potential for species restoration and reintroduction. However, global gaps currently exist in the representation of useful plants in such repositories (Khoury *et al.*, 2019).

Interdisciplinary research projects such as the UPFC, which unite a diverse range of knowledge bases through collaboration with local communities, alongside the involvement of researchers from multiple disciplines, are significant contributors to this process (UPFC, 2020). We believe that through sustained collaborative approaches, the conservation priorities presented in this study can act as a crucial starting point for useful plant conservation action in Colombia, and can enable participatory approaches to identify IPAs. While published conservation criteria and available conservation tools provide invaluable guidance for planning and prioritisation, stakeholder engagement is ultimately crucial to develop a realistic, practical, and sustainable network of sites that will result in effective conservation on the ground.

Participatory mapping to identify local conservation priorities in the Municipality of Becerril.
Mauricio Diazgranados

Acknowledgments

This work directly contributes to the 'Useful Plants and Fungi of Colombia' project, supported by a Professional Development & Engagement grant under the Newton-Caldas Fund partnership. The grant is funded by the UK Department for Business, Energy and Industrial Strategy (BEIS) and Minciencias, and delivered by the British Council. LK is supported by a studentship awarded by the Natural Environment Research Council [Grant Number NE/S007229/1].

References

Agencia Nacional de Tierras (2017). *Resguardos Indígenas*. Available at: https://data-agenciadetierras.opendata.arcgis.com/datasets/agenciadetierras::resguardos-ind%C3%ADgenas-1/about

Arponen A (2012) Prioritising species for conservation planning. *Biodiv Conserv,* 21:875-893.

Anderson S (2002) Identifying important plant areas: a site selection manual for Europe, and a basis for developing guidelines for other regions of the world. London, Plantlife International. 51p.

Arango JH (2017) Collective land tenure in Colombia: data and trends. *CIFOR Infobrief.* doi: 10.17528/cifor/006877

Beasley TM Schumacker Multiple regression approach to analyzing contingency tables: post hoc and planned comparison procedures. *J Exp Educ,* 64: 79-93.

Bernal H, García H, Quevedo G (2011) Pautas para el conocimiento, conservación y uso sostenible de las plantas medicinales nativas en Colombia. *Estrategia Nacional para la conservación de plantas. Ministerio de Ambiente Vivienda y Desarrollo Territorial.* Bogotá, Instituto de Investigación de Recursos Biológicos Alexander von Humboldt. 232p.

Bernal R, Gradstein SR, Celis M (2020) Catálogo de Plantas y Líquenes de Colombia. v1.1. Bogotá, Universidad Nacional de Colombia. 1499p.

Bivand R, Keitt T, Rowlingson B (2021) rgdal: Bindings for the 'Geospatial' Data Abstraction Library. R package version 1.5-23

Blasi C, Marignani M, Copiz R, Fipaldini M, Bonacquisti S, Del Vico E, Rosati L, Zavattero L (2011) Important plant areas in Italy: from data to mapping. *Biol Conserv,* 144:220-226.

Brehm JM, Maxted N, Martins-Loução MA, Ford-Lloyd BV (2010) New approaches for establishing conservation priorities for socio-economically important plant species. *Biodiv Conserv,* 19: 2715-2740

Bridge Colombia (2017) Preserving, restoring, and managing Colombian Biodiversity through responsible innovation. Available at: http://www.bridgecolombia.org/wp-content/uploads/2017/05/BRIDGE-Colombia-Workshop-Resolution-ENGLISH.pdf

Castellanos-Castro C, Sofrony C, Higuera D (2017) Plan de Acción de la Estrategia Nacional para la Conservación de Plantas de Colombia. Bogotá, Instituto de Investigación de Recursos Biológicos Alexander von Humboldt, Ministerio de Ambiente y Desarrollo Sostenible, Red Nacional de Jardines Botánicos de Colombia. 86p.

CBD (2012) The Global Strategy for Plant Conservation: 2011-2020. London, Botanic Gardens Conservation International. Available at: https://www.bgci.org/wp/wp-content/uploads/2019/04/GSPC_english.pdf

CBD (2021) First Draft of the Post-2020 Global Biodiversity Framework. Convention on Biological Diversity. Available at: https://www.cbd.int/article/draft-1-global-biodiversity-framework

Chamberlain S, Barve V, Mcglinn D, Oldoni D, Desmet P, Geffert L, Ram K (2021) rgbif: Interface to the Global Biodiversity Information Facility. R package version.

Chi X, Zhang Z, Xu X, Zhang X, Zhao Z, Liu Y, Wang Q, Wang H, Li Y, Yang G, Guo L, Tang Z, Huang L (2017) Threatened medicinal plants in China: distributions and conservation priorities. *Biol Conserv,* 210:89-95.

Cook FE (1995) Economic botany data collection standard. Prepared for the International Working Group on Taxonomic Databases for Plant Sciences (TDWG). London, Royal Botanic Gardens, Kew. Available at: https://www.tdwg.org/standards/economic-botany/

Corlett RT (2016) Plant diversity in a changing world: status, trends, and conservation needs. *Plant Diversity,* 38(1): 10-16. doi: 10.1016/j.pld.2016.01.001.

Corzo G, Ramírez W, Salamanca B, Londoño MC, Fonseca C, Castellanos C, Alcázar C, Lasso C, García H (2010) Planeación ambiental para la conservación de la biodiversidad en las áreas operativas de Ecopetrol localizadas en el Magdalena Medio y los Llanos Orientales. Bogotá, Instituto de Investigación de Recursos Biológicos Alexander von Humboldt y Ecopetrol. Available at: http://www.humboldt.org.co/es/component/k2/item/12-planeacion-ambiental-para-la-conservacion-ecopetrol

Cunningham AB (2001) *Applied Ethnobotany: People, Wild Plant Use and Conservation.* New York, Earthscan Publications. 300p.

Darbyshire I, Anderson S, Asatryan A, Byfield A, Cheek M, Clubbe C, Ghrabi Z, Harris T, Heatubun CD, Kalema J (2017) Important Plant Areas: revised selection criteria for a global approach to plant conservation. *Biodiv Conserv,* 26: 1767-1800.

Diazgranados M, Castellanos-Castro C (2018). Metodología para la identificación de áreas de importancia para plantas (AIP) en Colombia. Bogotá D.C.: Instituto de Investigación de Recursos Biológicos Alexander von Humboldt. http://repository.humboldt.org.co/handle/20.500.11761/35891

Diazgranados M, Allkin B, Black N, Cámara-Leret R, Canteiro C, Carretero J, Eastwood R, Hargreaves S, Hudson A, Milliken W, Nesbitt M, Ondo I, Patmore K, Pironon S, Turner R, Ulian T (2020) World Checklist of *Useful Plant Species*. KNB Data Repository. doi:10.5063/F1CV4G34.

Esri Inc (2021) ArcGIS Pro 2.8.0.

García Martínez H, Moreno LA, Londoño C, Sofrony C (2010) *Estrategia nacional para la conservación de plantas: actualización de los antecedentes normativos y políticos y revisión de avances*. Bogotá, Instituto de Investigación de Recursos Biológicos Alexander von Humboldt. Available at: http://www.humboldt.org.co/images/noticias/2010%20Estrategia%20Nacional%20Conservacion%20de%20Plantas.pdf

Garcia N, Galeano G, Salinas NR, Cardenas D, Saenz FC (2007) *Libro rojo de plantas de Colombia*. Volumen 5: Las magnoliáceas, las miristicáceas y las podocarpáceas. Serie Libros Rojos de Especies Amenazadas de Colombia. Bogotá, Instituto de Investigacion de Recursos Biologicos Alexander Von Humboldt. 236p.

GBIF Occurrence Download (2021) https://doi.org/10.15468/dl.cduqac. Accessed from R via rgbif (https://github.com/ropensci/rgbif) on 15 July 2021.

Hamidah M, Mohd Hasmadi I, Chua LSL, Lau KH, Faridah-Hanum I, Yong WSY, Pakhriazad HZ (2020) Towards identification of important plant areas (IPA) for Peninsular Malaysia. Methodology and future directions. *Heliyon* [online], 6: e04370.

Hamilton AC, Radford EA, Plantlife, Ethnobotanical Society (2007) Identification and conservation of important plant areas for medicinal plants in the Himalaya: project and workshop report. Salisbury, Wiltshire: Plantlife International. 48p.

IUCN (2008) *Nace el santuario de flora "Plantas medicionales Orito Ingi-Ande"* [Online]. Available: https://www.iucn.org/fr/node/1826 [Accessed 26 July 2021].

IUCN (2021) The IUCN Red List of Threatened Species. *Downloaded June 2021.* Version 2021-1. ed.

Khoury CK, Amariles D, Soto JS, Diaz MV, Sotelo S, Sosa CC, Ramirez-Villegas J, Achicanoy HA, Velasquez-Tibata J, Guarino L, Leon B, Navarro-Racines C, Castaneda-Alvarez, NP, Dempewolf H, Wiersema JH, Jarvis A (2019) Comprehensiveness of conservation of useful wild plants: an operational indicator for biodiversity and sustainable development targets. *Ecol Indic,* 98:420-429.

Kor L, Homewood K, Dawson TP, Diazgranados M (2021) Sustainability of wild plant use in the Andean Community of South America. *Ambio*, 50:1681–1697. doi: 10.1007/s13280-021-01529-7

MADS (2017) Plan de acción de biodiversidad para la implementación de la Política Nacional para la Gestión Integral de la Biodiversidad y sus Servicios Ecosistémicos 2016-2030. Bogotá, Ministerio de Ambiente y Desarrollo Sostenible. 69p.

MADS, IDEAM, IAvH, INVEMAR, IIAP, SINCHI, PNN & IGAC (2017) Mapa de Ecosistemas Continentales, Costeros y Marinos de Colombia a escala 1:100.000. Bogota, Colombia.

Margules CR, Sarkar S (2007) *Systematic Conservation Planning.* Cambridge: Cambridge University Press. 26p.

Meyer C, Weigelt P, Kreft H (2016) Multidimensional biases, gaps and uncertainties in global plant occurrence information. *Ecol Lett,* 19(8):992-1006.

Nic Lughadha E, Bachman SP, Leão TCC, Forest F, Halley JM, Moat J, Acedo C, Bacon KL, Brewer RFA, Gâteblé G, Gonçalves SC, Govaerts R, Hollingsworth PM, Krisai-Greilhuber I, Lirio EJ, Moore PGP, Negrão R, Onana JM, Rajaovelona LR, Razanajatovo H, Reich PB, Richards SL, Rivers MC, Cooper A, Iganci J, Lewis GP, Smidt EC, Antonelli A, Mueller GM, Walker BE (2020) Extinction risk and threats to plants and fungi. *Plants, People, Planet,* 2:389-408. doi: 10.1002/ppp3.10146

Orme CDL, Davies RG, Burgess M, Eigenbrod F, Pickup N, Olson VA, Webster AJ, Ding T-S, Rasmussen PC, Ridgely RS, Stattersfield AJ, Bennett PM, Blackburn TM, Gaston KJ, Owen IPF (2005) Global hotspots of species richness are not congruent with endemism or threat. *Nature,* 436: 1016-1019.

Palomo I, Montes C, Martín-López B, González JA, García-Llorente M, Alcorlo P, Mora MRG (2014) Incorporating the Social–Ecological Approach in Protected Areas in the Anthropocene. *BioScience,* 64:181-191.

Paniagua-Zambrana NY, Bussmann RW (2020) *Ethnobotany of the Andes,* Cham, Springer. doi: 10.1007/978-3-319-77093-2

POWO (2019) *Plants of the World Online.* Facilitated by the Royal Botanic Gardens, Kew. Available at: http://www.plantsoftheworldonline.org/

R Core Team (2020) R: a language and environment for statistical computing. Vienna, Austria: R Foundation for Statistical Computing.

Sharrock S (2020) Plant Conservation Report 2020: a review of progress in implementation of the Global Strategy for Plant Conservation 2011-2020. *Technical Series No. 95.* Montréal, Secretariat of the Convention on Biological Diversity and Botanic Gardens Conservation International.

Sweeney PW, Starly B, Morris PJ, Xu Y, Jones A, Radhakrishnan S, Grassa CJ, Davis CC (2018). Large-scale digitisation of herbarium specimens: Development and usage of an automated, high-throughput conveyor system. *Taxon, 67*: 165-178.

UN-WCMC (2014) *Megadiverse Countries.* Available at: https://www.biodiversitya-z.org/content/megadiverse-countries

UNEP-WCMC (2020) Protected Area Country Profile for Colombia from the World Database of Protected Areas. Available at: www.protectedplanet.net

UPFC (2020) *Useful Plants and Fungi of Colombia (UPFC).* Available at: https://www.kew.org/science/our-science/projects/useful-plants-and-fungi-colombia

Virapongse A, Brooks S, Metcalf EC, Zedalis M, Gosz J, Kliskey A, Alessa L (2016) A social-ecological systems approach for environmental management. *J Environ Manage,* 178:83-91.

Wickham H, François R, Henry L, Müller K (2021) dplyr: a grammar of data manipulation. R package version 1.0.6. ed.

Wilson KA, Carwardine J, Possingham HP (2009) Setting conservation priorities. *Ann NY Acad Sci,* 1162: 237-264. doi: 10.1111/j.1749-6632.2009.04149.x

Zizka A, Silvestro D, Andermann T, Azevedo J, Duarte Ritter C, Edler D, Farooq H, Herdean A, Ariza M, Scharn R, Svantesson S (2019) CoordinateCleaner: standardised cleaning of occurrence records from biological collection databases. *Methods Ecol Evol,* 10(5):744-751.

Espeletia killipii var. *chisacana* – medicinal endemic species from Colombia, occuring in Páramos.
Mauricio Diazgranados

Chapter 3

Endemic Medicinal Plants of Colombia

Henry Yesid Bernal M.[1]* & Catalina Mesa Sánchez[2]

[1] Ex-Profesor Asociado del Departamento de Biología, Facultad de Ciencias, Universidad Javeriana, Herbario de la Pontificia Universidad Javeriana (HPUJ), Bogotá, D.C., Colombia.

[2] Bióloga, M.Sc. en Gestión Ambiental, Facultad de Ciencias, Departamento de Biología, Pontificia Universidad Javeriana, Bogotá, D.C., Colombia.

***Corresponding author:** henryesid@gmail.com

Keywords: useful plants, ethnobotany, phytopharmaceuticals, phytomedicines, phytotherapeutic products

ABSTRACT

The medicinal species are highlighted from the Colombian biodiversity and its useful plants, including many native and a significant number of species endemic to the country. However, the medicinal species of Colombia are still underused or poorly known, with several of them already at risk of extinction. Medicinal species constitute a true natural heritage that requires concrete and innovative actions for their knowledge, conservation, and sustainable use. This study was carried out between 2014 and 2021, with the main goals to: a) review and update the inventory of medicinal plants used in Colombia (exotic and native species); b) specify which are the medicinal species endemic to Colombia; and c) evaluate the current state of traditional, scientific, and technological knowledge of the endemic species to Colombia used as medicinal and their potential contribution to the treatment of prevalent pathologies in the country. A total of 3,005 medicinal plants used in Colombia were identified and documented, of which 1,719 are native to this country (including 204 endemics), 558 exotic, and 728 species with an uncertain geographical distribution. The lack of knowledge on conservation and sustainable use of most of the Colombian endemic medicinal species was verified, and their traditional, scientific, and technological knowledge was characterised. Only three species have the traditional and scientific knowledge classified as 'acceptable': *Achyrocline bogotensis* (Asteraceae), *Espeletia killipii* (Asteraceae) and *Pentacalia corymbose* (Asteraceae). Another ten species have evidence of traditional knowledge classified as 'acceptable' and evidence of scientific knowledge classified as 'regular': *Berberis goudotii*, *Berberis rigidifolia* (Berberidaceae), *Cecropia telealba*, *Cecropia mutisiana* (Urticaceae), *Chromolaena barranquillensis* (Asteraceae), *Espeletia grandiflora*, *Espeletiopsis muiska* (Asteraceae), *Ocotea caparrapi* (Lauraceae), *Passiflora antioquiensis* (Passifloraceae) and *Senecio niveoaureus* (Asteraceae). Those 13 species present one or several mentions of uses for different pathologies with higher impact in Colombia.

RESUMEN

Dentro de la biodiversidad de Colombia y sus plantas útiles se destacan las especies medicinales, siendo muchas nativas y dentro de ellas un buen número son endémicas del país, pero son subutilizadas o poco conocidas y varias de ellas están en peligro de extinción. Se considera que estas especies medicinales constituyen un verdadero patrimonio natural que requiere de acciones concretas e innovadoras para su conocimiento, conservación y uso sostenible. Entre los años 2014 y 2021 se desarrolló esta investigación que tuvo como objetivos: a) revisar y actualizar el inventario de las plantas medicinales de uso en Colombia (exóticas y nativas); b) precisar cuáles son las plantas medicinales endémicas del país y c) evaluar el estado actual de los conocimientos tradicional, científico y tecnológico de las plantas medicinales endémicas de Colombia y su posible contribución para la atención de las patologías prevalentes en el país. En esta investigación se identificaron y documentaron 3.005 plantas medicinales de uso en Colombia de las cuales 1.719 son nativas de Colombia (204 endémicas), 558 son foráneas y 728 tienen distribución geográfica incierta. Se comprobó la falta de conocimientos para la conservación y uso sostenible de la mayoría de las especies medicinales endémicas de Colombia, y se logró la caracterización de sus conocimientos tradicional, científico y tecnológico. Solo 3 especies cuentan con conocimiento tradicional y científico clasificados como 'aceptable' *Achyrocline bogotensis* (Asteraceae), *Espeletia killipii* (Asteraceae) y *Pentacalia corymbosa* (Asteraceae), y 10 especies tienen evidencia de conocimiento tradicional clasificada como 'aceptable' y evidencia de conocimiento científico clasificada como 'regular' *Berberis goudotii* (Berberidaceae), *Berberis rigidifolia* (Berberidaceae), *Cecropia telealba* (Urticaceae), *Cecropia mutisiana* (Urticaceae), *Chromolaena barranquillensis* (Asteraceae), *Espeletia grandiflora* (Asteraceae), *Espeletiopsis muiska* (Asteraceae), *Ocotea caparrapi* (Lauraceae), *Passiflora antioquiensis* (Passifloraceae) y *Senecio niveoaureus* (Asteraceae). Estas 13 especies presentan uno o varias menciones de uso para las diferentes patologías de mayor impacto en Colombia.

INTRODUCTION

Since ancient times, all societies have resorted to plants as a source of medicines. According to the World Health Organization (WHO, 2021), 80% of the world's population use traditional medicine to meet their needs for primary health care. It is recognised that traditional medicine and medicinal plants are the main important ways to improve health for the planet's population. Several guidelines are found in a significant number of publications (WHO, 2021).

The conservation of biodiversity at the global level is a crucial issue for future possibilities for adapting living beings in a world of continuous change. The undoubted medicinal resource that plants constitute is under threat and is a treasure that cannot go to waste (Bernal, 2012). Many initiatives on biodiversity continue to register and count species, but there is a big gap in the associated traditional, scientific, and technological knowledge needed to conserve and use biodiversity sustainably.

The humid tropics in Latin American countries, especially Colombia, are a stripe of intense biological activity but low economic production of natural products. Several conditions can explain this contradiction, such as poor soils, excessive rainfall, plant and animal diseases, inadequate agricultural systems, physical structural deficiency, and lack of medium- and long-term investments. Paradoxically, Colombian biodiversity is considered megadiverse, and it is estimated that in terms of plant species, the country comprises between 30,000 and 41,000 species (Andrade, 2011). These figures place Colombia as one of the richest floras in the world in species number, which are mainly underutilised, underexploited, and poorly known in terms of their conservation and economic potentialities.

The great importance of environmental goods and services derived from biodiversity must be highlighted. The survival of humanity relies on biodiversity as a great source of food, fossil fuels, natural fibres, phytomedicines, biocides, natural colourants, soil productive capacity, water and air cycling, and many other environmental services. Likewise, biodiversity is the world's natural capital. In particular, the native and endemic species of megadiverse countries such as Colombia constitute its greatest biological heritage and represent fundamental options for equitable, democratic, and sustainable development (Bernal, 2012).

The main mortality-causing pathologies in Colombia are ischemic heart diseases, cerebrovascular diseases, chronic diseases of the lower respiratory tract, diabetes mellitus, acute respiratory infections, hypertensive diseases, malignant tumours of the stomach, trachea, bronchi, and lungs, and heart failure (Bernal *et al.*, 2011). Similarly, in descending order, the main morbidity-causing pathologies in Colombia are periodontal disease, dental caries, iron deficiency anaemia, asthma, trichuriasis by *Trichuris trichiura*, hypertensive heart disease, protein-calorie malnutrition, hookworm, abortion, unipolar major depression, and diarrheal diseases (Bernal *et al.*, 2011). The main pathologies on epidemiological surveillance in Colombia are classical dengue fever, haemorrhagic dengue fever, leishmaniasis, *Plasmodium vivax* and *P. falciparum* malaria, and tuberculosis (Bernal *et al.*, 2011).

In Colombia, important ethnobotanical, phytochemical, and pharmacological investigations have been carried out on the use of medicinal plants as a support treatment for several tropical pathologies. Regarding the inventory and the state of traditional, scientific, and technological knowledge of medicinal plants used in the country, it is worth mentioning the studies produced by: a) Enrique Pérez Arbeláez (1953–1966, 1978a,b); b) Victor Manuel Patiño (1967); c) Hernando García Barriga (1974, 1975 and 1992); d) José Joaquín Montes Giraldo (1981); e) Daniel J. González Patiño (1984); f) Mauro Hernández Mesa (1992); g) Ramiro Fonnegra *et al.* (Fonnegra & Jiménez, 1992, Fonnegra *et al.*, 2006); h) the Ibero-American Program of Science and Technology for Development (CYTED), Subprogram of Fine Pharmaceutical Chemistry and the International Organization of the Andrés Bello Agreement from the publication "Promising plant species from the Andrés Bello Agreement countries" (Bernal & Grupta 1995, ; Correa & Bernal, 1989–1995); and i) other Ibero-American medicinal plants studies (Bernal & Correa, 1995; Martínez *et al.*, 2000; Bernal & Jiménez, 2008). In 2008, the Ministry of Social Protection of Colombia published the first edition of the Colombian Vademecum of Medicinal Plants and updated guidelines, from which 127 plant species used as medicinal in Colombia were regulated (including 87 Neotropical species that are not native to Colombia and only two endemic species – *Achyrocline bogotensis* and *Cecropia mutisiana*) (Ministerio de Protección Social, 2008).

More recently, Bernal *et al.* (2011) published a book entitled 'Guidelines for Knowledge, conservation, and sustainable use of native medicinal plants of Colombia'. The inventory from this book was carried out by Vásquez & Bernal (2011), recording 2,404 medicinal plant species used in Colombia, including 1,870 native, 214 endemic, and 534 exotic species. In 2021, Bernal & Gómez prepared the second edition of the project "National collection of medicinal plants", which is a list of the endemic medicinal plants of Colombia and the native medicinal species of Colombia that have been prioritised for cultivation at the Quindío Botanical Garden (Calarca, Colombia).

The data and results presented here are part of a project developed between 2014 and 2021 from extensive previous studies on the medicinal plants of Colombia (Bernal & Mesa 2014). This chapter has as main goals: a) to review and update the inventory of medicinal plants used in Colombia (exotic and native); b) to specify which endemic species are used as medicines in Colombia; and c) to evaluate the current state of traditional, scientific, and technological knowledge on endemic species used as medicines and their possible contribution to the treatment of the prevalent pathologies in Colombia, according to data on mortality, morbidity, and diseases from epidemiological surveillance.

TABLE 1. National and international databases consulted.

Database name	URL
BHL Biodiversity Heritage Library	https://www.biodiversitylibrary.org/
BIOVIRTUAL (common names). Instituto de Ciencias Naturales-Universidad Nacional de Colombia	http://www.biovirtual.unal.edu.co/nombrescomunes
BIOVIRTUAL. Ejemplares de herbario Instituto de Ciencias Naturales-Universidad Nacional de Colombia	http://www.biovirtual.unal.edu.co/ICN/
Catálogo de plantas y líquenes de Colombia	http://catalogoplantasdecolombia.unal.edu.co/es
CBIF. Canadian Biodiversity Information Facility	https://www.cbif.gc.ca/acp
CEIBA. Catalogador de información biológica. Instituto de Investigación de Recursos Biológicos Alexander von Humboldt-Colombia	http://www.i2d.humboldt.org.co/ceiba
Colombia diversidad biótica. Instituto de Ciencias Naturales de la Universidad Nacional de Colombia	http://www.ciencias.bogota.unal.co/menu-principal/publicaciones/biblioteca-digital/Colombia-diversidad-biotica
CoL. Catalogue of Life	https://www.catalogueoflife.org
ColPlantA Colombian Resources for Plants made Accessible	https://colplanta.org
DR. DUCKE's. Phytochemical and Ethnobotanical Databases	http://www.ars-grin.gov/duke
EOL. Encyclopaedia of Life. National Museum of Natural History. Smithsonian	http://www.eol.org/
Flora de Colombia. Monografías del Instituto de Ciencias Naturales de la Universidad Nacional de Colombia	http://www.ciencias.bogota.unal.edu.co/menu-principal/publicaciones/biblioteca-digital/flora-de-colombia
Flora de la Real Expedición Botánica. Real Jardín Botánico de España	https://www.bibdigital.rjb.csic.es/medias
GBIF. Global Biodiversity Information Facility	http://www.gbif.org/species
Global Compositae database. International Compositae Alliance	https://compositae.org/
GRIN. Germplasm Resources Information Network	http://www.ars-grin.gov
Ildis database. International Legume Database and Information Service	http://www.ildis.org/
IPNI. International Plant Name Index	https://www.ipni.org/
IUCN-Red List. The IUCN red list of the threatened species	https://www.iucnredlist.org/species
ITIS. Integrated Taxonomic Information System	http://www.itis.gov
JSTOR Global Plants database	https://www.plants.jstor.org/
Kew herbarium catalogue. Royal Botanic Gardens, Kew	http://www.apps.kew.org/herbcat/
Kew Plant Information Portal. Royal Botanic Gardens, Kew	https://www.epic.kew.org/
Napralert Database. University of Illinois at Chicago	http://www.napralert.org/
NCBI. National Centre for Biotechnology Information. National Library of Medicine. National Institutes of Health. USA	https://www.ncbi.nlm.nih.gov/
NMNH. Smithsonian National Museum of Natural History	https://www.collections.nmnh.si.edu/
PL@NTUSE. Usos de las plantas	https://use.plantnet-project.org
RACCEFYN. Revista de la Academia Colombiana de Ciencias Exactas, Físicas y Naturales	https://raccefyn.co
REFLORA. Virtual Herbarium from the Flora of Brazil. Rio de Janeiro Botanic Garden.	http://reflora.jbrj.gov.br/
Science Direct	https://www.sciendedirect.com/
Scopus-Elsevier. Bibliographic database	http://www.scopues.com
SIAC. Sistema de información ambiental de Colombia	http://www.siac.gov.co/biodiversidad
SIB. Sistema de Información sobre la Biodiversidad de Colombia	http://www.dta.sibcolombia.net/inicio.htm
SPECIES 2000 and ITIS. Catalogue of life. Annual checklist. Integrated Taxonomic Information System	http://www.cataloguefllife.org/
The Plant List	http://www.theplantlist.org
TROPICOS. Vascular Plants of the Americas	http://www.tropicos.org/project/VPA
USDA-Plants Database. United States Department of Agriculture	http://www.plants.usda.gov
Useful Tropical Plants Database	https://www.tropical.the ferns.info
WCSP. World Checklist of Select Plant Families. Royal Botanic Gardens, Kew	https://www.wcsp.science.kew.org/
WCVP. The World Checklist of Vascular Plants. Royal Botanic Gardens, Kew	https://www.wcvp.science.kew.org/
WFO. World Flora Online Consortium	http://www.floraonline.org

MATERIALS AND METHODS

This study used a revised and updated version of the list of medicinal plants used in the country published by Vásquez & Bernal (2011, pages 165 to 231) as a starting point to identify the endemic medicinal plants of Colombia. Specialised bibliographic searches were carried out at the national and international levels covering the period from 2014 to 2021, and the corresponding databases were built.

A total of 13 Colombian herbaria were consulted: Herbarium of the Technological University of Chocó (CHOCO); Herbarium of the Eloy Valenzuela Botanical Garden (CDMB); Colombian Amazon Herbarium; Amazon Institute for Scientific Research, Sinchi (COAH); Colombian National Herbarium (COL); Herbarium of the University of Antioquia (HUA); Enrique Forero Herbarium of the University of the Amazon (HUAZ); Herbarium of the University of Córdoba (HUC); Herbarium of the Universidad Católica de Oriente (HUCO); Herbarium of the University of Quindío (HUQ); Herbarium of the Botanical Garden of Bogotá "José Celestino Mutis" (JBB); Herbarium of the Guillermo Piñeres Botanical Garden (JBGP); Herbarium of the Juan María Céspedes Garden. (TULV); and Herbario Forestal Gilberto Emilio Mahecha Vega of the District University (UDBC). Colombian and international databases were also consulted (Table 1), aiming to define the geographical distribution of medicinal species to Colombia, from which the species were grouped into three categories: a) restricted geographical distribution in one or two departments of Colombia, b) wide geographical distribution in three or more departments of Colombia, and c) uncertain geographic distribution in Colombia.

The current states of traditional, scientific, and technological knowledge of the endemic species of Colombia used as medicinal were based on the studies carried out by Mesa (2013). Theoretical variables were defined for each species to analyse the current state of knowledge (based on ethnobotanical, scientific, and technological evidence). Indicators were established for each of the variables according to the following variation scale: 5–7 (acceptable), 3–4 (regular), and 0–2 (scarce) with scores that allow the definition of the current state of knowledge for species (Tables 2, 3, and 4). The indicators of the evidence of traditional knowledge were: a) historically supported use based on reliable documentary references; b) pathologies prevented or treated by the used species; c) mentions of therapeutic uses in humans or animals; d) safety (no history of toxicity according to the ethnobotanical literature review); e) part(s) of the plant that are used for therapeutic purposes in humans or animals; f) traditional way of preparation; g) posology; h) dosage; i) topical administration route (external use); and j) systemic administration route (internal use) (Table 2).

The indicators of evidence of scientific knowledge were a) description and taxonomic diagnosis; b) morphology; c) geographical and altitudinal distribution; d) ecology; e) phytochemical profile; f) molecular characterisation; g) extraction and separation of active ingredients; h) identification of compounds and determination of their molecular structures; i) acute toxicity; j) subacute toxicity; k) chronic toxicity; l) subchronic toxicity; m) *in vitro* biological activities; n) biological activities in animal models; o) phase I clinical studies; p) phase II clinical studies; q) phase III clinical studies; and r) phase IV clinical studies (Table 3).

The indicators of evidence of technological knowledge were a) micropropagation; b) sexual propagation; c) asexual spread; d) soil preparation; e) planting and spacing; f) fertilisation

TABLE 2. Operationalisation for the theoretical variable 'ethnobotanical evidence' related to endemic species of Colombia used as medicinals.

Factor	Theoretical variable	Concept	Indicators	Scale
Traditional knowledge	Ethnobotanical evidence	Ethnobotanical evidence that an endemic plant species is used as medicinal in Colombia (in this study) is based on its cultural therapeutic use to treat or prevent diseases in humans or animals that mainly indigenous and local communities give to these species. It is based on the reliable bibliographic documentation available in the national and international literature.	• Use historically supported by reliable bibliographic references • Pathologies that are prevented or treated • Mentions of therapeutic uses in humans or animals • Safety (no history of toxicity according to the available botanical bibliography) • Part(s) of the plant that are used for therapeutic purposes in humans or animals • Traditional preparation method • Posology/dosage • Dose • Topical route of administration (external use) • Systemic route of administration (internal use)	• Acceptable • Regular • Scarce

TABLE 3. Operationalisation for the theoretical variable 'scientific evidence' related to endemic species of Colombia used as medicinals.

Factor	Theoretical variable	Concept	Indicators	Scale
Scientific knowledge	Scientific evidence	The scientific evidence related to endemic species of Colombia used as medicinals (in this study) is based on the current state of taxonomic, geographical, ecological, and phytochemical knowledge, toxicity, biological and pharmacological activities, and health studies (preclinical and clinical). It is based on the reliable bibliographic documentation available in the national and international literature.	**Taxonomic characterisation:** • Valid scientific name • Synonyms • Infraspecific categories • Cultivars • Taxonomic description. • Taxonomic diagnosis • Habit (life form) • Annual-perenniality data • Morphological studies • Molecular studies • Methods for the identification and taxonomic determination of species and infraspecific categories • Methods for the identification and taxonomic determination of cultivars **Geographical and ecological characterisation:** • Geographical distribution • Altitudinal distribution • Ecological studies (habitat, phenology, natural regeneration, populations, communities, plant-animal relationship, conservation status, conservation measures, etc.) **Phytochemical characterisation:** • Phytochemical workflow [extraction, fractionation, or separation, qualitative characterisation (preliminary or qualitative phytochemical workflow), isolation and purification (preparative or quantitative phytochemical workflow)] • Active compounds recognition • Identification of compounds and determination of molecular structures (structural interpretation) **Characterisation of innocuousness/safety:** • Acute toxicity • Subacute toxicity • Chronic toxicity • Subchronic toxicity • Contraindications and precautions • Interactions with other drugs **Preclinical characterisation:** • Biological and pharmacological activities *in vitro* • *In vivo* biological and pharmacological activities in animal models • Existence of studies on pharmacodynamics • Existence of studies on pharmacokinetics • Existence of studies on the kinetics **Clinical characterisation:** • Existence of Phase I sanitation/health studies • Existence of Phase II sanitation/health studies • Existence of Phase III sanitation/health studies • Existence of Phase IV sanitation/health studies	• Acceptable • Regular • Scarce

TABLE 4. Operationalisation for the theoretical variable 'technological evidence' related to endemic species of Colombia used as medicinals.

Factor	Theoretical variable	Concept	Indicators	Scale
Technological knowledge	Technological evidence	The technological evidence related to endemic species used as medicinal in Colombia (in this study) is based on the current state of agrotechnological knowledge and the technologies for preparing medicinal products made from them. It is based on the reliable bibliographic documentation available in the national and international literature.	**Agrotechnologies for organic and sustainable cultivation:** • Micropropagation • Sexual propagation • Asexual propagation • Soil preparation • Planting and spacing • Fertilisation and supply of nutrients • Organic farming and agroforestry (monoculture or polyculture) • Cultural work and plant care • Diseases and treatment • Harvest and collection of the part(s) of the plant used as medicine • Postharvest • Performance • Costs of organic and sustainable production **Technologies used in the elaboration of medicinal products:** • Technologies used in the extraction of plant raw material • Technologies used in the pharmaceutical formulation of medicinal products • Technologies for the use of plant-based excipients and auxiliary formulation to produce medicinal products • Quality control of medicinal products (thin-layer chromatography and high-performance liquid chromatography)	• Acceptable • Regular • Scarce

and nutrient supply; g) organic cultivation and agroforestry (monoculture or polyculture); h) cultural work and plant care; i) diseases and treatments; j) harvest and collection of parts for medicinal use; k) postharvest; l) performance; m) costs of organic and sustainable production; n) technologies for the extraction of vegetable raw materials; o) technologies for pharmaceutical formulation; p) technologies for the use of excipients and formulation aids; and q) quality control of medicinal products (Table 4).

The current state of traditional, scientific, and technological knowledge of the endemic species used as medicinals was based on the variables, indicators, scale, and scores indicated above, and defined according to five steps: 1) the endemic species used as medicinals were listed under 'acceptable', 'regular', or 'scarce' evidence for each of the three types of knowledge (traditional, scientific, and technological); 2) the endemic species used as medicinals that were associated with acceptable evidence of scientific and traditional knowledge were selected, and their geographical distributions in Colombian departments were presented; 3) species associated with acceptable evidence of traditional knowledge, but also with regular evidence of scientific knowledge, were selected and the same procedure was carried out; 4) species with acceptable evidence of scientific and traditional knowledge and acceptable or regular evidence of technological knowledge were analysed; and 5) a matrix was elaborated to compare pathologies of greater impact in the country in terms of mortality, morbidity and pathologies under epidemiological surveillance *versus* mentions of therapeutic uses considering the list of endemic medicinal species of Colombia.

RESULTS

In this study, we identified 3,005 plant species that have medicinal uses (2,847 spermatophytes and 158 species of mosses, hornworts, liverworts, or lycophytes) in Colombia, with 1,719 being native to this country (including 204 endemics), 558 exotics, and 728 with uncertain geographical distribution. Almost the totality of Colombian endemic species used as medicinals is represented by flowering plants (133 species) distributed into 50 families, with a single species of Lycophytes reported (*Selaginella rosea* – Selaginellaceae).

The endemic species to Colombia used as medicinals are represented mainly by 195 species (95.6%) of angiosperms distributed as follows: Compositae (Asteraceae, 52 species); Araliaceae (12 species); Gesneriaceae (nine species); Lamiaceae and Piperaceae (eight species each); Campanulaceae, Malvaceae, and Rubiaceae (seven species, each); Symplocaceae (six species); Caprifoliaceae and Fabaceae (five species each); Acanthaceae, Rosaceae, and Styracaceae (four species each); and Berberidaceae, Brassicaceae, Clusiaceae, Erythroxylaceae, Gentianaceae, Melastomataceae, and Urticaceae (three species each). The remaining families are represented by only two or a single species. Among the 204 endemic species of Colombia used as medicinal, 123 species (60.3%) are widely distributed (in three or more departments), 64 species (31.4%) have restricted distributions (in one or two departments), and 17 species (8.3%) have an uncertain distribution in the country. From the operationalisation and factors score for each theoretical variable, most species were considered as having "scarce" evidence of traditional or scientific knowledge (Figure 1A, B, respectively)

Only 26 (13%) endemic species used as medicinals in Colombia had their traditional knowledge defined as 'acceptable' (*Achyrocline bogotensis*, *Berberis goudotii*, *Berberis rigidifolia*, *Brachyotum strigosum*, *Cecropia mutisiana*, *Cecropia telealba*, *Chromolaena barranquillensis*, *Copaifera canime*, *Draba litamo*, *Espeletia grandiflora*, *Espeletia killipii*, *Espeletiopsis muiska*, *Justicia phytolaccoides*, *Ocotea caparrapi*, *Pachira speciosa*, *Passiflora antioquiensis*, *Pentacalia corymbosa*, *Peperomia garcia-barrigana*, *Peperomia putumayoensis*, *Psychotria boqueronensis*, *Rubus choachiensis*, *Senecio niveoaureus*, *Symplocos theiformis*, *Valeriana arborea*, *Vasconcellea goudotiana*, and *Verbena valerianoides*) (Figure 1A). Only three species had the_scientific knowledge factor qualified as 'acceptable' (*Achyrocline bogotensis*, *Espeletia killipii*, and *Pentacalia corymbosa*) (Figure 1B). The same three species were defined as presenting 'acceptable' evidence on both factors (scientific and traditional knowledge) (Table 5, Figure 2). An important result is that no endemic species used as a medicinal in Colombia, presents technological knowledge assessed as 'acceptable'.

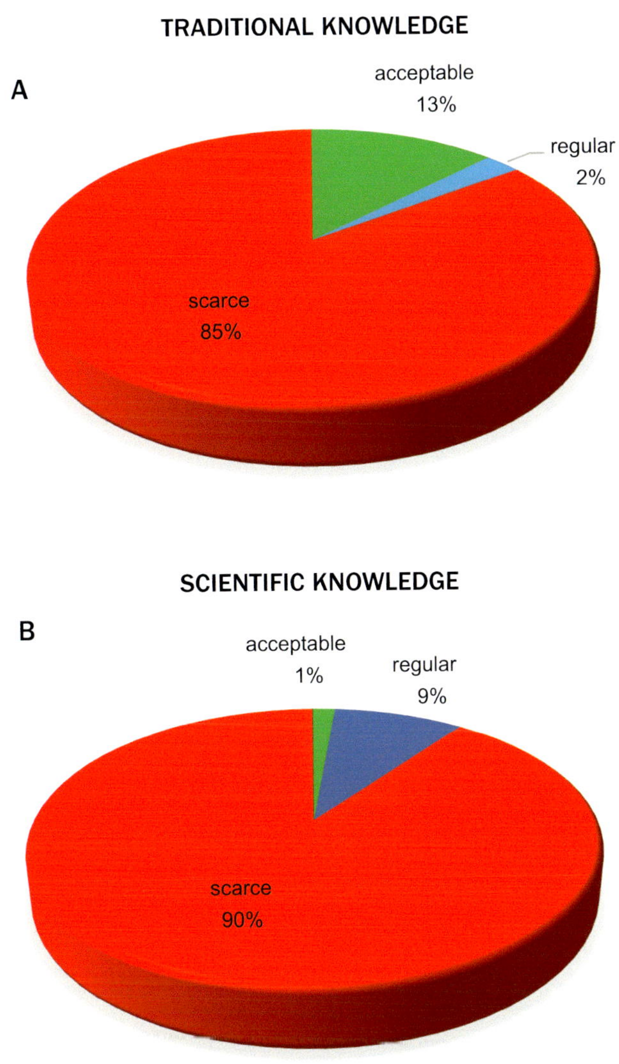

FIGURE 1. A Proportion of species endemic to Colombia used as medicinals, from which the traditional knowledge was defined as 'acceptable', 'regular' or 'scarce' according to ethnobotanical evidence. **B** Proportion of species endemic to Colombia used as medicinals, from which the scientific knowledge was defined as 'acceptable', 'regular' or 'scarce' according to scientific evidence.

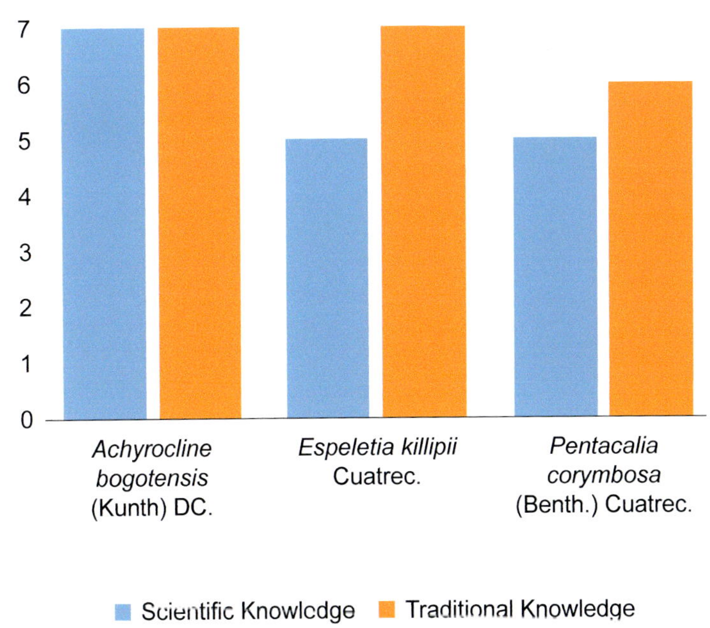

FIGURE 2. Species endemic to Colombia used as medicinals, presenting 'acceptable' evidence of traditional and scientific knowledge.

TABLE 5. Medicinal species endemic to Colombia with 'acceptable' traditional and scientific knowledge (*) and species with 'acceptable' traditional and 'regular' scientific knowledge (**). (***) m a.s.l. = metres above sea level; (****) Habitat Classification based on Cuatrecasas (1958).

No.	Scientific name/Family	Altitude (***) (m a.s.l.) and habitat (****)	Geographic distribution (Colombian Departments)
1	(*)***Achyrocline bogotensis*** (Kunth) DC. (Compositae-Asteraceae)	1,600 – 4,000 Sub-Andean Forest, Andean Forest, Subpáramo, Páramo, and Superpáramo	Boyacá, Cundinamarca, Magdalena, Nariño, Norte de Santander, and Santander
2	(*)***Espeletia killipii*** Cuatrec. (Compositae-Asteraceae)	2,120 – 4,010 Páramo, Subpáramo, and Andean Forest (occasional)	Cundinamarca and Caldas (?)
3	(*)***Pentacalia corymbosa*** (Benth.) Cuatrec. (Compositae-Asteraceae)	1,800 – 3,960 Andean Forest, Sub-Andean Forest, Subpáramo, and Páramo	Antioquia, Boyacá, Cauca, Cundinamarca, Meta, Norte de Santander, and Santander
4	(**)***Berberis goudotii*** Triana & Planch. (Berberidaceae)	1,900 – 3,960 Sub-Andean Forest, Andean Forest, Subpáramo, Páramo, and Superpáramo	Antioquia, Boyacá, Cundinamarca, and Meta
5	(**)***Berberis rigidifolia*** Kunth (Berberidaceae)	1,750 – 4,200 Sub-Andean Forest, Andean Forest, Subpáramo, Páramo, and Superpáramo	Boyacá, Cundinamarca, Norte de Santander, and Santander
6	(**)***Cecropia mutisiana*** Mildbr. (Urticaceae)	500 – 1,800 Lower Neotropical Forest and Sub-Andean Forest	Cauca, Caldas, Cundinamarca, Huila, Quindío, Risaralda, and Tolima
7	(**)***Cecropia telealba*** Cuatrec. (Urticaceae)	1,650 – 2,600 Sub-Andean Forest, Andean Forest	Caldas, Cauca, Risaralda, Valle del Cauca, and Quindío
8	(**)***Chromolaena barranquillensis***[1] (Hieron.) R.M. King & H. Rob. (Compositae-Asteraceae)	0 – 600 Lower Neotropical Forest	Atlántico, Bolívar, Chocó, Córdoba, La Guajira, Magdalena and San Andrés Isla
9	(**)***Espeletia grandiflora*** Humb. & Bonpl. (Compositae-Asteraceae)	2,120 – 4,300 Páramo and sometimes in the upper part of the Andean Forest	Boyacá, Cundinamarca, Chocó (?), Meta and Santander
10	(**)***Espeletiopsis muiska*** (Cuatrec.) Cuatrec. (Compositae-Asteraceae)	2,170 – 3,800 Subpáramo, Páramo and sometimes in the upper part of the Andean Forest	Boyacá, Cundinamarca, Norte de Santander and Santander
11	(**)***Ocotea caparrapi***[1] (Sand.-Groot ex Nates) Dugand (Lauraceae)	1,000 – 2,380 Sub-Andean Forest	Antioquia, Cundinamarca and Santander
12	***Passiflora antioquiensis*** H. Karst. (Passifloraceae)	1,800 – 3,000 Sub-Andean Forest, Andean Forest	Antioquia, Boyacá, Caldas, Cauca, Cundinamarca, Putumayo, Quindío, Risaralda, Tolima and Valle del Cauca
13	(*****Senecio niveoaureus*** Cuatrec. (Compositae-Asteraceae)	2,220 – 4,600 Páramo and sometimes in Andean Forest	Antioquia, Arauca, Boyacá, Caldas, Cauca, Cundinamarca, Huila, Meta, Nariño, Norte de Santander, Quindío, Santander, Tolima and Valle del Cauca

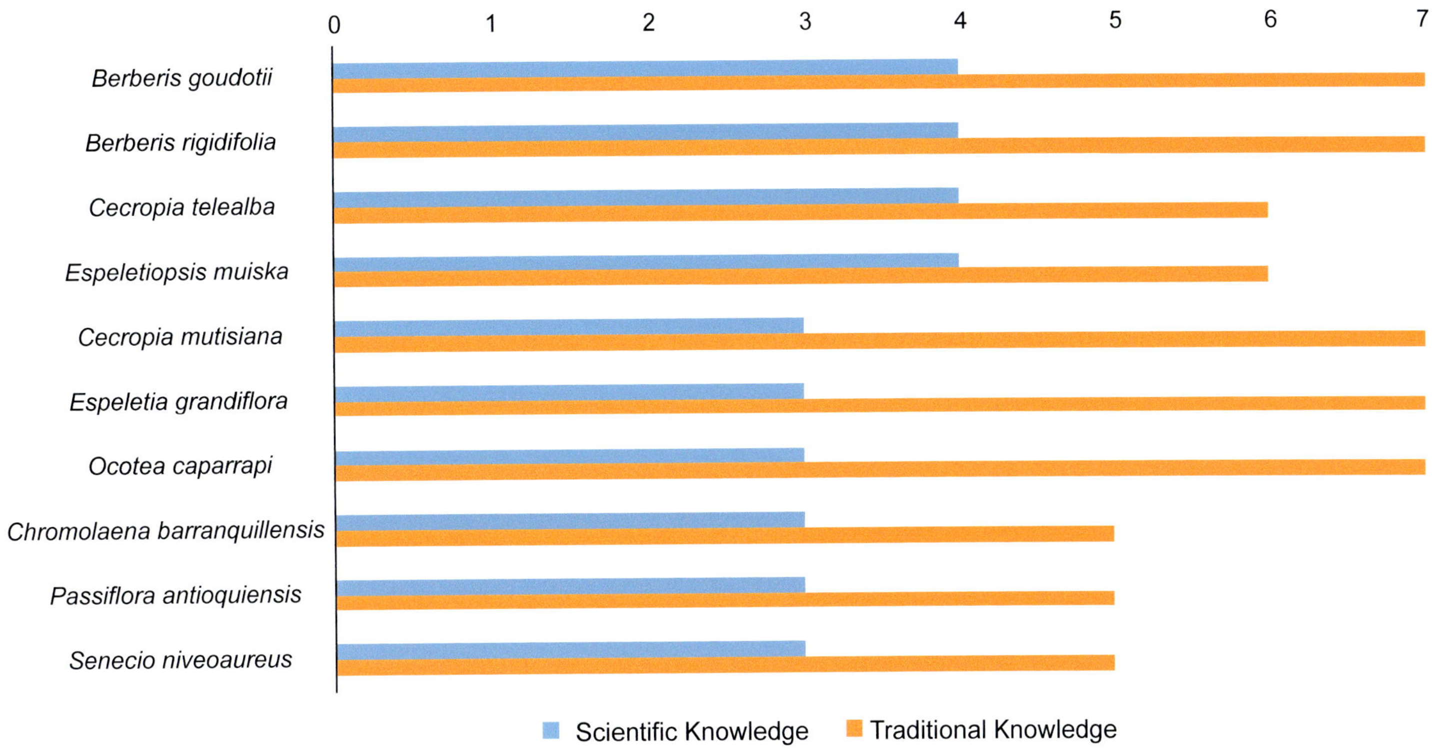

FIGURE 3. Colombian endemic species used as medicinals, presenting 'acceptable' of traditional knowledge evidence and 'regular' scientific knowledge evidence.

In terms of 'acceptable' traditional and 'regular' scientific knowledge, 10 species were recorded: (4) *Berberis goudotii*, (5) *Berberis rigidifolia*, (6) *Cecropia mutisiana*, (7) *Cecropia telealba*, (8) *Chromolaena barranquillensis*, (9) *Espeletia grandiflora*, (10) *Espeletiopsis muiska*, (11) *Ocotea caparrapi*[1], (12) *Passiflora antioquiensis*, and (13) *Senecio niveoaureus* (Table 5, Figure 3).

Whilst only five species presented regular traditional knowledge (*Clusia ellipticifolia*, *Clusia inesiana*, *Hesperomeles goudotiana*, *Salvia bogotensis*, and *Viburnum cornifolium*), 18 species presented regular scientific knowledge (*Baccharis bogotensis*, *Baccharis boyacensis*, *Berberis goudotii*, *Berberis rigidifolia*, *Calea peruviana*, *Cecropia goudotiana*, *Cecropia mutisiana*, *Cecropia telealba*, *Chromolaena barranquillensis*[1], *Chromolaena perglabra*, *Clusia ellipticifolia*, *Espeleptiosis muiska*, *Espeletia grandiflora*, *Ocotea caparrapi*[1], *Passiflora antioquiensis*, *Pentacalia ledifolia*, *Salvia bogotensis*, and *Senecio niveoaureus*), and only three species presented regular technological knowledge (*Achyrocline bogotensis*, *Passiflora antioquiensis*, and *Vasconcellea goudotiana*).

A matrix representing the relation between 28 prevalent pathologies in Colombia and those 13 species used as medicinals with acceptable traditional knowledge and acceptable or regular scientific knowledge is presented (Table 6). The results show, for example, that chronic lower respiratory diseases and acute respiratory infections, ranked, respectively, in third and the fifth place on the list of causes of mortality in Colombia, could be treated by at least by seven different species (1) *Achyrocline bogotensis*, (4) *Berberis goudotii*, (7) *Cecropia telealba*, (8) *Chromolaena barranquillensis*, (9) *Espeletia grandiflora*, (10) *Espeletiopsis muiska*, and (12) *Passiflora antioquiensis*.

Also, species such as *Pentacalia corymbosa* could treat various known diseases such as dental caries, diarrheal diseases, hypertensive diseases, leishmaniasis, *Plasmodium vivax* malaria, and periodontal disease (Table 6).

DISCUSSION

It is important to note that no other country is known for such a high number of native species with potential medicinal uses from the literature available on medicinal plants. The species represent just one of the hidden natural treasures of Colombia that should be broadly known, conserved, managed, and sustainably used by present and future generations. Several native plant species present in Colombia are recognised as having some type of medicinal activity. However, there

---

[1] *Ocotea caparrapi* and *Chromolaena barranquillensis* have been synonymised; however, for this chapter, these taxa do not follow the taxonomic criteria which proposed the synonymy of *Ocotea caparrapi* as *Mespilodaphne cymbarum* (*Ocotea cymbarum*) and *Chromolaena barranquillensis* as *Chromolaena odorata*. A different point of view has been prepared by the authors as a publication on the taxonomy, nomenclature, geographical distribution, and ecological status of these two species, which are recognized as exclusive to Colombia.

TABLE 6. Matrix of prevalent pathologies in Colombia and endemic medicinal species of Colombia with acceptable traditional knowledge and acceptable or regular scientific knowledge. Codes (present = 1 and absent = 0) represents mentions of any therapeutic use of each species for a prevalent disease in Colombia and shows the total of species that have uses mentioned for a certain pathology and the number of mentions of use for each species for these pathologies.

Prevalent pathologies/species	1	2	3	4	5	6	7	8	9	10	11	12	13	Total
MORTALITY														
Ischemic heart disease	0	0	0	1	1	0	0	0	0	0	0	0	0	2
Cerebrovascular diseases	0	0	0	0	0	0	0	0	0	0	0	0	0	0
Chronic lower respiratory diseases	1	0	0	1	0	0	1	1	1	1	0	1	0	7
Mellitus diabetes	0	0	0	1	0	0	0	0	0	0	1	0	0	2
Acute respiratory infections	1	0	0	1	0	0	1	1	1	1	0	1	0	7
Hypertensive diseases	0	1	1	0	0	0	0	0	0	0	0	0	0	2
Malignant stomach tumour	1	0	0	0	0	1	0	1	0	1	0	0	0	4
Malignant tumour of the trachea, bronchi, and lung	1	0	0	0	0	1	0	1	0	1	0	0	0	4
Kidney disease pain	1	0	0	0	0	0	0	0	0	0	0	0	0	1
Heart failure	0	0	0	1	1	0	0	0	0	0	0	0	0	2
MORBIDITY														
Periodontal disease	0	0	1	1	0	0	0	0	0	0	0	0	0	1
Dental caries	0	0	1	1	0	0	0	0	0	0	0	0	0	1
Iron deficiency anaemia	0	0	0	0	0	0	0	0	0	0	0	0	0	0
Asthma	0	0	0	0	1	0	1	1	1	0	0	0	0	4
Trichuriasis by *Trichuris trichiura*	0	0	0	0	0	0	0	0	0	0	0	0	0	0
Hypertensive heart disease	0	0	1	0	0	0	0	0	0	0	0	0	0	1
Protein-calorie malnutrition	0	0	0	0	0	0	0	0	0	0	0	0	0	0
Hookworm	0	0	0	0	0	0	0	0	0	0	0	0	0	0
Abortion	0	0	0	0	0	0	1	0	0	0	0	0	0	1
Unipolar major depression	0	0	0	0	0	0	0	0	0	0	0	0	0	0
Diarrheal diseases	0	0	1	0	0	0	0	0	0	1	0	0	0	2
EPIDEMIOLOGICAL SURVEILLANCE														
Malaria by *Plasmodium vivax*	0	1	1	0	0	0	0	0	0	0	0	0	0	2
Leishmaniasis	0	1	1	0	0	0	0	0	0	0	0	0	0	2
Classic Dengue	0	0	0	0	0	0	0	0	0	0	0	0	0	0
Malaria by *Plasmodium falciparum*	0	0	0	0	0	0	0	0	0	0	0	0	0	0
Tuberculosis	0	0	0	0	0	0	0	0	0	0	0	0	0	0
AIDS HIV	0	0	0	0	0	0	0	0	0	0	0	0	0	0
Dengue haemorrhagic fever	0	0	0	0	0	0	0	0	0	0	0	0	0	0
TOTAL	5	3	7	7	3	2	4	5	3	5	1	2	0	------

are no detailed publications for each species on the current state of traditional, scientific, and technological knowledge. Also, there is no published compilation to facilitate the identification and characterisation of species and with the potential to promote their use in primary health care. Likewise, there is also a lack of research, development, and innovation projects (I+D+i) (especially for endemic species) to allow the chemical-pharmaceutical synthesis or the sustainable use of raw materials to elaborate pharmaceutical preparations and phytotherapeutic products.

Achyrocline bogotensis, *Espeletia killipii*, and *Pentacalia corymbosa* are endemic plant species of Colombia used as medicinals that present the most extensive information on traditional and scientific knowledge, and used to treat and prevent some of the most prevalent pathologies within the country and in other tropical countries (see Table 6). However, it is still necessary to carry out studies focusing on these species regarding safety policies (toxicity) and in-depth investigations on efficacy policies for medicinal use, including phytochemical investigations, investigations of *in vivo* and *in vitro* biological activities and clinical studies. In the same way, it is still necessary to carry out new studies to expand the knowledge on agrotechnologies for organic and sustainable cultivation and on technologies for the elaboration of pharmaceutical preparations and phytotherapeutic products made from those three species.

Mentions of the traditional use of *Achyrocline bogotensis* are very varied, with most of the available information relating to conservation status and sustainable use. Regarding the evidence of traditional and scientific knowledge, the information on anti-inflammatory activity demonstrated from studies is remarkable. Several scientific investigations show that this species is also antioxidant, anticancer and has biocidal activities (García 1992; Pombo, 2003; Bernal *et al.*, 2011). This species also has trypanocidal activity against *Trypanosoma epimastigotes* and proven properties against *Staphylococcus aureus*, *Staphylococcus epidermidis*, *Staphylococcus pyogenes*, *Mycobacterium avium*, *Mycobacterium abscessus*, *Mycobacterium fortuitum*, *Mycobacterium chelonae*, *Microsporum gypseum*, and *Trichophyton mentagrophytes* (Bernal *et al.*, 2011; Baldisserra *et al.*, 2014). Evaluation of the extracts and fractions of leaves, stems and flowers demonstrated *in vitro* antiviral activity against a rotavirus (RRV strain) and an astrovirus (Yuc8 strain) (Téllez *et al.*, 2015). In the Colombian phytotherapeutic market, a product of *Achyrocline bogotensis* registered by the National Institute of Food and Drug Surveillance (Invima) (No. PFT2018-0000487-R1) presents sale conditions with medical formula as adjunct support in the treatment of mild inflammation of the lower urinary tract. Also, Téllez *et al.* (2015) developed a phytotherapeutic product to treat acute diarrheal disease on an industrial pilot scale obtained from the aerial parts of this species. The bioactive fraction and the formulation of the phytotherapeutic product were patented in 2017 (Patent No. 16-017515 Superintendency of Industry and Commerce Colombia). It is important to record the scientific findings on *Achyrocline bogotensis* because no other specific antiviral therapy reduces acute diarrheal illness. The World Health Organization (WHO, 2021) estimates that each year there are 1,700 million episodes of diarrhoea in children under five years in developing countries and 760,000 deaths due to acute diarrhoea related to 50–70% of cases with dehydration. For example, rotavirus alone produces about 500,000 deaths of children under five years old, and the protocols described by the WHO to treat this pathology are solely aimed at the rehydration of patients.

For *Espeletia killipii*, the small farmers of Colombia agree on their traditional medicinal uses, mainly for asthma and lung diseases, which seem to be endorsed by the scientific research that has been carried out (Torrenegra *et al.*, 1994; Bernal *et al.*, 2011). According to Torrenegra & Téllez (1996), the kaurenoic derivatives of this plant are toxic to the potato (*Solanum tuberosum*) white worm (*Premnotrypes vorax*). This species also presents marked antibacterial activity against *Bacillus subutilis*, *Salmonella tiphimurium*, *Staphylococcus aureus*, and *Streptococcus pneuminiae* (Bernal, in prep.).

In general, the traditional uses of *Pentacalia corymbosa* are not aligned with the results of the scientific investigations. In Colombian popular medicine, this plant is used as a vulnerary alexipharmaceutical and antisyphilitic. Several chemical compounds identified from this species are antioxidants (Uribe-Holguín, 2010), and they act against dermatophyte (*Trichophyton mentagrophytes*), have cytotoxic and antitumor action and are antifungal against *Botrytis cinerea*, *Fusarium oxysporum*, and *Trichoderma viride* (Torrenegra *et al.*, 2000).

From the ten endemic medicinal species in Colombia that have acceptable traditional knowledge and regular scientific knowledge (see Figure 3), the species from the genera *Berberis*, *Cecropia*, *Chromolaena*, and *Ocotea* stand out. *Berberis goudotii* and *Berberis rigidifolia* are used in popular medicine as a febrifuge, cutaneous leishmaniasis treatment, diaphoretic, haemostatic, treatment for pyorrhoea, purgative, remedy for throat infections, and tonic (García, 1974; Bernal *et al.*, 2011; Sequeda-Castañeda *et al.*, 2019). These species can play a significant role in the treatment or prevention of diseases that are important causes of morbidity in Colombia, since it has been proven that they contain chemical compounds that can act in a curative way in certain forms of cutaneous leishmaniasis which is a fairly frequent epidemiological surveillance disease in the country. More recently, a preliminary phytochemical analysis of *Berberis goudotii* determined that this species presents antimicrobial activity against bacteria associated with caries and periodontal disease (Sequeda-Castañeda *et al.*, 2019).

Cecropia mutisiana and *Cecropia telealba* are myrmecophilic species that are important for feeding birds and mammals in the Southern Neotropical rainforest. The Andean and sub-Andean forests of Colombia have many uses in traditional medicine (Bernal 2012). However, there are only a few scientific investigations to verify their biological activities (Sequeda-Castañeda *et al.*, 2015). The main uses in traditional medicine are as antiasthmatics, antibilious medicines, arthritis treatments, cardiotonics, chorea medicines, diuretics, hypoglycaemics, agents to improve nervous mobility of the body, and rheumatism remedies.

Cecropia mutisiana extracts have cardiotonic activity (Pinzón & Eraso, 1975). Something particular demonstrated by Zambrano-Ospina (2000) is that although the extract of the leaves is not effective against chronic seizures induced by pentylenetetrazol (PTZ), nor does it increase the latency time of seizures when used at normal doses and under test conditions; doses of 1,000 mg/kg protect 40% of the test animals from death.

From this group of 10 endemic medicinal species, *Chromolaena barranquillensis* has traditional uses mentioned as an antioxidant and cytotoxic plant, and as a treatment for boils, inflammations, and skin ulcers. There have been phytochemical studies on this species (Muñoz-Acevedo *et al.*, 2011). The antioxidant activity of the flowers and leaves of this species was also verified (Muñoz-Acevedo *et al.*, 2011). *Ocotea caparrapi* is a tree up to 30 m high that grows between 1,000 and 2,380 m a.s.l. in the sub-Andean forests of Colombia. Its oil, called "caparrapi oil", has a very broad historical record of traditional uses: as an alexipharmic, alexiteric, anticancer agent, antidote against wasp and scorpion stings, antivenom, antirheumatic, antiviral, mosquito repellent, cosmetic, anti-inflammatory, vulnerary, and veterinary medicine, as well as as a treatment for gonorrhoea, hair loss, bruises, diarrhoea, epigastric pain, eczema, skin diseases, erysipelas, intermittent fevers, malarial fevers, herpes, hydrophobia, ant (diseases of the mule and horse hoof), leprosy, lepromas, epithelial lesions, leucorrhoea, snake and rabid dog bites, insect bites, stingray stings, scabies, ringworm, tumours, ulcers, atonic ulcers, warts, and bronchial, skin and lung infections (extensive literature reviewed by Bernal, in prep.). This species has many phytochemical studies carried out by various authors between 1950 and 2010, and there are some basic studies on skin cancers (epitheliomas) (Bernal, in prep.).

This research corroborated the lack of knowledge on the conservation and sustainable use of medicinal species in Colombia, especially endemic species. All the endemic species to Colombia used as medicinals lack clinical studies or, at the very least, preclinical studies (including *in vitro* and *in vivo* biological and pharmacological activities, pharmacodynamics, pharmacokinetics, and kinetics). Most of these species have few scientific and technological studies. Only *Achryrocline bogotensis*, *Espeletiopsis muiska*, and *Vasconcellea goudotiana* have advances in this type of knowledge. The species *Achyrocline bogotensis*, *Ocotea caparrapi* and *Salvia bogotensis* present preclinical studies of biological and pharmacological activity *in vitro* and *in vivo*. The species *Berberis rigidifolia* and *Clusia ellipticifolia* have preclinical studies of biological and pharmacological activity *in vivo*; whilst the species *Cecropia goudotiana*, *Cecropia telealba*, *Chromolaena perglabra*, *Espeletia grandiflora*, *Espeletia killipii*, *Espeletiopsis muiska*, *Passiflora antioquiensis*, *Pentacalia corymbosa*, *Pentacalia ledifolia*, *Senecio niveoaureus*, and *Vasconcellea goudotiana* have preclinical biological and pharmacological studies of activity. In general, all endemic medicinal species of Colombia show insufficient knowledge about agrotechnologies for their organic and sustainable cultivation and studies on agroforestry in polyculture. Likewise, there are no technological studies on raw materials and elaboration of medicinal products from these species.

The native medicinal species of Colombia, especially the endemic species, constitute a true natural heritage that requires concrete and innovative actions to develop knowledge, conservation, and sustainable use. Several actions are currently taking place at the Alexander von Humboldt Biological Resources Research Institute, whose next publication will document the updated inventory and monographs on the current state of traditional, scientific, and technological knowledge on the main endemic medicinal plants in Colombia (Bernal, in prep.). Also, there are great efforts and initiatives from the Quindío Botanical Garden to establish on its premises the "National Collection of Medicinal Plants", supported by the National Botanical Gardens Network of Colombia, The Quindío Botanical Garden also and which seeks to establish a living reference collection for the country's native medicinal plants with an emphasis on endemic species (Bernal & Gómez, 2021).

In general, the native medicinal species of Colombia (especially the South American and Andean species present in the country and several of its endemic species) have great potential to prevent or treat the prevalent pathologies in Colombia and surely in other mainly tropical countries. However, scientific, and technological studies must be continuous to allow the conservation and sustainable use of medicinal species.

We strongly recommend the enrichment of programs and projects on research, development and innovation (R + D + I) which aim: i) to conserve and take advantage of species sustainably and profitably, either through the route of chemical-pharmaceutical synthesis for the generation of international exchange for patenting, or through the manufacture of medicinal products that serve to improve social security systems in health; ii) to reduce imports of phytotherapeutic products, and iii) to generate international exchange for exports of medicinal products made from native medicinal species to Colombia, in particular from those species that are to this country, which by their very nature present differential competitive advantages in the market of medicinal products at national and international levels.

Acknowledgements

The authors would like to thank Hernando García Barriga (q.e.p.d), Professor-Investigator at the Institute of Natural Sciences of the National University of Colombia; Elizabeth Hodson de Jaramillo, Former Director of the Department of Biology of the Pontificia Universidad Javeriana, and Hernando García Martínez, Director of the "Alexander von Humboldt" Institute for Research on Biological Resources, who have studied, and have always supported the scientific research, on the flora of Colombia and in particular the native medicinal plants of the country, and have been able to collect Colombian plants and put into practice a phrase by Enrique Pérez Arbeláez, who in 1956 pointed out that "a plant or animal cannot be considered as a natural resource, until when, as result of the investigation, be put at the service of humans".

References

Andrade C (2011) Estado del conocimiento de la biodiversidad en Colombia y sus amenazas. Consideraciones para fortalecer la interacción ciencia-política. *Rev. Acad. Colomb. Cienc. Exac. Fís y Nat.* 35 (137): 491-508.

Baldissera MD, Oliveira CB, Zimmermann CEP, Boligon AA, Athayde ML, Bolzan LP, Vaucher RA, Santurio JM, Sagrillo MR, da Silva AS, Monteiro SG (2014) *In vitro* trypanocidal activity of macela (*Achyrocline satureioides*) extracts against *Trypanosoma evansi*. *Korean J. Parasitol*. 52(3):311–315. doi: 10.3347/kjp.2014.52.3.311

Bernal, H. (in prep.). Especies medicinales endémicas de Colombia. Estado actual de conocimientos tradicionales, científicos y tecnológicos. Bogotá, Instituto de Investigación de Recursos Biológicos Alexander von Humboldt. 200 p.

Bernal H (2012). La biodiversidad como expresión humana. In: *La Javeriana explora el futuro de Colombia*. eds. FJ Liévano, G Hernández and SJ Castellanos Ramírez, pp. 133-140. Editorial Pontificia Universidad Javeriana.

Bernal H, Correa J (1989-1998) *Plantas medicinales de Colombia. In: Especies vegetales promisorias de los países del Convenio Andrés Bello*. ed. Secretaría Ejecutiva del Convenio Andrés Bello. Vols. 4, 6, 7, 8, 10, 12. Editora Guadalupe Ltda.

Bernal H, Gómez A (2021) *Plantas medicinales nativas de Colombia*. In: Proyecto colección nacional de plantas medicinales. 2nd edition. Calarca, Jardín Botánico del Quindío. 173p.

Bernal H, Jiménez L (2008) Plantas medicinales de Colombia. In: MP Gupta (ed) *Plantas medicinales iberoamericanas*. Bogotá, Editorial Presencia. 560p.

Bernal H, Mesa C (2014) Plantas medicinales endémicas de Colombia. 476 registros. Metadatos. Available at: http://ipt.sibcolombia.net/sib/resource.do?r=puj_002

Bernal H, García H, Quevedo G (comp.), Guerra Editores (ed) (2011) *Pautas para el conocimiento, conservación y uso sostenible de las plantas medicinales nativas de Colombia*. Bogotá, Instituto de Investigaciones de Recursos Biológicos Alexander von Humboldt. 232 p.

Bernal HY & Grupta M (1995) 270 plantas medicinales iberoamericanas. Secretaría Ejecutiva del Convenio Andrés Bello. 620 Pp. ISBN 10958-9206-50-6

Correa J, Bernal H (1989-1995) Plantas medicinales de Colombia. In: *Especies vegetales promisorias de los países del Convenio Andrés Bello*. ed. Secretaría Ejecutiva del Convenio Andrés Bello. Vols. 1, 3, 5, 7, 9, 11. Editora Guadalupe Ltda.

Cuatrecasas J (1958) Aspectos de la vegetaciòn natural de Colombia. *Rev. Acad. Col. Cienc. Ex. Fìs. Nat.* 10(40):221-268.

Fonnegra, R. & Jiménez, S. (1992). Plantas medicinales aprobadas en Colombia. Primera edición. Medellín, Colombia, Editorial Universidad de Antioquia. 368 p.

Fonnegra, R. & Jiménez, S. (2006). Plantas medicinales aprobadas en Colombia. 2nd edition. Editorial Universidad de Antioquia. Medellín, Colombia. 370p. ISBN 139789586559997

García Barriga H (1974) Flora medicinal de Colombia. Botánica médica. Bogotá, Instituto de Ciencias Naturales. Imprenta Nacional. 561p.

García Barriga H (1975) Flora medicinal de Colombia. Botánica médica. Vols. 2, 3. Bogotá, Imprenta Nacional.

García Barriga H (1992) Flora medicinal de Colombia. Botánica médica. 2nd edition. Vol 1, 2. Bogotá, Tercer Mundo Editores.

González Patiño D (1984) Utilización terapéutica de nuestras plantas medicinales. Un resumen de materia médica. Bogotá, Tercer Mundo Editores. 326p.

Hernández Mesa M (1992) Plantas colombianas su aplicación medicinal. Bogotá, Editorial Presencia. 290p.

Martínez J, Bernal H, Cáceres A (eds) (2000) Fundamentos de agrotecnología de cultivo de plantas medicinales iberoamericanas. Bogotá, Quebecor Impreandes. 524p.

Mesa C (2013) Plantas medicinales endémicas de Colombia y su potencial para el tratamiento de las patologías de mayor impacto en el país. Degree diss., Facultad de Ciencias, Pontificia Universidad Javeriana. Bogotá. 117p.

Ministerio de Protección Social (2008) Vademécum colombiano de plantas medicinales. Bogotá, Arte y Sistemas Integrados Ltda. 241p.

Muñoz-Acevedo A, Álvarez Carrillo VC, Niño Porras ME (2011) Chemical characterization of the volatile fractions and essential oils of leaves and flowers of Chromolaena barranquillensis from Sabanalarga (Atlántico, Colombia). *Bol Latinoam Caribe Plant Med Aromat* 10(6): 581-589.

Montes Giraldo J (1981) Medicina popular en Colombia. Vegetales y otras sustancias usadas como remedios. Yerbabuena, Instituto Caro y Cuervo.

Patiño V (1967) Plantas cultivadas y animales domésticos en América Equinoccial. Cali, Imprenta Departamental. 572p.

Pérez Arbeláez E (1953-1966) Recursos naturales de Colombia: su génesis, su medida, su aprovechamiento, conservación y renovación. Vol. 1, 2. Bogotá, Imprenta del Banco de la República. 334p.

Pérez Arbeláez E (1978a) Plantas medicinales y venenosas de Colombia. Estudio Botanico, étnico, Farmaceutico, veterinario y forense. ed. Medellín, Colombia. 295 p.

Pérez Arbeláez E (1978b) Plantas útiles de Colombia. 4th edition. Bogotá, Litografía Arco. 831 p.

Pombo LM (2003) Caracterización fitoquímica y farmacológica del extracto de la planta vira-vira (*Achyrocline bogotensis* H.B.K.). *Documenta Clínica* 16:11-18.

Pinzón R, Eraso S (1975) Análisis fitoquímico y farmacológico del guarumo (Cecropia mutisiana). Bogotá, Colombia: Universidad Nacional de Colombia.

Sequeda-Castañeda L, Muñoz-Realpe C, Celis-Zambrano C, Gutiérrez-Prieto S, Luengas-Caicedo P, Gamboa F (2019) Preliminary phytochemical analysis of *Berberis goudotii* Triana & Planch. ex Wedd. (Berberidaceae) with anticariogenic and antiperiodontal activities. *Scientia Pharmaceutica*, 87(1): 2. doi: 10.3390/scipharm87010002

Sequeda-Castañeda LG, Célis C, Ortíz-Ardila AE (2015) Phytochemical and therapeutic use of *Cecropia mutisiana* Mildbr. (Urticaceae) an endemic plantfrom Colombia. *Harmacolog Onl.*, 3:62–64.

Téllez MA, Téllez AN, Vélez F, Ulloa JC (2015) In vitro antiviral activity against rotavirus and astrovirus infection exerted by substances obtained from *Achyrocline bogotensis* (Kunth) DC. (Compositae). *BMC Complement Altern Med*, 15(1):1–11 doi: 10.1186/s12906-015-0949-0

Torrenegra RD, Pedrozo J, Téllez AN, Cabeza G, Granados A, Méndez D (2000) Chemistry and antifungal activity of *Pentacalia corymbosa* (Asteraceae - Senecioneae). *Rev Latinoam de Química*, 28: 31–34.

Torrenegra RD, Téllez AN (1996). Phytochemistry of *Espeletia killipii* Cuatr. and gibberellic activity of some of the isolated compounds. *Rev Latinoamer Quím*, 24, 2–6.

Torrenegra RD, Téllez AAN, García G (1994) Quimica de especies del género *Espeletia*: *Espeletia killipii* - *Espeletia tunjana*. *Rev Colomb de Química*, 23(2): 29–35

Uribe-Holguín C (2010) Evaluación de la actividad antioxidante de las hojas de *Pentacalia corymbosa* y *Pentacalia nitida* (Asterales: Asteraceae). Pontificia Universidad Javeriana.

Vásquez C, Bernal H (2011) Plantas medicinales de Colombia: origen, uso tradicional, eficacia, seguridad y aplicabilidad desde el sistema de salud. In: *Pautas para el conocimiento, conservación y uso sostenible de las plantas medicinales en Colombia*. 45-130pp. Bogotá, Alianza Ediprint Ltda., Instituto de Investigaciones de Recursos Biológicos Alexander von Humboldt.

WHO (World Health Organization) (2021) Temas de salud. Mortalidad y morbilidad. Available at: https://www.who.int/home/search. Accessed in June 2021.

Zambrano-Ospina AM (2000) Evaluación de la actividad ansiolítica y anticonvulsivante de las hojas de *Cecropia mutisiana*. Bogotá, Universidad Nacional de Colombia.

Boronía (confit aubergine with mashed plantain, crispy chickpeas and edible flowers and leaves); recipe from the Celele restaurant, in Cartagena, Colombia). .
Mauricio Diazgranados

Chapter 4

The hidden food basket of Latin America: an overview of Colombian edible plant diversity and its distribution

Benedetta Gori[1]*, Tiziana Ulian[1], Henry Yesid Bernal[2] & Mauricio Diazgranados[1]*

[1] Royal Botanic Gardens, Kew

[2] Ex-Profesor Asociado del Departamento de Biología, Facultad de Ciencias, Universidad Javeriana, Herbario de la Pontificia Universidad Javeriana (HPUJ), Bogotá, D.C., Colombia. henryesid@gmail.com

***Corresponding authors:** b.gori@kew.org, m.diazgranados@kew.org

Keywords: Edible plants, Colombia, biogeography, conservation.

ABSTRACT

Colombia is the second most biodiverse country globally, hosting more than 30,000 plant species and an extraordinary diversity of natural ecosystems. More than 7,000 of these species have reported human uses, and 3,806 of them are food plants. However, despite the high availability of natural resources, Colombia is characterised by widespread poverty and food insecurity. With growing attention being paid to the economic and nutritional benefits of neglected and underutilised edible plants, the present chapter explores the unrivalled diversity of edible plants in Colombia, providing a basis for future research on priority edible species for conservation and sustainable use in the country. It is focused on four fundamental questions: What does Colombian edible plant diversity look like? What is the distribution of Colombian edible species diversity across different bioregions? What is the conservation state of Colombian edible plant diversity *in-situ* and *ex-situ*? And, finally, what are the underutilised edible species that have the potential to improve the food security of the local communities and to be commercialised in national and international markets? Results show that the incredible biocultural richness of Colombian edible plants currently receives inadequate conservation actions, both *in-situ* and *ex-situ*. There are significant gaps in knowledge regarding the conservation status of some of the most important edible plant genera. The biogeographic analysis highlights the "Tropical Andes" as a diversity hotspot for edible plants, emphasising the urgent need for the preservation of this habitat. Further research is critical to understanding the complex interactions between people, food and the natural environment. At the same time, conservation efforts need to consider the intrinsic value of edible plant diversity and to identify its full potential for socio-economic development at the local and national levels.

RESUMEN

Colombia es el segundo país con mayor biodiversidad del mundo. Alberga más de 30.000 especies de plantas, así como una extraordinaria diversidad de ecosistemas naturales. Más de 7.400 especies presentan registros de usos humanos, y 3.806 de ellas son plantas alimenticias. Sin embargo, a pesar de la alta disponibilidad de recursos naturales, Colombia se caracteriza por una pobreza e inseguridad alimentaria generalizadas. Siguiendo la creciente atención a los beneficios económicos y nutricionales de las plantas comestibles desatendidas y subutilizadas, el presente capítulo explora la diversidad inigualable de plantas comestibles en Colombia, con el objetivo de proporcionar una base para futuras investigaciones sobre especies comestibles prioritarias para la conservación y el uso sostenible en el país. Este estudo se centra en cuatro preguntas fundamentales: ¿cómo es la diversidad de plantas comestibles colombianas? ¿cuál es la distribución de la diversidad de especies comestibles colombianas en diferentes bioregiones? ¿cuál es el estado de conservación *in-situ* y *ex-situ* de la diversidad de plantas comestibles colombianas? y ¿cuáles son las principales especies alimenticias subutilizadas de Colombia, que pueden ser usadas para mejorar la seguridad alimentaria de las comunidades locales y que tienen potencialidades económicas en los mercados nacionales e internacionales? Los resultados muestran que la increíble riqueza biocultural de las plantas comestibles colombianas recibe en la actualidad acciones de conservación inadecuadas, tanto *in situ* como *ex-situ*, y existen importantes lacunas en el conocimiento sobre el estado de conservación de algunos de los géneros de plantas comestibles más importantes. El análisis biogeográfico destaca las bioregiones "Andes" y "Páramo" como puntos críticos de diversidad para las plantas comestibles, enfatizando la urgente necesidad de su preservación. Es fundamental realizar más investigaciones para comprender las complejas interacciones entre los pueblos, los alimentos y el medio ambiente natural, mientras que los esfuerzos de conservación deben considerar el valor intrínseco de la diversidad de plantas comestibles e identificar todo su potencial para el desarrollo socioeconómico a nivel local y nacional.

INTRODUCTION

Food represents one of the deepest forms of interaction between humans and the environment. It lies at the foundation of human experience, shaping our relationship with non-human living beings and embedding forms of intangible cultural legacy. According to the World Checklist of Useful Plant Species (Diazgranados *et al.*, 2020), more than 7,000 plants are currently known to be edible. Many of them are part of the traditional gastronomic heritage of human populations and have the potential to support food security and develop sustainable agriculture around the world (Ulian *et al.*, 2020). However, only ten species contribute to almost all human caloric intake nowadays (FAO, 2015). This incongruous trend was triggered by the green revolution (Wolff, 2004), which started to replace traditional landraces and wild species with a restricted assortment of modern commercial hybrids by favouring yield production (Paludosi *et al.*, 2011). The downsides of such direction were manifold: on the one hand, people started to diminish their interest in local edible plants while progressively decreasing their attention on the wellbeing of the ecosystems hosting them, which led to their degradation (Kalamandeen *et al.*, 2018). On the other hand, increasing pressure on a narrow portion of natural resources, together with unsustainable cultivation and harvesting practices, resulted in a rapid depletion of natural plant populations (Pilgrim *et al.*, 2008; Nogueira *et al.*, 2011).

In response to this trend, international policy frameworks combining biodiversity protection and sustainable development have gained attention in recent decades (FAO, 2019). Efforts have increased focus on the revitalisation and greater use of what Borelli *et al.* (2020) defined as "orphan crops" and wild edible plants (Borelli *et al.*, 2020). Over the past few decades – more than 50 years after the green revolution – "Neglected and Underutilised Species" (NUS) have been shown to hold crucial importance for building sustainable livelihoods and mitigating environmental deterioration (Hunter *et al.*, 2019; Borelli *et al.*, 2020; Ulian *et al.*, 2020). They are also known to hold critical biocultural values, as they are linked to local agricultural and culinary traditions and practices, symbolising the organic relationship between nature and culture. Growing evidence has demonstrated that the value that peoples give to local plant resources can play a crucial role in their engagement in conservation and sustainable management (Brehm *et al.*, 2010; N'Danikou *et al.*, 2011). "Conservation-through-use" approaches – aimed at encouraging nature conservation through the sustainable use of its resources – are increasingly being applied in conservation programs worldwide (Kor *et al.*, 2021).

Colombia is one of the world's "megadiverse" countries (Renjifo *et al.*, 2020). More specifically, it is the second most biodiverse country on the planet, hosting 10% of the world's total biodiversity (Clerici *et al.*, 2019) and bringing together an unequalled number of distinct natural ecosystems and human cultures. However, despite its unrivalled biocultural richness, Colombia is nowadays characterised by widespread poverty and food insecurity (Hurtado-Bermudez *et al.*, 2020). While local communities had historically taken extensive advantage of local plant diversity, such diversity has been increasingly neglected over the past decades and replaced by high yielding commercial varieties (Grau & Aide, 2008). In addition, new land uses are now causing landscape conversion, driven by export-oriented industrial and agricultural policies and market conditions (Grau & Aide 2008; Boron *et al.*, 2016).

Considering these circumstances, investigating, and understanding the complex interactions between people, food, and their environment (Pieroni *et al.*, 2016) acquires crucial importance for formulating targeted and effective conservation and development activities at the local level. Within Colombian edible plants, numerous NUS hold the potential to address environmental degradation while creating sustainable livelihoods and boosting a new green growth, although being unknown outside of the country (Padulosi *et al.* 2011; Ulian *et al.* 2020). By combining biodiversity informatics and taxonomic approaches, this chapter reviews Colombian edible plant diversity, estimating its biogeographical distribution, highlighting biodiversity hotspots, and providing the basis for future efforts towards the revitalisation and conservation of edible plants in Colombia.

MATERIALS AND METHODS

Methods used for this study followed the approach employed by Diazgranados *et al.* (2020) for the World Checklist of Useful Plant Species. The Checklist added information from three large international datasets to the first State of the World's Plants report published in 2016 (RBG Kew, 2016). To date, the World Checklist of Useful Plant Species is the most comprehensive piece of work on useful plants at a global scale, containing key taxonomic and ethnobotanical information on 40,292 species. Among all of the plants classified as "useful", 7,039 are classified under the category of "Human Food", according to the Economic Botany Data Collection Standard categorisation (Cook, 1995). The same taxonomic backbone of the World Checklist of Useful Plant Species was adopted to reconcile taxonomically species from a combination of ten new Colombian and international datasets and publications, in addition to data from the Checklist (cf. Annotated Checklist of Useful Plants of Colombia). The new datasets were cleaned using RStudio4.1.0 v. (R Core Team, 2021). Taxon names were reconciled to Plants of the World Online (POWO, 2021) and, occasionally, to Tropicos (Missouri Botanical Gardens, 2021) using the "Plyr" and "Dplyr" packages (Wickham, 2011; Wickham & Francois, 2021). Higher taxonomy information was obtained from the Global Biodiversity Information Facility (GBIF, 2021). Data on species edibility was retrieved from the cleaned datasets. Data on the *in-situ* conservation actions for edible species were obtained from the IUCN Red List of Threatened Species (IUCN, 2021). PlantSearch (BGCI, 2021), a platform developed by Botanic Gardens Conservation International that contains taxon-level data from worldwide gene and seed banks, was used to gather data on *ex-situ* conservation actions.

Occurrences of edible species were downloaded from GBIF (2021) through the "RGbif" package (Chamberlain *et al.*, 2021) and cleaned using the "ShinyCCleaner" package (Ondo, in prep.) in RStudio. Filters for removing occurrences recorded within urban centres and institutions, areas (e.g., Botanic gardens and *ex-situ* repositories) were also applied to remove points in the sea and the centre of Colombia (i.e., centroid points). Moreover, latitude and longitude points with fewer than three decimals were removed, as were occurrences with equal latitude and longitude and occurrences with either latitude or longitude equal to zero. Mapping of edible species across Colombian departments and ecoregions was carried out in R 4.1.0 (R Core Team, 2021). Heatmaps showing species richness per department and grid were developed using the "ggplot2" package (Wickham, 2016). Grid analysis to measure species richness by ecoregions was carried out using the "rgdal" package (Bivand *et al.*, 2021), *raster* (Hijmans & van Etten, 2012), and *sp* (Bivand *et al.*, 2013). ArcGIS pro 2.8.1 was also employed to carry out biodiversity quantification. SDMtoolbox (Brown *et al.* 2017) – a python-based toolbox for spatial analysis – was used to obtain several biodiversity metrics such as species richness (i.e., the sum of species per cell), weighted endemism (i.e., the sum of the proportion of species' ranges found in a given grid cell – emphasising areas that are rich in species with restricted distributional ranges) and corrected weighted endemism (i.e., weighted endemism divided by the total number of species in a cell – emphasising areas that are rich in species that have restricted ranges, but that are not necessarily species-rich), employing a geographic resolution of 0.1 degrees.

RESULTS AND DISCUSSION

Overview of Colombian edible plants

The current work resulted in the cataloguing of 7,205 useful species found in Colombia, among which 3,805 were categorised as edible (i.e., characterised by a history of consumption by human populations). If compared to the data showcased by the World Checklist of Useful Plants (Diazgranados *et al.*, 2020), this number acquires crucial significance: in fact, the proportions of useful species and edible species at the global level (respectively, 40,292 and 7,039 in number) are remarkably lower than those specific to Colombia. According to the Economic Botany Data Collection Standard (Cook, 1995), at the global level, only 17.4% of useful plants have been recorded as "Human food" (Diazgranados *et al.*, 2020). In Colombia, this percentage rises to 53.6%, making this country a global reservoir for edible plant diversity. The distribution of Colombian edible species across different departments is showcased in Figures 1 and 2.

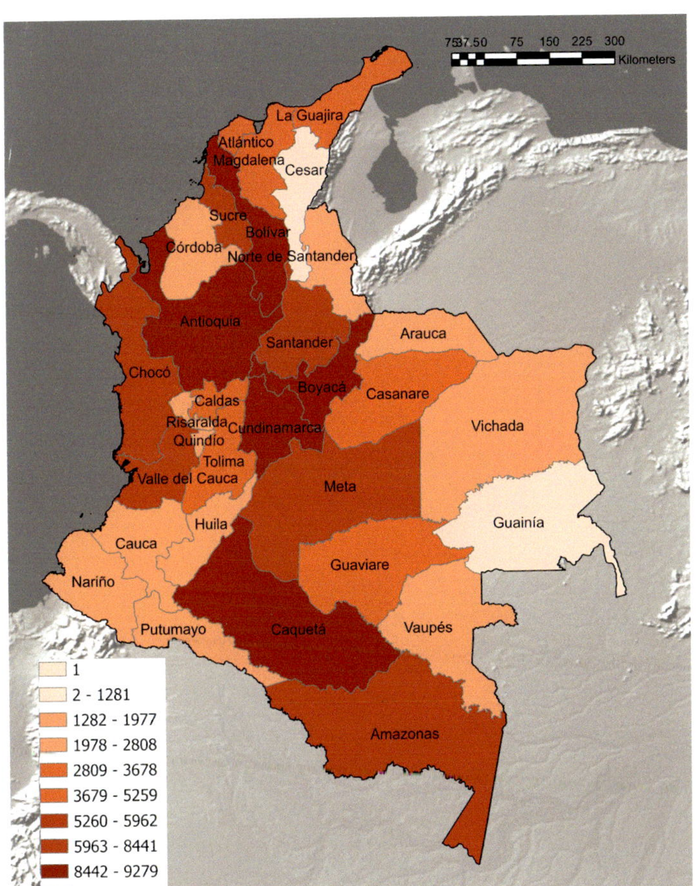

FIGURE 1. Richness of Colombian edible plant species per department.

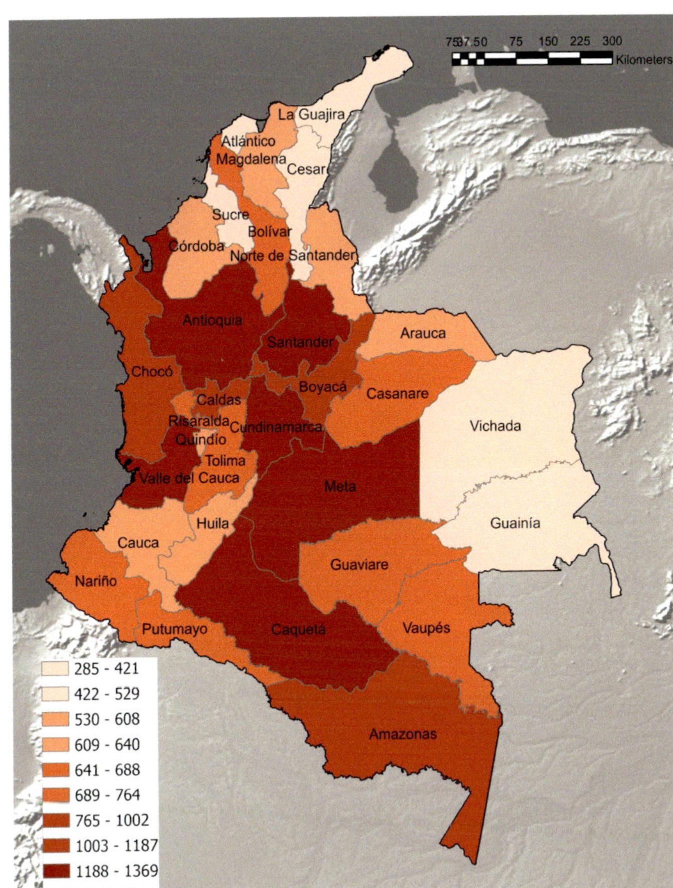

FIGURE 2. Richness of observations of Colombian edible plants by department.

Based on the clean georeferenced records for Colombian edible plants, the departments of Antioquia and Meta stand out for species richness. However, these are also the departments where most occurrences were recorded, and linear regressions revealed a significant correlation ($p<0.001$) between species richness and the number of georeferenced records (Figure 2). Therefore, it is possible to argue that other departments may not stand out simply because of a lack of adequate on-site investigation and the scarcity of georeferenced records collected.

Colombian edible plants are represented by 219 families and 1,389 genera. The most important families (i.e., those containing the highest number of species), taking into account both native and non-native species, include Fabaceae (119 genera, 351 species), Asteraceae (86 genera, 136 species), Poaceae (72 genera, 140 species), Arecaceae (55 genera, 180 species), and Rubiaceae (52 genera, 137 species). This scenario partly reflects the global trend highlighted by the World Checklist. However, if only native species are considered, families such as Melastomataceae (14 genera, 119 species) and Malvaceae (35 genera, 91 species) obtain greater relevance, followed by Moraceae (18 genera, 74 species), Annonaceae (14 genera, 66 species), and Myrtaceae (11 genera, 61 species).

Important genera, including both native and non-native species, comprise *Inga* (84 spp.), *Passiflora* (73 spp.), *Miconia* (63 spp.) *Solanum* (61 spp.), *Pouteria* (54 spp.), *Protium* (33 spp.), *Annona* (32 spp.), and *Bactris* (28 spp.). Genera such as *Ficus, Diospyros*, and *Garcinia*, known to be among the most relevant edible plants at the global level (Diazgranados *et al.*, 2020), are not significantly rich in edible species in Colombia. On the other hand, genera such as *Passiflora, Inga, Bactris*, and *Pouteria* are characterised by a high number of edible species and may represent a new frontier for ethnobotanical and bromatological studies in the country.

Colombian edible plants comprehend a great variety of growth forms, from trees to herbs, climbers, and epiphytes. Trees constitute the most dominant habit, with more than 1,500 species, followed by herbs, shrubs, and climbers. Growth habit highly reflects the biogeographic distribution of edible plants across various ecoregions (Figure 3).

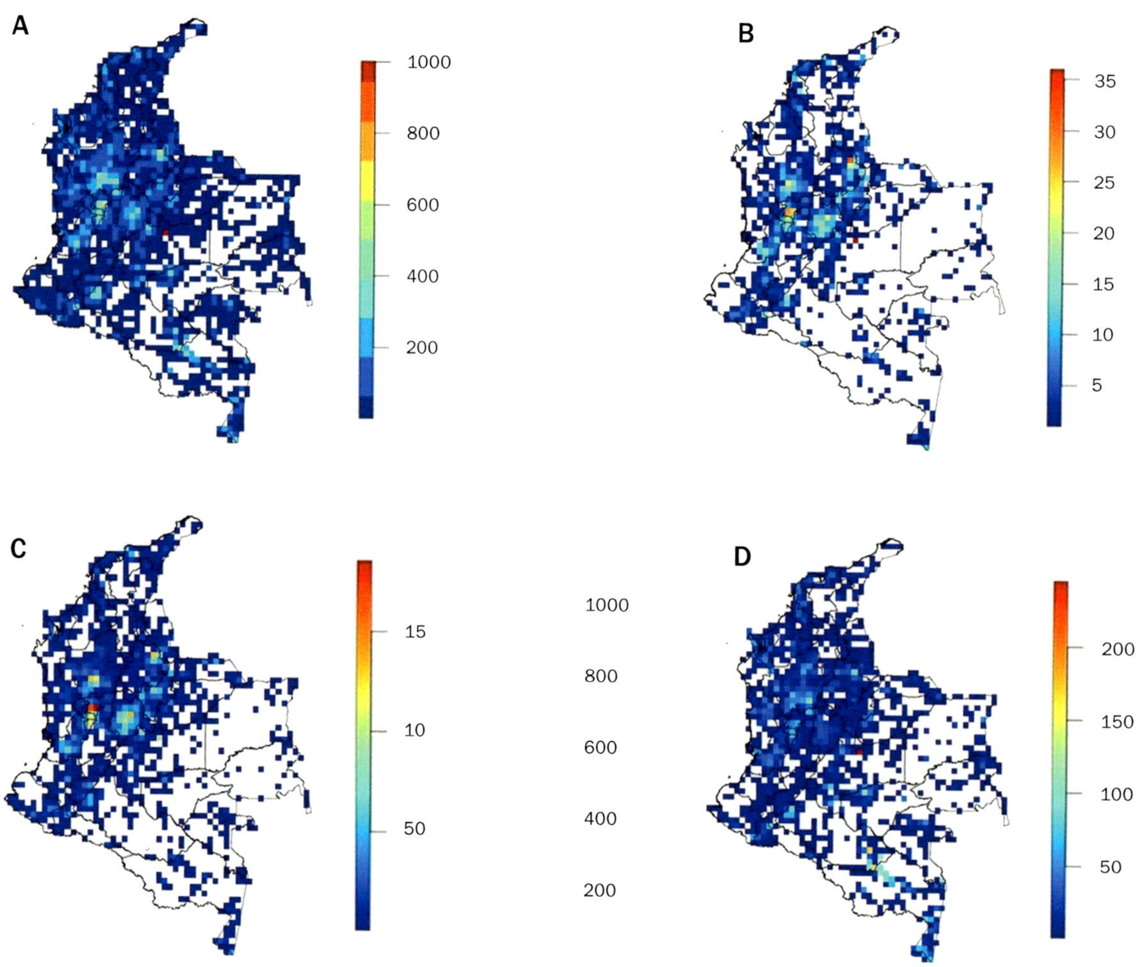

FIGURE 3. Distribution of Colombian edible plants by growing habit: **A** tree species, **B** shrub species, **C** herb species and **D** climbing species. The four distributions show relatively congruous trends, with high species concentrations across the Andean region.

Some of the most important tree genera include *Inga* (64 tree spp.), *Pouteria* (49 spp.), *Miconia* (39 spp.), *Protium* (32 spp.), *Annona* (27 spp.), *Ficus* (21 spp.), *Casearia* (21 spp.), and *Matisia* (20 spp.). On the other hand, important herbs include *Solanum* (30 herb spp.), *Cyperus* (16 spp.), *Miconia* (14 spp.), *Oxalis* (11 spp.), and *Eragrostis* (11 spp.). Predominant genera for shrub species include *Miconia* (54 shrub spp.), *Solanum* (33 spp.), *Bactris* (25 spp.), *Casearia* (20 spp.), *Piper* (15 spp.), and *Senna* (15 spp.). Finally, the most important genera for edible climbers comprise *Passiflora* (64 climbing spp.), *Ipomoea* (15 spp.), *Dioscorea* (15 spp.), *Paullinia* (14 spp.), and *Solanum* (12 spp.).

Examples of important genera

Passiflora *(Passifloraceae)*

Several species are characterised by great economic potential, but only eight of them are currently cultivated (*P. antioquiensis, P. caerulea, P. edulis, P. ligularis, P. mixta, P. quadrangularis, P. tarminiana,* and *P. tripartite*). Other species, including *P. vitifolia*, are wildly harvested for local consumption or sale in local markets. There is substantial morphological variation within the genus (e.g., in the colour and shapeof the fruits and flowers). The fruits can be eaten raw, cooked, or even used to make drinks, such as *P. antioquiensis*. The pulp is very aromatic, and flowers and leaves are also edible sometimes. In the case of *P. foetida*, the leaves are cooked and used as an ingredient in soups. The pulp of fruits is very variable in terms of taste: it can be sweet (e.g., *P. ligularis*), juicy and acid-flavoured (e.g., *P. coccinea*), or aromatic and mildly biting (e.g., *P. cumbalensis*). Finally, it is important to mention *P. mollissima*, which is traditionally eaten in various departments of Colombia and shows a great versatility of use. In fact, it can be consumed and marketed in the form of juice, sorbet, ice cream, or jam.

Solanum *(Solanaceae)*

With 61 known edible species, *Solanum* is the third most important genus in Colombia for food plants. In addition to some of the most economically valuable plants that the world has seen, such as *S. lycopersicum* (tomato), *Nicotiana tabacum* (tobacco), and *S. tuberosum* (potato), this family includes an outstanding number of edible plants, many of which hold lesser economic importance and are exclusively known, grown, and consumed at the local level. In Colombia, these are especially present in the Andean region and are still largely unexplored from a taxonomic, agronomic, and bromatological perspective. Examples include *S. cajanumense*, a fast-growing evergreen shrub whose golden-yellow fruits are usually eaten fresh when fully ripe, *S. capsicoides*, whose poisonous fruits can be eaten when roasted or cooked, and *S. pectinatum*, whose pale orange fruit, characterised by a sweet-acidic flavour, is delicious when cooked with sugar (Food Plants International, 2021). Only 14 out of 61 species are currently being cultivated. 51 species out of 61 are native to the country, and only one of them is endemic.

Inga *(Fabaceae)*

The edible fruit of this genus is very popular throughout South America, where it is commonly gathered from the wild and often cultivated (Lorenzi *et al.*, 2000). *Inga edulis* is the best known and most consumed species. Both the seeds and the white and jelly pulp surrounding them within the fruit's seedpod can be eaten. The pulp is characterised by a sweet and highly aromatic taste (Martin *et al.*, 1987), and it is usually eaten raw. The seeds are eaten cooked, usually boiled, or roasted, like in *I. ilta.* When young, they can also be eaten raw, blanched and salted, and added to salads (Food Plants International, 2021). *Inga* trees are commonly planted in coffee or cacao plantations to provide shade to the surrounding environment, and fruits are often sold in local markets (Leon, 1966). Out of 84 species found in Colombia, 76 are native to the country, and none of them is endemic. Only six species are currently being cultivated (*I. densiflora, I. edulis, I. feuillei, I. ornate, I. spectabilis*, and *I. vera*).

Bactris *(Arecaceae)*

Bactris is a genus of multi-purpose spiny palm trees native to Latin America, the most popular of which is *B. gasipaes*, commonly known as peach palm. In total, 28 species of *Bactris* are known to be edible in Colombia, but only *B. gasipaes* is currently cultivated. All 28 species are native to Colombia, and only one species – *B. chocoensis* – is endemic. The fruit of most species is inedible when raw (Blombery & Rodd, 1992). They are usually boiled in salted water for 30 to 60 minutes and eaten as a vegetable (Food Plants International, 2021). The pulp is characterised by a nutty flavour and a floury texture, and by its remarkably high nutritious qualities resulting from its high protein and carbohydrate contents. Fruits can also be made into flour and baked into bread, cakes, and other processed foods (Food Plants International, 2021; Blombery & Rodd, 1992). Seeds, such as those of *B. major* and *B. gasipes*, can be consumed raw (Wickens, 1995) as nuts. The palm heart of some species (e.g., *B. riparia* and *B. corossilla*) is also eaten raw, in salads, or cooked. Finally, fruits of *B. guineensis* can be fermented and used to produce a drink, known in Colombia as *Chicha de Corozo* (Chízmar, 2009; Rodriguez, 2019).

Vasconcellea *(Caricaceae)*

This genus contains 21 species, which were for a long time classified within the genus *Carica*. They are of great importance as genetic resources (wild relatives) for the improvement of the papaya (*Carica papaya*) and for their edible fruits. Additionally, they can be of great importance for producing papain enzyme and latex, which, as raw materials, are promising resources for food and non-food agro-industrial processes. Colombia has eight species popularly known as "tall papayas", of which three are endemic (*Vasconcellea goudotiana, Vasconcellea longiflora* and *Vasconcellea sphaerocarpa*). Several ethnobotanical studies carried out in the southern Colombian Andes show that *V. cundinamarcensis* and *V. goudotiana* are very often reported for their edible and medicinal uses, for both humans and animals (Fuertes, 2019). In the Andes, the "highland papayas" (*Vasconcellea* spp.) are mainly consumed fresh, toasted, juiced, as jams, sometimes preserved or cooked in sauces, or as fillings for cakes and pickles. These plants, therefore, possess great potential for use. They are usually grown on small Andean farms for home consumption for their tasty and aromatic fruits.

Conservation state and current gaps

Out of 3,806 Colombian edible plants, only 1,742 have been assessed by the International Union for Conservation of Nature. Out of these, 1,532 have been categorised as LC (Least Concern), 38 as NT (Near Threatened), 69 as VU (Vulnerable), 44 as EN (Endangered), 16 as CR (Critically Endangered), one as EW (Extinct In the Wild), and 40 as DD (Data Deficient). Additionally, two species have been categorised as EX (Extinct). Among the most threatened species, there are some endemic species, such as *Melicoccus antioquensis* (Sapindaceae) and *Panopsis hernandezii* (Proteaceae), as well introduced ones, such as *Eugenia cotinifolia* (Myrtaceae), *Araucaria angustifolia* (Araucariaceae), *Hyophorbe verschaffeltii* (Arecaceae), and *Lathyrus odoratus* (Fabaceae). The most common threats to these species include agricultural expansion and biological resource use. In fact, the degradation of habitats in which these species could be found in the past is mainly due to activities related to the expansion of the agricultural frontier (e.g., forest clearance for cattle ranching and crops cultivation) (IUCN, 2021). What is more, the usefulness of these plants makes them particularly subject to overexploitation and unsustainable harvesting. A significant portion of Colombian edible species boasts uses of other types, for example, having wood that is used for construction or other plant parts that are used as medicines. For instance, in the case *of Araucaria angustifolia*, in addition to more than 3,000 tons of fruits and seeds collected every year for human consumption, resource depletion can be attributed to its massive exploitation for timber (IUCN, 2021). The sustainable harvesting and use of natural resources remain key issues for determining their conservation: while overexploitation represents the second major driver of plant extinction, the sustainable use of local natural capital can be vital for the development of future livelihoods for indigenous peoples (Kor *et al.*, 2021).

As displayed by the significant portion of edible species whose conservation status has not yet been assessed, significant gaps in the knowledge required to support the protection of these species can be identified. For instance, out of the 73 Colombian edible species belonging to the genus *Passiflora*, none have been assessed by the IUCN. The same trend can be found for other genera, such as *Bactris:* only three out of 28 species present in Colombia have been assessed. Therefore, we can conclude that the conservation status of two of the richest genera for edible plants in Colombia is unknown.

Ex-situ conservation

The comparison between the Colombian edible plant dataset and PlantSearch showed a total match of 3,042 species. However, 547 species (17.9% of the total) are not currently kept in any collection (Total gardens = 0). Therefore, the number of edible species included in the present study and currently maintained within the *ex-situ* collections of the BGCI global consortium is 2,495. In this case, 82.1% of all edible species found in Colombia are currently preserved in at least one *ex-situ* collection. Categorical analysis results are summarised, with 379 useful plant taxa conserved in only one BGCI botanic garden's living collection, 780 taxa conserved within 2–10 collections, 290 taxa within 11–20 collections, 344 taxa within 2–40 collections, and 702 taxa within more than 40 collections each (Table 1).

TABLE 1. *Ex-situ* conservation of Colombian edible plants.

No. of *ex-situ* collections	No. of edible species	% of the total
0	547	17.9
1	379	12.4
2–10	780	25.6
11–20	290	9.5
21–40	344	11.3
>40	702	23

Results show that important genera for edible species, such as *Bactris*, *Inga*, and *Pouteria*, lack adequate *ex-situ* conservation. For instance, out of 28 edible *Bactris* species, one is currently not conserved in any *ex-situ* repository globally, and only two species are currently conserved in more than 15 repositories. Out of 84 edible species of *Inga* found in Colombia, 59 are not currently conserved in any *ex-situ* repository, and only one species – *Inga edulis* – is conserved in more than 15 repositories. Finally, out of 54 edible *Pouteria* species, 42 are not conserved in any *ex-situ* collection, and only two species – *P. caimito* and the introduced *P. sapota* – are maintained in more than 15 of them.

Promising edible species and diversity hotspots

The spatial analysis results carried out on the clean dataset, containing 221,838 georeferenced records for 3,220 Colombian edible species, revealed several hotspots for edible species diversity in the northern and north-western Andean region (Figure 4). The corrected weighted endemism distribution pattern for Colombian edible species emphasises areas that are characterised by a high proportion of species with restricted distributional ranges. Therefore, these areas are not only remarkably high in edible species richness but also stand out as diversity hotspots for rare and endemic species. Colombian Andes

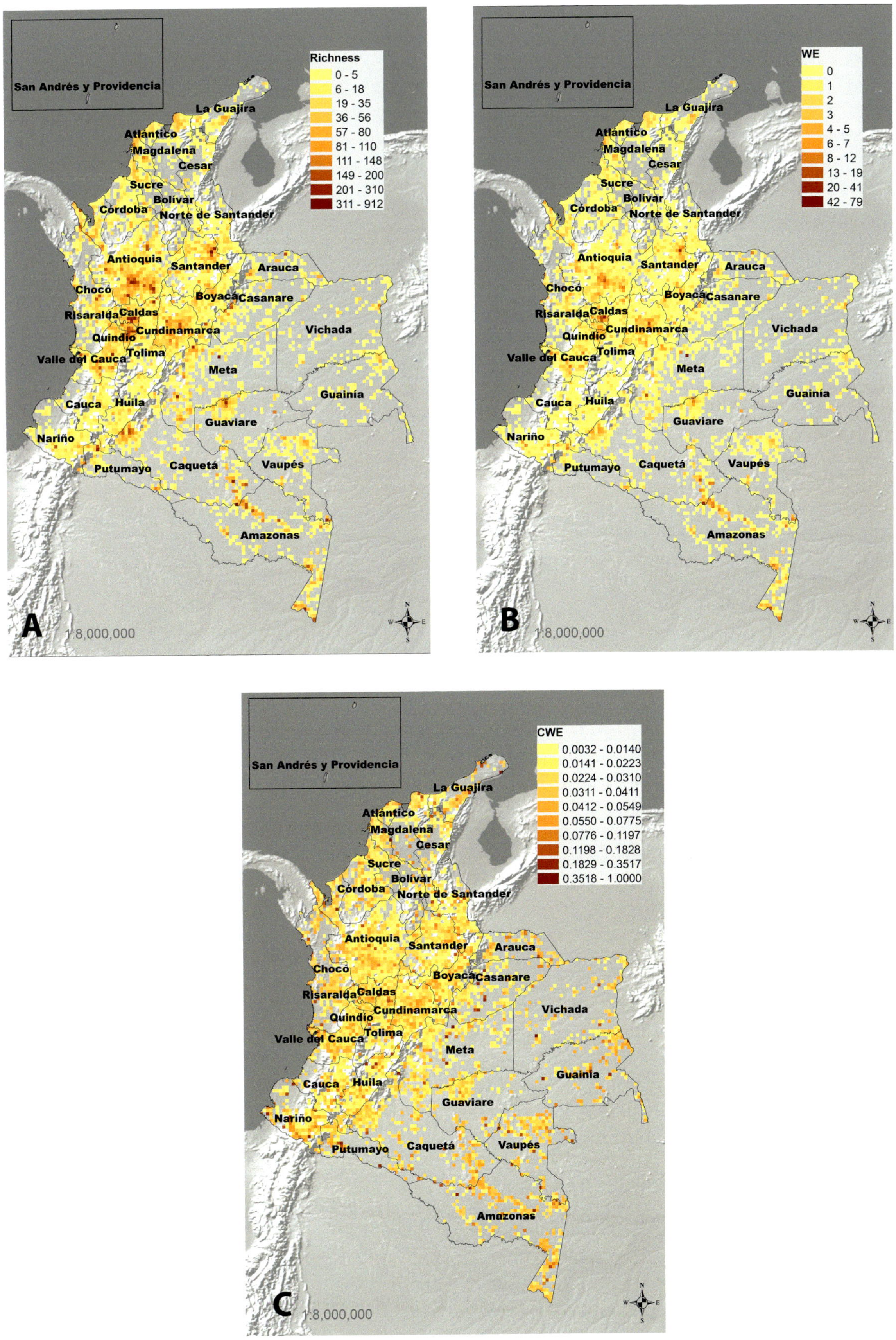

FIGURE 4. Diversity patterns of Colombian edible plants. **A** Species richness map, showing the Tropical Andes as the richest area. **B** Weighted endemism map. **C** Corrected weighted endemism, with the Andes standing out, together with the tropical rain forest and Sub-Andean rainforest of the Nariño department, the Savanna between the Amazonas and Caqueta departments, and the Paramo of la Guajira.

form part of the Tropical Andes, extending across the north of Chile, Argentina, Bolivia, Peru, Ecuador, Colombia, and Venezuela for over 1.5 million km^2, between the latitudinal range of 11° N to 30° S, and with an elevational range of approximately 500 to 6,000 m a.s.l. (Bax & Francesconi, 2019). According to Myers *et al.* (2020), the Tropical Andes support approximately 45,000 plant species, with nearly half of them endemic to the Andean ecoregion (Meyers *et al.*, 2020).

In particular, a significantly high density of narrow-distribution edible plants has been recorded in the northern part of the Huila department. Huila is considered one of the richest regions of Colombia in terms of plant biodiversity due to its great variety of thermal floors, from Páramo zones to extensive areas of tropical dry forest (Figure 5). At least 120,000 hectares of the department are localised within the Páramo area, with elevations ranging between 2,900 and 5,000 m a.s.l. Notably, the area has been described as one of the richest in the country in terms of the diversity of the genus *Passiflora* (Ocampo, 2013).

Another interesting hotspot was identified within the high Guajira department, in the proximity of the Macuira National Park. The Serrania de Macuira, located in the extreme north of La Guajira's department, is considered of great environmental importance. In fact, their evergreen cloud forest ecosystem, although localised outside of the Andean mountains, appears incredibly similar to the Andean forests located near the Páramo line (altitudes above 2,700 m a.s.l.). Despite this similarity, they are located at only 550 m a.s.l. This ecosystem functions as a water reserve for the surrounding desert region and supports a unique floral diversity. Furthermore, as showcased by the present study, it represents a national reservoir for unique edible plants.

Therefore, the present biogeographic analysis highlights Andean ecosystems as important repositories for singular edible plants, which are highly adapted to specific environmental conditions and characterised by narrow distributions. The number of food species present in the Colombian ecosystems that are of low and regular altitudes is very high compared to that in high Andean forests and Páramo, which are characterised by more extreme weather patterns. Therefore, to guarantee their conservation and use in the long term by local populations, it is critical to register the most important species found within these ecosystems, their culinary uses, and their ecological value. Páramo exhibits an unparalleled flora originating from sophisticated evolutionary processes over the past 3 million years and is characterised by a great proportion of endemic species (Londono *et al.*, 2014). However, due to increasing human-driven alterations of natural ecosystems and the progressive impact of climatic variations, ecosystems such as the Tropical Andes are nowadays recognised as one of the most critically threatened ecoregions in the tropics (Noh *et al.*, 2020). This scenario constitutes a particularly serious hazard to the native endemic edible species that inhabit the region (Figure 6), characterised by extremely specific habitat needs. In light of this, Colombian Andes are expected to lose a significant proportion of their native plant diversity due to environmental degradation (Brooks *et al.*, 2002; Noh *et al.*, 2020). This scenario would represent a significant loss not only for the biological heritage that is a feature of the region but also for its socio-cultural heritage, embedded in local traditional agricultural practices and gastronomy. High-altitude ecosystems and their edible flora must therefore be protected and yet further investigated from an ethnobotanical perspective to deepen the evolutionary and socio-cultural practices that underlie a remarkable gastronomic heritage.

Finally, based on the obtained results, it is key to showcase the most promising underutilised and neglected species (native and non-native) that are present and easily adapted to different Colombian ecosystems. Such species present significant competitive advantages compared to other cultivated species, which could greatly improve local communities' economies and, consequently, the country's bioeconomy. For this purpose, Colombia's most important food species for the warm, cold, and temperate thermal floors are summarised in Tables 2 and 3.

CONCLUSIONS

In conclusion, the showcased results highlight consistent discrepancies between the diversity and richness of Colombian edible plants and their current conservation. Both wild and domesticated edible plants have played a significant role in the subsistence of human communities across the country. Their preservation and valorisation, today, is more crucial than ever. After more than six decades of internal conflict, Colombia is currently going through an important time in the country's development, which will determine the fate of its natural resources in the coming years. However, agricultural expansion and urban development lead to deforestation and habitat loss, causing unprecedented levels of biodiversity loss nationwide. Local underutilised edible plants hold great potential for developing new livelihoods and consolidating a readjusted bioeconomy based on the sustainable use of local plant diversity.

The present chapter represents a first review of the Colombian edible diversity from a taxonomic, biogeographical and conservationist point of view. Taxonomic investigation revealed new plant groups worth deepening from an ethnobotanical and commercial perspective, laying the foundation for future efforts to revitalise underutilised edible species. Additionally, the spatial analysis highlighted the biodiversity hotspots in the country. Further analysis considering simultaneously the extinction risk assessments and the *in-situ* and *ex-situ* conservation actions in place for the most important edible plant groups could facilitate pinpointed conservation endeavours.

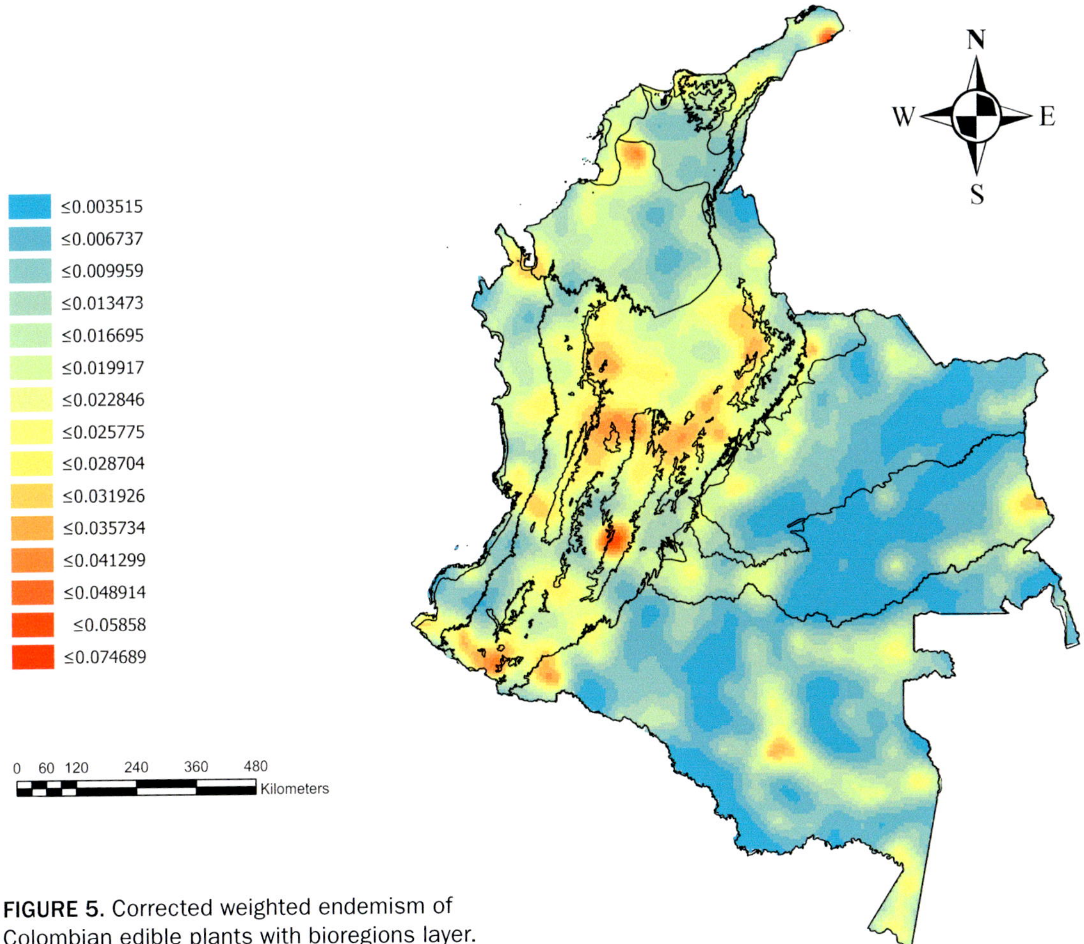

FIGURE 5. Corrected weighted endemism of Colombian edible plants with bioregions layer.

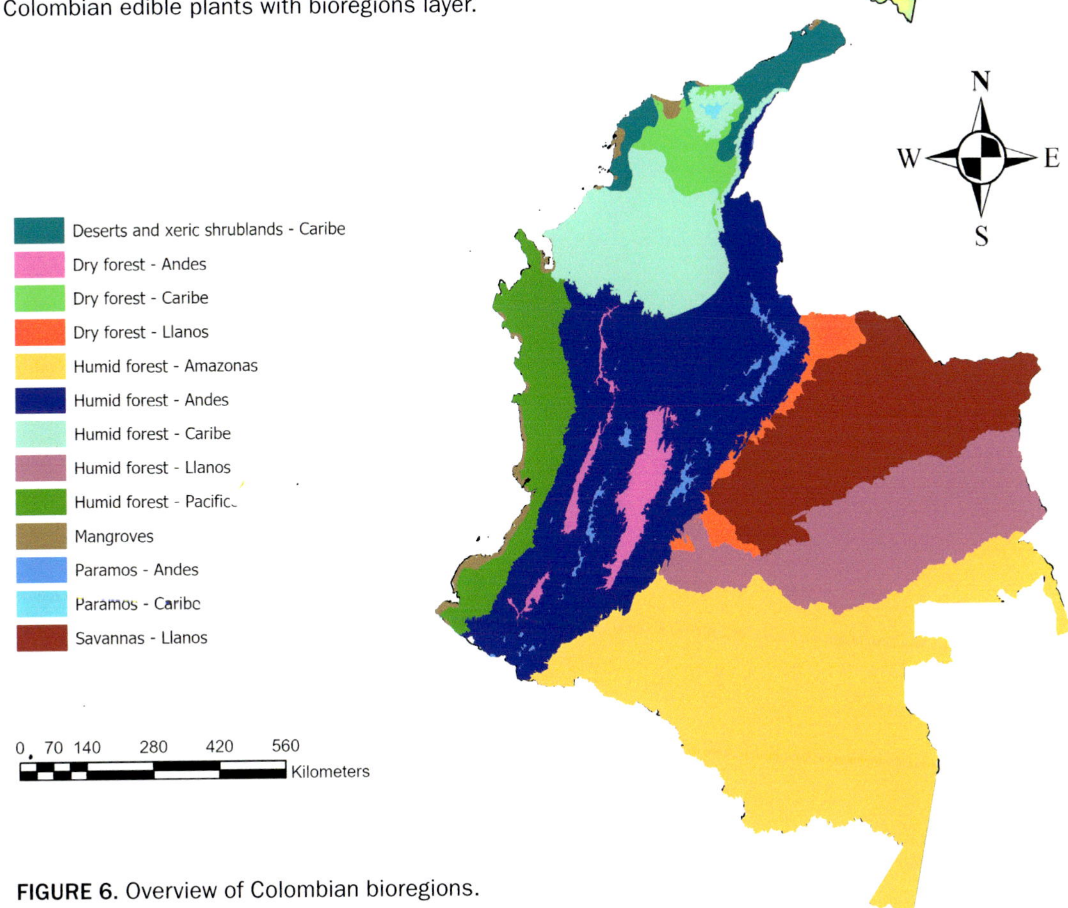

FIGURE 6. Overview of Colombian bioregions.

TABLE 2. Native and non-native edible species suitable for cold thermal floors in Colombia. (*) Altitudinal range (m a.s.l.); (**) Habitat Classification based on Cuatrecasas (1958).

Scientific name	Family	Common name(s)	Habit	Altitudinal range (*) and habitat (**)	Main uses
Amaranthus caudatus L.	Amaranthaceae	Amaranto, coime, achita	Herb 1.5 – 2.0 m	2,400 – 3,500 Andean Forest, Subpáramo and Páramo	Human food (seeds) and animal forage (the entire plant)
Cavendishia bracteata (Ruiz & Pav. Ex J.St.-Hil.) Hoerold	Ericaceae	Uva de anís, uva de monte, asnalulo	Shrub 0.8 – 3.0 m	2,400 – 3,200 Andean Forest and Subpáramo	Human food (fruits)
Chalybea macrocarpa (Uribe) Morales P. & Penneys	Melastomataceae	Churumbelo	Small tree 5 – 12 m	2,400 – 2,900 Andean Forest	Human food (fruits)
Chenopodium pallidicaule Aellen	Amaranthaceae	Cañahua, kañihua	Herb 0.2 – 0.8 m	3,500 – 4,200 Páramo and Superpáramo	Human food (seeds) and animal forage (stems)
Chenopodium quinoa Willd.	Amaranthaceae	Quinua, quinoa	Herb 1.0 – 2.5 m	2,400 – 4,000 Andean Forest, Páramo, Subpáramo and Superpáramo	Human food (seeds)
Gaylussacia buxifolia Kunth	Ericaceae	Agracillo, mosquito, uvo	Shrub 0.5 – 1.5 m	2,500 – 3,500 Andean Forest, Subpáramo and Páramo	Human food (fruits)
Hesperomeles goudotiana (Decne) Killip	Rosaceae	Mortiño, cerote, manzano, motemote	Small tree or shrub 2 – 8 m	2,500 – 3,500 Andean Forest, Subpáramo and Páramo	Human and animal food (fruits)
Juglans neotropica Diels	Juglandaceae	Nogal, cedro nogal, cedro negro, cedro bogotano, tocte	Tree 15 – 45 m	2,400 – 3,100 Andean Forest and Subpáramo	Human food (seeds)
Lupinus mutabilis Sweet	Fabaceae (Leguminosae)	Lupino, tarwi, altramuz, chocho	Herb 1.8 – 2.0 m	2,400 – 3,800 Andean Forest, Subpáramo and Páramo	Human food (seeds)
Macleania rupestris (Kunth) A.C. Sm.	Ericaceae	Uva camarona, uvita, uvilla, chaquilulo, otros	Shrub 0.6 – 2.5 m	2,400 – 4,100 Andean Forest, Páramo, Subpáramo and Superpáramo	Human food (fruits)
Miconia squamulosa Triana	Melastomataceae	Tuno esmeraldo, tunacòn	Small tree or shrub 2 – 8 m	2,400 – 3,200 Andean Forest and Subpáramo	Human food (fruits)
Mirabilis expansa (Ruiz & Pav.) Standley	Nyctaginaceae	Mauka, chago, arricòn, miso, taso, pega pega	Herb 0.8 – 1.0 m	2,400 – 3,500 Andean Forest, Subpáramo and páramo	Human food (leaves and tuberous roots) and animal food (leaves, stems, and roots)
Myrcianthes leucoxyla (Ortega) McVaugh	Myrtaceae	Arrayán, guayabo silvestre mirto, champo	Tree 10 – 15 m	2,400 – 3,300 Andean Forest and Subpáramo	Human food (fruits)

Scientific name	Family	Common name(s)	Habit	Altitudinal range (*) and habitat (**)	Main uses
Oxalis tuberosa Molina	Oxalidaceae	Oca, oka, papa oca, ibia, cubio	Herb 0.2 – 0.3 m	2,400 – 3,900 Andean Forest, Subpáramo and Páramo	Human food (tubers)
Passiflora antioquiensis H. Karst.	Passifloraceae	Curuba, curuba antioqueña, curuba quiteña	Vine 8 – 10 m	1,800 – 2,700 Andean and Sub-Andean Forest	Human food (fruits)
Passiflora mollissima (Kunth) Bailey	Passifloraceae	Curuba, tacso, tumbo, poroporo	Woody climber	1,800 – 3,000 Andean and Sub-Andean Forest	Human food (fruits)
Physalis peruviana L.	Solanaceae	Uchuva, guchuva, uvilla, aguaymato, vejigón	Herb 0.3 – 1.0 m	2,400 – 3,300 Andean and Sub-Andean Forest	Human food (fruits)
Plutarchia guascensis (Cuatrec.) A.C. Sm.	Ericaceae	Azabache, coronilla, corales, coralitos, uvo, uvito de guasca	Shrub 0.5 – 2.5 m	2,800 – 4,200 Andean Forest, Páramo, Subpáramo and Superpáramo	Human food (fruits)
Prestoea acuminata (Willd.) H.E. Moore	Arecaceae (Palmae)	Palmito de altura	Tree 13 – 15 m	2,400 – 2,650 Andean Forest	Human food (buds, palm heart)
Rubus nubigenus Kunth	Rosaceae	Morón de páramo, mora gigante, mora de oso	Climbing shrub 1 – 4 m	2,600 – 3,200 Andean Forest and Subpáramo	Human food (fruits)
Smallanthus sonchifolius (Poepp. & Endl.) H. Robinson	Compositae (Asteraceae)	Yacón, llacòn, aricona, jícama, jíquima	Herb 1.5 – 2.5 m	2,400 – 3,000 Andean Forest	Human food (tubers)
Solanum vestissimum Dunal	Solanaceae	Lulo de tierra fría, toronjo, tumo, cuque	Shrub 2 – 5 m	2,400 – 3,000 Andean Forest and Subpáramo	Human food (fruits)
Tropaeolum tuberosum Ruiz & Pav.	Tropaeolaceae	Mashua, añu, isaño, cubio, majua, otros	Crawler 0.8 – 1.0 m	2,400 – 3,900 Andean Forest, Subpáramo and Páramo	Human and animal food (tubers)
Ugni myricoides (Kunth) O. Berg	Myrtaceae	Arrayancillo, arrayán blanco, mirto	Shrub 0.5 – 1.5 m	2,900 – 3,400 Andean Forest, Subpáramo and Páramo	Human food (fruits)
Ullucus tuberosus Caldas	Basellaceae	Ulluco, melloco	Herb 0.3 – 1.2 m	2,400 – 3,400 Andean Forest, Subpáramo and Páramo	Human food (tubers)
Vaccinium meridionale Swartz	Ericaceae	Agraz, mortiño, uva de monte, congama	Shrub 1 – 3 m	2,400 – 3,000 Andean Forest and Subpáramo	Human food (fruits)
Vasconcellea cundinamarcensis V.M. Badillo	Caricaceae	Chilacùan, chihualcan, higuillo, tapaculo	Small tree or shrub 8 – 10 m	2,400 – 3,300 Andean Forest and Subpáramo	Human food (fruits)
Vasconcellea goudotiana Triana & Planch.	Caricaceae	Papayuela, papaya de tierra fría, tapacho, tapaculo	Small tree or shrub 1 – 6 m	2,640 – 3,575 Andean Forest, Subpáramo and Páramo	Human food (fruits)
Vicia andicola Kunth	Fabaceae (Leguminosae)	Haba forrajera, alverjilla, haka, mullu, shintu	Herb 0.4 – 0.5 m	2,400 – 4,200 Andean Forests, Páramo, Subpáramo and Superpáramo	Animal food (forage)

TABLE 3. Native and non-native food species suitable for the warm and temperate thermal floors of Colombia. (*) Altitudinal range (m a.s.l.); (**) Habitat classification based on Cuatrecasas (1958).

Scientific name	Family	Common name(s)	Habit	Altitudinal range (*) and habitat (**)	Main uses
Amaranthus dubius Mart. ex Thell.	Amaranthaceae	Bledo, bledo blanco, yuyo, amaranto	Herb 0.8 – 1.0 m	200 – 1,850 Lower Neotropical rainforest and Sub-Andean Forest	Human food (seeds and leaves) and animal food (leaves)
Anacardium excelsum (Bertero & Balb.) Skeels	Anacardiaceae	Marañón, merey, caracolí, cahu, aspavé	Tree 20 – 25 m	20 – 1,200 Lower Neotropical rainforest and Sub-Andean Forest	Human food (fruit peduncle and seeds)
Anacardium occidentale L.	Anacardiaceae	Anacardo, marañón, merey, caju	Small tree 3 – 10 m	0 – 1,300 Lower Neotropical rainforest and Sub-Andean Forest	Human food (seeds)
Artocarpus altilis (Parkinson) Fosberg	Moraceae	Árbol del pan, fruta del pan, frutipan	Tree 15 – 20 m	80 – 1,550 Lower Neotropical rainforest and Sub-Andean Forest	Human food (fruits)
Bactris gasipaes Kunth	Arecaceae (Palmae)	Chontaduro, pejibaye, pijuayo, cachipay	Tree 7 – 20 m	0 – 1,500 Lower Neotropical rainforest and Sub-Andean Forest	Human food (fruits)
Bixa orellana L.	Bixaceae	Achiote, onoto, bija, urucú	Small tree or shrub 2 – 6 m	0 – 1,200 Lower Neotropical rainforest and Sub-Andean Forest	Human food
Borojoa patinoi Cuatrec.	Rubiaceae	Borojó, burijo, burojò	Small tree 3 – 5 m	0 – 600 Lower Neotropical rainforest	Human food (fruits)
Cajanus cajan (L.) Millsp.	Fabaceae (Leguminosae)	Guandú, Guandul, fríjol de palo, quinchoncho	Shrub 1 – 3 m	100 – 900 Lower Neotropical rainforest	Human food (seeds)
Canavalia ensiformis (L.) DC.	Fabaceae (Leguminosae)	Haba criolla, haba de burro, fríjol espada	Herb 0.6 – 1.3 m	40 – 900 Lower Neotropical rainforest	Human food (seeds) and animal food (fruits, seeds and forage)
Canna indica L.	Cannaceae	Achira, sagú, chisgua	Herb 0.4 – 2.5 m	0 – 2,400 Lower Neotropical rainforest and Sub-Andean Forest	Human food (roots)
Caryodendron orinocense Karsten	Euphorbiaceae	Inchi, cacay, nuez llanera e iracana	Tree 15 – 25 m	0 – 1,200 Lower Neotropical rainforest and Sub-Andean Forest	Human and animal food (seeds)
Erythrina edulis (L.) DC.	Fabaceae (Leguminosae)	Balú, chachafruto, sachafruto, frijol nopas, poroto	Tree 6 – 10 m	1,400 – 2,400 Sub-Andean Forest	Human food (seeds) and Animal food (fruits and leaves), forage plant

Scientific name	Family	Common name(s)	Habit	Altitudinal range (*) and habitat (**)	Main uses
Eugenia stipitata McVaugh	Myrtaceae	Arazá, guayabo amazónico	Small tree or shrub 5 – 10 m	50 – 500 Lower Neotropical rainforest	Human food (fruits)
Ipomoea batatas (L.) Lam.	Convolvulaceae	Camote, batata, chaco, papa dulce.	Creeping or climbing grass	0 – 2.400 Lower Neotropical rainforest and Sub-Andean Forest	Human food (roots) and animal food (roots and leaves)
Maranta arundinacea L.	Marantaceae	Arrurruz, maranta, yuquilla, sagú	Herb 0.8 – 1.0 m	0 – 800 Lower Neotropical rainforest	Human food (roots)
Mauritia flexuosa L.f.	Arecaceae (Palmae)	Moriche, aguaje, buritì, morete	Tree 20 – 35 m	0 – 900 Lower Neotropical rainforest	Human and animal food (fruits)
Myrciaria dubia (Kunth) McVaugh	Myrtaceae	Camu camu	Shrub 4 – 8 m	0 – 300 Lower Neotropical rainforest	Human food (fruits)
Pachyrhizus ahipa (Wedd.) Parodi	Fabaceae (Leguminosae)	Ahipa, ajipa, jacatupé	Herb 30 – 60 cm	0 – 2,400 Lower Neotropical rainforest and Sub-Andean Forest	Human food (tuberous roots, pods and seeds)
Plukenetia volubilis L.	Euphorbiaceae	Sacha-inchi, sacha-maní, maní del inca	Climbing grass	0 – 2,000 Lower Neotropical rainforest and Sub-Andean Forest	Human food (seeds)
Pouteria lucuma (Ruíz & Pav.)	Sapotaceae	Lúcuma, fruta de oro, oro de los incas	Tree 13 – 15 m	1,500 – 2,400 Sub-Andean Forest	Human food (fruits)
Prosopis juliflora (Sw.) DC.	Fabaceae (Leguminosae)	Algarroba, trupillo, mezquite, cují yaque	Tree 10 – 15 m	0 – 2,100 Lower Neotropical rainforest and Sub-Andean Forest	Human food (fruits and seeds) and animal food (fruits, seed, leaves, forage)
Sechium edule (Jacq.) Sw.	Cucurbitaceae	Guatila, chocho, chayote, papa de pobre	Creeper	1,150 – 2,500	Human and animal food (fruits and roots)
Solanum sessiliflorum Dunal	Solanaceae	Cocona, tupiro, topiro	Herb 0.8 – 2.0 m	0 – 1,000 Lower Neotropical rainforest	Human food (fruits)
Theobroma bicolor Humb. & Bonpl.	Malvaceae	Pataxte, bacao, maraca, Mocambo de Chocó	Small tree 7 – 12 m	20 – 1,000 Lower Neotropical rainforest	Human food (fruits and seeds)
Theobroma grandiflorum (Willd. ex Spreng.) K. Schum.	Malvaceae	Copoasù, copoazù, cacao blanco	Tree 14 – 18 m	120 – 1,000 Lower Neotropical rainforest	Human food (fruits and seeds)
Trichanthera gigantea (Humb. & Bonpl.) Nees	Acanthaceae	Nacedero, naranjillo, cajeto	Tree 10 – 12 m	100 – 2,150 Lower Neotropical rainforest and Sub-Andean Forest	Animal food (tender leaves and stems as fodder)

Acknowledgements

Many people contributed to the production of this chapter. The authors wish to acknowledge the help provided by the 'Useful Plants and Fungi of Colombia' project team. The assistance provided by Tiziana Cossu, Laura Kor, Ian Ondo, Carolina Tovar and David Hammond was greatly appreciated. Special thanks go to Kaitalin White for the valuable technical and administrative support and dedicated time.

References

Bax V, Francesconi W (2019) Conservation gaps and priorities in the Tropical Andes biodiversity hotspot: Implications for the expansion of protected areas. J. Environ. Manage. 232 (15): 387–396. doi: 10.1016/j.jenvman.2018.11.086.

BGCI (2021) PlantSearch online database. Botanic Gardens Conservation International. Richmond, U.K. Available at https://tools.bgci.org/plant_search.php.

Bivand R, Keitt T, Rowlingson B, Pebesma E, Summer M, Hijmans R, Baston D, Rouault E, Warmerdam F, Ooms J, Rundel C (2021) Rgdal: Bindings for the "Geospatial" Data Abstraction Library. R Package Version 1.5-23. Available at: https://cran.r-project.org/web/packages/rgdal/index.html

Bivand RS, Pebesma E, Gomez-Rubio V (2013) Applied spatial data analysis with R, Second edition. New York, Springer. Available at: https://asdar-book.org/

Blombery A, Rodd T (1992) *Palms of the World. Their Cultivation, Care and Landscape Use*. London, Angus and Robertson. 199p.

Borelli T, Hunter D, Padulosi S, Amaya N, Meldrum G, de Oliveira Beltrame DM, Samarasinghe G, Wasike VW, Güner B, Tan A, Koreissi Dembélé Y, Lochetti G, Sidibé A, Tartanac F (2020) Local solutions for sustainable food systems: the contribution of orphan crops and wild edible species. *Agron. J,* 10 (2): 231.

Boron V, Payán E, McMillan D, Tzanopulos J (2016) Achieving sustainable development in rural areas in Colombia: future scenarios for biodiversity conservation under land-use change. *Land Use Policy*, 59: 27-37.

Brehm JM, Maxted N, Martin-Louçao MA, Frod-Lloyd BV (2010) New approaches for establishing conservation priorities for socio-economically important plant species. *Biodivers Conserv*, 19 (9): 2715-2740.

Brooks MT, Mittermeier RA, Mittermeier CG, Da Fonseca GAB, Rylands AB, Konstat, WP, Flick P, Pilgrim J, Oldfield S, Magin G, Hilton-Taylor C (2002) Habitat loss and extinction in the hotspots of biodiversity. *Biodivers Conserv*, 16 (4): 909-923.

Brown JL, Bennett J, French CM (2017) SDMtoolbox: the next generation Python-based GIS toolkit for landscape genetic, biogeographic and species distribution model analyses. PeerJ [online] 5:e4095. doi: 10.7717/peerj.4095.

Chamberlain S, Oldoni D, Barve V, Desmet P, Geffert L, Mcglinn D, Ram K (2021) Rgbif: Interface to the Global "Biodiversity" Information Facility API. R Package Version 3.6.0.

Chízmar C (2009) *Plantas comestibles de Centroamérica. Santo Domingo de Heredia*. Costa Rica: Instituto Nacional de Biodiversidad, INBio, 360p.

Clerici N, Salazar C, Pardo-Díaz C, Jiggins CD, Linares M (2019) Peace in Colombia is a critical moment for Neotropical connectivity and conservation: save the northern Andes–Amazon biodiversity bridge. *Conserv Lett,* 12: 1-7. doi: 10.1111/conl.12594

Cook EMF (1995) *Economic Botany Data Collection Standard.* London, Royal Botanic Gardens, Kew, 146p.

Cuatrecasas J (1958) Aspectos de la vegetaciòn natural de Colombia. Rev. Acad. Acad. Col. Cienc. Ex. Fìs. Nat. 10(40):221-268.

Diazgranados M, Allkin B, Black N, Cámara-Leret R, Canteiro C, Carretero J, Eastwood R, Hargreaves S, Hudson A, Milliken W, Nesbitt M, Ondo I, Patmore K, Pironon, S, Turner R, Ulian T (2020) *World Checklist of Useful Plant Species*. London, Royal Botanic Gardens, Kew. 689p.

FAO (2015) *The Second Report on the State of The World's Plant Genetic Resources for Food and Agriculture.* Rome, FAO Commission on Genetic Resources for Food and Agriculture Organization of United Nations. 370p.

FAO (2019) *The State of the World's Biodiversity for Food and Agriculture*. J. Bélanger & D. Pilling (eds.). Rome, FAO Commission on Genetic Resources for Food and Agriculture Assessments. 572 p.

Food Plants International (2001). Retrieved on 14 March 2021 at Articles & Books - Food Plants International. Available at: https://foodplantsinternational.com/articles/

French B (2019) Food Plants International database of edible plants of the world, a free resource for all. *Acta Hortic*, 1241: 1-6.

Fuertes CCA (2019) Diversidad, distribución y uso del género Vasconcellea (Caricaceae) en el sur de los Andes colombianos. Dissertation. Universidad del Cauca. Facultad de Ciencias Agrarias. Departamento de Ciencias Agropecuarias. Popayán, Colombia. 114 p.

GBIF (2021) GBIF Backbone Taxonomy. Available at: https://www.gbif.org/ [Accessed in April 2021].

Grau HR, Aide M (2008) Globalisation and land-use transitions in Latin America. *Ecol Soc*, 13 (2): 1-16

Hijmans RJ, van Etten J (2012) Raster: Geographic analysis and modelling with raster data. R package version 2.0-12. Available at: http://CRAN.R-project.org/package=raster

Hunter D, Borelli T, Beltrame DMO, Oliveira CNS, Coradin L, Wasike WV, Wasilwa L, Mwai J, Manjella A, Samarasinghe GWL, Madhujith T, Nadeeshani HVH, Tan A (2019) The potential of neglected and underutilised species for improving diets and nutrition. *Planta* 250: 709–729.

Hurtado-Bermudez LJ, Velez-Torres I, Mendez F (2020) No land for food: prevalence of food insecurity in ethnic communities enclosed by sugarcane monocrop in Colombia. *Int J of Public Health*, 65: 1087-1096.

IUCN (2021) *The IUCN Red List of Threatened Species. Version 2021-1.* Available at: https://www.iucnredlist.org [Downloaded on 24 June 2021].

Kalamandeen M, Gloor E, Mitchard E (2018) Pervasive raise of small-scale deforestation in Amazonia. *Sci Rep*, 8 (1600): 1-10.

Kor L, Homewood K, Dawson TP, Diazgranados M (2021) Sustainability of wild plant use in the Andean Community of South America. *Ambio,* 50: 1681-1697.

León J (1966) Central American and West Indian species of *Inga* (Leguminosae). *Ann Mo Bot Gard*, 53 (3):265-359.

Londono C, Cleef A, Madrinan S (2014) Angiosperm flora and biogeography of the páramo region of Colombia, Northern Andes. *Flora*, 209: 81–87.

Lorenzi H, Bacher L, Lacerda M, Sartori S (2000) *Brazilian Fruits and Cultivated Exotics*. Nova Odessa, Instituto Plantarum de Estudos da Flora LTDA. 672p.

Martin FW, Campbell CW, Rubertbe RM (1987) *Perennial Edible Fruits of the Tropics: An Inventory*. U.S. Dept. of Agriculture, Agricultural Research Service. 247p.

Meyers N, Mittermeier R, Mittermeier CG, Kent J (2020) Biodiversity hotspot for conservation priority. *Nature*, 403 (6772): 853-8

Missouri Botanical Gardens (2021) Tropicos.org. Available at: https://tropicos.org [Accessed on: April 2021].

N'Danikou S, Achigan-Dako EG, Wong JLG (2011) Eliciting local values of wild edible plants in Southern Benin to identify priority species for conservation. *Economic Botany,* 65, 381-395.

Nogueira S, Nogueira-Filho S (2011) Wildlife farming: an alternative to unsustainable hunting and deforestation in Neotropical forests? *Biodiv Conserv*, 20: 1385-1397.

Noh JK, Echeveirra C, Kleemann J, Koo H, Fürst C, Cuenc P (2020) Warning about conservation status of forest ecosystems in tropical Andes: National Assessment based on IUCN criteria. *PLoS ONE* [online], 15(8): e0237877.

Ocampo J (2013) Diversidad y distribución de las Passifloraceae en el departamento del Huila en Colombia. *Acta Biol Colomb.*, 18 (3): 511-516.

Ondo I, *et al.* (in prep.) ShinyCCleaner: an interactive app for cleaning species occurrence records.

Padulosi S, Heywood V, Hunter D, Jarvis A (2011) Underutilised species and climate change: current status and outlook.. In: Shyam SY, Redden RJ, Hatfield JL, Lotze-Campen H, Hall AE eds., *Crop Adaptation to Climate Change*. pp. 507-517. Blackwell Publishing.

Pieroni A, Pawera L, Shah GM (2016) Gastronomic ethnobiology. In: Albuquerque UP, Alves RN, eds. *Introduction to Ethnobiology*. pp. 53-62. Springer.

Pilgrim SE, Collen LC, Smith DJ, Pretty JN (2008) Ecological knowledge is lost in wealthier communities and countries. *Environ Sci Technol.* 42 (4): 1004-9.

POWO (2021) Plants of the World Online. Facilitated by the Royal Botanic Gardens, Kew. Available at: http://www.plantsoftheworldonline.org/ [Accessed in April 2021].

RBG, Kew (2016) State of the World's Plants. Available at: https://stateoftheworldsplants.com [Accessed in April 2021]

R Core Team (2021) R: a language and environment for statistical computing. R Foundation for Statistical Computing. Available at: https://www.R-project.org/

Renjifo LM, Amaya-Villareal AM, Butchart SHM (2020) Tracking extinction risk trends and patterns in a megadiverse country: a Red List Index for birds in Colombia. *PLoS ONE*, 15(1), 1-19.

Rodriguez LMG (2019) Estudio de crecimiento y de producción de frutos de *Bactris guineesis* (güiscoyol) en Sistemas Agroforestales como potencial de desarrollo en la Región Chorotega. Costa Rica, Universidad Técnica Nacional Investigación y transferencia. doi: 10.13140/RG.2.2.11970.22725

Ulian T, Diazgranados M, Pironon S, Padulosi S, Liu U, Davies L, Howes M-JR, Borrell JS, *et al* (2020) Unlocking plant resources to support food security and promote sustainable agriculture. *Plants, People, Planet*, 12 (5): 421-445.

Wickens GE (1995) *Edible Nuts: Non-Wood Forest Products*, Handbook 5. Rome, FAO - Food and Agriculture Organization of the United Nations. 150p.

Wickham H (2016) ggplot2: elegant graphics for data analysis. New York, Springer-Verlag. ISBN 978-3-319-24277-4. Available at: https://ggplot2.tidyverse.org

Wickham H (2011) The Split-Apply-Combine strategy for data analysis. *J Stat Softw*, 40 (1): 1–29.

Wickham H, Francois R (2021) Dplyr: A Grammar of Data Manipulation. R Package Version 0.4.3. Available at: https://dplyr.tidyverse.org

Wolff F (2004) Legal factors driving agrobiodiversity loss. *Elni Review,* 1: 1-11.

Leaf beetle (Chrysomelidae, Coleoptera).
Carolina Castellanos-Castro

Chapter 5

Insecticide Plants of Colombia

Jennyfer Andrea Aldana-Mejía[1], Henry Yesid Bernal M.[2]* & Alba Nohemí Téllez A.[3]

[1] Doutoranda em Ciências Farmacêuticas, Laboratório de Farmacognosia, Faculdade de Ciências Farmacêuticas, Universidade de São Paulo, Campus de Ribeirão Preto, Ribeirão Preto, São Paulo, Brasil. j.aldana@usp.br; j.aldana@javeriana.edu.co

[2] Ex-Profesor Asociado del Departamento de Biología, Facultad de Ciencias, Universidad Javeriana, Herbario de la Pontificia Universidad Javeriana (HPUJ), Bogotá, D.C., Colombia. henryesid@gmail.com

[3] Profesor Titular del Departamento de Química, Facultad de Ciencias, Universidad Javeriana, Grupo de Investigación en Fitoquímica (GIFUJ), Pontificia Universidad Javeriana, Carrera 7 No. 43–82, Laboratorio 206, Edificio Carlos Ortiz, S.J. Bogotá, D.C., Colombia. ntellez@javeriana.edu.co

***Corresponding author:** henryesid@gmail.com

Keywords: insectostatic, insectifugal, insectivorous plants, phytoecdysteroids, biocides, repellants, chemosterilisers.

ABSTRACT

Several plant species from Colombia are recognised as having some insecticide activity, but no publication has detailed this use to date. This study aimed to identify and characterise the plant species of Colombia that are used as insecticides, assessing the state of their traditional and scientific knowledge, and emphasising the native species of the country. A literature review of insecticide plants in Colombia allowed us to recognise 16 mechanisms of action and to classify them into four major categories: i) species that provide chemical pesticides; ii) insectostatic species; iii) insecticide species by mechanical means; iv) and species with unknown insecticide action. The main mechanisms of action reported were repellent (22.3%), followed by 'insecticide by unknown means' type (21.2%). In total, 632 species that are used as 'insecticides' were identified in Colombia, of which spermatophytes represent 98.9% and only 1.1% are lycophytes and ferns. The families Fabaceae, Asteraceae (Compositae), and Solanaceae present the largest number of reports on insecticide species. A total of 408 native species of Colombia used as insecticides were arranged into seven groups according to their geographical origin, including 11 endemic species: *Ageratina ampla*, *Ageratina vacciniaefolia*, *Berberis monguiensis*, *Berberis samacana*, *Berberis tabiensis*, *Chromolaena barranquillensis*, *Espeletia killipii*, *Lourteigia microphylla*, *Mandevilla mollisima*, *Ocotea caparrapi*, and *Ryania speciosa* var. *chocoensis*. Regarding the endemic species, eight (72.7%) present a wide geographic distribution in the country, and three species (27.3%) have restricted geographic distribution in the country, mainly in the departments of Boyacá, Cundinamarca, and Caldas. None of those Colombian endemic species presented sufficient evidence of scientific knowledge, and only a single species has sufficient evidence of traditional knowledge regarding its insecticide action.

RESUMEN

Numerosas especies vegetales presentes en el territorio colombiano se reconocen con algún tipo de actividad insecticida; sin embargo, no existen publicaciones que detallen el estado actual de este uso. Esta investigación se orientó a la identificación y caracterización de las especies de Colombia utilizadas como insecticidas y al establecimiento del estado actual de sus conocimientos tradicional y científico, con énfasis en las especies nativas del país. Una revisión bibliográfica de las plantas insecticidas de Colombia permitió reconocer 16 mecanismos de acción, y clasificarlos en cuatro grandes categorías, definidas como: especies proveedoras de plaguicidas químicos, especies insectostáticas, especies insecticidas por medios mecánicos y especies sin medio de acción conocido. El principal mecanismo de acción reportado fue el de repelente (22,3%), seguido por insecticidas por medios desconocidos (21,2%). Se identificaron 632 especies utilizadas en Colombia como insecticidas, de las cuales el 98,9% son espermatofitos y el 1,1% son musgos, hornworts, hepáticas o licofitas. Las familias Fabaceae, Asteraceae (Compositae), Solanaceae presentaron el mayor número de especies reportadas. Las 408 especies nativas de Colombia utilizadas como insecticidas se ordenaron en siete grupos de acuerdo con su origen geográfico, destacándose la presencia de 11 especies endémicas del país que son: *Ageratina ampla*, *Ageratina vacciniaefolia*, *Berberis monguiensis*, *Berberis samacana*, *Berberis tabiensis*, *Chromolaena barranquillensis*, *Espeletia killipii*, *Lourteigia microphylla*, *Mandevilla mollisima*, *Ocotea caparrapi* y *Ryania speciosa* var. *chocoensis*. De estas especies 8 (72,7%) son de distribución geográfica amplia en el país y 3 (27,3%) especies tienen distribución geográfica restringida en el país, con prevalencia en los departamentos de Boyacá, Cundinamarca y Caldas. Ninguna de estas especies endémicas de Colombia presentó suficiente evidencia de conocimientos científicos y una sola especie posee suficiente evidencia de conocimientos tradicionales para su utilización como insecticida.

INTRODUCTION

Research and field collection of plants in Colombia have been carried out for more than 200 years for national and international herbaria or botanical gardens. Nowadays, the country holds more than 30,000 registered plant species, ranking second in the world in terms of plant richness (Andrade, 2011; SIB Colombia, 2020). Many of these plant species are considered important due to the environmental services they provide upon which human societies depend. Those useful or promising plants species have applications in different fields. Three main aspects lead to the great variety of useful plants in Colombia: the great variety of ecosystems, the fantastic plant species richness, and the diversity of human groups represented by nearly 100 different ethnic groups, including indigenous Afro-descendants, mestizos, and other traditional groups. The Flora of Colombia (native and exotic) is used in almost all daily activities: as artisanal resources, biocides, coagulants, colourings, construction materials, dyes, essential oils, flocculants, food, fuel or industrial oils, honey, housing products, insecticides, insectostatics, insects traps or catalysts, medicines, oilseeds, repellents, tanning agents, timber, waxes, and wrapping, and in cultural activities, magical-religious activities, and dentistry, among other uses. A strong example of useful plants of Colombia is the medicinal species identified and documented in recent studies carried out by Bernal & Mesa (Chapter 3).

Ethnobotanical knowledge has been used as a starting point for developing new products (Ribeiro *et al.*, 2017). Terms such as “bioprospecting” have been applied to define the search for new medicinal products (Garnatje *et al.*, 2017). In general, many species that are considered useful and from which production chains have been derived (such as biocides, food, and textiles) have come from the link between traditional knowledge and its scientific application. Ethnobotanical bioprospecting for the identification of bioactive compounds has become an important tool that combines ethnobotanical, phytochemical, phylogenetic, and molecular data. From the traditional knowledge, it has been possible to validate scientific predictions about the biological potential of plants (Garnatje *et al.*, 2017).

The need for food and raw materials on a large scale, guaranteeing the profitability and economic stability of agroindustries, has generated dependence on synthetic chemicals for pest protection, causing an increase in environmental and health issues or a decrease in yield (Bonilla, 1991). Insecticides derived mainly from pyrethrum, rotenone, ryanodines, nicotine, neem extracts, and some essential oils (garlic, rosemary, eucalyptus, among others) have taken hold of the market, especially in developing countries (Isman, 2006; Isman, 2020). However, the agricultural use of plants as a protection resource against insects has been documented for more than two millennia in communities in ancient China, Egypt, Greece and India (Isman, 2006). Nowadays, due to the application of this traditional knowledge, several botanical products have been brought to the market to control and manage agriculturally important pests. The use of plants or their chemical compounds constitutes an environmentally safer alternative, which could be combined with other techniques and methods (Cerón, 1996). It can be used as a complementary strategy in regulating insect pests, even if the full replacement of pesticides is impossible (Cerón, 1996).

In general, an insecticide is a substance or compound that, due to the nature of its chemical structure, is used to kill or exert a biocidal action on insects (Lawrence, 2003; Silva *et al.*, 2002). However, most plant species used to protect or control harmful pests do not exhibit any insecticide effect in the strict sense. Their mechanism of action is due to other phenomena (repellency, deterrence of feeding, among others) that also have a protective effect against insects (Silva *et al.*, 2002). Under the classical perspective of pest control, the plant species that present insecticidal actions due to their physical or mechanical characters are not considered to be insecticide species.

In this present study, the conventional concept of ‘insecticide species’ is adapted based on the literature review, aiming to include other mechanisms of insecticide action. Here, the reviewed definition of ‘insecticide plant’ includes all plant species or their infraspecific categories (subspecies, variety, subvariety, form, or subform) with biological actions that mediate or interfere directly in the physiological and behavioural processes of insects, preventing, mitigating, or eradicating the damage that the insects produce on other plants, humans and/or animals.

Several plant species from Colombia are recognised as presenting insecticide activity (by chemical or mechanical means) or insectostatic action. However, there are no publications detailing the existing knowledge of this use. Therefore, this study was oriented toward identifying and characterising the traditional and scientific knowledge relating to species used as insecticides in Colombia, emphasising the native species of the country. Finally, this study aims to help the discovery of new and better options for pest control.

MATERIALS AND METHODS

The information gathered on plants used as insecticides in Colombia was initially based on searches using keywords related to the subject in different bibliographic sources. The subsequent step was conceptualising the terms that refer to the control and management of insects and the role of plants in these fields.

From the initial review on ‘plant species with biological activity against insects’ was possible to identify two differentiated groups in the literature: (i) ‘insecticide species’, which are those that exert a lethal effect on the target insect, similar to traditional chemical pesticides; and (ii) ‘insectostatic species’, which act on the behaviour or development of insects, exerting a long-term effect (Table 1).

The information was searched from libraries, documentation centres and national and international databases specialised in the use and exploitation of plant resources, from the perspective of pest control and management, using keywords related to the mechanisms of action found. The classification and conceptualisation of the terminology of the mechanisms of action of plants used as insecticides in Colombia were based on the literature (Table 1). The mechanisms were divided into four main groups and sixteen subgroups. In many cases, the mechanism was not defined, and this led us to consider the mention of use as "insecticide by unknown means" as an independent category. It is important to note that this is a classification made by this study, which we put forward as a proposal for further in-depth future research.

TABLE 1. Classification of the mechanisms of action of insecticide plants of Colombia.

Plants Supplying Insecticides or Chemical Pesticides Plant species presenting a source of molecules or chemical substances that cause a toxic or lethal effect for insects (Pedigo *et al.*, 2021; Vergara & Madrigal, 1994)	
Insecticide plants (by ingestion)	Plants producing chemical compounds causing a toxic effect on the digestive system of insects (Abubakar *et al.*, 2019; Alonso, 1999; Pedigo *et al.*, 2021)
Insecticide plants (by touch)	Plants producing insect-killing chemical compounds in direct contact with insects (Alonso, 1999; Pedigo *et al.*, 2021)
Fumigant insecticide plants	Plants with insecticide chemical compounds acting on target pests in the vapour or gas phase, leading to death (Abubakar *et al.*, 2019; Pedigo *et al.*, 2021; Rajendran & Sriranjini, 2008; Spochacz *et al.*, 2018)
Phototoxic insecticide plants	Plants with photosensitive chemical compounds that can accumulate in insects and, when exposed to visible radiation, become phototoxic products, or cause lethal photochemical reactions (Ahmed *et al.*, 2018; Pascual, 1996; Spochacz *et al.*, 2018)
Insectostatic Plants Plant species with chemical compounds that do not act directly and immediately as lethal agents but can contribute to the death or control of insects in the long term, intervening in their behaviour and/or development (Enrique *et al.*, 2008; Montesino *et al.*, 2009; Silva *et al.*, 2002)	
Repellent or insect-repellent plants	Plants with chemical compounds capable of making insects move away from the surface or structure from which they emanate due to the repulsion or irritation that they cause (Vergara & Madrigal, 1994; Enrique *et al.*, 2008; Abubakar *et al.*, 2019)
Feed-regulating plants	Plants with chemical compounds affecting the insect feeding process at any stage, either by inhibiting or interrupting it (Pascual, 1996; Abubakar *et al.*, 2019; Enrique *et al.*, 2008)
RIG (Regulating Insect Growth)	Plants with chemical compounds intervening in the functioning of the hormonal systems of insects, resulting in the alteration of physiological processes related to body growth (Abubakar *et al.*, 2019; Pedigo *et al.*, 2021)
RIG plants, seedling inducers: phytoecdysteroids	Plants with chemical compounds similar to ecdysones or hormones that induce seedlings, causing acceleration or advancement of seedlings, intervention in the formation of oocytes, generation of malformations, involvement in embryonic development and sexual immaturity of the adult male (Pascual, 1996; Primo, 1991; Vergara & Madrigal, 1994)
RIG plants, seedling inhibitors: anti-hormones	Plants with chemical structures capable of preventing the ecdysis process from taking place, causing the lengthening of the development phases and the subsequent death of the individual without the fulfilment of its normal life cycle (Pascual, 1996; Primo, 1991; Vergara & Madrigal, 1994).
RIG plants, metamorphosis inducers: juvenile antihormones	Plants with antagonist compounds of the juvenile hormone (JH), capable of generating effects such as early metamorphosis, the appearance of adults, sterility of females, reduction of fertility in males, interruption of embryogenesis and inhibition of sexual pheromone (Pascual & Robledo, 1998; Silva-Aguayo, 2021; Vergara & Madrigal, 1994)

RIG plants, metamorphosis inhibitors: 'phytojuvenoids'	Plants with 'phytojuvenoid-type' compounds, which fulfil roles analogous to those of the juvenile hormones of insects, and therefore impede metamorphosis and sterility by suspending embryogenesis and gametogenesis processes (Pascual, 1996; Primo, 1991; Silva-Aguayo, 2021)
IGI plants (Inhibiting the Growth of Insects)	Plants with chemical molecules capable of inhibiting chitin synthesis in insects, preventing them from completing ecdysis (Vergara & Madrigal, 1994)
Plants regulating the oviposition of insects	Plants with compounds capable of suppressing or interrupting oviposition through processes that cut off the needed perception of plant-insect stimuli (but not from any hormonal functioning intervention) (Montesino *et al.*, 2009)
Chemosterilizing plants	Plants with chemical compounds that exert a definitive sterilising action on target insects (Carrero, 1996; Pedigo *et al.*, 2021; Vergara & Madrigal, 1994)
Insecticide Plants by Mechanical Means Species with particular morphological adaptations acting as adherent traps that impede insect movement and lead to death by starvation or that are digested as a complementary activity in the acquisition of nutrients (García-Saavedra *et al.*, 2007)	
Insectivorous plants	Plant species using morphological and anatomical adaptations to carry out the mechanical process of trapping and digesting insects using proteolytic enzymes (Alvarado *et al.*, 2011; Fernández-Pérez, 1964; García-Saavedra *et al.*, 2007; Lawrence *et al.*, 2003)
Plants with sticky traps	Plant species with morphological characteristics including special structures which work as 'sticky traps' where insects can pose, fall, and die (LoPresti *et al.*, 2015)
Insecticides Plants by Unkown Means Plants reported with insecticide action, without specifying the type of biological action, nor how they exert it	

A database was built aiming to identify, compile, and analyse the existing traditional and scientific knowledge associated with the species used in Colombia as 'insecticide' and 'insectostatic' plants, containing the following information for each species: i) kingdom, ii) sub-kingdoms, iii) division (phylum), iv) class, v) order, vi) botanical family, vii) valid scientific name, viii) synonyms, ix) common names, x) geographical distribution, xi) insecticide action mechanisms, xii) bibliographic references, xiii) indicators of ethnobotanical knowledge, and xiv) indicators of scientific knowledge.

The geographical distribution for each species was organised into four large groups: 1) wild plants, 2) cultivated plants (native or exotic), 3) naturalised plants, and 4) species with uncertain geographic distribution. The wild (non-cultivated) species were organised into three groups: 1) native, indigenous, or autochthonous plants of Colombia; 2) non-native, introduced, or exotic plants, and 3) cosmopolitan plants. The native species used as insecticide plants in Colombia were organised into seven groups: 1) endemic species of Colombia (E); 2) native species of the Neotropics (N) with presence in Colombia; 3) native species of Mesoamerica (M) with presence in Colombia; 4) species native to South America (S) with presence in Colombia (in two or more non-Andean South American countries); 5) species from Mesoamerica and South America, present in Colombia (M + S); 6) Mesoamerican and Andean species present in Colombia (M + A), and 7) native plants of the Andean region (A) present in Colombia (in two or more Andean countries: Colombia, Ecuador, Peru, Venezuela, and Bolivia).

Parameters included in semi-structured interviews from ethnobotanical studies on useful plants, such as type of use, used organs, type of use, preparation and dosage (Espinosa-Jiménez *et al.*, 2021; Ahmad *et al.*, 2014; Gu *et al.*, 2020), were used define the information compiled on traditional knowledge. Available information on the spectrum of action and possible toxicity were added to the above parameters. The compilation of the state of scientific knowledge was based on the search for information on characterisation, efficacy, and safety in the use of insecticide plants. The search included literature on the characterisation of insecticide plants (taxonomic, systematic, biogeographical, and ecological studies), efficacy analysis of insecticide plants (phytochemical and biological activity studies), and safety analysis (toxicity studies).

The indicators of traditional knowledge related to each species were: a) insecticide actions (mention of insecticide use), b) part(s) used, c) traditional preparation mode, d) route of administration (topic or systemic), e) method of use (dose), f) method of use (posology), g) spectrum of action (low 1–3 pests, medium 4–6 pests and high 7 or more pests) and h) toxicity (low, medium, or high).

The indicators of scientific knowledge related to each species were: a) geographical distribution, b) altitudinal distribution, c) ecological characterisation, d) morphological characterisation, e) molecular characterisation, f) taxonomic description and diagnosis, g) methods for identification and taxonomic determination, h) toxicity tests in insects (LD_{50} and concentrations), and i) tests of

insecticide use efficacy (phytochemical investigations and biological activities investigations). In the phytochemical investigations, the phytochemical profiles (extraction, fractionation or separation, qualitative characterisation, isolation, and purification) and the identification of the substances (structural elucidation) were recorded. Regarding the investigation of biological activities, *in vivo* activities (adults, eggs, larvae, pupae, and nymphs) and spectrum of action (low 1–3 pests, medium 4–6 pests and high 7 or more pests) were recorded.

For each indicator, scores were assigned according to the absence (0) or presence (1) of bibliographic information to conduct a quantitative assessment of each level of knowledge. Considering that the indicators 'action spectrum', 'toxicity', and 'phytochemical profile' were subdivided into several levels according to the complexity of the information, the scoring scale was extended. It could also range from zero (no information) to 4 (high spectrum of action, high toxicity/ LD_{50}, and isolation/purification of active substances). Some ethnobotanical (traditional form of preparation, route of administration, method of use, and toxicity) and scientific indicators (toxicity tests on insects) were considered 'not applicable' in the quantification of data for plant species classified as insecticides by mechanical means.

A scale of scores was established to determine the level (low, medium, or high) of each species' state of ethnobotanical knowledge. The scale was differentiated according to the category of insecticide use. For example, for plants exerting non-mechanical mechanisms of action, within a maximum range of 13, it was determined that a score of 0–4 indicates a low state of knowledge; 5–9 intermediate, and 10–13 high. For insectivorous plants and 'sticky trap' species, within a maximum range of five, it was defined that a total score between 0–1 represents a low state of knowledge, between 2–3 medium, and 4–5 high.

At the scientific level, for species with non-mechanical mechanisms, within a maximum range of 34, it was defined that scores between 0–11 represent a low level of knowledge, 12–23 medium, and 24–34 high. For insectivorous plants and 'sticky trap' species, within a maximum range of 31, it was defined that scores between 0–11 represent a low state of knowledge, 12–22 intermediate, and 23–31 high.

The identification and characterisation of the insecticide plants of Colombia were carried out based on the review of 432 bibliographic references on the subject, from which a corresponding list was constructed. Owing to its length, only the basic bibliography is presented in this publication. For each species presenting information on insecticide uses in Colombia, a search was carried out from the national herbaria of the country that presents online data accessible from the Global Biodiversity Information Facility (GBIF) (Table 2).

National and international databases were also consulted regarding species nomenclature, geographical distribution, altitudinal distribution, and bibliography (Table 3).

TABLE 2. Consulted Colombian herbaria from the Global Biodiversity Information Facility (GBIF).

Herbarium	Acronym	City, department	Web site
Colombian Amazon Herbarium related to the Amazon Institute for Scientific Research, Sinchi	COAH (Sinchi)	Bogotá, Bogotá D.C.	https://sinchi.org.co/coah
Colombian National Herbarium	COL	Bogotá, Bogotá D.C.	http://www.biovirtual.unal.edu.co/es/colecciones/search/plants/
Herbarium of the University of Antioquia	HUA	Medellín, Antioquia	https://www.udea.edu.co/wps/portal/udea/web/inicio/unidades-academicas/ciencias-exactas-naturales/herbario#:~:text=El%20herbario%20de%20la%20Universidad,y%20otras%20regiones%20de%20Colombia
Universidad Católica de Oriente Herbarium	HUCO	Medellín, Antioquia	https://www.uco.edu.co/herbariouco/Paginas/Catalogo-Virtual.aspx
Botanical Garden of Bogotá "José Celestino Mutis" Herbarium	JBB	Bogotá, Bogotá D.C.	http://herbario.jbb.gov.co/especimen/simple
Guillermo Piñeres Botanical Garden Herbarium	JBGP	Cartagena, Bolívar	http://www.jbgp.org.co/el-herbario/
Juan María Céspedes Garden Herbarium	TULV	Tuluá, Valle del Cauca	https://www.inciva.gov.co/patrimonio-turistico/jard-iacute-n-bot-aacute-nico-juan-maria-c-eacute-spedes
Herbario Forestal Gilberto Emilio Mahecha Vega of the District University	UDBC	Bogotá, Bogotá D.C.	http://herbario.udistrital.edu.co/herbario/index.php?option=com_content&view=article&id=4&Itemid=3

TABLE 3. National and international databases consulted as online resources.

Database	Web site
BHL. Biodiversity Heritage Library	https://www.biodiversitylibrary.org/
BIOVIRTUAL (Common names). Institute of Natural Sciences – National University of Colombia	http://www.biovirtual.unal.edu.co/nombrescomunes/es/
BIOVIRTUAL (Herbarium specimens). Institute of Natural Sciences – National University of Colombia	http://www.biovirtual.unal.edu.co/es/
Catalogue of plants and lichens of Colombia	http://catalogoplantasdecolombia.unal.edu.co/es/
CBIF. Canadian Biodiversity Information Facility	https://www.cbif. gc.ca/acp
CEIBA. Cataloguer of Biological Information – Instituto de Investigación de Recursos Biológicos Alexander von Humboldt-Colombia	http://www.i2d.humboldt.org.co/ceiba/
Colombia Biotic Diversity, Institute of Natural Sciences, Museum of Natural History of the National University of Colombia	http://colombiadiversidadbiotica.com/Sitio_web/Bienvenida.html
CoL. Catalogue of Life	https://www.catalogueoflife.org/
ColPlantA. Colombian Resources for Plants	https://colplanta.org/
DR. DUCKE's. Phytochemical and Ethnobotanical Databases	https://phytochem.nal.usda.gov/phytochem/search
EOL. Encyclopedia of Life. National Museum of Natural History, Smithsonian	https://eol.org/
Flora Of Colombia. Monographs of the Institute of Natural Sciences of the National University of Colombia.	http://ciencias.bogota.unal.edu.co/menu-principal/publicaciones/biblioteca-digital/flora-de-colombia/
Flora of the Royal Botanical Expedition. Royal Botanical Garden of Spain	https://bibdigital.rjb.csic.es/
GBIF. Global Biodiversity Information Facility	https://www.gbif.org/species/search
Global Compositae Database. International Compositae Alliance	https://compositae.org/
GRIN. Germplasm Resources Information Network	https://www.ars-grin.gov/
ILDIS DATABASE. International Legume Database and Information Service	http://www.ildis.org/
IPNI. International Plant Name Index	https://www.ipni.org/
IUCN-Red List. The IUCN red list of the threatened species	https://www.iucnredlist.org/
ITIS. Integrated Taxonomic Information System	https://www.itis.gov/
JSTOR Global Plants Database	https://plants.jstor.org/
Kew Herbarium Catalogue. Royal Botanic Gardens, Kew	http://apps.kew.org/herbcat/navigator.do
Kew Plant Information Portal. Napralert Database. University of Illinois at Chicago	https://www.napralert.org/
NCBI. National Center for Biotechnology Information. National Library of Medicine, National Institutes of Health, USA	https://www.ncbi.nlm.nih.gov/
NMNH. Smithsonian National Museum of Natural History	https://collections.nmnh.si.edu/search/botany/
PL@NTUSE. Plant uses	https://uses.plantnet-project.org/en/Main_Page
RACCEFYN. Journal of the Colombian Academy of Exact, Physical and Natural Sciences	https://raccefyn.co/index.php/raccefyn

Database	Web site
REFLORA. Virtual herbarium of the flora of Brazil. Rio de Janeiro Botanical Garden	http://reflora.jbrj.gov.br/reflora/herbarioVirtual/ConsultaPublicoHVUC/ConsultaPublicoHVUC.do
Science Direct	https://www.sciendedirect.com/
SCOPUS-Elsevier. Bibliographic database	http://www.scopus.com
SIAC. Colombia's environmental information system	http://www.siac.gov.co/biodiversidad
SIB. Colombia's Biodiversity Information System	https://sibcolombia.net/
SPECIES 2000 and ITIS. Catalogue of Life. Annual checklist. Integrated Taxonomic Information System	http://www.catalogueoflife.org/annual-checklist/2019/info/about
The Plant List	http://www.theplantlist.org/
Tropicos. Vascular Plants of the Americas	http://legacy.tropicos.org/project/VPA
USDA-Plants Database. United States Department of Agriculture	http://www.plants.usda.gov
Useful Tropical Plants Database	http://tropical.theferns.info/
WCSP. World Checklist of Select Plant Families. Royal Botanic Gardens Kew	https://wcsp.science.kew.org/qsearch.do
WCVP. The World Checklist of Vascular Plants. Royal Botanic Gardens Kew	https://wcvp.science.kew.org/
WFO. World Flora Online Consortium	http://www.worldfloraonline.org/

RESULTS AND DISCUSSION

From the literature search, a total of 632 plant species with reports of insecticide activity were identified as presenting occurrence in Colombia, including 625 species (98.9%) of spermatophytes and seven species (1.1%) of lycophytes and ferns. The complete list of species and associated information on insecticide plants in Colombia has been added to the Col*Plant*A portal (https://colplanta.org), Royal Botanical Garden, Kew. The species recorded as insecticide belong to 44 orders, distributed in 117 botanical families. The highest percentages of insecticide useful species were found among the families: Asteraceae (Compositae), Euphorbiaceae, Lamiaceae, Leguminosae (Fabaceae), and Solanaceae. From the 117 botanical families identified, 41 families (35%) presented a single species with associated information as 'insecticide use'.

The non-monocots flowering plants represented 562 species distributed within 94 botanical families. The highest percentages of species with insecticide action were represented by Leguminosae (Fabaceae) with 92 species, Asteraceae (Compositae) with 66 species, Solanaceae with 30 species, and Lamiaceae with 28 species (Figure 1a). The 60 species of monocots reported as insecticides were distributed in 18 botanical families, with Poaceae (Gramineae) standing out with 16 species, Araceae with ten, Arecaceae (Palmae) with eight, and Zingiberaceae with five (Figure 1b). Three species of gymnosperms with presence in Colombia were registered.

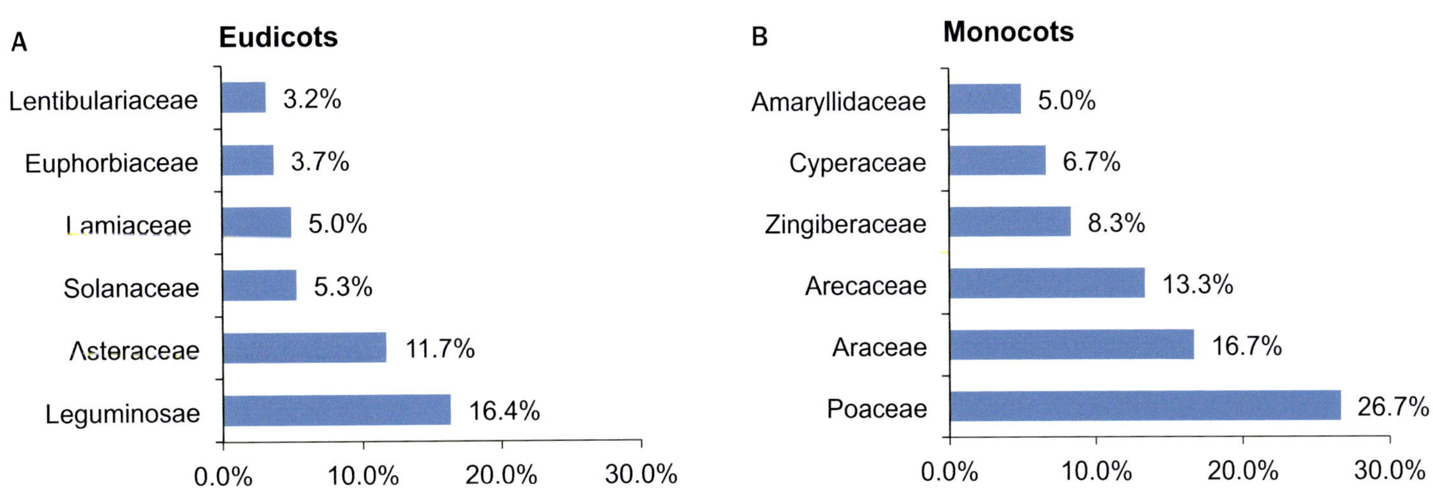

FIGURE 1. Taxonomic coverage of the main botanical families of **A** Eudicots and **B** Monocots, as percentage of species number/family.

Seven insecticide species representing lycophytes and ferns were identified, distributed in six botanical families (Figure 2). Lycopodiaceae is the only family reported with more than one record, including two species (28.6%), the other species are ferns, one species (14.3%) from Equisetales, and four species (57.2%) of Polypodiales.

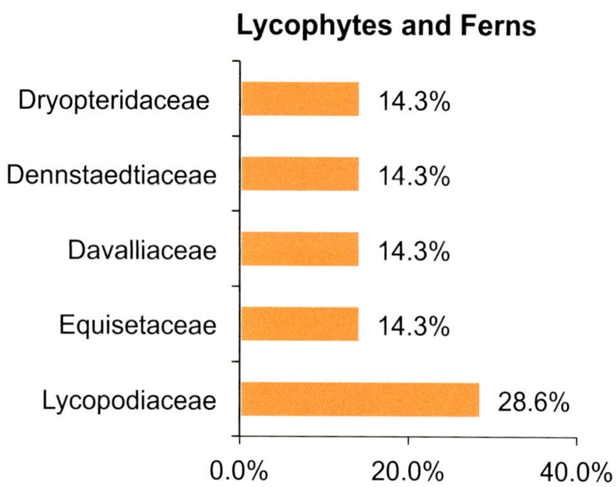

FIGURE 2. Taxonomic coverage of on the main botanical families of lycophyte and ferns, in percentage of species number/family.

Considering the 632 species reported as insecticides in Colombia, the lack of information on the particular mechanism exerted by the species on the 'target' insects category is highly prevalent any level of knowledge, being the second most frequently mentioned category in the literature (Figure 3). The categories 'repellent insecticide' and 'feeding regulation insecticide' were the most common and third most common categories. However, it was found that the categories comprising specific biological actions that have been studied at the scientific level (for example, insecticide related to 'phototoxic', 'inducer of metamorphosis', 'inducer of moulting', 'chemosterilating', and 'growth regulator') are those that have the least reports among the native species of Colombia.

From the total number of species used as 'insecticide' plants in Colombia, 408 species are considered native to the country, including 405 (99.3%) spermatophytes and three (0.7%) ferns and lycophytes. The native species belong to 38 orders, distributed in 90 botanical families. The native species to Colombia grouped by geographical distribution are represented by 222 (54.8%) Neotropical species, 77 (19.0%) Mesoamerican and South American species, 39 (9.6%) Andean species, 28 (6.9%) Mesoamerican species, 27 (6.7%) South American species, 11 (2.7%) species endemic to Colombia, as well as one (0.2%) Mesoamerican and Andean native species (Figure 4).

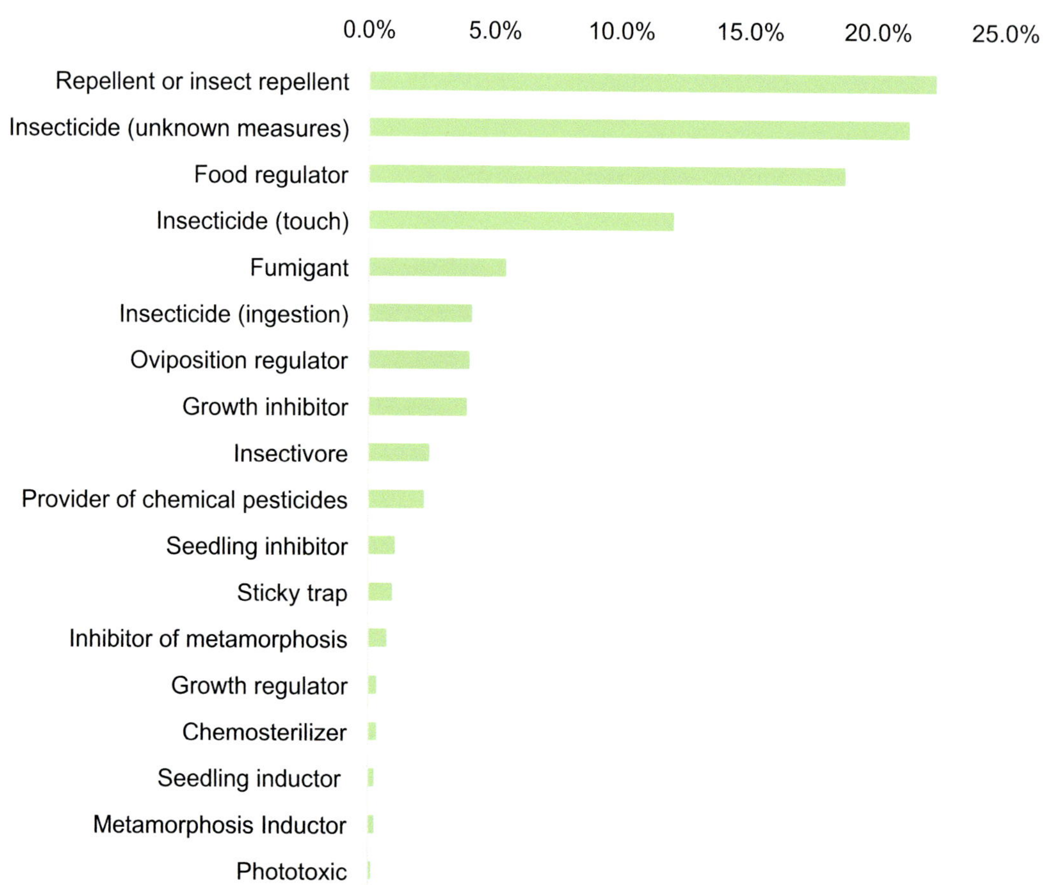

FIGURE 3. Registered mechanisms of action for insecticidal plants with occurrence in Colombia.

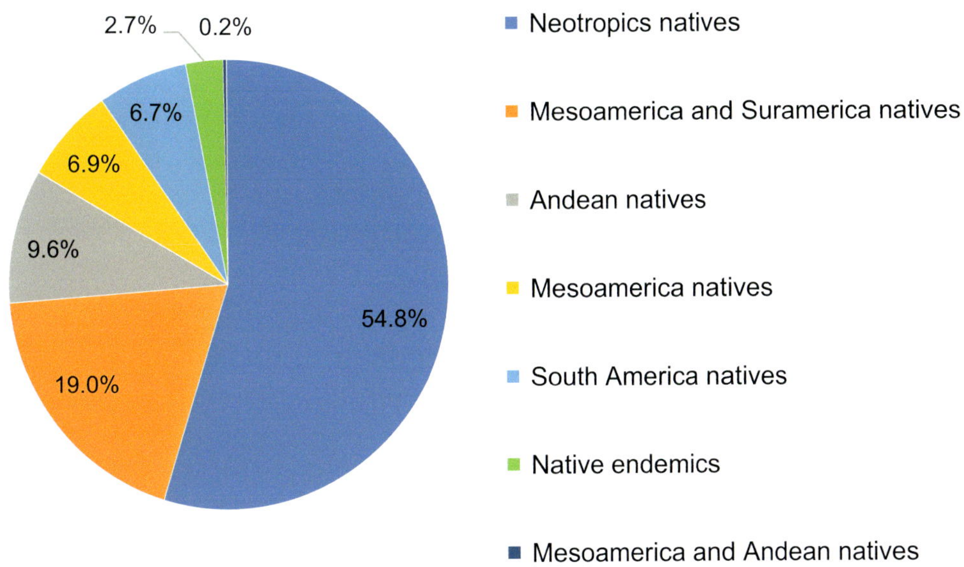

FIGURE 4. Geographical distribution of the native wild insecticide species of Colombia.

Among the native species used as 'insecticides', 11 species were identified as endemic to Colombia: *Ageratina ampla*, *Ageratina vacciniaefolia*, *Berberis monguiensis*, *Berberis samacana*, *Berberis tabiensis*, *Chromolaena barranquillensis*, *Espeletia killipii*, *Lourteigia microphylla*, *Mandevilla mollisima*, *Ocotea caparrapi*, and *Ryania speciosa* var. *chocoensis* (Table 4). The majority of species (72.7%) present a wide geographic distribution in Colombia, and only three species (27.3%) have restricted geographic distribution in the country, with prevalence especially in the departments of Boyacá, Cundinamarca, and Caldas (Table 4).

Among the native species of Colombia used as 'insecticide', none presented high levels of traditional and scientific knowledge about their insecticide activity. However, the endemic species *Ageratina ampla* is notable for presenting a high level of traditional knowledge, and it is also included in the category of insecticides by mechanical means, acting as a 'sticky trap' for insects, one of the least reported mechanisms reported in this study (Table 4).

TABLE 4. List of species endemic to Colombia. [1]WGDS = Wide geographic distribution species; RGDS = Restricted georaphic distribution species; m a.s.l. (m) = above sea level (metres). [2]H = High; I = Intermediate; L = Low.

No.	Scientific name and family	Synonyms	Common names	Altitudinal range (m a.s.l.) and habitat	Geographical distribution and departments[1]	Mechanism of action	State of knowledge[2]	
							Traditional	Scientific
1	***Ageratina ampla*** (Benth.) R.M. King & H. Rob. Compositae (Asteraceae)	*Eupatorium amplum* Benth.	Jarillo, jarrillo	1,750 – 3,400 Sub-Andean forest, Andean forest and Sub-páramo	WGDS Boyacá, Caldas, Cundinamarca, Meta, Risaralda, Tolima and Valle del Cauca	Sticky trap	H	L
2	***Ageratina vacciniaefolia*** (Benth.) R.M. King & H. Rob. Compositae (Asteraceae)	*Eupatorium confertifolium* Klatt *Eupatorium vacciniaefolium* Benth.		2,120 – 3,600 Sub-Andean forest, Andean forest, Sub-páramo and Páramo	WGDS Arauca, Boyacá, Casanare, Cauca, Cundinamarca, Meta and Santander	Insecticide (unknown means)	L	L

No.	Scientific name and family	Synonyms	Common names	Altitudinal range (m a.s.l.) and habitat	Geographical distribution and departments[1]	Mechanism of action	State of knowledge[2]	
							Traditional	Scientific
3	***Berberis monguiensis*** L.A. Camargo Berberidaceae			3,120 Sub-páramo	RGDS Boyacá	Growth inhibitor, feed regulator	L	L
4	***Berberis samacana*** L.A. Camargo Berberidaceae			2,425 – 2,980 Andean forest	RGDS Boyacá	Feed regulator	L	L
5	***Berberis tabiensis*** L.A. Camargo Berberidaceae		Uña de gato	1,750 – 3,670 Sub-Andean forest, Andean forest, Sub-páramo and Páramo	WGDS Antioquia, Boyacá, Chocó, Cundinamarca, La Guajira and Santander	Insecticide (by touch)	L	I
6	***Chromolaena barranquillensis*** (Hieron.) R.M. King & H.Rob. Compositae (Asteraceae)	*Eupatorium barranquillense* Hieron	Falso guaco, saragüay, rosita vieja, santa cruz, rosa vieja, santa maría	0 – 600 Lower Neotropical Rainforest	WGDS Atlántico, Bolívar, Chocó, Córdoba, La Guajira, Magdalena and San Andrés Isla	Repellent	L	L
7	***Espeletia killipii*** Cuatrec. Compositae (Asteraceae)	*Espeletia killipii* Cuatrec. var. *chicasana* Cuatrec. *Espeletia killipii* Cuatrec. var. *killipii*	Frailejón, frailejón de Chisacá	2,120 – 4,010 Páramo, Sub-páramo and sometimes found in the Andean forest	RGDS Cundinamarca and Caldas	Chemical pesticide supplier, feed regulator	L	I
8	***Lourteigia microphylla*** (L.f.) R.M.King & H.Rob. Compositae (Asteraceae)	*Conoclinium microphyllum* DC. *Eupatorium microphyllum* L.f.	Chilca, jarilla, patinegra, salvia, trébol aromatizador	1,990 – 3,600 Andes	WGDS Boyacá, Cundinamarca, Magdalena, Tolima	Feed regulator	L	L
9	***Mandevilla mollissima*** (Kunth) K.Schum. Apocynaceae	*Echites mollissimus* Kunth *Exothostemon mollissimum* (Kunth) G.Don	Bejuco amarillo	880 – 2,000 Andes	WGDS Antioquia, Boyacá, Caldas, Cauca, Cundinamarca, Huila, Magdalena, Nariño, Putumayo, Santander, Tolima, Valle	Feed regulator	L	L

No.	Scientific name and family	Synonyms	Common names	Altitudinal range (m a.s.l.) and habitat	Geographical distribution and departments[1]	Mechanism of action	State of knowledge[2]	
							Traditional	Scientific
10	***Ocotea caparrapi*** (Sand.-Groot ex Nates) Dugand Lauraceae	*Nectandra caparrapi* Sand.-Groot ex Nates. *Nectandra oleifera* Posada Arango ex Nates. *Oreodaphne oleifera* Posada Arango.	Aceite de caparrapí, aceite caparrapí real, aceite de palo, aceituno, aguarrás, aymendron, amacey, caparrapí, calaba, canelo, canelo real hoja chica, canelo real caparrapí, comino, laurel canelo, palo de aceite, palo de caparrapí	1,000 – 2,380 Sub-Andean forests	WGDS Antioquia, Cundinamarca and Santander	Repellent	I	L
11	***Ryania speciosa*** M.Vahl. var. ***chocoensis*** (Triana & Planch.) Monach. Salicaceae	*Ryania chocoensis* Triana & Planch. *Patrisa chocoensis* (Triana & Plach.) Warb. *Patrisia chocoensis* (Triana & Plach.) Warb.		10 – 550 Lower Neotropical Rainforest	WGDS Antioquia, Bolívar, Cauca, Chocó, Santander, Tolima and Valle del Cauca	Insecticide (by touch)	L	L

From the total of native species of Colombia used as 'insecticides', only *Nicotiana tabacum* was associated with a high state of traditional knowledge and *Ocotea caparrapi*, an endemic species used as an 'insecticide' plant, presented an intermediate level of traditional knowledge. According to García (1992), a repellent action against mosquitoes is produced from burning the leaves of *Ocotea caparrapi.* However, this species is mostly recognised as medicinal, and there are no reports on the specific dose, dosage, and toxicity, neither any indication of use as an 'insecticide'. Further endemic species of Colombia were mainly classified as 'low' as regards the level of traditional knowledge, and only a few records of use as an 'insecticide' were found for *Ageratina vacciniaefolia*, *Chromolaena barranquillensis*, and *Ryania speciosa var. chocoensis*.

It is important to note that no research linking traditional and scientific knowledge for the endemic species of Colombia used as 'insecticide' was found. The species that presented the highest scores for traditional knowledge had 'low' scientific knowledge. Indeed, none of the endemic species presented a high level of scientific knowledge, and only *Berberis tabiensis* and *Espeletia killipii* presented intermediate levels. For example, there is evidence that the dichloromethane fraction of a stem extract from *Berberis tabiensis* is rich in the alkaloid tabienin B, and that it exhibits activity as a 'contact insecticide' against *Culex quinquefasciatus* (Diptera: Culicidae) larvae, with an EC_{50} (effective concentration) of 78 mg/µL (Núñez, 2010). On the other hand, an extract of the leaves of *E. killipii* in petroleum ether, from which two of the grandifloric acid and kaurenal diterpenes were isolated and identified, showed insecticide activity against *Premnotrypes vorax* (Coleoptera: Curculionidae), an important pest of the potato – *Solanum tuberosum* (Marín, 1997). Acosta & Torrenegra (2011) also reported *E. killipii* as a regulating plant for feeding Lepidoptera larvae (Pterygote Subclass). In this case, the plant produces a leaf exudate rich in grandifloric acid, which has an antifeedant effect on insects, as a defence mechanism the attack by lepidopteran larvae (Acosta & Torrenegra 2011). A study on insecticide and antifungal activities and chemical constituents of the essential oils of *Ocotea longifolia* and *O. macrophylla* recorded a significant fumigant activity against *Sitophilus zeamais* (LC_{50} 280, 5 µL/L air) (Prieto *et al.*, 2010). Other species such as *Ryania speciosa* var. *chocoensis,* although presenting a low state of scientific knowledge with a lack of phytochemical characterisation, presented biological activity *in vivo* against *Toxoptera citricidus* (Hemiptera: Aphididae), with LC_{50} (lethal concentration) of 236 mg/L and 68 mg/L of the ethanolic extracts from stems and leaves after 36 hours of contact (Méndez *et al.*, 2011). On the other hand, species with a high state of scientific knowledge, such as *Annona* cherimola*, *Annona squamosa*, and *Ageratum conyzoides*, have intermediate to low levels of traditional knowledge. These results show that there are still gaps in the links between the different states of knowledge about native species that are used as insecticides.

The study of insecticide plants has grown over the past few years. Since 1980, more than 20% of the publications on pesticides have been related to chemical pesticides made from plant resources (Isman & Grieneisen, 2014). Recently, there has also been increased pressure worldwide for the production of organic products as people are more aware of the risk of pesticides poisoning food and water and affecting ecosystems. Not only for countries in development but for any country, it has been very important to search for better methods of pest control that overcome the need for chemical pesticides (especially those targeting a wide range of pests) but do not present negative effects for the environment and human health (Amoabeng *et al.*, 2014; Isman, 2020).

Although the mechanisms of action of insecticide plants are diverse, the literature review shows that studies have mainly focused on the potential of some plant secondary metabolites with biological activity at a scientific level. The use of plant extracts that have insecticide activity is a highly explored alternative due to the possibility of synergistic effects between the active substances, the reduction in resistance to isolated active compounds, the easy access to the plant material, the simple extraction processes and the good cost-benefit when compared with conventional pesticides (Silva *et al.*, 2002; Amoabeng *et al.*, 2014; Miresmailli & Isman, 2014).

A peculiarity of this study is the inclusion of the category of plants used as insecticides by mechanical means (insectivorous plants and 'sticky traps'). Those species are still poorly studied compared to other categories. An important example of the importance of understanding and prioritising native insecticide plants of Colombia is the endemic species *Ageratina ampla* (Compositae-Asteraceae), which is reported as a 'sticky trap' for insects and presents sufficient evidence of traditional knowledge. The recognition of this species as an auxiliary/complementary method in the control of insects should be explored, considering that the mechanism of action is free from the disadvantages seen for conventional pesticides.

Likewise, the recognition of endemic species that belong to botanical genera or families which already had secondary metabolites identified as insecticides is also a way forward in the search for alternative pesticides. For example, alkaloids such as ryanodine isolated from species of the genus *Ryania*, berberine and other alkaloids from *Berberis*, and acetogenins identified in species of genus *Annona*, are natural compounds with high potential for product development due to their biocidal capacity (Isman 2020; Méndez *et al.*, 2011; Moreira *et al.*, 2007; Rattan 2010).

CONCLUSIONS

The analysis of the insecticide mechanisms shows that the insecticide actions known from traditional knowledge (repellency, fumigant activity, and toxicity by ingestion) have been brought forward for evaluation and scientific verification. The results showed that these actions are among the categories presenting the highest number of mentions in the literature. In general, the lack of knowledge on the mechanisms of action of many of these species is predominant, indicating an unclear validation of some of these insecticide uses.

There is a predominant lack of knowledge on insecticide evidence for Colombian plant species, especially endemic species, at traditional and scientific knowledge levels. Scientific knowledge is more lacking than traditional knowledge, and the identification and prioritisation of useful species in projects of research, development, and innovation (R + D + I) still need to be deepened to allow the chemical synthesis or the sustainable use of plants as raw material to produce insecticide preparations or products. Likewise, there are still other barriers to be overcome in the development of plant-based insecticides, such as the lack of information on geographic characterisation, management and cultivation; the standardisation of the chemical composition of raw materials for the manufacture of insecticide products; production at large-scale; and the endorsement of controlling agencies to regulate the use and commercialisation (Isman, 2020; Miresmailli & Isman, 2014), as well as the study and alternative use of insecticide/insectifugal species that have non-traditional or unknown insecticide mechanisms.

Chapter's endnote: It is important to note that for this study, we did not follow the taxonomic criteria of some authors or databases (such as the World Checklist of Selected Plant Families or Plants of the World Online) for several of the insecticide species native to Colombia. As a result of our research, we found that the synonymies of *Ocotea caparrapi* with *Mespilodaphne cymbarum* (*Ocotea cymbarum*) and of *Chromolaena barranquillensis* with *Chromolaena odorata* were not applicable, as we found those species pairs to be divergent. Publications are in preparation to define the taxonomic, nomenclatural, geographical, and ecological distribution status of these species, which we consider to be endemic to Colombia.

Acknowledgements

The authors would like to thank Colciencias for funding this research through the Young Researchers and Innovators program "Virginia Gutiérrez de Pineda", Convocatoria Nacional No. 525–2011, and the Pontificia Universidad Javeriana, Bogota for allowing the development of research on its University campus.

References

Abubakar Y, Tijjani H, Egbuna, C, Adetunji C, Kala S, Kryeziu T, Patrick-Iwuanyanwu K (2019) Chapter 3. Pesticides, history, and classification. In: Egbuna C, Sawicka B, eds., *Natural Remedies for Pest, Disease and Weed Control. Part I: Green approach to pest and disease control.* Academic Press pp. 29–42.

Acosta N, Torrenegra R (2011) Extracción, separación e identificación de la sustancia activa producida por *Espeletia killipii* como defensa ante el ataque de larvas de lepidópteros. *Rev. Fac. Cienc. Básicas*, 7(1):126133.

Acero L (2005) Plantas útiles de la cuenca del Orinoco. 2nd Edition. Bogotá, Zona Ediciones. 605p.

Ahmad M, Sultana S, Fazl-i-Hadi S, Ben Hadda T, Rashid S, Zafar M, Khan MA, Khan MPZ, Yaseen G (2014) An ethnobotanical study of medicinal plants in high mountainous region of Chail valley (District Swat-Pakistan). *J Ethnobiol Ethnomed*, 10(1):36.

Ahmed S, El-Rahman S, Kader M (2018) Field evaluation of some photosensitisers and nanocomposites against cotton leaf worm, *Spodoptera littoralis* (Bois.) (Lepidoptera: Noctuidae). *Middle East J Appl Sci*, 8(4):1471–1479.

Alvarado J, Vásquez H, Delgado G, Trevisan D, Horna O, Pereira J, Rojas C (2011) Actividad inhibitoria de plantas *in vitro* de *Drosera capillaris* sobre *Mycobacterium tuberculosis*. *Rev. Peru Biol.*, 17(3):353–358.

Alonso O (1999) Los insecticidas botánicos: una opción ecológica para el control de plagas. Matanzas: Estación Experimental de Indio Hatuey (Cuba). *Pastos forrajes* [online], 22(1):3-13.

Amoabeng B, Gurr G, Gitau C, Stevenson P (2014) Cost: benefit analysis of botanical insecticide use in cabbage: implications for smallholder farmers in developing countries. *Crop Protection*, 57:71–76.

Andrade G (2011) Estado del conocimiento de la biodiversidad en Colombia y sus amenazas. Consideraciones para fortalecer la interacción ciencia-política. *Rev. Acad. Colomb. Cienc. Exactas Fis Nat*, 35(137):491–507.

Bonilla B (1991) Los extractos vegetales, una alternativa al uso de plaguicidas: aplicación de extractos de *Bidens pilosa* (Compositae-Asteraceae) y *Ruta graveolens* en cultivos hidropónicos de *Fragaria vesca* (Rosaceae) y *Capsicum annuum* (Solanaceae). In: Ríos M, Pedersen HB, eds. *Las plantas y el hombre: memorias del Primer Simposio Ecuatoriano de Etnobotánica y Botánica Ecónomica*. Quito, Herbario QCA-P.U.C.E. pp. 81–88.

Carrero JM (1996) *Lucha integrada contra las plagas agrícolas y forestales*. Libros MP, ed. Madrid, Mundi Prensa Libros. 276p.

Cerón P (1996) Plantas usadas en el control de artrópodos. In: Nates BC, Cerón P, Hernández E, eds. *Las plantas y el Territorio: Clasificaciones, usos y concepciones en los Andes Colombianos*. Quito, Ediciones Abya-Yala. pp. 80–94.

Enrique L Una P, Celis Á, Mendoza C, Pachón M, Cardona J, Delgado W, Cuca E (2008) Extractos vegetales utilizados como biocontroladores con énfasis en la familia Piperaceae. Una revisión. *Agron Colomb*, 26(1):97–106.

Espinosa-Jiménez J, Fonseca-Cortés A, Franco-Londoño C, Ramírez-Ortiz N, Rey-Gómez F, Rojas-Arias L, Romero-Gómez N, Riveros-Loaiza LM (2021) *Plantas usadas por la comunidad rural de la vereda Bradamonte, Sibaté*. Bogotá, Universidad Nacional de Colombia. 179p.

Fernández-Pérez A (1964) Plantas insectívoras, I Lentibulariaceas de Colombia y Perú. *Caldasia*, 9 (41):5–84.

García H (1992) *Flora medicinal de Colombia. Botánica Médica*. Vol I. 2nd edition. Bogotá, Tercer Mundo Editores. 559p.

García Saavedra M, López A, Mallo DA, Martínez OT (2007) Plantas carnívoras. *Rebigo*, 2:7–14.

Garnatje T, Peñuelas J, Vallès J (2017) Ethnobotany, phylogeny, and 'omics' for human health and food security. *Trends Plant Sci*, 22(3):187–191.

Gu W, Hao X, Wang Z, Zhang J, Huang L, Pei S (2020) Ethnobotanical study on medicinal plants from the Dragon Boat Festival herbal markets of Qianxinan, southwestern Guizhou, China. *Plant Diversity*, 42(6): 427–433.

Isman M (2006) Botanical insecticides, deterrents, and repellents in modern agriculture and an increasingly regulated world. *Ann Rev Entomol*, 51(1):45–66.

Isman M (2020) Botanical insecticides in the twenty-first century-fulfilling their promise? *Ann Rev Entomol*, 65:233–249.

Isman M, Grieneisen M (2014) Botanical insecticide research: many publications, limited useful data. *Trends Plant Sci*, 19(3):140–145.

Lawrence E (ed.) (2003) *Diccionario Akal de Términos biológicos*. Madrid, Ediciones Akal. 341p.

LoPresti E, Pearse I, Charles G (2015) The siren song of a sticky plant: Columbines provision mutualist arthropods by attracting and killing passerby insects. *Ecology*, 96(11):2862–2869.

Marín M (1997) Evaluación de la toxicidad de las hojas de la *Espeletia killipii* sobre el gusano blanco (*Pemnotrypes vorax*) que afecta tubérculos de papa (*Solanum tuberosum*). Trabajo de grado. Pontificia Universidad Javeriana, Facultad de Ciencias. Bogotá, Colombia. 83p.

Méndez A, González R, Castillo C (2011) Utilización de *Ryania speciosa* var. *chocoensis* (Triana y Planch) para el control de *Toxoptera citricidus* (Kirkaldy, 1907) (Hemiptera: Aphididae) en condiciones de laboratorio. *Fitosanidad*, 15(4):237-240.

Miresmailli S, Isman M (2014) Botanical insecticides inspired by plant-herbivore chemical interactions. *Trends Plant Sci*, 19(1):29–35.

Moreira M, Picanço M, Barbosa L, Guedes R, Barros E, Campos M (2007) Compounds from *Ageratum conyzoides*: isolation, structural elucidation and insecticide activity. *Pest Manage Sci*, 63:615–621.

Montesino M, López H, Hernández J, Zayas E (2009) Insecticidas botánicos como alternativas para el manejo de plagas en sistemas agroforestales. *Actaf*, 1:24.

Núñez A (2010) Estudio fitoquímico del extracto alcaloidal de la especie nativa *Berberis tabiensis* (Berberidaceae). Thesis. Universidad Nacional de Colombia, Facultad de Ciencias, Departamento de Química, Bogotá, Colombia. 94p.

Pascual M (1996) *Plaguicidas naturales de origen vegetal: estado actual de la investigación*. Madrid, Instituto Nacional de Investigación y Technologia Agraria y Alimentaria. 35p.

Pascual M, Robledo A (1998) Screening for anti-insect activity in Mediterranean plants. *Ind Crop Prod*, 8(3):183–194.

Pedigo L, Rice M, Krell R (2021) *Entomology and Pest Management*. 7th edition. New Jersey, Waveland Press. 58p.

Prieto J, Pabón L, Patiño O, Delgado W, Cuca L (2010) Constituyentes químicos, actividad insecticida y antifúngica de los aceites esenciales de hojas de dos especies del género *Ocotea* (Lauraceae). *Rev. Colomb. Quím*, 39(2):199-209.

Primo Y (1991) *Ecología química. Nuevos métodos de lucha contra insectos*. Madrid, Ediciones Mundi-Prensa. 191p.

Rajendran S, Sriranjini V (2008) Plant products as fumigants for stored-product insect control. *J Stored Prod Res*, 44(2):126–135.

Rattan R (2010) Mechanism of action of insecticide secondary metabolites of plant origin. *Crop Prot*, 29(9):913-920.

Ribeiro RV, Bieski IGC, Balogun SO, Martins DT de O (2017) Ethnobotanical study of medicinal plants used by Ribeirinhos in the North Araguaia microregion, Mato Grosso, Brazil. *J Ethnopharmacol*, 205:69–102.

SIB Colombia (2020) Biodiversidad en Cifras: número de especies de plantas registrados en el SIB Colombia. Consultado a través del SIB Colombia. Available at: https://cifras.biodiversidad.co/ [Accessed September 2021]

Silva-Aguayo G (2021) Botanical Insecticides. In: Radcliffe EB, Hutchison WD (eds). *Radcliffe's IPM World Textbook*. Minnesota, University of Minnesota. Available at: https://ipmworld.umn.edu/silva-aguayo-botanical [Accessed in June 2021]

Silva G, Lagunes A, Rodríguez C, Rodríguez D (2002) Insecticidas vegetales: una vieja y nueva alternativa para el manejo de plagas. *Manejo Integr. Plagas Agroecología (Costa Rica)*, 66:4–12.

Spochacz M, Chowański S, Walkowiak K, Szymczak M, Adamski Z (2018) Plant-derived substances used against beetles – pests of stored crops and food – and their mode of action: a review. *Compr Rev Food Sci F*, 17(5):1339–1366.

Vergara RRY, Madrigal CA (1994) Estado actual y futuro de los extractos de plantas para el control de plagas. In: S. C. de Entomología, ed. Memorias XXI Congreso Sociedad Colombiana de Entomología. Medellín, Sociedad Colombiana de Entomología. pp. 252–272.

Symphonia globulifera L.f (Clusiaceae) has several uses in Colombia, such as Environmental, Fuel, Human Food, Material and Medicinal.
Dairon Cárdenas-López

Chapter 6

Useful Plants of the Colombian Amazon

Dairon Cárdenas-López†[1], Nicolás Castaño-Arboleda[1]*, Nórida Marín-Canchala[1] & Andrés A. Barona-Colmenares[1]

[1] Herbario Amazónico Colombiano COAH, Instituto Amazónico de Investigaciones Científicas SINCHI.

***Corresponding author:** ncastano@sinchi.org.co

Keywords: food, fuel, medicinal, ornamental, and timber

ABSTRACT

This study presents a synthesis of the current state of knowledge about the Useful Plants of Colombia, based on information gathered from field expeditions and records added from the Colombian Amazon Herbarium (Herbario Amazónico Colombiano (COAH)) database, which is managed by the Amazonian Scientific Research Institute (Instituto Amazónico de Investigaciones Científicas SINCHI). The useful plants were identified and classified according to their different categories of uses, their origin (native or endemic), and their extinction risk assessments. In total, 2,286 useful plants were recorded in the Colombian Amazon, representing 177 botanical families and 896 genera. Among the 12 categories of uses identified, the medicinal (883) and food (690) plants presented the highest number of useful species. Challenges to achieving the sustainable use of useful species in the Colombian Amazon are discussed from the economic botany perspective.

RESUMEN

Se presenta una síntesis del conocimiento de las plantas útiles de la Amazonia colombiana, a partir de los registros de la base de datos del Herbario Amazónico Colombiano (COAH) del Instituto Amazónico de Investigaciones Científicas SINCHI e información colectada en campo. Se registran 2.286 especies útiles, pertenecientes a 896 **géneros y 17**7 familias y se identifican especies útiles nativas, endémicas, introducidas y con alguna categoría de amenaza. Se reconocen 12 categorías de uso en las que medicinal (883) y alimenticia (690) son las categorías que registran mayor número de especies. Se discuten los retos para alcanzar el aprovechamiento sostenible de las especies útiles en la Amazonia colombiana bajo la perspectiva de la botánica económica.

INTRODUCTION

The Colombian Amazon is an excellent expression of tropical forest biodiversity. Such high diversity has been attributed to the geological history resulting from the different tectonic movements that resulted in the Andean uplift, which involved different tectonic movements, changes in the direction of rivers, and sedimentation processes (Hoorn *et al*., 2010). Differences among tree communities have been suggested for geological units of the Colombian Amazon forest, the border of Orinoco savannas and the eastern border of the Andes (Cárdenas *et al*., 2017). The Guyana shield is represented by extensions of savannas (northeast part) and rocky outcrops (e.g., tepuis; as Serranía de Chiribiquete, Serranía de Naquén, Serranía de Araracuara, and some others distributed in the departments of Vaupés, Guainía, and Guaviare). Hernández (1984) estimates that more than 50% of Colombian living forms are represented there. According to Infante-Betancour & Rangel (2018), in the Colombian Amazon lowlands (Amazonia *sensu stricto*), there are 7,288 plant species recorded, whereas the Colombian Amazon Herbarium (COAH) holds records for 8,329 documented plant species (7,535 vascular and 794 non-vascular plants). Also, The Worldwide Fund for Nature (WWF) has identified several diverse ecoregions in the Amazon, such as the "Napo River Humid Forests", the richest biota in the world, the western arc of the Amazon, and in particular the areas near the Andes foothills (Dinerstein *et al*., 1995; Cardoso *et al*., 2017). The Amazonian forests are considered a warehouse for multiple natural products, including a great variety of foods, medicinal supplies, and craft production items; they also represent an important refuge for plant species that constitutes a genetic bank for humanity (Leyva, 1998). However, a critical point in making the most of this high diversity is the low density of individuals presented by most plant species (Gentry, 1989).

The first records on the use of Colombian Amazon plant species come from the notes of several naturalists about the local use of species, indigenous names for plants species, mythology and superstitions, and other aspects of indigenous life. The German Karl Friedrich Philipp von Martius was the first botanist to work in the Colombian Amazon, arriving by the Amazon River and crossing from the Caquetá River to the Araracuara rapids around 1820 (Schultes & Raffauf, 1994). His ethnobotanical notes were recorded in the monumental *Flora Brasiliensis* (Martius, 1829). For the 20th century, it is important to highlight several studies developed by Richard Evans Schultes (1970, 1979, 1983) and Schultes & Raffauf

(1990, 1994), which provided valuable information on phytochemical and pharmacological aspects of Amazonian plants, also enhancing the richness of plant diversity and the lack of awareness of the possibilities for taking advantage of these resources. Enrique Pérez-Arbeláez's masterpiece "Plantas Útiles de Colombia" was published for the first time in 1935, with many of the Amazonian plant species known at that time included. Likewise, stand out studies by J. M. Idrobo and A. Fernández were carried out in the 1950s using botanical collections with recorded information on plant uses in indigenous communities, mainly in the Igara-Paraná River sector.

Later, within the framework of the Amazon Grammetric Radar Project (Colombian mapping Project – PRORADAM), Acero (1979) produced an essential contribution to the knowledge of the plant resources of the region, reporting 100 species with timber potential, whilst Pabón (1982) started the publication of the magazine "Colombia Amazónica", reporting 327 useful plant species from the same region. On the other hand, Glenbosky (1983) carried out an inventory of medicinal plants used by the indigenous Tikuna of the Amazonian trapezium region, recording 184 species. Also, in 1983, la Rotta published a study describing the information recorded on 128 plant species used by the Andoque people, and later, la Rotta *et al.* (1987) carried out a similar study with the Miraña people from the middle Caquetá River.

More recently, the Flora team of the Amazonian Scientific Research Institute (SINCHI) has carried out floristic inventories in the Colombian Amazon, obtaining an important reference collection and valuable information on the use of vegetation in several localities of the Amazonas, Guaviare, Putumayo, Guainía, and Vaupés departments (Cárdenas *et al.*, 2002, 2012; Cárdenas & López, 2000; Cárdenas & Politis, 2000; Castaño *et al.*, 2007). This chapter presents a synthesis of the knowledge on useful species in the Colombian Amazon Forest.

MATERIALS AND METHODS

Study area

Several delimitations have been proposed for the Colombian Amazon through the years, based on different approaches and following biogeographic, hydrographic and/or political-administrative criteria (Domínguez, 1985; Rudas-Ll., 2009). The delimitation could include the Amazon and the Orinoco River basins, with the Vichada River as its northern limit. Physiognomic, climatic, and edaphic aspects were also included (Herrera, 1999; Domínguez, 1985; Rudas-Ll., 2009), as was the boundary of the Orinoco and Amazon River basins (Hernández-C. *et al.*, 1992). From the east and south, the boundaries were defined by the neighbouring countries of Brazil, Ecuador, Peru, and Venezuela. To the west and north-

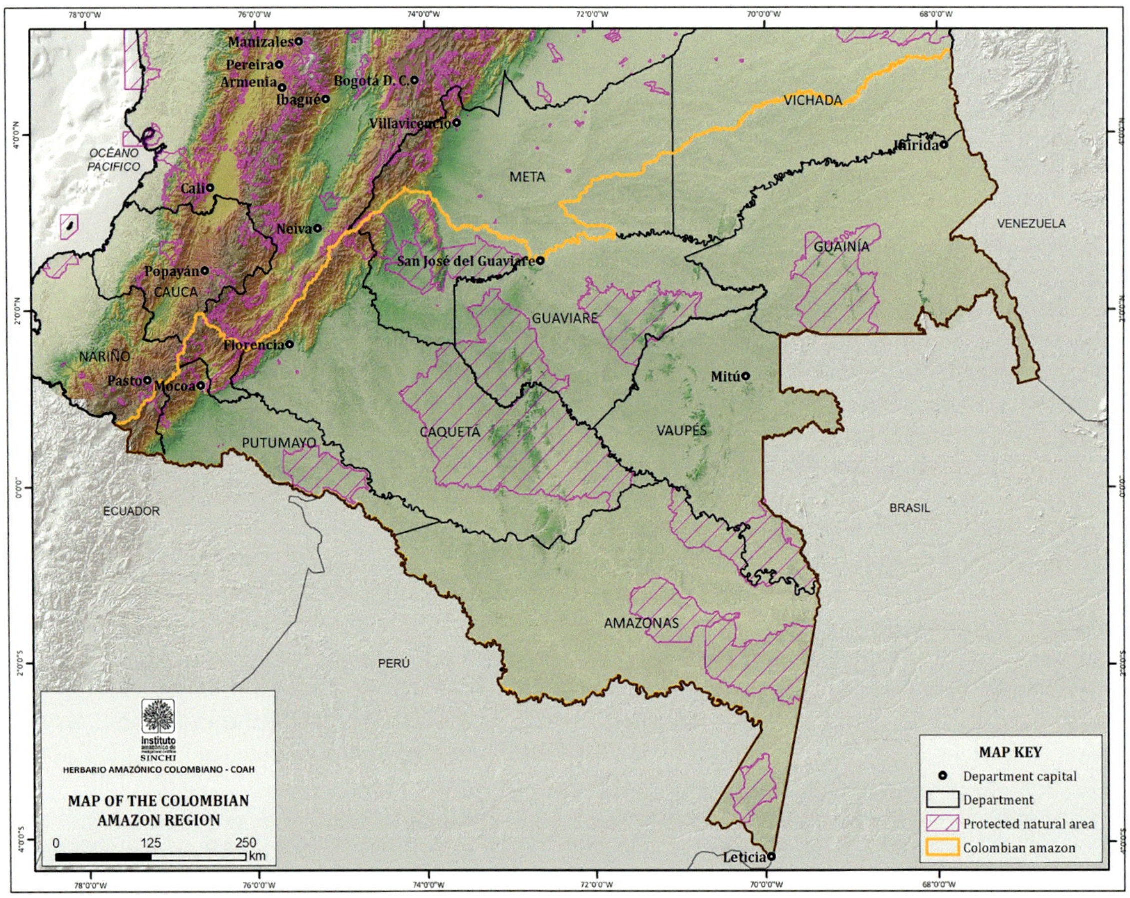

FIGURE 1. Distribution map of the Colombian Amazon, highlighting its departments and conservation units.

west, the Amazon basin borders the Andes foothills (Andean Amazonian transition), the rocky outcrops of the Guiana Shield, the Amazon *sensu stricto* or Amazonian lowlands, and the Serranía de La Macarena (Rangel-Ch & Infante-Betancour 2018). The western limit for the Colombian Amazon is the elevation of 1,000 m of the eastern slope of the eastern mountain range (Cordillera oriental), where elements of the Amazonian tropical humid forest predominate (Figure 1).

It is also essential to take into account that from the political point of view, the Amazon Forest covers the southern part of the Vichada department; the southeast of the Meta department; the entire territory of the departments of Amazonas, Caquetá, Guainía, Guaviare, Putumayo and Vaupés, as well as the Cauca Boot and the Amazonian slopes of Nariño department (which includes the middle part of the rivers Guamuez, Sucio, San Miguel, and Aguarico) (Salazar & Riaño 2016).

Methods

The potential uses of plant species were defined based on the information gathered from existing biological records in the database of COAH herbarium, obtained since 1983, and data from the vegetation surveys carried out in different projects at SINCHI. All species for which there are botanical records in the herbarium were considered, including those cultivated in orchards, gardens, and fields in the region. It is noteworthy that in the Amazon piedmont, many of the species are documented from photographic records of species growing in the residential gardens and orchards of settlers coming from the country's interior. Subsequently, all the scientific names were validated using the Taxonomic Name Resolution Service tool v4.0 (Boyle, 2013). The records with taxonomic issues were reviewed, and those accepted in the Catalogue of Plants and Lichens of Colombia were considered (Bernal *et al.*, 2016). Also, priority was given to names assigned recently by expert botanists. The record of threatened species was defined according to the categories assigned in two national resolutions (Resolución 1912/2017) (MADS, 2017), which establish the lists of wild threatened species of Colombia. The identification of endemic species was based on the Catalogue of Plants and Lichens of Colombia (Bernal *et al.*, 2016) and supplemented with new records added to the COAH collection. The categories of use are those used on the COAH database, and they are not mutually exclusive, as follows:

- **Construction:** species used in the structure, roofing, floors, and mooring of houses.
- **Cultural:** species used in rituals and activities related to indigenous ancestral culture.
- **Dye:** species that produce dyes or natural colours used in foods or crafts.
- **Food:** Forest and cultivated species, used as edibles.
- **Forage:** plant species used to feed animals.
- **Fuel:** species used as firewood and for lighting.
- **Handcrafts:** species used as dyes, fibres for basketry, pulp for handcrafted paper, wood for splint, trays, and containers.
- **Medicinal:** species with curative and preventive properties, according to local populations.
- **Ornamental:** species employed (or potentially employed) as an ornament and in the decoration of spaces.
- **Psychotropic:** species that produce effects on the nervous system.
- **Timber:** species used for house construction in the region and as commercial lumber.
- **Toxic:** species used as poisons for hunting, or which are recognized as harmful to humans.

RESULTS AND DISCUSSION

The useful plants of the Colombian Amazon are represented by 2,289 species belonging to 896 genera and 177 families. Of these species, 2,211 correspond to angiosperms, 14 gymnosperms, 56 ferns, and five lycophytes. Some families have marked dominance: 60% of the registered speciesare concentrated within the 30 richest families. Among the most diverse families, we can highlight those that have important uses as timber (Lecythidaceae and Meliaceae), ornamentals (Orchidaceae), food and handcrafts (Arecaceae), or species with multiple uses as timber, food, and medicinals (Rubiaceae, Fabaceae) (Table 1).

The genera with the highest species richness were *Piper* (Piperaceae) (49 spp.), *Inga* (Fabaceae) (41 spp.), *Miconia* (Melastomataceae) (38 spp.), *Protium* (Burseraceae) (32 spp.), and *Solanum* (Solanaceae) (27 spp.) (Figure 2). In contrast to what was observed for families, the richness of useful species at the generic level was more heterogeneous. The 30 most diverse genera account for only 22.8% of the species, while about 74% of the genera have only one or two useful species.

TABLE 1. Botanical families with the highest number of reports of useful plants in the Colombian Amazon.

Family	Species	Genera
Fabaceae	210	78
Rubiaceae	107	45
Arecaceae	80	31
Melastomataceae	72	22
Annonaceae	70	14
Malvaceae	64	25
Euphorbiaceae	62	27
Araceae	62	18
Asteraceae	58	38
Apocynaceae	58	24
Piperaceae	58	2
Lauraceae	56	10
Moraceae	55	18

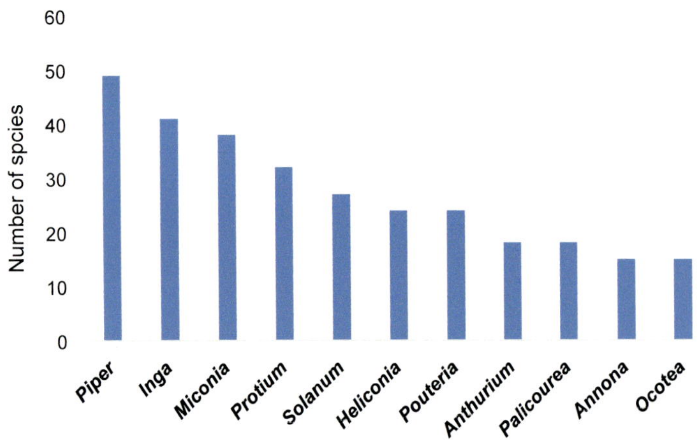

FIGURE 2. Plant genera with the highest number of useful species in the Colombian Amazon.

Useful species of the Colombian Amazon by category of use

The species with potential uses recorded in the Colombian Amazon were classified according to 12 categories. They corresponded to 2,286 species representing 28% of the total number of species registered in this region (8,329 species). In this case, the Colombian Amazon presents a high proportion of species compared to the number of useful species for the entire country, considering that Pérez-Arbeláez (1978), in his publication "Plantas Útiles de Colombia", considered the existence of about 1,850 species of useful plants in Colombia in the middle of the 20th century. On the other hand, the 252 species recorded by the 'Useful Plants of the Orinoco Basin' is very low if we consider that it includes species from cold, temperate, and warm zones of the Orinoco River basin in Colombia (Acero, 2005). However, it is important to highlight that for the Peruvian Amazon, 3,140 useful species were reported (Brack, 1993), representing a larger proportion of useful species than that reported in the present study for the Colombian Amazon. The regional richness of useful plants of the Colombian Amazon is extensive considering that it could represent about 50% of the useful plants recorded for entire South American countries. For example, 5,172 useful plant species were reported for Ecuador as the product of a several-year initiative with the participation of about 40 researchers (de la Torre *et al.*, 2008). Those results show the importance of the Amazon for those countries where indigenous communities most widely use those species.

The categories with the largest number of species were medicinals with 884 species, foods with 691 species including those introduced, and timber with 402 species, including a few introduced species (Table 2).

A list of species by category is presented below, starting with the categories including the largest number of useful species:

- **Construction:** 211 species were recorded, including those used in building structures [many genera of Annonaceae (*Duguetia, Guatteria, Oxandra,* and *Xylopia*); Lauraceae (*Aniba*); Lecythidaceae (*Eschweilera* and *Grias*); and Myristicaceae (*Osteophloeum* and *Virola*)], roofing [some Annonaceae and Arecaceae (*Iriartea deltoidea* (bombona), *Lepidocaryum tenue* (puy), and *Mauritia flexuosa* (canangucha)], floors (several families), and the moorings of houses [such as Araceae (*Heteropsis*)].
- **Cultural:** this category includes 65 species that are important in local communities and represents the contextual basis for each culture. Several useful plants are key elements that identify each ethnic group (Garibaldi & Turner, 2004). They include, for example, yagé – *Banisteriopsis caapi* (Malpighiaceae), coca – *Erythroxylum coca* (Erythroxylaceae), tobacco – *Nicotiana tabacum* (Solanaceae), ortiga – *Urera baccifera* (Urticaceae), as well as many other plants used in rituals and daily activities depending on the different cultural backgrounds (Schultes & Raffauf, 1994).
- **Dyes/colourants:** 39 species have been used as a dye, with a few used in rituals and others in the preparation of typical foods, such as the broadly used saino – *Goupia glabra* (Goupiaceae), *Picramnia sellowi* (Picramniaceae), azafrán de tierra – *Escobedia grandiflora* (Orobanchaceae), huito – *Genipa americana* (Rubiaceae), and achiote – *Bixa orellana* (Bixaceae).
- **Food:** 690 species were recorded, including important forest species such as asaí – *Euterpe precatoria* (Arecaceae), madroño – *Garcinia madruno* (Clusiaceae), canangucha – *Mauritia flexuosa* (Arecaceae), milpesos – *Oenocarpus bataua* (Arecaceae), and cucuy – *Macoubea guianensis* (Apocynaceae). The presence of a significant number of introduced species is also highlighted (84 species) such as guanábana – *Annona muricata*

TABLE 2. Numbers of useful plant species in the Colombian Amazon by category of use.

Category of use	Species
Medicinal	883
Food	690
Timber	401
Ornamental	392
Fuel	275
Handcrafts	233
Construction	211
Forage	81
Toxic	77
Cultural	65
Dye	39
Psychotropic	22

(Annonaceae), coco – *Cocos nucifera* (Arecaceae), coffee – *Coffea arabica* (Rubiaceae), mango – *Mangifera indica* (Anacardiaceae), and pomarrosa - *Syzygium malaccense* (Myrtaceae). Invasive species in this category include the african palm – *Elaeis guineensis* (Arecaceae) with a high potential of invasion in the Amazon region (Cárdenas *et al.*, 2011). Still, its use is currently promoted by government agencies. However, it is known that it is contributing to the transformation of natural areas and consequently impacting the natural populations of native species, some of them threatened.

An important part of the food category corresponds to the Non-Conventional Edible Plants (PANCs) – defined as native, not cultivated, and underutilized species (Marín, in press), which correspond to 605 registered species, representing a significant proportion of species included in the food category.

- **Forage:** this category includes 81 species used to feed animals, which mainly represent the families Poaceae (20 species), Fabaceae (17 species), Cyperaceae (10 species), and Asteraceae (eight species). Some of them have been incorporated by the extensive livestock management in the Amazon region, such as: yaguará – *Hyparrhenia rufa* (Poaceae), guayacana – *Imperata brasiliensis* (Poaceae), caminadora – *Rottboellia cochinchinensis* (Poaceae), and several species from braquiarias – *Urochloa* (Poaceae). A few other species that are widely used as forage in Amazon region are: mataratón – *Gliricidia sepium* (Fabaceae), leucaena – *Leucaena leucocephala* (Fabaceae), botón de oro – *Tithonia diversifolia* (Asteraceae), and quiebrabarrigo - *Trichanthera gigantea* (Acanthaceae).
- **Fuel:** 343 species from the Colombian Amazon are used as fuel, given the need to generate fire to carry out daily activities such as lighting the darkness of a *maloca* at night, preparing food or baking ceramics, among others. Previously, 295 species were recorded as being used as fuel in the Colombian Amazon region (Cárdenas *et al.*, 2012), and 226 species were recorded as fuels in Ecuador (Palacios, 2008). Both studies emphasized that these species present special characteristics, such as suitability as firewood, carbon content and the presence of exudates that make them good fuels or combustion initiators (Palacios, 2008, Cárdenas *et al.*, 2012). The species most used as fuel in the Colombian Amazon region are cucharo – *Arawakia weddelliana* (Clusiaceae), costillo - *Aspidosperma excelsum* (Apocynaceae), turí - *Eschweilera rufifolia* (Lecythidaceae), chaparro or motilón - *Hieronyma alchorneoides* (Phyllanthaceae), palo negro - *Piptocoma discolor* (Asteraceae), anime or copal – *Protium apiculatum* (Burseraceae), chimbe or bizcocho - *Pseudosenefeldera inclinata* (Euphorbiaceae), and seal or carate - *Vismia baccifera* (Hypericaceae).
- **Handcrafts:** 233 species were recorded in this category. Linares (1994) reported only 248 species in a preliminary inventory of the plants used to make handicrafts in Colombia, a number lower than expected considering the variety of ecosystems and landscapes in Colombia. A total of 114 species form a very important component of the artisanal species of Colombia, vegetable fibres that are extracted to produce traditional Colombian handcrafts (*'Artesanías de Colombia'*), a fundamental basis of a national trade (Linares *et al.*, 2008). The main species in this category are: **fibre sources:** cumare - *Astrocaryum chambira* (Arecaceae), papelillo - *Couratari guianensis* (Lecythidaceae), yanchama - *Ficus maxima* (Moraceae), chiquichiqui - *Leopoldinia piassaba* (Arecaceae), palo de algodón – *Pseudobombax munguba* (Malvaceae); **wood carving:** palisangre - *Brosimum rubescens* (Moraceae), granadillo - *Platymiscium pinnatum* (Fabaceae), and capinurí - *Pseudolmedia laevis* (Moraceae); **seed sources:** lágrimas de Job – *Coix lacryma-jobi* (Poaceae) and chochos - *Ormosia* spp. (Fabaceae); **fruits used to make containers:** totumo - *Crescentia amazonica* (Bignoniaceae); and **fibres sources for basketry:** different species of yaré – *Heteropsis* spp. (Araceae) are very popular throughout the region.
- **Medicinals:** 883 species were recorded, with the following species standing out: andiroba - *Carapa guianensis* (Meliaceae), copaiba - *Copaifera reticulata* (Fabaceae), sangre drago - *Croton lechleri* (Euphorbiaceae), chuchuhuasa - *Maytenus laevis* (Celastraceae), bálsamo - *Myrocarpus venezuelensis* (Fabaceae), curarina - *Potalia resinífera* (Gentianaceae), caraño – *Trattinnickia burserifolia* (Burseraceae), uña de gato - *Uncaria tomentosa* (Rubiaceae). This category also includes some introduced species (95 spp.), such as paico - *Chenopodium ambrosioides*[1] (Amaranthaceae), limón - *Citrus limon* (Rutaceae), totumo - *Crescentia cujete* (Bignoniaceae), and balsamina – *Momordica charantia* (Cucurbitaceae).

Several of those species are shared with the Peruvian Amazon, where Brack (1993) reported 1,044 species with medicinal uses. Likewise, in Ecuador, 3,118 species were documented as "plants used to cure, alleviate, and treat human diseases" (de la Torre *et al.*, 2008). In the Amazon, the medicinal species represent an important resource used by local communities, and the region presents a very high number of medicinal species according to the traditional knowledge, even though in most cases, there is still a lack of phytochemical and clinical studies documenting their true medicinal indications and effects (Taylor 2005).

Studies by SINCHI found tannins, saponins, steroids, and triterpenoids, as well as terpenic lactones and coumarins, in caraño - *Trattinnickia burserifolia* (Burseraceae) (Cárdenas *et al.*, 2002). In the Amazon region, this species is used for

[1] *Chenopodium ambrosioides* is accepted by Tropicos and Catálogo de plantas y líquenes de Colombia (Bernal *et al.*, 2016).

kidney diseases and as an antibiotic for skin problems (da Paz *et al.*, 2004, Guimarães *et al.*, 2014). This species is also used by the eastern Nukak (indigenous communities) from the Colombian Amazon for kidney diseases and as antibiotics (Cárdenas & Politis, 2000). For *tigre guasca* o *carare* – *Aristolochia ruiziana* (Aristolochiaceae), this study evidenced the presence of alkaloids in low quantity, steroids, and free triterpenoids (Cárdenas *et al.*, 2002). In the region of Mocoa, this species is used to relieve cramps (Cárdenas *et al.*, 2002). The same uses are reported from the scientific literature and attributed to its marked antispasmodic properties (against muscle spasms), which are potentially related to the presence of alkaloids and free triterpenoids (Cárdenas *et al.*, 2002). The study also found flavonoids, leucoanthocyanidins, and tannins for *chimbe* – *Senefeldera inclinata* (Euphorbiaceae), besides a small number of steroids and triterpenoids (Cárdenas *et al.*, 2002). Schultes and Raffauf (1990) have reported the direct use of *S. inclinata* scratched bark by the *Achuar* (*Jibaros Mayna*) – an indigenous ethnicity from Peru – as a treatment for toothaches. Alkaloids, tannins, saponins, and terpenic lactones were found in curarina – *Potalia resinifera* (Gentianaceae) (Cárdenas *et al.*, 2002). In Putumayo and other regions of the Amazon basin, this plant is used in plasters as a type of palliative for snakebite (Schultes & Raffauf, 1990). However, no pharmacological studies have been found to explain this activity.

- **Ornamentals**: 392 species from the Colombian Amazon were recorded in this category of use, including 189 introduced species from several different countries. In this category, all species with current or potential ornamental uses were included. Although most of the native species are not used in the region for this purpose in the interior market of Colombia, there is an important demand for ornamental plants. This group includes many palms, heliconias, bromeliads, zamias, and orchids, in addition to specific species such as *Goeppertia lanata* (Marantaceae), *Guacamaya superba* (Rapateaceae), *Schizaea elegans* (Schizaeaceae), *Schoenocephalium martianum* and *Schoenocephalium teretifolim* (Rapateaceae), *Victoria amazonica* (Nymphaeaceae), and *Vellozia tubiflora* (Velloziaceae). This group of species could constitute an economic alternative for the region involving a program of propagation of promising species.

This category has been influenced by the establishment of populated urban centres, farms, farmsteads, and dwellings in the Amazon region, as species were gradually brought from different parts of the world to recreate and decorate the environment with plants that adorned the places of origin of the new settlers. In this process, several species have been introduced, such as gold cup – *Allamanda cathartica* (Apocynaceae), araucaria – *Araucaria columnaris* (Araucariaceae), bismarkia – *Bismarckia nobilis* (Arecaceae), flanboyan – *Delonix regia* (Fabaceae), royal palm – *Roystonea regia* (Arecaceae), and african tulipan – *Spathodea campanulata* (Bignoniaceae), among others.

- **Psychotropic:** 22 species were recorded, such as yopo – *Anadenanthera peregrina* (Fabaceae), yagé – *Banisteriopsis caapi* (Malpighiaceae), coca – *Erythroxylum coca* (Erythroxylaceae), mamitavea – *Iryanthera crassifolia* (Myristicaceae), and tobacco – *Nicotiana tabacum* (Solanaceae), several of those 22 species have been identified and used by different indigenous peoples of the Amazon (Schultes & Raffauf, 1994).
- **Timber:** 401 species were included in this category, a number close to the 468 forest species studied in the diagnosis of Flor de Agosto (Perú) and Tarapacá (Colombia) forests (Unda, 1993). Several of those species are among the 30 most-mobilized timber species from the natural forests of Colombia (Blanco, 2020). They are ranked among highly among the 875 species reported as timber by the Regional Autonomous Corporations of Colombia and Sustainable Development (Cárdenas & Salinas, 2007). Some of the most important species used as timber are cachicamo – *Calophyllum brasiliense* (Calophyllaceae), andiroba – *Carapa guianensis* (Meliaceae), abarco amazónico – *Cariniana domestica* (Lecythidaceae), palo de arco – *Handroanthus serratifolius* (Bignoniaceae), itauba – *Mezilaurus itauba*[2] (Lauraceae), acapú – *Minquartia guianensis* (Olacaceae), achapo – *Cedrelinga cateniformis* and granadillo – *Platymiscium pinnatum* (Fabaceae). Some of those species were assessed as threatened according to the National List of threatened species (Cárdenas & Salinas 2007), such as the Critically Endangered (CR) palorosa – *Aniba rosodora* (Lauraceae), and the Endangered (EN) cedro – *Cedrela odorata* (Meliaceae), and canelo de los andaquíes – *Mespilodaphne quixos* (Lauraceae). There are also 14 species used as timber that have been introduced into the Colombian Amazon, such as melina – *Gmelina arborea* (Lamiaceae), ciprés – *Hesperocyparis lusitanica* (Cupressaceae), kaya – *Khaya anthotheca* (Meliaceae), pino caribea – *Pinus caribaea* (Pinaceae), and teca – *Tectona grandis* (Lamiaceae), some of them incorporated into agroforestry in the Guaviare and Caquetá departments.
- **Toxic:** 77 toxicspecies have been recorded, including especially the curares [*Abuta grandifolia*, *Curarea tecunarum* (Menispermaceae), and *Strychnos toxifera* (Loganiaceae)]. Species with toxic properties are recognized under the name of *curare*, although their extract is drawn from different species and mainly used in hunting arrows (pers. obs.). Several cultivated species were recorded, including those from the

[2] Identified by Henk Van der Werff and accepted by Tropicos

fields and small orchards of indigenous communities, and especially those predominantly incorporated in agroforestry arrangements such as the **food species**: anón amazónico - *Annona mucosa*, guanábana - *Annona muricata* (Annonaceae), maní - *Arachis hypogaea* (Fabaceae), chontaduro - *Bactris gasipaes* (Arecaceae), papaya - *Carica papaya* (Caricaceae), ahuyama - *Cucurbita maxima* (Cucurbitaceae), yuca - *Manihot esculenta* (Euphorbiaceae), aguacate – *Persea americana* (Lauraceae), cocona - *Solanum sessiliflorum* (Solanaceae), copoasú - *Theobroma grandiflorum* (Malvaceae), and maíz - *Zea mays* (Poaceae). Likewise, some recorded cultivated species are exclusively medicinal, such as albahaca - *Ocimum basilicum* (Lamiaceae), or psychotropic, such as coca - *Erythroxylum coca* (Erythroxylaceae) and tobacco - *Nicotiana tabacum* (Solanaceae). Among the cultivated species, there are also species from other continents, incorporated in the near and remote past into the crops of the American indigenous peoples, such as mafafa - *Colocasia esculenta* (Araceae) from South Asia, plátano - *Musa paradisiaca* (Musaceae) from Asia, caña de azúcar - *Saccharum officinarum* (Poaceae) from New Guinea, and several species of citrus – *Citrus* spp. also from Asia (Purseglove, 1968).

Endemic Useful Species

The reports of Colombian endemic species are important to ensure the protection of those species that are only found in the country. And therefore, a protection strategy that guarantees their conservation is needed. Between the recorded useful plants of the Colombian Amazon, there are nine endemic species, including *Guatteria araracuarae* (Annonaceae) registered only in the region of Araracuara and used by the natives to make artisanal fishing tools; *Astrocaryum cuatrecasasianum*[3] (Arecaceae) registered in the Amazonian piedmont between the departments of Caquetá and Putumayo, used by the local indigenous people as a food product; *Ouratea chiribiquetensis* (Ochnaceae) recorded exclusively in the Natural National Park Serranía de Chiribiquete, where it is used as fuel (firewood) by the local communities; and *Schoenocephalium martianum* (Rapateaceae) recorded in the region of Araracuara and used as ornamental.

Threatened Useful Species

Among the useful plants recorded for the Colombian Amazon, there are 19 species assessed in categories of risk of extinction (according to the IUCN criteria) and included on the National Resolution nº 1912 de 2017 produced by the Ministry of Environment and Sustainable Development. Five of these species were assessed as Critically Endangered (CR) [e.g., *Roystonea oleracea* (Arecaceae), *Aniba perutilis* (Lauraceae), and *Aniba rosodora* (Lauraceae)]; seven as Endangered (EN) mainly due to the overexploitation of wood: these included *Cedrela odorata* (Meliaceae), one of the species with the highest demand in the world (Cárdenas & Salinas 2007); *Pachira quinata*[3] (Malvaceae); and *Elaeis oleifera* (Arecaceae), not very common in Amazonia, but mainly affected at the national level by fragmentation of the natural habitat. Also, seven species were assessed as Vulnerable (VU), including *Caryocar amygdaliferum* (Caryocaraceae), *Podocarpus guatemalensis* (Podocarpaceae), *Syagrus sancona* (Arecaceae), and *Zamia amazonum* (Zamiaceae).

Among the threatened species at the national level, there are some introduced species that are considered to be transplants in the Amazon region. These include species assessed as Critically Endangered (CR), such as mahogany - *Swietenia macrophylla* (Meliaceae), abarco - *Cariniana pyriformis* (Lecythidaceae), and guayacán de bola – *Bulnesia arborea*[3] (Zygophyllaceae), a timber species assessed as Endangered (EN) in the Colombian Amazon and used as ornamental (Cárdenas & Salinas, 2007). The Brazil nut – *Bertholletia excelsa* (Lecythidaceae) is a food species assessed as Vulnerable (VU). Although there are very few records in Colombia, these are found in the surroundings of human settlements, which is why they are surmised to be introduced individuals as natural populations of the species have never been reported in the national territory.

Introduced Useful Species

Biological invasions are regarded as the second cause of biodiversity loss and one of the five main drivers of change globally (McNeely *et al.*, 2001). Any species introduction can become a biological invasion as some of the introduced species generate changes in the composition of species, displacement of native species and loss of biodiversity (Cárdenas *et al.* 2011). In this study, 323 introduced useful species were recorded, grouped in 85 families and 238 genera. This is a large number considering that 202 introduced species were reported previously for the region by Infante-Betancour & Rangel-Ch. (2018), and only 274 introduced plant species were recorded for the entire country according to Cárdenas *et al.* (2011). The families with the highest number of useful introduced plants were Fabaceae (26 species), Poaceae (21 species), Euphorbiaceae (17 species), and Lamiaceae (14 species). The main genera presenting the highest number of useful introduced species are *Citrus* (Rutaceae) and *Euphorbia* (Euphorbiaceae) (six species each), *Hibiscus* (Malvaceae) and *Jatropha* (Euphorbiaceae) (five species each), and *Allium* (Amaryllidaceae), *Canna* (Cannaceae), *Ficus* (Moraceae), *Musa* (Musaceae), *Pelargonium* (Geraniaceae), and *Urochloa* (Poaceae) (four species each). The categories of uses with the highest number of species introduced to the Colombian Amazon were Ornamental (191 species), Medicinal (102 species), Food (87 species), and Forage (19 species).

[3] Species accepted by Bernal *et al.*, 2016

Useful species in large landscapes of the Colombian Amazon

The following section presents the useful species related to large landscapes of the Colombian Amazon (sensu Duivenvoorden & Lips 1993). Some useful species are exclusive to the **Terraces of the great rivers**, where the greatest human activity of indigenous communities is concentrated. These include marañón de monte – *Anacardium giganteum* (Anacardiaceae), uflé – *Anaueria brasiliensis*[4] (Lauraceae), barbasco – *Caryocar nuciferum* (Caryocaraceae), madroño – *Garcinia madruno* (Clusiaceae), siringa – *Hevea spruceana* (Euphorbiaceae), juansoquillo – *Parahancornia oblonga* (Apocynaceae), toñeka – *Parinari montana* (Chrysobalanaceae), and juansoco de alacrán – *Rhigospira quadrangularis* (Apocynaceae).

Also, unique useful species were recorded for the **dissected surfaces of sedimentary origin**, such as cuchara caspi – *Ambelania occidentalis* (Apocynaceae), palo rosa – *Aniba rosodora* (Lauraceae), cachicamo o lagarto caspi – *Calophyllum brasiliense* (Calophyllaceae), sangre drago – *Croton lechleri* (Euphorbiaceae), charapilla – *Dipteryx odorata* (Fabaceae), chicle – *Lacmellea edulis* (Apocynaceae), *Picramnia latifolia* (Picramniaceae), and umarí – *Poraqueiba sericea* (Metteniusaceae).

The **Alluvial floodplains of the Andean and Amazonian rivers** also have species that only occur there, including *Anthodiscus amazonicus* (Caryocaraceae), costillo – *Aspidosperma excelsum* (Apocynaceae), yagé – *Banisteriopsis caapi* (Malpighiaceae), lagarto o yacaré – *Calophyllum longifolium* (Clusiaceae), copaiba – *Copaifera reticulata* (Fabaceae), curare – *Curarea tecunarum* (Menispermaceae), oreja de chimbe – *Erisma uncinatum* (Vocysiaceae), palo de arco – *Handroanthus serratifolius* (Bignoniaceae), catahua – *Hura crepitans* (Euphorbiaceae), palo de algodón – *Pseudobombax munguba* (Malvaceae), *Omphalea diandra* (Euphorbiaceae), and uña de gato – *Uncaria tomentosa* (Rubiaceae).

The rocky areas of sedimentary and igneous-metamorphic origin, recognized as the **rocky outcrops** in the northern part of the studied area, have relatively few records of only 13% (85 species) of the total useful species registered. Some of the useful species recorded for these ecosystems are *Acanthella sprucei* (Melastomataceae), *Aechmea rubiginosa* (Bromeliaceae), *Epidendrum nocturnum* (Orchidaceae), *Ouratea chiribiquetensis* (Ochnaceae), *Pachira fuscolepidota* (Malvaceae), *Parahancornia surrogata* (Apocynaceae), *Molongum lucidum* (Apocynaceae), *Navia acaulis* (Bromeliaceae), *Schoenocephalium martianum* (Rapateaceae), and *Vellozia tubiflora* (Velloziaceae).

On the other hand, there are species with wide distribution in different landscapes such as cagüi – *Caryocar glabrum* (Caryocaraceae), fariñero – *Clathrotropis macrocarpa* (Fabaceae), surba o juansoco – *Couma macrocarpa* (Apocynaceae), fono – *Eschweilera coriacea* (Lecythidaceae), asaí – *Euterpe precatoria* (Arecaceae), *Guatteria decurrens* (Annonaceae), milpesos – *Oenocarpus bataua* (Arecaceae), and curarina – *Potalia resinifera* (Gentianaceae). All of them are characterised by their in high densities within the different landscapes, and some are included among the 227 most abundant species in the Amazon rainforest (Ter Steege *et al.*, 2013).

[4] Species accepted by Bernal *et al.*, 2016

CONCLUSIONS

In general terms, the knowledge on the useful plants of the Colombian Amazon has been tackled from an ethnobotanical perspective, which is the science focused on the relationship between plants and humans. However, it is important to approach this knowledge from the perspective of economic botany, the science dedicated to useful plants, which considers taxonomic, pharmacological, ecological, and economic aspects, allowing us to know the precise identification, potentialities for forest production, and supply and demand conditions for products in the Colombian markets (Ricker & Daly, 1997).

It is also necessary to address this knowledge from the bromatological and phytochemical point of view to determine the species' properties. In this sense, it is important to indicate that although there is a high potential related to medicinal plant species, progress has yet to be made in identifying biochemical compounds, bromatological analyses, and other fields to support the reports of their safe use. Likewise, and ideally, for all species, it is necessary to evaluate the supply of resources to ensure their sustainable use, which must be the basis of any management plans that support their commercial use.

Acknowledgements

The authors thank Sonia Sua-Tunjano for supporting the database construction and for debugging and elaborating the distribution map.

References

Acero L (1979) *Principales Plantas Útiles de la Amazonia Colombiana. Proyecto Radargramétrico del Amazonas*. Bogotá, IGAC. 590p.

Acero L (2005) *Plantas Útiles de la Cuenca del Orinoco*. Bogotá, BP Exploration Company. 605p.

Blanco J (2020) *Caracterización de las 30 especies forestales maderables más movilizadas en Colombia provenientes del bosque natural*. Bogotá, Organización de las Naciones Unidas para la Alimentación y la Agricultura, Ministerio de Ambiente y Desarrollo Sostenible y Unión Europea. 84p.

Bernal R, Gradstein S, Celis M (eds) (2016) *Catálogo de plantas y líquenes de Colombia*. Bogotá, Instituto de Ciencias Naturales, Universidad Nacional de Colombia. Available at: http://catalogoplantasdecolombia.unal.edu.co/es (Accessed in May 2021).

Boyle B, Hopkins N, Lu Z, Garay JAR, Mozzherin D, Rees T, Matasci N, Narro ML, Piel WL, McKay SJ, Lowry S, Freeland C, Peet RK, Enquist BJ (2013) The taxonomic name resolution service: an online tool for automated standardization of plant names. *BMC Bioinformatics* [online], 14: 16. doi: 10.1186/1471-2105-14-16

Brack EA (1993) *Plantas nativas, comunidades humanas y desarrollo en el Perú*. Based Conservation Workshop. Airlie, Virginia.

Cárdenas D, López R (2000) *Plantas Útiles de la Amazonia Colombiana - Departamento del Amazonas. Perspectivas de los Productos Forestales no Maderables*. Bogotá, Instituto Amazónico de Investigaciones Científicas SINCHI. 133p.

Cárdenas D, Politis G (2000) *Movilidad, Territorialidad, Etnobotánica y Manejo del Bosque de los Nukak Orientales*. Bogotá, Instituto SINCHI and UNIANDES. 105p.

Cárdenas D, Marín C, Suárez S (2002) *Plantas útiles de Lagarto Cocha y Serranía de Churumbelo en el departamento de Putumayo*. Bogotá, Instituto Amazónico de Investigaciones Científicas SINCHI. 40p.

Cárdenas D, Salinas N (2007) *Libro Rojo de Plantas de Colombia. Volumen 4. Especies maderables amenazadas Primera parte. Serie Libros Rojos de Especies Amenazadas de Colombia*. Bogotá, Instituto Amazónico de Investigaciones Científicas SINCHI-Ministerio de Ambiente, Vivienda y Desarrollo Territorial, 232p.

Cárdenas D, Castaño N, Cárdenas-Toro J (2011) *Plantas introducidas, establecidas e invasoras en la Amazonia colombiana*. Bogotá, Instituto Amazónico de Investigaciones Científicas SINCHI. 154p.

Cárdenas D, Marín N, Castaño N (2012) Plantas alimenticias no convencionales en Amazonia colombiana y anotaciones sobre otras plantas alimenticias. *Rev Colomb Amaz*, 5:58-81.

Cárdenas D, Duque AJ, Duivenvoorden J, Gonzalez-Caro JS (2017). Asymmetrical niche determinism across geological units shapes phylogenetic tree communities in the Colombian Amazonia. *Perspect. Plant Ecol. Evol. Syst*, 28: 1-9. doi: 10.1016/j.ppees.2017.06.001

Cardoso D, Särkinen T, Alexander S, Amorim AM, Bittrich V, Cellis M, Daly DC, Fiaschi P, Funk VA, Giacomin LL, Goldenberg R, Heiden G, Iganci J, Kelloff CI, Knapp S, Cavalcante de Lima H, Machado AFP, Santos RM, Mello-Silva R, Michelangeli FA, Mitchell J, Moonlight P, Moraes PLR, Mori SA, Nunes TS, Pennington TD, Pirani JR, Prance GT, Queiroz LP, Rapini A, Riina R, Rincon CAV, Roque N, Shimizu G, Sobral M, Stehman JR, Stevens WD, Taylor CM, Trovo M, VAN DEN Berg C, van der Werff H, Viana PL, Zartman CE, Forzza R (2017). Amazon plant diversity revealed by a taxonomically verified species list. *PNAS* [online], 114(40):10695–10700. doi: 10.1073/pnas.1706756114

Castaño N, Cárdenas D y Rodríguez E (2007) *Ecología, Aprovechamiento y Manejo de nueve especies de Plantas del departamento de Amazonas, generadoras de Productos Maderables y no Maderables*. Bogotá, Instituto Amazónico de Investigaciones Científicas SINCHI-Corporación para el Desarrollo Sostenible del Sur de la Amazonia CORPOAMAZONIA. 266p.

da Paz Lima M, de Campos Braga PA, Lopes Macedo M, da Silva MFGF, Gilberto Ferreira A, Fernandes JB, Vieira PC (2004) Phytochemistry of *Trattinnickia burserifolia*, *T. rhoifolia*, and *Dacryodes hopkinsii*: chemosystematic implications. *J. Braz. Chem. Soc*, 15 (3): 385-394.

Dinerstein E, Olson D, Graham D, Webster A, Primm S, Bookbinder M, Ledec G (1995) *Una evaluación del estado de conservación de las eco-regiones terrestres de América Latina y el Caribe*. Washington, Banco Mundial. 135p.

de la Torre L, Navarrete H, Muriel P, Macía M, Balslev H eds. (2008) *Enciclopedia de las plantas útiles del Ecuador*. Quito, Herbario QCA de la Escuela de Ciencias Biológicas de la Pontificia Universidad Católica del Ecuador-Herbario AAU del Departamento de Ciencias Biológicas de la Universidad de Aarhus. 949p.

Domínguez C (1985) *Amazonia Colombiana*. Visión general. Bogotá, Banco Popular. 274p.

Duivenvoorden JF, Lips JM (1993) *Landscape Ecology of the Middle Caquetá Basin; Explanatory Notes to the Maps*. Bogotá, Tropenbos Colombia. 301p.

Garibaldi A, Turner N (2004) Cultural keystone species: implications for ecological conservation and restoration. *Ecol Soc* [online], 9(3):1. Available at: http://www.ecologyandsociety.org/vol9/iss3/art1/

Gentry AH (1989) Diversidad florística y fitogeografía de la Amazonia. In: INDERENA, ed. *Memorias del Simposio Internacional: Investigación y Manejo de la Amazonia*. 65-70pp. Bogotá, Instituto Nacional de los Recursos Naturales renovables y del Ambiente INDERENA.

Glenbosky L (1983) *Ethnobotany of the Tukuna Indians Amazonas, Colombia*. Bogotá, Universidad Nacional de Colombia. 92p.

Guimarães de Oliveira G, Bezerra da Silva HE, Salgado Vital MJ, Nascimento FC, de Melo Filho AA, Carvalho Rabelo SA (2014) Chemical constituents and antimicrobial activity of essential oil of *Trattinnickia burserifolia*, Tepequém, Roraima, Brazil. *Planta Med*, 80: LP86. doi: 10.1055/s-0034-1395117

Hernández CJ (1984) *Vistazo General sobre la Protección de la Naturaleza en Colombia*. Bogotá, Instituto Nacional de los Recursos Naturales renovables y del Ambiente INDERENA. 80p.

Hernández-C J, Hurtado A, Ortíz R, Walschburger T (1992) Unidades biogeográficas de Colombia. In: G Halffter (comp.). *La diversidad biológica de Iberoamérica I. Acta Zoológica Mexicana*, (n.s): pp. 105-173.

Herrera J (1999) Geología y características geomorfopedológicas. In: Botero P, ed. *Paisajes fisiográficos de la Orinoquía- Amazonia (ORAM) Colombia*. pp. 105-134. Bogotá, Instituto Geográfico Agustín Codazzi IGAC.

Hoorn C, Wesselingh FP, ter Steege H, Bermudez MA, Mora A, Sevink J, Sanmartín I, Sanchez-Meseguer A, Anderson CL, Figueiredo JP, Jaramillo C, Riff D, Negri FR, Hooghiemstra H, Lundberg J, Stadler T, Särkinen T, Antonelli A (2010) Amazonia through time: Andean uplift, climate change, landscape evolution, and biodiversity. *Science*, 330: 927–931. doi: 10.1126/science.1194585

Infante-Betancour J., Rangel-Ch O (2018) Riqueza y Diversidad de las Plantas con Flores de la Amazonía Colombiana: Su Significado En Un Contexto Global. In: Rangel-Ch JO ed. (2018) *La riqueza vegetal de la Amazonía de Colombia*. Bogotá, Fundación NATURA. 1029p.

Linares C (1994) Inventario preliminar de las plantas utilizadas para elaborar artesanías en Colombia. *Univ Sci*, 2 (1):7-43.

Linares C, Galeano G, García N, Figueroa Y (2008) *Fibras Vegetales Utilizadas en Artesanías en Colombia*. Bogotá, Artesanías de Colombia S.A.-Instituto de Ciencias Naturales Universidad Nacional de Colombia. 328p.

la Rotta C (1983) *Observaciones etnobotánicas sobre algunas especies utilizadas por la comunidad indígena Andoque*. Bogotá, Corporación Araracuara. 117p.

la Rotta C, Miraña P, Miraña M, Miraña B, Miraña M, Yucuna N (1987) *Observaciones etnobotánicas sobre algunas especies utilizadas por la comunidad indígena Miraña*. Bogotá, WWF-Fondo FEN Colombia. 381p.

Leyva P (ed.) (1998) *El Medio Ambiente en Colombia*. Bogotá, Instituto de Hidrología, Meteorología, y Estudios ambientales IDEAM. 495p.

MADS – Ministerio de Ambiente y Desarrollo Sostenible (2017) Resolución 1912 de 2017. Bogotá.

McNeely JA, Mooney H, Neville L, Schei P, Waage J (2001) *Global Strategy on Invasive Alien Species*. Gland, Switzerland and Cambridge, UK, IUCN. 50p.

Martius, KFP von (1829) Flora Brasiliensis: seu Enumeratio plantarum in Brasilia tam sua sponte quam accedente cultura provenientium. 2, vol. Sumptibus J. G Cottae.

Pabón M (1982) Botánica económica de la Amazonia Colombiana. *Col Amaz*, 1(1): 9-30.

Pérez-Arbeláez E (1978) *Plantas útiles de Colombia*. 4th edition. Bogotá, Litografía Arco. 831p.

Palacios W (2008) Plantas combustibles. In: L de la Torre, H Navarrete, P Muriel, M Macía, H Balslev eds. *Enciclopedia de las Plantas Útiles del Ecuador*. pp. 83-85. Quito and Aarhus, Herbario QCA y Herbario AAU.

Purseglove J (ed.) (1968) *Tropical crops. Dicotyledons 2*. New York, John Wiley and Sons Inc. 332p.

Rangel-Ch. O, Infante-Betancour J (2018) Aspectos generales sobre la Amazonia de Colombia. In: O Rangel-Ch. *et al.* eds. *Colombia Diversidad Biótica XVII. Región de la Serranía de Manacacías (Meta) Orinoquía colombiana*. pp. 1-20. Bogotá, Instituto de Ciencias Naturales Universidad Nacional de Colombia.

Ricker M, J Daly D (eds.) (1997) *Botánica Económica en Bosques Tropicales. Principios y Métodos para su Estudio y Aprovechamiento*. Ciudad de México, Editorial Diana México. 293p.

Rudas-Ll A (2009) Unidades ecogeográficas y su relación con la diversidad vegetal de la Amazonia Colombiana. Thesis. Instituto de Ciencias Naturales, Universidad Nacional de Colombia. Bogotá. 147p.

Salazar C, Riaño E (2016) *Perfiles urbanos en la Amazonia colombiana*. Bogotá, Instituto Amazónico de Investigaciones Científicas SINCHI. 216p.

Schultes R (1970) De plantis toxicaris e Mundo Novo tropicale commentationes, VII. Several ethnotoxicological notes from the Colombian Amazon. *Bot. Mus. Leafl.* 22(10): 345-352.

Schultes R (1979) The Amazonia as a source of new economic plants. *Econ. Bot.* 33: 259-266.

Schultes R (1983) De plantis toxicariis e Mundo Novo tropicale commentationes, XXXI. Further ethnopharmacological notes on Malpighiaceous plants of the Northwest Amazon. *Bot. Mus. Leafl.* 29(2):133-137.

Schultes R, Raffauf R (1990) *The Healing Forest: Medicinal and Toxic Plants of the Northwest Amazonia*. Portland, Dioscorides Press. 500p.

Schultes R, Raffauf R (1994) *El bejuco del alma. Los médicos tradicionales de la Amazonia colombiana, sus plantas y sus rituales*. Medellín, Universidad de Antioquia. 283p.

Taylor L (2005) *The Healing Power of Rainforest Herbs. A guide to Understanding and Using Herbal Medicinals*. New York, Square One Publishers. 520p.

ter Steege H, Pitman NC, Sabatier D, Baralotto C, Salomao RP, Guevara JE, Phillips OL, Castilho CV, Magnusson WE *et al.* (2013). Hyperdominance in the Amazonian tree flora. *Science* [online], 342 (6156). doi: 10.1126/science.1243092

Unda A (1993) *Manejo integral y sostenido de los bosques de Tarapacá y Flor de Agosto: Estudio de prefactibilidad*. Bogotá, OEA. 173p.

Snake carved from the twisted stem of a liana of *Entada polystachya*, collected in Antioquia, Colombia and sent by William R. Jervis to Kew in the 1850s.
Economic Botany Collection, EBC 59030

Chapter 7

Colombian plants in Kew's Economic Botany Collection

Hernando Echeverri-Sanchez[1,2], Fabio Ávila[1], Natalí Sánchez-Garzón[1] & Mark Nesbitt[1]*

[1] Royal Botanic Gardens Kew, Richmond, Surrey, TW9 3AE, UK

[2] Department of Anthropology, University College London, 14 Taviton St, Bloomsbury, London WC1H 0BW, UK

***Corresponding author:** m.nesbitt@kew.org

Keywords: ethnobotany, ethnobiology, museum, history, empire

ABSTRACT

Kew's Economic Botany Collection was first founded as the Museum of Economic Botany in 1847. Among a total collection of 100,000 plant raw materials and plant products, it contains 320 specimens from Colombia. Recent research has enabled botanical names to be updated and provenance data retrieved through archival research. The Colombian specimens are best understood as representing changes in the scientific relationship between Colombia and Kew from the mid-nineteenth century to today. We identify three stages in bioprospecting. In the nineteenth century, medicines, dyes, and other commercial materials were sent to Kew for evaluation and subsequent display, often by British residents in Colombia. The Royal Pharmaceutical Society Museum, now part of the Kew collection, also includes medicinal plants from Colombia. In the early twentieth century, the Colombian government employed Morley Thomas Dawe to explore Colombia's agricultural and botanical potential, leading to the donation of 70 specimens to Kew. Later in the twentieth century, Colombian botanists donated timber specimens. In the twenty-first century, the Colombia Bio project (between Colombia and Kew) led to the acquisition of new objects, emphasising the development of sustainable livelihoods and the preservation of biocultural knowledge.

RESUMEN

La Colección de Botánica Económica de Kew fue fundada como Museo de Botánica Económica en 1847. Con un total de 100.000 materias primas vegetales y productos vegetales en la colección, 320 especímenes son de Colombia. Investigaciones recientes han permitido actualizar los nombres botánicos y recuperar datos de procedencia mediante la investigación de archivos. Los especímenes colombianos representan cambios en la relación científica entre Colombia y Kew desde mediados del siglo XIX hasta la actualidad. Identificamos tres etapas en la bioprospección. En el siglo XIX, diversos medicamentos, tintes y otros materiales comerciales se enviaron a Kew para su evaluación, a menudo por parte de residentes británicos en Colombia. El Museo de la Real Sociedad Farmacéutica, ahora parte de la colección Kew, también incluye plantas medicinales de Colombia. A principios del siglo XX, el gobierno colombiano empleó a Morley Thomas Dawe para explorar el potencial agrícola y botánico de Colombia, lo que llevó a la donación de 70 especímenes a Kew. Posteriormente en el siglo XX se evidencia algún contacto con botánicos colombianos a través de donaciones de madera. En el siglo XXI, el proyecto Colombia Bio condujo a una intensa interacción entre Kew y Colombia, y la adquisición de objetos que enfatizan el desarrollo de medios de vida sostenibles y la preservación del conocimiento biocultural.

INTRODUCTION

The Economic Botany Collection (EBC) at the Royal Botanic Gardens, Kew (Kew) is a large, wide-ranging biocultural collection of over 100,000 materials and plant products spanning food products, tools, artefacts, medicines, poisons, wood, ornaments and textiles, among other objects, covering all aspects of plant uses in human life. The EBC is the successor to Kew's Museum of Economic Botany, founded in 1847. Such museums of economic botany were widespread in the second half of the nineteenth century, connecting producers of raw materials with manufacturers. Economic botany collections are identifiable by their emphasis on usefulness to humans, by their distinctive combination of raw materials and products, and by the consistent application of botanical (Latin) names to the specimens (Cornish & Nesbitt, 2014; Nesbitt & Cornish, 2016).

Economic botany was a fundamental part of the colonial system. By the 19th century, vegetable products were considered strategic elements, and oversight of their trade was a state priority (Drayton, 2000). As different colonial and commercial powers sought to control the economic potential of tropical biodiversity, they funded the creation and establishment of museums of economic botany. It is likely that similar materials, such as cinchona barks, were held by the 'Botanical House' in Bogotá: the base of the Real Expedición Botánica del Nuevo Reino de Granada. These specimens were removed to Madrid in 1816 (Díaz-Piedrahíta, 1986).

Although Colombia, with its recognised biological wealth, seems like a strategic source for economic botany specimens, it is surprisingly under-represented in the EBC. Unlike other countries of the region such as Mexico

(850 items; Rico Arce *et al.*, 2013) and Brazil (2,700 items), the Colombian component of the EBC is only 320 items (Echeverri-Sanchez *et al.*, in press). As part of the larger Colombia Bio project, the authors of this chapter have catalogued, digitised, and studied these items and uploaded them to the Plants of the World Online (http://www.plantsoftheworldonline.org/) and Col*Plant*A platforms (https://colplanta.org/). Updated records are also available on the EBC database, searchable online at: https://www.kew.org/science/collections-and-resources/collections/economic-botany-collection.

The Colombian element in EBC is small in number and is uneven in coverage of the many ways in which plants are used in the country. Although the EBC underrepresents the enormous botanical diversity of the country, it offers a useful perspective on British scientific and commercial interest in Colombia from the Victorian period to the current day.

BRITISH – COLOMBIAN RELATIONS

As a government institution since 1840, Kew's relationship with Colombia reflected the official stance of the British government. In the second half of the nineteenth century, this was characterised by a tentative interest in the country's natural wealth but little or no interest in formal administrative control. The British banking industry sought to make loans to the new nations of South America, establishing a debt system that, on many occasions, exerted a profound influence on these countries. This type of 'soft' power or, as historians now term it, 'informal empire', was considered more effective than formal political and economic control (Cain & Hopkins, 2016). However, British powers represented economic interests in the region, especially in countries with the greatest economic potential, such as Argentina, Brazil, and Chile (Thompson, 2008; Cain & Hopkins, 2016). Ensuring debt servicing of British loans and opening the Latin economies to English investors was London's main policy in South America.

In Colombia, informal political and economic control over the territory was rarely plausible. However, although Britain was one of the greatest investors in the young Andean nation, political instability, rurality, and geographical obstacles made it unattractive for English banks (Bushnell, 1993). Colombia was constantly trying to pay off large debts throughout its history and was quickly considered too risky for international investment (Deas, 1993, 2008). At the same time, the country had a relatively small luxury goods market and its rugged geography limited transportation. Foreign companies that tried to enter the domestic market often went bankrupt (Deas, 1993). Even large infrastructure investments, such as the Colombian rail network, failed. For these reasons, Colombia was one of the few countries in Latin American that did not produce large English fortunes, never being at the centre of British military and colonial interests.

The only product that could generate wealth in the country was gold, and as a result, several mines were granted by the Colombian government to pay interest on loans. These mines would be English cultural enclaves until the middle of the 20th century (Mejía Rivera, 2016; González Rodríguez, 2005). People working in the mines were one of the main sources of early collections in the EBC. They housed British doctors and engineers interested in local history and with the ability and interest to communicate directly with Kew. A search to find new medicines or industrial products was of great interest to local British enterprises. Scientists, merchants, and diplomats would send materials to Kew to explore their potential as a commodity.

As the geopolitics of empire and trade changed during the 20th century, objects collected for the EBC reflect these changes. Founded in part as a tool to classify and control the biodiversity of tropical colonies, the purpose of such institutions has transformed profoundly. The differences between the objects collected at the beginning of the 20th century and those at the end of the century demonstrate the gradual change. When the British colonial project ended, these scientific institutions began to focus on documenting and understanding biocultural diversity to enable its preservation and use in local livelihoods. As such, items collected in the past 40 years are focused on preserving biocultural diversity, with fuller documentation of the techniques and knowledge associated with them.

THE NINETEENTH CENTURY: SEARCHING FOR COMMODITIES

By the time the EBC was constituted in the 1840s, the mercantilist trade dynamics of the nineteenth century had led to a focus on Europe as an importer of raw materials, and an exporter of manufactured goods, with an emphasis on trade within a country's imperial territories. For countries outside the imperial system, entering this trade network required commodities and internal wealth that would incentivise foreign investment, which often came with heavy political and economic interventionism. The young South American nations were not part of any global imperial system at the time, following their recent independence from Spain and Portugal. They were often integrated into the north Atlantic trade network through a specific commodity or through banking and financial transactions. Throughout the history of the continent, several products caused exponential economic growth for local communities. Most of these products were raw resources produced from the local biodiversity, such as rubber, quinine, sugar, tobacco, alongside other lesser-known products, such as dyes and tropical fruits.

But Colombia was not part of the British colonial sphere, so there was no interest in investing in national agricultural production. Instead, the country was understood as a possible source of resources that could break the existing monopolies, such as Brazil with rubber and Peru with quinine. The race to find species that produce quinine and rubber, evidenced in the collection, demonstrates the interest in these products in the British markets. At the same time, it highlights the efforts of the Colombian government to look for new products for global trade, leading in the early twentieth century to financing European experts to visit Colombia and analyse the local biodiversity (see Morley Thomas Dawe, below)

Unlike other Latin American nations, Colombia was relatively poor. Its economy was based mostly on gold extraction, with smaller subsistence farming throughout the Andean region and larger plantations on the coast. In the absence of important export products such as guano in Peru, cocoa in Venezuela, or rubber in Brazil, Colombia had little to offer to international markets. Those products that gained a certain international value had a limited effect on the national economy, thanks to mismanagement and civil conflicts, as in the cases of the tobacco boom of 1850 (Bushnell, 1993) and quinine in the 1860s (Tirado Mejida, 1971). It was only with coffee in the late 19th century that Colombia effectively entered world trade. This market was directed towards the United States rather than Great Britain but began consolidating Colombia's influence throughout the continent.

MEDICINES

Some specimens highlight a local interest in finding alternative products that could generate income. An early example of this is Té de Mutis (*Symplocos theiformis*), with several samples in the EBC collection. The earliest specimen (Catalogue number EBC 50445, Figure 1a) was donated to Kew in 1893 by London's Pharmaceutical Society. This donation was an early attempt – involving the Spanish botanist Jose Celestino Mutis, resident in Bogotá – to commercialise an alternative to the lucrative tea (*Camellia sinensis*) trade.

However, as a biodiverse region, there was much interest in the potential of local medicines found in the forest and jungles of the new nation. Several items in the collection, given by W.R. Jervis, demonstrate the role of British imperialism and trade in the formation of botanical medicine bioprospecting. Jervis was probably William R. Jervis, a doctor in the English mines of Marmato (Mejía Rivera, 2016; González Rodríguez, 2005). Though we have little information on him, Jervis was a proficient collector with over 30 items collected between 1825 and 1850. He was one of several British citizens who emigrated to Antioquia during the first half of the nineteenth century, when a small population of foreigners managed the British-owned gold mines.

Jervis collected several specimens of cinchona bark (*Cinchona* spp., Figure 1b) and balsam (*Myroxylon* sp. and *Myrospermum* sp.), among other medicinal plants. The two examples of curare (*Strychnos toxifera*) from Colombia (EBC 49123 and EBC 49125) were sent to Kew by Jervis in 1853. Curare is used to poison darts used with blowguns or sometimes with arrows for hunting animals. It is prepared from the bark of *Strychnos* stems, mixed with a plant gum (usually from *Couma macrocarpa* latex) so that it will stick to the dart or arrow. *Strychnos* and other tubocurarine-containing plants are used throughout the Amazon basin. It is unclear how Jervis obtained this specimen, which might have come from the Amazon region (far from Marmato). Nevertheless, several *Strychnos* species grow in Andean and inter-Andean valleys. The 1850s were when curare played an important role in understanding the usefulness of paralysis-inducing drugs in medical treatment.

The EBC contains about 1,000 examples of *Cinchona* bark but only 32 specimens from Colombia. *Cinchona* bark

A

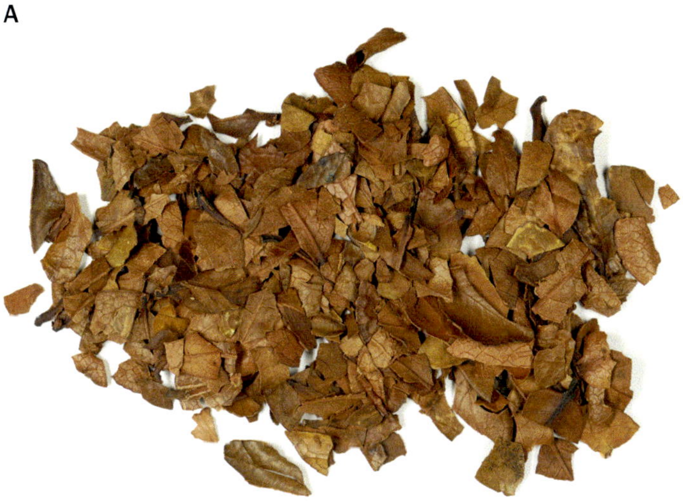

B

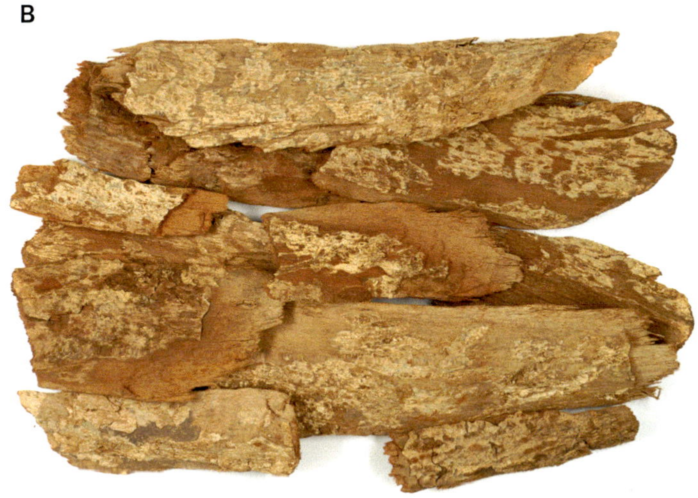

FIGURE 1. Medicines. **A** Té de Mutis (leaves of *Symplocos theiformis*), donated to Kew in 1893 by the Pharmaceutical Society of Great Britain (EBC 50445). **B** Bark of *Cinchona pitayensis* (Wedd.) Wedd., from Antioquia, Yarumal, sent to Kew by W.R. Jervis, 1854 (EBC 52393).

A

B

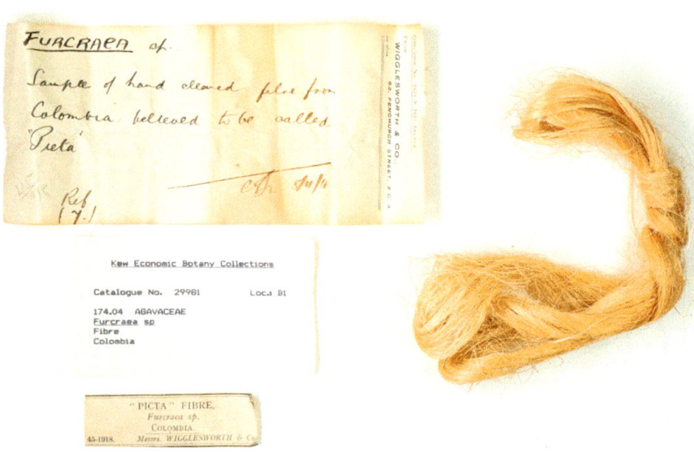

C

FIGURE 2. A Materials. Top. Leaves, twigs, and dyed fabric of '*Bignonia tinctoria*' (this is an unplaced name according to Plants of the World Online; it is probably related to *Fridericia chica* (Bonpl.) L.G. Lohmann), given by Robert Thomson to the Pharmaceutical Society of Great Britain c. 1890 (EBC 47638). **B** Fibre from leaves of *Furcraea* sp., given to Kew as 'picta' by London brokers Messrs. Wigglesworth & Co in 1919 (EBC 29981). **C** Bag waterproofed with *Castilla elastica* Cerv. Sent to Kew in 1919 by Morley Thomas Dawe, described by him as 'a rubber sack made by the Indians of that region [Darien Country]. They buy the material in the coastal towns or villages and waterproof it in the forests with the latex of *Castilla elastica*' (EBC 43012).

was the only effective treatment for malaria in much of the world from 1630 to 1940. It was used as powdered bark until the extraction of the quinine alkaloids, which was perfected in the 1820s (Walker & Nesbitt, 2019; Rønsted *et al.*, 2020). Since Peru had established a near-monopoly over *Cinchona* bark exports, Colombia had little effect on the international market (Crawford, 2016). However, there was a smaller-scale *Cinchona* bark industry in Colombia, which was characterised by a series of short-lived booms and busts over three periods: 1849–1852, 1867–1873, and 1877–1882 (Sastoque, 2011; Sandoval & Echandia, 1986; Tirado Mejida, 1971). Most of the quinine was harvested from the wild, and the samples found in the collection represent species common in Colombian cloud forests such as *Cinchona pitayensis* and *Cinchona pubescens*. Most of these specimens were donated to Kew by the Royal Pharmacological Society, which collected many samples from the London trade. From the 32 samples of the quinine-producing plants catalogued as Colombian, they represent five species of *Cinchona*, three species of *Ciliosemina*, and one each of the species of *Carapichea* and *Ladenbergia* (incl. *Cascarilla*). Colombia never had a sustained quinine industry, and when production shifted to Asia from the 1860s, Latin America lost its role in producing this important medicine.

MATERIALS

Dyes were of great interest to early Colombian and British bio prospectors. An interesting example is the *chica* dyes (*Fridericia chica*) held by the Pharmaceutical Society of Great Britain and transferred to Kew in 1983 (EBC 47638, Figure 2a). These were brought to London by Robert Thomson, a British gardener who visited Bogotá in the 1880s-1890s (Cendales Paredes, 2012). Even though this dye produced a beautiful red colour, correspondence in the Kew archives highlights that the colour does not last. This fact was the reason it was never a commercial alternative to cochineal dyes.

Rubber was another neotropical product of great interest for the British empire. Although there are many rubber products in the EBC, there are no Colombian items made with *Hevea brasiliensis*. Instead, five objects are made of the rubber 'substitute' known as caoutchouc or Indian rubber (*Castilla elastica*, Figure 2c). Since *H. brasiliensis* only grew wild in the Amazon basin, until the development of Asian plantations late in the nineteenth century, Brazil had a monopoly overexploiting this important commodity. Due to its strategic value, there was a constant incentive to search for alternatives that could compete against the Brazilian industry. Although *C. elastica* has been used as a source of rubber by indigenous people since pre-Colombian times, it was never a realistic alternative due to its lower quality and production limitations (Hosler *et al.*, 1999).

There was also much interest in materials to make strong ropes, which were considered strategic in an age before plastics. In the early twentieth century, the British naturalist and collector Morley Thomas Dawe had a particular interest in the fique or *Furcraea* genus plants, the source of excellent rope materials (Figure 2b). In 1916 he wrote, "There are probably few vegetable products, the subject of domestic industry, particularly in the fibre world, which have so wide an application in everyday use as fique in Colombia" (Dawe, 1916). Even though the production of fique in Colombia was mostly to supply a growing internal industry, the trade potential of this product was not lost to Dawe, who, in that same bulletin, highlights the cost of production and export to ports such as New York.

TIMBER

A key factor in Colombia's international trade was its access to transport. This fact is particularly the case with bulk materials such as timber. As late as 1916, its railways were poorly developed, with the main transport route being the Magdalena River. An easy connection between the country's Pacific and Atlantic coasts awaited the opening of the Panama Canal in 1914 (Simmons, 1916). Simmons reported an extremely challenging environment for timber exports, with stands of two desirable timber trees, Spanish cedar (*Cedrela odorata*) and Colombia mahogany (*Swietenia macrophylla*), so scattered as to be economically inviable – at least for export.

The wood component (xylarium) of the EBC contains about 42,000 specimens, with only 77 from Colombia. The history of the xylarium falls into two parts: collecting timber species from 1847 to the 1930s and then a broadening of interest to all woody species to support broad taxonomic surveys of wood anatomy (Cornish *et al.* 2014). The Colombian woods fall into two groups: a first period (1890–1920), mostly of Dawe's collections, and a second one (1960–1980) based on academic cooperation with the Universidad Nacional of Colombia (Figure 3).

The first group contains three important timbers even today: Andean cedar (cedro de Bogotá, *Cedrela montana*), native pine (pino chaquiro, *Prumnopitys montana*), and laurel (comino crespo or chachajo, *Aniba perutilis*). Santiago Cortés (1897) highlighted these species as important for the timber industry, and all are now categorised as at threat of extinction.

The second group of woody plants represented in the collection (and currently used in the country) belongs to 1960–1980 accessions. These items include commonly used plants in rural areas and are highly harvested throughout the country. Examples of this group are *Couma macrocarpa* (juansoco), *Terminalia amazonia* (arenillo), *Cariniana pyriformis* (abarco), and *Bulnesia arborea* (guayacán), which are also threatened (BGCI, 2021). Nowadays, according to MADS (2021), it is estimated that nearly 21,528 cubic metres of these trees were harvested between January 2020 and March 2021.

RICHARD SPRUCE'S ETHNOBOTANICAL COLLECTIONS

The scope of Kew's Museum of Economic Botany was, from its foundation, defined as 'all kinds of useful and curious Vegetable Products, which neither the living plants of the Garden nor the specimens in the Herbarium could exhibit'

(Hooker, 1855: 3; Cornish, 2012). In other words, not all specimens were purely utilitarian. This scope is reflected in the c. 350 ethnobotanical specimens collected from indigenous peoples in the Amazon region by the botanist Richard Spruce, mainly in the years 1850–1855 (Martins *et al.*, 2021). Some of these are clearly of economic potential, as in the fruits or the caapi vine (*Banisteriopsis caapi*), a component of ayahuasca, which was of medicinal interest in Europe. Spruce also collected other items of less obvious usefulness in a European context, such as ritual weapons or clothing, with the same scrupulous level of attention to their botanical identity.

Most of Spruce's ethnobotanical specimens were collected in the Rio Negro region of the northwest Amazon, along the Rio Uaupés, probably into Colombia. These Colombian specimens include poisoned arrows with shafts made from grass (EBC 38648, *Gynerium sagittatum*) and a palm leaf (*Mauritiella armata*, EBC 35883). He also collected items on the border between Venezuela and Colombia.

A

B

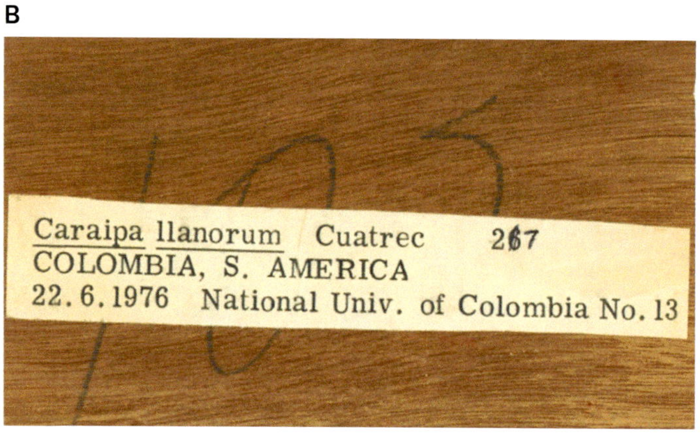

FIGURE 3. Wood specimens. **A** *Aniba perutilis* Hemsl., trade names 'Comino liso, 'Plain Comino'. Sent to Kew in 1893 by William Gordon, British Vice-Consul, Medellin (EBC 14468). **B** *Caraipa llanorum* Cuatrec. Sent to Kew in 1976 by the National University of Colombia (EBC 1130).

THE TWENTIETH CENTURY: COLOMBIA'S NATIONAL AGENDA

One of the most interesting themes drawn from the EBC is the transformation of British global science: from the commercial focus of the late colonial period to a postcolonial scientific interest in biodiversity. Botanists, diplomats, and experts would send items to Kew that reflected their interests and what they believed was the interest of Kew as an imperial scientific institute. From the late twentieth century onwards, these sources dry up, and donations are made almost exclusively by professional botanists operating within the Convention on Biological Diversity parameters, working in a collaborative and conservation-based framework for science.

The work of Morley Thomas Dawe, who was a consultant for the Colombian Ministry of Agriculture, is emblematic of the shift from individual or commercial collectors to the state agency, albeit still working within a framework of bioprospecting. Dawe travelled the country between 1915 and 1920, drawing on his expertise as a trained botanist and long-time collaborator with Kew and exploring Colombia's botanical potential (Figures 2c and 4). Though Colombia had a policy of highlighting its botanical potential on the global stage since its independence (del Castillo, 2010), it relied mostly on foreign experts.

Due to his pre-existing relationship with Kew, Dawe sent over 70 items (representing a quarter of the entire Colombian EBC collection). His expeditions were mostly in the northern and western regions of Colombia, traversing the different ecosystems of the Caribbean, Andean, and Pacific regions. He later published lengthy reports (such as Dawe 1918, 1921a, 1921b). This scenario highlights the mediating role he had between the UK and Colombia. As with the fique collections, these studies were fundamentally economic, looking for commercial and agricultural alternatives to benefit Colombia and the UK.

The lack of Colombian specimens (except for timber) during the rest of the twentieth century reflects both a wider decline in acquisitions during this time, as the development of the Museum of Economic Botany ceased to be a priority for Kew, and a parallel lack of engagement of Kew botanists with Colombia. Colombia entered the commercial sphere of the United States, and American botanists played a major role in the exploration of the Colombian flora.

Typical of the sporadic acquisitions of Colombian items during this period was the basketry purchased by Kew botanist Lauren Gardiner at the World Palm Symposium, held in 2015 in the municipality of Montenegro, Quindío

FIGURE 4. Model with figures and utensils made from corky outgrowths of the bark of 'hobor' tree, *Spondias mombin*. Sent by Morley Thomas Dawe in 1916 (EBC 37950).

Department (Figure 5). The Symposium included an organised market for attendees to buy traditional crafts and other palm products. This reflected the shift during the 20th century from an emphasis on exporting raw materials to benefit large corporations or manufacturing industries overseas, to projects that emphasise high-value products, the retention of income in local communities in Colombia, and the sustainable use of biological resources.

THE TWENTY-FIRST CENTURY: THE COLOMBIA BIO PROJECT

In the 21st century, Kew has developed a new relationship with Colombian researchers through collaboration with the national Colombia Bio project. This Colombian government initiative is based on the sustainable use of local biodiversity. Kew has collaborated with Colombia Bio in several projects, of which the 'Biodiversity Expeditions' in Boyacá provided several items for the collection. Although there are many similarities between these expeditions and those conducted by Dawe one hundred years earlier, Colombia Bio is based in a well-established community of botanical researchers in Colombia. The items collected during the Colombia Bio expeditions, such as the sample of the basketry in the Boyacá area (EBC 100447-100449), demonstrate a different vision and perspective: one of self-reflection and identification of the country's own ecological and cultural heritage.

CONCLUSIONS

Our initial study of the Colombian component of the EBC – and there is much more to be done – demonstrates the importance of such case studies of countries that (during our study period) did not form part of one of the European empires. We emphasise the complex routes by which specimens reached Kew, with the strong involvement of commercial interests and actors within Colombian society in the nineteenth century. While economic botany in the nineteenth century clearly continues to function as a tool for extracting wealth from South America, even after independence, this is a complex process in which Kew's engagement was fitful.

From the early twentieth century onwards, the main actors are the Colombian government and, later, Colombian scientists. In both the early twentieth and twenty-first centuries, the emphasis shifts towards supporting local economic development. Now, we can see an explicit focus on conservation and ethnobotanical methodologies as the bridge between biodiversity conservation and its use. The Economic Botany Collection is thus an archive both of traditional knowledge (for detailed case studies, see Echeverri-Sanchez *et al.*, 2021) and an archive of Colombia and Kew's changing engagement with the world.

FIGURE 5. Bag woven by Reinaldo da Silva, of the Tikuna people of the Colombian Amazon, with fibres from the unexpanded leaves of the chambira palm, *Astrocaryum chambira* Burret. This palm, the most important source of fibre in western Amazonia, was in process of domestication by the time of the European conquest (Garcia *et al.* 2015). Bought by Lauren Gardiner at the World Palm Symposium, Montenegro, Quindío, Colombia, 2015 (EBC 99453).

Acknowledgements

We thank Frances Cook for curatorial assistance in the EBC, Ellie Graves and Federico Fabriani for advice on plant identification, and Carmen Amalia Díaz Peña and William Milliken for helpful comments on the manuscript.

References

BGCI (2021) GlobalTree Portal. Available at: https://www.bgci.org/resources/bgci-databases/global-tree-assessment-portal/

Bushnell D (1993) *The making of modern Colombia: a nation in spite of itself*. Berkeley, University of California Press. 335p.

Cain PJ, Hopkins AG (2016) *British imperialism: 1688-2015*. 3rd Edition. London, Routledge. 794p.

Cendales Paredes C (2012) Robert Thomson (1840-1908): a Kew gardener in Bogotá, Colombia. J *Gard Hist*, 40: 239-52.

Cornish C (2012) Useful and curious: a totem pole at Kew's timber museum. *J Mus Ethnogr*, 25:138-151.

Cornish C, Gasson P, Nesbitt M (2014) The wood collection (xylarium) of the Royal Botanic Gardens, Kew. *IAWA J*, 35:85-104.

Cornish C, Nesbitt M (2014) Historical perspectives on western ethnobotanical collections. In: J. Salick, K. Konchar, M. Nesbitt eds. *Curating biocultural collections: a handbook*. pp. 271-293. London, Royal Botanic Gardens, Kew.

Cortés S (1897) *Flora de Colombia*. Vol. 1. Bogotá, Samper Matiz. 288p.

Crawford MJ (2016) *The Andean wonder drug: Cinchona bark and imperial science in the Spanish Atlantic,* 1630–1800. Pittsburgh, University of Pittsburgh Press. 284p.

Dawe MT (1916) Fique: *Furcraea gigantea*. Bulletin of Miscellaneous Information. *Kew Bull,* 1916(7), 169-170.

Dawe MT (1918) *Account of a journey down the Magdalena River, through the Magdalena Province and the Peninsula of Guajira (Colombia)*. Bogotá, Ministry of Agriculture and Commerce.

Dawe MT (1921a) *Account of a journey through the western portion of Colombia, showing the possibilities of the economic development of the districts visited*. London, Colombian Bureau of Information and Trade Propaganda. 28p.

Dawe MT (1921b) An account of a journey from Tolima to the Pacific Coast, through the Departments of Caldas and El Valle (Republic of Colombia). In: MT Dawe ed. *Account of a journey through the western portion of Colombia, showing the possibilities of the economic development of the districts visited*. London, Colombian Bureau of Information and Trade Propaganda. 28p.

Deas M (1993) *Del poder y la gramática: y otros ensayos sobre historia, política y literatura colombianas*. Santafé de Bogotá, Tercer Mundo Editores. 346p.

Deas M (2008) Weapons of the weak? Colombia and foreign powers in the nineteenth century. In: *Informal empire in Latin America: Culture, commerce and capital*, ed. M. Brown, pp. 173-86. Oxford, Wiley Blackwell.

del Castillo L (2010) La Gran Colombia de la Gran Bretaña: la importancia del lugar en la producción de imágenes nacionales, 1819 - 1830. *Araucaria,* 12(24): 124-149.

Díaz-PiedrahÍta S (1986) Las compuestas de la Real Expedición Botánica del Nuevo Reino de Granada (1783-1816). *Caldasia*, 14: 371-393.

Drayton RH (2000) *Nature's government: Science, imperial Britain, and the 'improvement' of the world*. New Haven, Yale University Press. 346p.

Echeverri-Sanchez H, Sánchez-Garzón N, Nesbitt M, Ávila, F, Diazgranados M (2021) Enviados directo a Londres: Los objetos colombianos en la Colección de Botánica Económica del Real Jardín Botánico de Kew. *Apuntes* Pontificia Universidad Javeriana, 34:doi.org/10.11144/Javeriana.apu34.edlo

Garcia N, Galeano G, Mesa L, Castano N, Balslev H, Bernal R (2015) Management of the palm *Astrocaryum chambira* Burret (Arecaceae) in northwest Amazon. *Acta Bot Bras*, 29: 45-57.

González Rodríguez AL (2005) Educación y práctica médicas en Antioquia. Antecedentes históricos de la fundación de la Escuela de Medicina de la Universidad de Antioquia. *Iatreia*, 18: 332-343.

Hooker WJ (1855) *Museum of Economic Botany: Or, a popular guide to the useful and remarkable vegetable products of the Museum of the Royal Gardens of Kew, London*. London, Royal Botanic Gardens, Kew. 11p.

Hosler D, Burkett SL, Tarkanian M J (1999). Prehistoric polymers: rubber processing in ancient Mesoamerica. *Science*, 284: 1988-1991.

MADS - Ministerio de Ambiente y Desarrollo Sostenible (2021) Base de datos de movilizaciones de madera con salvoconducto forestal en Colombia. Bogotá, Dirección de Bosques, Biodiversidad y Servicios Ecosistémicos.

Martins L, Fonseca-Kruel V, Cabalzar A, Lima Azevedo D, Milliken W, Nesbitt M, Scholz, A (2021) *A maloca entre artefatos e plantas: guia da coleção Rio Negro de Richard Spruce em Londres*. São Paulo, Instituto Socioambiental. Available at: https://acervo.socioambiental.org/acervo/publicacoes-isa/maloca-entre-artefatos-e-plantas-guia-da-colecao-rio-negro-de-richard-spruce

Mejía Rivera O (2016) *Historia de la medicina en el eje cafetero 1865-1965*. Manizalez, Universidad de Caldas. 440p.

Nesbitt M, Cornish C (2016) Seeds of industry and empire: Economic botany collections between nature and culture. *J Mus Ethnogr*, 29:53-70.

Rico Arce M de L, Nesbitt M, Rodriguez A, Hiscock C (2013) Mexican Plant Materials Housed in the Economic Botany Collection of The Royal Botanic Gardens, Kew. *Ibugana*, 5:3-38.

Rønsted N, Allasi Canales N, Hansen TG, Cornett C, Walker K, Driver F, Antonelli A, Maldonado C, Nesbitt M, Barnes C (2020) Historical chemical annotations of *Cinchona* bark collections are comparable to results from current day state-of-the-art High-Pressure Liquid Chromatography technologies. *J Ethnopharmacol*, 249:112375. doi:10.1016/j.jep.2019.112375

Sandoval Y, Echandía C (1986) La historia de la quina desde una perspectiva regional: Colombia, 1850-1882. *ACHSC*, 13-14:153-187.

Sastoque E (2011) Tabaco, quina y añil en el siglo XIX: Bonanzas efímeras. *Credencial Historia*, 255: 8-16.

Simmons RE (1916) *Lumber markets of the west and north coasts of South America*. Washington, Department of Commerce. 162p.

Thompson A (2008) Afterword: Informal empire: Past, present and future. In *Informal empire in Latin America: Culture, commerce and capital*, ed. M. Brown, pp. 229-241. Oxford, Wiley Blackwell.

Tirado Mejia A (1971) *Introduccion a la historia economica de Colombia*. Bogotá, Universidad Nacional de Colombia. 320p.

Walker K, Nesbitt M (2019) *Just the tonic: the natural history of tonic water.* London, Kew Publishing. 140p.

CENIVAM Garden.
Elena Stashenko

Chapter 8

Bioeconomy: a sustainable and responsible option for making the most of Colombian plant resources

Elizabeth Hodson de Jaramillo[1]* & Elena Stashenko[2]*

[1] Professor Emeritus, Pontificia Universidad Javeriana, Bogotá, Colombia. ehodson8@outlook.com
[2] Laureate Professor, General Director CENIVAM – Industrial University of Santander, Colombia. elenastashenko@gmail.com

***Corresponding authors:** ehodson8@outlook.com, elenastashenko@gmail.com

Keywords: circular bioeconomy, biodiversity, sustainability, SDGs, value-added bioproducts

ABSTRACT

Environmental deterioration, degradation of natural resources, and climate change are the greatest current challenges our planet faces. The environmental and social imperative of meeting the Sustainable Development Goals in a more environmentally friendly economy is not an option, but it is humanity's greatest responsibility today. The most obvious and clever response to this challenge is through the circular bioeconomy model for the sustainable production and use of resources and their transformation into value-added bioproducts. The bioeconomy strategies focus on value chains with integrated cycles, which increase efficiency and recycling through products and co-products in different biological systems. For a megadiverse country such as Colombia, sustainable bioeconomy requires a completely different way of producing: circular instead of linear, and using science intensively to add value to biological resources and biological processes. Basic concepts of circular bioeconomy, value chains and biotechnology are given, and some examples of bioeconomy-focused biotechnological developments in the Research Centre for Agro-industrialisation of Tropical Aromatic and Medicinal Plant Species CENIVAM are presented.

RESUMEN

El deterioro ambiental, la degradación de recursos naturales y el cambio climático son el mayor reto actual. El imperativo ambiental y social de atender los ODS en una economía más respetuosa con el ambiente no es una opción, es la mayor responsabilidad actual de la humanidad. Probablemente la respuesta más obvia e inteligente a este reto es a través del modelo de bioeconomía circular para la producción y aprovechamiento sostenible de los recursos y su transformación en bioproductos con valor agregado. Las estrategias de la bioeconomía se basan en cadenas de valor con ciclos secuenciales integrados, lo cual incrementa la eficiencia y reciclaje a través de productos y co-productos en diferentes sistemas biológicos. Para un país megadiverso como Colombia, la bioeconomía sostenible requiere una forma totalmente diferente de producir: sistemas circulares en vez de lineales y el uso intensivo de conocimiento para dar valor agregado a los recursos y procesos biológicos. Se presentan conceptos básicos de bioeconomía circular, cadenas de valor y biotecnología y se plantean algunos ejemplos de desarrollos biotecnológicos con enfoque de bioeconomía obtenidos en el Centro de Investigación para la Agroindustrialización de Especies Aromáticas y Medicinales Tropicales CENIVAM.

INTRODUCTION

The world is facing systemic climate, biodiversity, social, economic and health crises (IACGB, 2020). Biodiversity loss and land use are drivers correlated with zoonotic diseases such as COVID-19 and other emerging infectious diseases (Dasgupta, 2021). Environmental deterioration, degradation of natural resources, and climate change are the utmost current challenge. The global economic system must undergo radical transformation processes to improve sustainability (Lewandowski, 2018). The greatest endowment humankind has, and depends on, is nature that provides food, water, shelter, recreation, and resources to improve health and well-being, while maintaining nutrient cycles, soil, and water health, regulating climate, and increasing resilience. Inadequate human management has led to severe negative impacts, degradation, and decline of natural resources, challenging nature's productivity, resilience, and adaptability, and threatening global economies. Ecosystem health and integrity have been affected by the extraction of natural resources, including the waste generated from this process (Dasgupta, 2021). The wise use of Science and Technology (S&T) to face social and environmental threats is an ethical and social imperative. Scientific innovations play a crucial role in maintaining healthy soils and water, and in protecting or restoring natural habitats, conserving biodiversity, and increasing agrobiodiversity in local communities (von Braun *et al.*, 2021).

The fulfilment of the Sustainable Development Goals (SDGs) in an economy that is more respectful of the environment is not an option, but it is the greatest current responsibility of humanity. It is fundamental to promote human, environmental, social, and economic health. Probably the clearest and smartest approach to this challenge is through a circular bioeconomy model. Circular bioeconomies aim to use resources sustainably by mitigating waste streams through their transformation into side-line products that have added value such as food, feed, biobased products, biomaterials, and bioenergy. The goal is to produce more with less through waste recovery and to make use of all available S&T developments to harness the benefits of biodiversity by promoting resilient value chains (Misión Internacional de Sabios, 2021). Transformation to a sustainable system requires changes in food production and processing (the major driver of terrestrial biodiversity decline), changes in decarbonising our energy systems (the principal contributor to anthropogenic climate change), and drastically altering patterns and behaviours in production and consumption to better align with both environmental objectives and the entire global supply chain network (Dasgupta, 2021).

The concept of bioeconomy as an alternative path to sustainable development rests in its potential for renovation. A circular bioeconomy has a major role in re-aligning growth with the biosphere (IACGB, 2020). As stated by the European Commission in their updated Bioeconomy Strategy (European Commission, 2018), *"We live in a world of limited resources [...]. We must improve and innovate the way we produce and consume food, products, and materials within healthy ecosystems through a sustainable bioeconomy."*

The bioeconomy includes, in addition to the traditional agriculture, forestry, fisheries and aquaculture sectors, several downstream processing and service industries, such as food, paper, textiles, building and construction, chemistry, cosmetics, and biopharma. Key converging technologies, such as bio-, nano- and information technologies, are crucial to the knowledge-based bioeconomy, which applies biobased processes and principles in industrial applications for the obtention of value-added bioproducts (IACGB, 2020). The bioeconomy embraces principles of renewability and circularity and provides an opportunity to create a system based on sustainable economic growth, using science to add value to biological resources and biological processes while reducing resource consumption and protecting and regenerating ecosystem (IACGB-GBS 2020). The objective of a cascading and circular economy is to increase resource efficiency, reduce demand for fresh materials, and add knowledge and technology to transform and create new industrial value, create jobs, and improve the welfare of the local community (Trigo *et al*., 2021).

BIOECONOMY IN COLOMBIA – WHY?

Colombia's immense biodiversity constitutes the most exceptional asset. Worldwide, it ranks first in orchids, second in general plants, and third in palms. The conservation and sustainable use of its resources represent its development driver. The main proposal of the International Experts Mission 2019 (Misión Internacional de Sabios, 2021) is that the country's conceptual and political framework for the country's sustainable development is the bioeconomy, with a territorial approach within the framework of the SDGs. The Ministry of Science, Technology, and Innovation launched the Colombia Bioeconomy Mission (December 2020) and opened a call for proposals in bioeconomy for the development of new technologies and products based on the use of biomass, biodiversity, and ecosystem services, and additionally to encourage sustainable agriculture in regions (https://minciencias.gov.co/convocatorias/innovacion-y-productividad/convocatoria-para-el-apoyo-programas-y-proyectos-idi-que).

One of the goals is to consolidate science and innovation as the axis of growth in the generation of progress and social equity. This goal must beadjusted to the characteristics, resources, conditions and capacities of each region in the country so that the adoption of the circular bioeconomy model promotes job creation, income growth, and the well-being of the entire community. The bioeconomy path chosen by each region must be defined in participatory bottom-up processes to adjust to local conditions and requirements and to contribute to common prosperity. Therefore, the main objective of this chapter is to present a review of the fundamentals of the circular bioeconomy, to promote the sustainable management of biodiversity in Colombia and to present, as a case study in agribusiness, some of the advances that have been obtained in the Centre for Agro-industrialisation of Aromatic Plant Species and Tropical Medicines CENIVAM of the Industrial University of Santander.

THE CIRCULAR BIOECONOMY – WHAT IS IT?

The circular bioeconomy is considered the efficient management of biobased renewable resources to sustainably produce, technologically process and trade resources from land, fisheries and aquaculture, and to transform them into bioproducts while promoting new industries and jobs (Tan & Lamers, 2021). The circular bioeconomy model for territorial efficient conversion of resources into value added products is considered the best choice for achieving SDGs. This model directly covers at least 14 of the 17 SDGs, making important contributions to the health and nutrition of a growing population, to the sustainable provision of food, energy, water and raw materials, and to soil protection, climate, and environmental protection (IACGB-GBS 2020). This concept excludes the false dichotomy between productive development and sustainability. According to FAO (2021), sustainable and circular bioeconomic development implies balancing different SDGs because strategies are cross-cutting and multi-sectoral, and consequently face greater challenges than sustainable development strategies that target just a single sector. As stated by FAO (2021): "The bioeconomy sought unlocks the potential of nature to provide bioresources, bioprocesses and biodiversity and respects the planetary boundaries".

The circular bioeconomy has also been proposed as the new strategy for reactivation and reindustrialisation (IACGB, 2020), strengthening the eco-efficiency of processes and the use of recycled carbon to reduce current dependencies on fossil carbon. It is transversal, regional, and interdisciplinary (Misión Internacional de Sabios, 2021). The circular bioeconomy includes all kinds of biomaterial streams with innumerable application paths, and reconciles local and global integration by promoting the creation of diversified, innovative, and decentralised value-chains and adapting old ones (IACGB-GBS 2020).

This model is a disruptive process that needs new knowledge-based processes for innovation towards the eco-efficient circular sustainable management of renewable resources. The goal is to reposition agriculture and the management of rural areas as a part of a balanced development process that generates employment by giving added value to the primary uses of resources and waste by-products. The bioeconomy is the opportunity for territorial development and a complement to existing productive activities, diversifying and integrating them into value chains and networks that look forward to make responsible use of natural resources and locally generated waste in parallel, to maintain an appropriate balance with nature. In this system, the value of products, materials and resources is maintained in the economy for as long as possible, and waste generation is reduced to a minimum (Misión Internacional de Sabios, 2021). It not only adopts circularity by cascading materials and co-products into several uses in different value streams, but also increases the efficiency in each step of the value chain and minimises the residues generated (Tan & Lamers, 2021).

SUSTAINABLE, RESILIENT VALUE CHAINS – THE EFFICIENT MODEL

Bioeconomy is not inherently sustainable because it is based on natural resources; it requires the articulated integration of a sustainable supply chain, sustainable, appropriate transformation processes, and sustainable high-value bioproducts, biomaterials or bioenergy with appropriate trade systems (Tan & Lamers, 2021). The circular bioeconomy focuses on the efficient and sustainable valuation of biomass in comprehensive multi-product production chains, using waste and optimising the value of biomass in a cascade (Trigo *et al.*, 2021). We can look to future scientific and technological developments to increase the portfolio of bioproducts developed from the local biodiversity, keeping with a circular economy approach (Hodson *et al.*, 2021). The greatest potential of the circular bioeconomy lies in its wide diversity and ability to adjust to territorial conditions and resources, promoting unique developments that make it possible to establish niches in global markets and to promote local and regional initiatives that create synergies among the stakeholders (IACGB-GBS, 2020).

Nevertheless, to promote the advances and bioscience innovations and to direct them to being a factor of sustainability is not enough. It is fundamental to adapt them to local conditions, resources, and capacities, to make them available and accessible to the regional community, especially smallholders, and to use them to enhance local and traditional knowledge (von Braun *et al.*, 2021).

This approach reconciles local and global integration because it needs to define policies to strengthen clever international and local value chains, which requires a wise balance of global and local objectives. There is a need to maintain permanent investments in bioeconomy development to establish new, innovative, diversified, and decentralised value-chains and adjust old ones to ensure the sustainability, resilience, and adaptability of the selected supply chains. The bioeconomy maintenance lies in its diversity, adaptability, and close interactions with local and rural communities (IACGB-GBS, 2020).

In terms of sustainability and resilience, using a greater range of plant diversity is critically important. Numerous plant species have not been domesticated and show successful traits for cultivation, such as adaptation to stresses and high nutritional content. They offer new possibilities to improve food and nutritional safety and as raw materials for innumerable processes and bioproducts. They contain significant genetic diversity to increase agrobiodiversity and contribute to resilience and more sustainable agriculture to in the face of climate change (von Wettberg *et al.*, 2020). The value chain is a business efficiency model that makes it possible to find comparative and competitive advantages based on the particular activities of each system. For the bioeconomy approach, these value chains include natural ingredients used for several industries: food, feed, cosmetics, nutritional supplements, tropical flowers and foliage, plant extracts, sustainable agricultural products, and ecotourism, among many others. The prioritised value chains contribute in different ways to the conservation of biodiversity and territorial development.

BIOTECHNOLOGY – A KEY TO CREATING BIOECONOMIC VALUE

The bioeconomic model is the convergence of science towards biotechnology, focusing intently on the management of natural resources for the industrial use, transformation, and manipulation of biomass. It uses the present latent value of this biomass to achieve new growth and social benefits in thefuture (renewable fuels, bioplastics, agricultural inputs, and medicinal products). Biotechnology is the technological instrument of the bioeconomy. Innovation is essential when modifying inefficient, polluting and resource-based industrial processes to improve efficiency and respect for the environment (Misión Internacional de Sabios 2021). Biotechnology supports the development of bio-based industries to facilitate the sustainability of the sector and to reduce environmental impacts.

A sustainable bioeconomy requires sustainable supply chains and appropriate transforming biotechnologies for processing renewable resources into high-value-added bioproducts, materials and bioenergy (Tan & Lamers, 2021). Biotechnology is one of the keys enabling the circular bioeconomy to function, offering solutions for efficient and sustainable production of plant and microbial biomass for bioenergy production; for bioremediation and environmental protection; for transformation through green chemistry; and as an integral part of agri-food processes in value chains among other processes (Trigo *et al.*, 2021). In addition, biotechnology also brings solutions to improve post-harvest processing and thus optimise the bioeconomy sector, as a whole. For instance, in the forestry sector, biotechnology offers solutions to reduce the lignin content of trees or crops, thereby reducing the highly polluting processing of the wood in the paper or biofuel industry (Lewandowski, 2018). Biotechnology plays a central role in these value chains and technological transformation processes. It plays a vital role in numerous biochemical processes such as the fermentation of residues for animal feed or bioenergy, production of biotechnological fertilisers or soil amendments, microbial acidification in silage, enzymatic saccharification of starch and lignocellulose, obtention of food and feed additives, and production of supplements, vitamins, amino acids, biopolymers, cosmetics, and pharmaceutical bioproducts. Engineering the composting and residual biomass transformation by microorganisms is one example of the crucial value added by applying biotechnology to the ongoing issue of residue management (Kircher, 2021). Biotechnology applies science to disrupt business-as-usual processes by focusing on value added through renewability, circularity, and cascaded resource use. This 'biologisation' of the economy comprises social and high-tech innovations and the development of completely new solutions, services, and products that combine sustainability with increased consumer benefit and thus enable new lifestyle concepts (IACGB-GBS, 2020). Every present-day economic system considers sustainability essential. It mimics biological systems in what nature has done for millions of years: being dynamically balanced and sustainable. The central core of this model is the integration of all the steps of the processes in value chains or networks and not only processes of value-adding at each step. These chains must take advantage of cascade technologies to optimise processes across multiple products and must generate circularity (Misión Internacional de Sabios, 2021).

AGROINDUSTRY IN COLOMBIA

Colombian agriculture faces a series of very sensitive challenges to its future productivity, fundamentally related to the dramatic decline of the rural population (55% in 1955 and currently barely reaching 24%), its progressive ageing and the lack of labour, accompanied by few incentives for young people to stay in rural areas and associate their future with the countryside. Agro-industry in Colombia focuses on a basket of staple and cash crops; including many grains (rice, sorghum, corn, wheat, barley, quinoa, and others), oils and oilseeds (palm oil, coconut oil, cashew, soy), crops for beverages (coffee, cocoa, tea), fresh vegetables (potatoes, onions, bananas, yams, cassava, beans, peas, carrots, etc.), wood products, rubber and fibres (fique fibre plant, cotton), fresh fruits (pineapple, avocado, blackberry, mango, *lulo*, passion fruit, guava, banana, citrus, etc.), fresh aromatic spices (bay leaf, turmeric, ginger, basil, thyme, rosemary, etc.), and cut flowers (rose, carnation, and others).

Although not all the land in Colombia is arable, the current area occupied by these crops is less than 9%, and large areas remain that could support the introduction of promising new crops. While banana cultivation occupies 915,987 ha (DANE, 2014), that of aromatic plants is only 65,846 ha (0.78% of the total cultivated area). Agricultural agro-industrial projects in large areas can cultivate aromatic plants to obtain essential oils (EOs). Still, considering the topography and agricultural infrastructure of the country, production from smallholdings – through associations or cooperatives of growers and their collective work as small farm owners – is paramount to achieving socio-economic and environmental sustainability. This could prevent the displacement of growers and consequently preserve family farms. In many cases, men must abandon their families and move to other country regions as day labourers (for example, to collect coffee). However, by having new productive projects on their farms, they could obtain additional income, which will allow families to remain together.

THE EXAMPLE OF CENIVAM

The scientific research carried out by the Centre for Research on Tropical Aromatic and Medicinal Plant Species (CENIVAM) based at the Industrial University of Santander, UIS (Bucaramanga, Colombia) is focused on the transfer of scientific and technological knowledge to the field through the development of an Agro-industry of EOs and natural ingredients from aromatic and medicinal plants cultivated in small farms, by growers united in associations. The development route that needs to be covered is still quite long, as illustrated by the import/export activities of Colombia in the EO market, which represented in 2019 a mere 0.0058% of the US$ 5,6 billion-dollar global market (International Trade Center, 2021).

CENIVAM is a multidisciplinary research network of groups from several Colombian public and private universities that joined efforts to study Colombia's biodiversity, focusing on aromatic and medicinal plants, their EOs and extracts. The approach performs research activities under three major investigation blocks: bioprospection, biotransformation, and bioactivity. The long-term expected results are developing both natural ingredients and final commercial products that use their biological properties. Under a permit from Colombia's Ministry of Environment and Sustainable Development, more than 25 botanical outings and expeditions were organised to obtain plant material from different regions in the country. Verbenaceae (350 accessions), Asteraceae (241 accessions), Labiatae (235 accessions), Piperaceae (139 accessions), and Melastomataceae (27 accessions) are the main plant families studied, thus far yielding over a thousand EOs and extracts for further chemical and biochemical characterisation.

During the past fifteen years, CENIVAM has conducted multiple studies on native and introduced aromatic and medicinal plants, their secondary metabolites, EOs and extracts, their biological activities, and their applications in final products. For the instrumental analysis of volatile

FIGURE 1. *Lippia origanoides* and the Chicamocha river.
Elena Stashenko

fractions, EOs, and extracts, gas chromatography (GC) is used in one-dimensional or two-dimensional versions (GCxGC), in capillary columns of different polarities, using universal, selective, or very specific detection systems (Choi, 2003). High-resolution mass spectrometry detection systems (HR-TOF, Orbitrap) coupled to gas or liquid chromatographs are nowadays an excellent alternative for exact mass determination (elemental composition analysis), and for the selective and very sensitive detection of secondary metabolites in complex floral scent mixtures (Stashenko & Martínez, 2007, 2008), plant volatiles, EOs, and extracts obtained with organic solvents or supercritical fluids (CO_2).

This research network has carried out more than 5,000 biological tests during 2005-2020 on EOs and extracts obtained from more than a thousand botanical accessions collected in the field in different regions of Colombia. Around 45-47% of EOs or extracts possessed some interesting biological activity. Since the cytotoxicity of EOs and extracts may prevent their use in some pharmaceutical applications, this aspect is normally included among the first assays of the bioactivity investigation. Acute toxicity tests using brine shrimp with *Artemia franciscana* for a large set of EOs studied in CENIVAM showed that over 30% of these oils did not have any degree of toxicity (Olivero *et al.*, 2009). The virucidal activity of some essential oils against dengue (Meneses *et al.*, 2009a,b; Ocazionez *et al.*, 2010) and yellow fever (Meneses *et al.*, 2009a,b) was studied at the Research Center for Tropical Illnesses, CINTROP. Of the EOs tested, 83% were active against the dengue virus, and 66% showed positive results in inactivating the yellow fever virus. These results are a very interesting contribution, especially considering that there are not many data in the literature on the EO activity against yellow fever and related viruses. Chagas and leishmania are additional tropical diseases that deserve attention. Bioactivity assays against

Trypanosoma cruzi (epimastigotes and amastigotes) and *Leishmania chagasi* (promastigotes and amastigotes) were carried out at CINTROP using cell lineages commonly employed for these purposes (Vero and THP cells). From the results, 48% of the EOs examined were active against *T. cruzi*. The EOs were also tested against *L. chagasi* (promastigotes and amastigotes) and 19% were found to be active (Escobar *et al.*, 2010). Antifungal (Mesa-Arango *et al.*, 2010) and antibacterial (Caceres *et al.*, 2020) activities have also been extensively studied.

An important consideration to keep in mind during the study of secondary metabolites from aromatic and medicinal plants is the existence of plant chemotypes when plants with identical morphologies have different secondary metabolic profiles. Notable examples are *Lippia alba* and *L. origanoides*, two Verbenaceae species that have attracted lots of attention due to their biological properties. Seven chemotypes have been reported for *L. alba*, three found in Colombia (Lopez *et al.*, 2011; Mesa-Arango *et al.*, 2010). Up to five different chemotypes have been distinguished for *L. origanoides*, three of them present in Colombia (Stashenko *et al.*, 2010; Sarrazin *et al.*, 2015). Only two of the *L. origanoides* chemotypes are rich in thymol and carvacrol. Their high content confers antimicrobial properties to their EOs to make them important ingredients for a successful commercial chicken feed product. Mountain oregano, the common name used for *L. origanoides*, is a species that constitutes a very good example of a tropical plant with many potential applications in the pharmaceutical and food industries. It is a very good source of flavonoids (Stashenko *et al.*, 2013; Arias *et al.*, 2020), which may explain its activity against breast cancer (Raman *et al.*, 2017, 2018), as well as the efficacy of its EO as an insect repellent that may be used to protect grains from beetle attack during storage (Alcala-Orozco *et al.*, 2019; Caballero-Gallardo *et al.*, 2012). The antigenotoxicity of *L. origanoides* EOs (Vicuña *et al.*, 2010) makes them important ingredients of photoprotective cosmetics, which may as well take advantage of their antioxidant properties (Stashenko *et al.*, 2008). An example of circular economy application is the patent for a methodology that makes integral use of *L. origanoides* by producing its EO, fractions enriched in phenylpropanoids, extract, and hydrosol, in a combined process (Stashenko *et al.*, 2019). The post-distillation biomass is then used as biofuel and composting material, while the hydrosol can be used in pest control.

Various pilot research projects in the field have been financed in the past 15 years by Colciencias (currently Minciencias, Ministry of Science, Technology, and Innovation), the Ministry of Agriculture and Rural Development, the Governments of Santander and Arauca, and the Fund of the General System of Royalties (SGR, UIS - Government of Santander). More than 200 rural families have been involved in aromatic plant cultivation and their distillation to produce EOs.

The combined knowledge of chemical composition and biological activity serves as the basis for the sustainable use of biodiversity in developing new consumer products for the cosmetics, hygiene, food, and pharmaceutical industries. Pilot EO production units have been implemented in some municipalities in Santander, Arauca and Cundinamarca in Colombia, where farmer associations have been trained in good agricultural practices, post-harvest treatments, and the operation of stills designed at UIS for EO production using either hydrodistillation or steam distillation. Thanks to these pilot units, farmers have begun producing *Cymbopogon nardus*, *C. martinii*, and *Lippia origanoides* EOs. New developments have started to extend the cultivation and EO production to additional species, such as *Cananga odorata* (ylang ylang), *Pogostemon cablin* (patchouli), *Vanilla planifolia* (vanilla), *Lippia alba*, *Rosmarinus officinalis* (rosemary), *Thymus vulgaris* (thyme), and *Pelargonium graveolens* (geranium). In addition, many promising aromatic and medicinal plants are under investigation, and the intense study of their biological and chemical profiles should lead to the design of new Colombian products based on natural ingredients.

CONCLUSIONS — PROSPECTS AND CHALLENGES TO A COLOMBIAN BIOECONOMY

The sustainable development of the bioeconomy requires solid scientific support for all the value chain steps. This was the main justification for presenting the CENIVAM example, which shows contextualised developments of final commercial products for the essential oils agro-industry in which the circular economy principles are implemented. Crossing bioprospecting results with those from analytical chemistry and molecular biology (bioactivity assays) led to the detection of promising vegetal species to be used as ingredients for new consumer products. Field projects have established the conditions for biomass production under good agricultural practices and for their rural processing to afford essential oils, extracts, biocontrol solutions, composting material, and biofuel. The modular agro-industrial projects that currently in place in several Colombian municipalities serve as a reference for developing new enterprises that add value to the crops, diversify the rural economy, and creating career opportunities for farm workers and professionals while using resources comprehensively with low waste generation. Multiple challenges remain from diverse areas, such as establishing the commercial chains with reinvestment into the countryside, the achievement of origin recognition, road improvement, internet coverage, water supplies, and stable legislation.

Acknowledgements

Economic support from the Colombian General System of Royalties and the Colombian Ministry for Science, Technology, and Innovation is gratefully acknowledged.

References

Alcala-Orozco M, Caballero-Gallardo K, Stashenko E, Olivero-Verbel J (2019) Repellent and fumigant actions of the essential oils from

Elettaria cardamomum (L.) Maton, *Salvia officinalis* (L.) Linnaeus, and *Lippia origanoides* (V.) Kunth against *Tribolium castaneum* and *Ulomoides dermestoides*. *J Essent Oil-Bear Plants*, 22 (1):18-30. doi: 10.1080/0972060X.2019.1585966

Arias J, Mejía J, Córdoba Y, Martínez J, Stashenko E, del Valle J (2020) Optimisation of flavonoids extraction from *Lippia graveolens* and *Lippia origanoides* chemotypes with ethanol-modified supercritical CO_2 after steam distillation. *Ind Crops Prod*, 146:112170. doi: 10.1016/j.indcrop.2020.112170

Caballero-Gallardo K, Olivero-Verbel J, Stashenko E (2012) Repellency and toxicity of essential oils from *Cymbopogon martinii*, *Cymbopogon flexuosus* and *Lippia origanoides* cultivated in Colombia against *Tribolium castaneum*. *J Stored Prod Res*, 50:62-65. doi: 10.1016/j.jspr.2012.05.002

Cáceres M, Hidalgo W, Stashenko E, Torres R, Ortiz C (2020) Essential oils of aromatic plants with antibacterial, anti-biofilm and anti-quorum sensing activities against pathogenic bacteria. *Antibiotics*, 9:147. doi: 10.3390/antibiotics9040147

Choi HS (2003) Characterisation of *Citrus unshiu* (*C. unshiu* Marcov. Forma *Miyagawa-wase*) blossom aroma by solid-phase microextraction in conjunction with an electronic nose. *J Agric Food Chem*, 51:418-423.

DANE - Departamento Administrativo Nacional de Estadística (2014) Censo Nacional Agropecuario. Available at: https://www.dane.gov.co/index.php/estadisticas-por-tema/agropecuario/censo-nacional-agropecuario-2014

Dasgupta P (2021) The Economics of Biodiversity: The Dasgupta Review. London, HM Treasury. Available at: https://www.gov.uk/government/publications/final-report-the-economics-of-biodiversity-the-dasgupta-review

Escobar P, Leal S, Herrera L, Martínez J, Stashenko E (2010) Chemical composition and antiprotozoal activites of Colombian *Lippia* spp. essential oils and their major components. *Mem Inst Oswaldo Cruz*, 105:184-190.

European Commission (2018) A sustainable bioeconomy for Europe: strengthening the connection between economy, society and the environment Updated Bioeconomy Strategy. Directorate-General for Research and Innovation. doi: 10.2777/792130

FAO (2021) Aspirational principles and criteria for a sustainable bioeconomy. Rome, FAO Available at: www.fao.org/3/cb3706en/cb3706en.pdf

Hodson E, Nigli U, Kitajima K, Lal R, Sadoff C (2021) Boost nature positive production. A paper on Action Track 3 from the Scientific Group of the UN Food Systems Summit.

IACGB - GBS Global Bioeconomy Summit (2020) Expanding the Sustainable Bioeconomy –Vision and Way Forward. Communiqué of the Global Bioeconomy Summit 2020. Available at: https://gbs2020.net/wp-content/uploads/2020/11/GBS2020_IACGB-Communique.pdf

IACGB - International Advisory Council on Global Bioeconomy (2020) Global Bioeconomy Policy Report (IV): a decade of bioeconomy policy development around the world. Available at: https://knowledge4policy.ec.europa.eu/publication/global-bioeconomy-policy-report-iv-decade-bioeconomy-policy-development-around-world_en

International Trade Center (2021) International trade statistics. Available at: https://www.intracen.org/itc/market-info-tools/trade-statistics/

Kircher M (2021) Bioeconomy – present status and future needs of industrial value chains. *N Biotechnol*, 60: 96-104. doi: 10.1016/j.nbt.2020.09.005 .

Lewandowski I (ed.) (2018) *Bioeconomy: Shaping the Transition to a Sustainable, Biobased Economy*. Cham, Springer Nature. doi: 10.1007/978-3-319-68152-8

López M, Stashenko E, Fuentes J (2011) Chemical composition and antigenotoxic properties of *Lippia alba* essential oils. *Genet Mol Biol*, 34 (3), 479-488. doi: 10.1590/S1415-47572011005000030

Meneses R, Ocazionez R, Martínez J, Stashenko E (2009a) Inhibitory of essential oils obtained from plants grown in Colombia on yellow fever virus replication in vitro. *Ann Clin Microbiol Antimicrob*, 8:1-6.

Meneses R, Torres F, Stashenko E, Ocazionez R (2009b) Aceites esenciales de plantas colombianas inactivan el virus del dengue y el virus de la fiebre amarilla. *Rev Univ Ind Santander*, 41: 236-243.

Mesa-Arango C, Betancur-Galvis L, Montiel J, Bueno J, Baena A, Duran D, Martínez J, Stashenko E (2010) Antifungal activity and chemical composition of the essential oils of *Lippia alba* (Miller) N.E Brown grown in different regions of Colombia. *J Essent Oil Res*, 22:568-574. doi: 10.1080/10412905.2010.9700402

Misión Internacional de Sabios (2021) *Ciencia y Tecnología, fundamento de la Bioeconomía. Propuestas del Foco Biotecnología, Bioeconomía y Medio Ambiente*. Vol 3. Bogotá, Colección Misión Internacional de Sabios 2019. Ediciones Uniandes. 208p.

Ocazionez R, Meneses R, Torres F, Stashenko E (2010) Virucidal activity of Colombian *Lippia* essential oils on dengue virus replication in vitro. *Mem Inst Oswaldo Cruz*, 105: 304-309.

Olivero-Verbel J, Güette-Fernandez J, Stashenko E (2009) Acute toxity against *Artemia franciscana* of essential oils isolated from plants of the genus *Lippia* and *Piper* collected in Colombia. *BLACPMA* 8: 419-427.

Raman V, Aryal U, Hedrick V, Ferreira R, Fuentes J, Stashenko E, Levy M, Levy M, Camarillo I (2018) Proteomic analysis reveals that an extract of the plant *Lippia origanoides* suppresses mitochondrial metabolism in triple-negative breast cancer cells. *J Proteome Res*, 17 (10): 3370-3383. doi: 10.1021/acs. jproteome.8b00255

Raman V, Fuentes J, Stashenko E, Levy M, Levy M, Camarillo I (2017) *Lippia origanoides* extract induces cell cycle arrest and apoptosis and suppresses NF-kB signalling in triple-negative breast cancer cells. *Int J Oncol*, 51(6):1801-1808. doi: 10.3892/ijo.2017.4169

Sarrazin S, Andrade da Silva L, de Assunção A, Oliveira R, Perez V, da Silva R, Stashenko E, Maia J, Mourão R (2015) Antimicrobial and seasonal evaluation of the carvacrol-chemotype oil from *Lippia origanoides* Kunth. *Molecules*, 20:1860-1871. doi: 10.3390/molecules20021860

Stashenko E, Martínez J (2007) Sampling volatile compounds from natural products with headspace/solid-phase micro-extraction. *J Biochem Biophys Methods* 70: 235-242. doi: 10.1016/j.jbbm.2006.08.011.

Stashenko E, Martínez J (2008) Sampling flower scent for chromatographic analysis. *J Sep Sci*, 31:2022-2031 doi: 10.1002/jssc.200800151.

Stashenko E, Ruiz C, Muñoz A, Castañeda M, Martínez J (2008) Composition and antioxidant activity of essential oils of *Lippia origanoides* H.B.K. grown in Colombia. *Nat Prod Commun*, 3 (4):563-566. doi: 10.1177/1934578X0800300417

Stashenko E, Ruíz C, Arias G, Durán D, Salgar W, Cala M, Martínez J (2010) *Lippia origanoides* chemotype differentiation based on essential oil GC-MS analysis. *J Sep Sci*, 33 (1): 93-103. doi: 10.1002/jssc.200900452

Stashenko E, Martínez J, Cala M, Durán D, Caballero D (2013) Chromatographic and mass spectrometric characterisation of essential oils and extracts from *Lippia* (Verbenaceae) aromatic plants. *J Sep Sci*, 36:192-202. doi: 10.1002/jssc.201200877

Stashenko E, Martínez J, Durán D, Arias A, Córdoba Y, Mejía J, Tavera C (2019) Patent of invention WO2018122654A1: method for making full use of *Lippia origanoides*.

Tan E, Lamers P (2021) Circular Bioeconomy Concepts—A Perspective. *Front Sustain*, 2:701509. doi: 10.3389/frsus.2021.701509

Trigo E, Chavarria H, Pray C, Smyth SJ, Torroba A, Wesseler J, Zilberman D, Martinez JF (2021) *The Bioeconomy and Food Systems Transformation*. FSS Brief by Partners of Scientific Group. Available at: https://sc-fss2021.org/wp-content/uploads/2021/03/FSS_Brief_Bioeconomy_and_Food_Systems_Transformation.pdf

Vicuña G, Stashenko E, Fuentes J (2010) Chemical composition of the *Lippia origanoides* essential oils and their antigenotoxicity against bleomycin-induced DNA damage. *Fitoterapia*, 81(5): 343-349. doi: 10.1016/j.fitote.2009.10.008

von Braun J, Afsana K, Fresco LO, Hassan M (2021) Science for Transformation of Food Systems: Opportunities for the UN Food Systems Summit. Available at: https://sc-fss2021.org/wp-content/uploads/2021/07/Scientific_Group_Strategic_Paper_revised_July-22-2021.pdf

von Wettberg E, Davis TM, Smy' kal P (2020) Editorial: Wild Plants as Source of New Crops. *Front Plant Sci*, 11:591554. doi: 10.3389/fpls.2020.591554

Naidi—*Euterpe precatoria* Mart. (Arecaceae)—Infructescence.
Juan Carlos Copete

Chapter 9

Sustainable value chains and development pathways for natural ingredients in Colombia: the case of naidí (*Euterpe oleracea* Mart.)

Mabel Tatiana Rojas[1]*, Tiziana Ulian[2]*, Carlos Alberto Cortés[1], Germán Eduardo Torres-Morales[1], David Hammond[2], Felipe García[1] & Mauricio Diazgranados[2]

[1] Instituto de Investigación de Recursos Biológicos Alexander von Humboldt, Bogotá, Colombia.
[2] Royal Botanic Gardens, Kew.

***Corresponding authors:** mrojas@humboldt.org.co, t.ulian@kew.org

Keywords: biodiversity, bioeconomy, *Euterpe*, green growth, value-chains, sustainable development

ABSTRACT

Colombia is the second most biodiverse country globally, potentially having a comparative advantage over less biodiverse countries in shifting to a more sustainable, biobased economy. However, Colombia accounts for only 0.27% of natural ingredients' exports in Latin America; hence its potential is still locked. As a case study, we applied a value-chain analysis to the fruit of the naidí palm (*Euterpe oleracea*) growing along Colombia's Pacific Coast. We conducted semi-structured interviews involving community-based organisations, private companies, NGOs, international organisations, and universities to assess the full range of activities and relationships involved in its value creation. We found that the factors characterising its value chain consist of trust relationships between small-scale producers, a low distribution of profits among local collectors, limited profitability supported by external subsidies, and low regional and national demand. This value chain could be improved in the short- and medium-term by increasing the capacity of cold chains and storage centres, fostering economic incentives for implementing agroforestry systems, developing strategies for promoting traditional diets by using its natural ingredients, and improving technology and business capabilities to enhance its commercialisation.

RESUMEN

Colombia es el segundo país con más biodiversidad a nivel mundial, y potencialmente tiene una ventaja comparativa sobre los países con menor biodiversidad para cambiar a una economía de base biológica más sostenible. Sin embargo, Colombia representa solo el 0,27% de las exportaciones de ingredientes naturales en América Latina; de ahí que su potencial aún esté bloqueado. Como caso de estudio, aplicamos un análisis de la cadena de valor al fruto de la palma naidí (*Euterpe oleracea*) que crece a lo largo de la costa del Pacífico de Colombia. Realizamos entrevistas semiestructuradas en las que participaron organizaciones comunitarias, empresas privadas, ONGs, organizaciones internacionales y universidades, para evaluar la gama completa de actividades y relaciones involucradas en su creación de valor. Encontramos que los factores que caracterizan su cadena de valor son: las relaciones de confianza entre los pequeños productores, una baja distribución de utilidades entre los recolectores locales, una rentabilidad limitada sustentada en subsidios externos, y una baja demanda regional y nacional. Esto podría mejorarse en el corto y mediano plazo aumentando la capacidad de las cadenas de frío y los centros de almacenamiento, fomentando incentivos económicos para la implementación de sistemas agroforestales, desarrollando estrategias para promover dietas tradicionales utilizando sus ingredientes naturales, y mejorando la tecnología y las capacidades comerciales mejorando su comercialización.

INTRODUCTION

Colombia is the second most biodiverse country globally, hosting around 10% of the planet's biodiversity (Rincón Bermúdez *et al.*, 2009). It has at least 28,947 vascular plants, of which ca. 7,472 have a documented use (Negrao *et al.*, 2022). Of these, 79% have medicinal uses, 32% are harvested for materials (including personal care products and textiles), 28% provide environmental services (soil improvers, ornamentals, plants for erosion, and pollution control), and 26% are sources for human food (Diazgranados *et al.*, 2021). This high diversity of useful plants represents a potential comparative advantage over other countries, as it can be considered an asset in developing Colombia's bioeconomy.

Development of the Colombian bioeconomy and Green Growth

Recently, the concept of bioeconomy has received much attention at the international level as a viable system for transforming goods and services to overcome social, economic, techno-scientific, energetic, and environmental sustainability challenges (Staffas *et al.*, 2013; Moñux *et al.*, 2018). In 2018, more than 40 countries had policies or strategies in place to develop their bioeconomies based on the transformation of goods and services from the use of biological resources (Dietz *et al*., 2018). These were led by strategic priorities within each country, economic specialisation, biomass allocation, historical investments in research and development and labour productivity (Bracco *et al*., 2018). For Colombia, bioeconomic development has been identified as a priority to achieve the 17 UN Sustainable Development Goals (SDGs). Similarly, the Organisation for Economic Co-operation and Development (OECD) and the Economic Commission for Latin America and the Caribbean (ECLAC) have encouraged the country to design an agenda to unlock the potential of its biodiversity for sustainable development (OECD, 2009; Albrecht *et al*., 2015). Currently, the government is implementing a national bioeconomy through a Green Growth strategy (CONPES, 2018), taking advantage of the nation's large natural capital. After developing the Green Business Strategy, the concept of bioeconomy was promoted by the Green Growth Mission of the National Planning Department (Departamento Nacional de Planeación, DNP) in 2015. This Mission has identified that the development of a national bioeconomy is fundamental for achieving sustainable growth, defining it as *"a model that efficiently and sustainably manages biodiversity and biomass to generate new value-added products, processes, and services, based on knowledge and innovation"* (DNP, 2018, p. 28).

It is expected that by 2030 the bioeconomy will account for 10% of the Colombian Gross Domestic Product (GDP). Around 100,000 new jobs will be created in different sub-sectors, including biocommerce, natural ingredients, generation of new materials, biorefineries, bioenergy, agri-food systems, and advanced biotechnology (Misión Internacional de Sabios, 2019).

Connections between bioeconomy and natural ingredients

Growing a bioeconomy in Colombia involves identifying and producing natural ingredients via biotech and non-biotech pathways as part of a process of value-creation and innovation that earmarks the replacement of synthetic compounds with those of natural origin (Biointropic, 2019). Supplying natural ingredients is thus essential in differentiating Colombian products in global value chains (ONUDI, 2015), where the application of sustainable business models is increasingly needed to maintain market access and consumer interest. In this study, we adopted the following definition of natural ingredients: *"The solid or liquid raw material extracted from the physical, and sometimes simple chemical processing of animal sources, plants and other living, native organisms, respecting the principles of Biotrade, used in the development of products for therapeutic purposes, of hygiene or beauty of the body and of products that through its ingestion nourish the organism"* (Fondo Biocomercio, 2009, p. 14). These are classified into nine groups by the Fondo Biocomercio (2009): 1) tannins or dyestuffs; 2) active ingredients for therapeutic purposes; 3) condiments-spices and value added-fruits, sweeteners, thickeners, and flavourings; 4) essential oils; 5) fats and waxes; 6) natural flavouring substances; 7) saps, gums, resins, and oleoresins; 8) juice, pulps, extracts, juice concentrates; and 9) flours and starches.

Status of natural ingredients in Colombia

According to Rojas *et al*. (2021), Colombia exported a total of 650 tons of natural ingredients worth 3.4 million dollars in 2019. The group with the largest share of the country's exports of natural ingredients (49%) was juices and plant extracts. However, the group with the highest level of business dynamism between 2014-2019 was natural colourings, with a Compound Annual Growth Rate (CAGR) of 11.95%. In parallel, the country imported 5,552 tons of natural ingredients worth 60 million dollars (Rojas *et al*., 2021). This scenario means that the country has a trade deficit of 57.1 million dollars, which is significantly higher than expected when considering the rich country's biodiversity and potential for producing natural ingredients, reflecting the lack of development in the natural ingredients sector.

Colombia accounts for only 0.27% of Latin America's exports of natural ingredients (Rojas *et al*., 2021). Hence, its potential remains unfulfilled in helping to meet the Colombian government's commitment to Green Growth. Papellet *et al*. (2020) argue that this is due to a lack of knowledge of technical product standards and regulations, and the species management requirements needed to satisfy consumer demand.

Partly this is due to a lack of information regarding the biology and ecology of species of interest and to limited research in biotechnology, such as bioprospecting (Centro Red de Innovación, 2013; Biointropic, 2019). This gap is reflected in the small number of bio-innovative companies. As of 2018, the country had only 90 identified green businesses and 84 registered bioproducts (DNP, 2018), with even fewer established value chains in the country involving local communities. Achiote (*Bixa orellana*), naidí (*Euterpe oleracea*), Jagua (*Genipa americana*), and sacha inchi (*Plukenetia volubilis*) are exemplary exceptions (Biointropic, 2019).

Studies suggest an absence of institutional leadership, the inefficient use of natural capital, low investment in research and development, and regulatory barriers as the main causes hindering the development of bioeconomic initiatives in the country (DNP, 2018; Moñux *et al*., 2018). Administrative inefficiencies in obtaining sustainable harvest permits for non-timber forest products and contracts to access genetic resources for commercial purposes may also play a particularly important role (Biointropic, 2019; Papell *et al*., 2020). In this context, it is essential that biodiversity becomes a differentiating factor in Colombia's bioeconomy to overcome socio-economic challenges at the regional and national levels.

This chapter aims to assess the value chains of natural ingredients in Colombia and to explore the potential for developing Colombia's bioeconomy and supporting sustainable livelihoods. The objective is to provide a framework for developing sustainable value chains of natural ingredients and to create pathways for sustainable development in Colombia. To do this, we applied a value chain analysis to the fruit of the naidí palm (*Euterpe oleracea*). We examined limitations, opportunities, and success factors weighing on sustainable value chain development for natural ingredients in Colombia. This work has been carried out in the framework of the 'Useful Plants and Fungi of Colombia' project, led by the Royal Botanic Gardens, Kew, in partnership with the Alexander von Humboldt Biological Resources Research Institute (see Chapter 1).

MATERIAL AND METHODS

A value chain is understood as a strategic network among several organisations. According to Hobbs *et al.* (2000), this approach differs from a supply chain in several ways: value chain participants have a long-term strategic vision; recognise interdependence and align working together to solve common problems, sharing risks and benefits; value chains are demand-driven, not supply-driven, and thus focus on consumer needs. Value chains do not exist by themselves but are part of the socio-economic and institutional system, and therefore external forces affect the chains. Efficiency will transcend production processes and entail information flows among the actors in the chain and the ability of support services and enabling conditions to overcome challenges (Lundy *et al.*, 2007). Within this broader context, paying mind to internal factors, such as adaptive schemes of governance and social, human and natural capital that underpin livelihoods, will increase resilience to exogenous shocks and improve sustainability (Ellis & Allison, 2001).

Study area and actors

We conducted a value chain analysis of the processed fruit of the naidí palm (*Euterpe oleracea*) growing along Colombia's Pacific Coast. The analysis was carried out in the municipalities of Buenaventura (Valle del Cauca), Bajo Baudó (Chocó) and Vigía del Fuerte (Antioquia). We documented the source of natural ingredients (from raw material to the final product), the processes for capturing and creating value, and emerging interactions and relationships of autonomy and interdependence among

TABLE 1. Direct actors involved and interviewed for the value chain analysis of naidí palm fruit products.

Actor	Description
Naidí Pacífico SAS	Community-based company comprising 7 community councils (CC): two of them (CC Cajambre and CC Bahía Málaga) are in Buenaventura (Valle del Cauca), and the rest (CC Sivirú, CC Piliza, CC San Andrés de Usaragá, CC Guineal and CC Concosta) in the municipality of Bajo Baudó, comprising the Departments of Chocó and Cauca. This company was established in 2015, with sales and production management needing training and creating a legally constituted business model.
Planeta CHB SAS	A company created in February 2016, with ten partners belonging to the municipalities of Vigía del Fuerte (Antioquia) and Bojayá (Chocó), adjacent to the Atrato River. Of the ten partners, seven communities are represented, and three of these have been part of the organisational process of local Community Councils.
Corpocampo	A wholesale company focused on selling natural and functional ingredients from Non-timber Forest Products on international markets. The naidí-based products benefit 1,200 families in Guapi (Cauca), Buenaventura (Valle del Cauca), Tumaco (Nariño) and Puerto Asís (Putumayo), who manage 15,600 hectares of forest with sustainable agroforestry models. This company won the Norwegian 'Business for Peace' award in 2018, is a member of the 'Business Call to Action' initiative led by the United Nations Development Program (UNDP), and was nominated for recognition by them.
Fondo Acción	Colombian private fund with 20 years of experience in sustainable investments in environmental and childhood projects.
MUCHOCOL	A retail distribution company that eliminates intermediaries and reaches consumers directly.
Universidad de Los Andes Faculty of Economics	University that offers a broad research background on collective property in the Colombian Pacific and on governance schemes in this territory.
Partnerships for Forests (P4F)	Programme funded in Latin America by the British Government (Department for Business, Energy & Industrial Strategy [BEIS]). This initiative aims to protect the world's tropical forests through productive and commercial projects that combat deforestation and that, in turn, highlight the value of natural forests.

Source: Adapted from Rojas *et al.* (2021).

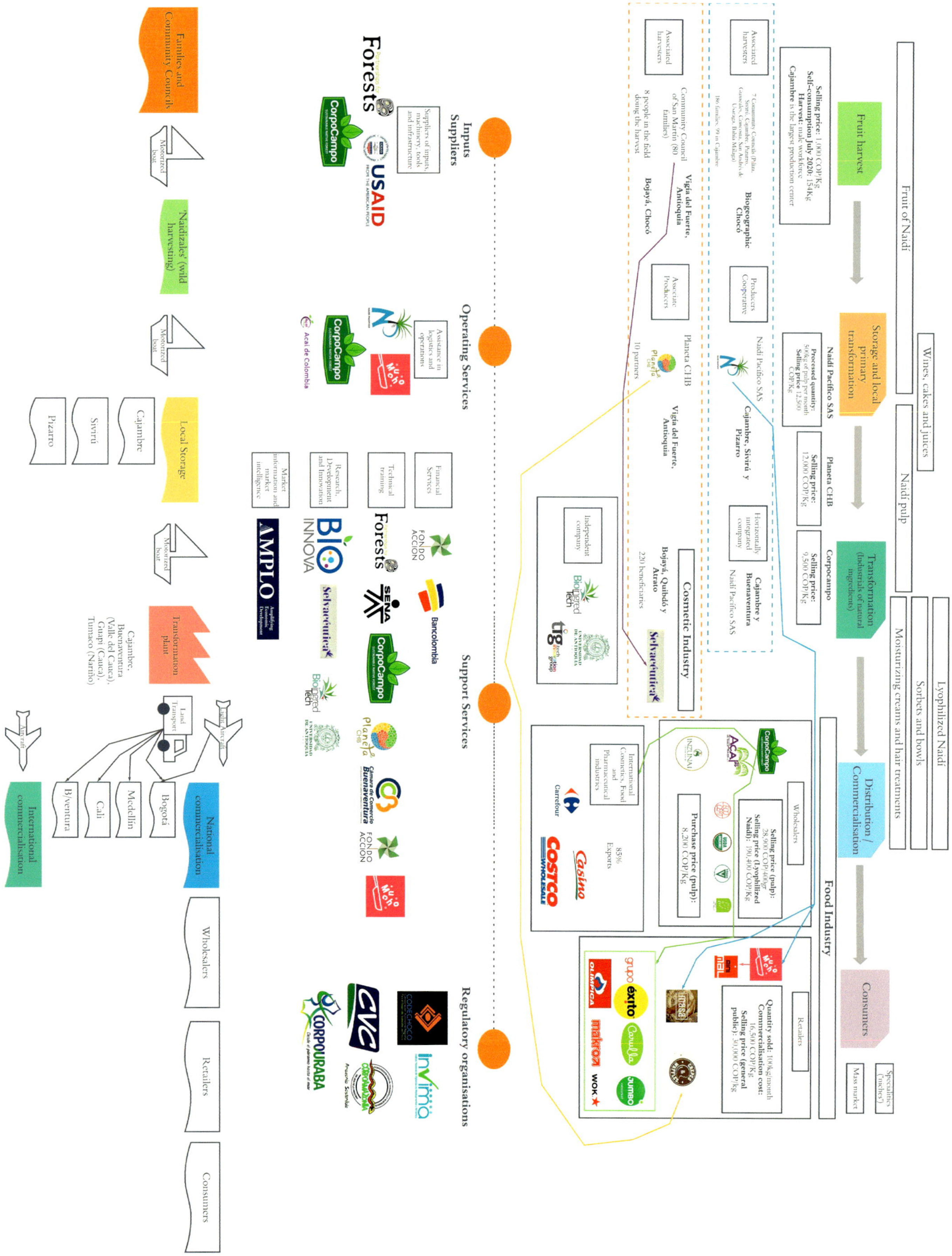

FIGURE 1. Mapping the value chain of the natural ingredients of the naidí palm.

stakeholders. A distinction is made between the different direct actors in the chain, those involved in the harvesting, processing, and marketing of products. In parallel, we consider indirect actors, who are not exposed directly to risk attached to the product and may provide ancillary services to chain actors in different links, and regulatory agencies (CIAT, 2014). Then, we explore some of the value chain's complexity and opportunities, which could help the development and commercialisation of natural ingredients, assessing challenges and opportunities. This approach can be viewed as a tool for identifying constraints, key public policy issues, and assistance interventions (Belcher & Schreckenberg, 2007; Rich *et al.*, 2009).

Data collection and analysis

We conducted seven semi-structured interviews with direct and indirect actors in the value chain (Table 1; https://figshare.com/s/07441b0d412923c79d2a). We refer to direct actors as those directly involved in the harvesting, production and commercialisation of goods in the value chain, and conversely, to indirect actors as those who provide support and operational services to direct actors. We explored whether the distribution of the value of the products was equitable along the value chain; that is, whether the costs and benefits of investments are shared across actors in each link. Simple commercialisation margins and consumer price ratios (Marshall *et al.*, 2006) were taken as a reference for measuring benefit distribution and equity within the value chain.

First, the calculation of the commercialisation margin was obtained from the price received by each player in the value chain, from harvesters to consumers, using the following equation:

$$\text{Commercialisation Margin} = ((P_S - P_P)/ P_C)*100$$
(Equation 1)

where P_S is the sale price, P_P is the purchase price, and P_C is the final retail price.

Commercialisation margin depends on the final consumer price, where the numerator of the equation reflects the gross income of each link (sale less purchase price). The commercialisation margin reflects the proportion of the gross income per link and its share on the final price. Alternatively, the proportion of the final price taken by the different actors in the value chain (Marshall *et al.*, 2006) can be calculated from information on the product's final price (Equation 2).

$$\text{Consumer Price Ratio} = (P_S / P_C)*100$$
(Equation 2)

RESULTS

Fruit of the naidí palm is transformed first into frozen pulp and then into the lyophilised powder used in the food and the cosmetic industry (Figure 1). We recorded a diversity of health care products, such as body lotions and shampoos. Furthermore, within the value chain, we assessed the direct stakeholders, their geographical location, the flows of products and their relationships (black arrows), as well as physical and monetary units of exchange and production (boxes with black frames). In parallel, indirect stakeholders, despite not taking direct action within the stages of production, transformation, and commercialisation, act as "enablers" of the entire chain (Figure 1). Finally, we looked at the transport used in each link (Figure 1).

Actor assessment

Link 1: Fruit harvest

In the municipalities of Bajo Baudó (Chocó) and Buenaventura (Valle del Cauca), naidí fruit is wild harvested from a total area of 56,000 hectares. This activity supports the livelihoods of 186 households who harvest the raw material for self-consumption and local sale. Currently, the Community Council of Río Cajambre concentrates most of the production and families (totalling 99), as well as the storage and processing of naidí pulp. The harvest of the fruit is organised by families and the Community Councils that own Naidí Pacífico SAS. This community-based company schedules the dates to receive the fruit, notifying families to take them to the collection centres, paying approximately 1,000 pesos per kilogram. Transport to the interior of the collective territory is 100% by motorboat, owned by the Community Councils and through Naidí Pacifico SAS, the boat circulates in the harvest areas.

In parallel, Planeta CHB SAS already has granted harvesting permits in the northern part of Chocó and Urabá antioqueño, harvesting naidí fruit from around 26,000 hectares of 'naidizales', benefiting 50 families. Planeta CHB also pays around 1,000 pesos per kilogram of fresh fruit. For this company, wild harvest is not sufficient to meet the demand. Hence this is complemented with the cultivation of the palm in agroforestry systems. On the other hand, Corpocampo buys a kilogram of naidí's fresh fruit for 1,000-1,300 pesos; price depends on the cleanliness, selection, the degree of ripeness of the fruit among other qualities. They also commit harvesters to generate conservation agreements and 'zero deforestation'; for example, this company requires that a percentage of fruits are left to guarantee food for other species that naturally disperse the seeds of the naidí palm, which could represent about 20% of the forest production.

Harvesting is carried out mainly by men since it is time-consuming being in the 'naidizales'. The soil is very soft, and harvesters work all day with the water at the waist and abundant insects and moisture. It is also an activity that requires climbing the palms with a safety team (harness), removing the bunches, and lowering them to the floor on a string. On the floor, there is usually an assistant who receives the bunch and takes it to a basket. Naidí is a species that blooms and fruits all year round, with peaks between January-February and May-July, offering a harvest of around 4 tons/peak of naidí pulp (data only available for Buenaventura and Bajo Baudó).

Link 2: Storage and local primary transformation

Primary transformation is undertaken in the territory by Naidí Pacífico SAS, which ensures that the naidí harvest can be stored all year round thanks to the installed capacity for harvesting, processing, and freezing. Between January and July 2020, Naidí Pacífico SAS managed to produce 3.6 tons of pulp, mainly from the Community Council of Río Cajambre (Buenaventura). However, in 2020, not even 80% of the harvested fresh fruit produced by this Community Council was transformed into frozen pulp. In parallel, there are two other collection and storage centres in Sivirú (Chocó) and Pizarro (Nariño), with smaller freezers. Soon, with the Partnerships for Forests (P4F) project and *Fondo Acción*, these two collection centres also plan to start processing plants. Moreover, an industrial scaling phase should soon be underway, with in 10 tons of pulp production forecasted for 2021. Naidí Pacifico SAS sells the kilogram placed in Buenaventura at 12,500 pesos; however, if the client wants the product in other cities of Colombia, an additional value (transport and packaging) is charged. In parallel, there are other local alternatives for processing the naidí fruit. In the Community Council of Río Cajambre, '*arrechón*' or naidí wine is produced by independent producers. The production and distribution of this product are already legalised, participating in the farmers market of Universidad Santiago de Cali and the music festival of the Pacific *'Petronio* Álvarez' (Cali, Valle del Cauca).

For Planeta CHB SAS, the processing factory of the company is in the urban centre of Vigía del Fuerte (Antioquia). There, the mature fruit of the naidí, the tasks of pulping, packing, and freezing are carried out. In this process, upright freezers are available, which, due to space limitations, have a total freezing capacity of 3.6 tons of pulp. Most of the employed staff are women. Planeta CHB sells a kilogram of frozen pulp to Bogotá and Medellín for 12,000 pesos, and locally, for 10,000 pesos. Likewise, Corpocampo's processing plants have been built in Buenaventura (Valle del Cauca), Tumaco (Nariño), and La Hormiga (Putumayo) since the Guapi (Cauca) plant had to close due to public order issues.

Link 3: Commercialisation

Local commercialisation is concentrated in the municipality of Buenaventura (see Table 2), which inherited the traditional use of the species for daily consumption as 'pepiao', an exotic traditional recipe that soaks the palm fruit in warm water for manual maceration until a liquid of thick consistency is obtained. Finally, local inhabitants add milk powder, sugar or honey and a pinch of salt. In this part of the country, several independent producers transform the pulp into juices, 'bolis' (traditional frozen juices and ice creams), cakes, and wines.

At the national level (Table 3), the naidí pulp is commercialised by retailers in Bogotá and Medellín, where it occupies a niche market, characterised by high-end products (e.g., organic food and novel foods). In parallel, wholesalers focus on selling the lyophilised naidí in international markets.

TABLE 2. Local-level commercialisation circuits.

Community-based companies	
Naidí Pacífico SAS	Planeta CHB SAS
The product (pulp) is transported by motorised boat from Cajambre to Buenaventura (the distribution centre), where the company's main demand comes from hotels, restaurants, and cafes. This distributor type is very important for Naidí Pacifico SAS since 80% of demand is concentrated in this sector. According to Fondo Acción, in 2020, 20–25% of the pulp was consumed in Buenaventura, by people and restaurants. In this municipality, transport is easier as it is distributed by motorbike, door to door. In Cali, they rely on family or friends to distribute the product since it is only two hours by land.	In Medio Atrato (both Vigía del Fuerte and Bojayá), local communities have no reports of historical consumption. However, Planeta CHB SAS sells the frozen pulp to local communities that come to the transformation plant and have a local dispatch system. As in the case of Naidí Pacífico SAS, this has allowed Planeta CHB SAS to recover sales in this time of pandemic.

Source: Own elaboration based on Rojas *et al.* (2021).

TABLE 3. National commercialisation circuits.

Community-based companies		Retailers	Wholesalers
Naidí Pacífico SAS	Planeta CHB SAS	MUCHOCOL	Corpocampo
This company sells approximately 500 kilograms of frozen pulp per month through its sales division, mainly to Bogotá, with several customers, among which are retail distributors such as MUCHOCOL and a restaurant called 'Masa'.	Since 2017, this company has delivered frozen naidí pulp to Medellín by light aircraft, which courier services distribute. Each aircraft can carry 5,000 plastic bags packed in different sizes (200 g, 500 g, 1 kg and 5 kg). Moreover, in April 2020, they were willing to expand their market to the city of Cali; the company had everything ready, but the restrictions of the pandemic did not allow it. In the short term, Planeta CHB's goal is to have a distribution point in Medellin, Bogotá, and Urabá. At this point, they have not been able to have a retail distribution channel in supermarkets and speciality stores because they are still organising all the information required on the labels and the brand.	MUCHOCOL buys approximately 100 kilograms per month of frozen pulp from Naidí Pacífico SAS, distributed equally to Medellín and Bogotá. The total is distributed as follows: 60 kg in presentations of 500 g and 40 kg in presentations of 100 g. Before the pandemic, MUCHOCOL bought the frozen pulp in Buenaventura and made shipments in cargo planes. However, during 2020, they have sought alternatives to make shipments from Buenaventura, making an agreement with the Magdalena SA fleet to distribute to Bogotá (12 hours journey). This company buys pulp from Naidi Pacifico SAS, which in Bogotá represents a cost of 16,500 pesos per kilogram, which is 73% higher than the purchase price of Corpocampo, and 101% higher than the purchase price at Inzunai. MUCHOCOL adds packaging and logistics costs to the selling price until the producer has an income higher than the minimum wage. In turn, MUCHOCOL sells frozen pulp for 20,444 pesos/kg to members and 30,000/kg to the general public.	Corpocampo has specialised retailers in Bogotá, Barranquilla, Cartagena, Cali, Medellín, Bucaramanga and Ibagué. Before the pandemic, they distributed their product line in 100 restaurants; now, they distribute their products directly to consumers through their website and WhatsApp. In addition, it is a wholesale company that distributes to other wholesalers such as Éxito, Jumbo, Olímpica and Makro. In 2014, they began exporting a range of products, concentrating on the European (France), Mexican, Chilean, Japanese and Lebanese markets. In this sense, Corpocampo is one of the few actors in the chain that has issued organic certifications (ECOCERT and USDA Organic). In turn, Corpocampo has the most diverse and value-added product line. They have a line of sorbets, frozen pulp (24,083 pesos/kg), lyophilised powder (190,400 pesos/kg), capsules, jams and jellies, as well as honey from *Melipona* colonies and related products.

Source: Own elaboration based on Rojas *et al.* (2021).

Link 4: Consumers

a. MUCHOCOL: establishes its demand through campaigns of responsible consumption in social networks to strengthen or gain consumers' loyalty. Within their discourse, they emphasise social justice: "Communities are represented on a plate."

b. P4F: given that consolidated producers such as Naidí Pacifico SAS or Planeta CHB SAS have not yet come to produce under economies of scale, niche customers (specialities) are willing to pay a premium for the social and environmental history of these products.

c. Naidí Pacífico SAS claims that consumption trends in large cities are based on the naidí bowl (pulp with a variety of nuts and fresh fruits) and fruit juice.

d. Corpocampo affirm the importance of the consumer in the value chains and how these actors can influence defining and encouraging different living methods for the peasant, indigenous and Afro-Colombian communities that harvest naidí fruit.

Supporting actors

1. Inputs, machinery, tools and infrastructure

    a. US Agency for International Development (USAID): through the project 'Territorios de oportunidad Guapi-Timbiquí', they offer machinery, tools, and infrastructure.

    b. P4F: make investments in equipment, and these are concentrated in the Community Council of Río Cajambre (Buenaventura). Emphasis is placed on expanding freezing capacity in Buenaventura (Valle del Cauca) and Pizarro (Nariño).

    c. Corpocampo: supplies organic fertilisers and plant material (seedlings and seeds).

2. Logistics and operations

    a. MUCHOCOL: assume shipping costs (packaging and distribution) of 4,000 pesos/kg to distribute frozen naidí pulp from Buenaventura to the main cities of Colombia.

    a. Corpocampo: guarantee fixed purchase prices and the collection of the naidí fruit 'door to door' from harvesters.

3. Finance, technical and research

    a. MUCHOCOL and Bancolombia: 'venda y aprenda' platform to strengthen digital markets, business, and personal finance issues; technological development was an incentive for producers to commercialise their products, generate their marketing and expand to other domestic markets. Training in invoicing and maintenance of production and sales records, as well as in matters of verification of equipment and transformation processes.

    b. Fondo Acción: this institution has worked since 2016 in the Colombian Pacific under a USAID umbrella project (BIOREDD), developing the entire carbon bond line as an incentive to conserve 'naidizales'. In this region, Fondo Acción supports not only Naidí Pacifico SAS but mainly the Community Councils within the REDD+ projects, in components of governance and productive reconversion, among others.

    c. Corpocampo: empowers communities to make better use of the forest. Ensures that when the fruit is harvested, there is no palm heart extraction, and when the peaks of production are low, the palm growth is managed. This company has a field team dedicated to these tasks.

    d. P4F: funded the exchange of experiences between Naidí Pacífico SAS and Belém do Pará (Brazil), the State with the highest production and consumption of naidí in the world. This organisation also hired these experts to go to the Colombian Pacific with the aim that Naidí Pacifico SAS could optimise the harvesting processes (including technical data sheets), pulp processing, operations logistics, investments in equipment, and human capital. In parallel, P4F outsourced a consultancy to design a participatory marketing strategy, business plan, and training in social media management.

    e. AMPLO: a consultancy company hired by P4F that develops projects and attracts capital to fund them, as inputs to the business plan for naidí pulp. Supports and determines the selling prices of the pulp.

    f. Chamber of commerce of Buenaventura: according to Fondo Acción, this actor has been training leaders and the community in general to formulate business plans.

    g. Planeta CHB SAS: trains communities in protocols for the sustainable harvest and management of wild species. They established an agreement with the National Learning Service (SENA) in Apartadó (Antioquia) to carry out forestry training, emphasising good harvesting and processing practices. Two hundred twenty people from 34 communities have been trained.

    h. USAID: business plan development for naidí in 2016. However, interviews with actors in the value chain state that the business plan, specifically for the lyophilisate of the pulp made by Naidi Pacifico SAS, was never carried out.

    i. Selvacéutica: a technology-based company that produces NTFPs from the Chocó Biogeographic region for the natural cosmetics and phytotherapy subsectors. It has a line of hair products (nutrition and growth treatment) made from naidí fruit. The company has a community innovation component that contributes to developing competitive biodiversity-based products through the implementation of R&D and knowledge management strategies.

TABLE 4. Quantitative value chain analysis seeks to analyse the distribution of (monetary) value in each link of the value chain for natural ingredients of the naidí palm, in this case, the frozen pulp.

Value chain link	Price (COP $) / kg		Gross income (COP $)	Consumer price ratio (%) PULP	Commercialisation margins (%) PULP
Harvest and/or gathering	Purchase	0	1,000	3.3%	3.3%
	Sales	1,000			
Storage and processing (Naidí Pacífico SAS)	Purchase	1,000	11,500	41.7%	38.3%
	Sales	12,500			
Retail (MUCHOCOL)	Purchase	16,500	13,500	55.0%	45.0%
	Sales	30,000			

Source: Rojas *et al.* (2021).

4. Regulatory
   a. Corpoamazonía: Corpocampo states that this institution has enabled processes in Putumayo, as has Vision Amazonia (an initiative from the Ministry of Environment), to facilitate the acquisition of permits for the sustainable harvest and use of Non-Timber Forest Products (NTFP).
   b. Corporación autónoma del Valle del Cauca (CVC): technical and logistical support to Naidí Pacífico SAS in obtaining permits to sustainably use and harvest NTFP for commercial purposes.
   c. CODECHOCÓ y CorpoUrabá: technical and logistical support to Planeta CHB SAS in obtaining harvesting permits.

Quantitative analysis of the value chain

Quantitative value chain analysis seeks to analyse the distribution of (monetary) value in each link (Marshall *et al.*, 2006). We made this analysis for the value chain of the natural ingredients of the naidí palm, considering that in the case of the frozen pulp, retailers such as MUCHOCOL fix the product's final price. In this regard, the small-scale harvesters capture only 3.33% of the price of the final product (kilogram of frozen pulp), representing a low distribution of profit to harvesters (Table 4). However, small-scale harvesters and Community Councils own Naidí Pacífico SAS, which manages to obtain a 41.67% share in the final price. Future research should explore how dividends and profits are distributed within the Community Councils to consider a more robust approach. Finally, and as MUCHOCOL stated in the interview, the highest margins of commercialisation, and therefore of profits, are concentrated in the retail link (45%).

The purchase price of the naidí fruit shows that any hidden costs of harvesting to communities have not been included. It was assumed that this purchase price was $0 (*sensu* Valderrama, 2011). However, no associated costs were considered related to the request for sustainable harvest permits or investments in labour, equipment, and tools. Also not accounted for are the diverse intrinsic and non-monetary values of the naidí fruit for local communities. Additionally, the gross income (difference between the purchase and sale price in each link) may not reflect the costing reality of how the naidí fruit is transformed into pulp since approximately 3 kg of fruit is needed to generate 1 kg of pulp.

DISCUSSION

Governance of the value chain

The Colombian Pacific is intimately intertwined with the first links of the value chain. In this region of Colombia, 90% of the population is Afro-descendant, 60% of its inhabitants live in urban areas, and 40% are along some 240 rivers (Grueso *et al.*, 1997). Although Afro-descendant communities have inhabited these lands for centuries, the right of formal ownership via Community Councils[1] was not recognised as the highest authority over these territories until the Constitution of 1991 (Peña *et al.*, 2017). The ownership of the land in this region of Colombia belongs to the Community Councils and not to individuals. However, in some cases, within the collective property, there might exist private property boundaries. In addition, within the Community Councils, like property ownership, decision-making is undertaken collectively by the community. In the context of the naidí value chain, for Naidí Pacifico SAS, the regulation of collective ownership of the land has allowed biodiversity-based products to be harvested, transformed, and commercialised. This fact is also supported by Decree

[1] Currently in Colombia there are 176 Community Councils with collective titles, representing 5% of the national territory.

690 of 2021, stating that only the owners of the land or those acting on response to a request made by authorisation of the owners have a valid right to sustainably use wild flora and NTFP. This means that third parties that are not part of the Community Councils cannot extract biodiversity-based products from these territories for commercial purposes. In this regard, both Naidí Pacífico SAS and Planeta CHB SAS were founded by members of different Community Councils. Therefore, the articulation of processes and the approval of commercial initiatives by the communities has been inclusive from the beginning. Besides, the request for sustainable harvest permits for wild flora to the corresponding local environmental authorities and the socialisation of productive alternatives are developed more organically.

Two key aspects can be highlighted by further analysing the governance of naidí's value chain. For example, the main suppliers of pulp are characterised by operating under a community-based business system. This kind of structure makes relationships and trust fundamental, revealing a horizontal integration. It means that the first links in the naidí value chain can be characterised by being relational: internal transactions are made possible by high levels of trust. However, in the next links of the value chain, where the transformation of the naidí fruit into a natural ingredient occurs, local producers begin to rely heavily on different commercialisation channels (retailers or wholesalers). Conversely, Naidí Pacífico SAS and Planeta CHB depend on very few buyers to satisfy their demand. This type of relationship makes community-based vulnerable to potential volatilities, such as the effects of COVID-19, because they do not have a broader demand. In addition, this also causes them to relate with distributors in a vertical or managerial manner due to their low bargaining power.

Livelihoods engaged in the value chain

The fruit of the naidí palm has great importance not only for local economies but also for the livelihoods of rural communities that inhabit the Chocó biogeographic region. According to the interviews undertaken, the cultural roots around the fruit are part of the essence of the people who inhabit the forest. Since childhood, local communities have eaten the fruit without its being transformed into pulp in the form of 'pepiao' (cooked naidí fruit mixed with powdered milk and sugar), traditional in local consumption and relevant for food security. The fruit has also been consumed historically and traditionally in juices and in the preparation of 'viche', which is a fermented drink based on the naidí palm's fruit. Similarly, the consumption of naidí in several communities has been used for medicinal purposes by the 'comadronas' (midwives) for the consumption of pregnant women and to treat anaemia and joint pains.

The Community Councils that makeup Naidí Pacifico SAS know very well the properties of the fruit (high content of antioxidants and vitamins), as well as the positive effects of adding it to the diet for personal health, being one of the main products of the family basket during harvest peaks. The sustainable use of the natural ingredients of the naidí palm can be considered substitutes for illicit crops through the guaranteed purchase of raw materials and the generation of complementary income in agroforestry systems, which contribute to the food security of the farms.

Additionally, in the municipalities where Planeta CHB is located (Vigía del Fuerte and Bojayá), the naidí palm has been used to construct housing and in traditional fishing techniques. However, neither the heart's palm nor pulp has been part of the diet in this region, despite the more than 150,000 hectares in these two municipalities of natural forest dominated by naidí. Faced with this situation, the company has started a social appropriation process so that, in hand with the sustainable harvest, local inhabitants begin to consume it. Notwithstanding, other livelihoods of the Medio Atrato region cannot be denied. Income from the different traditional economic activities such as agriculture in the Atrato River basin, fishing in the rainy season, mining in the upper part of the Atrato River, and timber harvesting are also of great importance in this area because they are used to purchase food such as bananas, meat, oil, soft drinks, and potable water from the north of Urabá.

Limitations

Colombia has approximately 55 million hectares of forest. However, according to the Partnership for Forest (P4F) (Table 1), less than 1% is used sustainably. This situation is not different for the Colombian Pacific region, which has historically been very vulnerable in terms of limited infrastructure, limited geographic access and public services, and has a high level of poverty (Grueso *et al.*, 1997). These, among other elements, represent limitations for developing the NTFP value chain, and in the case of naidí, they are summarised below.

- **High logistics costs:** the lack of road infrastructure and the difficulty in accessing the naidizales (mainly by motorboat) make some areas impossible to use. These situations make transportation costs in the Colombian Pacific very high, making the products that come from Chocó more expensive.
- **Supply of public services:** in municipalities such as Cajambre, public services are limited, which increases the production costs of maintaining the cold chain (temperature control to maintain product quality and safety) and makes it difficult to invest in other areas such as Good Agricultural Practices certifications (GAP).
- **Incomplete population data for the species:** projections cannot be made to guarantee that the process of use and management of the palms are sustainable and that the practices are standardised.
- **Local knowledge about the palm:** for Corpocampo, one of the great limitations in the use of the naidí is the lack of knowledge of the local populations about the palm. For example, they do not know its value and, therefore, they cut the palm as a means of subsistence.

- **Armed conflict:** some groups still charge bribes (called "vaccines") to extract raw materials from the port of Buenaventura, which directly affects the distribution channels.
- **Pressure from other economic activities:** the extraction of wood, a practice carried out for more than 200 years according to Planeta CHB, and illegal mining in the south of the Atrato river, among others, put pressure on the use of the naidí palm.
- **Limited domestic market:** the low cultural appropriation at local and national levels means that the production of companies such as Planeta CHB is not continuous over time, which makes it difficult to maintain contracted personnel throughout the year, forcing the employer to hire new workers each time, and repeat training them.
- **High certification costs:** due to the conservationist vision of forest management, use protocols and permits for the use of NTFPs could cost between 25 and 30 thousand dollars (in P4F estimates). This amount is not affordable for local communities. The same happens with other regulations, such as the work at height regulations.
- **Discontinuity of cooperation projects:** Subsidies and incentives remain essential to guarantee value chains because the commercialisation volumes are insufficient. Without cooperation projects, community-based companies had difficulties providing technical training, which not only affects production but ends up demotivating the people involved, and new personnel need to be hired.
- **COVID-19 pandemic:** deeply affected community enterprises, many of which already depended on a single retailer or wholesaler. People working both in the harvest and in the plant were left without the opportunity to work.

Success factors

Although the value chain for naidí pulp is still in the process of consolidation and commercialisation, several success factors are highlighted here that can be replicable for other natural ingredients. For instance, the differential factor has been the human factor, i.e., the relationship of trust between small-scale harvesters and food producers. This has allowed the creation of community-based organisations under joint-stock companies. In this regard, the Naidí Pacifico SAS is a successful example of how a community-based company, owned by Community Councils (hence families), contributes to local economic growth and territorial development. In addition, the inclusion of diverse segments of the population (youth, women, elderly, and ex-combatants) has been vital to boosting this value chain. Other actors in the value chain, such as Planeta CHB, argue that the needs of local inhabitants must be prioritised. Then later, it is possible to create conservation awareness on the resource since it has value, and, therefore, people take care of it.

On the other hand, Corpocampo, despite not being a community-based company, also relies on local communities for the provision of the naidí pulp and canned palm hearts. This company has encouraged the constant support of local communities to generate awareness of the forest's intrinsic value and to promote social inclusion. Hence, working with local communities is an enabling condition of the value chain that, in the case of Corpocampo, generates sustainability in the supply of raw materials and supports the monitoring and maintenance of the harvest areas. In addition, the company establishes conservation agreements with farmers to undertake forest enrichment activities in degraded areas. This system not only encourages the naidí value chain from its raw material but also gives a sense of belonging to the employees, helping to preserve nature and sustain their livelihoods. The scale and scope of Corpocampo show how incentives within the value chain can foster collaborations within local communities and can create a clear and common purpose in relation to the commercialisation of the naidí palm.

Most of the community-based companies addressed in this study emphasised that for these initiatives to be successful, they must offer concrete alternatives of work and complementary income to families at risk of falling into illegal or extractive economies. Therefore, families and individuals should be offered alternatives that are equally profitable, allow the conservation of biodiversity, and to help protect the forest. In this sense, it is essential that downstream value-chain actors downstream value-chain actors guarantee the supply of raw materials. In addition, the assistance of NGOs, national educational services, companies, national public programmes and international cooperation for harvesting, processing, scaling-up and marketing products and biocultural assets are key contributing factors for the sustainability of the value chain. Finally, another successful factor has been large companies leading the market development of the products. In this regard, this kind of company has the financial assets and resources necessary to develop sustained long-term actions and the capacity to share risks with small-scale producers.

Opportunities

In general, diverse opportunities were identified for scaling up the value chain of the naidí pulp by the interviewed actors. Some of these opportunities include the following approaches or milestones:

- **Demand-led approach:** in the traditional development discourse, assistance programs have usually focused on technical training for producers, forgetting demand consolidation; that is, market creation. Therefore, community-based companies such as Naidí Pacífico SAS and Planeta CHB must take advantage of the market development process that Corpocampo and Crepes and Waffles have done in the national and international markets.
- **Fund training:** it is important to implement agro-biodiverse systems and to develop business capacities in the territories to increase the bargaining power of local communities to access funding. Agro-biodiverse systems such as Agroforestry employing the naidí palm

as a pioneer species for natural regeneration are key with other cash crops that support food security.

- **Development of logistic chains:** subsidise the expansion of cold chains and storage centres to increase production capacity under the maximum yield established by the protocols for sustainably harvesting the fruits of the naidí palm.
- **Reach economies of scale:** not only focus on niche markets, but develop productive and technological capacities to reduce price, reach more customers and be more competitive in the market.
- **Community-based monitoring:** although project execution time is relatively short, monitoring is an important activity to incorporate environmental authorities and the community to generate alternative sources of income and sustainability in the initiatives.
- **Further efforts in research:** implement demonstration plots to determine potential and improve the management of the naidí palm and its NTFP.
- **Development of local markets and social appropriation of the species:** more educational processes are needed to socialise the nutritional virtues of naidí and pulp consumption. For instance, roadshows can be an alternative way of informing rural and urban communities about the naidí palm. In this regard, the promotion of gastronomic tourism around this species is also key.
- **Diversify products and clients:** not relying on a single customer for distribution and diversifying production, for instance, the use of bioproducts of the naidí palm such as fungi that grow in the compost and fermentation of the shell of the palm hearts or residues of the fruit after the pulping process.
- **Foray into the international market:** the transformation of naidí pulp to functional ingredients such as lyophilised powder has great potential for exports to countries such as the United States (State of California), Mexico, and Chile. Therefore, Procolombia should deliver aid to small (community-based) producers and entrepreneurs to access international markets.

CONCLUSIONS

The historical incidence of the armed conflict, extractive and illegal economies, a lack of public services and limited access to some areas weigh on the value chain of the naidí palm in Colombia. Until alternatives to dignified survival are guaranteed, local inhabitants will not be able to decide how to sustain their livelihoods, especially in areas where food dependence and poverty are very high. The natural ingredients of the naidí palm offer a real economic alternative for the local communities and have the potential to relieve the pressure from other extractive and illegal economic activities, to provide additional income and to diversify the food security of the inhabitants of the 'naidizales'. However, the pathway for commercial recognition and integration still has a long way to go, as the value chains of the naidí and other NTFPs have not yet been fully appreciated by the national government or incorporated into the country's business policy.

Conversely, the lack of market expansion and the low cultural ownership of the naidí palm at a national level have been decisive factors in generating higher sales, and consequently, better working conditions, or in the search for certifications. Ultimately, the demand is challenging, as the harvest area is approximately 300 thousand hectares, which is still not all currently accessed because commercialisation is still low. Moreover, our analysis revealed that the value chain of the natural ingredients from the naidí palm is not profitable yet unless it receives aid or subsidies. Small-scale producers depend highly on external funds, mainly through international cooperation. Thus, their income generation remains economically unsustainable. More studies are needed to improve the quality control standards of products and to carry out long-term economic, financial and sustainability analyses. Similarly, it is still impossible to achieve a fair and equitable distribution of profits without an in-depth analysis of production costs and considering the scale of businesses. For example, companies like Corpocampo operate under economies of scale, while other companies operate at a small scale. The higher prices may reflect market inefficiencies.

However, it is possible to take actions in the short- and medium-term, such as: 1) upgrading the expansion of cold chains and storage centres; 2) fostering incentives for the implementation of agroforestry systems and further investigation on hidden harvesting costs; 3) developing strategies for promoting local diets and cultural acceptance of the species, including investigating the cultural practices; and 4) transfering technology and business capabilities for the commercialisation of natural ingredients of the naidí palm. At the same time, complementing the harvesting of the naidí from the wild in the forest with its cultivation through the development of agroforestry systems would go a long way towards stimulating bioeconomic growth (Papell *et al.* 2020). More in general, companies in Colombia should be encouraged to focus on the cultivation of native species rather than non-native major crops.

Finally, to boost sustainable value chains of natural ingredients in Colombia, the focus should be concentrated on small landscape units to generate diverse mechanisms of self-consumption and complementary sources of income derived from the high diversity of native plant species. Likewise, it would be a mistake to prioritise only the value chain of the naidí pulp while ignoring that families traditionally consume a variety of products, such as *chontaduro, borojó, vanilla,* and *milpesos*, and rely on other productive activities, such as fishing, the legal and sustainable use of wood and ecotourism. In addition, the Covid-19 pandemic highlighted several innovative uses of the naidí pulp in diverse economies, promoting local commercialisation and consumption. In this regard, community livelihoods are compatible with sustainable value chains of natural ingredients, as these fit within the traditional local practices.

References

Albrecht K, von Braun J, Ettling S (2015) Bioeconomía: nuevas oportunidades para la agricultura. *CEPAL - Serie Desarrollo Productivo* 200. 47p.

Belcher B, Schreckenberg K (2007) Commercialisation of non-timber forest products: a reality check. *Development Policy Review*, 25: pp. 355–377. doi: 10.1111/j.1467-7679.2007.00374.x.

Biointropic (2019) Tendencias de ingredientes naturales para la industria cosmética. Available at: https://www.ccc.org.co/inc/uploads/2019/03/7.-Tendencia-de-Ingredientes-Naturales-en-la-Industria-Cosmética-Biointropic.pdf.

Bracco S, Calicioglu O, Gomez San Juan M, Flammini A (2018) Assessing the Contribution of Bioeconomy to the Total Economy: A Review of National Frameworks. *Sustainability*, 10(6): 1698. doi: 10.3390/su10061698

Centro Red de Innovación (2013) Desarrollo tecnológico de la industria de ingredientes naturales de aplicación cosmética y de aseo a partir del aprovechamiento de la biodiversidad del país. 20p.

CIAT (2014) Metodología LINK: una guía participativa para modelos empresariales incluyentes con pequeños agricultores Versión 2.0 Available at: https://cgspace.cgiar.org/handle/10568/49607.

CONPES (2018) Política de Crecimiento Verde. Bogotá, Consejo Nacional de Política Económica y Social (Documento CONPES 3934). Available at: https://colaboracion.dnp.gov.co/CDT/Conpes/Económicos/3934.pdf.

Diazgranados *et al.*, 2021. Rincón Bermúdez *et al.*, 2009. Stockholm Environment Institute, 2020.

Dietz T, Börner J, Förster J, von Braun J (2018). Governance of the Bioeconomy: A Global Comparative Study of National Bioeconomy Strategies. *Sustainability*, 10(9):3190. doi: 10.3390/su10093190

DNP (2018) Política de Crecimiento Verde (Documentos CONPES 3934), Departamento Nacional de Planeación. Bogotá, D.C: Departamento Nacional de Planeación. 114p. Available at: https://colaboracion.dnp.gov.co/CDT/Conpes/Económicos/3934.pdf.

Ellis F, Allison EH (2001) The livelihoods approach and management of small-scale fisheries. *Mar Policy*, 25(5):377–388.

Fondo de Biocomercio (2009) La cadena de valor de los ingredientes naturales del Biocomercio para las industrias farmacéutica, alimentaria y cosmética. Bogotá, Universidad de Bogotá Jorge Tadeo Lozano. 221p.

Grueso L, Rosero C, Escobar A (1997) El proceso organizativo de comunidades negras en Colombia. *Ecol Polít*, 14:47–64.

Hobbs J, Cooney A, Fulton M (2000). Value Chains in the Agri-Food Sector: What are they? How do the work? Are they for me? Department of Agricultural Economics, University of Saskatchewan, Saskatoon.

Lundy M, Gottret MV, Ostertag C (2007) Participatory Market Chain Analysis for Smallholder Producers, Good Practice Guide 4. Cali, Centro Internacional de Agricultura Tropical. 130p.

Marshall E, Rushton J, Schreckenberg K (2006) Practical tools for researching successful NTFP commercialisation: a methods manual. *CEPFOR report*, 72p.

Marshall E, Schreckenberg K, Newton AC (2006) Commercialisation of Non-timber Forest Products: Factors Influencing Success. Lessons Learned from Mexico and Bolivia and Policy Implications for Decision-makers. Cambridge, UNEP WCMC. 140p. doi: 10.15684/formath.10.219.

Misión Internacional de Sabios (2019) Informe de La Misión Internacional de Sabios 2019 Por La Educación, La Ciencia, La Tecnología y La Innovación. Misión Internacional de Sabios 2019. Gobierno de Colombia. 33p.

Moñux D *et al.* (2018) Estudio sobre la Bioeconomia como fuente de nuevas industrias basadas en el capital natural de Colombia No.1240667. Fase I., BIOINTROPIC 120p.

OECD (2009) The Bioeconomy to 2030 - Main Findings and Conclusions. Organización para la Cooperación y el Desarrollo Económico. Paris, OECD International Futures Project. 18p. Available at: https://www.oecd.org/futures/long-termtechnologicalsocietalchallenges/42837897.pdf

ONUDI (2015) Dinámica comercial y económica del sector cosméticos e ingredientes naturales en Colombia. Available at: https://fdocuments.net/download/dinamica-comercial-y-economica-del-sector-cosmeticos-e-ingredientes-naturales

Papell M, Domínguez JD, Medina F (2020) Estudio de oferta y demanda nacional de ingredientes naturales, en Presentación enviada por Toro, Camila (2020), mensaje para participantes de Mesa de Ingredientes Naturales, 18 de junio. Bogotá, Colombia: Cluster Development. 49p.

Peña X, Vélez MA, Cárdenas JC (2017) Collective Property Leads to Household Investments: Lessons From Land Titling in Afro-Colombian Communities. *World Dev*, 97: 27–48. doi: 10.1016/j.worlddev.2017.03.025.

Rich KM *et al.* (2009) Concepts, applications, and extensions of value chain analysis to livestock systems in developing countries. *International Association of Agricultural Economists Conference*, 1:16–22. doi: 10.21125/edulearn.2017.2707.

Rincón Bermúdez S, Salcedo JB, Calderón JT (2009) Lineamientos guía para la evaluación de criterios de biodiversidad en los estudios ambientales requeridos para licenciamiento ambiental. Instituto Alexander von Humboldt, Instituto de Estudios Ambientales - Universidad Nacional de Colombia. Opciones Gráficas Editores Ltda. 126pp. ISBN: 978-958-8343-40-2.

Rojas T, Cortés C, Noguera M, Ulian T, Diazgranados M (2021) Evaluación del estado de los desarrollos bioeconomicos colombianos en plantas y hongos. Royal Botanic Gardens, Kew; Instituto de Investigaciones en Recursos Biológicos Alexander von Humbold. 125p. doi: 10.34885/201

Staffas L, Gustavsson M, McCormick K (2013) Strategies and policies for the bioeconomy and bio-based economy: an analysis of official national approaches. *Sustainability*, 5(6):2751–2769. doi: 10.3390/su5062751.

Valderrama N (2011) *Value Chain Investigations on Four Colombian Palm Species*. MSc. Thesis. School of Forest Science and Resource Management, Technische Universität München. 113p.

Guacamaya superba (Flor de Inírida de Invierno) in the white sand savannas of Inírida.
Mateo Fernandez Lucero

Chapter 10

A taxonomic summary of useful plants in Colombia

Mauricio Diazgranados[1]*

[1] Royal Botanic Gardens, Kew, UK

***Corresponding author:** m.diazgranados@kew.org

Keywords: biodiversity, bioprospecting, economic botany, ethnobotany, richness.

ABSTRACT

As part of the Useful Plants and Fungi of Colombia project, we compiled a comprehensive checklist of useful plants for this country. This chapter provides notes on the taxonomic coverage of this checklist, highlighting the richest plant groups in useful plant species, categories of use, and phylogenetic coverage. It also provides notes on the evolutionary origin of species and comparisons with the Colombian flora as a whole. Potential taxonomic gaps (i.e., missing use reports for plant groups containing useful species) are also identified. The checklist of useful plants comprises 7,472 species, 2,140 genera, and 258 families, representing all major plant groups. More than 50% of the species belong to 18 families. Most of the species are native (78.6%), and 6.4% are endemic to Colombia. At least 14% of the useful species are commercially cultivated, and 8.6% are introduced and naturalised. The richest families in useful plant species are Fabaceae (665 spp.), Asteraceae (418), and Poaceae (338). The richest genera in useful plant species are *Solanum* (88), *Inga* (84), and *Passiflora* (78). The categories of use with the most associated plant species are medicines (5,108 spp.), human food (3,806), and materials (2,363). Before this project, Colombia did not have any updated national list of useful plants, with only a few lists of species by use category (i.e., medicinal, edible, and fibres), often with conflicting taxonomy and incomplete geographic coverage. The annotated checklist published in this volume provides records for more than 4,000 useful species not included in any of the previous lists, integrating all use categories with updated taxonomy, covering the country's geographic extent, and linking the records with the expanded species profiles in Col*Plant*A.

RESUMEN

Como parte del proyecto Plantas y Hongos Útiles de Colombia, un equipo multinacional de investigadores compiló una lista de verificación completa de plantas útiles para este país. Este capítulo proporciona notas sobre la cobertura taxonómica de esta lista de verificación, incluyendo menciones de los taxones más ricos y categorías de uso, cobertura filogenética, notas sobre el origen de las especies y comparaciones con la flora colombiana. También se identifican posibles vacíos taxonómicos (es decir, reportes de uso faltantes de taxones que contienen especies útiles). La lista de verificación de plantas útiles comprende 7.472 especies de plantas con usos documentados, clasificadas en 2.140 géneros y 258 familias, distribuidas equitativamente en la filogenia de plantas. Más del 50% de las especies útiles se clasifican en 18 familias. La mayoría de las especies reportadas son nativas (78,6%) y el 6,4% son endémicas de Colombia. Al menos el 14,4% de las especies útiles se cultivan comercialmente y el 8,6% son introducidas y naturalizadas. Las familias más ricas en especies son Fabaceae (665 spp.), Asteraceae (418) y Poaceae (338), mientras que los géneros más ricos son *Solanum* (88), *Inga* (84) y *Passiflora* (78). Las categorías de uso con más especies son medicamentos (5,108 spp.), alimentación humana (3,806) y materiales (2,363). Antes de este proyecto, Colombia no contaba con ningún listado nacional actualizado de plantas útiles; en cambio, había algunas listas de especies por categoría de uso (es decir, medicinales, comestibles, fibras), a menudo con taxonomía conflictiva y cobertura geográfica incompleta. La lista de verificación anotada publicada en este volumen proporciona registros para más de 4,000 especies útiles no incluidas en ninguna de las listas anteriores, integrando todas las categorías de uso, con una taxonomía revisada, cubriendo la extensión geográfica del país y vinculando los registros con los perfiles expandidos de especies en Col*Plant*A.

INTRODUCTION

Colombia is one of the world's most species-rich countries (Mittermeier *et al.*, 1997; Myers *et al.*, 2000; Raven *et al.*, 2020), holding a broad range of continental ecosystems, from tropical rainforests to mangroves, dry forests, savannas, deserts, mountain forests, and páramos (Galeano, 2016; MADS *et al.*, 2017). From sea level to perennial snow, the large range of elevations is reflected in the wide spectrum of climates and environments (Hooghiemstra & Flantua, 2019). Its tropical position grants high solar energy throughout the year, with reduced climate seasonality, mainly in terms of rainfall variation (Hurtado Montoya & Mesa Sánchez, 2015). The local biota relies on contrasting climate patterns with coasts on the Atlantic and Pacific oceans (Galeano, 2016).

TABLE 1. Diversity of plant species across four datasets: Plants of the World (PW), Useful Plants of the World (WCUP), Checklist of Plants of Colombia (CPC) and Checklist of Useful Plants of Colombia (CUPC). Both the CPC and the CUPC were developed by the Useful Plants and Fungi of Colombia project.

CATEGORY	PW	WCUP	%WCUP/PW	CPC	CUPC	%CUPC/CPC	%CUPC/ WCUP
Divisions	8	6	75.0	6	5	83.3	83.3
Classes	42	14	33.3	17	13	76.5	92.9
Orders	215	101	47.0	107	77	72.0	76.2
Families	952	433	45.5	394	258	65.5	59.6
Genera	14,717	6,737	45.8	3,583	2,140	59.7	31.8
Species	397,692	40,283	10.1	28,947	7,472	25.8	18.6

Colombia's geographic position at the confluence of South and Central/North America through the Isthmus of Panama facilitated past intercontinental dispersal of terrestrial animals and plants (Bacon *et al.*, 2015; O'Dea *et al.*, 2016), generating complex biotas such as the hyperdiverse flora of Chocó (Pérez-Escobar *et al.*, 2019). The above factors, together with its complex topography, including the three main Andean Cordilleras and numerous isolated mountain systems (Bernal, 2016), create a perfect scenario for speciation processes, resulting in very high biodiversity and endemism (Diazgranados & Barber, 2017; Flantua *et al.*, 2019; Rangel-Ch, 2015; Raven *et al.*, 2020). Colombia ranks second in the world in overall biodiversity and third in plant richness (Mittermeier *et al.*, 1997; Raven *et al.*, 2020), with at least 28,947 species based on Col*Plant*A (http://colplanta.org/), of which 93.7% are native and over 32.7% are endemic.

Colombia is also ethnically and culturally very rich, with mixed origins from the European colonisation, African descendants, immigrants from the Middle East, and the native indigenous inhabitants (DANE, 2019). At least 115 indigenous groups, with almost two million people, i.e., 4.4% of the country's population, have survived colonisation in the country (DANE, 2019). All of this contributes to a diverse cultural heritage manifested in how Colombians interact with nature and use plants. The history of colonisation is probably one of the reasons to explain the very long social conflicts that this country has had for centuries. Currently, Colombia remains one of the world's most unequal countries, where 42.5% of the population are under the line of monetary poverty, and 15.1% are under extreme monetary poverty, with inequality predominant in secondary cities and rural areas in conflict-affected regions (DANE, 2021). In 2019, the Royal Botanic Gardens, Kew, in collaboration with the Instituto de Investigación de Recursos Biológicos Alexander von Humboldt, began the Useful Plants and Fungi of Colombia Project, which aims to empower that country to turn its plant and fungal diversity into an economic resource for sustainable development, by documenting and conducting research into useful plants and fungi, broadly disseminating this knowledge, and promoting and supporting markets for high-value products based on their sustainable use (https://www.kew.org/upfc; https://in-colombia.org/).

Efforts to catalogue the useful plants of Colombia's territory can be traced back to the Chronicles of the Indies in the sixteenth century (Pardo-Tomás & López Terrada, 1993). During the Royal Botanical Expedition to New Granada (1783–1816), headed by José Celestino Mutis, detailed annotations of plant uses were added to the illustrations of species (Piedrahita, 2008). Humboldt's expeditions to the Americas (1799–1804) also included mentions of uses of plants (Humboldt & Bonpland, 1809 [1808]). In 1852, José Jerónimo Triana was appointed by the Colombian Government to create the first official inventory of useful plants for the country (Piedrahita & Lourteig, 1989). But it was Enrique Pérez Arbeláez, a century later, who cemented this path with his book "Useful Plants of Colombia", containing almost 2,000 species (Pérez Arbeláez, 1956). Since then, several publications have emerged, most focused either on specific use categories such as medicinal (Bernal *et al.*, 2011; García Barriga, 1975), edible (García, 2011; IAVH, 2014; López & García, 2021), fibres (Cadena *et al.*, 2007; Linares *et al.*, 2008) and timber (Cárdenas López & López Camacho, 2000; López Camacho *et al.*, 2014), or geographic areas [e.g., Amazon (Cárdenas López & López Camacho, 2000), Orinoco (Acero, 2005; Diazgranados & Moreno, 2020), páramos (Duarte-Abadía & Parra-Ortega, 2015)]. However, these sources often lack taxonomic consistency, and there was no recent consolidated list. Thus, as part of this project, a multinational team of researchers compiled a comprehensive checklist of useful plants for this country, based on previous databases and available literature (see the detailed methodology for the assembly, consolidation, and data cleaning in chapter 12). This chapter aims to provide a summary of the taxonomic coverage of this checklist, including indications of the richest taxa and categories of use, phylogenetic coverage, notes on the origin of species, and brief comparisons with the Colombian flora.

MATERIALS AND METHODS

This chapter analyses the taxonomic breadth of the Annotated Checklist of Useful Plants of Colombia (CUPC hereafter), available in chapter 12. The checklist resulted from the compilation of several data sources, taxonomic reconciliation, classification of reported uses, and data normalisation (for detailed methods, please refer to chapter 12). The CUPC was compared against the Checklist of Plants of Colombia (CPC hereafter), being updated as part of this project and available through Col*Plant*A (http://colplanta.org/). The CPC comprises 28,947 species, classified into 3,583 genera and 394 families (Table 1), ca. 3,000 more species than are currently reported by the Colombian Catalogue of Plants and Lichens, with 26,134 species (Bernal *et al.*, 2019). Results were also compared with the World Checklist of Useful Plants (WCUP), produced by RBG Kew, the most comprehensive checklist that exists today (Diazgranados *et al.*, 2020). It comprises 40,292 taxon names (including a few organisms frequently misidentified as plants, such as brown algae and nostoc), classified into three kingdoms (Plantae with 40,283 species, Chromista with eight species, and Bacteria with one species), six divisions/phyla, 14 classes, 101 orders, 433 families, and 6,737 genera (Diazgranados *et al.*, 2020).

The statistical analyses and figures were elaborated in R v.4.1.0, mainly using *ggplot2* v.3.3.5 (Wickham *et al.*, 2016) for graphical representations, and *dplyr* v.1.0.7 (Wickham *et al.*, 2021) for efficiently manipulating datasets and summarising results. The categories of use were mapped onto a modified phylogenetic tree of vascular plants (Gastauer and Meira, 2017), using the packages *ggtree* v.3.0.2 (Yu *et al.*, 2017), *ggplot2*, and *treeio* 3.13 (Wang *et al.*, 2020) to visualise how useful taxa are distributed across the plant phylogeny and to compute the richness of useful species by clades. Considering that species often have multiple uses, a chord diagram was calculated using *circlize* v.0.4.13 (Gu *et al.*, 2014) to reveal the network of inter-relationships among multiple use categories. Since some use categories could be correlated with each other, *corrplot* v.0.90 (Wei *et al.*, 2017) was used to calculate the correlogram with the Pearson correlation coefficient (ρ), which measures the strength and direction of the linear relationship between two variables (Hauke & Kossowski, 2011).

FIGURE 1. Treemap of the richness of useful species of Colombia by families. Red polygon: families comprising more than 50% of the useful species. Yellow polygon: families comprising more than 75% of the useful species. Green polygon: families (69) containing only one or two useful species. The numbers of species per family are indicated in chapter 12.

RESULTS

Richness by taxa and species origin

The CUPC comprises 7,472 species of useful plants, (7,472 species when including a few organisms frequently misidentified as plants, such as brown algae and nostoc) classified into 2,140 genera, 258 families, 77 orders, 13 classes, and five divisions (Table 1). When comparing these numbers to the CPC (28,947 spp.), almost one in four Colombian plant species has at least one reported use.

Also, 59.7% of the genera and 65.5% of the families in the CPC contain useful plant species. More than 50% of the species are classified within 18 families, with 69 families containing only one or two useful species (Figure 1). Most of the useful species in Colombia are native (5,881 spp., 78.6%), and 481 species (6.4%) are endemic to this country. At least 1,077 species (14.4%) are commercially cultivated, and 643 (8.6%) are introduced and naturalised. These numbers are relevant considering that only 677 species (2.3% of the total) in the CPC have been identified as naturalised. In other words, 95% of the naturalised species in Colombia have reported uses. We could not report the origin for 12.8% of the species, because they are either cryptogenic species or presumed to be exotic. Still, we do not have confirmation of their naturalised status. Most families (155) comprise just native (incl. endemic) species (i.e., with no introduced species), with Annonaceae (102 spp.), Piperaceae (102), and Sapindaceae (73) being the richest ones. Three families include only introduced useful species: Balsaminaceae (5 spp.), Musaceae (7), and Sphenocleaceae (1).

Overall, the most species-rich families in terms of useful species are Fabaceae (665 species), Asteraceae (418), Poaceae (338), Rubiaceae (285), and Malvaceae (221). At the same time, the richest genera are *Solanum* (88), *Inga* (84), *Passiflora* (78), *Miconia* (77), and *Piper* (67) (Table 2), which all have mainly native species, including several endemics to the country.

TABLE 2. Top species-rich families and genera in the Checklist of Useful Plants of Colombia (CUPC) and the checklist of plants of Colombia (CPC).

Top 10 most species-rich families				Top 10 most species-rich genera			
CPC		CUPC		CPC		CUPC	
Orchidaceae	4,436	Fabaceae	665	*Epidendrum*	558	*Solanum*	88
Asteraceae	1,405	Asteraceae	418	*Stelis*	450	*Inga*	84
Fabaceae	1,317	Poaceae	338	*Piper*	429	*Passiflora*	78
Rubiaceae	1,276	Rubiaceae	285	*Miconia*	414	*Miconia*	77
Melastomataceae	995	Malvaceae	221	*Anthurium*	392	*Piper*	67
Araceae	919	Arecaceae	214	*Lepanthes*	323	*Maxillaria*	62
Poaceae	919	Orchidaceae	195	*Palicourea*	313	*Pouteria*	54
Piperaceae	719	Solanaceae	171	*Maxillaria*	288	*Ficus*	46
Polypodiaceae	614	Melastomataceae	162	*Peperomia*	287	*Ipomoea*	43
Bromeliaceae	553	Euphorbiaceae	149	*Philodendron*	225	*Peperomia*	42

TABLE 3. Top endemic-species-rich families and genera in the Checklist of Useful Plants of Colombia (CUPC) and the Checklist of Plants of Colombia (CPC).

Top 10 most endemic species-rich families				Top 10 most endemic species-rich genera			
CPC		CUPC		CPC		CUPC	
Orchidaceae	2,097	Asteraceae	67	*Stelis*	323	*Espeletia*	29
Asteraceae	637	Melastomataceae	40	*Epidendrum*	238	*Clidemia*	17
Melastomataceae	474	Orchidaceae	27	*Lepanthes*	235	*Passiflora*	11
Araceae	403	Fabaceae	24	*Anthurium*	213	*Oreopanax*	9
Rubiaceae	295	Arecaceae	18	*Piper*	187	*Leandra*	8
Piperaceae	282	Araceae	17	*Miconia*	158	*Magnolia*	8
Lejeuneaceae	258	Malvaceae	16	*Pleurothallis*	114	*Symplocos*	7
Fabaceae	211	Araliaceae	14	*Masdevallia*	104	*Matisia*	7
Acanthaceae	208	Passifloraceae	11	*Peperomia*	95	*Miconia*	7
Bromeliaceae	186	Piperaceae	11	*Maxillaria*	94	*Solanum*	6

TABLE 4. Top naturalised-species-rich families and genera in the Checklist of Useful Plants of Colombia (CUPC) and the Checklist of Plants of Colombia (CPC).

Top 10 most naturalised species-rich families				Top 10 most naturalised species-rich genera			
CPC		CUPC		CPC		CUPC	
Poaceae	160	Poaceae	153	*Ipomoea*	12	*Eragrostis*	12
Fabaceae	93	Fabaceae	86	*Eragrostis*	12	*Ipomoea*	10
Asteraceae	80	Asteraceae	77	*Digitaria*	10	*Digitaria*	10
Brassicaceae	24	Brassicaceae	24	*Bambusa*	9	*Bambusa*	9
Solanaceae	20	Solanaceae	20	*Senna*	8	*Crotalaria*	8
Lamiaceae	19	Lamiaceae	19	*Crotalaria*	8	*Urochloa*	8
Rubiaceae	16	Rubiaceae	15	*Urochloa*	8	*Senna*	7
Convolvulaceae	15	Convolvulaceae	13	*Cenchrus*	7	*Cyperus*	7
Cucurbitaceae	13	Polygonaceae	12	*Cyperus*	7	*Setaria*	7
Polygonaceae	12	Cucurbitaceae	12	*Setaria*	7	*Trifolium*	6

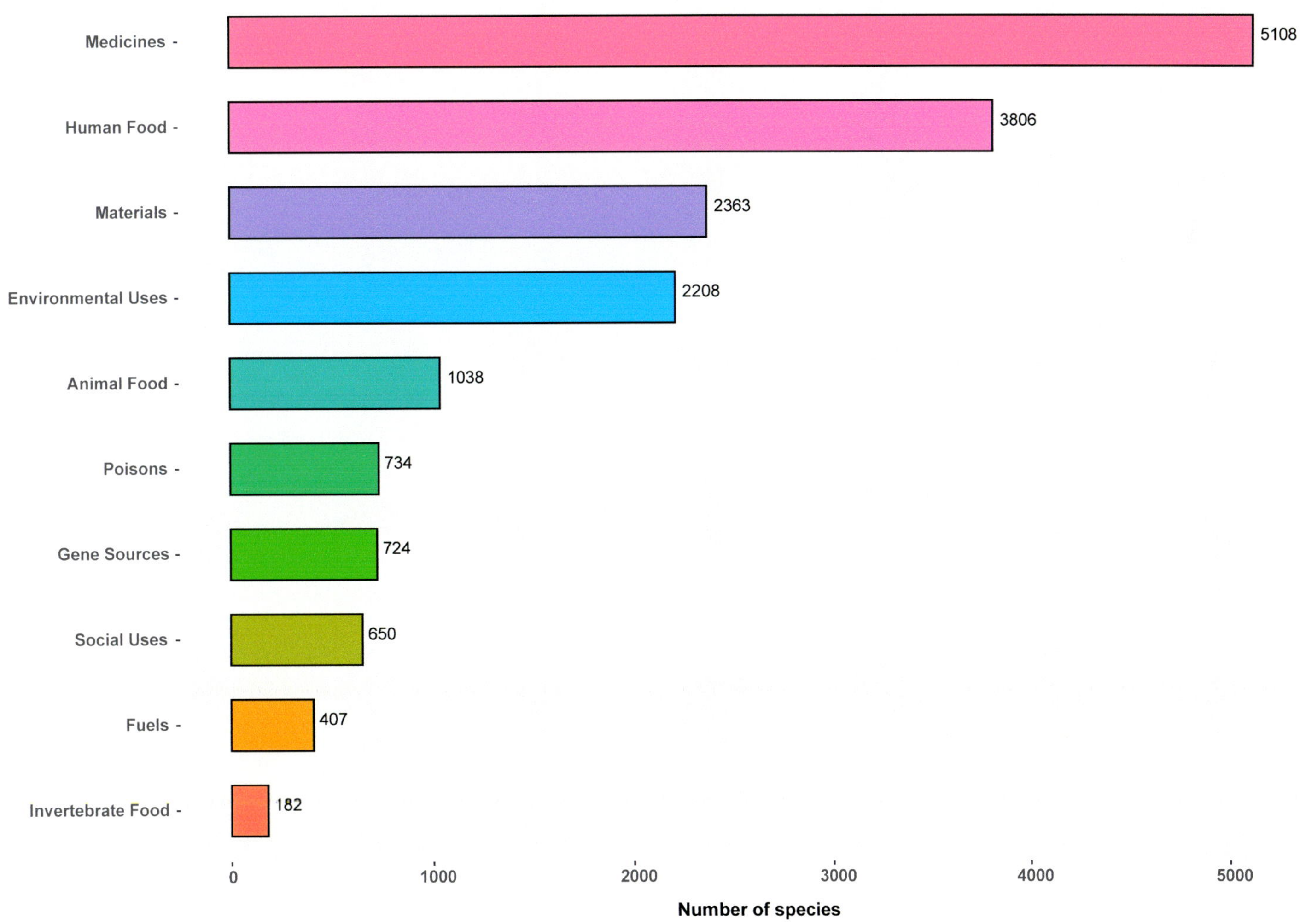

FIGURE 2. Number of plant species per category of use in the CUPC.

The highest number of useful endemic species is found within the families Asteraceae, Melastomataceae, and Orchidaceae (67, 40, and 27 spp., respectively) and within the genera *Espeletia*, *Clidemia*, and *Passiflora* (29, 17, and 11 spp., respectively) (Table 3).

Most of the top 10 richest genera in useful endemic species are not within the 50 richest genera in the CPC. Indeed, some occupy discrete positions in the ranking: *Matisia* (52 spp., ranked in 81st position, with 21 useful spp. of which seven are endemic), *Magnolia* (42, 109th, 13, 8), *Symplocos* (34, 149th, 12, 7), and *Leandra* (28, 210th, 8, 8). As for the naturalised species, the families that contain more useful species are Poaceae, Fabaceae, and Asteraceae (153, 86, and 77 species, respectively), and the genera are *Eragrostis*, *Ipomoea*, and *Digitaria* (12, 10, and 10 species, respectively) (Table 4).

Categories of use

The categories of use with most species are medicines (5,108 spp.), human food (3,806), materials (2,363), and environmental uses (2,208), the latter including ornamental species (Figure 2).

Useful taxa are fairly distributed across the plant phylogeny, with most families having many types of use (Figure 3). Monilophytes (i.e., ferns and horsetails) contribute with the fewest useful taxa (178 spp., 2.4%), whereas Rosids [Vitales + Eurosids (Fabids + Malvids)] is the richest among the large clades (2,795 spp., 37.4%). Within Rosids, Fabids are the richest subclade (1,809 spp., 24.2%): Fabales (688 spp.), Malpighiales (647), Rosales (291), Cucurbitales (76), Oxalidales (58), Celastrales (31), Zygophyllales (9), and Fagales (9). Asterids [Cornales + Ericales + Eurasterids (Lamiids + Campanulids)] is the second richest large clade (2,204 spp., 29.5%). Within Asterids, Lamiids (1,299 spp., 17.4%) is the richest subclade: Lamiales (509 spp.), Gentianales (474), Solanales (250), Boraginales (58), Icacinales (6), and Metteniusales (2).

Half of the species have more than one type of use (Figure 4). For instance, some species used as human food can also have social uses or are sources of materials and poisons. Uses in all ten categories are reported by 28 species [incl. 22 cultivated (e.g., *Arachis hypogaea*, *Zea mays*), 11 naturalised (e.g., *Cocos nucifera*, *Persea americana*), and eight natives (e.g., *Anacardium occidentale* and *Spondias mombin*)].

With at least four types of use, some endemic species are quite versatile, *e.g.*, *Hesperomeles goudotiana* (5 categories of use), *Astrocaryum malybo* (4), and *Passiflora antioquiensis* (4). There are no significant correlations between categories of use (-0.13 ≤ ρ ≤ 0.36), with human food and medicinal plants potentially having any other type of use (Figure 5). The cladogram shows the highest correlations between animal food and fuel, social use and materials, and poisons and gene sources.

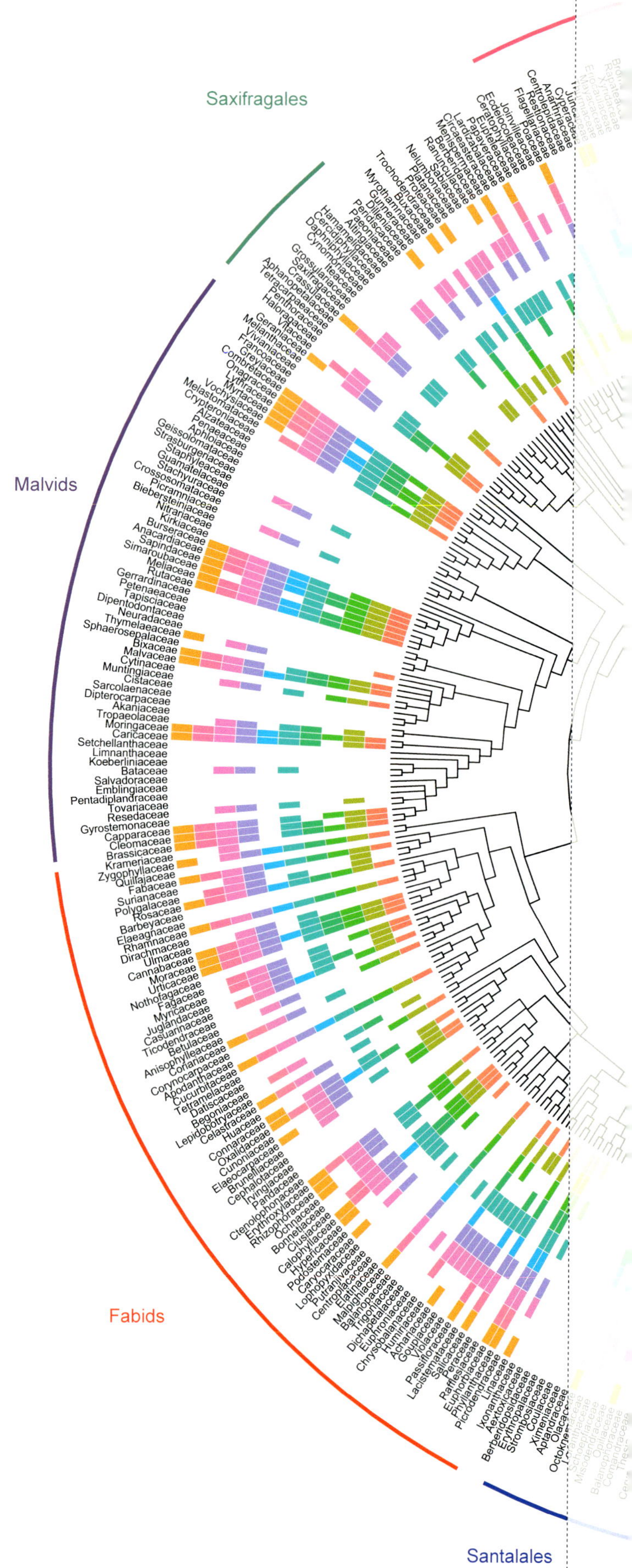

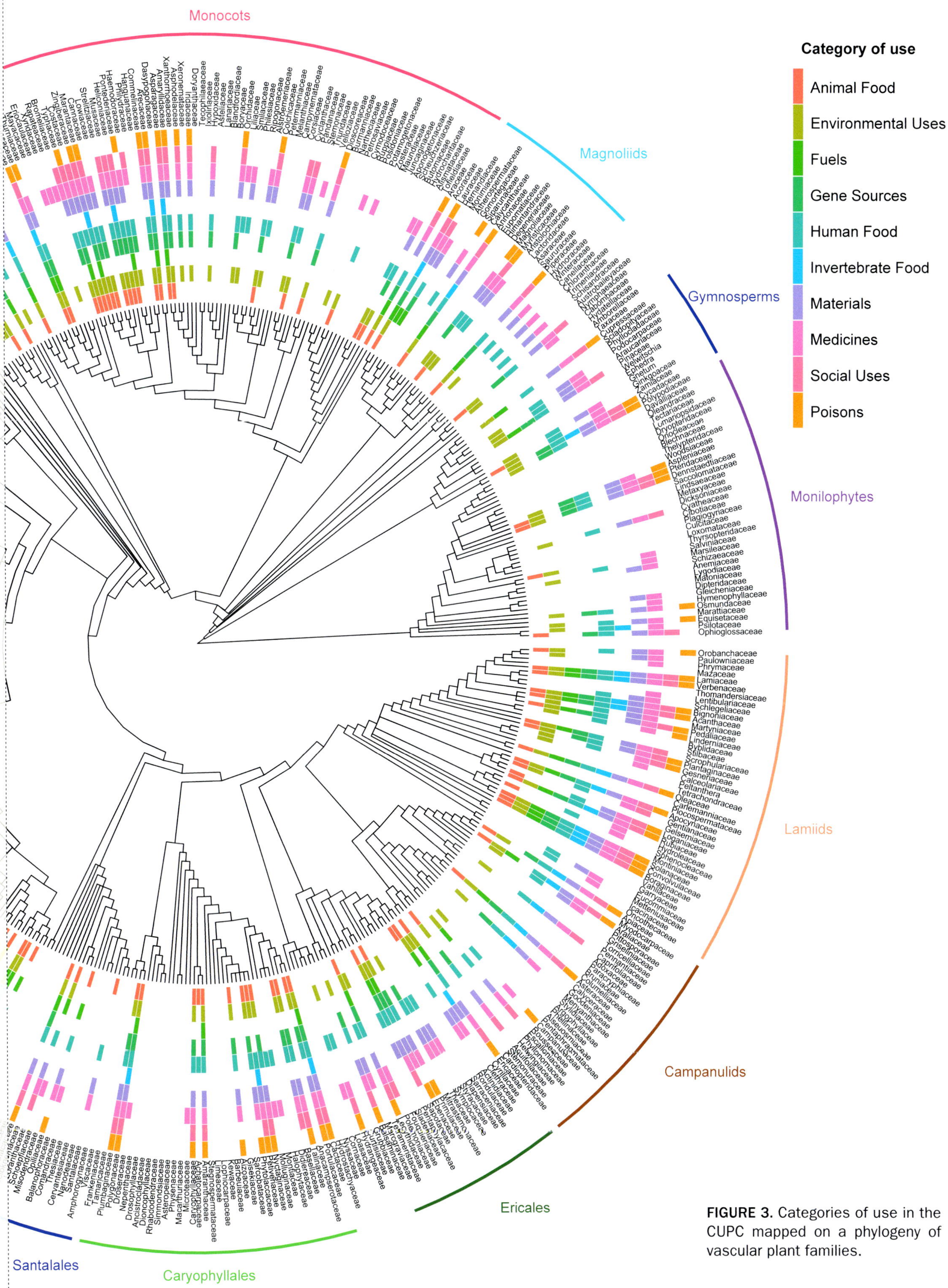

FIGURE 3. Categories of use in the CUPC mapped on a phylogeny of vascular plant families.

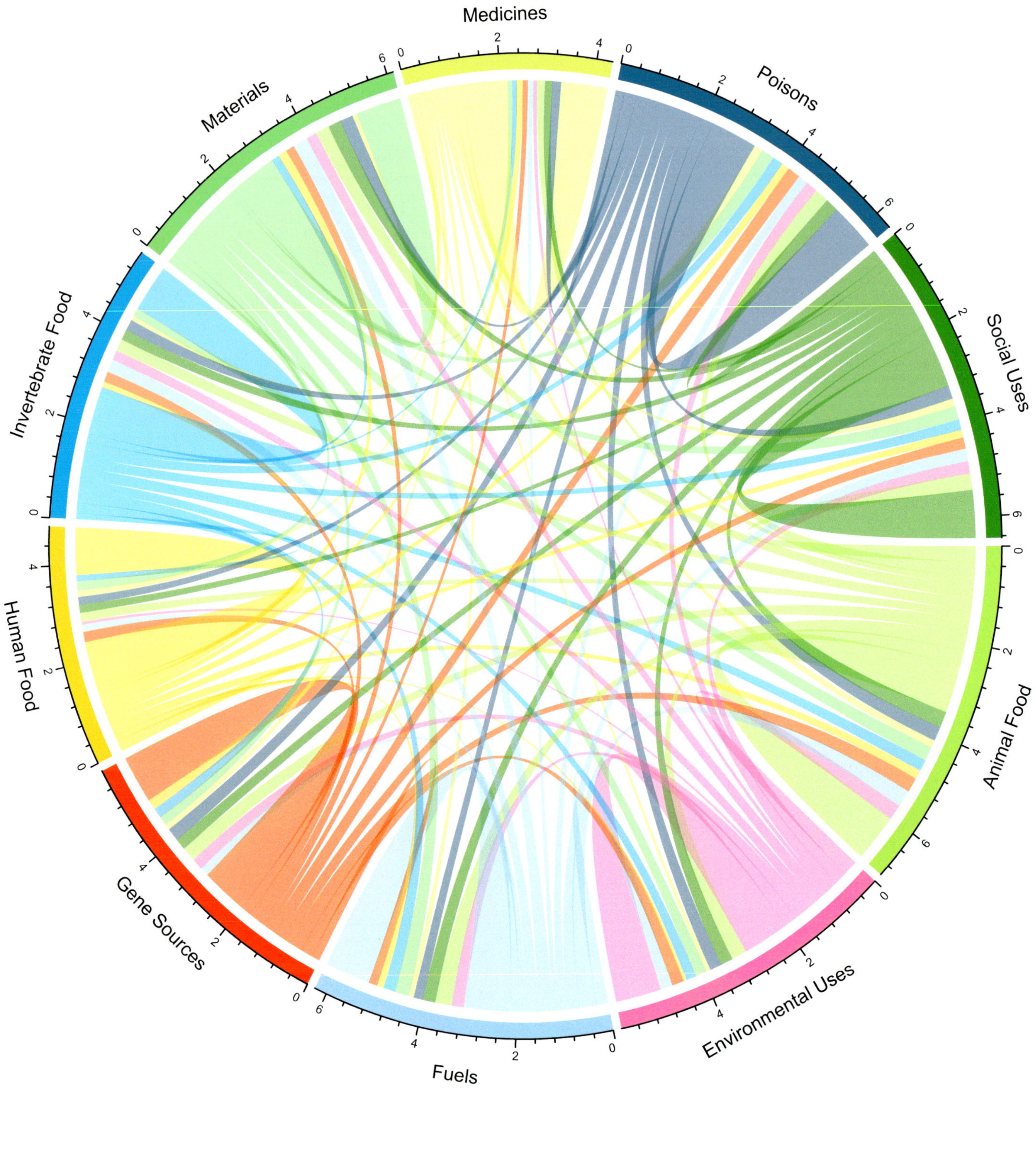

FIGURE 4. Chord diagram of categories of use for species in the CUPC.

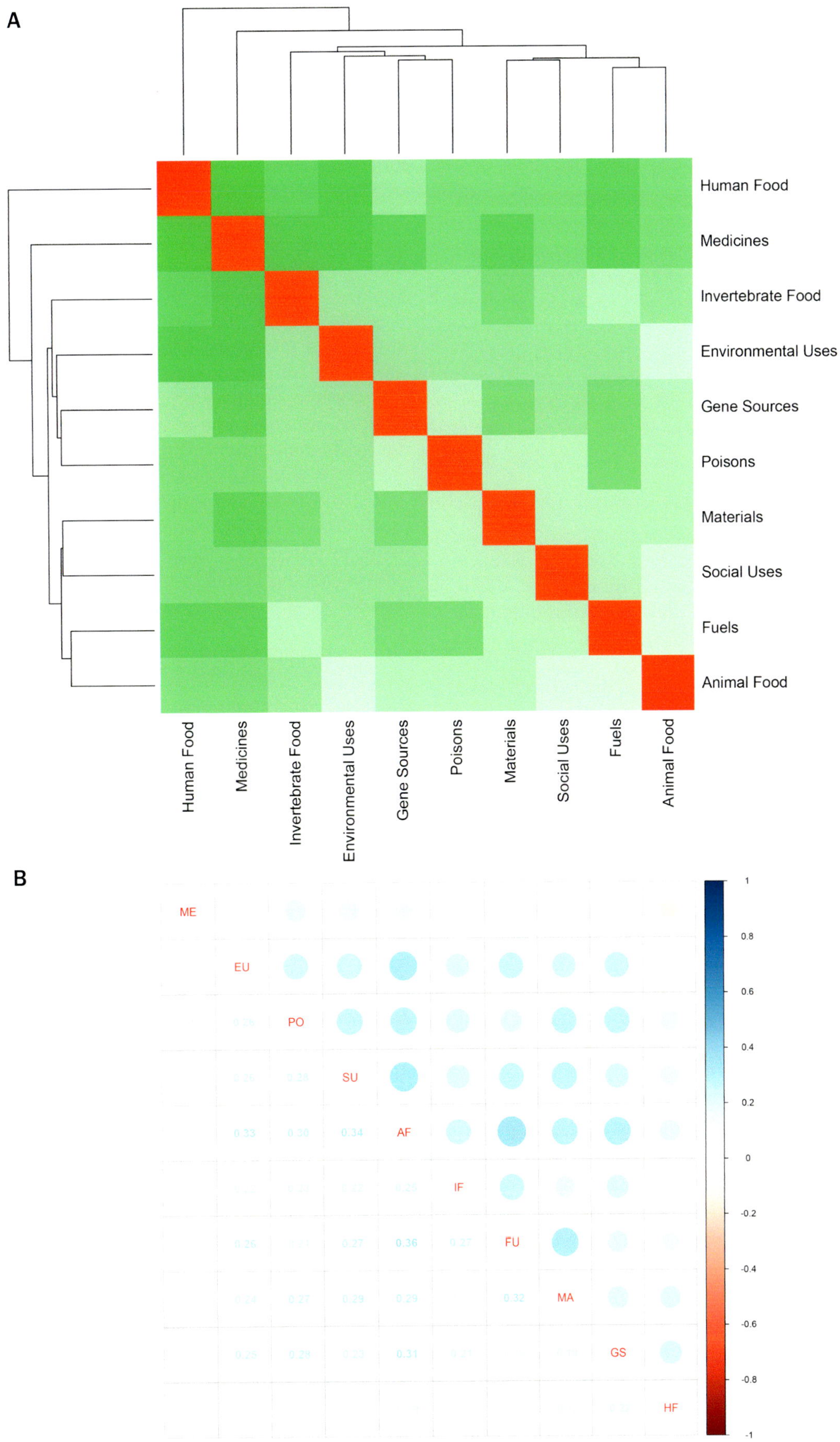

FIGURE 5. Correlogram of categories of use for species in the CUPC. **A** cladogram; **B** Pearson correlation coefficients.

DISCUSSION

Before this project, Colombia did not have an updated national list of useful plants. The available lists were not recent, and were focused on specific use categories or geographic areas. The largest species lists always included fewer than 2,600 species [e.g., 2,567 species in IAVH (2014), 2,404 in Bernal *et al.* (2011), fewer than 2,000 in Pérez Arbeláez (1956), 703 in López & García (2021)], but most of the species were repeated across the publications. By counting species in the CUPC that are not found in any of those important datasets, we estimate that the annotated checklist published in this volume (7,472 spp.) provides records for more than 4,000 useful species that were not included in any of the previous lists. However, the exact number is difficult to estimate. For the first time, the country now has a species list that integrates all use categories.

When comparing the number of useful species to the CPC, one in four Colombian plant species has uses reported. Other biodiverse, well-studied countries have a similar richness of useful plants. For instance, for Perú, there are at least 5,000 wild species with known uses (Brack, 1999), and for Ecuador, 5,172 species, which is three of every ten species reported for that country (De la Torre *et al.*, 2008). Unfortunately, only a few countries have comprehensive checklists of useful plants, and vast gaps in ethnobotanical information persist in the region (Cámara-Leret *et al.*, 2014), precluding us from making conclusions about these numbers.

Most useful species are native, but a significant proportion of them were introduced and are currently naturalised (8.6%), often transforming the natural ecosystems. Several of these species are considered invasive (e.g., various species of *Acacia*, *Eucalyptus*, and *Pinus*), which should be a major concern. Previous studies have reported at least 597 introduced species in Colombia, 42 of which are considered high risk of invasion (Cárdenas-López *et al.*, 2017). The CUPC cites uses for 95% of the naturalised species in Colombia and for 90% of the species with a high risk of invasion. Stunningly, one-third of the species with a high risk of invasion are considered species of high economic interest as well (Cárdenas-López *et al.*, 2017). A review on the possible pathways of introduction reveals both accidental and intentional introductions, such as deliberate release (e.g., ecological restoration), escape from captivity (e.g., ornamental garden plants), contaminants of commodities (e.g., weed seeds), and stowaways on transport vectors (Cárdenas-López *et al.*, 2017). Regardless of their use, biological invasions are deemed to be one of the greatest causes of biodiversity loss (Pyšek *et al.*, 2020), and the use, propagation, and transport of invasive species are prohibited in Colombia (e.g., Law 13 of 1990; Decree 2256 of 1991; Law 599 of 2000).

Sixty-five percent of the species (4,844) had uses reported in the WCUP, of which 1,571 species had no reported uses in the country but are used elsewhere. For instance, there are 1,022 species with medicinal uses in other parts of the world (MPNS, 2021), 101 useful species reported in tropical Africa (PROTA, 2020), 97 in Asia (Jansen *et al.*, 1991), 71 in arid and semi-arid lands of the world (RBG Kew, 1999), and 35 in New Guinea (Cámara-Leret & Dennehy, 2019). This significant proportion of neglected or underutilised useful species in Colombia represents an opportunity to further develop the bioeconomy and local practices of sustainable use.

Only 18 families contain more than 50% of the useful species in Colombia. This scenario is not surprising, considering that those families are in general species-rich, abundant, and sources of major crops [e.g., Fabaceae (beans), Asteraceae (sunflowers), Poaceae (cereals), Arecaceae (palms), Rubiaceae (coffee), and Solanaceae (tomatoes, potatoes)]. Fabaceae are the most used plants in Colombia, with hundreds of species used as a source of food and materials distributed across the elevation gradient in all regions. It is also one of the most important families of the lowlands in the neotropics (Gentry, 1988). Asteraceae ranks second in richness of useful species, with its medicinal use being the most important use category by far. The family becomes dominant towards the high-elevation ecosystems (e.g., in páramos) (Rangel-Ch., 2000). Perhaps that is the reason for the common belief that high mountains are the natural pharmacies of Colombians. Poaceae, in the third position, is particularly important as animal food, with most of the current pastures in the country belonging to species introduced in the 20th century, dramatically transforming the native ecosystems. However, many native species of Poaceae still maintain their traditional importance in producing materials (mainly fibres) for handicrafts and construction (Cadena *et al.*, 2007; Linares *et al.*, 2008).

Some of the richest genera also include major crops [e.g., *Solanum* (tomatoes, potatoes), *Passiflora* (passionfruit)], but many others, with hundreds or tens of useful species, are rarely commercialised. The percentage of endemic useful species is relatively low compared to the overall endemics in the Colombian flora (6.4% vs 32.7%, respectively). This scenario is probably because most of the endemic species are still poorly known and often recently described. They are considered rare or grow in remote areas from which ethnobotanical information is still scanty. Ranks of the most diverse families and genera in the CUPC (Table 2) are relatively consistent with the Colombian flora, listing most of these taxa in the top 25 positions of the CPC, except for *Ipomea* and *Pouteria* (in positions 49 and 53 in the CPC, respectively). Some remarkably species-rich genera with several potential ornamental species do not make it to the top positions in the CUPC: three genera of orchids, i.e., *Epidendrum* (558 spp.), *Stelis* (450), and *Lepanthes* (323); and two of aroids, i.e., *Anthurium* (392) and *Philodendron* (225) (Table 2). This scenario could be for various reasons, including information gaps on the local use of plants; lack of knowledge on the taxonomic diversity and potential of the flora; the natural history of species (rare species, difficult to find or harvest species, or difficult to propagate species); and current regulations restricting commercialisation of species in CITES lists, including most wild orchids in Colombia (Betancur *et al.*, 2015).

Results highlight some groups with comparatively low percentages of useful species in the Colombian flora. For instance, Orchidaceae, by far the most diverse family in the country (4,436 spp.) and with the largest genera, ranks in 7th position, with only 195 reported useful species. However, according to CITES, between 1975 and 2012 Colombia exported at least 1,352 species (excluding unidentified or exotic species), mainly to eight countries: Australia, Canada, France, Germany, Japan, Switzerland, United Kingdom and United States of America (Betancur *et al.*, 2015). At the same time, Colombia's national conservation plan for orchids promotes the artificial propagation and commercialisation of native species and their reintroduction into the natural ecosystems (Betancur *et al.*, 2015). Understanding that most of the Colombian species of orchids are under threat because of land-use change and wild harvesting (Calderón Sáenz, 2007), some recent initiatives in the country are promoting the artificial propagation and sustainable commercialisation of native species of orchids as a way to increase household income and improve the livelihoods of local communities (Castellanos-Castro & Torres-Morales, 2018).

The categories of use with more species are medicines (5,108 spp.), human food (3,806), materials (2,363), and environmental uses (2,208) (Figure 2). This scenario is consistent with the top four categories in the World Checklist of Useful Plants (Diazgranados *et al.*, 2020): medicines (26,662 species), materials (13,663), environmental uses (8,983), and human food (7,039). Also, it is not surprising to see medicinal use as the most frequent use for species in Colombia. Numerous studies have focused on this topic (see Chapter 3), and there are guidelines for the sustainable use of prioritised species (Bernal *et al.*, 2011). The report of 5,108 species with medicinal use doubled the 2,404 species previously reported (Bernal *et al.*, 2011). When matching the CPC list with the latest available version of the Medicinal Plant Names Services list (MPNS, 2021), we obtained 3,898 species with medicinal use, with only 75 species not found in any of the other sources we used to compile the WCUP.

It is of particular interest to draw attention to the relatively large diversity of plants reported as sources of human food (3,806 spp.), representing 50.1% of the world's total number of food plants identified to date (7,039) by Ulian *et al.* (2020), considering that the latter did not include the CUPC database. Also, this represented a significant leap from previous reports, e.g., 703 wild species with edible fruits (López & García, 2021), or 399 species used for human feeding in Colombia (García, 2011). As described in chapter 4, this striking diversity of edible plants evokes calling the country the "hidden food basket of Latin America".

A visual inspection of the distribution of useful taxa across the phylogeny (Figure 3) does not allow a clear identification of patterns, with most families having many types of use (Figure 3). Rosids and Asterids are the clades that concentrate most of the diversity of useful plants (66.9%). Detailed analyses, considering morphological traits, categories of use, ecosystem dominance and richness, are needed to understand these results.

Data limitations and potential taxonomic gaps

Despite the enormous effort of the team to compile and categorise the available information on plant uses, some caveats are worth highlighting:

- There are still numerous sources and records that have not been added to the CUPC because of access limitations, permits, taxonomic problems or limited time.
- Ethnobotanical research is still needed in many biodiverse areas of the country.
- There is a strong research bias towards certain geographic areas and ecosystems (e.g., Andean Forest) and certain taxonomic groups (e.g., vascular plants).
- Interpretation of the categories is not always straightforward. Reports of use often had to be discarded because of misinterpretations (e.g., a wild plant pollinated by bees should not be categorised as "invertebrate food" if it has not been intentionally planted with that purpose).

These data limitations can explain some potentially missing taxa. For instance, there are 30 plant families in Colombia (CPC) that have uses reported in the WCUP, but not in the CUPC (e.g., Cardiopteridaceae: 17 spp. in the WCUP / 4 spp. in the CPC; Elatinaceae: 6 / 3; Lentibulariaceae: 17 / 38; Sabiaceae: 17 / 27; and Velloziaceae: 10 / 1). Also, circa 250 genera found in Colombia are reported to include useful species in the WCUP but not in the CUPC (e.g., *Dichapetalum*: 24 spp. in the WCUP / 10 spp. in the CPC; *Homalomena*: 16 / 9; *Meliosma*: 13 / 23; *Merremia*: 10 / 9; *Utricularia*: 13 / 32; and *Vriesea*: 7 / 31). It is possible that with further work, uses within these clades could be added to updated versions of the CUPC.

Another limitation is that the database does not include frequency or intensity of use, which is only available for a small proportion of the species. Therefore, we cannot describe how much or how often taxa are used. However, it is quite likely that most species are rarely used, especially by people in urban areas.

Last, the database does not reflect the geographic or temporal scales of use. Plant uses vary over time and geographically (Cámara-Leret *et al.*, 2014), depending on cultural changes, society needs and opportunities, local regulations, availability, market behaviour and industrial developments. In the 18th century, frailejones (*Espeletia* spp.) were largely used to extract turpentine, to construct shelters in the páramos, and for medicinal and social purposes (Cuatrecasas, 2013). Today, those uses of frailejones are not permitted, and people risk jail or a significant fine if frailejones, now with a recognised and protected environmental use, are harvested as a source of materials. The African oil palm (*Elaeis guineensis* Jacq.) was introduced to Colombia in 1945 by the United Fruit Company (Aguilera-Díaz, 2002), but it only became an economically important species in recent decades. Today,

Colombia is the world's fourth oil palm producer, first in the Americas, with more than 500,000 ha of oil palm crops in 161 municipalities and 21 departments (Lain & Ocampo-Peñuela, 2019). Therefore, changes of plant use through time and space are relevant to fully understand the dynamic interaction of societies with nature.

CONCLUSIONS

This chapter provides an overview of the taxonomic coverage of the CUPC, with notes on the species origin, and mentions of the most diverse taxa, species richness by category of use, phylogenetic coverage, and data limitations. With 7,472 species of useful plants, classified in 2,140 genera and 258 families, Colombia's rich diversity stands out compared to that of other countries and the WCUP as a natural reservoir of useful species. Also, this reflects, at least in part, this project's efforts to compile, digitise, and categorise the available information published in this volume and accessible to the community through the Col*Plant*A portal. Most of the reported species in the CUPC are native (78.6%), but an important proportion is introduced and naturalised (8.6%), some of them with high risk of becoming invasive. The country must pay special attention to those species, as they can be harmful to the native ecosystems and species, yielding negative outcomes in the long run. Circa 21% of useful species had no reports of plant use in the country but are used elsewhere. This significant proportion of neglected or underutilised useful species in Colombia represents an opportunity to further develop the bioeconomy and local practices of sustainable use. The clades Rosids and Asterids are the most useful species-rich large clades. Within them, Fabids and Lamiids concentrate most of the diversity. At the family level, Fabaceae (665 spp.), Asteraceae (418) and Poaceae (339) are the richest, while *Solanum* (88), *Inga* (84) and *Passiflora* (78) are the richest genera. The categories of use with most species are medicines (5,108 spp.), human food (3,806), materials (2,363), and environmental uses (2,208). Using plants to develop this country's local and national bioeconomy may contribute to ending poverty and hunger, achieving food security, improving nutrition and health, and combating climate change, among other benefits. Therefore, research on economic botany and ethnobotany should continue to complete the inventory of the useful plants of Colombia, filling taxonomic and geographic gaps of information. Also, more studies are needed to investigate temporal and geographic patterns of plant uses and to continue identifying priority species to promote the country's sustainable development.

References

Acero L (2005) *Plantas útiles de la cuenca del Orinoco.* Bogotá, Ecopetrol y Corporinoquia. Editorial Panamericana Formas e Impresos S.A. 605p.

Aguilera-Díaz MM (2002) Palma africana en la Costa Caribe: un semillero de empresas solidarias. *Documentos de Trabajo Sobre Economía Regional y Urbana; No. 30.* doi: 10.32468/dtseru.30

Bacon CD, Silvestro D, Jaramillo C, Smith BT, Chakrabarty P, Antonelli A (2015). Biological evidence supports an early and complex emergence of the Isthmus of Panama. *Proc Natl Acad Sci USA,* 112:6110-6115.

Bernal HY, Quevedo Sánchez GF, García Martínez H (2011) *Pautas para el conocimiento, conservación y uso sostenible de las plantas medicinales nativas en Colombia: Estrategia nacional para la conservación de plantas* Bogotá, Instituto de Investigación de Recursos Biológicos Alexander von Humboldt. 232p.

Bernal R (2016) Geografía de Colombia. In: Bernal R, Gradstein S, Celis M (eds) *Catálogo de plantas y líquenes de Colombia.* Bogotá, Colombia: Universidad Nacional de Colombia, Facultad de Ciencias, Instituto de Ciencias Naturales. 3060p.

Bernal R, Gradstein S, Celis M (eds) (2019) *Catálogo de plantas y líquenes de Colombia.* Instituto de Ciencias Naturales, Universidad Nacional de Colombia. Available at: http://catalogoplantasdecolombia.unal.edu.co/

Betancur J, Sarmiento H, Toro-González L, Valencia J (2015) *Plan para el estudio y la conservación de las orquídeas en Colombia.* Bogotá, Ministerio de Ambiente y Desarrollo Sostenible. Universidad Nacional de Colombia. 336p.

Brack A (1999) Diccionario enciclopédico de plantas útiles del Peru (Text in Spanish). *CBC-Centro de Estudios Regionales Andinos "Bartolomé de las Casas". Peru: Cuzco,* 550.

Cadena C, Diazgranados M, Bernal HY (2007) Plantas útiles para la elaboración de artesanías de la comunidad indígena monifue amena (Amazonas, Colombia). *Univ Sci,* 12:97-116.

Calderón Sáenz E (2007) Libro rojo de plantas de Colombia. Vol. 6. Orquídeas, primera parte. 832p.

Cámara-Leret R, Dennehy Z (2019) Information gaps in indigenous and local knowledge for science-policy assessments. *Nat Sustain,* 2:736-741.

Cámara-Leret R, Paniagua-Zambrana N, Balslev H, Macía MJ (2014) Ethnobotanical knowledge is vastly under-documented in northwestern South America. *PLoS One,* e85794. doi:10.1371/journal.pone.0085794

Cárdenas-López D, Baptiste MP, Castaño N (eds) (2017) *Plantas exóticas con alto potencial de invasión en Colombia.* Bogota, Instituto de Investigación de Recursos Biológicos Alexander von Humboldt. 295p.

Cárdenas López D, López Camacho R (2000) *Plantas útiles de la Amazonia Colombiana-Departamento del Amazonas: perspectivas de los productos forestales no maderables.* Bogotá, SINCHI, Instituto Amazónico de Investigaciones Científicas, Ministerio del Medio Ambiente. 133p.

Castellanos-Castro C, Torres-Morales G (2018) *Orquídeas de Cundinamarca: conservación y aprovechamiento sostenible.* Bogotá, Instituto de Investigación de Recursos Biológicos Alexander von Humboldt, Pontificia Universidad Javeriana, Jardín Botánico de Bogotá "José Celestino Mutis", Corporación Colombiana de Investigación Agropecuaria Corpoica, Gobernación de Cundinamarca. 328p.

Cuatrecasas J (2013) *A systematic study of the subtribe Espeletiinae.* New York, The New York Botanical Garden. 689p.

DANE (2019) Población Indígena de Colombia. Online. Departamento Administrativo Nacional de Estadística (DANE). Available at: https://www.dane.gov.co/files/investigaciones/boletines/grupos-etnicos/presentacion-grupos-etnicos-2019.pdf

DANE (2021) Información Pobreza monetaria nacional 2020. Departamento Administrativo Nacional de Estadística (DANE). Available at: https://www.dane.gov.co/index.php/estadisticas-por-tema/pobreza-y-condiciones-de-vida/pobreza-monetaria

De La Torre L, Navarrete H, Muriel P, Macía MJ, Balslev H (eds) (2008) *Enciclopedia de las Plantas Útiles del Ecuador (con extracto de datos).* Quito & Aarhus, Herbario QCA de la Escuela de Ciencias Biológicas de la Pontificia. 322p.

Diazgranados M, Allkin B, Black N, Cámara-Leret R, Canteiro C, Carretero J, Eastwood R, Hargreaves S, Hudson A, Milliken W, Nesbitt M, Ondo I, Patmore K, Pironon S, Turner R, Ulian T (2020) World Checklist of Useful Plant Species. In: Royal Botanic Gardens, Kew (ed). Knowledge Network for Biocomplexity. 689p. doi: 10.5063/F1CV4G34

Diazgranados M, Barber JC (2017) Geography shapes the phylogeny of frailejones (Espeletiinae Cuatrec., Asteraceae): a remarkable example of recent rapid radiation in sky islands. *PeerJ,* 5: e2968. doi: 10.7717/peerj.2968

Diazgranados M, Moreno N (2020) Plantas útiles de la Reserva Natural Bojonawi (Escudo Guayanés), Orinoquia, Vichada, Colombia. In: Lasso CA, Trujillo F, Morales-Betancourt MA (eds). *Biodiversidad de la Reserva Natural Bojonawi, Vichada, Colombia: río Orinoco y planicie de inundación*. Bogotá, Instituto de Investigación de Recursos Biológicos Alexander von Humboldt. 559p.

Duarte- Abadía B, Parra-Ortega SL (2015) Plantas del páramo y sus usos para el buen vivir: Páramos de Guerrero y Rabanal. *Bogota,* Instituto de Investigacion de Recursos Biologicos Alexander von Humboldt. 60p.

Flantua SGA, O'dea A, Onstein RE, Giraldo C, Hooghiemstra H (2019) The flickering connectivity system of the north Andean páramos. *J Biogeogr,* 46:1808-1825.

Galeano G (2016) Vegetación Natural de Colombia. In: Bernal R, Gradstein SR, Celis M (eds) *Catálogo de plantas y líquenes de Colombia.* Bogotá, Universidad Nacional de Colombia, Facultad de Ciencias, Instituto de Ciencias Naturales.

García Barriga H (1975) *Flora Medicinal de Colombia, Tomos I–III.* Bogotá, Instituto de Ciencias Naturales, Universidad Nacional de Colombia.

García N (2011) *Plantas nativas empleadas en alimentación en Colombia*. Bogotá, Instituto de Investigación de Recursos Biológicos Alexander von Humboldt. 52p. Available at: http://hdl.handle.net/20.500.11761/31275

Gastauer M, Meira JA (2017) Updated angiosperm family tree for analysing phylogenetic diversity and community structure. *Acta Bot Bras,* 31:191-198.

Gentry AH (1988) Changes in Plant Community Diversity and Floristic Composition on Environmental and Geographical Gradients. *Ann Mo Bot Gard,* 75:1-34.

Gu Z, Gu L, Eils R, Schlesner M, Brors B (2014) Circlize implements and enhances circular visualisation in R. *Bioinformatics,* 30:2811-2812.

Hauke J, Kossowski T (2011) Comparison of values of Pearson's and Spearman's correlation coefficient on the same sets of data. *Quaest Geogr,* 30:87-93.

Hooghiemstra H, Flantua S (2019) Colombia in the Quaternary: An overview of environmental and climatic change. In: Gómez Tapias J, Pinilla-Pachon AO (eds). *The Geology of Colombia.* Bogotá, Servicio Geológico Colombiano.

Humboldt A, Bonpland A (1809 [1808]) *Plantae Aequinoctiales recueillies au Mexique: dans l'île de Cuba, dans les provinces de Caracas, de Cumana et de Barcelone, aux Andes de la Nouvelle Grenade, de Quito et du Pérou, et sur les bords du rio-Negro de Orénoque et de la rivière des Amazones*. Paris, F. Schoell. 600p.

Hurtado Montoya AF, Mesa Sánchez ÓJ (2015) Cambio climático y variabilidad espacio-temporal de la precipitación en Colombia. *Revista EIA*, 131-150.

IAVH (2014) Plantas alimenticias y medicinales nativas de Colombia. In: (Iavh) IDIDRBaVH (ed). 2567 registros, aportados por: Castellanos C, Valderrama N, Bernal Y, García N. doi: 10.15472/0bg7iq

Jansen P, Lemmens R, Oyen L, Siemonsma J, Stavast F, Van Valkenburg J (1991) Plant Resources of South-East Asia (PROSEA). *Basic list of species and commodity grouping. Final version.* Pudoc, Wageningen. 372p.

Lain E, Ocampo-Peñuela N (2019) Contexto actual del impacto ambiental de la palma de aceite en Colombia. *Rev Palmas,* 40:79-88.

Linares EL, Galeano G, García N, Figueroa Y (2008) *Fibras vegetales empleadas en artesanías en Colombia.* Bogotá, Artesanías de Colombia S.A., Ministerio de Comercio, Industria y Turismo, Instituto de Ciencias Naturales de la Universidad Nacional de Colombia. 330p.

López Camacho R, Pulido Rodríguez NE, González Martínez RO, Nieto Vargas JE, Vásquez MY (2014) *Maderas. Especies comercializadas en el territorio CAR: Guía para su identificación.* Bogota, Corporación Autónoma Regional de Cundinamarca-CAR. 120p.

López D, García N (2021) Frutos silvestres comestibles de Colombia: diversidad y perspectivas de uso. *Biota Colombiana,* 22:16-55.

MADS, IDEAM, IAVH, INVEMAR, IIAP, SINCHI, PNN, IGAC (2017) *Mapa de Ecosistemas Continentales, Costeros y Marinos de Colombia a escala 1:100.000.* Bogota, Colombia. IDEAM - Instituto de Hidrología, Meteorología y Estudios Ambientales.

Mittermeier RA, Mittermeier CG, Gil PR (1997) *Megadiversity: Earth's Biologically Wealthiest Nations*. CEMEX. 501p.

MPNS (2021) Medicinal Plant Names Services V10. In: Royal Botanic Gardens, Kew (ed). Available at: https://mpns.science.kew.org

Myers N, Mittermeier RA, Mittermeier CG, Da Fonseca GA, Kent J (2000) Biodiversity hotspots for conservation priorities. *Nature,* 403: 853-858.

O'dea A, Lessios HA, Coates AG, Eytan RI, Restrepo-Moreno SA, Cione AL, Collins LS, Queiroz AD, Farris DW, Norris RD, Stallard RF, Woodburne MO, Aguilera O, Aubry MP, Berggren WA, Budd AF, Cozzuol MA, Coppard SE, Duque-Caro H, Finnegan S, Gasparini GM, Grossman EL, Johnson KG, Keigwin LD, Knowlton N, Leigh EG, Leonard-Pingel JS, Marko PB, Pyenson ND, Rachello-Dolmen PG, Soibelzon E, Soibelzon L, Todd JA, Vermeij GJ, Jackson JBC (2016) Formation of the Isthmus of Panama. *Science Advances,* 2:e1600883.

Pardo-Tomás J, López Terrada ML (1993) *Las primeras noticias sobre plantas americanas en las relaciones de viajes y Crónicas de Indias (1493-1553).* Valencia, Insituto de Estudios Documentales e Historicos sobre la ciencia, Universitat de Valencia. 361p

Pérez-Escobar OA, Lucas E, Jaramillo C, Monro A, Morris SK, Bogarín D, Greer D, Dodsworth S, Aguilar-Cano J, Sanchez Meseguer A, Antonelli A (2019) The origin and diversification of the hyperdiverse flora in the Chocó Biogeographic Region. *Front Plant Sci,* 10:1328. doi: 10.3389/fpls.2019.01328

Pérez Arbeláez E (1956) *Plantas útiles de Colombia*. Madrid, Sucesores de Rivadeneyra. 831p.

Piedrahita SD (2008) *Mutis y el movimiento ilustrado en la Nueva Granada.* Bogotá, Universidad de América. 204p.

Piedrahita SD, Lourteig A (1989) *Génesis de una flora*. Bogotá, Academia Colombiana de Ciencias Exactas, Físicas y Naturales. 334p.

PROTA (2020) *Plant Resources of Tropical Africa (PROTA4U)* Ressources végétales de l'Afrique tropicale. Wageningen, FAO. Available at: https://www.prota4u.org/

Pyšek P, Hulme PE, Simberloff D, Bacher S, Blackburn TM, Carlton JT, Dawson W, Essl F, Foxcroft LC, Genovesi P, Jeschke JM, Kühn I, Liebhold AM, Mandrak NE, Meyerson LA, Pauchard A, Pergl J, Roy HE, Seebens H, Van Kleunen M, Vilà M, Wingfield MJ, Richardson DM (2020) Scientists' warning on invasive alien species. *Biol Rev,* 95:1511-1534.

Rangel-Ch JO (2015) La biodiversidad de Colombia: significado y distribución regional. *Rev Acad Colomb Cienc Exactas Fis Nat,* 39:176–200.

Rangel-Ch O (2000) Biodiversidad en la región del páramo: con especial referencia a Colombia. In: Rangel-Ch. O (ed) *La región de vida paramuna.* Bogotá, Instituto de Ciencias Naturales - Instituto Alexander von Humboldt. 902p.

Raven PH, Gereau RE, Phillipson PB, Chatelain C, Jenkins CN, Ulloa Ulloa C (2020) The distribution of biodiversity richness in the tropics. *Sci Adv,* 6:eabc6228.

RBG Kew (1999) *Survey of Economic Plants for Arid and Semi-Arid Lands (SEPASAL) database* [Online]. Royal Botanic Gardens, Kew. Available at: http://sftp.kew.org/pub/data-repositories/sepasal/

Ulian T, Diazgranados M, Pironon S, Padulosi S, Liu U, Davies L, Howes MJR, Borrell JS, Ondo I, Pérez-Escobar OA (2020) Unlocking plant resources to support food security and promote sustainable agriculture. *Plants, People, Planet,* 2:421-445.

Wang LG, Lam TTY, Xu S, Dai Z, Zhou L, Feng T, Guo P, Dunn CW, Jones BR, Bradley T (2020) Treeio: an R package for phylogenetic tree input and output with richly annotated and associated data. *Mol Biol Evol,* 37:599-603.

Wei T, Simko V, Levy M, Xie Y, Jin Y, Zemla J (2017) Package 'corrplot'. *Statistician,* 56:e24.

Wickham H, Chang W, Wickham MH (2016) Package 'ggplot2'. *Create Elegant Data Visualisations Using the Grammar of Graphics.* 2:1-189.

Wickham H, François R, Henry L, Müller K (2021) *dplyr: A Grammar of Data Manipulation.* Available at: https://CRAN.R-project.org/package=dplyr

Yu G, Smith DK, Zhu H, Guan Y, Lam TTY (2017) ggtree: an r package for visualisation and annotation of phylogenetic trees with their covariates and other associated data. *Methods Ecol Evol,* 8: 28-36.

Moon Island in the medium-zone of the Guaviare river.

Chapter 11

Notes on the geographic distribution of the useful plants of Colombia

Mauricio Diazgranados[1]* & Laura Kor[1]

[1] Royal Botanic Gardens, Kew.

*Corresponding author: m.diazgranados@kew.org

Keywords: biogeography, georeferenced records, beta diversity, plant diversity, weighted endemism.

ABSTRACT

Colombia ranks second in the world for species richness and has high cultural diversity, reflected in the range of ethnic groups and native languages it supports. This rich biocultural diversity has resulted in thousands of documented plant uses, with ecological and social importance. In this chapter, we characterise the geographic distribution of the species listed in the checklist of Useful Plants of Colombia (UPC) by describing distribution patterns at varying biological and geographical scales. In total, the 7,472 plant species with documented uses are represented by 435,230 publicly available georeferenced records. Of the 13 bioregions assessed, the Andean humid forests bioregion has by far the largest number of records, species, genera, and families. The humid forests of the Amazonia and Caribe followed in terms of species richness. Antioquia Department has the largest number of useful species, followed by Cundinamarca and Valle del Cauca, all of which have a remarkable proportion of non-native species. At the municipality level, Solano (Caquetá), Manizales (Caldas), and Leticia (Amazonas) report the highest species richness. However, records are unevenly distributed across both taxonomic groups and geographic areas. Just ten species account for 12.6% of all occurrences, while large and ecologically diverse bioregions such as the Llanos Orientales support very few records concentrated around urban centres. Within all use categories, species richness decreased with elevation. We discuss the potential drivers and implications of these data and knowledge gaps. This characterisation of our current understanding of the distribution of useful plants can be used as a baseline to target future research efforts, supporting plant conservation planning and efforts across Colombia.

RESUMEN

Colombia ocupa el segundo lugar en el mundo en el número de especies que contiene y posee una alta diversidad cultural, reflejada en la variedad de grupos étnicos y lenguas nativas que preserva. Esta rica diversidad biocultural ha dado lugar a miles de usos de plantas documentados, con importancia tanto ecológica como social. En este capítulo, caracterizamos la distribución geográfica de las especies enumeradas en la lista de verificación de Plantas Útiles de Colombia (UPC) mediante la descripción de patrones de ocurrencia en diversas escalas biológicas y geográficas. En total, las 7.472 especies de plantas con usos documentados están representadas por 435.230 registros georreferenciados disponibles públicamente. De las 13 biorregiones evaluadas, el bosque húmedo de los Andes fue, por mucho, la biorregión con el mayor número de registros, especies, géneros y familias. Le siguieron los bosques húmedos de la Amazonia y el Caribe en términos de riqueza de especies. Antioquia tiene la mayor cantidad de especies útiles, seguido de Cundinamarca y Valle del Cauca, aunque incluyen una proporción notable de especies no nativas. A nivel municipal, Solano (Caquetá), Manizales (Caldas) y Leticia (Amazonas) reportan la mayor riqueza de especies. Sin embargo, los registros están distribuidos de manera desigual entre grupos taxonómicos y áreas geográficas. Solo 10 especies representan el 12,6% de todas las ocurrencias, mientras que las bioregiones grandes y ecológicamente diversas como los Llanos Orientales mantienen muy pocos registros, que en cambio se concentran alrededor de los centros urbanos. Dentro de todas las categorías de uso, la riqueza de especies disminuyó con la elevación. Discutimos los predictores potenciales y las implicaciones de estos datos, así como los vacíos de conocimiento. Esta caracterización de nuestra comprensión actual de la distribución de plantas útiles se puede utilizar como línea de base para orientar los esfuerzos de investigación futuros, apoyando así la planificación y los esfuerzos de conservación de plantas en Colombia.

INTRODUCTION

The world is facing unprecedented environmental challenges, with consequences for the natural environment and the livelihoods of millions of people (Cardinale *et al.*, 2012). These changes are particularly striking in megadiverse countries such as Colombia. After decades of armed conflict, Colombia is experiencing rapid economic growth but concurrently has one of the world's highest deforestation rates (Clerici *et al.*, 2020). Large areas of forest, many still poorly explored by scientists, are being transformed into pastures for livestock or severely damaged by mining and other land-use changes. With this loss of biodiversity, nature's potential contribution to people is also being lost. Therefore, scientists are in a race against time to document Colombia's biological diversity

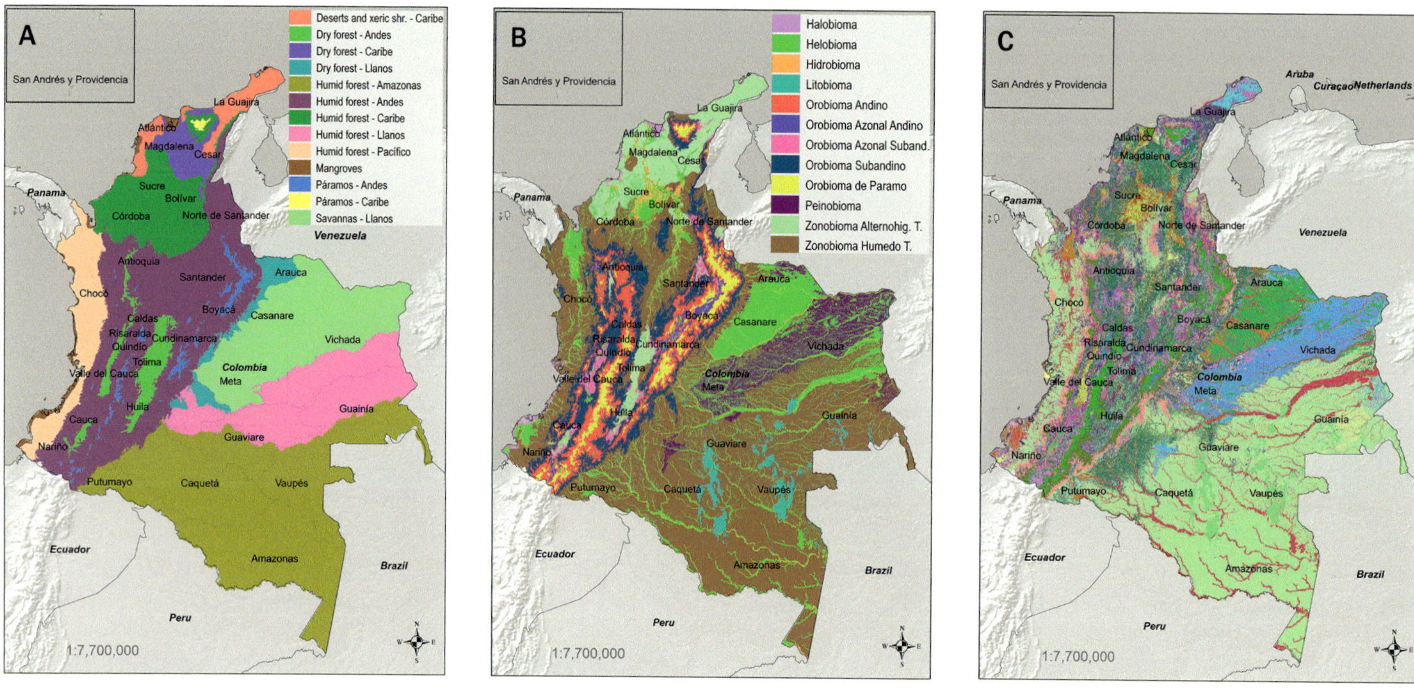

FIGURE 1. Biological regionalisation of Colombia. **A** Bioregions used in this chapter (13 classes); **B** macro-biomes (12 classes); **C** general terrestrial ecosystems (93 classes).

and its uses before they disappear. One of the fundamental questions in this race is understanding "what grows where".

Colombia is one of the most biodiverse countries on earth (Corzo *et al*., 2010; Raven *et al*., 2020), hosting at least 28,947 plant species (24% endemic; >1,000 threatened), belonging to 3,583 genera and 395 families. The country ranks second in the world for its diversity of plants and is the centre of origin of tomatoes, peppers, potatoes, chillies, and many other crops and their wild relatives (Diazgranados *et al*., 2020a; Khoury *et al*., 2016). The country's fungal diversity is also impressive, with at least 7,140 species belonging to 1,763 genera and 448 families. Given that the observed ratio between fungi and plant species in well-studied areas is 9.8:1, Colombia could harbour up to 300,000 species of fungi, representing 9% of the global diversity (Gaya *et al*., 2021). Colombia's ecosystem diversity is also very rich, with five continental ecological regions, 13 bioregions, at least 93 general ecosystems (MADS *et al*., 2017), and 23 climate provinces, from hot-arid to extremely cold and super humid (Figure 1).

As part of the Useful Plants and Fungi of Colombia project (UPFC), a multinational team of researchers led by the Royal Botanic Gardens, Kew, with the collaboration of the Instituto de Investigaciones en Recursos Biológicos Alexander von Humboldt, has been compiling, digitising, and georeferencing reported plant uses, as a first step in evaluating the status of knowledge on useful plants in the country. This effort has resulted in this publication, which presents a comprehensive annotated checklist of the plants with reported uses in Colombia, including 7,472 plant species (481 endemics; 250 threatened with extinction), belonging to 2,140 genera and 258 families.

Our life and well-being rely on the services provided by plants and fungi (Raven, 2021). This fact is true not just for regulating services at the landscape and ecosystem levels, such as soil enrichment, water regulation, and carbon storage, but also through more direct uses such as food, medicines, natural fibres, fuel, building materials, cosmetics, cultural resources, and more (Millennium Ecosystem Assessment, 2005). The preservation and sustainable use of this biodiversity are therefore essential to human well-being and economic prosperity. However, a crucial first step in protecting this diversity *in situ* is understanding its geographical distribution.

Useful plant species across bioregions

A first comprehensive assessment of the importance of different Colombian bioregions for useful plant species and the quality of available distribution data was undertaken by Bystriakova *et al*. (2021) during the first phase of the UPFC. The authors assembled a dataset of georeferenced collection localities from online databases of all vascular plants of Colombia (522,257 unique observations of 23,961 species), including useful plant species (197,166 unique distribution records of 3,870 species). The distribution of records per bioregion was investigated using a map with 14 generalised ecosystem types (bioregions). Gaps in the completeness and coverage of records in geographic and environmental space were identified to assess the reliability of species number estimates.

It was found that the largest numbers of useful plant species occurred in the Andean moist forests bioregion (13,909 species of vascular plants; 2,781 species of useful plants), followed by the Amazon (6,650; 1,612), Pacific (5,168; 1,336), Llanos (2,938; 1,151), and Caribbean moist forests (3,055; 1,166) (Bystriakova *et al.*, 2021). In addition, results showed that medicinal use was the most common useful attribute across all bioregions, followed by materials, environmental uses, and human food (Figure 2). The proportion of well-surveyed 10×10 km grid cells (*i.e.*, with ≥ 25 observation records of useful plants) was less than 50% of the total number of surveyed cells in all bioregions, except for the Andean páramo. The lowest survey coverage was observed in the three dry bioregions: Caribbean deserts and xeric shrublands, Llanos, and Caribbean dry forests. Bystriakova *et al.* (2021) highlighted that the sampling effort for useful plants of Colombia has been inadequate in the majority of the bioregions. The authors suggested prioritising future research on useful plants in the Amazon and Llanos (Orinoquia) regions, where survey completeness and coverage appeared to be less adequate than that in other regions.

Chapter aims and background

Building on the biogeographical analyses undertaken by Bystriakova *et al.* (2021), this chapter aims to describe the geographic distribution of the useful plants of Colombia (UPC) using a significantly larger dataset of compiled species (7,472 versus 3,870) published in this book. Results presented here are primarily descriptive, giving context to other chapters and offering a baseline upon which the checklist can be further improved. For detailed analyses on completeness and coverage of records in geographic and environmental space, see Bystriakova *et al.* (2021).

In this chapter, we aimed to describe the distribution, richness, species density, and other relevant characteristics of useful plant species at a range of biological and geographical scales. Biologically, we assessed unique georeferenced records of herbarium specimens (hereafter occurrences) of all plants *sensu lato* (Viridiplantae or "green plants") at the taxonomic level of species, genera, and families. Our geographic analyses considered bioregions (13), departments (32), municipalities (1,122), and grid cells (10×10 km) in recognition of the importance of both ecological and political factors in understanding and managing biodiversity.

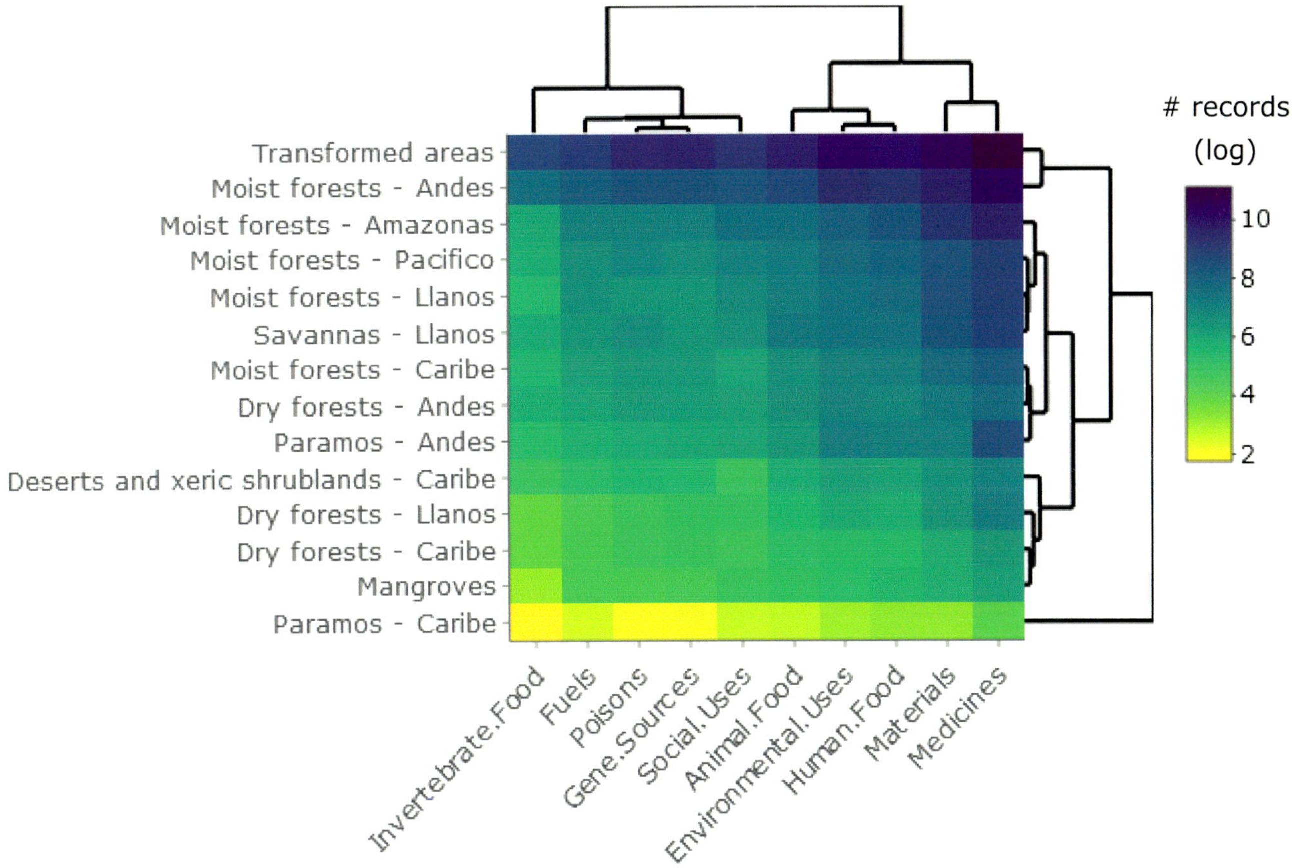

FIGURE 2. Number of records of useful plants (log-transformed) of Colombia per category of use (x-axis) and bioregion (y-axis). Dendrograms show clusters based on the number of records. From Bystriakova *et al.* (2021).

Species richness and endemism are two fundamental measures in conservation planning and studies on community ecology and biogeography (Gotelli *et al.*, 2009; Gotelli & Colwell, 2011). Despite this, they remain difficult variables to determine accurately, with various definitions, indices, and models developed for their calculation. Considering the descriptive scope of this chapter, we focused on absolute values of species richness (and the richness of higher taxonomic ranks) based on occurrence data rather than the application of predictive models. Even within this scope, many considerations are involved in measures of species richness, most simply defined as the number of species in an assemblage.

While the species-area relationship has long been used to explain plant richness at various scales, other factors such as vegetation types, elevational belts, and their interaction have also been shown to be important determinants (Kreft & Jetz, 2007). Meanwhile, the absolute value of a species' range is often used as an important determinant in the definition of endemism, an important indicator of concentrations of unique biodiversity. However, this has its limitations, with numerous alternative methods developed based on the relative range sizes of different taxa. For instance, 'weighted endemism' calculates the level of endemism of a particular grid cell based on its species richness inversely weighted by the range of each species. This value can be corrected to account for the often observed correlation between species richness and endemism, resulting in the 'corrected weighted endemism' index (Crisp *et al.*, 2001). Determining these variables for useful plant species in Colombia, and recognising both their importance and limitations, would significantly contribute to characterising our current understanding of their distribution, thereby supporting conservation efforts and the identification of further research required to fill knowledge and data gaps.

Another interesting biogeographical question is how species richness of UPC is distributed along elevation gradients. In Colombian folklore, while páramos are considered "the natural pharmacies" of the Andes, the middle elevations (*e.g.*, Andean and Sub-Andean forest) are referred to as the "food pantries of the cities". Additionally, Colombia's highest human population density occurs in the Andean region, at middle and high elevations (DANE, 2015). However, studying the relationship between richness and elevation is a complex subject due to the strong correlation between area and elevation, data limitations (*e.g.*, sampling effort) and various environmental and ecological factors affecting the species distributions. Considering these caveats and acknowledging the need for further detailed studies, we undertook preliminary descriptive analyses on the number of species by use category along elevation gradients.

MATERIALS AND METHODS

Analyses were based on the full checklist of UPC, published in this book and complied from relevant databases, reports, and scientific literature. The UPC comprises 7,472 species of green plants, classified within 2,140 genera and 258 families. Uses in the UPC were grouped into the ten categories applied in the World Checklist of Useful Plant Species (Diazgranados *et al.*, 2020b), based on the 'Level 1' categories of the Economic Botany Data Collection Standard (EBDCS) (Cook, 1995). Native, endemic, cultivated, and naturalised species were identified using the Colombian Catalogue of Plants and Lichens (CCPL) (Bernal *et al.*, 2020), the Plants of the World Online (POWO, 2019), and other reviewed sources.

Georeferenced occurrence records were downloaded from GBIF on 15 July 2021 (GBIF Occurrence Download, 2021), using the package *rgbif* v3.5.2 (Chamberlain & Boettiger, 2017) in R v.4.1.1 (R Core Team, 2020). We cleaned these data with the *shinyCCleaner* tool (Ondo *et al.*, unpub.) by removing records that were: with coordinates outside Colombia; duplicated; assigned to country's centroids or institutions; non-terrestrial; rounded to less than one decimal place; with equal or zero longitudes and latitudes; and within 1 km of the capital. Because of the importance of uses associated with some naturalised and cultivated species, which have been incorporated into the cultural practices of communities, we carried out all the analyses on all useful plants regardless of their origin.

Datasets for analyses and spatial mapping were prepared using the package *dplyr* v.1.0.6 (Wickham *et al.*, 2021). Spatial analyses were performed in R v.4.1.1 and ArcGIS Pro 2.8.0 (Esri Inc., 2021). Initially, we used the Colombian Ecosystem map (MADS *et al.*, 2017), containing five continental ecological regions, 12 general biomes, 30 synthesised ecosystems, and 93 general ecosystems. On a scale of 1:100,000, the map was too detailed with too many categories to use in our biome and ecosystem analyses and too coarse with too few categories to be used at the continental region levels. Therefore, we used a map of bioregions produced by the spatial project team, combining the map of Terrestrial Ecoregions of the World (Olson *et al.*, 2001) with the five continental ecological regions in Colombia (Bystriakova *et al.*, 2021). This map resulted in 13 discrete bioregions: deserts and xeric shrublands – Caribe, dry forests – Andes, dry forests – Caribe, dry forests – Llanos, humid forests – Amazonas, humid forests – Andes, humid forests – Caribe, humid forests – Llanos, humid forests – Pacific, mangroves, páramos – Andes, páramos – Caribe, and savannas – Llanos (Figure 1). Bystriakova *et al.* (2021) created a fourteenth 'bioregion' by merging the cells with transformed natural land. However, we did not include this category here for three reasons: 1) cells (10×10 km) of transformed vegetation generally include patches of natural vegetation, ecological corridors, and even nature reserves, where the native useful plants can be still found; 2) even in transformed areas, several local useful plants are cultivated, in orchards, gardens, riverbanks, and along roads; and 3) flora of transformed areas may differ considerably among bioregions because of contrasting environmental conditions; for example, introduced pastures for grazing cattle in the Caribbean region are floristically completely different to those in the high plains of the Andes. In addition, we did not carry out analyses of survey completeness and coverage, which were comprehensively undertaken by Bystriakova *et al.* (2021) and beyond the descriptive scope of this chapter.

We used the shapefiles of political divisions for Colombia from the GADM database v. 3.6 (https://gadm.org/). Grid analyses (10×10 km) were used to measure species density. The numbers of occurrences, species, genera, and families by bioregions, departments, and municipalities were computed using the R packages *rgdal* v1.5–23 (Bivand *et al.*, 2021), *raster* (Hijmans, 2021), and *sp* (Bivand *et al.*, 2013). Complete datasets with the results are accessible in Figshare (https://doi.org/10.6084/m9.figshare.19123157.v1).

To analyse broad-scale patterns of useful plant diversity, we calculated estimates of species richness (SR), weighted endemism (WE), and corrected weighted endemism (CWE) using the SDMtoolbox 2.4 (Brown 2014; Brown *et al.* 2017), a python-based ArcGIS toolbox for spatial analyses. The SR is the sum of unique species per cell, the WE corresponds to the sum of the reciprocal of the total number of cells each species in a grid cell is found in, and the CWE is simply the WE divided by the total species number in a cell. The WE and the CWE emphasise areas with a high proportion of species with restricted ranges, but in the latter, the areas are not necessarily species-rich. The SR, WE, and CWE were estimated using grid cells of 10×10 km. The resulting raster layers were classified using Natural Breaks (Jenks) with ten classes.

Studying the relationship between richness and elevation is a complex subject due to the strong correlation between area and elevation. At higher elevations, geographical area is restricted, therefore resulting in lower species richness. This relationship is not usually linear but often unimodal, with a mid-elevation peak (Rahbek, 1995). Species with moderate to large elevational ranges are more likely to overlap at mid-elevation than at the extremes, resulting in a phenomenon termed the "mid-domain effect" (MDE). MDE is attributed to geometrically constrained stochastic processes affecting the underlying range-size frequency distribution (Colwell, 2008). However, there may be multiple additional factors, such as area availability and environmental variables, that affect the species richness-elevation relationship and result in concentrations around midpoint attractors (Colwell *et al.*, 2016). Recognising the methodological challenges and data limitations, we undertook a simple descriptive analysis of useful species richness using category along elevation to provide an initial picture of the potential patterns.

To explore useful plant richness along elevation gradients, we based our analyses on the elevation ranges of each species reported in the Colombian Catalogue of Plants and Lichens, which were classified by experts of each group (Bernal *et al.*, 2020). The total elevation range (*i.e.*, range between the minimum and maximum elevations reported for all species) was divided every 100 m. Each 100 m band was transformed into a bin using the R package *linbin* (Welty *et al.*, 2015). The number of species occurring in each bin was counted with the R package *data.table* 1.14.0 (Dowle *et al.*, 2021). Results were visualised with the R package *ggplot2* (Wickham *et al.*, 2016). We also compared the empirical curve of species number along the elevation with the computed curve assuming a 'mid-domain effect" (MDE). The empirical richness means of computed richness, standard deviation, and Veech's statistic (*D*) (Veech, 2000) were obtained using 1,000 randomisations for 6,581 species with elevation ranges, at 50 sampling sites (from 0 to 5,000 m a.s.l.), using Range Model5Win (Colwell, 2008; Colwell *et al.*, 2016). The null distribution was estimated using the empirical range frequency distribution with random midpoints (box 5 in Range Model5Win). Veech's statistic measures the mean displacement of the simulated curve from the average random curve, treating MDE as a falsifiable hypothesis (Colwell, 2008). Using a Wilcoxon Test in R, we tested for differences between the empirical richness and the computed richness means.

RESULTS

Taxonomic coverage and species origin

The cleaned database of georeferenced records (occurrences) included 435,230 occurrences, belonging to 6,320 (85%) species, 1,936 (91%) genera, and 248 (96%) families. Occurrences were unevenly distributed across species, with the top ten species accounting for 12.6% of the total number of records (*Acacia decurrens* with 10,962 occurrences; *Eucalyptus globulus* 9,578; *Pinus radiata* 9,353; *Gliricidia sepium* 6,596; *Cedrela angustifolia* 4,588; *Guazuma ulmifolia* 4,484; *Quercus humboldtii* 3,187; *Pinus patula* 3,094; and *Tapirira guianensis* with 2,895 records). In total, 1,161 useful species did not have any adequate georeferenced records, 442 species had only one occurrence, 339 only two occurrences, and 2,172 had ten or fewer occurrences. The list used for the analyses included 6,897 (92%) species with confirmed presence in at least one Colombian department, and 6,581 (89%) species showed elevation range information. The majority of useful plants of Colombia are native (at least 5,881 species, 78.6%), including 481 endemic species to Colombia (8.2% of the native species). There are at least 643 naturalised species and 1,077 cultivated species reported. However, information about species origin is still unclear for 442 species.

Bioregions

Of the 13 bioregions assessed, the Andes region as a whole (including dry forests, humid forests, and páramos) supported the highest numbers of individual occurrences, species, genera, and families (Figure 3 A–D). The Orinoquia (including the dry forests and savannas in the Llanos Orientales) supported the fewest useful plant records, followed by the Caribbean dry forests and páramos, and the mangroves (Figure 3 A). The humid forests in the Andes held 52.4% of all the occurrences for the country, five times more than the Amazon region, which is the largest by area (Table 1). The Caribbean páramos, located in the Sierra Nevada de Santa Marta, had only 177 occurrences.

The humid forests in the Andes supported occurrence records for 82.3% of the useful species in Colombia, representing almost double the species richness of the Amazon region (5,236 compared to 2,848 species) (Table 1). It was also the richest in terms of genera (90.4%) and families (95.2%). The dry forests in the Andes had more genera and families than the Amazon region (1,028 and 186 versus 1,012 and 179, respectively).

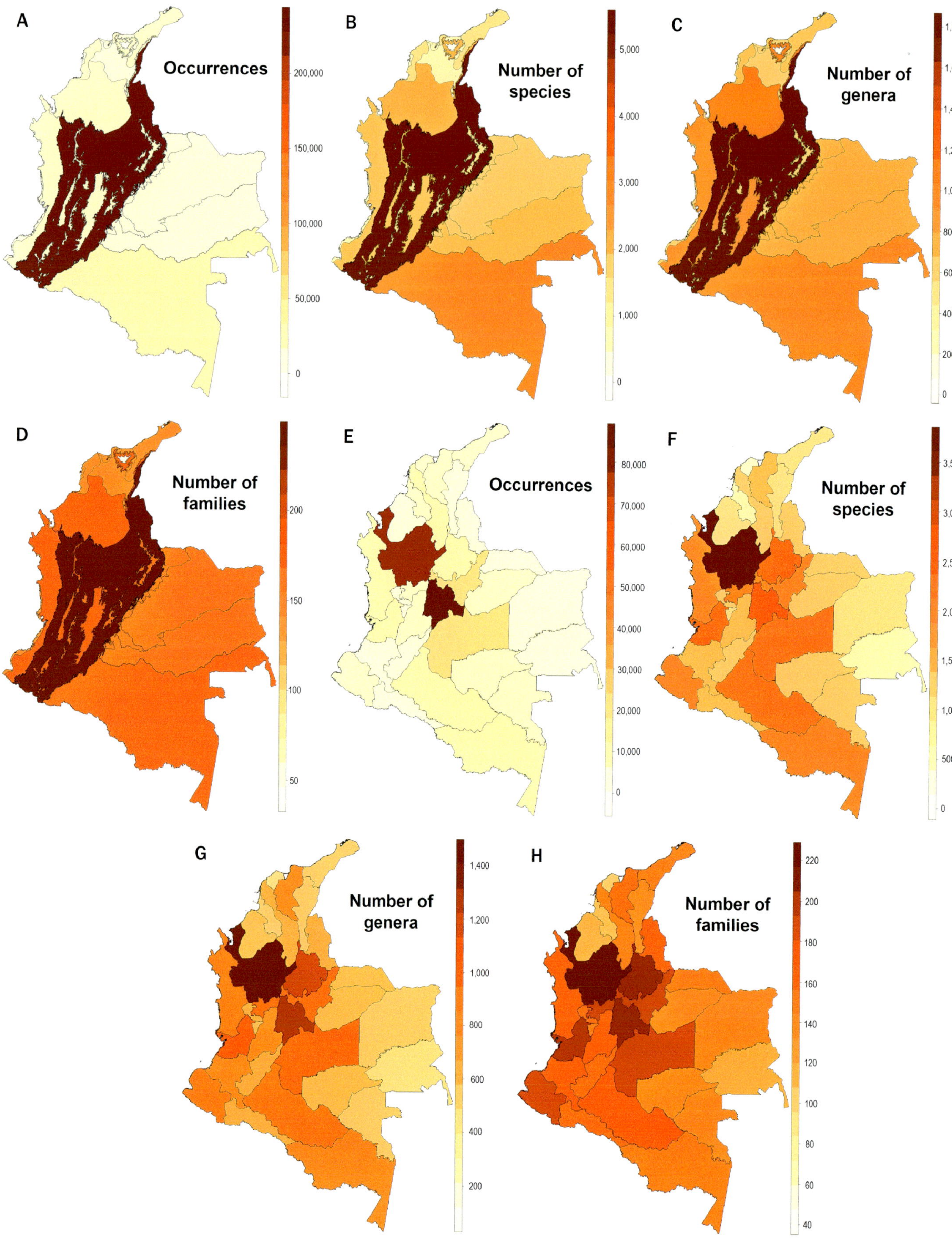

FIGURE 3. Occurrences, species, genera and family richness of useful plants of Colombia by bioregions **A–D** and department **E–H**.

TABLE 1. Occurrences, species, genera, and family richness of useful plants of Colombia per bioregion. Pale green: biomes with the highest overall species richness.

Biome region	Area (100 km²)	Occurrences (%)	Species (%)	Genera (%)	Families (%)
Deserts & xeric shrublands – Caribe	273.24	27,006 (6.2)	1,368 (21.6)	746 (38.5)	151 (60.9)
Dry forests – Andes	291.34	31,189 (7.2)	2,204 (34.9)	1,028 (53.1)	186 (75)
Dry forests – Caribe	248.42	3,660 (0.8)	942 (14.9)	575 (29.7)	141 (56.9)
Dry forests – Llanos	245.33	7,539 (1.7)	1,668 (26.4)	815 (42.1)	159 (64.1)
Humid forests – Amazonas	3215.99	41,260 (9.5)	2,848 (45.1)	1,012 (52.3)	179 (72.2)
Humid forests – Andes	2559.94	228,047 (52.4)	5,236 (82.8)	1,750 (90.4)	236 (95.2)
Humid forests – Caribe	862.46	20,494 (4.7)	2,398 (37.9)	1,060 (54.8)	188 (75.8)
Humid forests – Llanos	1324.20	16,136 (3.7)	2,026 (32.1)	839 (43.3)	159 (64.1)
Humid forests – Pacifico	582.37	16,737 (3.8)	2,145 (33.9)	938 (48.5)	183 (73.8)
Mangroves	85.81	7,980 (1.8)	1,175 (18.6)	673 (34.8)	142 (57.3)
Paramos – Andes	142.30	17,658 (4.1)	1,326 (21.0)	634 (32.7)	171 (69.0)
Paramos – Caribe	12.39	177 (0.1)	93 (1.5)	77 (4.0)	46 (18.5)
Savannas – Llanos	1525.94	15,936 (3.7)	1,716 (27.1)	802 (41.4)	164 (66.1)

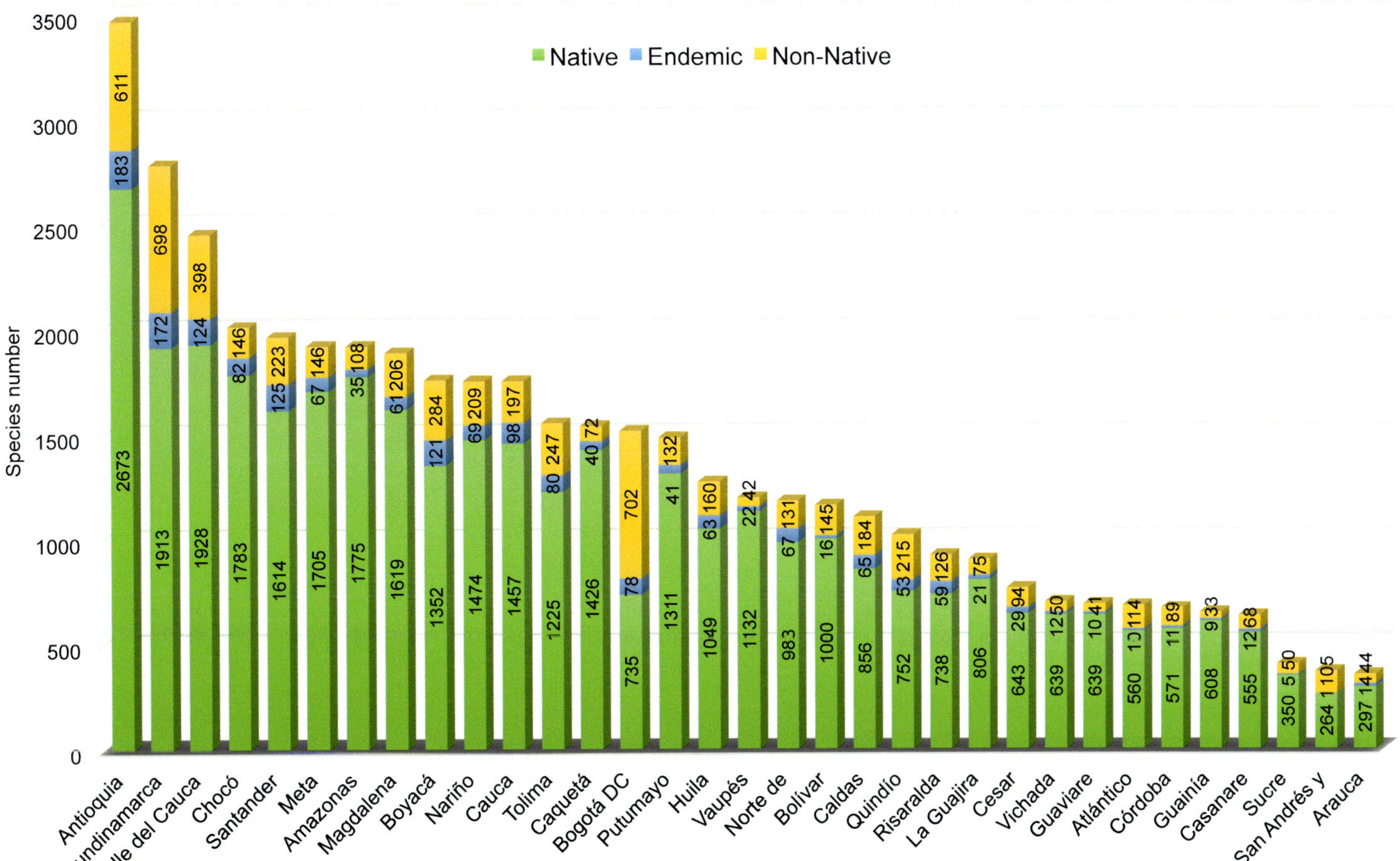

FIGURE 4. Species richness of useful plants of Colombia by departments.

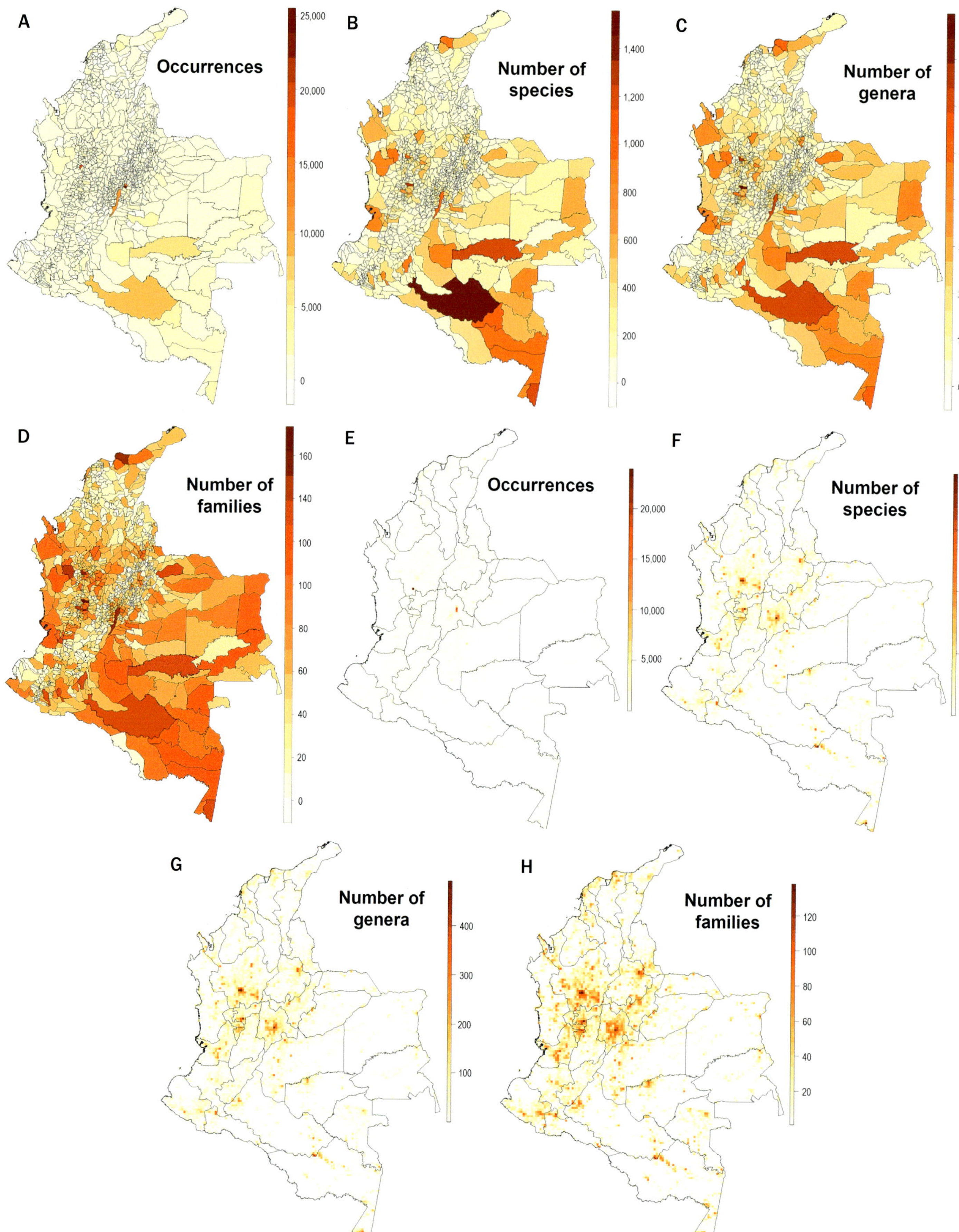

FIGURE 5. Occurrences, species, genera and family richness of useful plants of Colombia by municipality **A–D** and grid (10×10 km) **E–H.**

TABLE 2. Top 10 municipalities by overall species richness of useful plants (pale green).

Name	Area (100 km²)	Occurrences (%)	Species (%)	Genera (%)	Families (%)
Solano (Caquetá)	496.8	9,425 (2.2)	1,455 (23.0)	589 (30.4)	138 (55.6)
Manizales (Caldas)	4.4	4,057 (0.9)	1,321 (20.9)	745 (38.5)	162 (65.3)
Leticia (Amazonas)	41.7	3,692 (0.8)	1,236 (19.6)	558 (28.8)	134 (54.0)
San José del Guaviare (Guav.)	242.1	6,582 (1.5)	1,212 (19.2)	596 (30.8)	131 (52.8)
Medellín (Antioquia)	3.6	4,887 (1.1)	1,131 (17.9)	667 (34.5)	157 (63.3)
Santafé de Bogotá (Cund.)	16.5	13,400 (3.1)	1,107 (17.5)	608 (31.4)	153 (61.7)
Puerto Santander (Amazonas)	90.4	3,867 (0.9)	1,083 (17.1)	476 (24.6)	124 (50.0)
Florencia (Caquetá)	20.2	2,390 (0.5)	990 (15.7)	539 (27.8)	135 (54.4)
Urrao (Antioquia)	26.4	3,125 (0.7)	978 (15.5)	527 (27.2)	146 (58.9)
Santa Marta (D. Esp.) (Mag.)	23.9	2,316 (0.5)	960 (15.2)	584 (30.2)	151 (60.9)

Departments

Antioquia and Cundinamarca (including Bogotá DC) were the richest departments in terms of individual occurrences, species, genera, and families (Figure 3 E–H), whilst the poorest departments were in the Eastern side of the country. In terms of individual occurrences, Antioquia and Cundinamarca each had over 80,000 records, four times more than Boyacá, in the third position. Eight of the 32 departments had fewer than 5,000 records each. This scenario translated to species richness of 3,616, 2,783, and 2,587 species, respectively, across the top three richest departments. Meanwhile, Sucre, San Andrés y Providencia, and Arauca supported the lowest levels of species richness, with 405, 370, and 355 species, respectively (Figure 4). Antioquia, Cundinamarca, and Bogotá DC had the largest numbers of non-native species, all with more than 600 (Figure 4).

A similar pattern was observed for the richness of genera (Figure 3 G), with Antioquia, Cundinamarca, and Santander containing occurrence records for the greatest diversity of genera (1,405, 1,221, and 1,167, respectively), while Guanía, Atlántico, and San Andrés y Providencia had the lowest genus richness (426, 413, and 118 genera). Results at the family level showed the same departments in the top and bottom positions (Figure 3 H).

Municipalities

Most municipalities displayed relatively low numbers of occurrences compared to Guatavita (Cundinamarca department; 23,883), Jericó (Antioquia; 19,459), Sesquilé (Cundinamarca; 14,856), Santafé de Bogotá (Cundinamarca; 13,400), and Solano (Caquetá; 9,425 occurrences) (Figure 5 A). Excluding Solano, these are relatively small municipalities (the first three being less than 200 km^2 in size). Solano, however, is the largest municipality in the country, at over 43,000 km^2, making it larger than 23 of the departments. Twenty municipalities had no occurrences, 27 had one or two, and 137 (13%) had fewer than 10. The top 10 richest municipalities in species number are shown in Table 2 and Figure 5 B. Solano, Manizales (Caldas), and Leticia (Amazonas) are the richest in terms of species (1,455, 1,321, and 1,236 species, respectively). Three large cities, Manizales (Caldas), Bogotá DC, and Medellín (Antioquia), had the highest genus (745, 667 and 608, respectively) and family richness (162, 153 and 157, respectively).

Grid analyses and endemicity

Analyses by grid cells (10×10 km) showed gaps in occurrences in the Amazon and Orinoco (Llanos) regions (south-east of Colombia). Meanwhile, a few cells with particularly high occurrence numbers corresponded to the areas around the cities of Bogotá and Medellín (Figure 5 E). Five species-rich areas were clearly identified, all associated with the areas of influence of large cities: Bogotá (Cundinamarca), Medellín (Antioquia), the coffee belt (Manizales-Pereira-Armenia), Bucaramanga (Santander), and Cali (Valle del Cauca) (Figure 5 F). The same pattern was repeated for the richness of genera (Figure 5 G). The map showing family richness displayed some additional hotspots in Santa Marta (Magdalena), Leticia (Amazonas), Pasto (Nariño), and Florencia (Caquetá) (Figure 5 H).

Bogotá and Medellín's areas supported the highest diversity of species, followed by the coffee belt (Caldas-Risaralda-Quindío), Santander, and the Valle del Cauca (Figure 6 A). The weighted endemism map also displayed the same hotspots (Figure 6 B). When correcting by the species number (i.e., corrected weighted endemism), several hotspots of high endemic value emerged in areas such as northern Huila, Caquetá, Vaupés, Vichada, and Casanare (Figure 6 C).

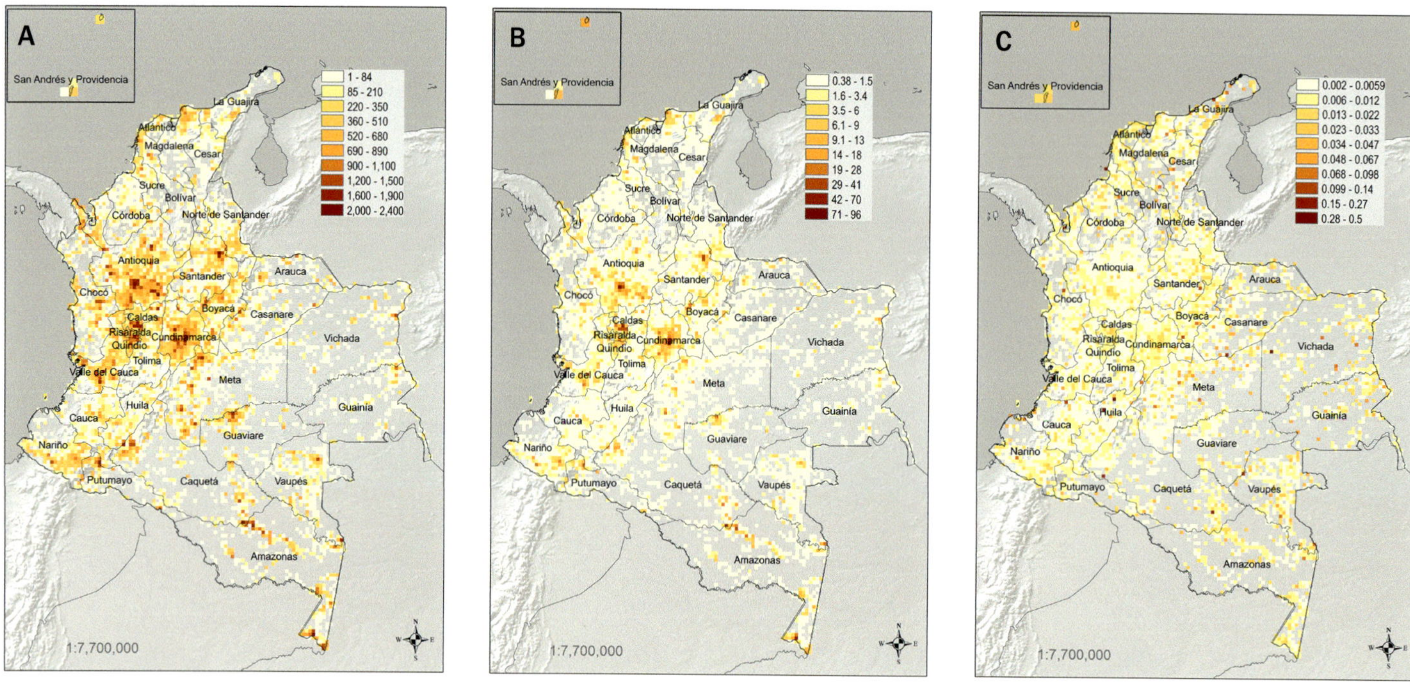

FIGURE 6. Biological regionalisation of Colombia. **A** Bioregions used in this chapter (13 classes); **B** macro-biomes (12 classes); **C** general terrestrial ecosystems (93 classes).

Elevation

Species richness along elevation gradients showed a strong inverse relationship between the number of useful species and increasing elevation in all categories of use (Figure 7 A). A linear regression of all useful plants along elevation was highly significant ($R^2 = 0.6$; p-value < 0.0001). Analyses by use types showed that plant species with medicinal uses were the most numerous at any elevation and reached the highest elevations. The diversity of plants with environmental use and gene sources was relatively constant between 100 m and 1,500 m before gradually decreasing as elevation increased beyond this point. Very few species with uses other than medicinal occurred above 4,000 m, with only medicinal plants reaching the highest reported elevations of almost 5,000 m (*Cerastium arvense*, *Hypochaeris sessiliflora* and *Stevia lucida*). We observed a weak mid-domain effect (Figure 7 B): for the observed curves, $D = 523.44$; for 1,000 null curves, mean $D = 18.4$, range of $D = 8.92–40.31$. However, we cannot reject the null hypothesis of no differences between the means of the observed distribution and the computed distribution based on 1,000 replications (p-value = 0.65).

DISCUSSION

Based on publicly available species occurrence data, this chapter presents a geographic description of the current state of knowledge on the useful plants of Colombia. As part of this process, we have characterised useful plant richness at varying biological and geographical scales and highlighted areas of high endemism. These analyses complement and add to that of Bystriakova *et al.* (2021) and can be used as a baseline upon which an improved understanding of the useful plants of Colombia is built over time. We used a larger dataset of georeferenced occurrences of useful plants (435,230 occurrences, belonging to 6,320 species and 248 families, compared to 197,166 occurrences of 3,870 species and 228 families in Bystriakova *et al.* (2021)), made possible through the additional work undertaken to increase the comprehensiveness of the UPC. In contrast to Bystriakova *et al.* (2021), we did not compare the distribution of all vascular plants with that of useful plants or estimate survey completeness and coverage. However, we present counts of species by department and municipality and a first descriptive view on the number of species along elevation. However, there are limitations involved in the data used and analyses undertaken. In this section, we provide context for the patterns observed, discuss the data limitations involved, and outline potential next steps to further our understanding of the useful plants of Colombia.

Origin of useful plants in Colombia

The checklist of UPC drew on numerous information sources to classify the origin of useful species in Colombia. The proportion of naturalised (8.6%) to native useful species (78.6%) was relatively high, compared to 2.3% and 93.6%, respectively, for all plant species in Colombia. In fact, 643 of the 677 naturalised species have uses reported, suggesting that their utility may be a potential cause for their introduction to natural habitats. A number of these naturalised useful species have become highly invasive both in Colombia and globally (Cárdenas-López *et al.*, 2017; Turbelin *et al.*, 2017), reflected in the higher number of occurrences (e.g.,

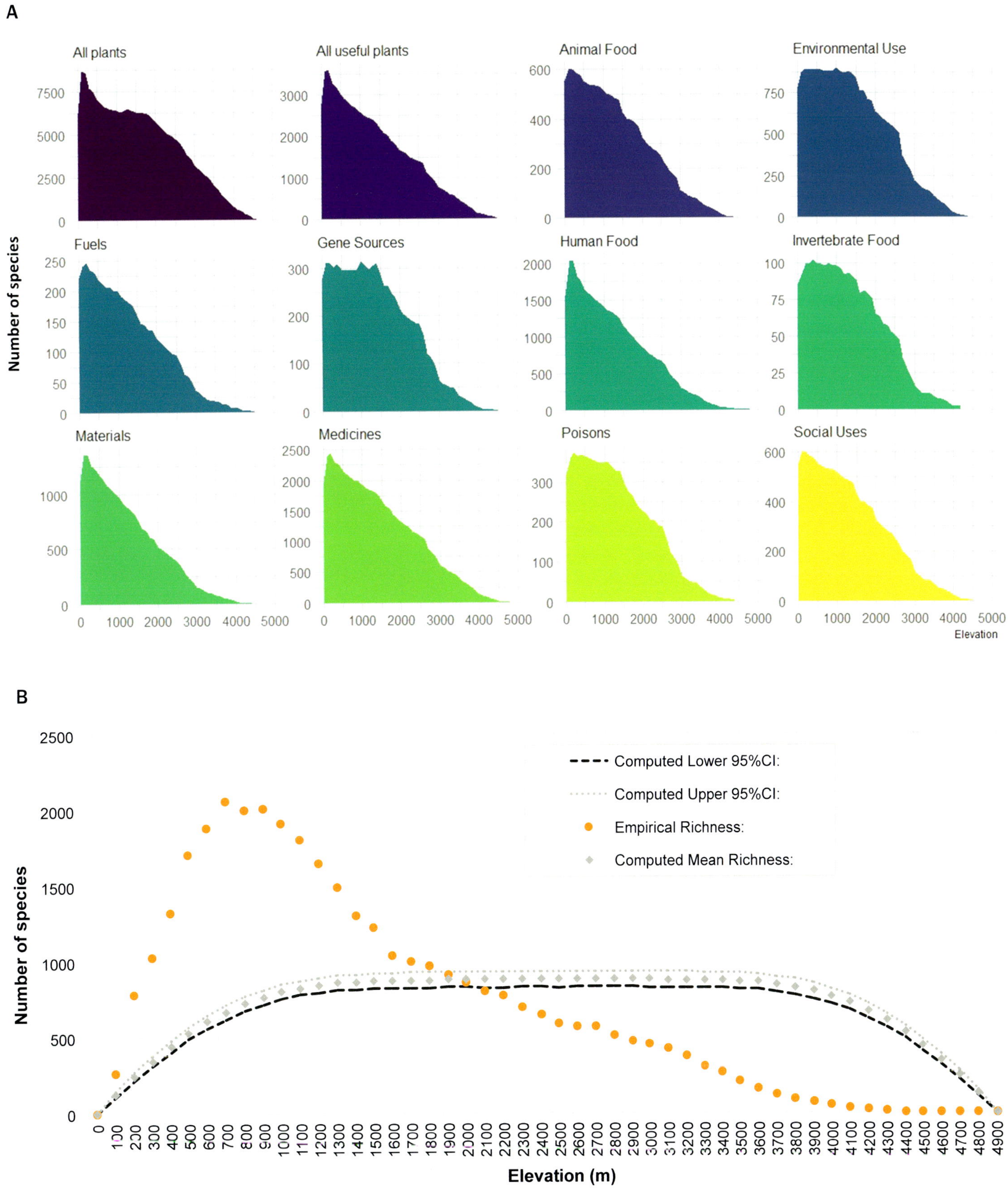

FIGURE 7. Richness of useful plants of Colombia along elevation. **A** Richness of all plants, all useful plants and species by level 1 category of use. **B** Empirical richness of useful plants compared to the mid-domain effect (MDE) null-model predictions.

Acacia decurrens, *Eucalyptus globulus*, *Pinus radiata*, and *P. patula*). This result is an important consideration as invasive species are a major driver of biodiversity loss globally, with huge associated livelihood and economic costs (Millennium Ecosystem Assessment, 2005). This scenario poses questions about how best to manage useful plant species, especially introduced and naturalised species, in a way that sustainably maximises their benefits while causing minimal impacts on the wider environment.

Distribution of occurrence records

Occurrence records were unevenly distributed across species. This result is a common pattern in distribution data of this kind, where the number of singletons (species represented by one individual) and doubletons (species represented by two individuals) often represent a high proportion of species (Gotelli & Colwell, 2011), in our case accounting for 10.4% of UPC. Occurrences were also unevenly distributed across bioregions. Despite the large land area of the llanos Orientales (Orinoquia region), this area's dry forests and savanna bioregions supported a relatively low number of occurrence records for useful plants. This result is consistent with previous species richness estimates for Colombia (Rangel-Ch, 2015). Still, it is likely to be partly due to under-sampling, as reported by Bystriakova *et al*. (2021). The estuary nature of large parts of the llanos causes it to be flooded for several months a year, affecting road connectivity. This is an important factor, with a global study on the patterns of plant diversity and floristic knowledge highlighting flooded grasslands and flooded savannas as especially lacking adequate data (Kier *et al*., 2005).

Historical and security factors are also likely to have affected sampling efforts. Traditionally, there has been more focus on sampling in tropical rainforests, such as remote areas of the Amazon region (Redford *et al*., 1990), than in the Orinoquia region. Meanwhile, some of the areas in the Orinoquia region are well-known corridors of illegal activities (e.g., "Corredor del Sarare-Cubará", "Corredor de Sácama-Tame-Saravena"). They have had a long history of violence hampering scientific research. Similar circumstances occur in other poorly studied bioregions, such as the Pacific mangroves and the Caribbean dry forests and páramos. The areas with the highest records remain near large cities, which are likely to have been most intensively surveyed as they are more easily accessible with better supporting infrastructure and safety conditions.

This pattern of uneven occurrence distribution is also visible at the department level (Figure 2). Departments such as Guanía and Vichada in the Orinoco region show relatively low numbers of occurrences and species. Their capital cities, Puerto Inírida and Puerto Carreño, are only accessible by plane, with infrequent, relatively costly flights. These areas therefore remain hard to access for biology students and researchers. Meanwhile, despite supporting a wide range of habitats and a large elevation range from sea level to over 5,000 m, the Arauca department shows very poor species diversity based on occurrence records. This result is likely to be because conflict has impeded research in the area for decades. A similar situation is seen in the departments of Norte de Santander, Cesar, Cauca, and Nariño. The latter, bordering Ecuador, has many characteristics which would be expected to make it one of the most biodiverse departments in Colombia. Geographically, it supports a range of biomes, including humid forests of the Amazon and the Pacific, Andean humid and dry forests, isolated areas of páramos, and mangroves, with large areas continuing to be covered by natural vegetation and protected as part of indigenous reserves (MADS *et al*., 2017). The cultural and ethnic diversity of the department is also notable, with the 2018 census identifying 15.7% of its population as belonging to indigenous groups and 17.8% as Afro-Colombians (DANE, 2018). However, the combination of years of conflict and its distance from the main cities of Colombia has limited the scientific exploration of its rich flora and its uses.

Results at the municipality level can be explained by considering their unequal land areas and the influence of urban centres. Very large municipalities or small municipalities near large urban centres tend to have a higher diversity of useful plants. In the second case, several of the richest municipalities by area (*e.g.*, Tenjo, Sasaima, Tabio) are within city regions or urban ecological footprints (Ortega-Montoya & Johari, 2020). Last, human population density is also an important factor. While 29 municipalities in the Amazon and Orinoco regions support fewer than one inhabitant per km^2, 25 municipalities, mainly in the Andes, have at least 1 million inhabitants per km^2 (DANE, 2018). This scenario produces disproportionate reports of plant uses across the territory.

The uneven distribution of sampling effort and the factors contributing to it are common limitations associated with the use of occurrence data, and they have knock-on effects for subsequent analyses and for conservation and management decisions (Gotelli & Colwell, 2011). While accounting for these data limitations through modelling approaches was beyond this chapter's scope, they should be recognised when interpreting the presented data and statistics on species richness and endemism.

Biogeographic characteristics of useful plant richness

In terms of overall species, genera, and family richness, the humid forests in the Andes were by far the richest bioregion, followed by the humid forests in the Amazon and Caribe (Table 1). These results coincide with the findings reported by Bystriakova *et al*. (2021). The regional importance of the Andes in terms of biodiversity, ahead of the Amazon region, has been indicated numerous times. It is possibly due to the higher variety of niches resulting from the area's diverse climates and soils (Rangel-Ch & Aguilar, 1995). Another reason is the population size of the region. All departments within the Andean region's combined population are over 26 million (55% of the country's population) (DANE, 2018). This scenario is important because useful plants can only be recognised where there is interaction with human communities, i.e., when they are used and reported.

The mangrove and páramos bioregions have a noteworthy diversity of useful plants, despite their very small size compared to other bioregions (Table 1). Mangroves are known to support the production of many goods of economic interest, including wood for firewood and charcoal, timber for construction of dwellings, furniture, boats and fishing gear, tannins for the textile industry, food, and medicinal extracts (Bandaranayake, 1998, García & Polanía, 2007). In the páramos, numerous studies have reported plants' uses and the ecosystem services they provide (De La Torre *et al*., 2008, Diazgranados *et al*., 2021, Duarte & Parra, 2015). Páramos are often referred to as "the pharmacies of the Andes" due to the abundance of a few important medicinal species of Asteraceae, a family supporting many species known to have local medicinal uses (Simpson, 2009). Additionally, páramo grasses are used to thatch roofs of shelters, while *Espeletia* species were traditionally used as a source of turpentine, for construction, as well as medicinally (Cuatrecasas, 2013). Perhaps the concentration of these useful species in smaller areas and the greater number of studies undertaken due to relatively good accessibility make these bioregions significant for the presence of useful plant species.

Based on the analyses of survey completeness and coverage, Bystriakova *et al*. (2021) suggested prioritising future research on useful plants in the Amazon and Llanos (Orinoquia) regions, for which sampling appeared to be less adequate compared to that in other regions. Our results support this suggestion, with the humid forests in the Amazon, Savannas and Llanos being the least surveyed areas (Figures 5 and 6). These regions have the lowest human population density, with large proportions of their area remaining understudied (Kier *et al*., 2005). These two factors may explain the results, and further ethnobotanical studies targeting such habitats are likely to unveil a much higher diversity of useful plants in them.

The analysis of corrected weighted endemism reveals several hotspots (seen as dark brown cells in Figure 6 C) of high endemic value in areas such as northern Huila, Amazonia (Amazonas, Caquetá, and Vaupés), and Orinoquia (Vichada and Casanare). Those hotspots might coincide with well-surveyed small areas, such as localised studies in nature reserves or around some urban centres. Also, they may reflect the importance of the unique diversity of useful plants of these regions. None of those hotspots can be seen in departments such as Cundinamarca and Antioquia.

We undertook a simple descriptive analysis of useful species richness using category along elevation, uslng equal elevation range size bands (every 100 m). This analysis suggested a general trend of decreasing useful plant species richness with increasing elevation across all categories of use, except for environmental uses and gene sources, where species numbers remain relatively constant between 100 and 1,500 m (Figure 7 A). This result contradicts the common belief that more useful species of certain use types can be found at higher elevations, with the highest species richness found to occur below 1,000 m elevation. In addition, the distribution of the number of useful species along elevation supported a weak MDE. The available data did not provide explanatory variables (climate, ecological conditions, and sampling effort) to address this relationship in detail, which was beyond the scope of this chapter. However, we suggest further exploring the relationship between useful plant richness and elevation, considering sampling across elevation bands of equal area, and including predictors such as climate and human population density.

CONCLUSIONS

The geographic analyses of the checklist of useful plants presented here reveal both important distributional patterns and gaps in data, which can be used to prioritise further work. The uneven distribution of sampling effort is evident. This scenario is expected in a country like Colombia, where several areas are only now becoming accessible to researchers while others remain difficult to study. At the bioregion level, the humid forests of the Andes were found to support the richest diversity of useful plants. More studies are required in areas such as the Orinoco savannas and forests, for which the information is still limited compared to that for other areas. At the department and municipality levels, size, population density, geolocation, conflict situation, and knowledge availability could explain the results. At the grid level, the observed hotspots of richness can often be explained by the vicinity of large urban centres, highlighting the potential bias in the data available to date. Colombia's inequality is evident not just in social indicators but also in terms of the distribution of scientific knowledge on the diversity of useful plants. We hope that future research can focus on filling in the identified gaps to decrease current biases in data and knowledge, thereby revealing more representative geographic patterns of useful plants in Colombia.

Acknowledgements

This work directly contributes to the 'Useful Plants and Fungi of Colombia' project, supported by a Professional Development and Engagement grant under the Newton-Caldas Fund partnership. The grant is funded by the UK Department for Business, Energy and Industrial Strategy (BEIS) and Minciencias and delivered by the British Council. LK is supported by a studentship awarded by the Natural Environment Research Council [Grant Number NE/S007229/1].

References

Bandaranayake WM (1998) Traditional and medicinal uses of mangroves. *Mangroves Salt Marshes,* 2:133-148.

Bernal R, Gradstein SR. Celis M (2020) *Catálogo de Plantas y Líquenes de Colombia*. Bogotá, Universidad Nacional de Colombia. doi: 10.15472/7avdhn

Bivand R, Keitt T, Rowlingson B (2021) *rgdal: Bindings for the 'Geospatial' Data Abstraction Library. R package version 1.5-23.* Available at: https://CRAN.R-project.org/package=rgdal

Bivand R, Pebesma E, Gómez-Rubio V (2013) *Applied spatial data: analysis with R.* New York, Springer. 405p.

Brown JL (2014) SDMtoolbox: a python-based GIS toolkit for landscape genetic, biogeographic and species distribution model analyses. Methods Ecol Evol, 5: 694-700. doi: 10.1111/2041-210X.12200.

Brown JL, Bennett JR, French CM (2017) SDMtoolbox 2.0: The next generation Python-based GIS toolkit for landscape genetic, biogeographic and species distribution model analyses. PeerJ, 2017(12). doi: 10.7717/peerj.4095.

Bystriakova N, Tovar C, Monro A, Moat J, Hendrigo P, Carretero J, Torres-Morales G, Diazgranados M (2021) Colombia's bioregions as a source of useful plants. *PLoS One,* 16: e0256457.

Cárdenas-López D, Baptiste MP, Castano N (eds) (2017) *Plantas exóticas con alto potencial de invasión en Colombia.* Bogota, Instituto de Investigación de Recursos Biológicos Alexander von Humboldt.

Cardinale BJ, Duffy JE, Gonzalez A, Hooper DU, Perrings C, Venail P, Narwani A, Mace GM, Tilman D, Wardle DA (2012) Biodiversity loss and its impact on humanity. *Nature,* 486, 59-67.

Chamberlain SA, Boettiger C (2017) R Python, and Ruby clients for GBIF species occurrence data.

Clerici N, Armenteras D, Kareiva P, Botero R, Ramírez-Delgado JP, Forero-Medina G, Ochoa J, Pedraza C, Schneider L, Lora C, Gómez C, Linares M, Hirashiki C, Biggs D (2020) Deforestation in Colombian protected areas increased during post-conflict periods. *Scientific Reports,* 10, 4971.

Colwell RK (2008) RangeModel: tools for exploring and assessing geometric constraints on species richness (the mid-domain effect) along transects. *Ecography,* 31, 4-7.

Colwell RK, Gotelli NJ, Ashton LA, Beck J, Brehm G, Fayle TM, Fiedler K, Forister ML, Kessler M, Kitching RL (2016) Midpoint attractors and species richness: modelling the interaction between environmental drivers and geometric constraints. *Ecology Letters,* 19, 1009-1022.

Cook F (1995) Economic botany data collection standards, prepared for the international working group on taxonomic database for plant science (TDWG). *Richmond,* Royal Botanic Gardens, Kew.

Corzo G, Ramírez W, Salamanca B, Londono MC, Fonseca C, Castellanos C, Alcázar C, Lasso C, García H (2010) Planeación ambiental para la conservación de la biodiversidad en las áreas operativas de Ecopetrol localizadas en el Magdalena Medio y los Llanos Orientales. Bogotá D.C., Colombia, Instituto de Investigación de Recursos Biológicos Alexander von Humboldt y Ecopetrol.

Crisp MD, Laffan S, Linder HP, Monro A (2001) Endemism in the Australian flora. *Journal of Biogeography,* 28, 183–198.

Cuatrecasas J (2013) *A systematic study of the subtribe Espeletiinae,* New York, USA, The New York Botanical Garden.

DANE (2015) *Cuentas Departamentales,* https://www.dane.gov.co, Departamento Administrativo Nacional de Estadística de Colombia.

DANE (2018) *Censo Nacional de Población y Vivienva* [Online]. Gobierno de Colombia. Available: https://www.dane.gov.co/index.php/estadisticas-por-tema/demografia-y-poblacion/censo-nacional-de-poblacion-y-vivenda-2018/cuantos-somos [Accessed 6 August 2021].

De La Torre L, Navarrete H, Muriel P, Macía MJ, Balslev H (2008) *Enciclopedia de las Plantas Útiles del Ecuador (con extracto de datos),* Herbario QCA de la Escuela de Ciencias Biológicas de la Pontificia ....

Diazgranados M, Allkin B, Ávila F, Baker W, Bishop D, Black N, Bystriakova N, Carretero J, Castellanos-Castro C, Cely M, Colville L, Cossu T, Davies L, Díaz A, Dsouza J, García F, Gaya E, Graves E, Green L, Haigh A, Hammond DS, Hendrigo P, Herrera A, Hillebrecht W, Howes MJ, Iggulden D, Kersey P, Kor L, Mattana E, Milliken W, Mira M, Moat J, Monro A, Morley J, Murphy V, Nesbitt M, Ondo I, Parker L, Phillips S, Piquer M, Pritchard HW, Reis P, Rojas T, Ruff J, Stone M, Tanimoto M, Torres G, Tovar C, Turner R, Taylor R, Utteridge T, Vargas N, Weech MH, White K, Wilkin P, Williams J, Ulian T (2020a) *ColPlantA: Colombian resources for Plants made Accessible. 2nd edition.,* Richmond, Royal Botanic Gardens, Kew.

Diazgranados M, Allkin B, Black N, Cámara-Leret R, Canteiro C, Carretero J, Eastwood R, Hargreaves S, Hudson A, Milliken W, Nesbitt M, Ondo I, Patmore K, Pironon S, Turner R, Ulian T (2020b) World Checklist of Useful Plant Species. Royal Botanic Gardens, Kew, Knowledge Network for Biocomplexity.

Diazgranados M, Tovar C, Etherington TR, Rodriguez-Zorro PA, Castellanos-Castro C, Galvis Rueda M, Flantua SGA (2021) Ecosystem services show variable responses to future climate conditions in the Colombian paramos. *PeerJ,* 9, e11370.

Dowle M, Srinivasan A, Gorecki J, Chirico M, Stetsenko P, Short T, Lianoglou S, Antonyan E, Bonsch M, Parsonage H, Ritchie S, Ren K, Tan X, Saporta R, Seiskari O, Dong X, Lang M, Iwasaki W, Wenchel S, Broman K, Schmidt T, Arenburg D, Smith E, Cocquemas F, Gomez M, Chataignon P, Blaser N, Selivanov D, Riabushenko A, Lee C, Groves D, Possenriede D, Parages F, Toth D, Yaramaz-David M, Perumal A, Sams J, Morgan M, Quinn M, @Javrucebo, @Marc-Outins, Storey R, Saraswat M, Jacob M, Schubmehl M, Vaughan D, Hocking T, Silvestri L, Barrett T, Hester J, Damico A, Freundt S, Simons D, Andrade ESD, Miller C, Meldgaard JP, Tlapak V, Ushey K, Eddelbuettel D, Schwen B (2021) *data.table 1.14.1.* Available at https://rdatatable.gitlab.io/data.table/.

Duarte B, Parra S (2015) Plantas del páramo y sus usos para el buen vivir: Páramos de Guerrero y Rabanal. Bogota, Colombia: Instituto de Investigacion de Recursos Biologicos Alexander von Humboldt.

Esri Inc. (2021). *ArcGIS Pro 2.8.0* [Online]. Available: https://www.esri.com/en-us/arcgis/products/arcgis-pro/overview.

García C, Polanía J (2007) Marco conceptual para productos no maderables del bosque en manglares de Colombia. *Gestión y Ambiente,* 10, 169-178.

Gaya E, Vasco-Palacios AM, Vargas-Estupinán N, Lücking R, Carretero J, Sanjuan T, Moncada B, Allkin B, Bolanos-Rojas AC, Castellanos-Castro C, Coca LF, Corrales A, Cossu T, Davis L, Dsouza J, Dufat A, Franco-Molano AE, García F, Gómez-Montoya N, González-Cuellar FE, Hammond D, Herrera A, Jaramillo-Ciro MM, Lasso-Benavides C, Mira MP, Morley J, Motato-Vásquez V, Nino-Fernández Y, Ortiz-Moreno ML, Pena-Canón ER, Ramírez-Castrillón M, Rojas T, Ruff J, Simijaca D, Sipman HJM, Soto-Medina E, Torres G, Torres-Andrade PA, Ulian T, White K, Diazgranados M (2021) *ColFungi: Colombian resources for Fungi Made Accessible,* Richmond, UK, Royal Botanic Gardens, Kew.

GBIF Occurrence Download (2021). https://doi.org/10.15468/dl.cduqac. Accessed from R via rgbif (https://github.com/ropensci/rgbif) on 15 July 2021.

Gotelli NJ, Anderson MJ, Arita HT, Chao A, Colwell RK, Connolly SR, Currie DJ, Dunn RR, Graves GR, Green JL, Grytnes JA, Jiang YH, Jetz W, Lyons SK, Mccain CM, Magurran AE, Rahbek C, Rangel TFLVB, Soberón J, Webb CO, Willig MR (2009) Patterns and causes of species richness: a general simulation model for macroecology. *Ecology Letters,* 12, 873–886.

Gotelli NJ, Colwell RK (2011) Estimating species richness. *Biological diversity: frontiers in measurement and assessment,* 12, 35. Oxford, Oxford University Press.

Hijmans RJ (2021) *raster: Geographic Data Analysis and Modeling. R package version 3.4-10.* Available: https://CRAN.R-project.org/package=raster.

Khoury CK, Achicanoy HA, Bjorkman AD, Navarro-Racines C, Guarino L, Flores-Palacios X, Engels JMM, Wiersema JH, Dempewolf H, Sotelo S, Ramírez-Villegas J, Castaneda-Álvarez NP, Fowler C, Jarvis A, Rieseberg LH, Struik PC (2016) Origins of food crops connect countries worldwide. *Proceedings of the Royal Society B: Biological Sciences,* 283, 20160792.

Kier G, Mutke J, Dinerstein E, Ricketts TH, Küper W, Kreft H, Barthlott W (2005) Global patterns of plant diversity and floristic knowledge. *Journal of Biogeography,* 32, 1107-1116.

Kreft H, Jetz W (2007) Global patterns and determinants of vascular plant diversity. *Proceedings of the National Academy of Sciences of the United States of America,* 104, 5925–5930.

MADS, IDEAM, IAVH, INVEMAR, IIAP, SINCHI, PNN, IGAC (2017) *Mapa de Ecosistemas Continentales, Costeros y Marinos de Colombia a escala 1:100.000.* Bogota, D.C., Colombia.

Millennium Ecosystem Assessment (2005). Ecosystems and human well-being. Washington, DC.

Olson DM, Dinerstein E, Wikramanayake ED, Burgess ND, Powell GV, Underwood EC, D'amico JA, Itoua I, Strand HE, Morrison JC (2001) Terrestrial Ecoregions of the World: A New Map of Life on Earth, A new global map of terrestrial ecoregions provides an innovative tool for conserving biodiversity. *BioScience,* 51, 933-938.

Ondo I, Zizka A, Pironon S, Antonelli A (unpub.) ShinyCCleaner: an interactive app for cleaning species occurrence records.

Ortega-Montoya CY, Johari A (2020) Urban Ecological Footprints. *Sustainable Cities and Communities*, 812-824.

POWO (2019). Plants of the World Online. Facilitated by the Royal Botanic Gardens, Kew.

R Core Team (2020) *R: A language and environment for statistical computing.* Vienna, Austria: R Foundation for Statistical Computing. Available: https://www.R-project.org/.

Rahbek C (1995) The elevational gradient of species richness: a uniform pattern? *Ecography*, 200-205.

Rangel-Ch JO (2015) La biodiversidad de Colombia: significado y distribución regional. *Revista de la Academia Colombiana de Ciencias Exactas, Físicas y Naturales,* 39, 176–200.

Rangel-Ch JO, Aguilar M (1995) Una aproximación sobre la diversidad climática en las regiones naturales de Colombia. *Diversidad Biótica I. Instituto de Ciencias Naturales-Universidad Nacional de Colombia-Inderena, Bogotá*, 25-77.

Raven PH (2021) Plants make our existence possible. *PLANTS, PEOPLE, PLANET*, 2–6.

Raven PH, Gereau RE, Phillipson PB, Chatelain C, Jenkins CN, Ulloa Ulloa C (2020) The distribution of biodiversity richness in the tropics. *Science Advances,* 6, eabc6228.

Redford KH, Taber A, Simonetti JA (1990) There is more to biodiversity than the tropical rain forests. *Conservation Biology,* 4, 328–330.

Simpson B (2009) Economic importance of Compositae. *In:* Funk VA, Susanna A, Stuessy TF, Bayer RJ (eds) *Systematics, evolution, and biogeography of Compositae.* Washington, D.C., USA: Smithsonian Institution.

Turbelin AJ, Malamud BD, Francis RA (2017) Mapping the global state of invasive alien species: patterns of invasion and policy responses. *Global Ecology and Biogeography,* 26, 78–92.

Veech JA (2000) A null model for detecting non-random patterns of species richness along spatial gradients. *Ecology,* 81, 1143-1149.

Welty EZ, Torgersen CE, Brenkman SJ, Duda JJ, Armstrong JB (2015) Multiscale analysis of river networks using the R package linbin. *North American Journal of Fisheries Management,* 35, 802-809.

Wickham H, Chang W, Wickham MH (2016) Package 'ggplot2'. *Create Elegant Data Visualisations Using the Grammar of Graphics. Version,* 2, 1-189.

Wickham H, François R, Henry L, Müller K (2021) *dplyr: A Grammar of Data Manipulation.* Available: https://CRAN.R-project.org/package=dplyr.

Specimen of *Cattleya trianae* Linden & Rchb.f., collected by F. C. Lehmann, in around 1901. It is Colombia's national flower (also known as Flor de Mayo, or the "Christmas orchid") and is from one of the identified 29 endemic, threatened species digitised by the UPFC digitisation project officers Priscila dos Reis and Ellie Graves.
Digital Collection - RBG Kew

Chapter 12

Annotated checklist of useful plants of Colombia

Mauricio Diazgranados[1]*, Tiziana Cossu[1], Laura Kor[1], Benedetta Gori[1], Germán Torres-Morales[2], Alejandra Aguilar-Giraldo[2], Daniel Jiménez-Pastrana[2], Lina Isabel Guevara-Ruiz[2], Julia Carretero[1], Joaquim de Souza[1], Henry Agudelo[3], Fabio Ávila[4] & Raquel Negrao[1]

[1] Royal Botanic Gardens, Kew.
[2] Instituto de Investigación de Recursos Biológicos Alexander von Humboldt.
[3] Instituto de Ciencias Naturales de Colombia, Universidad Nacional de Colombia.
[4] Jardín Botánico de Bogotá "José Celestino Mutis".

***Corresponding author:** m.diazgranados@kew.org

Keywords: edible plants, ethnobotany, food plants, fuel plants, medicinal plants, plant uses

ABSTRACT

A comprehensive checklist of the useful plants of Colombia is presented in this chapter. This checklist is one of the core outputs of the Useful Plants and Fungi of Colombia (UPFC) project, produced by a multinational team of researchers. The list includes supraspecific taxa, accepted species and authors, common names, species origin (when available), geographic information (habitats, regions, departments, elevation range), conservation status (assessments of the accepted name), and level 1 of the category of use. The taxonomic coverage and statistics are provided in Chapter 10. Notes on the geographic distribution are provided in Chapter 11. Analyses on the conservation status are included in Chapter 2. At the end of the catalogue, indexes of synonyms, families and genera, and common names are also provided. For full details on the species, readers should scan the QR codes or visit the species profiles in http://colplanta.org/, which include morphological descriptions, geographic distribution and maps, synonymy, notes on uses, links to herbarium specimens, images of live plants, illustrations, economic botany items and anatomical laminae, bibliography and additional sources.

RESUMEN

En este capítulo se presenta una lista de verificación completa de plantas útiles de Colombia. Éste es uno de los resultados centrales del proyecto Plantas y Hongos Útiles de Colombia (UPFC), desarrollado por un equipo multinacional de investigadores. La lista incluye taxones supraespecíficos, especies aceptadas y autores, nombres comunes, origen de las especies (si estaba disponible), información geográfica (hábitats, regiones, departamentos, rango de elevación), estado de conservación (evaluaciones de riesgo de extinction para nombres aceptados), y nivel 1 de la categoría de uso. La cobertura taxonómica y las estadísticas de diversidad son presentadas en el capítulo 10. En el capítulo 11 se presenta una descripción general de la distribución geográfica. Los análisis sobre el estado de conservación de plantas útiles se incluyen en el capítulo 2. Al final del catálogo, también se proporcionan índices de sinónimos, familias y géneros, y nombres comunes. Para obtener detalles completos sobre la especie, los lectores pueden escanear los códigos QR o visitar los perfiles de especies en http://colplanta.org/, que incluyen descripciones morfológicas, distribución geográfica y mapas, sinonimia, notas sobre usos, enlaces a especímenes de herbario, imágenes de plantas vivas, ilustraciones, elementos de botánica económica y láminas anatómicas, bibliografía y fuentes adicionales.

INTRODUCTION

The elaboration of the checklist of useful plants of Colombia followed five main steps: 1. identification of useful species, digitisation, and compilation of key datasets and sources; 2. taxonomic reconciliation; 3. classification of uses; 4. annotation of species descriptors; and 5. database normalisation. All plant species with a reported current or historical human use in published literature, in Colombia or elsewhere, were included regardless of origin or domestication status. Plants with no reported uses (even if they are informally known to be used or if they could be used) were excluded. Additionally, the checklist did not include infraspecific categories (e.g., subspecies, varieties, forms, and cultivars).

1. Identification, digitisation, and compilation of key datasets and sources

A thorough identification process of datasets and sources was carried out over this project (Nov. 2019–Feb. 2022). Initially, we matched the species from the World Checklist of Useful Plants (WCUP; Diazgranados *et al.*, 2020) with the updated Checklist of Plants of Colombia (CPC), available from Col*Plant*A (http://colplanta.org/). The 13 large datasets used to prepare the WCUP contributed a significant amount of information (in parentheses: the source and the number of Colombian species with reported uses):

1. Kew's Economic Botany Collection (EcBot; https://ecbot.science.kew.org/; 1,591 species);
2. European Red List of Medicinal Plants (Allen *et al.*, 2014; 83 species);
3. Germplasm Resources Information Network from the United States Department of Agriculture (GRIN; https://www.ars-grin.gov; 2,109 species);
4. Medicinal Plant Names Services v.10 (MPNS; http://mpns.kew.org/; 3,898 species);
5. Useful Plants of New Guinea (Cámara-Leret & Dennehy, 2019; 455 species);
6. Palms of the World Online (http://www.palmweb.org; 121 species);
7. Plant Resources of South-East Asia – PROSEA (Jansen *et al.*, 1991; 812 species);
8. Plant Resources for Tropical Africa (https://www.prota4u.org/database; 922 species);
9. Survey of Economic Plants for Arid and Semi-Arid Lands (SEPASAL; http://apps.kew.org/sepasalweb/sepaweb; 417 species);
10. Useful Plant Project (UPP; Ulian *et al.*, 2017; 253 species);
11. Useful Plants of West Tropical Africa (UPWTA; Burkill, 1995; 688 species);
12. Malaria and Fever in Latin America dataset (Malaria; Milliken, 1997; 626 species);
13. Crop Wild Relatives (CWR; https://www.cwrdiversity.org; Dempewolf *et al.*, 2014; 291 species).

We also identified and digitised data from over 1,500 sources reporting plant uses, most from Colombia, such as scientific journal publications, books, technical reports, theses, and databases (the full list is available in the bibliography of Col*Plant*A).

We recognise that some of the species from the checklist do not have reported uses in the country but have uses reported elsewhere, as indicated in Chapter 10. However, we decided that it was important to include them to call attention to the potential and breadth of uses of the Colombian flora. Those taxa can be considered neglected and underutilised species (NUS), and the incorporation of their uses may generate new venues for the improvement of livelihoods of local communities, supporting the further development of the country's bioeconomy (Hunter *et al.*, 2019; Padulosi *et al.*, 2013; Ulian *et al.*, 2020).

2. Taxonomic reconciliation

The nomenclature primarily follows the taxonomic backbone of the International Plant Name Index (IPNI, 2021) and the World Checklist of Vascular Plants v.5.0 (WCVP, 2021), also displayed in Plants of the World Online (POWO, 2021). For a few unresolved taxa (<0.1%) such as algae, which are not available in IPNI nor WCVP, we used Tropicos 3.0.2 (TROPICOS, 2021) and AlgaeBase (Guiry & Guiry, 2021). For the higher taxonomy (order level and above), we used the taxonomic backbone of the GBIF species matching tool (GBIF, 2021).

The reconciliation of scientific names was done primarily in R version 4.1.0, using the packages *plyr* version 1.8.5 (Wickham & Wickham, 2020), *rgdal* version 1.4-8 (Bivand *et al.*, 2015), *dplyr* version 1.0.6 (Wickham *et al.*, 2021), and *doBy* version 4.6-3 (Højsgaard *et al.*, 2019). Misspelt species names from the sources were corrected whenever possible, and mismatched names were discarded. Synonymy and homonymy were resolved, and illegitimate and invalid names were excluded. IPNI Life Sciences Identifier (LSID or IPNI ID) were assigned to most names (99.9%), except for 22 accepted names that were not in the current version of IPNI (15 in Tropicos, 3 in AlgaeBase, 1 in USDA, 1 in The Plant List).

3. Classification of uses

The categorisation of plant uses followed a simplified version of the level 1 states proposed by Cook (1995) in the Economic Botany Data Collections Standard, as defined in the WCUP (Diazgranados *et al.*, 2020):

- **Animal Food (AF):** Forage and fodder for vertebrate animals only;
- **Environmental Uses (EU):** Examples include intercrops and nurse crops, ornamentals, barrier hedges, shade plants, windbreaks, soil improvers, plants for revegetation and erosion control, wastewater purifiers, indicators of the presence of metals, pollution, or underground water;
- **Fuels (FU):** Wood, charcoal, petroleum substitutes, fuel alcohols, etc. – separated from MATERIALS because of their importance;
- **Gene Sources (GS):** Wild relatives of major crops which may possess traits associated with biotic or abiotic resistance and may be valuable for breeding programs;
- **Human Food (HF):** Food, including beverages, for humans only;
- **Invertebrate Food (IF):** Only plants eaten by invertebrates useful to humans, such as silkworms, lac insects and edible grubs, are covered here. This group does not include bee-pollinated wild species;
- **Materials (MA):** Woods, fibres, cork, cane, tannins, latex, resins, gums, waxes, oils, lipids, etc. and their derived products;
- **Medicines (ME):** Both human and veterinary;
- **Poisons (PO):** Plants that are poisonous to vertebrates and invertebrates, both accidentally and usefully, e.g., for hunting and fishing;

- **Social Uses (SU):** Plants used for social purposes, which are not definable as food or medicines, for instance, masticatories, smoking materials, narcotics, hallucinogens and psychoactive drugs, contraceptives and abortifacients, and plants with ritual or religious significance.

4. Annotation of species-related information

Common names from the original sources and from the database of common names of Colombian plants (Bernal *et al.*, 2017) were added to the species list. Indications of origin (i.e., endemic, native, introduced, naturalised) were obtained from POWO's database, from the Catalogue of Plants and Lichens of Colombia (CPLC; Bernal *et al.*, 2019), and from a literature review. We combined the geographic information in the CPLC with the cleaned geographic data in GBIF (used in Chapter 11) to determine the presence in regions and departments. The geographic dataset was also used to extract IUCN habitats. Global conservation status was retrieved from the IUCN Red List portal (IUCN, 2020) and matched to the CPC using the R package *dplyr* version 1.0.6 (Wickham *et al.*, 2021), taking into account only the assessments related to the accepted name. National assessments were based on data from the IUCN SSC Colombia Plant Specialist Group. Habitat information was extracted from the global map of terrestrial habitat types (Jung *et al.*, 2020), using the R packages *sp* (Bivand *et al.*, 2013), *raster* (Hijmans *et al.*, 2013), *dplyr* (Wickham *et al.*, 2021), and the available georeferenced records (Chapter 11). Last, the CITES status was obtained from the checklist of the CITES Species portal (https://checklist.cites.org/). QR codes linking the Col*Plant*A species profiles were created for each species from its Permanent URLs using the R package QRcode version 0.1.3 (Teh, 2015).

5. Database normalisation

The data were structured, normalised, and formatted according to the standards agreed for this catalogue and space limitations. Please see the list of abbreviations used below.

Example of plant record

6174 *Miconia formicoheterophylla*
Michelang. (Melastomataceae)
IPNI: 77207843-1
Habit: Shrub, Subshrub
Use: HF
Origin: Endemic Region: Amazonia, Andean
Dept.: AMA, ANT, CAQ, CAU, MET, PUT
Elev.: 150–1420 m a.s.l.
IUCN Habitat code: 1,3,14
Conservation status: (IUCN) NE

Abbreviations and codes used

IPNI number: IPNI Life Sciences Identifier (LSID) code. Each name is given a unique LSID code, which can solve misspelt names or homonymy. Therefore, we very much encourage the use of LSID codes, which can be seen at the end of the Col*Plant*A species profile URLs (e.g., http://powo.science.kew.org/taxon/urn:lsid:ipni.org:names:**665697-1**?site=colplanta), and can be retrieved using Kew's Names Reconciliation Service (http://namematch.science.kew.org/).

Level 1 categories of use: Animal Food (AF), Environmental Uses (EU), Fuels (FU), Gene Sources (GS), Human Food (HF), Invertebrate Food (IF), Materials (MA), Medicines (ME), Poisons (PO), Social Uses (SU).

IUCN Habitat codes: Forest (1), Savanna (2), Shrubland (3), Grassland (4), Wetlands (inland) (5), Rocky Areas (e.g., inland cliffs, mountain peaks) (6), Desert (8), Artificial – Terrestrial (14). Numbers follow the classification of habitats according to the IUCN global map of terrestrial habitat types (Jung *et al.*, 2020). Missing numbers are habitats not found for useful plants of Colombia. Detailed information on the species habitats may be available in the online species profiles and the respective references.

Colombian departments: Amazonas (AMA), Antioquia (ANT), Arauca (ARA), Atlántico (ATL), Bogotá D.C. (BOG), Bolívar (BOL), Boyacá (BOY), Caldas (CAL), Caquetá (CAQ), Casanare (CAS), Cauca (CAU), Cesar (CES), Chocó (CHO), Córdoba (COR), Cundinamarca (CUN), Guainía (GUA), Guaviare (GUV), Huila (HUI), La Guajira (LAG), Magdalena (MAG), Meta (MET), Nariño (NAR), Norte de Santander (NSA), Putumayo (PUT), Quindío (QUI), Risaralda (RIS), San Andrés y Providencia (SAP), Santander (SAN), Sucre (SUC), Tolima (TOL), Valle del Cauca (VAC), Vaupés (VAU), and Vichada (VID).

Conservation status: International and national conservation status were based on the IUCN Red List (IUCN, 2021) and the "Colombian Plants Specialist Group" CPSG (i.e., the list of national extinction risk assessments carried out by the IUCN Colombian Plant Specialist Group). The assessments were considered only when related to the currently accepted names or when the name was unambiguous (for more detailed information, see Chapter 2). CITES status was presented when available.

TAXONOMIC CLASSIFICATION

Kingdoms: Plantae (7,471 species) and Chromista (1).
Divisions/Phyla: Bryophyta (5 species), Marchantiophyta (2), Ochrophyta (1), Rhodophyta (2), and Tracheophyta (7,462).

Classes: Bryopsida (1 species), Cycadopsida (10), Florideophyceae (2), Ginkgoopsida (1), Gnetopsida (5), Lycopodiopsida (32), Magnoliophyta/Angiospermae (7,210), Marchantiopsida (2), Phaeophyceae (1), Pinopsida (26), Polypodiopsida (178), and Sphagnopsida (4).

Orders: Alismatales (160 species), Apiales (82), Aquifoliales (11), Arecales (214), Asparagales (298), Asterales (449), Austrobaileyales (1), Boraginales (58), Brassicales (100), Buxales (3), Canellales (2), Caryophyllales (254), Celastrales (31), Ceratophyllales (1), Chloranthales (15), Commelinales (36), Cornales (6), Crossosomatales (1), Cucurbitales (76), Cyatheales (15), Cycadales (10), Dilleniales (14), Dioscoreales (29), Dipsacales (32), Equisetales (5), Ericales (320), Escalloniales (5), Fabales (688), Fagales (9), Fucales (1), Gentianales (474), Geraniales (15), Ginkgoales (1), Gleicheniales (5), Gnetales (5), Gracilariales (1), Gunnerales (4), Hymenophyllales (6), Icacinales (6), Lamiales (509), Laurales (101), Liliales (18), Lunulariales (1), Lycopodiales (12), Magnoliales (156), Malpighiales (647), Malvales (232), Marchantiales (1), Metteniusales (2), Myrtales (368), Nymphaeales (10), Ophioglossales (4), Osmundales (2), Oxalidales (58), Pandanales (18), Picramniales (8), Pinales (26), Piperales (133), Plocamiales (1), Poales (485), Polypodiales (129), Polytrichales (1), Proteales (16), Psilotales (2), Ranunculales (70), Rosales (291), Salviniales (5), Santalales (42), Sapindales (251), Saxifragales (39), Schizaeales (5), Selaginellales (20), Solanales (250), Sphagnales (4), Vitales (11), Zingiberales (101), and Zygophyllales (9).

Families: Acanthaceae (73 species), Achariaceae (17), Achatocarpaceae (1), Actinidiaceae (16), Aizoaceae (11), Alismataceae (14), Alstroemeriaceae (7), Altingiaceae (1), Amaranthaceae (56), Amaryllidaceae (36), Anacardiaceae (31), Annonaceae (105), Apiaceae (34), Apocynaceae (128), Aquifoliaceae (10), Araceae (137), Araliaceae (46), Araucariaceae (4), Arecaceae (214), Aristolochiaceae (24), Asparagaceae (32), Asphodelaceae (9), Aspleniaceae (25), Asteraceae (418), Balanophoraceae (3), Balsaminaceae (5), Basellaceae (4), Bataceae (1), Begoniaceae (27), Berberidaceae (9), Betulaceae (1), Bignoniaceae (82), Bixaceae (6), Bonnetiaceae (1), Boraginaceae (58), Brassicaceae (46), Bromeliaceae (47), Brunelliaceae (5), Burseraceae (58), Buxaceae (3), Cabombaceae (1), Cactaceae (47), Calceolariaceae (3), Calophyllaceae (11), Campanulaceae (29), Cannabaceae (6), Cannaceae (5), Capparaceae (19), Caprifoliaceae (21), Caricaceae (14), Caryocaraceae (10), Caryophyllaceae (30), Casuarinaceae (2), Celastraceae (31), Ceratophyllaceae (1), Chloranthaceae (15), Chrysobalanaceae (50), Cleomaceae (12), Clethraceae (3), Clusiaceae (55), Colchicaceae (1), Combretaceae (18), Commelinaceae (26), Connaraceae (6), Convolvulaceae (77), Coriariaceae (1), Costaceae (20), Crassulaceae (31), Cucurbitaceae (48), Cunoniaceae (9), Cupressaceae (9), Cyatheaceae (15), Cycadaceae (2), Cyclanthaceae (16), Cyperaceae (81), Cyrillaceae (1), Cystopteridaceae (1), Dennstaedtiaceae (6), Dichapetalaceae (5), Dilleniaceae (14), Dioscoreaceae (29), Droseraceae (1), Ebenaceae (8), Elaeocarpaceae (12), Equisetaceae (5), Ericaceae (61), Eriocaulaceae (3), Erythroxylaceae (26), Escalloniaceae (5), Euphorbiaceae (149), Fabaceae (665), Fagaceae (1), Fucaceae (1), Gentianaceae (30), Geraniaceae (15), Gesneriaceae (58), Ginkgoaceae (1), Gleicheniaceae (5), Gnetaceae (5), Goupiaceae (1), Gracilariaceae (1), Grossulariaceae (1), Gunneraceae (4), Haemodoraceae (4), Haloragaceae (1), Hamamelidaceae (1), Heliconiaceae (18), Hernandiaceae (5), Humiriaceae (12), Hydrangeaceae (1), Hydrocharitaceae (4), Hydroleaceae (1), Hymenophyllaceae (6), Hypericaceae (21), Hypoxidaceae (3), Icacinaceae (2), Iridaceae (23), Juglandaceae (2), Juncaceae (9), Krameriaceae (2), Lacistemataceae (2), Lamiaceae (145), Lauraceae (81), Lecythidaceae (45), Lentibulariaceae (1), Liliaceae (1), Linaceae (2), Linderniaceae (7), Loasaceae (5), Loganiaceae (30), Loranthaceae (10), Lunulariaceae (1), Lycopodiaceae (12), Lythraceae (21), Magnoliaceae (13), Malpighiaceae (51), Malvaceae (221), Marantaceae (29), Marcgraviaceae (5), Marchantiaceae (1), Marsileaceae (1), Martyniaceae (2), Melastomataceae (162), Meliaceae (38), Menispermaceae (37), Menyanthaceae (2), Metteniusaceae (7), Microteaceae (1), Molluginaceae (2), Monimiaceae (2), Montiaceae (3), Moraceae (113), Moringaceae (1), Muntingiaceae (2), Musaceae (7), Myricaceae (3), Myristicaceae (38), Myrtaceae (119), Nelumbonaceae (1), Nyctaginaceae (19), Nymphaeaceae (9), Ochnaceae (22), Olacaceae (11), Oleaceae (19), Onagraceae (34), Ophioglossaceae (4), Opiliaceae (2), Orchidaceae (195), Orobanchaceae (10), Osmundaceae (2), Oxalidaceae (26), Pandanaceae (1), Papaveraceae (9), Passifloraceae (88), Paulowniaceae (1), Pedaliaceae (1), Pentaphylacaceae (4), Peraceae (3), Petiveriaceae (8), Phrymaceae (1), Phyllanthaceae (27), Phytolaccaceae (9), Picramniaceae (8), Pinaceae (8), Piperaceae (109), Pittosporaceae (2), Plantaginaceae (43), Platanaceae (1), Plocamiaceae (1), Plumbaginaceae (5), Poaceae (339), Podocarpaceae (5), Podostemaceae (1), Polemoniaceae (3), Polygalaceae (22), Polygonaceae (51), Polypodiaceae (52), Polytrichaceae (1), Pontederiaceae (6), Portulacaceae (4), Potamogetonaceae (4), Primulaceae (47), Proteaceae (14), Psilotaceae (2), Pteridaceae (45), Putranjivaceae (1), Ranunculaceae (15), Rapateaceae (1), Resedaceae (2), Rhamnaceae (18), Rhizophoraceae (4), Rosaceae (79), Rubiaceae (286), Ruppiaceae (1), Rutaceae (34), Sabiaceae (1), Salicaceae (56), Salviniaceae (4), Santalaceae (15), Sapindaceae (78), Sapotaceae (96), Saxifragaceae (4), Schisandraceae (1), Schizaeaceae (5), Schlegeliaceae (4), Schoepfiaceae (1), Scrophulariaceae (9), Selaginellaceae (20), Simaroubaceae (12), Siparunaceae (13), Smilacaceae (9), Solanaceae (171), Sphagnaceae (4), Sphenocleaceae (1), Staphyleaceae (1), Strelitziaceae (4), Styracaceae (10), Surianaceae (1), Symplocaceae (12), Talinaceae (2), Tetrameristaceae (1), Theaceae (3), Thymelaeaceae (3), Triuridaceae (1), Tropaeolaceae (5), Typhaceae (3), Ulmaceae (6), Urticaceae (69), Verbenaceae (50), Viburnaceae (11), Violaceae (31), Vitaceae (11), Vochysiaceae (14), Winteraceae (2), Xyridaceae (2), Zamiaceae (8), Zingiberaceae (18), and Zygophyllaceae (7).

References

Allen D, Bilz M, Leaman DJ, Miller RM, Timoshyna A, Window J (2014) European Red List of medicinal plants. *Publications Office of the European Union, Luxembourg,* 10, 907382.

Bernal R, Galeano G, Rodríguez A, Sarmiento H, Gutiérrez M (2017) Nombres Comunes de las Plantas de Colombia. http://www.biovirtual.unal.edu.co/nombrescomunes/.

Bernal R, Gradstein S, Celis M (eds) (2019) *Catálogo de plantas y líquenes de Colombia.* http://catalogoplantasdecolombia.unal.edu.co/. Instituto de Ciencias Naturales, Universidad Nacional de Colombia.

Bivand R, Keitt T, Rowlingson B, Pebesma E, Sumner M, Hijmans R, Rouault E, Bivand MR (2015) *Package 'rgdal'*. Bindings for the Geospatial Data Abstraction Library. https://cran.r-project.org/web/packages/rgdal/index.html.

Bivand R, Pebesma E, Gómez-Rubio V (2013) *Applied spatial data: analysis with R.* New York, Springer.

Burkill HM (1995) *The useful plants of west tropical Africa,* Vols. 1-3. Richmond, Royal Botanic Gardens, Kew.

Cámara-Leret R, Dennehy Z (2019) Information gaps in indigenous and local knowledge for science-policy assessments. *Nature Sustainability,* 2, 736-741.

Cook F (1995) *Economic botany data collection standards, prepared for the international working group on taxonomic database for plant science (TDWG).* Richmond, Royal Botanic Gardens, Kew.

Dempewolf H, Eastwood RJ, Guarino L, Khoury CK, Müller JV, Toll J (2014) Adapting agriculture to climate change: a global initiative to collect, conserve, and use crop wild relatives. *Agroecology and Sustainable Food Systems,* 38, 369-377.

Diazgranados M, Allkin B, Black N, Cámara-Leret R, Canteiro C, Carretero J, Eastwood R, Hargreaves S, Hudson A, Milliken W, Nesbitt M, Ondo I, Patmore K, Pironon S, Turner R, Ulian T (2020) *World Checklist of Useful Plant Species*. Knowledge Network for Biocomplexity. https://knb.ecoinformatics.org/view doi:10.5063/F1CV4G34. Richmond, Royal Botanic Gardens, Kew.

GBIF (2021) *GBIF species matching tool*. Facilitated by the Global Biodiversity Information Facility. https://www.gbif.org/tools/species-lookup/.

Guiry MD, Guiry GM (2021) *AlgaeBase*. World-wide electronic publication, National University of Ireland, Galway. https://www.algaebase.org.

Hijmans RJ, Van Etten J, Mattiuzzi M, Sumner M, Greenberg J, Lamigueiro O, Bevan A, Racine E, Shortridge A (2013) *Raster package in R*. Version 3.4-13. https://cran.r-project.org/.

Hojsgaard S, Halekoh U, Hojsgaard MS, Imports M (2019) *Package 'doBy'*. https://cran.r-project.org/.

Hunter D, Borelli T, Beltrame DMO, Oliveira CNS, Coradin L, Wasike VW, Wasilwa L, Mwai J, Manjella A, Samarasinghe GWL, Madhujith T, Nadeeshani HVH, Tan A, Ay ST, Güzelsoy N, Lauridsen N, Gee E, Tartanac F (2019) The potential of neglected and underutilised species for improving diets and nutrition. *Planta,* 250, 709-729.

IPNI (2021) *International Plant Names Index*. Facilitated by the The Royal Botanic Gardens, Kew, Harvard University Herbaria & Libraries and Australian National Botanic Gardens. http://www.ipni.org/.

IUCN (2020) The IUCN Red List of Threatened Species. Version 2021-1. http://www.iucnredlist.org.

Jansen P, Lemmens R, Oyen L, Siemonsma J, Stavast F, Van Valkenburg J (1991) Plant Resources of South-East Asia (PROSEA). *Basic list of species and commodity grouping. Final version. Pudoc, Wageningen.*

Jung M, Dahal PR, Butchart SHM, Donald PF, De Lamo X, Lesiv M, Kapos V, Rondinini C, Visconti P (2020) A global map of terrestrial habitat types. *Scientific Data,* 7, 256.

Milliken W (1997) *Plants for malaria, plants for fever: Medicinal species in Latin America, a bibliographic survey. Richmond,* Royal Botanic Gardens, Kew.

Padulosi S, Thompson J, Rudebjer P (2013) *Fighting poverty, hunger and malnutrition with neglected and underutilised species: needs, challenges and the way forward*, Bioversity International.

POWO (2021) *Plants of the World Online*. Facilitated by the Royal Botanic Gardens, Kew. http://www.plantsoftheworldonline.org/.

Teh V (2015) *QRcode Generator for R (0.1.3).* https://doi.org/10.5281/zenodo.5040088.

TROPICOS (2021) *Tropicos.org*. Facilitated by the Missouri Botanical Garden. https://www.tropicos.org/.

Ulian T, Diazgranados M, Pironon S, Padulosi S, Liu U, Davies L, Howes MJR, Borrell JS, Ondo I, Pérez-Escobar OA (2020) Unlocking plant resources to support food security and promote sustainable agriculture. *Plants, People, Planet,* 2, 421-445.

Ulian T, Sacandé M, Hudson A, Mattana E (2017) Conservation of indigenous plants to support community livelihoods: the MGU–Useful Plants Project. *Journal of Environmental Planning and Management,* 60, 668-683.

WCVP (2021) World Checklist of Vascular Plants, version 2.0. Facilitated by the Royal Botanic Gardens, Kew (ed.). http://wcvp.science.kew.org/.

Wickham H, François R, Henry L, Müller K (2021) *dplyr: a Grammar of Data Manipulation*. https://CRAN.R-project.org/package=dplyr.

Wickham H, Wickham MH (2020) *Package 'plyr'*. https://cran.r-project.org/.

Pacific forest.
Mar Palanca and Mateo Fernandez Lucer

Marchantiophyta Marchantiopsida Lunulariales

1 *Lunularia cruciata* (L.) Dumort. ex Lindb. (Lunulariaceae)
IPNI: 622149
Use: EU **Origin:** Endemic, Cultivated **Region:** Andean **Dept.:** CUN **Elev.:** 2650–2650 m a.s.l. **Conservation status: (IUCN)** NE

Marchantiales
2 *Marchantia polymorpha* L. (Marchantiaceae)
IPNI: 622330
Common name: paragüitas **Use:** ME **Origin:** Endemic **Region:** Andean **Dept.:** BOY, CAU, CES, CUN, MAG, NAR, NSA, SAN **Elev.:** 1600–4700 m a.s.l. **IUCN Habitat code:** 1,3,4,6,14 **Conservation status: (IUCN)** NE

Bryophyta Bryopsida Polytrichales
3 *Polytrichum juniperinum* Hedw. (Polytrichaceae)
IPNI: 579099
Use: ME **Origin:** Endemic **Region:** Andean **Dept.:** ANT, ARA, BOY, CAL, CAQ, CAU, CHO, CUN, MAG, MET, NAR, NSA, PUT, RIS, SAN, TOL, VAC **Elev.:** 950–4680 m a.s.l. **IUCN Habitat code:** 1,2,3,4,5,6,8,14 **Conservation status: (IUCN)** NE

Sphagnopsida Sphagnales
4 *Sphagnum cuspidatum* Ehrh. ex Hoffm. (Sphagnaceae)
IPNI: 604247
Use: EU, MA, ME **Origin:** Endemic **Region:** Andean **Dept.:** ANT, BOY, CAL, CAU, CUN, HUI, MET, NAR, NSA, SAN, VAC **Elev.:** 2340-4030 m.a.s.l. **IUCN Habitat code:** 1,3,4,8,14 **Conservation status: (IUCN)** NE **(CPSG)** LC

5 *Sphagnum magellanicum* Brid. (Sphagnaceae)
IPNI: 611262
Use: ME **Origin:** Endemic **Region:** Andean **Dept.:** ANT, ARA, BOY, CAL, CAS, CAU, CES, CHO, CUN, HUI, MAG, MET, NAR, NSA, PUT, QUI, RIS, SAN, TOL, VAC **Elev.:** 1450-3850 m.a.s.l. **IUCN Habitat code:** 1,3,4,6,8,14 **Conservation status: (IUCN)** NE **(CPSG)** LC

6 *Sphagnum meridense* (Hampe) Müll. Hal. (Sphagnaceae)
IPNI: 611352
Use: MA, ME **Origin:** Endemic **Region:** Andean **Dept.:** ANT, BOY, CAQ, CAU, CUN, HUI, MAG, NAR, PUT, RIS, SAN, TOL **Elev.:** 1220-3650 m.a.s.l. **IUCN Habitat code:** 1,3,4,6,8,14 **Conservation status: (IUCN)** NE **(CPSG)** LC

7 *Sphagnum subsecundum* Nees (Sphagnaceae)
IPNI: 612658
Use: ME **Origin:** Endemic **Region:** Andean **Dept.:** ANT, CUN, SAN **Elev.:** 2310-3500 m.a.s.l. **IUCN Habitat code:** 1,4,14 **Conservation status: (IUCN)** NE **(CPSG)** NT

Tracheophyta Lycopodiopsida Lycopodiales

8 *Huperzia attenuata* (Spring) Trevis. (Lycopodiaceae)
IPNI: 124842-2
Habit: Herb **Use:** ME **Origin:** Native **Region:** Andean **Dept.:** BOY, CUN, NAR, RIS **Elev.:** 2800–3800 m a.s.l. **IUCN Habitat code:** 1,4,14 **Conservation status: (IUCN)** NE

9 *Huperzia crassa* (Humb. & Bonpl. ex Willd.) Rothm. (Lycopodiaceae)
IPNI: 124844-2
Common name: cacho de venao, colchón de pobre, piligan **Habit:** Herb **Use:** ME **Origin:** Native **Region:** Andean **Dept.:** ARA, BOY, CAL, CHO, MAG, MET, NAR, PUT, RIS, SAN, TOL **Elev.:** 3350–4700 m a.s.l. **IUCN Habitat code:** 1,3,4,6,8,14 **Conservation status: (IUCN)** NE

10 *Huperzia dichotoma* (Jacq.) Trevis. (Lycopodiaceae)
IPNI: 1090207-2
Habit: Herb **Use:** ME **Origin:** Native **Region:** Andean, Pacific **Dept.:** ANT, CHO **Elev.:** 0–700 m a.s.l. **IUCN Habitat code:** 1,3 **Conservation status: (IUCN)** NE

11 *Huperzia saururus* (Lam.) Trevis. (Lycopodiaceae)
IPNI: 60447820-2
Habit: Herb **Use:** MA, ME **Origin:** Native **Region:** Andean **Dept.:** CHO, MAG **Elev.:** 3800–3850 m a.s.l. **IUCN Habitat code:** 4 **Conservation status: (IUCN)** NE

12 *Lycopodiella alopecuroides* (L.) Cranfill (Lycopodiaceae)
IPNI: 146929-2
Use: ME **IUCN Habitat code:** 1,3,4,14 **Conservation status: (IUCN)** NE **(CPSG)** Potential LC

13 *Lycopodiella caroliniana* (L.) Pic.Serm. (Lycopodiaceae)
IPNI: 146931-2
Habit: Herb **Use:** SU **Origin:** Native **Region:** Amazonia, Andean, Pacific **Dept.:** AMA, ANT, CAQ, CHO, GUA, NAR, SAN, VAU **Elev.:** 50–2500 m a.s.l. **IUCN Habitat code:** 1,4,14 **Conservation status: (IUCN)** NE

14 *Lycopodiella cernua* (L.) Pic.Serm. (Lycopodiaceae)
IPNI: 1054587-2
Common name: abrecaminos, cacho de venado, caminadera, colchon de pobre, colchón de pobre, gateadera, uña de gato **Habit:** Herb **Use:** EU, HF, MA, ME, SU **Origin:** Native **Region:** Amazonia, Andean, Guiana Shield, Caribbean, Orinoquia, Pacific **Dept.:** AMA, ANT, ARA, BOY, CAL, CAQ, CAU, CES, CHO, CUN, GUA, HUI, LAG, MAG, MET, NAR, NSA, QUI, RIS, SAP, SAN, TOL, VAC, VAU, VID **Elev.:** 50-3000 m.a.s.l. **IUCN Habitat code:** 1,2,3,4,5,14 **Conservation status: (IUCN)** NE **(CPSG)** Potential LC

15 *Lycopodiella pendulina* (Hook.) B.Øllg. (Lycopodiaceae)
IPNI: 279987-2
Habit: Herb **Use:** ME **Origin:** Native **Region:** Andean **Dept.:** ANT, BOY, CAU, CUN, NAR, PUT **Elev.:** 2100–3600 m a.s.l. **IUCN Habitat code:** 1,3,4,8,14 **Conservation status: (IUCN)** NE

16 *Lycopodium clavatum* L. (Lycopodiaceae)
IPNI: 304307-2
Common name: cacho de venado, cacho de venao, caminadera, colchón de pobre, gateadera, lycopodio **Habit:** Herb **Use:** AF, EU, FU, HF, MA, ME, SU **Origin:** Native **Region:** Andean, Caribbean **Dept.:** ANT, BOG, BOY, CAL, CAS, CAU, CES, CUN, HUI, LAG, MAG, NAR, NSA, RIS, SAN, TOL, VAC **Elev.:** 1050-4200 m.a.s.l. **IUCN Habitat code:** 1,2,3,4,6,8,14 **Conservation status: (IUCN)** NE **(CPSG)** Potential LC

17 *Lycopodium complanatum* L. (Lycopodiaceae)
IPNI: 147262-2
Common name: caminadera **Use:** ME, PO, SU **Dept.:** BOG **IUCN Habitat code:** 1,4,14 **Conservation status: (IUCN)** NE

18 *Lycopodium jussiaei* Desv. (Lycopodiaceae)
IPNI: 147420-2
Common name: cacho de venado **Habit:** Herb **Use:** ME **Origin:** Native **Region:** Andean **Dept.:** ANT, BOY, CAL, CAU, CHO, CUN, HUI, MAG, MET, NAR, NSA, PUT, QUI, RIS, SAN, TOL, VAC **Elev.:** 1000-3700 m.a.s.l. **IUCN Habitat code:** 1,3,4,8,14 **Conservation status: (IUCN)** NE **(CPSG)** Potential LC

19 *Lycopodium thyoides* Willd. (Lycopodiaceae)
IPNI: 147665-2
Common name: cacho de venado, caminadera, gateadera, helecho gateadero **Habit:** Herb **Use:** ME **Origin:** Native **Region:** Andean, Caribbean **Dept.:** ANT, BOY, CAL, CAU, CES, CHO, CUN, HUI, MAG, MET, NAR, NSA, PUT, QUI, RIS, SAN, TOL, VAC **Elev.:** 1700-4300 m.a.s.l. **Conservation status: (IUCN)** NE **(CPSG)** Potential LC

Selaginellales

20 *Selaginella amazonica* Spring (Selaginellaceae)
IPNI: 231969-2
Habit: Herb **Use:** ME **Origin:** Native **Region:** Amazonia, Pacific **Dept.:** AMA, CAQ, CAU, GUA, VAU **Elev.:** 150–250 m a.s.l. **IUCN Habitat code:** 1,2,3,5,14 **Conservation status: (IUCN)** NE

21 *Selaginella articulata* (Kunze) Spring (Selaginellaceae)
IPNI: 77186609-1
Habit: Herb **Use:** ME **Origin:** Native **Region:** Amazonia, Pacific **Dept.:** ANT, CHO, PUT, VAC **Elev.:** 0–1400 m a.s.l. **IUCN Habitat code:** 1,3,14 **Conservation status: (IUCN)** NE

22 *Selaginella asperula* Mart. ex Spring (Selaginellaceae)
IPNI: 77186584-1
Common name: busukobenaeji, little tree mano de sapo **Habit:** Herb **Use:** ME **Origin:** Native **Region:** Amazonia, Andean, Guiana Shield, Orinoquia **Dept.:** AMA, BOG, CAQ, CAU, GUA, GUV, MET, NAR, VAU, VID **Elev.:** 0–2000 m a.s.l. **IUCN Habitat code:** 1,3,4,5,14 **Conservation status: (IUCN)** NE

23 *Selaginella convoluta* (Arn.) Spring (Selaginellaceae)
IPNI: 77142531-1
Habit: Herb **Use:** ME **Origin:** Native **Region:** Andean, Caribbean **Dept.:** CES, MAG, NSA **Elev.:** 450–1900 m a.s.l. **IUCN Habitat code:** 1,2,14 **Conservation status: (IUCN)** NE

24 *Selaginella diffusa* (C.Presl) Spring (Selaginellaceae)
IPNI: 77186646-1
Habit: Herb **Use:** ME **Origin:** Native **Region:** Amazonia, Andean, Orinoquia, Pacific **Dept.:** ANT, BOY, CAL, CAQ, CAU, CES, CHO, CUN, HUI, MAG, MET, NAR, NSA, PUT, QUI, SAN, TOL, VAC **Elev.:** 200-2800 m.a.s.l. **IUCN Habitat code:** 1,2,3,14 **Conservation status: (IUCN)** NE **(CPSG)** Potential LC

25 *Selaginella erythropus* (Mart.) Spring (Selaginellaceae)
IPNI: 30243959-2
Habit: Herb **Use:** EU **Origin:** Native **Region:** Pacific **Dept.:** ANT, CAL, CHO, CUN, HUI, NAR, NSA, TOL, VAC **Elev.:** 150–2000 m a.s.l. **IUCN Habitat code:** 1,2,3,4,14 **Conservation status: (IUCN)** NE

26 *Selaginella exaltata* (Kunze) Spring (Selaginellaceae)
IPNI: 77186589-1
Common name: yerba de loro **Habit:** Herb **Use:** ME **Origin:** Native **Region:** Amazonia, Pacific **Dept.:** AMA, ANT, CAQ, CAU, CHO, GUV, NAR, PUT, VAC **Elev.:** 0–400 m

a.s.l. **IUCN Habitat code:** 1,3,5,14 **Conservation status: (IUCN)** NE

27 *Selaginella flabellata* (L.) Spring (Selaginellaceae) **IPNI:** 30226649-2 **Habit:** Herb **Use:** ME **Origin:** Native **Region:** Amazonia, Pacific **Dept.:** AMA, ANT, CAQ, CHO, VAC **Elev.:** 200–900 m a.s.l. **IUCN Habitat code:** 1,14 **Conservation status: (IUCN)** NE

28 *Selaginella flagellata* Spring (Selaginellaceae) **IPNI:** 232097-2 **Habit:** Herb **Use:** ME **Origin:** Native **Region:** Amazonia, Andean, Pacific **Dept.:** AMA, ANT, BOY, CAQ, MAG, MET, QUI, SAN **Elev.:** 200–1800 m a.s.l. **IUCN Habitat code:** 1,3,14 **Conservation status: (IUCN)** NE

29 *Selaginella geniculata* (C.Presl) Spring (Selaginellaceae) **IPNI:** 77186554-1 **Habit:** Herb **Use:** ME **Origin:** Native **Region:** Andean, Pacific **Dept.:** ANT, BOY, CAL, CAQ, CAU, CHO, CUN, HUI, NAR, PUT, QUI, RIS, TOL, VAC **Elev.:** 0-3400 m.a.s.l. **IUCN Habitat code:** 1,3,5,14 **Conservation status: (IUCN)** NE **(CPSG)** Potential LC

30 *Selaginella lingulata* Spring (Selaginellaceae) **IPNI:** 232147-2 **Common name:** doradilla **Habit:** Herb **Use:** ME **Origin:** Native **Region:** Andean, Pacific **Dept.:** ANT, BOY, CAU, CHO, CUN, NAR, PUT, RIS, SAN, TOL, VAC **Elev.:** 50–3500 m a.s.l. **IUCN Habitat code:** 1,3,5,14 **Conservation status: (IUCN)** NE

31 *Selaginella pallescens* (C.Presl) Spring (Selaginellaceae) **IPNI:** 30016320-2 **Habit:** Herb **Use:** ME **Origin:** Native **Region:** Amazonia, Andean, Caribbean **Dept.:** CES, GUA, LAG, MAG **Elev.:** 700–1200 m a.s.l. **IUCN Habitat code:** 1 **Conservation status: (IUCN)** NE

32 *Selaginella parkeri* (Hook. & Grev.) Spring (Selaginellaceae) **IPNI:** 77186636-1 **Habit:** Herb **Use:** ME **Origin:** Native **Region:** Amazonia **Dept.:** AMA, CAQ, PUT, VAU **Elev.:** 100–400 m a.s.l. **IUCN Habitat code:** 1,3,4,5,14 **Conservation status: (IUCN)** NE

33 *Selaginella plana* (Desv.) Hieron. (Selaginellaceae) **IPNI:** 77186707-1 **Habit:** Herb **Use:** AF, HF, MA, ME **Origin:** Naturalised **Region:** Pacific **Dept.:** NAR, TOL, VAC **Elev.:** 0–450 m a.s.l. **IUCN Habitat code:** 14 **Conservation status: (IUCN)** NE

34 *Selaginella rosea* Alston (Selaginellaceae) **IPNI:** 232255-2 **Common name:** yerba del carpintero **Habit:** Herb **Use:** ME **Origin:** Endemic **Region:** Andean **Dept.:** ANT, TOL **Elev.:** 1600–2500 m a.s.l. **IUCN Habitat code:** 1,3,14 **Conservation status: (IUCN)** NE

35 *Selaginella schultesii* Alston ex Crabbe & Jermy (Selaginellaceae) **IPNI:** 232288-2 **Habit:** Herb **Use:** ME **Origin:** Native **Region:** Amazonia **Dept.:** AMA, VAU **Elev.:** 200–700 m a.s.l. **IUCN Habitat code:** 1,5 **Conservation status: (IUCN)** NE

36 *Selaginella sellowii* Hieron. (Selaginellaceae) **IPNI:** 232292-2 **Habit:** Herb **Use:** ME **Origin:** Native **Region:** Andean **Dept.:** BOG, BOY, CUN, HUI, NAR, SAN **Elev.:** 700–2700 m a.s.l. **Conservation status: (IUCN)** NE

37 *Selaginella speciosa* A.Br. (Selaginellaceae) **IPNI:** 77186694-1 **Habit:** Herb **Use:** ME **Origin:** Native **Region:** Amazonia, Andean, Pacific **Dept.:** AMA, ANT, CAL, CAQ, CAU, CHO, CUN, PUT, TOL, VAC **Elev.:** 100–1800 m a.s.l. **Conservation status: (IUCN)** NE

38 *Selaginella stellata* Spring (Selaginellaceae) **IPNI:** 77186626-1 **Habit:** Herb **Use:** ME **Origin:** Native **Region:** Amazonia **Dept.:** VAU **Elev.:** 300–350 m a.s.l. **IUCN Habitat code:** 1,5 **Conservation status: (IUCN)** NE

39 *Selaginella willdenowii* (Desv.) Baker (Selaginellaceae) **IPNI:** 60471496-2 **Habit:** Herb **Use:** EU, HF, ME **Origin:** Naturalised **Region:** Andean **Dept.:** VAC **Elev.:** 1350–1350 m a.s.l. **IUCN Habitat code:** 14 **Conservation status: (IUCN)** NE

Polypodiopsida Cyatheales

40 *Alsophila cuspidata* (Kunze) D.S.Conant (Cyatheaceae) **IPNI:** 9954-2 **Use:** HF **Origin:** Native **Region:** Amazonia, Andean, Guiana Shield, Pacific **Dept.:** ANT, CAU, CHO, MET, NSA, PUT, RIS, SAN, VAC **Elev.:** 50-1600 m.a.s.l. **IUCN Habitat code:** 1,3,5,14 **Conservation status: (IUCN)** NE **(CPSG)** LC

41 *Alsophila erinacea* (H.Karst.) D.S.Conant (Cyatheaceae) **IPNI:** 9971-2 **Common name:** helecho estacoso **Use:** ME **Origin:** Native **Region:** Andean, Pacific **Dept.:** ANT, BOY, CAU, CHO, CUN, NAR, PUT, SAN, VAC **Elev.:** 500-2600 m.a.s.l. **IUCN Habitat code:** 1,3,14 **Conservation status: (IUCN)** NE **(CPSG)** LC

42 *Alsophila firma* (Baker) D.S.Conant (Cyatheaceae) **IPNI:** 9976-2 **Use:** ME **Origin:** Native **Region:** Andean **Dept.:** NAR **Elev.:** 1300-1300 m.a.s.l. **IUCN Habitat code:** 1 **Conservation status: (IUCN)** NE **(CPSG)** LC

43 *Cyathea arborea* (L.) Sm. (Cyatheaceae) **IPNI:** 17071820-1 **Use:** EU **Origin:** Native **Region:** Caribbean **Dept.:** LAG **Elev.:** 600-650 m.a.s.l. **IUCN Habitat code:** 3 **Conservation status: (IUCN)** NE **(CPSG)** VU

44 *Cyathea aterrima* Domin (Cyatheaceae) **IPNI:** 73117-2 **Use:** ME **Origin:** Native **Region:** Amazonia, Andean, Guiana Shield, Pacific **Dept.:** CAQ, CHO, GUV, MAG, MET, SAN, VAC, VAU **Elev.:** 250-1800 m.a.s.l. **IUCN Habitat code:** 1,4,14 **Conservation status: (IUCN)** NE **(CPSG)** LC

45 *Cyathea bicrenata* Liebm. (Cyatheaceae) **IPNI:** 73125-2 **Common name:** Palo de la vida, Pelma **Use:** HF **Origin:** Native **Region:** Amazonia, Andean, Pacific **Dept.:** AMA, ANT, CAL, CHO, HUI, NAR, TOL, VAC **Elev.:** 0-2500 m.a.s.l. **IUCN Habitat code:** 1,3,14 **Conservation status: (IUCN)** NE **(CPSG)** LC

46 *Cyathea caracasana* (Klotzsch) Domin (Cyatheaceae) **IPNI:** 73146-2 **Common name:** helecho, helecho macho, palma boba **Use:** EU, MA, ME, SU **Origin:** Native **Region:** Andean, Guiana Shield, Caribbean, Pacific **Dept.:** ANT, BOG, BOY, CAL, CAU, CHO, CUN, HUI, LAG, MAG, MET, NAR, NSA, PUT, RIS, SAN, TOL, VAC **Elev.:** 1200-4200 m.a.s.l. **IUCN Habitat code:** 1,2,3,4,14 **Conservation status: (IUCN)** NE **(CPSG)** Potential LC

47 *Cyathea fulva* (M.Martens & Galeotti) Fée (Cyatheaceae) **IPNI:** 17072730-1 **Use:** MA **Origin:** Native **Region:** Andean **Dept.:** ANT, BOY, CAL, CAU, CUN, NSA, QUI **Elev.:** 1800–3300 m a.s.l. **IUCN Habitat code:** 1,2,3,14 **Conservation status: (IUCN)** NE

48 *Cyathea lasiosora* Domin (Cyatheaceae) **IPNI:** 73267-2 **Common name:** helecho arborescente, tacis, ùjùmedù **Use:** ME **Origin:** Native **Region:** Amazonia, Andean, Orinoquia **Dept.:** AMA, ANT, CAQ, CAU, MET, NAR, PUT, VAC, VAU **Elev.:** 100-1800 m.a.s.l. **IUCN Habitat code:** 1,3,4,5,14 **Conservation status: (IUCN)** NE **(CPSG)** LC

49 *Cyathea microdonta* (Desv.) Domin (Cyatheaceae) **IPNI:** 73288-2 **Common name:** lecho, palma boba, trichi o tricho **Use:** HF, ME **Origin:** Native **Region:** Andean, Orinoquia, Pacific **Dept.:** ANT, CAQ, CAS, CAU, CHO, CUN, MET, NAR, VAC, VID **Elev.:** 0-2000 m.a.s.l. **IUCN Habitat code:** 1,3,4,5,14 **Conservation status: (IUCN)** NE **(CPSG)** LC

50 *Cyathea tryonorum* (Riba) Lellinger (Cyatheaceae) **IPNI:** 275500-2 **Use:** ME **Origin:** Native **Region:** Andean **Dept.:** ANT, BOY, CAU, CUN, SAN, VAC **Elev.:** 1650-2900 m.a.s.l. **IUCN Habitat code:** 1,3,14 **Conservation status: (IUCN)** NE **(CPSG)** LC

51 *Dicksonia karsteniana* (Klotzsch) T.Moore (Cyatheaceae) **IPNI:** 1190416-2 **Use:** ME **Origin:** Native **Conservation status: (IUCN)** NE **(CPSG)** LC

52 *Dicksonia sellowiana* (C.Presl) Hook. (Cyatheaceae) **IPNI:** 275823-2 **Use:** MA, ME **Origin:** Native **Region:** Andean **Dept.:** ANT, BOG, BOY, CAL, CAU, CES, CUN, HUI, MAG, NAR, NSA, PUT, QUI, RIS, SAN, TOL, VAC **Elev.:** 1500-3800 m.a.s.l. **IUCN Habitat code:** 1,3,4,5,8,14 **Conservation status: (IUCN)** NE **(CPSG)** Potential LC

53 *Lophosoria quadripinnata* (J.F.Gmel.) C.Chr. (Cyatheaceae) **IPNI:** 143624-2 **Common name:** boba, helecho de árbol, lecho **Use:** HF, ME **Origin:** Native **Region:** Andean, Guiana Shield, Pacific **Dept.:** ANT, BOG, BOY, CAL, CAQ, CAU, CHO, CUN, HUI, MAG, MET, NAR, NSA, PUT, RIS, SAN, TOL, VAC **Elev.:** 100-3510 m.a.s.l. **IUCN Habitat code:** 1,3,4,5,14 **Conservation status: (IUCN)** NE **(CPSG)** LC

54 *Sphaeropteris quindiuensis* (H.Karst.) R.M.Tryon (Cyatheaceae) **IPNI:** 241965-2 **Use:** ME **Origin:** Native **Region:** Andean **Dept.:** ANT, BOY, CAL, CUN, HUI, NAR, QUI, RIS, SAN, TOL **Elev.:** 750-3000 m.a.s.l. **IUCN Habitat code:** 1,3,4,14 **Conservation status: (IUCN)** NE **(CPSG)** LC

Equisetales

55 *Equisetum arvense* L. (Equisetaceae) **IPNI:** 300073-2 **Use:** HF, MA, ME, PO **Conservation status: (IUCN)** LC

56 *Equisetum bogotense* Kunth (Equisetaceae) **IPNI:** 329489-2 **Common name:** canutillo, cola de caballo, equisetum, herb de conejo, juntalli, tembladera, tembladera pequeña, yerba de chivo **Habit:** Herb **Use:** ME **Origin:** Native **Region:** Andean **Dept.:** ANT, ARA, BOG, BOY, CAQ, CAU, CES, CHO, CUN, HUI, MAG, MET, NAR, NSA, PUT, QUI, RIS, SAN, TOL, VAC **Elev.:** 1500-3800 m.a.s.l. **IUCN Habitat code:** 1,2,3,4,5,14 **Conservation status: (IUCN)** NE **(CPSG)** Potential LC

57 *Equisetum giganteum* L. (Equisetaceae) **IPNI:** 320302-2 **Common name:** canutillo, cola de caballo, rabo de mula **Habit:** Herb **Use:** ME **Origin:** Native **Region:** Amazonia, Andean, Caribbean **Dept.:** ANT, BOG, BOY, CES, CHO, CUN, GUA, HUI, NAR, PUT, QUI, TOL, VAC **Elev.:** 600–3000 m a.s.l. **IUCN Habitat code:** 1,2,3,14 **Conservation status: (IUCN)** LC

58 *Equisetum hyemale* L. (Equisetaceae) **IPNI:** 92939-2 **Use:** EU, MA, ME **Region:** Andean **Dept.:** BOG **Conservation status: (IUCN)** LC

59 *Equisetum myriochaetum* Schltdl. & Cham. (Equisetaceae) **IPNI:** 77060052-1 **Habit:** Herb **Use:** ME **Origin:** Native **Region:** Andean **Dept.:** ANT, CHO, CUN, TOL, VAC **Elev.:** 1400–2200 m a.s.l. **IUCN Habitat code:** 1,14 **Conservation status: (IUCN)** NE

Gleicheniales

60 *Dicranopteris flexuosa* (Schrad.) Underw. (Gleicheniaceae) **IPNI:** 79839-2 **Common name:** helecho pategallina **Habit:** Herb **Use:** ME **Origin:** Native **Region:** Amazonia, Andean, Guiana Shield, Pacific **Dept.:** ANT, BOY, CAQ, CAU, CES, CHO, CUN, GUA, MET, SAN, VAC, VAU **Elev.:** 100-2600 m.a.s.l. **IUCN Habitat code:** 1,2,3,14 **Conservation status: (IUCN)** NE **(CPSG)** Potential LC

61 *Gleichenella pectinata* (Willd.) Ching (Gleicheniaceae) **IPNI:** 110733-2 **Habit:** Herb **Use:** ME **Origin:** Native **Region:** Amazonia, Andean, Guiana Shield, Caribbean, Orinoquia, Pacific **Dept.:** AMA, ANT, BOY, CAQ, CAU, CES, CHO, CUN, GUV, HUI, MAG, MET, NAR, QUI, SAN, TOL, VAC, VID **Elev.:** 0-1950 m.a.s.l. **IUCN Habitat code:** 1,3,4,14 **Conservation status: (IUCN)** NE **(CPSG)** Potential LC

62 *Sticherus bifidus* (Willd.) Ching (Gleicheniaceae) **IPNI:** 245596-2 **Common name:** helecho pata de gallina, helecho pategallina **Habit:** Herb **Use:** MA **Origin:** Native **Region:** Andean, Pacific **Dept.:** ANT, BOG, BOY, CAL, CAQ, CAU, CHO, CUN, HUI, MAG, MET, NAR, NSA, PUT, SAN, VAC **Elev.:** 50-2450 m.a.s.l. **IUCN Habitat code:** 1,2,3,4,14 **Conservation status: (IUCN)** NE **(CPSG)** Potential LC

63 *Sticherus nudus* (Moritz ex Reichardt) Nakai (Gleicheniaceae) **IPNI:** 245632-2 **Habit:** Herb **Use:** MA **Origin:** Native **Region:** Andean **Dept.:** ANT, BOG, BOY, CAU, CUN, HUI, PUT, VAC **Elev.:** 1700–2500 m a.s.l. **IUCN Habitat code:** 1,3,14 **Conservation status: (IUCN)** NE

64 *Sticherus rubiginosus* (Mett.) Nakai (Gleicheniaceae) **IPNI:** 245647-2 **Common name:** helecho gallinero, pata de gallina **Habit:** Herb **Use:** MA **Origin:** Native **Region:** Andean, Pacific **Dept.:** ANT, BOG, BOY, CAQ, CAU, CHO, CUN, HUI, MAG, NAR, NSA, SAN, TOL, VAC **Elev.:** 1250-2800 m.a.s.l. **IUCN Habitat code:** 1,3,14 **Conservation status: (IUCN)** NE **(CPSG)** Potential LC

Hymenophyllales

65 *Hymenophyllum fragile* (Hedw.) C.V.Morton (Hymenophyllaceae) **IPNI:** 125766-2 **Habit:** Herb **Use:** ME **Origin:** Native **Region:** Andean, Pacific **Dept.:** ANT, BOY, CHO, CUN, MAG, RIS, VAC **Elev.:** 1450–2800 m a.s.l. **IUCN Habitat code:** 1,3,14 **Conservation status: (IUCN)** NE

66 *Hymenophyllum fucoides* (Sw.) Sw. (Hymenophyllaceae) **IPNI:** 1009149-2 **Habit:** Herb **Use:** ME **Origin:** Native **Region:** Andean **Dept.:** ANT, BOG, BOY, CAU, CES, CHO, CUN, HUI, MAG, NSA, PUT, RIS, SAN, TOL, VAC **Elev.:** 400-4300 m.a.s.l. **IUCN Habitat code:** 1,3,4,14 **Conservation status: (IUCN)** NE **(CPSG)** Potential LC

67 *Trichomanes crispum* L. (Hymenophyllaceae) **IPNI:** 257306-2 **Habit:** Herb **Use:** ME **Origin:** Native **Region:** Amazonia, Andean, Guiana Shield, Caribbean, Orinoquia, Pacific **Dept.:** AMA, ANT, BOY, CAU, CHO, CUN, MET, NAR, NSA, SAN, TOL, VAC **Elev.:** 0–2050 m a.s.l. **Conservation status: (IUCN)** NE

68 *Trichomanes diversifrons* (Bory) Mett. (Hymenophyllaceae) **IPNI:** 17228970-1 **Common name:** churrusco **Habit:** Herb **Use:** ME **Origin:** Native **Region:** Amazonia, Andean, Guiana Shield, Caribbean, Pacific **Dept.:** AMA, ANT, CAU, CHO, MET, NAR, SAN, TOL **Elev.:** 0–1000 m a.s.l. **IUCN Habitat code:** 1,3,4,5,14 **Conservation status: (IUCN)** NE

69 *Trichomanes elegans* Rich. (Hymenophyllaceae) **IPNI:** 17229030-1 **Common name:** loro, navegadora **Habit:** Herb **Use:** ME **Origin:** Native **Region:** Amazonia, Andean, Guiana Shield, Pacific **Dept.:** AMA, ANT, BOY, CAQ, CAU, CHO, CUN, GUA, MET, NAR, NSA, PUT, RIS, SAN, VAC, VAU **Elev.:** 0–1500 m a.s.l. **IUCN Habitat code:** 1,3,4,5,14 **Conservation status: (IUCN)** NE

70 *Trichomanes pinnatum* Hedw. (Hymenophyllaceae) **IPNI:** 320804-2 **Common name:** helecho de pescado, rabo de chucha **Habit:** Herb **Use:** ME **Origin:** Native **Region:** Amazonia, Andean, Guiana Shield, Caribbean, Orinoquia, Pacific **Dept.:** AMA, ANT, BOY, CAQ, CAU, CES, CHO, COR, CUN, MET, NAR, NSA, PUT, SAN, TOL, VAC, VAU, VID **Elev.:** 0-1000 m.a.s.l. **IUCN Habitat code:** 1,2,3,4,5,14 **Conservation status: (IUCN)** NE **(CPSG)** Potential LC

Ophioglossales

71 *Botrychium schaffneri* Underw. (Ophioglossaceae) **IPNI:** 35345-2 **Habit:** Herb **Use:** ME **Origin:** Native **Region:** Andean **Dept.:** BOY, MAG, MET, SAN **Elev.:** 3250–4000 m a.s.l. **IUCN Habitat code:** 4,6 **Conservation status: (IUCN)** NE

72 *Botrychium underwoodianum* Maxon (Ophioglossaceae) **IPNI:** 35373-2 **Habit:** Herb **Use:** ME **Origin:** Native **Conservation status: (IUCN)** NE

73 *Botrychium virginianum* (L.) Sw. (Ophioglossaceae) **IPNI:** 1106719-2 **Habit:** Herb **Use:** HF, ME **Origin:** Native **Region:** Andean **Dept.:** BOY, CUN, MAG, SAN **Elev.:** 1800–3300 m a.s.l. **IUCN Habitat code:** 1,3,4,14 **Conservation status: (IUCN)** NE

74 *Ophioglossum reticulatum* L. (Ophioglossaceae) **IPNI:** 174612-2 **Habit:** Herb **Use:** AF, GS, HF, MA, ME, SU **Origin:** Native **Region:** Andean **Dept.:** ANT, BOG, BOY, CAU, CUN, SAN, TOL, VAC **Elev.:** 1500–3850 m a.s.l. **IUCN Habitat code:** 14 **Conservation status: (IUCN)** LC

Osmundales

75 *Osmunda regalis* L. (Osmundaceae) **IPNI:** 305721-2 **Common name:** helecho de espiga **Habit:** Herb **Use:** EU, GS, HF, MA, ME, PO **Origin:** Native **Region:** Andean **Dept.:** ANT, CAU, CUN, HUI, SAN **Elev.:** 1600–2700 m a.s.l. **IUCN Habitat code:** 1,14 **Conservation status: (IUCN)** LC

76 *Osmundastrum cinnamomeum* (L.) C.Presl (Osmundaceae) **IPNI:** 60447721-2 **Habit:** Herb **Use:** EU, ME **Origin:** Native **Region:** Andean **Dept.:** ANT, BOY, HUI, SAN **Elev.:** 1800–2200 m a.s.l. **IUCN Habitat code:** 1,14 **Conservation status: (IUCN)** NE

Polypodiales

77 *Asplenium auritum* Sw. (Aspleniaceae) **IPNI:** 22222-2 **Habit:** Herb **Use:** ME **Origin:** Native **Region:** Amazonia, Andean, Guiana Shield, Orinoquia, Pacific **Dept.:** ANT, BOG, BOY, CAL, CAQ, CAU, CHO, CUN, HUI, MAG, MET, NAR, NSA, PUT, QUI, RIS, SAN, TOL, VAU, VID **Elev.:** 0-2820 m.a.s.l. **IUCN Habitat code:** 1,3,4,5,14 **Conservation status: (IUCN)** NE **(CPSG)** Potential LC

78 *Asplenium bipartitum* Bory ex Willd. (Aspleniaceae) **IPNI:** 17039970-1 **Use:** ME **Conservation status: (IUCN)** NE

79 *Asplenium cuneatum* Lam. (Aspleniaceae) **IPNI:** 328699-2 **Habit:** Herb **Use:** ME **Origin:** Native **Region:** Amazonia, Andean **Dept.:** AMA, ANT, NSA **Elev.:** 200–600 m a.s.l. **IUCN Habitat code:** 1,3,14 **Conservation status: (IUCN)** NE

80 *Asplenium formosum* Willd. (Aspleniaceae) **IPNI:** 22417-2 **Habit:** Herb **Use:** EU **Origin:** Native **Region:** Andean, Guiana Shield, Caribbean, Orinoquia, Pacific **Dept.:** ANT, BOY, CAL, CAU, CES, CHO, CUN, LAG, MAG, MET, NAR, TOL **Elev.:** 50–1300 m a.s.l. **IUCN Habitat code:** 1,2,3,14 **Conservation status: (IUCN)** LC

81 *Asplenium monanthes* L. (Aspleniaceae) **IPNI:** 22558-2 **Habit:** Herb **Use:** ME **Origin:** Native **Region:** Andean, Guiana Shield **Dept.:** ANT, ARA, BOG, BOY, CAL, CAU, CUN, MAG, MET, NAR, NSA, RIS, SAN, TOL **Elev.:** 1050-4400 m.a.s.l. **IUCN Habitat code:** 1,3,4,8,14 **Conservation status: (IUCN)** NE **(CPSG)** Potential LC

82 *Asplenium praemorsum* Sw. (Aspleniaceae) **IPNI:** 328509-2 **Habit:** Herb **Use:** ME **Origin:** Native **Region:** Andean **Dept.:** ANT, BOG, BOY, CAU, CES, CUN, HUI, MAG, NAR, NSA, QUI, SAN **Elev.:** 1200–3050 m a.s.l. **IUCN Habitat code:** 1,3,4,14 **Conservation status: (IUCN)** NE

83 *Asplenium radicans* L. (Aspleniaceae) **IPNI:** 22662-2 **Habit:** Herb **Use:** EU, ME **Origin:** Native **Region:** Andean, Guiana Shield, Pacific **Dept.:** ANT, BOG, BOY, CAL, CAQ, CAS, CAU, CES, CHO, CUN, HUI, MAG, MET, NAR, NSA, PUT, QUI, RIS, SAN, TOL, VAC **Elev.:** 100-3500 m.a.s.l. **IUCN Habitat code:** 1,3,4,5,14 **Conservation status: (IUCN)** NE **(CPSG)** Potential LC

84 *Asplenium serratum* L. (Aspleniaceae) **IPNI:** 22751-2 **Common name:** doradilla, pencón **Habit:** Herb **Use:** ME **Origin:** Native **Region:** Amazonia, Andean, Guiana Shield, Caribbean, Orinoquia, Pacific **Dept.:** AMA, ANT, CAQ, CAU, CHO, GUA, LAG, MAG, MET, NSA, PUT, RIS, SAN, VAC, VAU, VID **Elev.:** 40-900 m.a.s.l. **IUCN Habitat code:** 1,2,3,4,5,14 **Conservation status: (IUCN)** NE **(CPSG)** Potential LC

85 *Athyrium asplenioides* Desv. (Aspleniaceae) **IPNI:** 17054310-1 **Use:** ME **Region:** Andean **Dept.:** BOG **Conservation status: (IUCN)** NE

86 *Athyrium filix-femina* (L.) Roth (Aspleniaceae) **IPNI:** 30055316-2 **Common name:** helecho macho **Habit:** Herb **Use:** EU, HF, ME **Origin:** Native **Region:** Andean **Dept.:** ANT, BOG, CUN **Elev.:** 2300–3100 m a.s.l. **IUCN Habitat code:** 1,4 **Conservation status: (IUCN)** NE

87 *Blechnum brasiliense* Desv. (Aspleniaceae) **IPNI:** 17058320-1 **Habit:** Herb **Use:** EU **Origin:** Native, Cultivated **Region:** Andean **Dept.:** ANT, HUI **Elev.:** 850–1100 m a.s.l. **IUCN Habitat code:** 3 **Conservation status: (IUCN)** NE

88 *Blechnum cordatum* (Desv.) Hieron. (Aspleniaceae) **IPNI:** 33533-2 **Habit:** Herb **Use:** HF, ME **Origin:** Native **Region:** Andean, Caribbean, Pacific **Dept.:** ANT, BOL, BOY, CAL, CAQ, CAU, CHO, CUN, HUI, LAG, NAR, PUT, RIS, SAN, TOL, VAC **Elev.:** 200-3100 m.a.s.l. **IUCN Habitat code:** 1,3,4,8,14 **Conservation status: (IUCN)** NE **(CPSG)** Potential LC

89 *Blechnum loxense* (Kunth) Hook. ex Salomon (Aspleniaceae) **IPNI:** 77186436-1 **Habit:** Herb **Use:** EU **Origin:** Native **Region:** Andean **Dept.:** ANT, BOY, CAL, CAU, CES, CHO, CUN, HUI, MAG, MET, NAR, PUT, RIS, SAN, TOL **Elev.:** 2600-3900 m.a.s.l. **IUCN Habitat code:** 1,3,4,5,8,14 **Conservation status: (IUCN)** NE **(CPSG)** Potential LC

90 *Blechnum occidentale* L. (Aspleniaceae) **IPNI:** 33632-2 **Habit:** Herb **Use:** HF, ME **Origin:** Native **Region:** Andean, Guiana Shield, Caribbean, Pacific **Dept.:** ANT, BOG, BOY, CAL, CAU, CES, CHO, CUN, HUI, LAG, MAG, MET, NAR, NSA, QUI, RIS, SAN, TOL, VAC **Elev.:** 50-2850 m.a.s.l. **IUCN Habitat code:** 1,2,3,4,14 **Conservation status: (IUCN)** NE **(CPSG)** Potential LC

91 *Diplazium expansum* Willd. (Aspleniaceae) **IPNI:** 17086820-1 **Habit:** Herb **Use:** HF **Origin:** Native **Region:** Andean **Dept.:** ANT, BOY, CUN **Elev.:** 2000–2800 m a.s.l. **IUCN Habitat code:** 1,3 **Conservation status: (IUCN)** NE

92 *Diplazium subserratum* (Blume) T.Moore (Aspleniaceae) **IPNI:** 17088960-1 **Use:** HF **Conservation status: (IUCN)** NE

93 *Hymenasplenium obtusifolium* (L.) L.Regalado & Prada (Aspleniaceae) **IPNI:** 60458652-2 **Habit:** Herb **Use:** ME **Origin:** Native **Region:** Andean **Dept.:** CUN, MAG, NSA **Elev.:** 900–1600 m a.s.l. **IUCN Habitat code:** 1,2,14 **Conservation status: (IUCN)** NE

94 *Macrothelypteris torresiana* (Gaudich.) Ching (Aspleniaceae) **IPNI:** 1019960-2 **Habit:** Herb **Use:** EU, ME **Origin:** Naturalised **Region:** Andean, Guiana Shield, Caribbean, Orinoquia, Pacific **Dept.:** ANT, BOY, CAL, CAU, CHO, CUN, HUI, LAG, MAG, MET, NAR, NSA, PUT, QUI, RIS, SAN, TOL, VAC **Elev.:** 0–2500 m a.s.l. **IUCN Habitat code:** 1,3,14 **Conservation status: (IUCN)** NE

95 *Telmatoblechnum serrulatum* (Rich.) Perrie, D.J.Ohlsen & Brownsey (Aspleniaceae) **IPNI:** 77142412-1 **Habit:** Herb **Use:** HF **Origin:** Native **Region:** Amazonia, Guiana Shield, Caribbean, Orinoquia, Pacific **Dept.:** ANT, CAQ, CHO, COR, GUA, MET, NAR, SAN, VAU, VID **Elev.:** 0–550 m a.s.l. **IUCN Habitat code:** 1,3,4,5,14 **Conservation status: (IUCN)** NE

96 *Thelypteris dentata* (Forssk.) E.P.St.John (Aspleniaceae) **IPNI:** 251308-2 **Habit:** Herb **Use:** MA, ME, PO **Origin:** Naturalised **Region:** Amazonia, Andean, Guiana Shield, Caribbean, Orinoquia, Pacific **Dept.:** AMA, ANT, CAQ, CHO, CUN, HUI, MAG, MET, TOL, VAC **Elev.:** 50–1700 m a.s.l. **IUCN Habitat code:** 1,5,14 **Conservation status: (IUCN)** NE

97 *Thelypteris gongylodes* (Schkuhr) Small (Aspleniaceae) **IPNI:** 251353-2 **Use:** GS, HF, ME **Origin:** Native **IUCN Habitat code:** 1,3 **Conservation status: (IUCN)** NE

98 *Thelypteris interrupta* (Willd.) K.Iwats. (Aspleniaceae) **IPNI:** 17458930-1 **Habit:** Herb **Use:** HF, MA, ME **Origin:** Native **Region:** Andean, Caribbean, Pacific **Dept.:** ANT, BOL, CUN, VAC **Elev.:** 100–2400 m a.s.l. **Conservation status: (IUCN)** NE

99 *Thelypteris kunthii* (Desv.) C.V.Morton (Aspleniaceae) **IPNI:** 251400-2 **Habit:** Herb **Use:** ME **Origin:** Endemic **Region:** Andean, Caribbean **Dept.:** ATL, BOY, HUI, LAG **Elev.:** 100–1500 m a.s.l. **IUCN Habitat code:** 1,3,14 **Conservation status: (IUCN)** NE

100 *Thelypteris nicaraguensis* (E.Fourn.) C.V.Morton (Aspleniaceae) **IPNI:** 252198-2 **Habit:** Herb **Use:** ME **Origin:** Native **Region:** Pacific **Dept.:** CHO **Elev.:** 100–100 m a.s.l. **Conservation status: (IUCN)** NE

101 *Thelypteris opulenta* (Kaulf.) Fosberg (Aspleniaceae) **IPNI:** 17460390-1 **Habit:** Herb **Use:** HF, ME **Origin:** Naturalised **Region:** Amazonia, Guiana Shield, Orinoquia, Pacific **Dept.:** AMA, ANT, CAQ, CAU, CHO, MET, NAR, PUT, TOL, VAC **Elev.:** 0–500 m a.s.l. **IUCN Habitat code:** 1,14 **Conservation status: (IUCN)** NE

102 *Cystopteris fragilis* (L.) Bernh. (Cystopteridaceae) **IPNI:** 30013429-2 **Habit:** Herb **Use:** EU, GS, HF, ME **Origin:** Native **Region:** Andean **Dept.:** ANT, ARA, BOG, BOY, CAL, CAU, CUN, HUI, MAG, MET, NAR, NSA, PUT, RIS, SAN, TOL **Elev.:**

1400-4500 m.a.s.l. **IUCN Habitat code:** 1,3,4,14 **Conservation status: (IUCN)** NE **(CPSG)** Potential LC

103 *Dennstaedtia cicutaria* (Sw.) T.Moore (Dennstaedtiaceae) **IPNI:** 17081410-1 **Habit:** Herb **Use:** EU **Origin:** Native **Region:** Andean, Pacific **Dept.:** ANT, CAL, CHO, CUN, HUI, MAG, MET, NAR, NSA, SAN, TOL **Elev.:** 0–2600 m a.s.l. **IUCN Habitat code:** 1,2,3,5,14 **Conservation status: (IUCN)** NE

104 *Hypolepis hostilis* (Kunze) C.Presl (Dennstaedtiaceae) **IPNI:** 17126480-1 **Habit:** Herb **Use:** HF **Origin:** Native **Region:** Amazonia, Andean, Guiana Shield, Pacific **Dept.:** AMA, ANT, CAQ, CAU, MET, NAR, NSA, QUI, SAN, VAU **Elev.:** 100–1900 m a.s.l. **IUCN Habitat code:** 1,14 **Conservation status: (IUCN)** NE

105 *Hypolepis repens* (L.) C.Presl (Dennstaedtiaceae) **IPNI:** 1031307-2 **Habit:** Herb **Use:** HF, ME **Origin:** Native **Region:** Andean, Pacific **Dept.:** ANT, CAQ, CHO, NAR, NSA, VAC **Elev.:** 0–1650 m a.s.l. **IUCN Habitat code:** 1,3,14 **Conservation status: (IUCN)** NE

106 *Pteridium aquilinum* (L.) Kuhn (Dennstaedtiaceae) **IPNI:** 17210060-1 **Common name:** caramo, helecho de marrano, helecho liso, helecho marranero, milile, mililí, pata de cuervo **Use:** AF, GS, HF, MA, ME, PO, SU **Dept.:** BOG **IUCN Habitat code:** 1,3,4,5,14 **Conservation status: (IUCN)** NE

107 *Pteridium arachnoideum* (Kaulf.) Maxon (Dennstaedtiaceae) **IPNI:** 213798-2 **Common name:** helecho de marrano, helecho marranero **Habit:** Herb **Use:** ME **Origin:** Native **Region:** Amazonia, Andean, Caribbean, Orinoquia, Pacific **Dept.:** AMA, ANT, BOG, BOY, CAQ, CAU, CES, CHO, CUN, HUI, MAG, MET, NAR, NSA, SAN, VAU **Elev.:** 240-3200 m.a.s.l. **IUCN Habitat code:** 1,3,14 **Conservation status: (IUCN)** NE **(CPSG)** Potential LC

108 *Pteridium caudatum* Maxon (Dennstaedtiaceae) **IPNI:** 213799-2 **Habit:** Herb **Use:** HF, ME, PO **Origin:** Native **Region:** Amazonia, Andean, Caribbean, Pacific **Dept.:** AMA, ANT, BOG, BOY, CAL, CAQ, CUN, HUI, SAP, SAN **Elev.:** 0–2600 m a.s.l. **IUCN Habitat code:** 1,2,3,4,5,8,14 **Conservation status: (IUCN)** NE

109 *Campyloneurum albopunctatissimum* (Lellinger) Christenh. (Polypodiaceae) **IPNI:** 77183020-1 **Habit:** Herb **Use:** ME **Origin:** Native **Region:** Andean, Guiana Shield, Pacific **Dept.:** ANT, CAU, CHO, CUN, MET, NAR, SAN **Elev.:** 500–3050 m a.s.l. **Conservation status: (IUCN)** NE

110 *Campyloneurum angustifolium* Fée (Polypodiaceae) **IPNI:** 17063000-1 **Common name:** calaguala, pecosa **Habit:** Herb **Use:** EU, ME **Origin:** Native **Region:** Amazonia, Andean, Guiana Shield **Dept.:** AMA, ANT, BOG, BOY, CAL, CAS, CAU, CES, CUN, HUI, MAG, MET, NAR, QUI, RIS, SAN, VAC **Elev.:** 450-3500 m.a.s.l. **IUCN Habitat code:** 1,3,4,5,14 **Conservation status: (IUCN)** NE **(CPSG)** Potential LC

111 *Campyloneurum crassifolium* (L.) Christenh. (Polypodiaceae) **IPNI:** 77183236-1 **Common name:** maiku jat **Habit:** Herb **Use:** EU, ME **Origin:** Native **Region:** Amazonia, Andean, Guiana Shield, Caribbean, Orinoquia, Pacific **Dept.:** AMA, ANT, BOL, BOY, CAQ, CAU, CHO, CUN, HUI, MET, PUT, QUI, RIS, SAN, TOL, VAC, VAU **Elev.:** 0-3300 m.a.s.l. **Conservation status: (IUCN)** NE **(CPSG)** Potential LC

112 *Campyloneurum densifolium* (Hieron.) Lellinger (Polypodiaceae) **IPNI:** 274004-2 **Habit:** Herb **Use:** ME **Origin:** Native **Region:** Andean **Dept.:** ANT, BOG, BOY, CAU, CES, CUN, MAG, NAR, QUI, RIS, SAN, TOL **Elev.:** 1200–3400 m a.s.l. **IUCN Habitat code:** 1,3,4,8,14 **Conservation status: (IUCN)** NE

113 *Campyloneurum phyllitidis* (L.) C.Presl (Polypodiaceae) **IPNI:** 17063380-1 **Common name:** helecha **Habit:** Herb **Use:** EU, ME **Origin:** Native **Region:** Amazonia, Andean, Guiana Shield, Caribbean, Orinoquia, Pacific **Dept.:** AMA, ANT, BOY, CAL, CAQ, CAU, CES, CHO, COR, CUN, HUI, LAG, MAG, MET, NAR, NSA, TOL, VAC, VAU, VID **Elev.:** 0-2800 m.a.s.l. **IUCN Habitat code:** 1,2,3,5,14 **Conservation status: (IUCN)** NE **(CPSG)** Potential LC

114 *Cyclopeltis semicordata* (Sw.) J.Sm. (Polypodiaceae) **IPNI:** 1148359-2 **Habit:** Herb **Use:** ME **Origin:** Native **Region:** Andean, Guiana Shield, Caribbean, Pacific **Dept.:** ANT, BOL, BOY, CAL, CES, CHO, CUN, LAG, MAG, MET, NSA, SAN, SUC, TOL **Elev.:** 0-1600 m.a.s.l. **IUCN Habitat code:** 1,2,3,5,14 **Conservation status: (IUCN)** NE **(CPSG)** Potential LC

115 *Davallia canariensis* (L.) Sm. (Polypodiaceae) **IPNI:** 17078300-1 **Common name:** helecho encaje, rabo de mico **Habit:** Herb **Use:** EU **Origin:** Cultivated **Dept.:** ANT, BOG **IUCN Habitat code:** 1,14 **Conservation status: (IUCN)** NE

116 *Davallia solida* (G.Forst.) Sw. (Polypodiaceae) **IPNI:** 17080730-1 **Habit:** Herb **Use:** ME **Origin:** Cultivated **Region:** Andean **Dept.:** CUN **Elev.:** 2500–2600 m a.s.l. **Conservation status: (IUCN)** NE

117 *Didymochlaena truncatula* (Sw.) J.Sm. (Polypodiaceae) **IPNI:** 17084790-1 **Habit:** Herb **Use:** EU, ME **Origin:** Native **Region:** Amazonia, Andean, Caribbean, Pacific **Dept.:** AMA, ANT, CAU, CES, CHO, CUN, LAG, MAG, NAR, NSA, PUT, QUI, RIS, SAN, VAC **Elev.:** 50-2350 m.a.s.l. **IUCN Habitat code:** 1,2,3,4,5,14 **Conservation status: (IUCN)** LC **(CPSG)** Potential LC

118 *Dryopteris fengiana* (Ching) ined. (Polypodiaceae) **IPNI:** 77205539-1 **Habit:** Herb **Use:** EU **Origin:** Cultivated **Region:** Andean **Dept.:** TOL **Elev.:** 1300–1300 m a.s.l. **Conservation status: (IUCN)** NE

119 *Dryopteris wallichiana* (Spreng.) Hyl. (Polypodiaceae) **IPNI:** 17392950-1 **Common name:** helecho de la suerte, helecho macho **Habit:** Herb **Use:** EU, ME, SU **Origin:** Native **Region:** Andean **Dept.:** ANT, BOG, BOY, CAL, CAU, CES, CUN, MAG, MET, NAR, NSA, QUI, RIS, SAN, TOL **Elev.:** 350-3600 m.a.s.l. **IUCN Habitat code:** 1,3,4,8,14 **Conservation status: (IUCN)** NE **(CPSG)** Potential LC

120 *Elaphoglossum gayanum* (Fée) T.Moore (Polypodiaceae) **IPNI:** 17103530-1 **Habit:** Herb **Use:** ME **Origin:** Native **Region:** Andean **Dept.:** ANT, ARA, BOG, BOY, CAL, CAU, CES, CUN, MAG, MET, NAR, NSA, SAN **Elev.:** 1700-4300 m.a.s.l. **IUCN Habitat code:** 1,3,4,8,14 **Conservation status: (IUCN)** NE **(CPSG)** Potential LC

121 *Elaphoglossum metallicum* Mickel (Polypodiaceae) **IPNI:** 276374-2 **Habit:** Herb **Use:** EU **Origin:** Native **Region:** Andean **Dept.:** ANT **Elev.:** 2800–2800 m a.s.l. **IUCN Habitat code:** 14 **Conservation status: (IUCN)** NE

122 *Elaphoglossum minutum* (Pohl ex Fée) T.Moore (Polypodiaceae) **IPNI:** 17104440-1 **Habit:** Herb **Use:** ME **Origin:** Native **Region:** Andean **Dept.:** ANT, BOG, CAU, CUN, MAG, NAR **Elev.:** 2800–3650 m a.s.l. **Conservation status: (IUCN)** NE

123 *Elaphoglossum petiolatum* Urb. (Polypodiaceae) **IPNI:** 89270-2 **Habit:** Herb **Use:** ME **Origin:** Native **Region:** Andean **Dept.:** CHO, TOL **Elev.:** 1900–3000 m a.s.l. **IUCN Habitat code:** 1,14 **Conservation status: (IUCN)** NE

124 *Lomariopsis japurensis* (C.Mart.) J.Sm. (Polypodiaceae) **IPNI:** 17442740-1 **Common name:** bagre pajo, charapa huasca **Habit:** Herb **Use:** ME **Origin:** Native **Region:** Amazonia, Andean **Dept.:** AMA, ANT, CAQ, MET, PUT **Elev.:** 200–1000 m a.s.l. **IUCN Habitat code:** 1,3,5,14 **Conservation status: (IUCN)** NE

125 *Microgramma lycopodioides* (L.) Copel. (Polypodiaceae) **IPNI:** 160142-2 **Habit:** Herb **Use:** HF, ME **Origin:** Native **Region:** Amazonia, Andean, Guiana Shield, Caribbean, Pacific **Dept.:** AMA, ANT, BOY, CAL, CAU, CES, CHO, CUN, HUI, LAG, MAG, MET, NAR, PUT, SAN, TOL, VAC, VAU **Elev.:** 0-1600 m.a.s.l. **IUCN Habitat code:** 1,3,4,5,14 **Conservation status: (IUCN)** NE **(CPSG)** Potential LC

126 *Microgramma nitida* (J.Sm.) A.R.Sm. (Polypodiaceae) **IPNI:** 17444940-1 **Habit:** Herb **Use:** ME **Origin:** Native **Region:** Caribbean **Dept.:** SAP **Conservation status: (IUCN)** NE

127 *Microgramma percussa* (Cav.) de la Sota (Polypodiaceae) **IPNI:** 17546710-1 **Common name:** calaguala, helecho de bejuco, herb de lagartija, planta de sanguijuela **Habit:** Herb **Use:** ME **Origin:** Native **Region:** Amazonia, Andean, Guiana Shield, Caribbean, Pacific **Dept.:** AMA, ANT, BOY, CAL, CAQ, CES, CHO, CUN, HUI, LAG, MAG, MET, NAR, NSA, PUT, RIS, SAN, TOL, VAC, VID **Elev.:** 0-2600 m.a.s.l.

IUCN Habitat code: 1,2,3,4,5,14 **Conservation status: (IUCN)** NE **(CPSG)** Potential LC

128 *Microgramma vacciniifolia* (Langsd. & Fisch.) Copel. (Polypodiaceae) **IPNI:** 17406250-1 **Habit:** Herb **Use:** ME **Origin:** Native **Region:** Caribbean **Dept.:** LAG, MAG **Elev.:** 350–450 m a.s.l. **Conservation status: (IUCN)** NE

129 *Microsorum punctatum* Copel. (Polypodiaceae) **IPNI:** 17341250-1 **Common name:** manoetigre **Habit:** Herb **Use:** EU, HF, ME **Origin:** Cultivated, Naturalised **Dept.:** ANT **IUCN Habitat code:** 14 **Conservation status: (IUCN)** NE

130 *Nephrolepis biserrata* (Sw.) Schott (Polypodiaceae) **IPNI:** 17160710-1 **Common name:** helecho cola de pez **Habit:** Herb **Use:** EU, GS, HF, MA, ME, PO, SU **Origin:** Native **Region:** Amazonia, Andean, Caribbean, Pacific **Dept.:** AMA, ANT, CAQ, CHO, COR, CUN, MET, SAP, SAN, VAC, VID **Elev.:** 100–1150 m a.s.l. **IUCN Habitat code:** 1,3,4,5,14 **Conservation status: (IUCN)** NE

131 *Nephrolepis brownii* (Desv.) Hovenkamp & Miyam. (Polypodiaceae) **IPNI:** 77069554-1 **Habit:** Herb **Use:** ME **Origin:** Naturalised **Region:** Amazonia, Andean, Pacific **Dept.:** ANT, CAU, CHO, RIS, VAC, VAU **Elev.:** 0–2000 m a.s.l. **Conservation status: (IUCN)** NE

132 *Nephrolepis cordifolia* (L.) C.Presl (Polypodiaceae) **IPNI:** 17160730-1 **Habit:** Herb **Use:** EU, HF, ME, SU **Origin:** Cultivated, Naturalised **Region:** Amazonia, Andean, Pacific **Dept.:** ANT, BOG, BOY, CAL, CAU, CHO, CUN, HUI, MAG, NSA, PUT, SAN, TOL, VAC, VAU **Elev.:** 0–2800 m a.s.l. **IUCN Habitat code:** 1,3,14 **Conservation status: (IUCN)** NE

133 *Nephrolepis exaltata* (L.) Schott (Polypodiaceae) **IPNI:** 17160830-1 **Common name:** aliento de niño, encaje, espárrago fino, helecho churco, helecho crespo, llanto de niño, sueño, sueño de niño **Habit:** Herb **Use:** EU, ME **Origin:** Native **Region:** Caribbean, Pacific **Dept.:** ANT, BOG, CUN, LAG, VAC **Elev.:** 250–2500 m a.s.l. **IUCN Habitat code:** 1,2,3,14 **Conservation status: (IUCN)** NE

134 *Nephrolepis pectinata* (Willd.) Schott (Polypodiaceae) **IPNI:** 17161100-1 **Habit:** Herb **Use:** EU **Origin:** Native **Region:** Andean, Pacific **Dept.:** ANT, BOG, BOY, CAQ, CHO, CUN, NAR, NSA, PUT, SAN **Elev.:** 50–2470 m a.s.l. **IUCN Habitat code:** 1,3,14 **Conservation status: (IUCN)** NE

135 *Nephrolepis undulata* J.Sm. (Polypodiaceae) **IPNI:** 17161310-1 **Habit:** Herb **Use:** HF, ME, SU **Origin:** Native **Region:** Andean **Dept.:** ANT, BOY, CHO, HUI, MAG **Elev.:** 650–2450 m a.s.l. **IUCN Habitat code:** 1,3 **Conservation status: (IUCN)** LC

136 *Oleandra articulata* (Sw.) C.Presl (Polypodiaceae) **IPNI:** 1138771-2 **Common name:** yerba de ojo **Habit:** Herb **Use:** ME **Origin:** Native **Region:** Amazonia, Andean, Pacific **Dept.:** AMA, ANT, CAU, CHO, MET, NAR, NSA, PUT, VAC, VAU **Elev.:** 0–1800 m a.s.l. **IUCN Habitat code:** 1,3,5,14 **Conservation status: (IUCN)** NE

137 *Oleandra pilosa* Hook. (Polypodiaceae) **IPNI:** 173662-2 **Habit:** Herb **Use:** ME **Origin:** Native **Region:** Amazonia, Andean, Guiana Shield, Pacific **Dept.:** AMA, ANT, BOY, CAQ, CAU, CHO, CUN, MAG, MET, NSA, SAN, VAC, VAU **Elev.:** 500-1900 m.a.s.l. **IUCN Habitat code:** 1,3,4,5,14 **Conservation status: (IUCN)** NE **(CPSG)** Potential LC

138 *Phlebodium areolatum* (Willd.) J.Sm. (Polypodiaceae) **IPNI:** 17175380-1 **Common name:** calaguala **Habit:** Herb **Use:** ME **Origin:** Native **Region:** Andean **Dept.:** ANT, BOY, CAU, CUN, MAG, NSA, SAN **Elev.:** 800–2300 m a.s.l. **IUCN Habitat code:** 1,3,14 **Conservation status: (IUCN)** NE

139 *Phlebodium aureum* (L.) J.Sm. (Polypodiaceae) **IPNI:** 17175390-1 **Common name:** calaguala, calahuala, cartagenero, helecho costeño, helecho cuero, polipodio **Habit:** Herb **Use:** EU, HF, MA, ME **Origin:** Native **Region:** Andean **Dept.:** ANT, BOG **Elev.:** 450–2000 m a.s.l. **IUCN Habitat code:** 1,2,3,5,14 **Conservation status: (IUCN)** NE

140 *Phlebodium decumanum* (Willd.) J.Sm. (Polypodiaceae) **IPNI:** 17175400-1 **Common name:** copudo, rabo de ardita **Habit:** Herb **Use:** EU, ME **Origin:** Native **Region:** Amazonia, Andean, Caribbean, Orinoquia, Pacific **Dept.:** AMA, ANT, BOL, BOY, CAQ, CES, CHO, COR, CUN, MAG, MET, NAR, NSA, SAN, TOL, VAU, VID **Elev.:** 100-1200 m.a.s.l. **IUCN Habitat code:** 1,2,3,4,14 **Conservation status: (IUCN)** NE **(CPSG)** Potential LC

141 *Platycerium bifurcatum* (Cav.) C.Chr. (Polypodiaceae) **IPNI:** 17177460-1 **Common name:** cacho de venado, cachoevenado, cuerno, helecho cuerno **Habit:** Herb **Use:** EU, ME **Origin:** Cultivated, Naturalised **Dept.:** ANT, BOG **IUCN Habitat code:** 1,3,14 **Conservation status: (IUCN)** NE

142 *Platycerium elephantotis* Schweinf. (Polypodiaceae) **IPNI:** 17177470-1 **Habit:** Herb **Use:** EU, ME **Origin:** Cultivated, Naturalised **Region:** Andean **Dept.:** QUI, VAC **Conservation status: (IUCN)** NE

143 *Pleopeltis balaonensis* (Hieron.) A.R.Sm. (Polypodiaceae) **IPNI:** 77141144-1 **Habit:** Herb **Use:** ME **Origin:** Native **Conservation status: (IUCN)** NE

144 *Pleopeltis macrocarpa* (Willd.) Kaulf. (Polypodiaceae) **IPNI:** 17448650-1 **Common name:** calaguala **Habit:** Herb **Use:** ME, PO **Origin:** Native **Region:** Andean, Guiana Shield **Dept.:** ANT, BOG, BOY, CAL, CAU, CES, CUN, HUI, MAG, MET, NAR, NSA, PUT, RIS, SAN, TOL, VAC **Elev.:** 1400-3600 m.a.s.l. **IUCN Habitat code:** 1,2,3,4,8,14 **Conservation status: (IUCN)** NE **(CPSG)** Potential LC

145 *Pleopeltis polypodioides* (L.) E.G.Andrews & Windham (Polypodiaceae) **IPNI:** 1039912-2 **Habit:** Herb **Use:** ME **Origin:** Native **Region:** Amazonia, Andean, Guiana Shield, Caribbean **Dept.:** ANT, BOG, BOY, CAQ, CES, CUN, MAG, MET, NSA, TOL **Elev.:** 250–3000 m a.s.l. **IUCN Habitat code:** 1,3,4,14 **Conservation status: (IUCN)** NE

146 *Polybotrya botryoides* (Baker) C.Chr. (Polypodiaceae) **IPNI:** 17180780-1 **Habit:** Herb **Use:** ME **Origin:** Native **Region:** Andean **Dept.:** ANT **Elev.:** 1800–2000 m a.s.l. **IUCN Habitat code:** 1,14 **Conservation status: (IUCN)** NE

147 *Polybotrya osmundacea* Humb. & Bonpl. ex Willd. (Polypodiaceae) **IPNI:** 30017613-2 **Habit:** Herb **Use:** MA **Origin:** Native **Region:** Amazonia, Andean, Caribbean, Pacific **Dept.:** AMA, ANT, BOY, CAU, CHO, CUN, MAG, MET, PUT, RIS, SAN, VAC **Elev.:** 100-2500 m.a.s.l. **IUCN Habitat code:** 1,2,3,4,5,14 **Conservation status: (IUCN)** NE **(CPSG)** Potential LC

148 *Polybotrya serratifolia* (Fée) Klotzsch (Polypodiaceae) **IPNI:** 17181340-1 **Habit:** Herb **Use:** ME **Origin:** Native **Region:** Andean, Orinoquia **Dept.:** ANT, MET **Elev.:** 500–1500 m a.s.l. **IUCN Habitat code:** 1 **Conservation status: (IUCN)** NE

149 *Polypodium fraternum* Schltdl. & Cham. (Polypodiaceae) **IPNI:** 17189260-1 **Use:** ME **Origin:** Native **IUCN Habitat code:** 14 **Conservation status: (IUCN)** NE

150 *Polypodium vulgare* L. (Polypodiaceae) **IPNI:** 300720-2 **Common name:** helecho peine, polipodio **Use:** EU, GS, HF, ME **IUCN Habitat code:** 3,14 **Conservation status: (IUCN)** NE

151 *Polystichum montevidense* Rosenst. (Polypodiaceae) **IPNI:** 208026-2 **Habit:** Herb **Use:** ME **Origin:** Native **Region:** Andean **Dept.:** ANT, BOG, BOY, CAL, CAU, CUN, HUI, NAR, SAN, VAC **Elev.:** 2100–4050 m a.s.l. **IUCN Habitat code:** 3,4 **Conservation status: (IUCN)** NE

152 *Rumohra adiantiformis* (G.Forst.) Ching (Polypodiaceae) **IPNI:** 224543-2 **Common name:** helecho cuero, helecho de cuero **Habit:** Herb **Use:** EU, ME **Origin:** Native, Cultivated **Dept.:** BOG **IUCN Habitat code:** 3,14 **Conservation status: (IUCN)** NE

153 *Serpocaulon attenuatum* (Humb. & Bonpl. ex Willd.) A.R.Sm. (Polypodiaceae) **IPNI:** 77075087-1 **Habit:** Herb **Use:** ME **Origin:** Native **Region:** Caribbean **Dept.:** CES **Elev.:** 100–100 m a.s.l. **Conservation status: (IUCN)** NE

154 *Serpocaulon levigatum* (Cav.) A.R.Sm. (Polypodiaceae) **IPNI:** 77075105-1 **Common name:** calaguala **Habit:** Herb **Use:** ME **Origin:** Native **Region:** Andean, Guiana Shield, Caribbean, Pacific **Dept.:** ANT, BOG, BOY, CAL, CAU, CES, CHO, CUN, HUI, LAG, MAG, NAR, NSA, PUT, QUI, RIS, SAN, VAC **Elev.:** 200-2850 m.a.s.l. **IUCN Habitat code:** 1,2,3,4,14 **Conservation status: (IUCN)** NE **(CPSG)** Potential LC

155 *Serpocaulon maritimum* (Hieron.) A.R.Sm. (Polypodiaceae) **IPNI:** 77075108-1 **Habit:** Herb **Use:** ME **Origin:** Native **Region:** Andean, Caribbean, Pacific **Dept.:** ANT, CAU, CHO, LAG, NAR, VAC **Elev.:** 0–1650 m a.s.l. **IUCN Habitat code:** 1,3,14 **Conservation status: (IUCN)** NE

156 *Serpocaulon triseriale* (Sw.) A.R.Sm. (Polypodiaceae) **IPNI:** 77075127-1 **Common name:** helecho, nujubujupiali **Habit:** Herb **Use:** ME **Origin:** Native **Region:** Amazonia, Andean, Guiana Shield, Orinoquia, Pacific **Dept.:** AMA, ANT, BOG, BOY, CAQ, CAU, CHO, CUN, GUA, HUI, MAG, MET, NAR, SAN, VAC, VAU **Elev.:** 40-2800 m.a.s.l. **IUCN Habitat code:** 1,2,3,4,5,14 **Conservation status: (IUCN)** NE **(CPSG)** Potential LC

157 *Tectaria heracleifolia* Underw. (Polypodiaceae) **IPNI:** 249899-2 **Habit:** Herb **Use:** EU, ME **Origin:** Native **Region:** Amazonia, Andean, Caribbean **Dept.:** AMA, ANT, BOL, BOY, CAL, CUN, LAG, MAG **Elev.:** 50–1300 m a.s.l. **IUCN Habitat code:** 1,3,14 **Conservation status: (IUCN)** NE

158 *Tectaria incisa* Cav. (Polypodiaceae) **IPNI:** 17226640-1 **Habit:** Herb **Use:** EU, ME **Origin:** Native **Region:** Amazonia, Andean, Guiana Shield, Caribbean, Orinoquia, Pacific **Dept.:** AMA, ANT, ATL, BOL, BOY, CAL, CAQ, CES, CHO, COR, CUN, GUV, LAG, MAG, MET, NAR, NSA, PUT, QUI, RIS, SAN, TOL, VAC, VAU **Elev.:** 0-1900 m.a.s.l. **IUCN Habitat code:** 1,2,3,4,5,14 **Conservation status: (IUCN)** NE **(CPSG)** Potential LC

159 *Tectaria mexicana* (Fée) C.V.Morton (Polypodiaceae) **IPNI:** 249914-2 **Habit:** Herb **Use:** ME **Origin:** Native **Region:** Andean, Pacific **Dept.:** ANT, CHO, MAG, NAR **Elev.:** 0–700 m a.s.l. **IUCN Habitat code:** 1,2,3,5,14 **Conservation status: (IUCN)** NE

160 *Triplophyllum funestum* (Kunze) Holttum (Polypodiaceae) **IPNI:** 286436-2 **Habit:** Herb **Use:** ME **Origin:** Native **Region:** Amazonia, Guiana Shield, Pacific **Dept.:** AMA, ANT, CAQ, CHO, GUA, GUV, MET, NAR, SAN, VAC, VAU, VID **Elev.:** 0-1000 m.a.s.l. **IUCN Habitat code:** 1,3,5,14 **Conservation status: (IUCN)** NE **(CPSG)** Potential LC

161 *Acrostichum aureum* L. (Pteridaceae) **IPNI:** 319061-2 **Common name:** carranconcha, chicharrón, chiguamacho, corocilla, corozo de ciénaga, helecho chiguamacho, helecho mano de tigre, madrevieja, matatigre, racancha, ranconcha **Habit:** Herb **Use:** AF, EU, GS, HF, MA, ME **Origin:** Native **Region:** Andean, Caribbean, Pacific **Dept.:** ANT, ATL, BOL, CAU, CHO, COR, MAG, NAR, SUC, VAC **Elev.:** 0–950 m a.s.l. **IUCN Habitat code:** 1,2,3,4,5,14 **Conservation status: (IUCN)** LC

162 *Acrostichum danaeifolium* Langsd. & Fisch. (Pteridaceae) **IPNI:** 17002360-1 **Habit:** Herb **Use:** HF, ME **Origin:** Native **Region:** Andean, Caribbean, Pacific **Dept.:** ANT, CHO, LAG, MAG, VAC **Elev.:** 0–500 m a.s.l. **IUCN Habitat code:** 1,2,3,4,5,14 **Conservation status: (IUCN)** LC

163 *Adiantum anceps* Maxon & C.V.Morton (Pteridaceae) **IPNI:** 4862-2 **Habit:** Herb **Use:** EU **Origin:** Native **Conservation status: (IUCN)** NE

164 *Adiantum capillus-veneris* L. (Pteridaceae) **IPNI:** 325030-2 **Common name:** capilera, cilantrillo, culantrillo, helecho cilantro, helecho granizo **Use:** EU, GS, HF, ME, SU **Dept.:** BOG **IUCN Habitat code:** 1,3,14 **Conservation status: (IUCN)** LC

165 *Adiantum concinnum* Humb. & Bonpl. ex Willd. (Pteridaceae) **IPNI:** 315077-2 **Common name:** cilantrillo, culantrillo **Habit:** Herb **Use:** EU, ME **Origin:** Native **Region:** Andean, Guiana Shield, Caribbean, Pacific **Dept.:** BOG, BOY, CAL, CAU, CUN, HUI, LAG, MAG, MET, NAR, NSA, VAC **Elev.:** 250–2000 m a.s.l. **IUCN Habitat code:** 1,2,3,14 **Conservation status: (IUCN)** NE

166 *Adiantum deflectens* Mart. (Pteridaceae) **IPNI:** 17009750-1 **Habit:** Herb **Use:** HF **Origin:** Native **Region:** Caribbean, Pacific **Dept.:** LAG, VAC **Elev.:** 700–700 m a.s.l. **IUCN Habitat code:** 1 **Conservation status: (IUCN)** NE

167 *Adiantum latifolium* Lam. (Pteridaceae) **IPNI:** 271633-2 **Common name:** culantrillo negro, helechillo, helecho de diablo, helecho de perfume **Habit:** Herb **Use:** ME **Origin:** Native **Region:** Amazonia, Andean, Guiana Shield, Orinoquia, Pacific **Dept.:** AMA, ANT, BOY, CAQ, CAU, CHO, CUN, GUV, MET, NAR, NSA, PUT, VAC, VAU, VID **Elev.:** 100-1500 m.a.s.l. **IUCN Habitat code:** 1,2,3,4,5,14 **Conservation status: (IUCN)** NE **(CPSG)** Potential LC

168 *Adiantum lucidum* (Cav.) Sw. (Pteridaceae) **IPNI:** 17011270-1 **Common name:** nabugui **Habit:** Herb **Use:** ME **Origin:** Native **Region:** Andean, Caribbean, Pacific **Dept.:** ANT, ATL, CHO, COR, MAG **Elev.:** 0–300 m a.s.l. **IUCN Habitat code:** 1,2,3,4,14 **Conservation status: (IUCN)** NE

169 *Adiantum macrophyllum* Sw. (Pteridaceae) **IPNI:** 17011380-1 **Habit:** Herb **Use:** EU **Origin:** Native **Region:** Andean, Guiana Shield, Caribbean, Pacific **Dept.:** ANT, CAL, CHO, CUN, LAG, MAG, MET, NAR, NSA, RIS, SAN, VAC **Elev.:** 0-1600 m.a.s.l. **IUCN Habitat code:** 1,2,3,5,14 **Conservation status: (IUCN)** NE **(CPSG)** Potential LC

170 *Adiantum orbignyanum* Mett. (Pteridaceae) **IPNI:** 319723-2 **Habit:** Herb **Use:** ME **Origin:** Native **Region:** Andean **Dept.:** BOY, CAU **Elev.:** 1900–2200 m a.s.l. **IUCN Habitat code:** 1 **Conservation status: (IUCN)** NE

171 *Adiantum philippense* L. (Pteridaceae) **IPNI:** 325052-2 **Habit:** Herb **Use:** ME **Origin:** Native **Region:** Andean, Caribbean **Dept.:** BOY, MAG **Elev.:** 450–650 m a.s.l. **IUCN Habitat code:** 1,14 **Conservation status: (IUCN)** NE

172 *Adiantum poiretii* Wikstr. (Pteridaceae) **IPNI:** 17012320-1 **Habit:** Herb **Use:** EU, ME **Origin:** Native **Region:** Andean **Dept.:** BOG, BOY, CUN, NAR **Elev.:** 1500–3000 m a.s.l. **IUCN Habitat code:** 1,14 **Conservation status: (IUCN)** NE

173 *Adiantum raddianum* C.Presl (Pteridaceae) **IPNI:** 17012580-1 **Common name:** culantrillo **Habit:** Herb **Use:** EU, ME **Origin:** Native **Region:** Andean **Dept.:** ANT, BOG, BOY, CAL, CAU, CUN, HUI, MAG, NAR, QUI, SAN, TOL, VAC **Elev.:** 1000-3450 m.a.s.l. **IUCN Habitat code:** 1,2,3,4,14 **Conservation status: (IUCN)** NE **(CPSG)** Potential LC

174 *Adiantum tenerum* Sw. (Pteridaceae) **IPNI:** 17013290-1 **Habit:** Herb **Use:** EU, ME **Origin:** Native **Region:** Andean, Caribbean **Dept.:** BOY, CES, CUN, MAG **Elev.:** 800–1650 m a.s.l. **IUCN Habitat code:** 1,3,14 **Conservation status: (IUCN)** NE

175 *Adiantum tetraphyllum* Humb. & Bonpl. ex Willd. (Pteridaceae) **IPNI:** 315261-2 **Common name:** culantrillo **Habit:** Herb **Use:** ME **Origin:** Native **Region:** Amazonia, Andean, Caribbean, Pacific **Dept.:** ANT, BOY, CAL, CAU, CHO, CUN, MAG, MET, NSA, PUT, RIS, TOL, VAC, VAU **Elev.:** 50-1800 m.a.s.l. **IUCN Habitat code:** 1,2,3,4,5,14 **Conservation status: (IUCN)** NE **(CPSG)** Potential LC

176 *Adiantum tinctum* T.Moore (Pteridaceae) **IPNI:** 17013420-1 **Use:** ME **Origin:** Native **Conservation status: (IUCN)** NE

177 *Adiantum villosum* L. (Pteridaceae) **IPNI:** 315262-2 **Habit:** Herb **Use:** ME **Origin:** Native **Region:** Andean, Caribbean, Pacific **Dept.:** LAG, MAG, NAR **Elev.:** 850–850 m a.s.l. **IUCN Habitat code:** 1,2,3,14 **Conservation status: (IUCN)** NE

178 *Adiantum wilsonii* Hook. (Pteridaceae) **IPNI:** 17013890-1 **Habit:** Herb **Use:** ME **Origin:** Native **Region:** Pacific **Dept.:** ANT, CHO, SAN **Elev.:** 0–500 m a.s.l. **IUCN Habitat code:** 1,3 **Conservation status: (IUCN)** NE

179 *Ceratopteris pteridoides* (Hook.) Hieron. (Pteridaceae) **IPNI:** 17242590-1 **Common name:** helecho acuático **Habit:** Herb, Aquatic **Use:** EU, HF, ME **Origin:** Native **Region:** Amazonia, Caribbean, Orinoquia, Pacific **Dept.:** AMA, ANT, ARA, ATL, BOL, CHO, COR, SAN, SUC, VID **Elev.:** 0–150 m a.s.l. **IUCN Habitat code:** 1,2,3,4,5,14 **Conservation status: (IUCN)** NE

180 *Ceratopteris thalictroides* (L.) Brongn. (Pteridaceae) **IPNI:** 17064190-1 **Habit:** Herb, Aquatic **Use:** EU, GS, HF, ME **Origin:** Native **Region:** Caribbean, Orinoquia, Pacific **Dept.:** ANT, BOL, CHO, MET **Elev.:** 50–450 m a.s.l. **IUCN Habitat code:** 1,5,14 **Conservation status: (IUCN)** LC

181 *Hemionitis chlorophylla* (Sw.) Christenh. (Pteridaceae) **IPNI:** 77182519-1 **Habit:** Herb **Use:** ME **Origin:** Native **Conservation status: (IUCN)** NE

182 *Hemionitis concolor* (Langsd. & Fisch.) Christenh. (Pteridaceae) **IPNI:** 77182373-1 **Habit:** Herb **Use:** ME **Origin:** Native **Region:** Caribbean **Dept.:** CES, MAG **Elev.:** 50–50 m a.s.l. **Conservation status: (IUCN)** NE

183 *Hemionitis farinosa* (Forssk.) Christenh. (Pteridaceae) **IPNI:** 77182808-1 **Habit:** Herb **Use:** ME **Origin:** Native **Region:** Andean **Dept.:** CUN **Elev.:** 1900–3000 m a.s.l. **Conservation status: (IUCN)** NE

184 *Hemionitis marginata* (Kunth) Christenh. (Pteridaceae) **IPNI:** 77182463-1 **Habit:** Herb **Use:** ME **Origin:** Native **Region:** Andean **Dept.:** ANT, BOY, CAU, CUN, NAR **Elev.:** 1800–3800 m a.s.l. **Conservation status: (IUCN)** NE

185 *Hemionitis microphylla* (Sw.) Christenh. (Pteridaceae) **IPNI:** 77183177-1 **Habit:** Herb **Use:** EU **Origin:** Native **Region:** Andean, Caribbean **Dept.:** ANT, BOY, CAL, CES, CUN, HUI, LAG, MAG, NAR, SAN, VAC **Elev.:** 250–1600 m a.s.l. **Conservation status: (IUCN)** NE

186 *Hemionitis myriophylla* (Desv.) Christenh. (Pteridaceae) **IPNI:** 77183191-1 **Habit:** Herb **Use:** ME **Origin:** Native **Region:** Andean **Dept.:** BOY, CUN, MAG, NAR, SAN **Elev.:** 1000–3700 m a.s.l. **Conservation status: (IUCN)** NE

187 *Hemionitis nivea* (Poir.) Christenh. (Pteridaceae) **IPNI:** 77182941-1 **Habit:** Herb **Use:** ME **Origin:** Native **Conservation status: (IUCN)** NE

188 *Hemionitis obducta* (Mett.) Christenh. (Pteridaceae) **IPNI:** 77182303-1 **Habit:** Herb **Use:** ME **Origin:** Native **Region:** Caribbean **Dept.:** ANT, CES, CUN, HUI, MAG **Elev.:** 150–600 m a.s.l. **Conservation status: (IUCN)** NE

189 *Hemionitis ovata* (Desv.) Christenh. (Pteridaceae) **IPNI:** 77183106-1 **Habit:** Herb **Use:** ME **Origin:** Native **Region:** Andean **Dept.:** ANT, BOY, CES, CUN, NAR, PUT, SAN **Elev.:** 1000–2600 m a.s.l. **Conservation status: (IUCN)** NE

190 *Hemionitis ternifolia* (Cav.) Christenh. (Pteridaceae) **IPNI:** 77182326-1 **Habit:** Herb **Use:** ME **Origin:** Native **Region:** Andean **Dept.:** BOY, CUN, MAG, NAR, SAN, VAC **Elev.:** 2000–3350 m a.s.l. **Conservation status: (IUCN)** NE

191 *Jamesonia bogotensis* H.Karst. (Pteridaceae) **IPNI:** 17127250-1 **Habit:** Herb **Use:** EU, MA **Origin:** Native **Region:** Andean **Dept.:** ANT, BOG, BOL, BOY, CUN, NAR, RIS, SAN **Elev.:** 2950–4400 m a.s.l. **IUCN Habitat code:** 1,3,4,8,14 **Conservation status: (IUCN)** NE

192 *Jamesonia flexuosa* (Kunth) Christenh. (Pteridaceae) **IPNI:** 77110633-1 **Common name:** helechillo negro **Habit:** Herb **Use:** ME **Origin:** Native **Region:** Andean **Dept.:** ANT, BOG, BOL, BOY, CAL, CAU, CHO, CUN, HUI, MAG, MET, NAR, NSA, PUT, QUI, RIS, SAN, TOL **Elev.:** 1100–3700 m a.s.l. **IUCN Habitat code:** 1,3,4,8,14 **Conservation status: (IUCN)** NE

193 *Pityrogramma calomelanos* (L.) Link (Pteridaceae) **IPNI:** 17345640-1 **Common name:** helecho negro, yerba de la Virgen, yerba de la virgen **Habit:** Herb **Use:** EU, MA, ME, PO, SU **Origin:** Native **Region:** Amazonia, Andean, Guiana Shield, Caribbean, Orinoquia, Pacific **Dept.:** AMA, ANT, BOL, BOY, CAL, CAQ, CAU, CHO, CUN, GUA, GUV, HUI, LAG, MAG, MET, NAR, NSA, PUT, QUI, RIS, SAP, SAN, TOL, VAC, VAU, VID **Elev.:** 0-2700 m.a.s.l. **IUCN Habitat code:** 1,2,3,4,5,14 **Conservation status: (IUCN)** NE **(CPSG)** Potential LC

194 *Pityrogramma ebenea* (L.) Proctor (Pteridaceae) **IPNI:** 200935-2 **Common name:** helecho blanco **Habit:** Herb **Use:** ME **Origin:** Native **Region:** Amazonia, Andean, Pacific **Dept.:** ANT, BOY, CAL, CAQ, CAS, CAU, CHO, CUN, HUI, NAR, NSA, PUT, SAN, TOL, VAC **Elev.:** 200-3200 m.a.s.l. **IUCN Habitat code:** 1,3,14 **Conservation status: (IUCN)** NE **(CPSG)** Potential LC

195 *Polytaenium guayanense* (Hieron.) Alston (Pteridaceae) **IPNI:** 17355560-1 **Common name:** calaguala, yamichuí **Habit:** Herb **Use:** ME **Origin:** Native **Region:** Amazonia, Andean, Guiana Shield, Pacific **Dept.:** AMA, ANT, CAQ, CHO, GUV, HUI, MET, NSA **Elev.:** 100–1700 m a.s.l. **IUCN Habitat code:** 1,3,4,5,14 **Conservation status: (IUCN)** NE

196 *Polytaenium lineatum* J.Sm. (Pteridaceae) **IPNI:** 17209080-1 **Habit:** Herb **Use:** ME **Origin:** Native **Region:** Andean, Caribbean **Dept.:** ANT, CAL, CES, CUN, HUI, LAG, MAG, NAR, NSA, RIS, SAN, TOL **Elev.:** 1500-2800 m.a.s.l. **IUCN Habitat code:** 1,3,14 **Conservation status: (IUCN)** NE **(CPSG)** Potential LC

197 *Pteris biaurita* L. (Pteridaceae) **IPNI:** 213840-2 **Habit:** Herb **Use:** HF, ME **Origin:** Native **Region:** Andean, Guiana Shield, Caribbean, Orinoquia **Dept.:** BOY, CAQ, COR, CUN, MAG, MET **Elev.:** 50–1300 m a.s.l. **IUCN Habitat code:** 1,2,3,14 **Conservation status: (IUCN)** NE

198 *Pteris cretica* L. (Pteridaceae) **IPNI:** 17211870-1 **Habit:** Herb **Use:** EU, HF, ME **Origin:** Naturalised **Region:** Amazonia, Andean **Dept.:** AMA, BOG, CUN, SAN **Elev.:** 200–2700 m a.s.l. **IUCN Habitat code:** 1,3,5,14 **Conservation status: (IUCN)** NE

199 *Pteris grandifolia* L. (Pteridaceae) **IPNI:** 313710-2 **Habit:** Herb **Use:** EU, ME **Origin:** Native **Region:** Andean, Caribbean **Dept.:** ANT, ATL, BOY, CAL, CUN, HUI, TOL **Elev.:** 350–1800 m a.s.l. **IUCN Habitat code:** 1,3,14 **Conservation status: (IUCN)** NE

200 *Pteris quadriaurita* Retz. (Pteridaceae) **IPNI:** 17215900-1 **Habit:** Herb **Use:** EU, ME **Origin:** Native **Region:** Andean, Pacific **Dept.:** ANT, BOY, CAL, CAU, CES, CUN, HUI, MAG, VAC **Elev.:** 900–2700 m a.s.l. **IUCN Habitat code:** 1,2,3,14 **Conservation status: (IUCN)** NE

201 *Pteris tripartita* Sw. (Pteridaceae) **IPNI:** 17217170-1 **Habit:** Herb **Use:** EU, ME **Origin:** Naturalised **Region:** Amazonia, Pacific **Dept.:** AMA, CAQ, CHO, PUT **Elev.:** 50–700 m a.s.l. **IUCN Habitat code:** 1,3,5 **Conservation status: (IUCN)** NE

202 *Pteris vittata* L. (Pteridaceae) **IPNI:** 17217500-1 **Habit:** Herb **Use:** EU, HF, ME, PO **Origin:** Naturalised **Region:** Andean **Dept.:** ANT, VAC **Elev.:** 1500–1750 m a.s.l. **IUCN Habitat code:** 1,2,3,14 **Conservation status: (IUCN)** LC

203 *Pterozonium reniforme* (Mart.) Fée (Pteridaceae) **IPNI:** 17217830-1 **Habit:** Herb **Use:** ME **Origin:** Native **Region:** Amazonia, Guiana Shield **Dept.:** AMA, CAQ, VAU **Elev.:** 100–650 m a.s.l. **IUCN Habitat code:** 1,4,14 **Conservation status: (IUCN)** NE

204 *Vittaria graminifolia* Kaulf. (Pteridaceae) **IPNI:** 266216-2 **Habit:** Herb **Use:** ME **Origin:** Native **Region:** Andean, Guiana Shield, Orinoquia, Pacific **Dept.:** ANT, BOG, BOY, CAU, CES, CHO, CUN, LAG, MAG, MET, NAR, NSA, PUT, SAN, VAC **Elev.:** 100-3200 m.a.s.l. **IUCN Habitat code:** 1,2,3,4,14 **Conservation status: (IUCN)** NE **(CPSG)** Potential LC

205 *Vittaria lineata* (L.) Sm. (Pteridaceae) **IPNI:** 17234040-1 **Habit:** Herb **Use:** ME **Origin:** Native **Region:** Amazonia, Andean, Caribbean, Pacific **Dept.:** AMA, ANT, BOG, BOY, CAQ, CAU, CHO, LAG, MET, NAR, NSA, SAN, VAC **Elev.:** 0-3450 m.a.s.l. **IUCN Habitat code:** 1,2,3,4,5,14 **Conservation status: (IUCN)** NE **(CPSG)** Potential LC

Psilotales

206 *Psilotum complanatum* Sw. (Psilotaceae)
IPNI: 114220-3
Habit: Herb **Use:** ME **Origin:** Native **Dept.:** BOG **Conservation status: (IUCN)** NE

207 *Psilotum nudum* (L.) P.Beauv. (Psilotaceae)
IPNI: 212103-2
Habit: Herb **Use:** EU, HF, IF, ME **Origin:** Native **Region:** Andean, Pacific **Dept.:** ANT, CHO, VAC **Elev.:** 0–1500 m a.s.l. **IUCN Habitat code:** 1,14 **Conservation status: (IUCN)** NE

Salviniales

208 *Marsilea polycarpa* Hook. & Grev. (Marsileaceae)
IPNI: 280439-2
Common name: trébol de agua **Habit:** Herb **Use:** HF, ME **Origin:** Native **Region:** Caribbean **Dept.:** ATL, BOL, CAS, COR, HUI **Elev.:** 0–650 m a.s.l. **IUCN Habitat code:** 1,2,3,4,5,14 **Conservation status: (IUCN)** NE

209 *Azolla caroliniana* Willd. (Salviniaceae)
IPNI: 17056910-1
Use: EU **Origin:** Native **IUCN Habitat code:** 5,14 **Conservation status: (IUCN)** NE

210 *Azolla filiculoides* Lam. (Salviniaceae)
IPNI: 17056950-1
Common name: alfombra de agua, azola **Habit:** Herb, Aquatic **Use:** AF, EU, ME **Origin:** Native **Region:** Amazonia, Andean, Caribbean **Dept.:** AMA, ANT, BOG, BOY, COR, CUN, HUI, MET, RIS, VAU **Elev.:** 50–4200 m a.s.l. **IUCN Habitat code:** 1,3,4,5,8,14 **Conservation status: (IUCN)** NE

211 *Azolla microphylla* Kaulf. (Salviniaceae)
IPNI: 17057000-1
Habit: Herb, Aquatic **Use:** EU **Origin:** Native **Region:** Amazonia, Pacific **Dept.:** AMA, CHO **Elev.:** 100–100 m a.s.l. **IUCN Habitat code:** 1,5 **Conservation status: (IUCN)** LC

212 *Salvinia auriculata* Aubl. (Salviniaceae)
IPNI: 17219800-1
Habit: Herb, Aquatic **Use:** EU **Origin:** Native **Region:** Amazonia, Caribbean, Orinoquia, Pacific **Dept.:** AMA, ANT, ATL, CHO, COR, MET, TOL, VAC, VID **Elev.:** 50–400 m a.s.l. **IUCN Habitat code:** 1,2,3,4,5,14 **Conservation status: (IUCN)** NE

Schizaeales

213 *Anemia adiantifolia* (L.) Sw. (Schizaeaceae)
IPNI: 1031990-2
Habit: Herb **Use:** EU **Origin:** Native **Conservation status: (IUCN)** NE

214 *Anemia phyllitidis* (L.) Sw. (Schizaeaceae)
IPNI: 1202412-2
Habit: Herb **Use:** EU, ME **Origin:** Native **Region:** Andean, Guiana Shield, Caribbean **Dept.:** ANT, BOY, CAL, CAU, CES, CUN, LAG, MAG, QUI, RIS, SAN, VAC **Elev.:** 0-2500 m.a.s.l. **IUCN Habitat code:** 1,2,3,14 **Conservation status: (IUCN)** NE **(CPSG)** Potential LC

215 *Lygodium venustum* Sw. (Schizaeaceae)
IPNI: 17143510-1
Common name: bejuco alambre **Habit:** Climbing **Use:** ME **Origin:** Native **Region:** Andean, Guiana Shield, Caribbean, Orinoquia, Pacific **Dept.:** ANT, ATL, BOL, BOY, CAS, CAU, CES, CHO, COR, CUN, HUI, LAG, MAG, MET, NSA, SAP, SUC, TOL, VAC, VID **Elev.:** 0-2750 m.a.s.l. **IUCN Habitat code:** 1,2,3,4,5,14 **Conservation status: (IUCN)** NE **(CPSG)** Potential LC

216 *Lygodium volubile* Sw. (Schizaeaceae)
IPNI: 17143520-1
Habit: Climbing **Use:** MA, ME **Origin:** Native **Region:** Amazonia, Andean **Dept.:** AMA, ANT, CAQ, MAG, NSA **Elev.:** 200–700 m a.s.l. **IUCN Habitat code:** 1,3,4,14 **Conservation status: (IUCN)** NE

217 *Schizaea elegans* (Vahl) Sw. (Schizaeaceae)
IPNI: 17220460-1
Common name: helecho de madremonte, la diabla, nolocrea macho, sombrillita **Habit:** Herb **Use:** ME **Origin:** Native **Region:** Amazonia, Andean, Guiana Shield, Caribbean, Pacific **Dept.:** AMA, ANT, BOL, CAQ, CAS, CAU, CES, CHO, COR, GUA, GUV, MAG, MET, NAR, NSA, PUT, SAN, TOL, VAC **Elev.:** 50-1400 m.a.s.l. **IUCN Habitat code:** 1,2,3,4,5,14 **Conservation status: (IUCN)** NE **(CPSG)** Potential LC

Cycadopsida Cycadales

218 *Cycas revoluta* Thunb. (Cycadaceae)
IPNI: 328823-2
Common name: cícara, palma de sagú, palma funeral **Habit:** Shrub **Use:** EU, GS, HF, MA, ME **Origin:** Cultivated **Region:** Andean, Caribbean **Dept.:** ANT, BOG, CUN, QUI, SAP, TOL **Elev.:** 500–1600 m a.s.l. **IUCN Habitat code:** 1,3,14 **Conservation status: (IUCN)** LC

219 *Cycas rumphii* Miq. (Cycadaceae)
IPNI: 326820-2
Common name: cica, palma cica, palma sagú **Habit:** Shrub **Use:** EU, GS, HF, IF, MA, ME, PO, SU **Origin:** Cultivated **Region:** Andean, Caribbean **Dept.:** ANT, MET, QUI, TOL **Elev.:** 500–2000 m a.s.l. **IUCN Habitat code:** 14 **Conservation status: (IUCN)** NT

220 *Zamia amazonum* D.W.Stev. (Zamiaceae)
IPNI: 20009732-1
Common name: quiripa, yuca de monte **Habit:** Herb **Use:** HF **Origin:** Native **Region:** Amazonia **Dept.:** AMA, VAU **Elev.:** 100-500 m.a.s.l. **IUCN Habitat code:** 1 **Conservation status: (IUCN)** NT **(CPSG)** VU

221 *Zamia chigua* Seem. (Zamiaceae)
IPNI: 297256-1
Common name: chigua, chigua macho, helecho **Habit:** Shrub **Use:** AF, EU, HF **Origin:** Endemic **Region:** Pacific **Dept.:** CHO, VAC **Elev.:** 30-250 m.a.s.l. **IUCN Habitat code:** 1,5,14 **Conservation status: (IUCN)** NT **(CPSG)** NT

222 *Zamia hymenophyllidia* D.W.Stev. (Zamiaceae)
IPNI: 20009735-1
Habit: Herb **Use:** HF **Origin:** Native **Region:** Amazonia **Dept.:** AMA **Elev.:** 100-250 m.a.s.l. **IUCN Habitat code:** 1 **Conservation status: (IUCN)** CR **(CPSG)** EN

223 *Zamia melanorrhachis* D.W.Stev. (Zamiaceae)
IPNI: 20009736-1
Common name: corocita, corocito, helecho, ibaracú **Habit:** Herb **Use:** HF, ME **Origin:** Endemic **Dept.:** ANT, COR, SAN **Elev.:** 30-500 m.a.s.l. **IUCN Habitat code:** 1,3,14 **Conservation status: (IUCN)** EN **(CPSG)** EN

224 *Zamia muricata* Willd. (Zamiaceae)
IPNI: 270552-2
Habit: Herb **Use:** EU **Origin:** Native **Region:** Caribbean **Dept.:** BOY, LAG, SAN **Elev.:** 0-700 m.a.s.l. **IUCN Habitat code:** 1,4 **Conservation status: (IUCN)** NT **(CPSG)** EN

225 *Zamia obliqua* A.Braun (Zamiaceae)
IPNI: 297347-1
Common name: chigua, maicito, zamia **Habit:** Shrub **Use:** HF **Origin:** Native **Region:** Pacific **Dept.:** ANT, CHO, VAC **Elev.:** 0-600 m.a.s.l. **IUCN Habitat code:** 1,3 **Conservation status: (IUCN)** NT **(CPSG)** VU

226 *Zamia poeppigiana* Mart. & Eichler (Zamiaceae)
IPNI: 270556-2
Use: HF **IUCN Habitat code:** 1 **Conservation status: (IUCN)** NT

227 *Zamia roezlii* Regel ex Linden (Zamiaceae)
IPNI: 297372-1
Common name: chigua **Habit:** Shrub **Use:** EU, HF **Origin:** Native **Region:** Pacific **Dept.:** CHO, NAR, VAC **Elev.:** 0-200 m.a.s.l. **IUCN Habitat code:** 1 **Conservation status: (IUCN)** NT **(CPSG)** VU

Ginkgoopsida Ginkgoales

228 *Ginkgo biloba* L. (Ginkgoaceae)
IPNI: 262125-1
Common name: ginkgo biloba **Use:** EU, HF, MA, ME **Conservation status: (IUCN)** EN

Gnetopsida Gnetales

229 *Gnetum leyboldii* Tul. (Gnetaceae)
IPNI: 111368-2
Common name: plátano, purgante de picón **Habit:** Liana **Use:** HF **Origin:** Native **Region:** Amazonia, Orinoquia, Pacific **Dept.:** AMA, ANT, CAQ, CHO, GUV, MET, PUT, VAU, VID **Elev.:** 40–770 m a.s.l. **IUCN Habitat code:** 1,3,4,5,14 **Conservation status: (IUCN)** LC

230 *Gnetum nodiflorum* Brongn. (Gnetaceae)
IPNI: 111373-2
Habit: Climbing **Use:** HF **Origin:** Native **Region:** Amazonia, Guiana Shield **Dept.:** AMA, CAQ, GUA, MET, VAU, VID **Elev.:** 100–250 m a.s.l. **IUCN Habitat code:** 1,3,4,14 **Conservation status: (IUCN)** LC

231 *Gnetum paniculatum* Spruce ex Benth. (Gnetaceae)
IPNI: 111375-2
Common name: reventilla **Habit:** Climbing **Use:** HF **Origin:** Native **Region:** Amazonia **Dept.:** AMA, GUA **Elev.:** 150–240 m a.s.l. **IUCN Habitat code:** 1 **Conservation status: (IUCN)** LC

232 *Gnetum schwackeanum* Taub. ex Schenck (Gnetaceae) **IPNI:** 111377-2 **Habit:** Climbing **Use:** HF **Origin:** Native **Region:** Amazonia, Guiana Shield, Orinoquia **Dept.:** AMA, CAQ, GUA, VID **Elev.:** 100–250 m a.s.l. **IUCN Habitat code:** 1,3 **Conservation status: (IUCN)** LC

233 *Gnetum urens* (Aubl.) Blume (Gnetaceae) **IPNI:** 383578-1 **Habit:** Climbing **Use:** HF **Origin:** Native **Region:** Amazonia, Guiana Shield **Dept.:** AMA, CAQ, PUT, VID **Elev.:** 120–300 m a.s.l. **IUCN Habitat code:** 1,3,5 **Conservation status: (IUCN)** LC

Pinopsida Pinales

234 *Araucaria angustifolia* (Bertol.) Kuntze (Araucariaceae) **IPNI:** 17518-2 **Common name:** araucaria, araucaria crespa **Habit:** Tree **Use:** EU, HF, MA, ME **Origin:** Cultivated **Region:** Andean **Dept.:** ANT, BOG, CAL, CUN, QUI, RIS **Elev.:** 1200–2750 m a.s.l. **IUCN Habitat code:** 1,14 **Conservation status: (IUCN)** CR

235 *Araucaria araucana* (Molina) K.Koch (Araucariaceae) **IPNI:** 261681-1 **Habit:** Tree **Use:** EU, HF, MA, ME **Origin:** Cultivated **Region:** Andean **Dept.:** ANT, BOG **Elev.:** 2000–2500 m a.s.l. **IUCN Habitat code:** 1,3 **Conservation status: (IUCN)** EN

236 *Araucaria columnaris* (G.Forst.) Hook. (Araucariaceae) **IPNI:** 261688-1 **Use:** EU, HF, MA, ME **Region:** Andean **Dept.:** BOG **IUCN Habitat code:** 1,3,14 **Conservation status: (IUCN)** LC

237 *Araucaria heterophylla* (Salisb.) Franco (Araucariaceae) **IPNI:** 676669-1 **Habit:** Tree **Use:** EU, HF, MA, ME **Origin:** Cultivated **Region:** Andean, Caribbean **Dept.:** ANT, ATL, BOG, CUN, HUI, SAP, TOL **Elev.:** 10–2750 m a.s.l. **IUCN Habitat code:** 3,14 **Conservation status: (IUCN)** VU

238 *Chamaecyparis lawsoniana* (A.Murray bis) Parl. (Cupressaceae) **IPNI:** 261839-1 **Use:** EU, MA, ME **Region:** Andean **Dept.:** BOG **IUCN Habitat code:** 14 **Conservation status: (IUCN)** NT

239 *Cryptomeria japonica* (Thunb. ex L.f.) D.Don (Cupressaceae) **IPNI:** 261870-1 **Habit:** Tree **Use:** EU, HF, MA, ME **Origin:** Cultivated **Region:** Andean **Dept.:** ANT, BOG, CUN **Elev.:** 1500–2600 m a.s.l. **IUCN Habitat code:** 14 **Conservation status: (IUCN)** NT

240 *Cupressus sempervirens* L. (Cupressaceae) **IPNI:** 261974-1 **Common name:** ciprés, pino **Habit:** Tree **Use:** EU, FU, MA, ME, PO **Origin:** Cultivated **Region:** Andean **Dept.:** ANT, BOG, CUN **Elev.:** 1300–2700 m a.s.l. **IUCN Habitat code:** 1,3,14 **Conservation status: (IUCN)** LC

241 *Hesperocyparis arizonica* (Greene) Bartel (Cupressaceae) **IPNI:** 60451546-2 **Habit:** Tree **Use:** EU, FU, MA, ME, SU **Origin:** Cultivated **Region:** Andean **Dept.:** ANT **Elev.:** 1500–2000 m a.s.l. **Conservation status: (IUCN)** NE

242 *Hesperocyparis lusitanica* (Mill.) Bartel (Cupressaceae) **IPNI:** 60451554-2 **Common name:** pino común **Habit:** Tree **Use:** EU, HF, MA, ME **Origin:** Cultivated **Region:** Andean **Dept.:** ANT, BOG, CHO, CUN, QUI, RIS, TOL, VAC **Elev.:** 900–3000 m a.s.l. **IUCN Habitat code:** 3,14 **Conservation status: (IUCN)** NE

243 *Hesperocyparis macrocarpa* (Hartw.) Bartel (Cupressaceae) **IPNI:** 60451558-2 **Habit:** Tree **Use:** EU, FU, MA, ME, PO **Origin:** Cultivated **Region:** Andean **Dept.:** BOY **Elev.:** 3600–3600 m a.s.l. **Conservation status: (IUCN)** NE

244 *Platycladus orientalis* (L.) Franco (Cupressaceae) **IPNI:** 677124-1 **Common name:** pino, pino libro **Habit:** Tree **Use:** EU, HF, MA, ME **Origin:** Cultivated **Region:** Andean, Caribbean **Dept.:** BOG, CUN, SAP **Elev.:** 2620–2620 m a.s.l. **IUCN Habitat code:** 3,14 **Conservation status: (IUCN)** NT

245 *Sequoia sempervirens* (D.Don) Endl. (Cupressaceae) **IPNI:** 263741-1 **Habit:** Tree **Use:** EU, MA, ME **Origin:** Cultivated **Region:** Andean **Dept.:** ANT, BOG **Elev.:** 2475–2475 m a.s.l. **IUCN Habitat code:** 14 **Conservation status: (IUCN)** EN

246 *Thuja occidentalis* L. (Cupressaceae) **IPNI:** 263956-1 **Common name:** pino, tuya, tuya del Canadá **Habit:** Tree **Use:** EU, HF, MA, ME **Origin:** Cultivated **Region:** Andean **Dept.:** ANT, VAC **Elev.:** 900–2100 m a.s.l. **IUCN Habitat code:** 1,14 **Conservation status: (IUCN)** LC

247 *Pinus ayacahuite* Ehrenb. ex Schltdl. (Pinaceae) **IPNI:** 314973-2 **Use:** MA **Region:** Andean **Dept.:** BOG **Conservation status: (IUCN)** LC

248 *Pinus caribaea* Morelet (Pinaceae) **IPNI:** 314859-2 **Common name:** pino **Habit:** Tree **Use:** EU, GS, HF, MA, ME **Origin:** Cultivated **Region:** Andean, Caribbean **Dept.:** ANT, CAS, MET, VID **Elev.:** 40–2300 m a.s.l. **IUCN Habitat code:** 3,14 **Conservation status: (IUCN)** LC

249 *Pinus montezumae* Lamb. (Pinaceae) **IPNI:** 263125-1 **Use:** MA, ME **Region:** Andean **Dept.:** BOG **IUCN Habitat code:** 3,14 **Conservation status: (IUCN)** LC

250 *Pinus mugo* Turra (Pinaceae) **IPNI:** 677083-1 **Use:** EU, HF, MA, ME **Region:** Andean **Dept.:** BOG **Conservation status: (IUCN)** LC

251 *Pinus patula* Schiede ex Schltdl. & Cham. (Pinaceae) **IPNI:** 314961-2 **Common name:** barbas de chivo, pino, pino espátula, pino pátula **Habit:** Tree **Use:** EU, FU, HF, MA, ME **Origin:** Cultivated **Region:** Andean **Dept.:** ANT, BOG, CAL, CAU, CUN, HUI, NAR, NSA, PUT, QUI, RIS, SAN, TOL, VAC **Elev.:** 2000–3000 m a.s.l. **IUCN Habitat code:** 1,3,4,5,14 **Conservation status: (IUCN)** LC

252 *Pinus pinaster* Aiton (Pinaceae) **IPNI:** 263216-1 **Common name:** pino marítimo, pino negral **Use:** EU, HF, MA, ME **Dept.:** BOG **IUCN Habitat code:** 14 **Conservation status: (IUCN)** LC

253 *Pinus radiata* D.Don (Pinaceae) **IPNI:** 30020085-2 **Common name:** pino, pino candelabro, pino ocarpo **Habit:** Tree **Use:** EU, HF, MA, ME **Origin:** Cultivated **Region:** Andean **Dept.:** BOG, CUN **Elev.:** 2600–3000 m a.s.l. **IUCN Habitat code:** 1,3,5,14 **Conservation status: (IUCN)** EN

254 *Pinus sylvestris* L. (Pinaceae) **IPNI:** 263353-1 **Common name:** pino **Use:** EU, FU, HF, MA, ME, SU **Conservation status: (IUCN)** LC

255 *Pectinopitys harmsiana* (Pilg.) C.N.Page (Podocarpaceae) **IPNI:** 77203787-1 **Common name:** chaquiro, pino, pino colombiano, pino negro **Habit:** Tree **Use:** EU **Origin:** Native, Cultivated **Region:** Andean **Dept.:** CAU, MAG, QUI, RIS **Elev.:** 1800–2200 m a.s.l. **Conservation status: (IUCN)** NE

256 *Podocarpus guatemalensis* Standl. (Podocarpaceae) **IPNI:** 204582-2 **Common name:** ají, chaquiro, chaquiro dulce, pino, pino amarillo, pino chaquiro, pino colombiano **Habit:** Tree **Use:** MA **Origin:** Native **Region:** Andean, Pacific **Dept.:** ANT, CAQ, CAU, CHO, NAR, PUT, SAN **Elev.:** 0-1200 m.a.s.l. **IUCN Habitat code:** 1,3,5,14 **Conservation status: (IUCN)** LC **(CPSG)** VU

257 *Podocarpus oleifolius* D.Don (Podocarpaceae) **IPNI:** 263553-1 **Common name:** ají, chaquiro, hayuelo, pino, pino amarillo, pino chaquiro, pino colombiano, pino criollo, pino de peña, pino hayuelo, pino real, pino romerillo caleño, pino romerón **Habit:** Tree **Use:** HF, MA, ME **Origin:** Native **Region:** Andean **Dept.:** ANT, BOG, BOY, CAL, CAU, CES, CHO, CUN, HUI, LAG, MAG, MET, NAR, NSA, PUT, QUI, RIS, SAN, TOL, VAC **Elev.:** 1900-3800 m.a.s.l. **IUCN Habitat code:** 1,3,4,14 **Conservation status: (IUCN)** LC **(CPSG)** VU

258 *Podocarpus parlatorei* Pilg. (Podocarpaceae) **IPNI:** 204598-2 **Use:** MA, ME **Region:** Andean **Dept.:** BOG **Conservation status: (IUCN)** NT

259 *Retrophyllum rospigliosii* (Pilg.) C.N.Page (Podocarpaceae) **IPNI:** 946471-1 **Common name:** chaquiro, pino, pino chaquiro, pino colombiano, pino de Pacho, pino de montaña, pino hayuelo, pino montañero, pino negro, pino romerillo, pino romerón **Habit:** Tree **Use:** EU **Origin:** Native, Cultivated **Region:** Andean **Dept.:** ANT, BOG, CUN, HUI,

MAG, NSA, QUI, SAN **Elev.:** 1200–3750 m a.s.l. **Conservation status: (IUCN)** VU

Magnoliopsida Nymphaeales

260 *Cabomba caroliniana* A.Gray (Cabombaceae) **IPNI:** 273581-2 **Common name:** majate **Habit:** Herb, Aquatic **Use:** EU, ME **Origin:** Cultivated **Region:** Andean **Dept.:** ANT **Elev.:** 1500–1500 m a.s.l. **IUCN Habitat code:** 2,4 **Conservation status: (IUCN)** NE

261 *Nymphaea amazonum* Mart. & Zucc. (Nymphaeaceae) **IPNI:** 605491-1 **Habit:** Herb, Aquatic **Use:** HF, ME **Origin:** Native **Region:** Andean, Caribbean **Dept.:** CAU, COR, CUN, MAG, TOL, VAC **Elev.:** 6–1600 m a.s.l. **IUCN Habitat code:** 1,2,14 **Conservation status: (IUCN)** NE

262 *Nymphaea ampla* (Salisb.) DC. (Nymphaeaceae) **IPNI:** 605494-1 **Habit:** Herb, Aquatic **Use:** HF, ME **Origin:** Native **Region:** Andean, Caribbean, Pacific **Dept.:** CHO, COR, CUN, MAG **Elev.:** 100–1200 m a.s.l. **IUCN Habitat code:** 1,2,3,5,14 **Conservation status: (IUCN)** NE

263 *Nymphaea elegans* Hook. (Nymphaeaceae) **IPNI:** 171178-2 **Common name:** loto **Habit:** Herb, Aquatic **Use:** HF, ME **Origin:** Naturalised **Region:** Andean **Dept.:** ANT, CUN **Elev.:** 1600–2125 m a.s.l. **IUCN Habitat code:** 3,14 **Conservation status: (IUCN)** NE

264 *Nymphaea glandulifera* Rodschied (Nymphaeaceae) **IPNI:** 281434-2 **Habit:** Herb, Aquatic **Use:** EU, MA **Origin:** Native **Region:** Pacific **Dept.:** CHO **Elev.:** 150–150 m a.s.l. **IUCN Habitat code:** 1 **Conservation status: (IUCN)** NE

265 *Nymphaea jamesoniana* Planch. (Nymphaeaceae) **IPNI:** 605588-1 **Habit:** Herb, Aquatic **Use:** HF **Origin:** Native **Region:** Caribbean, Pacific **Dept.:** ANT, ATL, CHO **Elev.:** 20–200 m a.s.l. **IUCN Habitat code:** 14 **Conservation status: (IUCN)** NE

266 *Nymphaea lotus* L. (Nymphaeaceae) **IPNI:** 605604-1 **Habit:** Herb, Aquatic **Use:** AF, EU, GS, HF, MA, ME, SU **Origin:** Cultivated, Naturalised **Region:** Andean, Pacific **Dept.:** ANT, CUN, HUI **Elev.:** 110–1600 m a.s.l. **IUCN Habitat code:** 1,2,3,14 **Conservation status: (IUCN)** LC

267 *Nymphaea pulchella* DC. (Nymphaeaceae) **IPNI:** 605675-1 **Common name:** torta **Habit:** Herb, Aquatic **Use:** ME **Origin:** Native **Region:** Andean, Caribbean, Pacific **Dept.:** ANT, CHO, SUC **Elev.:** 0–180 m a.s.l. **IUCN Habitat code:** 2,14 **Conservation status: (IUCN)** NE

268 *Nymphaea rudgeana* G.Mey. (Nymphaeaceae) **IPNI:** 281442-2 **Habit:** Herb, Aquatic **Use:** ME **Origin:** Native **Region:** Caribbean **Dept.:** MAG **Elev.:** 50–50 m a.s.l. **Conservation status: (IUCN)** NE

269 *Victoria amazonica* (Poepp.) Klotzsch (Nymphaeaceae) **IPNI:** 605776-1 **Common name:** victoria regia **Habit:** Herb, Aquatic **Use:** EU, HF, MA **Origin:** Native **Region:** Amazonia **Dept.:** AMA **Elev.:** 180–360 m a.s.l. **IUCN Habitat code:** 1,5,14 **Conservation status: (IUCN)** NE

Austrobaileyales

270 *Illicium verum* Hook.f. (Schisandraceae) **IPNI:** 554553-1 **Common name:** anís estrella, anís estrellado **Use:** HF, MA, ME **Conservation status: (IUCN)** NE

Canellales

271 *Drimys granadensis* L.f. (Winteraceae) **IPNI:** 84100-2 **Common name:** ají, ají de páramo, bocadillo, canela de páramo, canelo, canelo de páramo, canelón, cupis, palo de ají, quijón, quinón, quirón, quiñón **Habit:** Small tree, Tree **Use:** EU, HF, ME **Origin:** Native **Region:** Andean **Dept.:** ANT, BOG, BOY, CAL, CAQ, CAU, CUN, HUI, NAR, NSA, PUT, QUI, RIS, SAN, TOL, VAC **Elev.:** 1800–3900 m a.s.l. **IUCN Habitat code:** 1,3,4,5,8,14 **Conservation status: (IUCN)** NE

272 *Drimys winteri* J.R.Forst. & G.Forst. (Winteraceae) **IPNI:** 330748-2 **Common name:** ajicillo, ajizuelo, ají, ají de páramo, boighe, boique, canela, canelo, canelo de monte, canelo de páramo, cascarilla, cupis, foiye, fuñe, fuñe boighe, palo de ají, quinón **Use:** EU, HF, MA, ME **IUCN Habitat code:** 1,4 **Conservation status: (IUCN)** LC

Chloranthales

273 *Hedyosmum angustifolium* (Ruiz & Pav.) Solms (Chloranthaceae) **IPNI:** 167832-1 **Habit:** Tree **Use:** HF **Origin:** Native **Region:** Andean **Dept.:** NAR **Elev.:** 900–2990 m a.s.l. **IUCN Habitat code:** 1,4 **Conservation status: (IUCN)** LC

274 *Hedyosmum anisodorum* Todzia (Chloranthaceae) **IPNI:** 277965-2 **Habit:** Shrub, Small tree, Tree **Use:** HF, ME **Origin:** Native **Region:** Andean **Dept.:** HUI, NAR, PUT **Elev.:** 1000–2800 m a.s.l. **Conservation status: (IUCN)** LC

275 *Hedyosmum bonplandianum* Kunth (Chloranthaceae) **IPNI:** 277969-2 **Common name:** aguanoso, almizcle, anisillo, canelo, canelón, chiflachifla, chocó, colchón de pobre, gallinazo, granicillo, granizo, guayusa, hurón, llorón, motilón, planta de soldado, planta de soledad, planta del soldado, silbador, silbasilba, silbato, silbo silbo, si **Habit:** Shrub, Small tree, Tree **Use:** HF, ME **Origin:** Native **Region:** Andean **Dept.:** ANT, BOG, CAU, CHO, QUI, RIS, VAC **Elev.:** 900–2400 m a.s.l. **IUCN Habitat code:** 1,3,4,14 **Conservation status: (IUCN)** LC

276 *Hedyosmum crenatum* Occhioni (Chloranthaceae) **IPNI:** 118320-2 **Common name:** aguanoso, granizo, silbasilba, silva silva **Habit:** Shrub, Subshrub, Tree **Use:** ME **Origin:** Native **Region:** Andean **Dept.:** ANT, BOG, BOY, CUN, LAG, MAG, MET, NSA, SAN **Elev.:** 1990–3700 m a.s.l. **IUCN Habitat code:** 1,3,4,14 **Conservation status: (IUCN)** NE

277 *Hedyosmum cuatrecazanum* Occhioni (Chloranthaceae) **IPNI:** 297228-2 **Common name:** canelo, canelón, chuchuguasa, granizo, guayusa **Habit:** Shrub, Small tree, Tree **Use:** HF **Origin:** Native **Region:** Andean **Dept.:** ANT, BOY, CAL, CAU, CUN, HUI, MAG, NAR, PUT, QUI, RIS, SAN, VAC **Elev.:** 1400-3700 m.a.s.l. **IUCN Habitat code:** 1,3,4,14 **Conservation status: (IUCN)** LC **(CPSG)** Potential LC

278 *Hedyosmum cumbalense* H.Karst. (Chloranthaceae) **IPNI:** 167850-1 **Common name:** granizo **Habit:** Shrub, Small tree, Tree **Use:** HF, ME **Origin:** Native **Region:** Andean **Dept.:** ANT, BOG, CAU, HUI, NAR, PUT, QUI, RIS, VAC **Elev.:** 1950–3800 m a.s.l. **IUCN Habitat code:** 1,3,4,8,14 **Conservation status: (IUCN)** LC

279 *Hedyosmum goudotianum* Solms (Chloranthaceae) **IPNI:** 277980-2 **Common name:** chiflador, granizo, silva silva, silvador, árnica de árbol **Habit:** Shrub, Small tree, Tree **Use:** HF, ME **Origin:** Native **Region:** Andean **Dept.:** ANT, BOG, BOY, CAL, CAU, CHO, CUN, HUI, MET, NAR, NSA, PUT, QUI, RIS, SAN, TOL, VAC **Elev.:** 1050-3300 m.a.s.l. **IUCN Habitat code:** 1,3,4,8,14 **Conservation status: (IUCN)** LC **(CPSG)** Potential LC

280 *Hedyosmum luteynii* Todzia (Chloranthaceae) **IPNI:** 277988-2 **Common name:** borracho, cashco, granecillo, granizo, guayusa andina, majagua, pururug colorado, tarqui **Habit:** Shrub, Small tree, Tree **Use:** HF, ME **Origin:** Native **Region:** Andean **Dept.:** CAU, HUI, NAR, QUI, RIS, TOL, VAC **Elev.:** 2600–3600 m a.s.l. **IUCN Habitat code:** 1,3,4,5,14 **Conservation status: (IUCN)** NE

281 *Hedyosmum maximum* (Kuntze) K.Schum. (Chloranthaceae) **IPNI:** 118340-2 **Use:** HF **IUCN Habitat code:** 14 **Conservation status: (IUCN)** VU

282 *Hedyosmum mexicanum* C.Cordem. (Chloranthaceae) **IPNI:** 167876-1 **Use:** HF **Origin:** Native **Conservation status: (IUCN)** VU

283 *Hedyosmum racemosum* (Ruiz & Pav.) G.Don (Chloranthaceae) **IPNI:** 167884-1 **Common name:** canelo, chiflichifli, chuchuguasa, granizo, guayusa, hojas de granizo, palo de agua, paloepiedra, pito, silbosilbo **Habit:** Shrub, Small tree, Tree **Use:** HF **Origin:** Native **Region:** Amazonia, Andean, Guiana Shield, Orinoquia **Dept.:** ANT, BOG, BOY, CAL, CAQ, CAU, CES, CUN, GUV, HUI, MAG, MET, NAR, NSA, PUT, QUI, RIS, SAN, TOL, VAC **Elev.:** 250-3350 m.a.s.l. **IUCN Habitat code:** 1,2,3,4,8,14 **Conservation status: (IUCN)** LC **(CPSG)** Potential LC

284 *Hedyosmum scaberrimum* Standl. (Chloranthaceae) **IPNI:** 167886-1 **Common name:** granizo, rodilla de pollo, zancazanca **Habit:** Shrub, Tree **Use:** HF **Origin:** Native **Region:** Andean, Pacific **Dept.:** ANT, CHO, NAR, VAC **Elev.:** 0–

2150 m a.s.l. **IUCN Habitat code:** 1,3,5,14 **Conservation status: (IUCN)** LC

285 *Hedyosmum scabrum* (Ruiz & Pav.) Solms (Chloranthaceae) **IPNI:** 167887-1 **Habit:** Shrub, Tree **Use:** MA **Origin:** Native **Region:** Andean **Dept.:** NAR, PUT **Elev.:** 2090–3130 m a.s.l. **IUCN Habitat code:** 1,14 **Conservation status: (IUCN)** LC

286 *Hedyosmum sprucei* Solms (Chloranthaceae) **IPNI:** 277998-2 **Common name:** granizo **Habit:** Shrub, Small tree **Use:** HF **Origin:** Native **Region:** Andean **Dept.:** CAU, PUT **Elev.:** 630–1050 m a.s.l. **IUCN Habitat code:** 1,3,14 **Conservation status: (IUCN)** LC

287 *Hedyosmum translucidum* Cuatrec. (Chloranthaceae) **IPNI:** 118351-2 **Common name:** aguaquín, granizo, silbasilba **Habit:** Shrub, Small tree, Tree **Use:** HF, ME **Origin:** Native **Region:** Andean **Dept.:** ANT, BOG, BOY, CAQ, CAU, CUN, HUI, MET, NAR, NSA, PUT, VAC **Elev.:** 2000–3100 m a.s.l. **IUCN Habitat code:** 1,3,4,14 **Conservation status: (IUCN)** LC

Laurales

288 *Gyrocarpus americanus* Jacq. (Hernandiaceae) **IPNI:** 170637-1 **Common name:** banco, volador **Habit:** Tree **Use:** FU, MA, ME, SU **Origin:** Native **Region:** Caribbean **Dept.:** ANT, ATL, BOL, COR, LAG, MAG, SAN, SUC **Elev.:** 5–1000 m a.s.l. **IUCN Habitat code:** 1,2,3,4,5,14 **Conservation status: (IUCN)** LC

289 *Hernandia didymantha* Donn.Sm. (Hernandiaceae) **IPNI:** 120657-2 **Common name:** aguacatón, banco, musana, palo blanco, tambor **Habit:** Tree **Use:** MA **Origin:** Native **Region:** Pacific **Dept.:** ANT, CHO **Elev.:** 0–100 m a.s.l. **IUCN Habitat code:** 1,3,5,14 **Conservation status: (IUCN)** NT

290 *Hernandia sonora* L. (Hernandiaceae) **IPNI:** 316357-2 **Common name:** hernandia **Use:** MA, ME **Conservation status: (IUCN)** NE

291 *Sparattanthelium amazonum* Mart. (Hernandiaceae) **IPNI:** 170915-1 **Habit:** Shrub, Climbing **Use:** ME **Origin:** Native **Dept.:** ANT **Elev.:** 120–600 m a.s.l. **IUCN Habitat code:** 1,3,14 **Conservation status: (IUCN)** NE

292 *Sparattanthelium glabrum* Rusby (Hernandiaceae) **IPNI:** 241044-2 **Common name:** canelón **Habit:** Liana **Use:** ME **Origin:** Native **Region:** Amazonia **Dept.:** AMA, COR, CUN, NAR **Elev.:** 200–750 m a.s.l. **IUCN Habitat code:** 1,14 **Conservation status: (IUCN)** NE

293 *Aiouea montana* (Sw.) R.Rohde (Lauraceae) **IPNI:** 77173628-1 **Common name:** aguacatillo, amarillo laurel, chambolán de lo frío, chaviaco, comino, guacharaco, gualpante, jigua, jigua laurel, laurel, laurel colorado, muena, opachiro, queso **Habit:** Tree **Use:** ME **Origin:** Native **Region:** Amazonia, Andean, Caribbean, Pacific **Dept.:** AMA, ANT, BOL, CAL, CAU, CUN, HUI, LAG, MAG, NAR, QUI, RIS, VAC **Elev.:** 0-2600 m.a.s.l. **IUCN Habitat code:** 1,2,3,4,5,14 **Conservation status: (IUCN)** LC **(CPSG)** Potential LC

294 *Aniba affinis* (Meisn.) Mez (Lauraceae) **IPNI:** 13808-2 **Habit:** Tree **Use:** MA **Origin:** Native **Region:** Amazonia, Guiana Shield **Dept.:** CAQ **Elev.:** 100–500 m a.s.l. **Conservation status: (IUCN)** LC

295 *Aniba burchellii* Kosterm. (Lauraceae) **IPNI:** 13815-2 **Habit:** Tree **Use:** MA **Origin:** Native **Region:** Amazonia, Guiana Shield **Dept.:** VAU **Elev.:** 100–200 m a.s.l. **IUCN Habitat code:** 1 **Conservation status: (IUCN)** LC

296 *Aniba canelilla* (Kunth) Mez (Lauraceae) **IPNI:** 13816-2 **Common name:** azafrás, canela de Quijos, canela muena, canelo, canelo de Andaquíes, cáscara, cáscara preciosa, sasafrás, tuabe **Habit:** Tree **Use:** HF, MA, ME, PO **Origin:** Native **Region:** Amazonia **Dept.:** CAQ **Elev.:** 400–400 m a.s.l. **IUCN Habitat code:** 1,14 **Conservation status: (IUCN)** LC

297 *Aniba coto* (Rusby) Kosterm. (Lauraceae) **IPNI:** 13822-2 **Common name:** cordillero, laurel, laurel comino **Habit:** Tree **Use:** MA, ME **Origin:** Native **Region:** Andean **Dept.:** QUI, TOL, VAC **Elev.:** 1200–2600 m a.s.l. **IUCN Habitat code:** 1,3,4,14 **Conservation status: (IUCN)** NE

298 *Aniba hostmanniana* (Nees) Mez (Lauraceae) **IPNI:** 13841-2 **Common name:** amarillo, canelo **Habit:** Tree **Use:** MA, PO **Origin:** Native **Region:** Amazonia, Andean **Dept.:** ANT, MET **Elev.:** 100–1200 m a.s.l. **IUCN Habitat code:** 1,3,14 **Conservation status: (IUCN)** LC

299 *Aniba panurensis* (Meisn.) Mez (Lauraceae) **IPNI:** 13863-2 **Common name:** amarillo, amarillo oloroso, comino, laurel, laurel oloroso, loiro, mediocomino, miratava, palo rosa, yema de huevo **Habit:** Tree **Use:** MA, ME **Origin:** Native **Region:** Amazonia, Andean, Guiana Shield, Orinoquia **Dept.:** AMA, CUN, MET, TOL, VAU, VID **Elev.:** 200–1000 m a.s.l. **IUCN Habitat code:** 1,3,4,5,14 **Conservation status: (IUCN)** LC

300 *Aniba perutilis* Hemsl. (Lauraceae) **IPNI:** 13869-2 **Common name:** caucho, chachajo, comino, comino crespo, comino real, laurel, laurel comino, medio comino, mediocomino, tuno **Habit:** Tree **Use:** MA, ME, SU **Origin:** Native **Region:** Andean **Dept.:** ANT, HUI, MET, SAN, VAC **Elev.:** 100-2600 m.a.s.l. **IUCN Habitat code:** 1,3,14 **Conservation status: (IUCN)** VU **(CPSG)** CR

301 *Aniba puchury-minor* (Mart.) Mez (Lauraceae) **IPNI:** 13872-2 **Common name:** amarillo corazón negro, canelo, guacharaco, jigua negro, laurel, laurel comino, tuabe, tuabe canelo, tuave **Habit:** Tree **Use:** MA, ME **Origin:** Native **Region:** Amazonia, Andean, Pacific **Dept.:** AMA, ANT, BOY, CAQ, MET, NAR, QUI, SAN, TOL, VAC **Elev.:** 200–2000 m a.s.l. **IUCN Habitat code:** 1,2,3,5,14 **Conservation status: (IUCN)** LC

302 *Aniba riparia* (Nees) Mez (Lauraceae) **IPNI:** 13878-2 **Common name:** canelo, comino **Habit:** Tree **Use:** MA, ME **Origin:** Native **Region:** Amazonia, Andean **Dept.:** AMA, ANT **Elev.:** 100–800 m a.s.l. **IUCN Habitat code:** 1,3,14 **Conservation status: (IUCN)** LC

303 *Aniba rosodora* Ducke (Lauraceae) **IPNI:** 13880-2 **Common name:** palo de rosa **Habit:** Tree **Use:** HF, MA, ME, PO **Origin:** Native **Region:** Amazonia **Dept.:** AMA **Elev.:** 100–500 m a.s.l. **Conservation status: (IUCN)** EN

304 *Beilschmiedia alloiophylla* (Rusby) Kosterm. (Lauraceae) **IPNI:** 3010840-1 **Habit:** Tree **Use:** MA, ME, PO **Origin:** Native **Region:** Andean, Pacific **Dept.:** MAG, VAC **Elev.:** 100–1900 m a.s.l. **IUCN Habitat code:** 1,14 **Conservation status: (IUCN)** LC

305 *Beilschmiedia brasiliensis* (Kosterm.) Kosterm. (Lauraceae) **IPNI:** 31042-2 **Common name:** abacachirana, aguacatillo, emege, emehe, uflé, uflé-aguacatillo, yurajicù **Habit:** Tree **Use:** HF, MA, ME **Origin:** Native **Region:** Amazonia **Dept.:** AMA **Elev.:** 100–200 m a.s.l. **Conservation status: (IUCN)** NE

306 *Beilschmiedia pendula* (Sw.) Hemsl. (Lauraceae) **IPNI:** 1057986-2 **Common name:** aguacate cimarrón, laurel **Habit:** Tree **Use:** MA, PO **Origin:** Native **Region:** Andean **Dept.:** ANT, NAR **Elev.:** 0–1700 m a.s.l. **IUCN Habitat code:** 1,3,14 **Conservation status: (IUCN)** NE

307 *Beilschmiedia sulcata* (Ruiz & Pav.) Kosterm. (Lauraceae) **IPNI:** 31062-2 **Use:** HF, ME **Origin:** Native **IUCN Habitat code:** 1,3,14 **Conservation status: (IUCN)** NE

308 *Beilschmiedia tovarensis* (Klotzsch & H.Karst. ex Meisn.) Sachiko Nishida (Lauraceae) **IPNI:** 1116029-2 **Common name:** aguacatillo, amarillo tara, laurel **Habit:** Tree **Use:** HF, ME **Origin:** Native **Region:** Andean **Dept.:** ANT, BOY, CAL, CAU, MAG, MET, VAC **Elev.:** 500–3000 m a.s.l. **IUCN Habitat code:** 1,3,4,14 **Conservation status: (IUCN)** LC

309 *Caryodaphnopsis cogolloi* van der Werff (Lauraceae) **IPNI:** 274159-2 **Common name:** yumbé, yumbé aguacatillo y zumbí **Habit:** Tree **Use:** MA **Origin:** Endemic **Dept.:** ANT **Elev.:** 300-500 m.a.s.l. **IUCN Habitat code:** 1,3,14 **Conservation status: (IUCN)** EN **(CPSG)** EN

310 *Cassytha filiformis* L. (Lauraceae) **IPNI:** 463202-1 **Common name:** bejuco de rema **Habit:** Herb, Parasitic **Use:** HF, MA, ME, PO, SU **Origin:** Native **Region:** Amazonia, Guiana Shield, Caribbean, Orinoquia, Pacific **Dept.:** ANT, CAQ, GUA, GUV, VAU, VID **Elev.:** 0–

1500 m a.s.l. **IUCN Habitat code:** 1,2,3,4,14 **Conservation status: (IUCN)** NE

311 *Chlorocardium venenosum* (Kosterm. & Pinkley) Rohwer, H.G.Richt. & van der Werff (Lauraceae)
IPNI: 274529-2
Habit: Tree **Use:** ME, PO **Origin:** Native **Region:** Amazonia, Pacific **Dept.:** AMA, CAQ, PUT, VAC **Elev.:** 0–300 m a.s.l. **IUCN Habitat code:** 1,3,14 **Conservation status: (IUCN)** LC

312 *Cinnamomum camphora* (L.) J.Presl (Lauraceae)
IPNI: 463336-1
Common name: alcanfor **Use:** EU, GS, HF, MA, ME, PO **Conservation status: (IUCN)** NE

313 *Cinnamomum verum* J.Presl (Lauraceae)
IPNI: 463752-1
Common name: canela, canelo **Habit:** Tree **Use:** EU, FU, HF, MA, ME **Origin:** Cultivated **Region:** Amazonia, Orinoquia **Dept.:** CAQ, CAS, TOL **Elev.:** 200–800 m a.s.l. **IUCN Habitat code:** 1,3,14 **Conservation status: (IUCN)** NE

314 *Damburneya martinicensis* (Mez) Trofimov (Lauraceae)
IPNI: 77158859-1
Habit: Tree **Use:** MA **Origin:** Native **Region:** Pacific **Dept.:** CHO **Elev.:** 0–100 m a.s.l. **IUCN Habitat code:** 1,2,14 **Conservation status: (IUCN)** LC

315 *Damburneya purpurea* (Ruiz & Pav.) Trofimov (Lauraceae)
IPNI: 77158867-1
Common name: amarillo, laurel
Habit: Tree **Use:** MA, ME **Origin:** Native **Region:** Andean, Pacific **Dept.:** ANT, CAL, CAU, CHO, CUN, HUI, LAG, MAG, NSA, QUI, VAC **Elev.:** 100–2400 m a.s.l. **IUCN Habitat code:** 1,2,3,4,8,14 **Conservation status: (IUCN)** LC

316 *Endlicheria anomala* (Nees) Mez (Lauraceae)
IPNI: 464481-1
Habit: Tree **Use:** MA, ME **Origin:** Native **Region:** Amazonia, Andean, Orinoquia **Dept.:** AMA, CAQ, MET, VAC, VAU, VID **Elev.:** 100–1400 m a.s.l. **IUCN Habitat code:** 1,3,4,5,14 **Conservation status: (IUCN)** LC

317 *Endlicheria arenosa* Chanderb. (Lauraceae)
IPNI: 60440993-2
Habit: Tree **Use:** MA, ME, PO **Origin:** Native **Region:** Amazonia **Dept.:** AMA **Elev.:** 100–400 m a.s.l. **IUCN Habitat code:** 1,4 **Conservation status: (IUCN)** LC

318 *Endlicheria canescens* Chanderbali (Lauraceae)
IPNI: 307011-2
Habit: Tree **Use:** MA **Origin:** Native **Region:** Amazonia, Andean **Dept.:** AMA, PUT **Elev.:** 100–1700 m a.s.l. **IUCN Habitat code:** 1,3 **Conservation status: (IUCN)** LC

319 *Endlicheria citriodora* van der Werff (Lauraceae)
IPNI: 276551-2
Habit: Tree **Use:** MA, ME **Origin:** Native **Region:** Amazonia **Dept.:** AMA, VAU **Elev.:** 100–200 m a.s.l. **IUCN Habitat code:** 1 **Conservation status: (IUCN)** VU

320 *Endlicheria dictifarinosa* C.K.Allen (Lauraceae)
IPNI: 90945-2
Habit: Tree **Use:** HF **Origin:** Native **Region:** Guiana Shield, Orinoquia **Dept.:** GUA, VID **Elev.:** 80–120 m a.s.l. **IUCN Habitat code:** 1 **Conservation status: (IUCN)** LC

321 *Endlicheria multiflora* (Miq.) Mez (Lauraceae)
IPNI: 90972-2
Use: MA **Origin:** Native **IUCN Habitat code:** 3 **Conservation status: (IUCN)** LC

322 *Endlicheria paniculata* (Spreng.) J.F.Macbr. (Lauraceae)
IPNI: 90974-2
Habit: Tree **Use:** MA **Origin:** Native **Region:** Andean **Dept.:** CUN, HUI, MAG, MET **Elev.:** 100–2000 m a.s.l. **IUCN Habitat code:** 1,14 **Conservation status: (IUCN)** LC

323 *Endlicheria sericea* Nees (Lauraceae)
IPNI: 464535-1
Common name: amarillo comino, guara, laurel amarillo, moena **Habit:** Tree **Use:** MA, ME **Origin:** Native **Region:** Amazonia, Andean, Pacific **Dept.:** AMA, CAQ, MET, NAR, PUT **Elev.:** 100–500 m a.s.l. **IUCN Habitat code:** 1,3,4,14 **Conservation status: (IUCN)** NE

324 *Laurus nobilis* L. (Lauraceae)
IPNI: 465049-1
Common name: canelón, laurel, laurel real **Use:** EU, HF, MA, ME **Dept.:** BOG **IUCN Habitat code:** 1,3,14 **Conservation status: (IUCN)** LC

325 *Licaria armeniaca* (Nees) Kosterm. (Lauraceae)
IPNI: 465248-1
Common name: laurel **Habit:** Tree **Use:** HF, MA **Origin:** Native **Region:** Amazonia **Dept.:** AMA **Elev.:** 60–200 m a.s.l. **IUCN Habitat code:** 1,3,14 **Conservation status: (IUCN)** LC

326 *Licaria cannella* (Meisn.) Kosterm. (Lauraceae)
IPNI: 139921-2
Common name: canela, canelo, comino baboso, diablo **Habit:** Tree **Use:** HF, MA **Origin:** Native **Region:** Amazonia, Andean, Guiana Shield, Pacific **Dept.:** AMA, ANT, BOY, CAQ, CHO, GUA, NAR, VAC, VAU **Elev.:** 0–2000 m a.s.l. **Conservation status: (IUCN)** LC

327 *Licaria chrysophylla* (Meisn.) Kosterm. (Lauraceae)
IPNI: 465260-1
Habit: Tree **Use:** MA **Origin:** Native **Region:** Amazonia **Dept.:** CAQ **Elev.:** 200–200 m a.s.l. **IUCN Habitat code:** 1 **Conservation status: (IUCN)** LC

328 *Licaria guianensis* Aubl. (Lauraceae)
IPNI: 465275-1
Common name: palo de guara
Habit: Tree **Use:** HF, MA **Origin:** Native **Region:** Amazonia, Andean **Dept.:** AMA, ANT **Elev.:** 100–700 m a.s.l. **IUCN Habitat code:** 1,3 **Conservation status: (IUCN)** LC

329 *Licaria macrophylla* (A.C.Sm.) Kosterm. (Lauraceae)
IPNI: 465280-1
Habit: Tree **Use:** MA **Origin:** Native **Region:** Amazonia **Dept.:** AMA, CAQ **Elev.:** 400–800 m a.s.l. **IUCN Habitat code:** 1,3 **Conservation status: (IUCN)** LC

330 *Licaria triandra* (Sw.) Kosterm. (Lauraceae)
IPNI: 465312-1
Common name: canelo, jigua canelo, laurel, palo de manteca, tuabe canelo **Habit:** Tree **Use:** HF, MA, PO **Origin:** Native **Region:** Andean, Pacific **Dept.:** CHO, VAC **Elev.:** 200–900 m a.s.l. **IUCN Habitat code:** 1,3,14 **Conservation status: (IUCN)** LC

331 *Mespilodaphne cymbarum* (Kunth) Trofimov (Lauraceae)
IPNI: 77198011-1
Common name: aceite de caparrapí, aceituno, aguarrás, amacey, canelo real caparrapí, canelo real hoja chica, chimar, chimar-karoa, comino, karoa, laurel canelo, palo de aceite, palo de caparrapí, sasafrás, zazafrás **Habit:** Tree **Use:** MA, ME **Origin:** Native **Region:** Amazonia, Guiana Shield **Dept.:** AMA, GUA, VAU **Elev.:** 100–300 m a.s.l. **IUCN Habitat code:** 1,3,14 **Conservation status: (IUCN)** NE

332 *Mespilodaphne quixos* (Lam.) Rohwer (Lauraceae)
IPNI: 77198017-1
Common name: canelo, canelo de Andaquíes, canelo de los Andaquíes, espingo **Habit:** Tree **Use:** HF, MA, ME, PO **Origin:** Native **Region:** Amazonia **Dept.:** AMA, CAQ, PUT **IUCN Habitat code:** 8 **Conservation status: (IUCN)** NE **(CPSG)** EN

333 *Mespilodaphne veraguensis* (Meisn.) Rohwer (Lauraceae)
IPNI: 77198019-1
Habit: Tree **Use:** MA, ME **Origin:** Native **Region:** Andean, Pacific **Dept.:** RIS, VAC **Elev.:** 0–1000 m a.s.l. **IUCN Habitat code:** 3,5,14 **Conservation status: (IUCN)** NE

334 *Mezilaurus ita-uba* (Meisn.) Taub. ex Mez (Lauraceae)
IPNI: 158649-2
Common name: comino, itauba, parature, parature negro **Habit:** Tree **Use:** HF, MA **Origin:** Native **Region:** Amazonia, Guiana Shield **Dept.:** AMA, CAQ, GUA, VAU **Elev.:** 50–200 m a.s.l. **Conservation status: (IUCN)** VU

335 *Nectandra acutifolia* (Ruiz & Pav.) Mez (Lauraceae)
IPNI: 167746-2
Common name: amarillo, amarillo baboso, amarillo tolonchero, jigua amarillo, jigua negro, jigua palialte, laurel, laurel jigua, tolonchero **Habit:** Tree **Use:** SU **Origin:** Native **Region:** Andean, Pacific **Dept.:** ANT, BOY, CAU, CHO, VAC **Elev.:** 100–2300 m a.s.l. **IUCN Habitat code:** 1,3,14 **Conservation status: (IUCN)** LC

336 *Nectandra amazonum* Nees (Lauraceae)
IPNI: 466626-1
Common name: chulo, laurel, tinto
Habit: Tree **Use:** ME **Origin:** Native **Region:** Amazonia **Dept.:** AMA **Elev.:** 50–300 m a.s.l. **IUCN Habitat code:** 1,3,4,5,14 **Conservation status: (IUCN)** LC

337 *Nectandra cuspidata* Nees & Mart. (Lauraceae)
IPNI: 281229-2
Common name: amarillo, laurel, laurel mierda, lengua de venado, moena **Habit:** Tree **Use:** FU, HF, MA, ME, PO **Origin:** Native **Region:** Amazonia, Andean, Guiana Shield, Orinoquia **Dept.:** ANT, BOL, BOY, CAL, CAQ, GUA, GUV, MAG, MET, QUI, SAN, TOL, VAU, VID **Elev.:** 200–1300 m a.s.l. **IUCN Habitat code:** 1,2,3,4,5,14 **Conservation status: (IUCN)** LC

338 *Nectandra discolor* (Kunth) Nees (Lauraceae)
IPNI: 1203274-2
Common name: laurel mierda **Habit:** Tree **Use:** MA **Origin:** Native **Region:** Andean **Dept.:** BOY, CAU, CUN, HUI, MAG, QUI, RIS, VAC **Elev.:** 1500–2300 m a.s.l. **IUCN Habitat code:** 1,3,14 **Conservation status: (IUCN)** LC

339 *Nectandra globosa* (Aubl.) Mez (Lauraceae)
IPNI: 167822-2
Common name: aguacatillo, incibe, jigua blanco, jigua negro, laurel, laurel negro **Habit:** Tree **Use:** MA, ME **Origin:** Native **Region:** Amazonia **Dept.:** AMA, MET **Elev.:** 250–400 m a.s.l. **IUCN Habitat code:** 1,3,4,5,14 **Conservation status: (IUCN)** LC

340 *Nectandra hihua* (Ruiz & Pav.) Rohwer (Lauraceae)
IPNI: 974281-1
Common name: laurel baboso **Habit:** Tree **Use:** MA, ME **Origin:** Native **Region:** Amazonia, Andean, Guiana Shield **Dept.:** AMA, ANT, CAQ, HUI, LAG, MAG, MET, VAU **Elev.:** 0–1000 m a.s.l. **IUCN Habitat code:** 1,2,3,14 **Conservation status: (IUCN)** LC

341 *Nectandra laurel* Klotzsch ex Nees (Lauraceae)
IPNI: 281234-2
Common name: amargo, laurel, laurel amargo, laurel amarillo, laurel llanero **Habit:** Tree **Use:** HF **Origin:** Native **Region:** Andean **Dept.:** ANT, BOG, CAU, LAG, MAG, QUI, VAC **Elev.:** 1500–2500 m a.s.l. **IUCN Habitat code:** 1,3,5,14 **Conservation status: (IUCN)** LC

342 *Nectandra lineata* (Kunth) Rohwer (Lauraceae)
IPNI: 974282-1
Common name: jigua amarillo, palialte **Habit:** Tree **Use:** MA, ME **Origin:** Native **Region:** Andean, Pacific **Dept.:** ANT, CAL, CUN, MAG, QUI, TOL, VAC **Elev.:** 20–2300 m a.s.l. **IUCN Habitat code:** 1,2,3,14 **Conservation status: (IUCN)** LC

343 *Nectandra membranacea* (Sw.) Griseb. (Lauraceae)
IPNI: 466773-1
Common name: amarillo, guacharaco, laurel, laurel amarillo, laurelillo **Use:** ME **Origin:** Native **IUCN Habitat code:** 1,2,3,4,5,14 **Conservation status: (IUCN)** LC

344 *Nectandra pichurim* (Kunth) Mez (Lauraceae)
IPNI: 1171177-2
Common name: ají de monte, laurel, palo anís **Habit:** Tree **Use:** HF, ME **Origin:** Native **Region:** Amazonia, Guiana Shield, Orinoquia **Dept.:** AMA, ARA, VAU **Elev.:** 20–100 m a.s.l. **IUCN Habitat code:** 1,2,3,14 **Conservation status: (IUCN)** LC

345 *Nectandra turbacensis* (Kunth) Nees (Lauraceae)
IPNI: 466885-1
Common name: ajicillo, ají, ají de monte, laurel, laurel ají, laurel mierda, lomo de babilla **Habit:** Tree **Use:** MA **Origin:** Native **Region:** Andean, Caribbean **Dept.:** ATL, BOL, CHO, CUN, LAG, MAG, PUT, RIS, SAP, VAC **Elev.:** 30–1400 m a.s.l. **IUCN Habitat code:** 1,2,3,4,14 **Conservation status: (IUCN)** LC

346 *Nectandra villosa* Nees & Mart. (Lauraceae)
IPNI: 466894-1
Common name: amarillo, amarillo real, balsero, guacharaco, jigua laurel, jigua negro, laurel **Habit:** Tree **Use:** MA, ME **Origin:** Native **Region:** Andean **Dept.:** ANT, CAU, CHO, CUN, LAG, NSA, PUT, QUI, TOL, VAC **Elev.:** 900–2300 m a.s.l. **Conservation status: (IUCN)** NE

347 *Ocotea aciphylla* (Nees & Mart.) Mez (Lauraceae)
IPNI: 171499-2
Common name: amarillo, comino, jigua laurel, laurel, laurel baboso, laurel blanco, laurel canelo, laurel comino, laurel oloroso, medio amarillo, medio comino, mediocomino, oloroso, palo de achiote **Habit:** Tree **Use:** HF, MA, ME **Origin:** Native **Region:** Amazonia, Andean **Dept.:** AMA, ANT, CAQ, CAU, PUT, VAC, VAU **Elev.:** 50–600 m a.s.l. **IUCN Habitat code:** 1,3,4,5,14 **Conservation status: (IUCN)** LC

348 *Ocotea acutangula* (Miq.) Mez (Lauraceae)
IPNI: 171503-2
Use: MA **Origin:** Native **IUCN Habitat code:** 1 **Conservation status: (IUCN)** NE

349 *Ocotea aurantiodora* (Ruiz & Pav.) Mez (Lauraceae)
IPNI: 171526-2
Common name: aguarrás amarillo, amarillo baboso, arenillo, canelo cenizo, chachajo, comino real, jigua laurel, laurel, laurel comino, laurel comino de caquetá, laurel guacharaco, matatigre, medio comino, tigrito **Habit:** Tree **Use:** MA, ME **Origin:** Native **Region:** Amazonia, Andean, Guiana Shield, Caribbean, Orinoquia **Dept.:** AMA, ANT, BOL, CAQ, CHO, GUV, MET, QUI, SAN, TOL, VAC, VID **Elev.:** 200–1600 m a.s.l. **IUCN Habitat code:** 1,3,4,5,14 **Conservation status: (IUCN)** LC

350 *Ocotea bofo* Kunth (Lauraceae)
IPNI: 467161-1
Common name: amarillo, laurel, laurel amarillo, laurel baboso, laurel baboso blanco **Habit:** Tree **Use:** MA, ME **Origin:** Native **Region:** Amazonia, Guiana Shield, Orinoquia, Pacific **Dept.:** AMA, ANT, BOL, CAQ, CHO, GUV, VAU, VID **Elev.:** 50–200 m a.s.l. **IUCN Habitat code:** 1,2,3,4,5,14 **Conservation status: (IUCN)** LC

351 *Ocotea caesariata* van der Werff (Lauraceae)
IPNI: 77167373-1
Common name: aguacate macho, amarillo, escobo, espadero real, laurel, laurel aliso, laurel blanco, peralejo, petaquín, susca, susque **Habit:** Tree **Use:** HF **Origin:** Native **Region:** Andean **Dept.:** ANT, BOG, BOY, CAL, CES, CUN, NSA, QUI, SAN **Elev.:** 2250–3900 m a.s.l. **IUCN Habitat code:** 1,3,4,14 **Conservation status: (IUCN)** NE

352 *Ocotea caracasana* (Nees) Mez (Lauraceae)
IPNI: 171568-2
Habit: Tree **Use:** MA **Origin:** Native **Conservation status: (IUCN)** NE

353 *Ocotea floribunda* (Sw.) Mez (Lauraceae)
IPNI: 1146025-2
Common name: laurel, laurel molinillo **Habit:** Tree **Use:** HF, MA **Origin:** Native **Region:** Amazonia, Pacific **Dept.:** ANT, MET, VAC **Elev.:** 100–1800 m a.s.l. **IUCN Habitat code:** 1,3,14 **Conservation status: (IUCN)** NE

354 *Ocotea glomerata* (Nees) Mez (Lauraceae)
IPNI: 171672-2
Use: MA **Origin:** Native **Conservation status: (IUCN)** LC

355 *Ocotea guianensis* Aubl. (Lauraceae)
IPNI: 467336-1
Common name: laurel, laurel pajita, laurel tabaidá, mestizo, terciopelo **Habit:** Tree **Use:** MA, ME **Origin:** Native **Region:** Andean, Guiana Shield **Dept.:** ANT, BOL, GUA, MAG, SAN, VAC **Elev.:** 50–1800 m a.s.l. **IUCN Habitat code:** 1,2,3,4,5,6,14 **Conservation status: (IUCN)** LC

356 *Ocotea heterochroma* Mez & Sodiro (Lauraceae)
IPNI: 171684-2
Common name: jigua blanco, laurel chaquiro **Habit:** Tree **Use:** HF **Origin:** Native **Region:** Andean **Dept.:** BOG, CAU, QUI, TOL, VAC **Elev.:** 2500-3500 m.a.s.l. **IUCN Habitat code:** 1,3,4,14 **Conservation status: (IUCN)** NE **(CPSG)** NT

357 *Ocotea insularis* (Meisn.) Mez (Lauraceae)
IPNI: 171697-2
Common name: jigua, jigua blanco, jigua pava, laurel, laurel blanco **Habit:** Tree **Use:** HF **Origin:** Native **Region:** Andean, Pacific **Dept.:** ANT, CHO, NAR, QUI, VAC **Elev.:** 50–2000 m a.s.l. **IUCN Habitat code:** 1,3,14 **Conservation status: (IUCN)** LC

358 *Ocotea leptobotra* (Ruiz & Pav.) Mez (Lauraceae)
IPNI: 171727-2
Common name: baboso, chachajo, comino, jigua baboso, jigua negro, jigua pava, laurel, orejemula **Habit:** Tree **Use:** MA, ME **Origin:** Native **Region:** Amazonia, Andean, Orinoquia, Pacific **Dept.:** AMA, ANT, CAQ, CAS, CHO, MET, NAR, VAC, VID **Elev.:** 0–1400 m a.s.l. **IUCN Habitat code:** 3 **Conservation status: (IUCN)** NE

359 *Ocotea macrophylla* Kunth (Lauraceae)
IPNI: 326508-2
Common name: aceituno, jigua, jigua amarillo, jigua negro, laurel **Habit:** Tree **Use:** ME, PO **Origin:** Native **Region:** Andean **Dept.:** ANT, CUN, QUI, VAC **Elev.:** 1000–1500 m a.s.l. **IUCN Habitat code:** 1,3,4,14 **Conservation status: (IUCN)** LC

360 *Ocotea oblonga* (Meisn.) Mez (Lauraceae)
IPNI: 171797-2
Common name: agey+, amarillo del blanco, chaquiro, chimbusa, iabeai, jigua, jigua sombrillo, laurel amarillo, laurel vava, planta del pajaro azul, puike, tamajecua **Habit:** Tree **Use:** HF, MA, ME **Origin:** Native **Region:**

Amazonia, Andean, Orinoquia **Dept.:** AMA, ANT, CAQ, VAC **Elev.:** 200–1600 m a.s.l. **IUCN Habitat code:** 1,3,4,5,14 **Conservation status: (IUCN)** LC

361 *Ocotea puberula* (Rich.) Nees (Lauraceae)
IPNI: 1032261-2
Common name: gurapo, laurel
Habit: Tree **Use:** MA, ME **Origin:** Native **Region:** Amazonia, Andean, Pacific **Dept.:** AMA, ANT, CHO, QUI, RIS **Elev.:** 100–1500 m a.s.l. **IUCN Habitat code:** 1,3,14 **Conservation status: (IUCN)** LC

362 *Ocotea schomburgkiana* (Nees) Mez (Lauraceae)
IPNI: 171883-2
Habit: Tree **Use:** MA **Origin:** Native **Region:** Orinoquia **Dept.:** VID **Elev.:** 120–120 m a.s.l. **IUCN Habitat code:** 1,3,4 **Conservation status: (IUCN)** LC

363 *Ocotea splendens* (Meisn.) Baill. (Lauraceae)
IPNI: 171904-2
Habit: Tree **Use:** MA **Origin:** Native **Region:** Guiana Shield **Dept.:** GUA, VAU **Elev.:** 0–300 m a.s.l. **IUCN Habitat code:** 1,3,14 **Conservation status: (IUCN)** LC

364 *Persea americana* Mill. (Lauraceae)
IPNI: 325643-2
Common name: aguacate, aguacate Lorena, aguacate búho, aguacate corneto, aguacate cura, aguacate de anís, aguacate de danta, aguacate de fantasma, aguacate de monte, aguacate de perico, aguacate macho, aguacate mariquiteño, aguacate negro, aguacate papelillo, avocado, **Habit:** Tree **Use:** AF, EU, FU, GS, HF, IF, MA, ME, PO, SU **Origin:** Cultivated, Naturalised **Region:** Amazonia, Andean, Guiana Shield, Caribbean, Orinoquia, Pacific **Dept.:** AMA, ANT, ATL, BOG, BOL, CAL, CAU, CES, CHO, COR, CUN, HUI, MAG, NAR, NSA, PUT, QUI, RIS, SAP, SAN, TOL, VAC **Elev.:** 0–2500 m a.s.l. **IUCN Habitat code:** 1,2,3,14 **Conservation status: (IUCN)** LC

365 *Persea areolatocostae* (C.K.Allen) van der Werff (Lauraceae)
IPNI: 282532-2
Habit: Tree **Use:** HF **Origin:** Native **Region:** Andean **Dept.:** ANT, VAC **Elev.:** 1000–2000 m a.s.l. **IUCN Habitat code:** 1,3,14 **Conservation status: (IUCN)** LC

366 *Persea caerulea* (Ruiz & Pav.) Mez (Lauraceae)
IPNI: 190613-2
Common name: aguacate cimarrón, aguacate de monte, aguacate macho, aguacate morado, aguacatillo, aguacatillo colorado, aguacato blanco, amarillo, chichacún, cuco, curomacho, curumacho, gourda, laurel, laurel bongo, rosicenizo **Habit:** Tree **Use:** GS, HF, ME **Origin:** Native **Region:** Andean **Dept.:** ANT, CAL, LAG, MAG, VAC **Elev.:** 1000–2500 m a.s.l. **IUCN Habitat code:** 1,2,3,5,14 **Conservation status: (IUCN)** LC

367 *Persea cuneata* Meisn. (Lauraceae)
IPNI: 190632-2
Common name: aguacarillo, aguacatillo, amarillo jigua, laurel
Habit: Tree **Use:** HF **Origin:** Native **Region:** Andean **Dept.:** ANT, CHO, VAC **Elev.:** 1200–2500 m a.s.l. **IUCN Habitat code:** 1,2,3,5,14 **Conservation status: (IUCN)** LC

368 *Persea ferruginea* Kunth (Lauraceae)
IPNI: 468017-1
Common name: aguacatillo colorado, jigua, laurel **Habit:** Tree **Use:** HF **Origin:** Native **Region:** Andean **Dept.:** ANT, BOG, CUN, HUI **Elev.:** 2500–4000 m a.s.l. **IUCN Habitat code:** 1,3,4,14 **Conservation status: (IUCN)** NE

369 *Persea schiedeana* Nees (Lauraceae)
IPNI: 190745-2
Habit: Tree **Use:** GS, HF, MA, ME **Origin:** Native **Region:** Andean **Dept.:** CHO **Elev.:** 1500–2000 m a.s.l. **IUCN Habitat code:** 14 **Conservation status: (IUCN)** EN

370 *Pleurothyrium parviflorum* Ducke (Lauraceae)
IPNI: 203826-2
Habit: Tree **Use:** ME **Origin:** Native **IUCN Habitat code:** 1 **Conservation status: (IUCN)** LC

371 *Pleurothyrium vasquezii* van der Werff (Lauraceae)
IPNI: 301916-2
Habit: Tree **Use:** MA, ME **Origin:** Native **Region:** Amazonia **Dept.:** AMA **Elev.:** 240–240 m a.s.l. **Conservation status: (IUCN)** LC

372 *Rhodostemonodaphne grandis* (Mez) Rohwer (Lauraceae)
IPNI: 284250-2
Common name: amarillo **Habit:** Tree **Use:** HF **Origin:** Native **Region:** Amazonia, Andean **Dept.:** AMA, PUT **Elev.:** 100–1500 m a.s.l. **IUCN Habitat code:** 1,3 **Conservation status: (IUCN)** LC

373 *Sextonia rubra* (Mez) van der Werff (Lauraceae)
IPNI: 999503-1
Use: MA **Origin:** Native **IUCN Habitat code:** 1 **Conservation status: (IUCN)** LC

374 *Mollinedia ovata* Ruiz & Pav. (Monimiaceae)
IPNI: 582116-1
Common name: carchilla **Habit:** Shrub, Tree **Use:** ME **Origin:** Native **Region:** Andean **Dept.:** ANT, BOG **Elev.:** 900–1500 m a.s.l. **IUCN Habitat code:** 1,3,4,14 **Conservation status: (IUCN)** LC

375 *Peumus boldus* Molina (Monimiaceae)
IPNI: 191376-2
Common name: boldo **Use:** FU, GS, HF, MA, ME **IUCN Habitat code:** 14 **Conservation status: (IUCN)** LC

376 *Siparuna aspera* (Ruiz & Pav.) A.DC. (Siparunaceae)
IPNI: 582223-1
Habit: Shrub, Small tree, Tree **Use:** ME **Origin:** Native **Region:** Andean, Pacific **Dept.:** ANT, CAU, CHO, CUN, NAR, QUI, RIS, VAC **Elev.:** 70–2500 m a.s.l. **IUCN Habitat code:** 1,2,3,14 **Conservation status: (IUCN)** LC

377 *Siparuna cuspidata* (Tul.) A.DC. (Siparunaceae)
IPNI: 582248-1
Habit: Shrub, Small tree, Tree **Use:** HF, MA **Origin:** Native **Region:** Amazonia, Andean, Guiana Shield, Pacific **Dept.:** AMA, ANT, BOY, CHO, HUI, MET, VAC, VAU **Elev.:** 80–2500 m a.s.l. **IUCN Habitat code:** 1,3,5,14 **Conservation status: (IUCN)** LC

378 *Siparuna decipiens* (Tul.) A.DC. (Siparunaceae)
IPNI: 582253-1
Common name: chucho, juan viejo, moyejo, sal de raspar, uamao, yemahuevo **Habit:** Shrub, Small tree, Tree **Use:** MA, ME **Origin:** Native **Region:** Amazonia, Andean **Dept.:** AMA, CAQ, GUV, QUI, VAU **Elev.:** 100–1560 m a.s.l. **IUCN Habitat code:** 1,3,4,5,14 **Conservation status: (IUCN)** LC

379 *Siparuna echinata* (Kunth) A.DC. (Siparunaceae)
IPNI: 582257-1
Common name: amapa, capitiú, limoncillo, malagre, napanga, raposapanga, surapanga **Habit:** Shrub, Small tree, Tree, Climbing **Use:** ME **Origin:** Native **Region:** Andean **Dept.:** ANT, CAU, CHO, HUI, NAR, PUT, QUI, RIS, TOL, VAC **Elev.:** 1300–3300 m a.s.l. **IUCN Habitat code:** 1,3,4,8,14 **Conservation status: (IUCN)** NE

380 *Siparuna guajalitensis* S.S.Renner & Hausner (Siparunaceae)
IPNI: 306947-2
Habit: Shrub **Use:** HF **Origin:** Native **Region:** Andean **Dept.:** NAR **Elev.:** 1800-1800 m.a.s.l. **IUCN Habitat code:** 1 **Conservation status: (IUCN)** VU **(CPSG)** VU

381 *Siparuna guianensis* Aubl. (Siparunaceae)
IPNI: 582285-1
Common name: ajicillo, ajícillo, gallinazo panga, limoncillo, limón hediondo, romadizo, tree de chucha, tumatumale, árbol de chucha **Habit:** Shrub, Small tree, Tree **Use:** MA, ME **Origin:** Native **Region:** Amazonia, Andean, Guiana Shield, Caribbean, Orinoquia, Pacific **Dept.:** AMA, ANT, BOY, CAQ, CAS, CAU, CHO, COR, CUN, GUA, LAG, MET, SAN, TOL, VAC, VAU, VID **Elev.:** 0–1450 m a.s.l. **IUCN Habitat code:** 1,2,3,4,5,14 **Conservation status: (IUCN)** LC

382 *Siparuna harlingii* S.S.Renner & Hausner (Siparunaceae)
IPNI: 982480-1
Common name: chucha panga, chuchapanga, chucho **Habit:** Shrub, Tree **Use:** HF, ME **Origin:** Native **Region:** Andean **Dept.:** CAQ, PUT **Elev.:** 800–1420 m a.s.l. **IUCN Habitat code:** 1,3,14 **Conservation status: (IUCN)** NE

383 *Siparuna pachyantha* A.C.Sm. (Siparunaceae)
IPNI: 237245-2
Habit: Small tree, Tree **Use:** ME **Origin:** Native **Region:** Amazonia **Dept.:** AMA **Elev.:** 100–120 m a.s.l. **IUCN Habitat code:** 1 **Conservation status: (IUCN)** LC

384 *Siparuna pauciflora* (Beurl.) A.DC. (Siparunaceae)
IPNI: 582342-1
Habit: Shrub, Tree **Use:** ME **Origin:** Native **Region:** Pacific **Dept.:** ANT, CHO **Elev.:** 0–300 m a.s.l. **IUCN Habitat code:** 1,3,5,14 **Conservation status: (IUCN)** LC

385 *Siparuna petiolaris* (Kunth) A.DC. (Siparunaceae)
IPNI: 582347-1
Common name: limón de monte, mono **Habit:** Tree **Use:** MA **Origin:** Endemic **Region:** Andean **Dept.:** ANT, MAG, QUI **Elev.:** 1700-1750 m.a.s.l. **IUCN Habitat code:** 1,14 **Conservation status: (IUCN)** LC **(CPSG)** LC

386 *Siparuna poeppigii* (Tul.) A.DC. (Siparunaceae) **IPNI:** 582354-1 **Habit:** Shrub **Use:** MA, ME **Origin:** Native **Region:** Guiana Shield **Dept.:** CAQ **Elev.:** 450–500 m a.s.l. **IUCN Habitat code:** 1 **Conservation status: (IUCN)** LC

387 *Siparuna sessiliflora* (Kunth) A.DC. (Siparunaceae) **IPNI:** 582373-1 **Common name:** chuchapanga, chuchapanga blanca, chuchapanga negra, chunduy, limoncillo, limón, limón cimarrón, limón de monte, mono, pepino mataculebra, semana santa **Habit:** Shrub, Small tree, Tree **Use:** MA, ME, PO **Origin:** Native **Region:** Amazonia, Orinoquia, Pacific **Dept.:** AMA, ANT, CAL, COR, TOL **Elev.:** 60–1220 m a.s.l. **IUCN Habitat code:** 1,2,3,4,5,14 **Conservation status: (IUCN)** LC

388 *Siparuna thecaphora* (Poepp. & Endl.) A.DC. (Siparunaceae) **IPNI:** 582392-1 **Common name:** cascabel, chucho, güeva de arriera, limoncillo, limoncillo de monte, purichingo, romadizo **Habit:** Shrub, Small tree, Tree, Climbing **Use:** MA, ME, PO **Origin:** Native **Region:** Amazonia, Andean, Guiana Shield, Pacific **Dept.:** AMA, ANT, BOG, CAQ, CAU, CHO, CUN, GUA, MAG, MET, PUT, SAN, VAC, VID **Elev.:** 0–2200 m a.s.l. **IUCN Habitat code:** 1,3,4,5,14 **Conservation status: (IUCN)** LC

Magnoliales

389 *Anaxagorea acuminata* (Dunal) A.St.-Hil. ex A.DC. (Annonaceae) **IPNI:** 72023-1 **Use:** ME **IUCN Habitat code:** 1,14 **Conservation status: (IUCN)** LC

390 *Anaxagorea brevipes* Benth. (Annonaceae) **IPNI:** 72027-1 **Habit:** Small tree, Tree **Use:** ME **Origin:** Native **Region:** Amazonia, Andean **Dept.:** AMA, ANT, CAQ **Elev.:** 90–600 m a.s.l. **IUCN Habitat code:** 1,3,5,14 **Conservation status: (IUCN)** LC

391 *Anaxagorea dolichocarpa* Sprague & Sandwith (Annonaceae) **IPNI:** 12508-2 **Common name:** carguero **Habit:** Subshrub, Small tree, Tree **Use:** ME **Origin:** Native **Region:** Amazonia, Caribbean, Pacific **Dept.:** AMA, ANT, BOL, COR **Elev.:** 150–300 m a.s.l. **IUCN Habitat code:** 1,3,5,14 **Conservation status: (IUCN)** LC

392 *Anaxagorea petiolata* R.E.Fr. (Annonaceae) **IPNI:** 12520-2 **Habit:** Small tree **Use:** ME **Origin:** Native **Region:** Amazonia **Dept.:** VAU **Elev.:** 250–250 m a.s.l. **IUCN Habitat code:** 1 **Conservation status: (IUCN)** LC

393 *Anaxagorea rufa* Timmerman (Annonaceae) **IPNI:** 272121-2 **Common name:** cãrucù, carguero, carguero de pava, carguero de ratón, carguero negro **Habit:** Small tree **Use:** ME **Origin:** Native **Region:** Amazonia, Pacific **Dept.:** AMA, GUA, VAC, VAU **Elev.:** 100–450 m a.s.l. **IUCN Habitat code:** 1,3,4 **Conservation status: (IUCN)** LC

394 *Annona acuminata* Saff. (Annonaceae) **IPNI:** 14218-2 **Habit:** Small tree, Tree **Use:** HF **Origin:** Native **Region:** Pacific **Dept.:** ANT, CHO **Elev.:** 0–600 m a.s.l. **IUCN Habitat code:** 1,3,14 **Conservation status: (IUCN)** NE

395 *Annona amazonica* R.E.Fr. (Annonaceae) **IPNI:** 14221-2 **Habit:** Tree **Use:** HF **Origin:** Native **Region:** Amazonia, Orinoquia **Dept.:** ANT, CAQ, MET, VAU **Elev.:** 250–800 m a.s.l. **IUCN Habitat code:** 1,3,5,14 **Conservation status: (IUCN)** LC

396 *Annona ambotay* Aubl. (Annonaceae) **IPNI:** 72148-1 **Common name:** casima, casima casima, casimana **Habit:** Shrub, Tree, Climbing **Use:** HF, MA, ME **Origin:** Native **Region:** Amazonia **Dept.:** AMA, CAQ, CAU, GUV, PUT, VAU **Elev.:** 100–850 m a.s.l. **IUCN Habitat code:** 1,3,14 **Conservation status: (IUCN)** LC

397 *Annona cercocarpa* Saff. (Annonaceae) **IPNI:** 14238-2 **Habit:** Tree **Use:** HF **Origin:** Endemic **Region:** Pacific **Dept.:** BOL, CHO **Elev.:** 100-300 m.a.s.l. **IUCN Habitat code:** 1,5,14 **Conservation status: (IUCN)** VU **(CPSG)** VU

398 *Annona cherimola* Mill. (Annonaceae) **IPNI:** 927235-1 **Common name:** anón, anón de monte, bichichinejo, casima, casimana, catreche, chirimorriñón, chirimoya, chirimoya de monte, chirimoyo, mocuya **Habit:** Tree **Use:** GS, HF, MA, ME, PO **Origin:** Native **Region:** Andean, Caribbean **Dept.:** ANT, BOG, CAL, CAU, CUN, QUI, SAP, VAC **Elev.:** 300–2500 m a.s.l. **IUCN Habitat code:** 1,2,3,14 **Conservation status: (IUCN)** LC

399 *Annona cherimolioides* Triana & Planch. (Annonaceae) **IPNI:** 300526-2 **Common name:** guanábana de monte, guanábano de monte, ubre de vaca **Habit:** Tree **Use:** HF **Origin:** Endemic **Region:** Amazonia, Andean **Dept.:** ANT, CAL, CUN, NSA, QUI, VAC **Elev.:** 900–2450 m a.s.l. **IUCN Habitat code:** 1,3,14 **Conservation status: (IUCN)** LC

400 *Annona cordifolia* (Szyszyl.) Poepp. ex Maas & Westra (Annonaceae) **IPNI:** 72182-1 **Habit:** Shrub, Liana **Use:** HF, ME **Origin:** Native **Region:** Amazonia **Dept.:** AMA **Elev.:** 100–100 m a.s.l. **IUCN Habitat code:** 1,3 **Conservation status: (IUCN)** LC

401 *Annona cuspidata* (Mart.) H.Rainer (Annonaceae) **IPNI:** 77080775-1 **Common name:** anonilla **Habit:** Tree **Use:** HF **Origin:** Native **Region:** Amazonia **Dept.:** AMA, CAQ **Elev.:** 100–300 m a.s.l. **IUCN Habitat code:** 1,3 **Conservation status: (IUCN)** LC

402 *Annona duckei* Diels (Annonaceae) **IPNI:** 14257-2 **Common name:** imbira blanco, nabueno **Habit:** Tree **Use:** HF, ME **Origin:** Native **Region:** Amazonia, Andean **Dept.:** AMA, PUT **Elev.:** 100–1450 m a.s.l. **IUCN Habitat code:** 1,3,14 **Conservation status: (IUCN)** LC

403 *Annona edulis* (Triana & Planch.) H.Rainer (Annonaceae) **IPNI:** 77080778-1 **Common name:** chirimoyo de monte, majagua **Habit:** Shrub, Tree **Use:** HF **Origin:** Native **Region:** Amazonia, Andean, Guiana Shield, Orinoquia **Dept.:** AMA, ARA, CAQ, GUV, HUI, MET, PUT, VAU **Elev.:** 100–1400 m a.s.l. **IUCN Habitat code:** 1,2,3,4,5,14 **Conservation status: (IUCN)** NE

404 *Annona exsucca* DC. (Annonaceae) **IPNI:** 72201-1 **Habit:** Shrub, Small tree, Tree **Use:** HF, MA **Origin:** Native **Region:** Orinoquia **Dept.:** VID **Elev.:** 100–100 m a.s.l. **IUCN Habitat code:** 1,3,4,5,14 **Conservation status: (IUCN)** LC

405 *Annona foetida* Mart. (Annonaceae) **IPNI:** 72204-1 **Common name:** nabueno **Habit:** Tree **Use:** HF **Origin:** Native **Region:** Amazonia **Dept.:** CAQ, VAU **Elev.:** 200–300 m a.s.l. **IUCN Habitat code:** 1,14 **Conservation status: (IUCN)** LC

406 *Annona glabra* L. (Annonaceae) **IPNI:** 72208-1 **Common name:** anón cienaguero, anón de cerdo, anón de riñón, anón liso, anón pelado, chirimoya, guanabanito de playa, guanábana arisca, guanábana cimarrona, guanábana de pozo, guanábano de pozo, pelón **Habit:** Tree **Use:** AF, EU, GS, HF, MA, ME, PO **Origin:** Native **Region:** Caribbean, Orinoquia, Pacific **Dept.:** ANT, ATL, CAU, CES, CHO, COR, LAG, MAG, MET **Elev.:** 0–300 m a.s.l. **IUCN Habitat code:** 1,2,3,5,14 **Conservation status: (IUCN)** LC

407 *Annona hayesii* Saff. (Annonaceae) **IPNI:** 14275-2 **Habit:** Tree **Use:** HF **Origin:** Native **Region:** Caribbean **Dept.:** CES **Elev.:** 0–250 m a.s.l. **IUCN Habitat code:** 1,3,5,14 **Conservation status: (IUCN)** NE

408 *Annona hypoglauca* Mart. (Annonaceae) **IPNI:** 72224-1 **Common name:** anón de charapa, anón de pescado, anón enredadera **Habit:** Shrub, Climbing **Use:** HF **Origin:** Native **Region:** Amazonia **Dept.:** AMA, CAQ, PUT, VID **Elev.:** 100–300 m a.s.l. **IUCN Habitat code:** 1,3,5,14 **Conservation status: (IUCN)** LC

409 *Annona jahnii* Saff. (Annonaceae) **IPNI:** 14285-2 **Common name:** anón de monte, guanabanito, makukula, manirita, manirito **Habit:** Shrub, Tree **Use:** HF, ME **Origin:** Native **Region:** Guiana Shield, Orinoquia **Dept.:** ARA, CAS, MET **Elev.:** 150–500 m a.s.l. **IUCN Habitat code:** 1,2,3,4,14 **Conservation status: (IUCN)** LC

410 *Annona montana* Macfad. (Annonaceae) **IPNI:** 72257-1 **Common name:** anonilla, anón de caimán, chirimoyo, guanábana, guanábano brasilero, guanábano cimarrona, sirimbo **Habit:** Tree **Use:** AF, GS, HF, MA, ME **Origin:** Native **Region:** Amazonia, Andean, Caribbean, Pacific **Dept.:** ANT, CAL, CHO, LAG, VAU

Elev.: 50–1800 m a.s.l. **IUCN Habitat code:** 1,2,3,14 **Conservation status: (IUCN)** LC

411 *Annona mucosa* Jacq. (Annonaceae)
IPNI: 72259-1
Common name: biribá **Habit:** Small tree, Tree **Use:** EU, FU, HF, MA, ME **Origin:** Native **Region:** Amazonia, Andean, Caribbean, Pacific **Dept.:** AMA, ANT, CAQ, CES, CHO, GUV, MAG, NSA, PUT, SAN, VAC, VAU **Elev.:** 0–1500 m a.s.l. **IUCN Habitat code:** 1,2,3,4,5,14 **Conservation status: (IUCN)** LC

412 *Annona muricata* L. (Annonaceae)
IPNI: 14308-2
Common name: anoncillo, anón, anón de corcho, anón de espino, catuche, guanabanilla, guanábana, guanábana cimarrona, guanábana de pozo, guanábano **Habit:** Tree **Use:** AF, EU, FU, GS, HF, IF, MA, ME, PO, SU **Origin:** Native **Region:** Amazonia, Andean, Caribbean, Orinoquia, Pacific **Dept.:** AMA, ANT, BOY, CAS, CAU, CHO, COR, CUN, HUI, MAG, MET, PUT, QUI, RIS, SAN, SUC, TOL, VAC, VAU **Elev.:** 100–2000 m a.s.l. **IUCN Habitat code:** 1,2,3,4,5,14 **Conservation status: (IUCN)** LC

413 *Annona neoinsignis* H.Rainer (Annonaceae)
IPNI: 77080806-1
Habit: Shrub, Tree **Use:** HF **Origin:** Native **Dept.:** AMA **Elev.:** 100–100 m a.s.l. **IUCN Habitat code:** 1 **Conservation status: (IUCN)** LC

414 *Annona paludosa* Aubl. (Annonaceae)
IPNI: 72271-1
Use: HF **IUCN Habitat code:** 3 **Conservation status: (IUCN)** LC

415 *Annona papilionella* (Diels) H.Rainer (Annonaceae)
IPNI: 77080793-1
Common name: nabueno **Habit:** Tree **Use:** HF **Origin:** Native **Region:** Amazonia, Andean, Caribbean, Pacific **Dept.:** AMA, ANT, BOL, CHO **Elev.:** 0–1500 m a.s.l. **IUCN Habitat code:** 1,2,3,14 **Conservation status: (IUCN)** LC

416 *Annona punicifolia* Triana & Planch. (Annonaceae)
IPNI: 72288-1
Common name: guanabanito **Habit:** Tree **Use:** HF, ME **Origin:** Native **Region:** Caribbean **Dept.:** BOL, CES, MAG **Elev.:** 0–50 m a.s.l. **IUCN Habitat code:** 2,5,14 **Conservation status: (IUCN)** NE

417 *Annona purpurea* Moc. & Sessé ex Dunal (Annonaceae)
IPNI: 72289-1
Common name: cabezona, catuche, gallina gorda, guanabana cimarrona, guanabanilla, guanabanillo, guanabarjo, guanacona, guanacón, guanábana, guanábana cabezona, guanábana de cabeza negra, guanábana de monte, guanábana matimbá, guanábana ñeca, guanábano, guanábano pun, **Habit:** Tree **Use:** GS, HF, ME **Origin:** Native **Region:** Caribbean, Pacific **Dept.:** ANT, BOL, CES, CHO, COR, LAG, SUC **Elev.:** 300–800 m a.s.l. **IUCN Habitat code:** 1,2,14 **Conservation status: (IUCN)** LC

418 *Annona quinduensis* Kunth (Annonaceae)
IPNI: 72292-1
Common name: anón cimarrón, anón de monte, chirimoya de monte, chirimoya montañera, chirimoyo de monte, guanábana de monte, guanábano **Habit:** Tree **Use:** HF **Origin:** Endemic **Region:** Andean **Dept.:** ANT, CAL, CAU, CUN, NAR, QUI, RIS, VAC **Elev.:** 750–2600 m a.s.l. **IUCN Habitat code:** 1,3,14 **Conservation status: (IUCN)** LC

419 *Annona rensoniana* (Standl.) H.Rainer (Annonaceae)
IPNI: 77080795-1
Common name: churumuyuí **Habit:** Tree **Use:** HF **Origin:** Native **Region:** Andean, Pacific **Dept.:** ANT, CAL, CAU, CHO, CUN, HUI, MAG, QUI, TOL, VAC **Elev.:** 300–1500 m a.s.l. **IUCN Habitat code:** 1,2,3,5,14 **Conservation status: (IUCN)** LC

420 *Annona reticulata* L. (Annonaceae)
IPNI: 72295-1
Common name: anón pelón, anón silvestre, candango, chirimoya roja, corazón, guanábana de pozo **Habit:** Tree **Use:** EU, GS, HF, MA, ME, PO **Origin:** Native **Region:** Andean, Pacific **Dept.:** ANT, CUN, MAG, SAN, TOL **Elev.:** 500–1900 m a.s.l. **IUCN Habitat code:** 1,3,14 **Conservation status: (IUCN)** LC

421 *Annona rufinervis* (Triana & Planch.) H.Rainer (Annonaceae)
IPNI: 77080796-1
Common name: guanábana de monte **Habit:** Tree **Use:** HF, ME **Origin:** Endemic **Region:** Andean **Dept.:** ANT, TOL **Elev.:** 300-1000 m.a.s.l. **Conservation status: (IUCN)** EN **(CPSG)** EN

422 *Annona scandens* Diels ex Pilg. (Annonaceae)
IPNI: 14345-2
Common name: anón de monte **Habit:** Climbing **Use:** HF **Origin:** Native **Region:** Amazonia **Dept.:** CAQ, VAU **Elev.:** 200–300 m a.s.l. **IUCN Habitat code:** 1 **Conservation status: (IUCN)** VU

423 *Annona sericea* Dunal (Annonaceae)
IPNI: 72310-1
Habit: Tree **Use:** HF, MA **Origin:** Native **Region:** Guiana Shield, Caribbean **Dept.:** BOL, MET **Elev.:** 150–600 m a.s.l. **IUCN Habitat code:** 1 **Conservation status: (IUCN)** LC

424 *Annona spraguei* Saff. (Annonaceae)
IPNI: 14352-2
Common name: guanabanito **Habit:** Tree **Use:** HF **Origin:** Native **Region:** Caribbean, Pacific **Dept.:** ANT, CHO, COR **Elev.:** 0–250 m a.s.l. **IUCN Habitat code:** 1,3,5,14 **Conservation status: (IUCN)** VU

425 *Annona squamosa* L. (Annonaceae)
IPNI: 72319-1
Common name: anona, anoncito, anón, anón borroñoso, anón caucano, anón de verruga, anón morado, anón rosado, anón verrugoso, riñón, saramuyo **Habit:** Shrub **Use:** AF, EU, FU, GS, HF, MA, ME, PO, SU **Origin:** Native **Region:** Andean, Caribbean **Dept.:** ANT, CAL, CUN, SAN, SUC, TOL, VAC **Elev.:** 200–1300 m a.s.l. **Conservation status: (IUCN)** LC

426 *Annona tenuiflora* Mart. (Annonaceae)
IPNI: 72324-1
Habit: Shrub, Tree **Use:** ME **Origin:** Native **Region:** Amazonia, Andean, Pacific **Dept.:** ANT, CAQ, SAN **Elev.:** 0–500 m a.s.l. **IUCN Habitat code:** 1,3 **Conservation status: (IUCN)** LC

427 *Bocageopsis multiflora* (Mart.) R.E.Fr. (Annonaceae)
IPNI: 34051-2
Common name: carguero, chucho, envira, imbira, majagüillo, tablón **Habit:** Tree **Use:** MA **Origin:** Native **Region:** Amazonia, Orinoquia **Dept.:** AMA, CAQ, GUA, VAU, VID **Elev.:** 100–400 m a.s.l. **IUCN Habitat code:** 1,3,4,5,14 **Conservation status: (IUCN)** LC

428 *Cananga odorata* (Lam.) Hook.f. & Thomson (Annonaceae)
IPNI: 72580-1
Common name: cadmia, cadmio, camay, cananga, flor de amor, ilang ilang, ilanga ilang, ilanilán, ilán ilán **Habit:** Small tree **Use:** AF, EU, FU, HF, IF, MA, ME, SU **Origin:** Cultivated **Region:** Andean, Orinoquia, Pacific **Dept.:** ANT, BOY, CAL, CAQ, CUN, MET, QUI, SAN, VAC, VID **Elev.:** 100–2220 m a.s.l. **IUCN Habitat code:** 1,2,3,4,14 **Conservation status: (IUCN)** LC

429 *Cremastosperma cauliflorum* R.E.Fr. (Annonaceae)
IPNI: 69238-2
Habit: Small tree, Tree **Use:** ME **Origin:** Native **Region:** Amazonia **Dept.:** AMA, PUT **Elev.:** 100–230 m a.s.l. **IUCN Habitat code:** 1,3,14 **Conservation status: (IUCN)** LC

430 *Cremastosperma gracilipes* R.E.Fr. (Annonaceae)
IPNI: 69239-2
Habit: Shrub, Tree **Use:** HF **Origin:** Native **Region:** Amazonia **Dept.:** AMA, CAQ **Elev.:** 0–500 m a.s.l. **IUCN Habitat code:** 1,3,14 **Conservation status: (IUCN)** LC

431 *Cremastosperma megalophyllum* R.E.Fr. (Annonaceae)
IPNI: 69245-2
Habit: Shrub, Tree **Use:** HF **Origin:** Native **Region:** Amazonia **Dept.:** AMA **Elev.:** 100–300 m a.s.l. **IUCN Habitat code:** 1,14 **Conservation status: (IUCN)** LC

432 *Cymbopetalum brasiliense* (Vell.) Benth. ex Baill. (Annonaceae)
IPNI: 72706-1
Habit: Small tree **Use:** HF, ME **Origin:** Native **Region:** Amazonia, Orinoquia **Dept.:** MET, PUT **Elev.:** 500–600 m a.s.l. **IUCN Habitat code:** 1,3,5 **Conservation status: (IUCN)** LC

433 *Desmopsis panamensis* (B.L.Rob.) Saff. (Annonaceae)
IPNI: 78729-2
Common name: yaya **Habit:** Small tree, Tree **Use:** ME **Origin:** Native **Region:** Pacific **Dept.:** ANT, CHO **Elev.:** 15–500 m a.s.l. **IUCN Habitat code:** 1,3,5,14 **Conservation status: (IUCN)** LC

434 *Diclinanona tessmannii* Diels (Annonaceae)
IPNI: 79677-2
Common name: carguero de puerco **Habit:** Tree **Use:** HF, ME **Origin:** Native **Region:** Amazonia **Dept.:** AMA, CAQ **Elev.:** 100–300 m a.s.l. **IUCN Habitat code:** 1,5 **Conservation status: (IUCN)** LC

435 *Duguetia cauliflora* R.E.Fr. (Annonaceae)
IPNI: 85691-2
Common name: cargero, carguero, pajarito **Habit:** Tree **Use:** HF **Origin:** Native **Region:** Amazonia, Orinoquia **Dept.:** AMA, CAQ, VAU, VID **Elev.:** 15–250 m a.s.l. **IUCN Habitat code:** 1,3,4 **Conservation status: (IUCN)** LC

436 *Duguetia confusa* Maas (Annonaceae) **IPNI:** 305889-2 **Common name:** torrijao **Habit:** Small tree **Use:** HF **Origin:** Native **Region:** Pacific **Dept.:** ANT, CHO, VAC **Elev.:** 40–50 m a.s.l. **IUCN Habitat code:** 1,3,5 **Conservation status: (IUCN)** LC

437 *Duguetia flagellaris* Huber (Annonaceae) **IPNI:** 85700-2 **Common name:** ají de guara, imbira blanco **Habit:** Tree **Use:** HF, MA, ME **Origin:** Native **Region:** Amazonia **Dept.:** AMA, ANT, CAQ, VAU **Elev.:** 100–900 m a.s.l. **IUCN Habitat code:** 1,4,14 **Conservation status: (IUCN)** LC

438 *Duguetia hadrantha* (Diels) R.E.Fr. (Annonaceae) **IPNI:** 72868-1 **Habit:** Shrub, Tree **Use:** HF, ME **Origin:** Native **Region:** Amazonia **Dept.:** AMA **Elev.:** 100–300 m a.s.l. **Conservation status: (IUCN)** LC

439 *Duguetia latifolia* R.E.Fr. (Annonaceae) **IPNI:** 85710-2 **Common name:** anoncillo, anón de guara, míyacù **Habit:** Shrub, Tree **Use:** HF **Origin:** Native **Region:** Amazonia **Dept.:** AMA, CAQ **Elev.:** 100–300 m a.s.l. **IUCN Habitat code:** 1,3 **Conservation status: (IUCN)** LC

440 *Duguetia lepidota* (Miq.) Pulle (Annonaceae) **IPNI:** 1077481-2 **Habit:** Tree **Use:** HF **Origin:** Native **Region:** Orinoquia **Dept.:** VID **Elev.:** 150–200 m a.s.l. **IUCN Habitat code:** 1,3 **Conservation status: (IUCN)** LC

441 *Duguetia macrophylla* R.E.Fr. (Annonaceae) **IPNI:** 85715-2 **Common name:** carguero, perfume de tigre, tortuga caspi **Habit:** Small tree **Use:** HF **Origin:** Native **Region:** Amazonia, Orinoquia **Dept.:** AMA, CAQ, GUA, VAU, VID **Elev.:** 100–300 m a.s.l. **IUCN Habitat code:** 1,3,5,14 **Conservation status: (IUCN)** LC

442 *Duguetia odorata* (Diels) J.F.Macbr. (Annonaceae) **IPNI:** 72891-1 **Common name:** carguero, chucho, palo de piedra de rebalse **Habit:** Tree **Use:** HF, ME **Origin:** Native **Region:** Amazonia, Orinoquia **Dept.:** AMA, CAQ, MET, VAU **Elev.:** 100–450 m a.s.l. **IUCN Habitat code:** 1,3,4,5,14 **Conservation status: (IUCN)** LC

443 *Duguetia quitarensis* Benth. (Annonaceae) **IPNI:** 85728-2 **Common name:** ají, nagüí, nagüí chirimoyo **Habit:** Tree **Use:** HF, MA, ME **Origin:** Native **Region:** Amazonia, Orinoquia **Dept.:** AMA, CAQ, MET, VAU **Elev.:** 200–600 m a.s.l. **IUCN Habitat code:** 1,3,5,14 **Conservation status: (IUCN)** LC

444 *Duguetia riparia* Huber (Annonaceae) **IPNI:** 85731-2 **Common name:** carguero **Habit:** Tree **Use:** HF, MA **Origin:** Native **Region:** Amazonia **Dept.:** AMA, CAQ **Elev.:** 150–300 m a.s.l. **IUCN Habitat code:** 1,5 **Conservation status: (IUCN)** LC

445 *Duguetia spixiana* Mart. (Annonaceae) **IPNI:** 309665-2 **Common name:** falso biribá, piño **Habit:** Tree **Use:** HF **Origin:** Native **Region:** Amazonia, Caribbean, Pacific **Dept.:** AMA, CAQ, CHO, LAG, PUT, VAU **Elev.:** 100–350 m a.s.l. **IUCN Habitat code:** 1,3 **Conservation status: (IUCN)** LC

446 *Duguetia stenantha* R.E.Fr. (Annonaceae) **IPNI:** 72913-1 **Common name:** barbásco, ekuku, ekuray, jabotí, mamo de tigre, mano de tigre, naranjito, noairay, noay, ziinijava **Habit:** Tree **Use:** HF, ME **Origin:** Native **Region:** Amazonia **Dept.:** AMA, CAQ, VAU **Elev.:** 100–400 m a.s.l. **IUCN Habitat code:** 1,3,14 **Conservation status: (IUCN)** LC

447 *Duguetia surinamensis* R.E.Fr. (Annonaceae) **IPNI:** 85740-2 **Habit:** Tree **Use:** MA **Origin:** Native **Dept.:** SAN **Elev.:** 200–200 m a.s.l. **IUCN Habitat code:** 1 **Conservation status: (IUCN)** LC

448 *Duguetia uniflora* (DC.) Mart. (Annonaceae) **IPNI:** 72920-1 **Habit:** Tree **Use:** HF **Origin:** Native **Region:** Amazonia **Dept.:** GUA **Elev.:** 100–100 m a.s.l. **IUCN Habitat code:** 1 **Conservation status: (IUCN)** LC

449 *Duguetia vallicola* J.F.Macbr. (Annonaceae) **IPNI:** 85746-2 **Habit:** Tree **Use:** HF **Origin:** Native **Region:** Caribbean, Pacific **Dept.:** BOL, CAL, CHO, SAN **Elev.:** 100–200 m a.s.l. **IUCN Habitat code:** 1,2,3,14 **Conservation status: (IUCN)** LC

450 *Ephedranthus amazonicus* R.E.Fr. (Annonaceae) **IPNI:** 91148-2 **Common name:** anón negro, carguero, indira, palo de arco, palo de imbira, palo de indira **Habit:** Tree **Use:** MA **Origin:** Native **Region:** Amazonia **Dept.:** AMA, CAQ **Elev.:** 100–300 m a.s.l. **IUCN Habitat code:** 1,3,5 **Conservation status: (IUCN)** LC

451 *Fusaea longifolia* (Aubl.) Saff. (Annonaceae) **IPNI:** 106440-2 **Common name:** anón, chirimoyo, yaya, yaya negra **Habit:** Tree **Use:** HF, MA **Origin:** Native **Region:** Amazonia, Andean, Caribbean, Pacific **Dept.:** AMA, ANT, CAQ, CHO, COR, SAN, VAU **Elev.:** 50–530 m a.s.l. **IUCN Habitat code:** 1,3,4,5,14 **Conservation status: (IUCN)** LC

452 *Guatteria amplifolia* Triana & Planch. (Annonaceae) **IPNI:** 73306-1 **Common name:** aporrejao, cargadera, cargadero, cargadero pita **Habit:** Tree **Use:** MA, ME **Origin:** Native **Region:** Andean, Pacific **Dept.:** ANT, CHO **Elev.:** 45–1350 m a.s.l. **IUCN Habitat code:** 1,3,5,14 **Conservation status: (IUCN)** LC

453 *Guatteria decurrens* R.E.Fr. (Annonaceae) **IPNI:** 114219-2 **Common name:** carguero, limoncillo, palo de buho, palo de búho, árbol de alivio **Habit:** Tree **Use:** ME **Origin:** Native **Region:** Amazonia **Dept.:** AMA **Elev.:** 200–300 m a.s.l. **IUCN Habitat code:** 1 **Conservation status: (IUCN)** LC

454 *Guatteria discolor* R.E.Fr. (Annonaceae) **IPNI:** 114224-2 **Use:** ME **Origin:** Native **IUCN Habitat code:** 1,3,14 **Conservation status: (IUCN)** LC

455 *Guatteria duckeana* R.E.Fr. (Annonaceae) **IPNI:** 114228-2 **Habit:** Small tree **Use:** ME **Origin:** Native **Region:** Amazonia **Dept.:** GUV, VAU **Elev.:** 200–600 m a.s.l. **IUCN Habitat code:** 1,3,4,14 **Conservation status: (IUCN)** LC

456 *Guatteria dura* R.E.Fr. (Annonaceae) **IPNI:** 114231-2 **Habit:** Tree **Use:** ME **Origin:** Native **Region:** Amazonia **Dept.:** AMA, VAU **Elev.:** 100–300 m a.s.l. **IUCN Habitat code:** 1,4,5,14 **Conservation status: (IUCN)** LC

457 *Guatteria foliosa* Benth. (Annonaceae) **IPNI:** 73416-1 **Common name:** caanuci, carguero negro **Habit:** Tree **Use:** ME **Origin:** Native **Region:** Amazonia **Dept.:** AMA, CAQ, VAU **Elev.:** 100–400 m a.s.l. **IUCN Habitat code:** 1,3,4,5,14 **Conservation status: (IUCN)** LC

458 *Guatteria guianensis* (Aubl.) R.E.Fr. (Annonaceae) **IPNI:** 114255-2 **Habit:** Tree **Use:** HF, ME **Origin:** Native **Region:** Amazonia, Andean **Dept.:** AMA, ANT, VAU **Elev.:** 200–500 m a.s.l. **IUCN Habitat code:** 1,3,4,14 **Conservation status: (IUCN)** LC

459 *Guatteria hirsuta* Ruiz & Pav. (Annonaceae) **IPNI:** 73441-1 **Common name:** cargadero, garrapato, guasca amarilla, palancabueno **Habit:** Tree **Use:** HF, MA **Origin:** Native **Region:** Amazonia, Andean, Orinoquia, Pacific **Dept.:** AMA, ANT, CAU, CHO, CUN, HUI, MET, NSA, SAN, VAC **Elev.:** 0–2000 m a.s.l. **IUCN Habitat code:** 1,2,3,4,14 **Conservation status: (IUCN)** LC

460 *Guatteria maypurensis* Kunth (Annonaceae) **IPNI:** 73503-1 **Habit:** Small tree **Use:** HF **Origin:** Native **Region:** Amazonia, Orinoquia **Dept.:** MET, VAU, VID **Elev.:** 50–120 m a.s.l. **IUCN Habitat code:** 1,3,4,14 **Conservation status: (IUCN)** LC

461 *Guatteria megalophylla* Diels (Annonaceae) **IPNI:** 114300-2 **Common name:** anón de monte, carguero, carguero de rebalse, carguero negro, golondrino **Habit:** Tree **Use:** ME **Origin:** Native **Region:** Amazonia, Andean **Dept.:** AMA, ANT, CAQ, PUT, VAU **Elev.:** 100–700 m a.s.l. **IUCN Habitat code:** 1,3,5,14 **Conservation status: (IUCN)** LC

462 *Guatteria punctata* (Aubl.) R.A.Howard (Annonaceae) **IPNI:** 912648-1 **Common name:** caanuci, cargadero, cargadero negro, carguero, cueronegro, garrapato, garrapato lanudo, garrapato negro, nagüí **Habit:** Small tree, Tree **Use:** HF, MA, ME **Origin:** Native **Region:** Amazonia,

Andean, Caribbean, Orinoquia, Pacific **Dept.:** AMA, ANT, BOL, CAL, CAQ, CAU, CHO, CUN, GUA, MET, NAR, SAN, VAC, VAU **Elev.:** 5–2000 m a.s.l. **IUCN Habitat code:** 1,3,4,5,14 **Conservation status: (IUCN)** LC

463 *Guatteria schomburgkiana* Mart. (Annonaceae) **IPNI:** 73626-1 **Common name:** caanuci, carguero **Habit:** Small tree, Tree **Use:** HF, MA, ME **Origin:** Native **Region:** Amazonia, Orinoquia **Dept.:** CAQ, CAS, MET, VAU, VID **Elev.:** 100–140 m a.s.l. **Conservation status: (IUCN)** LC

464 *Guatteria ucayalina* Huber (Annonaceae) **IPNI:** 114422-2 **Habit:** Small tree, Tree **Use:** ME **Origin:** Native **Region:** Amazonia, Andean, Pacific **Dept.:** AMA, ANT, BOL, BOY, CAQ, CHO, GUA, HUI, MET, NSA, VAU **Elev.:** 100–1800 m a.s.l. **IUCN Habitat code:** 1,3,4,5,14 **Conservation status: (IUCN)** LC

465 *Klarobelia anomala* (R.E.Fr.) Chatrou (Annonaceae) **IPNI:** 1003565-1 **Common name:** cargadero, yaya **Habit:** Tree **Use:** HF **Origin:** Native **Region:** Pacific **Dept.:** ANT, CHO **Elev.:** 0–200 m a.s.l. **IUCN Habitat code:** 1,2,3,5,14 **Conservation status: (IUCN)** LC

466 *Mosannona papillosa* Chatrou (Annonaceae) **IPNI:** 1003586-1 **Habit:** Tree **Use:** HF **Origin:** Native **Region:** Amazonia **Dept.:** PUT **Elev.:** 330–330 m a.s.l. **Conservation status: (IUCN)** LC

467 *Oxandra asbeckii* (Pulle) R.E.Fr. (Annonaceae) **IPNI:** 178617-2 **Use:** MA **Origin:** Native **IUCN Habitat code:** 1 **Conservation status: (IUCN)** LC

468 *Oxandra espintana* (Spruce ex Benth.) Baill. (Annonaceae) **IPNI:** 1013110-1 **Habit:** Tree **Use:** HF **Origin:** Native **Region:** Andean, Caribbean **Dept.:** HUI, RIS, SAP, TOL **Elev.:** 160-700 m.a.s.l. **IUCN Habitat code:** 1,3,14 **Conservation status: (IUCN)** LC **(CPSG)** CR

469 *Oxandra euneura* Diels (Annonaceae) **IPNI:** 178619-2 **Common name:** carguero, cuerinegro, vara de pescar **Habit:** Small tree **Use:** ME **Origin:** Native **Region:** Amazonia **Dept.:** AMA, CAQ, VAU **Elev.:** 100–300 m a.s.l. **IUCN Habitat code:** 1,4,14 **Conservation status: (IUCN)** LC

470 *Oxandra riedeliana* R.E.Fr. (Annonaceae) **IPNI:** 178638-2 **Use:** HF **Origin:** Native **IUCN Habitat code:** 1 **Conservation status: (IUCN)** LC

471 *Oxandra xylopioides* Diels (Annonaceae) **IPNI:** 178642-2 **Common name:** carguero, carguero de hormiga, carguero de pescado, cuerinegro, golondrino **Habit:** Small tree, Tree **Use:** HF, MA, ME **Origin:** Native **Region:** Amazonia **Dept.:** AMA, CAQ, GUA, VAU **Elev.:** 10–300 m a.s.l. **IUCN Habitat code:** 1,3,4,14 **Conservation status: (IUCN)** LC

472 *Polyalthia persicifolia* (Hook.f. & Thomson) Bedd. (Annonaceae) **IPNI:** 74692-1 **Habit:** Tree **Use:** ME **Origin:** Native **Region:** Andean, Orinoquia **Dept.:** MET, NAR, NSA, SAN **Elev.:** 1150–3500 m a.s.l. **Conservation status: (IUCN)** NE

473 *Porcelia mediocris* N.A.Murray (Annonaceae) **IPNI:** 975800-1 **Habit:** Tree **Use:** HF **Origin:** Native **Region:** Amazonia, Guiana Shield **Dept.:** MET, PUT **Elev.:** 450–450 m a.s.l. **IUCN Habitat code:** 3 **Conservation status: (IUCN)** LC

474 *Pseudomalmea boyacana* (J.F.Macbr.) Chatrou (Annonaceae) **IPNI:** 1003592-1 **Common name:** anón de monte **Habit:** Shrub, Tree **Use:** ME **Origin:** Endemic **Region:** Andean **Dept.:** ANT, BOY, CAL **Elev.:** 170-600 m.a.s.l. **IUCN Habitat code:** 1,2,3,14 **Conservation status: (IUCN)** LC **(CPSG)** LC

475 *Pseudoxandra sclerocarpa* Maas (Annonaceae) **IPNI:** 283917-2 **Common name:** frisolo, garrapato **Habit:** Tree **Use:** ME **Origin:** Endemic **Region:** Andean **Dept.:** ANT **Elev.:** 550-770 m.a.s.l. **IUCN Habitat code:** 1,3 **Conservation status: (IUCN)** VU **(CPSG)** VU

476 *Ruizodendron ovale* (Ruiz & Pav.) R.E.Fr. (Annonaceae) **IPNI:** 224360-2 **Habit:** Tree **Use:** HF **Origin:** Native **Region:** Amazonia, Guiana Shield **Dept.:** AMA, MET **Elev.:** 200–220 m a.s.l. **IUCN Habitat code:** 1 **Conservation status: (IUCN)** LC

477 *Tetrameranthus duckei* R.E.Fr. (Annonaceae) **IPNI:** 251707-2 **Habit:** Tree **Use:** MA **Origin:** Native **Region:** Amazonia **Dept.:** VAU **Elev.:** 260–280 m a.s.l. **IUCN Habitat code:** 1 **Conservation status: (IUCN)** LC

478 *Unonopsis floribunda* Diels (Annonaceae) **IPNI:** 259502-2 **Common name:** carguero, carguero negro **Habit:** Small tree, Tree **Use:** HF, ME **Origin:** Native **Region:** Amazonia, Pacific **Dept.:** AMA **Elev.:** 5–300 m a.s.l. **IUCN Habitat code:** 1,3,5,14 **Conservation status: (IUCN)** LC

479 *Unonopsis guatterioides* (A.DC.) R.E.Fr. (Annonaceae) **IPNI:** 259508-2 **Common name:** carguero de rebalse, carguero negro, imbira negra **Habit:** Shrub, Tree **Use:** ME **Origin:** Native **Region:** Amazonia **Dept.:** AMA, PUT **Elev.:** 150–200 m a.s.l. **IUCN Habitat code:** 1,3,5 **Conservation status: (IUCN)** LC

480 *Unonopsis spectabilis* Diels (Annonaceae) **IPNI:** 259528-2 **Common name:** anón de monte, espintana **Habit:** Shrub, Tree **Use:** HF, ME **Origin:** Native **Region:** Amazonia **Dept.:** AMA, VAU **Elev.:** 100–300 m a.s.l. **IUCN Habitat code:** 1,3 **Conservation status: (IUCN)** LC

481 *Unonopsis veneficiorum* (Mart.) R.E.Fr. (Annonaceae) **IPNI:** 259534-2 **Common name:** carguero, imbira **Habit:** Shrub, Tree **Use:** HF, ME, PO **Origin:** Native **Region:** Amazonia **Dept.:** AMA, GUA **Elev.:** 100–300 m a.s.l. **IUCN Habitat code:** 1,3,5,14 **Conservation status: (IUCN)** NE

482 *Xylopia amazonica* R.E.Fr. (Annonaceae) **IPNI:** 269768-2 **Common name:** cuerinegro, espintana, espintaza, tablón **Habit:** Tree **Use:** ME **Origin:** Native **Region:** Amazonia, Andean **Dept.:** AMA, ANT, SAN **Elev.:** 100–800 m a.s.l. **IUCN Habitat code:** 1,2,3,4,14 **Conservation status: (IUCN)** LC

483 *Xylopia aromatica* (Lam.) Mart. (Annonaceae) **IPNI:** 76067-1 **Common name:** achón, azotacaballo, cocopi mataefinae, cujicabro, escobillo, fruta de burro, fruta de capuchino, fruta del capuchino, frutaeburro, fruteburro, fruto de burra, fruto de burro, gallito, guaruchí, hachón, kokopi mata efinae, listón, látigo, majagüillo, mala **Habit:** Tree **Use:** AF, EU, FU, HF, MA, ME, SU **Origin:** Native **Region:** Amazonia, Andean, Caribbean, Orinoquia, Pacific **Dept.:** AMA, ANT, BOL, BOY, CAL, CAQ, CUN, GUA, GUV, HUI, LAG, MET, NSA, SAN, SUC, TOL, VAU, VID **Elev.:** 40–1220 m a.s.l. **IUCN Habitat code:** 1,2,3,4,5,14 **Conservation status: (IUCN)** LC

484 *Xylopia benthamii* R.E.Fr. (Annonaceae) **IPNI:** 269770-2 **Habit:** Small tree, Tree **Use:** ME **Origin:** Native **Region:** Amazonia, Andean **Dept.:** CAQ, GUV, NAR **Elev.:** 320–1325 m a.s.l. **IUCN Habitat code:** 1,3,5,14 **Conservation status: (IUCN)** LC

485 *Xylopia cuspidata* Diels (Annonaceae) **IPNI:** 269786-2 **Habit:** Small tree **Use:** ME **Origin:** Native **Region:** Amazonia **Dept.:** AMA, CAQ **Elev.:** 100–300 m a.s.l. **IUCN Habitat code:** 1,3 **Conservation status: (IUCN)** LC

486 *Xylopia discreta* (L.f.) Sprague & Hutch. (Annonaceae) **IPNI:** 76119-1 **Common name:** escubilla pimentona, escubillo, escubillo negro **Habit:** Tree **Use:** HF, MA, ME **Origin:** Native **Region:** Andean **Dept.:** ANT, NSA, SAN, TOL **Elev.:** 0–650 m a.s.l. **IUCN Habitat code:** 1,3,4,14 **Conservation status: (IUCN)** LC

487 *Xylopia emarginata* Mart. (Annonaceae) **IPNI:** 76126-1 **Common name:** carguero, carguero de hormiga, escobilla, palo astilloso **Habit:** Tree **Use:** MA, ME **Origin:** Native **Region:** Amazonia, Andean, Orinoquia **Dept.:** AMA, ANT, CAQ, GUA, MET, SAN **Elev.:** 90–600 m a.s.l. **IUCN Habitat code:** 1,2,3,4,5,14 **Conservation status: (IUCN)** LC

488 *Xylopia frutescens* Aubl. (Annonaceae) **IPNI:** 76133-1 **Common name:** escobillo, yaya **Habit:** Small tree, Tree **Use:** HF, MA, ME **Origin:** Native **Region:** Andean **Dept.:** ANT **Elev.:** 70–1000 m a.s.l. **IUCN Habitat code:** 1,2,3,5,14 **Conservation status: (IUCN)** LC

489 *Xylopia macrantha* Planch. & Triana (Annonaceae)
IPNI: 76181-1
Common name: achón sembe, azotacaballo, fruta de capuchino, malagueto, malagueño macho, quinaquina, rayado **Habit:** Tree **Use:** ME **Origin:** Native **Region:** Pacific **Dept.:** ANT, CHO, NAR, VAC **Elev.:** 10–160 m a.s.l. **IUCN Habitat code:** 1,3,5,14 **Conservation status: (IUCN)** LC

490 *Xylopia nitida* Dunal (Annonaceae)
IPNI: 76202-1
Common name: pelador **Habit:** Small tree **Use:** MA **Origin:** Native **Region:** Amazonia **Dept.:** AMA, CAQ **Elev.:** 100-200 m.a.s.l. **IUCN Habitat code:** 1,14 **Conservation status: (IUCN)** LC **(CPSG)** Potential LC

491 *Xylopia parviflora* Spruce (Annonaceae)
IPNI: 76217-1
Common name: carguero de hormiga **Habit:** Small tree **Use:** ME **Origin:** Native **Region:** Amazonia **Dept.:** AMA, CAQ, VAU **Elev.:** 200–650 m a.s.l. **IUCN Habitat code:** 1,3,14 **Conservation status: (IUCN)** LC

492 *Xylopia polyantha* R.E.Fr. (Annonaceae)
IPNI: 269821-2
Common name: burilico, buriligo, cargadero negro, escobillo, escubillo, malaguito, yaya blanca, yaya negra **Habit:** Tree **Use:** MA **Origin:** Native **Region:** Caribbean, Pacific **Dept.:** BOL, CHO, GUV, SAN, VAC **Elev.:** 0–900 m a.s.l. **IUCN Habitat code:** 1,2,3,4,5,14 **Conservation status: (IUCN)** LC

493 *Xylopia sericea* A.St.-Hil. (Annonaceae)
IPNI: 76256-1
Habit: Tree **Use:** HF, ME **Origin:** Native **Region:** Amazonia, Pacific **Dept.:** ANT, BOL, GUA, VAC, VAU **Elev.:** 50–500 m a.s.l. **IUCN Habitat code:** 1,3,5,14 **Conservation status: (IUCN)** LC

494 *Magnolia × soulangeana* Soul.-Bod. (Magnoliaceae)
IPNI: 60468302-2
Habit: Tree **Use:** EU **Origin:** Cultivated **Region:** Andean **Dept.:** ANT, CUN **Elev.:** 2600–2740 m a.s.l. **Conservation status: (IUCN)** NE

495 *Magnolia calimaensis* (Lozano) Govaerts (Magnoliaceae)
IPNI: 991406-1
Common name: almanegra del calima **Habit:** Tree **Use:** MA **Origin:** Endemic **Region:** Pacific **Dept.:** VAC **Elev.:** 0-100 m.a.s.l. **IUCN Habitat code:** 1 **Conservation status: (IUCN)** CR **(CPSG)** CR

496 *Magnolia cespedesii* (Triana & Planch.) Govaerts (Magnoliaceae)
IPNI: 991410-1
Common name: anón, cape grande **Habit:** Tree **Use:** ME **Origin:** Endemic **Region:** Andean **Dept.:** CUN **Elev.:** 1000-1000 m.a.s.l. **IUCN Habitat code:** 1,3 **Conservation status: (IUCN)** CR **(CPSG)** CR

497 *Magnolia chocoensis* (Lozano) Govaerts (Magnoliaceae)
IPNI: 991411-1
Common name: copachí, molinillo **Habit:** Tree **Use:** MA **Origin:** Endemic **Region:** Andean **Dept.:** CHO, RIS **Elev.:** 1450-1900 m.a.s.l. **IUCN Habitat code:** 1 **Conservation status: (IUCN)** EN **(CPSG)** EN

498 *Magnolia denudata* Desr. (Magnoliaceae)
IPNI: 554678-1
Use: EU, MA, ME **Region:** Andean **Dept.:** BOG **IUCN Habitat code:** 14 **Conservation status: (IUCN)** LC

499 *Magnolia espinalii* (Lozano) Govaerts (Magnoliaceae)
IPNI: 991415-1
Common name: hojarasco **Habit:** Tree **Use:** MA **Origin:** Endemic **Region:** Andean **Dept.:** ANT **Elev.:** 1800-2400 m.a.s.l. **IUCN Habitat code:** 1,14 **Conservation status: (IUCN)** CR **(CPSG)** CR

500 *Magnolia gilbertoi* (Lozano) Govaerts (Magnoliaceae)
IPNI: 991417-1
Common name: capachí, cañabrava, cañabravo, copachí, hojaras de gilberto, molinillo **Habit:** Tree **Use:** MA **Origin:** Endemic **Region:** Andean **Dept.:** QUI, RIS, VAC **Elev.:** 1800-3300 m.a.s.l. **IUCN Habitat code:** 1,3,14 **Conservation status: (IUCN)** EN **(CPSG)** EN

501 *Magnolia grandiflora* L. (Magnoliaceae)
IPNI: 554723-1
Common name: magnolia, magnolia blanca, magnolio **Habit:** Tree **Use:** EU, HF, MA, ME, PO **Origin:** Cultivated **Dept.:** ANT, BOG, BOY, CAL, CAU, CUN, NSA, QUI, RIS, SAN, TOL, VAC **Elev.:** 1500–2600 m a.s.l. **IUCN Habitat code:** 1,3,5,14 **Conservation status: (IUCN)** LC

502 *Magnolia hernandezii* (Lozano) Govaerts (Magnoliaceae)
IPNI: 991421-1
Common name: copachí, guanábano de monte, molinillo, molinillo del río cauca **Habit:** Tree **Use:** MA **Origin:** Endemic **Region:** Andean **Dept.:** ANT, QUI, RIS, VAC **Elev.:** 1700-2600 m.a.s.l. **IUCN Habitat code:** 1,3,14 **Conservation status: (IUCN)** EN **(CPSG)** EN

503 *Magnolia liliiflora* Desr. (Magnoliaceae)
IPNI: 554759-1
Use: EU, GS, ME **Origin:** Cultivated **Region:** Andean **Dept.:** CAL **Elev.:** 2200–2200 m a.s.l. **Conservation status: (IUCN)** DD

504 *Magnolia sambuensis* (Pittier) Govaerts (Magnoliaceae)
IPNI: 991437-1
Common name: charambirá, copachí, molinillo **Habit:** Tree **Use:** MA **Origin:** Native **Region:** Andean, Pacific **Dept.:** ANT, CHO, COR **Elev.:** 0–1000 m a.s.l. **IUCN Habitat code:** 1,3,14 **Conservation status: (IUCN)** NT

505 *Magnolia silvioi* (Lozano) Govaerts (Magnoliaceae)
IPNI: 991440-1
Common name: fruta de molinillo, guanabanadillo, guanabanillo, guanábana de monte, guanábano de monte **Habit:** Tree **Use:** MA **Origin:** Endemic **Region:** Andean **Dept.:** ANT **Elev.:** 400-1550 m.a.s.l. **IUCN Habitat code:** 1,3,14 **Conservation status: (IUCN)** EN **(CPSG)** EN

506 *Magnolia urraoensis* (Lozano) Govaerts (Magnoliaceae)
IPNI: 991441-1
Common name: alma negra, almanegra, almanegra de urrao, gallinaza, gallinazo **Habit:** Tree **Use:** MA **Origin:** Endemic **Region:** Andean **Dept.:** ANT **Elev.:** 1800-2400 m.a.s.l. **IUCN Habitat code:** 1,14 **Conservation status: (IUCN)** EN **(CPSG)** EN

507 *Compsoneura anoriensis* Janovec & A.K.Neill (Myristicaceae)
IPNI: 326808-2
Habit: Tree **Use:** HF **Origin:** Endemic **Dept.:** ANT **Elev.:** 400-900 m.a.s.l. **IUCN Habitat code:** 1,3,14 **Conservation status: (IUCN)** EN **(CPSG)** EN

508 *Compsoneura atopa* (A.C.Sm.) A.C.Sm. (Myristicaceae)
IPNI: 64003-2
Common name: castaño, castaño hojicolorado, castaño querré, garrucho, jerecú, josebé, manteco, nuánamo colorao, nuánamo sangregallo **Habit:** Small tree, Tree **Use:** HF, ME **Origin:** Native **Region:** Pacific **Dept.:** ANT, CAU, CHO, SAN, VAC **Elev.:** 0-800 m.a.s.l. **IUCN Habitat code:** 1,3,5,14 **Conservation status: (IUCN)** LC **(CPSG)** LC

509 *Compsoneura capitellata* (Poepp. ex A.DC.) Warb. (Myristicaceae)
IPNI: 64004-2
Common name: otobo, tatibuco **Habit:** Tree **Use:** HF **Origin:** Native **Region:** Amazonia, Guiana Shield, Pacific **Dept.:** AMA, ANT, CAQ, CHO, PUT, SAN, VAC, VAU **Elev.:** 50–800 m a.s.l. **IUCN Habitat code:** 1,3,5,14 **Conservation status: (IUCN)** LC

510 *Compsoneura claroensis* Janovec & A.K.Neill (Myristicaceae)
IPNI: 326809-2
Common name: güeba, mazorca, palo papa, soto sangre **Habit:** Tree **Use:** MA **Origin:** Endemic **Dept.:** ANT **Elev.:** 500-800 m.a.s.l. **IUCN Habitat code:** 1,3,14 **Conservation status: (IUCN)** LC **(CPSG)** LC

511 *Compsoneura cuatrecasasii* A.C.Sm. (Myristicaceae)
IPNI: 64006-2
Common name: castaña, castaño, chucha, cuángare, jigüa **Habit:** Tree **Use:** HF, ME **Origin:** Endemic **Region:** Andean, Pacific **Dept.:** CAQ, CAU, CHO, PUT, VAC **Elev.:** 0-1420 m.a.s.l. **IUCN Habitat code:** 1,3,5,14 **Conservation status: (IUCN)** LC **(CPSG)** VU

512 *Compsoneura sprucei* (A.DC.) Warb. (Myristicaceae)
IPNI: 64010-2
Common name: sangretoro **Habit:** Shrub, Tree **Use:** HF **Origin:** Native **Region:** Amazonia, Guiana Shield, Orinoquia, Pacific **Dept.:** AMA, ANT, CAQ, CHO, GUV, MET, PUT, SAN, VAU **Elev.:** 100–700 m a.s.l. **IUCN Habitat code:** 1,3,4,5,14 **Conservation status: (IUCN)** LC

513 *Iryanthera coriacea* Ducke (Myristicaceae)
IPNI: 130201-2
Habit: Tree **Use:** ME **Origin:** Native **Region:** Amazonia, Guiana Shield, Orinoquia **Dept.:** AMA, CAQ, GUV **Elev.:** 100–250 m a.s.l. **IUCN Habitat code:** 1 **Conservation status: (IUCN)** NE

514 *Iryanthera crassifolia* A.C.Sm. (Myristicaceae)
IPNI: 130202-2
Common name: cabo de hacha, cumala, espintana, mamito **Habit:** Shrub, Small tree, Tree **Use:** HF **Origin:** Native **Region:** Amazonia, Guiana Shield, Pacific **Dept.:** AMA, CAQ, GUA, VAC, VAU **Elev.:** 50–250 m a.s.l. **IUCN Habitat code:** 1,3,5,14 **Conservation status: (IUCN)** LC

515 *Iryanthera elliptica* Ducke (Myristicaceae) **IPNI:** 130205-2 **Common name:** mamita, mamito, otobo **Habit:** Tree **Use:** HF **Origin:** Native **Region:** Amazonia, Guiana Shield, Pacific **Dept.:** AMA, CAQ, VAC, VAU **Elev.:** 100–200 m a.s.l. **IUCN Habitat code:** 1,3,14 **Conservation status: (IUCN)** LC

516 *Iryanthera hostmannii* (Benth.) Warb. (Myristicaceae) **IPNI:** 585810-1 **Common name:** aace, bita, bolenillo, cabo de hacha, cabo de indio, cabo de judío, carnevaca, cumala, cuángare, dovida, foua, guacamayo caspi, guacharaca, jumo, mamilla, mamita, mamita mediana, mamito, molenillo, molenillo colorao, molinillo, nuámano, nuánamo sangregall **Habit:** Shrub, Small tree, Tree **Use:** HF, ME **Origin:** Native **Region:** Amazonia, Pacific **Dept.:** AMA, ANT, BOL, CAQ, CAU, CHO, COR, GUA, GUV, MET, NAR, SAN, VAC, VAU **Elev.:** 0–800 m a.s.l. **IUCN Habitat code:** 1,3,4,5,14 **Conservation status: (IUCN)** LC

517 *Iryanthera juruensis* Warb. (Myristicaceae) **IPNI:** 130210-2 **Common name:** cabo de hacha, carguero, cuángare, mamita, mamito, sangregallo, sangretoro **Habit:** Shrub, Small tree, Tree **Use:** HF, ME **Origin:** Native **Region:** Amazonia, Guiana Shield, Pacific **Dept.:** AMA, CAQ, CAU, CHO, COR, GUA, MET, NAR, VAC, VAU **Elev.:** 5–430 m a.s.l. **IUCN Habitat code:** 1,3,4,5,14 **Conservation status: (IUCN)** LC

518 *Iryanthera laevis* Markgr. (Myristicaceae) **IPNI:** 130212-2 **Common name:** cabo de hacha, cumala, duricke, mamita, mamito, mimito, poná, pucuna caspi, puná **Habit:** Tree **Use:** HF, ME **Origin:** Native **Region:** Amazonia, Guiana Shield, Orinoquia **Dept.:** AMA, CAQ, GUA, GUV, MET, VAU, VID **Elev.:** 100–800 m a.s.l. **IUCN Habitat code:** 1,3,4,5,14 **Conservation status: (IUCN)** NE

519 *Iryanthera lancifolia* Ducke (Myristicaceae) **IPNI:** 130213-2 **Common name:** cabo de hacha, charapilla, dekuyukai, kiriko, mamita, miirui, muuru, niba johuimeo, nivaoue, noikai, ofoma terikai, sangretoro, séhéñeede, tree camarón o cabo de hacha **Habit:** Tree **Use:** HF, MA, ME, SU **Origin:** Native **Region:** Amazonia, Guiana Shield, Pacific **Dept.:** AMA, CAQ, CHO, PUT, VAC, VAU **Elev.:** 5–900 m a.s.l. **IUCN Habitat code:** 1,3,5,14 **Conservation status: (IUCN)** LC

520 *Iryanthera macrophylla* Warb. (Myristicaceae) **IPNI:** 585818-1 **Common name:** cabo de indio, corobá, cuángare, mamito, sangregallina **Habit:** Tree **Use:** HF, MA, ME **Origin:** Native **Region:** Amazonia **Dept.:** AMA, CAQ, VAU **Elev.:** 200–220 m a.s.l. **IUCN Habitat code:** 1,5,14 **Conservation status: (IUCN)** LC

521 *Iryanthera megistophylla* A.C.Sm. (Myristicaceae) **IPNI:** 130217-2 **Common name:** cabo de indio, cuángare, cuángare cholo, nuánamo sangregallo **Habit:** Tree **Use:** ME **Origin:** Endemic **Region:** Amazonia, Pacific **Dept.:** AMA, ANT, CAU, CHO, NAR, VAC, VAU **Elev.:** 2-770 m.a.s.l. **IUCN Habitat code:** 1,3,14 **Conservation status: (IUCN)** LC **(CPSG)** VU

522 *Iryanthera paraensis* Huber (Myristicaceae) **IPNI:** 130222-2 **Common name:** maduraverde, mamito del pequeño, mamito pequeño, neoduricu, zoquete, árbol de borugo **Habit:** Tree **Use:** HF, MA **Origin:** Native **Region:** Amazonia, Guiana Shield, Orinoquia **Dept.:** AMA, GUV, MET, SAN, VAU **Elev.:** 100–850 m a.s.l. **IUCN Habitat code:** 1,3,14 **Conservation status: (IUCN)** LC

523 *Iryanthera polyneura* Ducke (Myristicaceae) **IPNI:** 130223-2 **Common name:** cumala, mamita, mamita de cangrejo, mamita grande, mamita mediana, mamito **Habit:** Tree **Use:** HF **Origin:** Native **Region:** Amazonia **Dept.:** AMA, CAQ, PUT, VAU **Elev.:** 50–250 m a.s.l. **IUCN Habitat code:** 1,3,5,14 **Conservation status: (IUCN)** LC

524 *Iryanthera tricornis* Ducke (Myristicaceae) **IPNI:** 130227-2 **Common name:** cabo de hacha, cabo hacha, mamita, mamito, nuánamo, palo de tabaco, palo mamito, sangretoro, trescáscaras, virola **Habit:** Small tree, Tree **Use:** HF **Origin:** Native **Region:** Amazonia, Guiana Shield **Dept.:** AMA, CAQ, SAN, VAU **Elev.:** 200–275 m a.s.l. **IUCN Habitat code:** 1,3,4,5,14 **Conservation status: (IUCN)** LC

525 *Myristica fragrans* Houtt. (Myristicaceae) **IPNI:** 586076-1 **Common name:** nuez moscada, nutmeg **Use:** EU, HF, MA, ME, PO, SU **Conservation status: (IUCN)** DD

526 *Osteophloeum platyspermum* (Spruce ex A.DC.) Warb. (Myristicaceae) **IPNI:** 586380-1 **Common name:** aguamiel, aguanoso, cabo de hacha, caracolí, chispare, chucha, cumala, cuángare, cuángare chucha, cuángare de loma, fierro amarillo, kimula flojo, madre de agua, mamita, mamita blanco, mamito, mamito blanco, marupa blanco, marupá, marupá blanco, miroco, n **Habit:** Tree **Use:** AF, HF, MA, ME **Origin:** Native **Region:** Amazonia, Guiana Shield, Pacific **Dept.:** AMA, CAQ, CHO, NAR, PUT, VAC, VAU **Elev.:** 2–770 m a.s.l. **IUCN Habitat code:** 1,3,5,14 **Conservation status: (IUCN)** LC

527 *Otoba acuminata* (Standl.) A.H.Gentry (Myristicaceae) **IPNI:** 586382-1 **Common name:** Santa María siete venas, cuángare, mamito, mamo, nuánamo, ocoba, otiba, otoba, otobo, sangre de pescado **Habit:** Shrub, Tree **Use:** HF, ME **Origin:** Native **Region:** Caribbean **Dept.:** ANT, VAC **Elev.:** 0-410 m.a.s.l. **IUCN Habitat code:** 1,5,14 **Conservation status: (IUCN)** NT **(CPSG)** VU

528 *Otoba gracilipes* (A.C.Sm.) A.H.Gentry (Myristicaceae) **IPNI:** 586384-1 **Common name:** cuángare, cuángare de loma, cuángare de mangual, cuángare otobo, hojicenizoso, otobo pajarito, sangregallo, tula pueta, tulapueta, urugá **Habit:** Tree **Use:** MA, ME **Origin:** Native **Region:** Pacific **Dept.:** ANT, CAU, CHO, GUA, NAR, VAC **Elev.:** 0-540 m.a.s.l. **IUCN Habitat code:** 1,3,5,14 **Conservation status: (IUCN)** NE **(CPSG)** NT

529 *Otoba novogranatensis* Moldenke (Myristicaceae) **IPNI:** 177341-2 **Common name:** cuángare, otoba, otobo, sangregallina, sebillo, soto, soto sangre **Habit:** Small tree, Tree **Use:** HF, MA, ME **Origin:** Native **Region:** Amazonia, Andean, Pacific **Dept.:** AMA, ANT, BOL, CAQ, CHO, HUI, NAR, NSA, PUT, RIS, VAC **Elev.:** 0–2150 m a.s.l. **IUCN Habitat code:** 1,3,5,14 **Conservation status: (IUCN)** LC

530 *Otoba parvifolia* (Markgr.) A.H.Gentry (Myristicaceae) **IPNI:** 586389-1 **Common name:** aguanillo, cumala colorada, cumala roja, cumalá blanca, cuángare, cuángare blanco, gabui, mamilla, naiquí, otoba, otobo, sangre de toro, sangregallina, sangretoro, sangrevaca, ucucal **Habit:** Small tree, Tree **Use:** HF, ME **Origin:** Native **Region:** Amazonia, Andean, Guiana Shield, Pacific **Dept.:** AMA, ANT, CAQ, CHO, MET, NAR, PUT, SAN, VAC **Elev.:** 10–1100 m a.s.l. **IUCN Habitat code:** 1,3,14 **Conservation status: (IUCN)** NE

531 *Virola albidiflora* Ducke (Myristicaceae) **IPNI:** 267133-2 **Common name:** po-ham-bó-raek, sangretoro **Habit:** Tree **Use:** ME **Origin:** Native **Region:** Amazonia **Dept.:** VAU **Elev.:** 100–100 m a.s.l. **IUCN Habitat code:** 1 **Conservation status: (IUCN)** NE

532 *Virola calophylla* (Spruce) Warb. (Myristicaceae) **IPNI:** 267140-2 **Common name:** are-de-yé, falsa sangretoro, harecha, mamita, mamita roja, mamito, reventillo, sangre toro, sangretoro, sangretoro de montaña, tierra amarillo, yaegaseii, yakee, yató **Habit:** Shrub, Small tree, Tree **Use:** MA, ME, PO, SU **Origin:** Native **Region:** Amazonia, Guiana Shield, Pacific **Dept.:** AMA, ANT, CAQ, CHO, GUA, MET, NAR, PUT, VAC, VAU **Elev.:** 50–750 m a.s.l. **IUCN Habitat code:** 1,3,4,5,14 **Conservation status: (IUCN)** LC

533 *Virola calophylloidea* Markgr. (Myristicaceae) **IPNI:** 267141-2 **Use:** MA, ME, PO, SU **IUCN Habitat code:** 1 **Conservation status: (IUCN)** NE

534 *Virola carinata* (Spruce ex Benth.) Warb. (Myristicaceae) **IPNI:** 267142-2 **Common name:** bon am, cuajo, cuajó, curiaci, enviroleira, kumula blanca, lasil me meju, nat sin ñemé, sangre de toro, ucuuba **Habit:** Tree **Use:** HF, MA, ME **Origin:** Native **Region:** Amazonia, Guiana Shield, Orinoquia, Pacific **Dept.:** AMA, CAQ, MET, PUT, VAU **Elev.:** 100–450 m a.s.l. **IUCN Habitat code:** 1,3,4,5,14 **Conservation status: (IUCN)** LC

535 *Virola duckei* A.C.Sm. (Myristicaceae) **IPNI:** 267151-2 **Common name:** carne de vaca, dogonpapoe, dogonpapohue, dunpahue, familia de kajao-jufaifimuo, jufaifim +o, juufaifim, mamita, pepiadero, sangre de toro blanco, sangretoro blanco, ucuúba **Habit:** Tree **Use:** HF, MA, ME **Origin:** Native **Region:** Amazonia, Pacific **Dept.:** AMA, CAQ, PUT, VAC, VAU **Elev.:** 100–300 m a.s.l. **IUCN Habitat code:** 1,3,14 **Conservation status: (IUCN)** LC

536 *Virola elongata* (Benth.) Warb. (Myristicaceae) **IPNI:** 267153-2 **Common name:** cabo de hacha, carnevaca, cuajo, cujupa, cumala, cumula blanca, falsa sangretoro, falso sangretoro, fierro amarillo, guarutata, hacha cabo, mamita, mamita pequeña, mamito, mamito de hoja ancha, maranae, nuánamo, nuánamo puntelanza, otobo, rase ñemee, sang **Habit:** Shrub, Small tree, Tree **Use:** HF, MA, ME, SU **Origin:** Native **Region:** Amazonia, Andean, Guiana Shield, Orinoquia, Pacific **Dept.:** AMA, ANT, BOL, CAQ, CHO, COR, GUA, GUV, MET, NAR, PUT, SAN, TOL, VAC, VAU, VID **Elev.:** 0–1420 m a.s.l. **IUCN Habitat code:** 1,3,4,5,14 **Conservation status: (IUCN)** LC

537 *Virola flexuosa* A.C.Sm. (Myristicaceae) **IPNI:** 267157-2 **Common name:** aire, cumala, nuámano, nuánamo, nuánamo sangregallo, sangre de gallo, sangre de pescado, sangretoro, soto, ucucal, virola **Habit:** Small tree, Tree **Use:** MA, ME **Origin:** Native **Region:** Amazonia, Guiana Shield, Orinoquia, Pacific **Dept.:** AMA, ANT, BOL, CAQ, CHO, GUA, GUV, MET, PUT, VAC, VAU **Elev.:** 50–1100 m a.s.l. **IUCN Habitat code:** 1,3,5,14 **Conservation status: (IUCN)** LC

538 *Virola koschnyi* Warb. (Myristicaceae) **IPNI:** 267165-2 **Habit:** Tree **Use:** HF, MA, ME **Origin:** Native **Region:** Pacific **Dept.:** CHO **IUCN Habitat code:** 5 **Conservation status: (IUCN)** LC

539 *Virola loretensis* A.C.Sm. (Myristicaceae) **IPNI:** 267168-2 **Common name:** mamita de boa, sangretoro copé **Habit:** Small tree, Tree **Use:** HF **Origin:** Native **Region:** Amazonia, Pacific **Dept.:** AMA, ANT, CAQ, CAU, CHO, NAR, VAC, VAU **Elev.:** 2–250 m a.s.l. **IUCN Habitat code:** 1,5,14 **Conservation status: (IUCN)** LC

540 *Virola pavonis* (A.DC.) A.C.Sm. (Myristicaceae) **IPNI:** 267190-2 **Common name:** cazuelo, fierro amarillo, mamita, mamita de cananguchal, sangretoro **Habit:** Small tree, Tree **Use:** ME **Origin:** Native **Region:** Amazonia, Guiana Shield **Dept.:** AMA, CAQ, GUA, PUT, VAU **Elev.:** 10–306 m a.s.l. **IUCN Habitat code:** 1,3,4,5,14 **Conservation status: (IUCN)** LC

541 *Virola peruviana* (A.DC.) Warb. (Myristicaceae) **IPNI:** 267191-2 **Common name:** carne de vaca, carnevaca, guapa, huapa, mamita, sangre de toro, sangretoro, tree del toro **Habit:** Tree **Use:** MA, ME **Origin:** Native **Region:** Amazonia, Andean, Guiana Shield, Orinoquia **Dept.:** AMA, ANT, GUA, GUV, MET, NAR, PUT, VAU **Elev.:** 100–1100 m a.s.l. **IUCN Habitat code:** 1,3,4,5,14 **Conservation status: (IUCN)** LC

542 *Virola sebifera* Aubl. (Myristicaceae) **IPNI:** 267199-2 **Common name:** bijo, carne de bagre, cascaraeyuca, cenizoso, chalviande, cuángare, cuángare de loma, cuángare punta de lanza, guanabanillo, lacre, sangre de toro, sangretoro, sangretoro bajito, sebillo, sebo, sota, soto, yaya sangre de pescado **Habit:** Tree **Use:** HF, MA, ME **Origin:** Native **Region:** Amazonia, Andean, Guiana Shield, Caribbean, Orinoquia, Pacific **Dept.:** AMA, ANT, CAQ, CAU, CHO, COR, GUA, GUV, LAG, MAG, MET, SAN, TOL, VAC, VAU, VID **Elev.:** 2–1330 m a.s.l. **IUCN Habitat code:** 1,3,4,5,14 **Conservation status: (IUCN)** LC

543 *Virola surinamensis* (Rol. ex Rottb.) Warb. (Myristicaceae) **IPNI:** 267202-2 **Common name:** cuajo, cumalá blanca, cumalá blanco, mamito, nuámano, nuánamo, sangretoro, sangretoro de bajo, sebo, ucucal **Habit:** Tree **Use:** HF, MA, ME **Origin:** Native **Region:** Amazonia, Pacific **Dept.:** AMA, ANT, CAQ, CHO, VAC, VAU **Elev.:** 20–400 m a.s.l. **IUCN Habitat code:** 1,2,3,4,5,14 **Conservation status: (IUCN)** EN

544 *Virola theiodora* (Spruce ex Benth.) Warb. (Myristicaceae) **IPNI:** 267203-2 **Common name:** gua, gua-roo'-ta-ta, roo', ta **Habit:** Tree **Use:** ME, PO, SU **Origin:** Native **Region:** Amazonia **Dept.:** AMA **Elev.:** 100–120 m a.s.l. **IUCN Habitat code:** 1 **Conservation status: (IUCN)** LC

Piperales

545 *Aristolochia acutifolia* Duch. (Aristolochiaceae) **IPNI:** 272643-2 **Common name:** carare **Habit:** Climbing **Use:** ME **Origin:** Native **Region:** Amazonia, Andean, Guiana Shield, Orinoquia **Dept.:** AMA, CAQ, CAU, PUT, VAU, VID **Elev.:** 170–800 m a.s.l. **IUCN Habitat code:** 1,3,4,14 **Conservation status: (IUCN)** NE

546 *Aristolochia anguicida* Jacq. (Aristolochiaceae) **IPNI:** 316428-2 **Common name:** capitana, contracapitana, curarín, gavilana, guaco, halconcito, waraara **Habit:** Climbing **Use:** HF, ME **Origin:** Native **Region:** Andean, Caribbean **Dept.:** ATL, BOL, CES, COR, LAG, MAG, SAN, SUC **Elev.:** 0–900 m a.s.l. **IUCN Habitat code:** 1,2,3,4,14 **Conservation status: (IUCN)** NE

547 *Aristolochia arborescens* L. (Aristolochiaceae) **IPNI:** 92743-1 **Use:** ME **Origin:** Native **Conservation status: (IUCN)** NE

548 *Aristolochia bicolor* Ule ex Pilg. (Aristolochiaceae) **IPNI:** 19406-2 **Use:** ME **Conservation status: (IUCN)** NE

549 *Aristolochia constricta* Griseb. (Aristolochiaceae) **IPNI:** 92853-1 **Common name:** bejuco carare, capitana, carare, carey, oreja de tigre, zaragoza **Habit:** Climbing **Use:** ME **Origin:** Native **Region:** Amazonia, Andean, Guiana Shield, Pacific **Dept.:** AMA, ANT, CHO, MET, NAR, PUT, SAN, VAC **Elev.:** 0–1100 m a.s.l. **Conservation status: (IUCN)** NE

550 *Aristolochia cordiflora* Mutis ex Kunth (Aristolochiaceae) **IPNI:** 92858-1 **Common name:** bejuco carare, bejuco del sol, bejuco guaco, capitana, carare, contracapitana de Mompós, corazón de Jesús, flor de Mompós, flor de alcatraz, flor de palomita, gallito, guaco, rasca, zaragoza **Habit:** Climbing **Use:** EU, ME **Origin:** Native **Region:** Andean, Pacific **Dept.:** ANT, CAU, CHO, CUN, NAR, PUT, SAN, TOL, VAC **Elev.:** 10–2000 m a.s.l. **IUCN Habitat code:** 1,3,14 **Conservation status: (IUCN)** NE

551 *Aristolochia didyma* S.Moore (Aristolochiaceae) **IPNI:** 19457-2 **Habit:** Climbing **Use:** MA, ME **Origin:** Native **Region:** Pacific **Dept.:** NAR **Elev.:** 215–215 m a.s.l. **Conservation status: (IUCN)** NE

552 *Aristolochia grandiflora* Sw. (Aristolochiaceae) **IPNI:** 92978-1 **Common name:** aristoloquia, bejuco carare, capitana, contracapitana, contracapitana de mompós, flor de alcatraz de mompós, gallito, guaco, oreja de elefante, yansaquitup, zaragoza **Habit:** Climbing **Use:** EU, ME **Origin:** Native, Cultivated **Region:** Amazonia, Caribbean, Pacific **Dept.:** ANT, CAQ, CHO, CUN, PUT **Elev.:** 0–600 m a.s.l. **IUCN Habitat code:** 1,2,3,14 **Conservation status: (IUCN)** NE

553 *Aristolochia guentheri* O.C.Schmidt (Aristolochiaceae) **IPNI:** 19489-2 **Common name:** carare, curarina de selva, curarina vegetal, zarzaparrilla **Habit:** Climbing **Use:** ME **Origin:** Native **Region:** Amazonia **Dept.:** AMA, CAQ **Elev.:** 100–265 m a.s.l. **IUCN Habitat code:** 1,5 **Conservation status: (IUCN)** NE

554 *Aristolochia inflata* Kunth (Aristolochiaceae) **IPNI:** 93027-1 **Common name:** bejuco carare, guaco **Habit:** Climbing **Use:** EU, ME **Origin:** Native **Region:** Caribbean **Dept.:** BOL, CES, COR, CUN, HUI, MAG, TOL, VAC **Elev.:** 20–1100 m a.s.l. **IUCN Habitat code:** 1,2,3,14 **Conservation status: (IUCN)** NE

555 *Aristolochia iquitensis* O.C.Schmidt (Aristolochiaceae) **IPNI:** 19503-2 **Habit:** Climbing **Use:** ME **Origin:** Native **Region:** Amazonia **Dept.:** AMA, CAQ **Elev.:** 100–250 m a.s.l. **IUCN Habitat code:** 1 **Conservation status: (IUCN)** NE

556 *Aristolochia klugii* O.C.Schmidt (Aristolochiaceae) **IPNI:** 272658-2 **Common name:** bejuco carare, bejuco carey, capitana, contracapitana, curarina, peuma huasca, tigre guasca verde **Habit:** Climbing **Use:** ME **Origin:** Native **Region:** Amazonia **Dept.:** AMA, CAQ, PUT **Elev.:** 325–420 m a.s.l. **IUCN Habitat code:** 1,14 **Conservation status: (IUCN)** NE

557 *Aristolochia leuconeura* Linden (Aristolochiaceae) **IPNI:** 272659-2 **Common name:** guaco **Habit:** Climbing **Use:** ME **Origin:** Native **Region:** Amazonia, Caribbean, Pacific **Dept.:** AMA, ANT, CAL, CHO, PUT, TOL **Elev.:** 100–500 m a.s.l. **IUCN Habitat code:** 1,2,3,5,14 **Conservation status: (IUCN)** NE

558 *Aristolochia littoralis* Parodi (Aristolochiaceae) **IPNI:** 93074-1 **Common name:** cisne, corazón de Jesús, gallito, gallo **Habit:** Climbing **Use:** EU, ME **Origin:** Native, Cultivated **Region:** Andean, Orinoquia **Dept.:** ANT, CAL, CAU, MAG, MET, VAC **Elev.:** 450–2150 m a.s.l. **IUCN Habitat code:** 14 **Conservation status: (IUCN)** NE

559 *Aristolochia maxima* Jacq. (Aristolochiaceae) **IPNI:** 316426-2 **Common name:** batatillo, bejuco canasta, bejuco carare, bejuco de Guayaquil, bejuco de canastillo, bejuco guasagil, bejuco guayaquil, canastilla, capitana, capitana de corazón, carare, cesticas, contracapitana, contraprieta, gallito, gallitos, gallo, guaco, guasca, gua **Habit:** Climbing **Use:** HF, ME **Origin:** Native **Region:** Andean, Guiana Shield, Caribbean, Orinoquia **Dept.:** ANT, BOG, BOY, CAL, CES, CUN, HUI, LAG, MAG, MET, RIS, SAN, SUC, TOL, VAC **Elev.:** 30–1800 m a.s.l. **IUCN Habitat code:** 1,2,3,5,14 **Conservation status: (IUCN)** NE

560 *Aristolochia nummulariifolia* Kunth (Aristolochiaceae) **IPNI:** 93167-1 **Common name:** curarina, dokobiijerojero, frutemato, mato, pepaemato **Habit:** Herb, Subshrub **Use:** ME **Origin:** Native **Region:** Orinoquia **Dept.:** ARA, CAS, CUN, MET **Elev.:** 100–500 m a.s.l. **Conservation status: (IUCN)** NE

561 *Aristolochia odoratissima* L. (Aristolochiaceae) **IPNI:** 316342-2 **Common name:** bejuco del sol, capitana, halconcito, zaragoza **Habit:** Climbing **Use:** ME **Origin:** Native **Region:** Caribbean, Orinoquia, Pacific **Dept.:** ANT, BOL, CES, LAG, MAG, MET, NAR, NSA, SAN, SUC **Elev.:** 0–450 m a.s.l. **IUCN Habitat code:** 1,2,3,4,14 **Conservation status: (IUCN)** NE

562 *Aristolochia pentandra* Jacq. (Aristolochiaceae) **IPNI:** 93219-1 **Habit:** Climbing **Use:** ME **Origin:** Native **Region:** Caribbean **Dept.:** SAP **Elev.:** 1–1 m a.s.l. **Conservation status: (IUCN)** NE

563 *Aristolochia pilosa* Kunth (Aristolochiaceae) **IPNI:** 93231-1 **Common name:** bejuco de Guayaquil, bejuco de zaragoza, gallito, gallo, guaco, guaco negro, niniche, tumbues, zaragoza **Habit:** Climbing **Use:** HF, ME **Origin:** Native **Region:** Amazonia, Andean, Caribbean, Orinoquia, Pacific **Dept.:** AMA, ANT, CAQ, CES, CHO, COR, CUN, GUV, HUI, MAG, MET, NSA, PUT, SAN, TOL, VAC **Elev.:** 0–1850 m a.s.l. **IUCN Habitat code:** 1,2,3,8,14 **Conservation status: (IUCN)** NE

564 *Aristolochia ringens* Vahl (Aristolochiaceae) **IPNI:** 93285-1 **Common name:** astroloja, bejuco carare, buche de pavo, capitana, capitana de corazón, contracapitana, flor de pato, gallito, gallo, guaco, guaco blanco, guascua, pajarito, pajaritos verdes, patico, paticos, pato, zaragoza **Habit:** Climbing **Use:** EU, ME **Origin:** Native, Cultivated **Region:** Andean, Guiana Shield, Caribbean, Orinoquia **Dept.:** AMA, ANT, BOL, CAL, CAU, CES, CUN, MAG, MET, NSA, SAN, SUC, TOL, VAC **Elev.:** 100–2150 m a.s.l. **Conservation status: (IUCN)** NE

565 *Aristolochia ruiziana* (Klotzsch) Duch. (Aristolochiaceae) **IPNI:** 93298-1 **Common name:** carare, oquis, tigre guasca, tigre guasca amarillo, tigre guasca morado **Habit:** Climbing **Use:** ME **Origin:** Native **Region:** Amazonia, Orinoquia **Dept.:** AMA, CAQ, CAS, PUT **Elev.:** 400–580 m a.s.l. **IUCN Habitat code:** 1,3,5,14 **Conservation status: (IUCN)** NE

566 *Aristolochia schultzeana* O.C.Schmidt (Aristolochiaceae) **IPNI:** 272665-2 **Habit:** Climbing **Use:** ME **Origin:** Native **Region:** Amazonia, Pacific **Dept.:** NAR, PUT **Elev.:** 0–325 m a.s.l. **IUCN Habitat code:** 1,14 **Conservation status: (IUCN)** NE

567 *Aristolochia tonduzii* O.C.Schmidt (Aristolochiaceae) **IPNI:** 19637-2 **Common name:** contra, zaragoza almendra **Habit:** Climbing **Use:** ME **Origin:** Native **Region:** Amazonia, Pacific **Dept.:** AMA, ANT, CAU, CHO, NAR, SAN, VAC **Elev.:** 0–500 m a.s.l. **IUCN Habitat code:** 1,3 **Conservation status: (IUCN)** NE

568 *Aristolochia trilobata* L. (Aristolochiaceae) **IPNI:** 319092-2 **Habit:** Climbing **Use:** ME **Origin:** Native **Region:** Pacific **Dept.:** ANT, CHO **Conservation status: (IUCN)** NE

569 *Peperomia acuminata* Ruiz & Pav. (Piperaceae) **IPNI:** 677918-1 **Common name:** cordoncillo **Habit:** Herb **Use:** HF, ME **Origin:** Native **Region:** Andean **Dept.:** ANT, BOY, CAU, CHO, CUN, HUI, MET, NAR, NSA, PUT, QUI, RIS, SAN, VAC **Elev.:** 1800-3600 m.a.s.l. **IUCN Habitat code:** 1,2,3,4,5,14 **Conservation status: (IUCN)** NE **(CPSG)** Potential LC

570 *Peperomia alata* Ruiz & Pav. (Piperaceae) **IPNI:** 677941-1 **Habit:** Herb, Epiphyte **Use:** ME **Origin:** Native **Region:** Andean, Guiana Shield **Dept.:** ANT, CAU, CHO, CUN, GUA, MAG, NSA, QUI, RIS, SAN, TOL, VAC **Elev.:** 100-2900 m.a.s.l. **IUCN Habitat code:** 1,3,5,14 **Conservation status: (IUCN)** NE **(CPSG)** Potential LC

571 *Peperomia angustata* Kunth (Piperaceae) **IPNI:** 677986-1 **Habit:** Herb, Epiphyte **Use:** HF **Origin:** Native **Region:** Amazonia, Guiana Shield, Caribbean, Pacific **Dept.:** AMA, ANT, BOL, BOY, CAQ, CES, CHO, MAG, NSA, SAN, VAU **Elev.:** 50–500 m a.s.l. **IUCN Habitat code:** 1,2,3,14 **Conservation status: (IUCN)** NE

572 *Peperomia argyreia* (Hook.f.) É.Morren (Piperaceae) **IPNI:** 678021-1 **Common name:** conchas, congas, cucharitas **Habit:** Herb **Use:** EU, ME, SU **Origin:** Cultivated **Region:** Andean **Dept.:** BOG, CUN **Elev.:** 2600–2600 m a.s.l. **IUCN Habitat code:** 14 **Conservation status: (IUCN)** NE

573 *Peperomia bethaniana* Trel. & Yunck. (Piperaceae) **IPNI:** 282355-2 **Common name:** siempreviva **Habit:** Herb **Use:** ME **Origin:** Endemic **Region:** Andean **Dept.:** CHO, CUN **Elev.:** 671–2700 m a.s.l. **IUCN Habitat code:** 1,3 **Conservation status: (IUCN)** NE

574 *Peperomia blanda* (Jacq.) Kunth (Piperaceae) **IPNI:** 678107-1 **Habit:** Herb, Epiphyte **Use:** ME **Origin:** Native **Region:** Andean **Dept.:** ANT, BOG, BOY, CAL, CAU, CES, CUN, HUI, LAG, MAG, NAR, NSA, QUI, SAN, VAC **Elev.:** 900-2985 m.a.s.l. **IUCN Habitat code:** 1,2,3,14 **Conservation status: (IUCN)** NE **(CPSG)** Potential LC

575 *Peperomia caperata* Yunck. (Piperaceae) **IPNI:** 678190-1 **Use:** EU **Region:** Andean **Dept.:** BOG **Conservation status: (IUCN)** NE

576 *Peperomia crispa* Sodiro (Piperaceae) **IPNI:** 678363-1 **Habit:** Herb **Use:** ME **Origin:** Native **Region:** Andean **Dept.:** TOL **Elev.:** 2600–2600 m a.s.l. **IUCN Habitat code:** 1 **Conservation status: (IUCN)** VU

577 *Peperomia dimota* Trel. & Yunck. (Piperaceae) **IPNI:** 290058-2 **Habit:** Herb **Use:** ME **Origin:** Endemic **Region:** Andean **Dept.:** NAR **Elev.:** 600-600 m.a.s.l. **Conservation status: (IUCN)** NE **(CPSG)** Potential LC

578 *Peperomia elsana* Trel. & Yunck. (Piperaceae) **IPNI:** 678527-1 **Habit:** Herb **Use:** ME **Origin:** Endemic **Region:** Andean **Dept.:** CHO, VAC **Elev.:** 975–975 m a.s.l. **Conservation status: (IUCN)** NE

579 *Peperomia emarginella* (Sw. ex Wikstr.) C.DC. (Piperaceae) **IPNI:** 678531-1 **Habit:** Herb, Epiphyte **Use:** ME **Origin:** Native **Region:** Amazonia, Andean, Guiana Shield, Pacific **Dept.:** ANT, CAQ, CAU, CHO, GUV, MAG, MET, NAR, NSA, PUT, SAN, VAC, VAU **Elev.:** 0-2000 m.a.s.l. **IUCN Habitat code:** 1,3,5,14 **Conservation status: (IUCN)** NE **(CPSG)** Potential LC

580 *Peperomia emarginulata* C.DC. (Piperaceae) **IPNI:** 678532-1 **Use:** EU **Origin:** Native **Dept.:** BOG **IUCN Habitat code:** 1 **Conservation status: (IUCN)** NE

581 *Peperomia fraseri* C.DC. (Piperaceae) **IPNI:** 678641-1 **Use:** EU, HF, SU **Origin:** Native **IUCN Habitat code:** 1 **Conservation status: (IUCN)** NE

582 *Peperomia galioides* Kunth (Piperaceae) **IPNI:** 678659-1 **Common name:** andaquí de loma, canelón, cuyanguilla, cuyanguillo, siempreviva **Habit:** Herb **Use:** ME **Origin:** Native **Region:** Andean **Dept.:** ANT, BOG, BOY, CAU, CES, CUN, HUI, MAG, MET, NAR, SAN, VAC **Elev.:** 1300-3700 m.a.s.l. **IUCN Habitat code:** 1,3,4,14 **Conservation status: (IUCN)** NE **(CPSG)** Potential LC

583 *Peperomia garcia-barrigana* Trel. & Yunck. (Piperaceae) **IPNI:** 678661-1 **Common name:** siempreviva **Habit:** Herb **Use:** ME **Origin:** Endemic **Region:** Andean **Dept.:** CUN **Elev.:** 1500–1600 m a.s.l. **Conservation status: (IUCN)** NE

584 *Peperomia glabella* (Sw.) A.Dietr. (Piperaceae) **IPNI:** 30001267-2 **Habit:** Herb, Epiphyte **Use:** ME **Origin:** Native **Region:** Amazonia, Andean, Orinoquia, Pacific **Dept.:** AMA, ANT, BOG, CAQ, CAU, CHO, GUV, MAG, MET, PUT, RIS, SAN, VAC **Elev.:** 10-2500 m.a.s.l. **Conservation status: (IUCN)** NE **(CPSG)** Potential LC

585 *Peperomia ilaloensis* Sodiro (Piperaceae)
IPNI: 678842-1
Common name: congona perfumada **Habit:** Herb **Use:** ME **Origin:** Native **Region:** Andean **Dept.:** BOG, BOY, CAL, CAU, CUN, LAG, PUT, QUI, RIS, SAN **Elev.:** 1600–3000 m a.s.l. **IUCN Habitat code:** 1,3,14 **Conservation status: (IUCN)** NE

586 *Peperomia inaequalifolia* Ruiz & Pav. (Piperaceae)
IPNI: 678848-1
Common name: canelón, congona **Habit:** Herb **Use:** ME **Origin:** Native, Cultivated **Region:** Andean **Dept.:** ANT, BOY, CES, CUN, RIS **Elev.:** 1700–2500 m a.s.l. **IUCN Habitat code:** 1,14 **Conservation status: (IUCN)** NE

587 *Peperomia macrostachyos* (Vahl) A.Dietr. (Piperaceae)
IPNI: 30130843-2
Common name: curcuncha, golocera, quereme, taracua **Habit:** Herb, Epiphyte **Use:** ME, SU **Origin:** Native **Region:** Amazonia, Andean, Pacific **Dept.:** AMA, ANT, CAQ, CAU, CHO, GUA, GUV, MET, PUT, SAN, VAC, VAU, VID **Elev.:** 0-830 m.a.s.l. **IUCN Habitat code:** 1,3,4,5,14 **Conservation status: (IUCN)** NE **(CPSG)** Potential LC

588 *Peperomia macrotricha* C.DC. (Piperaceae)
IPNI: 679042-1
Common name: quereme contra **Habit:** Herb **Use:** SU **Origin:** Endemic **Region:** Andean, Pacific **Dept.:** ANT **Elev.:** 0–1500 m a.s.l. **IUCN Habitat code:** 1,3,14 **Conservation status: (IUCN)** NE

589 *Peperomia maculosa* (L.) Hook. (Piperaceae)
IPNI: 679043-1
Common name: cordoncillo, pimiento, pipilongo **Habit:** Herb, Shrub **Use:** HF **Origin:** Native, Cultivated **Region:** Amazonia, Andean, Orinoquia, Pacific **Dept.:** ANT, BOY, CAQ, CAS, CHO, CUN, GUV, HUI, SAN, TOL, VAC **Elev.:** 250–1920 m a.s.l. **IUCN Habitat code:** 1,3,14 **Conservation status: (IUCN)** NE

590 *Peperomia magnoliifolia* (Jacq.) A.Dietr. (Piperaceae)
IPNI: 679047-1
Habit: Herb, Epiphyte **Use:** MA, ME, SU **Origin:** Native **Region:** Andean, Caribbean, Pacific **Dept.:** ANT, CHO, GUV, LAG, MAG, VAC **Elev.:** 0–1400 m a.s.l. **IUCN Habitat code:** 1,2,3,5,14 **Conservation status: (IUCN)** NE

591 *Peperomia martiana* Miq. (Piperaceae)
IPNI: 679080-1
Habit: Herb **Use:** ME **Origin:** Native **Region:** Andean **Dept.:** ANT, VAC **Elev.:** 760–760 m a.s.l. **IUCN Habitat code:** 3 **Conservation status: (IUCN)** NE

592 *Peperomia maypurensis* Kunth (Piperaceae)
IPNI: 679095-1
Habit: Herb, Epiphyte **Use:** EU **Origin:** Native **Region:** Orinoquia **Dept.:** VID **Elev.:** 80–80 m a.s.l. **IUCN Habitat code:** 3 **Conservation status: (IUCN)** NE

593 *Peperomia nitida* Dahlst. (Piperaceae)
IPNI: 189209-2
Habit: Herb, Epiphyte **Use:** SU **Origin:** Cultivated **Region:** Andean **Dept.:** ANT **Elev.:** 1470–2000 m a.s.l. **Conservation status: (IUCN)** NE

594 *Peperomia obtusifolia* (L.) A.Dietr. (Piperaceae)
IPNI: 679269-1
Common name: hoja de piedra, hoja de tunda **Habit:** Herb, Epiphyte **Use:** EU, HF, ME, SU **Origin:** Native, Cultivated **Region:** Amazonia, Andean, Guiana Shield **Dept.:** ANT, BOG, CAL, CAQ, CAU, CES, CHO, CUN, GUV, HUI, MAG, RIS, SAN, TOL, VAC, VAU **Elev.:** 0-1920 m.a.s.l. **IUCN Habitat code:** 1,2,3,5,14 **Conservation status: (IUCN)** NE **(CPSG)** Potential LC

595 *Peperomia pellucida* (L.) Kunth (Piperaceae)
IPNI: 679378-1
Common name: celedonia, doncella, neralla, palitaria, sacha yuyu, shushucu, yerba de jabotí **Habit:** Herb **Use:** EU, GS, HF, MA, ME, SU **Origin:** Native, Cultivated **Region:** Amazonia, Andean, Guiana Shield, Pacific **Dept.:** AMA, ANT, ATL, BOL, CAQ, CAS, CAU, CES, CHO, CUN, GUA, GUV, MAG, MET, PUT, RIS, SAN, SUC, VAC **Elev.:** 0-3500 m.a.s.l. **IUCN Habitat code:** 1,2,3,4,5,14 **Conservation status: (IUCN)** NE **(CPSG)** Potential LC

596 *Peperomia pereskiifolia* (Jacq.) Kunth (Piperaceae)
IPNI: 679404-1
Habit: Herb, Epiphyte **Use:** EU, HF **Origin:** Native **Region:** Andean **Dept.:** ANT, BOG, BOY, CAU, CUN, HUI, RIS, SAN, VAC **Elev.:** 350–2350 m a.s.l. **IUCN Habitat code:** 1,3,14 **Conservation status: (IUCN)** NE

597 *Peperomia putumayoensis* Trel. & Yunck. (Piperaceae)
IPNI: 679587-1
Common name: payaco, payaco-vinan, vinan **Habit:** Herb **Use:** ME **Origin:** Endemic **Region:** Amazonia **Dept.:** CAU, PUT **Elev.:** 100–250 m a.s.l. **IUCN Habitat code:** 1,3 **Conservation status: (IUCN)** NE

598 *Peperomia quadrangularis* (J.V.Thomps.) A.Dietr. (Piperaceae)
IPNI: 679591-1
Habit: Herb **Use:** MA **Origin:** Native **Region:** Amazonia, Andean, Guiana Shield, Caribbean, Orinoquia **Dept.:** ANT, CAQ, CAS, CES, MAG, MET **Elev.:** 400-2000 m.a.s.l. **IUCN Habitat code:** 1,3,4,14 **Conservation status: (IUCN)** NE **(CPSG)** Potential LC

599 *Peperomia quadrifolia* (L.) Kunth (Piperaceae)
IPNI: 679597-1
Common name: clavo de huerta **Habit:** Herb **Use:** HF **Origin:** Native **Region:** Andean **Dept.:** ANT, BOY, CAU, CES, CUN, LAG, NSA, SAN, VAC **Elev.:** 1690–3500 m a.s.l. **IUCN Habitat code:** 1,2,3,4,14 **Conservation status: (IUCN)** NE

600 *Peperomia rotundata* Kunth (Piperaceae)
IPNI: 679705-1
Habit: Herb **Use:** ME **Origin:** Native **Region:** Andean **Dept.:** ANT, BOG, BOY, CAU, CHO, CUN, MAG, MET, NAR, PUT, QUI, SAN, TOL, VAC **Elev.:** 740-3500 m.a.s.l. **IUCN Habitat code:** 1,3,4,8,14 **Conservation status: (IUCN)** NE **(CPSG)** Potential LC

601 *Peperomia rotundifolia* (L.) Kunth (Piperaceae)
IPNI: 189588-2
Common name: berrito caminante, congona arisca, cuartillito, cuartillo, perejil, platillo, pusana, queredora, siempreviva, yerba del cuartillo **Habit:** Herb, Epiphyte **Use:** HF, MA, ME, PO, SU **Origin:** Native **Region:** Amazonia, Andean, Orinoquia, Pacific **Dept.:** AMA, ANT, BOY, CAQ, CAU, CHO, CUN, GUV, HUI, MAG, MET, NAR, NSA, PUT, SAN, VAC, VAU **Elev.:** 0-1190 m.a.s.l. **IUCN Habitat code:** 1,2,3,5,14 **Conservation status: (IUCN)** NE **(CPSG)** Potential LC

602 *Peperomia serpens* (Sw.) G.Don (Piperaceae)
IPNI: 679841-1
Habit: Herb, Epiphyte **Use:** MA, ME **Origin:** Native **Region:** Amazonia, Andean, Guiana Shield, Pacific **Dept.:** AMA, ANT, CAQ, CAU, CHO, GUV, MET, NAR, NSA, PUT, SAN, VAC **Elev.:** 0–1000 m a.s.l. **IUCN Habitat code:** 1,3,14 **Conservation status: (IUCN)** NE

603 *Peperomia striata* Ruiz & Pav. (Piperaceae)
IPNI: 679909-1
Habit: Herb, Epiphyte **Use:** ME **Origin:** Native **Region:** Andean, Pacific **Dept.:** ANT, CHO, COR, CUN, HUI, NAR, NSA, PUT, SAN, VAC **Elev.:** 200–2425 m a.s.l. **IUCN Habitat code:** 1,3,14 **Conservation status: (IUCN)** NE

604 *Peperomia subspathulata* Yunck. (Piperaceae)
IPNI: 679961-1
Common name: canelo, canelón, clavo, congona, siempreviva **Habit:** Herb **Use:** ME **Origin:** Native **Region:** Andean **Dept.:** BOG, CAU **Elev.:** 3400–3450 m a.s.l. **IUCN Habitat code:** 3,14 **Conservation status: (IUCN)** NE

605 *Peperomia tenuipes* Trel. (Piperaceae)
IPNI: 189825-2
Habit: Herb **Use:** ME **Origin:** Native **Region:** Andean **Dept.:** ANT, CAL, CAU, NSA, PUT, RIS, SAN, TOL, VAC **Elev.:** 1210–2500 m a.s.l. **IUCN Habitat code:** 1,3,14 **Conservation status: (IUCN)** NE

606 *Peperomia tetraphylla* (G.Forst.) Hook. & Arn. (Piperaceae)
IPNI: 1068802-2
Habit: Herb, Epiphyte **Use:** EU, ME **Origin:** Native **Region:** Andean **Dept.:** ANT, BOG, CAU, CES, CHO, CUN, MAG, MET, NSA, PUT, QUI, RIS, SAN, VAC **Elev.:** 1200-3000 m.a.s.l. **IUCN Habitat code:** 1,2,3,4,14 **Conservation status: (IUCN)** NE **(CPSG)** Potential LC

607 *Peperomia trifolia* (L.) A.Dietr. (Piperaceae)
IPNI: 1090908-2
Use: ME **Conservation status: (IUCN)** NE

608 *Peperomia tristachya* Kunth (Piperaceae)
IPNI: 680084-1
Common name: corazón de hombre, cuartillo, moneda **Habit:** Herb **Use:** SU **Origin:** Native **Region:** Andean **Dept.:** ANT, CHO, CUN **Elev.:** 1341–1780 m a.s.l. **IUCN Habitat code:** 1,14 **Conservation status: (IUCN)** NE

609 *Peperomia tutunendoana* Trel. & Yunck. (Piperaceae)
IPNI: 680114-1
Habit: Herb, Epiphyte **Use:** EU, ME **Origin:** Native **Region:** Andean, Pacific **Dept.:** ANT, CHO, CUN **Elev.:** 150–1300 m a.s.l. **IUCN Habitat code:** 1 **Conservation status: (IUCN)** NE

610 *Peperomia velutina* Linden & André (Piperaceae)
IPNI: 680160-1
Habit: Herb **Use:** EU **Origin:** Native **Region:** Andean **Dept.:** MAG **Elev.:** 780–780 m a.s.l. **IUCN Habitat code:** 2,14 **Conservation status: (IUCN)** NE

611 *Piper aduncum* L. (Piperaceae)
IPNI: 680296-1
Common name: anisillo, chavica, cordoncillo, cordoncillo blanco, herb de canotillo, herb del soldado, matico, monte del soldado, nudillo, pipilongo, sabanagrande, tapa hueco **Habit:** Shrub **Use:** AF, EU, FU, HF, MA, ME, PO **Origin:** Native **Region:** Amazonia, Andean, Guiana Shield, Pacific **Dept.:** ANT, BOY, CAQ, CAS, CAU, CHO, COR, CUN, GUV, HUI, LAG, MAG, MET, NSA, PUT, QUI, RIS, SAN, TOL, VAC, VAU **Elev.:** 0–2600 m a.s.l. **IUCN Habitat code:** 1,2,3,4,5,8,14 **Conservation status: (IUCN)** LC

612 *Piper aequale* Vahl (Piperaceae)
IPNI: 680300-1
Common name: cordoncillo **Habit:** Shrub **Use:** ME **Origin:** Native **Region:** Amazonia, Andean, Orinoquia **Dept.:** AMA, ANT, BOY, CAL, CAQ, CAS, CAU, CHO, CUN, HUI, MAG, MET, PUT, QUI, RIS, SAN, TOL, VAC **Elev.:** 0-2400 m.a.s.l. **IUCN Habitat code:** 1,2,3,4,5,14 **Conservation status: (IUCN)** LC **(CPSG)** Potential LC

613 *Piper amalago* L. (Piperaceae)
IPNI: 680359-1
Common name: cordoncillo, cordoncillo de hoja ancha, cordoncillo grande, deshinchadora, rollera **Habit:** Shrub, Tree **Use:** HF, MA, ME, PO, SU **Origin:** Native **Region:** Andean, Caribbean **Dept.:** ANT, ATL, BOL, CAL, CUN, LAG, MAG, NSA, SAP, SAN, VAC **Elev.:** 0–1250 m a.s.l. **IUCN Habitat code:** 1,2,3,14 **Conservation status: (IUCN)** LC

614 *Piper anonifolium* (Kunth) Steud. (Piperaceae)
IPNI: 1003227-1
Common name: cordoncillo **Habit:** Shrub **Use:** MA **Origin:** Native **Region:** Guiana Shield **Dept.:** CAQ, MET **Elev.:** 250–740 m a.s.l. **IUCN Habitat code:** 1,3,14 **Conservation status: (IUCN)** NE

615 *Piper arboreum* Aubl. (Piperaceae)
IPNI: 680435-1
Common name: cordoncillo, desvanecedora, pipilongo **Habit:** Shrub, Small tree, Tree **Use:** AF, HF, MA, ME, PO, SU **Origin:** Native **Region:** Amazonia, Andean, Guiana Shield, Caribbean, Pacific **Dept.:** AMA, ANT, CAQ, CAS, CAU, CHO, CUN, GUA, GUV, LAG, MAG, MET, NSA, PUT, SAN, SUC, VAU, VID **Elev.:** 0-1700 m.a.s.l. **IUCN Habitat code:** 1,2,3,4,5,14 **Conservation status: (IUCN)** NE **(CPSG)** Potential LC

616 *Piper arieianum* C.DC. (Piperaceae)
IPNI: 197302-2
Use: ME, SU **IUCN Habitat code:** 1,3,5,14 **Conservation status: (IUCN)** NE

617 *Piper augustum* Rudge (Piperaceae)
IPNI: 680507-1
Common name: cordoncillo, hoja de vaso, teñir de negro **Habit:** Shrub **Use:** ME **Origin:** Native **Region:** Amazonia, Andean, Pacific **Dept.:** AMA, ANT, CAQ, CAU, CHO, GUV, MET, NAR, PUT, RIS, VAC **Elev.:** 0–1600 m a.s.l. **IUCN Habitat code:** 1,3,4,5,14 **Conservation status: (IUCN)** LC

618 *Piper auritum* Kunth (Piperaceae)
IPNI: 680520-1
Common name: anisillo, canelón, hoja de Santa María, hoja de la estrella, santa maría, santamarianí, santamarianís, santamaría, santamaría de anís, señora maría anís **Habit:** Herb, Shrub **Use:** AF, HF, MA, ME, PO, SU **Origin:** Native **Region:** Andean, Pacific **Dept.:** ANT, BOL, CAL, CAU, CHO, CUN, NAR, QUI, RIS, SAN, VAC **Elev.:** 0-1500 m.a.s.l. **IUCN Habitat code:** 1,2,3,4,5,14 **Conservation status: (IUCN)** LC **(CPSG)** Potential LC

619 *Piper barbatum* Kunth (Piperaceae)
IPNI: 680559-1
Common name: cordoncillo, pipilongo, pipote **Habit:** Shrub **Use:** EU, HF, ME, PO **Origin:** Native **Region:** Andean **Dept.:** BOG, CAU, CUN, NAR, NSA, PUT, SAN, VAC **Elev.:** 200–3500 m a.s.l. **IUCN Habitat code:** 1,3,4,14 **Conservation status: (IUCN)** NE

620 *Piper bartlingianum* (Miq.) C.DC. (Piperaceae)
IPNI: 680574-1
Common name: santamaría **Habit:** Shrub **Use:** MA **Origin:** Native **Region:** Guiana Shield **Dept.:** AMA, GUA, GUV, VAU **Elev.:** 75–630 m a.s.l. **IUCN Habitat code:** 1,14 **Conservation status: (IUCN)** NE

621 *Piper brachypodon* (Benth.) C.DC. (Piperaceae)
IPNI: 680665-1
Habit: Shrub, Climbing **Use:** ME **Origin:** Native **Region:** Pacific **Dept.:** ANT, CAU, CHO, NAR, VAC **Elev.:** 0–500 m a.s.l. **IUCN Habitat code:** 1,3,4,5,14 **Conservation status: (IUCN)** NE

622 *Piper bredemeyeri* J.Jacq. (Piperaceae)
IPNI: 680680-1
Common name: anisillo, cordoncillo **Habit:** Shrub **Use:** ME **Origin:** Native **Region:** Andean, Caribbean, Pacific **Dept.:** ANT, CAL, CAU, CHO, LAG, MAG, MET, SAN, VAC **Elev.:** 40–1800 m a.s.l. **IUCN Habitat code:** 1,2,3,4,14 **Conservation status: (IUCN)** NE

623 *Piper calceolarium* C.DC. (Piperaceae)
IPNI: 680737-1
Common name: cordoncillo, desvanecedora **Habit:** Herb, Shrub **Use:** EU, FU, ME **Origin:** Native **Region:** Andean **Dept.:** ANT, CAU, CHO, CUN, QUI, RIS, VAC **Elev.:** 1570–2600 m a.s.l. **IUCN Habitat code:** 1,3,14 **Conservation status: (IUCN)** NE

624 *Piper callosum* Ruiz & Pav. (Piperaceae)
IPNI: 680748-1
Habit: Herb, Epiphyte **Use:** HF, MA, ME **Origin:** Native **Region:** Pacific **Dept.:** CHO, COR **Elev.:** 50–500 m a.s.l. **IUCN Habitat code:** 1 **Conservation status: (IUCN)** NE

625 *Piper carpunya* Ruiz & Pav. (Piperaceae)
IPNI: 680810-1
Habit: Shrub, Climbing **Use:** HF, ME **Origin:** Native **Region:** Andean **Dept.:** ANT, NAR **Elev.:** 900–2380 m a.s.l. **IUCN Habitat code:** 1,3,14 **Conservation status: (IUCN)** NE

626 *Piper chavicoides* (Miq.) C.DC. (Piperaceae)
IPNI: 680871-1
Habit: Climbing **Use:** ME **Origin:** Native **Region:** Amazonia **Dept.:** AMA **Elev.:** 140–140 m a.s.l. **IUCN Habitat code:** 5 **Conservation status: (IUCN)** NE

627 *Piper chiadoense* Yunck. (Piperaceae)
IPNI: 680875-1
Habit: Shrub, Climbing **Use:** ME, PO **Origin:** Endemic **Region:** Pacific **Dept.:** ANT **Elev.:** 20–350 m a.s.l. **IUCN Habitat code:** 1,3,5,14 **Conservation status: (IUCN)** NE

628 *Piper cililimbum* Yunck. (Piperaceae)
IPNI: 197589-2
Common name: alacrán, cordoncillo, lengua de picón, tabaco de alacrán **Habit:** Shrub **Use:** ME **Origin:** Native **Region:** Amazonia, Guiana Shield **Dept.:** AMA, CAQ, PUT **Elev.:** 80–500 m a.s.l. **IUCN Habitat code:** 1,3,14 **Conservation status: (IUCN)** NE

629 *Piper confertinodum* (Trel. & Yunck.) M.A.Jaram. & Callejas (Piperaceae)
IPNI: 77061871-1
Common name: yerba de culebra **Habit:** Subshrub **Use:** ME **Origin:** Native **Region:** Pacific **Dept.:** CHO, VAC **Elev.:** 0–275 m a.s.l. **IUCN Habitat code:** 1,3,4,14 **Conservation status: (IUCN)** NE

630 *Piper consanguineum* (Kunth) Steud. (Piperaceae)
IPNI: 680988-1
Common name: cordoncillo **Habit:** Shrub **Use:** EU, ME **Origin:** Native **Region:** Amazonia, Caribbean **Dept.:** AMA, CAQ, MAG **Elev.:** 80–250 m a.s.l. **IUCN Habitat code:** 1,2,3,5,14 **Conservation status: (IUCN)** NE

631 *Piper contraverrugosum* (Cuatrec.) R.Bernal (Piperaceae)
IPNI: 77146701-1
Common name: contraverrugosa, herb de la verrugosa, yerba de la verrugosa **Habit:** Subshrub **Use:** ME **Origin:** Endemic **Region:** Pacific **Dept.:** CHO, COR, VAC **Elev.:** 0–600 m a.s.l. **Conservation status: (IUCN)** NE

632 *Piper coruscans* Kunth (Piperaceae)
IPNI: 681027-1
Common name: cordoncillo, danda kidai **Habit:** Shrub **Use:** ME, PO **Origin:** Native **Region:** Amazonia, Andean, Pacific **Dept.:** AMA, ANT, CAQ, CAU, CHO, GUV, MET, PUT, SAN, VAC **Elev.:** 0-1450 m.a.s.l. **IUCN Habitat code:** 1,3,4,5,14 **Conservation status: (IUCN)** NE **(CPSG)** Potential LC

633 *Piper crassinervium* Kunth (Piperaceae)
IPNI: 681041-1
Common name: cordoncillo, cordoncillo de hoja pequeña **Habit:** Shrub **Use:** AF, EU, MA, ME, PO **Origin:** Native **Region:** Amazonia, Andean, Pacific **Dept.:** ANT, CAQ, CAU, CHO, HUI, MET, NAR, NSA, PUT, QUI, RIS, SAN, TOL, VAC **Elev.:** 100-2420 m.a.s.l. **IUCN Habitat code:** 1,2,3,5,14 **Conservation status: (IUCN)** LC **(CPSG)** Potential LC

634 *Piper cuspidilimbum* C.DC. (Piperaceae)
IPNI: 197721-2
Common name: malaire kirú **Habit:** Shrub **Use:** ME **Origin:** Native

Region: Pacific **Dept.:** NAR **Elev.:** 100–100 m a.s.l. **Conservation status: (IUCN)** NE

635 *Piper darienense* C.DC. (Piperaceae) **IPNI:** 681128-1 **Common name:** cualecuina **Habit:** Herb, Shrub **Use:** ME **Origin:** Native **Region:** Caribbean, Pacific **Dept.:** ANT, BOL, CHO, MAG **Elev.:** 10–600 m a.s.l. **IUCN Habitat code:** 1,2,3,5,14 **Conservation status: (IUCN)** NE

636 *Piper demeraranum* (Miq.) C.DC. (Piperaceae) **IPNI:** 681157-1 **Common name:** cordoncillo, pintura negra, teñir de negro **Habit:** Herb, Subshrub **Use:** HF, MA **Origin:** Native **Region:** Amazonia, Guiana Shield **Dept.:** AMA, ANT, CAQ, GUA, GUV, MET, SAN, VAU, VID **Elev.:** 100–1200 m a.s.l. **IUCN Habitat code:** 1,3,4,5,14 **Conservation status: (IUCN)** NE

637 *Piper dilatatum* Rich. (Piperaceae) **IPNI:** 681193-1 **Common name:** cordoncillo **Habit:** Shrub **Use:** HF, MA, ME **Origin:** Native **Region:** Amazonia, Andean, Pacific **Dept.:** CAQ, CAS, CHO, GUV, MET, PUT, VAC **Elev.:** 0–1000 m a.s.l. **Conservation status: (IUCN)** NE

638 *Piper divaricatum* G.Mey. (Piperaceae) **IPNI:** 681224-1 **Common name:** cordoncillo, malaire **Habit:** Shrub **Use:** GS, MA, ME **Origin:** Native **Region:** Amazonia **Dept.:** AMA **Elev.:** 80–230 m a.s.l. **IUCN Habitat code:** 1,5,14 **Conservation status: (IUCN)** NE

639 *Piper dumosum* Rudge (Piperaceae) **IPNI:** 681251-1 **Common name:** cordoncillo **Habit:** Shrub, Small tree **Use:** ME **Origin:** Native **Region:** Amazonia **Dept.:** AMA, CAQ, GUV, MET, PUT **Elev.:** 80–250 m a.s.l. **IUCN Habitat code:** 1,3,14 **Conservation status: (IUCN)** NE

640 *Piper ecuadorense* Sodiro (Piperaceae) **IPNI:** 681264-1 **Habit:** Shrub **Use:** ME **Origin:** Native **Region:** Andean **Dept.:** CAU, QUI, TOL **Elev.:** 2100–2750 m a.s.l. **IUCN Habitat code:** 1,3 **Conservation status: (IUCN)** NE

641 *Piper erythroxyloides* R.E.Schult. & García-Barr. (Piperaceae) **IPNI:** 197884-2 **Common name:** achikaira, chankira **Use:** ME, SU **Origin:** Endemic **IUCN Habitat code:** 14 **Conservation status: (IUCN)** NE

642 *Piper grande* Vahl (Piperaceae) **IPNI:** 681559-1 **Habit:** Shrub, Small tree **Use:** ME **Origin:** Native **Region:** Andean, Pacific **Dept.:** ANT, BOL, CAL, CHO, COR, MAG, NSA, PUT, RIS, SAN, TOL, VAC **Elev.:** 10-2140 m.a.s.l. **IUCN Habitat code:** 1,2,3,5,14 **Conservation status: (IUCN)** LC **(CPSG)** Potential LC

643 *Piper heterophyllum* Ruiz & Pav. (Piperaceae) **IPNI:** 681651-1 **Habit:** Shrub **Use:** ME **Origin:** Native **Region:** Amazonia **Dept.:** AMA **Elev.:** 120–120 m a.s.l. **IUCN Habitat code:** 1 **Conservation status: (IUCN)** NE

644 *Piper hispidum* Sw. (Piperaceae) **IPNI:** 681674-1 **Common name:** cordoncillo, cordoncillo blanco **Habit:** Shrub **Use:** EU, HF, MA, ME, PO, SU **Origin:** Native **Region:** Amazonia, Andean, Guiana Shield, Pacific **Dept.:** AMA, ANT, CAQ, CAS, CAU, CHO, CUN, GUV, MAG, MET, NAR, NSA, PUT, RIS, SAN, VAC, VAU **Elev.:** 0-1800 m.a.s.l. **Conservation status: (IUCN)** LC **(CPSG)** Potential LC

645 *Piper holtonii* C.DC. (Piperaceae) **IPNI:** 681685-1 **Common name:** cordoncillo, raicillo **Habit:** Herb, Shrub, Subshrub **Use:** ME **Origin:** Native **Region:** Andean, Pacific **Dept.:** ANT, CAL, CAU, CES, CUN, HUI, RIS, TOL, VAC **Elev.:** 420-2700 m.a.s.l. **IUCN Habitat code:** 1,2,3,5,14 **Conservation status: (IUCN)** NE **(CPSG)** Potential LC

646 *Piper hostmannianum* (Miq.) C.DC. (Piperaceae) **IPNI:** 681692-1 **Habit:** Shrub **Use:** MA, ME **Origin:** Native **Region:** Amazonia, Guiana Shield **Dept.:** AMA, CAQ, PUT, VAU **Elev.:** 80–400 m a.s.l. **IUCN Habitat code:** 1,3,5,14 **Conservation status: (IUCN)** LC

647 *Piper immutatum* Trel. (Piperaceae) **IPNI:** 198173-2 **Habit:** Shrub **Use:** HF **Origin:** Native **Region:** Amazonia, Andean, Pacific **Dept.:** CAQ, CAU, CHO, PUT **Elev.:** 0–1420 m a.s.l. **IUCN Habitat code:** 1,3,14 **Conservation status: (IUCN)** NE

648 *Piper lacunosum* Kunth (Piperaceae) **IPNI:** 681864-1 **Common name:** cordoncillo, desvanecedora **Habit:** Shrub, Small tree **Use:** ME **Origin:** Native **Region:** Andean **Dept.:** ANT, BOG, CAL, CHO, CUN, HUI, NAR, PUT, QUI, VAC **Elev.:** 600–2800 m a.s.l. **IUCN Habitat code:** 1,3,4,14 **Conservation status: (IUCN)** LC

649 *Piper laevigatum* Kunth (Piperaceae) **IPNI:** 681876-1 **Common name:** cordoncillo **Habit:** Shrub **Use:** MA, ME **Origin:** Native **Region:** Amazonia, Orinoquia **Dept.:** AMA, ANT, CAQ, CAS, CUN, GUV, MET, PUT, SAN **Elev.:** 80–300 m a.s.l. **IUCN Habitat code:** 1,4,5,14 **Conservation status: (IUCN)** LC

650 *Piper macerispicum* Trel. & Yunck. (Piperaceae) **IPNI:** 682055-1 **Habit:** Shrub, Climbing **Use:** HF **Origin:** Native **Region:** Amazonia, Guiana Shield **Dept.:** AMA, CAQ, MET, PUT, VAU **Elev.:** 110–400 m a.s.l. **IUCN Habitat code:** 1,3,5,14 **Conservation status: (IUCN)** NE

651 *Piper maranyonense* Trel. (Piperaceae) **IPNI:** 198388-2 **Common name:** cordoncillo, nenakirú **Habit:** Shrub **Use:** HF, ME **Origin:** Native **Region:** Amazonia, Andean, Pacific **Dept.:** AMA, CAQ, CHO, MET, PUT **Elev.:** 200–600 m a.s.l. **IUCN Habitat code:** 1,3,14 **Conservation status: (IUCN)** NE

652 *Piper marginatum* Jacq. (Piperaceae) **IPNI:** 682138-1 **Common name:** Santa María de anís, aguardientillo, anisillo, anís, corazón de la Virgen, cordoncillo, cordoncillo blanco, cordoncillo negro, curadientes, deshinchadora, hoja de raya, santamaría, santamaría negra **Habit:** Shrub, Small tree **Use:** HF, MA, ME, PO **Origin:** Native **Region:** Amazonia, Andean, Guiana Shield, Caribbean, Orinoquia, Pacific **Dept.:** ANT, ARA, ATL, BOL, CAL, CAQ, CAS, CHO, COR, CUN, GUA, GUV, HUI, LAG, MAG, MET, NAR, NSA, SAN, SUC, TOL, VAU, VID **Elev.:** 0–2350 m a.s.l. **IUCN Habitat code:** 1,2,3,4,5,14 **Conservation status: (IUCN)** NE

653 *Piper marsupiiferum* Trel. (Piperaceae) **IPNI:** 198403-2 **Habit:** Herb **Use:** ME **Origin:** Native **Region:** Amazonia **Dept.:** PUT **Elev.:** 400–400 m a.s.l. **IUCN Habitat code:** 1,3,5 **Conservation status: (IUCN)** NE

654 *Piper multiplinervium* C.DC. (Piperaceae) **IPNI:** 682301-1 **Common name:** bejuco de cáncer, cordoncillo, cordoncillo de bejuco **Habit:** Liana, Climbing **Use:** ME, PO **Origin:** Native **Region:** Andean, Pacific **Dept.:** ANT, BOL, CHO, NAR, SAN, VAC **Elev.:** 0–900 m a.s.l. **IUCN Habitat code:** 1,3,5,14 **Conservation status: (IUCN)** NE

655 *Piper munchanum* C.DC. (Piperaceae) **IPNI:** 682305-1 **Common name:** chicopia chiguave **Habit:** Shrub, Small tree **Use:** ME, SU **Origin:** Native **Region:** Andean, Pacific **Dept.:** ANT, BOY, CHO, LAG, MAG, NSA, SAN, VAC **Elev.:** 0–2240 m a.s.l. **IUCN Habitat code:** 1,2,3,4,14 **Conservation status: (IUCN)** NE

656 *Piper nigrispicum* C.DC. (Piperaceae) **IPNI:** 198546-2 **Habit:** Shrub **Use:** MA **Origin:** Native **Region:** Amazonia **Dept.:** CAQ **Elev.:** 250–250 m a.s.l. **IUCN Habitat code:** 1,3,4 **Conservation status: (IUCN)** NE

657 *Piper nigrum* L. (Piperaceae) **IPNI:** 682369-1 **Common name:** pimentero, pimienta, pimienta blanca, pimienta de Castilla, pimienta negra, pimienta picante **Habit:** Climbing **Use:** AF, EU, GS, HF, MA, ME, PO **Origin:** Cultivated **Region:** Andean **Dept.:** ANT, TOL **Elev.:** 500–1500 m a.s.l. **IUCN Habitat code:** 1 **Conservation status: (IUCN)** NE

658 *Piper nubigenum* (Kunth) Kunth ex Steud. (Piperaceae) **IPNI:** 682399-1 **Habit:** Shrub **Use:** AF,ME **Origin:** Native **Region:** Andean **Dept.:** BOY, CUN, NAR, TOL, VAC **Elev.:** 2500–3170 m a.s.l. **IUCN Habitat code:** 1,3,4,14 **Conservation status: (IUCN)** NE

659 *Piper obliquum* Ruiz & Pav. (Piperaceae) **IPNI:** 682429-1 **Common name:** cordoncillo, cordoncillo hojancha, desvanecedera, rodilla de pollo, santamaría **Habit:** Shrub **Use:** ME **Origin:** Native **Region:** Amazonia, Andean, Orinoquia, Pacific **Dept.:** AMA, ANT, CAQ, CAS, CAU, CHO, CUN, GUV, MET, NSA, PUT, QUI, SAN, VAC, VAU **Elev.:** 80–3000 m a.s.l. **IUCN**

Habitat code: 1,2,3,4,5,14 **Conservation status: (IUCN)** NE

660 *Piper obtusilimbum* C.DC. (Piperaceae) **IPNI:** 198599-2 **Common name:** airon, cordoncillo, desvanecedera, desvanecedora, dolor de muelas, hierba maldita, hoja para la viruela **Habit:** Herb **Use:** ME **Origin:** Native **Region:** Amazonia, Guiana Shield **Dept.:** CAQ, GUV, PUT, VAU **Elev.:** 100–630 m a.s.l. **IUCN Habitat code:** 1,3,14 **Conservation status: (IUCN)** LC

661 *Piper peltatum* L. (Piperaceae) **IPNI:** 316334-2 **Common name:** Santa María, anisillo, contraverrugosa, cordoncillo, cordoncillo rayo, curillo, hoja de raya, hoja de rayo, indio viejo, pato, platillo, plaza, raya, rayo, santa maría boba, santamaría, santamaría boba, señora maría boba, tapoca **Habit:** Herb, Shrub, Subshrub **Use:** EU, HF, MA, ME, PO **Origin:** Native **Region:** Amazonia, Andean, Guiana Shield, Caribbean, Pacific **Dept.:** AMA, ANT, BOL, CAQ, CAU, CHO, COR, CUN, GUV, LAG, MAG, MET, NAR, NSA, PUT, RIS, SAN, SUC, TOL, VAC, VAU **Elev.:** 0-2200 m.a.s.l. **IUCN Habitat code:** 1,2,3,4,5,14 **Conservation status: (IUCN)** NE **(CPSG)** Potential LC

662 *Piper pseudofuligineum* C.DC. (Piperaceae) **IPNI:** 682899-1 **Habit:** Shrub **Use:** ME **Origin:** Native **Region:** Andean, Caribbean, Pacific **Dept.:** CHO, CUN, HUI, MAG, NSA, SAN, VAC **Elev.:** 0–1800 m a.s.l. **IUCN Habitat code:** 1,2,3,5,14 **Conservation status: (IUCN)** LC

663 *Piper pubiovarium* Yunck. (Piperaceae) **IPNI:** 198926-2 **Use:** ME **IUCN Habitat code:** 1,2,3,14 **Conservation status: (IUCN)** NE

664 *Piper pulchrum* C.DC. (Piperaceae) **IPNI:** 682973-1 **Common name:** tree de la verrugosa, árbol de la verrugosa **Habit:** Shrub, Tree **Use:** ME, PO **Origin:** Native **Region:** Andean **Dept.:** ANT, CAU, HUI, NAR, VAC **Elev.:** 500-2440 m.a.s.l. **IUCN Habitat code:** 1,3,5,14 **Conservation status: (IUCN)** LC **(CPSG)** LC

665 *Piper reticulatum* L. (Piperaceae) **IPNI:** 198988-2 **Habit:** Shrub, Tree **Use:** MA, ME **Origin:** Native **Region:** Amazonia, Andean, Caribbean, Pacific **Dept.:** AMA, ANT, BOL, BOY, CAL, CAU, CHO, LAG, MAG, MET, NSA, SUC, TOL, VAC **Elev.:** 0-1150 m.a.s.l. **IUCN Habitat code:** 1,2,3,5,14 **Conservation status: (IUCN)** LC **(CPSG)** Potential LC

666 *Piper schiedeanum* Steud. (Piperaceae) **IPNI:** 683316-1 **Common name:** cordoncillo **Habit:** Shrub, Small tree **Use:** MA, ME **Origin:** Native **Region:** Andean, Pacific **Dept.:** ANT, BOY, CAL, CHO, MET, VAC **Elev.:** 100–1500 m a.s.l. **IUCN Habitat code:** 2 **Conservation status: (IUCN)** LC

667 *Piper soledadense* Trel. (Piperaceae) **IPNI:** 199210-2 **Habit:** Shrub **Use:** ME **Origin:** Native **Region:** Amazonia, Guiana Shield, Pacific **Dept.:** AMA, CAQ, GUV, PUT, VAU **Elev.:** 100–250 m a.s.l. **IUCN Habitat code:** 1,3,5,14 **Conservation status: (IUCN)** NE

668 *Piper sphaeroides* C.DC. (Piperaceae) **IPNI:** 199219-2 **Common name:** cordoncillo **Habit:** Herb, Shrub **Use:** ME **Origin:** Endemic **Region:** Andean **Dept.:** ANT, CHO, NAR, TOL **Elev.:** 800–2330 m a.s.l. **IUCN Habitat code:** 1,3,14 **Conservation status: (IUCN)** NE

669 *Piper sternii* Yunck. (Piperaceae) **IPNI:** 199236-2 **Habit:** Subshrub **Use:** ME **Origin:** Native **Region:** Andean, Pacific **Dept.:** ANT, CHO **Elev.:** 20–1000 m a.s.l. **IUCN Habitat code:** 1,3,5,14 **Conservation status: (IUCN)** NE

670 *Piper terryae* Standl. (Piperaceae) **IPNI:** 199390-2 **Common name:** yerba de la verrugosa **Habit:** Herb **Use:** ME **Origin:** Native **Region:** Andean **Dept.:** ANT, CHO **Elev.:** 600–1360 m a.s.l. **IUCN Habitat code:** 1,3,14 **Conservation status: (IUCN)** NE

671 *Piper trianae* C.DC. (Piperaceae) **IPNI:** 683758-1 **Common name:** hoja redonda **Habit:** Subshrub **Use:** ME **Origin:** Native **Region:** Andean **Dept.:** CAU, NAR **Elev.:** 1400–2500 m a.s.l. **IUCN Habitat code:** 1,4,5,14 **Conservation status: (IUCN)** NE

672 *Piper tricuspe* (Miq.) C.DC. (Piperaceae) **IPNI:** 683774-1 **Common name:** costeña, costeño, hoja de mano, hoja de tresdedos, hoja mano, hojamano, sueldoconsueldo, tres dedos, tresdedos **Habit:** Shrub **Use:** ME **Origin:** Native, Cultivated **Region:** Pacific **Dept.:** CAU, CHO, NAR, VAC **Elev.:** 0–215 m a.s.l. **IUCN Habitat code:** 1,3,14 **Conservation status: (IUCN)** NE

673 *Piper trigonum* C.DC. (Piperaceae) **IPNI:** 683782-1 **Common name:** cordoncillo, curachimici, desinflamador, lengua de pichón, malaire kirú, santamaría **Habit:** Herb, Shrub **Use:** ME **Origin:** Native **Region:** Amazonia, Andean, Pacific **Dept.:** AMA, ANT, CAQ, CAU, CHO, GUV, MET, NAR, NSA, SAN, VAC, VAU **Elev.:** 0-718 m.a.s.l. **IUCN Habitat code:** 1,3,5,14 **Conservation status: (IUCN)** NE **(CPSG)** Potential LC

674 *Piper tuberculatum* Jacq. (Piperaceae) **IPNI:** 683806-1 **Common name:** cordoncillo, cordoncillo chinibo, santamaría **Habit:** Shrub, Small tree **Use:** EU, HF, MA, ME, PO, SU **Origin:** Native **Region:** Amazonia, Andean, Caribbean, Orinoquia, Pacific **Dept.:** AMA, ANT, ATL, BOL, BOY, CAQ, CAS, CAU, CES, CHO, COR, CUN, GUV, HUI, LAG, MAG, MET, NAR, NSA, PUT, SAN, TOL, VAC **Elev.:** 0-2200 m.a.s.l. **Conservation status: (IUCN)** NE **(CPSG)** Potential LC

675 *Piper umbellatum* L. (Piperaceae) **IPNI:** 316339-2 **Common name:** Santa María, anisillo, congo, contraverrugosa, cordoncillo, hoja de rayo, santa maría macho, santamaría, uña de caballo **Habit:** Herb, Shrub **Use:** GS, HF, MA, ME **Origin:** Native **Region:** Amazonia, Andean, Guiana Shield **Dept.:** AMA, ANT, ATL, CAL, CAQ, CAU, CHO, CUN, GUA, MAG, NSA, PUT, SAN, VAC **Elev.:** 0-2200 m.a.s.l. **IUCN Habitat code:** 1,2,3,5,14 **Conservation status: (IUCN)** NE **(CPSG)** Potential LC

676 *Piper vitaceum* Yunck. (Piperaceae) **IPNI:** 199557-2 **Habit:** Hemiepiphyte, Climbing **Use:** MA **Origin:** Native **Region:** Amazonia **Dept.:** AMA, CAQ **Elev.:** 80–120 m a.s.l. **IUCN Habitat code:** 1 **Conservation status: (IUCN)** NE

Liliopsida Alismatales

677 *Alisma plantago-aquatica* L. (Alismataceae) **IPNI:** 30019518-2 **Use:** HF, MA, ME **Conservation status: (IUCN)** LC

678 *Aquarius cordifolius* (L.) Christenh. & Byng (Alismataceae) **IPNI:** 77182733-1 **Habit:** Herb, Aquatic **Use:** EU **Origin:** Native **Region:** Caribbean **Dept.:** LAG, MAG **Elev.:** 50–50 m a.s.l. **Conservation status: (IUCN)** NE

679 *Aquarius floribundus* (Seub.) Christenh. & Byng (Alismataceae) **IPNI:** 77183476-1 **Habit:** Herb, Aquatic **Use:** ME **Origin:** Native **Dept.:** VAC **Elev.:** 1000–1000 m a.s.l. **Conservation status: (IUCN)** NE

680 *Aquarius grisebachii* (Small) Christenh. & Byng (Alismataceae) **IPNI:** 77183004-1 **Habit:** Herb, Aquatic **Use:** EU **Origin:** Native **Region:** Orinoquia, Pacific **Dept.:** BOY, CHO, MET **Elev.:** 40–40 m a.s.l. **Conservation status: (IUCN)** NE

681 *Aquarius horizontalis* (Rataj) Christenh. & Byng (Alismataceae) **IPNI:** 77183523-1 **Common name:** comida de morrocoy **Habit:** Herb, Aquatic **Use:** EU **Origin:** Native **Region:** Amazonia, Guiana Shield **Dept.:** AMA, GUV **Elev.:** 100–280 m a.s.l. **Conservation status: (IUCN)** NE

682 *Aquarius paniculatus* (Micheli) Christenh. & Byng (Alismataceae) **IPNI:** 77182731-1 **Common name:** lirio de agua **Habit:** Herb, Aquatic **Use:** HF, ME **Origin:** Native **Region:** Caribbean, Orinoquia **Dept.:** ANT, ARA, ATL, BOL, COR, MAG, SUC **Elev.:** 5–250 m a.s.l. **Conservation status: (IUCN)** NE

683 *Helanthium bolivianum* (Rusby) Lehtonen & Myllys (Alismataceae) **IPNI:** 77089409-1 **Habit:** Herb, Aquatic **Use:** EU **Origin:** Native **Region:** Amazonia, Guiana Shield, Pacific **Dept.:** ANT, GUV, MET, TOL, VAC, VAU **Elev.:** 0–280 m a.s.l. **IUCN Habitat code:** 1,14 **Conservation status: (IUCN)** NE

684 *Helanthium tenellum* (Mart. ex Schult.f.) J.G.Sm. (Alismataceae) **IPNI:** 118649-2 **Habit:** Herb, Aquatic **Use:** EU **Origin:** Native **Region:** Orinoquia **Dept.:** BOY, MET, SAN, VID **Elev.:** 130–500 m a.s.l. **IUCN Habitat code:** 1,2,3,14 **Conservation status: (IUCN)** NE

685 *Hydrocleys nymphoides* (Humb. & Bonpl. ex Willd.) Buchenau (Alismataceae)
IPNI: 58399-1
Common name: falso loto, loto
Habit: Herb, Aquatic **Use:** ME **Origin:** Native **Region:** Andean, Caribbean **Dept.:** ANT, HUI, MAG **Elev.:** 400–1900 m a.s.l. **IUCN Habitat code:** 1,2,3,14 **Conservation status: (IUCN)** NE

686 *Limnocharis flava* (L.) Buchenau (Alismataceae)
IPNI: 58404-1
Common name: buchón, gracia, hoja de buitre, sapito **Habit:** Herb, Aquatic **Use:** AF, EU, HF, ME **Origin:** Native **Region:** Amazonia, Andean, Caribbean, Pacific **Dept.:** AMA, ANT, ATL, BOG, CHO, CUN, GUV, MAG, NAR, SAN, SUC, VAC **Elev.:** 5–950 m a.s.l. **IUCN Habitat code:** 1,2,3,4,5,14 **Conservation status: (IUCN)** NE

687 *Sagittaria guayanensis* Kunth (Alismataceae)
IPNI: 58497-1
Habit: Herb, Aquatic **Use:** EU, GS, HF, ME **Origin:** Native **Region:** Amazonia, Andean, Guiana Shield, Caribbean, Orinoquia **Dept.:** ANT, CAQ, CAS, CAU, CES, MAG, MET, SAN, TOL **Elev.:** 100–600 m a.s.l. **IUCN Habitat code:** 1,2,3,4,5,14 **Conservation status: (IUCN)** LC

688 *Sagittaria lancifolia* L. (Alismataceae)
IPNI: 58516-1
Habit: Herb, Aquatic **Use:** ME **Origin:** Native **Region:** Caribbean, Pacific **Dept.:** ANT, ATL, CAU, CHO, COR, MAG, NAR **Elev.:** 0–100 m a.s.l. **IUCN Habitat code:** 2,5 **Conservation status: (IUCN)** NE

689 *Sagittaria latifolia* Willd. (Alismataceae)
IPNI: 225280-2
Habit: Herb, Aquatic **Use:** EU, HF, ME **Origin:** Native **Region:** Amazonia, Andean, Caribbean, Orinoquia **Dept.:** ANT, MAG, VAC **Elev.:** 200–1130 m a.s.l. **IUCN Habitat code:** 1,3,4 **Conservation status: (IUCN)** LC

690 *Sagittaria rhombifolia* Cham. (Alismataceae)
IPNI: 58564-1
Habit: Herb, Aquatic **Use:** HF, ME **Origin:** Native **Region:** Amazonia, Guiana Shield **Dept.:** GUV **Elev.:** 240–240 m a.s.l. **IUCN Habitat code:** 1,2,3,4,14 **Conservation status: (IUCN)** NE

691 *Aglaonema commutatum* Schott (Araceae)
IPNI: 84050-1
Common name: aglomena, cafeto, cajeto, cucaracho, dólar **Habit:** Herb **Use:** EU, ME, SU **Origin:** Cultivated **Region:** Andean, Caribbean, Orinoquia **Dept.:** ANT, BOG, CAL, CUN, MET, RIS, SAP, SUC, TOL, VAC **Elev.:** 100–1600 m a.s.l. **IUCN Habitat code:** 1,3,14 **Conservation status: (IUCN)** NE

692 *Aglaonema costatum* N.E.Br. (Araceae)
IPNI: 84052-1
Habit: Herb **Use:** EU, HF, ME, SU **Origin:** Cultivated **IUCN Habitat code:** 1,14 **Conservation status: (IUCN)** NE

693 *Aglaonema modestum* Schott ex Engl. (Araceae)
IPNI: 84081-1
Habit: Herb **Use:** EU, ME **Origin:** Cultivated **Region:** Andean **Dept.:** NAR **Elev.:** 1700–1700 m a.s.l. **IUCN Habitat code:** 1 **Conservation status: (IUCN)** NE

694 *Alocasia cucullata* (Lour.) G.Don (Araceae)
IPNI: 84151-1
Common name: huevo verraco, terciopelo **Use:** EU, HF, ME **IUCN Habitat code:** 3 **Conservation status: (IUCN)** NE

695 *Alocasia macrorrhizos* (L.) G.Don (Araceae)
IPNI: 60444565-2
Habit: Herb **Use:** AF, EU, GS, HF, MA, ME, PO, SU **Origin:** Cultivated **Region:** Caribbean, Pacific **Dept.:** CAU, CHO, SAP **Elev.:** 0–200 m a.s.l. **IUCN Habitat code:** 1,14 **Conservation status: (IUCN)** NE

696 *Anthurium affine* Schott (Araceae)
IPNI: 272224-2
Habit: Herb **Use:** ME **Origin:** Native **Region:** Amazonia **Dept.:** CAQ **Elev.:** 500–500 m a.s.l. **Conservation status: (IUCN)** NE

697 *Anthurium alatum* Engl. (Araceae)
IPNI: 15067-2
Habit: Herb, Epiphyte **Use:** ME **Origin:** Endemic **Region:** Andean, Pacific **Dept.:** ANT, CAL, CHO, QUI, RIS, VAC **Elev.:** 220-2300 m.a.s.l. **IUCN Habitat code:** 1,3,14 **Conservation status: (IUCN)** NE **(CPSG)** LC

698 *Anthurium andraeanum* Linden ex André (Araceae)
IPNI: 84549-1
Common name: anturio, anturio rojo, capotillo colorado **Habit:** Herb, Epiphyte, Climbing **Use:** EU, ME, PO **Origin:** Native, Cultivated **Region:** Andean **Dept.:** ANT, BOG, CHO, NAR, RIS, VAC **Elev.:** 600–2650 m a.s.l. **IUCN Habitat code:** 1,2,3,4,14 **Conservation status: (IUCN)** NE

699 *Anthurium atropurpureum* R.E.Schult. & Maguire (Araceae)
IPNI: 15105-2
Common name: ciempatas, col de monte **Habit:** Herb, Epiphyte **Use:** ME **Origin:** Native **Region:** Amazonia, Andean, Orinoquia **Dept.:** AMA, CAQ, CAS, MET, NAR, PUT, VAU **Elev.:** 100-1700 m.a.s.l. **IUCN Habitat code:** 1,3,4,5,14 **Conservation status: (IUCN)** NE **(CPSG)** Potential LC

700 *Anthurium bonplandii* G.S.Bunting (Araceae)
IPNI: 15128-2
Common name: canilla de grulla
Habit: Herb, Epiphyte, Hemiepiphyte, Climbing **Use:** ME **Origin:** Native **Region:** Amazonia, Guiana Shield, Orinoquia **Dept.:** AMA, CAQ, GUA, GUV, MET, VAU, VID **Elev.:** 65-700 m.a.s.l. **IUCN Habitat code:** 1,2,3,4,5,14 **Conservation status: (IUCN)** LC **(CPSG)** Potential LC

701 *Anthurium breviscapum* Kunth (Araceae)
IPNI: 84611-1
Common name: col de monte **Habit:** Herb, Epiphyte, Hemiepiphyte, Climbing **Use:** HF **Origin:** Native **Region:** Amazonia, Andean, Orinoquia **Dept.:** AMA, CAQ, CES, HUI, MAG, MET, NAR, PUT, VAC, VAU **Elev.:** 100–3000 m a.s.l. **IUCN Habitat code:** 1,3,4,14 **Conservation status: (IUCN)** NE

702 *Anthurium crassinervium* (Jacq.) Schott (Araceae)
IPNI: 84707-1
Common name: abrazapalo, anturio, hoja de piedra, hojas de corazón **Habit:** Herb, Epiphyte, Climbing **Use:** ME **Origin:** Native **Region:** Andean, Caribbean, Orinoquia **Dept.:** BOY, CAU, CES, CUN, LAG, MAG, MET, NAR, RIS, SAN, VID **Elev.:** 100–3000 m a.s.l. **IUCN Habitat code:** 1,2,3,14 **Conservation status: (IUCN)** NE

703 *Anthurium crystallinum* Linden & André (Araceae)
IPNI: 84713-1
Common name: anturio **Habit:** Herb, Epiphyte **Use:** EU **Origin:** Native **Region:** Andean, Pacific **Dept.:** ANT, CHO, RIS, TOL **Elev.:** 500–1000 m a.s.l. **IUCN Habitat code:** 1,3,14 **Conservation status: (IUCN)** NE

704 *Anthurium eminens* Schott (Araceae)
IPNI: 84777-1
Habit: Herb, Epiphyte, Hemiepiphyte, Climbing **Use:** MA, ME **Origin:** Native **Region:** Amazonia, Andean, Guiana Shield, Pacific **Dept.:** AMA, ANT, CAQ, CHO, CUN, GUA, MET, PUT **Elev.:** 40–1600 m a.s.l. **IUCN Habitat code:** 1,3,5,14 **Conservation status: (IUCN)** NE

705 *Anthurium ernesti* Engl. (Araceae)
IPNI: 15267-2
Habit: Herb, Epiphyte **Use:** HF **Origin:** Native **Region:** Amazonia **Dept.:** AMA, MET, PUT **Elev.:** 100–350 m a.s.l. **Conservation status: (IUCN)** NE

706 *Anthurium giganteum* Engl. (Araceae)
IPNI: 15299-2
Habit: Herb, Epiphyte **Use:** HF **Origin:** Native **Region:** Andean **Dept.:** VAC **Elev.:** 1600–1600 m a.s.l. **IUCN Habitat code:** 3 **Conservation status: (IUCN)** NE

707 *Anthurium glaucospadix* Croat (Araceae)
IPNI: 272290-2
Common name: anturio, col de monte **Habit:** Herb, Epiphyte **Use:** ME **Origin:** Endemic **Region:** Andean **Dept.:** CUN, HUI, MET, NAR, RIS, SAN, TOL, VAC **Elev.:** 140-3000 m.a.s.l. **IUCN Habitat code:** 1,2,3,14 **Conservation status: (IUCN)** LC **(CPSG)** LC

708 *Anthurium gracile* (Rudge) Lindl. (Araceae)
IPNI: 84849-1
Habit: Herb, Epiphyte **Use:** ME **Origin:** Native **Region:** Amazonia, Andean, Guiana Shield, Caribbean, Orinoquia, Pacific **Dept.:** AMA, ANT, BOL, CAQ, CAS, CAU, CHO, COR, CUN, GUA, LAG, MET, NAR, PUT, RIS, SAN, VAC, VAU **Elev.:** 10-2300 m.a.s.l. **IUCN Habitat code:** 1,3,4,5,14 **Conservation status: (IUCN)** NE **(CPSG)** Potential LC

709 *Anthurium infectorium* R.E.Schult. (Araceae)
IPNI: 15361-2
Habit: Herb, Epiphyte **Use:** ME **Origin:** Endemic **Region:** Guiana Shield **Dept.:** VAU **Elev.:** 200–244 m a.s.l. **IUCN Habitat code:** 1 **Conservation status: (IUCN)** NE

710 *Anthurium lancea* Sodiro (Araceae)
IPNI: 15381-2
Habit: Herb, Epiphyte, Climbing **Use:** HF **Origin:** Native **Region:** Andean, Pacific **Dept.:** CAU, CHO, NAR, VAC

Elev.: 300–1800 m a.s.l. **IUCN Habitat code:** 1,3,14 **Conservation status: (IUCN)** NE

711 *Anthurium marmoratum* Sodiro (Araceae) **IPNI:** 15456-2 **Common name:** terciopelo **Habit:** Herb, Hemiepiphyte **Use:** ME **Origin:** Native **Region:** Andean **Dept.:** VAC **Elev.:** 1450–1450 m a.s.l. **IUCN Habitat code:** 1,14 **Conservation status: (IUCN)** NE

712 *Anthurium nymphaeifolium* K.Koch & C.D.Bouché (Araceae) **IPNI:** 85127-1 **Common name:** chupapanga, corazón, guaje, tinracha **Habit:** Herb, Epiphyte, Climbing **Use:** GS **Origin:** Native **Region:** Andean, Pacific **Dept.:** ANT, BOY, CAU, CES, CHO, CUN, HUI, MAG, NAR, NSA, PUT, RIS, SAN, TOL, VAC **Elev.:** 100-3750 m.a.s.l. **IUCN Habitat code:** 1,2,3,4,14 **Conservation status: (IUCN)** NE **(CPSG)** Potential LC

713 *Anthurium obtusum* (Engl.) Grayum (Araceae) **IPNI:** 1001003-1 **Habit:** Herb, Epiphyte, Climbing **Use:** HF **Origin:** Native **Region:** Amazonia, Andean, Pacific **Dept.:** AMA, ANT, BOY, CAL, CAQ, CHO, CUN, GUA, HUI, MAG, MET, NAR, PUT, RIS, VAC, VAU **Elev.:** 0-1940 m.a.s.l. **IUCN Habitat code:** 1,3,4,5,14 **Conservation status: (IUCN)** NE **(CPSG)** Potential LC

714 *Anthurium oxycarpum* Poepp. (Araceae) **IPNI:** 272336-2 **Habit:** Herb, Epiphyte, Climbing **Use:** ME **Origin:** Native **Region:** Amazonia **Dept.:** AMA, CAQ **Elev.:** 100–300 m a.s.l. **IUCN Habitat code:** 1,14 **Conservation status: (IUCN)** NE

715 *Anthurium palmatum* (L.) Schott (Araceae) **IPNI:** 85170-1 **Use:** ME **Origin:** Native **Conservation status: (IUCN)** NE

716 *Anthurium pentaphyllum* (Aubl.) G.Don (Araceae) **IPNI:** 85201-1 **Common name:** miyaibo **Habit:** Herb, Epiphyte, Hemiepiphyte, Climbing **Use:** ME **Origin:** Native **Region:** Amazonia, Andean, Guiana Shield, Caribbean, Orinoquia, Pacific **Dept.:** AMA, ATL, CAQ, CAS, CHO, MAG, MET, PUT, RIS, SUC **Elev.:** 20–1280 m a.s.l. **IUCN Habitat code:** 1,2,3,4,5,14 **Conservation status: (IUCN)** NE

717 *Anthurium sagittatum* (Sims) G.Don (Araceae) **IPNI:** 85321-1 **Habit:** Herb **Use:** HF **Origin:** Native **Region:** Amazonia, Andean **Dept.:** AMA, ANT, CAQ, NAR, PUT **Elev.:** 110–1700 m a.s.l. **IUCN Habitat code:** 1,3,4,14 **Conservation status: (IUCN)** LC

718 *Anthurium scandens* (Aubl.) Engl. (Araceae) **IPNI:** 85333-1 **Common name:** anturio **Habit:** Herb, Epiphyte, Hemiepiphyte, Climbing **Use:** HF, ME **Origin:** Native **Region:** Amazonia, Andean, Pacific **Dept.:** ANT, BOY, CAL, CAQ, CAU, CHO, CUN, HUI, LAG, MAG, MET, NAR, NSA, PUT, QUI, RIS, SAN, TOL, VAC, VAU **Elev.:** 5-2800 m.a.s.l. **IUCN Habitat code:** 1,2,3,4,5,14 **Conservation status: (IUCN)** NE **(CPSG)** Potential LC

719 *Anthurium uleanum* Engl. (Araceae) **IPNI:** 15792-2 **Common name:** afuera, col, maceto **Habit:** Herb, Epiphyte, Hemiepiphyte, Climbing **Use:** ME **Origin:** Native **Region:** Amazonia, Andean **Dept.:** AMA, CAQ, MET, VAU **Elev.:** 60–1200 m a.s.l. **IUCN Habitat code:** 1,3,5,14 **Conservation status: (IUCN)** NE

720 *Anthurium vanderknaapii* Croat (Araceae) **IPNI:** 77084919-1 **Habit:** Herb **Use:** EU **Origin:** Endemic, Cultivated **Region:** Andean **Dept.:** CAL, VAC **Elev.:** 500–2000 m a.s.l. **Conservation status: (IUCN)** NE

721 *Anthurium veitchii* Mast. (Araceae) **IPNI:** 85473-1 **Habit:** Herb, Epiphyte **Use:** EU **Origin:** Endemic, Cultivated **Region:** Andean, Pacific **Dept.:** ANT, CHO **Elev.:** 300–1510 m a.s.l. **Conservation status: (IUCN)** NE

722 *Anthurium versicolor* Sodiro (Araceae) **IPNI:** 15813-2 **Common name:** guaje **Habit:** Herb, Epiphyte, Hemiepiphyte, Climbing **Use:** HF, ME **Origin:** Native **Region:** Andean, Pacific **Dept.:** ANT, CAU, CHO, CUN, NAR, QUI, RIS, VAC **Elev.:** 220–3000 m a.s.l. **IUCN Habitat code:** 1,3,14 **Conservation status: (IUCN)** NE

723 *Anthurium warocqueanum* T.Moore (Araceae) **IPNI:** 85496-1 **Common name:** piel de tigre, terciopelo **Habit:** Herb, Epiphyte, Hemiepiphyte, Climbing **Use:** EU **Origin:** Endemic, Cultivated **Region:** Andean, Pacific **Dept.:** ANT, CHO, QUI, RIS, VAC **Elev.:** 0-2250 m.a.s.l. **Conservation status: (IUCN)** NE **(CPSG)** LC

724 *Anthurium watermaliense* L.H.Bailey & Nash (Araceae) **IPNI:** 77177182-1 **Common name:** anturio negro **Habit:** Herb **Use:** EU **Origin:** Native, Cultivated **Region:** Andean **Dept.:** RIS **Elev.:** 2500–2500 m a.s.l. **Conservation status: (IUCN)** NE

725 *Caladium bicolor* (Aiton) Vent. (Araceae) **IPNI:** 30010088-2 **Common name:** boricá, corazón, corazón de Jesús, corazón de jesús, corazón herido, corazóno herido, jaspeado, lágrimas de María, lágrimas de maría, morrocoy, oreja de mula, pañoleta, pañueleta, penitente, platillo, platino, platino rosado, rascadera, rascaderilla, rayo **Habit:** Herb **Use:** EU, GS, HF, ME, PO, SU **Origin:** Native, Cultivated **Region:** Amazonia, Andean, Caribbean, Orinoquia, Pacific **Dept.:** AMA, ANT, BOY, CAQ, CAS, CAU, CHO, CUN, MAG, MET, NAR, QUI, RIS, SAP, SAN, TOL, VAC, VAU **Elev.:** 30-2300 m.a.s.l. **IUCN Habitat code:** 1,2,3,4,5,14 **Conservation status: (IUCN)** NE **(CPSG)** Potential LC

726 *Caladium lindenii* (André) Madison (Araceae) **IPNI:** 40313-2 **Habit:** Herb **Use:** EU, HF **Origin:** Native **Region:** Andean, Pacific **Dept.:** ANT, CAL, CHO, VAC **Elev.:** 40–1030 m a.s.l. **IUCN Habitat code:** 1,3,5,14 **Conservation status: (IUCN)** NE

727 *Caladium macrotites* Schott (Araceae) **IPNI:** 86379-1 **Common name:** anapa, rascadera **Habit:** Herb **Use:** SU **Origin:** Native **Region:** Amazonia, Orinoquia **Dept.:** CAS, MET, VID **Elev.:** 100–500 m a.s.l. **IUCN Habitat code:** 1,2,3,4,5,14 **Conservation status: (IUCN)** NE

728 *Colocasia esculenta* (L.) Schott (Araceae) **IPNI:** 1170772-2 **Common name:** achín, bore morado, chicul, chime, chonque, chonte, chunque, chícalo, chícol, chícolo, cun, mafafa, mafafa de charapa, mafafa de murciélago, malanga, papa, papa cun, papachina, pipa, rascadera, rascadera blanca, rascadera rosada, taro, yota, ñame yota **Habit:** Herb **Use:** AF, EU, GS, HF, MA, ME, PO, SU **Origin:** Cultivated, Naturalised **Region:** Amazonia, Andean, Caribbean, Orinoquia, Pacific **Dept.:** AMA, ANT, BOG, BOY, CAQ, CHO, MAG, PUT, VAC **Elev.:** 5–2200 m a.s.l. **IUCN Habitat code:** 1,2,3,5,14 **Conservation status: (IUCN)** LC

729 *Dieffenbachia bowmanni* H.J.Veitch (Araceae) **IPNI:** 86846-1 **Use:** ME **Origin:** Native **Conservation status: (IUCN)** NE

730 *Dieffenbachia cannifolia* Engl. (Araceae) **IPNI:** 86850-1 **Habit:** Herb **Use:** ME **Origin:** Native **Region:** Amazonia **Dept.:** AMA, PUT **Elev.:** 88–150 m a.s.l. **IUCN Habitat code:** 1,14 **Conservation status: (IUCN)** NE

731 *Dieffenbachia costata* Klotzsch ex Schott (Araceae) **IPNI:** 86857-1 **Habit:** Herb **Use:** ME **Origin:** Native **Region:** Guiana Shield **Dept.:** MET **Elev.:** 500–500 m a.s.l. **IUCN Habitat code:** 1,4 **Conservation status: (IUCN)** NE

732 *Dieffenbachia longispatha* Engl. & K.Krause (Araceae) **IPNI:** 80133-2 **Common name:** bala de zaino, hoja de chucha, hoja de saíno, igalán, matapuerco, rascadera, saíno, tronco de zaino **Habit:** Herb **Use:** ME **Origin:** Native **Region:** Andean, Guiana Shield, Orinoquia, Pacific **Dept.:** AMA, ANT, CHO, CUN, MAG, MET, VAU **Elev.:** 5–1500 m a.s.l. **IUCN Habitat code:** 1,3,4,14 **Conservation status: (IUCN)** NE

733 *Dieffenbachia oerstedii* Schott (Araceae) **IPNI:** 86896-1 **Habit:** Herb **Use:** ME **Origin:** Native **Region:** Andean **Dept.:** NAR **Elev.:** 1700–1700 m a.s.l. **IUCN Habitat code:** 1,3,14 **Conservation status: (IUCN)** NE

734 *Dieffenbachia parlatorei* Linden & André (Araceae) **IPNI:** 86899-1 **Common name:** platanillo **Habit:** Herb **Use:** ME **Origin:** Native **Region:** Andean, Pacific **Dept.:** ANT, CAL, CAU, CHO, CUN, MAG, MET, RIS, TOL, VAC **Elev.:** 20–1000 m a.s.l. **IUCN Habitat code:** 1,3,5,14 **Conservation status: (IUCN)** NE

735 *Dieffenbachia seguine* (Jacq.) Schott (Araceae) **IPNI:** 86911-1 **Common name:** ajo de monte, calla, carateja, caña cimarrona, caña de puente, caña muda, chicapoy,

chucha, chula, chulo, col del diablo, cucarachero, cucaracho, gota de aceite, jaspe, manjó, mármol, nieve tropical, pentagrama, picatón, rascaor, rocío, saíno, wáshington, **Habit:** Herb **Use:** EU, HF, ME, PO **Origin:** Native, Cultivated **Region:** Amazonia, Andean, Caribbean **Dept.:** BOG, CAQ, CUN, MAG, MET, VAC, VAU, VID **Elev.:** 213–2600 m a.s.l. **IUCN Habitat code:** 1,2,3,4,5,14 **Conservation status: (IUCN)** NE

736 *Dracontium croatii* G.H.Zhu (Araceae)
IPNI: 303133-2
Common name: naipe, palo de la equis, papayuelo **Habit:** Herb **Use:** ME **Origin:** Native **Region:** Andean **Dept.:** NAR **Elev.:** 660–1300 m a.s.l. **IUCN Habitat code:** 1 **Conservation status: (IUCN)** EN

737 *Dracontium spruceanum* (Schott) G.H.Zhu (Araceae)
IPNI: 991397-1
Common name: chupadera, coronilla, mapaná, palo de la equis, papa de culebra, papaculebra, papaindia **Habit:** Herb **Use:** HF, ME **Origin:** Native **Region:** Amazonia, Andean, Guiana Shield, Orinoquia, Pacific **Dept.:** AMA, ANT, BOY, CAQ, CAU, CHO, MET, NAR, PUT, SAN, VAC, VAU **Elev.:** 0-1900 m.a.s.l. **IUCN Habitat code:** 1,2,3,4,5,14 **Conservation status: (IUCN)** NE **(CPSG)** Potential LC

738 *Epipremnum pinnatum* (L.) Engl. (Araceae)
IPNI: 87046-1
Habit: Herb, Hemiepiphyte **Use:** AF, EU, HF, MA, ME **Origin:** Cultivated, Naturalised **Region:** Amazonia **Dept.:** VAC **Elev.:** 100–1000 m a.s.l. **IUCN Habitat code:** 1,2,3,4,14 **Conservation status: (IUCN)** NE

739 *Heteropsis ecuadorensis* Sodiro (Araceae)
IPNI: 121107-2
Common name: daápi, mamirí, mamure, nanio, waja, yai'h-meá, yaré, yaucasi **Habit:** Herb, Epiphyte, Climbing **Use:** MA **Origin:** Native **Region:** Amazonia, Guiana Shield, Pacific **Dept.:** AMA, ANT, CAQ, CHO, GUA **Elev.:** 10-520 m.a.s.l. **IUCN Habitat code:** 1 **Conservation status: (IUCN)** NE **(CPSG)** Potential LC

740 *Heteropsis flexuosa* (Kunth) G.S.Bunting (Araceae)
IPNI: 121108-2
Common name: bejuco burro, bejuco cangrejo, bejuco de lagartija, bejuco de mamure, bejuco mamure, bejuco yaré, mamure, pamitano, yaré, yurape **Habit:** Herb, Epiphyte, Climbing **Use:** HF, MA, ME, SU **Origin:** Native **Region:** Amazonia, Guiana Shield, Orinoquia **Dept.:** AMA, CAQ, GUV, VAU, VID **Elev.:** 160-500 m.a.s.l. **IUCN Habitat code:** 1,5,14 **Conservation status: (IUCN)** NE **(CPSG)** Potential LC

741 *Heteropsis linearis* A.C.Sm. (Araceae)
IPNI: 121110-2
Common name: intetano, omachi **Habit:** Herb, Epiphyte, Climbing **Use:** MA **Origin:** Native **Region:** Amazonia **Dept.:** AMA **Elev.:** 160–160 m a.s.l. **IUCN Habitat code:** 1 **Conservation status: (IUCN)** NE

742 *Heteropsis oblongifolia* Kunth (Araceae)
IPNI: 87123-1
Common name: bejuco de costilla de chimbe, bejuco de costillas de chimbe, bejuco yaré, ochaiteiba, pamitano, piquigua, pirigua, pitigua, quibru negro, yaré **Habit:** Herb, Hemiepiphyte, Climbing **Use:** MA, ME **Origin:** Native **Region:** Amazonia, Pacific **Dept.:** AMA, ANT, CAQ, CHO, VAU **Elev.:** 20-500 m.a.s.l. **IUCN Habitat code:** 1,3,5,14 **Conservation status: (IUCN)** NE **(CPSG)** Potential LC

743 *Heteropsis spruceana* Schott (Araceae)
IPNI: 87130-1
Common name: acui huipo, bejuco yaré, jaolu, jinollu, tamishi, tedbai, tun, yaré, yaucasi **Habit:** Herb, Epiphyte, Climbing **Use:** MA **Origin:** Native **Region:** Amazonia, Guiana Shield, Orinoquia **Dept.:** AMA, CAQ, GUA, VAU, VID **Elev.:** 80–550 m a.s.l. **IUCN Habitat code:** 1,3,4,5 **Conservation status: (IUCN)** NE

744 *Heteropsis steyermarkii* G.S.Bunting (Araceae)
IPNI: 278149-2
Common name: bejuco yaré **Habit:** Herb, Climbing **Use:** MA **Origin:** Native **Region:** Amazonia, Guiana Shield **Dept.:** AMA, CAQ, GUA **Elev.:** 100–100 m a.s.l. **IUCN Habitat code:** 1,5 **Conservation status: (IUCN)** NE

745 *Heteropsis tenuispadix* G.S.Bunting (Araceae)
IPNI: 278150-2
Common name: yaré **Habit:** Herb, Epiphyte, Hemiepiphyte, Climbing **Use:** MA, ME **Origin:** Native **Region:** Amazonia, Orinoquia **Dept.:** AMA, CAQ, GUV, VAU **Elev.:** 50–240 m a.s.l. **IUCN Habitat code:** 1 **Conservation status: (IUCN)** NE

746 *Lemna aequinoctialis* Welw. (Araceae)
IPNI: 526159-1
Habit: Herb, Aquatic **Use:** EU, HF, ME **Origin:** Native **Region:** Amazonia, Caribbean **Dept.:** AMA, ANT, ATL, BOL, CES, COR, CUN, HUI, MAG, SAN, SUC, TOL **Elev.:** 0-1000 m.a.s.l. **IUCN Habitat code:** 1,2,3,4,14 **Conservation status: (IUCN)** LC **(CPSG)** Potential LC

747 *Lemna gibba* L. (Araceae)
IPNI: 526178-1
Habit: Herb, Aquatic **Use:** AF, HF **Origin:** Native **Region:** Andean **Dept.:** BOG, CUN **Elev.:** 2550–2650 m a.s.l. **IUCN Habitat code:** 1,3,14 **Conservation status: (IUCN)** LC

748 *Lemna minor* L. (Araceae)
IPNI: 526194-1
Use: AF,ME **Region:** Andean **Dept.:** BOG **IUCN Habitat code:** 1,2,3,14 **Conservation status: (IUCN)** LC

749 *Lemna valdiviana* Phil. (Araceae)
IPNI: 279392-2
Habit: Herb, Aquatic **Use:** HF **Origin:** Native **Region:** Andean, Caribbean **Dept.:** ATL, BOL, CES, COR, CUN **Elev.:** 0–2700 m a.s.l. **IUCN Habitat code:** 1,14 **Conservation status: (IUCN)** NE

750 *Monstera adansonii* Schott (Araceae)
IPNI: 87464-1
Common name: abalazo, abalazos, abrazapalo, balazo, balazos, bejuco arauato, bejuco tripa de pollo, ceriman, costilla de Adán, folla furada, herb de puerco, hoja de piedra, piragua, umbe **Habit:** Herb, Hemiepiphyte, Climbing **Use:** EU, HF, ME **Origin:** Native **Region:** Amazonia, Andean, Guiana Shield, Caribbean, Orinoquia, Pacific **Dept.:** AMA, ANT, ARA, ATL, BOL, BOY, CAL, CAQ, CAS, CES, CHO, COR, CUN, GUV, HUI, LAG, MAG, MET, NAR, PUT, RIS, SAN, TOL, VAC, VAU, VID **Elev.:** 0-2500 m.a.s.l. **IUCN Habitat code:** 1,2,3,4,5,14 **Conservation status: (IUCN)** NE **(CPSG)** Potential LC

751 *Monstera deliciosa* Liebm. (Araceae)
IPNI: 87478-1
Common name: abalazo, abalazos, balazo, balazos, ceriman, hojadillo, hojarrota, milbalazos, ojalillo, piña anona, piñanona, uñedanta **Habit:** Herb, Hemiepiphyte, Climbing **Use:** EU, HF, ME, PO, SU **Origin:** Cultivated **Region:** Andean **Dept.:** ANT, BOG, CUN, MET, NAR **Elev.:** 450–2620 m a.s.l. **IUCN Habitat code:** 1,2,3,14 **Conservation status: (IUCN)** NE

752 *Monstera dubia* (Kunth) Engl. & K.Krause (Araceae)
IPNI: 163921-2
Common name: bobo, peseta **Habit:** Herb, Hemiepiphyte, Climbing **Use:** ME **Origin:** Native **Region:** Amazonia, Andean, Guiana Shield, Caribbean, Orinoquia, Pacific **Dept.:** ANT, CAQ, CES, CHO, LAG, MAG, MET, NSA, PUT, SAN **Elev.:** 0-1600 m.a.s.l. **IUCN Habitat code:** 1,2,3,5,14 **Conservation status: (IUCN)** NE **(CPSG)** Potential LC

753 *Monstera lechleriana* Schott (Araceae)
IPNI: 87509-1
Habit: Herb, Hemiepiphyte, Climbing **Use:** ME **Origin:** Native **Region:** Amazonia, Andean, Orinoquia, Pacific **Dept.:** AMA, ANT, CAQ, CAU, CHO, NAR, PUT, VAC, VID **Elev.:** 250–2100 m a.s.l. **IUCN Habitat code:** 1,3,4,14 **Conservation status: (IUCN)** NE

754 *Monstera pinnatipartita* Schott (Araceae)
IPNI: 315164-2
Common name: nujü wukarto, nujübosonae, tjalatjala **Habit:** Herb, Hemiepiphyte, Climbing **Use:** EU, ME, SU **Origin:** Native **Region:** Andean, Caribbean, Orinoquia, Pacific **Dept.:** ANT, BOL, BOY, CAQ, CES, CHO, MAG, MET, NAR, RIS, TOL, VAC **Elev.:** 0-1950 m.a.s.l. **IUCN Habitat code:** 1,2,3,4,5,14 **Conservation status: (IUCN)** NE **(CPSG)** Potential LC

755 *Montrichardia arborescens* (L.) Schott (Araceae)
IPNI: 87563-1
Common name: anica, aninga, arracacho, bore de raya, chupalla, chupayo, rascadera, rábano **Habit:** Herb, Aquatic **Use:** HF, MA, ME **Origin:** Native **Region:** Amazonia, Guiana Shield, Orinoquia, Pacific **Dept.:** AMA, ANT, GUA, MET, VAU, VID **Elev.:** 0–500 m a.s.l. **IUCN Habitat code:** 1,2,3,4,5,14 **Conservation status: (IUCN)** NE

756 *Montrichardia linifera* (Arruda) Schott (Araceae)
IPNI: 87565-1
Common name: bore de raya, chupalla **Habit:** Herb **Use:** HF, MA, ME **Origin:** Native **Region:** Amazonia, Caribbean, Pacific **Dept.:** AMA, ANT, CAQ, CAS, CHO, COR, MET, PUT **Elev.:** 40–350 m a.s.l. **IUCN Habitat code:** 1,2,3,4,5,14 **Conservation status: (IUCN)** NE

757 *Philodendron anisotomum* Schott (Araceae)
IPNI: 192963-2
Habit: Herb, Climbing **Use:** ME **Origin:** Native **Region:** Amazonia, Andean, Pacific **Dept.:** AMA, CHO, NAR **Elev.:** 100–1700 m a.s.l. **Conservation status: (IUCN)** NE

758 *Philodendron colombianum* R.E.Schult. (Araceae)
IPNI: 193000-2
Habit: Herb, Epiphyte **Use:** ME **Origin:** Native **Region:** Amazonia, Andean **Dept.:** AMA, CAQ, NAR, PUT,

VAU **Elev.:** 250–1500 m a.s.l. **IUCN Habitat code:** 1,3,14 **Conservation status: (IUCN)** NE

759 *Philodendron craspedodromum* R.E.Schult. (Araceae) **IPNI:** 193001-2 **Habit:** Herb **Use:** ME **Origin:** Endemic **Region:** Guiana Shield **Dept.:** VAU **Elev.:** 450–450 m a.s.l. **Conservation status: (IUCN)** NE

760 *Philodendron dyscarpium* R.E.Schult. (Araceae) **IPNI:** 193028-2 **Common name:** do, jepe, jepe-kutata, kutata, ká, o, o-do-ká, odoká **Habit:** Herb **Use:** ME **Origin:** Native **Region:** Guiana Shield **Dept.:** AMA, VAU **Elev.:** 180–700 m a.s.l. **IUCN Habitat code:** 1,3,4,5 **Conservation status: (IUCN)** NE

761 *Philodendron erubescens* K.Koch & Augustin (Araceae) **IPNI:** 87759-1 **Habit:** Herb, Climbing **Use:** EU **Origin:** Endemic, Cultivated **Region:** Caribbean **Dept.:** LAG **Elev.:** 520–520 m a.s.l. **IUCN Habitat code:** 1,3,14 **Conservation status: (IUCN)** NE

762 *Philodendron fragrantissimum* (Hook.) G.Don (Araceae) **IPNI:** 87769-1 **Common name:** cachetona **Habit:** Herb, Epiphyte, Hemiepiphyte, Climbing **Use:** ME **Origin:** Native **Region:** Amazonia, Andean, Guiana Shield, Caribbean, Orinoquia, Pacific **Dept.:** AMA, ANT, BOL, CAQ, CAU, CHO, LAG, MET, SAN, VAU **Elev.:** 10–1780 m a.s.l. **IUCN Habitat code:** 1,2,3,4,5,14 **Conservation status: (IUCN)** NE

763 *Philodendron gloriosum* André (Araceae) **IPNI:** 87777-1 **Habit:** Herb **Use:** EU **Origin:** Endemic **Region:** Andean, Orinoquia **Dept.:** CUN, MET **Elev.:** 350-700 m.a.s.l. **IUCN Habitat code:** 1,3,5,14 **Conservation status: (IUCN)** VU **(CPSG)** VU

764 *Philodendron hederaceum* (Jacq.) Schott (Araceae) **IPNI:** 87797-1 **Common name:** abrazapalo, camote de sargotín, chaldé, masipapa, matapuerco, mimbre, quemacarate, sanguijuela, tundushé **Habit:** Herb, Epiphyte, Hemiepiphyte, Climbing **Use:** EU, ME, PO **Origin:** Native **Region:** Andean, Caribbean, Orinoquia, Pacific **Dept.:** ANT, ATL, BOL, CAS, CES, CHO, CUN, LAG, MAG, MET, NAR, NSA, RIS, SUC, VAC **Elev.:** 0-1650 m.a.s.l. **IUCN Habitat code:** 1,2,3,4,5,14 **Conservation status: (IUCN)** NE **(CPSG)** Potential LC

765 *Philodendron inaequilaterum* Liebm. (Araceae) **IPNI:** 87815-1 **Habit:** Herb, Epiphyte, Hemiepiphyte, Climbing **Use:** ME **Origin:** Native **Region:** Amazonia, Andean, Guiana Shield, Caribbean, Orinoquia, Pacific **Dept.:** AMA, ANT, CAU, CES, CHO, CUN, MET, NAR, NSA, PUT, RIS, SAN, VAC **Elev.:** 0-2400 m.a.s.l. **IUCN Habitat code:** 1,3,4,5,14 **Conservation status: (IUCN)** NE **(CPSG)** Potential LC

766 *Philodendron insigne* Schott (Araceae) **IPNI:** 87819-1 **Habit:** Herb, Epiphyte **Use:** ME **Origin:** Native **Region:** Amazonia, Guiana Shield **Dept.:** AMA, GUA, VAU **Elev.:** 160–580 m a.s.l. **IUCN Habitat code:** 1,3,4,5 **Conservation status: (IUCN)** NE

767 *Philodendron lindenii* Schott (Araceae) **IPNI:** 87852-1 **Common name:** abrazapalo **Habit:** Herb, Climbing **Use:** ME **Origin:** Native **Region:** Andean, Caribbean **Dept.:** ATL, CES, MAG **Elev.:** 20–1333 m a.s.l. **IUCN Habitat code:** 1 **Conservation status: (IUCN)** NE

768 *Philodendron linnaei* Kunth (Araceae) **IPNI:** 87858-1 **Common name:** solimón **Habit:** Herb, Climbing **Use:** ME **Origin:** Native **Region:** Amazonia, Guiana Shield, Pacific **Dept.:** AMA, CAQ, CHO, GUA, VAU **Elev.:** 100–500 m a.s.l. **IUCN Habitat code:** 1,3,4 **Conservation status: (IUCN)** NE

769 *Philodendron longirrhizum* M.M.Mora & Croat (Araceae) **IPNI:** 77087061-1 **Common name:** bejuco tripa de perro, bejuco tripaeperro, estiladera, tripa de perro, tripaeperro **Habit:** Herb, Epiphyte, Climbing **Use:** MA **Origin:** Native **Region:** Andean **Dept.:** ANT, CAL, CHO, CUN, HUI, QUI, RIS, VAC **Elev.:** 1200–2400 m a.s.l. **IUCN Habitat code:** 1,3,14 **Conservation status: (IUCN)** NE

770 *Philodendron melanochrysum* Linden & André (Araceae) **IPNI:** 87882-1 **Habit:** Herb, Climbing **Use:** EU **Origin:** Endemic **Region:** Andean **Dept.:** ANT **Elev.:** 800–800 m a.s.l. **IUCN Habitat code:** 1 **Conservation status: (IUCN)** NE

771 *Philodendron ornatum* Schott (Araceae) **IPNI:** 87918-1 **Common name:** corazón **Habit:** Herb, Hemiepiphyte, Climbing **Use:** EU **Origin:** Native **Region:** Amazonia, Andean **Dept.:** AMA, ANT, CAQ, MET **Elev.:** 200–1330 m a.s.l. **IUCN Habitat code:** 1,3,4,5,14 **Conservation status: (IUCN)** NE

772 *Philodendron pedatum* (Hook.) Kunth (Araceae) **IPNI:** 87928-1 **Habit:** Herb, Climbing **Use:** EU, MA, ME **Origin:** Native **Region:** Andean **Dept.:** CAQ, CHO, MET **Elev.:** 350–600 m a.s.l. **IUCN Habitat code:** 1,14 **Conservation status: (IUCN)** NE

773 *Philodendron quinquenervium* Miq. (Araceae) **IPNI:** 87967-1 **Common name:** bejuco chinche, ponowaiinboto, wanalijorobarii **Habit:** Herb, Hemiepiphyte, Climbing **Use:** ME **Origin:** Native **Region:** Amazonia, Andean **Dept.:** AMA, ANT, CAQ, GUV, MET, PUT, SAN, VID **Elev.:** 100–1450 m a.s.l. **IUCN Habitat code:** 1,3,4,14 **Conservation status: (IUCN)** NE

774 *Philodendron radiatum* Schott (Araceae) **IPNI:** 314452-2 **Common name:** balazo, pata de danta **Habit:** Herb, Climbing **Use:** ME **Origin:** Native **Dept.:** ANT **Elev.:** 300–300 m a.s.l. **IUCN Habitat code:** 1,14 **Conservation status: (IUCN)** NE

775 *Philodendron remifolium* R.E.Schult. (Araceae) **IPNI:** 193163-2 **Habit:** Herb **Use:** HF **Origin:** Native **Region:** Amazonia, Guiana Shield **Dept.:** VAU **Elev.:** 198–350 m a.s.l. **IUCN Habitat code:** 1 **Conservation status: (IUCN)** NE

776 *Philodendron sagittifolium* Liebm. (Araceae) **IPNI:** 87989-1 **Habit:** Herb, Hemiepiphyte, Climbing **Use:** ME **Origin:** Native **Region:** Andean, Pacific **Dept.:** ANT, CHO, MAG, MET, NAR, SAN **Elev.:** 0–1700 m a.s.l. **IUCN Habitat code:** 1,3,4,5,14 **Conservation status: (IUCN)** NE

777 *Philodendron subhastatum* K.Krause (Araceae) **IPNI:** 289907-2 **Common name:** hojarrota **Habit:** Herb, Epiphyte, Hemiepiphyte, Climbing **Use:** MA **Origin:** Native **Region:** Andean, Pacific **Dept.:** ANT, CAU, CHO, CUN, NAR, VAC **Elev.:** 5–2300 m a.s.l. **IUCN Habitat code:** 1,3,5,14 **Conservation status: (IUCN)** NE

778 *Philodendron tripartitum* (Jacq.) Schott (Araceae) **IPNI:** 88045-1 **Common name:** anturio de tres dedos, anturio tresdedos, guaral, pata de danta, tres dedo, tres dedos, tresdedos **Habit:** Herb, Epiphyte, Hemiepiphyte, Climbing **Use:** HF, ME **Origin:** Native **Region:** Amazonia, Andean, Guiana Shield, Orinoquia, Pacific **Dept.:** AMA, ANT, CAQ, CAU, CHO, CUN, HUI, MET, NAR, RIS, SAN, VAC **Elev.:** 0–1700 m a.s.l. **IUCN Habitat code:** 1,3,4,5,14 **Conservation status: (IUCN)** NE

779 *Philodendron uleanum* Engl. (Araceae) **IPNI:** 193204-2 **Habit:** Herb **Use:** ME **Origin:** Native **Region:** Andean **Dept.:** MET **Elev.:** 800–800 m a.s.l. **IUCN Habitat code:** 1 **Conservation status: (IUCN)** NE

780 *Philodendron victoriae* G.S.Bunting (Araceae) **IPNI:** 282674-2 **Common name:** burikui **Habit:** Herb, Climbing **Use:** ME **Origin:** Native **Region:** Amazonia **Dept.:** GUV **Elev.:** 200–200 m a.s.l. **IUCN Habitat code:** 1 **Conservation status: (IUCN)** NE

781 *Philodendron wadedavisii* Croat (Araceae) **IPNI:** 77075661-1 **Common name:** bejuco chinche **Habit:** Herb **Use:** MA **Origin:** Native **Region:** Guiana Shield **Dept.:** AMA, VAU **Elev.:** 200–700 m a.s.l. **IUCN Habitat code:** 1,5 **Conservation status: (IUCN)** NE

782 *Philodendron wendlandii* Schott (Araceae) **IPNI:** 314476-2 **Habit:** Herb **Use:** GS **Origin:** Native **Region:** Andean **Dept.:** TOL, VAC **Elev.:** 415–600 m a.s.l. **IUCN Habitat code:** 1 **Conservation status: (IUCN)** NE

783 *Philodendron wurdackii* G.S.Bunting (Araceae) **IPNI:** 282675-2 **Common name:** hikatatano **Habit:** Herb, Epiphyte, Hemiepiphyte **Use:** MA, ME **Origin:** Native **Region:** Amazonia, Orinoquia **Dept.:** AMA, MET **Elev.:** 110–500

m a.s.l. **IUCN Habitat code:** 1,3,4,14 **Conservation status: (IUCN)** NE

784 *Pistia stratiotes* L. (Araceae) **IPNI:** 317684-2 **Common name:** buchona, flor de totumo, hidrópica, lechuga, lechuga de agua, lechuga de río, lechugilla, lechuguilla, murure, oreja de mula, oreja de mulo, repollito de agua, sirena **Habit:** Herb, Aquatic **Use:** AF, EU, FU, GS, HF, MA, ME **Origin:** Native, Cultivated **Region:** Amazonia, Andean, Caribbean, Orinoquia, Pacific **Dept.:** AMA, ANT, ARA, BOL, CAU, CHO, COR, CUN, HUI, MAG, VAC, VAU **Elev.:** 0-2000 m.a.s.l. **IUCN Habitat code:** 1,2,3,4,5,14 **Conservation status: (IUCN)** LC **(CPSG)** Potential LC

785 *Raphidophora pertusa* (Roxb.) Schott (Araceae) **IPNI:** 88567-1 **Use:** HF, ME **Conservation status: (IUCN)** NE

786 *Rhodospatha latifolia* Poepp. (Araceae) **IPNI:** 88624-1 **Habit:** Herb, Climbing **Use:** HF **Origin:** Native **Region:** Amazonia, Andean, Pacific **Dept.:** ANT, CAQ, CAU, CHO, PUT, RIS, SAN **Elev.:** 0–2350 m a.s.l. **IUCN Habitat code:** 1,3,14 **Conservation status: (IUCN)** NE

787 *Rhodospatha moritziana* Schott (Araceae) **IPNI:** 88631-1 **Habit:** Herb, Climbing **Use:** HF **Origin:** Native **Region:** Andean, Orinoquia, Pacific **Dept.:** ANT, CAU, CHO, MET, NSA, VAC **Elev.:** 5–1900 m a.s.l. **IUCN Habitat code:** 1,4,5,14 **Conservation status: (IUCN)** NE

788 *Rhodospatha oblongata* Poepp. (Araceae) **IPNI:** 88634-1 **Common name:** pos, tusilla **Habit:** Herb, Epiphyte, Hemiepiphyte, Climbing **Use:** ME **Origin:** Native **Region:** Amazonia, Andean, Orinoquia, Pacific **Dept.:** CAQ, CAU, CHO, COR, MET, NAR, VAC, VAU **Elev.:** 5–1800 m a.s.l. **IUCN Habitat code:** 1,4,5,14 **Conservation status: (IUCN)** NE

789 *Spathiphyllum cannifolium* (Dryand. ex Sims) Schott (Araceae) **IPNI:** 88955-1 **Common name:** jimaboto, oreja de burro, tibabana **Habit:** Herb **Use:** EU, ME **Origin:** Native, Cultivated **Region:** Amazonia, Andean, Guiana Shield, Orinoquia **Dept.:** AMA, BOG, CAQ, CAS, CUN, GUA, GUV, MET, NSA, PUT, SAN, VAU, VID **Elev.:** 40-1900 m.a.s.l. **IUCN Habitat code:** 1,3,4,14 **Conservation status: (IUCN)** NE **(CPSG)** Potential LC

790 *Spathiphyllum floribundum* (Linden & André) N.E.Br. (Araceae) **IPNI:** 88963-1 **Common name:** espatifilo **Habit:** Herb **Use:** EU, ME **Origin:** Native **Region:** Andean, Caribbean **Dept.:** ANT, BOL, BOY, CAL, COR, CUN, NSA, SAN **Elev.:** 40–1300 m a.s.l. **IUCN Habitat code:** 1,3,5,14 **Conservation status: (IUCN)** NE

791 *Spathiphyllum friedrichsthalii* Schott (Araceae) **IPNI:** 88964-1 **Common name:** canoita, garza, hoja de garza, niño de cuna, pos **Habit:** Herb **Use:** HF, ME **Origin:** Native, Cultivated **Region:** Pacific **Dept.:** ANT, CAU, CHO, NAR, VAC **Elev.:** 0–400 m a.s.l. **IUCN Habitat code:** 1,3,5,14 **Conservation status: (IUCN)** NE

792 *Spathiphyllum lanceifolium* (Jacq.) Schott (Araceae) **IPNI:** 88983-1 **Habit:** Herb **Use:** ME **Origin:** Native **Region:** Andean **Dept.:** ANT, CAL, CUN, SAN, TOL **Elev.:** 700–1200 m a.s.l. **IUCN Habitat code:** 1,3,4 **Conservation status: (IUCN)** NE

793 *Spathiphyllum phryniifolium* Schott (Araceae) **IPNI:** 88999-1 **Habit:** Herb **Use:** HF **Origin:** Native **Region:** Andean, Pacific **Dept.:** ANT, CHO, SAN **Elev.:** 50–1830 m a.s.l. **IUCN Habitat code:** 1,3,14 **Conservation status: (IUCN)** NE

794 *Spathiphyllum wallisii* Regel (Araceae) **IPNI:** 89011-1 **Common name:** anturio garcita, anturio garza, garcita, garza **Habit:** Herb **Use:** HF, ME **Origin:** Native, Cultivated **Dept.:** ANT **Elev.:** 1465–1465 m a.s.l. **IUCN Habitat code:** 1,2,3,14 **Conservation status: (IUCN)** NE

795 *Spirodela polyrhiza* (L.) Schleid. (Araceae) **IPNI:** 526250-1 **Habit:** Herb, Aquatic **Use:** AF,ME **Origin:** Native **Region:** Caribbean **Dept.:** COR, CUN, HUI, MAG **Elev.:** 100–200 m a.s.l. **Conservation status: (IUCN)** LC

796 *Spirodela punctata* (G.Mey.) C.H.Thomps. (Araceae) **IPNI:** 242868-2 **Habit:** Herb, Aquatic **Use:** AF **Origin:** Native **Region:** Amazonia, Andean, Caribbean **Dept.:** AMA, ANT, ATL, BOG, BOL, BOY, COR, CUN, GUA, HUI, SAN **Elev.:** 0–3020 m a.s.l. **IUCN Habitat code:** 1,2,3,4,5,14 **Conservation status: (IUCN)** NE

797 *Stenospermation ammiticum* G.S.Bunting (Araceae) **IPNI:** 244978-2 **Habit:** Herb, Epiphyte **Use:** HF **Origin:** Native **Region:** Guiana Shield **Dept.:** GUA, VAU **Elev.:** 200–250 m a.s.l. **IUCN Habitat code:** 1,4,5 **Conservation status: (IUCN)** NE

798 *Stenospermation amomifolium* (Poepp.) Schott (Araceae) **IPNI:** 89035-1 **Habit:** Herb, Epiphyte, Climbing **Use:** ME **Origin:** Native **Region:** Amazonia **Dept.:** AMA, CAQ **Elev.:** 100–500 m a.s.l. **IUCN Habitat code:** 1,3,5,14 **Conservation status: (IUCN)** NE

799 *Syngonium angustatum* Schott (Araceae) **IPNI:** 89127-1 **Habit:** Herb, Climbing **Use:** EU, ME **Origin:** Native **Region:** Caribbean **Dept.:** SAP **Elev.:** 200–250 m a.s.l. **IUCN Habitat code:** 1 **Conservation status: (IUCN)** NE

800 *Syngonium podophyllum* Schott (Araceae) **IPNI:** 89152-1 **Common name:** airo kajo, singonio, ñanta juju **Habit:** Herb, Epiphyte, Hemiepiphyte, Climbing **Use:** EU, HF, ME, SU **Origin:** Native **Region:** Amazonia, Andean, Caribbean, Orinoquia, Pacific **Dept.:** AMA, ANT, BOG, CAU, CHO, CUN, HUI, MAG, MET, NAR, PUT, QUI, RIS, SAP, SAN **Elev.:** 0-2420 m.a.s.l. **IUCN Habitat code:** 1,2,3,4,14 **Conservation status: (IUCN)** NE **(CPSG)** Potential LC

801 *Thaumatophyllum bipinnatifidum* (Schott ex Endl.) Sakur., Calazans & Mayo (Araceae) **IPNI:** 77178497-1 **Use:** EU, ME, PO **IUCN Habitat code:** 1,2,14 **Conservation status: (IUCN)** NE

802 *Thaumatophyllum solimoesense* (A.C.Sm.) Sakur., Calazans & Mayo (Araceae) **IPNI:** 77178490-1 **Common name:** bejuco burro, bejuco de sarna, bejuco panzeburro, bejuco silvestre, foei, foeí, panzaeburro, piragua, pirawa, sibari huipo, ñodku **Habit:** Herb **Use:** MA, ME **Origin:** Native **Region:** Amazonia, Orinoquia **Dept.:** AMA, CAQ, SAN, VID **Elev.:** 160–300 m a.s.l. **Conservation status: (IUCN)** NE

803 *Thaumatophyllum spruceanum* Schott (Araceae) **IPNI:** 89184-1 **Common name:** memeo, panza de burro **Habit:** Herb, Epiphyte, Hemiepiphyte **Use:** MA **Origin:** Native **Region:** Amazonia, Guiana Shield **Dept.:** AMA, CAQ, GUA **Elev.:** 50–160 m a.s.l. **Conservation status: (IUCN)** NE

804 *Urospatha antisylleptica* R.E.Schult. (Araceae) **IPNI:** 261700-2 **Common name:** hepe, hepe-kuu-tata, kuu, tata **Habit:** Herb **Use:** ME **Origin:** Endemic **Region:** Amazonia **Dept.:** AMA, GUA, VAU **Elev.:** 240–580 m a.s.l. **IUCN Habitat code:** 1 **Conservation status: (IUCN)** NE

805 *Urospatha caudata* (Poepp.) Schott (Araceae) **IPNI:** 89280-1 **Use:** ME **IUCN Habitat code:** 1 **Conservation status: (IUCN)** NE

806 *Urospatha sagittifolia* (Rudge) Schott (Araceae) **IPNI:** 89296-1 **Common name:** talla **Habit:** Herb **Use:** HF, ME **Origin:** Native **Region:** Amazonia, Guiana Shield, Orinoquia **Dept.:** AMA, CAQ, GUA, GUV, PUT, VAU, VID **Elev.:** 5–700 m a.s.l. **IUCN Habitat code:** 1,3,4,5,14 **Conservation status: (IUCN)** NE

807 *Wolffia globosa* (Roxb.) Hartog & Plas (Araceae) **IPNI:** 1135607-2 **Habit:** Herb, Aquatic **Use:** AF, HF, ME **Origin:** Naturalised **Region:** Caribbean **Dept.:** ATL, BOL, COR **Elev.:** 0–200 m a.s.l. **Conservation status: (IUCN)** LC

808 *Xanthosoma belophyllum* (Willd.) Kunth (Araceae) **IPNI:** 89310-1 **Habit:** Herb **Use:** GS, HF, ME **Origin:** Native **Region:** Guiana Shield **IUCN Habitat code:** 1 **Conservation status: (IUCN)** NE

809 *Xanthosoma bilineatum* Rusby (Araceae) **IPNI:** 269302-2 **Habit:** Herb **Use:** GS **Origin:** Endemic **Region:** Andean **IUCN Habitat code:** 1,2 **Conservation status: (IUCN)** NE

810 *Xanthosoma brasiliense* (Desf.) Engl. (Araceae) **IPNI:** 269304-2 **Habit:** Herb **Use:** EU, GS, HF, ME **Origin:** Native, Cultivated **Dept.:** GUA,

VID **Elev.:** 100–100 m a.s.l. **IUCN Habitat code:** 1 **Conservation status: (IUCN)** NE

811 *Xanthosoma caracu* K.Koch & C.D.Bouché (Araceae) **IPNI:** 89318-1 **Use:** GS, HF **Conservation status: (IUCN)** NE

812 *Xanthosoma daguense* Engl. (Araceae) **IPNI:** 269311-2 **Common name:** bore, rascadera **Habit:** Herb **Use:** GS **Origin:** Native **Region:** Andean, Pacific **Dept.:** MAG **Elev.:** 750–1200 m a.s.l. **IUCN Habitat code:** 1,3,5,14 **Conservation status: (IUCN)** NE

813 *Xanthosoma flavomaculatum* Engl. (Araceae) **IPNI:** 269313-2 **Common name:** boya **Habit:** Herb **Use:** GS **Origin:** Endemic **Region:** Andean **IUCN Habitat code:** 1,2,3,14 **Conservation status: (IUCN)** NE

814 *Xanthosoma guttatum* Croat & D.C.Bay (Araceae) **IPNI:** 77083952-1 **Habit:** Herb **Use:** GS **Origin:** Endemic **Region:** Pacific **Dept.:** ANT, CAU, CHO, NAR, VAC **Elev.:** 0-2100 m.a.s.l. **IUCN Habitat code:** 1 **Conservation status: (IUCN)** EN **(CPSG)** EN

815 *Xanthosoma hebetatum* Croat & D.C.Bay (Araceae) **IPNI:** 77083953-1 **Habit:** Herb **Use:** GS **Origin:** Endemic **Region:** Pacific **Dept.:** CAU, CUN, TOL **Elev.:** 1000-1600 m.a.s.l. **IUCN Habitat code:** 1 **Conservation status: (IUCN)** EN **(CPSG)** EN

816 *Xanthosoma helleborifolium* (Jacq.) Schott (Araceae) **IPNI:** 89333-1 **Common name:** bore, rascadera **Habit:** Herb **Use:** GS, HF, ME **Origin:** Native **Region:** Amazonia, Andean, Caribbean, Pacific **Dept.:** VAC **Elev.:** 30–50 m a.s.l. **IUCN Habitat code:** 1,3,14 **Conservation status: (IUCN)** LC

817 *Xanthosoma hylaeae* Engl. & K.Krause (Araceae) **IPNI:** 269320-2 **Habit:** Herb **Use:** GS **Origin:** Native **Region:** Amazonia, Andean **Dept.:** ANT, BOY, CAU, CES, CHO, CUN, MAG, MET, PUT, VAC **Elev.:** 0–1800 m a.s.l. **IUCN Habitat code:** 1,3,5,14 **Conservation status: (IUCN)** LC

818 *Xanthosoma latestigmatum* Bogner & E.G.Gonç. (Araceae) **IPNI:** 77071637-1 **Habit:** Herb **Use:** GS **Origin:** Native **Region:** Andean **Dept.:** NAR **Elev.:** 1820–1820 m a.s.l. **Conservation status: (IUCN)** DD

819 *Xanthosoma mexicanum* Liebm. (Araceae) **IPNI:** 89347-1 **Common name:** mafafa **Habit:** Herb **Use:** GS, ME **Origin:** Native **Region:** Andean **Dept.:** CUN, VAC **Elev.:** 1300–1600 m a.s.l. **IUCN Habitat code:** 1,2,3,14 **Conservation status: (IUCN)** NE

820 *Xanthosoma narinoense* Bogner & L.P.Hannon (Araceae) **IPNI:** 77086185-1 **Habit:** Herb **Use:** GS **Origin:** Endemic **Region:** Andean **Conservation status: (IUCN)** CR **(CPSG)** CR

821 *Xanthosoma paradoxum* (Bogner & Mayo) Bogner (Araceae) **IPNI:** 77071638-1 **Habit:** Herb **Use:** GS **Origin:** Endemic **Region:** Orinoquia, Pacific **Dept.:** ANT, BOY, CUN, HUI, MAG, MET, NAR, PUT, SAN, TOL **Elev.:** 100-1700 m.a.s.l. **IUCN Habitat code:** 1 **Conservation status: (IUCN)** DD **(CPSG)** DD

822 *Xanthosoma robustum* Schott (Araceae) **IPNI:** 89369-1 **Habit:** Herb **Use:** EU, GS, HF, ME **Origin:** Cultivated **Dept.:** BOY, CHO, MET **Elev.:** 100–500 m a.s.l. **IUCN Habitat code:** 1,3,14 **Conservation status: (IUCN)** NE

823 *Xanthosoma sagittifolium* (L.) Schott (Araceae) **IPNI:** 89373-1 **Common name:** alín, batata, bore, chunga, mafafa, mafafa blanca, mafafa mija, mafafa morada, mafafa pata de gallineta, malanga, malangay, mija, nipachire, ocumo, otó, rascadera, rascadera morada, rascadera negra, rascadera rosada, yota, yota de picón, yota de sapito, y **Habit:** Herb **Use:** AF, EU, GS, HF, MA, ME, PO **Origin:** Native, Cultivated **Region:** Amazonia, Andean, Pacific **Dept.:** BOY, CAQ **Elev.:** 300–1100 m a.s.l. **IUCN Habitat code:** 1,3,5,14 **Conservation status: (IUCN)** NE

824 *Xanthosoma striatipes* (Kunth & C.D.Bouché) Madison (Araceae) **IPNI:** 909842-1 **Common name:** guaral **Habit:** Herb **Use:** GS, HF, ME **Origin:** Native **Region:** Guiana Shield, Orinoquia **IUCN Habitat code:** 1,2,3,4,14 **Conservation status: (IUCN)** LC

825 *Xanthosoma trichophyllum* K.Krause (Araceae) **IPNI:** 269348-2 **Habit:** Herb **Use:** GS **Origin:** Native **Region:** Amazonia **IUCN Habitat code:** 1 **Conservation status: (IUCN)** NE

826 *Xanthosoma undipes* (K.Koch & C.D.Bouché) K.Koch (Araceae) **IPNI:** 89384-1 **Habit:** Herb **Use:** EU, GS, HF, ME, PO **Origin:** Native **Region:** Andean, Pacific **Dept.:** BOY, MET, VID **Elev.:** 200–230 m a.s.l. **Conservation status: (IUCN)** LC

827 *Zantedeschia aethiopica* (L.) Spreng. (Araceae) **IPNI:** 89403-1 **Common name:** cartucho, cartucho amarillo, cartucho blanco, rascador **Habit:** Herb **Use:** EU, HF, ME, PO **Origin:** Naturalised **Region:** Andean **Dept.:** AMA, BOG, BOY, CUN, PUT, RIS **Elev.:** 100–2000 m a.s.l. **IUCN Habitat code:** 1,3,5,14 **Conservation status: (IUCN)** LC

828 *Elodea densa* (Planch.) Casp. (Hydrocharitaceae) **IPNI:** 431710-1 **Habit:** Herb, Aquatic **Use:** EU **Origin:** Naturalised **Region:** Andean **Dept.:** ANT, BOY, CUN **Elev.:** 1480–2246 m a.s.l. **Conservation status: (IUCN)** NE

829 *Hydrilla verticillata* (L.f.) Royle (Hydrocharitaceae) **IPNI:** 114224-3 **Habit:** Herb, Aquatic **Use:** AF, EU, HF, ME **Origin:** Naturalised **Dept.:** CUN **Elev.:** 250–250 m a.s.l. **IUCN Habitat code:** 14 **Conservation status: (IUCN)** LC

830 *Hydrocharis spongia* Bosc (Hydrocharitaceae) **IPNI:** 431804-1 **Use:** EU **Conservation status: (IUCN)** NE

831 *Vallisneria americana* Michx. (Hydrocharitaceae) **IPNI:** 431972-1 **Habit:** Herb, Aquatic **Use:** EU, HF **Origin:** Native **Region:** Andean **Dept.:** ANT, VAC **Elev.:** 980–1480 m a.s.l. **Conservation status: (IUCN)** LC

832 *Potamogeton crispus* L. (Potamogetonaceae) **IPNI:** 323116-2 **Habit:** Herb, Aquatic **Use:** HF, ME **Origin:** Naturalised **Dept.:** BOG, VAC **Elev.:** 1000–1000 m a.s.l. **IUCN Habitat code:** 4,14 **Conservation status: (IUCN)** LC

833 *Potamogeton illinoensis* Morong (Potamogetonaceae) **IPNI:** 208837-2 **Habit:** Herb, Aquatic **Use:** EU **Origin:** Native **Region:** Andean **Dept.:** BOY, CAU, CUN, MET, NAR, PUT **Elev.:** 2550–3370 m a.s.l. **IUCN Habitat code:** 4,14 **Conservation status: (IUCN)** NE

834 *Stuckenia pectinata* (L.) Börner (Potamogetonaceae) **IPNI:** 77099639-1 **Habit:** Herb, Aquatic **Use:** AF, HF, ME **Origin:** Native **Region:** Andean **Dept.:** BOG, NAR **Conservation status: (IUCN)** LC

835 *Zannichellia palustris* L. (Potamogetonaceae) **IPNI:** 603593-1 **Use:** ME **Origin:** Native **Conservation status: (IUCN)** LC

836 *Ruppia maritima* L. (Ruppiaceae) **IPNI:** 224565-2 **Common name:** pasto de agua **Habit:** Herb, Aquatic **Use:** EU, ME **Origin:** Native **Region:** Caribbean **Dept.:** LAG **Elev.:** 0–0 m a.s.l. **IUCN Habitat code:** 8 **Conservation status: (IUCN)** LC

Arecales

837 *Acoelorraphe wrightii* (Griseb. & H.Wendl.) H.Wendl. ex Becc. (Arecaceae) **IPNI:** 663912-1 **Habit:** Shrub, Caespitose palm **Use:** EU, ME **Origin:** Native **Region:** Caribbean **Dept.:** SAP **Elev.:** 200-350 m.a.s.l. **Conservation status: (IUCN)** NE **(CPSG)** VU

838 *Acrocomia aculeata* (Jacq.) Lodd. ex R.Keith (Arecaceae) **IPNI:** 663914-1 **Common name:** amolada, amolado, amolao, chonta, corozo, corozo amolado, corozo de amolao, corozo de vino, corozo grande, corozo redondo, cuesco, jabara, palma chonta, palma de corozo, palma tamaco, tamaca, tamaco **Habit:** Tree, Solitary palm

Use: AF, EU, HF, MA, ME **Origin:** Native **Region:** Andean, Caribbean, Orinoquia **Dept.:** ANT, ARA, ATL, BOL, BOY, CAL, CAS, CES, CUN, HUI, LAG, MAG, MET, NSA, SAN, SUC, VAC **Elev.:** 0-1300 m.a.s.l. **IUCN Habitat code:** 1,2,3,4,14 **Conservation status: (IUCN)** NE **(CPSG)** LC

839 *Adonidia merrillii* (Becc.) Becc. (Arecaceae)
IPNI: 663990-1
Common name: palma de Manila, palma manila **Habit:** Shrub, Tree **Use:** EU, HF, MA, ME, SU **Origin:** Cultivated **Dept.:** ANT, CUN **Elev.:** 0–1900 m a.s.l. **IUCN Habitat code:** 1,3,14 **Conservation status: (IUCN)** VU

840 *Aiphanes deltoidea* Burret (Arecaceae)
IPNI: 7897-2
Habit: Shrub, Subshrub, Caespitose palm **Use:** HF **Origin:** Native **Region:** Amazonia **Dept.:** AMA **Elev.:** 100-300 m.a.s.l. **IUCN Habitat code:** 1 **Conservation status: (IUCN)** NE **(CPSG)** LC

841 *Aiphanes gelatinosa* H.E.Moore (Arecaceae)
IPNI: 7906-2
Habit: Shrub, Caespitose palm, Solitary palm **Use:** HF **Origin:** Native **Region:** Andean **Dept.:** NAR **Elev.:** 1200–1200 m a.s.l. **IUCN Habitat code:** 1 **Conservation status: (IUCN)** NE

842 *Aiphanes hirsuta* Burret (Arecaceae)
IPNI: 7908-2
Common name: cirquí, corocito de monte, corozo de monte **Habit:** Shrub, Caespitose palm **Use:** HF **Origin:** Native **Region:** Andean, Pacific **Dept.:** ANT, CAU, CHO, NAR, RIS, VAC **Elev.:** 100-2200 m.a.s.l. **IUCN Habitat code:** 1,3,14 **Conservation status: (IUCN)** LC **(CPSG)** NT

843 *Aiphanes horrida* (Jacq.) Burret (Arecaceae)
IPNI: 7909-2
Common name: chascaraiza, chascaral, chascaraza, corozo, corozo chiquito, corozo colorado, corozo rojo, mararave, mararay, palma corozo, palma de mararay, palma mararay, pujamo **Habit:** Tree, Solitary palm **Use:** EU, HF, MA, ME, SU **Origin:** Native, Cultivated **Region:** Andean, Orinoquia **Dept.:** ANT, ARA, CAL, CAS, CAU, CUN, MET, QUI, RIS, SAN, TOL, VAC **Elev.:** 100-1700 m.a.s.l. **IUCN Habitat code:** 1,2,3,14 **Conservation status: (IUCN)** NE **(CPSG)** LC

844 *Aiphanes lindeniana* (H.Wendl.) H.Wendl. (Arecaceae)
IPNI: 664018-1
Common name: albarico **Habit:** Shrub, Solitary palm **Use:** HF **Origin:** Native **Region:** Andean **Dept.:** NSA, SAN **Elev.:** 1700-2200 m.a.s.l. **IUCN Habitat code:** 1,2,3,14 **Conservation status: (IUCN)** VU **(CPSG)** EN

845 *Aiphanes linearis* Burret (Arecaceae)
IPNI: 7914-2
Common name: chascaray, chirca, chonta, cirquí, corocito de agua, corozo de agüita **Habit:** Shrub, Tree, Caespitose palm **Use:** HF, ME **Origin:** Endemic **Region:** Andean **Dept.:** ANT, VAC **Elev.:** 1800-2600 m.a.s.l. **IUCN Habitat code:** 1,2,14 **Conservation status: (IUCN)** LC **(CPSG)** LC

846 *Aiphanes macroloba* Burret (Arecaceae)
IPNI: 7916-2
Common name: chirquía **Habit:** Shrub, Caespitose palm **Use:** HF **Origin:** Native **Region:** Andean, Pacific **Dept.:** ANT, CHO, NAR **Elev.:** 100-1400 m.a.s.l. **IUCN Habitat code:** 1,3 **Conservation status: (IUCN)** NE **(CPSG)** NT

847 *Aiphanes ulei* (Dammer) Burret (Arecaceae)
IPNI: 7926-2
Common name: chontilla **Habit:** Shrub, Subshrub, Solitary palm **Use:** HF, MA, ME, SU **Origin:** Native **Region:** Amazonia, Andean **Dept.:** AMA, CAQ, PUT **Elev.:** 90-2200 m.a.s.l. **IUCN Habitat code:** 1,3,14 **Conservation status: (IUCN)** LC **(CPSG)** LC

848 *Ammandra decasperma* O.F.Cook (Arecaceae)
IPNI: 11322-2
Common name: antá, cabecilla, cabecita, palma de cabeza, tagua, tagua blanca, tagua de monte, vena, venita, yarina, yarina blanca, yumal, yume, ñume, ñume blanco **Habit:** Acaulescent palm **Use:** EU, HF, MA, ME **Origin:** Native **Region:** Amazonia, Pacific **Dept.:** ANT, BOY, CAU, CHO, PUT, VAC **Elev.:** 0-200 m.a.s.l. **IUCN Habitat code:** 1,3,5,14 **Conservation status: (IUCN)** NE **(CPSG)** LC

849 *Archontophoenix alexandrae* (F.Muell.) H.Wendl. & Drude (Arecaceae)
IPNI: 664083-1
Common name: alejandra, palma Alejandra, palma reina Alejandra, reina Alejandra **Habit:** Tree, Solitary palm **Use:** EU, HF, ME **Origin:** Cultivated **Dept.:** ANT, BOG, CUN, VAC **Elev.:** 500–1600 m a.s.l. **IUCN Habitat code:** 1,14 **Conservation status: (IUCN)** LC

850 *Archontophoenix cunninghamiana* (H.Wendl.) H.Wendl. & Drude (Arecaceae)
IPNI: 134734-3
Common name: palma Popayán, palma botella, palma payanesa, payanesa **Habit:** Tree **Use:** EU, HF **Origin:** Cultivated **Dept.:** ANT, BOG, CUN, VAC **Elev.:** 500–2700 m a.s.l. **IUCN Habitat code:** 1,3,14 **Conservation status: (IUCN)** LC

851 *Areca catechu* L. (Arecaceae)
IPNI: 664107-1
Common name: catechu **Habit:** Tree **Use:** EU, FU, HF, IF, MA, ME, PO, SU **Origin:** Cultivated **Dept.:** CUN, VAC **Elev.:** 400–1600 m a.s.l. **Conservation status: (IUCN)** NE

852 *Asterogyne martiana* (H.Wendl.) H.Wendl. ex Hemsl. (Arecaceae)
IPNI: 664278-1
Common name: cola de gallo, cortadera, panda, panga, quituá, rabihorcao, rabo de zorro **Habit:** Shrub, Solitary palm **Use:** HF, MA **Origin:** Native **Region:** Andean, Pacific **Dept.:** ANT, CAU, CHO, NAR, RIS, VAC **Elev.:** 0-1100 m.a.s.l. **IUCN Habitat code:** 1,3,5,14 **Conservation status: (IUCN)** NE **(CPSG)** LC

853 *Astrocaryum acaule* Mart. (Arecaceae)
IPNI: 664283-1
Common name: coco de tigre, corocito, corozo, cubarro, espina, kãhíka tsiambo **Habit:** Shrub, Acaulescent palm **Use:** HF, MA, ME **Origin:** Native **Region:** Guiana Shield, Orinoquia **Dept.:** ARA, GUA, MET, VAU, VID **Elev.:** 100-450 m.a.s.l. **IUCN Habitat code:** 1,3,4,5,14 **Conservation status: (IUCN)** NE **(CPSG)** LC

854 *Astrocaryum aculeatum* G.Mey. (Arecaceae)
IPNI: 326535-2
Common name: albarico montañero, chonta, chuchana, coco, coquillo, coquito, cubarra, cubarrillo, cubarro, cumarillo, cumaré, espina, macana, macanilla, macanillo, palma de cerrillo, palma macana, tucumá **Habit:** Shrub, Tree, Caespitose palm **Use:** AF, EU, FU, HF, MA, ME, SU **Origin:** Native, Cultivated **Region:** Amazonia, Guiana Shield, Orinoquia **Dept.:** AMA, CAQ, GUA, MET, PUT, VAU, VID **Elev.:** 100-350 m.a.s.l. **IUCN Habitat code:** 1,3,4,5,14 **Conservation status: (IUCN)** NE **(CPSG)** LC

855 *Astrocaryum alatum* Loomis (Arecaceae)
IPNI: 25851-2
Use: ME **Origin:** Native **Conservation status: (IUCN)** NT

856 *Astrocaryum chambira* Burret (Arecaceae)
IPNI: 25854-2
Common name: behtáñó, betaño, betañő, beto, betoñi, chambira, coco, coco chambira, corombolo, cumare, dumestri, hiimasa, hókiki, kaméla, komtë?gn, kumaki, kumali, kumaliboto, kumaliphé, kumaliñu, kumari, kumaria, mũĩũká, matáigahiba, naaĩ, naĩ, nãi, néhe, palma coco, **Habit:** Tree, Solitary palm **Use:** AF, EU, FU, HF, MA, ME, SU **Origin:** Native **Region:** Amazonia, Guiana Shield, Orinoquia **Dept.:** AMA, CAQ, CAS, GUA, GUV, MET, PUT, VAU, VID **Elev.:** 100-500 m.a.s.l. **IUCN Habitat code:** 1,3,5,14 **Conservation status: (IUCN)** NE **(CPSG)** LC

857 *Astrocaryum ciliatum* F.Kahn & B.Millán (Arecaceae)
IPNI: 984104-1
Common name: coco, coco de monte, coco de puerco, coco peludo, coquillo, cumare de guara **Habit:** Acaulescent palm **Use:** HF **Origin:** Native **Region:** Amazonia, Guiana Shield **Dept.:** AMA, CAQ **Elev.:** 100-300 m.a.s.l. **IUCN Habitat code:** 1,5,14 **Conservation status: (IUCN)** NE **(CPSG)** LC

858 *Astrocaryum ferrugineum* F.Kahn & B.Millán (Arecaceae)
IPNI: 984103-1
Common name: güicungo, huiconga, huicongo, huicungo **Habit:** Tree, Solitary palm **Use:** MA **Origin:** Native **Region:** Amazonia **Dept.:** AMA **Elev.:** 100-300 m.a.s.l. **IUCN Habitat code:** 1 **Conservation status: (IUCN)** NE **(CPSG)** LC

859 *Astrocaryum jauari* Mart. (Arecaceae)
IPNI: 664323-1
Common name: albarico, colomboro, corombolo, guara, güiridima, hiabetoñi, huiririma, macanilla, yavarí **Habit:** Tree, Caespitose palm **Use:** HF, MA **Origin:** Native **Region:** Amazonia, Guiana Shield, Orinoquia **Dept.:** AMA, ARA, CAQ, CAS, GUA, GUV, MET, PUT, VAU, VID **Elev.:** 40-200 m.a.s.l. **IUCN Habitat code:** 1,2,3,4,5 **Conservation status: (IUCN)** NE **(CPSG)** LC

860 *Astrocaryum macrocalyx* Burret (Arecaceae)
IPNI: 25873-2
Common name: coco de puerco **Habit:** Tree, Solitary palm **Use:** HF, MA, ME **Origin:** Native **Region:** Amazonia **Dept.:** AMA **Elev.:** 300-300 m.a.s.l. **IUCN Habitat code:** 1,3 **Conservation status: (IUCN)** LC **(CPSG)** LC

861 *Astrocaryum malybo* H.Karst. (Arecaceae)
IPNI: 272909-2
Common name: alúnka, anchamba, anchambe, ancharma, bobil, bobila, chingale, chingalé, enchama, enchamba, enchambe, lanceta, lanzeta, malibu, malibú, mobil, palma bobila, palma chingalé, palma de estera, palma enchama, palma estera, palma lanceta, palma malibú, palma r **Habit:** Shrub, Acaulescent palm **Use:** HF, MA, ME **Origin:** Endemic **Region:** Caribbean, Pacific **Dept.:** ANT, BOL, CES, CHO, COR, CUN, SAN, SUC, TOL **Elev.:** 0–750 m a.s.l. **IUCN Habitat code:** 1,2,3,5,14 **Conservation status: (IUCN)** NE

862 *Astrocaryum murumuru* Mart. (Arecaceae)
IPNI: 664335-1
Use: HF **IUCN Habitat code:** 1,3,14
Conservation status: (IUCN) NE

863 *Astrocaryum sciophilum* (Miq.) Pulle (Arecaceae)
IPNI: 25885-2
Use: EU, HF, MA, ME, SU
Conservation status: (IUCN) NE

864 *Astrocaryum standleyanum* L.H.Bailey (Arecaceae)
IPNI: 25887-2
Common name: chunga, guinul, güerre, güinul, güéguerre, güérregue, higuatá, higuá, huerre, huéguerre, huérregue, palma güinul, werre, wigirbü, wéger **Habit:** Tree, Solitary palm **Use:** AF, EU, FU, HF, MA, ME, SU **Origin:** Native **Region:** Pacific **Dept.:** ANT, CAU, CHO, NAR, VAC **Elev.:** 0-200 m.a.s.l. **IUCN Habitat code:** 1,3,5,14 **Conservation status: (IUCN)** LC **(CPSG)** LC

865 *Astrocaryum triandrum* Galeano, R.Bernal & F.Kahn (Arecaceae)
IPNI: 272910-2
Common name: cabecenegro, cabecinegro, guatinajo, guatinajo hembra, mobil **Habit:** Shrub, Tree, Solitary palm **Use:** MA **Origin:** Endemic **Dept.:** ANT, BOY, CAL, SAN **Elev.:** 160-500 m.a.s.l. **IUCN Habitat code:** 1,3,14 **Conservation status: (IUCN)** EN **(CPSG)** EN

866 *Astrocaryum tucuma* Mart. (Arecaceae)
IPNI: 326537-2
Use: HF, MA **Origin:** Native
Conservation status: (IUCN) NE

867 *Astrocaryum urostachys* Burret (Arecaceae)
IPNI: 25891-2
Common name: chuchana **Habit:** Tree, Solitary palm **Use:** AF, FU, HF, MA, ME, SU **Origin:** Native **Region:** Amazonia **Dept.:** CAQ, PUT **Elev.:** 100–700 m a.s.l. **IUCN Habitat code:** 1,3 **Conservation status: (IUCN)** LC

868 *Attalea allenii* H.E.Moore (Arecaceae)
IPNI: 26553-2
Common name: corozo pajarito, taparín, táparo **Habit:** Acaulescent palm **Use:** HF, MA, SU **Origin:** Native **Region:** Pacific **Dept.:** ANT, BOL, CHO, COR, VAC **Elev.:** 0-100 m.a.s.l. **IUCN Habitat code:** 1,3,5,14 **Conservation status: (IUCN)** NE **(CPSG)** LC

869 *Attalea amygdalina* Kunth (Arecaceae)
IPNI: 664411-1
Common name: almendrón, palma almendrón, palma táparo, táparo **Habit:** Acaulescent palm **Use:** EU, HF, ME, SU **Origin:** Endemic **Region:** Andean **Dept.:** ANT, CAL, QUI, VAC **Elev.:** 1000–1600 m a.s.l. **IUCN Habitat code:** 1,3,14 **Conservation status: (IUCN)** NE

870 *Attalea butyracea* (Mutis ex L.f.) Wess.Boer (Arecaceae)
IPNI: 1172851-2
Common name: barie, buha tuuko, canambo, chapaja, coroza, corozo, corozo de marrano, corozo de puerco, corozo de vaca, corúa, cuesco, curumuta, curúa, ikutiñu, kuluuala, kurua, kóchbot, mĩã phã pú, mabái, mapanaré, nulyi, pa pa, palma corúa, palma cuesco, palma de cue **Habit:** Tree, Solitary palm **Use:** AF, EU, FU, HF, MA, ME, SU **Origin:** Native **Region:** Amazonia, Caribbean, Orinoquia **Dept.:** AMA, ANT, ARA, ATL, BOL, BOY, CAL, CAQ, CAS, CES, COR, CUN, GUA, GUV, HUI, LAG, MAG, MET, NSA, PUT, QUI, RIS, SAN, SUC, TOL, VAC, VAU, VID **Elev.:** 0-1000 m.a.s.l. **IUCN Habitat code:** 1,2,3,5,14 **Conservation status: (IUCN)** NE **(CPSG)** LC

871 *Attalea cohune* Mart. (Arecaceae)
IPNI: 26558-2
Common name: mamarro?n, mamarrón, palma de cuesco, palma real **Habit:** Tree, Solitary palm **Use:** AF, EU, FU, HF, MA, ME **Origin:** Native **Dept.:** ANT, BOY, CAL, SAN **Elev.:** 150-400 m.a.s.l. **IUCN Habitat code:** 1,3,14 **Conservation status: (IUCN)** NE **(CPSG)** EN

872 *Attalea colenda* (O.F.Cook) Balslev & A.J.Hend. (Arecaceae)
IPNI: 272952-2
Common name: palm real de tumaco, palma real **Habit:** Tree, Solitary palm **Use:** AF, EU, HF, MA, ME **Origin:** Native **Region:** Pacific **Dept.:** NAR **Elev.:** 0-200 m.a.s.l. **IUCN Habitat code:** 1,5,14 **Conservation status: (IUCN)** NE **(CPSG)** EN

873 *Attalea cuatrecasasiana* (Dugand) A.J.Hend., Galeano & R.Bernal (Arecaceae)
IPNI: 77177001-1
Common name: corozo, palma corozo, táparo, táparo calimeño **Habit:** Acaulescent palm **Use:** HF **Origin:** Endemic **Region:** Pacific **Dept.:** CAU, CHO, NAR, VAC **Elev.:** 0–100 m a.s.l. **Conservation status: (IUCN)** NE

874 *Attalea fairchildensis* (Glassman) Zona (Arecaceae)
IPNI: 20008131-1
Use: MA **Origin:** Endemic
Conservation status: (IUCN) NE

875 *Attalea iguadummat* de Nevers (Arecaceae)
IPNI: 272953-2
Habit: Acaulescent palm **Use:** HF **Origin:** Native **Region:** Pacific **Dept.:** CHO **Elev.:** 100-200 m.a.s.l. **IUCN Habitat code:** 1 **Conservation status: (IUCN)** NE **(CPSG)** VU

876 *Attalea insignis* (Mart.) Drude (Arecaceae)
IPNI: 26571-2
Common name: coco, palma coco, palma de ramo, palma yagua, yagua **Habit:** Acaulescent palm **Use:** AF, HF, MA, ME, SU **Origin:** Native **Region:** Amazonia, Guiana Shield, Orinoquia **Dept.:** AMA, CAQ, CAS, GUV, MET, VAU **Elev.:** 100-600 m.a.s.l. **IUCN Habitat code:** 1,3,4,5,14 **Conservation status: (IUCN)** NE **(CPSG)** LC

877 *Attalea luetzelburgii* (Burret) Wess.Boer (Arecaceae)
IPNI: 961936-1
Common name: curúa **Habit:** Acaulescent palm **Use:** HF, MA, SU **Origin:** Native **Region:** Guiana Shield **Dept.:** GUA, VAU **Elev.:** 200-500 m.a.s.l. **IUCN Habitat code:** 1,14 **Conservation status: (IUCN)** NE **(CPSG)** LC

878 *Attalea magdalenica* (Dugand) Zona (Arecaceae)
IPNI: 20008141-1
Use: HF **Origin:** Endemic
Conservation status: (IUCN) NE

879 *Attalea maripa* (Aubl.) Mart. (Arecaceae)
IPNI: 664452-1
Common name: abacú, bacú, chichire, cucurita, cucurito, flecha, guajo, güichire, güichirre, huichira, inayá, marija, palma crespa, palma cucurita, palma de sal, palma de virote, palma huichire, palma inayá, palma real **Habit:** Tree, Solitary palm **Use:** FU, HF, MA, ME **Origin:** Native **Region:** Amazonia, Guiana Shield, Orinoquia **Dept.:** AMA, ARA, CAQ, CAS, GUA, GUV, MET, PUT, VAU, VID **Elev.:** 40-600 m.a.s.l. **IUCN Habitat code:** 1,3,4,5,14 **Conservation status: (IUCN)** NE **(CPSG)** LC

880 *Attalea microcarpa* Mart. (Arecaceae)
IPNI: 319686-2
Common name: ẽmiyapochĩabo, mabaco **Habit:** Acaulescent palm **Use:** HF, MA **Origin:** Native **Region:** Amazonia, Guiana Shield **Dept.:** AMA, GUA, VAU, VID **Elev.:** 80-300 m.a.s.l. **IUCN Habitat code:** 1,3,4,5 **Conservation status: (IUCN)** NE **(CPSG)** LC

881 *Attalea nucifera* H.Karst. (Arecaceae)
IPNI: 319677-2
Common name: almendrón, cuesco, mangué **Habit:** Acaulescent palm **Use:** AF, HF, MA, ME **Origin:** Endemic **Dept.:** ANT, BOL, CUN, NSA, SAN, TOL **Elev.:** 200–800 m a.s.l. **IUCN Habitat code:** 1,2,3,14 **Conservation status: (IUCN)** NE

882 *Attalea phalerata* Mart. ex Spreng. (Arecaceae)
IPNI: 26580-2
Common name: chapaja de bajo **Habit:** Tree **Use:** AF, EU, FU, HF, MA, ME, SU **Origin:** Native **Region:** Amazonia **Dept.:** AMA **Elev.:** 100-100 m.a.s.l. **IUCN Habitat code:** 1 **Conservation status: (IUCN)** NE **(CPSG)** LC

883 *Attalea plowmanii* (Glassman) Zona (Arecaceae)
IPNI: 20008146-1
Common name: chapaja **Habit:** Acaulescent palm **Use:** HF, MA **Origin:** Native **Region:** Amazonia **Dept.:** AMA **Elev.:** 100-300 m.a.s.l. **IUCN Habitat code:** 1,14 **Conservation status: (IUCN)** NE **(CPSG)** LC

884 *Attalea racemosa* Spruce (Arecaceae)
IPNI: 664463-1
Common name: coco, coco mabaco, emumeke **Habit:** Acaulescent palm **Use:** HF **Origin:** Native **Region:** Amazonia, Guiana Shield, Orinoquia **Dept.:** AMA, CAQ, GUA, GUV, VAU, VID **Elev.:** 80-400 m.a.s.l. **IUCN Habitat code:** 1,3,5,14 **Conservation status: (IUCN)** NE **(CPSG)** LC

885 *Attalea septuagenata* Dugand (Arecaceae)
IPNI: 26587-2
Common name: cujita **Habit:** Tree **Use:** HF, MA, ME **Origin:** Endemic **Region:** Amazonia **Dept.:** AMA **Elev.:** 100-100 m.a.s.l. **IUCN Habitat code:** 3 **Conservation status: (IUCN)** LC **(CPSG)** LC

886 *Bactris acanthocarpa* Mart. (Arecaceae)
IPNI: 664489-1
Common name: chontaduro de los peces, chontaduro de monte, chontilla, espina **Habit:** Shrub, Acaulescent palm **Use:** AF, HF, MA, ME **Origin:** Native **Region:** Amazonia, Guiana Shield **Dept.:** AMA, CAQ, GUA, GUV, PUT, VAU, VID **Elev.:** 75-300 m.a.s.l. **IUCN Habitat code:** 1,3,4,5,14 **Conservation status: (IUCN)** NE **(CPSG)** LC

887 *Bactris balanophora* Spruce (Arecaceae)
IPNI: 664513-1
Common name: albarico, chontaduro falso, chontilla, cubaro, cubarro **Habit:** Shrub, Caespitose palm **Use:** HF, MA, ME, PO **Origin:** Native **Region:** Amazonia, Guiana Shield **Dept.:** AMA, GUA, VAU **Elev.:** 100-600 m.a.s.l. **Conservation status: (IUCN)** NE **(CPSG)** LC

888 *Bactris barronis* L.H.Bailey (Arecaceae)
IPNI: 28392-2
Common name: chascaray, chascarrá, chontadurillo, jingapá, lata, lata montan?era, lata montañera, maquenque **Habit:** Shrub, Tree, Caespitose palm **Use:** HF, MA, SU **Origin:** Native **Region:** Pacific **Dept.:** ANT, BOL, BOY, CAL, CHO, SAN, VAC **Elev.:** 0-700 m.a.s.l. **IUCN Habitat code:** 1,3,5,14 **Conservation status: (IUCN)** NE **(CPSG)** LC

889 *Bactris bidentula* Spruce (Arecaceae)
IPNI: 664519-1
Common name: chontadurillo, chontaduro de pescado, coquito espinoso de lago, cubarro **Habit:** Shrub **Use:** AF, HF **Origin:** Native **Region:** Amazonia, Guiana Shield, Orinoquia **Dept.:** AMA, ARA, GUA, VAU, VID **Elev.:** 80-200 m.a.s.l. **IUCN Habitat code:** 1,3,4,5 **Conservation status: (IUCN)** NE **(CPSG)** LC

890 *Bactris bifida* Mart. (Arecaceae)
IPNI: 664520-1
Common name: cola de pato **Habit:** Shrub, Subshrub, Caespitose palm **Use:** HF **Origin:** Native **Region:** Amazonia **Dept.:** AMA **Elev.:** 80–220 m a.s.l. **IUCN Habitat code:** 1,3,14 **Conservation status: (IUCN)** NE

891 *Bactris brongniartii* Mart. (Arecaceae)
IPNI: 664527-1
Common name: albarico, cachipay montañero, caña negra, chacarra de agua, chacarrá, chascarrá, chascarrás, cubaro, cubarra, cubarro, espina, mararay, palma cubarro, palma de cubarro, uvita de gallinaza, uvita gallinaza **Habit:** Shrub, Caespitose palm **Use:** HF, MA **Origin:** Native **Region:** Amazonia, Caribbean, Orinoquia, Pacific **Dept.:** AMA, ANT, BOL, CES, CHO, GUA, GUV, MAG, MET, SAN **Elev.:** 10-240 m.a.s.l. **IUCN Habitat code:** 1,2,3,5,14 **Conservation status: (IUCN)** NE **(CPSG)** LC

892 *Bactris chocoensis* R.Bernal, Galeano, Copete & Cámara-Leret (Arecaceae)
IPNI: 77137250-1
Habit: Shrub, Caespitose palm **Use:** HF **Origin:** Endemic **Region:** Pacific **Dept.:** CHO **Elev.:** 60–200 m a.s.l. **IUCN Habitat code:** 14 **Conservation status: (IUCN)** NE

893 *Bactris coloradonis* L.H.Bailey (Arecaceae)
IPNI: 28417-2
Common name: chacarrá, chacarrás, corozo, lata prieta, palma de tunda, pelo de zaino **Habit:** Shrub, Caespitose palm **Use:** HF, MA **Origin:** Native **Region:** Andean, Pacific **Dept.:** ANT, CAU, CHO, COR, NAR, SAN, VAC **Elev.:** 5-1000 m.a.s.l. **IUCN Habitat code:** 1,3,5,14 **Conservation status: (IUCN)** NE **(CPSG)** LC

894 *Bactris concinna* Mart. (Arecaceae)
IPNI: 664555-1
Habit: Shrub, Caespitose palm **Use:** HF **Origin:** Native **Region:** Amazonia, Guiana Shield **Dept.:** AMA, GUA, PUT, VAU **Elev.:** 80-250 m.a.s.l. **IUCN Habitat code:** 1,3,14 **Conservation status: (IUCN)** NE **(CPSG)** LC

895 *Bactris corossilla* H.Karst. (Arecaceae)
IPNI: 664560-1
Common name: barba de tigre, chontilla, coquito, cubarro, cubarro pintón **Habit:** Shrub, Caespitose palm **Use:** AF, HF, MA, ME **Origin:** Native **Region:** Amazonia, Andean, Guiana Shield, Orinoquia **Dept.:** AMA, ARA, CAQ, CAS, GUA, GUV, MET, NSA, PUT, VAU **Elev.:** 90-1400 m.a.s.l. **IUCN Habitat code:** 1,3,4,14 **Conservation status: (IUCN)** NE **(CPSG)** LC

896 *Bactris elegans* Schaedtler (Arecaceae)
IPNI: 664582-1
Common name: chonta **Habit:** Shrub, Caespitose palm **Use:** HF **Origin:** Native **Region:** Amazonia **Dept.:** AMA, CAQ **Elev.:** 80-300 m.a.s.l. **IUCN Habitat code:** 1,5 **Conservation status: (IUCN)** NE **(CPSG)** LC

897 *Bactris fissifrons* Mart. (Arecaceae)
IPNI: 664596-1
Common name: chontilla, chontilla sitanó, coquito, coquito espinoso, cubarro, cubarru de espinas, yachoro **Habit:** Shrub, Caespitose palm **Use:** HF **Origin:** Native **Region:** Amazonia, Guiana Shield **Dept.:** AMA, GUA, PUT, VAU **Elev.:** 80-300 m.a.s.l. **IUCN Habitat code:** 1,3,4,5,14 **Conservation status: (IUCN)** LC **(CPSG)** LC

898 *Bactris gasipaes* Kunth (Arecaceae)
IPNI: 28453-2
Common name: bobi, cachipae, cachipay, cachipá, chichagua, chima, chonta, chonta de castilla, chontadura, chontaduro, chontaduro amarillo, chontaduro colorado, chontaduro de pescado, chontaduro mantecoso, chontaduro rojo, chontaduro sin pepa, macana, macanilla, maquen **Habit:** Tree, Caespitose palm **Use:** AF, EU, FU, HF, MA, ME, SU **Origin:** Native, Cultivated **Region:** Amazonia, Andean, Guiana Shield, Caribbean, Orinoquia, Pacific **Dept.:** AMA, ANT, ARA, BOL, BOY, CAL, CAQ, CAS, CAU, CHO, COR, CUN, GUA, GUV, LAG, MAG, MET, NAR, NSA, PUT, QUI, RIS, SAN, SUC, TOL, VAC, VAU **Elev.:** 0-1500 m.a.s.l. **IUCN Habitat code:** 1,2,3,5,14 **Conservation status: (IUCN)** NE **(CPSG)** VU

899 *Bactris glandulosa* Oerst. (Arecaceae)
IPNI: 664607-1
Habit: Shrub, Caespitose palm **Use:** HF **Origin:** Native **Region:** Andean, Pacific **Dept.:** ANT, CHO, COR **Elev.:** 150-1000 m.a.s.l. **Conservation status: (IUCN)** LC **(CPSG)** LC

900 *Bactris guineensis* (L.) H.E.Moore (Arecaceae)
IPNI: 28460-2
Common name: castilla, caña de lata, cañabrava, chonta, corocito, corozo, corozo de lata, corozo hembra, gallinaza, guevotigre, güevoetigre, higuero de lata, lata, lata de corozo, lata hembra, lata sabanera, palma de corozo, palma de lata, palma lata, palma lata monta **Habit:** Shrub, Caespitose palm **Use:** EU, HF, MA, ME **Origin:** Native **Region:** Caribbean, Orinoquia **Dept.:** ANT, ARA, ATL, BOL, CES, COR, LAG, MAG, SUC **Elev.:** 0-300 m.a.s.l. **IUCN Habitat code:** 1,2,3,4,5,14 **Conservation status: (IUCN)** NE **(CPSG)** NT

901 *Bactris halmoorei* A.J.Hend. (Arecaceae)
IPNI: 1017212-1
Habit: Shrub, Caespitose palm **Use:** AF, HF, MA **Origin:** Native **Region:** Amazonia **Dept.:** AMA **Elev.:** 80-150 m.a.s.l. **IUCN Habitat code:** 1 **Conservation status: (IUCN)** NE **(CPSG)** LC

902 *Bactris hirta* Mart. (Arecaceae)
IPNI: 664621-1
Common name: chontaduro de grillo, chontaduro de monte, chontaduro de rana, chontaduro de rana de rastrojo, coquillo, coquito, espina **Habit:** Shrub, Subshrub, Caespitose palm **Use:** HF **Origin:** Native **Region:** Amazonia, Guiana Shield **Dept.:** AMA, CAQ, GUA, GUV, PUT, VAU, VID **Elev.:** 80-500 m.a.s.l. **IUCN Habitat code:** 1,3,5,14 **Conservation status: (IUCN)** NE **(CPSG)** LC

903 *Bactris killipii* Burret (Arecaceae)
IPNI: 28479-2
Habit: Subshrub, Caespitose palm **Use:** MA **Origin:** Native **Region:** Amazonia **Dept.:** AMA, VAU **Elev.:** 80-300 m.a.s.l. **IUCN Habitat code:** 1 **Conservation status: (IUCN)** NE **(CPSG)** LC

904 *Bactris macroacantha* Mart. (Arecaceae)
IPNI: 664663-1
Common name: chontilla coco, coco, wajuñe **Habit:** Shrub, Caespitose palm **Use:** HF **Origin:** Native **Region:** Amazonia **Dept.:** AMA, VAU **Elev.:** 80-300 m.a.s.l. **IUCN Habitat code:** 1,3,5 **Conservation status: (IUCN)** NE **(CPSG)** LC

905 *Bactris major* Jacq. (Arecaceae)
IPNI: 664668-1
Common name: albarica, albarico, chonta, chontadura, cocorote, corozo de gallinazo, corozo de lata, cubaro, cubarro, lata, lata arroyera, lata de arroyo, lata de gallinazo, lata de playa, lata macho, lata playera, palma chonta, uvita de tigre, uvita gallinaza **Habit:** Shrub, Caespitose palm **Use:** HF, MA, ME **Origin:** Native **Region:** Caribbean, Orinoquia **Dept.:** ANT, ARA, ATL, BOL, CAS, CES, COR, HUI, LAG, MAG, MET, NSA, SUC, TOL **Elev.:** 0-300 m.a.s.l. **IUCN Habitat code:** 1,2,3,4,5,14 **Conservation status: (IUCN)** NE **(CPSG)** LC

906 *Bactris maraja* Mart. (Arecaceae)
IPNI: 28503-2
Common name: buru-moj-quere, cachipay montañero, chacarrá, chascarray, chascarrá, chirquí, chonta, chontadurillo, chontaduro de búho, chontaduro de pescado, chontilla, cubarra, cubarro, espina, lata, lata montañera, lata sabanera, tewpede **Habit:** Shrub, Caespitose palm **Use:** AF, FU, HF, MA,

ME **Origin:** Native **Region:** Amazonia, Andean, Guiana Shield, Orinoquia, Pacific **Dept.:** AMA, ANT, BOL, CAQ, CAU, CHO, COR, GUA, GUV, MET, NAR, NSA, PUT, SAN, VAC, VAU, VID **Elev.:** 0-1500 m.a.s.l. **IUCN Habitat code:** 1,3,4,5,14 **Conservation status: (IUCN)** NE **(CPSG)** LC

907 *Bactris martiana* A.J.Hend. (Arecaceae) **IPNI:** 1017218-1 **Common name:** nejilla **Habit:** Shrub, Tree, Caespitose palm **Use:** HF **Origin:** Native **Region:** Amazonia **Dept.:** AMA, PUT **Elev.:** 80-300 m.a.s.l. **IUCN Habitat code:** 1,3 **Conservation status: (IUCN)** NE **(CPSG)** LC

908 *Bactris obovata* H.Wendl. ex Schaedtler (Arecaceae) **IPNI:** 664697-1 **Common name:** chontaduro del diablo **Habit:** Shrub, Caespitose palm **Use:** HF **Origin:** Native **Region:** Pacific **Dept.:** ANT, CAU, CHO, COR, NAR, VAC **Elev.:** 0-700 m.a.s.l. **IUCN Habitat code:** 1,3,4,5,14 **Conservation status: (IUCN)** NE **(CPSG)** NT

909 *Bactris pilosa* H.Karst. (Arecaceae) **IPNI:** 664717-1 **Common name:** chacarrá, chonta, chúntaro, lata, lata blanca, lata macho, lata montañera, lata peluda, pelo de saíno **Habit:** Shrub, Caespitose palm **Use:** HF **Origin:** Native **Region:** Caribbean, Orinoquia, Pacific **Dept.:** ANT, ARA, BOL, CAL, CES, CHO, COR, CUN, MAG, NSA, SAN, SUC, TOL **Elev.:** 0-1000 m.a.s.l. **IUCN Habitat code:** 1,2,3,5,14 **Conservation status: (IUCN)** NE **(CPSG)** NT

910 *Bactris riparia* Mart. (Arecaceae) **IPNI:** 664736-1 **Common name:** chontadurillo, chontaduro de pescado, chontilla **Habit:** Shrub, Tree **Use:** HF **Origin:** Native **Region:** Amazonia **Dept.:** AMA, CAQ, PUT, VAU **Elev.:** 80-300 m.a.s.l. **IUCN Habitat code:** 1,5 **Conservation status: (IUCN)** NE **(CPSG)** LC

911 *Bactris schultesii* (L.H.Bailey) Glassman (Arecaceae) **IPNI:** 28556-2 **Common name:** chontilla **Habit:** Subshrub, Caespitose palm **Use:** HF, MA **Origin:** Native **Region:** Amazonia, Andean **Dept.:** AMA, CAQ, CAU, PUT **Elev.:** 120-1200 m.a.s.l. **IUCN Habitat code:** 1,5 **Conservation status: (IUCN)** NE **(CPSG)** LC

912 *Bactris setulosa* H.Karst. (Arecaceae) **IPNI:** 664746-1 **Common name:** chacarrá, chacarrá macho, chacarrás, chapí, chascarrial, chascarrá, jingapá, macana, piganá, pinganá **Habit:** Tree, Caespitose palm **Use:** EU, HF, MA, ME **Origin:** Native **Region:** Andean, Pacific **Dept.:** ANT, ARA, BOY, CAL, CAQ, CAU, CHO, CUN, MAG, MET, NAR, NSA, PUT, SAN, VAC **Elev.:** 0-1800 m.a.s.l. **IUCN Habitat code:** 1,3,5,14 **Conservation status: (IUCN)** NT **(CPSG)** LC

913 *Bactris simplicifrons* Mart. (Arecaceae) **IPNI:** 664750-1 **Common name:** chontadurillo, chontadurillo liso, chontaduro de lagartija, chontaduro de rana de rastrojo, espina, palma de grulla **Habit:** Shrub, Subshrub, Caespitose palm **Use:** HF **Origin:** Native **Region:** Amazonia, Guiana Shield, Orinoquia, Pacific **Dept.:** AMA, CAQ, COR, GUA, GUV, PUT, VAU, VID **Elev.:** 80–400 m a.s.l. **Conservation status: (IUCN)** NE

914 *Bactris sphaerocarpa* Trail (Arecaceae) **IPNI:** 664754-1 **Habit:** Shrub, Subshrub, Caespitose palm **Use:** HF **Origin:** Native **Region:** Amazonia **Dept.:** AMA **Elev.:** 300-800 m.a.s.l. **IUCN Habitat code:** 1 **Conservation status: (IUCN)** NE **(CPSG)** LC

915 *Bismarckia nobilis* Hildebrandt & H.Wendl. (Arecaceae) **IPNI:** 664859-1 **Common name:** bismarckia, palma Bismarck, palma bismarckia **Habit:** Tree **Use:** EU, HF, MA **Origin:** Cultivated **Dept.:** ANT, CUN, MET, QUI, TOL, VAC **Elev.:** 0–1600 m a.s.l. **IUCN Habitat code:** 3,14 **Conservation status: (IUCN)** LC

916 *Caryota mitis* Lour. (Arecaceae) **IPNI:** 665616-1 **Common name:** cola de pescado, mariposa, palma mariposa, palma mariposa chica **Habit:** Shrub, Tree **Use:** EU, FU, GS, HF, MA, ME, PO, SU **Origin:** Cultivated **Region:** Caribbean **Dept.:** ANT, CUN, SAP, TOL, VAC **Elev.:** 0–2600 m a.s.l. **IUCN Habitat code:** 14 **Conservation status: (IUCN)** LC

917 *Caryota urens* L. (Arecaceae) **IPNI:** 665636-1 **Common name:** cola de pescado, mariposa, palma cola de pescado, palma mariposa, palma mariposa grande **Habit:** Tree, Solitary palm **Use:** AF, EU, FU, HF, MA, ME, PO, SU **Origin:** Cultivated **Dept.:** ANT, CUN, QUI, RIS, TOL, VAC **Elev.:** 0–2600 m a.s.l. **IUCN Habitat code:** 1,14 **Conservation status: (IUCN)** LC

918 *Ceroxylon alpinum* Bonpl. ex DC. (Arecaceae) **IPNI:** 30119294-2 **Common name:** chonta, palma de cera, palma real **Habit:** Tree, Solitary palm **Use:** AF, HF, MA, ME, SU **Origin:** Native **Region:** Andean **Dept.:** ANT, CAL, CUN, QUI, RIS, VAC **Elev.:** 1400-2000 m.a.s.l. **IUCN Habitat code:** 1,3,14 **Conservation status: (IUCN)** EN **(CPSG)** EN

919 *Ceroxylon ceriferum* (H.Karst.) Pittier (Arecaceae) **IPNI:** 53436-2 **Common name:** palma de cera, palma de ramo **Habit:** Tree, Solitary palm **Use:** HF **Origin:** Native **Region:** Andean **Dept.:** CES, LAG, MAG **Elev.:** 1900-2800 m.a.s.l. **IUCN Habitat code:** 1 **Conservation status: (IUCN)** NE **(CPSG)** EN

920 *Ceroxylon parvifrons* (Engel) H.Wendl. (Arecaceae) **IPNI:** 665695-1 **Common name:** palma de ramo, ramo **Habit:** Tree **Use:** FU, HF, MA, SU **Origin:** Native **Region:** Andean **Dept.:** ANT, BOY, CAL, CAU, HUI, NAR, NSA, PUT, QUI, RIS, TOL, VAC **Elev.:** 1900-3200 m.a.s.l. **IUCN Habitat code:** 1,3,4,14 **Conservation status: (IUCN)** NE **(CPSG)** VU

921 *Ceroxylon quindiuense* (H.Karst.) H.Wendl. (Arecaceae) **IPNI:** 665697-1 **Common name:** chonta, palma de cera, palma de ramo **Habit:** Tree, Solitary palm **Use:** AF, EU, FU, HF, MA, ME, SU **Origin:** Native **Region:** Andean **Dept.:** ANT, BOG, BOY, CAL, CAQ, CUN, MET, NSA, PUT, QUI, RIS, SAN, TOL, VAC **Elev.:** 1550-3100 m.a.s.l. **IUCN Habitat code:** 1,3,4,14 **Conservation status: (IUCN)** VU **(CPSG)** EN

922 *Ceroxylon sasaimae* Galeano (Arecaceae) **IPNI:** 300985-2 **Common name:** palma de cera, palma de ramo, palma real **Habit:** Tree, Solitary palm **Use:** ME, SU **Origin:** Endemic **Region:** Andean **Dept.:** ANT, CUN **Elev.:** 1350-1800 m.a.s.l. **IUCN Habitat code:** 1,2,3,14 **Conservation status: (IUCN)** CR **(CPSG)** EN

923 *Ceroxylon ventricosum* Burret (Arecaceae) **IPNI:** 53445-2 **Common name:** palma real **Habit:** Tree, Solitary palm **Use:** AF, EU, FU, HF, MA, SU **Origin:** Native **Region:** Andean **Dept.:** CAU, NAR, PUT **Elev.:** 1600-3000 m.a.s.l. **IUCN Habitat code:** 1 **Conservation status: (IUCN)** NE **(CPSG)** EN

924 *Ceroxylon vogelianum* (Engel) H.Wendl. (Arecaceae) **IPNI:** 665703-1 **Common name:** chile, chonta, gallinazo, palma de cera, palma de ramo, palma negra, palma ramo, palmito real **Habit:** Tree, Solitary palm **Use:** EU, HF, MA, SU **Origin:** Native **Region:** Andean **Dept.:** ANT, BOY, CAL, CAQ, CAU, CUN, HUI, NAR, NSA, PUT, QUI, RIS, SAN, VAC **Elev.:** 1900-3000 m.a.s.l. **IUCN Habitat code:** 1,2,3,14 **Conservation status: (IUCN)** NE **(CPSG)** VU

925 *Chamaedorea elegans* Mart. (Arecaceae) **IPNI:** 54865-2 **Common name:** linda, palma bella **Habit:** Herb, Subshrub, Solitary palm **Use:** EU, HF **Origin:** Cultivated **Dept.:** ANT, CUN, VAC **Elev.:** 0–2600 m a.s.l. **IUCN Habitat code:** 1,3 **Conservation status: (IUCN)** NE

926 *Chamaedorea ernesti-augusti* H.Wendl. (Arecaceae) **IPNI:** 54867-2 **Use:** EU **Region:** Andean **Dept.:** BOG **Conservation status: (IUCN)** NE

927 *Chamaedorea linearis* (Ruiz & Pav.) Mart. (Arecaceae) **IPNI:** 54901-2 **Common name:** caña de San Pablo, caña de víbora, chanul, inguanul, inguanul grande, macana, matamba, palmiche, palmilla, turumuto **Habit:** Shrub, Solitary palm **Use:** EU, HF, ME **Origin:** Native **Region:** Andean, Pacific **Dept.:** ANT, CAL, CAQ, CAU, CHO, HUI, LAG, MAG, NAR, NSA, PUT, QUI, RIS, SAN, TOL, VAC **Elev.:** 300-2800 m.a.s.l. **IUCN Habitat code:** 1,3,5,14 **Conservation status: (IUCN)** NE **(CPSG)** NT

928 *Chamaedorea pauciflora* Mart. (Arecaceae) **IPNI:** 54928-2 **Habit:** Herb, Subshrub, Solitary palm **Use:** EU, HF, MA, ME, SU **Origin:** Native **Region:** Amazonia **Dept.:** AMA, CAQ, PUT **Elev.:** 80-410 m.a.s.l. **IUCN Habitat code:** 1,3,5,14 **Conservation status: (IUCN)** NE **(CPSG)** LC

929 *Chamaedorea pinnatifrons* (Jacq.) Oerst. (Arecaceae) **IPNI:** 665826-1 **Common name:** caña de San Pablo, caña de víbora, carmaná, caña de San Pablo, caña de molinillo, caña de víbora, colegallo, iguanul, inguanul, inguanul pequeño, labicá, mataculebra, mataculebra verde, matamba, molenillo, molinilla, molinillo, palma de culebra, palma de **Habit:** Subshrub, Solitary palm **Use:** AF, EU, HF, MA, ME, SU **Origin:** Native **Region:** Amazonia, Andean, Pacific **Dept.:** AMA, ANT, BOY, CAL,

CAQ, CAS, CAU, CHO, COR, CUN, HUI, LAG, MAG, MET, NAR, NSA, PUT, QUI, RIS, SAN, TOL, VAC **Elev.:** 30-2500 m.a.s.l. **IUCN Habitat code:** 1,2,3,4,5,14 **Conservation status: (IUCN)** LC **(CPSG)** LC

930 *Chamaedorea seifrizii* Burret (Arecaceae) **IPNI:** 54954-2 **Common name:** palma bambú **Habit:** Shrub, Subshrub, Solitary palm **Use:** EU **Origin:** Cultivated **Dept.:** ANT, CUN, TOL, VAC **Elev.:** 0–2600 m a.s.l. **Conservation status: (IUCN)** NE

931 *Chamaedorea tepejilote* Liebm. (Arecaceae) **IPNI:** 54964-2 **Common name:** molinillo **Habit:** Shrub, Solitary palm **Use:** HF, ME **Origin:** Native **Region:** Andean, Pacific **Dept.:** ANT, CHO, QUI, VAC **Elev.:** 0-1500 m.a.s.l. **IUCN Habitat code:** 1,3,14 **Conservation status: (IUCN)** NE **(CPSG)** NT

932 *Chamaerops humilis* L. (Arecaceae) **IPNI:** 665917-1 **Habit:** Shrub **Use:** AF, EU, FU, GS, HF, MA, ME **Origin:** Cultivated **Dept.:** BOG, CUN **Elev.:** 2600–2600 m a.s.l. **Conservation status: (IUCN)** LC

933 *Chelyocarpus repens* F.Kahn & K.Mejia (Arecaceae) **IPNI:** 274480-2 **Habit:** Subshrub, Solitary palm **Use:** HF, MA **Origin:** Native **Region:** Amazonia **Dept.:** AMA **Elev.:** 90-90 m.a.s.l. **IUCN Habitat code:** 1 **Conservation status: (IUCN)** NE **(CPSG)** LC

934 *Chelyocarpus ulei* Dammer (Arecaceae) **IPNI:** 55986-2 **Habit:** Shrub, Solitary palm **Use:** AF, FU, HF, MA, SU **Origin:** Native **Region:** Amazonia **Dept.:** AMA, CAQ **Elev.:** 300-300 m.a.s.l. **IUCN Habitat code:** 1,5 **Conservation status: (IUCN)** NE **(CPSG)** LC

935 *Coccothrinax argentata* (Jacq.) L.H.Bailey (Arecaceae) **IPNI:** 62341-2 **Common name:** pactá, palma de plata, palmicha, wild pop thatch **Habit:** Shrub, Small tree, Solitary palm **Use:** EU, MA **Origin:** Native **Region:** Caribbean **Dept.:** SAP **Elev.:** 0-300 m.a.s.l. **IUCN Habitat code:** 1 **Conservation status: (IUCN)** LC **(CPSG)** EN

936 *Cocos nucifera* L. (Arecaceae) **IPNI:** 666160-1 **Common name:** coca, coco, coco caballero, cocotera, cocotero, coquera, coquero, mata de coco, palm de coco, palma cocotera, palma de coco, palo de coco, típico **Habit:** Tree, Solitary palm **Use:** AF, EU, FU, GS, HF, IF, MA, ME, PO, SU **Origin:** Cultivated, Naturalised **Region:** Amazonia, Andean, Guiana Shield, Caribbean, Orinoquia, Pacific **Dept.:** AMA, ANT, ARA, ATL, BOL, BOY, CAL, CAQ, CAS, CAU, CES, CHO, COR, CUN, GUA, GUV, HUI, LAG, MAG, MET, NAR, NSA, PUT, QUI, RIS, SAP, SAN, SUC, TOL, VAC, VAU, VID **Elev.:** 0–1500 m a.s.l. **IUCN Habitat code:** 1,2,3,5,14 **Conservation status: (IUCN)** NE

937 *Copernicia tectorum* (Kunth) Mart. (Arecaceae) **IPNI:** 666287-1 **Common name:** palma chará, palma saraza, palma sará, palmicha, palmiche, páutta, sará **Habit:** Shrub, Tree, Solitary palm **Use:** AF, HF, MA **Origin:** Native **Region:** Caribbean **Dept.:** ATL, BOL, CES, LAG, MAG, SUC **Elev.:** 0-100 m.a.s.l. **IUCN Habitat code:** 1,2,14 **Conservation status: (IUCN)** LC **(CPSG)** NT

938 *Cryosophila kalbreyeri* (Dammer ex Burret) Dahlgren (Arecaceae) **IPNI:** 666356-1 **Common name:** barbasco, barbasco de escoba, nupa, palma barbasco, palma escoba **Habit:** Shrub, Tree, Solitary palm **Use:** MA **Origin:** Native **Region:** Andean, Caribbean, Pacific **Dept.:** ANT, BOL, BOY, CHO, COR, SUC **Elev.:** 0-1200 m.a.s.l. **IUCN Habitat code:** 1,2,3,14 **Conservation status: (IUCN)** NT **(CPSG)** VU

939 *Cyrtostachys renda* Blume (Arecaceae) **IPNI:** 666411-1 **Common name:** palma panameña, palma pintalabios, palma roja **Habit:** Shrub, Solitary palm **Use:** EU, MA, ME **Origin:** Cultivated **Region:** Andean, Caribbean **Dept.:** ANT, CUN, MET, QUI, VAC **Elev.:** 400–1500 m a.s.l. **Conservation status: (IUCN)** NE

940 *Desmoncus cirrhifer* A.H.Gentry & Zardini (Arecaceae) **IPNI:** 275791-2 **Common name:** guasca matamba, jaladanta, matamba, piyandé **Habit:** Climbing **Use:** HF **Origin:** Native **Region:** Pacific **Dept.:** ANT, CAU, CHO, NAR, VAC **Elev.:** 0-800 m.a.s.l. **IUCN Habitat code:** 1,3,14 **Conservation status: (IUCN)** NE **(CPSG)** LC

941 *Desmoncus giganteus* A.J.Hend. (Arecaceae) **IPNI:** 981498-1 **Common name:** pendejón **Habit:** Liana, Climbing **Use:** AF, HF, MA, ME, SU **Origin:** Native **Region:** Amazonia **Dept.:** AMA, CAQ, GUA, PUT, VAU **Elev.:** 200-600 m.a.s.l. **IUCN Habitat code:** 1,2,3 **Conservation status: (IUCN)** NE **(CPSG)** LC

942 *Desmoncus horridus* Splitg. ex Mart. (Arecaceae) **IPNI:** 666606-1 **Habit:** Climbing **Use:** MA **Origin:** Native **Region:** Amazonia, Guiana Shield, Orinoquia **Dept.:** ARA, CAS, GUA, VAU **Elev.:** 100-500 m.a.s.l. **IUCN Habitat code:** 1,3,14 **Conservation status: (IUCN)** NE **(CPSG)** LC

943 *Desmoncus mitis* Mart. (Arecaceae) **IPNI:** 666632-1 **Common name:** atajadanta, bejuco alcalde, jalapatrás, yachitara, yasitara **Habit:** Climbing **Use:** AF, HF, MA, ME, SU **Origin:** Native **Region:** Amazonia **Dept.:** AMA, PUT **Elev.:** 80-400 m.a.s.l. **IUCN Habitat code:** 1,3,5,14 **Conservation status: (IUCN)** NE **(CPSG)** LC

944 *Desmoncus orthacanthos* Mart. (Arecaceae) **IPNI:** 666637-1 **Common name:** atajadanta, camahua, matamba, tirapatrás **Use:** AF, HF, MA, ME, SU **IUCN Habitat code:** 1,2,3,4,5,14 **Conservation status: (IUCN)** NE

945 *Desmoncus polyacanthos* Mart. (Arecaceae) **IPNI:** 666648-1 **Common name:** atajadanta, diente de culebra, namemo, yachitara, yashitara, yasitara **Habit:** Climbing **Use:** AF, HF, MA, SU **Origin:** Native **Region:** Amazonia, Guiana Shield, Orinoquia **Dept.:** AMA, ARA, GUA, MET, NSA, PUT, VID **Elev.:** 80-300 m.a.s.l. **IUCN Habitat code:** 1,2,3,4,5,14 **Conservation status: (IUCN)** NE **(CPSG)** LC

946 *Dictyocaryum lamarckianum* (Mart.) H.Wendl. (Arecaceae) **IPNI:** 666671-1 **Common name:** barrigona, barrigona blanca, bombona, chicón, chontula, corozo, mapora, palma bombona, tagua **Habit:** Tree, Solitary palm **Use:** AF, EU, HF, MA, SU **Origin:** Native **Region:** Andean **Dept.:** ANT, BOY, CAQ, CAU, CHO, HUI, MAG, MET, RIS, VAC **Elev.:** 400-2100 m.a.s.l. **IUCN Habitat code:** 1,3,4,14 **Conservation status: (IUCN)** NE **(CPSG)** LC

947 *Dictyocaryum ptarianum* (Steyerm.) H.E.Moore & Steyerm. (Arecaceae) **IPNI:** 79985-2 **Common name:** bombona, bombona falsa, maripa **Habit:** Tree, Solitary palm **Use:** HF, MA **Origin:** Native **Region:** Amazonia, Guiana Shield **Dept.:** AMA **Elev.:** 100-300 m.a.s.l. **IUCN Habitat code:** 1 **Conservation status: (IUCN)** NE **(CPSG)** LC

948 *Dypsis decaryi* (Jum.) Beentje & J.Dransf. (Arecaceae) **IPNI:** 985350-1 **Common name:** palma triangular, triangularis **Habit:** Tree, Solitary palm **Use:** EU, GS, HF, MA **Origin:** Cultivated **Dept.:** ANT, CUN, TOL, VAC **Elev.:** 500–1500 m a.s.l. **Conservation status: (IUCN)** VU

949 *Dypsis lutescens* (H.Wendl.) Beentje & J.Dransf. (Arecaceae) **IPNI:** 985360-1 **Common name:** areca, palma areca **Habit:** Shrub **Use:** EU, ME **Origin:** Cultivated **Dept.:** AMA, ANT, ARA, ATL, BOG, BOL, BOY, CAL, CAQ, CAS, CAU, CES, CHO, COR, CUN, GUA, GUV, HUI, LAG, MAG, MET, NAR, NSA, PUT, QUI, RIS, SAP, SAN, SUC, TOL, VAC, VAU, VID **Elev.:** 0–2600 m a.s.l. **IUCN Habitat code:** 1,3,14 **Conservation status: (IUCN)** NT

950 *Dypsis madagascariensis* D.T.Fish (Arecaceae) **IPNI:** 77211330-1 **Habit:** Tree, Solitary palm **Use:** EU, GS, HF, MA, ME **Origin:** Endemic, Cultivated **Dept.:** ANT, CUN, VAC **Elev.:** 500–1500 m a.s.l. **Conservation status: (IUCN)** LC

951 *Elaeis guineensis* Jacq. (Arecaceae) **IPNI:** 666802-1 **Common name:** palma africana, palma africana de aceite, palma de aceite **Habit:** Tree, Solitary palm **Use:** AF, EU, FU, GS, HF, MA, ME, PO, SU **Origin:** Cultivated **Dept.:** AMA, ANT, BOL, CAQ, CAS, CES, CHO, COR, CUN, MAG, MET, NAR, NSA, PUT, QUI, RIS, SAN, SUC, TOL, VAC **Elev.:** 100–1500 m a.s.l. **IUCN Habitat code:** 1,2,3,4,14 **Conservation status: (IUCN)** LC

952 *Elaeis oleifera* (Kunth) Cortés (Arecaceae) **IPNI:** 88982-2 **Common name:** coroza, corozo, corozo de manteca, nolí, palama ñoli, palma de corozo, palma de nolí, palma de ñolí, palma nolí, palma ñolí, palmaeñolí, yolí, ñolí **Habit:** Shrub, Tree, Solitary palm **Use:** AF, EU, FU, GS, HF, MA, ME **Origin:** Native **Region:** Caribbean, Pacific **Dept.:** ANT, BOL, BOY, CAL, CES, CHO, COR, MAG, SAN, SUC, TOL **Elev.:** 0-300 m.a.s.l. **IUCN Habitat code:** 1,2,3,5,14 **Conservation status: (IUCN)** NE **(CPSG)** EN

953 *Euterpe catinga* Wallace (Arecaceae) **IPNI:** 317294-2 **Common name:** asaí, asaí de sabana, asaí falso, asaí rojo, asaí

sabanero, guasaí de sabana, guasaí pequeño, manaca, manaquita, najoemimoeñu **Habit:** Tree, Caespitose palm **Use:** HF, MA, ME **Origin:** Native **Region:** Amazonia, Guiana Shield **Dept.:** AMA, CAQ, GUA, GUV, VAU **Elev.:** 80-350 m.a.s.l. **IUCN Habitat code:** 1,3,4,5,14 **Conservation status: (IUCN)** NE **(CPSG)** LC

954 *Euterpe oleracea* Mart. (Arecaceae) **IPNI:** 666941-1 **Common name:** acaí, asaí, asaí de Pará, bambil, cansin, manaca brasilera, maquenco, maquenque, murrapho, murrapo, naidi, naidí, nãhõemimueñi, palma murrapo, palma naidí, palma triste, palmicha, palmiche, pará, tapafrío, tarábwa mĩhĩ, tarapãnẽ, wasái do pará **Habit:** Tree, Caespitose palm **Use:** AF, EU, HF, MA, ME **Origin:** Native **Region:** Pacific **Dept.:** ANT, BOY, CAU, CHO, COR, NAR, SAN, VAC **Elev.:** 0-200 m.a.s.l. **IUCN Habitat code:** 1,2,3,4,5,14 **Conservation status: (IUCN)** NE **(CPSG)** LC

955 *Euterpe precatoria* Mart. (Arecaceae) **IPNI:** 666949-1 **Common name:** asaí, asaí verdadero, báha, cecilia, cecilio, chapín, cicilio, cuyara, divá, emimoe, fiífia, guasay, guasaí, guypani, hinkató, huasaí, inibue,kerantë´´gn, macana, maizpepe, malakala, mana, manaca, manaco, manakai, manakaiboto, manakái, manaqué, manáka **Habit:** Tree, Solitary palm **Use:** AF, EU, FU, HF, MA, ME, PO, SU **Origin:** Native **Region:** Amazonia, Guiana Shield, Orinoquia **Dept.:** AMA, ARA, CAQ, CAS, GUA, GUV, MET, NSA, PUT, VAU, VID **Elev.:** 0-450 m.a.s.l. **Conservation status: (IUCN)** NE **(CPSG)** LC

956 *Geonoma brongniartii* Mart. (Arecaceae) **IPNI:** 667027-1 **Common name:** sampablo **Habit:** Herb, Shrub, Subshrub, Solitary palm **Use:** MA **Origin:** Native **Region:** Amazonia, Andean, Guiana Shield, Orinoquia **Dept.:** AMA, CAQ, CAU, MET, PUT **Elev.:** 75-1200 m.a.s.l. **IUCN Habitat code:** 1,3,5,14 **Conservation status: (IUCN)** NE **(CPSG)** LC

957 *Geonoma camana* Trail (Arecaceae) **IPNI:** 667031-1 **Habit:** Shrub, Subshrub, Solitary palm **Use:** AF, HF, MA, SU **Origin:** Native **Region:** Amazonia **Dept.:** AMA, CAQ, PUT, VAU **Elev.:** 80-300 m.a.s.l. **IUCN Habitat code:** 1,5 **Conservation status: (IUCN)** NE **(CPSG)** LC

958 *Geonoma chococola* Wess.Boer (Arecaceae) **IPNI:** 667044-1 **Common name:** calbón, zanca de pava **Habit:** Shrub, Solitary palm **Use:** MA **Origin:** Native **Region:** Andean, Pacific **Dept.:** ANT, CAU, CHO, NAR, VAC **Elev.:** 0-700 m.a.s.l. **IUCN Habitat code:** 1,5,14 **Conservation status: (IUCN)** NE **(CPSG)** LC

959 *Geonoma cuneata* H.Wendl. ex Spruce (Arecaceae) **IPNI:** 667052-1 **Common name:** caló, hoja de tortuga, palmilla, panga negra, panguita, puy de paloma, rabihorcao, revolcado **Habit:** Herb, Shrub, Acaulescent palm, Solitary palm **Use:** AF, HF, MA, ME, SU **Origin:** Native **Region:** Andean, Pacific **Dept.:** ANT, CAU, CHO, COR, NAR, VAC **Elev.:** 0-1500 m.a.s.l. **IUCN Habitat code:** 1,2,3,5,14 **Conservation status: (IUCN)** NE **(CPSG)** LC

960 *Geonoma deversa* (Poit.) Kunth (Arecaceae) **IPNI:** 667064-1 **Common name:** bolenillo, chontilla, guaguarín, guagüera, mataculebra, mataculebra blanco, molinillo, palmillo, pata de venado, puy de paloma, sampablo, sampaulo, tataba **Habit:** Shrub, Caespitose palm **Use:** EU, HF, MA, ME, SU **Origin:** Native **Region:** Amazonia, Orinoquia, Pacific **Dept.:** AMA, ANT, ARA, CAQ, CAS, CAU, CHO, COR, GUA, GUV, MET, NAR, PUT, SAN, VAC, VAU, VID **Elev.:** 0-1200 m.a.s.l. **IUCN Habitat code:** 1,3,4,5,14 **Conservation status: (IUCN)** NE **(CPSG)** LC

961 *Geonoma interrupta* (Ruiz & Pav.) Mart. (Arecaceae) **IPNI:** 667119-1 **Common name:** barayá, cola de gallo, colegallo, cortadera, maraya, marayón, palma blanca, palmicha, palmiche, panguita, patevenado, rabo de gallo, rabo de zorro, sampablo, sampedrito, verduguillo **Habit:** Shrub, Solitary palm **Use:** AF, HF, MA **Origin:** Native **Region:** Andean, Pacific **Dept.:** ANT, BOY, CAU, CHO, MAG, MET, NSA, SAN, VAC **Elev.:** 20-1540 m.a.s.l. **IUCN Habitat code:** 1,2,3,5,14 **Conservation status: (IUCN)** NE **(CPSG)** LC

962 *Geonoma leptospadix* Trail (Arecaceae) **IPNI:** 667140-1 **Habit:** Shrub, Caespitose palm **Use:** MA **Origin:** Native **Region:** Amazonia **Dept.:** AMA, CAQ, GUV, PUT, VAU **Elev.:** 100-500 m.a.s.l. **IUCN Habitat code:** 1,3,5 **Conservation status: (IUCN)** NE **(CPSG)** LC

963 *Geonoma longepedunculata* Burret (Arecaceae) **IPNI:** 108914-2 **Habit:** Subshrub **Use:** MA **Origin:** Native **Region:** Amazonia, Guiana Shield **Dept.:** MET, PUT **Elev.:** 200-500 m.a.s.l. **IUCN Habitat code:** 1,3,14 **Conservation status: (IUCN)** NE **(CPSG)** LC

964 *Geonoma macrostachys* Mart. (Arecaceae) **IPNI:** 667158-1 **Common name:** horquetilla, sampablo, ubí **Habit:** Herb, Subshrub, Acaulescent palm **Use:** AF, HF, MA, ME, SU **Origin:** Native **Region:** Amazonia, Orinoquia **Dept.:** AMA, CAQ, PUT, VAU **Elev.:** 80-600 m.a.s.l. **IUCN Habitat code:** 1,3,5,14 **Conservation status: (IUCN)** NE **(CPSG)** LC

965 *Geonoma maxima* (Poit.) Kunth (Arecaceae) **IPNI:** 667168-1 **Common name:** chontilla, falso puy, osopanga, puy falso **Habit:** Shrub, Caespitose palm **Use:** AF, EU, HF, MA, SU **Origin:** Native **Region:** Amazonia, Guiana Shield, Pacific **Dept.:** AMA, ANT, CAQ, CHO, COR, GUA, GUV, NSA, PUT, SAN, VAU, VID **Elev.:** 80-360 m.a.s.l. **IUCN Habitat code:** 1,3,4,5,14 **Conservation status: (IUCN)** NE **(CPSG)** LC

966 *Geonoma multisecta* (Burret) Burret (Arecaceae) **IPNI:** 108938-2 **Habit:** Subshrub, Solitary palm **Use:** MA **Origin:** Native **Region:** Amazonia, Andean **Dept.:** CAQ, PUT **Elev.:** 200-1000 m.a.s.l. **IUCN Habitat code:** 1,3,14 **Conservation status: (IUCN)** NE **(CPSG)** LC

967 *Geonoma orbignyana* Mart. (Arecaceae) **IPNI:** 667191-1 **Common name:** huasipanga, maquenque, maraya, palma de San Pablo, palma de corona, palma puy, palma ñonce, palmicha, sampablo, verduguillo **Habit:** Shrub, Subshrub, Solitary palm **Use:** HF, MA, SU **Origin:** Native **Region:** Andean **Dept.:** ANT, BOY, CAL, CAQ, CAU, CES, CHO, CUN, HUI, LAG, MAG, MET, NAR, NSA, PUT, QUI, RIS, SAN, TOL, VAC **Elev.:** 700-2900 m.a.s.l. **IUCN Habitat code:** 1,3,14 **Conservation status: (IUCN)** NE **(CPSG)** NT

968 *Geonoma poeppigiana* Mart. (Arecaceae) **IPNI:** 667216-1 **Common name:** osopanga **Habit:** Shrub, Solitary palm **Use:** HF, MA, ME, SU **Origin:** Native **Region:** Amazonia **Dept.:** AMA, CAQ, PUT **Elev.:** 80-250 m.a.s.l. **IUCN Habitat code:** 1,4,5,14 **Conservation status: (IUCN)** NE **(CPSG)** LC

969 *Geonoma stricta* (Poit.) Kunth (Arecaceae) **IPNI:** 667262-1 **Common name:** sampaulo **Habit:** Subshrub, Caespitose palm **Use:** AF, MA, ME, SU **Origin:** Native **Region:** Amazonia, Guiana Shield, Pacific **Dept.:** AMA, ANT, CAQ, CHO, PUT, VAU **Elev.:** 80-1000 m.a.s.l. **IUCN Habitat code:** 1,3,4,5,14 **Conservation status: (IUCN)** NE **(CPSG)** LC

970 *Geonoma triglochin* Burret (Arecaceae) **IPNI:** 108998-2 **Common name:** chontilla, cola de morada, sampablito, sampablo **Habit:** Shrub, Solitary palm **Use:** MA, ME **Origin:** Native **Region:** Andean **Dept.:** CAQ, CAU, PUT **Elev.:** 700-1500 m.a.s.l. **IUCN Habitat code:** 1,3,14 **Conservation status: (IUCN)** NE **(CPSG)** NT

971 *Geonoma undata* Klotzsch (Arecaceae) **IPNI:** 667287-1 **Common name:** cananá, carmaná, chalar, lindona, osapanga, palma colepato, palma macuma, palmicha, palmiche, palmicho **Habit:** Shrub, Small tree, Tree, Solitary palm **Use:** HF, MA **Origin:** Native **Region:** Andean **Dept.:** ANT, BOG, BOY, CAL, CAQ, CAU, CES, CHO, CUN, HUI, LAG, MAG, MET, NAR, NSA, PUT, QUI, RIS, SAN, TOL, VAC **Elev.:** 1200-3200 m.a.s.l. **IUCN Habitat code:** 1,3,4,14 **Conservation status: (IUCN)** NE **(CPSG)** NT

972 *Howea forsteriana* (F.Muell.) Becc. (Arecaceae) **IPNI:** 667434-1 **Common name:** kencia, kentia **Habit:** Tree, Solitary palm **Use:** EU **Origin:** Cultivated **Dept.:** ANT, CUN, VAC **Elev.:** 0–2600 m a.s.l. **IUCN Habitat code:** 14 **Conservation status: (IUCN)** VU

973 *Hyophorbe verschaffeltii* H.Wendl. (Arecaceae) **IPNI:** 667456-1 **Common name:** cemento, palma de cemento **Habit:** Shrub, Solitary palm **Use:** EU, HF, MA **Origin:** Cultivated **Dept.:** ANT, CUN, VAC **Elev.:** 0–1600 m a.s.l. **IUCN Habitat code:** 14 **Conservation status: (IUCN)** CR

974 *Hyospathe elegans* Mart. (Arecaceae) **IPNI:** 667463-1 **Common name:** chontilla, guagualín, hoja de capa, mataculebra, palma blanca, palma chontilla, palma de verano, palmito, tumapá **Habit:** Shrub, Caespitose palm **Use:** HF, MA, ME **Origin:** Native **Region:** Amazonia, Andean, Pacific **Dept.:** AMA, ANT, ARA, BOY, CAQ, CAS, CAU, CHO, CUN, GUA, GUV, MET, NAR, PUT, RIS, VAC, VAU **Elev.:** 40-1900 m.a.s.l. **IUCN Habitat code:** 1,3,4,5,14 **Conservation status: (IUCN)** NE **(CPSG)** LC

975 *Iriartea deltoidea* Ruiz & Pav. (Arecaceae)
IPNI: 667594-1
Common name: aayae, ar, arrá, ayako, bahtápohkoño, barrigona, barrigona negra, barrigonilla, barriguda, bombo, bombona, bumbuhe, butsiboto, bëi', cachuda, cachuda barriguda, cachudo, choapo, chonta, chonta negra, chuapo, corneto, eiñaphí, etá, fegona, hayaku, hiaigin, **Habit:** Tree, Solitary palm **Use:** AF, EU, FU, HF, MA, ME, SU **Origin:** Native **Region:** Amazonia, Andean, Guiana Shield, Orinoquia, Pacific **Dept.:** AMA, ANT, CAQ, CAU, CHO, COR, GUA, GUV, MET, NAR, PUT, VAC, VAU **Elev.:** 0-1350 m.a.s.l. **IUCN Habitat code:** 1,3,4,5,14 **Conservation status:** **(IUCN)** LC **(CPSG)** LC

976 *Iriartella setigera* (Mart.) H.Wendl. (Arecaceae)
IPNI: 667622-1
Common name: bodoquera, boloquera, bombona, caraiñu, chontica, mavé, pachuba, pachubita, palma de cerbatana, pimpí, pona, ponilla, yaripa, yupatí, zanconcita **Habit:** Shrub, Small tree, Caespitose palm **Use:** MA, ME **Origin:** Native **Region:** Amazonia, Guiana Shield **Dept.:** AMA, CAQ, GUA, GUV, VAU, VID **Elev.:** 100-800 m.a.s.l. **IUCN Habitat code:** 1,3,4,5,14 **Conservation status:** **(IUCN)** NE **(CPSG)** LC

977 *Iriartella stenocarpa* Burret (Arecaceae)
IPNI: 130009-2
Habit: Shrub, Caespitose palm **Use:** MA, ME **Origin:** Native **Region:** Amazonia **Dept.:** AMA **Elev.:** 100-250 m.a.s.l. **IUCN Habitat code:** 1,5 **Conservation status:** **(IUCN)** NE **(CPSG)** LC

978 *Itaya amicorum* H.E.Moore (Arecaceae)
IPNI: 130738-2
Common name: bombona, bombonaje, bomboná, caraná bravo, iraca **Habit:** Shrub, Solitary palm **Use:** HF, MA, SU **Origin:** Native **Region:** Amazonia **Dept.:** AMA **Elev.:** 100-250 m.a.s.l. **IUCN Habitat code:** 1,5 **Conservation status:** **(IUCN)** LC **(CPSG)** LC

979 *Klopstockia utilis* H.Karst. (Arecaceae)
IPNI: 667716-1
Use: HF **Origin:** Endemic
Conservation status: **(IUCN)** NE

980 *Leopoldinia piassaba* Wallace (Arecaceae)
IPNI: 60447541-2
Common name: chiquichiqui, chìkichìki, fibra, hañõn, malama, marama, maramapé [palm], maramathe [fruit], maramá, maráma, palma chiquichiqui, palma de fibra, piasaba, sikisikiboto **Habit:** Tree, Solitary palm **Use:** AF, EU, HF, MA, ME, SU **Origin:** Native **Region:** Guiana Shield **Dept.:** GUA, VID **Elev.:** 80-200 m.a.s.l. **IUCN Habitat code:** 1,4,5 **Conservation status:** **(IUCN)** NE **(CPSG)** NT

981 *Leopoldinia pulchra* Mart. (Arecaceae)
IPNI: 667794-1
Common name: palmalito, palmarito, palmiche, palmito, yará **Habit:** Shrub, Caespitose palm **Use:** HF, MA **Origin:** Native **Region:** Guiana Shield **Dept.:** GUA, VID **Elev.:** 50-200 m.a.s.l. **IUCN Habitat code:** 1,4,5,14 **Conservation status:** **(IUCN)** NE **(CPSG)** LC

982 *Lepidocaryum tenue* Mart. (Arecaceae)
IPNI: 667804-1
Common name: aheko, ahi, bibimiku háheku, caraná, cigao háheku, deene háhe, diku háheku, dikuháhe imoháhe, ereri, hota mohe, háheku, iiamaiki, irapay, kaarú, karugiri, kotu, muhí, muĩña, muy, ngupe, niya háheku, potamé, pui espinoso, puy, puy de pava, puy de rajar, pu **Habit:** Shrub, Caespitose palm **Use:** AF, HF, MA, ME, SU **Origin:** Native **Region:** Amazonia, Guiana Shield **Dept.:** AMA, CAQ, GUA, VAU **Elev.:** 80-300 m.a.s.l. **IUCN Habitat code:** 1,3,5,14 **Conservation status:** **(IUCN)** NE **(CPSG)** LC

983 *Licuala grandis* (T.Moore) H.Wendl. (Arecaceae)
IPNI: 667880-1
Common name: licuala **Habit:** Shrub **Use:** EU, MA, ME **Origin:** Cultivated **Dept.:** ANT, CUN, VAC **Elev.:** 0–1500 m a.s.l. **IUCN Habitat code:** 1,2 **Conservation status:** **(IUCN)** NE

984 *Livistona chinensis* (Jacq.) R.Br. ex Mart. (Arecaceae)
IPNI: 668021-1
Common name: palma abanico, palma abanico de la China **Habit:** Tree, Solitary palm **Use:** EU, HF, MA, ME **Origin:** Cultivated **Dept.:** ANT, CUN, VAC **Elev.:** 0–1600 m a.s.l. **IUCN Habitat code:** 14 **Conservation status:** **(IUCN)** NE

985 *Manicaria martiana* Burret (Arecaceae)
IPNI: 152345-2
Common name: ubí **Habit:** Shrub, Caespitose palm **Use:** HF **Origin:** Native **Region:** Amazonia, Guiana Shield **Dept.:** AMA, VAU **Elev.:** 80-250 m.a.s.l. **IUCN Habitat code:** 1,3 **Conservation status:** **(IUCN)** NE **(CPSG)** LC

986 *Manicaria saccifera* Gaertn. (Arecaceae)
IPNI: 668116-1
Common name: ampiapúne,bohsomuhĩ́, bohsomũĩ, busú, cabecinegro, jiclilla, jicra, jigra, jícara, jíquera, kaĩngát, kaĩntokëgn, kéd, mũã pũrí,mekwáibak, muichiãbo, paja, palm de jícara, palma de jícara, palma de jíquera, palma jícara, sanagua, sanaguas, tṹkür, taah **Habit:** Tree, Caespitose palm **Use:** AF, FU, HF, MA, ME, SU **Origin:** Native **Region:** Amazonia, Guiana Shield, Pacific **Dept.:** AMA, ANT, CAQ, CAU, CHO, GUA, NAR, VAC, VAU **Elev.:** 0-250 m.a.s.l. **IUCN Habitat code:** 1,3,4,5,14 **Conservation status:** **(IUCN)** NE **(CPSG)** LC

987 *Mauritia carana* Wallace (Arecaceae)
IPNI: 319735-2
Common name: canangucha de sabana, canangucha falsa, cananguche falso, canangucho de sabana, caraná, china **Habit:** Tree, Solitary palm **Use:** HF, MA, SU **Origin:** Native **Region:** Guiana Shield **Dept.:** AMA, CAQ, GUA, GUV, VAU **Elev.:** 100-250 m.a.s.l. **IUCN Habitat code:** 1,3,4,5,14 **Conservation status:** **(IUCN)** CD **(CPSG)** LC

988 *Mauritia flexuosa* L.f. (Arecaceae)
IPNI: 668158-1
Common name: aguaje, canangucha, canangucha macho, cananguche, canangucho, dauri [red fruit], diwita, eú, hiaukiña, híduiki, idéwi, inhoboto, inohoboto, inóho, itebi, itewi, itewída, iyõ, iñéhe, ínẽ̀káhá, ínẽ́'ẽ̀tsóó'ò, kanangucho, kanongocho, kinena, konta, konéna, k **Habit:** Tree, Solitary palm **Use:** AF, EU, FU, HF, MA, ME, SU **Origin:** Native **Region:** Amazonia, Guiana Shield **Dept.:** AMA, ARA, CAQ, CAS, GUA, GUV, MET, PUT, VAU, VID **Elev.:** 40-450 m.a.s.l. **IUCN Habitat code:** 1,2,3,4,5,14 **Conservation status:** **(IUCN)** NE **(CPSG)** LC

989 *Mauritiella aculeata* (Kunth) Burret (Arecaceae)
IPNI: 154896-2
Common name: caranaí, cu?va, morichito **Habit:** Tree, Caespitose palm **Use:** AF, HF, MA, ME, SU **Origin:** Native **Region:** Guiana Shield **Dept.:** GUA, GUV, VAU, VID **Elev.:** 100-250 m.a.s.l. **IUCN Habitat code:** 1,4,5,14 **Conservation status:** **(IUCN)** NE **(CPSG)** LC

990 *Mauritiella armata* (Mart.) Burret (Arecaceae)
IPNI: 154897-2
Common name: cahuaia, cahuayo, cananguchilla, cananguchillo, cananguchillo de sabana, canangucho bravo, cuvañu, moriche macho **Habit:** Tree, Caespitose palm **Use:** AF, EU, FU, HF, MA **Origin:** Native **Region:** Amazonia, Guiana Shield, Orinoquia **Dept.:** AMA, ARA, CAQ, CAS, GUA, GUV, MET, PUT, VAU, VID **Elev.:** 50-500 m.a.s.l. **IUCN Habitat code:** 1,3,4,5,14 **Conservation status:** **(IUCN)** NE **(CPSG)** LC

991 *Mauritiella macroclada* (Burret) Burret (Arecaceae)
IPNI: 154903-2
Common name: miralsol, mirasol, quitasol **Habit:** Tree **Use:** EU, HF, MA, ME **Origin:** Native **Region:** Andean, Pacific **Dept.:** ANT, CAU, CHO, VAC **Elev.:** 0-100 m.a.s.l. **IUCN Habitat code:** 1,3,5,14 **Conservation status:** **(IUCN)** LC **(CPSG)** LC

992 *Mauritiella pumila* (Wallace) Burret (Arecaceae)
IPNI: 154909-2
Habit: Shrub, Caespitose palm **Use:** HF **Origin:** Native **Region:** Guiana Shield **Dept.:** GUA **Elev.:** 100-150 m.a.s.l. **IUCN Habitat code:** 1,4 **Conservation status:** **(IUCN)** NE **(CPSG)** LC

993 *Oenocarpus bacaba* Mart. (Arecaceae)
IPNI: 668509-1
Common name: bacaba, boriabea, ibacaba, milpesillo, milpesillos, patabá de guara, seje pequeño **Habit:** Tree, Solitary palm **Use:** HF, MA, ME **Origin:** Native **Region:** Guiana Shield **Dept.:** AMA, CAQ, GUA, VAU **Elev.:** 60-300 m.a.s.l. **IUCN Habitat code:** 1,3,4,14 **Conservation status:** **(IUCN)** NE **(CPSG)** LC

994 *Oenocarpus balickii* F.Kahn (Arecaceae)
IPNI: 281514-2
Common name: milpesillo **Habit:** Tree, Solitary palm **Use:** HF **Origin:** Native **Region:** Amazonia, Guiana Shield **Dept.:** AMA, GUA, VAU **Elev.:** 80-200 m.a.s.l. **IUCN Habitat code:** 1,3 **Conservation status:** **(IUCN)** NE **(CPSG)** LC

995 *Oenocarpus bataua* Mart. (Arecaceae)
IPNI: 668510-1
Common name: arú, ataiboto, babayhu igai, bareu puori, bateí, batú, bokohañu, boreyabeñu, bórúá, chapil, cõja, come, coomeji, coroba, coroiba, dobómaña, duebokohañu, duibokane, düü, gibiño, guakaria, gõsa, haadió kuume, hebukanu, hiidíko paña, ibe, iidikuumee, isoi, **Habit:** Tree, Solitary palm **Use:** AF, EU, FU, HF, MA, ME, SU **Origin:** Native **Region:** Amazonia, Andean, Guiana Shield, Orinoquia, Pacific **Dept.:** AMA, ANT, ARA, BOL, BOY, CAL, CAQ, CAS, CAU, CHO, COR, CUN, GUA, GUV, MET, NAR, NSA, PUT, RIS, SAN, VAC, VAU, VID **Elev.:** 0-1000 m.a.s.l. **IUCN Habitat code:** 1,3,4,5,14 **Conservation status:** **(IUCN)** NE **(CPSG)** LC

996 *Oenocarpus circumtextus* Mart. (Arecaceae)
IPNI: 668516-1
Common name: milpesillo, milpesillo de sabana, milpesillos **Habit:** Shrub, Solitary palm **Use:** HF, ME **Origin:** Native **Region:** Amazonia **Dept.:** AMA **Elev.:** 300–300 m a.s.l. **IUCN Habitat code:** 1 **Conservation status: (IUCN)** VU

997 *Oenocarpus makeru* R.Bernal, Galeano & A.J.Hend. (Arecaceae)
IPNI: 959997-1
Common name: makeru **Habit:** Tree, Solitary palm **Use:** HF, ME **Origin:** Endemic **Region:** Amazonia **Dept.:** AMA **Elev.:** 100-150 m.a.s.l. **IUCN Habitat code:** 1 **Conservation status: (IUCN)** LC **(CPSG)** LC

998 *Oenocarpus mapora* H.Karst. (Arecaceae)
IPNI: 668532-1
Use: AF, EU, FU, HF, MA, ME, SU **Origin:** Native **IUCN Habitat code:** 1,3,4,5,14 **Conservation status: (IUCN)** LC

999 *Oenocarpus minor* Mart. (Arecaceae)
IPNI: 668534-1
Common name: bacaba, chapil, chapililla, chapingo, chapín, cocoromo, colemula, dompedrito, manguengue, manquenque, maquenco, maquenque, milpesilla, milpesillo, milpesillo de rebalse, milpesillos, palma dompedrito, palma maquenque, pusuy, sampedrito, seje pequeño **Habit:** Tree, Caespitose palm **Use:** AF, HF, MA, ME **Origin:** Native **Region:** Amazonia, Andean, Guiana Shield, Orinoquia, Pacific **Dept.:** AMA, ANT, ARA, BOY, CAL, CAQ, CAS, CAU, CES, CHO, COR, MET, NAR, NSA, PUT, RIS, SAN, TOL, VAC, VAU, VID **Elev.:** 0-1000 m.a.s.l. **IUCN Habitat code:** 1,2,3,4,5,14 **Conservation status: (IUCN)** NE **(CPSG)** LC

1000 *Parajubaea cocoides* Burret (Arecaceae)
IPNI: 182952-2
Common name: coco, coco cumbe **Habit:** Tree, Solitary palm **Use:** EU, HF, MA **Origin:** Native, Cultivated **Region:** Andean **Dept.:** BOG, CUN, NAR **Elev.:** 1500–3000 m a.s.l. **IUCN Habitat code:** 14 **Conservation status: (IUCN)** NE

1001 *Phoenix canariensis* H.Wildpret (Arecaceae)
IPNI: 668909-1
Common name: palma de Canarias, palma de dátil de las Canarias, palma fénix **Habit:** Tree, Solitary palm **Use:** EU, GS, HF, ME **Origin:** Cultivated **Dept.:** ANT, BOG, BOY, CUN **Elev.:** 1500–2600 m a.s.l. **IUCN Habitat code:** 3,14 **Conservation status: (IUCN)** LC

1002 *Phoenix dactylifera* L. (Arecaceae)
IPNI: 668912-1
Common name: datilera, dátil, palm de dátil, palma datilera, palma de dátil **Habit:** Tree, Solitary palm **Use:** AF, EU, FU, GS, HF, MA, ME, SU **Origin:** Cultivated **Dept.:** ANT, BOG **Elev.:** 1500–1500 m a.s.l. **IUCN Habitat code:** 1,14 **Conservation status: (IUCN)** NE

1003 *Phoenix reclinata* Jacq. (Arecaceae)
IPNI: 668943-1
Common name: palma fénix **Habit:** Tree **Use:** AF, EU, FU, GS, HF, IF, MA, ME, SU **Origin:** Cultivated **Dept.:** ANT, BOG, CUN, TOL, VAC **Elev.:** 0–2600 m a.s.l. **Conservation status: (IUCN)** LC

1004 *Phoenix roebelenii* O'Brien (Arecaceae)
IPNI: 668945-1
Common name: palma robeleni, palma robelini, rebelén, rebelín, robelini **Habit:** Shrub, Solitary palm **Use:** EU, GS, HF **Origin:** Cultivated **Dept.:** ANT, BOG, CUN, TOL, VAC **IUCN Habitat code:** 1,3,14 **Conservation status: (IUCN)** NE

1005 *Pholidostachys dactyloides* H.E.Moore (Arecaceae)
IPNI: 193854-2
Common name: carmaná, chalar, chalá, rabo de gallo, santainés, yalar **Habit:** Shrub, Solitary palm **Use:** HF, MA **Origin:** Native **Region:** Andean, Pacific **Dept.:** ANT, CAU, CHO, NAR, RIS, VAC **Elev.:** 5-1500 m.a.s.l. **IUCN Habitat code:** 1,3,14 **Conservation status: (IUCN)** NE **(CPSG)** LC

1006 *Pholidostachys synanthera* (Mart.) H.E.Moore (Arecaceae)
IPNI: 193857-2
Common name: chalar, puy de pava, sampablo **Habit:** Shrub, Solitary palm **Use:** AF, HF, MA, ME, SU **Origin:** Native **Region:** Amazonia **Dept.:** AMA, ANT, CAQ, CHO, NAR, VAC **Elev.:** 0-1625 m.a.s.l. **IUCN Habitat code:** 1,2,3,5,14 **Conservation status: (IUCN)** NE **(CPSG)** LC

1007 *Phytelephas macrocarpa* Ruiz & Pav. (Arecaceae)
IPNI: 668996-1
Common name: antá, cabeza de negro, chipati, cuca, ngumũ chi, palma cabeza de negro, pátiu, sagu, taaudau, tagua, yarina, ãtá **Habit:** Shrub, Solitary palm **Use:** AF, EU, FU, HF, MA, ME, SU **Origin:** Native **Region:** Amazonia, Andean, Pacific **Dept.:** AMA, ANT, BOY, CAU, CES, CHO, COR, CUN, HUI, NSA, SAN, TOL, VAC **Elev.:** 0-1700 m.a.s.l. **IUCN Habitat code:** 1,3,4,14 **Conservation status: (IUCN)** NE **(CPSG)** NT

1008 *Phytelephas tenuicaulis* (Barfod) A.J.Hend. (Arecaceae)
IPNI: 981519-1
Common name: tagua, yarina **Habit:** Shrub, Caespitose palm **Use:** AF, EU, FU, HF, MA, ME **Origin:** Native **Region:** Amazonia **Dept.:** PUT **Elev.:** 280-350 m.a.s.l. **IUCN Habitat code:** 1,14 **Conservation status: (IUCN)** NE **(CPSG)** NT

1009 *Phytelephas tumacana* O.F.Cook (Arecaceae)
IPNI: 195814-2
Common name: tagua, tagua de tumaco **Habit:** Shrub, Tree, Solitary palm **Use:** EU, HF, MA, ME **Origin:** Endemic **Region:** Pacific **Dept.:** NAR **Elev.:** 0-200 m.a.s.l. **IUCN Habitat code:** 1,5 **Conservation status: (IUCN)** EN **(CPSG)** EN

1010 *Prestoea acuminata* (Willd.) H.E.Moore (Arecaceae)
IPNI: 210057-2
Common name: buitsaha, murrapho, palma colorada, palmicha, palmiche, palmiche colorado, palmicho, palmito **Habit:** Shrub, Tree, Caespitose palm **Use:** AF, EU, HF, MA, ME **Origin:** Native **Region:** Andean **Dept.:** ANT, BOY, CAL, CAQ, CES, CHO, CUN, HUI, LAG, MAG, MET, NAR, NSA, PUT, QUI, RIS, SAN, TOL, VAC **Elev.:** 1500-2650 m.a.s.l. **IUCN Habitat code:** 1,2,3,4,14 **Conservation status: (IUCN)** NE **(CPSG)** LC

1011 *Prestoea decurrens* H.E.Moore (Arecaceae)
IPNI: 210062-2
Common name: cecilia, chapilde, chichiburrú **Habit:** Shrub, Tree, Caespitose palm **Use:** HF, MA **Origin:** Native **Region:** Andean, Pacific **Dept.:** ANT, CAU, CHO, COR, NAR, RIS, VAC **Elev.:** 0-1500 m.a.s.l. **IUCN Habitat code:** 1,3,5,14 **Conservation status: (IUCN)** NE **(CPSG)** LC

1012 *Prestoea ensiformis* (Ruiz & Pav.) H.E.Moore (Arecaceae)
IPNI: 210063 2
Common name: hoja de aguacero, lindona, rabihorcao **Habit:** Shrub, Caespitose palm **Use:** EU, FU, HF, MA **Origin:** Native **Region:** Andean, Pacific **Dept.:** ANT, BOY, CAL, CAU, CHO, NAR, RIS, VAC **Elev.:** 450-1750 m.a.s.l. **IUCN Habitat code:** 1,3,14 **Conservation status: (IUCN)** NE **(CPSG)** NT

1013 *Prestoea longepetiolata* (Oerst.) H.E. Moore (Arecaceae)
IPNI: 669239-1
Habit: Shrub, Caespitose palm **Use:** HF **Origin:** Endemic **Region:** Andean **Dept.:** NSA, SAN **Elev.:** 470-1750 m.a.s.l. **IUCN Habitat code:** 1 **Conservation status: (IUCN)** NE **(CPSG)** NE

1014 *Prestoea schultzeana* (Burret) H.E.Moore (Arecaceae)
IPNI: 210069-2
Habit: Shrub, Caespitose palm **Use:** AF, FU, HF, MA, ME, SU **Origin:** Native **Region:** Amazonia, Andean **Dept.:** AMA, CAU, PUT **Elev.:** 100-1100 m.a.s.l. **IUCN Habitat code:** 1,5 **Conservation status: (IUCN)** NE **(CPSG)** LC

1015 *Pritchardia pacifica* Seem. & H.Wendl. (Arecaceae)
IPNI: 669288-1
Common name: abanico, palma abanico **Habit:** Tree, Solitary palm **Use:** EU, HF **Origin:** Cultivated **Dept.:** ANT, CUN, TOL, VAC **Elev.:** 0–1600 m a.s.l. **IUCN Habitat code:** 14 **Conservation status: (IUCN)** NE

1016 *Ptychosperma elegans* (R.Br.) Blume (Arecaceae)
IPNI: 669378-1
Common name: palma solitaria **Habit:** Shrub, Tree **Use:** EU, HF, MA **Origin:** Cultivated **Dept.:** ANT, BOL, CUN, TOL, VAC **Elev.:** 0–1600 m a.s.l. **Conservation status: (IUCN)** LC

1017 *Ptychosperma macarthurii* (H.Wendl. ex H.J.Veitch) H.Wendl. ex Hook.f. (Arecaceae)
IPNI: 55245-3
Habit: Tree **Use:** EU, HF, MA, ME, PO **Origin:** Cultivated **Dept.:** BOL, CUN, VAC **Elev.:** 0–1600 m a.s.l. **Conservation status: (IUCN)** NE

1018 *Raphia taedigera* (Mart.) Mart. (Arecaceae)
IPNI: 669539-1
Common name: palma pángana, panga, pángama, pángana, yupatí **Habit:** Tree, Caespitose palm **Use:** HF, MA **Origin:** Native **Region:** Pacific **Dept.:** ANT, CHO **Elev.:** 0-30 m.a.s.l. **IUCN Habitat code:** 1,5 **Conservation status: (IUCN)** NE **(CPSG)** LC

1019 *Rhapis excelsa* (Thunb.) A.Henry (Arecaceae)
IPNI: 669577-1
Common name: palma rafis, rapis **Habit:** Shrub **Use:** EU, MA, ME **Origin:** Cultivated **Dept.:** ANT, CUN, TOL, VAC **Elev.:** 0–1600 m a.s.l. **IUCN Habitat code:** 1,14 **Conservation status: (IUCN)** NE

1020 *Roystonea oleracea* (Jacq.) O.F.Cook (Arecaceae) **IPNI:** 222805-2 **Common name:** mapora, palma mapora **Habit:** Tree, Solitary palm **Use:** EU, HF, MA, ME **Origin:** Native, Cultivated **Region:** Orinoquia **Dept.:** ARA, CAS, MET **Elev.:** 150-300 m.a.s.l. **IUCN Habitat code:** 1,3,4,14 **Conservation status:** **(IUCN)** NE **(CPSG)** VU

1021 *Roystonea regia* (Kunth) O.F.Cook (Arecaceae) **IPNI:** 222809-2 **Common name:** palma botella, palma real, palma real de Cuba **Habit:** Tree, Solitary palm **Use:** AF, EU, HF, MA, ME **Origin:** Cultivated **Dept.:** ANT, ATL, BOL, BOY, COR, CUN, HUI, MAG, MET, NSA, QUI, RIS, SAN, TOL, VAC **Elev.:** 0–1800 m a.s.l. **IUCN Habitat code:** 1,3,14 **Conservation status:** **(IUCN)** LC

1022 *Sabal mauritiiformis* (H.Karst.) Griseb. & H.Wendl. (Arecaceae) **IPNI:** 669714-1 **Common name:** calicá, chilangué, palma amarga, palma redonda, palmiche, palmicho, palmito **Habit:** Tree, Solitary palm **Use:** AF, HF, MA, ME, SU **Origin:** Native **Region:** Caribbean, Orinoquia **Dept.:** ANT, ATL, BOL, CAS, CES, CHO, COR, CUN, LAG, MAG, SAP, SUC, TOL, VAC **Elev.:** 0-1000 m.a.s.l. **IUCN Habitat code:** 1,2,3,14 **Conservation status:** **(IUCN)** NE **(CPSG)** NT

1023 *Saribus rotundifolius* (Lam.) Blume (Arecaceae) **IPNI:** 669832-1 **Common name:** palma abanico **Habit:** Tree, Solitary palm **Use:** EU, HF, MA, ME **Origin:** Cultivated **Dept.:** ANT, CUN, VAC **Elev.:** 0–2600 m a.s.l. **Conservation status:** **(IUCN)** NE

1024 *Serenoa repens* (W.Bartram) Small (Arecaceae) **IPNI:** 941782-1 **Common name:** saw palmeto **Use:** EU, HF, MA, ME **Dept.:** BOG **Conservation status:** **(IUCN)** NE

1025 *Socratea exorrhiza* (Mart.) H.Wendl. (Arecaceae) **IPNI:** 1200702-2 **Common name:** ẽeñá, anaku tsatsavó, araco, araque, arra sĩi, bonbon, cachuda zancona, choapo, chonta, chonta pambil, chuapo, crespa, dorida, ebáñi maihóbariño, eta, etá irakü, éwani kíhiñi, gualte macho, guasaina, gõháño, himuntḗgn, hira, hiriwa, huba, huru, hurudã, h **Habit:** Tree, Solitary palm **Use:** AF, EU, FU, HF, MA, ME, SU **Origin:** Native **Region:** Amazonia, Andean, Guiana Shield, Orinoquia, Pacific **Dept.:** AMA, ANT, ARA, BOL, CAQ, CAS, CAU, CHO, COR, GUA, GUV, MET, NAR, NSA, PUT, RIS, SAN, VAC, VAU, VID **Elev.:** 0-1150 m.a.s.l. **IUCN Habitat code:** 1,3,4,5,14 **Conservation status:** **(IUCN)** NE **(CPSG)** LC

1026 *Socratea rostrata* Burret (Arecaceae) **IPNI:** 238168-2 **Common name:** gualte patudo, maquenque, rallador **Habit:** Tree, Solitary palm **Use:** HF, MA **Origin:** Native **Region:** Andean **Dept.:** ANT, CAQ, CAU, CHO, NAR, PUT, RIS, VAC **Elev.:** 500-1800 m.a.s.l. **IUCN Habitat code:** 1,14 **Conservation status:** **(IUCN)** NE **(CPSG)** LC

1027 *Syagrus inajai* (Spruce) Becc. (Arecaceae) **IPNI:** 246946-2 **Use:** HF, MA, ME **IUCN Habitat code:** 1,3,4 **Conservation status:** **(IUCN)** NE

1028 *Syagrus orinocensis* (Spruce) Burret (Arecaceae) **IPNI:** 246957-2 **Common name:** churrubay, churruhay, coco, coquillo, maispepe, palma churrubay **Habit:** Tree, Solitary palm **Use:** HF, MA **Origin:** Native **Region:** Guiana Shield, Orinoquia **Dept.:** AMA, CAQ, CAS, GUA, GUV, MET, VAU, VID **Elev.:** 40-600 m.a.s.l. **IUCN Habitat code:** 1,3,4,5,14 **Conservation status:** **(IUCN)** NE **(CPSG)** LC

1029 *Syagrus romanzoffiana* (Cham.) Glassman (Arecaceae) **IPNI:** 246966-2 **Common name:** coco plumoso, palma reina **Habit:** Tree, Solitary palm **Use:** EU, HF, MA, ME **Origin:** Cultivated **Dept.:** ANT, CUN **Elev.:** 0–2600 m a.s.l. **IUCN Habitat code:** 14 **Conservation status:** **(IUCN)** NE

1030 *Syagrus sancona* (Kunth) H.Karst. (Arecaceae) **IPNI:** 670066-1 **Common name:** chiragua, chonta, palma de sarare, palma real, palma sarare, palma zancona, quirache, sarare, saray, zancona **Habit:** Tree, Solitary palm **Use:** AF, EU, HF, MA, SU **Origin:** Native, Cultivated **Region:** Amazonia, Andean, Guiana Shield, Orinoquia, Pacific **Dept.:** ANT, ARA, CAQ, CAS, CUN, MET, NSA, PUT, QUI, RIS, TOL, VAC **Elev.:** 200-1200 m.a.s.l. **IUCN Habitat code:** 1,3,14 **Conservation status:** **(IUCN)** NE **(CPSG)** VU

1031 *Syagrus smithii* (H.E.Moore) Glassman (Arecaceae) **IPNI:** 246971-2 **Common name:** churrubay coco **Habit:** Shrub, Solitary palm **Use:** HF, MA **Origin:** Native **Region:** Amazonia **Dept.:** AMA **Elev.:** 100-250 m.a.s.l. **IUCN Habitat code:** 1,5 **Conservation status:** **(IUCN)** LC **(CPSG)** LC

1032 *Syagrus weddelliana* (H.Wendl.) Becc. (Arecaceae) **IPNI:** 246979-2 **Common name:** coquito, palma de coquito **Habit:** Shrub, Solitary palm **Use:** EU **Origin:** Cultivated **Dept.:** ANT **Elev.:** 1500–1700 m a.s.l. **Conservation status:** **(IUCN)** NE

1033 *Synechanthus warscewiczianus* H.Wendl. (Arecaceae) **IPNI:** 248082-2 **Common name:** guaguarín, jincha, palmilla **Habit:** Shrub, Caespitose palm **Use:** HF, MA, SU **Origin:** Native **Region:** Andean, Pacific **Dept.:** ANT, CAU, CHO, COR, NAR, RIS, VAC **Elev.:** 0-1700 m.a.s.l. **IUCN Habitat code:** 1,3,5,14 **Conservation status:** **(IUCN)** NE **(CPSG)** LC

1034 *Thrinax parviflora* Sw. (Arecaceae) **IPNI:** 670156-1 **Common name:** palma abanico **Habit:** Tree, Solitary palm **Use:** EU, MA **Origin:** Cultivated **Dept.:** ANT, CUN, SAP **Elev.:** 0–1500 m a.s.l. **Conservation status:** **(IUCN)** NE

1035 *Trachycarpus fortunei* (Hook.) H.Wendl. (Arecaceae) **IPNI:** 670182-1 **Common name:** palma de viento, palma molino de viento chino **Habit:** Tree, Solitary palm **Use:** EU, HF, MA, ME **Origin:** Cultivated **Dept.:** ANT **Elev.:** 1500–1800 m a.s.l. **IUCN Habitat code:** 14 **Conservation status:** **(IUCN)** NE

1036 *Washingtonia filifera* (Rafarin) H.Wendl. ex de Bary (Arecaceae) **IPNI:** 267788-2 **Common name:** palma abanico, palma de abanico, palma washingtonia, palma wasintonia, washingtonia, wasintonia **Habit:** Tree, Solitary palm **Use:** EU, FU, GS, HF, MA, ME **Origin:** Cultivated **Dept.:** ANT, BOG, BOY, CUN, TOL, VAC **Elev.:** 0–2600 m a.s.l. **Conservation status:** **(IUCN)** LC

1037 *Washingtonia robusta* H.Wendl. (Arecaceae) **IPNI:** 670278-1 **Common name:** palma washingtonia, washingtonia **Habit:** Tree, Solitary palm **Use:** EU, HF, MA **Origin:** Cultivated **Dept.:** ANT, BOY, CUN, TOL, VAC **Elev.:** 0–2600 m a.s.l. **IUCN Habitat code:** 3,14 **Conservation status:** **(IUCN)** NE

1038 *Welfia regia* H.Wendl. (Arecaceae) **IPNI:** 670285-1 **Common name:** amarga, amargo, cuasbil, gender, mona, palma amarga, palma amargo, palma mona, palma sanjuán, sanjuán, trupa **Habit:** Tree, Solitary palm **Use:** HF, MA, ME **Origin:** Native **Region:** Andean, Pacific **Dept.:** ANT, BOY, CAU, CHO, NAR, RIS, SAN, VAC **Elev.:** 0-1200 m.a.s.l. **IUCN Habitat code:** 1,3,5,14 **Conservation status:** **(IUCN)** NE **(CPSG)** LC

1039 *Wettinia aequalis* (O.F.Cook & Doyle) R.Bernal (Arecaceae) **IPNI:** 983907-1 **Habit:** Shrub, Tree, Solitary palm **Use:** HF, MA **Origin:** Native **Region:** Andean, Pacific **Dept.:** ANT, CAU, CHO, NAR, RIS, VAC **Elev.:** 0-800 m.a.s.l. **IUCN Habitat code:** 1,3,14 **Conservation status:** **(IUCN)** NE **(CPSG)** LC

1040 *Wettinia augusta* Poepp. & Endl. (Arecaceae) **IPNI:** 670291-1 **Habit:** Shrub, Tree **Use:** HF, MA, ME, SU **Origin:** Native **Region:** Amazonia **Dept.:** AMA **Elev.:** 100-250 m.a.s.l. **IUCN Habitat code:** 1,5 **Conservation status:** **(IUCN)** NE **(CPSG)** LC

1041 *Wettinia disticha* (R.Bernal) R.Bernal (Arecaceae) **IPNI:** 983909-1 **Common name:** macana, macana abanico **Habit:** Tree, Solitary palm **Use:** AF, MA **Origin:** Endemic **Region:** Andean **Dept.:** ANT, CAU, CHO, RIS, VAC **Elev.:** 1700-2250 m.a.s.l. **IUCN Habitat code:** 1,3 **Conservation status:** **(IUCN)** LC **(CPSG)** LC

1042 *Wettinia drudei* (O.F.Cook & Doyle) A.J.Hend. (Arecaceae) **IPNI:** 981485-1 **Common name:** pachuba **Habit:** Shrub, Caespitose palm **Use:** MA **Origin:** Native **Region:** Amazonia, Andean **Dept.:** AMA, CAQ, PUT, VAU **Elev.:** 70-1100 m.a.s.l. **IUCN Habitat code:** 1,3,5,14 **Conservation status:** **(IUCN)** LC **(CPSG)** LC

1043 *Wettinia fascicularis* (Burret) H.E.Moore & J.Dransf. (Arecaceae) **IPNI:** 268487-2 **Common name:** chonta de mico, macana, maquenque, palma chonta **Habit:** Shrub, Tree **Use:** AF, HF, MA **Origin:** Native **Region:** Andean **Dept.:** ANT, BOL, CAQ, CAU, CHO, HUI, MET, PUT, RIS, VAC **Elev.:** 1200-2200 m.a.s.l. **IUCN Habitat code:** 1,3,14 **Conservation status:** **(IUCN)** LC **(CPSG)** LC

1044 *Wettinia hirsuta* Burret (Arecaceae)
IPNI: 268488-2
Common name: lata, macana, maquenque, mazorca, palm mazorca, palma mazorca, tablemina, zancaraño **Habit:** Shrub, Tree **Use:** EU, MA **Origin:** Endemic **Region:** Andean **Dept.:** ANT, BOY, COR, SAN **Elev.:** 200-1350 m.a.s.l. **IUCN Habitat code:** 1,2,3,14 **Conservation status: (IUCN)** EN **(CPSG)** VU

1045 *Wettinia kalbreyeri* (Burret) R.Bernal (Arecaceae)
IPNI: 983910-1
Common name: crespa, guaira bombón, gualte, gualte bola, macana, palma bolillo **Habit:** Tree, Solitary palm **Use:** AF, HF, MA **Origin:** Native **Region:** Andean **Dept.:** ANT, CAL, CAQ, CAU, CHO, HUI, NAR, PUT, QUI, RIS, VAC **Elev.:** 400-2500 m.a.s.l. **IUCN Habitat code:** 1,3,14 **Conservation status: (IUCN)** LC **(CPSG)** LC

1046 *Wettinia maynensis* Spruce (Arecaceae)
IPNI: 670297-1
Common name: corunta, curunta, gualte, palma curunta, urunta **Habit:** Tree, Solitary palm **Use:** AF, EU, FU, HF, MA, ME **Origin:** Native **Region:** Amazonia, Andean **Dept.:** CAQ, CAU, PUT **Elev.:** 200-1200 m.a.s.l. **IUCN Habitat code:** 1,14 **Conservation status: (IUCN)** NE **(CPSG)** LC

1047 *Wettinia oxycarpa* Galeano & R.Bernal (Arecaceae)
IPNI: 268490-2
Common name: gualte upián **Habit:** Tree, Solitary palm **Use:** MA **Origin:** Native **Region:** Andean, Pacific **Dept.:** ANT, CHO, NAR, VAC **Elev.:** 400-1300 m.a.s.l. **IUCN Habitat code:** 1 **Conservation status: (IUCN)** NE **(CPSG)** NT

1048 *Wettinia quinaria* (O.F.Cook & Doyle) Burret (Arecaceae)
IPNI: 268491-2
Common name: chorr, gualte, guarnul, macana, maquenque, meme, memé, murrapo, palma memé **Habit:** Tree, Solitary palm **Use:** AF, EU, FU, HF, MA, SU **Origin:** Native **Region:** Andean, Pacific **Dept.:** ANT, CAU, CHO, NAR, RIS, VAC **Elev.:** 0-1100 m.a.s.l. **IUCN Habitat code:** 1,3,5,14 **Conservation status: (IUCN)** NE **(CPSG)** LC

1049 *Wettinia radiata* (O.F.Cook & Doyle) R.Bernal (Arecaceae)
IPNI: 1169616-2
Common name: baqueta, chaguaitera, crespa, maquenque, sapa **Habit:** Shrub, Tree **Use:** MA **Origin:** Native **Region:** Pacific **Dept.:** ANT, CAU, CHO, RIS, VAC **Elev.:** 0-1000 m.a.s.l. **IUCN Habitat code:** 1,3,5,14 **Conservation status: (IUCN)** NE **(CPSG)** LC

1050 *Wodyetia bifurcata* A.K.Irvine (Arecaceae)
IPNI: 903383-1
Common name: cola de zorro **Habit:** Tree, Solitary palm **Use:** EU **Origin:** Cultivated **Dept.:** ANT **Elev.:** 1500 1600 m a.s.l. **Conservation status: (IUCN)** CD

Asparagales

1051 *Agapanthus africanus* (L.) Hoffmanns. (Amaryllidaceae)
IPNI: 527258-1
Use: EU, ME **Region:** Andean **Dept.:** BOG **IUCN Habitat code:** 1,14 **Conservation status: (IUCN)** NE

1052 *Agapanthus praecox* Willd. (Amaryllidaceae)
IPNI: 527283-1
Habit: Herb **Use:** EU, HF, ME **Origin:** Cultivated **Region:** Andean **Dept.:** ANT, BOG, BOY, CUN, QUI **Elev.:** 1500–2600 m a.s.l. **IUCN Habitat code:** 1,14 **Conservation status: (IUCN)** NE

1053 *Allium ampeloprasum* L. (Amaryllidaceae)
IPNI: 332079-2
Habit: Herb **Use:** GS, HF, MA, ME **Origin:** Cultivated **Region:** Andean **Dept.:** BOG, CUN **Elev.:** 2600–2600 m a.s.l. **Conservation status: (IUCN)** LC

1054 *Allium cepa* L. (Amaryllidaceae)
IPNI: 527795-1
Common name: cebolla, cebolla blanca, cebolla cabezona, cebolla de a ciento, cebolla de huevo, cebolla morada, cebolla peruana, papa cebolla **Habit:** Herb **Use:** AF, EU, GS, HF, MA, ME, PO, SU **Origin:** Cultivated, Naturalised **Region:** Andean **Dept.:** ANT, BOG, CUN, NAR **Elev.:** 1400–2700 m a.s.l. **IUCN Habitat code:** 1,3,14 **Conservation status: (IUCN)** NE

1055 *Allium fistulosum* L. (Amaryllidaceae)
IPNI: 528046-1
Common name: cebolla, cebolla común, cebolla de rabo, cebolla de rama, cebolla junca, cebolla junco, cebolla larga, cebollín **Habit:** Herb **Use:** AF, EU, GS, HF, MA, ME **Origin:** Cultivated **Region:** Andean **Dept.:** ANT, BOG, CUN, NAR, RIS **Elev.:** 1600–2700 m a.s.l. **IUCN Habitat code:** 1,2,3,8,14 **Conservation status: (IUCN)** NE

1056 *Allium obliquum* L. (Amaryllidaceae)
IPNI: 528506-1
Use: GS, HF, ME **IUCN Habitat code:** 14 **Conservation status: (IUCN)** NE

1057 *Allium sativum* L. (Amaryllidaceae)
IPNI: 528796-1
Common name: ajo, ajo macho **Habit:** Herb **Use:** AF, EU, GS, HF, MA, ME, PO, SU **Origin:** Cultivated **Region:** Andean **Dept.:** ANT, BOG, BOY, CUN, NAR **Elev.:** 1500–2600 m a.s.l. **IUCN Habitat code:** 1,3,4,14 **Conservation status: (IUCN)** NE

1058 *Allium schoenoprasum* L. (Amaryllidaceae)
IPNI: 528823-1
Common name: cebollín, cibulé **Habit:** Herb **Use:** GS, HF, ME, PO **Origin:** Cultivated **Region:** Andean **Dept.:** BOG, CUN **Elev.:** 2600–2600 m a.s.l. **IUCN Habitat code:** 1,14 **Conservation status: (IUCN)** LC

1059 *Clivia miniata* (Lindl.) Verschaff. (Amaryllidaceae)
IPNI: 994513-1
Common name: doce apóstoles, tulipán **Habit:** Herb **Use:** EU, ME, SU **Origin:** Cultivated **Region:** Andean **Dept.:** BOG, CUN **Elev.:** 2600–2600 m a.s.l. **IUCN Habitat code:** 14 **Conservation status: (IUCN)** NE

1060 *Crinum × powellii* Baker (Amaryllidaceae)
IPNI: 64029-1
Use: ME **Region:** Andean **Dept.:** BOG **Conservation status: (IUCN)** NE

1061 *Crinum americanum* L. (Amaryllidaceae)
IPNI: 319172-2
Common name: narcizo **Use:** EU, ME **IUCN Habitat code:** 14 **Conservation status: (IUCN)** NE

1062 *Crinum bulbispermum* (Burm.f.) Milne-Redh. & Schweick. (Amaryllidaceae)
IPNI: 63828-1
Habit: Herb **Use:** EU, HF, ME, PO **Origin:** Cultivated **Region:** Caribbean **Dept.:** SAP **Elev.:** 0–50 m a.s.l. **IUCN Habitat code:** 1,3 **Conservation status: (IUCN)** NE

1063 *Crinum erubescens* L.f. ex Aiton (Amaryllidaceae)
IPNI: 30032850-2
Common name: tulipán **Habit:** Herb **Use:** ME **Origin:** Native, Cultivated **Region:** Amazonia, Pacific **Dept.:** AMA, CHO **Elev.:** 0–200 m a.s.l. **IUCN Habitat code:** 1,3,4,5,14 **Conservation status: (IUCN)** NE

1064 *Crinum jagus* (J.Thomps.) Dandy (Amaryllidaceae)
IPNI: 63934-1
Habit: Herb **Use:** EU, MA, ME **Origin:** Cultivated **Region:** Pacific **Dept.:** ANT **Elev.:** 220–220 m a.s.l. **IUCN Habitat code:** 14 **Conservation status: (IUCN)** LC

1065 *Crinum zeylanicum* (L.) L. (Amaryllidaceae)
IPNI: 64125-1
Use: EU, MA, ME, PO **Origin:** Naturalised **IUCN Habitat code:** 1,3 **Conservation status: (IUCN)** NE

1066 *Hippeastrum elegans* (Spreng.) H.E.Moore (Amaryllidaceae)
IPNI: 122772-2
Habit: Herb **Use:** EU, GS **Origin:** Native **Region:** Andean **Dept.:** CUN, MAG, MET, TOL **Elev.:** 500–1800 m a.s.l. **IUCN Habitat code:** 1,2,3 **Conservation status: (IUCN)** NE

1067 *Hippeastrum puniceum* (Lam.) Voss (Amaryllidaceae)
IPNI: 122815-2
Common name: azucena roja, cebolleta colorada, lirio rojo, lirio rosado **Habit:** Herb **Use:** EU, GS, ME, PO **Origin:** Native **Region:** Andean, Guiana Shield, Orinoquia **Dept.:** ANT, CUN, MET, SAN, TOL **Elev.:** 100–3500 m a.s.l. **IUCN Habitat code:** 1,2,14 **Conservation status: (IUCN)** NE

1068 *Hippeastrum vittatum* (L'Hér.) Herb. (Amaryllidaceae)
IPNI: 65185-1
Use: GS, ME **Region:** Andean **Dept.:** BOG **IUCN Habitat code:** 3,14 **Conservation status: (IUCN)** NE

1069 *Hymenocallis caribaea* (L.) Herb. (Amaryllidaceae)
IPNI: 65212-1
Common name: lirio, planta de arroyo, varita de San José, varita de san José **Use:** EU, ME **Dept.:** SAP **Conservation status: (IUCN)** NE

1070 *Hymenocallis littoralis* (Jacq.) Salisb. (Amaryllidaceae)
IPNI: 30048918-2
Common name: azucena, azucena chirosa, cebolleta, cebolleta blanca, cebollina, chirosa, flor de pila, lirio, sirena, vara de San José, varita de San José **Habit:** Herb **Use:** EU, ME **Origin:** Native **Region:** Caribbean

Dept.: BOL, SAP **Elev.:** 0–300 m a.s.l. **IUCN Habitat code:** 1,2,3,5,14 **Conservation status: (IUCN)** NE

1071 *Nothoscordum gracile* (Aiton) Stearn (Amaryllidaceae)
IPNI: 130726-3
Habit: Herb **Use:** EU, HF **Origin:** Native, Cultivated **Region:** Andean **Dept.:** ANT **Elev.:** 1460–1460 m a.s.l. **Conservation status: (IUCN)** NE

1072 *Tulbaghia alliacea* L.f. (Amaryllidaceae)
IPNI: 542757-1
Habit: Herb **Use:** GS, HF, ME **Origin:** Cultivated **Region:** Andean **Dept.:** BOG, CUN **Elev.:** 2600–2600 m a.s.l. **Conservation status: (IUCN)** NE

1073 *Tulbaghia simmleri* Beauverd (Amaryllidaceae)
IPNI: 542797-1
Use: ME **Region:** Andean **Dept.:** BOG **Conservation status: (IUCN)** NE

1074 *Tulbaghia violacea* Harv. (Amaryllidaceae)
IPNI: 542801-1
Common name: cebollina **Habit:** Herb **Use:** HF, ME **Origin:** Cultivated **Region:** Andean **Dept.:** ANT, BOG, CUN, QUI, VAC **Elev.:** 1200–2600 m a.s.l. **IUCN Habitat code:** 3,14 **Conservation status: (IUCN)** NE

1075 *Urceolina × grandiflora* (Planch. & Linden) Traub (Amaryllidaceae)
IPNI: 261579-2
Use: HF, ME **Origin:** Native **IUCN Habitat code:** 14 **Conservation status: (IUCN)** NE

1076 *Urceolina amazonica* (Linden ex Planch.) Christenh. & Byng (Amaryllidaceae)
IPNI: 77182693-1
Use: EU, ME **Region:** Andean **Dept.:** BOG **Conservation status: (IUCN)** NE

1077 *Urceolina candida* (Planch. & Linden) Traub (Amaryllidaceae)
IPNI: 261572-2
Common name: cebollita **Habit:** Herb **Use:** EU **Origin:** Native **Region:** Amazonia, Andean, Orinoquia **Dept.:** AMA, MET, PUT **Elev.:** 100-1100 m.a.s.l. **Conservation status: (IUCN)** NE **(CPSG)** DD

1078 *Urceolina formosa* (Meerow) Christenh. & Byng (Amaryllidaceae)
IPNI: 77183240-1
Common name: ajo de perdiz, sachacebolla **Habit:** Herb **Use:** EU **Origin:** Native, Cultivated **Region:** Amazonia, Andean **Dept.:** AMA, CAQ, CAU **Elev.:** 100–1800 m a.s.l. **Conservation status: (IUCN)** NE

1079 *Urceolina sanderi* (Baker) Traub (Amaryllidaceae)
IPNI: 261595-2
Habit: Herb **Use:** EU **Origin:** Endemic **Region:** Andean, Pacific **Dept.:** CAU, CHO, NAR, RIS, VAC **Elev.:** 0-1000 m.a.s.l. **Conservation status: (IUCN)** NE **(CPSG)** DD

1080 *Urceolina subedentata* (Baker) Traub (Amaryllidaceae)
IPNI: 67017-1
Common name: azucena, billete, cebolla de perdiz, cebolleta, cebolleta blanca, cebollina, estrella de Belén, estrella del Caquetá, lirio, reina del Caquetá, varita de San José, vida del hombre **Habit:** Herb **Use:** EU **Origin:** Endemic **Region:** Andean **Dept.:** CAU, CUN, VAC **Elev.:** 600-2000 m.a.s.l. **Conservation status: (IUCN)** NE **(CPSG)** VU

1081 *Zephyranthes albiella* Traub (Amaryllidaceae)
IPNI: 270910-2
Habit: Herb **Use:** EU **Origin:** Native, Cultivated **Region:** Andean, Caribbean **Dept.:** BOG, BOL, CUN **Elev.:** 350–350 m a.s.l. **Conservation status: (IUCN)** NE

1082 *Zephyranthes carinata* Herb. (Amaryllidaceae)
IPNI: 270933-2
Common name: cebolleta, jacinto, lirio **Habit:** Herb **Use:** EU, ME **Origin:** Native **Region:** Andean **Dept.:** ANT, BOG, BOY, CUN, SAN **Elev.:** 1700–2600 m a.s.l. **IUCN Habitat code:** 1,3,14 **Conservation status: (IUCN)** NE

1083 *Zephyranthes puertoricensis* Traub (Amaryllidaceae)
IPNI: 270983-2
Habit: Herb **Use:** EU **Origin:** Native, Cultivated **Region:** Caribbean **Dept.:** BOL, TOL **Elev.:** 0–200 m a.s.l. **Conservation status: (IUCN)** NE

1084 *Zephyranthes robusta* (Herb.) Baker (Amaryllidaceae)
IPNI: 270986-2
Habit: Herb **Use:** EU **Origin:** Cultivated, Naturalised **Region:** Andean **Dept.:** ANT **Elev.:** 1470–1470 m a.s.l. **Conservation status: (IUCN)** NE

1085 *Zephyranthes rosea* Lindl. (Amaryllidaceae)
IPNI: 67293-1
Habit: Herb **Use:** EU, MA, ME **Origin:** Native, Cultivated **Region:** Andean, Caribbean **Dept.:** BOG, BOL, CUN, SAP **Elev.:** 0–0 m a.s.l. **IUCN Habitat code:** 1,14 **Conservation status: (IUCN)** NE

1086 *Zephyranthes tubispatha* (L'Hér.) Herb. (Amaryllidaceae)
IPNI: 67314-1
Habit: Herb **Use:** EU, ME **Origin:** Naturalised **Dept.:** CUN **Elev.:** 2650–3000 m a.s.l. **Conservation status: (IUCN)** NE

1087 *Agave americana* L. (Asparagaceae)
IPNI: 319063-2
Common name: agave, cabuya, cabuya azul, cabuya de México, cabuya montua, chaguar, cocuy, fique, henequén, maguey, magué, motua, penca, penca de cabuya, pita, yucatán **Habit:** Herb **Use:** EU, HF, IF, MA, ME, PO, SU **Origin:** Cultivated, Naturalised **Region:** Andean **Dept.:** ANT, BOG **Elev.:** 1500–1500 m a.s.l. **IUCN Habitat code:** 2,3,5,14 **Conservation status: (IUCN)** LC

1088 *Agave amica* (Medik.) Thiede & Govaerts (Asparagaceae)
IPNI: 77153983-1
Common name: azaritos, azucena de la costa, narciso, nardo, varita de San José **Habit:** Herb **Use:** EU, GS, HF, MA, ME **Origin:** Cultivated **Region:** Andean **Dept.:** CUN **Elev.:** 2500–2500 m a.s.l. **IUCN Habitat code:** 14 **Conservation status: (IUCN)** NE

1089 *Agave angustifolia* Haw. (Asparagaceae)
IPNI: 327760-2
Common name: palma montañera, penca **Use:** EU, HF, MA, ME **Origin:** Cultivated **Region:** Andean, Caribbean **Dept.:** ANT **Elev.:** 1500–1500 m a.s.l. **IUCN Habitat code:** 3,14 **Conservation status: (IUCN)** LC

1090 *Agave attenuata* Salm-Dyck (Asparagaceae)
IPNI: 61824-1
Use: EU **Region:** Andean **Dept.:** BOG **IUCN Habitat code:** 14 **Conservation status: (IUCN)** LC

1091 *Agave boldinghiana* Trel. (Asparagaceae)
IPNI: 6386-2
Habit: Herb **Use:** EU **Origin:** Native, Cultivated **Region:** Caribbean **Dept.:** LAG **Elev.:** 50–50 m a.s.l. **Conservation status: (IUCN)** VU

1092 *Agave cocui* Trel. (Asparagaceae)
IPNI: 6425-2
Habit: Herb **Use:** HF, ME **Origin:** Native, Cultivated **Region:** Andean, Caribbean **Dept.:** BOY, MAG, NSA, SAN **Elev.:** 400–1000 m a.s.l. **IUCN Habitat code:** 2,14 **Conservation status: (IUCN)** LC

1093 *Albuca bracteata* (Thunb.) J.C.Manning & Goldblatt (Asparagaceae)
IPNI: 77097737-1
Habit: Herb **Use:** ME **Origin:** Cultivated **Region:** Andean **Dept.:** BOG, BOY, CUN **Elev.:** 2600–2600 m a.s.l. **Conservation status: (IUCN)** NE

1094 *Asparagus aethiopicus* L. (Asparagaceae)
IPNI: 530994-1
Use: EU, HF, ME **Region:** Andean **Dept.:** BOG **IUCN Habitat code:** 14 **Conservation status: (IUCN)** LC

1095 *Asparagus densiflorus* (Kunth) Jessop (Asparagaceae)
IPNI: 531072-1
Common name: castillo, espárrago, espárrago de jardín, helecho de Castilla, rabo de zorro **Habit:** Herb **Use:** EU, GS, ME, SU **Origin:** Cultivated **Region:** Andean **Dept.:** BOG, BOL, CAL, CUN, RIS **Elev.:** 1000–2600 m a.s.l. **IUCN Habitat code:** 1,3,14 **Conservation status: (IUCN)** NE

1096 *Asparagus officinalis* L. (Asparagaceae)
IPNI: 531229-1
Common name: espárrago **Habit:** Herb **Use:** EU, GS, HF, ME, PO **Origin:** Cultivated **Region:** Andean **Dept.:** ANT, CUN **Elev.:** 1500–2600 m a.s.l. **IUCN Habitat code:** 1,3,14 **Conservation status: (IUCN)** LC

1097 *Asparagus setaceus* (Kunth) Jessop (Asparagaceae)
IPNI: 531312-1
Common name: amor de madre, brisa, brisa de Rin, brisa del río, brisas del Rin, bruma, bruma del mar, ensueño, espuma de mar, espárrago, manto de la Virgen, pino, rin, sombra, sombra de mar **Habit:** Herb **Use:** AF, EU, GS, HF, ME, PO, SU **Origin:** Cultivated **Region:** Andean, Caribbean, Orinoquia, Pacific **Dept.:** ANT, ARA, BOG, BOY, CAL, CAS, CAU, CUN, HUI, MAG, MET, NAR, NSA, RIS, SAN, SUC, TOL, VAC **Elev.:** 100–2600 m a.s.l. **IUCN Habitat code:** 1,3,14 **Conservation status: (IUCN)** NE

1098 *Asparagus virgatus* Baker (Asparagaceae)
IPNI: 531381-1
Common name: trífer **Habit:** Herb **Use:** AF, EU, GS, HF, ME, SU **Origin:** Cultivated **Region:** Andean **Dept.:**

CUN **Elev.:** 2600–2600 m a.s.l. **Conservation status: (IUCN)** NE

1099 *Chlorophytum comosum* (Thunb.) Jacques (Asparagaceae)
IPNI: 532810-1
Common name: cinta, cintilla, lazo de amor, madre moderna **Habit:** Herb **Use:** EU, HF, ME, SU **Origin:** Cultivated **Region:** Andean **Dept.:** ANT, BOG **Elev.:** 1480–1500 m a.s.l. **IUCN Habitat code:** 1,3,14 **Conservation status: (IUCN)** NE

1100 *Cordyline australis* (G.Forst.) Endl. (Asparagaceae)
IPNI: 533544-1
Use: EU, HF, MA, ME **Region:** Andean **Dept.:** BOG **IUCN Habitat code:** 14 **Conservation status: (IUCN)** NE

1101 *Cordyline fruticosa* (L.) A.Chev. (Asparagaceae)
IPNI: 533580-1
Common name: apararrayo, berlinda, capaelarreina, caraca, carey, carey morado, carey verde, caña de adorno, cañaelaindia, cinta, espada, flor de casa, flor de difunto, hoja de palma, linda palmera, palma carey, palma de Cristo, palma de San José, palma de carey, palma **Habit:** Shrub, Small tree **Use:** AF, EU, FU, GS, HF, IF, MA, ME, SU **Origin:** Cultivated **Region:** Amazonia, Caribbean **Dept.:** BOG, CAQ, PUT, SAP, VAC **Elev.:** 200–1000 m a.s.l. **IUCN Habitat code:** 1,2,3,4,14 **Conservation status: (IUCN)** LC

1102 *Cordyline rubra* Otto & A.Dietr. (Asparagaceae)
IPNI: 97540-3
Habit: Shrub **Use:** HF, ME, SU **Origin:** Cultivated **Dept.:** VAC **Elev.:** 930–930 m a.s.l. **IUCN Habitat code:** 1,3,14 **Conservation status: (IUCN)** LC

1103 *Dracaena angustifolia* (Medik.) Roxb. (Asparagaceae)
IPNI: 534103-1
Use: AF, EU, HF, MA, ME, SU **Conservation status: (IUCN)** NE

1104 *Dracaena braunii* Engl. (Asparagaceae)
IPNI: 534129-1
Habit: Herb **Use:** EU **Origin:** Cultivated **Region:** Andean, Orinoquia **Dept.:** ANT, ARA, CUN **Elev.:** 100–2600 m a.s.l. **Conservation status: (IUCN)** NE

1105 *Dracaena fragrans* (L.) Ker Gawl. (Asparagaceae)
IPNI: 534207-1
Common name: carey, palmita roja, sangre de carey, tronco **Habit:** Shrub **Use:** EU, HF, MA, ME, SU **Origin:** Cultivated **Region:** Andean, Pacific **Dept.:** ANT, BOG, CUN, QUI **Elev.:** 0–2000 m a.s.l. **IUCN Habitat code:** 3,14 **Conservation status: (IUCN)** LC

1106 *Dracaena hyacinthoides* (L.) Mabb. (Asparagaceae)
IPNI: 77164234-1
Common name: culebrilla, curarina, cáñamo de África, mapaná **Habit:** Herb **Use:** AF, EU, HF, MA, ME, SU **Origin:** Cultivated **Region:** Andean **Dept.:** CUN, SAN **Conservation status: (IUCN)** NE

1107 *Dracaena reflexa* Lam. (Asparagaceae)
IPNI: 534338-1
Common name: canción de la India **Habit:** Herb **Use:** EU, GS, HF, MA, ME **Origin:** Cultivated **Region:** Andean, Caribbean **Dept.:** ANT, BOG, CUN, QUI, SAP **Elev.:** 1500–2600 m a.s.l. **IUCN Habitat code:** 3,14 **Conservation status: (IUCN)** LC

1108 *Dracaena trifasciata* (Prain) Mabb. (Asparagaceae)
IPNI: 77164235-1
Common name: calavera, cinta, cola de gato, cola de pescado, corbatín, culebra, culebrero, culebrina, culebro, curarina, espada, espada de Bolívar, espada de Judas, lengua de suegra, lenguaesuegra, lengüevaca, mapanare, mapaná, mejicano, penca, pitahaya, rabo de caimá **Habit:** Herb **Use:** EU, MA, ME, SU **Origin:** Cultivated **Region:** Andean, Caribbean, Orinoquia, Pacific **Dept.:** ANT, ARA, ATL, BOY, CAS, CAU, CES, CHO, COR, CUN, HUI, LAG, MET, NAR, NSA, RIS, SAP, SAN, SUC, TOL, VAC **Elev.:** 100–2000 m a.s.l. **Conservation status: (IUCN)** NE

1109 *Furcraea andina* Trel. (Asparagaceae)
IPNI: 106421-2
Use: HF **Conservation status: (IUCN)** NE

1110 *Furcraea cabuya* Trel. (Asparagaceae)
IPNI: 106423-2
Common name: cabuya, cocuiza, fique, fique americano, fique canalito, fique común, fique criollo, fique de Castilla, fique espada, fique espino, fique espinoso, fique fique, fique hembrita, fique hembro, fique liso, fique macho, fique pelón, fique perolero, fique punt **Habit:** Herb **Use:** AF, EU, MA, ME, PO **Origin:** Native, Cultivated **Region:** Andean **Dept.:** BOG, BOY, CUN, MAG, NAR **Elev.:** 1600–2400 m a.s.l. **IUCN Habitat code:** 1,3,14 **Conservation status: (IUCN)** NE

1111 *Furcraea foetida* (L.) Haw. (Asparagaceae)
IPNI: 64429-1
Common name: cabuya, cabuya común, cabuya hembra, cabuya sin espinas, fique, fique hembra, pita **Habit:** Herb **Use:** AF, EU, FU, HF, MA, ME, PO **Origin:** Native, Cultivated **Region:** Andean **Dept.:** CUN **Elev.:** 1750–2100 m a.s.l. **IUCN Habitat code:** 1,14 **Conservation status: (IUCN)** NE

1112 *Furcraea hexapetala* (Jacq.) Urb. (Asparagaceae)
IPNI: 105442-2
Common name: cabuya, cabuya con espinas, cabuya macho, fique, fique macho, maguey **Habit:** Herb **Use:** EU, MA, ME, PO **Origin:** Cultivated **Region:** Andean **Dept.:** BOG, BOY, CUN **Elev.:** 1580–2700 m a.s.l. **IUCN Habitat code:** 1,14 **Conservation status: (IUCN)** LC

1113 *Furcraea selloana* K.Koch (Asparagaceae)
IPNI: 64452-1
Common name: fique rayado, maguey, motua **Habit:** Herb **Use:** EU, MA, ME **Origin:** Native, Cultivated **Region:** Andean **Dept.:** MAG, VAC **Elev.:** 900–1300 m a.s.l. **Conservation status: (IUCN)** NE

1114 *Ornithogalum thyrsoides* Jacq (Asparagaceae)
IPNI: 539581-1
Common name: estrella de Belén, estrella de David **Habit:** Herb **Use:** EU, ME, PO **Origin:** Cultivated **Region:** Andean **Dept.:** BOG, CUN **Elev.:** 2600–2600 m a.s.l. **IUCN Habitat code:** 1 **Conservation status: (IUCN)** NE

1115 *Ruscus aculeatus* L. (Asparagaceae)
IPNI: 540443-1
Common name: brusco **Habit:** Herb **Use:** EU, HF, MA, ME **Origin:** Cultivated **Region:** Andean **Dept.:** ANT **Elev.:** 1500–2000 m a.s.l. **IUCN Habitat code:** 14 **Conservation status: (IUCN)** NE

1116 *Ruscus hypoglossum* L. (Asparagaceae)
IPNI: 540452-1
Common name: rusco **Habit:** Shrub **Use:** EU, HF, ME **Origin:** Cultivated **Region:** Andean **Dept.:** BOG, CUN **Elev.:** 2600–2600 m a.s.l. **Conservation status: (IUCN)** NE

1117 *Yucca baccata* Torr. (Asparagaceae)
IPNI: 270284-2
Use: MA, ME **Region:** Andean **Dept.:** BOG **Conservation status: (IUCN)** LC

1118 *Yucca gigantea* Lem. (Asparagaceae)
IPNI: 543700-1
Habit: Tree **Use:** EU, HF, ME **Origin:** Cultivated **Region:** Andean **Dept.:** ANT, BOG, CUN **Elev.:** 1500–2600 m a.s.l. **IUCN Habitat code:** 1,3,14 **Conservation status: (IUCN)** NE

1119 *Aloe arborescens* Mill. (Asphodelaceae)
IPNI: 529214-1
Habit: Herb, Shrub **Use:** EU, GS, HF, ME **Origin:** Cultivated **Region:** Andean **Dept.:** BOG, CUN **Elev.:** 2600–2600 m a.s.l. **IUCN Habitat code:** 14 **Conservation status: (IUCN)** LC

1120 *Aloe vera* (L.) Burm.f. (Asphodelaceae)
IPNI: 530017-1
Common name: aloe, penca de sábila, penca sábila, sábila, zábila **Habit:** Herb **Use:** EU, GS, HF, MA, ME, PO, SU **Origin:** Cultivated **Dept.:** ANT, BOG, BOY, CAL, CUN, HUI **Elev.:** 1300–2600 m a.s.l. **IUCN Habitat code:** 1,2,3,14 **Conservation status: (IUCN)** NE

1121 *Haworthiopsis fasciata* (Willd.) G.D.Rowley (Asphodelaceae)
IPNI: 77138009-1
Use: EU, ME **Region:** Andean **Dept.:** BOG **Conservation status: (IUCN)** NE

1122 *Hemerocallis citrina* Baroni (Asphodelaceae)
IPNI: 536319-1
Use: HF, ME **Region:** Andean **Dept.:** BOG **Conservation status: (IUCN)** NE

1123 *Hemerocallis fulva* (L.) L. (Asphodelaceae)
IPNI: 536335-1
Use: EU, HF, ME **Region:** Andean **Dept.:** BOG **IUCN Habitat code:** 14 **Conservation status: (IUCN)** NE

1124 *Hemerocallis lilioasphodelus* L. (Asphodelaceae)
IPNI: 536346-1
Common name: azucena amarilla, lirio, lirio amarillo, lirio de los pantanos **Habit:** Herb **Use:** EU, HF, ME **Origin:** Cultivated **Region:** Andean **Dept.:** ANT, BOG, BOY **Elev.:** 1000–2600 m a.s.l. **IUCN Habitat code:** 1,3 **Conservation status: (IUCN)** NE

1125 *Kniphofia × praecox* Baker (Asphodelaceae) **IPNI:** 537075-1 **Habit:** Herb **Use:** ME **Origin:** Cultivated **Region:** Andean **Dept.:** ANT **Elev.:** 2400–2400 m a.s.l. **IUCN Habitat code:** 1 **Conservation status: (IUCN)** NE

1126 *Kniphofia uvaria* (L.) Oken (Asphodelaceae) **IPNI:** 537112-1 **Common name:** llama, llamas, llamitas, rabo de runcho **Habit:** Herb **Use:** EU, ME **Origin:** Cultivated **Region:** Andean **Dept.:** ANT, BOG, CUN **Elev.:** 2000–2600 m a.s.l. **IUCN Habitat code:** 1,3,4,14 **Conservation status: (IUCN)** NE

1127 *Phormium tenax* J.R.Forst. & G.Forst. (Asphodelaceae) **IPNI:** 539926-1 **Common name:** economía, lino, lino enano, lino morado **Habit:** Herb **Use:** EU, HF, MA, ME **Origin:** Cultivated **Region:** Andean **Dept.:** ANT, BOG, CUN, QUI, RIS **Elev.:** 2000–2850 m a.s.l. **IUCN Habitat code:** 1,3,4,14 **Conservation status: (IUCN)** NE

1128 *Curculigo capitulata* (Lour.) Kuntze (Hypoxidaceae) **IPNI:** 64141-1 **Habit:** Herb **Use:** EU, GS, HF, MA, ME, SU **Origin:** Cultivated, Naturalised **Region:** Andean, Orinoquia **Dept.:** ANT, CAS, COR, CUN, GUV, MET, QUI, VID **Elev.:** 1000–1500 m a.s.l. **IUCN Habitat code:** 1 **Conservation status: (IUCN)** NE

1129 *Curculigo scorzonerifolia* (Lam.) Baker (Hypoxidaceae) **IPNI:** 64173-1 **Habit:** Herb **Use:** ME **Origin:** Native **Region:** Guiana Shield, Caribbean, Orinoquia **Dept.:** CAS, COR, CUN, GUV, MET, VID **Elev.:** 70–850 m a.s.l. **IUCN Habitat code:** 1,2,3,4 **Conservation status: (IUCN)** NE

1130 *Hypoxis decumbens* L. (Hypoxidaceae) **IPNI:** 65385-1 **Common name:** tiririca **Habit:** Herb **Use:** HF **Origin:** Native **Region:** Andean **Dept.:** ANT, BOG, BOY, CHO, COR, CUN, HUI, MAG, NAR, PUT **Elev.:** 1400–2800 m a.s.l. **IUCN Habitat code:** 1,3,4,14 **Conservation status: (IUCN)** NE

1131 *Crocosmia × crocosmiiflora* (Lemoine) N.E.Br. (Iridaceae) **IPNI:** 436427-1 **Use:** ME **Origin:** Naturalised **Dept.:** BOG **IUCN Habitat code:** 1,2,3,4,5,8,14 **Conservation status: (IUCN)** NE

1132 *Crocosmia aurea* (Pappe ex Hook.) Planch. (Iridaceae) **IPNI:** 436425-1 **Use:** EU, GS, HF, MA, ME **Region:** Andean **Dept.:** BOG **Conservation status: (IUCN)** NE

1133 *Crocosmia pottsii* (Baker) N.E.Br. (Iridaceae) **IPNI:** 436437-1 **Use:** ME **Origin:** Naturalised **IUCN Habitat code:** 1,3 **Conservation status: (IUCN)** NE

1134 *Dietes grandiflora* N.E.Br. (Iridaceae) **IPNI:** 436870-1 **Use:** EU **Region:** Andean **Dept.:** BOG **IUCN Habitat code:** 14 **Conservation status: (IUCN)** NE

1135 *Eleutherine bulbosa* (Mill.) Urb. (Iridaceae) **IPNI:** 89940-2 **Common name:** juari corobeabo, nolocrea **Habit:** Herb **Use:** HF, MA, ME **Origin:** Native **Region:** Amazonia, Andean **Dept.:** CAU, NAR, PUT **Elev.:** 300–1400 m a.s.l. **IUCN Habitat code:** 1,4 **Conservation status: (IUCN)** NE

1136 *Freesia leichtlinii* (G.L.Mey.) J.C.Manning & Goldblatt (Iridaceae) **IPNI:** 437022-1 **Use:** EU **Region:** Andean **Dept.:** BOG **Conservation status: (IUCN)** NE

1137 *Freesia refracta* (Jacq.) Klatt (Iridaceae) **IPNI:** 437030-1 **Common name:** fresia, lirio, nardo **Habit:** Herb **Use:** EU, ME **Origin:** Cultivated **Region:** Andean **Dept.:** ANT, BOG, BOY, CAL, CAU, CUN, QUI, RIS, SAN **Elev.:** 2000–2500 m a.s.l. **Conservation status: (IUCN)** NE

1138 *Gladiolus communis* L. (Iridaceae) **IPNI:** 437372-1 **Habit:** Herb **Use:** EU, ME **Origin:** Cultivated **Region:** Andean **Dept.:** ANT, BOG, BOY, CAL, CAU, CUN, QUI, RIS, SAN **Elev.:** 2000–3000 m a.s.l. **IUCN Habitat code:** 1,3,4,14 **Conservation status: (IUCN)** NE

1139 *Herbertia pulchella* Sweet (Iridaceae) **IPNI:** 278134-2 **Use:** HF **Origin:** Native **Conservation status: (IUCN)** NE

1140 *Iris × germanica* L. (Iridaceae) **IPNI:** 438637-1 **Habit:** Herb **Use:** ME **Origin:** Cultivated **Region:** Andean **Dept.:** ANT, BOY, CAL, CAU, CUN, QUI, RIS, SAN **Elev.:** 2000–3000 m a.s.l. **IUCN Habitat code:** 1,14 **Conservation status: (IUCN)** NE

1141 *Iris japonica* Thunb. (Iridaceae) **IPNI:** 438741-1 **Use:** ME **Region:** Andean **Dept.:** BOG **IUCN Habitat code:** 3 **Conservation status: (IUCN)** NE

1142 *Orthrosanthus chimboracensis* (Kunth) Baker (Iridaceae) **IPNI:** 440470-1 **Common name:** amargoso, aneiba, carrizo, escobo, espadilla, esterilla, esterillo, estrellita, fibra de María, lirio de páramo, paja garnona, raíz de cepillos **Habit:** Herb **Use:** EU, MA, ME **Origin:** Native **Region:** Andean **Dept.:** ANT, BOG, BOY, CAU, CUN, MAG, NAR, PUT, SAN **Elev.:** 2200-4000 m.a.s.l. **IUCN Habitat code:** 1,2,3,4,5,8,14 **Conservation status: (IUCN)** NE **(CPSG)** Potential LC

1143 *Orthrosanthus monadelphus* Ravenna (Iridaceae) **IPNI:** 176894-2 **Common name:** sebujá **Habit:** Herb **Use:** ME **Origin:** Native **Region:** Andean **Dept.:** BOG, BOY, CAU, CUN, MAG, SAN **Elev.:** 2600–3500 m a.s.l. **IUCN Habitat code:** 1,3,4,14 **Conservation status: (IUCN)** NE

1144 *Sisyrinchium chilense* Hook. (Iridaceae) **IPNI:** 330256-2 **Common name:** espadilla **Habit:** Herb **Use:** ME **Origin:** Native **Region:** Andean **Dept.:** BOG, BOY, CUN, MET, NAR, SAN **Elev.:** 2500-3660 m.a.s.l. **IUCN Habitat code:** 1,3,4,14 **Conservation status: (IUCN)** NE **(CPSG)** Potential LC

1145 *Sisyrinchium convolutum* Nocca (Iridaceae) **IPNI:** 285111-2 **Common name:** espadilla **Habit:** Herb **Use:** EU, ME **Origin:** Native **Region:** Andean **Dept.:** ANT, BOG, BOY, CAL, CAU, CUN, MET, NAR, NSA, PUT, SAN, TOL, VAC **Elev.:** 1900–4000 m a.s.l. **IUCN Habitat code:** 1,4,14 **Conservation status: (IUCN)** NE

1146 *Sisyrinchium micranthum* Cav. (Iridaceae) **IPNI:** 330260-2 **Common name:** cortadera, espadilla **Habit:** Herb **Use:** ME, PO **Origin:** Native **Region:** Andean **Dept.:** ANT, BOG, BOY, CAL, CAQ, CAU, CUN, HUI, MAG, NAR, NSA, PUT, QUI, RIS, SAN, TOL, VAC **Elev.:** 1000-4500 m.a.s.l. **IUCN Habitat code:** 1,3,4,8,14 **Conservation status: (IUCN)** NE **(CPSG)** Potential LC

1147 *Sisyrinchium palmifolium* L. (Iridaceae) **IPNI:** 237727-2 **Habit:** Herb **Use:** EU, ME **Origin:** Native **Region:** Andean **Dept.:** ANT, BOY, MAG, NSA, SAN **Elev.:** 1750–3900 m a.s.l. **IUCN Habitat code:** 1,3,4,14 **Conservation status: (IUCN)** NE

1148 *Sisyrinchium tinctorium* Kunth (Iridaceae) **IPNI:** 441310-1 **Common name:** espadilla, esterillo, fito-fito, fitofito, ispaguilla, lirio de monte **Habit:** Herb **Use:** ME **Origin:** Native **Region:** Andean **Dept.:** ANT, BOG, BOY, CAL, CAU, CUN, HUI, LAG, MAG, NAR, NSA, PUT, SAN **Elev.:** 1400-4500 m.a.s.l. **IUCN Habitat code:** 1,3,4,14 **Conservation status: (IUCN)** NE **(CPSG)** Potential LC

1149 *Sisyrinchium vaginatum* Spreng. (Iridaceae) **IPNI:** 237808-2 **Common name:** espadilla **Habit:** Herb **Use:** ME **Origin:** Native **Region:** Andean, Guiana Shield, Orinoquia **Dept.:** BOG, CAS, GUV, MET, VID **Elev.:** 80–500 m a.s.l. **IUCN Habitat code:** 1,2,3,4,14 **Conservation status: (IUCN)** NE

1150 *Tigridia pavonia* (L.f.) Redouté (Iridaceae) **IPNI:** 441553-1 **Common name:** flor de la maravilla, flor de un día, lirio, maravilla, tulipán **Habit:** Herb **Use:** EU, HF, ME **Origin:** Naturalised **Region:** Andean **Dept.:** ANT, BOG, BOY, CUN, NAR, PUT, QUI, RIS, VAC **Elev.:** 1000–2800 m a.s.l. **IUCN Habitat code:** 1,2,3,14 **Conservation status: (IUCN)** NE

1151 *Trimezia longifolia* (Link & Otto) Christenh. & Byng (Iridaceae)
IPNI: 77182480-1
Use: SU **Origin:** Native **Conservation status: (IUCN)** NE

1152 *Trimezia martinicensis* (Jacq.) Herb. (Iridaceae)
IPNI: 441642-1
Common name: malamadre **Habit:** Herb **Use:** EU, MA, ME **Origin:** Native, Cultivated **Region:** Andean **Dept.:** ANT, CUN, QUI **Elev.:** 1200–2500 m a.s.l. **IUCN Habitat code:** 2,3,14 **Conservation status: (IUCN)** NE

1153 *Watsonia borbonica* (Pourr.) Goldblatt (Iridaceae)
IPNI: 944889-1
Common name: flor de mayo, garzovia, palma, sanjosé, varita de San José, varsovia, vida del hombre **Habit:** Herb **Use:** EU **Origin:** Cultivated **Region:** Andean **Dept.:** ANT, BOG, BOY, CAL, CAU, CUN, QUI, RIS, SAN **Elev.:** 2500–3000 m a.s.l. **IUCN Habitat code:** 1,14 **Conservation status: (IUCN)** NE

1154 *Aa maderoi* Schltr. (Orchidaceae)
IPNI: 302812-2
Habit: Herb **Use:** ME **Origin:** Native **Region:** Andean **Dept.:** BOG, CAU, CUN, HUI **Elev.:** 2800–3720 m a.s.l. **IUCN Habitat code:** 4 **Conservation status: (IUCN)** NE

1155 *Acineta superba* (Kunth) Rchb.f. (Orchidaceae)
IPNI: 614722-1
Common name: calavera, calaveras **Habit:** Herb, Epiphyte **Use:** EU **Origin:** Native **Region:** Andean **Dept.:** ANT, CUN, MAG, NSA **Elev.:** 1800–2353 m a.s.l. **IUCN Habitat code:** 2,4,14 **Conservation status: (IUCN)** NE

1156 *Anathallis brevipes* (H.Focke) Pridgeon & M.W.Chase (Orchidaceae)
IPNI: 1147281-2
Habit: Herb, Epiphyte **Use:** EU **Origin:** Native, Cultivated **Region:** Orinoquia **Dept.:** MET **Elev.:** 300–300 m a.s.l. **Conservation status: (IUCN)** NE

1157 *Anguloa uniflora* Ruiz & Pav. (Orchidaceae)
IPNI: 616523-1
Common name: cuna de Venus **Use:** EU **IUCN Habitat code:** 1 **Conservation status: (IUCN)** NE

1158 *Aspidogyne debilis* (Lindl.) Meneguzzo (Orchidaceae)
IPNI: 77125687-1
Habit: Herb **Use:** EU **Origin:** Native **Region:** Andean **Dept.:** NAR **Conservation status: (IUCN)** NE

1159 *Aspidogyne querceticola* (Lindl.) Meneguzzo (Orchidaceae)
IPNI: 77125705-1
Habit: Herb **Use:** EU **Origin:** Native **Region:** Andean **Dept.:** ANT, CHO, QUI **Elev.:** 600–1450 m a.s.l. **IUCN Habitat code:** 1,14 **Conservation status: (IUCN)** NE

1160 *Bletia campanulata* Lex. (Orchidaceae)
IPNI: 617639-1
Habit: Herb, Epiphyte **Use:** ME **Origin:** Native **Region:** Andean, Orinoquia **Dept.:** CAU, HUI, VID **Elev.:** 120–1700 m a.s.l. **Conservation status: (IUCN)** NE

1161 *Bletia catenulata* Ruiz & Pav. (Orchidaceae)
IPNI: 617644-1
Habit: Herb, Epiphyte **Use:** EU **Origin:** Native **Region:** Andean **Elev.:** 1500–1500 m a.s.l. **Conservation status: (IUCN)** NE

1162 *Bletia purpurea* (Lam.) A.DC. (Orchidaceae)
IPNI: 1218040-2
Habit: Herb, Epiphyte **Use:** EU, ME **Origin:** Native **Region:** Andean **Dept.:** ANT, MAG **Elev.:** 620–1700 m a.s.l. **IUCN Habitat code:** 1,3,14 **Conservation status: (IUCN)** NE

1163 *Brassavola cucullata* (L.) R.Br. (Orchidaceae)
IPNI: 618026-1
Habit: Herb, Epiphyte **Use:** EU **Origin:** Native **Region:** Andean, Caribbean, Pacific **Dept.:** ANT, ATL, BOL, CES, CHO, CUN, GUA, MAG, SAN, TOL **Elev.:** 50–1200 m a.s.l. **IUCN Habitat code:** 3 **Conservation status: (IUCN)** NE

1164 *Brassavola martiana* Lindl. (Orchidaceae)
IPNI: 325518-2
Common name: herb cebolla, hierba cebolla **Habit:** Herb, Epiphyte **Use:** MA **Origin:** Native **Region:** Amazonia **Dept.:** AMA, VAU **Elev.:** 180–450 m a.s.l. **Conservation status: (IUCN)** NE

1165 *Catasetum barbatum* (Lindl.) Lindl. (Orchidaceae)
IPNI: 1092236-2
Habit: Herb, Epiphyte **Use:** ME **Origin:** Native **Region:** Amazonia **Dept.:** VAU **Elev.:** 200–220 m a.s.l. **Conservation status: (IUCN)** LC

1166 *Catasetum discolor* (Lindl.) Lindl. (Orchidaceae)
IPNI: 621811-1
Habit: Herb, Epiphyte **Use:** MA **Origin:** Native **Region:** Amazonia, Orinoquia **Dept.:** AMA, GUA, GUV, VAU **Elev.:** 180–500 m a.s.l. **IUCN Habitat code:** 1,3,4,14 **Conservation status: (IUCN)** NE

1167 *Catasetum macrocarpum* Rich. ex Kunth (Orchidaceae)
IPNI: 325522-2
Common name: cebolleta, cebolleta de pegar **Habit:** Herb, Epiphyte **Use:** EU, MA **Origin:** Native **Region:** Amazonia, Guiana Shield **Dept.:** VAU **Elev.:** 50–200 m a.s.l. **IUCN Habitat code:** 1,3,14 **Conservation status: (IUCN)** NE

1168 *Catasetum saccatum* Lindl. (Orchidaceae)
IPNI: 621924-1
Habit: Herb, Epiphyte **Use:** EU **Origin:** Native **Region:** Amazonia, Andean, Orinoquia **Dept.:** MET **Elev.:** 120–1000 m a.s.l. **IUCN Habitat code:** 4 **Conservation status: (IUCN)** NE

1169 *Cattleya dowiana* Bateman & Rchb.f. (Orchidaceae)
IPNI: 622063-1
Common name: carleya amarilla, catleya de Dabeiba, catleya de oro, catleya dorada **Habit:** Herb, Epiphyte **Use:** EU **Origin:** Native, Cultivated **Region:** Andean, Pacific **Dept.:** ANT, CHO, COR, RIS **Elev.:** 50–1600 m a.s.l. **Conservation status: (IUCN)** NE

1170 *Cattleya loddigesii* Lindl. (Orchidaceae)
IPNI: 622144-1
Common name: flor de mayo **Use:** EU **Conservation status: (IUCN)** NE

1171 *Cattleya maxima* Lindl. (Orchidaceae)
IPNI: 622161-1
Use: EU **Origin:** Native **Conservation status: (IUCN)** NE

1172 *Cattleya mendelii* Dombrain (Orchidaceae)
IPNI: 70028975-1
Habit: Herb, Epiphyte **Use:** EU **Origin:** Endemic, Cultivated **Region:** Andean **Dept.:** BOY, NSA, SAN **Elev.:** 1200-1800 m.a.s.l. **Conservation status: (IUCN)** NE **(CPSG)** DD

1173 *Cattleya percivaliana* (Rchb.f.) O'Brien (Orchidaceae)
IPNI: 622189-1
Habit: Herb, Epiphyte **Use:** EU **Origin:** Native **Region:** Orinoquia **Dept.:** ARA, NSA **Elev.:** 1300-2000 m.a.s.l. **Conservation status: (IUCN)** NE **(CPSG)** DD

1174 *Cattleya quadricolor* B.S.Williams (Orchidaceae)
IPNI: 622208-1
Habit: Herb, Epiphyte **Use:** EU **Origin:** Endemic **Region:** Andean **Dept.:** QUI, RIS, VAC **Elev.:** 950-1700 m.a.s.l. **IUCN Habitat code:** 1,3,14 **Conservation status: (IUCN)** NE **(CPSG)** EN

1175 *Cattleya schroederae* (Rchb.f.) Sander (Orchidaceae)
IPNI: 50291-2
Habit: Herb, Epiphyte **Use:** EU **Origin:** Endemic, Cultivated **Region:** Andean, Orinoquia **Dept.:** ARA, BOY, CAS, CUN, MET **Elev.:** 500-1600 m.a.s.l. **Conservation status: (IUCN)** NE **(CPSG)** EN

1176 *Cattleya trianae* Linden & Rchb.f. (Orchidaceae)
IPNI: 622257-1
Common name: carleya blanca, catleya, catleya blanca, flor de San Juan, flor de mayo, lirio, lirio de mayo, mayo, parásita, tulipán **Habit:** Herb, Epiphyte **Use:** EU, ME **Origin:** Endemic, Cultivated **Region:** Andean **Dept.:** CUN, HUI, TOL **Elev.:** 900-1800 m.a.s.l. **IUCN Habitat code:** 1,3,4,14 **Conservation status: (IUCN)** NE **(CPSG)** EN

1177 *Cattleya violacea* (Kunth) Lindl. (Orchidaceae)
IPNI: 622273-1
Habit: Herb, Epiphyte **Use:** EU **Origin:** Native, Cultivated **Region:** Amazonia, Guiana Shield, Orinoquia **Dept.:** AMA, ARA, CAQ, CAS, GUA, GUV, MET, PUT, VAU, VID **Elev.:** 200–646 m a.s.l. **Conservation status: (IUCN)** NE

1178 *Cattleya warscewiczii* Rchb.f. (Orchidaceae)
IPNI: 622282-1
Habit: Herb, Epiphyte **Use:** EU **Origin:** Endemic **Region:** Andean **Dept.:** ANT, CAL, COR, RIS **Elev.:** 300-1600 m.a.s.l. **IUCN Habitat code:** 1,4,14 **Conservation status: (IUCN)** NE **(CPSG)** EN

1179 *Caularthron bicornutum* (Hook.) Raf. (Orchidaceae) **IPNI:** 622297-1 **Habit:** Herb, Epiphyte **Use:** EU **Origin:** Native **Region:** Guiana Shield, Orinoquia **Dept.:** GUA, VAU, VID **Elev.:** 100–250 m a.s.l. **IUCN Habitat code:** 1,3,5,14 **Conservation status: (IUCN)** NE

1180 *Cischweinfia colombiana* Garay (Orchidaceae) **IPNI:** 59191-2 **Common name:** orquídea **Habit:** Herb, Epiphyte **Use:** EU **Origin:** Endemic **Region:** Andean **Dept.:** ANT **Elev.:** 1050–1535 m a.s.l. **Conservation status: (IUCN)** NE

1181 *Coccineorchis cernua* (Lindl.) Garay (Orchidaceae) **IPNI:** 61935-2 **Habit:** Herb **Use:** EU **Origin:** Native, Cultivated **Region:** Andean **Dept.:** ATL, BOY, CAU, CHO, CUN, RIS, VAC **Elev.:** 1800–3200 m a.s.l. **Conservation status: (IUCN)** NE

1182 *Coryanthes elegantium* Linden & Rchb.f. (Orchidaceae) **IPNI:** 624302-1 **Habit:** Herb, Epiphyte **Use:** MA **Origin:** Native **Region:** Pacific **Dept.:** ANT, CHO, NAR **Elev.:** 100-100 m.a.s.l. **Conservation status: (IUCN)** NE **(CPSG)** NT

1183 *Cyclopogon elatus* (Sw.) Schltr. (Orchidaceae) **IPNI:** 73688-2 **Habit:** Herb, Epiphyte **Use:** ME **Origin:** Native **Region:** Andean **Dept.:** ANT, BOG, CUN, NAR, SAN, VAC **Elev.:** 1000–2926 m a.s.l. **IUCN Habitat code:** 1,3,14 **Conservation status: (IUCN)** NE

1184 *Cycnoches chlorochilon* Klotzsch (Orchidaceae) **IPNI:** 73806-2 **Common name:** caucana, cisne, pelícano, sinnoches **Habit:** Herb, Epiphyte **Use:** EU **Origin:** Native **Region:** Andean **Dept.:** MAG **Elev.:** 1250-1500 m.a.s.l. **IUCN Habitat code:** 3,14 **Conservation status: (IUCN)** NE **(CPSG)** LC

1185 *Cyrtochilum halteratum* (Lindl.) Kraenzl. (Orchidaceae) **IPNI:** 75479-2 **Habit:** Herb, Epiphyte **Use:** EU **Origin:** Native **Region:** Andean **Dept.:** ANT, QUI **Elev.:** 2630–2700 m a.s.l. **IUCN Habitat code:** 1,2,3,4,14 **Conservation status: (IUCN)** LC

1186 *Cyrtochilum ramosissimum* (Lindl.) Dalström (Orchidaceae) **IPNI:** 1021207-1 **Common name:** americana **Habit:** Herb, Epiphyte, Terrestrial **Use:** EU **Origin:** Native **Region:** Andean **Dept.:** ANT, BOG, NAR **Elev.:** 1900–3600 m a.s.l. **IUCN Habitat code:** 1,3,4,14 **Conservation status: (IUCN)** NE

1187 *Cyrtochilum revolutum* (Lindl.) Dalström (Orchidaceae) **IPNI:** 1021208-1 **Common name:** aguadija **Habit:** Herb, Epiphyte **Use:** HF, ME **Origin:** Native **Region:** Andean **Dept.:** BOG, BOY, CAU, CUN, NSA **Elev.:** 3000-3700 m.a.s.l. **IUCN Habitat code:** 1,3,4,14 **Conservation status: (IUCN)** NE **(CPSG)** VU

1188 *Cyrtopodium andersonii* (Lamb. ex Andrews) R.Br. (Orchidaceae) **IPNI:** 1173554-2 **Habit:** Herb, Epiphyte **Use:** MA **Origin:** Native **Region:** Orinoquia **Dept.:** VID **Elev.:** 100–100 m a.s.l. **IUCN Habitat code:** 1,4,14 **Conservation status: (IUCN)** NE

1189 *Cyrtopodium punctatum* (L.) Lindl. (Orchidaceae) **IPNI:** 1147853-2 **Habit:** Herb, Epiphyte, Lithophytes **Use:** EU, ME **Origin:** Native **Region:** Andean **Dept.:** CAU, CUN, MAG **Elev.:** 600–1500 m a.s.l. **IUCN Habitat code:** 1,3 **Conservation status: (IUCN)** NE

1190 *Dichaea morrisii* Fawc. & Rendle (Orchidaceae) **IPNI:** 79276-2 **Habit:** Herb, Epiphyte **Use:** ME **Origin:** Native **Region:** Andean **Dept.:** ANT, CUN, MAG, NAR **Elev.:** 939–2280 m a.s.l. **IUCN Habitat code:** 1,3,14 **Conservation status: (IUCN)** NE

1191 *Dracula bella* (Rchb.f.) Luer (Orchidaceae) **IPNI:** 83961-2 **Habit:** Herb, Epiphyte **Use:** EU **Origin:** Endemic **Region:** Andean **Dept.:** ANT, CHO, RIS **Elev.:** 1700-2440 m.a.s.l. **IUCN Habitat code:** 1 **Conservation status: (IUCN)** NE **(CPSG)** VU

1192 *Dracula chimaera* (Rchb.f.) Luer (Orchidaceae) **IPNI:** 83967-2 **Habit:** Herb, Epiphyte **Use:** EU **Origin:** Endemic **Region:** Andean **Dept.:** ANT, CAU, CHO, VAC **Elev.:** 1300-2200 m.a.s.l. **IUCN Habitat code:** 1,14 **Conservation status: (IUCN)** NE **(CPSG)** NT

1193 *Elleanthus aurantiacus* (Lindl.) Rchb.f. (Orchidaceae) **IPNI:** 630704-1 **Habit:** Herb **Use:** ME **Origin:** Native **Region:** Andean **Dept.:** ANT, ARA, BOG, BOY, CAL, CAU, CUN, HUI, MAG, MET, NAR, PUT, QUI, RIS, SAN, TOL, VAC **Elev.:** 1600-3600 m.a.s.l. **IUCN Habitat code:** 1,3,4,8,14 **Conservation status: (IUCN)** NE **(CPSG)** LC

1194 *Elleanthus kermesinus* (Lindl.) Rchb.f. (Orchidaceae) **IPNI:** 630759-1 **Common name:** vara de alcalde **Habit:** Herb **Use:** EU **Origin:** Native **Region:** Andean **Dept.:** NSA, TOL **Elev.:** 1200–1500 m a.s.l. **IUCN Habitat code:** 1,4,8 **Conservation status: (IUCN)** NE

1195 *Epidendrum × elongatum* Jacq. (Orchidaceae) **IPNI:** 631770-1 **Habit:** Herb, Epiphyte **Use:** EU **Origin:** Native **Region:** Andean **Dept.:** BOY, CAL, CAU, CES, CHO, CUN, HUI, MAG, NAR, PUT, RIS, SAN, TOL, VAC **Elev.:** 1500–3500 m a.s.l. **IUCN Habitat code:** 1,2,3,4,5,14 **Conservation status: (IUCN)** NE

1196 *Epidendrum amplexigastrium* Hágsater & Dodson (Orchidaceae) **IPNI:** 318038-2 **Habit:** Herb, Epiphyte **Use:** EU **Origin:** Native, Cultivated **Region:** Andean **Conservation status: (IUCN)** NE

1197 *Epidendrum arachnoglossum* Rchb.f. ex André (Orchidaceae) **IPNI:** 631350-1 **Habit:** Herb **Use:** EU **Origin:** Native **Region:** Andean **Dept.:** BOG, CAU, CUN, NAR **Elev.:** 2500-3000 m.a.s.l. **IUCN Habitat code:** 1,3,14 **Conservation status: (IUCN)** NE **(CPSG)** DD

1198 *Epidendrum ciliare* L. (Orchidaceae) **IPNI:** 631578-1 **Habit:** Herb, Epiphyte **Use:** EU **Origin:** Native **Region:** Andean, Orinoquia **Dept.:** ANT, BOY, CUN, MET, NSA, TOL **Elev.:** 500–1800 m a.s.l. **IUCN Habitat code:** 2,3,14 **Conservation status: (IUCN)** NE

1199 *Epidendrum cinnabarinum* Salzm. ex Lindl. (Orchidaceae) **IPNI:** 631580-1 **Habit:** Herb **Use:** EU **Origin:** Cultivated **Dept.:** PUT **Elev.:** 500–500 m a.s.l. **Conservation status: (IUCN)** NE

1200 *Epidendrum ellisii* Rolfe (Orchidaceae) **IPNI:** 91520-2 **Habit:** Herb, Epiphyte **Use:** EU **Origin:** Endemic, Cultivated **Region:** Andean **Conservation status: (IUCN)** NE

1201 *Epidendrum erosum* Ames & C.Schweinf. (Orchidaceae) **IPNI:** 91528-2 **Common name:** epidendro, jagalana **Habit:** Herb, Epiphyte **Use:** ME **Origin:** Native **Region:** Andean **Dept.:** ANT, BOG, CAU, CUN, MAG, NSA **Elev.:** 2160–3685 m a.s.l. **IUCN Habitat code:** 1,3,4,8,14 **Conservation status: (IUCN)** NE

1202 *Epidendrum fimbriatum* Kunth (Orchidaceae) **IPNI:** 631826-1 **Common name:** pajarito blanco **Habit:** Herb, Epiphyte **Use:** EU, ME **Origin:** Native **Region:** Andean **Dept.:** ANT, BOG, BOY, CAL, CAQ, CAU, CHO, CUN, HUI, LAG, NAR, NSA, PUT, QUI, SAN, TOL, VAC **Elev.:** 1070-3800 m.a.s.l. **IUCN Habitat code:** 1,2,3,4,5,8,14 **Conservation status: (IUCN)** NE **(CPSG)** Potential LC

1203 *Epidendrum friderici-guilielmi* Warsz. ex Lindl. (Orchidaceae) **IPNI:** 631856-1 **Use:** EU **Conservation status: (IUCN)** NE

1204 *Epidendrum ibaguense* Kunth (Orchidaceae) **IPNI:** 633050-1 **Common name:** pajaritos **Habit:** Herb, Epiphyte **Use:** EU, ME **Origin:** Native **Region:** Andean **Dept.:** ANT, CAU, CUN, HUI, NSA, SAN, TOL **Elev.:** 1300–1750 m a.s.l. **IUCN Habitat code:** 1,2,3,4,14 **Conservation status: (IUCN)** NE

1205 *Epidendrum ilense* Dodson (Orchidaceae) **IPNI:** 91670-2 **Habit:** Herb, Epiphyte **Use:** EU **Origin:** Cultivated **Conservation status: (IUCN)** NE

1206 *Epidendrum jamiesonis* Rchb.f. (Orchidaceae) **IPNI:** 632050-1 **Habit:** Herb **Use:** ME **Origin:** Native **Region:** Andean **Dept.:** NAR **Elev.:**

2600–3100 m a.s.l. **Conservation status: (IUCN)** NE

1207 *Epidendrum leeanum* (Rchb.f.) Hágsater (Orchidaceae) **IPNI:** 77100813-1 **Habit:** Herb, Epiphyte **Use:** EU **Origin:** Native, Cultivated **Region:** Orinoquia **Dept.:** BOY, MET **Elev.:** 100–400 m a.s.l. **Conservation status: (IUCN)** NE

1208 *Epidendrum nocturnum* Jacq. (Orchidaceae) **IPNI:** 30010607-2 **Habit:** Herb, Epiphyte **Use:** EU **Origin:** Native **Region:** Amazonia, Andean, Guiana Shield, Caribbean, Orinoquia, Pacific **Dept.:** AMA, ANT, CAQ, CHO, GUA, GUV, NAR, VAC, VAU **Elev.:** 0–2500 m a.s.l. **IUCN Habitat code:** 1,3,4,5,14 **Conservation status: (IUCN)** NE

1209 *Epidendrum paniculatum* Ruiz & Pav. (Orchidaceae) **IPNI:** 632433-1 **Habit:** Herb, Epiphyte **Use:** ME **Origin:** Native **Region:** Andean **Dept.:** CAU **Elev.:** 2000–2000 m a.s.l. **IUCN Habitat code:** 1,3,4,14 **Conservation status: (IUCN)** DD

1210 *Epidendrum pristes* Rchb.f. (Orchidaceae) **IPNI:** 632551-1 **Habit:** Herb, Epiphyte **Use:** EU **Origin:** Endemic, Cultivated **Region:** Andean **Conservation status: (IUCN)** NE

1211 *Epidendrum radicans* Pav. ex Lindl. (Orchidaceae) **IPNI:** 632612-1 **Habit:** Herb, Epiphyte **Use:** EU, ME **Origin:** Native **Region:** Andean **Dept.:** ANT, CAL, CAU, RIS, TOL, VAC **Elev.:** 1000–2400 m a.s.l. **IUCN Habitat code:** 1,3,4,14 **Conservation status: (IUCN)** NE

1212 *Epidendrum secundum* Jacq. (Orchidaceae) **IPNI:** 92159-2 **Common name:** araña, melena, pajarito amarillo **Habit:** Herb, Epiphyte **Use:** EU, ME **Origin:** Native **Region:** Andean **Dept.:** ANT **Elev.:** 2000–2000 m a.s.l. **IUCN Habitat code:** 1,3,4,5,14 **Conservation status: (IUCN)** NE

1213 *Epidendrum shigenobui* Hágsater (Orchidaceae) **IPNI:** 318117-2 **Habit:** Herb, Epiphyte **Use:** EU **Origin:** Endemic, Cultivated **Region:** Andean **Dept.:** RIS **Conservation status: (IUCN)** NE

1214 *Epidendrum strobiliferum* Rchb.f. (Orchidaceae) **IPNI:** 92214-2 **Habit:** Herb, Epiphyte **Use:** ME **Origin:** Native **Region:** Amazonia, Orinoquia, Pacific **Dept.:** ANT, MET **Elev.:** 400–490 m a.s.l. **IUCN Habitat code:** 1,3,14 **Conservation status: (IUCN)** NE

1215 *Epidendrum summerhayesii* Hágsater (Orchidaceae) **IPNI:** 302760-2 **Habit:** Herb, Epiphyte **Use:** EU **Origin:** Native, Cultivated **Region:** Andean **Dept.:** ANT **Elev.:** 1200–1800 m a.s.l. **Conservation status: (IUCN)** NE

1216 *Epidendrum wallisii* Rchb.f. (Orchidaceae) **IPNI:** 633025-1 **Habit:** Herb, Epiphyte **Use:** EU **Origin:** Native **Region:** Andean **Dept.:** ANT, CHO, VAC **Elev.:** 500–2100 m a.s.l. **IUCN Habitat code:** 1,14 **Conservation status: (IUCN)** NE

1217 *Epidendrum xanthinum* Lindl. (Orchidaceae) **IPNI:** 633042-1 **Use:** EU **IUCN Habitat code:** 1,2,3,5,14 **Conservation status: (IUCN)** NE

1218 *Epidendrum zipaquiranum* Schltr. (Orchidaceae) **IPNI:** 92363-2 **Habit:** Herb **Use:** EU **Origin:** Native **Region:** Andean **Dept.:** ARA, BOY, CUN, MET, NSA, PUT, SAN **Elev.:** 3000–3500 m a.s.l. **IUCN Habitat code:** 1,3,4,5,14 **Conservation status: (IUCN)** NE

1219 *Eriopsis rutidobulbon* Hook. (Orchidaceae) **IPNI:** 634223-1 **Habit:** Herb, Epiphyte **Use:** EU **Origin:** Native **Region:** Andean **Dept.:** ANT, BOY **Elev.:** 1200–1200 m a.s.l. **Conservation status: (IUCN)** NE

1220 *Erycina pusilla* (L.) N.H.Williams & M.W.Chase (Orchidaceae) **IPNI:** 1021256-1 **Common name:** pajaritos **Habit:** Herb, Epiphyte **Use:** ME **Origin:** Native **Region:** Amazonia, Andean, Pacific **Dept.:** ANT, CHO, VAU **Elev.:** 50–1500 m a.s.l. **IUCN Habitat code:** 1,3,4,5,14 **Conservation status: (IUCN)** NE

1221 *Gomesa radicans* (Rchb.f.) M.W.Chase & N.H.Williams (Orchidaceae) **IPNI:** 77101300-1 **Habit:** Herb, Epiphyte **Use:** EU **Origin:** Cultivated **Dept.:** VAC **Elev.:** 500–1000 m a.s.l. **Conservation status: (IUCN)** NE

1222 *Gongora portentosa* Linden & Rchb.f. (Orchidaceae) **IPNI:** 636033-1 **Common name:** torito **Habit:** Herb, Epiphyte **Use:** EU **Origin:** Native **Region:** Andean **Dept.:** BOY, CAQ, CUN, MET **Elev.:** 800–1000 m a.s.l. **IUCN Habitat code:** 3 **Conservation status: (IUCN)** LC

1223 *Habenaria conopsea* Rchb.f. (Orchidaceae) **IPNI:** 637060-1 **Habit:** Herb **Use:** HF, ME **Origin:** Endemic **Region:** Andean **Conservation status: (IUCN)** NE

1224 *Habenaria floribunda* Lindl. (Orchidaceae) **IPNI:** 637263-1 **Habit:** Herb **Use:** ME **Origin:** Native **Region:** Andean **Dept.:** ANT **Elev.:** 2000–2000 m a.s.l. **IUCN Habitat code:** 1,2,14 **Conservation status: (IUCN)** NE

1225 *Huntleya meleagris* Lindl. (Orchidaceae) **IPNI:** 639074-1 **Common name:** belmania **Habit:** Herb, Epiphyte **Use:** EU **Origin:** Native **Region:** Andean **Dept.:** ANT, RIS, VAC **Elev.:** 500–1000 m a.s.l. **IUCN Habitat code:** 1,14 **Conservation status: (IUCN)** NE

1226 *Laelia colombiana* J.M.H.Shaw (Orchidaceae) **IPNI:** 77090555-1 **Habit:** Herb, Epiphyte **Use:** EU **Origin:** Endemic, Cultivated **Conservation status: (IUCN)** NE

1227 *Laelia gloriosa* (Rchb.f.) L.O.Williams (Orchidaceae) **IPNI:** 134675-2 **Use:** EU **Conservation status: (IUCN)** NE **(CPSG)** LC

1228 *Liparis nervosa* (Thunb.) Lindl. (Orchidaceae) **IPNI:** 641061-1 **Habit:** Herb **Use:** EU, MA, ME **Origin:** Native **Region:** Amazonia, Andean, Guiana Shield **Dept.:** ANT, SAN, VAU **Elev.:** 180–1250 m a.s.l. **IUCN Habitat code:** 1,3,14 **Conservation status: (IUCN)** NE

1229 *Lockhartia acuta* (Lindl.) Rchb.f. (Orchidaceae) **IPNI:** 641704-1 **Habit:** Herb, Epiphyte **Use:** EU **Origin:** Native, Cultivated **Region:** Amazonia, Andean, Caribbean, Orinoquia **Dept.:** MAG, MET **Elev.:** 30–500 m a.s.l. **Conservation status: (IUCN)** NE

1230 *Lockhartia longifolia* (Lindl.) Schltr. (Orchidaceae) **IPNI:** 142885-2 **Habit:** Herb, Epiphyte **Use:** EU **Origin:** Native, Cultivated **Region:** Andean, Pacific **Dept.:** ANT, CAU, CHO, MAG, VAC **Elev.:** 1400–2200 m a.s.l. **Conservation status: (IUCN)** NE

1231 *Masdevallia caudata* Lindl. (Orchidaceae) **IPNI:** 642726-1 **Common name:** bandera de cola, colibrí **Habit:** Herb, Epiphyte **Use:** EU **Origin:** Native **Region:** Andean **Dept.:** BOG, CUN, NSA, QUI, SAN **Elev.:** 1670–2800 m.a.s.l. **IUCN Habitat code:** 1,4,14 **Conservation status: (IUCN)** NE **(CPSG)** EN

1232 *Masdevallia coccinea* Linden ex Lindl. (Orchidaceae) **IPNI:** 642746-1 **Common name:** bandera, bandera colorada, banderitas **Habit:** Herb, Epiphyte **Use:** EU **Origin:** Endemic **Region:** Andean **Dept.:** SAN **Elev.:** 2900–2900 m.a.s.l. **IUCN Habitat code:** 1,4,8,14 **Conservation status: (IUCN)** NE **(CPSG)** EN

1233 *Masdevallia coriacea* Lindl. (Orchidaceae) **IPNI:** 642756-1 **Habit:** Herb, Epiphyte **Use:** EU **Origin:** Native **Region:** Andean **Dept.:** BOG, BOY, CUN, NSA, PUT **Elev.:** 2600–3350 m.a.s.l. **IUCN Habitat code:** 1,3,4,14 **Conservation status: (IUCN)** NE **(CPSG)** LC

1234 *Masdevallia fragrans* Woolward (Orchidaceae) **IPNI:** 642833-1 **Habit:** Herb, Epiphyte **Use:** EU **Origin:** Endemic, Cultivated **Region:** Andean **Dept.:** ANT **Elev.:** 3000–3500 m a.s.l. **Conservation status: (IUCN)** NE

1235 *Masdevallia macrura* Rchb.f. (Orchidaceae) **IPNI:** 642941-1 **Habit:** Herb, Epiphyte **Use:** EU **Origin:** Endemic **Region:** Andean **Dept.:** ANT, CHO **Elev.:** 1850–2600 m.a.s.l. **IUCN**

Habitat code: 1 **Conservation status: (IUCN)** NE **(CPSG)** EN

1236 *Masdevallia parvula* Schltr. (Orchidaceae) **IPNI:** 154214-2 **Habit:** Herb, Epiphyte **Use:** EU **Origin:** Native, Cultivated **Region:** Andean **Dept.:** ANT **Elev.:** 2000–2500 m a.s.l. **Conservation status: (IUCN)** NE

1237 *Masdevallia rosea* Lindl. (Orchidaceae) **IPNI:** 643089-1 **Habit:** Herb, Epiphyte **Use:** EU **Origin:** Native **Region:** Andean **Dept.:** NAR **Elev.:** 2550-3200 m.a.s.l. **Conservation status: (IUCN)** NE **(CPSG)** NT

1238 *Masdevallia schlimii* Linden ex Lindl. (Orchidaceae) **IPNI:** 643106-1 **Habit:** Herb, Epiphyte **Use:** EU **Origin:** Native **Region:** Andean **Dept.:** NSA **Elev.:** 2400-3200 m.a.s.l. **IUCN Habitat code:** 1,14 **Conservation status: (IUCN)** NE **(CPSG)** VU

1239 *Masdevallia trochilus* Linden & André (Orchidaceae) **IPNI:** 643178-1 **Common name:** colibrí, tominejo **Habit:** Herb, Epiphyte **Use:** EU **Origin:** Native **Region:** Andean **Dept.:** ANT, CAU **Elev.:** 1750-2500 m.a.s.l. **IUCN Habitat code:** 1 **Conservation status: (IUCN)** NE **(CPSG)** VU

1240 *Masdevallia velifera* Rchb.f. (Orchidaceae) **IPNI:** 643196-1 **Habit:** Herb, Epiphyte **Use:** EU **Origin:** Endemic **Region:** Andean **Dept.:** ANT **Elev.:** 2130-2130 m.a.s.l. **Conservation status: (IUCN)** NE **(CPSG)** EN

1241 *Masdevallia wendlandiana* Rchb.f. (Orchidaceae) **IPNI:** 154376-2 **Habit:** Herb, Epiphyte **Use:** EU **Origin:** Native, Cultivated **Region:** Amazonia, Andean **Dept.:** CAU, VAU **Elev.:** 150–1800 m a.s.l. **Conservation status: (IUCN)** NE

1242 *Maxillaria aciantha* Rchb.f. (Orchidaceae) **IPNI:** 643235-1 **Habit:** Herb, Epiphyte **Use:** EU **Origin:** Native, Cultivated **Region:** Pacific **Dept.:** CHO **IUCN Habitat code:** 1,5,14 **Conservation status: (IUCN)** NE

1243 *Maxillaria alba* (Hook.) Lindl. (Orchidaceae) **IPNI:** 643250-1 **Habit:** Herb, Epiphyte **Use:** EU **Origin:** Native, Cultivated **Region:** Amazonia, Andean, Orinoquia, Pacific **Dept.:** ANT, CHO, GUA, MAG, MET **Elev.:** 0–1850 m a.s.l. **IUCN Habitat code:** 1 **Conservation status: (IUCN)** NE

1244 *Maxillaria bicallosa* (Rchb.f.) Garay (Orchidaceae) **IPNI:** 154952-2 **Habit:** Herb, Epiphyte **Use:** EU **Origin:** Native, Cultivated **Region:** Amazonia, Andean, Pacific **Dept.:** ANT, CAQ, CHO **Elev.:** 300–1375 m a.s.l. **Conservation status: (IUCN)** NE

1245 *Maxillaria carinulata* Rchb.f. (Orchidaceae) **IPNI:** 643344-1 **Habit:** Herb, Epiphyte **Use:** EU **Origin:** Native, Cultivated **Region:** Andean, Pacific **Dept.:** ANT, CHO, PUT **Elev.:** 0–1670 m a.s.l. **Conservation status: (IUCN)** NE

1246 *Maxillaria cordyline* (Rchb.f.) Dodson (Orchidaceae) **IPNI:** 984957-1 **Habit:** Herb **Use:** EU **Origin:** Native, Cultivated **Region:** Andean **Dept.:** ANT, CUN **Elev.:** 1600–2500 m a.s.l. **Conservation status: (IUCN)** NE

1247 *Maxillaria crassifolia* (Lindl.) Rchb.f. (Orchidaceae) **IPNI:** 643387-1 **Habit:** Herb, Epiphyte **Use:** EU **Origin:** Native, Cultivated **Region:** Amazonia, Andean, Guiana Shield, Caribbean, Orinoquia, Pacific **Dept.:** CAL, CES, CHO, HUI, MAG, MET, VAU **Elev.:** 100–1000 m a.s.l. **Conservation status: (IUCN)** NE

1248 *Maxillaria cryptobulbon* Carnevali & J.T.Atwood (Orchidaceae) **IPNI:** 305047-2 **Habit:** Herb, Epiphyte **Use:** EU **Origin:** Native, Cultivated **Region:** Pacific **Dept.:** CHO **Conservation status: (IUCN)** NE

1249 *Maxillaria desvauxiana* Rchb.f. (Orchidaceae) **IPNI:** 643410-1 **Habit:** Herb, Epiphyte **Use:** EU **Origin:** Native, Cultivated **Region:** Amazonia, Andean, Guiana Shield **Dept.:** AMA, CAL, RIS, VAU **Elev.:** 150–2250 m a.s.l. **Conservation status: (IUCN)** NE

1250 *Maxillaria discolor* (Lodd. ex Lindl.) Rchb. (Orchidaceae) **IPNI:** 643414-1 **Habit:** Herb, Epiphyte **Use:** EU **Origin:** Native, Cultivated **Region:** Amazonia, Guiana Shield, Orinoquia, Pacific **Dept.:** AMA, ANT, CHO, GUA, GUV, VAU **Elev.:** 0–1000 m a.s.l. **Conservation status: (IUCN)** NE

1251 *Maxillaria ecuadorensis* Schltr. (Orchidaceae) **IPNI:** 155039-2 **Habit:** Herb, Epiphyte **Use:** EU **Origin:** Native, Cultivated **Region:** Andean **Dept.:** ANT, CHO, NAR, VAC **Elev.:** 1800–2700 m a.s.l. **Conservation status: (IUCN)** NE

1252 *Maxillaria egertoniana* (Bateman ex Lindl.) Molinari (Orchidaceae) **IPNI:** 77155180-1 **Habit:** Herb, Epiphyte **Use:** EU **Origin:** Native, Cultivated **Region:** Amazonia, Andean, Caribbean, Pacific **Dept.:** CHO, LAG, MAG **Elev.:** 20–712 m a.s.l. **Conservation status: (IUCN)** NE

1253 *Maxillaria fractiflexa* Rchb.f. (Orchidaceae) **IPNI:** 643449-1 **Habit:** Herb, Epiphyte **Use:** EU **Origin:** Native, Cultivated **Region:** Andean **Dept.:** ANT, CAU, CHO, NAR **Elev.:** 1700–2150 m a.s.l. **Conservation status: (IUCN)** NE

1254 *Maxillaria fulgens* (Rchb.f.) L.O.Williams (Orchidaceae) **IPNI:** 155059-2 **Habit:** Herb, Epiphyte **Use:** EU **Origin:** Native, Cultivated **Region:** Andean, Pacific **Dept.:** CHO, VAC **Elev.:** 0–1150 m a.s.l. **IUCN Habitat code:** 1,5,14 **Conservation status: (IUCN)** NE

1255 *Maxillaria grandiflora* (Kunth) Lindl. (Orchidaceae) **IPNI:** 643478-1 **Common name:** quijadas **Habit:** Herb, Epiphyte **Use:** EU **Origin:** Endemic, Cultivated **Region:** Andean **Dept.:** CAU, MET, NAR, PUT **Elev.:** 2000–3200 m a.s.l. **IUCN Habitat code:** 1,3,4,14 **Conservation status: (IUCN)** NE

1256 *Maxillaria grandis* Rchb.f. (Orchidaceae) **IPNI:** 643480-1 **Habit:** Herb, Epiphyte **Use:** EU **Origin:** Native, Cultivated **Region:** Andean **Conservation status: (IUCN)** NE

1257 *Maxillaria guareimensis* Rchb.f. (Orchidaceae) **IPNI:** 643483-1 **Habit:** Herb, Epiphyte **Use:** EU **Origin:** Native, Cultivated **Region:** Andean **Dept.:** ANT **Elev.:** 1590–1590 m a.s.l. **Conservation status: (IUCN)** NE

1258 *Maxillaria hedwigiae* Hamer & Dodson (Orchidaceae) **IPNI:** 902533-1 **Habit:** Herb, Epiphyte **Use:** EU **Origin:** Native, Cultivated **Region:** Andean, Pacific **Dept.:** CHO **Elev.:** 500–500 m a.s.l. **Conservation status: (IUCN)** NE

1259 *Maxillaria hennisiana* Schltr. (Orchidaceae) **IPNI:** 155084-2 **Habit:** Herb, Epiphyte **Use:** EU **Origin:** Native, Cultivated **Region:** Andean, Pacific **Dept.:** CHO, RIS, VAC **Elev.:** 0–540 m a.s.l. **Conservation status: (IUCN)** NE

1260 *Maxillaria huebschii* Rchb.f. (Orchidaceae) **IPNI:** 643514-1 **Habit:** Herb, Epiphyte **Use:** EU **Origin:** Endemic, Cultivated **Region:** Andean **Dept.:** CAU, NAR **Elev.:** 2200–3000 m a.s.l. **Conservation status: (IUCN)** NE

1261 *Maxillaria inaudita* Rchb.f. (Orchidaceae) **IPNI:** 643519-1 **Habit:** Herb, Epiphyte **Use:** EU **Origin:** Native, Cultivated **Region:** Andean **Conservation status: (IUCN)** NE

1262 *Maxillaria insolita* Dressler (Orchidaceae) **IPNI:** 280607-2 **Habit:** Herb, Epiphyte **Use:** EU **Origin:** Native, Cultivated **Region:** Andean **Dept.:** RIS **Elev.:** 1500–2000 m a.s.l. **Conservation status: (IUCN)** NE

1263 *Maxillaria irrorata* Rchb.f. (Orchidaceae) **IPNI:** 643525-1 **Habit:** Herb, Epiphyte **Use:** EU **Origin:** Native, Cultivated **Region:** Andean **Dept.:** NAR **Elev.:** 2600–2600 m a.s.l. **Conservation status: (IUCN)** NE

1264 *Maxillaria lehmannii* Rchb.f. (Orchidaceae) **IPNI:** 322889-2 **Habit:** Herb, Epiphyte **Use:** EU **Origin:** Native, Cultivated **Region:** Andean **Elev.:** 2000–2000 m a.s.l. **Conservation status: (IUCN)** NE

1265 *Maxillaria lepidota* Lindl. (Orchidaceae)
IPNI: 643558-1
Habit: Herb, Epiphyte **Use:** EU **Origin:** Native, Cultivated **Region:** Andean **Dept.:** ANT, CAU, CHO, CUN, NAR, PUT, SAN **Elev.:** 1350–2700 m a.s.l. **Conservation status: (IUCN)** NE

1266 *Maxillaria leptosepala* Hook. (Orchidaceae)
IPNI: 643559-1
Habit: Herb, Epiphyte **Use:** EU **Origin:** Native, Cultivated **Region:** Pacific **Dept.:** CHO **Conservation status: (IUCN)** NE

1267 *Maxillaria longipetala* Ruiz & Pav. (Orchidaceae)
IPNI: 155147-2
Habit: Herb, Epiphyte **Use:** EU **Origin:** Native, Cultivated **Region:** Andean **Conservation status: (IUCN)** NE

1268 *Maxillaria longissima* Lindl. (Orchidaceae)
IPNI: 643585-1
Habit: Herb, Epiphyte **Use:** EU **Origin:** Native, Cultivated **Region:** Andean **Dept.:** ANT, CAU, NAR, VAC **Elev.:** 1350–2750 m a.s.l. **Conservation status: (IUCN)** NE

1269 *Maxillaria luteoalba* Lindl. (Orchidaceae)
IPNI: 643588-1
Habit: Herb, Epiphyte **Use:** EU **Origin:** Native, Cultivated **Region:** Andean, Orinoquia **Dept.:** ANT, BOY, CAS, CHO, CUN, MET, NSA **Elev.:** 490–1750 m a.s.l. **Conservation status: (IUCN)** NE

1270 *Maxillaria lutescens* Scheidw. (Orchidaceae)
IPNI: 643592-1
Common name: jadzami jinguaque
Habit: Herb, Epiphyte **Use:** EU, ME **Origin:** Native **Region:** Amazonia, Andean, Guiana Shield, Orinoquia, Pacific **Dept.:** ANT, CHO, GUV, MAG, NSA, VID **Elev.:** 10–1850 m a.s.l. **Conservation status: (IUCN)** NE

1271 *Maxillaria magliana* Molinari (Orchidaceae)
IPNI: 77155011-1
Habit: Herb, Epiphyte **Use:** EU **Origin:** Native, Cultivated **Region:** Andean **Dept.:** CAU **Elev.:** 1200–1200 m a.s.l. **Conservation status: (IUCN)** NE

1272 *Maxillaria maleolens* Schltr. (Orchidaceae)
IPNI: 155157-2
Habit: Herb, Epiphyte **Use:** EU **Origin:** Native, Cultivated **Region:** Pacific **Dept.:** CHO **Elev.:** 0–100 m a.s.l. **Conservation status: (IUCN)** NE

1273 *Maxillaria miniata* (Lindl.) L.O.Williams (Orchidaceae)
IPNI: 155165-2
Habit: Herb, Epiphyte **Use:** EU **Origin:** Native, Cultivated **Region:** Andean **Dept.:** MAG, MET, NSA, SAN **Elev.:** 1300–2800 m a.s.l. **Conservation status: (IUCN)** NE

1274 *Maxillaria nasuta* Rchb.f. (Orchidaceae)
IPNI: 643645-1
Habit: Herb, Epiphyte **Use:** EU **Origin:** Native, Cultivated **Region:** Andean **Dept.:** BOY, CUN, SAN **Elev.:** 750–2500 m a.s.l. **Conservation status: (IUCN)** NE

1275 *Maxillaria neophylla* Rchb.f. (Orchidaceae)
IPNI: 643647-1
Habit: Herb, Epiphyte **Use:** EU **Origin:** Native, Cultivated **Region:** Andean **Dept.:** ANT, RIS **Elev.:** 500–2500 m a.s.l. **Conservation status: (IUCN)** NE

1276 *Maxillaria niesseniae* Christenson (Orchidaceae)
IPNI: 327410-2
Habit: Herb, Epiphyte **Use:** EU **Origin:** Endemic, Cultivated **Region:** Andean **Conservation status: (IUCN)** NE

1277 *Maxillaria nigrescens* Lindl. (Orchidaceae)
IPNI: 643650-1
Habit: Herb, Epiphyte **Use:** EU **Origin:** Native, Cultivated **Region:** Andean **Dept.:** BOY **Elev.:** 1700–1700 m a.s.l. **IUCN Habitat code:** 3,14 **Conservation status: (IUCN)** NE

1278 *Maxillaria obtusa* (Lindl.) Molinari (Orchidaceae)
IPNI: 77155190-1
Habit: Herb, Epiphyte **Use:** EU **Origin:** Native, Cultivated **Region:** Amazonia, Guiana Shield, Orinoquia, Pacific **Dept.:** CAQ, CHO, GUV **Elev.:** 0–650 m a.s.l. **Conservation status: (IUCN)** NE

1279 *Maxillaria parkeri* Hook. (Orchidaceae)
IPNI: 643684-1
Habit: Herb, Epiphyte **Use:** EU **Origin:** Native, Cultivated **Region:** Amazonia, Guiana Shield **Dept.:** CAQ, VAU **Elev.:** 150–250 m a.s.l. **Conservation status: (IUCN)** NE

1280 *Maxillaria porrecta* Lindl. (Orchidaceae)
IPNI: 643731-1
Habit: Herb, Epiphyte **Use:** EU **Origin:** Native, Cultivated **Region:** Amazonia, Andean, Pacific **Dept.:** ANT, CAU, CHO, NAR, NSA, PUT **Elev.:** 0–2200 m a.s.l. **Conservation status: (IUCN)** NE

1281 *Maxillaria procurrens* Lindl. (Orchidaceae)
IPNI: 643736-1
Habit: Herb, Epiphyte **Use:** EU **Origin:** Native, Cultivated **Region:** Andean **Dept.:** ANT, BOG, BOY, CAS, CUN, MAG **Elev.:** 1800–2800 m a.s.l. **Conservation status: (IUCN)** NE

1282 *Maxillaria pseudoreichenheimiana* Dodson (Orchidaceae)
IPNI: 896587-1
Habit: Herb, Epiphyte **Use:** EU **Origin:** Native, Cultivated **Region:** Andean **Dept.:** CHO, RIS **Elev.:** 1350–2473 m a.s.l. **Conservation status: (IUCN)** NE

1283 *Maxillaria pulla* Linden & Rchb.f. (Orchidaceae)
IPNI: 643743-1
Habit: Herb, Epiphyte **Use:** EU **Origin:** Native, Cultivated **Region:** Andean **Dept.:** NSA **Conservation status: (IUCN)** NE

1284 *Maxillaria reichenheimiana* Endrés & Rchb.f. (Orchidaceae)
IPNI: 643763-1
Habit: Herb, Epiphyte **Use:** EU **Origin:** Native, Cultivated **Region:** Andean **Elev.:** 800–1200 m a.s.l. **Conservation status: (IUCN)** NE

1285 *Maxillaria ringens* Rchb.f. (Orchidaceae)
IPNI: 318680-2
Habit: Herb, Epiphyte **Use:** EU **Origin:** Native, Cultivated **Region:** Pacific **Dept.:** CHO **Elev.:** 80–80 m a.s.l. **Conservation status: (IUCN)** NE

1286 *Maxillaria ruberrima* (Lindl.) Garay (Orchidaceae)
IPNI: 155262-2
Habit: Herb, Epiphyte **Use:** EU **Origin:** Native, Cultivated **Region:** Andean **Dept.:** ANT, BOG, BOY, CAU, MAG, SAN **Elev.:** 800–2700 m a.s.l. **Conservation status: (IUCN)** NE

1287 *Maxillaria rubioi* Dodson (Orchidaceae)
IPNI: 305358-2
Habit: Herb, Epiphyte **Use:** EU **Origin:** Native, Cultivated **Region:** Pacific **Dept.:** VAC **Elev.:** 50–100 m a.s.l. **Conservation status: (IUCN)** NE

1288 *Maxillaria rufescens* Lindl. (Orchidaceae)
IPNI: 643781-1
Habit: Herb, Epiphyte **Use:** EU **Origin:** Native, Cultivated **Region:** Andean, Guiana Shield, Pacific **Dept.:** CHO, CUN, GUA, HUI, NSA **Elev.:** 0–1750 m a.s.l. **Conservation status: (IUCN)** NE

1289 *Maxillaria sophronitis* (Rchb.f.) Garay (Orchidaceae)
IPNI: 155290-2
Habit: Herb, Epiphyte **Use:** EU **Origin:** Native, Cultivated **Region:** Andean, Guiana Shield **Dept.:** GUA, MAG **Elev.:** 650–750 m a.s.l. **Conservation status: (IUCN)** NE

1290 *Maxillaria splendens* Poepp. & Endl. (Orchidaceae)
IPNI: 643821-1
Habit: Herb, Epiphyte **Use:** EU **Origin:** Native, Cultivated **Region:** Andean **Dept.:** NAR **Elev.:** 1600–1600 m a.s.l. **Conservation status: (IUCN)** NE

1291 *Maxillaria striata* Rolfe (Orchidaceae)
IPNI: 155303-2
Habit: Herb, Epiphyte **Use:** EU **Origin:** Native, Cultivated **Region:** Andean **Dept.:** PUT **Elev.:** 700–700 m a.s.l. **Conservation status: (IUCN)** NE

1292 *Maxillaria subrepens* (Rolfe) Schuit. & M.W.Chase (Orchidaceae)
IPNI: 77149732-1
Habit: Herb, Epiphyte **Use:** EU **Origin:** Native, Cultivated **Region:** Amazonia, Guiana Shield, Orinoquia **Dept.:** GUV, VAU **Elev.:** 184–500 m a.s.l. **Conservation status: (IUCN)** NE

1293 *Maxillaria superflua* Rchb.f. (Orchidaceae)
IPNI: 643845-1
Habit: Herb, Epiphyte **Use:** EU **Origin:** Native, Cultivated **Region:** Amazonia **Dept.:** AMA, VAU **Elev.:** 115–500 m a.s.l. **Conservation status: (IUCN)** NE

1294 *Maxillaria tenuibulba* Christenson (Orchidaceae)
IPNI: 320703-2
Habit: Herb, Epiphyte **Use:** EU **Origin:** Endemic, Cultivated **Region:** Andean **Conservation status: (IUCN)** NE

1295 *Maxillaria tenuis* C.Schweinf. (Orchidaceae) **IPNI:** 155317-2 **Habit:** Herb, Epiphyte **Use:** EU **Origin:** Native, Cultivated **Region:** Amazonia, Guiana Shield **Dept.:** GUV, VAU **Elev.:** 180–500 m a.s.l. **Conservation status: (IUCN)** NE

1296 *Maxillaria uncata* Lindl. (Orchidaceae) **IPNI:** 155331-2 **Habit:** Herb, Epiphyte **Use:** EU **Origin:** Native, Cultivated **Region:** Amazonia, Guiana Shield, Orinoquia, Pacific **Dept.:** AMA, ANT, CHO, GUV, VAC **Elev.:** 10–700 m a.s.l. **Conservation status: (IUCN)** NE

1297 *Maxillaria valenzuelana* (A.Rich.) Nash (Orchidaceae) **IPNI:** 155338-2 **Habit:** Herb, Epiphyte **Use:** EU **Origin:** Native, Cultivated **Region:** Andean **Dept.:** ANT, VAC **Elev.:** 1050–1100 m a.s.l. **Conservation status: (IUCN)** NE

1298 *Maxillaria variabilis* Bateman ex Lindl. (Orchidaceae) **IPNI:** 30044409-2 **Habit:** Herb, Epiphyte **Use:** EU **Origin:** Native, Cultivated **Region:** Andean **Conservation status: (IUCN)** NE

1299 *Maxillaria venusta* Linden & Rchb.f. (Orchidaceae) **IPNI:** 643893-1 **Habit:** Herb, Epiphyte **Use:** EU **Origin:** Native, Cultivated **Region:** Andean **Dept.:** NSA **Elev.:** 1800–1800 m a.s.l. **Conservation status: (IUCN)** NE

1300 *Maxillaria villosa* (Barb.Rodr.) Cogn. (Orchidaceae) **IPNI:** 155347-2 **Habit:** Herb, Epiphyte **Use:** EU **Origin:** Native, Cultivated **Region:** Amazonia, Andean **Dept.:** AMA, CAL **Elev.:** 240–1000 m a.s.l. **Conservation status: (IUCN)** NE

1301 *Maxillaria violaceopunctata* Rchb.f. (Orchidaceae) **IPNI:** 643899-1 **Habit:** Herb, Epiphyte **Use:** EU **Origin:** Native, Cultivated **Region:** Amazonia, Pacific **Dept.:** CHO, VAU **Elev.:** 0–120 m a.s.l. **Conservation status: (IUCN)** NE

1302 *Maxillaria virguncula* Rchb.f. (Orchidaceae) **IPNI:** 643901-1 **Habit:** Herb, Epiphyte **Use:** EU **Origin:** Native, Cultivated **Region:** Andean **Dept.:** CAU, CUN, PUT **Elev.:** 1500–1800 m a.s.l. **Conservation status: (IUCN)** NE

1303 *Maxillaria wojii* Christenson (Orchidaceae) **IPNI:** 1016159-1 **Habit:** Herb, Epiphyte **Use:** EU **Origin:** Endemic, Cultivated **Region:** Andean **Conservation status: (IUCN)** NE

1304 *Miltoniopsis vexillaria* (Rchb.f.) God.-Leb. (Orchidaceae) **IPNI:** 161453-2 **Common name:** josefina, josefita, lirio, mayo, parásita, tulipá **Habit:** Herb, Epiphyte **Use:** EU **Origin:** Native **Region:** Andean **Dept.:** ANT, CAL, CUN, HUI, NAR, RIS, VAC **Elev.:** 1060-1950 m.a.s.l. **IUCN Habitat code:** 1,3,14 **Conservation status: (IUCN)** NE **(CPSG)** VU

1305 *Mormodes buccinator* Lindl. (Orchidaceae) **IPNI:** 644946-1 **Habit:** Herb, Epiphyte **Use:** EU **Origin:** Native **Region:** Amazonia, Andean, Orinoquia **Dept.:** BOY, MET **Elev.:** 450–2550 m a.s.l. **IUCN Habitat code:** 1 **Conservation status: (IUCN)** NE

1306 *Neomoorea wallisii* (Rchb.f.) Schltr. (Orchidaceae) **IPNI:** 168902-2 **Habit:** Herb, Epiphyte **Use:** EU **Origin:** Native, Cultivated **Region:** Andean, Pacific **Dept.:** ANT, CHO **Elev.:** 50–1470 m a.s.l. **Conservation status: (IUCN)** NE

1307 *Oeceoclades maculata* (Lindl.) Lindl. (Orchidaceae) **IPNI:** 1073117-2 **Habit:** Herb, Epiphyte **Use:** EU, ME, SU **Origin:** Naturalised **Region:** Andean **Dept.:** ANT, MAG, NSA **Elev.:** 500–1200 m a.s.l. **IUCN Habitat code:** 1,2,3,5,14 **Conservation status: (IUCN)** LC

1308 *Oncidium alexandrae* (Bateman) M.W.Chase & N.H.Williams (Orchidaceae) **IPNI:** 77101415-1 **Habit:** Herb, Epiphyte **Use:** EU, GS **Origin:** Native **Region:** Andean **Dept.:** BOY, CAU, CUN, HUI, NAR, PUT, SAN **Elev.:** 1700-3200 m.a.s.l. **IUCN Habitat code:** 1,3,14 **Conservation status: (IUCN)** NE **(CPSG)** EN

1309 *Oncidium sphacelatum* Lindl. (Orchidaceae) **IPNI:** 323062-2 **Use:** EU **IUCN Habitat code:** 3,14 **Conservation status: (IUCN)** NE

1310 *Peristeria elata* Hook. (Orchidaceae) **IPNI:** 650144-1 **Common name:** espiritusanto, paloma, vara de San José **Habit:** Herb **Use:** EU, ME **Origin:** Native **Region:** Andean, Pacific **Dept.:** ANT, CAL, TOL, VAC **Elev.:** 0–1120 m a.s.l. **IUCN Habitat code:** 1,3,14 **Conservation status: (IUCN)** NE

1311 *Pescatoria lalindei* (Rchb.f.) Dressler ex P.A.Harding (Orchidaceae) **IPNI:** 60440679-2 **Habit:** Herb, Epiphyte **Use:** EU **Origin:** Endemic **Region:** Andean **Dept.:** SAN **Elev.:** 1000–2000 m a.s.l. **IUCN Habitat code:** 14 **Conservation status: (IUCN)** NE

1312 *Phragmipedium caudatum* (Lindl.) Rolfe (Orchidaceae) **IPNI:** 194355-2 **Common name:** viejo barbudo, zapato de Venus **Habit:** Herb **Use:** EU, GS **Origin:** Native **Region:** Andean **Dept.:** ANT, VAC **Elev.:** 2000-2500 m.a.s.l. **Conservation status: (IUCN)** EN **(CPSG)** DD

1313 *Phragmipedium lindenii* (Lindl.) Dressler & N.H.Williams (Orchidaceae) **IPNI:** 1119571-2 **Common name:** zapato de Venus **Habit:** Herb **Use:** EU **Origin:** Native **Region:** Andean **Dept.:** ANT, BOY, CHO, CUN, NSA, PUT, SAN, VAC **Elev.:** 1500-2100 m.a.s.l. **IUCN Habitat code:** 1,3,14 **Conservation status: (IUCN)** LC **(CPSG)** LC

1314 *Pleurothallis baudoensis* Luer & R.Escobar (Orchidaceae) **IPNI:** 316910-2 **Habit:** Herb, Epiphyte **Use:** EU **Origin:** Endemic, Cultivated **Region:** Andean, Pacific **Dept.:** CHO, VAC **Elev.:** 400–1000 m a.s.l. **Conservation status: (IUCN)** NE

1315 *Polystachya concreta* (Jacq.) Garay & H.R.Sweet (Orchidaceae) **IPNI:** 654273-1 **Habit:** Herb, Epiphyte **Use:** EU, ME **Origin:** Native **Region:** Andean, Pacific **Dept.:** CAU, CHO, MAG **Elev.:** 0–1600 m a.s.l. **IUCN Habitat code:** 1,3,4,14 **Conservation status: (IUCN)** NE

1316 *Prosthechea cochleata* (L.) W.E.Higgins (Orchidaceae) **IPNI:** 1001247-1 **Habit:** Herb, Epiphyte **Use:** HF **Origin:** Native **Region:** Andean, Guiana Shield **Dept.:** CUN, MAG, MET **Elev.:** 350-1200 m.a.s.l. **IUCN Habitat code:** 1,14 **Conservation status: (IUCN)** NE **(CPSG)** LC

1317 *Prosthechea fragrans* (Sw.) W.E.Higgins (Orchidaceae) **IPNI:** 1001253-1 **Common name:** espíritu santo **Habit:** Herb, Epiphyte **Use:** EU, MA **Origin:** Native **Dept.:** CHO **Elev.:** 0–0 m a.s.l. **IUCN Habitat code:** 1,3,4,5,14 **Conservation status: (IUCN)** NE

1318 *Psychopsis krameriana* (Rchb.f.) H.G.Jones (Orchidaceae) **IPNI:** 212378-2 **Common name:** mariposa **Habit:** Herb, Epiphyte **Use:** EU **Origin:** Native **Region:** Andean, Pacific **Dept.:** CHO, CUN, VAC **Elev.:** 200-1000 m.a.s.l. **IUCN Habitat code:** 1,3 **Conservation status: (IUCN)** NE **(CPSG)** NT

1319 *Psychopsis papilio* (Lindl.) H.G.Jones (Orchidaceae) **IPNI:** 212379-2 **Common name:** mariposa **Habit:** Herb, Epiphyte **Use:** EU **Origin:** Native **Region:** Andean **Elev.:** 800-1200 m.a.s.l. **Conservation status: (IUCN)** NE **(CPSG)** DD

1320 *Restrepia antennifera* Kunth (Orchidaceae) **IPNI:** 655513-1 **Common name:** zancudo **Habit:** Herb, Epiphyte **Use:** EU **Origin:** Native **Region:** Andean **Dept.:** ANT, CAU, NAR, NSA **Elev.:** 1700-3500 m.a.s.l. **IUCN Habitat code:** 3,14 **Conservation status: (IUCN)** NE **(CPSG)** LC

1321 *Restrepia elegans* H.Karst. (Orchidaceae) **IPNI:** 655533-1 **Habit:** Herb, Epiphyte **Use:** EU **Origin:** Native **Region:** Andean **Dept.:** SAN **IUCN Habitat code:** 1,3,4 **Conservation status: (IUCN)** NE

1322 *Rodriguezia granadensis* (Lindl.) Rchb.f. (Orchidaceae) **IPNI:** 655792-1 **Common name:** gallita, matatotumo, quereme **Habit:** Herb, Epiphyte **Use:** EU **Origin:** Endemic **Region:** Andean **Dept.:** ANT, CAU, CUN, NAR, NSA, TOL **Elev.:** 1350–2500 m a.s.l. **IUCN Habitat code:** 1,3,5,14 **Conservation status: (IUCN)** NE

1323 *Rodriguezia lanceolata* Ruiz & Pav. (Orchidaceae)
IPNI: 655797-1
Common name: gallita, matatotumo, quereme **Habit:** Herb, Epiphyte **Use:** EU, ME **Origin:** Native **Region:** Amazonia, Andean, Pacific **Dept.:** ANT, CHO, CUN, GUA, NSA, QUI, VAC, VAU **Elev.:** 250-1400 m.a.s.l. **IUCN Habitat code:** 1,3,5,14 **Conservation status: (IUCN)** NE **(CPSG)** LC

1324 *Scaphyglottis prolifera* (R.Br.) Cogn. (Orchidaceae)
IPNI: 228831-2
Habit: Herb, Epiphyte **Use:** HF **Origin:** Native **Region:** Amazonia, Andean, Guiana Shield, Pacific **Dept.:** ANT, CAU, CHO, MAG, PUT, VAU **Elev.:** 5–1000 m a.s.l. **IUCN Habitat code:** 1,3,4,5,14 **Conservation status: (IUCN)** NE

1325 *Sobralia decora* Bateman (Orchidaceae)
IPNI: 657877-1
Habit: Herb **Use:** EU **Origin:** Native **Region:** Amazonia, Andean, Guiana Shield, Pacific **Dept.:** CHO, CUN, NAR, VAU **Elev.:** 50–1550 m a.s.l. **IUCN Habitat code:** 1,2,3,14 **Conservation status: (IUCN)** NE

1326 *Sobralia fragrans* Lindl. (Orchidaceae)
IPNI: 657888-1
Common name: aguacolda, sobralia **Habit:** Herb **Use:** EU **Origin:** Native **Region:** Amazonia, Andean, Guiana Shield, Orinoquia, Pacific **Dept.:** ANT, CAQ, CHO, MET **Elev.:** 200–630 m a.s.l. **IUCN Habitat code:** 1,3,14 **Conservation status: (IUCN)** NE

1327 *Sobralia macrophylla* Rchb.f. (Orchidaceae)
IPNI: 657915-1
Habit: Herb **Use:** EU **Origin:** Native **Region:** Amazonia, Andean, Guiana Shield, Pacific **Dept.:** AMA, ANT, CAQ, CAU, CHO, PUT, VAC, VAU **Elev.:** 0–1500 m a.s.l. **IUCN Habitat code:** 1,3,5,14 **Conservation status: (IUCN)** NE

1328 *Stanhopea jenischiana* F.Kramer ex Rchb.f. (Orchidaceae)
IPNI: 243490-2
Habit: Herb, Epiphyte **Use:** HF **Origin:** Native **Region:** Andean **Dept.:** CAU, HUI **Elev.:** 900–1400 m a.s.l. **IUCN Habitat code:** 1 **Conservation status: (IUCN)** NE

1329 *Stanhopea tigrina* Bateman ex Lindl. (Orchidaceae)
IPNI: 658613-1
Common name: cucarrón **Use:** EU, HF, ME **Dept.:** RIS **Conservation status: (IUCN)** NE

1330 *Stenorrhynchos speciosum* (Jacq.) Rich. (Orchidaceae)
IPNI: 659300-1
Habit: Herb, Epiphyte **Use:** EU **Origin:** Native **Region:** Andean **Dept.:** ANT, BOG, BOY, CAL, CUN, HUI, LAG, NAR, NSA, VAC **Elev.:** 1450–3900 m a.s.l. **IUCN Habitat code:** 1,3,4,14 **Conservation status: (IUCN)** NE

1331 *Sudamerlycaste fulvescens* (Hook.) Archila (Orchidaceae)
IPNI: 77096738-1
Habit: Herb **Use:** EU **Origin:** Native **Region:** Andean **Dept.:** BOY, CUN, NSA, SAN **Elev.:** 1700–2000 m a.s.l. **Conservation status: (IUCN)** NE

1332 *Sudamerlycaste gigantea* (Lindl.) Archila (Orchidaceae)
IPNI: 77096741-1
Habit: Herb **Use:** EU **Origin:** Native **Region:** Andean **IUCN Habitat code:** 1,14 **Conservation status: (IUCN)** NE

1333 *Telipogon nervosus* (L.) Druce (Orchidaceae)
IPNI: 659882-1
Habit: Herb, Epiphyte **Use:** EU **Origin:** Native **Region:** Andean **Dept.:** BOG, BOY, CUN, SAN **Elev.:** 2200–3300 m a.s.l. **IUCN Habitat code:** 1,3,4,5,14 **Conservation status: (IUCN)** NE

1334 *Trichocentrum lanceanum* (Lindl.) M.W.Chase & N.H.Williams (Orchidaceae)
IPNI: 1021269-1
Habit: Herb, Epiphyte **Use:** MA **Origin:** Native **Region:** Amazonia, Guiana Shield **Dept.:** GUA **Elev.:** 150–150 m a.s.l. **Conservation status: (IUCN)** NE

1335 *Trichopilia fragrans* (Lindl.) Rchb.f. (Orchidaceae)
IPNI: 660593-1
Habit: Herb, Epiphyte **Use:** EU **Origin:** Native **Region:** Andean **Dept.:** ANT, CAU, MAG, NSA **Elev.:** 1150–1800 m a.s.l. **IUCN Habitat code:** 1,14 **Conservation status: (IUCN)** NE

1336 *Vanilla appendiculata* Rolfe (Orchidaceae)
IPNI: 262538-2
Use: GS **Origin:** Native **IUCN Habitat code:** 1 **Conservation status: (IUCN)** NE

1337 *Vanilla bicolor* Lindl. (Orchidaceae)
IPNI: 661161-1
Use: EU **Origin:** Native **IUCN Habitat code:** 1,3,4,5 **Conservation status: (IUCN)** NE

1338 *Vanilla columbiana* Rolfe (Orchidaceae)
IPNI: 262547-2
Common name: vainilla **Habit:** Herb, Climbing **Use:** ME **Origin:** Endemic **Dept.:** CUN, VAC **Elev.:** 40–1000 m a.s.l. **IUCN Habitat code:** 1,14 **Conservation status: (IUCN)** NE

1339 *Vanilla guianensis* Splitg. (Orchidaceae)
IPNI: 661200-1
Common name: vainilla **Use:** HF, ME **Origin:** Native **IUCN Habitat code:** 1,4 **Conservation status: (IUCN)** NE

1340 *Vanilla helleri* A.D.Hawkes (Orchidaceae)
IPNI: 262563-2
Use: GS **Origin:** Native **Dept.:** CHO **Conservation status: (IUCN)** DD

1341 *Vanilla mexicana* Mill. (Orchidaceae)
IPNI: 661225-1
Use: EU **Origin:** Native **IUCN Habitat code:** 1,3 **Conservation status: (IUCN)** NE

1342 *Vanilla odorata* C.Presl (Orchidaceae)
IPNI: 661231-1
Common name: vainilla, vainillo **Habit:** Herb, Climbing **Use:** GS, HF, MA, ME, PO, SU **Origin:** Native **Region:** Amazonia, Andean, Orinoquia, Pacific **Dept.:** AMA, ANT, CAU, MET **Elev.:** 150–2000 m a.s.l. **IUCN Habitat code:** 1,3,14 **Conservation status: (IUCN)** EN

1343 *Vanilla palmarum* (Salzm. ex Lindl.) Lindl. (Orchidaceae)
IPNI: 661237-1
Common name: platanillo, vainillón **Habit:** Herb, Climbing **Use:** EU, HF **Origin:** Native **Region:** Amazonia **Elev.:** 200–200 m a.s.l. **IUCN Habitat code:** 1,3,14 **Conservation status: (IUCN)** NE

1344 *Vanilla phaeantha* Rchb.f. (Orchidaceae)
IPNI: 661245-1
Use: GS **Origin:** Native **IUCN Habitat code:** 2 **Conservation status: (IUCN)** EN

1345 *Vanilla planifolia* Andrews (Orchidaceae)
IPNI: 262578-2
Common name: bejuco vainilla, bejuco vainillo, bejuquillo, bejuquito, cuyanquillo, vainilla, vainillo, verde vainilla **Habit:** Herb, Climbing **Use:** EU, GS, HF, MA, ME **Origin:** Native, Cultivated **Region:** Andean, Caribbean, Pacific **Dept.:** ANT, BOL, CHO, CUN, MAG, NAR, SUC, VAC **Elev.:** 0–730 m a.s.l. **IUCN Habitat code:** 1,2,3,4,14 **Conservation status: (IUCN)** EN

1346 *Vanilla pompona* Schiede (Orchidaceae)
IPNI: 661259-1
Habit: Herb, Climbing **Use:** EU, GS, HF, ME **Origin:** Native **Region:** Andean, Caribbean **Dept.:** CUN, MAG, SUC, TOL **Elev.:** 184–1700 m a.s.l. **IUCN Habitat code:** 1,2,3,14 **Conservation status: (IUCN)** EN

1347 *Vanilla ribeiroi* Hoehne (Orchidaceae)
IPNI: 262585-2
Habit: Herb, Climbing **Use:** GS **Origin:** Native **Region:** Amazonia **Dept.:** VAU **Elev.:** 150–150 m a.s.l. **Conservation status: (IUCN)** NE

1348 *Warczewiczella marginata* Rchb. f. (Orchidaceae)
IPNI: 661384-1
Use: EU **Origin:** Endemic **IUCN Habitat code:** 14 **Conservation status: (IUCN)** NE

Commelinales

1349 *Aneilema umbrosum* (Vahl) Kunth (Commelinaceae)
IPNI: 171758-1
Common name: siempreviva **Habit:** Herb **Use:** ME **Origin:** Native **Region:** Amazonia, Andean, Pacific **Dept.:** AMA, ANT, CHO, MET, NAR, VAC **Elev.:** 150–1350 m a.s.l. **IUCN Habitat code:** 1,3,5,14 **Conservation status: (IUCN)** NE

1350 *Callisia fragrans* (Lindl.) Woodson (Commelinaceae)
IPNI: 42441-2
Habit: Herb **Use:** EU, ME **Origin:** Cultivated **Region:** Andean **Dept.:** ANT **Elev.:** 1470–1470 m a.s.l. **IUCN Habitat code:** 14 **Conservation status: (IUCN)** NE

1351 *Callisia gracilis* (Kunth) D.R.Hunt (Commelinaceae)
IPNI: 42443-2
Habit: Herb **Use:** ME **Origin:** Native **Region:** Andean **Dept.:** ANT, HUI, NAR, VAC **Elev.:** 700–2600 m a.s.l. **IUCN Habitat code:** 1,2,3,14 **Conservation status: (IUCN)** LC

1352 *Callisia procumbens* (Willd.) ined. (Commelinaceae)
IPNI: 77205602-1
Common name: codillo, nudillo, siempreviva, suelda, suelda consuelda, sueldaconsuelda **Habit:** Climbing **Use:** ME **Origin:** Native **Region:** Andean **Dept.:** ANT, BOY, CAQ, CUN **Elev.:** 1000–3000 m a.s.l. **Conservation status: (IUCN)** NE

1353 *Callisia repens* (Jacq.) L. (Commelinaceae)
IPNI: 171826-1
Common name: suelda **Habit:** Herb **Use:** EU, ME **Origin:** Native **Region:** Andean, Caribbean **Dept.:** ANT, CAU, MAG, SAN, VAC **Elev.:** 60–2640 m a.s.l. **IUCN Habitat code:** 1,2,14 **Conservation status: (IUCN)** NE

1354 *Callisia serrulata* (Vahl) Christenh. & Byng (Commelinaceae)
IPNI: 77184055-1
Common name: chicoria, conejera, siempreviva, siempreviva chocoana **Habit:** Herb **Use:** HF, ME **Origin:** Native **Region:** Andean, Caribbean, Orinoquia, Pacific **Dept.:** ANT, BOY, CAL, CHO, CUN, LAG, MAG, MET, NAR, QUI, RIS, VAC **Elev.:** 0-1380 m.a.s.l. **Conservation status: (IUCN)** NE **(CPSG)** Potential LC

1355 *Commelina benghalensis* L. (Commelinaceae)
IPNI: 332090-2
Habit: Herb **Use:** AF, GS, HF, MA, ME, PO **Origin:** Cultivated **Region:** Andean **Dept.:** ANT **Elev.:** 1480–1500 m a.s.l. **Conservation status: (IUCN)** LC

1356 *Commelina clavata* C.B.Clarke (Commelinaceae)
IPNI: 171972-1
Use: HF, ME **IUCN Habitat code:** 14 **Conservation status: (IUCN)** LC

1357 *Commelina diffusa* Burm.f. (Commelinaceae)
IPNI: 172020-1
Common name: barquito morado, barquito morao, canutillo, canutillo morado, jalechepil, ojito de Santa Lucía **Habit:** Herb **Use:** AF, EU, HF, MA, ME, SU **Origin:** Naturalised **Region:** Andean, Orinoquia, Pacific **Dept.:** ANT, BOG, BOY, CAS, CHO, CUN, MAG, MET, SAN, VAC **Elev.:** 120–2550 m a.s.l. **IUCN Habitat code:** 1,2,3,4,5,14 **Conservation status: (IUCN)** LC

1358 *Commelina elliptica* Kunth (Commelinaceae)
IPNI: 172043-1
Use: HF **Origin:** Native **Conservation status: (IUCN)** NE

1359 *Commelina erecta* L. (Commelinaceae)
IPNI: 172048-1
Common name: cadillo, cohitre, ojera, ojito de Santa Lucía, ojo de Santa Lucía, panameña, santalucía, siempreviva, suelda con suelda, sueldaconsuelda, sueldaquesuelda, sueldasuelda **Habit:** Herb **Use:** AF, EU, HF, ME, SU **Origin:** Native **Region:** Andean, Caribbean, Pacific **Dept.:** ANT, ATL, BOG, BOL, CAL, CHO, COR, CUN, HUI, MAG, SAP, SUC, TOL, VAC **Elev.:** 5-2260 m.a.s.l. **IUCN Habitat code:** 1,2,3,4,5,14 **Conservation status: (IUCN)** LC **(CPSG)** Potential LC

1360 *Commelina obliqua* Vahl (Commelinaceae)
IPNI: 63900-2
Common name: sueldaconsuelda **Habit:** Herb **Use:** HF, ME **Origin:** Native **Region:** Andean **Dept.:** CUN, SAN **Elev.:** 1500–1800 m a.s.l. **IUCN Habitat code:** 1,2,3,14 **Conservation status: (IUCN)** NE

1361 *Commelina tuberosa* L. (Commelinaceae)
IPNI: 302798-2
Habit: Herb **Use:** EU, HF, ME **Origin:** Native **Dept.:** TOL **Elev.:** 480–480 m a.s.l. **IUCN Habitat code:** 14 **Conservation status: (IUCN)** NE

1362 *Dichorisandra hexandra* (Aubl.) C.B.Clarke (Commelinaceae)
IPNI: 79497-2
Habit: Herb, Climbing **Use:** ME, SU **Origin:** Native **Region:** Amazonia, Andean, Guiana Shield, Caribbean, Orinoquia, Pacific **Dept.:** AMA, ANT, BOY, CAS, CHO, GUV, MAG, MET, NAR, PUT, QUI, RIS, SAN, TOL, VAC, VAU **Elev.:** 10-1800 m.a.s.l. **IUCN Habitat code:** 1,2,3,4,5,14 **Conservation status: (IUCN)** NE **(CPSG)** Potential LC

1363 *Dichorisandra musaica* K.Koch & Linden (Commelinaceae)
IPNI: 172628-1
Habit: Liana **Use:** ME **Origin:** Native **Region:** Andean, Pacific **Dept.:** CHO, VAC **Elev.:** 100–1800 m a.s.l. **IUCN Habitat code:** 1,3 **Conservation status: (IUCN)** NE

1364 *Dichorisandra ulei* J.F.Macbr. (Commelinaceae)
IPNI: 79534-2
Habit: Herb **Use:** ME **Origin:** Native **Region:** Amazonia **Dept.:** AMA, ANT, CAQ, CAU, PUT, VAC **Elev.:** 100–800 m a.s.l. **IUCN Habitat code:** 1,3,5,14 **Conservation status: (IUCN)** NE

1365 *Geogenanthus ciliatus* G.Brückn. (Commelinaceae)
IPNI: 108816-2
Common name: aparoje, jaqui, oreja de negro, turu **Use:** ME **IUCN Habitat code:** 1,4,14 **Conservation status: (IUCN)** NE

1366 *Gibasis pellucida* (M.Martens & Galeotti) D.R.Hunt (Commelinaceae)
IPNI: 109899-2
Habit: Herb **Use:** EU **Origin:** Cultivated **Region:** Andean **Dept.:** CUN **Elev.:** 2600–2600 m a.s.l. **IUCN Habitat code:** 14 **Conservation status: (IUCN)** NE

1367 *Murdannia nudiflora* (L.) Brenan (Commelinaceae)
IPNI: 172873-1
Habit: Herb **Use:** AF, HF, MA, ME, SU **Origin:** Naturalised **Region:** Andean, Caribbean, Orinoquia, Pacific **Dept.:** ANT, ATL, CAS, CES, CHO, MAG, MET, SAN, TOL, VAC **Elev.:** 60–500 m a.s.l. **IUCN Habitat code:** 1,2,3,4,14 **Conservation status: (IUCN)** NE

1368 *Tinantia erecta* (Jacq.) Fenzl (Commelinaceae)
IPNI: 173146-1
Common name: sueldaconsuelda **Habit:** Herb **Use:** HF, ME **Origin:** Native **Region:** Andean **Dept.:** BOG, CUN, MAG **Elev.:** 1230–2600 m a.s.l. **IUCN Habitat code:** 1,2,3,14 **Conservation status: (IUCN)** NE

1369 *Tradescantia crassifolia* Cav. (Commelinaceae)
IPNI: 173264-1
Use: ME **Region:** Andean **Dept.:** BOG **Conservation status: (IUCN)** NE

1370 *Tradescantia fluminensis* Vell. (Commelinaceae)
IPNI: 30090792-2
Use: EU, ME **Region:** Andean **Dept.:** BOG **IUCN Habitat code:** 14 **Conservation status: (IUCN)** NE

1371 *Tradescantia pallida* (Rose) D.R.Hunt (Commelinaceae)
IPNI: 256037-2
Habit: Herb **Use:** EU, ME **Origin:** Cultivated **Region:** Andean, Caribbean **Dept.:** ANT, ATL, BOG, CUN, SAP **Elev.:** 50–1500 m a.s.l. **IUCN Habitat code:** 1,3,14 **Conservation status: (IUCN)** NE

1372 *Tradescantia spathacea* Sw. (Commelinaceae)
IPNI: 173471-1
Common name: barco, boul di mas, canoa, croto, croto palmita, flotilla de gaitán, magueyito, palmita **Use:** EU, HF, ME, PO **IUCN Habitat code:** 1,2,5,14 **Conservation status: (IUCN)** NE

1373 *Tradescantia zanonia* (L.) Sw. (Commelinaceae)
IPNI: 1191409-2
Common name: cascajo flor, cañagria, cañagua, cinta, rascadera, suelda **Habit:** Herb, Subshrub **Use:** ME **Origin:** Native **Region:** Amazonia, Andean, Orinoquia, Pacific **Dept.:** AMA, ANT, BOG, BOY, CAL, CAS, CAU, CHO, CUN, HUI, MAG, MET, NAR, PUT, QUI, RIS, SAN, TOL, VAC **Elev.:** 30-2850 m.a.s.l. **IUCN Habitat code:** 1,2,3,5,14 **Conservation status: (IUCN)** NE **(CPSG)** Potential LC

1374 *Tradescantia zebrina* Bosse (Commelinaceae)
IPNI: 30049896-2
Common name: cohitre morado, cordoncillo morado, cucaracha, judío errante, maravilla, morado, panameña, plateada, sueldaconsuelda, trancadera, víbora, zebra **Habit:** Herb **Use:** EU, ME **Origin:** Native **Region:** Pacific **Dept.:** ANT, BOG, CHO **Elev.:** 60–1500 m a.s.l. **IUCN Habitat code:** 1,2,3,5,14 **Conservation status: (IUCN)** NE

1375 *Anigozanthos flavidus* Redouté (Haemodoraceae)
IPNI: 429575-1
Common name: garra de canguro **Habit:** Herb **Use:** EU, GS, HF **Origin:** Naturalised **Region:** Andean **Dept.:** BOG, CUN **Elev.:** 2600–2600 m a.s.l. **Conservation status: (IUCN)** NE

1376 *Anigozanthos pulcherrimus* Hook. (Haemodoraceae)
IPNI: 429585-1
Use: EU **Region:** Andean **Dept.:** BOG **Conservation status: (IUCN)** NE

1377 *Schiekia orinocensis* (Kunth) Meisn. (Haemodoraceae)
IPNI: 429893-1
Common name: cebolleta sabanera, cebollín **Habit:** Herb **Use:** ME **Origin:** Native **Region:** Guiana Shield, Orinoquia **Dept.:** AMA, CAS, GUA, GUV, MET, VAU, VID **Elev.:** 80–600 m a.s.l. **IUCN Habitat code:** 1,3,4,5,14 **Conservation status: (IUCN)** NE

1378 *Xiphidium caeruleum* Aubl. (Haemodoraceae)
IPNI: 304977-2
Common name: cebolleta, cebolleta blanca, cola de babilla, cola de paloma, helecho, jabón, manito de Dios, manito de dios, mano de Dios, mano de ángel, manoediós, padilla, palm bruja, placo, tres dedos,

tresdedos **Habit:** Herb **Use:** ME, SU **Origin:** Native **Region:** Amazonia, Andean, Guiana Shield, Caribbean, Orinoquia, Pacific **Dept.:** AMA, ANT, BOL, BOY, CAL, CAQ, CAS, CAU, CHO, COR, CUN, GUA, GUV, LAG, MAG, MET, NAR, PUT, QUI, SAN, TOL, VAC, VAU **Elev.:** 0-2100 m.a.s.l. **IUCN Habitat code:** 1,2,3,4,5,14 **Conservation status: (IUCN)** NE **(CPSG)** Potential LC

1379 *Heteranthera reniformis* Ruiz & Pav. (Pontederiaceae) **IPNI:** 698080-1 **Common name:** berro de laguna, buche de gallina, violeta de agua **Habit:** Herb, Aquatic **Use:** HF, ME **Origin:** Native **Region:** Andean, Caribbean, Orinoquia, Pacific **Dept.:** ANT, BOY, CAQ, CAS, CHO, HUI, MET, NAR, PUT, SAN, TOL, VAC **Elev.:** 0–1850 m a.s.l. **IUCN Habitat code:** 1,2,3,4,5,14 **Conservation status: (IUCN)** NE

1380 *Pontederia azurea* Sw. (Pontederiaceae) **IPNI:** 698138-1 **Common name:** balata, batata, bora, buchón, cebolleta, chupalla, jacinto de agua, lechuguilla, nomeolvides, oreja de mula, tapón, taruya **Habit:** Herb, Aquatic **Use:** HF, ME **Origin:** Native, Cultivated **Region:** Amazonia, Andean, Caribbean, Orinoquia, Pacific **Dept.:** ANT, ARA, BOL, CAQ, CAS, CES, CHO, COR, CUN, GUV, HUI, MAG, PUT, SAN, VAU **Elev.:** 40–1000 m a.s.l. **IUCN Habitat code:** 2,5 **Conservation status: (IUCN)** NE

1381 *Pontederia cordata* L. (Pontederiaceae) **IPNI:** 283610-2 **Common name:** oreja de mulo **Habit:** Herb, Aquatic **Use:** EU, HF, ME **Origin:** Native, Cultivated **Region:** Amazonia, Guiana Shield, Caribbean, Pacific **Dept.:** BOL, CHO, GUV, MAG, PUT, VAU **Elev.:** 20–325 m a.s.l. **IUCN Habitat code:** 1,2,3,4,5,14 **Conservation status: (IUCN)** NE

1382 *Pontederia crassipes* Mart. (Pontederiaceae) **IPNI:** 310928-2 **Common name:** batata, bora, buchón, buchón de agua, cebolleta, jacinto de agua, lechuguilla, oreja de mula, oreja de mulo, orejemula, pato, taruya **Habit:** Herb, Aquatic **Use:** AF, EU, FU, GS, HF, MA, ME **Origin:** Native, Cultivated **Region:** Amazonia, Andean, Caribbean, Pacific **Dept.:** AMA, ANT, ATL, BOL, CAQ, CES, CHO, COR, CUN, GUA, GUV, HUI, MET, NAR, PUT, QUI, VAC **Elev.:** 5–2600 m a.s.l. **IUCN Habitat code:** 1,2,3,5,14 **Conservation status: (IUCN)** NE

1383 *Pontederia diversifolia* (Vahl) M.Pell. & C.N.Horn (Pontederiaceae) **IPNI:** 77188085-1 **Habit:** Herb, Aquatic **Use:** EU,PO **Origin:** Native **Region:** Orinoquia, Pacific **Dept.:** CAQ, CAU, CHO, MET, NSA, PUT, SAN, VID **Elev.:** 20–180 m a.s.l. **Conservation status: (IUCN)** NE

1384 *Pontederia paniculata* Spreng. (Pontederiaceae) **IPNI:** 698175-1 **Use:** ME **Origin:** Native **Conservation status: (IUCN)** NE

Dioscoreales

1385 *Dioscorea acanthogene* Rusby (Dioscoreaceae) **IPNI:** 80658-2 **Habit:** Climbing **Use:** HF, ME **Origin:** Native **Region:** Pacific **Dept.:** ANT, BOL, BOY, CAL, CHO, CUN, HUI, SAN, SUC, TOL **Elev.:** 10–910 m a.s.l. **IUCN Habitat code:** 1,2,3,14 **Conservation status: (IUCN)** NE

1386 *Dioscorea alata* L. (Dioscoreaceae) **IPNI:** 317625-1 **Common name:** cabeza de negro, chilma, tabena, ñame, ñame congo, ñame morado, ñampí **Habit:** Climbing **Use:** AF, EU, GS, HF, MA, ME, PO, SU **Origin:** Cultivated, Naturalised **Region:** Amazonia, Andean, Caribbean **Dept.:** AMA, ANT, BOL, COR, MAG, SAP **Elev.:** 0–1500 m a.s.l. **IUCN Habitat code:** 1,3,5,14 **Conservation status: (IUCN)** NE

1387 *Dioscorea amaranthoides* C.Presl (Dioscoreaceae) **IPNI:** 323907-2 **Habit:** Climbing **Use:** HF, ME **Origin:** Native **Region:** Amazonia, Andean **Dept.:** AMA, ANT, BOY **Elev.:** 90–1000 m a.s.l. **IUCN Habitat code:** 1,2,3,4,14 **Conservation status: (IUCN)** NE

1388 *Dioscorea amazonum* Mart. ex Griseb. (Dioscoreaceae) **IPNI:** 324013-2 **Habit:** Climbing **Use:** HF, ME **Origin:** Native **Region:** Amazonia **Dept.:** AMA, CAQ, GUA, MET, PUT, VAU **Elev.:** 0–800 m a.s.l. **IUCN Habitat code:** 1,3,4,14 **Conservation status: (IUCN)** NE

1389 *Dioscorea bulbifera* L. (Dioscoreaceae) **IPNI:** 323908-2 **Common name:** papa de aire, ñame papa, ñame volador **Habit:** Climbing **Use:** AF, EU, GS, HF, MA, ME, PO, SU **Origin:** Cultivated, Naturalised **Region:** Andean, Guiana Shield **Dept.:** ANT, BOL, CAU, COR, VAU **Elev.:** 34–1630 m a.s.l. **IUCN Habitat code:** 1,2,3 **Conservation status: (IUCN)** NE

1390 *Dioscorea cayenensis* Lam. (Dioscoreaceae) **IPNI:** 317788-1 **Habit:** Climbing **Use:** GS, HF, MA, ME, SU **Origin:** Cultivated **Region:** Amazonia, Guiana Shield, Caribbean **Dept.:** AMA, BOL, COR, GUA, SUC **Elev.:** 100–120 m a.s.l. **IUCN Habitat code:** 1,2,3,4,5,14 **Conservation status: (IUCN)** NE

1391 *Dioscorea coriacea* Humb. & Bonpl. ex Willd. (Dioscoreaceae) **IPNI:** 324018-2 **Common name:** guatamo **Habit:** Climbing **Use:** MA, ME **Origin:** Native **Region:** Andean **Dept.:** ANT, BOG, BOY, CAU, CES, CHO, CUN, HUI, MAG, NAR, PUT, QUI, RIS, TOL, VAC **Elev.:** 1200–3900 m a.s.l. **IUCN Habitat code:** 1,2,3,4,14 **Conservation status: (IUCN)** NE

1392 *Dioscorea dodecaneura* Vell. (Dioscoreaceae) **IPNI:** 323912-2 **Habit:** Climbing **Use:** HF, ME **Origin:** Native **Region:** Amazonia **Conservation status: (IUCN)** NE

1393 *Dioscorea esculenta* (Lour.) Burkill (Dioscoreaceae) **IPNI:** 317946-1 **Habit:** Climbing **Use:** GS, HF, MA, ME, SU **Origin:** Cultivated **Elev.:** 0–100 m a.s.l. **Conservation status: (IUCN)** NE

1394 *Dioscorea flaccida* R.Knuth (Dioscoreaceae) **IPNI:** 80807-2 **Habit:** Climbing **Use:** ME **Origin:** Endemic **Region:** Andean **Dept.:** CAU, CUN, NAR, TOL **Elev.:** 1100–2500 m a.s.l. **IUCN Habitat code:** 1,3,4,14 **Conservation status: (IUCN)** NE

1395 *Dioscorea glandulosa* (Griseb.) Klotzsch ex Kunth (Dioscoreaceae) **IPNI:** 324020-2 **Habit:** Climbing **Use:** HF, ME **Origin:** Native **Region:** Andean **Dept.:** BOY, CUN, QUI **Elev.:** 420–1800 m a.s.l. **IUCN Habitat code:** 1,2,3,14 **Conservation status: (IUCN)** NE

1396 *Dioscorea guianensis* R.Knuth (Dioscoreaceae) **IPNI:** 80836-2 **Common name:** ñame de espina, ñame espina, ñame espino, ñame espinoso **Habit:** Climbing **Use:** HF **Origin:** Native **Region:** Guiana Shield **Dept.:** GUA, GUV, VAU **Elev.:** 150–200 m a.s.l. **IUCN Habitat code:** 1,3 **Conservation status: (IUCN)** NE

1397 *Dioscorea hastatissima* Rusby (Dioscoreaceae) **IPNI:** 80843-2 **Habit:** Climbing **Use:** ME **Origin:** Endemic **Region:** Andean **Dept.:** ANT, MAG **Elev.:** 300–1350 m a.s.l. **IUCN Habitat code:** 2,14 **Conservation status: (IUCN)** NE

1398 *Dioscorea lehmannii* Uline (Dioscoreaceae) **IPNI:** 80894-2 **Common name:** zarzaparrilla **Habit:** Climbing **Use:** MA, ME **Origin:** Native **Region:** Andean **Dept.:** ANT, BOG, CUN **Elev.:** 2000–3000 m a.s.l. **IUCN Habitat code:** 1,3,4,14 **Conservation status: (IUCN)** NE

1399 *Dioscorea matagalpensis* Uline (Dioscoreaceae) **IPNI:** 80927-2 **Use:** ME **IUCN Habitat code:** 14 **Conservation status: (IUCN)** NE

1400 *Dioscorea meridensis* Kunth (Dioscoreaceae) **IPNI:** 318260-1 **Common name:** zarzaparrilla **Habit:** Climbing **Use:** MA, ME **Origin:** Native **Region:** Andean **Dept.:** ANT, CAL, CHO, CUN **Elev.:** 500–2500 m a.s.l. **IUCN Habitat code:** 1,3,14 **Conservation status: (IUCN)** NE

1401 *Dioscorea mexicana* Scheidw. (Dioscoreaceae) **IPNI:** 318262-1 **Habit:** Climbing **Use:** HF, ME **Origin:** Native **Dept.:** BOL **Elev.:** 400–400 m a.s.l. **IUCN Habitat code:** 2,3 **Conservation status: (IUCN)** NE

1402 *Dioscorea olfersiana* Klotzsch ex Griseb. (Dioscoreaceae) **IPNI:** 318343-1 **Use:** HF, ME **IUCN Habitat code:** 14 **Conservation status: (IUCN)** NE

1403 *Dioscorea pallens* Schltdl. (Dioscoreaceae) **IPNI:** 318367-1 **Use:** HF, ME **IUCN Habitat code:** 1,14 **Conservation status: (IUCN)** NE

1404 *Dioscorea pennellii* R.Knuth (Dioscoreaceae) **IPNI:** 80993-2 **Habit:** Climbing **Use:** ME **Origin:** Endemic **Region:** Andean **Dept.:** ANT,

CAL, CAU, RIS **Elev.:** 1500–2600 m a.s.l. **Conservation status: (IUCN)** NE

1405 *Dioscorea pilosiuscula* Bertero ex Spreng. (Dioscoreaceae) **IPNI:** 318417-1 **Common name:** masoko, ñame huitoto **Habit:** Climbing **Use:** ME **Origin:** Native **Region:** Amazonia, Guiana Shield, Pacific **Dept.:** AMA, CHO, MET **Elev.:** 0–500 m a.s.l. **IUCN Habitat code:** 1,2,14 **Conservation status: (IUCN)** NE

1406 *Dioscorea piperifolia* Humb. & Bonpl. ex Willd. (Dioscoreaceae) **IPNI:** 323920-2 **Habit:** Climbing **Use:** EU, HF, ME **Origin:** Native **Region:** Andean **Dept.:** BOG, CAU, MAG, SAN **Elev.:** 600–2100 m a.s.l. **IUCN Habitat code:** 1,2,3,4,5,14 **Conservation status: (IUCN)** NE

1407 *Dioscorea polygonoides* Humb. & Bonpl. ex Willd. (Dioscoreaceae) **IPNI:** 323921-2 **Common name:** guaco blanco **Habit:** Climbing **Use:** AF, HF, ME **Origin:** Native **Region:** Andean, Caribbean, Pacific **Dept.:** ANT, BOL, BOY, CAL, CAU, CES, CHO, CUN, HUI, LAG, MAG, MET, NSA, RIS, SUC, VAC **Elev.:** 10–2000 m a.s.l. **Conservation status: (IUCN)** NE

1408 *Dioscorea santanderensis* R.Knuth (Dioscoreaceae) **IPNI:** 81065-2 **Common name:** bejuco **Habit:** Climbing **Use:** MA, ME **Origin:** Endemic **Region:** Andean **Dept.:** ANT, SAN **Elev.:** 60–1300 m a.s.l. **IUCN Habitat code:** 1,3,5,14 **Conservation status: (IUCN)** NE

1409 *Dioscorea sericea* R.Knuth (Dioscoreaceae) **IPNI:** 81074-2 **Habit:** Climbing **Use:** ME **Origin:** Native **Region:** Andean **Dept.:** CUN, PUT, SUC, VAC **Elev.:** 950-2200 m.a.s.l. **IUCN Habitat code:** 3,14 **Conservation status: (IUCN)** LC **(CPSG)** LC

1410 *Dioscorea stegelmanniana* R.Knuth (Dioscoreaceae) **IPNI:** 81088-2 **Common name:** cipó pequeño **Use:** ME **IUCN Habitat code:** 4 **Conservation status: (IUCN)** NE

1411 *Dioscorea trifida* L.f. (Dioscoreaceae) **IPNI:** 323927-2 **Common name:** batata, chisca, paje, papayote, pekuoino, racu, tabena, ñame, ñame blanco, ñame común, ñame de cotudo, ñame de milpesillo, ñame de monte, ñame negro, ñame redondito, ñame redondo, ñamu **Habit:** Climbing **Use:** AF, GS, HF, ME **Origin:** Native, Cultivated **Region:** Amazonia, Andean, Guiana Shield, Orinoquia, Pacific **Dept.:** AMA, ANT, BOY, CHO, GUA, GUV, MET, NAR, TOL, VAU **Elev.:** 5–1120 m a.s.l. **IUCN Habitat code:** 1,3,5,14 **Conservation status: (IUCN)** NE

1412 *Dioscorea trifoliata* Kunth (Dioscoreaceae) **IPNI:** 324037-2 **Habit:** Climbing **Use:** HF, ME **Origin:** Native **Region:** Caribbean **Dept.:** LAG, MAG **Elev.:** 10–150 m a.s.l. **IUCN Habitat code:** 1 **Conservation status: (IUCN)** NE

1413 *Dioscorea villosa* L. (Dioscoreaceae) **IPNI:** 318778-1 **Common name:** raíz de china, ñame silvestre **Use:** HF, ME **IUCN Habitat code:** 3,14 **Conservation status: (IUCN)** NE

Liliales

1414 *Alstroemeria aurea* Graham (Alstroemeriaceae) **IPNI:** 62457-1 **Common name:** astromelia **Habit:** Herb **Use:** EU, GS, HF, ME **Origin:** Cultivated **Region:** Andean **Dept.:** ANT, BOG, BOY, CAL, CAU, CUN, NAR, QUI, RIS, SAN **Elev.:** 1800–2700 m a.s.l. **IUCN Habitat code:** 1,3,14 **Conservation status: (IUCN)** NE

1415 *Alstroemeria ligtu* L. (Alstroemeriaceae) **IPNI:** 62542-1 **Use:** EU, GS, MA, ME **Region:** Andean **Dept.:** BOG **Conservation status: (IUCN)** NE

1416 *Bomarea carderi* Mast. (Alstroemeriaceae) **IPNI:** 63253-1 **Common name:** cortapicos **Habit:** Climbing **Use:** ME **Origin:** Native **Region:** Andean, Pacific **Dept.:** ANT, BOY, CAL, CAU, CHO, CUN, NAR, NSA, QUI, RIS, SAN, VAC **Elev.:** 300-3400 m.a.s.l. **IUCN Habitat code:** 1,3,4,8,14 **Conservation status: (IUCN)** NE **(CPSG)** Potential LC

1417 *Bomarea edulis* (Tussac) Herb. (Alstroemeriaceae) **IPNI:** 63283-1 **Common name:** papaguasca **Habit:** Climbing **Use:** HF, ME **Origin:** Native **Region:** Amazonia, Andean, Caribbean, Orinoquia, Pacific **Dept.:** ANT, CES, CUN, LAG, MAG, MET, QUI, RIS **Elev.:** 200-2900 m.a.s.l. **IUCN Habitat code:** 1,3,4,6,14 **Conservation status: (IUCN)** NE **(CPSG)** Potential LC

1418 *Bomarea hirsuta* (Kunth) Herb. (Alstroemeriaceae) **IPNI:** 63318-1 **Common name:** cortapicos, pecosa **Habit:** Climbing **Use:** AF **Origin:** Native **Region:** Andean **Dept.:** ANT, BOG, BOY, CAL, CAU, CUN, HUI, MET, NAR, PUT, QUI, RIS, VAC **Elev.:** 1900-3800 m.a.s.l. **IUCN Habitat code:** 1,3,4,8,14 **Conservation status: (IUCN)** NE **(CPSG)** Potential LC

1419 *Bomarea multiflora* (L.f.) Mirb. (Alstroemeriaceae) **IPNI:** 63359-1 **Common name:** bejuco cortapicos, bejuco pecoso, cortapico, cortapicos, papaguasca, papasacha, peca, pecosa, pecosita, petacas, petaquita, rompeplato, rompeplatos, tetona **Habit:** Climbing **Use:** ME **Origin:** Native **Region:** Andean **Dept.:** ANT, BOG, BOY, CAL, CAU, CUN, HUI, LAG, NAR, NSA, PUT, QUI, RIS, SAN, TOL, VAC **Elev.:** 1700-3800 m.a.s.l. **IUCN Habitat code:** 1,3,4,5,8,14 **Conservation status: (IUCN)** NE **(CPSG)** Potential LC

1420 *Bomarea setacea* (Ruiz & Pav.) Herb. (Alstroemeriaceae) **IPNI:** 63411-1 **Common name:** chimpamplena, pecas, tetón **Habit:** Climbing **Use:** ME **Origin:** Native **Region:** Andean **Dept.:** ANT, BOG, BOY, CAL, CAU, CHO, CUN, HUI, MET, NAR, NSA, PUT, QUI, RIS, SAN, TOL, VAC **Elev.:** 1900-4000 m.a.s.l. **IUCN Habitat code:** 1,3,4,14 **Conservation status: (IUCN)** LC **(CPSG)** Potential LC

1421 *Gloriosa superba* L. (Colchicaceae) **IPNI:** 535953-1 **Habit:** Herb **Use:** EU, HF, MA, ME, PO, SU **Origin:** Cultivated **Region:** Andean **Dept.:** BOG, CUN **Elev.:** 2600–2600 m a.s.l. **Conservation status: (IUCN)** LC

1422 *Lilium candidum* L. (Liliaceae) **IPNI:** 537512-1 **Common name:** azucena, azucena de Obando, azucena de Quito, lirio, nardo **Habit:** Herb **Use:** EU, HF, ME **Origin:** Cultivated **Region:** Andean **Dept.:** BOG, BOY, CUN, NAR **Elev.:** 2600–2700 m a.s.l. **IUCN Habitat code:** 1 **Conservation status: (IUCN)** NE

1423 *Smilax cuspidata* Duhamel (Smilacaceae) **IPNI:** 541388-1 **Use:** ME **IUCN Habitat code:** 1,3,4,14 **Conservation status: (IUCN)** NE

1424 *Smilax domingensis* Willd. (Smilacaceae) **IPNI:** 541406-1 **Common name:** bejuco canasto, bejuco negro, china, guayacana, lindachina, uña de gato, zarzaparrilla **Habit:** Liana, Climbing **Use:** HF, MA, ME **Origin:** Native **Region:** Amazonia, Andean, Caribbean, Pacific **Dept.:** AMA, ANT, BOG, BOL, BOY, CAL, CAQ, CAU, CHO, COR, CUN, GUA, GUV, HUI, MAG, NAR, NSA, PUT, QUI, RIS, SAN, TOL, VAC, VAU **Elev.:** 0-3320 m.a.s.l. **IUCN Habitat code:** 1,2,3,4,5,14 **Conservation status: (IUCN)** NE **(CPSG)** Potential LC

1425 *Smilax longifolia* Rich. (Smilacaceae) **IPNI:** 315317-2 **Habit:** Liana, Climbing **Use:** HF, ME **Origin:** Native **Region:** Amazonia **Dept.:** AMA, PUT **Elev.:** 200–360 m a.s.l. **IUCN Habitat code:** 1,5 **Conservation status: (IUCN)** NE

1426 *Smilax officinalis* Kunth (Smilacaceae) **IPNI:** 541685-1 **Common name:** uña de gato, zarzaparrilla **Habit:** Liana, Climbing **Use:** HF, MA, ME **Origin:** Native **Region:** Amazonia, Andean, Caribbean, Pacific **Dept.:** ANT, BOL, CAQ, CHO, COR, CUN, PUT, QUI, VAC **Elev.:** 48–1500 m a.s.l. **IUCN Habitat code:** 1,2,3 **Conservation status: (IUCN)** NE

1427 *Smilax purhampuy* Ruiz (Smilacaceae) **IPNI:** 541778-1 **Common name:** diente de perro **Habit:** Liana, Climbing **Use:** ME **Origin:** Native **Region:** Andean, Caribbean **Dept.:** ANT, CAL, CES, CUN, HUI, NSA, QUI, RIS, SAN, VAC **Elev.:** 66–2450 m a.s.l. **IUCN Habitat code:** 1,3,4,14 **Conservation status: (IUCN)** NE

1428 *Smilax schomburgkiana* Kunth (Smilacaceae) **IPNI:** 541837-1 **Habit:** Liana, Climbing **Use:** ME **Origin:** Native **Region:** Amazonia **Dept.:** AMA **Elev.:** 100–100 m a.s.l. **IUCN Habitat code:** 1 **Conservation status: (IUCN)** NE

1429 *Smilax siphilitica* Humb. & Bonpl. ex Willd. (Smilacaceae) **IPNI:** 541854-1 **Common name:** bejuco espuelo, china, jalapatrás, uña de gato, zarcilla, zarzaparrilla, zarzilla **Habit:** Liana, Climbing **Use:** HF, ME **Origin:** Native **Region:**

Amazonia, Andean, Guiana Shield, Pacific **Dept.:** AMA, ANT, CAQ, CAU, CHO, GUA, GUV, PUT, VAU **Elev.:** 0–1900 m a.s.l. **IUCN Habitat code:** 1,3,4,5,14 **Conservation status: (IUCN)** NE

1430 *Smilax spinosa* Mill. (Smilacaceae) **IPNI:** 541862-1 **Common name:** chinese root, colmillo de perro, raíz china, uña de gato, zarzaparrilla **Habit:** Liana, Climbing **Use:** HF, ME **Origin:** Native **Region:** Amazonia, Andean, Caribbean, Orinoquia, Pacific **Dept.:** ANT, ATL, BOG, BOL, BOY, CAQ, CAS, CAU, CES, CHO, COR, CUN, HUI, LAG, MAG, MET, PUT, QUI, SAP, SAN, SUC, TOL, VAC, VAU **Elev.:** 0–2500 m a.s.l. **IUCN Habitat code:** 1,2,3,4,5,14 **Conservation status: (IUCN)** NE

1431 *Smilax tomentosa* Kunth (Smilacaceae) **IPNI:** 541909-1 **Common name:** adorote, atacorral, bejuco de uñegato, bejuco espuelón, china, guayacana, uña de gato, zarzaparrilla de páramo **Habit:** Liana, Climbing **Use:** MA **Origin:** Native **Region:** Andean, Guiana Shield **Dept.:** ANT, BOG, BOY, CAL, CAQ, CAU, CHO, CUN, HUI, MET, NAR, NSA, PUT, QUI, RIS, SAN, TOL, VAC **Elev.:** 1700-3680 m.a.s.l. **IUCN Habitat code:** 1,3,4,14 **Conservation status: (IUCN)** NE **(CPSG)** Potential LC

Pandanales

1432 *Asplundia alata* Harling (Cyclanthaceae) **IPNI:** 22877-2 **Habit:** Herb, Hemiepiphyte **Use:** EU **Origin:** Native **Region:** Amazonia, Pacific **Dept.:** AMA, ANT, CHO, PUT **Elev.:** 0–1100 m a.s.l. **IUCN Habitat code:** 1,3 **Conservation status: (IUCN)** NE

1433 *Asplundia moritziana* (Klotzsch) Harling (Cyclanthaceae) **IPNI:** 22935-2 **Common name:** cairasco **Habit:** Herb, Hemiepiphyte **Use:** MA **Origin:** Native **Region:** Andean **Dept.:** CAL, CUN, PUT, SAN **Elev.:** 330–2500 m a.s.l. **IUCN Habitat code:** 1,3 **Conservation status: (IUCN)** NE

1434 *Asplundia sarmentosa* Galeano & R.Bernal (Cyclanthaceae) **IPNI:** 272788-2 **Common name:** cestillo, cestillo hembra **Habit:** Herb, Hemiepiphyte, Climbing **Use:** HF, MA **Origin:** Endemic **Region:** Andean **Dept.:** ANT, QUI, RIS **Elev.:** 2140–2440 m a.s.l. **IUCN Habitat code:** 1,3,4,14 **Conservation status: (IUCN)** NE

1435 *Asplundia urophylla* Harling (Cyclanthaceae) **IPNI:** 22966-2 **Common name:** cestillo **Habit:** Herb, Hemiepiphyte, Climbing **Use:** MA **Origin:** Endemic **Region:** Andean **Dept.:** ANT, VAC **Elev.:** 1900–2100 m a.s.l. **IUCN Habitat code:** 1,14 **Conservation status: (IUCN)** NE

1436 *Asplundia utilis* (Oerst.) Harling (Cyclanthaceae) **IPNI:** 22967-2 **Habit:** Herb, Hemiepiphyte **Use:** HF **Origin:** Native **Region:** Andean, Pacific **Dept.:** ANT, CHO, VAC **Elev.:** 0–2330 m a.s.l. **IUCN Habitat code:** 1,14 **Conservation status: (IUCN)** NE

1437 *Asplundia xiphophylla* Harling (Cyclanthaceae) **IPNI:** 297504-1 **Common name:** bejuco de lagartija, sal de lagartija **Habit:** Herb, Hemiepiphyte **Use:** HF, ME **Origin:** Native **Region:** Amazonia **Dept.:** AMA, VAU **Elev.:** 150–250 m a.s.l. **IUCN Habitat code:** 1,5 **Conservation status: (IUCN)** NE

1438 *Carludovica drudei* Mast. (Cyclanthaccae) **IPNI:** 297529-1 **Habit:** Herb **Use:** HF, ME **Origin:** Native **Region:** Andean, Pacific **Dept.:** ANT, CHO, NAR **Elev.:** 0–1000 m a.s.l. **IUCN Habitat code:** 1,5 **Conservation status: (IUCN)** NE

1439 *Carludovica palmata* Ruiz & Pav. (Cyclanthaceae) **IPNI:** 297578-1 **Common name:** bacuma, caraná, cestillo, chulquín, cubeo, dócoro, horquetilla, iraca, jipijapa, jiraca, joropo, lucaica, lucaina, lucateba, lucua, macuma, mismiche, murraco, murrapa, murrapo, nacuma, naguala, napa, paja, palma iraca, palmicha, palmiche, palmicho, panga, **Habit:** Herb **Use:** EU, GS, HF, MA, ME, SU **Origin:** Native **Region:** Amazonia, Andean, Guiana Shield, Caribbean, Orinoquia, Pacific **Dept.:** AMA, ANT, ARA, ATL, BOL, BOY, CAL, CAQ, CAS, CAU, CES, CHO, COR, CUN, HUI, LAG, MAG, MET, NAR, NSA, PUT, QUI, RIS, SAN, SUC, TOL, VAC, VAU **Elev.:** 0-1300 m.a.s.l. **IUCN Habitat code:** 1,2,3,5,14 **Conservation status: (IUCN)** LC **(CPSG)** Potential LC

1440 *Cyclanthus bipartitus* Poit. ex A.Rich. (Cyclanthaceae) **IPNI:** 30336798-2 **Common name:** calzoncillo, coca, cola de pato, colepato, contraconga, hoja de lapa, horqueta, mamira, mella, melliza, misiraca, pángala, rajarraja, sordera, tapamuerto, tijereta, tornillo, totora, yoyo **Habit:** Herb **Use:** EU, HF, MA, ME **Origin:** Native **Region:** Amazonia, Andean, Guiana Shield, Caribbean, Orinoquia, Pacific **Dept.:** AMA, ANT, BOL, CAL, CAQ, CAU, CHO, COR, CUN, GUA, GUV, HUI, LAG, MAG, MET, NAR, NSA, PUT, QUI, RIS, SAN, TOL, VAC, VAU **Elev.:** 10-2020 m.a.s.l. **IUCN Habitat code:** 1,3,4,5,14 **Conservation status: (IUCN)** NE **(CPSG)** Potential LC

1441 *Dicranopygium grandifolium* Harling (Cyclanthaceae) **IPNI:** 79912-2 **Common name:** palmiche **Habit:** Herb **Use:** ME **Origin:** Native **Region:** Andean, Pacific **Dept.:** CHO, RIS, VAC **Elev.:** 0–1000 m a.s.l. **IUCN Habitat code:** 1,14 **Conservation status: (IUCN)** NE

1442 *Dicranopygium yacu-sisa* Harling (Cyclanthaceae) **IPNI:** 79945-2 **Use:** ME **Origin:** Native **Conservation status: (IUCN)** NE

1443 *Evodianthus funifer* (Poit.) Lindm. (Cyclanthaceae) **IPNI:** 102548-2 **Common name:** ancla, bejuco cestillo, bejuco potré, chopuré, chópure, contrapito, sajalagua, yaré **Habit:** Herb, Hemiepiphyte, Climbing **Use:** MA **Origin:** Native **Region:** Amazonia, Andean, Caribbean, Orinoquia, Pacific **Dept.:** AMA, ANT, CAQ, CAU, CHO, COR, CUN, NAR, PUT, RIS, SAN, VAC, VAU **Elev.:** 0-1500 m.a.s.l. **IUCN Habitat code:** 1,3,5,14 **Conservation status: (IUCN)** NE **(CPSG)** Potential LC

1444 *Ludovia lancifolia* Brongn. (Cyclanthaceae) **IPNI:** 297687-1 **Common name:** lagarto **Habit:** Herb, Hemiepiphyte **Use:** ME **Origin:** Native **Region:** Amazonia **Dept.:** AMA, CAQ, VAU **Elev.:** 100–500 m a.s.l. **Conservation status: (IUCN)** NE

1445 *Sphaeradenia crocea* Harling (Cyclanthaceae) **IPNI:** 241558-2 **Common name:** palmiche **Habit:** Herb, Epiphyte **Use:** ME **Origin:** Native **Region:** Andean, Pacific **Dept.:** CAU, CHO, VAC **Elev.:** 0–1100 m a.s.l. **IUCN Habitat code:** 1,5 **Conservation status: (IUCN)** NE

1446 *Sphaeradenia garciae* Harling (Cyclanthaceae) **IPNI:** 241565-2 **Habit:** Herb, Hemiepiphyte **Use:** EU, MA, ME **Origin:** Native **Region:** Andean **Dept.:** ANT, CHO, RIS, VAC **Elev.:** 600–2000 m a.s.l. **IUCN Habitat code:** 1,3,14 **Conservation status: (IUCN)** NE

1447 *Thoracocarpus bissectus* (Vell.) Harling (Cyclanthaceae) **IPNI:** 252718-2 **Common name:** ancla, bejuco cestillo, bejuco potré, chópure, cuangata, potré, puño huasca, yaré **Habit:** Herb, Hemiepiphyte, Climbing **Use:** HF, MA **Origin:** Native **Region:** Amazonia, Andean, Guiana Shield, Orinoquia, Pacific **Dept.:** AMA, ANT, CAQ, CAU, CHO, MAG, MET, NAR, PUT, RIS, SAN, VAC, VID **Elev.:** 10-1450 m.a.s.l. **IUCN Habitat code:** 1,3,4,5,14 **Conservation status: (IUCN)** NE **(CPSG)** Potential LC

1448 *Pandanus tectorius* Parkinson ex Du Roi (Pandanaceae) **IPNI:** 895770-1 **Habit:** Tree **Use:** AF, EU, HF, MA, ME, SU **Origin:** Cultivated **Region:** Andean, Caribbean, Pacific **Dept.:** ANT, CAU, QUI, SAP **Elev.:** 0–1500 m a.s.l. **Conservation status: (IUCN)** LC

1449 *Sciaphila purpurea* Benth. (Triuridaceae) **IPNI:** 836332-1 **Common name:** levantapalos, yerba del comején **Habit:** Herb, Saprophyte **Use:** ME **Origin:** Native **Region:** Amazonia, Andean, Guiana Shield, Pacific **Dept.:** AMA, ANT, CAL, SAN, VAC, VAU **Elev.:** 50–1354 m a.s.l. **IUCN Habitat code:** 1,3,5,14 **Conservation status: (IUCN)** NE

Poales

1450 *Aechmea bracteata* (Sw.) Griseb. (Bromeliaceae) **IPNI:** 121570-1 **Habit:** Herb, Epiphyte **Use:** HF **Origin:** Native **Region:** Caribbean **Dept.:** BOL, MAG, SAN **Elev.:** 100-420 m.a.s.l. **Conservation status: (IUCN)** NE **(CPSG)** LC

1451 *Aechmea bromeliifolia* (Rudge) Baker ex Benth. & Hook.f. (Bromeliaceae) **IPNI:** 121574-1 **Habit:** Herb, Epiphyte **Use:** HF **Origin:** Native **Region:** Andean, Guiana Shield **Dept.:** MAG, MET, NSA, VAU **Elev.:** 150-1400 m.a.s.l. **IUCN Habitat code:** 1 **Conservation status: (IUCN)** NE **(CPSG)** LC

1452 *Aechmea chantinii* (Carrière) Baker (Bromeliaceae) **IPNI:** 5385-2 **Habit:** Herb, Epiphyte, Terrestrial **Use:** EU **Origin:** Native **Region:** Andean, Guiana Shield **Dept.:** AMA, CAQ, PUT, VAU **Elev.:** 200-660 m.a.s.l. **IUCN Habitat code:** 1,2,3,14 **Conservation status: (IUCN)** NE **(CPSG)** LC

1453 *Aechmea corymbosa* (Mart. ex Schult. & Schult.f.) Mez (Bromeliaceae) **IPNI:** 5405-2 **Common name:** sacha piña **Habit:** Herb, Epiphyte **Use:** HF **Origin:** Native **Region:** Amazonia, Guiana Shield **Dept.:** AMA, CAQ, GUA, GUV, PUT, VAU **Elev.:** 66-600 m.a.s.l. **IUCN Habitat code:** 1,3,4,5,14 **Conservation status: (IUCN)** NE **(CPSG)** LC

1454 *Aechmea hoppii* (Harms) L.B.Sm. (Bromeliaceae) **IPNI:** 5475-2 **Habit:** Herb **Use:** HF **Origin:** Native **Region:** Amazonia, Andean, Macarena, Guayana **Dept.:** AMA, ANT, BOY, CAL, CAQ, CAS, CAU, CHO, COR, CUN, GUA, MAG, MET, NAR, NSA, PUT, SAN, TOL, VAC, VAU, VID **Elev.:** 92 – 815 m a.s.l. **Conservation status: (IUCN)** NE

1455 *Aechmea magdalenae* (André) André ex Baker (Bromeliaceae) **IPNI:** 5516-2 **Common name:** pita, pita floja, piña silvestre, piñuela, piñuela silvestre **Habit:** Herb **Use:** EU, HF, MA, ME, SU **Origin:** Native **Region:** Andean, Caribbean, Pacific **Dept.:** ANT, BOL, CHO, CUN, MAG, QUI, TOL, VAC **Elev.:** 0-1400 m.a.s.l. **IUCN Habitat code:** 1,3,5,14 **Conservation status: (IUCN)** NE **(CPSG)** LC

1456 *Aechmea mexicana* Baker (Bromeliaceae) **IPNI:** 271722-2 **Habit:** Herb, Epiphyte **Use:** HF **Origin:** Native **Region:** Andean **Dept.:** ANT, BOY, CHO, CUN, VAU **Elev.:** 300-1265 m.a.s.l. **IUCN Habitat code:** 1,3,14 **Conservation status: (IUCN)** NE **(CPSG)** LC

1457 *Aechmea nudicaulis* (L.) Griseb. (Bromeliaceae) **IPNI:** 121752-1 **Habit:** Herb, Epiphyte **Use:** HF **Origin:** Native **Region:** Pacific **Dept.:** ANT, NAR **Elev.:** 50-800 m.a.s.l. **IUCN Habitat code:** 1,5 **Conservation status: (IUCN)** NE **(CPSG)** LC

1458 *Aechmea penduliflora* André (Bromeliaceae) **IPNI:** 5555-2 **Habit:** Herb, Epiphyte **Use:** HF **Origin:** Native **Region:** Amazonia, Andean, Guiana Shield **Dept.:** AMA, ANT, BOY, CAQ, CAU, CHO, CUN, GUA, MET, NAR, NSA, PUT, VAU **Elev.:** 97-1440 m.a.s.l. **IUCN Habitat code:** 1,3,5,14 **Conservation status: (IUCN)** NE **(CPSG)** LC

1459 *Aechmea rubiginosa* Mez (Bromeliaceae) **IPNI:** 271755-2 **Common name:** piña de monte, piñuela **Habit:** Herb **Use:** HF **Origin:** Native **Region:** Amazonia, Andean, Guiana Shield, Orinoquia **Dept.:** AMA, CAQ, GUV, MET, VAU, VID **Elev.:** 100-1350 m.a.s.l. **IUCN Habitat code:** 1,3,4,5,14 **Conservation status: (IUCN)** NE **(CPSG)** LC

1460 *Aechmea tillandsioides* (Mart. ex Schult. & Schult.f.) Baker (Bromeliaceae) **IPNI:** 121847-1 **Habit:** Herb, Epiphyte **Use:** HF **Origin:** Native **Region:** Amazonia, Andean, Guiana Shield, Orinoquia, Pacific **Dept.:** AMA, ANT, BOY, CAL, CAQ, CAU, CHO, COR, GUA, GUV, MET, NSA, PUT, SAN, TOL, VAU **Elev.:** 0-1100 m.a.s.l. **IUCN Habitat code:** 1,3,5,14 **Conservation status: (IUCN)** NE **(CPSG)** LC

1461 *Aechmea zebrina* L.B.Sm. (Bromeliaceae) **IPNI:** 271782-2 **Habit:** Herb, Epiphyte **Use:** EU **Origin:** Native **Region:** Amazonia **Dept.:** PUT **Elev.:** 200-350 m.a.s.l. **Conservation status: (IUCN)** NE **(CPSG)** EN

1462 *Ananas comosus* (L.) Merr. (Bromeliaceae) **IPNI:** 12322-2 **Common name:** cóijibo, piña, piña banqueña, piña brava, piña de agua, piña de borugo, piña de charapa, piña de chucha, piña de cumare, piña de cuzumbo, piña de danta, piña de guacamayo, piña de iguana, piña de monte, piña de omima, piña de panguana, piña de paujil, pi **Habit:** Herb **Use:** EU, GS, HF, MA, ME **Origin:** Native, Cultivated **Region:** Amazonia, Andean, Guiana Shield, Caribbean, Orinoquia, Pacific **Dept.:** AMA, ANT, ARA, ATL, BOL, BOY, CAL, CAQ, CAS, CAU, CES, CHO, COR, CUN, GUA, GUV, HUI, LAG, MAG, MET, NAR, NSA, PUT, QUI, RIS, SAP, SAN, SUC, TOL, VAC, VAU, VID **Elev.:** 0–2000 m a.s.l. **IUCN Habitat code:** 1,2,3,4,5,14 **Conservation status: (IUCN)** NE

1463 *Bromelia balansae* Mez (Bromeliaceae) **IPNI:** 36836-2 **Habit:** Herb **Use:** EU, HF, MA, ME **Origin:** Native **Region:** Guiana Shield, Orinoquia **Dept.:** ARA, CAS, GUV, MET, VID **Elev.:** 69-520 m.a.s.l. **IUCN Habitat code:** 1,3,4,14 **Conservation status: (IUCN)** NE **(CPSG)** LC

1464 *Bromelia chrysantha* Jacq. (Bromeliaceae) **IPNI:** 122289-1 **Common name:** chivichivi, mayita, mayita chivichivi, piñuela, piñuela de tierra, piñuelo, tukupe **Habit:** Herb **Use:** AF, HF, MA, ME **Origin:** Native **Region:** Caribbean, Pacific **Dept.:** ATL, BOL, CES, CHO, LAG, MAG, NSA, SAN **Elev.:** 0-1005 m.a.s.l. **IUCN Habitat code:** 1,2,4,14 **Conservation status: (IUCN)** NE **(CPSG)** LC

1465 *Bromelia karatas* L. (Bromeliaceae) **IPNI:** 328361-2 **Common name:** chihichuine, curucujurro, curujujul, cusuca, dodake, dorokolle, piñuela **Habit:** Herb **Use:** EU, HF, MA, ME **Origin:** Native **Region:** Amazonia, Andean, Caribbean **Dept.:** ANT, CAQ, CAU, COR, HUI, MAG, NSA, PUT, SAN, TOL, VAC **Elev.:** 30-1800 m.a.s.l. **IUCN Habitat code:** 1,2,3,4,5,14 **Conservation status: (IUCN)** NE **(CPSG)** LC

1466 *Bromelia nidus-puellae* (André) André ex Mez (Bromeliaceae) **IPNI:** 122363-1 **Habit:** Herb **Use:** HF, ME **Origin:** Endemic **Dept.:** CUN, VAC **Elev.:** 400-950 m.a.s.l. **IUCN Habitat code:** 14 **Conservation status: (IUCN)** EN **(CPSG)** CR

1467 *Bromelia pinguin* L. (Bromeliaceae) **IPNI:** 319113-2 **Common name:** besa, chibichibe, chicorrón, chocorrón, curujul, cusuca, haya, maya, pinguin, piñuela, piñuelo **Habit:** Herb **Use:** HF, MA, ME **Origin:** Native **Region:** Andean, Caribbean, Orinoquia **Dept.:** ATL, BOL, CAS, CES, LAG, MAG, SAP **Elev.:** 0-1900 m.a.s.l. **IUCN Habitat code:** 1,2,5,14 **Conservation status: (IUCN)** NE **(CPSG)** LC

1468 *Bromelia trianae* Mez (Bromeliaceae) **IPNI:** 77177315-1 **Common name:** pita, piñuela, piñuelo **Habit:** Herb **Use:** HF, ME **Origin:** Endemic **Dept.:** CUN, TOL **Elev.:** 700-900 m.a.s.l. **Conservation status: (IUCN)** NE **(CPSG)** CR

1469 *Catopsis berteroniana* (Schult. & Schult.f.) Mez (Bromeliaceae) **IPNI:** 50112-2 **Habit:** Herb, Epiphyte **Use:** EU **Origin:** Native **Region:** Pacific **Dept.:** ANT, CHO **Elev.:** 100-610 m.a.s.l. **IUCN Habitat code:** 1,3 **Conservation status: (IUCN)** NE **(CPSG)** LC

1470 *Greigia collina* L.B.Sm. (Bromeliaceae) **IPNI:** 113406-2 **Habit:** Herb **Use:** HF **Origin:** Endemic **Region:** Andean **Dept.:** CUN, SAN **Elev.:** 2400-3250 m.a.s.l. **IUCN Habitat code:** 1,4,14 **Conservation status: (IUCN)** VU **(CPSG)** VU

1471 *Greigia mulfordii* L.B.Sm. (Bromeliaceae) **IPNI:** 113413-2 **Habit:** Herb **Use:** HF **Origin:** Native **Region:** Andean **Dept.:** BOG, CUN, PUT **Elev.:** 3450-3670 m.a.s.l. **IUCN Habitat code:** 1,3,4 **Conservation status: (IUCN)** LC **(CPSG)** DD

1472 *Greigia sodiroana* Mez (Bromeliaceae) **IPNI:** 113421-2 **Habit:** Herb **Use:** HF **Origin:** Native **Region:** Andean **Dept.:** BOG, NAR **Elev.:** 3050-3050 m.a.s.l. **IUCN Habitat code:** 1 **Conservation status: (IUCN)** LC **(CPSG)** LC

1473 *Greigia stenolepis* L.B.Sm. (Bromeliaceae) **IPNI:** 113423-2 **Common name:** cardón, piñuela **Habit:** Herb **Use:** EU, HF **Origin:** Native **Region:** Andean **Dept.:** BOG, BOY, CUN, NSA **Elev.:** 2500-3720 m.a.s.l. **IUCN Habitat code:** 1,3,4,14 **Conservation status: (IUCN)** LC **(CPSG)** LC

1474 *Greigia vulcanica* André (Bromeliaceae) **IPNI:** 113426-2 **Common name:** piñuela **Habit:** Herb **Use:** HF **Origin:** Native **Region:** Andean **Dept.:** ANT, CAL, NAR, PUT, QUI, TOL, VAC **Elev.:** 3100-3600 m.a.s.l. **IUCN Habitat code:** 1,4,14 **Conservation status: (IUCN)** LC **(CPSG)** LC

1475 *Guzmania lingulata* (L.) Mez (Bromeliaceae) **IPNI:** 115161-2 **Common name:** cardo, chupaila **Habit:** Herb, Epiphyte, Terrestrial **Use:** EU **Origin:** Native, Cultivated **Region:** Amazonia, Andean, Guiana Shield, Caribbean, Pacific **Dept.:** AMA, ANT, BOY, CAL, CAQ, CES, CHO, COR, CUN, GUA, GUV, LAG, MAG, MET, NAR, NSA, QUI, RIS, SAN, VAC, VAU **Elev.:** 0–1650 m a.s.l. **IUCN Habitat code:** 1,2,3,5,14 **Conservation status: (IUCN)** NE

1476 *Guzmania monostachia* (L.) Rusby ex Mez (Bromeliaceae) **IPNI:** 115178-2 **Common name:** quiche **Habit:** Herb, Epiphyte, Terrestrial **Use:** EU **Origin:** Native **Region:** Amazonia, Andean, Caribbean, Orinoquia, Pacific **Dept.:** ANT, BOY, CAL, CAQ, CAS, CAU, CES, CHO, CUN, HUI, LAG, MAG, MET, NAR, NSA, QUI, SAN, TOL, VAC **Elev.:** 0-1800 m.a.s.l. **IUCN Habitat code:** 1,3,14 **Conservation status:** **(IUCN)** NE **(CPSG)** LC

1477 *Guzmania musaica* (Linden & André) Mez (Bromeliaceae) **IPNI:** 115185-2 **Common name:** chupalla, guiche, quiche **Habit:** Herb, Epiphyte, Terrestrial **Use:** EU **Origin:** Native **Region:** Andean, Pacific **Dept.:** ANT, CAU, CHO, NAR, NSA, VAC **Elev.:** 0-1200 m.a.s.l. **IUCN Habitat code:** 1,3,5 **Conservation status:** **(IUCN)** NE **(CPSG)** LC

1478 *Guzmania sanguinea* (André) André ex Mez (Bromeliaceae) **IPNI:** 115224-2 **Common name:** chupalla, quiche, quiche morado **Habit:** Herb, Epiphyte, Terrestrial **Use:** ME **Origin:** Native **Region:** Caribbean, Pacific **Dept.:** BOG, CHO, LAG, NAR, VAC **Elev.:** 100-650 m.a.s.l. **Conservation status:** **(IUCN)** NE **(CPSG)** LC

1479 *Lutheria splendens* (Brongn.) Barfuss & W.Till (Bromeliaceae) **IPNI:** 77158988-1 **Common name:** cardo, cardos, quiche, quiche morado **Habit:** Herb **Use:** EU, ME **Origin:** Native, Cultivated **Region:** Caribbean **Dept.:** BOG, LAG **Elev.:** 550–750 m a.s.l. **Conservation status:** **(IUCN)** NE

1480 *Neoregelia carolinae* (Beer) L.B.Sm. (Bromeliaceae) **IPNI:** 169003-2 **Use:** EU **Region:** Andean **Dept.:** BOG **Conservation status:** **(IUCN)** NE

1481 *Pitcairnia andreana* Linden (Bromeliaceae) **IPNI:** 123813-1 **Habit:** Herb **Use:** EU **Origin:** Endemic **Dept.:** CHO **IUCN Habitat code:** 1 **Conservation status:** **(IUCN)** NE **(CPSG)** DD

1482 *Pitcairnia heterophylla* (Lindl.) Beer (Bromeliaceae) **IPNI:** 123989-1 **Habit:** Herb **Use:** EU **Origin:** Native **Region:** Andean **Dept.:** ANT, BOY, CAU, CUN, MAG, NSA **Elev.:** 1000-1800 m.a.s.l. **IUCN Habitat code:** 1,2,14 **Conservation status:** **(IUCN)** NE **(CPSG)** LC

1483 *Pitcairnia pungens* Kunth (Bromeliaceae) **IPNI:** 124143-1 **Common name:** cardo **Habit:** Herb **Use:** ME **Origin:** Native **Region:** Andean **Dept.:** BOG, BOY, CAU, CUN, HUI, NAR, QUI, RIS, VAC **Elev.:** 1800-3000 m.a.s.l. **IUCN Habitat code:** 1,14 **Conservation status:** **(IUCN)** LC **(CPSG)** LC

1484 *Puya goudotiana* Mez (Bromeliaceae) **IPNI:** 214589-2 **Common name:** cardón, puya **Habit:** Herb **Use:** EU **Origin:** Endemic **Region:** Andean **Dept.:** BOG, BOY, CUN, NSA, SAN **Elev.:** 2760-3550 m.a.s.l. **IUCN Habitat code:** 1,2,3,4,8,14 **Conservation status:** **(IUCN)** LC **(CPSG)** LC

1485 *Puya hamata* L.B.Sm. (Bromeliaceae) **IPNI:** 214597-2 **Common name:** cardo, cardo de páramo, puya **Habit:** Herb **Use:** HF **Origin:** Native **Region:** Andean **Dept.:** CAU, NAR, PUT, TOL, VAC **Elev.:** 2500-3750 m.a.s.l. **IUCN Habitat code:** 1,4,14 **Conservation status:** **(IUCN)** LC **(CPSG)** LC

1486 *Puya santosii* Cuatrec. (Bromeliaceae) **IPNI:** 214682-2 **Common name:** cardón **Habit:** Herb **Use:** AF **Origin:** Endemic **Region:** Andean **Dept.:** ARA, BOG, BOY, CUN, MET **Elev.:** 2800-4285 m.a.s.l. **IUCN Habitat code:** 1,3,4,5,8,14 **Conservation status:** **(IUCN)** LC **(CPSG)** LC

1487 *Puya trianae* Baker (Bromeliaceae) **IPNI:** 214698-2 **Common name:** cardón, puya **Habit:** Herb **Use:** AF, EU **Origin:** Native **Region:** Andean **Dept.:** ANT, ARA, BOG, BOY, CUN, QUI, SAN, TOL, VAC **Elev.:** 3000-4100 m.a.s.l. **IUCN Habitat code:** 1,3,4,14 **Conservation status:** **(IUCN)** LC **(CPSG)** LC

1488 *Racinaea tetrantha* (Ruiz & Pav.) M.A.Spencer & L.B.Sm. (Bromeliaceae) **IPNI:** 970621-1 **Habit:** Herb, Epiphyte **Use:** EU **Origin:** Native **Region:** Andean **Dept.:** ANT, BOG, BOY, CAL, CAU, CES, CHO, CUN, HUI, MAG, NAR, NSA, PUT, QUI, RIS, SAN, TOL, VAC **Elev.:** 1280-3900 m.a.s.l. **IUCN Habitat code:** 1,3,4,8,14 **Conservation status:** **(IUCN)** NE **(CPSG)** LC

1489 *Tillandsia complanata* Benth. (Bromeliaceae) **IPNI:** 125000-1 **Habit:** Herb, Epiphyte, Terrestrial **Use:** HF **Origin:** Native **Region:** Andean **Dept.:** ANT, BOG, BOY, CAL, CAU, CES, CUN, HUI, LAG, MAG, NAR, NSA, PUT, QUI, RIS, SAN, TOL, VAC **Elev.:** 800-3650 m.a.s.l. **IUCN Habitat code:** 1,3,4,14 **Conservation status:** **(IUCN)** NE **(CPSG)** LC

1490 *Tillandsia incarnata* Kunth (Bromeliaceae) **IPNI:** 253931-2 **Habit:** Herb, Epiphyte, Terrestrial **Use:** EU **Origin:** Native **Region:** Andean **Dept.:** BOG, BOY, CUN, NAR, NSA, SAN **Elev.:** 760-3100 m.a.s.l. **IUCN Habitat code:** 1,3,4,5,14 **Conservation status:** **(IUCN)** NE **(CPSG)** LC

1491 *Tillandsia recurvata* (L.) L. (Bromeliaceae) **IPNI:** 1194855-2 **Common name:** guincho, guinchos **Habit:** Herb, Epiphyte, Terrestrial **Use:** EU, HF, MA, ME **Origin:** Native **Region:** Andean, Guiana Shield, Caribbean, Orinoquia **Dept.:** ANT, ARA, BOG, BOY, CAL, CAQ, CAS, CAU, CES, CUN, HUI, LAG, MAG, MET, NAR, NSA, QUI, SAN, VAC, VID **Elev.:** 50-2700 m.a.s.l. **IUCN Habitat code:** 1,2,3,4,14 **Conservation status:** **(IUCN)** NE **(CPSG)** LC

1492 *Tillandsia rubella* Baker (Bromeliaceae) **IPNI:** 254243-2 **Habit:** Herb, Epiphyte **Use:** HF **Origin:** Native **Region:** Andean **Dept.:** NAR **Elev.:** 3011–3280 m a.s.l. **Conservation status:** **(IUCN)** NE

1493 *Tillandsia tenuifolia* L. (Bromeliaceae) **IPNI:** 125713-1 **Habit:** Herb, Epiphyte **Use:** EU **Origin:** Native **Region:** Andean, Caribbean **Dept.:** CES, MAG, NSA **Elev.:** 350–2500 m a.s.l. **IUCN Habitat code:** 1,2 **Conservation status:** **(IUCN)** NE

1494 *Tillandsia turneri* Baker (Bromeliaceae) **IPNI:** 254385-2 **Habit:** Herb, Epiphyte **Use:** EU **Origin:** Native **Region:** Andean **Dept.:** BOG, BOY, CUN, NSA, SAN **Elev.:** 2600-3600 m.a.s.l. **IUCN Habitat code:** 1,3,4,14 **Conservation status:** **(IUCN)** NE **(CPSG)** LC

1495 *Tillandsia usneoides* (L.) L. (Bromeliaceae) **IPNI:** 1169665-2 **Common name:** barba de palo, barba de viejo, barbas, barbas de palo, barbas de viejo, barbechivo, cuarque, lana, melena, musgo, musgo blanco, salvaje, salvajina, yerbarrucia **Habit:** Herb, Epiphyte, Terrestrial **Use:** EU, MA, ME, SU **Origin:** Native **Region:** Andean, Guiana Shield, Caribbean, Pacific **Dept.:** ANT, ATL, BOG, BOL, BOY, CAL, CAU, CES, CUN, HUI, LAG, MAG, MET, NAR, NSA, QUI, SAN, SUC, TOL, VAC **Elev.:** 15-2850 m.a.s.l. **IUCN Habitat code:** 1,2,3,5,14 **Conservation status:** **(IUCN)** LC **(CPSG)** LC

1496 *Tillandsia utriculata* L. (Bromeliaceae) **IPNI:** 254406-2 **Common name:** gallogiro **Use:** ME **IUCN Habitat code:** 1,2,14 **Conservation status:** **(IUCN)** NE

1497 *Bolboschoenus maritimus* (L.) Palla (Cyperaceae) **IPNI:** 298094-1 **Habit:** Herb, Aquatic **Use:** EU, GS, HF, MA, ME, PO **Origin:** Native **Region:** Caribbean, Pacific **Dept.:** ANT, SUC **Elev.:** 0–0 m a.s.l. **IUCN Habitat code:** 1 **Conservation status:** **(IUCN)** LC

1498 *Bulbostylis capillaris* (L.) Kunth ex C.B.Clarke (Cyperaceae) **IPNI:** 1186511-2 **Habit:** Herb **Use:** ME **Origin:** Native **Region:** Amazonia, Andean, Guiana Shield, Caribbean **Dept.:** ANT, CES, GUV, MAG, MET, VAU **Elev.:** 30–2180 m a.s.l. **IUCN Habitat code:** 1,2,3,4,5,14 **Conservation status:** **(IUCN)** LC

1499 *Bulbostylis hispidula* (Vahl) R.W.Haines (Cyperaceae) **IPNI:** 912172-1 **Use:** AF, HF, MA, ME **Origin:** Native **Dept.:** CES **Conservation status:** **(IUCN)** LC

1500 *Bulbostylis junciformis* (Kunth) C.B.Clarke (Cyperaceae) **IPNI:** 38049-2 **Habit:** Herb **Use:** ME **Origin:** Native **Region:** Andean, Guiana Shield, Orinoquia **Dept.:** ANT, BOY, CAQ, CAS, CUN, GUV, MAG, MET, SAN, VAU, VID **Elev.:** 100–2800 m a.s.l. **IUCN Habitat code:** 1,2,3,4,14 **Conservation status:** **(IUCN)** NE

1501 *Bulbostylis lanata* (Kunth) Lindm. (Cyperaceae) **IPNI:** 30060098-2 **Common name:** cabeza de negro **Habit:** Herb **Use:** ME **Origin:** Native **Region:** Amazonia, Guiana Shield, Orinoquia **Dept.:** AMA, GUA, MET, VAU, VID **Elev.:** 80–

300 m a.s.l. **IUCN Habitat code:** 1,2,3,4,5,14 **Conservation status: (IUCN)** NE

1502 *Carex hamata* Sw. (Cyperaceae) **IPNI:** 46311-2 **Common name:** cortadera **Habit:** Herb **Use:** ME **Origin:** Native **Region:** Andean **Dept.:** ANT, BOG, BOY, CAL, CAU, CES, CHO, CUN, HUI, LAG, MAG, NAR, PUT, QUI, RIS, SAN, TOL, VAC **Elev.:** 910-3650 m.a.s.l. **IUCN Habitat code:** 1,3,4,14 **Conservation status: (IUCN)** NE **(CPSG)** Potential LC

1503 *Carex lasiocarpa* Ehrh. (Cyperaceae) **IPNI:** 30041747-2 **Use:** ME **Region:** Andean **Dept.:** BOG **Conservation status: (IUCN)** LC

1504 *Carex pichinchensis* Kunth (Cyperaceae) **IPNI:** 301577-1 **Common name:** cortadera, paja **Habit:** Herb **Use:** EU **Origin:** Native **Region:** Andean **Dept.:** ANT, ARA, BOG, BOY, CAL, CAU, CES, CUN, MAG, MET, NAR, NSA, QUI, TOL **Elev.:** 1325-4500 m.a.s.l. **IUCN Habitat code:** 1,2,3,4,5,6,8,14 **Conservation status: (IUCN)** NE **(CPSG)** Potential LC

1505 *Cladium mariscus* (L.) Pohl (Cyperaceae) **IPNI:** 303489-1 **Common name:** cortadera **Habit:** Herb **Use:** EU, FU, HF, MA, ME **Origin:** Native **Region:** Pacific **Dept.:** CHO, VAC **Elev.:** 0–500 m a.s.l. **IUCN Habitat code:** 3 **Conservation status: (IUCN)** LC

1506 *Cyperus aggregatus* (Willd.) Endl. (Cyperaceae) **IPNI:** 303704-1 **Common name:** acaneboto, cresta de gallo, pasto víbora **Habit:** Herb, Shrub **Use:** ME **Origin:** Native **Region:** Andean, Guiana Shield, Caribbean, Orinoquia, Pacific **Dept.:** ANT, ARA, BOG, BOY, CAQ, CHO, CUN, GUA, LAG, MET, NAR, SAN, VAC, VID **Elev.:** 0-1000 m.a.s.l. **IUCN Habitat code:** 1,2,3,4,14 **Conservation status: (IUCN)** NE **(CPSG)** Potential LC

1507 *Cyperus alternifolius* L. (Cyperaceae) **IPNI:** 303729-1 **Common name:** coquito, cortadera **Habit:** Herb, Aquatic **Use:** AF, EU, HF, MA, ME, SU **Origin:** Naturalised **Region:** Andean **Dept.:** ANT, CUN, GUA, TOL, VAC **Elev.:** 900–2900 m a.s.l. **IUCN Habitat code:** 1,3,14 **Conservation status: (IUCN)** LC

1508 *Cyperus amabilis* Vahl (Cyperaceae) **IPNI:** 303735-1 **Habit:** Herb **Use:** AF,ME **Origin:** Native **Region:** Amazonia, Andean, Caribbean, Orinoquia **Dept.:** AMA, CES, HUI, MAG, VID **Elev.:** 100-1130 m.a.s.l. **IUCN Habitat code:** 3,4 **Conservation status: (IUCN)** LC **(CPSG)** Potential LC

1509 *Cyperus articulatus* L. (Cyperaceae) **IPNI:** 303822-1 **Common name:** coco-chunduro, cocochunduro, herb de arroz, hierba de arroz, junco, paripiri **Habit:** Herb, Aquatic **Use:** EU, HF, MA, ME, PO, SU **Origin:** Native, Cultivated **Region:** Andean, Guiana Shield, Caribbean, Orinoquia **Dept.:** ANT, ARA, ATL, BOL, CAL, COR, GUA, LAG, MAG, MET, TOL, VAC, VAU **Elev.:** 6-2000 m.a.s.l. **IUCN Habitat code:** 1,2,3,5,14 **Conservation status: (IUCN)** LC **(CPSG)** Potential LC

1510 *Cyperus brevifolius* (Rottb.) Hassk. (Cyperaceae) **IPNI:** 30313989-2 **Common name:** espadilla, fosforito **Habit:** Herb, Climbing **Use:** AF,ME **Origin:** Native **Region:** Andean, Caribbean, Orinoquia, Pacific **Dept.:** ANT, ARA, ATL, BOY, CAS, CAU, CHO, CUN, HUI, MAG, MET, NAR, PUT, VAC, VID **Elev.:** 0-2850 m.a.s.l. **IUCN Habitat code:** 1,2,3,4,14 **Conservation status: (IUCN)** NE **(CPSG)** Potential LC

1511 *Cyperus canus* J.Presl & C.Presl (Cyperaceae) **IPNI:** 304039-1 **Habit:** Herb **Use:** MA **Origin:** Native **Region:** Caribbean **Dept.:** ANT, MAG **Elev.:** 300–300 m a.s.l. **Conservation status: (IUCN)** NE

1512 *Cyperus chalaranthus* J.Presl & C.Presl (Cyperaceae) **IPNI:** 304087-1 **Habit:** Herb **Use:** ME **Origin:** Native **Region:** Andean, Pacific **Dept.:** ANT, CHO **Elev.:** 0–960 m a.s.l. **IUCN Habitat code:** 1,3 **Conservation status: (IUCN)** NE

1513 *Cyperus compressus* L. (Cyperaceae) **IPNI:** 304155-1 **Common name:** coquito, paja de pisco **Habit:** Herb **Use:** AF, EU, HF, MA, ME **Origin:** Native **Region:** Andean, Caribbean, Orinoquia **Dept.:** ANT, ARA, ATL, BOL, CAL, CAU, COR, CUN, MAG, MET, SAP, TOL, VID **Elev.:** 0-2700 m.a.s.l. **IUCN Habitat code:** 1,2,3,4,14 **Conservation status: (IUCN)** LC **(CPSG)** Potential LC

1514 *Cyperus cyperinus* (Retz.) Valck.Sur. (Cyperaceae) **IPNI:** 304256-1 **Habit:** Herb **Use:** ME **Origin:** Naturalised **Region:** Caribbean **Dept.:** BOG, LAG, MAG **Elev.:** 200–500 m a.s.l. **IUCN Habitat code:** 14 **Conservation status: (IUCN)** NE

1515 *Cyperus difformis* L. (Cyperaceae) **IPNI:** 304311-1 **Habit:** Herb **Use:** AF, GS, HF, ME **Origin:** Naturalised **Dept.:** VAU **Conservation status: (IUCN)** LC

1516 *Cyperus digitatus* Roxb. (Cyperaceae) **IPNI:** 304315-1 **Common name:** cortadera, piripiri **Habit:** Herb **Use:** MA, ME **Origin:** Native **Region:** Caribbean, Orinoquia, Pacific **Dept.:** ANT, ARA, ATL, CAU, COR, MAG, SAN, VAC **Elev.:** 5-950 m.a.s.l. **IUCN Habitat code:** 1,2,3,5,14 **Conservation status: (IUCN)** LC **(CPSG)** Potential LC

1517 *Cyperus distans* L.f. (Cyperaceae) **IPNI:** 301880-2 **Habit:** Herb **Use:** AF, HF, MA, ME **Origin:** Native **Region:** Andean, Orinoquia **Dept.:** SAN, VAC, VID **Elev.:** 970-1630 m.a.s.l. **IUCN Habitat code:** 1,2,3,4,5,14 **Conservation status: (IUCN)** LC **(CPSG)** Potential LC

1518 *Cyperus entrerianus* Boeckeler (Cyperaceae) **IPNI:** 74636-2 **Use:** ME **Origin:** Native **Conservation status: (IUCN)** NE

1519 *Cyperus eragrostis* Lam. (Cyperaceae) **IPNI:** 304426-1 **Use:** AF, EU, HF **Origin:** Native **IUCN Habitat code:** 4,14 **Conservation status: (IUCN)** LC **(CPSG)** Potential LC

1520 *Cyperus esculentus* L. (Cyperaceae) **IPNI:** 300668-2 **Common name:** chingo, chufa, coquito, corocillo, corocito **Habit:** Herb **Use:** AF, EU, FU, GS, HF, IF, MA, ME, SU **Origin:** Native **Region:** Andean, Caribbean, Orinoquia **Dept.:** ANT, ARA, ATL, BOG, CUN, HUI, LAG, MAG, MET, TOL, VAC **Elev.:** 230–1550 m a.s.l. **IUCN Habitat code:** 1,3,14 **Conservation status: (IUCN)** LC

1521 *Cyperus flavescens* L. (Cyperaceae) **IPNI:** 74698-2 **Habit:** Herb **Use:** ME **Origin:** Native **Region:** Andean **Dept.:** BOY, CHO, CUN, VAC **Elev.:** 1000–2820 m a.s.l. **Conservation status: (IUCN)** LC

1522 *Cyperus giganteus* Vahl (Cyperaceae) **IPNI:** 304595-1 **Habit:** Herb **Use:** EU, MA, ME **Origin:** Native **Region:** Caribbean, Pacific **Dept.:** ATL, CHO, COR, MAG **Elev.:** 6-8 m.a.s.l. **IUCN Habitat code:** 1,2,14 **Conservation status: (IUCN)** NE **(CPSG)** Potential LC

1523 *Cyperus haspan* L. (Cyperaceae) **IPNI:** 304701-1 **Common name:** chandur, cortadera, paja de pantano **Habit:** Herb, Aquatic **Use:** AF, EU, FU, HF, MA, ME **Origin:** Native **Region:** Andean, Guiana Shield, Orinoquia **Dept.:** ANT, BOG, BOL, BOY, CAU, CHO, CUN, GUA, HUI, MET, SAN, TOL, VAC, VAU, VID **Elev.:** 65-2551 m.a.s.l. **IUCN Habitat code:** 1,2,3,4,5,14 **Conservation status: (IUCN)** LC **(CPSG)** Potential LC

1524 *Cyperus hermaphroditus* (Jacq.) Standl. (Cyperaceae) **IPNI:** 74768-2 **Habit:** Herb **Use:** AF,ME **Origin:** Native **Region:** Andean, Caribbean, Orinoquia, Pacific **Dept.:** ANT, ATL, BOG, BOY, CHO, CUN, MET, NSA, PUT **Elev.:** 0–2900 m a.s.l. **IUCN Habitat code:** 1,2,3,14 **Conservation status: (IUCN)** NE

1525 *Cyperus hortensis* (Salzm. ex Steud.) Dorr (Cyperaceae) **IPNI:** 77144505-1 **Common name:** botoncillo, espadilla, fósforo **Habit:** Herb **Use:** HF, MA, ME **Origin:** Native **Region:** Amazonia, Andean, Caribbean, Orinoquia, Pacific **Dept.:** AMA, ANT, BOY, CAS, CHO, CUN, GUA, GUV, HUI, MAG, MET, NAR, NSA, PUT, RIS, SAN, TOL, VAC **Elev.:** 5–3000 m a.s.l. **IUCN Habitat code:** 1,2,3,5,14 **Conservation status: (IUCN)** LC

1526 *Cyperus imbricatus* Retz. (Cyperaceae) **IPNI:** 304793-1 **Habit:** Herb **Use:** AF, MA **Origin:** Native **Region:** Amazonia, Andean, Orinoquia **Dept.:** AMA, ARA, ATL, BOG, CAS, MET, VAC **Elev.:** 5-400 m.a.s.l. **Conservation status: (IUCN)** LC **(CPSG)** Potential LC

1527 *Cyperus iria* L. (Cyperaceae) **IPNI:** 304844-1 **Habit:** Herb **Use:** AF, MA, ME **Origin:** Naturalised **Region:** Caribbean, Orinoquia, Pacific **Dept.:** ARA, ATL,

COR, HUI, MET, NAR, NSA, TOL **Elev.:** 6–230 m a.s.l. **IUCN Habitat code:** 1,2,3,14 **Conservation status: (IUCN)** LC

1528 *Cyperus laevigatus* L. (Cyperaceae) **IPNI:** 304938-1 **Habit:** Herb **Use:** AF, HF, MA, ME **Origin:** Native **Region:** Pacific **Dept.:** VAC **Elev.:** 100–100 m a.s.l. **Conservation status: (IUCN)** LC

1529 *Cyperus laxus* Lam. (Cyperaceae) **IPNI:** 304973-1 **Common name:** chundul, coquito, cortadera, macano, paja cortadera **Habit:** Herb **Use:** EU, HF, ME **Origin:** Native **Region:** Amazonia, Andean, Guiana Shield, Caribbean, Orinoquia, Pacific **Dept.:** AMA, ANT, ARA, BOL, BOY, CAQ, CAS, CAU, CHO, COR, CUN, GUA, MAG, MET, NAR, PUT, SAN, TOL, VAC, VAU, VID **Elev.:** 0-2200 m.a.s.l. **IUCN Habitat code:** 1,2,3,4,5,14 **Conservation status: (IUCN)** NE **(CPSG)** Potential LC

1530 *Cyperus ligularis* L. (Cyperaceae) **IPNI:** 305019-1 **Common name:** cortacorta **Habit:** Herb **Use:** AF, EU, HF, MA, ME **Origin:** Native **Region:** Guiana Shield, Caribbean, Pacific **Dept.:** ANT, ATL, BOL, CHO, COR, HUI, LAG, MAG, MET, NAR, SAP, SAN, SUC, TOL, VAC, VAU **Elev.:** 0-460 m.a.s.l. **IUCN Habitat code:** 1,2,3,4,5,14 **Conservation status: (IUCN)** NE **(CPSG)** Potential LC

1531 *Cyperus longifolius* Poir. (Cyperaceae) **IPNI:** 305041-1 **Use:** MA **Conservation status: (IUCN)** NE

1532 *Cyperus lucidus* R.Br. (Cyperaceae) **IPNI:** 305060-1 **Use:** MA, SU **IUCN Habitat code:** 14 **Conservation status: (IUCN)** NE

1533 *Cyperus luzulae* (L.) Retz. (Cyperaceae) **IPNI:** 74851-2 **Common name:** cabezoncillo, chandul, chandur, chundul, chundún, cigarrilla, coquito, cortadera, espadillón, estrella, horquetilla, pasto estrella, piripiri **Habit:** Herb **Use:** EU, ME, SU **Origin:** Native **Region:** Amazonia, Andean, Guiana Shield, Caribbean, Orinoquia, Pacific **Dept.:** AMA, ANT, ARA, ATL, BOG, BOL, BOY, CAL, CAQ, CAS, CAU, CES, CHO, COR, CUN, GUA, GUV, HUI, MAG, MET, NAR, NSA, PUT, QUI, RIS, SAP, SAN, SUC, TOL, VAC, VAU, VID **Elev.:** 0-1500 m.a.s.l. **IUCN Habitat code:** 1,2,3,4,5,14 **Conservation status: (IUCN)** NE **(CPSG)** Potential LC

1534 *Cyperus mindorensis* (Steud.) Huygh (Cyperaceae) **IPNI:** 77140538-1 **Habit:** Herb **Use:** ME **Origin:** Naturalised **Region:** Andean **Dept.:** CUN **Elev.:** 300–2660 m a.s.l. **Conservation status: (IUCN)** NE

1535 *Cyperus niger* Ruiz & Pav. (Cyperaceae) **IPNI:** 74915-2 **Common name:** chundur de gente **Habit:** Herb **Use:** SU **Origin:** Native **Region:** Andean **Dept.:** ANT, BOG, BOY, CUN, HUI, NAR, NSA, PUT, QUI, RIS, SAN, VAC **Elev.:** 1300-3496 m.a.s.l. **IUCN Habitat code:** 1,2,3,8,14 **Conservation status: (IUCN)** NE **(CPSG)** Potential LC

1536 *Cyperus obtusatus* (J.Presl & C.Presl) Mattf. & Kük. (Cyperaceae) **IPNI:** 74934-2 **Habit:** Herb **Use:** MA, ME **Origin:** Native **Region:** Andean, Caribbean, Pacific **Dept.:** ANT, CAU, CHO, COR, MAG, NAR, SAP, VAC **Elev.:** 0–10 m a.s.l. **IUCN Habitat code:** 5 **Conservation status: (IUCN)** NE

1537 *Cyperus odoratus* L. (Cyperaceae) **IPNI:** 74939-2 **Common name:** coquito, cortadera, funche, pasto estrella, tresfilos **Habit:** Herb, Aquatic **Use:** HF, ME, SU **Origin:** Native **Region:** Amazonia, Andean, Guiana Shield, Caribbean, Orinoquia, Pacific **Dept.:** AMA, ANT, ARA, ATL, BOG, BOL, BOY, CAQ, CAS, CES, CHO, COR, CUN, GUA, HUI, MAG, MET, NAR, NSA, PUT, SAN, TOL, VAC, VAU **Elev.:** 0-1900 m.a.s.l. **IUCN Habitat code:** 1,2,3,4,5,14 **Conservation status: (IUCN)** NE **(CPSG)** Potential LC

1538 *Cyperus papyrus* L. (Cyperaceae) **IPNI:** 330934-2 **Common name:** minipapiro, papiro, papiro enano **Habit:** Herb **Use:** AF, EU, FU, HF, MA, ME, SU **Origin:** Cultivated, Naturalised **Region:** Andean **Dept.:** ANT, BOG, BOY, CUN, QUI, TOL **Elev.:** 1200–2820 m a.s.l. **IUCN Habitat code:** 1,2,3,5,14 **Conservation status: (IUCN)** LC

1539 *Cyperus polystachyos* Rottb. (Cyperaceae) **IPNI:** 75004-2 **Common name:** hierba de perro, paja de pisco **Habit:** Herb **Use:** AF **Origin:** Native **Region:** Pacific **Dept.:** NAR, TOL, VAC **Elev.:** 0–220 m a.s.l. **IUCN Habitat code:** 2,5,14 **Conservation status: (IUCN)** LC

1540 *Cyperus prolixus* Kunth (Cyperaceae) **IPNI:** 305627-1 **Common name:** chundur, cortadera, danta chondur, junco **Habit:** Herb **Use:** HF, MA, ME **Origin:** Native **Region:** Andean, Orinoquia **Dept.:** ANT, CUN, PUT, TOL, VAC, VID **Elev.:** 110–2200 m a.s.l. **IUCN Habitat code:** 1,3,4,5,14 **Conservation status: (IUCN)** NE

1541 *Cyperus rotundus* L. (Cyperaceae) **IPNI:** 305797-1 **Common name:** chivasa, coco, coquillo, coquito, corocillo, cortadera, junco de coquito **Habit:** Herb **Use:** AF, EU, GS, HF, MA, ME, PO **Origin:** Naturalised **Region:** Caribbean **Dept.:** ATL, BOL, CES, COR, CUN, MAG, MET, SAP, SUC, TOL, VAC **Elev.:** 0–1000 m a.s.l. **IUCN Habitat code:** 3 **Conservation status: (IUCN)** LC

1542 *Cyperus sesquiflorus* (Torr.) Mattf. & Kük. (Cyperaceae) **IPNI:** 75123-2 **Common name:** espadilla **Habit:** Herb **Use:** AF, EU, MA, ME **Origin:** Native **Region:** Amazonia, Andean, Guiana Shield, Caribbean, Orinoquia, Pacific **Dept.:** ANT, ARA, BOG, BOY, CAQ, CAU, CHO, CUN, GUA, GUV, HUI, MAG, MET, NAR, PUT, SAN, VAC, VAU **Elev.:** 50–2600 m a.s.l. **IUCN Habitat code:** 1,2,3,4,14 **Conservation status: (IUCN)** LC

1543 *Cyperus sphacelatus* Rottb. (Cyperaceae) **IPNI:** 305967-1 **Habit:** Herb **Use:** AF, MA **Origin:** Native **Region:** Amazonia, Andean, Guiana Shield, Caribbean, Orinoquia **Dept.:** AMA, ANT, ATL, CAQ, CAS, COR, MET, QUI, VAU **Elev.:** 120-1500 m.a.s.l. **IUCN Habitat code:** 1,2,3,4,5,14 **Conservation status: (IUCN)** NE **(CPSG)** Potential LC

1544 *Cyperus squarrosus* L. (Cyperaceae) **IPNI:** 324890-2 **Habit:** Herb **Use:** HF **Origin:** Native **Region:** Caribbean **Dept.:** ATL **Elev.:** 10–10 m a.s.l. **Conservation status: (IUCN)** LC

1545 *Cyperus tenuis* Sw. (Cyperaceae) **IPNI:** 75178-2 **Habit:** Herb **Use:** MA, ME **Origin:** Native **Region:** Amazonia, Andean, Caribbean, Orinoquia, Pacific **Dept.:** ANT, BOG, BOY, CAQ, CHO, CUN, MAG, MET, RIS, SAN **Elev.:** 10–2890 m a.s.l. **IUCN Habitat code:** 1,2,3,5,14 **Conservation status: (IUCN)** LC

1546 *Cyperus thyrsiflorus* Jungh. (Cyperaceae) **IPNI:** 75184-2 **Habit:** Herb **Use:** ME **Origin:** Native **Region:** Andean, Caribbean **Dept.:** MAG, NAR, VAC **Elev.:** 0–1700 m a.s.l. **IUCN Habitat code:** 1,3 **Conservation status: (IUCN)** NE

1547 *Cyperus tibialis* (Poit. ex Ledeb.) Govaerts (Cyperaceae) **IPNI:** 1009998-1 **Use:** ME **Origin:** Native **IUCN Habitat code:** 1,5 **Conservation status: (IUCN)** NE

1548 *Eleocharis acicularis* (L.) Roem. & Schult. (Cyperaceae) **IPNI:** 1169169-2 **Common name:** cebolleta de pantano, mara, piso **Habit:** Herb **Use:** EU **Origin:** Native **Region:** Andean **Dept.:** BOG, BOY, CUN, MET, NAR, PUT **Elev.:** 2300–3725 m a.s.l. **IUCN Habitat code:** 1,3,4,14 **Conservation status: (IUCN)** LC

1549 *Eleocharis acutangula* (Roxb.) Schult. (Cyperaceae) **IPNI:** 157356-3 **Habit:** Herb, Aquatic **Use:** MA **Origin:** Native **Region:** Andean, Guiana Shield, Orinoquia, Pacific **Dept.:** ANT, CHO, MET, SAN, VAU **Elev.:** 40–2480 m a.s.l. **IUCN Habitat code:** 1,2,3,4,14 **Conservation status: (IUCN)** LC

1550 *Eleocharis elegans* (Kunth) Roem. & Schult. (Cyperaceae) **IPNI:** 306803-1 **Common name:** cebolleta de pantano, cola de caballo, junco, junco papuche, totorilla **Habit:** Herb, Aquatic **Use:** MA, ME **Origin:** Native **Region:** Amazonia, Andean, Caribbean, Orinoquia, Pacific **Dept.:** AMA, ANT, ARA, ATL, BOL, BOY, CAS, CAU, CHO, COR, CUN, HUI, LAG, MAG, MET, NAR, PUT, SAN, SUC, TOL, VAC **Elev.:** 10-2200 m.a.s.l. **IUCN Habitat code:** 1,2,3,4,5,14 **Conservation status: (IUCN)** NE **(CPSG)** Potential LC

1551 *Eleocharis geniculata* (L.) Roem. & Schult. (Cyperaceae) **IPNI:** 1029981-2 **Common name:** cebolleta de pantano, junco, mara, pasto, raíz de mara **Use:** ME **Origin:** Native **Region:** Andean, Caribbean, Pacific **Dept.:** BOG, CAU, CHO, HUI, LAG, MAG, MET, NSA, SAP, VAC **Elev.:** 0–1300 m a.s.l. **IUCN Habitat code:** 1,2,3,4,5,14 **Conservation status: (IUCN)** LC

1552 *Eleocharis interstincta* (Vahl) Roem. & Schult. (Cyperaceae) **IPNI:** 60444117-2 **Common name:** jon, junco, junco de seda, yundul **Habit:** Herb **Use:** MA, ME **Origin:** Native **Region:** Andean, Guiana Shield, Caribbean, Orinoquia, Pacific **Dept.:** ANT, ARA, ATL, CAU, CHO, HUI, NAR, SAN, VAC, VAU **Elev.:** 10-1100 m.a.s.l. **IUCN Habitat code:** 1,2,3,4,5,14 **Conservation status: (IUCN)** NE **(CPSG)** Potential LC

1553 *Fimbristylis aestivalis* (Retz.) Vahl (Cyperaceae) **IPNI:** 307664-1 **Habit:** Herb **Use:** ME **Origin:** Naturalised **Region:** Guiana Shield, Orinoquia **Dept.:** MET, VAU **Elev.:** 350–400 m a.s.l. **IUCN Habitat code:** 4 **Conservation status: (IUCN)** NE

1554 *Fimbristylis argentea* (Rottb.) Vahl (Cyperaceae) **IPNI:** 307702-1 **Habit:** Herb **Use:** ME **Origin:** Naturalised **Region:** Caribbean, Orinoquia **Dept.:** ATL, COR, VID **Elev.:** 5–100 m a.s.l. **Conservation status: (IUCN)** LC

1555 *Fimbristylis complanata* (Retz.) Link (Cyperaceae) **IPNI:** 1111946-2 **Habit:** Herb **Use:** MA, ME **Origin:** Native **Region:** Andean, Orinoquia **Dept.:** ANT, ARA, BOG, BOY, CAS, CUN, MAG, MET, NAR, NSA, RIS, VAC **Elev.:** 50-2160 m.a.s.l. **IUCN Habitat code:** 1,2,3,4,14 **Conservation status: (IUCN)** LC **(CPSG)** Potential LC

1556 *Fimbristylis dichotoma* (L.) Vahl (Cyperaceae) **IPNI:** 307853-1 **Common name:** arrocillo, barba de indio, coquito, espartillo, maicillo, yerba fina **Habit:** Herb **Use:** AF, EU, IF, MA, ME **Origin:** Native **Region:** Amazonia, Andean, Guiana Shield, Caribbean, Orinoquia, Pacific **Dept.:** AMA, ANT, ARA, BOL, BOY, CAL, CAQ, CAS, CAU, CES, CHO, COR, CUN, HUI, MAG, MET, NAR, PUT, QUI, SAN, SUC, VAC, VAU, VID **Elev.:** 0-2800 m.a.s.l. **IUCN Habitat code:** 1,2,3,4,14 **Conservation status: (IUCN)** LC **(CPSG)** Potential LC

1557 *Fimbristylis littoralis* Gaudich. (Cyperaceae) **IPNI:** 308035-1 **Habit:** Herb **Use:** AF, SU **Origin:** Native **Region:** Amazonia, Caribbean, Orinoquia, Pacific **Dept.:** AMA, ANT, ARA, ATL, BOL, BOY, CAS, CAU, CHO, COR, HUI, MAG, MET, NAR, NSA, PUT, SAN, TOL, VAC, VAU **Elev.:** 0-900 m.a.s.l. **IUCN Habitat code:** 1,2,3,4,14 **Conservation status: (IUCN)** LC **(CPSG)** Potential LC

1558 *Fimbristylis ovata* (Burm.f.) J.Kern (Cyperaceae) **IPNI:** 104972-2 **Habit:** Herb **Use:** AF,ME **Origin:** Native **Region:** Caribbean **Dept.:** MAG, TOL **Elev.:** 152–350 m a.s.l. **IUCN Habitat code:** 3 **Conservation status: (IUCN)** LC

1559 *Fimbristylis spadicea* (L.) Vahl (Cyperaceae) **IPNI:** 308274-1 **Common name:** juncia **Habit:** Herb **Use:** MA **Origin:** Native **Region:** Caribbean, Pacific **Dept.:** ANT, ATL, BOL, CHO, COR, MAG, SUC, VAC **Elev.:** 0-150 m.a.s.l. **IUCN Habitat code:** 1,5 **Conservation status: (IUCN)** NE **(CPSG)** Potential LC

1560 *Fimbristylis squarrosa* Vahl (Cyperaceae) **IPNI:** 308286-1 **Habit:** Herb **Use:** ME **Origin:** Native **Region:** Caribbean **Dept.:** CES **Elev.:** 50–50 m a.s.l. **Conservation status: (IUCN)** LC

1561 *Fuirena umbellata* Rottb. (Cyperaceae) **IPNI:** 308552-1 **Common name:** cortadera **Habit:** Herb, Aquatic **Use:** AF, EU, HF, MA, ME **Origin:** Native **Region:** Orinoquia, Pacific **Dept.:** ANT, CAU, CHO, COR, HUI, MAG, MET, NAR, SAN, TOL, VAC, VID **Elev.:** 0-500 m.a.s.l. **IUCN Habitat code:** 1,2,3,4,5,14 **Conservation status: (IUCN)** LC **(CPSG)** Potential LC

1562 *Isolepis cernua* (Vahl) Roem. & Schult. (Cyperaceae) **IPNI:** 309238-1 **Habit:** Herb **Use:** EU, ME **Origin:** Native **Region:** Andean **Dept.:** BOG, CAL, CAU, VAC **Elev.:** 3550–4100 m a.s.l. **IUCN Habitat code:** 4,14 **Conservation status: (IUCN)** LC

1563 *Rhynchospora blepharophora* (J.Presl & C.Presl) H.Pfeiff. (Cyperaceae) **IPNI:** 220187-2 **Common name:** estrella, estrellita, estrellita de Japón, flor de grama, flor de sabana, fosforito, tote, totecito **Habit:** Herb **Use:** ME **Origin:** Native **Region:** Andean, Orinoquia **Dept.:** ANT, ARA, BOG, BOY, CAQ, CUN, HUI, MAG, MET, TOL, VID **Elev.:** 220–3180 m a.s.l. **IUCN Habitat code:** 1,2,3,4,14 **Conservation status: (IUCN)** NE

1564 *Rhynchospora cephalotes* (L.) Vahl (Cyperaceae) **IPNI:** 312067-1 **Common name:** paja judía **Habit:** Herb **Use:** ME **Origin:** Native **Region:** Amazonia, Andean, Guiana Shield, Caribbean, Orinoquia, Pacific **Dept.:** AMA, ANT, ARA, BOG, CAQ, CAS, CAU, CHO, COR, CUN, GUA, GUV, MAG, MET, SAP, SAN, TOL, VAC, VAU, VID **Elev.:** 0-1700 m.a.s.l. **IUCN Habitat code:** 1,2,3,4,5,14 **Conservation status: (IUCN)** NE **(CPSG)** Potential LC

1565 *Rhynchospora colorata* (L.) H.Pfeiff. (Cyperaceae) **IPNI:** 311841-1 **Use:** AF,ME, SU **Region:** Andean **Dept.:** BOG **IUCN Habitat code:** 1,2,3,14 **Conservation status: (IUCN)** NE

1566 *Rhynchospora corymbosa* (L.) Britton (Cyperaceae) **IPNI:** 220271-2 **Common name:** barba de indio **Habit:** Herb **Use:** EU, MA, ME, SU **Origin:** Native **Region:** Andean, Orinoquia, Pacific **Dept.:** ANT, ARA, BOG, CAU, CHO, CUN, HUI, MET, PUT, SAN, VAC, VAU, VID **Elev.:** 5-1900 m.a.s.l. **IUCN Habitat code:** 1,3,4,14 **Conservation status: (IUCN)** LC **(CPSG)** Potential LC

1567 *Rhynchospora nervosa* (Vahl) Boeckeler (Cyperaceae) **IPNI:** 312330-1 **Common name:** Santa Juana, akanepona, chivasa, coquito blanco, estrella, estrella tote, estrellita, herb de estrella, hierba de estrella, horquetilla, mecanito, pisco, purgante de perro, tote, totecito, totes, totesito **Habit:** Herb **Use:** MA, ME, SU **Origin:** Native **Dept.:** ANT, BOG, BOY, CAU, CUN, HUI, MAG, MET, QUI, SAN, TOL, VAC, VAU **Elev.:** 240-3000 m.a.s.l. **IUCN Habitat code:** 1,2,3,4,5,14 **Conservation status: (IUCN)** NE **(CPSG)** Potential LC

1568 *Rhynchospora polystachys* (Turrill) H.Pfeiff. (Cyperaceae) **IPNI:** 220575-2 **Common name:** totes **Habit:** Herb **Use:** ME **Origin:** Native **Region:** Amazonia, Andean, Orinoquia, Pacific **Dept.:** ANT, CAU, CHO, CUN, MET, NAR, RIS, TOL, VAC **Elev.:** 40-2100 m.a.s.l. **IUCN Habitat code:** 1,14 **Conservation status: (IUCN)** NE **(CPSG)** Potential LC

1569 *Rhynchospora pura* (Nees) Griseb. (Cyperaceae) **IPNI:** 312402-1 **Common name:** totes **Habit:** Herb **Use:** ME **Origin:** Native **Region:** Amazonia, Andean, Guiana Shield, Caribbean, Orinoquia, Pacific **Dept.:** AMA, ANT, ARA, BOL, BOY, CAQ, CAS, CAU, CHO, COR, CUN, GUA, GUV, HUI, MAG, MET, NAR, NSA, PUT, SAN, SUC, VAC, VID **Elev.:** 0-3420 m.a.s.l. **IUCN Habitat code:** 1,3,4,5,14 **Conservation status: (IUCN)** NE **(CPSG)** Potential LC

1570 *Schoenoplectus californicus* (C.A.Mey.) Soják (Cyperaceae) **IPNI:** 229810-2 **Common name:** junco, junco redondo, totora **Habit:** Herb, Aquatic **Use:** EU, HF, MA, ME **Origin:** Native **Region:** Andean, Pacific **Dept.:** BOG, BOY, CAU, CUN, NAR, PUT, VAC **Elev.:** 0–2090 m a.s.l. **IUCN Habitat code:** 1,3,4,5,14 **Conservation status: (IUCN)** NE

1571 *Schoenoplectus tabernaemontani* (C.C.Gmel.) Palla (Cyperaceae) **IPNI:** 1019164-1 **Use:** EU, HF, ME **Region:** Andean **Dept.:** BOG **IUCN Habitat code:** 14 **Conservation status: (IUCN)** LC

1572 *Scleria bracteata* Cav. (Cyperaceae) **IPNI:** 314601-1 **Common name:** cortadera, curíbano, lambedera **Habit:** Herb, Climbing **Use:** ME **Origin:** Native **Region:** Andean, Caribbean, Orinoquia, Pacific **Dept.:** ANT, CAU, CES, HUI, MAG, MET, SAN, VAC **Elev.:** 250–1700 m a.s.l. **IUCN Habitat code:** 1,3,4,5,14 **Conservation status: (IUCN)** LC

1573 *Scleria hirtella* Sw. (Cyperaceae) **IPNI:** 329993-2 **Common name:** algofar, curîbano, gengibrillo, raíz de resfrío **Use:** MA, ME **Origin:** Native **IUCN Habitat code:** 1 **Conservation status: (IUCN)** LC

1574 *Scleria macrophylla* J.Presl & C.Presl (Cyperaceae) **IPNI:** 314849-1 **Common name:** cortadera, ponetukuerenae **Habit:** Herb, Climbing **Use:** ME **Origin:** Native **Region:** Amazonia, Andean, Caribbean, Orinoquia, Pacific **Dept.:** AMA, ANT, CAQ, CHO, MAG, MET, PUT, VAC, VAU, VID **Elev.:** 20–1900 m a.s.l. **IUCN Habitat code:** 1,3,4,5,14 **Conservation status: (IUCN)** LC

1575 *Scleria melanomphala* Kunth (Cyperaceae) **IPNI:** 314862-1 **Habit:** Herb, Climbing **Use:** ME **Origin:** Native **Region:** Orinoquia **Dept.:** MET **Elev.:** 410–410 m a.s.l. **IUCN Habitat code:** 1 **Conservation status: (IUCN)** LC

1576 *Scleria robusta* Camelb. & Goetgh. (Cyperaceae) **IPNI:** 320885-2 **Habit:** Herb, Climbing **Use:** HF **Origin:** Native **Region:** Amazonia, Guiana Shield **Dept.:** CAQ, VAU **Elev.:** 200–500 m a.s.l. **IUCN Habitat code:** 1,4 **Conservation status: (IUCN)** LC

1577 *Scleria secans* (L.) Urb. (Cyperaceae) **IPNI:** 230730-2 **Common name:** cortadera **Habit:** Herb, Climbing **Use:** ME, SU **Origin:** Native **Region:** Amazonia, Andean, Guiana Shield, Caribbean, Orinoquia, Pacific **Dept.:** AMA, ANT, BOY, CAQ, CAS, CAU, CHO, CUN, LAG, MAG, MET, NSA, SAP, SAN, VAC, VAU, VID **Elev.:** 0-2000 m.a.s.l. **IUCN Habitat code:** 1,2,3,5,14 **Conservation status: (IUCN)** LC **(CPSG)** Potential LC

1578 *Eriocaulon setaceum* L. (Eriocaulaceae) **IPNI:** 335722-1 **Habit:** Herb, Aquatic **Use:** ME **Origin:** Native **Region:** Guiana Shield **Dept.:** GUV, MET **Elev.:** 195–450 m a.s.l. **IUCN Habitat code:** 1,3,4,14 **Conservation status: (IUCN)** LC

1579 *Paepalanthus alpinus* Körn. (Eriocaulaceae) **IPNI:** 336004-1 **Common name:** cardita, cardoncillo **Habit:** Herb **Use:** EU **Origin:** Endemic **Region:** Andean **Dept.:** ANT, BOG, BOY, CAU, CUN, HUI, NSA, SAN **Elev.:** 1900–3600 m a.s.l. **IUCN Habitat code:** 1,3,4,8,14 **Conservation status: (IUCN)** NE

1580 *Paepalanthus dendroides* (Kunth) Kunth (Eriocaulaceae) **IPNI:** 336162-1 **Common name:** cardoncillo espinoso **Habit:** Herb, Aquatic **Use:** EU **Origin:** Native **Region:** Andean **Dept.:** ANT, BOG, BOY, CAS, CAU, CUN, NSA, SAN **Elev.:** 1300–3575 m a.s.l. **IUCN Habitat code:** 1,3,4,14 **Conservation status: (IUCN)** NE

1581 *Juncus balticus* Willd. (Juncaceae) **IPNI:** 324783-2 **Habit:** Herb **Use:** EU, HF, ME **Origin:** Native **Region:** Andean **Dept.:** NAR **Elev.:** 2890–2890 m a.s.l. **Conservation status: (IUCN)** LC

1582 *Juncus bufonius* L. (Juncaceae) **IPNI:** 30079279-2 **Common name:** junco **Habit:** Herb **Use:** AF, EU, ME **Origin:** Native **Region:** Andean **Dept.:** ANT, BOG, BOY, CAL, CUN, NSA, QUI, RIS, SAN, TOL **Elev.:** 1100–3700 m a.s.l. **IUCN Habitat code:** 1,3,4,14 **Conservation status: (IUCN)** LC

1583 *Juncus dudleyi* Wiegand (Juncaceae) **IPNI:** 132109-2 **Habit:** Herb **Use:** ME **Origin:** Naturalised **Region:** Andean **Dept.:** BOG, BOY, CUN, SAN, VAC **Elev.:** 2600–3900 m a.s.l. **IUCN Habitat code:** 4,14 **Conservation status: (IUCN)** LC

1584 *Juncus echinocephalus* Balslev (Juncaceae) **IPNI:** 132111-2 **Common name:** junco **Habit:** Herb **Use:** AF, MA **Origin:** Native **Region:** Andean **Dept.:** ANT, BOG, BOY, CAU, CUN, MET, NAR, NSA, PUT, RIS, SAN, TOL **Elev.:** 2550-4200 m.a.s.l. **IUCN Habitat code:** 1,3,4,14 **Conservation status: (IUCN)** NE **(CPSG)** Potential LC

1585 *Juncus effusus* L. (Juncaceae) **IPNI:** 442917-1 **Common name:** esparto, junco **Habit:** Herb **Use:** AF, EU, FU, GS, HF, MA, ME, PO **Origin:** Native **Region:** Andean **Dept.:** ANT, BOG, BOY, CAL, CAU, CUN, HUI, LAG, MAG, NAR, NSA, PUT, QUI, RIS, SAN, TOL **Elev.:** 1500-3700 m.a.s.l. **IUCN Habitat code:** 1,3,4,5,14 **Conservation status: (IUCN)** LC **(CPSG)** Potential LC

1586 *Juncus microcephalus* Kunth (Juncaceae) **IPNI:** 443213-1 **Habit:** Herb **Use:** HF **Origin:** Native **Region:** Andean **Dept.:** ANT, BOG, BOY, CAU, CUN, HUI, NAR, NSA, PUT, RIS, SAN, VAC **Elev.:** 1500-3600 m.a.s.l. **IUCN Habitat code:** 1,3,4,5,14 **Conservation status: (IUCN)** NE **(CPSG)** Potential LC

1587 *Juncus pallescens* Lam. (Juncaceae) **IPNI:** 443313-1 **Habit:** Herb **Use:** ME **Origin:** Native **Region:** Andean **Dept.:** NAR **Elev.:** 2530–2530 m a.s.l. **IUCN Habitat code:** 1 **Conservation status: (IUCN)** NE

1588 *Juncus ramboi* Barros (Juncaceae) **IPNI:** 132285-2 **Common name:** esparto **Habit:** Herb **Use:** MA **Origin:** Native **Region:** Andean **Dept.:** BOG, BOY, CUN **Elev.:** 900–1600 m a.s.l. **IUCN Habitat code:** 1,4,14 **Conservation status: (IUCN)** NE

1589 *Juncus tenuis* Willd. (Juncaceae) **IPNI:** 304453-2 **Habit:** Herb **Use:** ME **Origin:** Naturalised **Region:** Andean **Dept.:** ANT, BOG, BOY, CAU, CHO, CUN, NSA, PUT, QUI, RIS, SAN, VAC **Elev.:** 1300–2850 m a.s.l. **IUCN Habitat code:** 1,2,3,4,8,14 **Conservation status: (IUCN)** NE

1590 *Acroceras zizanioides* (Kunth) Dandy (Poaceae) **IPNI:** 3538-2 **Common name:** grama macho, nudillo, pasto nudillo, pasto nulludo **Habit:** Herb **Use:** AF, EU, HF, MA, ME, SU **Origin:** Native **Region:** Amazonia, Andean, Guiana Shield, Caribbean, Orinoquia, Pacific **Dept.:** AMA, ANT, ARA, BOL, CAQ, CAS, CAU, CHO, CUN, GUA, GUV, HUI, LAG, MAG, MET, NAR, PUT, QUI, SAN, TOL, VAC, VAU **Elev.:** 0-1500 m.a.s.l. **IUCN Habitat code:** 1,2,3,4,5,14 **Conservation status: (IUCN)** LC **(CPSG)** Potential LC

1591 *Agrostis capillaris* L. (Poaceae) **IPNI:** 30197865-2 **Habit:** Herb **Use:** AF, EU, GS **Origin:** Cultivated, Naturalised **IUCN Habitat code:** 1,4,8,14 **Conservation status: (IUCN)** NE

1592 *Agrostis gigantea* Roth (Poaceae) **IPNI:** 385783-1 **Habit:** Herb **Use:** AF, EU,PO **Origin:** Cultivated, Naturalised **Conservation status: (IUCN)** NE

1593 *Agrostis perennans* (Walter) Tuck. (Poaceae) **IPNI:** 1128639-2 **Common name:** paja, pasto de gallina, pasto gallina **Habit:** Herb **Use:** HF **Origin:** Native **Region:** Andean **Dept.:** ANT, BOG, BOY, CAL, CAQ, CAU, CES, CUN, HUI, MAG, MET, NAR, PUT, QUI, RIS, SAN, TOL, VAC **Elev.:** 1500-4500 m.a.s.l. **IUCN Habitat code:** 1,3,4,14 **Conservation status: (IUCN)** LC **(CPSG)** Potential LC

1594 *Agrostis stolonifera* L. (Poaceae) **IPNI:** 386353-1 **Common name:** cundidora, grama blanca, yerba cundidora **Habit:** Herb **Use:** EU, HF **Origin:** Cultivated, Naturalised **Region:** Andean **Dept.:** BOG, BOY, CUN **Elev.:** 2500–3000 m a.s.l. **Conservation status: (IUCN)** LC

1595 *Aira caryophyllea* L. (Poaceae) **IPNI:** 386633-1 **Use:** AF **Origin:** Native **Region:** Andean **Conservation status: (IUCN)** NE

1596 *Alopecurus aequalis* Sobol. (Poaceae) **IPNI:** 387005-1 **Habit:** Herb **Use:** HF, ME **Origin:** Native **Region:** Andean **Dept.:** BOG, BOY, CAL, CAU, CUN **Elev.:** 2500–4500 m a.s.l. **IUCN Habitat code:** 4,14 **Conservation status: (IUCN)** LC

1597 *Alopecurus pratensis* L. (Poaceae) **IPNI:** 387176-1 **Habit:** Herb **Use:** AF, EU, HF, ME **Origin:** Naturalised **Region:** Andean **Dept.:** BOG, CUN **Elev.:** 2500–3500 m a.s.l. **Conservation status: (IUCN)** NE

1598 *Andropogon bicornis* L. (Poaceae) **IPNI:** 12758-2 **Common name:** cola de caballo, cola de vaca, cola de venado, cola de zorro, colizorro, kamuwe, paimbo, paja, paja de zorro, paja montosa, pajaezorro, pasto elefante, pasto puntero, pasto rabo de zorro, puntero, rabo de gato, rabo de vaca, rabo de zorro, raboevaca, rabo **Habit:** Herb **Use:** EU, HF, MA, ME **Origin:** Native **Region:** Amazonia, Andean, Guiana Shield, Caribbean, Orinoquia, Pacific **Dept.:** AMA, ANT, ARA, BOL, BOY, CAL, CAQ, CAS, CAU, CHO, COR, CUN, GUA, GUV, HUI, MAG, MET, NAR, NSA, PUT, QUI, RIS, SAP, SAN, TOL, VAC, VAU, VID **Elev.:** 0-2500 m.a.s.l. **IUCN Habitat code:** 1,2,3,4,5,14 **Conservation status: (IUCN)** NE **(CPSG)** Potential LC

1599 *Andropogon gayanus* Kunth (Poaceae) **IPNI:** 387915-1 **Common name:** carimagua **Habit:** Herb **Use:** AF, EU, HF, MA, ME, SU **Origin:** Naturalised **Elev.:** 1000–1500 m a.s.l. **IUCN Habitat code:** 1,3,14 **Conservation status: (IUCN)** NE

1600 *Andropogon glomeratus* (Walter) Britton, Sterns & Poggenb. (Poaceae) **IPNI:** 12850-2 **Habit:** Herb **Use:** AF, EU, ME **Origin:** Native **Region:** Andean, Caribbean **Dept.:** ANT, BOL, CES, LAG, MAG, SAP **Elev.:** 0–1000 m a.s.l. **IUCN Habitat code:** 1,14 **Conservation status: (IUCN)** NE

1601 *Andropogon leucostachyus* Kunth (Poaceae)
IPNI: 12920-2
Common name: cola de vaca, paimbo, pluma de garza, rabo de gato, rabo de vaca **Habit:** Herb **Use:** ME **Origin:** Native **Region:** Amazonia, Andean, Guiana Shield, Caribbean, Orinoquia, Pacific **Dept.:** ANT, ARA, BOL, BOY, CAL, CAQ, CAS, CAU, CES, CHO, CUN, GUA, GUV, HUI, LAG, MAG, MET, NAR, NSA, RIS, SAN, TOL, VAC, VAU, VID **Elev.:** 0-2500 m.a.s.l. **IUCN Habitat code:** 1,2,3,4,5,14 **Conservation status: (IUCN)** NE **(CPSG)** Potential LC

1602 *Andropogon virgatus* Desv. (Poaceae)
IPNI: 13112-2
Habit: Herb **Use:** ME **Origin:** Native **Region:** Andean, Caribbean, Orinoquia **Dept.:** ARA, CAS, CES, MET, SAN, VID **Elev.:** 0–500 m a.s.l. **IUCN Habitat code:** 1,2,3,4,14 **Conservation status: (IUCN)** NE

1603 *Anthephora hermaphrodita* (L.) Kuntze (Poaceae)
IPNI: 14931-2
Common name: araña, carrizo agrio **Habit:** Herb **Use:** EU **Origin:** Native **Region:** Amazonia, Andean, Caribbean, Orinoquia, Pacific **Dept.:** ANT, ATL, BOL, CAL, CAS, CAU, CES, COR, HUI, LAG, MAG, NAR, NSA, SAP, SAN, SUC, TOL, VAC **Elev.:** 0-1500 m.a.s.l. **IUCN Habitat code:** 2,5,14 **Conservation status: (IUCN)** NE **(CPSG)** Potential LC

1604 *Anthoxanthum odoratum* L. (Poaceae)
IPNI: 389034-1
Common name: alpistillo, coluda, díctamo, grama, grama olorosa, oloroso, paja, paja aromática, pasto de olor, pasto dulce, rabo de zorro **Habit:** Herb **Use:** HF, MA, ME, PO **Origin:** Naturalised **Region:** Andean **Dept.:** ANT, BOG, BOY, CAL, CAU, CES, CHO, CUN, HUI, LAG, MAG, MET, NAR, NSA, PUT, QUI, RIS, SAN, TOL, VAC **Elev.:** 1800–4500 m a.s.l. **IUCN Habitat code:** 1,3,4,5,8,14 **Conservation status: (IUCN)** NE

1605 *Anthoxanthum redolens* (Vahl) P.Royen (Poaceae)
IPNI: 389043-1
Common name: etamo real **Use:** ME **IUCN Habitat code:** 1 **Conservation status: (IUCN)** NE

1606 *Aristida adscensionis* L. (Poaceae)
IPNI: 319759-2
Habit: Herb **Use:** AF, EU, HF, MA, ME, SU **Origin:** Native **Region:** Andean **Dept.:** BOY, CES, CUN, LAG, MAG, NAR, VAC **Elev.:** 500–2500 m a.s.l. **IUCN Habitat code:** 1,2,3,14 **Conservation status: (IUCN)** NE

1607 *Arrhenatherum elatius* (L.) P.Beauv. ex J.Presl & C.Presl (Poaceae)
IPNI: 389923-1
Habit: Herb **Use:** AF, EU, HF, ME **Origin:** Cultivated, Naturalised **Region:** Andean **Dept.:** BOG, CUN **Elev.:** 2500–3000 m a.s.l. **Conservation status: (IUCN)** NE

1608 *Arundinella hispida* (Kunth ex Willd.) Kuntze (Poaceae)
IPNI: 21052-2
Habit: Herb **Use:** ME **Origin:** Native **Region:** Andean, Caribbean, Orinoquia, Pacific **Dept.:** ANT, ARA, BOY, CAL, CAS, CAU, CUN, HUI, MAG, MET, NAR, NSA, QUI, RIS, SAN, TOL, VAC, VID **Elev.:** 0-2000 m.a.s.l. **IUCN Habitat code:** 1,3,4,14 **Conservation status: (IUCN)** NE **(CPSG)** Potential LC

1609 *Arundo donax* L. (Poaceae)
IPNI: 390837-1
Common name: carrizo, caña, caña de Castilla, caña de queco, caña de tierra fría, cañabrava, cañabrava de Castilla, cañapopa, cañote, chin, chin de Castilla, juco, lata, lata de Castilla, popo, popo de chumbe, tute **Habit:** Herb **Use:** AF, EU, FU, HF, MA, ME, SU **Origin:** Cultivated, Naturalised **Region:** Andean **Dept.:** ANT, BOG, BOY, CAL, CAQ, CAU, CUN, HUI, NAR, QUI, RIS, SAN, TOL, VAC **Elev.:** 500–3000 m a.s.l. **IUCN Habitat code:** 1,2,3,5,14 **Conservation status: (IUCN)** LC

1610 *Aulonemia queko* Goudot (Poaceae)
IPNI: 26916-2
Common name: carrizo, flauta, sorbetana, tunda **Habit:** Small tree **Use:** ME **Origin:** Native **Region:** Andean **Dept.:** CAU, CUN, HUI, PUT, QUI, TOL, VAC **Elev.:** 2330–3000 m a.s.l. **IUCN Habitat code:** 1,14 **Conservation status: (IUCN)** NE

1611 *Aulonemia trianae* (Munro) McClure (Poaceae)
IPNI: 26923-2
Common name: chusco, cira, cira junca, micay, queco **Habit:** Climbing **Use:** MA **Origin:** Native **Region:** Andean **Dept.:** ANT, BOG, BOY, CAU, CUN, NSA, VAC **Elev.:** 2600–3200 m a.s.l. **IUCN Habitat code:** 1,3,4,14 **Conservation status: (IUCN)** NE

1612 *Avena fatua* L. (Poaceae)
IPNI: 391443-1
Common name: avena, avena cimarrona, avena silvestre **Habit:** Herb **Use:** AF, GS, HF, ME, PO **Origin:** Naturalised **Region:** Andean **Dept.:** BOG, BOY, CUN, NAR **Elev.:** 2000–3000 m a.s.l. **IUCN Habitat code:** 3 **Conservation status: (IUCN)** LC

1613 *Avena sativa* L. (Poaceae)
IPNI: 391732-1
Common name: avena, cebada santafereña **Habit:** Herb **Use:** AF, EU, GS, HF, MA, ME, PO **Origin:** Naturalised **Region:** Andean **Dept.:** BOG, BOY, CUN, NAR **Elev.:** 2000–3000 m a.s.l. **IUCN Habitat code:** 1,3,14 **Conservation status: (IUCN)** NE

1614 *Axonopus aureus* P.Beauv. (Poaceae)
IPNI: 321761-2
Common name: paja ciempiés, pasto sabana **Habit:** Herb **Use:** ME **Origin:** Native **Region:** Amazonia, Andean, Guiana Shield, Caribbean, Orinoquia **Dept.:** AMA, ANT, ARA, BOY, CAS, CES, GUA, GUV, HUI, MAG, MET, NSA, SAN, VID **Elev.:** 0-2500 m.a.s.l. **IUCN Habitat code:** 1,2,3,4,5,14 **Conservation status: (IUCN)** NE **(CPSG)** Potential LC

1615 *Axonopus compressus* (Sw.) P.Beauv. (Poaceae)
IPNI: 392087-1
Common name: barbacoa, caminacompié, cañamazo, cuchipén, gateadora morada, grama blanca, grama criolla, grama dulce, grama extranjera de hoja ancha, grama horquetilla, grama natural, gramalote, gramilla rastrera, gramínea de sabana, hojiancha, nudillo, panceburro, pan **Habit:** Herb **Use:** AF, EU, ME **Origin:** Native, Cultivated **Region:** Amazonia, Andean, Guiana Shield, Caribbean, Orinoquia, Pacific **Dept.:** AMA, ANT, ARA, ATL, BOG, BOL, BOY, CAL, CAQ, CAS, CAU, CES, CHO, COR, CUN, GUA, GUV, HUI, LAG, MAG, MET, NAR, NSA, PUT, QUI, RIS, SAP, SAN, SUC, TOL, VAC, VID **Elev.:** 0-3000 m.a.s.l. **IUCN Habitat code:** 1,2,3,4,5,14 **Conservation status: (IUCN)** NE **(CPSG)** Potential LC

1616 *Axonopus fissifolius* (Raddi) Kuhlm. (Poaceae)
IPNI: 27338-2
Common name: bacaria, gallina, pasto alfombra, pata de gallina **Habit:** Herb **Use:** AF, EU, ME **Origin:** Native **Region:** Amazonia, Andean, Guiana Shield, Caribbean, Orinoquia **Dept.:** AMA, ANT, ARA, BOG, BOY, CAQ, GUA, MAG, MET, NAR, RIS, TOL, VAC, VID **Elev.:** 0-2000 m.a.s.l. **IUCN Habitat code:** 1,2,4,14 **Conservation status: (IUCN)** NE **(CPSG)** Potential LC

1617 *Axonopus scoparius* (Flüggé) Kuhlm. (Poaceae)
IPNI: 27415-2
Common name: capín venezuela, gramalote, gramalote blanco, gramalote colorado, imperial, macael, micay, micay blanco, pasto chigüiro, pasto de trenza, pasto imperial, pasto micael, pasto micay, pasto negro, telembí, yerba micay **Habit:** Herb **Use:** AF,ME **Origin:** Native, Cultivated **Region:** Amazonia, Andean, Guiana Shield, Caribbean, Orinoquia, Pacific **Dept.:** AMA, ANT, BOG, BOL, BOY, CAL, CAQ, CAS, CAU, CES, CHO, CUN, GUA, GUV, HUI, MAG, MET, NAR, NSA, PUT, QUI, RIS, SAN, TOL, VAC, VAU **Elev.:** 0-3000 m.a.s.l. **IUCN Habitat code:** 1,3,4,5,14 **Conservation status: (IUCN)** LC **(CPSG)** Potential LC

1618 *Bambusa beecheyana* Munro (Poaceae)
IPNI: 392260-1
Habit: Herb **Use:** EU, HF, ME **Origin:** Cultivated, Naturalised **Conservation status: (IUCN)** NE

1619 *Bambusa eutuldoides* McClure (Poaceae)
IPNI: 392318-1
Habit: Tree **Use:** HF, MA **Origin:** Cultivated, Naturalised **Dept.:** CES, CUN, GUV, MET, QUI, TOL, VAC **Elev.:** 500–1500 m a.s.l. **IUCN Habitat code:** 3 **Conservation status: (IUCN)** NE

1620 *Bambusa longispiculata* Gamble (Poaceae)
IPNI: 392396-1
Use: HF, MA **Origin:** Naturalised **Dept.:** VAC **Conservation status: (IUCN)** NE

1621 *Bambusa multiplex* (Lour.) Raeusch. ex Schult.f. (Poaceae)
IPNI: 1097548-2
Habit: Tree **Use:** EU, HF, MA, ME **Origin:** Cultivated, Naturalised **Region:** Andean **Dept.:** CUN, QUI, TOL, VAC **Elev.:** 800–1500 m a.s.l. **IUCN Habitat code:** 1,14 **Conservation status: (IUCN)** NE

1622 *Bambusa oldhamii* Munro (Poaceae)
IPNI: 392445-1
Habit: Herb **Use:** EU, HF, ME **Origin:** Cultivated, Naturalised **Region:** Andean **Dept.:** VAC **Elev.:** 1000–1500 m a.s.l. **Conservation status: (IUCN)** NE

1623 *Bambusa textilis* McClure (Poaceae)
IPNI: 392542-1
Use: EU, HF, MA, ME **Origin:** Naturalised **Conservation status: (IUCN)** NE

1624 *Bambusa tulda* Roxb. (Poaceae)
IPNI: 392551-1
Habit: Tree **Use:** EU, HF, MA, ME **Origin:** Cultivated, Naturalised

Region: Andean Dept.: CAU, QUI, TOL, VAC Elev.: 800–1300 m a.s.l. Conservation status: (IUCN) NE

1625 *Bambusa tuldoides* Munro (Poaceae)
IPNI: 392553-1
Use: EU, HF, MA, ME Origin: Naturalised Dept.: CAU Conservation status: (IUCN) NE

1626 *Bambusa vulgaris* Schrad. ex J.C.Wendl. (Poaceae)
IPNI: 392574-1
Common name: bambú, bambú de la india, bambú de tallos amarillos, guadua amarilla Habit: Tree Use: AF, EU, FU, GS, HF, MA, ME, PO, SU Origin: Cultivated, Naturalised Region: Andean, Caribbean Dept.: AMA, ANT, BOG, BOY, CAL, CAU, CES, CHO, COR, CUN, GUV, HUI, MAG, MET, QUI, RIS, SAP, SAN, SUC, TOL, VAC, VID Elev.: 100–1500 m a.s.l. IUCN Habitat code: 1,2,3,14 Conservation status: (IUCN) NE

1627 *Bothriochloa barbinodis* (Lag.) Herter (Poaceae)
IPNI: 35184-2
Habit: Herb Use: AF, EU Origin: Native Region: Andean Dept.: CAL, CAU, CUN, NAR, TOL Elev.: 0–2000 m a.s.l. Conservation status: (IUCN) NE

1628 *Bothriochloa bladhii* (Retz.) S.T.Blake (Poaceae)
IPNI: 392701-1
Habit: Herb Use: AF, EU, MA, ME Origin: Naturalised Region: Andean Dept.: BOY, TOL Elev.: 500–2500 m a.s.l. Conservation status: (IUCN) NE

1629 *Bothriochloa ischaemum* (L.) Keng (Poaceae)
IPNI: 1191417-2
Habit: Herb Use: AF, EU Origin: Naturalised Region: Andean, Caribbean Dept.: MAG, SAN, SUC Elev.: 0–2000 m a.s.l. Conservation status: (IUCN) NE

1630 *Bothriochloa pertusa* (L.) A.Camus (Poaceae)
IPNI: 65406-3
Common name: caminadora, caminera, cebo de Cuba, cocuyina, colosoana, colosuana, colusuana, cucuyina, kikuyo, olorosa, pastoangletón, pasto cordobés Habit: Herb Use: AF, EU, ME Origin: Naturalised Region: Andean, Caribbean Dept.: ANT, ATL, BOL, BOY, CAL, CAU, CES, CUN, HUI, LAG, MAG, SUC, TOL, VAC Elev.: 0–1500 m a.s.l. IUCN Habitat code: 1,2,3,14 Conservation status: (IUCN) NE

1631 *Bothriochloa saccharoides* (Sw.) Rydb. (Poaceae)
IPNI: 35205-2
Habit: Herb Use: ME Origin: Native Region: Andean Dept.: ANT, BOG, BOY, CUN, HUI, MET, NAR, SAN, VAC Elev.: 500–3000 m a.s.l. IUCN Habitat code: 14 Conservation status: (IUCN) NE

1632 *Bouteloua aristidoides* (Kunth) Griseb. (Poaceae)
IPNI: 1049662-2
Habit: Herb Use: AF, EU, HF Origin: Native Region: Caribbean Dept.: LAG Elev.: 0–500 m a.s.l. IUCN Habitat code: 1 Conservation status: (IUCN) NE

1633 *Bouteloua curtipendula* (Michx.) Torr. (Poaceae)
IPNI: 1198286-2
Habit: Herb Use: AF, EU, MA Origin: Native Region: Andean Dept.: BOY, CAL, HUI, NAR, QUI, SAN, TOL, VAC Elev.: 500–2500 m a.s.l. IUCN Habitat code: 1,2,14 Conservation status: (IUCN) NE

1634 *Bouteloua simplex* Lag. (Poaceae)
IPNI: 328564-2
Habit: Herb Use: ME Origin: Native Region: Andean Dept.: BOG, BOY, CAL, CUN Elev.: 2000–3000 m a.s.l. IUCN Habitat code: 14 Conservation status: (IUCN) NE

1635 *Briza maxima* L. (Poaceae)
IPNI: 393293-1
Common name: ilusión Habit: Herb Use: EU, HF, ME Origin: Naturalised Region: Andean Dept.: ANT, BOY, MAG, NAR Elev.: 2000–3000 m a.s.l. IUCN Habitat code: 14 Conservation status: (IUCN) NE

1636 *Briza minor* L. (Poaceae)
IPNI: 393299-1
Common name: corazoncito, corazoncitos, ilusión, realidad Habit: Herb Use: AF, EU, ME Origin: Naturalised Region: Andean Dept.: ANT, BOG, BOY, CAL, CUN, NAR, PUT Elev.: 2500–3500 m a.s.l. IUCN Habitat code: 1,3,14 Conservation status: (IUCN) NE

1637 *Bromus berteroanus* Colla (Poaceae)
IPNI: 393559-1
Habit: Herb Use: AF Origin: Native Region: Andean Conservation status: (IUCN) NE

1638 *Bromus carinatus* Hook. & Arn. (Poaceae)
IPNI: 30065892-2
Use: AF, EU, ME Region: Andean Dept.: BOG IUCN Habitat code: 1 Conservation status: (IUCN) NE

1639 *Bromus catharticus* Vahl (Poaceae)
IPNI: 37001-2
Common name: cebadilla, cebadilla criolla, criolla, hierba de perro, maleza de perro, paja de perro, pajaeperro, pasto blanco, pasto rescate, triguillo, vendeaguja de montaña, yerba de perro Habit: Herb Use: AF, EU, HF, ME Origin: Native Region: Andean Dept.: ANT, BOG, BOY, CAL, CAU, CUN, HUI, MAG, MET, NAR, NSA, PUT, QUI, RIS, TOL, VAC Elev.: 1000–4000 m a.s.l. IUCN Habitat code: 1,3,4,14 Conservation status: (IUCN) NE

1640 *Bromus pitensis* Kunth (Poaceae)
IPNI: 393995-1
Habit: Herb Use: ME Origin: Native Region: Andean Dept.: ANT, BOG, BOY, CAL, CAU, CUN, HUI, MET, NAR, NSA, QUI, RIS, SAN, TOL, VAC Elev.: 2500–4000 m a.s.l. IUCN Habitat code: 3,4,8,14 Conservation status: (IUCN) NE

1641 *Bromus secalinus* L. (Poaceae)
IPNI: 394095-1
Habit: Herb Use: AF, HF Origin: Naturalised Conservation status: (IUCN) NE

1642 *Catabrosa aquatica* (L.) P.Beauv. (Poaceae)
IPNI: 395086-1
Habit: Herb Use: HF, ME Origin: Native Region: Andean Conservation status: (IUCN) LC

1643 *Cenchrus americanus* (L.) Morrone (Poaceae)
IPNI: 77105978-1
Habit: Herb Use: AF, HF, ME Origin: Naturalised Dept.: BOG Conservation status: (IUCN) NE

1644 *Cenchrus brownii* Roem. & Schult. (Poaceae)
IPNI: 395204-1
Common name: abrojo, cadillo, cadillo carretón, garretón Habit: Herb Use: ME Origin: Native Region: Andean, Guiana Shield, Caribbean, Orinoquia Dept.: ANT, ATL, BOL, CAL, COR, CUN, HUI, LAG, MAG, RIS, SAP, SAN, TOL, VAC, VID Elev.: 0–1000 m a.s.l. IUCN Habitat code: 1,2,3,4,5,14 Conservation status: (IUCN) NE

1645 *Cenchrus ciliaris* L. (Poaceae)
IPNI: 395214-1
Common name: pasto bufel Habit: Herb Use: AF, EU, GS, HF, MA, ME, PO Origin: Naturalised Region: Andean, Caribbean Dept.: ANT, CES, CUN, HUI, LAG, MAG, SAN, TOL, VAC Elev.: 0–1000 m a.s.l. IUCN Habitat code: 1,2,3,5,14 Conservation status: (IUCN) LC

1646 *Cenchrus clandestinus* (Hochst. ex Chiov.) Morrone (Poaceae)
IPNI: 77105989-1
Habit: Herb Use: AF, EU, ME Origin: Naturalised Region: Andean Dept.: ANT, BOG, BOY, CAL, CAQ, CAU, CES, CHO, CUN, HUI, MAG, MET, NAR, NSA, PUT, QUI, RIS, SAN, TOL, VAC Elev.: 1500–4000 m a.s.l. IUCN Habitat code: 1,3,4,5,14 Conservation status: (IUCN) LC

1647 *Cenchrus echinatus* L. (Poaceae)
IPNI: 316518-2
Common name: abrojo, cachorro, cadillo, cadillo arroz, cadillo carretón, cadillo carretón morado, cadillo correntino, cadillo de paja, cadillo tigre, carretón, carretón morado, garretón, pega-pega, pegapega, purula, roseta Habit: Herb Use: AF, HF, ME Origin: Native Region: Andean, Guiana Shield, Caribbean, Orinoquia Dept.: ANT, ATL, BOL, CAL, CAS, CAU, HUI, MAG, NAR, QUI, SAP, SAN, SUC, TOL, VAC, VID Elev.: 0-1500 m.a.s.l. IUCN Habitat code: 1,2,3,5,14 Conservation status: (IUCN) NE (CPSG) Potential LC

1648 *Cenchrus longisetus* M.C.Johnst. (Poaceae)
IPNI: 51255-2
Habit: Herb Use: AF, EU, ME Origin: Cultivated, Naturalised Region: Andean, Caribbean Dept.: ANT, BOL, BOY, CAL, CES, CUN, HUI, NSA, QUI, RIS, SAN, TOL, VAC Elev.: 0–3000 m a.s.l. IUCN Habitat code: 14 Conservation status: (IUCN) NE

1649 *Cenchrus myosuroides* Kunth (Poaceae)
IPNI: 51261-2
Habit: Herb Use: AF, EU, GS, ME Origin: Native Conservation status: (IUCN) NE

1650 *Cenchrus purpureus* (Schumach.) Morrone (Poaceae)
IPNI: 77106033-1
Habit: Herb Use: AF, EU, FU, GS, HF, IF, MA, ME, SU Origin: Cultivated, Naturalised Region: Andean, Guiana Shield, Caribbean, Orinoquia, Pacific Dept.: ANT, ARA, BOL, CAL, CAS, CAU, CHO, CUN, HUI, MET, NAR, SAN, TOL, VAC, VID Elev.: 0–3000 m a.s.l. IUCN Habitat code: 1,2,3,4,14 Conservation status: (IUCN) LC

1651 *Cenchrus setaceus* (Forssk.) Morrone (Poaceae)
IPNI: 77106046-1
Habit: Herb **Use:** AF, EU, MA **Origin:** Cultivated, Naturalised **Region:** Andean, Caribbean, Orinoquia **Dept.:** ANT, BOG, BOY, CAL, CES, CUN, MET, NAR, NSA, QUI, RIS, SAN, TOL, VAC **Elev.:** 0–3000 m a.s.l. **IUCN Habitat code:** 14 **Conservation status: (IUCN)** LC

1652 *Cenchrus spinifex* Cav. (Poaceae)
IPNI: 395325-1
Habit: Herb **Use:** HF, ME **Origin:** Native **Region:** Andean, Pacific **Dept.:** HUI, NAR **Elev.:** 0–1000 m a.s.l. **Conservation status: (IUCN)** NE

1653 *Chascolytrum monandrum* (Hack.) Essi, Longhi-Wagner & Souza-Chies (Poaceae)
IPNI: 77116853-1
Habit: Herb **Use:** ME **Origin:** Native **Region:** Andean **Dept.:** ANT, CAL, CAU, NAR **Elev.:** 2500–3500 m a.s.l. **IUCN Habitat code:** 1,14 **Conservation status: (IUCN)** NE

1654 *Chloris barbata* Sw. (Poaceae)
IPNI: 1168070-2
Common name: paja blanca, pasto blanco **Habit:** Herb **Use:** AF,ME **Origin:** Naturalised **Region:** Andean, Guiana Shield, Caribbean, Orinoquia, Pacific **Dept.:** ARA, ATL, BOL, BOY, CUN, HUI, LAG, MAG, MET, NAR, SAP, SAN, TOL, VAC, VID **Elev.:** 0-2500 m.a.s.l. **IUCN Habitat code:** 1,2,3,14 **Conservation status: (IUCN)** NE **(CPSG)** Potential LC

1655 *Chloris gayana* Kunth (Poaceae)
IPNI: 395891-1
Common name: cosia **Habit:** Herb **Use:** AF, EU, MA, ME **Origin:** Naturalised **Region:** Andean **Dept.:** ANT, BOG, BOY, CAL, CUN, TOL, VAC **Elev.:** 500–2500 m a.s.l. **IUCN Habitat code:** 1,3,14 **Conservation status: (IUCN)** NE

1656 *Chloris pycnothrix* Trin. (Poaceae)
IPNI: 395997-1
Habit: Herb **Use:** AF **Origin:** Native **Dept.:** VAC **Conservation status: (IUCN)** NE

1657 *Chloris virgata* Sw. (Poaceae)
IPNI: 396063-1
Habit: Herb **Use:** AF, EU, HF, ME **Origin:** Native **Region:** Andean **Dept.:** CUN, HUI, VAC **Elev.:** 500–2500 m a.s.l. **IUCN Habitat code:** 1,14 **Conservation status: (IUCN)** NE

1658 *Chrysopogon aciculatus* (Retz.) Trin. (Poaceae)
IPNI: 396110-1
Habit: Herb **Use:** AF, EU, MA, ME **Origin:** Naturalised **Region:** Andean **Dept.:** ANT, SAN **Elev.:** 0–1000 m a.s.l. **Conservation status: (IUCN)** NE

1659 *Chrysopogon zizanioides* (L.) Roberty (Poaceae)
IPNI: 396213-1
Common name: mandrágora, vetiver **Habit:** Herb **Use:** AF, EU, HF, MA, ME, PO, SU **Origin:** Cultivated, Naturalised **Region:** Andean **Dept.:** CUN, TOL **Elev.:** 0–2000 m a.s.l. **IUCN Habitat code:** 14 **Conservation status: (IUCN)** NE

1660 *Chusquea scandens* Kunth (Poaceae)
IPNI: 1135523-2
Common name: carrizo, chusque **Habit:** Shrub, Climbing **Use:** EU, MA **Origin:** Native **Region:** Andean **Dept.:** ANT, BOG, CAU, CUN, HUI, NAR, NSA, QUI, SAN, TOL, VAC **Elev.:** 2500–3500 m a.s.l. **IUCN Habitat code:** 1,3,4,8,14 **Conservation status: (IUCN)** NE

1661 *Chusquea tessellata* Munro (Poaceae)
IPNI: 274648-2
Common name: carrizo, chusque, chusquejón **Habit:** Shrub, Subshrub **Use:** EU, MA **Origin:** Native **Region:** Andean **Dept.:** ANT, ARA, BOG, BOL, BOY, CAU, CES, CUN, HUI, MAG, MET, NSA, PUT, QUI, RIS, SAN, TOL, VAC **Elev.:** 2800–4350 m a.s.l. **IUCN Habitat code:** 1,2,3,4,8,14 **Conservation status: (IUCN)** NE

1662 *Coix lacryma-jobi* L. (Poaceae)
IPNI: 30100521-2
Common name: camándula, cuenta de milagro, cuentas de milagro, lágrimas de Job, lágrimas de San Pedro, milagro, moisés, ojo de Dios, sampedrito, sorgo, sukumusha, trigo, trigo amazónico, trigo brasilero, trigo peruano, yerba de milagro **Habit:** Herb **Use:** AF, EU, GS, HF, IF, MA, ME, SU **Origin:** Naturalised **Region:** Amazonia, Andean, Guiana Shield, Caribbean, Orinoquia, Pacific **Dept.:** AMA, ANT, ARA, CAQ, CAU, CES, CHO, CUN, GUA, GUV, HUI, LAG, MET, NAR, PUT, RIS, TOL, VAC, VAU **Elev.:** 0–2000 m a.s.l. **IUCN Habitat code:** 1,2,3,8,14 **Conservation status: (IUCN)** NE

1663 *Cortaderia jubata* (Lemoine) Stapf (Poaceae)
IPNI: 66302-2
Habit: Herb **Use:** EU, ME **Origin:** Native **Region:** Andean **Dept.:** BOY, CUN **Elev.:** 2000–3000 m a.s.l. **IUCN Habitat code:** 3,14 **Conservation status: (IUCN)** NE

1664 *Cortaderia nitida* (Kunth) Pilg. (Poaceae)
IPNI: 66306-2
Common name: carrizo, cortadera, cortadero, pajita del Niño Dios, pajita del pesebre, rabo de zorro, sixe **Habit:** Herb **Use:** ME **Origin:** Native **Region:** Andean **Dept.:** ANT, BOG, BOY, CAL, CAU, CHO, CUN, HUI, MET, NAR, PUT, RIS, SAN, TOL, VAC **Elev.:** 2500–4000 m a.s.l. **IUCN Habitat code:** 1,3,4,5,8,14 **Conservation status: (IUCN)** NE

1665 *Cortaderia selloana* (Schult. & Schult.f.) Asch. & Graebn. (Poaceae)
IPNI: 66317-2
Common name: penacho, pluma, plumón **Habit:** Herb **Use:** AF, EU, HF, ME **Origin:** Cultivated, Naturalised **Region:** Andean **Dept.:** ANT, BOG, BOY, CAL, CUN, QUI, RIS, SAN, VAC **Elev.:** 1000–3000 m a.s.l. **IUCN Habitat code:** 1,3,14 **Conservation status: (IUCN)** NE

1666 *Cymbopogon caesius* (Hook. & Arn.) Stapf (Poaceae)
IPNI: 396888-1
Use: AF, EU, HF, MA, ME, PO
Conservation status: (IUCN) NE

1667 *Cymbopogon citratus* (DC.) Stapf (Poaceae)
IPNI: 396896-1
Common name: caña limonaria, citronela, fever grass, guayacana, herb de limón, hierba de limón, hierba limón, hierba luisa, limonaria, limoncillo, limonera, marialuisa, molojillo criollo, pachulí, paja de limón, paja limón, sitwonel, toronjil de paja, té limón, yerba **Habit:** Herb **Use:** AF, EU, GS, HF, MA, ME, PO, SU **Origin:** Cultivated, Naturalised **Region:** Amazonia, Andean, Caribbean, Orinoquia, Pacific **Dept.:** AMA, ANT, BOG, BOL, CAL, CAU, CES, CHO, CUN, HUI, MAG, MET, RIS, SAP, TOL, VAC **Elev.:** 0–3000 m a.s.l. **IUCN Habitat code:** 1,2,3,14 **Conservation status: (IUCN)** NE

1668 *Cymbopogon nardus* (L.) Rendle (Poaceae)
IPNI: 396972-1
Common name: citronela, limoncillo **Habit:** Herb **Use:** AF, EU, HF, MA, ME **Origin:** Cultivated, Naturalised **Region:** Andean **Dept.:** CAL, CUN **Elev.:** 1500–2000 m a.s.l. **IUCN Habitat code:** 1,2,3,14 **Conservation status: (IUCN)** NE

1669 *Cymbopogon winterianus* Jowitt ex Bor (Poaceae)
IPNI: 397039-1
Habit: Herb **Use:** EU, HF, MA, ME **Origin:** Cultivated, Naturalised **Region:** Andean **Dept.:** ANT **Elev.:** 1500–2000 m a.s.l. **Conservation status: (IUCN)** NE

1670 *Cynodon dactylon* (L.) Pers. (Poaceae)
IPNI: 397065-1
Common name: argentina, barrenillo, bermuda, cruceta, crítica, grama, grama de gallina, grama dulce, gramilla brava, gramilla colorada, paja de la Virgen, pasto argentina, pasto argentino, pasto bermuda, pasto de gallina, pata de perdiz, pelo de vieja, tejedora, yerba **Habit:** Herb **Use:** AF, EU, HF, MA, ME, PO, SU **Origin:** Naturalised **Region:** Amazonia, Andean, Guiana Shield, Caribbean, Orinoquia, Pacific **Dept.:** AMA, ANT, ARA, ATL, BOL, BOY, CAL, CAQ, CAS, CAU, CES, CHO, COR, CUN, GUA, GUV, HUI, LAG, MAG, MET, NAR, NSA, PUT, QUI, RIS, SAP, SAN, SUC, TOL, VAC, VAU, VID **Elev.:** 0–3000 m a.s.l. **IUCN Habitat code:** 1,2,3,4,14 **Conservation status: (IUCN)** NE

1671 *Cynodon nlemfuensis* Vanderyst (Poaceae)
IPNI: 126195-3
Common name: enredalocas, estrella, estrella africana, pasto estrella, tumbabobos, tumbaviejos **Habit:** Herb **Use:** AF, EU, MA, ME, PO **Origin:** Naturalised **Region:** Andean, Orinoquia, Pacific **Dept.:** ANT, CAU, CHO, CUN, HUI, MET, QUI, RIS, TOL, VAC **Elev.:** 0–2000 m a.s.l. **IUCN Habitat code:** 1,2,3,4,14 **Conservation status: (IUCN)** NE

1672 *Dactylis glomerata* L. (Poaceae)
IPNI: 397309-1
Common name: bermuda, dorchoro, grama, ochor, ochoro, orchoro, orochoa, pasto argentina, pasto azul, pasto ochor, pasto ochoro, pasto orchoro, pasto orillo, pastoazul, raigrás **Habit:** Herb **Use:** AF, EU, HF, ME, PO **Origin:** Naturalised **Region:** Andean **Dept.:** ANT, BOG, BOY, CAL, CAQ, CES, CUN, HUI, LAG, MAG, NAR, NSA, PUT, QUI, RIS, SAN, TOL, VAC **Elev.:** 2000–4500 m a.s.l. **IUCN Habitat code:** 1,3,4,8,14 **Conservation status: (IUCN)** NE

1673 *Dactyloctenium aegyptium* (L.) Willd. (Poaceae)
IPNI: 397387-1
Common name: grama egipcia, mindaca, mindacá, paja tresdedos, tresdedos, yerba de Egipto **Habit:** Herb **Use:** AF, EU, GS, HF, MA, ME, PO **Origin:** Naturalised **Region:** Amazonia, Andean, Guiana Shield, Caribbean, Orinoquia **Dept.:** AMA, ANT, ARA, ATL, BOG, BOL, CAL, CAS, CAU, CES, COR, CUN, HUI, LAG, MAG, MET, NAR, RIS, SAP, SAN, SUC, TOL, VAC, VID **Elev.:** 0–3000 m a.s.l. **IUCN Habitat code:** 1,2,3,5,14 **Conservation status: (IUCN)** NE

1674 *Dendrocalamus asper* (Schult.f.) Backer (Poaceae) **IPNI:** 397862-1 **Habit:** Herb **Use:** EU, GS, HF, MA, ME **Origin:** Cultivated, Naturalised **Region:** Andean **Dept.:** VAC **Elev.:** 1000–1500 m a.s.l. **Conservation status: (IUCN)** NE

1675 *Dichanthelium acuminatum* (Sw.) Gould & C.A.Clark (Poaceae) **IPNI:** 79328-2 **Habit:** Herb **Use:** ME **Origin:** Native **Region:** Amazonia, Andean, Guiana Shield, Pacific **Dept.:** ANT, CAL, CAU, CHO, CUN, HUI, MAG, NAR, RIS, SAN, TOL, VAU **Elev.:** 0–3000 m a.s.l. **IUCN Habitat code:** 1,2,3,14 **Conservation status: (IUCN)** NE

1676 *Dichanthelium sphaerocarpon* (Elliott) Gould (Poaceae) **IPNI:** 79391-2 **Habit:** Herb **Use:** ME **Origin:** Native **Region:** Andean **Dept.:** ANT, BOG, CAL, CAU, CUN, TOL **Elev.:** 1500–3000 m a.s.l. **Conservation status: (IUCN)** NE

1677 *Dichanthelium strigosum* (Muhl. ex Elliott) Freckmann (Poaceae) **IPNI:** 79395-2 **Habit:** Herb **Use:** ME **Origin:** Native **Region:** Andean **Dept.:** CAU, MAG **Elev.:** 2000–2500 m a.s.l. **Conservation status: (IUCN)** NE

1678 *Dichanthium annulatum* (Forssk.) Stapf (Poaceae) **IPNI:** 398528-1 **Common name:** climacuna, ángleton **Habit:** Herb **Use:** AF, EU, GS, HF, MA, ME **Origin:** Naturalised **Region:** Andean, Caribbean, Orinoquia **Dept.:** ANT, ARA, ATL, BOY, CAL, COR, CUN, HUI, SAP, SUC, TOL, VAC, VID **Elev.:** 0–1500 m a.s.l. **IUCN Habitat code:** 1,2,3,14 **Conservation status: (IUCN)** NE

1679 *Dichanthium aristatum* (Poir.) C.E.Hubb. (Poaceae) **IPNI:** 398529-1 **Common name:** angle, pasto ángleton, solana, ángleto, ángleton, ánglica, ánglico, ánglito, ángliton **Habit:** Herb **Use:** AF, EU **Origin:** Naturalised **Region:** Andean, Caribbean **Dept.:** ANT, BOG, CAU, COR, CUN, HUI, NAR, NSA, TOL, VAC **Elev.:** 0–1500 m a.s.l. **IUCN Habitat code:** 2,3,14 **Conservation status: (IUCN)** NE

1680 *Dichanthium caricosum* (L.) A.Camus (Poaceae) **IPNI:** 398532-1 **Habit:** Herb **Use:** AF, EU, ME **Origin:** Naturalised **Conservation status: (IUCN)** NE

1681 *Diectomis fastigiata* (Sw.) P.Beauv. (Poaceae) **IPNI:** 398608-1 **Habit:** Herb **Use:** AF, MA, ME **Origin:** Native **Region:** Andean, Caribbean, Orinoquia **Dept.:** CAS, HUI, MAG, MET, NAR, TOL, VID **Elev.:** 0–1500 m a.s.l. **IUCN Habitat code:** 2,4 **Conservation status: (IUCN)** NE

1682 *Digitaria abyssinica* (Hochst. ex A.Rich.) Stapf (Poaceae) **IPNI:** 1209084-2 **Habit:** Herb **Use:** AF, EU, GS, HF, ME **Origin:** Naturalised **Region:** Andean **Dept.:** BOG, CUN, HUI, QUI, VAC **Elev.:** 1000–2000 m a.s.l. **Conservation status: (IUCN)** NE

1683 *Digitaria andicola* Gir.-Cañas (Poaceae) **IPNI:** 77066998-1 **Habit:** Herb **Use:** GS **Origin:** Endemic **Region:** Andean **Dept.:** BOY, CUN, NAR, VAC **Elev.:** 1000–3000 m a.s.l. **IUCN Habitat code:** 14 **Conservation status: (IUCN)** LC

1684 *Digitaria argillacea* (Hitchc. & Chase) Fernald (Poaceae) **IPNI:** 398643-1 **Habit:** Herb **Use:** AF, GS **Origin:** Native **Region:** Andean, Caribbean **Dept.:** MAG, NSA **Elev.:** 0–1000 m a.s.l. **Conservation status: (IUCN)** NE

1685 *Digitaria bicornis* (Lam.) Roem. & Schult. (Poaceae) **IPNI:** 1218147-2 **Common name:** coneja, gaudín, guaduín, guardarrocío, guardarrocío colchado **Habit:** Herb **Use:** GS **Origin:** Naturalised **Region:** Amazonia, Andean, Guiana Shield, Caribbean, Orinoquia, Pacific **Dept.:** ANT, ARA, ATL, BOL, CAL, CAS, CHO, COR, CUN, GUV, HUI, LAG, MAG, MET, NSA, RIS, SAP, SAN, SUC, TOL, VAC, VAU, VID **Elev.:** 0–2000 m a.s.l. **IUCN Habitat code:** 1,2,3,4,5,14 **Conservation status: (IUCN)** NE

1686 *Digitaria californica* (Benth.) Henrard (Poaceae) **IPNI:** 80220-2 **Common name:** pasto plateado, rabo de zorro **Habit:** Herb **Use:** AF, EU, GS **Origin:** Native **Region:** Andean, Caribbean **Dept.:** BOY, CUN, HUI, LAG, NAR, NSA, SAN, TOL, VAC **Elev.:** 0–2000 m a.s.l. **IUCN Habitat code:** 1,2,3,14 **Conservation status: (IUCN)** LC

1687 *Digitaria cardenasiana* Gir.-Cañas (Poaceae) **IPNI:** 60427627-2 **Habit:** Herb **Use:** GS **Origin:** Endemic **Region:** Andean, Orinoquia **Dept.:** BOY, MET, SAN, VID **Elev.:** 0-2500 m.a.s.l. **IUCN Habitat code:** 1,4,14 **Conservation status: (IUCN)** EN **(CPSG)** EN

1688 *Digitaria ciliaris* (Retz.) Koeler (Poaceae) **IPNI:** 1038558-2 **Habit:** Herb **Use:** AF, EU, GS, HF, ME **Origin:** Naturalised **Region:** Amazonia, Andean, Guiana Shield, Caribbean, Orinoquia **Dept.:** AMA, ANT, BOL, BOY, CAL, CAQ, CAU, CES, COR, CUN, GUV, HUI, MAG, MET, NAR, NSA, RIS, SAN, TOL, VAC **Elev.:** 0-3000 m.a.s.l. **IUCN Habitat code:** 1,2,3,4,14 **Conservation status: (IUCN)** NE **(CPSG)** Potential LC

1689 *Digitaria curvinervis* (Hack.) Fernald (Poaceae) **IPNI:** 80233-2 **Habit:** Herb **Use:** GS **Origin:** Native **Conservation status: (IUCN)** LC

1690 *Digitaria dioica* Killeen & Rúgolo (Poaceae) **IPNI:** 301826-2 **Habit:** Herb **Use:** GS **Origin:** Native **Region:** Guiana Shield, Orinoquia **Dept.:** GUV, MET, VID **Elev.:** 0–500 m a.s.l. **IUCN Habitat code:** 1,3,4 **Conservation status: (IUCN)** LC

1691 *Digitaria eriantha* Steud. (Poaceae) **IPNI:** 398792-1 **Common name:** pangola, pasto pangola **Habit:** Herb **Use:** AF, EU, GS, MA, ME **Origin:** Naturalised **Region:** Andean, Caribbean, Orinoquia **Dept.:** ANT, ATL, COR, CUN, MAG, MET, SAP, SAN, TOL, VAC **Elev.:** 0–2000 m a.s.l. **IUCN Habitat code:** 1,3,4,14 **Conservation status: (IUCN)** NE

1692 *Digitaria fragilis* (Steud.) Luces (Poaceae) **IPNI:** 80253-2 **Habit:** Herb **Use:** GS **Origin:** Native **Region:** Caribbean **Dept.:** MAG **Elev.:** 0–500 m a.s.l. **Conservation status: (IUCN)** LC

1693 *Digitaria fuscescens* (J.Presl) Henrard (Poaceae) **IPNI:** 80254-2 **Habit:** Herb **Use:** EU, GS **Origin:** Naturalised **Region:** Amazonia, Guiana Shield, Orinoquia **Dept.:** AMA, CAS, MET, VID **Elev.:** 0–500 m a.s.l. **IUCN Habitat code:** 1,4,14 **Conservation status: (IUCN)** NE

1694 *Digitaria horizontalis* Willd. (Poaceae) **IPNI:** 329014-2 **Common name:** cebadilla, coneja, grama alta, guardarrocío, pasto conejo **Habit:** Herb **Use:** AF, GS, HF, ME **Origin:** Native **Region:** Amazonia, Andean, Guiana Shield, Caribbean, Orinoquia, Pacific **Dept.:** AMA, ATL, BOL, CAL, CAS, CES, CUN, GUV, HUI, MAG, MET, NAR, NSA, QUI, RIS, SAN, TOL, VAC, VAU, VID **Elev.:** 0-3000 m.a.s.l. **IUCN Habitat code:** 1,3,4,5,14 **Conservation status: (IUCN)** LC **(CPSG)** Potential LC

1695 *Digitaria insularis* (L.) Mez ex Ekman (Poaceae) **IPNI:** 398888-1 **Common name:** cabezona, cola de zorro, hierba de zorra, rabo de zorro **Habit:** Herb **Use:** GS, ME **Origin:** Native **Region:** Amazonia, Andean, Guiana Shield, Caribbean, Orinoquia, Pacific **Dept.:** AMA, ANT, BOL, BOY, CAL, CAQ, CAU, CES, CHO, CUN, GUV, HUI, LAG, MAG, MET, NAR, NSA, RIS, SAP, SAN, TOL, VAC, VAU, VID **Elev.:** 0-2500 m.a.s.l. **IUCN Habitat code:** 1,2,3,14 **Conservation status: (IUCN)** LC **(CPSG)** Potential LC

1696 *Digitaria lehmanniana* Henrard (Poaceae) **IPNI:** 80276-2 **Habit:** Herb **Use:** GS **Origin:** Native **Region:** Andean **Dept.:** CAU **Elev.:** 1500–2000 m a.s.l. **Conservation status: (IUCN)** LC

1697 *Digitaria leucites* (Trin.) Henrard (Poaceae) **IPNI:** 80280-2 **Habit:** Herb **Use:** GS **Origin:** Native **Region:** Andean **Dept.:** CUN **Elev.:** 1500–2000 m a.s.l. **Conservation status: (IUCN)** LC

1698 *Digitaria longiflora* (Retz.) Pers. (Poaceae) **IPNI:** 398933-1 **Habit:** Herb **Use:** AF, EU, GS, HF, MA, ME **Origin:** Naturalised **Region:** Caribbean **Dept.:** CES **Elev.:** 0–500 m a.s.l. **Conservation status: (IUCN)** NE

1699 *Digitaria nuda* Schumach. (Poaceae) **IPNI:** 329018-2 **Habit:** Herb **Use:** AF, GS, HF, ME **Origin:** Naturalised **Region:** Pacific **Dept.:** VAC **Elev.:** 0–500 m a.s.l. **Conservation status: (IUCN)** LC

1700 *Digitaria sacchariflora* (Raddi) Henrard (Poaceae) **IPNI:** 1050107-2 **Habit:** Herb **Use:** GS **Origin:** Native **Region:** Andean **Dept.:** ANT, CAL, CUN, HUI, SAN, TOL, VAC **Elev.:** 0–

2000 m a.s.l. **IUCN Habitat code:** 2,14 **Conservation status: (IUCN)** LC

1701 *Digitaria sanguinalis* (L.) Scop. (Poaceae)
IPNI: 399115-1
Common name: azulita, cebadilla, gramilla, gramilla zacate, guadín, guardarrocío, guardasereno, hierba de conejo, paja colchón, pasto chato, pasto de cuaresma, pata de gallina, pata de gallo, pategallina **Habit:** Herb **Use:** AF, GS, HF, ME **Origin:** Naturalised **Region:** Andean, Caribbean **Dept.:** ANT, ATL, CAQ, CAU, CHO, COR, CUN, HUI, MAG, NAR, SAN, TOL, VAC **Elev.:** 0–2500 m a.s.l. **IUCN Habitat code:** 1,2,3,14 **Conservation status: (IUCN)** NE

1702 *Digitaria sellowii* (C.Muell.) Henrard (Poaceae)
IPNI: 80323-2
Habit: Herb **Use:** GS **Origin:** Native **Region:** Amazonia, Andean **Dept.:** AMA, ANT, BOY, MAG, TOL **Elev.:** 0–2000 m a.s.l. **IUCN Habitat code:** 1 **Conservation status: (IUCN)** NE

1703 *Digitaria setigera* Roth (Poaceae)
IPNI: 1203032-2
Habit: Herb **Use:** AF, GS, ME **Origin:** Naturalised **Region:** Andean, Caribbean **Dept.:** ANT, QUI, SAP **Elev.:** 0–1500 m a.s.l. **IUCN Habitat code:** 3,14 **Conservation status: (IUCN)** NE

1704 *Digitaria violascens* Link (Poaceae)
IPNI: 399226-1
Habit: Herb **Use:** AF, GS, ME **Origin:** Naturalised **Region:** Amazonia, Andean, Guiana Shield, Caribbean, Orinoquia, Pacific **Dept.:** AMA, ANT, BOY, CAQ, CAU, CES, CHO, CUN, HUI, MAG, MET, NSA, QUI, RIS, SAN, TOL, VAC, VID **Elev.:** 0-2500 m.a.s.l. **IUCN Habitat code:** 3,14 **Conservation status: (IUCN)** NE **(CPSG)** Potential LC

1705 *Diplachne fusca* (L.) P.Beauv. ex Roem. & Schult. (Poaceae)
IPNI: 60458632-2
Habit: Herb **Use:** AF, EU, FU, MA, ME **Origin:** Native **Region:** Andean, Caribbean **Dept.:** ATL, BOL, LAG, MAG, SUC **Elev.:** 0–500 m a.s.l. **IUCN Habitat code:** 2 **Conservation status: (IUCN)** NE

1706 *Disakisperma dubium* (Kunth) P.M.Peterson & N.Snow (Poaceae)
IPNI: 60458636-2
Habit: Herb **Use:** AF **Origin:** Native **Region:** Andean **Dept.:** NAR **Elev.:** 1500–2000 m a.s.l. **Conservation status: (IUCN)** NE

1707 *Distichlis spicata* (L.) Greene (Poaceae)
IPNI: 82343-2
Common name: pasto de salitre
Habit: Herb **Use:** AF, EU, HF, ME **Origin:** Native **Region:** Andean **Dept.:** BOY **Elev.:** 2000–2500 m a.s.l. **Conservation status: (IUCN)** LC

1708 *Echinochloa colonum* (L.) Link (Poaceae)
IPNI: 399655-1
Common name: arrocillo, grama salada, hierba de pisingo, liendra de puerco, liendre de puerco, paja de arroz, paja de burro, paja de pato, paja salada, pasto colorado, pasto de cuaresma, pasto de pisingo, pasto janeiro, pasto liendre de puerco, taebũ **Habit:** Herb **Use:** AF, EU, GS, HF, MA, ME, PO, SU **Origin:** Naturalised **Region:** Amazonia, Andean, Guiana Shield, Caribbean, Orinoquia, Pacific **Dept.:** AMA, ANT, ARA, ATL, BOL, BOY, CAL, CAQ, CAS, CAU, CES, CHO, COR, CUN, GUV, HUI, LAG, MAG, MET, NAR, NSA, PUT, QUI, RIS, SAP, SAN, SUC, TOL, VAC, VAU, VID **Elev.:** 0-2000 m.a.s.l. **IUCN Habitat code:** 14 **Conservation status: (IUCN)** NE **(CPSG)** Potential LC

1709 *Echinochloa crus-galli* (L.) P.Beauv. (Poaceae)
IPNI: 399660-1
Habit: Herb **Use:** AF, EU, GS, HF, ME **Origin:** Naturalised **Region:** Andean, Caribbean **Dept.:** BOY, HUI, MAG, QUI, RIS, TOL, VAC **Elev.:** 0–2500 m a.s.l. **IUCN Habitat code:** 1,3,5 **Conservation status: (IUCN)** LC

1710 *Echinochloa crus-pavonis* (Kunth) Schult. (Poaceae)
IPNI: 30149942-2
Common name: alemán, barba de indio, barba roja, gramalote, liendra de puerco, liendre de puerco, moco de pavo, mono alemán, pasto alemán, pasto colorado, pata de gallo, peludo **Habit:** Herb **Use:** AF, GS, HF, ME **Origin:** Naturalised **Region:** Andean, Guiana Shield, Caribbean **Dept.:** ANT, BOL, CAU, COR, CUN, HUI, NSA, SAN, TOL, VAC, VID **Elev.:** 0–2500 m a.s.l. **IUCN Habitat code:** 1,2,3,4,5,14 **Conservation status: (IUCN)** LC

1711 *Echinochloa polystachya* (Kunth) Hitchc. (Poaceae)
IPNI: 88097-2
Common name: alemana, alemán, angola, canutillo, gramalote, pasto alemán, pasto de yulo **Habit:** Herb **Use:** AF, GS **Origin:** Native **Region:** Amazonia, Andean, Guiana Shield, Caribbean, Orinoquia **Dept.:** AMA, ANT, ARA, BOL, CAS, COR, GUV, HUI, SAN, SUC, TOL, VAU, VID **Elev.:** 0–500 m a.s.l. **IUCN Habitat code:** 1,2,3,5,14 **Conservation status: (IUCN)** LC

1712 *Echinolaena inflexa* (Poir.) Chase (Poaceae)
IPNI: 88264-2
Common name: pasto dominícola, pasto dulce **Habit:** Herb **Use:** AF **Origin:** Native **Region:** Andean, Guiana Shield, Orinoquia **Dept.:** MET, SAN, VID **Elev.:** 0–1500 m a.s.l. **IUCN Habitat code:** 1,2,3,4,14 **Conservation status: (IUCN)** NE

1713 *Eleusine indica* (L.) Gaertn. (Poaceae)
IPNI: 74675-3
Common name: cola de burro, espartillo, grama, grama de horqueta, grama de matojo, grama de moño, gramablanca, herb dulce, hierba dulce, horquetilla, matojo dulce, mogote, paja de burro, pasto estrada, pata de gallina, pata de gallo, pataegallina, pategallina, rabo de **Habit:** Herb **Use:** AF, EU, GS, HF, MA, ME, PO, SU **Origin:** Naturalised **Region:** Amazonia, Andean, Guiana Shield, Caribbean, Orinoquia, Pacific **Dept.:** AMA, ANT, ARA, ATL, BOG, BOL, BOY, CAL, CAQ, CAS, CAU, CES, CHO, COR, CUN, GUA, GUV, HUI, LAG, MAG, MET, NAR, NSA, PUT, QUI, RIS, SAP, SAN, SUC, TOL, VAC, VAU, VID **Elev.:** 0–3000 m a.s.l. **IUCN Habitat code:** 1,2,3,4,5,14 **Conservation status: (IUCN)** LC

1714 *Eleusine tristachya* (Lam.) Lam. (Poaceae)
IPNI: 30127333-2
Habit: Herb **Use:** GS, HF, ME **Origin:** Naturalised **Region:** Caribbean **Dept.:** BOL **Elev.:** 0–500 m a.s.l. **IUCN Habitat code:** 14 **Conservation status: (IUCN)** LC

1715 *Elionurus ciliaris* Kunth (Poaceae)
IPNI: 400052-1
Habit: Herb **Use:** AF, MA, SU **Origin:** Native **IUCN Habitat code:** 3,4,14 **Conservation status: (IUCN)** NE

1716 *Elionurus muticus* (Spreng.) Kuntze (Poaceae)
IPNI: 90411-2
Common name: leche de miel, pasto de estero **Habit:** Herb **Use:** AF, MA, ME **Origin:** Native **Region:** Andean, Orinoquia **Dept.:** BOG, HUI, MET, TOL, VID **Elev.:** 0–1000 m a.s.l. **Conservation status: (IUCN)** NE

1717 *Elymus cordilleranus* Davidse & R.W.Pohl (Poaceae)
IPNI: 961772-1
Habit: Herb **Use:** GS **Origin:** Native **Region:** Andean **Dept.:** ANT, BOG, BOY, CAL, CUN, NAR, RIS **Elev.:** 2500–4000 m a.s.l. **IUCN Habitat code:** 3,4,14 **Conservation status: (IUCN)** NE

1718 *Enneapogon desvauxii* P.Beauv. (Poaceae)
IPNI: 400626-1
Habit: Herb **Use:** AF, EU, HF, IF **Origin:** Native **Conservation status: (IUCN)** NE

1719 *Eragrostis atrovirens* (Desf.) Trin. ex Steud. (Poaceae)
IPNI: 400813-1
Habit: Herb **Use:** AF,ME **Origin:** Naturalised **Region:** Guiana Shield, Orinoquia, Pacific **Dept.:** VAC, VID **Elev.:** 0–500 m a.s.l. **IUCN Habitat code:** 1,3,14 **Conservation status: (IUCN)** LC

1720 *Eragrostis bahiensis* Schult. (Poaceae)
IPNI: 93057-2
Habit: Herb **Use:** AF,ME **Origin:** Native **Region:** Amazonia, Andean, Orinoquia, Pacific **Dept.:** ANT, ARA, BOY, CAL, CAU, CHO, MET, PUT, VAC **Elev.:** 0–1500 m a.s.l. **IUCN Habitat code:** 1,2,3,4,14 **Conservation status: (IUCN)** NE

1721 *Eragrostis barrelieri* Daveau (Poaceae)
IPNI: 322547-2
Habit: Herb **Use:** AF, HF **Origin:** Naturalised **Region:** Andean **Dept.:** ANT **Elev.:** 1500–2000 m a.s.l. **Conservation status: (IUCN)** NE

1722 *Eragrostis cilianensis* (All.) Vignolo ex Janch. (Poaceae)
IPNI: 1104599-2
Habit: Herb **Use:** AF, EU, HF, MA, ME, PO **Origin:** Naturalised **Region:** Andean, Guiana Shield, Caribbean **Dept.:** BOG, CES, CUN, HUI, MAG, NAR, NSA, SAN, TOL, VAC, VID **Elev.:** 0–3000 m a.s.l. **IUCN Habitat code:** 1,3,14 **Conservation status: (IUCN)** NE

1723 *Eragrostis ciliaris* (L.) R.Br. (Poaceae)
IPNI: 1156566-2
Habit: Herb **Use:** AF, EU, GS, HF, MA, ME, SU **Origin:** Naturalised **Region:** Andean, Guiana Shield, Caribbean, Orinoquia, Pacific **Dept.:** ANT, ARA, ATL, BOL, CAS, CES, COR, CUN, HUI, LAG, MAG, MET, NAR, NSA, SAP, SAN, TOL, VAC, VID **Elev.:** 0–1500 m a.s.l. **IUCN Habitat code:** 1,2,3,4,5,14 **Conservation status: (IUCN)** NE

1724 *Eragrostis curvula* (Schrad.) Nees (Poaceae)
IPNI: 1169955-2
Common name: pasto llorón **Habit:** Herb **Use:** AF, EU, GS, HF, ME **Origin:** Cultivated, Naturalised **Region:** Andean **Dept.:** ANT, VAC **Elev.:** 500–2500 m a.s.l. **Conservation status: (IUCN)** NE

1725 *Eragrostis gangetica* (Roxb.) Steud. (Poaceae)
IPNI: 401103-1
Habit: Herb **Use:** AF, EU, HF, ME **Origin:** Naturalised **Region:** Guiana Shield, Orinoquia **Dept.:** VID **Elev.:** 0–500 m a.s.l. **Conservation status: (IUCN)** LC

1726 *Eragrostis intermedia* Hitchc. (Poaceae)
IPNI: 93115-2
Habit: Herb **Use:** AF **Origin:** Native **Region:** Andean **Dept.:** BOG, CUN **Elev.:** 2000–3000 m a.s.l. **Conservation status: (IUCN)** NE

1727 *Eragrostis japonica* (Thunb.) Trin. (Poaceae)
IPNI: 401189-1
Habit: Herb **Use:** AF,ME **Origin:** Naturalised **Region:** Amazonia, Andean, Caribbean, Orinoquia **Dept.:** AMA, ATL, CAS, CUN, MAG, MET, TOL, VID **Elev.:** 0–500 m a.s.l. **IUCN Habitat code:** 1,3,4,5,14 **Conservation status: (IUCN)** LC

1728 *Eragrostis maypurensis* (Kunth) Steud. (Poaceae)
IPNI: 1147585-2
Habit: Herb **Use:** HF **Origin:** Native **Region:** Amazonia, Andean, Guiana Shield, Orinoquia, Pacific **Dept.:** AMA, ANT, ARA, BOY, CAS, CAU, GUA, GUV, HUI, MAG, MET, NAR, SAN, TOL, VAC, VID **Elev.:** 0-1500 m.a.s.l. **IUCN Habitat code:** 1,2,3,4,5,14 **Conservation status: (IUCN)** NE **(CPSG)** Potential LC

1729 *Eragrostis mexicana* (Hornem.) Link (Poaceae)
IPNI: 1141212-2
Habit: Herb **Use:** HF **Origin:** Native **Region:** Andean **Dept.:** ANT, BOY, CAU, CUN, HUI, NAR, QUI, RIS, SAN, TOL, VAC **Elev.:** 1000–3000 m a.s.l. **IUCN Habitat code:** 1,2,14 **Conservation status: (IUCN)** NE

1730 *Eragrostis mokensis* Pilg. (Poaceae)
IPNI: 322530-2
Habit: Herb **Use:** AF **Origin:** Naturalised **Region:** Andean **Dept.:** CAU **Elev.:** 1500–2000 m a.s.l. **Conservation status: (IUCN)** NE

1731 *Eragrostis pectinacea* (Michx.) Nees (Poaceae)
IPNI: 1015280-2
Habit: Herb **Use:** HF **Origin:** Native **Region:** Amazonia, Andean, Guiana Shield, Caribbean, Orinoquia, Pacific **Dept.:** ANT, BOG, BOY, CAL, CAQ, CAS, CAU, CES, CHO, CUN, HUI, MET, NSA, QUI, RIS, SAP, SAN, TOL, VAC, VID **Elev.:** 0-3000 m.a.s.l. **IUCN Habitat code:** 3,4,14 **Conservation status: (IUCN)** NE **(CPSG)** Potential LC

1732 *Eragrostis pilosa* (L.) P.Beauv. (Poaceae)
IPNI: 1180620-2
Habit: Herb **Use:** AF, GS, HF, ME **Origin:** Naturalised **Region:** Amazonia, Andean, Guiana Shield, Caribbean, Orinoquia, Pacific **Dept.:** ANT, ARA, BOL, CAS, HUI, LAG, MAG, MET, NAR, QUI, SAN, TOL, VAC, VAU, VID **Elev.:** 0–2000 m a.s.l. **Conservation status: (IUCN)** NE

1733 *Eragrostis prolifera* (Sw.) Steud. (Poaceae)
IPNI: 1075146-2
Habit: Herb **Use:** EU, MA **Origin:** Native **Region:** Caribbean **Dept.:** ATL, BOL, SAP, SUC **Elev.:** 0–500 m a.s.l. **IUCN Habitat code:** 2,5 **Conservation status: (IUCN)** LC

1734 *Eragrostis secundiflora* J.Presl (Poaceae)
IPNI: 322572-2
Habit: Herb **Use:** AF, HF, ME **Origin:** Native **Region:** Andean, Orinoquia **Dept.:** CES, MET, SAN **Elev.:** 0–2000 m a.s.l. **IUCN Habitat code:** 3,5 **Conservation status: (IUCN)** NE

1735 *Eragrostis tenella* (L.) P.Beauv. ex Roem. & Schult. (Poaceae)
IPNI: 1088685-2
Common name: grama ilusión, paja ilusión, pasto amor **Habit:** Herb **Use:** AF, HF, ME **Origin:** Naturalised **Region:** Amazonia, Andean, Guiana Shield, Caribbean, Orinoquia, Pacific **Dept.:** AMA, ANT, ARA, ATL, BOL, BOY, CAL, CAQ, CAS, CAU, CES, CHO, COR, CUN, GUA, GUV, HUI, LAG, MAG, MET, NAR, NSA, PUT, QUI, RIS, SAP, SAN, SUC, TOL, VAC, VAU, VID **Elev.:** 0–2000 m a.s.l. **IUCN Habitat code:** 14 **Conservation status: (IUCN)** NE

1736 *Eragrostis tenuifolia* (A.Rich.) Hochst. ex Steud. (Poaceae)
IPNI: 1064189-2
Habit: Herb **Use:** ME **Origin:** Naturalised **Region:** Amazonia, Andean, Guiana Shield, Caribbean, Orinoquia **Dept.:** ANT, BOG, BOY, CAQ, CAU, CES, CUN, HUI, LAG, MAG, NAR, NSA, QUI, RIS, SAN, TOL, VAC, VID **Elev.:** 0–3000 m a.s.l. **IUCN Habitat code:** 1,3,14 **Conservation status: (IUCN)** NE

1737 *Eragrostis viscosa* (Retz.) Trin. (Poaceae)
IPNI: 401703-1
Habit: Herb **Use:** AF, EU, MA, ME **Origin:** Naturalised **Region:** Andean, Caribbean **Dept.:** LAG, MAG, NSA, VAC **Elev.:** 0–2000 m a.s.l. **Conservation status: (IUCN)** NE

1738 *Eremocaulon aureofimbriatum* Soderstr. & Londoño (Poaceae)
IPNI: 316770-2
Habit: Herb **Use:** EU **Origin:** Cultivated **Region:** Andean **Dept.:** QUI **Elev.:** 1500–2000 m a.s.l. **Conservation status: (IUCN)** NE

1739 *Eriochloa aristata* Vasey (Poaceae)
IPNI: 94589-2
Habit: Herb **Use:** HF **Origin:** Native **Conservation status: (IUCN)** NE

1740 *Eriochloa procera* (Retz.) C.E.Hubb. (Poaceae)
IPNI: 402044-1
Habit: Herb **Use:** AF,ME **Origin:** Naturalised **Region:** Caribbean **Dept.:** COR **Elev.:** 0–500 m a.s.l. **IUCN Habitat code:** 5 **Conservation status: (IUCN)** LC

1741 *Eriochloa punctata* (L.) Ham. (Poaceae)
IPNI: 1021229-2
Common name: alparada, parada **Habit:** Herb **Use:** ME **Origin:** Native **Region:** Andean, Guiana Shield, Caribbean, Orinoquia **Dept.:** ANT, BOL, CAS, CUN, MAG, MET, SAN, TOL, VAC, VID **Elev.:** 0–1500 m a.s.l. **IUCN Habitat code:** 1,2,3,14 **Conservation status: (IUCN)** NE

1742 *Euclasta condylotricha* (Hochst. ex Steud.) Stapf (Poaceae)
IPNI: 97916-2
Habit: Herb **Use:** AF **Origin:** Native **Dept.:** MAG **Conservation status: (IUCN)** NE

1743 *Festuca bromoides* L. (Poaceae)
IPNI: 402472-1
Habit: Herb **Use:** AF **Origin:** Naturalised **Region:** Andean **Dept.:** ANT, BOY, CAL, CUN, HUI, NAR, NSA, QUI, RIS, SAN, TOL, VAC **Elev.:** 2000–3500 m a.s.l. **Conservation status: (IUCN)** NE

1744 *Festuca glauca* Vill. (Poaceae)
IPNI: 20008857-1
Common name: festuca azul, pasto azul, pasto chino **Habit:** Herb **Use:** EU **Origin:** Cultivated, Naturalised **Region:** Andean **Dept.:** BOG, CUN **Elev.:** 2000–3000 m a.s.l. **IUCN Habitat code:** 14 **Conservation status: (IUCN)** NE

1745 *Festuca myuros* L. (Poaceae)
IPNI: 403134-1
Habit: Herb **Use:** AF, EU, ME **Origin:** Naturalised **Region:** Andean **Dept.:** BOY, CAL, CUN, HUI, NAR, QUI, RIS, SAN, TOL, VAC **Elev.:** 2500–4000 m a.s.l. **IUCN Habitat code:** 14 **Conservation status: (IUCN)** NE

1746 *Festuca rubra* L. (Poaceae)
IPNI: 30010036-2
Habit: Herb **Use:** AF, EU,PO **Origin:** Naturalised **Region:** Andean **Dept.:** CAL, CUN **Elev.:** 2500–4000 m a.s.l. **Conservation status: (IUCN)** NE

1747 *Gigantochloa verticillata* (Willd.) Munro (Poaceae)
IPNI: 404020-1
Habit: Herb **Use:** EU, HF, MA, ME, SU **Origin:** Cultivated, Naturalised **Region:** Andean **Dept.:** VAC **Elev.:** 1000–1500 m a.s.l. **IUCN Habitat code:** 1 **Conservation status: (IUCN)** NE

1748 *Glyceria fluitans* (L.) R.Br. (Poaceae)
IPNI: 404100-1
Habit: Herb **Use:** HF, ME **Origin:** Naturalised **Region:** Andean **Dept.:** ANT, BOG, CAL, CUN, NAR **Elev.:** 2500–3000 m a.s.l. **IUCN Habitat code:** 4 **Conservation status: (IUCN)** LC

1749 *Guadua amplexifolia* J.Presl (Poaceae)
IPNI: 113747-2
Common name: cañabrava, cañaguadua, cañaguauda, guadua, guadúa, guauda, guaúda **Habit:** Tree **Use:** HF, MA **Origin:** Native **Region:** Caribbean, Orinoquia **Dept.:** ANT, ATL, BOL, CAS, CES, LAG **Elev.:** 10–300 m a.s.l. **IUCN Habitat code:** 1,2,3,14 **Conservation status: (IUCN)** NE

1750 *Guadua angustifolia* Kunth (Poaceae)
IPNI: 1074014-2
Common name: bambú, bambú guadua, bambú nativo, caña, caña hueca fina, cañuela, guabra, guada, guadua, guadua cebolla, guadua lisa, guadua macana, guadua macha, guadua rayada, guadúa, guafa, guafilla, guauda, guaudua, guaudúa, guaúda, yaripa **Habit:** Tree **Use:** AF, EU, FU, HF, MA, ME, SU **Origin:** Native **Region:** Andean **Dept.:** ANT, BOY, CAL,

CAQ, CAU, CHO, CUN, GUV, HUI, MET, NAR, NSA, PUT, QUI, RIS, SAN, TOL, VAC **Elev.:** 10-2250 m.a.s.l. **IUCN Habitat code:** 1,2,3,4,5,14 **Conservation status: (IUCN)** NE **(CPSG)** Potential LC

1751 *Gymnopogon spicatus* (Spreng.) Kuntze (Poaceae) **IPNI:** 115792-2 **Habit:** Herb **Use:** ME **Origin:** Native **Region:** Andean, Caribbean **Dept.:** MAG, SAN, TOL **Elev.:** 0–1500 m a.s.l. **IUCN Habitat code:** 14 **Conservation status: (IUCN)** NE

1752 *Gynerium sagittatum* (Aubl.) P.Beauv. (Poaceae) **IPNI:** 404486-1 **Common name:** berada, carrizo, caña amarga, caña de Castilla, caña de flecha, caña india, cañablanca, cañaboba, cañabrava, cañabrava playera, cañafina, cañaflecha, cañaflete, cañamansa, cañamenuda, cañapángala, chacla, chalare, chin, chin de Castilla, chulquín, flecha, **Habit:** Herb **Use:** EU, HF, MA, ME **Origin:** Native **Region:** Amazonia, Andean, Guiana Shield, Caribbean, Orinoquia, Pacific **Dept.:** AMA, ANT, ARA, ATL, BOL, BOY, CAL, CAQ, CAS, CAU, CES, CHO, COR, CUN, GUA, GUV, HUI, LAG, MAG, MET, NAR, NSA, PUT, QUI, RIS, SAN, SUC, TOL, VAC, VAU, VID **Elev.:** 0-2000 m.a.s.l. **IUCN Habitat code:** 1,3,5,14 **Conservation status: (IUCN)** NE **(CPSG)** Potential LC

1753 *Hackelochloa granularis* (L.) Kuntze (Poaceae) **IPNI:** 116848-2 **Habit:** Herb **Use:** AF, EU, HF, ME **Origin:** Naturalised **Region:** Andean, Guiana Shield, Caribbean, Orinoquia, Pacific **Dept.:** CAL, CHO, COR, CUN, HUI, MAG, MET, NSA, SAN, TOL, VAC, VID **Elev.:** 0–1500 m a.s.l. **IUCN Habitat code:** 1,3,5,14 **Conservation status: (IUCN)** NE

1754 *Hemarthria altissima* (Poir.) Stapf & C.E.Hubb. (Poaceae) **IPNI:** 404684-1 **Habit:** Herb **Use:** AF, HF, MA **Origin:** Naturalised **IUCN Habitat code:** 1 **Conservation status: (IUCN)** LC

1755 *Heteropogon contortus* (L.) P.Beauv. ex Roem. & Schult. (Poaceae) **IPNI:** 404772-1 **Habit:** Herb **Use:** AF, EU, IF, MA, ME, PO **Origin:** Native **Region:** Andean **Dept.:** ANT, BOY, CES, CUN, HUI, MAG, MET, NAR, SAN, TOL **Elev.:** 1000–3000 m a.s.l. **IUCN Habitat code:** 1,2,4,14 **Conservation status: (IUCN)** NE

1756 *Heteropogon melanocarpus* (Elliott) Benth. (Poaceae) **IPNI:** 404802-1 **Habit:** Herb **Use:** AF **Origin:** Native **Conservation status: (IUCN)** NE

1757 *Holcus lanatus* L. (Poaceae) **IPNI:** 405034-1 **Common name:** avena choclo, falsa poa, pasto lanoso, pasto lanudo, pasto poa, poa, poa azul, saboyá **Habit:** Herb **Use:** AF, EU, ME, PO **Origin:** Naturalised **Region:** Andean **Dept.:** ANT, BOG, BOY, CAL, CAU, CUN, HUI, LAG, MAG, NAR, NSA, PUT, QUI, RIS, SAN, TOL, VAC **Elev.:** 1500–4000 m a.s.l. **IUCN Habitat code:** 1,3,4,5,8,14 **Conservation status: (IUCN)** NE

1758 *Holcus mollis* L. (Poaceae) **IPNI:** 123530-2 **Habit:** Herb **Use:** EU **Origin:** Cultivated, Naturalised **Region:** Andean **Dept.:** BOG, CUN **Elev.:** 2500–3000 m a.s.l. **Conservation status: (IUCN)** NE

1759 *Homolepis glutinosa* (Sw.) Zuloaga & Soderstr. (Poaceae) **IPNI:** 278239-2 **Common name:** cola de caballo **Habit:** Herb **Use:** PO **Origin:** Native **Region:** Amazonia, Andean **Dept.:** AMA, ANT, CAL, CAU, CUN, HUI, MAG, NAR, RIS, SAN, VAC **Elev.:** 0–2000 m a.s.l. **IUCN Habitat code:** 1,3,14 **Conservation status: (IUCN)** NE

1760 *Hordeum muticum* J.Presl (Poaceae) **IPNI:** 278265-2 **Habit:** Herb **Use:** GS **Origin:** Native **Region:** Andean **Dept.:** BOY **Conservation status: (IUCN)** LC

1761 *Hordeum vulgare* L. (Poaceae) **IPNI:** 405391-1 **Common name:** cebada, cebada anchuela **Habit:** Herb **Use:** AF, EU, FU, GS, HF, MA, ME, PO, SU **Origin:** Cultivated, Naturalised **Region:** Andean **Dept.:** BOG, BOY, CUN **Elev.:** 2500–3000 m a.s.l. **IUCN Habitat code:** 3,4,14 **Conservation status: (IUCN)** NE

1762 *Hymenachne amplexicaulis* (Rudge) Nees (Poaceae) **IPNI:** 405429-1 **Common name:** canutillo, churro, gramalote, pasto muczuro, yerbalancha **Habit:** Herb **Use:** AF, EU, FU, HF, ME **Origin:** Native **Region:** Amazonia, Andean, Guiana Shield, Caribbean, Orinoquia, Pacific **Dept.:** AMA, ANT, ARA, BOL, CAS, CAU, CES, CHO, COR, GUV, LAG, MAG, MET, NAR, PUT, SAN, SUC, TOL, VAC, VAU, VID **Elev.:** 0-1000 m.a.s.l. **IUCN Habitat code:** 1,2,3,4,5,14 **Conservation status: (IUCN)** NE **(CPSG)** Potential LC

1763 *Hymenachne donacifolia* (Raddi) Chase (Poaceae) **IPNI:** 125410-2 **Habit:** Herb **Use:** ME **Origin:** Native **Region:** Amazonia, Andean, Guiana Shield, Caribbean, Orinoquia **Dept.:** ANT, CAQ, HUI, MAG, MET, PUT, SAN, TOL, VID **Elev.:** 0–1000 m a.s.l. **Conservation status: (IUCN)** NE

1764 *Hymenachne grandis* (Hitchc. & Chase) Zuloaga (Poaceae) **IPNI:** 77200715-1 **Habit:** Herb **Use:** ME **Origin:** Native **Region:** Caribbean **Dept.:** BOL, SAN, VAC **Elev.:** 0–500 m a.s.l. **Conservation status: (IUCN)** NE

1765 *Hyparrhenia rufa* (Nees) Stapf (Poaceae) **IPNI:** 405583-1 **Common name:** araguá, faragua, pasto puntero, pasto yaraguá, puntero, uribe, yaguará, yaguará uribe, yaraguá, yaraguá Uribe, yaraguá legítimo, yaraguá peludo **Habit:** Herb **Use:** AF, EU, HF, MA **Origin:** Naturalised **Region:** Amazonia, Andean, Guiana Shield, Caribbean, Orinoquia, Pacific **Dept.:** AMA, ANT, ARA, ATL, BOL, BOY, CAL, CAQ, CAS, CAU, CES, CHO, COR, CUN, GUA, GUV, HUI, LAG, MAG, MET, NAR, NSA, PUT, QUI, RIS, SAN, SUC, TOL, VAC, VAU, VID **Elev.:** 0–2000 m a.s.l. **IUCN Habitat code:** 1,2,3,4,5,14 **Conservation status: (IUCN)** NE

1766 *Hyperthelia dissoluta* (Nees ex Steud.) Clayton (Poaceae) **IPNI:** 126405-2 **Habit:** Herb **Use:** AF, MA **Origin:** Cultivated, Naturalised **Dept.:** MAG **Conservation status: (IUCN)** NE

1767 *Ichnanthus pallens* (Sw.) Munro ex Benth. (Poaceae) **IPNI:** 405749-1 **Common name:** guaduilla **Habit:** Herb **Use:** AF,ME **Origin:** Native **Region:** Amazonia, Andean, Guiana Shield, Caribbean, Orinoquia, Pacific **Dept.:** AMA, ANT, CAL, CAQ, CAU, CHO, CUN, GUV, HUI, LAG, MAG, MET, NAR, NSA, PUT, SAN, TOL, VAC, VAU, VID **Elev.:** 0-1500 m.a.s.l. **IUCN Habitat code:** 1,2,3,5,14 **Conservation status: (IUCN)** NE **(CPSG)** Potential LC

1768 *Imperata brasiliensis* Trin. (Poaceae) **IPNI:** 405808-1 **Common name:** guayacana, víbora **Habit:** Herb **Use:** ME **Origin:** Native **Region:** Andean, Guiana Shield, Caribbean, Orinoquia **Dept.:** BOL, BOY, COR, MET, VID **Elev.:** 0–1000 m a.s.l. **IUCN Habitat code:** 1,2,3,4,5,14 **Conservation status: (IUCN)** NE

1769 *Imperata contracta* Hitchc. (Poaceae) **IPNI:** 405817-1 **Common name:** guayacana, sauye, suaye, vendeaguja, vendiaguja **Habit:** Herb **Use:** ME **Origin:** Native **Region:** Amazonia, Andean, Guiana Shield, Orinoquia, Pacific **Dept.:** ANT, CAQ, CHO, HUI, MET, NAR, SAN, TOL, VAC, VID **Elev.:** 0–1500 m a.s.l. **Conservation status: (IUCN)** NE

1770 *Ischaemum rugosum* Salisb. (Poaceae) **IPNI:** 406281-1 **Common name:** arrocillo, caminadera, caminadora **Habit:** Herb **Use:** AF, HF, ME, SU **Origin:** Naturalised **Region:** Andean, Caribbean, Orinoquia **Dept.:** ANT, CAL, CAU, CUN, MAG, MET, NAR, NSA, TOL, VAC, VID **Elev.:** 0–1000 m a.s.l. **IUCN Habitat code:** 1,3,4,14 **Conservation status: (IUCN)** LC

1771 *Lagurus ovatus* L. (Poaceae) **IPNI:** 406799-1 **Use:** EU **Region:** Andean **Dept.:** BOG **Conservation status: (IUCN)** NE

1772 *Lamarckia aurea* (L.) Moench (Poaceae) **IPNI:** 406803-1 **Habit:** Herb **Use:** EU, ME **Origin:** Cultivated, Naturalised **Region:** Andean **Dept.:** VAC **Elev.:** 500–1000 m a.s.l. **Conservation status: (IUCN)** NE

1773 *Lasiacis maculata* (Aubl.) Urb. (Poaceae) **IPNI:** 135707-2 **Common name:** carrizo, cola de caballo **Habit:** Herb **Use:** ME **Origin:** Native **Region:** Amazonia, Andean, Guiana Shield, Caribbean, Orinoquia, Pacific **Dept.:** AMA, ANT, BOL, BOY, CAQ, CAU, CHO, CUN, GUV, HUI, LAG, MAG, MET, NAR, NSA, QUI, RIS, SAP, TOL, VAC, VID **Elev.:** 0-2000 m.a.s.l. **IUCN Habitat code:** 1,2,3,4,14 **Conservation status: (IUCN)** NE **(CPSG)** Potential LC

1774 *Lasiacis procerrima* (Hack.) Hitchc. ex Chase (Poaceae) **IPNI:** 135714-2 **Common name:** carrizo, carrizo blanco, cañuela, cañuela trepadora **Habit:** Herb **Use:** HF, ME **Origin:** Native **Region:** Amazonia, Andean, Guiana Shield, Caribbean, Orinoquia, Pacific **Dept.:** AMA, ANT, ARA, ATL, BOL, BOY, CAL, CAQ, CAS, CAU, CES, CHO, COR, CUN, GUA, GUV, HUI, LAG, MAG, MET, NAR, NSA, PUT, QUI, RIS, SAN, SUC, TOL, VAC, VAU, VID **Elev.:** 0-2000 m.a.s.l. **IUCN Habitat code:** 1,3,5,14 **Conservation status: (IUCN)** NE **(CPSG)** Potential LC

1775 *Lasiacis ruscifolia* (Kunth) Hitchc. ex Chase (Poaceae) **IPNI:** 406853-1 **Common name:** chusquito **Habit:** Herb **Use:** AF, EU, MA, ME **Origin:** Native **Region:** Amazonia, Andean, Guiana Shield, Caribbean, Orinoquia, Pacific **Dept.:** ANT, ATL, BOL, CAL, CAQ, CUN, MAG, MET, NAR, PUT, SAP, SUC, TOL, VAC **Elev.:** 0-2000 m.a.s.l. **IUCN Habitat code:** 1,2,3,14 **Conservation status: (IUCN)** NE **(CPSG)** Potential LC

1776 *Lasiacis sloanei* (Griseb.) Hitchc. (Poaceae) **IPNI:** 135721-2 **Habit:** Herb **Use:** ME **Origin:** Native **Region:** Andean **Dept.:** CUN, MAG, MET, NSA, RIS **Elev.:** 0–1500 m a.s.l. **Conservation status: (IUCN)** NE

1777 *Leersia hexandra* Sw. (Poaceae) **IPNI:** 137026-2 **Common name:** arrocillo, hierba de arroz, lambedora, pasto de agua, pasto lamedora **Habit:** Herb **Use:** AF, EU, GS, HF, MA, ME **Origin:** Native **Region:** Amazonia, Andean, Guiana Shield, Caribbean, Orinoquia, Pacific **Dept.:** AMA, ANT, ARA, ATL, BOG, BOL, CAL, CAS, CHO, COR, CUN, HUI, MAG, MET, QUI, RIS, SAN, SUC, TOL, VAC **Elev.:** 0-3000 m.a.s.l. **IUCN Habitat code:** 1,2,3,4,5,14 **Conservation status: (IUCN)** LC **(CPSG)** Potential LC

1778 *Leptochloa panicea* (Retz.) Ohwi (Poaceae) **IPNI:** 407138-1 **Habit:** Herb **Use:** AF **Origin:** Native **Region:** Andean, Caribbean, Orinoquia **Dept.:** ANT, ATL, BOL, CAS, CAU, CES, COR, CUN, HUI, LAG, MAG, SAN, TOL, VAC **Elev.:** 0–2000 m a.s.l. **IUCN Habitat code:** 5,14 **Conservation status: (IUCN)** LC

1779 *Leptochloa pluriflora* (E.Fourn.) P.M.Peterson & N.Snow (Poaceae) **IPNI:** 60458641-2 **Habit:** Herb **Use:** AF **Origin:** Native **Region:** Andean **Dept.:** BOY, CUN, SAN **Elev.:** 0–1500 m a.s.l. **Conservation status: (IUCN)** NE

1780 *Lolium arundinaceum* (Schreb.) Darbysh. (Poaceae) **IPNI:** 973267-1 **Common name:** cañuela, paja de páramo **Habit:** Herb **Use:** AF, EU, HF, ME, PO **Origin:** Naturalised **Region:** Andean **Dept.:** CUN, NAR **Elev.:** 2500–3000 m a.s.l. **IUCN Habitat code:** 3,4,14 **Conservation status: (IUCN)** NE

1781 *Lolium multiflorum* Lam. (Poaceae) **IPNI:** 407486-1 **Common name:** ballico, lolium, raigrás, raigrás anual, raigrás inglés, raigrás italiano, reigrás **Habit:** Herb **Use:** AF, EU, HF, ME, PO **Origin:** Naturalised **Region:** Andean **Dept.:** ANT, BOG, BOY, CAL, CAU, CUN, HUI, NAR, QUI, RIS, TOL, VAC **Elev.:** 1500–3500 m a.s.l. **IUCN Habitat code:** 3,4,14 **Conservation status: (IUCN)** NE

1782 *Lolium perenne* L. (Poaceae) **IPNI:** 407493-1 **Common name:** ballico, reigrás **Habit:** Herb **Use:** AF, EU, GS, HF, ME, PO **Origin:** Naturalised **Region:** Andean **Dept.:** ANT, BOG, BOY, CAL, CAU, CUN, HUI, NAR, NSA, PUT, QUI, RIS, SAN, TOL, VAC **Elev.:** 2000–3500 m a.s.l. **IUCN Habitat code:** 3,14 **Conservation status: (IUCN)** NE

1783 *Lolium temulentum* L. (Poaceae) **IPNI:** 407516-1 **Common name:** ballico, borrachuelo, cizaña, cizaño, joyo, ray grass inglés, yoyo **Habit:** Herb **Use:** AF, GS, HF, ME, PO **Origin:** Naturalised **Region:** Andean **Dept.:** BOG, BOY, CUN, NSA **Elev.:** 2000–3000 m a.s.l. **Conservation status: (IUCN)** NE

1784 *Louisiella elephantipes* (Nees ex Trin.) Zuloaga (Poaceae) **IPNI:** 77147487-1 **Common name:** gramalote, gramalote grueso **Habit:** Herb **Use:** GS **Origin:** Native **Region:** Caribbean, Orinoquia **Dept.:** ATL, BOL, BOY, CAL, COR, HUI, MET, TOL **Elev.:** 0–500 m a.s.l. **IUCN Habitat code:** 2,5 **Conservation status: (IUCN)** NE

1785 *Luziola peruviana* J.F.Gmel. (Poaceae) **IPNI:** 289770-2 **Habit:** Herb **Use:** HF **Origin:** Native **Region:** Amazonia, Andean, Orinoquia, Pacific **Dept.:** ARA, CAS, CAU, MET, PUT, VAC **Elev.:** 0–1500 m a.s.l. **IUCN Habitat code:** 2,3,4 **Conservation status: (IUCN)** NE

1786 *Megathyrsus maximus* (Jacq.) B.K.Simon & S.W.L.Jacobs (Poaceae) **IPNI:** 60426565-2 **Common name:** castilla, guinea, guinea pajarito, guinedón, hierba india, india, indio, pasto castilla, pasto chigüiro, pasto guinea, pasto indio, saboya, yerba de Guinea, yerba india, zaina **Habit:** Herb **Use:** AF, EU, ME **Origin:** Naturalised **Region:** Amazonia, Andean, Guiana Shield, Caribbean, Orinoquia, Pacific **Dept.:** AMA, ANT, ARA, ATL, BOL, BOY, CAL, CAQ, CAS, CAU, CES, CHO, COR, CUN, GUA, GUV, HUI, MAG, MET, NAR, NSA, PUT, QUI, RIS, SAP, SAN, SUC, TOL, VAC, VAU, VID **Elev.:** 0–2000 m a.s.l. **IUCN Habitat code:** 1,2,3,14 **Conservation status: (IUCN)** NE

1787 *Melinis minutiflora* P.Beauv. (Poaceae) **IPNI:** 408232-1 **Common name:** chopín, chopín gordura, chupín, contraincendio, gordura, hierba del melado, pasto chopín, pasto gordura, pasto yaraguá, yaguará, yaraguá, yaraguá Uribe, yaraguá falso, yaraguá peluda **Habit:** Herb **Use:** AF, EU, MA, ME **Origin:** Naturalised **Region:** Amazonia, Andean, Guiana Shield, Caribbean, Orinoquia, Pacific **Dept.:** ANT, BOY, CAL, CAS, CAU, CES, CUN, GUA, GUV, HUI, MAG, MET, NAR, NSA, PUT, QUI, RIS, SAN, TOL, VAC, VAU **Elev.:** 0–3000 m a.s.l. **IUCN Habitat code:** 1,3,4,14 **Conservation status: (IUCN)** NE

1788 *Melinis repens* (Willd.) Zizka (Poaceae) **IPNI:** 2945817-1 **Common name:** grama seda, paja brillante, paja rosada, pajillita rosada, pasto brillante, pasto natal, pasto rosado **Habit:** Herb **Use:** AF, EU, HF, MA, ME **Origin:** Naturalised **Region:** Andean, Orinoquia **Dept.:** ANT, ARA, BOY, CAL, CES, CUN, HUI, LAG, MAG, NAR, NSA, RIS, SAN, TOL, VAC **Elev.:** 0–2500 m a.s.l. **IUCN Habitat code:** 1,2,3,4,14 **Conservation status: (IUCN)** NE

1789 *Melocanna baccifera* (Roxb.) Kurz (Poaceae) **IPNI:** 408271-1 **Use:** HF, MA, ME **Origin:** Naturalised **IUCN Habitat code:** 14 **Conservation status: (IUCN)** NE

1790 *Microchloa indica* (L.f.) P.Beauv. (Poaceae) **IPNI:** 408465-1 **Habit:** Herb **Use:** AF,ME **Origin:** Naturalised **Region:** Caribbean **Dept.:** MAG **Elev.:** 0–500 m a.s.l. **IUCN Habitat code:** 1,14 **Conservation status: (IUCN)** NE

1791 *Microchloa kunthii* Desv. (Poaceae) **IPNI:** 408467-1 **Habit:** Herb **Use:** ME **Origin:** Native **Region:** Andean **Dept.:** BOG, BOY, CAL, CAU, CUN, HUI, NAR, SAN, TOL, VAC **Elev.:** 0–3000 m a.s.l. **Conservation status: (IUCN)** NE

1792 *Muhlenbergia microsperma* (DC.) Kunth (Poaceae) **IPNI:** 409081-1 **Habit:** Herb **Use:** HF **Origin:** Native **Region:** Andean **Dept.:** BOY, MAG, SAN **Elev.:** 500–2000 m a.s.l. **IUCN Habitat code:** 2 **Conservation status: (IUCN)** NE

1793 *Muhlenbergia rigida* (Kunth) Kunth (Poaceae) **IPNI:** 60445990-2 **Habit:** Herb **Use:** EU **Origin:** Native **Region:** Andean **Dept.:** BOY, NAR **Elev.:** 1500–2500 m a.s.l. **Conservation status: (IUCN)** NE

1794 *Nassella neesiana* (Trin. & Rupr.) Barkworth (Poaceae) **IPNI:** 948410-1 **Habit:** Herb **Use:** PO **Origin:** Native **Region:** Andean **Dept.:** CUN **Elev.:** 2500–3000 m a.s.l. **Conservation status: (IUCN)** NE

1795 *Nassella tenuissima* (Trin.) Barkworth (Poaceae) **IPNI:** 281184-2 **Common name:** avenilla **Habit:** Herb **Use:** EU **Origin:** Cultivated, Naturalised **Region:** Andean **Dept.:** ANT, CUN **Elev.:** 1500–3000 m a.s.l. **Conservation status: (IUCN)** NE

1796 *Ocellochloa stolonifera* (Poir.) Zuloaga & Morrone (Poaceae) **IPNI:** 77105897-1 **Habit:** Herb **Use:** MA **Origin:** Native **Region:** Amazonia, Andean, Guiana Shield, Orinoquia, Pacific **Dept.:** AMA, CHO, MET, PUT, VAC, VAU, VID **Elev.:** 0–1500 m a.s.l. **IUCN Habitat code:** 1,3,4,5,14 **Conservation status: (IUCN)** NE

1797 *Olyra latifolia* L. (Poaceae) **IPNI:** 316519-2 **Common name:** arrocillo, carricillo, carrizo, cañabrava, guaira sacha, jiwa jiwa, jiwaboto, pitillo **Habit:** Herb, Climbing **Use:** AF, EU, HF, MA, ME, SU **Origin:** Native **Region:** Amazonia, Andean, Guiana Shield, Caribbean, Orinoquia, Pacific **Dept.:** AMA, ANT, BOL, CAL, CAQ, CAS, CAU, CHO, COR, CUN, GUV, LAG, MAG, MET, NAR, PUT, QUI, RIS, SAP, SAN, TOL, VAC, VAU, VID **Elev.:** 0-1300 m.a.s.l. **IUCN Habitat**

code: 1,2,3,4,5,14 **Conservation status: (IUCN)** NE **(CPSG)** Potential LC

1798 *Oplismenus burmanni* (Retz.) P.Beauv. (Poaceae)
IPNI: 30453766-2
Common name: pelillo **Habit:** Herb **Use:** AF,ME, SU **Origin:** Naturalised **Region:** Amazonia, Andean, Guiana Shield, Caribbean, Orinoquia, Pacific **Dept.:** ANT, ATL, BOG, BOL, BOY, CAS, CAU, CES, CHO, CUN, GUV, HUI, LAG, MAG, MET, NSA, QUI, SAP, SAN, TOL, VAC **Elev.:** 0-3000 m.a.s.l. **IUCN Habitat code:** 1,2,3,4,5,14 **Conservation status: (IUCN)** NE **(CPSG)** Potential LC

1799 *Oplismenus compositus* (L.) P.Beauv. (Poaceae)
IPNI: 409823-1
Habit: Herb **Use:** AF,ME **Origin:** Native **IUCN Habitat code:** 1,14 **Conservation status: (IUCN)** NE

1800 *Oplismenus hirtellus* (L.) P.Beauv. (Poaceae)
IPNI: 1004002-1
Habit: Herb **Use:** AF, EU, MA, ME **Origin:** Native **Region:** Andean **Dept.:** ANT, MAG, NAR, TOL, VAC **Elev.:** 500–2500 m a.s.l. **IUCN Habitat code:** 1,2,3,4,14 **Conservation status: (IUCN)** NE

1801 *Oryza grandiglumis* (Döll) Prodoehl (Poaceae)
IPNI: 176926-2
Habit: Herb **Use:** GS **Origin:** Native **Region:** Amazonia, Orinoquia **Dept.:** AMA, ARA, CAS, MET, VID **Elev.:** 0–500 m a.s.l. **IUCN Habitat code:** 1,3 **Conservation status: (IUCN)** LC

1802 *Oryza latifolia* Desv. (Poaceae)
IPNI: 316809-2
Common name: arrocillo, arroz, arroz cimarrón **Habit:** Herb **Use:** GS, HF, ME **Origin:** Native **Region:** Amazonia, Andean, Guiana Shield, Caribbean, Orinoquia, Pacific **Dept.:** AMA, ANT, BOL, CAS, CHO, MAG, MET, NAR, PUT, SAN, VID **Elev.:** 0–1000 m a.s.l. **IUCN Habitat code:** 1,2,3,4,5,6,14 **Conservation status: (IUCN)** LC

1803 *Oryza rufipogon* Griff. (Poaceae)
IPNI: 410115-1
Habit: Herb **Use:** AF, GS, HF, ME, SU **Origin:** Naturalised **IUCN Habitat code:** 1,2,3,4 **Conservation status: (IUCN)** LC

1804 *Oryza sativa* L. (Poaceae)
IPNI: 316812-2
Common name: arroz, arroz blanco, arroz chino, arroz colorado, arroz fortuna, arroz garrapata, arroz llanero, arroz lucinda, arroz macho, arroz mira, arroz miramono, arroz monito, arroz monolaya, arroz rojo, arroz rongoy, arroz sanjerónimo, mechudo, pipón desgranador **Habit:** Herb **Use:** AF, EU, FU, GS, HF, MA, ME, SU **Origin:** Cultivated, Naturalised **Region:** Amazonia, Andean, Orinoquia, Pacific **Dept.:** AMA, ANT, BOG, CAS, CHO, CUN, HUI, MAG, MET, NSA, SAN, TOL, VAC **Elev.:** 0–2000 m a.s.l. **IUCN Habitat code:** 1,2,3,4,5,14 **Conservation status: (IUCN)** NE

1805 *Panicum antidotale* Retz. (Poaceae)
IPNI: 410542-1
Habit: Herb **Use:** AF, EU, HF, ME **Origin:** Naturalised **Region:** Andean, Caribbean **Dept.:** ATL, CAU, COR, CUN, MAG, SAN, TOL, VAC **Elev.:** 0–1000 m a.s.l. **IUCN Habitat code:** 1,2,3,14 **Conservation status: (IUCN)** NE

1806 *Panicum aquaticum* Poir. (Poaceae)
IPNI: 410555-1
Habit: Herb **Use:** GS **Origin:** Native **Dept.:** AMA **Conservation status: (IUCN)** NE

1807 *Panicum dichotomiflorum* Michx. (Poaceae)
IPNI: 325362-2
Habit: Herb **Use:** GS, ME **Origin:** Native **Region:** Amazonia, Caribbean, Orinoquia **Dept.:** AMA, ARA, CAS, CES, MET **Elev.:** 0–500 m a.s.l. **IUCN Habitat code:** 3,4,14 **Conservation status: (IUCN)** NE

1808 *Panicum hirticaule* J.Presl (Poaceae)
IPNI: 411695-1
Habit: Herb **Use:** HF **Origin:** Native **Region:** Orinoquia **Dept.:** VID **Elev.:** 0–500 m a.s.l. **IUCN Habitat code:** 14 **Conservation status: (IUCN)** NE

1809 *Panicum hirtum* Lam. (Poaceae)
IPNI: 411699-1
Habit: Herb **Use:** AF, GS, HF, ME, SU **Origin:** Native **Region:** Orinoquia **Dept.:** MET **Elev.:** 0–500 m a.s.l. **IUCN Habitat code:** 1,3,4,14 **Conservation status: (IUCN)** NE

1810 *Panicum miliaceum* L. (Poaceae)
IPNI: 412217-1
Common name: millo **Habit:** Herb **Use:** AF, GS, HF, MA, ME **Origin:** Cultivated, Naturalised **Region:** Andean **Dept.:** VAC **Elev.:** 1000–1500 m a.s.l. **IUCN Habitat code:** 1,3,14 **Conservation status: (IUCN)** NE

1811 *Panicum millegrana* Poir. (Poaceae)
IPNI: 412226-1
Common name: caminacompié, canajá, gateadora blanca, grama, pasto amargo **Habit:** Herb **Use:** ME **Origin:** Native **Region:** Andean, Guiana Shield, Caribbean **Dept.:** CAL, CUN, HUI, MAG, TOL, VID **Elev.:** 0–1500 m a.s.l. **IUCN Habitat code:** 1,2 **Conservation status: (IUCN)** LC

1812 *Panicum olyroides* Kunth (Poaceae)
IPNI: 412456-1
Common name: paja lanosa **Habit:** Herb **Use:** GS **Origin:** Native **Region:** Guiana Shield, Orinoquia **Dept.:** CAS, MET, VID **Elev.:** 0–500 m a.s.l. **IUCN Habitat code:** 1,2,3,4,14 **Conservation status: (IUCN)** NE

1813 *Panicum rudgei* Roem. & Schult. (Poaceae)
IPNI: 412983-1
Common name: paja de tigre, paja mona, paja tigre, pasaganado, pasto sabana, rabo de zorro **Habit:** Herb **Use:** GS **Origin:** Native **Region:** Amazonia, Andean, Orinoquia **Dept.:** ANT, CAQ, CAS, GUA, MAG, MET, NSA, SAN, TOL, VAU, VID **Elev.:** 0–1500 m a.s.l. **IUCN Habitat code:** 1,2,3,4,5,14 **Conservation status: (IUCN)** NE

1814 *Panicum stramineum* Hitchc. & Chase (Poaceae)
IPNI: 182558-2
Habit: Herb **Use:** GS **Origin:** Native **Dept.:** CUN **Elev.:** 0–500 m a.s.l. **Conservation status: (IUCN)** LC

1815 *Panicum trichanthum* Nees (Poaceae)
IPNI: 413376-1
Habit: Herb **Use:** ME **Origin:** Native **Region:** Amazonia, Andean, Caribbean, Pacific **Dept.:** ANT, ATL, BOY, CAL, CAQ, CHO, COR, CUN, HUI, MAG, NSA, RIS, SAN, TOL **Elev.:** 0–1500 m a.s.l. **IUCN Habitat code:** 1,2,3,14 **Conservation status: (IUCN)** NE

1816 *Panicum trichoides* Sw. (Poaceae)
IPNI: 324582-2
Common name: guardasireno, ilusión, ilusión de hojita ancha, ilusión golondrina, leguminosa, paja churcada **Habit:** Herb **Use:** AF,ME **Origin:** Native **Region:** Amazonia, Andean, Guiana Shield, Caribbean, Orinoquia, Pacific **Dept.:** AMA, ANT, ATL, CAL, CAQ, CAS, CAU, CES, CHO, COR, GUV, HUI, LAG, MAG, MET, NAR, QUI, SAP, SAN, SUC, TOL, VAC **Elev.:** 0-1500 m.a.s.l. **IUCN Habitat code:** 1,2,3,4,5,14 **Conservation status: (IUCN)** NE **(CPSG)** Potential LC

1817 *Panicum tricholaenoides* Steud. (Poaceae)
IPNI: 413385-1
Habit: Herb **Use:** GS, ME **Origin:** Native **Region:** Orinoquia **Dept.:** CAS, MET **Elev.:** 0–500 m a.s.l. **IUCN Habitat code:** 3,4 **Conservation status: (IUCN)** NE

1818 *Pappophorum pappiferum* (Lam.) Kuntze (Poaceae)
IPNI: 182865-2
Habit: Herb **Use:** ME **Origin:** Native **Region:** Andean, Caribbean **Dept.:** BOY, CAL, CUN, HUI, MAG, NAR, SAN, TOL, VAC **Elev.:** 0–2000 m a.s.l. **IUCN Habitat code:** 14 **Conservation status: (IUCN)** NE

1819 *Paramochloa effusa* (Kunth) P.M.Peterson, Soreng, Romasch. & Barberá (Poaceae)
IPNI: 77206801-1
Common name: cañuela, espartillo, esparto, güinche, paja, paja blanca, paja de empacar, paja de empajar, paja de páramo, paja de ratón, pajas **Habit:** Herb **Use:** EU, MA **Origin:** Endemic **Region:** Andean **Dept.:** ANT, ARA, BOG, BOY, CAL, CAU, CES, CHO, CUN, HUI, LAG, MAG, MET, NAR, NSA, PUT, QUI, RIS, SAN, TOL, VAC **Elev.:** 2500–4500 m a.s.l. **IUCN Habitat code:** 1,2,3,4,6,8,14 **Conservation status: (IUCN)** NE

1820 *Paspalum bonplandianum* Flüggé (Poaceae)
IPNI: 413896-1
Common name: pasto cañitas **Habit:** Herb **Use:** ME **Origin:** Native **Region:** Andean **Dept.:** BOG, BOY, CAL, CUN, HUI, NAR **Elev.:** 2500–4000 m a.s.l. **IUCN Habitat code:** 1,3,4,5,14 **Conservation status: (IUCN)** NE

1821 *Paspalum conjugatum* P.J.Bergius (Poaceae)
IPNI: 30011315-2
Common name: camina con pie, churro, grama, grama andadora, grama blanca, grama común, horquetilla, jengibrillo, paja de panela, pasto amargo, pasto horqueta, sabana, yerba agria **Habit:** Herb **Use:** AF, EU, HF, MA, ME, SU **Origin:** Native **Region:** Amazonia, Andean, Guiana Shield, Caribbean, Orinoquia, Pacific **Dept.:** AMA, ANT, ARA, ATL, BOL, BOY, CAL, CAQ, CAS, CAU, CES, CHO, COR, CUN, GUA, GUV, HUI, LAG, MAG, MET, NAR, NSA, PUT, QUI, RIS, SAP, SAN, SUC, TOL, VAC, VAU, VID **Elev.:** 0-2000 m.a.s.l. **IUCN Habitat code:** 1,2,3,4,5,14 **Conservation status: (IUCN)** LC **(CPSG)** Potential LC

1822 *Paspalum convexum* Flüggé (Poaceae)
IPNI: 414002-1
Common name: paja zancona **Habit:** Herb **Use:** AF,ME **Origin:** Native **Region:** Andean, Caribbean, Orinoquia **Dept.:** ANT, CES, MET, VID **Elev.:** 0–1000 m a.s.l. **IUCN Habitat code:** 14 **Conservation status: (IUCN)** NE

1823 *Paspalum denticulatum* Trin. (Poaceae)
IPNI: 414056-1
Habit: Herb **Use:** EU **Origin:** Native **Region:** Andean **Dept.:** CUN, NAR **Elev.:** 2500–3000 m a.s.l. **Conservation status: (IUCN)** NE

1824 *Paspalum dilatatum* Poir. (Poaceae)
IPNI: 414070-1
Common name: gramalote **Habit:** Herb **Use:** AF, EU, ME, PO **Origin:** Naturalised **IUCN Habitat code:** 1,3,14 **Conservation status: (IUCN)** NE

1825 *Paspalum distichum* L. (Poaceae)
IPNI: 184275-2
Common name: grama blanca **Habit:** Herb **Use:** AF, EU, GS, HF, ME, PO **Origin:** Native **Region:** Caribbean, Orinoquia, Pacific **Dept.:** ANT, MET, TOL **Elev.:** 0–500 m a.s.l. **IUCN Habitat code:** 1,14 **Conservation status: (IUCN)** LC

1826 *Paspalum fimbriatum* Kunth (Poaceae)
IPNI: 414155-1
Habit: Herb **Use:** EU **Origin:** Native **Region:** Andean **Dept.:** ANT, BOG, BOY, CUN, HUI, QUI, SAN, TOL, VAC **Elev.:** 500–2500 m a.s.l. **IUCN Habitat code:** 1,3,14 **Conservation status: (IUCN)** NE

1827 *Paspalum foliiforme* S.Denham (Poaceae)
IPNI: 60440769-2
Common name: paja hoz **Habit:** Herb **Use:** ME **Origin:** Native **Region:** Amazonia, Andean, Guiana Shield, Orinoquia **Dept.:** ANT, GUA, GUV, HUI, MET, SAN, TOL, VID **Elev.:** 0–2000 m a.s.l. **IUCN Habitat code:** 3,4,14 **Conservation status: (IUCN)** NE

1828 *Paspalum gardnerianum* Nees (Poaceae)
IPNI: 414188-1
Habit: Herb **Use:** EU **Origin:** Native **Region:** Orinoquia **Dept.:** VID **Elev.:** 0–500 m a.s.l. **IUCN Habitat code:** 1,4 **Conservation status: (IUCN)** NE

1829 *Paspalum intermedium* Munro ex Morong (Poaceae)
IPNI: 184394-2
Common name: maciga **Habit:** Herb **Use:** MA **Origin:** Native **Region:** Amazonia, Guiana Shield, Orinoquia **Dept.:** AMA, MET, VID **Elev.:** 0–500 m a.s.l. **IUCN Habitat code:** 1,3,14 **Conservation status: (IUCN)** NE

1830 *Paspalum maritimum* Trin. (Poaceae)
IPNI: 414411-1
Habit: Herb **Use:** AF, EU **Origin:** Native **Conservation status: (IUCN)** NE

1831 *Paspalum millegranum* Schrad. (Poaceae)
IPNI: 414431-1
Habit: Herb **Use:** AF, MA **Origin:** Native **Region:** Andean **Dept.:** ANT, BOY, CAQ, CUN, HUI, NSA, PUT, QUI, SAN, TOL, VAC **Elev.:** 1000–2000 m a.s.l. **Conservation status: (IUCN)** NE

1832 *Paspalum notatum* Flüggé (Poaceae)
IPNI: 414477-1
Common name: bahía, grama de trenza, grama trencilla, grama trenza, gramilla blanca, pasto bahía, pasto macana, pasto trencilla, trencita, trenza **Habit:** Herb **Use:** AF, EU, ME **Origin:** Cultivated, Naturalised **Region:** Amazonia, Andean, Guiana Shield, Caribbean, Orinoquia, Pacific **Dept.:** AMA, ANT, ARA, BOY, CAL, CAS, CAU, CES, CHO, CUN, GUV, MAG, MET, QUI, RIS, SAN, TOL, VAC, VAU, VID **Elev.:** 0-2500 m.a.s.l. **IUCN Habitat code:** 1,2,3,4,14 **Conservation status: (IUCN)** NE **(CPSG)** Potential LC

1833 *Paspalum paniculatum* L. (Poaceae)
IPNI: 184526-2
Common name: paja brava, paja de camino, paja del camino **Habit:** Herb **Use:** EU, ME **Origin:** Native **Region:** Andean, Caribbean, Orinoquia, Pacific **Dept.:** ANT, BOG, BOL, CAL, CAU, CHO, COR, CUN, HUI, MAG, MET, NAR, NSA, PUT, QUI, RIS, SAP, SAN, TOL, VAC **Elev.:** 0–2000 m a.s.l. **IUCN Habitat code:** 1,2,3,4,5,14 **Conservation status: (IUCN)** NE

1834 *Paspalum plicatulum* Michx. (Poaceae)
IPNI: 184571-2
Common name: pasto negro **Habit:** Herb **Use:** AF **Origin:** Native **Region:** Amazonia, Andean, Guiana Shield, Caribbean, Orinoquia, Pacific **Dept.:** AMA, ANT, BOY, CAS, CAU, CES, CHO, CUN, GUV, MAG, MET, SAN, TOL, VAC, VID **Elev.:** 0-2000 m.a.s.l. **IUCN Habitat code:** 1,2,3,4,5,14 **Conservation status: (IUCN)** NE **(CPSG)** Potential LC

1835 *Paspalum racemosum* Lam. (Poaceae)
IPNI: 414634-1
Habit: Herb **Use:** EU **Origin:** Native **Region:** Andean **Dept.:** ANT, NAR **Elev.:** 1500–2000 m a.s.l. **Conservation status: (IUCN)** NE

1836 *Paspalum repens* P.J.Bergius (Poaceae)
IPNI: 30475806-2
Common name: canutillo, churri, churrichurri, gramalote, nudillo acuático, trenza acuática **Habit:** Herb **Use:** AF **Origin:** Native **Region:** Amazonia, Andean, Guiana Shield, Caribbean, Orinoquia, Pacific **Dept.:** AMA, ANT, ARA, BOL, CAS, CAU, CES, CHO, COR, GUA, GUV, MAG, MET, SAN, VAC, VAU, VID **Elev.:** 0-1500 m.a.s.l. **Conservation status: (IUCN)** LC **(CPSG)** Potential LC

1837 *Paspalum scrobiculatum* L. (Poaceae)
IPNI: 414702-1
Common name: paja peluda **Habit:** Herb **Use:** AF, EU, GS, HF, MA, ME, PO, SU **Origin:** Naturalised **Region:** Andean **Dept.:** ANT, TOL **Elev.:** 0–1000 m a.s.l. **Conservation status: (IUCN)** LC

1838 *Paspalum urvillei* Steud. (Poaceae)
IPNI: 414828-1
Habit: Herb **Use:** AF,ME **Origin:** Naturalised **Conservation status: (IUCN)** NE

1839 *Paspalum vaginatum* Sw. (Poaceae)
IPNI: 184742-2
Habit: Herb **Use:** AF, EU, ME **Origin:** Native **Region:** Caribbean, Pacific **Dept.:** ANT, ATL, MAG, SAP, VAC **Elev.:** 0–500 m a.s.l. **IUCN Habitat code:** 3 **Conservation status: (IUCN)** LC

1840 *Paspalum virgatum* L. (Poaceae)
IPNI: 414854-1
Common name: cabezona, gramalote, maciega, marciega, pajón, remolina **Habit:** Herb **Use:** EU, MA **Origin:** Native **Region:** Amazonia, Andean, Guiana Shield, Caribbean, Orinoquia, Pacific **Dept.:** AMA, ANT, ARA, BOL, BOY, CAL, CAU, CHO, CUN, GUA, GUV, MAG, MET, NAR, RIS, SAN, TOL, VAC, VID **Elev.:** 0-2000 m.a.s.l. **IUCN Habitat code:** 1,2,3,5,14 **Conservation status: (IUCN)** NE **(CPSG)** Potential LC

1841 *Phalaris angusta* Nees ex Trin. (Poaceae)
IPNI: 311946-2
Habit: Herb **Use:** ME **Origin:** Native **Conservation status: (IUCN)** NE

1842 *Phalaris aquatica* L. (Poaceae)
IPNI: 415560-1
Habit: Herb **Use:** AF,ME, PO **Origin:** Cultivated, Naturalised **Dept.:** BOG **IUCN Habitat code:** 3,14 **Conservation status: (IUCN)** NE

1843 *Phalaris arundinacea* L. (Poaceae)
IPNI: 415571-1
Common name: alpiste de forraje, pasto brasilero, pasto cinta **Habit:** Herb **Use:** AF, EU, HF, MA, ME, PO **Origin:** Cultivated, Naturalised **Region:** Andean **Dept.:** ANT, BOG, BOY, CAU, CUN, HUI, NAR, NSA, PUT, SAN, TOL, VAC **Elev.:** 2000–4000 m a.s.l. **IUCN Habitat code:** 1,3,4,5,14 **Conservation status: (IUCN)** LC

1844 *Phalaris canariensis* L. (Poaceae)
IPNI: 415588-1
Common name: alpiste, alpiste de canarios, pasto alpiste **Habit:** Herb **Use:** AF, HF, MA, ME **Origin:** Naturalised **Region:** Andean **Dept.:** ANT, BOG **Elev.:** 1000–2000 m a.s.l. **Conservation status: (IUCN)** NE

1845 *Phalaris minor* Retz. (Poaceae)
IPNI: 415632-1
Common name: alpiste, pasto romano **Habit:** Herb **Use:** AF, GS, HF, ME, PO **Origin:** Naturalised **Region:** Andean **Dept.:** BOG, BOY, CAL, CUN **Elev.:** 2000–3000 m a.s.l. **Conservation status: (IUCN)** NE

1846 *Pharus lappulaceus* Aubl. (Poaceae)
IPNI: 415708-1
Habit: Herb **Use:** HF, ME **Origin:** Native **Region:** Andean, Caribbean, Orinoquia, Pacific **Dept.:** ANT, CHO, CUN, MAG, MET, QUI, TOL, VAC **Elev.:** 400–1900 m a.s.l. **IUCN Habitat code:** 1,3,5,14 **Conservation status: (IUCN)** NE

1847 *Pharus latifolius* L. (Poaceae)
IPNI: 316520-2
Common name: camaperro **Habit:** Herb **Use:** ME **Origin:** Native **Region:** Amazonia, Andean, Guiana Shield, Caribbean, Pacific **Dept.:** AMA, ANT, CHO, COR, LAG, MAG, MET, PUT, SAN, VAC, VAU **Elev.:** 100-1700 m.a.s.l. **IUCN Habitat code:** 1,2,3,4,5,14 **Conservation status: (IUCN)** LC **(CPSG)** Potential LC

1848 *Phragmites australis* (Cav.) Trin. ex Steud. (Poaceae) **IPNI:** 30337627-2 **Habit:** Herb **Use:** AF, EU, FU, GS, HF, MA, ME, PO **Origin:** Native **Region:** Caribbean **Dept.:** LAG **Elev.:** 0–500 m a.s.l. **Conservation status: (IUCN)** LC

1849 *Phyllostachys aurea* (André) Rivière & C.Rivière (Poaceae) **IPNI:** 415983-1 **Common name:** bambú, guaduilla **Habit:** Small tree **Use:** EU, GS, HF, MA, ME **Origin:** Cultivated, Naturalised **Dept.:** ANT, BOG, CAL, CAU, CUN, HUI, QUI, RIS, TOL, VAC **Elev.:** 800–1500 m a.s.l. **IUCN Habitat code:** 1, 2, 14 **Conservation status: (IUCN)** NE

1850 *Phyllostachys meyeri* McClure (Poaceae) **IPNI:** 416023-1 **Habit:** Small tree **Use:** HF, MA **Origin:** Cultivated, Naturalised **Region:** Andean **Dept.:** CAL, QUI, RIS, TOL, VAC **Elev.:** 1000–1800 m a.s.l. **Conservation status: (IUCN)** NE

1851 *Piresia leptophylla* Soderstr. (Poaceae) **IPNI:** 199907-2 **Habit:** Herb **Use:** ME **Origin:** Native **Region:** Amazonia **Dept.:** AMA **Elev.:** 100–100 m a.s.l. **Conservation status: (IUCN)** NE

1852 *Pleioblastus fortunei* (Van Houtte) Nakai (Poaceae) **IPNI:** 416271-1 **Use:** EU **Origin:** Naturalised **Elev.:** 500–1000 m a.s.l. **Conservation status: (IUCN)** LC

1853 *Poa annua* L. (Poaceae) **IPNI:** 320035-2 **Common name:** pasto gallina, pata de gallina, piojillo, poa, yerba de gallina, zorrito **Habit:** Herb **Use:** AF, EU, ME **Origin:** Naturalised **Region:** Andean **Dept.:** ANT, ARA, BOG, BOY, CAL, CAQ, CAU, CES, CUN, HUI, LAG, MAG, NAR, NSA, PUT, QUI, RIS, SAN, TOL, VAC **Elev.:** 1000–4000 m a.s.l. **IUCN Habitat code:** 1,3,4,14 **Conservation status: (IUCN)** LC

1854 *Poa pratensis* L. (Poaceae) **IPNI:** 417792-1 **Use:** AF, EU, HF, ME **IUCN Habitat code:** 4 **Conservation status: (IUCN)** LC

1855 *Poa trivialis* L. (Poaceae) **IPNI:** 320346-2 **Use:** AF, EU, ME **Origin:** Naturalised **IUCN Habitat code:** 4 **Conservation status: (IUCN)** NE

1856 *Polypogon monspeliensis* (L.) Desf. (Poaceae) **IPNI:** 1048074-2 **Habit:** Herb **Use:** AF, EU, HF, ME **Origin:** Naturalised **Region:** Andean **Dept.:** BOG, BOY, CUN **Elev.:** 1000–3000 m a.s.l. **IUCN Habitat code:** 14 **Conservation status: (IUCN)** LC

1857 *Polytrias indica* (Houtt.) Veldkamp (Poaceae) **IPNI:** 963150-1 **Habit:** Herb **Use:** AF, EU, ME **Origin:** Cultivated, Naturalised **Region:** Andean, Caribbean **Dept.:** ATL, CUN, VAC **Elev.:** 0–1500 m a.s.l. **Conservation status: (IUCN)** LC

1858 *Pseudechinolaena polystachya* (Kunth) Stapf (Poaceae) **IPNI:** 211125-2 **Common name:** coneja, yerba de conejo **Habit:** Herb **Use:** AF, EU, ME, SU **Origin:** Native **Region:** Amazonia, Andean, Guiana Shield, Caribbean, Orinoquia, Pacific **Dept.:** ANT, BOG, BOY, CAL, CAQ, CAU, CHO, CUN, GUV, HUI, MAG, MET, NAR, NSA, PUT, QUI, RIS, SAN, TOL, VAC **Elev.:** 0-2500 m.a.s.l. **Conservation status: (IUCN)** NE **(CPSG)** Potential LC

1859 *Rhipidocladum racemiflorum* (Steud.) McClure (Poaceae) **IPNI:** 219234-2 **Habit:** Climbing **Use:** EU, ME **Origin:** Native **Region:** Andean, Caribbean **Dept.:** ANT, BOL, CAL, CAU, CES, CUN, MAG, QUI, RIS, TOL, VAC **Elev.:** 150-1750 m.a.s.l. **IUCN Habitat code:** 1,3,14 **Conservation status: (IUCN)** NE **(CPSG)** Potential LC

1860 *Rhytachne rottboellioides* Desv. (Poaceae) **IPNI:** 325559-2 **Habit:** Herb **Use:** AF, MA **Origin:** Naturalised **Region:** Guiana Shield, Orinoquia **Dept.:** MET, VID **Elev.:** 0–500 m a.s.l. **Conservation status: (IUCN)** LC

1861 *Rottboellia cochinchinensis* (Lour.) Clayton (Poaceae) **IPNI:** 909653-1 **Common name:** caminadora, paja brava, pasto trejos **Habit:** Herb **Use:** AF, HF, MA, ME **Origin:** Naturalised **Region:** Andean, Orinoquia **Dept.:** ANT, ARA, BOY, CAL, CAS, HUI, MET, QUI, RIS, SAN, TOL, VAC, VID **Elev.:** 0–1500 m a.s.l. **IUCN Habitat code:** 1,2,3,14 **Conservation status: (IUCN)** NE

1862 *Rugoloa pilosa* (Sw.) Zuloaga (Poaceae) **IPNI:** 77144198-1 **Common name:** caminacompié, canajá, gateadora blanca, grama, pasto amargo **Habit:** Herb **Use:** HF, ME **Origin:** Native **Region:** Amazonia, Andean, Guiana Shield, Caribbean, Orinoquia, Pacific **Dept.:** AMA, ANT, ARA, BOL, CAL, CAQ, CAS, CAU, CHO, CUN, GUA, GUV, HUI, LAG, MAG, MET, NAR, NSA, PUT, QUI, SAP, TOL, VAC, VAU, VID **Elev.:** 0-1500 m.a.s.l. **IUCN Habitat code:** 1,3,14 **Conservation status: (IUCN)** NE **(CPSG)** Potential LC

1863 *Saccharum angustifolium* (Nees) Trin. (Poaceae) **IPNI:** 419853-1 **Habit:** Herb **Use:** GS **Origin:** Native **Region:** Andean, Orinoquia **Dept.:** CAL, HUI, MET, TOL **Elev.:** 0–1500 m a.s.l. **IUCN Habitat code:** 14 **Conservation status: (IUCN)** LC

1864 *Saccharum officinarum* L. (Poaceae) **IPNI:** 419977-1 **Common name:** canapoy, caña, caña amarilla, caña canapoy, caña criolla, caña de azúcar, caña de azúcar altacumbre, caña de azúcar amarilla, caña de azúcar amarilla rayuna, caña de azúcar americana, caña de azúcar azucarera, caña de azúcar azúcar, caña de azúcar bella, **Habit:** Herb **Use:** AF, EU, FU, GS, HF, IF, MA, ME, SU **Origin:** Cultivated, Naturalised **Region:** Amazonia, Andean, Guiana Shield, Caribbean, Orinoquia, Pacific **Dept.:** AMA, ANT, ARA, ATL, BOL, BOY, CAL, CAQ, CAS, CAU, CES, CHO, COR, CUN, GUA, GUV, HUI, LAG, MAG, MET, NAR, NSA, PUT, QUI, RIS, SAP, SAN, SUC, TOL, VAC, VAU, VID **Elev.:** 0–2500 m a.s.l. **IUCN Habitat code:** 1,3,14 **Conservation status: (IUCN)** NE

1865 *Saccharum villosum* Steud. (Poaceae) **IPNI:** 420046-1 **Habit:** Herb **Use:** GS **Origin:** Native **Region:** Orinoquia **Dept.:** ANT, MET **Elev.:** 0–500 m a.s.l. **IUCN Habitat code:** 4,14 **Conservation status: (IUCN)** LC

1866 *Sacciolepis myuros* (Lam.) Chase (Poaceae) **IPNI:** 225088-2 **Habit:** Herb **Use:** ME **Origin:** Native **Region:** Andean, Guiana Shield, Orinoquia **Dept.:** CAS, MET, VID **Elev.:** 0–1000 m a.s.l. **Conservation status: (IUCN)** NE

1867 *Schizachyrium brevifolium* (Sw.) Nees ex Buse (Poaceae) **IPNI:** 1126606-2 **Habit:** Herb **Use:** AF, EU, ME **Origin:** Native **Region:** Andean, Guiana Shield, Caribbean, Orinoquia **Dept.:** ANT, BOY, CAL, CAS, CUN, MAG, MET, NSA, QUI, RIS, SAN, VAC, VID **Elev.:** 0–2000 m a.s.l. **IUCN Habitat code:** 1,2,3,4,14 **Conservation status: (IUCN)** LC

1868 *Schizachyrium cirratum* (Hack.) Wooton & Standl. (Poaceae) **IPNI:** 420874-1 **Use:** AF **Origin:** Native **Conservation status: (IUCN)** NE

1869 *Schizachyrium condensatum* (Kunth) Nees (Poaceae) **IPNI:** 420877-1 **Common name:** grama azul, paja cola de mula, pasto espiga, pasto fino **Habit:** Herb **Use:** ME **Origin:** Native **Region:** Andean, Orinoquia **Dept.:** ANT, CAU, HUI, MAG, MET, SAN, TOL **Elev.:** 0–2000 m a.s.l. **IUCN Habitat code:** 1,3,4,14 **Conservation status: (IUCN)** NE

1870 *Schizachyrium maclaudii* (Jacq.-Fél.) S.T.Blake (Poaceae) **IPNI:** 229330-2 **Habit:** Herb **Use:** AF **Origin:** Naturalised **Region:** Caribbean **Dept.:** MAG **Elev.:** 0–500 m a.s.l. **Conservation status: (IUCN)** NE

1871 *Schizachyrium microstachyum* (Desv.) Roseng., B.R.Arrill. & Izag. (Poaceae) **IPNI:** 420918-1 **Habit:** Herb **Use:** ME **Origin:** Native **Region:** Andean, Caribbean, Orinoquia, Pacific **Dept.:** ANT, BOL, CAL, CHO, CUN, HUI, MET, SAN, TOL **Elev.:** 0–3000 m a.s.l. **IUCN Habitat code:** 1,14 **Conservation status: (IUCN)** NE

1872 *Schizachyrium sanguineum* (Retz.) Alston (Poaceae) **IPNI:** 420951-1 **Common name:** pasto aguja, pasto simple, puntero del llano **Habit:** Herb **Use:** AF, MA, ME **Origin:** Native **Region:** Andean, Guiana Shield, Caribbean, Orinoquia **Dept.:** ANT, BOG, BOY, CAL, CAU, CUN, GUV, HUI, MAG, MET, NAR, TOL, VID **Elev.:** 0–2500 m a.s.l. **IUCN Habitat code:** 1,3,4,14 **Conservation status: (IUCN)** NE

1873 *Secale cereale* L. (Poaceae) **IPNI:** 421164-1 **Common name:** centeno **Habit:** Herb **Use:** AF, EU, GS, HF, MA, ME, PO **Origin:** Cultivated, Naturalised **Region:** Andean **Dept.:** BOG, CUN,

NAR **Elev.:** 2500–3000 m a.s.l. **IUCN Habitat code:** 14 **Conservation status: (IUCN)** NE

1874 *Setaria adhaerens* (Forssk.) Chiov. (Poaceae) **IPNI:** 421428-1 **Habit:** Herb **Use:** ME **Origin:** Naturalised **Conservation status: (IUCN)** NE

1875 *Setaria barbata* (Lam.) Kunth (Poaceae) **IPNI:** 1172068-2 **Habit:** Herb **Use:** AF, EU, HF, ME **Origin:** Naturalised **Conservation status: (IUCN)** NE

1876 *Setaria geminata* (Forssk.) Veldkamp (Poaceae) **IPNI:** 979871-1 **Habit:** Herb **Use:** AF, HF, ME **Origin:** Naturalised **Region:** Caribbean **Dept.:** BOL, COR, MAG **Elev.:** 0–500 m a.s.l. **IUCN Habitat code:** 2,4 **Conservation status: (IUCN)** NE

1877 *Setaria grisebachii* E.Fourn. (Poaceae) **IPNI:** 319034-2 **Habit:** Herb **Use:** AF,ME **Origin:** Native **Region:** Caribbean **Dept.:** MAG **Elev.:** 0–500 m a.s.l. **IUCN Habitat code:** 14 **Conservation status: (IUCN)** NE

1878 *Setaria italica* (L.) P.Beauv. (Poaceae) **IPNI:** 1207294-2 **Habit:** Herb **Use:** AF, HF, ME **Origin:** Cultivated, Naturalised **Conservation status: (IUCN)** NE

1879 *Setaria liebmannii* E.Fourn. (Poaceae) **IPNI:** 319037-2 **Habit:** Herb **Use:** AF **Origin:** Native **Region:** Caribbean **Dept.:** MAG **Elev.:** 0–500 m a.s.l. **Conservation status: (IUCN)** NE

1880 *Setaria macrostachya* Kunth (Poaceae) **IPNI:** 235234-2 **Habit:** Herb **Use:** AF **Origin:** Native **Region:** Andean, Caribbean **Dept.:** BOY, CAL, HUI, MAG, SAN, TOL, VAC **Elev.:** 0–1500 m a.s.l. **IUCN Habitat code:** 1,2,3,4,5,14 **Conservation status: (IUCN)** NE

1881 *Setaria palmifolia* (J.Koenig) Stapf (Poaceae) **IPNI:** 421711-1 **Common name:** pasto de palma **Habit:** Herb **Use:** AF, EU, HF, MA, ME, SU **Origin:** Naturalised **Region:** Andean, Caribbean, Pacific **Dept.:** ANT, CUN, HUI, MAG, NAR, NSA, PUT, TOL, VAC **Elev.:** 0–2000 m a.s.l. **IUCN Habitat code:** 1,3,14 **Conservation status: (IUCN)** NE

1882 *Setaria parviflora* (Poir.) Kerguélen (Poaceae) **IPNI:** 943811-1 **Common name:** cola de zorro, deshollinador, gusanillo, gusanito, limpiafrascos, munchira roja, pajilla blanca, rabo de gato, rabo de zorra, tusillo **Habit:** Herb **Use:** AF, EU, HF, MA, ME, SU **Origin:** Native **Region:** Amazonia, Andean, Guiana Shield, Caribbean, Orinoquia, Pacific **Dept.:** AMA, ANT, ARA, ATL, BOG, BOL, BOY, CAS, CAU, CES, CHO, COR, CUN, GUA, GUV, HUI, LAG, MAG, MET, NAR, NSA, PUT, QUI, RIS, SAN, SUC, TOL, VAC, VAU, VID **Elev.:** 0-3000 m.a.s.l. **IUCN Habitat code:** 1,2,3,4,5,14 **Conservation status: (IUCN)** LC **(CPSG)** Potential LC

1883 *Setaria pumila* (Poir.) Roem. & Schult. (Poaceae) **IPNI:** 421764-1 **Habit:** Herb **Use:** AF, GS, HF, MA, ME, PO **Origin:** Naturalised **IUCN Habitat code:** 3,14 **Conservation status: (IUCN)** NE

1884 *Setaria sulcata* Raddi (Poaceae) **IPNI:** 421828-1 **Common name:** palmera, raboemico **Habit:** Herb **Use:** ME **Origin:** Native **Region:** Amazonia, Andean, Guiana Shield, Orinoquia, Pacific **Dept.:** ANT, CAS, CAU, CHO, CUN, MAG, MET, NAR, PUT, QUI, TOL, VAU, VID **Elev.:** 0–3000 m a.s.l. **IUCN Habitat code:** 1,14 **Conservation status: (IUCN)** NE

1885 *Setaria verticillata* (L.) P.Beauv. (Poaceae) **IPNI:** 421857-1 **Habit:** Herb **Use:** AF, GS, HF, MA, ME, PO, SU **Origin:** Naturalised **Dept.:** CUN**Conservation status: (IUCN)** NE

1886 *Setaria vulpiseta* (Lam.) Roem. & Schult. (Poaceae) **IPNI:** 1182687-2 **Common name:** rabo de mico **Habit:** Herb **Use:** AF, EU, HF, ME **Origin:** Native **Region:** Amazonia, Andean, Guiana Shield, Caribbean, Orinoquia, Pacific **Dept.:** ANT, BOL, CAQ, CAS, CHO, MAG, MET, PUT, SAN, VAU **Elev.:** 0–1000 m a.s.l. **Conservation status: (IUCN)** NE

1887 *Sorghastrum incompletum* (J.Presl) Nash (Poaceae) **IPNI:** 240818-2 **Habit:** Herb **Use:** AF **Origin:** Native **Dept.:** ARA**Conservation status: (IUCN)** NE

1888 *Sorghastrum stipoides* (Kunth) Nash (Poaceae) **IPNI:** 240834-2 **Habit:** Herb **Use:** AF, MA **Origin:** Naturalised **Region:** Andean **Dept.:** CAU, MET, NAR, TOL **Elev.:** 500–2000 m a.s.l. **IUCN Habitat code:** 3 **Conservation status: (IUCN)** LC

1889 *Sorghum arundinaceum* (Desv.) Stapf (Poaceae) **IPNI:** 422078-1 **Use:** AF, EU, FU, GS, HF, MA, ME, PO **Origin:** Naturalised **Conservation status: (IUCN)** LC

1890 *Sorghum bicolor* (L.) Moench (Poaceae) **IPNI:** 422090-1 **Common name:** maíz, maíz de Guinea, maíz millo, mijo, sorgo, sorgo forrajero, sudán **Habit:** Herb **Use:** AF, EU, FU, GS, HF, MA, ME, PO, SU **Origin:** Naturalised **Region:** Andean, Caribbean **Dept.:** ANT, ATL, BOG, BOY, CAL, CAU, COR, CUN, HUI, MAG, TOL, VAC **Elev.:** 0–2500 m a.s.l. **IUCN Habitat code:** 1,3,14 **Conservation status: (IUCN)** NE

1891 *Sorghum halepense* (L.) Pers. (Poaceae) **IPNI:** 422139-1 **Common name:** canutillo, grama china, maicillo, pasto Johnson, pasto johnson, pasto ruso, sorgo, sorguillo **Habit:** Herb **Use:** AF, GS, HF, MA, ME, PO, SU **Origin:** Naturalised **Region:** Andean, Caribbean **Dept.:** ANT, ATL, BOG, BOY, CAL, CAU, COR, CUN, HUI, NSA, QUI, RIS, SAP, SAN, SUC, TOL, VAC **Elev.:** 0–2500 m a.s.l. **IUCN Habitat code:** 2,3,14 **Conservation status: (IUCN)** NE

1892 *Sporobolus giganteus* Nash (Poaceae) **IPNI:** 242986-2 **Use:** AF, HF, ME **Region:** Andean **Dept.:** BOG**Conservation status: (IUCN)** NE

1893 *Sporobolus indicus* (L.) R.Br. (Poaceae) **IPNI:** 1207790-2 **Common name:** castillera, costillera, cutupén, espartillo, grama, grama dura, gramote, guayacán, llano, paja de burro, paja de grama, paja dulce, paja puyita, pasto negro, rabo de burro, rabo de mula, raboeburro, sabana **Habit:** Herb **Use:** AF, HF, MA, ME **Origin:** Native **Region:** Amazonia, Andean, Caribbean, Orinoquia, Pacific **Dept.:** ANT, ARA, BOG, BOL, BOY, CAL, CAQ, CAU, COR, CUN, HUI, LAG, MAG, NAR, PUT, QUI, RIS, SAN, TOL, VAC **Elev.:** 0-3000 m.a.s.l. **IUCN Habitat code:** 1,3,4,14 **Conservation status: (IUCN)** NE **(CPSG)** Potential LC

1894 *Sporobolus junceus* (P.Beauv.) Kunth (Poaceae) **IPNI:** 1103484-2 **Common name:** grama blanca **Use:** ME **Dept.:** BOG**IUCN Habitat code:** 1,14 **Conservation status: (IUCN)** NE

1895 *Sporobolus pyramidalis* P.Beauv. (Poaceae) **IPNI:** 422898-1 **Common name:** sabana de Castilla **Habit:** Herb **Use:** AF, EU, HF, MA, ME, SU **Origin:** Native **Region:** Amazonia, Andean, Guiana Shield, Caribbean, Orinoquia, Pacific **Dept.:** AMA, ANT, ARA, ATL, BOL, BOY, CAL, CAQ, CAS, CAU, CES, CHO, CUN, GUA, GUV, HUI, LAG, MAG, MET, NSA, PUT, QUI, RIS, SAP, SAN, TOL, VAC, VAU, VID **Elev.:** 0-2000 m.a.s.l. **IUCN Habitat code:** 1,2,3,4,14 **Conservation status: (IUCN)** NE **(CPSG)** Potential LC

1896 *Sporobolus pyramidatus* (Lam.) Hitchc. (Poaceae) **IPNI:** 243026-2 **Habit:** Herb **Use:** ME **Origin:** Native **Region:** Caribbean **Dept.:** ATL, BOL, LAG, MAG **Elev.:** 0–500 m a.s.l. **IUCN Habitat code:** 1,2,5,14 **Conservation status: (IUCN)** NE

1897 *Sporobolus tenuissimus* (Mart. ex Schrank) Kuntze (Poaceae) **IPNI:** 243044-2 **Common name:** pelo de vieja **Habit:** Herb **Use:** EU **Origin:** Native **Region:** Amazonia, Andean, Guiana Shield, Caribbean, Orinoquia, Pacific **Dept.:** AMA, ANT, ARA, ATL, BOL, BOY, CAL, CAQ, CAS, CAU, CES, CHO, COR, CUN, GUV, HUI, LAG, MAG, MET, NSA, PUT, QUI, RIS, SAP, SAN, TOL, VAC, VID **Elev.:** 0-2000 m.a.s.l. **IUCN Habitat code:** 4,14 **Conservation status: (IUCN)** NE **(CPSG)** Potential LC

1898 *Sporobolus virginicus* (L.) Kunth (Poaceae) **IPNI:** 423010-1 **Habit:** Herb **Use:** AF, EU, HF, ME **Origin:** Native **Region:** Caribbean **Dept.:** ATL, BOL, LAG, MAG, SAP, SUC **Elev.:** 0–500 m a.s.l. **IUCN Habitat code:** 1,2,5 **Conservation status: (IUCN)** NE

1899 *Stapfochloa elata* (Desv.) P.M.Peterson (Poaceae) **IPNI:** 60469006-2 **Habit:** Herb **Use:** AF,ME **Origin:** Native **Region:** Andean, Caribbean, Orinoquia **Dept.:** ATL, BOL, CAL, CUN, HUI, LAG, MAG, MET, NAR, SAN, TOL, VAC **Elev.:** 0–1500 m a.s.l. **Conservation status: (IUCN)** NE

1900 *Steinchisma laxum* (Sw.) Zuloaga (Poaceae)
IPNI: 1156911-2
Habit: Herb **Use:** ME **Origin:** Native **Region:** Amazonia, Andean, Guiana Shield, Caribbean, Orinoquia, Pacific **Dept.:** AMA, ANT, ARA, BOY, CAL, CAQ, CAS, CAU, CES, CHO, COR, CUN, GUA, GUV, HUI, MAG, MET, NAR, NSA, QUI, RIS, SAN, TOL, VAC, VAU, VID **Elev.:** 0-2500 m.a.s.l. **IUCN Habitat code:** 1,2,3,4,5,14 **Conservation status: (IUCN)** NE **(CPSG)** Potential LC

1901 *Stenotaphrum secundatum* (Walter) Kuntze (Poaceae)
IPNI: 423051-1
Common name: hedionda, yerba de San Agustín **Habit:** Herb **Use:** AF, EU, GS, HF, ME **Origin:** Cultivated, Naturalised **Region:** Amazonia, Andean, Caribbean, Pacific **Dept.:** ANT, CAL, CAQ, CHO, CUN, NAR, NSA, SAP, SAN, VAC **Elev.:** 0–3000 m a.s.l. **IUCN Habitat code:** 1,3,14 **Conservation status: (IUCN)** NE

1902 *Trachypogon spicatus* (L.f.) Kuntze (Poaceae)
IPNI: 424391-1
Common name: paja lisa, saeta, saeta lisa **Habit:** Herb **Use:** AF, MA, ME, PO, SU **Origin:** Native **Region:** Amazonia, Andean, Guiana Shield, Caribbean, Orinoquia **Dept.:** ANT, ARA, BOG, BOY, CAS, CES, CUN, GUA, GUV, MAG, MET, TOL, VAC, VAU, VID **Elev.:** 0-1500 m.a.s.l. **IUCN Habitat code:** 1,2,3,4,14 **Conservation status: (IUCN)** NE **(CPSG)** Potential LC

1903 *Tragus berteronianus* Schult. (Poaceae)
IPNI: 256481-2
Habit: Herb **Use:** AF, MA, ME **Origin:** Naturalised **Region:** Andean, Caribbean, Orinoquia **Dept.:** ANT, BOY, CAL, CAS, CES, HUI, LAG, MAG, NSA, SAN, TOL, VAC **Elev.:** 0–2500 m a.s.l. **IUCN Habitat code:** 2,3 **Conservation status: (IUCN)** NE

1904 *Tripsacum andersonii* J.R.Gray (Poaceae)
IPNI: 258687-2
Common name: guatemala, pasto guatemala, pasto maíz **Habit:** Herb **Use:** AF, EU, GS **Origin:** Native **IUCN Habitat code:** 1,3,4,14 **Conservation status: (IUCN)** LC

1905 *Tripsacum australe* Cutler & E.S.Anderson (Poaceae)
IPNI: 258688-2
Common name: arroz silvestre, cañarote, gramalote **Habit:** Herb **Use:** GS **Origin:** Native **Region:** Andean, Guiana Shield, Orinoquia **Dept.:** BOY, CUN, GUV, HUI, MET, TOL **Elev.:** 0–1500 m a.s.l. **IUCN Habitat code:** 1,14 **Conservation status: (IUCN)** LC

1906 *Tripsacum cundinamarce* De Wet & Timothy (Poaceae)
IPNI: 258691-2
Habit: Herb **Use:** GS **Origin:** Endemic **Region:** Andean **Dept.:** CUN **Elev.:** 1000–1500 m a.s.l. **Conservation status: (IUCN)** NE

1907 *Tripsacum dactyloides* (L.) L. (Poaceae)
IPNI: 424990-1
Habit: Herb **Use:** AF, EU, GS, HF **Origin:** Native **Dept.:** BOGIUCN **Habitat code:** 1,2,3,4,14 **Conservation status: (IUCN)** NE

1908 *Tripsacum laxum* Nash (Poaceae)
IPNI: 287719-2
Habit: Herb **Use:** AF, GS, ME **Origin:** Naturalised **Region:** Andean **Dept.:** BOY, CUN, VAC **Elev.:** 500–1500 m a.s.l. **IUCN Habitat code:** 3,14 **Conservation status: (IUCN)** LC

1909 *Trisetum spicatum* (L.) K.Richt. (Poaceae)
IPNI: 258855-2
Habit: Herb **Use:** AF, HF **Origin:** Native **Region:** Andean **Dept.:** BOG, CAL, CUN **Elev.:** 3500–4500 m a.s.l. **IUCN Habitat code:** 4,14 **Conservation status: (IUCN)** NE

1910 *Triticum aestivum* L. (Poaceae)
IPNI: 332110-2
Common name: marengo, trigo, trigo almequemé, trigo argentino, trigo arroceño, trigo barcino, trigo bolapicota, trigo bolarraspa, trigo bolas, trigo bonza, trigo cartagua, trigo cañablanca, trigo centeno, trigo colorao, trigo cuanca, trigo diacó, trigo florel marengo, **Habit:** Herb **Use:** AF, GS, HF, MA, ME, PO, SU **Origin:** Cultivated, Naturalised **Region:** Andean **Dept.:** ANT, BOG, BOY, CUN, NAR **Elev.:** 2000–3000 m a.s.l. **IUCN Habitat code:** 1,3,4,14 **Conservation status: (IUCN)** NE

1911 *Urochloa arrecta* (Hack.) Morrone & Zuloaga (Poaceae)
IPNI: 972082-1
Common name: braquipará **Habit:** Herb **Use:** AF, EU, MA, PO **Origin:** Cultivated, Naturalised **Region:** Andean **Dept.:** ANT **Elev.:** 1500–2000 m a.s.l. **IUCN Habitat code:** 2,14 **Conservation status: (IUCN)** NE

1912 *Urochloa brizantha* (A.Rich.) R.D.Webster (Poaceae)
IPNI: 943736-1
Common name: bacaria lisa, braquiara amargo, braquiaria, pasto alambre **Habit:** Herb **Use:** AF, EU, HF, IF, ME **Origin:** Naturalised **Region:** Amazonia, Andean, Caribbean, Orinoquia **Dept.:** ANT, ARA, BOY, CAL, CAQ, COR, HUI, MET, TOL, VAC **Elev.:** 0–2500 m a.s.l. **IUCN Habitat code:** 1,2,3,4,14 **Conservation status: (IUCN)** NE

1913 *Urochloa dictyoneura* (Fig. & De Not.) Veldkamp (Poaceae)
IPNI: 992521-1
Common name: pasto dulce, pasto llanero **Habit:** Herb **Use:** AF, EU **Origin:** Naturalised **Region:** Orinoquia **Dept.:** ARA, MET, VAC **Elev.:** 0–500 m a.s.l. **IUCN Habitat code:** 1,3,4,14 **Conservation status: (IUCN)** NE

1914 *Urochloa distachya* (L.) T.Q.Nguyen (Poaceae)
IPNI: 426250-1
Habit: Herb **Use:** AF, EU, ME **Origin:** Naturalised **Region:** Caribbean **Dept.:** ATL, CUN, LAG, SAP, TOL **Elev.:** 0–500 m a.s.l. **IUCN Habitat code:** 5 **Conservation status: (IUCN)** NE

1915 *Urochloa eminii* (Mez) Davidse (Poaceae)
IPNI: 974586-1
Common name: bracaria, braquiaria, braquiaria común, brecharia, mulato, pasto amargo, pasto braquiaria, pasto peludo, peludo **Habit:** Herb **Use:** AF, HF, ME, PO **Origin:** Naturalised **Region:** Amazonia, Andean, Guiana Shield, Caribbean, Orinoquia **Dept.:** ANT, ARA, CAL, CAQ, CAS, COR, CUN, HUI, MET, PUT, TOL, VAC, VID **Elev.:** 0–2500 m a.s.l. **Conservation status: (IUCN)** NE

1916 *Urochloa fusca* (Sw.) B.F.Hansen & Wunderlin (Poaceae)
IPNI: 1156554-2
Common name: grama de horqueta, granadilla, grano de oro **Habit:** Herb **Use:** AF **Origin:** Native **Region:** Amazonia, Andean, Caribbean, Orinoquia **Dept.:** ANT, ARA, ATL, BOL, CAQ, CHO, COR, CUN, HUI, LAG, MAG, TOL, VAC **Elev.:** 0–1000 m a.s.l. **IUCN Habitat code:** 1,2,3,14 **Conservation status: (IUCN)** NE

1917 *Urochloa mollis* (Sw.) Morrone & Zuloaga (Poaceae)
IPNI: 972086-1
Common name: paja de Páez, paja de laguna, paja de pay, pajaepaje, pajaepay, pará, pasto de laguna, yerba de Pará **Habit:** Herb **Use:** ME **Origin:** Native **Region:** Andean, Caribbean, Orinoquia **Dept.:** ANT, ARA, ATL, BOL, CAS, CAU, CUN, HUI, MAG, NAR, SAN, TOL, VAC **Elev.:** 0-1000 m.a.s.l. **IUCN Habitat code:** 1,2,14 **Conservation status: (IUCN)** NE **(CPSG)** Potential LC

1918 *Urochloa mutica* (Forssk.) T.Q.Nguyen (Poaceae)
IPNI: 426266-1
Common name: admirable, braquipará, mirable, paja de pay, paja pará, paja páez, pará, pasto de laguna, pasto janeiro, pasto pará, yaraguá **Habit:** Herb **Use:** AF, EU, ME, SU **Origin:** Naturalised **Region:** Amazonia, Andean, Guiana Shield **Dept.:** ANT, BOY, CAL, CUN, HUI, MAG, MET, PUT, TOL, VAC, VID **Elev.:** 0–1500 m a.s.l. **IUCN Habitat code:** 1,2,3,14 **Conservation status: (IUCN)** NE

1919 *Urochloa plantaginea* (Link) R.D.Webster (Poaceae)
IPNI: 286530-2
Common name: braquiaria amargo, braquiaria dulce, pasto braquiaria **Habit:** Herb **Use:** AF **Origin:** Native **Region:** Andean **Dept.:** ANT **Elev.:** 1500–2000 m a.s.l. **IUCN Habitat code:** 1,14 **Conservation status: (IUCN)** NE

1920 *Urochloa polystachya* (Kunth) Mabb. (Poaceae)
IPNI: 77164258-1
Common name: janeiro, pasto janeiro, yerba janeiro **Habit:** Herb **Use:** AF, HF, ME **Origin:** Native **Region:** Andean, Pacific **Dept.:** ANT, BOY, CAU, QUI, VAC **Elev.:** 0–1500 m a.s.l. **Conservation status: (IUCN)** NE

1921 *Urochloa reptans* (L.) Stapf (Poaceae)
IPNI: 426280-1
Common name: mindaca **Habit:** Herb **Use:** AF,ME **Origin:** Naturalised **Region:** Andean, Caribbean **Dept.:** ANT, ATL, BOL, COR, CUN, HUI, MAG, SUC, TOL **Elev.:** 0–2000 m a.s.l. **IUCN Habitat code:** 2,3,14 **Conservation status: (IUCN)** NE

1922 *Urochloa subquadripara* (Trin.) R.D.Webster (Poaceae)
IPNI: 943744-1
Use: ME **Origin:** Naturalised **Conservation status: (IUCN)** NE

1923 *Zea diploperennis* Iltis, Doebley & R.Guzmán (Poaceae)
IPNI: 306962-2
Habit: Herb **Use:** GS, HF **Origin:** Cultivated, Naturalised **Region:** Andean **Dept.:** CAL, CAU, CUN **Elev.:** 2000–3000 m a.s.l. **Conservation status: (IUCN)** EN

1924 *Zea luxurians* (Durieu & Asch.) R.M.Bird (Poaceae)
IPNI: 270861-2
Common name: teosinte **Habit:** Herb **Use:** AF, GS, HF, ME **Origin:** Cultivated, Naturalised **Region:** Andean **Dept.:** VAC **Elev.:** 1000–1500 m a.s.l. **Conservation status: (IUCN)** VU

1925 *Zea mays* L. (Poaceae)
IPNI: 426810-1
Common name: cangil, canguil, capi, capí, choclo, chulpe, granizo, maicillo, maiki, mayi, maíz, maíz Yucatán, maíz azulito, maíz berrendo, maíz blanco, maíz brisa, maíz cacho de buey, maíz capio, maíz cariaco, maíz cariaco amarillo, maíz cariaco rayado, maíz cariaco r **Habit:** Herb **Use:** AF, EU, FU, GS, HF, IF, MA, ME, PO, SU **Origin:** Cultivated, Naturalised **Region:** Amazonia, Andean, Guiana Shield, Caribbean, Orinoquia, Pacific **Dept.:** AMA, ANT, ARA, ATL, BOG, BOL, BOY, CAL, CAQ, CAS, CAU, CES, CHO, CUN, GUA, GUV, HUI, LAG, MAG, MET, NAR, NSA, PUT, QUI, RIS, SAN, SUC, TOL, VAC, VAU, VID **Elev.:** 0–3000 m a.s.l. **IUCN Habitat code:** 1,2,3,4,5,6,8,14 **Conservation status: (IUCN)** LC

1926 *Zoysia matrella* (L.) Merr. (Poaceae)
IPNI: 1214917-2
Habit: Herb **Use:** AF, EU, ME **Origin:** Cultivated, Naturalised **Region:** Amazonia, Andean, Caribbean, Orinoquia, Pacific **Dept.:** AMA, ANT, ATL, BOL, CAQ, CES, CHO, COR, HUI, LAG, MAG, NSA, QUI, RIS, SAP, SAN, SUC, TOL, VAC **Elev.:** 0–2000 m a.s.l. **IUCN Habitat code:** 1,14 **Conservation status: (IUCN)** NE

1927 *Zuloagaea bulbosa* (Kunth) Bess (Poaceae)
IPNI: 77076421-1
Habit: Herb **Use:** ME **Origin:** Native **Region:** Andean **Dept.:** BOG, CUN, NAR, VAC **Elev.:** 2500–3000 m a.s.l. **Conservation status: (IUCN)** NE

1928 *Rapatea spectabilis* Pilg. (Rapateaceae)
IPNI: 217816-2
Common name: yerba de zaíno **Habit:** Herb **Use:** ME **Origin:** Native **Region:** Amazonia, Pacific **Dept.:** AMA, CHO, VAC **Elev.:** 50–500 m a.s.l. **IUCN Habitat code:** 1,5,14 **Conservation status: (IUCN)** NE

1929 *Sparganium americanum* Nutt. (Typhaceae)
IPNI: 836744-1
Habit: Herb, Aquatic **Use:** HF **Origin:** Native **Region:** Andean **Dept.:** ANT **Elev.:** 1850–1850 m a.s.l. **Conservation status: (IUCN)** LC

1930 *Typha domingensis* Pers. (Typhaceae)
IPNI: 836837-1
Common name: enea **Habit:** Herb, Aquatic **Use:** AF, EU, FU, GS, HF, MA, ME, SU **Origin:** Native **Region:** Andean, Caribbean, Pacific **Dept.:** ANT, BOG, BOY, COR, MAG, SAN **Elev.:** 10–2000 m a.s.l. **IUCN Habitat code:** 2,3,5,14 **Conservation status: (IUCN)** LC

1931 *Typha latifolia* L. (Typhaceae)
IPNI: 836870-1
Common name: anea, caña de la pasión, enea, espadaña, junco, junco de estera, junco de la pasión, totora **Habit:** Herb, Aquatic **Use:** AF, EU, GS, HF, MA, ME, PO **Origin:** Native **Region:** Andean **Dept.:** BOG, SAN **Elev.:** 1680–1680 m a.s.l. **IUCN Habitat code:** 1,3,5,14 **Conservation status: (IUCN)** LC

1932 *Xyris jupicai* Rich. (Xyridaceae)
IPNI: 315094-2
Common name: junco sudadero **Habit:** Herb **Use:** ME **Origin:** Native **Region:** Andean, Guiana Shield, Orinoquia, Pacific **Dept.:** ANT, CAS, CAU, CHO, GUA, MET, SAN, VAC, VAU, VID **Elev.:** 40–2200 m a.s.l. **IUCN Habitat code:** 1,2,3,4,5,14 **Conservation status: (IUCN)** NE

1933 *Xyris savanensis* Miq. (Xyridaceae)
IPNI: 319776-2
Habit: Herb **Use:** ME **Origin:** Native **Region:** Amazonia, Guiana Shield, Orinoquia **Dept.:** AMA, CAQ, CAS, GUA, VAU, VID **Elev.:** 240–275 m a.s.l. **IUCN Habitat code:** 1,2,3,4,14 **Conservation status: (IUCN)** NE

Zingiberales

1934 *Canna glauca* L. (Cannaceae)
IPNI: 796016-1
Common name: achira, achirilla, buatensmesh **Habit:** Herb **Use:** EU, GS, HF, MA, ME **Origin:** Native **Region:** Caribbean, Pacific **Dept.:** CHO, MAG, VAC **Elev.:** 50–1000 m a.s.l. **IUCN Habitat code:** 2,14 **Conservation status: (IUCN)** NE

1935 *Canna indica* L. (Cannaceae)
IPNI: 319130-2
Common name: achera, achila, achira, achira blanca, achira bugueña, achira colorada, achira criolla, achira morada, achira natagueña, achira oscura, achira roja, achira verde, achirilla, adura, bandera, bandera amarilla, bandera de Uribe, bijao rosario, brigua, buria, **Habit:** Herb **Use:** AF, EU, HF, MA, ME **Origin:** Native, Cultivated **Region:** Amazonia, Andean, Guiana Shield, Caribbean, Pacific **Dept.:** AMA, ANT, BOG, CAL, CAQ, CAU, CHO, COR, CUN, HUI, MAG, MET, NAR, NSA, PUT, QUI, SAP, SAN, VAC, VAU **Elev.:** 20-2620 m.a.s.l. **IUCN Habitat code:** 1,2,3,4,5,14 **Conservation status: (IUCN)** NE **(CPSG)** Potential LC

1936 *Canna jaegeriana* Urb. (Cannaceae)
IPNI: 44425-2
Common name: achira, achirilla, bijao de agua, chisgua **Habit:** Herb **Use:** HF **Origin:** Native **Region:** Andean **Dept.:** ANT, BOY, PUT, SAN **Elev.:** 600–2500 m a.s.l. **IUCN Habitat code:** 1,2,3,14 **Conservation status: (IUCN)** NE

1937 *Canna paniculata* Ruiz & Pav. (Cannaceae)
IPNI: 315579-2
Habit: Herb **Use:** HF **Origin:** Native **Region:** Caribbean, Orinoquia **Dept.:** LAG, MET **Elev.:** 300–300 m a.s.l. **Conservation status: (IUCN)** NE

1938 *Canna tuerckheimii* Kraenzl. (Cannaceae)
IPNI: 44441-2
Habit: Herb **Use:** HF, ME **Origin:** Native **Region:** Andean, Pacific **Dept.:** CHO, NAR **Elev.:** 730–730 m a.s.l. **IUCN Habitat code:** 1 **Conservation status: (IUCN)** NE

1939 *Costus amazonicus* (Loes.) J.F.Macbr. (Costaceae)
IPNI: 66817-2
Common name: caña, cañagria, cañaguate **Habit:** Herb **Use:** HF **Origin:** Native **Region:** Amazonia, Andean, Guiana Shield **Dept.:** AMA, CAQ, PUT, VAU **Elev.:** 100–1100 m a.s.l. **IUCN Habitat code:** 1,3,14 **Conservation status: (IUCN)** NE

1940 *Costus arabicus* L. (Costaceae)
IPNI: 796267-1
Common name: cañabrava, cañagria **Habit:** Herb **Use:** HF, ME **Origin:** Native **Region:** Amazonia, Guiana Shield, Caribbean, Orinoquia **Dept.:** AMA, ARA, BOL, CAS, CES, GUA, GUV, MET, VAU, VID **Elev.:** 35–320 m a.s.l. **IUCN Habitat code:** 1,2,3,4,5,14 **Conservation status: (IUCN)** NE

1941 *Costus comosus* (Jacq.) Roscoe (Costaceae)
IPNI: 796280-1
Common name: cañagria **Habit:** Herb **Use:** ME **Origin:** Native **Region:** Andean, Orinoquia **Dept.:** BOY, CAQ, CAS, CUN, MAG, MET **Elev.:** 230–1080 m a.s.l. **IUCN Habitat code:** 1,3,14 **Conservation status: (IUCN)** NE

1942 *Costus erythrocoryne* K.Schum. (Costaceae)
IPNI: 66844-2
Habit: Herb **Use:** HF **Origin:** Native **Region:** Amazonia **Dept.:** AMA **Elev.:** 100–110 m a.s.l. **IUCN Habitat code:** 1 **Conservation status: (IUCN)** NE

1943 *Costus erythrophyllus* Loes. (Costaceae)
IPNI: 66845-2
Habit: Herb **Use:** HF, ME **Origin:** Native **Region:** Amazonia, Andean, Orinoquia **Dept.:** BOY, CUN, MET, PUT **Elev.:** 250–1800 m a.s.l. **IUCN Habitat code:** 1,3,5 **Conservation status: (IUCN)** NE

1944 *Costus guanaiensis* Rusby (Costaceae)
IPNI: 796308-1
Common name: cañabrava, cañagria **Habit:** Herb **Use:** AF, HF, ME **Origin:** Native **Region:** Amazonia, Andean, Guiana Shield, Caribbean, Orinoquia, Pacific **Dept.:** ANT, ATL, BOL, CAQ, CAS, CAU, CHO, GUV, LAG, MAG, MET, NAR, NSA, PUT, QUI, VAC, VAU **Elev.:** 0-1880 m.a.s.l. **IUCN Habitat code:** 1,2,3,4,14 **Conservation status: (IUCN)** NE **(CPSG)** Potential LC

1945 *Costus laevis* Ruiz & Pav. (Costaceae)
IPNI: 796315-1
Common name: cañaguate, cañeja, guira, güira, mocuillibua **Habit:** Herb **Use:** EU, ME **Origin:** Native **Region:** Andean, Caribbean, Orinoquia, Pacific **Dept.:** ANT, ARA, BOY, CAL, CAQ, CAU, CHO, CUN, HUI, MAG, MET, NAR, NSA, PUT, QUI, SAN, TOL, VAC **Elev.:** 0-2000 m.a.s.l. **IUCN Habitat code:** 1,2,3,5,14 **Conservation status: (IUCN)** NE **(CPSG)** Potential LC

1946 *Costus lasius* Loes. (Costaceae)
IPNI: 66866-2
Common name: cañagria, cañagria menuda, cañaguate, cañaguatos, de flor naranja **Habit:** Herb **Use:** ME **Origin:** Native **Region:** Amazonia, Andean, Guiana Shield, Orinoquia, Pacific **Dept.:** AMA, ANT, CAQ, CAU, CHO, MET, PUT, SAN, TOL, VAU **Elev.:** 5–1900 m a.s.l. **IUCN Habitat code:** 1,3,14 **Conservation status: (IUCN)** LC

1947 *Costus lima* K.Schum. (Costaceae)
IPNI: 66870-2
Common name: mocuillibua **Habit:** Herb **Use:** ME **Origin:** Native **Region:** Pacific **Dept.:** ANT, CAU, CHO, MAG, NAR, VAC **Elev.:** 0–950 m a.s.l. **IUCN Habitat code:** 1,3,14 **Conservation status: (IUCN)** LC

1948 *Costus longibracteolatus* Maas (Costaceae) **IPNI:** 66873-2 **Habit:** Herb **Use:** ME **Origin:** Native **Region:** Amazonia, Andean, Guiana Shield **Dept.:** AMA, CAQ, PUT **Elev.:** 50–1250 m a.s.l. **IUCN Habitat code:** 1 **Conservation status: (IUCN)** NE

1949 *Costus productus* Gleason ex Maas (Costaceae) **IPNI:** 66892-2 **Habit:** Herb **Use:** HF **Origin:** Native **Region:** Amazonia **Dept.:** PUT **Elev.:** 230–250 m a.s.l. **Conservation status: (IUCN)** NE

1950 *Costus pulverulentus* C.Presl (Costaceae) **IPNI:** 796364-1 **Common name:** caña agria **Habit:** Herb **Use:** HF, ME **Origin:** Native **Region:** Andean, Guiana Shield, Caribbean, Orinoquia, Pacific **Dept.:** ANT, ARA, BOL, BOY, CAL, CAU, CHO, LAG, MAG, MET, NAR, NSA, RIS, SAN, TOL, VAC **Elev.:** 10-900 m.a.s.l. **IUCN Habitat code:** 1,3,5,14 **Conservation status: (IUCN)** NE **(CPSG)** Potential LC

1951 *Costus scaber* Ruiz & Pav. (Costaceae) **IPNI:** 796376-1 **Common name:** caña, caña selvática, cañafuerte, cañagria, cañaguate, flor candelabro, pendaya, tilia **Habit:** Herb **Use:** HF, ME **Origin:** Native **Region:** Amazonia, Andean, Guiana Shield, Caribbean, Orinoquia, Pacific **Dept.:** AMA, ANT, ARA, BOL, BOY, CAQ, CAS, CAU, CHO, COR, CUN, GUA, LAG, MAG, MET, NAR, NSA, PUT, RIS, SAN, VAC, VAU, VID **Elev.:** 0–1700 m a.s.l. **IUCN Habitat code:** 1,2,3,4,5,14 **Conservation status: (IUCN)** NE

1952 *Costus spicatus* (Jacq.) Sw. (Costaceae) **IPNI:** 329964-2 **Common name:** cañagria, tilia **Use:** HF, ME **IUCN Habitat code:** 1,2,14 **Conservation status: (IUCN)** NE

1953 *Costus spiralis* (Jacq.) Roscoe (Costaceae) **IPNI:** 796387-1 **Common name:** cañagria, cañaguate, cañejo **Habit:** Herb **Use:** HF, ME **Origin:** Native **Region:** Amazonia, Andean, Guiana Shield, Caribbean, Orinoquia, Pacific **Dept.:** AMA, ANT, ARA, BOL, BOY, CAQ, CAS, CAU, CES, CHO, CUN, GUA, GUV, MAG, MET, NSA, PUT, QUI, SAN, VAU, VID **Elev.:** 55-1630 m.a.s.l. **IUCN Habitat code:** 1,3,4,5,14 **Conservation status: (IUCN)** NE **(CPSG)** Potential LC

1954 *Costus villosissimus* Jacq. (Costaceae) **IPNI:** 796399-1 **Common name:** cañagria, cañaguatos, cañalanuda, cañeja, pendaya **Habit:** Herb **Use:** EU, MA, ME **Origin:** Native **Region:** Andean, Caribbean, Orinoquia, Pacific **Dept.:** ANT, BOY, CAL, CAS, CAU, CHO, CUN, LAG, MAG, MET, NAR, PUT, SAN, VAC, VID **Elev.:** 0-1300 m.a.s.l. **IUCN Habitat code:** 1,2,3,14 **Conservation status: (IUCN)** LC **(CPSG)** Potential LC

1955 *Costus woodsonii* Maas (Costaceae) **IPNI:** 66932-2 **Habit:** Herb **Use:** HF **Origin:** Native **Region:** Pacific **Dept.:** CHO, VAC **Elev.:** 0–130 m a.s.l. **IUCN Habitat code:** 1,14 **Conservation status: (IUCN)** LC

1956 *Dimerocostus strobilaceus* Kuntze (Costaceae) **IPNI:** 80399-2 **Common name:** cagaloduro, cañagria, cañaguate, cañapángala, cañeja, pángala, vara de mico **Habit:** Herb **Use:** EU, HF, ME **Origin:** Native **Region:** Amazonia, Andean, Guiana Shield, Orinoquia **Dept.:** AMA, ANT, ARA, BOY, CAL, CAQ, CAS, CES, CHO, COR, CUN, MET, NAR, NSA, PUT, SAN, VAC **Elev.:** 0–1500 m a.s.l. **IUCN Habitat code:** 1,2,3,5,14 **Conservation status: (IUCN)** NE

1957 *Tapeinochilos ananassae* (Hassk.) K.Schum. (Costaceae) **IPNI:** 873025-1 **Habit:** Herb **Use:** HF, ME, SU **Origin:** Cultivated **Region:** Andean **Dept.:** ANT, CAL, CUN, QUI, SAN, VAC **Elev.:** 1000–1500 m a.s.l. **IUCN Habitat code:** 1 **Conservation status: (IUCN)** NE

1958 *Heliconia acuminata* A.Rich. (Heliconiaceae) **IPNI:** 796910-1 **Common name:** platanillo, sororokinha **Habit:** Herb **Use:** ME **Origin:** Native **Region:** Amazonia, Guiana Shield, Orinoquia **Dept.:** AMA, CAQ, GUA, MET, PUT, VAU, VID **Elev.:** 50–1300 m a.s.l. **IUCN Habitat code:** 1,3,4,5,14 **Conservation status: (IUCN)** NE

1959 *Heliconia bihai* (L.) L. (Heliconiaceae) **IPNI:** 796925-1 **Common name:** bihao, bihaos, bijao, bijao de puerco, bijaos, maurú, murrapo, platanillo **Habit:** Herb **Use:** EU, GS, HF, MA, ME **Origin:** Native **Region:** Andean, Caribbean **Dept.:** LAG, MAG, NSA, SUC **Elev.:** 0–1700 m a.s.l. **IUCN Habitat code:** 1,2,3,4,14 **Conservation status: (IUCN)** NE

1960 *Heliconia curtispatha* Petersen (Heliconiaceae) **IPNI:** 119344-2 **Common name:** platanillo rojo **Habit:** Herb **Use:** ME **Origin:** Native **Region:** Pacific **Dept.:** ANT, CHO, COR, NAR, VAC **Elev.:** 0–350 m a.s.l. **IUCN Habitat code:** 1,3,14 **Conservation status: (IUCN)** NE

1961 *Heliconia episcopalis* Vell. (Heliconiaceae) **IPNI:** 119355-2 **Common name:** platanillo **Habit:** Herb **Use:** EU **Origin:** Native **Region:** Amazonia, Guiana Shield, Caribbean, Orinoquia **Dept.:** AMA, ANT, ARA, BOL, CAQ, CAS, CUN, GUV, MET, NSA, PUT, SAP, SAN, SUC **Elev.:** 50–600 m a.s.l. **IUCN Habitat code:** 1,2,3,5,14 **Conservation status: (IUCN)** NE

1962 *Heliconia hirsuta* L.f. (Heliconiaceae) **IPNI:** 796953-1 **Common name:** bijao, cañabrava, churca, cuchilla, gualba, guargo, isira, isirra, murrapo, muu, pipiboto, platanillo, platanillo fosforito, tacana **Habit:** Herb **Use:** EU, HF, MA, ME **Origin:** Native **Region:** Amazonia, Andean, Guiana Shield, Caribbean, Orinoquia, Pacific **Dept.:** AMA, ANT, ARA, BOL, BOY, CAL, CAQ, CAS, CAU, CES, CHO, COR, CUN, GUA, GUV, HUI, LAG, MAG, MET, NAR, NSA, PUT, QUI, RIS, SAN, TOL, VAC, VAU, VID **Elev.:** 30–1680 m a.s.l. **IUCN Habitat code:** 1,2,3,4,5,14 **Conservation status: (IUCN)** NE

1963 *Heliconia latispatha* Benth. (Heliconiaceae) **IPNI:** 796959-1 **Common name:** ave del paraíso, flor pajarito, lengua de vaca, platanilla, platanillo **Habit:** Herb **Use:** EU, HF **Origin:** Native **Region:** Amazonia, Andean, Guiana Shield, Caribbean, Orinoquia, Pacific **Dept.:** ANT, ARA, ATL, BOL, BOY, CAL, CAQ, CAS, CAU, CHO, COR, CUN, GUV, LAG, MAG, MET, NAR, NSA, QUI, RIS, SAN, SUC, TOL, VAC **Elev.:** 0-1810 m.a.s.l. **IUCN Habitat code:** 1,2,3,5,14 **Conservation status: (IUCN)** NE **(CPSG)** Potential LC

1964 *Heliconia lentiginosa* Abalo & G.Morales (Heliconiaceae) **IPNI:** 278062-2 **Common name:** heliconia **Habit:** Herb **Use:** EU **Origin:** Endemic **Region:** Andean **Dept.:** ANT **Elev.:** 300–1150 m a.s.l. **IUCN Habitat code:** 1,3 **Conservation status: (IUCN)** NE

1965 *Heliconia longa* (Griggs) H.J.P.Winkl. (Heliconiaceae) **IPNI:** 119402-2 **Habit:** Herb **Use:** EU **Origin:** Native **Region:** Andean, Pacific **Dept.:** ANT, CAL, CHO, COR, SAN **Elev.:** 50–1800 m a.s.l. **Conservation status: (IUCN)** NE

1966 *Heliconia longiflora* R.R.Sm. (Heliconiaceae) **IPNI:** 119403-2 **Habit:** Herb **Use:** ME **Origin:** Native **Region:** Caribbean, Pacific **Dept.:** ANT, CHO, COR, NAR, RIS, SUC, VAC **Elev.:** 0–800 m a.s.l. **IUCN Habitat code:** 1,3,14 **Conservation status: (IUCN)** NE

1967 *Heliconia marginata* (Griggs) Pittier (Heliconiaceae) **IPNI:** 119411-2 **Common name:** bijagüillo, bijao, platanillo **Habit:** Herb **Use:** AF **Origin:** Native **Region:** Amazonia, Guiana Shield, Caribbean, Orinoquia, Pacific **Dept.:** AMA, ANT, ARA, BOL, BOY, CAQ, CAS, CHO, COR, CUN, GUV, LAG, MAG, MET, QUI, SAN, TOL, VID **Elev.:** 5-400 m.a.s.l. **IUCN Habitat code:** 1,2,3,4,5,14 **Conservation status: (IUCN)** NE **(CPSG)** Potential LC

1968 *Heliconia mariae* Hook.f. (Heliconiaceae) **IPNI:** 796965-1 **Common name:** lengua de vaca, tacana, tapamuerto **Habit:** Herb **Use:** EU, HF **Origin:** Native **Region:** Caribbean, Pacific **Dept.:** ANT, ATL, CHO, COR **Elev.:** 0–520 m a.s.l. **IUCN Habitat code:** 1,3,14 **Conservation status: (IUCN)** NE

1969 *Heliconia metallica* Planch. & Linden ex Hook. (Heliconiaceae) **IPNI:** 796967-1 **Common name:** hoja de sal, lengua de vaca, platanillo **Habit:** Herb **Use:** EU **Origin:** Native **Region:** Andean, Caribbean, Pacific **Dept.:** ANT, ATL, BOL, CHO, COR, MAG, SAN **Elev.:** 0–900 m a.s.l. **IUCN Habitat code:** 1,2,3,5,14 **Conservation status: (IUCN)** NE

1970 *Heliconia pogonantha* Cufod. (Heliconiaceae) **IPNI:** 119444-2 **Habit:** Herb **Use:** HF **Origin:** Native **Region:** Andean, Pacific **Dept.:** ANT, CHO, COR **Elev.:** 0–1700 m a.s.l. **IUCN Habitat code:** 1,14 **Conservation status: (IUCN)** NE

1971 *Heliconia psittacorum* L.f. (Heliconiaceae) **IPNI:** 796972-1 **Common name:** heliconia, platanillo **Habit:** Herb **Use:** EU, HF, MA, ME **Origin:** Native **Region:** Amazonia, Andean, Guiana Shield, Orinoquia **Dept.:** AMA, ARA, CAQ, CAS, GUA, GUV, MET, PUT, VAU, VID **Elev.:** 80-

1000 m.a.s.l. **IUCN Habitat code:** 1,2,3,4,14 **Conservation status: (IUCN)** NE **(CPSG)** Potential LC

1972 *Heliconia rostrata* Ruiz & Pav. (Heliconiaceae) **IPNI:** 796978-1 **Common name:** platanillo **Habit:** Herb **Use:** EU, HF **Origin:** Native **Region:** Amazonia, Andean, Guiana Shield **Dept.:** CAQ, CAU, MET, PUT **Elev.:** 140–900 m a.s.l. **IUCN Habitat code:** 1,2,3,14 **Conservation status: (IUCN)** NE

1973 *Heliconia spathocircinata* Aristeg. (Heliconiaceae) **IPNI:** 119496-2 **Common name:** heliconia enana, platanillo **Habit:** Herb **Use:** EU **Origin:** Native **Region:** Amazonia, Andean, Guiana Shield, Caribbean, Orinoquia, Pacific **Dept.:** AMA, ANT, CAL, CAQ, CAS, CHO, COR, CUN, GUV, MET, NAR, PUT, RIS, VAC, VAU **Elev.:** 0-1620 m.a.s.l. **IUCN Habitat code:** 1,3,4,5,14 **Conservation status: (IUCN)** NE **(CPSG)** Potential LC

1974 *Heliconia stricta* Huber (Heliconiaceae) **IPNI:** 119505-2 **Common name:** heliconia, platanillo **Habit:** Herb **Use:** EU **Origin:** Native **Region:** Amazonia, Andean, Guiana Shield, Orinoquia, Pacific **Dept.:** AMA, ANT, ARA, BOY, CAL, CAQ, CAS, CAU, CHO, CUN, GUA, GUV, LAG, MAG, MET, NAR, NSA, PUT, SAN, TOL, VAC, VAU **Elev.:** 0-1700 m.a.s.l. **IUCN Habitat code:** 1,3,4,5,14 **Conservation status: (IUCN)** NE **(CPSG)** Potential LC

1975 *Heliconia wagneriana* Petersen (Heliconiaceae) **IPNI:** 119524-2 **Common name:** guarderiana **Habit:** Herb **Use:** EU, MA, ME, SU **Origin:** Native **Region:** Andean, Caribbean, Pacific **Dept.:** ANT, CHO, CUN, NAR, RIS, SAP, SAN, TOL, VAC **Elev.:** 0-1420 m.a.s.l. **IUCN Habitat code:** 1,2,3,4,5,14 **Conservation status: (IUCN)** NE **(CPSG)** Potential LC

1976 *Goeppertia bachemiana* (É.Morren) Borchs. & S.Suárez (Marantaceae) **IPNI:** 60458927-2 **Habit:** Herb **Use:** EU **Origin:** Endemic, Cultivated **Region:** Andean **Dept.:** CUN **Elev.:** 2600–2600 m a.s.l. **Conservation status: (IUCN)** NE

1977 *Calathea crotalifera* S.Watson (Marantaceae) **IPNI:** 273648-2 **Common name:** bihao, bihao negro, bijao, bijao mantagorda, cuero, jupe, mantagorda **Habit:** Herb **Use:** HF **Origin:** Native **Region:** Amazonia, Andean, Guiana Shield, Caribbean, Orinoquia, Pacific **Dept.:** ANT, BOY, CAL, CAS, CAU, CHO, COR, GUV, LAG, MAG, MET, NAR, NSA, PUT, RIS, VAC **Elev.:** 5-1850 m.a.s.l. **IUCN Habitat code:** 1,3,5,14 **Conservation status: (IUCN)** NE **(CPSG)** Potential LC

1978 *Calathea latrinotecta* H.Kenn. (Marantaceae) **IPNI:** 273661-2 **Habit:** Herb **Use:** ME **Origin:** Native **Region:** Amazonia, Andean **Dept.:** CAU, PUT **Elev.:** 100–760 m a.s.l. **IUCN Habitat code:** 1,3 **Conservation status: (IUCN)** VU

1979 *Calathea lutea* (Aubl.) E.Mey. ex Schult. (Marantaceae) **IPNI:** 60458866-2 **Common name:** biao, bihao, bihao blanco, bijao, calatea, hoja de sal, hojablanca **Habit:** Herb **Use:** EU, HF, MA, ME, SU **Origin:** Native, Cultivated **Region:** Amazonia, Andean, Guiana Shield, Caribbean, Orinoquia, Pacific **Dept.:** ANT, BOL, BOY, CAL, CAQ, CAS, CAU, CES, CHO, CUN, MAG, MET, NAR, NSA, PUT, QUI, RIS, SAN, SUC, TOL, VAC **Elev.:** 0-2200 m.a.s.l. **IUCN Habitat code:** 1,2,3,4,5,14 **Conservation status: (IUCN)** NE **(CPSG)** Potential LC

1980 *Goeppertia cyclophora* (Baker) Borchs. & S.Suárez (Marantaceae) **IPNI:** 60459273-2 **Habit:** Herb **Use:** HF **Origin:** Native **Region:** Amazonia, Guiana Shield, Orinoquia **Dept.:** CAQ, GUA, GUV, MET, VAU, VID **Elev.:** 100–700 m a.s.l. **IUCN Habitat code:** 1,3,4,14 **Conservation status: (IUCN)** NE

1981 *Goeppertia latifolia* (Willd. ex Link) Borchs. & S.Suárez (Marantaceae) **IPNI:** 60459373-2 **Common name:** bijao cuero, bijao de agua **Habit:** Herb **Use:** HF, ME **Origin:** Native **Region:** Amazonia, Andean, Guiana Shield, Caribbean, Orinoquia, Pacific **Dept.:** ANT, BOL, BOY, CAS, CES, CHO, CUN, GUV, LAG, MAG, MET, VAC **Elev.:** 15–1000 m a.s.l. **IUCN Habitat code:** 1,2,3,14 **Conservation status: (IUCN)** NE

1982 *Goeppertia lietzei* (É.Morren) Saka (Marantaceae) **IPNI:** 77174968-1 **Habit:** Herb **Use:** EU **Origin:** Cultivated **Region:** Andean **Dept.:** CUN **Elev.:** 1680–1680 m a.s.l. **Conservation status: (IUCN)** NE

1983 *Goeppertia loeseneri* (J.F.Macbr.) Borchs. & S.Suárez (Marantaceae) **IPNI:** 77128944-1 **Common name:** color azul, tinta verde **Habit:** Herb **Use:** MA, ME **Origin:** Native **Region:** Amazonia **Dept.:** AMA **Elev.:** 100–150 m a.s.l. **IUCN Habitat code:** 1 **Conservation status: (IUCN)** NE

1984 *Goeppertia macrosepala* (K.Schum.) Borchs. & S.Suárez (Marantaceae) **IPNI:** 60461039-2 **Habit:** Herb **Use:** HF, MA **Origin:** Native **Region:** Pacific **Dept.:** AMA, ANT, NAR, VAC **Elev.:** 5–300 m a.s.l. **Conservation status: (IUCN)** NE

1985 *Goeppertia majestica* (Linden) Borchs. & S.Suárez (Marantaceae) **IPNI:** 60461040-2 **Habit:** Herb **Use:** EU **Origin:** Native **Region:** Amazonia, Andean **Dept.:** NSA, PUT **Elev.:** 230–1100 m a.s.l. **IUCN Habitat code:** 4 **Conservation status: (IUCN)** NE

1986 *Goeppertia micans* (L.Mathieu) Borchs. & S.Suárez (Marantaceae) **IPNI:** 60461053-2 **Habit:** Herb **Use:** ME **Origin:** Native **Region:** Amazonia, Andean, Guiana Shield, Orinoquia, Pacific **Dept.:** AMA, ANT, CAQ, CHO, MET, NSA, SAN, VAU **Elev.:** 100–600 m a.s.l. **Conservation status: (IUCN)** NE

1987 *Goeppertia ornata* (Linden) Borchs. & S.Suárez (Marantaceae) **IPNI:** 60461169-2 **Common name:** pincel **Habit:** Herb **Use:** EU, ME **Origin:** Native **Region:** Orinoquia **Dept.:** MET **Elev.:** 595–595 m a.s.l. **IUCN Habitat code:** 1,3,14 **Conservation status: (IUCN)** NE

1988 *Goeppertia panamensis* (Rowlee) Borchs. & S.Suárez (Marantaceae) **IPNI:** 60461175-2 **Habit:** Herb **Use:** HF, ME **Origin:** Native **Region:** Guiana Shield, Orinoquia, Pacific **Dept.:** CHO, MET, VAU **Elev.:** 23–500 m a.s.l. **IUCN Habitat code:** 1,2,3 **Conservation status: (IUCN)** NE

1989 *Goeppertia paucifolia* (H.Kenn.) Borchs. & S.Suárez (Marantaceae) **IPNI:** 60461176-2 **Habit:** Herb **Use:** EU **Origin:** Cultivated **Region:** Andean **Dept.:** CAQ **Elev.:** 800–800 m a.s.l. **Conservation status: (IUCN)** NE

1990 *Goeppertia propinqua* (Poepp. & Endl.) Borchs. & S.Suárez (Marantaceae) **IPNI:** 60461303-2 **Habit:** Herb **Use:** ME **Origin:** Native **Region:** Amazonia, Andean, Guiana Shield, Orinoquia, Pacific **Dept.:** AMA, ANT, CAS, CHO, GUV, MET, VAU **Elev.:** 40–1410 m a.s.l. **Conservation status: (IUCN)** NE

1991 *Goeppertia roseopicta* (Linden ex Lem.) Borchs. & S.Suárez (Marantaceae) **IPNI:** 60461319-2 **Habit:** Herb **Use:** ME **Origin:** Native **Region:** Amazonia **Dept.:** AMA, CAQ, PUT **Elev.:** 110–250 m a.s.l. **Conservation status: (IUCN)** NE

1992 *Goeppertia zebrina* (Sims) Nees (Marantaceae) **IPNI:** 796810-1 **Common name:** mano de tigre, pizarro **Habit:** Herb **Use:** EU, MA **Origin:** Cultivated **Dept.:** CUN **Elev.:** 1900–2100 m a.s.l. **IUCN Habitat code:** 1,3,14 **Conservation status: (IUCN)** NE

1993 *Ischnosiphon arouma* (Aubl.) Körn. (Marantaceae) **IPNI:** 797103-1 **Common name:** aria, balay, balayo, chocolatillo, chocolatí, conga, congo, guaruma, guarumo, guarumá, huarumá, jingurú, juajuá, lengua de vaca, tirita, yaruma **Habit:** Herb **Use:** HF, MA, ME **Origin:** Native **Region:** Amazonia, Andean, Guiana Shield, Caribbean, Orinoquia, Pacific **Dept.:** AMA, ANT, CAQ, CHO, COR, GUV, MET, PUT, VAC, VAU, VID **Elev.:** 20-850 m.a.s.l. **IUCN Habitat code:** 1,3,4,5,14 **Conservation status: (IUCN)** NE **(CPSG)** Potential LC

1994 *Ischnosiphon inflatus* L.Andersson (Marantaceae) **IPNI:** 130281-2 **Common name:** chocolatillo **Habit:** Herb **Use:** MA **Origin:** Native **Region:** Andean, Pacific **Dept.:** CAU, CHO **Elev.:** 20–760 m a.s.l. **IUCN Habitat code:** 1,3,5,14 **Conservation status: (IUCN)** NE

1995 *Ischnosiphon rotundifolius* (Poepp. & Endl.) Körn. (Marantaceae) **IPNI:** 797124-1 **Habit:** Herb **Use:** ME **Origin:** Native **Region:** Amazonia, Guiana Shield **Dept.:** AMA, MET, VAU **Elev.:** 200–600 m a.s.l. **IUCN Habitat code:** 1,3 **Conservation status: (IUCN)** NE

1996 *Maranta amplifolia* K.Schum. (Marantaceae) **IPNI:** 152845-2 **Habit:** Herb **Use:** EU **Origin:** Native, Cultivated **Region:** Amazonia **Dept.:**

AMA **Elev.:** 220–220 m a.s.l. **Conservation status: (IUCN)** NE

1997 *Maranta arundinacea* L. (Marantaceae) **IPNI:** 280408-2 **Common name:** achira, amaranta, arruruz, bijao de agua, bordoncillo, guapo, guate, jamachipeke, juajuá, maranta, platanillo de ciénaga, pusanga, sago, sagú, sulú, yunka oqa, yuquilla **Habit:** Herb **Use:** AF, EU, HF, MA, ME, SU **Origin:** Native, Cultivated **Region:** Amazonia, Andean, Guiana Shield, Caribbean, Orinoquia **Dept.:** AMA, ANT, BOL, CAS, CES, HUI, MAG, PUT, SUC, VAU **Elev.:** 0–800 m a.s.l. **IUCN Habitat code:** 1,2,3,4,5,14 **Conservation status: (IUCN)** NE

1998 *Maranta gibba* Sm. (Marantaceae) **IPNI:** 280409-2 **Habit:** Herb **Use:** HF **Origin:** Native **Region:** Amazonia, Andean, Caribbean, Orinoquia, Pacific **Dept.:** ANT, CUN, MAG, MET, PUT, QUI, SAN **Elev.:** 50–1800 m a.s.l. **IUCN Habitat code:** 1,2,3,14 **Conservation status: (IUCN)** NE

1999 *Maranta leuconeura* É.Morren (Marantaceae) **IPNI:** 797361-1 **Common name:** caracol, huella de tigre, mano de tigre **Habit:** Herb **Use:** EU **Origin:** Cultivated, Naturalised **Region:** Andean **Dept.:** BOG, CUN **Elev.:** 2600–2600 m a.s.l. **IUCN Habitat code:** 1,3,14 **Conservation status: (IUCN)** NE

2000 *Maranta ruiziana* Körn. (Marantaceae) **IPNI:** 797449-1 **Habit:** Herb **Use:** HF, MA **Origin:** Native, Cultivated **Region:** Amazonia, Guiana Shield **Dept.:** AMA, VAU **Elev.:** 180–250 m a.s.l. **IUCN Habitat code:** 1 **Conservation status: (IUCN)** NE

2001 *Myrosma cannifolia* L.f. (Marantaceae) **IPNI:** 797633-1 **Habit:** Herb **Use:** HF **Origin:** Native **Region:** Amazonia, Orinoquia **Dept.:** CAS, PUT, VID **Elev.:** 100–200 m a.s.l. **IUCN Habitat code:** 1,14 **Conservation status: (IUCN)** NE

2002 *Stromanthe jacquinii* (Roem. & Schult.) H.Kenn. & Nicolson (Marantaceae) **IPNI:** 246147-2 **Common name:** alia, bijagüillo, bijao de palito, congo, paja tetera, palma tetera, tacana, tetera **Habit:** Herb **Use:** EU, HF, MA, ME **Origin:** Native **Region:** Andean, Caribbean, Orinoquia, Pacific **Dept.:** ANT, BOL, BOY, CAQ, CES, CHO, COR, CUN, LAG, MAG, MET, NAR, SAN, TOL, VAC **Elev.:** 80-1800 m.a.s.l. **IUCN Habitat code:** 1,2,3,5,14 **Conservation status: (IUCN)** NE **(CPSG)** Potential LC

2003 *Stromanthe macrochlamys* (Woodson & Standl.) H.Kenn. & Nicolson (Marantaceae) **IPNI:** 285700-2 **Use:** HF **Origin:** Native **IUCN Habitat code:** 1 **Conservation status: (IUCN)** NE

2004 *Thalia geniculata* L. (Marantaceae) **IPNI:** 798255-1 **Common name:** bihao, bijao, bijao de fardo, bocachica, lengüevaca, platanillo **Habit:** Herb **Use:** AF, EU, GS, HF, MA, ME, PO, SU **Origin:** Native **Region:** Guiana Shield, Caribbean, Orinoquia, Pacific **Dept.:** ANT, ARA, ATL, BOL, CAL, CHO, COR, GUV, MAG, MET, SAN **Elev.:** 6-400 m.a.s.l. **Conservation status: (IUCN)** LC **(CPSG)** Potential LC

2005 *Ensete ventricosum* (Welw.) Cheesman (Musaceae) **IPNI:** 584810-1 **Common name:** plátano de Abisinia **Habit:** Herb **Use:** AF, EU, GS, HF, IF, MA, ME **Origin:** Cultivated **Region:** Andean **Dept.:** ANT, BOG, CUN **Elev.:** 1500–2600 m a.s.l. **IUCN Habitat code:** 1,3,14 **Conservation status: (IUCN)** LC

2006 *Musa × paradisiaca* L. (Musaceae) **IPNI:** 797595-1 **Common name:** africano, ahogacachaco, artillero, banano, barranquillo, bellaco, blanco, bogotano, cachaco, cajaró, cajeto, cambul negro, cambur sopero, cambure negro, camburú, carrión, chileno, chimilo, chino, chirineto, cholo, choncho, choncho plátano, chopo, colecaba **Use:** AF, EU, HF, IF, MA, ME, SU **Dept.:** BOGIUCN **Habitat code:** 1,2,3,14 **Conservation status: (IUCN)** NE

2007 *Musa acuminata* Colla (Musaceae) **IPNI:** 797527-1 **Common name:** africano, andaquí, asentao, banano, banano cerenano, banano enano, banano indio, banano indio rojo, banano maduraverde, banano manchoso, banano manzano, barranquillo, bocadillo, bámbaro, cacique, cajeto, camaní, cambur, cambur guajibo, cera, chengue, chic **Habit:** Herb **Use:** AF, GS, HF, MA, ME, SU **Origin:** Cultivated **Region:** Amazonia, Andean, Guiana Shield, Caribbean, Orinoquia, Pacific **Dept.:** AMA, ANT, ARA, ATL, BOL, BOY, CAL, CAQ, CAS, CAU, CES, CHO, COR, CUN, GUA, GUV, HUI, LAG, MAG, MET, NAR, NSA, PUT, QUI, RIS, SAP, SAN, SUC, TOL, VAC, VAU, VID **Elev.:** 0–1800 m a.s.l. **IUCN Habitat code:** 1,3,14 **Conservation status: (IUCN)** LC

2008 *Musa balbisiana* Colla (Musaceae) **IPNI:** 797536-1 **Common name:** cachaco, cuatrofilos, pacífico, popocho **Use:** AF, GS, HF, MA, ME **IUCN Habitat code:** 1,2,14 **Conservation status: (IUCN)** LC

2009 *Musa coccinea* Andrews (Musaceae) **IPNI:** 797554-1 **Common name:** banano amazónico, colicero, platanillo, plátano guineo **Habit:** Herb **Use:** EU, GS, HF, MA, ME **Origin:** Cultivated **Region:** Andean **Dept.:** ANT, CUN, PUT **Elev.:** 1000–1500 m a.s.l. **IUCN Habitat code:** 1,3,14 **Conservation status: (IUCN)** EN

2010 *Musa ornata* Roxb. (Musaceae) **IPNI:** 584973-1 **Habit:** Herb **Use:** EU, GS, HF, MA, ME **Origin:** Cultivated, Naturalised **Region:** Andean **Dept.:** ANT **Elev.:** 0–500 m a.s.l. **IUCN Habitat code:** 1,14 **Conservation status: (IUCN)** LC

2011 *Musa velutina* H.Wendl. & Drude (Musaceae) **IPNI:** 797625-1 **Common name:** plátano ornamental **Habit:** Herb **Use:** EU, GS, HF, ME **Origin:** Cultivated **Region:** Andean **Dept.:** ANT **Elev.:** 0–1500 m a.s.l. **IUCN Habitat code:** 1,2,3,5,14 **Conservation status: (IUCN)** NE

2012 *Phenakospermum guyannense* (A.Rich.) Endl. ex Miq. (Strelitziaceae) **IPNI:** 585013-1 **Common name:** ejedje, koko, meta], platanillo, platanillo de monte, platanillo macho, platano de monte, plátano de monte, tarriago, terriago en San Martín, turriago **Habit:** Tree **Use:** EU, HF, MA, ME, SU **Origin:** Native **Region:** Amazonia, Guiana Shield, Orinoquia **Dept.:** AMA, CAQ, GUA, GUV, MET, VAU **Elev.:** 65–150 m a.s.l. **IUCN Habitat code:** 1,3,4,5,14 **Conservation status: (IUCN)** LC

2013 *Ravenala madagascariensis* Sonn. (Strelitziaceae) **IPNI:** 60466138-2 **Common name:** palma del viajero, árbol del viajero **Habit:** Tree **Use:** AF, EU, GS, HF, MA, ME **Origin:** Cultivated **Region:** Andean, Caribbean **Dept.:** ANT, CUN, QUI, RIS, SAP, SAN, TOL, VAC **Elev.:** 0–1600 m a.s.l. **IUCN Habitat code:** 14 **Conservation status: (IUCN)** LC

2014 *Strelitzia nicolai* Regel & Körn. (Strelitziaceae) **IPNI:** 798185-1 **Habit:** Shrub **Use:** EU, GS, HF, ME **Origin:** Cultivated **Region:** Andean **Dept.:** ANT, CUN, QUI, RIS, TOL, VAC **Elev.:** 0–1500 m a.s.l. **Conservation status: (IUCN)** NE

2015 *Strelitzia reginae* Banks (Strelitziaceae) **IPNI:** 798194-1 **Common name:** ave del paraíso **Habit:** Herb **Use:** EU, HF, MA, ME **Origin:** Cultivated **Region:** Andean, Caribbean **Dept.:** ANT, ATL, BOL, CAL, CUN, MAG, QUI, RIS, TOL, VAC **Elev.:** 0–2600 m a.s.l. **Conservation status: (IUCN)** NE

2016 *Alpinia purpurata* (Vieill.) K.Schum. (Zingiberaceae) **IPNI:** 795383-1 **Common name:** sanjuanico, yinye, yínyer **Habit:** Herb **Use:** EU, HF, ME **Origin:** Cultivated, Naturalised **Region:** Amazonia, Andean, Guiana Shield, Caribbean, Orinoquia, Pacific **Dept.:** ANT, ATL, CAQ, CAU, CHO, MAG, PUT, QUI, RIS, SAP, VAC **Elev.:** 0–1500 m a.s.l. **IUCN Habitat code:** 1,2,3,14 **Conservation status: (IUCN)** NE

2017 *Alpinia zerumbet* (Pers.) B.L.Burtt & R.M.Sm. (Zingiberaceae) **IPNI:** 872083-1 **Habit:** Herb **Use:** EU, GS, HF, MA, ME **Origin:** Cultivated **Region:** Andean, Caribbean **Dept.:** ANT, SAP, VAC **Elev.:** 1200–1500 m a.s.l. **IUCN Habitat code:** 1,2,3,14 **Conservation status: (IUCN)** DD

2018 *Curcuma longa* L. (Zingiberaceae) **IPNI:** 796451-1 **Common name:** azafrán, azafrán cimarrón, curcuma, cúrcuma, guisador **Habit:** Herb **Use:** AF, GS, HF, MA, ME, PO, SU **Origin:** Cultivated **Region:** Amazonia, Andean, Pacific **Dept.:** AMA, GUV, QUI **Elev.:** 0–1200 m a.s.l. **IUCN Habitat code:** 1,2,3,4,14 **Conservation status: (IUCN)** DD

2019 *Elettaria cardamomum* (L.) Maton (Zingiberaceae) **IPNI:** 796556-1 **Common name:** cardamomo **Habit:** Herb **Use:** EU, GS, HF, MA, ME **Origin:** Cultivated **Region:** Andean **Dept.:** ANT, QUI, VAC **Elev.:** 1300–1400 m a.s.l. **IUCN Habitat code:** 1,3,14 **Conservation status: (IUCN)** NE

2020 *Etlingera elatior* (Jack) R.M.Sm. (Zingiberaceae) **IPNI:** 942355-1 **Common name:** bastón de obispo, bastón del emperador, bastón del rey, bastón imperial **Habit:** Herb **Use:** AF, EU, GS, HF, MA, ME, SU **Origin:** Cultivated **Region:** Amazonia, Andean, Pacific **Dept.:** AMA, ANT, QUI **Elev.:** 0–1600 m a.s.l. **IUCN Habitat code:** 1,2,3,4,14 **Conservation status: (IUCN)** DD

2021 *Hedychium coccineum* Buch.-Ham. ex Sm. (Zingiberaceae) **IPNI:** 796834-1 **Habit:** Herb **Use:** EU, HF, ME **Origin:** Cultivated **Region:** Andean **Dept.:** QUI **Elev.:** 1600–1600 m a.s.l. **IUCN Habitat code:** 1,3 **Conservation status: (IUCN)** NE

2022 *Hedychium coronarium* J.Koenig (Zingiberaceae) **IPNI:** 796836-1 **Common name:** ajengible, ajengibre, angélica, bengible, bijagüillo, conga, congo, heliotropo, hidotropo, hiliotropo, hilotropo, inopropio, jengibre, jilotropos, lirio, matandrea, quereme de las quebradas **Habit:** Herb **Use:** EU, HF, MA, ME **Origin:** Naturalised **Region:** Amazonia, Andean, Guiana Shield, Caribbean, Orinoquia, Pacific **Dept.:** AMA, ANT, BOG, BOY, CAL, CAQ, CAU, CHO, CUN, LAG, MAG, MET, NAR, PUT, QUI, RIS, SAN, TOL, VAC **Elev.:** 0–1800 m a.s.l. **IUCN Habitat code:** 1,2,3,4,5,14 **Conservation status: (IUCN)** DD

2023 *Renealmia alpinia* (Rottb.) Maas (Zingiberaceae) **IPNI:** 218583-2 **Common name:** abebe, bodó, canopio, cañagria cimarrona, cañuela, chisgua, conopio, huitillo, istri, matandrea, misiraca, naikú, palma surca, pintura negra, platanilla, platanillo, platanillo de altura, sarandago gigante, sarandango, sirindango, surca, surco **Habit:** Herb **Use:** HF, MA, ME **Origin:** Native **Region:** Amazonia, Andean, Guiana Shield, Orinoquia, Pacific **Dept.:** AMA, ANT, ARA, CAL, CAQ, CAS, CHO, CUN, GUA, MET, NSA, PUT, SAN, VAC, VAU, VID **Elev.:** 0-2500 m.a.s.l. **IUCN Habitat code:** 1,3,5,14 **Conservation status: (IUCN)** NE **(CPSG)** Potential LC

2024 *Renealmia aromatica* (Aubl.) Griseb. (Zingiberaceae) **IPNI:** 798014-1 **Common name:** San Juanito, abebe, abebe grande, achira de monte, bagala, conopio, mata andrea, matandrea, matandrea blanca, platanillo, sanjuanito, wewena **Habit:** Herb **Use:** EU, HF, MA, ME **Origin:** Native **Region:** Andean, Caribbean, Orinoquia, Pacific **Dept.:** ANT, CAS, CAU, CHO, CUN, LAG, MAG, PUT, RIS, SAN, TOL, VAC **Elev.:** 50-2000 m.a.s.l. **IUCN Habitat code:** 1,2,3,14 **Conservation status: (IUCN)** NE **(CPSG)** Potential LC

2025 *Renealmia breviscapa* (Poepp. & Endl.) Poepp. & Endl. (Zingiberaceae) **IPNI:** 798020-1 **Common name:** matandrea **Habit:** Herb **Use:** HF **Origin:** Native **Region:** Amazonia, Andean, Guiana Shield **Dept.:** AMA, CAQ, CAU, MET, PUT, VAU **Elev.:** 100–1580 m a.s.l. **IUCN Habitat code:** 1,14 **Conservation status: (IUCN)** NE

2026 *Renealmia cernua* (Sw. ex Roem. & Schult.) J.F.Macbr. (Zingiberaceae) **IPNI:** 218597-2 **Common name:** jengibre, matandrea, platanillo **Habit:** Herb **Use:** EU **Origin:** Native **Region:** Amazonia, Andean, Guiana Shield, Caribbean, Orinoquia, Pacific **Dept.:** ANT, ARA, BOL, BOY, CAL, CAQ, CAS, CAU, CES, CHO, CUN, GUV, HUI, MAG, MET, NSA, PUT, QUI, SAN, TOL, VAC **Elev.:** 0-2000 m.a.s.l. **IUCN Habitat code:** 1,2,3,4,5,14 **Conservation status: (IUCN)** NE **(CPSG)** Potential LC

2027 *Renealmia floribunda* K.Schum. (Zingiberaceae) **IPNI:** 218616-2 **Common name:** matacolibrí, sanapechos **Habit:** Herb **Use:** HF, MA **Origin:** Native **Region:** Amazonia, Guiana Shield **Dept.:** AMA, CAQ, GUA, GUV, VAU **Elev.:** 100–280 m a.s.l. **IUCN Habitat code:** 1,3,4,5,14 **Conservation status: (IUCN)** NE

2028 *Renealmia mexicana* Klotzsch ex Petersen (Zingiberaceae) **IPNI:** 218639-2 **Habit:** Herb **Use:** HF **Origin:** Native **Region:** Andean **Dept.:** ANT, MAG **Elev.:** 700–1750 m a.s.l. **IUCN Habitat code:** 1 **Conservation status: (IUCN)** NE

2029 *Renealmia monosperma* Miq. (Zingiberaceae) **IPNI:** 798069-1 **Common name:** jengibre, jengibre de monte **Habit:** Herb **Use:** ME **Origin:** Native **Region:** Amazonia, Andean, Guiana Shield **Dept.:** AMA, CAQ, GUV, MET, PUT, VAU **Elev.:** 200–620 m a.s.l. **IUCN Habitat code:** 1,3,5,14 **Conservation status: (IUCN)** NE

2030 *Renealmia nicolaioides* Loes. (Zingiberaceae) **IPNI:** 218643-2 **Common name:** jengibre, sarandango **Habit:** Herb **Use:** HF **Origin:** Native **Region:** Amazonia, Andean, Guiana Shield, Orinoquia **Dept.:** AMA, ANT, CAQ, MET, PUT, SAN, VAU **Elev.:** 100–1870 m a.s.l. **IUCN Habitat code:** 1,2,3,5,14 **Conservation status: (IUCN)** NE

2031 *Renealmia thyrsoidea* (Ruiz & Pav.) Poepp. & Endl. (Zingiberaceae) **IPNI:** 798093-1 **Common name:** ala de panguana, berebere, chici ordinario, coroba, platanillo, vira **Habit:** Herb **Use:** HF, ME **Origin:** Native **Region:** Amazonia, Andean, Guiana Shield, Orinoquia **Dept.:** AMA, ANT, BOY, CAQ, CAS, CAU, CHO, CUN, GUV, HUI, MAG, MET, NAR, NSA, PUT, SAN, TOL, VAC, VAU, VID **Elev.:** 100-2250 m.a.s.l. **IUCN Habitat code:** 1,3,4,5,14 **Conservation status: (IUCN)** NE **(CPSG)** Potential LC

2032 *Zingiber officinale* Roscoe (Zingiberaceae) **IPNI:** 798372-1 **Common name:** ajengible, ajengibre, ajingible, anjengibre, anjigible, jengibre **Habit:** Herb **Use:** AF, EU, FU, GS, HF, IF, MA, ME, PO, SU **Origin:** Cultivated **Region:** Amazonia, Andean, Caribbean, Pacific **Dept.:** AMA, BOL, CHO, QUI **Elev.:** 0–1500 m a.s.l. **IUCN Habitat code:** 1,3,14 **Conservation status: (IUCN)** DD

2033 *Zingiber spectabile* Griff. (Zingiberaceae) **IPNI:** 798388-1 **Common name:** maraca **Habit:** Herb **Use:** HF, ME **Origin:** Cultivated **Region:** Amazonia, Andean, Pacific **Dept.:** ANT, QUI, VAC **Elev.:** 0–1500 m a.s.l. **IUCN Habitat code:** 1,2,3,14 **Conservation status: (IUCN)** DD

Ceratophyllales

2034 *Ceratophyllum demersum* L. (Ceratophyllaceae) **IPNI:** 30006047-2 **Common name:** candelabro de agua **Habit:** Herb, Aquatic **Use:** EU, HF, ME **Origin:** Native **Region:** Caribbean, Pacific **Dept.:** ANT, COR **Elev.:** 6–30 m a.s.l. **IUCN Habitat code:** 1,2,5,14 **Conservation status: (IUCN)** LC

Apiales

2035 *Ammi majus* L. (Apiaceae) **IPNI:** 837486-1 **Habit:** Herb **Use:** EU, HF, ME, PO **Origin:** Naturalised **Region:** Andean **Dept.:** BOY, CUN **Elev.:** 2030–2400 m a.s.l. **IUCN Habitat code:** 3,14 **Conservation status: (IUCN)** NE

2036 *Anethum graveolens* L. (Apiaceae) **IPNI:** 837530-1 **Common name:** aneldo, eneldo **Habit:** Herb **Use:** EU, GS, HF, MA, ME **Origin:** Cultivated **Region:** Andean **Dept.:** ANT, BOG, CUN, HUI, MAG, VAC **Elev.:** 1000–2600 m a.s.l. **IUCN Habitat code:** 1,2,3,14 **Conservation status: (IUCN)** NE

2037 *Angelica archangelica* L. (Apiaceae) **IPNI:** 837560-1 **Common name:** angélica **Use:** EU, HF, IF, MA, ME **IUCN Habitat code:** 3 **Conservation status: (IUCN)** NE

2038 *Apium graveolens* L. (Apiaceae) **IPNI:** 838067-1 **Common name:** apio, apio criollo, apio de páramo, apio grande, apio pequeño **Habit:** Herb **Use:** AF, GS, HF, MA, ME, PO **Origin:** Cultivated **Region:** Andean **Dept.:** ANT, BOG, CUN, HUI, VAC **Elev.:** 1700–2700 m a.s.l. **IUCN Habitat code:** 1,2,3,4,14 **Conservation status: (IUCN)** LC

2039 *Arracacia tolucensis* (Kunth) Hemsl. (Apiaceae) **IPNI:** 838268-1 **Habit:** Herb **Use:** ME **Origin:** Native **Region:** Andean **Dept.:** CES, LAG, MAG **Elev.:** 3100–3750 m a.s.l. **IUCN Habitat code:** 14 **Conservation status: (IUCN)** NE

2040 *Arracacia xanthorrhiza* Bancr. (Apiaceae) **IPNI:** 838272-1 **Common name:** apio, arracacha, arracacha amarilla, arracacha de tarro, arrecate, racacha, sarracacha, zanahoria blanca **Habit:** Herb **Use:** HF, ME **Origin:** Native, Cultivated **Region:** Andean **Dept.:** ANT, BOG, BOY, CAU, NSA, PUT, VAC **Elev.:** 1200–2250 m a.s.l. **IUCN Habitat code:** 1,3,8,14 **Conservation status: (IUCN)** NE

2041 *Azorella biloba* (Schltdl.) Wedd. (Apiaceae) **IPNI:** 838552-1 **Habit:** Herb **Use:** ME **Origin:** Native **Region:** Andean **Dept.:** NAR **Elev.:** 2800–4000 m a.s.l. **Conservation status: (IUCN)** LC

2042 *Azorella crenata* (Ruiz & Pav.) Pers. (Apiaceae) **IPNI:** 838571-1 **Common name:** achicoria **Habit:** Herb **Use:** HF **Origin:** Native **Region:** Andean **Dept.:** ANT, BOG, BOY, CAL, CAU, CUN, HUI, MAG, MET, NAR, QUI, RIS, TOL, VAC

Elev.: 2900-4600 m.a.s.l. **IUCN Habitat code:** 1,3,4,8,14 **Conservation status: (IUCN)** LC **(CPSG)** Potential LC

2043 *Bupleurum rotundifolium* L. (Apiaceae)
IPNI: 38361-2
Habit: Herb **Use:** EU, HF, ME **Origin:** Cultivated **Region:** Andean **Dept.:** ANT **Elev.:** 2200–2200 m a.s.l. **IUCN Habitat code:** 14 **Conservation status: (IUCN)** NE

2044 *Centella asiatica* (L.) Urb. (Apiaceae)
IPNI: 1197718-2
Common name: berro macho, centella asiatica **Habit:** Herb **Use:** EU, GS, HF, MA, ME, PO, SU **Origin:** Naturalised **Region:** Andean **Dept.:** ANT, CAU, RIS **Elev.:** 1500–2610 m a.s.l. **IUCN Habitat code:** 1,3,14 **Conservation status: (IUCN)** LC

2045 *Centella erecta* (L.f.) Fernald (Apiaceae)
IPNI: 51383-2
Habit: Herb **Use:** ME **Origin:** Native **Region:** Andean **Dept.:** ANT **Elev.:** 2370–2370 m a.s.l. **IUCN Habitat code:** 1,3,14 **Conservation status: (IUCN)** NE

2046 *Chaerophyllum andicola* (Kunth) K.F.Chung (Apiaceae)
IPNI: 77085100-1
Common name: apio de páramo **Habit:** Herb **Use:** HF **Origin:** Native **Region:** Andean **Dept.:** BOG, BOY, CAL, CAU, CUN, MET, NAR, QUI, SAN, TOL **Elev.:** 3100–4500 m a.s.l. **IUCN Habitat code:** 1,4,6,8,14 **Conservation status: (IUCN)** NE

2047 *Conium maculatum* L. (Apiaceae)
IPNI: 840668-1
Common name: cicuta, humaria, perla **Habit:** Herb **Use:** EU, HF, ME, PO **Origin:** Naturalised **Region:** Andean **Dept.:** ANT, BOG, CAU, CUN **Elev.:** 1500–3045 m a.s.l. **IUCN Habitat code:** 1,3,4,5,14 **Conservation status: (IUCN)** NE

2048 *Coriandrum sativum* L. (Apiaceae)
IPNI: 840760-1
Common name: bragueta de viejo, cilantro, cilantro castilla, cilantro de Castilla, cilantro de paisa, culantro, culantro de burro, culantro doméstico **Habit:** Herb **Use:** AF, EU, GS, HF, MA, ME, SU **Origin:** Cultivated **Region:** Amazonia, Andean **Dept.:** AMA, ANT, BOG, CUN, QUI, VAC **Elev.:** 100–2850 m a.s.l. **IUCN Habitat code:** 1,2,3,4,14 **Conservation status: (IUCN)** NE

2049 *Cuminum cyminum* L. (Apiaceae)
IPNI: 840882-1
Common name: comino, planta de cominos **Use:** EU, GS, HF, MA, ME **Conservation status: (IUCN)** NE

2050 *Cyclospermum leptophyllum* (Pers.) Sprague ex Britton & P.Wilson (Apiaceae)
IPNI: 840899-1
Common name: anisillo, apio de páramo, culantrillo, fumaria, jumaria **Habit:** Herb **Use:** AF, EU, HF, ME **Origin:** Native **Region:** Andean **Dept.:** ANT, BOG, BOY, CUN, NAR, NSA, PUT, QUI, SAN, VAC **Elev.:** 1700-3600 m.a.s.l. **IUCN Habitat code:** 1,2,3,14 **Conservation status: (IUCN)** NE **(CPSG)** Potential LC

2051 *Daucus carota* L. (Apiaceae)
IPNI: 841063-1
Common name: zanahoria, zanahoria amarilla **Habit:** Herb **Use:** AF, GS, HF, MA, ME, PO **Origin:** Cultivated **Region:** Andean **Dept.:** BOG, CUN **Elev.:** 2550–2850 m a.s.l. **IUCN Habitat code:** 1,3,4,14 **Conservation status: (IUCN)** NE

2052 *Daucus montanus* Humb. & Bonpl. ex Schult. (Apiaceae)
IPNI: 841167-1
Common name: cilantrillo, fumaria **Habit:** Herb **Use:** HF, ME **Origin:** Native **Region:** Andean **Dept.:** ANT, BOG, BOY, CAU, CES, CUN, HUI, MET, NAR, NSA, PUT, SAN, TOL **Elev.:** 1800-3960 m.a.s.l. **IUCN Habitat code:** 1,3,4,14 **Conservation status: (IUCN)** NE **(CPSG)** Potential LC

2053 *Eryngium alpinum* L. (Apiaceae)
IPNI: 841634-1
Common name: cardo **Habit:** Herb **Use:** EU **Origin:** Cultivated **Region:** Andean **Dept.:** BOG, CUN **Elev.:** 2600–2600 m a.s.l. **Conservation status: (IUCN)** NT

2054 *Eryngium ebracteatum* Lam. (Apiaceae)
IPNI: 841769-1
Habit: Herb **Use:** HF, ME **Origin:** Native **Region:** Andean, Orinoquia **Dept.:** ANT, BOG, CAS, CAU, HUI, MET, NAR, TOL, VAC **Elev.:** 130–1600 m a.s.l. **IUCN Habitat code:** 1,3,14 **Conservation status: (IUCN)** NE

2055 *Eryngium foetidum* L. (Apiaceae)
IPNI: 841793-1
Common name: apio de monte, chadwoin beni, chilangua, chillangua, chiluanga, cilantrillo, cilantro, cilantro ancho, cilantro chicorio, cilantro cimarrón, cilantro común, cilantro de indio, cilantro de la tierra, cilantro de monte, cilantro de sabana, cilantro montañer **Habit:** Herb **Use:** AF, HF, MA, ME, PO **Origin:** Native, Cultivated **Region:** Amazonia, Andean, Guiana Shield, Caribbean, Orinoquia, Pacific **Dept.:** AMA, ANT, BOG, BOL, BOY, CAL, CAQ, CAU, CES, CHO, COR, CUN, GUA, HUI, MET, NAR, NSA, QUI, SAN, VAC, VAU **Elev.:** 5-1600 m.a.s.l. **IUCN Habitat code:** 1,2,3,4,5,14 **Conservation status: (IUCN)** NE **(CPSG)** Potential LC

2056 *Eryngium humboldtii* F.Delaroche (Apiaceae)
IPNI: 841840-1
Common name: cardoncillo, cardón **Habit:** Herb **Use:** ME **Origin:** Native **Region:** Andean **Dept.:** ANT, BOG, BOY, CAU, CES, CUN, LAG, MET, NSA, QUI, SAN **Elev.:** 1990–4100 m a.s.l. **IUCN Habitat code:** 1,3,4,5,8,14 **Conservation status: (IUCN)** NE

2057 *Eryngium humile* Cav. (Apiaceae)
IPNI: 841842-1
Common name: estrella, yerba gorda **Habit:** Herb **Use:** ME **Origin:** Native **Region:** Andean **Dept.:** ANT, BOG, BOY, CAL, CAU, CUN, MET, NAR, NSA, PUT, RIS, SAN, TOL **Elev.:** 2400-4500 m.a.s.l. **IUCN Habitat code:** 1,3,4,8,14 **Conservation status: (IUCN)** NE **(CPSG)** Potential LC

2058 *Eryngium paniculatum* Cav. & Dombey ex F.Delaroche (Apiaceae)
IPNI: 841951-1
Common name: cardo **Use:** HF, ME **Conservation status: (IUCN)** NE

2059 *Foeniculum vulgare* Mill. (Apiaceae)
IPNI: 842680-1
Common name: eneldo, hinojo, hinojo dulce **Habit:** Herb **Use:** AF, EU, GS, HF, MA, ME **Origin:** Cultivated, Naturalised **Region:** Andean **Dept.:** ANT, BOG, BOY, CAU, CUN, HUI **Elev.:** 1350–2850 m a.s.l. **IUCN Habitat code:** 1,2,3,4,14 **Conservation status: (IUCN)** NE

2060 *Lilaeopsis macloviana* (Gand.) A.W.Hill (Apiaceae)
IPNI: 140089-2
Habit: Herb **Use:** HF **Origin:** Native **Region:** Andean **Dept.:** BOY, CUN, MET **Elev.:** 3300–3900 m a.s.l. **IUCN Habitat code:** 4 **Conservation status: (IUCN)** NE

2061 *Neonelsonia acuminata* (Benth.) J.M.Coult. & Rose (Apiaceae)
IPNI: 168905-2
Common name: anocuchuelo, arracacha de monte, arracachilla **Habit:** Herb **Use:** ME **Origin:** Native **Region:** Andean **Dept.:** ANT, BOG, CAU, MET, NAR, PUT, QUI, RIS, TOL **Elev.:** 2180-3510 m.a.s.l. **IUCN Habitat code:** 1,3,4,5,8,14 **Conservation status: (IUCN)** NE **(CPSG)** Potential LC

2062 *Niphogeton glaucescens* (Kunth) J.F.Macbr. (Apiaceae)
IPNI: 170074-2
Common name: apio, apio de monte **Habit:** Herb **Use:** ME **Origin:** Native **Region:** Andean **Dept.:** ANT, BOG, BOY, CAL, CAU, CUN, MET, TOL **Elev.:** 2400–4100 m a.s.l. **IUCN Habitat code:** 1,3,4,5,8,14 **Conservation status: (IUCN)** NE

2063 *Niphogeton ternata* (Willd. ex Schult.) Mathias & Constance (Apiaceae)
IPNI: 170084-2
Common name: anamú, apio, apio de monte, culantrillo, perejil **Habit:** Herb **Use:** ME **Origin:** Native **Region:** Andean **Dept.:** ANT, BOG, BOY, CAL, CAU, CUN, HUI, MET, NAR, NSA, PUT, RIS, SAN, VAC **Elev.:** 2230-4290 m.a.s.l. **IUCN Habitat code:** 1,3,4,8,14 **Conservation status: (IUCN)** NE **(CPSG)** Potential LC

2064 *Osmorhiza mexicana* Griseb. (Apiaceae)
IPNI: 845560-1
Habit: Herb **Use:** ME **Origin:** Native **Region:** Andean **Dept.:** CAL, TOL **Elev.:** 3500–3730 m a.s.l. **Conservation status: (IUCN)** NE

2065 *Petroselinum crispum* (Mill.) Fuss (Apiaceae)
IPNI: 60442790-2
Common name: alpichala, perejil, perejil crespo, perejil liso **Habit:** Herb **Use:** GS, HF, MA, ME **Origin:** Cultivated **Region:** Andean **Dept.:** ANT, BOG, CUN, PUT, VAC **Elev.:** 1000–2900 m a.s.l. **IUCN Habitat code:** 1,2,3,4,14 **Conservation status: (IUCN)** NE

2066 *Pimpinella anisum* L. (Apiaceae)
IPNI: 846658-1
Common name: anís, anís común, anís verde **Habit:** Herb **Use:** AF, GS, HF, MA, ME **Origin:** Cultivated **Region:** Andean **Dept.:** ANT, NAR **Elev.:** 1900–3000 m a.s.l. **IUCN Habitat code:** 14 **Conservation status: (IUCN)** NE

2067 *Sanicula liberta* Cham. & Schltdl. (Apiaceae) **IPNI:** 847852-1 **Habit:** Herb **Use:** ME **Origin:** Native **Region:** Andean **Dept.:** ANT, BOY, CAL, CUN, HUI, MAG, NAR, NSA, PUT, QUI, RIS, SAN, TOL, VAC **Elev.:** 1700-3200 m.a.s.l. **IUCN Habitat code:** 1,3,14 **Conservation status: (IUCN)** NE **(CPSG)** Potential LC

2068 *Visnaga daucoides* Gaertn. (Apiaceae) **IPNI:** 849956-1 **Habit:** Herb **Use:** HF, MA, ME, PO **Origin:** Naturalised **Region:** Andean **Dept.:** CUN **Elev.:** 1600–1720 m a.s.l. **IUCN Habitat code:** 14 **Conservation status: (IUCN)** NE

2069 *Aralia excelsa* (Griseb.) J.Wen (Araliaceae) **IPNI:** 20012157-1 **Habit:** Shrub, Tree **Use:** HF, MA **Origin:** Native **Region:** Amazonia, Andean, Caribbean **Dept.:** BOL, CAQ, CES, COR, MET **Elev.:** 34–1400 m a.s.l. **IUCN Habitat code:** 1,2,3,14 **Conservation status: (IUCN)** LC

2070 *Dendropanax arboreus* (L.) Decne. & Planch. (Araliaceae) **IPNI:** 1133937-2 **Common name:** aguacatón, cambusil, cascabel, cascajero, cincodedos, guayusa verde, lembo, pategallina, platero **Habit:** Tree **Use:** HF, MA, ME **Origin:** Native **Region:** Amazonia, Andean, Guiana Shield, Caribbean, Orinoquia, Pacific **Dept.:** AMA, ANT, CAQ, CAS, CAU, CES, CHO, CUN, GUV, HUI, LAG, MAG, MET, NAR, PUT, QUI, RIS, SAN, VAC, VID **Elev.:** 0–2600 m a.s.l. **IUCN Habitat code:** 1,2,3,4,5,14 **Conservation status: (IUCN)** NE

2071 *Dendropanax querceti* Donn.Sm. (Araliaceae) **IPNI:** 77666-2 **Habit:** Shrub **Use:** ME **Origin:** Native **Region:** Andean **Dept.:** CHO **Elev.:** 1200–1650 m a.s.l. **IUCN Habitat code:** 1 **Conservation status: (IUCN)** LC

2072 *Didymopanax morototoni* (Aubl.) Decne. & Planch. (Araliaceae) **IPNI:** 90355-1 **Common name:** beno, cafeto, cajete, cajeto, guacharaco, guarumero, guarumón, laurel baboso, mano de león, mano de oso, mano de tigre, manodioso, manuelión, pata de gallina, patagallina, pategallina, sachauva, torcazo, tortolito, tostado, tumbamaco, tórtolo, yagrume, ya **Habit:** Shrub, Tree **Use:** MA, ME **Origin:** Native **Region:** Amazonia, Andean, Guiana Shield, Caribbean, Orinoquia **Dept.:** AMA, ANT, BOL, CAQ, CAU, CES, CHO, COR, CUN, GUA, GUV, HUI, LAG, MET, SAN, VAC, VID **Elev.:** 0–2140 m a.s.l. **IUCN Habitat code:** 1,3,14 **Conservation status: (IUCN)** LC

2073 *Fatsia japonica* (Thunb.) Decne. & Planch. (Araliaceae) **IPNI:** 90496-1 **Habit:** Shrub **Use:** EU, MA, ME **Origin:** Cultivated **Region:** Andean **Dept.:** BOG**Conservation status: (IUCN)** NE

2074 *Hedera canariensis* Willd. (Araliaceae) **IPNI:** 90665-1 **Habit:** Climbing **Use:** EU, ME **Origin:** Cultivated **Dept.:** CUN **Elev.:** 2700–2700 m a.s.l. **Conservation status: (IUCN)** NE

2075 *Hedera helix* L. (Araliaceae) **IPNI:** 90723-1 **Common name:** hiedra, yedra **Habit:** Climbing **Use:** EU, MA, ME, PO **Origin:** Cultivated **Dept.:** BOG, CUN **Elev.:** 2700–2700 m a.s.l. **IUCN Habitat code:** 1,3,14 **Conservation status: (IUCN)** NE

2076 *Heptapleurum actinophyllum* (Endl.) Lowry & G.M.Plunkett (Araliaceae) **IPNI:** 77213393-1 **Common name:** cheflera **Habit:** Tree **Use:** AF, EU, FU, MA, ME, SU **Origin:** Endemic, Cultivated **Region:** Andean **Dept.:** ANT, BOG **Elev.:** 1000–1000 m a.s.l. **IUCN Habitat code:** 1,3,14 **Conservation status: (IUCN)** LC

2077 *Heptapleurum arboricola* Hayata (Araliaceae) **IPNI:** 90845-1 **Habit:** Shrub **Use:** EU, ME, SU **Origin:** Cultivated **Region:** Andean **Dept.:** ANT, BOY **Elev.:** 2775–2775 m a.s.l. **Conservation status: (IUCN)** NE

2078 *Hydrocotyle bonariensis* Comm. ex Lam. (Araliaceae) **IPNI:** 125117-2 **Habit:** Herb **Use:** EU, HF, MA, ME **Origin:** Native **Dept.:** VAC **Elev.:** 1000–1000 m a.s.l. **IUCN Habitat code:** 2,3 **Conservation status: (IUCN)** LC

2079 *Hydrocotyle bonplandii* A.Rich. (Araliaceae) **IPNI:** 843403-1 **Common name:** berros, chupana, chuparán, oreja, paragüitas, plegadera, sombrerito de agua, sombrerito de sapo **Habit:** Herb **Use:** ME **Origin:** Native **Region:** Andean **Dept.:** ANT, BOG, BOY, CAU, CHO, CUN, HUI, MET, NAR, NSA, PUT, QUI, RIS, SAN, TOL **Elev.:** 1300-4000 m.a.s.l. **IUCN Habitat code:** 1,3,4,5,8,14 **Conservation status: (IUCN)** NE **(CPSG)** Potential LC

2080 *Hydrocotyle grossulariifolia* Rusby (Araliaceae) **IPNI:** 125145-2 **Habit:** Herb **Use:** ME **Origin:** Native **Region:** Andean **Dept.:** MAG **Elev.:** 1600–3700 m a.s.l. **IUCN Habitat code:** 1,3,4,14 **Conservation status: (IUCN)** NE

2081 *Hydrocotyle gunnerifolia* Wedd. (Araliaceae) **IPNI:** 843519-1 **Common name:** chupana, oreja de ratón, plegadera, sombrerito **Habit:** Herb **Use:** ME **Origin:** Native **Region:** Andean **Dept.:** ANT, BOG, BOY, CAU, CUN, MET, NAR, NSA, QUI, RIS, SAN, TOL **Elev.:** 2570-4290 m.a.s.l. **IUCN Habitat code:** 1,3,4,14 **Conservation status: (IUCN)** NE **(CPSG)** Potential LC

2082 *Hydrocotyle hederacea* Mathias (Araliaceae) **IPNI:** 125149-2 **Habit:** Herb **Use:** ME **Origin:** Native **Region:** Andean **Dept.:** BOG, BOY, CAU, CUN, MAG, NSA, QUI, SAN, TOL **Elev.:** 2600–4000 m a.s.l. **IUCN Habitat code:** 1,3,4 **Conservation status: (IUCN)** NE

2083 *Hydrocotyle humboldtii* A.Rich. (Araliaceae) **IPNI:** 843540-1 **Common name:** chupana **Habit:** Herb **Use:** ME **Origin:** Native **Region:** Andean **Dept.:** ANT, BOG, BOY, CAQ, CAU, CUN, HUI, MAG, NAR, PUT, QUI, RIS, TOL, VAC **Elev.:** 1200-3200 m.a.s.l. **IUCN Habitat code:** 1,3,4,14 **Conservation status: (IUCN)** NE **(CPSG)** Potential LC

2084 *Hydrocotyle lehmannii* Mathias (Araliaceae) **IPNI:** 125166-2 **Habit:** Herb **Use:** ME **Origin:** Endemic **Region:** Andean **Dept.:** CAQ, CAU, HUI, NAR, PUT, VAC **Elev.:** 1900–3600 m a.s.l. **IUCN Habitat code:** 1,5,14 **Conservation status: (IUCN)** NE

2085 *Hydrocotyle leucocephala* Cham. & Schltdl. (Araliaceae) **IPNI:** 843574-1 **Common name:** chupana, golondrina, oreja de ratón **Habit:** Herb **Use:** EU, ME **Origin:** Native **Region:** Andean, Pacific **Dept.:** ANT, BOG, CAU, CHO, CUN, QUI, RIS, VAC **Elev.:** 0-2700 m.a.s.l. **IUCN Habitat code:** 1,3,4,14 **Conservation status: (IUCN)** NE **(CPSG)** Potential LC

2086 *Hydrocotyle mexicana* Schltdl. & Cham. (Araliaceae) **IPNI:** 843596-1 **Habit:** Herb **Use:** HF, ME **Origin:** Native **Region:** Andean **Dept.:** BOY, CUN, NSA, SAN, TOL **Elev.:** 2200–3400 m a.s.l. **IUCN Habitat code:** 3,4,8,14 **Conservation status: (IUCN)** NE

2087 *Hydrocotyle ranunculoides* L.f. (Araliaceae) **IPNI:** 843671-1 **Common name:** chupana, oreja de ratón, paraguas, sombrerito de agua **Habit:** Herb **Use:** HF, ME **Origin:** Native **Region:** Andean **Dept.:** ANT, BOG, BOY, CAU, CUN, QUI, RIS **Elev.:** 1500-3800 m.a.s.l. **IUCN Habitat code:** 1,3,4,14 **Conservation status: (IUCN)** LC **(CPSG)** Potential LC

2088 *Hydrocotyle sphenoloba* Wedd. (Araliaceae) **IPNI:** 843705-1 **Habit:** Herb **Use:** ME **Origin:** Endemic **Region:** Andean **Dept.:** BOY, CAL, CAU, CUN **Elev.:** 2000–2900 m a.s.l. **IUCN Habitat code:** 3 **Conservation status: (IUCN)** NE

2089 *Hydrocotyle tenerrima* Rose ex Mathias (Araliaceae) **IPNI:** 125218-2 **Habit:** Herb **Use:** ME **Origin:** Native **Region:** Andean **Dept.:** BOG, CUN, QUI, SAN **Elev.:** 2000–2870 m a.s.l. **IUCN Habitat code:** 1,3,4,14 **Conservation status: (IUCN)** NE

2090 *Hydrocotyle umbellata* L. (Araliaceae) **IPNI:** 843746-1 **Common name:** chupana, orejitas, paraguas de sapo, plegadera, redonditas de agua, sombrerito de agua **Habit:** Herb **Use:** EU, HF, ME **Origin:** Native **Region:** Andean, Caribbean, Pacific **Dept.:** ANT, ATL, BOY, CAU, CHO, COR, NAR, NSA, PUT, QUI, RIS, SAN, TOL, VAC **Elev.:** 0-2090 m.a.s.l. **IUCN Habitat code:** 1,2,3,4,5,14 **Conservation status: (IUCN)** LC **(CPSG)** Potential LC

2091 *Hydrocotyle verticillata* Thunb. (Araliaceae) **IPNI:** 843759-1 **Habit:** Herb **Use:** ME **Origin:** Native **Region:** Caribbean **Dept.:** ATL, BOG, HUI **Elev.:** 10–300 m a.s.l. **IUCN Habitat code:** 3,14 **Conservation status: (IUCN)** NE

2092 *Oreopanax bogotensis* Cuatrec. (Araliaceae) **IPNI:** 314126-2 **Common name:** amarillo, higuerillo, higuerón, mano de oso **Habit:** Shrub, Tree **Use:** AF, EU, MA **Origin:** Native **Region:** Andean **Dept.:** ANT, ARA, BOG, BOY, CAL, CAU, CUN, HUI, MET, NAR, QUI, SAN **Elev.:** 2050–4117 m a.s.l. **IUCN Habitat code:** 1,3,4,5,14 **Conservation status: (IUCN)** NE

2093 *Oreopanax capitatus* (Jacq.) Decne. & Planch. (Araliaceae) **IPNI:** 30050876-2 **Common name:** candelero, pata de gallina, tinajero **Habit:** Tree, Hemiepiphyte **Use:** MA, ME **Origin:** Native **Region:** Andean, Caribbean, Pacific **Dept.:** ANT, BOL, BOY, CES, CHO, CUN, LAG, MAG, NAR, NSA, QUI, RIS, SAN, VAC **Elev.:** 5–4000 m a.s.l. **IUCN Habitat code:** 1,3,4,14 **Conservation status: (IUCN)** LC

2094 *Oreopanax catalpifolius* (Willd. ex Schult.) Decne. & Planch. (Araliaceae) **IPNI:** 91160-1 **Habit:** Shrub, Tree **Use:** ME **Origin:** Native **Region:** Andean **Dept.:** ANT, BOY, CUN **Elev.:** 2200–2950 m a.s.l. **IUCN Habitat code:** 1 **Conservation status: (IUCN)** NE

2095 *Oreopanax gargantae* Cuatrec. (Araliaceae) **IPNI:** 176183-2 **Common name:** higuerón **Habit:** Tree **Use:** ME **Origin:** Endemic **Region:** Andean **Dept.:** BOY, NSA **Elev.:** 2300-3600 m.a.s.l. **IUCN Habitat code:** 14 **Conservation status: (IUCN)** CR **(CPSG)** CR

2096 *Oreopanax glabrifolius* Cuatrec. (Araliaceae) **IPNI:** 176184-2 **Habit:** Shrub, Tree **Use:** MA **Origin:** Endemic **Region:** Andean, Pacific **Dept.:** ANT, BOY, CAL, QUI, RIS, TOL **Elev.:** 100-3050 m.a.s.l. **IUCN Habitat code:** 1,2,3,4,8,14 **Conservation status: (IUCN)** LC **(CPSG)** LC

2097 *Oreopanax hederaceus* Cuatrec. (Araliaceae) **IPNI:** 176187-2 **Habit:** Shrub, Epiphyte **Use:** ME **Origin:** Endemic **Region:** Andean **Dept.:** ANT, CAU, CES, NSA, QUI, RIS, SAN **Elev.:** 1500–2300 m a.s.l. **IUCN Habitat code:** 1 **Conservation status: (IUCN)** NE

2098 *Oreopanax incisus* (Willd. ex Schult.) Decne. & Planch. (Araliaceae) **IPNI:** 91222-1 **Common name:** candelero, chiviná, guache, mano de oso, pata de gallina **Habit:** Tree **Use:** AF, EU, MA, ME **Origin:** Native **Region:** Andean, Orinoquia **Dept.:** ANT, BOG, BOY, CAL, CAS, CAU, CES, CUN, HUI, QUI, RIS, SAN, TOL **Elev.:** 200–3750 m a.s.l. **IUCN Habitat code:** 1,3,4,5,8,14 **Conservation status: (IUCN)** NE

2099 *Oreopanax integrifolius* Cuatrec. (Araliaceae) **IPNI:** 176190-2 **Habit:** Tree **Use:** ME **Origin:** Endemic **Region:** Andean **Dept.:** HUI **Elev.:** 2200-2300 m.a.s.l. **IUCN Habitat code:** 1,14 **Conservation status: (IUCN)** LC **(CPSG)** LC

2100 *Oreopanax mutisianus* (Kunth) Decne. & Planch. (Araliaceae) **IPNI:** 91261-1 **Common name:** mano de oso, yuco **Habit:** Small tree, Tree **Use:** EU, ME **Origin:** Endemic **Region:** Andean **Dept.:** BOG, BOY, CUN, MET **Elev.:** 2700-3700 m.a.s.l. **IUCN Habitat code:** 1,3,4,8,14 **Conservation status: (IUCN)** LC **(CPSG)** LC

2101 *Oreopanax pallidus* Cuatrec. (Araliaceae) **IPNI:** 176219-2 **Common name:** cincodedos **Habit:** Shrub, Small tree **Use:** ME **Origin:** Endemic **Region:** Andean **Dept.:** CAL, CUN, NSA, TOL **Elev.:** 1700-2866 m.a.s.l. **IUCN Habitat code:** 1,3,14 **Conservation status: (IUCN)** NT **(CPSG)** NT

2102 *Oreopanax ruizanus* Cuatrec. (Araliaceae) **IPNI:** 176230-2 **Common name:** cincodedos **Habit:** Shrub, Tree **Use:** ME **Origin:** Endemic **Region:** Andean **Dept.:** CAL, CHO, HUI, RIS, TOL **Elev.:** 2150-4000 m.a.s.l. **IUCN Habitat code:** 1,3,4 **Conservation status: (IUCN)** EN **(CPSG)** EN

2103 *Oreopanax sucrensis* Cuatrec. (Araliaceae) **IPNI:** 176246-2 **Habit:** Small tree, Tree **Use:** ME **Origin:** Endemic **Region:** Andean **Dept.:** CAQ **Elev.:** 1000-1000 m.a.s.l. **IUCN Habitat code:** 1 **Conservation status: (IUCN)** CR **(CPSG)** CR

2104 *Oreopanax trianae* Decne. & Planch. ex Harms (Araliaceae) **IPNI:** 91329-1 **Common name:** cincodedos **Habit:** Tree **Use:** ME **Origin:** Endemic **Region:** Andean **Dept.:** ANT, CUN **Elev.:** 1730-2300 m.a.s.l. **IUCN Habitat code:** 1 **Conservation status: (IUCN)** VU **(CPSG)** VU

2105 *Panax ginseng* C.A.Mey. (Araliaceae) **IPNI:** 91472-1 **Common name:** aralea, ginseng **Use:** HF, MA, ME **IUCN Habitat code:** 14 **Conservation status: (IUCN)** NE

2106 *Plerandra elegantissima* (H.J.Veitch ex Mast.) Lowry, G.M.Plunkett & Frodin (Araliaceae) **IPNI:** 77131172-1 **Use:** EU **Region:** Andean **Dept.:** BOG**IUCN Habitat code:** 3 **Conservation status: (IUCN)** EN

2107 *Polyscias cumingiana* (C.Presl) Fern.-Vill. (Araliaceae) **IPNI:** 91749-1 **Common name:** árnica **Habit:** Shrub **Use:** EU, HF, ME **Origin:** Cultivated **Region:** Amazonia, Pacific **Dept.:** CHO, PUT, VAC **Elev.:** 500–1050 m a.s.l. **Conservation status: (IUCN)** NE

2108 *Polyscias fruticosa* (L.) Harms (Araliaceae) **IPNI:** 847351-1 **Habit:** Shrub **Use:** EU, HF, ME, SU **Origin:** Cultivated **Region:** Andean, Caribbean **Dept.:** ATL, BOL, TOL **Elev.:** 10–1170 m a.s.l. **IUCN Habitat code:** 5 **Conservation status: (IUCN)** NE

2109 *Polyscias guilfoylei* (W.Bull) L.H.Bailey (Araliaceae) **IPNI:** 91773-1 **Common name:** cana de vieja, millonario blanco, ovejitos, pestaña de vieja, pestañaevieja **Habit:** Shrub **Use:** EU, HF, ME, PO, SU **Origin:** Cultivated **Region:** Andean **Elev.:** 1000–2000 m a.s.l. **IUCN Habitat code:** 3,14 **Conservation status: (IUCN)** NE

2110 *Polyscias scutellaria* (Burm.f.) Fosberg (Araliaceae) **IPNI:** 91817-1 **Common name:** millonaria, millonario, millón **Habit:** Shrub **Use:** AF, EU, HF, IF, ME, SU **Origin:** Cultivated **Region:** Andean **Dept.:** BOG, RIS **Elev.:** 1749–1749 m a.s.l. **Conservation status: (IUCN)** NE

2111 *Sciodaphyllum bogotense* (Cuatrec.) Lowry, G.M.Plunkett & M.M.Mora (Araliaceae) **IPNI:** 77210463-1 **Common name:** yuco **Habit:** Tree **Use:** EU, MA **Origin:** Endemic **Region:** Andean **Dept.:** ANT, BOY, CAU, CHO, CUN, HUI, MET, NSA, RIS, SAN, VAC **Elev.:** 2500-3950 m.a.s.l. **IUCN Habitat code:** 1,3,4,5,8,14 **Conservation status: (IUCN)** NE **(CPSG)** LC

2112 *Sciodaphyllum sphaerocoma* Benth. (Araliaceae) **IPNI:** 92546-1 **Common name:** mano de oso, pategallina **Habit:** Shrub, Tree, Epiphyte, Climbing **Use:** ME **Origin:** Native **Region:** Andean, Pacific **Dept.:** ANT, BOY, CAU, CES, CHO, NAR, PUT, RIS, SAN, VAC **Elev.:** 0–2400 m a.s.l. **IUCN Habitat code:** 1 **Conservation status: (IUCN)** NE

2113 *Sciodaphyllum trianae* Planch. & Linden ex Marchal (Araliaceae) **IPNI:** 230167-2 **Common name:** candelabro, pata de gallina, pategallina **Habit:** Shrub **Use:** EU, MA **Origin:** Endemic **Region:** Andean **Dept.:** ANT, BOY, RIS, SAN, TOL **Elev.:** 2000-3300 m.a.s.l. **IUCN Habitat code:** 1,3,14 **Conservation status: (IUCN)** LC **(CPSG)** LC

2114 *Tetrapanax papyrifer* (Hook.) K.Koch (Araliaceae) **IPNI:** 86417-3 **Habit:** Shrub **Use:** EU, HF, MA, ME **Origin:** Cultivated **Region:** Andean **Dept.:** BOG, CUN **Elev.:** 2200–2600 m a.s.l. **IUCN Habitat code:** 1,5,14 **Conservation status: (IUCN)** LC

2115 *Pittosporum tobira* (Thunb.) W.T.Aiton (Pittosporaceae) **IPNI:** 684716-1 **Use:** EU, ME **Region:** Andean **Dept.:** BOG**IUCN Habitat code:** 14 **Conservation status: (IUCN)** NE

2116 *Pittosporum undulatum* Vent. (Pittosporaceae) **IPNI:** 684734-1 **Common name:** blanquillo, jazmín, jazmín australiano, laurel huesito **Habit:** Small tree **Use:** EU, HF, MA **Origin:** Cultivated **Region:** Andean **Dept.:** ANT, BOG, CUN **Elev.:** 1500–3100 m a.s.l. **IUCN Habitat code:** 1,2,3,5,14 **Conservation status: (IUCN)** NE

Aquifoliales

2117 *Ilex danielis* Killip & Cuatrec. (Aquifoliaceae) **IPNI:** 127793-2 **Common name:** huesito, palo negro **Habit:** Shrub, Small tree **Use:** ME

Origin: Native **Region:** Andean **Dept.:** ANT **Elev.:** 2200–2800 m a.s.l. **IUCN Habitat code:** 1,3,14 **Conservation status: (IUCN)** NE

2118 *Ilex guayusa* Loes. (Aquifoliaceae) **IPNI:** 127854-2 **Common name:** aguayusa, guayusa, guañusa, huayusa, wayusa **Habit:** Tree **Use:** HF, MA, ME, SU **Origin:** Native **Region:** Andean **Dept.:** ANT, BOY, PUT, RIS, VAC **Elev.:** 200–2000 m a.s.l. **IUCN Habitat code:** 1,3,14 **Conservation status: (IUCN)** LC

2119 *Ilex guianensis* (Aubl.) Kuntze (Aquifoliaceae) **IPNI:** 127855-2 **Habit:** Shrub, Small tree, Tree **Use:** HF, MA **Origin:** Native **Region:** Andean, Guiana Shield **Dept.:** ANT, CAQ, LAG **Elev.:** 0–1000 m a.s.l. **IUCN Habitat code:** 1,3,4,5 **Conservation status: (IUCN)** LC

2120 *Ilex kunthiana* Triana (Aquifoliaceae) **IPNI:** 30054025-2 **Common name:** palo mulato **Habit:** Shrub, Small tree, Tree **Use:** AF, EU, ME **Origin:** Native **Region:** Andean **Dept.:** ANT, BOG, BOY, CUN, HUI, MAG, NSA, QUI, SAN, TOL **Elev.:** 1990–3950 m a.s.l. **IUCN Habitat code:** 1,2,3,4,6,8,14 **Conservation status: (IUCN)** LC

2121 *Ilex laureola* Triana (Aquifoliaceae) **IPNI:** 83432-1 **Common name:** ají, palo de ají **Habit:** Small tree, Tree **Use:** ME **Origin:** Native **Region:** Andean, Guiana Shield, Caribbean **Dept.:** ANT, BOL, CAQ, CUN, HUI, MET, PUT, SAN, TOL, VAU **Elev.:** 20–1670 m a.s.l. **IUCN Habitat code:** 1,3,4,5,14 **Conservation status: (IUCN)** NE

2122 *Ilex laurina* Kunth (Aquifoliaceae) **IPNI:** 83435-1 **Common name:** algodón, cardenillo, piedro **Habit:** Small tree, Tree **Use:** GS, ME **Origin:** Native **Region:** Andean, Guiana Shield **Dept.:** ANT, BOY, CAU, CHO, CUN, MAG, MET, NAR, NSA, SAN, VAC **Elev.:** 1300–3750 m a.s.l. **IUCN Habitat code:** 1,2,3,5,14 **Conservation status: (IUCN)** NE

2123 *Ilex myricoides* Kunth (Aquifoliaceae) **IPNI:** 83545-1 **Common name:** cabo de hacha **Habit:** Shrub, Small tree **Use:** ME **Origin:** Native **Region:** Andean **Dept.:** BOY, CAU, MAG, NAR, PUT, QUI, SAN, VAC **Elev.:** 1500–3300 m a.s.l. **IUCN Habitat code:** 1,3,4 **Conservation status: (IUCN)** LC

2124 *Ilex petiolaris* Benth. (Aquifoliaceae) **IPNI:** 306411-2 **Habit:** Tree **Use:** MA **Origin:** Native **Region:** Amazonia, Orinoquia **Dept.:** AMA, VID **Elev.:** 300–300 m a.s.l. **IUCN Habitat code:** 1 **Conservation status: (IUCN)** NE

2125 *Ilex retusa* Klotzsch ex Reissek (Aquifoliaceae) **IPNI:** 83691-1 **Habit:** Small tree **Use:** GS **Origin:** Native **Region:** Guiana Shield **Dept.:** GUA, VAU **Elev.:** 730–830 m a.s.l. **IUCN Habitat code:** 1 **Conservation status: (IUCN)** LC

2126 *Ilex uniflora* Benth. (Aquifoliaceae) **IPNI:** 308592-2 **Common name:** horcaperros **Habit:** Small tree, Tree **Use:** ME **Origin:** Native **Region:** Andean **Dept.:** CAU, NAR, QUI, RIS, SAN, VAC **Elev.:** 2600–3800 m a.s.l. **IUCN Habitat code:** 1,3,4,14 **Conservation status: (IUCN)** EN

2127 *Dendrobangia boliviana* Rusby (Metteniusaceae) **IPNI:** 77582-2 **Common name:** aceituno, aguadulce, arenillo, marcelo **Habit:** Small tree, Tree **Use:** HF, MA **Origin:** Native **Region:** Amazonia, Andean **Dept.:** AMA, ANT, BOL, COR, NAR, PUT, SAN, VAC, VAU **Elev.:** 50–2100 m a.s.l. **IUCN Habitat code:** 1,3,5,14 **Conservation status: (IUCN)** LC

Asterales

2128 *Acanthospermum australe* (Loefl.) Kuntze (Asteraceae=Compositae) **IPNI:** 1692-2 **Common name:** abrojo, cáncer de loma, tapecue, yerba del cáncer **Habit:** Subshrub **Use:** ME **Origin:** Native **Region:** Andean **Dept.:** ARA, BOY, CAL, CAS, CAU, CES, CUN, MAG, MET, NAR, SAN, VID **Elev.:** 950–1800 m a.s.l. **IUCN Habitat code:** 1,3,4,5,14 **Conservation status: (IUCN)** NE

2129 *Acanthospermum hispidum* DC. (Asteraceae=Compositae) **IPNI:** 173718-1 **Common name:** abrojo, cacharro, cachito de cabro, cachoecabro, cizaña, espuela de caballero, espuela de caballo, guajirilla, pecado mortal, pecao mortal, yerba del pecado mortal **Habit:** Herb **Use:** AF, HF, MA, ME **Origin:** Native **Region:** Andean **Dept.:** ANT, CAU, CUN, HUI, LAG, NAR, SAN, TOL, VAC **Elev.:** 350–2070 m a.s.l. **IUCN Habitat code:** 1,2,3,5,14 **Conservation status: (IUCN)** NE

2130 *Achillea millefolium* L. (Asteraceae=Compositae) **IPNI:** 2294-2 **Common name:** cola de reina, colchón de pobre, flor de pluma, hierba del carpintero, manzanilla de los montes, mielenrama, mil de rama, milderama, milefolio, milenrama, yarrow, yerba de carpintero, yerba del carpintero **Habit:** Herb **Use:** AF, EU, GS, HF, MA, ME, PO, SU **Origin:** Cultivated, Naturalised **Region:** Andean **Dept.:** ANT, BOG, CUN, MET, PUT, TOL **Elev.:** 2220–3500 m a.s.l. **IUCN Habitat code:** 1,3,4,14 **Conservation status: (IUCN)** LC

2131 *Achyrocline alata* DC. (Asteraceae=Compositae) **IPNI:** 174368-1 **Common name:** viravira, viravirón **Habit:** Herb, Subshrub **Use:** HF, ME **Origin:** Native **Region:** Andean **Dept.:** ANT, BOG, BOY, CAL, CAU, CES, CUN, MAG, NAR, PUT, QUI, TOL **Elev.:** 1680–4100 m a.s.l. **IUCN Habitat code:** 1,3,4,8,14 **Conservation status: (IUCN)** NE

2132 *Achyrocline bogotensis* DC. (Asteraceae=Compositae) **IPNI:** 174373-1 **Common name:** cenizo, lechuguilla, suso, vira-vira, viravira, viravirón **Habit:** Herb **Use:** ME **Origin:** Endemic **Region:** Andean **Dept.:** BOG, BOY, CUN, MAG, NAR, NSA, QUI, SAN **Elev.:** 1600–4000 m a.s.l. **IUCN Habitat code:** 1,3,4,14 **Conservation status: (IUCN)** NE

2133 *Achyrocline lehmannii* Hieron. (Asteraceae=Compositae) **IPNI:** 2556-2 **Common name:** botón de oro, vira-vira, viravira **Habit:** Herb **Use:** ME **Origin:** Native **Region:** Andean **Dept.:** ANT, ARA, BOG, BOL, BOY, CUN, MAG, NSA, SAN **Elev.:** 2650–4300 m a.s.l. **IUCN Habitat code:** 1,3,4,8,14 **Conservation status: (IUCN)** NE

2134 *Achyrocline saturejoides* (Lam.) DC. (Asteraceae=Compositae) **IPNI:** 174407-1 **Common name:** huira-huira, juamblanco, juan blanco, kea-kea, marcela, marcela hembra, suso, vejiga de lobo, venadillo, vira-vira, viravilona, viravira, yerba de chivo **Habit:** Herb, Shrub, Subshrub **Use:** HF, MA, ME **Origin:** Native **Region:** Andean **Dept.:** ANT, BOY, CAU, CES, CHO, CUN, HUI, LAG, MAG, NAR, PUT, QUI, TOL, VAC **Elev.:** 1250–4000 m a.s.l. **Conservation status: (IUCN)** NE

2135 *Achyrocline tomentosa* Rusby (Asteraceae=Compositae) **IPNI:** 2571-2 **Use:** ME **Region:** Andean **Dept.:** BOG**IUCN Habitat code:** 14 **Conservation status: (IUCN)** NE

2136 *Acmella alba* (L'Hér.) R.K.Jansen (Asteraceae=Compositae) **IPNI:** 271528-2 **Common name:** botoncillo, botón de oro, llambu **Habit:** Herb **Use:** HF, ME **Origin:** Native **Region:** Andean **Dept.:** ANT, CUN, MAG, NAR, PUT **Elev.:** 900–1800 m a.s.l. **IUCN Habitat code:** 1,2 **Conservation status: (IUCN)** NE

2137 *Acmella brachyglossa* Cass. (Asteraceae=Compositae) **IPNI:** 174530-1 **Common name:** botoncillo, botoncillo hembra, guaca, yuyo quemado **Habit:** Herb, Subshrub **Use:** HF, ME **Origin:** Native **Region:** Andean, Guiana Shield **Dept.:** AMA, ANT, ARA, BOY, CAL, CAS, CHO, CUN, HUI, MAG, MET, NAR, NSA, VAC **Elev.:** 250–2600 m a.s.l. **IUCN Habitat code:** 1,2,3,4,5,8,14 **Conservation status: (IUCN)** NE

2138 *Acmella ciliata* (Kunth) Cass. (Asteraceae=Compositae) **IPNI:** 174534-1 **Common name:** botoncillo, botoncillo de Castilla, botón de oro, campanilla de oro, chisacá, guaca, guaca negra, risacá, yuyito, yuyo, yuyo de flor amarilla, yuyo quemado, yuyo quemao **Habit:** Herb **Use:** HF, ME **Origin:** Native **Region:** Andean **Dept.:** AMA, ANT, BOG, BOY, CAL, CAQ, CAU, CES, CHO, CUN, HUI, LAG, MAG, MET, NAR, NSA, PUT, SAN, TOL, VAC **Elev.:** 500-2670 m.a.s.l. **IUCN Habitat code:** 1,2,3,5,14 **Conservation status: (IUCN)** NE **(CPSG)** Potential LC

2139 *Acmella filipes* (Greenm.) R.K.Jansen (Asteraceae=Compositae) **IPNI:** 271533-2 **Use:** ME **Conservation status: (IUCN)** NE

2140 *Acmella oleracea* (L.) R.K.Jansen (Asteraceae=Compositae) **IPNI:** 914778-1 **Habit:** Herb **Use:** EU, GS, HF, ME **Origin:** Naturalised **Region:** Andean **Dept.:** MAG **Elev.:** 1300–1300 m a.s.l. **IUCN Habitat code:** 3,14 **Conservation status: (IUCN)** NE

2141 *Acmella paniculata* (Wall. ex DC.) R.K.Jansen (Asteraceae=Compositae) **IPNI:** 914780-1 **Habit:** Herb **Use:** HF, ME, PO **Origin:** Naturalised **Region:** Pacific **Dept.:** CHO **Elev.:** 100–100 m a.s.l. **IUCN Habitat code:** 1,4 **Conservation status: (IUCN)** LC

2142 *Acmella radicans* (Jacq.) R.K.Jansen (Asteraceae=Compositae) **IPNI:** 271548-2 **Common name:** guaca, guasca, yuyo quemado **Habit:** Herb **Use:** ME **Origin:** Native **Region:** Andean, Caribbean **Dept.:** ANT, CAL, CAU, CUN, MAG, NSA, PUT **Elev.:** 0–1650 m a.s.l. **IUCN Habitat code:** 1,14 **Conservation status: (IUCN)** NE

2143 *Acmella repens* (Walter) Rich. (Asteraceae=Compositae) **IPNI:** 174552-1 **Common name:** botoncillo, chisacá, guaca, quemadera, risacá, yuyo, yuyo quemado **Habit:** Herb **Use:** ME **Origin:** Native **Region:** Andean **Dept.:** ANT, BOY, CUN **Elev.:** 2100–2100 m a.s.l. **IUCN Habitat code:** 1,14 **Conservation status: (IUCN)** NE

2144 *Acmella uliginosa* (Sw.) Cass. (Asteraceae=Compositae) **IPNI:** 174556-1 **Common name:** botón de oro, yuyo quemado **Habit:** Herb **Use:** GS, HF, ME **Origin:** Native **Region:** Andean **Dept.:** BOG, BOY, CUN, HUI, MAG, VAC **Elev.:** 49–1700 m a.s.l. **IUCN Habitat code:** 1,2 **Conservation status: (IUCN)** LC

2145 *Adenostemma fosbergii* R.M.King & H.Rob. (Asteraceae=Compositae) **IPNI:** 4432-2 **Habit:** Subshrub **Use:** ME **Origin:** Native **Region:** Andean **Dept.:** CHO, HUI, NAR **Elev.:** 1600–2000 m a.s.l. **IUCN Habitat code:** 1 **Conservation status: (IUCN)** NE

2146 *Adenostemma lavenia* (L.) Kuntze (Asteraceae=Compositae) **IPNI:** 175003-1 **Common name:** doñajuana **Habit:** Herb, Subshrub **Use:** AF, HF, ME **Origin:** Native **Region:** Andean, Pacific **Dept.:** CHO, HUI, NAR, VAC **Elev.:** 10–700 m a.s.l. **IUCN Habitat code:** 1,3,5,14 **Conservation status: (IUCN)** NE

2147 *Adenostemma platyphyllum* Cass. (Asteraceae=Compositae) **IPNI:** 175016-1 **Common name:** aralén, doñajuana, hierba de cabro, mamajuana, tuhui yasi **Habit:** Herb **Use:** ME **Origin:** Native **Region:** Andean, Pacific **Dept.:** ANT, CAL, CAU, CHO, COR, HUI, NAR, PUT, QUI, RIS, TOL, VAC **Elev.:** 0–2600 m a.s.l. **IUCN Habitat code:** 1,3,5,14 **Conservation status: (IUCN)** NE

2148 *Adenostemma viscosum* J.R.Forst. & G.Forst. (Asteraceae=Compositae) **IPNI:** 175030-1 **Use:** EU, HF, MA, ME, SU **Origin:** Naturalised **Conservation status: (IUCN)** LC

2149 *Aequatorium verrucosum* (Wedd.) S.Díaz & Cuatrec. (Asteraceae=Compositae) **IPNI:** 970931-1 **Habit:** Small tree, Tree **Use:** ME **Origin:** Endemic **Region:** Andean **Dept.:** CAL, CAU, HUI, QUI, TOL, VAC **Elev.:** 3075-3480 m.a.s.l. **IUCN Habitat code:** 1,3,4,14 **Conservation status: (IUCN)** LC **(CPSG)** LC

2150 *Aetheolaena doryphylla* (Cuatrec.) B.Nord. (Asteraceae=Compositae) **IPNI:** 6086-2 **Common name:** singuinaria cabé **Habit:** Herb, Subshrub, Liana **Use:** ME **Origin:** Native **Region:** Andean **Dept.:** CES, LAG, MAG **Elev.:** 2400–4400 m a.s.l. **Conservation status: (IUCN)** NE

2151 *Aetheolaena patens* (Kunth) B.Nord. (Asteraceae=Compositae) **IPNI:** 6096-2 **Habit:** Herb **Use:** ME **Origin:** Native **Region:** Andean **Dept.:** CAL, CAU, CHO, CUN, NAR, RIS, TOL **Elev.:** 2438–4300 m a.s.l. **IUCN Habitat code:** 1,3,4,8,14 **Conservation status: (IUCN)** NE

2152 *Ageratina altissima* (L.) R.M.King & H.Rob. (Asteraceae=Compositae) **IPNI:** 6815-2 **Use:** ME, PO **IUCN Habitat code:** 14 **Conservation status: (IUCN)** NE

2153 *Ageratina angustifolia* (Kunth) R.M.King & H.Rob. (Asteraceae=Compositae) **IPNI:** 6821-2 **Common name:** chilca **Habit:** Subshrub **Use:** ME **Origin:** Endemic **Region:** Andean **Dept.:** BOY, CUN, TOL **Elev.:** 2050–3575 m a.s.l. **IUCN Habitat code:** 1,4,14 **Conservation status: (IUCN)** NE

2154 *Ageratina asclepiadea* (L.f.) R.M.King & H.Rob. (Asteraceae=Compositae) **IPNI:** 271815-2 **Common name:** amargoso, amarguero, guasco, zorro **Habit:** Shrub, Small tree **Use:** ME **Origin:** Endemic **Region:** Andean **Dept.:** BOG, BOY, CUN, TOL **Elev.:** 1800–3575 m a.s.l. **IUCN Habitat code:** 1,3,4,5,14 **Conservation status: (IUCN)** NE

2155 *Ageratina baccharoides* (Kunth) R.M.King & H.Rob. (Asteraceae=Compositae) **IPNI:** 6833-2 **Habit:** Shrub, Subshrub **Use:** ME **Origin:** Endemic **Region:** Andean **Dept.:** BOG, BOY, CAU, CES, CUN, NSA, SAN **Elev.:** 1700–4000 m a.s.l. **IUCN Habitat code:** 1,3,4,14 **Conservation status: (IUCN)** NE

2156 *Ageratina pichinchensis* (Kunth) R.M.King & H.Rob. (Asteraceae=Compositae) **IPNI:** 6988-2 **Common name:** manrubio, marrubio, yerba de chivo blanca **Habit:** Subshrub, Climbing **Use:** ME **Origin:** Native **Region:** Andean **Dept.:** ANT, BOG, BOY, CAL, CAU, CES, CUN, HUI, MAG, NAR, NSA, PUT, QUI, RIS, TOL, VAC **Elev.:** 470-4500 m.a.s.l. **IUCN Habitat code:** 1,3,4,14 **Conservation status: (IUCN)** NE **(CPSG)** Potential LC

2157 *Ageratina pseudochilca* (Benth.) R.M.King & H.Rob. (Asteraceae=Compositae) **IPNI:** 6999-2 **Habit:** Shrub **Use:** ME **Origin:** Native **Region:** Andean **Dept.:** BOG, CUN, NAR **Elev.:** 2600–3500 m a.s.l. **Conservation status: (IUCN)** NE

2158 *Ageratina tinifolia* (Kunth) R.M.King & H.Rob. (Asteraceae=Compositae) **IPNI:** 7050-2 **Common name:** amargo, amargoso, amarguero negro, ayubara, chilca amarga, chilco, chilco amargo, salvia amarga **Habit:** Shrub, Subshrub, Small tree **Use:** EU, ME **Origin:** Native **Region:** Andean **Dept.:** ANT, ARA, BOG, BOY, CAL, CAU, CES, CHO, CUN, HUI, MET, NAR, NSA, PUT, QUI, RIS, SAN, TOL, VAC **Elev.:** 1350-3900 m.a.s.l. **IUCN Habitat code:** 1,3,4,5,8,14 **Conservation status: (IUCN)** NE **(CPSG)** Potential LC

2159 *Ageratina vacciniifolia* (Benth.) R.M.King & H.Rob. (Asteraceae=Compositae) **IPNI:** 7058-2 **Habit:** Shrub **Use:** ME **Origin:** Endemic **Region:** Andean **Dept.:** ARA, BOG, BOY, CAS, CAU, CUN, MET, SAN **Elev.:** 2120–3800 m a.s.l. **IUCN Habitat code:** 1,3,4,14 **Conservation status: (IUCN)** NE

2160 *Ageratum conyzoides* L. (Asteraceae=Compositae) **IPNI:** 7086-2 **Common name:** barba de chivo, berba de chivo, celestina azul, chamiza, chiva, chivatera, chivaza, chivita, chivo, chivo macho, copetona, copo de nieve, empella, estancadera, flor de octubre, florazul, florota, hierba de Santa Lucía, hierba de chivo, hierba de chucho, l **Habit:** Herb, Subshrub **Use:** AF, EU, GS, HF, IF, MA, ME, PO, SU **Origin:** Naturalised **Region:** Amazonia, Andean, Caribbean, Orinoquia, Pacific **Dept.:** AMA, ANT, BOG, BOY, CAL, CAQ, CAS, CAU, CHO, CUN, HUI, MAG, MET, NAR, NSA, PUT, QUI, RIS, SAN, TOL, VAC **Elev.:** 0-3420 m.a.s.l. **IUCN Habitat code:** 1,2,3,4,5,14 **Conservation status: (IUCN)** NE **(CPSG)** Potential LC

2161 *Ageratum houstonianum* Mill. (Asteraceae=Compositae) **IPNI:** 175500-1 **Common name:** manrubio, marrubio, yerba de chivo **Habit:** Herb, Shrub, Subshrub **Use:** EU, MA, ME **Origin:** Naturalised **Region:** Andean **Dept.:** ANT, BOG, BOY, CAU, CHO, CUN, HUI, MAG, NAR, NSA, SAN, VAC **Elev.:** 1300–2625 m a.s.l. **IUCN Habitat code:** 1,2,3,14 **Conservation status: (IUCN)** NE

2162 *Alloispermum caracasanum* (Kunth) H.Rob. (Asteraceae=Compositae) **IPNI:** 9082-2 **Common name:** bejuco blanco, carrasposa, jarilla **Habit:** Herb, Climbing **Use:** ME **Origin:** Native **Region:** Andean **Dept.:** ANT, BOG, BOY, CAL, CAU, CES, CUN, HUI, LAG, MAG, NAR, NSA, QUI, SAN, TOL, VAC **Elev.:** 800-4000 m.a.s.l. **IUCN Habitat code:** 1,2,3,4,14 **Conservation status: (IUCN)** NE **(CPSG)** Potential LC

2163 *Ambrosia arborescens* Mill. (Asteraceae=Compositae) **IPNI:** 176055-1 **Common name:** altamisa, altamisa grande, artamisa, marco **Habit:** Herb **Use:** ME, SU **Origin:** Native **Region:** Andean **Dept.:** BOG, BOY, CAU, CUN, HUI, NAR **Elev.:** 266–3025 m a.s.l. **IUCN Habitat code:** 1,8,14 **Conservation status: (IUCN)** NE

2164 *Ambrosia artemisiifolia* L. (Asteraceae=Compositae) **IPNI:** 176058-1 **Common name:** altamisa, artemisa, artemisia **Habit:** Herb **Use:** HF, MA, ME, PO, SU **Origin:** Native **Region:** Caribbean **Dept.:** CUN, MAG, NAR **Elev.:** 0–500 m a.s.l. **IUCN Habitat code:** 1,2,3,14 **Conservation status: (IUCN)** NE

2165 *Ambrosia artemisioides* Meyen & Walp. (Asteraceae=Compositae) **IPNI:** 176059-1 **Common name:** marju **Use:** HF, ME **Dept.:** CAU**Conservation status:** (IUCN) NE

2166 *Ambrosia cumanensis* Kunth (Asteraceae=Compositae) **IPNI:** 11053-2 **Common name:** altamisa, altamisa blanca, altamisa chiquita, altamisa negra, amargosa, ambrosia silvestre, artamisa, artemisa, artemisia, atrapabruja, maki, markju, ruda **Habit:** Herb **Use:** HF, ME, SU **Origin:** Native **Region:** Amazonia, Andean, Caribbean, Orinoquia **Dept.:** AMA, ANT, ATL, BOL, BOY, CAL, CAQ, CAU, CES, CHO, CUN, MAG, MET, NAR, NSA, QUI, SAN, TOL, VAC **Elev.:** 10–3000 m a.s.l. **IUCN Habitat code:** 1,2,3,4,5,14 **Conservation status:** (IUCN) NE

2167 *Ambrosia peruviana* All. (Asteraceae=Compositae) **IPNI:** 60466716-2 **Use:** ME **Region:** Andean **Dept.:** BOG**IUCN Habitat code:** 1,2,3,5,14 **Conservation status:** (IUCN) NE

2168 *Anaphalis margaritacea* (L.) Benth. & Hook.f. (Asteraceae=Compositae) **IPNI:** 30173179-2 **Use:** EU, HF, ME **IUCN Habitat code:** 1 **Conservation status:** (IUCN) NE

2169 *Argyranthemum frutescens* (L.) Sch.Bip. (Asteraceae=Compositae) **IPNI:** 178745-1 **Common name:** margarita, margarita amarilla, margarita blanca **Habit:** Herb **Use:** EU, HF **Origin:** Cultivated, Naturalised **Region:** Andean **Dept.:** ANT, BOG, BOY, CUN **Elev.:** 2620–2800 m a.s.l. **IUCN Habitat code:** 1,3,4,14 **Conservation status:** (IUCN) NE

2170 *Aristeguietia lamiifolia* (Kunth) R.M.King & H.Rob. (Asteraceae=Compositae) **IPNI:** 19109-2 **Common name:** chilca, ciro, gurrubo, matico, pega, pegajosa, puliz, salvia, salvia macho, tola **Habit:** Shrub **Use:** ME **Origin:** Native **Region:** Andean, Orinoquia **Dept.:** BOY, MET, NAR, PUT, VID **Elev.:** 170–3000 m a.s.l. **IUCN Habitat code:** 1,3,14 **Conservation status:** (IUCN) NE

2171 *Artemisia absinthium* L. (Asteraceae=Compositae) **IPNI:** 300106-2 **Common name:** ajenjo, altamisa, artemisa, curahígado **Habit:** Herb, Subshrub **Use:** HF, MA, ME, PO **Origin:** Cultivated, Naturalised **Region:** Andean **Dept.:** BOG, BOY, CAL, CAQ, CUN, QUI, RIS, SAN, VAC **Elev.:** 950–2700 m a.s.l. **IUCN Habitat code:** 1,2,3,4,14 **Conservation status:** (IUCN) NE

2172 *Artemisia dracunculus* L. (Asteraceae=Compositae) **IPNI:** 307848-2 **Use:** HF, MA, ME **Region:** Andean **Dept.:** BOG**Conservation status:** (IUCN) NE

2173 *Artemisia sodiroi* Hieron. (Asteraceae=Compositae) **IPNI:** 20757-2 **Common name:** alcanfor, alhucema, artemisa **Habit:** Herb **Use:** HF, ME **Origin:** Native **Region:** Andean **Dept.:** CUN **Elev.:** 2500–2680 m a.s.l. **IUCN Habitat code:** 1,14 **Conservation status:** (IUCN) NE

2174 *Artemisia verlotiorum* Lamotte (Asteraceae=Compositae) **IPNI:** 180264-1 **Use:** HF, ME **Origin:** Naturalised **Conservation status:** (IUCN) NE

2175 *Artemisia vulgaris* L. (Asteraceae=Compositae) **IPNI:** 20812-2 **Common name:** ajenjo, ajenjo rucio, ajenjol, altamisa, altamisa negra, altamisa pastusa, artemisa **Habit:** Herb **Use:** HF, MA, ME, PO, SU **Origin:** Naturalised **Region:** Andean **Dept.:** ANT, BOG, CAU, MET, NAR, QUI, RIS **Elev.:** 450–1690 m a.s.l. **IUCN Habitat code:** 1,3,5,14 **Conservation status:** (IUCN) NE

2176 *Aster indamellus* Grierson (Asteraceae=Compositae) **IPNI:** 181476-1 **Use:** ME **Region:** Andean **Dept.:** BOG**Conservation status:** (IUCN) NE

2177 *Austroeupatorium inulifolium* (Kunth) R.M.King & H.Rob. (Asteraceae=Compositae) **IPNI:** 27149-2 **Common name:** almoraduz, aromático, balsilla, cadillo, chicharrón, chilca amarga, chilco, gavilana, indioviejo blanco, jarilla, marubio, pegapega, pulisa, salvia, salvia amarga, salvia blanca, salvia de Castilla, salvia de hora, salvio, salvio amargo **Habit:** Herb, Shrub, Subshrub **Use:** AF, EU, FU, MA, ME **Origin:** Native **Region:** Andean, Caribbean **Dept.:** ANT, BOL, BOY, CAL, CAS, CAU, CES, CHO, CUN, HUI, MAG, MET, NAR, NSA, QUI, RIS, SAN, TOL, VAC **Elev.:** 0-2700 m.a.s.l. **IUCN Habitat code:** 1,2,3,5,14 **Conservation status:** (IUCN) NE **(CPSG)** Potential LC

2178 *Ayapana amygdalina* (Lam.) R.M.King & H.Rob. (Asteraceae=Compositae) **IPNI:** 27442-2 **Common name:** amígdala, centaura **Habit:** Herb, Shrub **Use:** HF, MA, ME **Origin:** Native **Region:** Amazonia, Andean, Orinoquia **Dept.:** AMA, ANT, BOG, BOY, CAQ, CAS, CAU, CUN, GUV, MAG, MET, VID **Elev.:** 100–1310 m a.s.l. **IUCN Habitat code:** 1,2,3,4,5,14 **Conservation status:** (IUCN) NE

2179 *Baccharis antioquensis* Killip & Cuatrec. (Asteraceae=Compositae) **IPNI:** 27812-2 **Habit:** Shrub **Use:** MA, ME **Origin:** Endemic **Region:** Andean **Dept.:** ANT, VAC **Elev.:** 2400–3000 m a.s.l. **IUCN Habitat code:** 1,3,14 **Conservation status:** (IUCN) NE

2180 *Baccharis bogotensis* Kunth (Asteraceae=Compositae) **IPNI:** 300387-2 **Common name:** chilca, chilco, ciro **Habit:** Shrub **Use:** EU, FU, ME **Origin:** Endemic **Region:** Andean **Dept.:** BOG, BOY, CES, CUN, MET, NSA, SAN **Elev.:** 1950–4100 m a.s.l. **IUCN Habitat code:** 1,3,4,5,8,14 **Conservation status:** (IUCN) NE

2181 *Baccharis caespitosa* Pers. (Asteraceae=Compositae) **IPNI:** 183170-1 **Use:** ME **Region:** Andean **Dept.:** BOG**Conservation status:** (IUCN) NE

2182 *Baccharis decussata* (Klatt) Hieron. (Asteraceae=Compositae) **IPNI:** 27895-2 **Common name:** almanga, barzalita, mochila vieja **Habit:** Subshrub, Climbing **Use:** ME **Origin:** Native **Region:** Andean **Dept.:** ANT, BOY, CAU, CUN, HUI, MET, NAR, NSA, PUT, VAC **Elev.:** 1100–3000 m a.s.l. **IUCN Habitat code:** 1,3,14 **Conservation status:** (IUCN) NE

2183 *Baccharis genistelloides* (Lam.) Pers. (Asteraceae=Compositae) **IPNI:** 183309-1 **Common name:** araucaria de páramo, charara, chilca, chilco, jatum kimsa nkuchu, muqu-muqu, nuchu-kuchu, nudo-nudo, quimsa-kuchu, tres espigas **Habit:** Herb, Shrub **Use:** ME **Origin:** Native **Region:** Andean **Dept.:** ANT, CAL, CAU, CUN, NAR, PUT, RIS, TOL **Elev.:** 2600–4300 m a.s.l. **IUCN Habitat code:** 1,3,4,6,8,14 **Conservation status:** (IUCN) NE

2184 *Baccharis latifolia* Pers. (Asteraceae=Compositae) **IPNI:** 183399-1 **Common name:** algodoncillo, auraucho, barsalito, barzalito, buéntsamo, chilca, chilca blanca, chilca dulce, chilca morada, chilca negra, chilca rucia, chilca verde, chilco, chilco blanco, chilco común, chilco de teñir, chilco grande, chilco negro, chilquita, chiquito, **Habit:** Herb, Shrub, Tree **Use:** EU, MA, ME, SU **Origin:** Native **Region:** Andean **Dept.:** ANT, BOG, BOY, CAL, CAU, CHO, CUN, HUI, LAG, MAG, MET, NAR, NSA, PUT, QUI, RIS, SAN, TOL, VAC **Elev.:** 1400-4000 m.a.s.l. **IUCN Habitat code:** 1,3,4,5,8,14 **Conservation status:** (IUCN) LC **(CPSG)** Potential LC

2185 *Baccharis macrantha* Kunth (Asteraceae=Compositae) **IPNI:** 303714-2 **Common name:** chilco, ciro, cirro **Habit:** Shrub, Subshrub **Use:** AF, EU, MA **Origin:** Native **Region:** Andean **Dept.:** ARA, BOG, BOY, CAL, CAU, CES, CHO, CUN, HUI, LAG, MET, NAR, PUT, TOL **Elev.:** 1700–4100 m a.s.l. **IUCN Habitat code:** 1,3,4,5,8,14 **Conservation status:** (IUCN) NE

2186 *Baccharis nitida* Pers. (Asteraceae=Compositae) **IPNI:** 183511-1 **Common name:** chilca, chilca blanca, chilco, chilco blanco, salvio **Habit:** Shrub, Tree **Use:** ME **Origin:** Native **Region:** Andean, Caribbean **Dept.:** ANT, BOG, BOL, BOY, CAL, CAQ, CAU, CES, CHO, CUN, HUI, LAG, MAG, MET, NAR, NSA, PUT, QUI, RIS, SAN, TOL, VAC **Elev.:** 40-3600 m.a.s.l. **IUCN Habitat code:** 1,2,3,4,14 **Conservation status:** (IUCN) LC **(CPSG)** Potential LC

2187 *Baccharis pedunculata* (Mill.) Cabrera (Asteraceae=Compositae) **IPNI:** 28074-2 **Common name:** chilca **Habit:** Herb, Shrub **Use:** MA, ME **Origin:** Native **Region:** Andean **Dept.:** ANT, BOY, CAU, CHO, CUN, HUI, NAR, NSA, PUT, QUI, SAN, TOL, VAC **Elev.:** 100-3200 m.a.s.l. **IUCN Habitat code:** 1,2,3,4,14 **Conservation status:** (IUCN) NE **(CPSG)** Potential LC

2188 *Baccharis pentlandii* DC. (Asteraceae=Compositae) **IPNI:** 183580-1 **Habit:** Shrub **Use:** ME **Origin:** Native **Region:** Andean **Dept.:** NAR **Elev.:** 3000–3000 m a.s.l. **Conservation status:** (IUCN) NE

2189 *Baccharis prunifolia* Kunth (Asteraceae=Compositae) **IPNI:** 183631-1 **Common name:** almoraús, chilca, chilco, chilco de páramo, chirco, ciro **Habit:** Shrub, Small tree **Use:** FU **Origin:** Native **Region:** Andean **Dept.:** ANT, ARA, BOG, BOY, CAL, CAU, CES, CUN, LAG, MAG, MET, NAR, NSA, SAN, TOL **Elev.:** 1700-4500 m.a.s.l. **IUCN Habitat code:** 1,2,3,4,8,14 **Conservation status: (IUCN)** NE **(CPSG)** Potential LC

2190 *Baccharis revoluta* Kunth (Asteraceae=Compositae) **IPNI:** 183680-1 **Common name:** chinchimali, chinchimalí, conejo, romerillo, sanalotodo, viravira, zorro **Habit:** Subshrub **Use:** ME **Origin:** Native **Region:** Andean **Dept.:** BOG, BOY, CAL, CUN, TOL **Elev.:** 3700–3700 m a.s.l. **IUCN Habitat code:** 1,3,4,14 **Conservation status: (IUCN)** NE

2191 *Baccharis sagittalis* (Less.) DC. (Asteraceae=Compositae) **IPNI:** 30159244-2 **Use:** HF, ME **Origin:** Native **Conservation status: (IUCN)** NE

2192 *Baccharis salicifolia* (Ruiz & Pav.) Pers. (Asteraceae=Compositae) **IPNI:** 183704-1 **Common name:** chilca, chilco **Habit:** Shrub **Use:** ME **Origin:** Native **Region:** Andean **Dept.:** ANT, BOG, CAS, CAU, HUI, MET, NAR, PUT, TOL, VAC **Elev.:** 400–3500 m a.s.l. **IUCN Habitat code:** 1,3 **Conservation status: (IUCN)** NE

2193 *Baccharis teindalensis* Kunth (Asteraceae=Compositae) **IPNI:** 183793-1 **Common name:** amapola silvestre, cadillo de huerta, chipaca, masequia, papunga, pecunga **Habit:** Shrub **Use:** ME **Origin:** Native **Region:** Andean **Dept.:** BOG, CAU, CUN, NAR **Elev.:** 2800–3700 m a.s.l. **IUCN Habitat code:** 4,14 **Conservation status: (IUCN)** NE

2194 *Baccharis tricuneata* Pers. (Asteraceae=Compositae) **IPNI:** 183814-1 **Common name:** algodoncillo, asawín, chilca de teñir, chilka, china-thula, escobito, hirwa-kowa, marucha, mosca, mosquita, sanalotodo, susumuco, tayanqa **Habit:** Shrub **Use:** ME **Origin:** Native **Region:** Andean **Dept.:** ANT, ARA, BOG, BOY, CAL, CUN, MAG, MET, NSA, QUI, RIS, SAN, TOL, VAC **Elev.:** 1850-4400 m.a.s.l. **IUCN Habitat code:** 1,3,4,5,8,14 **Conservation status: (IUCN)** NE **(CPSG)** Potential LC

2195 *Baccharis trinervis* Pers. (Asteraceae=Compositae) **IPNI:** 183821-1 **Common name:** arnica, barzalita, barzalito, barzalito de monte, chaparral de bestia, chilca, chilca blanca, chilco, chilopo, chilquilla, chilquita, culeco, gavilana, gusanero, lengua de gato, lengüegato, machucha, machucho, mandaguasca, manzanillo, marucha, maruchenga **Habit:** Herb, Shrub, Small tree, Liana, Climbing **Use:** AF, HF, ME **Origin:** Native **Region:** Andean, Caribbean **Dept.:** ANT, ATL, BOG, BOL, BOY, CAL, CAQ, CAS, CAU, CES, CHO, CUN, GUV, HUI, LAG, MAG, MET, NAR, NSA, PUT, QUI, RIS, SAN, SUC, TOL, VAC **Elev.:** 0-3000 m.a.s.l. **IUCN Habitat code:** 1,2,3,4,5,14 **Conservation status: (IUCN)** NE **(CPSG)** Potential LC

2196 *Badilloa steetzii* (B.L.Rob.) R.M.King & H.Rob. (Asteraceae=Compositae) **IPNI:** 28627-2 **Habit:** Shrub **Use:** ME **Origin:** Native **Region:** Andean **Dept.:** SAN **Elev.:** 2600–3000 m a.s.l. **Conservation status: (IUCN)** NE

2197 *Baltimora geminata* (Brandegee) Stuessy (Asteraceae=Compositae) **IPNI:** 28847-2 **Use:** ME **Origin:** Native **Dept.:** MAG**Conservation status: (IUCN)** NE

2198 *Barnadesia arborea* Kunth (Asteraceae=Compositae) **IPNI:** 273073-2 **Use:** EU, ME **Dept.:** CUN**Conservation status: (IUCN)** NE

2199 *Barnadesia spinosa* L.f. (Asteraceae=Compositae) **IPNI:** 184250-1 **Common name:** abrojo, aromo, corono, cuasco, erizo, espina, espino, guagüilla, guasco, guntsiangre, palosanto, roseto **Habit:** Shrub, Small tree **Use:** EU, ME **Origin:** Native **Region:** Andean **Dept.:** ANT, BOG, BOY, CAL, CAU, CUN, NAR, PUT, QUI, SAN, VAC **Elev.:** 1950–4100 m a.s.l. **IUCN Habitat code:** 1,2,3,4,8,14 **Conservation status: (IUCN)** NE

2200 *Bellis perennis* L. (Asteraceae=Compositae) **IPNI:** 184409-1 **Common name:** pascua, pascualito, pascuitas, risueña **Habit:** Herb **Use:** EU, HF, ME **Origin:** Naturalised **Region:** Andean **Dept.:** ANT, BOG, NAR **Elev.:** 2600–2800 m a.s.l. **IUCN Habitat code:** 3,14 **Conservation status: (IUCN)** NE

2201 *Bidens alba* (L.) DC. (Asteraceae=Compositae) **IPNI:** 184711-1 **Habit:** Herb **Use:** ME **Origin:** Native **Region:** Andean, Caribbean **Dept.:** ANT, ATL, BOG, BOY, CAL, CAQ, CAU, CHO, CUN, HUI, MAG, MET, NAR, NSA, PUT, QUI, RIS, SAP, SAN, TOL, VAC, VAU **Elev.:** 60–3000 m a.s.l. **IUCN Habitat code:** 1,3,4,14 **Conservation status: (IUCN)** NE

2202 *Bidens andicola* Kunth (Asteraceae=Compositae) **IPNI:** 184720-1 **Habit:** Herb **Use:** HF, ME **Origin:** Native **Region:** Andean **Dept.:** ANT, BOG, BOY, CAU, CUN, LAG, MAG, NAR, PUT, QUI, SAN, TOL, VAC **Elev.:** 2000-4500 m.a.s.l. **IUCN Habitat code:** 1,3,4,8,14 **Conservation status: (IUCN)** NE **(CPSG)** Potential LC

2203 *Bidens bipinnata* L. (Asteraceae=Compositae) **IPNI:** 184769-1 **Common name:** masiquía **Habit:** Shrub **Use:** AF, EU, GS, HF, IF, ME, PO **Origin:** Naturalised **Region:** Pacific **Dept.:** BOG, CHO, CUN **Elev.:** 450–450 m a.s.l. **IUCN Habitat code:** 1,3,4,5,14 **Conservation status: (IUCN)** NE

2204 *Bidens cynapiifolia* Kunth (Asteraceae=Compositae) **IPNI:** 32394-2 **Common name:** cadillo, cadillo de indio, chipaca, chisacá, duarte, masiquía, pacunga, pupunga, tsalikali **Habit:** Shrub, Subshrub **Use:** ME, SU **Origin:** Native **Region:** Andean, Guiana Shield, Caribbean, Orinoquia **Dept.:** ATL, BOG, BOL, CUN, GUV, MAG, MET, NAR, NSA, SUC, VAC **Elev.:** 25–3300 m a.s.l. **IUCN Habitat code:** 1,3,4,14 **Conservation status: (IUCN)** NE

2205 *Bidens laevis* (L.) Britton, Sterns & Poggenb. (Asteraceae=Compositae) **IPNI:** 32497-2 **Common name:** botoncillo, chipaca, guaca negra, masiquía **Habit:** Herb **Use:** EU, GS, HF, ME **Origin:** Native **Region:** Andean **Dept.:** BOG, BOY, CUN **Elev.:** 2500–3500 m a.s.l. **IUCN Habitat code:** 1,3,4,5,14 **Conservation status: (IUCN)** NE

2206 *Bidens pilosa* L. (Asteraceae=Compositae) **IPNI:** 32564-2 **Common name:** abujua, amapola silvestre, amorseco, cacho de cabra, cadillo, cadillo de huerta, carapacunga, caripacunga, chipa, chipaca, chipaco, mandigüiche, masequía, masiquiara, masiquía, pacunga, papunga, pegapega, picuyo, tapunga, venturosa **Habit:** Herb **Use:** AF, EU, GS, HF, IF, MA, ME, PO, SU **Origin:** Native **Region:** Andean, Caribbean, Pacific **Dept.:** ANT, BOG, BOY, CAL, CAQ, CAU, CHO, CUN, HUI, MAG, MET, NAR, NSA, PUT, QUI, SAP, SAN, TOL, VAC **Elev.:** 0-3100 m.a.s.l. **IUCN Habitat code:** 1,2,3,4,5,14 **Conservation status: (IUCN)** NE **(CPSG)** Potential LC

2207 *Bidens reptans* G.Don (Asteraceae=Compositae) **IPNI:** 185190-1 **Habit:** Herb, Shrub, Subshrub, Climbing **Use:** HF, ME **Origin:** Native **Region:** Andean **Dept.:** BOG, BOY, CAS, CES, CUN, HUI, MAG, MET, NSA, TOL **Elev.:** 500–2880 m a.s.l. **IUCN Habitat code:** 3,4,14 **Conservation status: (IUCN)** NE

2208 *Bidens riparia* Kunth (Asteraceae=Compositae) **IPNI:** 185198-1 **Habit:** Herb **Use:** ME **Origin:** Native **Region:** Andean, Caribbean **Dept.:** ANT, CAU, MAG, PUT, SAN **Elev.:** 275–1350 m a.s.l. **IUCN Habitat code:** 1,2 **Conservation status: (IUCN)** NE

2209 *Bidens subalternans* DC. (Asteraceae=Compositae) **IPNI:** 185264-1 **Habit:** Herb **Use:** HF, ME **Origin:** Native **Region:** Andean **Dept.:** CUN, MAG, MET **Elev.:** 1500–2500 m a.s.l. **IUCN Habitat code:** 3,14 **Conservation status: (IUCN)** NE

2210 *Bidens triplinervia* Kunth (Asteraceae=Compositae) **IPNI:** 185294-1 **Common name:** chipaca, chipaco, duarte, masiquía, mijí, pacunga, pakuash, rosa amarilla común, sunila, ñachag **Habit:** Herb **Use:** ME **Origin:** Native **Region:** Andean **Dept.:** ANT, BOG, BOY, CAL, CAU, CES, CUN, LAG, MAG, MET, NAR, NSA, PUT, QUI, SAN, TOL, VAC **Elev.:** 470-4500 m.a.s.l. **IUCN Habitat code:** 1,3,4,5,14 **Conservation status: (IUCN)** NE **(CPSG)** Potential LC

2211 *Calea berteroana* DC. (Asteraceae=Compositae) **IPNI:** 187576-1 **Common name:** amargosita, cabezona, carrasposa, carrasquillo, chicharrón, fostón **Habit:** Shrub **Use:** ME **Origin:** Native **Region:** Andean, Caribbean **Dept.:** CAU, CUN, MAG, MET, NSA, SAN, TOL **Elev.:** 150–2100 m a.s.l. **Conservation status: (IUCN)** NE

2212 *Calea jamaicensis* (L.) L. (Asteraceae=Compositae) **IPNI:** 1157019-2 **Habit:** Herb, Shrub, Climbing **Use:** ME **Origin:** Native **Region:** Andean, Caribbean **Dept.:** ANT, CAL, CAU, CHO, CUN, HUI, NAR, NSA, SAP, SAN, TOL, VAC **Elev.:** 0-1400 m.a.s.l. **IUCN Habitat code:** 1,3,14 **Conservation status: (IUCN)** NE **(CPSG)** Potential LC

2213 *Calea peruviana* Benth. ex S.F.Blake (Asteraceae=Compositae) **IPNI:** 41956-2 **Common name:** amargosita, cabezona, carrasposa, carrasquillo, chicharrón **Habit:** Herb, Shrub, Subshrub **Use:** ME **Origin:** Native **Region:** Andean **Dept.:** BOG, BOY, CUN, MET, SAN, TOL **Elev.:** 300–3100 m a.s.l. **IUCN Habitat code:** 1,3,4,14 **Conservation status: (IUCN)** NE

2214 *Calea prunifolia* Kunth (Asteraceae=Compositae) **IPNI:** 187732-1 **Common name:** amargosita, cabezona, carrasposa, carrasposa de tierra caliente, chicharrón, concha, jarilla **Habit:** Liana **Use:** ME **Origin:** Native **Region:** Pacific **Dept.:** CHO, NAR **Elev.:** 20–20 m a.s.l. **Conservation status: (IUCN)** NE

2215 *Calea sessiliflora* Less. (Asteraceae=Compositae) **IPNI:** 187759-1 **Common name:** carrasposa, chicharrón, conchita **Habit:** Herb, Shrub **Use:** ME **Origin:** Endemic **Region:** Andean **Dept.:** ANT, CAL, CAU, CUN, HUI, NAR, NSA, PUT, QUI, SAN, TOL, VAC **Elev.:** 500–2150 m a.s.l. **IUCN Habitat code:** 1,2,3,14 **Conservation status: (IUCN)** NE

2216 *Calea ternifolia* Kunth (Asteraceae=Compositae) **IPNI:** 187778-1 **Common name:** carrasposa **Habit:** Herb **Use:** ME **Origin:** Native **Region:** Andean **Dept.:** CUN, TOL **Elev.:** 1900–2100 m a.s.l. **IUCN Habitat code:** 14 **Conservation status: (IUCN)** NE

2217 *Calendula officinalis* L. (Asteraceae=Compositae) **IPNI:** 187894-1 **Common name:** botón de oro, caléndula, flor de muerto, maravilla, rosa amarilla común **Habit:** Herb **Use:** AF, EU, HF, MA, ME **Origin:** Cultivated, Naturalised **Region:** Andean **Dept.:** ANT, BOG, BOY, CAL, CUN, NSA, QUI, RIS, SAN, TOL, VAC **Elev.:** 1350–2800 m a.s.l. **IUCN Habitat code:** 1,2,3,4,14 **Conservation status: (IUCN)** NE

2218 *Campuloclinium macrocephalum* DC. (Asteraceae=Compositae) **IPNI:** 188196-1 **Use:** ME **Origin:** Native **Conservation status: (IUCN)** NE

2219 *Carthamus tinctorius* L. (Asteraceae=Compositae) **IPNI:** 324467-2 **Common name:** alazor, azafrancillo, azafrán bastardo, cártamo **Habit:** Herb **Use:** AF, EU, GS, HF, IF, MA, ME **Origin:** Cultivated, Naturalised **Region:** Andean **Dept.:** BOG, CUN, TOL **Elev.:** 1300–2560 m a.s.l. **IUCN Habitat code:** 14 **Conservation status: (IUCN)** NE

2220 *Centaurea cyanus* L. (Asteraceae=Compositae) **IPNI:** 190310-1 **Common name:** albarina, ambarinas, azulejos, azurinas, botón de soltero **Habit:** Herb **Use:** EU, IF, ME **Origin:** Cultivated, Naturalised **Region:** Andean **Dept.:** BOG, CUN **Elev.:** 2620–2620 m a.s.l. **IUCN Habitat code:** 14 **Conservation status: (IUCN)** NE

2221 *Centratherum punctatum* Cass. (Asteraceae=Compositae) **IPNI:** 51409-2 **Common name:** botoncillo, prontoalivio, siempreviva **Habit:** Shrub **Use:** EU, ME **Origin:** Native **Region:** Amazonia, Andean, Guiana Shield, Caribbean, Orinoquia, Pacific **Dept.:** AMA, ANT, ARA, CAL, CAQ, CAS, CAU, CES, CHO, GUV, NAR, QUI, TOL, VAC **Elev.:** 0-1700 m.a.s.l. **IUCN Habitat code:** 1,2,3,14 **Conservation status: (IUCN)** NE **(CPSG)** Potential LC

2222 *Chamaemelum nobile* (L.) All. (Asteraceae=Compositae) **IPNI:** 192489-1 **Common name:** manzanilla, manzanilla matricaria, manzanilla romana, marimoña, matricaria **Habit:** Herb **Use:** EU, HF, MA, ME **Origin:** Naturalised **Region:** Andean **Dept.:** ANT, BOG, CUN **Elev.:** 2620–2620 m a.s.l. **IUCN Habitat code:** 1,3,4,14 **Conservation status: (IUCN)** LC

2223 *Chaptalia nutans* (L.) Pol. (Asteraceae=Compositae) **IPNI:** 192653-1 **Common name:** amargón, arracachuelo, balsilla, baratsui naekono, capitana, chicoria, contragavilana, contrahierba, corro, diente de león, herb amarga, lechugilla, lechuguilla, lechuguita, lengua de vaca, mano de lagarto, mano de tigre, pipita, punuyoronaekono, salvia g **Habit:** Herb **Use:** HF, MA, ME, SU **Origin:** Native **Region:** Andean, Caribbean, Orinoquia, Pacific **Dept.:** ANT, BOY, CAL, CAQ, CAS, CAU, CES, CHO, CUN, HUI, MAG, MET, NAR, PUT, QUI, RIS, SAN, TOL, VAC **Elev.:** 50-2000 m.a.s.l. **IUCN Habitat code:** 1,2,3,4,5,14 **Conservation status: (IUCN)** NE **(CPSG)** Potential LC

2224 *Chromolaena bullata* (Klatt) R.M.King & H.Rob. (Asteraceae=Compositae) **IPNI:** 57366-2 **Habit:** Shrub **Use:** ME **Origin:** Endemic **Region:** Andean **Dept.:** ANT, BOG, BOY, CUN, VAC **Elev.:** 1900–3200 m a.s.l. **IUCN Habitat code:** 1,3,4,14 **Conservation status: (IUCN)** NE

2225 *Chromolaena ivifolia* (L.) R.M.King & H.Rob. (Asteraceae=Compositae) **IPNI:** 57410-2 **Habit:** Shrub **Use:** ME **Origin:** Native **Region:** Andean, Orinoquia **Dept.:** CAS, CAU, CUN, MAG, MET, RIS **Elev.:** 300–2551 m a.s.l. **IUCN Habitat code:** 1,4 **Conservation status: (IUCN)** NE

2226 *Chromolaena laevigata* (Lam.) R.M.King & H.Rob. (Asteraceae=Compositae) **IPNI:** 57414-2 **Common name:** balsilla, chilca, chilco **Habit:** Shrub **Use:** MA, ME **Origin:** Native **Region:** Andean, Guiana Shield, Orinoquia **Dept.:** ANT, BOY, CAL, CAQ, CAS, CAU, CUN, GUV, HUI, LAG, MET, NSA, QUI, SAN, TOL, VAC, VID **Elev.:** 240-1800 m.a.s.l. **IUCN Habitat code:** 1,2,3,14 **Conservation status: (IUCN)** NE **(CPSG)** Potential LC

2227 *Chromolaena odorata* (L.) R.M.King & H.Rob. (Asteraceae=Compositae) **IPNI:** 57443-2 **Common name:** balsilla, cadillo, jarilla, lang chat, morada, rompe zaragüey, rozavieja, salvia, salvia amarga, salvia blanca, salvia morada, sanalotodo, varejón, verbena arisca, verbena blanca **Habit:** Shrub **Use:** AF, EU, HF, MA, ME, PO **Origin:** Native **Region:** Amazonia, Andean, Guiana Shield, Caribbean, Orinoquia, Pacific **Dept.:** AMA, ANT, ATL, BOL, BOY, CAL, CAQ, CAS, CHO, COR, CUN, LAG, MAG, MET, NAR, NSA, PUT, RIS, SAP, SAN, SUC, TOL, VAC, VAU, VID **Elev.:** 5-2430 m.a.s.l. **IUCN Habitat code:** 1,2,3,4,5,14 **Conservation status: (IUCN)** NE **(CPSG)** Potential LC

2228 *Chromolaena perglabra* (B.L.Rob.) R.M.King & H.Rob. (Asteraceae=Compositae) **IPNI:** 57461-2 **Common name:** jarilla **Habit:** Shrub **Use:** ME **Origin:** Endemic **Region:** Andean **Dept.:** BOG, BOL, BOY, CUN, SAN **Elev.:** 1990–2850 m a.s.l. **IUCN Habitat code:** 1,3,14 **Conservation status: (IUCN)** NE

2229 *Chromolaena scabra* (L.f.) R.M.King & H.Rob. (Asteraceae=Compositae) **IPNI:** 57486-2 **Common name:** amarguero, salvio amargo **Habit:** Shrub **Use:** ME **Origin:** Native **Region:** Andean **Dept.:** ANT, BOG, BOY, CAL, CAU, CUN, HUI, MAG, NAR, PUT, QUI, RIS, SAN, TOL, VAC **Elev.:** 1680–3100 m a.s.l. **IUCN Habitat code:** 1,3,5,14 **Conservation status: (IUCN)** NE

2230 *Chromolaena squalida* (DC.) R.M.King & H.Rob. (Asteraceae=Compositae) **IPNI:** 57489-2 **Habit:** Shrub **Use:** MA, ME **Origin:** Native **Region:** Guiana Shield, Orinoquia **Dept.:** GUV, MET, SAN **Elev.:** 235–650 m a.s.l. **IUCN Habitat code:** 1,3,14 **Conservation status: (IUCN)** NE

2231 *Chromolaena tacotana* (Klatt) R.M.King & H.Rob. (Asteraceae=Compositae) **IPNI:** 57497-2 **Common name:** balsilla, salvia negra **Habit:** Shrub **Use:** ME **Origin:** Endemic **Region:** Andean **Dept.:** ANT, BOG, BOL, BOY, CAU, CUN, HUI, MAG, NAR, NSA, SAN, TOL, VAC **Elev.:** 600–2800 m a.s.l. **IUCN Habitat code:** 1,2,3,14 **Conservation status: (IUCN)** NE

2232 *Chrysanthemum × morifolium* (Ramat.) Hemsl. (Asteraceae=Compositae) **IPNI:** 77111328-1 **Habit:** Herb, Subshrub **Use:** HF, ME **Origin:** Cultivated, Naturalised **Region:** Andean **Dept.:** ANT **Elev.:** 2200–2200 m a.s.l. **IUCN Habitat code:** 1,3,14 **Conservation status: (IUCN)** NE

2233 *Chrysanthemum indicum* L. (Asteraceae=Compositae) **IPNI:** 193510-1 **Use:** EU, GS, MA, ME **Region:** Andean **Dept.:** BOG**Conservation status: (IUCN)** NE

2234 *Chryselium gnaphalioides* (Kunth) Urtubey & S.E.Freire (Asteraceae=Compositae) **IPNI:** 77197189-1 **Habit:** Herb **Use:** ME **Origin:** Native **Region:** Andean **Dept.:** ANT, ARA, BOG, BOY, CAL, CAU, CES, CUN, MAG, MET, NAR, NSA,

RIS, SAN, TOL **Elev.:** 1940–4300 m a.s.l. **IUCN Habitat code:** 1 **Conservation status: (IUCN)** NE

2235 *Chuquiraga jussieui* J.F.Gmel. (Asteraceae=Compositae) **IPNI:** 194359-1 **Common name:** cartagena, chiquiragua, chuquiraga cuasco, chuquiragua, vela de páramo, velita de páramo **Habit:** Shrub **Use:** ME **Origin:** Native **Region:** Andean **Dept.:** NAR, QUI, TOL, VAC **Elev.:** 3000–3750 m a.s.l. **IUCN Habitat code:** 1,3,4,8,14 **Conservation status: (IUCN)** NE

2236 *Cichorium intybus* L. (Asteraceae=Compositae) **IPNI:** 194533-1 **Common name:** achicoria **Use:** AF, GS, HF, ME, PO, SU **IUCN Habitat code:** 1,14 **Conservation status: (IUCN)** NE

2237 *Cirsium echinatum* (Desf.) DC. (Asteraceae=Compositae) **IPNI:** 195292-1 **Habit:** Herb **Use:** EU **Origin:** Cultivated, Naturalised **Region:** Andean **Dept.:** BOG, CAU, CUN **Elev.:** 1750–2600 m a.s.l. **Conservation status: (IUCN)** NE

2238 *Cirsium mexicanum* DC. (Asteraceae=Compositae) **IPNI:** 195679-1 **Habit:** Herb **Use:** HF, ME **Origin:** Native **Region:** Andean **Dept.:** CHO **Elev.:** 0–1800 m a.s.l. **Conservation status: (IUCN)** NE

2239 *Cirsium vulgare* (Savi) Ten. (Asteraceae=Compositae) **IPNI:** 196191-1 **Habit:** Herb **Use:** EU, GS, HF, ME **Origin:** Cultivated, Naturalised **Region:** Andean **Dept.:** BOG, CUN, NAR, QUI **Elev.:** 2500–3500 m a.s.l. **IUCN Habitat code:** 1,3,4,8,14 **Conservation status: (IUCN)** NE

2240 *Cissampelopsis corymbosa* (Wall. ex DC.) C.Jeffrey & Y.L.Chen (Asteraceae=Compositae) **IPNI:** 913807-1 **Use:** ME **Dept.:** BOY**Conservation status: (IUCN)** NE

2241 *Clibadium eggersii* Hieron. (Asteraceae=Compositae) **IPNI:** 60769-2 **Common name:** barbasco **Habit:** Herb, Shrub **Use:** ME **Origin:** Native **Region:** Amazonia, Andean **Dept.:** AMA, ANT, BOY, CAU, CHO, MET, NAR, PUT, RIS, VAC **Elev.:** 0–2500 m a.s.l. **IUCN Habitat code:** 1,3,5,14 **Conservation status: (IUCN)** NE

2242 *Clibadium surinamense* L. (Asteraceae=Compositae) **IPNI:** 319147-2 **Common name:** barbasco, barbasco amarillo, barbasco de hoja, canilla de mula, catalina, conami, gamboa, guaca, juque, lavaplatos, lengua de vaca, limpiaculo, manrubio, maprara, mastranto, matagusanos, papagayo, pintadilla, reventador, salvia, salvia amarga, salvio, sal **Habit:** Herb, Shrub, Subshrub, Tree **Use:** MA, ME, PO **Origin:** Native **Region:** Amazonia, Andean, Guiana Shield, Caribbean, Orinoquia, Pacific **Dept.:** AMA, ANT, ARA, BOL, BOY, CAL, CAQ, CAS, CAU, CHO, COR, CUN, GUA, GUV, LAG, MAG, MET, NAR, PUT, QUI, RIS, SAN, TOL, VAC, VAU **Elev.:** 0-2800 m.a.s.l. **IUCN Habitat code:** 1,2,3,5,14 **Conservation status: (IUCN)** LC **(CPSG)** Potential LC

2243 *Clibadium sylvestre* (Aubl.) Baill. (Asteraceae=Compositae) **IPNI:** 1050569-2 **Common name:** almoraduz, barbasco, barbasco amarillo, barbasco de lagartija, barbasco huitoto, guaca, huaca, matapescado **Habit:** Herb, Shrub, Tree, Climbing **Use:** HF, MA, ME, PO **Origin:** Native **Region:** Amazonia, Andean, Caribbean, Pacific **Dept.:** AMA, ANT, BOL, CAQ, CAS, CHO, CUN, GUV, MAG, NAR, PUT, VAC, VAU **Elev.:** 0–2700 m a.s.l. **IUCN Habitat code:** 1,3,14 **Conservation status: (IUCN)** LC

2244 *Clibadium terebinthinaceum* (Sw.) DC. (Asteraceae=Compositae) **IPNI:** 1006328-2 **Common name:** panal **Habit:** Shrub, Subshrub **Use:** ME **Origin:** Native **Region:** Andean, Caribbean, Pacific **Dept.:** ANT, CHO, MAG, NAR, PUT, VAC **Elev.:** 20–2550 m a.s.l. **IUCN Habitat code:** 1,3,14 **Conservation status: (IUCN)** NE

2245 *Conocliniopsis prasiifolia* (DC.) R.M.King & H.Rob. (Asteraceae=Compositae) **IPNI:** 64237-2 **Common name:** yerba de chivo **Habit:** Shrub **Use:** ME **Origin:** Native **Region:** Andean **Dept.:** ANT, CAL, CAS, CUN, HUI, LAG, SAN, TOL **Elev.:** 40–1805 m a.s.l. **IUCN Habitat code:** 1,2,3,14 **Conservation status: (IUCN)** NE

2246 *Cosmos bipinnatus* Cav. (Asteraceae=Compositae) **IPNI:** 198320-1 **Common name:** cadillo, correo, guatemala, maravilla, maría, musgo florido, pamplonita **Habit:** Herb **Use:** EU, HF, MA, ME **Origin:** Naturalised **Region:** Andean **Dept.:** ANT, BOG, CUN, VAC **Elev.:** 1088–2620 m a.s.l. **Conservation status: (IUCN)** NE

2247 *Cosmos caudatus* Kunth (Asteraceae=Compositae) **IPNI:** 198324-1 **Common name:** correo, rosa amarilla **Habit:** Herb **Use:** EU, HF, MA, ME **Origin:** Native **Region:** Amazonia, Andean, Caribbean, Pacific **Dept.:** AMA, ANT, CAU, CHO, CUN, MAG, NAR, PUT, SAN, VAC **Elev.:** 5–2200 m a.s.l. **IUCN Habitat code:** 1,2,3,14 **Conservation status: (IUCN)** NE

2248 *Cosmos sulphureus* Cav. (Asteraceae=Compositae) **IPNI:** 198368-1 **Common name:** cadillo, correo **Habit:** Herb **Use:** EU, HF, MA, ME **Origin:** Naturalised **Region:** Andean **Dept.:** ANT, CHO, VAC **Elev.:** 540–540 m a.s.l. **IUCN Habitat code:** 1,2,3,14 **Conservation status: (IUCN)** NE

2249 *Crepis capillaris* (L.) Wallr. (Asteraceae=Compositae) **IPNI:** 199753-1 **Habit:** Herb **Use:** HF **Origin:** Naturalised **Region:** Andean **Dept.:** BOG, CUN **Elev.:** 2600–2600 m a.s.l. **IUCN Habitat code:** 14 **Conservation status: (IUCN)** NE

2250 *Critonia morifolia* (Mill.) R.M.King & H.Rob. (Asteraceae=Compositae) **IPNI:** 69558-2 **Habit:** Herb, Shrub **Use:** ME **Origin:** Native **Region:** Andean, Guiana Shield **Dept.:** ANT, ATL, BOY, CAL, CES, CUN, GUA, HUI, LAG, MAG, MET, NAR, QUI, SAN, TOL, VAC **Elev.:** 100-2200 m.a.s.l. **IUCN Habitat code:** 1,2,3,5,14 **Conservation status: (IUCN)** LC **(CPSG)** Potential LC

2251 *Critonia quadrangularis* (DC.) R.M.King & H.Rob. (Asteraceae=Compositae) **IPNI:** 69565-2 **Use:** ME **IUCN Habitat code:** 1,3,14 **Conservation status: (IUCN)** NE

2252 *Critoniella acuminata* (Kunth) R.M.King & H.Rob. (Asteraceae=Compositae) **IPNI:** 69573-2 **Common name:** Santa María, almoraduz, apio de monte, chilca, chupadera, cutuma, esmeraldo, gerillo, jarilla, margaritón de monte, patinegra, patinegro, quebrantadera, salvio, santamaría, trébol aromatizador, tumbaguay **Habit:** Shrub, Tree, Climbing **Use:** ME **Origin:** Native **Region:** Andean **Dept.:** ANT, BOY, CAQ, CAU, CES, CHO, CUN, HUI, MAG, QUI, RIS, SAN, TOL, VAC **Elev.:** 1000-4000 m.a.s.l. **IUCN Habitat code:** 1,2,3,4,14 **Conservation status: (IUCN)** LC **(CPSG)** Potential LC

2253 *Critoniopsis bogotana* (Cuatrec.) H.Rob. (Asteraceae=Compositae) **IPNI:** 200446-1 **Common name:** amarguero amarillo **Habit:** Tree **Use:** ME **Origin:** Native **Region:** Andean **Dept.:** BOG, CAU, CUN, MET, NSA, SAN **Elev.:** 2500–3045 m a.s.l. **IUCN Habitat code:** 1,3,4,14 **Conservation status: (IUCN)** NE

2254 *Critoniopsis glandulata* (Cuatrec.) H.Rob. (Asteraceae=Compositae) **IPNI:** 200452-1 **Habit:** Tree **Use:** ME **Origin:** Native **Region:** Andean **Dept.:** BOY, CUN, NSA, SAN **Elev.:** 1750-2700 m.a.s.l. **IUCN Habitat code:** 1,2,3,4,14 **Conservation status: (IUCN)** VU **(CPSG)** VU

2255 *Culcitium canescens* Bonpl. (Asteraceae=Compositae) **IPNI:** 200642-1 **Habit:** Herb **Use:** ME **Origin:** Native **Region:** Andean **Dept.:** ARA, BOG, BOY, CAL, CAU, CUN, MET, NAR, NSA, RIS, SAN, TOL, VAC **Elev.:** 3100–4710 m a.s.l. **IUCN Habitat code:** 1,3,4,6,8,14 **Conservation status: (IUCN)** NE

2256 *Curio radicans* (L.f.) P.V.Heath (Asteraceae=Compositae) **IPNI:** 1011411-1 **Habit:** Herb **Use:** ME **Origin:** Cultivated **Region:** Andean **Dept.:** BOG, CUN, VAC **Elev.:** 2560–2820 m a.s.l. **Conservation status: (IUCN)** NE

2257 *Cyanthillium cinereum* (L.) H.Rob. (Asteraceae=Compositae) **IPNI:** 962413-1 **Common name:** moradita, tisaca **Habit:** Herb **Use:** AF, HF, MA, ME **Origin:** Naturalised **Region:** Andean, Caribbean, Pacific **Dept.:** BOL, BOY, CUN, LAG, NAR, SAP, TOL, VAC **Elev.:** 0–1000 m a.s.l. **IUCN Habitat code:** 1,2,14 **Conservation status: (IUCN)** NE

2258 *Cynara cardunculus* L. (Asteraceae=Compositae) **IPNI:** 200876-1 **Common name:** alcachofa, cardo **Habit:** Herb **Use:** AF, EU, GS, HF, MA, ME **Origin:** Cultivated, Naturalised **Region:** Andean **Dept.:** ANT, BOG, CUN, NSA, RIS, SAN, VAC **Elev.:** 2200–2200 m a.s.l. **IUCN Habitat code:** 1,14 **Conservation status: (IUCN)** NE

2259 *Cyrtocymura scorpioides* (Lam.) H.Rob. (Asteraceae=Compositae) **IPNI:** 944993-1 **Common name:** cerbatana, pebetera, varejón de caballo **Habit:** Climbing **Use:** ME **Origin:** Native **Region:** Andean, Caribbean, Orinoquia **Dept.:** ANT, BOL, BOY, CAS, CES, CHO, HUI, LAG, MAG, MET, NSA, VAC **Elev.:** 500–2195 m a.s.l. **IUCN Habitat code:** 1,2,3,4,14 **Conservation status: (IUCN)** NE

2260 *Dahlia imperialis* Roezl ex Ortgies (Asteraceae=Compositae) **IPNI:** 200959-1 **Common name:** dalia, dalia silvestre **Habit:** Herb, Subshrub **Use:** AF, EU, HF, ME **Origin:** Native **Region:** Andean **Dept.:** BOG, BOY, CAU, CUN, HUI, MET, NAR, RIS, TOL **Elev.:** 1600–3500 m a.s.l. **IUCN Habitat code:** 1,3,14 **Conservation status: (IUCN)** NE

2261 *Dahlia pinnata* Cav. (Asteraceae=Compositae) **IPNI:** 30027297-2 **Common name:** dalia **Habit:** Herb **Use:** EU, GS, HF, MA, ME **Origin:** Naturalised **Region:** Andean **Dept.:** ANT, BOG, CAU, CUN, QUI, VAC **Elev.:** 1300–2800 m a.s.l. **IUCN Habitat code:** 1,2,3,5,14 **Conservation status: (IUCN)** NE

2262 *Delairea odorata* Lem. (Asteraceae=Compositae) **IPNI:** 201136-1 **Common name:** hiedra amarilla **Habit:** Herb **Use:** EU, ME, PO **Origin:** Naturalised **Region:** Andean **Dept.:** ANT, BOG **Elev.:** 1700–1700 m a.s.l. **IUCN Habitat code:** 3,14 **Conservation status: (IUCN)** NE

2263 *Dendrophorbium lloense* (Hieron. ex Sodiro) C.Jeffrey (Asteraceae=Compositae) **IPNI:** 963541-1 **Common name:** colla, santamaría **Habit:** Shrub, Small tree, Tree **Use:** HF **Origin:** Native **Region:** Andean **Dept.:** CAU, HUI, NAR **Elev.:** 1350–3800 m a.s.l. **IUCN Habitat code:** 1 **Conservation status: (IUCN)** NE

2264 *Dimorphotheca ecklonis* DC. (Asteraceae=Compositae) **IPNI:** 201468-1 **Use:** EU, MA, ME **Region:** Andean **Dept.:** BOG**Conservation status: (IUCN)** NE

2265 *Dimorphotheca jucunda* E.Phillips (Asteraceae=Compositae) **IPNI:** 201475-1 **Use:** ME **Region:** Andean **Dept.:** BOG**Conservation status: (IUCN)** NE

2266 *Eclipta prostrata* (L.) L. (Asteraceae=Compositae) **IPNI:** 1186403-2 **Common name:** antoniodíaz, botoncillo, botón blanco, chisacá, clavo de pozo, congolala, herb de clavo, hierba de clavo, juan eliberto, palo de agua, pimienta, vangolalá, yerba de laguna **Habit:** Herb **Use:** AF, EU, GS, HF, MA, ME **Origin:** Native **Region:** Amazonia, Andean, Caribbean **Dept.:** AMA, ANT, ATL, BOL, BOY, CAL, CAS, CAU, CES, CHO, COR, CUN, GUA, HUI, LAG, MAG, MET, NAR, NSA, PUT, QUI, SAP, SAN, SUC, TOL, VAC **Elev.:** 0-1900 m.a.s.l. **IUCN Habitat code:** 1,2,3,4,5,14 **Conservation status: (IUCN)** LC **(CPSG)** Potential LC

2267 *Egletes viscosa* (L.) Less. (Asteraceae=Compositae) **IPNI:** 1037395-2 **Habit:** Herb **Use:** ME **Origin:** Native **Region:** Amazonia, Orinoquia **Dept.:** AMA, BOY, CAS, LAG **Elev.:** 140–150 m a.s.l. **IUCN Habitat code:** 1,2,3,5 **Conservation status: (IUCN)** NE

2268 *Eirmocephala brachiata* (Benth.) H.Rob. (Asteraceae=Compositae) **IPNI:** 944995-1 **Common name:** arroz llanero, indioviejo, mano de tigre, tabaquillo, yabaquillo **Habit:** Shrub **Use:** AF, EU, ME **Origin:** Native **Region:** Andean, Orinoquia **Dept.:** ANT, BOY, CAL, CAQ, CAS, CAU, CUN, HUI, MET, NSA, SAN, VAC **Elev.:** 100–2200 m a.s.l. **IUCN Habitat code:** 1,2,3,5,14 **Conservation status: (IUCN)** NE

2269 *Elephantopus mollis* Kunth (Asteraceae=Compositae) **IPNI:** 202942-1 **Common name:** chicoria, colaepuerco, coleburro, estancadera, frenteloro, hierba de puerco, oreja de puerco, orejemula, pastel, quebradura, rabo de puerco, rabo de zorro, raboemarrano, suelda, suelda consuelda, suelda de marrano, suelda marranera, sueldaconsuelda, sueld **Habit:** Shrub **Use:** HF, ME **Origin:** Native **Region:** Andean, Guiana Shield, Orinoquia, Pacific **Dept.:** AMA, ANT, BOL, BOY, CAL, CAS, CAU, CES, CHO, CUN, HUI, MAG, MET, NAR, NSA, QUI, RIS, SAN, TOL, VAC, VAU, VID **Elev.:** 0-2500 m.a.s.l. **IUCN Habitat code:** 1,2,3,4,5,14 **Conservation status: (IUCN)** NE **(CPSG)** Potential LC

2270 *Elephantopus tomentosus* L. (Asteraceae=Compositae) **IPNI:** 202967-1 **Common name:** chicoria, hierba de caballo, lengua de vaca, suelda, suelda chicoria, suelda de sabana, suelda espigada, sueldaconsuelda, sueldaconsuelda mayor, totumo, yerba de caballo **Habit:** Herb **Use:** ME **Origin:** Naturalised **Region:** Andean **Dept.:** AMA, CAQ, CUN, MAG, MET **Elev.:** 152–1780 m a.s.l. **IUCN Habitat code:** 1,3 **Conservation status: (IUCN)** NE

2271 *Eleutheranthera ruderalis* (Sw.) Sch.Bip. (Asteraceae=Compositae) **IPNI:** 202984-1 **Common name:** verdolaga **Habit:** Herb **Use:** ME **Origin:** Native **Region:** Andean, Caribbean, Orinoquia, Pacific **Dept.:** ANT, ATL, BOL, BOY, CAS, CAU, CES, CHO, CUN, HUI, MAG, MET, SAP, TOL, VAC **Elev.:** 100-2400 m.a.s.l. **IUCN Habitat code:** 1,2,5,14 **Conservation status: (IUCN)** NE **(CPSG)** Potential LC

2272 *Emilia coccinea* (Sims) G.Don (Asteraceae=Compositae) **IPNI:** 203038-1 **Common name:** borlita, borlitas, carrillo, diente de león rojo, lechuguilla, pincel, pincel de amor, yerba socialista **Habit:** Herb **Use:** AF, EU, GS, HF, ME, SU **Origin:** Naturalised **Region:** Andean, Orinoquia **Dept.:** ANT, BOG, CAQ, CAS, CAU, CHO, CUN, GUV, HUI, MET, NAR, NSA, QUI, SAN, TOL, VAC **Elev.:** 350–3000 m a.s.l. **IUCN Habitat code:** 1,2,3,14 **Conservation status: (IUCN)** NE

2273 *Emilia fosbergii* Nicolson (Asteraceae=Compositae) **IPNI:** 90491-2 **Common name:** borlita, clavelillo, miona, pincelito, yerba socialista, árnica de montaña **Habit:** Herb **Use:** HF, ME **Origin:** Native **Region:** Andean **Dept.:** ANT, BOL, BOY, CAL, GUA, VAC **Elev.:** 350–1600 m a.s.l. **Conservation status: (IUCN)** NE

2274 *Emilia sonchifolia* (L.) DC. (Asteraceae=Compositae) **IPNI:** 203080-1 **Common name:** borla, borlita, borlitas, cerraja, clavelito, coneja, mapolilla, pincel, pincelito, yerba socialista **Habit:** Herb **Use:** EU, GS, HF, ME, SU **Origin:** Naturalised **Region:** Andean, Caribbean **Dept.:** ANT, CAL, CAQ, CAS, CAU, CHO, CUN, HUI, MAG, MET, NAR, NSA, PUT, QUI, SAP, SAN, TOL, VAC **Elev.:** 0–1800 m a.s.l. **IUCN Habitat code:** 1,2,3,4,5,14 **Conservation status: (IUCN)** NE

2275 *Enydra fluctuans* Lour. (Asteraceae=Compositae) **IPNI:** 203207-1 **Common name:** chipaca **Habit:** Herb **Use:** GS, HF, ME **Origin:** Naturalised **Region:** Caribbean **Dept.:** BOL, COR **Elev.:** 30–30 m a.s.l. **IUCN Habitat code:** 1 **Conservation status: (IUCN)** LC

2276 *Erechtites hieraciifolius* (L.) Raf. ex DC. (Asteraceae=Compositae) **IPNI:** 1152557-2 **Common name:** isimalinae **Habit:** Herb **Use:** HF, ME **Origin:** Native **Region:** Amazonia, Andean, Orinoquia, Pacific **Dept.:** AMA, ANT, BOY, CAQ, CAS, CAU, CHO, CUN, HUI, LAG, MAG, MET, NAR, NSA, PUT, RIS, SAN, TOL, VAC, VAU, VID **Elev.:** 0-2700 m.a.s.l. **IUCN Habitat code:** 1,2,3,4,14 **Conservation status: (IUCN)** NE **(CPSG)** Potential LC

2277 *Erechtites valerianifolius* (Link ex Spreng.) DC. (Asteraceae=Compositae) **IPNI:** 203365-1 **Habit:** Herb **Use:** HF, ME **Origin:** Native **Region:** Andean **Dept.:** ANT, BOG, BOY, CAL, CAQ, CAU, CHO, CUN, HUI, MAG, MET, NAR, NSA, PUT, QUI, RIS, SAN, TOL, VAC **Elev.:** 400-3400 m.a.s.l. **IUCN Habitat code:** 1,3,4,14 **Conservation status: (IUCN)** NE **(CPSG)** Potential LC

2278 *Erigeron bonariensis* L. (Asteraceae=Compositae) **IPNI:** 93603-2 **Common name:** juamparao, varejón, venadillo, yerba de caballo **Habit:** Herb **Use:** AF, EU, HF, IF, MA, ME, PO **Origin:** Native **Region:** Andean, Caribbean, Orinoquia **Dept.:** AMA, ANT, BOG, BOY, CAL, CAS, CAU, CES, CHO, CUN, GUA, HUI, LAG, MAG, MET, NAR, NSA, PUT, QUI, RIS, SAN, TOL, VAC **Elev.:** 100–3900 m a.s.l. **IUCN Habitat code:** 1,2,3,4,5,14 **Conservation status: (IUCN)** NE

2279 *Erigeron canadensis* L. (Asteraceae=Compositae) **IPNI:** 203653-1 **Common name:** juamparao, pericón, venadillo **Habit:** Herb **Use:** FU, HF, MA, ME, PO **Origin:** Native **Region:** Andean, Caribbean **Dept.:** ANT, BOG, CAS, CUN, LAG, MAG, NAR, SAP, SAN **Elev.:** 76–3250 m a.s.l. **IUCN Habitat code:** 1,3,5,14 **Conservation status: (IUCN)** NE

2280 *Erigeron floribundus* (Kunth) Sch.Bip. (Asteraceae=Compositae) **IPNI:** 60473415-2 **Use:** AF,ME **Origin:** Native **IUCN Habitat code:** 1,3,14 **Conservation status: (IUCN)** NE

2281 *Erigeron karvinskianus* DC. (Asteraceae=Compositae) **IPNI:** 204028-1 **Habit:** Herb **Use:** AF,ME **Origin:** Native **Region:** Andean **Dept.:** ANT, BOG, BOY, CAL, CUN, NSA, QUI, SAN, TOL **Elev.:** 1500–3150 m a.s.l. **IUCN Habitat code:** 1,2,3,4,14 **Conservation status:** **(IUCN)** NE

2282 *Erigeron sumatrensis* Retz. (Asteraceae=Compositae) **IPNI:** 204572-1 **Common name:** yaguache **Habit:** Herb **Use:** HF, IF, MA, ME, SU **Origin:** Native **Region:** Andean **Dept.:** ANT, BOG, BOL, CAU, CUN, HUI, MAG, NAR, PUT, SAN, VAC **Elev.:** 50–4000 m a.s.l. **IUCN Habitat code:** 1,3,14 **Conservation status:** **(IUCN)** NE

2283 *Erigeron trihecatactis* S.F.Blake (Asteraceae=Compositae) **IPNI:** 94203-2 **Common name:** venadillo **Habit:** Herb **Use:** ME **Origin:** Native **Region:** Andean **Dept.:** BOG, BOY, CAL, CUN, NSA, QUI, SAN **Elev.:** 2100–3900 m a.s.l. **IUCN Habitat code:** 1,3,4,8,14 **Conservation status:** **(IUCN)** NE

2284 *Erigeron uliginosus* Benth. (Asteraceae=Compositae) **IPNI:** 204638-1 **Common name:** venadillo **Habit:** Herb **Use:** ME **Origin:** Native **Region:** Andean **Dept.:** BOG, BOY, CAL, CAU, CES, CUN, HUI, LAG, MAG, MET, NAR, NSA, RIS, SAN, TOL, VAC **Elev.:** 2300–4500 m a.s.l. **IUCN Habitat code:** 1,3,4,5,8,14 **Conservation status:** **(IUCN)** NE

2285 *Espeletia × almorzana* Cuatrec. (Asteraceae=Compositae) **IPNI:** 289835-2 **Common name:** frailejón **Use:** ME **Origin:** Endemic **IUCN Habitat code:** 4 **Conservation status:** **(IUCN)** NE

2286 *Espeletia × bogotensis* Cuatrec. (Asteraceae=Compositae) **IPNI:** 97630-2 **Common name:** frailejón **Use:** ME **Origin:** Endemic **IUCN Habitat code:** 4 **Conservation status:** **(IUCN)** NE

2287 *Espeletia × garcibarrigae* Cuatrec. (Asteraceae=Compositae) **IPNI:** 97672-2 **Common name:** frailejón **Use:** MA, ME **Origin:** Endemic **IUCN Habitat code:** 4,14 **Conservation status:** **(IUCN)** NE

2288 *Espeletia arbelaezii* Cuatrec. (Asteraceae=Compositae) **IPNI:** 97618-2 **Habit:** Herb, Shrub **Use:** EU **Origin:** Endemic **Region:** Andean **Dept.:** BOY **Elev.:** 3000-3300 m.a.s.l. **IUCN Habitat code:** 1,4,8,14 **Conservation status:** **(IUCN)** VU **(CPSG)** VU

2289 *Espeletia argentea* Bonpl. (Asteraceae=Compositae) **IPNI:** 205061-1 **Common name:** frailejón **Habit:** Herb, Shrub **Use:** ME **Origin:** Endemic **Region:** Andean **Dept.:** BOG, BOY, CAL, CES, CUN, MET, SAN **Elev.:** 2600-3800 m.a.s.l. **IUCN Habitat code:** 1,3,4,5,14 **Conservation status:** **(IUCN)** LC **(CPSG)** LC

2290 *Espeletia barclayana* Cuatrec. (Asteraceae=Compositae) **IPNI:** 97628-2 **Common name:** frailejón **Habit:** Herb, Shrub **Use:** ME **Origin:** Endemic **Region:** Andean **Dept.:** BOG, BOY, CUN **Elev.:** 3100-3785 m.a.s.l. **IUCN Habitat code:** 1,3,4,5,8,14 **Conservation status:** **(IUCN)** NT **(CPSG)** NT

2291 *Espeletia boyacensis* Cuatrec. (Asteraceae=Compositae) **IPNI:** 97631-2 **Common name:** frailejón, frailejón plateado **Habit:** Herb, Shrub **Use:** EU, ME **Origin:** Endemic **Region:** Andean **Dept.:** BOY, CUN, MAG, SAN **Elev.:** 2345-3900 m.a.s.l. **IUCN Habitat code:** 1,3,4,8,14 **Conservation status:** **(IUCN)** LC **(CPSG)** LC

2292 *Espeletia caldasii* Cuatrec. (Asteraceae=Compositae) **IPNI:** 97641-2 **Common name:** frailejoncito, frailejón **Habit:** Herb, Shrub **Use:** ME **Origin:** Endemic **Region:** Andean **Dept.:** NSA, SAN **Elev.:** 3200-3600 m.a.s.l. **Conservation status:** **(IUCN)** NE **(CPSG)** CR

2293 *Espeletia chocontana* Cuatrec. (Asteraceae=Compositae) **IPNI:** 97646-2 **Common name:** frailejón **Habit:** Herb, Shrub **Use:** EU **Origin:** Endemic **Region:** Andean **Dept.:** ARA, BOG, BOY, CUN **Elev.:** 2760-3700 m.a.s.l. **IUCN Habitat code:** 1,4,5,14 **Conservation status:** **(IUCN)** EN **(CPSG)** EN

2294 *Espeletia congestiflora* Cuatrec. (Asteraceae=Compositae) **IPNI:** 97651-2 **Common name:** frailejón, incana **Habit:** Herb, Shrub **Use:** ME **Origin:** Endemic **Region:** Andean **Dept.:** BOY, CUN, NSA, SAN **Elev.:** 2500-4000 m.a.s.l. **IUCN Habitat code:** 1,3,4,8,14 **Conservation status:** **(IUCN)** LC **(CPSG)** LC

2295 *Espeletia corymbosa* Bonpl. (Asteraceae=Compositae) **IPNI:** 97655-2 **Common name:** frailejón **Habit:** Herb, Shrub **Use:** ME **Origin:** Endemic **Region:** Andean **Dept.:** BOG, BOY, CUN, MET, NSA **Elev.:** 2000-4000 m.a.s.l. **IUCN Habitat code:** 1,4,14 **Conservation status:** **(IUCN)** NE **(CPSG)** LC

2296 *Espeletia curialensis* Cuatrec. (Asteraceae=Compositae) **IPNI:** 97661-2 **Habit:** Herb, Shrub **Use:** ME **Origin:** Endemic **Region:** Andean **Dept.:** ARA, BOY, CAS **Elev.:** 2900-3650 m.a.s.l. **IUCN Habitat code:** 4,14 **Conservation status:** **(IUCN)** LC **(CPSG)** LC

2297 *Espeletia discoidea* Cuatrec. (Asteraceae=Compositae) **IPNI:** 97662-2 **Common name:** frailejón **Habit:** Herb, Shrub **Use:** ME, SU **Origin:** Endemic **Region:** Andean **Dept.:** BOY, SAN **Elev.:** 3200-3910 m.a.s.l. **IUCN Habitat code:** 4,8,14 **Conservation status:** **(IUCN)** EN **(CPSG)** EN

2298 *Espeletia garciae* Cuatrec. (Asteraceae=Compositae) **IPNI:** 97671-2 **Common name:** frailejón **Habit:** Herb, Shrub **Use:** EU, MA **Origin:** Endemic **Region:** Andean **Dept.:** BOY, CUN, SAN **Elev.:** 1990-3640 m.a.s.l. **IUCN Habitat code:** 1,4 **Conservation status:** **(IUCN)** NE **(CPSG)** VU

2299 *Espeletia glandulosa* Cuatrec. (Asteraceae=Compositae) **IPNI:** 97673-2 **Common name:** frailejón **Habit:** Herb **Use:** ME **Origin:** Endemic **Region:** Andean **Dept.:** BOY, CUN, NSA, SAN **Elev.:** 1990-3650 m.a.s.l. **IUCN Habitat code:** 1,4,8,14 **Conservation status:** **(IUCN)** NE **(CPSG)** VU

2300 *Espeletia grandiflora* Bonpl. (Asteraceae=Compositae) **IPNI:** 205109-1 **Common name:** frailejón, orejas de burro, tabaqueros, tabaquillos, trementinos **Habit:** Herb, Shrub **Use:** EU, MA, ME **Origin:** Endemic **Region:** Andean **Dept.:** ANT, BOG, BOY, CUN, MET, SAN **Elev.:** 2120-4100 m.a.s.l. **IUCN Habitat code:** 1,3,4,5,8,14 **Conservation status:** **(IUCN)** LC **(CPSG)** LC

2301 *Espeletia incana* Cuatrec. (Asteraceae=Compositae) **IPNI:** 97697-2 **Common name:** frailejón **Habit:** Herb, Shrub **Use:** EU, MA, ME **Origin:** Endemic **Region:** Andean **Dept.:** BOY, SAN **Elev.:** 2900-3900 m.a.s.l. **IUCN Habitat code:** 1,4,14 **Conservation status:** **(IUCN)** VU **(CPSG)** VU

2302 *Espeletia jaramilloi* S.Díaz (Asteraceae=Compositae) **IPNI:** 97702-2 **Habit:** Herb, Shrub **Use:** ME **Origin:** Endemic **Region:** Andean **Dept.:** BOY **Elev.:** 3480-3750 m.a.s.l. **IUCN Habitat code:** 4,14 **Conservation status:** **(IUCN)** VU **(CPSG)** VU

2303 *Espeletia jimenez-quesadae* Cuatrec. (Asteraceae=Compositae) **IPNI:** 97703-2 **Common name:** frailejón negro **Habit:** Herb, Shrub **Use:** EU **Origin:** Endemic **Region:** Andean **Dept.:** BOY, SAN, VAC **Elev.:** 2200-3850 m.a.s.l. **IUCN Habitat code:** 4 **Conservation status:** **(IUCN)** NE **(CPSG)** VU

2304 *Espeletia killipii* Cuatrec. (Asteraceae=Compositae) **IPNI:** 97705-2 **Common name:** frailejón **Habit:** Herb, Shrub **Use:** ME **Origin:** Endemic **Region:** Andean **Dept.:** BOG, BOY, CUN **Elev.:** 2120-4010 m.a.s.l. **IUCN Habitat code:** 1,3,4,14 **Conservation status:** **(IUCN)** LC **(CPSG)** LC

2305 *Espeletia lopezii* Cuatrec. (Asteraceae=Compositae) **IPNI:** 97712-2 **Common name:** frailejón, frailejón perrito **Habit:** Herb, Shrub **Use:** ME **Origin:** Endemic **Region:** Andean **Dept.:** ARA, BOY, CUN **Elev.:** 3100-4550 m.a.s.l. **IUCN Habitat code:** 1,2,3,4,14 **Conservation status:** **(IUCN)** LC **(CPSG)** LC

2306 *Espeletia muiska* Cuatrec. (Asteraceae=Compositae) **IPNI:** 97726-2 **Common name:** frailejón **Habit:** Herb, Shrub **Use:** EU, ME **Origin:** Endemic **Region:** Andean **Dept.:** BOY, NSA, SAN **Elev.:** 2700-3810 m.a.s.l. **IUCN Habitat code:** 1,3,4,8,14 **Conservation status:** **(IUCN)** NE **(CPSG)** LC

2307 *Espeletia nemekenei* Cuatrec. (Asteraceae=Compositae) **IPNI:** 97732-2 **Habit:** Herb, Shrub **Use:** ME **Origin:** Endemic **Region:** Andean **Dept.:** ARA, BOY, SAN **Elev.:** 2850-3500 m.a.s.l. **IUCN Habitat code:** 1,4,8,14 **Conservation status: (IUCN)** EN **(CPSG)** EN

2308 *Espeletia occulta* S.F.Blake (Asteraceae=Compositae) **IPNI:** 97736-2 **Common name:** frailejón, nabalá, tabaquillo **Habit:** Tree **Use:** ME **Origin:** Endemic **Region:** Andean **Dept.:** CES, LAG, MAG, NSA **Elev.:** 2700-4000 m.a.s.l. **IUCN Habitat code:** 1,4,14 **Conservation status: (IUCN)** NE **(CPSG)** CR

2309 *Espeletia oswaldiana* S.Díaz (Asteraceae=Compositae) **IPNI:** 97737-2 **Habit:** Herb, Shrub **Use:** ME **Origin:** Endemic **Region:** Andean **Dept.:** BOY **Elev.:** 2500-3015 m.a.s.l. **IUCN Habitat code:** 1,4,14 **Conservation status: (IUCN)** CR **(CPSG)** CR

2310 *Espeletia pleiochasia* Cuatrec. (Asteraceae=Compositae) **IPNI:** 97751-2 **Common name:** frailejón **Habit:** Herb, Shrub **Use:** MA, ME **Origin:** Endemic **Region:** Andean **Dept.:** BOG, BOY, CUN, MAG, SAN **Elev.:** 2200-3624 m.a.s.l. **IUCN Habitat code:** 4,14 **Conservation status: (IUCN)** NE **(CPSG)** NT

2311 *Espeletia pycnophylla* Cuatrec. (Asteraceae=Compositae) **IPNI:** 97756-2 **Common name:** frailejón **Habit:** Shrub, Subshrub **Use:** ME **Origin:** Native **Region:** Andean **Dept.:** CAL, CAU, NAR, PUT, TOL **Elev.:** 2000-4250 m.a.s.l. **IUCN Habitat code:** 1,3,4,8,14 **Conservation status: (IUCN)** LC **(CPSG)** LC

2312 *Espeletia santanderensis* A.C.Sm. (Asteraceae=Compositae) **IPNI:** 97768-2 **Common name:** frailejón, tache **Habit:** Herb, Shrub, Subshrub **Use:** ME **Origin:** Endemic **Region:** Andean **Dept.:** NSA, SAN **Elev.:** 2500-4300 m.a.s.l. **IUCN Habitat code:** 4 **Conservation status: (IUCN)** NE **(CPSG)** LC

2313 *Espeletia tunjana* Cuatrec. (Asteraceae=Compositae) **IPNI:** 97794-2 **Common name:** frailejón **Habit:** Herb, Shrub **Use:** ME **Origin:** Endemic **Region:** Andean **Dept.:** BOY **Elev.:** 2000-3600 m.a.s.l. **IUCN Habitat code:** 1,3,4,8,14 **Conservation status: (IUCN)** EN **(CPSG)** EN

2314 *Espeletia uribei* Cuatrec. (Asteraceae=Compositae) **IPNI:** 97796-2 **Common name:** frailejón **Habit:** Shrub, Small tree **Use:** EU, ME, SU **Origin:** Endemic **Region:** Andean **Dept.:** BOG, BOY, CUN, HUI, MET **Elev.:** 1900-3700 m.a.s.l. **IUCN Habitat code:** 1,3,4,14 **Conservation status: (IUCN)** LC **(CPSG)** LC

2315 *Fleischmannia microstemon* (Cass.) R.M.King & H.Rob. (Asteraceae=Compositae) **IPNI:** 105109-2 **Habit:** Herb **Use:** SU **Origin:** Native **Region:** Amazonia, Andean **Dept.:** ANT, CAL, CAU, CES, CHO, CUN, GUV, HUI, LAG, MAG, MET, NSA, PUT, QUI, RIS, TOL, VAC **Elev.:** 20-2000 m.a.s.l. **IUCN Habitat code:** 1,3,4,5,14 **Conservation status: (IUCN)** NE **(CPSG)** Potential LC

2316 *Gaillardia pulchella* Foug. (Asteraceae=Compositae) **IPNI:** 329185-2 **Habit:** Herb **Use:** EU, GS, ME **Origin:** Naturalised **Region:** Andean **Dept.:** ANT, BOG **Elev.:** 2550–2550 m a.s.l. **IUCN Habitat code:** 1,14 **Conservation status: (IUCN)** NE

2317 *Galinsoga parviflora* Cav. (Asteraceae=Compositae) **IPNI:** 208554-1 **Common name:** caripacunga, chagraquigua, chipaca, cominillo rosado, consuelda menor, guasca, guascas, hierba de puerco, melosita, pacunga, pajaritos, pan con queso, yerba de puerco, yerba pajarilla, yerbaepuerco, yuyo **Habit:** Herb **Use:** AF, EU, GS, HF, ME, PO **Origin:** Native **Region:** Andean **Dept.:** ANT, BOG, BOY, CUN, NAR, PUT, SAN, TOL **Elev.:** 900–3100 m a.s.l. **IUCN Habitat code:** 1,2,3,14 **Conservation status: (IUCN)** NE

2318 *Galinsoga quadriradiata* Ruiz & Pav. (Asteraceae=Compositae) **IPNI:** 208557-1 **Common name:** cadillo de bestia, cominillo rosado, estrellita, florecilla, guasca, guascas, hierba de puerco, pacoyuyo, pacunga pequeña, pan con queso, pan de queso, yerba de puerco, yuyo **Habit:** Herb **Use:** GS, HF, ME **Origin:** Native **Region:** Andean **Dept.:** ANT, BOG, BOY, CAU, CUN, HUI, LAG, MAG, NAR, PUT, RIS, SAN, TOL, VAC **Elev.:** 40-4500 m.a.s.l. **IUCN Habitat code:** 1,3,14 **Conservation status: (IUCN)** NE **(CPSG)** Potential LC

2319 *Gamochaeta americana* (Mill.) Wedd. (Asteraceae=Compositae) **IPNI:** 1213296-2 **Common name:** cenizo, chubilla, chupana, don alonso, donalonso, lechugilla, lechuguilla, meloso, mirabira colorada, oreja de ratón, palomita, pascuala, peralonso, tusa, vira-vira, viravira **Habit:** Herb **Use:** ME **Origin:** Native **Region:** Andean **Dept.:** ANT, BOY, CAL, CUN, MAG, MET, NAR, NSA, PUT, SAN, TOL, VAC **Elev.:** 1000-4100 m.a.s.l. **IUCN Habitat code:** 1,3,4,8,14 **Conservation status: (IUCN)** NE **(CPSG)** Potential LC

2320 *Gamochaeta coarctata* (Willd.) Kerguélen (Asteraceae=Compositae) **IPNI:** 943810-1 **Habit:** Herb **Use:** ME **Origin:** Native **Region:** Andean **Dept.:** ANT, BOG, CUN, NAR **Elev.:** 2300–2300 m a.s.l. **IUCN Habitat code:** 1,3,4,5,14 **Conservation status: (IUCN)** NE

2321 *Gamochaeta purpurea* (L.) Cabrera (Asteraceae=Compositae) **IPNI:** 107379-2 **Common name:** espadilla blanca, lechuguilla, vira-vira, viravira **Habit:** Herb **Use:** HF, ME **Origin:** Native **Region:** Andean **Dept.:** ANT, BOG, BOY, CAL, CAU, CES, CUN, HUI, MAG, MET, NAR, NSA, RIS, SAN, TOL **Elev.:** 1750–4300 m a.s.l. **Conservation status: (IUCN)** NE

2322 *Gamochaeta simplicicaulis* (Willd. ex Spreng.) Cabrera (Asteraceae=Compositae) **IPNI:** 107382-2 **Habit:** Herb **Use:** ME **Origin:** Native **Region:** Andean **Dept.:** ANT **Elev.:** 1840–1840 m a.s.l. **IUCN Habitat code:** 1 **Conservation status: (IUCN)** NE

2323 *Gazania krebsiana* Less. (Asteraceae=Compositae) **IPNI:** 208693-1 **Use:** AF, EU, ME **Region:** Andean **Dept.:** BOG**IUCN Habitat code:** 14 **Conservation status: (IUCN)** NE

2324 *Gazania linearis* (Thunb.) Druce (Asteraceae=Compositae) **IPNI:** 208700-1 **Use:** EU, ME **Region:** Andean **Dept.:** BOG**IUCN Habitat code:** 14 **Conservation status: (IUCN)** NE

2325 *Gazania pectinata* (Thunb.) Hartweg (Asteraceae=Compositae) **IPNI:** 941423-1 **Habit:** Herb **Use:** ME **Origin:** Cultivated **Region:** Andean **Dept.:** BOG, CUN **Elev.:** 2620–2620 m a.s.l. **Conservation status: (IUCN)** NE

2326 *Gazania rigens* (L.) Gaertn. (Asteraceae=Compositae) **IPNI:** 208720-1 **Use:** EU **Region:** Andean **Dept.:** BOG**IUCN Habitat code:** 1,3,14 **Conservation status: (IUCN)** NE

2327 *Gerbera jamesonii* Bolus (Asteraceae=Compositae) **IPNI:** 208893-1 **Common name:** gérbera, margarita africana **Habit:** Herb **Use:** EU **Origin:** Cultivated, Naturalised **Region:** Andean **Dept.:** ANT, BOG, CUN **Elev.:** 1680–1700 m a.s.l. **IUCN Habitat code:** 3,14 **Conservation status: (IUCN)** NE

2328 *Glebionis coronaria* (L.) Cass. ex Spach (Asteraceae=Compositae) **IPNI:** 1013480-1 **Use:** EU, GS, HF, MA, ME **Region:** Andean **Dept.:** BOG**IUCN Habitat code:** 1 **Conservation status: (IUCN)** NE

2329 *Glebionis segetum* (L.) Fourr. (Asteraceae=Compositae) **IPNI:** 209007-1 **Habit:** Herb **Use:** EU, HF, ME **Origin:** Naturalised **Region:** Andean **Dept.:** ANT **Elev.:** 1800–1800 m a.s.l. **IUCN Habitat code:** 1,3,14 **Conservation status: (IUCN)** NE

2330 *Gynura aurantiaca* (Blume) DC. (Asteraceae=Compositae) **IPNI:** 211219-1 **Common name:** manto del padre **Habit:** Herb **Use:** EU, HF, ME **Origin:** Naturalised **Region:** Andean **Dept.:** ANT, NAR, VAC **Elev.:** 1100–1100 m a.s.l. **IUCN Habitat code:** 1 **Conservation status: (IUCN)** NE

2331 *Hebeclinium macrophyllum* (L.) DC. (Asteraceae=Compositae) **IPNI:** 1092582-2 **Common name:** aralén, carate de mariposa, catalina, flor de mariposa, hoja de mariposa, pata de gallina **Habit:** Herb, Shrub **Use:** MA, ME **Origin:** Native **Region:** Amazonia, Andean, Caribbean, Orinoquia, Pacific **Dept.:** AMA, ANT, BOL, BOY, CAL, CAQ, CAU, CES, CHO, COR, CUN, GUV, HUI, MAG, MET, NAR, NSA, PUT, QUI, RIS, TOL, VAC, VAU **Elev.:** 0–3000 m a.s.l. **IUCN Habitat code:** 1,3,4,5,14 **Conservation status: (IUCN)** NE

2332 *Helianthus annuus* L. (Asteraceae=Compositae) **IPNI:** 119003-2 **Common name:** girasol, maravilla, mirasol **Habit:** Herb **Use:** AF, EU, FU, GS, HF, IF, MA, ME, PO, SU **Origin:**

Cultivated, Naturalised **Region:** Andean **Dept.:** ANT, BOG, BOY, CUN **Elev.:** 2630–2630 m a.s.l. **IUCN Habitat code:** 1,2,3,4,14 **Conservation status: (IUCN)** LC

2333 *Helichrysum luteoalbum* (L.) Rchb. (Asteraceae=Compositae) **IPNI:** 212994-1 **Habit:** Herb **Use:** AF, HF, MA, ME **Origin:** Naturalised **Region:** Andean **IUCN Habitat code:** 14 **Conservation status: (IUCN)** NE

2334 *Helichrysum orientale* (L.) Gaertn. (Asteraceae=Compositae) **IPNI:** 213129-1 **Common name:** donalonso, viravira **Use:** ME, SU **IUCN Habitat code:** 1,14 **Conservation status: (IUCN)** NE

2335 *Helichrysum petiolare* Hilliard & B.L.Burtt (Asteraceae=Compositae) **IPNI:** 213170-1 **Use:** EU, ME **Region:** Andean **Dept.:** BOG**IUCN Habitat code:** 14 **Conservation status: (IUCN)** NE

2336 *Heliopsis buphthalmoides* (Jacq.) Dunal (Asteraceae=Compositae) **IPNI:** 213524-1 **Common name:** botoncillo, botoncillo amarillo, botoncillo blanco, botoncillo de Castilla, botón de oro, chisacá, chuchoquemao, chuco, gamboa, guaca, guaca cimarrona, guaca de marrano, guaquilla, guasca, munuaniitabukuerunae, quemadera, risacá, salvia, yuya, yuyo, yuyo **Habit:** Herb, Subshrub **Use:** HF, ME **Origin:** Native **Region:** Andean, Pacific **Dept.:** ANT, BOG, BOL, BOY, CAS, CAU, CHO, CUN, NAR, NSA, SAN, VAC **Elev.:** 0–2200 m a.s.l. **IUCN Habitat code:** 1,2,3,14 **Conservation status: (IUCN)** NE

2337 *Heliopsis helianthoides* (L.) Sweet (Asteraceae=Compositae) **IPNI:** 213534-1 **Common name:** botoncillo, botoncillo macho, floranta, heralda, margarita amarilla **Habit:** Herb **Use:** EU, HF, ME **Origin:** Naturalised **Region:** Amazonia, Andean **Dept.:** AMA, ANT, BOY, CAQ, CAU, CUN, HUI, LAG, MET, NAR, NSA, PUT, QUI, SAN, TOL, VAC **Elev.:** 40–2850 m a.s.l. **IUCN Habitat code:** 1,2,3,5,14 **Conservation status: (IUCN)** NE

2338 *Heterosperma achaetum* S.F.Blake (Asteraceae=Compositae) **IPNI:** 121378-2 **Common name:** anisillo **Habit:** Herb **Use:** ME **Origin:** Endemic **Region:** Andean **Dept.:** BOG, BOY, CUN **Elev.:** 1800–3000 m a.s.l. **IUCN Habitat code:** 1,14 **Conservation status: (IUCN)** NE

2339 *Hypochaeris chillensis* (Kunth) Britton (Asteraceae=Compositae) **IPNI:** 126428-2 **Habit:** Herb **Use:** HF, ME **Origin:** Native **Region:** Andean **Dept.:** BOG, BOY, CUN **Elev.:** 2500–2824 m a.s.l. **IUCN Habitat code:** 1 **Conservation status: (IUCN)** NE

2340 *Hypochaeris elata* Griseb. (Asteraceae=Compositae) **IPNI:** 225495-1 **Habit:** Herb **Use:** ME **Origin:** Native **Region:** Andean **Dept.:** ANT, BOY, CUN, NSA **Elev.:** 2400–3600 m a.s.l. **Conservation status: (IUCN)** NE

2341 *Hypochaeris glabra* L. (Asteraceae=Compositae) **IPNI:** 225504-1 **Habit:** Herb **Use:** HF, ME **Origin:** Naturalised **Region:** Andean **Dept.:** CUN **Elev.:** 1600–1700 m a.s.l. **Conservation status: (IUCN)** NE

2342 *Hypochaeris microcephala* (Sch.Bip.) Cabrera (Asteraceae=Compositae) **IPNI:** 126475-2 **Use:** ME **Origin:** Native **Conservation status: (IUCN)** NE

2343 *Hypochaeris radicata* L. (Asteraceae=Compositae) **IPNI:** 225575-1 **Common name:** achicoria, centella, cerraja, chicoria, diente de león, hierba de chancho, margarita, margaritón, penacho, trompetilla **Habit:** Herb **Use:** HF, ME **Origin:** Naturalised **Region:** Andean **Dept.:** ANT, BOG, BOY, CAU, CUN, HUI, LAG, MAG, NAR, NSA, QUI, RIS, SAN, TOL **Elev.:** 900-4500 m.a.s.l. **IUCN Habitat code:** 1,3,4,8,14 **Conservation status: (IUCN)** NE **(CPSG)** Potential LC

2344 *Hypochaeris sessiliflora* Kunth (Asteraceae=Compositae) **IPNI:** 317594-2 **Common name:** achicora, achicoria, ayak pilli, berros, chchillamatu, chicoria, chicoria blanca, chicuria cucui, chikku chikku **Habit:** Herb **Use:** EU, HF, ME **Origin:** Native **Region:** Andean **Dept.:** ANT, ARA, BOG, BOY, CAL, CAS, CAU, CES, CUN, LAG, MAG, MET, NAR, NSA, RIS, SAN, TOL **Elev.:** 2100-4900 m.a.s.l. **IUCN Habitat code:** 1,3,4,5,6,8,14 **Conservation status: (IUCN)** NE **(CPSG)** Potential LC

2345 *Hypochaeris taraxacoides* Ball (Asteraceae=Compositae) **IPNI:** 296211-2 **Common name:** mamito **Habit:** Herb **Use:** HF **Origin:** Native **Region:** Andean **Dept.:** BOG, CUN, MET **Elev.:** 3550–3720 m a.s.l. **IUCN Habitat code:** 4 **Conservation status: (IUCN)** NE

2346 *Ichthyothere terminalis* S.F.Blake (Asteraceae=Compositae) **IPNI:** 127605-2 **Common name:** barbasco, díctamo real, galicosa **Habit:** Herb, Shrub, Subshrub **Use:** MA, ME, PO **Origin:** Native **Region:** Andean, Orinoquia **Dept.:** BOY, CAQ, CAS, CUN, GUA, GUV, MAG, MET, NSA, SAN, VAU, VID **Elev.:** 65–2500 m a.s.l. **IUCN Habitat code:** 1,2,3,4,14 **Conservation status: (IUCN)** NE

2347 *Isocarpha oppositifolia* (L.) Cass. (Asteraceae=Compositae) **IPNI:** 1008406-1 **Habit:** Herb **Use:** ME **Origin:** Native **Region:** Andean, Caribbean **Dept.:** BOL, CES, LAG, MAG **Elev.:** 50–1130 m a.s.l. **Conservation status: (IUCN)** NE

2348 *Jacobaea maritima* (L.) Pelser & Meijden (Asteraceae=Compositae) **IPNI:** 77075194-1 **Common name:** cineraria **Habit:** Herb **Use:** EU, ME, PO **Origin:** Cultivated, Naturalised **Region:** Andean **Dept.:** ANT, BOG, CUN **Elev.:** 1500–2600 m a.s.l. **IUCN Habitat code:** 3,14 **Conservation status: (IUCN)** NE

2349 *Jungia paniculata* A.Gray (Asteraceae=Compositae) **IPNI:** 226804-1 **Common name:** manito de oso, mano de león, manoeleón, manoelión, manolión, manuelión **Habit:** Herb, Subshrub **Use:** ME **Origin:** Native **Region:** Andean **Dept.:** BOG, BOY, CUN, SAN **Elev.:** 2350–3300 m a.s.l. **IUCN Habitat code:** 1,3,4,8,14 **Conservation status: (IUCN)** NE

2350 *Lactuca sativa* L. (Asteraceae=Compositae) **IPNI:** 228239-1 **Common name:** lechuga, lechuga arrepollada, lechuga crespa **Habit:** Herb **Use:** AF, GS, HF, MA, ME, PO **Origin:** Cultivated, Naturalised **Region:** Andean **Dept.:** ANT, BOG, BOY, CUN **Elev.:** 2514–2890 m a.s.l. **IUCN Habitat code:** 1,3,14 **Conservation status: (IUCN)** NE

2351 *Laennecia filaginoides* DC. (Asteraceae=Compositae) **IPNI:** 228375-1 **Common name:** venadillo **Habit:** Herb **Use:** ME **Origin:** Native **Region:** Andean **Dept.:** BOG, BOY, CUN, NAR **Elev.:** 1990–3180 m a.s.l. **IUCN Habitat code:** 4,14 **Conservation status: (IUCN)** NE

2352 *Laennecia gnaphalioides* Cass. (Asteraceae=Compositae) **IPNI:** 228376-1 **Habit:** Herb **Use:** ME **Origin:** Native **Region:** Andean **Dept.:** CUN **Elev.:** 2650–3000 m a.s.l. **IUCN Habitat code:** 1 **Conservation status: (IUCN)** NE

2353 *Laennecia sophiifolia* (Kunth) G.L.Nesom (Asteraceae=Compositae) **IPNI:** 947840-1 **Common name:** plumaria chiquita **Habit:** Herb **Use:** ME **Origin:** Native **Region:** Andean **Dept.:** BOY, CUN, NSA **Elev.:** 2200–3600 m a.s.l. **Conservation status: (IUCN)** NE

2354 *Lagascea mollis* Cav. (Asteraceae=Compositae) **IPNI:** 134795-2 **Habit:** Herb **Use:** ME **Origin:** Native **Region:** Caribbean **Dept.:** ANT, COR, CUN, HUI, MAG, NSA, TOL, VAC **Elev.:** 12–1050 m a.s.l. **Conservation status: (IUCN)** NE

2355 *Lapsana communis* L. (Asteraceae=Compositae) **IPNI:** 228732-1 **Habit:** Herb **Use:** HF, ME **Origin:** Cultivated, Naturalised **Region:** Andean **Dept.:** CUN **Elev.:** 500–500 m a.s.l. **IUCN Habitat code:** 14 **Conservation status: (IUCN)** NE

2356 *Lasiocephalus ovatus* D.F.K.Schltdl. (Asteraceae=Compositae) **IPNI:** 228818-1 **Habit:** Herb **Use:** ME **Origin:** Native **Region:** Andean **Dept.:** NAR **Elev.:** 3200–4500 m a.s.l. **IUCN Habitat code:** 4 **Conservation status: (IUCN)** NE

2357 *Launaea intybacea* (Jacq.) Beauverd (Asteraceae=Compositae) **IPNI:** 229022-1 **Common name:** cerraja **Habit:** Herb **Use:** AF, HF, ME **Origin:** Naturalised **Region:** Caribbean **Dept.:** ATL, BOL, LAG **Elev.:** 100–120 m a.s.l. **IUCN Habitat code:** 2,3,14 **Conservation status: (IUCN)** NE

2358 *Lepidaploa canescens* (Kunth) H.Rob. (Asteraceae=Compositae) **IPNI:** 948489-1 **Common name:** maruchonga, salvia blanca, salvión, tabaquillo, vara santa, varajón, varasanta, varejón, varejón recio **Habit:** Subshrub, Liana, Climbing **Use:** ME **Origin:** Native **Region:** Andean, Caribbean, Pacific **Dept.:** ANT, BOG, BOL, BOY, CAL, CAU, CHO, CUN, HUI, MAG, MET, NAR, NSA, QUI, RIS, SAN, TOL, VAC **Elev.:** 70-2800 m.a.s.l. **IUCN Habitat code:** 1,2,3,5,14 **Conservation status: (IUCN)** NE **(CPSG)** Potential LC

2359 *Lepidaploa karstenii* (Sch.Bip.) H.Rob. (Asteraceae=Compositae) **IPNI:** 948527-1 **Common name:** huesito, varito **Habit:** Subshrub **Use:** ME **Origin:** Native **Region:** Andean **Dept.:** BOG, BOY, CUN, SAN **Elev.:** 1650–3600 m a.s.l. **IUCN Habitat code:** 1,3,4,14 **Conservation status: (IUCN)** NE

2360 *Lepidaploa lehmannii* (Hieron.) H.Rob. (Asteraceae=Compositae) **IPNI:** 948528-1 **Common name:** salvión, varejón **Habit:** Subshrub **Use:** ME **Origin:** Native **Region:** Andean, Pacific **Dept.:** ANT, CAU, CHO, CUN, MAG, NAR, NSA, QUI, VAC **Elev.:** 50–2500 m a.s.l. **IUCN Habitat code:** 1,2,3,14 **Conservation status: (IUCN)** NE

2361 *Lessingianthus rubricaulis* (Bonpl.) H.Rob. (Asteraceae=Compositae) **IPNI:** 962014-1 **Habit:** Shrub **Use:** PO **Origin:** Native **Dept.:** ANT, CUN, MET **Elev.:** 300–600 m a.s.l. **IUCN Habitat code:** 1,14 **Conservation status: (IUCN)** NE

2362 *Leucanthemum maximum* (Ramond) DC. (Asteraceae=Compositae) **IPNI:** 230047-1 **Common name:** margarita **Habit:** Herb **Use:** EU, ME **Origin:** Cultivated **Region:** Andean **Dept.:** ANT, BOG, CAL, CUN, TOL **Elev.:** 2133–2620 m a.s.l. **IUCN Habitat code:** 3 **Conservation status: (IUCN)** NE

2363 *Leucanthemum vulgare* Lam. (Asteraceae=Compositae) **IPNI:** 30296198-2 **Common name:** margarita, margarita de jardín, margaritón **Habit:** Herb **Use:** EU, HF, MA, ME **Origin:** Cultivated, Naturalised **Region:** Andean **Dept.:** BOG, CAL, CES, CUN **Elev.:** 2100–3400 m a.s.l. **IUCN Habitat code:** 1,2,3,4,14 **Conservation status: (IUCN)** NE

2364 *Liabum igniarium* Less. (Asteraceae=Compositae) **IPNI:** 230435-1 **Common name:** Santa Juana, santamaría **Habit:** Liana **Use:** FU, ME **Origin:** Native **Region:** Andean **Dept.:** CAU, CUN, MAG, NAR **Elev.:** 1800–3000 m a.s.l. **IUCN Habitat code:** 1,3,14 **Conservation status: (IUCN)** NE

2365 *Liatris spicata* (L.) Willd. (Asteraceae=Compositae) **IPNI:** 230634-1 **Use:** EU, ME **Region:** Andean **Dept.:** BOG**Conservation status: (IUCN)** NE

2366 *Linochilus anactinotus* (Wedd.) Saldivia & O.M.Vargas (Asteraceae=Compositae) **IPNI:** 77203458-1 **Common name:** balguerajalé **Habit:** Shrub, Subshrub **Use:** ME **Origin:** Endemic **Region:** Andean **Dept.:** LAG, MAG **Elev.:** 3450-4430 m.a.s.l. **IUCN Habitat code:** 2,4,14 **Conservation status: (IUCN)** NE **(CPSG)** LC

2367 *Linochilus cyparissias* (Wedd.) Saldivia & O.M.Vargas (Asteraceae=Compositae) **IPNI:** 77203471-1 **Common name:** romero ninyucué **Habit:** Shrub **Use:** ME **Origin:** Endemic **Region:** Andean **Dept.:** LAG, MAG **Elev.:** 3450-4000 m.a.s.l. **Conservation status: (IUCN)** NE **(CPSG)** LC

2368 *Linochilus nevadensis* (Cuatrec.) Saldivia & O.M.Vargas (Asteraceae=Compositae) **IPNI:** 77203490-1 **Common name:** cascujalé **Habit:** Subshrub **Use:** ME **Origin:** Endemic **Region:** Andean **Dept.:** MAG **Elev.:** 3260–4700 m a.s.l. **Conservation status: (IUCN)** NE

2369 *Linochilus revolutus* (S.F.Blake) Saldivia & O.M.Vargas (Asteraceae=Compositae) **IPNI:** 77203500-1 **Common name:** romerillo, romero, romero de monte, romero de páramo, romero negro **Habit:** Shrub, Tree **Use:** AF, EU, FU, HF, MA, ME **Origin:** Endemic **Region:** Andean **Dept.:** ANT, BOG, BOY, CAL, CUN, MAG, MET, NSA, QUI, SAN **Elev.:** 2551–4200 m a.s.l. **IUCN Habitat code:** 1,2,3,4,6,8,14 **Conservation status: (IUCN)** NE

2370 *Linochilus rosmarinifolius* Benth. (Asteraceae=Compositae) **IPNI:** 230916-1 **Common name:** romerillo, romero, romero blanco, romero de monte, romero de páramo, romero negro **Habit:** Tree **Use:** EU, FU, HF, MA, ME **Origin:** Native **Region:** Andean **Dept.:** ANT, BOG, BOY, CES, CHO, CUN, LAG, MAG, MET, NSA, SAN, TOL, VAC **Elev.:** 2000–3900 m a.s.l. **IUCN Habitat code:** 1,3,4,5,8,14 **Conservation status: (IUCN)** NE

2371 *Linochilus weddellii* (S.F.Blake) Saldivia & O.M.Vargas (Asteraceae=Compositae) **IPNI:** 77203515-1 **Common name:** yayajalé **Habit:** Shrub **Use:** ME **Origin:** Endemic **Region:** Andean **Dept.:** LAG, MAG **Elev.:** 3100–4400 m a.s.l. **IUCN Habitat code:** 4,14 **Conservation status: (IUCN)** NE

2372 *Loricaria ilinissae* (Benth.) Cuatrec. (Asteraceae=Compositae) **IPNI:** 143805-2 **Habit:** Shrub **Use:** ME **Origin:** Native **Region:** Andean **Dept.:** NAR **Elev.:** 3500–4450 m a.s.l. **Conservation status: (IUCN)** NE

2373 *Loricaria thuyoides* (Lam.) Sch.Bip. (Asteraceae=Compositae) **IPNI:** 231180-1 **Common name:** romero **Habit:** Shrub, Subshrub **Use:** ME **Origin:** Native **Region:** Andean **Dept.:** CAU, NAR, VAC **Elev.:** 3000–4500 m a.s.l. **IUCN Habitat code:** 1,3,4,5,8,14 **Conservation status: (IUCN)** NE

2374 *Lourteigia ballotifolia* (Kunth) R.M.King & H.Rob. (Asteraceae=Compositae) **IPNI:** 144056-2 **Habit:** Herb **Use:** ME **Origin:** Native **Region:** Andean **Dept.:** ANT, BOG, BOY, CAU, CUN, MET, SAN, TOL, VAC **Elev.:** 1800–3750 m a.s.l. **IUCN Habitat code:** 1,3,4,14 **Conservation status: (IUCN)** NE

2375 *Lourteigia humilis* (Benth.) R.M.King & H.Rob. (Asteraceae=Compositae) **IPNI:** 144058-2 **Common name:** hierba de los Uribe, sanalotodo, venturosa, venturosa morada, yerba de los Uribe, yerba de los uribes **Habit:** Herb **Use:** ME **Origin:** Endemic **Region:** Andean **Dept.:** BOG, BOY, CUN, MET, NSA, SAN **Elev.:** 980–3750 m a.s.l. **IUCN Habitat code:** 1,3,4,14 **Conservation status: (IUCN)** NE

2376 *Lourteigia stoechadifolia* (L.f.) R.M.King & H.Rob. (Asteraceae=Compositae) **IPNI:** 144063-2 **Common name:** viravira, viroviro blanco, yerba blanca **Habit:** Herb **Use:** ME **Origin:** Native **Region:** Andean **Dept.:** BOG, BOY, CES, CUN, LAG, MAG, NSA, SAN, VAC **Elev.:** 2531–3490 m a.s.l. **IUCN Habitat code:** 1,3,4,5,14 **Conservation status: (IUCN)** NE

2377 *Lycoseris mexicana* Cass. (Asteraceae=Compositae) **IPNI:** 231341-1 **Common name:** cascabel de tierra caliente, yerba de Santa María **Habit:** Shrub, Subshrub, Climbing **Use:** ME **Origin:** Endemic **Region:** Andean **Dept.:** ANT, CAL, CUN, HUI, QUI, TOL, VAC **Elev.:** 200–2551 m a.s.l. **IUCN Habitat code:** 1,3,4,14 **Conservation status: (IUCN)** NE

2378 *Lycoseris triplinervia* Less. (Asteraceae=Compositae) **IPNI:** 231347-1 **Common name:** buenastardes, mapola, rozavieja **Habit:** Herb, Shrub **Use:** EU, ME **Origin:** Native **Region:** Andean, Caribbean, Orinoquia, Pacific **Dept.:** ANT, ATL, BOL, BOY, CAS, CAU, CHO, COR, MAG, NAR, SAN, SUC **Elev.:** 10–1030 m a.s.l. **IUCN Habitat code:** 1,2,3,4,5,14 **Conservation status: (IUCN)** NE

2379 *Matricaria chamomilla* L. (Asteraceae=Compositae) **IPNI:** 154715-2 **Common name:** camomilla, manzanilla, manzanilla chiquita, manzanilla cimarrona, manzanilla común, manzanilla de Castilla, manzanilla dulce, manzanilla loca, manzanilla matricaria, manzanilla romana **Habit:** Herb **Use:** AF, HF, MA, ME, SU **Origin:** Naturalised **Region:** Andean **Dept.:** ANT, BOG, BOY, CUN, HUI, MAG, NAR, PUT, RIS, SAN, TOL, VAC **Elev.:** 1700–3600 m a.s.l. **IUCN Habitat code:** 1,3,4,8,14 **Conservation status: (IUCN)** NE

2380 *Melampodium divaricatum* (Rich.) DC. (Asteraceae=Compositae) **IPNI:** 156179-2 **Common name:** botón de oro, chisacá, dormidera, drosella, floramarilla, yuyo **Habit:** Herb **Use:** AF, HF, MA, ME **Origin:** Native **Region:** Andean, Caribbean, Orinoquia, Pacific **Dept.:** ANT, BOL, CAL, CAS, CHO, COR, CUN, HUI, LAG, MAG, NAR, SAP, TOL, VAC **Elev.:** 10-1700 m.a.s.l. **IUCN Habitat code:** 1,2,3,5,14 **Conservation status: (IUCN)** NE **(CPSG)** Potential LC

2381 *Melanthera nivea* (L.) Small (Asteraceae=Compositae) **IPNI:** 156327-2 **Common name:** botón blanco, cariaquito blanco, carraquillo, carrasposa, chucharrón, conchita blanca, panal, salvia, yuyo de sábalo **Habit:** Herb, Shrub, Subshrub **Use:** AF,ME **Origin:** Native **Region:** Amazonia, Andean, Caribbean, Pacific **Dept.:** AMA, ANT, ATL, BOL, BOY, CAL, CAQ, CHO, CUN, HUI, LAG, MAG, NAR, NSA, RIS, SAP, SAN, TOL, VAC **Elev.:** 0-2600 m.a.s.l. **IUCN Habitat code:** 1,2,3,5,14 **Conservation status:** **(IUCN)** NE **(CPSG)** Potential LC

2382 *Mikania aschersonii* Hieron. (Asteraceae=Compositae) **IPNI:** 160759-2 **Habit:** Climbing **Use:** ME **Origin:** Native **Region:** Andean **Dept.:** ANT, BOY, CAU, CHO, CUN, VAC **Elev.:** 465–3660 m a.s.l. **IUCN Habitat code:** 1,3,14 **Conservation status:** **(IUCN)** NE

2383 *Mikania banisteriae* DC. (Asteraceae=Compositae) **IPNI:** 232743-1 **Common name:** bejuco sapo, murapo, terciopelo **Habit:** Liana, Climbing **Use:** EU, MA, ME, PO **Origin:** Native **Region:** Amazonia, Andean, Guiana Shield, Orinoquia **Dept.:** AMA, ANT, BOY, CAL, CAQ, CAS, CAU, CHO, CUN, HUI, LAG, MAG, MET, NAR, NSA, PUT, QUI, RIS, SAN, TOL, VAC, VAU, VID **Elev.:** 40-3000 m.a.s.l. **IUCN Habitat code:** 1,2,3,4,5,8,14 **Conservation status:** **(IUCN)** NE **(CPSG)** Potential LC

2384 *Mikania caldasana* B.L.Rob. (Asteraceae=Compositae) **IPNI:** 160806-2 **Habit:** Climbing **Use:** ME **Origin:** Endemic **Region:** Andean **Dept.:** BOY, QUI **Elev.:** 3000–3000 m a.s.l. **IUCN Habitat code:** 1,3,14 **Conservation status:** **(IUCN)** NE

2385 *Mikania clematidiflora* Rusby (Asteraceae=Compositae) **IPNI:** 160833-2 **Habit:** Climbing **Use:** ME **Origin:** Native **Region:** Andean **Dept.:** ANT, HUI, MAG, NSA, TOL **Elev.:** 1370–2700 m a.s.l. **IUCN Habitat code:** 1,2,3,4,14 **Conservation status:** **(IUCN)** NE

2386 *Mikania congesta* DC. (Asteraceae=Compositae) **IPNI:** 232821-1 **Common name:** bejuco de chivo, guaco **Habit:** Climbing **Use:** MA, ME **Origin:** Native **Region:** Amazonia, Andean, Guiana Shield, Caribbean, Orinoquia, Pacific **Dept.:** AMA, ANT, BOL, CAQ, CAS, CAU, CHO, COR, CUN, LAG, MAG, MET, TOL, VAC, VID **Elev.:** 50–1720 m a.s.l. **IUCN Habitat code:** 1,2,3,4,5,14 **Conservation status:** **(IUCN)** NE

2387 *Mikania cordifolia* (L.f.) Willd. (Asteraceae=Compositae) **IPNI:** 232826-1 **Common name:** batata de sapo venenoso, bejuco de chivo, guaco, vidita mía **Habit:** Climbing **Use:** HF, ME **Origin:** Native **Region:** Amazonia, Andean, Caribbean **Dept.:** AMA, ANT, ATL, BOL, CAL, CAQ, CAS, CAU, CHO, CUN, MAG, MET, NAR, TOL **Elev.:** 10-1800 m.a.s.l. **IUCN Habitat code:** 1,2,3,5,14 **Conservation status:** **(IUCN)** NE **(CPSG)** Potential LC

2388 *Mikania flabellata* Rusby (Asteraceae=Compositae) **IPNI:** 160902-2 **Habit:** Liana **Use:** ME **Origin:** Native **Region:** Andean **Dept.:** BOY, CUN, MAG, NSA **Elev.:** 1676–2650 m a.s.l. **IUCN Habitat code:** 1 **Conservation status:** **(IUCN)** NE

2389 *Mikania granulata* B.L.Rob. (Asteraceae=Compositae) **IPNI:** 160925-2 **Habit:** Liana **Use:** ME **Origin:** Native **Region:** Andean **Dept.:** BOG, CAL, CAU, CUN, PUT, SAN, TOL, VAC **Elev.:** 1650–3100 m a.s.l. **IUCN Habitat code:** 1,14 **Conservation status:** **(IUCN)** NE

2390 *Mikania guaco* Bonpl. (Asteraceae=Compositae) **IPNI:** 232913-1 **Common name:** bejuco de culebra, capitana, contracapitana, guaco, guaco morado, hierba capitana **Habit:** Climbing **Use:** ME **Origin:** Native **Region:** Amazonia, Andean, Pacific **Dept.:** AMA, ANT, BOY, CAL, CHO, CUN, HUI, LAG, MAG, NAR, PUT, QUI, SAN **Elev.:** 10-3350 m.a.s.l. **IUCN Habitat code:** 1,3,4,14 **Conservation status:** **(IUCN)** NE **(CPSG)** Potential LC

2391 *Mikania hookeriana* DC. (Asteraceae=Compositae) **IPNI:** 232937-1 **Habit:** Liana, Climbing **Use:** MA, ME **Origin:** Native **Region:** Andean, Caribbean, Pacific **Dept.:** ANT, BOY, CAQ, CAU, CES, CHO, MAG, MET, NAR, PUT, QUI, SAN, VAC **Elev.:** 0–2900 m a.s.l. **IUCN Habitat code:** 1,3,5,14 **Conservation status:** **(IUCN)** NE

2392 *Mikania lehmannii* Hieron. (Asteraceae=Compositae) **IPNI:** 161007-2 **Common name:** guaco **Habit:** Climbing **Use:** ME **Origin:** Endemic **Region:** Andean **Dept.:** ANT, BOG, BOY, SAN **Elev.:** 1900–2600 m a.s.l. **IUCN Habitat code:** 1,14 **Conservation status:** **(IUCN)** NE

2393 *Mikania lindleyana* DC. (Asteraceae=Compositae) **IPNI:** 233001-1 **Habit:** Climbing **Use:** MA **Origin:** Native **Region:** Amazonia **Dept.:** PUT **Elev.:** 325–325 m a.s.l. **IUCN Habitat code:** 1 **Conservation status:** **(IUCN)** NE

2394 *Mikania micrantha* Kunth (Asteraceae=Compositae) **IPNI:** 161051-2 **Common name:** bejuco de chivo, bejuco guaco, cernilla, guaco blanco, guaquito, mata cuy, o'mochacho', tee-shai-pee-ka **Habit:** Liana, Climbing **Use:** AF, HF, ME **Origin:** Native **Region:** Amazonia, Andean, Caribbean, Orinoquia, Pacific **Dept.:** AMA, ANT, ATL, BOL, BOY, CAL, CAQ, CAS, CAU, CES, CHO, COR, CUN, HUI, MAG, MET, NAR, PUT, QUI, SAN, TOL, VAC, VAU, VID **Elev.:** 0-2900 m.a.s.l. **IUCN Habitat code:** 1,2,3,4,5,14 **Conservation status:** **(IUCN)** NE **(CPSG)** Potential LC

2395 *Mikania parviflora* (Aubl.) H.Karst. (Asteraceae=Compositae) **IPNI:** 233110-1 **Habit:** Climbing **Use:** MA, ME **Origin:** Native **Region:** Amazonia, Andean, Pacific **Dept.:** AMA, ANT, CAQ, CES, CHO, MET, VAC, VAU **Elev.:** 0–900 m a.s.l. **IUCN Habitat code:** 1,5,14 **Conservation status:** **(IUCN)** NE

2396 *Mikania psilostachya* DC. (Asteraceae=Compositae) **IPNI:** 233144-1 **Habit:** Climbing **Use:** ME **Origin:** Native **Region:** Amazonia, Andean, Guiana Shield, Caribbean, Orinoquia, Pacific **Dept.:** AMA, ANT, ARA, BOL, BOY, CAQ, CAS, CHO, GUA, GUV, LAG, MET, NAR, PUT, SAN, TOL, VAU **Elev.:** 10-1200 m.a.s.l. **IUCN Habitat code:** 1,2,3,4,5,14 **Conservation status:** **(IUCN)** NE **(CPSG)** Potential LC

2397 *Mikania selloi* Spreng. (Asteraceae=Compositae) **IPNI:** 233209-1 **Common name:** yavímajuna **Use:** ME **IUCN Habitat code:** 1 **Conservation status:** **(IUCN)** NE

2398 *Mikania sylvatica* Klatt (Asteraceae=Compositae) **IPNI:** 161279-2 **Habit:** Liana, Climbing **Use:** ME **Origin:** Native **Region:** Andean **Dept.:** ANT, BOG, BOY, CAL, CAU, CUN, HUI, LAG, MAG, NAR, QUI, RIS, TOL, VAC **Elev.:** 1200–3150 m a.s.l. **IUCN Habitat code:** 1,3,14 **Conservation status:** **(IUCN)** NE

2399 *Mikania vitifolia* DC. (Asteraceae=Compositae) **IPNI:** 161327-2 **Common name:** jarapijinae **Habit:** Climbing **Use:** ME **Origin:** Native **Region:** Andean **Dept.:** ANT, BOY, CUN, HUI, MAG, MET, NAR, NSA, PUT **Elev.:** 400–2073 m a.s.l. **IUCN Habitat code:** 1,3,14 **Conservation status:** **(IUCN)** NE

2400 *Milleria quinqueflora* L. (Asteraceae=Compositae) **IPNI:** 315761-2 **Common name:** doña juana, doñajuana **Habit:** Herb, Shrub **Use:** ME **Origin:** Native **Region:** Andean, Caribbean, Pacific **Dept.:** ATL, BOL, CHO, CUN, LAG, MAG, NSA **Elev.:** 0–2280 m a.s.l. **IUCN Habitat code:** 1,14 **Conservation status:** **(IUCN)** NE

2401 *Montanoa ovalifolia* K.Koch (Asteraceae=Compositae) **IPNI:** 233594-1 **Common name:** acapón, arboloco, camargo, jiquimilla, jome, jícama, jíquina, pauche, treeoco de bogotá, ucapón, upa, upacón, upo **Habit:** Small tree, Tree **Use:** EU, FU, HF, ME **Origin:** Native **Region:** Andean **Dept.:** BOG, BOY, CUN, NAR, NSA, SAN **Elev.:** 1500–3200 m a.s.l. **IUCN Habitat code:** 1,3,4,14 **Conservation status:** **(IUCN)** NE

2402 *Montanoa quadrangularis* Sch.Bip. (Asteraceae=Compositae) **IPNI:** 233605-1 **Common name:** arboloco, balso, caya negra, cocua, colla negra, pauche, pauche blanco, treeoco **Habit:** Shrub, Small tree, Tree **Use:** MA, ME, SU **Origin:** Native **Region:** Andean **Dept.:** ANT, BOG, BOY, CAL, CAQ, CUN, HUI, MET, QUI, TOL, VAC **Elev.:** 1200–2800 m a.s.l. **IUCN Habitat code:** 1,3,14 **Conservation status:** **(IUCN)** NE

2403 *Monticalia abietina* (Willd. ex Wedd.) C.Jeffrey (Asteraceae=Compositae) **IPNI:** 963566-1 **Common name:** panque, panque romero, pino cansabrazo, romero, romero chiquito, romero de monte **Habit:** Shrub **Use:** ME **Origin:** Native **Region:** Andean **Dept.:** BOG, BOY, CES, CUN, MET, NAR, NSA, SAN, TOL **Elev.:** 1750–4200 m a.s.l. **IUCN Habitat code:** 1,3,4,8,14 **Conservation status:** **(IUCN)** NE

2404 *Monticalia corymbosa* (Benth.) C.Jeffrey (Asteraceae=Compositae) **IPNI:** 963576-1 **Common name:** basgüín, chitacá, guaguín huashüin, guasgüín, guasquín, hierba blanca, huashuín, panque, romerillo, romero de monte, yerba blanca **Habit:** Shrub,

Subshrub **Use:** ME **Origin:** Endemic **Region:** Andean **Dept.:** BOG, BOY, CUN, MET, SAN **Elev.:** 1800–4000 m a.s.l. **IUCN Habitat code:** 1,2,3,4,5,14 **Conservation status: (IUCN)** NE

2405 *Monticalia guicanensis* (Cuatrec.) C.Jeffrey (Asteraceae=Compositae) **IPNI:** 963584-1 **Common name:** basgüín, basquín, guasgüín, guasquín **Habit:** Shrub, Subshrub **Use:** ME **Origin:** Endemic **Region:** Andean **Dept.:** ARA, BOY **Elev.:** 3500-4525 m.a.s.l. **IUCN Habitat code:** 3,4,6 **Conservation status: (IUCN)** NE **(CPSG)** LC

2406 *Monticalia ledifolia* (Kunth) C.Jeffrey (Asteraceae=Compositae) **IPNI:** 963586-1 **Common name:** chitacá, guasgüín, güiesquín, hierba cana, hierba de Santa María, hierba santa, huasguin, huasgüín, romerillo, romerito, romerito de hoja ancha, romero, romero de monte, yerba blanca **Habit:** Herb, Shrub **Use:** AF,ME **Origin:** Native **Region:** Andean **Dept.:** ANT, BOG, BOY, CAS, COR, CUN, MET, NSA, SAN **Elev.:** 2200–4100 m a.s.l. **IUCN Habitat code:** 1,2,3,4,5,8,14 **Conservation status: (IUCN)** NE

2407 *Monticalia leioclada* (Cuatrec.) C.Jeffrey (Asteraceae=Compositae) **IPNI:** 963587-1 **Use:** ME **Origin:** Endemic **IUCN Habitat code:** 4 **Conservation status: (IUCN)** NE

2408 *Monticalia nitida* (Kunth) C.Jeffrey (Asteraceae=Compositae) **IPNI:** 963597-1 **Common name:** basgüín, guasgüín, guasquín, huasgüín **Habit:** Shrub **Use:** ME **Origin:** Endemic **Region:** Andean **Dept.:** BOG, CAU, CUN, MET **Elev.:** 2700–3685 m a.s.l. **IUCN Habitat code:** 1,3,4,8,14 **Conservation status: (IUCN)** NE

2409 *Monticalia pulchella* (Kunth) C.Jeffrey (Asteraceae=Compositae) **IPNI:** 963604-1 **Common name:** buba, guasgüín, helecho, panque, romerillo, romerillo blanco, romero, romero blanco, romero de monte, romero de páramo, vieja **Habit:** Shrub, Subshrub, Small tree **Use:** MA **Origin:** Endemic **Region:** Andean **Dept.:** ANT, BOG, BOY, CAU, CUN, MET, NSA, SAN, TOL, VAC **Elev.:** 1900–3850 m a.s.l. **IUCN Habitat code:** 1,3,4,8,14 **Conservation status: (IUCN)** NE

2410 *Monticalia vaccinioides* (Kunth) C.Jeffrey (Asteraceae=Compositae) **IPNI:** 963618-1 **Common name:** chilquilla menuda, chitón, guasgüís blanco, maíz tostado, romerillo rucio, sanalotodo, tangue, teta de bruja **Habit:** Shrub **Use:** ME **Origin:** Native **Region:** Andean **Dept.:** ANT, ARA, BOG, BOY, CAL, CAS, CAU, CES, CHO, CUN, HUI, MAG, MET, NAR, NSA, PUT, QUI, RIS, SAN, TOL, VAC **Elev.:** 3000-3990 m.a.s.l. **IUCN Habitat code:** 1,3,4,6,8,14 **Conservation status: (IUCN)** NE **(CPSG)** Potential LC

2411 *Munnozia hastifolia* (Poepp.) H.Rob. & Brettell (Asteraceae=Compositae) **IPNI:** 164998-2 **Common name:** santamaría, árnica, árnica santamaría **Habit:** Herb, Shrub, Climbing **Use:** ME **Origin:** Native **Region:** Andean **Dept.:** ANT, BOG, BOY, CAL, CAQ, CAU, CHO, CUN, MAG, NAR, PUT, QUI, RIS, TOL, VAC **Elev.:** 440–2700 m a.s.l. **IUCN Habitat code:** 1,3,4,14 **Conservation status: (IUCN)** NE

2412 *Mutisia clematis* L.f. (Asteraceae=Compositae) **IPNI:** 233864-1 **Common name:** bejuco clavellino, clavellina, clavellina de bejuco, clavellino, clavellino de bejuco, zarcillejo **Habit:** Liana **Use:** EU **Origin:** Endemic **Region:** Andean **Dept.:** ANT, BOG, BOY, CAL, CUN, HUI, MET, NSA, QUI, SAN, TOL, VAC **Elev.:** 1990–3900 m a.s.l. **IUCN Habitat code:** 1,3,4,8,14 **Conservation status: (IUCN)** NE

2413 *Neurolaena lobata* (L.) Cass. (Asteraceae=Compositae) **IPNI:** 234386-1 **Common name:** almizclillo, capitana, contra, contragavilana, gavilana, inaciabi, indio viejo, mano de lagrato, mano de tigre, mostaza, patelancha, salvia gavilana, tainde, tres puntas, valeriana, varejón blanco, venadillo, wareara, yerba de cáncer **Habit:** Herb, Shrub **Use:** MA, ME **Origin:** Native **Region:** Amazonia, Andean, Pacific **Dept.:** AMA, ANT, BOL, BOY, CAL, CAQ, CAU, CHO, GUV, LAG, MAG, MET, NAR, NSA, PUT, QUI, RIS, SAN, TOL, VAC **Elev.:** 0-1950 m.a.s.l. **IUCN Habitat code:** 1,3,5,14 **Conservation status: (IUCN)** NE **(CPSG)** Potential LC

2414 *Oligactis sessiliflora* (Kunth) H.Rob. & Brettell (Asteraceae=Compositae) **IPNI:** 173688-2 **Habit:** Climbing **Use:** ME **Origin:** Native **Region:** Andean **Dept.:** ANT, BOG, BOY, CUN, NSA, SAN **Elev.:** 1780–2940 m a.s.l. **IUCN Habitat code:** 1,3,4,14 **Conservation status: (IUCN)** NE

2415 *Onoseris drakeana* André (Asteraceae=Compositae) **IPNI:** 235410-1 **Common name:** árnica **Habit:** Shrub **Use:** ME **Origin:** Endemic **Region:** Andean, Pacific **Dept.:** CAU, NAR **Elev.:** 500–2600 m a.s.l. **IUCN Habitat code:** 14 **Conservation status: (IUCN)** NE

2416 *Onoseris onoseroides* (Kunth) B.L.Rob. (Asteraceae=Compositae) **IPNI:** 174455-2 **Common name:** hoja de Santa María, hoja de santamaría, papelillo, planta de Santa María, planta de santamaría, santamaría **Habit:** Shrub **Use:** ME **Origin:** Native **Region:** Andean, Orinoquia **Dept.:** ANT, CAL, CAS, CES, CUN, MAG, MET **Elev.:** 350–1900 m a.s.l. **IUCN Habitat code:** 1,3,14 **Conservation status: (IUCN)** NE

2417 *Onoseris purpurea* (L.f.) S.F.Blake (Asteraceae=Compositae) **IPNI:** 77181848-1 **Common name:** altamisa, amapola de monte, herb de santamaría, hierba de Santa María, hoja de Santa María, hoja de santamaría, juamblanco, santamaría, árnica **Habit:** Herb, Shrub **Use:** ME **Origin:** Endemic **Region:** Andean **Dept.:** CAS, CUN, HUI, QUI, TOL, VAC **Elev.:** 300–2720 m a.s.l. **IUCN Habitat code:** 1,2,3,14 **Conservation status: (IUCN)** NE

2418 *Onoseris silvatica* Greenm. (Asteraceae=Compositae) **IPNI:** 174463-2 **Common name:** santamaría **Habit:** Shrub **Use:** ME **Origin:** Native **Region:** Andean, Guiana Shield, Orinoquia **Dept.:** BOY, CAQ, CAS, CUN, MET **Elev.:** 400–1250 m a.s.l. **IUCN Habitat code:** 1,3,14 **Conservation status: (IUCN)** NE

2419 *Orthopappus angustifolius* (Sw.) Gleason (Asteraceae=Compositae) **IPNI:** 176829-2 **Habit:** Herb **Use:** ME **Origin:** Native **Region:** Andean, Guiana Shield, Orinoquia **Dept.:** ANT, BOY, CAQ, CAU, CES, CUN, GUV, MAG, MET, NAR, SAN, TOL, VID **Elev.:** 100–1900 m a.s.l. **IUCN Habitat code:** 1,2,3,4,14 **Conservation status: (IUCN)** NE

2420 *Pacourina edulis* Aubl. (Asteraceae=Compositae) **IPNI:** 236351-1 **Common name:** lechuga **Habit:** Shrub **Use:** HF, ME **Origin:** Native **Region:** Caribbean, Orinoquia **Dept.:** ANT, ARA, BOL, COR, CUN, MAG, SUC, TOL **Elev.:** 25–300 m a.s.l. **Conservation status: (IUCN)** NE

2421 *Parthenium hysterophorus* L. (Asteraceae=Compositae) **IPNI:** 236510-1 **Habit:** Herb **Use:** ME, PO **Origin:** Native **Region:** Andean **Dept.:** CUN, VAC **Elev.:** 2600–2600 m a.s.l. **IUCN Habitat code:** 1,3,14 **Conservation status: (IUCN)** NE

2422 *Pectis elongata* Kunth (Asteraceae=Compositae) **IPNI:** 236598-1 **Common name:** comino, comino silvestre, pimento, yerba de limón **Habit:** Herb **Use:** HF, MA, ME **Origin:** Native **Region:** Amazonia, Andean **Dept.:** AMA, ANT, CAQ, CAU, CES, HUI, LAG, MAG, MET, TOL, VAC **Elev.:** 150–1225 m a.s.l. **IUCN Habitat code:** 1,2,3,14 **Conservation status: (IUCN)** NE

2423 *Pectis linearis* La Llave (Asteraceae=Compositae) **IPNI:** 236637-1 **Common name:** corazoncillo, culantrillo, mentolín, tomillo, tomillo oloroso **Habit:** Herb **Use:** MA, ME **Origin:** Native **Region:** Andean, Pacific **Dept.:** ANT, CAL, CAU, CHO, CUN, MAG, NAR, SAN, TOL, VAC **Elev.:** 0–1280 m a.s.l. **IUCN Habitat code:** 1,2,14 **Conservation status: (IUCN)** NE

2424 *Pectis linifolia* L. (Asteraceae=Compositae) **IPNI:** 236639-1 **Common name:** anisillo, comino **Habit:** Herb **Use:** ME **Origin:** Native **Region:** Caribbean **Dept.:** BOL, LAG, MAG **Elev.:** 10–200 m a.s.l. **IUCN Habitat code:** 2,3,14 **Conservation status: (IUCN)** NE

2425 *Pectis prostrata* Cav. (Asteraceae=Compositae) **IPNI:** 236681-1 **Habit:** Herb **Use:** ME **Origin:** Native **Dept.:** CES **Elev.:** 100–600 m a.s.l. **Conservation status: (IUCN)** NE

2426 *Pentacalia decomposita* (Sch.Bip. ex Hieron.) Cuatrec. (Asteraceae=Compositae) **IPNI:** 909415-1 **Habit:** Liana **Use:** ME **Origin:** Native **Region:** Andean **Dept.:** NSA, SAN, VAC **Elev.:** 3000–3900 m a.s.l. **IUCN Habitat code:** 1,3 **Conservation status: (IUCN)** NE

2427 *Pentanema squarrosum* (L.) D.Gut.Larr., Santos-Vicente, Anderb., E.Rico & M.M.Mart.Ort. (Asteraceae=Compositae) **IPNI:** 60478388-2 **Habit:** Subshrub **Use:** ME **Origin:** Naturalised **Region:** Andean **Dept.:** CUN **Elev.:** 500–500 m a.s.l. **Conservation status: (IUCN)** NE

2428 *Perezia multiflora* (Bonpl.) Less. (Asteraceae=Compositae) **IPNI:** 237026-1 **Common name:** escorzonera **Habit:** Herb **Use:** ME **Origin:** Native **Region:** Andean **Dept.:** CAU, NAR **Elev.:** 3000–4300 m a.s.l. **IUCN Habitat code:** 1 **Conservation status: (IUCN)** NE

2429 *Pericallis × hybrida* (Regel) B.Nord. (Asteraceae=Compositae) **IPNI:** 1164364-2 **Habit:** Subshrub **Use:** EU **Origin:** Naturalised **Region:** Andean **Dept.:** ANT **Elev.:** 1500–1500 m a.s.l. **IUCN Habitat code:** 14 **Conservation status: (IUCN)** NE

2430 *Pericallis cruenta* Webb & Berthel. (Asteraceae=Compositae) **IPNI:** 237108-1 **Use:** EU **Dept.:** BOG**IUCN Habitat code:** 4 **Conservation status: (IUCN)** NE

2431 *Piptocoma acuminata* (Humb., Bonpl. & Kunth) Pruski (Asteraceae=Compositae) **IPNI:** 988439-1 **Habit:** Tree **Use:** ME **Origin:** Native **Region:** Andean **Dept.:** CAS, NSA **Elev.:** 1300–1800 m a.s.l. **IUCN Habitat code:** 1,3,14 **Conservation status: (IUCN)** NE

2432 *Piptocoma discolor* (Kunth) Pruski (Asteraceae=Compositae) **IPNI:** 988437-1 **Common name:** aliso, aliso negro, amorseco, cacique, cenizo, chasmiande, congo, gallinaza, gallinazo, gallinazo blanco, gallinazo negro, guayacán, mulata, mulato, palo negro, palo quinde, quinde, sauce, saucito, susacá **Habit:** Small tree, Tree **Use:** AF,ME **Origin:** Native **Region:** Amazonia, Andean, Guiana Shield, Orinoquia, Pacific **Dept.:** AMA, ANT, ARA, BOY, CAL, CAQ, CAS, CAU, CHO, CUN, GUA, GUV, HUI, MET, NAR, NSA, PUT, SAN, TOL, VAC, VAU **Elev.:** 12-2700 m.a.s.l. **IUCN Habitat code:** 1,2,3,4,5,14 **Conservation status: (IUCN)** LC **(CPSG)** Potential LC

2433 *Plagiocheilus bogotensis* Wedd. (Asteraceae=Compositae) **IPNI:** 238661-1 **Common name:** arracachuela **Habit:** Herb **Use:** ME **Origin:** Native **Region:** Andean **Dept.:** BOG, BOY, CAU, CUN, NAR, PUT, RIS **Elev.:** 2200–3750 m a.s.l. **Conservation status: (IUCN)** NE

2434 *Pluchea carolinensis* (Jacq.) G.Don (Asteraceae=Compositae) **IPNI:** 238803-1 **Use:** ME **Origin:** Native **IUCN Habitat code:** 3 **Conservation status: (IUCN)** LC

2435 *Pluchea foetida* (L.) DC. (Asteraceae=Compositae) **IPNI:** 238826-1 **Common name:** venadillo **Use:** ME **IUCN Habitat code:** 1,3,4,8,14 **Conservation status: (IUCN)** NE

2436 *Pluchea odorata* (L.) Cass. (Asteraceae=Compositae) **IPNI:** 1099254-2 **Habit:** Herb, Subshrub **Use:** MA, ME **Origin:** Native **Region:** Andean, Caribbean **Dept.:** ATL, BOY, CHO, COR, LAG, MAG **Elev.:** 20–1300 m a.s.l. **IUCN Habitat code:** 2,5,14 **Conservation status: (IUCN)** NE

2437 *Pluchea sagittalis* (Lam.) Cabrera (Asteraceae=Compositae) **IPNI:** 203916-2 **Habit:** Herb **Use:** MA, ME **Origin:** Native **Region:** Amazonia, Andean **Dept.:** GUV, MET, VAC **Elev.:** 200–1050 m a.s.l. **IUCN Habitat code:** 3,14 **Conservation status: (IUCN)** NE

2438 *Podachaenium eminens* (Lag.) Sch.Bip. (Asteraceae=Compositae) **IPNI:** 238923-1 **Use:** EU **Origin:** Native **Conservation status: (IUCN)** LC

2439 *Porophyllum ruderale* (Jacq.) Cass. (Asteraceae=Compositae) **IPNI:** 208446-2 **Common name:** chiata, chipaca, chivata, chucha, flor de muerto, gallinaza, gallinaza de azotea, gallinazo, hierba de chulo, hierba de gallinazo, hierbaechulo, maturranga, purranga, ruda, ruda de gallina, ruda de gallinazo, ruda gallinaza, venadillo, verdolaga, yerba de **Habit:** Herb, Subshrub **Use:** HF, MA, ME, SU **Origin:** Native **Region:** Andean, Caribbean, Orinoquia, Pacific **Dept.:** ANT, ATL, BOL, BOY, CAL, CAU, CES, CHO, CUN, HUI, LAG, MAG, MET, NAR, NSA, QUI, SAN, TOL, VAC **Elev.:** 0-2000 m.a.s.l. **IUCN Habitat code:** 1,2,3,5,14 **Conservation status: (IUCN)** NE **(CPSG)** Potential LC

2440 *Pseudelephantopus spicatus* (Juss. ex Aubl.) C.F.Baker (Asteraceae=Compositae) **IPNI:** 239804-1 **Common name:** amargón, amorseco, chicoria, cola de burro, escobilla blanca, herb de caballo, herb de golpe, herb de marrano, hierba de golpe, hierba de marrano, hierba golpe, rabo de puerco, sanfrancisco, suelda, suelda-consuelda, sueldaconsuelda, yerba de caballo, yer **Habit:** Herb **Use:** ME **Origin:** Native **Region:** Amazonia, Andean, Orinoquia, Pacific **Dept.:** AMA, ANT, BOY, CAQ, CAS, CAU, CHO, CUN, GUV, MAG, MET, NAR, NSA, PUT, QUI, RIS, SAN, TOL, VAC, VID **Elev.:** 50-2600 m.a.s.l. **IUCN Habitat code:** 1,2,3,4,5,14 **Conservation status: (IUCN)** NE **(CPSG)** Potential LC

2441 *Pseudelephantopus spiralis* (Less.) Cronquist (Asteraceae=Compositae) **IPNI:** 211400-2 **Common name:** amargón, amorseco, cadillo, chicoria, consuelda, escoba, escoba de San Antonio, escobilla real, falsa verbena, hierba de marrano, hierba golpe, oreja de mula, suelda, suelda de sabana, suelda de tierra, suelda marranera, sueldaconsuelda, totumo, tutuma, y **Habit:** Shrub **Use:** ME **Origin:** Native **Region:** Andean, Caribbean, Orinoquia, Pacific **Dept.:** AMA, ANT, BOY, CAL, CAS, CAU, CHO, CUN, HUI, LAG, MET, NAR, PUT, QUI, SAN, TOL, VAC **Elev.:** 0-2700 m.a.s.l. **IUCN Habitat code:** 1,2,3,4,5,14 **Conservation status: (IUCN)** NE **(CPSG)** Potential LC

2442 *Pseudoconyza viscosa* (Mill.) D'Arcy (Asteraceae=Compositae) **IPNI:** 211317-2 **Habit:** Herb **Use:** EU, HF, MA, ME **Origin:** Native **Region:** Caribbean **Dept.:** ATL, BOL, BOY, MAG **Elev.:** 10–200 m a.s.l. **Conservation status: (IUCN)** NE

2443 *Pseudognaphalium cheiranthifolium* (Lam.) Hilliard & B.L.Burtt (Asteraceae=Compositae) **IPNI:** 909620-1 **Common name:** killi-killi, mull kketo-kketo, vira-vira, viravira **Habit:** Herb **Use:** HF, ME **Origin:** Native **Region:** Andean **Dept.:** BOG, CAL, CUN, MAG, NSA **Elev.:** 2500–4200 m a.s.l. **IUCN Habitat code:** 1,3,4 **Conservation status: (IUCN)** NE

2444 *Pseudognaphalium cymatoides* (DC.) Anderb. (Asteraceae=Compositae) **IPNI:** 283846-2 **Common name:** viravira **Use:** ME **Dept.:** BOG**IUCN Habitat code:** 1,4,14 **Conservation status: (IUCN)** NE

2445 *Pseudognaphalium dombeyanum* (DC.) Anderb. (Asteraceae=Compositae) **IPNI:** 283847-2 **Habit:** Herb **Use:** ME **Origin:** Native **Region:** Andean **Dept.:** ANT, BOY, CAL, CUN, MAG, NAR, NSA, PUT, QUI, TOL, VAC **Elev.:** 3100–3450 m a.s.l. **IUCN Habitat code:** 14 **Conservation status: (IUCN)** NE

2446 *Pseudognaphalium dysodes* (Spreng.) S.E.Freire, N.Bayón & C.Monti (Asteraceae=Compositae) **IPNI:** 77142354-1 **Common name:** suso, vira, vira-vira **Use:** ME **Dept.:** BOG**IUCN Habitat code:** 1,4,14 **Conservation status: (IUCN)** NE

2447 *Pseudognaphalium elegans* Kartesz (Asteraceae=Compositae) **IPNI:** 1039426-2 **Common name:** don alonso, donalonso, vira-vira, viravira **Habit:** Herb **Use:** ME **Origin:** Native **Region:** Andean **Dept.:** ANT, BOY, CAU, CUN, MAG, NSA, TOL, VAC **Elev.:** 480–3900 m a.s.l. **IUCN Habitat code:** 3,14 **Conservation status: (IUCN)** NE

2448 *Pseudognaphalium gaudichaudianum* (DC.) Anderb. (Asteraceae=Compositae) **IPNI:** 283850-2 **Common name:** vira, vira-vira **Habit:** Herb **Use:** ME **Origin:** Native **Region:** Andean **Dept.:** BOG, CAU, CUN, TOL **Elev.:** 2650–2927 m a.s.l. **IUCN Habitat code:** 1,3,4,8,14 **Conservation status: (IUCN)** NE

2449 *Pseudognaphalium lanuginosum* (Kunth) Anderb. (Asteraceae=Compositae) **IPNI:** 963030-1 **Habit:** Herb **Use:** ME **Origin:** Native **Region:** Andean **Dept.:** BOG, BOY, CUN, MAG, RIS, SAN, TOL **Elev.:** 2600–4500 m a.s.l. **IUCN Habitat code:** 1,4,14 **Conservation status: (IUCN)** NE

2450 *Pseudognaphalium meridanum* (Aristeg.) Anderb. (Asteraceae=Compositae) **IPNI:** 283867-2 **Habit:** Herb **Use:** ME **Origin:** Native **Region:** Andean **Dept.:** BOG, BOY, CUN, LAG, MAG **Elev.:** 2560–4500 m a.s.l. **IUCN Habitat code:** 1,4,14 **Conservation status: (IUCN)** NE

2451 *Pseudognaphalium obtusifolium* (L.) Hilliard & B.L.Burtt (Asteraceae=Compositae) **IPNI:** 909626-1 **Habit:** Herb **Use:** ME **Origin:** Naturalised **Region:** Andean **Dept.:** BOG, CUN **Elev.:** 1500–2900 m a.s.l. **IUCN Habitat code:** 4,14 **Conservation status: (IUCN)** NE

2452 *Pseudognaphalium paramorum* (S.F.Blake) M.O.Dillon (Asteraceae=Compositae) **IPNI:** 77154313-1 **Habit:** Herb **Use:** ME **Origin:** Native **Region:** Andean **Dept.:** ANT, BOG, BOY, CUN, LAG, MAG, MET, SAN **Elev.:** 2800–3900 m

a.s.l. **IUCN Habitat code:** 1,3,4,14 **Conservation status: (IUCN)** NE

2453 *Pseudognaphalium pellitum* (Kunth) Anderb. (Asteraceae=Compositae) **IPNI:** 963040-1 **Habit:** Herb **Use:** ME **Origin:** Endemic **Region:** Andean **Dept.:** BOY, CUN, VAC **Elev.:** 2700–3950 m a.s.l. **IUCN Habitat code:** 4 **Conservation status: (IUCN)** NE

2454 *Pseudognaphalium viravira* (Molina) Anderb. (Asteraceae=Compositae) **IPNI:** 283893-2 **Use:** ME **Region:** Andean **Dept.:** BOGIUCN **Habitat code:** 3 **Conservation status: (IUCN)** NE

2455 *Pseudognaphalium viscosum* (Kunth) Anderb. (Asteraceae=Compositae) **IPNI:** 283895-2 **Common name:** kamomy, vira vira, viravira **Habit:** Herb **Use:** ME **Origin:** Native **Region:** Andean **Dept.:** CUN **Elev.:** 2500–3500 m a.s.l. **IUCN Habitat code:** 14 **Conservation status: (IUCN)** NE

2456 *Pseudogynoxys bogotensis* (Spreng.) Cuatrec. (Asteraceae=Compositae) **IPNI:** 211359-2 **Common name:** agualibia, agualivia, amapola del camino, patá **Habit:** Herb, Climbing **Use:** ME **Origin:** Native **Region:** Andean **Dept.:** ANT, BOY, CAU, CUN, SAN, TOL **Elev.:** 700–1800 m a.s.l. **IUCN Habitat code:** 1,3,14 **Conservation status: (IUCN)** NE

2457 *Pseudogynoxys chenopodioides* (Kunth) Cabrera (Asteraceae=Compositae) **IPNI:** 211361-2 **Habit:** Herb, Climbing **Use:** EU, ME **Origin:** Native **Region:** Andean, Caribbean **Dept.:** ANT, MAG, NAR, TOL **Elev.:** 10–2000 m a.s.l. **IUCN Habitat code:** 1,3,14 **Conservation status: (IUCN)** NE

2458 *Pseudogynoxys cummingii* (Benth. ex Oerst.) H.Rob. & Cuatrec. (Asteraceae=Compositae) **IPNI:** 211365-2 **Habit:** Climbing **Use:** EU, ME **Origin:** Native **Region:** Andean **Dept.:** MAG **Elev.:** 600–600 m a.s.l. **Conservation status: (IUCN)** NE

2459 *Pterocaulon alopecuroides* (Lam.) DC. (Asteraceae=Compositae) **IPNI:** 240147-1 **Common name:** pomo, venadillo grande **Habit:** Herb **Use:** ME **Origin:** Native **Region:** Andean, Guiana Shield **Dept.:** ANT, BOG, BOY, CAQ, CUN, MET, RIS, TOL **Elev.:** 280–2620 m a.s.l. **IUCN Habitat code:** 1,3,14 **Conservation status: (IUCN)** NE

2460 *Pterocaulon virgatum* (L.) DC. (Asteraceae=Compositae) **IPNI:** 30022662-2 **Common name:** lechuguilla, peralonso, santaura, tusa, venadillo grande, viravira **Habit:** Herb **Use:** ME **Origin:** Native **Region:** Andean **Dept.:** ANT, MET, TOL, VAC **Elev.:** 740–1200 m a.s.l. **IUCN Habitat code:** 1,3,4,6,14 **Conservation status: (IUCN)** NE

2461 *Rhodanthe chlorocephala* (Turcz.) Paul G.Wilson (Asteraceae=Compositae) **IPNI:** 969106-1 **Habit:** Herb **Use:** EU **Origin:** Cultivated, Naturalised **Region:** Andean **Dept.:** BOG, CUN **Elev.:** 2620–2620 m a.s.l. **Conservation status: (IUCN)** NE

2462 *Rhodanthe manglesii* Lindl. (Asteraceae=Compositae) **IPNI:** 241451-1 **Common name:** botón de oro, inmortal, siempreviva, siemprevivas **Habit:** Herb **Use:** EU **Origin:** Cultivated, Naturalised **Region:** Andean **Dept.:** ANT, BOG, CUN **Elev.:** 1540–2620 m a.s.l. **IUCN Habitat code:** 14 **Conservation status: (IUCN)** NE

2463 *Rolandra fruticosa* (L.) Kuntze (Asteraceae=Compositae) **IPNI:** 221674-2 **Habit:** Herb **Use:** ME **Origin:** Native **Region:** Amazonia, Andean, Guiana Shield, Orinoquia, Pacific **Dept.:** AMA, CAS, CHO, CUN, GUA, GUV, MET, VAC, VID **Elev.:** 50–770 m a.s.l. **IUCN Habitat code:** 1,3,4,5,14 **Conservation status: (IUCN)** NE

2464 *Rudbeckia laciniata* L. (Asteraceae=Compositae) **IPNI:** 241716-1 **Habit:** Herb **Use:** EU, HF, ME, PO **Origin:** Naturalised **Region:** Andean **Dept.:** CUN **Elev.:** 1300–1300 m a.s.l. **Conservation status: (IUCN)** NE

2465 *Salmea scandens* (L.) DC. (Asteraceae=Compositae) **IPNI:** 241903-1 **Common name:** barbasco, duermeboca, saltafuera **Habit:** Shrub, Subshrub **Use:** ME **Origin:** Native **Region:** Amazonia, Andean, Caribbean **Dept.:** ATL, BOL, MAG, MET **Elev.:** 10–610 m a.s.l. **IUCN Habitat code:** 1 **Conservation status: (IUCN)** NE

2466 *Santolina chamaecyparissus* L. (Asteraceae=Compositae) **IPNI:** 241927-1 **Common name:** pino brasilero **Habit:** Herb **Use:** EU, HF, ME **Origin:** Cultivated, Naturalised **Region:** Andean **Dept.:** CUN **Elev.:** 2500–3000 m a.s.l. **IUCN Habitat code:** 3,14 **Conservation status: (IUCN)** NE

2467 *Schkuhria pinnata* (Lam.) Kuntze ex Thell. (Asteraceae=Compositae) **IPNI:** 242907-1 **Habit:** Herb **Use:** AF, HF, MA, ME, SU **Origin:** Native **Region:** Andean **Dept.:** BOG, CUN, NAR **Elev.:** 2650–3300 m a.s.l. **IUCN Habitat code:** 14 **Conservation status: (IUCN)** NE

2468 *Sclerocarpus africanus* Jacq. (Asteraceae=Compositae) **IPNI:** 242967-1 **Habit:** Herb **Use:** HF, ME **Origin:** Naturalised **Dept.:** MAG **Elev.:** 200–200 m a.s.l. **IUCN Habitat code:** 2 **Conservation status: (IUCN)** NE

2469 *Sclerocarpus baranguillae* (Spreng.) S.F.Blake (Asteraceae=Compositae) **IPNI:** 230784-2 **Use:** ME **Origin:** Native **IUCN Habitat code:** 2 **Conservation status: (IUCN)** NE

2470 *Sclerocarpus divaricatus* (Benth.) Benth. & Hook.f. ex Hemsl. (Asteraceae=Compositae) **IPNI:** 242973-1 **Common name:** mapola **Habit:** Herb **Use:** ME **Origin:** Native **Region:** Caribbean **Dept.:** ATL, COR, MAG **Elev.:** 50–100 m a.s.l. **IUCN Habitat code:** 2,14 **Conservation status: (IUCN)** NE

2471 *Senecio formosoides* Cuatrec. (Asteraceae=Compositae) **IPNI:** 233107-2 **Common name:** árnica **Habit:** Herb **Use:** ME **Origin:** Native **Region:** Andean **Dept.:** ARA, BOG, BOY, CAU, CES, CUN, LAG, MET, NSA, SAN **Elev.:** 3000–4425 m a.s.l. **IUCN Habitat code:** 1,3,4,14 **Conservation status: (IUCN)** NE

2472 *Senecio formosus* Kunth (Asteraceae=Compositae) **IPNI:** 245177-1 **Common name:** extraña de páramo, falsa árnica, suelda consuelda, sueldaconsuelda, tabacote, tabacón, tabaquillo, árnica, árnica colombiana, árnica de Bogotá, árnica de bogotá, árnica de flor morada, árnica de la tierra, árnica de monte, árnica de párampo, árnica de tie **Habit:** Herb, Subshrub **Use:** ME, PO **Origin:** Native **Region:** Andean **Dept.:** ANT, ARA, BOG, BOY, CAL, CAU, CES, CUN, HUI, LAG, MAG, MET, NAR, NSA, QUI, SAN, TOL, VAC **Elev.:** 1025–4500 m a.s.l. **IUCN Habitat code:** 1,3,4,6,8,14 **Conservation status: (IUCN)** NE

2473 *Senecio garcibarrigae* Cuatrec. (Asteraceae=Compositae) **IPNI:** 233136-2 **Common name:** árnica **Habit:** Herb **Use:** ME **Origin:** Endemic **Region:** Andean **Dept.:** BOG, BOY, CUN, MET **Elev.:** 1500–4000 m a.s.l. **IUCN Habitat code:** 1,3,4,14 **Conservation status: (IUCN)** NE

2474 *Senecio macroglossus* DC. (Asteraceae=Compositae) **IPNI:** 246196-1 **Habit:** Herb **Use:** EU, ME **Origin:** Cultivated, Naturalised **Region:** Andean **Dept.:** CUN **Elev.:** 2600–2600 m a.s.l. **IUCN Habitat code:** 14 **Conservation status: (IUCN)** NE

2475 *Senecio madagascariensis* Poir. (Asteraceae=Compositae) **IPNI:** 246209-1 **Habit:** Herb **Use:** HF, PO **Origin:** Naturalised **Region:** Andean **Dept.:** BOG, BOY, CUN **Elev.:** 2050–2880 m a.s.l. **IUCN Habitat code:** 1,3,4,5,14 **Conservation status: (IUCN)** NE

2476 *Senecio niveoaureus* Cuatrec. (Asteraceae=Compositae) **IPNI:** 233596-2 **Common name:** blanquizco, rurruco, árnica blanca, árnica ceniza **Habit:** Herb **Use:** EU, ME **Origin:** Endemic **Region:** Andean **Dept.:** ANT, ARA, BOG, BOY, CAL, CAU, CUN, HUI, MET, NSA, SAN, TOL, VAC **Elev.:** 3000–4600 m a.s.l. **IUCN Habitat code:** 1,3,4,6,8,14 **Conservation status: (IUCN)** NE

2477 *Senecio tamoides* DC. (Asteraceae=Compositae) **IPNI:** 247873-1 **Use:** ME **Origin:** Naturalised **Dept.:** BOGIUCN **Habitat code:** 14 **Conservation status: (IUCN)** NE

2478 *Senecio vulgaris* L. (Asteraceae=Compositae) **IPNI:** 248251-1 **Common name:** cardo, cerrajilla, flor de paraguas, yuyito **Habit:** Herb **Use:** HF, ME, PO **Origin:** Naturalised **Region:** Andean **Dept.:** BOG, BOY, CUN, NAR, PUT, SAN, TOL **Elev.:** 2100–3500 m a.s.l. **IUCN Habitat code:** 1,3,4,5,14 **Conservation status: (IUCN)** NE

2479 *Sigesbeckia jorullensis* Kunth (Asteraceae=Compositae) **IPNI:** 249077-1 **Common name:** pacunga, pausa, pegajosa, pegapega, trompetilla **Habit:** Herb, Shrub, Climbing **Use:** ME **Origin:** Native **Region:** Andean **Dept.:** ANT, BOG, BOY, CAU, CES, CUN, HUI, LAG, MAG, MET, NAR, PUT, QUI, RIS, SAN, TOL, VAC **Elev.:** 1025-3860 m.a.s.l. **IUCN Habitat code:** 1,3,4,14 **Conservation status: (IUCN)** NE **(CPSG)** Potential LC

2480 *Sigesbeckia orientalis* L. (Asteraceae=Compositae) **IPNI:** 249084-1 **Common name:** escorzonera, pacunga, pausa, trompetilla **Habit:** Shrub, Small tree **Use:** HF, MA, ME, PO **Origin:** Naturalised **Region:** Andean **Dept.:** BOG, CUN **Elev.:** 3000–3500 m a.s.l. **Conservation status: (IUCN)** NE

2481 *Silybum marianum* (L.) Gaertn. (Asteraceae=Compositae) **IPNI:** 249211-1 **Common name:** cardo blanco, cardo de María, cardo mariano, cardosanto, poma **Habit:** Herb **Use:** HF, IF, ME, PO **Origin:** Naturalised **Region:** Andean **Dept.:** BOG, CUN, NAR, SAN **Elev.:** 2500–4000 m a.s.l. **IUCN Habitat code:** 3,14 **Conservation status: (IUCN)** LC

2482 *Smallanthus cocuyensis* (Cuatrec.) Vitali (Asteraceae=Compositae) **IPNI:** 77137314-1 **Common name:** treeoco **Habit:** Tree **Use:** ME **Origin:** Endemic **Region:** Andean **Dept.:** BOY **Elev.:** 2750-2750 m.a.s.l. **Conservation status: (IUCN)** LC **(CPSG)** LC

2483 *Smallanthus fruticosus* (Benth.) H.Rob. (Asteraceae=Compositae) **IPNI:** 237495-2 **Habit:** Small tree, Tree **Use:** HF, ME **Origin:** Native **Region:** Andean **Dept.:** CUN, NAR **Elev.:** 3000–3200 m a.s.l. **IUCN Habitat code:** 4,14 **Conservation status: (IUCN)** NE

2484 *Smallanthus pyramidalis* (Triana) H.Rob. (Asteraceae=Compositae) **IPNI:** 237506-2 **Common name:** anime, arboloco, camargo, colla, escorzonera, jiquimillo, pauche, treeoco **Habit:** Small tree, Tree **Use:** EU, HF, MA, ME **Origin:** Native **Region:** Andean **Dept.:** ANT, BOG, BOY, CAU, CUN, NAR, NSA, PUT, QUI, TOL, VAC **Elev.:** 1700–3500 m a.s.l. **IUCN Habitat code:** 1,2,3,5,8,14 **Conservation status: (IUCN)** NE

2485 *Smallanthus riparius* (Kunth) H.Rob. (Asteraceae=Compositae) **IPNI:** 237508-2 **Common name:** arboloco, camargo, cumaná, gamboa, jiquimilla, pauche **Habit:** Herb, Shrub **Use:** ME **Origin:** Native **Region:** Andean **Dept.:** ANT, BOG, BOY, CAL, CAU, CUN, HUI, NAR, NSA, SAN, TOL, VAC **Elev.:** 100–2900 m a.s.l. **IUCN Habitat code:** 1,2,3,14 **Conservation status: (IUCN)** NE

2486 *Smallanthus sonchifolius* (Poepp.) H.Rob. (Asteraceae=Compositae) **IPNI:** 237510-2 **Common name:** arboloco, colla, escorzonera, jiquimilla, jícama, jíquima, pipilanga, yacón **Habit:** Shrub, Small tree **Use:** HF, ME **Origin:** Native **Region:** Andean **Dept.:** BOG, BOY, CAU, CUN **Elev.:** 1700–2700 m a.s.l. **IUCN Habitat code:** 3,4,8,14 **Conservation status: (IUCN)** NE

2487 *Solidago chilensis* Meyen (Asteraceae=Compositae) **IPNI:** 249473-1 **Use:** EU, ME **Origin:** Naturalised **IUCN Habitat code:** 3,14 **Conservation status: (IUCN)** NE

2488 *Solidago gigantea* Aiton (Asteraceae=Compositae) **IPNI:** 249575-1 **Habit:** Herb, Subshrub **Use:** EU, HF, IF, ME **Origin:** Naturalised **Region:** Andean **Dept.:** ANT, CHO **Elev.:** 860–1450 m a.s.l. **IUCN Habitat code:** 1,3,14 **Conservation status: (IUCN)** NE

2490 *Soliva anthemifolia* (Juss.) R.Br. (Asteraceae=Compositae) **IPNI:** 249968-1 **Common name:** yerba de las almorranas **Habit:** Herb **Use:** HF, ME **Origin:** Native **Region:** Andean **Dept.:** BOG, BOY, CUN, NAR, VAC **Elev.:** 2600–2700 m a.s.l. **IUCN Habitat code:** 3,14 **Conservation status: (IUCN)** NE

2491 *Sonchus asper* (L.) Hill (Asteraceae=Compositae) **IPNI:** 1003017-2 **Common name:** cerraja lechosa, cerraja áspera **Habit:** Herb **Use:** GS, HF, MA, ME **Origin:** Naturalised **Region:** Andean **Dept.:** ANT, BOG, CUN, NAR, TOL **Elev.:** 1800–4100 m a.s.l. **IUCN Habitat code:** 1,2,3,4,5,8,14 **Conservation status: (IUCN)** NE

2492 *Sonchus oleraceus* L. (Asteraceae=Compositae) **IPNI:** 250268-1 **Common name:** achicoria, canayuyo, cerraja, cerraja espinosa, chicoria, diente de león, escarola, lechosa, lechuga amarilla, lechuguilla, yerba de sapo **Habit:** Herb **Use:** AF, GS, HF, MA, ME **Origin:** Naturalised **Region:** Andean **Dept.:** ANT, BOG, BOY, CAU, CES, CHO, CUN, HUI, MAG, NAR, PUT, QUI, TOL, VAC **Elev.:** 580-3900 m.a.s.l. **Conservation status: (IUCN)** NE **(CPSG)** Potential LC

2493 *Sphagneticola trilobata* (L.) Pruski (Asteraceae=Compositae) **IPNI:** 1093589-2 **Common name:** botoncillo, botoncillo de monte, botón de oro, guaco, pata de lancha **Habit:** Herb **Use:** EU, MA, ME, PO **Origin:** Native **Region:** Amazonia, Andean, Caribbean, Pacific **Dept.:** AMA, ANT, ATL, CAQ, CAU, CHO, COR, GUA, MAG, NAR, QUI, SAP, VAC **Elev.:** 0-1465 m.a.s.l. **IUCN Habitat code:** 1,2,3,4,5,14 **Conservation status: (IUCN)** NE **(CPSG)** Potential LC

2494 *Spilanthes urens* Jacq. (Asteraceae=Compositae) **IPNI:** 250890-1 **Common name:** dormidera **Habit:** Herb **Use:** ME **Origin:** Native **Region:** Caribbean, Orinoquia **Dept.:** ANT, ATL, BOL, LAG, MAG, SAP, SUC, VID **Elev.:** 0–250 m a.s.l. **IUCN Habitat code:** 1,2,3,5,14 **Conservation status: (IUCN)** NE

2495 *Stevia caracasana* DC. (Asteraceae=Compositae) **IPNI:** 251284-1 **Use:** ME **Origin:** Native **IUCN Habitat code:** 3,14 **Conservation status: (IUCN)** NE

2496 *Stevia collina* Gardner (Asteraceae=Compositae) **IPNI:** 251300-1 **Use:** ME **Dept.:** BOY**Conservation status: (IUCN)** NE

2497 *Stevia lucida* Lag. (Asteraceae=Compositae) **IPNI:** 251437-1 **Common name:** chilca, chilca blanca, chupadera, donduada, jarilla, jarilla blanca, jarillo, jarrilla **Habit:** Herb, Shrub, Subshrub **Use:** ME **Origin:** Native **Region:** Andean **Dept.:** BOG, BOY, CAU, CES, CUN, LAG, MAG, NAR, NSA, QUI, SAN, VAC **Elev.:** 250–4900 m a.s.l. **IUCN Habitat code:** 1,3,4,5,14 **Conservation status: (IUCN)** NE

2498 *Stevia ovata* Willd. (Asteraceae=Compositae) **IPNI:** 30284853-2 **Common name:** jarilla jumarunna **Habit:** Herb, Shrub **Use:** ME **Origin:** Native **Region:** Andean **Dept.:** ANT, BOG, BOY, CAU, CUN, LAG, MAG, NSA, SAN **Elev.:** 1600–3900 m a.s.l. **IUCN Habitat code:** 1,3,4,14 **Conservation status: (IUCN)** NE

2499 *Stevia pallida* Hieron. (Asteraceae=Compositae) **IPNI:** 293446-2 **Habit:** Herb **Use:** ME **Origin:** Endemic **Region:** Andean **Dept.:** BOG, CUN, PUT, SAN **Elev.:** 3200–3700 m a.s.l. **IUCN Habitat code:** 1 **Conservation status: (IUCN)** NE

2500 *Stevia rebaudiana* (Bertoni) Bertoni (Asteraceae=Compositae) **IPNI:** 245468-2 **Common name:** estevia **Use:** HF, ME **IUCN Habitat code:** 5 **Conservation status: (IUCN)** NE

2501 *Stevia serrata* Cav. (Asteraceae=Compositae) **IPNI:** 251571-1 **Common name:** chipatoria **Habit:** Herb **Use:** HF, ME **Origin:** Native **Region:** Andean **Dept.:** BOG, BOY, CUN, NAR, NSA **Elev.:** 1800–3700 m a.s.l. **IUCN Habitat code:** 4,14 **Conservation status: (IUCN)** NE

2502 *Stevia suaveolens* Lag. (Asteraceae=Compositae) **IPNI:** 251582-1 **Habit:** Herb **Use:** ME **Origin:** Native **Region:** Andean **Dept.:** MAG, NSA **Elev.:** 760–760 m a.s.l. **Conservation status: (IUCN)** NE

2503 *Struchium sparganophorum* (L.) Kuntze (Asteraceae=Compositae) **IPNI:** 1059151-2 **Common name:** pata de babilla **Habit:** Herb **Use:** AF, EU, HF, MA, ME, SU **Origin:** Native **Region:** Amazonia, Caribbean, Orinoquia, Pacific **Dept.:** AMA, CES, CHO, MET **Elev.:** 5–350 m a.s.l. **IUCN Habitat code:** 1,2,5,14 **Conservation status: (IUCN)** NE

2504 *Symphyotrichum laeve* (L.) Á.Löve & D.Löve (Asteraceae=Compositae) **IPNI:** 247114-2

Habit: Herb **Use:** EU, ME **Origin:** Cultivated, Naturalised **Region:** Andean **Dept.:** ANT, BOG **Elev.:** 500–500 m a.s.l. **IUCN Habitat code:** 1 **Conservation status: (IUCN)** NE

2505 *Symphyotrichum subulatum* (Michx.) G.L.Nesom (Asteraceae=Compositae) **IPNI:** 981909-1 **Habit:** Herb **Use:** ME **Origin:** Cultivated, Naturalised **Region:** Andean **Dept.:** ANT, BOG, CAU, CUN, MAG, NAR, VAC **Elev.:** 1900–2800 m a.s.l. **IUCN Habitat code:** 1,3,14 **Conservation status: (IUCN)** NE

2506 *Synedrella nodiflora* (L.) Gaertn. (Asteraceae=Compositae) **IPNI:** 252024-1 **Common name:** cerbatana, espinillo, floramarilla, venturosa, yerba de gallinaza, yerba de gallinazo, yuyo **Habit:** Herb **Use:** AF, GS, HF, MA, ME, SU **Origin:** Native **Region:** Andean, Pacific **Dept.:** ANT, ARA, CAS, CAU, CHO, CUN, HUI, LAG, MAG, MET, NAR, QUI, SAN, TOL, VAC **Elev.:** 0–1900 m a.s.l. **IUCN Habitat code:** 1,2,3,5,14 **Conservation status: (IUCN)** NE

2507 *Tagetes apetala* Posada-Ar. (Asteraceae=Compositae) **IPNI:** 252069-1 **Common name:** gallinaza, ruda de gallinaza, ruda gallinaza, yerba de gallinazo **Habit:** Herb **Use:** ME, PO **Origin:** Endemic **Region:** Andean **Dept.:** ANT, CAU, SAN **Elev.:** 2350–2400 m a.s.l. **IUCN Habitat code:** 1,14 **Conservation status: (IUCN)** NE

2508 *Tagetes erecta* L. (Asteraceae=Compositae) **IPNI:** 252092-1 **Common name:** amapola, amapola amarilla, amapola de jardín, chinchimalí, clavelina, flor de muerto, flor de muerto o amapola, flora amarilla, floramarilla, hoja de difunto, marigold, maringol, merigol, osa de muerto, rosa, rosa amarilla, rosa amarilla común, rosa de mu **Habit:** Herb, Shrub **Use:** AF, EU, HF, MA, ME, PO, SU **Origin:** Cultivated, Naturalised **Region:** Andean, Orinoquia, Pacific **Dept.:** AMA, ANT, BOG, BOY, CAL, CAQ, CAS, CAU, CHO, CUN, MET, NAR, PUT, RIS, SAN, TOL, VAC **Elev.:** 0-3000 m.a.s.l. **IUCN Habitat code:** 1,2,3,4,14 **Conservation status: (IUCN)** NE **(CPSG)** Potential LC

2509 *Tagetes filifolia* Lag. (Asteraceae=Compositae) **IPNI:** 252097-1 **Common name:** anisillo, anís anís, anís cimarrón, anís del monte, anís silvestre, anís verde, kitha anís, pampa anís, tomillo, tunilla, tunillo **Habit:** Herb **Use:** EU, HF, ME **Origin:** Native **Region:** Andean **Dept.:** ANT, BOG, BOY, CAU, CUN, NAR **Elev.:** 1450–2850 m a.s.l. **IUCN Habitat code:** 1,3,5,14 **Conservation status: (IUCN)** NE

2510 *Tagetes lucida* Cav. (Asteraceae=Compositae) **IPNI:** 252121-1 **Common name:** herb de San Juan, herb de san juan, iyá, jolomocox, pericón **Use:** EU, HF, MA, ME, PO **IUCN Habitat code:** 14 **Conservation status: (IUCN)** NE

2511 *Tagetes micrantha* Cav. (Asteraceae=Compositae) **IPNI:** 252130-1 **Habit:** Herb **Use:** HF, ME **Origin:** Native **Region:** Andean **Dept.:** BOY **Elev.:** 2320–2320 m a.s.l. **IUCN Habitat code:** 14 **Conservation status: (IUCN)** NE

2512 *Tagetes minuta* L. (Asteraceae=Compositae) **IPNI:** 252132-1 **Common name:** ruda gallinaza **Habit:** Herb **Use:** EU, HF, MA, ME, PO **Origin:** Naturalised **Region:** Andean **Dept.:** CAL, NAR **Elev.:** 2455–2455 m a.s.l. **IUCN Habitat code:** 1,3,14 **Conservation status: (IUCN)** NE

2513 *Tagetes tenuifolia* Cav. (Asteraceae=Compositae) **IPNI:** 252177-1 **Common name:** ruda de arado **Habit:** Herb **Use:** EU, HF, ME **Origin:** Native **Region:** Andean, Caribbean **Dept.:** BOY, MAG **Elev.:** 300–2130 m a.s.l. **Conservation status: (IUCN)** NE

2514 *Tagetes terniflora* Kunth (Asteraceae=Compositae) **IPNI:** 60461729-2 **Common name:** amayuyo, clavelina, flor de muerto, ruda, ruda de Castilla, ruda de Quito, ruda de castilla, ruda dulce **Habit:** Herb **Use:** HF, ME **Origin:** Native **Region:** Andean **Dept.:** ANT, BOG, BOY, CUN, HUI, RIS **Elev.:** 1400–2600 m a.s.l. **IUCN Habitat code:** 1,3,14 **Conservation status: (IUCN)** NE

2515 *Tagetes verticillata* Lag. & Rodr. (Asteraceae=Compositae) **IPNI:** 252181-1 **Common name:** amapola, asnayuyo, flor de muerto, gallinazo común, ruda, ruda de arao, ruda gallinaza, ruda silvestre **Habit:** Herb **Use:** ME **Origin:** Native **Region:** Andean **Dept.:** ANT, BOY, CAS, CUN, HUI, MAG, NSA, RIS, SAN, TOL, VAC **Elev.:** 900–2300 m a.s.l. **IUCN Habitat code:** 1,2,3,14 **Conservation status: (IUCN)** NE

2516 *Tagetes zypaquirensis* Bonpl. (Asteraceae=Compositae) **IPNI:** 252183-1 **Common name:** chinchimalí, ruda **Habit:** Herb **Use:** EU, MA, ME **Origin:** Native **Region:** Andean **Dept.:** BOG, BOY, CUN, NAR **Elev.:** 2600–3750 m a.s.l. **IUCN Habitat code:** 4,5,14 **Conservation status: (IUCN)** NE

2517 *Tanacetum cinerariifolium* (Trevis.) Sch.Bip. (Asteraceae=Compositae) **IPNI:** 252275-1 **Common name:** pelitre, piletre, piretro **Habit:** Subshrub **Use:** EU, MA, ME, PO **Origin:** Cultivated, Naturalised **Region:** Andean **Dept.:** NAR **Elev.:** 2820–2820 m a.s.l. **Conservation status: (IUCN)** LC

2518 *Tanacetum parthenium* (L.) Sch.Bip. (Asteraceae=Compositae) **IPNI:** 252460-1 **Common name:** altamisa, manzanilla, manzanilla grande, manzanilla matricaria, manzanillón, matricaria, matricaria doble **Habit:** Herb **Use:** EU, HF, MA, ME **Origin:** Naturalised **Region:** Andean **Dept.:** ANT, BOG, BOY, CAL, CAU, CES, CUN, MAG, NAR, SAN **Elev.:** 1280–3300 m a.s.l. **IUCN Habitat code:** 1,2,3,4,14 **Conservation status: (IUCN)** NE

2519 *Tanacetum vulgare* L. (Asteraceae=Compositae) **IPNI:** 252568-1 **Use:** EU, HF, MA, ME, PO **Origin:** Naturalised **IUCN Habitat code:** 14 **Conservation status: (IUCN)** NE

2520 *Taraxacum officinale* F.H.Wigg. (Asteraceae=Compositae) **IPNI:** 1003018-2 **Common name:** achicoria, amargón, chicoria, chicoria amarilla, colmillo de león, diente de león **Habit:** Herb **Use:** AF, HF, IF, ME **Origin:** Endemic **Region:** Andean **Dept.:** ANT, BOG, BOY, CAL, CUN, HUI, NAR, PUT, QUI, RIS, SAN, TOL, VAC **Elev.:** 1100–4300 m a.s.l. **IUCN Habitat code:** 1,2,3,4,8,14 **Conservation status: (IUCN)** NE

2521 *Tessaria integrifolia* Ruiz & Pav. (Asteraceae=Compositae) **IPNI:** 30283369-2 **Common name:** aliso, aliso calentano, aliso de tierra caliente, aliso playero, barredera, cenizo, cenizo playero, erizo, sauce, sauce blanco, sauce bobo, sauce de playa, sauce playero **Habit:** Shrub, Small tree, Tree **Use:** HF, MA, ME **Origin:** Native **Region:** Amazonia, Andean, Caribbean, Orinoquia **Dept.:** AMA, ANT, ARA, ATL, BOL, BOY, CAL, CAQ, CAS, CAU, CHO, COR, CUN, GUV, HUI, LAG, MAG, MET, NSA, PUT, QUI, SAN, TOL, VAC, VID **Elev.:** 0-2050 m.a.s.l. **IUCN Habitat code:** 1,2,3,5,14 **Conservation status: (IUCN)** LC **(CPSG)** Potential LC

2522 *Tilesia baccata* (L.) Pruski (Asteraceae=Compositae) **IPNI:** 993359-1 **Common name:** newutjuebarunae'r, tree lengua de juagar **Habit:** Herb, Shrub, Liana, Climbing **Use:** HF, MA, ME **Origin:** Native **Region:** Amazonia, Andean, Orinoquia, Pacific **Dept.:** AMA, ANT, BOG, BOL, BOY, CAL, CAQ, CAS, CAU, CHO, CUN, HUI, LAG, MAG, MET, NAR, PUT, QUI, RIS, SAN, TOL, VAC, VID **Elev.:** 0–2650 m a.s.l. **IUCN Habitat code:** 1,2,3,4,5,14 **Conservation status: (IUCN)** NE

2523 *Tithonia diversifolia* (Hemsl.) A.Gray (Asteraceae=Compositae) **IPNI:** 255747-1 **Common name:** amargón, botondioro, botón de oro, girasol, mano de Dios, margaritón, mirasol, yerba de bruja, árnica **Habit:** Shrub **Use:** AF, EU, FU, HF, IF, MA, ME **Origin:** Cultivated, Naturalised **Region:** Andean, Caribbean **Dept.:** ANT, BOG, BOL, BOY, CAQ, CAU, CHO, CUN, GUV, NAR, PUT, QUI, SAN, TOL, VAC **Elev.:** 100–2000 m a.s.l. **IUCN Habitat code:** 1,2,3,4,5,14 **Conservation status: (IUCN)** NE

2524 *Tithonia rotundifolia* (Mill.) S.F.Blake (Asteraceae=Compositae) **IPNI:** 254724-2 **Habit:** Herb **Use:** EU, IF, ME **Origin:** Cultivated, Naturalised **Dept.:** CAU, VAC **Elev.:** 1050–1050 m a.s.l. **IUCN Habitat code:** 1,2,3,14 **Conservation status: (IUCN)** NE

2525 *Tragopogon dubius* Scop. (Asteraceae=Compositae) **IPNI:** 255993-1 **Use:** ME **Region:** Andean **Dept.:** BOG**Conservation status: (IUCN)** NE

2526 *Tragopogon porrifolius* L. (Asteraceae=Compositae) **IPNI:** 256109-1 **Use:** EU, HF, ME **Region:** Andean **Dept.:** BOG**IUCN Habitat code:** 14 **Conservation status: (IUCN)** NE

2527 *Tridax procumbens* L. (Asteraceae=Compositae) **IPNI:** 257735-2 **Common name:** botoncillo, cadillo chisacá, cadillo de chipaca, chipaca, chisacá **Habit:** Herb **Use:** AF, EU, GS, HF, MA, ME **Origin:** Native **Region:** Andean, Caribbean, Orinoquia, Pacific **Dept.:** ANT, ATL, BOL, BOY, CAL, CAU, CHO, COR, CUN, HUI, LAG, MAG, NAR, NSA, QUI, SAP, SAN, TOL, VAC **Elev.:** 5-2000 m.a.s.l. **IUCN Habitat code:** 1,2,3,5,14 **Conservation status: (IUCN)** NE **(CPSG)** Potential LC

2528 *Trixis divaricata* (Kunth) Spreng. (Asteraceae=Compositae) **IPNI:** 1207471-2 **Common name:** chucha, palo de santamaría **Habit:** Herb **Use:** ME **Origin:** Native **Region:** Andean **Dept.:** ANT, CAL, CUN, MAG, NSA, SAN **Elev.:** 1350–1600 m a.s.l. **IUCN Habitat code:** 3,14 **Conservation status: (IUCN)** NE

2529 *Trixis inula* Crantz (Asteraceae=Compositae) **IPNI:** 256729-1 **Common name:** barelita, chucha, curanina, mallorquín, mapola, varejón, árnica de monte **Habit:** Herb, Shrub, Climbing **Use:** ME **Origin:** Native **Region:** Andean, Caribbean **Dept.:** ANT, ATL, BOG, BOL, BOY, CAL, CES, COR, CUN, HUI, LAG, MAG, NSA, SAN, TOL **Elev.:** 100-2130 m.a.s.l. **IUCN Habitat code:** 1,2,3,5,14 **Conservation status: (IUCN)** LC **(CPSG)** Potential LC

2530 *Unxia camphorata* L.f. (Asteraceae=Compositae) **IPNI:** 257035-1 **Habit:** Herb **Use:** HF, MA, ME **Origin:** Native **Region:** Andean, Guiana Shield, Orinoquia **Dept.:** BOY, GUA, GUV, MAG, MET, VAU, VID **Elev.:** 60–1170 m a.s.l. **IUCN Habitat code:** 1,3,4,14 **Conservation status: (IUCN)** NE

2531 *Verbesina encelioides* (Cav.) Benth. & Hook.f. ex A.Gray (Asteraceae=Compositae) **IPNI:** 30147294-2 **Habit:** Herb **Use:** IF, ME, PO **Origin:** Native **Region:** Andean **Dept.:** CAU, VAC **Elev.:** 1000–1000 m a.s.l. **IUCN Habitat code:** 14 **Conservation status: (IUCN)** NE

2532 *Verbesina turbacensis* Kunth (Asteraceae=Compositae) **IPNI:** 257696-1 **Common name:** cerbatana, manito de Dios, mano de Dios **Habit:** Herb, Subshrub **Use:** ME **Origin:** Native **Region:** Caribbean, Orinoquia **Dept.:** ATL, BOL, COR, MAG, MET, NSA, SAP **Elev.:** 30–230 m a.s.l. **IUCN Habitat code:** 2,3,14 **Conservation status: (IUCN)** LC

2533 *Vernonanthura brasiliana* (L.) H.Rob. (Asteraceae=Compositae) **IPNI:** 969180-1 **Common name:** estoraque, olivón, palo de agua, pebetera, varejón, varote **Habit:** Subshrub **Use:** ME **Origin:** Native **Region:** Andean, Orinoquia **Dept.:** ANT, ARA, CAS, CAU, CUN, HUI, MAG, MET, NAR, TOL, VAC, VID **Elev.:** 170–1700 m a.s.l. **IUCN Habitat code:** 1,2,3,4,14 **Conservation status: (IUCN)** NE

2534 *Vernonanthura patens* (Kunth) H.Rob. (Asteraceae=Compositae) **IPNI:** 969218-1 **Common name:** almoraduz, balsilla, estoraque, indio viejo, indioviejo, lengüevaco, matapaja, olivón, palotal, salvión, tucutuco, varejón, varejón blanco, varejón de Castilla, volantona, yasmande, yasmiande, ñasmande **Habit:** Shrub **Use:** EU, ME **Origin:** Native **Region:** Andean, Caribbean, Orinoquia, Pacific **Dept.:** AMA, ANT, CAL, CAQ, CAS, CAU, CES, CHO, COR, CUN, GUV, HUI, MAG, MET, NAR, PUT, QUI, RIS, SAN, TOL, VAC **Elev.:** 0-2100 m.a.s.l. **IUCN Habitat code:** 1,2,3,4,5,14 **Conservation status: (IUCN)** LC **(CPSG)** Potential LC

2535 *Vernonanthura phosphorica* (Vell.) H.Rob. (Asteraceae=Compositae) **IPNI:** 1038787-2 **Use:** HF, ME **Dept.:** COR**Conservation status: (IUCN)** NE

2536 *Viguiera dentata* (Cav.) Spreng. (Asteraceae=Compositae) **IPNI:** 259782-1 **Habit:** Herb **Use:** HF, ME **Origin:** Naturalised **Region:** Andean **Dept.:** HUI **Elev.:** 1000–1000 m a.s.l. **IUCN Habitat code:** 14 **Conservation status: (IUCN)** NE

2537 *Wedelia acapulcensis* Kunth (Asteraceae=Compositae) **IPNI:** 30439090-2 **Use:** ME **Origin:** Native **IUCN Habitat code:** 3 **Conservation status: (IUCN)** NE

2538 *Wedelia fruticosa* Jacq. (Asteraceae=Compositae) **IPNI:** 60468294-2 **Common name:** botón de oro, brusca sabanera, dosdeoros **Use:** PO **Origin:** Native **IUCN Habitat code:** 1,2,3,4,14 **Conservation status: (IUCN)** NE

2539 *Werneria pygmaea* Gillies ex Hook. & Arn. (Asteraceae=Compositae) **IPNI:** 260433-1 **Common name:** chicuria **Habit:** Herb **Use:** ME **Origin:** Native **Region:** Andean **Dept.:** ARA, BOG, BOY, CAL, CAS, CHO, CUN, LAG, MAG, NAR, NSA, RIS, SAN, TOL, VAC **Elev.:** 3140–4850 m a.s.l. **IUCN Habitat code:** 1,3,4,6,8,14 **Conservation status: (IUCN)** NE

2540 *Xanthium spinosum* L. (Asteraceae=Compositae) **IPNI:** 260892-1 **Common name:** casamarucha **Habit:** Subshrub **Use:** HF, ME, PO **Origin:** Native **Region:** Andean **Dept.:** NAR, VAC **Elev.:** 2000–2560 m a.s.l. **Conservation status: (IUCN)** NE

2541 *Xanthium strumarium* L. (Asteraceae=Compositae) **IPNI:** 260893-1 **Common name:** cadillo, cardo **Habit:** Herb, Shrub **Use:** HF, ME **Origin:** Naturalised **Region:** Andean **Dept.:** ANT, CES, NSA, SAN **Elev.:** 500–1600 m a.s.l. **IUCN Habitat code:** 3,14 **Conservation status: (IUCN)** NE

2542 *Xerochrysum bracteatum* (Vent.) Tzvelev (Asteraceae=Compositae) **IPNI:** 962307-1 **Common name:** botón de oro, botón de plata, botón dorado, inmortal, inmortales, siempreviva **Habit:** Herb **Use:** EU, HF, IF, MA, ME, SU **Origin:** Naturalised **Region:** Andean **Dept.:** ANT, BOG, CUN **Elev.:** 1800–3180 m a.s.l. **IUCN Habitat code:** 3,14 **Conservation status: (IUCN)** NE

2543 *Youngia japonica* (L.) DC. (Asteraceae=Compositae) **IPNI:** 1124129-2 **Habit:** Herb **Use:** HF, ME **Origin:** Naturalised **Region:** Andean **Dept.:** CAU, CES, MAG, QUI, SAN, TOL, VAC **Elev.:** 1200–2300 m a.s.l. **IUCN Habitat code:** 1,3,14 **Conservation status: (IUCN)** NE

2544 *Zinnia elegans* Jacq. (Asteraceae=Compositae) **IPNI:** 261331-1 **Use:** EU, HF, ME **Origin:** Naturalised **IUCN Habitat code:** 1,2,3,14 **Conservation status: (IUCN)** NE

2545 *Zinnia peruviana* (L.) L. (Asteraceae=Compositae) **IPNI:** 261363-1 **Common name:** amapola, buenastardes, caléndula, cartagena, clavellina, clavelón, colegiala, colegiales, colegianta, colegiante, cuarenta días, cuarentana, cuarentena, estable, insignia, insinia, montenegro, perpetua, petronita, portal, rudón, sanfrancisco, sangre de Ad **Habit:** Herb **Use:** AF, EU, ME **Origin:** Native, Cultivated **Region:** Andean **Dept.:** ANT, BOL, BOY, CAQ, CHO, COR, CUN, QUI, SAN, VAC **Elev.:** 100–2130 m a.s.l. **IUCN Habitat code:** 1,14 **Conservation status: (IUCN)** NE

2546 *Burmeistera vulgaris* E.Wimm. (Campanulaceae) **IPNI:** 38509-2 **Habit:** Herb, Shrub **Use:** HF **Origin:** Native **Region:** Andean **Dept.:** CHO, NSA **Elev.:** 1200–3130 m a.s.l. **IUCN Habitat code:** 1,14 **Conservation status: (IUCN)** NE

2547 *Campanula medium* L. (Campanulaceae) **IPNI:** 140670-1 **Common name:** campana, campanillas, campanitas, campanola, campánula, faroles, farolillo, farolito, farolitos **Habit:** Herb **Use:** EU, HF, ME **Origin:** Cultivated **Region:** Andean **Dept.:** BOG, CUN, VAC **Elev.:** 1090–2620 m a.s.l. **IUCN Habitat code:** 1,14 **Conservation status: (IUCN)** NE

2548 *Centropogon coccineus* (Hook.) Regel ex B.D.Jacks. (Campanulaceae) **IPNI:** 77083518-1 **Common name:** zarcillejo **Habit:** Herb **Use:** ME **Origin:** Native **Region:** Andean, Pacific **Dept.:** ANT, CHO **Elev.:** 0–1400 m a.s.l. **IUCN Habitat code:** 3,14 **Conservation status: (IUCN)** NE

2549 *Centropogon cornutus* (L.) Druce (Campanulaceae) **IPNI:** 51553-2 **Common name:** gallina de pobre, gallito rosado, lechero, narizlargo, zarcillejo **Habit:** Herb, Shrub, Subshrub, Climbing **Use:** HF, ME **Origin:** Native **Region:** Amazonia, Andean, Pacific **Dept.:** AMA, ANT, CAL, CAQ, CHO, CUN, HUI, MAG, MET, NAR, NSA, PUT, QUI, SAN, TOL, VAC **Elev.:** 30-2780 m.a.s.l. **IUCN Habitat code:** 1,2,3,5,14 **Conservation status: (IUCN)** NE **(CPSG)** Potential LC

2550 *Centropogon erianthus* (Benth.) Benth. & Hook.f. ex Drake (Campanulaceae) **IPNI:** 141468-1 **Habit:** Herb, Subshrub **Use:** HF **Origin:** Native **Region:** Andean **Dept.:** ANT, BOY, CAL, MAG, NAR **Elev.:** 1700–3100 m a.s.l. **IUCN Habitat code:** 1,3,4,8,14 **Conservation status: (IUCN)** NE

2551 *Centropogon lehmannii* Zahlbr. (Campanulaceae) **IPNI:** 51673-2 **Common name:** chicle, chicle fino, gallo, huevo de gallo, piojito, piojo **Habit:** Herb, Shrub, Climbing **Use:** HF **Origin:** Endemic **Region:** Andean **Dept.:** CAU, CUN, HUI, NAR, QUI, TOL, VAC **Elev.:** 1600–2800 m a.s.l. **IUCN Habitat code:** 1,3,14 **Conservation status: (IUCN)** NE

2552 *Centropogon lianeus* E.Wimm. (Campanulaceae) **IPNI:** 51680-2 **Habit:** Climbing **Use:** HF **Origin:** Endemic **Region:** Andean **Dept.:** ANT

Elev.: 2000–3050 m a.s.l. **IUCN Habitat code:** 1,14 **Conservation status: (IUCN)** NE

2553 *Centropogon solanifolius* Benth. (Campanulaceae)
IPNI: 141634-1
Habit: Herb, Shrub, Climbing **Use:** ME **Origin:** Native **Region:** Andean **Dept.:** ANT, CAL, CHO, CUN, HUI, NAR, NSA, QUI, RIS, TOL, VAC **Elev.:** 900–3200 m a.s.l. **IUCN Habitat code:** 1,3,14 **Conservation status: (IUCN)** NE

2554 *Diastatea micrantha* (Kunth) McVaugh (Campanulaceae)
IPNI: 79151-2
Habit: Shrub **Use:** HF **Origin:** Native **Region:** Andean **Dept.:** ANT, CUN, MAG, SAN **Elev.:** 2600–3724 m a.s.l. **IUCN Habitat code:** 1,2,3,14 **Conservation status: (IUCN)** NE

2555 *Hippobroma longiflora* (L.) G.Don (Campanulaceae)
IPNI: 1133957-2
Common name: azucena, borraja, estrella, estrella polar, estrellita, jazmin chino, jazmin silvestre de estrella, jazmín, jazmín chino, revientacaballos, tabaco macho **Habit:** Herb **Use:** ME, PO **Origin:** Naturalised **Region:** Amazonia, Andean, Guiana Shield, Caribbean, Orinoquia, Pacific **Dept.:** AMA, ANT, BOL, CAL, CHO, CUN, HUI, MAG, MET, NAR, NSA, PUT, QUI, SAP, SAN, TOL, VAC, VAU **Elev.:** 0-2000 m.a.s.l. **IUCN Habitat code:** 1,2,3,5,14 **Conservation status: (IUCN)** NE **(CPSG)** Potential LC

2556 *Lobelia acuminata* Sw. (Campanulaceae)
IPNI: 142908-1
Use: ME **Conservation status: (IUCN)** NE

2557 *Lobelia andrewsii* Lammers (Campanulaceae)
IPNI: 1020602-1
Habit: Herb **Use:** EU **Origin:** Cultivated **Region:** Andean **Dept.:** ANT **Elev.:** 1620–1620 m a.s.l. **Conservation status: (IUCN)** NE

2558 *Lobelia cardinalis* L. (Campanulaceae)
IPNI: 142115-2
Habit: Herb **Use:** EU, HF, ME, PO **Origin:** Native **Region:** Andean **Dept.:** ANT **Elev.:** 2600–2600 m a.s.l. **Conservation status: (IUCN)** LC

2559 *Lobelia erinus* L. (Campanulaceae)
IPNI: 143240-1
Common name: avispitas, zulia
Habit: Herb **Use:** EU, HF, ME **Origin:** Cultivated **Region:** Andean **Dept.:** ANT, BOG, CUN **Elev.:** 2150–2300 m a.s.l. **IUCN Habitat code:** 3,14 **Conservation status: (IUCN)** NE

2560 *Lobelia laxiflora* Kunth (Campanulaceae)
IPNI: 142260-2
Common name: zarcillejo **Habit:** Herb, Shrub **Use:** ME **Origin:** Native **Region:** Andean **Dept.:** CAU, NAR **Elev.:** 1400–2700 m a.s.l. **IUCN Habitat code:** 1,14 **Conservation status: (IUCN)** NE

2561 *Lobelia nana* Kunth (Campanulaceae)
IPNI: 143584-1
Habit: Shrub **Use:** HF **Origin:** Native **Region:** Andean **Dept.:** SAN **Elev.:** 2700–2700 m a.s.l. **IUCN Habitat code:** 1,4 **Conservation status: (IUCN)** NE

2562 *Lobelia tenera* Kunth (Campanulaceae)
IPNI: 143931-1
Common name: avispita, avispitas, palomita, palomitas, teresita, zulia
Habit: Herb **Use:** ME **Origin:** Native **Region:** Andean **Dept.:** ANT, BOG, BOY, CAU, CUN **Elev.:** 1830–3750 m a.s.l. **IUCN Habitat code:** 1,3,4,8,14 **Conservation status: (IUCN)** NE

2563 *Siphocampylus benthamianus* Walp. (Campanulaceae)
IPNI: 145082-1
Habit: Climbing **Use:** ME **Origin:** Endemic **Region:** Andean **Dept.:** ANT, BOG, CAU, CUN, HUI, QUI, TOL, VAC **Elev.:** 1500–3950 m a.s.l. **IUCN Habitat code:** 1,3,4,14 **Conservation status: (IUCN)** NE

2564 *Siphocampylus bogotensis* E.Wimm. (Campanulaceae)
IPNI: 236693-2
Habit: Herb **Use:** ME **Origin:** Endemic **Region:** Andean **Dept.:** BOG, CUN, NSA **Elev.:** 1700–3000 m a.s.l. **IUCN Habitat code:** 1,4,14 **Conservation status: (IUCN)** NE

2565 *Siphocampylus columnae* (L.f.) G.Don (Campanulaceae)
IPNI: 145124-1
Common name: cascabel **Habit:** Shrub **Use:** ME **Origin:** Native **Region:** Andean **Dept.:** ANT, BOG, BOY, CAL, CUN, HUI **Elev.:** 2100–4100 m a.s.l. **IUCN Habitat code:** 1,3,4,8,14 **Conservation status: (IUCN)** NE

2566 *Siphocampylus ellipticus* (Willd. ex Schult.) Vatke (Campanulaceae)
IPNI: 145166-1
Habit: Climbing **Use:** ME **Origin:** Endemic **Region:** Andean **Dept.:** NSA, SAN **Elev.:** 2400–3100 m a.s.l. **IUCN Habitat code:** 1,3,4,14 **Conservation status: (IUCN)** NE

2567 *Siphocampylus giganteus* (Cav.) G.Don (Campanulaceae)
IPNI: 145189-1
Common name: caucho, chicle
Habit: Shrub **Use:** ME **Origin:** Native **Region:** Andean **Dept.:** ANT, CAU, NAR, PUT **Elev.:** 2200–3600 m a.s.l. **IUCN Habitat code:** 1,3,4,8,14 **Conservation status: (IUCN)** NE

2568 *Siphocampylus lucifer* E.Wimm. (Campanulaceae)
IPNI: 236827-2
Habit: Herb, Subshrub, Climbing **Use:** ME **Origin:** Native **Region:** Andean **Dept.:** CAS, CES, CUN, MAG, NSA, SAN **Elev.:** 845–2640 m a.s.l. **IUCN Habitat code:** 1,2,3,14 **Conservation status: (IUCN)** NE

2569 *Siphocampylus mirabilis* E.Wimm. (Campanulaceae)
IPNI: 236841-2
Habit: Subshrub, Climbing **Use:** ME **Origin:** Endemic **Region:** Andean **Dept.:** CAL **Elev.:** 3400–3400 m a.s.l. **IUCN Habitat code:** 1,3,4 **Conservation status: (IUCN)** NE

2570 *Siphocampylus niveus* (Willd. ex Schult.) Vatke (Campanulaceae)
IPNI: 145280 1
Habit: Shrub **Use:** ME **Origin:** Endemic **Region:** Andean **Dept.:** CAU, NAR **Elev.:** 1700–3380 m a.s.l. **IUCN Habitat code:** 1,4,14 **Conservation status: (IUCN)** NE

2571 *Siphocampylus polyphyllus* Planch. (Campanulaceae)
IPNI: 145312-1
Habit: Shrub **Use:** ME **Origin:** Native **Region:** Andean **Dept.:** BOY, CES, CUN, MAG, MET, NSA **Elev.:** 580–3100 m a.s.l. **IUCN Habitat code:** 1,3,14 **Conservation status: (IUCN)** NE

2572 *Siphocampylus retrorsus* (Willd. ex Schult.) Vatke (Campanulaceae)
IPNI: 145335-1
Common name: zarcillejo **Habit:** Epiphyte, Climbing **Use:** ME **Origin:** Endemic **Region:** Andean **Dept.:** ANT, BOG, BOY, CAU, CUN, HUI, PUT, SAN, TOL **Elev.:** 1000–3500 m a.s.l. **IUCN Habitat code:** 1,3,4,14 **Conservation status: (IUCN)** NE

2573 *Triodanis perfoliata* (L.) Nieuwl. (Campanulaceae)
IPNI: 258447-2
Habit: Herb **Use:** ME **Origin:** Native **Region:** Andean **Dept.:** CAU, CUN **Elev.:** 2300–2700 m a.s.l. **IUCN Habitat code:** 14 **Conservation status: (IUCN)** NE

2574 *Nymphoides humboldtiana* (Kunth) Kuntze (Menyanthaceae)
IPNI: 171243-2
Use: ME **Origin:** Native **IUCN Habitat code:** 1,2,3,4,5,14 **Conservation status: (IUCN)** NE

2575 *Nymphoides indica* (L.) Kuntze (Menyanthaceae)
IPNI: 1182548-2
Common name: hoja de raya, loto, loto silvestre, patodiagua, taruya raya **Habit:** Herb, Aquatic **Use:** EU, GS, HF, IF, MA, ME **Origin:** Native **Region:** Andean, Caribbean, Orinoquia, Pacific **Dept.:** ANT, ATL, CAS, CAU, CHO, COR, MET, RIS, VID **Elev.:** 5–1125 m a.s.l. **Conservation status: (IUCN)** LC

Boraginales

2576 *Borago officinalis* L. (Boraginaceae)
IPNI: 113618-1
Common name: borraja, borraja blanca **Habit:** Herb, Subshrub **Use:** EU, GS, HF, IF, MA, ME, PO **Origin:** Cultivated, Naturalised **Region:** Andean **Dept.:** ANT, BOG, BOY, CUN, SAN, TOL, VAC **Elev.:** 1500–3600 m a.s.l. **IUCN Habitat code:** 1,3,4,14 **Conservation status: (IUCN)** NE

2577 *Bourreria mollis* Standl. (Boraginaceae)
IPNI: 35505-2
Habit: Tree **Use:** ME **Origin:** Native **Dept.:** ANT **Elev.:** 60–90 m a.s.l. **IUCN Habitat code:** 3,14 **Conservation status: (IUCN)** NE

2578 *Bourreria succulenta* Jacq. (Boraginaceae)
IPNI: 1088317-2
Habit: Tree **Use:** HF, MA, ME **Origin:** Native **Region:** Caribbean **Dept.:** SAP **Elev.:** 80–80 m a.s.l. **Conservation status: (IUCN)** LC

2579 *Cordia alliodora* (Ruiz & Pav.) Oken (Boraginaceae)
IPNI: 113943-1
Common name: ajo de montaña, ajo-ajo, barcino del amarillo, bálsamo, canalete, canalete de humo, canalete prieto, cedrito, guásimo vegetal, jigua laurel, laurel, laurel blanco, laurel de puná, laurel negro, laurelito, mo, moho, moncro, mu, murrapo,

móncoro, nogal, nog **Habit:** Tree **Use:** AF, EU, HF, IF, MA, ME, SU **Origin:** Native **Region:** Andean, Caribbean, Pacific **Dept.:** ANT, ARA, BOL, BOY, CAL, CES, CHO, CUN, HUI, LAG, MAG, NAR, NSA, PUT, QUI, SAN, SUC, TOL, VAC **Elev.:** 1-1800 m.a.s.l. **IUCN Habitat code:** 1,2,3,4,5,14 **Conservation status: (IUCN)** LC **(CPSG)** Potential LC

2580 *Cordia bicolor* A.DC. (Boraginaceae) **IPNI:** 274895-2 **Common name:** cacho de venado, carbonero, guácimo, guásimo negro, muñeco, palo de agua **Habit:** Small tree, Tree **Use:** HF, ME **Origin:** Native **Region:** Andean, Guiana Shield, Caribbean, Pacific **Dept.:** AMA, ANT, ARA, BOL, CES, CHO, LAG, MET, NSA, VID **Elev.:** 70-1400 m.a.s.l. **IUCN Habitat code:** 1,2,3,4,5,14 **Conservation status: (IUCN)** LC **(CPSG)** Potential LC

2581 *Cordia bogotensis* Benth. (Boraginaceae) **IPNI:** 274898-2 **Common name:** gomo, mano de oso, verde y negro **Habit:** Shrub, Small tree, Tree **Use:** ME **Origin:** Native **Region:** Andean, Pacific **Dept.:** ANT, BOG, CUN, HUI, NSA, QUI, RIS, TOL, VAC **Elev.:** 30-2340 m.a.s.l. **IUCN Habitat code:** 1,2,3,14 **Conservation status: (IUCN)** LC **(CPSG)** LC

2582 *Cordia collococca* L. (Boraginaceae) **IPNI:** 114058-1 **Common name:** arato, caujaro, muñeco **Habit:** Small tree, Tree **Use:** HF, MA, ME **Origin:** Native **Region:** Amazonia, Caribbean, Orinoquia, Pacific **Dept.:** AMA, ANT, ARA, ATL, BOL, BOY, LAG, MAG, NAR, SAP, SAN, TOL **Elev.:** 10-1036 m.a.s.l. **IUCN Habitat code:** 1,2,3,4,5,14 **Conservation status: (IUCN)** LC **(CPSG)** Potential LC

2583 *Cordia cymosa* (Donn.Sm.) Standl. (Boraginaceae) **IPNI:** 65279-2 **Common name:** falso guascayuca, pacocito **Habit:** Shrub, Small tree, Tree **Use:** HF **Origin:** Native **Region:** Andean, Pacific **Dept.:** ANT, CHO, NAR, VAC **Elev.:** 200–1900 m a.s.l. **IUCN Habitat code:** 1,3,14 **Conservation status: (IUCN)** LC

2584 *Cordia dentata* Poir. (Boraginaceae) **IPNI:** 927551-1 **Common name:** biyuyo, cajuaro, canjara, caujaro, cuajaro, gomo, gomo blanco, gomoso, hobito, jobito, koshot, murciélago, saúco, uva, uvero, uvillo, uvita, uvita mocosa, uvita pegajosa, uvito de murciélago, uvito mocoso **Habit:** Shrub, Small tree, Tree **Use:** HF, ME **Origin:** Native **Region:** Caribbean, Orinoquia **Dept.:** ANT, ATL, BOL, CAS, CES, COR, CUN, HUI, LAG, MAG, NSA, SUC, TOL, VAC **Elev.:** 0-1000 m.a.s.l. **IUCN Habitat code:** 1,2,3,4,5,14 **Conservation status: (IUCN)** LC **(CPSG)** Potential LC

2585 *Cordia diversifolia* Pav. ex A.DC. (Boraginaceae) **IPNI:** 274931-2 **Habit:** Shrub **Use:** HF **Origin:** Native **Dept.:** SAN **Elev.:** 300–300 m a.s.l. **IUCN Habitat code:** 3 **Conservation status: (IUCN)** LC

2586 *Cordia exaltata* Lam. (Boraginaceae) **IPNI:** 114135-1 **Common name:** anoncillo **Habit:** Small tree, Tree **Use:** MA **Origin:** Native **Region:** Amazonia **Dept.:** AMA, CAQ, GUA **Elev.:** 80–80 m a.s.l. **IUCN Habitat code:** 1,3,4,14 **Conservation status: (IUCN)** LC

2587 *Cordia gerascanthus* L. (Boraginaceae) **IPNI:** 65321-2 **Use:** HF, MA, ME **IUCN Habitat code:** 1,2,3,5,14 **Conservation status: (IUCN)** NE

2588 *Cordia kingstoniana* J.S.Mill. (Boraginaceae) **IPNI:** 322119-2 **Habit:** Small tree **Use:** ME **Origin:** Native **Region:** Andean **Dept.:** VAC **Elev.:** 1500–1500 m a.s.l. **IUCN Habitat code:** 1,3,14 **Conservation status: (IUCN)** LC

2589 *Cordia lasiocalyx* Pittier (Boraginaceae) **IPNI:** 274981-2 **Habit:** Shrub **Use:** MA **Origin:** Native **Region:** Pacific **Dept.:** CHO **Elev.:** 100–100 m a.s.l. **IUCN Habitat code:** 1,14 **Conservation status: (IUCN)** NE

2590 *Cordia lucidula* I.M.Johnst. (Boraginaceae) **IPNI:** 274992-2 **Common name:** arato, carguero balsudo de rebalse, jigua laurel **Habit:** Small tree, Tree **Use:** MA **Origin:** Native **Region:** Amazonia, Caribbean, Pacific **Dept.:** AMA, ANT, BOL, CHO **Elev.:** 34–415 m a.s.l. **IUCN Habitat code:** 1,3,14 **Conservation status: (IUCN)** LC

2591 *Cordia lutea* Lam. (Boraginaceae) **IPNI:** 274993-2 **Common name:** biyuyo, caujaro, gomo, gomo amarillo, murciélago, ubre de vaca, ubrevaco, uvito **Habit:** Herb, Shrub, Small tree, Tree **Use:** HF, ME **Origin:** Native **Region:** Andean, Caribbean **Dept.:** ANT, BOL, CUN, NAR, TOL, VAC **Elev.:** 100–1500 m a.s.l. **IUCN Habitat code:** 1,3,14 **Conservation status: (IUCN)** LC

2592 *Cordia nodosa* Lam. (Boraginaceae) **IPNI:** 114394-1 **Common name:** babasaino, bastón del diablo, bolas de gato, canalete, cucubo, goma, gomo, peludo, tumbatoro, tumbatoro peludo, turbemono, turmemono, turmetoro, vara de humo **Habit:** Shrub, Small tree, Tree **Use:** AF, FU, HF, MA, ME **Origin:** Native **Region:** Amazonia, Andean, Orinoquia **Dept.:** AMA, ANT, BOL, BOY, CAQ, CAS, CUN, GUA, GUV, MET, PUT, SAN, VAU, VID **Elev.:** 50-1400 m.a.s.l. **IUCN Habitat code:** 1,2,3,4,5,14 **Conservation status: (IUCN)** LC **(CPSG)** Potential LC

2593 *Cordia panamensis* L.Riley (Boraginaceae) **IPNI:** 275016-2 **Common name:** bretaño, candelero, cunula, cuzú, guácimo peludo, muñeco blanco, nogal, quesero **Habit:** Herb, Shrub, Small tree, Tree **Use:** HF **Origin:** Native **Region:** Amazonia, Andean, Pacific **Dept.:** AMA, ANT, BOL, CAL, CAQ, CAS, CAU, CES, CHO, CUN, MET, NAR, PUT, QUI, SAN, VAC **Elev.:** 2-2200 m.a.s.l. **IUCN Habitat code:** 1,2,3,5,14 **Conservation status: (IUCN)** LC **(CPSG)** Potential LC

2594 *Cordia protracta* I.M.Johnst. (Boraginaceae) **IPNI:** 275031-2 **Common name:** combito, muñeco, muñeco prieto, tumbatoro **Habit:** Shrub, Small tree, Tree **Use:** ME **Origin:** Native **Region:** Andean, Pacific **Dept.:** ANT, CHO, CUN, TOL **Elev.:** 10–2300 m a.s.l. **IUCN Habitat code:** 1,3,5,14 **Conservation status: (IUCN)** EN

2595 *Cordia sebestena* L. (Boraginaceae) **IPNI:** 65551-2 **Common name:** San Joaquín, bonitel colorado, camalete, carambolí, caujaro, gomo, hoja de pulir, nomeolvides, sanagustín, sanjoaquín, sebestén **Habit:** Shrub, Small tree, Tree **Use:** EU, HF, MA, ME **Origin:** Native, Cultivated **Region:** Caribbean **Dept.:** ATL, BOL, COR, CUN, LAG, NSA, SAP, SUC **Elev.:** 0–600 m a.s.l. **IUCN Habitat code:** 1,2,3,5,14 **Conservation status: (IUCN)** LC

2596 *Cordia sericicalyx* A.DC. (Boraginaceae) **IPNI:** 275053-2 **Habit:** Small tree, Tree **Use:** HF, MA **Origin:** Native **Region:** Andean, Pacific **Dept.:** CAU, CUN, MET, PUT **Elev.:** 0–1725 m a.s.l. **IUCN Habitat code:** 1,3,4,14 **Conservation status: (IUCN)** LC

2597 *Cordia tetrandra* Aubl. (Boraginaceae) **IPNI:** 275071-2 **Common name:** baboso, canalete de río, candelero, cariaco blanco, caujaro, dinde, pardillo blanco, quitasol **Habit:** Shrub, Small tree, Tree **Use:** HF, MA, ME **Origin:** Native **Region:** Caribbean, Orinoquia **Dept.:** ANT, ATL, BOL, CAL, CAS, COR, MAG **Elev.:** 10–750 m a.s.l. **IUCN Habitat code:** 1,2,3,4,5,14 **Conservation status: (IUCN)** LC

2598 *Cordia toqueve* Aubl. (Boraginaceae) **IPNI:** 275076-2 **Common name:** chicharrón **Habit:** Small tree **Use:** HF **Origin:** Native **Region:** Caribbean, Orinoquia **Dept.:** CAS, NSA, SUC **Elev.:** 120–625 m a.s.l. **IUCN Habitat code:** 1,2,14 **Conservation status: (IUCN)** LC

2599 *Cynoglossum amabile* Stapf & J.R.Drumm. (Boraginaceae) **IPNI:** 115031-1 **Common name:** borraja, cieloazul, miosotis **Habit:** Herb, Shrub **Use:** EU, HF, ME **Origin:** Naturalised **Region:** Andean **Dept.:** ANT, BOG, BOY, CAL, CAU, CUN, HUI, NAR, NSA, PUT, SAN, VAC **Elev.:** 1088–2750 m a.s.l. **IUCN Habitat code:** 1,3,4,14 **Conservation status: (IUCN)** NE

2600 *Euploca procumbens* (Mill.) Diane & Hilger (Boraginaceae) **IPNI:** 60430582-2 **Common name:** alacrán, rabo de alacrán **Habit:** Herb, Shrub **Use:** ME **Origin:** Native **Region:** Caribbean, Orinoquia, Pacific **Dept.:** ANT, ATL, BOL, BOY, CAS, CHO, COR, MAG, MET, SAN **Elev.:** 10–140 m a.s.l. **IUCN Habitat code:** 1,2,3,4,5,14 **Conservation status: (IUCN)** LC

2601 *Heliotropium angiospermum* Murray (Boraginaceae) **IPNI:** 116662-1 **Common name:** alacrán, hierba alacrán, jeyutse pal, mulato, rabo de alacran, rabo de alacrán, verbena, verbena blanca, verbena criolla, verbenilla **Habit:** Herb **Use:** IF, ME, PO **Origin:** Native **Region:** Andean, Caribbean **Dept.:** ANT, ATL, BOL, BOY, CES, CUN, HUI, LAG, MAG, NSA, SAP, SAN, SUC, TOL, VAC **Elev.:** 0-2300 m.a.s.l. **IUCN Habitat code:** 1,2,3,4,5,14 **Conservation status: (IUCN)** NE **(CPSG)** Potential LC

2602 *Heliotropium angustiflorum* (Ruiz & Pav.) Govaerts (Boraginaceae) **IPNI:** 77192212-1 **Habit:** Shrub, Tree, Liana, Climbing **Use:** ME **Origin:** Native **Region:**

Amazonia, Andean, Caribbean, Pacific **Dept.:** ANT, BOL, CAQ, CHO, GUV, SAN, SUC **Elev.:** 200–1900 m a.s.l. **Conservation status: (IUCN)** NE

2603 *Heliotropium arborescens* L. (Boraginaceae) **IPNI:** 116670-1 **Common name:** heliotoropo de olor a vainilla, heliotropo, vainilla **Habit:** Shrub **Use:** EU, HF, MA, ME, PO **Origin:** Native, Cultivated **Region:** Andean **Dept.:** ANT, BOG, VAC **Elev.:** 1200–1600 m a.s.l. **IUCN Habitat code:** 1,3,14 **Conservation status: (IUCN)** NE

2604 *Heliotropium curassavicum* L. (Boraginaceae) **IPNI:** 116784-1 **Habit:** Herb **Use:** HF, ME, PO **Origin:** Native **Region:** Caribbean **Dept.:** ATL, BOL, LAG, MAG, SAP **Elev.:** 0-32 m.a.s.l. **IUCN Habitat code:** 1,2,5,14 **Conservation status: (IUCN)** LC **(CPSG)** Potential LC

2605 *Heliotropium cuspidatum* (Kunth) Feuillet (Boraginaceae) **IPNI:** 77203945-1 **Common name:** borrajo, maíz de gallo **Habit:** Shrub, Subshrub, Small tree, Climbing **Use:** ME **Origin:** Native **Region:** Amazonia, Andean, Caribbean, Pacific **Dept.:** AMA, ANT, ARA, BOL, BOY, CAQ, CAS, CAU, CES, CHO, CUN, HUI, MET, NAR, PUT, SAN **Elev.:** 0-1500 m.a.s.l. **Conservation status: (IUCN)** NE **(CPSG)** Potential LC

2606 *Heliotropium europaeum* L. (Boraginaceae) **IPNI:** 116833-1 **Use:** ME, PO **Region:** Andean **Dept.:** BOG**IUCN Habitat code:** 1 **Conservation status: (IUCN)** NE

2607 *Heliotropium glabrum* (L.) Feuillet (Boraginaceae) **IPNI:** 77210153-1 **Common name:** frutillo, muñeco **Habit:** Tree **Use:** ME **Origin:** Endemic **Region:** Amazonia, Andean, Guiana Shield, Caribbean **Dept.:** ANT, BOY, CAL, CAQ, CUN, LAG, MAG, MET, SUC, VAC **Elev.:** 0–2800 m a.s.l. **IUCN Habitat code:** 1,2,3 **Conservation status: (IUCN)** LC

2608 *Heliotropium indicum* L. (Boraginaceae) **IPNI:** 116938-1 **Common name:** alacrana, alacrán, borrajón, cordoncillo, jeyutse, rabo de alacrán, rabo de caimán, rabo de zorro, tiriaodes, verbena, verbena blanca, verbena morada, verbena rabo de alacrán **Habit:** Herb, Shrub, Subshrub **Use:** AF, HF, MA, ME, PO, SU **Origin:** Naturalised **Region:** Amazonia, Andean, Guiana Shield, Caribbean, Orinoquia, Pacific **Dept.:** AMA, ANT, ARA, ATL, BOL, CAL, CAS, CAU, CES, CHO, COR, CUN, GUA, LAG, MAG, MET, NSA, SAP, SAN, SUC, TOL, VAC **Elev.:** 0-1675 m.a.s.l. **IUCN Habitat code:** 1,2,3,4,5,14 **Conservation status: (IUCN)** NE **(CPSG)** Potential LC

2609 *Heliotropium verdcourtii* Craven (Boraginaceae) **IPNI:** 77069564-1 **Common name:** fricgaplatos, lágrimas de San Pedro, maquionso, niguo **Habit:** Shrub, Subshrub, Small tree, Tree, Liana, Climbing **Use:** AF, HF, ME **Origin:** Native **Region:** Amazonia, Andean, Caribbean, Pacific **Dept.:** AMA, ANT, ATL, BOL, BOY, CAL, CAQ, CAU, CES, CHO, CUN, HUI, LAG, MAG, NAR, PUT, QUI, RIS, SAP, SAN, TOL, VAC **Elev.:** 0–2000 m a.s.l. **IUCN Habitat code:** 1,2,3,14 **Conservation status: (IUCN)** NE

2610 *Lennoa madreporoides* Lex. (Boraginaceae) **IPNI:** 526306-1 **Habit:** Herb, Parasitic **Use:** HF **Origin:** Native **Region:** Caribbean **Dept.:** MAG **Elev.:** 25–25 m a.s.l. **IUCN Habitat code:** 2 **Conservation status: (IUCN)** NE

2611 *Lithospermum officinale* L. (Boraginaceae) **IPNI:** 118163-1 **Common name:** caracolito **Habit:** Subshrub **Use:** HF, ME **Origin:** Naturalised **Region:** Andean **Dept.:** CUN **Elev.:** 2700–2700 m a.s.l. **IUCN Habitat code:** 14 **Conservation status: (IUCN)** NE

2612 *Myosotis alpestris* F.W.Schmidt (Boraginaceae) **IPNI:** 118892-1 **Use:** EU **Region:** Andean **Dept.:** BOG**IUCN Habitat code:** 14 **Conservation status: (IUCN)** NE

2613 *Myriopus maculatus* (Jacq.) Feuillet (Boraginaceae) **IPNI:** 60449244-2 **Habit:** Herb, Shrub, Tree, Liana, Climbing **Use:** ME **Origin:** Native **Region:** Amazonia, Andean, Caribbean, Pacific **Dept.:** AMA, ANT, ATL, CAQ, CHO, CUN, HUI, LAG, MAG, MET, NSA, QUI, SAN, TOL, VAC **Elev.:** 5-2250 m.a.s.l. **IUCN Habitat code:** 1,2,3,4,14 **Conservation status: (IUCN)** NE **(CPSG)** Potential LC

2614 *Myriopus psilostachya* (Kunth) Diane & Hilger (Boraginaceae) **IPNI:** 60430568-2 **Habit:** Shrub **Use:** ME **Origin:** Native **Region:** Amazonia, Andean **Dept.:** NSA, PUT, VAC **Elev.:** 325–2500 m a.s.l. **IUCN Habitat code:** 1,14 **Conservation status: (IUCN)** NE

2615 *Myriopus volubilis* (L.) Small (Boraginaceae) **IPNI:** 119329-1 **Habit:** Shrub, Small tree, Liana, Climbing **Use:** ME **Origin:** Native **Region:** Andean, Caribbean **Dept.:** ANT, ATL, BOL, BOY, LAG, MAG, NAR, SAN, SUC, VAC **Elev.:** 0–1550 m a.s.l. **IUCN Habitat code:** 1,2,3,14 **Conservation status: (IUCN)** NE

2616 *Symphytum officinale* L. (Boraginaceae) **IPNI:** 120815-1 **Common name:** cofrey, confrey **Use:** AF, EU, GS, HF, MA, ME, PO **Dept.:** BOG**IUCN Habitat code:** 1,3,14 **Conservation status: (IUCN)** NE

2617 *Tournefortia brevilobata* K.Krause (Boraginaceae) **IPNI:** 255461-2 **Habit:** Shrub, Small tree, Tree **Use:** ME **Origin:** Native **Region:** Andean **Dept.:** ANT, CAU, MAG, NAR **Elev.:** 1550–3300 m a.s.l. **IUCN Habitat code:** 1,4,14 **Conservation status: (IUCN)** NE

2618 *Tournefortia fuliginosa* Kunth (Boraginaceae) **IPNI:** 120988-1 **Common name:** buandn-doch, guásimo, palo monte, verdenegro **Habit:** Shrub, Subshrub, Small tree, Tree **Use:** HF, ME **Origin:** Native **Region:** Andean **Dept.:** ANT, CAU, CUN, HUI, LAG, MAG, NAR, NSA, PUT, QUI, RIS, SAN, TOL, VAC **Elev.:** 1380-3700 m.a.s.l. **IUCN Habitat code:** 1,3,4,5,8,14 **Conservation status: (IUCN)** NE **(CPSG)** Potential LC

2619 *Tournefortia gnaphalodes* (L.) R.Br. ex Roem. & Schult. (Boraginaceae) **IPNI:** 120998-1 **Common name:** tabaco pescador **Habit:** Shrub, Subshrub **Use:** MA, ME **Origin:** Native **Region:** Caribbean **Dept.:** LAG, SAP **Elev.:** 0–20 m a.s.l. **IUCN Habitat code:** 5 **Conservation status: (IUCN)** NE

2620 *Tournefortia macrostachya* Rusby (Boraginaceae) **IPNI:** 255500-2 **Habit:** Shrub, Subshrub, Small tree, Tree **Use:** ME **Origin:** Native **Region:** Andean **Dept.:** ANT, BOG, CES, LAG, MAG, PUT, QUI **Elev.:** 1420–3000 m a.s.l. **IUCN Habitat code:** 1,3,14 **Conservation status: (IUCN)** NE

2621 *Tournefortia polystachya* Ruiz & Pav. (Boraginaceae) **IPNI:** 121103-1 **Common name:** calchilla, friegaplatos **Habit:** Shrub, Small tree, Tree **Use:** AF, EU, FU, MA, ME **Origin:** Native **Region:** Andean **Dept.:** ANT, BOY, CAL, CUN, MAG, NSA, PUT, QUI, RIS, VAC **Elev.:** 1500–3150 m a.s.l. **IUCN Habitat code:** 1,2,3,4,5,14 **Conservation status: (IUCN)** LC

2622 *Tournefortia scabrida* Kunth (Boraginaceae) **IPNI:** 121133-1 **Common name:** heliotropo, mayorquín, verdenegro, viravira blanca **Habit:** Shrub, Small tree, Tree **Use:** HF, ME **Origin:** Native **Region:** Andean, Caribbean **Dept.:** ANT, BOY, CAL, CAQ, CAU, CHO, CUN, HUI, NAR, QUI, RIS, SAN, TOL, VAC **Elev.:** 1000-3200 m.a.s.l. **IUCN Habitat code:** 1,3,4,14 **Conservation status: (IUCN)** LC **(CPSG)** Potential LC

2623 *Varronia acuta* (Pittier) Borhidi (Boraginaceae) **IPNI:** 286671-2 **Common name:** guasco, guácimo, guásimo, mallorquín, palonegro, salvio negro **Habit:** Shrub, Small tree, Tree **Use:** ME **Origin:** Endemic **Region:** Andean **Dept.:** ANT, CAL, CAU, CHO, CUN, MET, NAR, NSA, QUI, RIS, VAC **Elev.:** 970-3150 m.a.s.l. **IUCN Habitat code:** 1,3,14 **Conservation status: (IUCN)** EN **(CPSG)** EN

2624 *Varronia bullata* L. (Boraginaceae) **IPNI:** 318022-2 **Common name:** maiztostado, palo quemado, rasgarropa, salvia de monte **Habit:** Shrub, Subshrub, Small tree, Tree **Use:** ME **Origin:** Native **Region:** Andean, Caribbean **Dept.:** ANT, ATL, BOL, CAL, CUN, HUI, LAG, MAG, NSA, SAP, SAN, TOL **Elev.:** 10–1005 m a.s.l. **IUCN Habitat code:** 2 **Conservation status: (IUCN)** LC

2625 *Varronia curassavica* Jacq. (Boraginaceae) **IPNI:** 262640-2 **Common name:** basura prieta, cariaquito negro, cariaqyuito, guasca negra, guasco, guásimo, majao negro, mulato, negrito, palo negro, palonegro, pata de gallina, penda, salvia negra, salvio, salvio negro, tuni esmeraldo **Habit:** Shrub, Subshrub **Use:** AF, HF, IF, ME **Origin:** Native **Region:** Andean, Caribbean, Pacific **Dept.:** ANT, ARA, ATL, BOG, BOL, BOY, CES, CHO, COR, CUN, LAG, MAG, NAR, NSA, SAP, SAN, TOL, VAC **Elev.:** 0–2400 m a.s.l. **IUCN Habitat code:** 1,2,3,4,5,14 **Conservation status: (IUCN)** NE

2626 *Varronia cylindrostachya* Ruiz & Pav. (Boraginaceae) **IPNI:** 262641-2 **Common name:** guasca negra, mallorquín, mulato, palonegro, salvia negra, salvio, salvio blanco, salvio negro, tuniesmeraldo, verdenegro **Habit:** Shrub, Subshrub, Small tree, Tree **Use:** MA, ME **Origin:** Native **Region:** Andean **Dept.:** ANT, BOG, BOY, CAQ, CAU, CHO, CUN, HUI, MET, NAR, NSA, PUT, QUI, RIS, SAN, TOL, VAC **Elev.:** 520-3400 m.a.s.l. **Conservation status: (IUCN)** NE **(CPSG)** Potential LC

2627 *Varronia dichotoma* Ruiz & Pav. (Boraginaceae) **IPNI:** 121424-1 **Common name:** escobo, mallorquí, morita **Habit:** Herb, Shrub **Use:** ME **Origin:** Native **Region:** Andean, Pacific **Dept.:** ANT, BOG, BOL, BOY, CAL, CAQ, CAU, CES, CHO, CUN, MET, NAR, NSA, PUT, RIS, SAN, VAC **Elev.:** 20-3820 m.a.s.l. **IUCN Habitat code:** 1,3,4,5,14 **Conservation status: (IUCN)** NE **(CPSG)** Potential LC

2628 *Varronia grandiflora* Desv. (Boraginaceae) **IPNI:** 121434-1 **Use:** ME **Origin:** Native **Conservation status: (IUCN)** NE

2629 *Varronia inermis* (Mill.) Borhidi (Boraginaceae) **IPNI:** 946177-1 **Use:** MA, ME **Origin:** Native **Conservation status: (IUCN)** NE

2630 *Varronia linnaei* (Stearn) J.S.Mill. (Boraginaceae) **IPNI:** 77084349-1 **Common name:** romperropa **Habit:** Shrub, Subshrub, Small tree **Use:** ME **Origin:** Native **Region:** Andean, Caribbean, Orinoquia **Dept.:** ANT, BOL, BOY, CAS, CHO, CUN, HUI, MET, RIS, TOL, VAC **Elev.:** 3–2410 m a.s.l. **IUCN Habitat code:** 1,3,14 **Conservation status: (IUCN)** NE

2631 *Varronia polycephala* Lam. (Boraginaceae) **IPNI:** 262677-2 **Common name:** arquina, brazo de negro, maiztostado, mallorquín, palo quemado, rasgarropa, salvia de montesurundé, varraco, verdenegro **Habit:** Herb, Shrub, Subshrub, Small tree **Use:** ME **Origin:** Native **Region:** Amazonia, Andean, Caribbean **Dept.:** ANT, BOY, CAL, CAQ, CES, COR, CUN, HUI, MAG, MET, PUT, QUI, SAN, TOL, VAC **Elev.:** 31-2440 m.a.s.l. **IUCN Habitat code:** 1,2,3,4,5,14 **Conservation status: (IUCN)** LC **(CPSG)** Potential LC

2632 *Varronia spinescens* (L.) Borhidi (Boraginaceae) **IPNI:** 946226-1 **Common name:** bejuco choco, bejuco nacundo, bejuco negro, brazonegro, calentana, cansamulato, cuerito rubio, gomo, guasca, guasco, guácimo, mallorquín, mortiño, mulato, nacancio, nacansio, naquinso, naquinzo, naquiunso, naquiunzo, natiunsa, natiunza, rasgarropa, rompe **Habit:** Herb, Shrub, Subshrub, Tree, Climbing **Use:** ME **Origin:** Native **Region:** Amazonia, Andean, Orinoquia, Pacific **Dept.:** AMA, ANT, BOG, BOY, CAL, CAQ, CAU, CHO, CUN, HUI, MAG, MET, NAR, NSA, PUT, QUI, RIS, SAN, SUC, TOL, VAC **Elev.:** 0-2900 m.a.s.l. **Conservation status: (IUCN)** LC **(CPSG)** Potential LC

2633 *Wigandia urens* (Ruiz & Pav.) Kunth (Boraginaceae) **IPNI:** 114242-3 **Common name:** borragón, borrajón, caracasuna, pringamoza **Habit:** Shrub **Use:** EU, ME, PO, SU **Origin:** Native **Region:** Andean **Dept.:** ANT, CES, CUN, HUI, LAG, MAG, NAR, NSA, SAN, TOL, VAC **Elev.:** 250–1400 m a.s.l. **IUCN Habitat code:** 1,2,3,14 **Conservation status: (IUCN)** LC

Brassicales

2634 *Batis maritima* L. (Bataceae) **IPNI:** 319368-2 **Common name:** bicho, platanillo, platanito, platanito salado, shot-shot, varilla de vidrio **Habit:** Herb **Use:** HF, MA, ME **Origin:** Native **Region:** Caribbean **Dept.:** ANT, ATL, BOL, COR, LAG, MAG, SUC **Elev.:** 0–10 m a.s.l. **IUCN Habitat code:** 1,2,5,14 **Conservation status: (IUCN)** NE

2635 *Barbarea verna* (Mill.) Asch. (Brassicaceae) **IPNI:** 1109261-2 **Use:** GS, HF, ME **Region:** Andean **Dept.:** BOGIUCN **Habitat code:** 14 **Conservation status: (IUCN)** DD

2636 *Brassica juncea* (L.) Czern. (Brassicaceae) **IPNI:** 60442520-2 **Habit:** Herb **Use:** AF, EU, FU, GS, HF, MA, ME, PO, SU **Origin:** Naturalised **Region:** Andean **Dept.:** ANT, BOG, CAU, MAG, SAN, VAC **Elev.:** 1750–3250 m a.s.l. **IUCN Habitat code:** 1,3,14 **Conservation status: (IUCN)** NE

2637 *Brassica napus* L. (Brassicaceae) **IPNI:** 279419-1 **Common name:** canola, nabo **Habit:** Herb **Use:** AF, EU, FU, GS, HF, MA, ME, PO **Origin:** Naturalised **Region:** Andean **Dept.:** BOG, CUN, PUT **Elev.:** 2200–2600 m a.s.l. **IUCN Habitat code:** 3,4,14 **Conservation status: (IUCN)** NE

2638 *Brassica nigra* (L.) W.D.J.Koch (Brassicaceae) **IPNI:** 279422-1 **Common name:** mostaza, mostaza negra **Habit:** Herb **Use:** EU, GS, HF, ME, PO **Origin:** Naturalised **Region:** Andean **Dept.:** BOG, BOY, CUN **Elev.:** 2400–2600 m a.s.l. **IUCN Habitat code:** 1,3,14 **Conservation status: (IUCN)** LC

2639 *Brassica oleracea* L. (Brassicaceae) **IPNI:** 279435-1 **Common name:** col rizada, cáliz **Habit:** Herb **Use:** AF, EU, GS, HF, MA, ME, PO **Origin:** Cultivated **Region:** Amazonia, Andean **Dept.:** AMA, ANT, BOG, CUN, PUT, RIS **Elev.:** 0–2620 m a.s.l. **IUCN Habitat code:** 1,2,3,4,14 **Conservation status: (IUCN)** DD

2640 *Brassica rapa* L. (Brassicaceae) **IPNI:** 279485-1 **Common name:** alpiste, alpiste de canarios, alpistillo, colinabo, mostacilla, mostacilla silvestre, mostaza, mostaza de canarios, nabo, nabo amarillo, nabo de canarios, rebancá, rábano de canarios **Habit:** Herb **Use:** AF, EU, FU, GS, HF, MA, ME, PO **Origin:** Naturalised **Region:** Andean **Dept.:** ANT, BOG, BOY, CUN, NAR, PUT, RIS, VAC **Elev.:** 1000–4200 m a.s.l. **IUCN Habitat code:** 1,2,3,4,14 **Conservation status: (IUCN)** NE

2641 *Cakile lanceolata* (Willd.) O.E.Schulz (Brassicaceae) **IPNI:** 118574-3 **Habit:** Herb **Use:** HF **Origin:** Native **Region:** Caribbean **Dept.:** BOL **Elev.:** 0–0 m a.s.l. **Conservation status: (IUCN)** NE

2642 *Capsella bursa-pastoris* (L.) Medik. (Brassicaceae) **IPNI:** 30092589-2 **Common name:** bosa de pastor, calzoncitos, calzón, empanaditas, maleticas, pan con queso, pan y quesito **Habit:** Herb **Use:** AF, GS, HF, MA, ME, PO **Origin:** Naturalised **Region:** Andean **Dept.:** BOG, BOY, CAU, CUN, NAR, SAN **Elev.:** 2200–4100 m a.s.l. **IUCN Habitat code:** 1,3,4,14 **Conservation status: (IUCN)** NE

2643 *Cardamine bonariensis* Juss. ex Pers. (Brassicaceae) **IPNI:** 280189-1 **Common name:** berro de Castilla, berros **Habit:** Herb **Use:** HF, ME **Origin:** Native **Region:** Andean **Dept.:** ANT, BOG, BOY, CAL, CAU, CUN, HUI, MAG, MET, NAR, PUT, QUI, RIS, SAN, TOL, VAC **Elev.:** 1700-4300 m.a.s.l. **IUCN Habitat code:** 1,3,4,5,6,8,14 **Conservation status: (IUCN)** NE **(CPSG)** Potential LC

2644 *Cardamine flexuosa* With. (Brassicaceae) **IPNI:** 45189-2 **Habit:** Herb **Use:** HF, ME **Origin:** Naturalised **Region:** Andean **Dept.:** ANT, BOG, CAL, CUN, QUI, RIS, VAC **Elev.:** 990–2600 m a.s.l. **IUCN Habitat code:** 1,2,3,4,14 **Conservation status: (IUCN)** NE

2645 *Cardamine hirsuta* L. (Brassicaceae) **IPNI:** 280380-1 **Habit:** Herb **Use:** HF, ME **Origin:** Naturalised **Region:** Andean **Dept.:** ANT, BOG, CAL, CUN, QUI, SAN **Elev.:** 1500–3250 m a.s.l. **IUCN Habitat code:** 1,3,14 **Conservation status: (IUCN)** NE

2646 *Cardamine obliqua* Hochst. ex A.Rich. (Brassicaceae) **IPNI:** 280530-1 **Habit:** Herb **Use:** HF **Origin:** Native **Region:** Andean **Dept.:** CAL, CAU, NAR, TOL **Elev.:** 2150–3000 m a.s.l. **Conservation status: (IUCN)** NE

2647 *Cardamine ovata* Benth. (Brassicaceae) **IPNI:** 280543-1 **Habit:** Herb **Use:** ME **Origin:** Native **Region:** Andean **Dept.:** ANT, BOG, BOY, CAU, CUN, HUI, MAG, MET, NAR, NSA, QUI, RIS, SAN, TOL, VAC **Elev.:** 1200-3900 m.a.s.l. **IUCN Habitat code:** 1,3,4,14 **Conservation status: (IUCN)** NE **(CPSG)** Potential LC

2648 *Coincya tournefortii* (Gouan) Alcaraz, T.E.Díaz, Rivas Mart. & Sánchez-Gómez (Brassicaceae) **IPNI:** 999093-1 **Habit:** Herb **Use:** GS, HF, ME **Origin:** Naturalised **Region:** Andean **Dept.:** CUN **Elev.:** 2600–2600 m a.s.l. **Conservation status: (IUCN)** NE

2649 *Draba cheiranthoides* Hook.f. (Brassicaceae) **IPNI:** 282475-1 **Use:** ME **Origin:** Endemic **IUCN Habitat code:** 4,8 **Conservation status: (IUCN)** NE

2650 *Draba litamo* L.Uribe (Brassicaceae) **IPNI:** 83668-2 **Common name:** dictamo, litamo, litamo real, litamo rial, lítamo, lítamo real, lítamo rial **Habit:** Herb **Use:** ME **Origin:** Endemic **Region:** Andean **Dept.:** BOY **Elev.:** 3600–4800 m a.s.l. **IUCN Habitat code:** 4,8 **Conservation status:** **(IUCN)** NE

2651 *Draba pachythyrsa* Triana & Planch. (Brassicaceae) **IPNI:** 282916-1 **Habit:** Herb **Use:** ME **Origin:** Endemic **Region:** Andean **Dept.:** CAL, RIS, TOL **Elev.:** 4000–4400 m a.s.l. **IUCN Habitat code:** 4 **Conservation status:** **(IUCN)** NE

2652 *Draba pamplonensis* Planch. & Linden (Brassicaceae) **IPNI:** 282923-1 **Habit:** Herb **Use:** ME **Origin:** Endemic **Region:** Andean **Dept.:** NSA, SAN **Elev.:** 3850–3850 m a.s.l. **IUCN Habitat code:** 1,4 **Conservation status:** **(IUCN)** NE

2653 *Draba pennell-hazenii* O.E.Schulz (Brassicaceae) **IPNI:** 83794-2 **Habit:** Herb **Use:** ME **Origin:** Endemic **Region:** Andean **Dept.:** CAL, RIS **Elev.:** 3700–4500 m a.s.l. **IUCN Habitat code:** 1,4,14 **Conservation status:** **(IUCN)** NE

2654 *Eruca vesicaria* (L.) Cav. (Brassicaceae) **IPNI:** 1137625-2 **Habit:** Herb **Use:** AF, FU, GS, HF, MA, ME, PO **Origin:** Naturalised **Region:** Andean **Dept.:** ANT, BOG, CUN **Elev.:** 1910–2600 m a.s.l. **IUCN Habitat code:** 1,14 **Conservation status:** **(IUCN)** NE

2655 *Erysimum × cheiri* (L.) Crantz (Brassicaceae) **IPNI:** 283703-1 **Common name:** alhelí, alhelí amarillo **Habit:** Herb **Use:** EU, ME **Origin:** Cultivated **Region:** Andean **Dept.:** CUN **Elev.:** 2600–2600 m a.s.l. **IUCN Habitat code:** 1,14 **Conservation status:** **(IUCN)** NE

2656 *Iberis amara* L. (Brassicaceae) **IPNI:** 324660-2 **Habit:** Herb **Use:** EU, HF, ME **Origin:** Cultivated **Region:** Andean **Dept.:** BOG, CUN **Elev.:** 2600–2600 m a.s.l. **Conservation status:** **(IUCN)** NE

2657 *Iberis umbellata* L. (Brassicaceae) **IPNI:** 324671-2 **Common name:** carraspique, lila, lilas **Habit:** Herb **Use:** EU, ME **Origin:** Cultivated **Region:** Andean **Dept.:** BOG, CUN **Elev.:** 2600–2600 m a.s.l. **IUCN Habitat code:** 3,14 **Conservation status:** **(IUCN)** NE

2658 *Lepidium bipinnatifidum* Desv. (Brassicaceae) **IPNI:** 137955-2 **Common name:** carraspique, chichera, chinchira, maiztostado, mastuerzo, maíz tostado **Habit:** Herb **Use:** GS, ME **Origin:** Native **Region:** Andean **Dept.:** ANT, BOG, BOY, CAL, CAU, CUN, NAR, NSA, PUT, QUI, SAN, TOL, VAC **Elev.:** 1500-3750 m.a.s.l. **IUCN Habitat code:** 1,3,4,14 **Conservation status:** **(IUCN)** NE **(CPSG)** Potential LC

2659 *Lepidium didymum* L. (Brassicaceae) **IPNI:** 286148-1 **Habit:** Herb **Use:** AF, GS, HF, ME **Origin:** Naturalised **Region:** Andean **Dept.:** BOG, CUN **Elev.:** 2600–2600 m a.s.l. **IUCN Habitat code:** 14 **Conservation status:** **(IUCN)** NE

2660 *Lepidium spinosum* Ard. (Brassicaceae) **IPNI:** 286421-1 **Use:** GS **Region:** Andean **Dept.:** BOG**Conservation status:** **(IUCN)** NE

2661 *Lepidium virginicum* L. (Brassicaceae) **IPNI:** 138164-2 **Common name:** mastuerzo **Habit:** Herb **Use:** HF, ME **Origin:** Naturalised **Region:** Andean **Dept.:** ANT, SAN **Elev.:** 1500–1800 m a.s.l. **IUCN Habitat code:** 1,2,3,14 **Conservation status:** **(IUCN)** NE

2662 *Lobularia maritima* (L.) Desv. (Brassicaceae) **IPNI:** 1085072-2 **Common name:** arroz, bellísima, brisas del mar, colación, golondrina, guardaparque blanco, llovizna de enero, miel de angelita, nieve, orgullo, pañuelito, reseda, reseda blanca, rocío, siempreviva **Habit:** Herb **Use:** EU, HF, ME **Origin:** Naturalised **Region:** Andean **Dept.:** ANT, BOG, BOY, CUN, NAR, NSA, QUI, RIS **Elev.:** 1500–2640 m a.s.l. **IUCN Habitat code:** 1,3,14 **Conservation status:** **(IUCN)** NE

2663 *Lunaria annua* L. (Brassicaceae) **IPNI:** 286647-1 **Common name:** moneda del Papa, panderitas **Habit:** Herb **Use:** EU, HF **Origin:** Cultivated **Region:** Andean **Dept.:** BOG, CUN **Elev.:** 2600–2600 m a.s.l. **IUCN Habitat code:** 14 **Conservation status:** **(IUCN)** NE

2664 *Matthiola incana* (L.) W.T.Aiton (Brassicaceae) **IPNI:** 286910-1 **Common name:** alhelí, alhelí blanco, alhelí morado **Habit:** Herb **Use:** EU, HF, ME **Origin:** Cultivated, Naturalised **Region:** Andean **Dept.:** ANT, BOG, CUN **Elev.:** 1500–2600 m a.s.l. **IUCN Habitat code:** 3,14 **Conservation status:** **(IUCN)** NE

2665 *Nasturtium officinale* W.T.Aiton (Brassicaceae) **IPNI:** 287534-1 **Common name:** berro, berros, mastruerzo, mastuerzo, mastuerzo de agua **Habit:** Herb **Use:** EU, GS, HF, ME **Origin:** Naturalised **Region:** Andean **Dept.:** ANT, BOG, BOY, CAL, CAU, CUN, MAG, NAR, NSA, QUI, RIS, SAN, VAC **Elev.:** 1500–3700 m a.s.l. **IUCN Habitat code:** 1,3,4,8,14 **Conservation status:** **(IUCN)** LC

2666 *Raphanus raphanistrum* L. (Brassicaceae) **IPNI:** 217826-2 **Common name:** mostaza, rabanit, rabanito colorado, rebancá, rabanillo, rábano, rábano blanco, rábano maleza, rábano morado, rábano negro, rábano silvestre **Habit:** Herb **Use:** GS, HF, ME, PO **Origin:** Cultivated, Naturalised **Region:** Andean **Dept.:** ANT, BOG, BOY, CAU, CUN, NAR, RIS **Elev.:** 1760–2900 m a.s.l. **IUCN Habitat code:** 1,3,14 **Conservation status:** **(IUCN)** NE

2667 *Rapistrum rugosum* (L.) All. (Brassicaceae) **IPNI:** 1061021-2 **Habit:** Herb **Use:** AF, GS, HF, ME, PO **Origin:** Naturalised **Region:** Andean **Dept.:** BOG, CUN **Elev.:** 2600–2600 m a.s.l. **IUCN Habitat code:** 14 **Conservation status:** **(IUCN)** NE

2668 *Rorippa dubia* (Pers.) H.Hara (Brassicaceae) **IPNI:** 288735-1 **Habit:** Herb **Use:** GS, HF, ME **Origin:** Naturalised **Region:** Andean **Dept.:** BOG, CUN, QUI **Elev.:** 1200–2600 m a.s.l. **Conservation status:** **(IUCN)** NE

2669 *Rorippa eggersii* (O.E.Schulz) J.F.Macbr. (Brassicaceae) **IPNI:** 288736-1 **Habit:** Herb **Use:** GS **Origin:** Native **Region:** Andean **Dept.:** ANT **Elev.:** 1500–1500 m a.s.l. **Conservation status:** **(IUCN)** NE

2670 *Rorippa indica* (L.) Hiern (Brassicaceae) **IPNI:** 222144-2 **Common name:** alpiste, nabo **Habit:** Herb **Use:** GS, HF, ME **Origin:** Naturalised **Region:** Andean **Dept.:** ANT, BOG, CAL, CAU, CUN, HUI, NAR, QUI, VAC **Elev.:** 900–2000 m a.s.l. **IUCN Habitat code:** 3,14 **Conservation status:** **(IUCN)** NE

2671 *Rorippa mandonii* (E.Fourn.) Mart.-Laborde (Brassicaceae) **IPNI:** 284416-2 **Common name:** berro **Habit:** Herb **Use:** GS **Origin:** Native **Region:** Andean **Dept.:** ANT, BOG, CAL, CAU, CUN, NAR, VAC **Elev.:** 960–2900 m a.s.l. **IUCN Habitat code:** 1,3,14 **Conservation status:** **(IUCN)** NE

2672 *Rorippa mexicana* (Moc., Sessé & Cerv. ex DC.) Standl. & Steyerm. (Brassicaceae) **IPNI:** 222160-2 **Use:** GS **Region:** Andean **Dept.:** BOG**IUCN Habitat code:** 14 **Conservation status:** **(IUCN)** NE

2673 *Rorippa nana* (Schltdl.) J.F.Macbr. (Brassicaceae) **IPNI:** 222164-2 **Habit:** Herb **Use:** GS **Origin:** Native **Region:** Andean **Dept.:** BOG, CUN, RIS **Elev.:** 3700–4000 m a.s.l. **IUCN Habitat code:** 4 **Conservation status:** **(IUCN)** NE

2674 *Rorippa palustris* (L.) Besser (Brassicaceae) **IPNI:** 288669-1 **Common name:** berro de huerta, berros, mastuerzo, mastuerzo de huerta **Habit:** Herb **Use:** GS, HF, ME **Origin:** Naturalised **Region:** Andean, Orinoquia **Dept.:** ANT, BOY, CAS, CAU, CUN, QUI, SAN **Elev.:** 110–2170 m a.s.l. **IUCN Habitat code:** 1,14 **Conservation status:** **(IUCN)** NE

2675 *Rorippa pinnata* (Sessé & Moc.) Rollins (Brassicaceae) **IPNI:** 222187-2 **Habit:** Herb **Use:** GS **Origin:** Native **Region:** Andean **Dept.:** ANT, BOG, BOY, CUN, RIS, VAC **Elev.:** 1560–3220 m a.s.l. **IUCN Habitat code:** 3,14 **Conservation status:** **(IUCN)** NE

2676 *Sinapis alba* L. (Brassicaceae) **IPNI:** 288952-1 **Common name:** alpiste, mostacilla, mostaza, mostaza amarilla, mostaza blanca, nabo silvestre **Habit:** Herb **Use:** AF, EU, FU, GS, HF, ME, PO, SU **Origin:** Naturalised **Region:** Andean **Dept.:** BOG, CUN **Elev.:** 2600–2600 m a.s.l. **IUCN Habitat code:** 3,14 **Conservation status: (IUCN)** NE

2677 *Sisymbrium altissimum* L. (Brassicaceae) **IPNI:** 289144-1 **Habit:** Herb **Use:** HF, ME **Origin:** Naturalised **Region:** Andean **Dept.:** CUN **Elev.:** 2600–2600 m a.s.l. **Conservation status: (IUCN)** NE

2678 *Sisymbrium irio* L. (Brassicaceae) **IPNI:** 237940-2 **Habit:** Herb **Use:** HF, ME **Origin:** Naturalised **Region:** Andean **Dept.:** CUN **Elev.:** 2600–2600 m a.s.l. **Conservation status: (IUCN)** NE

2679 *Sisymbrium officinale* (L.) Scop. (Brassicaceae) **IPNI:** 289589-1 **Habit:** Herb **Use:** HF, ME **Origin:** Naturalised **Region:** Andean **Dept.:** BOG, CUN, TOL **Elev.:** 1800–2900 m a.s.l. **IUCN Habitat code:** 5,14 **Conservation status: (IUCN)** NE

2680 *Thlaspi arvense* L. (Brassicaceae) **IPNI:** 290477-1 **Habit:** Herb **Use:** FU, HF, ME, PO **Origin:** Naturalised **Region:** Andean **Dept.:** BOG, CUN **Elev.:** 2600–2600 m a.s.l. **IUCN Habitat code:** 14 **Conservation status: (IUCN)** NE

2681 *Crateva tapia* L. (Capparaceae) **IPNI:** 316363-2 **Common name:** amor de mico, cachimonda, huevo de burro, limón toco, mamón de puerco, naranjillo, naranjito, naranjuelo, naranjuelo blanco, pepa de sábalo, sorrocloco, toco, toco negro, totojando, zorrocloco **Habit:** Small tree, Tree **Use:** AF, EU, FU, HF, MA, ME **Origin:** Native **Region:** Amazonia, Caribbean, Orinoquia, Pacific **Dept.:** AMA, ATL, BOL, CAQ, CHO, CUN, HUI, LAG, MAG, NSA, PUT, VAC, VAU, VID **Elev.:** 5-1500 m.a.s.l. **Conservation status: (IUCN)** LC **(CPSG)** Potential LC

2682 *Morisonia americana* L. (Capparaceae) **IPNI:** 316366-2 **Common name:** naranjito, naranjuelo, níspero de saíno, pancayate, pangalleta, toco, totumito **Habit:** Shrub, Small tree, Tree, Liana **Use:** AF, HF, MA, ME **Origin:** Native **Region:** Caribbean **Dept.:** ANT, ATL, BOL, CES, CUN, LAG, MAG, SAN **Elev.:** 5-950 m.a.s.l. **IUCN Habitat code:** 1,2,3,4,14 **Conservation status: (IUCN)** LC **(CPSG)** Potential LC

2683 *Morisonia amplissima* (Lam.) Christenh. & Byng (Capparaceae) **IPNI:** 77184082-1 **Habit:** Shrub, Tree, Liana **Use:** HF, MA **Origin:** Native **Region:** Amazonia, Andean, Caribbean, Pacific **Dept.:** AMA, ANT, ATL, BOL, CUN, HUI, NAR, TOL, VAC **Elev.:** 30–1200 m a.s.l. **Conservation status: (IUCN)** NE

2684 *Morisonia crotonoides* (Kunth) Christenh. & Byng (Capparaceae) **IPNI:** 77183834-1 **Use:** ME **Origin:** Native **Conservation status: (IUCN)** NE

2685 *Morisonia detonsa* (Triana & Planch.) Christenh. & Byng (Capparaceae) **IPNI:** 77183701-1 **Common name:** copijuique **Habit:** Shrub, Small tree, Tree, Liana, Climbing **Use:** ME **Origin:** Native **Region:** Amazonia, Andean, Guiana Shield, Pacific **Dept.:** AMA, ANT, BOL, CAQ, CHO, CUN, MET, PUT, QUI, RIS, SAN, VAC, VAU **Elev.:** 50–1890 m a.s.l. **Conservation status: (IUCN)** NE

2686 *Morisonia flexuosa* L. (Capparaceae) **IPNI:** 316361-2 **Common name:** arara, olot **Habit:** Shrub, Small tree, Tree, Liana, Climbing **Use:** MA, ME **Origin:** Native **Region:** Caribbean **Dept.:** ATL, BOL, LAG, MAG **Elev.:** 0–120 m a.s.l. **Conservation status: (IUCN)** NE

2687 *Morisonia frondosa* (Jacq.) Christenh. & Byng (Capparaceae) **IPNI:** 77183570-1 **Common name:** concha prieta, contraprieta, guanabanito, olivo negro, sincogollo **Habit:** Shrub, Small tree, Tree **Use:** AF, FU, ME **Origin:** Native **Region:** Andean, Guiana Shield, Caribbean, Orinoquia, Pacific **Dept.:** ANT, ATL, BOL, BOY, CAQ, CES, CHO, CUN, LAG, MAG, MET, SAP, SAN, SUC, TOL **Elev.:** 20–1700 m a.s.l. **Conservation status: (IUCN)** NE

2688 *Morisonia indica* (L.) ined. (Capparaceae) **IPNI:** 77217338-1 **Common name:** auso, funita, naranjuelo, olivo, olivo macho, olivo santo, pachaca, palo blanco, tablón, tinto, tunito **Habit:** Shrub, Subshrub, Small tree, Tree **Use:** ME **Origin:** Endemic **Region:** Andean, Caribbean **Dept.:** ANT, ATL, BOL, CUN, LAG, MAG, SUC, TOL **Elev.:** 0–1350 m a.s.l. **Conservation status: (IUCN)** NE

2689 *Morisonia linearis* (Jacq.) Christenh. & Byng (Capparaceae) **IPNI:** 77184255-1 **Common name:** contraarará, kateesua, lengua de venado, medialuna, oivo, olivo, olivo macho **Habit:** Shrub, Small tree, Tree **Use:** AF, FU, HF, MA, ME, SU **Origin:** Native **Region:** Caribbean **Dept.:** ATL, BOL, LAG, MAG **Elev.:** 10–245 m a.s.l. **Conservation status: (IUCN)** NE

2690 *Morisonia nemorosa* (Jacq.) Christenh. & Byng (Capparaceae) **IPNI:** 77183738-1 **Common name:** calabazuelo, guevo de burro, güeva de burro, toco **Habit:** Shrub, Small tree, Tree **Use:** AF, EU, FU, MA, PO **Origin:** Native **Region:** Caribbean **Dept.:** ATL, BOL, LAG, MAG **Elev.:** 0–500 m a.s.l. **Conservation status: (IUCN)** NE

2691 *Morisonia oblongifolia* Britton (Capparaceae) **IPNI:** 164357-2 **Use:** HF **IUCN Habitat code:** 1,3,14 **Conservation status: (IUCN)** LC

2692 *Morisonia odoratissmia* (Jacq.) Christenh. & Byng (Capparaceae) **IPNI:** 77184072-1 **Common name:** bejuco inglés, fruta de burro, fruta de zorro, fruteloro, fruto de zorro, kapuchiru, naranjillo, naranjuelo, olivo, olivo hembra, olivo santo, palo de burro, tabloncillo, tablón, tinto **Habit:** Shrub, Small tree, Tree **Use:** EU, HF, ME, SU **Origin:** Native, Cultivated **Region:** Andean, Caribbean **Dept.:** ANT, ATL, BOL, COR, CUN, HUI, LAG, MAG, SAP, SUC, TOL, VAC **Elev.:** 5-1320 m.a.s.l. **Conservation status: (IUCN)** NE **(CPSG)** Potential LC

2693 *Morisonia osmantha* (Diels) Christenh. & Byng (Capparaceae) **IPNI:** 77184046-1 **Habit:** Shrub, Small tree, Tree **Use:** HF **Origin:** Native **Region:** Amazonia, Andean, Pacific **Dept.:** AMA, ANT, SAN **Elev.:** 20–1000 m a.s.l. **Conservation status: (IUCN)** NE

2694 *Morisonia pachaca* (Kunth) Christenh. & Byng (Capparaceae) **IPNI:** 77184114-1 **Common name:** cotorrero, jilrawal, naranjito, naranjuelo, toco negro **Habit:** Small tree, Tree **Use:** ME **Origin:** Native **Region:** Caribbean **Dept.:** ANT, ATL, BOL, LAG, MAG, NSA **Elev.:** 0–610 m a.s.l. **Conservation status: (IUCN)** NE

2695 *Morisonia pittieri* (Standl.) Christenh. & Byng (Capparaceae) **IPNI:** 77184198-1 **Habit:** Shrub, Small tree, Tree **Use:** ME **Origin:** Native **Region:** Caribbean, Pacific **Dept.:** ANT, BOL, CHO **Elev.:** 0–1000 m a.s.l. **Conservation status: (IUCN)** NE

2696 *Morisonia pulcherrima* (Jacq.) Christenh. & Byng (Capparaceae) **IPNI:** 77183588-1 **Common name:** huevo burro, huevoburro, jantukuru **Habit:** Shrub, Small tree **Use:** MA, ME, PO **Origin:** Native **Region:** Caribbean **Dept.:** ATL, BOL, MAG **Elev.:** 5–200 m a.s.l. **Conservation status: (IUCN)** NE

2697 *Morisonia sessilis* (Banks ex DC.) Christenh. & Byng (Capparaceae) **IPNI:** 77183680-1 **Common name:** bola de perro, cruceto, sietenudos **Habit:** Shrub, Small tree **Use:** MA **Origin:** Native **Region:** Caribbean **Dept.:** ATL, BOL, CUN, LAG, MAG, SUC **Elev.:** 5–700 m a.s.l. **Conservation status: (IUCN)** NE

2698 *Morisonia sola* (J.F.Macbr.) Christenh. & Byng (Capparaceae) **IPNI:** 77184186-1 **Common name:** cambuayo, clavo de agua **Habit:** Shrub, Small tree, Tree **Use:** ME **Origin:** Native **Region:** Amazonia, Andean, Guiana Shield, Caribbean, Orinoquia **Dept.:** ANT, CAQ, CES, MET, PUT, VAU **Elev.:** 200–1670 m a.s.l. **Conservation status: (IUCN)** NE

2699 *Morisonia verrucosa* (Jacq.) Christenh. & Byng (Capparaceae) **IPNI:** 77183644-1 **Habit:** Shrub, Small tree, Tree, Liana **Use:** MA **Origin:** Native **Region:** Caribbean **Dept.:** ATL, BOL, CES, HUI, LAG, MAG, SAN, SUC **Elev.:** 5–500 m a.s.l. **Conservation status: (IUCN)** NE

2700 *Carica papaya* L. (Caricaceae) **IPNI:** 30011248-2 **Common name:** chamburo, fruta bomba, lechosa, lechoso, mamaubo, mamona, melón, papaya, papaya calentana, papaya hawaiana, papaya hembra, papaya macho, papaya maradol, papaya melona, papayo, zapote **Habit:** Shrub, Small tree, Tree **Use:** AF, EU, GS, HF, IF, MA, ME, PO, SU **Origin:** Native, Cultivated **Region:** Amazonia, Andean, Orinoquia, Pacific **Dept.:** AMA, ANT, CHO, CUN, GUV, HUI, MAG, MET, NAR, TOL, VAU **Elev.:** 60–2160 m a.s.l. **IUCN Habitat code:** 1,2,3,4,5,14 **Conservation status: (IUCN)** DD

2701 *Jacaratia chocoensis* A.H.Gentry & Forero (Caricaceae) **IPNI:** 130967-2 **Habit:** Tree **Use:** GS **Origin:** Native **Region:** Andean **Dept.:** ANT, CHO **Elev.:** 630–1500 m a.s.l. **IUCN Habitat code:** 1 **Conservation status: (IUCN)** NE

2702 *Jacaratia digitata* (Poepp. & Endl.) Solms (Caricaceae) **IPNI:** 130970-2 **Common name:** barril, ceiba boba, papaya, papayuelo **Habit:** Tree **Use:** GS, HF, ME **Origin:** Native **Region:** Amazonia, Andean, Guiana Shield **Dept.:** AMA, MET, NAR, PUT **Elev.:** 120–1670 m a.s.l. **IUCN Habitat code:** 1,3,14 **Conservation status: (IUCN)** LC

2703 *Jacaratia spinosa* (Aubl.) A.DC. (Caricaceae) **IPNI:** 30104966-2 **Habit:** Shrub, Tree **Use:** GS, HF, ME **Origin:** Native **Region:** Andean, Pacific **Dept.:** ANT, CHO **Elev.:** 10–980 m a.s.l. **IUCN Habitat code:** 1,3,5,14 **Conservation status: (IUCN)** LC

2704 *Vasconcellea × pentagona* (Heilborn) Mabb. (Caricaceae) **IPNI:** 77164259-1 **Common name:** babaco, chamburo, papayuela **Use:** HF, ME **Origin:** Native **Conservation status: (IUCN)** NE

2705 *Vasconcellea cauliflora* (Jacq.) A.DC. (Caricaceae) **IPNI:** 675769-1 **Common name:** asparcuarguat, papaya de monte, papayo de montaña, papayuela, papayuelo, tapaculo **Habit:** Shrub, Small tree **Use:** GS, HF, ME **Origin:** Native **Region:** Andean, Guiana Shield, Caribbean, Pacific **Dept.:** ANT, ATL, BOL, CHO, LAG, MAG, MET, SUC, VAC **Elev.:** 100–2050 m a.s.l. **IUCN Habitat code:** 1,2,3,5,14 **Conservation status: (IUCN)** NE

2706 *Vasconcellea crassipetala* (V.M.Badillo) V.M.Badillo (Caricaceae) **IPNI:** 1015560-1 **Habit:** Subshrub, Small tree **Use:** GS, HF **Origin:** Native **Region:** Andean **Dept.:** ANT, CAU, HUI, QUI, RIS **Elev.:** 550–2600 m a.s.l. **IUCN Habitat code:** 1,3,14 **Conservation status: (IUCN)** NE

2707 *Vasconcellea goudotiana* Triana & Planch. (Caricaceae) **IPNI:** 675773-1 **Common name:** chupico, papaya del monte, papaya silvestre, papayo, papayote, papayuela, papayuelo, tapaculo **Habit:** Small tree, Tree **Use:** GS, HF, ME **Origin:** Native, Cultivated **Region:** Amazonia, Andean, Caribbean, Pacific **Dept.:** ANT, BOY, CAU, CUN, HUI, MAG, NSA, PUT, TOL, VAC **Elev.:** 20–2200 m a.s.l. **IUCN Habitat code:** 1,3,5,8,14 **Conservation status: (IUCN)** NE

2708 *Vasconcellea longiflora* (V.M.Badillo) V.M.Badillo (Caricaceae) **IPNI:** 1015565-1 **Habit:** Small tree **Use:** GS, HF **Origin:** Endemic **Region:** Andean **Dept.:** ANT, CAU, NAR, VAC **Elev.:** 1100-1710 m.a.s.l. **IUCN Habitat code:** 1 **Conservation status: (IUCN)** LC **(CPSG)** LC

2709 *Vasconcellea microcarpa* (Jacq.) A.DC. (Caricaceae) **IPNI:** 675778-1 **Common name:** higo de monte, papayo, papayuelo **Habit:** Small tree, Tree **Use:** GS, HF, ME **Origin:** Native **Region:** Amazonia, Andean, Guiana Shield, Orinoquia **Dept.:** AMA, ANT, CAL, CAS, CHO, CUN, MET, NAR **Elev.:** 100–2250 m a.s.l. **IUCN Habitat code:** 1,3,8,14 **Conservation status: (IUCN)** LC

2710 *Vasconcellea monoica* (Desf.) A.DC. (Caricaceae) **IPNI:** 675779-1 **Use:** GS, HF **Conservation status: (IUCN)** NE

2711 *Vasconcellea pubescens* A.DC. (Caricaceae) **IPNI:** 675782-1 **Common name:** cammon, chilacuan, chilacuán, chilguacán, chilhuacán, chilicuan, higuillo, papacú, papaya de olor, papaya de tierra fría, papaya silvestre, papayito de los andes, papayo, papayote, papayuela, papayuela maligna, papayuelo, tapaculo, tapacú, tapaya de altur **Habit:** Tree **Use:** AF, EU, GS, HF, ME **Origin:** Native, Cultivated **Region:** Andean **Dept.:** ANT, BOG, BOY, CAU, CUN, MET, NAR, PUT, SAN **Elev.:** 1100–3600 m a.s.l. **IUCN Habitat code:** 1,3,14 **Conservation status: (IUCN)** NE

2712 *Vasconcellea pulchra* (V.M.Badillo) V.M.Badillo (Caricaceae) **IPNI:** 1015572-1 **Habit:** Small tree **Use:** GS, HF **Origin:** Native **Region:** Andean **Dept.:** NAR **Elev.:** 1750–1750 m a.s.l. **IUCN Habitat code:** 14 **Conservation status: (IUCN)** NT

2713 *Vasconcellea sphaerocarpa* (García-Barr. & Hern.Cam.) V.M.Badillo (Caricaceae) **IPNI:** 1015573-1 **Habit:** Small tree **Use:** GS, HF, ME **Origin:** Endemic **Region:** Andean **Dept.:** CUN, SAN, VAC **Elev.:** 700-2050 m.a.s.l. **IUCN Habitat code:** 1,2,3,14 **Conservation status: (IUCN)** VU **(CPSG)** VU

2714 *Cleome aculeata* L. (Cleomaceae) **IPNI:** 316362-2 **Habit:** Herb **Use:** ME **Origin:** Native **Dept.:** TOL, VAC **Elev.:** 600–980 m a.s.l. **IUCN Habitat code:** 1,2,3,14 **Conservation status: (IUCN)** NE

2715 *Cleome anomala* Kunth (Cleomaceae) **IPNI:** 146919-1 **Common name:** centella, mismia, yerba de murciélago **Habit:** Herb, Shrub, Small tree **Use:** ME **Origin:** Native **Region:** Andean **Dept.:** ANT, BOG, BOY, CAU, CUN, HUI, NAR, PUT, QUI, RIS, VAC **Elev.:** 1490–3380 m a.s.l. **IUCN Habitat code:** 1,3,4,5,14 **Conservation status: (IUCN)** NE

2716 *Cleome arborea* Kunth (Cleomaceae) **IPNI:** 146924-1 **Habit:** Herb, Shrub, Subshrub **Use:** ME **Origin:** Native **Region:** Andean **Dept.:** ANT, CUN **Elev.:** 1000–1400 m a.s.l. **IUCN Habitat code:** 1,2,3,4,5,14 **Conservation status: (IUCN)** NE

2717 *Cleome gigantea* L. (Cleomaceae) **IPNI:** 147057-1 **Common name:** mismia, ámbar **Habit:** Herb, Subshrub **Use:** MA **Origin:** Native **Region:** Andean, Guiana Shield **Dept.:** ANT, CUN, HUI, MET, NAR, PUT, TOL **Elev.:** 350–1700 m a.s.l. **IUCN Habitat code:** 1,3,5,14 **Conservation status: (IUCN)** NE

2718 *Cleome glandulosa* Ruiz & Pav. ex DC. (Cleomaceae) **IPNI:** 147061-1 **Habit:** Herb, Shrub **Use:** MA **Origin:** Native **Region:** Andean **Dept.:** CES, QUI **Elev.:** 2300–3963 m a.s.l. **Conservation status: (IUCN)** NE

2719 *Cleome gynandra* L. (Cleomaceae) **IPNI:** 60518-2 **Common name:** barba de chivo, cabellos de ángel **Habit:** Herb **Use:** AF, EU, GS, HF, MA, ME, PO **Origin:** Naturalised **Region:** Caribbean **Dept.:** ATL, LAG, MAG, SAN **Elev.:** 100–800 m a.s.l. **IUCN Habitat code:** 2,14 **Conservation status: (IUCN)** NE

2720 *Cleome houtteana* Schltdl. (Cleomaceae) **IPNI:** 147105-1 **Habit:** Herb, Shrub **Use:** EU, MA, ME **Origin:** Naturalised **Region:** Andean, Caribbean **Dept.:** ANT, CUN, HUI, LAG **Elev.:** 40–2200 m a.s.l. **IUCN Habitat code:** 1,14 **Conservation status: (IUCN)** NE

2721 *Cleome parviflora* Kunth (Cleomaceae) **IPNI:** 147226-1 **Common name:** cabello de ángel, jaramago, rajateta, yerba garza **Habit:** Herb **Use:** ME **Origin:** Native **Region:** Amazonia, Andean, Orinoquia, Pacific **Dept.:** AMA, ANT, ARA, CAQ, CAU, CHO, MET, SAN, VAC **Elev.:** 5–1320 m a.s.l. **IUCN Habitat code:** 1,3,4,5,14 **Conservation status: (IUCN)** NE

2722 *Cleome pilosa* Benth. (Cleomaceae) **IPNI:** 147242-1 **Habit:** Herb, Subshrub **Use:** ME **Origin:** Native **Region:** Andean, Caribbean, Orinoquia, Pacific **Dept.:** ANT, BOY, CAL, CHO, CUN, MAG, MET, NSA, RIS **Elev.:** 100–1700 m a.s.l. **IUCN Habitat code:** 1,2,3 **Conservation status: (IUCN)** NE

2723 *Cleome speciosa* Raf. (Cleomaceae) **IPNI:** 147320-1 **Common name:** barba de sapo, cabellito de ángel, cabello de ángel, caracol, caracol de jardín, rabo de mico **Habit:** Herb, Subshrub **Use:** EU, HF, ME **Origin:** Naturalised **Region:** Amazonia, Andean, Caribbean, Pacific **Dept.:** ANT, BOY, CAL, CAU, CHO, CUN, MET, NAR, NSA, PUT, SAN, VAC **Elev.:** 60–1740 m a.s.l. **Conservation status: (IUCN)** NE

2724 *Cleome spinosa* Jacq. (Cleomaceae) **IPNI:** 60589-2 **Common name:** arañagato, barba de chivo, barba de sapo, cabello de ángel, cabellos de ángel, centella, desbaratabaile, desdicha, flor de muerto, flor del diablo, garcita, jazmín del río, mismia, pitopito, rompepollera **Habit:** Herb, Shrub **Use:** EU, GS, HF, MA, ME, PO **Origin:** Native **Region:** Amazonia, Andean, Caribbean, Pacific **Dept.:** AMA, ANT, ATL, BOL, CAS, CES, CHO, COR, CUN, HUI, LAG, MAG, MET, NAR, QUI, SAP, SAN, TOL, VAC **Elev.:** 0-1800 m.a.s.l. **IUCN Habitat code:** 1,2,3,4,5,14 **Conservation status: (IUCN)** NE **(CPSG)** Potential LC

2725 *Cleome viscosa* L. (Cleomaceae) **IPNI:** 320133-2 **Habit:** Herb **Use:** AF, EU, GS, HF, MA, ME, SU **Origin:** Naturalised **Region:** Caribbean **Dept.:** SAP **Elev.:** 0–300 m a.s.l. **IUCN Habitat code:** 14 **Conservation status: (IUCN)** NE

2726 *Moringa oleifera* Lam. (Moringaceae) **IPNI:** 584736-1 **Common name:** aceite, aceite de Ben, bambou-bananier, ben, ben olifere, benzolive, chagualo, cucharo, cucharo colorado, espadero, goma, graines benné, jacinto, ángela, árbol vela **Habit:** Small tree **Use:** AF, EU, FU, GS, HF, IF, MA, ME, PO, SU **Origin:** Cultivated **Region:** Caribbean **Dept.:** ANT, BOL, COR, LAG, MAG, SAP, VAC **Elev.:** 35–1050 m a.s.l. **IUCN Habitat code:** 1,2,3,14 **Conservation status: (IUCN)** LC

2727 *Reseda luteola* L. (Resedaceae) **IPNI:** 715535-1 **Use:** HF, MA, ME **IUCN Habitat code:** 14 **Conservation status: (IUCN)** NE

2728 *Reseda odorata* L. (Resedaceae) **IPNI:** 715569-1 **Use:** EU, HF, MA, ME **Dept.:** BOG**IUCN Habitat code:** 14 **Conservation status: (IUCN)** NE

2729 *Tropaeolum magnificum* Sparre (Tropaeolaceae) **IPNI:** 260036-2 **Habit:** Climbing **Use:** EU **Origin:** Native, Cultivated **Region:** Andean **Dept.:** ANT **Elev.:** 2200–2200 m a.s.l. **Conservation status: (IUCN)** VU

2730 *Tropaeolum majus* L. (Tropaeolaceae) **IPNI:** 310974-2 **Common name:** antioqueña, cachaco, cachaco de muladar, cahaco de muladar, capuchina, caracucho, caracucho grande, coqueta, espuela de galán, llagas de Cristo, malva vagabunda, malva vagamunda **Habit:** Climbing **Use:** EU, GS, HF, MA, ME **Origin:** Cultivated, Naturalised **Region:** Andean **Dept.:** ANT, BOG, CUN, PUT, QUI, TOL **Elev.:** 2850–2900 m a.s.l. **IUCN Habitat code:** 1,2,3,8,14 **Conservation status: (IUCN)** NE

2731 *Tropaeolum peltophorum* Benth. (Tropaeolaceae) **IPNI:** 308535-2 **Common name:** flor de pajarito, pajarito **Habit:** Climbing **Use:** EU **Origin:** Native **Region:** Andean **Dept.:** CAU, VAC **Elev.:** 1750–2350 m a.s.l. **IUCN Habitat code:** 1 **Conservation status: (IUCN)** NE

2732 *Tropaeolum peregrinum* L. (Tropaeolaceae) **IPNI:** 377401-1 **Common name:** pajarito amarillo **Habit:** Climbing **Use:** EU, HF **Origin:** Cultivated **Region:** Andean **Dept.:** BOG, CUN **Elev.:** 2600–2600 m a.s.l. **Conservation status: (IUCN)** NE

2733 *Tropaeolum tuberosum* Ruiz & Pav. (Tropaeolaceae) **IPNI:** 310353-2 **Common name:** cubia, cubia cimarrona, cubias, cubio, cubios, majua, mashua, nabio, nabios, nabo, nabos **Habit:** Climbing **Use:** EU, HF, ME **Origin:** Native, Cultivated **Region:** Andean **Dept.:** BOG, BOY, CAU, CUN, NAR, RIS, TOL **Elev.:** 2800–3600 m a.s.l. **IUCN Habitat code:** 1,3,4,14 **Conservation status: (IUCN)** NE

Buxales

2734 *Buxus microphylla* Siebold & Zucc. (Buxaceae) **IPNI:** 340303-1 **Habit:** Shrub **Use:** EU, MA, ME **Origin:** Cultivated **Region:** Andean **Dept.:** BOG, CUN, QUI **Elev.:** 1500–2600 m a.s.l. **Conservation status: (IUCN)** LC

2735 *Buxus sempervirens* L. (Buxaceae) **IPNI:** 340317-1 **Use:** EU, MA, ME, PO **Region:** Andean **Dept.:** BOG**Conservation status: (IUCN)** LC

2736 *Styloceras laurifolium* (Willd.) Kunth (Buxaceae) **IPNI:** 356634-1 **Common name:** cachitos, cacho de venado, limoncillo de cachos, limoncito de cachos, naranjillo, tijereto, totumito **Habit:** Tree **Use:** HF **Origin:** Native **Region:** Andean **Dept.:** BOG, CES, CUN, LAG, QUI, RIS **Elev.:** 2000-2785 m.a.s.l. **IUCN Habitat code:** 1,3,14 **Conservation status: (IUCN)** NE **(CPSG)** Potential LC

Caryophyllales

2737 *Achatocarpus nigricans* Triana (Achatocarpaceae) **IPNI:** 58636-1 **Common name:** cocacá, como, dulomoco, limonacho, pepito, pino manteco, rabo de iguana, tintorera, totocal, totocá **Habit:** Shrub, Tree **Use:** AF, HF **Origin:** Native **Region:** Caribbean **Dept.:** ANT, ATL, BOL, CUN, HUI, LAG, MAG, SAP, SUC, TOL, VAC **Elev.:** 10–1300 m a.s.l. **IUCN Habitat code:** 1,2,3,4,14 **Conservation status: (IUCN)** LC

2738 *Carpobrotus acinaciformis* (L.) L.Bolus (Aizoaceae) **IPNI:** 360773-1 **Use:** EU, ME **Region:** Andean **Dept.:** BOG**Conservation status: (IUCN)** NE

2739 *Carpobrotus edulis* (L.) N.E.Br. (Aizoaceae) **IPNI:** 1200427-2 **Use:** AF, EU, GS, HF, MA, ME **Region:** Andean **Dept.:** BOG**Conservation status: (IUCN)** NE

2740 *Delosperma cooperi* (Hook.f.) L.Bolus (Aizoaceae) **IPNI:** 361479-1 **Use:** EU **Region:** Andean **Dept.:** BOG**Conservation status: (IUCN)** NE

2741 *Lampranthus aureus* (L.) N.E.Br. (Aizoaceae) **IPNI:** 362165-1 **Use:** EU **Region:** Andean **Dept.:** BOG**Conservation status: (IUCN)** NE

2742 *Lampranthus multiradiatus* (Jacq.) N.E.Br. (Aizoaceae) **IPNI:** 362274-1 **Use:** EU, ME **Region:** Andean **Dept.:** BOG**Conservation status: (IUCN)** NE

2743 *Lampranthus productus* (Haw.) N.E.Br. (Aizoaceae) **IPNI:** 362301-1 **Common name:** clavel chino, clavel de día **Habit:** Herb **Use:** EU **Origin:** Cultivated **Region:** Andean **Dept.:** CUN **Elev.:** 2600–2600 m a.s.l. **Conservation status: (IUCN)** NE

2744 *Lampranthus spectabilis* (Haw.) N.E.Br. (Aizoaceae) **IPNI:** 362323-1 **Use:** EU **Region:** Andean **Dept.:** BOG**IUCN Habitat code:** 14 **Conservation status: (IUCN)** NE

2745 *Malephora crocea* (Jacq.) Schwantes (Aizoaceae) **IPNI:** 1011059-2 **Use:** EU **Region:** Andean **Dept.:** BOG**IUCN Habitat code:** 14 **Conservation status: (IUCN)** NE

2746 *Mesembryanthemum cordifolium* L.f. (Aizoaceae) **IPNI:** 322566-2 **Common name:** escarcha, rocío, tristeza **Habit:** Herb **Use:** EU, ME **Origin:** Naturalised **Region:** Andean **Dept.:** CUN **Elev.:** 2500–2600 m a.s.l. **IUCN Habitat code:** 1,14 **Conservation status: (IUCN)** NE

2747 *Sesuvium portulacastrum* (L.) L. (Aizoaceae) **IPNI:** 235145-2 **Common name:** chupahuevo, verdolaga **Habit:** Herb **Use:** AF, EU, GS, HF, ME **Origin:** Native **Region:** Caribbean, Pacific **Dept.:** ANT, ATL, BOL, CHO, LAG, MAG **Elev.:** 0–2 m a.s.l. **IUCN Habitat code:** 1,2,3,5,14 **Conservation status: (IUCN)** NE

2748 *Trianthema portulacastrum* L. (Aizoaceae) **IPNI:** 322565-2 **Common name:** araña de perro, verdolaga **Habit:** Herb **Use:** AF, EU, GS, HF, ME, PO **Origin:** Native **Region:** Caribbean **Dept.:** ANT, LAG **Elev.:** 0–120 m a.s.l. **IUCN Habitat code:** 1,2,3,5,14 **Conservation status: (IUCN)** NE

2749 *Achyranthes aspera* L. (Amaranthaceae) **IPNI:** 2468-2 **Common name:** abrojo, cadillo, cadillo de mazorca, canilla de golero, chiche de borugo, chichegato, chimboeborugo, cocúa, cola de caimán, descancel, discancer, discancer grande, iscanser, kuamna, mazorca, mevoycontigo, pegapega, perico, pinga de gato, rabo de caimán, r **Habit:** Herb **Use:** AF, EU, GS, HF, MA, ME, SU **Origin:** Naturalised **Region:** Andean, Caribbean, Orinoquia, Pacific **Dept.:** ANT, ATL, BOL, CAL, CAU, CES, CHO, COR, CUN, HUI, MAG, MET, NAR, SAP, SUC, TOL, VAC **Elev.:** 0–1600 m a.s.l. **IUCN Habitat code:** 1,2,3,4,5,14 **Conservation status: (IUCN)** NE

2750 *Alternanthera albotomentosa* Suess. (Amaranthaceae) **IPNI:** 10265-2 **Common name:** abrojo, abrojo blanco, amorseco, arabia, bacalao, botoncito blanco, cordoncillo, discancer chiquito, tapón, tripa de pollo **Habit:** Herb **Use:** ME **Origin:** Native **Region:** Andean, Caribbean,

Orinoquia, Pacific **Dept.:** ANT, ATL, BOL, BOY, CHO, COR, CUN, HUI, LAG, MAG, MET, NSA, QUI, SAN, SUC, TOL, VAC **Elev.:** 0-2500 m.a.s.l. **IUCN Habitat code:** 1,2,3,5,14 **Conservation status: (IUCN)** NE **(CPSG)** Potential LC

2751 *Alternanthera bettzickiana* (Regel) G.Nicholson (Amaranthaceae) **IPNI:** 10291-2 **Common name:** manito de Dios, té, té verde **Habit:** Herb **Use:** EU, HF, ME **Origin:** Cultivated **Region:** Andean, Caribbean, Orinoquia, Pacific **Dept.:** BOG, BOL, CHO, CUN, MAG, NSA, QUI, SAN, VAC, VAU **Elev.:** 0–2600 m a.s.l. **IUCN Habitat code:** 1,2,3,5,14 **Conservation status: (IUCN)** NE

2752 *Alternanthera brasiliana* (L.) Kuntze (Amaranthaceae) **IPNI:** 10294-2 **Common name:** escancé gigante, iscancel chiquito, penicilina, veterina **Habit:** Herb **Use:** EU, HF, ME **Origin:** Native **Region:** Andean, Caribbean **Dept.:** BOG, BOL, BOY, CAU, CES, HUI, MAG, NSA, VAC **Elev.:** 0–2500 m a.s.l. **IUCN Habitat code:** 1,2,3,4,5,14 **Conservation status: (IUCN)** NE

2753 *Alternanthera caracasana* Kunth (Amaranthaceae) **IPNI:** 327344-2 **Common name:** abrojo, almorrana, verdolaga blanca **Habit:** Herb **Use:** EU **Origin:** Native **Region:** Andean **Dept.:** BOG, BOY, CUN, NAR, QUI, SAN, TOL **Elev.:** 500–2800 m a.s.l. **IUCN Habitat code:** 1,3,8,14 **Conservation status: (IUCN)** NE

2754 *Alternanthera ficoidea* (L.) P.Beauv. (Amaranthaceae) **IPNI:** 1078972-2 **Common name:** descansé menudo, guardaparque, sanguinaria, siempreviva **Use:** EU, HF, ME **Origin:** Native **Dept.:** BOG**IUCN Habitat code:** 1,4 **Conservation status: (IUCN)** NE

2755 *Alternanthera halimifolia* (Lam.) Standl. ex Pittier (Amaranthaceae) **IPNI:** 10362-2 **Common name:** coralillo, schasha **Use:** ME **Origin:** Native **Dept.:** BOG**IUCN Habitat code:** 1,2,3,5,14 **Conservation status: (IUCN)** NE

2756 *Alternanthera laguroides* (Standl.) Standl. (Amaranthaceae) **IPNI:** 10377-2 **Use:** ME **IUCN Habitat code:** 1,2,14 **Conservation status: (IUCN)** NE

2757 *Alternanthera lanceolata* (Benth.) Schinz (Amaranthaceae) **IPNI:** 10378-2 **Common name:** borrachera, borrachera del yagé, borrachero, cancer blanco, cancer morado, cáncer blanco, cáncer morado, descancel, descancé, descancé de Castilla, descansel, discancer chiquito, discancer grande, escancel, escansel, escansé morado, iscancer, iscanser, la **Habit:** Herb, Climbing **Use:** ME **Origin:** Native **Region:** Amazonia, Andean, Caribbean, Pacific **Dept.:** AMA, ANT, BOG, BOL, BOY, CAL, CAU, CUN, HUI, NAR, PUT, QUI, RIS, SAN, TOL **Elev.:** 0-3500 m.a.s.l. **IUCN Habitat code:** 1,3,5,14 **Conservation status: (IUCN)** NE **(CPSG)** Potential LC

2758 *Alternanthera paronychioides* A.St.-Hil. (Amaranthaceae) **IPNI:** 327343-2 **Common name:** abrojo, descansé, inosadilla, pampera, tripa de pollo, tripapollo, yerba de ojo **Habit:** Herb **Use:** HF, ME **Origin:** Native **Region:** Andean, Caribbean, Orinoquia **Dept.:** ANT, ATL, BOL, BOY, CAL, COR, MAG, VID **Elev.:** 0–1500 m a.s.l. **IUCN Habitat code:** 2,3,5,14 **Conservation status: (IUCN)** NE

2759 *Alternanthera philoxeroides* (Mart.) Griseb. (Amaranthaceae) **IPNI:** 1024156-2 **Use:** AF, EU, HF, ME **Origin:** Native **IUCN Habitat code:** 14 **Conservation status: (IUCN)** NE

2760 *Alternanthera porrigens* (Jacq.) Kuntze (Amaranthaceae) **IPNI:** 10447-2 **Common name:** cartagena **Habit:** Herb **Use:** ME **Origin:** Native **Region:** Andean, Caribbean **Dept.:** BOG, BOY, CAU, CES, CUN, MAG, NAR, NSA, PUT, SAN **Elev.:** 500–3500 m a.s.l. **IUCN Habitat code:** 1,2,3,4,14 **Conservation status: (IUCN)** NE

2761 *Alternanthera pubiflora* (Benth.) Kuntze (Amaranthaceae) **IPNI:** 10450-2 **Common name:** abrojo, abrojo rojo, almorrana, alpargatera, alternatera, amaranto, arancel, bacalao, escances morado, pegapega, sanguinaria, vinotinto **Habit:** Herb, Climbing **Use:** ME **Origin:** Native **Region:** Andean, Caribbean, Pacific **Dept.:** ATL, BOG, BOL, BOY, CHO, CUN, LAG, NAR, QUI, VAC **Elev.:** 0–3000 m a.s.l. **IUCN Habitat code:** 1,2,3,14 **Conservation status: (IUCN)** NE

2762 *Alternanthera pungens* Kunth (Amaranthaceae) **IPNI:** 327342-2 **Common name:** abrojo, almorrana, verdolaga, vidriosa **Habit:** Herb **Use:** EU, HF, MA, ME **Origin:** Native **Region:** Andean **Dept.:** ANT, BOG, BOY, CAU, CUN, HUI, NAR, TOL, VAC **Elev.:** 300–2000 m a.s.l. **IUCN Habitat code:** 1,2,3,14 **Conservation status: (IUCN)** NE

2763 *Alternanthera sessilis* (L.) R.Br. ex DC. (Amaranthaceae) **IPNI:** 1069792-2 **Common name:** abrojo, alpargatera, amaranto, botoncillo, descancé, escansel, iscansé, moradilla, pampera, pimpollo **Habit:** Herb **Use:** AF, EU, GS, HF, MA, ME, SU **Origin:** Native **Region:** Andean, Caribbean, Orinoquia, Pacific **Dept.:** ANT, BOL, CAU, CHO, COR, CUN, HUI, MAG, MET, NAR, QUI, TOL, VAC **Elev.:** 0–2000 m a.s.l. **IUCN Habitat code:** 1,2,3,5,14 **Conservation status: (IUCN)** LC

2764 *Amaranthus atropurpureus* Roxb. (Amaranthaceae) **IPNI:** 59409-1 **Use:** AF, HF, ME **Conservation status: (IUCN)** NE

2765 *Amaranthus australis* (A.Gray) J.D.Sauer (Amaranthaceae) **IPNI:** 10611-2 **Common name:** bleo **Habit:** Liana **Use:** HF **Origin:** Native **Region:** Caribbean, Pacific **Dept.:** CHO, COR, MAG **Elev.:** 15–100 m a.s.l. **IUCN Habitat code:** 1 **Conservation status: (IUCN)** NE

2766 *Amaranthus blitum* L. (Amaranthaceae) **IPNI:** 10619-2 **Common name:** amaranto espinoso, bledo **Habit:** Herb **Use:** AF, EU, GS, HF, MA, ME **Origin:** Naturalised **Region:** Andean, Pacific **Dept.:** ANT, BOG, CHO, CUN, NAR, QUI, TOL, VAC **Elev.:** 80–1900 m a.s.l. **IUCN Habitat code:** 1,3,14 **Conservation status: (IUCN)** NE

2767 *Amaranthus caudatus* L. (Amaranthaceae) **IPNI:** 316347-2 **Common name:** abanico, achita, amaranto, ataco, ataco morado, bledo, bleo, borlas, coime, cresta de gallo, felpa, kiwicha, kiwincha, moco de pavo **Habit:** Shrub **Use:** AF, EU, GS, HF, MA, ME, SU **Origin:** Cultivated, Naturalised **Region:** Andean, Caribbean **Dept.:** BOG, BOL, CAL, CHO, CUN, HUI, MET, NAR, QUI **Elev.:** 400–2600 m a.s.l. **IUCN Habitat code:** 1,3,14 **Conservation status: (IUCN)** NE

2768 *Amaranthus crassipes* Schltdl. (Amaranthaceae) **IPNI:** 327385-2 **Common name:** bledo **Habit:** Herb **Use:** HF **Origin:** Native **Region:** Caribbean **Dept.:** ATL, LAG, MAG **Elev.:** 5–30 m a.s.l. **IUCN Habitat code:** 1,2,3,14 **Conservation status: (IUCN)** NE

2769 *Amaranthus cruentus* L. (Amaranthaceae) **IPNI:** 10631-2 **Use:** AF, EU, GS, HF, MA, ME **Origin:** Naturalised **IUCN Habitat code:** 2,14 **Conservation status: (IUCN)** NE

2770 *Amaranthus hybridus* L. (Amaranthaceae) **IPNI:** 10648-2 **Common name:** bejuco reuma, bledo, bledo chico, bleo, cadillo, ortiga morada **Habit:** Herb **Use:** AF, EU, GS, HF, MA, ME, SU **Origin:** Native **Region:** Andean, Orinoquia **Dept.:** ANT, BOG, BOL, BOY, CAL, CAU, CHO, CUN, HUI, NAR, PUT, QUI, SAN, TOL, VAC **Elev.:** 400-2700 m.a.s.l. **IUCN Habitat code:** 1,2,3,4,5,14 **Conservation status: (IUCN)** NE **(CPSG)** Potential LC

2771 *Amaranthus hypochondriacus* L. (Amaranthaceae) **IPNI:** 10665-2 **Use:** AF, EU, GS, HF, MA, ME, PO **Region:** Andean **Dept.:** BOG**Conservation status: (IUCN)** NE

2772 *Amaranthus spinosus* L. (Amaranthaceae) **IPNI:** 10711-2 **Common name:** bledo, bledo blanco, bledo espinoso, bledo macho, cáncer de laguna, cáncer espinoso, escansé espinoso, husquilete, husquilla, rabo de gallo, siempreviva **Habit:** Herb, Subshrub **Use:** AF, GS, HF, MA, ME, PO, SU **Origin:** Native **Region:** Andean, Caribbean, Orinoquia, Pacific **Dept.:** ANT, ATL, BOL, CAL, CAU, CHO, COR, HUI, MAG, MET, NAR, NSA, PUT, QUI, SAN, TOL, VAC **Elev.:** 0-2000 m.a.s.l. **IUCN Habitat code:** 1,2,3,4,5,14 **Conservation status: (IUCN)** NE **(CPSG)** Potential LC

2773 *Amaranthus tortuosus* Hornem. (Amaranthaceae) **IPNI:** 59693-1 **Common name:** amaranto, amaranto blanco, atakco, bledo, bledo blanco, bledo chico, bledo colorado, bledo colorao, bledo de espina, bledo de puerco, bledo hembra, bledo liso, bledo maleza, bledo rojizo, bleo, bleo de puerco, calalú, gleo, lirgua, pira, yuyo, yuyo hembra **Habit:** Herb **Use:** AF, EU, GS, HF,

MA, ME **Origin:** Native **Region:** Amazonia, Andean, Guiana Shield, Caribbean, Orinoquia, Pacific **Dept.:** AMA, ANT, ARA, ATL, BOL, BOY, CAL, CAS, CAU, CES, CHO, COR, CUN, HUI, LAG, MAG, MET, NAR, NSA, QUI, RIS, SAP, SAN, SUC, TOL, VAC **Elev.:** 0–2000 m a.s.l. **IUCN Habitat code:** 1,2,3,4,5,14 **Conservation status: (IUCN)** NE

2774 *Amaranthus tricolor* L. (Amaranthaceae) **IPNI:** 327386-2 **Common name:** ala de loro, amaranto, cucaracha **Habit:** Herb **Use:** EU, GS, HF, MA, ME, PO, SU **Origin:** Cultivated **Region:** Andean **Dept.:** QUI **Elev.:** 1560–1560 m a.s.l. **IUCN Habitat code:** 14 **Conservation status: (IUCN)** NE

2775 *Amaranthus viridis* L. (Amaranthaceae) **IPNI:** 316349-2 **Common name:** amaranto, ataco, bledo, bledo blanco, bledo chico, bledo manso, bleo, bleo blanco, bleo negro, hierba de puerco **Habit:** Herb **Use:** AF, GS, HF, MA, ME **Origin:** Native **Region:** Andean, Caribbean, Orinoquia **Dept.:** ANT, ATL, BOL, CAS, GUV, HUI, LAG, MAG, MET, QUI, TOL, VAC, VAU **Elev.:** 0–1600 m a.s.l. **IUCN Habitat code:** 1,2,3,14 **Conservation status: (IUCN)** NE

2776 *Beta vulgaris* L. (Amaranthaceae) **IPNI:** 164505-1 **Common name:** remolacha **Habit:** Herb **Use:** AF, EU, FU, GS, HF, MA, ME, PO **Origin:** Cultivated **Region:** Andean **Dept.:** ANT, BOG, BOY, CAL, CUN, NAR, TOL, VAC **Elev.:** 1000–2600 m a.s.l. **IUCN Habitat code:** 1,3,4,14 **Conservation status: (IUCN)** NE

2777 *Blutaparon portulacoides* (A.St.-Hil.) Mears (Amaranthaceae) **IPNI:** 34030-2 **Habit:** Herb **Use:** EU **Origin:** Native **Region:** Caribbean **Dept.:** LAG **Elev.:** 0–20 m a.s.l. **IUCN Habitat code:** 2 **Conservation status: (IUCN)** NE

2778 *Blutaparon vermiculare* (L.) Mears (Amaranthaceae) **IPNI:** 34033-2 **Habit:** Herb **Use:** AF, EU, GS, HF, ME, SU **Origin:** Native **Region:** Caribbean, Pacific **Dept.:** ATL, BOL, MAG, NAR, SAP **Elev.:** 0–20 m a.s.l. **IUCN Habitat code:** 2,5,14 **Conservation status: (IUCN)** NE

2779 *Celosia argentea* L. (Amaranthaceae) **IPNI:** 316350-2 **Common name:** abanico, borlas, celosia, cresta de gallo, crestas, crestegallo, manitas, moco de pavo, mocoepavo, moño de seda, plumosa, suspiro, terciopelo, tusa **Habit:** Herb **Use:** AF, EU, GS, HF, MA, ME, PO, SU **Origin:** Cultivated, Naturalised **Region:** Amazonia, Andean, Caribbean, Orinoquia, Pacific **Dept.:** AMA, ANT, BOG, CHO, CUN, GUA, MAG, MET, NAR, PUT, QUI, RIS, SAP, TOL, VAC, VAU **Elev.:** 100–1500 m a.s.l. **IUCN Habitat code:** 1,2,3,4,5,14 **Conservation status: (IUCN)** LC

2780 *Celosia grandifolia* Moq. (Amaranthaceae) **IPNI:** 59885-1 **Habit:** Shrub **Use:** ME **Origin:** Native **Region:** Andean **Dept.:** HUI, VAC **Elev.:** 1900–2100 m a.s.l. **Conservation status: (IUCN)** NE

2781 *Celosia virgata* Jacq. (Amaranthaceae) **IPNI:** 60011-1 **Habit:** Herb **Use:** ME **Origin:** Native **Region:** Andean, Caribbean **Dept.:** CAU, CUN, HUI, LAG, MAG, NAR, TOL, VAC **Elev.:** 340–1300 m a.s.l. **IUCN Habitat code:** 2,3,14 **Conservation status: (IUCN)** NE

2782 *Chamissoa altissima* (Jacq.) Kunth (Amaranthaceae) **IPNI:** 60077-1 **Common name:** bejuco de guacharaca, cadillo, lyann panye, pinta bollo, pintabollo, plumilla, siempreviva **Habit:** Climbing **Use:** AF, HF, MA, ME, PO **Origin:** Native **Region:** Amazonia, Andean, Guiana Shield, Caribbean, Orinoquia, Pacific **Dept.:** AMA, ANT, ATL, BOG, BOL, CAQ, CAS, CES, CHO, CUN, GUV, HUI, LAG, MAG, MET, NSA, PUT, QUI, RIS, SAN, SUC, TOL, VAC, VID **Elev.:** 0-2500 m.a.s.l. **IUCN Habitat code:** 1,2,3,4,5,14 **Conservation status: (IUCN)** NE **(CPSG)** Potential LC

2783 *Chenopodiastrum murale* (L.) S.Fuentes, Uotila & Borsch (Amaranthaceae) **IPNI:** 77121007-1 **Common name:** cenizo **Habit:** Herb **Use:** AF, EU, GS, HF, ME **Origin:** Naturalised **Region:** Andean **Dept.:** BOG, CUN, NSA **Elev.:** 2300–2900 m a.s.l. **IUCN Habitat code:** 3,14 **Conservation status: (IUCN)** NE

2784 *Chenopodium album* L. (Amaranthaceae) **IPNI:** 55993-2 **Habit:** Herb **Use:** AF, EU, GS, HF, MA, ME, PO **Origin:** Naturalised **Region:** Andean **Dept.:** BOG, CUN **Elev.:** 2700–2700 m a.s.l. **IUCN Habitat code:** 1,14 **Conservation status: (IUCN)** NE

2785 *Chenopodium hircinum* Schrad. (Amaranthaceae) **IPNI:** 164974-1 **Use:** GS, HF, ME **Conservation status: (IUCN)** NE

2786 *Chenopodium pallidicaule* Aellen (Amaranthaceae) **IPNI:** 56201-2 **Common name:** cañahua **Use:** HF **Dept.:** CUN**Conservation status: (IUCN)** NE

2787 *Chenopodium petiolare* Kunth (Amaranthaceae) **IPNI:** 165126-1 **Common name:** cenizo, cenizo blanco, quinua verde **Habit:** Herb **Use:** AF, HF, ME **Origin:** Native **Region:** Andean **Dept.:** BOG, BOY, CUN, NAR **Elev.:** 2100–2650 m a.s.l. **IUCN Habitat code:** 3,14 **Conservation status: (IUCN)** NE

2788 *Chenopodium quinoa* Willd. (Amaranthaceae) **IPNI:** 165175-1 **Common name:** achita, alpaquina, alpaquinoa, alpaquinua, arrocillo, arroz del perú, canigua, canihua, cenizo, jiura, liccha, pasca, quinoa, quinua, quinua cimarrona, suba **Habit:** Herb **Use:** GS, HF, ME **Origin:** Cultivated **Region:** Andean **Dept.:** BOG, BOY, CUN, PUT **Elev.:** 2200–2800 m a.s.l. **IUCN Habitat code:** 3,8,14 **Conservation status: (IUCN)** NE

2789 *Cyathula achyranthoides* (Kunth) Moq. (Amaranthaceae) **IPNI:** 60139-1 **Common name:** alacrán, cadillo, liendra de puerco, pega pega, pulmonaria, sanalototal **Habit:** Herb **Use:** EU, HF, MA, ME **Origin:** Native **Region:** Amazonia, Andean, Caribbean, Orinoquia, Pacific **Dept.:** AMA, ANT, ATL, BOY, CAQ, CHO, CUN, HUI, MAG, MET, NAR, NSA, QUI, SAN, VAC, VAU **Elev.:** 0–1400 m a.s.l. **IUCN Habitat code:** 1,2,3,5,8,14 **Conservation status: (IUCN)** NE

2790 *Cyathula prostrata* (L.) Blume (Amaranthaceae) **IPNI:** 60185-1 **Common name:** abrojo, abugua, alacrán, bledo blanco, bledo de puerco, cadillo, cadillo de bejuco, cadillo de bolsa, cadillo sacalotodo, chichiborugo, chichoborugo, chorrillo, cucua kjun, hierba de arco, hierba de golpe, hoja cucua, marranera, pega pega, pegahueso, pega **Habit:** Herb **Use:** AF, GS, HF, ME, SU **Origin:** Native **Region:** Amazonia, Andean, Caribbean, Orinoquia, Pacific **Dept.:** AMA, ANT, CAL, CAQ, CAS, CAU, CHO, CUN, GUA, HUI, MAG, MET, NAR, PUT, QUI, RIS, SAN, TOL, VAC **Elev.:** 0–1900 m a.s.l. **IUCN Habitat code:** 1,3,4,5,14 **Conservation status: (IUCN)** NE

2791 *Dysphania ambrosioides* (L.) Mosyakin & Clemants (Amaranthaceae) **IPNI:** 1198060-2 **Common name:** ambrosía, apazote, camatai, chemen contra, epasota, epazota, hierba sagrada, paico, pasota, pasote, pazota, pazote, semen contra, shipiwai, simen contra, wormweed, yerba santa, yerbasanta **Habit:** Herb **Use:** AF, GS, HF, MA, ME, PO, SU **Origin:** Native **Region:** Amazonia, Andean, Caribbean **Dept.:** AMA, ANT, ATL, BOG, BOY, CAU, CUN, HUI, MAG, NAR, NSA, PUT, SAN, VAC **Elev.:** 50–3600 m a.s.l. **IUCN Habitat code:** 1,2,3,5,14 **Conservation status: (IUCN)** NE

2792 *Dysphania anthelmintica* (L.) Mosyakin & Clemants (Amaranthaceae) **IPNI:** 1090820-2 **Use:** ME **Origin:** Native **IUCN Habitat code:** 14 **Conservation status: (IUCN)** NE

2793 *Gomphrena globosa* L. (Amaranthaceae) **IPNI:** 327333-2 **Common name:** San Diego, abanico, alumbra a las once, barbegallo, botón, camarón, cartagena, colocación, don diego morado, inmortal, inmortal blanca, inmortales, manto de cristo, para todo, perpetuas, sanagustín, siempreviva, terciopelo, viravira, yerbavirgen **Habit:** Herb **Use:** EU, GS, HF, MA, ME, SU **Origin:** Native, Cultivated **Region:** Amazonia, Andean, Caribbean, Orinoquia **Dept.:** AMA, BOG, BOY, CAQ, CAS, CAU, CUN, MAG, MET, NAR, PUT, QUI, SAP, SAN, VAC **Elev.:** 27–2600 m a.s.l. **IUCN Habitat code:** 1,2,3,4,5,14 **Conservation status: (IUCN)** NE

2794 *Gomphrena serrata* L. (Amaranthaceae) **IPNI:** 112097-2 **Common name:** abrojo, arrasa con todo, perpetua **Habit:** Herb **Use:** ME **Origin:** Native **Region:** Andean, Caribbean, Orinoquia, Pacific **Dept.:** ANT, ATL, BOY, CAL, CAU, CES, CUN, HUI, MAG, MET, NAR, QUI, TOL, VAC **Elev.:** 25-2570 m.a.s.l. **IUCN Habitat code:** 1,2,3,14 **Conservation status: (IUCN)** NE **(CPSG)** Potential LC

2795 *Guilleminea densa* (Willd. ex Schult.) Moq. (Amaranthaceae) **IPNI:** 1183159-2 **Habit:** Herb **Use:** HF, ME **Origin:** Native **Region:** Andean **Dept.:** NAR, VAC **Elev.:** 1100–2600 m a.s.l. **IUCN Habitat code:** 1,14 **Conservation status: (IUCN)** NE

2796 *Heterostachys ritteriana* (Moq.) Ung.-Sternb. (Amaranthaceae) **IPNI:** 165886-1 **Common name:** hierba de vidrio, yerba de vidrio, yerbaevidrio **Habit:** Herb **Use:** HF **Origin:** Native **Region:** Caribbean **Dept.:** LAG **Elev.:** 0–10 m a.s.l. **IUCN Habitat code:** 2,3 **Conservation status: (IUCN)** NE

2797 *Iresine diffusa* Humb. & Bonpl. ex Willd. (Amaranthaceae) **IPNI:** 129900-2 **Common name:** alverjón, amaranto, barbas de viejo, brasileña, calambombo, capitana, capitana macho, cáncer, cáncer sol, descancer común, descancé, descancé común, descancé gigante, escudilla, ilusión, lucero, macán borrachero, molleja, moradita, patepava, pinta amarill **Habit:** Herb, Climbing **Use:** EU, GS, HF, MA, ME, SU **Origin:** Native, Cultivated **Region:** Amazonia, Andean, Caribbean, Orinoquia, Pacific **Dept.:** ANT, ATL, BOG, BOL, BOY, CAL, CAQ, CAU, CES, CHO, CUN, HUI, LAG, MAG, MET, NAR, NSA, PUT, QUI, RIS, SAP, SAN, SUC, TOL, VAC **Elev.:** 0-3500 m.a.s.l. **IUCN Habitat code:** 1,2,3,4,5,8,14 **Conservation status: (IUCN)** NE **(CPSG)** Potential LC

2798 *Iresine latifolia* (M.Martens & Galeotti) Benth. & Hook.f. (Amaranthaceae) **IPNI:** 60789-1 **Habit:** Climbing **Use:** ME **Origin:** Native **Region:** Andean, Caribbean, Pacific **Dept.:** ATL, BOL, CAU, CHO **Elev.:** 6–2920 m a.s.l. **Conservation status: (IUCN)** NE

2799 *Iresine rhizomatosa* Standl. (Amaranthaceae) **IPNI:** 129956-2 **Use:** ME **Origin:** Native **Conservation status: (IUCN)** NE

2800 *Pedersenia argentata* (Mart.) Holub (Amaranthaceae) **IPNI:** 1002675-1 **Common name:** amaranto, cana de vieja, lagaña de vieja, plumilla **Habit:** Climbing **Use:** ME **Origin:** Native **Region:** Andean, Caribbean **Dept.:** ATL, BOL, BOY, CAU, CUN, MAG, VAC **Elev.:** 50–1800 m a.s.l. **IUCN Habitat code:** 1,14 **Conservation status: (IUCN)** NE

2801 *Pfaffia glomerata* (Spreng.) Pedersen (Amaranthaceae) **IPNI:** 191400-2 **Common name:** corango sempreviva **Use:** HF, ME **Origin:** Native **IUCN Habitat code:** 1,14 **Conservation status: (IUCN)** NE

2802 *Pfaffia iresinoides* (Kunth) Spreng. (Amaranthaceae) **IPNI:** 61055-1 **Common name:** madre de la yuca, plumaria cimarrona, pulmonaria **Habit:** Shrub **Use:** ME **Origin:** Native **Region:** Amazonia, Andean, Caribbean, Orinoquia **Dept.:** AMA, ANT, ATL, BOG, BOY, CAQ, CAS, CES, CUN, GUA, LAG, MAG, MET, NSA, SAN, TOL **Elev.:** 0-2500 m.a.s.l. **IUCN Habitat code:** 1,2,3,14 **Conservation status: (IUCN)** NE **(CPSG)** Potential LC

2803 *Pfaffia tuberosa* (Spreng.) Hicken (Amaranthaceae) **IPNI:** 1219488-2 **Habit:** Herb **Use:** ME **Origin:** Native **Region:** Amazonia, Andean, Orinoquia **Dept.:** CAQ, CAS, CUN, GUV, MET, TOL, VID **Elev.:** 150–1800 m a.s.l. **IUCN Habitat code:** 1,3,14 **Conservation status: (IUCN)** NE

2804 *Spinacia oleracea* L. (Amaranthaceae) **IPNI:** 167384-1 **Common name:** espinaca **Habit:** Herb **Use:** GS, HF, ME, PO **Origin:** Cultivated **Region:** Andean **Dept.:** BOG, BOY, CUN **Elev.:** 2600–2600 m a.s.l. **IUCN Habitat code:** 1,3,14 **Conservation status: (IUCN)** NE

2805 *Anredera cordifolia* (Ten.) Steenis (Basellaceae) **IPNI:** 14549-2 **Common name:** insulina, sachaulluco **Habit:** Climbing **Use:** EU, HF, ME **Origin:** Native **Region:** Amazonia, Andean, Caribbean **Dept.:** ANT, BOG, CAU, COR, MET, NAR, QUI **Elev.:** 100–1700 m a.s.l. **IUCN Habitat code:** 1,3,14 **Conservation status: (IUCN)** NE

2806 *Anredera vesicaria* (Lam.) C.F.Gaertn. (Basellaceae) **IPNI:** 1153747-2 **Habit:** Climbing **Use:** EU, HF, ME **Origin:** Native **Region:** Caribbean **Dept.:** ATL, BOL **Elev.:** 0–300 m a.s.l. **IUCN Habitat code:** 14 **Conservation status: (IUCN)** NE

2807 *Basella alba* L. (Basellaceae) **IPNI:** 164286-1 **Common name:** espinaca, espinaca de enredadera, gallinita **Habit:** Herb, Climbing **Use:** AF, EU, GS, HF, MA, ME, SU **Origin:** Cultivated, Naturalised **Region:** Andean, Caribbean, Pacific **Dept.:** CUN, NAR, QUI, SAP, VAC **Elev.:** 0–1500 m a.s.l. **IUCN Habitat code:** 1,2,3,14 **Conservation status: (IUCN)** NE

2808 *Ullucus tuberosus* Caldas (Basellaceae) **IPNI:** 167633-1 **Common name:** chincheno, chuagua, chugua, chuguas, colluco, lisas, melloco, olloco, papas lisas, ruba, rubas, ruhua, ulloco, ulluco, ullucu, ulluma **Habit:** Herb, Climbing **Use:** HF, ME **Origin:** Cultivated, Naturalised **Region:** Andean **Dept.:** BOG, BOY, CAU, CUN, HUI, NAR, PUT **Elev.:** 2250–3400 m a.s.l. **IUCN Habitat code:** 1,4,14 **Conservation status: (IUCN)** NE

2809 *Acanthocereus tetragonus* (L.) Hummelinck (Cactaceae) **IPNI:** 1646-2 **Common name:** cardo cuatrofilos, cardo pitajayo, cardo pitaya, cardón, pitahaya, pitajaya, pitaya, pitaya amarilla, pitaya roja, pitayo **Habit:** Shrub **Use:** AF, EU, HF, ME **Origin:** Native **Region:** Andean, Caribbean **Dept.:** ATL, BOL, CES, COR, CUN, HUI, MAG, SAP, SAN **Elev.:** 0–1300 m a.s.l. **IUCN Habitat code:** 1,2,3,5,14 **Conservation status: (IUCN)** LC

2810 *Austrocylindropuntia cylindrica* (Lam.) Backeb. (Cactaceae) **IPNI:** 128942-1 **Habit:** Shrub **Use:** EU, SU **Origin:** Native, Cultivated **Region:** Andean **Dept.:** BOG, CUN **Elev.:** 2600–2600 m a.s.l. **IUCN Habitat code:** 3,14 **Conservation status: (IUCN)** NT

2811 *Austrocylindropuntia subulata* (Muehlenpf.) Backeb. (Cactaceae) **IPNI:** 27133-2 **Habit:** Tree **Use:** EU, HF **Origin:** Native **Region:** Andean **Dept.:** CUN, NAR **Elev.:** 2300–2900 m a.s.l. **IUCN Habitat code:** 14 **Conservation status: (IUCN)** LC

2812 *Borzicactus sepium* (Kunth) Britton & Rose (Cactaceae) **IPNI:** 35149-2 **Habit:** Shrub **Use:** HF **Origin:** Native, Cultivated **Region:** Andean **Dept.:** NAR **Elev.:** 2300–2900 m a.s.l. **Conservation status: (IUCN)** NE

2813 *Brasiliopuntia brasiliensis* (Willd.) A.Berger (Cactaceae) **IPNI:** 129102-1 **Habit:** Tree **Use:** ME **Origin:** Naturalised **Region:** Andean **Dept.:** BOY **Elev.:** 1100–1100 m a.s.l. **IUCN Habitat code:** 3,14 **Conservation status: (IUCN)** LC

2814 *Cereus fricii* Backeb. (Cactaceae) **IPNI:** 52982-2 **Common name:** cardón peludo **Habit:** Small tree **Use:** HF, ME **Origin:** Native **Region:** Caribbean **Dept.:** MAG **Elev.:** 5–150 m a.s.l. **Conservation status: (IUCN)** VU

2815 *Cereus hexagonus* (L.) Mill. (Cactaceae) **IPNI:** 130283-1 **Common name:** cacto, cactus, cardón, cirio, yaurero **Habit:** Tree **Use:** AF, EU, HF, ME **Origin:** Native **Region:** Andean **Dept.:** CAU, LAG, NAR, SAN **Elev.:** 180–1900 m a.s.l. **IUCN Habitat code:** 1,2,3,14 **Conservation status: (IUCN)** LC

2816 *Cereus pterogonus* Lem. (Cactaceae) **IPNI:** 130602-1 **Use:** HF **Origin:** Endemic **Conservation status: (IUCN)** NE

2817 *Cereus repandus* (L.) Mill. (Cactaceae) **IPNI:** 130627-1 **Common name:** cardón higo **Habit:** Tree **Use:** AF, EU, FU, HF, MA, ME **Origin:** Native **Region:** Caribbean **Dept.:** ATL, BOL, LAG, MAG **Elev.:** 0–200 m a.s.l. **IUCN Habitat code:** 1,2,14 **Conservation status: (IUCN)** LC

2818 *Cylindropuntia caribaea* (Britton & Rose) F.M.Knuth (Cactaceae) **IPNI:** 73852-2 **Common name:** guasábara **Habit:** Shrub **Use:** HF **Origin:** Native **Region:** Andean **Dept.:** SAN **Elev.:** 500–1400 m a.s.l. **IUCN Habitat code:** 2 **Conservation status: (IUCN)** LC

2819 *Cylindropuntia tunicata* (Lehm.) F.M.Knuth (Cactaceae) **IPNI:** 73905-2 **Habit:** Shrub **Use:** ME **Origin:** Naturalised **Region:** Andean **Dept.:** BOY **Elev.:** 2100–2100 m a.s.l. **IUCN Habitat code:** 14 **Conservation status: (IUCN)** LC

2820 *Epiphyllum hookeri* Haw. (Cactaceae) **IPNI:** 132926-1 **Habit:** Herb, Epiphyte **Use:** HF, ME **Origin:** Native **Region:** Caribbean, Pacific **Dept.:** ANT, CHO, MAG **Elev.:** 0–700 m a.s.l. **IUCN Habitat code:** 1,3,5,14 **Conservation status: (IUCN)** LC

2821 *Epiphyllum phyllanthus* (L.) Haw. (Cactaceae) **IPNI:** 132939-1 **Common name:** cactus, cactus de flor roja, calaguala, calanguala, caraguala, cola de caimán, flor de baile, rabiguano, rabo de iguana **Habit:** Herb, Epiphyte **Use:** HF, ME **Origin:** Native **Region:** Amazonia, Andean, Guiana Shield, Caribbean, Pacific **Dept.:** AMA, ANT, BOY, CAL, CHO, CUN, GUA, HUI, LAG, MET, NAR, NSA, QUI, SAN, SUC, TOL, VAC **Elev.:** 0–1700 m a.s.l. **IUCN**

Habitat code: 1,2,3,4,5,14 **Conservation status: (IUCN)** LC

2822 *Leuenbergeria bleo* (Kunth) Lodé (Cactaceae)
IPNI: 60461844-2
Common name: bledochupa, bleo, bleo de chupa, bleochupa, chupa, chupa melón, chupamelón, corona de espinas, disciplinas, guacamacho falso, guamacho extranjero, pipichuelo, putarsaniko, rungumá, tuna, ñajú **Habit:** Subshrub, Small tree **Use:** EU, HF, ME, PO **Origin:** Native, Cultivated **Region:** Caribbean **Dept.:** ANT, BOL, CHO, COR, MAG **Elev.:** 0–300 m a.s.l. **IUCN Habitat code:** 1,2,3,14 **Conservation status: (IUCN)** NE

2823 *Leuenbergeria guamacho* (F.A.C.Weber) Lodé (Cactaceae)
IPNI: 60461845-2
Common name: algarrobito, chupachupa, guacamacho, guamache, guamachito, guamacho, mokochira, supi, supire, suspiro **Habit:** Shrub, Small tree **Use:** AF, HF, MA, ME **Origin:** Native **Region:** Caribbean, Orinoquia **Dept.:** ATL, BOL, CAS, CES, LAG, MAG **Elev.:** 0–200 m a.s.l. **IUCN Habitat code:** 1,2,14 **Conservation status: (IUCN)** NE

2824 *Mammillaria columbiana* Salm-Dyck (Cactaceae)
IPNI: 314810-2
Habit: Herb **Use:** EU **Origin:** Native **Region:** Andean **Dept.:** BOY, CAU, CUN, HUI, SAN **Elev.:** 1100–2700 m a.s.l. **IUCN Habitat code:** 1,14 **Conservation status: (IUCN)** LC

2825 *Mammillaria mammillaris* (L.) H.Karst. (Cactaceae)
IPNI: 134797-1
Habit: Herb **Use:** HF **Origin:** Native **Region:** Andean **Dept.:** SAN **Elev.:** 500–1200 m a.s.l. **IUCN Habitat code:** 2,14 **Conservation status: (IUCN)** LC

2826 *Melocactus curvispinus* Pfeiff. (Cactaceae)
IPNI: 305646-2
Common name: cabecinegro, cabeza de indio, cabeza de negro, cactus, gorro de obispo **Habit:** Herb **Use:** EU, HF, ME **Origin:** Native **Region:** Andean, Caribbean **Dept.:** BOY, CUN, HUI, LAG, MAG, NSA, TOL, VAC **Elev.:** 350–1500 m a.s.l. **IUCN Habitat code:** 1,2,3,14 **Conservation status: (IUCN)** LC

2827 *Melocactus matanzanus* León (Cactaceae)
IPNI: 156850-2
Habit: Herb **Use:** EU **Origin:** Native **Region:** Orinoquia **Dept.:** VID **Elev.:** 100–100 m a.s.l. **IUCN Habitat code:** 1 **Conservation status: (IUCN)** EN

2828 *Opuntia caracassana* Salm-Dyck (Cactaceae)
IPNI: 136428-1
Common name: cacto, higo, tuna, tuna guajira, tunito, tuno **Habit:** Shrub **Use:** EU, HF, ME **Origin:** Native **Region:** Andean, Caribbean **Dept.:** ATL, BOL, LAG, MAG, NSA, SAN **Elev.:** 0–1200 m a.s.l. **IUCN Habitat code:** 1,2,5,14 **Conservation status: (IUCN)** LC

2829 *Opuntia cochenillifera* (L.) Mill. (Cactaceae)
IPNI: 1008816-2
Common name: nopal **Habit:** Shrub **Use:** AF, EU, HF, IF, MA, ME **Origin:** Cultivated, Naturalised **Region:** Caribbean **Dept.:** MAG **Elev.:** 0–50 m a.s.l. **IUCN Habitat code:** 2,3,5,14 **Conservation status: (IUCN)** NE

2830 *Opuntia dillenii* (Ker Gawl.) Haw. (Cactaceae)
IPNI: 136534-1
Use: AF, EU, HF, MA, ME, SU **Origin:** Naturalised **IUCN Habitat code:** 1,2,3,14 **Conservation status: (IUCN)** NE

2831 *Opuntia elatior* Mill. (Cactaceae)
IPNI: 136561-1
Common name: arepo, higo, tuna, tunas **Habit:** Shrub **Use:** AF, EU, HF, ME **Origin:** Native **Region:** Andean, Caribbean **Dept.:** ANT, BOY, CUN, HUI, LAG, SAN, VAC **Elev.:** 0–2100 m a.s.l. **IUCN Habitat code:** 1,3,14 **Conservation status: (IUCN)** LC

2832 *Opuntia ficus-indica* (L.) Mill. (Cactaceae)
IPNI: 1151735-2
Common name: cacto, chumbera, higo, higo chumbo, nopal, penca de Castilla, tuna **Habit:** Shrub, Small tree **Use:** AF, EU, FU, GS, HF, MA, ME, SU **Origin:** Cultivated, Naturalised **Region:** Andean **Dept.:** ANT, BOG, BOY, CUN, SAN **Elev.:** 1300–2600 m a.s.l. **IUCN Habitat code:** 2,3,14 **Conservation status: (IUCN)** DD

2833 *Opuntia microdasys* (Lehm.) Pfeiff. (Cactaceae)
IPNI: 136806-1
Use: EU,PO **Region:** Andean **Dept.:** BOG**Conservation status: (IUCN)** LC

2834 *Opuntia pittieri* Britton & Rose (Cactaceae)
IPNI: 175432-2
Common name: higo, tuno **Habit:** Small tree **Use:** HF **Origin:** Endemic **Region:** Andean **Dept.:** BOY, CUN, HUI, VAC **Elev.:** 900–2600 m a.s.l. **IUCN Habitat code:** 1,2,3,14 **Conservation status: (IUCN)** DD

2835 *Opuntia quitensis* F.A.C.Weber (Cactaceae)
IPNI: 175475-2
Habit: Shrub **Use:** HF **Origin:** Native **Region:** Andean **Dept.:** NAR **Elev.:** 1900–1900 m a.s.l. **IUCN Habitat code:** 14 **Conservation status: (IUCN)** LC

2836 *Opuntia schumannii* F.A.C.Weber ex A.Berger (Cactaceae)
IPNI: 175540-2
Common name: penca, tuna, tuna de la sabana, tuno, tuno o tunas **Habit:** Shrub **Use:** HF, ME **Origin:** Native **Region:** Andean **Dept.:** BOG, BOY, CES, CUN, NSA, SAN **Elev.:** 1200–2600 m a.s.l. **IUCN Habitat code:** 2,3,14 **Conservation status: (IUCN)** VU

2837 *Opuntia stricta* (Haw.) Haw. (Cactaceae)
IPNI: 137078-1
Habit: Shrub **Use:** EU, HF, ME **Origin:** Naturalised **Region:** Andean **Dept.:** SAN **Elev.:** 500–1200 m a.s.l. **Conservation status: (IUCN)** LC

2838 *Opuntia tuna* (L.) Mill. (Cactaceae)
IPNI: 137125-1
Habit: Shrub **Use:** AF, EU, HF, ME **Origin:** Native **Region:** Caribbean **Dept.:** MAG, NSA **Elev.:** 240–240 m a.s.l. **Conservation status: (IUCN)** NE

2839 *Pereskia aculeata* Mill. (Cactaceae)
IPNI: 137452-1
Common name: guamacho serrano **Habit:** Shrub **Use:** EU, HF, MA, ME **Origin:** Native **Region:** Caribbean **Dept.:** LAG **Elev.:** 600–600 m a.s.l. **IUCN Habitat code:** 14 **Conservation status: (IUCN)** LC

2840 *Pilosocereus lanuginosus* (L.) Byles & G.D.Rowley (Cactaceae)
IPNI: 196706-2
Habit: Shrub, Small tree **Use:** HF **Origin:** Native **Region:** Andean, Caribbean **Dept.:** CAU, HUI, LAG, MAG, NAR, VAC **Elev.:** 0–900 m a.s.l. **IUCN Habitat code:** 1,2,3,5,14 **Conservation status: (IUCN)** LC

2841 *Praecereus euchlorus* (F.A.C.Weber ex K.Schum.) N.P.Taylor (Cactaceae)
IPNI: 1001185-1
Habit: Shrub **Use:** HF **Origin:** Native **Region:** Andean, Caribbean **Dept.:** CUN, HUI, NSA, SAN **Elev.:** 400–1200 m a.s.l. **IUCN Habitat code:** 1,3,14 **Conservation status: (IUCN)** LC

2842 *Pseudorhipsalis amazonica* (K.Schum.) Ralf Bauer (Cactaceae)
IPNI: 60426912-2
Common name: calaguala, caraguala, cola de caballo, cola de caimán **Habit:** Herb, Epiphyte **Use:** ME **Origin:** Native **Region:** Amazonia, Andean, Guiana Shield, Caribbean, Pacific **Dept.:** AMA, ANT, BOL, CAQ, CES, CHO, COR, MAG, NSA, PUT, SAN, VAU **Elev.:** 20–1200 m a.s.l. **IUCN Habitat code:** 1,3,5,14 **Conservation status: (IUCN)** LC

2843 *Pseudorhipsalis ramulosa* (Salm-Dyck) Barthlott (Cactaceae)
IPNI: 963240-1
Habit: Herb, Epiphyte **Use:** EU **Origin:** Native **Region:** Andean **Dept.:** ANT **Elev.:** 1000–1840 m a.s.l. **IUCN Habitat code:** 1,3,14 **Conservation status: (IUCN)** LC

2844 *Rhipsalis baccifera* (J.S.Muell.) Stearn (Cactaceae)
IPNI: 219250-2
Common name: disciplina, disciplinas, lluvia de perlas, parásita **Habit:** Herb, Epiphyte **Use:** AF, EU, HF, ME **Origin:** Native **Region:** Andean, Guiana Shield, Caribbean, Orinoquia **Dept.:** ANT, BOY, CAQ, CAU, CES, CHO, CUN, GUA, HUI, LAG, MAG, MET, NSA, QUI, SAN, TOL, VAC **Elev.:** 100–1500 m a.s.l. **Conservation status: (IUCN)** LC

2845 *Rhipsalis micrantha* (Kunth) DC. (Cactaceae)
IPNI: 138328-1
Habit: Herb, Epiphyte **Use:** HF **Origin:** Native **Region:** Andean, Orinoquia **Dept.:** ANT, ARA, CAU, RIS **Elev.:** 500–2000 m a.s.l. **IUCN Habitat code:** 1,3,14 **Conservation status: (IUCN)** LC

2846 *Rhodocactus grandifolius* (Haw.) F.M.Knuth (Cactaceae)
IPNI: 219420-2
Common name: cardón **Use:** EU, HF, ME **Conservation status: (IUCN)** NE

2847 *Schlumbergera truncata* (Haw.) Moran (Cactaceae)
IPNI: 229597-2
Use: EU, GS, HF, ME **Dept.:** BOG**IUCN Habitat code:** 1,3,14 **Conservation status: (IUCN)** VU

2848 *Selenicereus costaricensis* (F.A.C.Weber) S.Arias & N.Korotkova ex Hammel (Cactaceae)
IPNI: 77177683-1
Common name: pitahaya **Habit:** Herb, Climbing **Use:** EU, HF, ME, SU **Origin:** Native, Cultivated **Region:** Caribbean **Dept.:**

MAG **Elev.:** 5–50 m a.s.l. **IUCN Habitat code:** 1,3 **Conservation status: (IUCN)** LC

2849 *Selenicereus grandiflorus* (L.) Britton & Rose (Cactaceae) **IPNI:** 138488-1 **Use:** EU, HF, MA, ME **Conservation status: (IUCN)** LC

2850 *Selenicereus megalanthus* (K.Schum. ex Vaupel) Moran (Cactaceae) **IPNI:** 232386-2 **Common name:** pitahaya, pitaya **Habit:** Herb, Climbing **Use:** EU, HF, ME **Origin:** Native, Cultivated **Region:** Andean, Guiana Shield **Dept.:** ANT, CUN, QUI, VAC **Elev.:** 90–1800 m a.s.l. **IUCN Habitat code:** 1,14 **Conservation status: (IUCN)** NE

2851 *Selenicereus monacanthus* (Lem.) D.R.Hunt (Cactaceae) **IPNI:** 77163180-1 **Common name:** cardón marica, makibosoto, pitaya **Habit:** Herb, Epiphyte **Use:** HF, MA, ME, SU **Origin:** Native **Region:** Andean, Caribbean, Orinoquia **Dept.:** BOL, CAS, LAG, MAG, SAN, VAC **Elev.:** 0–2000 m a.s.l. **IUCN Habitat code:** 1,2,3,5,14 **Conservation status: (IUCN)** NE

2852 *Selenicereus ocamponis* (Salm-Dyck) D.R.Hunt (Cactaceae) **IPNI:** 77163181-1 **Use:** HF **Conservation status: (IUCN)** NE

2853 *Selenicereus undatus* (Haw.) D.R.Hunt (Cactaceae) **IPNI:** 77163184-1 **Common name:** cacto, pitahaya roja, pitaya, pitaya roja **Habit:** Herb, Epiphyte, Climbing **Use:** EU, GS, HF, ME **Origin:** Cultivated, Naturalised **Region:** Andean **Dept.:** ANT, CUN, HUI, VAC **Elev.:** 1000–1500 m a.s.l. **IUCN Habitat code:** 1,2,3,5,14 **Conservation status: (IUCN)** NE

2854 *Stenocereus griseus* (Haw.) Buxb. (Cactaceae) **IPNI:** 244634-2 **Common name:** canelón, cardón, cardón de arenales, cardón guajiro, cardón real, yasu, yotojoro, yotójoro **Habit:** Shrub, Small tree **Use:** EU, HF, MA, ME **Origin:** Native **Region:** Andean, Caribbean **Dept.:** BOY, CAU, HUI, LAG, MAG, NAR, SAN **Elev.:** 0–1200 m a.s.l. **IUCN Habitat code:** 1,2,3,5,14 **Conservation status: (IUCN)** LC

2855 *Trichocereus macrogonus* (Salm-Dyck) Riccob. (Cactaceae) **IPNI:** 257107-2 **Habit:** Shrub **Use:** EU, ME, SU **Origin:** Naturalised **Region:** Andean **Dept.:** BOG **Elev.:** 1900–2000 m a.s.l. **Conservation status: (IUCN)** NE

2856 *Arenaria lanuginosa* (Michx.) Rohrb. (Caryophyllaceae) **IPNI:** 30040285-2 **Common name:** abrojillo, abrojita, abrojito, clavito kirukise **Habit:** Herb **Use:** ME **Origin:** Native **Region:** Andean **Dept.:** ANT, BOG, BOY, CAU, CHO, CUN, HUI, MAG, QUI, RIS, SAN, TOL **Elev.:** 1200–3400 m a.s.l. **IUCN Habitat code:** 1,2,3,4,14 **Conservation status: (IUCN)** NE

2857 *Atocion armeria* (L.) Raf. (Caryophyllaceae) **IPNI:** 77093328-1 **Common name:** alelí, amor fino, cascabel, gatico, gaticos **Habit:** Herb **Use:** EU **Origin:** Naturalised **Region:** Andean **Dept.:** ANT, CUN **Elev.:** 2400–2400 m a.s.l. **Conservation status: (IUCN)** NE

2858 *Cardionema ramosissima* (Weinm.) A.Nelson & J.F.Macbr. (Caryophyllaceae) **IPNI:** 45444-2 **Habit:** Herb **Use:** ME **Origin:** Native **Region:** Andean **Dept.:** BOY, CUN **Elev.:** 2600–3125 m a.s.l. **IUCN Habitat code:** 3,14 **Conservation status: (IUCN)** NE

2859 *Cerastium arvense* L. (Caryophyllaceae) **IPNI:** 329786-2 **Habit:** Herb **Use:** ME **Origin:** Native **Region:** Andean **Dept.:** ANT, BOG, BOY, CUN, MAG, MET, SAN, TOL **Elev.:** 2500–4900 m a.s.l. **IUCN Habitat code:** 1,3,4,8,14 **Conservation status: (IUCN)** LC

2860 *Cerastium fontanum* Baumg. (Caryophyllaceae) **IPNI:** 152255-1 **Habit:** Herb **Use:** ME **Origin:** Naturalised **Region:** Andean **Dept.:** BOG, BOY, CAL, CAU, CUN, RIS, TOL **Elev.:** 2470–4050 m a.s.l. **Conservation status: (IUCN)** NE

2861 *Cerastium glomeratum* Thuill. (Caryophyllaceae) **IPNI:** 152274-1 **Habit:** Herb **Use:** ME **Origin:** Naturalised **Region:** Andean **Dept.:** BOG, CES, CUN, MAG, NAR **Elev.:** 2600–3700 m a.s.l. **IUCN Habitat code:** 1,3,4,8,14 **Conservation status: (IUCN)** NE

2862 *Cerastium holosteoides* Fr. (Caryophyllaceae) **IPNI:** 152312-1 **Use:** HF, ME **Origin:** Naturalised **IUCN Habitat code:** 1,3,4,8 **Conservation status: (IUCN)** NE

2863 *Dianthus armeria* L. (Caryophyllaceae) **IPNI:** 310755-2 **Common name:** clavellina **Habit:** Herb **Use:** EU **Origin:** Cultivated **Region:** Andean **Dept.:** BOG **Elev.:** 2600–2600 m a.s.l. **Conservation status: (IUCN)** NE

2864 *Dianthus barbatus* L. (Caryophyllaceae) **IPNI:** 306674-2 **Common name:** clavel, clavel chino, clavellina, clavellina pintada, macetilla **Habit:** Herb **Use:** EU, HF, ME **Origin:** Cultivated **Region:** Andean **Dept.:** ANT, BOG **Elev.:** 2500–2500 m a.s.l. **IUCN Habitat code:** 1,3,14 **Conservation status: (IUCN)** NE

2865 *Dianthus caryophyllus* L. (Caryophyllaceae) **IPNI:** 153205-1 **Common name:** clavel, clavel reventón, clavel rojo, clavellina, clavellina doble, clavellino **Habit:** Herb **Use:** EU, GS, HF, MA, ME **Origin:** Cultivated **Region:** Andean **Dept.:** ANT, BOG, CUN **Elev.:** 1400–2600 m a.s.l. **IUCN Habitat code:** 1,3,14 **Conservation status: (IUCN)** NE

2866 *Dianthus chinensis* L. (Caryophyllaceae) **IPNI:** 153215-1 **Use:** EU, ME **Region:** Andean **Dept.:** BOG**IUCN Habitat code:** 1,3,14 **Conservation status: (IUCN)** NE

2867 *Dianthus gratianopolitanus* Vill. (Caryophyllaceae) **IPNI:** 153408-1 **Use:** EU **Region:** Andean **Dept.:** BOG**Conservation status: (IUCN)** NE

2868 *Dianthus plumarius* L. (Caryophyllaceae) **IPNI:** 153717-1 **Use:** EU, GS, ME **Region:** Andean **Dept.:** BOG**IUCN Habitat code:** 14 **Conservation status: (IUCN)** NE

2869 *Drymaria cordata* (L.) Willd. ex Schult. (Caryophyllaceae) **IPNI:** 30172747-2 **Common name:** celedonia, celidonia, encaje, golondrina, hoja de cáncer, peledillo, poleo macho **Habit:** Herb **Use:** AF, EU, GS, HF, MA, ME, SU **Origin:** Native **Region:** Amazonia, Andean, Caribbean, Orinoquia, Pacific **Dept.:** AMA, ANT, BOG, BOY, CAS, CAU, CHO, CUN, HUI, MAG, MET, NAR, NSA, PUT, SAN, TOL, VAC **Elev.:** 0-3100 m.a.s.l. **IUCN Habitat code:** 1,2,3,4,5,14 **Conservation status: (IUCN)** NE **(CPSG)** Potential LC

2870 *Drymaria glandulosa* Bartl. (Caryophyllaceae) **IPNI:** 154139-1 **Habit:** Herb **Use:** ME **Origin:** Native **Dept.:** BOY**Conservation status: (IUCN)** NE

2871 *Drymaria ovata* Willd. ex Schult. (Caryophyllaceae) **IPNI:** 154166-1 **Habit:** Herb **Use:** ME **Origin:** Native **Region:** Andean **Dept.:** ANT, HUI **Elev.:** 1900–2500 m a.s.l. **IUCN Habitat code:** 1,2,4,14 **Conservation status: (IUCN)** NE

2872 *Drymaria villosa* Schltdl. & Cham. (Caryophyllaceae) **IPNI:** 154201-1 **Habit:** Herb **Use:** AF, HF, ME **Origin:** Native **Region:** Andean **Dept.:** ANT, BOG, CUN, HUI, MAG, RIS, TOL **Elev.:** 1950–2600 m a.s.l. **IUCN Habitat code:** 1,3,4,14 **Conservation status: (IUCN)** NE

2873 *Gypsophila elegans* M.Bieb. (Caryophyllaceae) **IPNI:** 154453-1 **Common name:** gasa **Habit:** Herb **Use:** EU **Origin:** Cultivated, Naturalised **Region:** Andean **Dept.:** ANT, BOG, CUN **Elev.:** 2500–2500 m a.s.l. **IUCN Habitat code:** 3,14 **Conservation status: (IUCN)** NE

2874 *Paronychia communis* Cambess. (Caryophyllaceae) **IPNI:** 435017-1 **Habit:** Herb **Use:** ME **Origin:** Native **Region:** Andean **Dept.:** CUN **Elev.:** 2600–2600 m a.s.l. **Conservation status: (IUCN)** NE

2875 *Sagina procumbens* L. (Caryophyllaceae) **IPNI:** 302082-2 **Habit:** Herb **Use:** EU **Origin:** Naturalised **Region:** Andean **Dept.:** ANT, BOY, NAR **Elev.:** 3200–3660 m

a.s.l. **IUCN Habitat code:** 1,3,4,14 **Conservation status: (IUCN)** NE

2876 *Scleranthus annuus* L. (Caryophyllaceae) **IPNI:** 300494-2 **Habit:** Herb **Use:** ME **Origin:** Naturalised **Region:** Andean **Dept.:** CUN **Elev.:** 2650–2650 m a.s.l. **IUCN Habitat code:** 14 **Conservation status: (IUCN)** NE

2877 *Silene coronaria* (L.) Clairv. (Caryophyllaceae) **IPNI:** 60461121-2 **Common name:** cruz de Malta **Use:** EU, ME **Origin:** Cultivated **Region:** Andean **Dept.:** ANT, BOG **Elev.:** 2500–2500 m a.s.l. **IUCN Habitat code:** 14 **Conservation status: (IUCN)** NE

2878 *Silene gallica* L. (Caryophyllaceae) **IPNI:** 332106-2 **Common name:** argentina, botijita, botijitas, calabacillo, cascabel, choroticos, forastera, pichincha **Habit:** Herb **Use:** ME **Origin:** Naturalised **Region:** Andean **Dept.:** ANT, BOG, BOY, CUN, NAR **Elev.:** 2150–3040 m a.s.l. **IUCN Habitat code:** 1,3,14 **Conservation status: (IUCN)** NE

2879 *Spergula arvensis* L. (Caryophyllaceae) **IPNI:** 158725-1 **Common name:** abrejillo, abrojillo, abrojita, abrojito, anisillo, gasa, gasilla, linacilla, miona **Habit:** Herb **Use:** AF, HF, ME **Origin:** Naturalised **Region:** Andean **Dept.:** ANT, BOG, BOY, CUN, NAR **Elev.:** 2350–3600 m a.s.l. **IUCN Habitat code:** 1,3,4,14 **Conservation status: (IUCN)** NE

2880 *Spergularia marina* (L.) Besser (Caryophyllaceae) **IPNI:** 241336-2 **Habit:** Herb **Use:** ME **Origin:** Naturalised **Region:** Andean **Dept.:** CUN **Elev.:** 2680–2680 m a.s.l. **IUCN Habitat code:** 14 **Conservation status: (IUCN)** LC

2881 *Spergularia rubra* (L.) J.Presl & C.Presl (Caryophyllaceae) **IPNI:** 158983-1 **Use:** AF, EU, ME **Region:** Andean **Dept.:** BOG**Conservation status: (IUCN)** NE

2882 *Spergularia villosa* (Pers.) Cambess. (Caryophyllaceae) **IPNI:** 159015-1 **Use:** ME **Region:** Andean **Dept.:** BOG**IUCN Habitat code:** 14 **Conservation status: (IUCN)** NE

2883 *Stellaria cuspidata* Willd. ex D.F.K.Schltdl. (Caryophyllaceae) **IPNI:** 244185-2 **Habit:** Herb **Use:** ME **Origin:** Native **Region:** Andean **Dept.:** ANT, BOG, BOY, CAU, CUN, HUI, MET, RIS, TOL **Elev.:** 820–3820 m a.s.l. **IUCN Habitat code:** 1,3,4,14 **Conservation status: (IUCN)** NE

2884 *Stellaria media* (L.) Vill. (Caryophyllaceae) **IPNI:** 159330-1 **Common name:** berros, pajarera, yerba de pájaro **Habit:** Herb **Use:** AF, EU, GS, HF, MA, ME, PO **Origin:** Naturalised **Region:** Andean **Dept.:** ANT, BOG, CUN, NAR, QUI, SAN **Elev.:** 2300–3700 m a.s.l. **IUCN Habitat code:** 1,3,4,14 **Conservation status: (IUCN)** NE

2885 *Stellaria ovata* Willd. ex D.F.K.Schltdl. (Caryophyllaceae) **IPNI:** 159379-1 **Habit:** Herb **Use:** HF **Origin:** Native **Region:** Andean **Dept.:** ANT, BOG, CAQ, CHO, CUN, HUI, VAC **Elev.:** 560–3115 m a.s.l. **IUCN Habitat code:** 1,3,4,8,14 **Conservation status: (IUCN)** NE

2886 *Drosera capillaris* Poir. (Droseraceae) **IPNI:** 321807-1 **Habit:** Herb, Carnivorous **Use:** ME **Origin:** Native **Region:** Guiana Shield, Orinoquia **Dept.:** AMA, CAQ, VID **Elev.:** 250–400 m a.s.l. **IUCN Habitat code:** 1,3,4,14 **Conservation status: (IUCN)** NE

2887 *Microtea debilis* Sw. (Microteaceae) **IPNI:** 676291-1 **Habit:** Herb **Use:** ME **Origin:** Native **Region:** Amazonia, Guiana Shield, Caribbean, Orinoquia, Pacific **Dept.:** AMA, ANT, BOL, CHO, GUA, MAG, MET, SAP, VAC **Elev.:** 100–900 m a.s.l. **IUCN Habitat code:** 1,2,3,5,14 **Conservation status: (IUCN)** NE

2888 *Glinus radiatus* (Ruiz & Pav.) Rohrb. (Molluginaceae) **IPNI:** 1008757-2 **Habit:** Herb **Use:** ME **Origin:** Native **Region:** Caribbean **Dept.:** ATL, BOL, COR, MAG **Elev.:** 5–80 m a.s.l. **IUCN Habitat code:** 1,2,5,14 **Conservation status: (IUCN)** NE

2889 *Mollugo verticillata* L. (Molluginaceae) **IPNI:** 280992-2 **Common name:** tomillo, verdolaga de césped **Habit:** Herb **Use:** HF, ME **Origin:** Native **Region:** Caribbean **Dept.:** ATL, BOG, LAG, MAG, SUC, TOL **Elev.:** 5–440 m a.s.l. **IUCN Habitat code:** 1,2,3,5,14 **Conservation status: (IUCN)** NE

2890 *Calandrinia acaulis* Kunth (Montiaceae) **IPNI:** 698305-1 **Common name:** chicoria ñinguicue **Habit:** Herb **Use:** HF, ME **Origin:** Native **Region:** Andean **Dept.:** BOG, BOY, CAL, CUN, LAG, MAG, TOL **Elev.:** 3400–4400 m a.s.l. **IUCN Habitat code:** 3,4,8,14 **Conservation status: (IUCN)** NE

2891 *Calandrinia ciliata* (Ruiz & Pav.) DC. (Montiaceae) **IPNI:** 698357-1 **Common name:** moradita **Habit:** Herb **Use:** HF **Origin:** Native **Region:** Andean **Dept.:** BOG, BOY, CES, NAR **Elev.:** 2824–3750 m a.s.l. **IUCN Habitat code:** 4,14 **Conservation status: (IUCN)** NE

2892 *Montia fontana* L. (Montiaceae) **IPNI:** 311155-2 **Habit:** Herb **Use:** HF **Origin:** Native **Region:** Andean **Dept.:** BOG, CAL, CAU, CUN **Elev.:** 3410–4500 m a.s.l. **IUCN Habitat code:** 1,4 **Conservation status: (IUCN)** LC

2893 *Allionia incarnata* L. (Nyctaginaceae) **IPNI:** 8606-2 **Habit:** Herb **Use:** ME **Origin:** Native **Region:** Caribbean **Dept.:** LAG, SAN **Elev.:** 0–500 m a.s.l. **IUCN Habitat code:** 2,14 **Conservation status: (IUCN)** NE

2894 *Boerhavia coccinea* Mill. (Nyctaginaceae) **IPNI:** 326873-2 **Common name:** hierba centella, pega pega **Habit:** Herb **Use:** AF, GS, HF, ME **Origin:** Native **Region:** Caribbean, Pacific **Dept.:** BOL, CHO, COR, CUN, MAG, VAC **Elev.:** 4–1000 m a.s.l. **IUCN Habitat code:** 1,2,3,14 **Conservation status: (IUCN)** NE

2895 *Boerhavia diffusa* L. (Nyctaginaceae) **IPNI:** 326899-2 **Common name:** lagaña de perro, pegajosa, pegajosita, rodilla de pollo **Habit:** Herb **Use:** AF, GS, HF, ME, PO, SU **Origin:** Native **Region:** Caribbean **Dept.:** ANT, ATL, BOL, CHO, COR, CUN, HUI, LAG, MAG, NAR, NSA, SAN, TOL, VAC **Elev.:** 0–1000 m a.s.l. **IUCN Habitat code:** 1,2,3,5,14 **Conservation status: (IUCN)** NE

2896 *Boerhavia erecta* L. (Nyctaginaceae) **IPNI:** 319111-2 **Common name:** rodilla de pollo **Habit:** Herb **Use:** AF, HF, ME **Origin:** Native **Region:** Caribbean **Dept.:** ANT, ATL, BOL, CAU, CES, CUN, MAG, NAR, SAP, SAN, SUC, TOL, VAC **Elev.:** 5-1088 m.a.s.l. **IUCN Habitat code:** 1,2,3,14 **Conservation status: (IUCN)** NE **(CPSG)** Potential LC

2897 *Bougainvillea glabra* Choisy (Nyctaginaceae) **IPNI:** 35444-2 **Common name:** brisa, buganvil, buganvil anaranjado, buganvilia, cimbidio, crenitaria, curazao, curazao rojo, flor de verano, pompadur, primavera, trinitaria, veranera, verano **Habit:** Shrub, Liana, Climbing **Use:** EU, MA, ME **Origin:** Cultivated **Region:** Andean, Caribbean **Dept.:** ANT, ATL, BOG, BOL, CUN **Elev.:** 0–2500 m a.s.l. **IUCN Habitat code:** 1,2,3,4,14 **Conservation status: (IUCN)** LC

2898 *Bougainvillea peruviana* Bonpl. (Nyctaginaceae) **IPNI:** 604352-1 **Habit:** Shrub, Liana, Climbing **Use:** EU **Origin:** Cultivated **Region:** Andean, Pacific **Dept.:** HUI **Elev.:** 0–1500 m a.s.l. **Conservation status: (IUCN)** NE

2899 *Bougainvillea spectabilis* Willd. (Nyctaginaceae) **IPNI:** 604358-1 **Common name:** bellísima, buganvil, curazao, primavera, trinitaria, veranera, veranera morada **Habit:** Shrub, Liana, Climbing **Use:** EU, HF, MA, ME **Origin:** Cultivated **Region:** Andean, Caribbean **Dept.:** ANT, BOG, MAG **Elev.:** 200–2260 m a.s.l. **IUCN Habitat code:** 1,3,4,14 **Conservation status: (IUCN)** NE

2900 *Commicarpus scandens* (L.) Standl. (Nyctaginaceae) **IPNI:** 63937-2 **Common name:** rodillo de pollo, shaashal **Habit:** Herb, Climbing **Use:** ME **Origin:** Native **Region:** Caribbean **Dept.:** HUI, MAG **Elev.:** 3–620 m a.s.l. **IUCN Habitat code:** 1,2,14 **Conservation status: (IUCN)** NE

2901 *Guapira cuspidata* (Heimerl) Lundell (Nyctaginaceae) **IPNI:** 113926-2 **Habit:** Small tree, Tree **Use:** MA **Origin:** Native **Region:** Orinoquia **Dept.:** MET, VID **Elev.:** 100–350 m a.s.l. **IUCN Habitat code:** 1,3 **Conservation status: (IUCN)** LC

2902 *Guapira fragrans* (Dum.Cours.) Little (Nyctaginaceae) **IPNI:** 113934-2 **Habit:** Tree **Use:** MA **Origin:** Native **Region:** Caribbean **Dept.:** ATL, BOL, MAG **Elev.:** 75–100 m a.s.l. **IUCN Habitat code:** 14 **Conservation status: (IUCN)** NE

2903 *Mirabilis jalapa* L. (Nyctaginaceae) **IPNI:** 162591-2 **Common name:** bella de noche, buenas tardes, buenastardes, centella, diego, don diego de noche, dondiego, dondiego de noche, donjuán de noche, flor de Panamá, flor de tarde, jalapa, jazmín de México, jazmín de tarde, maravilla, tabaquillo, trompetilla **Habit:** Herb **Use:** EU, HF, MA, ME, PO, SU **Origin:** Naturalised **Region:** Andean, Pacific **Dept.:** ANT, BOG, BOY, CHO, CUN, QUI, SAN, TOL, VAC **Elev.:** 100–1780 m a.s.l. **IUCN Habitat code:** 1,2,3,4,14 **Conservation status: (IUCN)** NE

2904 *Neea divaricata* Poepp. & Endl. (Nyctaginaceae) **IPNI:** 604755-1 **Common name:** palometa huaiyú, palometa huelto **Habit:** Shrub, Small tree, Tree **Use:** ME **Origin:** Native **Region:** Amazonia, Andean **Dept.:** AMA, ANT, BOL, GUV, PUT, RIS, VAC **Elev.:** 80–1950 m a.s.l. **IUCN Habitat code:** 1,2,3,4,5,14 **Conservation status: (IUCN)** LC

2905 *Neea floribunda* Poepp. & Endl. (Nyctaginaceae) **IPNI:** 604762-1 **Common name:** garopa **Habit:** Shrub, Tree **Use:** HF, ME **Origin:** Native **Region:** Amazonia, Andean **Dept.:** AMA, ANT, PUT **Elev.:** 100–1850 m a.s.l. **IUCN Habitat code:** 1,3,4,14 **Conservation status: (IUCN)** LC

2906 *Neea macrophylla* Poepp. & Endl. (Nyctaginaceae) **IPNI:** 604776-1 **Habit:** Shrub, Small tree, Tree **Use:** HF, ME **Origin:** Native **Region:** Amazonia, Andean **Dept.:** AMA, ANT, PUT **Elev.:** 100–1500 m a.s.l. **IUCN Habitat code:** 1,3,5,14 **Conservation status: (IUCN)** LC

2907 *Neea parviflora* Poepp. & Endl. (Nyctaginaceae) **IPNI:** 604789-1 **Common name:** bollo de puerco, cajeto **Habit:** Tree **Use:** HF **Origin:** Native **Region:** Amazonia, Pacific **Dept.:** AMA, CHO, PUT **Elev.:** 20–130 m a.s.l. **IUCN Habitat code:** 1,3,5,14 **Conservation status: (IUCN)** LC

2908 *Neea psychotrioides* Donn.Sm. (Nyctaginaceae) **IPNI:** 168038-2 **Habit:** Shrub, Subshrub, Small tree **Use:** ME **Origin:** Native **Region:** Pacific **Dept.:** ANT, CAU, CHO **Elev.:** 70–1000 m a.s.l. **IUCN Habitat code:** 1 **Conservation status: (IUCN)** LC

2909 *Neea verticillata* Ruiz & Pav. (Nyctaginaceae) **IPNI:** 604820-1 **Common name:** arenoso **Habit:** Shrub, Small tree **Use:** ME **Origin:** Native **Region:** Amazonia **Dept.:** AMA, CAQ **Elev.:** 100–200 m a.s.l. **IUCN Habitat code:** 1,5 **Conservation status: (IUCN)** NE

2910 *Pisonia aculeata* L. (Nyctaginaceae) **IPNI:** 326897-2 **Common name:** adorote, aruñagato, diente de perro, mancha mano, uña de gato, uña de perro **Habit:** Shrub, Small tree, Liana **Use:** EU, MA, ME **Origin:** Native **Region:** Andean, Caribbean **Dept.:** ANT, ATL, BOL, MAG, VAC **Elev.:** 10–1000 m a.s.l. **IUCN Habitat code:** 1,2,3,5,14 **Conservation status: (IUCN)** LC

2911 *Salpianthus purpurascens* (Cav. ex Lag.) Hook. & Arn. (Nyctaginaceae) **IPNI:** 605135-1 **Habit:** Herb, Subshrub **Use:** HF, ME **Origin:** Native **Region:** Caribbean **Elev.:** 0–200 m a.s.l. **Conservation status: (IUCN)** NE

2912 *Hilleria latifolia* (Lam.) H.Walter (Petiveriaceae) **IPNI:** 122647-2 **Habit:** Herb **Use:** AF, EU, HF, MA, ME, PO, SU **Origin:** Native **Region:** Amazonia, Andean **Dept.:** CUN, NSA **Elev.:** 1000–1200 m a.s.l. **Conservation status: (IUCN)** NE

2913 *Ledenbergia peruviana* O.C.Schmidt (Petiveriaceae) **IPNI:** 136986-2 **Habit:** Shrub, Small tree **Use:** EU **Origin:** Native, Cultivated **Region:** Andean **Dept.:** BOG, CUN **Elev.:** 2600–2900 m a.s.l. **Conservation status: (IUCN)** NE

2914 *Petiveria alliacea* L. (Petiveriaceae) **IPNI:** 323288-2 **Common name:** anamú, apacín, ave, barbasco, chimú, fimercarra, guinea hen, herb hedionda, ipacina, jazmillo, koujourouk, lancetilla, mapurite, mapurito, mikura, mucura, mucuracá, namú, palindera, pisajachu, raíz de pipí, samut apal, sunikila, urgat, zorrillo **Habit:** Herb, Subshrub **Use:** AF, HF, MA, ME, PO, SU **Origin:** Native, Cultivated **Region:** Amazonia, Andean, Caribbean, Orinoquia, Pacific **Dept.:** AMA, ANT, BOG, BOL, CAL, CAQ, CAU, CES, CHO, COR, CUN, LAG, MAG, MET, NAR, PUT, RIS, SAP, SAN, SUC, TOL, VAC **Elev.:** 5-1780 m.a.s.l. **IUCN Habitat code:** 1,2,3,14 **Conservation status: (IUCN)** NE **(CPSG)** Potential LC

2915 *Rivina humilis* L. (Petiveriaceae) **IPNI:** 221479-2 **Common name:** ajicillo, azafrán, carmín, coral, coralillo, fruta de pava, guacamaya, pimpín **Habit:** Herb, Subshrub **Use:** AF, EU, HF, IF, MA, ME, PO **Origin:** Native **Region:** Andean, Caribbean, Pacific **Dept.:** ANT, ATL, BOL, CAL, CHO, COR, CUN, HUI, LAG, MAG, SAP, SAN, SUC, TOL, VAC **Elev.:** 0-1580 m.a.s.l. **IUCN Habitat code:** 1,2,3,14 **Conservation status: (IUCN)** NE **(CPSG)** Potential LC

2916 *Seguieria americana* L. (Petiveriaceae) **IPNI:** 676493-1 **Common name:** barbero, rabo de iguana **Use:** MA, ME, PO **IUCN Habitat code:** 1,2,14 **Conservation status: (IUCN)** NE

2917 *Seguieria macrophylla* Benth. (Petiveriaceae) **IPNI:** 676510-1 **Common name:** rabo de iguana **Habit:** Liana **Use:** ME **Origin:** Native **Region:** Caribbean, Orinoquia **Dept.:** ARA, ATL, BOL, LAG, MET, VID **Elev.:** 210–360 m a.s.l. **IUCN Habitat code:** 1,2,3,4 **Conservation status: (IUCN)** NE

2918 *Trichostigma octandrum* (L.) H.Walter (Petiveriaceae) **IPNI:** 257665-2 **Common name:** bejuco de burro, bejuco negro **Habit:** Shrub, Liana **Use:** AF, HF, MA, ME **Origin:** Native **Region:** Amazonia, Andean, Caribbean, Orinoquia, Pacific **Dept.:** ANT, ATL, BOL, CAQ, CHO, CUN, MET, QUI, VAC **Elev.:** 10–2200 m a.s.l. **IUCN Habitat code:** 1,2,3,14 **Conservation status: (IUCN)** LC

2919 *Trichostigma peruvianum* (Moq.) H.Walter (Petiveriaceae) **IPNI:** 257667-2 **Habit:** Shrub **Use:** HF **Origin:** Native **Region:** Amazonia **Dept.:** CAQ **Elev.:** 300–300 m a.s.l. **IUCN Habitat code:** 1 **Conservation status: (IUCN)** NE

2920 *Phytolacca americana* L. (Phytolaccaceae) **IPNI:** 323290-2 **Use:** EU, GS, HF, MA, ME, PO **Dept.:** BOG**IUCN Habitat code:** 1,14 **Conservation status: (IUCN)** NE

2921 *Phytolacca bogotensis* Kunth (Phytolaccaceae) **IPNI:** 323291-2 **Common name:** altasara, cargamanta, cargamanto, guaba, hierba de culebra, jaboncillo, mantavieja, maíz de lobo, maíz de perro, tintamorada, yerba de culebra **Habit:** Herb **Use:** MA, ME **Origin:** Native **Region:** Andean **Dept.:** ANT, BOG, BOY, CAU, CES, CUN, HUI, NAR, QUI, RIS, SAN, TOL **Elev.:** 1560-3600 m.a.s.l. **IUCN Habitat code:** 1,3,4,5,6,8,14 **Conservation status: (IUCN)** NE **(CPSG)** Potential LC

2922 *Phytolacca dioica* L. (Phytolaccaceae) **IPNI:** 676348-1 **Common name:** guaba, yerba de culebra **Habit:** Tree **Use:** AF, EU, GS, HF, MA, ME, PO **Origin:** Native **Region:** Andean **Dept.:** ANT, BOG, QUI, RIS, TOL **Elev.:** 1700–2400 m a.s.l. **IUCN Habitat code:** 1,14 **Conservation status: (IUCN)** NE

2923 *Phytolacca icosandra* L. (Phytolaccaceae) **IPNI:** 323292-2 **Common name:** altasara, amole, atusara, bella sombra, cargamanta, caricillo, congora, conguerán, guaba, hierba de culebra, jaboncillo, matavieja, maíz de perro, saúco, yerba de culebra **Habit:** Herb **Use:** HF, MA, ME, PO **Origin:** Native **Region:** Andean **Dept.:** ANT, BOG, BOY, CAU, CES, CUN, HUI, MAG, MET, QUI, RIS, TOL **Elev.:** 1200-3800 m.a.s.l. **IUCN Habitat code:** 1,3,4,5,14 **Conservation status: (IUCN)** NE **(CPSG)** Potential LC

2924 *Phytolacca octandra* L. (Phytolaccaceae) **IPNI:** 323293-2 **Habit:** Herb **Use:** EU, GS, HF, ME **Origin:** Native **Region:** Andean **Dept.:** ANT, BOY, CAU, NAR, NSA **Elev.:** 1700-2800 m.a.s.l. **Conservation status: (IUCN)** NE **(CPSG)** Potential LC

2925 *Phytolacca rivinoides* Kunth & C.D.Bouché (Phytolaccaceae) **IPNI:** 676397-1 **Common name:** airambo, altusa, altusara morado, antusara, atonsora, atusara, calamanta, cargamanta, cargamanto, carurú, guaba, guagua, jaboncillo, juíbaro, lechuguilla, lusara, wakarabiijiiterenae, yerba de culebra **Habit:** Herb, Subshrub **Use:** EU, HF, MA, ME, PO **Origin:** Native **Region:** Amazonia, Andean, Guiana Shield, Orinoquia, Pacific **Dept.:** AMA, ANT, BOG, CAQ, CAU, CHO, GUV,

LAG, MAG, MET, NAR, PUT, SAN, VAC, VAU **Elev.:** 50-2900 m.a.s.l. **Conservation status: (IUCN)** NE **(CPSG)** Potential LC

2926 *Phytolacca rugosa* A.Braun & C.D.Bouché (Phytolaccaceae) **IPNI:** 676398-1 **Common name:** guaba, guabo, jaboncillo **Habit:** Herb, Subshrub **Use:** ME **Origin:** Native **Region:** Andean **Dept.:** ANT, BOG, CAU, CUN, MAG, PUT, SAN **Elev.:** 2000–3600 m a.s.l. **IUCN Habitat code:** 1,3,4,14 **Conservation status: (IUCN)** NE

2927 *Phytolacca sanguinea* H.Walter (Phytolaccaceae) **IPNI:** 195836-2 **Common name:** guaba, jaboncillo **Habit:** Herb **Use:** ME **Origin:** Native **Region:** Andean **Dept.:** ANT, CAU, CUN, NAR, PUT, SAN, TOL, VAC **Elev.:** 2200–3600 m a.s.l. **IUCN Habitat code:** 1,3,4,14 **Conservation status: (IUCN)** NE

2928 *Phytolacca thyrsiflora* Fenzl ex J.A.Schmidt (Phytolaccaceae) **IPNI:** 30307474-2 **Habit:** Herb **Use:** HF **Origin:** Native **Region:** Guiana Shield **Dept.:** VAU **Elev.:** 120–120 m a.s.l. **IUCN Habitat code:** 1 **Conservation status: (IUCN)** NE

2929 *Armeria maritima* (Mill.) Willd. (Plumbaginaceae) **IPNI:** 19793-2 **Common name:** rosa del cielo **Habit:** Herb **Use:** EU, HF, ME **Origin:** Cultivated **Region:** Andean **Dept.:** BOG, BOY, CUN **Elev.:** 2600–2800 m a.s.l. **Conservation status: (IUCN)** NE

2930 *Limonium sinuatum* (L.) Mill. (Plumbaginaceae) **IPNI:** 686979-1 **Common name:** estatis **Habit:** Herb **Use:** EU, GS, HF, ME **Origin:** Cultivated **Region:** Andean **Dept.:** ANT, BOG, CUN **Elev.:** 1000–2620 m a.s.l. **IUCN Habitat code:** 14 **Conservation status: (IUCN)** NE

2931 *Plumbago auriculata* Lam. (Plumbaginaceae) **IPNI:** 687056-1 **Common name:** azulina, bellaemilia, centella, floroazul, jazmín azul, pegosa **Habit:** Shrub **Use:** AF, EU, HF, MA, ME, PO, SU **Origin:** Cultivated **Region:** Andean, Caribbean, Orinoquia **Dept.:** ANT, BOG, CAL, CAU, CUN, MAG, MET, QUI, SAP, TOL, VAC **Elev.:** 50–2600 m a.s.l. **IUCN Habitat code:** 1,2,3,4,14 **Conservation status: (IUCN)** NE

2932 *Plumbago caerulea* Kunth (Plumbaginaceae) **IPNI:** 687057-1 **Habit:** Herb **Use:** ME **Origin:** Native **Region:** Andean **Dept.:** BOY **Elev.:** 2210–2500 m a.s.l. **IUCN Habitat code:** 3 **Conservation status: (IUCN)** NE

2933 *Plumbago zeylanica* L. (Plumbaginaceae) **IPNI:** 687109-1 **Common name:** bella emilia, centella, pegosa **Habit:** Climbing **Use:** EU, HF, MA, ME, PO **Origin:** Native **Region:** Caribbean **Dept.:** ANT, BOL, BOY, CUN, HUI, LAG, MAG, SUC, TOL, VAC **Elev.:** 300–2350 m a.s.l. **Conservation status: (IUCN)** NE

2934 *Antigonon guatimalense* Meisn. (Polygonaceae) **IPNI:** 693090-1 **Common name:** bellísima **Habit:** Climbing **Use:** EU **Origin:** Cultivated **Region:** Andean **Dept.:** ANT, NSA **Elev.:** 700–1800 m a.s.l. **IUCN Habitat code:** 14 **Conservation status: (IUCN)** NE

2935 *Antigonon leptopus* Hook. & Arn. (Polygonaceae) **IPNI:** 15860-2 **Common name:** bella, bellita, bellísima, bellísimo, cadena de amor, recuerdo **Habit:** Climbing **Use:** AF, EU, GS, HF, MA, ME **Origin:** Cultivated, Naturalised **Region:** Caribbean, Pacific **Dept.:** ANT, ATL, BOL, CHO, CUN, MAG, SAP **Elev.:** 0–200 m a.s.l. **IUCN Habitat code:** 1,2,3,4,14 **Conservation status: (IUCN)** NE

2936 *Coccoloba acuminata* Kunth (Polygonaceae) **IPNI:** 693560-1 **Common name:** laurel, maizcocho, maiztostao **Habit:** Shrub, Small tree **Use:** HF, ME **Origin:** Native **Region:** Caribbean, Orinoquia, Pacific **Dept.:** ANT, BOL, CAU, CHO, LAG, MAG, MET, TOL **Elev.:** 0–500 m a.s.l. **IUCN Habitat code:** 1,2,3,5,14 **Conservation status: (IUCN)** LC

2937 *Coccoloba ascendens* Duss ex Lindau (Polygonaceae) **IPNI:** 693572-1 **Habit:** Tree, Liana **Use:** HF **Origin:** Native **Region:** Orinoquia, Pacific **Dept.:** ANT, CHO, MET **Elev.:** 0–500 m a.s.l. **IUCN Habitat code:** 1,3,4,5,14 **Conservation status: (IUCN)** NE

2938 *Coccoloba caracasana* Meisn. (Polygonaceae) **IPNI:** 693597-1 **Common name:** hojaancha, hojarasco, hueso, uvero, uvero macho, uvero uvita, uvita roja **Habit:** Tree **Use:** AF, HF **Origin:** Native **Region:** Amazonia, Caribbean, Orinoquia **Dept.:** BOL, CAS, GUV, MET **Elev.:** 0–500 m a.s.l. **IUCN Habitat code:** 1,2,3,4,5,14 **Conservation status: (IUCN)** NE

2939 *Coccoloba coronata* Jacq. (Polygonaceae) **IPNI:** 693617-1 **Common name:** juan garrote, vara de piedra **Habit:** Small tree **Use:** HF, MA **Origin:** Native **Region:** Andean, Caribbean, Orinoquia, Pacific **Dept.:** ANT, BOL, BOY, CHO, LAG, MAG, MET, SAN **Elev.:** 0–600 m a.s.l. **IUCN Habitat code:** 1,2,3,5,14 **Conservation status: (IUCN)** LC

2940 *Coccoloba densifrons* Mart. ex Meisn. (Polygonaceae) **IPNI:** 693638-1 **Common name:** carneasada, guacharaco, rompehacha, vino **Habit:** Small tree **Use:** HF **Origin:** Native **Region:** Amazonia, Caribbean, Orinoquia **Dept.:** AMA, ANT, BOL, MET **Elev.:** 100–500 m a.s.l. **IUCN Habitat code:** 1,2,3,5,14 **Conservation status: (IUCN)** LC

2941 *Coccoloba excelsa* Benth. (Polygonaceae) **IPNI:** 693654-1 **Common name:** palo diablo **Habit:** Small tree, Liana, Climbing **Use:** HF **Origin:** Native **Region:** Amazonia, Pacific **Dept.:** AMA, ANT, GUA **Elev.:** 100–300 m a.s.l. **IUCN Habitat code:** 1,3,4,14 **Conservation status: (IUCN)** LC

2942 *Coccoloba latifolia* Poir. (Polygonaceae) **IPNI:** 693712-1 **Common name:** palo de bejuco **Habit:** Tree **Use:** HF **Origin:** Native **Region:** Amazonia **Dept.:** AMA **Elev.:** 200–200 m a.s.l. **IUCN Habitat code:** 1,2,3,4,5,14 **Conservation status: (IUCN)** NE

2943 *Coccoloba mollis* Casar. (Polygonaceae) **IPNI:** 693754-1 **Common name:** arepo **Habit:** Small tree, Tree **Use:** HF, MA **Origin:** Native **Region:** Amazonia, Andean, Caribbean, Orinoquia, Pacific **Dept.:** AMA, ANT, CAQ, CAS, COR, MET, NAR, VID **Elev.:** 100–600 m a.s.l. **Conservation status: (IUCN)** LC

2944 *Coccoloba nitida* Kunth (Polygonaceae) **IPNI:** 693777-1 **Habit:** Tree, Liana **Use:** HF, ME **Origin:** Native **Region:** Amazonia **Dept.:** AMA **Elev.:** 100–100 m a.s.l. **IUCN Habitat code:** 1,2,3,4 **Conservation status: (IUCN)** NE

2945 *Coccoloba obovata* Kunth (Polygonaceae) **IPNI:** 693787-1 **Common name:** buchegallino, hueso negro, juan garrote, mortin, mortín, uvillo **Habit:** Tree **Use:** HF, ME **Origin:** Native **Region:** Andean, Caribbean, Pacific **Dept.:** ANT, BOL, CES, CHO, HUI, MAG, QUI, SUC, TOL **Elev.:** 80–1000 m a.s.l. **IUCN Habitat code:** 1,2,3,5,14 **Conservation status: (IUCN)** LC

2946 *Coccoloba obtusifolia* Jacq. (Polygonaceae) **IPNI:** 693788-1 **Common name:** corralero, tacaloa **Habit:** Small tree **Use:** AF, FU, MA **Origin:** Native **Region:** Andean, Caribbean **Dept.:** ANT, ATL, BOL, LAG, MAG **Elev.:** 0–700 m a.s.l. **IUCN Habitat code:** 1,2,3,5,14 **Conservation status: (IUCN)** NE

2947 *Coccoloba ovata* Benth. (Polygonaceae) **IPNI:** 693795-1 **Habit:** Shrub, Small tree **Use:** HF **Origin:** Native **Region:** Amazonia, Guiana Shield, Orinoquia **Dept.:** AMA, VAU, VID **Elev.:** 0–200 m a.s.l. **Conservation status: (IUCN)** LC

2948 *Coccoloba uvifera* (L.) L. (Polygonaceae) **IPNI:** 693918-1 **Common name:** agraz, buchegallina, corralero, güino, juan garrote, koulit, micongo, murta, quino, uva, uva caleta, uva de la playa, uva de mar, uva de playa, uva playa, uvero, uvero de playa, uvilla, uvita de playa, uvita playera, uvito, uvito de playa, uvo de mar, uvo **Habit:** Tree **Use:** EU, FU, GS, HF, MA, ME **Origin:** Native, Cultivated **Region:** Caribbean, Pacific **Dept.:** ANT, BOL, MAG, VAC **Elev.:** 20–900 m a.s.l. **IUCN Habitat code:** 1,2,3,4,5,14 **Conservation status: (IUCN)** LC

2949 *Enneatypus ramiflorus* (C.A.Mey.) Roberty & Vautier (Polygonaceae) **IPNI:** 693983-1 **Common name:** cargamuchacho, cocaemico, cóndor, guarapero, palo prieto, pastelillo, shoponoy, volador **Habit:** Tree **Use:** EU, FU, HF, MA, ME **Origin:** Native **Region:** Caribbean **Dept.:** ATL, BOL, LAG, MAG, SAN **Elev.:** 0–300 m a.s.l. **Conservation status: (IUCN)** NE

2950 *Fagopyrum esculentum* Moench (Polygonaceae) **IPNI:** 694526-1 **Common name:** trigo sarraceno **Use:** AF, EU, GS, HF, IF, MA, ME, PO, SU **IUCN Habitat code:** 3,4,14 **Conservation status: (IUCN)** NE

2951 *Muehlenbeckia platyclada* (F.Muell.) Meisn. (Polygonaceae) **IPNI:** 694774-1 **Common name:** cinta, mamajuanita **Habit:** Shrub **Use:** EU, HF, ME, SU **Origin:** Cultivated **Region:** Andean **Dept.:** ANT **Elev.:** 1500–2500 m a.s.l. **IUCN Habitat code:** 14 **Conservation status: (IUCN)** NE

2952 *Muehlenbeckia tamnifolia* (Kunth) Meisn. (Polygonaceae) **IPNI:** 694785-1 **Common name:** bejuco chivo, bejuco coloradito, bejuco colorado, bejuco mollantín, chibo, chipaque, chío, coronillo, mamajuana, mamajuanita, mocantia, mollantín, mollentín, murupacha de frío, pajarito, shiatushi **Habit:** Liana, Climbing **Use:** HF, ME **Origin:** Native **Region:** Andean **Dept.:** ANT, BOG, BOY, CAQ, CAU, CES, CUN, HUI, LAG, MAG, NAR, PUT, VAC **Elev.:** 390-3900 m.a.s.l. **IUCN Habitat code:** 1,3,4,5,8,14 **Conservation status: (IUCN)** NE **(CPSG)** Potential LC

2953 *Muehlenbeckia volcanica* (Benth.) Endl. (Polygonaceae) **IPNI:** 1022243-2 **Habit:** Subshrub **Use:** HF, ME **Origin:** Native **Region:** Andean **Dept.:** BOG, BOY, CAL, CAU, CUN, MET, RIS, TOL **Elev.:** 2900–4000 m a.s.l. **IUCN Habitat code:** 1,4,8 **Conservation status: (IUCN)** NE

2954 *Persicaria acuminata* (Kunth) M.Gómez (Polygonaceae) **IPNI:** 694896-1 **Common name:** fumo bravo, hierba caimán, tabaquillo, yaco **Habit:** Herb, Subshrub **Use:** AF, EU, HF, MA, ME, SU **Origin:** Native **Region:** Andean, Orinoquia **Dept.:** ANT, ARA **Elev.:** 200–2100 m a.s.l. **IUCN Habitat code:** 1,2,3,4,5,14 **Conservation status: (IUCN)** NE

2955 *Persicaria capitata* (Buch.-Ham. ex D.Don) H.Gross (Polygonaceae) **IPNI:** 60433515-2 **Habit:** Herb **Use:** EU, ME **Origin:** Naturalised **Region:** Andean **Dept.:** ANT, CUN **Elev.:** 1000–2600 m a.s.l. **IUCN Habitat code:** 1,3,14 **Conservation status: (IUCN)** NE

2956 *Persicaria glabra* (Willd.) M.Gómez (Polygonaceae) **IPNI:** 1005755-2 **Common name:** barbasco **Habit:** Herb, Subshrub **Use:** HF, ME **Origin:** Native **Region:** Caribbean **Dept.:** COR, VAC **Elev.:** 6–1000 m a.s.l. **IUCN Habitat code:** 1,2,3,5 **Conservation status: (IUCN)** NE

2957 *Persicaria hispida* (Kunth) M.Gómez (Polygonaceae) **IPNI:** 694999-1 **Common name:** barbasco, tabaquillo **Habit:** Herb **Use:** ME, PO **Origin:** Native **Region:** Caribbean, Pacific **Dept.:** ANT, BOL, COR **Elev.:** 5–120 m a.s.l. **IUCN Habitat code:** 1,2,3,5,14 **Conservation status: (IUCN)** NE

2958 *Persicaria hydropiper* (L.) Delarbre (Polygonaceae) **IPNI:** 77069479-1 **Use:** HF, ME, PO **Region:** Andean **Dept.:** BOG**Conservation status: (IUCN)** LC

2959 *Persicaria hydropiperoides* (Michx.) Small (Polygonaceae) **IPNI:** 190811-2 **Common name:** barbasco, barbasquillo, gualola, herb de sapo, hierba de sapo, tabaquillo **Habit:** Herb **Use:** MA, ME **Origin:** Native **Region:** Andean, Orinoquia, Pacific **Dept.:** ANT, BOG, BOY, MET **Elev.:** 200–3200 m a.s.l. **IUCN Habitat code:** 1,2,3,4,5,14 **Conservation status: (IUCN)** LC

2960 *Persicaria lapathifolia* (L.) Delarbre (Polygonaceae) **IPNI:** 77069480-1 **Common name:** barbasco, barbasquillo, catay, corazón herido, hierba de sapo, picantilla, picantillo **Use:** EU, HF, ME, PO **Conservation status: (IUCN)** LC

2961 *Persicaria maculosa* Gray (Polygonaceae) **IPNI:** 695052-1 **Use:** ME, PO **Region:** Andean **Dept.:** BOG**IUCN Habitat code:** 1,3,14 **Conservation status: (IUCN)** LC

2962 *Persicaria meisneriana* (Cham. & Schltdl.) M.Gómez (Polygonaceae) **IPNI:** 1050850-2 **Common name:** barbasco, yerba de sapo **Habit:** Herb **Use:** ME **Origin:** Native **Region:** Andean **Dept.:** CAU, CUN **Elev.:** 1700–2500 m a.s.l. **IUCN Habitat code:** 4 **Conservation status: (IUCN)** NE

2963 *Persicaria nepalensis* (Meisn.) H.Gross (Polygonaceae) **IPNI:** 60433514-2 **Common name:** amaranto, barbasco, barbasquillo, bledo, bogotana, catay, corazón herido, frijolillo, hierba de sapo, malavecina, orejas del diablo, picantilla, picantillo **Habit:** Herb **Use:** AF, EU, HF, MA, ME, SU **Origin:** Naturalised **Region:** Andean **Dept.:** ANT, HUI **Elev.:** 1500–2400 m a.s.l. **IUCN Habitat code:** 1,3,4,14 **Conservation status: (IUCN)** NE

2964 *Persicaria posumbu* (Buch.-Ham. ex D.Don) H.Gross (Polygonaceae) **IPNI:** 695113-1 **Use:** ME **Region:** Andean **Dept.:** BOG**Conservation status: (IUCN)** NE

2965 *Persicaria punctata* (Elliott) Small (Polygonaceae) **IPNI:** 190855-2 **Common name:** barbasco, barbasquillo, catay, corazón herido, hierba de sapo, picantilla, picantillo, yerba de sapo **Habit:** Herb **Use:** HF, ME **Origin:** Native **Region:** Andean, Pacific **Dept.:** ANT, BOY, CHO, CUN, MAG, MET, SAN, VAC **Elev.:** 700–2400 m a.s.l. **IUCN Habitat code:** 1,2,3,5,8,14 **Conservation status: (IUCN)** NE

2966 *Persicaria segetum* (Kunth) Small (Polygonaceae) **IPNI:** 190869-2 **Common name:** barbasco, barbasco rosado, corazón herido, envidia, gloria, gualola, hierba de sapo **Habit:** Herb **Use:** EU, ME **Origin:** Native **Region:** Andean, Caribbean **Dept.:** ANT, BOG, BOL, COR, CUN **Elev.:** 6–1600 m a.s.l. **IUCN Habitat code:** 1,2,3,4,14 **Conservation status: (IUCN)** NE

2967 *Polygonum aviculare* L. (Polygonaceae) **IPNI:** 695422-1 **Habit:** Herb **Use:** AF, HF, MA, ME, PO **Origin:** Naturalised **Region:** Andean **Dept.:** BOG, CUN **Elev.:** 2600–2600 m a.s.l. **IUCN Habitat code:** 4,14 **Conservation status: (IUCN)** NE

2968 *Polygonum patulum* M.Bieb. (Polygonaceae) **IPNI:** 696222-1 **Habit:** Herb **Use:** HF, ME **Origin:** Naturalised **Region:** Andean **Dept.:** CUN **Elev.:** 2600–2600 m a.s.l. **IUCN Habitat code:** 14 **Conservation status: (IUCN)** NE

2969 *Rheum officinale* Baill. (Polygonaceae) **IPNI:** 696768-1 **Common name:** ruibarbo **Habit:** Herb **Use:** EU, HF, ME, SU **Origin:** Cultivated **Region:** Andean **Dept.:** ANT **Elev.:** 2000–2500 m a.s.l. **Conservation status: (IUCN)** NE

2970 *Rheum palmatum* L. (Polygonaceae) **IPNI:** 696772-1 **Common name:** raíz de ruibarbo, ruibarbo, ruibarbo del tíbet **Habit:** Herb **Use:** EU, GS, HF, ME **Origin:** Naturalised **Region:** Andean **Dept.:** BOG, CUN **Elev.:** 2600–2600 m a.s.l. **Conservation status: (IUCN)** NE

2971 *Rheum rhaponticum* L. (Polygonaceae) **IPNI:** 696786-1 **Habit:** Herb **Use:** HF, ME, PO **Origin:** Naturalised **Region:** Andean **Dept.:** BOG, CUN **Elev.:** 2560–2560 m a.s.l. **Conservation status: (IUCN)** NE

2972 *Rumex acetosella* L. (Polygonaceae) **IPNI:** 224391-2 **Common name:** acedera, acederilla, acederita, barrabacillo, barrabasillo, barrabás, envidia, fredolina, lamparilla, lengüilla, romacilla, sangre de toro, sangretoro, toritos, yerba envidia **Habit:** Herb **Use:** HF, ME, PO **Origin:** Naturalised **Region:** Andean, Caribbean **Dept.:** ANT, BOG, BOY, CAU, CES, CUN, SUC **Elev.:** 300–3900 m a.s.l. **IUCN Habitat code:** 1,2,3,4,8,14 **Conservation status: (IUCN)** NE

2973 *Rumex conglomeratus* Murray (Polygonaceae) **IPNI:** 697003-1 **Common name:** bijuacá, lengua de vaca, romaza **Habit:** Herb **Use:** HF, ME **Origin:** Naturalised **Region:** Andean **Dept.:** BOG, BOY, CAL **Elev.:** 2000–3000 m a.s.l. **IUCN Habitat code:** 1,3,5,14 **Conservation status: (IUCN)** NE

2974 *Rumex crispus* L. (Polygonaceae) **IPNI:** 224413-2 **Common name:** arracachuela, arracachuelo, barbasco, barrabás, bijuacá, chueca, lengua de vaca, lengüevaca, paciencia, ribarbo, riobarbo, romacillo, romaza, romazo, ruibarbo, ruibarbo de huerta **Habit:** Herb **Use:** AF, EU, GS, HF, MA, ME, PO **Origin:** Naturalised **Region:** Andean **Dept.:** ANT, BOG **Elev.:** 2500–2500 m a.s.l. **IUCN Habitat code:** 1,2,3,4,5,8,14 **Conservation status: (IUCN)** NE

2975 *Rumex obtusifolius* L. (Polygonaceae)
IPNI: 310739-2
Common name: acedera, acederita, bijuacá, lengua de vaca, lengüevaca, paciencia, romaza, rumaza, sangre de toro, sangretoro **Habit:** Herb **Use:** HF, ME **Origin:** Naturalised **Region:** Andean **Dept.:** ANT, BOG, NAR, SAN, VAC **Elev.:** 1500–3000 m a.s.l. **IUCN Habitat code:** 1,3,4,14 **Conservation status: (IUCN)** NE

2976 *Rumex patientia* L. (Polygonaceae)
IPNI: 697388-1
Habit: Herb **Use:** HF, ME **Origin:** Naturalised **Region:** Andean **Dept.:** BOG, CUN **Elev.:** 2690–2690 m a.s.l. **IUCN Habitat code:** 14 **Conservation status: (IUCN)** NE

2977 *Symmeria paniculata* Benth. (Polygonaceae)
IPNI: 697710-1
Common name: mangle, mangle bobo, mangle rojo **Habit:** Tree **Use:** HF **Origin:** Native **Region:** Amazonia, Caribbean, Orinoquia **Dept.:** AMA, BOL, GUA, VAU **Elev.:** 5–120 m a.s.l. **IUCN Habitat code:** 1,2,3,4,5,14 **Conservation status: (IUCN)** NE

2978 *Triplaris americana* L. (Polygonaceae)
IPNI: 697785-1
Common name: barba de mono, barrabás, guacamayo, guayabo, guayabo zancón, hormiguero, palo de María, palo de Santa María, palo maría, palo santo, palosanto, tangarana, varasanta, zancón **Habit:** Tree **Use:** EU, MA, ME **Origin:** Native **Region:** Amazonia, Andean, Caribbean, Orinoquia, Pacific **Dept.:** AMA, ANT, ATL, CHO, MAG, MET, PUT, SAN, TOL, VAU **Elev.:** 0–2000 m a.s.l. **IUCN Habitat code:** 1,2,3,4,5,14 **Conservation status: (IUCN)** LC

2979 *Triplaris caracasana* Cham. (Polygonaceae)
IPNI: 30350551-2
Habit: Tree **Use:** ME **Origin:** Native **Region:** Caribbean **Dept.:** MAG **Elev.:** 200–200 m a.s.l. **Conservation status: (IUCN)** NE

2980 *Triplaris cumingiana* Fisch. & C.A.Mey. (Polygonaceae)
IPNI: 697802-1
Common name: varasanta **Habit:** Tree **Use:** MA **Origin:** Native **Region:** Andean, Caribbean, Orinoquia **Dept.:** ANT, BOL, BOY, CES, CUN, HUI, LAG, MAG, MET, SAN, SUC, VAC **Elev.:** 300–1000 m a.s.l. **Conservation status: (IUCN)** LC

2981 *Triplaris melaenodendron* (Bertol.) Standl. & Steyerm. (Polygonaceae)
IPNI: 258609-2
Common name: varasanta **Habit:** Tree **Use:** MA **Region:** Andean, Pacific **Dept.:** ANT, CUN, MAG, TOL, VAC **Elev.:** 0-1000 m.a.s.l. **IUCN Habitat code:** 1,3,14 **Conservation status: (IUCN)** LC **(CPSG)** LC

2982 *Triplaris peruviana* Fisch. & C.A.Mey. (Polygonaceae)
IPNI: 286414-2
Common name: cajao-amo, tangarana **Use:** ME **IUCN Habitat code:** 1 **Conservation status: (IUCN)** NE

2983 *Triplaris purdiei* Meisn. (Polygonaceae)
IPNI: 286418-2
Habit: Tree **Use:** EU **Origin:** Native **Region:** Caribbean **Dept.:** ATL, BOL, LAG, MAG **Elev.:** 0-100 m.a.s.l. **IUCN Habitat code:** 2,14 **Conservation status: (IUCN)** VU **(CPSG)** VU

2984 *Triplaris weigeltiana* (Rchb.) Kuntze (Polygonaceae)
IPNI: 258621-2
Common name: varablanca **Habit:** Tree **Use:** MA, ME **Origin:** Native **Region:** Amazonia **Dept.:** AMA **Elev.:** 0–100 m a.s.l. **Conservation status: (IUCN)** LC

2985 *Portulaca oleracea* L. (Portulacaceae)
IPNI: 323270-2
Common name: anu wan u, paliaría, verdolaga, yerbepollo **Habit:** Herb **Use:** AF, EU, GS, HF, MA, ME, PO, SU **Origin:** Naturalised **Region:** Amazonia, Andean, Guiana Shield, Caribbean, Orinoquia, Pacific **Dept.:** AMA, ANT, BOG, BOL, CAQ, CAS, CHO, COR, CUN, GUA, GUV, LAG, MAG, MET, NAR, PUT, SAP, TOL, VAC **Elev.:** 0–2550 m a.s.l. **IUCN Habitat code:** 1,2,3,4,5,14 **Conservation status: (IUCN)** LC

2986 *Portulaca pilosa* L. (Portulacaceae)
IPNI: 316354-2
Common name: atarraya, bellaalasonce, bellaonce, beso de Gaitán, bilasonce, billalasonce, brillaalasonce, brillalasonce, buenastardes, clavel chino, diez del día, flor de las once, hierba de vidrio, hierba del vidrio, pechiche, pedorrera, portulaca, sol de mediodía, s **Habit:** Herb **Use:** EU, HF, ME **Origin:** Native **Region:** Andean, Guiana Shield, Caribbean, Orinoquia **Dept.:** ANT, ATL, BOL, CUN, GUA, HUI, LAG, MAG, NAR, SAP, TOL, VAC, VAU, VID **Elev.:** 0–1420 m a.s.l. **IUCN Habitat code:** 1,2,3,4,5,14 **Conservation status: (IUCN)** NE

2987 *Portulaca rubricaulis* Kunth (Portulacaceae)
IPNI: 323271-2
Habit: Herb **Use:** ME **Origin:** Native **Region:** Caribbean, Orinoquia **Dept.:** ATL, BOL, CAS, MAG, SUC, VID **Elev.:** 0–100 m a.s.l. **Conservation status: (IUCN)** NE

2988 *Portulaca umbraticola* Kunth (Portulacaceae)
IPNI: 323272-2
Habit: Herb **Use:** HF, ME **Origin:** Native **Region:** Caribbean **IUCN Habitat code:** 1,3,14 **Conservation status: (IUCN)** NE

2989 *Talinum fruticosum* (L.) Juss. (Talinaceae)
IPNI: 1182764-2
Common name: aclaraboyos, espinaca de filipinas, espinaca de java, espinaca de surinam, hierba de sapo, lechuga platanera, lustrosa grande, verdolaga, verdolaga francés, verdolaga montañera **Habit:** Herb **Use:** AF, EU, GS, HF, MA, ME, PO, SU **Origin:** Native **Region:** Andean, Guiana Shield, Pacific **Dept.:** ANT, ATL, BOL, CHO, CUN, GUA, LAG, MAG, SUC, TOL, VAC **Elev.:** 5–1580 m a.s.l. **IUCN Habitat code:** 1,2,3,4,5,14 **Conservation status: (IUCN)** NE

2990 *Talinum paniculatum* (Jacq.) Gaertn. (Talinaceae)
IPNI: 1007146-2
Common name: cueroesapo, hoja de arroz, lechuga de platanal, portulaca de playa, verdolaga, verdolaga de playa, verdolago **Habit:** Herb **Use:** EU, GS, HF, ME **Origin:** Native **Region:** Andean, Caribbean **Dept.:** ANT, BOL, CAQ, CAU, CUN, HUI, LAG, NSA, PUT, QUI, SAP, TOL, VAC **Elev.:** 120-1670 m.a.s.l. **IUCN Habitat code:** 1,3 **Conservation status: (IUCN)** NE **(CPSG)** Potential LC

Celastrales

2991 *Cheiloclinium anomalum* Miers (Celastraceae)
IPNI: 430947-1
Habit: Liana **Use:** HF **Origin:** Native **Region:** Amazonia **Dept.:** VAU **Elev.:** 0–300 m a.s.l. **IUCN Habitat code:** 1,14 **Conservation status: (IUCN)** NE

2992 *Cheiloclinium articulatum* (A.C.Sm.) A.C.Sm. (Celastraceae)
IPNI: 55842-2
Habit: Liana **Use:** ME **Origin:** Native **Region:** Amazonia **Dept.:** AMA **Elev.:** 0–300 m a.s.l. **IUCN Habitat code:** 1 **Conservation status: (IUCN)** NE

2993 *Cheiloclinium cognatum* (Miers) A.C.Sm. (Celastraceae)
IPNI: 55844-2
Habit: Shrub, Small tree **Use:** HF **Origin:** Native **Region:** Amazonia, Andean, Guiana Shield, Pacific **Dept.:** ANT, CHO, NAR, NSA, PUT, SAN, VAU **Elev.:** 0–900 m a.s.l. **IUCN Habitat code:** 1,3,4,5,14 **Conservation status: (IUCN)** LC

2994 *Cheiloclinium hippocrateoides* (Peyr.) A.C.Sm. (Celastraceae)
IPNI: 55849-2
Habit: Liana **Use:** HF, ME **Origin:** Native **Region:** Amazonia, Guiana Shield **Dept.:** AMA, VAU **Elev.:** 0–400 m a.s.l. **IUCN Habitat code:** 1,3,4,14 **Conservation status: (IUCN)** NE

2995 *Crossopetalum rhacoma* Crantz (Celastraceae)
IPNI: 60458018-2
Habit: Shrub **Use:** HF **Origin:** Native **Region:** Caribbean **Dept.:** SAP, SUC **Elev.:** 0–100 m a.s.l. **Conservation status: (IUCN)** LC

2996 *Cuervea kappleriana* (Miq.) A.C.Sm. (Celastraceae)
IPNI: 72123-2
Habit: Liana **Use:** HF, MA **Origin:** Native **Region:** Amazonia **Dept.:** AMA, CAQ, CUN **Elev.:** 0–600 m a.s.l. **IUCN Habitat code:** 1,2,4,14 **Conservation status: (IUCN)** NE

2997 *Euonymus europaeus* L. (Celastraceae)
IPNI: 160876-1
Use: EU, MA, ME, PO **Region:** Andean **Dept.:** BOG**Conservation status: (IUCN)** LC

2998 *Euonymus japonicus* Thunb. (Celastraceae)
IPNI: 160944-1
Habit: Shrub, Small tree, Liana **Use:** EU, HF, ME **Origin:** Cultivated **Region:** Andean **Dept.:** BOG, CAL, CUN, NSA **Elev.:** 1800–2600 m a.s.l. **IUCN Habitat code:** 14 **Conservation status: (IUCN)** NE

2999 *Hippocratea volubilis* L. (Celastraceae)
IPNI: 161653-1
Common name: bejuco alambre, malibú **Habit:** Liana **Use:** HF, ME **Origin:** Native **Region:** Amazonia, Andean, Caribbean **Dept.:** AMA, ATL, MAG, VAC, VAU, VID **Elev.:** 0–1800 m a.s.l. **IUCN Habitat code:** 1,2,3,4,5,14 **Conservation status: (IUCN)** NE

3000 *Hylenaea comosa* (Sw.) Miers (Celastraceae) **IPNI:** 161663-1 **Habit:** Small tree, Liana **Use:** HF **Origin:** Native **Region:** Amazonia **Dept.:** VAU **Elev.:** 0–200 m a.s.l. **IUCN Habitat code:** 1 **Conservation status: (IUCN)** LC

3001 *Hylenaea praecelsa* (Miers) A.C.Sm. (Celastraceae) **IPNI:** 125377-2 **Habit:** Liana **Use:** HF **Origin:** Native **Region:** Pacific **Dept.:** ANT, CHO **Elev.:** 10–220 m a.s.l. **IUCN Habitat code:** 1,5,14 **Conservation status: (IUCN)** NE

3002 *Maytenus amazonica* Mart. ex Reissek (Celastraceae) **IPNI:** 161805-1 **Common name:** chuchuguasa **Habit:** Tree **Use:** ME **Origin:** Native **Region:** Amazonia, Guiana Shield **Dept.:** AMA, CAQ, GUA **Elev.:** 200–630 m a.s.l. **IUCN Habitat code:** 1 **Conservation status: (IUCN)** NE

3003 *Maytenus guyanensis* Klotzsch ex Reissek (Celastraceae) **IPNI:** 161926-1 **Common name:** chichuá **Habit:** Tree **Use:** MA, ME **Origin:** Native **Region:** Amazonia, Orinoquia **Dept.:** CAQ, CAS, GUV, MET, VID **Elev.:** 35–350 m a.s.l. **IUCN Habitat code:** 1,3,5 **Conservation status: (IUCN)** LC

3004 *Maytenus laevis* Reissek (Celastraceae) **IPNI:** 161953-1 **Common name:** chuchuaza, chuchugasi, chuchuguacha, chuchuguasa, chuchuguasi, chuchuhuasa, chuchuhuasha, chuchuuaso, chugchuguao, coemeni **Habit:** Tree **Use:** HF, ME, PO **Origin:** Native **Region:** Amazonia **Dept.:** AMA, CAQ, PUT, VAU **Elev.:** 250–400 m a.s.l. **IUCN Habitat code:** 1,3 **Conservation status: (IUCN)** NE

3005 *Maytenus macrocarpa* (Ruiz & Pav.) Briq. (Celastraceae) **IPNI:** 155542-2 **Common name:** cajón, carguero rosado, chuchuguache, chuchuguasa **Habit:** Shrub, Small tree, Tree **Use:** HF, MA, ME, SU **Origin:** Native **Region:** Amazonia, Andean **Dept.:** AMA, ANT, GUA, PUT, VAU **Elev.:** 300–2600 m a.s.l. **IUCN Habitat code:** 1,3,14 **Conservation status: (IUCN)** NE

3006 *Maytenus sieberiana* Krug & Urb. (Celastraceae) **IPNI:** 155604-2 **Common name:** camarón, corocito **Habit:** Tree **Use:** AF, FU, ME **Origin:** Native **Region:** Amazonia, Caribbean **Dept.:** ANT, ATL, BOL, CHO, GUV, LAG, MAG, SAN **Elev.:** 0-600 m.a.s.l. **IUCN Habitat code:** 14 **Conservation status: (IUCN)** NE **(CPSG)** LC

3007 *Maytenus verticillata* (Ruiz & Pav.) DC. (Celastraceae) **IPNI:** 162118-1 **Habit:** Shrub, Small tree **Use:** ME **Origin:** Native **Region:** Andean **Dept.:** CAU, HUI, NAR **Elev.:** 2300–3600 m a.s.l. **IUCN Habitat code:** 1,3,4,14 **Conservation status: (IUCN)** LC

3008 *Peritassa compta* Miers (Celastraceae) **IPNI:** 162329-1 **Common name:** huevo de gato **Use:** HF **IUCN Habitat code:** 1,2,3,4,5,14 **Conservation status: (IUCN)** NE

3009 *Peritassa laevigata* (Hoffmanns. ex Link) A.C.Sm. (Celastraceae) **IPNI:** 190399-2 **Common name:** huevo de gato **Habit:** Shrub, Liana **Use:** HF **Origin:** Native **Region:** Amazonia, Guiana Shield, Orinoquia, Pacific **Dept.:** AMA, ANT, CHO, GUA, MET, VAU, VID **Elev.:** 0–770 m a.s.l. **Conservation status: (IUCN)** LC

3010 *Peritassa pruinosa* (Seem.) A.C.Sm. (Celastraceae) **IPNI:** 190402-2 **Habit:** Shrub, Liana **Use:** HF **Origin:** Native **Region:** Caribbean **Dept.:** MAG **Elev.:** 0–100 m a.s.l. **IUCN Habitat code:** 1,3 **Conservation status: (IUCN)** NE

3011 *Prionostemma aspera* (Lam.) Miers (Celastraceae) **IPNI:** 162403-1 **Habit:** Liana **Use:** HF **Origin:** Native **Region:** Guiana Shield, Orinoquia, Pacific **Dept.:** ANT, MET, SAN **Elev.:** 10–800 m a.s.l. **IUCN Habitat code:** 1,2,3,4,14 **Conservation status: (IUCN)** LC

3012 *Pristimera celastroides* (Kunth) A.C.Sm. (Celastraceae) **IPNI:** 210308-2 **Habit:** Shrub **Use:** ME, PO **Origin:** Native **Region:** Caribbean **Dept.:** BOL, LAG, SUC **Elev.:** 0–300 m a.s.l. **IUCN Habitat code:** 2,14 **Conservation status: (IUCN)** NE

3013 *Salacia cordata* (Miers) Mennega (Celastraceae) **IPNI:** 964785-1 **Habit:** Liana **Use:** HF, ME **Origin:** Native **Region:** Amazonia, Pacific **Dept.:** AMA, ANT, CHO, NAR, RIS, VAC, VAU **Elev.:** 100–550 m a.s.l. **IUCN Habitat code:** 1,3,14 **Conservation status: (IUCN)** LC

3014 *Salacia elliptica* (Mart.) G.Don (Celastraceae) **IPNI:** 162581-1 **Habit:** Small tree, Liana **Use:** HF, MA, ME **Origin:** Native **Region:** Guiana Shield, Orinoquia **Dept.:** AMA, ANT, BOY, CAS, MET, SAN **Elev.:** 70–1200 m a.s.l. **IUCN Habitat code:** 1,4,14 **Conservation status: (IUCN)** LC

3015 *Salacia gigantea* Loes. (Celastraceae) **IPNI:** 431218-1 **Common name:** caimillo, caimo, cajeto, chuchuguasa de rebalse, ojo de buey, palo de cemento **Habit:** Shrub, Small tree **Use:** HF **Origin:** Native **Region:** Amazonia, Pacific **Dept.:** AMA, CHO, VAC **Elev.:** 50–180 m a.s.l. **IUCN Habitat code:** 1,3 **Conservation status: (IUCN)** LC

3016 *Salacia impressifolia* (Miers) A.C.Sm. (Celastraceae) **IPNI:** 225455-2 **Habit:** Shrub, Liana **Use:** HF **Origin:** Native **Region:** Amazonia, Orinoquia **Dept.:** AMA, MET, VAU, VID **Elev.:** 100–450 m a.s.l. **IUCN Habitat code:** 1,3,4,5 **Conservation status: (IUCN)** LC

3017 *Salacia insignis* A.C.Sm. (Celastraceae) **IPNI:** 225457-2 **Common name:** alcanzaviejo, caparo opedoamo, güeva de tigre **Habit:** Liana **Use:** HF, ME **Origin:** Native **Region:** Amazonia, Guiana Shield **Dept.:** AMA, GUV, PUT, VAU **Elev.:** 0–200 m a.s.l. **IUCN Habitat code:** 1 **Conservation status: (IUCN)** LC

3018 *Salacia juruana* Loes. (Celastraceae) **IPNI:** 225458-2 **Habit:** Small tree **Use:** HF **Origin:** Native **Region:** Amazonia **Dept.:** AMA, VAU **Elev.:** 0–300 m a.s.l. **IUCN Habitat code:** 1,14 **Conservation status: (IUCN)** LC

3019 *Salacia macrantha* A.C.Sm. (Celastraceae) **IPNI:** 225467-2 **Habit:** Shrub, Small tree **Use:** HF **Origin:** Native **Region:** Amazonia, Guiana Shield, Caribbean **Dept.:** AMA, ANT, BOL, CAQ, MET, VAU **Elev.:** 100–800 m a.s.l. **IUCN Habitat code:** 1,3,5,14 **Conservation status: (IUCN)** LC

3020 *Schaefferia frutescens* Jacq. (Celastraceae) **IPNI:** 162760-1 **Common name:** limoncillo **Habit:** Shrub **Use:** MA **Origin:** Native **Region:** Caribbean **Dept.:** ATL, BOL, CAU, CUN, HUI, LAG, MAG, NAR, NSA, VAC **Elev.:** 0–1100 m a.s.l. **IUCN Habitat code:** 1,2,3,14 **Conservation status: (IUCN)** LC

3021 *Tontelea attenuata* Miers (Celastraceae) **IPNI:** 162852-1 **Habit:** Shrub, Liana, Climbing **Use:** HF **Origin:** Native **Region:** Amazonia, Guiana Shield **Dept.:** AMA, CAQ, GUA, MET, PUT **Elev.:** 150–1000 m a.s.l. **IUCN Habitat code:** 1,3,5,14 **Conservation status: (IUCN)** LC

Cornales

3022 *Hydrangea macrophylla* (Thunb.) Ser. (Hydrangeaceae) **IPNI:** 791637-1 **Common name:** berengüela, hortensia **Habit:** Herb **Use:** EU, HF, ME, PO, SU **Origin:** Cultivated **Region:** Andean **Dept.:** ANT, BOG, CUN, QUI, RIS, VAC **Elev.:** 1200–2830 m a.s.l. **IUCN Habitat code:** 1,3,4,14 **Conservation status: (IUCN)** NE

3023 *Aosa grandis* (Standl.) R.H.Acuña & Weigend (Loasaceae) **IPNI:** 77188059-1 **Habit:** Herb **Use:** ME **Origin:** Native **Region:** Andean, Pacific **Dept.:** ANT **Elev.:** 150–150 m a.s.l. **Conservation status: (IUCN)** NE

3024 *Gronovia scandens* L. (Loasaceae) **IPNI:** 545139-1 **Common name:** ortiga, pegapega de bejuco **Habit:** Climbing **Use:** ME **Origin:** Native **Region:** Andean **Dept.:** CUN, MAG, NAR, VAC **Elev.:** 30–1500 m a.s.l. **IUCN Habitat code:** 1,2,3,14 **Conservation status: (IUCN)** NE

3025 *Mentzelia aspera* L. (Loasaceae) **IPNI:** 545374-1 **Common name:** cadillo, pegapega **Habit:** Herb, Climbing **Use:** HF, ME **Origin:** Native **Dept.:** ANT, ATL, CAU, CUN, MAG, NAR, SAN, SUC, TOL, VAC **Elev.:** 1–1800 m a.s.l. **IUCN Habitat code:** 1,2,3,5,14 **Conservation status: (IUCN)** NE

3026 *Mentzelia scabra* Kunth (Loasaceae) **IPNI:** 545481-1 **Habit:** Shrub, Subshrub **Use:** ME **Origin:** Native **Dept.:** CES, NAR **Elev.:** 1800–2200 m a.s.l. **IUCN Habitat code:** 1,14 **Conservation status: (IUCN)** NE

3027 *Nasa triphylla* (Juss.) Weigend (Loasaceae) **IPNI:** 1014841-1 **Common name:** ortiga, picapica **Habit:** Herb, Shrub **Use:** ME **Origin:** Native **Dept.:** ANT, CAL, CAQ, CUN, HUI, MAG, NAR, NSA, QUI, RIS, SAN, TOL, VAC **Elev.:** 1100-2800 m.a.s.l. **IUCN Habitat code:** 1,3,14 **Conservation status:** **(IUCN)** NE **(CPSG)** Potential LC

Crossosomatales

3028 *Turpinia occidentalis* (Sw.) G.Don (Staphyleaceae) **IPNI:** 785554-1 **Common name:** ahuyano, almendro, bois pilori, guacamure, mantequillo, palo santo de montaña, rodillón, vetiao **Habit:** Shrub, Small tree, Tree **Use:** HF, MA, ME **Origin:** Native **Region:** Amazonia, Andean, Pacific **Dept.:** AMA, ANT, BOL, BOY, CAU, CHO, CUN, HUI, LAG, MAG, PUT, QUI, RIS, SAN, TOL, VAC **Elev.:** 10-3200 m.a.s.l. **IUCN Habitat code:** 1,2,3,5,14 **Conservation status:** **(IUCN)** LC **(CPSG)** Potential LC

Cucurbitales

3029 *Begonia boliviensis* A.DC. (Begoniaceae) **IPNI:** 104291-1 **Common name:** ala de ángel, begonia, corazón, sombrero de obispo **Habit:** Herb **Use:** GS, HF **Origin:** Cultivated **Region:** Andean **Dept.:** BOY, CUN, NSA**IUCN Habitat code:** 1 **Conservation status:** **(IUCN)** NE

3030 *Begonia bowerae* Ziesenh. (Begoniaceae) **IPNI:** 30424-2 **Common name:** begonia de hoja **Habit:** Herb **Use:** EU **Origin:** Cultivated **Region:** Andean **Dept.:** BOG, BOY, RIS**Conservation status:** **(IUCN)** NE

3031 *Begonia cardiocarpa* Liebm. (Begoniaceae) **IPNI:** 104371-1 **Habit:** Herb **Use:** EU **Origin:** Cultivated **Dept.:** BOL **Elev.:** 130–130 m a.s.l. **Conservation status:** **(IUCN)** NE

3032 *Begonia coccinea* Hook. (Begoniaceae) **IPNI:** 315561-2 **Common name:** abayao, alas de ángel, alegría, begonia **Habit:** Herb **Use:** EU, GS **Origin:** Cultivated **Region:** Andean **Dept.:** BOG, CAL, NSA, RIS**IUCN Habitat code:** 14 **Conservation status:** **(IUCN)** NE

3033 *Begonia crassicaulis* Lindl. (Begoniaceae) **IPNI:** 104473-1 **Habit:** Herb, Epiphyte **Use:** EU **Origin:** Cultivated **Region:** Andean **Dept.:** CAL **Elev.:** 1300–1300 m a.s.l. **Conservation status:** **(IUCN)** NE

3034 *Begonia cucullata* Willd. (Begoniaceae) **IPNI:** 315562-2 **Common name:** alegría, alitas de ángel, arete, begonia, begonia chiquita, begonia sencilla, cerera, conchitas, corazones partidos, corazón, ojo de pichón, pajarito, pata de pichón, pichón, pichón de ángel, porcelana **Habit:** Herb, Subshrub **Use:** EU, GS, HF, ME **Origin:** Cultivated, Naturalised **Region:** Andean, Pacific **Dept.:** ANT, BOG, CAL, CAU, CHO, CUN, PUT, VAC **Elev.:** 500–3000 m a.s.l. **IUCN Habitat code:** 1,2,3,14 **Conservation status:** **(IUCN)** NE

3035 *Begonia fischeri* Schrank (Begoniaceae) **IPNI:** 315565-2 **Common name:** begonia de pantano **Habit:** Herb **Use:** ME **Origin:** Native **Region:** Andean **Dept.:** AMA, ANT, BOG, BOL, BOY, CAL, CAQ, CAS, CAU, CHO, CUN, HUI, LAG, MAG, MET, NSA, PUT, RIS, SAN, VAC **Elev.:** 30–2700 m a.s.l. **IUCN Habitat code:** 1,2,3,5,14 **Conservation status:** **(IUCN)** NE

3036 *Begonia fuchsioides* Hook. (Begoniaceae) **IPNI:** 104715-1 **Use:** EU **Origin:** Native **Region:** Andean, Pacific, Santa Marta **Dept.:** ANT, BOY, CAL, CAU, CES, CHO, CUND, MAG, NOR, QUI, RIS, SAN, TOL, VALLE **Elev.:** 1150–3200 m a.s.l. **IUCN Habitat code:** 1,3,14 **Conservation status:** **(IUCN)** NE

3037 *Begonia glabra* Aubl. (Begoniaceae) **IPNI:** 104759-1 **Common name:** antra blanca, antra blanco, pichón **Habit:** Herb, Hemiepiphyte, Climbing **Use:** HF, ME **Origin:** Native **Region:** Amazonia, Andean, Guiana Shield, Caribbean, Pacific **Dept.:** AMA, ANT, CAL, CAQ, CAU, CHO, CUN, MAG, MET, NSA, PUT, VAC **Elev.:** 0-1930 m.a.s.l. **IUCN Habitat code:** 1,3,14 **Conservation status:** **(IUCN)** NE **(CPSG)** Potential LC

3038 *Begonia heracleifolia* Schltdl. & Cham. (Begoniaceae) **IPNI:** 104828-1 **Common name:** pata de lancho, patelanche **Habit:** Herb **Use:** EU, HF, ME **Origin:** Cultivated **Dept.:** ATL, VAC **Elev.:** 1000–1000 m a.s.l. **Conservation status:** **(IUCN)** NE

3039 *Begonia hirsuta* Aubl. (Begoniaceae) **IPNI:** 104848-1 **Habit:** Herb **Use:** ME **Origin:** Native **Region:** Caribbean, Pacific **Dept.:** ATL, CHO, MAG **Elev.:** 0–250 m a.s.l. **IUCN Habitat code:** 1,3 **Conservation status:** **(IUCN)** NE

3040 *Begonia hirtella* Link (Begoniaceae) **IPNI:** 104855-1 **Habit:** Herb **Use:** HF, ME **Origin:** Native, Cultivated **Region:** Andean **Dept.:** ANT, CUN, PUT, TOL **Elev.:** 50–1280 m a.s.l. **IUCN Habitat code:** 1,3,14 **Conservation status:** **(IUCN)** NE

3041 *Begonia humilis* Aiton (Begoniaceae) **IPNI:** 30121655-2 **Use:** ME **Origin:** Native **Conservation status:** **(IUCN)** NE

3042 *Begonia longirostris* Benth. (Begoniaceae) **IPNI:** 105087-1 **Habit:** Herb, Subshrub **Use:** ME **Origin:** Native **Region:** Andean **Dept.:** ANT, CAL, CAQ, NAR, PUT, VAC **Elev.:** 750–2500 m a.s.l. **IUCN Habitat code:** 1,3,14 **Conservation status:** **(IUCN)** NE

3043 *Begonia maculata* Raddi (Begoniaceae) **IPNI:** 105130-1 **Common name:** pata de pichón, pataepichón, patepichona **Habit:** Climbing **Use:** EU, GS, ME **Origin:** Cultivated **Region:** Andean **Dept.:** ANT, CUN, TOL **Elev.:** 1580–1700 m a.s.l. **IUCN Habitat code:** 1,3,14 **Conservation status:** **(IUCN)** NE

3044 *Begonia manicata* Brongn. (Begoniaceae) **IPNI:** 105149-1 **Common name:** borla de San Pedro **Habit:** Herb **Use:** EU, HF **Origin:** Cultivated **Conservation status:** **(IUCN)** NE

3045 *Begonia masoniana* Irmsch. ex Ziesenh. (Begoniaceae) **IPNI:** 105166-1 **Common name:** begonia cruz de hierro, tapete **Habit:** Herb **Use:** EU, SU **Origin:** Cultivated **Region:** Andean **Dept.:** CAL, CUN**Conservation status:** **(IUCN)** NE

3046 *Begonia maurandiae* A.DC. (Begoniaceae) **IPNI:** 105168-1 **Common name:** sacha tos de guasca **Habit:** Herb, Hemiepiphyte, Climbing **Use:** ME **Origin:** Native **Region:** Andean **Dept.:** ANT, CAL, CAU, CUN, NAR, PUT, RIS, VAC **Elev.:** 1300–3300 m a.s.l. **IUCN Habitat code:** 1,3,14 **Conservation status:** **(IUCN)** NE

3047 *Begonia minor* Jacq. (Begoniaceae) **IPNI:** 105214-1 **Habit:** Herb **Use:** HF **Origin:** Cultivated **Region:** Andean, Caribbean **Dept.:** CES, NSA, RIS **Elev.:** 100–2000 m a.s.l. **Conservation status:** **(IUCN)** NE

3048 *Begonia nelumbiifolia* Schltdl. & Cham. (Begoniaceae) **IPNI:** 273181-2 **Habit:** Herb **Use:** EU, HF, ME **Origin:** Native **Region:** Caribbean, Pacific **Dept.:** ANT, BOL, CHO **Elev.:** 100–400 m a.s.l. **IUCN Habitat code:** 1,14 **Conservation status:** **(IUCN)** NE

3049 *Begonia novogranatae* A.DC. (Begoniaceae) **IPNI:** 105280-1 **Habit:** Herb **Use:** EU **Origin:** Native, Cultivated **Region:** Andean **Dept.:** BOY, CUN **Elev.:** 800–1500 m a.s.l. **Conservation status:** **(IUCN)** NE

3050 *Begonia pustulata* Liebm. (Begoniaceae) **IPNI:** 105525-1 **Habit:** Herb **Use:** EU **Origin:** Cultivated **Dept.:** VAC**Conservation status:** **(IUCN)** NE

3051 *Begonia rex* Putz. (Begoniaceae) **IPNI:** 105573-1 **Common name:** begonia, begonia de salón, begonia regia, caracol, matrimonio, rucia **Habit:** Herb **Use:** EU, GS, HF, ME **Origin:** Cultivated **Region:** Andean, Caribbean **Dept.:** BOG, CES, NSA, RIS, SAN **Elev.:** 100–2000 m a.s.l. **IUCN Habitat code:** 3,14 **Conservation status:** **(IUCN)** NE

3052 *Begonia rosacea* Putz. (Begoniaceae) **IPNI:** 105604-1 **Habit:** Herb **Use:** EU **Origin:** Endemic, Cultivated **Region:** Andean, Orinoquia **Dept.:** BOG, BOY, MET, VID **Elev.:** 500–2150 m a.s.l. **Conservation status:** **(IUCN)** NE

3053 *Begonia rossmanniae* A.DC. (Begoniaceae) **IPNI:** 105608-1 **Habit:** Herb, Shrub, Epiphyte, Climbing **Use:** ME **Origin:** Native **Region:** Amazonia, Andean **Dept.:** AMA, CAQ, NAR, PUT **Elev.:** 100–1200 m a.s.l. **IUCN Habitat code:** 1,3,14 **Conservation status: (IUCN)** NE

3054 *Begonia semiovata* Liebm. (Begoniaceae) **IPNI:** 105712-1 **Common name:** begonia, chulco, churco, churquillo **Habit:** Herb, Shrub, Epiphyte **Use:** ME **Origin:** Native **Region:** Amazonia, Andean, Pacific **Dept.:** AMA, ANT, CHO, MET, NAR, SAN, VAC **Elev.:** 0–900 m a.s.l. **IUCN Habitat code:** 1,2,3,5,14 **Conservation status: (IUCN)** NE

3055 *Begonia sericoneura* Liebm. (Begoniaceae) **IPNI:** 105718-1 **Habit:** Herb **Use:** EU **Origin:** Native, Cultivated **Region:** Andean, Pacific **Dept.:** ANT, CAU, CUN, NAR, QUI **Elev.:** 300–2100 m a.s.l. **Conservation status: (IUCN)** NE

3056 *Coriaria ruscifolia* L. (Coriariaceae) **IPNI:** 271369-1 **Common name:** barbasco, chanchi, chanchí, curtidera, falso helecho, mortiño, mortiño borrachero, mortiño zumaque, reventadera, sancia, sancie, sansá, sansí, sansú, teñidera, tinta, tisis, uvilla, zeu, zumaque **Habit:** Shrub **Use:** HF, MA, ME, PO, SU **Origin:** Native **Region:** Andean **Dept.:** ANT, BOG, BOY, CAL, CAU, CHO, CUN, HUI, LAG, MAG, NAR, PUT, QUI, RIS, TOL, VAC **Elev.:** 900-3600 m.a.s.l. **IUCN Habitat code:** 1,3,4,8,14 **Conservation status: (IUCN)** NE **(CPSG)** Potential LC

3057 *Cayaponia capitata* Cogn. ex Harms (Cucurbitaceae) **IPNI:** 50564-2 **Habit:** Climbing **Use:** HF **Origin:** Native **Region:** Amazonia **Dept.:** MET, PUT **Elev.:** 350–350 m a.s.l. **IUCN Habitat code:** 1 **Conservation status: (IUCN)** NE

3058 *Cayaponia glandulosa* (Mart.) Cogn. (Cucurbitaceae) **IPNI:** 291801-1 **Habit:** Climbing **Use:** ME **Origin:** Native **Region:** Andean, Caribbean, Pacific **Dept.:** ANT, BOY, CAU, CHO, GUV, HUI, MAG, MET, NAR, SAN, SUC, TOL, VAC **Elev.:** 50-2360 m.a.s.l. **IUCN Habitat code:** 1,2,3,14 **Conservation status: (IUCN)** NE **(CPSG)** Potential LC

3059 *Cayaponia granatensis* Cogn. (Cucurbitaceae) **IPNI:** 291805-1 **Habit:** Climbing **Use:** HF **Origin:** Native **Region:** Orinoquia, Pacific **Dept.:** ANT, BOY, CHO, MET **Elev.:** 30–600 m a.s.l. **IUCN Habitat code:** 1,3,4,5,14 **Conservation status: (IUCN)** NE

3060 *Cayaponia kathematophora* R.E.Schult. (Cucurbitaceae) **IPNI:** 50583-2 **Habit:** Climbing **Use:** MA **Origin:** Endemic **Region:** Amazonia **Dept.:** AMA **Elev.:** 180–180 m a.s.l. **IUCN Habitat code:** 1 **Conservation status: (IUCN)** NE

3061 *Cayaponia macrocalyx* Harms (Cucurbitaceae) **IPNI:** 50589-2 **Habit:** Climbing **Use:** HF **Origin:** Native **Region:** Amazonia, Andean, Pacific **Dept.:** AMA, ANT, CAQ, CHO, MET **Elev.:** 20–1020 m a.s.l. **IUCN Habitat code:** 1,5 **Conservation status: (IUCN)** NE

3062 *Cayaponia ophthalmica* R.E.Schult. (Cucurbitaceae) **IPNI:** 320916-2 **Common name:** jo, mun, mun-teka, teka, tsun, tsun-jo **Habit:** Climbing **Use:** ME **Origin:** Native **Region:** Amazonia, Andean **Dept.:** ANT, BOL, GUV **Elev.:** 700–1650 m a.s.l. **IUCN Habitat code:** 1,3 **Conservation status: (IUCN)** NE

3063 *Cayaponia podantha* Cogn. (Cucurbitaceae) **IPNI:** 291841-1 **Habit:** Climbing **Use:** HF, ME **Origin:** Native **Region:** Andean, Caribbean, Orinoquia **Dept.:** ATL, COR, CUN, MAG, MET **Elev.:** 10–1700 m a.s.l. **IUCN Habitat code:** 2,3,5 **Conservation status: (IUCN)** NE

3064 *Cayaponia racemosa* (Mill.) Cogn. (Cucurbitaceae) **IPNI:** 291845-1 **Habit:** Climbing **Use:** ME **Origin:** Native **Region:** Andean, Caribbean **Dept.:** ANT, ATL, CHO, COR, MAG **Elev.:** 5–1700 m a.s.l. **IUCN Habitat code:** 1,2,3 **Conservation status: (IUCN)** NE

3065 *Cayaponia selysioides* C.Jeffrey (Cucurbitaceae) **IPNI:** 50620-2 **Habit:** Climbing **Use:** HF **Origin:** Native **Region:** Amazonia **Dept.:** PUT **Elev.:** 200–200 m a.s.l. **IUCN Habitat code:** 1 **Conservation status: (IUCN)** NE

3066 *Ceratosanthes palmata* (L.) Urb. (Cucurbitaceae) **IPNI:** 52543-2 **Habit:** Climbing **Use:** HF, ME **Origin:** Native **Region:** Caribbean **Dept.:** ATL, LAG **Elev.:** 0–100 m a.s.l. **IUCN Habitat code:** 2 **Conservation status: (IUCN)** NE

3067 *Citrullus lanatus* (Thunb.) Matsum. & Nakai (Cucurbitaceae) **IPNI:** 291938-1 **Common name:** melón de agua, patilla, patía, sandia, sandía **Habit:** Climbing **Use:** AF, EU, FU, GS, HF, MA, ME, PO, SU **Origin:** Cultivated, Naturalised **Region:** Amazonia, Andean, Caribbean, Orinoquia **Dept.:** AMA, ANT, ATL, BOL, CAS, CUN, NSA, SAP, VAC **Elev.:** 5–1500 m a.s.l. **IUCN Habitat code:** 1,2,3,14 **Conservation status: (IUCN)** NE

3068 *Coccinia grandis* (L.) Voigt (Cucurbitaceae) **IPNI:** 291972-1 **Habit:** Climbing **Use:** AF, GS, HF, IF, ME **Origin:** Naturalised **Region:** Caribbean **Dept.:** BOL, MAG **Elev.:** 0–0 m a.s.l. **IUCN Habitat code:** 14 **Conservation status: (IUCN)** NE

3069 *Cucumis anguria* L. (Cucurbitaceae) **IPNI:** 292141-1 **Common name:** cocombro, cohombro, concombro, melón de golero, patilla de golero, patilla de monte, patilla de ratón, pepinillo, pepinito de monte, pepino cimarrón, **Habit:** Climbing **Use:** AF, GS, HF, MA, ME, PO **Origin:** Naturalised **Region:** Caribbean **Dept.:** ANT, ATL, BOL, CUN, LAG, MAG, SAN, SUC, VAC **Elev.:** 25–1040 m a.s.l. **IUCN Habitat code:** 1,2,3,14 **Conservation status: (IUCN)** NE

3070 *Cucumis dipsaceus* Ehrenb. ex Spach (Cucurbitaceae) **IPNI:** 292180-1 **Common name:** esponjilla, limoncillo, meloncillo, melón de culebra, pepino del diablo **Habit:** Climbing **Use:** AF, EU, GS, HF, MA, ME, PO **Origin:** Naturalised **Region:** Andean, Caribbean, Pacific **Dept.:** ANT, BOL, CHO, CUN, HUI, LAG, MAG, TOL, VAC **Elev.:** 5–1200 m a.s.l. **IUCN Habitat code:** 1,2,3,4,14 **Conservation status: (IUCN)** NE

3071 *Cucumis melo* L. (Cucurbitaceae) **IPNI:** 292238-1 **Common name:** bejuco de sapo, calabacilla, calabazo, coladera, esponja, estopa, estropajo, jaboncillo de campo, lavaplato, limpión, melocotón, meloncillo, melón, melón de conejo, melón de gallinazo, melón de golero, pepino cimarrón, pepino de monte, quingombó, serville **Habit:** Climbing **Use:** AF, EU, FU, GS, HF, IF, MA, ME, PO, SU **Origin:** Cultivated, Naturalised **Region:** Amazonia, Andean, Caribbean, Pacific **Dept.:** AMA, ANT, ATL, BOL, CAL, CAQ, CHO, COR, CUN, HUI, LAG, MAG, NAR, QUI, RIS, SAP, SUC, TOL, VAC **Elev.:** 5–2350 m a.s.l. **IUCN Habitat code:** 1,2,3,14 **Conservation status: (IUCN)** NE

3072 *Cucumis sativus* L. (Cucurbitaceae) **IPNI:** 292296-1 **Common name:** cocombro, cohombro, pepino, pepino cocombro, pepino cohombro **Habit:** Climbing **Use:** AF, GS, HF, MA, ME, PO **Origin:** Cultivated, Naturalised **Region:** Andean, Caribbean **Dept.:** ANT, BOG, CUN **Elev.:** 30–2120 m a.s.l. **IUCN Habitat code:** 1,2,3,4,14 **Conservation status: (IUCN)** NE

3073 *Cucurbita ficifolia* Bouché (Cucurbitaceae) **IPNI:** 292369-1 **Common name:** bolo, calabaza, mejicano, victoria, vitoria, vitoriera **Habit:** Climbing **Use:** AF, EU, GS, HF, ME **Origin:** Cultivated, Naturalised **Region:** Andean **Dept.:** ANT, BOG, CAU, CUN, MET, NAR, PUT, SAN **Elev.:** 550–3000 m a.s.l. **IUCN Habitat code:** 1,3,14 **Conservation status: (IUCN)** NE

3074 *Cucurbita maxima* Duchesne (Cucurbitaceae) **IPNI:** 30006538-2 **Common name:** ahuyama, ahuyama burrera, ahuyama candelaria, ahuyama cintura, ahuyama peruana, auyama, auyama blanca, auyama candelaria, auyama caturra, auyama común, auyama corazón de res, auyama lagartija, auyama pirul, auyama poteca, gallina, huyama, uyama, uyamera, **Habit:** Climbing **Use:** AF, EU, GS, HF, IF, MA, ME, SU **Origin:** Cultivated **Region:** Amazonia, Andean, Caribbean **Dept.:** AMA, ANT, ATL, BOG, BOL, BOY, CAU, COR, CUN, MAG, NAR, NSA, QUI, SAP, SAN, TOL, VAC, VID **Elev.:** 20–1750 m a.s.l. **IUCN Habitat code:** 1,2,3,5,14 **Conservation status: (IUCN)** NE

3075 *Cucurbita moschata* Duchesne (Cucurbitaceae) **IPNI:** 320034-2 **Common name:** ahuyama, ahuyama chica, auyama, churí, gallina, jiromon, jiroumou, joumou, kalawasu, lakayote, uyama, zapallo, zapallo marranero, zapayo, zácata **Habit:** Climbing **Use:** AF, EU, GS, HF, ME **Origin:** Cultivated, Naturalised **Region:** Caribbean **Dept.:** ANT, ATL, BOL, CAS, CAU, CES, CUN, NAR, PUT, TOL, VAC **Elev.:** 0–2200 m a.s.l. **IUCN Habitat code:** 1,2,3,4,5,14 **Conservation status: (IUCN)** NE

3076 *Cucurbita pepo* L. (Cucurbitaceae) **IPNI:** 292416-1 **Common name:** auyama, bellota, berenjena, bichuela, bitora, bitoria, bitoriera, bolito, bolo, bolos, borombolo, calabacita, calabacín, calabaza, calabaza biche, calabaza bonetera, calabaza chura, calabaza churita, calabaza dulce, calabaza en bichuela, calabaza guagua, **Habit:** Climbing **Use:** AF, EU, FU, GS, HF, IF, MA, ME, PO **Origin:** Cultivated **Region:** Andean **Dept.:** ANT, BOG, CUN, NSA **Elev.:** 500–2700 m a.s.l. **IUCN Habitat code:** 1,3,4,5,14 **Conservation status: (IUCN)** LC

3077 *Cyclanthera brachybotrys* (Poepp. & Endl.) Cogn. (Cucurbitaceae) **IPNI:** 292472-1 **Habit:** Climbing **Use:** HF **Origin:** Native, Cultivated **Region:** Andean **Dept.:** CAU, LAG **Elev.:** 2100–3600 m a.s.l. **IUCN Habitat code:** 1,3,4,14 **Conservation status: (IUCN)** NE

3078 *Cyclanthera brachystachya* (DC.) Cogn. (Cucurbitaceae) **IPNI:** 292473-1 **Common name:** chochilla, cochilla, pepino diablito **Habit:** Climbing **Use:** GS, HF, ME **Origin:** Native, Cultivated **Region:** Andean **Dept.:** ANT, BOG, CAU, CUN **Elev.:** 1080–1950 m a.s.l. **Conservation status: (IUCN)** NE

3079 *Cyclanthera carthagenensis* (Jacq.) H.Schaef. & S.S.Renner (Cucurbitaceae) **IPNI:** 77110356-1 **Habit:** Climbing **Use:** HF, ME, PO **Origin:** Native **Region:** Amazonia, Andean, Orinoquia, Pacific **Dept.:** ANT, CHO, CUN, MAG, MET, PUT **Elev.:** 1–1650 m a.s.l. **IUCN Habitat code:** 1,2,3,4,5,14 **Conservation status: (IUCN)** NE

3080 *Cyclanthera cordifolia* Cogn. (Cucurbitaceae) **IPNI:** 292478-1 **Common name:** achocha cimarrona, salchucha **Habit:** Climbing **Use:** HF **Origin:** Native **Region:** Andean **Dept.:** CUN, NAR **Elev.:** 1800–2200 m a.s.l. **IUCN Habitat code:** 1 **Conservation status: (IUCN)** NE

3081 *Cyclanthera pedata* (L.) Schrad. (Cucurbitaceae) **IPNI:** 292512-1 **Common name:** achocha, achocho, achoncha, alchocha, alchucha, archucha, avichucha, caigua, chauchilla, pepino, pepino crespo, pepino de picar, pepino de rellenar, pepino de relleno, pepino hueco **Habit:** Climbing **Use:** GS, HF, MA, ME **Origin:** Native, Cultivated **Region:** Andean **Dept.:** ANT, BOG, CUN, NAR **Elev.:** 1700–2080 m a.s.l. **IUCN Habitat code:** 1,3,14 **Conservation status: (IUCN)** NE

3082 *Doyerea emetocathartica* Grosourdy (Cucurbitaceae) **IPNI:** 592558-1 **Habit:** Climbing **Use:** ME **Origin:** Native **Region:** Caribbean **Dept.:** ATL, BOL, LAG **Elev.:** 5–20 m a.s.l. **IUCN Habitat code:** 1 **Conservation status: (IUCN)** NE

3083 *Echinopepon racemosus* (Steud.) C.Jeffrey (Cucurbitaceae) **IPNI:** 88306-2 **Habit:** Climbing **Use:** ME **Origin:** Native **Region:** Andean **Dept.:** SAN **Elev.:** 1500–1800 m a.s.l. **Conservation status: (IUCN)** NE

3084 *Fevillea cordifolia* L. (Cucurbitaceae) **IPNI:** 104422-2 **Common name:** aviata, aviatá, calabacita, calabacito, colombo, gabilla, gavilla, habilla, jabilla, javilla, mantecona, marianita, nacha, necha, neche, nicha, olla de mono, papa de cruz, pasta, pepa de cruz, pepa de culebra, secúa, tanta jabilla, tutumilla **Habit:** Climbing **Use:** FU, HF, MA, ME **Origin:** Native, Cultivated **Region:** Amazonia, Andean, Caribbean, Pacific **Dept.:** AMA, ANT, ATL, BOL, CHO, CUN, MAG, MET, PUT, VAU **Elev.:** 5–1700 m a.s.l. **IUCN Habitat code:** 1,2,3,5,14 **Conservation status: (IUCN)** NE

3085 *Gurania bignoniacea* (Poepp. & Endl.) C.Jeffrey (Cucurbitaceae) **IPNI:** 114819-2 **Common name:** bien te veo, bienteveo **Habit:** Climbing **Use:** ME **Origin:** Native **Region:** Amazonia, Andean, Orinoquia, Pacific **Dept.:** AMA, ANT, CHO, MET, NAR, NSA, QUI, VAC, VAU **Elev.:** 5–1100 m a.s.l. **IUCN Habitat code:** 1,3,4,14 **Conservation status: (IUCN)** NE

3086 *Gurania eriantha* (Poepp. & Endl.) Cogn. (Cucurbitaceae) **IPNI:** 292789-1 **Common name:** tonchuré **Habit:** Climbing **Use:** ME **Origin:** Native **Region:** Amazonia, Andean, Orinoquia, Pacific **Dept.:** AMA, ANT, CAQ, CHO, CUN, HUI, MET, PUT, VAU **Elev.:** 110–1700 m a.s.l. **IUCN Habitat code:** 1,3,14 **Conservation status: (IUCN)** NE

3087 *Gurania lobata* (L.) Pruski (Cucurbitaceae) **IPNI:** 1010655-1 **Habit:** Climbing **Use:** ME **Origin:** Native **Region:** Amazonia, Andean, Pacific **Dept.:** AMA, ANT, BOY, CAL, CAQ, CAU, CHO, CUN, HUI, MAG, MET, NAR, NSA, PUT, QUI, RIS, SAN, TOL, VAC **Elev.:** 10-2460 msnm **IUCN Habitat code:** 1,3,5,14 **Conservation status: (IUCN)** NE **(CPSG)** Potential LC

3088 *Gurania makoyana* (Lem.) Cogn. (Cucurbitaceae) **IPNI:** 292814-1 **Habit:** Climbing **Use:** ME **Origin:** Native **Region:** Andean, Pacific **Dept.:** ANT, CHO **Elev.:** 10–1560 m a.s.l. **IUCN Habitat code:** 1,3,5,14 **Conservation status: (IUCN)** NE

3089 *Lagenaria siceraria* (Molina) Standl. (Cucurbitaceae) **IPNI:** 134809-2 **Common name:** bangaña, bangaño, bangaño de bejuco, bola, cabazo, calabacera, calabacito, calabacito amargo, calabaza, calabaza de guasca, calabaza vinatera, calabazo, calabazo amargoso, calabazo bangaño, calabazo blanco, calabazo cinturo, calabazo de bangaño, calabazo **Habit:** Climbing **Use:** AF, EU, GS, HF, MA, ME, PO, SU **Origin:** Cultivated **Region:** Amazonia, Andean, Caribbean, Orinoquia, Pacific **Dept.:** AMA, ANT, BOL, CAS, CHO, CUN, GUA, MAG, VAC, VAU **Elev.:** 50–1760 m a.s.l. **IUCN Habitat code:** 1,2,3,4,14 **Conservation status: (IUCN)** NE

3090 *Luffa acutangula* (L.) Roxb. (Cucurbitaceae) **IPNI:** 105280-3 **Use:** EU, GS, HF, MA, ME, PO **IUCN Habitat code:** 14 **Conservation status: (IUCN)** NE

3091 *Luffa aegyptiaca* Mill. (Cucurbitaceae) **IPNI:** 293060-1 **Use:** HF, ME **Origin:** Naturalised **IUCN Habitat code:** 1,2,3,4,5,14 **Conservation status: (IUCN)** NE

3092 *Luffa operculata* (L.) Cogn. (Cucurbitaceae) **IPNI:** 293097-1 **Common name:** esponjilla, mochilita, muchilita **Habit:** Climbing **Use:** HF, MA, ME **Origin:** Native **Region:** Amazonia, Andean, Caribbean, Pacific **Dept.:** AMA, ANT, ATL, BOL, CHO, COR, CUN, LAG, MAG **Elev.:** 2–1300 m a.s.l. **IUCN Habitat code:** 1,2,5,14 **Conservation status: (IUCN)** NE

3093 *Luffa sepium* (G.Mey.) C.Jeffrey (Cucurbitaceae) **IPNI:** 1076251-2 **Common name:** chacarita, chacarrita, chauchilla, esponja, esponjilla, esponjillas, esponjillas indígenas, mochila, mochila vieja, mochilita, mochilitas, muchilita, purga de Juan Páez **Use:** ME **Origin:** Native **IUCN Habitat code:** 1,2,4,5,14 **Conservation status: (IUCN)** NE

3094 *Melothria dulcis* Wunderlin (Cucurbitaceae) **IPNI:** 157061-2 **Habit:** Climbing **Use:** HF **Origin:** Native **Region:** Andean, Pacific **Dept.:** ANT, CHO, NAR **Elev.:** 0–1000 m a.s.l. **IUCN Habitat code:** 1,3 **Conservation status: (IUCN)** NE

3095 *Melothria pendula* L. (Cucurbitaceae) **IPNI:** 293293-1 **Common name:** fruta de culebra, guaquillo, muchila, patillita, pepa de culebra, pepinillo de monte, pepinillo silvestre, sandillita **Habit:** Climbing **Use:** HF, ME **Origin:** Native **Region:** Amazonia, Andean, Caribbean, Pacific **Dept.:** AMA, ANT, BOL, BOY, CAL, CAQ, CAU, CHO, CUN, HUI, LAG, MAG, MET, NAR, NSA, PUT, QUI, RIS, SAN, TOL, VAC **Elev.:** 3-2100 msnm **IUCN Habitat code:** 1,2,3,4,5,14 **Conservation status: (IUCN)** NE **(CPSG)** Potential LC

3096 *Melothria scabra* Naudin (Cucurbitaceae) **IPNI:** 293320-1 **Habit:** Climbing **Use:** HF, ME **Origin:** Native **Region:** Andean **Dept.:** ANT, BOY, TOL **Elev.:** 750–1090 m a.s.l. **IUCN Habitat code:** 1,3,14 **Conservation status: (IUCN)** NE

3097 *Melothria sphaerocarpa* (Cogn.) H.Schaef. & S.S.Renner (Cucurbitaceae) **IPNI:** 77110571-1 **Common name:** bejuquito maraquero, calabacillo, taparito maraquero, tarrali, tarralí **Habit:** Climbing **Use:** AF, FU, GS, HF, MA, ME, PO, SU **Origin:** Native **Region:** Andean, Pacific **Dept.:** ANT, BOY, CHO, CUN, VAC **Elev.:** 90–1760 m a.s.l. **IUCN Habitat code:** 1,3,5,14 **Conservation status: (IUCN)** NE

3098 *Melothria trilobata* Cogn. (Cucurbitaceae) **IPNI:** 293334-1 **Common name:** esponjilla, patilla de monte, patillita de monte, pepino de monte **Habit:** Climbing **Use:** AF, HF **Origin:** Native **Region:** Andean, Caribbean, Pacific **Dept.:** ANT, BOL, BOY, CAL, CHO, CUN, LAG, MAG, MET, RIS, SAN, SUC, VAC **Elev.:** 20-1500 msnm **IUCN Habitat code:** 1,2,3,4,5,14 **Conservation status: (IUCN)** NE **(CPSG)** Potential LC

3099 *Momordica charantia* L. (Cucurbitaceae) **IPNI:** 293413-1 **Common name:** alchucha, archucha, archurchis, balsamina, balsamino, balsamira, balsamín, bejuco de cojé, bejuco de culebra, bocado de culebra, bocaoeculebra, chauchilla, cojé, cundeamor, fruto de la culebra, maravilla, marsamila, mavilla, pepinillo, pepino, pepino de c **Habit:** Climbing **Use:** AF, EU, GS, HF, MA, ME, PO **Origin:** Naturalised **Region:** Amazonia, Andean, Caribbean, Orinoquia, Pacific **Dept.:** AMA, ANT, ARA, ATL, BOL, CAL, CAS, CES, CHO, COR, CUN, GUA, HUI, LAG, MAG, NSA, QUI, SAP, SAN, SUC, TOL, VAC, VID **Elev.:** 5–1500 m a.s.l. **Conservation status: (IUCN)** NE

3100 *Psiguria triphylla* (Miq.) C.Jeffrey (Cucurbitaceae) **IPNI:** 211990-2 **Common name:** bejuco almendro **Habit:** Climbing **Use:** HF **Origin:** Native **Region:** Amazonia, Andean, Orinoquia, Pacific **Dept.:** BOY, CAQ, CHO, MET, NAR, VAC **Elev.:** 20–1800 m a.s.l. **IUCN Habitat code:** 1,3,14 **Conservation status: (IUCN)** NE

3101 *Sicana odorifera* (Vell.) Naudin (Cucurbitaceae) **IPNI:** 293902-1 **Common name:** cagua, melocotón, melón, melón de olor, pepino, pepino de olor, susuca **Habit:** Climbing **Use:** EU, HF, ME **Origin:** Cultivated, Naturalised **Region:** Amazonia, Andean, Caribbean **Dept.:** ANT, PUT, SUC, TOL **Elev.:** 100–1300 m a.s.l. **IUCN Habitat code:** 1,3 **Conservation status: (IUCN)** NE

3102 *Sicyos edulis* Jacq. (Cucurbitaceae) **IPNI:** 293946-1 **Common name:** achocha, alchonchas, ayota, bellota, biota, chayota, chota, choyota, cidra, cidrapapa, cidrayota, guasquila, guasquilla, guatila, guatilla, huisquila, huisquilla, papa berlín, papa de pobre, papaepobre, papepobre, pepilla, rosquil, sayota, sidra, yota **Habit:** Climbing **Use:** AF, EU, GS, HF, MA, ME **Origin:** Cultivated, Naturalised **Region:** Andean **Dept.:** ANT, CUN, PUT, VAC **Elev.:** 1150–2500 m a.s.l. **Conservation status: (IUCN)** NE

3103 *Sicyos parviflorus* Willd. (Cucurbitaceae) **IPNI:** 294002-1 **Habit:** Climbing **Use:** HF **Origin:** Native **Conservation status: (IUCN)** NE

3104 *Sicyos polyacanthos* Cogn. (Cucurbitaceae) **IPNI:** 294010-1 **Use:** HF, ME **Origin:** Native **Conservation status: (IUCN)** NE

Dilleniales

3105 *Curatella americana* L. (Dilleniaceae) **IPNI:** 321019-2 **Common name:** carne de fiambre, chaparro, chaparro bobo, chaparro de agua, chaparro sabanero, chaparrón, curata, guayabillo, gusanero, lija vegetal, manteco yuco, paraejo, parica, pedralejo, peralejo, peralejo blanco, peralejo macho **Habit:** Shrub, Small tree, Tree **Use:** HF, MA, ME **Origin:** Native **Region:** Amazonia, Guiana Shield, Orinoquia **Dept.:** BOL, CAQ, GUV, HUI, MAG, MET, SUC, TOL, VID **Elev.:** 90–1000 m a.s.l. **IUCN Habitat code:** 1,2,3,4,5,14 **Conservation status: (IUCN)** NE

3106 *Davilla kunthii* A.St.-Hil. (Dilleniaceae) **IPNI:** 316577-1 **Common name:** bejuco chaparro, bejuco de agua, bejuco de candela, bejuco quemador, bejuco tomé **Habit:** Shrub, Liana, Climbing **Use:** HF, ME **Origin:** Native **Region:** Amazonia, Andean, Guiana Shield, Caribbean, Orinoquia, Pacific **Dept.:** ANT, ARA, BOL, CAQ, CAU, CHO, CUN, GUA, LAG, MAG, MET, NSA, PUT, SAP, SAN, TOL, VAC, VID **Elev.:** 0-1700 msnm **IUCN Habitat code:** 1,2,3,4,5,14 **Conservation status: (IUCN)** NE **(CPSG)** Potential LC

3107 *Davilla nitida* (Vahl) Kubitzki (Dilleniaceae) **IPNI:** 77025-2 **Common name:** bejuco chaparro, bejuco de agua, bejuco haparro, bejuco rayado, chaparro, koumboto, nishé, tomé **Habit:** Shrub, Liana, Climbing **Use:** HF, ME **Origin:** Native **Region:** Amazonia, Andean, Guiana Shield, Caribbean, Orinoquia, Pacific **Dept.:** ANT, BOL, BOY, CAL, CAS, CAU, CHO, GUV, LAG, MET, NAR, TOL, VAC, VAU, VID **Elev.:** 0-2200 msnm **IUCN Habitat code:** 1,2,3,4,5,14 **Conservation status: (IUCN)** NE **(CPSG)** Potential LC

3108 *Davilla rugosa* Poir. (Dilleniaceae) **IPNI:** 328431-2 **Common name:** bejuco de agua, bejuco quemador **Habit:** Climbing **Use:** MA, ME **Origin:** Native **Region:** Amazonia, Guiana Shield, Caribbean, Orinoquia, Pacific **Dept.:** CHO, MAG, MET, VAU, VID **Elev.:** 140–500 m a.s.l. **IUCN Habitat code:** 1,3,14 **Conservation status: (IUCN)** NE

3109 *Dillenia indica* L. (Dilleniaceae) **IPNI:** 316684-1 **Habit:** Tree **Use:** EU, FU, GS, HF, MA, ME, PO **Origin:** Cultivated **Region:** Caribbean **Dept.:** ATL, HUI, TOL **Elev.:** 250–250 m a.s.l. **Conservation status: (IUCN)** LC

3110 *Doliocarpus dentatus* (Aubl.) Standl. (Dilleniaceae) **IPNI:** 82916-2 **Common name:** aguarrás, bejuco agraz, bejuco de agua, bejuco rosa, bejuco tomé **Habit:** Shrub, Liana, Climbing **Use:** AF, HF, ME **Origin:** Native **Region:** Amazonia, Andean, Guiana Shield, Caribbean, Orinoquia, Pacific **Dept.:** AMA, ANT, BOL, BOY, CAQ, CES, CHO, GUA, GUV, MAG, MET, NSA, PUT, SAP, SAN, VAU **Elev.:** 10-2000 msnm **IUCN Habitat code:** 1,2,3,4,5,14 **Conservation status: (IUCN)** NE **(CPSG)** Potential LC

3111 *Doliocarpus macrocarpus* Mart. ex Eichler (Dilleniaceae) **IPNI:** 316807-1 **Habit:** Liana **Use:** HF **Origin:** Native **Region:** Amazonia **Dept.:** SAN, VAU **Elev.:** 0–300 m a.s.l. **IUCN Habitat code:** 1,3 **Conservation status: (IUCN)** NE

3112 *Doliocarpus major* J.F.Gmel. (Dilleniaceae) **IPNI:** 30327947-2 **Common name:** bejuco agraz **Habit:** Shrub, Liana, Climbing **Use:** HF, MA, ME **Origin:** Native **Region:** Amazonia, Pacific **Dept.:** AMA, ANT, CAQ, CHO **Elev.:** 10–690 m a.s.l. **IUCN Habitat code:** 1,4,5,14 **Conservation status: (IUCN)** NE

3113 *Doliocarpus nitidus* (Triana) Triana & Planch. (Dilleniaceae) **IPNI:** 316812-1 **Common name:** bejuco tomé, tomé **Habit:** Climbing **Use:** ME **Origin:** Native **Region:** Pacific **Dept.:** ANT, CHO, CUN, HUI, SAN, TOL **Elev.:** 45–500 m a.s.l. **IUCN Habitat code:** 1,3,14 **Conservation status: (IUCN)** NE

3114 *Doliocarpus olivaceus* Sprague & R.O.Williams ex G.E.Hunter (Dilleniaceae) **IPNI:** 316817-1 **Habit:** Liana **Use:** ME **Origin:** Native **Region:** Andean, Pacific **Dept.:** ANT, CHO **Elev.:** 10–1460 m a.s.l. **IUCN Habitat code:** 1,3,14 **Conservation status: (IUCN)** NE

3115 *Pinzona coriacea* Mart. & Zucc. (Dilleniaceae) **IPNI:** 317113-1 **Common name:** bejuco agraz, bejuco chaparro, bejuco de agua, bejuco de monte, bejuco mora dulce **Habit:** Liana, Climbing **Use:** HF, ME **Origin:** Native **Region:** Amazonia, Andean, Caribbean, Orinoquia, Pacific **Dept.:** AMA, ANT, BOL, CAQ, CHO, COR, MET, SAN, VID **Elev.:** 50–950 m a.s.l. **IUCN Habitat code:** 1,3,5,14 **Conservation status: (IUCN)** NE

3116 *Tetracera portobellensis* Beurl. (Dilleniaceae) **IPNI:** 317470-1 **Common name:** agraz, bejuco chaparro, bejuco de agua, bejuco tomé **Habit:** Shrub, Liana **Use:** HF, MA **Origin:** Native **Region:** Andean, Caribbean, Pacific **Dept.:** ANT, BOL, BOY, CUN, HUI, SAN **Elev.:** 400–1350 m a.s.l. **IUCN Habitat code:** 2 **Conservation status: (IUCN)** NE

3117 *Tetracera volubilis* L. (Dilleniaceae) **IPNI:** 332108-2 **Common name:** ambe mora, bejuco chaparro, bejuco colorado, bejuco de agua, bejuco tomé, tomé **Habit:** Liana **Use:** HF, MA, ME **Origin:** Native **Region:** Amazonia, Caribbean, Orinoquia, Pacific **Dept.:** AMA, ANT, ATL, BOY, CHO, COR, MAG, MET, SAP, VID **Elev.:** 10–250 m a.s.l. **IUCN Habitat code:** 1,2,5,14 **Conservation status: (IUCN)** NE

3118 *Tetracera willdenowiana* Steud. (Dilleniaceae) **IPNI:** 317513-1 **Common name:** bejuco chaparro, bejuco de agua **Habit:** Climbing **Use:** HF **Origin:** Native **Region:** Amazonia, Guiana Shield, Caribbean, Orinoquia, Pacific **Dept.:** ANT, CHO, MAG, MET, SAN, VAU **Elev.:** 100–700 m a.s.l. **IUCN Habitat code:** 1,3,14 **Conservation status: (IUCN)** NE

Dipsacales

3119 *Dipsacus fullonum* L. (Caprifoliaceae) **IPNI:** 319260-1 **Common name:** carda, cardencha, cardencho, cardo, cardol, flor del cardo **Habit:** Herb **Use:** EU, HF, IF, MA, ME, SU **Origin:** Naturalised **Region:** Andean **Dept.:** BOG, BOY, CUN **Elev.:** 2000–2650 m a.s.l. **IUCN Habitat code:** 14 **Conservation status: (IUCN)** NE

3120 *Dipsacus sativus* (L.) Honck. (Caprifoliaceae) **IPNI:** 319292-1 **Common name:** carda, cardencha, cardo, cardol, cardón, karkimchu, kartusan, kopajhiri **Use:** MA, ME **Conservation status: (IUCN)** NE

3121 *Lomelosia caucasica* (M.Bieb.) Greuter & Burdet (Caprifoliaceae) **IPNI:** 915212-1 **Common name:** escabiosa **Habit:** Herb **Use:** EU **Origin:** Cultivated **Region:** Andean **Dept.:** CUN **Elev.:** 2600–2600 m a.s.l. **Conservation status: (IUCN)** NE

3122 *Lonicera japonica* Thunb. (Caprifoliaceae) **IPNI:** 148850-1 **Common name:** lonicera **Habit:** Shrub, Climbing **Use:** EU, HF, MA, ME **Origin:** Cultivated, Naturalised **Region:** Andean **Dept.:** ANT, BOG, CUN, VAC **Elev.:** 1000–2620 m a.s.l. **IUCN Habitat code:** 1,3,14 **Conservation status: (IUCN)** NE

3123 *Scabiosa atropurpurea* L. (Caprifoliaceae) **IPNI:** 319717-1 **Common name:** alfiletero, escabiosa, viuda **Habit:** Herb **Use:** EU, ME **Origin:** Cultivated **Region:** Andean **Dept.:** CUN **Elev.:** 2600–2600 m a.s.l. **IUCN Habitat code:** 14 **Conservation status: (IUCN)** NE

3124 *Valeriana arborea* Killip & Cuatrec. (Caprifoliaceae) **IPNI:** 262131-2 **Common name:** valeriana **Habit:** Shrub, Tree **Use:** MA, ME **Origin:** Endemic **Region:** Andean **Dept.:** BOG, BOY, CUN, MET, SAN **Elev.:** 3400-4500 msnm **IUCN Habitat code:** 3,4 **Conservation status: (IUCN)** LC **(CPSG)** LC

3125 *Valeriana chaerophylloides* Sm. (Caprifoliaceae) **IPNI:** 859680-1 **Habit:** Herb **Use:** ME **Origin:** Native **Region:** Amazonia, Andean, Caribbean **Dept.:** ANT, CAU, CUN, HUI, LAG, NAR, PUT, QUI, SAN **Elev.:** 40–2200 m a.s.l. **IUCN Habitat code:** 1,2,3,14 **Conservation status: (IUCN)** NE

3126 *Valeriana clematitis* Kunth (Caprifoliaceae) **IPNI:** 286620-2 **Habit:** Shrub, Liana, Climbing **Use:** ME **Origin:** Native **Region:** Andean **Dept.:** ANT, BOG, BOY, CAU, CHO, CUN, HUI, MAG, MET, NAR, NSA, PUT, QUI, RIS, SAN, TOL, VAC **Elev.:** 950-3820 msnm **IUCN Habitat code:** 1,3,4,8,14 **Conservation status: (IUCN)** NE **(CPSG)** Potential LC

3127 *Valeriana decussata* Ruiz & Pav. (Caprifoliaceae) **IPNI:** 859734-1 **Habit:** Subshrub **Use:** ME **Origin:** Native **Region:** Andean **Elev.:** 2650–3350 m a.s.l. **IUCN Habitat code:** 14 **Conservation status: (IUCN)** NE

3128 *Valeriana karstenii* Briq. (Caprifoliaceae) **IPNI:** 859894-1 **Common name:** cascujalé **Habit:** Shrub **Use:** ME **Origin:** Endemic **Region:** Andean **Dept.:** CES, MAG **Elev.:** 2800–4500 m a.s.l. **IUCN Habitat code:** 4 **Conservation status: (IUCN)** NE

3129 *Valeriana microphylla* Kunth (Caprifoliaceae) **IPNI:** 286639-2 **Habit:** Shrub **Use:** ME **Origin:** Native **Region:** Andean **Dept.:** CAU, NAR, VAC **Elev.:** 2800–4500 m a.s.l. **IUCN Habitat code:** 1,3,4,5,8,14 **Conservation status: (IUCN)** NE

3130 *Valeriana officinalis* L. (Caprifoliaceae) **IPNI:** 860012-1 **Common name:** valeriana **Habit:** Herb **Use:** HF, MA, ME **Origin:** Cultivated **Dept.:** BOG, CUN **Elev.:** 2600–2800 m a.s.l. **IUCN Habitat code:** 1,14 **Conservation status: (IUCN)** NE

3131 *Valeriana pilosa* Ruiz & Pav. (Caprifoliaceae) **IPNI:** 286641-2 **Common name:** anamú, valeriana **Habit:** Herb **Use:** ME **Origin:** Native **Region:** Andean **Dept.:** ANT, BOG, BOY, CAL, CAU, CUN, MAG, MET, NAR, RIS, SAN, TOL, VAC **Elev.:** 2650-4800 msnm **IUCN Habitat code:** 1,3,4,5,8,14 **Conservation status: (IUCN)** NE **(CPSG)** Potential LC

3132 *Valeriana quindiensis* Killip (Caprifoliaceae) **IPNI:** 262323-2 **Habit:** Shrub, Climbing **Use:** ME **Origin:** Endemic **Region:** Andean **Dept.:** CAL, CAU, CES, NSA, QUI, RIS, TOL **Elev.:** 2000–4000 m a.s.l. **IUCN Habitat code:** 1,4 **Conservation status: (IUCN)** NE

3133 *Valeriana rigida* Ruiz & Pav. (Caprifoliaceae) **IPNI:** 286650-2 **Habit:** Herb **Use:** ME **Origin:** Native **Region:** Andean **Dept.:** BOY, NAR, NSA, SAN **Elev.:** 3350–4500 m a.s.l. **IUCN Habitat code:** 4,8,14 **Conservation status: (IUCN)** NE

3134 *Valeriana rubra* L. (Caprifoliaceae) **IPNI:** 860141-1 **Use:** EU, ME **Region:** Andean **Dept.:** BOG**Conservation status: (IUCN)** NE

3135 *Valeriana scandens* L. (Caprifoliaceae) **IPNI:** 30039915-2 **Habit:** Herb, Climbing **Use:** ME **Origin:** Native **Region:** Andean **Dept.:** ANT, BOY, CAU, CHO, NAR, QUI, SAN, VAC **Elev.:** 825–2000 m a.s.l. **IUCN Habitat code:** 1,3,4,14 **Conservation status: (IUCN)** NE

3136 *Valeriana sorbifolia* Kunth (Caprifoliaceae) **IPNI:** 286655-2 **Habit:** Herb **Use:** HF, ME **Origin:** Native **Region:** Andean **Dept.:** ANT, CUN, MAG, MET, NSA, QUI **Elev.:** 600–2700 m a.s.l. **IUCN Habitat code:** 1,3,14 **Conservation status: (IUCN)** NE

3137 *Valeriana tatamana* Killip (Caprifoliaceae) **IPNI:** 262378-2 **Habit:** Herb **Use:** ME **Origin:** Native **Region:** Andean **Dept.:** RIS **Elev.:** 3400–3700 m a.s.l. **IUCN Habitat code:** 1 **Conservation status: (IUCN)** NE

3138 *Valeriana triphylla* Kunth (Caprifoliaceae) **IPNI:** 860264-1 **Common name:** valeriana **Habit:** Subshrub **Use:** ME **Origin:** Native **Region:** Andean **Dept.:** BOG, BOY, CUN **Elev.:** 3000–4100 m a.s.l. **IUCN Habitat code:** 1,3,4,14 **Conservation status: (IUCN)** NE

3139 *Valeriana urticifolia* Kunth (Caprifoliaceae) **IPNI:** 286659-2 **Habit:** Herb **Use:** ME **Origin:** Native **Region:** Andean **Dept.:** CAU, VAC **Elev.:** 1700–2720 m a.s.l. **IUCN Habitat code:** 3,14 **Conservation status: (IUCN)** NE

3140 *Sambucus canadensis* L. (Viburnaceae) **IPNI:** 321978-2 **Common name:** bajman, sacatsun, sauco **Use:** EU, HF, MA, ME, PO **Origin:** Native **Dept.:** BOG**IUCN Habitat code:** 1,3,14 **Conservation status: (IUCN)** NE

3141 *Sambucus mexicana* C.Presl ex DC. (Viburnaceae) **IPNI:** 227173-2 **Common name:** saúco **Use:** HF, ME, SU **IUCN Habitat code:** 1,2,3,4,5,14 **Conservation status: (IUCN)** NE

3142 *Sambucus nigra* L. (Viburnaceae) **IPNI:** 30122169-2 **Common name:** carbunquera, sabuguera, salbuguera, sauco, sauco blanco, sauco de Castilla, sauco de monte, saúco **Habit:** Shrub, Small tree **Use:** AF, EU, HF, MA, ME, PO **Origin:** Cultivated **Region:** Andean **Dept.:** ANT, BOG, BOY, CAL, CAU, CUN, HUI, MAG, MET, NAR, PUT, QUI, SAN, VAC **Elev.:** 700-2700 msnm **Conservation status: (IUCN)** LC **(CPSG)** Potential LC

3143 *Sambucus peruviana* Kunth (Viburnaceae) **IPNI:** 321979-2 **Common name:** carbunquera, sabuguera, salbuguera, sauco, sauco blanco, sauco de castilla, sauco de monte, saúco **Use:** HF, ME **Origin:** Native **Dept.:** BOG**IUCN Habitat code:** 1,3,14 **Conservation status: (IUCN)** LC

3144 *Viburnum anabaptista* Graebn. (Viburnaceae) **IPNI:** 264840-2 **Common name:** palo boroqueño, pita, pitá, sauco de montaña, sauco de monte **Habit:** Shrub **Use:** ME **Origin:** Native **Region:** Andean **Dept.:** ANT, BOG, QUI **Elev.:** 2100–2850 m a.s.l. **IUCN Habitat code:** 1,3,14 **Conservation status: (IUCN)** NE

3145 *Viburnum glabratum* Kunth (Viburnaceae) **IPNI:** 149702-1 **Common name:** bodoquero, chuguacá, sauco, saúco **Habit:** Tree **Use:** ME **Origin:** Native **Region:** Andean **Dept.:** CUN, HUI, TOL **Elev.:** 1000–1700 m a.s.l. **IUCN Habitat code:** 1,3,14 **Conservation status: (IUCN)** NE

3146 *Viburnum lasiophyllum* Benth. (Viburnaceae) **IPNI:** 149770-1 **Common name:** manzano **Habit:** Shrub, Small tree **Use:** ME **Origin:** Native **Region:** Andean **Dept.:** ANT, BOG, CHO, CUN, HUI, RIS **Elev.:** 1200–2845 m a.s.l. **IUCN Habitat code:** 1,3,14 **Conservation status: (IUCN)** NE

3147 *Viburnum pichinchense* Benth. (Viburnaceae) **IPNI:** 30356705-2 **Common name:** cabo de hacha, doblador, pelotillo, pita, sauco montañero **Habit:** Shrub, Small tree **Use:** ME **Origin:** Native **Region:** Andean **Dept.:** BOG, CAU, CUN, NSA, PUT, QUI, RIS, TOL **Elev.:** 1300–3600 m a.s.l. **IUCN Habitat code:** 1,2,3,4,5,14 **Conservation status: (IUCN)** NE

3148 *Viburnum tinoides* L.f. (Viburnaceae)
IPNI: 149971-1
Common name: chucua, garrocho, juco, pita, sauco de monte **Habit:** Shrub, Small tree, Tree, Climbing **Use:** EU, ME **Origin:** Endemic **Region:** Andean **Dept.:** ANT, BOG, BOY, CUN, LAG, NAR, QUI, RIS, SAN, VAC **Elev.:** 920–3724 m a.s.l. **IUCN Habitat code:** 1,2,3,4,5,8,14 **Conservation status: (IUCN)** NE

3149 *Viburnum toronis* Killip & A.C.Sm. (Viburnaceae)
IPNI: 264984-2
Common name: cabo de hacha, juco, pita, sauco de monte **Habit:** Shrub, Small tree, Tree **Use:** ME **Origin:** Native **Region:** Andean **Dept.:** ANT, BOG, BOY, CAL, CUN, HUI, MAG, NAR, QUI, RIS, SAN, VAC **Elev.:** 1100–3650 m a.s.l. **IUCN Habitat code:** 1,2,3,4,5,14 **Conservation status: (IUCN)** NE

3150 *Viburnum triphyllum* Benth. (Viburnaceae)
IPNI: 149985-1
Common name: chuque, garrocho, juco, sauco montañero **Habit:** Shrub, Small tree, Tree **Use:** AF, EU, FU, MA, ME **Origin:** Native **Region:** Andean **Dept.:** ANT, BOG, BOY, CAU, CUN, LAG, NAR, RIS, SAN, VAC **Elev.:** 1700–3200 m a.s.l. **IUCN Habitat code:** 1,3,4,5,8,14 **Conservation status: (IUCN)** NE

Ericales

3151 *Saurauia aromatica* R.E.Schult. (Actinidiaceae)
IPNI: 228008-2
Common name: moquillo **Habit:** Tree **Use:** HF **Origin:** Endemic **Region:** Andean **Dept.:** CAL, CAQ, CAU, HUI, PUT, VAC **Elev.:** 500-2600 msnm **IUCN Habitat code:** 1,3,14 **Conservation status: (IUCN)** LC **(CPSG)** LC

3152 *Saurauia brachybotrys* Turcz. (Actinidiaceae)
IPNI: 829301-1
Common name: dulomoco, moquillo **Habit:** Shrub, Small tree, Tree **Use:** HF, ME **Origin:** Native **Region:** Andean **Dept.:** ANT, CAL, CAU, CHO, HUI, NAR, PUT, QUI, RIS, TOL, VAC **Elev.:** 671–2500 m a.s.l. **IUCN Habitat code:** 1,2,3,4,14 **Conservation status: (IUCN)** NE

3153 *Saurauia bullosa* Wawra (Actinidiaceae)
IPNI: 228016-2
Common name: dulomoco **Habit:** Tree **Use:** HF, ME **Origin:** Native **Region:** Andean **Dept.:** ANT, CAL, CAU, CUN, HUI, NAR, PUT, QUI, SAN, TOL, VAC **Elev.:** 2200–3600 m a.s.l. **IUCN Habitat code:** 1,3,4,14 **Conservation status: (IUCN)** LC

3154 *Saurauia chocoensis* Soejarto (Actinidiaceae)
IPNI: 317977-2
Common name: camita **Habit:** Shrub, Tree **Use:** HF, MA **Origin:** Endemic **Region:** Andean, Pacific **Dept.:** ANT, CHO **Elev.:** 50-1670 msnm **IUCN Habitat code:** 1,3,14 **Conservation status: (IUCN)** VU **(CPSG)** VU

3155 *Saurauia cuatrecasana* R.E.Schult. (Actinidiaceae)
IPNI: 228033-2
Common name: dulumoco, moquillo **Use:** HF **Origin:** Endemic **IUCN Habitat code:** 1,3,14 **Conservation status: (IUCN)** LC

3156 *Saurauia excelsa* Willd. (Actinidiaceae)
IPNI: 829326-1
Habit: Small tree **Use:** HF **Origin:** Native **Region:** Andean **Dept.:** ANT, CAS, MAG, NAR, NSA, SAN **Elev.:** 420–2150 m a.s.l. **IUCN Habitat code:** 1,2 **Conservation status: (IUCN)** NE

3157 *Saurauia laevigata* Triana & Planch. (Actinidiaceae)
IPNI: 228054-2
Common name: chupachupa, dulomoco **Habit:** Tree **Use:** HF **Origin:** Native **Region:** Andean **Dept.:** ANT, CUN, TOL **Elev.:** 600-1900 msnm **IUCN Habitat code:** 1,3,5,14 **Conservation status: (IUCN)** LC **(CPSG)** LC

3158 *Saurauia lehmannii* Hieron. (Actinidiaceae)
IPNI: 228056-2
Habit: Tree **Use:** HF **Origin:** Native **Region:** Andean **Dept.:** ANT **Elev.:** 1650-1810 msnm **IUCN Habitat code:** 1,3,14 **Conservation status: (IUCN)** NT **(CPSG)** NT

3159 *Saurauia omichlophila* R.E.Schult. (Actinidiaceae)
IPNI: 228075-2
Common name: dulomoco blanco, moquillo, moquillo de páramo **Habit:** Shrub **Use:** HF **Origin:** Endemic **Region:** Andean **Dept.:** CAU, CHO, NAR, PUT **Elev.:** 1700–3800 m a.s.l. **IUCN Habitat code:** 1,3,4,5,8,14 **Conservation status: (IUCN)** NE

3160 *Saurauia prainiana* Buscal. (Actinidiaceae)
IPNI: 228097-2
Habit: Shrub, Small tree, Tree **Use:** HF **Origin:** Native **Region:** Andean **Dept.:** ANT, VAC **Elev.:** 1000–2100 m a.s.l. **IUCN Habitat code:** 1,3,14 **Conservation status: (IUCN)** LC

3161 *Saurauia pseudoleucocarpa* Buscal. (Actinidiaceae)
IPNI: 228105-2
Habit: Shrub, Tree **Use:** HF **Origin:** Native **Region:** Andean, Pacific **Dept.:** CAU, CHO, VAC **Elev.:** 0–830 m a.s.l. **IUCN Habitat code:** 1,3,5,14 **Conservation status: (IUCN)** NE

3162 *Saurauia pulchra* Sprague (Actinidiaceae)
IPNI: 228115-2
Common name: dulumoco, moquillo **Habit:** Small tree, Tree **Use:** HF, ME **Origin:** Endemic **Region:** Andean **Dept.:** CAU, HUI **Elev.:** 1300-2300 msnm **IUCN Habitat code:** 1 **Conservation status: (IUCN)** VU **(CPSG)** VU

3163 *Saurauia scabra* (Kunth) D.Dietr. (Actinidiaceae)
IPNI: 1183910-2
Common name: dulomoco, moco, moquillo **Habit:** Tree **Use:** AF, HF **Origin:** Native **Region:** Andean **Dept.:** BOG, BOY, CAL, CAQ, CAU, CUN, HUI, MET, RIS, SAN, TOL **Elev.:** 1400–2800 m a.s.l. **IUCN Habitat code:** 1,3,4,14 **Conservation status: (IUCN)** NE

3164 *Saurauia tomentosa* (Kunth) Spreng. (Actinidiaceae)
IPNI: 829436-1
Common name: moquillo **Habit:** Tree **Use:** HF **Origin:** Native **Region:** Andean **Dept.:** CAU, HUI, NAR, NSA, PUT, SAN, VAC **Elev.:** 1200–3200 m a.s.l. **IUCN Habitat code:** 1,3,8,14 **Conservation status: (IUCN)** NE

3165 *Saurauia ursina* Triana & Planch. (Actinidiaceae)
IPNI: 228162-2
Common name: aguanoso, cargagua, churimo, dulomoco, dulumoco, lulumoco, moco, moquillo **Habit:** Shrub, Small tree, Tree **Use:** HF, MA **Origin:** Endemic **Region:** Andean **Dept.:** ANT, BOG, CUN, RIS, SAN **Elev.:** 700–2700 m a.s.l. **IUCN Habitat code:** 1,3,4,8,14 **Conservation status: (IUCN)** NE

3166 *Saurauia yasicae* Loes. (Actinidiaceae)
IPNI: 228173-2
Habit: Tree **Use:** HF **Origin:** Native **Region:** Andean, Caribbean, Pacific **Dept.:** ANT, CHO, CUN, LAG, MAG, TOL **Elev.:** 0–2000 m a.s.l. **IUCN Habitat code:** 1,2,3,5,14 **Conservation status: (IUCN)** LC

3167 *Impatiens balsamina* L. (Balsaminaceae)
IPNI: 373978-1
Common name: caracucha, caracucho, caracucho blanco, caracuchos, primavera **Habit:** Herb **Use:** EU, GS, HF, MA, ME, PO **Origin:** Cultivated **Region:** Amazonia, Andean, Caribbean **Dept.:** ANT, BOG, BOY, CAL, CAQ, CAU, CES, COR, CUN, HUI, MET, NAR, NSA, PUT, QUI, RIS, SAN, SUC, TOL, VAC **Elev.:** 50–2100 m a.s.l. **IUCN Habitat code:** 1,2,3,4,14 **Conservation status: (IUCN)** NE

3168 *Impatiens hawkeri* W.Bull (Balsaminaceae)
IPNI: 374127-1
Use: EU, HF, ME, SU **Region:** Andean **Dept.:** BOG**IUCN Habitat code:** 1,3,14 **Conservation status: (IUCN)** NE

3169 *Impatiens noli-tangere* L. (Balsaminaceae)
IPNI: 30208246-2
Use: HF, ME **IUCN Habitat code:** 1,3,14 **Conservation status: (IUCN)** LC

3170 *Impatiens sodenii* Engl. & Warb. (Balsaminaceae)
IPNI: 374351-1
Use: EU, ME **Region:** Andean **Dept.:** BOG**IUCN Habitat code:** 1,3,14 **Conservation status: (IUCN)** NE

3171 *Impatiens walleriana* Hook.f. (Balsaminaceae)
IPNI: 374419-1
Common name: alegría, amor ardiente, amor encarnado, bellahelena, besito, besito de jardín, besito de monte, besitos, beso de antioqueño, beso de novia, besos, buenastardes, canutillo, caracucho, ojo de gato, pensamiento, primavera, princesa, violeta **Habit:** Herb **Use:** EU, HF, ME **Origin:** Cultivated, Naturalised **Region:** Amazonia, Andean, Caribbean **Dept.:** AMA, ANT, BOG, BOY, CAL, CAQ, CES, CHO, CUN, HUI, MAG, MET, NAR, NSA, PUT, QUI, RIS, SAN, TOL, VAC **Elev.:** 200–2800 m a.s.l. **IUCN Habitat code:** 1,2,3,4,14 **Conservation status: (IUCN)** NE

3172 *Clethra fagifolia* Kunth (Clethraceae)
IPNI: 327589-1
Common name: ahuyamo, ajicillo, carbonero, cargagua, chiriguaco, estoraque negro, guayabo colorao, madredeagua, manzano montañero **Habit:** Shrub, Tree **Use:** AF, EU, FU **Origin:** Native **Region:** Andean **Dept.:** ANT, BOG, BOY, CAU, CUN, HUI, MAG, MET, NAR, NSA, PUT, QUI, RIS, SAN, TOL, VAC **Elev.:** 1500-3420 msnm **IUCN Habitat code:** 1,2,3,4,8,14 **Conservation status: (IUCN)** NE **(CPSG)** Potential LC

3173 *Clethra fimbriata* Kunth (Clethraceae) **IPNI:** 327593-1 **Common name:** auyamo, azafrán, colorado, istoraque, manzano, patevaca **Habit:** Shrub **Use:** AF, EU, FU, MA, ME **Origin:** Native **Region:** Andean **Dept.:** BOG, BOY, CAL, CES, CUN, MAG, NSA, QUI, SAN **Elev.:** 1830–3700 m a.s.l. **IUCN Habitat code:** 1,3,4,8,14 **Conservation status: (IUCN)** LC

3174 *Clethra mexicana* DC. (Clethraceae) **IPNI:** 327610-1 **Common name:** níspero macho **Habit:** Small tree, Tree **Use:** MA **Origin:** Native **Region:** Andean **Dept.:** ANT, BOG, BOY, LAG, MAG, VAC **Elev.:** 350–2900 m a.s.l. **IUCN Habitat code:** 1,2,3,4,14 **Conservation status: (IUCN)** LC

3175 *Cyrilla racemiflora* L. (Cyrillaceae) **IPNI:** 319176-2 **Habit:** Small tree, Tree **Use:** EU, IF, MA **Origin:** Native **Region:** Guiana Shield **Dept.:** CAQ, GUA, VAU **Elev.:** 200–900 m a.s.l. **IUCN Habitat code:** 1,4,14 **Conservation status: (IUCN)** LC

3176 *Diospyros artanthifolia* Mart. ex Miq. (Ebenaceae) **IPNI:** 322093-1 **Habit:** Small tree, Tree **Use:** HF **Origin:** Native **Region:** Amazonia, Andean, Pacific **Dept.:** AMA, ANT, CAQ, CHO, GUV, MET, NSA, PUT, VAC, VAU **Elev.:** 10–900 m a.s.l. **IUCN Habitat code:** 1,3,4,5,14 **Conservation status: (IUCN)** LC

3177 *Diospyros capreifolia* Mart. ex Hiern (Ebenaceae) **IPNI:** 322202-1 **Common name:** yemahuevo, yemehuevo **Habit:** Small tree, Tree **Use:** MA **Origin:** Native **Region:** Amazonia, Guiana Shield **Dept.:** AMA, CAQ, VAU **Elev.:** 100–300 m a.s.l. **IUCN Habitat code:** 1 **Conservation status: (IUCN)** NE

3178 *Diospyros discolor* Willd. (Ebenaceae) **IPNI:** 322328-1 **Habit:** Tree **Use:** EU, FU, GS, HF, MA, ME **Origin:** Cultivated **Region:** Andean, Pacific **Dept.:** ANT, CHO **Elev.:** 100–1500 m a.s.l. **IUCN Habitat code:** 1,14 **Conservation status: (IUCN)** NE

3179 *Diospyros inconstans* Jacq. (Ebenaceae) **IPNI:** 322526-1 **Common name:** juandediós, limoncillo, limpiadiente, quende **Habit:** Shrub, Small tree, Tree **Use:** HF, MA, ME **Origin:** Native, Cultivated **Region:** Caribbean **Dept.:** ANT, ATL, BOL, CES, MAG, SUC **Elev.:** 5–500 m a.s.l. **IUCN Habitat code:** 1,2,3,5,14 **Conservation status: (IUCN)** NE

3180 *Diospyros myrmecocarpa* Mart. ex Miq. (Ebenaceae) **IPNI:** 322748-1 **Habit:** Shrub, Small tree, Tree **Use:** HF **Origin:** Native **Region:** Amazonia **Dept.:** AMA, CAQ, GUV, VAU **Elev.:** 100–350 m a.s.l. **IUCN Habitat code:** 1,4,5,14 **Conservation status: (IUCN)** NE

3181 *Diospyros nigra* (J.F.Gmel.) Perr. (Ebenaceae) **IPNI:** 322764-1 **Common name:** zapote negro **Habit:** Tree **Use:** EU, GS, HF, MA, ME, PO **Origin:** Native, Cultivated **Region:** Andean **Dept.:** ANT, CUN **Elev.:** 20–1600 m a.s.l. **IUCN Habitat code:** 1,2,5,14 **Conservation status: (IUCN)** LC

3182 *Diospyros sericea* A.DC. (Ebenaceae) **IPNI:** 323004-1 **Habit:** Small tree, Tree **Use:** MA **Origin:** Native **Region:** Amazonia **Dept.:** AMA, ANT, GUV, MET **Elev.:** 70–900 m a.s.l. **IUCN Habitat code:** 1,3,5,14 **Conservation status: (IUCN)** NE

3183 *Diospyros tetrasperma* Sw. (Ebenaceae) **IPNI:** 323099-1 **Habit:** Tree **Use:** HF **Origin:** Native **Region:** Caribbean **Dept.:** SAP **Elev.:** 0–125 m a.s.l. **Conservation status: (IUCN)** NE

3184 *Anthopterus wardii* Ball (Ericaceae) **IPNI:** 858211-1 **Habit:** Shrub, Epiphyte, Hemiepiphyte **Use:** HF, ME **Origin:** Native **Region:** Andean, Pacific **Dept.:** ANT, CHO, NAR, VAC **Elev.:** 0–2130 m a.s.l. **IUCN Habitat code:** 1,3,5,14 **Conservation status: (IUCN)** LC

3185 *Bejaria aestuans* Mutis (Ericaceae) **IPNI:** 317630-2 **Common name:** angucha, anguche, angucho, arrayán, azalea de monte, carbonerito, carbonero, carbonero de frío, carbonero de monte, carbonero de tierra fría, carbón, carbón mosco, cepa, cepo, clavelito, clavellino, coronito, cucharo, encillo, fragua, garrocho, guinda, ma **Habit:** Shrub, Small tree **Use:** EU, HF, ME **Origin:** Native **Region:** Andean **Dept.:** ANT, BOG, BOY, CUN, HUI, LAG, MAG, NAR, SAN, VAC, VAU **Elev.:** 800–3900 m a.s.l. **IUCN Habitat code:** 1,2,3,4,5,8,14 **Conservation status: (IUCN)** LC

3186 *Bejaria ledifolia* Bonpl. (Ericaceae) **IPNI:** 327001-1 **Common name:** pegamoso **Use:** ME **Dept.:** BOG**IUCN Habitat code:** 14 **Conservation status: (IUCN)** NE

3187 *Bejaria mathewsii* Fielding & Gardner (Ericaceae) **IPNI:** 327003-1 **Common name:** angucho, angujo, carbonero, fragua, mosquito, pegamosco, rosadito **Habit:** Shrub, Small tree **Use:** ME **Origin:** Native **Region:** Andean **Dept.:** ANT, BOG, BOY, CAL, CAU, CUN, HUI, MET, NAR, QUI, TOL, VAC **Elev.:** 1300–3230 m a.s.l. **IUCN Habitat code:** 1,3,4,14 **Conservation status: (IUCN)** LC

3188 *Bejaria resinosa* Mutis ex L.f. (Ericaceae) **IPNI:** 327015-1 **Common name:** angucha, angucho, angujo, azalea del monte, carbonero, carbón, clavelillo, matamosca, mosco, mosquito, pega pega, pegamosca, pegamosco, pegamosco de arbolito, pegamoscos, pegapega, rosadito, uva de monte, uva monte **Habit:** Shrub, Small tree **Use:** AF, EU, ME **Origin:** Native **Region:** Andean **Dept.:** ANT, ARA, BOG, BOY, CAU, CUN, NSA, SAN, VAC **Elev.:** 1750–3900 m a.s.l. **IUCN Habitat code:** 1,3,4,5,8,14 **Conservation status: (IUCN)** LC

3189 *Cavendishia adenophora* Mansf. (Ericaceae) **IPNI:** 50373-2 **Common name:** quereme, quereme rosa, uvito **Habit:** Shrub, Epiphyte **Use:** HF **Origin:** Endemic **Region:** Andean, Pacific **Dept.:** ANT, CAU, CHO, RIS, VAC **Elev.:** 500-2420 msnm **IUCN Habitat code:** 1,3,5,14 **Conservation status: (IUCN)** NT **(CPSG)** LC

3190 *Cavendishia bracteata* (Ruiz & Pav. ex J.St.-Hil.) Hoerold (Ericaceae) **IPNI:** 60458979-2 **Common name:** anizo, botón-botón, cacagüito, chaquilulo, coral, esmeraldo, gualicón, joyapa, maycha, mozgal, muyaña, muñuño, pichaque, puchato, pucsato, quemadero, quinoy, sagalita, salapa, siete cueros, sietecueros, tulla, tullá, uva blanca, uva de anis, uva de anís, **Habit:** Shrub, Epiphyte, Hemiepiphyte **Use:** AF, EU, FU, HF, MA, ME, SU **Origin:** Native **Region:** Andean **Dept.:** ANT, BOG, BOY, CAL, CAU, CHO, CUN, HUI, LAG, MAG, MET, NAR, NSA, PUT, QUI, RIS, SAN, TOL, VAC **Elev.:** 1000-3820 msnm **IUCN Habitat code:** 1,2,3,4,5,8,14 **Conservation status: (IUCN)** LC **(CPSG)** Potential LC

3191 *Cavendishia divaricata* A.C.Sm. (Ericaceae) **IPNI:** 50414-2 **Common name:** uva, zanca de mula **Habit:** Shrub, Epiphyte **Use:** HF **Origin:** Endemic **Region:** Andean **Dept.:** CAU, CUN, HUI **Elev.:** 1200-2800 msnm **IUCN Habitat code:** 1,3,14 **Conservation status: (IUCN)** NT **(CPSG)** NT

3192 *Cavendishia dulcis* Luteyn (Ericaceae) **IPNI:** 50418-2 **Habit:** Shrub **Use:** HF **Origin:** Endemic **Region:** Andean **Dept.:** VAC **Elev.:** 1890–1890 m a.s.l. **IUCN Habitat code:** 1,14 **Conservation status: (IUCN)** NE

3193 *Cavendishia nitida* (Kunth) A.C.Sm. (Ericaceae) **IPNI:** 50474-2 **Common name:** uva camarona **Habit:** Shrub, Hemiepiphyte **Use:** HF, ME **Origin:** Endemic **Region:** Andean **Dept.:** ANT, BOG, CAL, CUN, HUI, QUI, RIS, SAN, TOL **Elev.:** 1300-3000 msnm **IUCN Habitat code:** 1,2,3,4,5,14 **Conservation status: (IUCN)** LC **(CPSG)** LC

3194 *Cavendishia pubescens* (Kunth) Hemsl. (Ericaceae) **IPNI:** 858236-1 **Common name:** camerona, quemadera, quereme, uva, uva de monte, uvita, uvita de monte, uvito, uvito de barrancos, uvito de monte, uvito noble, uvo, uvo de monte **Habit:** Shrub, Epiphyte **Use:** AF, EU, HF, ME **Origin:** Native **Region:** Andean **Dept.:** ANT, BOY, CAL, CHO, CUN, HUI, NSA, QUI, RIS, SAN, TOL, VAC **Elev.:** 1000-3000 msnm **IUCN Habitat code:** 1,2,3,4,5,14 **Conservation status: (IUCN)** LC **(CPSG)** Potential LC

3195 *Cavendishia quereme* (Kunth) Benth. & Hook.f. (Ericaceae) **IPNI:** 1064712-2 **Common name:** quereme **Habit:** Shrub, Epiphyte **Use:** ME, SU **Origin:** Native, Cultivated **Region:** Andean, Pacific **Dept.:** ANT, CHO, NAR, VAC **Elev.:** 300–2350 m a.s.l. **IUCN Habitat code:** 1,3,14 **Conservation status: (IUCN)** NE

3196 *Cavendishia tarapotana* (Meisn.) Benth. & Hook.f. (Ericaceae) **IPNI:** 858240-1 **Habit:** Shrub, Epiphyte, Hemiepiphyte, Liana **Use:** HF **Origin:** Native **Region:** Andean **Dept.:** ANT, CAU, CHO, MET, NAR, NSA, RIS, SAN, VAC **Elev.:** 1000–2000 m a.s.l.

IUCN Habitat code: 1,3,5,14 **Conservation status: (IUCN)** NE

3197 *Ceratostema alatum* (Hoerold) Sleumer (Ericaceae)
IPNI: 52551-2
Habit: Shrub **Use:** HF **Origin:** Native **Region:** Andean **Dept.:** NAR **Elev.:** 3050–3200 m a.s.l. **Conservation status: (IUCN)** NE

3198 *Disterigma acuminatum* (Kunth) Nied. (Ericaceae)
IPNI: 82286-2
Common name: piquisique **Habit:** Shrub, Epiphyte, Climbing **Use:** HF **Origin:** Native **Region:** Andean **Dept.:** ANT, CAL, CAU, CUN, HUI, NAR, PUT, QUI, RIS, SAN, TOL, VAC **Elev.:** 2000-3700 msnm **IUCN Habitat code:** 1,3,4,14 **Conservation status: (IUCN)** NE **(CPSG)** Potential LC

3199 *Disterigma alaternoides* (Kunth) Nied. (Ericaceae)
IPNI: 82288-2
Common name: capulí silvestre, totiadera **Habit:** Shrub, Epiphyte, Climbing **Use:** HF **Origin:** Native **Region:** Andean **Dept.:** ANT, BOG, BOY, CAL, CAS, CAU, CHO, CUN, HUI, MET, NAR, NSA, PUT, QUI, RIS, SAN, TOL, VAC **Elev.:** 1500-3600 msnm **IUCN Habitat code:** 1,3,4,5,14 **Conservation status: (IUCN)** LC **(CPSG)** Potential LC

3200 *Disterigma balslevii* Luteyn (Ericaceae)
IPNI: 275983-2
Habit: Shrub, Epiphyte, Climbing **Use:** HF **Origin:** Native **Region:** Andean **Dept.:** CAU, NAR, PUT **Elev.:** 2500-3700 msnm **IUCN Habitat code:** 1,4,8,14 **Conservation status: (IUCN)** LC **(CPSG)** LC

3201 *Disterigma dumontii* Luteyn (Ericaceae)
IPNI: 82296-2
Common name: chaquilulo **Habit:** Shrub, Epiphyte, Climbing **Use:** HF **Origin:** Native **Region:** Andean, Pacific **Dept.:** NAR **Elev.:** 300–1900 m a.s.l. **IUCN Habitat code:** 1 **Conservation status: (IUCN)** NE

3202 *Disterigma empetrifolium* (Kunth) Nied. (Ericaceae)
IPNI: 327994-1
Habit: Shrub, Climbing **Use:** HF **Origin:** Native **Region:** Andean **Dept.:** ANT, ARA, BOG, BOY, CAL, CAS, CAU, CUN, HUI, LAG, MET, NAR, NSA, PUT, QUI, RIS, SAN, TOL, VAC **Elev.:** 2500-4500 msnm **IUCN Habitat code:** 1,2,3,4,5,6,8,14 **Conservation status: (IUCN)** LC **(CPSG)** Potential LC

3203 *Disterigma humboldtii* (Klotzsch) Nied. (Ericaceae)
IPNI: 82301-2
Common name: reventadera **Habit:** Shrub, Epiphyte, Climbing **Use:** HF **Origin:** Native **Region:** Andean **Dept.:** ANT, BOG, BOY, CHO, CUN, VAC **Elev.:** 1800–3100 m a.s.l. **IUCN Habitat code:** 1,3,4,14 **Conservation status: (IUCN)** NE

3204 *Erica lusitanica* Rudolphi (Ericaceae)
IPNI: 329044-1
Use: EU **Region:** Andean **Dept.:** BOG**Conservation status: (IUCN)** LC

3205 *Gaultheria anastomosans* (Mutis ex L.f.) Kunth (Ericaceae)
IPNI: 330459-1
Common name: borrachera, borrachero, cacao de perro, chirriadera, maíz de perro, mortiño, mortiño borrachero, mortiño falso, reventadera, totiadera, totiadera blanca, té del canadá, uvito, veneno **Habit:** Shrub, Subshrub **Use:** AF, EU, ME **Origin:** Native **Region:** Andean **Dept.:** ANT, ARA, BOG, BOY, CAL, CAU, CES, CHO, CUN, HUI, MAG, MET, NSA, QUI, RIS, SAN, TOL, VAC **Elev.:** 2500-4100 msnm **IUCN Habitat code:** 1,3,4,5,6,8,14 **Conservation status: (IUCN)** NE **(CPSG)** Potential LC

3206 *Gaultheria erecta* Vent. (Ericaceae)
IPNI: 330527-1
Common name: chichaja, mortiño, pachín, pata de gallina, pegamosco, reventadera, uva, uvo **Habit:** Herb, Subshrub, Epiphyte **Use:** HF, MA, ME **Origin:** Native **Region:** Andean **Dept.:** ANT, ARA, BOG, BOY, CAL, CAU, CES, CHO, CUN, HUI, LAG, MAG, MET, NAR, NSA, PUT, QUI, RIS, SAN, TOL, VAC **Elev.:** 2000-3700 msnm **IUCN Habitat code:** 1,2,3,4,5,8,14 **Conservation status: (IUCN)** LC **(CPSG)** Potential LC

3207 *Gaultheria foliolosa* Benth. (Ericaceae)
IPNI: 330535-1
Habit: Shrub, Small tree **Use:** HF **Origin:** Native **Region:** Andean **Dept.:** ANT, CAL, CAQ, CAU, HUI, MET, NAR, PUT, QUI, RIS, TOL, VAC **Elev.:** 2400-3800 msnm **IUCN Habitat code:** 1,3,4,8,14 **Conservation status: (IUCN)** NE **(CPSG)** Potential LC

3208 *Gaultheria glomerata* (Cav.) Sleumer (Ericaceae)
IPNI: 107641-2
Common name: pata de gallo **Habit:** Shrub, Subshrub **Use:** HF **Origin:** Native **Region:** Andean **Dept.:** ANT, CAU, MAG, NAR, NSA, TOL **Elev.:** 2600–4100 m a.s.l. **IUCN Habitat code:** 1,3,4,8,14 **Conservation status: (IUCN)** LC

3209 *Gaultheria hapalotricha* A.C.Sm. (Ericaceae)
IPNI: 107644-2
Habit: Shrub **Use:** HF **Origin:** Native **Region:** Andean **Dept.:** BOG, CUN, MET, NSA, SAN **Elev.:** 2800–3600 m a.s.l. **IUCN Habitat code:** 1,3,4,14 **Conservation status: (IUCN)** NE

3210 *Gaultheria insipida* Benth. (Ericaceae)
IPNI: 330570-1
Habit: Shrub **Use:** HF **Origin:** Native **Region:** Andean **Dept.:** ANT, BOY, CAL, CAU, CHO, CUN, HUI, NAR, PUT, QUI, RIS, TOL, VAC **Elev.:** 1800-3200 msnm **IUCN Habitat code:** 1,3,4,5,8,14 **Conservation status: (IUCN)** NE **(CPSG)** Potential LC

3211 *Gaultheria lanigera* Hook. (Ericaceae)
IPNI: 330584-1
Habit: Shrub **Use:** AF **Origin:** Native **Region:** Andean **Dept.:** BOG, BOY, CUN **Elev.:** 3200–3600 m a.s.l. **IUCN Habitat code:** 1,3,4,14 **Conservation status: (IUCN)** NE

3212 *Gaultheria myrsinoides* Kunth (Ericaceae)
IPNI: 330618-1
Common name: bichachá, borrachera, borrachero, chirriadera, chirridera, cimarrón, colorado, gateadero, guayabilla negra, macha-macha, maíz de perro, moridera, mortiño, mortiño cimarrón, mortiño venenoso, reventadera, uvito de monte **Habit:** Shrub, Subshrub **Use:** AF, HF, ME **Origin:** Native **Region:** Andean **Dept.:** ANT, ARA, BOY, CAL, CAU, CES, CUN, HUI, LAG, MAG, MET, NAR, NSA, PUT, QUI, RIS, SAN, TOL, VAC **Elev.:** 2000-4525 msnm **IUCN Habitat code:** 1,2,3,4,6,8,14 **Conservation status: (IUCN)** NE **(CPSG)** Potential LC

3213 *Gaultheria rigida* Kunth (Ericaceae)
IPNI: 330676-1
Common name: camarero, cimarrón, reventadera, uvito, uvito de monte, uvo, uvo falso **Habit:** Shrub **Use:** AF, HF **Origin:** Native **Region:** Andean **Dept.:** BOG, BOY, CAU, CUN, MET, NSA, SAN **Elev.:** 2700–3500 m a.s.l. **IUCN Habitat code:** 1,3,4,14 **Conservation status: (IUCN)** NE

3214 *Gaultheria sclerophylla* Cuatrec. (Ericaceae)
IPNI: 107704-2
Common name: reventadora, totiadora, uva camarona, uva de anís **Habit:** Shrub **Use:** HF **Origin:** Native **Region:** Andean **Dept.:** BOG, BOY, CAL, CAU, CUN, NAR, QUI, TOL, VAC **Elev.:** 3200–4000 m a.s.l. **IUCN Habitat code:** 1,3,4,6,8,14 **Conservation status: (IUCN)** NE

3215 *Gaultheria shallon* Pursh (Ericaceae)
IPNI: 317388-2
Use: AF, EU, HF, ME **Region:** Andean **Dept.:** BOG**Conservation status: (IUCN)** NE

3216 *Gaylussacia buxifolia* Kunth (Ericaceae)
IPNI: 858406-1
Common name: mosquito, uvo **Habit:** Shrub **Use:** EU, HF **Origin:** Native **Region:** Andean **Dept.:** ANT, BOG, BOY, CUN, LAG, MAG, SAN **Elev.:** 1990–3500 m a.s.l. **IUCN Habitat code:** 1,3,4,5,14 **Conservation status: (IUCN)** NE

3217 *Macleania bullata* Yeo (Ericaceae)
IPNI: 148680-2
Common name: chaquilulo, chupa quinde, chupequinde **Habit:** Shrub, Epiphyte **Use:** HF **Origin:** Native **Region:** Andean **Dept.:** CAU, NAR **Elev.:** 1500–2100 m a.s.l. **IUCN Habitat code:** 1,3,4,14 **Conservation status: (IUCN)** NE

3218 *Macleania hirtiflora* (Benth.) A.C.Sm. (Ericaceae)
IPNI: 148696-2
Common name: mechá, uvo **Use:** HF, ME **IUCN Habitat code:** 1,3,4,5,8,14 **Conservation status: (IUCN)** NE

3219 *Macleania rupestris* (Kunth) A.C.Sm. (Ericaceae)
IPNI: 148722-2
Common name: cacagüito, chaquilulo, hualicón llucho, joyapa, tamadero, uva, uva camarera, uva camarona, uva de perro, uvito, uvo **Habit:** Shrub, Epiphyte, Climbing **Use:** AF, EU, HF, MA, ME **Origin:** Native **Region:** Andean **Dept.:** ANT, BOG, BOY, CAL, CAU, CUN, HUI, LAG, MAG, NAR, NSA, PUT, QUI, SAN, TOL, VAC **Elev.:** 2000-4100 msnm **Conservation status: (IUCN)** LC **(CPSG)** Potential LC

3220 *Macleania smithiana* Luteyn (Ericaceae)
IPNI: 317635-2
Habit: Shrub, Epiphyte **Use:** HF **Origin:** Native **Region:** Andean, Pacific **Dept.:** ANT, CAU, CHO, NAR,

RIS, VAC **Elev.:** 0–1850 m a.s.l. **IUCN Habitat code:** 1,3,5,14 **Conservation status: (IUCN)** NE

3221 *Monotropa uniflora* L. (Ericaceae) **IPNI:** 281016-2 **Common name:** pipa de indio **Habit:** Herb **Use:** HF, ME **Origin:** Native **Region:** Andean **Dept.:** ANT, BOY, CUN, HUI, QUI, SAN, VAC **Elev.:** 1200–2700 m a.s.l. **IUCN Habitat code:** 1,3,14 **Conservation status: (IUCN)** NE

3222 *Plutarchia guascensis* (Cuatrec.) A.C.Sm. (Ericaceae) **IPNI:** 204021-2 **Common name:** agráz, chorotas, coronilla, uvito, uvito de Guasca **Habit:** Shrub **Use:** HF, ME **Origin:** Endemic **Region:** Andean **Dept.:** BOG, BOY, CUN, SAN **Elev.:** 2800-4180 msnm **IUCN Habitat code:** 1,3,4,14 **Conservation status: (IUCN)** VU **(CPSG)** EN

3223 *Psammisia aberrans* A.C.Sm. (Ericaceae) **IPNI:** 211010-2 **Common name:** quereme **Habit:** Shrub, Epiphyte **Use:** HF **Origin:** Native **Region:** Andean, Pacific **Dept.:** ANT, NAR, VAC **Elev.:** 350–2800 m a.s.l. **IUCN Habitat code:** 1,3,14 **Conservation status: (IUCN)** NE

3224 *Psammisia caloneura* A.C.Sm. (Ericaceae) **IPNI:** 211014-2 **Common name:** chaquilulo **Habit:** Shrub, Climbing **Use:** HF **Origin:** Native **Region:** Andean, Pacific **Dept.:** NAR, RIS, VAC **Elev.:** 250–1800 m a.s.l. **IUCN Habitat code:** 1,3 **Conservation status: (IUCN)** NE

3225 *Psammisia falcata* (Kunth) Klotzsch (Ericaceae) **IPNI:** 858617-1 **Habit:** Shrub **Use:** HF **Origin:** Endemic **Region:** Andean **Dept.:** BOG, BOY, CUN, HUI, SAN, TOL, VAC **Elev.:** 1800–3500 m a.s.l. **IUCN Habitat code:** 1,14 **Conservation status: (IUCN)** NE

3226 *Psammisia graebneriana* Hoerold (Ericaceae) **IPNI:** 211039-2 **Habit:** Shrub, Epiphyte, Liana **Use:** HF **Origin:** Native **Region:** Andean **Dept.:** ANT, BOG, BOY, CAU, CUN, HUI, MET, NAR, PUT, RIS, TOL **Elev.:** 2000–3400 m a.s.l. **IUCN Habitat code:** 1,3,4,5,8,14 **Conservation status: (IUCN)** NE

3227 *Psammisia hookeriana* Klotzsch (Ericaceae) **IPNI:** 858621-1 **Habit:** Shrub, Climbing **Use:** HF **Origin:** Native **Region:** Andean **Dept.:** ANT, BOY, CAQ **Elev.:** 1128–2500 m a.s.l. **IUCN Habitat code:** 1 **Conservation status: (IUCN)** NE

3228 *Psammisia macrophylla* (Kunth) Klotzsch (Ericaceae) **IPNI:** 858626-1 **Common name:** uva camarona, uvo **Habit:** Shrub, Tree, Hemiepiphyte **Use:** ME **Origin:** Endemic **Region:** Andean, Pacific **Dept.:** ANT, CAL, CAU, CHO, CUN, HUI, MAG, NAR, RIS, TOL, VAC **Elev.:** 1500–3700 m a.s.l. **IUCN Habitat code:** 1,3,5,14 **Conservation status: (IUCN)** NE

3229 *Psammisia sodiroi* Hoerold (Ericaceae) **IPNI:** 211066-2 **Habit:** Shrub, Subshrub, Epiphyte **Use:** HF **Origin:** Native **Region:** Andean **Dept.:** ANT, HUI, NAR **Elev.:** 1500–2800 m a.s.l. **IUCN Habitat code:** 1,3,14 **Conservation status: (IUCN)** NE

3230 *Psammisia ulbrichiana* Hoerold (Ericaceae) **IPNI:** 211070-2 **Habit:** Shrub, Liana **Use:** HF **Origin:** Native **Region:** Andean **Dept.:** ANT, HUI **Elev.:** 1220–2070 m a.s.l. **IUCN Habitat code:** 1,3,14 **Conservation status: (IUCN)** NE

3231 *Rhododendron indicum* (L.) Sweet (Ericaceae) **IPNI:** 1137397-2 **Use:** EU **Region:** Andean **Dept.:** BOG**IUCN Habitat code:** 1,3,14 **Conservation status: (IUCN)** NE

3232 *Rhododendron simsii* Planch. (Ericaceae) **IPNI:** 333359-1 **Common name:** azalea, biflora, biflora reina de Hungría, miflora, milflora, rododendro **Use:** EU, GS, HF, ME **Origin:** Cultivated **Region:** Andean **Dept.:** ANT, BOG, VAC **Elev.:** 1000–1000 m a.s.l. **IUCN Habitat code:** 1,3,14 **Conservation status: (IUCN)** NE

3233 *Satyria breviflora* Hoerold (Ericaceae) **IPNI:** 227968-2 **Common name:** uvito de monte **Habit:** Shrub, Epiphyte, Liana **Use:** HF **Origin:** Endemic **Region:** Andean **Dept.:** ANT, BOY, CAL, CAU, CUN, QUI, TOL **Elev.:** 1250-3300 msnm **IUCN Habitat code:** 1,3,14 **Conservation status: (IUCN)** DD **(CPSG)** DD

3234 *Satyria panurensis* (Benth. ex Meisn.) Hook.f. ex Nied. (Ericaceae) **IPNI:** 77163666-1 **Common name:** poprillo **Habit:** Shrub, Epiphyte, Liana **Use:** HF **Origin:** Native **Region:** Andean, Guiana Shield, Orinoquia **Dept.:** ANT, CAQ, CHO, CUN, MET, PUT, VAU **Elev.:** 0–2700 m a.s.l. **IUCN Habitat code:** 1,3,5,14 **Conservation status: (IUCN)** NE

3235 *Satyria warszewiczii* Klotzsch (Ericaceae) **IPNI:** 858655-1 **Use:** HF **Origin:** Native **IUCN Habitat code:** 1,3,4,14 **Conservation status: (IUCN)** NE

3236 *Sphyrospermum buxifolium* Poepp. & Endl. (Ericaceae) **IPNI:** 858679-1 **Common name:** aguadulce, uvito, uvito de monte **Habit:** Shrub, Epiphyte, Climbing **Use:** HF, ME **Origin:** Native **Region:** Andean, Pacific **Dept.:** ANT, BOG, BOY, CAL, CAQ, CAU, CHO, CUN, HUI, MAG, MET, NAR, NSA, PUT, QUI, RIS, SAN, TOL, VAC **Elev.:** 5-3600 msnm **IUCN Habitat code:** 1,2,3,4,5,14 **Conservation status: (IUCN)** NE **(CPSG)** Potential LC

3237 *Themistoclesia dependens* (Benth.) A.C.Sm. (Ericaceae) **IPNI:** 252411-2 **Habit:** Shrub, Epiphyte **Use:** HF **Origin:** Native **Region:** Andean **Dept.:** ANT, CAU, RIS **Elev.:** 2500–3360 m a.s.l. **IUCN Habitat code:** 1,4,14 **Conservation status: (IUCN)** NE

3238 *Thibaudia floribunda* Kunth (Ericaceae) **IPNI:** 858728-1 **Common name:** chaquilulo, coral, uva de monte, uvo, uvo de monte **Habit:** Shrub, Epiphyte, Hemiepiphyte **Use:** HF **Origin:** Native **Region:** Andean, Pacific **Dept.:** ANT, BOG, BOY, CAL, CAU, CHO, CUN, HUI, NAR, PUT, QUI, RIS, SAN, TOL, VAC **Elev.:** 200-4180 msnm **IUCN Habitat code:** 1,3,4,8,14 **Conservation status: (IUCN)** LC **(CPSG)** Potential LC

3239 *Thibaudia grantii* A.C.Sm. (Ericaceae) **IPNI:** 334071-1 **Common name:** chorontas, chorota, chorotas, uvo de monte **Habit:** Liana **Use:** HF, ME **Origin:** Endemic **Region:** Andean **Dept.:** BOG, CUN, MET **Elev.:** 2800-3460 msnm **IUCN Habitat code:** 1,3,4 **Conservation status: (IUCN)** LC **(CPSG)** LC

3240 *Thibaudia parvifolia* (Benth.) Hoerold (Ericaceae) **IPNI:** 252657-2 **Habit:** Shrub **Use:** HF **Region:** Andean **Dept.:** BOG, CAU, NAR, PUT **Elev.:** 2900-3520 msnm **IUCN Habitat code:** 1,3,4,8,14 **Conservation status: (IUCN)** LC **(CPSG)** LC

3241 *Vaccinium corymbosum* L. (Ericaceae) **IPNI:** 261823-2 **Use:** EU, GS, HF, MA, ME **Region:** Andean **Dept.:** BOG**Conservation status: (IUCN)** NE

3242 *Vaccinium crenatum* (G.Don) Sleumer (Ericaceae) **IPNI:** 261842-2 **Use:** EU, HF **IUCN Habitat code:** 4 **Conservation status: (IUCN)** NE

3243 *Vaccinium floribundum* Kunth (Ericaceae) **IPNI:** 858930-1 **Common name:** agras, agraz, agráz, calabatá, chivaco, mortiño, mortiño amargo, mortiño de árbol, mortiño falso, mortiño venenoso, reventadera **Habit:** Shrub **Use:** AF, EU, FU, HF, MA, ME, SU **Origin:** Native **Region:** Andean **Dept.:** ANT, BOG, BOY, CAU, CES, CHO, CUN, HUI, LAG, MAG, MET, NAR, NSA, PUT, QUI, SAN, TOL, VAC **Elev.:** 1600-4600 msnm **IUCN Habitat code:** 1,2,3,4,5,8,14 **Conservation status: (IUCN)** LC **(CPSG)** Potential LC

3244 *Vaccinium meridionale* Sw. (Ericaceae) **IPNI:** 859020-1 **Common name:** agraz, agraz coronel, agrás, calabatá, guarapita, mortiño, teta de bruja **Habit:** Shrub **Use:** AF, EU, HF, MA, ME, PO **Origin:** Native **Region:** Andean **Dept.:** ANT, BOG, BOY, CUN, HUI, LAG, MAG, NAR, SAN **Elev.:** 1850–3800 m a.s.l. **IUCN Habitat code:** 1,3,4,14 **Conservation status: (IUCN)** NE

3245 *Allantoma lineata* (Mart. ex O.Berg) Miers (Lecythidaceae) **IPNI:** 590593-1 **Habit:** Tree **Use:** HF **Origin:** Native **Region:** Guiana Shield, Orinoquia **Dept.:** GUA, VID **Elev.:** 95–110 m a.s.l. **Conservation status: (IUCN)** NE

3246 *Allantoma pluriflora* S.A.Mori, Ya Y.Huang & Prance (Lecythidaceae) **IPNI:** 77094248-1 **Common name:** acaño de aceite, papelillo, tabarique, tabarí **Habit:**

Tree **Use:** MA **Origin:** Native **Region:** Amazonia, Guiana Shield **Dept.:** AMA, CAQ, VAU **Elev.:** 250-250 msnm **IUCN Habitat code:** 1,5,14 **Conservation status: (IUCN)** NE **(CPSG)** LC

3247 *Bertholletia excelsa* Bonpl. (Lecythidaceae) **IPNI:** 591270-1 **Common name:** castaña, castaño, castaño de Brasil, castaño de paca, castaño del Marañón, matamatá, matamatá blanco, matamatá de altura, nuez de Pará, nuez del Brasil, nuez del brazil **Habit:** Tree **Use:** GS, HF, MA, ME **Origin:** Native, Cultivated **Region:** Amazonia, Orinoquia **Dept.:** AMA, MET, VAU **Elev.:** 80–480 m a.s.l. **IUCN Habitat code:** 1,14 **Conservation status: (IUCN)** VU

3248 *Cariniana pyriformis* Miers (Lecythidaceae) **IPNI:** 592127-1 **Common name:** abarco, chibugá, cocoabarco, papelillo, piloncillo **Habit:** Tree **Use:** EU, HF, MA **Origin:** Native **Region:** Pacific **Dept.:** ANT, BOY, CHO, COR, NSA, SAN **Elev.:** 30-770 msnm **IUCN Habitat code:** 1,2,3,5,14 **Conservation status: (IUCN)** NT **(CPSG)** CR

3249 *Couratari guianensis* Aubl. (Lecythidaceae) **IPNI:** 592371-1 **Common name:** cabuyo, carguero, coco blanco, coco cabuyo, coco manteco, fono, guasca, guasca peo, olleto, ollita, papelillo, piloncillo, pona **Habit:** Tree **Use:** HF, MA **Origin:** Native **Region:** Pacific **Dept.:** ANT, BOL, CHO, SAN, VAC **Elev.:** 50-770 msnm **IUCN Habitat code:** 1,3,5,14 **Conservation status: (IUCN)** VU **(CPSG)** LC

3250 *Couratari stellata* A.C.Sm. (Lecythidaceae) **IPNI:** 67276-2 **Common name:** carguero negro, colorado, fono hediondo, fono negro, fono podrido, guasca nato pedo, guasca peo, guasco, matamatá, matamatá blanco, matamatá colorado, palo de gusano **Habit:** Tree **Use:** MA **Origin:** Native **Region:** Amazonia **Dept.:** AMA, VAU **Elev.:** 100-100 msnm **IUCN Habitat code:** 1,14 **Conservation status: (IUCN)** NE **(CPSG)** VU/NT

3251 *Couroupita guianensis* Aubl. (Lecythidaceae) **IPNI:** 592385-1 **Common name:** bala de cañón, cabuyo, castaña de macaco, castaño, chupo, clavelino, coco de mono, coco picho, cocopicho, guasco, mamey hediondo, maraco, muco, mucurutu, palo fara, tabarí, taparón **Habit:** Tree **Use:** EU, HF, MA, ME **Origin:** Native **Region:** Amazonia, Orinoquia **Dept.:** AMA, CAQ, CAS, GUV, MET, PUT, VAU, VID **Elev.:** 200-560 msnm **IUCN Habitat code:** 1,2,3,4,5,14 **Conservation status: (IUCN)** LC **(CPSG)** LC

3252 *Couroupita nicaraguarensis* DC. (Lecythidaceae) **IPNI:** 592389-1 **Habit:** Tree **Use:** MA **Origin:** Native **Dept.:** ANT **Elev.:** 70-100 msnm **IUCN Habitat code:** 14 **Conservation status: (IUCN)** LC **(CPSG)** VU

3253 *Eschweilera alata* A.C.Sm. (Lecythidaceae) **IPNI:** 97363-2 **Common name:** cabo de hacha, carguero, carguero de gavilán, palo de gavilán **Habit:** Tree **Use:** MA **Origin:** Native **Region:** Amazonia **Dept.:** AMA **Elev.:** 190-190 msnm **IUCN Habitat code:** 1,3 **Conservation status: (IUCN)** NE **(CPSG)** LC

3254 *Eschweilera amazonica* R.Knuth (Lecythidaceae) **IPNI:** 97368-2 **Common name:** fono blanco, fono colorado, turí **Use:** MA **IUCN Habitat code:** 4 **Conservation status: (IUCN)** LC

3255 *Eschweilera caudiculata* R.Knuth (Lecythidaceae) **IPNI:** 97378-2 **Common name:** guasco colorado, tete **Habit:** Small tree, Tree **Use:** HF **Origin:** Native **Region:** Andean, Pacific **Dept.:** ANT, CAU, CUN, MAG, NAR, VAC **Elev.:** 50-2240 msnm **IUCN Habitat code:** 1,3,14 **Conservation status: (IUCN)** LC **(CPSG)** LC

3256 *Eschweilera chartaceifolia* S.A.Mori (Lecythidaceae) **IPNI:** 939647-1 **Habit:** Tree **Use:** HF **Origin:** Native **Region:** Amazonia **Dept.:** CAQ **Elev.:** 200-200 msnm **IUCN Habitat code:** 1,2,3 **Conservation status: (IUCN)** NE **(CPSG)** LC

3257 *Eschweilera coriacea* (DC.) S.A.Mori (Lecythidaceae) **IPNI:** 947902-1 **Common name:** achi, achí, cabuyo, carguero, carguero blanco, carguero colorado, carguero de gavilán, carguero de guacamaya, carguero de lombriz, carguero de mojarra, carguero lombriz, carguero negro, cazuelo, coco, coco cazuelo, coco mojagua, cuyo seuyodopirai, doejue- **Habit:** Small tree, Tree **Use:** HF, MA, ME **Origin:** Native **Region:** Amazonia, Guiana Shield, Pacific **Dept.:** AMA, ANT, BOL, BOY, GUA, SAN, VAC, VAU **Elev.:** 50-900 msnm **IUCN Habitat code:** 1,3,5,14 **Conservation status: (IUCN)** LC **(CPSG)** LC

3258 *Eschweilera itayensis* R.Knuth (Lecythidaceae) **IPNI:** 97409-2 **Common name:** carguero, carguero de araña, carguero de gavilán, carguero de guacamayo, carguero guacamaya, carguero popay, copay, ducu, fono blanco, fono colorado, matamatá, popay de rebalse **Habit:** Shrub, Small tree, Tree **Use:** HF **Origin:** Native **Region:** Amazonia **Dept.:** AMA **Elev.:** 100-220 msnm **IUCN Habitat code:** 1,5 **Conservation status: (IUCN)** LC **(CPSG)** LC

3259 *Eschweilera parvifolia* Mart. ex DC. (Lecythidaceae) **IPNI:** 592624-1 **Common name:** carguero, carguero de mojarra, fono blanco, fono colorado, huasca, matamatá, palo de gavilán, palo de ratón, popay, popay carguero, popay de muchilero **Habit:** Tree **Use:** HF **Origin:** Native **Region:** Amazonia, Guiana Shield, Orinoquia **Dept.:** AMA, GUA, MET, VAU, VID **Elev.:** 85-200 msnm **IUCN Habitat code:** 1,3,4,5,14 **Conservation status: (IUCN)** LC **(CPSG)** LC

3260 *Eschweilera pedicellata* (Rich.) S.A.Mori (Lecythidaceae) **IPNI:** 276907-2 **Common name:** matamatá, matamatá blanco **Habit:** Tree **Use:** MA **Origin:** Native **Region:** Amazonia, Guiana Shield **Dept.:** AMA, GUA **Elev.:** 100–100 m a.s.l. **IUCN Habitat code:** 1,5 **Conservation status: (IUCN)** LC

3261 *Eschweilera reversa* Pittier (Lecythidaceae) **IPNI:** 97452-2 **Common name:** cargadero blanco, coco, coco blanco, coco cabuyo, coco colorado, coco cristal, coco manteco, coco redondo, guaco, guasca, guasca colorada, guasca negra, guasco, olleto **Habit:** Shrub, Tree **Use:** MA **Origin:** Native **Region:** Andean, Pacific **Dept.:** ANT, CAU, CHO, SAN, VAC **Elev.:** 0-1400 msnm **IUCN Habitat code:** 1,3,5,14 **Conservation status: (IUCN)** NE **(CPSG)** LC

3262 *Eschweilera rufifolia* S.A.Mori (Lecythidaceae) **IPNI:** 939654-1 **Common name:** carguero, chibillo, fono, fono carguero, fono medicines, mata mata, matamatá, matamatá pequeño, palo de gavilán, palo gavilán, popay, turi **Habit:** Small tree, Tree **Use:** ME **Origin:** Native **Region:** Amazonia **Dept.:** AMA, CAQ **Elev.:** 100-200 msnm **IUCN Habitat code:** 1,3,5,14 **Conservation status: (IUCN)** NE **(CPSG)** LC

3263 *Grias cauliflora* L. (Lecythidaceae) **IPNI:** 596618-1 **Common name:** pacó **Habit:** Tree **Use:** HF, MA **Origin:** Native **Region:** Pacific **Dept.:** ANT, CHO **Elev.:** 0–250 m a.s.l. **IUCN Habitat code:** 1,3,5,14 **Conservation status: (IUCN)** LC

3264 *Grias haughtii* R.Knuth (Lecythidaceae) **IPNI:** 113436-2 **Common name:** coco espárrago, membrillo, pacó, tapaculo **Habit:** Tree **Use:** HF, ME **Origin:** Endemic **Region:** Andean, Pacific **Dept.:** ANT, CHO, NAR, SAN, VAC **Elev.:** 50-1300 msnm **IUCN Habitat code:** 1,3,5,14 **Conservation status: (IUCN)** LC **(CPSG)** LC

3265 *Grias neuberthii* J.F.Macbr. (Lecythidaceae) **IPNI:** 113443-2 **Common name:** cocoro, mortecino, pelotillo, pepeguará **Habit:** Tree **Use:** HF, ME **Origin:** Native **Region:** Amazonia **Dept.:** CAQ, CAU, MET, PUT **Elev.:** 230-610 msnm **IUCN Habitat code:** 1,3,14 **Conservation status: (IUCN)** LC **(CPSG)** LC

3266 *Grias peruviana* Miers (Lecythidaceae) **IPNI:** 113444-2 **Habit:** Tree **Use:** HF, ME **Origin:** Native **Region:** Amazonia **Dept.:** AMA **Elev.:** 200-220 msnm **IUCN Habitat code:** 1,14 **Conservation status: (IUCN)** NE **(CPSG)** VU/NT

3267 *Gustavia angustifolia* Benth. (Lecythidaceae) **IPNI:** 114913-2 **Habit:** Tree **Use:** HF, ME **Origin:** Native **Region:** Pacific **Dept.:** VAC **Elev.:** 20-20 msnm **IUCN Habitat code:** 1,14 **Conservation status: (IUCN)** NE **(CPSG)** EN

3268 *Gustavia augusta* L. (Lecythidaceae) **IPNI:** 114915-2 **Common name:** coco mono, mortecino, palo hediondo, rabo pelado **Habit:** Tree **Use:** HF, MA, ME **Origin:** Native **Region:** Amazonia, Guiana Shield, Orinoquia **Dept.:** AMA, MET, VID **Elev.:** 100-500 msnm **IUCN Habitat code:** 1,2,3,4,5,14 **Conservation status: (IUCN)** LC **(CPSG)** LC

3269 *Gustavia dodsonii* S.A.Mori (Lecythidaceae) **IPNI:** 114926 2 **Habit:** Tree **Use:** HF **Origin:** Native **Region:** Pacific **Dept.:** NAR **Elev.:** 150–150 m a.s.l. **Conservation status: (IUCN)** EN

3270 *Gustavia dubia* (Kunth) O.Berg (Lecythidaceae) **IPNI:** 596736-1 **Common name:** almanegra, mulamuerta **Habit:** Tree **Use:** HF **Origin:** Native **Region:** Pacific **Dept.:** ANT, BOL, CHO, COR, SAN, VAC **Elev.:** 20-600 msnm **IUCN Habitat code:** 1,3,5,14 **Conservation status: (IUCN)** NE **(CPSG)** VU

3271 *Gustavia gracillima* Miers (Lecythidaceae) **IPNI:** 114943-2 **Common name:** confite, membrillo **Habit:** Tree **Use:** ME **Origin:** Endemic **Region:** Andean, Pacific **Dept.:** ANT, CAL, NAR **Elev.:** 330-1200 msnm **Conservation status: (IUCN)** VU **(CPSG)** VU

3272 *Gustavia hexapetala* (Aubl.) Sm. (Lecythidaceae) **IPNI:** 596743-1 **Common name:** fara, matamatá blanco, mortecino, níspero, palo de sal **Habit:** Tree **Use:** HF, MA, ME **Origin:** Native **Region:** Amazonia, Guiana Shield, Orinoquia **Dept.:** AMA, CAQ, GUA, MET, VAU **Elev.:** 25-490 msnm **IUCN Habitat code:** 1,3,4,5,14 **Conservation status: (IUCN)** LC **(CPSG)** LC

3273 *Gustavia longifolia* Poepp. ex O.Berg (Lecythidaceae) **IPNI:** 114952-2 **Habit:** Tree **Use:** FU, HF, ME **Origin:** Native **Region:** Amazonia, Guiana Shield **Dept.:** AMA, CAQ, MET, PUT, VAU **Elev.:** 150-500 msnm **IUCN Habitat code:** 1,5,14 **Conservation status: (IUCN)** NE **(CPSG)** LC

3274 *Gustavia macarenensis* Philipson (Lecythidaceae) **IPNI:** 114954-2 **Habit:** Tree **Use:** HF **Origin:** Native **Region:** Amazonia, Guiana Shield **Dept.:** AMA, MET **Elev.:** 100-1400 msnm **IUCN Habitat code:** 1,14 **Conservation status: (IUCN)** LC **(CPSG)** VU/NT

3275 *Gustavia nana* Pittier (Lecythidaceae) **IPNI:** 469407-1 **Common name:** membrillo, membrillo hembra, pacó **Habit:** Tree **Use:** HF, MA, ME **Origin:** Native **Region:** Pacific **Dept.:** ANT, BOY, CHO, COR, VAC **Elev.:** 10-600 msnm **IUCN Habitat code:** 1,3,5,14 **Conservation status: (IUCN)** NE **(CPSG)** VU

3276 *Gustavia occidentalis* Cuatrec. (Lecythidaceae) **IPNI:** 114963-2 **Common name:** chupo **Habit:** Tree **Use:** HF **Origin:** Endemic **Region:** Andean, Pacific **Dept.:** VAC **Elev.:** 50-1500 msnm **IUCN Habitat code:** 1,14 **Conservation status: (IUCN)** NE **(CPSG)** EN

3277 *Gustavia petiolata* S.A.Mori (Lecythidaceae) **IPNI:** 114964-2 **Habit:** Tree **Use:** HF **Origin:** Endemic **Region:** Pacific **Dept.:** ANT, CHO, NAR **Elev.:** 50-780 msnm **IUCN Habitat code:** 1,3,14 **Conservation status: (IUCN)** EN **(CPSG)** VU

3278 *Gustavia poeppigiana* O.Berg (Lecythidaceae) **IPNI:** 114966-2 **Common name:** chicho borugo, matamatá, mortecino, palomuerto **Habit:** Tree **Use:** HF **Origin:** Native **Region:** Amazonia, Guiana Shield, Orinoquia **Dept.:** AMA, CAQ, MET, VAU, VID **Elev.:** 100-1000 msnm **IUCN Habitat code:** 1,3,4,5,14 **Conservation status: (IUCN)** LC **(CPSG)** LC

3279 *Gustavia speciosa* (Kunth) DC. (Lecythidaceae) **IPNI:** 596758-1 **Common name:** chupa, chupo, chupo de montaña, chupo de monte, cocomuerto, flor de muerto, hediondo, mulamuerta **Habit:** Tree **Use:** HF, ME **Origin:** Native, Cultivated **Region:** Andean, Pacific **Dept.:** ANT, BOL, CHO, COR, CUN, MAG, RIS, TOL, VAC **Elev.:** 50–1600 m a.s.l. **IUCN Habitat code:** 1,3,5,14 **Conservation status: (IUCN)** NE

3280 *Gustavia superba* (Kunth) O.Berg (Lecythidaceae) **IPNI:** 323304-2 **Common name:** bolero, canutillo, chocolatil, chupo, fara, membrillo, membrillo hembra, membrillo macho, mortecino, paco, pacó, pacó de monte, tupu, vaga local, wild mango **Habit:** Tree **Use:** AF, EU, HF, MA, ME, PO **Origin:** Native **Region:** Pacific **Dept.:** ANT, BOL, CHO, COR, MAG, NAR, VAC **Elev.:** 0-1300 msnm **IUCN Habitat code:** 1,2,3,5,14 **Conservation status: (IUCN)** LC **(CPSG)** LC

3281 *Lecythis ampla* Miers (Lecythidaceae) **IPNI:** 597240-1 **Common name:** coco cristal, coco mono, salero, úrtugo **Habit:** Tree **Use:** HF, ME **Origin:** Native **Region:** Pacific **Dept.:** ANT, BOL, BOY, CHO, SAN **Elev.:** 760-914 msnm **IUCN Habitat code:** 1,3,5,14 **Conservation status: (IUCN)** NE **(CPSG)** NT

3282 *Lecythis chartacea* O.Berg (Lecythidaceae) **IPNI:** 597250-1 **Common name:** carguero, fono blanco **Habit:** Tree **Use:** HF, MA **Origin:** Native **Region:** Amazonia **Dept.:** AMA, GUV, VAU **Elev.:** 100-300 msnm **IUCN Habitat code:** 1,4,5,14 **Conservation status: (IUCN)** LC **(CPSG)** LC

3283 *Lecythis lanceolata* Poir. (Lecythidaceae) **IPNI:** 597275-1 **Common name:** olla de mono, olleto **Use:** EU, HF, MA, ME **Conservation status: (IUCN)** CD **(CPSG)** LC

3284 *Lecythis mesophylla* S.A.Mori (Lecythidaceae) **IPNI:** 136951-2 **Common name:** cococristal **Habit:** Tree **Use:** MA **Origin:** Native **Region:** Caribbean **Dept.:** ANT, ATL, SAN **Elev.:** 50-770 msnm **IUCN Habitat code:** 1,3,14 **Conservation status: (IUCN)** NE **(CPSG)** VU

3285 *Lecythis minor* Jacq. (Lecythidaceae) **IPNI:** 597290-1 **Common name:** carguero, coco, coco cristal, coco de mono, coco mono, cocomono, cocuelo, coquillo, coquito, indio pelo, olla de mono, olleto, ollita de mona, ollita de mono, so ulua **Habit:** Shrub, Tree **Use:** AF, FU, HF, MA, ME, SU **Origin:** Native **Region:** Caribbean, Pacific **Dept.:** ANT, ATL, BOL, CHO, COR, LAG, MAG, SAN, SUC, VAC **Elev.:** 0–280 m a.s.l. **IUCN Habitat code:** 1,2,3,4,5,14 **Conservation status: (IUCN)** LC

3286 *Lecythis ollaria* Loefl. (Lecythidaceae) **IPNI:** 77176926-1 **Common name:** coco cuna, olla de mono, olla mono, salero **Use:** HF, MA, ME, PO **IUCN Habitat code:** 1,3,5,14 **Conservation status: (IUCN)** LC

3287 *Lecythis pisonis* Cambess. (Lecythidaceae) **IPNI:** 597315-1 **Common name:** almendro, cabeza de churuco, carguero, castaño, huesillo, nomana, olla de mono **Habit:** Tree **Use:** EU, HF, MA, ME **Origin:** Native **Region:** Amazonia **Dept.:** CAQ **Elev.:** 280-280 msnm **Conservation status: (IUCN)** NE **(CPSG)** LC

3288 *Lecythis tuyrana* Pittier (Lecythidaceae) **IPNI:** 136980-2 **Common name:** olla de mono, olleta de mono, olleto **Habit:** Tree **Use:** HF, MA, ME **Origin:** Native **Region:** Pacific **Dept.:** ANT, BOL, BOY, COR, SAN **Elev.:** 10-220 msnm **IUCN Habitat code:** 1,2,3,5,14 **Conservation status: (IUCN)** NE **(CPSG)** VU

3289 *Lecythis zabucajo* Aubl. (Lecythidaceae) **IPNI:** 597352-1 **Common name:** castaña **Use:** HF, MA, ME **IUCN Habitat code:** 1 **Conservation status: (IUCN)** LC

3290 *Marcgravia brownei* (Triana & Planch.) Krug & Urb. (Marcgraviaceae) **IPNI:** 152975-2 **Common name:** uva montañera **Use:** MA **Origin:** Native **Region:** Andean **Dept.:** ANT, BOY, CAL, CAU, CES, CHO, CUN, HUI, MAG, NAR, NSA, PUT, QUI, RIS, VAC **Elev.:** 1000-3000 msnm **IUCN Habitat code:** 1,3,14 **Conservation status: (IUCN)** NE **(CPSG)** Potential LC

3291 *Marcgravia myriostigma* Triana & Planch. (Marcgraviaceae) **IPNI:** 829069-1 **Use:** ME **Origin:** Endemic **Region:** Caribbean, Pacific **Dept.:** ANT, CHO **Elev.:** 0–2000 m a.s.l. **IUCN Habitat code:** 1,14 **Conservation status: (IUCN)** NE

3292 *Marcgraviastrum mixtum* (Triana & Planch.) Bedell (Marcgraviaceae) **IPNI:** 997443-1 **Use:** MA **Origin:** Native **Region:** Andean, Guiana Shield, Orinoquia, Pacific **Dept.:** ANT, BOY, CAQ, CAU, CHO, CUN, HUI, MET, NAR, PUT, QUI, SAN, TOL, VAC **Elev.:** 0-2000 msnm **IUCN Habitat code:** 1,3,14 **Conservation status: (IUCN)** NE **(CPSG)** Potential LC

3293 *Souroubea corallina* (Mart.) de Roon (Marcgraviaceae) **IPNI:** 829489-1 **Use:** ME **Origin:** Native **Region:** Amazonia, Guiana Shield **Dept.:** AMA, CAQ, VAU **Elev.:** 0–500 m a.s.l. **IUCN Habitat code:** 1,3,14 **Conservation status: (IUCN)** NE

3294 *Souroubea guianensis* Aubl. (Marcgraviaceae) **IPNI:** 829493-1 **Common name:** chilapo, mandibá, oreja de mula **Use:** ME **Origin:** Native **Region:** Amazonia, Guiana Shield **Dept.:** AMA, CAQ, GUA, VAU **Elev.:** 0–500 m a.s.l. **IUCN Habitat code:** 1,3,4,14 **Conservation status: (IUCN)** NE

3295 *Freziera calophylla* Triana & Planch. (Pentaphylacaceae) **IPNI:** 828818-1 **Habit:** Small tree, Tree **Use:** HF **Origin:** Native **Region:** Andean,

Pacific **Dept.:** ANT, CHO, SAN, VAC **Elev.:** 0–2000 m a.s.l. **IUCN Habitat code:** 1,3,14 **Conservation status: (IUCN)** LC

3296 *Freziera tomentosa* (Ruiz & Pav.) Tul. (Pentaphylacaceae) **IPNI:** 106032-2 **Common name:** cerezo, cerezo blanco, cerezo de monte, chantre, mandul **Habit:** Shrub, Small tree, Tree **Use:** HF **Origin:** Native **Region:** Andean **Dept.:** ANT, CAU, NAR, RIS, VAC **Elev.:** 1500-2800 msnm **IUCN Habitat code:** 1,14 **Conservation status: (IUCN)** NT **(CPSG)** EN

3297 *Freziera verrucosa* (Hieron.) Kobuski (Pentaphylacaceae) **IPNI:** 106037-2 **Habit:** Small tree **Use:** HF **Origin:** Native **Region:** Andean **Dept.:** ANT, CAU, CHO **Elev.:** 1350–2800 m a.s.l. **IUCN Habitat code:** 1,3,4,14 **Conservation status: (IUCN)** LC

3298 *Ternstroemia macrocarpa* Triana & Planch. (Pentaphylacaceae) **IPNI:** 829623-1 **Common name:** corregidor **Habit:** Small tree, Tree **Use:** MA **Origin:** Endemic **Region:** Andean **Dept.:** ANT, CUN, NSA, QUI, RIS **Elev.:** 2500–2770 m a.s.l. **IUCN Habitat code:** 1,3,14 **Conservation status: (IUCN)** NE

3299 *Cantua buxifolia* Juss. ex Lam. (Polemoniaceae) **IPNI:** 688721-1 **Use:** EU, ME **Region:** Andean **Dept.:** BOG**Conservation status: (IUCN)** LC

3300 *Cobaea scandens* Cav. (Polemoniaceae) **IPNI:** 61922-2 **Common name:** campana, funeraria, zapato de obispo **Habit:** Liana, Climbing **Use:** EU, ME, SU **Origin:** Cultivated, Naturalised **Region:** Andean **Dept.:** ANT, BOG, BOY, CAQ, CUN, MET, PUT **Elev.:** 1750–2700 m a.s.l. **IUCN Habitat code:** 1,3,4,14 **Conservation status: (IUCN)** NE

3301 *Phlox drummondii* Hook. (Polemoniaceae) **IPNI:** 193463-2 **Common name:** anelvia, aurora, aurorita, aurorita crespa, bella, besitos, beso, jazmín, reina, simpática, viuda **Habit:** Herb **Use:** EU **Origin:** Cultivated **Region:** Andean, Caribbean **Dept.:** ANT, BOG, BOY, CAL, HUI, MAG, MET, NSA, QUI, RIS, SAN, TOL, VAC **Elev.:** 100–2600 m a.s.l. **IUCN Habitat code:** 3,14 **Conservation status: (IUCN)** NE

3302 *Ardisia bartletii* Lundell (Primulaceae) **IPNI:** 18048-2 **Habit:** Shrub, Tree **Use:** HF **Origin:** Native **Region:** Caribbean, Pacific **Dept.:** BOL, CHO, SUC **Elev.:** 150–180 m a.s.l. **Conservation status: (IUCN)** NE

3303 *Ardisia colombiana* Lundell (Primulaceae) **IPNI:** 18093-2 **Habit:** Small tree **Use:** HF **Origin:** Native **Region:** Pacific **Dept.:** CHO **Elev.:** 0–20 m a.s.l. **IUCN Habitat code:** 1 **Conservation status: (IUCN)** NE

3304 *Ardisia compressa* Kunth (Primulaceae) **IPNI:** 586863-1 **Habit:** Shrub, Subshrub, Tree **Use:** HF **Origin:** Native **IUCN Habitat code:** 1,14 **Conservation status: (IUCN)** LC

3305 *Ardisia fimbrillifera* Lundell (Primulaceae) **IPNI:** 18143-2 **Habit:** Tree **Use:** HF **Origin:** Native **Region:** Pacific **Dept.:** CHO **Elev.:** 0–520 m a.s.l. **IUCN Habitat code:** 1 **Conservation status: (IUCN)** LC

3306 *Ardisia guianensis* (Aubl.) Mez (Primulaceae) **IPNI:** 18168-2 **Common name:** ajicito del diablo, cadillo, fresco, manglillo, mote **Habit:** Shrub, Subshrub, Small tree, Tree **Use:** HF, MA, ME **Origin:** Native **Region:** Amazonia, Andean, Guiana Shield, Caribbean, Pacific **Dept.:** AMA, ANT, CAQ, CAU, CES, CHO, HUI, MAG, MET, NAR, NSA, PUT, RIS, VAC, VAU **Elev.:** 20-2440 msnm **IUCN Habitat code:** 1,2,3,5,14 **Conservation status: (IUCN)** LC **(CPSG)** Potential LC

3307 *Ardisia manglillo* Cuatrec. (Primulaceae) **IPNI:** 18218-2 **Common name:** chingongo, manglillo **Habit:** Tree **Use:** HF, ME **Origin:** Endemic **Region:** Pacific **Dept.:** VAC **Elev.:** 5-100 msnm **IUCN Habitat code:** 1,14 **Conservation status: (IUCN)** LC **(CPSG)** LC

3308 *Ardisia perinsignis* Lundell (Primulaceae) **IPNI:** 18275-2 **Habit:** Small tree, Tree **Use:** ME **Origin:** Native **Region:** Andean, Caribbean, Pacific **Dept.:** BOL, CHO, CUN, SUC **Elev.:** 100–1320 m a.s.l. **IUCN Habitat code:** 1,2,14 **Conservation status: (IUCN)** NE

3309 *Ardisia sapida* Cuatrec. (Primulaceae) **IPNI:** 18325-2 **Common name:** ayuelo, chanud, charmolán, charmuelán, guayabillo, hayuelo **Habit:** Tree **Use:** HF, ME **Origin:** Endemic **Region:** Andean **Dept.:** BOG, CUN, VAC **Elev.:** 1800-2200 msnm **IUCN Habitat code:** 1,3,14 **Conservation status: (IUCN)** NT **(CPSG)** NT

3310 *Bonellia frutescens* (Mill.) B.Ståhl & Källersjö (Primulaceae) **IPNI:** 60430434-2 **Common name:** barbasco, barbasco de púa, barrabás, hincahinca, manca mulo, mancamulo, ruutka, sarniclo, sarnisclo, sarnícalo **Habit:** Shrub, Small tree, Tree **Use:** HF, ME **Origin:** Native **Region:** Caribbean **Dept.:** ATL, BOL, CES, LAG, MAG **Elev.:** 0–500 m a.s.l. **IUCN Habitat code:** 1,2,4,14 **Conservation status: (IUCN)** NE

3311 *Clavija costaricana* Pittier (Primulaceae) **IPNI:** 59800-2 **Habit:** Shrub, Small tree **Use:** HF **Origin:** Native **Region:** Pacific **Dept.:** ANT, BOL, CHO, RIS **Elev.:** 80–600 m a.s.l. **IUCN Habitat code:** 1,14 **Conservation status: (IUCN)** LC

3312 *Clavija eggersiana* Mez (Primulaceae) **IPNI:** 274721-2 **Use:** HF **IUCN Habitat code:** 14 **Conservation status: (IUCN)** NE

3313 *Clavija fusca* B.Ståhl (Primulaceae) **IPNI:** 274722-2 **Habit:** Shrub, Small tree **Use:** HF **Origin:** Native **Region:** Pacific **Dept.:** ANT, CHO **Elev.:** 250–1550 m a.s.l. **IUCN Habitat code:** 1,3,14 **Conservation status: (IUCN)** LC

3314 *Clavija harlingii* B.Ståhl (Primulaceae) **IPNI:** 274723-2 **Habit:** Shrub **Use:** HF **Origin:** Native **Region:** Amazonia **Dept.:** AMA, CAQ **Elev.:** 200–550 m a.s.l. **IUCN Habitat code:** 1 **Conservation status: (IUCN)** NE

3315 *Clavija latifolia* (Willd. ex Roem. & Schult.) K.Koch (Primulaceae) **IPNI:** 587931-1 **Common name:** huevo de hicotea, huevo de morroco, huevo de morrocoy, huevoemorrocoy, membrillo **Habit:** Shrub, Small tree **Use:** AF, HF **Origin:** Endemic **Region:** Caribbean **Dept.:** ANT, ATL, BOL, CAL, CES, MAG, SAN **Elev.:** 20-500 msnm **IUCN Habitat code:** 1,2,14 **Conservation status: (IUCN)** NE **(CPSG)** LC

3316 *Clavija lehmannii* Mez (Primulaceae) **IPNI:** 311309-2 **Common name:** durazno de monte **Habit:** Shrub, Small tree **Use:** HF **Origin:** Endemic **Region:** Andean **Dept.:** ANT, CAU, TOL, VAC **Elev.:** 870-1560 msnm **IUCN Habitat code:** 1,14 **Conservation status: (IUCN)** LC **(CPSG)** LC

3317 *Clavija macrocarpa* Ruiz & Pav. (Primulaceae) **IPNI:** 311911-2 **Use:** HF **IUCN Habitat code:** 14 **Conservation status: (IUCN)** NE

3318 *Clavija membranacea* Mez (Primulaceae) **IPNI:** 309153-2 **Habit:** Shrub, Small tree **Use:** HF **Origin:** Native **Region:** Andean, Pacific **Dept.:** CAU, CHO, NAR, VAC **Elev.:** 5–1500 m a.s.l. **IUCN Habitat code:** 1,2 **Conservation status: (IUCN)** NE

3319 *Clavija mezii* Pittier (Primulaceae) **IPNI:** 59824-2 **Common name:** huevo de morrocoy, huevo morroco, huevo morrocoy **Habit:** Shrub, Small tree **Use:** HF, ME **Origin:** Native **Region:** Caribbean, Pacific **Dept.:** ANT, BOL, CHO, COR **Elev.:** 10–400 m a.s.l. **IUCN Habitat code:** 1,2,3,5,14 **Conservation status: (IUCN)** LC

3320 *Clavija ornata* D.Don (Primulaceae) **IPNI:** 587939-1 **Common name:** manzano, membrillo **Habit:** Shrub, Small tree **Use:** EU, HF **Origin:** Native **Region:** Amazonia, Andean, Caribbean **Dept.:** CAQ, GUV, MET, NSA, PUT **Elev.:** 250–1500 m a.s.l. **IUCN Habitat code:** 1,2,3,5,14 **Conservation status: (IUCN)** NE

3321 *Clavija weberbaueri* Mez (Primulaceae) **IPNI:** 308892-2 **Common name:** marañón, muopicoroboado, sanango **Habit:** Shrub, Small tree **Use:** HF, ME **Origin:** Native **Region:** Amazonia **Dept.:** AMA, CAQ, PUT, VAU **Elev.:** 100–1400 m a.s.l. **IUCN Habitat code:** 1,3,5,14 **Conservation status: (IUCN)** LC

3322 *Cybianthus amplus* (Mez) G.Agostini (Primulaceae) **IPNI:** 73449-2 **Habit:** Shrub, Tree **Use:** HF **Origin:** Native **Region:** Amazonia, Andean, Guiana Shield **Dept.:** AMA, CAQ, GUA, VAU **Elev.:** 100–2150 m a.s.l. **IUCN Habitat code:** 1,3,4,5,14 **Conservation status: (IUCN)** LC

3323 *Cybianthus guyanensis* (A.DC.) Miq. (Primulaceae) **IPNI:** 1166848-2 **Habit:** Shrub, Small tree **Use:** HF **Origin:** Native **Region:** Amazonia, Guiana Shield **Dept.:** AMA, GUA **Elev.:** 100–350 m a.s.l. **IUCN Habitat code:** 1,14 **Conservation status: (IUCN)** LC

3324 *Cybianthus kayapii* (Lundell) Pipoly (Primulaceae) **IPNI:** 1006171-1 **Habit:** Shrub, Subshrub, Tree **Use:** HF **Origin:** Native **Region:** Amazonia, Andean, Pacific **Dept.:** AMA, ANT, CHO, NAR **Elev.:** 110–1750 m a.s.l. **IUCN Habitat code:** 1,3,14 **Conservation status: (IUCN)** LC

3325 *Cybianthus poeppigii* Mez (Primulaceae) **IPNI:** 73545-2 **Habit:** Shrub, Small tree, Tree **Use:** HF **Origin:** Native **Region:** Amazonia, Andean, Pacific **Dept.:** ANT, CAU, CHO, CUN, PUT, VAC **Elev.:** 50–2000 m a.s.l. **IUCN Habitat code:** 1,3,5,14 **Conservation status: (IUCN)** NE

3326 *Cybianthus schlimii* (Hook.f.) G.Agostini (Primulaceae) **IPNI:** 73559-2 **Common name:** palo sin rama, palosinrama **Habit:** Shrub, Small tree, Tree **Use:** ME **Origin:** Native **Region:** Amazonia, Andean, Orinoquia, Pacific **Dept.:** ANT, BOL, CAS, CHO, CUN, MET, NAR, PUT, SAN, VAC **Elev.:** 5–2600 m a.s.l. **IUCN Habitat code:** 1,3,5,14 **Conservation status: (IUCN)** LC

3327 *Cybianthus spicatus* (Kunth) G.Agostini (Primulaceae) **IPNI:** 73568-2 **Habit:** Shrub, Tree **Use:** HF **Origin:** Native **Region:** Guiana Shield, Orinoquia **Dept.:** CAQ, GUA, VID **Elev.:** 100–730 m a.s.l. **IUCN Habitat code:** 1,4,5,14 **Conservation status: (IUCN)** LC

3328 *Cybianthus sprucei* (Hook.f.) G.Agostini (Primulaceae) **IPNI:** 73569-2 **Common name:** chocó **Habit:** Shrub, Subshrub, Tree **Use:** HF **Origin:** Native **Region:** Andean, Pacific **Dept.:** ANT, CAU, CHO, NAR, VAC **Elev.:** 0–1800 m a.s.l. **IUCN Habitat code:** 1,14 **Conservation status: (IUCN)** LC

3329 *Cyclamen persicum* Mill. (Primulaceae) **IPNI:** 700765-1 **Common name:** ciclamen, violeta de los Alpes **Habit:** Herb **Use:** EU, HF, ME, PO **Origin:** Cultivated **Region:** Andean **Dept.:** ANT, BOG, BOY, CUN **Elev.:** 2000–2600 m a.s.l. **Conservation status: (IUCN)** NE

3330 *Geissanthus longistamineus* (A.C.Sm.) Pipoly (Primulaceae) **IPNI:** 975391-1 **Common name:** capulí, negrito **Habit:** Shrub, Small tree, Tree **Use:** HF, ME **Origin:** Native **Region:** Andean, Pacific **Dept.:** ANT, CAU, CHO, NAR **Elev.:** 80–2100 m a.s.l. **IUCN Habitat code:** 1,5 **Conservation status: (IUCN)** NE

3331 *Jacquinia armillaris* Jacq. (Primulaceae) **IPNI:** 312066-2 **Common name:** barbasco **Habit:** Shrub, Small tree **Use:** MA, ME, PO **Origin:** Native **Region:** Caribbean **Dept.:** ATL, BOL, LAG, MAG **Elev.:** 0–100 m a.s.l. **IUCN Habitat code:** 1,2,14 **Conservation status: (IUCN)** NE

3332 *Lysimachia arvensis* (L.) U.Manns & Anderb. (Primulaceae) **IPNI:** 77100386-1 **Habit:** Herb **Use:** AF, HF, MA, ME, PO **Origin:** Naturalised **Region:** Andean **Dept.:** BOG, BOY, CUN, NAR, SAN **Elev.:** 800–3100 m a.s.l. **IUCN Habitat code:** 1,3,4,14 **Conservation status: (IUCN)** NE

3333 *Lysimachia minima* (L.) U.Manns & Anderb. (Primulaceae) **IPNI:** 77100404-1 **Habit:** Herb **Use:** PO **Origin:** Naturalised **Region:** Andean, Caribbean **Dept.:** BOY, CES, CUN **Elev.:** 100–2600 m a.s.l. **Conservation status: (IUCN)** NE

3334 *Myrsine coriacea* (Sw.) R.Br. ex Roem. & Schult. (Primulaceae) **IPNI:** 1028287-2 **Common name:** aguaqu, ceibito, chagualo, cuchara, cucharero, cucharillo blanco, cucharo, cucharo blanco, cucharo pequeño, cucharro, espadero, espadero blanco, espadero de hoja chica, espadero rubio, garrocho, manteco, morochillo, mortiño, platanillo, trementina, tremen **Habit:** Shrub, Subshrub, Small tree, Tree **Use:** HF, MA, ME **Origin:** Native **Region:** Andean, Caribbean, Pacific **Dept.:** ANT, BOG, BOY, CAL, CAU, CES, CHO, CUN, HUI, MAG, MET, NAR, NSA, PUT, QUI, RIS, SAN, TOL, VAC **Elev.:** 820–3360 m a.s.l. **IUCN Habitat code:** 1,2,3,4,5,8,14 **Conservation status: (IUCN)** NE

3335 *Myrsine cubana* A.DC. (Primulaceae) **IPNI:** 589091-1 **Common name:** espadero **Use:** EU, ME **Conservation status: (IUCN)** LC

3336 *Myrsine dependens* (Ruiz & Pav.) Spreng. (Primulaceae) **IPNI:** 1119456-2 **Common name:** arrayán, cucharillo, cucharo, espadero, maiztostado, maiztostao, maíz tostado **Habit:** Shrub, Small tree, Tree **Use:** EU **Origin:** Native **Region:** Andean **Dept.:** ANT, BOG, BOY, CAL, CAU, CES, CHO, CUN, HUI, LAG, MAG, MET, NAR, NSA, PUT, QUI, RIS, SAN, TOL, VAC **Elev.:** 2500-3800 msnm **IUCN Habitat code:** 1,3,4,8,14 **Conservation status: (IUCN)** LC **(CPSG)** Potential LC

3337 *Myrsine guianensis* (Aubl.) Kuntze (Primulaceae) **IPNI:** 166565-2 **Common name:** cascarillo, chagualito, chagualo, chaguato, changuelito, cucharero, cucharo, espadero, garrucho, mortiño, quiebramachete, rapabarbo **Habit:** Tree **Use:** HF, MA **Origin:** Native **Region:** Amazonia, Andean **Dept.:** ANT, BOG, BOY, CAU, CUN, HUI, MAG, NAR, NSA, PUT, SAN, TOL, VAC **Elev.:** 0–3000 m a.s.l. **IUCN Habitat code:** 1,2,3,4,5,8,14 **Conservation status: (IUCN)** NE

3338 *Myrsine latifolia* (Ruiz & Pav.) Spreng. (Primulaceae) **IPNI:** 1033305-2 **Common name:** chagualo, cucharo, cucharo blanco, espadero colorado, guayabo, trementino **Habit:** Shrub, Tree **Use:** ME **Origin:** Native **Region:** Andean **Dept.:** ANT, BOG, CUN, SAN, VAC **Elev.:** 1300–2640 m a.s.l. **IUCN Habitat code:** 1,3,4,14 **Conservation status: (IUCN)** NE

3339 *Primula obconica* Hance (Primulaceae) **IPNI:** 702340-1 **Common name:** primavera, primaveras, prímula, victoria **Habit:** Herb **Use:** EU, ME, PO **Origin:** Cultivated **Region:** Andean **Dept.:** ANT, BOG, BOY, CAL, CAU, CUN, NAR, NSA, RIS, SAN, TOL **Elev.:** 250–2600 m a.s.l. **IUCN Habitat code:** 14 **Conservation status: (IUCN)** NE

3340 *Primula praenitens* Ker Gawl. (Primulaceae) **IPNI:** 702446-1 **Use:** EU, GS, ME **Region:** Andean **Dept.:** BOG**Conservation status: (IUCN)** NE

3341 *Primula veris* L. (Primulaceae) **IPNI:** 702751-1 **Use:** EU, HF, ME **Region:** Andean **Dept.:** BOG**Conservation status: (IUCN)** NE

3342 *Stylogyne ardisioides* (Kunth) Mez (Primulaceae) **IPNI:** 247441-2 **Use:** HF **IUCN Habitat code:** 1,3,4,14 **Conservation status: (IUCN)** LC

3343 *Stylogyne longifolia* (Mart. ex Miq.) Mez (Primulaceae) **IPNI:** 247465-2 **Common name:** almidón, arrayán de río, cucharo, madroño, manglillo, ojo de tucano, ojo de tucano picón, uvo **Habit:** Shrub, Small tree, Tree **Use:** HF, ME **Origin:** Native **Region:** Amazonia, Guiana Shield, Orinoquia, Pacific **Dept.:** AMA, ANT, CAQ, CAS, CHO, CUN, GUA, MET, NSA, PUT, VAC, VAU, VID **Elev.:** 20-1740 msnm **IUCN Habitat code:** 1,2,3,4,5,14 **Conservation status: (IUCN)** LC **(CPSG)** Potential LC

3344 *Stylogyne micrantha* (Kunth) Mez (Primulaceae) **IPNI:** 247469-2 **Common name:** cucharo **Habit:** Shrub, Tree **Use:** HF, ME **Origin:** Native **Region:** Andean, Caribbean, Orinoquia **Dept.:** ANT, BOL, BOY, LAG, MET, QUI, SUC, VID **Elev.:** 10–370 m a.s.l. **IUCN Habitat code:** 1,2,3,4,14 **Conservation status: (IUCN)** LC

3345 *Stylogyne nigricans* (A.DC.) Mez (Primulaceae) **IPNI:** 247472-2 **Habit:** Shrub, Small tree, Tree **Use:** HF **Origin:** Native **Region:** Amazonia, Guiana Shield **Dept.:** AMA, VAU **Elev.:** 160–250 m a.s.l. **IUCN Habitat code:** 1 **Conservation status: (IUCN)** NE

3346 *Stylogyne orinocensis* (Kunth) Mez (Primulaceae) **IPNI:** 247474-2 **Use:** HF **Origin:** Native **IUCN Habitat code:** 3,4 **Conservation status: (IUCN)** LC

3347 *Stylogyne serpentina* Mez (Primulaceae)
IPNI: 247483-2
Use: HF **Origin:** Native **Conservation status: (IUCN)** NE

3348 *Stylogyne turbacensis* (Kunth) Mez (Primulaceae)
IPNI: 247490-2
Common name: calabomba, chagualo, cucharo, fruta de pava, guaba, mortiño, patica de paloma **Habit:** Shrub, Subshrub, Small tree **Use:** AF,ME **Origin:** Native **Region:** Andean, Guiana Shield, Caribbean, Orinoquia, Pacific **Dept.:** ANT, ATL, BOL, CAU, CES, CHO, CUN, GUV, LAG, MAG, MET, SAN, SUC, TOL **Elev.:** 30-1700 msnm **IUCN Habitat code:** 1,2,3,4,14 **Conservation status: (IUCN)** LC **(CPSG)** Potential LC

3349 *Chromolucuma rubriflora* Ducke (Sapotaceae)
IPNI: 307657-2
Habit: Tree **Use:** HF **Origin:** Native **Dept.:** ANT **Elev.:** 470–800 m a.s.l. **IUCN Habitat code:** 1 **Conservation status: (IUCN)** LC

3350 *Chrysophyllum amazonicum* T.D.Penn. (Sapotaceae)
IPNI: 941286-1
Common name: caimitillo, caimo, nashau, puirasado, pun, pun-nashau, quinilla colorada **Habit:** Tree **Use:** HF, ME **Origin:** Native **Region:** Amazonia **Dept.:** AMA, PUT, VAU **Elev.:** 0–400 m a.s.l. **IUCN Habitat code:** 1,3,14 **Conservation status: (IUCN)** LC

3351 *Chrysophyllum argenteum* Jacq. (Sapotaceae)
IPNI: 311591-2
Common name: caimitillo, caimito, caimito blanco, caimito fundipaila, caimito jiowoo-mesese-e, caimo, caimo cerezo, caimo de monte, caimo liso de monte, guanabanillo, higo, higuerón, islugu, jujulinae, manzano, níspero caimito, papabü emalinae, ponni inna, quina, tawa **Habit:** Tree **Use:** HF, MA, ME **Origin:** Native **Region:** Amazonia, Andean, Pacific **Dept.:** AMA, ANT, CAQ, CHO, CUN, MAG, MET, NAR, NSA, PUT, RIS, SAN, VAC **Elev.:** 10–1600 m a.s.l. **IUCN Habitat code:** 1,2,3,4,5,14 **Conservation status: (IUCN)** LC

3352 *Chrysophyllum bombycinum* T.D.Penn. (Sapotaceae)
IPNI: 274581-2
Common name: caimo, palo de cangrejo, teta churuco, yugo **Habit:** Tree **Use:** HF **Origin:** Native **Region:** Amazonia, Guiana Shield **Dept.:** AMA, CAQ, GUA **Elev.:** 110–400 m a.s.l. **IUCN Habitat code:** 1,3,5 **Conservation status: (IUCN)** NT

3353 *Chrysophyllum cainito* L. (Sapotaceae)
IPNI: 304384-2
Common name: caimitillo, caimito, caimito arisco, caimito blanco, caimito de montaña, caimito de monte, caimito morado, caimo, caimo blanco, caimo colorado, caimo de monte, caimo morado, caimo verde, madura, madura verde, maduraverde, maduroverde **Habit:** Tree **Use:** EU, GS, HF, MA, ME, SU **Origin:** Cultivated, Naturalised **Region:** Andean, Guiana Shield, Pacific **Dept.:** AMA, ANT, ATL, CUN, GUA, HUI, MAG, NAR, QUI, TOL, VAC **Elev.:** 10–1560 m a.s.l. **IUCN Habitat code:** 1,2,3,5,14 **Conservation status: (IUCN)** NE

3354 *Chrysophyllum colombianum* (Aubrév.) T.D.Penn. (Sapotaceae)
IPNI: 948084-1
Habit: Tree **Use:** HF **Origin:** Native **Region:** Amazonia, Andean, Pacific **Dept.:** AMA, ANT, CAQ, SAN, VAC **Elev.:** 100–1200 m a.s.l. **IUCN Habitat code:** 1,3,14 **Conservation status: (IUCN)** LC

3355 *Chrysophyllum lucentifolium* Cronquist (Sapotaceae)
IPNI: 57770-2
Habit: Tree **Use:** HF **Origin:** Native **Region:** Amazonia, Caribbean **Dept.:** ANT, ATL, GUV, MET, PUT, SUC **Elev.:** 20–1400 m a.s.l. **IUCN Habitat code:** 1,2,3 **Conservation status: (IUCN)** LC

3356 *Chrysophyllum manaosense* (Aubrév.) T.D.Penn. (Sapotaceae)
IPNI: 948085-1
Habit: Tree **Use:** HF **Origin:** Native **Region:** Amazonia **Dept.:** AMA, CAQ, VAU **Elev.:** 0–350 m a.s.l. **IUCN Habitat code:** 1,3,5 **Conservation status: (IUCN)** LC

3357 *Chrysophyllum pomiferum* (Eyma) T.D.Penn. (Sapotaceae)
IPNI: 948086-1
Common name: cacho, caimito, caimo, caimo amarillo, caimo morado, punteamarillo, puntegranado **Habit:** Tree **Use:** HF, MA **Origin:** Native **Region:** Andean, Guiana Shield, Pacific **Dept.:** CHO, SAN, VAU **Elev.:** 0–700 m a.s.l. **IUCN Habitat code:** 1,3,14 **Conservation status: (IUCN)** LC

3358 *Chrysophyllum prieurii* A.DC. (Sapotaceae)
IPNI: 57816-2
Common name: caimo, caimo amarillo, caimo de lombriz, caimo hembra, chicle, palo de casabe de trueno, palo largo, palo loro **Habit:** Tree **Use:** HF **Origin:** Native **Region:** Amazonia, Andean, Guiana Shield, Pacific **Dept.:** AMA, ANT, BOY, CAQ, CHO, VAU **Elev.:** 100–1850 m a.s.l. **IUCN Habitat code:** 1,3,14 **Conservation status: (IUCN)** LC

3359 *Chrysophyllum sanguinolentum* (Pierre) Baehni (Sapotaceae)
IPNI: 57826-2
Common name: bafacona, cagao de lombriz, caimo, caimo balata, caimo churuco, caimo de churruco, caimo de churuco, caimo de monte, chicle, palo de cangrejo, siringa de sol, yugo chicle **Habit:** Small tree, Tree **Use:** HF **Origin:** Native **Region:** Amazonia, Guiana Shield **Dept.:** AMA, CAQ, GUA, PUT, VAU **Elev.:** 160–420 m a.s.l. **IUCN Habitat code:** 1,3,5,14 **Conservation status: (IUCN)** LC

3360 *Chrysophyllum superbum* T.D.Penn. (Sapotaceae)
IPNI: 274588-2
Common name: caimillo, caimo negro **Habit:** Tree **Use:** HF **Origin:** Native **Region:** Amazonia, Guiana Shield **Dept.:** AMA, CAQ **Elev.:** 50–400 m a.s.l. **IUCN Habitat code:** 1 **Conservation status: (IUCN)** CR

3361 *Chrysophyllum venezuelanense* (Pierre) T.D.Penn. (Sapotaceae)
IPNI: 948088-1
Common name: caimo, caimo de monte, zapote de monte **Habit:** Tree **Use:** HF **Origin:** Native **Region:** Amazonia, Andean, Caribbean **Dept.:** ANT, ATL, BOL, BOY, LAG, MET **Elev.:** 325–1710 m a.s.l. **IUCN Habitat code:** 1,3,14 **Conservation status: (IUCN)** LC

3362 *Ecclinusa bullata* T.D.Penn. (Sapotaceae)
IPNI: 276174-2
Common name: nalga de araña pollera **Habit:** Tree **Use:** HF **Origin:** Native **Region:** Amazonia, Guiana Shield **Dept.:** AMA, CAQ, GUA **Elev.:** 120–240 m a.s.l. **IUCN Habitat code:** 1 **Conservation status: (IUCN)** LC

3363 *Ecclinusa guianensis* Eyma (Sapotaceae)
IPNI: 86806-2
Common name: cáscara de cangrejo, picaci **Habit:** Tree **Use:** HF, MA **Origin:** Native **Region:** Amazonia, Andean, Guiana Shield **Dept.:** AMA, ANT, CAQ **Elev.:** 250–700 m a.s.l. **IUCN Habitat code:** 1,3,5 **Conservation status: (IUCN)** LC

3364 *Ecclinusa lanceolata* (Mart. & Eichler) Pierre (Sapotaceae)
IPNI: 86808-2
Common name: caimito popa, caimo, caimo de monte, caucho, mamey **Habit:** Tree **Use:** HF **Origin:** Native **Region:** Amazonia, Guiana Shield, Orinoquia, Pacific **Dept.:** AMA, ANT, ARA, CAQ, MET, NSA, PUT, VAC, VAU **Elev.:** 50–900 m a.s.l. **IUCN Habitat code:** 1,3,14 **Conservation status: (IUCN)** LC

3365 *Ecclinusa ramiflora* Mart. (Sapotaceae)
IPNI: 786784-1
Common name: balata, barbacona, caimito, caimito leche, caimo **Habit:** Tree **Use:** HF **Origin:** Native **Region:** Amazonia, Pacific **Dept.:** AMA, ANT, VAC **Elev.:** 50–200 m a.s.l. **IUCN Habitat code:** 1,14 **Conservation status: (IUCN)** LC

3366 *Elaeoluma glabrescens* (Mart. & Eichler) Aubrév. (Sapotaceae)
IPNI: 89008-2
Common name: caimitillo balso, chicle, palo de churuco **Habit:** Tree **Use:** HF **Origin:** Native **Region:** Guiana Shield **Dept.:** AMA, CAQ, GUA, GUV, VAU **Elev.:** 100–700 m a.s.l. **IUCN Habitat code:** 1,3,5,14 **Conservation status: (IUCN)** LC

3367 *Elaeoluma schomburgkiana* (Miq.) Baill. (Sapotaceae)
IPNI: 89010-2
Common name: waiarima **Use:** HF **IUCN Habitat code:** 1,4,14 **Conservation status: (IUCN)** LC

3368 *Manilkara bidentata* (A.DC.) A.Chev. (Sapotaceae)
IPNI: 1032531-2
Common name: balata, balata roja, balato, caimito, caimo, caimo colorao, caucho rey, chicle, falsa balata, masarandi, masarandua, nisperillo, níspero, palo de chicle, purgo, quinilla, trapichero, yugo de charapa **Habit:** Tree **Use:** HF, MA, ME **Origin:** Native **Region:** Amazonia, Guiana Shield, Pacific **Dept.:** AMA, CAQ, CHO, GUA, GUV, NAR, SAN, VAC, VAU **Elev.:** 15–240 m a.s.l. **IUCN Habitat code:** 1,3,5,14 **Conservation status: (IUCN)** NE

3369 *Manilkara chicle* (Pittier) Gilly (Sapotaceae)
IPNI: 152591-2
Habit: Tree **Use:** HF, MA, ME **Origin:** Native **Region:** Andean, Caribbean **Dept.:** ATL, CHO, MAG, SUC **Elev.:** 30–1100 m a.s.l. **IUCN Habitat code:** 1,2,5,14 **Conservation status: (IUCN)** LC

3370 *Manilkara huberi* (Ducke) Standl. (Sapotaceae)
IPNI: 152609-2
Use: HF, MA, ME **IUCN Habitat code:** 14 **Conservation status: (IUCN)** NE

3371 *Manilkara zapota* (L.) P.Royen (Sapotaceae)
IPNI: 152641-2
Common name: chicle, mamey colorado, níspero, níspero de anís, sapote, sapote mamey, zapote **Habit:** Tree **Use:** EU, FU, GS, HF, MA, ME, SU **Origin:** Native, Cultivated **Region:** Andean, Caribbean **Dept.:** ANT, ATL, BOL, CES, COR, CUN, MAG, SAP, SUC, TOL, VAC **Elev.:** 0–1600 m a.s.l. **IUCN Habitat code:** 1,2,3,14 **Conservation status: (IUCN)** NE

3372 *Micropholis acutangula* (Ducke) Eyma (Sapotaceae)
IPNI: 160358-2
Habit: Tree **Use:** HF **Origin:** Native **Region:** Amazonia **Dept.:** CAQ **Elev.:** 250–250 m a.s.l. **IUCN Habitat code:** 1,14 **Conservation status: (IUCN)** NE

3373 *Micropholis crotonoides* (Pierre) Pierre (Sapotaceae)
IPNI: 160368-2
Common name: jigua piedra **Habit:** Small tree, Tree **Use:** HF **Origin:** Native **Region:** Amazonia, Andean, Pacific **Dept.:** CAQ, CHO, MAG, VAC **Elev.:** 20–2600 m a.s.l. **IUCN Habitat code:** 1,3,14 **Conservation status: (IUCN)** LC

3374 *Micropholis egensis* (A.DC.) Pierre (Sapotaceae)
IPNI: 160374-2
Common name: caimitillo **Habit:** Tree **Use:** HF, MA **Origin:** Native **Region:** Amazonia, Pacific **Dept.:** AMA, ANT, CAQ, COR, PUT **Elev.:** 100–900 m a.s.l. **IUCN Habitat code:** 1,2,3,5,14 **Conservation status: (IUCN)** LC

3375 *Micropholis gardneriana* (A.DC.) Pierre (Sapotaceae)
IPNI: 787792-1
Use: MA **Origin:** Native **IUCN Habitat code:** 1 **Conservation status: (IUCN)** LC

3376 *Micropholis guyanensis* (A.DC.) Pierre (Sapotaceae)
IPNI: 160386-2
Common name: balata, balata blanca, breo amarillo, caimillo, caimitillo, caimito, caimito silbador, caimo, caimo cascabel, caimo colorado, caimo de monte, capure, capure blanco, capure negro, capuri balachina, lengua de perro, pa, palo de cangrejo, palo de inflamación **Habit:** Tree **Use:** HF, MA, ME **Origin:** Native **Region:** Amazonia, Guiana Shield, Pacific **Dept.:** AMA, ANT, BOL, CAQ, CHO, CUN, GUA, GUV, MET, NAR, PUT, SAN, VAC, VAU **Elev.:** 5–1500 m a.s.l. **IUCN Habitat code:** 1,3,5,14 **Conservation status: (IUCN)** LC

3377 *Micropholis melinoniana* Pierre (Sapotaceae)
IPNI: 160393-2
Common name: caimito mora, canilla de vieja blanco **Habit:** Tree **Use:** HF **Origin:** Native **Region:** Amazonia, Guiana Shield, Pacific **Dept.:** AMA, BOY, CAQ, NAR, VAU **Elev.:** 100–500 m a.s.l. **IUCN Habitat code:** 1,5,14 **Conservation status: (IUCN)** LC

3378 *Micropholis obscura* T.D.Penn. (Sapotaceae)
IPNI: 280898-2
Common name: palo de erizo **Habit:** Tree **Use:** HF **Origin:** Native **Region:** Guiana Shield **Dept.:** AMA, CAQ, GUA, VAU **Elev.:** 100–250 m a.s.l. **IUCN Habitat code:** 1 **Conservation status: (IUCN)** NE

3379 *Micropholis venulosa* (Mart. & Eichler) Pierre (Sapotaceae)
IPNI: 160417-2
Common name: aguacatillo, balata blanca, caimitillo, caimito, caimito rucio, caimo, culefierro, dobonare, hobo, mamón de mico, mare, mazarandua, nisperillo, níspero, palo de inflamación **Habit:** Tree **Use:** HF, MA **Origin:** Native **Region:** Amazonia, Guiana Shield, Pacific **Dept.:** AMA, ANT, CAQ, CHO, GUA, GUV, MET, PUT, VAC, VAU **Elev.:** 0–1000 m a.s.l. **IUCN Habitat code:** 1,3,4,5,14 **Conservation status: (IUCN)** LC

3380 *Pouteria amygdalicarpa* (Pittier) T.D.Penn. (Sapotaceae)
IPNI: 948062-1
Common name: caimito, caimo vela **Habit:** Tree **Use:** HF **Origin:** Native **Region:** Pacific **Dept.:** BOY, CHO, VAC **Elev.:** 50–250 m a.s.l. **IUCN Habitat code:** 1,5,14 **Conservation status: (IUCN)** LC

3381 *Pouteria arguacoensium* (H.Karst.) Baehni (Sapotaceae)
IPNI: 209690-2
Common name: manzano **Habit:** Tree **Use:** HF, ME **Origin:** Endemic **Region:** Andean **Dept.:** MAG **Elev.:** 1300-2300 msnm **IUCN Habitat code:** 3 **Conservation status: (IUCN)** LC (CPSG) LC

3382 *Pouteria baehniana* Monach. (Sapotaceae)
IPNI: 209695-2
Common name: lechero de plomo, yúcucù **Habit:** Tree **Use:** HF **Origin:** Native **Region:** Andean, Pacific **Dept.:** ANT, CUN, MAG, NSA, VAC **Elev.:** 120–2500 m a.s.l. **IUCN Habitat code:** 1,3,14 **Conservation status: (IUCN)** LC

3383 *Pouteria bangii* (Rusby) T.D.Penn. (Sapotaceae)
IPNI: 948066-1
Common name: caimo **Habit:** Tree **Use:** HF **Origin:** Native **Region:** Guiana Shield, Orinoquia **Dept.:** MET, NSA **Elev.:** 100–1200 m a.s.l. **IUCN Habitat code:** 1,3,14 **Conservation status: (IUCN)** LC

3384 *Pouteria bilocularis* (H.J.P.Winkl.) Baehni (Sapotaceae)
IPNI: 209700-2
Habit: Tree **Use:** HF **Origin:** Native **Region:** Amazonia, Pacific **Dept.:** AMA, CAQ, VAC, VAU **Elev.:** 5–200 m a.s.l. **IUCN Habitat code:** 1,3,5,14 **Conservation status: (IUCN)** LC

3385 *Pouteria bracteata* T.D.Penn. (Sapotaceae)
IPNI: 283713-2
Habit: Tree **Use:** HF **Origin:** Native **Region:** Andean **Dept.:** VAC **Elev.:** 900–1200 m a.s.l. **IUCN Habitat code:** 1 **Conservation status: (IUCN)** EN

3386 *Pouteria caimito* (Ruiz & Pav.) Radlk. (Sapotaceae)
IPNI: 788845-1
Common name: abiorama, abiu, abiurama, buburi, cafuche, caimitillo, caimito, caimito amarillo, caimito de monte, caimo, caimo de borugo, caimo de charapa, caimo de monte, caimo de sol, caimo maduroverde, caimo pequeñito, cari?ca, cauje, emali, emalinae, jificona, madu **Habit:** Tree **Use:** AF, HF, MA, ME **Origin:** Native, Cultivated **Region:** Amazonia, Andean, Guiana Shield, Orinoquia, Pacific **Dept.:** AMA, ANT, CAQ, CAS, CHO, GUA, GUV, HUI, MET, NAR, PUT, TOL, VAU **Elev.:** 0–1700 m a.s.l. **IUCN Habitat code:** 1,2,3,4,5,14 **Conservation status: (IUCN)** LC

3387 *Pouteria campechiana* (Kunth) Baehni (Sapotaceae)
IPNI: 209714-2
Use: EU, GS, HF, MA, ME
Conservation status: (IUCN) LC

3388 *Pouteria cayennensis* (A.DC.) Eyma (Sapotaceae)
IPNI: 209723-2
Habit: Tree **Use:** HF **Origin:** Native **Region:** Amazonia, Guiana Shield **Dept.:** AMA, CAQ **Elev.:** 240–240 m a.s.l. **IUCN Habitat code:** 1,14 **Conservation status: (IUCN)** LC

3389 *Pouteria cladantha* Sandwith (Sapotaceae)
IPNI: 209727-2
Common name: caimito, caimo, yugo de pescado **Habit:** Tree **Use:** HF, MA **Origin:** Native **Region:** Amazonia, Guiana Shield **Dept.:** AMA, ANT, CAQ, GUA, GUV, VAU **Elev.:** 80–800 m a.s.l. **IUCN Habitat code:** 1,3,5,14 **Conservation status: (IUCN)** LC

3390 *Pouteria collina* (Little) T.D.Penn. (Sapotaceae)
IPNI: 1048092-2
Common name: caimito **Habit:** Small tree, Tree **Use:** HF **Origin:** Native **Dept.:** NAR, VAC **Elev.:** 0–650 m a.s.l. **IUCN Habitat code:** 1,14 **Conservation status: (IUCN)** VU

3391 *Pouteria coriacea* (Pierre) Pierre (Sapotaceae)
IPNI: 209732-2
Habit: Tree **Use:** HF **Origin:** Native **Region:** Amazonia, Guiana Shield **Dept.:** AMA, VAU **Elev.:** 100–100 m a.s.l. **IUCN Habitat code:** 1,3,14 **Conservation status: (IUCN)** LC

3392 *Pouteria cuspidata* (A.DC.) Baehni (Sapotaceae)
IPNI: 209738-2
Common name: caimitillo, caimo amarillo, caimo sapo, palo de almidón, popay, popay de sapo **Habit:** Tree **Use:** HF **Origin:** Native **Region:** Amazonia, Guiana Shield, Pacific **Dept.:** AMA, ANT, CAQ, CHO, COR, GUA, MET, PUT, SAN, VAC, VAU **Elev.:** 0–550 m a.s.l. **IUCN Habitat code:** 1,2,3,5,14 **Conservation status: (IUCN)** LC

3393 *Pouteria durlandii* (Standl.) Baehni (Sapotaceae)
IPNI: 788920-1
Common name: caimitillo, níspero de monte, sapotillo **Habit:** Tree **Use:** EU, FU, HF, MA **Origin:** Native **Region:** Amazonia, Caribbean **Dept.:** AMA, ANT, CAQ, CES, PUT **Elev.:** 0–500 m a.s.l. **IUCN Habitat code:** 1,2,3,14 **Conservation status: (IUCN)** LC

3394 *Pouteria egregia* Sandwith (Sapotaceae)
IPNI: 209757-2
Common name: caimo **Habit:** Tree **Use:** HF, MA **Origin:** Native **Region:** Guiana Shield **Dept.:** CAQ, MET **Elev.:** 250–800 m a.s.l. **IUCN Habitat code:** 1,4,14 **Conservation status: (IUCN)** LC

3395 *Pouteria elegans* (A.DC.) Baehni (Sapotaceae)
IPNI: 209758-2
Common name: palo sapo, reventillo **Habit:** Tree **Use:** HF, ME **Origin:** Native **Region:** Amazonia, Guiana Shield, Orinoquia **Dept.:** AMA, GUA, MET, VAU, VID **Elev.:** 80–700 m a.s.l. **IUCN Habitat code:** 1,3,4,5 **Conservation status: (IUCN)** LC

3396 *Pouteria engleri* Eyma (Sapotaceae) **IPNI:** 291747-2 **Habit:** Tree **Use:** HF, MA **Origin:** Native **Region:** Pacific **Dept.:** VAC **Elev.:** 0–250 m a.s.l. **IUCN Habitat code:** 1 **Conservation status: (IUCN)** NE

3397 *Pouteria ephedrantha* (A.C.Sm.) T.D.Penn. (Sapotaceae) **IPNI:** 948081-1 **Habit:** Tree **Use:** HF **Origin:** Native **Region:** Amazonia **Dept.:** AMA **Elev.:** 100–100 m a.s.l. **IUCN Habitat code:** 1 **Conservation status: (IUCN)** NE

3398 *Pouteria filipes* Eyma (Sapotaceae) **IPNI:** 209766-2 **Use:** HF **IUCN Habitat code:** 1 **Conservation status: (IUCN)** LC

3399 *Pouteria franciscana* Baehni (Sapotaceae) **IPNI:** 209771-2 **Common name:** caimo **Habit:** Tree **Use:** HF **Origin:** Native **Region:** Amazonia **Dept.:** CAQ **Elev.:** 240–350 m a.s.l. **IUCN Habitat code:** 1,4 **Conservation status: (IUCN)** LC

3400 *Pouteria glauca* T.D.Penn. (Sapotaceae) **IPNI:** 283727-2 **Common name:** popay de sapo **Habit:** Tree **Use:** HF **Origin:** Native **Region:** Amazonia **Dept.:** AMA, CAQ, GUV **Elev.:** 100–250 m a.s.l. **IUCN Habitat code:** 1,3 **Conservation status: (IUCN)** VU

3401 *Pouteria glomerata* (Miq.) Radlk. (Sapotaceae) **IPNI:** 788965-1 **Common name:** arrayán, cocosapo **Habit:** Tree **Use:** HF **Origin:** Native **Region:** Amazonia, Andean, Caribbean **Dept.:** AMA, ANT, BOL, CAQ, CAU, CUN, MAG, PUT, SAN, SUC **Elev.:** 80–1300 m a.s.l. **IUCN Habitat code:** 1,2,3,4,5,14 **Conservation status: (IUCN)** NE

3402 *Pouteria gomphiifolia* (Mart. ex Miq.) Radlk. (Sapotaceae) **IPNI:** 788967-1 **Habit:** Tree **Use:** HF **Origin:** Native **Region:** Amazonia, Guiana Shield **Dept.:** GUA, VAU **Elev.:** 100–250 m a.s.l. **IUCN Habitat code:** 1,2 **Conservation status: (IUCN)** LC

3403 *Pouteria gongrijpii* Eyma (Sapotaceae) **IPNI:** 209784-2 **Habit:** Tree **Use:** HF, MA **Origin:** Native **Region:** Amazonia **Dept.:** AMA **Elev.:** 80–240 m a.s.l. **IUCN Habitat code:** 1 **Conservation status: (IUCN)** LC

3404 *Pouteria guianensis* Aubl. (Sapotaceae) **IPNI:** 788972-1 **Common name:** caimillo fino, caimito, caimito colorado, caimo, caimo amarillo, caimo negro, caimón, leche de cabra, mantequilla, noyocaricaci **Habit:** Tree **Use:** HF, MA **Origin:** Native **Region:** Amazonia, Guiana Shield, Orinoquia, Pacific **Dept.:** AMA, ANT, CAQ, CAS, CHO, COR, GUV, MET, PUT, VAC, VAU, VID **Elev.:** 80–500 m a.s.l. **IUCN Habitat code:** 1,2,3,4,14 **Conservation status: (IUCN)** LC

3405 *Pouteria juruana* K.Krause (Sapotaceae) **IPNI:** 788995-1 **Habit:** Tree **Use:** HF **Origin:** Native **Region:** Amazonia **Dept.:** CAQ, PUT **Elev.:** 100–250 m a.s.l. **IUCN Habitat code:** 1 **Conservation status: (IUCN)** EN

3406 *Pouteria krukovii* (A.C.Sm.) Baehni (Sapotaceae) **IPNI:** 209797-2 **Habit:** Tree **Use:** HF **Origin:** Native **Region:** Amazonia, Orinoquia **Dept.:** AMA, ANT, MET **Elev.:** 100–500 m a.s.l. **IUCN Habitat code:** 1,3,5,14 **Conservation status: (IUCN)** VU

3407 *Pouteria laevigata* (Mart.) Radlk. (Sapotaceae) **IPNI:** 789009-1 **Common name:** kunda, siringa de pescado, tree de churuco, árbol de churuco **Habit:** Tree **Use:** HF **Origin:** Native **Region:** Amazonia, Guiana Shield **Dept.:** AMA, CAQ, VAU **Elev.:** 80–250 m a.s.l. **IUCN Habitat code:** 1,4,5,14 **Conservation status: (IUCN)** LC

3408 *Pouteria lucuma* (Ruiz & Pav.) Kuntze (Sapotaceae) **IPNI:** 209804-2 **Common name:** caimitillo, caimo, lucmo, lúcuma, lúcumo, maco, maco serrano, mediacara, mediacaro, pepo, sapote, zapote **Habit:** Tree **Use:** HF, ME **Origin:** Native **Region:** Andean **Dept.:** ANT, CAU, HUI, NAR, PUT, SAN, TOL, VAC **Elev.:** 1500–3000 m a.s.l. **IUCN Habitat code:** 1,3,14 **Conservation status: (IUCN)** LC

3409 *Pouteria lucumifolia* (Reissek ex Maxim.) T.D.Penn. (Sapotaceae) **IPNI:** 948065-1 **Common name:** maco **Habit:** Tree **Use:** HF **Origin:** Native **Region:** Amazonia, Guiana Shield **Dept.:** CAU, GUA, HUI, PUT **Elev.:** 120–200 m a.s.l. **IUCN Habitat code:** 1 **Conservation status: (IUCN)** NE

3410 *Pouteria macrocarpa* (Mart.) D.Dietr. (Sapotaceae) **IPNI:** 789045-1 **Common name:** anchique **Habit:** Tree **Use:** HF **Origin:** Native **Region:** Andean, Guiana Shield **Dept.:** BOY, MET **Elev.:** 500–1800 m a.s.l. **IUCN Habitat code:** 1,3,5,14 **Conservation status: (IUCN)** VU

3411 *Pouteria macrophylla* (Lam.) Eyma (Sapotaceae) **IPNI:** 209808-2 **Habit:** Tree **Use:** HF, MA **Origin:** Native **Region:** Andean **Dept.:** AMA, PUT **Elev.:** 1300–1400 m a.s.l. **Conservation status: (IUCN)** LC

3412 *Pouteria maguirei* (Aubrév.) T.D.Penn. (Sapotaceae) **IPNI:** 948078-1 **Common name:** caimo, caimo de monte **Habit:** Tree **Use:** HF **Origin:** Native **Region:** Amazonia, Guiana Shield **Dept.:** AMA, CAQ, GUA **Elev.:** 120–300 m a.s.l. **IUCN Habitat code:** 1,3 **Conservation status: (IUCN)** NT

3413 *Pouteria multiflora* (A.DC.) Eyma (Sapotaceae) **IPNI:** 209824-2 **Common name:** caimito, caimo, lengua de vaca, lengüevaca, ojo de venado **Habit:** Tree **Use:** HF, MA, ME **Origin:** Native **Region:** Andean, Pacific **Dept.:** AMA, ANT, BOL, CHO, CUN, PUT, RIS, SAN **Elev.:** 100–2000 m a.s.l. **Conservation status: (IUCN)** NE

3414 *Pouteria nemorosa* Baehni (Sapotaceae) **IPNI:** 209829-2 **Habit:** Tree **Use:** HF **Origin:** Native **Region:** Pacific **Dept.:** ANT **Elev.:** 180–400 m a.s.l. **Conservation status: (IUCN)** VU

3415 *Pouteria oblanceolata* Pires (Sapotaceae) **IPNI:** 209832-2 **Common name:** caimo morado **Habit:** Tree **Use:** HF **Origin:** Native **Region:** Amazonia **Dept.:** AMA, VAU **Elev.:** 100–240 m a.s.l. **IUCN Habitat code:** 1 **Conservation status: (IUCN)** LC

3416 *Pouteria petiolata* T.D.Penn. (Sapotaceae) **IPNI:** 283741-2 **Habit:** Tree **Use:** HF **Origin:** Native **Region:** Amazonia **Dept.:** CAQ **Elev.:** 250–250 m a.s.l. **Conservation status: (IUCN)** VU

3417 *Pouteria plicata* T.D.Penn. (Sapotaceae) **IPNI:** 283742-2 **Habit:** Tree **Use:** HF **Origin:** Native **Region:** Amazonia, Guiana Shield **Dept.:** AMA, CAQ, GUV, VAU **Elev.:** 80–250 m a.s.l. **IUCN Habitat code:** 1,3,5,14 **Conservation status: (IUCN)** LC

3418 *Pouteria procera* (Mart.) K.Hammer (Sapotaceae) **IPNI:** 943696-1 **Habit:** Tree **Use:** HF, MA **Origin:** Native **Region:** Amazonia, Caribbean **Dept.:** MET, NSA **Elev.:** 380–400 m a.s.l. **IUCN Habitat code:** 1 **Conservation status: (IUCN)** LC

3419 *Pouteria pubescens* (Aubrév. & Pellegr.) T.D.Penn. (Sapotaceae) **IPNI:** 948067-1 **Habit:** Tree **Use:** HF **Origin:** Native **Region:** Amazonia **Dept.:** AMA, CAQ **Elev.:** 120–250 m a.s.l. **IUCN Habitat code:** 1 **Conservation status: (IUCN)** VU

3420 *Pouteria putamen-ovi* T.D.Penn. (Sapotaceae) **IPNI:** 283744-2 **Common name:** caimo **Habit:** Tree **Use:** HF, ME **Origin:** Native **Region:** Amazonia **Dept.:** CAQ, PUT **Elev.:** 300–400 m a.s.l. **IUCN Habitat code:** 1 **Conservation status: (IUCN)** VU

3421 *Pouteria reticulata* (Engl.) Eyma (Sapotaceae) **IPNI:** 209864-2 **Habit:** Tree **Use:** HF **Origin:** Native **Region:** Amazonia, Orinoquia, Pacific **Dept.:** AMA, CAQ, CAS, MET, VAC, VAU **Elev.:** 30–250 m a.s.l. **IUCN Habitat code:** 1,3,14 **Conservation status: (IUCN)** LC

3422 *Pouteria retinervis* T.D.Penn. (Sapotaceae) **IPNI:** 283746-2 **Habit:** Tree **Use:** HF **Origin:** Native **Region:** Amazonia, Guiana Shield **Dept.:** AMA, CAQ **Elev.:** 100–250 m a.s.l. **IUCN Habitat code:** 1 **Conservation status: (IUCN)** CD

3423 *Pouteria sapota* (Jacq.) H.E.Moore & Stearn (Sapotaceae) **IPNI:** 1110421-2 **Common name:** galán de noche, mamey, mamey colorado, mamey rojo, mamey sapote, mamey zapote, níspero, sapote, sapote colorado, sapote costeño, sapote cuna, sapote de carne, zapote, zapote colorado, zapote costeño, zapote cuna, zapote de carne **Habit:** Tree **Use:** HF, ME **Origin:** Cultivated, Naturalised **Region:** Caribbean, Pacific **Dept.:** ATL, BOL, CES, CHO, MAG, NAR, SUC, VAC **Elev.:** 80–110 m a.s.l. **IUCN Habitat code:** 1,2,3,14 **Conservation status: (IUCN)** LC

3424 *Pouteria sclerocarpa* (Pittier) Cronquist (Sapotaceae) **IPNI:** 209883-2 **Habit:** Tree **Use:** HF **Origin:** Native **Region:** Pacific **Dept.:** CHO **Elev.:** 0–150 m a.s.l. **Conservation status: (IUCN)** NT

3425 *Pouteria simulans* Monach. (Sapotaceae) **IPNI:** 209890-2 **Common name:** pendare **Habit:** Tree **Use:** HF **Origin:** Native **Region:** Amazonia **Dept.:** CAQ **Elev.:** 200–250 m a.s.l. **IUCN Habitat code:** 1 **Conservation status: (IUCN)** LC

3426 *Pouteria speciosa* (Ducke) Baehni (Sapotaceae) **IPNI:** 209892-2 **Common name:** palo yuca **Use:** HF, MA **Origin:** Native **Dept.:** VAU**Conservation status: (IUCN)** NE

3427 *Pouteria stipitata* Cronquist (Sapotaceae) **IPNI:** 209900-2 **Common name:** caimito sabanero **Use:** AF, EU, FU, HF, ME **Origin:** Native **IUCN Habitat code:** 1 **Conservation status: (IUCN)** LC

3428 *Pouteria surumuensis* Baehni (Sapotaceae) **IPNI:** 209907-2 **Use:** HF **Origin:** Native **IUCN Habitat code:** 1,3,14 **Conservation status: (IUCN)** LC

3429 *Pouteria torta* (Mart.) Radlk. (Sapotaceae) **IPNI:** 789233-1 **Common name:** caimillo, caimitillo, caimito, caimito de rebalse, caimito pelao, caimo, caimo de lombriz, caimo de monte, caimo peludo, caimo zanca de venado, casaco, piaste, piaste caimito, sapote de monte, sapotillo, zapote de monte, zapotillo **Habit:** Tree **Use:** HF, MA, ME **Origin:** Native **Region:** Amazonia, Andean, Pacific **Dept.:** AMA, ANT, BOL, CAQ, CHO, GUA, GUV, NAR, QUI, SAN, VAC, VAU **Elev.:** 0–2440 m a.s.l. **IUCN Habitat code:** 1,3,4,5,14 **Conservation status: (IUCN)** LC

3430 *Pouteria trilocularis* Cronquist (Sapotaceae) **IPNI:** 209912-2 **Common name:** caimo **Habit:** Tree **Use:** HF **Origin:** Native **Region:** Amazonia, Andean, Pacific **Dept.:** AMA, ANT, CAQ, CHO, MET, VAC **Elev.:** 0–1900 m a.s.l. **IUCN Habitat code:** 1,3,14 **Conservation status: (IUCN)** LC

3431 *Pouteria ucuqui* Pires & R.E.Schult. (Sapotaceae) **IPNI:** 209914-2 **Common name:** caimito, k-do-fru, k-na-pla, kajae, kajáe, kugna, pa?coro, quicugau, sopo-ka, sopoca, trapichero, ucuquí, yucu, yugo **Habit:** Tree **Use:** FU, HF, MA, ME **Origin:** Native **Region:** Guiana Shield **Dept.:** AMA, GUA, VAU **Elev.:** 100–250 m a.s.l. **IUCN Habitat code:** 1,5 **Conservation status: (IUCN)** NE

3432 *Pouteria venosa* (Mart.) Baehni (Sapotaceae) **IPNI:** 209920-2 **Habit:** Tree **Use:** HF **Origin:** Native **Region:** Guiana Shield, Orinoquia **Dept.:** ARA, CAQ, VID **Elev.:** 40–250 m a.s.l. **IUCN Habitat code:** 1,4,14 **Conservation status: (IUCN)** LC

3433 *Pouteria vernicosa* T.D.Penn. (Sapotaceae) **IPNI:** 283759-2 **Common name:** caimillo, caimo de monte **Habit:** Tree **Use:** HF **Origin:** Native **Region:** Amazonia, Guiana Shield **Dept.:** AMA, CAQ, VAU **Elev.:** 100–250 m a.s.l. **IUCN Habitat code:** 1 **Conservation status: (IUCN)** LC

3434 *Pradosia caracasana* (Pittier) T.D.Penn. (Sapotaceae) **IPNI:** 948100-1 **Habit:** Tree **Use:** HF **Origin:** Native **Dept.:** ANT **Elev.:** 330–330 m a.s.l. **IUCN Habitat code:** 1,3 **Conservation status: (IUCN)** NE

3435 *Pradosia cochlearia* (Lecomte) T.D.Penn. (Sapotaceae) **IPNI:** 948096-1 **Habit:** Tree **Use:** HF **Origin:** Native **Region:** Amazonia **Dept.:** AMA, CAQ **Elev.:** 100–240 m a.s.l. **IUCN Habitat code:** 1,2,3,14 **Conservation status: (IUCN)** NE

3436 *Pradosia colombiana* (Standl.) T.D.Penn. ex T.J.Ayers & Boufford (Sapotaceae) **IPNI:** 77176719-1 **Common name:** huevo de burro, joveroso, madera de hierro, mamón de tigre **Habit:** Tree **Use:** FU, HF, MA **Origin:** Native **Region:** Caribbean **Dept.:** ATL, BOL, CES, LAG, MAG, SUC **Elev.:** 0–550 m a.s.l. **IUCN Habitat code:** 1,2,3,14 **Conservation status: (IUCN)** LC

3437 *Pradosia mutisii* Cronquist (Sapotaceae) **IPNI:** 209946-2 **Use:** HF **Origin:** Native **IUCN Habitat code:** 1,3,14 **Conservation status: (IUCN)** EX

3438 *Pradosia schomburgkiana* (A.DC.) Cronquist (Sapotaceae) **IPNI:** 209950-2 **Habit:** Tree **Use:** HF **Origin:** Native **Region:** Amazonia, Andean, Guiana Shield **Dept.:** AMA, ANT, CAQ, GUA, PUT, VAU **Elev.:** 120–1600 m a.s.l. **IUCN Habitat code:** 1,3,4,5 **Conservation status: (IUCN)** LC

3439 *Pradosia subverticillata* Ducke (Sapotaceae) **IPNI:** 309845-2 **Common name:** caimo danto **Habit:** Tree **Use:** HF **Origin:** Native **Region:** Amazonia **Dept.:** CAQ, MET **Elev.:** 200–400 m a.s.l. **IUCN Habitat code:** 1 **Conservation status: (IUCN)** VU

3440 *Sarcaulus brasiliensis* (A.DC.) Eyma (Sapotaceae) **IPNI:** 227688-2 **Common name:** caimito, semilla lisa **Habit:** Tree **Use:** HF, ME **Origin:** Native **Region:** Amazonia, Orinoquia, Pacific **Dept.:** AMA, ANT, ARA, CHO, GUV **Elev.:** 10–350 m a.s.l. **IUCN Habitat code:** 1,2,3,14 **Conservation status: (IUCN)** LC

3441 *Sideroxylon celastrinum* (Kunth) T.D.Penn. (Sapotaceae) **IPNI:** 948032-1 **Habit:** Shrub, Small tree, Tree **Use:** HF **Origin:** Native **Region:** Caribbean **Dept.:** ATL, BOL, MAG **Elev.:** 0–500 m a.s.l. **IUCN Habitat code:** 2,4 **Conservation status: (IUCN)** LC

3442 *Sideroxylon obtusifolium* (Roem. & Schult.) T.D.Penn. (Sapotaceae) **IPNI:** 1008737-2 **Common name:** brujo, caimitico, caimito, cruceta, espina de Cristo, pasito, periquera, waliruwain **Habit:** Shrub, Tree **Use:** HF, MA, ME **Origin:** Native **Region:** Caribbean **Dept.:** ATL, BOL, COR, CUN, HUI, LAG, MAG, NSA, SUC **Elev.:** 0–1100 m a.s.l. **IUCN Habitat code:** 1,2,3,4,14 **Conservation status: (IUCN)** LC

3443 *Sideroxylon persimile* (Hemsl.) T.D.Penn. (Sapotaceae) **IPNI:** 948021-1 **Common name:** espino de brujo, guayacán de bola **Habit:** Shrub **Use:** HF **Origin:** Native **Region:** Caribbean **Dept.:** ATL, BOL, MAG **Elev.:** 0–100 m a.s.l. **IUCN Habitat code:** 2 **Conservation status: (IUCN)** LC

3444 *Styrax argenteus* C.Presl (Styracaceae) **IPNI:** 327046-2 **Use:** HF **IUCN Habitat code:** 1 **Conservation status: (IUCN)** LC

3445 *Styrax guyanensis* A.DC. (Styracaceae) **IPNI:** 327042-2 **Common name:** jigua blanco **Habit:** Tree **Use:** ME **Origin:** Native **Region:** Amazonia, Guiana Shield, Orinoquia **Dept.:** AMA, GUA, GUV, VAU, VID **Elev.:** 100–300 m a.s.l. **Conservation status: (IUCN)** LC

3446 *Styrax microphyllus* Perkins (Styracaceae) **IPNI:** 247679-2 **Habit:** Shrub **Use:** ME **Origin:** Endemic **Region:** Andean **Dept.:** ANT, NSA **Elev.:** 1600–2000 m a.s.l. **IUCN Habitat code:** 2 **Conservation status: (IUCN)** NE

3447 *Styrax pavonii* A.DC. (Styracaceae) **IPNI:** 826890-1 **Common name:** estoraque **Habit:** Shrub, Tree **Use:** ME **Origin:** Native **Region:** Andean **Dept.:** ANT, CAL, CAU, CES, CHO, HUI, NSA, QUI, RIS, TOL, VAC **Elev.:** 450–3300 m a.s.l. **IUCN Habitat code:** 1,3,14 **Conservation status: (IUCN)** LC

3448 *Styrax pentlandianus* J.Rémy (Styracaceae) **IPNI:** 327039-2 **Common name:** estoraque **Habit:** Tree **Use:** ME **Origin:** Native **Region:** Andean **Dept.:** ANT, CES, HUI, NSA **Elev.:** 420–2300 m a.s.l. **IUCN Habitat code:** 1 **Conservation status: (IUCN)** NE

3449 *Styrax pohlii* A.DC. (Styracaceae) **IPNI:** 826900-1 **Common name:** estoraque **Habit:** Tree **Use:** HF, MA, ME **Origin:** Native **Region:** Orinoquia **Dept.:** MET, VID **Elev.:** 200–300 m a.s.l. **IUCN Habitat code:** 1 **Conservation status: (IUCN)** NE

3450 *Styrax rigidifolius* Idrobo & R.E.Schult. (Styracaceae) **IPNI:** 247710-2 **Habit:** Shrub **Use:** ME **Origin:** Endemic **Region:** Guiana Shield **Dept.:** AMA, CAQ, MET, VAU **Elev.:** 100–640 m a.s.l. **IUCN Habitat code:** 1,4,14 **Conservation status: (IUCN)** NE

3451 *Styrax schultzei* Perkins (Styracaceae) **IPNI:** 247713-2 **Habit:** Tree **Use:** ME **Origin:** Endemic **Region:** Andean **Dept.:** CES, MAG **Elev.:** 1800-3000 m.a.s.l. **Conservation status: (IUCN)** CR **(CPSG)** CR

3452 *Styrax tomentosus* Bonpl. (Styracaceae) **IPNI:** 826958-1 **Common name:** benjuí, estoraque, pastilla **Use:** ME **IUCN Habitat code:** 1,3,4,14 **Conservation status: (IUCN)** NE

3453 *Styrax trichocalyx* Perkins (Styracaceae) **IPNI:** 247727-2 **Habit:** Tree **Use:** ME **Origin:** Native **Region:** Andean **Dept.:** ANT, BOG, BOY, CAU, CES, CUN, HUI, MAG, SAN **Elev.:** 1400–3000 m a.s.l. **IUCN Habitat code:** 1,3,14 **Conservation status: (IUCN)** NE

3454 *Symplocos arechea* L'Hér. (Symplocaceae) **IPNI:** 827005-1 **Habit:** Tree **Use:** ME **Origin:** Native **Region:** Andean **Dept.:** CAU, NSA **Elev.:** 900–1700 m a.s.l. **IUCN Habitat code:** 1 **Conservation status: (IUCN)** LC

3455 *Symplocos ecuadorensis* Little (Symplocaceae) **IPNI:** 247908-2 **Habit:** Tree **Use:** ME **Origin:** Native **Region:** Pacific **Dept.:** NAR **Elev.:** 50–50 m a.s.l. **IUCN Habitat code:** 14 **Conservation status: (IUCN)** NE

3456 *Symplocos lutescens* Brand (Symplocaceae) **IPNI:** 247957-2 **Habit:** Shrub, Small tree **Use:** ME **Origin:** Endemic **Region:** Andean **Dept.:** CES **Elev.:** 2800–2940 m a.s.l. **IUCN Habitat code:** 1 **Conservation status: (IUCN)** NE

3457 *Symplocos martinicensis* Jacq. (Symplocaceae) **IPNI:** 827186-1 **Habit:** Tree **Use:** MA **Origin:** Native **Region:** Andean, Pacific **Dept.:** CHO **Elev.:** 200–2150 m a.s.l. **Conservation status: (IUCN)** NE

3458 *Symplocos nivalis* Linden ex Brand (Symplocaceae) **IPNI:** 307879-2 **Common name:** murtá **Habit:** Shrub, Tree **Use:** ME **Origin:** Endemic **Region:** Andean **Dept.:** LAG, MAG **Elev.:** 2800-4000 m a.s.l. **IUCN Habitat code:** 2 **Conservation status: (IUCN)** LC **(CPSG)** LC

3459 *Symplocos pichindensis* Cuatrec. (Symplocaceae) **IPNI:** 307456-2 **Habit:** Tree **Use:** ME **Origin:** Endemic **Region:** Andean **Dept.:** VAC **Elev.:** 1800-2380 m a.s.l. **Conservation status: (IUCN)** CR **(CPSG)** CR

3460 *Symplocos sararensis* Cuatrec. (Symplocaceae) **IPNI:** 248023-2 **Habit:** Shrub, Small tree **Use:** ME **Origin:** Endemic **Region:** Andean **Dept.:** BOY, NSA, SAN **Elev.:** 1930–2300 m a.s.l. **IUCN Habitat code:** 1,14 **Conservation status: (IUCN)** NE

3461 *Symplocos serrulata* Bonpl. (Symplocaceae) **IPNI:** 827299-1 **Common name:** fruta de pava, fruto de pava, palo yanaconas, pepa de pava, symploco **Habit:** Tree **Use:** HF, ME **Origin:** Native **Region:** Andean **Dept.:** ANT, CAU, CUN, HUI, NSA, QUI, TOL, VAC **Elev.:** 1200–2550 m a.s.l. **IUCN Habitat code:** 1,2,3,14 **Conservation status: (IUCN)** NE

3462 *Symplocos suaveolens* Klotzsch ex Brand (Symplocaceae) **IPNI:** 248033-2 **Use:** ME **IUCN Habitat code:** 1,14 **Conservation status: (IUCN)** NE

3463 *Symplocos theiformis* (L.f.) Oken (Symplocaceae) **IPNI:** 827837-1 **Common name:** calabacito, palo blanco, té de Bogotá, té de bogotá **Habit:** Shrub, Small tree **Use:** HF, ME **Origin:** Endemic **Region:** Andean **Dept.:** ANT, BOG, BOY, CUN, SAN **Elev.:** 2450–3724 m a.s.l. **IUCN Habitat code:** 1,3,4,8,14 **Conservation status: (IUCN)** LC

3464 *Symplocos trianae* Brand (Symplocaceae) **IPNI:** 305493-2 **Habit:** Tree **Use:** ME **Origin:** Endemic **Region:** Orinoquia **Dept.:** CAS, MET **Elev.:** 250-400 m a.s.l. **IUCN Habitat code:** 3 **Conservation status: (IUCN)** EN **(CPSG)** EN

3465 *Symplocos venulosa* Cuatrec. (Symplocaceae) **IPNI:** 248055-2 **Habit:** Tree **Use:** ME **Origin:** Endemic **Region:** Andean **Dept.:** CUN, HUI, MET, PUT **Elev.:** 2200-2950 m.a.s.l. **IUCN Habitat code:** 1,3,4,14 **Conservation status: (IUCN)** VU **(CPSG)** VU

3466 *Pelliciera rhizophorae* Planch. & Triana (Tetrameristaceae) **IPNI:** 829218-1 **Common name:** comedero, mangle comedero, mangle de Buenaventura, mangle picudo, mangle piñuelo, piñuelo, piñuelo picudo **Habit:** Tree **Use:** HF **Origin:** Native **Region:** Caribbean, Pacific **Dept.:** BOL, CAU, CHO, NAR, VAC **Elev.:** 0-0 m.a.s.l. **IUCN Habitat code:** 1,3,5,14 **Conservation status: (IUCN)** VU **(CPSG)** Potential LC

3467 *Camellia japonica* L. (Theaceae) **IPNI:** 828524-1 **Common name:** camelia **Habit:** Shrub, Small tree **Use:** EU, GS, HF, MA, ME **Origin:** Cultivated **Region:** Andean **Dept.:** ANT, BOG **Elev.:** 2300–2400 m a.s.l. **IUCN Habitat code:** 1,3,14 **Conservation status: (IUCN)** LC

3468 *Camellia sinensis* (L.) Kuntze (Theaceae) **IPNI:** 828548-1 **Common name:** té, té de china **Habit:** Shrub, Small tree **Use:** GS, HF, ME **Origin:** Cultivated **Region:** Andean **Dept.:** CUN **Elev.:** 1300–1750 m a.s.l. **IUCN Habitat code:** 1,3,14 **Conservation status: (IUCN)** DD

3469 *Gordonia fruticosa* (Schrad.) H.Keng (Theaceae) **IPNI:** 112597-2 **Common name:** candelo, carretillo, chilco, chisgo, cogote, impar, impar colorado, liscano, lizcano, manzanillo, mata de borugo, quimulá, rey, roso **Habit:** Shrub, Small tree, Tree **Use:** MA, ME **Origin:** Native **Region:** Andean **Dept.:** ANT, BOG, CAU, CHO, CUN, HUI, LAG, MAG, MET, NAR, PUT, QUI, RIS, SAN, TOL, VAC **Elev.:** 1000-3600 m.a.s.l. **IUCN Habitat code:** 1,2,3,4,14 **Conservation status: (IUCN)** NE **(CPSG)** Potential LC

Escalloniales

3470 *Escallonia discolor* Vent. (Escalloniaceae) **IPNI:** 791015-1 **Common name:** chilca, chilco dorado **Habit:** Small tree, Tree **Use:** MA **Origin:** Endemic, Cultivated **Region:** Andean **Dept.:** BOY, CUN, NSA, SAN **Elev.:** 2100-3300 m.a.s.l. **IUCN Habitat code:** 1,3,4,5,14 **Conservation status: (IUCN)** NT **(CPSG)** NT

3471 *Escallonia myrtilloides* L.f. (Escalloniaceae) **IPNI:** 791067-1 **Common name:** chilco, chilco colorado, chilco de páramo, cochinito, cuasa, mangel, pagoda, pagüis, rodamonte, sombrerito, tiba, tobo, tíbar, tíber **Habit:** Shrub, Small tree **Use:** HF, ME **Origin:** Native **Region:** Andean **Dept.:** ANT, BOG, BOY, CAL, CAU, CES, CHO, CUN, HUI, LAG, MAG, MET, NAR, NSA, PUT, QUI, RIS, SAN, TOL, VAC **Elev.:** 2500-3900 m.a.s.l. **IUCN Habitat code:** 1,3,4,5,6,8,14 **Conservation status: (IUCN)** NE **(CPSG)** Potential LC

3472 *Escallonia paniculata* (Ruiz & Pav.) Schult. (Escalloniaceae) **IPNI:** 791071-1 **Common name:** carbón, chiclo colorado, chilco, chilco colorado, cuasa, rodamonte, tobo, tíbar **Habit:** Shrub, Small tree, Tree **Use:** EU, FU, MA, ME, SU **Origin:** Native **Region:** Andean **Dept.:** ANT, BOG, BOY, CAL, CAU, CHO, CUN, HUI, LAG, MAG, MET, NAR, NSA, QUI, RIS, SAN, TOL, VAC **Elev.:** 1500-3800 m.a.s.l. **IUCN Habitat code:** 1,2,3,4,5,14 **Conservation status: (IUCN)** LC **(CPSG)** Potential LC

3473 *Escallonia pendula* (Ruiz & Pav.) Pers. (Escalloniaceae) **IPNI:** 791075-1 **Common name:** loqueto, mangle **Habit:** Shrub, Small tree **Use:** EU, FU, MA **Origin:** Native, Cultivated **Region:** Andean **Dept.:** BOG, BOY, CUN, NSA **Elev.:** 1450–2660 m a.s.l. **IUCN Habitat code:** 1,2,3,4,5,14 **Conservation status: (IUCN)** NE

3474 *Escallonia rubra* (Ruiz & Pav.) Pers. (Escalloniaceae) **IPNI:** 791101-1 **Use:** EU, ME **Region:** Andean **Dept.:** BOG**IUCN Habitat code:** 1,2,5 **Conservation status: (IUCN)** NE

Fabales

3475 *Abarema auriculata* (Benth.) Barneby & J.W.Grimes (Fabaceae) **IPNI:** 989358-1 **Habit:** Tree **Use:** HF **Origin:** Native **Region:** Amazonia **Dept.:** AMA **Elev.:** 0–150 m a.s.l. **IUCN Habitat code:** 1,5 **Conservation status: (IUCN)** LC

3476 *Abarema jupunba* (Willd.) Britton & Killip (Fabaceae)
IPNI: 469538-1
Common name: aguacerito, arenillo de monte, aserrín, barbasquillo, campanillo, capitán, carbonero, carbonero amarillo, chirrinchao, chocho, esmeralda, guacamayo blanco, llovizno, palmillo, rayo, zorro **Habit:** Shrub, Tree **Use:** EU, MA, ME, SU **Origin:** Native **Region:** Amazonia, Andean, Guiana Shield, Orinoquia, Pacific **Dept.:** AMA, ANT, CAQ, CHO, GUA, GUV, MET, NAR, PUT, SAN, VAC, VID **Elev.:** 40-1100 m.a.s.l. **IUCN Habitat code:** 1,2,3,4,5,14 **Conservation status: (IUCN)** LC **(CPSG)** Potential LC

3477 *Abarema killipii* (Britton & Rose) Barneby & J.W.Grimes (Fabaceae)
IPNI: 1096490-2
Common name: chocho azul **Habit:** Tree **Use:** HF **Origin:** Native **Region:** Andean **Dept.:** CAL, QUI, RIS, TOL, VAC **Elev.:** 1700–2500 m a.s.l. **IUCN Habitat code:** 1,3,14 **Conservation status: (IUCN)** VU

3478 *Abarema laeta* (Benth.) Barneby & J.W.Grimes (Fabaceae)
IPNI: 989371-1
Common name: fierrito **Habit:** Shrub, Small tree, Tree **Use:** HF, ME **Origin:** Native **Region:** Amazonia, Andean, Guiana Shield, Orinoquia **Dept.:** AMA, BOY, CAQ, MET, PUT, VAU **Elev.:** 200-1670 m.a.s.l. **IUCN Habitat code:** 1,3,14 **Conservation status: (IUCN)** LC **(CPSG)** CR

3479 *Abarema leucophylla* (Spruce ex Benth.) Barneby & J.W.Grimes (Fabaceae)
IPNI: 989368-1
Habit: Shrub, Small tree **Use:** HF **Origin:** Native **Region:** Guiana Shield **Dept.:** GUA, VAU **Elev.:** 75–300 m a.s.l. **IUCN Habitat code:** 1,3,4,5 **Conservation status: (IUCN)** LC

3480 *Abrus melanospermus* Hassk. (Fabaceae)
IPNI: 469600-1
Habit: Herb, Climbing **Use:** HF, ME **Origin:** Naturalised **Region:** Amazonia, Guiana Shield **Dept.:** GUV, MET, VAU **Elev.:** 290–290 m a.s.l. **Conservation status: (IUCN)** NE

3481 *Abrus precatorius* L. (Fabaceae)
IPNI: 469605-1
Common name: bejuco de peonía, brujitos, chochitas, chochito de indio, chochitos, chochitos de indio, chocho, chocho pequeño, chocho trepador, chochos, chochos de bejuco, chochos de pinta negra, jequirite, ojito de Santa Lucía, peonia, peonía, peronilla, pianida, pion **Habit:** Herb, Shrub, Climbing **Use:** AF, EU, HF, MA, ME, PO, SU **Origin:** Naturalised **Region:** Amazonia, Andean, Caribbean **Dept.:** AMA, ANT, BOL, GUV, MAG, MET, PUT, SAP, TOL, VAC, VAU **Elev.:** 5–2500 m a.s.l. **IUCN Habitat code:** 1,2,3,5,14 **Conservation status: (IUCN)** NE

3482 *Acacia baileyana* F.Muell. (Fabaceae)
IPNI: 469798-1
Common name: acacia **Habit:** Tree **Use:** AF, EU, FU, HF, IF, ME **Origin:** Cultivated, Naturalised **Region:** Andean **Dept.:** BOG, BOY, CUN **Elev.:** 2560–3050 m a.s.l. **IUCN Habitat code:** 3,5,14 **Conservation status: (IUCN)** NE

3483 *Acacia dealbata* Link (Fabaceae)
IPNI: 470130-1
Common name: acacia **Habit:** Tree **Use:** AF, EU, FU, GS, HF, IF, MA, ME **Origin:** Cultivated **Region:** Andean **Dept.:** BOG, BOY, CUN, NSA **Elev.:** 1800–3000 m a.s.l. **IUCN Habitat code:** 1,2,3,14 **Conservation status: (IUCN)** NE

3484 *Acacia decurrens* (J.C.Wendl.) Willd. (Fabaceae)
IPNI: 470138-1
Common name: acacia, acacia ceniza, acacia gris, acacia negra, uña de gato **Habit:** Tree **Use:** AF, EU, FU, GS, HF, MA, ME **Origin:** Cultivated, Naturalised **Region:** Andean **Dept.:** ANT, BOG, CAL, CUN, HUI, MAG, VAC **Elev.:** 1000–3000 m a.s.l. **IUCN Habitat code:** 1,2,3,4,5,14 **Conservation status: (IUCN)** NE

3485 *Acacia longifolia* (Andrews) Willd. (Fabaceae)
IPNI: 470782-1
Common name: acacia **Habit:** Shrub, Tree **Use:** EU, GS, HF, MA, ME **Origin:** Cultivated, Naturalised **Region:** Andean **Dept.:** BOG, CUN **Elev.:** 2600–2670 m a.s.l. **IUCN Habitat code:** 1,14 **Conservation status: (IUCN)** NE

3486 *Acacia mangium* Willd. (Fabaceae)
IPNI: 470839-1
Common name: acacio **Habit:** Tree **Use:** AF, EU, FU, HF, MA, ME **Origin:** Cultivated **Region:** Andean, Caribbean **Dept.:** ANT, LAG, QUI **Elev.:** 0–1450 m a.s.l. **IUCN Habitat code:** 1,2,3,5,14 **Conservation status: (IUCN)** LC

3487 *Acacia melanoxylon* R.Br. (Fabaceae)
IPNI: 470873-1
Common name: acacia, acacia de tierra fría, acacia gris, acacia japonesa, acacia negra **Habit:** Tree **Use:** AF, EU, FU, HF, MA, ME, PO **Origin:** Cultivated, Naturalised **Region:** Andean **Dept.:** ANT, BOG, BOY, CUN, SAN, VAC **Elev.:** 1500–3200 m a.s.l. **IUCN Habitat code:** 1,3,4,5,14 **Conservation status: (IUCN)** NE

3488 *Acacia podalyriifolia* A.Cunn. ex G.Don (Fabaceae)
IPNI: 76522-3
Use: EU, ME **Region:** Andean **Dept.:** BOG**Conservation status: (IUCN)** NE

3489 *Acacia retinodes* Schltdl. (Fabaceae)
IPNI: 471313-1
Common name: acacia, acacia alemana, acacia amarilla, acacia japonesa **Habit:** Shrub, Tree **Use:** EU, HF, MA, PO **Origin:** Naturalised **Region:** Andean **Dept.:** BOG, CUN **Elev.:** 2560–2800 m a.s.l. **IUCN Habitat code:** 14 **Conservation status: (IUCN)** NE

3490 *Acacia saligna* (Labill.) H.L.Wendl. (Fabaceae)
IPNI: 471383-1
Use: AF, EU, FU, HF, MA **Region:** Andean **Dept.:** BOG**Conservation status: (IUCN)** LC

3491 *Acacia verticillata* (L'Hér.) Willd. (Fabaceae)
IPNI: 471751-1
Common name: acacia, ambuque **Habit:** Shrub, Small tree **Use:** HF **Origin:** Naturalised **Region:** Andean **Dept.:** BOG, CUN, VAC **Elev.:** 1025–2620 m a.s.l. **IUCN Habitat code:** 1,3,14 **Conservation status: (IUCN)** NE

3492 *Acaciella angustissima* (Mill.) Britton & Rose (Fabaceae)
IPNI: 712-2
Habit: Shrub, Tree **Use:** AF, EU, HF, MA, ME **Origin:** Native **Region:** Andean **Dept.:** ANT, BOG, BOY, CAL, CAU, CHO, CUN, LAG, MAG, NAR, NSA, SAN, VAC **Elev.:** 700-2500 m.a.s.l. **IUCN Habitat code:** 1,2,3,4,14 **Conservation status: (IUCN)** LC **(CPSG)** Potential LC

3493 *Acaciella glauca* (L.) L.Rico (Fabaceae)
IPNI: 77077786-1
Habit: Tree **Use:** EU **Origin:** Native **Region:** Caribbean **Dept.:** BOG, MAG **Elev.:** 0–50 m a.s.l. **Conservation status: (IUCN)** NE

3494 *Acaciella villosa* (Sw.) Britton & Rose (Fabaceae)
IPNI: 762-2
Use: EU, ME **Origin:** Native **Conservation status: (IUCN)** LC

3495 *Acrocarpus fraxinifolius* Wight & Arn. (Fabaceae)
IPNI: 472105-1
Common name: mundani, árbol del dinero **Habit:** Tree **Use:** AF, EU, FU, HF, IF, MA, ME **Origin:** Cultivated **Region:** Andean **Dept.:** ANT, QUI, TOL **Elev.:** 1100–1400 m a.s.l. **Conservation status: (IUCN)** NE

3496 *Adenanthera pavonina* L. (Fabaceae)
IPNI: 317691-2
Common name: chocho **Habit:** Tree **Use:** AF, EU, FU, GS, HF, IF, MA, ME, PO, SU **Origin:** Cultivated **Region:** Caribbean, Pacific **Dept.:** CAU, HUI, LAG **Elev.:** 0–600 m a.s.l. **IUCN Habitat code:** 1,14 **Conservation status: (IUCN)** LC

3497 *Aeschynomene americana* L. (Fabaceae)
IPNI: 316404-2
Common name: adormidera, adormidera dulce, cadillo, cargarrocío, cujicillo, dormidera, dormidera cimarrona, dormideras, pega pega, pegapega, pulmón, rudu huequi aho, zarza **Habit:** Herb, Shrub, Climbing **Use:** AF, EU, ME **Origin:** Native **Region:** Amazonia, Andean, Caribbean, Orinoquia, Pacific **Dept.:** AMA, ANT, ARA, ATL, BOY, CAL, CAU, CHO, COR, CUN, GUV, HUI, MAG, MET, NAR, NSA, PUT, SAN, TOL, VAC **Elev.:** 15-1900 m.a.s.l. **IUCN Habitat code:** 1,2,3,4,5,14 **Conservation status: (IUCN)** NE **(CPSG)** Potential LC

3498 *Aeschynomene evenia* C.Wright (Fabaceae)
IPNI: 472720-1
Common name: cadillo **Habit:** Herb **Use:** AF **Origin:** Native **Region:** Andean, Caribbean, Orinoquia, Pacific **Dept.:** ARA, ATL, CAS, LAG, MET, NAR, RIS, VAC **Elev.:** 0–1050 m a.s.l. **IUCN Habitat code:** 2,3,4,14 **Conservation status: (IUCN)** NE

3499 *Aeschynomene rudis* Benth. (Fabaceae)
IPNI: 472933-1
Habit: Herb, Subshrub, Tree **Use:** ME **Origin:** Native **Region:** Amazonia, Andean, Orinoquia, Pacific **Dept.:** ANT, BOL, BOY, CAQ, CHO, COR, MAG, MET, NAR, SAN, TOL, VAC **Elev.:** 0-1850 m.a.s.l. **IUCN Habitat code:** 1,2,3,14 **Conservation status: (IUCN)** NE **(CPSG)** Potential LC

3500 *Aeschynomene sensitiva* Sw. (Fabaceae)
IPNI: 472949-1
Habit: Herb, Shrub, Tree **Use:** EU, MA, ME **Origin:** Native **Region:** Amazonia, Andean, Orinoquia, Pacific **Dept.:** AMA, ANT, ARA, BOL, CAU, CHO, COR, MAG, MET, NAR, PUT, QUI, SAN, TOL, VAC **Elev.:** 20-2000 m.a.s.l. **IUCN Habitat code:** 1,2,3,4,5,14 **Conservation status: (IUCN)** NE **(CPSG)** Potential LC

3501 *Aeschynomene villosa* Poir. (Fabaceae)
IPNI: 6008-2
Habit: Herb, Subshrub **Use:** AF,ME **Origin:** Native **Region:** Andean, Pacific **Dept.:** ANT, CAU, CHO, CUN, NAR, VAC **Elev.:** 420–1500 m a.s.l. **IUCN Habitat code:** 1,3,14 **Conservation status: (IUCN)** NE

3502 *Albizia carbonaria* Britton (Fabaceae)
IPNI: 8016-2
Common name: baranoa, barboquillo, bayeto, buche, capitán, carbonero, carbonero blanco, carbonero de sombrío, carbonero gigante, dormilón, galapo, gigante, guacamayo, guamuche, guango muche, jalapo, mataguacuco, muche, mucheblanco, pisquín **Habit:** Tree **Use:** AF, EU, FU, MA, ME **Origin:** Native **Region:** Andean, Caribbean, Pacific **Dept.:** ANT, CAL, CHO, CUN, MAG, NSA, SAP, TOL, VAC **Elev.:** 10–1800 m a.s.l. **IUCN Habitat code:** 1,2,3,4,5,14 **Conservation status: (IUCN)** LC

3503 *Albizia lebbeck* (L.) Benth. (Fabaceae)
IPNI: 99109-3
Common name: bayeto, carbonero, carbonero de sombrío, carbonero dormilón, carita, dormilón, guamuche, guango, muche, mucheblanco, pisquín **Habit:** Tree **Use:** AF, EU, FU, HF, MA, ME, PO **Origin:** Cultivated, Naturalised **Region:** Andean, Caribbean, Orinoquia **Dept.:** ANT, ATL, COR, LAG, MAG, MET **Elev.:** 0–1000 m a.s.l. **IUCN Habitat code:** 1,2,3,14 **Conservation status: (IUCN)** LC

3504 *Albizia multiflora* (Kunth) Barneby & J.W.Grimes (Fabaceae)
IPNI: 989448-1
Habit: Tree **Use:** MA **Origin:** Native **Region:** Andean **Elev.:** 0–1800 m a.s.l. **IUCN Habitat code:** 5 **Conservation status: (IUCN)** NE

3505 *Albizia niopoides* (Spruce ex Benth.) Burkart (Fabaceae)
IPNI: 8032-2
Common name: bayeto, guacamayo, jarijana **Habit:** Tree **Use:** AF, EU, FU, MA, ME **Origin:** Native **Region:** Caribbean **Dept.:** ATL, BOL, LAG, MAG **Elev.:** 0–150 m a.s.l. **IUCN Habitat code:** 1,2,3,14 **Conservation status: (IUCN)** LC

3506 *Aldina heterophylla* Benth. (Fabaceae)
IPNI: 473437-1
Common name: bujeque, sangretoro **Use:** ME **IUCN Habitat code:** 1,4 **Conservation status: (IUCN)** NE

3507 *Aldina latifolia* Benth. (Fabaceae)
IPNI: 473440-1
Common name: palo de sangre **Habit:** Small tree, Tree **Use:** HF **Origin:** Native **Region:** Amazonia, Guiana Shield **Dept.:** GUA, VAU **Elev.:** 120–250 m a.s.l. **IUCN Habitat code:** 1 **Conservation status: (IUCN)** LC

3508 *Alysicarpus vaginalis* (L.) DC. (Fabaceae)
IPNI: 473569-1
Common name: orejita de ratón **Habit:** Herb **Use:** AF, EU, HF, MA, ME **Origin:** Naturalised **Region:** Andean, Caribbean, Pacific **Dept.:** ANT, ATL, BOL, COR, MAG, NAR, SAN, VAC **Elev.:** 0–1000 m a.s.l. **IUCN Habitat code:** 1,2,5,14 **Conservation status: (IUCN)** NE

3509 *Anadenanthera colubrina* (Vell.) Brenan (Fabaceae)
IPNI: 12264-2
Common name: nopo **Use:** ME, SU **Conservation status: (IUCN)** LC

3510 *Anadenanthera peregrina* (L.) Speg. (Fabaceae)
IPNI: 12268-2
Common name: carripaco, cohoba, cojoba, dopa, dopamagerná, dopanae, dopane, guahibo, lomo de caimán, niopo, yompa, yopa, yopo, yoto, yumpa, yupa **Habit:** Tree **Use:** EU, MA, ME, PO, SU **Origin:** Native **Region:** Andean, Guiana Shield, Orinoquia **Dept.:** ARA, CAL, CAS, MET, TOL, VID **Elev.:** 100–770 m a.s.l. **IUCN Habitat code:** 1,3,4,14 **Conservation status: (IUCN)** LC

3511 *Ancistrotropis peduncularis* (Fawc. & Rendle) A.Delgado (Fabaceae)
IPNI: 60457232-2
Common name: frijolillo, frijolillo nativo, zaragoza **Habit:** Herb, Climbing **Use:** ME **Origin:** Native **Region:** Amazonia, Andean, Caribbean, Orinoquia, Pacific **Dept.:** ANT, BOL, CAL, CAQ, CAU, CHO, CUN, MAG, MET, NAR, SAN, VAU, VID **Elev.:** 0-2000 m.a.s.l. **IUCN Habitat code:** 1,2,3,4,14 **Conservation status: (IUCN)** NE **(CPSG)** Potential LC

3512 *Andira inermis* (W.Wright) DC. (Fabaceae)
IPNI: 1075415-2
Common name: ajunado, ají, almendrillo, amargo, angelina, angelino, arroz, arroz con coco, avinge, bagatá, bolombolo, bongo de agua, cabo de oro, cacaco, cacho cabra, chirai, ciruelo de playa, coco, cocu, congo, congo de agua, conguito, culnicre, dividive, frito, guax **Habit:** Shrub, Small tree, Tree **Use:** AF, EU, FU, GS, HF, IF, MA, ME, PO, SU **Origin:** Native **Region:** Amazonia, Andean, Caribbean, Orinoquia, Pacific **Dept.:** AMA, ANT, CAL, CAQ, CAU, CHO, COR, GUA, LAG, MAG, MET, NAR, QUI, SAN, SUC, TOL, VAC **Elev.:** 0-2000 m.a.s.l. **IUCN Habitat code:** 1,2,3,5,14 **Conservation status: (IUCN)** LC **(CPSG)** Potential LC

3513 *Andira surinamensis* (Bondt) Splitg. ex Pulle (Fabaceae)
IPNI: 12607-2
Common name: arenoso, chochos, congo **Habit:** Small tree, Tree **Use:** MA, ME **Origin:** Native **Region:** Amazonia, Guiana Shield, Orinoquia **Dept.:** AMA, GUV, MET, VAU, VID **Elev.:** 200–220 m a.s.l. **IUCN Habitat code:** 1,3,4,5,14 **Conservation status: (IUCN)** LC

3514 *Andira taurotesticulata* R.T.Penn. (Fabaceae)
IPNI: 20012192-1
Common name: almanegra, almendrillo, amargo, cañabravo, coco de monte, palo pilón, peloto, turmemono **Habit:** Small tree, Tree **Use:** AF, EU, MA, ME **Origin:** Native **Region:** Andean **Dept.:** ANT, BOY, CUN, MET, QUI, SAN **Elev.:** 500–2100 m a.s.l. **IUCN Habitat code:** 1,3,5,14 **Conservation status: (IUCN)** LC

3515 *Apuleia leiocarpa* (Vogel) J.F.Macbr. (Fabaceae)
IPNI: 16766-2
Common name: cobre, guacamayo, guayacán, jero, maqui, naune, tigrito **Habit:** Tree **Use:** HF, MA, ME **Origin:** Native **Region:** Amazonia, Guiana Shield, Caribbean, Orinoquia **Dept.:** AMA, ARA, BOL, GUV, MET, VID **Elev.:** 20–200 m a.s.l. **IUCN Habitat code:** 1,3,4,5,14 **Conservation status: (IUCN)** LC

3516 *Arachis hypogaea* L. (Fabaceae)
IPNI: 318562-2
Common name: cacahuate, cacahuete, maní, maní blanco, maní blanco de picón, maní de picón, maní rojo, planta de maní, tiga, yacana **Habit:** Herb, Climbing **Use:** AF, EU, FU, GS, HF, IF, MA, ME, PO, SU **Origin:** Cultivated, Naturalised **Region:** Amazonia, Andean **Dept.:** AMA, ANT, CAL, CAQ, CAU, CES, CUN, PUT **Elev.:** 280–1850 m a.s.l. **IUCN Habitat code:** 1,3,5,14 **Conservation status: (IUCN)** NE

3517 *Arachis pintoi* Krapov. & W.C.Greg. (Fabaceae)
IPNI: 318518-2
Common name: maní, maní de huerta, maní forrajero **Habit:** Herb **Use:** AF, EU, GS **Origin:** Naturalised **Region:** Orinoquia **Dept.:** BOG, GUV, PUT **Elev.:** 200–200 m a.s.l. **IUCN Habitat code:** 1,2,3,4,14 **Conservation status: (IUCN)** NE

3518 *Arachis prostrata* Benth. (Fabaceae)
IPNI: 474941-1
Habit: Herb **Use:** GS **Origin:** Cultivated **Dept.:** VAC **Elev.:** 900–900 m a.s.l. **IUCN Habitat code:** 3 **Conservation status: (IUCN)** NE

3519 *Arachis repens* Handro (Fabaceae)
IPNI: 17315-2
Habit: Herb **Use:** EU, GS, HF **Origin:** Cultivated **Dept.:** VAC **Elev.:** 1000–1000 m a.s.l. **IUCN Habitat code:** 14 **Conservation status: (IUCN)** NE

3520 *Balizia pedicellaris* (DC.) Barneby & J.W.Grimes (Fabaceae)
IPNI: 989349-1
Habit: Tree **Use:** MA **Origin:** Native **Region:** Guiana Shield, Pacific **Dept.:** CAQ, VAU **Elev.:** 0–800 m a.s.l. **IUCN Habitat code:** 1,14 **Conservation status: (IUCN)** NE

3521 *Bauhinia aculeata* L. (Fabaceae)
IPNI: 316399-2
Common name: casco de vaca, pata de vaca, patevaca **Habit:** Shrub, Tree **Use:** MA, ME **Origin:** Native **Region:** Caribbean **Dept.:** ATL, BOL, MAG, NSA, SAN, SUC **Elev.:** 0–350 m a.s.l. **IUCN Habitat code:** 1,2,14 **Conservation status: (IUCN)** LC

3522 *Bauhinia forficata* Link (Fabaceae)
IPNI: 481315-1
Use: HF, MA, ME **Conservation status: (IUCN)** LC

3523 *Bauhinia longicuspis* Spruce ex Benth. (Fabaceae)
IPNI: 30050541-2
Common name: pezuño **Habit:** Shrub, Tree, Climbing **Use:** ME **Origin:** Native **Region:** Amazonia, Andean, Guiana Shield, Orinoquia **Dept.:** AMA, CUN, GUV, MET, SAN, VAU **Elev.:** 120–600 m a.s.l. **IUCN**

Habitat code: 1,3,4,5,14 **Conservation status: (IUCN)** LC

3524 *Bauhinia longifolia* (Bong.) Steud. (Fabaceae)
IPNI: 481456-1
Use: ME **Origin:** Native **Conservation status: (IUCN)** LC

3525 *Bauhinia monandra* Kurz (Fabaceae)
IPNI: 30129-2
Common name: casco de buey, casco de vaca, cocla, pata de vaca, patebuey **Habit:** Shrub, Tree **Use:** EU, HF, MA, ME **Origin:** Cultivated, Naturalised **Region:** Andean, Caribbean, Pacific **Dept.:** ANT, CHO, MAG, VAC **Elev.:** 0–1550 m a.s.l. **IUCN Habitat code:** 1,3,14 **Conservation status: (IUCN)** DD

3526 *Bauhinia picta* (Kunth) DC. (Fabaceae)
IPNI: 481565-1
Common name: bejuco escalero, cairo, casco de buey, casco de vaca, cascoevaca, cocla, pata de buey, pata de vaca, patebuey, patevaca, pat'evaca, urrapia **Habit:** Tree **Use:** EU, ME **Origin:** Native, Cultivated **Region:** Andean **Dept.:** ANT, BOG, BOY, CAL, CAU, CHO, CUN, HUI, QUI, SAN, VAC **Elev.:** 100–1700 m a.s.l. **IUCN Habitat code:** 1,2,3,4,14 **Conservation status: (IUCN)** LC

3527 *Bauhinia purpurea* L. (Fabaceae)
IPNI: 481605-1
Common name: casco de buey, casco de vaca, patebuey, patevaca, pate'vaca, pie de cabra **Use:** AF, EU, FU, GS, HF, MA, ME **IUCN Habitat code:** 1,2,3,14 **Conservation status: (IUCN)** LC

3528 *Bauhinia tarapotensis* Benth. (Fabaceae)
IPNI: 481718-1
Common name: bejuco cadeno, casco de vaca, escalero, pata de vaca, patevaca, zanco de mula **Habit:** Shrub, Tree **Use:** ME **Origin:** Native **Region:** Amazonia, Guiana Shield, Pacific **Dept.:** AMA, CAQ, CAU, CHO, GUV, MET, PUT **Elev.:** 200–600 m a.s.l. **IUCN Habitat code:** 1,3,5,14 **Conservation status: (IUCN)** LC

3529 *Bauhinia ungulata* L. (Fabaceae)
IPNI: 316241-2
Common name: pata de vaca, patevaca **Habit:** Shrub, Tree **Use:** FU, MA, ME **Origin:** Native **Region:** Amazonia, Andean, Caribbean, Orinoquia, Pacific **Dept.:** AMA, CAQ, CES, CHO, COR, MAG, MET, SAN, VID **Elev.:** 50–900 m a.s.l. **IUCN Habitat code:** 1,2,3,4,5,14 **Conservation status: (IUCN)** LC

3530 *Bauhinia variegata* L. (Fabaceae)
IPNI: 481756-1
Common name: casco de buey, casco de vaca, flor rosada, pata de buey, pata de vaca, patevaca, árbol orquídea **Habit:** Shrub, Tree **Use:** AF, EU, FU, GS, HF, MA, ME, PO **Origin:** Cultivated, Naturalised **Region:** Andean **Dept.:** ANT, BOG, BOY, CAL, CAU, CUN, NAR, VAC **Elev.:** 1025–2400 m a.s.l. **IUCN Habitat code:** 1,2,3,14 **Conservation status: (IUCN)** LC

3531 *Biancaea decapetala* (Roth) O.Deg. (Fabaceae)
IPNI: 481887-1
Habit: Shrub **Use:** EU, HF, MA, ME **Origin:** Cultivated **Region:** Andean **Dept.:** CAU **Elev.:** 1580–1580 m a.s.l. **IUCN Habitat code:** 1,3 **Conservation status: (IUCN)** LC

3532 *Bowdichia virgilioides* Kunth (Fabaceae)
IPNI: 482144-1
Common name: alcornoco, alcornoque, alcornoque sabanero, algarrobo, capote, cornoco, mandingo, omonoes, sucupira, yatomay, yatonay, yatornay, yatorona **Habit:** Shrub, Tree **Use:** MA, ME **Origin:** Native **Region:** Amazonia, Andean, Guiana Shield, Orinoquia **Dept.:** BOY, CAS, HUI, MET, VID **Elev.:** 90–1120 m a.s.l. **IUCN Habitat code:** 1,2,3,4,5,14 **Conservation status: (IUCN)** LC

3533 *Brownea ariza* Benth. (Fabaceae)
IPNI: 482406-1
Common name: amapolo, arizal, arizá, clavellino, flor de la cruz, florisanto, monterillo, palo cruz, palo de candela, palo de cruz, palo de la cruz, palo de rosa, palo rosa, palocruz, palocruz hembra, palosanto, rosa de cruz, rosa de la cruz, rosa de monte, roso de mo **Habit:** Shrub, Tree **Use:** EU, HF, MA, ME **Origin:** Native, Cultivated **Region:** Amazonia, Andean, Guiana Shield, Caribbean, Orinoquia, Pacific **Dept.:** AMA, ANT, ARA, BOY, CAQ, CAS, CUN, GUV, HUI, MAG, MET, NSA, PUT, SUC, TOL, VAC, VAU **Elev.:** 0–1700 m a.s.l. **IUCN Habitat code:** 1,2,3,4,5,14 **Conservation status: (IUCN)** NE

3534 *Brownea coccinea* Jacq. (Fabaceae)
IPNI: 482413-1
Common name: arizá, palo de cruz, roso **Habit:** Shrub, Tree **Use:** EU, MA, ME **Origin:** Native, Cultivated **Region:** Amazonia, Andean, Guiana Shield, Caribbean, Orinoquia, Pacific **Dept.:** AMA, CAS, GUA, NAR, NSA, SAN, VAU **Elev.:** 50–700 m a.s.l. **IUCN Habitat code:** 1,2,3,5,14 **Conservation status: (IUCN)** LC

3535 *Brownea grandiceps* Jacq. (Fabaceae)
IPNI: 482418-1
Common name: anisillo, ariza, arizá, caipe, palo de candela, palo de cruz, palo rosa, palocruz, rosa de cruz, rosa de montaña, rosa de venezuela, roso de monte **Habit:** Small tree, Tree **Use:** EU, MA, ME, SU **Origin:** Native, Cultivated **Region:** Amazonia, Andean, Guiana Shield, Orinoquia, Pacific **Dept.:** AMA, ANT, CAQ, CHO, CUN, HUI, MET, PUT, TOL, VAC, VAU, VID **Elev.:** 40–1570 m a.s.l. **Conservation status: (IUCN)** LC

3536 *Brownea longipedicellata* Huber (Fabaceae)
IPNI: 37410-2
Common name: ají de barbasco, palo de cruz **Habit:** Tree **Use:** ME **Origin:** Native **Region:** Amazonia **Dept.:** AMA, CAQ **Elev.:** 250–250 m a.s.l. **IUCN Habitat code:** 1,3 **Conservation status: (IUCN)** NE

3537 *Brownea macrophylla* Linden ex Mast. (Fabaceae)
IPNI: 482431-1
Common name: arizal, arizá, cruceto, palo de cruz, palocruz **Habit:** Small tree, Tree **Use:** EU, MA, ME **Origin:** Native **Region:** Amazonia, Andean, Caribbean, Pacific **Dept.:** AMA, ANT, CAQ, CAS, CHO, CUN, MAG, TOL **Elev.:** 20–1350 m a.s.l. **IUCN Habitat code:** 1,2,3,5,14 **Conservation status: (IUCN)** LC

3538 *Brownea multijuga* Britton & Killip (Fabaceae)
IPNI: 37413-2
Common name: clavellino **Habit:** Tree **Use:** ME **Origin:** Native **Region:** Caribbean, Pacific **Dept.:** ANT, CHO, SAN **Elev.:** 40–500 m a.s.l. **IUCN Habitat code:** 1,3,14 **Conservation status: (IUCN)** NE

3539 *Brownea negrensis* Benth. (Fabaceae)
IPNI: 482433-1
Common name: arizá amapolo, cruceto, palo cruz, palo de cruz, palocruz, árbol de la cruz **Habit:** Small tree, Tree **Use:** ME **Origin:** Native **Region:** Amazonia, Guiana Shield, Orinoquia **Dept.:** AMA, ANT, CAQ, GUA, GUV, MET, PUT, VAU **Elev.:** 150–500 m a.s.l. **IUCN Habitat code:** 1,3,4,14 **Conservation status: (IUCN)** NE

3540 *Brownea rosa-de-monte* P.J.Bergius (Fabaceae)
IPNI: 482440-1
Common name: arisa, ariza, arizá, clavellino, florisanto, garrapato, palo cruz, palo de cruz, palocruz, palocruz macho **Habit:** Shrub, Tree **Use:** HF, ME **Origin:** Native **Region:** Amazonia, Caribbean, Orinoquia, Pacific **Dept.:** ANT, CAQ, CES, CHO, COR, CUN, LAG, MAG, MET, NSA, SAN, TOL, VAC **Elev.:** 0-1100 m.a.s.l. **Conservation status: (IUCN)** LC **(CPSG)** Potential LC

3541 *Brownea similis* R.S.Cowan (Fabaceae)
IPNI: 37420-2
Common name: maveci, palo cruz **Habit:** Tree **Use:** ME **Origin:** Native **Region:** Amazonia **Dept.:** AMA, VAU **Elev.:** 150–250 m a.s.l. **IUCN Habitat code:** 1,3 **Conservation status: (IUCN)** NE

3542 *Brownea stenantha* Britton & Killip (Fabaceae)
IPNI: 37422-2
Common name: arizal, arizá, clavellino, florisanto **Habit:** Tree **Use:** ME **Origin:** Endemic **Region:** Caribbean, Pacific **Dept.:** ANT, BOL, CHO, COR, MAG, SUC **Elev.:** 0-500 m.a.s.l. **IUCN Habitat code:** 1,2,3,5,14 **Conservation status: (IUCN)** LC **(CPSG)** LC

3543 *Browneopsis excelsa* Pittier (Fabaceae)
IPNI: 37425-2
Common name: cruceto, palo de cruz, palocruz **Habit:** Tree **Use:** MA **Origin:** Native **Region:** Pacific **Dept.:** ANT, CHO **Elev.:** 30–450 m a.s.l. **IUCN Habitat code:** 1,3,14 **Conservation status: (IUCN)** VU

3544 *Caesalpinia cassioides* Willd. (Fabaceae)
IPNI: 482664-1
Common name: clavellino **Habit:** Shrub **Use:** MA **Origin:** Native **Region:** Andean **Dept.:** CAU, HUI, VAC **Elev.:** 600–900 m a.s.l. **IUCN Habitat code:** 1,2,3,5,14 **Conservation status: (IUCN)** LC

3545 *Caesalpinia pulcherrima* (L.) Sw. (Fabaceae)
IPNI: 1164658-2
Common name: acacia, acacia de jardín, acacio, angelita, angelito, angelito clavellino, bellísima, bermudera, cabellos de ángel, camarón, clavelillo, clavelino, clavellina, clavellino, dividivi, flor de muerto, flor de pavo, flor de ángel, florito, gallito, guacamaya, **Habit:** Shrub, Tree **Use:** EU, GS, HF, MA, ME, PO **Origin:** Cultivated, Naturalised **Region:** Amazonia, Andean, Guiana Shield, Caribbean, Orinoquia, Pacific **Dept.:** AMA, ANT, ATL, BOL, BOY, CAL, CAS, CES, CHO, COR, CUN, GUV, LAG, MAG, MET, PUT, SAP, SAN, TOL, VAC, VID **Elev.:** 0–1700 m a.s.l. **IUCN Habitat code:** 1,2,3,5,14 **Conservation status: (IUCN)** LC

3546 *Cajanus cajan* (L.) Huth (Fabaceae)
IPNI: 1152177-2
Common name: cascabelillos, chinchoncho, choncho, frisol de palo, frisol guandús, frĭjol guandul, frĭjol guanduz, frĭjol guandó, frĭjol guandú, frĭjol guandús, frĭjol quinchancho, frĭjol quinchoncho, gondu, guandul, guandumbo, guanduz, guandú, guandús, gungo, quinchon **Habit:** Shrub **Use:** AF, EU, FU, GS, HF, IF, MA, ME, PO **Origin:** Cultivated, Naturalised **Region:** Andean, Caribbean **Dept.:** ANT, CAL, CAU, CUN, LAG, MAG, NAR, NSA, QUI, SAN, TOL **Elev.:** 50–2000 m a.s.l. **IUCN Habitat code:** 1,2,3,14 **Conservation status: (IUCN)** NE

3547 *Calliandra angustifolia* Spruce ex Benth. (Fabaceae)
IPNI: 42066-2
Common name: carbón **Habit:** Shrub, Tree **Use:** ME **Origin:** Native **Region:** Amazonia, Andean, Pacific **Dept.:** AMA, CAU, CHO, NAR, PUT, VAC **Elev.:** 40–2100 m a.s.l. **IUCN Habitat code:** 1,14 **Conservation status: (IUCN)** LC

3548 *Calliandra falcata* Benth. (Fabaceae)
IPNI: 42154-2
Habit: Shrub **Use:** EU **Origin:** Native, Cultivated **Region:** Andean, Orinoquia **Dept.:** MET, VAC **Elev.:** 500–1080 m a.s.l. **Conservation status: (IUCN)** NE

3549 *Calliandra guildingii* Benth. (Fabaceae)
IPNI: 42179-2
Common name: guamilla **Habit:** Tree **Use:** ME **Origin:** Native **Region:** Amazonia, Andean, Guiana Shield **Dept.:** AMA, ANT **Elev.:** 400–1850 m a.s.l. **IUCN Habitat code:** 1,3,14 **Conservation status: (IUCN)** LC

3550 *Calliandra haematocephala* Hassk. (Fabaceae)
IPNI: 42180-2
Common name: carbonero, clavellino **Habit:** Shrub, Small tree **Use:** EU, ME **Origin:** Cultivated **Region:** Andean, Caribbean, Orinoquia, Pacific **Dept.:** ANT, BOG, CAU, CUN, HUI, MAG, MET, PUT, TOL, VAC **Elev.:** 500–1600 m a.s.l. **IUCN Habitat code:** 1,2,3,5,14 **Conservation status: (IUCN)** NE

3551 *Calliandra laxa* (Willd.) Benth. (Fabaceae)
IPNI: 1161746-2
Habit: Tree **Use:** MA **Origin:** Native **Region:** Guiana Shield, Pacific **Dept.:** CHO, VAU **Elev.:** 200–250 m a.s.l. **IUCN Habitat code:** 1 **Conservation status: (IUCN)** LC

3552 *Calliandra magdalenae* (Bertero ex DC.) Benth. (Fabaceae)
IPNI: 1051245-2
Common name: carbonero, rayado **Habit:** Shrub, Tree **Use:** ME, SU **Origin:** Native **Region:** Andean, Caribbean **Dept.:** ANT, BOL, CAL, CES, CUN, HUI, LAG, MAG, SAN, TOL **Elev.:** 200–1550 m a.s.l. **IUCN Habitat code:** 1,2,3,14 **Conservation status: (IUCN)** LC

3553 *Calliandra pittieri* Standl. (Fabaceae)
IPNI: 42259-2
Common name: carbonerito, carbonero, carbonero morado, carbonero rojo, clavellino, quebrajacho, quebrajancho, quiebracho, quiebrajacho **Habit:** Shrub, Tree **Use:** ME **Origin:** Native **Region:** Andean, Guiana Shield, Caribbean, Pacific **Dept.:** ANT, ATL, BOG, BOY, CAL, CAQ, CAU, CHO, COR, CUN, HUI, NAR, RIS, SAN, TOL, VAC **Elev.:** 100-2600 m.a.s.l. **IUCN Habitat code:** 1,2,3,14 **Conservation status: (IUCN)** LC **(CPSG)** Potential LC

3554 *Calliandra purdiei* Benth. (Fabaceae)
IPNI: 42269-2
Common name: carbonero, carbonero morado, carbón, churimo, florito, pichindecillo, quebrajacho **Habit:** Shrub, Small tree **Use:** MA, ME **Origin:** Native **Region:** Andean, Caribbean, Orinoquia, Pacific **Dept.:** ANT, BOG, CES, CUN, MET, NSA, QUI, SAN, TOL, VAC **Elev.:** 40–1600 m a.s.l. **IUCN Habitat code:** 1,2,3,14 **Conservation status: (IUCN)** NE

3555 *Calliandra riparia* Pittier (Fabaceae)
IPNI: 483259-1
Common name: carbonero, dormidero **Habit:** Shrub, Small tree **Use:** EU **Origin:** Native, Cultivated **Region:** Andean, Caribbean, Orinoquia, Pacific **Dept.:** ANT, ATL, BOL, CES, CHO, COR, CUN, HUI, LAG, MAG, MET, NSA, TOL **Elev.:** 5-1330 m.a.s.l. **IUCN Habitat code:** 1,2,3,4,14 **Conservation status: (IUCN)** LC **(CPSG)** Potential LC

3556 *Calliandra surinamensis* Benth. (Fabaceae)
IPNI: 316281-2
Common name: acacia, clavellino, dormidero, dormilón **Habit:** Shrub, Tree **Use:** ME **Origin:** Native **Region:** Amazonia, Andean, Guiana Shield, Caribbean, Orinoquia **Dept.:** ANT, BOY, CAQ, CAS, CUN, LAG, MET, NSA, PUT, VAU **Elev.:** 250–1780 m a.s.l. **IUCN Habitat code:** 1,3,4,14 **Conservation status: (IUCN)** LC

3557 *Calliandra trinervia* Benth. (Fabaceae)
IPNI: 42330-2
Common name: carbonero bayeto, carbonero rojo, casco de vaca **Habit:** Shrub, Small tree **Use:** ME **Origin:** Native **Region:** Amazonia, Andean, Guiana Shield **Dept.:** AMA, ANT, BOG, CAL, CAQ, CAU, CUN, PUT, VAC, VAU **Elev.:** 120–2200 m a.s.l. **IUCN Habitat code:** 1,3,14 **Conservation status: (IUCN)** LC

3558 *Calopogonium caeruleum* (Benth.) C.Wright (Fabaceae)
IPNI: 60471356-2
Common name: bejuco de carnero, calopo, caracucha, frijolito azul, frĭjol de tubérculo **Habit:** Climbing **Use:** HF **Origin:** Native **Region:** Amazonia, Andean, Caribbean, Orinoquia, Pacific **Dept.:** AMA, ANT, ARA, ATL, BOL, CAU, CHO, COR, CUN, HUI, MAG, MET, TOL, VAC **Elev.:** 0-1700 m.a.s.l. **IUCN Habitat code:** 1,2,3,4,5,14 **Conservation status: (IUCN)** NE **(CPSG)** Potential LC

3559 *Calopogonium mucunoides* Desv. (Fabaceae)
IPNI: 483437-1
Common name: bejuco de ortiga, calopo, kudzú, rabo de iguana **Habit:** Liana, Climbing **Use:** AF, EU, HF, ME **Origin:** Native **Region:** Amazonia, Andean, Caribbean, Orinoquia, Pacific **Dept.:** AMA, ANT, ARA, ATL, CAL, CAS, CAU, CHO, COR, CUN, GUA, MAG, MET, NAR, PUT, RIS, SAP, SAN, TOL, VAC, VID **Elev.:** 0-2500 m.a.s.l. **IUCN Habitat code:** 1,2,3,4,5,14 **Conservation status: (IUCN)** NE **(CPSG)** Potential LC

3560 *Campsiandra angustifolia* Spruce ex Benth. (Fabaceae)
IPNI: 483500-1
Common name: amanague, guacapurano, huacapurana, lavallaga **Habit:** Tree **Use:** HF, ME **Origin:** Native **Region:** Amazonia, Guiana Shield, Caribbean **Dept.:** AMA, GUA, VAU **Elev.:** 140–180 m a.s.l. **IUCN Habitat code:** 1,3,5,14 **Conservation status: (IUCN)** LC

3561 *Campsiandra comosa* Benth. (Fabaceae)
IPNI: 307054-2
Common name: chigo, huacapurana, lavallaga, palo de bufeo, árbol de orejas **Habit:** Tree **Use:** HF, MA, ME **Origin:** Native **Region:** Amazonia, Orinoquia **Dept.:** AMA, GUA, VAU, VID **Elev.:** 7–265 m a.s.l. **IUCN Habitat code:** 1,3,4,5 **Conservation status: (IUCN)** LC

3562 *Campsiandra implexicaulis* Stergios (Fabaceae)
IPNI: 307062-2
Habit: Tree **Use:** HF **Origin:** Native **Region:** Amazonia, Orinoquia **Dept.:** VAU, VID**IUCN Habitat code:** 1,4,5 **Conservation status: (IUCN)** NE

3563 *Campsiandra laurifolia* Benth. (Fabaceae)
IPNI: 44114-2
Use: ME **Origin:** Native **IUCN Habitat code:** 1,5 **Conservation status: (IUCN)** NE

3564 *Campsiandra steyermarkiana* Stergios (Fabaceae)
IPNI: 307067-2
Habit: Small tree **Use:** EU **Origin:** Native, Cultivated **Region:** Amazonia, Orinoquia **Dept.:** AMA, VID**Conservation status: (IUCN)** NE

3565 *Canavalia brasiliensis* Mart. ex Benth. (Fabaceae)
IPNI: 483623-1
Common name: frijolón, frĭjol machete, haba **Habit:** Liana, Climbing **Use:** HF, ME **Origin:** Native **Region:** Amazonia, Andean, Caribbean, Pacific **Dept.:** AMA, ANT, ATL, BOL, CHO, CUN, LAG, MAG, SAN, SUC, VAC **Elev.:** 0–1800 m a.s.l. **IUCN Habitat code:** 1,2,3,5,14 **Conservation status: (IUCN)** NE

3566 *Canavalia ensiformis* (L.) DC. (Fabaceae)
IPNI: 1209500-2
Common name: canavalia, canavalia Saldaña, frijol blanco, frijolón, frisol, frĭjol blanco, frĭjol burro, frĭjol cacao, frĭjol canavalia, frĭjol de playa, frĭjol de puerco, frĭjol extranjero, frĭjol guamo, frĭjol machete, frĭjol mataarriera, frĭjol morro, frĭjol veneno **Habit:** Climbing **Use:** AF, EU, HF, MA, ME, PO **Origin:** Native, Cultivated **Region:** Andean, Orinoquia **Dept.:** ANT, CAL, CUN, MET, PUT, QUI, VAC **Elev.:** 450–1400 m a.s.l. **IUCN Habitat code:** 1,3,14 **Conservation status: (IUCN)** NE

3567 *Canavalia rosea* (Sw.) DC. (Fabaceae)
IPNI: 30187778-2
Common name: frĭjol **Habit:** Herb, Liana, Climbing **Use:** AF, EU, HF, MA, ME, PO, SU **Origin:** Native **Region:** Caribbean, Pacific **Dept.:** ANT, ATL, BOL, CAU, CHO, LAG, MAG, NAR, SAP **Elev.:** 0-200 m.a.s.l. **IUCN Habitat code:** 1,2,3,5,14 **Conservation status: (IUCN)** NE **(CPSG)** Potential LC

3568 *Cassia fistula* L. (Fabaceae)
IPNI: 484507-1
Common name: acacia, acacia lluvia de oro, cañafístola, cañafístula, cañafístulo, cañandonga, cañandunga, chorro de oro, floramarillo, golden shower, lluvia de oro, lluvia de verano, matarratón extranjero **Habit:** Tree **Use:** AF, EU, FU, HF, MA, ME, PO, SU **Origin:** Cultivated, Naturalised **Region:** Amazonia, Andean, Caribbean, Orinoquia **Dept.:** AMA, ANT, ATL, CAS, COR, CUN, MAG, TOL, VAC **Elev.:** 0–1470 m a.s.l. **IUCN Habitat code:** 1,2,3,5,14 **Conservation status: (IUCN)** LC

3569 *Cassia grandis* L.f. (Fabaceae)
IPNI: 484607-1
Common name: capote, carao, cayandonga, cañadonga, cañafisto, cañafistula, cañaflota, cañaflote, cañafístol, cañafístola, cañafístola delgada, cañafístola gruesa, cañafístula, cañafístula burrero, cañafístula delgada, cañafístula gruesa, cañafístulo, cañafístulo burre **Habit:** Tree **Use:** AF, EU, FU, HF, MA, ME, SU **Origin:** Native, Cultivated **Region:** Guiana Shield, Caribbean, Orinoquia, Pacific **Dept.:** ANT, BOL, CAS, CAU, CHO, GUV, LAG, MAG, MET, TOL, VAC **Elev.:** 10–1200 m a.s.l. **IUCN Habitat code:** 1,2,3,4,14 **Conservation status: (IUCN)** LC

3570 *Cassia javanica* L. (Fabaceae)
IPNI: 484704-1
Common name: acacia brasileña, acacia rosada, casia de Java, casia rosada, cañafístulo macho **Habit:** Tree **Use:** EU, HF, MA, ME **Origin:** Cultivated, Naturalised **Region:** Andean, Caribbean **Dept.:** COR, SAN **Elev.:** 100–1000 m a.s.l. **IUCN Habitat code:** 1,3,14 **Conservation status: (IUCN)** LC

3571 *Cassia leiandra* Benth. (Fabaceae)
IPNI: 48715-2
Common name: cañafístol, cañafístola, cañafístole, cañafístolo, cañafístula, cañafístula delgada, cañafístula pequeña, cañafístulo, cañafístulo amarillo, cañafístulo pequeño, sietecueros **Habit:** Shrub, Tree **Use:** HF, ME **Origin:** Native **Region:** Caribbean, Orinoquia **Dept.:** ANT, VID **Elev.:** 50–100 m a.s.l. **IUCN Habitat code:** 1,3,4,14 **Conservation status: (IUCN)** LC

3572 *Cassia moschata* Kunth (Fabaceae)
IPNI: 484879-1
Habit: Tree **Use:** AF, HF, MA, ME **Origin:** Native **Region:** Andean, Caribbean, Orinoquia **Dept.:** ANT, BOL, BOY, CAS, COR, CUN, MET, VID **Elev.:** 50–1200 m a.s.l. **IUCN Habitat code:** 1,2,3,4,5,14 **Conservation status: (IUCN)** LC

3573 *Cassia spruceana* Benth. (Fabaceae)
IPNI: 485255-1
Use: ME **Origin:** Native **IUCN Habitat code:** 1 **Conservation status: (IUCN)** LC

3574 *Cedrelinga cateniformis* (Ducke) Ducke (Fabaceae)
IPNI: 51040-2
Common name: acahapo rosado, achapo, achapo blanco, achapo colorado, achapo rojo, achapo rosado, achuapo, arenoso, cedro achapo, cedro mure, cedrorana, iachapo, latape de altura, mure, seique, taecillo, tornillo, yacayaca **Habit:** Tree **Use:** EU, FU, MA, ME **Origin:** Native **Region:** Amazonia, Guiana Shield **Dept.:** AMA, CAQ, PUT, VAU, VID **Elev.:** 50–250 m a.s.l. **IUCN Habitat code:** 1,3,5,14 **Conservation status: (IUCN)** LC

3575 *Cenostigma pluviosum* (DC.) Gagnon & G.P.Lewis (Fabaceae)
IPNI: 77158081-1
Habit: Tree **Use:** ME **Origin:** Cultivated **Region:** Andean **Dept.:** ANT **Elev.:** 1480–1480 m a.s.l. **IUCN Habitat code:** 1,14 **Conservation status: (IUCN)** NE

3576 *Centrolobium paraense* Tul. (Fabaceae)
IPNI: 485534-1
Common name: amarillo, araribis, balaúste, balaústre, balaústre colorado, cartur, colorado, guayacán hobo, guayaquil **Habit:** Tree **Use:** EU, FU, MA **Origin:** Native **Region:** Caribbean, Pacific **Dept.:** ANT, ATL, CHO, COR, GUV, SAN **Elev.:** 40–700 m a.s.l. **IUCN Habitat code:** 1,2,3,5,14 **Conservation status: (IUCN)** NE

3577 *Centrosema acutifolium* Benth. (Fabaceae)
IPNI: 485540-1
Habit: Herb **Use:** AF **Origin:** Native **Region:** Caribbean, Orinoquia **Dept.:** ANT, MET **Elev.:** 220–220 m a.s.l. **IUCN Habitat code:** 1,2,3,4,5,14 **Conservation status: (IUCN)** NE

3578 *Centrosema brasilianum* (L.) Benth. (Fabaceae)
IPNI: 485555-1
Habit: Liana, Climbing **Use:** AF,ME **Origin:** Native **Region:** Caribbean, Orinoquia, Pacific **Dept.:** ARA, CHO, MAG **Elev.:** 40–130 m a.s.l. **IUCN Habitat code:** 1,3,14 **Conservation status: (IUCN)** NE

3579 *Centrosema macrocarpum* Benth. (Fabaceae)
IPNI: 485597-1
Common name: frijolillo **Habit:** Herb, Climbing **Use:** AF, EU, GS, HF **Origin:** Native, Cultivated **Region:** Andean, Caribbean, Orinoquia **Dept.:** ANT, ARA, CAL, CHO, CUN, GUV, MAG, MET, NSA, VAC, VID **Elev.:** 70–2300 m a.s.l. **IUCN Habitat code:** 1,2,3,4,5,14 **Conservation status: (IUCN)** LC

3580 *Centrosema molle* Mart. ex Benth. (Fabaceae)
IPNI: 485600-1
Habit: Climbing **Use:** AF, EU **Origin:** Native **Region:** Orinoquia **Dept.:** SAN, VID **Elev.:** 140–500 m a.s.l. **IUCN Habitat code:** 1,2,3,4,14 **Conservation status: (IUCN)** NE

3581 *Centrosema plumieri* (Turpin ex Pers.) Benth. (Fabaceae)
IPNI: 485609-1
Common name: bejuco pito, fríjol de hormiga, gallito, pirigallo, pito, platanito, zaragoza **Habit:** Herb, Climbing **Use:** EU, HF, MA, ME, PO **Origin:** Native **Region:** Andean, Guiana Shield, Caribbean, Pacific **Dept.:** ANT, ATL, BOL, BOY, CAL, CAU, CHO, COR, CUN, HUI, MAG, MET, NSA, PUT, RIS, SAP, SAN, SUC, TOL, VAC **Elev.:** 10-2000 m.a.s.l. **IUCN Habitat code:** 1,2,3,5,14 **Conservation status: (IUCN)** NE **(CPSG)** Potential LC

3582 *Centrosema pubescens* Benth. (Fabaceae)
IPNI: 485611-1
Common name: centrosema, frijolito, gallito **Habit:** Climbing **Use:** AF, EU, HF, ME **Origin:** Native **Region:** Amazonia, Andean, Caribbean, Orinoquia, Pacific **Dept.:** ANT, ATL, BOG, BOL, CAL, CAQ, CAS, CAU, CES, CHO, COR, CUN, GUV, HUI, LAG, MAG, MET, NAR, NSA, QUI, RIS, SAN, TOL, VAC **Elev.:** 0-2100 m.a.s.l. **IUCN Habitat code:** 1,2,3,4,5,14 **Conservation status: (IUCN)** NE **(CPSG)** Potential LC

3583 *Centrosema sagittatum* (Humb. & Bonpl. ex Willd.) Brandegee (Fabaceae)
IPNI: 485618-1
Habit: Herb, Climbing **Use:** EU **Origin:** Native **Region:** Guiana Shield, Caribbean **Dept.:** ATL, MAG, MET **Elev.:** 20–450 m a.s.l. **IUCN Habitat code:** 1,2,14 **Conservation status: (IUCN)** NE

3584 *Centrosema virginianum* (L.) Benth. (Fabaceae)
IPNI: 51897-2
Common name: bejuco de sonajera, boca de monte, frijolillo silvestre, frijolito, sonajera **Habit:** Herb, Climbing **Use:** AF, EU, ME **Origin:** Native **Region:** Andean, Pacific **Dept.:** ANT, BOY, CAU, CHO, CUN, HUI, MET, NAR, NSA, SAN, TOL, VAC **Elev.:** 40-2600 m.a.s.l. **IUCN Habitat code:** 1,2,3,4,14 **Conservation status: (IUCN)** NE **(CPSG)** Potential LC

3585 *Chamaecrista absus* (L.) H.S.Irwin & Barneby (Fabaceae)
IPNI: 54310-2
Habit: Shrub **Use:** AF, HF, ME, PO **Origin:** Native **Region:** Andean **Dept.:** HUI **Elev.:** 880–880 m a.s.l. **IUCN Habitat code:** 14 **Conservation status: (IUCN)** LC

3586 *Chamaecrista desvauxii* (Collad.) Killip (Fabaceae)
IPNI: 54460-2
Common name: alverjilla, cargarrocío, guardarocío, guardarrocío, pensamiento **Habit:** Herb, Shrub **Use:** ME **Origin:** Native **Region:** Amazonia, Andean, Guiana Shield, Orinoquia **Dept.:** AMA, BOY, CAL, CAQ, CAS, CAU, CUN, GUA, GUV, HUI, MET, SAN, VAU, VID **Elev.:** 100-2200 m.a.s.l. **IUCN Habitat code:** 1,2,3,4,5,14 **Conservation status: (IUCN)** LC **(CPSG)** Potential LC

3587 *Chamaecrista diphylla* (L.) Greene (Fabaceae)
IPNI: 54479-2
Habit: Herb, Subshrub **Use:** ME **Origin:** Native **Region:** Guiana Shield, Orinoquia **Dept.:** CAS, GUV, MET, VID **Elev.:** 100–900 m a.s.l. **IUCN Habitat code:** 1,2,3,4,5,14 **Conservation status: (IUCN)** NE

3588 *Chamaecrista flexuosa* (L.) Greene (Fabaceae)
IPNI: 54508-2
Habit: Herb, Subshrub **Use:** ME **Origin:** Native **Region:** Guiana Shield, Caribbean, Orinoquia **Dept.:** CAS, CES, GUA, LAG, MAG, MET, VID **Elev.:** 50–300 m a.s.l. **IUCN Habitat code:** 1,2,4 **Conservation status: (IUCN)** NE

3589 *Chamaecrista kunthiana* (Schltdl. & Cham.) H.S.Irwin & Barneby (Fabaceae)
IPNI: 54568-2
Habit: Herb **Use:** ME **Origin:** Native **Region:** Orinoquia **Dept.:** VIDIUCN **Habitat code:** 2,3,4 **Conservation status: (IUCN)** NE

3590 *Chamaecrista nictitans* (L.) Moench (Fabaceae)
IPNI: 30485908-2
Common name: adormidera, chilinchile, frijolillo, hiebra de gallinazo, planchito **Habit:** Herb, Subshrub **Use:** EU, ME **Origin:** Native **Region:** Amazonia, Andean, Caribbean, Orinoquia, Pacific **Dept.:** ANT, ARA, ATL, BOY, CAL, CAQ, CAS, CAU, CHO, CUN, GUA, HUI, LAG, MAG, MET, NSA, SAN, SUC, TOL, VAC, VAU, VID **Elev.:** 0-2500 m.a.s.l. **IUCN Habitat code:** 1,2,3,4,5,14 **Conservation status: (IUCN)** LC **(CPSG)** Potential LC

3591 *Chamaecrista rotundifolia* (Pers.) Greene (Fabaceae)
IPNI: 54723-2
Common name: zarza menor **Habit:** Herb, Shrub **Use:** AF,ME **Origin:** Native **Region:** Andean, Orinoquia **Dept.:** ANT, ARA, CUN, GUA, HUI, MET, TOL, VID **Elev.:** 100–1250 m a.s.l. **IUCN Habitat code:** 1,3,4,5,14 **Conservation status: (IUCN)** NE

3592 *Chloroleucon mangense* (Jacq.) Britton & Rose (Fabaceae) **IPNI:** 56985-2 **Common name:** algarrobilla, algarrobillo, amarillo, angarilla, angarillo, arbolito, atupa, cachito, carbonero, chiminango, espinito blanco, gallinazo, hoyo de zorro, payandé, pino blanco, rampayuco, raspayuco, vainillo, vivaseca, árbol de tomar **Habit:** Tree **Use:** AF, MA, ME **Origin:** Native **Region:** Caribbean **Dept.:** ATL, BOL, COR, CUN, HUI, LAG, MAG, TOL **Elev.:** 0–1290 m a.s.l. **IUCN Habitat code:** 1,2,3,4,5,14 **Conservation status: (IUCN)** LC

3593 *Cicer arietinum* L. (Fabaceae) **IPNI:** 486336-1 **Common name:** garbanzo **Use:** AF, GS, HF, MA, ME **Origin:** Naturalised **IUCN Habitat code:** 1,4,14 **Conservation status: (IUCN)** NE

3594 *Clathrotropis brachypetala* (Tul.) Kleinh. (Fabaceae) **IPNI:** 59774-2 **Common name:** sapán, vaino **Habit:** Tree **Use:** MA, ME **Origin:** Native **Region:** Amazonia, Guiana Shield **Dept.:** GUA, GUV **Elev.:** 120–200 m a.s.l. **IUCN Habitat code:** 1,2,3,5,14 **Conservation status: (IUCN)** LC

3595 *Clathrotropis brunnea* Amshoff (Fabaceae) **IPNI:** 59775-2 **Common name:** sangre de toro, sapan, sapán **Habit:** Tree **Use:** MA **Origin:** Endemic **Dept.:** ANT, BOL, SAN **Elev.:** 100-750 m.a.s.l. **IUCN Habitat code:** 1,3,14 **Conservation status: (IUCN)** NE **(CPSG)** EN

3596 *Clathrotropis glaucophylla* R.S.Cowan (Fabaceae) **IPNI:** 59777-2 **Habit:** Tree **Use:** ME **Origin:** Native **Region:** Amazonia, Guiana Shield **Dept.:** AMA, CAQ, GUA **Elev.:** 75–75 m a.s.l. **IUCN Habitat code:** 1,5 **Conservation status: (IUCN)** LC

3597 *Clathrotropis macrocarpa* Ducke (Fabaceae) **IPNI:** 59779-2 **Common name:** barbasco, bizcocho, cabari, fariñero, fariñoso, manana, manzana, masano, palo venenoso, paya, tabaco, tabaquillo, tepaci, tepá **Habit:** Shrub, Tree **Use:** HF, MA, ME **Origin:** Native **Region:** Amazonia, Guiana Shield, Orinoquia **Dept.:** AMA, CAQ, GUA, GUV, VAU, VID **Elev.:** 100–250 m a.s.l. **IUCN Habitat code:** 1,3,5,14 **Conservation status: (IUCN)** LC

3598 *Clianthus puniceus* (G.Don) Sol. ex Lindl. (Fabaceae) **IPNI:** 486457-1 **Habit:** Shrub, Tree **Use:** EU **Origin:** Cultivated **Region:** Andean **Dept.:** BOG, CAU, CUN, NAR, TOL **Elev.:** 1300–3300 m a.s.l. **IUCN Habitat code:** 1,3,14 **Conservation status: (IUCN)** EN

3599 *Clitoria arborescens* R.Br. (Fabaceae) **IPNI:** 486481-1 **Habit:** Climbing **Use:** MA **Origin:** Native **Region:** Amazonia, Orinoquia, Pacific **Dept.:** AMA, CAS, CHO, VAC **Elev.:** 50–1000 m a.s.l. **IUCN Habitat code:** 1,2,5,14 **Conservation status: (IUCN)** LC

3600 *Clitoria fairchildiana* R.A.Howard (Fabaceae) **IPNI:** 291409-2 **Common name:** cañafisto, sombrero **Habit:** Tree **Use:** EU **Origin:** Native, Cultivated **Region:** Amazonia, Orinoquia **Dept.:** AMA, GUV **Elev.:** 100–100 m a.s.l. **IUCN Habitat code:** 1,14 **Conservation status: (IUCN)** LC

3601 *Clitoria javitensis* (Kunth) Benth. (Fabaceae) **IPNI:** 486540-1 **Common name:** chimbillo bravo largo, newütjümujüyorotonae **Habit:** Liana, Climbing **Use:** EU, HF, ME, SU **Origin:** Native **Region:** Amazonia, Andean, Guiana Shield, Caribbean, Orinoquia, Pacific **Dept.:** AMA, ANT, ARA, CAQ, CAS, CHO, CUN, GUA, GUV, MET, PUT, TOL, VAU **Elev.:** 0-780 m.a.s.l. **IUCN Habitat code:** 1,3,4,5,14 **Conservation status: (IUCN)** NE **(CPSG)** Potential LC

3602 *Clitoria ternatea* L. (Fabaceae) **IPNI:** 486606-1 **Common name:** azulejo, bejuco conchita, campanilla, campanita, conchitas, doncenón, flormarina, frijolillo silvestre, papito, patico azul, sapotillo, zapatico de la Virgen, zapatico de la reina, zapatico de reina **Habit:** Herb, Shrub, Climbing **Use:** AF, EU, GS, HF, MA, ME, PO **Origin:** Naturalised **Region:** Andean, Caribbean, Orinoquia, Pacific **Dept.:** ANT, BOL, CAL, CAS, CHO, COR, CUN, HUI, MET, NAR, QUI, SAN, TOL, VAC **Elev.:** 20–1500 m a.s.l. **IUCN Habitat code:** 1,2,3,4,14 **Conservation status: (IUCN)** NE

3603 *Cochliasanthus caracalla* (L.) Trew (Fabaceae) **IPNI:** 60471746-2 **Habit:** Climbing **Use:** EU, HF **Origin:** Native, Cultivated **Region:** Andean, Guiana Shield, Caribbean, Orinoquia, Pacific **Dept.:** ANT, ATL, BOL, BOY, CAS, CHO, CUN, MAG, MET, SUC, TOL **Elev.:** 20–1300 m a.s.l. **IUCN Habitat code:** 1,2,3,14 **Conservation status: (IUCN)** NE

3604 *Cojoba arborea* (L.) Britton & Rose (Fabaceae) **IPNI:** 62882-2 **Common name:** acacia, carbonero, carbonero dormilón, carbonero negro, carbonero zorro, cojoba, dormilón, mimosa **Habit:** Tree **Use:** EU, IF, MA, ME, PO **Origin:** Native, Cultivated **Region:** Andean **Dept.:** ANT, CAL, CUN, VAC **Elev.:** 140–1500 m a.s.l. **IUCN Habitat code:** 1,3,14 **Conservation status: (IUCN)** LC

3605 *Cojoba rufescens* (Benth.) Britton & Rose (Fabaceae) **IPNI:** 62892-2 **Common name:** guamo **Habit:** Tree **Use:** ME **Origin:** Native **Region:** Andean, Caribbean, Pacific **Dept.:** ANT, ATL, BOL, CHO, CUN, SAN **Elev.:** 0–1800 m a.s.l. **IUCN Habitat code:** 1,2,3,5,14 **Conservation status: (IUCN)** LC

3606 *Colutea arborescens* L. (Fabaceae) **IPNI:** 486805-1 **Use:** AF, EU, GS, HF, MA, ME **Region:** Andean **Dept.:** BOG**Conservation status: (IUCN)** NE

3607 *Copaifera canime* Harms (Fabaceae) **IPNI:** 64898-2 **Common name:** aceite, campano, canime, copaiba, copaiva, copayero, majagua, quiebracha **Habit:** Tree **Use:** ME, SU **Origin:** Endemic **Region:** Caribbean **Dept.:** ANT, MAG **Elev.:** 50–300 m a.s.l. **IUCN Habitat code:** 1,2,3,5,14 **Conservation status: (IUCN)** LC

3608 *Copaifera guianensis* Desf. (Fabaceae) **IPNI:** 487006-1 **Habit:** Tree **Use:** ME **Origin:** Native **Region:** Amazonia **Dept.:** AMA **Elev.:** 100–100 m a.s.l. **Conservation status: (IUCN)** LC

3609 *Copaifera officinalis* L. (Fabaceae) **IPNI:** 64940-2 **Common name:** aceite, aceite canime, amansamujer, bálsamo de copaiba, campano, copaiba, copaibí, copayero, copaíba, palo de aceite, quebrahacha, tree de aceite, trementino, árbol de aceite **Habit:** Tree **Use:** AF, EU, MA, ME, PO **Origin:** Native **Region:** Amazonia, Orinoquia **Dept.:** CAS, VAU **Elev.:** 200–500 m a.s.l. **IUCN Habitat code:** 1,3,4,14 **Conservation status: (IUCN)** NE

3610 *Copaifera pubiflora* Benth. (Fabaceae) **IPNI:** 487038-1 **Common name:** aceite, bálsamo de canime, cabima, canime, copaiba, copaibí, maramo, moramo, palo aceite, palo de aceite, tree de copaiba, tunoso, árbol de la copaiba **Habit:** Tree **Use:** MA, ME **Origin:** Native **Region:** Andean, Orinoquia **Dept.:** ARA, CAS, CAU, MET **Elev.:** 130–1100 m a.s.l. **IUCN Habitat code:** 1,2,3,4,5,14 **Conservation status: (IUCN)** LC

3611 *Copaifera reticulata* Ducke (Fabaceae) **IPNI:** 64945-2 **Habit:** Tree **Use:** HF, MA, ME **Origin:** Native **Region:** Amazonia **Dept.:** AMA, GUA, VAU **Elev.:** 120–200 m a.s.l. **IUCN Habitat code:** 1,3 **Conservation status: (IUCN)** LC

3612 *Coulteria mollis* Kunth (Fabaceae) **IPNI:** 67201-2 **Common name:** piula, yaguaro **Habit:** Shrub, Tree **Use:** EU, MA, ME **Origin:** Native **Region:** Caribbean **Dept.:** ATL, LAG, MAG **Elev.:** 10–400 m a.s.l. **IUCN Habitat code:** 1,2,4 **Conservation status: (IUCN)** DD

3613 *Coursetia caribaea* (Jacq.) Lavin (Fabaceae) **IPNI:** 275194-2 **Common name:** yerba de caballo, zuná **Habit:** Herb, Shrub **Use:** AF **Origin:** Native **Region:** Andean, Caribbean **Dept.:** ATL, BOL, BOY, CUN, HUI, LAG, MAG, NAR, NSA, SAN, SUC, TOL, VAC **Elev.:** 5-2600 m.a.s.l. **IUCN Habitat code:** 1,2,3,5,14 **Conservation status: (IUCN)** LC **(CPSG)** Potential LC

3614 *Coursetia ferruginea* (Kunth) Lavin (Fabaceae) **IPNI:** 944647-1 **Common name:** bálsamo, bálsamo hediondo, macuiro, macurutú, matarratón cimarrón, ramoncillo, ramón de conejo **Habit:** Shrub, Tree **Use:** MA **Origin:** Native **Region:** Guiana Shield, Caribbean, Orinoquia **Dept.:** ATL, BOL, COR, LAG, MAG, MET, TOL **Elev.:** 0–1000 m a.s.l. **IUCN Habitat code:** 1,2,3,14 **Conservation status: (IUCN)** LC

3615 *Crotalaria agatiflora* Schweinf. (Fabaceae) **IPNI:** 487769-1 **Common name:** crotalaria, pajarito, zapatico **Habit:** Shrub **Use:** AF, EU, ME, PO **Origin:** Cultivated, Naturalised **Region:** Andean **Dept.:** BOG, BOY, CUN **Elev.:** 2600–2820 m a.s.l. **IUCN Habitat code:** 1,3,14 **Conservation status: (IUCN)** LC

3616 *Crotalaria brevidens* Benth. (Fabaceae) **IPNI:** 487901-1 **Habit:** Subshrub **Use:** AF, EU, HF, MA, ME **Origin:** Cultivated, Naturalised **Region:** Andean **Dept.:** ANT, CAL, CUN **Elev.:** 1240–1560 m a.s.l. **IUCN Habitat code:** 1,3 **Conservation status:** **(IUCN)** NE

3617 *Crotalaria incana* L. (Fabaceae) **IPNI:** 316245-2 **Common name:** bicho, campana amarilla, cargadita, cascabel, cascabeles, cascabelillo, cascabelito, cascabelitos, chilinchín, chochitos, empanaditas, hierba de la luna, maraquitas, quebraplatos, tote, yerba de caballo, yerba de chulo, yerba de sapo **Habit:** Herb, Subshrub **Use:** EU, HF, MA, ME, PO **Origin:** Native **Region:** Amazonia, Andean, Caribbean, Orinoquia **Dept.:** AMA, ANT, ATL, BOL, BOY, CAL, CAQ, CES, COR, CUN, HUI, LAG, MAG, NAR, NSA, QUI, SAN, SUC, TOL, VAC, VID **Elev.:** 0-2600 m.a.s.l. **IUCN Habitat code:** 1,2,3,4,5,14 **Conservation status:** **(IUCN)** NE **(CPSG)** Potential LC

3618 *Crotalaria juncea* L. (Fabaceae) **IPNI:** 322601-2 **Common name:** cascabel, cascabelito, crotalaria, maraquitas **Habit:** Herb, Subshrub **Use:** AF, EU, HF, MA, ME, PO, SU **Origin:** Naturalised **Region:** Andean **Dept.:** ANT, CAL, CUN, MAG, TOL, VAC **Elev.:** 50–2000 m a.s.l. **IUCN Habitat code:** 1,3,14 **Conservation status:** **(IUCN)** NE

3619 *Crotalaria lanceolata* E.Mey. (Fabaceae) **IPNI:** 322602-2 **Habit:** Herb, Subshrub **Use:** AF, EU, HF, ME **Origin:** Cultivated, Naturalised **Region:** Andean, Pacific **Dept.:** ANT, CAL, VAC **Elev.:** 30–1550 m a.s.l. **IUCN Habitat code:** 1,3,5,14 **Conservation status:** **(IUCN)** LC

3620 *Crotalaria maypurensis* Kunth (Fabaceae) **IPNI:** 322603-2 **Habit:** Herb, Subshrub **Use:** ME **Origin:** Native **Region:** Andean, Guiana Shield, Caribbean, Orinoquia **Dept.:** GUA, GUV, MAG, MET, SAN **Elev.:** 75–1400 m a.s.l. **IUCN Habitat code:** 1,4,5,14 **Conservation status:** **(IUCN)** NE

3621 *Crotalaria micans* Link (Fabaceae) **IPNI:** 322606-2 **Common name:** cascabel, cascabeles, cascabelito, cascabelitos, chochitos, chucho, crotalaria, guache **Habit:** Herb, Shrub **Use:** AF, EU, HF, ME **Origin:** Native, Cultivated **Region:** Amazonia, Andean, Orinoquia **Dept.:** ANT, BOY, CAL, CAU, CES, CUN, HUI, MAG, NAR, NSA, PUT, QUI, SAN, TOL, VAC, VAU, VID **Elev.:** 200–3000 m a.s.l. **IUCN Habitat code:** 1,2,3,4,14 **Conservation status:** **(IUCN)** LC

3622 *Crotalaria nitens* Kunth (Fabaceae) **IPNI:** 488559-1 **Common name:** cacabel, cascabel, cascabeles, cascabelito, cascabelitos, cascabelitos, chuchos, guachecito, guachecitos, maraco, quiebraplatos, quiebrapuro **Habit:** Herb, Shrub, Subshrub **Use:** ME **Origin:** Native **Region:** Amazonia, Andean, Guiana Shield, Caribbean, Orinoquia, Pacific **Dept.:** AMA, ANT, BOL, BOY, CAL, CAQ, CAU, CHO, CUN, HUI, LAG, MET, NAR, NSA, PUT, RIS, SAN, TOL, VAC, VAU **Elev.:** 0-2700 m.a.s.l. **IUCN Habitat code:** 1,2,3,4,5,14 **Conservation status:** **(IUCN)** LC **(CPSG)** Potential LC

3623 *Crotalaria pallida* Aiton (Fabaceae) **IPNI:** 322608-2 **Common name:** brusquillo, cascabel, cascabeles, cascabelito, cascabelitos, crotalaria **Habit:** Herb, Shrub **Use:** AF, EU, HF, MA, ME, PO, SU **Origin:** Naturalised **Region:** Amazonia, Andean, Caribbean, Orinoquia, Pacific **Dept.:** ANT, CAL, CAQ, CAS, CAU, CHO, CUN, GUA, HUI, LAG, MET, QUI, SAN, TOL, VAC, VAU, VID **Elev.:** 5-1850 m.a.s.l. **IUCN Habitat code:** 1,2,3,4,5,14 **Conservation status:** **(IUCN)** NE **(CPSG)** Potential LC

3624 *Crotalaria paulina* Schrank (Fabaceae) **IPNI:** 488624-1 **Habit:** Shrub **Use:** EU **Origin:** Native, Cultivated **Region:** Andean **Dept.:** ANT, CAL, QUI, TOL, VAC **Elev.:** 1260–2780 m a.s.l. **IUCN Habitat code:** 1,3,14 **Conservation status:** **(IUCN)** NE

3625 *Crotalaria pilosa* Mill. (Fabaceae) **IPNI:** 488660-1 **Common name:** guache, herb de venado, hierba de venado, jomokwei **Habit:** Herb **Use:** ME **Origin:** Native **Region:** Andean, Guiana Shield, Caribbean, Orinoquia **Dept.:** ANT, BOY, CAS, CAU, CES, CUN, HUI, MAG, MET, SAN, TOL **Elev.:** 100-1730 m.a.s.l. **IUCN Habitat code:** 1,2,3,4,5,14 **Conservation status:** **(IUCN)** NE **(CPSG)** Potential LC

3626 *Crotalaria retusa* L. (Fabaceae) **IPNI:** 322611-2 **Common name:** alorva, cascabel, cascabelito, isiraachon, maraquita, mariposa **Habit:** Herb, Subshrub **Use:** AF, EU, HF, MA, ME, PO **Origin:** Naturalised **Region:** Andean, Caribbean, Pacific **Dept.:** ANT, ARA, ATL, BOL, CAS, CES, CHO, COR, CUN, LAG, MAG, NAR, NSA, SAP, SAN, TOL, VAC **Elev.:** 0–2450 m a.s.l. **IUCN Habitat code:** 1,2,3,5,14 **Conservation status:** **(IUCN)** NE

3627 *Crotalaria sagittalis* L. (Fabaceae) **IPNI:** 318914-2 **Common name:** cascabelito, cascabelitos, guache, maraquitas **Habit:** Herb **Use:** HF, ME, PO **Origin:** Native **Region:** Amazonia, Andean, Caribbean, Orinoquia, Pacific **Dept.:** AMA, ANT, BOY, CAL, CAS, CHO, CUN, GUV, HUI, LAG, MAG, MET, NAR, SAN, TOL, VAC, VID **Elev.:** 0-2600 m.a.s.l. **Conservation status:** **(IUCN)** NE **(CPSG)** Potential LC

3628 *Crotalaria spectabilis* Roth (Fabaceae) **IPNI:** 322612-2 **Common name:** crotalaria **Habit:** Shrub **Use:** EU, HF, MA, ME, PO **Origin:** Cultivated, Naturalised **Region:** Andean, Pacific **Dept.:** CAL, VAC **Elev.:** 30–1400 m a.s.l. **IUCN Habitat code:** 1,3,14 **Conservation status:** **(IUCN)** NE

3629 *Crotalaria stipularia* Desv. (Fabaceae) **IPNI:** 488910-1 **Use:** ME **Origin:** Native **IUCN Habitat code:** 2 **Conservation status:** **(IUCN)** NE

3630 *Crotalaria verrucosa* L. (Fabaceae) **IPNI:** 489042-1 **Common name:** cascabel, cascabelito, chochito **Habit:** Herb **Use:** EU, ME, PO **Origin:** Cultivated, Naturalised **Region:** Caribbean **Dept.:** ATL, BOL, SAP **Elev.:** 1–200 m a.s.l. **IUCN Habitat code:** 2,4,14 **Conservation status:** **(IUCN)** NE

3631 *Crotalaria vitellina* Ker Gawl. (Fabaceae) **IPNI:** 322616-2 **Common name:** guache **Habit:** Herb, Shrub **Use:** AF, HF, PO **Origin:** Native **Region:** Andean, Caribbean **Dept.:** CES, LAG, MAG **Elev.:** 30–1830 m a.s.l. **IUCN Habitat code:** 1,2,14 **Conservation status:** **(IUCN)** NE

3632 *Crudia amazonica* Spruce ex Benth. (Fabaceae) **IPNI:** 489094-1 **Habit:** Shrub, Tree **Use:** MA, ME **Origin:** Native **Region:** Amazonia **Dept.:** AMA, GUA, VAU **Elev.:** 200–250 m a.s.l. **IUCN Habitat code:** 1,5 **Conservation status:** **(IUCN)** LC

3633 *Ctenodon brasilianus* (Poir.) D.B.O.S.Cardoso, P.L.R.Moraes & H.C.Lima (Fabaceae) **IPNI:** 77211517-1 **Habit:** Herb, Shrub **Use:** ME **Origin:** Endemic **Region:** Amazonia, Andean, Guiana Shield, Caribbean, Orinoquia, Pacific **Dept.:** ANT, CAQ, CAU, CUN, HUI, LAG, MAG, MET, TOL, VAC, VID **Elev.:** 20-1800 m.a.s.l. **IUCN Habitat code:** 1,3,4,5,14 **Conservation status:** **(IUCN)** LC **(CPSG)** Potential LC

3634 *Ctenodon elegans* (Schltdl. & Cham.) D.B.O.S.Cardoso & A.Delgado (Fabaceae) **IPNI:** 77211590-1 **Habit:** Herb, Subshrub **Use:** EU **Origin:** Endemic **Region:** Andean, Caribbean, Orinoquia **Dept.:** ANT, BOY, CAL, CAS, CAU, CUN, HUI, LAG, MAG, MET, NAR, NSA, TOL, VAC, VID **Elev.:** 30-1800 m.a.s.l. **IUCN Habitat code:** 1,3,4,14 **Conservation status:** **(IUCN)** NE **(CPSG)** Potential LC

3635 *Ctenodon falcatus* (Poir.) D.B.O.S.Cardoso, P.L.R.Moraes & H.C.Lima (Fabaceae) **IPNI:** 77211592-1 **Common name:** amorseco, cortejo **Habit:** Herb, Subshrub, Climbing **Use:** AF,ME **Origin:** Endemic **Region:** Andean, Caribbean, Orinoquia **Dept.:** ANT, CAL, CAU, CUN, HUI, MET, SAN, TOL, VAC, VID **Elev.:** 180–2200 m a.s.l. **IUCN Habitat code:** 1,3,4,14 **Conservation status:** **(IUCN)** NE

3636 *Ctenodon fascicularis* (Schltdl. & Cham.) A.Delgado (Fabaceae) **IPNI:** 77211890-1 **Common name:** megua **Habit:** Herb, Shrub **Use:** ME **Origin:** Endemic **Region:** Andean, Caribbean **Dept.:** ATL, BOL, BOY, CES, CUN, HUI, MAG **Elev.:** 0–2250 m a.s.l. **IUCN Habitat code:** 1,2,3,14 **Conservation status:** **(IUCN)** NE

3637 *Cynometra bauhiniifolia* Benth. (Fabaceae) **IPNI:** 489367-1 **Common name:** algarrobo, angolito, mangle duro **Habit:** Shrub, Tree **Use:** HF **Origin:** Native **Region:** Amazonia, Guiana Shield **Dept.:** ANT, GUV, VAU **Elev.:** 250–500 m a.s.l. **IUCN Habitat code:** 1,3,4,5,14 **Conservation status:** **(IUCN)** LC

3638 *Cynometra marginata* Benth. (Fabaceae) **IPNI:** 489448-1 **Common name:** algarrobo **Habit:** Shrub, Tree **Use:** HF **Origin:** Native **Region:** Amazonia **Dept.:** AMA, GUA, VAU **Elev.:** 100–100 m a.s.l. **IUCN Habitat code:** 1,4,5 **Conservation status:** **(IUCN)** LC

3639 *Dalbergia brownei* (Jacq.) Schinz (Fabaceae)
IPNI: 75919-2
Common name: bejuco de penda blanca, bejuco penda blanca, bejuco pende, penda blanca **Habit:** Shrub, Liana, Climbing **Use:** ME **Origin:** Native **Region:** Amazonia, Caribbean, Pacific **Dept.:** ANT, ATL, CAQ, CAU, CHO, MAG, NAR, SAP, VAC, VAU **Elev.:** 0–200 m a.s.l. **IUCN Habitat code:** 1,2,3,5,14 **Conservation status: (IUCN)** LC

3640 *Dalbergia ecastaphyllum* (L.) Taub. (Fabaceae)
IPNI: 75934-2
Habit: Shrub **Use:** EU, MA, ME **Origin:** Native **Region:** Amazonia, Guiana Shield, Caribbean, Pacific **Dept.:** ANT, CAQ, CHO, COR, VAU **Elev.:** 0–600 m a.s.l. **IUCN Habitat code:** 1,2,3,4,5,14 **Conservation status: (IUCN)** LC

3641 *Dalbergia foliosa* (Benth.) T.S.Ralph (Fabaceae)
IPNI: 1006519-1
Habit: Shrub, Small tree **Use:** MA **Origin:** Native **Region:** Amazonia, Guiana Shield **Dept.:** AMA, GUA, VAU **Elev.:** 80–240 m a.s.l. **IUCN Habitat code:** 1,5 **Conservation status: (IUCN)** NE

3642 *Dalbergia frutescens* (Vell.) Britton (Fabaceae)
IPNI: 75939-2
Common name: barbacobo **Habit:** Shrub, Tree, Liana, Climbing **Use:** MA, ME **Origin:** Native **Region:** Amazonia, Andean, Caribbean **Dept.:** AMA, ANT, CAQ, COR, CUN **Elev.:** 100–1320 m a.s.l. **IUCN Habitat code:** 1,3,14 **Conservation status: (IUCN)** NE

3643 *Dalbergia inundata* Spruce ex Benth. (Fabaceae)
IPNI: 75954-2
Habit: Shrub, Small tree **Use:** MA **Origin:** Native **Region:** Orinoquia **Dept.:** AMA, GUA, VID **Elev.:** 65–260 m a.s.l. **IUCN Habitat code:** 1,3,4,5 **Conservation status: (IUCN)** NE

3644 *Dalbergia ovalis* (L.) P.L.R.Moraes & L.P.Queiroz (Fabaceae)
IPNI: 77117114-1
Common name: uña de gavilán **Habit:** Shrub, Liana, Climbing **Use:** MA, ME **Origin:** Native **Region:** Amazonia, Andean, Guiana Shield, Caribbean, Orinoquia, Pacific **Dept.:** AMA, ANT, CAQ, CAS, CHO, COR, CUN, GUA, HUI, MET, SUC, TOL, VAU **Elev.:** 0-1900 m.a.s.l. **IUCN Habitat code:** 1,2,3,5,14 **Conservation status: (IUCN)** NE **(CPSG)** Potential LC

3645 *Dalbergia riparia* (Mart. ex Benth.) Benth. (Fabaceae)
IPNI: 490433-1
Habit: Liana **Use:** MA **Origin:** Native **Region:** Amazonia **Dept.:** AMA **Elev.:** 200–200 m a.s.l. **IUCN Habitat code:** 1,3 **Conservation status: (IUCN)** NE

3646 *Dalea cliffortiana* Willd. (Fabaceae)
IPNI: 76085-2
Habit: Herb **Use:** ME **Origin:** Native **Region:** Andean **Dept.:** SAN **Elev.:** 1000–1800 m a.s.l. **IUCN Habitat code:** 2,14 **Conservation status: (IUCN)** NE

3647 *Dalea coerulea* (L.f.) Schinz & Thell. (Fabaceae)
IPNI: 76086-2
Common name: arizá, carica, chiripique, chivato, gusanillo, palo de cruz, pasto unca, pispura, pispure, pispusa, pulguito, roso de monte, unca **Habit:** Herb, Shrub **Use:** ME **Origin:** Native **Region:** Andean **Dept.:** ANT, BOG, BOY, CAU, CUN, NAR **Elev.:** 500–3100 m a.s.l. **IUCN Habitat code:** 1,3,4,5,14 **Conservation status: (IUCN)** NE

3648 *Dalea foliolosa* (Aiton) Barneby (Fabaceae)
IPNI: 76146-2
Use: ME **Origin:** Native **Conservation status: (IUCN)** NE

3649 *Deguelia amazonica* Killip (Fabaceae)
IPNI: 77178-2
Use: ME **Origin:** Native **IUCN Habitat code:** 1,3,4,14 **Conservation status: (IUCN)** LC

3650 *Deguelia densiflora* (Benth.) A.M.G.Azevedo ex M.Sousa (Fabaceae)
IPNI: 60453072-2
Habit: Shrub, Tree, Climbing **Use:** ME **Origin:** Native **Region:** Amazonia, Caribbean, Orinoquia, Pacific **Dept.:** AMA, ANT, CHO, GUV, MET, VAU **Elev.:** 10–250 m a.s.l. **IUCN Habitat code:** 1,2,3,5,14 **Conservation status: (IUCN)** NE

3651 *Deguelia nitidula* (Benth.) A.M.G.Azevedo & R.A.Camargo (Fabaceae)
IPNI: 77143238-1
Common name: matarratón **Habit:** Small tree **Use:** MA, ME **Origin:** Native **Region:** Amazonia, Orinoquia **Dept.:** AMA, CAQ, VID **Elev.:** 180–220 m a.s.l. **IUCN Habitat code:** 1,3,4,14 **Conservation status: (IUCN)** NE

3652 *Deguelia picta* (Pittier) R.A.Camargo & A.M.G.Azevedo (Fabaceae)
IPNI: 77185872-1
Common name: barbasco, cube, majagua de gallina, timbo nocou **Habit:** Tree, Liana **Use:** MA, ME, PO **Origin:** Native **Region:** Orinoquia **Dept.:** BOY, CAS, SAN**IUCN Habitat code:** 1,2,3,5,14 **Conservation status: (IUCN)** NE

3653 *Deguelia rariflora* (Mart. ex Benth.) G.P.Lewis & Acev.-Rodr. (Fabaceae)
IPNI: 77116882-1
Use: ME **Origin:** Native **Conservation status: (IUCN)** NE

3654 *Deguelia scandens* Aubl. (Fabaceae)
IPNI: 491205-1
Habit: Small tree, Liana, Climbing **Use:** ME, PO **Origin:** Native **Region:** Amazonia **Dept.:** AMA, ANT, CAQ, VAU **Elev.:** 100–900 m a.s.l. **IUCN Habitat code:** 1,2,3,4,5,14 **Conservation status: (IUCN)** LC

3655 *Deguelia utilis* (A.C.Sm.) A.M.G.Azevedo (Fabaceae)
IPNI: 1005531-1
Habit: Shrub **Use:** ME, PO **Origin:** Native **Region:** Amazonia **Dept.:** AMA **Elev.:** 200–200 m a.s.l. **Conservation status: (IUCN)** NE

3656 *Delonix regia* (Bojer ex Hook.) Raf. (Fabaceae)
IPNI: 491231-1
Common name: abeto, acacia, acacia de Girardot, acacia roja, acacio, acacio extranjero, acacio ordinario, acacio rojo, calamayó, calambayó, clavelino, clavellino, flamboyán, flor de fuego, flor de pavo, flor de ángel, florito, gallino, jalamayó, josefino, matarratón, **Habit:** Tree **Use:** AF, EU, FU, HF, IF, MA, ME, PO **Origin:** Cultivated, Naturalised **Region:** Amazonia, Andean, Caribbean, Orinoquia, Pacific **Dept.:** AMA, ANT, ATL, BOL, CAS, COR, CUN, MET, SAP, TOL, VAC **Elev.:** 0–1540 m a.s.l. **IUCN Habitat code:** 1,2,3,4,5,14 **Conservation status: (IUCN)** LC

3657 *Desmanthus virgatus* (L.) Willd. (Fabaceae)
IPNI: 1125234-2
Common name: bicho, deshinchadera, frisolillo de llano, manzanillo, peinecillo, platanito filú, zarza **Habit:** Herb, Subshrub **Use:** AF, EU, HF, MA, ME **Origin:** Native **Region:** Andean, Caribbean, Pacific **Dept.:** ANT, ATL, BOL, BOY, COR, CUN, HUI, LAG, MAG, NAR, SAP, SAN, SUC, TOL, VAC **Elev.:** 0-2200 m.a.s.l. **IUCN Habitat code:** 1,2,3,5,14 **Conservation status: (IUCN)** NE **(CPSG)** Potential LC

3658 *Desmodium affine* Schltdl. (Fabaceae)
IPNI: 78342-2
Common name: amorseco **Habit:** Herb, Subshrub **Use:** ME **Origin:** Native **Region:** Andean, Caribbean **Dept.:** ANT, CHO, LAG, VAC **Elev.:** 200–1300 m a.s.l. **IUCN Habitat code:** 1,2,3,14 **Conservation status: (IUCN)** NE

3659 *Desmodium angustifolium* (Kunth) DC. (Fabaceae)
IPNI: 78351-2
Habit: Herb **Use:** AF, EU **Origin:** Native **Region:** Andean, Orinoquia **Dept.:** BOY, HUI, MAG **Elev.:** 380–1200 m a.s.l. **IUCN Habitat code:** 14 **Conservation status: (IUCN)** NE

3660 *Desmodium axillare* (Sw.) DC. (Fabaceae)
IPNI: 491595-1
Common name: cadillo, frijolillo, pegapega **Habit:** Herb, Subshrub, Climbing **Use:** ME **Origin:** Native **Region:** Amazonia, Andean, Guiana Shield, Caribbean, Orinoquia, Pacific **Dept.:** AMA, ANT, BOY, CAQ, CAU, CHO, CUN, GUV, LAG, MAG, MET, NAR, PUT, QUI, RIS, SAN, TOL, VAC **Elev.:** 0-3200 m.a.s.l. **IUCN Habitat code:** 1,3,5,14 **Conservation status: (IUCN)** NE **(CPSG)** Potential LC

3661 *Desmodium cajanifolium* (Kunth) DC. (Fabaceae)
IPNI: 1108524-2
Common name: amorseco, mermelada **Habit:** Herb, Subshrub, Tree **Use:** ME **Origin:** Native **Region:** Andean, Caribbean, Orinoquia, Pacific **Dept.:** ANT, BOY, CAL, CAU, CES, CHO, COR, CUN, GUV, HUI, MAG, MET, NSA, SAN, TOL, VAC, VID **Elev.:** 25-1700 m.a.s.l. **IUCN Habitat code:** 1,2,3,4,14 **Conservation status: (IUCN)** LC **(CPSG)** Potential LC

3662 *Desmodium distortum* (Aubl.) J.F.Macbr. (Fabaceae)
IPNI: 78415-2
Common name: amorseco **Habit:** Herb **Use:** AF,ME **Origin:** Native **Region:** Amazonia, Andean, Caribbean, Orinoquia **Dept.:** ANT, CAS, CUN, GUA, MAG **Elev.:** 150–1400 m a.s.l. **IUCN Habitat code:** 3,14 **Conservation status: (IUCN)** NE

3663 *Desmodium glabrum* (Mill.) DC. (Fabaceae)
IPNI: 1130757-2
Habit: Herb, Shrub, Subshrub **Use:** AF, EU, ME **Origin:** Native **Region:** Andean, Caribbean, Orinoquia **Dept.:** ATL, BOL, BOY, CHO, CUN, HUI, MAG, MET, TOL, VAC **Elev.:** 25–1600 m a.s.l. **IUCN Habitat code:** 1,2,3,5,14 **Conservation status: (IUCN)** NE

3664 *Desmodium incanum* (Sw.) DC. (Fabaceae) **IPNI:** 1015219-2 **Common name:** amor do campo, amor seco, amorseco, arroz con coco, cadillo, cadillo juancho, cadillo puya, empanadilla, pegadera, pegapega **Habit:** Herb, Shrub, Climbing **Use:** AF, HF, MA, ME **Origin:** Native **Region:** Amazonia, Andean, Caribbean, Orinoquia, Pacific **Dept.:** ANT, ARA, ATL, BOL, BOY, CAL, CAQ, CAU, CES, CHO, COR, CUN, HUI, LAG, MAG, NAR, NSA, PUT, QUI, RIS, SAP, SAN, SUC, TOL, VAC **Elev.:** 0-2200 m.a.s.l. **IUCN Habitat code:** 1,2,3,4,5,14 **Conservation status: (IUCN)** NE **(CPSG)** Potential LC

3665 *Desmodium intortum* (Mill.) Urb. (Fabaceae) **IPNI:** 134286-3 **Common name:** amor seco, amorseco, cadillo, empanaditas, guabillo, pega pega, pegapega **Habit:** Herb, Subshrub, Climbing **Use:** AF, EU, FU, ME **Origin:** Native **Region:** Andean, Caribbean, Orinoquia, Pacific **Dept.:** ANT, BOY, CAL, CAU, CHO, CUN, HUI, MAG, MET, NAR, QUI, SAN, TOL, VAC **Elev.:** 60-2200 m.a.s.l. **IUCN Habitat code:** 1,2,3,4,14 **Conservation status: (IUCN)** LC **(CPSG)** Potential LC

3666 *Desmodium molliculum* (Kunth) DC. (Fabaceae) **IPNI:** 1124646-2 **Common name:** amor seco, amorseco, amorseco del valle, pan, pega-pega, pegapega, pegasaya, sordé **Habit:** Herb, Climbing **Use:** ME **Origin:** Native **Region:** Andean **Dept.:** ANT, BOG, BOY, CAU, CUN, LAG, MAG, NAR, PUT, QUI, RIS, SAN, TOL, VAC **Elev.:** 1350-3420 m.a.s.l. **IUCN Habitat code:** 1,3,14 **Conservation status: (IUCN)** NE **(CPSG)** Potential LC

3667 *Desmodium orbiculare* Schltdl. (Fabaceae) **IPNI:** 491958-1 **Habit:** Herb, Subshrub **Use:** ME **Origin:** Native **Region:** Andean, Caribbean, Pacific **Dept.:** ANT, BOG, BOY, CAU, CHO, CUN, LAG, SAN, VAC **Elev.:** 40–1860 m a.s.l. **IUCN Habitat code:** 1,3,14 **Conservation status: (IUCN)** NE

3668 *Desmodium procumbens* (Mill.) C.L.Hitchc. (Fabaceae) **IPNI:** 78541-2 **Common name:** pegapega **Habit:** Herb, Shrub **Use:** AF, EU **Origin:** Native **Region:** Caribbean, Pacific **Dept.:** ANT, ATL, BOL, CHO, LAG, MAG **Elev.:** 10-550 m.a.s.l. **IUCN Habitat code:** 1,2,14 **Conservation status: (IUCN)** NE **(CPSG)** Potential LC

3669 *Desmodium scorpiurus* (Sw.) Desv. ex DC. (Fabaceae) **IPNI:** 135026-3 **Common name:** tregua **Habit:** Herb, Climbing **Use:** AF **Origin:** Native **Region:** Andean, Caribbean, Orinoquia, Pacific **Dept.:** ANT, ARA, ATL, BOL, BOY, CAL, CHO, COR, CUN, HUI, MAG, MET, SAN, TOL, VAC **Elev.:** 0-1460 m.a.s.l. **IUCN Habitat code:** 1,2,3,4,5,14 **Conservation status: (IUCN)** NE **(CPSG)** Potential LC

3670 *Desmodium sericophyllum* Schltdl. (Fabaceae) **IPNI:** 78576-2 **Common name:** amorseco **Habit:** Herb, Subshrub, Climbing **Use:** HF **Origin:** Native **Region:** Andean **Dept.:** BOY, CAU, CES, CUN, HUI, NAR, SAN **Elev.:** 1200–1380 m a.s.l. **IUCN Habitat code:** 3,14 **Conservation status: (IUCN)** NE

3671 *Desmodium tortuosum* (Sw.) DC. (Fabaceae) **IPNI:** 1028008-2 **Common name:** amorseco, cadillo, cadillo juancho, pegapega, robalgodón **Habit:** Herb, Shrub, Climbing **Use:** AF, EU, ME **Origin:** Native **Region:** Andean, Caribbean, Orinoquia **Dept.:** ANT, ATL, BOY, CAU, CUN, HUI, MAG, MET, NAR, SAN, TOL, VAC **Elev.:** 50–2000 m a.s.l. **IUCN Habitat code:** 1,2,3,5,14 **Conservation status: (IUCN)** NE

3672 *Desmodium uncinatum* (Jacq.) DC. (Fabaceae) **IPNI:** 1109494-2 **Common name:** amorseco, cadillo, lambedera, pegapega, zarza **Habit:** Herb, Shrub, Climbing **Use:** AF, HF, ME **Origin:** Native **Region:** Andean, Caribbean, Orinoquia **Dept.:** ANT, BOY, CAL, CAU, CHO, CUN, HUI, MAG, NAR, QUI, SAN, TOL, VAC, VID **Elev.:** 250–3300 m a.s.l. **IUCN Habitat code:** 1,3,5,14 **Conservation status: (IUCN)** NE

3673 *Dialium guianense* (Aubl.) Sandwith (Fabaceae) **IPNI:** 78900-2 **Common name:** abrojo, algarrobillo, angolito, borombolo, bálsamo, camaparu, canime, chicharrón, chocolatín de monte, corazón, corazón negro, culo de hierro, cumaricama, fique, frito, granadillo, guabillo, habillo, hueso, mano de tigre, paleta de perico, palo de pelejo, **Habit:** Tree **Use:** HF, MA, ME **Origin:** Native **Region:** Amazonia, Andean, Caribbean, Orinoquia, Pacific **Dept.:** AMA, ANT, BOL, CAQ, CHO, COR, GUV, MET, SAN, VAC, VAU **Elev.:** 5-1000 m.a.s.l. **IUCN Habitat code:** 1,2,3,4,5,14 **Conservation status: (IUCN)** LC **(CPSG)** Potential LC

3674 *Dicorynia paraensis* Benth. (Fabaceae) **IPNI:** 30086855-2 **Habit:** Tree **Use:** MA **Origin:** Native **Region:** Amazonia, Guiana Shield **Dept.:** AMA, CAQ, GUA, VAU **Elev.:** 100–120 m a.s.l. **IUCN Habitat code:** 1,3,14 **Conservation status: (IUCN)** LC

3675 *Dioclea guianensis* Benth. (Fabaceae) **IPNI:** 492636-1 **Common name:** alverja de monte **Habit:** Herb, Shrub, Climbing **Use:** SU **Origin:** Native **Region:** Amazonia, Andean, Orinoquia **Dept.:** AMA, ANT, BOL, CAU, CUN, GUV, MAG, MET, NSA, SAN, VAC, VAU, VID **Elev.:** 0-2000 m.a.s.l. **IUCN Habitat code:** 1,2,3,4,5,14 **Conservation status: (IUCN)** LC **(CPSG)** Potential LC

3676 *Dioclea virgata* (Rich.) Amshoff (Fabaceae) **IPNI:** 492690-1 **Habit:** Shrub, Liana, Climbing **Use:** ME **Origin:** Native **Region:** Amazonia, Andean, Caribbean, Orinoquia, Pacific **Dept.:** AMA, ANT, BOY, CAQ, CHO, COR, GUA, GUV, LAG, MET, NSA, SAN, VAC, VAU **Elev.:** 5-1800 m.a.s.l. **IUCN Habitat code:** 1,2,3,4,5,14 **Conservation status: (IUCN)** NE **(CPSG)** Potential LC

3677 *Diphysa americana* (Mill.) M.Sousa (Fabaceae) **IPNI:** 948330-1 **Habit:** Shrub **Use:** EU, FU, HF, MA, ME **Origin:** Native **Region:** Caribbean **Dept.:** SUC **Elev.:** 150–300 m a.s.l. **Conservation status: (IUCN)** LC

3678 *Diphysa carthagenensis* Jacq. (Fabaceae) **IPNI:** 492709-1 **Common name:** cacique, carate, chicharrón, vivaseca **Habit:** Shrub, Tree **Use:** FU, MA, ME **Origin:** Native **Region:** Caribbean **Dept.:** ATL, BOL, LAG, MAG, NSA **Elev.:** 5–60 m a.s.l. **IUCN Habitat code:** 1,2,3,14 **Conservation status: (IUCN)** LC

3679 *Diplotropis martiusii* Benth. (Fabaceae) **IPNI:** 492746-1 **Common name:** kò-ma-ma, palo amarillo, palo de pescado, árbol de fruta venenosa **Habit:** Small tree, Tree **Use:** HF, MA, ME **Origin:** Native **Region:** Amazonia **Dept.:** AMA, CAQ, GUA, GUV, VAU **Elev.:** 100–250 m a.s.l. **IUCN Habitat code:** 1,3,4,5 **Conservation status: (IUCN)** NE

3680 *Diplotropis purpurea* (Rich.) Amshoff (Fabaceae) **IPNI:** 492750-1 **Common name:** bálsamo negro, cacho, mascuriaga, sucupira, árbol de barbasco falso **Habit:** Tree **Use:** MA, ME **Origin:** Native **Region:** Amazonia, Andean, Caribbean, Orinoquia **Dept.:** AMA, ANT, CAQ, MET, VAC, VAU **Elev.:** 300–700 m a.s.l. **IUCN Habitat code:** 1,3,14 **Conservation status: (IUCN)** LC

3681 *Dipogon lignosus* (L.) Verdc. (Fabaceae) **IPNI:** 492756-1 **Habit:** Climbing **Use:** EU, HF, ME **Origin:** Cultivated, Naturalised **Region:** Andean **Dept.:** BOG, CUN **Elev.:** 2600–2600 m a.s.l. **Conservation status: (IUCN)** NE

3682 *Dipteryx micrantha* Harms (Fabaceae) **IPNI:** 492767-1 **Common name:** aceituno, charapilla, guacamayo, sarrapio **Habit:** Tree **Use:** HF, MA, ME **Origin:** Native **Region:** Amazonia, Guiana Shield, Orinoquia **Dept.:** AMA, CAQ, MET **Elev.:** 100–100 m a.s.l. **IUCN Habitat code:** 1,5 **Conservation status: (IUCN)** DD

3683 *Dipteryx odorata* (Aubl.) Forsyth f. (Fabaceae) **IPNI:** 492769-1 **Common name:** almendro, botagajo, botaguajo, charapilla, choibá, sarrapia, sarrapio, serrapia chibá, tonca, tonga, yape **Habit:** Tree **Use:** EU, HF, MA, ME, SU **Origin:** Native **Region:** Amazonia **Dept.:** AMA, CAQ **Elev.:** 240–240 m a.s.l. **IUCN Habitat code:** 1,3,14 **Conservation status: (IUCN)** DD

3684 *Dipteryx oleifera* Benth. (Fabaceae) **IPNI:** 492770-1 **Common name:** almendro, choibá, iguá **Habit:** Tree **Use:** EU, HF, MA, ME **Origin:** Native **Region:** Pacific **Dept.:** ANT, BOL, CHO, COR **Elev.:** 5-400 m.a.s.l. **IUCN Habitat code:** 1,3,5,14 **Conservation status: (IUCN)** LC **(CPSG)** VU

3685 *Dipteryx polyphylla* Huber (Fabaceae) **IPNI:** 82092-2 **Common name:** ovegue, palo de grulla, sarrapia, serrapia **Habit:** Tree **Use:** HF **Origin:** Native **Region:** Amazonia, Caribbean **Dept.:** AMA, CAQ **Elev.:** 40–300 m a.s.l. **IUCN Habitat code:** 1 **Conservation status: (IUCN)** NE

3686 *Dipteryx punctata* (S.F.Blake) Amshoff (Fabaceae) **IPNI:** 492778-1 **Common name:** sarrapia, sarrapia real **Habit:** Tree **Use:** HF, ME **Origin:** Native, Cultivated **Region:** Andean, Orinoquia **Dept.:** AMA, CAQ, CAS, CUN, GUA, GUV,

MET, TOL, VAC, VID **Elev.:** 150–400 m a.s.l. **IUCN Habitat code:** 1,3,4,14 **Conservation status:** **(IUCN)** LC

3687 *Dipteryx rosea* Spruce ex Benth. (Fabaceae) **IPNI:** 492781-1 **Common name:** sarrapia, sarrapio **Habit:** Tree **Use:** HF, ME **Origin:** Native **Region:** Amazonia, Orinoquia **Dept.:** VAU, VID **Elev.:** 100–350 m a.s.l. **IUCN Habitat code:** 1,4,14 **Conservation status:** **(IUCN)** NE

3688 *Dussia lehmannii* Harms (Fabaceae) **IPNI:** 86180-2 **Common name:** bagatá, chocho, coropo, embagatao, purga, tara **Habit:** Tree **Use:** MA **Origin:** Native **Region:** Andean, Pacific **Dept.:** ANT, CHO, NAR, VAC **Elev.:** 0–1400 m a.s.l. **IUCN Habitat code:** 1,3,14 **Conservation status:** **(IUCN)** LC

3689 *Entada gigas* (L.) Fawc. & Rendle (Fabaceae) **IPNI:** 1078643-2 **Common name:** bejuco de estribo, bejuco de garza, bejuco parta, congolo, cóngolo, pastillas de chocolate, riñoncitos **Habit:** Liana **Use:** EU, HF, MA, ME **Origin:** Native **Region:** Andean, Caribbean, Pacific **Dept.:** ANT, CHO, CUN, LAG, MAG, NAR, NSA, RIS, SAN **Elev.:** 30–1420 m a.s.l. **IUCN Habitat code:** 1,2,3,5,14 **Conservation status:** **(IUCN)** NE

3690 *Entada phaseoloides* (L.) Merr. (Fabaceae) **IPNI:** 493866-1 **Common name:** bejuco cadeno, bejuco parta, cobalongo, congolo, habo, ojo de buey, parta **Use:** EU, HF, IF, MA, ME, PO, SU **Dept.:** CAS**Conservation status:** **(IUCN)** NE

3691 *Entada polyphylla* Benth. (Fabaceae) **IPNI:** 91067-2 **Common name:** bejuco garcero, bejuco jabón, escoba de macaco, jipioca, jipió, ojo de buey, pashaco, sichaca **Habit:** Subshrub, Liana **Use:** ME **Origin:** Native **Region:** Amazonia, Andean, Guiana Shield, Orinoquia, Pacific **Dept.:** AMA, ARA, CAQ, CAU, GUV, MET, PUT **Elev.:** 100–1500 m a.s.l. **IUCN Habitat code:** 1,3,5,14 **Conservation status:** **(IUCN)** NE

3692 *Entada polystachya* (L.) DC. (Fabaceae) **IPNI:** 1009172-2 **Common name:** bejuco de garza, bejuco garcero, bejuco zorro, carpeta, guamillo, zapatillo **Habit:** Liana **Use:** ME **Origin:** Native **Region:** Caribbean, Pacific **Dept.:** ANT, ATL, BOL, CHO, COR, LAG, MAG **Elev.:** 7–295 m a.s.l. **IUCN Habitat code:** 1,2,3,5,14 **Conservation status:** **(IUCN)** LC

3693 *Enterolobium cyclocarpum* (Jacq.) Griseb. (Fabaceae) **IPNI:** 1037928-2 **Common name:** aguacerito, anjera, anjero, campanillo, caracaro, carita, carito, caro, caro-caro, carocaro, corotú, curutú, dormidero, dormilón, guacamayo, guanacaste, guarano blanco, hueso de pescado, jaboncillo, matojiro, orejera, orejero, orejero piñón, orejo, orejue **Habit:** Tree **Use:** AF, EU, FU, HF, IF, MA, ME, PO **Origin:** Native, Cultivated **Region:** Amazonia, Andean, Guiana Shield, Caribbean **Dept.:** AMA, ANT, ATL, BOY, COR, CUN, MAG, MET, SAN, SUC, VAC, VAU **Elev.:** 0-1200 m.a.s.l. **IUCN Habitat code:** 1,2,3,4,5,14 **Conservation status:** **(IUCN)** LC **(CPSG)** Potential LC

3694 *Enterolobium schomburgkii* (Benth.) Benth. (Fabaceae) **IPNI:** 493928-1 **Common name:** caracaro, dormidero, dormilón, hediondo, jaboncillo, orejero, resbalamono, revientatetas **Habit:** Tree **Use:** AF, EU, HF, MA **Origin:** Native **Region:** Guiana Shield, Orinoquia **Dept.:** GUV, MET, SAN, VID **Elev.:** 80–700 m a.s.l. **IUCN Habitat code:** 1,3,4,5,14 **Conservation status:** **(IUCN)** LC

3695 *Eperua falcata* Aubl. (Fabaceae) **IPNI:** 493934-1 **Habit:** Tree **Use:** MA, ME **Origin:** Native **Region:** Amazonia **Dept.:** VAU **Elev.:** 370–370 m a.s.l. **IUCN Habitat code:** 14 **Conservation status:** **(IUCN)** LC

3696 *Eperua purpurea* Benth. (Fabaceae) **IPNI:** 493943-1 **Common name:** alcanfor, cincuentacentavos, guanano, guapa, guayacán cavío, palo cincuentacentavos, palo pegante, pegante, yévaro, zapatero **Habit:** Tree **Use:** HF, MA **Origin:** Native **Region:** Amazonia, Guiana Shield **Dept.:** AMA, GUA, VAU **Elev.:** 120–320 m a.s.l. **IUCN Habitat code:** 1 **Conservation status:** **(IUCN)** LC

3697 *Eriosema diffusum* (Kunth) G.Don (Fabaceae) **IPNI:** 494021-1 **Habit:** Herb **Use:** ME **Origin:** Native **Region:** Andean **Dept.:** ANT, BOG, CAU, CES, CUN, MAG **Elev.:** 1200–2300 m a.s.l. **IUCN Habitat code:** 1,3,14 **Conservation status:** **(IUCN)** NE

3698 *Erythrina amazonica* Krukoff (Fabaceae) **IPNI:** 96551-2 **Habit:** Small tree, Tree **Use:** ME **Origin:** Native **Region:** Amazonia, Guiana Shield, Orinoquia **Dept.:** MET, PUT, VID **Elev.:** 225–310 m a.s.l. **IUCN Habitat code:** 1,14 **Conservation status:** **(IUCN)** LC

3699 *Erythrina berteroana* Urb. (Fabaceae) **IPNI:** 96559-2 **Common name:** coralillo, cresta de gallo, matacaimán, patsua, peronia, peronio, pionío, pito, piñón de cerca, plonío, poro **Habit:** Shrub, Small tree, Tree **Use:** AF, EU, HF, MA, ME, PO **Origin:** Native **Region:** Andean, Caribbean, Pacific **Dept.:** ANT, ATL, BOL, CES, CHO, LAG, MAG **Elev.:** 30–1210 m a.s.l. **IUCN Habitat code:** 1,2,3,14 **Conservation status:** **(IUCN)** NE

3700 *Erythrina corallodendron* L. (Fabaceae) **IPNI:** 316421-2 **Common name:** chochos, coraltree **Use:** EU, HF, MA, ME **IUCN Habitat code:** 1,2,3,14 **Conservation status:** **(IUCN)** NE

3701 *Erythrina costaricensis* Micheli (Fabaceae) **IPNI:** 96578-2 **Common name:** chocho, chochos, chucho, gachingo, seca, secá **Habit:** Shrub, Tree **Use:** ME **Origin:** Native **Region:** Andean, Caribbean, Pacific **Dept.:** ANT, CHO, CUN, HUI, MAG, MET, RIS **Elev.:** 10–1600 m a.s.l. **IUCN Habitat code:** 1,3,5,14 **Conservation status:** **(IUCN)** LC

3702 *Erythrina crista-galli* L. (Fabaceae) **IPNI:** 494400-1 **Common name:** cresta de gallo, crestagallo **Habit:** Shrub, Small tree **Use:** EU, FU, HF, MA, ME **Origin:** Cultivated **Region:** Andean **Dept.:** ANT, VAC **Elev.:** 1100–1500 m a.s.l. **IUCN Habitat code:** 1,3,14 **Conservation status:** **(IUCN)** LC

3703 *Erythrina edulis* Triana ex Micheli (Fabaceae) **IPNI:** 96586-2 **Common name:** balay, baluy, balú, cachimbo, calú, cantagallo, caporuto, chachafruta, chachafruto, chaforuto, chaporuto, chimbo poruto, colorado, frisol calu, frisol calú, fríjol de monte, fríjol gualuy, fríjol nopas, fríjol nopaz, fríjol nopás, güimo, güimó, ite, jite, **Habit:** Shrub, Tree **Use:** AF, EU, HF, MA, ME **Origin:** Native, Cultivated **Region:** Andean **Dept.:** ANT, BOY, CAU, CHO, CUN, HUI, MET, NAR, NSA, PUT, QUI, SAN, TOL, VAC **Elev.:** 750–2600 m a.s.l. **IUCN Habitat code:** 1,2,3,4,14 **Conservation status:** **(IUCN)** LC

3704 *Erythrina fusca* Lour. (Fabaceae) **IPNI:** 494431-1 **Common name:** amacise, amasisa, amasise, bucago, bucare, bucaro, búcare, búcaro, búcaro de pantano, cachimbo, cachingo, cantagallo, cantagallo colorado, cantagallos, chambul, chengue, chengué, chocha, chochas, cámbulo, cámbulo rosado, gachico, gallito de pantano, kasho **Habit:** Tree, Liana **Use:** AF, EU, FU, HF, IF, MA, ME, PO, SU **Origin:** Native **Region:** Amazonia, Andean, Caribbean, Pacific **Dept.:** AMA, ANT, ARA, ATL, CHO, COR, CUN, GUV, HUI, MAG, MET, PUT, SAN, TOL, VAC **Elev.:** 5-1560 m.a.s.l. **IUCN Habitat code:** 1,2,3,4,5,14 **Conservation status:** **(IUCN)** LC **(CPSG)** Potential LC

3705 *Erythrina poeppigiana* (Walp.) O.F.Cook (Fabaceae) **IPNI:** 96637-2 **Common name:** anaco, anaco rojo, barbatuscas, barbatusco, barbatuzcas, búcare, búcaro, cachimbo, cachingo, cachungo, ceibo, chocho, cámbulo, guaney, pito jigante, písamo, saibo, saivo **Habit:** Tree **Use:** AF, EU, HF, MA, ME, SU **Origin:** Native **Region:** Amazonia, Andean, Orinoquia, Pacific **Dept.:** ANT, BOY, CAL, CAS, CAU, CHO, CUN, GUV, HUI, MAG, MET, NSA, PUT, QUI, RIS, SAN, VAC **Elev.:** 0-2080 m.a.s.l. **IUCN Habitat code:** 1,2,3,5,14 **Conservation status:** **(IUCN)** LC **(CPSG)** Potential LC

3706 *Erythrina rubrinervia* Kunth (Fabaceae) **IPNI:** 494567-1 **Common name:** balú, bucare peonía, chochitos, chocho, chocho de árbol, chochos, chochos de tree, peronio, porotillo, poroto, sirigual, siriguay, siringay **Habit:** Shrub, Tree **Use:** EU, HF, MA, ME, PO, SU **Origin:** Native **Region:** Andean, Caribbean, Pacific **Dept.:** ANT, BOG, BOY, CAL, CAU, CES, CHO, CUN, HUI, LAG, MAG, NAR, PUT, QUI, RIS, SAN, TOL, VAC **Elev.:** 20-2700 m.a.s.l. **IUCN Habitat code:** 1,2,3,4,5,14 **Conservation status:** **(IUCN)** LC **(CPSG)** Potential LC

3707 *Erythrina ulei* Harms (Fabaceae) **IPNI:** 96661-2 **Habit:** Tree **Use:** ME **Origin:** Native **Region:** Andean **Dept.:** CAQ, PUT **Elev.:** 810–910 m a.s.l. **IUCN Habitat code:** 3 **Conservation status:** **(IUCN)** LC

3708 *Erythrina variegata* L. (Fabaceae) **IPNI:** 494614-1 **Common name:** caraqueño **Habit:** Tree **Use:** AF, EU, FU, HF, IF, MA, ME, PO, SU **Origin:** Cultivated **Region:** Caribbean **Dept.:** ANT, VAC **Elev.:** 20–900 m a.s.l. **Conservation status:** **(IUCN)** LC

3709 *Erythrina velutina* Willd. (Fabaceae) **IPNI:** 494616-1 **Common name:** pionía **Habit:** Tree **Use:** AF, EU, MA, ME **Origin:** Native **Region:** Caribbean **Dept.:** LAG, MAG **Elev.:** 10–150 m a.s.l. **IUCN Habitat code:** 2,3 **Conservation status:** **(IUCN)** NE

3710 *Flemingia strobilifera* (L.) W.T.Aiton (Fabaceae) **IPNI:** 495307-1 **Habit:** Shrub **Use:** AF, EU, HF, MA, ME, SU **Origin:** Naturalised **Region:** Pacific **Dept.:** ANT **Elev.:** 100–100 m a.s.l. **IUCN Habitat code:** 2,14 **Conservation status:** **(IUCN)** NE

3711 *Galactia marginalis* Benth. (Fabaceae) **IPNI:** 495473-1 **Use:** HF **Origin:** Native **IUCN Habitat code:** 2 **Conservation status:** **(IUCN)** NE

3712 *Galactia striata* (Jacq.) Urb. (Fabaceae) **IPNI:** 106712-2 **Common name:** frijolillo **Habit:** Herb, Liana, Climbing **Use:** EU, ME **Origin:** Native **Region:** Andean, Caribbean, Orinoquia **Dept.:** ANT, ATL, BOL, BOY, CAU, CES, COR, CUN, HUI, LAG, MAG, NAR, SAN, TOL, VAC, VID **Elev.:** 25-2000 m.a.s.l. **IUCN Habitat code:** 1,2,3,14 **Conservation status:** **(IUCN)** NE **(CPSG)** Potential LC

3713 *Genista canariensis* L. (Fabaceae) **IPNI:** 495939-1 **Use:** ME **Region:** Andean **Dept.:** BOG**IUCN Habitat code:** 1 **Conservation status:** **(IUCN)** LC

3714 *Genista monspessulana* (L.) L.A.S.Johnson (Fabaceae) **IPNI:** 496189-1 **Common name:** retama, retama blanca, retama de jardín, retamo, retamo blanco, retamo de jardín, retamo liso, retamo pequeño **Habit:** Herb, Shrub **Use:** EU, MA, ME **Origin:** Naturalised **Region:** Andean **Dept.:** ANT, BOG, CUN, TOL **Elev.:** 2000–3100 m a.s.l. **IUCN Habitat code:** 1,3,4,5,8,14 **Conservation status:** **(IUCN)** NE

3715 *Geoffroea spinosa* Jacq. (Fabaceae) **IPNI:** 496501-1 **Common name:** jigua, puy, seca, silbadera, silbadero **Habit:** Shrub, Tree **Use:** HF, MA, ME **Origin:** Native **Region:** Caribbean **Dept.:** ATL, BOL, HUI, LAG, VAC **Elev.:** 0–1300 m a.s.l. **IUCN Habitat code:** 1,2,5,14 **Conservation status:** **(IUCN)** LC

3716 *Gliricidia sepium* (Jacq.) Steud. (Fabaceae) **IPNI:** 496639-1 **Common name:** acacia, desnudo florecido, madre del cacao, madrecacao, matarratón, matarratón criollo, matarratón de Castilla, moutaraton, piñón florido, raborratón **Habit:** Shrub, Tree **Use:** AF, EU, FU, GS, HF, IF, MA, ME, PO, SU **Origin:** Native, Cultivated **Region:** Amazonia, Andean, Caribbean, Pacific **Dept.:** AMA, ANT, ATL, BOL, CAL, CAQ, CAU, CES, CHO, CUN, HUI, MAG, NAR, NSA, RIS, SAP, SAN, SUC, TOL, VAC **Elev.:** 5–1900 m a.s.l. **IUCN Habitat code:** 1,2,3,4,5,14 **Conservation status:** **(IUCN)** LC

3717 *Glycine max* (L.) Merr. (Fabaceae) **IPNI:** 60450240-2 **Common name:** arveja soya, frisol sojo, fríjol soya, haba soya, poroto, soja, soya **Habit:** Herb, Shrub **Use:** AF, EU, FU, GS, HF, MA, ME, PO **Origin:** Native, Cultivated **Region:** Amazonia **Dept.:** CAQ, TOL, VAC **Elev.:** 500–1100 m a.s.l. **IUCN Habitat code:** 1,3,14 **Conservation status:** **(IUCN)** NE

3718 *Grona adscendens* (Sw.) H.Ohashi & K.Ohashi (Fabaceae) **IPNI:** 77187388-1 **Common name:** amor de campo, amor seco, amorseca, amorseco, amorseco pequeño, cadillo, cadillo pipa, cargadita, garrapatilla, pega pega, pegaparuma, pegapega, pijao, prendeamor, seca **Habit:** Herb, Shrub, Liana, Climbing **Use:** AF, EU, HF, ME, SU **Origin:** Native **Region:** Amazonia, Andean, Caribbean, Orinoquia, Pacific **Dept.:** AMA, ANT, CAL, CAQ, CAS, CAU, CHO, CUN, GUA, HUI, LAG, MAG, MET, NAR, PUT, SAN, VAC, VAU, VID **Elev.:** 0-2600 m.a.s.l. **IUCN Habitat code:** 1,2,3,4,5,14 **Conservation status:** **(IUCN)** NE **(CPSG)** Potential LC

3719 *Grona barbata* (L.) H.Ohashi & K.Ohashi (Fabaceae) **IPNI:** 77187394-1 **Common name:** amorseco, bünü taju boto, cargadita, cáncer de loma, papabü natsaketsiwaü, pegapega, wepe taxusinae **Habit:** Herb, Shrub, Climbing **Use:** HF, ME, SU **Origin:** Native **Region:** Amazonia, Andean, Guiana Shield, Caribbean, Orinoquia **Dept.:** AMA, ANT, ARA, CAQ, CAS, CAU, COR, CUN, GUA, GUV, HUI, MAG, MET, NAR, SAN, TOL, VAC, VID **Elev.:** 25-1820 m.a.s.l. **IUCN Habitat code:** 1,2,3,4,14 **Conservation status:** **(IUCN)** NE **(CPSG)** Potential LC

3720 *Grona heterocarpos* (L.) H.Ohashi & K.Ohashi (Fabaceae) **IPNI:** 77187416-1 **Use:** ME **Origin:** Naturalised **Conservation status:** **(IUCN)** NE

3721 *Grona triflora* (L.) H.Ohashi & K.Ohashi (Fabaceae) **IPNI:** 77187443-1 **Common name:** amorseco, cadillo, pega pega **Habit:** Herb **Use:** AF, EU, HF, ME **Origin:** Native **Region:** Andean, Caribbean, Orinoquia, Pacific **Dept.:** ANT, CAU, CHO, MET, NAR, TOL, VAC **Elev.:** 100–1700 m a.s.l. **IUCN Habitat code:** 1,2,3,4,5,14 **Conservation status:** **(IUCN)** NE

3722 *Guilandina bonduc* L. (Fabaceae) **IPNI:** 114706-2 **Common name:** huevo'egato, mate **Habit:** Climbing **Use:** EU, HF, MA, ME, PO, SU **Origin:** Native **Region:** Caribbean, Pacific **Dept.:** ANT, BOL, CAU, CHO, MAG, NAR, SAP, VAC **Elev.:** 0–200 m a.s.l. **IUCN Habitat code:** 1,14 **Conservation status:** **(IUCN)** LC

3723 *Guilandina major* (Medik.) Small (Fabaceae) **IPNI:** 114718-2 **Habit:** Tree **Use:** EU, MA, ME **Origin:** Cultivated **Region:** Caribbean **Dept.:** ATL, MAG **Elev.:** 10–60 m a.s.l. **Conservation status:** **(IUCN)** NE

3724 *Haematoxylum brasiletto* H.Karst. (Fabaceae) **IPNI:** 497227-1 **Common name:** ataa, brasil, brasilete, brasileto, brasilito, brazil, brazilete, hala, palo brasil, palo de Brasil, palo sanguíneo campeche **Habit:** Shrub, Tree **Use:** FU, MA, ME, PO **Origin:** Native **Region:** Andean, Caribbean **Dept.:** BOL, LAG, MAG, SAN **Elev.:** 10–1200 m a.s.l. **IUCN Habitat code:** 1,2,3,4,14 **Conservation status:** **(IUCN)** LC

3725 *Haematoxylum campechianum* L. (Fabaceae) **IPNI:** 316401-2 **Common name:** brazilete, brazilito, bwa kanpéch, palo de brazil **Use:** EU, FU, HF, IF, MA, ME **Dept.:** MAG**Conservation status:** **(IUCN)** LC

3726 *Helicotropis linearis* (Kunth) A.Delgado (Fabaceae) **IPNI:** 60457240-2 **Common name:** punuyoro, yerba mara **Habit:** Herb **Use:** SU **Origin:** Native **Region:** Amazonia, Caribbean, Orinoquia **Dept.:** AMA, CAS, MET, SUC, VID **Elev.:** 140–600 m a.s.l. **IUCN Habitat code:** 1,2,3,4,5,14 **Conservation status:** **(IUCN)** NE

3727 *Heterostemon mimosoides* Desf. (Fabaceae) **IPNI:** 498316-1 **Common name:** palo de carguero **Habit:** Shrub, Tree **Use:** ME **Origin:** Native **Region:** Amazonia, Orinoquia **Dept.:** AMA, CAQ, GUA, VAU, VID **Elev.:** 0–550 m a.s.l. **IUCN Habitat code:** 1,3,5,14 **Conservation status:** **(IUCN)** LC

3728 *Hydrochorea corymbosa* (Rich.) Barneby & J.W.Grimes (Fabaceae) **IPNI:** 989341-1 **Habit:** Tree **Use:** MA **Origin:** Native **Region:** Amazonia, Guiana Shield, Orinoquia **Dept.:** AMA, GUA, MET, VAU **Elev.:** 150–250 m a.s.l. **IUCN Habitat code:** 1,2,3,4,5,14 **Conservation status:** **(IUCN)** LC

3729 *Hydrochorea marginata* (Spruce ex Benth.) Barneby & J.W.Grimes (Fabaceae) **IPNI:** 989342-1 **Habit:** Tree **Use:** HF **Origin:** Native **Region:** Amazonia, Guiana Shield **Dept.:** AMA, MET, VAU **Elev.:** 0–200 m a.s.l. **IUCN Habitat code:** 1,4,5 **Conservation status:** **(IUCN)** NE

3730 *Hymenaea courbaril* L. (Fabaceae) **IPNI:** 125420-2 **Common name:** algarroba, algarrobillo, algarrobo, carapa, copal, copal americano, corobore, corobore., guapinol, mangle duro, nazareno, pecueca, ámbar **Habit:** Tree **Use:** AF, EU, GS, HF, IF, MA, ME, SU **Origin:** Native, Cultivated **Region:** Amazonia, Andean, Caribbean, Orinoquia **Dept.:** AMA, ANT, ATL, BOL, BOY, CAU, LAG, MAG, MET, QUI, SAN, VID **Elev.:** 30–2000 m a.s.l. **IUCN Habitat code:** 1,2,3,4,5,14 **Conservation status:** **(IUCN)** LC

3731 *Hymenaea intermedia* Ducke (Fabaceae) **IPNI:** 125429-2 **Common name:** algarrobo **Habit:** Tree **Use:** HF, ME **Origin:** Native **Region:** Amazonia **Dept.:** AMA **Elev.:** 130–130 m a.s.l. **IUCN Habitat code:** 1,5 **Conservation status:** **(IUCN)** NE

3732 *Hymenaea martiana* (Hayne) Hayne (Fabaceae)
IPNI: 498789-1
Common name: algarrobo, basijo, chirivaña, cuero de marrano, guacamari **Habit:** Tree **Use:** HF, MA, ME **Origin:** Native **Region:** Amazonia, Pacific **Dept.:** ANT, VAU **Elev.:** 10–250 m a.s.l. **IUCN Habitat code:** 5,14 **Conservation status:** (IUCN) LC

3733 *Hymenaea oblongifolia* Huber (Fabaceae)
IPNI: 125434-2
Common name: algarroba, algarrobilla, algarrobillo, algarrobo, brasil, coca, coca de mico, eekegua, nazareno, palo de carapa, roble **Habit:** Tree **Use:** AF, EU, HF, MA, ME **Origin:** Native **Region:** Amazonia, Guiana Shield, Caribbean, Orinoquia, Pacific **Dept.:** AMA, ANT, CAQ, CHO, GUV, MET, PUT, VAC, VAU **Elev.:** 5-600 m.a.s.l. **IUCN Habitat code:** 1,3,14 **Conservation status:** (IUCN) LC (CPSG) NT

3734 *Hymenaea parvifolia* Huber (Fabaceae)
IPNI: 125440-2
Common name: algarrobo, azúcar hayo, azúcar huayo, copal, guama de altura **Habit:** Tree **Use:** HF, MA, ME **Origin:** Native **Region:** Amazonia **Dept.:** AMA, CAQ **Elev.:** 230–230 m a.s.l. **IUCN Habitat code:** 1 **Conservation status:** (IUCN) LC

3735 *Hymenaea reticulata* Ducke (Fabaceae)
IPNI: 125441-2
Common name: coca, coca de mico, pecuequilla **Use:** HF **Origin:** Native **IUCN Habitat code:** 1 **Conservation status:** (IUCN) NE

3736 *Hymenolobium excelsum* Ducke (Fabaceae)
IPNI: 125608-2
Habit: Tree **Use:** MA **Origin:** Native **Region:** Guiana Shield **Dept.:** GUA **Elev.:** 100–100 m a.s.l. **IUCN Habitat code:** 1,3,4 **Conservation status:** (IUCN) NE

3737 *Hymenolobium heterocarpum* Ducke (Fabaceae)
IPNI: 125611-2
Habit: Tree **Use:** HF **Origin:** Native **Region:** Amazonia **Dept.:** GUV **Elev.:** 240–240 m a.s.l. **IUCN Habitat code:** 1,4,14 **Conservation status:** (IUCN) NE

3738 *Hymenolobium petraeum* Ducke (Fabaceae)
IPNI: 125617-2
Common name: marimari **Habit:** Tree **Use:** MA **Origin:** Native **Region:** Amazonia, Orinoquia **Dept.:** AMA, VID **Elev.:** 90–150 m a.s.l. **IUCN Habitat code:** 1,3,4 **Conservation status:** (IUCN) LC

3739 *Indigofera hendecaphylla* Jacq. (Fabaceae)
IPNI: 499440-1
Habit: Herb **Use:** EU, MA, ME, PO **Origin:** Naturalised **Region:** Andean **Dept.:** ANT **Elev.:** 2500–2500 m a.s.l. **IUCN Habitat code:** 14 **Conservation status:** (IUCN) NE

3740 *Indigofera hirsuta* L. (Fabaceae)
IPNI: 499456-1
Common name: añil, púnciga **Habit:** Herb, Shrub **Use:** AF, EU, HF, MA, ME, PO **Origin:** Naturalised **Region:** Andean, Orinoquia **Dept.:** ARA, BOY, CAL, MET, TOL, VAC **Elev.:** 150–1000 m a.s.l. **IUCN Habitat code:** 1,2,3,4,5,14 **Conservation status:** (IUCN) NE

3741 *Indigofera lespedezioides* Kunth (Fabaceae)
IPNI: 499569-1
Common name: añil, añilón, barbasco **Habit:** Herb, Shrub **Use:** ME **Origin:** Native **Region:** Andean, Caribbean, Orinoquia **Dept.:** ANT, ARA, BOG, BOL, CAS, CAU, CES, CUN, HUI, MAG, MET, NAR, SAN, TOL, VAC, VID **Elev.:** 0-1900 m.a.s.l. **IUCN Habitat code:** 1,2,3,4,5,14 **Conservation status:** (IUCN) NE (CPSG) Potential LC

3742 *Indigofera microcarpa* Desv. (Fabaceae)
IPNI: 499658-1
Habit: Herb **Use:** AF, EU, ME **Origin:** Native **Region:** Caribbean **Dept.:** LAG **Elev.:** 100–100 m a.s.l. **Conservation status:** (IUCN) NE

3743 *Indigofera spicata* Forssk. (Fabaceae)
IPNI: 500027-1
Use: AF, EU, HF, MA, ME, PO **Origin:** Naturalised **IUCN Habitat code:** 1,14 **Conservation status:** (IUCN) NE

3744 *Indigofera suffruticosa* Mill. (Fabaceae)
IPNI: 500063-1
Common name: anilina, añil, añil indigo, añil silvestre, añilina, curí, guarrás, planta de guineíto, platanito, raíz colorada, wara atsu **Habit:** Subshrub **Use:** EU, HF, MA, ME, PO **Origin:** Native **Region:** Andean, Caribbean, Orinoquia, Pacific **Dept.:** AMA, ANT, ATL, BOL, BOY, CAQ, CAS, CAU, CHO, CUN, GUV, HUI, LAG, MAG, MET, NSA, QUI, SAP, SAN, TOL, VAC, VID **Elev.:** 0-2700 m.a.s.l. **IUCN Habitat code:** 1,2,3,4,5,14 **Conservation status:** (IUCN) NE (CPSG) Potential LC

3745 *Indigofera trita* L.f. (Fabaceae)
IPNI: 500135-1
Common name: aceite, arroz con coco, añil **Habit:** Herb **Use:** AF, EU, ME **Origin:** Naturalised **Region:** Andean, Caribbean **Dept.:** ANT, ATL, BOL, CAU, CUN, HUI, LAG, MAG, SUC, VAC, VAU **Elev.:** 40-1400 m.a.s.l. **IUCN Habitat code:** 1,2,3,5,14 **Conservation status:** (IUCN) LC (CPSG) Potential LC

3746 *Inga acreana* Harms (Fabaceae)
IPNI: 128353-2
Common name: guamo, guamo machete **Habit:** Small tree **Use:** HF, ME **Origin:** Native **Region:** Amazonia, Andean, Guiana Shield, Orinoquia, Pacific **Dept.:** AMA, ANT, CAQ, CHO, CUN, GUV, HUI, MET, NAR, PUT, TOL, VAC **Elev.:** 5-2450 m.a.s.l. **IUCN Habitat code:** 1,3,14 **Conservation status:** (IUCN) LC (CPSG) Potential LC

3747 *Inga acrocephala* Steud. (Fabaceae)
IPNI: 128354-2
Common name: boquiblanco, churimo, guama de churuco, guamillo, guamo, guamo chirimo, guamo churimo, guamo de guacamayo, guamo de mico, guamo machete, palo de pintura, palo de sábalo, palopintura **Habit:** Tree **Use:** HF **Origin:** Native **Region:** Amazonia, Andean, Guiana Shield, Orinoquia, Pacific **Dept.:** AMA, ANT, BOG, BOY, CAL, CAQ, CHO, CUN, GUV, MAG, MET, RIS, SAN, VAC, VAU **Elev.:** 30-2250 m.a.s.l. **IUCN Habitat code:** 1,3,4,5,14 **Conservation status:** (IUCN) LC (CPSG) Potential LC

3748 *Inga acuminata* Benth. (Fabaceae)
IPNI: 313763-2
Common name: guamo, guamo churimo **Habit:** Small tree **Use:** HF **Origin:** Native **Region:** Amazonia, Andean, Guiana Shield, Caribbean, Orinoquia, Pacific **Dept.:** ANT, BOG, BOL, BOY, CAL, CHO, CUN, GUV, MAG, MET, RIS, VAC, VAU **Elev.:** 100-2100 m.a.s.l. **IUCN Habitat code:** 1,3,4,5,14 **Conservation status:** (IUCN) LC (CPSG) Potential LC

3749 *Inga affinis* DC. (Fabaceae)
IPNI: 128358-2
Use: HF, MA, ME **Origin:** Native **IUCN Habitat code:** 1,3,4,5,14 **Conservation status:** (IUCN) NE

3750 *Inga alata* Benoist (Fabaceae)
IPNI: 128360-2
Habit: Tree **Use:** HF **Origin:** Native **Region:** Guiana Shield **Dept.:** MET, VAU **Elev.:** 100–100 m a.s.l. **IUCN Habitat code:** 1,14 **Conservation status:** (IUCN) LC

3751 *Inga alba* (Sw.) Willd. (Fabaceae)
IPNI: 1129244-2
Common name: carbonero, churimo, guamo, guamo caliente, guamo churimo, guamos, oti, otinae **Habit:** Tree **Use:** EU, FU, HF, MA, ME **Origin:** Native **Region:** Amazonia, Andean, Guiana Shield, Orinoquia, Pacific **Dept.:** AMA, ANT, CAQ, CHO, CUN, GUV, MET, VAU, VID **Elev.:** 90–1025 m a.s.l. **IUCN Habitat code:** 1,2,3,4,5,14 **Conservation status:** (IUCN) LC

3752 *Inga auristellae* Harms (Fabaceae)
IPNI: 128375-2
Common name: churimbo, guamito **Habit:** Tree **Use:** HF, MA **Origin:** Native **Region:** Amazonia, Guiana Shield, Orinoquia, Pacific **Dept.:** AMA, CAU, GUA, PUT **Elev.:** 80–600 m a.s.l. **IUCN Habitat code:** 1,14 **Conservation status:** (IUCN) LC

3753 *Inga bourgoni* (Aubl.) DC. (Fabaceae)
IPNI: 500286-1
Common name: guama de hoja grande, guamo machetero **Habit:** Small tree **Use:** HF **Origin:** Native **Region:** Amazonia, Andean, Guiana Shield **Dept.:** AMA, CAQ, PUT, VAU **Elev.:** 100–1420 m a.s.l. **IUCN Habitat code:** 1,2,3 **Conservation status:** (IUCN) NE

3754 *Inga brachyrhachis* Harms (Fabaceae)
IPNI: 128393-2
Common name: guamo **Habit:** Small tree, Tree **Use:** HF **Origin:** Native **Region:** Amazonia, Andean, Guiana Shield, Orinoquia **Dept.:** AMA, CAQ, GUV, MET, VAU **Elev.:** 90–1200 m a.s.l. **IUCN Habitat code:** 1,3,14 **Conservation status:** (IUCN) LC

3755 *Inga brachystachys* Ducke (Fabaceae)
IPNI: 288863-2
Common name: guamillo **Habit:** Tree **Use:** HF **Origin:** Native **Region:** Amazonia **Dept.:** AMA **Elev.:** 100–160 m a.s.l. **IUCN Habitat code:** 1,4 **Conservation status:** (IUCN) LC

3756 *Inga capitata* Desv. (Fabaceae)
IPNI: 128414-2
Common name: barbasco, churimo, guamo, guamo plancheto, palo de sábalo **Habit:** Small tree **Use:** HF, MA **Origin:** Native **Region:** Amazonia, Andean, Guiana Shield, Orinoquia **Dept.:** AMA, CAL, CAQ, CAS, CHO, GUV, MET, PUT, VAC, VAU **Elev.:** 50–1000 m a.s.l. **IUCN Habitat code:** 1,3,4,5,14 **Conservation status:** (IUCN) LC

3757 *Inga cayennensis* Sagot ex Benth. (Fabaceae)
IPNI: 128427-2
Common name: guamo **Habit:** Tree **Use:** HF **Origin:** Native **Region:** Amazonia, Andean, Guiana Shield, Pacific **Dept.:** ANT, BOY, CHO, VAC, VAU **Elev.:** 50–1800 m a.s.l. **IUCN Habitat code:** 1,3,5,14 **Conservation status: (IUCN)** LC

3758 *Inga chartacea* Poepp. (Fabaceae)
IPNI: 128432-2
Habit: Small tree **Use:** HF **Origin:** Native **Region:** Amazonia, Andean, Pacific **Dept.:** AMA, CAQ, VAC **Elev.:** 0–1000 m a.s.l. **IUCN Habitat code:** 1,3,5 **Conservation status: (IUCN)** LC

3759 *Inga chocoensis* Killip ex T.S.Elias (Fabaceae)
IPNI: 128436-2
Common name: guabo mico **Habit:** Tree **Use:** HF, MA **Origin:** Native **Region:** Andean, Pacific **Dept.:** CHO, VAC **Elev.:** 0–800 m a.s.l. **IUCN Habitat code:** 1,14 **Conservation status: (IUCN)** LC

3760 *Inga ciliata* C.Presl (Fabaceae)
IPNI: 128440-2
Common name: guama, guamo **Habit:** Small tree, Tree **Use:** HF **Origin:** Native **Region:** Amazonia, Andean, Guiana Shield **Dept.:** AMA, BOY, CAQ, CAS, PUT, VAC, VAU **Elev.:** 200–2600 m a.s.l. **IUCN Habitat code:** 1,3,14 **Conservation status: (IUCN)** LC

3761 *Inga cinnamomea* Spruce ex Benth. (Fabaceae)
IPNI: 128442-2
Habit: Small tree **Use:** HF **Origin:** Native **Region:** Amazonia, Andean, Guiana Shield, Pacific **Dept.:** AMA, ANT, BOG, CAQ, CAU, CHO, CUN, HUI, MET, NAR, RIS, VAC, VAU **Elev.:** 10-2450 m.a.s.l. **IUCN Habitat code:** 1,2,3,14 **Conservation status: (IUCN)** LC **(CPSG)** Potential LC

3762 *Inga cocleensis* Pittier (Fabaceae)
IPNI: 128446-2
Common name: churima, guabo, guamo, guamo blanco, guamo churimo, guamo lanoso **Habit:** Tree **Use:** HF **Origin:** Native **Region:** Amazonia, Andean, Guiana Shield, Caribbean, Orinoquia, Pacific **Dept.:** ANT, BOY, CAL, CAU, CHO, CUN, LAG, MET, NAR, SAN, TOL, VAC, VAU **Elev.:** 40-2700 m.a.s.l. **IUCN Habitat code:** 1,3,14 **Conservation status: (IUCN)** LC **(CPSG)** Potential LC

3763 *Inga cordatoalata* Ducke (Fabaceae)
IPNI: 128456-2
Common name: guamo **Habit:** Tree **Use:** HF **Origin:** Native **Region:** Amazonia, Guiana Shield **Dept.:** AMA, PUT, VAU **Elev.:** 160–600 m a.s.l. **IUCN Habitat code:** 1 **Conservation status: (IUCN)** LC

3764 *Inga coruscans* Humb. & Bonpl. ex Willd. (Fabaceae)
IPNI: 128458-2
Common name: churimito, churimo, churimo rosario, guamito de orilla, guamo, guamo churimo, guamo querré **Habit:** Small tree **Use:** HF **Origin:** Native **Region:** Andean, Caribbean, Pacific **Dept.:** ANT, BOG, BOY, CHO, CUN, HUI, MAG, NSA, VAC **Elev.:** 0–2200 m a.s.l. **IUCN Habitat code:** 1,2,3,4,5,14 **Conservation status: (IUCN)** LC

3765 *Inga cylindrica* (Vell.) Mart. (Fabaceae)
IPNI: 1090709-2
Common name: guamo, guamo pepaloro, guamo rosado, sangrito **Habit:** Tree **Use:** HF **Origin:** Native **Region:** Amazonia, Andean, Guiana Shield, Orinoquia, Pacific **Dept.:** AMA, ANT, CHO, GUV, MET **Elev.:** 50–600 m a.s.l. **IUCN Habitat code:** 1,3,14 **Conservation status: (IUCN)** LC

3766 *Inga densiflora* Benth. (Fabaceae)
IPNI: 128466-2
Common name: guamo, guamo machete, guamo macheto **Habit:** Tree **Use:** HF **Origin:** Native, Cultivated **Region:** Amazonia, Andean, Guiana Shield, Caribbean, Orinoquia, Pacific **Dept.:** AMA, ANT, CAL, CAU, CUN, HUI, MET, NAR, PUT, QUI, TOL, VAC **Elev.:** 20-2000 m.a.s.l. **IUCN Habitat code:** 1,3,14 **Conservation status: (IUCN)** LC **(CPSG)** Potential LC

3767 *Inga edulis* Mart. (Fabaceae)
IPNI: 128478-2
Common name: chumillo, churimo, cómenecù, guaba, guaba de bejuco, guaba del Putumayo, guaba manza, guabillo, guabo, guama, guamillo, guamillo de rastrojo, guamo, guamo bejuco, guamo chairus, guamo churimo, guamo de mico, guamo de osito hormiguero, guamo largo, guamo **Habit:** Tree **Use:** AF, EU, FU, GS, HF, MA, ME **Origin:** Native, Cultivated **Region:** Amazonia, Andean, Guiana Shield, Caribbean, Orinoquia, Pacific **Dept.:** AMA, ANT, BOL, CAL, CAQ, CAS, CAU, CHO, CUN, GUA, GUV, MAG, MET, NAR, PUT, RIS, SAN, TOL, VAC, VAU, VID **Elev.:** 0-2140 m.a.s.l. **IUCN Habitat code:** 1,2,3,4,5,14 **Conservation status: (IUCN)** LC **(CPSG)** Potential LC

3768 *Inga fastuosa* (Jacq.) Willd. (Fabaceae)
IPNI: 1040744-2
Common name: churimo, guamo, guamo de monte, guamo rabo de mono **Habit:** Small tree **Use:** HF **Origin:** Native **Region:** Amazonia, Andean, Guiana Shield, Pacific **Dept.:** AMA, ANT, CAL, CAQ, CAS, CAU, CHO, CUN, HUI, MET, NAR, QUI, RIS, VAC, VAU **Elev.:** 200-2400 m.a.s.l. **IUCN Habitat code:** 1,3,14 **Conservation status: (IUCN)** LC **(CPSG)** Potential LC

3769 *Inga fendleriana* Benth. (Fabaceae)
IPNI: 500432-1
Habit: Small tree **Use:** HF **Origin:** Native **Region:** Andean **Dept.:** RIS **Elev.:** 1700–1700 m a.s.l. **IUCN Habitat code:** 1,3 **Conservation status: (IUCN)** LC

3770 *Inga feuillei* DC. (Fabaceae)
IPNI: 128507-2
Common name: guabo **Habit:** Tree **Use:** HF **Origin:** Cultivated, Naturalised **Region:** Amazonia, Andean, Pacific **Dept.:** PUT, VAC **Elev.:** 0–2700 m a.s.l. **IUCN Habitat code:** 1,3 **Conservation status: (IUCN)** NE

3771 *Inga filiformis* N.Zamora (Fabaceae)
IPNI: 278805-2
Habit: Tree **Use:** HF **Origin:** Native **Region:** Pacific **Dept.:** CHO, VAC **Elev.:** 0–160 m a.s.l. **IUCN Habitat code:** 1,5 **Conservation status: (IUCN)** LC

3772 *Inga flexuosa* Schltdl. (Fabaceae)
IPNI: 500443-1
Habit: Tree **Use:** HF, ME **Origin:** Native **Region:** Andean **Dept.:** ANT, CUN, NSA, RIS **Elev.:** 750-2430 m.a.s.l. **IUCN Habitat code:** 1,3 **Conservation status: (IUCN)** NE **(CPSG)** Potential LC

3773 *Inga gracilifolia* Ducke (Fabaceae)
IPNI: 128520-2
Common name: guamo loro, palo de piedra **Habit:** Tree **Use:** HF **Origin:** Native **Region:** Amazonia, Orinoquia **Dept.:** AMA, VID **Elev.:** 0–100 m a.s.l. **IUCN Habitat code:** 1,4 **Conservation status: (IUCN)** LC

3774 *Inga gracilior* Sprague (Fabaceae)
IPNI: 128521-2
Common name: churimo **Habit:** Small tree, Tree **Use:** HF **Origin:** Native **Region:** Amazonia, Andean, Guiana Shield, Caribbean, Orinoquia, Pacific **Dept.:** AMA, ANT, BOY, CAQ, CAS, CAU, CHO, CUN, GUA, LAG, MAG, MET, PUT, QUI, RIS, TOL, VAU **Elev.:** 10-2200 m.a.s.l. **IUCN Habitat code:** 1,2,3,4,5,14 **Conservation status: (IUCN)** LC **(CPSG)** Potential LC

3775 *Inga hayesii* Benth. (Fabaceae)
IPNI: 128532-2
Common name: guamito **Habit:** Small tree, Tree **Use:** AF, HF **Origin:** Native **Region:** Caribbean, Pacific **Dept.:** ANT, BOL, CHO **Elev.:** 0–150 m a.s.l. **IUCN Habitat code:** 1,2,4,14 **Conservation status: (IUCN)** LC

3776 *Inga heterophylla* Willd. (Fabaceae)
IPNI: 128535-2
Common name: churimo **Habit:** Tree **Use:** HF, MA **Origin:** Native **Region:** Amazonia, Andean, Guiana Shield, Caribbean, Orinoquia **Dept.:** ANT, BOY, SAN, VAU, VID **Elev.:** 20-1500 m.a.s.l. **IUCN Habitat code:** 1,2,3,4,5,14 **Conservation status: (IUCN)** LC **(CPSG)** Potential LC

3777 *Inga ilta* T.D.Penn. (Fabaceae)
IPNI: 313755-2
Common name: guamo de semilla **Habit:** Tree **Use:** HF **Origin:** Native **Region:** Amazonia **Dept.:** AMA **Elev.:** 200–200 m a.s.l. **IUCN Habitat code:** 1 **Conservation status: (IUCN)** LC

3778 *Inga ingoides* (Rich.) Willd. (Fabaceae)
IPNI: 1131764-2
Common name: guamo cafetero, guamo chinibo, guamo de Popayán, guamo rabo de mico, guamo rabo de mono **Habit:** Tree **Use:** HF, ME **Origin:** Native **Region:** Amazonia, Andean **Dept.:** AMA, ANT, BOG, BOY, CUN, VAC **Elev.:** 20–2600 m a.s.l. **IUCN Habitat code:** 1,3,14 **Conservation status: (IUCN)** LC

3779 *Inga insignis* Kunth (Fabaceae)
IPNI: 128550-2
Use: HF **Origin:** Native **Conservation status: (IUCN)** LC

3780 *Inga involucrata* R.S.Cowan (Fabaceae)
IPNI: 128553-2
Habit: Small tree, Tree **Use:** HF **Origin:** Native **Region:** Pacific **Dept.:** ANT **Elev.:** 50–300 m a.s.l. **IUCN Habitat code:** 3 **Conservation status: (IUCN)** LC

3781 *Inga laevigata* M.Martens & Galeotti (Fabaceae)
IPNI: 500531-1
Common name: guamo de monte **Habit:** Tree **Use:** HF **Origin:** Native **Region:** Amazonia, Andean, Guiana

Shield, Pacific **Dept.:** AMA, ANT, CAQ, CHO, CUN, HUI, PUT, RIS, VAU **Elev.:** 20–2300 m a.s.l. **IUCN Habitat code:** 1,3,14 **Conservation status: (IUCN)** NE

3782 *Inga lallensis* Spruce ex Benth. (Fabaceae)
IPNI: 128564-2
Common name: churimo, guamo
Habit: Small tree, Tree **Use:** HF **Origin:** Native **Region:** Andean, Guiana Shield, Pacific **Dept.:** ANT, CHO, HUI, MET, PUT, VAC **Elev.:** 50–2800 m a.s.l. **IUCN Habitat code:** 1,3,14 **Conservation status: (IUCN)** LC

3783 *Inga lateriflora* Miq. (Fabaceae)
IPNI: 128568-2
Use: HF, MA, ME **IUCN Habitat code:** 1,14 **Conservation status: (IUCN)** LC

3784 *Inga laurina* (Sw.) Willd. (Fabaceae)
IPNI: 1199545-2
Common name: churimo, guamo, guamo churimo hojiamarilllo, guamo negrito **Habit:** Tree **Use:** EU, FU, HF, IF, MA, ME **Origin:** Native **Region:** Amazonia, Andean, Guiana Shield, Orinoquia, Pacific **Dept.:** AMA, ANT, GUA, NAR, VAC, VAU, VID **Elev.:** 6–1500 m a.s.l. **IUCN Habitat code:** 1,2,3,4,14 **Conservation status: (IUCN)** LC

3785 *Inga leptocarpa* T.D.Penn. (Fabaceae)
IPNI: 313764-2
Common name: guama de churuco
Habit: Small tree **Use:** HF **Origin:** Native **Region:** Amazonia, Andean, Pacific **Dept.:** AMA, CAQ, NAR, VAC, VAU **Elev.:** 90–1325 m a.s.l. **IUCN Habitat code:** 1 **Conservation status: (IUCN)** NE

3786 *Inga longiflora* Spruce ex Benth. (Fabaceae)
IPNI: 128586-2
Habit: Small tree **Use:** HF **Origin:** Native **Region:** Amazonia **Dept.:** CAQ, GUV **Elev.:** 0–240 m a.s.l. **IUCN Habitat code:** 1,14 **Conservation status: (IUCN)** NE

3787 *Inga lopadadenia* Harms (Fabaceae)
IPNI: 128591-2
Common name: guamo, guamo churimo **Habit:** Tree **Use:** HF **Origin:** Native **Region:** Amazonia, Andean **Dept.:** AMA, ANT, VAC **Elev.:** 100–1650 m a.s.l. **IUCN Habitat code:** 1,14 **Conservation status: (IUCN)** LC

3788 *Inga macrophylla* Humb. & Bonpl. ex Willd. (Fabaceae)
IPNI: 128596-2
Common name: guama machete, guamo, guamo hembra, guamo machete **Habit:** Tree **Use:** HF **Origin:** Native **Region:** Amazonia, Andean, Guiana Shield, Orinoquia, Pacific **Dept.:** AMA, ANT, CAQ, CUN, MET, NAR, VAC **Elev.:** 100–2000 m a.s.l. **IUCN Habitat code:** 1,2,3,5,14 **Conservation status: (IUCN)** LC

3789 *Inga manabiensis* T.D.Penn. (Fabaceae)
IPNI: 313749-2
Habit: Small tree, Tree **Use:** HF **Origin:** Native **Region:** Andean **Dept.:** VAC **Elev.:** 1000–1150 m a.s.l. **IUCN Habitat code:** 3 **Conservation status: (IUCN)** LC

3790 *Inga marginata* Willd. (Fabaceae)
IPNI: 128599-2
Common name: brijolito, churimito, churimo, guamo, guamo churimo, guamo churito, guamo negrito, guamo rosarito **Habit:** Tree **Use:** EU, HF, IF, MA, ME **Origin:** Native **Region:** Amazonia, Andean, Guiana Shield, Caribbean, Orinoquia, Pacific **Dept.:** AMA, ANT, BOG, CAL, CAQ, CAU, CES, CHO, CUN, GUV, HUI, LAG, MET, PUT, QUI, RIS, SAN, VAC, VAU **Elev.:** 0-2350 m.a.s.l. **IUCN Habitat code:** 1,3,5,14 **Conservation status: (IUCN)** LC **(CPSG)** Potential LC

3791 *Inga mortoniana* J.León (Fabaceae)
IPNI: 128626-2
Use: HF **Origin:** Native **Dept.:** CHO**Conservation status: (IUCN)** LC

3792 *Inga mucuna* Walp. (Fabaceae)
IPNI: 500629-1
Common name: guama, guamo cajeto **Habit:** Tree **Use:** HF **Origin:** Native **Region:** Andean, Caribbean, Pacific **Dept.:** ANT, CHO, MET **Elev.:** 25–830 m a.s.l. **IUCN Habitat code:** 1,3,5,14 **Conservation status: (IUCN)** LC

3793 *Inga multijuga* Benth. (Fabaceae)
IPNI: 128628-2
Common name: guama de churuco
Habit: Tree **Use:** HF **Origin:** Native **Region:** Amazonia, Andean, Guiana Shield, Caribbean, Pacific **Dept.:** AMA, ANT, CAQ, CHO, NAR, NSA, SAN, TOL, VAC **Elev.:** 0-2370 m.a.s.l. **IUCN Habitat code:** 1,3,4,5,14 **Conservation status: (IUCN)** LC **(CPSG)** Potential LC

3794 *Inga multinervis* T.D.Penn. (Fabaceae)
IPNI: 313745-2
Habit: Tree **Use:** HF **Origin:** Native **Region:** Amazonia, Andean, Pacific **Dept.:** CHO, NAR, PUT, VAC **Elev.:** 50–730 m a.s.l. **IUCN Habitat code:** 1 **Conservation status: (IUCN)** LC

3795 *Inga nobilis* Willd. (Fabaceae)
IPNI: 128633-2
Common name: cansamuelas, chimbilla, churimito, churimo, guabo, guamillo, guamita, guamito, guamo, guamo cansamuelas, guamo cerindo, guamo churimo, guamo de montaña, guamo mene, guamo rosario, guamo serindo, serindo colorado **Habit:** Tree **Use:** EU, FU, HF **Origin:** Native **Region:** Amazonia, Andean, Guiana Shield, Caribbean, Orinoquia, Pacific **Dept.:** AMA, ANT, BOL, BOY, CAL, CAQ, CAU, CES, CHO, COR, CUN, GUV, MAG, MET, NAR, NSA, PUT, QUI, SAN, VAU **Elev.:** 0-2890 m.a.s.l. **IUCN Habitat code:** 1,2,3,4,5,14 **Conservation status: (IUCN)** LC **(CPSG)** Potential LC

3796 *Inga obtusata* Spruce ex Benth. (Fabaceae)
IPNI: 128644-2
Habit: Small tree, Tree **Use:** HF **Origin:** Native **Region:** Guiana Shield **Dept.:** VAU **Elev.:** 0–100 m a.s.l. **IUCN Habitat code:** 1 **Conservation status: (IUCN)** LC

3797 *Inga oerstediana* Benth. (Fabaceae)
IPNI: 128647-2
Common name: guacuna, guamo blanco, guamo calzamuelas, guamo hojiancho, guamos **Habit:** Tree **Use:** FU, HF, ME **Origin:** Native **Region:** Amazonia, Andean, Guiana Shield, Caribbean, Pacific **Dept.:** AMA, ANT, BOY, CAL, CAU, CHO, CUN, HUI, LAG, MAG, MET, NAR, NSA, PUT, QUI, RIS, SAN, TOL, VAC **Elev.:** 60-2800 m.a.s.l. **IUCN Habitat code:** 1,3,5,14 **Conservation status: (IUCN)** LC **(CPSG)** Potential LC

3798 *Inga ornata* Kunth (Fabaceae)
IPNI: 500662-1
Common name: guamo, guamo bejuco, guamo mico, guamo santafereño **Habit:** Small tree, Tree **Use:** HF, ME **Origin:** Native, Cultivated **Region:** Andean, Caribbean **Dept.:** ANT, BOY, CAL, CAQ, CAU, CES, CUN, HUI, QUI, SAN, TOL, VAC **Elev.:** 800-2350 m.a.s.l. **IUCN Habitat code:** 1,2,3,14 **Conservation status: (IUCN)** LC **(CPSG)** Potential LC

3799 *Inga paraensis* Ducke (Fabaceae)
IPNI: 128661-2
Use: HF **IUCN Habitat code:** 14
Conservation status: (IUCN) LC

3800 *Inga pezizifera* Benth. (Fabaceae)
IPNI: 128673-2
Common name: guama, guama de antiguo, guamo, guamo blanco
Habit: Tree **Use:** HF, MA **Origin:** Native **Region:** Amazonia, Andean, Guiana Shield, Pacific **Dept.:** AMA, ANT, BOY, CAL, CAQ, CHO, VAC, VAU **Elev.:** 0–1850 m a.s.l. **IUCN Habitat code:** 1,3,4,5,14 **Conservation status: (IUCN)** LC

3801 *Inga pilosula* (Rich.) J.F.Macbr. (Fabaceae)
IPNI: 128675-2
Common name: chimbillo, guamilla, guamito de danta, guamito de muinanes, guamo, guamo de arqueado **Habit:** Small tree, Tree **Use:** HF, ME **Origin:** Native **Region:** Amazonia, Guiana Shield, Orinoquia, Pacific **Dept.:** AMA, CAQ, CHO, GUA, GUV, VAU **Elev.:** 100–380 m a.s.l. **Conservation status: (IUCN)** LC

3802 *Inga plumifera* Spruce ex Benth. (Fabaceae)
IPNI: 128680-2
Common name: guama, guama de barbasco, guama peluda, guamo, guamo de barbasco **Habit:** Small tree **Use:** HF **Origin:** Native **Region:** Amazonia, Guiana Shield **Dept.:** AMA, CAQ, GUV, VAU **Elev.:** 120–300 m a.s.l. **IUCN Habitat code:** 1,3,14 **Conservation status: (IUCN)** LC

3803 *Inga pruriens* Poepp. (Fabaceae)
IPNI: 128687-2
Common name: chimbillo, guamo
Habit: Tree **Use:** HF **Origin:** Native **Region:** Amazonia **Dept.:** AMA **Elev.:** 120–220 m a.s.l. **IUCN Habitat code:** 1 **Conservation status: (IUCN)** LC

3804 *Inga punctata* Willd. (Fabaceae)
IPNI: 128693-2
Common name: chimbillo, churimo de monte, churimo de orilla, guaba, guabo, guama, guamo, guamo borrachero **Habit:** Tree **Use:** EU, FU, HF, MA, ME **Origin:** Native **Region:** Amazonia, Andean, Guiana Shield, Caribbean, Orinoquia, Pacific **Dept.:** AMA, ANT, ARA, BOY, CAL, CAQ, CAS, CAU, CES, CHO, COR, CUN, GUV, HUI, MAG, MET, NAR, NSA, PUT, QUI, RIS, SAN, TOL, VAC **Elev.:** 0-2780 m.a.s.l. **IUCN Habitat code:** 1,2,3,4,5,14 **Conservation status: (IUCN)** LC **(CPSG)** Potential LC

3805 *Inga rubiginosa* (Rich.) DC. (Fabaceae)
IPNI: 1068738-2
Habit: Tree **Use:** HF **Origin:** Native **Region:** Guiana Shield, Caribbean, Orinoquia, Pacific **Dept.:** ANT, CHO, MET, VAU **Elev.:** 50–490 m a.s.l. **IUCN Habitat code:** 1,2,3,14 **Conservation status: (IUCN)** LC

3806 *Inga ruiziana* G.Don (Fabaceae) **IPNI:** 128721-2 **Common name:** guama, guamo, guamo churimbo, guamo de chupar **Habit:** Tree **Use:** HF, ME **Origin:** Native **Region:** Amazonia, Andean, Guiana Shield, Pacific **Dept.:** AMA, ANT, CAL, CAQ, CHO, CUN, NAR, PUT, RIS, VAC **Elev.:** 0–2150 m a.s.l. **IUCN Habitat code:** 1,3,5,14 **Conservation status: (IUCN)** LC

3807 *Inga saffordiana* Pittier (Fabaceae) **IPNI:** 128724-2 **Habit:** Small tree, Tree **Use:** HF **Origin:** Native **Region:** Andean, Pacific **Dept.:** ANT, CHO **Elev.:** 0–1250 m a.s.l. **IUCN Habitat code:** 1,3,5,14 **Conservation status: (IUCN)** LC

3808 *Inga sapindoides* Willd. (Fabaceae) **IPNI:** 128732-2 **Common name:** churimo, guamo, guamo blanco, guamo cansamuelas, guamo cheneve, guamo cuatrofilos **Habit:** Tree **Use:** HF **Origin:** Native **Region:** Andean, Guiana Shield, Orinoquia, Pacific **Dept.:** ANT, BOG, BOY, CAS, CHO, CUN, GUV, MET, NSA, QUI, RIS, VAC **Elev.:** 50–1600 m a.s.l. **IUCN Habitat code:** 1,3,5,14 **Conservation status: (IUCN)** LC

3809 *Inga sertulifera* DC. (Fabaceae) **IPNI:** 500808-1 **Common name:** churima, churimo **Habit:** Tree **Use:** HF, ME **Origin:** Native **Region:** Amazonia, Andean, Guiana Shield, Orinoquia, Pacific **Dept.:** AMA, ANT, CAQ, CAS, CAU, CHO, GUA, GUV, PUT, VAC, VID **Elev.:** 5-1420 m.a.s.l. **IUCN Habitat code:** 1,3,5,14 **Conservation status: (IUCN)** LC **(CPSG)** Potential LC

3810 *Inga setosa* G.Don (Fabaceae) **IPNI:** 128746-2 **Habit:** Small tree, Tree **Use:** HF **Origin:** Native **Region:** Andean, Caribbean, Pacific **Dept.:** CHO, CUN, HUI, MAG, MET, SAN, VAC **Elev.:** 350–2180 m a.s.l. **IUCN Habitat code:** 1,14 **Conservation status: (IUCN)** LC

3811 *Inga spectabilis* (Vahl) Willd. (Fabaceae) **IPNI:** 1122303-2 **Common name:** cajeto, churimba, guaba churimba, guabo macheto, guama coperuna, guama de churuco, guama macheta, guamo, guamo cafeto, guamo cajeto, guamo churito, guamo copero, guamo machete, guamo macheto, guamo rabo de mono, guamo santafereño **Habit:** Tree **Use:** EU, FU, HF, MA, ME **Origin:** Native, Cultivated **Region:** Amazonia, Andean, Guiana Shield, Caribbean, Orinoquia, Pacific **Dept.:** AMA, ANT, BOL, BOY, CAQ, CAS, CAU, CES, CHO, COR, CUN, GUV, HUI, MAG, MET, NAR, NSA, PUT, SAN, SUC, TOL, VAC **Elev.:** 0-2000 m.a.s.l. **IUCN Habitat code:** 1,2,3,5,14 **Conservation status: (IUCN)** LC **(CPSG)** Potential LC

3812 *Inga splendens* Willd. (Fabaceae) **IPNI:** 128749-2 **Habit:** Tree **Use:** HF, MA **Origin:** Native **Region:** Amazonia **Elev.:** 120–120 m a.s.l. **IUCN Habitat code:** 1,3,4,5,14 **Conservation status: (IUCN)** LC

3813 *Inga stenoptera* Benth. (Fabaceae) **IPNI:** 128764-2 **Common name:** chimbillo, guamo cupero, guamo de güio, guamo loro, to-to'-ko **Habit:** Tree **Use:** HF, ME **Origin:** Native **Region:** Amazonia, Andean, Guiana Shield, Caribbean, Orinoquia **Dept.:** AMA, BOL, CAQ, CUN, GUV, HUI, MET, PUT, VAU **Elev.:** 100–1200 m a.s.l. **Conservation status: (IUCN)** LC

3814 *Inga stipularis* DC. (Fabaceae) **IPNI:** 500843-1 **Habit:** Small tree, Tree **Use:** HF **Origin:** Native **Region:** Amazonia, Andean, Guiana Shield **Dept.:** CAQ, CAS, VAU **Elev.:** 200–600 m a.s.l. **IUCN Habitat code:** 1 **Conservation status: (IUCN)** LC

3815 *Inga striata* Benth. (Fabaceae) **IPNI:** 500844-1 **Habit:** Tree **Use:** EU, HF **Origin:** Native **Region:** Andean **Dept.:** NAR **Elev.:** 2000–2000 m a.s.l. **IUCN Habitat code:** 1,2,14 **Conservation status: (IUCN)** LC

3816 *Inga tenuicalyx* T.D.Penn. (Fabaceae) **IPNI:** 313754-2 **Use:** HF **Origin:** Native **IUCN Habitat code:** 1,14 **Conservation status: (IUCN)** VU

3817 *Inga tenuistipula* Ducke (Fabaceae) **IPNI:** 128784-2 **Habit:** Tree **Use:** HF **Origin:** Native **Region:** Amazonia, Andean, Guiana Shield, Orinoquia, Pacific **Dept.:** AMA, CAU, CHO, MET, PUT, VAC, VAU, VID **Elev.:** 0–1650 m a.s.l. **IUCN Habitat code:** 1,3,5,14 **Conservation status: (IUCN)** LC

3818 *Inga thibaudiana* DC. (Fabaceae) **IPNI:** 128789-2 **Common name:** guama de mico, guama de montaña, guamito silvestre, guamo, guamo de loma, guamo de montana, guamo de montaña, guamo loro **Habit:** Tree **Use:** HF **Origin:** Native **Region:** Amazonia, Andean, Guiana Shield, Orinoquia, Pacific **Dept.:** AMA, ANT, CAQ, CAU, CHO, COR, GUV, MET, NAR, PUT, SAN, VAC, VAU **Elev.:** 0-1550 m.a.s.l. **IUCN Habitat code:** 1,3,5,14 **Conservation status: (IUCN)** LC **(CPSG)** Potential LC

3819 *Inga tomentosa* Benth. (Fabaceae) **IPNI:** 128792-2 **Use:** HF **Origin:** Native **IUCN Habitat code:** 1,3 **Conservation status: (IUCN)** NE

3820 *Inga ulei* Harms (Fabaceae) **IPNI:** 128800-2 **Habit:** Tree **Use:** HF **Origin:** Native **Region:** Guiana Shield **Dept.:** GUA **Elev.:** 0–200 m a.s.l. **Conservation status: (IUCN)** LC

3821 *Inga umbellifera* (Vahl) Steud. ex DC. (Fabaceae) **IPNI:** 1035636-2 **Common name:** guama, guamo churimo **Habit:** Tree **Use:** HF **Origin:** Native **Region:** Amazonia, Andean, Guiana Shield, Orinoquia, Pacific **Dept.:** AMA, ANT, BOY, CAQ, CHO, CUN, GUA, GUV, MET, NAR, PUT, QUI, SAN, TOL, VAC, VAU **Elev.:** 20-1600 m.a.s.l. **IUCN Habitat code:** 1,3,5,14 **Conservation status: (IUCN)** LC **(CPSG)** Potential LC

3822 *Inga umbratica* Poepp. (Fabaceae) **IPNI:** 500893-1 **Habit:** Tree **Use:** HF **Origin:** Native **Region:** Pacific **Dept.:** CHO **Elev.:** 80–80 m a.s.l. **IUCN Habitat code:** 1,14 **Conservation status: (IUCN)** LC

3823 *Inga velutina* Willd. (Fabaceae) **IPNI:** 128810-2 **Habit:** Tree **Use:** HF, MA **Origin:** Native **Region:** Amazonia, Andean, Caribbean **Dept.:** AMA, ANT, COR, NAR **Elev.:** 50–1800 m a.s.l. **IUCN Habitat code:** 1,5 **Conservation status: (IUCN)** LC

3824 *Inga venusta* Standl. (Fabaceae) **IPNI:** 128812-2 **Habit:** Tree **Use:** HF **Origin:** Native **Region:** Amazonia, Andean, Pacific **Dept.:** AMA, ANT, BOY, CAU, CHO, NAR, PUT, RIS, VAC **Elev.:** 50–1800 m a.s.l. **IUCN Habitat code:** 1,3,14 **Conservation status: (IUCN)** LC

3825 *Inga vera* Willd. (Fabaceae) **IPNI:** 1006087-2 **Common name:** churimo, guabo, guamito, guamo, guamo racero, guamo santafereño, guarno arroyero, pepa de sierra **Habit:** Tree **Use:** EU, FU, HF, MA, ME **Origin:** Native, Cultivated **Region:** Amazonia, Andean, Guiana Shield, Caribbean, Orinoquia, Pacific **Dept.:** AMA, ANT, ARA, ATL, BOL, BOY, CAL, CAQ, CAS, CAU, CHO, COR, CUN, GUA, GUV, HUI, LAG, MAG, MET, NAR, NSA, RIS, SAN, SUC, TOL, VAC, VAU, VID **Elev.:** 10-2300 m.a.s.l. **IUCN Habitat code:** 1,2,3,4,5,14 **Conservation status: (IUCN)** LC **(CPSG)** Potential LC

3826 *Inga villosissima* Benth. (Fabaceae) **IPNI:** 128820-2 **Habit:** Tree **Use:** HF **Origin:** Native **Region:** Andean **Dept.:** ANT, BOY, CAL, CAU, CUN, HUI, NAR, QUI, RIS, VAC, VAU **Elev.:** 200–2500 m a.s.l. **IUCN Habitat code:** 1,2,3,14 **Conservation status: (IUCN)** NE

3827 *Inga vismiifolia* Poepp. (Fabaceae) **IPNI:** 500919-1 **Use:** HF **Origin:** Native **IUCN Habitat code:** 1,3,5,14 **Conservation status: (IUCN)** LC

3828 *Inga yacoana* J.F.Macbr. (Fabaceae) **IPNI:** 128830-2 **Habit:** Tree **Use:** HF **Origin:** Native **Region:** Amazonia, Guiana Shield **Dept.:** AMA, VAU **Elev.:** 100–100 m a.s.l. **IUCN Habitat code:** 1 **Conservation status: (IUCN)** LC

3829 *Inga yasuniana* T.D.Penn. (Fabaceae) **IPNI:** 313740-2 **Common name:** guama de boruga, guama de borugo, guamo de borugo **Habit:** Tree **Use:** HF **Origin:** Native **Region:** Pacific **Dept.:** CAU, CHO, VAC **Elev.:** 50–130 m a.s.l. **IUCN Habitat code:** 1 **Conservation status: (IUCN)** VU

3830 *Lablab purpureus* (L.) Sweet (Fabaceae) **IPNI:** 501411-1 **Common name:** alverjones, arroz con coco, arveja real, bugueño, frisol tropical, frisol tropical o mulato, frisoles, fríjol arveja real, fríjol bugueño, fríjol calentano, fríjol guisante, fríjol jacinto, fríjol peruano, fríjol risueño, fríjol sobaco, fríjol soya, guisa **Habit:** Herb, Climbing **Use:** AF, EU, GS, HF, MA, ME, PO **Origin:** Cultivated, Naturalised **Region:** Andean **Dept.:** ANT, ATL, BOY, CAL, CAU, CES, CUN, MAG, QUI **Elev.:**

100–2130 m a.s.l. **IUCN Habitat code:** 1,3,14 **Conservation status: (IUCN)** NE

3831 *Lachesiodendron viridiflorum* (Kunth) P.G.Ribeiro, L.P.Queiroz & Luckow (Fabaceae) **IPNI:** 77178320-1 **Common name:** bocachico, chicho, guacamayo **Habit:** Shrub, Tree **Use:** FU, MA, ME, PO **Origin:** Native **Region:** Caribbean **Dept.:** ATL, BOL, LAG, MAG, SAN **Elev.:** 20–1000 m a.s.l. **IUCN Habitat code:** 1,2,3,4,14 **Conservation status: (IUCN)** NE

3832 *Lathyrus magellanicus* Lam. (Fabaceae) **IPNI:** 501845-1 **Common name:** alverja de monte, alverjillo **Habit:** Herb, Climbing **Use:** HF, ME **Origin:** Native **Region:** Andean **Dept.:** ANT, BOY, CAU, CUN, NAR **Elev.:** 2800–3300 m a.s.l. **IUCN Habitat code:** 4,14 **Conservation status: (IUCN)** LC

3833 *Lathyrus nervosus* Lam. (Fabaceae) **IPNI:** 323132-2 **Use:** HF **Origin:** Native **Conservation status: (IUCN)** NE

3834 *Lathyrus odoratus* L. (Fabaceae) **IPNI:** 501909-1 **Common name:** cenón, doncenón, guisante de olor **Habit:** Herb, Climbing **Use:** EU, GS, HF, MA, ME, PO **Origin:** Cultivated **Region:** Andean **Dept.:** BOG, BOY, CUN **Elev.:** 2550–2620 m a.s.l. **IUCN Habitat code:** 14 **Conservation status: (IUCN)** CR

3835 *Lathyrus oleraceus* Lam. (Fabaceae) **IPNI:** 501912-1 **Common name:** alverja, arveja, arveja parda, arvejón blanco, criolla, guisante **Habit:** Herb, Climbing **Use:** AF, GS, HF, ME **Origin:** Cultivated, Naturalised **Region:** Andean **Dept.:** ANT, BOY, NSA, PUT, SAN, VAC **Elev.:** 1090–3100 m a.s.l. **Conservation status: (IUCN)** NE

3836 *Lathyrus sativus* L. (Fabaceae) **IPNI:** 501990-1 **Use:** AF, EU, GS, HF, ME, PO **Region:** Andean **Dept.:** BOG**IUCN Habitat code:** 14 **Conservation status: (IUCN)** NE

3837 *Lecointea amazonica* Ducke (Fabaceae) **IPNI:** 136896-2 **Common name:** carrapo **Habit:** Shrub, Tree **Use:** HF **Origin:** Native **Region:** Orinoquia **Dept.:** CAS**IUCN Habitat code:** 1,3 **Conservation status: (IUCN)** LC

3838 *Leptolobium nitens* Vogel (Fabaceae) **IPNI:** 502385-1 **Common name:** cangrio, palo de boa **Habit:** Small tree, Tree **Use:** MA, ME **Origin:** Native **Region:** Amazonia, Guiana Shield **Dept.:** CAQ, CAS, GUA, MET, VAU, VID **Elev.:** 90–200 m a.s.l. **IUCN Habitat code:** 1,2,3,4,5 **Conservation status: (IUCN)** LC

3839 *Leptolobium panamense* (Benth.) Sch.Rodr. & A.M.G.Azevedo (Fabaceae) **IPNI:** 77091925-1 **Use:** EU, MA, ME **Origin:** Native **Conservation status: (IUCN)** LC

3840 *Leptospron adenanthum* (G.Mey.) A.Delgado (Fabaceae) **IPNI:** 60457245-2 **Common name:** batatillo, frijolillo, platanillo **Habit:** Herb, Climbing **Use:** HF, ME **Origin:** Native, Cultivated **Region:** Andean, Guiana Shield, Caribbean, Orinoquia, Pacific **Dept.:** ARA, ATL, CAU, CHO, CUN, HUI, MAG, MET, PUT, VAC **Elev.:** 0–1300 m a.s.l. **IUCN Habitat code:** 1,2,3,5,14 **Conservation status: (IUCN)** NE

3841 *Leucaena leucocephala* (Lam.) de Wit (Fabaceae) **IPNI:** 138955-2 **Common name:** acacia blanca, acacia forrajera, carbonero, carbonero blanco, carbonero quiebrahacha, leucaena, panelo, ucaena **Habit:** Shrub, Small tree **Use:** AF, EU, FU, GS, HF, IF, MA, ME, PO, SU **Origin:** Cultivated, Naturalised **Region:** Andean, Guiana Shield, Caribbean, Orinoquia, Pacific **Dept.:** ANT, BOG, BOY, CAL, CES, CHO, COR, CUN, MAG, MET, NSA, QUI, SAP, SAN, SUC, TOL, VAC, VAU **Elev.:** 0-2200 m.a.s.l. **IUCN Habitat code:** 1,2,3,4,5,14 **Conservation status: (IUCN)** NE **(CPSG)** Potential LC

3842 *Libidibia coriaria* (Jacq.) Schltdl. (Fabaceae) **IPNI:** 1010207-2 **Common name:** baranó, dividivi, dividivi calentano, dividivi de tierra caliente, dm, dm-dm, ichil, ichí, libidivi **Habit:** Shrub, Tree **Use:** AF, EU, FU, HF, IF, MA, ME, SU **Origin:** Native **Region:** Caribbean, Orinoquia **Dept.:** ARA, ATL, BOL, CES, LAG, MAG, SUC **Elev.:** 0–350 m a.s.l. **IUCN Habitat code:** 1,2,4,5,14 **Conservation status: (IUCN)** LC

3843 *Libidibia glabrata* (Kunth) C.Cast. & G.P.Lewis (Fabaceae) **IPNI:** 77130395-1 **Habit:** Tree **Use:** HF, MA **Origin:** Native **Region:** Caribbean **Dept.:** LAG, MAG **Elev.:** 50–50 m a.s.l. **IUCN Habitat code:** 14 **Conservation status: (IUCN)** LC

3844 *Libidibia punctata* (Willd.) Britton (Fabaceae) **IPNI:** 139695-2 **Common name:** granadillo, morado, tanamé, ébano **Habit:** Tree **Use:** AF, EU, FU, MA, ME **Origin:** Native **Region:** Caribbean **Dept.:** ANT, ATL, BOL, LAG, MAG **Elev.:** 6-200 m.a.s.l. **IUCN Habitat code:** 1,2,3,14 **Conservation status: (IUCN)** NE **(CPSG)** VU

3845 *Lonchocarpus heptaphyllus* (Poir.) DC. (Fabaceae) **IPNI:** 502971-1 **Common name:** acrorutú, acurutú, barbasco, barbasco de bejuco, barbasco de costilla, barbasco de dormilón, barbasco del sol, barbasco fantasma, barbasco propio, barbasco sembrado, bejuco barbasco, eomu, índigo **Habit:** Tree, Climbing **Use:** MA, ME, PO **Origin:** Native **Region:** Amazonia, Guiana Shield, Caribbean, Pacific **Dept.:** AMA, ANT, CAQ, CHO, GUV, MAG, MET, PUT, SAP, VAU **Elev.:** 0–250 m a.s.l. **IUCN Habitat code:** 1,2,5 **Conservation status: (IUCN)** LC

3846 *Lonchocarpus macrophyllus* Kunth (Fabaceae) **IPNI:** 503009-1 **Common name:** barbasco, baurá, mataopez, matapez, palo de José, palo de josé **Habit:** Tree **Use:** MA, ME, PO **Origin:** Native **Region:** Caribbean **Dept.:** ANT **Elev.:** 0–250 m a.s.l. **IUCN Habitat code:** 1,2,3,5,14 **Conservation status: (IUCN)** LC

3847 *Lonchocarpus punctatus* Kunth (Fabaceae) **IPNI:** 503074-1 **Common name:** chocho, guayacán, meaparado, meaparao, miaparao **Habit:** Shrub, Tree **Use:** HF, PO **Origin:** Native **Region:** Caribbean **Dept.:** ATL, BOL, COR, LAG, MAG, VAC **Elev.:** 5–970 m a.s.l. **IUCN Habitat code:** 1,2,14 **Conservation status: (IUCN)** LC

3848 *Lonchocarpus sericeus* (Poir.) Kunth ex DC. (Fabaceae) **IPNI:** 143380-2 **Common name:** amasisa, baurá, carrapato, gachico, garrapato, kasho, kuaso, oakta, savonette **Habit:** Shrub, Tree **Use:** EU, FU, HF, MA, ME **Origin:** Native **Region:** Caribbean, Pacific **Dept.:** ANT, CAL, CHO, CUN **Elev.:** 70–500 m a.s.l. **IUCN Habitat code:** 1,2,3,14 **Conservation status: (IUCN)** LC

3849 *Lonchocarpus velutinus* Benth. (Fabaceae) **IPNI:** 503128-1 **Common name:** bollo limpio, macurutú **Habit:** Small tree, Tree **Use:** FU, MA **Origin:** Native **Region:** Caribbean **Dept.:** MAG, NSA **Elev.:** 250–250 m a.s.l. **IUCN Habitat code:** 1,2,3,14 **Conservation status: (IUCN)** LC

3850 *Lonchocarpus violaceus* (Jacq.) Kunth ex DC. (Fabaceae) **IPNI:** 503131-1 **Common name:** arepito, baurá **Habit:** Tree **Use:** MA, ME **Origin:** Native **Region:** Caribbean **Dept.:** BOL, LAG **Elev.:** 0–100 m a.s.l. **IUCN Habitat code:** 1,2,3 **Conservation status: (IUCN)** NE

3851 *Lotus corniculatus* L. (Fabaceae) **IPNI:** 30179739-2 **Habit:** Herb **Use:** AF, EU, HF, IF, ME, PO **Origin:** Naturalised **Region:** Andean **Dept.:** BOG, CUN **Elev.:** 2600–2600 m a.s.l. **IUCN Habitat code:** 1,14 **Conservation status: (IUCN)** NE

3852 *Lupinus albus* L. (Fabaceae) **IPNI:** 504069-1 **Common name:** altramuz, chocha, chocho de jardín, chochos de flor, chochos de hoja abanicada **Use:** AF, EU, GS, HF, IF, MA, ME, PO **Dept.:** BOG**IUCN Habitat code:** 1,3,14 **Conservation status: (IUCN)** LC

3853 *Lupinus alopecuroides* Desr. (Fabaceae) **IPNI:** 504088-1 **Common name:** borreguitas, chocho **Habit:** Herb **Use:** AF, EU, ME **Origin:** Native **Region:** Andean **Dept.:** ARA, BOY, CAL, CAU, CUN, MET, QUI, RIS, SAN, TOL **Elev.:** 2700–4500 m a.s.l. **IUCN Habitat code:** 1,2,3,4,6,8 **Conservation status: (IUCN)** NE

3854 *Lupinus arvensis* Benth. (Fabaceae) **IPNI:** 504161-1 **Habit:** Herb **Use:** GS **Origin:** Native **Region:** Andean **Dept.:** NAR **Elev.:** 3000–3000 m a.s.l. **Conservation status: (IUCN)** DD

3855 *Lupinus bogotensis* Benth. (Fabaceae) **IPNI:** 504211-1 **Common name:** altramuz, chochitos, chochos de flor, choco, lupino **Habit:** Herb, Shrub **Use:** AF, EU, GS, ME **Origin:** Native **Region:** Andean **Dept.:** ANT, BOG, BOY, CAL, CUN, MET, NSA **Elev.:** 1700–3700 m a.s.l. **IUCN**

Habitat code: 1,3,4,8,14 Conservation status: (IUCN) LC

3856 *Lupinus carrikeri* C.P.Sm. (Fabaceae)
IPNI: 144957-2
Common name: cuibangui jalé
Habit: Herb, Subshrub Use: ME Origin: Native Region: Andean Dept.: BOY, LAG, MAG Elev.: 3200–4500 m a.s.l. IUCN Habitat code: 4,14 Conservation status: (IUCN) NE

3857 *Lupinus hortorum* C.P.Sm. (Fabaceae)
IPNI: 145282-2
Common name: chocho Habit: Herb Use: EU Origin: Endemic, Cultivated Region: Andean Dept.: BOG, CUN Elev.: 2600–2600 m a.s.l. Conservation status: (IUCN) NE

3858 *Lupinus microphyllus* Desr. (Fabaceae)
IPNI: 504770-1
Habit: Herb Use: GS Origin: Native Region: Andean Dept.: CAL, QUI, RIS, TOL Elev.: 3650–4520 m a.s.l. IUCN Habitat code: 3,4,6 Conservation status: (IUCN) LC

3859 *Lupinus mutabilis* Sweet (Fabaceae)
IPNI: 504808-1
Common name: barbasco, chocho, chochos, tarwi, tauli, tauri, ullus
Habit: Herb, Subshrub Use: AF, EU, GS, HF, IF, ME, PO Origin: Native, Cultivated Region: Andean Dept.: ANT, BOG, BOY, CAL, CAU, CUN, SAN Elev.: 1700–3000 m a.s.l. IUCN Habitat code: 1,3,8,14 Conservation status: (IUCN) NE

3860 *Lupinus paniculatus* Desr. (Fabaceae)
IPNI: 504878-1
Common name: chocho Habit: Subshrub Use: GS Origin: Native Region: Andean Dept.: CUN Elev.: 2000–2000 m a.s.l. Conservation status: (IUCN) NE

3861 *Lupinus peruvianus* Ulbr. (Fabaceae)
IPNI: 145720-2
Habit: Herb, Subshrub Use: GS Origin: Native Region: Andean Dept.: BOG, CUN, HUI Elev.: 2700–3450 m a.s.l. IUCN Habitat code: 1 Conservation status: (IUCN) NE

3862 *Lupinus protrusus* C.P.Sm. (Fabaceae)
IPNI: 145775-2
Use: EU Origin: Endemic, Cultivated Region: Andean Dept.: BOG, CUN Elev.: 2600–2600 m a.s.l. Conservation status: (IUCN) NE

3863 *Lupinus pubescens* Benth. (Fabaceae)
IPNI: 504972-1
Common name: chochos Habit: Herb Use: GS, ME Origin: Native Region: Andean Dept.: BOG, CAU, CUN, NAR Elev.: 2500–3575 m a.s.l. IUCN Habitat code: 1,3,4,14 Conservation status: (IUCN) LC

3864 *Lupinus ramosissimus* Benth. (Fabaceae)
IPNI: 505001-1
Habit: Subshrub Use: GS Origin: Native Region: Andean Dept.: CAU, CUN Elev.: 3100–3100 m a.s.l. IUCN Habitat code: 14 Conservation status: (IUCN) LC

3865 *Lupinus revolutus* C.P.Sm. (Fabaceae)
IPNI: 145827-2
Habit: Herb Use: GS Origin: Native Region: Andean Dept.: CAU, HUI, NAR, PUT Elev.: 2950–3800 m a.s.l. Conservation status: (IUCN) NE

3866 *Lupinus rusbyanus* C.P.Sm. (Fabaceae)
IPNI: 145851-2
Use: GS Origin: Native Region: Andean Dept.: CUNConservation status: (IUCN) NE

3867 *Lupinus sarmentosus* Desr. (Fabaceae)
IPNI: 505050-1
Habit: Herb, Climbing Use: GS Origin: Native Region: Andean Dept.: NAR, QUI Elev.: 3200–4100 m a.s.l. IUCN Habitat code: 4,14 Conservation status: (IUCN) NE

3868 *Lupinus smithianus* Kunth (Fabaceae)
IPNI: 505094-1
Habit: Subshrub Use: GS Origin: Native Region: Andean Conservation status: (IUCN) DD

3869 *Lupinus tauris* Benth. (Fabaceae)
IPNI: 505172-1
Habit: Shrub Use: GS, HF Origin: Native Region: Andean Dept.: CAL, NAR Elev.: 3200–4100 m a.s.l. IUCN Habitat code: 1 Conservation status: (IUCN) NE

3870 *Machaerium arboreum* (Jacq.) Vogel (Fabaceae)
IPNI: 505373-1
Common name: arrayán blanco, cucaracho, látigo, purgación, sangregao Habit: Shrub, Tree, Liana Use: EU, FU, PO Origin: Native Region: Caribbean, Orinoquia, Pacific Dept.: ANT, ATL, BOL, CAS, CHO, LAG, MAG, MET, SUC Elev.: 5–700 m a.s.l. IUCN Habitat code: 1,2,3,4,5,14 Conservation status: (IUCN) LC

3871 *Machaerium capote* Triana ex Dugand (Fabaceae)
IPNI: 148477-2
Common name: anamos, ananios, capote, carbón, chucunda, dinde, duirá, granadillo, negrillo, perón, sietecueros, sietecáscaras, tachuelo Habit: Shrub, Tree Use: MA, ME Origin: Native Region: Andean, Caribbean, Pacific Dept.: ANT, ATL, BOL, CAL, CHO, COR, CUN, HUI, MAG, RIS, SAN, TOL, VAC Elev.: 2-1600 m a.s.l. IUCN Habitat code: 1,2,3,5,14 Conservation status: (IUCN) LC (CPSG) Potential LC

3872 *Machaerium inundatum* (Mart. ex Benth.) Ducke (Fabaceae)
IPNI: 148542-2
Habit: Small tree, Climbing Use: MA Origin: Native Region: Amazonia, Andean, Orinoquia Dept.: AMA, CUN, VID Elev.: 90–1320 m a.s.l. IUCN Habitat code: 1,3,4,14 Conservation status: (IUCN) LC

3873 *Machaerium isadelphum* (E.Mey.) Amshoff (Fabaceae)
IPNI: 148544-2
Common name: bejuco sangre,frente de toro pico de loro, rasguñagato, frente de toro, zarza Habit: Shrub, Tree, Climbing Use: ME Origin: Native Region: Amazonia, Caribbean, Pacific Dept.: ANT, ATL, BOL, CHO, COR, CUN, GUV, MAG, NSA, SAN Elev.: 10–1000 m a.s.l. IUCN Habitat code: 1,2,3,4,5,14 Conservation status: (IUCN) LC

3874 *Machaerium lunatum* (L.f.) Ducke (Fabaceae)
IPNI: 148561-2
Common name: uña de gato Habit: Shrub, Climbing Use: HF, MA, ME, PO Origin: Native Region: Orinoquia, Pacific Dept.: ANT, CAS, CHO Elev.: 0–100 m a.s.l. IUCN Habitat code: 1,4,5 Conservation status: (IUCN) LC

3875 *Macrolobium acaciifolium* (Benth.) Benth. (Fabaceae)
IPNI: 505635-1
Common name: arapara, arapari, arepito, barapaná, cincocentavos, cincuentacentavos, dormidera, dormidero, dormilón, guamillo, guamo rosado, guarango, parinari Habit: Tree Use: HF, ME Origin: Native Region: Amazonia, Guiana Shield, Orinoquia Dept.: AMA, CAQ, GUV, MET, PUT, VAU, VID Elev.: 10–900 m a.s.l. IUCN Habitat code: 1,3,4,5,14 Conservation status: (IUCN) LC

3876 *Macrolobium bifolium* (Aubl.) Pers. (Fabaceae)
IPNI: 30147160-2
Habit: Tree Use: ME Origin: Native Region: Amazonia Dept.: AMA, GUA Elev.: 100–150 m a.s.l. IUCN Habitat code: 1 Conservation status: (IUCN) LC

3877 *Macrolobium multijugum* (DC.) Benth. (Fabaceae)
IPNI: 505740-1
Habit: Shrub, Tree Use: ME Origin: Native Region: Amazonia, Orinoquia Dept.: AMA, CAQ, GUA, GUV, MET, VAU, VID Elev.: 80–500 m a.s.l. IUCN Habitat code: 1,2,3,4,5,14 Conservation status: (IUCN) LC

3878 *Macropsychanthus comosus* (G.Mey.) L.P.Queiroz & Snak (Fabaceae)
IPNI: 77212307-1
Common name: hoja de boa Habit: Herb, Liana, Climbing Use: EU, FU, GS, HF, MA, ME, PO, SU Origin: Endemic Region: Amazonia, Andean, Pacific Dept.: AMA, ANT, CAQ, CHO, CUN, NAR, VAC Elev.: 10–1080 m a.s.l. IUCN Habitat code: 1,3,14 Conservation status: (IUCN) NE

3879 *Macropsychanthus macrocarpus* (Huber) L.P.Queiroz & Snak (Fabaceae)
IPNI: 77212340-1
Habit: Shrub, Liana, Climbing Use: MA Origin: Endemic Region: Amazonia, Guiana Shield, Orinoquia, Pacific Dept.: CAQ, CHO, GUA, GUV, VAU Elev.: 5–810 m a.s.l. IUCN Habitat code: 1,14 Conservation status: (IUCN) NE

3880 *Macropsychanthus megacarpus* (Rolfe) L.P.Queiroz & Snak (Fabaceae)
IPNI: 77212324-1
Use: MA, ME, PO Origin: Endemic IUCN Habitat code: 3 Conservation status: (IUCN) NE

3881 *Macropsychanthus ucayalinus* (Harms) L.P.Queiroz & Snak (Fabaceae)
IPNI: 77212332-1
Habit: Climbing Use: ME Origin: Endemic Region: Amazonia, Andean Dept.: AMA, CAQ, PUT Elev.: 100–1000 m a.s.l. IUCN Habitat code: 1 Conservation status: (IUCN) NE

3882 *Macropsychanthus violaceus* (Mart. ex Benth.) L.P.Queiroz & Snak (Fabaceae)
IPNI: 77212334-1
Habit: Liana, Climbing Use: HF, ME Origin: Endemic Region: Amazonia,

Pacific **Dept.:** CHO, SAN **Elev.:** 80–200 m a.s.l. **IUCN Habitat code:** 1,2,5 **Conservation status: (IUCN)** NE

3883 *Macropsychanthus wilsonii* (Standl.) L.P.Queiroz & Snak (Fabaceae)
IPNI: 77212335-1
Habit: Liana, Climbing **Use:** ME **Origin:** Endemic **Region:** Andean, Caribbean **Dept.:** ANT, NAR **Elev.:** 0–1900 m a.s.l. **IUCN Habitat code:** 1,3,5 **Conservation status: (IUCN)** NE

3884 *Macroptilium atropurpureum* (DC.) Urb. (Fabaceae)
IPNI: 57174-3
Common name: frijolillo, siratro
Habit: Herb, Climbing **Use:** AF, EU, ME **Origin:** Native **Region:** Andean, Caribbean **Dept.:** ANT, ATL, CAU, CUN, HUI, MAG, MET, SAN, TOL, VAC **Elev.:** 10–2200 m a.s.l. **IUCN Habitat code:** 1,2,3,4,5,14 **Conservation status: (IUCN)** NE

3885 *Macroptilium erythroloma* (Mart. ex Benth.) Urb. (Fabaceae)
IPNI: 149061-2
Common name: frijolito **Habit:** Climbing **Use:** EU **Origin:** Cultivated, Naturalised **Region:** Andean **Dept.:** ANT, CUN **Elev.:** 850–1600 m a.s.l. **Conservation status: (IUCN)** NE

3886 *Macroptilium gracile* (Poepp. ex Benth.) Urb. (Fabaceae)
IPNI: 149063-2
Habit: Herb, Climbing **Use:** AF **Origin:** Native **Region:** Guiana Shield, Caribbean, Orinoquia **Dept.:** ATL, BOY, CAS, CES, MAG, MET, NSA, TOL, VAU, VID **Elev.:** 10–550 m a.s.l. **IUCN Habitat code:** 1,2,3,4,5,14 **Conservation status: (IUCN)** NE

3887 *Macroptilium lathyroides* (L.) Urb. (Fabaceae)
IPNI: 149066-2
Common name: chocho de sabana, frijolillo, moradita **Habit:** Herb, Climbing **Use:** AF, EU, GS, HF, ME, PO **Origin:** Native **Region:** Andean, Guiana Shield, Caribbean, Orinoquia **Dept.:** ANT, ATL, BOY, CAL, CAQ, CAU, CUN, GUV, HUI, MAG, MET, PUT, SAP, SAN, TOL, VAC **Elev.:** 10-1600 m.a.s.l. **IUCN Habitat code:** 1,2,3,4,5,14 **Conservation status: (IUCN)** NE **(CPSG)** Potential LC

3888 *Medicago lupulina* L. (Fabaceae)
IPNI: 506268-1
Common name: carretón de caballo, carretón de huerta **Habit:** Herb **Use:** AF, EU, HF, IF, ME, PO **Origin:** Naturalised **Region:** Andean **Dept.:** BOG, BOY, CUN **Elev.:** 2550–3600 m a.s.l. **IUCN Habitat code:** 3,4,14 **Conservation status: (IUCN)** NE

3889 *Medicago polymorpha* L. (Fabaceae)
IPNI: 60465997-2
Common name: cadillo, cadillo de perro, carretoncito, carretón, carretón cadillo, cuartillita, cuartillito, trébol cadillo **Habit:** Herb, Climbing **Use:** AF, EU, HF, ME, PO **Origin:** Naturalised **Region:** Andean **Dept.:** BOG, BOY, CAU, CUN, NAR, SAN, VAC **Elev.:** 2090–3200 m a.s.l. **IUCN Habitat code:** 14 **Conservation status: (IUCN)** NE

3890 *Medicago sativa* L. (Fabaceae)
IPNI: 30234961-2
Common name: alfalfa, alfalfa peruana, alfileres, alfilerillo, cadillo de tierra, ojo de buey, ojovenao, trébol de carretilla **Habit:** Herb **Use:** AF, EU, FU, GS, HF, IF, MA, ME, PO, SU **Origin:** Cultivated, Naturalised **Region:** Andean **Dept.:** ANT, BOG, BOY, CUN, NSA **Elev.:** 1150–2700 m a.s.l. **IUCN Habitat code:** 2,3,14 **Conservation status: (IUCN)** LC

3891 *Melilotus albus* Medik. (Fabaceae)
IPNI: 156656-2
Common name: carretón dulce blanco, trébol de olor, trébol dulce
Habit: Herb **Use:** AF, EU, HF, IF, MA, ME, PO **Origin:** Naturalised **Region:** Andean **Dept.:** BOG, BOY, CUN, NAR, SAN **Elev.:** 1990–2900 m a.s.l. **IUCN Habitat code:** 1,3,14 **Conservation status: (IUCN)** NE

3892 *Melilotus indicus* (L.) All. (Fabaceae)
IPNI: 1207360-2
Habit: Herb, Subshrub **Use:** AF, EU, HF, IF, ME, PO **Origin:** Naturalised **Region:** Andean **Dept.:** BOG, BOY, CUN, NSA **Elev.:** 700–2824 m a.s.l. **IUCN Habitat code:** 3,14 **Conservation status: (IUCN)** NE

3893 *Mimosa albida* Humb. & Bonpl. ex Willd. (Fabaceae)
IPNI: 507622-1
Common name: zarza, zarza de vega
Habit: Herb, Shrub, Subshrub, Climbing **Use:** ME **Origin:** Native **Region:** Andean, Caribbean, Orinoquia, Pacific **Dept.:** ANT, BOY, CAU, CES, CHO, HUI, LAG, MAG, MET, NAR, NSA, QUI, RIS, SAN, TOL, VAC **Elev.:** 40-2800 m.a.s.l. **IUCN Habitat code:** 1,2,3,4,5,14 **Conservation status: (IUCN)** LC **(CPSG)** Potential LC

3894 *Mimosa annularis* Spruce ex Benth. (Fabaceae)
IPNI: 507645-1
Habit: Shrub, Small tree **Use:** ME **Origin:** Native **Region:** Amazonia, Guiana Shield **Dept.:** AMA, CAQ, GUA, VAU **Elev.:** 0–300 m a.s.l. **IUCN Habitat code:** 14 **Conservation status: (IUCN)** NE

3895 *Mimosa arenosa* (Willd.) Poir. (Fabaceae)
IPNI: 507659-1
Habit: Shrub, Small tree, Tree **Use:** ME **Origin:** Native **Region:** Caribbean **Dept.:** BOL, LAG, MAG **Elev.:** 3–200 m a.s.l. **IUCN Habitat code:** 1,2,3,4,5,14 **Conservation status: (IUCN)** LC

3896 *Mimosa camporum* Benth. (Fabaceae)
IPNI: 507771-1
Common name: dormidera, salsa, zarza **Habit:** Herb, Shrub, Subshrub **Use:** ME **Origin:** Native **Region:** Andean, Guiana Shield, Caribbean, Orinoquia, Pacific **Dept.:** ATL, BOL, BOY, CAU, CUN, GUA, HUI, LAG, MAG, MET, SAN, SUC, TOL **Elev.:** 0-2000 m.a.s.l. **IUCN Habitat code:** 1,2,3,4,14 **Conservation status: (IUCN)** NE **(CPSG)** Potential LC

3897 *Mimosa candollei* R.Grether (Fabaceae)
IPNI: 1014032-1
Use: ME **Origin:** Native **IUCN Habitat code:** 3 **Conservation status: (IUCN)** NE

3898 *Mimosa casta* L. (Fabaceae)
IPNI: 313177-2
Common name: adormidera, cariño de suegra **Habit:** Shrub **Use:** ME **Origin:** Native **Region:** Guiana Shield, Caribbean, Orinoquia **Dept.:** CAQ, CAS, COR, GUV, MET **Elev.:** 20–500 m a.s.l. **IUCN Habitat code:** 1,2,3,5,14 **Conservation status: (IUCN)** NE

3899 *Mimosa debilis* Humb. & Bonpl. ex Willd. (Fabaceae)
IPNI: 507880-1
Habit: Subshrub, Climbing **Use:** ME **Origin:** Native **Region:** Orinoquia **Dept.:** MET **Elev.:** 250–405 m a.s.l. **IUCN Habitat code:** 14 **Conservation status: (IUCN)** NE

3900 *Mimosa diplotricha* C.Wright (Fabaceae)
IPNI: 312750-2
Habit: Herb, Subshrub, Climbing **Use:** AF, EU, ME **Origin:** Native **Region:** Amazonia, Andean, Pacific **Dept.:** AMA, ANT, BOY, CHO, CUN, HUI, NAR, TOL, VAC **Elev.:** 200–1200 m a.s.l. **IUCN Habitat code:** 1,3,5,14 **Conservation status: (IUCN)** NE

3901 *Mimosa distachya* Cav. (Fabaceae)
IPNI: 317695-2
Habit: Shrub **Use:** MA, ME **Origin:** Native **Region:** Caribbean, Orinoquia **Dept.:** CAS, LAG, MAG **Elev.:** 0–650 m a.s.l. **IUCN Habitat code:** 1,2,3,14 **Conservation status: (IUCN)** NE

3902 *Mimosa invisa* Mart. ex Colla (Fabaceae)
IPNI: 508165-1
Common name: zarza **Habit:** Shrub, Subshrub **Use:** AF, EU, ME **Origin:** Native **Region:** Andean **Dept.:** ANT, NSA **Elev.:** 100–1560 m a.s.l. **IUCN Habitat code:** 1,3,14 **Conservation status: (IUCN)** NE

3903 *Mimosa pigra* L. (Fabaceae)
IPNI: 313003-2
Common name: adormidera, dormidera, vergonzosa, zarza, zarza prieta **Habit:** Shrub, Subshrub **Use:** AF, EU, FU, HF, MA, ME, SU **Origin:** Native **Region:** Amazonia, Andean, Guiana Shield, Caribbean, Orinoquia, Pacific **Dept.:** AMA, ANT, ARA, ATL, BOL, BOY, CAQ, CAS, CAU, CHO, COR, CUN, GUA, HUI, MAG, MET, NSA, SAN, SUC, VAC **Elev.:** 0-1600 m.a.s.l. **IUCN Habitat code:** 1,2,3,4,5,14 **Conservation status: (IUCN)** LC **(CPSG)** Potential LC

3904 *Mimosa polyantha* Benth. (Fabaceae)
IPNI: 508515-1
Use: MA **Origin:** Native **IUCN Habitat code:** 3 **Conservation status: (IUCN)** LC

3905 *Mimosa polycarpa* Kunth (Fabaceae)
IPNI: 508517-1
Habit: Herb **Use:** ME **Origin:** Native **Region:** Orinoquia **Dept.:** CAS, CUN **Elev.:** 300–500 m a.s.l. **IUCN Habitat code:** 3,14 **Conservation status: (IUCN)** NE

3906 *Mimosa polydactyla* Humb. & Bonpl. ex Willd. (Fabaceae)
IPNI: 508520-1
Habit: Herb, Shrub **Use:** ME **Origin:** Native **Region:** Amazonia, Andean, Pacific **Dept.:** AMA, CHO, NAR, PUT **Elev.:** 80–1350 m a.s.l. **IUCN Habitat code:** 1,14 **Conservation status: (IUCN)** NE

3907 *Mimosa pudica* L. (Fabaceae)
IPNI: 313242-2
Common name: adormidera, biumía, cerrate puta, cierra tus puertas, cierrateciérrate, cierrateputa, ciérrate, dormidera, dormidera espinosa, dormilona, guamito, mimosa, rastrera, sensitiva, tsakeke, tsakenae, tsakewaül, vergonzosa **Habit:** Herb, Shrub, Subshrub, Liana **Use:** AF, EU, HF, MA, ME, SU **Origin:** Native **Region:** Amazonia, Andean, Guiana Shield, Caribbean, Orinoquia, Pacific **Dept.:**

AMA, ANT, ATL, BOL, BOY, CAL, CAS, CAU, CHO, COR, CUN, GUA, HUI, MAG, MET, NAR, NSA, PUT, QUI, RIS, SAP, SAN, SUC, TOL, VAC **Elev.:** 0-1780 m.a.s.l. **IUCN Habitat code:** 1,2,3,4,5,8,14 **Conservation status: (IUCN)** LC **(CPSG)** Potential LC

3908 *Mimosa scabrella* Benth. (Fabaceae)
IPNI: 508647-1
Use: EU, FU, HF, MA **Origin:** Naturalised **Dept.:** ANT**Conservation status: (IUCN)** NE

3909 *Mimosa sensitiva* L. (Fabaceae)
IPNI: 508671-1
Common name: zarza **Habit:** Shrub **Use:** EU **Origin:** Native **Region:** Caribbean **Dept.:** NSA **Elev.:** 120–120 m a.s.l. **IUCN Habitat code:** 14 **Conservation status: (IUCN)** NE

3910 *Mimosa somnians* Humb. & Bonpl. ex Willd. (Fabaceae)
IPNI: 508701-1
Common name: adormidera, adormidera rosada, dormidera, mimosa, zarza de vega **Habit:** Herb, Shrub, Subshrub, Small tree **Use:** ME **Origin:** Native **Region:** Amazonia, Andean, Caribbean, Orinoquia, Pacific **Dept.:** AMA, BOY, CAL, CAS, CAU, CUN, HUI, LAG, MET, QUI, SAN, TOL, VID **Elev.:** 40-1830 m.a.s.l. **IUCN Habitat code:** 1,2,3,4,14 **Conservation status: (IUCN)** NE **(CPSG)** Potential LC

3911 *Mimosa tenuiflora* (Willd.) Poir. (Fabaceae)
IPNI: 508775-1
Common name: carbonal **Habit:** Shrub, Tree **Use:** AF, FU, MA, ME, SU **Origin:** Native **Region:** Caribbean **Dept.:** LAG **Elev.:** 0–250 m a.s.l. **IUCN Habitat code:** 1,14 **Conservation status: (IUCN)** LC

3912 *Monopteryx angustifolia* Spruce ex Benth. (Fabaceae)
IPNI: 509092-1
Habit: Tree **Use:** HF **Origin:** Native **Region:** Amazonia **Dept.:** VAU **Elev.:** 200–200 m a.s.l. **IUCN Habitat code:** 1 **Conservation status: (IUCN)** LC

3913 *Monopteryx uaucu* Spruce ex Benth. (Fabaceae)
IPNI: 509095-1
Common name: abina, cincuentacentavos, creolino, hígado de diablo, hígado del diablo **Habit:** Tree **Use:** HF, MA **Origin:** Native **Region:** Amazonia **Dept.:** AMA, CAQ, VAU **Elev.:** 200–200 m a.s.l. **IUCN Habitat code:** 1,5 **Conservation status: (IUCN)** LC

3914 *Mora oleifera* (Triana ex Hemsl.) Ducke (Fabaceae)
IPNI: 164265-2
Common name: alcornoque, canime, mangle nato, nato, nato rojo **Habit:** Tree **Use:** MA, ME **Origin:** Native **Region:** Pacific **Dept.:** CAU, CHO, NAR, VAC **Elev.:** 0-100 m.a.s.l. **IUCN Habitat code:** 1,5,14 **Conservation status: (IUCN)** VU **(CPSG)** EN

3915 *Mucuna holtonii* (Kuntze) Moldenke (Fabaceae)
IPNI: 164648-2
Habit: Liana, Climbing **Use:** ME **Origin:** Native **Region:** Andean **Dept.:** ANT, BOL, CAL, QUI **Elev.:** 100–1680 m a.s.l. **IUCN Habitat code:** 1,2,3,14 **Conservation status: (IUCN)** NE

3916 *Mucuna mollis* (Kunth) DC. (Fabaceae)
IPNI: 509248-1
Common name: chocho, ojo de buey, ojo de vaca, ojo de venado, pica pica, picapica, quemador **Habit:** Liana **Use:** ME, SU **Origin:** Endemic **Region:** Amazonia, Andean, Pacific **Dept.:** AMA, ANT, CAU, CHO, COR, CUN, QUI, SAN, VAC **Elev.:** 200–1750 m a.s.l. **IUCN Habitat code:** 1,2,3,5,14 **Conservation status: (IUCN)** NE

3917 *Mucuna mutisiana* (Kunth) DC. (Fabaceae)
IPNI: 509253-1
Common name: chocho, cojón de fraile, congolo, ojo de buey, ojo de venado, ojo de zamuro, pica pica, picapica **Habit:** Liana **Use:** MA, ME, SU **Origin:** Native **Region:** Andean, Caribbean, Orinoquia, Pacific **Dept.:** ANT, BOY, CHO, CUN, HUI, LAG, MAG, NAR, NSA, QUI, RIS, SAN, TOL, VAC **Elev.:** 0-2300 m.a.s.l. **IUCN Habitat code:** 1,2,3,5,14 **Conservation status: (IUCN)** NE **(CPSG)** Potential LC

3918 *Mucuna pruriens* (L.) DC. (Fabaceae)
IPNI: 509275-1
Common name: bejuco picapica, congolo, cóngolo, frijol terciopelo, fríjol terciopelo, nescafé, ojo de buey, picapica **Habit:** Liana, Climbing **Use:** AF, EU, GS, HF, MA, ME, PO, SU **Origin:** Naturalised **Region:** Andean, Orinoquia, Pacific **Dept.:** ANT, BOL, BOY, CAL, CAU, CHO, GUA, GUV, HUI, MAG, MET, QUI, SAN, SUC, VAC **Elev.:** 80–1800 m a.s.l. **IUCN Habitat code:** 1,2,3,5,14 **Conservation status: (IUCN)** NE

3919 *Mucuna rostrata* Benth. (Fabaceae)
IPNI: 509285-1
Common name: congola, embisca, ojo de buey **Habit:** Liana **Use:** EU, ME **Origin:** Native **Region:** Amazonia, Andean, Orinoquia, Pacific **Dept.:** AMA, ANT, CHO, GUV, VAU, VID **Elev.:** 0–1000 m a.s.l. **IUCN Habitat code:** 1,14 **Conservation status: (IUCN)** NE

3920 *Mucuna sloanei* Fawc. & Rendle (Fabaceae)
IPNI: 291345-2
Common name: carnero, ojo de buey, ojo de venado, ojo de zamuro **Habit:** Liana **Use:** EU, HF, MA, ME **Origin:** Native **Region:** Amazonia, Andean, Guiana Shield, Orinoquia **Dept.:** CAQ, CUN, MET, NAR, PUT **Elev.:** 130–1300 m a.s.l. **IUCN Habitat code:** 1,2,3,14 **Conservation status: (IUCN)** NE

3921 *Mucuna urens* (L.) Medik. (Fabaceae)
IPNI: 1197939-2
Common name: cóngolo, ojo de buey, ojo de venado, picapica **Habit:** Liana **Use:** HF, ME, PO **Origin:** Native **Region:** Amazonia **Dept.:** VAU **Elev.:** 250–500 m a.s.l. **IUCN Habitat code:** 1,3,14 **Conservation status: (IUCN)** NE

3922 *Muellera broadwayi* (Urb.) M.Sousa (Fabaceae)
IPNI: 60458868-2
Habit: Shrub, Tree **Use:** ME **Origin:** Native **Region:** Caribbean **Dept.:** ATL, BOL, CES, LAG, MAG **Elev.:** 30–250 m a.s.l. **IUCN Habitat code:** 1,2 **Conservation status: (IUCN)** NE

3923 *Muellera fendleri* (Benth.) M.J.Silva & A.M.G.Azevedo (Fabaceae)
IPNI: 77118389-1
Common name: macurutú **Habit:** Tree **Use:** ME **Origin:** Native **Region:** Caribbean **Dept.:** MAG **Elev.:** 10–10 m a.s.l. **IUCN Habitat code:** 2 **Conservation status: (IUCN)** NE

3924 *Muellera monilis* (L.) M.J.Silva & A.M.G.Azevedo (Fabaceae)
IPNI: 77118402-1
Habit: Shrub, Small tree, Tree **Use:** ME **Origin:** Native **Region:** Amazonia, Caribbean, Orinoquia, Pacific **Dept.:** AMA, ANT, CHO, MET, NAR, VAC **Elev.:** 0–200 m a.s.l. **IUCN Habitat code:** 1,3,4,5,14 **Conservation status: (IUCN)** NE

3925 *Myrospermum frutescens* Jacq. (Fabaceae)
IPNI: 509434-1
Common name: balsamito, guatamare, mano de pilón **Habit:** Shrub, Small tree, Tree **Use:** AF, EU, FU, MA, ME **Origin:** Native **Region:** Caribbean **Dept.:** ATL, BOL, CES, COR, LAG, MAG, NSA, SUC **Elev.:** 0–350 m a.s.l. **IUCN Habitat code:** 1,2,3,14 **Conservation status: (IUCN)** LC

3926 *Myroxylon balsamum* (L.) Harms (Fabaceae)
IPNI: 166483-2
Common name: bálsamo, bálsamo blanco, bálsamo de Tolú, bálsamo de tolú, bálsamo del Perú, bálsamo rubio, bálsamo tolutiano, carano, caraña, cereipo, elemí, estoraque, guararo, olor, pidoquera, roble maría, sarrapio, tache, tolú, árbol de Tolú, árbol de caraña **Habit:** Shrub, Small tree, Tree **Use:** EU, HF, MA, ME, SU **Origin:** Native, Cultivated **Region:** Amazonia, Andean, Caribbean, Orinoquia **Dept.:** AMA, ANT, BOL, MAG, MET, PUT, VAC **Elev.:** 15–1600 m a.s.l. **IUCN Habitat code:** 1,2,3,4,5,14 **Conservation status: (IUCN)** NE

3927 *Myroxylon peruiferum* L.f. (Fabaceae)
IPNI: 509454-1
Common name: bálsamo **Use:** MA, ME **Origin:** Native **IUCN Habitat code:** 1 **Conservation status: (IUCN)** LC

3928 *Neptunia plena* (L.) Benth. (Fabaceae)
IPNI: 1056199-2
Common name: adormidera **Habit:** Herb, Shrub, Tree **Use:** EU, HF, MA, ME **Origin:** Native **Region:** Caribbean, Pacific **Dept.:** ANT, ATL, BOL, CHO, COR, LAG, MAG, SUC, TOL **Elev.:** 0–350 m a.s.l. **IUCN Habitat code:** 1,2,3,5,14 **Conservation status: (IUCN)** LC

3929 *Neptunia prostrata* (Lam.) Baill. (Fabaceae)
IPNI: 509612-1
Common name: dormidera acuática, neptunia, sensitiva acuática, sensitiva de agua, tripa de babilla, tripa de pollo, zarzón **Habit:** Herb, Aquatic **Use:** EU, GS, HF, ME **Origin:** Native **Region:** Amazonia, Caribbean, Orinoquia, Pacific **Dept.:** AMA, ARA, ATL, BOL, CES, CHO, COR, LAG **Elev.:** 5–150 m a.s.l. **IUCN Habitat code:** 1,2,3,5,14 **Conservation status: (IUCN)** NE

3930 *Neustanthus phaseoloides* (Roxb.) Benth. (Fabaceae)
IPNI: 509661-1
Common name: cusú, cuzú, kudzú, kudzú americano, kudzú tropical **Habit:** Herb, Climbing **Use:** AF, EU, HF, MA, ME, SU **Origin:** Naturalised **Region:** Caribbean, Orinoquia, Pacific **Dept.:** AMA, ANT, CAL, CAQ, CES, GUV, MET, VAC **Elev.:** 150–1500 m a.s.l. **IUCN Habitat code:** 1,2,3,14 **Conservation status: (IUCN)** NE

3931 *Nissolia vincentina* (Ker Gawl.) T.M.Moura & Fort.-Perez (Fabaceae) **IPNI:** 77191263-1 **Habit:** Herb, Climbing **Use:** AF **Origin:** Native **Region:** Caribbean **Dept.:** ATL, BOL, CUN, LAG, MAG, SAN **Elev.:** 50–600 m a.s.l. **IUCN Habitat code:** 1,2,3,4,5,14 **Conservation status: (IUCN)** NE

3932 *Ormosia amazonica* Ducke (Fabaceae) **IPNI:** 176344-2 **Common name:** cairuro, cairurú, chocho, peonio, peonío, pionía **Habit:** Small tree, Tree **Use:** MA, ME, SU **Origin:** Native **Region:** Amazonia, Orinoquia **Dept.:** AMA, CAQ, MET **Elev.:** 100–200 m a.s.l. **IUCN Habitat code:** 1,3,14 **Conservation status: (IUCN)** LC

3933 *Ormosia coarctata* Jacks. (Fabaceae) **IPNI:** 510681-1 **Habit:** Tree **Use:** MA **Origin:** Native **Region:** Amazonia, Orinoquia **Dept.:** BOY, CAQ, GUV **Elev.:** 500–500 m a.s.l. **IUCN Habitat code:** 1,3,5,14 **Conservation status: (IUCN)** NE

3934 *Ormosia coccinea* (Aubl.) Jacks. (Fabaceae) **IPNI:** 510682-1 **Common name:** cairurú, chocho, huaruro, huayro, tento malungo **Habit:** Shrub, Small tree, Tree **Use:** ME **Origin:** Native **Region:** Amazonia, Andean, Guiana Shield, Orinoquia **Dept.:** AMA, ANT, GUA, PUT, VAU, VID **Elev.:** 65–1420 m a.s.l. **IUCN Habitat code:** 1,3,4,14 **Conservation status: (IUCN)** LC

3935 *Ormosia colombiana* Rudd (Fabaceae) **IPNI:** 176354-2 **Common name:** cacho, chocho, peonía **Habit:** Tree **Use:** MA **Origin:** Endemic **Region:** Andean **Dept.:** ANT, CAQ, CUN, MAG, QUI, VAC **Elev.:** 1300-2310 m.a.s.l. **IUCN Habitat code:** 1,2,3,4,14 **Conservation status: (IUCN)** LC **(CPSG)** LC

3936 *Ormosia fastigiata* Tul. (Fabaceae) **IPNI:** 510703-1 **Common name:** chocho, macucú, sangretoro, tento, árbol guaruro **Habit:** Shrub, Tree **Use:** MA **Origin:** Native **Region:** Amazonia **Dept.:** GUA **Elev.:** 500–500 m a.s.l. **IUCN Habitat code:** 1 **Conservation status: (IUCN)** LC

3937 *Ormosia macrocalyx* Ducke (Fabaceae) **IPNI:** 176377-2 **Habit:** Tree **Use:** ME **Origin:** Native **Region:** Amazonia, Orinoquia **Dept.:** AMA, GUA, GUV **Elev.:** 150–150 m a.s.l. **IUCN Habitat code:** 1,3,4,14 **Conservation status: (IUCN)** LC

3938 *Ormosia macrophylla* Benth. (Fabaceae) **IPNI:** 510743-1 **Common name:** chocho, palo de mico, pionio, pionío **Habit:** Small tree, Tree **Use:** ME, PO, SU **Origin:** Native **Region:** Amazonia **Dept.:** AMA, CAQ, GUA, VAU **Elev.:** 100–730 m a.s.l. **IUCN Habitat code:** 1,3,4,14 **Conservation status: (IUCN)** LC

3939 *Ormosia nobilis* Tul. (Fabaceae) **IPNI:** 510761-1 **Common name:** chocho, guacharaco, peonía, pionía **Habit:** Tree **Use:** ME **Origin:** Native **Region:** Amazonia, Andean, Guiana Shield, Orinoquia, Pacific **Dept.:** AMA, ARA, CAQ, CHO, CUN, MET, PUT, VAU **Elev.:** 80–1420 m a.s.l. **IUCN Habitat code:** 1,3,4,14 **Conservation status: (IUCN)** LC

3940 *Ormosia paraensis* Ducke (Fabaceae) **IPNI:** 176389-2 **Common name:** chocho **Habit:** Tree **Use:** MA **Origin:** Native **Region:** Amazonia, Andean, Caribbean **Dept.:** AMA, ANT, SAN **Elev.:** 80–770 m a.s.l. **IUCN Habitat code:** 1,3,5,14 **Conservation status: (IUCN)** LC

3941 *Ormosia stipularis* Ducke (Fabaceae) **IPNI:** 176399-2 **Common name:** tento, árbol guaruto **Habit:** Shrub **Use:** MA **Origin:** Native **Region:** Amazonia **Dept.:** GUA **Elev.:** 500–500 m a.s.l. **Conservation status: (IUCN)** NE

3942 *Orphanodendron bernalii* Barneby & J.W.Grimes (Fabaceae) **IPNI:** 958253-1 **Common name:** almanegra, tirateté **Habit:** Tree **Use:** MA **Origin:** Endemic **Region:** Pacific **Dept.:** ANT **Elev.:** 150-350 m.a.s.l. **IUCN Habitat code:** 1,3,5,14 **Conservation status: (IUCN)** LC **(CPSG)** LC

3943 *Otholobium mexicanum* (L.f.) J.W.Grimes (Fabaceae) **IPNI:** 281771-2 **Common name:** amansapeon, amansapeón, culen, pulen, púlen, ruchica, rúchica, rúchico, rúchigo, tarta, té de Popayán, té de popayan, té de popayán **Habit:** Herb, Shrub **Use:** ME **Origin:** Native **Region:** Andean **Dept.:** ANT, BOG, BOY, CAL, CAU, CUN, MAG, NAR, NSA, PUT, QUI, SAN, TOL, VAC **Elev.:** 1800–4000 m a.s.l. **IUCN Habitat code:** 1,3,4,5,14 **Conservation status: (IUCN)** NE

3944 *Oxyrhynchus trinervius* (Donn.Sm.) Rudd (Fabaceae) **IPNI:** 179469-2 **Use:** MA **Origin:** Native **Conservation status: (IUCN)** NE

3945 *Pachyrhizus ferrugineus* (Piper) M.Sørensen (Fabaceae) **IPNI:** 945254-1 **Habit:** Climbing **Use:** GS, HF **Origin:** Native **Region:** Pacific **Dept.:** CHO **Elev.:** 40–450 m a.s.l. **IUCN Habitat code:** 1 **Conservation status: (IUCN)** NE

3946 *Pachyrhizus panamensis* R.T.Clausen (Fabaceae) **IPNI:** 179940-2 **Habit:** Climbing **Use:** GS **Origin:** Native **Region:** Caribbean **Dept.:** MAG **Elev.:** 100–100 m a.s.l. **Conservation status: (IUCN)** NE

3947 *Pachyrhizus tuberosus* (Lam.) Spreng. (Fabaceae) **IPNI:** 512027-1 **Common name:** fríjol de tubérculo, jícama, nupe **Habit:** Climbing **Use:** GS, HF, ME, PO **Origin:** Native **Region:** Andean **Dept.:** CAL **Elev.:** 1400–1400 m a.s.l. **IUCN Habitat code:** 1 **Conservation status: (IUCN)** NE

3948 *Paraserianthes lophantha* (Vent.) I.C.Nielsen (Fabaceae) **IPNI:** 912944-1 **Common name:** acacia, acacia bracatinga **Habit:** Tree **Use:** EU, FU, HF, MA, ME **Origin:** Cultivated, Naturalised **Region:** Andean, Orinoquia **Dept.:** BOG, BOY, CAL, CUN, MET **Elev.:** 100–3000 m a.s.l. **Conservation status: (IUCN)** LC

3949 *Parkia discolor* Spruce ex Benth. (Fabaceae) **IPNI:** 512191-1 **Common name:** araratucupé, araratucupí, aserrín, dormidero blanco **Habit:** Tree **Use:** HF **Origin:** Native **Region:** Amazonia, Guiana Shield, Orinoquia **Dept.:** AMA, CAQ, GUA, VAU, VID **Elev.:** 250–1000 m a.s.l. **IUCN Habitat code:** 1,3,4,5,14 **Conservation status: (IUCN)** LC

3950 *Parkia igneiflora* Ducke (Fabaceae) **IPNI:** 183268-2 **Common name:** guamo, guamo borugo, guamo negro, guarango **Habit:** Small tree **Use:** HF **Origin:** Native **Region:** Amazonia, Guiana Shield **Dept.:** AMA, GUA, VAU **Elev.:** 140–200 m a.s.l. **IUCN Habitat code:** 1,3,4,5 **Conservation status: (IUCN)** LC

3951 *Parkia multijuga* Benth. (Fabaceae) **IPNI:** 512214-1 **Common name:** dormilón, gatara, guama, guama de chucha, guamillo, guamo de pepa negra, guamo negro, guarango, guarango blanco, pachaco **Habit:** Tree **Use:** HF, MA, ME **Origin:** Native **Region:** Amazonia, Guiana Shield, Orinoquia **Dept.:** AMA, GUV, MET, PUT, VAU **Elev.:** 100–700 m a.s.l. **IUCN Habitat code:** 1,3,5,14 **Conservation status: (IUCN)** LC

3952 *Parkia nitida* Miq. (Fabaceae) **IPNI:** 512215-1 **Common name:** aserrín, dormidera, dormidero, dormilón, guama, guama negra, guamillo, guamo, guarango, guarango rayao, gurango rayao, paleta **Habit:** Tree **Use:** HF, MA, ME **Origin:** Native **Region:** Amazonia, Guiana Shield, Orinoquia, Pacific **Dept.:** AMA, CAQ, CHO, MET, PUT, VAU **Elev.:** 100–1200 m a.s.l. **IUCN Habitat code:** 1,3,14 **Conservation status: (IUCN)** LC

3953 *Parkia pectinata* (Humb. & Bonpl. ex Willd.) Benth. (Fabaceae) **IPNI:** 512223-1 **Common name:** araratucupí, dormidero, guama, guama de chucha, guamita, guamo **Habit:** Tree **Use:** HF **Origin:** Native **Region:** Amazonia, Guiana Shield **Dept.:** AMA, CAQ, VAU **Elev.:** 150–200 m a.s.l. **IUCN Habitat code:** 1,3,14 **Conservation status: (IUCN)** NE

3954 *Parkia velutina* Benoist (Fabaceae) **IPNI:** 282124-2 **Common name:** dormilón, guarango **Habit:** Tree **Use:** HF **Origin:** Native **Region:** Amazonia, Guiana Shield, Pacific **Dept.:** AMA, CAQ, VAC, VAU **Elev.:** 25–280 m a.s.l. **IUCN Habitat code:** 1,3,14 **Conservation status: (IUCN)** LC

3955 *Parkinsonia aculeata* L. (Fabaceae) **IPNI:** 512242-1 **Common name:** espinillo, espino, mapuja, palo de rayo, palo verde mexicano, paloverde, retaima, retama, retamo, retamo calentano, retamo espinoso, sauce, sauce espino, sauce espinoso, sauce guajiro, sulfatillo, turco, yaba, yabo, yayo **Habit:** Shrub, Tree **Use:** AF, EU, FU, GS, HF, IF, MA, ME **Origin:** Native, Cultivated **Region:** Andean, Caribbean, Orinoquia, Pacific **Dept.:** ANT, ATL, BOL, BOY, CAS, CUN, HUI, LAG, MAG, NAR, SAP, SAN, TOL, VAC **Elev.:** 0–2000 m a.s.l. **IUCN Habitat code:** 1,2,3,4,14 **Conservation status: (IUCN)** LC

3956 *Parkinsonia praecox* (Ruiz & Pav.) Hawkins (Fabaceae) **IPNI:** 1006943-1 **Common name:** brea, ciuca, cuica, cuico, cujaca, mapua, mapuso, palo verde, paloverde, quica, yabo **Habit:** Shrub, Tree **Use:** EU, HF, MA, ME **Origin:** Native **Region:** Andean, Caribbean **Dept.:** BOY, LAG, MAG, NSA, SAN **Elev.:** 0–2000 m a.s.l. **IUCN Habitat code:** 1,2,3,14 **Conservation status: (IUCN)** LC

3957 *Peltogyne paniculata* Benth. (Fabaceae) **IPNI:** 512994-1 **Common name:** algarrobito, brasilete, brasileto, cananeo, guacamari, guamo loro, mochilero, moradillo, muchilero, nazareno, tananeo, tanané **Habit:** Tree **Use:** MA **Origin:** Native **Region:** Caribbean **Dept.:** ANT, CES, SAN **Elev.:** 60-520 m.a.s.l. **IUCN Habitat code:** 1,2,3,5,14 **Conservation status: (IUCN)** NE **(CPSG)** NT

3958 *Peltogyne purpurea* Pittier (Fabaceae) **IPNI:** 187015-2 **Common name:** almanegra, azulito, brasilete, cananeo, nazareno **Habit:** Tree **Use:** MA **Origin:** Native **Region:** Caribbean **Dept.:** BOL, CES, MAG, SUC, TOL **Elev.:** 100-540 m.a.s.l. **IUCN Habitat code:** 1,2,14 **Conservation status: (IUCN)** NE **(CPSG)** VU

3959 *Peltogyne venosa* (Vahl) Benth. (Fabaceae) **IPNI:** 513006-1 **Habit:** Tree **Use:** MA **Origin:** Native **Region:** Guiana Shield **Dept.:** GUA **Elev.:** 120–120 m a.s.l. **IUCN Habitat code:** 1,4 **Conservation status: (IUCN)** NE

3960 *Peltophorum pterocarpum* (DC.) Backer ex K.Heyne (Fabaceae) **IPNI:** 513021-1 **Common name:** acacia **Habit:** Tree **Use:** AF, EU, FU, HF, MA, ME, SU **Origin:** Cultivated **Region:** Andean **Dept.:** ANT, TOL **Elev.:** 800–1500 m a.s.l. **IUCN Habitat code:** 3,14 **Conservation status: (IUCN)** NE

3961 *Pentaclethra macroloba* (Willd.) Kuntze (Fabaceae) **IPNI:** 188076-2 **Common name:** aserrín, capitancillo, dormilón, laurel, sserrín **Habit:** Tree **Use:** HF, MA, ME **Origin:** Native **Region:** Pacific **Dept.:** ANT, CAU, CHO, COR, NAR, VAC **Elev.:** 0–290 m a.s.l. **IUCN Habitat code:** 1,3,5,14 **Conservation status: (IUCN)** LC

3962 *Phaseolus × dumosus* Macfad. (Fabaceae) **IPNI:** 513799-1 **Common name:** botón de saco, cacha, cacha amarilla, cacha azul, cacha blanco, cachas, cachetón, calarqueño, carchas, carriel, carrielito, chiquinquireño, frijolito, frisol carriel, frisol de vida, frisol sapo, frĳol botón de saco, frĳol cacha, frĳol cacha amarilla, **Habit:** Climbing **Use:** GS, HF **Origin:** Naturalised **Region:** Andean **Dept.:** ANT, BOY, CAL, CAQ, CAU, CUN, HUI, MAG, MET, NAR, NSA, PUT, QUI, RIS, SAN, VAC **Elev.:** 1400–2800 m a.s.l. **IUCN Habitat code:** 1,2,3,4,14 **Conservation status: (IUCN)** NE

3963 *Phaseolus coccineus* L. (Fabaceae) **IPNI:** 513754-1 **Common name:** azulejo, bichuela, cachas, cachetón, cacho, frisol, frisol cachetón, frisol de monte, frĳol azulejo, frĳol bichuela, frĳol bichuela rojo, frĳol cachas, frĳol cachetón, frĳol cacho, frĳol de año, frĳol de enredadera, frĳol de todo el año, frĳol d **Habit:** Climbing **Use:** AF, EU, GS, HF, ME, PO **Origin:** Cultivated, Naturalised **Region:** Andean **Dept.:** ANT, BOG, CAL, CAU, CUN, PUT, QUI, VAC **Elev.:** 1500–2850 m a.s.l. **IUCN Habitat code:** 1,3,14 **Conservation status: (IUCN)** LC

3964 *Phaseolus filiformis* Benth. (Fabaceae) **IPNI:** 513814-1 **Habit:** Herb, Climbing **Use:** AF, EU, GS, HF **Origin:** Naturalised **Dept.:** VAC **Elev.:** 1000–1000 m a.s.l. **Conservation status: (IUCN)** LC

3965 *Phaseolus lunatus* L. (Fabaceae) **IPNI:** 513918-1 **Common name:** alverjo, arveja, bejuco de currucu, bejuco de jirigüelo, bejuco negro, blanco, cacha, cacha calentano, cacho, caratejo, carauta, carauta blanca, carauta silvestre, carcha, carcha rojo, carchas, carchita, carcho blanco, carcho rojo, carevaca, castillejo, e **Habit:** Climbing **Use:** AF, EU, GS, HF, ME, PO **Origin:** Native **Region:** Amazonia, Andean **Dept.:** ANT, BOY, CAL, CHO, CUN, GUA, HUI, MAG, QUI, VAC **Elev.:** 150–1875 m a.s.l. **IUCN Habitat code:** 1,2,3,4,14 **Conservation status: (IUCN)** LC

3966 *Phaseolus vulgaris* L. (Fabaceae) **IPNI:** 514191-1 **Common name:** aguadeño, ahumado, algarrobeño, algarrobo, amarillo de mata, américa, andino trepador, angulo, antioqueño, antioquia, argentina, argüello, azul, azulín, bala, bala rojo, balita, balitas, balín, bambaro, bambaro negro, barbuches, barcino, barrero, bayo, ba **Habit:** Herb, Climbing **Use:** AF, EU, GS, HF, ME, PO **Origin:** Cultivated, Naturalised **Region:** Amazonia, Andean, Orinoquia, Pacific **Dept.:** AMA, ANT, BOG, CAQ, CUN, HUI, MET, NAR, NSA, PUT, VAC, VAU **Elev.:** 30–2550 m a.s.l. **IUCN Habitat code:** 1,2,3,4,5,14 **Conservation status: (IUCN)** LC

3967 *Piptadenia pteroclada* Benth. (Fabaceae) **IPNI:** 514434-1 **Habit:** Tree **Use:** MA **Origin:** Native **Region:** Amazonia, Andean, Orinoquia **Dept.:** AMA, CAS, MET, PUT **Elev.:** 100–710 m a.s.l. **IUCN Habitat code:** 1,3,14 **Conservation status: (IUCN)** LC

3968 *Piptadenia retusa* (Jacq.) P.G.Ribeiro, Seigler & Ebinger (Fabaceae) **IPNI:** 77208976-1 **Common name:** dormilón, zarza, zarza colorada **Habit:** Shrub, Small tree, Tree **Use:** AF, MA, ME **Origin:** Endemic **Region:** Caribbean, Orinoquia **Dept.:** ATL, BOL, CUN, LAG, MAG, MET **Elev.:** 0–500 m a.s.l. **IUCN Habitat code:** 1,2,3,5,14 **Conservation status: (IUCN)** LC

3969 *Piptadenia uliginosa* Britton & Killip (Fabaceae) **IPNI:** 199687-2 **Habit:** Tree **Use:** ME **Origin:** Endemic **Region:** Caribbean **Dept.:** ATL, BOL **Elev.:** 80-100 m.a.s.l. **IUCN Habitat code:** 1 **Conservation status: (IUCN)** EN **(CPSG)** EN

3970 *Piscidia carthagenensis* Jacq. (Fabaceae) **IPNI:** 316422-2 **Common name:** arepo, barbasco, barbasco amarillo, baurá, borracho, capón, chicharrón, jebe, matapez **Habit:** Tree **Use:** FU, MA, ME, PO **Origin:** Native **Region:** Caribbean **Dept.:** ATL, BOL, SAP **Elev.:** 0–100 m a.s.l. **IUCN Habitat code:** 1,2,14 **Conservation status: (IUCN)** NE

3971 *Pithecellobium dulce* (Roxb.) Benth. (Fabaceae) **IPNI:** 1003508-2 **Common name:** acacia tunuda, azabache, buche, buche colorado, chanchán, chiminango, dinde, espino, gallinero, ojito de nena, payandé, payandé bobo, pintacanillo, pisca, tiraco **Habit:** Small tree, Tree **Use:** AF, EU, FU, GS, HF, IF, MA, ME, PO **Origin:** Native, Cultivated **Region:** Andean, Caribbean, Orinoquia, Pacific **Dept.:** ANT, ATL, BOL, BOY, CAL, CAS, CAU, CUN, HUI, LAG, MAG, NSA, QUI, SAN, SUC, TOL, VAC **Elev.:** 0-2130 m.a.s.l. **IUCN Habitat code:** 1,2,3,5,14 **Conservation status: (IUCN)** LC **(CPSG)** Potential LC

3972 *Pithecellobium hymenaeifolium* (Humb. & Bonpl. ex Willd.) Benth. (Fabaceae) **IPNI:** 1183336-2 **Common name:** barba de ángel, chiracó, pico de loro, uñas de tigre **Habit:** Shrub, Small tree **Use:** HF, MA **Origin:** Native **Region:** Caribbean **Dept.:** ATL, BOL, MAG, SUC **Elev.:** 0–180 m a.s.l. **IUCN Habitat code:** 1,2,5,14 **Conservation status: (IUCN)** LC

3973 *Pithecellobium lanceolatum* (Humb. & Bonpl. ex Willd.) Benth. (Fabaceae) **IPNI:** 514978-1 **Common name:** buche, buche blanco, carnefresca, cascoebuey, cañandronga, chiminango, espino, espinoso mono, gallinero, patevaca, payandé, payandé bobo, pinta mono, pintamono, tiraco, tiribuche, tiribuchi **Habit:** Shrub, Small tree, Tree **Use:** AF, FU, HF, MA, ME **Origin:** Native **Region:** Andean, Caribbean, Pacific **Dept.:** ANT, ATL, BOL, BOY, CAU, CES, COR, CUN, HUI, LAG, MAG, SAP, SAN, SUC, TOL, VAC **Elev.:** 0-1250 m.a.s.l. **IUCN Habitat code:** 1,2,3,5,14 **Conservation status: (IUCN)** LC **(CPSG)** Potential LC

3974 *Pithecellobium roseum* (Vahl) Barneby & J.W.Grimes (Fabaceae) **IPNI:** 996931-1 **Common name:** buche colorado, payandé clavo, tiraco **Habit:** Shrub, Tree **Use:** HF **Origin:** Native **Region:** Caribbean **Dept.:** ATL, BOL, LAG, MAG **Elev.:** 0–350 m a.s.l. **IUCN Habitat code:** 1,2,14 **Conservation status: (IUCN)** NE

3975 *Pithecellobium unguis-cati* (L.) Benth. (Fabaceae) **IPNI:** 515137-1 **Common name:** payandé clavo **Habit:** Shrub, Tree **Use:** EU, HF, IF, MA, ME, PO **Origin:** Native **Region:** Andean, Caribbean **Dept.:** ATL, BOL, COR, CUN, HUI, LAG, MAG, SAP, SAN, VAC **Elev.:** 0–1420 m a.s.l. **IUCN Habitat code:** 1,2,3,14 **Conservation status: (IUCN)** LC

3976 *Platymiscium pinnatum* (Jacq.) Dugand (Fabaceae) **IPNI:** 201767-2 **Common name:** aceituno montés, angelino, corazonfino, crucero, granadilla, granadillo, guayacán, guayacán trébol, palo de ardilla, palo de barbasco, peraco, roble, roble blanco, taray, tasajo, trébol, trébol negro, wajonae, wakonae **Habit:** Small tree, Tree **Use:** EU, FU, HF, MA, ME **Origin:** Native **Region:** Andean, Caribbean, Orinoquia **Dept.:** ANT, ATL, BOL, CAL, CES, LAG, MAG, MET, NSA, PUT, SAN, VAC **Elev.:** 5–1300 m a.s.l. **IUCN Habitat code:** 1,2,3,4,5,14 **Conservation status: (IUCN)** NE

3977 *Platymiscium stipulare* Benth. (Fabaceae)
IPNI: 515251-1
Habit: Tree **Use:** ME, SU **Origin:** Native **Region:** Amazonia **Dept.:** AMA, CAQ **Elev.:** 120–120 m a.s.l. **IUCN Habitat code:** 1,2,3,4,14 **Conservation status:** **(IUCN)** NE

3978 *Platymiscium trinitatis* Benth. (Fabaceae)
IPNI: 515253-1
Common name: garrapato **Habit:** Tree **Use:** EU, MA **Origin:** Native **Region:** Guiana Shield **Dept.:** GUV **Elev.:** 250–250 m a.s.l. **IUCN Habitat code:** 1,14 **Conservation status:** **(IUCN)** NE

3979 *Platypodium elegans* Vogel (Fabaceae)
IPNI: 515259-1
Common name: cajón, carbonero, combito, lomo de caimán, matarratón extranjero **Habit:** Tree **Use:** MA **Origin:** Native **Region:** Andean, Caribbean, Orinoquia, Pacific **Dept.:** ANT, ATL, BOL, CES, CHO, MAG, MET, SUC **Elev.:** 60–800 m a.s.l. **IUCN Habitat code:** 1,2,3,14 **Conservation status:** **(IUCN)** LC

3980 *Poeppigia procera* (Poepp. ex Spreng.) C.Presl (Fabaceae)
IPNI: 515473-1
Habit: Tree **Use:** HF **Origin:** Native **Region:** Caribbean, Orinoquia, Pacific **Dept.:** CHO, MAG, MET **Elev.:** 250–250 m a.s.l. **IUCN Habitat code:** 1,2,3,14 **Conservation status:** **(IUCN)** LC

3981 *Prioria copaifera* Griseb. (Fabaceae)
IPNI: 515766-1
Common name: aceite, aceite canime, amansa mujer, amansamujer, camivor, canime, cativo, cativo cucharo, cautivo, cucharo, trementino **Habit:** Tree **Use:** HF, MA, ME **Origin:** Native **Region:** Caribbean, Pacific **Dept.:** ANT, BOL, CHO, COR, MAG, SUC **Elev.:** 2-400 m.a.s.l. **IUCN Habitat code:** 1,3,5,14 **Conservation status:** **(IUCN)** LC **(CPSG)** EN

3982 *Prosopis juliflora* (Sw.) DC. (Fabaceae)
IPNI: 1002432-2
Common name: al pia, algarrobo, anchipia guaira, argentino, aromo, bayawond blan, chepesi, cují, cují negro, jpayandé, mancacaballo, mesquite, romero, trupillo, trupí **Habit:** Shrub, Tree **Use:** AF, EU, FU, GS, HF, IF, MA, ME, PO **Origin:** Native **Region:** Andean, Caribbean, Pacific **Dept.:** ATL, BOL, BOY, CAL, CUN, HUI, LAG, MAG, NAR, NSA, SAN, TOL, VAC **Elev.:** 0-2100 m.a.s.l. **IUCN Habitat code:** 1,2,3,4,5,14 **Conservation status:** **(IUCN)** NE **(CPSG)** Potential LC

3983 *Prosopis pallida* (Humb. & Bonpl. ex Willd.) Kunth (Fabaceae)
IPNI: 1175026-2
Common name: algarrobo, zarza **Habit:** Tree **Use:** AF, EU, FU, HF, IF, MA **Origin:** Native **Region:** Orinoquia **Dept.:** MET, VAC **Elev.:** 280–1050 m a.s.l. **Conservation status:** **(IUCN)** NE

3984 *Pseudopiptadenia psilostachya* (DC.) G.P.Lewis & M.P.Lima (Fabaceae)
IPNI: 962800-1
Common name: arenillo, cabiche blanco, caviche blanco, palo de pelusa, palo pelusa, saramán **Habit:** Tree **Use:** MA **Origin:** Native **Region:** Amazonia, Guiana Shield, Caribbean **Dept.:** AMA, BOL, VAU **Elev.:** 50–300 m a.s.l. **IUCN Habitat code:** 1,2,4,5,14 **Conservation status:** **(IUCN)** LC

3985 *Pseudosamanea guachapele* (Kunth) Harms (Fabaceae)
IPNI: 211561-2
Common name: cedro amarillo, falso samán, genízaro, higo amarillo, iguamarillo, iguá, iguá amarillo, jigua, masaguaro, naumo, nauno, pisquín, roble, roble amarillo, samán iguá, sanaguaro, tabaca, tobaco **Habit:** Tree **Use:** EU, FU, HF, MA **Origin:** Native **Region:** Andean, Caribbean, Orinoquia **Dept.:** ANT, ATL, CAU, CUN, HUI, MAG, MET, NAR, NSA, SAN, TOL, VAC **Elev.:** 50-1500 m.a.s.l. **IUCN Habitat code:** 1,2,3,5,14 **Conservation status:** **(IUCN)** LC **(CPSG)** Potential LC

3986 *Psoralea pinnata* L. (Fabaceae)
IPNI: 516230-1
Common name: pino azul **Habit:** Shrub **Use:** EU, HF, MA, ME **Origin:** Cultivated **Region:** Andean **Dept.:** BOG, CUN **Elev.:** 2050–3100 m a.s.l. **IUCN Habitat code:** 14 **Conservation status:** **(IUCN)** NE

3987 *Pterocarpus acapulcensis* Rose (Fabaceae)
IPNI: 214038-2
Common name: bollo blanco, drago, grau blanco, lagunero, ojo de venado, sangre, sangre de drago, sangrecorteza, sangregado, sangregao, sangregao blanco, sangro, tasajo, vacavieja **Habit:** Shrub, Tree **Use:** FU, MA, ME **Origin:** Native **Region:** Andean, Caribbean, Orinoquia **Dept.:** ATL, BOL, CAS, CES, LAG, MAG, NSA **Elev.:** 5–700 m a.s.l. **IUCN Habitat code:** 1,2,3,5,14 **Conservation status:** **(IUCN)** VU

3988 *Pterocarpus amazonum* (Mart. ex Benth.) Amshoff (Fabaceae)
IPNI: 214040-2
Common name: cumaca, guamo, manzanillo, oreja de guapore, palo de arco, palo de bejuco, palo de loro **Habit:** Small tree, Tree **Use:** HF, ME **Origin:** Native **Region:** Amazonia, Guiana Shield, Orinoquia **Dept.:** AMA, CAQ, GUA, MET **Elev.:** 10–340 m a.s.l. **IUCN Habitat code:** 1,3,5,14 **Conservation status:** **(IUCN)** NE

3989 *Pterocarpus officinalis* Jacq. (Fabaceae)
IPNI: 214060-2
Common name: bambudo, chiparo, churimbo, corazón negro, corcho, fruta sábalo, juana mestiza, lagunero, mucunana, palo de pollo, sabroso, sangre, sangre de drago, sangre drago, sangregallo, sangregao, sangregao blanco, suela, vacavieja **Habit:** Small tree, Tree **Use:** FU, MA, ME **Origin:** Native **Region:** Amazonia, Caribbean, Pacific **Dept.:** AMA, ANT, CAU, CHO, NAR, VAC **Elev.:** 0–100 m a.s.l. **IUCN Habitat code:** 1,2,3,5,14 **Conservation status:** **(IUCN)** NT

3990 *Pterocarpus rohrii* Vahl (Fabaceae)
IPNI: 516561-1
Common name: guayacán sabanero, sangre de gallo, sangregallina, sangregao, tablón **Habit:** Shrub, Small tree, Tree **Use:** MA, ME, PO **Origin:** Native **Region:** Amazonia, Guiana Shield, Caribbean, Orinoquia, Pacific **Dept.:** AMA, ANT, ATL, BOL, CES, CHO, GUV, MAG, MET, SUC **Elev.:** 15–1000 m a.s.l. **IUCN Habitat code:** 1,2,3,4,5,14 **Conservation status:** **(IUCN)** NE

3991 *Pterocarpus santalinoides* L'Hér. ex DC. (Fabaceae)
IPNI: 516568-1
Common name: palo de arco **Habit:** Tree **Use:** AF, EU, FU, HF, MA, ME **Origin:** Native **Region:** Amazonia **Dept.:** AMA, VAU **Elev.:** 180–180 m a.s.l. **IUCN Habitat code:** 1,3,14 **Conservation status:** **(IUCN)** LC

3992 *Pterogyne nitens* Tul. (Fabaceae)
IPNI: 516619-1
Habit: Tree **Use:** MA, ME **Origin:** Cultivated **Dept.:** VAC **Elev.:** 1000–1000 m a.s.l. **IUCN Habitat code:** 14 **Conservation status:** **(IUCN)** NT

3993 *Pueraria montana* (Lour.) Merr. (Fabaceae)
IPNI: 516708-1
Habit: Climbing **Use:** AF, EU, HF, MA, ME, SU **Origin:** Cultivated **Region:** Andean **Dept.:** CAL **Elev.:** 1250–1250 m a.s.l. **IUCN Habitat code:** 1,3,14 **Conservation status:** **(IUCN)** NE

3994 *Rhynchosia edulis* Griseb. (Fabaceae)
IPNI: 517240-1
Habit: Herb, Climbing **Use:** MA **Origin:** Native **Region:** Andean **Dept.:** ANT, BOL, HUI, SAN, TOL **Elev.:** 380–1075 m a.s.l. **IUCN Habitat code:** 1,2,3,5,14 **Conservation status:** **(IUCN)** NE

3995 *Rhynchosia minima* (L.) DC. (Fabaceae)
IPNI: 1197938-2
Common name: arroz con coco, cadillo, frijolillo dos pepas, platanito, rabo de ratón **Habit:** Herb, Liana, Climbing **Use:** AF, EU, HF, ME, PO, SU **Origin:** Native **Region:** Andean, Caribbean **Dept.:** ANT, ARA, ATL, BOL, BOY, CAL, COR, CUN, HUI, LAG, MAG, NSA, SAP, SAN, SUC, TOL, VAC **Elev.:** 0-1700 m.a.s.l. **IUCN Habitat code:** 1,2,3,5,14 **Conservation status:** **(IUCN)** LC **(CPSG)** Potential LC

3996 *Rhynchosia phaseoloides* (Sw.) DC. (Fabaceae)
IPNI: 1006990-2
Common name: bejuco de las peonias, chochos, peonia, pionia, zaragocilla **Habit:** Liana, Climbing **Use:** ME **Origin:** Native **Region:** Amazonia, Andean, Guiana Shield, Caribbean **Dept.:** ANT, CAQ, CHO, MET, NSA, QUI, SAN, TOL **Elev.:** 100–1900 m a.s.l. **IUCN Habitat code:** 1,3,5,14 **Conservation status:** **(IUCN)** NE

3997 *Rhynchosia precatoria* (Humb. & Bonpl. ex Willd.) DC. (Fabaceae)
IPNI: 517450-1
Use: MA **Origin:** Native **IUCN Habitat code:** 14 **Conservation status:** **(IUCN)** NE

3998 *Rhynchosia pyramidalis* (Lam.) Urb. (Fabaceae)
IPNI: 517469-1
Habit: Climbing **Use:** ME **Origin:** Native **Region:** Andean, Pacific **Dept.:** VAC **Elev.:** 50–1550 m a.s.l. **IUCN Habitat code:** 1,3 **Conservation status:** **(IUCN)** LC

3999 *Samanea saman* (Jacq.) Merr. (Fabaceae)
IPNI: 227126-2
Common name: algarrobillo, campano, campano bleo, campano hembra, campaño, cenicero, genízaro, guango, llovizno, samaguare, samán, sanaguaro, sanaguaré, sámano, urero, árbol de la lluvia, árbol de lluvia **Habit:** Tree **Use:** AF, EU, FU, GS, HF, MA, ME, PO, SU **Origin:** Native, Cultivated **Region:** Andean, Caribbean, Pacific **Dept.:** ANT, ATL, BOL, BOY, CAU, CES, COR, CUN, HUI, LAG, MAG, NAR, SAN, SUC, TOL, VAC **Elev.:** 0-1300 m.a.s.l. **IUCN Habitat code:** 1,2,3,4,5,14 **Conservation status:** **(IUCN)** LC **(CPSG)** Potential LC

4000 *Schizolobium parahyba* (Vell.) S.F.Blake (Fabaceae)
IPNI: 517988-1
Common name: arenillas, carambombo, catura, curaúbo, frijolillo, melote, tambolero, tambor, tamborero, tamboro, zorro **Habit:** Tree **Use:** EU, HF, MA, ME **Origin:** Native, Cultivated **Region:** Amazonia, Andean, Caribbean, Orinoquia **Dept.:** AMA, ANT, BOY, CAL, CHO, MET, QUI, RIS, SAN **Elev.:** 20–1400 m a.s.l. **IUCN Habitat code:** 1,2,3,5,14 **Conservation status: (IUCN)** LC

4001 *Schnella glabra* (Jacq.) Dugand (Fabaceae)
IPNI: 229758-2
Common name: bejuco cadena, bejuco de cadena, cadeno, casco de vaca, janami unüboto, jororo unubot, pata de buey, pata de vaca, patevaca **Habit:** Shrub, Liana, Climbing **Use:** MA, ME, SU **Origin:** Native **Region:** Amazonia, Andean, Guiana Shield, Caribbean, Orinoquia, Pacific **Dept.:** AMA, ATL, BOL, CES, CHO, COR, CUN, GUV, LAG, MAG, MET, NSA, PUT, SAN, SUC, TOL **Elev.:** 0-1200 m.a.s.l. **IUCN Habitat code:** 1,2,3,5,14 **Conservation status: (IUCN)** NE **(CPSG)** Potential LC

4002 *Schnella guianensis* (Aubl.) Wunderlin (Fabaceae)
IPNI: 77111925-1
Common name: bejuco cade bohobejuco, bejuco cadena, bejuco cadeno, bejuco de cadena marrón, bejuco escalera, bejuco negro, bohobejuco, escalera de mico, escalera de mono, hoja de venado, jorojoro, jororo unüboto, jororounüboto, suulishaya pa a **Habit:** Liana, Climbing **Use:** HF, MA, ME, SU **Origin:** Native **Region:** Amazonia, Andean, Guiana Shield, Caribbean, Orinoquia, Pacific **Dept.:** AMA, ANT, BOL, CAQ, CHO, GUA, GUV, MET, NAR, PUT, SAN, VAC, VAU, VID **Elev.:** 5-900 m.a.s.l. **IUCN Habitat code:** 1,2,3,4,5,14 **Conservation status: (IUCN)** NE **(CPSG)** Potential LC

4003 *Schnella herrerae* Britton & Rose (Fabaceae)
IPNI: 229759-2
Habit: Climbing **Use:** ME **Origin:** Native **Region:** Amazonia **Dept.:** AMA **Elev.:** 100–100 m a.s.l. **Conservation status: (IUCN)** NE

4004 *Schnella kunthiana* (Vogel) Wunderlin (Fabaceae)
IPNI: 77111927-1
Habit: Liana **Use:** ME **Origin:** Native **Region:** Orinoquia **Dept.:** CAS **Elev.:** 450–450 m a.s.l. **IUCN Habitat code:** 1,3 **Conservation status: (IUCN)** NE

4005 *Schnella microstachya* Raddi (Fabaceae)
IPNI: 229762-2
Habit: Liana, Climbing **Use:** ME **Origin:** Native **Region:** Amazonia, Pacific **Dept.:** AMA, ANT, CAQ, CHO **Elev.:** 0–520 m a.s.l. **IUCN Habitat code:** 1 **Conservation status: (IUCN)** NE

4006 *Schnella splendens* (Kunth) Benth. (Fabaceae)
IPNI: 518025-1
Use: ME **Origin:** Native **Conservation status: (IUCN)** NE

4007 *Senegalia martiusiana* (Steud.) Bocage & L.P.Queiroz (Fabaceae)
IPNI: 77085269-1
Habit: Shrub, Climbing **Use:** MA **Origin:** Native **Region:** Andean, Caribbean, Orinoquia **Dept.:** BOY, CES, MET **Elev.:** 100–1100 m a.s.l. **IUCN Habitat code:** 3 **Conservation status: (IUCN)** NE

4008 *Senegalia paniculata* (Willd.) Killip (Fabaceae)
IPNI: 234346-2
Habit: Shrub, Tree, Liana **Use:** ME **Origin:** Native **Region:** Amazonia, Andean, Guiana Shield, Caribbean, Orinoquia, Pacific **Dept.:** AMA, ANT, CAQ, CHO, MAG, MET, PUT, QUI, VAC, VAU **Elev.:** 400–830 m a.s.l. **IUCN Habitat code:** 1,2,3,14 **Conservation status: (IUCN)** NE

4009 *Senegalia polyphylla* (DC.) Britton & Rose (Fabaceae)
IPNI: 518289-1
Habit: Shrub, Tree **Use:** ME **Origin:** Native **Region:** Amazonia, Andean, Guiana Shield, Caribbean, Orinoquia, Pacific **Dept.:** AMA, ANT, ATL, BOL, CAU, CES, CHO, CUN, GUV, HUI, LAG, MAG, MET, NSA, SAN, SUC, TOL **Elev.:** 5-1400 m.a.s.l. **IUCN Habitat code:** 1,2,3,4,5,14 **Conservation status: (IUCN)** LC **(CPSG)** Potential LC

4010 *Senegalia tamarindifolia* (L.) Britton & Rose (Fabaceae)
IPNI: 234370-2
Habit: Shrub, Tree **Use:** EU **Origin:** Native, Cultivated **Region:** Caribbean **Dept.:** ATL, BOL, LAG, MAG **Elev.:** 1–150 m a.s.l. **Conservation status: (IUCN)** NE

4011 *Senna alata* (L.) Roxb. (Fabaceae)
IPNI: 1192167-2
Common name: babacú, bagua, bajagua, barajo, barbasco, bicho, bruscón, cacaona, dorancel, dorancé, favoranto, frisolato, frisolillos, frisolito, gabeche, galve, galve amarillo, galves, glaves, hierba de playa, kasyalata, lenguevaca, lengüevaca, lucutema, lucutena, mah **Habit:** Shrub, Small tree **Use:** AF, EU, GS, HF, MA, ME, PO, SU **Origin:** Native **Region:** Amazonia, Andean, Orinoquia, Pacific **Dept.:** ANT, BOL, CAL, CAU, CHO, GUA, MET, NAR, NSA, PUT, RIS, SAP, SAN, SUC, TOL, VAC, VID **Elev.:** 0-1000 m.a.s.l. **IUCN Habitat code:** 1,2,3,4,5,14 **Conservation status: (IUCN)** LC **(CPSG)** Potential LC

4012 *Senna alexandrina* Mill. (Fabaceae)
IPNI: 518323-1
Common name: hojas de sen, jalelo, sen, sen españa, sena alejandrina **Habit:** Herb, Subshrub **Use:** HF, ME **Origin:** Naturalised **Region:** Amazonia **Dept.:** AMA **Elev.:** 100–100 m a.s.l. **IUCN Habitat code:** 14 **Conservation status: (IUCN)** LC

4013 *Senna atomaria* (L.) H.S.Irwin & Barneby (Fabaceae)
IPNI: 234410-2
Common name: brucha macho, caranganito, chibato, chivato, cuchillito **Habit:** Shrub, Small tree, Tree **Use:** AF, FU, HF, MA, ME **Origin:** Native **Region:** Caribbean **Dept.:** ATL, BOL, CES, LAG, MAG, SAN **Elev.:** 5–400 m a.s.l. **IUCN Habitat code:** 1,2,3,4,14 **Conservation status: (IUCN)** LC

4014 *Senna bacillaris* (L.f.) H.S.Irwin & Barneby (Fabaceae)
IPNI: 234415-2
Common name: alcaparro, alcaparro pálido, amarillo, arracacho, bicho abejón, bicho bombito, bombito, burücunae, caranganito, chinchile, chivato, congüe, cuchillito, frijolero, frijolito, javiamena, matapasto bravo, platanito, tienderropa, yema de huevo **Habit:** Shrub, Subshrub, Small tree, Tree **Use:** AF, EU, MA, ME, SU **Origin:** Native **Region:** Amazonia, Andean, Orinoquia, Pacific **Dept.:** AMA, ANT, ATL, BOL, BOY, CAL, CAQ, CAS, CES, CHO, CUN, GUA, GUV, MAG, MET, NAR, PUT, SAN, SUC, TOL, VAU, VID **Elev.:** 5-1780 m.a.s.l. **IUCN Habitat code:** 1,2,3,4,5,14 **Conservation status: (IUCN)** LC **(CPSG)** Potential LC

4015 *Senna bicapsularis* (L.) Roxb. (Fabaceae)
IPNI: 518330-1
Common name: ahitera, alcaparro, algarrobo, biche, bicho, bicho bombito, bicho de café, bombito, bonguito, brusca, búho bombito, café de pobre, caranganito, chiki-chiki, chilchi, chilifruta, chilinchile, chiquichique, chivato, coijol, cuchillito, dandeleón, frijolillo **Habit:** Herb, Shrub, Small tree, Tree **Use:** AF, EU, FU, HF, MA, ME, PO **Origin:** Native **Region:** Amazonia, Andean, Caribbean, Orinoquia, Pacific **Dept.:** AMA, ATL, BOL, CAU, CUN, HUI, LAG, MAG, MET, PUT, TOL **Elev.:** 0-2400 m.a.s.l. **IUCN Habitat code:** 1,2,3,5,14 **Conservation status: (IUCN)** LC **(CPSG)** Potential LC

4016 *Senna dariensis* (Britton & Rose) H.S.Irwin & Barneby (Fabaceae)
IPNI: 911330-1
Common name: guabo tirisio, ictericio **Habit:** Herb, Shrub, Small tree, Tree **Use:** ME **Origin:** Native **Region:** Andean, Pacific **Dept.:** ANT, CAL, CHO, COR, NAR, VAC **Elev.:** 0–1600 m a.s.l. **IUCN Habitat code:** 1,2,3,5,14 **Conservation status: (IUCN)** LC

4017 *Senna fruticosa* (Mill.) H.S.Irwin & Barneby (Fabaceae)
IPNI: 234475-2
Common name: abejón, capacho, cañafístula, cigarrón, floramarillo, frijolillo, justo sin razón, matapasto, velero **Habit:** Shrub, Small tree, Tree **Use:** FU, HF, MA, ME **Origin:** Naturalised **Region:** Amazonia, Andean, Caribbean, Pacific **Dept.:** AMA, ANT, ATL, BOL, CAQ, CHO, CUN, MAG, PUT, SAN, VAC, VAU **Elev.:** 0–1760 m a.s.l. **IUCN Habitat code:** 1,2,3,4,14 **Conservation status: (IUCN)** LC

4018 *Senna hayesiana* (Britton & Rose) H.S.Irwin & Barneby (Fabaceae)
IPNI: 911344-1
Common name: abejín, contrarreuma **Habit:** Shrub, Subshrub, Small tree **Use:** ME **Origin:** Native **Region:** Andean, Caribbean, Orinoquia **Dept.:** ANT, BOG, BOL, CES, CHO, COR, MAG, MET, NSA, SAN, TOL **Elev.:** 5–1450 m a.s.l. **IUCN Habitat code:** 1,2,3,4,14 **Conservation status: (IUCN)** LC

4019 *Senna hirsuta* (L.) H.S.Irwin & Barneby (Fabaceae)
IPNI: 234491-2
Common name: ahitera, alcaparro, bicho, bicho de café, café furrusco, chilinchele, chunchu vainilla, dorancé, lumbaga, taka taka **Habit:** Herb, Subshrub, Tree **Use:** EU, HF, ME, PO, SU **Origin:** Native **Region:** Andean **Dept.:** ANT, BOG, BOY, CAL, CAU, CUN, HUI, SAN, TOL, VAC **Elev.:** 400-2200 m.a.s.l. **IUCN Habitat code:** 1,2,3,14 **Conservation status: (IUCN)** NE **(CPSG)** Potential LC

4020 *Senna italica* Mill. (Fabaceae)
IPNI: 518341-1
Common name: se in, sen **Habit:** Herb, Subshrub **Use:** AF, EU, GS, HF, MA, ME, PO, SU **Origin:** Naturalised **Region:** Caribbean **Dept.:** LAG **Elev.:** 52–52 m a.s.l. **IUCN Habitat code:** 1,2,3,4 **Conservation status: (IUCN)** NE

4021 *Senna latifolia* (G.Mey.) H.S.Irwin & Barneby (Fabaceae) **IPNI:** 911354-1 **Habit:** Climbing **Use:** ME **Origin:** Native **Region:** Amazonia **Dept.:** AMA **Elev.:** 100–100 m a.s.l. **IUCN Habitat code:** 1 **Conservation status: (IUCN)** LC

4022 *Senna macranthera* (DC. ex Collad.) H.S.Irwin & Barneby (Fabaceae) **IPNI:** 911359-1 **Common name:** alcaparro florón, alcaparro macho, cuala, florón **Habit:** Shrub, Tree **Use:** HF **Origin:** Native **Region:** Andean **Dept.:** BOY, CUN **Elev.:** 1650–2450 m a.s.l. **Conservation status: (IUCN)** LC

4023 *Senna macrophylla* (Kunth) H.S.Irwin & Barneby (Fabaceae) **IPNI:** 234523-2 **Common name:** frijolillo, mierdegato, tirisia **Habit:** Shrub, Small tree **Use:** ME **Origin:** Native **Region:** Amazonia, Andean, Orinoquia, Pacific **Dept.:** AMA, ANT, BOY, CAL, CAQ, CAU, CHO, CUN, GUV, HUI, MET, PUT, QUI, RIS, TOL, VAC, VAU, VID **Elev.:** 50-2000 m.a.s.l. **IUCN Habitat code:** 1,2,3,4,5,14 **Conservation status: (IUCN)** LC **(CPSG)** Potential LC

4024 *Senna multiglandulosa* (Jacq.) H.S.Irwin & Barneby (Fabaceae) **IPNI:** 234543-2 **Common name:** alcaparrito, alcaparro, alcaparro chico, alcaparro enano **Habit:** Shrub, Small tree, Tree **Use:** EU, HF, ME **Origin:** Native, Cultivated **Region:** Andean **Dept.:** ANT, BOG, BOY, CAU, CUN, NAR, PUT **Elev.:** 1500–3700 m a.s.l. **IUCN Habitat code:** 1,3,5,14 **Conservation status: (IUCN)** LC

4025 *Senna multijuga* (Rich.) H.S.Irwin & Barneby (Fabaceae) **IPNI:** 234544-2 **Common name:** cañafístulo de altura, guarango negro **Habit:** Shrub, Small tree, Tree **Use:** EU, MA, ME **Origin:** Native **Region:** Amazonia, Andean, Caribbean, Orinoquia **Dept.:** AMA, BOL, CAQ, CES, MET, PUT, VID **Elev.:** 225–2300 m a.s.l. **IUCN Habitat code:** 1,3,4,14 **Conservation status: (IUCN)** NE

4026 *Senna obtusifolia* (L.) H.S.Irwin & Barneby (Fabaceae) **IPNI:** 234560-2 **Common name:** bejuco de bicho, bicho, bicho bombito, bicho macho, brusca, burücunae, chilinchil, chilinchile, chirrinchado, chirrincho, comida de murciélago, frijolato, frisolato, hierba bicho de platanito, hierba bicho guinea, potra **Habit:** Herb, Shrub, Subshrub, Small tree **Use:** AF, EU, GS, HF, MA, ME, PO, SU **Origin:** Native **Region:** Amazonia, Andean, Caribbean, Orinoquia, Pacific **Dept.:** AMA, ANT, ATL, BOL, BOY, CAL, CAQ, CAU, CES, CHO, COR, CUN, HUI, LAG, MAG, MET, NSA, RIS, SAN, SUC, TOL, VAC **Elev.:** 0-2100 m.a.s.l. **IUCN Habitat code:** 1,2,3,4,5,14 **Conservation status: (IUCN)** LC **(CPSG)** Potential LC

4027 *Senna occidentalis* (L.) Link (Fabaceae) **IPNI:** 1117728-2 **Common name:** ahitera, aitera, alcaparro, alcoparro, bicho, bicho de cafe, bicho de café, bruca, brusca, burücunae, busaca, cafe furrusco, cafecillo, cafecito, cafelillo, café chilinchil, café churrusco, café de brusca, café de brusco, café furrusco, café moucha, chili **Habit:** Herb, Shrub, Subshrub, Small tree **Use:** EU, HF, MA, ME, PO **Origin:** Native **Region:** Amazonia, Orinoquia **Dept.:** ANT, ARA, ATL, BOL, BOY, CAL, CAQ, CAS, CAU, CES, CHO, CUN, LAG, MAG, MET, NAR, SAP, SAN, SUC, TOL, VAC, VAU, VID **Elev.:** 0–2350 m a.s.l. **IUCN Habitat code:** 1,2,3,4,5,14 **Conservation status: (IUCN)** NE

4028 *Senna pallida* (Vahl) H.S.Irwin & Barneby (Fabaceae) **IPNI:** 234569-2 **Common name:** bicho, bicho negro, bombito, chilinchil, chirrinchao, escobito, tienderropa **Habit:** Shrub, Subshrub, Small tree **Use:** ME **Origin:** Native **Region:** Andean, Caribbean **Dept.:** ANT, ATL, BOL, CAU, HUI, LAG, MAG, NAR, SAN **Elev.:** 3–1820 m a.s.l. **IUCN Habitat code:** 1,2,3,5,14 **Conservation status: (IUCN)** LC

4029 *Senna papillosa* (Britton & Rose) H.S.Irwin & Barneby (Fabaceae) **IPNI:** 234589-2 **Common name:** cambombulo **Habit:** Shrub, Small tree, Tree **Use:** HF **Origin:** Native **Region:** Andean, Guiana Shield, Caribbean, Orinoquia **Dept.:** ANT, BOL, CAU, CES, CUN, MAG, MET, NSA, QUI, VAC **Elev.:** 100–2350 m a.s.l. **IUCN Habitat code:** 1,2,3,4,14 **Conservation status: (IUCN)** LC

4030 *Senna pendula* (Humb. & Bonpl. ex Willd.) H.S.Irwin & Barneby (Fabaceae) **IPNI:** 911386-1 **Common name:** ahitera, alcaparro, bicho, bicho de café, bombito, brusca, café de pobre, chilinchile, chiquichique, coijol, dandeleón, frijolillo, furrusca, hediondilla, platanito, sen, yerba de gallinazo, yerba de murciélago **Habit:** Herb, Shrub, Small tree, Tree **Use:** EU, HF, ME **Origin:** Native **Region:** Andean, Orinoquia **Dept.:** ANT, BOY, CAL, CAS, CAU, CES, CUN, LAG, MET, NSA, SAN, VAC **Elev.:** 50-1940 m.a.s.l. **IUCN Habitat code:** 1,2,3,4,14 **Conservation status: (IUCN)** LC **(CPSG)** Potential LC

4031 *Senna pilifera* (Vogel) H.S.Irwin & Barneby (Fabaceae) **IPNI:** 911389-1 **Habit:** Herb, Subshrub **Use:** HF **Origin:** Native **Region:** Orinoquia **Dept.:** MET **Elev.:** 450–450 m a.s.l. **IUCN Habitat code:** 2 **Conservation status: (IUCN)** NE

4032 *Senna reticulata* (Willd.) H.S.Irwin & Barneby (Fabaceae) **IPNI:** 234636-2 **Common name:** acacia, alcaparro, bajagua, bajaguo, caballo, dorancel, dorancé, dorancé flor amarillo, floramarillo, francés, frentín, frisolato, galbecillo, galve, galvecillo, galves, galves arisco, galvillo, galvis, garve, guayabillo, jarra nasitsinae, labranced, maja **Habit:** Shrub, Small tree **Use:** HF, MA, ME, SU **Origin:** Native **Region:** Amazonia, Andean, Caribbean, Orinoquia, Pacific **Dept.:** AMA, ANT, ARA, ATL, BOL, CAL, CAQ, CAS, CAU, CES, CHO, COR, CUN, HUI, LAG, MAG, MET, NAR, QUI, RIS, SUC, TOL, VAC, VAU, VID **Elev.:** 0-2000 m.a.s.l. **IUCN Habitat code:** 1,2,3,4,5,14 **Conservation status: (IUCN)** LC **(CPSG)** Potential LC

4033 *Senna ruiziana* (G.Don) H.S.Irwin & Barneby (Fabaceae) **IPNI:** 911404-1 **Habit:** Shrub, Subshrub, Small tree, Tree **Use:** ME **Origin:** Native **Region:** Amazonia, Pacific **Dept.:** AMA, ANT, CAQ, NAR, PUT **Elev.:** 300–700 m a.s.l. **IUCN Habitat code:** 1,3,5,14 **Conservation status: (IUCN)** LC

4034 *Senna septemtrionalis* (Viv.) H.S.Irwin & Barneby (Fabaceae) **IPNI:** 234652-2 **Common name:** alcaparro **Habit:** Shrub **Use:** EU, HF, MA, ME, PO **Origin:** Naturalised **Region:** Andean **Dept.:** ANT, CAL, SAN **Elev.:** 1600–1931 m a.s.l. **IUCN Habitat code:** 1,14 **Conservation status: (IUCN)** NE

4035 *Senna siamea* (Lam.) H.S.Irwin & Barneby (Fabaceae) **IPNI:** 911410-1 **Common name:** abeto, acacia, acacia amarilla, acacia de Siam, acacio amarillo, acacio santandereano, alcaparro, carmín, casia amarilla, floramarillo, guamo de mico **Habit:** Tree **Use:** AF, EU, FU, GS, HF, IF, MA, ME, PO **Origin:** Cultivated, Naturalised **Region:** Amazonia, Andean, Caribbean, Pacific **Dept.:** ANT, BOL, CHO, COR, CUN, LAG, MAG, NSA, PUT, SAN, TOL, VAC **Elev.:** 0–1600 m a.s.l. **IUCN Habitat code:** 1,2,3,4,14 **Conservation status: (IUCN)** LC

4036 *Senna silvestris* (Vell.) H.S.Irwin & Barneby (Fabaceae) **IPNI:** 234653-2 **Common name:** alcaparrillo, carnevaca, floramarillo, yema de huevo **Habit:** Shrub, Small tree, Tree **Use:** EU, SU **Origin:** Native **Region:** Andean, Guiana Shield, Orinoquia **Dept.:** AMA, CAQ, CHO, COR, CUN, GUA, GUV, MET, NSA, VAU, VID **Elev.:** 0–1300 m a.s.l. **IUCN Habitat code:** 1,3,4,5,14 **Conservation status: (IUCN)** LC

4037 *Senna skinneri* (Benth.) H.S.Irwin & Barneby (Fabaceae) **IPNI:** 911412-1 **Habit:** Shrub **Use:** MA, ME **Origin:** Native **Region:** Caribbean **Dept.:** LAG **Elev.:** 150–150 m a.s.l. **IUCN Habitat code:** 2 **Conservation status: (IUCN)** LC

4038 *Senna sophera* (L.) Roxb. (Fabaceae) **IPNI:** 518364-1 **Common name:** ahitera, alcaparro, bicho, café de pobres, chilinchile **Habit:** Herb, Shrub, Subshrub **Use:** EU, HF, ME, SU **Origin:** Naturalised **Region:** Andean **Dept.:** ANT, BOY, CUN, NSA, PUT, RIS **Elev.:** 0–2130 m a.s.l. **IUCN Habitat code:** 1,2,3,14 **Conservation status: (IUCN)** NE

4039 *Senna spectabilis* (DC.) H.S.Irwin & Barneby (Fabaceae) **IPNI:** 234663-2 **Common name:** alcaparro, beleño, candelillo, cañafistulillo, cañafistulo macho, cañafístol, cañafístola macho, cañafístolo, cañafístula, cañafístula de caballo, cañafístula macho, cañafístulo, cañafístulo macho, floramarillo, floramarillo montero, fruta de murciélago, **Habit:** Small tree, Tree **Use:** EU, HF, MA, ME, PO **Origin:** Native **Region:** Andean **Dept.:** ANT, CAL, CAU, CUN, HUI, QUI, RIS, TOL, VAC **Elev.:** 160-1800 m.a.s.l. **IUCN Habitat code:** 1,2,3,14 **Conservation status: (IUCN)** LC **(CPSG)** Potential LC

4040 *Senna surattensis* (Burm.f.) H.S.Irwin & Barneby (Fabaceae) **IPNI:** 911423-1 **Habit:** Shrub **Use:** EU, HF, MA, ME **Origin:** Naturalised **Region:** Andean **Dept.:** CUN **Elev.:** 1350–1350 m a.s.l. **IUCN Habitat code:** 2,14 **Conservation status: (IUCN)** LC

4041 *Senna tora* (L.) Roxb. (Fabaceae) **IPNI:** 518369-1 **Common name:** bicho, brusca macho, brusca negra, chilinchile, comida de murciélago, potra **Use:** AF, EU, HF, MA, ME, PO **IUCN Habitat code:** 1,2,3,5,14 **Conservation status: (IUCN)** NE

4042 *Senna undulata* (Benth.) H.S.Irwin & Barneby (Fabaceae) **IPNI:** 234686-2 **Habit:** Shrub **Use:** HF, ME **Origin:** Native **Region:** Guiana Shield **Dept.:**

ANT, BOG, CUN, MAG, NSA, SAN, TOL **Elev.:** 40–1728 m a.s.l. **IUCN Habitat code:** 1,2,3,14 **Conservation status: (IUCN)** LC

4043 *Senna uniflora* (Mill.) H.S.Irwin & Barneby (Fabaceae)
IPNI: 234687-2
Habit: Herb **Use:** AF, HF, ME **Origin:** Native **Dept.:** HUI **Elev.:** 360–360 m a.s.l. **IUCN Habitat code:** 2,3,14 **Conservation status: (IUCN)** NE

4044 *Senna velutina* (Vogel) H.S.Irwin & Barneby (Fabaceae)
IPNI: 234691-2
Common name: alcaparro, alcaparro grande **Use:** ME **Dept.:** BOG **Conservation status: (IUCN)** NE

4045 *Senna viarum* (Little) H.S.Irwin & Barneby (Fabaceae)
IPNI: 234694-2
Common name: alcaparro, alcaparro de Bogotá, alcaparro grande **Habit:** Small tree, Tree **Use:** ME **Origin:** Native, Cultivated **Region:** Andean **Dept.:** BOG, CUN, NSA, TOL **Elev.:** 2000–2700 m a.s.l. **IUCN Habitat code:** 1,3,4,14 **Conservation status: (IUCN)** NE

4046 *Sesbania grandiflora* (L.) Poir. (Fabaceae)
IPNI: 518476-1
Common name: cantagallo blanco, gallito, pirigallo **Habit:** Small tree, Tree **Use:** AF, EU, FU, HF, MA, ME, SU **Origin:** Cultivated, Naturalised **Region:** Andean, Caribbean **Dept.:** ANT, ARA, COR, CUN, MAG, SAP, VAC **Elev.:** 0–1500 m a.s.l. **IUCN Habitat code:** 3,5,14 **Conservation status: (IUCN)** NE

4047 *Sesbania herbacea* (Mill.) McVaugh (Fabaceae)
IPNI: 284993-2
Habit: Herb, Shrub **Use:** EU, MA, PO **Origin:** Native **Region:** Andean, Caribbean, Pacific **Dept.:** ANT, CUN, VAC **Elev.:** 80–1000 m a.s.l. **IUCN Habitat code:** 1,2,3,5 **Conservation status: (IUCN)** NE

4048 *Sesbania sericea* (Willd.) Link (Fabaceae)
IPNI: 1015399-2
Habit: Herb, Shrub, Subshrub **Use:** AF, EU, FU, HF, MA, ME, PO **Origin:** Naturalised **Region:** Caribbean **Dept.:** ANT, ATL, BOL, MAG, SAN **Elev.:** 50–600 m a.s.l. **IUCN Habitat code:** 1,5 **Conservation status: (IUCN)** LC

4049 *Sophora tomentosa* L. (Fabaceae)
IPNI: 518954-1
Use: AF, EU, HF, MA, ME, PO **Origin:** Native **Conservation status: (IUCN)** LC

4050 *Spartium junceum* L. (Fabaceae)
IPNI: 519057-1
Common name: ginesta, retama, retama de olor, retama junca, retamo, retamo de olor, retamo junca, rubacá, talhui **Habit:** Shrub **Use:** EU, HF, MA, ME, PO **Origin:** Naturalised **Region:** Andean **Dept.:** BOG, BOL, BOY, CAU, CUN **Elev.:** 1900–3000 m a.s.l. **IUCN Habitat code:** 1,14 **Conservation status: (IUCN)** NE

4051 *Strongylodon macrobotrys* A.Gray (Fabaceae)
IPNI: 519595-1
Common name: jade, jade verdeazul **Habit:** Climbing **Use:** EU, HF, ME **Origin:** Cultivated **Region:** Andean **Dept.:** ANT, CAL, QUI **Elev.:** 1200–1500 m a.s.l. **IUCN Habitat code:** 1,14 **Conservation status: (IUCN)** NE

4052 *Stryphnodendron pulcherrimum* (Willd.) Hochr. (Fabaceae)
IPNI: 519643-1
Common name: guarango **Habit:** Small tree, Tree **Use:** HF **Origin:** Native **Region:** Guiana Shield **Dept.:** CAQ, GUA **Elev.:** 810–810 m a.s.l. **IUCN Habitat code:** 1,3,14 **Conservation status: (IUCN)** LC

4053 *Stylosanthes guianensis* (Aubl.) Sw. (Fabaceae)
IPNI: 1020451-2
Common name: alfalfa, alfalfa del Brasil, arco, cargadita, neblina **Habit:** Herb, Shrub, Subshrub **Use:** AF, EU, ME, PO **Origin:** Native **Region:** Andean, Caribbean, Orinoquia, Pacific **Dept.:** ANT, BOY, CAL, CAQ, CAS, CAU, CHO, CUN, GUV, HUI, LAG, MET, NAR, NSA, SAN, TOL, VAC, VID **Elev.:** 0-2500 m.a.s.l. **IUCN Habitat code:** 1,2,3,4,14 **Conservation status: (IUCN)** NE **(CPSG)** Potential LC

4054 *Stylosanthes hamata* (L.) Taub. (Fabaceae)
IPNI: 247544-2
Habit: Herb, Shrub, Subshrub **Use:** AF, EU, ME **Origin:** Native **Region:** Andean, Caribbean **Dept.:** ATL, BOL, CUN, HUI, LAG, MAG, SAN **Elev.:** 0–1000 m a.s.l. **IUCN Habitat code:** 1,2,4,5,14 **Conservation status: (IUCN)** NE

4055 *Stylosanthes humilis* Kunth (Fabaceae)
IPNI: 519681-1
Common name: neblina **Habit:** Herb, Shrub, Subshrub **Use:** AF, EU, HF **Origin:** Native **Region:** Andean, Caribbean **Dept.:** ANT, ATL, BOL, HUI, MAG, NSA, SAN, TOL, VAC **Elev.:** 0–1230 m a.s.l. **IUCN Habitat code:** 2,3,14 **Conservation status: (IUCN)** NE

4056 *Stylosanthes scabra* Vogel (Fabaceae)
IPNI: 304693-2
Habit: Herb, Shrub **Use:** AF,ME, PO **Origin:** Native **Region:** Andean **Dept.:** CAU, HUI, NAR, TOL, VAC **Elev.:** 450–1150 m a.s.l. **IUCN Habitat code:** 1,2,3,14 **Conservation status: (IUCN)** NE

4057 *Stylosanthes viscosa* (L.) Sw. (Fabaceae)
IPNI: 1176701-2
Habit: Herb **Use:** AF, EU, GS, ME, PO **Origin:** Native **Region:** Andean **Dept.:** ANT, CAS **Elev.:** 1080–1080 m a.s.l. **IUCN Habitat code:** 3,14 **Conservation status: (IUCN)** NE

4058 *Sulla coronaria* (L.) B.H.Choi & H.Ohashi (Fabaceae)
IPNI: 519725-1
Habit: Herb **Use:** AF, EU, HF, IF, ME **Origin:** Cultivated **Region:** Andean **Dept.:** SAN **Elev.:** 1800–1800 m a.s.l. **IUCN Habitat code:** 14 **Conservation status: (IUCN)** NE

4059 *Swainsona formosa* (G.Don) Joy Thomps. (Fabaceae)
IPNI: 948198-1
Use: AF, EU, HF, SU **Region:** Andean **Dept.:** BOG **Conservation status: (IUCN)** NE

4060 *Swartzia amplifolia* Harms (Fabaceae)
IPNI: 246570-2
Common name: café, cuña, escobillo, frisolillo, nuicuna, tirisia, zapatillo **Habit:** Shrub, Small tree, Tree **Use:** MA **Origin:** Native **Region:** Andean, Caribbean, Pacific **Dept.:** ANT, BOL, CHO, CUN, SAN, VAC **Elev.:** 0–1725 m a.s.l. **IUCN Habitat code:** 1,3,4,5,14 **Conservation status: (IUCN)** LC

4061 *Swartzia arborescens* (Aubl.) Pittier (Fabaceae)
IPNI: 519836-1
Common name: ají de barbasco **Habit:** Small tree, Tree **Use:** HF, MA, ME, PO **Origin:** Native **Region:** Amazonia, Guiana Shield **Dept.:** AMA, CAQ, MET, VAU **Elev.:** 100–400 m a.s.l. **IUCN Habitat code:** 1,3,4,5,14 **Conservation status: (IUCN)** LC

4062 *Swartzia benthamiana* Miq. (Fabaceae)
IPNI: 519841-1
Common name: guamo de bejuco, guamo de venado, palo de bufeo **Habit:** Shrub, Small tree, Tree **Use:** MA **Origin:** Native **Region:** Amazonia **Dept.:** AMA, CAQ, GUA, VAU **Elev.:** 100–250 m a.s.l. **IUCN Habitat code:** 1 **Conservation status: (IUCN)** LC

4063 *Swartzia leptopetala* Benth. (Fabaceae)
IPNI: 519927-1
Habit: Small tree, Tree **Use:** MA **Origin:** Native **Region:** Amazonia, Orinoquia **Dept.:** AMA, GUV, MET, SAN, VAU, VID **Elev.:** 90–300 m a.s.l. **IUCN Habitat code:** 1,2,3,4,5,14 **Conservation status: (IUCN)** LC

4064 *Swartzia macrophylla* Willd. ex Vogel (Fabaceae)
IPNI: 519937-1
Common name: guacharaco, macuná, mucuna, mucuná, rumbo **Habit:** Shrub, Small tree, Tree **Use:** MA **Origin:** Endemic **Region:** Andean, Caribbean, Pacific **Dept.:** ANT, CHO, CUN, SAN **Elev.:** 10-1400 m.a.s.l. **IUCN Habitat code:** 1,3,5,14 **Conservation status: (IUCN)** LC **(CPSG)** LC

4065 *Swartzia pinnata* (Vahl) Willd. (Fabaceae)
IPNI: 519989-1
Habit: Tree **Use:** EU, MA, ME **Origin:** Native **Region:** Andean **Dept.:** TOL **Elev.:** 850–850 m a.s.l. **IUCN Habitat code:** 1 **Conservation status: (IUCN)** NE

4066 *Swartzia polyphylla* DC. (Fabaceae)
IPNI: 519994-1
Habit: Small tree, Tree **Use:** ME **Origin:** Native **Region:** Amazonia **Dept.:** AMA, CAQ, GUA, VAU **Elev.:** 100–455 m a.s.l. **IUCN Habitat code:** 1,3,5,14 **Conservation status: (IUCN)** LC

4067 *Swartzia racemosa* Benth. (Fabaceae)
IPNI: 520000-1
Common name: algarrobillo, cacao, paleta de perico **Use:** HF **IUCN Habitat code:** 1 **Conservation status: (IUCN)** LC

4068 *Swartzia schomburgkii* Benth. (Fabaceae)
IPNI: 520015-1
Common name: bamba de alacrán, costillo, costillo blanco, costillo de mal olor, guama, palo de alacrán **Habit:** Small tree, Tree, Climbing **Use:** MA **Origin:** Native **Region:** Amazonia, Guiana Shield, Orinoquia

Dept.: AMA, CAQ, GUA, VAU, VID **Elev.:** 100–200 m a.s.l. **IUCN Habitat code:** 1,3,4,5,14 **Conservation status: (IUCN)** LC

4069 *Swartzia sericea* Vogel (Fabaceae)
IPNI: 520018-1
Common name: guamuche **Habit:** Tree **Use:** HF, ME **Origin:** Native **Region:** Guiana Shield **Dept.:** GUA **Elev.:** 80–120 m a.s.l. **IUCN Habitat code:** 1,5,14 **Conservation status: (IUCN)** LC

4070 *Swartzia simplex* (Sw.) Spreng. (Fabaceae)
IPNI: 520020-1
Common name: achotillo, arará, caballo caspi, caranuca, costillo, nagüí, naranja, negro caspi, palo de caballo, palo negrito, pernilla del monte, porotillo, pusilde, tete **Habit:** Shrub, Small tree, Tree **Use:** AF, EU, HF, MA, ME **Origin:** Native **Region:** Amazonia, Andean, Caribbean, Pacific **Dept.:** AMA, ANT, ATL, BOL, CAQ, CHO, GUA, NAR, SAN, TOL, VAC **Elev.:** 0-1250 m.a.s.l. **IUCN Habitat code:** 1,2,3,5,14 **Conservation status: (IUCN)** LC **(CPSG)** Potential LC

4071 *Tachigali cavipes* (Spruce ex Benth.) J.F.Macbr. (Fabaceae)
IPNI: 248955-2
Common name: caracha caspi, diablo, guamillo blanco, varasanta **Habit:** Shrub, Tree **Use:** ME **Origin:** Native **Region:** Amazonia, Guiana Shield, Orinoquia **Dept.:** AMA, CAQ, GUA, PUT, VAU, VID **Elev.:** 100–400 m a.s.l. **IUCN Habitat code:** 1,3,4,5,14 **Conservation status: (IUCN)** LC

4072 *Tachigali guianensis* (Benth.) Zarucchi & Herend. (Fabaceae)
IPNI: 974508-1
Habit: Tree **Use:** MA, ME **Origin:** Native **Region:** Amazonia, Guiana Shield **Dept.:** AMA, GUV, MET **Elev.:** 100–100 m a.s.l. **IUCN Habitat code:** 1,3,4,14 **Conservation status: (IUCN)** NE

4073 *Tachigali paniculata* Aubl. (Fabaceae)
IPNI: 520138-1
Common name: guamo hojiancho, kö-ma-na, negro, tangarana negro, varasanta **Habit:** Small tree, Tree **Use:** ME **Origin:** Native **Region:** Amazonia **Dept.:** AMA, VAU **Elev.:** 100–200 m a.s.l. **IUCN Habitat code:** 1,3,5,14 **Conservation status: (IUCN)** LC

4074 *Tachigali ptychophysca* Spruce ex Benth. (Fabaceae)
IPNI: 520143-1
Habit: Tree **Use:** ME **Origin:** Native **Region:** Amazonia **Dept.:** CAQ, VAU **Elev.:** 250–650 m a.s.l. **IUCN Habitat code:** 1 **Conservation status: (IUCN)** NE

4075 *Tachigali tinctoria* (Benth.) Zarucchi & Herend. (Fabaceae)
IPNI: 974515-1
Habit: Tree **Use:** MA **Origin:** Native **Region:** Orinoquia **Dept.:** VID **Elev.:** 115–115 m a.s.l. **IUCN Habitat code:** 1,3,4 **Conservation status: (IUCN)** NE

4076 *Tamarindus indica* L. (Fabaceae)
IPNI: 520167-1
Common name: tamarindo **Habit:** Tree **Use:** AF, EU, FU, GS, HF, IF, MA, ME, PO, SU **Origin:** Cultivated, Naturalised **Region:** Andean, Caribbean, Orinoquia, Pacific **Dept.:** ANT, ATL, BOL, CAL, CAS, CAU, COR, CUN, MAG, NSA, SAP, TOL, VAC **Elev.:** 0–1200 m a.s.l. **IUCN Habitat code:** 1,2,3,4,5,14 **Conservation status: (IUCN)** LC

4077 *Tara spinosa* (Molina) Britton & Rose (Fabaceae)
IPNI: 249443-2
Common name: dividivi, dividivi de tierra fría, guarango, menejùmecacù, tara, taya **Habit:** Shrub, Tree **Use:** EU, HF, MA, ME **Origin:** Native **Region:** Andean, Guiana Shield **Dept.:** AMA, ANT, BOY, CAU, CUN, NAR **Elev.:** 150–2750 m a.s.l. **IUCN Habitat code:** 1,2,3,4,14 **Conservation status: (IUCN)** NE

4078 *Taralea cordata* Ducke (Fabaceae)
IPNI: 249447-2
Habit: Shrub **Use:** ME **Origin:** Native **Region:** Amazonia **Dept.:** VAU **Elev.:** 200–200 m a.s.l. **IUCN Habitat code:** 1,4 **Conservation status: (IUCN)** LC

4079 *Taralea oppositifolia* Aubl. (Fabaceae)
IPNI: 520179-1
Common name: charapillo, oreja de pellejo, ovegue de rebalse, palo de ají, palo de pescado, palopescado, ponilla, sarrapio **Habit:** Shrub, Small tree, Tree **Use:** HF, ME **Origin:** Native **Region:** Amazonia, Guiana Shield **Dept.:** AMA, CAQ, GUA, VAU **Elev.:** 50–760 m a.s.l. **IUCN Habitat code:** 1,3,4,5,14 **Conservation status: (IUCN)** LC

4080 *Tephrosia adunca* Benth. (Fabaceae)
IPNI: 520366-1
Common name: barbasco **Habit:** Herb **Use:** ME **Origin:** Native **Region:** Orinoquia **Dept.:** MET **Elev.:** 150–450 m a.s.l. **IUCN Habitat code:** 14 **Conservation status: (IUCN)** NE

4081 *Tephrosia candida* DC. (Fabaceae)
IPNI: 520444-1
Common name: barbasco **Habit:** Herb **Use:** AF, EU, HF, ME, PO **Origin:** Cultivated, Naturalised **Dept.:** ANT, CAU, VAC **Elev.:** 1050–1050 m a.s.l. **IUCN Habitat code:** 14 **Conservation status: (IUCN)** NE

4082 *Tephrosia cinerea* (L.) Pers. (Fabaceae)
IPNI: 1071691-2
Common name: alverjillo, barbasco, barbasco de loma, barbasco de playa, barbasco de raíz, sen, yerba de barbasco **Habit:** Herb, Shrub, Subshrub **Use:** AF, MA, ME **Origin:** Native **Region:** Andean, Caribbean, Orinoquia, Pacific **Dept.:** ANT, ATL, BOL, CAU, CES, COR, CUN, HUI, LAG, MAG, MET, NSA, SAN, TOL, VAC **Elev.:** 0-1050 m.a.s.l. **IUCN Habitat code:** 1,2,3,5,14 **Conservation status: (IUCN)** NE **(CPSG)** Potential LC

4083 *Tephrosia purpurea* (L.) Pers. (Fabaceae)
IPNI: 520851-1
Common name: barbasco **Habit:** Subshrub **Use:** AF, EU, HF, MA, ME, PO **Origin:** Cultivated, Naturalised **Region:** Caribbean **Dept.:** MAG **Elev.:** 75–150 m a.s.l. **IUCN Habitat code:** 2,14 **Conservation status: (IUCN)** NE

4084 *Tephrosia sinapou* (Buc'hoz) A.Chev. (Fabaceae)
IPNI: 250423-2
Common name: ba-ki-rá, barbasco, barbasco de Castilla, barbasco de hoja, barbasco de loma, barbasco de raíz, barbasco de árbol, barbasco huaranga, barbascoerraíz, chilipique, cube, cube ordinario, guarango barbasco, sauce, timbo **Habit:** Herb, Shrub, Subshrub **Use:** EU, MA, ME, PO **Origin:** Native, Cultivated **Region:** Andean, Orinoquia, Pacific **Dept.:** ANT, BOY, CAS, CAU, CHO, MET, NAR, NSA, PUT, QUI, SAN, TOL, VAC, VID **Elev.:** 5–1800 m a.s.l. **IUCN Habitat code:** 1,2,3,4,5,14 **Conservation status: (IUCN)** NE

4085 *Tephrosia vogelii* Hook.f. (Fabaceae)
IPNI: 521002-1
Habit: Shrub **Use:** AF, EU, FU, HF, MA, ME, PO, SU **Origin:** Cultivated, Naturalised **Region:** Andean **Dept.:** BOG, CUN, QUI, SAN **Elev.:** 1200–2600 m a.s.l. **IUCN Habitat code:** 3,14 **Conservation status: (IUCN)** LC

4086 *Teramnus uncinatus* (L.) Sw. (Fabaceae)
IPNI: 521046-1
Common name: bejuquillo, uña de gato **Habit:** Herb, Climbing **Use:** MA **Origin:** Native **Region:** Andean, Caribbean, Orinoquia **Dept.:** ANT, CAL, CAU, CUN, HUI, MAG, NSA, PUT, SAN, VAC **Elev.:** 130–1900 m a.s.l. **IUCN Habitat code:** 1,2,3,14 **Conservation status: (IUCN)** NE

4087 *Trifolium alpestre* L. (Fabaceae)
IPNI: 522831-1
Use: GS, ME **Region:** Andean **Dept.:** BOG **Conservation status: (IUCN)** NE

4088 *Trifolium amabile* Kunth (Fabaceae)
IPNI: 522840-1
Common name: carretón, trébol **Habit:** Herb, Climbing **Use:** HF **Origin:** Native **Region:** Andean **Dept.:** BOG, BOY, CUN, TOL **Elev.:** 2500–3800 m a.s.l. **IUCN Habitat code:** 4,14 **Conservation status: (IUCN)** LC

4089 *Trifolium arvense* L. (Fabaceae)
IPNI: 522881-1
Use: AF, GS, HF, ME **Origin:** Naturalised **Dept.:** BOG **Conservation status: (IUCN)** NE

4090 *Trifolium campestre* Schreb. (Fabaceae)
IPNI: 522975-1
Common name: carretón amarillo, trébol **Habit:** Herb **Use:** AF, HF, PO **Origin:** Naturalised **Region:** Andean **Dept.:** BOG, BOY, CUN, VAC **Elev.:** 2000–3600 m a.s.l. **IUCN Habitat code:** 1,2,3,8,14 **Conservation status: (IUCN)** NE

4091 *Trifolium dubium* Sibth. (Fabaceae)
IPNI: 523099-1
Common name: carretón centella, trébol dorado **Habit:** Herb **Use:** AF, EU, IF, ME **Origin:** Naturalised **Region:** Andean **Dept.:** ANT, BOG, BOY, CAL, CUN, LAG, MAG, MET, NAR, VAC **Elev.:** 1990–3475 m a.s.l. **IUCN Habitat code:** 1,3,4,14 **Conservation status: (IUCN)** NE

4092 *Trifolium hybridum* L. (Fabaceae)
IPNI: 30240399-2
Common name: carretón, carretón híbrido, trébol híbrido **Habit:** Herb **Use:** AF, EU, HF, IF, ME, PO **Origin:** Cultivated, Naturalised **Dept.:** BOG, CUN **Elev.:** 2600–2600 m a.s.l. **IUCN Habitat code:** 1,14 **Conservation status: (IUCN)** NE

4093 *Trifolium pratense* L. (Fabaceae)
IPNI: 523575-1
Common name: carretón, carretón colorado, carretón rojo, trébol, trébol rojo **Habit:** Herb **Use:** AF, EU, GS, HF,

IF, ME, PO, SU **Origin:** Naturalised **Region:** Andean **Dept.:** ANT, BOG, BOY, CUN, LAG, MAG **Elev.:** 1750–3200 m a.s.l. **IUCN Habitat code:** 1,3,4,14 **Conservation status: (IUCN)** LC

4094 *Trifolium repens* L. (Fabaceae) **IPNI:** 523626-1 **Common name:** carretón, carretón blanco, trébol, trébol blanco **Habit:** Herb **Use:** AF, EU, GS, HF, IF, ME, PO, SU **Origin:** Naturalised **Region:** Andean **Dept.:** ANT, BOG, BOY, CAU, CUN, MAG, NAR, NSA, PUT, QUI, RIS, SAN, TOL **Elev.:** 1750–3750 m a.s.l. **IUCN Habitat code:** 1,2,3,4,8,14 **Conservation status: (IUCN)** NE

4095 *Ulex europaeus* L. (Fabaceae) **IPNI:** 524298-1 **Common name:** espino amarillo, espino de jardín, espino de oro, retama amarilla, retama espinosa, retamo espinoso **Habit:** Herb, Shrub **Use:** AF, EU, FU, HF, MA, ME **Origin:** Naturalised **Region:** Andean **Dept.:** ANT, BOG, CUN **Elev.:** 2000–3180 m a.s.l. **IUCN Habitat code:** 1,3,4,14 **Conservation status: (IUCN)** LC

4096 *Uribea tamarindoides* Dugand & Romero (Fabaceae) **IPNI:** 261663-2 **Common name:** granadillo, tamarindo, tamarindo de mico, tamarindo de montaña **Habit:** Tree **Use:** HF **Origin:** Native **Region:** Caribbean **Dept.:** ANT, BOL, CHO, CUN, MAG **Elev.:** 200–260 m a.s.l. **IUCN Habitat code:** 1,2,3,5,14 **Conservation status: (IUCN)** NE

4097 *Vachellia collinsii* (Saff.) Seigler & Ebinger (Fabaceae) **IPNI:** 60441206-2 **Common name:** cachito **Habit:** Shrub **Use:** ME **Origin:** Naturalised **Region:** Caribbean **Dept.:** ATL, BOL, CES, SAP **Elev.:** 20–350 m a.s.l. **IUCN Habitat code:** 1,2,14 **Conservation status: (IUCN)** LC

4098 *Vachellia farnesiana* (L.) Wight & Arn. (Fabaceae) **IPNI:** 1099713-2 **Common name:** aromo, coromo, cují, espino blanco, jinnuttureo, pelá, pelá de castilla, uña de cabra **Habit:** Shrub, Small tree **Use:** AF, EU, FU, GS, HF, IF, MA, ME, SU **Origin:** Native, Cultivated **Region:** Andean, Caribbean, Pacific **Dept.:** ANT, BOY, CAU, CUN, HUI, LAG, NAR, SAN, TOL, VAC **Elev.:** 0–2130 m a.s.l. **IUCN Habitat code:** 1,2,3,4,5,14 **Conservation status: (IUCN)** LC

4099 *Vachellia flexuosa* (Humb. & Bonpl. ex Willd.) Forero & C.Romero (Fabaceae) **IPNI:** 77100888-1 **Common name:** aromo, pelá **Habit:** Shrub, Tree **Use:** ME **Origin:** Endemic **Region:** Andean **Dept.:** CUN, HUI, TOL **Elev.:** 350-1300 m.a.s.l. **IUCN Habitat code:** 2,3 **Conservation status: (IUCN)** LC **(CPSG)** LC

4100 *Vachellia macracantha* (Humb. & Bonpl. ex Willd.) Seigler & Ebinger (Fabaceae) **IPNI:** 60441223-2 **Habit:** Shrub, Tree **Use:** AF, EU, FU, HF, MA, ME **Origin:** Native **Region:** Caribbean, Orinoquia **Dept.:** ATL, BOL, HUI, LAG, MAG, MET **Elev.:** 0–250 m a.s.l. **IUCN Habitat code:** 1,2,3,5,14 **Conservation status: (IUCN)** LC

4101 *Vachellia nilotica* (L.) P.J.H.Hurter & Mabb. (Fabaceae) **IPNI:** 77089275-1 **Habit:** Tree **Use:** AF, EU, FU, GS, HF, IF, MA, ME, SU **Origin:** Cultivated **Dept.:** ANT **Elev.:** 600–700 m a.s.l. **Conservation status: (IUCN)** LC

4102 *Vachellia pennatula* (Schltdl. & Cham.) Seigler & Ebinger (Fabaceae) **IPNI:** 60441228-2 **Habit:** Tree **Use:** AF, HF, ME **Origin:** Native **Region:** Andean **Dept.:** CAU, VAC **Elev.:** 537–1500 m a.s.l. **IUCN Habitat code:** 1,2,3,14 **Conservation status: (IUCN)** NE

4103 *Vachellia tortuosa* (L.) Seigler & Ebinger (Fabaceae) **IPNI:** 60441241-2 **Common name:** ambuca, aromo, cacho de cabra, murrai **Habit:** Shrub, Small tree **Use:** ME **Origin:** Native **Region:** Andean, Caribbean, Orinoquia **Dept.:** ANT, ATL, BOL, BOY, CAS, CES, CUN, HUI, LAG, MAG, VAC **Elev.:** 0–2144 m a.s.l. **IUCN Habitat code:** 1,2,4,5,14 **Conservation status: (IUCN)** NE

4104 *Vatairea fusca* (Ducke) Ducke (Fabaceae) **IPNI:** 262720-2 **Habit:** Tree **Use:** MA **Origin:** Native **Region:** Amazonia **Dept.:** AMA, VAU **Elev.:** 100–120 m a.s.l. **IUCN Habitat code:** 1 **Conservation status: (IUCN)** LC

4105 *Vatairea guianensis* Aubl. (Fabaceae) **IPNI:** 524446-1 **Common name:** gallinazo ordinario, guerra, marimari, palo de gallinazo, palo de grulla **Habit:** Tree **Use:** MA, ME **Origin:** Native **Region:** Amazonia **Dept.:** AMA, CAQ, VAU **Elev.:** 150–200 m a.s.l. **IUCN Habitat code:** 1,3,4,5 **Conservation status: (IUCN)** LC

4106 *Vataireopsis surinamensis* H.C.Lima (Fabaceae) **IPNI:** 262732-2 **Common name:** bálsamo **Habit:** Tree **Use:** ME **Origin:** Native **Region:** Amazonia, Guiana Shield **Dept.:** VAU **Elev.:** 350–350 m a.s.l. **IUCN Habitat code:** 1 **Conservation status: (IUCN)** LC

4107 *Vicia andicola* Kunth (Fabaceae) **IPNI:** 524519-1 **Common name:** alverjilla, arveja de monte, arvejilla, zarcillejo **Habit:** Herb, Climbing **Use:** ME **Origin:** Native **Region:** Andean **Dept.:** ANT, BOG, BOY, CAU, CUN, MET, NAR, NSA, SAN **Elev.:** 2000–4200 m a.s.l. **IUCN Habitat code:** 1,3,4,14 **Conservation status: (IUCN)** NE

4108 *Vicia benghalensis* L. (Fabaceae) **IPNI:** 524572-1 **Habit:** Herb, Climbing **Use:** AF, EU, IF, PO **Origin:** Cultivated **Region:** Andean **Dept.:** BOG, CUN, NAR **Elev.:** 2550–2700 m a.s.l. **IUCN Habitat code:** 4,14 **Conservation status: (IUCN)** NE

4109 *Vicia faba* L. (Fabaceae) **IPNI:** 524737-1 **Common name:** cargadita, haba **Habit:** Herb, Climbing **Use:** AF, EU, GS, HF, ME, PO **Origin:** Cultivated, Naturalised **Region:** Andean **Dept.:** ANT, BOG, BOY, CUN, TOL **Elev.:** 1000–3300 m a.s.l. **IUCN Habitat code:** 1,3,4,14 **Conservation status: (IUCN)** NE

4110 *Vicia graminea* Sm. (Fabaceae) **IPNI:** 524789-1 **Common name:** arveja del diablo **Habit:** Herb, Climbing **Use:** AF **Origin:** Native **Region:** Andean **Dept.:** BOG, BOY, CAU, CUN, NAR **Elev.:** 2700–3900 m a.s.l. **IUCN Habitat code:** 1,3,4,14 **Conservation status: (IUCN)** NE

4111 *Vicia lathyroides* L. (Fabaceae) **IPNI:** 524865-1 **Use:** GS **Region:** Andean **Dept.:** BOG **Conservation status: (IUCN)** LC

4112 *Vicia lens* (L.) Coss. & Germ. (Fabaceae) **IPNI:** 524872-1 **Common name:** lenteja **Habit:** Herb **Use:** AF, EU, GS, HF, ME **Origin:** Cultivated, Naturalised **Region:** Andean **Dept.:** BOY, CUN **Elev.:** 2600–2600 m a.s.l. **IUCN Habitat code:** 3,14 **Conservation status: (IUCN)** NE

4113 *Vicia sativa* L. (Fabaceae) **IPNI:** 525117-1 **Use:** AF, EU, GS, HF, ME, PO **Origin:** Naturalised **Dept.:** BOG **IUCN Habitat code:** 1,3,14 **Conservation status: (IUCN)** LC

4114 *Vicia villosa* Roth (Fabaceae) **IPNI:** 525259-1 **Habit:** Herb, Climbing **Use:** AF, EU, GS, HF, ME, PO **Origin:** Naturalised **Region:** Andean **Dept.:** ANT **Elev.:** 1000–1500 m a.s.l. **IUCN Habitat code:** 3 **Conservation status: (IUCN)** NE

4115 *Vigna angularis* (Willd.) Ohwi & H.Ohashi (Fabaceae) **IPNI:** 525291-1 **Common name:** frijolito, frĩjol arrocillo, frĩjol arroz, frĩjol pispo, pispo **Habit:** Climbing **Use:** AF, EU, GS, HF, MA, ME **Origin:** Cultivated **Region:** Andean **Dept.:** CAL, QUI, VAC **Elev.:** 1100–1400 m a.s.l. **IUCN Habitat code:** 1 **Conservation status: (IUCN)** LC

4116 *Vigna luteola* (Jacq.) Benth. (Fabaceae) **IPNI:** 525419-1 **Common name:** frisol de vaca, melusca **Habit:** Herb, Climbing **Use:** AF, EU, GS, HF, ME **Origin:** Native **Region:** Amazonia, Andean, Caribbean, Orinoquia, Pacific **Dept.:** AMA, ANT, BOL, CAL, CAS, CAU, CHO, CUN, GUA, SAP, SAN, VAC, VID **Elev.:** 0-1600 m.a.s.l. **IUCN Habitat code:** 1,2,3,4,5,14 **Conservation status: (IUCN)** LC **(CPSG)** Potential LC

4117 *Vigna marina* (Burm.) Merr. (Fabaceae) **IPNI:** 525428-1 **Habit:** Climbing **Use:** AF, EU, GS, HF, ME, SU **Origin:** Naturalised **Region:** Caribbean **Dept.:** ANT **Elev.:** 0–0 m a.s.l. **IUCN Habitat code:** 14 **Conservation status: (IUCN)** LC

4118 *Vigna racemosa* (G.Don) Hutch. & Dalziel ex Baker f. (Fabaceae) **IPNI:** 525491-1 **Common name:** naichinupa **Habit:** Climbing **Use:** AF **Origin:** Naturalised **Region:** Amazonia, Caribbean, Pacific **Dept.:** AMA, ANT, CHO **Elev.:** 75–100 m a.s.l. **IUCN Habitat code:** 1,3 **Conservation status: (IUCN)** LC

4119 *Vigna umbellata* (Thunb.) Ohwi & H.Ohashi (Fabaceae)
IPNI: 525554-1
Common name: frijol linaza, frijolito pijindí **Use:** AF, EU, GS, HF, MA, ME **IUCN Habitat code:** 1 **Conservation status: (IUCN)** NE

4120 *Vigna unguiculata* (L.) Walp. (Fabaceae)
IPNI: 1127257-2
Common name: cabecitanegra, caupí, chiclayo, frijol, fríjol batata, fríjol cabecenegra, fríjol cabecitanegra, fríjol cabezanegra, fríjol caupí, fríjol de cabeza negra, fríjol ojinegro, fríjol veleño **Habit:** Climbing **Use:** AF, EU, GS, HF, MA, ME **Origin:** Cultivated, Naturalised **Region:** Caribbean, Orinoquia **Dept.:** AMA, ANT, ATL, BOL, CAQ, CAS, CES, COR, CUN, LAG, MAG, SAN, VAC **Elev.:** 70–1000 m a.s.l. **IUCN Habitat code:** 1,2,3,5,8,14 **Conservation status: (IUCN)** NE

4121 *Vigna vexillata* (L.) A.Rich. (Fabaceae)
IPNI: 525561-1
Common name: bejuco de chivo, frijolillo, frijolillo rastrero **Habit:** Herb, Climbing **Use:** AF, EU, GS, HF, MA, ME **Origin:** Native, Cultivated **Region:** Amazonia, Andean, Caribbean, Pacific **Dept.:** AMA, ANT, CAL, CAQ, CHO, COR, CUN, HUI, MAG, NAR, PUT, SAN, VAC **Elev.:** 20–2000 m a.s.l. **IUCN Habitat code:** 1,2,3,4,5,14 **Conservation status: (IUCN)** NE

4122 *Zapoteca caracasana* (Jacq.) H.M.Hern. (Fabaceae)
IPNI: 287164-2
Common name: carbonero **Habit:** Shrub, Subshrub, Small tree **Use:** EU **Origin:** Native **Region:** Andean, Caribbean, Orinoquia **Dept.:** ANT, BOG, CES, CUN, LAG, MET, SAN, TOL, VAC **Elev.:** 0–1400 m a.s.l. **IUCN Habitat code:** 1,2,14 **Conservation status: (IUCN)** NE

4123 *Zapoteca formosa* (Kunth) H.M.Hern. (Fabaceae)
IPNI: 287168-2
Common name: bicho **Habit:** Herb, Shrub, Tree **Use:** AF **Origin:** Native **Region:** Andean, Caribbean **Dept.:** ATL, BOL, LAG, MAG, NSA **Elev.:** 0–600 m a.s.l. **IUCN Habitat code:** 1,2,3,14 **Conservation status: (IUCN)** LC

4124 *Zapoteca portoricensis* (Jacq.) H.M.Hern. (Fabaceae)
IPNI: 943976-1
Habit: Herb **Use:** EU, MA, ME **Origin:** Native **Region:** Andean, Orinoquia **Dept.:** CUN, MET **Elev.:** 0–1800 m a.s.l. **IUCN Habitat code:** 1 **Conservation status: (IUCN)** NE

4125 *Zornia curvata* Mohlenbr. (Fabaceae)
IPNI: 271272-2
Habit: Herb **Use:** MA **Origin:** Native **Region:** Andean **Dept.:** CAU, HUI, NSA, TOL, VAC **Elev.:** 500–1740 m a.s.l. **IUCN Habitat code:** 3 **Conservation status: (IUCN)** NE

4126 *Zornia diphylla* (L.) Pers. (Fabaceae)
IPNI: 30026493-2
Common name: alverjilla, barba de burro, cargadita, cargadita de frío, encarrugada, portadora, trencilla, urinaria **Habit:** Herb **Use:** AF,ME **Origin:** Naturalised **Region:** Andean, Caribbean, Orinoquia **Dept.:** ANT, BOY, CAU, CES, CUN, GUA, MET, NAR, TOL, VAU, VID **Elev.:** 65-2475 m.a.s.l. **IUCN Habitat code:** 1,2,3,4,14 **Conservation status: (IUCN)** NE **(CPSG)** Potential LC

4127 *Zornia latifolia* Sm. (Fabaceae)
IPNI: 30156577-2
Common name: cargadita, esterilla **Habit:** Herb **Use:** AF, EU, GS, HF, MA, ME **Origin:** Native **Region:** Amazonia, Andean, Orinoquia **Dept.:** AMA, ANT, BOY, CAQ, CAS, CAU, GUA, GUV, MET, SAN, TOL, VAC, VAU, VID **Elev.:** 100–1740 m a.s.l. **IUCN Habitat code:** 1,2,3,4,14 **Conservation status: (IUCN)** NE

4128 *Zornia leptophylla* (Benth.) Pittier (Fabaceae)
IPNI: 526041-1
Habit: Herb **Use:** ME **Origin:** Native **Region:** Guiana Shield **Dept.:** VAU **Elev.:** 350–350 m a.s.l. **IUCN Habitat code:** 3,14 **Conservation status: (IUCN)** NE

4129 *Zornia reticulata* Sm. (Fabaceae)
IPNI: 271325-2
Common name: alverjita, cargadita, sen de sabana, trencita **Habit:** Herb, Shrub **Use:** ME **Origin:** Native **Region:** Amazonia, Andean, Guiana Shield, Caribbean, Orinoquia, Pacific **Dept.:** AMA, ANT, CAL, CAS, CAU, CHO, CUN, HUI, LAG, MAG, MET, NAR, PUT, QUI, SAN, TOL, VAC, VAU, VID **Elev.:** 80-2800 m.a.s.l. **IUCN Habitat code:** 1,2,3,4,14 **Conservation status: (IUCN)** NE **(CPSG)** Potential LC

4130 *Zornia thymifolia* Kunth (Fabaceae)
IPNI: 526091-1
Common name: cargadita **Habit:** Herb **Use:** HF **Origin:** Native **Region:** Amazonia, Pacific **Dept.:** CAQ, CHO **Elev.:** 450–470 m a.s.l. **IUCN Habitat code:** 1,3,14 **Conservation status: (IUCN)** NE

4131 *Zygia andaquiensis* (C.Barbosa) L.Rico (Fabaceae)
IPNI: 962818-1
Habit: Tree **Use:** ME **Origin:** Endemic **Region:** Amazonia **Dept.:** CAQ **Elev.:** 150-300 m.a.s.l. **IUCN Habitat code:** 1 **Conservation status: (IUCN)** EN **(CPSG)** EN

4132 *Zygia cataractae* (Kunth) L.Rico (Fabaceae)
IPNI: 962821-1
Common name: carbón, cimbrapotro, guamo, palo boya **Habit:** Shrub, Tree **Use:** HF **Origin:** Native **Region:** Amazonia, Guiana Shield, Orinoquia **Dept.:** AMA, CAS, GUA, GUV, MET, VAU **Elev.:** 150–500 m a.s.l. **IUCN Habitat code:** 1,3,4,5,14 **Conservation status: (IUCN)** LC

4133 *Zygia coccinea* (G.Don) L.Rico (Fabaceae)
IPNI: 962823-1
Habit: Subshrub, Tree **Use:** HF **Origin:** Native **Region:** Amazonia, Andean **Dept.:** NAR, PUT **Elev.:** 250–1190 m a.s.l. **IUCN Habitat code:** 1,3,14 **Conservation status: (IUCN)** LC

4134 *Zygia inaequalis* (Humb. & Bonpl. ex Willd.) Pittier (Fabaceae)
IPNI: 271431-2
Common name: guamilla, guamo arroyero **Habit:** Shrub, Tree **Use:** AF, EU, FU, HF, MA **Origin:** Native **Region:** Amazonia, Andean, Guiana Shield, Caribbean, Orinoquia, Pacific **Dept.:** AMA, ANT, ARA, BOL, BOY, CAL, CAQ, CAS, CAU, CES, CHO, COR, GUA, GUV, MAG, MET, RIS, SAN, SUC, VAC, VAU, VID **Elev.:** 5-1500 m.a.s.l. **IUCN Habitat code:** 1,2,3,5,14 **Conservation status: (IUCN)** LC **(CPSG)** Potential LC

4135 *Zygia juruana* (Harms) L.Rico (Fabaceae)
IPNI: 962830-1
Common name: guama seco **Habit:** Tree **Use:** HF **Origin:** Native **Region:** Amazonia **Dept.:** AMA **Elev.:** 120–200 m a.s.l. **IUCN Habitat code:** 1,4 **Conservation status: (IUCN)** LC

4136 *Zygia latifolia* (L.) Fawc. & Rendle (Fabaceae)
IPNI: 271432-2
Common name: carbonero, churimo, cimbrapotro, guamilla, guamillo, guamo venenoso **Habit:** Shrub, Small tree, Tree **Use:** ME **Origin:** Native **Region:** Amazonia, Andean, Guiana Shield, Caribbean, Orinoquia, Pacific **Dept.:** AMA, CAL, CAS, CHO, LAG, MAG, MET, NAR, SAN, VAC, VAU **Elev.:** 0-1100 m.a.s.l. **Conservation status: (IUCN)** NE **(CPSG)** Potential LC

4137 *Zygia longifolia* (Humb. & Bonpl. ex Willd.) Britton & Rose (Fabaceae)
IPNI: 271434-2
Common name: achil, achí, amé, carbón, chiparo, chipero, chíparo, chípero, cimbrapotro, espino de mono, guamo de río, guamo macho, guamo playero, mané, payandé, pichindé, playero, suribio **Habit:** Shrub, Small tree, Tree **Use:** HF, ME, SU **Origin:** Native **Region:** Amazonia, Andean, Guiana Shield, Caribbean, Orinoquia, Pacific **Dept.:** ANT, BOL, CAL, CAQ, CES, CHO, COR, CUN, HUI, LAG, MAG, MET, NAR, PUT, SAN, TOL, VAC, VAU **Elev.:** 0-1700 m.a.s.l. **IUCN Habitat code:** 1,2,3,5,14 **Conservation status: (IUCN)** LC **(CPSG)** Potential LC

4138 *Zygia racemosa* (Ducke) Barneby & J.W.Grimes (Fabaceae)
IPNI: 996977-1
Habit: Small tree, Tree **Use:** MA **Origin:** Native **Region:** Amazonia, Guiana Shield **Dept.:** AMA, GUA **Elev.:** 0–600 m a.s.l. **IUCN Habitat code:** 1,3,5 **Conservation status: (IUCN)** LC

4139 *Asemeia acuminata* (Willd.) J.F.B.Pastore & J.R.Abbott (Polygalaceae)
IPNI: 77121649-1
Habit: Herb, Subshrub **Use:** ME **Origin:** Native **Region:** Andean, Pacific **Dept.:** CES, CHO, MAG, NAR **Elev.:** 100–1000 m a.s.l. **IUCN Habitat code:** 1 **Conservation status: (IUCN)** NE

4140 *Asemeia violacea* (Aubl.) J.F.B.Pastore & J.R.Abbott (Polygalaceae)
IPNI: 77121670-1
Habit: Herb **Use:** ME **Origin:** Native **Region:** Caribbean, Orinoquia **Dept.:** CUN, MAG, TOL, VID **Elev.:** 121–1000 m a.s.l. **IUCN Habitat code:** 1,2,3,4,14 **Conservation status: (IUCN)** NE

4141 *Bredemeyera floribunda* Willd. (Polygalaceae)
IPNI: 690122-1
Common name: rabo de iguana **Habit:** Small tree, Liana, Climbing **Use:** HF, ME **Origin:** Native **Region:** Caribbean **Dept.:** ANT, BOL, COR, MAG, SAN, SUC **Elev.:** 50–1000 m a.s.l. **IUCN Habitat code:** 1,2,3,14 **Conservation status: (IUCN)** LC

4142 *Caamembeca spectabilis* (DC.) J.F.B.Pastore (Polygalaceae)
IPNI: 77119197-1
Habit: Herb, Subshrub **Use:** ME **Origin:** Native **Region:** Amazonia, Andean **Dept.:** CAQ, VAU **Elev.:** 270–

1000 m a.s.l. **IUCN Habitat code:** 1,3,14 **Conservation status: (IUCN)** NE

4143 *Diclidanthera penduliflora* Mart. (Polygalaceae)
IPNI: 826364-1
Habit: Small tree, Liana **Use:** HF **Origin:** Native **Region:** Amazonia **Dept.:** AMA **Elev.:** 100–100 m a.s.l. **IUCN Habitat code:** 1,2,5,14 **Conservation status: (IUCN)** NE

4144 *Monnina aestuans* (L.f.) DC. (Polygalaceae)
IPNI: 690457-1
Common name: bodoquera, guaquito, ibilán, ingerta de shrub, injerta, rústico, sorbetana, tintillo, tinto, tinto macho **Habit:** Shrub, Small tree **Use:** ME **Origin:** Native **Region:** Andean **Dept.:** ANT, BOG, BOL, BOY, CAU, CES, CUN, MAG, NSA, SAN **Elev.:** 1700–4500 m a.s.l. **IUCN Habitat code:** 1,2,3,4,5,8,14 **Conservation status: (IUCN)** NE

4145 *Monnina bracteata* Chodat (Polygalaceae)
IPNI: 163536-2
Common name: pica pica, picapica **Habit:** Shrub, Subshrub **Use:** ME **Origin:** Native **Region:** Andean **Dept.:** NSA **Elev.:** 1400–2400 m a.s.l. **IUCN Habitat code:** 1,4,14 **Conservation status: (IUCN)** NE

4146 *Monnina cladostachya* Turcz. (Polygalaceae)
IPNI: 690494-1
Common name: tinto chiquito **Use:** ME **Origin:** Native **IUCN Habitat code:** 1,4 **Conservation status: (IUCN)** LC

4147 *Monnina crassifolia* (Bonpl.) Kunth (Polygalaceae)
IPNI: 690504-1
Habit: Shrub, Small tree **Use:** ME **Origin:** Native **Region:** Andean **Dept.:** BOG, CAU, CUN, NAR, NSA, PUT, QUI, SAN **Elev.:** 3100–4200 m a.s.l. **IUCN Habitat code:** 1,3,4,8,14 **Conservation status: (IUCN)** NE

4148 *Monnina fastigiata* (Bonpl.) DC. (Polygalaceae)
IPNI: 690538-1
Common name: sorbetana **Habit:** Shrub, Small tree **Use:** ME **Origin:** Endemic **Region:** Andean **Dept.:** ANT, BOG, BOY, CAL, CAU, CUN, MAG, NSA, PUT, QUI, RIS, SAN, TOL, VAC **Elev.:** 1340-3700 m.a.s.l. **IUCN Habitat code:** 1,3,4,5,14 **Conservation status: (IUCN)** NE **(CPSG)** Potential LC

4149 *Monnina padifolia* (Bonpl.) Freire-Fierro (Polygalaceae)
IPNI: 77143255-1
Habit: Herb, Shrub, Small tree, Tree **Use:** ME **Origin:** Endemic **Region:** Andean **Dept.:** ANT, BOG, BOL, BOY, CAL, CAU, CUN, HUI, MAG, NAR, NSA, PUT, QUI, RIS, SAN, TOL, VAC **Elev.:** 800-3950 m.a.s.l. **Conservation status: (IUCN)** NE **(CPSG)** LC

4150 *Monnina polystachya* Ruiz & Pav. (Polygalaceae)
IPNI: 163646-2
Habit: Herb, Shrub **Use:** ME **Origin:** Native **Region:** Andean **Dept.:** ANT, CAL, CAU, CUN, QUI **Elev.:** 1400–2500 m a.s.l. **IUCN Habitat code:** 1,3,8,14 **Conservation status: (IUCN)** NE

4151 *Monnina salicifolia* Ruiz & Pav. (Polygalaceae)
IPNI: 690654-1
Common name: eguielin, iguilan, iguilan phytolaca, tinto **Habit:** Shrub, Subshrub, Small tree **Use:** EU, HF, MA, ME **Origin:** Native **Region:** Andean **Dept.:** ANT, BOG, BOY, CAL, CAU, CUN, MAG, MET, NAR, PUT, SAN, VAC **Elev.:** 1900–4150 m a.s.l. **IUCN Habitat code:** 1,3,4,5,8,14 **Conservation status: (IUCN)** NE

4152 *Moutabea aculeata* (Ruiz & Pav.) Poepp. & Endl. (Polygalaceae)
IPNI: 690705-1
Common name: caparo apedomu **Habit:** Liana **Use:** HF, ME **Origin:** Native **Region:** Amazonia, Andean, Caribbean **Dept.:** AMA, BOL, PUT **Elev.:** 80–1350 m a.s.l. **IUCN Habitat code:** 1,3,5,14 **Conservation status: (IUCN)** LC

4153 *Moutabea guianensis* Aubl. (Polygalaceae)
IPNI: 690710-1
Common name: curare, ojo de venado, ojo de vendao **Habit:** Small tree, Liana **Use:** HF **Origin:** Native **Region:** Amazonia, Guiana Shield, Orinoquia **Dept.:** GUA, VID **Elev.:** 85–85 m a.s.l. **IUCN Habitat code:** 1,4,5,14 **Conservation status: (IUCN)** LC

4154 *Polygala appendiculata* Vell. (Polygalaceae)
IPNI: 691063-1
Use: ME **Origin:** Native **Conservation status: (IUCN)** NE

4155 *Polygala asperuloides* Kunth (Polygalaceae)
IPNI: 205398-2
Common name: menta, mentolín, sarpoleta, yerba de la Virgen **Habit:** Herb **Use:** ME **Origin:** Native **Region:** Andean, Orinoquia, Pacific **Dept.:** ANT, CAL, CHO, CUN, HUI, MET, NAR, TOL, VAC **Elev.:** 50–1800 m a.s.l. **IUCN Habitat code:** 1,3,14 **Conservation status: (IUCN)** NE

4156 *Polygala capillaris* E.Mey. ex Harv. (Polygalaceae)
IPNI: 1003233-1
Common name: rústica, sarpoleta **Use:** ME **Dept.:** BOG **IUCN Habitat code:** 1,3,14 **Conservation status: (IUCN)** NE

4157 *Polygala chinensis* L. (Polygalaceae)
IPNI: 691242-1
Habit: Herb **Use:** ME **Origin:** Native **Region:** Caribbean **Dept.:** MAG **Elev.:** 0–0 m a.s.l. **Conservation status: (IUCN)** NE

4158 *Polygala paniculata* L. (Polygalaceae)
IPNI: 330527-2
Common name: alegría del duende, angelita, angélica, canchalagua, chiegua, chinchimaní, colagoge, hoja de higo, ipeacuana, lanillo, mentol, sarpoleta, yerba de la leche **Habit:** Herb **Use:** HF, ME, SU **Origin:** Native **Region:** Andean, Caribbean, Pacific **Dept.:** ANT, BOG, BOY, CAU, CHO, CUN, HUI, MAG, MET, NAR, RIS, SAN, TOL, VAC **Elev.:** 0-3380 m.a.s.l. **IUCN Habitat code:** 1,2,3,4,5,14 **Conservation status: (IUCN)** NE **(CPSG)** Potential LC

4159 *Polygala trichosperma* Jacq. (Polygalaceae)
IPNI: 692483-1
Common name: sarpoleta **Habit:** Herb **Use:** ME **Origin:** Native **Region:** Orinoquia **Dept.:** MET, VID **Elev.:** 400–400 m a.s.l. **Conservation status: (IUCN)** NE

4160 *Securidaca diversifolia* (L.) S.F.Blake (Polygalaceae)
IPNI: 231560-2
Habit: Liana, Climbing **Use:** MA, ME **Origin:** Native **Region:** Andean, Caribbean, Pacific **Dept.:** ANT, BOL, CHO, MAG, NSA, SAN, SUC, VAC **Elev.:** 10–1700 m a.s.l. **IUCN Habitat code:** 1,2,3,4,14 **Conservation status: (IUCN)** LC

4161 *Suriana maritima* L. (Surianaceae)
IPNI: 814158-1
Common name: cedro playero **Habit:** Shrub **Use:** EU, MA, ME **Origin:** Native **Region:** Caribbean **Dept.:** BOL, LAG, SAP **Elev.:** 0–10 m a.s.l. **IUCN Habitat code:** 1,2,14 **Conservation status: (IUCN)** LC

Fagales

4162 *Alnus acuminata* Kunth (Betulaceae)
IPNI: 294882-1
Common name: aliso, alnus, cerezo, cerezo real, chaquero, chaquiro, fresno, lambrán **Habit:** Tree **Use:** AF, EU, FU, MA, ME **Origin:** Native, Cultivated **Region:** Andean **Dept.:** ANT, BOG, BOY, CAL, CAU, CUN, HUI, NAR, NSA, QUI, RIS, SAN, TOL, VAC **Elev.:** 1500–3400 m a.s.l. **IUCN Habitat code:** 1,3,4,5,8,14 **Conservation status: (IUCN)** LC

4163 *Casuarina cunninghamiana* Miq. (Casuarinaceae)
IPNI: 159845-1
Common name: casuarina, pino de Australia **Habit:** Tree **Use:** AF, EU, FU, HF, IF, MA **Origin:** Cultivated **Dept.:** BOG, VAC **Elev.:** 900–900 m a.s.l. **Conservation status: (IUCN)** LC

4164 *Casuarina equisetifolia* L. (Casuarinaceae)
IPNI: 159856-1
Common name: casuarina, falso pino, pino, pino australiano, pino de Australia, pino extranjero, pino falso australiano **Habit:** Tree **Use:** AF, EU, FU, HF, MA, ME, SU **Origin:** Cultivated **Region:** Andean, Caribbean **Dept.:** ANT, BOG, CAU, CUN, MAG, SAP **Elev.:** 5–2650 m a.s.l. **IUCN Habitat code:** 1,2,3,14 **Conservation status: (IUCN)** LC

4165 *Quercus humboldtii* Bonpl. (Fagaceae)
IPNI: 296268-1
Common name: roble, roble de tierra fría **Habit:** Tree **Use:** AF, EU, FU, HF, MA, ME, SU **Origin:** Native, Cultivated **Region:** Andean **Dept.:** ANT, BOG, BOY, CAL, CAQ, CAU, CHO, CUN, HUI, MET, NAR, QUI, RIS, SAN, TOL, VAC **Elev.:** 1400–3300 m a.s.l. **IUCN Habitat code:** 1,2,3,4,5,8,14 **Conservation status: (IUCN)** LC

4166 *Juglans neotropica* Diels (Juglandaceae)
IPNI: 131772-2
Common name: cedro grande, cedro negro, cedro nogal, nogal, nogal bogotano, nogal sabanero **Habit:** Tree **Use:** EU, GS, HF, MA, ME **Origin:** Native **Region:** Andean **Dept.:** ANT, BOG, CAL, CUN, PUT, RIS, SAN, VAC **Elev.:** 1600-3100 m.a.s.l. **IUCN Habitat code:** 1,3,14 **Conservation status: (IUCN)** EN **(CPSG)** EN

4167 *Juglans regia* L. (Juglandaceae) **IPNI:** 442427-1 **Common name:** nogal **Use:** EU, GS, HF, MA, ME **IUCN Habitat code:** 3,14 **Conservation status: (IUCN)** LC

4168 *Myrica cerifera* L. (Myricaceae) **IPNI:** 166220-2 **Habit:** Shrub, Small tree **Use:** HF, MA, ME **Origin:** Native **Region:** Caribbean **Dept.:** SAP **Elev.:** 300–310 m a.s.l. **Conservation status: (IUCN)** NE

4169 *Myrica parvifolia* Benth. (Myricaceae) **IPNI:** 326824-2 **Common name:** aromo, laurel, laurel de cera, laurel hojipequeño, laurel negro **Habit:** Shrub, Small tree, Tree **Use:** AF, EU, MA, ME, SU **Origin:** Native **Region:** Andean **Dept.:** ANT, BOY, CAS, CAU, CHO, CUN, NAR, NSA, PUT, QUI, RIS, SAN, VAC **Elev.:** 1600–3800 m a.s.l. **IUCN Habitat code:** 4,5,14 **Conservation status: (IUCN)** NE

4170 *Myrica pubescens* Humb. & Bonpl. ex Willd. (Myricaceae) **IPNI:** 326827-2 **Common name:** aromo, laurel, laurel de cera, olivo, olivo de cera **Habit:** Shrub, Small tree, Tree **Use:** MA, ME **Origin:** Native **Region:** Andean **Dept.:** ANT, BOY, CAL, CAQ, CAU, CES, CHO, CUN, HUI, LAG, MAG, NAR, NSA, PUT, QUI, RIS, SAN, TOL, VAC **Elev.:** 1700–3900 m a.s.l. **IUCN Habitat code:** 1,2,3,4,14 **Conservation status: (IUCN)** NE

Gentianales

4171 *Allamanda blanchetii* A.DC. (Apocynaceae) **IPNI:** 76468-1 **Common name:** trompeta **Habit:** Subshrub **Use:** EU, ME **Origin:** Cultivated **Region:** Caribbean **Dept.:** BOG, BOL, SAP **Elev.:** 0–10 m a.s.l. **IUCN Habitat code:** 2,3,14 **Conservation status: (IUCN)** NE

4172 *Allamanda cathartica* L. (Apocynaceae) **IPNI:** 319065-2 **Common name:** bejuco de sapo, campana de oro, capitana, copa de oro, copita de oro, flor de muerto, jazmín, jazmín amarillo **Habit:** Shrub, Subshrub, Climbing **Use:** EU, MA, ME, PO **Origin:** Native, Cultivated **Region:** Amazonia, Andean, Caribbean, Orinoquia, Pacific **Dept.:** AMA, ANT, ATL, CAL, CAS, CAU, CHO, CUN, MAG, MET, SAP, SAN, VAC, VAU **Elev.:** 0–1900 m a.s.l. **IUCN Habitat code:** 1,2,3,4,5,14 **Conservation status: (IUCN)** NE

4173 *Ambelania occidentalis* Zarucchi (Apocynaceae) **IPNI:** 272043-2 **Habit:** Small tree, Tree **Use:** HF **Origin:** Native **Region:** Amazonia **Dept.:** AMA **Elev.:** 100–200 m a.s.l. **IUCN Habitat code:** 1,5 **Conservation status: (IUCN)** NE

4174 *Anechites nerium* (Aubl.) Urb. (Apocynaceae) **IPNI:** 13196-2 **Habit:** Climbing **Use:** ME **Origin:** Native **Region:** Andean, Pacific **Dept.:** ANT, CHO, LAG **Elev.:** 45–600 m a.s.l. **IUCN Habitat code:** 1,3,14 **Conservation status: (IUCN)** NE

4175 *Asclepias curassavica* L. (Apocynaceae) **IPNI:** 94213-1 **Common name:** abortaganado, algodoncillo, algodón de seda, asesina, bailarina, bencenuco, cimarrona, flor de calentura, flor de seda, guajuro, guájaro, hierba de leche, jalapa, lechera, lombricera, lombricero, malcasada, mataganado, maturia, milkweed, niñomuerto, palom **Habit:** Herb, Subshrub **Use:** AF, EU, HF, IF, MA, ME, PO **Origin:** Native **Region:** Amazonia, Andean, Guiana Shield, Caribbean, Orinoquia, Pacific **Dept.:** AMA, ANT, BOL, BOY, CAL, CAU, CHO, CUN, HUI, LAG, MAG, MET, NAR, NSA, PUT, QUI, SAP, SAN, VAC **Elev.:** 0-2700 m.a.s.l. **IUCN Habitat code:** 1,2,3,4,5,14 **Conservation status: (IUCN)** NE **(CPSG)** Potential LC

4176 *Aspidosperma album* (Vahl) Benoist ex Pichon (Apocynaceae) **IPNI:** 77325-1 **Use:** ME **IUCN Habitat code:** 1 **Conservation status: (IUCN)** NE

4177 *Aspidosperma cuspa* (Kunth) S.F.Blake ex Pittier (Apocynaceae) **IPNI:** 77347-1 **Common name:** amargo, amargoso, carreto, chivato, cuspa, guacharaco, kalaipata, kalaitpana, malambito de perro, palo amargo, paloamargo, vara de piedra **Habit:** Shrub, Tree **Use:** ME **Origin:** Native **Region:** Andean, Caribbean **Dept.:** ANT, ATL, BOL, CES, CUN, HUI, MAG **Elev.:** 0–1805 m a.s.l. **IUCN Habitat code:** 1,2,3,5,14 **Conservation status: (IUCN)** LC

4178 *Aspidosperma desmanthum* Benth. ex Müll.Arg. (Apocynaceae) **IPNI:** 77353-1 **Common name:** cane, caney **Habit:** Tree **Use:** MA, ME **Origin:** Native **Region:** Amazonia, Andean, Guiana Shield, Caribbean, Pacific **Dept.:** AMA, ANT, BOL, GUA, MAG, PUT, VAC **Elev.:** 50–1490 m a.s.l. **IUCN Habitat code:** 1,2,3,4,5,14 **Conservation status: (IUCN)** LC

4179 *Aspidosperma excelsum* Benth. (Apocynaceae) **IPNI:** 77367-1 **Common name:** cabo de hacha, costillo, costillo blanco, costillo verdadero, costillocaspi, joetor+mu, juansoquillo, matú, remocaspi **Habit:** Small tree, Tree **Use:** MA, ME **Origin:** Native **Region:** Amazonia, Guiana Shield, Pacific **Dept.:** AMA, CAQ, CHO, COR, GUA, MET, PUT, VAC, VAU, VID **Elev.:** 50–500 m a.s.l. **IUCN Habitat code:** 1,3,4,5,14 **Conservation status: (IUCN)** LC

4180 *Aspidosperma megalocarpon* Müll.Arg. (Apocynaceae) **IPNI:** 77405-1 **Common name:** caimito canelo, carretillo, carreto, combo, contraveneno, costillo, costillo acanalado, costillo redondo, lechero, mameicillo, tomasuco, tumasuco **Habit:** Tree **Use:** HF, MA, ME **Origin:** Native **Region:** Caribbean, Pacific **Dept.:** ANT, BOL, CHO, MAG, SAN, SUC **Elev.:** 230-770 m.a.s.l. **IUCN Habitat code:** 1,2,3,4,5,14 **Conservation status: (IUCN)** NT **(CPSG)** NT

4181 *Aspidosperma myristicifolium* (Markgr.) Woodson (Apocynaceae) **IPNI:** 77411-1 **Common name:** remocaspi **Habit:** Tree **Use:** ME **Origin:** Native **Region:** Amazonia, Pacific **Dept.:** AMA, CHO **Elev.:** 100–300 m a.s.l. **IUCN Habitat code:** 1 **Conservation status: (IUCN)** LC

4182 *Aspidosperma parvifolium* A.DC. (Apocynaceae) **IPNI:** 315552-2 **Common name:** garlipo **Habit:** Tree **Use:** MA, ME **Origin:** Native **Region:** Andean **Dept.:** CES **Elev.:** 990–1400 m a.s.l. **IUCN Habitat code:** 1,2,14 **Conservation status: (IUCN)** NE

4183 *Aspidosperma polyneuron* Müll.Arg. (Apocynaceae) **IPNI:** 77433-1 **Common name:** amargo, carretillo, carreto, carreto de vela, chivato, comulá, costillo, cumulá, cuomulá, macuiro, piedeyerro, quimulá **Habit:** Small tree, Tree **Use:** AF, FU, MA, ME **Origin:** Native **Region:** Caribbean **Dept.:** ATL, BOL, BOY, CES, CUN, LAG, MAG, SUC, TOL **Elev.:** 50-350 m.a.s.l. **IUCN Habitat code:** 1,2,3,14 **Conservation status: (IUCN)** EN **(CPSG)** EN

4184 *Aspidosperma rigidum* Rusby (Apocynaceae) **IPNI:** 22058-2 **Common name:** cabo de hacha, costillo, pachipa **Habit:** Tree **Use:** ME **Origin:** Native **Region:** Amazonia, Guiana Shield, Pacific **Dept.:** AMA, CAQ, CHO, MET, NSA, PUT, SAN **Elev.:** 100–350 m a.s.l. **IUCN Habitat code:** 1,2,3,8,14 **Conservation status: (IUCN)** LC

4185 *Aspidosperma schultesii* Woodson (Apocynaceae) **IPNI:** 22062-2 **Common name:** matú, pelusa, popay de sapo **Habit:** Tree **Use:** ME **Origin:** Native **Region:** Guiana Shield **Dept.:** AMA, CAQ, COR, GUA **Elev.:** 200–274 m a.s.l. **IUCN Habitat code:** 1,3,4,5,14 **Conservation status: (IUCN)** LC

4186 *Aspidosperma spruceanum* Benth. ex Müll.Arg. (Apocynaceae) **IPNI:** 22069-2 **Common name:** brasil, caney, cané, carreto, carreto mameyón, costillo, costillo redondo, garrocho, lengua de venado, mamey, palo brasil, palo pilón, rascador, sangre de toro, sardinato **Habit:** Small tree, Tree **Use:** HF, MA **Origin:** Native **Region:** Amazonia, Guiana Shield, Orinoquia **Dept.:** GUA, GUV, VID **Elev.:** 100–125 m a.s.l. **IUCN Habitat code:** 1,2,3,4,5,14 **Conservation status: (IUCN)** LC

4187 *Blepharodon mucronatum* (Schltdl.) Decne. (Apocynaceae) **IPNI:** 94927-1 **Common name:** bejuco de pescado, bejuco sapo, pischicachilla **Habit:** Climbing **Use:** ME **Origin:** Native **Region:** Andean, Caribbean, Orinoquia, Pacific **Dept.:** ANT, BOL, CHO, COR, CUN, HUI, LAG, MAG, MET, SAN, TOL, VAC **Elev.:** 30-1760 m.a.s.l. **IUCN Habitat code:** 1,2,3,5,14 **Conservation status: (IUCN)** NE **(CPSG)** Potential LC

4188 *Calotropis gigantea* (L.) W.T.Aiton (Apocynaceae) **IPNI:** 95173-1 **Common name:** algodoncillo, algodón chino, algodón de seda, algodón higuerillo, capullo, lechoso, palomitas, palomitas de seda, vejigo, vejigón, árbol de seda **Habit:** Herb **Use:** EU, FU, HF, MA, ME, PO, SU **Origin:** Naturalised **Region:** Andean **Dept.:** CUN, HUI **Elev.:** 1100–1200 m a.s.l. **IUCN Habitat code:** 1,2,3,5,14 **Conservation status: (IUCN)** NE

4189 *Calotropis procera* (Aiton) W.T.Aiton (Apocynaceae) **IPNI:** 1004515-2 **Common name:** algodón, algodón andariego, algodón chino, algodón de Castilla, algodón de seda, algodón extranjero, algodón fino, algodón lechero, gallito, güevas de perro, lechero, palo de algodón, quiebraplatos, vagamundo, vejigo, vejigón **Habit:** Herb, Shrub, Subshrub, Small tree, Liana, Climbing **Use:** AF, EU, FU, HF, MA, ME, PO, SU **Origin:** Naturalised **Region:** Andean, Caribbean **Dept.:** ANT, ATL, BOL, BOY, CAS, CUN, HUI, MAG, SAN, TOL, VAC **Elev.:** 0–2720 m a.s.l. **IUCN Habitat code:** 3 **Conservation status: (IUCN)** NE

4190 *Carissa macrocarpa* (Eckl.) A.DC. (Apocynaceae)
IPNI: 77727-1
Habit: Shrub **Use:** EU, GS, HF, MA, ME **Origin:** Cultivated **Region:** Andean, Caribbean **Dept.:** ANT, BOL, VAC **Elev.:** 700–1500 m a.s.l. **IUCN Habitat code:** 14 **Conservation status: (IUCN)** LC

4191 *Cascabela thevetia* (L.) Lippold (Apocynaceae)
IPNI: 47927-2
Common name: amacay, amancai, barbasco, cabalonga, cabalongo, cachimolivo, cahimolivo, cascabel, castañeto, catape, catapes, catapis, caucho, cavadonga hembra, cobalonga, cobalongo, cojón de cabrito, cojón de perro, covalonga, covalongo, cucaracho, enebro, fey sezi, g **Habit:** Subshrub, Small tree, Tree **Use:** AF, EU, FU, HF, MA, ME, PO, SU **Origin:** Native, Cultivated **Region:** Andean, Caribbean, Pacific **Dept.:** ANT, BOL, CHO, CUN, NAR, PUT, SAP, SAN, VAC **Elev.:** 0–1500 m a.s.l. **IUCN Habitat code:** 1,2,3,4,14 **Conservation status: (IUCN)** LC

4192 *Catharanthus roseus* (L.) G.Don (Apocynaceae)
IPNI: 77880-1
Common name: alhelí, buenastardes, caraqueña, cartagena, chocolata, colegiala, corazón de Jesús, corte, cortejo, cortesita, cortés, margarita, moradita, nazarena, niñabonita, pensamiento, primavera, princesa, princesita, reina, sietecueros, vinca, viuda, viudita **Habit:** Herb, Shrub **Use:** AF, EU, GS, HF, MA, ME, PO **Origin:** Cultivated **Region:** Andean, Caribbean, Pacific **Dept.:** ANT, BOG, BOL, CAL, CAS, CAU, CUN, MAG, NAR, SAP, SAN, SUC, TOL, VAC **Elev.:** 100–2000 m a.s.l. **IUCN Habitat code:** 1,2,3,4,14 **Conservation status: (IUCN)** NE

4193 *Chloropetalum denticulatum* (Vahl) Morillo (Apocynaceae)
IPNI: 77206440-1
Common name: lecherote **Habit:** Climbing **Use:** ME **Origin:** Endemic **Region:** Amazonia, Orinoquia, Pacific **Dept.:** ANT, CHO, CUN, LAG, MAG, MET, PUT, QUI, SAN, TOL, VAC **Elev.:** 75–1000 m a.s.l. **IUCN Habitat code:** 1,3,14 **Conservation status: (IUCN)** NE

4194 *Couma catingae* Ducke (Apocynaceae)
IPNI: 67213-2
Common name: juansoco, juansoco de alacrán, juansoco de la sabana, juansoquillo, juansoquillo de alacrán, juansoquillo de alacrán pendare, pendare, sorba **Habit:** Tree **Use:** HF **Origin:** Native **Region:** Amazonia **Dept.:** AMA, CAQ, PUT, VAU **Elev.:** 100–300 m a.s.l. **IUCN Habitat code:** 1,4,5 **Conservation status: (IUCN)** NE

4195 *Couma macrocarpa* Barb.Rodr. (Apocynaceae)
IPNI: 67217-2
Common name: avichure, bolsa de mono, caimito lirio, caimito popa, chicle, jansoco, juan soco, juan solo, juansoco, juansoco corriente, juansoco de monte, juansoco propio, lechecaspi, lirio, pendare, pendare grande, pendaré, perillo, perillo negro, perillo rosado, per **Habit:** Tree **Use:** HF, MA, ME **Origin:** Native **Region:** Amazonia, Guiana Shield, Orinoquia, Pacific **Dept.:** AMA, ANT, BOL, CAQ, CHO, GUA, GUV, NAR, PUT, SAN, VAC, VAU, VID **Elev.:** 10-700 m.a.s.l. **IUCN Habitat code:** 1,3,4,5,14 **Conservation status: (IUCN)** LC **(CPSG)** Potential LC

4196 *Couma rigida* Müll.Arg. (Apocynaceae)
IPNI: 78175-1
Use: HF, MA **Conservation status: (IUCN)** NE

4197 *Couma utilis* (Mart.) Müll.Arg. (Apocynaceae)
IPNI: 78177-1
Common name: juan soco, juansoco, juansoco de rebalse, pendare, pendarito, sorba, sorva, sódanie **Habit:** Tree **Use:** HF, MA, ME **Origin:** Native **Region:** Amazonia, Orinoquia **Dept.:** MET, VAU, VID **Elev.:** 100–110 m a.s.l. **IUCN Habitat code:** 1,3,4,5 **Conservation status: (IUCN)** NE

4198 *Cryptostegia grandiflora* Roxb. ex R.Br. (Apocynaceae)
IPNI: 95962-1
Common name: veinte de julio **Habit:** Subshrub, Climbing **Use:** EU, MA, ME, PO **Origin:** Naturalised **Region:** Andean, Caribbean, Pacific **Dept.:** ATL, CAL, CHO, COR, TOL, VAC **Elev.:** 350–4200 m a.s.l. **IUCN Habitat code:** 2,3,4,14 **Conservation status: (IUCN)** NE

4199 *Cryptostegia madagascariensis* Bojer ex Decne. (Apocynaceae)
IPNI: 95963-1
Habit: Shrub **Use:** EU, MA, ME, PO **Origin:** Naturalised **Region:** Amazonia, Caribbean, Pacific **Dept.:** AMA, BOL, CHO **Elev.:** 10–50 m a.s.l. **IUCN Habitat code:** 1,2,14 **Conservation status: (IUCN)** NE

4200 *Cynanchum montevidense* Spreng. (Apocynaceae)
IPNI: 74165-2
Habit: Climbing **Use:** ME **Origin:** Native **Region:** Caribbean, Orinoquia, Pacific **Dept.:** ANT, BOL, CHO, COR, MAG, MET **Elev.:** 15–120 m a.s.l. **Conservation status: (IUCN)** NE

4201 *Echites umbellatus* Jacq. (Apocynaceae)
IPNI: 78821-1
Habit: Liana, Climbing **Use:** ME **Origin:** Native **Region:** Caribbean **Dept.:** SAP **Elev.:** 0–10 m a.s.l. **Conservation status: (IUCN)** NE

4202 *Fischeria stellata* (Vell.) E.Fourn. (Apocynaceae)
IPNI: 97427-1
Habit: Climbing **Use:** ME **Origin:** Native **Region:** Amazonia, Andean, Orinoquia **Dept.:** AMA, BOY, CUN, GUV, MET, NSA, PUT, TOL **Elev.:** 100–1600 m a.s.l. **IUCN Habitat code:** 1,3,14 **Conservation status: (IUCN)** NE

4203 *Forsteronia graciloides* Woodson (Apocynaceae)
IPNI: 105364-2
Habit: Liana, Climbing **Use:** ME **Origin:** Native **Region:** Guiana Shield **Dept.:** ANT, BOY, MET **Elev.:** 340–900 m a.s.l. **IUCN Habitat code:** 1,3 **Conservation status: (IUCN)** NE

4204 *Funastrum clausum* (Jacq.) Schltr. (Apocynaceae)
IPNI: 106366-2
Common name: bejuco de leche, bejuco de pescado, bejuco de sapo, bejuco lechoso, bejuco sapo, corrimiento, flor de cera, tisuapo **Habit:** Herb, Climbing **Use:** EU, HF, MA, ME, SU **Origin:** Native **Region:** Andean, Guiana Shield, Caribbean, Orinoquia, Pacific **Dept.:** ANT, ARA, ATL, BOL, BOY, CAU, CHO, COR, CUN, HUI, LAG, MAG, MET, NAR, NSA, PUT, SUC, TOL, VAC **Elev.:** 0-2000 m.a.s.l. **IUCN Habitat code:** 1,2,3,5,14 **Conservation status: (IUCN)** NE **(CPSG)** Potential LC

4205 *Funastrum glaucum* (Kunth) Schltr. (Apocynaceae)
IPNI: 106382-2
Common name: bejuco alambre, bejuco de leche, bejuco del diablo, bejuco sapo, corrimiento, ipecacuana, iwana alu **Habit:** Liana, Climbing **Use:** ME **Origin:** Native **Region:** Caribbean, Orinoquia **Dept.:** ANT, ATL, BOL, CAS, COR, CUN, HUI, LAG, MAG, NSA, SAN, SUC, TOL **Elev.:** 0-450 m.a.s.l. **IUCN Habitat code:** 1,2,3,14 **Conservation status: (IUCN)** NE **(CPSG)** Potential LC

4206 *Gomphocarpus fruticosus* (L.) W.T.Aiton (Apocynaceae)
IPNI: 97610-1
Common name: bolsa de lobo, bolsa de pastor, esponjilla, globitos, globos, vejigo, vejigón **Habit:** Subshrub **Use:** AF, EU, GS, HF, MA, ME **Origin:** Naturalised **Region:** Andean **Dept.:** ANT, BOY, NSA **Elev.:** 1540–2100 m a.s.l. **IUCN Habitat code:** 1,3,14 **Conservation status: (IUCN)** NE

4207 *Gomphocarpus physocarpus* E.Mey. (Apocynaceae)
IPNI: 97670-1
Common name: bombas, globito, globitos **Habit:** Subshrub **Use:** EU, MA, ME, PO, SU **Origin:** Cultivated, Naturalised **Region:** Andean **Dept.:** ANT, NAR, VAC **Elev.:** 950–2180 m a.s.l. **IUCN Habitat code:** 1,3,14 **Conservation status: (IUCN)** NE

4208 *Himatanthus articulatus* (Vahl) Woodson (Apocynaceae)
IPNI: 122688-2
Common name: bachaco, caimito plátano, caimo de plátano, caimo plátano, cauchillo, caucho lechudo, perillo blanco, platanillo, platanote, plátano **Habit:** Shrub, Small tree, Tree **Use:** FU, HF, MA, ME **Origin:** Native **Region:** Amazonia, Orinoquia, Pacific **Dept.:** AMA, ANT, BOL, CAQ, CAS, CHO, MET, SAN, SUC, VAC, VAU, VID **Elev.:** 20–900 m a.s.l. **IUCN Habitat code:** 1,2,3,4,5,14 **Conservation status: (IUCN)** LC

4209 *Himatanthus bracteatus* (A.DC.) Woodson (Apocynaceae)
IPNI: 122690-2
Use: HF, ME **IUCN Habitat code:** 1,4,14 **Conservation status: (IUCN)** LC

4210 *Himatanthus phagedaenicus* (Mart.) Woodson (Apocynaceae)
IPNI: 122695-2
Habit: Shrub, Small tree, Tree **Use:** ME **Origin:** Native **Region:** Amazonia **Dept.:** VAU **Elev.:** 100–200 m a.s.l. **IUCN Habitat code:** 1,3,4 **Conservation status: (IUCN)** LC

4211 *Himatanthus tarapotensis* (K.Schum. ex Markgr.) Plumel (Apocynaceae)
IPNI: 962262-1

Use: ME **Origin:** Native **IUCN Habitat code:** 1,3,4,5 **Conservation status:** **(IUCN)** LC

4212 *Hoya carnosa* (L.f.) R.Br. (Apocynaceae) **IPNI:** 107308-3 **Common name:** cera, cerilla, enredadera cera, flor de cera **Habit:** Climbing **Use:** EU, ME **Origin:** Cultivated **Dept.:** BOG **IUCN Habitat code:** 3,14 **Conservation status: (IUCN)** NE

4213 *Ibatia maritima* (Jacq.) Decne. (Apocynaceae) **IPNI:** 98854-1 **Common name:** bejuco'eleche **Habit:** Climbing **Use:** ME **Origin:** Native **Region:** Andean, Caribbean **Dept.:** ATL, BOL, BOY, MAG, NSA **Elev.:** 10–2250 m a.s.l. **IUCN Habitat code:** 1,2,14 **Conservation status: (IUCN)** NE

4214 *Lacmellea aculeata* (Ducke) Monach. (Apocynaceae) **IPNI:** 134481-2 **Habit:** Tree **Use:** HF, ME **Origin:** Native **Region:** Amazonia **Dept.:** AMA **Elev.:** 120–120 m a.s.l. **IUCN Habitat code:** 1,4 **Conservation status: (IUCN)** LC

4215 *Lacmellea arborescens* (Müll.Arg.) Markgr. (Apocynaceae) **IPNI:** 79531-1 **Common name:** chicle, pegapega **Habit:** Shrub, Tree **Use:** HF **Origin:** Native **Region:** Amazonia **Dept.:** AMA, CAQ, PUT, VAU **Elev.:** 180–250 m a.s.l. **IUCN Habitat code:** 1,3,14 **Conservation status: (IUCN)** LC

4216 *Lacmellea edulis* H.Karst. (Apocynaceae) **IPNI:** 79551-1 **Common name:** agua miel, chicle, chicle con espinas, chicle espinoso, juansoco con espinas, juansoquillo, lechemiel, lechero, lechudo, palo espino, pegapega **Habit:** Shrub, Small tree, Tree **Use:** HF, ME **Origin:** Native **Region:** Amazonia, Guiana Shield, Caribbean, Orinoquia **Dept.:** AMA, ANT, BOL, CAQ, CAS, GUV, MET, PUT, VAU **Elev.:** 20–1400 m a.s.l. **IUCN Habitat code:** 1,3,4,5,14 **Conservation status: (IUCN)** LC

4217 *Lacmellea floribunda* (Poepp.) Benth. & Hook.f. (Apocynaceae) **IPNI:** 79552-1 **Common name:** chicle, lirio, tacheulo, tree vaca, árbol vaca **Habit:** Tree **Use:** HF, ME **Origin:** Native **Region:** Amazonia, Guiana Shield **Dept.:** GUV, MET **Elev.:** 5–120 m a.s.l. **IUCN Habitat code:** 1,3,4,5,14 **Conservation status: (IUCN)** NE

4218 *Lacmellea foxii* (Stapf) Markgr. (Apocynaceae) **IPNI:** 134491-2 **Habit:** Tree **Use:** HF **Origin:** Native **Region:** Amazonia, Orinoquia **Dept.:** AMA, CAQ, VID **Elev.:** 200–300 m a.s.l. **IUCN Habitat code:** 1,4,5,14 **Conservation status: (IUCN)** NE

4219 *Lacmellea gracilis* (Müll.Arg.) Markgr. (Apocynaceae) **IPNI:** 79536-1 **Habit:** Tree **Use:** HF **Origin:** Native **Region:** Amazonia **Dept.:** AMA **Elev.:** 200–300 m a.s.l. **IUCN Habitat code:** 1,14 **Conservation status: (IUCN)** NE

4220 *Lacmellea oblongata* Markgr. (Apocynaceae) **IPNI:** 134498-2 **Habit:** Shrub, Small tree, Tree **Use:** HF **Origin:** Native **Region:** Amazonia **Dept.:** AMA **Elev.:** 110–300 m a.s.l. **IUCN Habitat code:** 1 **Conservation status: (IUCN)** LC

4221 *Lacmellea panamensis* (Woodson) Markgr. (Apocynaceae) **IPNI:** 134499-2 **Common name:** barbasco, barbasco lechoso, bigarro, escupidijo, lirio, perillo, tupirijo **Habit:** Small tree, Tree **Use:** HF, ME **Origin:** Native **Region:** Caribbean, Pacific **Dept.:** ANT, BOL, CHO, COR, NAR **Elev.:** 0–700 m a.s.l. **IUCN Habitat code:** 1,2,3,5,14 **Conservation status: (IUCN)** LC

4222 *Lacmellea ramosissima* (Müll.Arg.) Markgr. (Apocynaceae) **IPNI:** 79547-1 **Common name:** chicle, lechemiel, parajuansoco, pegapega, pendarito **Habit:** Shrub, Small tree, Tree **Use:** HF **Origin:** Native **Region:** Amazonia, Orinoquia **Dept.:** AMA, CAQ, VAU, VID **Elev.:** 90–810 m a.s.l. **IUCN Habitat code:** 1,3,4,5,14 **Conservation status: (IUCN)** LC

4223 *Lacmellea speciosa* Woodson (Apocynaceae) **IPNI:** 134506-2 **Common name:** caimito, caimito Pumarejo, perillo, popa, pumarejo, árbol vaca **Habit:** Tree **Use:** HF **Origin:** Native **Region:** Pacific **Dept.:** ANT, CHO, NAR, VAC **Elev.:** 5–880 m a.s.l. **IUCN Habitat code:** 1,14 **Conservation status: (IUCN)** LC

4224 *Macoubea guianensis* Aubl. (Apocynaceae) **IPNI:** 79852-1 **Common name:** carne de vaca, cocuy, costillero de sangre, cucuy, cucuy de monte, cucuye, güeva de toro, mamón, maucuari, palo morrocoy, pendare, pucarón, pukarón, turmas de toro, ucuye **Habit:** Tree **Use:** HF, MA, ME **Origin:** Native **Region:** Amazonia, Guiana Shield, Orinoquia **Dept.:** AMA, CAQ, GUA, PUT, VAU, VID **Elev.:** 80–300 m a.s.l. **IUCN Habitat code:** 1,3,4,14 **Conservation status: (IUCN)** LC

4225 *Macoubea sprucei* (Müll.Arg.) Markgr. (Apocynaceae) **IPNI:** 148754-2 **Common name:** costillo propio, cucuy **Habit:** Small tree, Tree **Use:** HF, ME **Origin:** Native **Region:** Amazonia, Guiana Shield **Dept.:** AMA, CAQ, VAU **Elev.:** 150–350 m a.s.l. **IUCN Habitat code:** 1 **Conservation status: (IUCN)** LC

4226 *Malouetia flavescens* (Willd. ex Roem. & Schult.) Müll.Arg. (Apocynaceae) **IPNI:** 79896-1 **Habit:** Shrub **Use:** HF, ME, PO **Origin:** Native **Region:** Amazonia **Dept.:** GUV, MET, VAU **Elev.:** 230–230 m a.s.l. **IUCN Habitat code:** 1,4,5 **Conservation status: (IUCN)** NE

4227 *Malouetia glandulifera* Miers (Apocynaceae) **IPNI:** 79898-1 **Common name:** palo de boya blanco **Habit:** Tree **Use:** ME **Origin:** Native **Region:** Amazonia, Guiana Shield **Dept.:** GUA **Elev.:** 100–130 m a.s.l. **IUCN Habitat code:** 1,4,5 **Conservation status: (IUCN)** NE

4228 *Malouetia naias* M.E.Endress (Apocynaceae) **IPNI:** 280308-2 **Common name:** cuchara, cucharacaspi, cullera, juansoco, molonga, molongo, palo de boa, palo de cuchara, siringa pegajosa **Habit:** Shrub, Tree **Use:** ME **Origin:** Native **Region:** Amazonia, Orinoquia **Dept.:** CAQ, CAS, GUV, VAU **Elev.:** 90–300 m a.s.l. **IUCN Habitat code:** 1,4,5,14 **Conservation status: (IUCN)** NE

4229 *Malouetia nitida* Spruce ex Müll.Arg. (Apocynaceae) **IPNI:** 79913-1 **Common name:** can, can-cay, cay, guachamaca, guachamacay, guachamacá, guachamacán **Use:** ME, PO **IUCN Habitat code:** 1,4 **Conservation status: (IUCN)** NE

4230 *Malouetia tamaquarina* (Aubl.) A.DC. (Apocynaceae) **IPNI:** 79926-1 **Common name:** cuchara caspi, cullera, guachamaca, molonga **Habit:** Tree **Use:** HF, ME **Origin:** Native **Region:** Amazonia, Orinoquia **Dept.:** AMA, VAU, VID **Elev.:** 90–125 m a.s.l. **IUCN Habitat code:** 1,3,4,5,14 **Conservation status: (IUCN)** LC

4231 *Mandevilla annulariifolia* Woodson (Apocynaceae) **IPNI:** 151923-2 **Habit:** Liana, Climbing **Use:** ME **Origin:** Native **Region:** Amazonia, Guiana Shield **Dept.:** AMA, GUA, VAU **Elev.:** 180–900 m a.s.l. **IUCN Habitat code:** 1,5 **Conservation status: (IUCN)** NE

4232 *Mandevilla boliviensis* (J.J.Veitch) Woodson (Apocynaceae) **IPNI:** 151935-2 **Habit:** Climbing **Use:** EU **Origin:** Native **Region:** Caribbean **Dept.:** LAG **Elev.:** 650–650 m a.s.l. **IUCN Habitat code:** 1,3 **Conservation status: (IUCN)** NE

4233 *Mandevilla hirsuta* (Rich.) K.Schum. (Apocynaceae) **IPNI:** 151979-2 **Habit:** Liana, Climbing **Use:** ME **Origin:** Native **Region:** Amazonia, Andean, Pacific **Dept.:** AMA, ANT, BOY, CAL, CAQ, CAS, CHO, COR, GUV, LAG, MET, NAR, NSA, PUT, SAN, VAC, VAU, VID **Elev.:** 5-2000 m.a.s.l. **IUCN Habitat code:** 1,3,4,5,14 **Conservation status: (IUCN)** NE **(CPSG)** Potential LC

4234 *Mandevilla javitensis* (Kunth) K.Schum. (Apocynaceae) **IPNI:** 151987-2 **Habit:** Subshrub, Climbing **Use:** ME **Origin:** Native **Region:** Amazonia **Dept.:** AMA, GUA, VAU **Elev.:** 120–400 m a.s.l. **IUCN Habitat code:** 1 **Conservation status: (IUCN)** NE

4235 *Mandevilla mollissima* (Kunth) K.Schum. (Apocynaceae) **IPNI:** 152016-2 **Common name:** bejuco amarillo, kulupaneekono **Habit:** Climbing **Use:** SU **Origin:** Endemic **Region:** Andean **Dept.:** ANT, BOY, CAL, CAU, CUN, HUI, MAG, NAR, PUT, SAN, TOL, VAC **Elev.:** 880-2000 m.a.s.l. **IUCN Habitat code:** 1,2,3,14 **Conservation status: (IUCN)** NE **(CPSG)** Potential LC

4236 *Mandevilla sanderi* (Hemsl.) Woodson (Apocynaceae) **IPNI:** 152049-2 **Common name:** diplodemia, mandevila **Use:** EU **Origin:** Cultivated **Region:** Andean **Dept.:** CUN **Elev.:** 2600–2600 m a.s.l. **Conservation status: (IUCN)** NE

4237 *Mandevilla scabra* (Hoffmanns. ex Roem. & Schult.) K.Schum. (Apocynaceae) **IPNI:** 152052-2

Habit: Liana **Use:** ME **Origin:** Native **Region:** Amazonia, Guiana Shield, Orinoquia **Dept.:** AMA, ARA, CAQ, GUA, GUV, PUT, VAU, VID **Elev.:** 150–760 m a.s.l. **IUCN Habitat code:** 1,2,3,4,5,14 **Conservation status: (IUCN)** NE

4238 *Mandevilla subcarnosa* (Benth.) Benth. & Hook.f. ex B.D.Jacks. (Apocynaceae) **IPNI:** 80081-1 **Habit:** Shrub, Subshrub, Liana **Use:** ME **Origin:** Native **Region:** Guiana Shield **Dept.:** CAQ **Elev.:** 400–650 m a.s.l. **IUCN Habitat code:** 1,14 **Conservation status: (IUCN)** LC

4239 *Mandevilla subsagittata* (Ruiz & Pav.) Woodson (Apocynaceae) **IPNI:** 152067-2 **Habit:** Climbing **Use:** ME **Origin:** Native **Region:** Amazonia, Andean, Caribbean, Orinoquia **Dept.:** ANT, BOG, CAL, CES, CUN, GUV, HUI, LAG, MAG, MET, NAR, NSA, PUT, QUI, RIS, SAN, TOL, VAC, VAU **Elev.:** 76-1850 m.a.s.l. **IUCN Habitat code:** 1,2,3,5,14 **Conservation status: (IUCN)** NE **(CPSG)** Potential LC

4240 *Marsdenia cundurango* Rchb.f. (Apocynaceae) **IPNI:** 99126-1 **Common name:** bejuco de sapo, belero, condurango, cundurango, matagente, orosús, orozús, velero **Habit:** Climbing **Use:** MA, ME **Origin:** Native **Region:** Andean **Dept.:** CUN, TOL **Elev.:** 900–1200 m a.s.l. **IUCN Habitat code:** 1,3,14 **Conservation status: (IUCN)** NE

4241 *Matelea obliqua* (Jacq.) Woodson (Apocynaceae) **IPNI:** 154570-2 **Use:** ME **Conservation status: (IUCN)** NE

4242 *Mesechites trifidus* (Jacq.) Müll.Arg. (Apocynaceae) **IPNI:** 30132318-2 **Common name:** bejuco mechocuán, guachamacá, mataviri **Habit:** Liana, Climbing **Use:** ME **Origin:** Native **Region:** Amazonia, Andean, Caribbean, Orinoquia, Pacific **Dept.:** AMA, ANT, ARA, ATL, BOL, BOY, CAQ, CHO, CUN, GUV, HUI, LAG, MAG, MET, PUT, QUI, RIS, SAN, TOL, VAC, VAU, VID **Elev.:** 0-1700 m.a.s.l. **IUCN Habitat code:** 1,2,3,4,5,14 **Conservation status: (IUCN)** NE **(CPSG)** Potential LC

4243 *Molongum lucidum* (Kunth) Zarucchi (Apocynaceae) **IPNI:** 945268-1 **Common name:** juansoquillo, mi-mi-wa-jo-ka-ki **Habit:** Shrub, Tree **Use:** HF **Origin:** Native **Region:** Amazonia, Guiana Shield **Dept.:** AMA, CAQ, GUV, VAU **Elev.:** 160–400 m a.s.l. **IUCN Habitat code:** 1,3,4,14 **Conservation status: (IUCN)** NE

4244 *Mucoa duckei* (Markgr.) Zarucchi (Apocynaceae) **IPNI:** 281069-2 **Common name:** caimo morado, cascotillo, falso juansoco, guayaba de danta, monachino **Habit:** Tree **Use:** HF, ME **Origin:** Native **Region:** Amazonia, Guiana Shield **Dept.:** CAQ, VAU **Elev.:** 200–300 m a.s.l. **IUCN Habitat code:** 1,3,4,5 **Conservation status: (IUCN)** NE

4245 *Neocouma ternstroemiacea* (Müll.Arg.) Pierre (Apocynaceae) **IPNI:** 168521-2 **Habit:** Tree **Use:** HF **Origin:** Native **Region:** Amazonia, Guiana Shield **Dept.:** AMA, CAQ **Elev.:** 250–300 m a.s.l. **IUCN Habitat code:** 1,5 **Conservation status: (IUCN)** NE

4246 *Nerium oleander* L. (Apocynaceae) **IPNI:** 80460-1 **Common name:** adelfa, adelfo, alejandría, amaranto, andelfo, azucena de La Habana, azuceno, azuceno de La Habana, berbería, clavel habano, delfo, flor de La Habana, habana, habano, laurel, laurel de Judea, laurel rosa, rosa de La Habana, yolanda, yucatán, ébano **Habit:** Shrub, Subshrub **Use:** EU, HF, MA, ME, PO **Origin:** Cultivated **Region:** Andean, Caribbean **Dept.:** ANT, ATL, BOG, CAU, CUN, MAG, QUI, SAP, VAC **Elev.:** 0–1760 m a.s.l. **IUCN Habitat code:** 1,2,3,4,14 **Conservation status: (IUCN)** LC

4247 *Odontadenia nitida* (Vahl) Müll.Arg. (Apocynaceae) **IPNI:** 1054185-2 **Habit:** Liana, Climbing **Use:** HF, ME **Origin:** Native **Region:** Amazonia, Guiana Shield, Orinoquia **Dept.:** AMA, CAQ, GUV, MET **Elev.:** 150–340 m a.s.l. **IUCN Habitat code:** 1,3,4,14 **Conservation status: (IUCN)** NE

4248 *Odontadenia puncticulosa* (Rich.) Pulle (Apocynaceae) **IPNI:** 291380-2 **Habit:** Liana, Climbing **Use:** ME **Origin:** Native **Region:** Amazonia, Andean, Pacific **Dept.:** AMA, ANT, CAQ, CHO, CUN, PUT, SAN, VAC **Elev.:** 5–1335 m a.s.l. **IUCN Habitat code:** 1,3,4,5,14 **Conservation status: (IUCN)** NE

4249 *Odontadenia semidigyna* (P.J.Bergius) J.F.Morales (Apocynaceae) **IPNI:** 77191232-1 **Common name:** paloma, palomilla **Habit:** Liana, Climbing **Use:** EU, MA, ME **Origin:** Native **Region:** Amazonia, Pacific **Dept.:** AMA, ANT, BOL, CHO, GUA, GUV, MET, SAN, VAC **Elev.:** 0–700 m a.s.l. **IUCN Habitat code:** 1,3,5,14 **Conservation status: (IUCN)** NE

4250 *Pacouria guianensis* Aubl. (Apocynaceae) **IPNI:** 80734-1 **Habit:** Climbing **Use:** HF **Origin:** Native **Region:** Guiana Shield **Dept.:** MET, VAU **Elev.:** 350–400 m a.s.l. **IUCN Habitat code:** 1,14 **Conservation status: (IUCN)** NE

4251 *Parahancornia fasciculata* (Poir.) Benoist (Apocynaceae) **IPNI:** 182946-2 **Common name:** amapá, naranjo podrido, tree de vaca **Use:** HF, MA, ME **Origin:** Native **IUCN Habitat code:** 1 **Conservation status: (IUCN)** LC

4252 *Parahancornia krukovii* Monach. (Apocynaceae) **IPNI:** 182947-2 **Common name:** rajan **Use:** HF, MA, ME **IUCN Habitat code:** 1,2,3,4,5,14 **Conservation status: (IUCN)** NE

4253 *Parahancornia oblonga* (Benth. ex Müll.Arg.) Monach. (Apocynaceae) **IPNI:** 182949-2 **Common name:** falso juansoco, juansoquillo, monachino, pendare, salivón **Habit:** Tree **Use:** HF **Origin:** Native **Region:** Amazonia, Guiana Shield, Orinoquia **Dept.:** AMA, CAQ, GUV, VID **Elev.:** 90–180 m a.s.l. **Conservation status: (IUCN)** NE

4254 *Parahancornia peruviana* Monach. (Apocynaceae) **IPNI:** 182950-2 **Common name:** juansoquillo **Habit:** Tree **Use:** HF **Origin:** Native **Region:** Amazonia **Dept.:** AMA **Elev.:** 150–200 m a.s.l. **IUCN Habitat code:** 1 **Conservation status: (IUCN)** NE

4255 *Parahancornia surrogata* Zarucchi (Apocynaceae) **IPNI:** 958560-1 **Common name:** juansoco, juansoco de la sabana, juansoquillo, pendare, pendarillo **Habit:** Shrub, Small tree, Tree **Use:** HF **Origin:** Native **Region:** Amazonia, Guiana Shield **Dept.:** AMA, CAQ, GUA, GUV, VAU **Elev.:** 120–635 m a.s.l. **IUCN Habitat code:** 1,3,4,5,14 **Conservation status: (IUCN)** LC

4256 *Plumeria alba* L. (Apocynaceae) **IPNI:** 81171-1 **Use:** EU, HF, MA, ME, PO **IUCN Habitat code:** 1,2,3,14 **Conservation status: (IUCN)** NE

4257 *Plumeria inodora* Jacq. (Apocynaceae) **IPNI:** 81220-1 **Common name:** alelí, amancayo, amaranto, azucena, azuceno, azuceno amarillo, azuceno blanco, estoraque, flor de la playa, flor de playa, floral, florón, frangipán, laurel de Dafne, lirio, mancayo, orejones **Habit:** Shrub, Tree **Use:** ME **Origin:** Native **Region:** Orinoquia **Dept.:** VID **Elev.:** 100–100 m a.s.l. **IUCN Habitat code:** 1,2,4,5 **Conservation status: (IUCN)** NE

4258 *Plumeria pudica* Jacq. (Apocynaceae) **IPNI:** 81268-1 **Common name:** azuceno, floral, florón **Habit:** Tree **Use:** MA **Origin:** Native **Region:** Caribbean **Dept.:** ATL, LAG, MAG **Elev.:** 150–250 m a.s.l. **IUCN Habitat code:** 1,2,3,4,5,14 **Conservation status: (IUCN)** LC

4259 *Plumeria rubra* L. (Apocynaceae) **IPNI:** 81275-1 **Common name:** amancayo, azuceno, azuceno colorado, azuceno de monte, azuceno rojo, flor de mayo, florón, frangipán, zapotillo **Habit:** Shrub, Small tree, Tree **Use:** AF, EU, GS, HF, MA, ME, PO, SU **Origin:** Native **Region:** Andean, Caribbean, Pacific **Dept.:** ANT, ATL, BOL, CES, CHO, LAG, SAP, SUC, VAC **Elev.:** 0–1500 m a.s.l. **IUCN Habitat code:** 1,2,3,5,14 **Conservation status: (IUCN)** LC

4260 *Prestonia mollis* Kunth (Apocynaceae) **IPNI:** 81387-1 **Use:** ME **Origin:** Native **Conservation status: (IUCN)** NE

4261 *Prestonia quinquangularis* (Jacq.) Spreng. (Apocynaceae) **IPNI:** 81405-1 **Habit:** Liana, Climbing **Use:** HF **Origin:** Native **Region:** Amazonia, Caribbean, Orinoquia, Pacific **Dept.:** AMA, ANT, ARA, ATL, BOL, COR, CUN, GUV, LAG, MET, QUI, RIS, SAN, TOL, VAC **Elev.:** 30-975 m.a.s.l. **IUCN Habitat code:** 1,2,3,5,14 **Conservation status: (IUCN)** NE **(CPSG)** Potential LC

4262 *Prestonia trifida* (Poepp.) Woodson (Apocynaceae) **IPNI:** 210147-2 **Habit:** Liana, Climbing **Use:** ME **Origin:** Native **Region:** Caribbean,

Pacific **Dept.:** ANT, BOL, CHO, COR, CUN, SAN **Elev.:** 0–790 m a.s.l. **IUCN Habitat code:** 1,2,3,5,14 **Conservation status: (IUCN)** NE

4263 *Rauvolfia leptophylla* A.S.Rao (Apocynaceae) **IPNI:** 217918-2 **Habit:** Shrub, Tree **Use:** ME **Origin:** Native **Region:** Andean, Caribbean, Pacific **Dept.:** ANT, BOY, CHO, LAG, MAG, NAR, PUT, RIS, TOL, VAC **Elev.:** 0–2285 m a.s.l. **IUCN Habitat code:** 1,2,3,5,14 **Conservation status: (IUCN)** NE

4264 *Rauvolfia ligustrina* Roem. & Schult. (Apocynaceae) **IPNI:** 81565-1 **Common name:** piñique piñique, piñiquepiñique, plan tana, solito, venenito **Habit:** Herb, Shrub **Use:** AF,ME, PO **Origin:** Native **Region:** Andean, Caribbean **Dept.:** ATL, BOL, COR, MAG, SAP, SUC, VAC **Elev.:** 10–865 m a.s.l. **IUCN Habitat code:** 1,2,3,14 **Conservation status: (IUCN)** LC

4265 *Rauvolfia littoralis* Rusby (Apocynaceae) **IPNI:** 217921-2 **Common name:** anguito, cruceto, solita **Habit:** Shrub, Small tree, Tree **Use:** ME **Origin:** Native **Region:** Caribbean, Pacific **Dept.:** ANT, ATL, BOL, CHO, MAG, SUC **Elev.:** 0–200 m a.s.l. **IUCN Habitat code:** 1,2,3,4,5,14 **Conservation status: (IUCN)** LC

4266 *Rauvolfia pachyphylla* Markgr. (Apocynaceae) **IPNI:** 217932-2 **Use:** ME **IUCN Habitat code:** 3 **Conservation status: (IUCN)** NE

4267 *Rauvolfia paraensis* Ducke (Apocynaceae) **IPNI:** 217933-2 **Habit:** Tree **Use:** HF **Origin:** Native **Region:** Amazonia **Dept.:** VAU **Conservation status: (IUCN)** NE

4268 *Rauvolfia tetraphylla* L. (Apocynaceae) **IPNI:** 81663-1 **Common name:** anguito, cerecillo, cruceto, fruta de pava, mirto, papa de culebra, pepa de culebra, piñique, piñiqui, solita, venenito **Habit:** Herb, Shrub, Small tree, Tree **Use:** MA, ME **Origin:** Native **Region:** Andean, Caribbean, Pacific **Dept.:** ATL, BOL, CAL, CES, CHO, COR, CUN, HUI, LAG, MAG, SAN, SUC, TOL, VAC **Elev.:** 3-1200 m.a.s.l. **IUCN Habitat code:** 1,2,3,4,5,6,14 **Conservation status: (IUCN)** NE **(CPSG)** Potential LC

4269 *Rauvolfia viridis* Roem. & Schult. (Apocynaceae) **IPNI:** 81670-1 **Common name:** cruceto, mata puerco, matapuerco, solita, solito, venenito **Habit:** Shrub **Use:** ME **Origin:** Native **Region:** Caribbean **Dept.:** ATL, BOL, MAG, SAP, SUC **Elev.:** 5–70 m a.s.l. **IUCN Habitat code:** 1,2,3,14 **Conservation status: (IUCN)** NE

4270 *Rhabdadenia biflora* (Jacq.) Müll.Arg. (Apocynaceae) **IPNI:** 81684-1 **Common name:** carrizo, lecho, lecho de mar, lechoso **Habit:** Liana, Climbing **Use:** ME **Origin:** Native **Region:** Caribbean, Pacific **Dept.:** ANT, ATL, CHO, MAG, NAR, SAP, SAN, SUC, VAC **Elev.:** 0–20 m a.s.l. **IUCN Habitat code:** 1,2,4,5,14 **Conservation status: (IUCN)** NE

4271 *Rhigospira quadrangularis* (Müll.Arg.) Miers (Apocynaceae) **IPNI:** 81716-1 **Common name:** caimo morado, cumarana, guayabo de danta, juansoco de alacrán, juansoco de danta, juansoquillo, mi-mi-wa-jo-ka-ki **Habit:** Tree **Use:** HF, MA **Origin:** Native **Region:** Amazonia **Dept.:** AMA, CHO, PUT, VAC, VAU **Elev.:** 5–300 m a.s.l. **IUCN Habitat code:** 1,4,5,14 **Conservation status: (IUCN)** LC

4272 *Ruehssia altissima* (Jacq.) F.Esp.Santo & Rapini (Apocynaceae) **IPNI:** 60478412-2 **Habit:** Liana, Climbing **Use:** MA, ME **Origin:** Native **Region:** Andean, Caribbean **Dept.:** ATL, BOL, CES, CUN, MAG **Elev.:** 0–1700 m a.s.l. **IUCN Habitat code:** 1,2,14 **Conservation status: (IUCN)** NE

4273 *Ruehssia macrophylla* (Humb. & Bonpl. ex Schult.) H.Karst. (Apocynaceae) **IPNI:** 100911-1 **Common name:** barbascos, bejuco de cáncer, palomitas **Habit:** Liana, Climbing **Use:** HF, ME, SU **Origin:** Native **Region:** Andean, Caribbean, Orinoquia **Dept.:** ANT, ATL, BOL, BOY, CAU, CUN, MAG, MET, NSA, SAP, TOL **Elev.:** 15–1600 m a.s.l. **IUCN Habitat code:** 1,2,3,5,14 **Conservation status: (IUCN)** NE

4274 *Spongiosperma macrophyllum* (Müll.Arg.) Zarucchi (Apocynaceae) **IPNI:** 945265-1 **Common name:** palo bálsamo **Habit:** Tree **Use:** ME **Origin:** Native **Region:** Amazonia, Guiana Shield **Dept.:** CAQ, VAU **Elev.:** 198–198 m a.s.l. **IUCN Habitat code:** 1,3,4,14 **Conservation status: (IUCN)** NE

4275 *Tabernaemontana alba* Mill. (Apocynaceae) **IPNI:** 81985-1 **Habit:** Tree **Use:** ME **Origin:** Native **Region:** Pacific **Dept.:** ANT **Elev.:** 30–80 m a.s.l. **IUCN Habitat code:** 1,3 **Conservation status: (IUCN)** LC

4276 *Tabernaemontana amygdalifolia* Jacq. (Apocynaceae) **IPNI:** 81996-1 **Common name:** cojón de fraile, cojón de gato, frailecillo, huevo de perro, jazmín de monte, kash kol, lirio macho **Habit:** Shrub, Tree **Use:** HF, ME **Origin:** Native **Region:** Andean, Caribbean, Pacific **Dept.:** ANT, ATL, BOL, CAL, CAU, CHO, CUN, LAG, MAG, NAR, SAN, SUC, TOL, VAC **Elev.:** 0-2600 m.a.s.l. **IUCN Habitat code:** 1,2,3,4,5,14 **Conservation status: (IUCN)** LC **(CPSG)** Potential LC

4277 *Tabernaemontana arborea* Rose (Apocynaceae) **IPNI:** 248839-2 **Common name:** mangle colorado **Habit:** Tree **Use:** MA **Origin:** Native **Region:** Pacific **Dept.:** ANT, CHO, VAC **Elev.:** 20–500 m a.s.l. **IUCN Habitat code:** 1,2,3,5,14 **Conservation status: (IUCN)** LC

4278 *Tabernaemontana columbiensis* (L.Allorge) Leeuwenb. (Apocynaceae) **IPNI:** 285800-2 **Habit:** Shrub, Tree **Use:** HF **Origin:** Native **Region:** Pacific **Dept.:** CAU, NAR, PUT, VAC **Elev.:** 0–75 m a.s.l. **IUCN Habitat code:** 1,3 **Conservation status: (IUCN)** LC

4279 *Tabernaemontana cymosa* Jacq. (Apocynaceae) **IPNI:** 82084-1 **Common name:** cojón de burro, cojón de cabrito, cojón de fraile, cojón de verraco, frailecillo, guacharaco, güevo de varraco, huevo de perro, huevo de varraco, huevo de verraco, huevos de verraco, jazmín de montaña, lechero, palo verraco, turma de perro, verraco **Habit:** Shrub, Small tree, Tree **Use:** ME **Origin:** Native **Region:** Caribbean, Pacific **Dept.:** ANT, ATL, BOL, CES, COR, LAG, MAG, NAR, SUC, VAC **Elev.:** 4–400 m a.s.l. **IUCN Habitat code:** 1,2,5,14 **Conservation status: (IUCN)** LC

4280 *Tabernaemontana divaricata* (L.) R.Br. ex Roem. & Schult. (Apocynaceae) **IPNI:** 82097-1 **Common name:** adelicia, azuceno, buquecito, huevo de gato, jazmín, jazmín de Malabar, malabar, malabarí, sanango, sanjosé, vengueverraco, venjués **Habit:** Shrub, Subshrub **Use:** EU, FU, HF, MA, ME, PO **Origin:** Cultivated **Region:** Amazonia, Caribbean, Pacific **Dept.:** ANT, ATL, CHO, CUN, HUI, SAP, SAN, TOL, VAC, VAU **Elev.:** 100–900 m a.s.l. **IUCN Habitat code:** 1,2,3,5,14 **Conservation status: (IUCN)** LC

4281 *Tabernaemontana grandiflora* Jacq. (Apocynaceae) **IPNI:** 82139-1 **Common name:** cofrón, cojoncito, cojón de fraile, cojón de puerco, hueva de verraco, jazmín malabar, lechoso, sapo **Habit:** Shrub, Small tree, Tree **Use:** MA, ME **Origin:** Native **Region:** Amazonia, Andean, Guiana Shield, Caribbean, Orinoquia, Pacific **Dept.:** AMA, ANT, ATL, BOL, BOY, CAL, CAS, CHO, COR, CUN, GUV, HUI, LAG, MAG, MET, SAN, SUC, TOL, VAC **Elev.:** 0-1500 m.a.s.l. **IUCN Habitat code:** 1,2,3,4,5,14 **Conservation status: (IUCN)** LC **(CPSG)** Potential LC

4282 *Tabernaemontana heterophylla* Vahl (Apocynaceae) **IPNI:** 82146-1 **Common name:** barbasco, barbascos, cojón, güeva peluda, lechoso **Habit:** Shrub, Small tree, Tree **Use:** HF, ME **Origin:** Native **Region:** Amazonia, Andean, Guiana Shield, Pacific **Dept.:** AMA, ANT, CAU, CHO, COR, CUN, GUV, MET, NSA, PUT, SAN, VAC, VAU **Elev.:** 50-1650 m.a.s.l. **IUCN Habitat code:** 1,3,4,5,14 **Conservation status: (IUCN)** LC **(CPSG)** Potential LC

4283 *Tabernaemontana litoralis* Kunth (Apocynaceae) **IPNI:** 285810-2 **Common name:** cojón de gato, huevos de perro, jazmín, jazmín de Calamar, jazmín de Malabar **Habit:** Shrub, Small tree **Use:** EU **Origin:** Native **Region:** Andean **Dept.:** ANT, QUI, VAC **Elev.:** 900–1500 m a.s.l. **IUCN Habitat code:** 1,3,14 **Conservation status: (IUCN)** LC

4284 *Tabernaemontana longipes* Donn.Sm. (Apocynaceae) **IPNI:** 248876-2 **Habit:** Shrub, Small tree, Tree **Use:** HF **Origin:** Native **Region:** Andean, Pacific **Dept.:** ANT, CAQ, CHO, SAN, VAC **Elev.:** 10–1800 m a.s.l. **IUCN Habitat code:** 1,3,5,14 **Conservation status: (IUCN)** LC

4285 *Tabernaemontana macrocalyx* Müll.Arg. (Apocynaceae) **IPNI:** 285812-2 **Common name:** caimo, pintura de zorra **Habit:** Shrub **Use:** HF **Origin:** Native **Region:** Amazonia, Guiana Shield **Dept.:** CAQ, GUA, PUT, VAU **Elev.:** 250–400 m a.s.l. **IUCN Habitat code:** 1,3,4,5,14 **Conservation status: (IUCN)** NE

4286 *Tabernaemontana markgrafiana* J.F.Macbr. (Apocynaceae) **IPNI:** 288916-2 **Common name:** huevo de berraco, jaen sananho, lagarto micunan, lechito, penoncahue, piangüita, sanangillo, sananho, uchu sananho **Habit:** Shrub, Small tree, Tree **Use:** HF, ME **Origin:** Native **Region:** Amazonia, Pacific **Dept.:** AMA, ANT, CAL, CHO, CUN, NAR, SAN, VAC **Elev.:** 20–900 m a.s.l. **IUCN Habitat code:** 1,3,5,14 **Conservation status: (IUCN)** LC

4287 *Tabernaemontana maxima* Markgr. (Apocynaceae) **IPNI:** 82212-1 **Habit:** Shrub, Small tree **Use:** HF **Origin:** Native **Region:** Amazonia, Pacific **Dept.:** ANT, SAN, VAC, VAU **Elev.:** 50–200 m a.s.l. **IUCN Habitat code:** 1 **Conservation status: (IUCN)** NE

4288 *Tabernaemontana palustris* Markgr. (Apocynaceae) **IPNI:** 248891-2 **Habit:** Shrub **Use:** ME **Origin:** Native **Region:** Amazonia, Guiana Shield **Dept.:** GUA, VAU **Elev.:** 95–183 m a.s.l. **IUCN Habitat code:** 1,4 **Conservation status: (IUCN)** NE

4289 *Tabernaemontana panamensis* (Markgr., Boiteau & L.Allorge) Leeuwenb. (Apocynaceae) **IPNI:** 248892-2 **Common name:** huevo de mono **Habit:** Shrub, Tree **Use:** HF **Origin:** Native **Region:** Andean, Pacific **Dept.:** ANT, CHO, NAR, VAC **Elev.:** 10–1200 m a.s.l. **IUCN Habitat code:** 1,5,14 **Conservation status: (IUCN)** LC

4290 *Tabernaemontana robinsonii* (Woodson) A.O.Simões & M.E.Endress (Apocynaceae) **IPNI:** 77105798-1 **Habit:** Shrub, Tree **Use:** ME **Origin:** Native **Region:** Caribbean, Pacific **Dept.:** CAU, CHO, LAG, VAC **Elev.:** 10–550 m a.s.l. **IUCN Habitat code:** 1,5,14 **Conservation status: (IUCN)** LC

4291 *Tabernaemontana sananho* Ruiz & Pav. (Apocynaceae) **IPNI:** 82327-1 **Common name:** barbasco, guachamaca de tree, guacharaco, jaen sananho, lagarto micunam, lechimullo, penoncahue, sanangillo, sanango, sananho, turmeperro, uchu sananho **Habit:** Shrub, Small tree, Tree **Use:** HF, ME **Origin:** Native **Region:** Amazonia, Andean, Guiana Shield, Caribbean **Dept.:** AMA, ANT, BOY, CAQ, GUV, MET, PUT, SUC **Elev.:** 100–1500 m a.s.l. **IUCN Habitat code:** 1,2,3,5,14 **Conservation status: (IUCN)** NE

4292 *Tabernaemontana siphilitica* (L.f.) Leeuwenb. (Apocynaceae) **IPNI:** 20001045-1 **Common name:** azúcar, cajón del amo, cojón de amo, gapuy, lechoso, sacha-sananjo, sanango, sanango de altura **Habit:** Shrub, Small tree **Use:** HF, ME **Origin:** Native **Region:** Amazonia, Caribbean, Orinoquia, Pacific **Dept.:** AMA, ARA, BOL, BOY, CAS, CHO, GUV, MET, PUT, SAN, VAU **Elev.:** 20–450 m a.s.l. **IUCN Habitat code:** 1,2,3,4,5,14 **Conservation status: (IUCN)** NE

4293 *Tabernaemontana undulata* Vahl (Apocynaceae) **IPNI:** 82371-1 **Habit:** Shrub **Use:** HF, MA, ME **Origin:** Native **Region:** Amazonia, Pacific **Dept.:** AMA, CHO, CUN, GUA, GUV, PUT, VAU **Elev.:** 195–450 m a.s.l. **IUCN Habitat code:** 1,3,14 **Conservation status: (IUCN)** LC

4294 *Tabernaemontana vanheurckii* Müll.Arg. (Apocynaceae) **IPNI:** 285821-2 **Habit:** Shrub, Tree **Use:** HF, ME **Origin:** Native **Region:** Pacific **Dept.:** VAC **Elev.:** 50–100 m a.s.l. **IUCN Habitat code:** 1,3,14 **Conservation status: (IUCN)** LC

4295 *Thevetia ahouai* (L.) A.DC. (Apocynaceae) **IPNI:** 82440-1 **Common name:** achulunga, bola de gato, bola de perro, bola de toro, cobalonga, covalonga, güeva de gato, habano, huevas de Adán, huevita de gato, huevito de gato, huevo de tigre, tomate de monte, tomate de árbol, tomate del diablo, tomatico de monte, tomatillo, tomatón **Habit:** Shrub, Subshrub, Small tree, Tree **Use:** HF, ME **Origin:** Native **Region:** Amazonia, Andean, Caribbean, Pacific **Dept.:** ANT, BOL, CHO, COR, GUV, LAG, MAG, SAN, SUC, VAC, VAU **Elev.:** 5–1320 m a.s.l. **IUCN Habitat code:** 1,2,3,4,5,14 **Conservation status: (IUCN)** LC

4296 *Vallesia glabra* (Cav.) Link (Apocynaceae) **IPNI:** 82668-1 **Habit:** Shrub **Use:** HF, MA, ME **Origin:** Native **Region:** Andean **Dept.:** VAC **Elev.:** 700–1500 m a.s.l. **IUCN Habitat code:** 14 **Conservation status: (IUCN)** LC

4297 *Vinca major* L. (Apocynaceae) **IPNI:** 82697-1 **Common name:** buenastardes, cortejo, pervinca, princesa, viuditas **Habit:** Herb, Liana, Climbing **Use:** EU, HF, ME **Origin:** Cultivated, Naturalised **Region:** Andean **Dept.:** ANT, BOG, CAL, CAU, CUN, QUI, SAN **Elev.:** 1950–2665 m a.s.l. **IUCN Habitat code:** 1,2,3,14 **Conservation status: (IUCN)** NE

4298 *Vinca minor* L. (Apocynaceae) **IPNI:** 82701-1 **Common name:** amor antioqueño, cortejo **Use:** EU, HF, ME **Dept.:** BOG **IUCN Habitat code:** 14 **Conservation status: (IUCN)** NE

4299 *Canscora diffusa* (Vahl) R.Br. ex Roem. & Schult. (Gentianaceae) **IPNI:** 366552-1 **Habit:** Herb, Liana **Use:** HF, ME **Origin:** Naturalised **Region:** Andean **Dept.:** BOG, BOY, CAU, CUN, MET, NAR, SAN, TOL **Elev.:** 3000–3750 m a.s.l. **IUCN Habitat code:** 1,4,14 **Conservation status: (IUCN)** NE

4300 *Centaurium erythraea* Rafn (Gentianaceae) **IPNI:** 326619-2 **Common name:** centaura, centaura menor **Habit:** Herb **Use:** EU, HF, ME **Origin:** Cultivated **Region:** Andean **Dept.:** NAR, PUT **Elev.:** 2000–3300 m a.s.l. **IUCN Habitat code:** 14 **Conservation status: (IUCN)** LC

4301 *Chelonanthus purpurascens* (Aubl.) Struwe, S.Nilsson & V.A.Albert (Gentianaceae) **IPNI:** 1002418-1 **Habit:** Herb **Use:** ME **Origin:** Native **Region:** Amazonia, Andean, Guiana Shield, Orinoquia, Pacific **Dept.:** AMA, BOY, CAQ, CUN, GUA, HUI, MET, PUT, TOL, VAC, VAU, VID **Elev.:** 100–3180 m.a.s.l. **IUCN Habitat code:** 1,2,3,4,14 **Conservation status: (IUCN)** NE **(CPSG)** Potential LC

4302 *Coutoubea ramosa* Aubl. (Gentianaceae) **IPNI:** 367033-1 **Common name:** piel de pescado, venadillo **Habit:** Herb, Shrub, Subshrub **Use:** ME **Origin:** Native **Region:** Amazonia, Guiana Shield, Orinoquia, Pacific **Dept.:** AMA, CAQ, CAS, CHO, GUA, GUV, MET, SAN, VAU, VID **Elev.:** 40–300 m a.s.l. **IUCN Habitat code:** 1,2,3,4,5,14 **Conservation status: (IUCN)** NE

4303 *Coutoubea spicata* Aubl. (Gentianaceae) **IPNI:** 367036-1 **Common name:** soldado parado **Habit:** Herb, Shrub **Use:** ME **Origin:** Native **Region:** Amazonia, Orinoquia **Dept.:** AMA, ANT, CAS, COR, CUN, GUV, MAG, MET, NSA, SAN, TOL, VID **Elev.:** 60-1000 m.a.s.l. **IUCN Habitat code:** 1,3,4,5,14 **Conservation status: (IUCN)** NE **(CPSG)** Potential LC

4304 *Eustoma exaltatum* (L.) Salisb. (Gentianaceae) **IPNI:** 367440-1 **Use:** ME **Origin:** Native **Conservation status: (IUCN)** NE

4305 *Gentiana prostrata* Wahlenb. (Gentianaceae) **IPNI:** 108504-2 **Use:** ME **Region:** Andean **Dept.:** BOG **Conservation status: (IUCN)** NE

4306 *Gentiana sedifolia* Kunth (Gentianaceae) **IPNI:** 317620-2 **Common name:** alegría de páramo, caaporopita, callpachina yuyu, cervate, dormilona, genciana, penga pienqa, penga-penga, ppenccacuc, siquita pacay, teresita, teresitas **Habit:** Herb **Use:** ME **Origin:** Native **Region:** Andean **Dept.:** ANT, ARA, BOY, CAL, CAU, CUN, MAG, MET, NAR, PUT, QUI, RIS, SAN, TOL, VAC **Elev.:** 2800-4700 m.a.s.l. **IUCN Habitat code:** 1,2,3,4,6,8,14 **Conservation status: (IUCN)** NE **(CPSG)** Potential LC

4307 *Gentianella amarella* (L.) Börner (Gentianaceae) **IPNI:** 108639-2 **Common name:** regencias **Use:** ME **Conservation status: (IUCN)** NE

4308 *Gentianella cerastioides* (Kunth) Fabris (Gentianaceae) **IPNI:** 108656-2 **Common name:** regencia **Habit:** Herb **Use:** ME **Origin:** Native **Region:** Andean **Dept.:** NAR, PUT **Elev.:** 2800–4700 m a.s.l. **IUCN Habitat code:** 8 **Conservation status: (IUCN)** NE

4309 *Gentianella corymbosa* (Kunth) Weaver & Rüdenberg (Gentianaceae) **IPNI:** 108661-2 **Common name:** parásita **Habit:** Herb **Use:** ME **Origin:** Native **Region:** Andean **Dept.:** ANT, BOG, BOY, CES, CUN, MAG **Elev.:** 2600–4500 m a.s.l. **IUCN Habitat code:** 1,3,4,8,14 **Conservation status: (IUCN)** NE

4310 *Gentianella dacrydioides* (Gilg) Weaver & Rüdenberg (Gentianaceae) **IPNI:** 108667-2 **Habit:** Subshrub **Use:** ME **Origin:** Native **Region:** Andean **Dept.:** CAU, HUI, NAR, PUT **Elev.:** 2500–4200 m a.s.l. **IUCN**

Habitat code: 1,3,4,6,8 **Conservation status: (IUCN)** NE

4311 *Gentianella nevadensis* (Gilg) Weaver & Rüdenberg (Gentianaceae)
IPNI: 108708-2
Habit: Herb **Use:** ME **Origin:** Native **Region:** Andean **Dept.:** BOG, BOY, CAU, CUN, MAG, NSA, SAN **Elev.:** 3100–4300 m a.s.l. **IUCN Habitat code:** 3,4 **Conservation status: (IUCN)** NE

4312 *Gentianella selaginifolia* (Griseb.) Fabris (Gentianaceae)
IPNI: 108732-2
Habit: Herb **Use:** ME **Origin:** Native **Region:** Andean **Dept.:** BOY, CAL, CAU, NAR, QUI, RIS, TOL, VAC **Elev.:** 3400–4500 m a.s.l. **IUCN Habitat code:** 1,3,4,14 **Conservation status: (IUCN)** NE

4313 *Goeppertia allouia* (Aubl.) Borchs. & S.Suárez (Gentianaceae)
IPNI: 60458762-2
Common name: agua bendita, bihao, dale dale, daledale, hoja de sal, lairén, lerene, lerén, sewi, yairén **Habit:** Herb **Use:** HF, ME **Origin:** Native **Region:** Amazonia, Guiana Shield, Orinoquia **Dept.:** AMA, MET, VAU **Elev.:** 200–460 m a.s.l. **IUCN Habitat code:** 1,3 **Conservation status: (IUCN)** NE

4314 *Halenia asclepiadea* (Kunth) G.Don (Gentianaceae)
IPNI: 369443-1
Common name: cachitos, dictamo real, don, díctamo, díctamo real, hierba de venado, yerba de venado **Habit:** Herb **Use:** AF,ME **Origin:** Native **Region:** Andean **Dept.:** BOG, BOY, CUN, MET, NSA, SAN **Elev.:** 2700–4000 m a.s.l. **IUCN Habitat code:** 1,3,4,5,8,14 **Conservation status: (IUCN)** NE

4315 *Halenia elata* Wedd. (Gentianaceae)
IPNI: 369463-1
Habit: Herb **Use:** HF **Origin:** Endemic **Region:** Andean **Dept.:** ANT, BOG, CAL, CAU, CES, CHO, HUI, LAG, MAG, NAR, QUI, RIS, TOL, VAC **Elev.:** 2800–4500 m a.s.l. **IUCN Habitat code:** 1,3,4,6,8,14 **Conservation status: (IUCN)** NE

4316 *Halenia weddelliana* Gilg (Gentianaceae)
IPNI: 117020-2
Habit: Herb **Use:** ME **Origin:** Native **Region:** Andean **Dept.:** BOG, CAU, HUI, NAR, PUT **Elev.:** 2700–4000 m a.s.l. **IUCN Habitat code:** 1,4,8,14 **Conservation status: (IUCN)** NE

4317 *Helia alata* (Aubl.) Kuntze (Gentianaceae)
IPNI: 118811-2
Common name: nucalé, yerba de adán **Use:** ME **IUCN Habitat code:** 1,14 **Conservation status: (IUCN)** NE

4318 *Irlbachia nemorosa* (Willd. ex Schult.) Merr. (Gentianaceae)
IPNI: 130185-2
Common name: flor de tierra **Habit:** Herb **Use:** ME **Origin:** Native **Region:** Amazonia, Guiana Shield **Dept.:** AMA, GUA, VAU **Elev.:** 75–415 m a.s.l. **IUCN Habitat code:** 1,3,4,5 **Conservation status: (IUCN)** NE

4319 *Potalia amara* Aubl. (Gentianaceae)
IPNI: 546882-1
Common name: amargoandré **Use:** HF, ME, SU **IUCN Habitat code:** 1,3,4,5,14 **Conservation status: (IUCN)** LC

4320 *Potalia chocoensis* Struwe & V.A.Albert (Gentianaceae)
IPNI: 77062953-1
Common name: amargoandré, yerbandrés **Habit:** Small tree **Use:** ME **Origin:** Endemic **Region:** Pacific **Dept.:** CHO **Elev.:** 50–60 m a.s.l. **IUCN Habitat code:** 1 **Conservation status: (IUCN)** NE

4321 *Potalia coronata* Struwe & V.A.Albert (Gentianaceae)
IPNI: 77062954-1
Common name: curarina, matiguaja **Habit:** Shrub, Small tree, Tree **Use:** ME **Origin:** Native **Region:** Amazonia, Guiana Shield **Dept.:** AMA, CAQ, PUT, VAU **Elev.:** 100–250 m a.s.l. **IUCN Habitat code:** 1,14 **Conservation status: (IUCN)** LC

4322 *Potalia crassa* Struwe & V.A.Albert (Gentianaceae)
IPNI: 77062955-1
Habit: Shrub **Use:** ME **Origin:** Endemic **Region:** Pacific **Dept.:** CHO **Elev.:** 70–70 m a.s.l. **Conservation status: (IUCN)** NE

4323 *Potalia elegans* Struwe & V.A.Albert (Gentianaceae)
IPNI: 317042-2
Common name: curarina, martiguaja, martiguaje **Habit:** Shrub, Small tree, Tree **Use:** ME **Origin:** Native **Region:** Guiana Shield **Dept.:** AMA, GUA, VAU **Elev.:** 120–250 m a.s.l. **IUCN Habitat code:** 1,4 **Conservation status: (IUCN)** NE

4324 *Potalia resinifera* Mart. (Gentianaceae)
IPNI: 546884-1
Common name: amargoandrés, barniz, curalina, curarina, maticuaja, matiguaja, matiguaje **Habit:** Shrub, Small tree **Use:** ME **Origin:** Native **Region:** Amazonia, Andean, Guiana Shield **Dept.:** AMA, ANT, CAQ, COR, MET, NAR, VAU, VID **Elev.:** 100–1600 m a.s.l. **IUCN Habitat code:** 1,3,4,5,14 **Conservation status: (IUCN)** LC

4325 *Schultesia brachyptera* Cham. (Gentianaceae)
IPNI: 370581-1
Habit: Herb **Use:** ME **Origin:** Native **Region:** Caribbean, Orinoquia **Dept.:** BOL, CAS, MET, TOL **Elev.:** 80–800 m a.s.l. **IUCN Habitat code:** 1,2,4,14 **Conservation status: (IUCN)** NE

4326 *Schultesia guianensis* (Aubl.) Malme (Gentianaceae)
IPNI: 230042-2
Habit: Herb **Use:** ME **Origin:** Native **Region:** Caribbean, Orinoquia **Dept.:** ARA, CAS, MAG, MET, TOL, VID **Elev.:** 20–560 m a.s.l. **IUCN Habitat code:** 1,2,3,4,14 **Conservation status: (IUCN)** NE

4327 *Voyria aphylla* (Jacq.) Pers. (Gentianaceae)
IPNI: 1167766-2
Habit: Herb, Saprophyte **Use:** ME **Origin:** Native **Region:** Amazonia, Andean, Guiana Shield, Pacific **Dept.:** ANT, CAQ, CAU, MAG, MET, SAN, VAC, VAU **Elev.:** 50–1850 m a.s.l. **IUCN Habitat code:** 1,2,3,5,14 **Conservation status: (IUCN)** NE

4328 *Xestaea lisianthoides* Griseb. (Gentianaceae)
IPNI: 371484-1
Habit: Herb **Use:** ME **Origin:** Native **Region:** Caribbean **Dept.:** BOL, MAG **Elev.:** 70–1000 m a.s.l. **Conservation status: (IUCN)** NE

4329 *Antonia ovata* Pohl (Loganiaceae)
IPNI: 545655-1
Habit: Tree **Use:** MA, ME **Origin:** Native **Region:** Guiana Shield **Dept.:** CAQ, GUV **Elev.:** 200–250 m a.s.l. **IUCN Habitat code:** 1,14 **Conservation status: (IUCN)** LC

4330 *Spigelia anthelmia* L. (Loganiaceae)
IPNI: 318920-2
Common name: caricia, guambi, guambia, gumbía, huambia, lagaña de gallina, lombricera, quiteria **Habit:** Herb **Use:** AF, HF, MA, ME, PO, SU **Origin:** Native **Region:** Andean, Caribbean, Orinoquia, Pacific **Dept.:** ANT, ATL, BOL, BOY, CAL, CAS, CES, CHO, CUN, HUI, LAG, MAG, MET, NAR, NSA, PUT, RIS, SAP, SAN, TOL, VAC **Elev.:** 0-2670 m.a.s.l. **IUCN Habitat code:** 1,2,3,4,5,14 **Conservation status: (IUCN)** NE **(CPSG)** Potential LC

4331 *Spigelia hamellioides* Kunth (Loganiaceae)
IPNI: 546941-1
Common name: lombricera **Habit:** Herb **Use:** ME **Origin:** Native **Region:** Amazonia, Andean, Guiana Shield, Caribbean, Pacific **Dept.:** AMA, ANT, BOY, CAQ, CAU, CES, CHO, CUN, MAG, MET, NAR, PUT, QUI, RIS, SAN, TOL, VAC, VAU **Elev.:** 75-1000 m.a.s.l. **Conservation status: (IUCN)** NE **(CPSG)** Potential LC

4332 *Spigelia humboldtiana* Cham. & Schltdl. (Loganiaceae)
IPNI: 546947-1
Habit: Herb, Subshrub **Use:** ME **Origin:** Native **Region:** Amazonia, Caribbean, Pacific **Dept.:** ANT, QUI, SAN **Elev.:** 80–300 m a.s.l. **IUCN Habitat code:** 1 **Conservation status: (IUCN)** NE

4333 *Spigelia pedunculata* Kunth (Loganiaceae)
IPNI: 546982-1
Common name: guambía, guamuco, huambía, huambía o yerba de huambía, quitería, yerba de huambía **Habit:** Herb, Subshrub **Use:** ME **Origin:** Native **Region:** Andean **Dept.:** ANT, CAL, CAU, CUN, HUI, MAG, NAR, NSA, PUT, QUI, RIS, SAN, TOL, VAC **Elev.:** 1500-2600 m.a.s.l. **IUCN Habitat code:** 1,3,4,14 **Conservation status: (IUCN)** NE **(CPSG)** Potential LC

4334 *Spigelia persicarioides* Ewan (Loganiaceae)
IPNI: 242225-2
Habit: Herb **Use:** ME **Origin:** Endemic **Region:** Pacific **Dept.:** CAU, VAC **Elev.:** 0–200 m a.s.l. **IUCN Habitat code:** 1,14 **Conservation status: (IUCN)** NE

4335 *Strychnos amazonica* Krukoff (Loganiaceae)
IPNI: 247278-2
Habit: Liana **Use:** ME **Origin:** Native **Region:** Amazonia **Dept.:** AMA, PUT **Elev.:** 300–300 m a.s.l. **IUCN Habitat code:** 1,3,5 **Conservation status: (IUCN)** NE

4336 *Strychnos brachiata* Ruiz & Pav. (Loganiaceae)
IPNI: 547077-1
Common name: sijipe **Habit:** Liana **Use:** ME **Origin:** Native **Region:** Amazonia **Dept.:** PUT **Elev.:** 380–380 m a.s.l. **IUCN Habitat code:** 4,5 **Conservation status: (IUCN)** NE

4337 *Strychnos bredemeyeri* (Schult.) Sprague & Sandwith (Loganiaceae)
IPNI: 547083-1
Habit: Liana **Use:** ME, PO **Origin:** Native **Region:** Amazonia **Dept.:** ARA, GUA, VID **Elev.:** 90–120 m a.s.l. **IUCN Habitat code:** 1,3,14 **Conservation status: (IUCN)** NE

4338 *Strychnos castelnaeana* Baill. (Loganiaceae)
IPNI: 547100-1
Habit: Liana **Use:** MA, ME **Origin:** Native **Region:** Amazonia **Dept.:** AMA **Elev.:** 120–120 m a.s.l. **Conservation status: (IUCN)** NE

4339 *Strychnos cogens* Benth. (Loganiaceae)
IPNI: 547125-1
Habit: Liana **Use:** ME **Origin:** Native **Region:** Amazonia, Guiana Shield, Pacific **Dept.:** AMA, CAQ, CHO, GUA, VAC **Elev.:** 0–200 m a.s.l. **IUCN Habitat code:** 1 **Conservation status: (IUCN)** NE

4340 *Strychnos darienensis* Seem. (Loganiaceae)
IPNI: 547151-1
Common name: bejuco grande **Habit:** Liana **Use:** ME **Origin:** Native **Region:** Amazonia, Pacific **Dept.:** AMA, CAU, VAC **Elev.:** 50–120 m a.s.l. **IUCN Habitat code:** 1,14 **Conservation status: (IUCN)** NE

4341 *Strychnos diaboli* Sandwith (Loganiaceae)
IPNI: 247300-2
Habit: Liana **Use:** MA, ME **Origin:** Native **Region:** Amazonia **Dept.:** AMA **Elev.:** 200–200 m a.s.l. **Conservation status: (IUCN)** NE

4342 *Strychnos erichsonii* R.H.Schomb. ex Progel (Loganiaceae)
IPNI: 547183-1
Common name: chuchuguasa macho, cuit, erequí, irirochí, limanajú, michichijo, palo de gusano, palo veneno **Habit:** Liana **Use:** HF, ME **Origin:** Native **Region:** Amazonia, Pacific **Dept.:** ANT, CAQ, CHO, CUN, GUA, VAC **Elev.:** 5–280 m a.s.l. **IUCN Habitat code:** 1,3,5 **Conservation status: (IUCN)** NE

4343 *Strychnos fendleri* Sprague & Sandwith (Loganiaceae)
IPNI: 247306-2
Habit: Small tree **Use:** ME **Origin:** Native **Region:** Caribbean **Dept.:** LAG **Elev.:** 100–100 m a.s.l. **Conservation status: (IUCN)** NE

4344 *Strychnos glabra* Sagot ex Progel (Loganiaceae)
IPNI: 547207-1
Habit: Liana **Use:** ME **Origin:** Native **Region:** Amazonia **Dept.:** AMA, VAU **Elev.:** 120–200 m a.s.l. **IUCN Habitat code:** 1,3,14 **Conservation status: (IUCN)** LC

4345 *Strychnos gubleri* G.Planch. (Loganiaceae)
IPNI: 547217-1
Common name: ong, ong-uué, uué **Habit:** Liana **Use:** ME **Origin:** Native **Region:** Amazonia, Guiana Shield, Orinoquia, Pacific **Dept.:** AMA, BOL, CHO, COR, MET, NAR, PUT, VAC, VAU **Elev.:** 0-300 m.a.s.l. **Conservation status: (IUCN)** NE **(CPSG)** Potential LC

4346 *Strychnos guianensis* (Aubl.) Mart. (Loganiaceae)
IPNI: 547220-1
Common name: curare de pescado, ya-hi-ae-o **Habit:** Liana **Use:** ME **Origin:** Native **Region:** Amazonia, Guiana Shield **Dept.:** AMA, CAQ, GUA, PUT, VAU, VID **Elev.:** 90-300 m.a.s.l. **IUCN Habitat code:** 1,3,4,5,14 **Conservation status: (IUCN)** NE **(CPSG)** Potential LC

4347 *Strychnos javariensis* Krukoff (Loganiaceae)
IPNI: 247314-2
Common name: veneno **Habit:** Liana **Use:** ME **Origin:** Native **Region:** Amazonia **Dept.:** AMA, CAQ, PUT **Elev.:** 200–300 m a.s.l. **IUCN Habitat code:** 1,5 **Conservation status: (IUCN)** NE

4348 *Strychnos jobertiana* Baill. (Loganiaceae)
IPNI: 547247-1
Common name: chii, iro, iro-chii **Habit:** Liana **Use:** HF, ME **Origin:** Native **Region:** Amazonia, Pacific **Dept.:** AMA, CHO, GUA, PUT, VAC, VAU **Elev.:** 100–300 m a.s.l. **IUCN Habitat code:** 1,4 **Conservation status: (IUCN)** LC

4349 *Strychnos mitscherlichii* M.R.Schomb. (Loganiaceae)
IPNI: 547341-1
Common name: veneno, yayu-ae-o **Habit:** Liana **Use:** MA, ME **Origin:** Native **Region:** Amazonia, Pacific **Dept.:** AMA, ANT, BOL, CHO, PUT, VAC, VAU **Elev.:** 5-250 m.a.s.l. **IUCN Habitat code:** 1,3,5,14 **Conservation status: (IUCN)** NE **(CPSG)** Potential LC

4350 *Strychnos panamensis* Seem. (Loganiaceae)
IPNI: 547390-1
Common name: bejuco del curare, curare, mataperros **Habit:** Liana **Use:** HF, ME **Origin:** Native **Region:** Andean, Guiana Shield, Caribbean, Pacific **Dept.:** ANT, CAL, CHO, GUA, MAG, SAN, VAC **Elev.:** 5–200 m a.s.l. **IUCN Habitat code:** 1,2,3,5,14 **Conservation status: (IUCN)** NE

4351 *Strychnos parviflora* Spruce ex Benth. (Loganiaceae)
IPNI: 547398-1
Habit: Liana **Use:** HF **Origin:** Native **Dept.:** AMA, CAQ **Elev.:** 120–200 m a.s.l. **IUCN Habitat code:** 1,5 **Conservation status: (IUCN)** NE

4352 *Strychnos peckii* B.L.Rob. (Loganiaceae)
IPNI: 547401-1
Common name: pico de paujil, veneno **Habit:** Shrub, Liana **Use:** HF, ME **Origin:** Native **Region:** Amazonia, Caribbean, Pacific **Dept.:** AMA, ANT, BOL, COR, GUA, MET, NAR, PUT, VAC, VAU **Elev.:** 80-1000 m.a.s.l. **IUCN Habitat code:** 1,3,5,14 **Conservation status: (IUCN)** NE **(CPSG)** Potential LC

4353 *Strychnos rondeletioides* Spruce ex Benth. (Loganiaceae)
IPNI: 547453-1
Habit: Liana **Use:** ME **Origin:** Native **Region:** Amazonia, Guiana Shield, Pacific **Dept.:** AMA, ANT, GUA, VAC, VAU **Elev.:** 60–900 m a.s.l. **IUCN Habitat code:** 1,3,14 **Conservation status: (IUCN)** LC

4354 *Strychnos solimoesana* Krukoff (Loganiaceae)
IPNI: 247356-2
Habit: Liana **Use:** ME **Origin:** Native **Region:** Amazonia **Elev.:** 120–120 m a.s.l. **Conservation status: (IUCN)** NE

4355 *Strychnos subcordata* Spruce ex Benth. (Loganiaceae)
IPNI: 547493-1
Common name: curare, veneno **Habit:** Liana **Use:** ME **Origin:** Native **Region:** Amazonia **Dept.:** AMA, PUT **Elev.:** 150–380 m a.s.l. **IUCN Habitat code:** 1,5 **Conservation status: (IUCN)** NE

4356 *Strychnos tarapotensis* Sprague & Sandwith (Loganiaceae)
IPNI: 247359-2
Use: AF, HF, ME **Origin:** Native **Region:** Amazonia, Pacific **Dept.:** CAU, CHO, CUN **Elev.:** 10–550 m a.s.l. **IUCN Habitat code:** 1,2,14 **Conservation status: (IUCN)** LC

4357 *Strychnos toxifera* R.H.Schomb. ex Lindl. (Loganiaceae)
IPNI: 547525-1
Common name: bejuco de maracure, curare, veneno **Habit:** Liana **Use:** MA, ME, PO **Origin:** Native **Region:** Amazonia, Orinoquia, Pacific **Dept.:** ANT, CHO, GUA, PUT, VID **Elev.:** 0–900 m a.s.l. **IUCN Habitat code:** 1,3,5,14 **Conservation status: (IUCN)** NE

4358 *Strychnos xinguensis* Krukoff (Loganiaceae)
IPNI: 247367-2
Use: ME **IUCN Habitat code:** 1,3,5,14 **Conservation status: (IUCN)** NE

4359 *Agouticarpa curviflora* (Dwyer) C.H.Perss. (Rubiaceae)
IPNI: 1011811-2
Habit: Shrub, Tree **Use:** HF **Origin:** Native **Region:** Amazonia, Andean, Pacific **Dept.:** AMA, CHO, MET, PUT **Elev.:** 0–1420 m a.s.l. **IUCN Habitat code:** 1,3 **Conservation status: (IUCN)** LC

4360 *Agouticarpa isernii* (Standl.) C.H.Perss. (Rubiaceae)
IPNI: 1095647-2
Habit: Tree **Use:** HF **Origin:** Native **Region:** Amazonia **Dept.:** AMA, PUT **Elev.:** 100–700 m a.s.l. **IUCN Habitat code:** 1,3,5,14 **Conservation status: (IUCN)** LC

4361 *Alibertia claviflora* K.Schum. (Rubiaceae)
IPNI: 8381-2
Common name: borojó, maogüa **Habit:** Shrub **Use:** HF, ME **Origin:** Cultivated **Region:** Amazonia **Dept.:** AMA **Elev.:** 250–250 m a.s.l. **IUCN Habitat code:** 1,3,14 **Conservation status: (IUCN)** LC

4362 *Alibertia duckeana* Delprete & C.H.Perss. (Rubiaceae)
IPNI: 77175918-1
Common name: borojó del Pacífico, borojó del pacífico **Habit:** Small tree **Use:** HF **Origin:** Native **Region:** Guiana Shield **Dept.:** VAU **Elev.:** 250–250 m a.s.l. **IUCN Habitat code:** 1 **Conservation status: (IUCN)** NE

4363 *Alibertia edulis* (Rich.) A.Rich. (Rubiaceae) **IPNI:** 743001-1 **Common name:** arrebianca, borojó, caimitillo, guayaba perurera, guayabito, guayabito montañero, pasita, pera, pera montañera, pera o perita, perilla, perita, pero, sulsul, toroncoyo, yaya de cajón **Habit:** Shrub, Small tree, Tree **Use:** HF, MA, ME **Origin:** Native **Region:** Amazonia, Andean, Caribbean, Orinoquia, Pacific **Dept.:** AMA, ANT, BOL, BOY, CAS, CAU, CES, CHO, COR, MAG, MET, SAP, VID **Elev.:** 0–1200 m a.s.l. **IUCN Habitat code:** 1,2,3,4,5,14 **Conservation status: (IUCN)** LC

4364 *Alibertia latifolia* (Benth.) K.Schum. (Rubiaceae) **IPNI:** 8399-2 **Habit:** Shrub **Use:** HF **Origin:** Native **Region:** Amazonia **Dept.:** AMA, MET **Elev.:** 110–120 m a.s.l. **IUCN Habitat code:** 1,3,4,5,14 **Conservation status: (IUCN)** LC

4365 *Alibertia patinoi* (Cuatrec.) Delprete & C.H.Perss. (Rubiaceae) **IPNI:** 77116169-1 **Common name:** borojó, borojó del pacífico, borojó montano **Habit:** Tree **Use:** HF, MA, ME, SU **Origin:** Native, Cultivated **Region:** Pacific **Dept.:** ANT, CHO, COR, GUA, GUV, NAR, PUT, VAC **Elev.:** 0–700 m a.s.l. **IUCN Habitat code:** 1,3,5,14 **Conservation status: (IUCN)** LC

4366 *Alibertia sorbilis* Huber ex Ducke (Rubiaceae) **IPNI:** 8417-2 **Use:** HF **IUCN Habitat code:** 1 **Conservation status: (IUCN)** NE

4367 *Alibertia verticillata* (Ducke) W.Schultze-Motel (Rubiaceae) **IPNI:** 943695-1 **Use:** HF **Origin:** Native **Conservation status: (IUCN)** NE

4368 *Alseis blackiana* Hemsl. (Rubiaceae) **IPNI:** 743041-1 **Common name:** aceituno, hueso, palo guitarro **Habit:** Small tree, Tree **Use:** MA **Origin:** Native **Region:** Pacific **Dept.:** ANT, BOL, CAL, CHO, SAN **Elev.:** 10–900 m a.s.l. **IUCN Habitat code:** 1,2,3,5,14 **Conservation status: (IUCN)** NE

4369 *Amaioua glomerulata* (Lam. ex Poir.) Delprete & C.H.Perss. (Rubiaceae) **IPNI:** 77124830-1 **Common name:** macanillo, palo sarna, vara de piedra, vara de sarna **Habit:** Shrub, Small tree, Tree **Use:** HF **Origin:** Native **Region:** Amazonia, Guiana Shield, Caribbean, Pacific **Dept.:** AMA, ANT, BOL, CHO, COR, GUA, GUV, NSA, SAP, SAN, TOL, VAC **Elev.:** 0–380 m a.s.l. **IUCN Habitat code:** 1,2,3,5,14 **Conservation status: (IUCN)** LC

4370 *Amaioua guianensis* Aubl. (Rubiaceae) **IPNI:** 743074-1 **Habit:** Shrub, Small tree, Tree **Use:** HF, MA **Origin:** Native **Region:** Amazonia, Andean, Guiana Shield, Caribbean, Orinoquia **Dept.:** AMA, ANT, CAQ, GUV, SAP, SAN, VAU, VID **Elev.:** 0–2300 m a.s.l. **IUCN Habitat code:** 1,3,4,14 **Conservation status: (IUCN)** LC

4371 *Amphidasya ambigua* (Standl.) Standl. (Rubiaceae) **IPNI:** 11534-2 **Common name:** guaquito, juambautista, promontorio **Habit:** Herb **Use:** ME **Origin:** Native **Region:** Andean, Pacific **Dept.:** CAU, CHO, NAR, VAC **Elev.:** 0–1000 m a.s.l. **IUCN Habitat code:** 1,3,5,14 **Conservation status: (IUCN)** NE

4372 *Appunia seibertii* Standl. (Rubiaceae) **IPNI:** 16747-2 **Habit:** Shrub, Small tree **Use:** ME **Origin:** Native **Region:** Caribbean, Pacific **Dept.:** ANT, CHO, COR, MAG **Elev.:** 20–600 m a.s.l. **IUCN Habitat code:** 1,2,3,4,14 **Conservation status: (IUCN)** NE

4373 *Appunia tenuiflora* (Benth.) B.D.Jacks. (Rubiaceae) **IPNI:** 1061341-2 **Habit:** Shrub **Use:** HF **Origin:** Native **Region:** Caribbean **Dept.:** CES **Elev.:** 100–100 m a.s.l. **IUCN Habitat code:** 1,4 **Conservation status: (IUCN)** NE

4374 *Arachnothryx reflexa* (Benth.) Planch. (Rubiaceae) **IPNI:** 743570-1 **Habit:** Shrub, Tree **Use:** HF **Origin:** Native **Region:** Andean **Dept.:** ANT, BOY, CAU, CUN, MAG, MET, NSA, SAN **Elev.:** 420-2300 msnm **IUCN Habitat code:** 1,2,3,4,5,14 **Conservation status: (IUCN)** LC **(CPSG)** LC

4375 *Arcytophyllum filiforme* (Ruiz & Pav.) Standl. (Rubiaceae) **IPNI:** 17992-2 **Habit:** Herb **Use:** ME **Origin:** Native **Region:** Andean **Dept.:** BOY, CAU, NAR **Elev.:** 3200–4100 m a.s.l. **IUCN Habitat code:** 1,3,4,14 **Conservation status: (IUCN)** LC

4376 *Arcytophyllum nitidum* (Kunth) Schltdl. (Rubiaceae) **IPNI:** 743608-1 **Common name:** chite, escobo, gurrubu, piojito, piojo, raberruncho, romero de páramo, sanalotodo, velillo, velito, venadillo **Habit:** Shrub, Subshrub **Use:** ME **Origin:** Native **Region:** Andean **Dept.:** ANT, ARA, BOG, BOY, CAU, CES, CHO, CUN, HUI, LAG, MAG, MET, NAR, NSA, PUT, SAN **Elev.:** 630–4500 m a.s.l. **IUCN Habitat code:** 1,3,4,5,6,8,14 **Conservation status: (IUCN)** NE

4377 *Arcytophyllum thymifolium* (Ruiz & Pav.) Standl. (Rubiaceae) **IPNI:** 18012-2 **Common name:** granicillo **Habit:** Shrub **Use:** HF **Origin:** Native **Region:** Andean **Dept.:** BOG, CAU, CUN, NAR **Elev.:** 2470–4500 m a.s.l. **IUCN Habitat code:** 1,3,8,14 **Conservation status: (IUCN)** NE

4378 *Augusta rivalis* (Benth.) J.H.Kirkbr. (Rubiaceae) **IPNI:** 998196-1 **Habit:** Herb **Use:** ME **Origin:** Native **Region:** Pacific **Dept.:** CHO **Elev.:** 150–150 m a.s.l. **Conservation status: (IUCN)** NE

4379 *Bertiera guianensis* Aubl. (Rubiaceae) **IPNI:** 744444-1 **Common name:** cafecillo, guamo negro, guayabillo, kaphensayo, yerba mora, yerbamora, yurune **Habit:** Shrub, Small tree, Tree **Use:** ME **Origin:** Native **Region:** Amazonia, Andean, Guiana Shield, Orinoquia, Pacific **Dept.:** AMA, ANT, BOL, BOY, CAQ, CAS, CAU, CHO, COR, GUV, LAG, MET, NAR, PUT, SAN, VAU, VID **Elev.:** 0–1350 m a.s.l. **IUCN Habitat code:** 1,2,3,4,5,14 **Conservation status: (IUCN)** LC

4380 *Botryarrhena pendula* Ducke (Rubiaceae) **IPNI:** 35222-2 **Common name:** bedecapîcù, cola de gurre, costillo **Habit:** Tree **Use:** HF **Origin:** Native **Region:** Amazonia, Guiana Shield **Dept.:** AMA, CAQ, GUV, VAU **Elev.:** 100–200 m a.s.l. **IUCN Habitat code:** 1,4 **Conservation status: (IUCN)** LC

4381 *Bouvardia longiflora* (Cav.) Kunth (Rubiaceae) **IPNI:** 745065-1 **Habit:** Shrub **Use:** EU, MA **Origin:** Cultivated **Region:** Andean **Dept.:** ANT **Elev.:** 2200–2200 m a.s.l. **Conservation status: (IUCN)** NE

4382 *Calycophyllum candidissimum* (Vahl) DC. (Rubiaceae) **IPNI:** 745241-1 **Common name:** alazano, ardeverde, guayabillo, guayabo, guayabo colorado, guayabo macanillo, guayabo macho, harino, palosol **Habit:** Tree **Use:** EU, FU, IF, MA, ME, SU **Origin:** Native, Cultivated **Region:** Andean, Caribbean, Pacific **Dept.:** ATL, BOL, CES, CHO, LAG, MAG, SUC **Elev.:** 0–1000 m a.s.l. **IUCN Habitat code:** 1,2,3,5,14 **Conservation status: (IUCN)** LC

4383 *Calycophyllum megistocaulum* (K.Krause) C.M.Taylor (Rubiaceae) **IPNI:** 969466-1 **Habit:** Small tree, Tree **Use:** ME **Origin:** Native **Region:** Amazonia, Guiana Shield **Dept.:** AMA, VAU **Elev.:** 120–250 m a.s.l. **IUCN Habitat code:** 1,3,14 **Conservation status: (IUCN)** LC

4384 *Calycophyllum obovatum* (Ducke) Ducke (Rubiaceae) **IPNI:** 43052-2 **Common name:** guayabete, matapalo, turmamico **Habit:** Tree **Use:** ME **Origin:** Native **Region:** Guiana Shield **Dept.:** GUA, VAU **Elev.:** 250–250 m a.s.l. **IUCN Habitat code:** 1,5 **Conservation status: (IUCN)** NE

4385 *Calycophyllum spruceanum* (Benth.) Hook.f. ex K.Schum. (Rubiaceae) **IPNI:** 745249-1 **Common name:** capirona, capirón, guayabete, huesito, resbalamono **Habit:** Tree **Use:** MA, ME **Origin:** Native **Region:** Amazonia **Dept.:** AMA, PUT **Elev.:** 100–380 m a.s.l. **IUCN Habitat code:** 1,3,5 **Conservation status: (IUCN)** NE

4386 *Capirona macrophylla* (Poepp.) Delprete (Rubiaceae) **IPNI:** 77210192-1 **Common name:** capirona, capirona de altura, cañegua, cañena, guayabete, guayabeto de altura, palo de indio, resbalamono, resbalamonos **Habit:** Tree **Use:** HF, MA, ME **Origin:** Endemic **Region:** Amazonia, Guiana Shield, Orinoquia **Dept.:** AMA, CAQ, GUV, MET, VID **Elev.:** 180–475 m a.s.l. **IUCN Habitat code:** 1,3,4,14 **Conservation status: (IUCN)** LC

4387 *Carapichea ipecacuanha* (Brot.) L.Andersson (Rubiaceae) **IPNI:** 1214950-2 **Common name:** ipeca, ipecacua raicilla, ipecacuana, ipecacuana de cartage ipeca, ipecacuana de cartagena, poajá, raicilla, raicilla gateadora, raíz de ipeca, raíz de montaña, raíz de oro **Habit:** Herb **Use:** MA, ME **Origin:** Native **Region:** Orinoquia, Pacific **Dept.:** ANT, BOL, CHO, COR, MET **Elev.:** 30–400 m a.s.l. **IUCN Habitat code:** 1,3,5,14 **Conservation status:** **(IUCN)** NE

4388 *Catunaregam spinosa* (Thunb.) Tirveng. (Rubiaceae) **IPNI:** 745904-1 **Common name:** mariangola **Use:** EU, FU, HF, IF, MA, ME, PO **IUCN Habitat code:** 1,14 **Conservation status:** **(IUCN)** NE

4389 *Chimarrhis glabriflora* Ducke (Rubiaceae) **IPNI:** 56520-2 **Habit:** Tree **Use:** MA **Origin:** Native **Region:** Amazonia **Dept.:** AMA, CAQ, MET, PUT **Elev.:** 120–400 m a.s.l. **IUCN Habitat code:** 1 **Conservation status:** **(IUCN)** LC

4390 *Chimarrhis hookeri* K.Schum. (Rubiaceae) **IPNI:** 56522-2 **Common name:** chichiburra **Use:** MA **Origin:** Native **IUCN Habitat code:** 1,3,5,14 **Conservation status:** **(IUCN)** LC

4391 *Chimarrhis turbinata* DC. (Rubiaceae) **IPNI:** 308173-2 **Use:** MA **Origin:** Native **IUCN Habitat code:** 1,3 **Conservation status:** **(IUCN)** LC

4392 *Chiococca alba* (L.) Hitchc. (Rubiaceae) **IPNI:** 56535-2 **Common name:** bejuco negro, bejuco yute, caica, cainica, chupachupa, crucecita, lágrimas de María, murciélago, raíz de muerciélago, raíz de murciélago, sampedro blanco **Habit:** Shrub, Liana, Climbing **Use:** HF, ME **Origin:** Native **Region:** Andean, Caribbean, Pacific **Dept.:** ANT, ATL, BOL, BOY, CAU, CES, CHO, COR, CUN, HUI, LAG, MAG, NAR, NSA, PUT, QUI, RIS, SAP, SAN, SUC, TOL, VAC **Elev.:** 0–2440 m a.s.l. **IUCN Habitat code:** 1,2,3,4,14 **Conservation status:** **(IUCN)** LC

4393 *Chione venosa* (Sw.) Urb. (Rubiaceae) **IPNI:** 56650-2 **Habit:** Small tree, Tree **Use:** MA, ME **Origin:** Native **Region:** Pacific **Dept.:** CHO, VAC **Elev.:** 0–500 m a.s.l. **IUCN Habitat code:** 1 **Conservation status:** **(IUCN)** NE

4394 *Chomelia paniculata* (Bartl. ex DC.) Steyerm. (Rubiaceae) **IPNI:** 57096-2 **Use:** ME **Origin:** Native **Conservation status:** **(IUCN)** LC

4395 *Chomelia spinosa* Jacq. (Rubiaceae) **IPNI:** 746662-1 **Common name:** chocolatillo, fruta de pava, fruta de pavo, frutaepava **Habit:** Shrub, Small tree, Tree **Use:** HF, ME, SU **Origin:** Native **Region:** Andean, Caribbean, Orinoquia **Dept.:** ANT, ATL, BOL, CAS, CES, GUV, LAG, MAG, MET, NSA, SAP, SUC, TOL **Elev.:** 0–1350 m a.s.l. **IUCN Habitat code:** 1,2,3,4,14 **Conservation status:** **(IUCN)** LC

4396 *Ciliosemina pedunculata* (H.Karst.) Antonelli (Rubiaceae) **IPNI:** 77065717-1 **Common name:** quina cuprea, quina punta de lanza, quina rosada **Habit:** Small tree, Tree **Use:** MA, ME **Origin:** Native **Region:** Andean, Guiana Shield, Orinoquia **Dept.:** AMA, ANT, CAQ, CUN, GUV, MET, NSA, SAN **Elev.:** 150–1700 m a.s.l. **IUCN Habitat code:** 1,3,5,14 **Conservation status:** **(IUCN)** LC

4397 *Ciliosemina purdieana* (Wedd.) Antonelli (Rubiaceae) **IPNI:** 77065718-1 **Common name:** cascarilla, quina cuprea, quina rosada **Habit:** Small tree, Tree **Use:** ME **Origin:** Endemic **Dept.:** ANT, BOL, SAN **Elev.:** 80-700 msnm **IUCN Habitat code:** 1,3,5,14 **Conservation status:** **(IUCN)** LC **(CPSG)** LC

4398 *Cinchona barbacoensis* H.Karst. (Rubiaceae) **IPNI:** 746708-1 **Common name:** cascarilla, quina **Habit:** Shrub, Small tree, Tree **Use:** ME **Origin:** Native **Region:** Andean, Pacific **Dept.:** ANT, CHO, NAR, RIS, VAC **Elev.:** 10–1540 m a.s.l. **IUCN Habitat code:** 1,14 **Conservation status:** **(IUCN)** NE

4399 *Cinchona calisaya* Wedd. (Rubiaceae) **IPNI:** 746718-1 **Use:** HF, MA, ME **IUCN Habitat code:** 1 **Conservation status:** **(IUCN)** LC

4400 *Cinchona lancifolia* Mutis (Rubiaceae) **IPNI:** 746806-1 **Habit:** Tree **Use:** MA, ME **Origin:** Native **Region:** Andean **Dept.:** ANT, BOY, CAU, CUN, HUI, MET, NSA, PUT, SAN **Elev.:** 1900–2900 m a.s.l. **IUCN Habitat code:** 1,3,5,14 **Conservation status:** **(IUCN)** NE

4401 *Cinchona officinalis* L. (Rubiaceae) **IPNI:** 58723-2 **Common name:** azuceno, cascarilla amarilla, cascarilla fina, cascarilla gris, funa, quina, quina anaranjada, quina anaranjada y cascarilla, quina blanca, quina canchosa, quina roja, quina rosada, quina tuna, quina tunita, quina túnica, quino, tunita **Use:** EU, HF, MA, ME **Origin:** Naturalised **Dept.:** BOG **IUCN Habitat code:** 1,3,4,14 **Conservation status:** **(IUCN)** NE

4402 *Cinchona pitayensis* (Wedd.) Wedd. (Rubiaceae) **IPNI:** 746868-1 **Common name:** cascarillo, quina, quina roja, quina roja de pitayó, quino **Habit:** Tree **Use:** MA, ME **Origin:** Native **Region:** Andean **Dept.:** BOG, CAU, NAR, PUT **Elev.:** 2400–3100 m a.s.l. **IUCN Habitat code:** 1,3,4,14 **Conservation status:** **(IUCN)** NE

4403 *Cinchona pubescens* Vahl (Rubiaceae) **IPNI:** 746872-1 **Common name:** azuceno, azuceno colorado, cascarilla, cascarilla amarga, cascarilla amarilla, cascarilla colorada, cascarilla gallinazo, cascarillo, cuas, flor de azar, juino, manglar, mazamorro, oreja de mula, palo, quin felpa, quina, quina amarilla, quina boliviana, q **Habit:** Tree **Use:** EU, GS, HF, MA, ME, SU **Origin:** Native, Cultivated **Region:** Andean **Dept.:** ANT, BOG, BOY, CAL, CAQ, CAU, CES, CHO, CUN, HUI, MAG, MET, NAR, NSA, PUT, QUI, RIS, SAN, TOL, VAC **Elev.:** 600–3300 m a.s.l. **IUCN Habitat code:** 1,2,3,4,14 **Conservation status:** **(IUCN)** LC

4404 *Coccocypselum aureum* (Spreng.) Cham. & Schltdl. (Rubiaceae) **IPNI:** 746936-1 **Habit:** Herb **Use:** HF **Origin:** Native **Region:** Amazonia, Andean, Pacific **Dept.:** AMA, NSA, TOL, VAC **Elev.:** 120–1200 m a.s.l. **IUCN Habitat code:** 3,4,14 **Conservation status:** **(IUCN)** NE

4405 *Coccocypselum condalia* Pers. (Rubiaceae) **IPNI:** 746945-1 **Habit:** Herb **Use:** HF **Origin:** Native **Region:** Andean **Dept.:** CAQ **Elev.:** 1000–1300 m a.s.l. **IUCN Habitat code:** 1 **Conservation status:** **(IUCN)** NE

4406 *Coccocypselum guianense* (Aubl.) K.Schum. (Rubiaceae) **IPNI:** 61955-2 **Habit:** Herb **Use:** ME **Origin:** Native **Region:** Andean, Orinoquia, Pacific **Dept.:** ANT, CHO, VID **Elev.:** 40–2200 m a.s.l. **IUCN Habitat code:** 1,3,4,14 **Conservation status:** **(IUCN)** NE

4407 *Coccocypselum hirsutum* Bartl. ex DC. (Rubiaceae) **IPNI:** 746963-1 **Common name:** cascujalé, mortiño de culebra **Habit:** Herb **Use:** ME **Origin:** Native **Region:** Amazonia, Andean, Orinoquia, Pacific **Dept.:** AMA, ANT, BOY, CAQ, CHO, CUN, HUI, MAG, MET, NAR, RIS, SAN, TOL, VAC, VID **Elev.:** 0–3300 m a.s.l. **IUCN Habitat code:** 1,2,3,4,8,14 **Conservation status:** **(IUCN)** NE

4408 *Coccocypselum lanceolatum* (Ruiz & Pav.) Pers. (Rubiaceae) **IPNI:** 315609-2 **Common name:** fruta de culebra, hojitas de sapo, mortiño de culebra, venturino **Habit:** Herb **Use:** HF **Origin:** Native **Region:** Andean, Caribbean, Pacific **Dept.:** ANT, BOY, CAQ, CAU, CES, CHO, CUN, HUI, MAG, NAR, NSA, QUI, RIS, SAN, TOL, VAC **Elev.:** 100-2800 msnm **IUCN Habitat code:** 1,2,3,4,5,14 **Conservation status:** **(IUCN)** NE **(CPSG)** Potential LC

4409 *Coffea arabica* L. (Rubiaceae) **IPNI:** 747038-1 **Common name:** arabio, arábigo, borbón, borbón amarillo, cafeto, cafeto arábico, café, café Colombia, café San Lorenzo, café San Ramón, café arábico, café borbón, café borbón amarillo, café común, café de Colombia, café enano, caturra, colombia, maragojito, sanramón, si **Habit:** Shrub, Tree **Use:** AF, EU, GS, HF, IF, MA, ME, SU **Origin:** Cultivated, Naturalised **Region:** Amazonia, Andean, Guiana Shield **Dept.:** AMA, ANT, BOG, BOY, CAL, CUN, GUA, HUI, MAG, MET, PUT, QUI, RIS, SAN, TOL, VAC, VAU **Elev.:** 200–2600 m a.s.l. **IUCN Habitat code:** 1,2,3,4,8,14 **Conservation status:** **(IUCN)** EN

4410 *Coffea liberica* W.Bull (Rubiaceae) **IPNI:** 747169-1 **Common name:** café excelso, café libérico **Habit:** Small tree **Use:** HF, ME **Origin:** Cultivated, Naturalised **Region:** Andean **Dept.:** ANT, CAL, CUN, QUI **Elev.:** 1300–1500 m a.s.l. **IUCN Habitat code:** 1,3,14 **Conservation status:** **(IUCN)** LC

4411 *Condaminea corymbosa* (Ruiz & Pav.) DC. (Rubiaceae) **IPNI:** 1010049-2 **Common name:** azuceno, cacao cimarrón, cascarillo, hojaancha, quino blanco **Habit:** Shrub, Small tree, Tree **Use:** ME **Origin:** Native **Region:** Andean, Pacific **Dept.:** ANT, BOG, CAL, CAQ, CAU, CHO, CUN,

HUI, MET, NAR, NSA, PUT, QUI, RIS, SAN, TOL, VAC **Elev.:** 0-2000 msnm **IUCN Habitat code:** 1,2,3,14 **Conservation status: (IUCN)** LC **(CPSG)** Potential LC

4412 *Cordiera hadrantha* (Standl.) C.H.Perss. & Delprete (Rubiaceae) **IPNI:** 77089908-1 **Use:** HF **Origin:** Native **IUCN Habitat code:** 1 **Conservation status: (IUCN)** NE

4413 *Cordiera myrciifolia* (K.Schum.) Perss. & Delprete (Rubiaceae) **IPNI:** 60436803-2 **Habit:** Shrub, Tree **Use:** HF **Origin:** Native **Region:** Amazonia, Orinoquia **Dept.:** ARA, GUV, MET **Elev.:** 100–260 m a.s.l. **IUCN Habitat code:** 1,3,4,5,14 **Conservation status: (IUCN)** LC

4414 *Cosmibuena grandiflora* (Ruiz & Pav.) Rusby (Rubiaceae) **IPNI:** 66725-2 **Common name:** azuceno, capecape, capecito, capenegro, cascarillo, mestiza, narciso **Habit:** Tree, Epiphyte **Use:** ME **Origin:** Native **Region:** Andean, Guiana Shield, Pacific **Dept.:** ANT, CAU, CHO, CUN, HUI, MET, NAR, NSA, SAN, TOL, VAC, VAU **Elev.:** 0–2200 m a.s.l. **IUCN Habitat code:** 1,2,3,4,14 **Conservation status: (IUCN)** LC

4415 *Coussarea amplifolia* C.M.Taylor (Rubiaceae) **IPNI:** 319914-2 **Habit:** Tree **Use:** HF **Origin:** Native **Region:** Amazonia **Dept.:** PUT **Elev.:** 0–300 m a.s.l. **IUCN Habitat code:** 1 **Conservation status: (IUCN)** NE

4416 *Coussarea antioquiana* C.M.Taylor (Rubiaceae) **IPNI:** 321109-2 **Habit:** Shrub, Small tree, Tree **Use:** HF **Origin:** Endemic **Region:** Andean, Pacific **Dept.:** ANT, COR **Elev.:** 0-1200 msnm **IUCN Habitat code:** 1,3 **Conservation status: (IUCN)** VU **(CPSG)** VU

4417 *Coussarea bernardii* Steyerm. (Rubiaceae) **IPNI:** 67399-2 **Common name:** brasil, brazil, tree de pava, árbol de pava **Habit:** Tree **Use:** HF **Origin:** Native **Region:** Andean **Dept.:** NSA **IUCN Habitat code:** 1,3 **Conservation status: (IUCN)** NE

4418 *Coussarea brevicaulis* K.Krause (Rubiaceae) **IPNI:** 747645-1 **Habit:** Shrub, Subshrub, Small tree **Use:** HF, ME **Origin:** Native **Region:** Amazonia, Andean **Dept.:** AMA, CAQ, PUT, VAU **Elev.:** 90–1000 m a.s.l. **IUCN Habitat code:** 1,5,14 **Conservation status: (IUCN)** NE

4419 *Coussarea cerroazulensis* Dwyer & M.V.Hayden (Rubiaceae) **IPNI:** 67403-2 **Common name:** matecillo, pajarito **Habit:** Shrub **Use:** HF **Origin:** Native **Region:** Pacific **Dept.:** BOY, CHO **Elev.:** 100–680 m a.s.l. **IUCN Habitat code:** 1,3 **Conservation status: (IUCN)** NE

4420 *Coussarea enneantha* Standl. (Rubiaceae) **IPNI:** 67412-2 **Common name:** matecillo **Habit:** Shrub, Tree **Use:** HF **Origin:** Native **Region:** Andean, Pacific **Dept.:** ANT, CAU, CHO, VAC **Elev.:** 0–1350 m a.s.l. **IUCN Habitat code:** 1,3,5,14 **Conservation status: (IUCN)** NE

4421 *Coussarea flava* Poepp. (Rubiaceae) **IPNI:** 747666-1 **Common name:** guevas de chucha, güevas de chucha **Habit:** Shrub **Use:** HF **Origin:** Native **Region:** Amazonia **Dept.:** CAQ **Elev.:** 250–250 m a.s.l. **IUCN Habitat code:** 1,14 **Conservation status: (IUCN)** NE

4422 *Coussarea garciae* Standl. (Rubiaceae) **IPNI:** 67416-2 **Common name:** café de monte, hierbadulce **Habit:** Shrub, Small tree, Tree **Use:** HF **Origin:** Native **Region:** Andean, Pacific **Dept.:** ANT, CHO, RIS **Elev.:** 0–1400 m a.s.l. **IUCN Habitat code:** 1,3,14 **Conservation status: (IUCN)** LC

4423 *Coussarea klugii* Steyerm. (Rubiaceae) **IPNI:** 67428-2 **Habit:** Tree **Use:** HF **Origin:** Native **Region:** Amazonia **Dept.:** AMA, CAQ **Elev.:** 100–100 m a.s.l. **IUCN Habitat code:** 1,14 **Conservation status: (IUCN)** DD

4424 *Coussarea latifolia* Standl. (Rubiaceae) **IPNI:** 67431-2 **Habit:** Shrub, Small tree **Use:** HF, ME **Origin:** Native **Region:** Andean, Pacific **Dept.:** CAU, CHO, NAR, VAC **Elev.:** 50–900 m a.s.l. **IUCN Habitat code:** 1 **Conservation status: (IUCN)** LC

4425 *Coussarea leptoloba* (Spreng. ex Benth. & Hook.f.) Müll.Arg. (Rubiaceae) **IPNI:** 747692-1 **Common name:** caimo de monte **Habit:** Shrub **Use:** HF **Origin:** Native **Region:** Amazonia, Andean, Guiana Shield **Dept.:** AMA, VAU **Elev.:** 300–300 m a.s.l. **IUCN Habitat code:** 1 **Conservation status: (IUCN)** NE

4426 *Coussarea resinosa* C.M.Taylor (Rubiaceae) **IPNI:** 319919-2 **Habit:** Shrub **Use:** HF **Origin:** Native **Region:** Amazonia **Dept.:** PUT **Elev.:** 0–300 m a.s.l. **IUCN Habitat code:** 1,3,8,14 **Conservation status: (IUCN)** LC

4427 *Coussarea tenuiflora* Standl. (Rubiaceae) **IPNI:** 67475-2 **Habit:** Tree **Use:** HF **Origin:** Native **Region:** Amazonia **Dept.:** PUT **Elev.:** 300–300 m a.s.l. **Conservation status: (IUCN)** NE

4428 *Coutarea hexandra* (Jacq.) K.Schum. (Rubiaceae) **IPNI:** 67491-2 **Common name:** amargo, azulejo, campanilla, chaparrito, hawapi para mujer, jazmín de mariangola, madiagola, mariangola, mariangola macho, mediagola, napaima, quina **Habit:** Shrub, Small tree, Tree **Use:** MA, ME **Origin:** Native **Region:** Andean, Caribbean, Pacific **Dept.:** ANT, ARA, ATL, BOL, CES, CHO, COR, CUN, HUI, LAG, MAG, NSA, SAN, SUC, TOL **Elev.:** 10-1728 msnm **IUCN Habitat code:** 1,2,3,14 **Conservation status: (IUCN)** LC **(CPSG)** Potential LC

4429 *Cruciata laevipes* Opiz (Rubiaceae) **IPNI:** 747959-1 **Habit:** Herb **Use:** HF, ME **Origin:** Naturalised **Region:** Andean **Dept.:** NSA **Elev.:** 2432–2432 m a.s.l. **Conservation status: (IUCN)** NE

4430 *Duroia bolivarensis* Steyerm. (Rubiaceae) **IPNI:** 86126-2 **Common name:** borojó **Habit:** Tree **Use:** HF **Origin:** Native **Region:** Amazonia, Guiana Shield **Dept.:** AMA, CAQ, VAU **Elev.:** 200–200 m a.s.l. **IUCN Habitat code:** 1,5,14 **Conservation status: (IUCN)** NE

4431 *Duroia duckei* Huber (Rubiaceae) **IPNI:** 86128-2 **Habit:** Tree **Use:** HF **Origin:** Native **Region:** Amazonia **Dept.:** AMA **Elev.:** 80–200 m a.s.l. **Conservation status: (IUCN)** NE

4432 *Duroia genipoides* Hook.f. ex K.Schum. (Rubiaceae) **IPNI:** 86133-2 **Common name:** pepa de monte **Habit:** Shrub, Tree **Use:** HF, ME **Origin:** Native **Region:** Amazonia, Guiana Shield **Dept.:** CAQ, GUA, VID **Elev.:** 85–250 m a.s.l. **IUCN Habitat code:** 1,3 **Conservation status: (IUCN)** LC

4433 *Duroia hirsuta* (Poepp.) K.Schum. (Rubiaceae) **IPNI:** 86135-2 **Common name:** amigo del diablo, babaimio, chacra, coropo, huevas de tigre, kaai, maoguatsegesesa, matamaleza, matamonte, matapalo, matatodo, palo de hormiga, palo de tigre, solimán, taruquillo, turma de mico, turmaemico, turmamico, turmemico, turmemono, turmequé **Habit:** Shrub, Small tree, Tree **Use:** EU, HF, MA, ME **Origin:** Native **Region:** Amazonia, Andean, Guiana Shield, Orinoquia, Pacific **Dept.:** AMA, BOY, CAQ, CAU, CHO, CUN, GUV, MET, PUT, SAN, VAU **Elev.:** 0–1465 m a.s.l. **IUCN Habitat code:** 1,3,4,5,14 **Conservation status: (IUCN)** NE

4434 *Duroia kotchubioides* Steyerm. (Rubiaceae) **IPNI:** 86137-2 **Use:** ME **Origin:** Native **Conservation status: (IUCN)** NE

4435 *Duroia maguirei* Steyerm. (Rubiaceae) **IPNI:** 86141-2 **Common name:** borojó **Habit:** Tree **Use:** HF **Origin:** Native **Region:** Amazonia, Guiana Shield **Dept.:** AMA, CAQ, GUV, VAU **Elev.:** 110–550 m a.s.l. **IUCN Habitat code:** 1,3,14 **Conservation status: (IUCN)** NE

4436 *Duroia micrantha* (Ladbr.) Zarucchi & J.H.Kirkbr. (Rubiaceae) **IPNI:** 276143-2 **Common name:** caimo de pescado, caruto, pintura de zorra **Habit:** Shrub, Tree **Use:** HF, MA **Origin:** Native **Region:** Amazonia, Guiana Shield, Orinoquia **Dept.:** AMA, CAQ, MET, VAU, VID **Elev.:** 90–400 m a.s.l. **IUCN Habitat code:** 1,2,3,4,5,14 **Conservation status: (IUCN)** LC

4437 *Duroia petiolaris* Spruce ex K.Schum. (Rubiaceae) **IPNI:** 86152-2 **Common name:** caimillo ordinario, pintura de zorra **Habit:** Tree **Use:** ME **Origin:** Native **Region:** Amazonia **Dept.:** AMA **Elev.:** 100–100 m a.s.l. **IUCN Habitat code:** 1,3 **Conservation status: (IUCN)** NE

4438 *Duroia saccifera* (Mart. ex Schult. & Schult.f.) K.Schum. (Rubiaceae)
IPNI: 60467403-2
Common name: palo de majiña **Habit:** Shrub, Small tree, Tree **Use:** HF, ME, PO **Origin:** Native **Region:** Amazonia, Guiana Shield **Dept.:** AMA, CAQ, GUA, VAU **Elev.:** 100–160 m a.s.l. **IUCN Habitat code:** 1,3,4,5,14 **Conservation status: (IUCN)** LC

4439 *Elaeagia karstenii* Standl. (Rubiaceae)
IPNI: 88958-2
Common name: copal, marfil **Habit:** Tree **Use:** ME **Origin:** Native **Region:** Andean **Dept.:** ANT, CHO, CUN, RIS **Elev.:** 950–1950 m a.s.l. **IUCN Habitat code:** 1,2,3,14 **Conservation status: (IUCN)** LC

4440 *Elaeagia pastoensis* L.E.Mora (Rubiaceae)
IPNI: 748734-1
Common name: barniz, barniz de Pasto, barniz de pasto, cera de rayo, lacre, mopamopa, rayo **Habit:** Shrub, Tree **Use:** HF, MA, ME **Origin:** Native **Region:** Andean, Pacific **Dept.:** ANT, CAU, CUN, PUT, VAC **Elev.:** 20–2350 m a.s.l. **IUCN Habitat code:** 1,3,14 **Conservation status: (IUCN)** VU

4441 *Elaeagia utilis* (Goudot) Wedd. (Rubiaceae)
IPNI: 748737-1
Common name: azuceno, barniz, barniz blanco, barniz de Pasto, barniz de pasto, canelón, cero, cero mopa, copal, lacre, marfil, mopa, mopamopa, quinocera, quinocero, resino, resume, resune, árbol de cera **Habit:** Shrub, Tree **Use:** ME **Origin:** Native **Region:** Andean **Dept.:** ANT, BOY, CAU, CHO, CUN, MET, NAR, RIS, SAN, TOL, VAC **Elev.:** 630–2500 m a.s.l. **IUCN Habitat code:** 1,3,4,14 **Conservation status: (IUCN)** LC

4442 *Emmeorhiza umbellata* (Spreng.) K.Schum. (Rubiaceae)
IPNI: 90501-2
Common name: ají **Habit:** Herb, Climbing **Use:** ME **Origin:** Native **Region:** Andean, Caribbean **Dept.:** ANT, BOY, CUN, HUI, MAG, MET, NSA, SAN, VAC **Elev.:** 300–2500 m a.s.l. **IUCN Habitat code:** 1,2,3,14 **Conservation status: (IUCN)** NE

4443 *Erithalis fruticosa* L. (Rubiaceae)
IPNI: 315444-2
Habit: Shrub **Use:** ME **Origin:** Native **Region:** Caribbean **Dept.:** SAP **Elev.:** 0–0 m a.s.l. **Conservation status: (IUCN)** LC

4444 *Ernodea littoralis* Sw. (Rubiaceae)
IPNI: 748814-1
Use: MA, ME **Origin:** Native **Conservation status: (IUCN)** LC

4445 *Eumachia microdon* (DC.) Delprete & J.H.Kirkbr. (Rubiaceae)
IPNI: 77154321-1
Habit: Shrub, Small tree, Tree, Climbing **Use:** ME **Origin:** Native **Region:** Andean, Caribbean **Dept.:** ANT, ATL, BOL, LAG, MAG, VAC **Elev.:** 10–1000 m a.s.l. **IUCN Habitat code:** 1,2,3,5,14 **Conservation status: (IUCN)** LC

4446 *Exostema caribaeum* (Jacq.) Schult. (Rubiaceae)
IPNI: 1125915-2
Common name: quina **Use:** MA, ME **Origin:** Native **Dept.:** SAP **Conservation status: (IUCN)** LC

4447 *Faramea anisocalyx* Poepp. (Rubiaceae)
IPNI: 749043-1
Habit: Tree **Use:** MA, ME **Origin:** Native **Region:** Amazonia **Dept.:** AMA, CAQ, PUT, VAU **Elev.:** 100–400 m a.s.l. **IUCN Habitat code:** 1,3,14 **Conservation status: (IUCN)** LC

4448 *Faramea capillipes* Müll.Arg. (Rubiaceae)
IPNI: 749070-1
Common name: cafetillo **Habit:** Shrub, Small tree, Tree **Use:** HF **Origin:** Native **Region:** Amazonia, Andean, Guiana Shield, Caribbean, Orinoquia, Pacific **Dept.:** AMA, ANT, BOL, CAQ, CHO, COR, GUV, SAN, VAU, VID **Elev.:** 20–1350 m a.s.l. **IUCN Habitat code:** 1,3,4,5,14 **Conservation status: (IUCN)** LC

4449 *Faramea glandulosa* Poepp. (Rubiaceae)
IPNI: 749120-1
Common name: simón **Habit:** Shrub, Small tree, Tree **Use:** ME **Origin:** Native **Region:** Amazonia, Andean, Pacific **Dept.:** AMA, ANT, BOY, CAQ, CAU, CHO, CUN, NAR, PUT, VAC **Elev.:** 5–2050 m a.s.l. **IUCN Habitat code:** 1,3,5,14 **Conservation status: (IUCN)** LC

4450 *Faramea latifolia* (Cham. & Schltdl.) DC. (Rubiaceae)
IPNI: 1039243-2
Habit: Small tree **Use:** MA **Origin:** Native **Region:** Andean **Dept.:** MAG **Elev.:** 1830–1830 m a.s.l. **Conservation status: (IUCN)** NE

4451 *Faramea multiflora* A.Rich. ex DC. (Rubiaceae)
IPNI: 749186-1
Common name: ara, cafetillo, guayabo, negrillo, uvo **Habit:** Shrub, Small tree, Tree **Use:** HF **Origin:** Native **Region:** Amazonia, Andean, Guiana Shield, Caribbean, Orinoquia, Pacific **Dept.:** AMA, ANT, BOL, BOY, CAQ, CAU, CHO, CUN, GUA, GUV, MET, NAR, NSA, PUT, SAN, VAU **Elev.:** 0-3043 msnm **IUCN Habitat code:** 1,2,3,5,14 **Conservation status: (IUCN)** LC **(CPSG)** Potential LC

4452 *Faramea occidentalis* (L.) A.Rich. (Rubiaceae)
IPNI: 1187456-2
Common name: comida de danta, huesito, jazmín, malibú, marfilito amarillo, palo de cucharo, piedrito, teack teca **Habit:** Shrub, Small tree, Tree **Use:** MA, ME **Origin:** Native **Region:** Amazonia, Andean, Guiana Shield, Caribbean, Orinoquia, Pacific **Dept.:** AMA, ANT, BOL, CAQ, CAS, CES, CHO, GUV, HUI, MAG, MET, NAR, SAP, SAN, VAC, VAU **Elev.:** 20–1000 m a.s.l. **IUCN Habitat code:** 1,2,3,4,5,14 **Conservation status: (IUCN)** NE

4453 *Faramea tamberlikiana* Müll.Arg. (Rubiaceae)
IPNI: 749271-1
Common name: hierba de danta **Habit:** Shrub, Small tree, Tree **Use:** HF **Origin:** Native **Region:** Amazonia, Andean, Caribbean, Orinoquia **Dept.:** AMA, ANT, BOL, CHO, VID **Elev.:** 0–1800 m a.s.l. **IUCN Habitat code:** 1,3,5,14 **Conservation status: (IUCN)** LC

4454 *Faramea torquata* Müll.Arg. (Rubiaceae)
IPNI: 749277-1
Common name: colmillo de danta, iwitsulinae, tsukunae, ziipaye **Habit:** Shrub, Small tree **Use:** HF, ME **Origin:** Native **Region:** Amazonia, Guiana Shield, Pacific **Dept.:** AMA, CAQ, CHO, GUA, VAC **Elev.:** 200–1800 m a.s.l. **IUCN Habitat code:** 1,3,5,14 **Conservation status: (IUCN)** LC

4455 *Ferdinandusa chlorantha* (Wedd.) Standl. (Rubiaceae)
IPNI: 103549-2
Common name: maní de perico, rabo de gurre **Habit:** Tree **Use:** ME **Origin:** Native **Region:** Amazonia, Guiana Shield **Dept.:** AMA, CAQ **Elev.:** 600–650 m a.s.l. **IUCN Habitat code:** 1,5,14 **Conservation status: (IUCN)** LC

4456 *Galianthe bogotensis* (Willd.) E.L.Cabral & Bacigalupo (Rubiaceae)
IPNI: 999397-1
Common name: botoncito, orégano, sanalotodo, totecito **Habit:** Subshrub **Use:** ME **Origin:** Endemic **Region:** Andean **Dept.:** BOG, BOY, CUN **Elev.:** 2200-3575 msnm **IUCN Habitat code:** 1,3,4,5,8,14 **Conservation status: (IUCN)** NE **(CPSG)** LC

4457 *Galium aparine* L. (Rubiaceae)
IPNI: 30007294-2
Habit: Herb **Use:** HF, MA, ME **Origin:** Naturalised **Region:** Andean **Dept.:** NAR **Elev.:** 2500–3000 m a.s.l. **Conservation status: (IUCN)** NE

4458 *Galium corymbosum* Ruiz & Pav. (Rubiaceae)
IPNI: 749770-1
Habit: Herb, Climbing **Use:** HF **Origin:** Native **Region:** Andean **Dept.:** ANT, BOG, BOY, CUN, NAR, RIS, TOL **Elev.:** 2000–4300 m a.s.l. **IUCN Habitat code:** 1,3,4,8,14 **Conservation status: (IUCN)** NE

4459 *Galium hypocarpium* (L.) Endl. ex Griseb. (Rubiaceae)
IPNI: 750043-1
Common name: ají de monte, bruja, brujita, brujita negra, brujo, castillo de monte, colorado, coral, coralito, coralitos, enredadera, herb de bruja, hierba de bruja, puca, raicilla, rojitos, rubia, yerba de bruja **Habit:** Herb, Epiphyte, Climbing **Use:** HF, ME **Origin:** Native **Region:** Andean **Dept.:** ANT, BOG, BOY, CAL, CAU, CES, CHO, CUN, HUI, LAG, MAG, MET, NAR, NSA, PUT, QUI, RIS, SAN, SUC, TOL, VAC **Elev.:** 700-4350 msnm **IUCN Habitat code:** 1,2,3,4,8,14 **Conservation status: (IUCN)** NE **(CPSG)** Potential LC

4460 *Galium mexicanum* Kunth (Rubiaceae)
IPNI: 750229-1
Common name: cepa de caballo **Habit:** Herb **Use:** ME **Origin:** Native **Region:** Andean **Dept.:** BOG, BOY, CUN, PUT **Elev.:** 2090–3100 m a.s.l. **IUCN Habitat code:** 1,4,8,14 **Conservation status: (IUCN)** NE

4461 *Gardenia jasminoides* J.Ellis (Rubiaceae)
IPNI: 751073-1
Habit: Shrub, Tree **Use:** EU, HF, MA, ME **Origin:** Cultivated **Region:** Andean, Guiana Shield **Dept.:** ANT, BOG, CAQ, CUN, HUI, VAU **Elev.:** 200–2620 m a.s.l. **IUCN Habitat code:** 1,2,3,14 **Conservation status: (IUCN)** NE

4462 *Genipa americana* L. (Rubiaceae)
IPNI: 751331-1
Common name: angelina, angelito, borojó, botellón, cabo de hacha, carate, careto, carreto, carupo, caruta, caruto, chibará, chipará, granado, guaitil, guanapay, guaricha irayol, guate, guayabo de tejado, huito, jago, jagua, jagua maluco, jaguarito, jaguo, jagüito, jal **Habit:** Shrub, Small tree, Tree **Use:** AF, EU, FU, HF, MA, ME, PO, SU **Origin:** Native **Region:** Amazonia, Andean, Guiana Shield, Caribbean, Orinoquia, Pacific **Dept.:** AMA, ANT, ARA, BOL, CAQ, CES, CHO, COR, CUN, GUV, HUI, LAG, MAG, MET, NAR, PUT, QUI, RIS, SAN, TOL, VAC, VAU, VID **Elev.:** 0-1850 msnm **IUCN Habitat code:** 1,2,3,4,5,14 **Conservation status: (IUCN)** NE **(CPSG)** Potential LC

4463 *Genipa spruceana* Steyerm. (Rubiaceae)
IPNI: 108120-2
Habit: Shrub, Small tree, Tree **Use:** HF, MA **Origin:** Native **Region:** Amazonia, Guiana Shield **Dept.:** AMA, CAQ **Elev.:** 100–250 m a.s.l. **IUCN Habitat code:** 1,3,4,14 **Conservation status: (IUCN)** LC

4464 *Geophila cordifolia* Miq. (Rubiaceae)
IPNI: 751420-1
Common name: pepita, sabañón doble **Habit:** Herb, Climbing **Use:** HF **Origin:** Native **Region:** Amazonia, Pacific **Dept.:** AMA, CAQ, CHO, MET, NAR, PUT **Elev.:** 100–1200 m a.s.l. **IUCN Habitat code:** 1,3,5,14 **Conservation status: (IUCN)** NE

4465 *Geophila macropoda* (Ruiz & Pav.) DC. (Rubiaceae)
IPNI: 1080914-2
Common name: carinta, morita
Habit: Herb, Climbing **Use:** EU **Origin:** Native **Region:** Amazonia, Andean, Caribbean, Pacific **Dept.:** AMA, BOL, CAQ, MET, NAR, PUT, RIS, VAC **Elev.:** 60–1040 m a.s.l. **IUCN Habitat code:** 1,3,4,5,14 **Conservation status: (IUCN)** NE

4466 *Geophila repens* (L.) I.M.Johnst. (Rubiaceae)
IPNI: 109024-2
Habit: Herb, Climbing **Use:** HF, ME **Origin:** Native **Region:** Amazonia, Andean, Guiana Shield, Orinoquia, Pacific **Dept.:** AMA, ANT, BOY, CAQ, CAS, CUN, GUA, MAG, MET, PUT, VAC, VAU **Elev.:** 10–1020 m a.s.l. **IUCN Habitat code:** 1,3,4,5,14 **Conservation status: (IUCN)** NE

4467 *Gonzalagunia bunchosioides* Standl. (Rubiaceae)
IPNI: 112490-2
Habit: Shrub **Use:** HF **Origin:** Native **Region:** Amazonia, Andean **Dept.:** CAU, CHO, MAG, PUT, VAC **Elev.:** 40–1950 m a.s.l. **IUCN Habitat code:** 1,14 **Conservation status: (IUCN)** LC

4468 *Gonzalagunia cornifolia* (Kunth) Standl. (Rubiaceae)
IPNI: 112495-2
Habit: Shrub, Tree **Use:** ME **Origin:** Native **Region:** Amazonia, Andean, Orinoquia, Pacific **Dept.:** AMA, ANT, ARA, BOL, CAL, CAQ, CAU, CES, CHO, COR, CUN, HUI, LAG, MAG, MET, NAR, NSA, PUT, RIS, SAN, VAC **Elev.:** 0-1800 msnm **IUCN Habitat code:** 1,2,3,4,5,14 **Conservation status: (IUCN)** NE **(CPSG)** Potential LC

4469 *Gonzalagunia dicocca* Cham. & Schltdl. (Rubiaceae)
IPNI: 751530-1
Habit: Shrub **Use:** HF **Origin:** Native **Region:** Andean **Dept.:** CUN, NAR, PUT **Elev.:** 1400–1890 m a.s.l. **IUCN Habitat code:** 1 **Conservation status: (IUCN)** NE

4470 *Gonzalagunia ovatifolia* (Donn.Sm.) B.L.Rob. (Rubiaceae)
IPNI: 112521-2
Habit: Shrub **Use:** HF **Origin:** Native **Region:** Pacific **Dept.:** CHO **Elev.:** 5–300 m a.s.l. **IUCN Habitat code:** 1,14 **Conservation status: (IUCN)** LC

4471 *Gonzalagunia panamensis* (Cav.) K.Schum. (Rubiaceae)
IPNI: 30107991-2
Common name: palito equis **Habit:** Shrub, Tree **Use:** ME **Origin:** Native **Region:** Andean, Caribbean, Pacific **Dept.:** ANT, BOY, CAU, CHO, LAG, MAG, RIS, SAN, VAC **Elev.:** 0–1860 m a.s.l. **IUCN Habitat code:** 1,3,5,14 **Conservation status: (IUCN)** NE

4472 *Gonzalagunia rudis* (Standl.) Standl. (Rubiaceae)
IPNI: 112531-2
Habit: Shrub, Subshrub **Use:** ME **Origin:** Native **Region:** Andean, Pacific **Dept.:** ANT, CHO, MAG, RIS, VAC **Elev.:** 50–2100 m a.s.l. **IUCN Habitat code:** 1,14 **Conservation status: (IUCN)** NE

4473 *Guettarda divaricata* (Humb. & Bonpl. ex Schult.) Standl. (Rubiaceae)
IPNI: 114524-2
Habit: Shrub, Tree **Use:** HF **Origin:** Native **Region:** Andean **Dept.:** LAG **Elev.:** 650–650 m a.s.l. **IUCN Habitat code:** 1 **Conservation status: (IUCN)** LC

4474 *Guettarda elliptica* Sw. (Rubiaceae)
IPNI: 114528-2
Habit: Shrub, Small tree **Use:** MA, ME **Origin:** Native **Region:** Caribbean, Pacific **Dept.:** BOL, SAP **Elev.:** 0–230 m a.s.l. **IUCN Habitat code:** 1,2,3,14 **Conservation status: (IUCN)** LC

4475 *Guettarda odorata* (Jacq.) Lam. (Rubiaceae)
IPNI: 752006-1
Habit: Shrub, Small tree **Use:** MA **Origin:** Native **Region:** Caribbean **Dept.:** ATL, BOL, CES **Elev.:** 100–100 m a.s.l. **IUCN Habitat code:** 1,2,4 **Conservation status: (IUCN)** NE

4476 *Guettarda scabra* (L.) Vent. (Rubiaceae)
IPNI: 30058289-2
Use: HF, MA **Origin:** Native **IUCN Habitat code:** 2 **Conservation status: (IUCN)** LC

4477 *Hamelia axillaris* Sw. (Rubiaceae)
IPNI: 117176-2
Common name: amargura, coralillo
Habit: Shrub, Subshrub, Small tree **Use:** ME **Origin:** Native **Region:** Amazonia, Andean, Guiana Shield, Caribbean, Pacific **Dept.:** AMA, ANT, CAL, CHO, CUN, GUV, LAG, MAG, MET, NAR, VAC **Elev.:** 0–2250 m a.s.l. **IUCN Habitat code:** 1,2,3,5,14 **Conservation status: (IUCN)** NE

4478 *Hamelia patens* Jacq. (Rubiaceae)
IPNI: 752194-1
Common name: ajicito, anavaco, bencenuco, buzunuco, canuto, chichipince, colorado, coral, coralillo, coralitas, coralito, corchito, cresta de gallo, cuetillo, herb de eripsepela, hoja de carpintero, koray, leoncico, leoncito, pata de totoco, quiebra olla, quiebraolla, **Habit:** Shrub, Small tree, Tree **Use:** EU, HF, MA, ME, PO **Origin:** Native **Region:** Amazonia, Andean, Guiana Shield, Caribbean, Orinoquia, Pacific **Dept.:** AMA, ANT, ARA, ATL, BOG, BOL, BOY, CAL, CAQ, CES, CHO, CUN, GUV, HUI, LAG, MAG, MET, NAR, NSA, PUT, QUI, RIS, SAP, SAN, SUC, TOL, VAC, VAU **Elev.:** 0-3700 msnm **IUCN Habitat code:** 1,2,3,5,14 **Conservation status: (IUCN)** LC **(CPSG)** Potential LC

4479 *Hexasepalum sarmentosum* (Sw.) Delprete & J.H.Kirkbr. (Rubiaceae)
IPNI: 77154343-1
Habit: Herb, Subshrub **Use:** ME **Origin:** Native **Region:** Andean, Pacific **Dept.:** ANT, CHO, HUI, MAG **Elev.:** 5–1400 m a.s.l. **IUCN Habitat code:** 1,2,3,4,5 **Conservation status: (IUCN)** NE

4480 *Hexasepalum serrulatum* (P.Beauv.) J.H.Kirkbr. & Delprete (Rubiaceae)
IPNI: 77154345-1
Habit: Herb **Use:** EU, MA, ME **Origin:** Native **Region:** Caribbean **Dept.:** ATL **Elev.:** 60–60 m a.s.l. **Conservation status: (IUCN)** NE

4481 *Hillia macrophylla* Standl. (Rubiaceae)
IPNI: 122668-2
Habit: Shrub, Subshrub, Tree **Use:** ME **Origin:** Native **Region:** Andean, Guiana Shield **Dept.:** ANT, BOY, CHO, CUN, MET, NAR, NSA, QUI, VAC **Elev.:** 620–2300 m a.s.l. **IUCN Habitat code:** 1,14 **Conservation status: (IUCN)** NE

4482 *Hoffmannia viridis* Rusby (Rubiaceae)
IPNI: 123345-2
Habit: Herb, Shrub, Subshrub **Use:** HF **Origin:** Native **Region:** Andean, Pacific **Dept.:** ANT, CAL, CHO, LAG, MAG, PUT, VAC **Elev.:** 100–1750 m a.s.l. **IUCN Habitat code:** 1,3,14 **Conservation status: (IUCN)** NE

4483 *Isertia coccinea* (Aubl.) J.F.Gmel. (Rubiaceae)
IPNI: 30281446-2
Common name: carne de vaca
Habit: Shrub, Small tree **Use:** HF, ME **Origin:** Native **Region:** Amazonia, Guiana Shield **Dept.:** AMA, CAQ, MET, PUT, VAU **Elev.:** 100–240 m a.s.l. **Conservation status: (IUCN)** LC

4484 *Isertia haenkeana* DC. (Rubiaceae)
IPNI: 753632-1
Common name: carutillo, coralillo, cuyarocrí, flor de mayo, guacharaco, pava, tabaquillo **Habit:** Shrub, Small tree, Tree **Use:** EU, ME **Origin:** Native **Region:** Amazonia, Andean, Pacific **Dept.:** AMA, ANT, BOL, BOY, CAL, CHO, GUV, MET, PUT, SAN, TOL **Elev.:** 0–2300 m a.s.l. **IUCN Habitat code:** 1,2,3,4,5,14 **Conservation status: (IUCN)** NE

4485 *Isertia hypoleuca* Benth. (Rubiaceae)
IPNI: 753635-1
Common name: capacira, tefé, tefé-capacira **Habit:** Shrub, Small tree, Tree **Use:** HF, ME **Origin:** Native **Region:** Amazonia, Guiana Shield **Dept.:** AMA, PUT **Elev.:** 100–160 m a.s.l. **IUCN Habitat code:** 1,3,14 **Conservation status: (IUCN)** LC

4486 *Isertia laevis* (Triana) B.M.Boom (Rubiaceae)
IPNI: 278892-2
Common name: asaquiro, asaquirú, azuceno, cruceto, guacharaco, guacharaco pavero, jaboncillo, jabonsillo, llorón, nazareno, quino, tabaquillo, vara

santa, varablanca, varasanta **Habit:** Shrub, Small tree, Tree **Use:** ME **Origin:** Native **Region:** Andean, Pacific **Dept.:** ANT, BOY, CAL, CAQ, CAS, CAU, CHO, CUN, HUI, MET, PUT, SAN, TOL, VAC **Elev.:** 0–2000 m a.s.l. **IUCN Habitat code:** 1,2,3,4,5,14 **Conservation status: (IUCN)** LC

4487 *Isertia parviflora* Vahl (Rubiaceae)
IPNI: 753641-1
Habit: Shrub, Small tree **Use:** MA **Origin:** Native **Region:** Guiana Shield **Dept.:** MET **Elev.:** 315–315 m a.s.l. **IUCN Habitat code:** 1,3,5,14 **Conservation status: (IUCN)** LC

4488 *Isertia pittieri* (Standl.) Standl. (Rubiaceae)
IPNI: 130336-2
Common name: azuceno, jaboncillo, jabón, mapán, mazamorra roja, mazamorro, palo mazamorra, platanillo **Habit:** Shrub, Small tree, Tree **Use:** ME **Origin:** Native **Region:** Andean, Pacific **Dept.:** ANT, CAU, CHO, NAR, VAC **Elev.:** 0–1700 m a.s.l. **IUCN Habitat code:** 1,3,5,14 **Conservation status: (IUCN)** NE

4489 *Isertia rosea* Spruce ex K.Schum. (Rubiaceae)
IPNI: 278894-2
Habit: Shrub, Small tree, Tree **Use:** ME **Origin:** Native **Region:** Amazonia, Guiana Shield **Dept.:** AMA, CAQ, GUA, MET, PUT, VAU, VID **Elev.:** 80–450 m a.s.l. **IUCN Habitat code:** 1,3,4,5,14 **Conservation status: (IUCN)** LC

4490 *Ixora casei* Hance (Rubiaceae)
IPNI: 753825-1
Common name: bouquet de novia, coral **Habit:** Shrub, Tree **Use:** ME **Origin:** Cultivated, Naturalised **Region:** Andean, Guiana Shield **Dept.:** CAL, GUV, VAC **Elev.:** 200–280 m a.s.l. **IUCN Habitat code:** 1 **Conservation status: (IUCN)** NE

4491 *Ixora chinensis* Lam. (Rubiaceae)
IPNI: 753834-1
Use: EU, HF, MA, ME **Origin:** Naturalised **IUCN Habitat code:** 1,3 **Conservation status: (IUCN)** NE

4492 *Ixora coccinea* L. (Rubiaceae)
IPNI: 753844-1
Common name: amor ardiente, amor de madre, buqué, buqué costeño, buqué de novia, coral, coral de la India, coral extranjero, coral gaitán, coral rojo, coral sencillo, coralillo, coralito, costeño, icaco, isora, jazmín rojo **Habit:** Shrub **Use:** EU, HF, MA, ME **Origin:** Cultivated, Naturalised **Region:** Amazonia, Andean, Caribbean, Pacific **Dept.:** AMA, ANT, ATL, BOL, CAU, COR, CUN, MAG, NAR, VAC **Elev.:** 0–1088 m a.s.l. **IUCN Habitat code:** 1,2,3,4,14 **Conservation status: (IUCN)** NE

4493 *Ixora finlaysoniana* Wall. ex G.Don (Rubiaceae)
IPNI: 753956-1
Common name: bouquet de novia, buqué de novia, jazmín de novio **Habit:** Shrub **Use:** EU, ME **Origin:** Cultivated, Naturalised **Region:** Amazonia, Andean, Caribbean **Dept.:** AMA, CUN, MAG, SAP, TOL, VAC **Elev.:** 0–1100 m a.s.l. **Conservation status: (IUCN)** NE

4494 *Ixora killipii* Standl. (Rubiaceae)
IPNI: 130853-2
Habit: Small tree, Tree **Use:** HF **Origin:** Native **Region:** Amazonia, Guiana Shield **Dept.:** AMA, MET, PUT **Elev.:** 100–400 m a.s.l. **IUCN Habitat code:** 1,3,5,14 **Conservation status: (IUCN)** LC

4495 *Ixora macrothyrsa* (Teijsm. & Binn.) N.E.Br. (Rubiaceae)
IPNI: 60459342-2
Common name: coral, coral doble, coralito **Habit:** Shrub **Use:** EU, ME **Origin:** Cultivated, Naturalised **Region:** Andean, Caribbean **Dept.:** ATL, CUN **IUCN Habitat code:** 14 **Conservation status: (IUCN)** NE

4496 *Kutchubaea micrantha* Steyerm. (Rubiaceae)
IPNI: 133852-2
Habit: Tree **Use:** HF **Origin:** Native **Region:** Amazonia **Dept.:** AMA **Elev.:** 100–250 m a.s.l. **IUCN Habitat code:** 1 **Conservation status: (IUCN)** NE

4497 *Kutchubaea sericantha* Standl. (Rubiaceae)
IPNI: 133858-2
Habit: Tree **Use:** HF **Origin:** Native **Region:** Amazonia **Dept.:** AMA **Elev.:** 100–300 m a.s.l. **IUCN Habitat code:** 1 **Conservation status: (IUCN)** NE

4498 *Ladenbergia carua* (Wedd.) Standl. (Rubiaceae)
IPNI: 754813-1
Common name: cascarilla, falsa quina **Use:** ME **Dept.:** CAQ **Conservation status: (IUCN)** LC

4499 *Ladenbergia macrocarpa* (Vahl) Klotzsch (Rubiaceae)
IPNI: 1215741-2
Common name: azuceno, cascarilla, cascarillo, falsa quina, lirio, quina, quina blanca, quina roja, quino **Habit:** Small tree, Tree **Use:** ME **Origin:** Native **Region:** Andean **Dept.:** ANT, BOG, BOY, CAL, CAU, CHO, CUN, HUI, MET, NAR, NSA, PUT, QUI, RIS, SAN, TOL, VAC **Elev.:** 800–3400 m a.s.l. **IUCN Habitat code:** 1,3,5,14 **Conservation status: (IUCN)** NE

4500 *Ladenbergia muzonensis* (Goudot) Standl. (Rubiaceae)
IPNI: 134610-2
Common name: quina **Habit:** Small tree, Tree **Use:** ME **Origin:** Native **Region:** Andean, Guiana Shield, Pacific **Dept.:** AMA, ANT, BOL, BOY, CAQ, CHO, NSA, PUT, SAN, VAC, VAU **Elev.:** 75–2000 m a.s.l. **IUCN Habitat code:** 1,3,14 **Conservation status: (IUCN)** LC

4501 *Ladenbergia oblongifolia* (Humb. ex Mutis) L.Andersson (Rubiaceae)
IPNI: 1106418-2
Common name: azahar macho, azuceno, azuceno cascarillo, brazil, carnefresca, cascarilla, cascarillo, cascarillo amarillo, cascarillo blanco, cascarillo hojigrande, guacamayo, quina, quina amarilla, quina roja, quina tuna, quino, quino rojo, reventador **Habit:** Tree **Use:** HF, MA, ME **Origin:** Native **Region:** Andean, Guiana Shield **Dept.:** ANT, BOL, BOY, CAQ, CAU, CES, CUN, GUV, HUI, NAR, PUT, QUI, RIS, TOL, VAC, VAU **Elev.:** 150–2340 m a.s.l. **IUCN Habitat code:** 1,3,4,5,14 **Conservation status: (IUCN)** LC

4502 *Macrocnemum roseum* (Ruiz & Pav.) Wedd. (Rubiaceae)
IPNI: 755513-1
Common name: azuceno, cascajero **Habit:** Small tree, Tree **Use:** MA, ME **Origin:** Native **Region:** Andean, Pacific **Dept.:** ANT, BOL, CAL, CAQ, CAU, CHO, HUI, QUI, VAC **Elev.:** 0–2700 m a.s.l. **IUCN Habitat code:** 1,2,3,5,14 **Conservation status: (IUCN)** LC

4503 *Manettia cordifolia* Mart. (Rubiaceae)
IPNI: 755652-1
Habit: Climbing **Use:** ME **Origin:** Native **Dept.:** ANT, SAN **Elev.:** 200–500 m a.s.l. **IUCN Habitat code:** 1,3 **Conservation status: (IUCN)** NE

4504 *Manettia reclinata* L. (Rubiaceae)
IPNI: 755756-1
Habit: Climbing **Use:** ME **Origin:** Native **Region:** Andean, Caribbean **Dept.:** ANT, BOL, CAL, CES, CHO, CUN, LAG, MAG, MET, SAN, TOL, VAC **Elev.:** 5–1385 m a.s.l. **IUCN Habitat code:** 1,2,3,14 **Conservation status: (IUCN)** NE

4505 *Mitracarpus hirtus* (L.) DC. (Rubiaceae)
IPNI: 756228-1
Common name: estrellitas **Habit:** Herb, Subshrub, Climbing **Use:** AF, EU, HF, ME, PO, SU **Origin:** Native **Region:** Andean, Orinoquia, Pacific **Dept.:** ANT, BOY, CAU, CHO, HUI, MET, VAC, VID **Elev.:** 0–1950 m a.s.l. **IUCN Habitat code:** 1,2,3,5,14 **Conservation status: (IUCN)** NE

4506 *Morinda citrifolia* L. (Rubiaceae)
IPNI: 756359-1
Common name: fey doulé, manzana de cerdo, manzana de puerco, monkey apple, noni, piñuela **Habit:** Shrub, Tree **Use:** EU, FU, GS, HF, MA, ME, SU **Origin:** Cultivated, Naturalised **Region:** Andean, Caribbean, Pacific **Dept.:** ANT, CHO, COR, QUI, SAP **Elev.:** 12–1200 m a.s.l. **IUCN Habitat code:** 1,2,3,5,14 **Conservation status: (IUCN)** NE

4507 *Morinda panamensis* Seem. (Rubiaceae)
IPNI: 756446-1
Habit: Shrub **Use:** HF, ME **Origin:** Native **Region:** Caribbean, Pacific **Dept.:** BOL, CHO, MAG **Elev.:** 10–100 m a.s.l. **IUCN Habitat code:** 1 **Conservation status: (IUCN)** LC

4508 *Morinda royoc* L. (Rubiaceae)
IPNI: 756470-1
Common name: piñita **Habit:** Shrub, Climbing **Use:** HF, ME **Origin:** Native **Region:** Caribbean **Dept.:** BOL, CHO, COR, LAG, MAG, SAP, SUC **Elev.:** 0–400 m a.s.l. **IUCN Habitat code:** 1,2,3,14 **Conservation status: (IUCN)** LC

4509 *Mussaenda erythrophylla* Schumach. & Thonn. (Rubiaceae)
IPNI: 756618-1
Common name: musaenda, musaendo, musandra **Use:** EU, HF, MA, ME, SU **IUCN Habitat code:** 1,14 **Conservation status: (IUCN)** LC

4510 *Nertera granadensis* (Mutis ex L.f.) Druce (Rubiaceae)
IPNI: 757486-1
Common name: coral, coral de agua, coralito, coralito de agua, coralitos, tomatillo, uvita de agua, yerba de sapo **Habit:** Herb, Climbing **Use:** AF, EU, HF, IF, ME, SU **Origin:** Native **Region:** Andean **Dept.:** ANT, ARA, BOG, BOY, CAL, CAU, CES, CHO, CUN, HUI, MAG, MET, NAR, NSA, PUT, QUI, RIS, SAN, TOL, VAC **Elev.:** 1300-4300 msnm **IUCN Habitat code:** 1,2,3,4,6,8,14 **Conservation status: (IUCN)** NE **(CPSG)** Potential LC

4511 *Notopleura uliginosa* (Sw.) Bremek. (Rubiaceae)
IPNI: 170751-2
Common name: azulejo **Habit:** Herb, Shrub **Use:** ME **Origin:** Native **Region:** Amazonia, Andean, Caribbean, Pacific **Dept.:** AMA, ANT, BOY, CAU, CES, CHO, HUI, MAG, NAR, PUT, RIS, VAC **Elev.:** 0–2286 m a.s.l. **IUCN Habitat code:** 1,2,3,5,14 **Conservation status: (IUCN)** NE

4512 *Oldenlandia corymbosa* L. (Rubiaceae)
IPNI: 757791-1
Habit: Herb **Use:** AF, HF, MA, ME, SU **Origin:** Naturalised **Region:** Amazonia, Caribbean, Orinoquia, Pacific **Dept.:** ARA, ATL, BOG, CAU, CHO, MET, NAR, TOL, VAC **Elev.:** 0–920 m a.s.l. **IUCN Habitat code:** 1,2,3,4,14 **Conservation status: (IUCN)** NE

4513 *Oldenlandia herbacea* (L.) Roxb. (Rubiaceae)
IPNI: 757892-1
Use: ME **Origin:** Naturalised **IUCN Habitat code:** 1,3,14 **Conservation status: (IUCN)** NE

4514 *Oldenlandia lancifolia* (Schumach.) DC. (Rubiaceae)
IPNI: 757942-1
Habit: Herb **Use:** HF, MA, ME **Origin:** Naturalised **Region:** Amazonia, Andean, Orinoquia, Pacific **Dept.:** AMA, ANT, CAQ, CAS, CHO, MET, PUT, SAN, TOL, VAC **Elev.:** 5–1500 m a.s.l. **IUCN Habitat code:** 1,2,3,4,5,14 **Conservation status: (IUCN)** LC

4515 *Pagamea coriacea* Spruce ex Benth. (Rubiaceae)
IPNI: 546862-1
Habit: Tree **Use:** ME **Origin:** Native **Region:** Amazonia, Andean, Guiana Shield **Dept.:** AMA, CAQ, GUA, VAU **Elev.:** 100–1800 m a.s.l. **IUCN Habitat code:** 1,3,4,5,14 **Conservation status: (IUCN)** LC

4516 *Pagamea macrophylla* Spruce ex Benth. (Rubiaceae)
IPNI: 546865-1
Common name: pintura de zorra **Habit:** Shrub, Small tree **Use:** ME **Origin:** Native **Region:** Amazonia, Guiana Shield **Dept.:** VAU **Elev.:** 200–200 m a.s.l. **IUCN Habitat code:** 1 **Conservation status: (IUCN)** NE

4517 *Pagamea plicata* Spruce ex Benth. (Rubiaceae)
IPNI: 546866-1
Common name: caimitillo **Habit:** Shrub, Small tree **Use:** HF **Origin:** Native **Region:** Amazonia, Guiana Shield **Dept.:** AMA, CAQ, GUA, VAU **Elev.:** 100–300 m a.s.l. **IUCN Habitat code:** 1,4,5,14 **Conservation status: (IUCN)** LC

4518 *Palicourea acuminata* (Benth.) Borhidi (Rubiaceae)
IPNI: 77115588-1
Common name: cordoncillo paramero **Habit:** Shrub **Use:** ME **Origin:** Native **Region:** Amazonia, Andean, Caribbean, Orinoquia, Pacific **Dept.:** AMA, ANT, BOY, CAL, CAQ, CAS, CAU, CES, CHO, CUN, HUI, MAG, MET, NAR, NSA, PUT, RIS, SAN, VAC, VAU **Elev.:** 0-2700 msnm **IUCN Habitat code:** 1,3,5,14 **Conservation status: (IUCN)** LC **(CPSG)** Potential LC

4519 *Palicourea alba* (Aubl.) Delprete & J.H.Kirkbr. (Rubiaceae)
IPNI: 60473597-2
Habit: Herb **Use:** ME **Origin:** Native **Region:** Amazonia **Dept.:** PUT **Elev.:** 0–300 m a.s.l. **IUCN Habitat code:** 1,14 **Conservation status: (IUCN)** NE

4520 *Palicourea angustifolia* Kunth (Rubiaceae)
IPNI: 319913-2
Common name: aguadulce, cafeto de monte, café, café de montaña, café de monte, cordoncillo, crucito, granicillo, sanalotodo, tinto **Habit:** Shrub, Small tree, Tree **Use:** HF **Origin:** Native **Region:** Andean, Guiana Shield **Dept.:** ANT, BOG, BOY, CAL, CAQ, CAS, CAU, CES, CHO, CUN, HUI, MAG, MET, NAR, NSA, PUT, QUI, RIS, SAN, TOL, VAC **Elev.:** 500-3600 msnm **IUCN Habitat code:** 1,2,3,4,5,8,14 **Conservation status: (IUCN)** LC **(CPSG)** Potential LC

4521 *Palicourea berteroana* (DC.) Borhidi (Rubiaceae)
IPNI: 77115593-1
Habit: Shrub, Small tree, Tree **Use:** HF **Origin:** Native **Region:** Andean, Caribbean, Pacific **Dept.:** ANT, CHO, LAG, QUI **Elev.:** 200–1880 m a.s.l. **IUCN Habitat code:** 1,3,5,14 **Conservation status: (IUCN)** NE

4522 *Palicourea blepharophora* (Standl.) Borhidi (Rubiaceae)
IPNI: 77161948-1
Habit: Shrub **Use:** ME **Origin:** Native **Region:** Andean **Dept.:** NSA **Elev.:** 1800–2200 m a.s.l. **Conservation status: (IUCN)** NE

4523 *Palicourea boqueronensis* (Wernham) C.M.Taylor (Rubiaceae)
IPNI: 77202779-1
Common name: amarilla, cacaíto, café de páramo, mirto, tinto **Habit:** Shrub **Use:** ME **Origin:** Endemic **Region:** Andean **Dept.:** BOG, BOY, CUN, NAR, SAN **Elev.:** 1800–2900 m a.s.l. **IUCN Habitat code:** 3,14 **Conservation status: (IUCN)** NE

4524 *Palicourea brachiata* (Sw.) Borhidi (Rubiaceae)
IPNI: 77115595-1
Common name: algodoncillo, chifla, morao **Habit:** Shrub, Small tree, Tree **Use:** ME **Origin:** Native **Region:** Amazonia, Andean, Guiana Shield, Caribbean, Orinoquia, Pacific **Dept.:** AMA, ANT, CAL, CAQ, CAU, CHO, CUN, GUA, GUV, HUI, LAG, MAG, MET, NAR, NSA, PUT, QUI, SAN, TOL, VAC, VAU **Elev.:** 10–2590 m a.s.l. **IUCN Habitat code:** 1,2,3,5,14 **Conservation status: (IUCN)** LC

4525 *Palicourea bracteocardia* (DC.) Delprete & J.H.Kirkbr. (Rubiaceae)
IPNI: 60473604-2
Common name: metsajakobejapanae, planta mano de danta **Habit:** Shrub **Use:** ME **Origin:** Native **Region:** Amazonia, Andean, Guiana Shield, Orinoquia **Dept.:** ANT, BOL, CAQ, CAS, CUN, GUV, MAG, MET, NSA, SAN, TOL, VAU, VID **Elev.:** 50–1000 m a.s.l. **IUCN Habitat code:** 1,2,3,4,5,14 **Conservation status: (IUCN)** NE

4526 *Palicourea colorata* (Hoffmanns. ex Willd.) Delprete & J.H.Kirkbr. (Rubiaceae)
IPNI: 77161054-1
Habit: Shrub, Subshrub **Use:** ME **Origin:** Native **Region:** Amazonia, Andean, Guiana Shield **Dept.:** AMA, ANT, BOL, BOY, CAL, CAQ, COR, GUA, MET, SAN, TOL, VAU **Elev.:** 50–800 m a.s.l. **IUCN Habitat code:** 1,3,14 **Conservation status: (IUCN)** NE

4527 *Palicourea crocea* (Sw.) Schult. (Rubiaceae)
IPNI: 759061-1
Common name: cascajero, caumano, flor de muerto, pata de picón **Habit:** Shrub, Subshrub, Small tree **Use:** ME, PO **Origin:** Native **Region:** Amazonia, Caribbean, Orinoquia, Pacific **Dept.:** AMA, ANT, BOG, BOL, CAL, CAQ, CES, CHO, COR, CUN, MAG, MET, PUT, QUI, VAC, VAU, VID **Elev.:** 5-400 msnm **IUCN Habitat code:** 1,2,3,4,5,14 **Conservation status: (IUCN)** NE **(CPSG)** Potential LC

4528 *Palicourea cuspidata* (Bredem. ex Schult.) C.M.Taylor (Rubiaceae)
IPNI: 77148083-1
Habit: Shrub, Small tree **Use:** ME **Origin:** Native **Region:** Andean, Pacific **Dept.:** ANT, CAU, COR, MAG, SAN, VAC **Elev.:** 100–2100 m a.s.l. **IUCN Habitat code:** 1,3,4,5,14 **Conservation status: (IUCN)** NE

4529 *Palicourea deflexa* (DC.) Borhidi (Rubiaceae)
IPNI: 77115599-1
Common name: comida de danta **Habit:** Shrub, Small tree, Tree **Use:** AF,ME **Origin:** Native **Region:** Amazonia, Andean, Guiana Shield, Caribbean, Orinoquia, Pacific **Dept.:** AMA, ANT, BOL, CAL, CAQ, CES, CHO, COR, CUN, GUA, GUV, LAG, MAG, MET, NSA, PUT, SAN, TOL, VAC, VAU **Elev.:** 20-2300 msnm **IUCN Habitat code:** 1,2,3,4,5,14 **Conservation status: (IUCN)** NE **(CPSG)** Potential LC

4530 *Palicourea egensis* (Müll.Arg.) Borhidi (Rubiaceae)
IPNI: 77161866-1
Habit: Shrub **Use:** ME **Origin:** Native **Region:** Amazonia, Guiana Shield, Orinoquia **Dept.:** AMA, CAQ, GUA, GUV, MET, VAU, VID **Elev.:** 160–635 m a.s.l. **IUCN Habitat code:** 1,4,14 **Conservation status: (IUCN)** NE

4531 *Palicourea elata* (Sw.) Borhidi (Rubiaceae)
IPNI: 77115600-1
Common name: boca de diablo **Habit:** Shrub, Small tree, Tree **Use:** ME **Origin:** Native **Region:** Amazonia, Andean, Pacific **Dept.:** AMA, ANT, CHO, NAR **Elev.:** 0–1450 m a.s.l. **IUCN Habitat code:** 1,3,5,14 **Conservation status: (IUCN)** LC

4532 *Palicourea galeottiana* M.Martens (Rubiaceae)
IPNI: 281967-2
Habit: Shrub, Small tree **Use:** HF **Origin:** Native **Region:** Andean **Dept.:** ANT, CHO **Elev.:** 1100–1354 m a.s.l. **IUCN Habitat code:** 1 **Conservation status: (IUCN)** LC

4533 *Palicourea glomerulata* (Donn.Sm.) Borhidi (Rubiaceae)
IPNI: 77115607-1
Habit: Shrub, Subshrub, Tree **Use:** ME **Origin:** Native **Region:** Pacific **Dept.:** ANT, CAU, CHO, VAC **Elev.:** 0–500 m a.s.l. **IUCN Habitat code:** 1,2 **Conservation status: (IUCN)** LC

4534 *Palicourea gracilenta* (Müll.Arg.) Delprete & J.H.Kirkbr. (Rubiaceae)
IPNI: 77161069-1
Common name: ubejuici **Habit:** Shrub, Subshrub **Use:** ME **Origin:** Native **Region:** Amazonia, Andean, Guiana Shield, Caribbean, Orinoquia, Pacific **Dept.:** AMA, ANT, ARA, CAQ, CHO, GUA, LAG, MET, SAN, VAU, VID **Elev.:** 60–1500 m a.s.l. **IUCN Habitat code:** 1,3 **Conservation status: (IUCN)** LC

4535 *Palicourea grandiflora* (Kunth) Standl. (Rubiaceae)
IPNI: 181012-2
Habit: Shrub, Subshrub, Small tree, Tree **Use:** ME, PO **Origin:** Native **Region:** Amazonia, Andean, Guiana Shield **Dept.:** AMA, ANT, CAQ, GUA, PUT, VAU **Elev.:** 75–1420 m a.s.l. **IUCN Habitat code:** 1,3,4,5,14 **Conservation status: (IUCN)** LC

4536 *Palicourea guianensis* Aubl. (Rubiaceae)
IPNI: 759121-1
Common name: cafetillo, cagalera, camacho, corpus, flor de mayo, floramarillo, jaboncillo, rodillón, taurutoto **Habit:** Shrub, Small tree, Tree **Use:** EU, MA, ME **Origin:** Native **Region:** Amazonia, Andean, Guiana Shield, Caribbean, Orinoquia, Pacific **Dept.:** AMA, ANT, BOL, BOY, CAL, CAQ, CAU, CES, CHO, COR, CUN, GUA, HUI, MAG, MET, NAR, PUT, RIS, SAN, TOL, VAC, VAU **Elev.:** 0-3150 msnm **IUCN Habitat code:** 1,2,3,4,5,14 **Conservation status: (IUCN)** NE **(CPSG)** Potential LC

4537 *Palicourea hebeclada* (DC.) Borhidi (Rubiaceae)
IPNI: 77115609-1
Habit: Shrub, Subshrub **Use:** ME **Origin:** Native **Region:** Andean, Caribbean, Pacific **Dept.:** ANT, BOL, CHO, MAG, RIS, SAN, VAC **Elev.:** 10–1460 m a.s.l. **IUCN Habitat code:** 1,2,3,5,14 **Conservation status: (IUCN)** NE

4538 *Palicourea hoffmannseggiana* (Willd. ex Schult.) Borhidi (Rubiaceae)
IPNI: 77115613-1
Common name: mortiño **Habit:** Herb, Subshrub **Use:** ME **Origin:** Native **Region:** Amazonia, Andean, Guiana Shield, Orinoquia, Pacific **Dept.:** AMA, ANT, BOL, CAS, CHO, CUN, GUA, GUV, MET, PUT, SAN, VAC, VID **Elev.:** 5–1450 m a.s.l. **IUCN Habitat code:** 1,3,4,5,14 **Conservation status: (IUCN)** LC

4539 *Palicourea humboldtiana* (Cham.) Delprete & J.H.Kirkbr. (Rubiaceae)
IPNI: 77161073-1
Habit: Herb, Shrub **Use:** ME **Origin:** Native **Region:** Amazonia, Guiana Shield, Pacific **Dept.:** AMA, CHO, GUA, VAU **Elev.:** 150–300 m a.s.l. **Conservation status: (IUCN)** NE

4540 *Palicourea iodotricha* (Müll.Arg.) Delprete & J.H.Kirkbr. (Rubiaceae)
IPNI: 77161076-1
Habit: Shrub **Use:** HF **Origin:** Native **Region:** Amazonia, Guiana Shield **Dept.:** AMA, CAQ, VAU **Elev.:** 100–400 m a.s.l. **Conservation status: (IUCN)** NE

4541 *Palicourea justiciifolia* (Rudge) Delprete & J.H.Kirkbr. (Rubiaceae)
IPNI: 77161077-1
Common name: pulsario caspi, sisipiyanae **Habit:** Herb, Shrub, Subshrub **Use:** ME **Origin:** Native **Region:** Amazonia, Andean, Guiana Shield, Orinoquia, Pacific **Dept.:** AMA, ANT, BOY, CAQ, CAS, CHO, CUN, GUA, GUV, MET, SAN, VAU **Elev.:** 0–1871 m a.s.l. **Conservation status: (IUCN)** NE

4542 *Palicourea lineariflora* Wernham (Rubiaceae)
IPNI: 181080-2
Common name: amargo, amarillo, muyo pava, muyo-pava, saca ojo, sacaojo, tinto **Habit:** Shrub, Small tree **Use:** ME **Origin:** Native **Region:** Amazonia, Andean, Orinoquia **Dept.:** BOG, BOY, CUN, MET, TOL, VAU **Elev.:** 180-3100 msnm **IUCN Habitat code:** 1,3,14 **Conservation status: (IUCN)** LC **(CPSG)** LC

4543 *Palicourea longicuspis* (Müll.Arg.) Delprete & J.H.Kirkbr. (Rubiaceae)
IPNI: 77161078-1
Habit: Shrub, Subshrub, Tree **Use:** HF **Origin:** Native **Region:** Amazonia, Andean, Guiana Shield, Pacific **Dept.:** AMA, ANT, CHO, COR, VAC, VAU **Elev.:** 0–1590 m a.s.l. **IUCN Habitat code:** 1,3,5,14 **Conservation status: (IUCN)** LC

4544 *Palicourea longirostris* (Rusby) Borhidi (Rubiaceae)
IPNI: 77161818-1
Common name: cafeto de monte **Habit:** Shrub, Small tree, Tree **Use:** ME **Origin:** Native **Region:** Andean, Pacific **Dept.:** ANT, BOY, CHO, HUI, MAG, NAR, RIS **Elev.:** 200–2300 m a.s.l. **IUCN Habitat code:** 1,3,14 **Conservation status: (IUCN)** LC

4545 *Palicourea macrobotrys* (Ruiz & Pav.) Schult. (Rubiaceae)
IPNI: 759203-1
Habit: Shrub, Tree **Use:** ME **Origin:** Native **Region:** Andean, Guiana Shield, Orinoquia **Dept.:** GUA, MET, VAU **Elev.:** 110–300 m a.s.l. **IUCN Habitat code:** 1,14 **Conservation status: (IUCN)** NE

4546 *Palicourea nigricans* K.Krause (Rubiaceae)
IPNI: 759234-1
Habit: Shrub, Small tree, Tree **Use:** HF **Origin:** Native **Region:** Amazonia, Guiana Shield **Dept.:** AMA, CAQ, GUA **Elev.:** 100–300 m a.s.l. **IUCN Habitat code:** 1,3 **Conservation status: (IUCN)** LC

4547 *Palicourea obovata* (Ruiz & Pav.) DC. (Rubiaceae)
IPNI: 759242-1
Habit: Shrub, Small tree **Use:** ME **Origin:** Native **Region:** Andean **Dept.:** CHO, HUI **Elev.:** 630–2100 m a.s.l. **Conservation status: (IUCN)** NE

4548 *Palicourea padifolia* (Willd. ex Schult.) C.M.Taylor & Lorence (Rubiaceae)
IPNI: 281971-2
Habit: Shrub, Small tree **Use:** ME **Origin:** Native **Region:** Andean **Dept.:** CHO **Elev.:** 800–2200 m a.s.l. **IUCN Habitat code:** 1,3,14 **Conservation status: (IUCN)** LC

4549 *Palicourea punicea* (Ruiz & Pav.) DC. (Rubiaceae)
IPNI: 759286-1
Use: ME **IUCN Habitat code:** 1 **Conservation status: (IUCN)** NE

4550 *Palicourea racemosa* (Aubl.) G.Nicholson (Rubiaceae)
IPNI: 77115627-1
Habit: Herb, Shrub, Subshrub **Use:** ME **Origin:** Native **Region:** Amazonia, Andean, Guiana Shield, Orinoquia, Pacific **Dept.:** AMA, ANT, BOY, CAQ, CHO, COR, GUV, MET, NAR, RIS, SAN, TOL, VAU, VID **Elev.:** 10–1850 m a.s.l. **IUCN Habitat code:** 1,2,3,4,5,14 **Conservation status: (IUCN)** LC

4551 *Palicourea rigida* Kunth (Rubiaceae)
IPNI: 759300-1
Common name: chaparrillo **Habit:** Shrub, Small tree **Use:** HF, ME **Origin:** Native **Region:** Amazonia, Andean, Guiana Shield, Orinoquia **Dept.:** ANT, BOY, CAQ, CAS, MAG, MET, NSA, SAN, VID **Elev.:** 100–3550 m a.s.l. **IUCN Habitat code:** 1,2,3,4,14 **Conservation status: (IUCN)** NE

4552 *Palicourea stenostachya* (Standl.) C.M.Taylor (Rubiaceae)
IPNI: 77160493-1
Common name: pintura de danta
Habit: Herb, Shrub, Subshrub **Use:** ME **Origin:** Native **Region:** Amazonia, Andean, Pacific **Dept.:** AMA, ANT, CAQ, HUI, NAR, PUT **Elev.:** 60–1980 m a.s.l. **IUCN Habitat code:** 1,3,14 **Conservation status: (IUCN)** NE

4553 *Palicourea subspicata* Huber (Rubiaceae)
IPNI: 181229-2
Habit: Shrub **Use:** HF **Origin:** Native **Region:** Amazonia **Dept.:** AMA, CAQ, MET, PUT **Elev.:** 90–400 m a.s.l. **IUCN Habitat code:** 1,3,4,14 **Conservation status: (IUCN)** NE

4554 *Palicourea tenerior* (Cham.) Delprete & J.H.Kirkbr. (Rubiaceae)
IPNI: 77161098-1
Use: ME **Origin:** Native **Conservation status: (IUCN)** NE

4555 *Palicourea tomentosa* (Aubl.) Borhidi (Rubiaceae)
IPNI: 77115635-1
Common name: abejorro, ají del diablo, beso de negra, boca caliente, boca de negro, boca de sapo, chochitas, dowatjiitajutokuerenae, flor de cagalera, flor de la cagalera, flor roja, gorretas, herb de la cagalera, labio de negra, labios de novia, lenguaediablo, lindon **Habit:** Shrub **Use:** EU, ME, PO, SU **Origin:** Native **Region:** Amazonia, Andean, Guiana Shield, Orinoquia, Pacific **Dept.:** AMA, ANT, BOY, CAQ, CAU, CHO, COR, CUN, GUA, HUI, LAG, MAG, MET, NAR, NSA, PUT, RIS, SAN, TOL, VAC, VAU, VID **Elev.:** 0-2500 msnm **IUCN Habitat code:** 1,2,3,4,5,14 **Conservation status: (IUCN)** NE **(CPSG)** Potential LC

4556 *Palicourea triphylla* DC. (Rubiaceae)
IPNI: 759353-1
Common name: falso tabaquillo
Habit: Shrub, Small tree, Tree **Use:** HF **Origin:** Native **Region:** Amazonia, Andean, Caribbean, Orinoquia, Pacific **Dept.:** AMA, ANT, CAQ, CHO, COR, GUV, LAG, MET, NSA, SAP, SAN, VAU, VID **Elev.:** 0–1900 m a.s.l. **IUCN Habitat code:** 1,2,3,4,5,14 **Conservation status: (IUCN)** NE

4557 *Palicourea violacea* (Aubl.) A.Rich. (Rubiaceae)
IPNI: 759372-1
Common name: cafetillo **Habit:** Shrub, Subshrub, Small tree, Tree **Use:** ME **Origin:** Native **Region:** Amazonia, Andean, Guiana Shield, Orinoquia, Pacific **Dept.:** AMA, ANT, BOY, CAQ, CAU, CHO, CUN, GUA, MAG, MET, NAR, NSA, TOL, VAC, VAU, VID **Elev.:** 0–1350 m a.s.l. **IUCN Habitat code:** 1,2,3,4,5,14 **Conservation status: (IUCN)** LC

4558 *Palicourea zevallosii* (C.M.Taylor) C.M.Taylor (Rubiaceae)
IPNI: 77153569-1
Habit: Shrub **Use:** HF **Origin:** Native **Region:** Amazonia, Andean **Dept.:** PUT **Elev.:** 180–700 m a.s.l. **IUCN Habitat code:** 1,5 **Conservation status: (IUCN)** NE

4559 *Pentagonia amazonica* (Ducke) L.Andersson & Rova (Rubiaceae)
IPNI: 77065394-1
Use: HF **Origin:** Native **IUCN Habitat code:** 1 **Conservation status: (IUCN)** LC

4560 *Pentagonia clementinensis* Cornejo (Rubiaceae)
IPNI: 77100312-1
Habit: Shrub, Tree **Use:** HF **Origin:** Native **Region:** Andean **Dept.:** NAR **Elev.:** 600–1600 m a.s.l. **Conservation status: (IUCN)** NE

4561 *Pentagonia grandiflora* Standl. (Rubiaceae)
IPNI: 188107-2
Habit: Shrub, Small tree, Tree **Use:** HF **Origin:** Native **Region:** Andean, Pacific **Dept.:** ANT, NAR, VAC **Elev.:** 5–1400 m a.s.l. **IUCN Habitat code:** 1,5,14 **Conservation status: (IUCN)** NE

4562 *Pentagonia macrophylla* Benth. (Rubiaceae)
IPNI: 760184-1
Common name: hoja de cuero, lagartijo, murciélago, papún, perico, raya, remedio de raya, tapaculo **Habit:** Shrub, Small tree, Tree **Use:** HF, MA **Origin:** Native **Region:** Amazonia, Andean, Pacific **Dept.:** ANT, BOL, CAQ, CAU, CHO, NAR, PUT, VAC **Elev.:** 0–1750 m a.s.l. **IUCN Habitat code:** 1,3,14 **Conservation status: (IUCN)** LC

4563 *Pentagonia parvifolia* Steyerm. (Rubiaceae)
IPNI: 188116-2
Common name: cascabel **Habit:** Herb, Small tree **Use:** HF **Origin:** Native **Region:** Amazonia, Andean, Pacific **Dept.:** AMA, CAU **Elev.:** 100–1250 m a.s.l. **IUCN Habitat code:** 1 **Conservation status: (IUCN)** LC

4564 *Pentagonia pinnatifida* Seem. (Rubiaceae)
IPNI: 760190-1
Common name: crestagallo, murciélago **Habit:** Shrub, Subshrub, Small tree **Use:** HF, ME **Origin:** Native **Region:** Pacific **Dept.:** ANT, CHO **Elev.:** 0–520 m a.s.l. **IUCN Habitat code:** 1,3,5,14 **Conservation status: (IUCN)** LC

4565 *Pentagonia spathicalyx* K.Schum. (Rubiaceae)
IPNI: 188124-2
Habit: Shrub, Small tree, Tree **Use:** HF **Origin:** Native **Region:** Amazonia, Andean **Dept.:** AMA, CAQ, PUT **Elev.:** 100–1130 m a.s.l. **IUCN Habitat code:** 1,3,5,14 **Conservation status: (IUCN)** LC

4566 *Pentagonia tinajita* Seem. (Rubiaceae)
IPNI: 760198-1
Habit: Shrub **Use:** HF **Origin:** Native **Region:** Andean **Dept.:** CHO **Elev.:** 710–880 m a.s.l. **IUCN Habitat code:** 1 **Conservation status: (IUCN)** NE

4567 *Pentagonia williamsii* Standl. (Rubiaceae)
IPNI: 188131-2
Use: HF **Origin:** Native **Conservation status: (IUCN)** NE

4568 *Pentagonia wurdackii* Steyerm. (Rubiaceae)
IPNI: 188133-2
Habit: Shrub, Small tree **Use:** HF **Origin:** Native **Region:** Amazonia, Andean **Dept.:** AMA, ANT, CAQ, PUT **Elev.:** 0–1300 m a.s.l. **IUCN Habitat code:** 1,3 **Conservation status: (IUCN)** NE

4569 *Pentas lanceolata* (Forssk.) Deflers (Rubiaceae)
IPNI: 760275-1
Common name: crispeta **Habit:** Herb, Shrub, Subshrub **Use:** EU, ME **Origin:** Cultivated, Naturalised **Region:** Andean **Dept.:** ANT, CUN, QUI, VAC **Elev.:** 1100–2500 m a.s.l. **IUCN Habitat code:** 1,2,3,4,14 **Conservation status: (IUCN)** NE

4570 *Pimentelia glomerata* Wedd. (Rubiaceae)
IPNI: 760466-1
Use: ME **Origin:** Native **IUCN Habitat code:** 1,3 **Conservation status: (IUCN)** NE

4571 *Platycarpum schultesii* Steyerm. (Rubiaceae)
IPNI: 201723-2
Habit: Shrub, Small tree **Use:** ME **Origin:** Native **Region:** Guiana Shield, Orinoquia **Dept.:** GUA, VAU, VID **Elev.:** 100–200 m a.s.l. **IUCN Habitat code:** 1,4,5 **Conservation status: (IUCN)** NE

4572 *Pogonopus speciosus* (Jacq.) K.Schum. (Rubiaceae)
IPNI: 204802-2
Common name: pacó amarillo **Habit:** Shrub, Small tree, Tree **Use:** MA, ME **Origin:** Native **Region:** Andean, Caribbean **Dept.:** CES, MAG **Elev.:** 50–1200 m a.s.l. **IUCN Habitat code:** 1,2,3,5,14 **Conservation status: (IUCN)** NE

4573 *Posoqueria latifolia* (Rudge) Schult. (Rubiaceae)
IPNI: 1079607-2
Common name: azuceno, azuceno de monte, borojocillo, borojó, cachaco, cafeto de monte, café, calabacillo, calabazo, churumbelo, clavo, dicharachero, falso borojó, fruto de pava, granizo, guayabo, guayabo de flor, jazmín de embarcadero, jazmín de monte, lombrirera, man **Habit:** Small tree, Tree **Use:** HF, MA, ME **Origin:** Native **Region:** Amazonia, Andean, Guiana Shield, Pacific **Dept.:** AMA, ANT, ARA, ATL, BOL, BOY, CAQ, CAU, CHO, HUI, MAG, MET, PUT, SAN, TOL, VAC **Elev.:** 0-1000 msnm **IUCN Habitat code:** 1,2,3,4,5,14 **Conservation status: (IUCN)** LC **(CPSG)** Potential LC

4574 *Posoqueria longiflora* Aubl. (Rubiaceae)
IPNI: 761025-1
Common name: bocavieja, fruta de mono, guayabo de mico, jazmín de embarcadero, jazmín de tree, jazmín de árbol **Habit:** Small tree, Tree **Use:** HF **Origin:** Native **Region:** Amazonia, Orinoquia **Dept.:** AMA, MET **Elev.:** 0–300 m a.s.l. **IUCN Habitat code:** 1,2,4,5,14 **Conservation status: (IUCN)** LC

4575 *Psychotria alba* Ruiz & Pav. (Rubiaceae)
IPNI: 212398-2
Use: ME **Origin:** Native **IUCN Habitat code:** 1,3,14 **Conservation status: (IUCN)** NE

4576 *Psychotria carthagenensis* Jacq. (Rubiaceae)
IPNI: 761655-1
Common name: ají de monte, cafeto de monte, caimito pelao, coto, flor de muerto, ipecacuana, raicilla **Habit:** Shrub, Small tree, Tree **Use:** AF, HF, MA, ME, SU **Origin:** Native **Region:** Amazonia, Andean, Guiana Shield, Caribbean, Orinoquia, Pacific **Dept.:** AMA, ANT, ATL, BOL, CAQ, CAS, CAU, CES, CHO, CUN, GUV, HUI, LAG, MAG, MET, RIS, SAN, SUC, TOL, VAC, VAU, VID **Elev.:** 5-2300 msnm **IUCN Habitat code:** 1,2,3,4,5,14 **Conservation status: (IUCN)** NE **(CPSG)** Potential LC

4577 *Psychotria ernestii* K.Krause (Rubiaceae)
IPNI: 212751-2
Common name: amargo pajarito **Habit:** Shrub, Small tree **Use:** ME **Origin:** Native **Region:** Amazonia **Dept.:** AMA, PUT **Elev.:** 90–325 m a.s.l. **IUCN Habitat code:** 1,3,4,5 **Conservation status: (IUCN)** LC

4578 *Psychotria fortuita* Standl. (Rubiaceae)
IPNI: 212798-2
Common name: cafeto de monte **Habit:** Shrub **Use:** ME **Origin:** Native **Region:** Andean **Dept.:** ANT, CAL, CUN, NSA, QUI, SAN, VAC **Elev.:** 1371–2780 m a.s.l. **IUCN Habitat code:** 1,3,14 **Conservation status: (IUCN)** LC

4579 *Psychotria grandis* Sw. (Rubiaceae)
IPNI: 762156-1
Common name: cafeto de monte **Habit:** Shrub, Small tree **Use:** ME **Origin:** Native **Region:** Andean, Caribbean, Pacific **Dept.:** ANT, CAU, CHO, MAG, VAC **Elev.:** 0–1080 m a.s.l. **IUCN Habitat code:** 1,3,5,14 **Conservation status: (IUCN)** LC

4580 *Psychotria guianensis* (Aubl.) Clos (Rubiaceae)
IPNI: 60473461-2
Habit: Small tree, Tree **Use:** MA **Origin:** Native **Region:** Amazonia, Guiana Shield, Pacific **Dept.:** CHO, VAU **Elev.:** 5–200 m a.s.l. **Conservation status: (IUCN)** NE

4581 *Psychotria horizontalis* Sw. (Rubiaceae)
IPNI: 762253-1
Common name: ají de monte **Habit:** Shrub **Use:** ME **Origin:** Native **Region:** Andean, Caribbean, Orinoquia, Pacific **Dept.:** ANT, BOL, CHO, COR, HUI, MAG, MET, NSA, SAP, SAN **Elev.:** 10–1500 m a.s.l. **IUCN Habitat code:** 1,2,3,4,5,14 **Conservation status: (IUCN)** NE

4582 *Psychotria marginata* Sw. (Rubiaceae)
IPNI: 213065-2
Habit: Shrub, Tree **Use:** ME **Origin:** Native **Region:** Andean, Caribbean, Orinoquia, Pacific **Dept.:** ANT, CHO, COR, CUN, LAG, MAG, MET, NAR, PUT, RIS, SAN **Elev.:** 0–1800 m a.s.l. **IUCN Habitat code:** 1,2,3,5,14 **Conservation status: (IUCN)** NE

4583 *Psychotria micrantha* Kunth (Rubiaceae)
IPNI: 762666-1
Common name: baba, cascajero, gorretas, gorro del diablo, mirto **Habit:** Herb, Shrub, Small tree, Tree **Use:** ME **Origin:** Native **Region:** Amazonia, Andean, Orinoquia, Pacific **Dept.:** ANT, ARA, CAL, CAQ, CHO, GUV, MAG, MET, NAR, PUT, RIS, TOL, VAC **Elev.:** 5–2250 m a.s.l. **IUCN Habitat code:** 1,3,5,14 **Conservation status: (IUCN)** LC

4584 *Psychotria nervosa* Sw. (Rubiaceae)
IPNI: 329821-2
Habit: Herb, Shrub **Use:** HF, ME **Origin:** Native **Region:** Andean, Caribbean, Pacific **Dept.:** ANT, ATL, BOL, CHO, CUN, HUI, LAG, MAG, NAR, SUC **Elev.:** 0–1020 m a.s.l. **Conservation status: (IUCN)** LC

4585 *Psychotria psychotriifolia* (Seem.) Standl. (Rubiaceae) **IPNI:** 213293-2 **Habit:** Herb, Shrub **Use:** ME **Origin:** Native **Region:** Amazonia, Andean, Orinoquia, Pacific **Dept.:** ANT, CHO, COR, GUV, MET, NAR **Elev.:** 0–900 m a.s.l. **IUCN Habitat code:** 1,3,5,14 **Conservation status: (IUCN)** LC

4586 *Psychotria remota* Benth. (Rubiaceae) **IPNI:** 763194-1 **Common name:** guaisulidia **Habit:** Shrub, Small tree **Use:** ME **Origin:** Native **Region:** Amazonia, Andean, Pacific **Dept.:** AMA, ANT, BOY, CAQ, CAU, CHO, GUA, NAR, PUT, VAU **Elev.:** 30–1100 m a.s.l. **IUCN Habitat code:** 1,3,4,5,14 **Conservation status: (IUCN)** LC

4587 *Psychotria saltatrix* C.M.Taylor (Rubiaceae) **IPNI:** 306284-2 **Habit:** Shrub, Small tree **Use:** ME **Origin:** Native **Region:** Andean **Dept.:** ANT, CHO, NAR, RIS, VAC **Elev.:** 600–4730 m a.s.l. **IUCN Habitat code:** 1,3,14 **Conservation status: (IUCN)** NE

4588 *Psychotria tenuifolia* Sw. (Rubiaceae) **IPNI:** 213510-2 **Habit:** Shrub, Subshrub **Use:** ME **Origin:** Native **Region:** Amazonia, Andean, Pacific **Dept.:** ANT, CHO, MET **Elev.:** 100–1200 m a.s.l. **IUCN Habitat code:** 1,3 **Conservation status: (IUCN)** NE

4589 *Psychotria viridis* Ruiz & Pav. (Rubiaceae) **IPNI:** 763751-1 **Common name:** chagropanga **Habit:** Shrub, Small tree **Use:** HF, ME **Origin:** Native, Cultivated **Region:** Amazonia, Andean, Guiana Shield, Orinoquia, Pacific **Dept.:** AMA, ANT, CAQ, CHO, MET, PUT, VAC **Elev.:** 10–1040 m a.s.l. **IUCN Habitat code:** 1,3,5,14 **Conservation status: (IUCN)** NE

4590 *Randia aculeata* L. (Rubiaceae) **IPNI:** 763958-1 **Common name:** cruceta, cruceta negra, crucetillo, cruceto, espinas de Cristo, espino totumito, guayaba perulera, guayabo perulero, maiztostado, maiztostao, mariangola, palo de cruz, tintero **Habit:** Shrub, Small tree, Tree **Use:** HF, MA, ME **Origin:** Native **Region:** Andean, Caribbean, Pacific **Dept.:** BOL, CHO, CUN, HUI, LAG, MAG, SAP, TOL **Elev.:** 0–1020 m a.s.l. **IUCN Habitat code:** 1,2,3,14 **Conservation status: (IUCN)** LC

4591 *Randia armata* (Sw.) DC. (Rubiaceae) **IPNI:** 1115268-2 **Common name:** cacho de venado, chocolate, cruceta negra, cruceto, cruceto real, manangola, mariangola, mariangola de espina, pilón, tachuelo, tuno **Habit:** Shrub, Small tree, Tree **Use:** AF, HF, MA, ME **Origin:** Native **Region:** Andean, Caribbean, Orinoquia, Pacific **Dept.:** ANT, ARA, ATL, BOL, BOY, CAS, CES, CHO, COR, CUN, HUI, LAG, MAG, MET, QUI, SAP, SAN, SUC, TOL, VAC **Elev.:** 0-2240 msnm **IUCN Habitat code:** 1,2,3,5,14 **Conservation status: (IUCN)** LC **(CPSG)** Potential LC

4592 *Randia calycina* Cham. (Rubiaceae) **IPNI:** 764017-1 **Habit:** Shrub, Tree **Use:** HF **Origin:** Native **Region:** Andean **Dept.:** BOY, CUN **Elev.:** 2000–2200 m a.s.l. **IUCN Habitat code:** 1,3,14 **Conservation status: (IUCN)** NE

4593 *Randia dioica* H.Karst. (Rubiaceae) **IPNI:** 764082-1 **Common name:** cajón, cruceto, mariangola **Habit:** Shrub, Small tree **Use:** FU, HF, ME **Origin:** Native **Region:** Caribbean **Dept.:** BOL, CES **Elev.:** 230–580 m a.s.l. **IUCN Habitat code:** 1,2,3,14 **Conservation status: (IUCN)** NE

4594 *Randia hebecarpa* Benth. (Rubiaceae) **IPNI:** 764142-1 **Habit:** Shrub, Small tree **Use:** HF **Origin:** Native **Region:** Andean **Dept.:** CAS **Elev.:** 710–710 m a.s.l. **IUCN Habitat code:** 2,3 **Conservation status: (IUCN)** NE

4595 *Randia hondensis* H.Karst. (Rubiaceae) **IPNI:** 764152-1 **Habit:** Shrub, Small tree, Tree **Use:** HF **Origin:** Native **Region:** Andean, Pacific **Dept.:** ANT, BOL, CES, CHO, MET **Elev.:** 20–2055 m a.s.l. **IUCN Habitat code:** 1,2,3,5,14 **Conservation status: (IUCN)** LC

4596 *Randia nitida* (Kunth) DC. (Rubiaceae) **IPNI:** 764273-1 **Use:** ME **Origin:** Native **Conservation status: (IUCN)** NE

4597 *Randia obcordata* S.Watson (Rubiaceae) **IPNI:** 217048-2 **Habit:** Shrub, Small tree **Use:** HF **Origin:** Native **Region:** Caribbean **Dept.:** LAG, MAG, SAP **Elev.:** 0–100 m a.s.l. **IUCN Habitat code:** 1,2,4,14 **Conservation status: (IUCN)** LC

4598 *Remijia amphithrix* Standl. (Rubiaceae) **IPNI:** 218543-2 **Habit:** Small tree **Use:** ME **Origin:** Endemic **Region:** Orinoquia **Dept.:** VID **Elev.:** 150–150 m a.s.l. **IUCN Habitat code:** 1,4 **Conservation status: (IUCN)** NE

4599 *Remijia macrophylla* (H.Karst.) Benth. & Hook.f. ex Flueck. (Rubiaceae) **IPNI:** 764577-1 **Common name:** quina, quina amarilla **Habit:** Shrub, Small tree **Use:** ME **Origin:** Endemic **Region:** Orinoquia **Dept.:** MET **Elev.:** 450–450 m a.s.l. **Conservation status: (IUCN)** NE

4600 *Remijia trianae* Wernham (Rubiaceae) **IPNI:** 218576-2 **Common name:** quino **Habit:** Small tree **Use:** ME **Origin:** Endemic **Region:** Orinoquia **Dept.:** MET **Elev.:** 390–400 m a.s.l. **IUCN Habitat code:** 14 **Conservation status: (IUCN)** NE

4601 *Remijia ulei* K.Krause (Rubiaceae) **IPNI:** 218577-2 **Habit:** Small tree, Tree **Use:** ME **Origin:** Native **Region:** Amazonia **Dept.:** AMA **Elev.:** 200–200 m a.s.l. **IUCN Habitat code:** 1 **Conservation status: (IUCN)** LC

4602 *Retiniphyllum schomburgkii* (Benth.) Müll.Arg. (Rubiaceae) **IPNI:** 764631-1 **Habit:** Shrub, Small tree **Use:** HF **Origin:** Native **Region:** Guiana Shield **Dept.:** GUA, VAU **Elev.:** 120–200 m a.s.l. **IUCN Habitat code:** 1,3,4,5,14 **Conservation status: (IUCN)** NE

4603 *Retiniphyllum truncatum* Müll.Arg. (Rubiaceae) **IPNI:** 764635-1 **Habit:** Shrub, Small tree **Use:** HF **Origin:** Native **Region:** Guiana Shield **Dept.:** AMA, GUA, VAU **Elev.:** 100–762 m a.s.l. **IUCN Habitat code:** 1,4,5 **Conservation status: (IUCN)** NE

4604 *Richardia scabra* L. (Rubiaceae) **IPNI:** 764679-1 **Common name:** botoncillo, ipecacuana blanca o amilácea, tapon, tapón **Habit:** Herb **Use:** HF, ME **Origin:** Native **Region:** Amazonia, Andean, Pacific **Dept.:** ANT, CAL, CAU, CUN, HUI, MET, PUT, SAN, TOL, VAC **Elev.:** 50–2100 m a.s.l. **IUCN Habitat code:** 1,2,3,4,5,14 **Conservation status: (IUCN)** NE

4605 *Ronabea emetica* (L.f.) A.Rich. (Rubiaceae) **IPNI:** 764727-1 **Common name:** ipeca de agua, ipeca falsa, lengua de venado, raicilla de agua, raicilla preta **Habit:** Herb, Shrub, Subshrub **Use:** ME **Origin:** Native **Region:** Pacific **Dept.:** ANT, CHO, SAN **Elev.:** 10–500 m a.s.l. **IUCN Habitat code:** 1,2,3,14 **Conservation status: (IUCN)** NE

4606 *Ronabea latifolia* Aubl. (Rubiaceae) **IPNI:** 764729-1 **Habit:** Shrub, Subshrub **Use:** ME **Origin:** Native **Region:** Amazonia, Andean, Pacific **Dept.:** ANT, CAU, CHO, GUA, MET, NAR, SAN, VAC, VAU **Elev.:** 0–2780 m a.s.l. **IUCN Habitat code:** 1,3,4,5,14 **Conservation status: (IUCN)** LC

4607 *Rosenbergiodendron densiflorum* (K.Schum.) Fagerl. (Rubiaceae) **IPNI:** 222507-2 **Habit:** Shrub **Use:** ME **Origin:** Native **Region:** Orinoquia **Dept.:** CAS, MET **Elev.:** 50–200 m a.s.l. **Conservation status: (IUCN)** NE

4608 *Rosenbergiodendron formosum* (Jacq.) Fagerl. (Rubiaceae) **IPNI:** 222508-2 **Common name:** mariangola, maría angola **Habit:** Shrub, Tree **Use:** EU, HF, MA, ME **Origin:** Native **Region:** Caribbean **Dept.:** ANT, ATL, BOL, CHO, CUN, LAG, MAG, SUC **Elev.:** 5–770 m a.s.l. **IUCN Habitat code:** 1,2,3,4,5,14 **Conservation status: (IUCN)** LC

4609 *Rosenbergiodendron longiflorum* (Ruiz & Pav.) Fagerl. (Rubiaceae) **IPNI:** 222509-2 **Habit:** Shrub, Small tree, Tree **Use:** HF, ME **Origin:** Native **Region:** Amazonia, Andean, Pacific **Dept.:** AMA, BOL, CAQ, CHO, CUN, GUV **Elev.:** 70–200 m a.s.l. **IUCN Habitat code:** 1,5,14 **Conservation status: (IUCN)** LC

4610 *Rudgea bracteata* J.H.Kirkbr. (Rubiaceae) **IPNI:** 223832-2 **Habit:** Shrub, Tree **Use:** HF **Origin:** Native **Region:** Amazonia **Dept.:** AMA, CAQ **Elev.:** 80–400 m a.s.l. **IUCN Habitat code:** 1 **Conservation status: (IUCN)** NE

4611 *Rudgea loretensis* Standl. (Rubiaceae)
IPNI: 223934-2
Common name: comida de danta, palo de sardina **Habit:** Shrub **Use:** HF **Origin:** Native **Region:** Amazonia **Dept.:** AMA, PUT **Elev.:** 100–300 m a.s.l. **IUCN Habitat code:** 1,3,5 **Conservation status: (IUCN)** NE

4612 *Rudgea panurensis* Müll.Arg. (Rubiaceae)
IPNI: 765569-1
Habit: Subshrub **Use:** HF **Origin:** Native **Region:** Amazonia, Guiana Shield **Dept.:** CAU, PUT, VAU **Elev.:** 100–200 m a.s.l. **IUCN Habitat code:** 1,3,5,14 **Conservation status: (IUCN)** LC

4613 *Rudgea pittieri* Standl. (Rubiaceae)
IPNI: 223977-2
Common name: zucarito **Habit:** Small tree, Tree **Use:** HF **Origin:** Native **Region:** Pacific **Dept.:** CAU, CHO, COR **Elev.:** 0–200 m a.s.l. **IUCN Habitat code:** 1 **Conservation status: (IUCN)** NE

4614 *Rudgea verticillata* (Ruiz & Pav.) Spreng. (Rubiaceae)
IPNI: 1114920-2
Habit: Shrub, Small tree **Use:** HF **Origin:** Native **Region:** Amazonia, Pacific **Dept.:** AMA, CHO **Elev.:** 0–100 m a.s.l. **IUCN Habitat code:** 1,3,14 **Conservation status: (IUCN)** LC

4615 *Rudgea viburnoides* (Cham.) Benth. (Rubiaceae)
IPNI: 1139608-2
Habit: Tree **Use:** HF, ME **Origin:** Native **Region:** Amazonia, Guiana Shield **Dept.:** AMA, CAQ, GUV, VAU **Elev.:** 110–250 m a.s.l. **IUCN Habitat code:** 1 **Conservation status: (IUCN)** NE

4616 *Sabicea amazonensis* Wernham (Rubiaceae)
IPNI: 224936-2
Habit: Climbing **Use:** HF **Origin:** Native **Region:** Amazonia, Guiana Shield **Dept.:** AMA, GUA, MET, VAU **Elev.:** 100–480 m a.s.l. **IUCN Habitat code:** 1,2,3,4,5,14 **Conservation status: (IUCN)** NE

4617 *Sabicea asperula* (Ball) Wernham (Rubiaceae)
IPNI: 224945-2
Common name: metisinajalé **Use:** ME **Origin:** Native **IUCN Habitat code:** 14 **Conservation status: (IUCN)** NE

4618 *Sabicea cinerea* Aubl. (Rubiaceae)
IPNI: 765830-1
Use: ME **Origin:** Native **Conservation status: (IUCN)** NE

4619 *Sabicea panamensis* Wernham (Rubiaceae)
IPNI: 224977-2
Habit: Shrub, Small tree, Liana, Climbing **Use:** HF, ME **Origin:** Native **Region:** Amazonia, Andean, Pacific **Dept.:** AMA, ANT, CHO, SAN, VAC **Elev.:** 0–1920 m a.s.l. **IUCN Habitat code:** 1,2,3,4,5,14 **Conservation status: (IUCN)** NE

4620 *Sabicea villosa* Willd. ex Schult. (Rubiaceae)
IPNI: 765972-1
Common name: curare ordinario, manzanilla, piña de abeja, piña de mono **Habit:** Liana, Climbing **Use:** HF, ME **Origin:** Native **Region:** Amazonia, Andean, Guiana Shield, Caribbean, Pacific **Dept.:** AMA, ANT, CAL, CAQ, CHO, CUN, GUV, HUI, MET, NAR, PUT, SAN, VAC, VAU, VID **Elev.:** 0–2800 m a.s.l. **IUCN Habitat code:** 1,2,3,4,5,14 **Conservation status: (IUCN)** NE

4621 *Schizocalyx bracteosa* Wedd. (Rubiaceae)
IPNI: 766156-1
Habit: Shrub, Tree **Use:** AF, EU, FU **Origin:** Native **Region:** Andean, Guiana Shield, Orinoquia **Dept.:** ANT, BOY, CAL, CAQ, CAS, GUV, MET, PUT, SAN, VAU **Elev.:** 150–1000 m a.s.l. **IUCN Habitat code:** 1,3,5,14 **Conservation status: (IUCN)** NE

4622 *Schradera andina* Steyerm. (Rubiaceae)
IPNI: 229942-2
Common name: ají, kidi **Habit:** Shrub **Use:** ME **Origin:** Native **Region:** Andean **Dept.:** ANT, CHO, PUT, VAC **Elev.:** 1400–2000 m a.s.l. **IUCN Habitat code:** 1,14 **Conservation status: (IUCN)** NE

4623 *Simira maxonii* (Standl.) Steyerm. (Rubiaceae)
IPNI: 236581-2
Habit: Tree **Use:** MA, ME **Origin:** Native **Region:** Andean **Dept.:** ANT **Elev.:** 900–1450 m a.s.l. **IUCN Habitat code:** 1 **Conservation status: (IUCN)** NE

4624 *Simira rubescens* (Benth.) Bremek. ex Steyerm. (Rubiaceae)
IPNI: 236591-2
Common name: brasil, paraguatán **Habit:** Shrub, Small tree, Tree **Use:** ME **Origin:** Native **Region:** Amazonia, Andean, Guiana Shield, Caribbean, Orinoquia **Dept.:** AMA, ANT, CAQ, CAS, GUV, LAG, SUC, VAU **Elev.:** 120–775 m a.s.l. **IUCN Habitat code:** 1,2,3,14 **Conservation status: (IUCN)** LC

4625 *Sipanea pratensis* Aubl. (Rubiaceae)
IPNI: 766491-1
Habit: Herb **Use:** ME **Origin:** Native **Region:** Orinoquia **Dept.:** MAG, MET, NSA **Elev.:** 100–300 m a.s.l. **IUCN Habitat code:** 1,3,4,5,14 **Conservation status: (IUCN)** NE

4626 *Spermacoce alata* Aubl. (Rubiaceae)
IPNI: 766585-1
Habit: Herb **Use:** ME **Origin:** Native **Region:** Amazonia, Andean, Guiana Shield, Orinoquia, Pacific **Dept.:** AMA, ANT, ARA, CAQ, CHO, GUA, HUI, MET, NAR, SAN, VAC **Elev.:** 60–2040 m a.s.l. **IUCN Habitat code:** 1,2,3,4,5,14 **Conservation status: (IUCN)** NE

4627 *Spermacoce capitata* Ruiz & Pav. (Rubiaceae)
IPNI: 317930-2
Common name: botoncillo, cordon de fraile, culegato, sanalotodo, shiiyo u muusa **Habit:** Herb, Shrub **Use:** ME **Origin:** Native **Region:** Amazonia, Andean, Guiana Shield, Caribbean, Orinoquia, Pacific **Dept.:** ANT, BOG, BOY, CAL, CAS, CAU, CES, COR, CUN, GUV, HUI, MAG, MET, NAR, NSA, SAN, TOL, VAC, VID **Elev.:** 25–2840 m a.s.l. **IUCN Habitat code:** 1,2,3,4,5,14 **Conservation status: (IUCN)** NE

4628 *Spermacoce confusa* Rendle (Rubiaceae)
IPNI: 766663-1
Common name: verbenilla **Habit:** Herb **Use:** ME **Origin:** Native **Region:** Andean **Dept.:** ANT, MAG, NAR, VAC **Elev.:** 550–2000 m a.s.l. **IUCN Habitat code:** 1,2,3,14 **Conservation status: (IUCN)** NE

4629 *Spermacoce densiflora* (DC.) Alain (Rubiaceae)
IPNI: 945476-1
Common name: hierba de tortero, tripa de pollo **Habit:** Herb **Use:** ME **Origin:** Native **Region:** Andean, Guiana Shield, Caribbean, Orinoquia, Pacific **Dept.:** ANT, ATL, CAS, LAG, MAG, MET, TOL, VAC **Elev.:** 50–1350 m a.s.l. **IUCN Habitat code:** 1,2,3,4,14 **Conservation status: (IUCN)** NE

4630 *Spermacoce latifolia* Aubl. (Rubiaceae)
IPNI: 766800-1
Common name: botoncillo, chupana, fruta de culebra, golondrina, manea blanca, menea blanca, totecito **Habit:** Herb, Subshrub **Use:** ME **Origin:** Native **Region:** Amazonia, Andean, Orinoquia, Pacific **Dept.:** AMA, ANT, CAQ, CHO, GUA, MET, VAU, VID **Elev.:** 50–1600 m a.s.l. **Conservation status: (IUCN)** NE

4631 *Spermacoce ocymifolia* Willd. (Rubiaceae)
IPNI: 766862-1
Habit: Herb, Shrub, Subshrub **Use:** ME **Origin:** Native **Region:** Amazonia, Andean, Guiana Shield, Orinoquia, Pacific **Dept.:** AMA, ANT, CAL, CAU, CHO, COR, CUN, GUA, HUI, LAG, MET, NAR, PUT, RIS, VAC, VAU, VID **Elev.:** 0-1850 msnm **IUCN Habitat code:** 1,2,3,4,5,14 **Conservation status: (IUCN)** NE **(CPSG)** Potential LC

4632 *Spermacoce ocymoides* Burm.f. (Rubiaceae)
IPNI: 766863-1
Use: ME, SU **Conservation status: (IUCN)** NE

4633 *Spermacoce prostrata* Aubl. (Rubiaceae)
IPNI: 766900-1
Habit: Herb **Use:** ME, SU **Origin:** Native **Region:** Amazonia, Andean, Caribbean, Orinoquia, Pacific **Dept.:** AMA, ANT, CAU, CHO, HUI, MAG, MET, NAR, VAC, VAU **Elev.:** 5–2400 m a.s.l. **IUCN Habitat code:** 1,3,4,14 **Conservation status: (IUCN)** NE

4634 *Spermacoce remota* Lam. (Rubiaceae)
IPNI: 319724-2
Habit: Herb, Shrub, Subshrub **Use:** ME **Origin:** Native **Region:** Amazonia, Andean, Caribbean, Orinoquia, Pacific **Dept.:** ANT, BOL, BOY, CHO, LAG, MET, NAR, RIS, SAN, VAC **Elev.:** 60–2300 m a.s.l. **IUCN Habitat code:** 1,2,3,4,5,14 **Conservation status: (IUCN)** NE

4635 *Spermacoce tenuior* L. (Rubiaceae)
IPNI: 241474-2
Common name: botoncillo, cansamozo, doncella negra, yerba de hierro **Habit:** Herb **Use:** ME **Origin:** Native **Region:** Amazonia, Andean, Guiana Shield, Caribbean, Orinoquia, Pacific **Dept.:** AMA, ANT, ATL, BOL, CAQ, CHO, CUN, HUI, LAG, MAG, MET, NAR, NSA, QUI, RIS, SAN, SUC, TOL, VAC **Elev.:** 10-2500 msnm **IUCN Habitat code:** 2,3,4,5,14 **Conservation status: (IUCN)** NE **(CPSG)** Potential LC

4636 *Spermacoce verticillata* L. (Rubiaceae)
IPNI: 767031-1
Common name: mosquita, totecito, yerba del toro **Habit:** Herb, Shrub, Subshrub **Use:** AF, MA, ME, PO **Origin:** Native **Region:** Andean, Caribbean, Orinoquia, Pacific **Dept.:** ANT, BOG, BOY, CHO, CUN, HUI, LAG,

MAG, MET, NAR, RIS, SAP, TOL, VAC, VID **Elev.:** 350–3100 m a.s.l. **IUCN Habitat code:** 1,2,3,4,5,14 **Conservation status: (IUCN)** NE

4637 *Stachyarrhena penduliflora* K.Schum. (Rubiaceae) **IPNI:** 243094-2 **Habit:** Shrub, Tree **Use:** HF **Origin:** Native **Region:** Amazonia, Guiana Shield **Dept.:** VAU **Elev.:** 50–400 m a.s.l. **IUCN Habitat code:** 1,4,5 **Conservation status: (IUCN)** LC

4638 *Stenostomum acreanum* (K.Krause) Achille & Delprete (Rubiaceae) **IPNI:** 77107342-1 **Habit:** Tree **Use:** HF, ME **Origin:** Native **Region:** Pacific **Dept.:** ANT, CHO, SAN **Elev.:** 20–500 m a.s.l. **IUCN Habitat code:** 1,3,5,14 **Conservation status: (IUCN)** LC

4639 *Tocoyena pittieri* (Standl.) Standl. (Rubiaceae) **IPNI:** 254985-2 **Habit:** Small tree, Tree **Use:** HF **Origin:** Native **Region:** Andean, Pacific **Dept.:** ANT, CAQ, CHO, VAC **Elev.:** 0–2150 m a.s.l. **IUCN Habitat code:** 1,3,14 **Conservation status: (IUCN)** VU

4640 *Uncaria guianensis* (Aubl.) J.F.Gmel. (Rubiaceae) **IPNI:** 768250-1 **Common name:** bejuco de anzuelo, bejuco mercade, buenamigo, cacho de venado, donguaudú, espinosa, garabato, jalapatrás, mercader, uña de gato, uña de águila **Habit:** Shrub, Liana, Climbing **Use:** ME **Origin:** Native **Region:** Amazonia, Andean, Guiana Shield, Pacific **Dept.:** AMA, ANT, BOL, CAQ, CHO, CUN, GUV, NSA, SAN, VAU **Elev.:** 20–750 m a.s.l. **IUCN Habitat code:** 1,2,3,4,14 **Conservation status: (IUCN)** NE

4641 *Uncaria tomentosa* (Willd. ex Schult.) DC. (Rubiaceae) **IPNI:** 768322-1 **Common name:** bejuco de agua, deixa, garabato, garabato amarillo, paotati-mosha, rangaya, sarmento, uncucha, unganangi, uña de gato, uña de gato de altura **Habit:** Liana **Use:** HF, MA, ME **Origin:** Native **Region:** Amazonia, Guiana Shield **Dept.:** CAQ, MET **Elev.:** 2–450 m a.s.l. **IUCN Habitat code:** 1,3,14 **Conservation status: (IUCN)** NE

4642 *Warszewiczia coccinea* (Vahl) Klotzsch (Rubiaceae) **IPNI:** 770293-1 **Common name:** barba de gallo, barba de gallo colorado, barbaegallo, barbagallo, barbas de gallo, barbegallo, capirona, carcoma, cola de guacamaya, cresta de gallo, crestegallo, cucutún, guacamaya, guacamaya roja, guacamayo, lengua de gallo, lirio colorado, palo de sol, **Habit:** Shrub, Small tree, Tree **Use:** ME **Origin:** Native **Region:** Amazonia, Andean, Guiana Shield, Orinoquia, Pacific **Dept.:** AMA, ANT, BOY, CAQ, CAS, CAU, CHO, COR, CUN, MET, NSA, PUT, QUI, RIS, SAN, TOL, VAU, VID **Elev.:** 0-1900 msnm **IUCN Habitat code:** 1,2,3,4,5,14 **Conservation status: (IUCN)** LC **(CPSG)** Potential LC

4643 *Warszewiczia peltata* Wedd. (Rubiaceae) **IPNI:** 770298-1 **Common name:** barba de gallo **Habit:** Tree **Use:** ME **Origin:** Endemic **Dept.:** CUN **IUCN Habitat code:** 1,3,4,14 **Conservation status: (IUCN)** NE

Geraniales

4644 *Erodium cicutarium* (L.) L'Hér. (Geraniaceae) **IPNI:** 372219-1 **Use:** AF, HF, MA, ME, PO **Region:** Andean **Dept.:** BOG **IUCN Habitat code:** 1,14 **Conservation status: (IUCN)** NE

4645 *Erodium moschatum* (L.) L'Hér. (Geraniaceae) **IPNI:** 372354-1 **Common name:** agujas, alfiler, alfileres, alfilerillo, almizclera, aujilla, puntero tenedor, relojito, yerba de almizcle **Habit:** Herb **Use:** HF, ME, PO **Origin:** Naturalised **Region:** Andean **Dept.:** BOG, BOY, CAU, CUN, NAR **Elev.:** 2551–3300 m a.s.l. **IUCN Habitat code:** 1,4,14 **Conservation status: (IUCN)** NE

4646 *Geranium bicknellii* Britton (Geraniaceae) **IPNI:** 109083-2 **Habit:** Herb **Use:** HF **Origin:** Naturalised **Region:** Andean **Dept.:** ANT, QUI **Elev.:** 2400–2700 m a.s.l. **IUCN Habitat code:** 1,2,3 **Conservation status: (IUCN)** NE

4647 *Geranium diffusum* Kunth (Geraniaceae) **IPNI:** 372978-1 **Common name:** geranio de páramo **Habit:** Herb **Use:** EU **Origin:** Native **Region:** Andean **Dept.:** ANT, BOY, CAL, CAU, MET, NAR, PUT, QUI, SAN, TOL, VAC **Elev.:** 2600–3900 m a.s.l. **IUCN Habitat code:** 1,3,4,8,14 **Conservation status: (IUCN)** NE

4648 *Geranium hernandesii* Moc. & Sessé ex DC. (Geraniaceae) **IPNI:** 373129-1 **Use:** ME **Region:** Andean **Dept.:** BOG **IUCN Habitat code:** 1,4 **Conservation status: (IUCN)** NE

4649 *Geranium himalayense* Klotzsch (Geraniaceae) **IPNI:** 373138-1 **Use:** EU, ME **Region:** Andean **Dept.:** BOG **IUCN Habitat code:** 3 **Conservation status: (IUCN)** NE

4650 *Geranium killipii* R.Knuth (Geraniaceae) **IPNI:** 109179-2 **Habit:** Herb **Use:** ME **Origin:** Native **Region:** Andean **Dept.:** BOG, BOY, CAL, CAU, CUN, NAR, NSA, PUT, SAN, TOL **Elev.:** 1600–3400 m a.s.l. **IUCN Habitat code:** 1,3,4,8,14 **Conservation status: (IUCN)** NE

4651 *Geranium lindenianum* Turcz. (Geraniaceae) **IPNI:** 327766-2 **Common name:** hierba del dedo, yerba del dedo, yerbamora **Habit:** Herb **Use:** MA **Origin:** Native **Region:** Andean **Dept.:** BOG, BOY, CES, LAG, SAN **Elev.:** 2700–3500 m a.s.l. **IUCN Habitat code:** 1,4,14 **Conservation status: (IUCN)** NE

4652 *Pelargonium × hybridum* (L.) L'Hér. (Geraniaceae) **IPNI:** 376335-1 **Use:** ME **Region:** Andean **Dept.:** BOG **Conservation status: (IUCN)** NE

4653 *Pelargonium capitatum* (L.) L'Hér. (Geraniaceae) **IPNI:** 375818-1 **Common name:** geranio de rosa **Habit:** Herb **Use:** EU, GS, HF, MA, ME **Origin:** Cultivated **Region:** Andean **Dept.:** BOG, CUN **Elev.:** 2620–2620 m a.s.l. **IUCN Habitat code:** 14 **Conservation status: (IUCN)** NE

4654 *Pelargonium cucullatum* (L.) L'Hér. (Geraniaceae) **IPNI:** 164626-3 **Common name:** geranio, geranio blanco, geranio rosado **Habit:** Herb **Use:** GS, ME **Origin:** Cultivated **Region:** Andean **Dept.:** BOG, CUN **Elev.:** 1780–2600 m a.s.l. **IUCN Habitat code:** 4 **Conservation status: (IUCN)** NE

4655 *Pelargonium graveolens* L'Hér. (Geraniaceae) **IPNI:** 376259-1 **Habit:** Herb **Use:** EU, HF, MA, ME **Origin:** Cultivated **Region:** Andean **Dept.:** BOG, CAU, CUN **Elev.:** 1700–2900 m a.s.l. **IUCN Habitat code:** 14 **Conservation status: (IUCN)** NE

4656 *Pelargonium odoratissimum* (L.) L'Hér. (Geraniaceae) **IPNI:** 376654-1 **Common name:** aroma, aromo, geranio, geranio de olor, malva de olor, malva olorosa, malvilla **Habit:** Herb **Use:** EU, HF, MA, ME **Origin:** Cultivated **Region:** Andean **Dept.:** ANT, BOG, BOY, MET **Elev.:** 1900–2517 m a.s.l. **IUCN Habitat code:** 1,3,14 **Conservation status: (IUCN)** NE

4657 *Pelargonium peltatum* (L.) L'Hér. (Geraniaceae) **IPNI:** 376728-1 **Common name:** cortejo, geranio, geranio blanco, geranio de yedra, novios, yedra, yedra geranio, yerda **Habit:** Herb **Use:** EU, GS, HF, MA, ME **Origin:** Cultivated **Region:** Andean **Dept.:** BOG, CUN **Elev.:** 1700–2800 m a.s.l. **IUCN Habitat code:** 1,3,14 **Conservation status: (IUCN)** NE

4658 *Pelargonium zonale* (L.) L'Hér. (Geraniaceae) **IPNI:** 377229-1 **Common name:** colegial, cortejo, geranio, geranio rojo, novio, novios, toronjil cimarrón **Habit:** Herb **Use:** EU, GS, HF, ME **Origin:** Cultivated **Region:** Andean **Dept.:** ANT, BOG, CUN **Elev.:** 1480–2620 m a.s.l. **IUCN Habitat code:** 1,2,3,4,14 **Conservation status: (IUCN)** NE

Gunnerales

4659 *Gunnera magellanica* Lam. (Gunneraceae) **IPNI:** 430163-1 **Common name:** hoja de pantano **Habit:** Herb **Use:** HF **Origin:** Native **Region:** Andean **Dept.:** CAL, CAU, CHO, NAR, VAC **Elev.:** 3000–4500 m a.s.l. **IUCN Habitat code:** 1,3,4,14 **Conservation status: (IUCN)** NE

4660 *Gunnera pilosa* Kunth (Gunneraceae) **IPNI:** 430181-1 **Common name:** hoja de pantano, hojas de pantano, mazorca de la parra, rascadera **Habit:** Herb **Use:** ME **Origin:** Native **Region:** Andean **Dept.:** ANT, BOG, BOY, CAL, CAU, CUN, MET, NAR, PUT, SAN, TOL, VAC **Elev.:** 1990–3800 m a.s.l. **IUCN Habitat code:** 1,3,4,14 **Conservation status: (IUCN)** NE

4661 *Gunnera schultesii* L.E.Mora (Gunneraceae) **IPNI:** 277820-2 **Common name:** mazorca de agua, parra de pantano **Habit:** Herb **Use:** EU, ME **Origin:** Endemic **Region:** Andean **Dept.:** CUN **Elev.:** 2840–3350 m a.s.l. **IUCN Habitat code:** 1,2,3,4,14 **Conservation status: (IUCN)** NE

4662 *Gunnera tinctoria* (Molina) Mirb. (Gunneraceae) **IPNI:** 430194-1 **Common name:** hoja ancha, hoja de pantano, mazorca de la parra, quitasol **Use:** EU, HF, ME **IUCN Habitat code:** 1,14 **Conservation status: (IUCN)** NE

Icacinales

4663 *Casimirella ampla* (Miers) R.A.Howard (Icacinaceae) **IPNI:** 963946-1 **Habit:** Shrub, Climbing **Use:** HF **Origin:** Native **Region:** Amazonia **Dept.:** AMA **Elev.:** 80–120 m a.s.l. **IUCN Habitat code:** 1 **Conservation status: (IUCN)** NE

4664 *Leretia cordata* Vell. (Icacinaceae) **IPNI:** 607989-1 **Common name:** bejuco de almidón, curarina, doctorina **Habit:** Liana, Climbing **Use:** ME **Origin:** Native **Region:** Amazonia, Pacific **Dept.:** AMA, ANT, CHO **Elev.:** 10–580 m a.s.l. **IUCN Habitat code:** 1,3,14 **Conservation status: (IUCN)** LC

4665 *Calatola costaricensis* Standl. (Metteniusaceae) **IPNI:** 41180-2 **Common name:** chuguacá, chulo, gallinazo, jaquillo, marfil, tinto **Habit:** Small tree, Tree **Use:** HF **Origin:** Native **Region:** Andean, Pacific **Dept.:** ANT, BOY, CAU, CHO, HUI, MAG, NAR, PUT, RIS, VAC **Elev.:** 280-2600 msnm **IUCN Habitat code:** 1,3,14 **Conservation status: (IUCN)** LC **(CPSG)** Potential LC

4666 *Poraqueiba guianensis* Aubl. (Metteniusaceae) **IPNI:** 608256-1 **Common name:** umarí de monte **Habit:** Shrub, Tree **Use:** HF, MA **Origin:** Native **Region:** Amazonia **Dept.:** VAU **Elev.:** 100–200 m a.s.l. **IUCN Habitat code:** 1 **Conservation status: (IUCN)** NE

4667 *Poraqueiba paraensis* Ducke (Metteniusaceae) **IPNI:** 208363-2 **Habit:** Shrub, Tree **Use:** HF **Origin:** Native **Region:** Amazonia, Guiana Shield **Dept.:** CAQ, VAU **Elev.:** 100–200 m a.s.l. **IUCN Habitat code:** 1,3,5,14 **Conservation status: (IUCN)** NE

4668 *Poraqueiba sericea* Tul. (Metteniusaceae) **IPNI:** 608257-1 **Common name:** guacure, guacure de danta, guacure de monte, guacure morado, guacure negro, guacure verde, guacuri, guacuri amarillo, guacuri anaranjado, guacuri de carne gruesa, guacuri grasoso, guacuri negro, humari, humarí, jitomagaro, mari, marí, umarí, umarí amarill **Habit:** Tree **Use:** HF, MA, ME **Origin:** Native, Cultivated **Region:** Amazonia, Guiana Shield **Dept.:** AMA, CAQ, GUA, VAU **Elev.:** 100–300 m a.s.l. **IUCN Habitat code:** 1,3,4,5,14 **Conservation status: (IUCN)** LC

Lamiales

4669 *Acanthus mollis* L. (Acanthaceae) **IPNI:** 44892-1 **Common name:** acanto **Habit:** Herb **Use:** EU, HF, ME **Origin:** Cultivated **Region:** Andean **Dept.:** ANT, BOG, CUN **Elev.:** 1483–2650 m a.s.l. **IUCN Habitat code:** 3,14 **Conservation status: (IUCN)** NE

4670 *Aphelandra aurantiaca* (Scheidw.) Lindl. (Acanthaceae) **IPNI:** 1053173-2 **Habit:** Herb, Subshrub **Use:** EU, ME **Origin:** Native **Region:** Amazonia, Andean **Dept.:** ANT, CAU, CHO, PUT, VAC **Elev.:** 100-1800 msnm **IUCN Habitat code:** 1,3,5,14 **Conservation status: (IUCN)** NE **(CPSG)** Potential LC

4671 *Aphelandra pilosa* Leonard (Acanthaceae) **IPNI:** 16278-2 **Common name:** borrachero montero, cajeto, diablo, nacedero, quiebrabarrigo **Habit:** Shrub **Use:** ME **Origin:** Native **Region:** Amazonia, Guiana Shield, Orinoquia **Dept.:** CAQ, CAS, GUV, MET, PUT, VAU **Elev.:** 200–1300 m a.s.l. **IUCN Habitat code:** 1,2,3,5,14 **Conservation status: (IUCN)** NE

4672 *Aphelandra scabra* (Vahl) Sm. (Acanthaceae) **IPNI:** 941389-1 **Common name:** cangreja, palotalo **Habit:** Shrub, Subshrub **Use:** ME **Origin:** Native **Region:** Guiana Shield, Caribbean, Orinoquia, Pacific **Dept.:** ANT, ARA, BOL, CHO, COR, GUV, MET, NSA, VAU, VID **Elev.:** 0–500 m a.s.l. **IUCN Habitat code:** 1,2,3,4,5,14 **Conservation status: (IUCN)** NE

4673 *Aphelandra squarrosa* Nees (Acanthaceae) **IPNI:** 315544-2 **Use:** EU **Region:** Andean **Dept.:** BOG **Conservation status: (IUCN)** NE

4674 *Aphelandra terryae* Standl. (Acanthaceae) **IPNI:** 16313-2 **Common name:** rumalda **Habit:** Shrub, Subshrub, Small tree **Use:** SU **Origin:** Native **Region:** Pacific **Dept.:** ANT, BOL, CHO, SAN, VAC **Elev.:** 0–550 m a.s.l. **IUCN Habitat code:** 1,14 **Conservation status: (IUCN)** NE

4675 *Asystasia gangetica* (L.) T.Anderson (Acanthaceae) **IPNI:** 45742-1 **Common name:** nevado **Habit:** Herb **Use:** AF, EU, FU, GS, HF, MA, ME, SU **Origin:** Cultivated **Dept.:** ANT, BOL, MAG, MET, SAN **Elev.:** 100–1500 m a.s.l. **IUCN Habitat code:** 1,2,3,4,14 **Conservation status: (IUCN)** NE

4676 *Avicennia germinans* (L.) L. (Acanthaceae) **IPNI:** 861125-1 **Common name:** bombito, comedera, comedero, comedero de iguana, iguanero, mangle, mangle blanco, mangle negro, mangle prieto, mangle salado, mangle salao, palo de sal, prieto **Habit:** Shrub, Small tree, Tree **Use:** FU, GS, HF, MA, ME **Origin:** Native **Region:** Caribbean, Pacific **Dept.:** ANT, ATL, BOL, CHO, LAG, MAG, NAR, SAP, SUC, VAC **Elev.:** 0-100 msnm **IUCN Habitat code:** 1,2,3,5,14 **Conservation status: (IUCN)** LC **(CPSG)** Potential LC

4677 *Barleria cristata* L. (Acanthaceae) **IPNI:** 45927-1 **Habit:** Herb, Subshrub **Use:** EU, HF, ME **Origin:** Cultivated **Region:** Caribbean **Dept.:** ATL, BOL **Elev.:** 10–100 m a.s.l. **IUCN Habitat code:** 14 **Conservation status: (IUCN)** NE

4678 *Barleria oenotheroides* Dum.Cours. (Acanthaceae) **IPNI:** 46114-1 **Habit:** Herb, Subshrub **Use:** EU **Origin:** Native **Region:** Caribbean **Dept.:** ATL, BOL, SAN **Elev.:** 350–350 m a.s.l. **IUCN Habitat code:** 1,2,14 **Conservation status: (IUCN)** NE

4679 *Bravaisia integerrima* (Spreng.) Standl. (Acanthaceae) **IPNI:** 36223-2 **Common name:** palo de agua, zancoaraño **Habit:** Tree **Use:** EU, MA, ME **Origin:** Native **Region:** Caribbean **Dept.:** ATL, BOL, MAG, SUC **Elev.:** 0–200 m a.s.l. **IUCN Habitat code:** 1,2,3,14 **Conservation status: (IUCN)** LC

4680 *Dianthera candelariae* (Oerst.) Hemsl. (Acanthaceae) **IPNI:** 1140399-2 **Common name:** calambombo **Habit:** Herb **Use:** MA, ME, SU **Origin:** Native **Region:** Amazonia, Andean **Dept.:** ANT, CAL, CAU, CUN, NAR, QUI, TOL, VAC, VAU **Elev.:** 50–2500 m a.s.l. **Conservation status: (IUCN)** NE

4681 *Dianthera comata* L. (Acanthaceae) **IPNI:** 78920-2 **Common name:** beringuera, chicaipe, chupador del pequeño, flor de la culebra, sueldaconsuelda, venenosa, yerba de chivo **Habit:** Herb **Use:** ME **Origin:** Native **Region:** Amazonia, Andean, Caribbean, Orinoquia, Pacific **Dept.:** AMA, ANT, BOL, CAU, CHO, COR, CUN, HUI, MAG, MET, NAR, SAN, TOL, VAC, VID **Elev.:** 0–2280 m a.s.l. **Conservation status: (IUCN)** NE

4682 *Dianthera pectoralis* (Jacq.) J.F.Gmel. (Acanthaceae) **IPNI:** 47290-1 **Common name:** amansaguapo, amansaguapos, amansajusticia, amansamacho, ancú, cariñito, carpintero, cerebril, chapantye, curia, curía, herb de San Antonio, herb de camarón, hierba de camarón, insulina, madre de yuca, mejorana, tila, tilo, tilo criollo, tilo cubano, técri **Habit:** Herb **Use:** HF, ME, SU **Origin:** Native **Region:** Amazonia, Caribbean, Pacific **Dept.:** AMA, ANT, BOL, CHO, MAG, MET, TOL, VAC, VAU **Elev.:** 50–600 m a.s.l. **Conservation status: (IUCN)** NE

4683 *Dianthera secunda* (Lam.) Griseb. (Acanthaceae) **IPNI:** 47308-1 **Common name:** curatodo, insulina, sangrito, singamochila **Habit:** Shrub, Subshrub **Use:** ME **Origin:** Native **Region:** Andean, Caribbean **Dept.:** ANT, CAU, CUN, LAG, MAG, MET, QUI, TOL **Elev.:** 50–2100 m a.s.l. **IUCN Habitat code:** 14 **Conservation status: (IUCN)** NE

4684 *Dicliptera sexangularis* (L.) Juss. (Acanthaceae) **IPNI:** 1114607-2 **Habit:** Herb **Use:** ME **Origin:** Native **Region:** Caribbean **Dept.:** BOL, LAG, MAG, SAP **Elev.:** 30–435 m a.s.l. **IUCN Habitat code:** 1,2,14 **Conservation status: (IUCN)** NE

4685 *Elytraria imbricata* (Vahl) Pers. (Acanthaceae) **IPNI:** 1167021-2 **Common name:** cilantro de monte, soldado parado **Habit:** Herb **Use:** HF, ME **Origin:** Native **Region:** Caribbean **Dept.:** ANT, ATL, BOL, COR, MAG, SAP, SUC, VAC **Elev.:** 10–500 m a.s.l. **IUCN Habitat code:** 1,2,14 **Conservation status: (IUCN)** NE

4686 *Eranthemum pulchellum* Andrews (Acanthaceae) **IPNI:** 48879-1 **Habit:** Subshrub **Use:** EU, HF, ME **Origin:** Cultivated **Dept.:** VAC **Elev.:** 1450–1450 m a.s.l. **Conservation status: (IUCN)** NE

4687 *Fittonia albivenis* (Lindl. ex Veitch) Brummitt (Acanthaceae) **IPNI:** 105033-2 **Common name:** abrecaminos, malamujer, parapalo, yerba de conejo, zaragoza **Habit:** Herb **Use:** EU, ME **Origin:** Native, Cultivated **Region:** Amazonia, Pacific **Dept.:** CAQ, CAU, PUT, VAC **Elev.:** 200–400 m a.s.l. **IUCN Habitat code:** 1,2,3,14 **Conservation status: (IUCN)** NE

4688 *Hygrophila costata* Nees (Acanthaceae) **IPNI:** 49672-1 **Common name:** chupador del grande, dadeató **Habit:** Herb **Use:** HF, ME **Origin:** Native **Region:** Amazonia, Andean, Caribbean **Dept.:** ANT, BOY, CAL, CAU, CHO, CUN, MAG, MET, NAR, PUT, SAN, TOL, VAC **Elev.:** 80-1850 msnm **IUCN Habitat code:** 1,2,3,4,5,14 **Conservation status: (IUCN)** NE **(CPSG)** Potential LC

4689 *Hygrophila guianensis* Nees (Acanthaceae) **IPNI:** 30085162-2 **Use:** ME **Origin:** Native **Conservation status: (IUCN)** NE

4690 *Hygrophila tyttha* Leonard (Acanthaceae) **IPNI:** 125371-2 **Common name:** cuchanguilla, huarme cuyanguilla **Habit:** Herb **Use:** ME, SU **Origin:** Native **Region:** Pacific **Dept.:** BOG, CHO, NAR, VAC **Elev.:** 10–280 m a.s.l. **IUCN Habitat code:** 1,14 **Conservation status: (IUCN)** NE

4691 *Hypoestes phyllostachya* Baker (Acanthaceae) **IPNI:** 49962-1 **Common name:** abrecaminos, destrancadera, estrancadera, pecas, pintor, polca **Habit:** Herb **Use:** EU **Origin:** Naturalised **Region:** Andean **Dept.:** ANT, BOG, CUN, HUI **Elev.:** 1250–1800 m a.s.l. **IUCN Habitat code:** 1,2,3,5,14 **Conservation status: (IUCN)** NE

4692 *Justicia brandegeeana* Wassh. & L.B.Sm. (Acanthaceae) **IPNI:** 289220-2 **Habit:** Herb **Use:** EU, ME **Origin:** Cultivated **Region:** Andean, Caribbean **Dept.:** ANT, BOG, MET **Elev.:** 450–1500 m a.s.l. **IUCN Habitat code:** 1,3,14 **Conservation status: (IUCN)** NE

4693 *Justicia carnea* Lindl. (Acanthaceae) **IPNI:** 132826-2 **Common name:** justicia, tango **Habit:** Herb, Subshrub **Use:** EU, ME **Origin:** Cultivated, Naturalised **Region:** Andean **Dept.:** ANT, BOG, CAU, VAC **Elev.:** 1950–2100 m a.s.l. **IUCN Habitat code:** 1,3,14 **Conservation status: (IUCN)** NE

4694 *Justicia carthaginensis* Jacq. (Acanthaceae) **IPNI:** 50468-1 **Common name:** amansaguapos, camarón, herb de la trinidad, hierba borriquera, yerba del susto **Habit:** Herb **Use:** ME, SU **Origin:** Native **Region:** Caribbean **Dept.:** ATL, BOL, MAG, SAP **Elev.:** 5–10 m a.s.l. **IUCN Habitat code:** 1,2 **Conservation status: (IUCN)** NE

4695 *Justicia idiogenes* Leonard (Acanthaceae) **IPNI:** 132939-2 **Habit:** Subshrub **Use:** ME **Origin:** Native **Region:** Amazonia **Dept.:** PUT **Elev.:** 230–250 m a.s.l. **IUCN Habitat code:** 1 **Conservation status: (IUCN)** NE

4696 *Justicia macarenensis* Leonard (Acanthaceae) **IPNI:** 132978-2 **Habit:** Herb **Use:** ME **Origin:** Endemic **Region:** Guiana Shield **Dept.:** MET **Elev.:** 450–770 m a.s.l. **IUCN Habitat code:** 1,3,14 **Conservation status: (IUCN)** NE

4697 *Justicia malacophylla* Leonard (Acanthaceae) **IPNI:** 132982-2 **Habit:** Shrub **Use:** SU **Origin:** Endemic **Region:** Andean **Dept.:** VAC **Elev.:** 2170–2170 m a.s.l. **Conservation status: (IUCN)** NE

4698 *Justicia phytolaccoides* Leonard (Acanthaceae) **IPNI:** 133044-2 **Common name:** aborrecedora, descancé, descansé **Habit:** Herb, Shrub, Subshrub **Use:** EU, ME **Origin:** Endemic **Region:** Andean, Pacific **Dept.:** ANT, CUN, SAN, TOL **Elev.:** 20–1120 m a.s.l. **IUCN Habitat code:** 1,2,3,14 **Conservation status: (IUCN)** NE

4699 *Justicia polygonoides* Kunth (Acanthaceae) **IPNI:** 51367-1 **Common name:** aborrecedora **Habit:** Herb **Use:** ME **Origin:** Native **Region:** Andean, Caribbean, Pacific **Dept.:** ANT, BOL, CAU, CHO, CUN, TOL, VAC **Elev.:** 0–2250 m a.s.l. **IUCN Habitat code:** 1,3,14 **Conservation status: (IUCN)** NE

4700 *Justicia spicigera* Schltdl. (Acanthaceae) **IPNI:** 133119-2 **Habit:** Shrub **Use:** MA, ME **Origin:** Native **Region:** Andean **Dept.:** ANT **Elev.:** 300–500 m a.s.l. **Conservation status: (IUCN)** NE

4701 *Justicia tinctoriella* Bennet & Raizada (Acanthaceae) **IPNI:** 907564-1 **Common name:** azul, pánciga, pánsiga **Habit:** Shrub, Subshrub **Use:** ME **Origin:** Cultivated **Region:** Andean **Dept.:** ANT, CAU, QUI **Elev.:** 500–1700 m a.s.l. **IUCN Habitat code:** 2,14 **Conservation status: (IUCN)** NE

4702 *Justicia xanthostachya* Leonard (Acanthaceae) **IPNI:** 133169-2 **Common name:** pulmonaria **Habit:** Shrub **Use:** ME **Origin:** Endemic **Region:** Andean **Dept.:** NSA **Elev.:** 700–900 m a.s.l. **IUCN Habitat code:** 1 **Conservation status: (IUCN)** NE

4703 *Lepidagathis alopecuroidea* (Vahl) R.Br. ex Griseb. (Acanthaceae) **IPNI:** 51951-1 **Habit:** Herb **Use:** ME, PO **Origin:** Native **Region:** Amazonia, Caribbean, Pacific **Dept.:** ANT, CAL, CES, CHO, MAG, MET, PUT, VAC, VAU **Elev.:** 500-650 msnm **IUCN Habitat code:** 1,2,3,4,14 **Conservation status: (IUCN)** NE **(CPSG)** Potential LC

4704 *Megaskepasma erythrochlamys* Lindau (Acanthaceae) **IPNI:** 155854-2 **Common name:** angelito **Habit:** Shrub **Use:** EU, MA **Origin:** Cultivated **Dept.:** BOG, CUN, NSA, SAN, VAC **Elev.:** 1500–2000 m a.s.l. **IUCN Habitat code:** 1,2,3,4,14 **Conservation status: (IUCN)** NE

4705 *Mendoncia aspera* Ruiz & Pav. (Acanthaceae) **IPNI:** 52381-1 **Habit:** Liana, Climbing **Use:** PO **Origin:** Native **Region:** Amazonia, Pacific **Dept.:** CAQ, CHO, PUT **Elev.:** 200–325 m a.s.l. **IUCN Habitat code:** 1,14 **Conservation status: (IUCN)** NE

4706 *Mendoncia retusa* Turrill (Acanthaceae) **IPNI:** 309179-2 **Habit:** Climbing **Use:** ME **Origin:** Native **Region:** Andean, Pacific **Dept.:** ANT, CHO **Elev.:** 0–600 m a.s.l. **IUCN Habitat code:** 1,3,14 **Conservation status: (IUCN)** NE

4707 *Nelsonia canescens* (Lam.) Spreng. (Acanthaceae) **IPNI:** 52558-1 **Habit:** Herb **Use:** AF, HF, MA, ME **Origin:** Naturalised **Region:** Caribbean **Dept.:** ATL, COR **Elev.:** 0–10 m a.s.l. **IUCN Habitat code:** 1 **Conservation status: (IUCN)** LC

4708 *Nicoteba betonica* (L.) Lindau (Acanthaceae) **IPNI:** 52612-1 **Common name:** jardín de seda **Habit:** Herb **Use:** AF, EU, IF, ME **Origin:** Naturalised **Region:** Andean **Dept.:** ANT, CUN **Elev.:** 1740–1830 m a.s.l. **Conservation status: (IUCN)** NE

4709 *Odontonema bracteolatum* (Jacq.) Kuntze (Acanthaceae) **IPNI:** 172440-2 **Habit:** Subshrub **Use:** ME **Origin:** Native **Region:** Andean, Caribbean **Dept.:** ATL, BOL, CES, MAG, NSA, SUC **Elev.:** 70–800 m a.s.l. **Conservation status: (IUCN)** NE

4710 *Odontonema tubaeforme* (Bertol.) Kuntze (Acanthaceae) **IPNI:** 172486-2 **Habit:** Subshrub **Use:** EU **Origin:** Native **Region:** Pacific **Dept.:** CHO **Elev.:** 0–20 m a.s.l. **IUCN Habitat code:** 1,2,14 **Conservation status: (IUCN)** NE

4711 *Pachystachys lutea* Nees (Acanthaceae) **IPNI:** 52814-1 **Common name:** boca de dragón, camarona, camarón, camarón amarillo, candelabro, coramina, velitas **Habit:** Shrub, Subshrub **Use:** EU **Origin:** Cultivated **Region:** Andean **Dept.:** ANT, BOG, QUI, VAC **Elev.:** 1700–1700 m a.s.l. **IUCN Habitat code:** 1,2,3,4,14 **Conservation status: (IUCN)** NE

4712 *Pachystachys spicata* (Ruiz & Pav.) Wassh. (Acanthaceae) **IPNI:** 943504-1 **Habit:** Subshrub **Use:** EU **Origin:** Native **Region:** Amazonia, Pacific **Dept.:** CAQ, MET, VAC **Elev.:** 230–600 m a.s.l. **IUCN Habitat code:** 1,3 **Conservation status: (IUCN)** NE

4713 *Pseuderanthemum maculatum* (G.Lodd.) I.M.Turner (Acanthaceae) **IPNI:** 77158889-1 **Habit:** Shrub, Small tree **Use:** EU, HF, ME **Origin:** Cultivated **Region:** Andean, Caribbean **Dept.:** ATL, BOL, CUN, TOL **Elev.:** 0–1000 m a.s.l. **Conservation status: (IUCN)** NE

4714 *Pulchranthus adenostachyus* (Lindau) V.M.Baum, Reveal & Nowicke (Acanthaceae) **IPNI:** 913240-1 **Habit:** Herb **Use:** ME **Origin:** Native **Region:** Amazonia **Dept.:** AMA, PUT **Elev.:** 100–250 m a.s.l. **IUCN Habitat code:** 1,3 **Conservation status: (IUCN)** NE

4715 *Ruellia biolleyi* Lindau (Acanthaceae) **IPNI:** 53850-1 **Habit:** Herb **Use:** ME **Origin:** Native **Region:** Pacific **Dept.:** CHO **Elev.:** 100–300 m a.s.l. **IUCN Habitat code:** 1 **Conservation status: (IUCN)** NE

4716 *Ruellia blechum* L. (Acanthaceae) **IPNI:** 224052-2 **Common name:** mazorca, mazorquilla **Habit:** Herb **Use:** AF,ME, SU **Origin:** Native **Region:** Andean, Caribbean, Orinoquia, Pacific **Dept.:** ANT, ATL, BOG, BOL, CAU, CHO, COR, CUN, HUI, MAG, MET, NAR, QUI, RIS, SAN, VAC **Elev.:** 0-2285 msnm **IUCN Habitat code:** 1,2,3,4,5,14 **Conservation status: (IUCN)** NE **(CPSG)** Potential LC

4717 *Ruellia brevifolia* (Pohl) C.Ezcurra (Acanthaceae) **IPNI:** 947317-1 **Use:** EU **Origin:** Native **IUCN Habitat code:** 14 **Conservation status: (IUCN)** NE

4718 *Ruellia chartacea* (T.Anderson) Wassh. (Acanthaceae) **IPNI:** 284442-2 **Habit:** Subshrub **Use:** ME **Origin:** Native **Region:** Amazonia **Dept.:** AMA, CAU, PUT **Elev.:** 200–300 m a.s.l. **IUCN Habitat code:** 1,2,3,14 **Conservation status: (IUCN)** LC

4719 *Ruellia ciliatiflora* Hook. (Acanthaceae) **IPNI:** 53920-1 **Common name:** ojerón **Habit:** Herb, Subshrub **Use:** ME **Origin:** Native **Region:** Caribbean **Dept.:** BOL, CUN, MAG, TOL, VAC **Elev.:** 300–1000 m a.s.l. **IUCN Habitat code:** 2,3 **Conservation status: (IUCN)** NE

4720 *Ruellia humboldtiana* (Nees) Lindau (Acanthaceae) **IPNI:** 54157-1 **Common name:** diablo, gota de tinta, rabo de gallo **Habit:** Herb, Shrub **Use:** ME **Origin:** Native **Region:** Amazonia, Orinoquia **Dept.:** CAS, CUN, GUV, MET, NSA, VAU, VID **Elev.:** 100–820 m a.s.l. **IUCN Habitat code:** 1,3,4,5,14 **Conservation status: (IUCN)** NE

4721 *Ruellia inundata* Kunth (Acanthaceae) **IPNI:** 54193-1 **Common name:** escobilla de ratón **Habit:** Herb, Subshrub **Use:** ME **Origin:** Native **Region:** Caribbean **Dept.:** ATL, BOL, MAG **Elev.:** 100–200 m a.s.l. **IUCN Habitat code:** 1,14 **Conservation status: (IUCN)** NE

4722 *Ruellia longifilamentosa* Lindau (Acanthaceae) **IPNI:** 224179-2 **Habit:** Shrub **Use:** ME **Origin:** Endemic **Region:** Andean **Dept.:** CAU, CUN **Elev.:** 1050–1550 m a.s.l. **IUCN Habitat code:** 1,3,14 **Conservation status: (IUCN)** NE

4723 *Ruellia obtusa* Nees (Acanthaceae) **IPNI:** 54363-1 **Common name:** oreja de mula, orejón, traquitraqui **Habit:** Herb **Use:** ME **Origin:** Endemic **Region:** Caribbean **Dept.:** ATL, BOL, COR, CUN, LAG, MAG, SAN, TOL **Elev.:** 5-500 msnm **IUCN Habitat code:** 1,2,3,14 **Conservation status: (IUCN)** NE **(CPSG)** LC

4724 *Ruellia paniculata* L. (Acanthaceae) **IPNI:** 54389-1 **Common name:** culantro, escobilla de ratón **Habit:** Herb, Subshrub **Use:** ME, SU **Origin:** Native **Region:** Caribbean **Dept.:** ATL, BOL, LAG, MAG **Elev.:** 0–100 m a.s.l. **IUCN Habitat code:** 1,2,14 **Conservation status: (IUCN)** NE

4725 *Ruellia simplex* C.Wright (Acanthaceae) **IPNI:** 54585-1 **Common name:** habano morado, toscana **Habit:** Herb **Use:** EU, ME **Origin:** Cultivated **Region:** Andean, Caribbean **Dept.:** ANT, BOL, BOY, CAU, QUI **Elev.:** 10–1500 m a.s.l. **IUCN Habitat code:** 1,3,14 **Conservation status: (IUCN)** NE

4726 *Ruellia terminalis* (Nees) Wassh. (Acanthaceae) **IPNI:** 1014800-1 **Common name:** chicoria **Habit:** Herb, Subshrub **Use:** ME **Origin:** Native **Region:** Amazonia, Andean, Pacific **Dept.:** ANT, CAL, CAQ, CAU, CHO, COR, MAG, NAR, SAN, VAC **Elev.:** 10–1400 m a.s.l. **IUCN Habitat code:** 1,2,3,14 **Conservation status: (IUCN)** NE

4727 *Ruellia tuberosa* L. (Acanthaceae) **IPNI:** 54671-1 **Common name:** chipaca, copetica, copialila, copialina, cortecorte, escopetica, explotaexplota, frito, oreja de mulo, tote, traquitraqui, triqui traque copia-lila, triqui-traque, triquitraque, trueno, yuca de puerco **Habit:** Herb **Use:** EU, HF, ME **Origin:** Native **Region:** Andean, Caribbean **Dept.:** ANT, BOL, BOY, COR, CUN, HUI, LAG, MAG, MET, NSA, SAP, SAN, SUC, TOL **Elev.:** 5-1900 msnm **IUCN Habitat code:** 1,2,3,5,14 **Conservation status: (IUCN)** NE **(CPSG)** Potential LC

4728 *Sanchezia macrocnemis* (Nees) Wassh. (Acanthaceae) **IPNI:** 1014801-1 **Common name:** rodilla de pollo **Habit:** Shrub **Use:** AF,ME **Origin:** Native **Region:** Amazonia, Pacific **Dept.:** ANT, BOL, CHO, MAG, MET, PUT, QUI, SAN, VAC, VAU **Elev.:** 0–1200 m a.s.l. **IUCN Habitat code:** 1,3,5,14 **Conservation status: (IUCN)** NE

4729 *Sanchezia oblonga* Ruiz & Pav. (Acanthaceae) **IPNI:** 54890-1 **Common name:** afelandra, sombra **Habit:** Shrub **Use:** EU, ME **Origin:** Native, Cultivated **Region:** Amazonia **Dept.:** AMA **Elev.:** 100–110 m a.s.l. **IUCN Habitat code:** 1,2,3,14 **Conservation status: (IUCN)** NE

4730 *Sanchezia parvibracteata* Sprague & Hutch. (Acanthaceae) **IPNI:** 227313-2 **Habit:** Shrub **Use:** EU, HF **Origin:** Cultivated, Naturalised **Region:** Amazonia **Dept.:** PUT **Elev.:** 270–310 m a.s.l. **IUCN Habitat code:** 3 **Conservation status: (IUCN)** NE

4731 *Sanchezia speciosa* Leonard (Acanthaceae) **IPNI:** 227328-2 **Habit:** Shrub **Use:** ME **Origin:** Native, Cultivated **Region:** Amazonia **Dept.:** PUT **IUCN Habitat code:** 1,14 **Conservation status: (IUCN)** NE

4732 *Stenandrium dulce* (Cav.) Nees (Acanthaceae) **IPNI:** 55411-1 **Common name:** violetilla **Habit:** Herb **Use:** ME **Origin:** Native **Region:** Andean **Dept.:** BOG, BOY, CUN **Elev.:** 2100–2680 m a.s.l. **IUCN Habitat code:** 4,14 **Conservation status: (IUCN)** NE

4733 *Stenostephanus leiorhachis* (Lindau) Hammel (Acanthaceae) **IPNI:** 60478486-2 **Habit:** Shrub **Use:** ME **Origin:** Native **Region:** Andean **Dept.:** ANT, BOY, CUN **Elev.:** 1000–1090 m a.s.l. **Conservation status: (IUCN)** NE

4734 *Strobilanthes alternata* (Burm.f.) Moylan ex J.R.I.Wood (Acanthaceae) **IPNI:** 77147688-1 **Habit:** Herb **Use:** EU, ME, SU **Origin:** Cultivated **Region:** Andean **Dept.:** ANT **Elev.:** 1500–1500 m a.s.l. **IUCN Habitat code:** 1,14 **Conservation status: (IUCN)** NE

4735 *Tetramerium nervosum* Nees (Acanthaceae) **IPNI:** 56148-1 **Habit:** Herb, Subshrub **Use:** ME **Origin:** Native **Region:** Andean, Caribbean **Dept.:** BOL, MAG, SAN, SUC **Elev.:** 120–600 m a.s.l. **IUCN Habitat code:** 1,2,3,4,14 **Conservation status: (IUCN)** NE

4736 *Thunbergia alata* Bojer ex Sims (Acanthaceae) **IPNI:** 56177-1 **Common name:** amor de madre, amor escondido, anteojo de poeta, bejuco de perdiz, colombiana, juanita, ojitos, ojo de poeta, susana, susana amarilla, susanita, susanita amarilla **Habit:** Climbing **Use:** AF, EU, GS, HF, MA, ME, PO **Origin:** Naturalised **Region:** Andean, Caribbean **Dept.:** ANT, BOG, CES, CUN, LAG, MET, QUI, SAN, VAC **Elev.:** 350–2600 m a.s.l. **IUCN Habitat code:** 1,2,3,4,14 **Conservation status: (IUCN)** NE

4737 *Thunbergia erecta* (Benth.) T.Anderson (Acanthaceae) **IPNI:** 56259-1 **Common name:** campanitas **Habit:** Shrub **Use:** EU, ME **Origin:** Cultivated **Region:** Andean **Dept.:** CUN, SAN **Elev.:** 750–1000 m a.s.l. **IUCN Habitat code:** 1,2,3,14 **Conservation status: (IUCN)** NE

4738 *Thunbergia fragrans* Roxb. (Acanthaceae) **IPNI:** 56270-1 **Habit:** Climbing **Use:** EU, MA, ME **Origin:** Cultivated, Naturalised **Region:** Andean **Dept.:** ANT, BOL, BOY, CAU, CHO, CUN, NAR, QUI, SAP, SAN, VAC **Elev.:** 8–1700 m a.s.l. **IUCN Habitat code:** 1,2,3,5,14 **Conservation status: (IUCN)** NE

4739 *Thunbergia grandiflora* Roxb. (Acanthaceae) **IPNI:** 117460-3 **Common name:** amor de madre, brasilera, carolia, carolina, emperatriz, quinceañera **Habit:** Liana **Use:** AF, EU, HF, ME **Origin:** Cultivated **Region:** Andean, Caribbean **Dept.:** ANT, ATL, BOG, BOL, CHO, CUN, NAR, VAC **Elev.:** 5–1700 m a.s.l. **IUCN Habitat code:** 1,2,3,14 **Conservation status: (IUCN)** NE

4740 *Thunbergia mysorensis* (Wight) T.Anderson (Acanthaceae) **IPNI:** 60447289-2 **Common name:** cabaret, calavera, carnaval, cortina africana, cortina china, zapatico **Habit:** Climbing **Use:** EU **Origin:** Cultivated **Region:** Andean **Dept.:** ANT, BOY, QUI **Elev.:** 1300–2100 m a.s.l. **IUCN Habitat code:** 1,3,14 **Conservation status: (IUCN)** NE

4741 *Trichanthera gigantea* (Bonpl.) Nees (Acanthaceae) **IPNI:** 56505-1 **Common name:** aro, cafetero, cafeto, cajeto, cajeto arco, cajeto-arco, carneazada, cenicero, cenicero suiban, chumbaguás, cochayuyo, cuchayuyo, cuchiyuyo, cuchuyuyo, fune, güibán, madre agua, madre de agua, madreagua, nacedera, nacedero, naceró, palo de agua, paloesal, **Habit:** Small tree, Tree **Use:** AF, EU, HF, ME **Origin:** Native **Region:** Amazonia, Andean, Pacific **Dept.:** ANT, BOL, BOY, CAL, CAQ, CAU, CHO, CUN, HUI, MET, NAR, NSA, PUT, QUI, RIS, SAN, TOL, VAC **Elev.:** 0-2000 msnm **IUCN Habitat code:** 1,2,3,4,5,8,14 **Conservation status: (IUCN)** LC **(CPSG)** Potential LC

4742 *Adenocalymma cladotrichum* (Sandwith) L.G.Lohmann (Bignoniaceae) **IPNI:** 77100436-1 **Common name:** guayacán, longapijo, palo de hueso **Habit:** Shrub, Small tree, Tree, Climbing **Use:** ME **Origin:** Native **Region:** Amazonia, Guiana Shield, Orinoquia **Dept.:** AMA, CAQ, GUV, MET, PUT, VAU **Elev.:** 120–500 m a.s.l. **IUCN Habitat code:** 1,3,4,14 **Conservation status: (IUCN)** LC

4743 *Adenocalymma flaviflorum* (Miq.) L.G.Lohmann (Bignoniaceae) **IPNI:** 77100438-1 **Habit:** Climbing **Use:** MA **Origin:** Native **Region:** Amazonia **Dept.:** AMA **Elev.:** 200–200 m a.s.l. **IUCN Habitat code:** 1,4,14 **Conservation status: (IUCN)** NE

4744 *Adenocalymma pseudopatulum* (A.H.Gentry) L.G.Lohmann (Bignoniaceae) **IPNI:** 77100440-1 **Habit:** Climbing **Use:** ME **Origin:** Native **Region:** Amazonia, Guiana Shield **Dept.:** AMA, GUV **Elev.:** 100–300 m a.s.l. **IUCN Habitat code:** 1,5,14 **Conservation status: (IUCN)** NE

4745 *Adenocalymma schomburgkii* (DC.) L.G.Lohmann (Bignoniaceae) **IPNI:** 77100441-1 **Common name:** bejuco mataganado **Habit:** Climbing **Use:** MA **Origin:** Native **Region:** Amazonia, Guiana Shield **Dept.:** AMA, CAQ, GUA, VAU **Elev.:** 70–300 m a.s.l. **IUCN Habitat code:** 1,14 **Conservation status: (IUCN)** NE

4746 *Amphilophium crucigerum* (L.) L.G.Lohmann (Bignoniaceae) **IPNI:** 77100444-1 **Common name:** bejuco canastilla, bejuco de chapolas, bejuco de cruz, bejuco peinemico, bejuquillo, erizo, garrotillo, peine, peine de mico, peinemico, rallo de mico **Habit:** Liana, Climbing **Use:** ME **Origin:** Native **Region:** Amazonia, Andean, Caribbean, Pacific **Dept.:** ANT, ATL, BOL, CAQ, CAU, CHO, CUN, LAG, MAG, NAR, SAN, SUC, VAC **Elev.:** 0-1690 msnm **IUCN Habitat code:** 1,2,3,4,14 **Conservation status: (IUCN)** NE **(CPSG)** Potential LC

4747 *Amphilophium magnoliifolium* (Kunth) L.G.Lohmann (Bignoniaceae) **IPNI:** 77100449-1 **Common name:** bejuco de rana **Habit:** Shrub, Liana, Climbing **Use:** MA **Origin:** Native **Region:** Amazonia, Orinoquia, Pacific **Dept.:** AMA, CAQ, GUA, VAU **Elev.:** 0–800 m a.s.l. **IUCN Habitat code:** 1,3,4,5,14 **Conservation status: (IUCN)** NE

4748 *Amphilophium paniculatum* (L.) Kunth (Bignoniaceae) **IPNI:** 108202-1 **Common name:** amarradiablo, bejuco de orote, bejuco de oroto, bejuco esquinero, bejuco sardinato, cascarillo, pico de pato, rabo de iguana **Habit:** Subshrub, Liana, Climbing **Use:** MA **Origin:** Native **Region:** Amazonia, Andean, Caribbean, Orinoquia, Pacific **Dept.:** AMA, ANT, ATL, BOL, BOY, CAQ, CAU, CES, CHO, CUN, MAG, NAR, QUI, RIS, SAN, SUC, TOL, VAC, VID **Elev.:** 0-2470 msnm **IUCN Habitat code:** 1,2,3,5,14 **Conservation status: (IUCN)** NE **(CPSG)** Potential LC

4749 *Amphitecna latifolia* (Mill.) A.H.Gentry (Bignoniaceae) **IPNI:** 108218-1 **Common name:** calabacillo, matecillo, matecito, totumillo, totumito **Habit:** Shrub, Tree **Use:** HF, MA, ME **Origin:** Native **Region:** Caribbean, Pacific **Dept.:** ANT, CAU, CHO, NAR, VAC **Elev.:** 0–200 m a.s.l. **IUCN Habitat code:** 1,3,5,14 **Conservation status: (IUCN)** NE

4750 *Anemopaegma puberulum* (Seibert) Miranda (Bignoniaceae) **IPNI:** 13609-2 **Habit:** Liana **Use:** MA **Origin:** Native **Region:** Andean **Dept.:** CAU, VAC **Elev.:** 1800–2000 m a.s.l. **IUCN Habitat code:** 1 **Conservation status: (IUCN)** NE

4751 *Anemopaegma robustum* Bureau & K.Schum. (Bignoniaceae) **IPNI:** 13613-2 **Habit:** Liana **Use:** MA **Origin:** Native **Region:** Guiana Shield **Dept.:** GUA **Elev.:** 120–120 m a.s.l. **Conservation status: (IUCN)** NE

4752 *Bignonia aequinoctialis* L. (Bignoniaceae) **IPNI:** 108550-1 **Common name:** campana, campana morada, chirriador **Habit:** Shrub, Liana **Use:** EU, ME **Origin:** Native **Region:** Amazonia, Andean, Caribbean, Orinoquia **Dept.:** AMA, ANT, ARA, ATL, BOL, BOY, CAL, CAQ, CAU, CES, CHO, COR, CUN, GUA, GUV, MAG, MET, NAR, NSA, SAN, SUC, TOL, VAC, VAU, VID **Elev.:** 0-2000 msnm **IUCN Habitat code:** 1,2,3,4,5,14 **Conservation status: (IUCN)** NE **(CPSG)** Potential LC

4753 *Bignonia corymbosa* Vent. (Bignoniaceae) **IPNI:** 60474996-2 **Common name:** barqui, barqui blanco, barqui de bejuco, barqui negro, bejuco de ajo, cometa, flor de pascua **Habit:** Shrub, Liana, Climbing **Use:** MA, ME **Origin:** Native **Region:** Amazonia, Andean, Guiana Shield, Caribbean, Orinoquia, Pacific **Dept.:** ANT, ATL, BOL, BOY, CAL, CAQ, CAU, CES, CHO, COR, LAG, MAG, MET, NAR, NSA, SAN, SUC, VAU, VID **Elev.:** 0-1000 msnm **IUCN Habitat code:** 1,2,3,4,5,14 **Conservation status: (IUCN)** NE **(CPSG)** Potential LC

4754 *Bignonia cuneata* (Dugand) L.G.Lohmann (Bignoniaceae) **IPNI:** 77100457-1 **Habit:** Liana **Use:** MA **Origin:** Native **Region:** Caribbean **Dept.:** ATL, BOL **Elev.:** 10–10 m a.s.l. **IUCN Habitat code:** 1,2,14 **Conservation status: (IUCN)** NE

4755 *Bignonia hyacinthina* (Standl.) L.G.Lohmann (Bignoniaceae) **IPNI:** 77100458-1 **Habit:** Liana, Climbing **Use:** HF, ME **Origin:** Native **Region:** Guiana Shield, Caribbean, Pacific **Dept.:** CHO, LAG, MET, VAC **Elev.:** 50–600 m a.s.l. **IUCN Habitat code:** 1 **Conservation status: (IUCN)** NE

4756 *Bignonia magnifica* W.Bull (Bignoniaceae) **IPNI:** 108858-1 **Habit:** Liana **Use:** EU, MA, ME **Origin:** Native, Cultivated **Region:** Andean, Caribbean **Dept.:** ATL, CES, MAG, NSA, RIS, SAN, VAC **Elev.:** 50–2150 m a.s.l. **IUCN Habitat code:** 1,2,3,14 **Conservation status: (IUCN)** NE

4757 *Bignonia neoheterophylla* L.G.Lohmann (Bignoniaceae) **IPNI:** 77141971-1 **Habit:** Liana, Climbing **Use:** MA **Origin:** Native **Region:** Caribbean **Dept.:** BOL, SUC **Elev.:** 100–350 m a.s.l. **IUCN Habitat code:** 1,2,14 **Conservation status: (IUCN)** NE

4758 *Bignonia nocturna* (Barb.Rodr.) L.G.Lohmann (Bignoniaceae) **IPNI:** 77100460-1 **Common name:** bihua junu, erchu chuá, sape mandur **Habit:** Liana, Climbing **Use:** MA, ME **Origin:** Native **Region:** Amazonia, Pacific **Dept.:** AMA, ANT, BOL, CHO, SAN **Elev.:** 20–770 m a.s.l. **IUCN Habitat code:** 1,2,3,14 **Conservation status: (IUCN)** NE

4759 *Callichlamys latifolia* (Rich.) K.Schum. (Bignoniaceae) **IPNI:** 42392-2 **Common name:** adorote, alpargate, bejuco de chucha, botecito, curare, jumeni tataboa, mataganao, totecito **Habit:** Tree, Liana, Climbing **Use:** HF, MA, ME **Origin:** Native **Region:** Amazonia, Andean, Caribbean, Pacific **Dept.:** AMA, ANT, BOL, BOY, CAQ, CES, CHO, COR, CUN, LAG, MAG, NAR, RIS, SAN, SUC, TOL, VAC **Elev.:** 33-1050 msnm **IUCN Habitat code:** 1,2,3,5,14 **Conservation status: (IUCN)** NE **(CPSG)** Potential LC

4760 *Catalpa longissima* (Jacq.) Dum.Cours. (Bignoniaceae) **IPNI:** 77176826-1 **Common name:** catalpa, crispetas **Habit:** Tree **Use:** EU, MA, ME **Origin:** Cultivated **Region:** Andean **Dept.:** ANT, VAC **Elev.:** 1000–1500 m a.s.l. **Conservation status: (IUCN)** NE

4761 *Crescentia cujete* L. (Bignoniaceae) **IPNI:** 319161-2 **Common name:** allita, caabaza, calabash tree, calabaza, calabazo, calabazo candongo, calabazo cimarrón, calabazo de palo, calabazo de pico, camasa, camazo, chumbo, chícaro, cubo, cucharito, cucharo, cuya, gourd, guan, güira, higüero, kalbas, mata de calabazo, mate, mat **Habit:** Shrub, Small tree, Tree **Use:** AF, EU, GS, HF, IF, MA, ME, PO, SU **Origin:** Native, Cultivated **Region:** Amazonia, Andean, Caribbean, Orinoquia, Pacific **Dept.:** AMA, ANT, ATL, BOL, BOY, CAS, CES, CHO, CUN, HUI, LAG, MAG, MET, NAR, NSA, PUT, QUI, SAN, SUC, TOL, VAC **Elev.:** 0–1370 m a.s.l. **IUCN Habitat code:** 1,2,3,4,5,14 **Conservation status: (IUCN)** LC

4762 *Cuspidaria inaequalis* (DC. ex Splitg.) L.G.Lohmann (Bignoniaceae) **IPNI:** 77100463-1 **Habit:** Liana, Climbing **Use:** HF, MA, ME **Origin:** Native **Region:** Amazonia **Dept.:** AMA, ANT, CAQ, MET, VAU **Elev.:** 120–600 m a.s.l. **IUCN Habitat code:** 1,3,5,14 **Conservation status: (IUCN)** NE

4763 *Delostoma integrifolium* D.Don (Bignoniaceae) **IPNI:** 109429-1 **Common name:** almanegra, cajeto, cajeto crecedor, calentano, crecedor, fresno, guarapo, guayacán morado, juagüito, molde, nacedero, navajuelo, terebinto **Habit:** Shrub, Small tree, Tree **Use:** ME **Origin:** Native **Region:** Andean **Dept.:** ANT, BOG, CAU, CUN, HUI, NAR, NSA, PUT, QUI, SAN, TOL, VAC **Elev.:** 1450–3200 m a.s.l. **IUCN Habitat code:** 1,3,4,14 **Conservation status: (IUCN)** LC

4764 *Dolichandra quadrivalvis* (Jacq.) L.G.Lohmann (Bignoniaceae) **IPNI:** 77100464-1 **Common name:** bejuco de cangrejo, bejuco de tierra, cangrejo, matacangrejo **Habit:** Liana **Use:** ME **Origin:** Native **Region:** Andean, Caribbean **Dept.:** ATL, BOL, CES, CUN, MAG, TOL **Elev.:** 20–1000 m a.s.l. **IUCN Habitat code:** 1,2 **Conservation status: (IUCN)** NE

4765 *Dolichandra uncata* (Andrews) L.G.Lohmann (Bignoniaceae) **IPNI:** 77100511-1 **Common name:** bejuco uñita, lagartija **Habit:** Liana, Climbing **Use:** HF, MA, ME **Origin:** Native **Region:** Andean, Caribbean, Orinoquia, Pacific **Dept.:** ANT, ATL, BOL, BOY, CES, CHO, COR, LAG, MAG, MET, SUC, VAC **Elev.:** 0–540 m a.s.l. **IUCN Habitat code:** 1,2,3,5,14 **Conservation status: (IUCN)** NE

4766 *Dolichandra unguis-cati* (L.) L.G.Lohmann (Bignoniaceae) **IPNI:** 77100466-1 **Common name:** bejuco uñita, lagartija, liana iguana, murciélago, pegapega, uña de gato **Habit:** Liana, Climbing **Use:** EU, HF, MA, ME **Origin:** Native **Region:** Andean, Guiana Shield, Caribbean, Orinoquia, Pacific **Dept.:** ANT, ATL, BOL, BOY, CAL, CES, CHO, COR, CUN, GUV, LAG, MAG, MET, SAP, SUC **Elev.:** 0-1660 msnm **IUCN Habitat code:** 1,2,3,5,14 **Conservation status: (IUCN)** NE **(CPSG)** Potential LC

4767 *Fridericia candicans* (Rich.) L.G.Lohmann (Bignoniaceae) **IPNI:** 77141989-1 **Common name:** bejuco culatero, bejuco de adorote, bejuco de culata, pintabollo macho, sueldilla **Habit:** Shrub, Liana, Climbing **Use:** ME **Origin:** Native **Region:** Amazonia, Andean, Guiana Shield, Caribbean, Orinoquia, Pacific **Dept.:** AMA, ANT, ARA, ATL, BOL, CAL, CAU, CES, CHO, COR, CUN, HUI, LAG, MAG, MET, NSA, PUT, SAN, SUC, TOL, VAC, VID **Elev.:** 20-2000 msnm **IUCN Habitat code:** 1,2,3,4,5,14 **Conservation status: (IUCN)** NE **(CPSG)** Potential LC

4768 *Fridericia chica* (Bonpl.) L.G.Lohmann (Bignoniaceae) **IPNI:** 77142115-1 **Common name:** barqui, bija, caballito, carajirú, chica, cipó cruz, cudio, curí, curí-huasca, hoja de teñir, hojita de teñir, koorí, makuri, neacurí, piranga, puca panga, puchicama, querabare, tail, tiza **Habit:** Liana **Use:** HF, MA, ME **Origin:** Native **Region:** Amazonia, Andean, Caribbean, Orinoquia, Pacific **Dept.:** AMA, ANT, BOL, CAQ, CHO, COR, GUV, LAG, MAG, MET, NAR, PUT, SAN, VAC, VAU **Elev.:** 20-1200 msnm **IUCN Habitat code:** 1,2,3,5,14 **Conservation status: (IUCN)** NE **(CPSG)** Potential LC

4769 *Fridericia conjugata* (Vell.) L.G.Lohmann (Bignoniaceae) **IPNI:** 77141996-1 **Common name:** bejuco calabacilla **Habit:** Liana, Climbing **Use:** AF **Origin:** Native **Region:** Caribbean, Orinoquia, Pacific **Dept.:** ATL, BOL, CAQ, CES, CHO, MAG, SAN, SUC, VID **Elev.:** 20–300 m a.s.l. **IUCN Habitat code:** 1,2,3,4,14 **Conservation status: (IUCN)** NE

4770 *Fridericia japurensis* (DC.) L.G.Lohmann (Bignoniaceae) **IPNI:** 77142016-1 **Common name:** ambar, bejuco blanco, bejuco pintapollo, chupa chupa, pinta bollo real **Habit:** Liana, Climbing **Use:** ME, PO **Origin:** Native **Region:** Amazonia, Caribbean, Pacific **Dept.:** AMA, BOL, CAQ, COR, NAR, VAU **Elev.:** 200–250 m a.s.l. **IUCN Habitat code:** 1,3,14 **Conservation status: (IUCN)** NE

4771 *Fridericia mollissima* (Kunth) L.G.Lohmann (Bignoniaceae) **IPNI:** 77142038-1 **Common name:** bejuco de chupachupa, bejuco pintabollo, chupa chupa, chupachupa, pintabollo real **Habit:** Liana, Climbing **Use:** ME **Origin:** Native **Region:** Caribbean **Dept.:** ANT, ATL, BOL, CES, CHO, COR, CUN, LAG, MAG, SUC **Elev.:** 5–460 m a.s.l. **IUCN Habitat code:** 1,2,3,5,14 **Conservation status: (IUCN)** NE

4772 *Fridericia pubescens* (L.) L.G.Lohmann (Bignoniaceae) **IPNI:** 77142059-1 **Common name:** bejuco culatero, bejuco de adorote, culatero, pintabollo colatero, pintabollo culatero, pintabollo macho **Habit:** Liana, Climbing **Use:** ME **Origin:** Native **Region:** Andean, Caribbean, Orinoquia **Dept.:** ATL, BOL, CES, CUN, HUI, LAG, MAG, MET, SUC, VID **Elev.:** 0–1320 m a.s.l. **IUCN Habitat code:** 1,2,3,4,14 **Conservation status: (IUCN)** NE

4773 *Godmania aesculifolia* (Kunth) Standl. (Bignoniaceae) **IPNI:** 111572-2 **Common name:** araguaney sabanero, cacho de chivo, cacho de vaca, cacho de venado, cacho de venado sabanero, potrillo, yuco **Habit:** Small tree **Use:** MA, ME **Origin:** Native **Region:** Andean, Orinoquia **Dept.:** ANT, CAS, CES, MET, VID **Elev.:** 170–1200 m a.s.l. **IUCN Habitat code:** 1,3,4,14 **Conservation status: (IUCN)** LC

4774 *Handroanthus barbatus* (E.Mey.) Mattos (Bignoniaceae) **IPNI:** 109687-1 **Common name:** apamate, ermaño caspi, jaico, palo de arco, palo mosquito **Habit:** Tree **Use:** MA, ME **Origin:** Native **Region:** Guiana Shield, Orinoquia **Dept.:** GUA, VID **Elev.:** 100–100 m a.s.l. **IUCN Habitat code:** 1,4,5 **Conservation status: (IUCN)** NE

4775 *Handroanthus billbergii* (Bureau & K.Schum.) S.O.Grose (Bignoniaceae) **IPNI:** 77085061-1 **Common name:** coralibe, puy **Habit:** Small tree, Tree **Use:** AF, EU, FU, MA, ME **Origin:** Native **Region:** Caribbean **Dept.:** ANT, ATL, BOL, CES, LAG, MAG **Elev.:** 5–200 m a.s.l. **IUCN Habitat code:** 1,2,3,4,14 **Conservation status: (IUCN)** NE

4776 *Handroanthus capitatus* (Bureau & K.Schum.) Mattos (Bignoniaceae) **IPNI:** 109688-1 **Habit:** Small tree **Use:** MA, ME **Origin:** Native **Region:** Amazonia **Dept.:** AMA **Elev.:** 230–230 m a.s.l. **IUCN Habitat code:** 1 **Conservation status: (IUCN)** NE

4777 *Handroanthus chrysanthus* (Jacq.) S.O.Grose (Bignoniaceae) **IPNI:** 77085066-1 **Common name:** abejón, araguaney, cañaguate, chicalá, coninae, floramarillo, guayacán, guayacán amarillo, guayacán polvillo, kawinanae, palo de arco, polvillo, roble amarillo **Habit:** Small tree, Tree **Use:** EU, HF, MA, ME, SU **Origin:** Native, Cultivated **Region:** Amazonia, Andean, Caribbean, Pacific **Dept.:** AMA, ANT, ATL, BOL, BOY, CAL, CAQ, CES, CHO, COR, CUN, MAG, MET, QUI, SAN, TOL, VAC **Elev.:** 5–2300 m a.s.l. **IUCN Habitat code:** 1,2,3,4,5,6,14 **Conservation status: (IUCN)** NE

4778 *Handroanthus chrysotrichus* (Mart. ex DC.) Mattos (Bignoniaceae) **IPNI:** 109691-1 **Habit:** Shrub **Use:** MA, ME **Origin:** Cultivated **Dept.:** VAC **Elev.:** 900–1000 m a.s.l. **IUCN Habitat code:** 1 **Conservation status: (IUCN)** NE

4779 *Handroanthus coralibe* (Standl.) S.O.Grose (Bignoniaceae) **IPNI:** 77085069-1 **Habit:** Tree **Use:** FU, MA **Origin:** Endemic **Region:** Caribbean **Dept.:** ATL, BOL, MAG **Elev.:** 10-200 msnm **IUCN Habitat code:** 1,2,14 **Conservation status: (IUCN)** VU **(CPSG)** VU

4780 *Handroanthus guayacan* (Seem.) S.O.Grose (Bignoniaceae) **IPNI:** 77085089-1 **Common name:** polvillo **Habit:** Shrub, Small tree, Tree **Use:** MA **Origin:** Native **Region:** Caribbean **Dept.:** ANT, CHO, COR, SAN **Elev.:** 50–900 m a.s.l. **IUCN Habitat code:** 1,2,3,5,14 **Conservation status: (IUCN)** NE

4781 *Handroanthus impetiginosus* (Mart. ex DC.) Mattos (Bignoniaceae) **IPNI:** 117330-2 **Habit:** Small tree, Tree **Use:** EU, HF, MA, ME **Origin:** Native **Region:** Andean, Caribbean **Dept.:** AMA, ATL, BOL, BOY, MAG, SAN, SUC **Elev.:** 5–1800 m a.s.l. **Conservation status: (IUCN)** LC

4782 *Handroanthus incanus* (A.H.Gentry) S.O.Grose (Bignoniaceae) **IPNI:** 77085090-1 **Habit:** Tree **Use:** HF **Origin:** Native **Region:** Amazonia **Dept.:** AMA **Elev.:**

120–120 m a.s.l. **IUCN Habitat code:** 1,3,14 **Conservation status: (IUCN)** NE

4783 *Handroanthus obscurus* (Bureau & K.Schum.) Mattos (Bignoniaceae) **IPNI:** 109700-1 **Habit:** Shrub **Use:** ME **Origin:** Native **Region:** Amazonia, Guiana Shield **Dept.:** AMA, CAQ, GUA, MET, VAU **Elev.:** 100–250 m a.s.l. **IUCN Habitat code:** 1,3,4,5 **Conservation status: (IUCN)** NE

4784 *Handroanthus ochraceus* (Cham.) Mattos (Bignoniaceae) **IPNI:** 117333-2 **Common name:** cañaguate, flor amarilla, guayacán, katsubarunae, palo de arco, to hua ri **Habit:** Small tree, Tree **Use:** MA, ME **Origin:** Native, Cultivated **Region:** Amazonia, Andean, Caribbean, Orinoquia **Dept.:** ANT, ATL, BOG, BOL, CAS, CES, COR, CUN, GUA, LAG, MAG, PUT, TOL, VAC, VID **Elev.:** 5–1850 m a.s.l. **IUCN Habitat code:** 1,2,3,5,14 **Conservation status: (IUCN)** NE

4785 *Handroanthus serratifolius* (Vahl) S.O.Grose (Bignoniaceae) **IPNI:** 77085096-1 **Common name:** acapro, acapro blanco, chicalá, curarire, flor amarillo, guayacan polvillo, guayacán, guayacán polvillo, pau darco, puy, vero **Habit:** Shrub, Small tree, Tree **Use:** EU, MA, ME **Origin:** Native **Region:** Amazonia, Caribbean, Orinoquia **Dept.:** AMA, ANT, ATL, CAL, CUN, NSA, PUT, SAN, VID **Elev.:** 80–350 m a.s.l. **IUCN Habitat code:** 1,2,3,14 **Conservation status: (IUCN)** NE

4786 *Incarvillea sinensis* Lam. (Bignoniaceae) **IPNI:** 109760-1 **Habit:** Herb **Use:** HF, ME **Origin:** Cultivated **Region:** Andean **Dept.:** ANT, CUN, NSA, TOL **Elev.:** 600–1750 m a.s.l. **Conservation status: (IUCN)** NE

4787 *Jacaranda caucana* Pittier (Bignoniaceae) **IPNI:** 130903-2 **Common name:** acacia, aceituno, caballito, cacao, caco, caro, chingalé, cornique, flormorado, gualanday, guayacán, palo de boba, piñón de oreja, riñón de oreja **Habit:** Small tree, Tree **Use:** MA, ME **Origin:** Native **Region:** Andean, Caribbean, Pacific **Dept.:** ANT, BOG, BOL, CAL, CAQ, CES, CHO, COR, CUN, HUI, MAG, MET, NAR, RIS, SAN, TOL, VAC **Elev.:** 20-1700 msnm **IUCN Habitat code:** 1,2,3,4,5,14 **Conservation status: (IUCN)** LC **(CPSG)** Potential LC

4788 *Jacaranda copaia* (Aubl.) D.Don (Bignoniaceae) **IPNI:** 109789-1 **Common name:** aspingo, caco, caco castañete, canalete, canalete común, caroba, caroba manaca, chingalé, copaia, escobillo, gallinazo, guabillo, gualanday, guawmoamú, hualanday, machaco, madura plátano, madurachontaduro, maduraplátano, papelillo, para-pará, pavito, pavo **Habit:** Small tree, Tree **Use:** FU, HF, MA, ME **Origin:** Native **Region:** Amazonia, Andean, Caribbean, Orinoquia, Pacific **Dept.:** AMA, ANT, BOL, BOY, CAQ, CHO, COR, CUN, MET, NAR, PUT, SAN, TOL, VAC, VAU, VID **Elev.:** 80-1450 msnm **IUCN Habitat code:** 1,3,4,5,14 **Conservation status: (IUCN)** LC **(CPSG)** Potential LC

4789 *Jacaranda glabra* (DC.) Bureau & K.Schum. (Bignoniaceae) **IPNI:** 130921-2 **Habit:** Tree **Use:** MA, ME, SU **Origin:** Native **Region:** Amazonia, Pacific **Dept.:** CAQ, CHO, PUT **Elev.:** 230–400 m a.s.l. **IUCN Habitat code:** 1,3,14 **Conservation status: (IUCN)** LC

4790 *Jacaranda hesperia* Dugand (Bignoniaceae) **IPNI:** 130925-2 **Common name:** chingalé, cunite, curnique, curnite, gualanday, pinguasí, pulga, tunisco, vainillo **Habit:** Small tree, Tree **Use:** EU, FU, MA, ME **Origin:** Endemic **Region:** Andean, Pacific **Dept.:** ANT, CAL, CHO, CUN, NAR, RIS, SAN, VAC **Elev.:** 0-960 msnm **IUCN Habitat code:** 1,2,3,5,14 **Conservation status: (IUCN)** LC **(CPSG)** LC

4791 *Jacaranda mimosifolia* D.Don (Bignoniaceae) **IPNI:** 130936-2 **Common name:** gualanday **Habit:** Tree **Use:** EU, FU, MA, ME, SU **Origin:** Cultivated **Region:** Andean **Dept.:** ANT **Elev.:** 1500–1500 m a.s.l. **IUCN Habitat code:** 2,3,14 **Conservation status: (IUCN)** VU

4792 *Jacaranda obtusifolia* Bonpl. (Bignoniaceae) **IPNI:** 109832-1 **Common name:** abey, caco, flormorado, gualanday, guarupa, pavito, árbol guaca **Habit:** Shrub, Small tree, Tree **Use:** EU, MA, ME **Origin:** Native **Region:** Amazonia, Andean, Guiana Shield, Orinoquia **Dept.:** AMA, ARA, BOY, CAQ, CAS, CUN, GUA, GUV, MET, PUT, VID **Elev.:** 200–1340 m a.s.l. **IUCN Habitat code:** 1,2,3,4,5,14 **Conservation status: (IUCN)** LC

4793 *Kigelia africana* (Lam.) Benth. (Bignoniaceae) **IPNI:** 109874-1 **Common name:** árbol de las salchichas, árbol salchicha, árbol vela **Habit:** Tree **Use:** AF, EU, FU, GS, HF, IF, MA, ME, PO, SU **Origin:** Cultivated **Region:** Andean **Dept.:** ANT **IUCN Habitat code:** 3,14 **Conservation status: (IUCN)** LC

4794 *Lundia erionema* DC. (Bignoniaceae) **IPNI:** 109921-1 **Habit:** Liana **Use:** ME **Origin:** Native **Region:** Guiana Shield **Dept.:** VAU **Elev.:** 180–700 m a.s.l. **IUCN Habitat code:** 1 **Conservation status: (IUCN)** NE

4795 *Mansoa alliacea* (Lam.) A.H.Gentry (Bignoniaceae) **IPNI:** 152672-2 **Common name:** ajo sacha, chachaajo **Habit:** Liana **Use:** HF, MA, ME **Origin:** Native, Cultivated **Region:** Caribbean **Dept.:** MAG, TOL **Elev.:** 375–375 m a.s.l. **IUCN Habitat code:** 1 **Conservation status: (IUCN)** NE

4796 *Mansoa hymenaea* (DC.) A.H.Gentry (Bignoniaceae) **IPNI:** 152677-2 **Common name:** ago, ajo de monte, bejuco, bejuco ajo, bejuco cebollo, bejuco de ajo, hosmeca, jumeca, olmeca, pedo de padre **Habit:** Climbing **Use:** HF, ME **Origin:** Native **Region:** Caribbean, Orinoquia **Dept.:** MAG, MET, VAC **Elev.:** 30–1050 m a.s.l. **IUCN Habitat code:** 1,2,3,14 **Conservation status: (IUCN)** NE

4797 *Mansoa standleyi* (Steyerm.) A.H.Gentry (Bignoniaceae) **IPNI:** 152684-2 **Habit:** Liana **Use:** ME **Origin:** Native **Region:** Amazonia, Caribbean, Orinoquia **Dept.:** AMA, MAG, MET, VID **Elev.:** 30–300 m a.s.l. **IUCN Habitat code:** 1,3,5 **Conservation status: (IUCN)** NE

4798 *Mansoa verrucifera* (Schltdl.) A.H.Gentry (Bignoniaceae) **IPNI:** 152688-2 **Common name:** bejuco ajo, bejuco de ajo **Habit:** Liana, Climbing **Use:** MA, ME **Origin:** Native **Region:** Amazonia, Caribbean, Pacific **Dept.:** ATL, BOL, CAQ, CES, CHO, CUN, MAG, SAP, SAN, SUC, TOL **Elev.:** 5-700 msnm **IUCN Habitat code:** 1,2,3,14 **Conservation status: (IUCN)** NE **(CPSG)** Potential LC

4799 *Martinella obovata* (Kunth) Bureau & K.Schum. (Bignoniaceae) **IPNI:** 153710-2 **Common name:** bejuco colorado, bejuco colorao, bejuco de ajo, bejuquillo de la hembra, cortalonso, orejón, yuquilla **Habit:** Liana, Climbing **Use:** ME **Origin:** Native **Region:** Amazonia, Andean, Guiana Shield, Caribbean, Orinoquia, Pacific **Dept.:** AMA, ANT, ATL, BOL, BOY, CAQ, CHO, COR, CUN, GUA, GUV, MAG, NAR, PUT, SAN, SUC, VAC, VAU, VID **Elev.:** 5-1430 msnm **IUCN Habitat code:** 1,2,3,4,5,14 **Conservation status: (IUCN)** NE **(CPSG)** Potential LC

4800 *Parmentiera aculeata* (Kunth) Seem. (Bignoniaceae) **IPNI:** 110241-1 **Habit:** Tree **Use:** HF, MA, ME **Origin:** Cultivated **Dept.:** VAC **Elev.:** 1100–1100 m a.s.l. **Conservation status: (IUCN)** LC

4801 *Parmentiera cereifera* Seem. (Bignoniaceae) **IPNI:** 110245-1 **Common name:** árbol de velas, árbol vela **Habit:** Tree **Use:** AF, EU, HF, MA, ME **Origin:** Cultivated **Region:** Caribbean **Dept.:** BOL **Elev.:** 10–10 m a.s.l. **IUCN Habitat code:** 14 **Conservation status: (IUCN)** EN

4802 *Parmentiera stenocarpa* Dugand & L.B.Sm. (Bignoniaceae) **IPNI:** 183296-2 **Common name:** esperanza, esperma, palo de vela, palo vela, platanillo, plátano de machin, sul maduno, vela, zulmatuno **Habit:** Shrub, Tree **Use:** EU, FU, HF, ME **Origin:** Endemic **Region:** Caribbean, Pacific **Dept.:** ANT, CHO **Elev.:** 0-220 msnm **IUCN Habitat code:** 1,3,5,14 **Conservation status: (IUCN)** LC **(CPSG)** LC

4803 *Pleonotoma jasminifolia* (Kunth) Miers (Bignoniaceae) **IPNI:** 110423-1 **Common name:** bejuco de puño, imuño, lombriz **Habit:** Liana, Climbing **Use:** ME **Origin:** Native **Region:** Amazonia, Andean, Guiana Shield, Orinoquia **Dept.:** AMA, CAQ, CUN, GUA, GUV, MET, VAU, VID **Elev.:** 140–1000 m a.s.l. **IUCN Habitat code:** 1,3,4,14 **Conservation status: (IUCN)** NE

4804 *Pleonotoma melioides* (S.Moore) A.H.Gentry (Bignoniaceae) **IPNI:** 202082-2 **Habit:** Liana, Climbing **Use:** ME **Origin:** Native **Region:** Amazonia **Dept.:** AMA, CAQ **Elev.:** 200–200 m a.s.l. **IUCN Habitat code:** 1 **Conservation status: (IUCN)** NE

4805 *Podranea ricasoliana* (Tanfani) Sprague (Bignoniaceae) **IPNI:** 110440-1 **Common name:** gaitana **Habit:** Liana, Climbing **Use:** EU **Origin:** Cultivated **Region:** Andean **Dept.:** ANT, BOY, MET, VAC **Elev.:** 500–2050 m a.s.l. **IUCN Habitat code:** 1,3,14 **Conservation status: (IUCN)** NE

4806 *Pyrostegia venusta* (Ker Gawl.) Miers (Bignoniaceae) **IPNI:** 110483-1 **Common name:** boliviana, cayena, jazmín de fuego, llamarada, lluvia de coral, lluvia de oro, siete de bastos, tango **Habit:** Climbing **Use:** EU, HF, MA, ME **Origin:** Native, Cultivated **Region:** Andean, Guiana Shield, Orinoquia **Dept.:** ANT, BOG, CAQ, CAU, MET, VAC, VID **Elev.:** 100–1900 m a.s.l. **IUCN Habitat code:** 1,3,4,5,14 **Conservation status: (IUCN)** NE

4807 *Roseodendron chryseum* (S.F.Blake) Miranda (Bignoniaceae) **IPNI:** 222517-2 **Common name:** cañahuate, roble amarillo **Habit:** Tree **Use:** EU, MA, ME **Origin:** Native, Cultivated **Region:** Caribbean **Dept.:** ANT, ATL, BOL, CES, LAG, SUC **Elev.:** 50–650 m a.s.l. **IUCN Habitat code:** 1,2,3,4,14 **Conservation status: (IUCN)** NT

4808 *Roseodendron donnell-smithii* (Rose) Miranda (Bignoniaceae) **IPNI:** 222518-2 **Habit:** Tree **Use:** MA **Origin:** Native **Region:** Caribbean **Dept.:** COR **Elev.:** 100–100 m a.s.l. **IUCN Habitat code:** 3 **Conservation status: (IUCN)** LC

4809 *Spathodea campanulata* P.Beauv. (Bignoniaceae) **IPNI:** 110661-1 **Common name:** barquitos, miona, pipí, roble africano, rosa china, tulipán, tulipán africano, tulipán de Gabón, tulipán rojo **Habit:** Tree **Use:** AF, EU, FU, GS, HF, MA, ME, PO, SU **Origin:** Cultivated **Region:** Amazonia, Andean **Dept.:** AMA, ANT, CAL, QUI, RIS, VAC **Elev.:** 200–1800 m a.s.l. **IUCN Habitat code:** 1,2,3,14 **Conservation status: (IUCN)** LC

4810 *Stizophyllum riparium* (Kunth) Sandwith (Bignoniaceae) **IPNI:** 246385-2 **Habit:** Liana, Climbing **Use:** ME **Origin:** Native **Region:** Amazonia, Andean, Caribbean, Pacific **Dept.:** AMA, ANT, BOL, BOY, CAQ, CHO, CUN, SAN, SUC, TOL, VAU **Elev.:** 50-800 msnm **IUCN Habitat code:** 1,2,3,5,14 **Conservation status: (IUCN)** NE **(CPSG)** Potential LC

4811 *Tabebuia heterophylla* (DC.) Britton (Bignoniaceae) **IPNI:** 248689-2 **Common name:** roble **Use:** EU, HF, MA, ME **IUCN Habitat code:** 1,3,14 **Conservation status: (IUCN)** LC

4812 *Tabebuia insignis* (Miq.) Sandwith (Bignoniaceae) **IPNI:** 248703-2 **Common name:** guayacán, palo de diablo **Habit:** Small tree, Tree **Use:** MA, ME **Origin:** Native **Region:** Amazonia, Orinoquia **Dept.:** AMA, CAQ, VAU, VID **Elev.:** 100–250 m a.s.l. **IUCN Habitat code:** 1,3,4,5 **Conservation status: (IUCN)** NE

4813 *Tabebuia palustris* Hemsl. (Bignoniaceae) **IPNI:** 110998-1 **Common name:** roble playero **Habit:** Shrub, Tree **Use:** ME **Origin:** Native **Region:** Pacific **Dept.:** CAU, CHO, NAR, VAC **Elev.:** 0–0 m a.s.l. **IUCN Habitat code:** 1,5 **Conservation status: (IUCN)** VU

4814 *Tabebuia rosea* (Bertol.) Bertero ex A.DC. (Bignoniaceae) **IPNI:** 111027-1 **Common name:** apamate, chicalá, flor morado, flor rosao, florblanca, florblanco, flormorado, flormorao, guayacán, guayacán lila, guayacán morado, guayacán rosado, minás, ocobo, palo colorado, polvillo, roble, roble de río, roble morado, roble sabanero, roso, urapo **Habit:** Small tree, Tree **Use:** EU, FU, HF, MA, ME **Origin:** Native, Cultivated **Region:** Amazonia, Andean, Guiana Shield, Pacific **Dept.:** AMA, ANT, ATL, BOL, CAU, CES, CHO, COR, CUN, GUA, LAG, MAG, NAR, NSA, SAN, SUC, VAC **Elev.:** 0–1700 m a.s.l. **IUCN Habitat code:** 1,2,3,4,5,14 **Conservation status: (IUCN)** LC

4815 *Tabebuia roseoalba* (Ridl.) Sandwith (Bignoniaceae) **IPNI:** 248789-2 **Habit:** Tree **Use:** MA **Origin:** Cultivated **Region:** Andean **Dept.:** ANT, VAC **Elev.:** 1000–1340 m a.s.l. **IUCN Habitat code:** 1,3,14 **Conservation status: (IUCN)** NE

4816 *Tanaecium bilabiatum* (Sprague) L.G.Lohmann (Bignoniaceae) **IPNI:** 77100467-1 **Habit:** Liana, Climbing **Use:** PO **Origin:** Native **Region:** Amazonia, Guiana Shield, Orinoquia **Dept.:** AMA, ARA, CAQ, GUA **Elev.:** 90–300 m a.s.l. **IUCN Habitat code:** 1 **Conservation status: (IUCN)** NE

4817 *Tanaecium exitiosum* Dugand (Bignoniaceae) **IPNI:** 249329-2 **Common name:** bejuco blanco, mataganado, mataganao **Habit:** Shrub, Liana **Use:** ME, PO **Origin:** Endemic **Dept.:** ANT, BOY, CAL, SAN **Elev.:** 0–500 m a.s.l. **IUCN Habitat code:** 1,3,14 **Conservation status: (IUCN)** NE

4818 *Tanaecium jaroba* Sw. (Bignoniaceae) **IPNI:** 111083-1 **Common name:** bejuco calabacilla, bejuco canastero, calabacillo, calabacillo prieto, calabaza **Habit:** Liana, Climbing **Use:** EU, MA, ME **Origin:** Native **Region:** Amazonia, Caribbean, Pacific **Dept.:** AMA, ANT, ATL, BOL, CAQ, LAG, MAG, SUC **Elev.:** 3–300 m a.s.l. **IUCN Habitat code:** 1,2,3,4,5,14 **Conservation status: (IUCN)** NE

4819 *Tanaecium xanthophyllum* (DC.) L.G.Lohmann (Bignoniaceae) **IPNI:** 77142096-1 **Common name:** curi guasca, curí, curí-guasca, guasca **Habit:** Liana **Use:** ME **Origin:** Native **Region:** Amazonia, Pacific **Dept.:** AMA, NAR, PUT **Elev.:** 100–270 m a.s.l. **IUCN Habitat code:** 1,3,5 **Conservation status: (IUCN)** NE

4820 *Tecoma stans* (L.) Juss. ex Kunth (Bignoniaceae) **IPNI:** 111284-1 **Common name:** amarillo, araguaney bobo, caballito, cabalongo extranjero, cañaguate, chicalá, chicalá polvillo, chirlobirlo, chirlobirlos, chirobiro, cucaracho, floramarilla, floramarillo, floro, fresnillo, fresno, guayacán chicalá, juaquín, quillotocto, santodomingo **Habit:** Shrub, Small tree, Tree **Use:** AF, EU, HF, IF, MA, ME, SU **Origin:** Native, Cultivated **Region:** Amazonia, Andean, Caribbean **Dept.:** AMA, ANT, ATL, BOG, BOL, BOY, CAU, CUN, HUI, MAG, NAR, QUI, SAN, TOL, VAC **Elev.:** 0–2800 m a.s.l. **IUCN Habitat code:** 1,2,3,4,5,14 **Conservation status: (IUCN)** LC

4821 *Tecomaria capensis* (Thunb.) Spach (Bignoniaceae) **IPNI:** 111333-1 **Common name:** bignonia, candela, llamarada **Habit:** Shrub **Use:** AF, EU, HF, MA, ME **Origin:** Cultivated **Region:** Andean **Dept.:** ANT, CUN, VAC **Elev.:** 900–2300 m a.s.l. **IUCN Habitat code:** 14 **Conservation status: (IUCN)** NE

4822 *Tynanthus panurensis* (Bureau) Sandwith (Bignoniaceae) **IPNI:** 259273-2 **Common name:** ambar, bejuco candela **Habit:** Climbing **Use:** ME **Origin:** Native **Region:** Amazonia **Dept.:** AMA, CAQ, PUT, VAU **Elev.:** 200–320 m a.s.l. **IUCN Habitat code:** 1,3,14 **Conservation status: (IUCN)** NE

4823 *Tynanthus polyanthus* (Bureau) Sandwith (Bignoniaceae) **IPNI:** 259274-2 **Habit:** Liana **Use:** ME **Origin:** Native **Region:** Orinoquia, Pacific **Dept.:** ANT, CHO, MET **Elev.:** 0–200 m a.s.l. **IUCN Habitat code:** 1 **Conservation status: (IUCN)** NE

4824 *Calceolaria dichotoma* Lam. (Calceolariaceae) **IPNI:** 800278-1 **Habit:** Herb **Use:** EU **Origin:** Native **Region:** Andean **Dept.:** CAU, CUN, NAR **Elev.:** 1300–2600 m a.s.l. **IUCN Habitat code:** 14 **Conservation status: (IUCN)** NE

4825 *Calceolaria mexicana* Benth. (Calceolariaceae) **IPNI:** 41487-2 **Common name:** redículo **Habit:** Herb **Use:** EU, ME **Origin:** Native **Region:** Andean **Dept.:** ANT, BOG, BOY, CAU, CES, CHO, CUN, MAG, NAR, NSA, QUI, SAN, TOL, VAC **Elev.:** 1700-3600 msnm **IUCN Habitat code:** 1,3,4,8,14 **Conservation status: (IUCN)** NE **(CPSG)** Potential LC

4826 *Calceolaria tripartita* Ruiz & Pav. (Calceolariaceae) **IPNI:** 41696-2 **Common name:** carrielito amarillo **Habit:** Herb **Use:** ME **Origin:** Native **Region:** Andean **Dept.:** ANT, BOG, CAL, CAQ, CAU, CUN, HUI, NAR, NSA, PUT, QUI, SAN, TOL, VAC **Elev.:** 1500-3380 msnm **IUCN Habitat code:** 1,3,4,14 **Conservation status: (IUCN)** NE **(CPSG)** Potential LC

4827 *Achimenes erecta* (Lam.) H.P.Fuchs (Gesneriaceae) **IPNI:** 2352-2 **Common name:** caracola **Habit:** Herb **Use:** EU, ME **Origin:** Native **Region:** Andean **Dept.:** ANT, CUN **Elev.:** 1700–2890 m a.s.l. **Conservation status: (IUCN)** NE

4828 *Achimenes longiflora* DC. (Gesneriaceae) **IPNI:** 377615-1 **Common name:** sinigual **Habit:** Herb **Use:** EU, ME **Origin:** Native, Cultivated **Region:** Andean **Dept.:** ANT, CUN, RIS **Elev.:** 1000–1725 m a.s.l. **IUCN Habitat code:** 1,14 **Conservation status: (IUCN)** NE

4829 *Besleria calantha* C.V.Morton (Gesneriaceae) **IPNI:** 31867-2 **Common name:** atajasangre **Habit:** Herb, Shrub **Use:** ME **Origin:** Endemic **Region:** Andean, Guiana Shield **Dept.:** ANT, CUN, HUI, MET **Elev.:** 1300–2300 m a.s.l. **IUCN Habitat code:** 1,14 **Conservation status: (IUCN)** NE

4830 *Besleria eriocalyx* C.V.Morton (Gesneriaceae) **IPNI:** 31912-2 **Habit:** Shrub **Use:** ME **Origin:** Endemic **Region:** Andean **Dept.:** HUI **Elev.:** 1800–2100 m a.s.l. **IUCN Habitat code:** 1,8 **Conservation status: (IUCN)** NE

4831 *Besleria fecunda* C.V.Morton (Gesneriaceae) **IPNI:** 31913-2 **Common name:** gota de sangre **Habit:** Shrub, Subshrub **Use:** ME **Origin:** Native **Region:** Andean **Dept.:** ANT, CAL, CAQ, SAN **Elev.:** 100–1700 m a.s.l. **IUCN Habitat code:** 1,3,5,14 **Conservation status: (IUCN)** NE

4832 *Besleria leucostoma* (Hook.) Hanst. (Gesneriaceae) **IPNI:** 378201-1 **Common name:** mata de conga **Habit:** Subshrub **Use:** ME **Origin:** Endemic **IUCN Habitat code:** 1 **Conservation status: (IUCN)** NE

4833 *Besleria nitens* Fritsch (Gesneriaceae) **IPNI:** 31989-2 **Common name:** mancha de sangre **Habit:** Herb, Shrub, Subshrub **Use:** ME **Origin:** Endemic **Region:** Andean **Dept.:** ANT, MAG, SAN, TOL, VAC **Elev.:** 200–1800 m a.s.l. **IUCN Habitat code:** 1,3,14 **Conservation status: (IUCN)** NE

4834 *Besleria reticulata* Fritsch (Gesneriaceae) **IPNI:** 32018-2 **Habit:** Herb, Shrub, Subshrub, Small tree **Use:** ME **Origin:** Native **Region:** Andean, Pacific **Dept.:** ANT, BOG, BOY, CAQ, CAU, CUN, HUI, MET, NSA, PUT, SAN, TOL, VAC **Elev.:** 20-3300 msnm **IUCN Habitat code:** 1,3,5,14 **Conservation status: (IUCN)** NE **(CPSG)** Potential LC

4835 *Besleria solanoides* Kunth (Gesneriaceae) **IPNI:** 378279-1 **Common name:** atajasangre, cafecillo **Habit:** Herb, Shrub, Subshrub, Small tree **Use:** ME **Origin:** Native **Region:** Amazonia, Andean, Pacific **Dept.:** ANT, BOY, CAL, CAU, CHO, CUN, HUI, MAG, MET, NAR, NSA, QUI, RIS, SAN, TOL, VAC, VAU **Elev.:** 0-3380 msnm **IUCN Habitat code:** 1,2,3,4,14 **Conservation status: (IUCN)** LC **(CPSG)** Potential LC

4836 *Chrysothemis friedrichsthaliana* (Hanst.) H.E.Moore (Gesneriaceae) **IPNI:** 58213-2 **Common name:** bellísima, cuna, hierba cuna **Habit:** Herb, Subshrub **Use:** ME **Origin:** Native **Region:** Andean, Pacific **Dept.:** ANT, BOL, CAU, CHO, NAR, RIS, VAC **Elev.:** 0–1000 m a.s.l. **IUCN Habitat code:** 1,3,5,14 **Conservation status: (IUCN)** NE

4837 *Chrysothemis pulchella* (Donn ex Sims) Decne. (Gesneriaceae) **IPNI:** 378695-1 **Habit:** Herb, Subshrub **Use:** EU **Origin:** Native, Cultivated **Region:** Andean, Guiana Shield, Orinoquia, Pacific **Dept.:** ANT, ARA, BOY, CES, CHO, CUN, MET, NAR, RIS, SAN **Elev.:** 70–1200 m a.s.l. **Conservation status: (IUCN)** NE

4838 *Codonanthopsis crassifolia* (H.Focke) Chautems & Mat.Perret (Gesneriaceae) **IPNI:** 77139499-1 **Common name:** yerba de hormiga mediana **Habit:** Herb, Epiphyte, Climbing **Use:** EU, ME **Origin:** Native **Region:** Amazonia, Andean, Guiana Shield, Pacific **Dept.:** AMA, ANT, BOY, CAQ, CAU, CHO, CUN, GUA, GUV, MET, NAR, RIS, SAN, VAC, VAU **Elev.:** 0–1100 m a.s.l. **IUCN Habitat code:** 1,3,4,5,14 **Conservation status: (IUCN)** NE

4839 *Columnea anisophylla* DC. (Gesneriaceae) **IPNI:** 63508-2 **Common name:** atajasangre **Habit:** Herb, Shrub, Epiphyte **Use:** ME **Origin:** Native **Region:** Andean, Pacific **Dept.:** ANT, BOY, CAL, CAQ, CAU, CHO, CUN, HUI, MAG, NSA, PUT, RIS, TOL, VAC **Elev.:** 30-2800 msnm **IUCN Habitat code:** 1,3,14 **Conservation status: (IUCN)** NE **(CPSG)** Potential LC

4840 *Columnea arguta* C.V.Morton (Gesneriaceae) **IPNI:** 63510-2 **Habit:** Epiphyte **Use:** EU **Origin:** Native **Region:** Andean **Dept.:** ANT **Elev.:** 1830–1830 m a.s.l. **IUCN Habitat code:** 1 **Conservation status: (IUCN)** NE

4841 *Columnea bilabiata* Seem. (Gesneriaceae) **IPNI:** 328486-2 **Common name:** sanguinaria **Habit:** Herb, Shrub, Epiphyte **Use:** ME **Origin:** Native **Region:** Andean, Pacific **Dept.:** ANT, CAL, CAU, CHO, NAR, RIS, VAC **Elev.:** 0–2000 m a.s.l. **IUCN Habitat code:** 1,3,5,14 **Conservation status: (IUCN)** NE

4842 *Columnea brenneri* (Wiehler) B.D.Morley (Gesneriaceae) **IPNI:** 378787-1 **Common name:** mordaza **Habit:** Shrub, Epiphyte **Use:** ME **Origin:** Native **Region:** Andean **Dept.:** CAU, PUT, VAC **Elev.:** 1350–1420 m a.s.l. **IUCN Habitat code:** 3 **Conservation status: (IUCN)** VU

4843 *Columnea calotricha* Donn.Sm. (Gesneriaceae) **IPNI:** 63523-2 **Habit:** Herb, Shrub, Epiphyte **Use:** ME **Origin:** Native **Region:** Pacific **Dept.:** ANT, VAC **Elev.:** 40–100 m a.s.l. **IUCN Habitat code:** 1 **Conservation status: (IUCN)** NE

4844 *Columnea dictyophylla* Donn.Sm. (Gesneriaceae) **IPNI:** 63557-2 **Habit:** Herb, Subshrub, Epiphyte **Use:** ME **Origin:** Native **Region:** Andean **Dept.:** ANT, CAU, RIS, VAC **Elev.:** 1890–3050 m a.s.l. **IUCN Habitat code:** 1,3,14 **Conservation status: (IUCN)** NE

4845 *Columnea dimidiata* (Benth.) Kuntze (Gesneriaceae) **IPNI:** 63559-2 **Common name:** ataja sangre, atajasangre, mancha de sangre, sangre de Cristo, sangre de cristo **Habit:** Herb, Shrub, Subshrub, Epiphyte **Use:** ME **Origin:** Native **Region:** Andean, Pacific **Dept.:** ANT, BOY, CAL, CAU, CHO, CUN, NAR, QUI, RIS, VAC **Elev.:** 0–3000 m a.s.l. **IUCN Habitat code:** 1,3,4,14 **Conservation status: (IUCN)** NE

4846 *Columnea eburnea* (Wiehler) L.P.Kvist & L.E.Skog (Gesneriaceae) **IPNI:** 969729-1 **Habit:** Herb, Epiphyte **Use:** HF **Origin:** Native **Region:** Andean, Pacific **Dept.:** CHO, NAR, RIS, VAC **Elev.:** 200–1400 m a.s.l. **IUCN Habitat code:** 1,14 **Conservation status: (IUCN)** NE

4847 *Columnea ericae* Mansf. (Gesneriaceae) **IPNI:** 378844-1 **Common name:** irisipela rosada, noiguita, rabo de tucán, árnica morada **Habit:** Subshrub, Epiphyte, Hemiepiphyte, Climbing **Use:** ME **Origin:** Native **Region:** Amazonia, Andean, Pacific **Dept.:** AMA, CAQ, CHO, HUI, PUT, VAU **Elev.:** 90–1800 m a.s.l. **IUCN Habitat code:** 1,3,14 **Conservation status: (IUCN)** NE

4848 *Columnea guttata* Poepp. (Gesneriaceae) **IPNI:** 63584-2 **Habit:** Herb, Shrub, Epiphyte, Climbing **Use:** ME **Origin:** Native **Region:** Andean **Dept.:** ANT, CAQ, CUN, HUI, NAR, PUT, VAC **Elev.:** 1350–1700 m a.s.l. **IUCN Habitat code:** 1 **Conservation status: (IUCN)** NE

4849 *Columnea kalbreyeriana* Mast. (Gesneriaceae) **IPNI:** 305407-2 **Common name:** caracola, hoja de sangre, sangre de Cristo, sangre de cristo, sanguinaria **Habit:** Herb, Shrub, Epiphyte, Climbing **Use:** ME, PO, SU **Origin:** Native **Region:** Andean, Caribbean, Pacific **Dept.:** ANT, BOL, CHO, CUN, MAG, NAR, SAN, VAC **Elev.:** 10–1750 m a.s.l. **IUCN Habitat code:** 1,3,14 **Conservation status: (IUCN)** NE

4850 *Columnea medicinalis* (Wiehler) L.E.Skog & L.P.Kvist (Gesneriaceae) **IPNI:** 999467-1 **Habit:** Herb, Subshrub, Epiphyte, Climbing **Use:** ME **Origin:** Native **Region:** Andean, Pacific **Dept.:** CAU, CHO, NAR **Elev.:** 80–2400 m a.s.l. **IUCN Habitat code:** 1,3,14 **Conservation status: (IUCN)** NE

4851 *Columnea microcalyx* Hanst. (Gesneriaceae) **IPNI:** 378924-1 **Habit:** Herb, Epiphyte **Use:** EU **Origin:** Native **Region:** Andean, Pacific **Dept.:** ANT, CHO **Elev.:** 0–1450 m a.s.l. **IUCN Habitat code:** 1,3,14 **Conservation status: (IUCN)** NE

4852 *Columnea microphylla* Klotzsch & Hanst. ex Oerst. (Gesneriaceae) **IPNI:** 378925-1 **Use:** EU **Origin:** Native **IUCN Habitat code:** 1,3,14 **Conservation status: (IUCN)** NE

4853 *Columnea parviflora* C.V.Morton (Gesneriaceae) **IPNI:** 63636-2 **Common name:** golondrina **Habit:** Epiphyte, Climbing **Use:** ME **Origin:** Native **Region:** Andean, Pacific **Dept.:** CHO, NAR, VAC **Elev.:** 20–2000 m a.s.l. **IUCN Habitat code:** 1,3,4,5,14 **Conservation status: (IUCN)** NE

4854 *Columnea picta* H.Karst. (Gesneriaceae) **IPNI:** 63643-2 **Common name:** jarekita, pico de loro, sanguinaria **Habit:** Herb, Epiphyte, Climbing **Use:** ME **Origin:**

Native **Region:** Andean, Pacific **Dept.:** ANT, CAL, CAU, CHO, NAR, QUI, RIS, VAC **Elev.:** 0–2860 m a.s.l. **IUCN Habitat code:** 1,3,5,14 **Conservation status: (IUCN)** NE

4855 *Columnea pulcherrima* C.V.Morton (Gesneriaceae) **IPNI:** 63653-2 **Common name:** guaquito, sanguinaria **Habit:** Herb, Shrub, Tree, Epiphyte **Use:** ME **Origin:** Endemic **Region:** Andean, Pacific **Dept.:** ANT, CHO, VAC **Elev.:** 0–2100 m a.s.l. **IUCN Habitat code:** 1,3,5,14 **Conservation status: (IUCN)** NE

4856 *Columnea purpurata* Hanst. (Gesneriaceae) **IPNI:** 378968-1 **Common name:** reina **Habit:** Herb, Shrub, Subshrub, Epiphyte **Use:** ME **Origin:** Native **Region:** Amazonia, Andean, Pacific **Dept.:** AMA, ANT, BOY, CHO, CUN, MAG, NAR, QUI, RIS, SAN, VAC **Elev.:** 0–2840 m a.s.l. **IUCN Habitat code:** 1,3,14 **Conservation status: (IUCN)** NE

4857 *Columnea sanguinea* (Pers.) Hanst. (Gesneriaceae) **IPNI:** 378985-1 **Common name:** atajasangre, mancha, sangre de Cristo **Habit:** Herb, Shrub, Epiphyte, Climbing **Use:** EU, ME **Origin:** Native **Region:** Amazonia, Andean, Orinoquia, Pacific **Dept.:** AMA, ANT, BOY, CAQ, CHO, CUN, MAG, MET, NSA, PUT, QUI, RIS, SAN, TOL, VAC, VAU **Elev.:** 100-2600 msnm **IUCN Habitat code:** 1,3,14 **Conservation status: (IUCN)** NE **(CPSG)** Potential LC

4858 *Columnea strigosa* Benth. (Gesneriaceae) **IPNI:** 379021-1 **Common name:** caracola, conejitos, gateadora, pico de loro amarillo, yerba de sangre **Habit:** Herb, Shrub, Epiphyte, Climbing **Use:** EU, ME, PO **Origin:** Native **Region:** Andean **Dept.:** ANT, BOG, BOY, CAL, CAQ, CAU, CHO, CUN, HUI, MET, NAR, NSA, PUT, QUI, RIS, SAN, TOL, VAC **Elev.:** 1100-4350 msnm **IUCN Habitat code:** 1,2,3,4,5,8,14 **Conservation status: (IUCN)** NE **(CPSG)** Potential LC

4859 *Columnea tenella* L.P.Kvist & L.E.Skog (Gesneriaceae) **IPNI:** 969025-1 **Habit:** Herb, Epiphyte, Climbing **Use:** HF **Origin:** Native **Region:** Andean, Pacific **Dept.:** ANT, CHO, NAR, RIS, VAC **Elev.:** 200–1550 m a.s.l. **IUCN Habitat code:** 1,14 **Conservation status: (IUCN)** NE

4860 *Drymonia coccinea* (Aubl.) Wiehler (Gesneriaceae) **IPNI:** 84424-2 **Common name:** arco, atajasangre, bejuco soldador, carpintero trenza, hierba de sarna, mancha, penitente **Habit:** Herb, Epiphyte, Liana, Climbing **Use:** ME **Origin:** Native **Region:** Amazonia, Andean, Guiana Shield, Orinoquia, Pacific **Dept.:** AMA, BOG, CAQ, CAU, CHO, GUV, MET, PUT, VAC, VAU **Elev.:** 0–1420 m a.s.l. **IUCN Habitat code:** 1,3,5,14 **Conservation status: (IUCN)** NE

4861 *Drymonia coriacea* (Oerst. ex Hanst.) Wiehler (Gesneriaceae) **IPNI:** 84425-2 **Habit:** Herb, Subshrub, Epiphyte, Hemiepiphyte **Use:** HF, ME **Origin:** Native **Region:** Amazonia, Andean, Pacific **Dept.:** AMA, ANT, CHO, NAR, VAC, VAU **Elev.:** 0–3140 m a.s.l. **IUCN Habitat code:** 1,3,14 **Conservation status: (IUCN)** NE

4862 *Drymonia serrulata* (Jacq.) Mart. (Gesneriaceae) **IPNI:** 380742-1 **Common name:** arrojasangre, desbaratadora, desbaratadora blanca, desbaratadora colorada, desbaratadora morada, pico de paujil, sombrerito del diablo **Habit:** Herb, Shrub, Epiphyte, Hemiepiphyte, Liana, Climbing **Use:** EU, ME **Origin:** Native **Region:** Amazonia, Andean, Guiana Shield, Caribbean, Orinoquia, Pacific **Dept.:** AMA, ANT, ARA, BOL, BOY, CAL, CAQ, CAS, CAU, CHO, CUN, LAG, MAG, MET, NAR, NSA, PUT, RIS, TOL, VAC, VAU **Elev.:** 0-3200 msnm **IUCN Habitat code:** 1,2,3,4,14 **Conservation status: (IUCN)** NE **(CPSG)** Potential LC

4863 *Episcia cupreata* (Hook.) Hanst. (Gesneriaceae) **IPNI:** 380772-1 **Common name:** antra roja, caracola, corbata de rico, corbatas, manta, tapete, tapiz **Habit:** Herb **Use:** EU **Origin:** Native **Region:** Amazonia, Andean, Guiana Shield, Orinoquia, Pacific **Dept.:** ANT, ARA, ATL, BOY, CAL, CAQ, CAS, CAU, CHO, CUN, GUV, MAG, MET, NAR, NSA, PUT, SAN, TOL, VAC, VAU **Elev.:** 0-2600 msnm **IUCN Habitat code:** 1,2,3,4,5,14 **Conservation status: (IUCN)** NE **(CPSG)** Potential LC

4864 *Episcia lilacina* Hanst. (Gesneriaceae) **IPNI:** 380794-1 **Common name:** estrella caracha roja **Habit:** Herb **Use:** EU, ME **Origin:** Native **Region:** Andean, Pacific **Dept.:** ANT, CHO **Elev.:** 0–700 m a.s.l. **IUCN Habitat code:** 1,3,14 **Conservation status: (IUCN)** NE

4865 *Episcia reptans* Mart. (Gesneriaceae) **IPNI:** 380813-1 **Common name:** añadidor, estrella caracha morada **Habit:** Herb, Epiphyte **Use:** EU, ME **Origin:** Native **Region:** Amazonia, Andean **Dept.:** AMA, ANT, CAQ, CUN, GUA, MET, NSA, PUT, SAN, VAU **Elev.:** 70–2000 m a.s.l. **IUCN Habitat code:** 1,3,4,5,14 **Conservation status: (IUCN)** NE

4866 *Gasteranthus corallinus* (Fritsch) Wiehler (Gesneriaceae) **IPNI:** 107520-2 **Habit:** Herb, Subshrub **Use:** ME **Origin:** Native **Region:** Amazonia, Andean, Pacific **Dept.:** AMA, CAU, NAR, PUT **Elev.:** 300–1900 m a.s.l. **IUCN Habitat code:** 1,3,14 **Conservation status: (IUCN)** NE

4867 *Gasteranthus pansamalanus* (Donn.Sm.) Wiehler (Gesneriaceae) **IPNI:** 107533-2 **Habit:** Herb, Subshrub **Use:** ME **Origin:** Native **Region:** Andean **Dept.:** BOY, CAU, HUI, NAR, PUT **Elev.:** 1300–2300 m a.s.l. **IUCN Habitat code:** 1,3,14 **Conservation status: (IUCN)** NE

4868 *Glossoloma ichthyoderma* (Hanst.) J.L.Clark (Gesneriaceae) **IPNI:** 60441766-2 **Common name:** vergüenza **Habit:** Herb, Shrub, Small tree, Epiphyte, Climbing **Use:** ME **Origin:** Native **Region:** Andean **Dept.:** ANT, BOG, BOY, CAL, CAQ, CAU, CES, CUN, HUI, MET, NAR, NSA, PUT, QUI, RIS, SAN, TOL, VAC **Elev.:** 1000-3900 msnm **IUCN Habitat code:** 1,3,4,6,14 **Conservation status: (IUCN)** NE **(CPSG)** Potential LC

4869 *Glossoloma panamense* (C.V.Morton) J.L.Clark (Gesneriaceae) **IPNI:** 60441770-2 **Common name:** cerveza, chupaflor, pavonilla **Habit:** Herb, Shrub, Subshrub, Epiphyte **Use:** ME **Origin:** Native **Region:** Andean, Pacific **Dept.:** ANT, CAU, CHO, NAR, QUI, RIS, VAC **Elev.:** 0–2300 m a.s.l. **IUCN Habitat code:** 1,3,5,14 **Conservation status: (IUCN)** NE

4870 *Gloxinia perennis* (L.) Druce (Gesneriaceae) **IPNI:** 381282-1 **Common name:** caracola, tosilla **Habit:** Herb, Subshrub **Use:** EU, ME **Origin:** Native, Cultivated **Region:** Amazonia, Andean, Caribbean, Orinoquia, Pacific **Dept.:** ATL, BOY, CAQ, CAS, CHO, CUN, MAG, MET, PUT, VAC **Elev.:** 350–1500 m a.s.l. **IUCN Habitat code:** 1,3,4,14 **Conservation status: (IUCN)** NE

4871 *Heppiella ulmifolia* (Kunth) Hanst. (Gesneriaceae) **IPNI:** 381388-1 **Common name:** caracol, caracola, moneda, sen de tierra **Habit:** Herb, Shrub, Subshrub, Climbing **Use:** ME **Origin:** Native **Region:** Andean **Dept.:** ANT, BOG, CAL, CAU, CUN, MET, NAR, PUT, QUI, RIS, TOL, VAC **Elev.:** 1600–4000 m a.s.l. **IUCN Habitat code:** 1,3,4,14 **Conservation status: (IUCN)** NE

4872 *Kohleria amabilis* (Planch. & Linden) Fritsch (Gesneriaceae) **IPNI:** 133755-2 **Habit:** Herb **Use:** EU **Origin:** Endemic **Region:** Andean **Dept.:** ANT, ARA, BOY, CUN, MAG, TOL **Elev.:** 250–1200 m a.s.l. **IUCN Habitat code:** 1,3,14 **Conservation status: (IUCN)** NE

4873 *Kohleria hirsuta* (Kunth) Regel (Gesneriaceae) **IPNI:** 381572-1 **Common name:** bejuco de riñoncito, caracola, oreja de perro, riñonada, riñoncito, tusa, tusilla, yerba de arco, yerbadiarco **Habit:** Herb, Shrub, Subshrub **Use:** EU, ME **Origin:** Native **Region:** Andean **Dept.:** ANT, BOG, BOY, CAL, CAS, CAU, CUN, HUI, MAG, MET, NAR, NSA, PUT, QUI, SAN, TOL **Elev.:** 100-2700 msnm **IUCN Habitat code:** 1,2,3,4,5,14 **Conservation status: (IUCN)** NE **(CPSG)** Potential LC

4874 *Kohleria spicata* (Kunth) Oerst. (Gesneriaceae) **IPNI:** 381628-1 **Common name:** caracola, caracola roja, doncella, gorrito rojo, gorro liberal, gotas de amor, gusanito, sen de la tierra, tusilla **Habit:** Herb, Subshrub **Use:** EU, ME **Origin:** Native **Region:** Andean, Pacific **Dept.:** ANT, BOL, CAL, CAU, CHO, CUN, HUI, MAG, MET, NAR, NSA, PUT, QUI, RIS, TOL, VAC **Elev.:** 0-2800 msnm **IUCN Habitat code:** 1,3,14 **Conservation status: (IUCN)** NE **(CPSG)** Potential LC

4875 *Kohleria tubiflora* (Cav.) Hanst. (Gesneriaceae) **IPNI:** 381638-1 **Common name:** caracola, pana **Habit:** Herb, Subshrub **Use:** ME **Origin:** Native **Region:** Andean, Guiana Shield, Caribbean, Orinoquia, Pacific **Dept.:** ANT, CAU, CES, CHO, COR, LAG, MAG, MET, NAR, NSA, QUI **Elev.:** 0-2600 msnm **IUCN Habitat code:** 1,2,3,14 **Conservation status: (IUCN)** NE **(CPSG)** Potential LC

4876 *Napeanthus apodemus* Donn.Sm. (Gesneriaceae) **IPNI:** 167091-2 **Common name:** chocó blanco **Habit:** Herb, Epiphyte **Use:** ME **Origin:** Native **Region:** Andean, Pacific **Dept.:** ANT, BOY, CAQ, CHO, MET, NAR, PUT, SAN, VAC **Elev.:** 18–1450 m a.s.l. **IUCN Habitat code:** 1,3,14 **Conservation status: (IUCN)** NE

4877 *Nautilocalyx antioquensis* Wiehler (Gesneriaceae)
IPNI: 306818-2
Common name: toje **Habit:** Herb **Use:** EU **Origin:** Endemic, Cultivated **Dept.:** ANT **Elev.:** 180–325 m a.s.l. **IUCN Habitat code:** 1,3 **Conservation status: (IUCN)** NE

4878 *Nautilocalyx lynchii* (Hook.f.) Sprague (Gesneriaceae)
IPNI: 167431-2
Use: EU **Origin:** Native **Conservation status: (IUCN)** NE

4879 *Reldia grandiflora* L.P.Kvist & L.E.Skog (Gesneriaceae)
IPNI: 284179-2
Habit: Herb, Subshrub **Use:** ME **Origin:** Endemic **Region:** Andean **Dept.:** CAQ, QUI, RIS, TOL **Elev.:** 1600–2750 m a.s.l. **IUCN Habitat code:** 1,3,4,14 **Conservation status: (IUCN)** NE

4880 *Reldia minutiflora* (L.E.Skog) L.P.Kvist & L.E.Skog (Gesneriaceae)
IPNI: 284180-2
Habit: Herb, Shrub **Use:** ME **Origin:** Native **Region:** Amazonia, Andean, Pacific **Dept.:** ANT, BOY, CAL, CAQ, CAU, CHO, CUN, PUT, QUI, RIS, TOL, VAC **Elev.:** 110-2800 msnm **IUCN Habitat code:** 1,3,14 **Conservation status: (IUCN)** NE **(CPSG)** Potential LC

4881 *Sinningia incarnata* (Aubl.) D.L.Denham (Gesneriaceae)
IPNI: 237093-2
Common name: tusilla **Habit:** Herb, Shrub, Subshrub **Use:** HF, ME **Origin:** Native **Region:** Amazonia, Andean, Guiana Shield, Orinoquia **Dept.:** AMA, ARA, BOY, CAS, CAU, CUN, GUV, MAG, MET, SAN, VAC, VID **Elev.:** 355-2640 msnm **IUCN Habitat code:** 1,3,4,5,14 **Conservation status: (IUCN)** NE **(CPSG)** Potential LC

4882 *Sinningia speciosa* (G.Lodd. ex Ker Gawl.) Hiern (Gesneriaceae)
IPNI: 382786-1
Use: EU **Region:** Andean **Dept.:** BOG **IUCN Habitat code:** 1 **Conservation status: (IUCN)** NE

4883 *Streptocarpus ionanthus* (H.Wendl.) Christenh. (Gesneriaceae)
IPNI: 77117375-1
Common name: violeta, violeta africana **Habit:** Herb **Use:** EU, ME **Origin:** Cultivated **Region:** Andean **Dept.:** CUN **Elev.:** 2600–2600 m a.s.l. **IUCN Habitat code:** 1 **Conservation status: (IUCN)** NE

4884 *Streptocarpus saxorum* Engl. (Gesneriaceae)
IPNI: 383021-1
Habit: Herb **Use:** EU **Origin:** Cultivated **Region:** Andean **Dept.:** BOY **Elev.:** 2200–2200 m a.s.l. **Conservation status: (IUCN)** LC

4885 *Aegiphila alba* Moldenke (Lamiaceae)
IPNI: 5660-2
Habit: Tree **Use:** HF **Origin:** Native **Region:** Andean, Pacific **Dept.:** ANT, BOG, BOL, CHO, CUN, VAC **Elev.:** 100–2170 m a.s.l. **IUCN Habitat code:** 1,3,14 **Conservation status: (IUCN)** LC

4886 *Aegiphila bogotensis* (Spreng.) Moldenke (Lamiaceae)
IPNI: 860809-1
Common name: mantequillo **Habit:** Tree **Use:** HF **Origin:** Native **Region:** Andean **Dept.:** ANT, BOG, CAU, CES, CUN, LAG, MAG, NAR, QUI, RIS, VAC **Elev.:** 2000-3400 msnm **IUCN Habitat code:** 1,3,14 **Conservation status: (IUCN)** LC **(CPSG)** LC

4887 *Aegiphila elata* Sw. (Lamiaceae)
IPNI: 860847-1
Common name: bejuco falso **Habit:** Shrub, Liana, Climbing **Use:** ME **Origin:** Native **Region:** Andean, Pacific **Dept.:** ANT, ARA, BOL, CAL, CHO, CUN, MET, VAC **Elev.:** 0–1000 m a.s.l. **IUCN Habitat code:** 1,3,5,14 **Conservation status: (IUCN)** NE

4888 *Aegiphila integrifolia* (Jacq.) B.D.Jacks. (Lamiaceae)
IPNI: 860891-1
Common name: abubuchique, guacamayo caspi, papelillo, santamaría, tabaco, tabaquillo **Habit:** Shrub, Tree **Use:** ME **Origin:** Native **Region:** Amazonia, Andean, Guiana Shield, Orinoquia, Pacific **Dept.:** AMA, ANT, CAQ, CAS, CHO, CUN, GUV, MET, PUT, SAN, TOL, VAC, VAU **Elev.:** 40-2000 msnm **IUCN Habitat code:** 1,2,3,4,5,14 **Conservation status: (IUCN)** LC **(CPSG)** LC

4889 *Aegiphila martinicensis* Jacq. (Lamiaceae)
IPNI: 860919-1
Habit: Shrub, Tree **Use:** HF, ME **Origin:** Native **Region:** Andean, Pacific **Dept.:** ANT, NSA **Elev.:** 280–280 m a.s.l. **IUCN Habitat code:** 1,14 **Conservation status: (IUCN)** NE

4890 *Aegiphila mollis* Kunth (Lamiaceae)
IPNI: 860928-1
Common name: cordillero, gavilán, mantequillo, tsematsemanea **Habit:** Shrub **Use:** ME **Origin:** Native **Region:** Amazonia, Andean, Caribbean, Orinoquia, Pacific **Dept.:** ANT, ATL, BOL, CAS, CUN, GUV, HUI, MAG, MET, QUI, TOL, VAC **Elev.:** 10-2900 msnm **IUCN Habitat code:** 1,2,3,4,14 **Conservation status: (IUCN)** LC **(CPSG)** LC

4891 *Aegiphila panamensis* Moldenke (Lamiaceae)
IPNI: 5786-2
Common name: mantequillo, ochorróneta **Habit:** Shrub, Small tree, Tree **Use:** ME **Origin:** Native **Region:** Amazonia, Andean, Pacific **Dept.:** AMA, ANT, BOL, BOY, CHO, GUV, MET, SAN, VAU **Elev.:** 5–1700 m a.s.l. **IUCN Habitat code:** 1,3,4,5,14 **Conservation status: (IUCN)** LC

4892 *Aegiphila peruviana* Turcz. (Lamiaceae)
IPNI: 860961-1
Habit: Shrub, Small tree **Use:** ME **Origin:** Native **Region:** Amazonia **Dept.:** AMA **Elev.:** 100–125 m a.s.l. **IUCN Habitat code:** 1,5 **Conservation status: (IUCN)** NE

4893 *Agastache rugosa* (Fisch. & C.A.Mey.) Kuntze (Lamiaceae)
IPNI: 444504-1
Habit: Herb **Use:** EU, HF, MA, ME **Origin:** Cultivated **Region:** Andean **Dept.:** CUN **Elev.:** 2600–2800 m a.s.l. **Conservation status: (IUCN)** NE

4894 *Ajuga genevensis* L. (Lamiaceae)
IPNI: 444576-1
Use: ME **Region:** Andean **Dept.:** BOG **Conservation status: (IUCN)** NE

4895 *Ajuga pyramidalis* L. (Lamiaceae)
IPNI: 444657-1
Common name: afuga, avispitas, remolacha **Habit:** Herb **Use:** ME **Origin:** Cultivated **Region:** Andean **Dept.:** ANT, BOG, BOY, CUN **Elev.:** 1400–2800 m a.s.l. **IUCN Habitat code:** 1,3,14 **Conservation status: (IUCN)** NE

4896 *Ajuga reptans* L. (Lamiaceae)
IPNI: 444664-1
Habit: Herb **Use:** EU, HF, ME **Origin:** Cultivated **Region:** Andean **Dept.:** BOG, BOY, CUN **Elev.:** 2500–2800 m a.s.l. **Conservation status: (IUCN)** NE

4897 *Callicarpa acuminata* Kunth (Lamiaceae)
IPNI: 861228-1
Common name: guacharaca **Habit:** Shrub, Tree **Use:** HF, ME, SU **Origin:** Native **Region:** Andean, Caribbean, Orinoquia, Pacific **Dept.:** ANT, CAL, CHO, COR, CUN, HUI, LAG, MET, RIS, SAP, TOL, VAC **Elev.:** 0-2100 msnm **IUCN Habitat code:** 1,2,3,14 **Conservation status: (IUCN)** LC **(CPSG)** LC

4898 *Cantinoa americana* (Aubl.) Harley & J.F.B.Pastore (Lamiaceae)
IPNI: 77120747-1
Habit: Herb **Use:** EU, GS, HF, MA, ME, PO, SU **Origin:** Native **Region:** Andean, Pacific **Dept.:** CAU, TOL, VAC **Elev.:** 600-2000 msnm **IUCN Habitat code:** 1,3,14 **Conservation status: (IUCN)** NE **(CPSG)** LC

4899 *Cantinoa colombiana* (Epling) Harley & J.F.B.Pastore (Lamiaceae)
IPNI: 77120750-1
Common name: cardosanto, mastranto **Habit:** Herb, Subshrub **Use:** ME **Origin:** Native **Region:** Andean, Guiana Shield, Orinoquia, Pacific **Dept.:** ANT, BOY, CAL, CAS, CAU, CUN, HUI, MAG, MET, QUI, RIS, SAN, TOL, VAC **Elev.:** 600–2700 m a.s.l. **IUCN Habitat code:** 1,2,3,14 **Conservation status: (IUCN)** NE

4900 *Cantinoa mutabilis* (Rich.) Harley & J.F.B.Pastore (Lamiaceae)
IPNI: 77120760-1
Common name: bretónica, clavillo, contrafuego, guaca, gusanero, gusanillo, mastranco, mastranto, verbena negra **Habit:** Herb, Subshrub **Use:** ME **Origin:** Native **Region:** Amazonia, Andean, Guiana Shield, Caribbean, Orinoquia, Pacific **Dept.:** AMA, ANT, ARA, ATL, BOL, BOY, CAL, CAQ, CAS, CAU, CES, CHO, COR, CUN, GUV, HUI, LAG, MAG, MET, NAR, NSA, PUT, QUI, RIS, SAN, TOL, VAC, VID **Elev.:** 0–1900 m a.s.l. **IUCN Habitat code:** 1,2,3,4,5,14 **Conservation status: (IUCN)** NE

4901 *Catoferia spicata* (Benth.) Benth. (Lamiaceae)
IPNI: 1150341-2
Habit: Subshrub **Use:** ME **Origin:** Native **Region:** Andean **Dept.:** CUN, NSA, SAN **Elev.:** 1000–2000 m a.s.l. **Conservation status: (IUCN)** NE

4902 *Clerodendrum chinense* (Osbeck) Mabb. (Lamiaceae) **IPNI:** 946989-1 **Habit:** Subshrub **Use:** EU, HF, ME **Origin:** Cultivated **Region:** Andean **Dept.:** ANT, CHO **Elev.:** 750–1560 m a.s.l. **IUCN Habitat code:** 1,14 **Conservation status: (IUCN)** LC

4903 *Clerodendrum emirnense* Bojer ex Hook. (Lamiaceae) **IPNI:** 862052-1 **Habit:** Subshrub, Climbing **Use:** EU **Origin:** Cultivated **Region:** Andean **Dept.:** CUN **Elev.:** 1750–2080 m a.s.l. **Conservation status: (IUCN)** NE

4904 *Clerodendrum paniculatum* L. (Lamiaceae) **IPNI:** 862301-1 **Habit:** Shrub **Use:** EU, HF, ME, SU **Origin:** Cultivated **Region:** Pacific **Dept.:** CHO, VAC **Elev.:** 1000–1000 m a.s.l. **IUCN Habitat code:** 1,14 **Conservation status: (IUCN)** NE

4905 *Clerodendrum speciosissimum* Van Geert ex Drapiez (Lamiaceae) **IPNI:** 50426038-2 **Habit:** Shrub **Use:** EU, ME **Origin:** Cultivated **Region:** Andean, Caribbean, Pacific **Dept.:** ANT, ATL, BOL, CHO **Elev.:** 0–1500 m a.s.l. **IUCN Habitat code:** 14 **Conservation status: (IUCN)** NE

4906 *Clerodendrum thomsoniae* Balf.f. (Lamiaceae) **IPNI:** 862437-1 **Common name:** corazón herido, gota de sangre, lágrimas de Obando **Habit:** Climbing **Use:** EU, ME **Origin:** Cultivated, Naturalised **Region:** Andean, Orinoquia **Dept.:** ANT, CAS, VAC **Elev.:** 100–1600 m a.s.l. **IUCN Habitat code:** 1,3,14 **Conservation status: (IUCN)** NE

4907 *Clerodendrum umbellatum* Poir. (Lamiaceae) **IPNI:** 862461-1 **Common name:** lágrimas de Obando **Habit:** Climbing **Use:** EU, MA, ME **Origin:** Cultivated, Naturalised **Region:** Andean, Caribbean **Dept.:** ANT, ATL, CAL, CAU **Elev.:** 6–1600 m a.s.l. **IUCN Habitat code:** 1,2,3 **Conservation status: (IUCN)** NE

4908 *Clinopodium brownei* (Sw.) Kuntze (Lamiaceae) **IPNI:** 61179-2 **Common name:** ajedrea, polea, poleo, poleo chiquito, poleo de castilla, poleo pequeño **Habit:** Herb **Use:** HF, ME **Origin:** Native **Region:** Amazonia, Andean, Caribbean, Orinoquia, Pacific **Dept.:** ANT, BOG, BOL, BOY, CAL, CAS, CAU, CES, CHO, CUN, HUI, MAG, MET, NAR, NSA, PUT, QUI, RIS, SAP, SAN, TOL, VAC **Elev.:** 300-3000 msnm **IUCN Habitat code:** 1,3,4,5,14 **Conservation status: (IUCN)** NE **(CPSG)** LC

4909 *Clinopodium nubigenum* (Kunth) Kuntze (Lamiaceae) **IPNI:** 61198-2 **Habit:** Subshrub **Use:** ME **Origin:** Native **Region:** Andean **Dept.:** ANT, BOG, BOY, CAL, CAU, CHO, CUN, HUI, MET, NAR, NSA, PUT, QUI, RIS, SAN, TOL, VAC **Elev.:** 3100–4300 m a.s.l. **IUCN Habitat code:** 1,4,8,14 **Conservation status: (IUCN)** NE

4910 *Clinopodium taxifolium* (Kunth) Govaerts (Lamiaceae) **IPNI:** 1009966-1 **Common name:** tusílago **Habit:** Subshrub **Use:** ME **Origin:** Native **Region:** Andean **Dept.:** LAG, MAG **Elev.:** 2000-3600 msnm **Conservation status: (IUCN)** NE **(CPSG)** VU

4911 *Clinopodium vimineum* (L.) Kuntze (Lamiaceae) **IPNI:** 61205-2 **Habit:** Subshrub **Use:** MA **Origin:** Cultivated **Region:** Andean **Dept.:** BOY, CUN, SAN **Elev.:** 2400–2900 m a.s.l. **Conservation status: (IUCN)** NE

4912 *Coleus amboinicus* Lour. (Lamiaceae) **IPNI:** 445902-1 **Common name:** guiso, orégano, orégano orejón, poleo orejón **Habit:** Herb, Subshrub **Use:** AF, EU, GS, HF, MA, ME, SU **Origin:** Cultivated, Naturalised **Region:** Andean, Caribbean, Pacific **Dept.:** ATL, BOG, BOL, BOY, CAL, CAU, CES, COR, CUN, GUA, HUI, MAG, NAR, QUI, RIS, SAP, SAN, TOL **Elev.:** 0–1700 m a.s.l. **IUCN Habitat code:** 1,2,14 **Conservation status: (IUCN)** NE

4913 *Coleus forsteri* (Benth.) A.J.Paton (Lamiaceae) **IPNI:** 77201075-1 **Use:** ME **Region:** Andean **Dept.:** BOG **Conservation status: (IUCN)** NE

4914 *Coleus paniculatus* Benth. (Lamiaceae) **IPNI:** 446113-1 **Use:** SU **Origin:** Endemic **Conservation status: (IUCN)** NE

4915 *Coleus scutellarioides* (L.) Benth. (Lamiaceae) **IPNI:** 446165-1 **Common name:** amaranto, begonia sencilla, borrachera de la morada, capaelarreina, carateja, carnefresca, coleo, coleos, cromos, gitana, gota de Cristo, manchada, manchados, nazareno, pampadur, pana, pana churca, panas, pompadur, radical, radio, simpática, sinvergüenza, **Habit:** Herb **Use:** EU, HF, MA, ME, SU **Origin:** Naturalised **Region:** Amazonia, Andean, Guiana Shield, Caribbean, Orinoquia, Pacific **Dept.:** AMA, ANT, ATL, BOG, BOL, BOY, CAL, CAQ, CAS, CAU, CHO, COR, CUN, GUV, HUI, MAG, NAR, NSA, PUT, QUI, RIS, SAP, SAN, TOL, VAC, VID **Elev.:** 0–2200 m a.s.l. **IUCN Habitat code:** 1,2,3,5,14 **Conservation status: (IUCN)** NE

4916 *Condea verticillata* (Jacq.) Harley & J.F.B.Pastore (Lamiaceae) **IPNI:** 77120819-1 **Common name:** albahaca cimarrona, mastrantillo, oreja micu albahaca cimarrona, oreja micuna, soro sacha **Habit:** Herb, Subshrub **Use:** ME, SU **Origin:** Native **Region:** Andean, Caribbean, Pacific **Dept.:** ANT, BOL, CAL, CAU, CHO, COR, CUN, HUI, MAG, NAR, QUI, SAP, SAN, TOL, VAC **Elev.:** 0-1000 msnm **IUCN Habitat code:** 1,2,3,5,14 **Conservation status: (IUCN)** NE **(CPSG)** LC

4917 *Congea tomentosa* Roxb. (Lamiaceae) **IPNI:** 862610-1 **Common name:** enredadera pava **Habit:** Liana, Climbing **Use:** EU, MA, ME **Origin:** Cultivated, Naturalised **Region:** Andean **Dept.:** ANT **Elev.:** 1500–1500 m a.s.l. **IUCN Habitat code:** 1,14 **Conservation status: (IUCN)** NE

4918 *Cornutia odorata* (Poepp.) Schauer (Lamiaceae) **IPNI:** 862634-1 **Common name:** aguadulce, romerillo, salvio **Habit:** Shrub, Tree **Use:** ME **Origin:** Native **Region:** Andean, Orinoquia, Pacific **Dept.:** ANT, BOL, CUN, MAG, MET, NAR, SUC, TOL **Elev.:** 80–2100 m a.s.l. **IUCN Habitat code:** 1,2,3,4,5,14 **Conservation status: (IUCN)** LC

4919 *Cornutia pyramidata* L. (Lamiaceae) **IPNI:** 319153-2 **Common name:** ahumapescao, arciniegas, culape **Habit:** Shrub, Small tree, Tree **Use:** EU, ME, SU **Origin:** Native **Region:** Andean, Pacific **Dept.:** ANT, CHO, CUN, LAG, MAG, NAR, VAC **Elev.:** 0-1800 msnm **IUCN Habitat code:** 1,2,3,14 **Conservation status: (IUCN)** LC **(CPSG)** LC

4920 *Eriope crassipes* Benth. (Lamiaceae) **IPNI:** 446895-1 **Habit:** Subshrub **Use:** ME **Origin:** Native **Region:** Amazonia, Andean, Guiana Shield, Orinoquia **Dept.:** ARA, BOY, CAS, CUN, GUV, MET, VAU, VID **Elev.:** 100–600 m a.s.l. **IUCN Habitat code:** 1,2,3,4,14 **Conservation status: (IUCN)** NE

4921 *Gmelina arborea* Roxb. ex Sm. (Lamiaceae) **IPNI:** 77159191-1 **Common name:** melina **Habit:** Tree **Use:** AF, EU, FU, HF, MA, ME **Origin:** Cultivated **Region:** Orinoquia, Pacific **Dept.:** ANT, GUV, MAG, SUC **Elev.:** 30–200 m a.s.l. **IUCN Habitat code:** 1,2,3,14 **Conservation status: (IUCN)** LC

4922 *Holmskioldia sanguinea* Retz. (Lamiaceae) **IPNI:** 863073-1 **Common name:** buganvil rojo, sombrero de la China, sombrilla japonesa, veranera **Habit:** Shrub, Climbing **Use:** EU, MA, ME **Origin:** Cultivated **Region:** Andean **Dept.:** ANT, BOG **Elev.:** 1500–1920 m a.s.l. **IUCN Habitat code:** 1,2,3,14 **Conservation status: (IUCN)** NE

4923 *Hyptidendron arboreum* (Benth.) Harley (Lamiaceae) **IPNI:** 278655-2 **Common name:** aguanoso, borrajo, gallinazo, gallinazo blanco, gallinazo negro, mulato **Habit:** Tree **Use:** MA **Origin:** Native **Region:** Andean **Dept.:** ANT, BOL, CAL, SAN **Elev.:** 450-1700 msnm **IUCN Habitat code:** 1,3,5,14 **Conservation status: (IUCN)** NE **(CPSG)** VU

4924 *Hyptis atrorubens* Poit. (Lamiaceae) **IPNI:** 447964-1 **Common name:** botoncillo, cartagena, herb de oveja, hierba de oveja, mastranto, yerbabuenilla **Habit:** Herb **Use:** HF, MA, ME **Origin:** Native **Region:** Amazonia, Andean, Guiana Shield, Caribbean, Orinoquia, Pacific **Dept.:** AMA, ANT, ARA, BOL, BOY, CAL, CAQ, CAS, CAU, CHO, CUN, GUV, HUI, MAG, MET, NAR, NSA, PUT, QUI, RIS, SAN, TOL, VAC, VID **Elev.:** 0–2300 m a.s.l. **IUCN Habitat code:** 1,2,3,4,14 **Conservation status: (IUCN)** NE

4925 *Hyptis brachiata* Briq. (Lamiaceae) **IPNI:** 126717-2 **Common name:** ajicillo, jujure, lavaplatos, mastranto **Habit:** Subshrub **Use:** ME **Origin:** Native **Region:** Amazonia, Andean, Guiana Shield, Orinoquia **Dept.:** ARA, BOY, CAQ, CAS, CUN, GUV, HUI, MAG, MET, NSA, SAN, VAU, VID **Elev.:** 250–2100 m a.s.l. **IUCN Habitat code:** 1,2,3,4,14 **Conservation status: (IUCN)** NE

4926 *Hyptis brevipes* Poit. (Lamiaceae)
IPNI: 447982-1
Habit: Herb **Use:** HF, MA, ME **Origin:** Native **Region:** Amazonia, Andean, Guiana Shield, Caribbean, Orinoquia, Pacific **Dept.:** AMA, ANT, ARA, BOY, CAL, CAQ, CAS, CAU, CES, CHO, CUN, GUV, HUI, MAG, MET, NAR, NSA, PUT, QUI, RIS, SAN, TOL, VAC **Elev.:** 0–1500 m a.s.l. **IUCN Habitat code:** 1,2,3,4,14 **Conservation status: (IUCN)** NE

4927 *Hyptis capitata* Jacq. (Lamiaceae)
IPNI: 447996-1
Common name: botoncillo, botón negro, cartagena, cuchilla de Perdomo, cuchilla de perdomo, cáncer, desinflamadora, escobilla arisca, lampazo, mastardo, mastranto, trampolín, tresbolas, trespelotas **Habit:** Herb **Use:** EU, HF, ME **Origin:** Native **Region:** Amazonia, Andean, Guiana Shield, Caribbean, Orinoquia, Pacific **Dept.:** ANT, ATL, BOY, CAL, CAS, CAU, CHO, CUN, GUV, HUI, LAG, MAG, MET, NAR, QUI, RIS, SAP, SAN, TOL, VAC **Elev.:** 0–2000 m a.s.l. **IUCN Habitat code:** 1,2,3,5,14 **Conservation status: (IUCN)** NE

4928 *Hyptis conferta* Pohl ex Benth. (Lamiaceae)
IPNI: 448022-1
Common name: mastranto **Habit:** Herb, Subshrub **Use:** ME **Origin:** Native **Region:** Amazonia, Andean, Guiana Shield, Orinoquia **Dept.:** AMA, ARA, BOY, CAQ, CAS, CUN, GUV, MAG, MET, VAU, VID **Elev.:** 0–700 m a.s.l. **IUCN Habitat code:** 1,2,3,4,14 **Conservation status: (IUCN)** NE

4929 *Hyptis dilatata* Benth. (Lamiaceae)
IPNI: 448057-1
Common name: moradita **Habit:** Herb **Use:** MA, ME **Origin:** Native **Region:** Amazonia, Andean, Guiana Shield, Orinoquia **Dept.:** ARA, BOY, CAQ, CAS, CUN, GUV, MET, VAU, VID **Elev.:** 100–850 m a.s.l. **IUCN Habitat code:** 1,2,3,4,5,14 **Conservation status: (IUCN)** NE

4930 *Hyptis hirsuta* Kunth (Lamiaceae)
IPNI: 448128-1
Habit: Subshrub **Use:** MA **Origin:** Native **Region:** Orinoquia **Dept.:** MET, VID **Elev.:** 200–500 m a.s.l. **IUCN Habitat code:** 1 **Conservation status: (IUCN)** NE

4931 *Hyptis lanceolata* Poir. (Lamiaceae)
IPNI: 448173-1
Habit: Herb **Use:** GS, HF, MA, ME **Origin:** Native **Region:** Guiana Shield, Orinoquia **Dept.:** ARA, CAS, MET **Elev.:** 400–600 m a.s.l. **IUCN Habitat code:** 2,3 **Conservation status: (IUCN)** NE

4932 *Hyptis lantanifolia* Poit. (Lamiaceae)
IPNI: 448177-1
Common name: carrasposa, herb de sapo, hierba de sapo, mastranto **Habit:** Herb **Use:** MA, ME, PO **Origin:** Native **Region:** Amazonia, Andean, Guiana Shield, Caribbean, Orinoquia **Dept.:** AMA, ANT, BOY, CAQ, CAS, CAU, CES, CUN, GUV, LAG, MAG, MET, NSA, SAN, VAC, VAU, VID **Elev.:** 200–1800 m a.s.l. **IUCN Habitat code:** 1,2,3,4,14 **Conservation status: (IUCN)** NE

4933 *Hyptis lutescens* Pohl ex Benth. (Lamiaceae)
IPNI: 448211-1
Habit: Herb, Subshrub **Use:** MA **Origin:** Native **Region:** Guiana Shield, Orinoquia **Dept.:** CUN, MET, SAN **Elev.:** 300–800 m a.s.l. **Conservation status: (IUCN)** NE

4934 *Hyptis obtusiflora* C.Presl ex Benth. (Lamiaceae)
IPNI: 127032-2
Common name: botoncillo, piojito, poleo de monte, tacotaco, tucutaco **Habit:** Subshrub **Use:** ME **Origin:** Native **Region:** Amazonia, Andean, Guiana Shield, Orinoquia, Pacific **Dept.:** AMA, ANT, BOY, CAL, CAQ, CAS, CAU, CHO, CUN, HUI, MET, NAR, NSA, PUT, QUI, RIS, VAC **Elev.:** 0–2000 m a.s.l. **IUCN Habitat code:** 1,2,3,4,5,14 **Conservation status: (IUCN)** NE

4935 *Hyptis recurvata* Poit. (Lamiaceae)
IPNI: 448358-1
Common name: betónica, hoja de boa, mastranto blanco, sarna de Castilla, siempreviva **Habit:** Herb **Use:** ME **Origin:** Native **Region:** Amazonia, Andean, Guiana Shield, Orinoquia, Pacific **Dept.:** AMA, ANT, BOY, CAL, CAQ, CAS, CAU, CHO, CUN, GUV, HUI, LAG, MAG, MET, NAR, NSA, QUI, RIS, SAN, TOL, VAC, VAU, VID **Elev.:** 0–2200 m a.s.l. **IUCN Habitat code:** 1,2,3,4,5,14 **Conservation status: (IUCN)** NE

4936 *Hyptis savannarum* Briq. (Lamiaceae)
IPNI: 127129-2
Common name: cuchilla de Perdomo, cuchilla de perdomo, perdomo **Habit:** Herb, Subshrub **Use:** ME **Origin:** Native **Region:** Andean, Guiana Shield, Orinoquia **Dept.:** BOY, CAS, CAU, CUN, HUI, MAG, NAR, SAN, TOL **Elev.:** 300–2000 m a.s.l. **IUCN Habitat code:** 1,3,14 **Conservation status: (IUCN)** NE

4937 *Hyptis sinuata* Pohl ex Benth. (Lamiaceae)
IPNI: 448405-1
Common name: abejón, albarrana, albarrana verde, alvarrana, alvarrana verde, cartagena **Habit:** Herb **Use:** ME **Origin:** Native **Region:** Andean **Dept.:** ANT, BOL, BOY, CAL, CAU, CHO, CUN, MET, NAR, QUI, RIS, SAN, TOL, VAC **Elev.:** 350–2200 m a.s.l. **IUCN Habitat code:** 1,2,3,14 **Conservation status: (IUCN)** NE

4938 *Lamium amplexicaule* L. (Lamiaceae)
IPNI: 448797-1
Habit: Herb **Use:** HF, ME, PO **Origin:** Naturalised **Region:** Andean **Dept.:** BOG, BOY, CUN, NAR **Elev.:** 2500–3000 m a.s.l. **IUCN Habitat code:** 3,14 **Conservation status: (IUCN)** NE

4939 *Lamium galeobdolon* (L.) L. (Lamiaceae)
IPNI: 60452701-2
Habit: Herb **Use:** EU, ME **Origin:** Cultivated **Region:** Andean **Dept.:** CUN **Elev.:** 2200–3000 m a.s.l. **Conservation status: (IUCN)** NE

4940 *Lavandula latifolia* Medik. (Lamiaceae)
IPNI: 449051-1
Use: EU, HF, IF, MA, ME **IUCN Habitat code:** 14 **Conservation status: (IUCN)** LC

4941 *Leonotis leonurus* (L.) R.Br. (Lamiaceae)
IPNI: 449133-1
Common name: oreja de león **Habit:** Shrub **Use:** EU, IF, ME, SU **Origin:** Cultivated **Region:** Andean **Dept.:** BOY, CUN **Elev.:** 2300–2800 m a.s.l. **IUCN Habitat code:** 1,14 **Conservation status: (IUCN)** NE

4942 *Leonotis nepetifolia* (L.) R.Br. (Lamiaceae)
IPNI: 127310-3
Common name: bretónica, cordón de fraile, cuchillo de los Perdomo, tresbolas **Habit:** Herb **Use:** AF, EU, HF, MA, ME, PO, SU **Origin:** Naturalised **Region:** Amazonia, Andean, Guiana Shield, Caribbean **Dept.:** ANT, BOG, BOL, BOY, CAU, CES, CUN, GUV, HUI, LAG, NAR, SAN, TOL, VAC **Elev.:** 500–1900 m a.s.l. **IUCN Habitat code:** 1,2,3,5,14 **Conservation status: (IUCN)** NE

4943 *Leonurus japonicus* Houtt. (Lamiaceae)
IPNI: 449195-1
Common name: abundancia **Habit:** Herb **Use:** HF, ME **Origin:** Naturalised **Region:** Amazonia, Andean, Caribbean **Dept.:** ANT, BOL, BOY, CES, COR, NSA, PUT, QUI, SAN, TOL **Elev.:** 400–2000 m a.s.l. **IUCN Habitat code:** 1,3,14 **Conservation status: (IUCN)** NE

4944 *Lepechinia betonicifolia* (Lam.) Epling (Lamiaceae)
IPNI: 137859-2
Habit: Shrub **Use:** ME **Origin:** Native **Region:** Andean **Dept.:** BOG, NAR **Elev.:** 1700-2700 msnm **IUCN Habitat code:** 3,8,14 **Conservation status: (IUCN)** NE **(CPSG)** VU

4945 *Lepechinia bullata* (Kunth) Epling (Lamiaceae)
IPNI: 137860-2
Common name: chirco, sacaojos, salvia negra, salvialugo, salvielugo, salvio, selam, tumbué, uva de perro **Habit:** Shrub, Small tree **Use:** ME **Origin:** Native **Region:** Andean, Guiana Shield **Dept.:** ANT, BOG, BOY, CAL, CAU, CES, CUN, HUI, MAG, NAR, NSA, PUT, QUI, RIS, SAN, TOL, VAC **Elev.:** 2000-3000 msnm **IUCN Habitat code:** 1,3,4,14 **Conservation status: (IUCN)** NE **(CPSG)** LC

4946 *Lepechinia conferta* (Benth.) Epling (Lamiaceae)
IPNI: 137869-2
Common name: salvia **Habit:** Shrub, Small tree **Use:** ME **Origin:** Native **Region:** Andean **Dept.:** BOG, BOY, CAU, CUN, NAR, NSA, SAN, VAC **Elev.:** 2300-3350 msnm **IUCN Habitat code:** 1,3,4,8,14 **Conservation status: (IUCN)** NE **(CPSG)** LC

4947 *Lepechinia salviifolia* (Kunth) Epling (Lamiaceae)
IPNI: 137894-2
Common name: salvia, salvio **Habit:** Shrub, Small tree **Use:** ME **Origin:** Native **Region:** Andean **Dept.:** BOG, BOY, CES, CUN, NSA, SAN **Elev.:** 2550–3700 m a.s.l. **IUCN Habitat code:** 1,3,4,5,14 **Conservation status: (IUCN)** NE

4948 *Lepechinia schiedeana* (Schltdl.) Vatke (Lamiaceae)
IPNI: 449287-1
Habit: Herb, Subshrub **Use:** ME **Origin:** Native **Region:** Andean **Dept.:** ANT, BOG, BOY, CUN, MAG, SAN **Elev.:** 2600–3700 m a.s.l. **IUCN Habitat code:** 1,3,4,14 **Conservation status: (IUCN)** NE

4949 *Lepechinia vulcanicola* J.R.I.Wood (Lamiaceae)
IPNI: 279529-2
Common name: salvia real silvestre **Habit:** Shrub **Use:** ME **Origin:** Endemic **Region:** Andean **Dept.:** BOG **IUCN Habitat code:** 1,3,4,8,14 **Conservation status:** (IUCN) NE

4950 *Marrubium vulgare* L. (Lamiaceae)
IPNI: 449990-1
Common name: manrubio blanco, marrubio **Habit:** Herb **Use:** EU, HF, IF, ME **Origin:** Naturalised **Region:** Andean **Dept.:** BOG, BOY, CUN, HUI, SAN, TOL **Elev.:** 1800–2900 m a.s.l. **IUCN Habitat code:** 1,3,14 **Conservation status:** (IUCN) NE

4951 *Marsypianthes chamaedrys* (Vahl) Kuntze (Lamiaceae)
IPNI: 449997-1
Common name: bolas de gato, cancerín, mastranto, menta, pegajosa **Habit:** Herb **Use:** ME **Origin:** Native **Region:** Amazonia, Andean, Guiana Shield, Caribbean, Orinoquia, Pacific **Dept.:** AMA, ANT, ARA, ATL, BOY, CAL, CAS, CAU, CES, CHO, COR, CUN, HUI, MAG, MET, NAR, NSA, QUI, RIS, SAN, TOL, VAC, VID **Elev.:** 50–1950 m a.s.l. **IUCN Habitat code:** 1,2,3,5,14 **Conservation status:** (IUCN) NE

4952 *Melissa officinalis* L. (Lamiaceae)
IPNI: 450084-1
Common name: samuda negra, toronjil **Habit:** Herb **Use:** EU, HF, IF, MA, ME **Origin:** Cultivated **Region:** Andean **Dept.:** BOG, BOY, CAU, CUN, HUI, SAN, VAC **Elev.:** 1200–2800 m a.s.l. **IUCN Habitat code:** 1,3,4,14 **Conservation status:** (IUCN) NE

4953 *Mentha × piperita* L. (Lamiaceae)
IPNI: 450969-1
Common name: menta, yerbabue menta, yerbabuena, yerbabuena lisa **Use:** HF, ME **Dept.:** BOG **IUCN Habitat code:** 1,3,4,14 **Conservation status:** (IUCN) NE

4954 *Mentha aquatica* L. (Lamiaceae)
IPNI: 450192-1
Use: EU, GS, HF, MA, ME **IUCN Habitat code:** 1 **Conservation status:** (IUCN) LC

4955 *Mentha arvensis* L. (Lamiaceae)
IPNI: 30026217-2
Common name: alubeno, menta **Use:** GS, HF, MA, ME **Dept.:** BOG **IUCN Habitat code:** 14 **Conservation status:** (IUCN) LC

4956 *Mentha longifolia* (L.) L. (Lamiaceae)
IPNI: 450735-1
Use: EU, GS, HF, IF, MA, ME, PO **Region:** Andean **Dept.:** BOG **Conservation status:** (IUCN) LC

4957 *Mentha pulegium* L. (Lamiaceae)
IPNI: 451022-1
Common name: menta, sígueme **Habit:** Herb **Use:** EU, HF, IF, MA, ME **Origin:** Cultivated **Region:** Amazonia, Andean **Dept.:** BOG, BOY, CUN, PUT **Elev.:** 400–2900 m a.s.l. **IUCN Habitat code:** 1,2,14 **Conservation status:** (IUCN) LC

4958 *Mentha spicata* L. (Lamiaceae)
IPNI: 451162-1
Habit: Herb **Use:** GS, HF, MA, ME **Origin:** Cultivated **Region:** Andean **Dept.:** BOG, BOY, CAL, CAU, CUN, NAR, QUI, SAN, TOL **Elev.:** 1500–3900 m a.s.l. **IUCN Habitat code:** 1,3,14 **Conservation status:** (IUCN) LC

4959 *Mentha suaveolens* Ehrh. (Lamiaceae)
IPNI: 451188-1
Common name: yerbabuena **Habit:** Herb **Use:** EU, GS, HF, ME **Origin:** Naturalised **Region:** Andean **Dept.:** ANT, BOG, BOY, CAL, CAU, CUN, NAR, QUI, SAN **Elev.:** 2000–3900 m a.s.l. **IUCN Habitat code:** 1,2,3,14 **Conservation status:** (IUCN) LC

4960 *Mesosphaerum pectinatum* (L.) Kuntze (Lamiaceae)
IPNI: 158242-2
Common name: bretónica, mastranto, salvia, sorbetana **Habit:** Herb, Subshrub **Use:** GS, HF, MA, ME, PO, SU **Origin:** Native **Region:** Andean, Guiana Shield, Caribbean, Orinoquia, Pacific **Dept.:** ANT, ATL, BOL, BOY, CAU, CHO, CUN, HUI, LAG, MAG, MET, NAR, QUI, RIS, SAN, VAC **Elev.:** 0-2400 msnm **IUCN Habitat code:** 1,2,3,5,14 **Conservation status:** (IUCN) NE (CPSG) LC

4961 *Mesosphaerum sidifolium* (L'Hér.) Harley & J.F.B.Pastore (Lamiaceae)
IPNI: 77121103-1
Common name: cartagena, mastranto, mastranto morado, prontoalivio **Habit:** Shrub, Subshrub **Use:** ME **Origin:** Native **Region:** Andean, Pacific **Dept.:** ANT, BOY, CAL, CAU, CHO, CUN, HUI, MAG, MET, NAR, QUI, RIS, SAN, TOL, VAC **Elev.:** 1000–2800 m a.s.l. **IUCN Habitat code:** 1,2,3,4,14 **Conservation status:** (IUCN) NE

4962 *Mesosphaerum suaveolens* (L.) Kuntze (Lamiaceae)
IPNI: 158297-2
Common name: albahaca de monte, alcanfor de monte, altamisa, botón morado, chavito, clavito, destrancadera, mastranco, mastrantillo, mastranto, mastranto de perro, mastranto de sabana, orégano, poleo, poleo morado, yerba de las muelas, yerba de reuma **Habit:** Herb **Use:** AF, EU, GS, HF, MA, ME, PO, SU **Origin:** Native **Region:** Andean, Guiana Shield, Caribbean, Orinoquia, Pacific **Dept.:** ANT, ATL, BOL, BOY, CAS, CAU, CES, CHO, COR, CUN, HUI, LAG, MAG, MET, RIS, SAP, SAN, SUC, TOL, VAC, VID **Elev.:** 700-1600 msnm **IUCN Habitat code:** 1,2,3,14 **Conservation status:** (IUCN) NE (CPSG) LC

4963 *Minthostachys mollis* (Benth.) Griseb. (Lamiaceae)
IPNI: 451957-1
Common name: poleo grande, tusilago chabama **Habit:** Subshrub **Use:** HF, ME **Origin:** Native **Region:** Andean **Dept.:** BOG, CAU, NAR, PUT, VAC **Elev.:** 2200-3900 msnm **IUCN Habitat code:** 1,2,3,4,5,8,14 **Conservation status:** (IUCN) NE (CPSG) LC

4964 *Moluccella laevis* L. (Lamiaceae)
IPNI: 451989-1
Use: EU **Region:** Andean **Dept.:** BOG **IUCN Habitat code:** 14 **Conservation status:** (IUCN) NE

4965 *Nepeta cataria* L. (Lamiaceae)
IPNI: 452333-1
Habit: Herb **Use:** EU, HF, ME **Origin:** Naturalised **Region:** Andean **Dept.:** CUN **Elev.:** 1800–2700 m a.s.l. **IUCN Habitat code:** 3,14 **Conservation status:** (IUCN) NE

4966 *Obtegomeria caerulescens* (Benth.) Doroszenko & P.D.Cantino (Lamiaceae)
IPNI: 1001364-1
Common name: tonmía **Habit:** Subshrub **Use:** ME **Origin:** Endemic **Region:** Andean **Dept.:** CES, MAG **Elev.:** 2900-4000 msnm **IUCN Habitat code:** 2,4,14 **Conservation status:** (IUCN) NE (CPSG) LC

4967 *Ocimum americanum* L. (Lamiaceae)
IPNI: 316218-2
Common name: albahaca, albahaca blanca, albahaca blanca del Santísimo, albahaca común, albahaca de anís, toronjil **Habit:** Herb, Subshrub **Use:** AF, EU, GS, HF, IF, MA, ME, PO, SU **Origin:** Naturalised **Region:** Amazonia, Andean, Caribbean, Pacific **Dept.:** ANT, ATL, BOG, BOL, CAL, CAU, CES, CHO, CUN, MAG, NAR, SAP, SAN, TOL, VAC **Elev.:** 0–1400 m a.s.l. **IUCN Habitat code:** 1,3,14 **Conservation status:** (IUCN) NE

4968 *Ocimum basilicum* L. (Lamiaceae)
IPNI: 452874-1
Common name: albahaca, albahaca blanca, albahaca chiracán, albahaca chirarán, albahaca cimarrona, albahaca común, albahaca de Castilla, albahaca del Santísimo, albahaca morada, albahaquilla, albajaca, arbaquita, basilik, chirarán, chirarán morado, cimarrona, cotorera, **Habit:** Herb **Use:** AF, EU, GS, HF, MA, ME, PO, SU **Origin:** Cultivated, Naturalised **Region:** Amazonia, Andean, Caribbean, Pacific **Dept.:** AMA, ANT, ATL, BOG, BOL, CAU, CHO, COR, CUN, MAG, NSA, QUI, SAP, SAN, TOL, VAC **Elev.:** 0–1700 m a.s.l. **IUCN Habitat code:** 1,2,3,4,14 **Conservation status:** (IUCN) NE

4969 *Ocimum campechianum* Mill. (Lamiaceae)
IPNI: 452886-1
Common name: abaca, albahaca, albahaca blanca, albahaca canela, albahaca de gallinaza, albahaca morada, albahaca toronjil, aruwalo, basil, gallinazo, huo-ca, toronjil **Habit:** Herb, Subshrub **Use:** HF, MA, ME, PO, SU **Origin:** Native **Region:** Amazonia, Andean, Guiana Shield, Caribbean, Orinoquia, Pacific **Dept.:** AMA, ANT, ATL, BOG, BOL, BOY, CAL, CAQ, CAS, CAU, CES, CHO, COR, CUN, GUA, GUV, HUI, LAG, MAG, NAR, NSA, PUT, QUI, SAP, SAN, TOL, VAC **Elev.:** 0-1450 msnm **IUCN Habitat code:** 1,2,3,5,14 **Conservation status:** (IUCN) NE (CPSG) LC

4970 *Ocimum gratissimum* L. (Lamiaceae)
IPNI: 452969-1
Common name: albahaca vaca, atiyayo, basen, fonbazen, vanne van **Habit:** Herb, Subshrub **Use:** AF, EU, HF, IF, MA, ME, PO, SU **Origin:** Cultivated **Region:** Amazonia, Andean, Pacific **Dept.:** CAQ, CAU, CHO, CUN, QUI, VAC **Elev.:** 300–1600 m a.s.l. **IUCN Habitat code:** 1,3,14 **Conservation status:** (IUCN) NE

4971 *Ocimum minimum* L. (Lamiaceae)
IPNI: 453031-1
Common name: toronjil **Use:** HF, ME **Origin:** Endemic **Dept.:** MAG **Conservation status:** (IUCN) NE

4972 *Ocimum tenuiflorum* L. (Lamiaceae)
IPNI: 453130-1
Common name: albahaca morada, toronjil, toronjil morado, toronjil silvestre **Habit:** Herb, Subshrub **Use:** HF, MA, ME, PO, SU **Origin:** Naturalised **Region:**

Andean, Caribbean, Pacific **Dept.:** ATL, BOL, CAU, CES, CHO, COR, LAG, MAG, SAP, SUC, VAC **Elev.:** 0–1500 m a.s.l. **IUCN Habitat code:** 1,2,3,14 **Conservation status: (IUCN)** NE

4973 *Origanum × majoricum* Cambess. (Lamiaceae) **IPNI:** 453305-1 **Use:** ME **Conservation status: (IUCN)** NE

4974 *Origanum majorana* L. (Lamiaceae) **IPNI:** 453303-1 **Common name:** mejorana, mejorana blanca, mejorana silvestre, orégano, orégano de España, orégano griego **Habit:** Herb, Subshrub **Use:** HF, MA, ME **Origin:** Cultivated **Region:** Andean, Pacific **Dept.:** ANT, BOG, BOY, CAU, CUN, HUI, MAG, SAN, TOL, VAC **Elev.:** 0–2500 m a.s.l. **IUCN Habitat code:** 1,2,3,4,14 **Conservation status: (IUCN)** NE

4975 *Origanum vulgare* L. (Lamiaceae) **IPNI:** 453395-1 **Common name:** orégano **Habit:** Herb **Use:** HF, IF, MA, ME **Origin:** Cultivated **Region:** Andean **Dept.:** ANT, BOG, BOY, CAS, CAU, CUN, SAN, VAC **Elev.:** 1500–2800 m a.s.l. **IUCN Habitat code:** 3,14 **Conservation status: (IUCN)** NE

4976 *Petitia domingensis* Jacq. (Lamiaceae) **IPNI:** 864000-1 **Common name:** capá **Habit:** Tree **Use:** MA **Origin:** Cultivated **Region:** Andean **Dept.:** ANT **Elev.:** 1500–1500 m a.s.l. **IUCN Habitat code:** 1 **Conservation status: (IUCN)** NE

4977 *Plectranthus verticillatus* (L.f.) Druce (Lamiaceae) **IPNI:** 454770-1 **Common name:** millonaria, sinvergüenza **Habit:** Herb, Subshrub **Use:** EU **Origin:** Cultivated **Region:** Andean **Dept.:** ANT, BOG, BOY, CAL, CUN, NAR, QUI, SAN, VAC **Elev.:** 700–2700 m a.s.l. **Conservation status: (IUCN)** NE

4978 *Prunella vulgaris* L. (Lamiaceae) **IPNI:** 455176-1 **Habit:** Herb **Use:** HF, ME **Origin:** Naturalised **Region:** Andean **Dept.:** ANT, BOG, BOY, CAL, HUI, MAG, NAR, NSA, PUT, QUI, RIS, SAN, TOL **Elev.:** 2200–3900 m a.s.l. **IUCN Habitat code:** 1,3,4,14 **Conservation status: (IUCN)** LC

4979 *Salvia amethystina* Sm. (Lamiaceae) **IPNI:** 30148459-2 **Common name:** mastranto grande, salvia **Habit:** Herb, Subshrub **Use:** ME **Origin:** Native **Region:** Andean **Dept.:** BOG, BOY, CAU, CUN, NSA, QUI, SAN, VAC **Elev.:** 2300-4000 msnm **IUCN Habitat code:** 1,3,4,8,14 **Conservation status: (IUCN)** NE **(CPSG)** EN

4980 *Salvia bogotensis* Benth. (Lamiaceae) **IPNI:** 455788-1 **Common name:** salvia, velero **Habit:** Shrub, Subshrub **Use:** ME **Origin:** Endemic **Region:** Andean **Dept.:** BOG, BOY, CUN, SAN **Elev.:** 1000-4000 msnm **IUCN Habitat code:** 3,4,14 **Conservation status: (IUCN)** NE **(CPSG)** LC

4981 *Salvia camarifolia* Benth. (Lamiaceae) **IPNI:** 455863-1 **Habit:** Subshrub **Use:** ME **Origin:** Endemic **Region:** Andean **Dept.:** BOG, CES, MAG, NSA **Elev.:** 400-1850 msnm **IUCN Habitat code:** 1,2,3,14 **Conservation status: (IUCN)** NE **(CPSG)** EN

4982 *Salvia carnea* Kunth (Lamiaceae) **IPNI:** 455889-1 **Common name:** salvia **Habit:** Herb **Use:** ME **Origin:** Native **Region:** Andean **Dept.:** ANT, BOG, BOY, CAL, CAS, CAU, CUN, MAG, MET, NAR, NSA, PUT, QUI, RIS, SAN, TOL, VAC **Elev.:** 2000-3900 msnm **IUCN Habitat code:** 1,3,4,5,8,14 **Conservation status: (IUCN)** NE **(CPSG)** LC

4983 *Salvia coccinea* Buc'hoz ex Etl. (Lamiaceae) **IPNI:** 455978-1 **Common name:** flor de Jericó **Habit:** Herb **Use:** AF, EU, HF, ME, PO **Origin:** Native **Region:** Amazonia, Andean, Orinoquia **Dept.:** ANT, BOG, CAU, CUN, HUI, MAG, MET, NAR, NSA, PUT, SAN, TOL, VAC **Elev.:** 500–1900 m a.s.l. **IUCN Habitat code:** 1,2,3,14 **Conservation status: (IUCN)** NE

4984 *Salvia hispanica* L. (Lamiaceae) **IPNI:** 456385-1 **Habit:** Herb **Use:** HF, MA, ME **Origin:** Native **Region:** Andean **Dept.:** ANT, BOG, BOY, CUN **Elev.:** 1800–2400 m a.s.l. **IUCN Habitat code:** 1,14 **Conservation status: (IUCN)** NE

4985 *Salvia lasiocephala* Hook. & Arn. (Lamiaceae) **IPNI:** 456526-1 **Habit:** Herb **Use:** ME **Origin:** Native **Region:** Andean, Guiana Shield, Caribbean, Orinoquia, Pacific **Dept.:** ANT, BOY, CAL, CAU, CES, CUN, MAG, MET, NAR, QUI, RIS, SAP, SAN, TOL, VAC **Elev.:** 400-1600 msnm **IUCN Habitat code:** 1,3,14 **Conservation status: (IUCN)** NE **(CPSG)** LC

4986 *Salvia leucantha* Cav. (Lamiaceae) **IPNI:** 226705-2 **Common name:** bella salvia, cordón de obispo, salvia, terciopelo **Habit:** Subshrub **Use:** ME **Origin:** Cultivated, Naturalised **Region:** Andean **Dept.:** ANT, BOG, BOL, BOY, CAL, CAU, CUN, HUI, RIS, SAN, TOL, VAC **Elev.:** 700–3400 m a.s.l. **IUCN Habitat code:** 1,2,3,14 **Conservation status: (IUCN)** NE

4987 *Salvia macrophylla* Benth. (Lamiaceae) **IPNI:** 456630-1 **Common name:** salvia china, salvia real **Habit:** Subshrub **Use:** ME **Origin:** Native **Region:** Andean **Dept.:** CAL, CAU, HUI, NAR, PUT, QUI, RIS, TOL, VAC **Elev.:** 1400–2300 m a.s.l. **IUCN Habitat code:** 1,3,14 **Conservation status: (IUCN)** NE

4988 *Salvia macrostachya* Kunth (Lamiaceae) **IPNI:** 456635-1 **Common name:** junjún, salviarrial **Habit:** Herb, Subshrub **Use:** ME **Origin:** Native **Region:** Andean **Dept.:** BOG, NAR **Elev.:** 2500-3100 msnm **IUCN Habitat code:** 4,8,14 **Conservation status: (IUCN)** NE **(CPSG)** CR

4989 *Salvia occidentalis* Sw. (Lamiaceae) **IPNI:** 456822-1 **Common name:** canasperro, cansaperro **Habit:** Herb **Use:** EU, ME **Origin:** Native **Region:** Andean, Caribbean, Pacific **Dept.:** ANT, ATL, BOL, BOY, CAL, CHO, COR, CUN, MAG, NAR, QUI, RIS, SAN, VAC **Elev.:** 50–1500 m a.s.l. **IUCN Habitat code:** 1,2,3,14 **Conservation status: (IUCN)** NE

4990 *Salvia officinalis* L. (Lamiaceae) **IPNI:** 456833-1 **Common name:** salvia **Habit:** Herb, Subshrub **Use:** EU, HF, MA, ME, PO **Origin:** Cultivated **Region:** Andean **Dept.:** BOG, BOY, CUN **Elev.:** 1600–2900 m a.s.l. **Conservation status: (IUCN)** LC

4991 *Salvia palifolia* Kunth (Lamiaceae) **IPNI:** 456870-1 **Common name:** agapaca, azapaca, banderilla, cadillo, cineraria, contrafuego, cueroesapo, golondrina, hierba de chucha, lusesteban, mangapaca, manguipaca, mantoevieja, mastranto, oreja de perro, parindera, salvia, salvia chiquita, washa washa, yerba de sapo **Habit:** Herb **Use:** ME **Origin:** Native **Region:** Andean **Dept.:** ANT, BOG, BOY, CAL, CAU, CES, CUN, HUI, MET, NAR, NSA, QUI, RIS, SAN, TOL, VAC **Elev.:** 1400-3500 msnm **IUCN Habitat code:** 1,2,3,4,14 **Conservation status: (IUCN)** NE **(CPSG)** LC

4992 *Salvia pauciserrata* Benth. (Lamiaceae) **IPNI:** 456908-1 **Common name:** salvia **Habit:** Subshrub **Use:** ME **Origin:** Native **Region:** Andean **Dept.:** ANT, BOG, BOY, CAL, CAS, CAU, CES, CUN, HUI, MET, NAR, NSA, PUT, QUI, RIS, SAN, TOL, VAC **Elev.:** 1500-3500 msnm **IUCN Habitat code:** 1,3,4,5,8,14 **Conservation status: (IUCN)** NE **(CPSG)** VU

4993 *Salvia rosmarinus* Spenn. (Lamiaceae) **IPNI:** 457138-1 **Common name:** romero, romero común **Habit:** Shrub, Subshrub **Use:** AF, EU, HF, IF, MA, ME, SU **Origin:** Cultivated **Region:** Andean **Dept.:** ANT, BOY, CAU, CUN, HUI, NAR, NSA, SAN, TOL, VAC **Elev.:** 1300–3000 m a.s.l. **IUCN Habitat code:** 1,2,3,4,14 **Conservation status: (IUCN)** NE

4994 *Salvia rubescens* Kunth (Lamiaceae) **IPNI:** 457145-1 **Common name:** barbasco, bretónica, dominico **Habit:** Herb **Use:** ME **Origin:** Native **Region:** Andean **Dept.:** BOG, BOY, CUN, MET, NSA, SAN **Elev.:** 600-3400 msnm **IUCN Habitat code:** 1,2,3,4,14 **Conservation status: (IUCN)** NE **(CPSG)** VU

4995 *Salvia rufula* Kunth (Lamiaceae) **IPNI:** 457156-1 **Common name:** salvia **Habit:** Shrub **Use:** ME **Origin:** Native **Region:** Andean **Dept.:** ANT, BOG, CAL, CAU, CUN, HUI, MAG, MET, QUI, RIS, SAN, TOL, VAC **Elev.:** 900-2900 msnm **IUCN Habitat code:** 1,3,4,5,14 **Conservation status: (IUCN)** NE **(CPSG)** EN

4996 *Salvia scutellarioides* Kunth (Lamiaceae) **IPNI:** 457216-1 **Common name:** contrafuego, cuero de sapo, mangapaca, manguepaque, manguipaque, mastranto, salvia **Habit:** Herb **Use:** ME **Origin:** Native **Region:** Andean **Dept.:** ANT, BOG, BOY, CAL, CAQ, CAU, CHO, CUN, HUI, MET, NAR, PUT, QUI, RIS, SAN, TOL, VAC **Elev.:** 1100-3200 msnm **IUCN Habitat code:** 1,3,4,14 **Conservation status: (IUCN)** NE **(CPSG)** LC

4997 *Salvia splendens* Sellow ex Nees (Lamiaceae) **IPNI:** 60466516-2 **Common name:** calzones del diablo, calzón del diablo, liberal, salviarroja, viudalegre, viudaloca **Habit:** Herb, Subshrub **Use:** EU, ME, SU **Origin:** Cultivated, Naturalised **Region:** Andean **Dept.:** ANT, BOG, BOY, CAL, CAU, CES, CUN, HUI, MAG, MET, NAR, NSA, PUT, QUI, RIS, SAN, TOL, VAC **Elev.:** 1400–3100 m a.s.l. **IUCN Habitat code:** 1,2,3,14 **Conservation status: (IUCN)** NE

4998 *Salvia tiliifolia* Vahl (Lamiaceae) **IPNI:** 457397-1 **Common name:** yerba clara **Habit:** Herb **Use:** HF, ME **Origin:** Native **Region:** Andean **Dept.:** ANT, BOY, CAL, CAS, CAU, CUN, HUI, MET, NAR, NSA, QUI, RIS, SAN, TOL, VAC **Elev.:** 1100-2400 msnm **IUCN Habitat code:** 1,2,3,5,14 **Conservation status: (IUCN)** NE **(CPSG)** LC

4999 *Salvia tortuosa* Kunth (Lamiaceae) **IPNI:** 457410-1 **Common name:** flor de quinde, quinde **Habit:** Shrub **Use:** ME **Origin:** Native **Region:** Andean **Dept.:** BOG, CAU, NAR, PUT, VAC **Elev.:** 1000-3400 msnm **IUCN Habitat code:** 1,2,3,8,14 **Conservation status: (IUCN)** NE **(CPSG)** LC

5000 *Salvia verbenaca* L. (Lamiaceae) **IPNI:** 457495-1 **Use:** ME **Region:** Andean **Dept.:** BOG **Conservation status: (IUCN)** NE

5001 *Satureja montana* L. (Lamiaceae) **IPNI:** 457736-1 **Common name:** orégano de tierra fría **Habit:** Subshrub **Use:** EU, HF, MA, ME **Origin:** Cultivated **Region:** Andean **Dept.:** ANT, CAU, CUN **Elev.:** 2500–3000 m a.s.l. **IUCN Habitat code:** 14 **Conservation status: (IUCN)** NE

5002 *Scutellaria agrestis* A.St.-Hil. ex Benth. (Lamiaceae) **IPNI:** 458104-1 **Common name:** coramina, moradilla, valeriana **Habit:** Herb **Use:** ME **Origin:** Native **Region:** Amazonia, Andean, Guiana Shield, Pacific **Dept.:** AMA, CAQ, CAU, NAR, PUT **Elev.:** 200–1200 m a.s.l. **IUCN Habitat code:** 1,3,5 **Conservation status: (IUCN)** NE

5003 *Scutellaria incarnata* Vent. (Lamiaceae) **IPNI:** 458352-1 **Common name:** alegría, contentica, cáncer morado silvestre, cáncer rojo, flor de quinde pequeño, fósforo, hierba alegre, pajacana, salvia **Habit:** Herb **Use:** EU, ME **Origin:** Native **Region:** Andean, Guiana Shield **Dept.:** ANT, BOG, BOY, CAL, CAQ, CAU, CUN, HUI, MAG, MET, NAR, NSA, PUT, QUI, RIS, SAN, TOL, VAC **Elev.:** 400–2400 m a.s.l. **IUCN Habitat code:** 1,2,3,5,14 **Conservation status: (IUCN)** NE

5004 *Stachys arvensis* (L.) L. (Lamiaceae) **IPNI:** 459319-1 **Habit:** Herb **Use:** ME, PO **Origin:** Naturalised **Region:** Andean **Dept.:** BOG, BOY, CUN **Elev.:** 2000–3000 m a.s.l. **IUCN Habitat code:** 14 **Conservation status: (IUCN)** NE

5005 *Stachys bogotensis* Kunth (Lamiaceae) **IPNI:** 459365-1 **Common name:** limpiafrascos, salviecita, salviecita de bogotá **Habit:** Herb **Use:** ME **Origin:** Endemic **Region:** Andean **Dept.:** ANT, BOG, BOY, CAU, CES, CUN, MET, NAR, NSA, SAN **Elev.:** 1700–3800 m a.s.l. **IUCN Habitat code:** 1,3,4,14 **Conservation status: (IUCN)** NE

5006 *Stachys byzantina* K.Koch (Lamiaceae) **IPNI:** 459393-1 **Use:** EU, ME **Region:** Andean **Dept.:** BOG **IUCN Habitat code:** 8 **Conservation status: (IUCN)** NE

5007 *Stachys elliptica* Kunth (Lamiaceae) **IPNI:** 30008354-2 **Common name:** cuychunchullo, guchunchullo **Habit:** Herb **Use:** ME **Origin:** Native **Region:** Andean **Dept.:** ARA, BOG, BOY, CUN, MET, NSA, SAN **Elev.:** 3450–4200 m a.s.l. **IUCN Habitat code:** 1,3,4,8,14 **Conservation status: (IUCN)** NE

5008 *Stachys gilliesii* Benth. (Lamiaceae) **IPNI:** 459558-1 **Common name:** hortiga blanca **Habit:** Herb **Use:** ME **Origin:** Native **Region:** Andean **Dept.:** ANT, ARA, BOG, BOY, CAL, CAU, CHO, COR, CUN, HUI, MAG, MET, NAR, PUT, QUI, RIS, SAN, TOL, VAC **Elev.:** 700–2400 m a.s.l. **IUCN Habitat code:** 1,3,4,5,14 **Conservation status: (IUCN)** NE

5009 *Stachys spathulata* Burch. ex Benth. (Lamiaceae) **IPNI:** 459988-1 **Use:** ME **Region:** Andean **Dept.:** BOG **IUCN Habitat code:** 1 **Conservation status: (IUCN)** NE

5010 *Tectona grandis* L.f. (Lamiaceae) **IPNI:** 864923-1 **Common name:** teca, teco **Habit:** Tree **Use:** EU, FU, HF, MA, ME, SU **Origin:** Cultivated **Region:** Andean, Caribbean **Dept.:** ANT, BOL, BOY, COR, MAG, QUI, RIS, SAP, SAN, VAC **Elev.:** 80–1200 m a.s.l. **IUCN Habitat code:** 1,2,3,4,5,14 **Conservation status: (IUCN)** NE

5011 *Teucrium fruticans* L. (Lamiaceae) **IPNI:** 460416-1 **Habit:** Shrub **Use:** AF, EU, HF, ME **Origin:** Cultivated **Region:** Andean **Dept.:** BOG, BOY, CUN **Elev.:** 2500–2700 m a.s.l. **Conservation status: (IUCN)** NE

5012 *Teucrium vesicarium* Mill. (Lamiaceae) **IPNI:** 460785-1 **Habit:** Herb **Use:** ME **Origin:** Native **Region:** Andean, Caribbean **Dept.:** ATL, BOL, COR, MAG **Elev.:** 0–500 m a.s.l. **IUCN Habitat code:** 5 **Conservation status: (IUCN)** NE

5013 *Thymus vulgaris* L. (Lamiaceae) **IPNI:** 461765-1 **Common name:** orégano, tomillo **Habit:** Subshrub **Use:** EU, HF, MA, ME, PO **Origin:** Cultivated **Region:** Andean **Dept.:** BOG, BOY, CUN, SAN, VAC **Elev.:** 2300–3800 m a.s.l. **IUCN Habitat code:** 1,3,4,14 **Conservation status: (IUCN)** LC

5014 *Vitex agnus-castus* L. (Lamiaceae) **IPNI:** 865568-1 **Habit:** Shrub **Use:** EU, HF, IF, MA, ME **Origin:** Cultivated **Region:** Amazonia **Dept.:** AMA **Elev.:** 100–100 m a.s.l. **IUCN Habitat code:** 1 **Conservation status: (IUCN)** DD

5015 *Vitex capitata* Vahl (Lamiaceae) **IPNI:** 865638-1 **Common name:** aceituno **Habit:** Shrub, Tree **Use:** HF, MA **Origin:** Native **Region:** Caribbean, Orinoquia, Pacific **Dept.:** ANT, BOL, CAS, COR, VID **Elev.:** 60–480 m a.s.l. **IUCN Habitat code:** 1,2,3,4,5,14 **Conservation status: (IUCN)** LC

5016 *Vitex compressa* Turcz. (Lamiaceae) **IPNI:** 865663-1 **Common name:** aceituno, iruwa **Habit:** Small tree, Tree **Use:** AF, EU, HF, MA, ME **Origin:** Native **Region:** Amazonia, Andean, Caribbean, Orinoquia **Dept.:** CAQ, CES, LAG, MAG, MET **Elev.:** 50–1200 m a.s.l. **IUCN Habitat code:** 1,2,3,5,14 **Conservation status: (IUCN)** LC

5017 *Vitex cooperi* Standl. (Lamiaceae) **IPNI:** 865667-1 **Use:** HF, MA **Origin:** Native **Dept.:** COR **Conservation status: (IUCN)** EN

5018 *Vitex cymosa* Bertero ex Spreng. (Lamiaceae) **IPNI:** 865678-1 **Common name:** aceituna, aceituno **Habit:** Shrub, Tree **Use:** AF, HF, MA, ME **Origin:** Native **Region:** Andean, Caribbean **Dept.:** BOL, CAL, LAG, MAG, SAP, SAN, VAC **Elev.:** 0–2350 m a.s.l. **IUCN Habitat code:** 1,2,3,14 **Conservation status: (IUCN)** LC

5019 *Vitex divaricata* Sw. (Lamiaceae) **IPNI:** 865688-1 **Habit:** Shrub, Tree **Use:** HF **Origin:** Native **Region:** Caribbean, Orinoquia **Dept.:** ANT, MET, TOL **Elev.:** 40–430 m a.s.l. **IUCN Habitat code:** 1,2,3,14 **Conservation status: (IUCN)** NE

5020 *Vitex flavens* Kunth (Lamiaceae) **IPNI:** 865718-1 **Common name:** aceituno **Habit:** Tree **Use:** AF, HF, MA **Origin:** Native **Region:** Caribbean **Dept.:** LAG, TOL **Elev.:** 200-280 msnm **IUCN Habitat code:** 3 **Conservation status: (IUCN)** LC **(CPSG)** LC

5021 *Vitex floridula* Duchass. & Walp. (Lamiaceae) **IPNI:** 865720-1 **Habit:** Tree **Use:** MA **Origin:** Native **Region:** Pacific **Dept.:** ANT, CHO **Elev.:** 0–500 m a.s.l. **IUCN Habitat code:** 3 **Conservation status: (IUCN)** NT

5022 *Vitex gaumeri* Greenm. (Lamiaceae) **IPNI:** 265966-2 **Habit:** Tree **Use:** MA, ME **Origin:** Native **Region:** Caribbean **Dept.:** SAP **Elev.:** 0–200 m a.s.l. **IUCN Habitat code:** 1 **Conservation status: (IUCN)** LC

5023 *Vitex gigantea* Kunth (Lamiaceae) **IPNI:** 865730-1 **Common name:** aceituno, ajicito, trúntago blanco **Habit:** Tree **Use:** HF **Origin:** Native **Dept.:** CUN, HUI **Elev.:** 350–400 m a.s.l. **IUCN Habitat code:** 3,4,14 **Conservation status: (IUCN)** LC

5024 *Vitex orinocensis* Kunth (Lamiaceae) **IPNI:** 865904-1 **Common name:** acapu, aceituno, calavero, flormorado, guarataro, guaretero, guataro, guayacancito, guayacán, nocuito, nocuito amarillo, nocuito blanco **Habit:** Tree **Use:** HF **Origin:** Native **Region:** Amazonia, Andean, Orinoquia, Pacific **Dept.:** AMA, ANT, ARA, CAS, CHO, CUN, MET, SAN **Elev.:** 140–1500 m a.s.l. **IUCN Habitat code:** 1,2,3,4,5,14 **Conservation status: (IUCN)** LC

5025 *Vitex parviflora* A.Juss. (Lamiaceae) **IPNI:** 865916-1 **Habit:** Tree **Use:** MA, ME, PO **Origin:** Cultivated, Naturalised **Region:** Andean **Dept.:** ANT **Elev.:** 1480–1500 m a.s.l. **Conservation status: (IUCN)** LC

5026 *Vitex pseudolea* Rusby (Lamiaceae) **IPNI:** 265993-2 **Habit:** Tree **Use:** HF **Origin:** Native **Region:** Amazonia **Dept.:** AMA, SAN **Elev.:** 100–500 m a.s.l. **Conservation status: (IUCN)** NE

5027 *Vitex stahelii* Moldenke (Lamiaceae) **IPNI:** 866014-1 **Use:** HF, MA **Origin:** Native **Dept.:** MET **Conservation status: (IUCN)** LC

5028 *Vitex triflora* Vahl (Lamiaceae) **IPNI:** 866038-1 **Common name:** aceituno, palo de arco **Habit:** Shrub, Tree **Use:** HF, ME **Origin:** Native **Region:** Caribbean **Dept.:** ANT, BOL, SUC **Elev.:** 200–700 m a.s.l. **IUCN Habitat code:** 1,2,3,14 **Conservation status: (IUCN)** NE

5029 *Volkameria aculeata* L. (Lamiaceae) **IPNI:** 866088-1 **Habit:** Shrub **Use:** EU, ME, PO **Origin:** Native **Region:** Andean, Caribbean **Dept.:** BOL, TOL **IUCN Habitat code:** 1,2,3,14 **Conservation status: (IUCN)** LC

5030 *Utricularia gibba* L. (Lentibulariaceae) **IPNI:** 526777-1 **Habit:** Herb, Aquatic **Use:** ME **Origin:** Native **Region:** Amazonia, Andean, Caribbean, Orinoquia, Pacific **Dept.:** AMA, ANT, BOG, BOY, CAU, CES, CHO, CUN, MET, SAN **Elev.:** 20-3850 msnm **Conservation status: (IUCN)** LC **(CPSG)** Potential LC

5031 *Bonnaya antipoda* (L.) Druce (Linderniaceae) **IPNI:** 799766-1 **Habit:** Herb **Use:** ME **Origin:** Native **Region:** Orinoquia **Dept.:** VID **Elev.:** 90–90 m a.s.l. **Conservation status: (IUCN)** LC

5032 *Lindernia rotundifolia* (L.) Alston (Linderniaceae) **IPNI:** 1061345-2 **Habit:** Herb, Semi-aquatic **Use:** ME **Origin:** Native **Region:** Amazonia **Dept.:** AMA **Elev.:** 90–90 m a.s.l. **Conservation status: (IUCN)** LC

5033 *Torenia asiatica* L. (Linderniaceae) **IPNI:** 810361-1 **Habit:** Herb **Use:** ME **Origin:** Cultivated **Region:** Pacific **Dept.:** CHO **Elev.:** 80–2000 m a.s.l. **Conservation status: (IUCN)** NE

5034 *Torenia crustacea* (L.) Cham. & Schltdl. (Linderniaceae) **IPNI:** 1189750-2 **Common name:** yerba de pollito **Habit:** Herb **Use:** HF, ME **Origin:** Native **Region:** Amazonia, Andean, Orinoquia, Pacific **Dept.:** AMA, ANT, CAS, CAU, CHO, GUV, MET, NAR, SAN, TOL, VAC **Elev.:** 0-1000 msnm **IUCN Habitat code:** 1,2,3,4,5,14 **Conservation status: (IUCN)** LC **(CPSG)** Potential LC

5035 *Torenia fournieri* Linden ex E.Fourn. (Linderniaceae) **IPNI:** 810406-1 **Common name:** copelia, gotelia, pensamiento, pensamiento de Bolívar **Habit:** Herb **Use:** EU, ME **Origin:** Cultivated **Region:** Andean, Caribbean, Orinoquia **Dept.:** ANT, BOY, CES, CHO, CUN, MET **Elev.:** 50–2600 m a.s.l. **IUCN Habitat code:** 1,3,14 **Conservation status: (IUCN)** NE

5036 *Torenia thouarsii* (Cham. & Schltdl.) Kuntze (Linderniaceae) **IPNI:** 255156-2 **Habit:** Herb **Use:** HF, ME **Origin:** Naturalised **Region:** Amazonia, Andean, Orinoquia **Dept.:** AMA, BOY, CAS, CUN, MET, NAR, PUT **Elev.:** 200–1400 m a.s.l. **IUCN Habitat code:** 1,14 **Conservation status: (IUCN)** LC

5037 *Vandellia diffusa* L. (Linderniaceae) **IPNI:** 262512-2 **Common name:** coneja, cuartillo, oreja de ratón, paraguay, yerba anís, yerba de pollo, yerba pollo **Habit:** Herb **Use:** HF, ME, PO **Origin:** Naturalised **Region:** Amazonia, Andean, Guiana Shield, Caribbean, Orinoquia **Dept.:** AMA, ANT, BOL, BOY, CAS, CAU, CES, CHO, CUN, GUA, GUV, MAG, MET, NAR, NSA, PUT, SAN, TOL, VAC, VAU, VID **Elev.:** 100-1500 msnm **IUCN Habitat code:** 1,2,3,4,5,14 **Conservation status: (IUCN)** LC **(CPSG)** Potential LC

5038 *Craniolaria annua* L. (Martyniaceae) **IPNI:** 675800-1 **Common name:** escorzonera, joural, yuca escorzonera **Habit:** Herb **Use:** HF, ME **Origin:** Native **Region:** Caribbean, Orinoquia **Dept.:** BOL, CES, MAG, SUC, VID **Elev.:** 50–100 m a.s.l. **IUCN Habitat code:** 2,14 **Conservation status: (IUCN)** NE

5039 *Martynia annua* L. (Martyniaceae) **IPNI:** 675840-1 **Common name:** escorzonera **Use:** EU, HF, MA, ME **IUCN Habitat code:** 4 **Conservation status: (IUCN)** NE

5040 *Chionanthus compactus* Sw. (Oleaceae) **IPNI:** 301087-2 **Habit:** Tree **Use:** MA **Origin:** Native **Region:** Caribbean, Orinoquia, Pacific **Dept.:** CAS, CAU, MAG, NSA, SAN **Elev.:** 0–460 m a.s.l. **IUCN Habitat code:** 1,2,14 **Conservation status: (IUCN)** NE

5041 *Chionanthus pubescens* Kunth (Oleaceae) **IPNI:** 300138-2 **Common name:** arupo **Habit:** Small tree **Use:** EU **Origin:** Cultivated **Region:** Andean **Dept.:** BOG, CUN, NAR **Elev.:** 2550–2600 m a.s.l. **Conservation status: (IUCN)** LC

5042 *Chionanthus virginicus* L. (Oleaceae) **IPNI:** 319141-2 **Use:** EU, MA, ME **Region:** Andean **Dept.:** BOG **Conservation status: (IUCN)** LC

5043 *Chrysojasminum humile* (L.) Banfi (Oleaceae) **IPNI:** 77144222-1 **Use:** EU, HF, MA, ME **Region:** Andean **Dept.:** BOG **Conservation status: (IUCN)** NE

5044 *Fraxinus longicuspis* Siebold & Zucc. (Oleaceae) **IPNI:** 609077-1 **Use:** ME **Region:** Andean **Dept.:** BOG **Conservation status: (IUCN)** LC

5045 *Fraxinus ornus* L. (Oleaceae) **IPNI:** 609128-1 **Use:** EU, MA, ME **Region:** Andean **Dept.:** BOG **IUCN Habitat code:** 14 **Conservation status: (IUCN)** LC

5046 *Fraxinus uhdei* (Wenz.) Lingelsh. (Oleaceae) **IPNI:** 105922-2 **Common name:** fresno, fresno urapán, urapán **Habit:** Tree **Use:** AF, EU, IF, MA, ME **Origin:** Cultivated **Region:** Andean **Dept.:** ANT, BOG, CUN **Elev.:** 1470–2800 m a.s.l. **IUCN Habitat code:** 1,3,14 **Conservation status: (IUCN)** LC

5047 *Jasminum azoricum* L. (Oleaceae) **IPNI:** 609354-1 **Use:** ME **Region:** Andean **Dept.:** BOG **Conservation status: (IUCN)** CR

5048 *Jasminum fluminense* Vell. (Oleaceae) **IPNI:** 609490-1 **Habit:** Climbing **Use:** AF, EU, FU, GS, HF, MA, ME **Origin:** Cultivated **Region:** Caribbean **Dept.:** ATL **Elev.:** 50–50 m a.s.l. **IUCN Habitat code:** 14 **Conservation status: (IUCN)** NE

5049 *Jasminum grandiflorum* L. (Oleaceae)
IPNI: 328130-2
Habit: Climbing **Use:** EU, HF, MA, ME **Origin:** Cultivated **Region:** Amazonia, Andean **Dept.:** AMA, ANT, BOG, CUN **Elev.:** 100–2620 m a.s.l. **IUCN Habitat code:** 3,14 **Conservation status: (IUCN)** NE

5050 *Jasminum multiflorum* (Burm.f.) Andrews (Oleaceae)
IPNI: 609641-1
Common name: jazmín estrellado
Habit: Climbing **Use:** EU, HF, MA, ME, SU **Origin:** Cultivated **Region:** Andean, Caribbean **Dept.:** ANT, BOG, COR, MAG, SAN **Elev.:** 20–1530 m a.s.l. **IUCN Habitat code:** 1,14 **Conservation status: (IUCN)** NE

5051 *Jasminum officinale* L. (Oleaceae)
IPNI: 609672-1
Use: EU, HF, MA, ME **Region:** Andean **Dept.:** BOG **Conservation status: (IUCN)** NE

5052 *Jasminum polyanthum* Franch. (Oleaceae)
IPNI: 609699-1
Use: EU, MA, ME **Region:** Andean **Dept.:** BOG **Conservation status: (IUCN)** NE

5053 *Jasminum sambac* (L.) Aiton (Oleaceae)
IPNI: 609755-1
Common name: jazmín, jazmín de Arabia, jazmín estrellado **Habit:** Shrub, Climbing **Use:** EU, HF, MA, ME **Origin:** Cultivated **Region:** Amazonia, Andean, Orinoquia **Dept.:** AMA, ANT, CAS, CUN **Elev.:** 100–1800 m a.s.l. **IUCN Habitat code:** 1,3,14 **Conservation status: (IUCN)** NE

5054 *Ligustrum japonicum* Thunb. (Oleaceae)
IPNI: 610001-1
Habit: Tree **Use:** EU, HF, MA, ME **Origin:** Cultivated **Region:** Andean **Dept.:** BOG, CUN **Elev.:** 2650–2650 m a.s.l. **IUCN Habitat code:** 14 **Conservation status: (IUCN)** NE

5055 *Ligustrum lucidum* W.T.Aiton (Oleaceae)
IPNI: 610017-1
Habit: Shrub, Small tree **Use:** AF, EU, HF, MA, ME **Origin:** Cultivated **Dept.:** BOG **IUCN Habitat code:** 3,14 **Conservation status: (IUCN)** LC

5056 *Ligustrum ovalifolium* Hassk. (Oleaceae)
IPNI: 610056-1
Use: EU, MA, ME, PO **Region:** Andean **Dept.:** BOG **IUCN Habitat code:** 14 **Conservation status: (IUCN)** LC

5057 *Ligustrum vulgare* L. (Oleaceae)
IPNI: 610130-1
Common name: aligustre, azahar de indio, azahar de novia, ligustro, trueno **Habit:** Tree **Use:** EU, MA, ME, PO **Origin:** Cultivated **Region:** Andean **Dept.:** ANT, BOG **Elev.:** 2100–2100 m a.s.l. **IUCN Habitat code:** 1,3,14 **Conservation status: (IUCN)** NE

5058 *Olea europaea* L. (Oleaceae)
IPNI: 610675-1
Common name: acebuche, aceituna, aceituno, olivo, olivo europeo **Habit:** Shrub, Small tree, Tree **Use:** AF, EU, FU, GS, HF, MA, ME, SU **Origin:** Cultivated **Region:** Andean **Dept.:** BOG, BOY, CUN **Elev.:** 1990–2620 m a.s.l. **IUCN Habitat code:** 1,14 **Conservation status: (IUCN)** NE

5059 *Agalinis lanceolata* (Ruiz & Pav.) D'Arcy (Orobanchaceae)
IPNI: 6185-2
Use: HF **Origin:** Native **Conservation status: (IUCN)** NE

5060 *Aureolaria levigata* (Raf.) Raf. (Orobanchaceae)
IPNI: 799460-1
Use: ME **Region:** Andean **Dept.:** BOG **Conservation status: (IUCN)** NE

5061 *Buchnera longifolia* Kunth (Orobanchaceae)
IPNI: 800009-1
Habit: Herb **Use:** ME **Origin:** Native **Region:** Andean, Orinoquia **Dept.:** ANT, CAU, CUN, HUI, MET, TOL **Elev.:** 150–1500 m a.s.l. **Conservation status: (IUCN)** NE

5062 *Castilleja arvensis* Schltdl. & Cham. (Orobanchaceae)
IPNI: 800871-1
Common name: peona, sanjuán, venadillo **Habit:** Herb **Use:** ME **Origin:** Native **Region:** Andean **Dept.:** ANT, BOG, BOY, CAL, CAQ, CAU, CHO, CUN, HUI, MET, NAR, PUT, QUI, RIS, SAN, VAC **Elev.:** 1000-3500 msnm **IUCN Habitat code:** 1,2,3,4,5,8,14 **Conservation status: (IUCN)** NE **(CPSG)** Potential LC

5063 *Castilleja fissifolia* L.f. (Orobanchaceae)
IPNI: 800957-1
Common name: cantimplora, chirlobirlos, liberal, trompetillo **Habit:** Herb **Use:** AF, EU **Origin:** Native **Region:** Andean **Dept.:** ANT, BOG, BOY, CAL, CAU, CES, CHO, CUN, LAG, MAG, MET, NAR, QUI, RIS, SAN, TOL, VAC **Elev.:** 2225-4500 msnm **IUCN Habitat code:** 1,2,3,4,5,6,8,14 **Conservation status: (IUCN)** NE **(CPSG)** Potential LC

5064 *Castilleja integrifolia* L.f. (Orobanchaceae)
IPNI: 801003-1
Habit: Herb **Use:** AF, EU **Origin:** Native **Region:** Andean **Dept.:** ANT, BOG, BOY, CES, CUN, LAG, NSA, SAN **Elev.:** 2580–3900 m a.s.l. **IUCN Habitat code:** 1,2,3,4,8,14 **Conservation status: (IUCN)** NE

5065 *Escobedia grandiflora* (L.f.) Kuntze (Orobanchaceae)
IPNI: 97540-2
Common name: azafrán, azafrán criollo, azafrán de Perú, azafrán de la tierra, azafrán de montaña, azafrán de tierra, color, nukcalá, raicilla, raíz de color, zafrán **Habit:** Herb **Use:** HF, MA, ME **Origin:** Native **Region:** Andean **Dept.:** ANT, BOY, CAL, CAU, CES, CHO, CUN, LAG, MAG, NAR, PUT, SAN, TOL, VAC **Elev.:** 500-3600 msnm **IUCN Habitat code:** 1,2,3,4,14 **Conservation status: (IUCN)** NE **(CPSG)** Potential LC

5066 *Lamourouxia virgata* Kunth (Orobanchaceae)
IPNI: 804205-1
Habit: Herb **Use:** HF, ME **Origin:** Native **Region:** Andean **Dept.:** BOG, BOY, CAU, CUN, NAR, NSA, SAN **Elev.:** 1800–3420 m a.s.l. **IUCN Habitat code:** 1,3,4,14 **Conservation status: (IUCN)** NE

5067 *Neobartsia santolinifolia* (Kunth) Uribe-Convers & Tank (Orobanchaceae)
IPNI: 77160774-1
Common name: singuinaria chabama **Habit:** Herb **Use:** ME **Origin:** Native **Region:** Andean **Dept.:** ANT, BOG, BOY, CAU, CES, CUN, MAG, MET, NAR, NSA, QUI, SAN **Elev.:** 2900–4400 m a.s.l. **IUCN Habitat code:** 1,3,4,8,14 **Conservation status: (IUCN)** NE

5068 *Orobanche minor* Sm. (Orobanchaceae)
IPNI: 53607-3
Habit: Herb, Parasitic **Use:** HF, ME, PO **Origin:** Naturalised **Region:** Andean **Dept.:** BOG, BOY, CUN **Elev.:** 2600–3100 m a.s.l. **IUCN Habitat code:** 1,4,14 **Conservation status: (IUCN)** NE

5069 *Sesamum indicum* L. (Pedaliaceae)
IPNI: 675971-1
Common name: ajonjolí, sésamo
Habit: Herb, Subshrub **Use:** AF, EU, FU, GS, HF, MA, ME, PO, SU **Origin:** Cultivated **Region:** Andean, Caribbean, Orinoquia, Pacific **Dept.:** ANT, ATL, BOL, BOY, CES, COR, CUN, HUI, MAG, MET, SUC, TOL **Elev.:** 150–1500 m a.s.l. **IUCN Habitat code:** 1,2,3,5,14 **Conservation status: (IUCN)** NE

5070 *Erythranthe glabrata* (Kunth) G.L.Nesom (Phrymaceae)
IPNI: 77120251-1
Habit: Herb, Semi-aquatic **Use:** ME **Origin:** Native **Region:** Andean **Dept.:** BOG, BOY, CAU, CUN, SAN **Elev.:** 2160–3700 m a.s.l. **IUCN Habitat code:** 1,3,4,5,8,14 **Conservation status: (IUCN)** NE

5071 *Angelonia angustifolia* Benth. (Plantaginaceae)
IPNI: 798919-1
Habit: Herb **Use:** EU, ME **Origin:** Native **Region:** Andean, Guiana Shield, Caribbean, Pacific **Dept.:** ANT, CAU, CHO, COR, VAU **Elev.:** 0–2000 m a.s.l. **IUCN Habitat code:** 1,3,5,14 **Conservation status: (IUCN)** NE

5072 *Angelonia salicariifolia* Bonpl. (Plantaginaceae)
IPNI: 798952-1
Common name: almizclado, almizclillo, angelón, cangalla, melosa, pilpe **Habit:** Herb **Use:** EU, ME **Origin:** Naturalised **Region:** Andean **Dept.:** ANT, ARA, BOY, CAS, CHO, CUN, HUI, TOL, VAC **Elev.:** 450–1400 m a.s.l. **IUCN Habitat code:** 1,2,3,14 **Conservation status: (IUCN)** NE

5073 *Antirrhinum majus* L. (Plantaginaceae)
IPNI: 799180-1
Common name: becerra, boca de dragón, bocas de dragón, boquelión, boquiabierto, boquiabiertos, dragón, guargüerón, jetiabierto, perrito, perritos **Habit:** Herb **Use:** EU, GS, HF, ME **Origin:** Cultivated **Region:** Andean **Dept.:** ANT, BOG, CUN **Elev.:** 2000–2600 m a.s.l. **IUCN Habitat code:** 1,4,8,14 **Conservation status: (IUCN)** NE

5074 *Bacopa axillaris* (Benth.) Standl. (Plantaginaceae)
IPNI: 28282-2
Habit: Herb, Aquatic **Use:** MA **Origin:** Native **Region:** Caribbean, Orinoquia **Dept.:** CES, MAG, MET **Elev.:** 100–450 m a.s.l. **Conservation status: (IUCN)** NE

5075 *Bacopa monnieri* (L.) Wettst. (Plantaginaceae) **IPNI:** 1072674-2 **Common name:** bacopa, herb de la gracia, herb del dinero, hisopo de agua costera **Habit:** Herb, Aquatic **Use:** EU, GS, HF, MA, ME **Origin:** Native **Region:** Amazonia, Pacific **Dept.:** AMA, CAU, VAC **Elev.:** 0–90 m a.s.l. **Conservation status: (IUCN)** LC

5076 *Callitriche terrestris* Raf. (Plantaginaceae) **IPNI:** 430053-1 **Habit:** Herb **Use:** ME **Origin:** Native **Region:** Andean **Dept.:** CUN, NSA, VAC **Elev.:** 1950–1950 m a.s.l. **IUCN Habitat code:** 4,14 **Conservation status: (IUCN)** LC

5077 *Conobea scoparioides* (Cham. & Schltdl.) Benth. (Plantaginaceae) **IPNI:** 801672-1 **Common name:** aromática, hierba de indio, hierba de sapo, pilpe, yerba de sapo **Habit:** Herb **Use:** MA, ME, SU **Origin:** Native **Region:** Amazonia, Andean, Pacific **Dept.:** ANT, CAU, CHO, NAR, PUT, RIS, VAC **Elev.:** 0–1500 m a.s.l. **IUCN Habitat code:** 1,3,5,14 **Conservation status: (IUCN)** NE

5078 *Cymbalaria muralis* G.Gaertn., B.Mey. & Scherb. (Plantaginaceae) **IPNI:** 325773-2 **Common name:** hiedra, hiedra de tiesto, hiedra de tiestos, yedra, yedrita **Habit:** Herb **Use:** EU, HF, ME **Origin:** Naturalised **Region:** Andean **Dept.:** BOG **Elev.:** 2000–2000 m a.s.l. **IUCN Habitat code:** 1,3,4,14 **Conservation status: (IUCN)** NE

5079 *Digitalis lanata* Ehrh. (Plantaginaceae) **IPNI:** 802028-1 **Common name:** guarguerón **Use:** EU, ME, PO **Region:** Andean **Dept.:** BOG **Conservation status: (IUCN)** NE

5080 *Digitalis purpurea* L. (Plantaginaceae) **IPNI:** 802077-1 **Common name:** dedalera, digital, digitalis, guarguerón o digital, guargüerones, guargüerón, guergüerón **Habit:** Herb **Use:** EU, HF, MA, ME, PO **Origin:** Naturalised **Region:** Andean **Dept.:** ANT, BOG, CUN, NAR, VAC **Elev.:** 2200–3600 m a.s.l. **IUCN Habitat code:** 1,2,3,4,5,8,14 **Conservation status: (IUCN)** NE

5081 *Gratiola bogotensis* Cortés ex Pennell (Plantaginaceae) **IPNI:** 113323-2 **Habit:** Herb **Use:** ME **Origin:** Native **Region:** Andean **Dept.:** ANT, BOG, BOY, CUN, MET, NAR, NSA, SAN, TOL **Elev.:** 2500–3700 m a.s.l. **IUCN Habitat code:** 1,2,3,4,5,14 **Conservation status: (IUCN)** NE

5082 *Gratiola peruviana* L. (Plantaginaceae) **IPNI:** 803515-1 **Use:** ME **Region:** Andean **Dept.:** BOG **Conservation status: (IUCN)** NE

5083 *Hydrotriche hottoniiflora* Zucc. (Plantaginaceae) **IPNI:** 803925-1 **Habit:** Herb, Aquatic **Use:** EU **Origin:** Cultivated **Region:** Andean **Dept.:** ANT, QUI **Elev.:** 1200–1500 m a.s.l. **Conservation status: (IUCN)** NT

5084 *Limosella australis* R.Br. (Plantaginaceae) **IPNI:** 329497-2 **Habit:** Herb **Use:** ME **Origin:** Native **Region:** Andean **Dept.:** ARA, BOG, CUN, MET, RIS, TOL **Elev.:** 3500–4250 m a.s.l. **IUCN Habitat code:** 3,4 **Conservation status: (IUCN)** NE

5085 *Lophospermum erubescens* D.Don (Plantaginaceae) **IPNI:** 279831-2 **Habit:** Climbing **Use:** EU **Origin:** Naturalised **Region:** Andean **Dept.:** ANT, CUN **Elev.:** 1500–2500 m a.s.l. **IUCN Habitat code:** 1,2,3,14 **Conservation status: (IUCN)** NE

5086 *Maurandella antirrhiniflora* (Humb. & Bonpl. ex Willd.) Rothm. (Plantaginaceae) **IPNI:** 805502-1 **Common name:** recuerdo **Habit:** Herb **Use:** EU **Origin:** Cultivated **Region:** Andean **Dept.:** ANT, CAU **Elev.:** 900–2000 m a.s.l. **IUCN Habitat code:** 14 **Conservation status: (IUCN)** NE

5087 *Maurandya barclayana* Lindl. (Plantaginaceae) **IPNI:** 280591-2 **Habit:** Climbing **Use:** EU **Origin:** Cultivated, Naturalised **Region:** Andean **Dept.:** ANT **Elev.:** 1500–2500 m a.s.l. **Conservation status: (IUCN)** NE

5088 *Mecardonia procumbens* (Mill.) Small (Plantaginaceae) **IPNI:** 155661-2 **Habit:** Herb **Use:** ME **Origin:** Native **Region:** Amazonia, Andean, Pacific **Dept.:** AMA, ANT, BOG, BOL, BOY, CAL, CAU, CHO, CUN, HUI, MAG, MET, NAR, NSA, SAN, TOL, VAC **Elev.:** 0-3000 msnm **IUCN Habitat code:** 1,2,3,4,5,14 **Conservation status: (IUCN)** NE **(CPSG)** Potential LC

5089 *Nuttallanthus canadensis* (L.) D.A.Sutton (Plantaginaceae) **IPNI:** 281417-2 **Habit:** Herb **Use:** EU **Origin:** Native **Region:** Andean **Dept.:** CUN, NAR **Elev.:** 2500–2600 m a.s.l. **IUCN Habitat code:** 14 **Conservation status: (IUCN)** NE

5090 *Penstemon hartwegii* Benth. (Plantaginaceae) **IPNI:** 808129-1 **Habit:** Herb **Use:** EU, GS **Origin:** Cultivated **Region:** Andean **Dept.:** BOG, CUN **Elev.:** 2600–2600 m a.s.l. **IUCN Habitat code:** 14 **Conservation status: (IUCN)** NE

5091 *Plantago australis* Lam. (Plantaginaceae) **IPNI:** 684911-1 **Common name:** llanten del cuichi, llantén, llantén de manga, llantén de monte, llantén de páramo, llantén del cuichi, mostaza **Habit:** Herb **Use:** HF, ME **Origin:** Native **Region:** Andean **Dept.:** ANT, BOG, BOY, CAU, CUN, HUI, MAG, MET, NAR, NSA, PUT, QUI, RIS, SAN, TOL, VAC **Elev.:** 1200–4100 m a.s.l. **IUCN Habitat code:** 1,3,4,8,14 **Conservation status: (IUCN)** NE

5092 *Plantago goudotiana* Decne. (Plantaginaceae) **IPNI:** 685196-1 **Common name:** llantén **Habit:** Herb **Use:** ME **Origin:** Endemic **Region:** Andean **Dept.:** BOG, CUN, SAN **Elev.:** 2500–2600 m a.s.l. **IUCN Habitat code:** 1,3,14 **Conservation status: (IUCN)** NE

5093 *Plantago lanceolata* L. (Plantaginaceae) **IPNI:** 321285-2 **Common name:** llantén **Habit:** Herb **Use:** AF, GS, HF, MA, ME **Origin:** Naturalised **Region:** Andean **Dept.:** BOG, CAU **Elev.:** 1590–1590 m a.s.l. **IUCN Habitat code:** 1,3,4,14 **Conservation status: (IUCN)** NE

5094 *Plantago linearis* Kunth (Plantaginaceae) **IPNI:** 685346-1 **Common name:** llantén **Habit:** Herb **Use:** ME **Origin:** Native **Region:** Andean **Dept.:** BOG, BOY, CAU, CUN, NSA, RIS **Elev.:** 2500–4300 m a.s.l. **IUCN Habitat code:** 1,3,4,14 **Conservation status: (IUCN)** NE

5095 *Plantago major* L. (Plantaginaceae) **IPNI:** 321286-2 **Common name:** chiracru, chiracu, llantén **Habit:** Herb **Use:** AF, EU, GS, HF, MA, ME **Origin:** Naturalised **Region:** Amazonia, Andean **Dept.:** AMA, ANT, BOG, BOY, CUN, HUI, MAG, MET, NAR, PUT, VAC **Elev.:** 100–2600 m a.s.l. **IUCN Habitat code:** 1,2,3,8,14 **Conservation status: (IUCN)** LC

5096 *Plantago monticola* Decne. (Plantaginaceae) **IPNI:** 685440-1 **Common name:** llantén **Use:** ME **IUCN Habitat code:** 1,3,4,5,8,14 **Conservation status: (IUCN)** NE

5097 *Plantago orbignyana* Steinh. ex Decne. (Plantaginaceae) **IPNI:** 201466-2 **Habit:** Herb **Use:** ME **Origin:** Native **Region:** Andean **Dept.:** BOG, NAR **Conservation status: (IUCN)** NE

5098 *Plantago sericea* Ruiz & Pav. (Plantaginaceae) **IPNI:** 685641-1 **Habit:** Herb **Use:** ME **Origin:** Native **Region:** Andean **Dept.:** BOG, BOY, CES, CUN, LAG, MAG, NSA, SAN **Elev.:** 2500–3900 m a.s.l. **Conservation status: (IUCN)** NE

5099 *Russelia equisetiformis* Schltdl. & Cham. (Plantaginaceae) **IPNI:** 808903-1 **Common name:** cabello de indio, cola de caballo, colaecaballo, coral, escoba, lluvia de fuego **Habit:** Herb **Use:** EU, ME **Origin:** Cultivated **Region:** Andean **Dept.:** ANT, CHO, CUN **Elev.:** 1500–2600 m a.s.l. **IUCN Habitat code:** 2,3,14 **Conservation status: (IUCN)** NE

5100 *Russelia sarmentosa* Jacq. (Plantaginaceae) **IPNI:** 808935-1 **Common name:** coralito **Habit:** Herb **Use:** EU, ME **Origin:** Native **Region:** Andean, Caribbean **Dept.:** ANT, COR, MAG **Elev.:** 450–920 m a.s.l. **IUCN Habitat code:** 1,2,3,14 **Conservation status: (IUCN)** NE

5101 *Scoparia dulcis* L. (Plantaginaceae) **IPNI:** 230902-2 **Common name:** amargo, añí, basuriña, berro, chamico, changa dulce, color, cominillo, culantrillo, escoba de Dios, escoba de castilla, escoba de paraguay, escoba dulce, escobilla, escobilla amarga, escobilla amarga granito de aníz, escobilla menuda, escobilla menudita, **Habit:** Herb **Use:** AF, EU, HF, MA, ME, PO, SU **Origin:** Native **Region:** Amazonia, Andean, Orinoquia, Pacific **Dept.:** ANT, BOL, CAU, CES, CHO, COR, CUN, GUA, HUI, MAG, MET, NAR, PUT, SUC, TOL,

VAC, VID **Elev.:** 0-1600 msnm **IUCN Habitat code:** 1,2,3,4,5,14 **Conservation status: (IUCN)** NE **(CPSG)** Potential LC

5102 *Stemodia durantifolia* (L.) Sw. (Plantaginaceae) **IPNI:** 328499-2 **Habit:** Herb **Use:** AF,ME, PO **Origin:** Native **Region:** Andean, Caribbean, Orinoquia, Pacific **Dept.:** ANT, ATL, BOL, CAS, HUI, MAG, SUC, VAC **Elev.:** 30-1250 msnm **IUCN Habitat code:** 1,2,3,5,14 **Conservation status: (IUCN)** NE **(CPSG)** Potential LC

5103 *Stemodia maritima* L. (Plantaginaceae) **IPNI:** 308488-2 **Common name:** yerba amarga **Habit:** Herb **Use:** ME **Origin:** Native **Region:** Caribbean **Dept.:** LAG **Elev.:** 300–300 m a.s.l. **IUCN Habitat code:** 2,3 **Conservation status: (IUCN)** NE

5104 *Stemodia verticillata* (Mill.) Hassl. (Plantaginaceae) **IPNI:** 244449-2 **Habit:** Herb **Use:** ME **Origin:** Native **Region:** Amazonia, Andean, Caribbean, Pacific **Dept.:** AMA, ANT, BOL, BOY, CAU, CES, CHO, COR, MAG, NAR, VAC **Elev.:** 0–600 m a.s.l. **IUCN Habitat code:** 1,2,14 **Conservation status: (IUCN)** NE

5105 *Veronica arvensis* L. (Plantaginaceae) **IPNI:** 811649-1 **Habit:** Herb **Use:** ME **Origin:** Naturalised **Region:** Andean **Dept.:** BOG, BOY, CAL, CUN, NAR, SAN **Elev.:** 2200–3500 m a.s.l. **IUCN Habitat code:** 1,3,4,14 **Conservation status: (IUCN)** NE

5106 *Veronica elliptica* G.Forst. (Plantaginaceae) **IPNI:** 811951-1 **Habit:** Herb **Use:** MA **Origin:** Cultivated **Dept.:** BOG **Conservation status: (IUCN)** NE

5107 *Veronica javanica* Blume (Plantaginaceae) **IPNI:** 812151-1 **Habit:** Herb **Use:** ME **Origin:** Naturalised **Region:** Andean **Dept.:** ANT, BOG, BOY, CAL, CUN **Elev.:** 2150–2600 m a.s.l. **Conservation status: (IUCN)** NE

5108 *Veronica odora* Hook.f. (Plantaginaceae) **IPNI:** 812400-1 **Habit:** Herb **Use:** EU **Origin:** Cultivated **Dept.:** BOG **Conservation status: (IUCN)** NE

5109 *Veronica peregrina* L. (Plantaginaceae) **IPNI:** 812460-1 **Habit:** Herb **Use:** ME **Origin:** Native **Region:** Andean **Dept.:** BOG, BOY, CAL, CAU, CUN, NAR, SAN **Elev.:** 2150–3200 m a.s.l. **IUCN Habitat code:** 1,3,4,14 **Conservation status: (IUCN)** LC

5110 *Veronica persica* Poir. (Plantaginaceae) **IPNI:** 30198355-2 **Common name:** azulita, golondrina, riñón morado, santamaría, violetilla **Habit:** Herb **Use:** HF, ME **Origin:** Naturalised **Region:** Andean **Dept.:** ANT, BOG, BOY, CUN, NAR, PUT, VAC **Elev.:** 2000–2700 m a.s.l. **IUCN Habitat code:** 1,3,5,14 **Conservation status: (IUCN)** NE

5111 *Veronica salicifolia* G.Forst. (Plantaginaceae) **IPNI:** 812616-1 **Habit:** Herb **Use:** ME **Origin:** Cultivated **Region:** Andean **Dept.:** BOG, BOY, CUN **Elev.:** 2600–2760 m a.s.l. **IUCN Habitat code:** 1,14 **Conservation status: (IUCN)** NE

5112 *Veronica serpyllifolia* L. (Plantaginaceae) **IPNI:** 812650-1 **Common name:** verónica **Habit:** Herb **Use:** HF, ME **Origin:** Naturalised **Region:** Andean **Dept.:** ANT, BOG, BOY, CAL, CAU, CUN, HUI, MAG, MET, NAR, PUT, QUI, RIS, SAN, TOL, VAC **Elev.:** 2200-4200 msnm **IUCN Habitat code:** 1,3,4,5,8,14 **Conservation status: (IUCN)** LC **(CPSG)** Potential LC

5113 *Veronica speciosa* R.Cunn. ex A.Cunn. (Plantaginaceae) **IPNI:** 812675-1 **Habit:** Herb **Use:** EU **Origin:** Cultivated **Region:** Andean **Dept.:** ANT, BOG **Elev.:** 2000–2500 m a.s.l. **Conservation status: (IUCN)** NE

5114 *Platanus occidentalis* L. (Platanaceae) **IPNI:** 685871-1 **Use:** EU, MA, ME **Region:** Andean **Dept.:** BOG **IUCN Habitat code:** 1 **Conservation status: (IUCN)** LC

5115 *Schlegelia chocoensis* A.H.Gentry (Schlegeliaceae) **IPNI:** 229505-2 **Habit:** Hemiepiphyte, Liana, Climbing **Use:** HF **Origin:** Native **Region:** Pacific **Dept.:** CHO, NAR, VAC **Elev.:** 30–300 m a.s.l. **IUCN Habitat code:** 1,3,14 **Conservation status: (IUCN)** NE

5116 *Schlegelia fastigiata* Schery (Schlegeliaceae) **IPNI:** 229555-2 **Habit:** Hemiepiphyte, Liana **Use:** HF **Origin:** Native **Region:** Andean, Pacific **Dept.:** ANT, NAR **Elev.:** 30–1320 m a.s.l. **IUCN Habitat code:** 1,3,5 **Conservation status: (IUCN)** LC

5117 *Schlegelia parviflora* (Oerst.) Monach. (Schlegeliaceae) **IPNI:** 229565-2 **Habit:** Small tree, Tree, Hemiepiphyte, Liana **Use:** MA **Origin:** Native **Region:** Amazonia, Andean, Pacific **Dept.:** AMA, ANT, BOY, CHO, LAG, MAG, NAR, VAC **Elev.:** 5–1800 m a.s.l. **IUCN Habitat code:** 1,3,5,14 **Conservation status: (IUCN)** LC

5118 *Schlegelia sulphurea* Diels (Schlegeliaceae) **IPNI:** 229576-2 **Habit:** Tree, Hemiepiphyte, Liana **Use:** ME **Origin:** Native **Dept.:** ANT **Elev.:** 400–900 m a.s.l. **IUCN Habitat code:** 1 **Conservation status: (IUCN)** NE

5119 *Alonsoa meridionalis* (L.f.) Kuntze (Scrophulariaceae) **IPNI:** 9536-2 **Common name:** caracolito, cascabel, cascabel del diablo, cascabelillo, cascabelito, choroticos, coconuco, colorado, diablo, napán blanco, zarallejo, zarcillejo **Habit:** Herb **Use:** EU, ME **Origin:** Native **Region:** Andean **Dept.:** ANT, BOG, BOY, CAU, CES, CUN, LAG, MAG, NAR, NSA, PUT, QUI, SAN, VAC **Elev.:** 2225-3900 msnm **IUCN Habitat code:** 1,3,4,5,8,14 **Conservation status: (IUCN)** NE **(CPSG)** Potential LC

5120 *Bontia daphnoides* L. (Scrophulariaceae) **IPNI:** 319112-2 **Use:** EU, HF, MA, ME **Origin:** Native **IUCN Habitat code:** 14 **Conservation status: (IUCN)** NE

5121 *Buddleja americana* L. (Scrophulariaceae) **IPNI:** 319122-2 **Common name:** gavilán, salvia blanca, tabaco de monte **Habit:** Shrub, Small tree **Use:** EU, HF, MA, ME, SU **Origin:** Native **Region:** Andean **Dept.:** ANT, BOG, BOY, CAL, CAU, CES, CUN, LAG, MAG, MET, NAR, NSA, PUT, SAN, TOL **Elev.:** 1190-3400 msnm **IUCN Habitat code:** 1,2,3,4,14 **Conservation status: (IUCN)** LC **(CPSG)** Potential LC

5122 *Buddleja davidii* Franch. (Scrophulariaceae) **IPNI:** 545734-1 **Common name:** copa de David, ramillete **Habit:** Shrub **Use:** EU, GS, ME **Origin:** Cultivated, Naturalised **Region:** Andean **Dept.:** ANT, BOG, PUT, VAC **Elev.:** 970–2200 m a.s.l. **IUCN Habitat code:** 1,3,4,14 **Conservation status: (IUCN)** NE

5123 *Buddleja incana* Ruiz & Pav. (Scrophulariaceae) **IPNI:** 545779-1 **Common name:** gavilán **Habit:** Shrub, Tree **Use:** EU, FU, HF, MA, ME **Origin:** Native **Region:** Andean **Dept.:** ANT, BOG, BOY, CUN, MET **Elev.:** 2500–3300 m a.s.l. **IUCN Habitat code:** 1,2,3,4,8,14 **Conservation status: (IUCN)** LC

5124 *Capraria biflora* L. (Scrophulariaceae) **IPNI:** 800773-1 **Common name:** año nuevo, escoba de monte, escoba peluda, guají, pericón, té criollo **Habit:** Herb **Use:** HF, MA, ME **Origin:** Native **Region:** Caribbean **Dept.:** ANT, ATL, BOL, CES, COR, CUN, HUI, MAG, SAP **Elev.:** 0–600 m a.s.l. **IUCN Habitat code:** 1,2,3,4,5,8,14 **Conservation status: (IUCN)** NE

5125 *Capraria peruviana* Benth. (Scrophulariaceae) **IPNI:** 800802-1 **Use:** ME **Origin:** Native **Conservation status: (IUCN)** NE

5126 *Leucophyllum frutescens* (Berland.) I.M.Johnst. (Scrophulariaceae) **IPNI:** 139187-2 **Common name:** frailejón **Habit:** Shrub **Use:** AF, EU, ME **Origin:** Cultivated **Region:** Caribbean **Dept.:** BOL, CUN, TOL **Elev.:** 0–500 m a.s.l. **IUCN Habitat code:** 14 **Conservation status: (IUCN)** NE

5127 *Verbascum virgatum* Stokes (Scrophulariaceae) **IPNI:** 770211-1 **Habit:** Herb **Use:** EU, ME **Origin:** Naturalised **Region:** Andean **Dept.:** BOG, BOY, CUN, HUI **Elev.:** 1750–2700 m a.s.l. **IUCN Habitat code:** 3,14 **Conservation status: (IUCN)** NE

5128 *Aloysia citrodora* Paláu (Verbenaceae) **IPNI:** 1015994-1 **Common name:** alegría, cedrón, cidrón, lemon verbena, luisa, luisa de Chile, saca ojo, yerba luisa, yerbaluisa, zorrillo **Habit:** Herb, Shrub **Use:** EU, HF, MA, ME **Origin:** Cultivated **Region:** Andean, Caribbean,

Orinoquia **Dept.:** ANT, BOY, CAL, CAU, CUN, MET, PUT, SAN, VAC **Elev.:** 250–2800 m a.s.l. **IUCN Habitat code:** 1,14 **Conservation status: (IUCN)** NE

5129 *Bouchea prismatica* (L.) Kuntze (Verbenaceae)
IPNI: 35430-2
Common name: arrocillo, granavi, rabo de zorro, verbena de Castilla, verbena de castilla, verbena negra **Habit:** Herb, Subshrub **Use:** ME **Origin:** Native **Region:** Andean, Caribbean **Dept.:** ANT, BOL, BOY, CAU, CUN, MAG, SAN, TOL, VAC **Elev.:** 5-2130 msnm **IUCN Habitat code:** 1,2,3,14 **Conservation status: (IUCN)** NE **(CPSG)** Potential LC

5130 *Citharexylum caudatum* L. (Verbenaceae)
IPNI: 330600-2
Habit: Shrub, Tree **Use:** EU, ME **Origin:** Native **Region:** Caribbean, Pacific **Dept.:** ANT, BOL, NAR, SAP **Elev.:** 300–310 m a.s.l. **IUCN Habitat code:** 1,2,3,14 **Conservation status: (IUCN)** LC

5131 *Citharexylum macrophyllum* Poir. (Verbenaceae)
IPNI: 861796-1
Use: HF, ME **Origin:** Native **IUCN Habitat code:** 1 **Conservation status: (IUCN)** LC

5132 *Citharexylum poeppigii* Walp. (Verbenaceae)
IPNI: 861827-1
Habit: Small tree, Tree **Use:** ME **Origin:** Native **Region:** Amazonia, Orinoquia, Pacific **Dept.:** CAS, CHO, MET, PUT, VAC **Elev.:** 100–600 m a.s.l. **IUCN Habitat code:** 1,3,4,14 **Conservation status: (IUCN)** LC

5133 *Citharexylum subflavescens* S.F.Blake (Verbenaceae)
IPNI: 59539-2
Common name: cajeto, caragay, cascarillo, garagay, palo guitarra, quimulá, uruapo **Habit:** Small tree, Tree **Use:** HF **Origin:** Native, Cultivated **Region:** Andean **Dept.:** BOG, CUN, QUI, TOL, VAC **Elev.:** 1100–2800 m a.s.l. **IUCN Habitat code:** 1,2,3,4,14 **Conservation status: (IUCN)** NE

5134 *Duranta erecta* L. (Verbenaceae)
IPNI: 330606-2
Common name: adonis, cucaná, cucas, espino, espino negro, fruta de tórtolas, garbancillo, guacharaco, guapante, navidad, totocal, totocul, totocá **Habit:** Shrub **Use:** EU, HF, MA, ME, PO **Origin:** Native **Region:** Andean, Caribbean **Dept.:** ANT, ATL, BOG, BOY, CAL, CAU, COR, CUN, SAN, TOL, VAC **Elev.:** 100-3140 msnm **IUCN Habitat code:** 1,2,3,5,14 **Conservation status: (IUCN)** LC **(CPSG)** Potential LC

5135 *Duranta mutisii* L.f. (Verbenaceae)
IPNI: 862753-1
Common name: cacho de venado, choca, cruceto, cucanao, cucaná, cucas, cucuná, cúcane, cúcano, espino, espino blanco, espino cruz, espino de garbanzo, espino negro, espino tabio, garbancillo, guapante, guapanto, guayabito **Habit:** Shrub, Tree, Climbing **Use:** AF, EU, MA, ME **Origin:** Native **Region:** Andean **Dept.:** ANT, BOG, BOY, CAL, CUN, HUI, NSA, QUI, SAN, VAC **Elev.:** 1300-3500 msnm **IUCN Habitat code:** 1,3,4,5,8,14 **Conservation status: (IUCN)** NE **(CPSG)** Potential LC

5136 *Duranta triacantha* Juss. (Verbenaceae)
IPNI: 862775-1
Common name: cruceto, espino, guapante **Habit:** Shrub **Use:** HF **Origin:** Native **Region:** Andean **Dept.:** VAC **Elev.:** 2900–2900 m a.s.l. **IUCN Habitat code:** 1 **Conservation status: (IUCN)** LC

5137 *Lantana × strigocamara* R.W.Sanders (Verbenaceae)
IPNI: 60444377-2
Use: ME **Origin:** Naturalised **Conservation status: (IUCN)** NE

5138 *Lantana achyranthifolia* Desf. (Verbenaceae)
IPNI: 324301-2
Habit: Herb, Subshrub **Use:** HF **Origin:** Native **Region:** Caribbean **Dept.:** CES **Elev.:** 260–400 m a.s.l. **IUCN Habitat code:** 1 **Conservation status: (IUCN)** NE

5139 *Lantana camara* L. (Verbenaceae)
IPNI: 325686-2
Common name: andalucía, bwa wa tau, cariaquita, cariaquito, carraquilla, carraquillo, carrasposa, carrasquillo, cinco negritos, colorada, conchita, curaverrugas, flor de sangre, gurupacha, huesito, julio, mabizou, macanillo, mastranto, mille fleurs, morupacha, murupac **Habit:** Herb, Shrub, Subshrub, Climbing **Use:** AF, EU, FU, GS, HF, IF, MA, ME, PO, SU **Origin:** Native **Region:** Amazonia, Andean, Guiana Shield, Caribbean, Orinoquia, Pacific **Dept.:** AMA, ANT, ATL, BOG, BOL, BOY, CAL, CAQ, CAS, CAU, CHO, COR, CUN, GUV, HUI, LAG, MAG, MET, NAR, NSA, PUT, QUI, SAP, SAN, TOL, VAC, VID **Elev.:** 0-2800 msnm **IUCN Habitat code:** 1,2,3,4,5,14 **Conservation status: (IUCN)** NE **(CPSG)** Potential LC

5140 *Lantana canescens* Kunth (Verbenaceae)
IPNI: 863236-1
Common name: albahaca toronjil, aliviadolor, mejorana, oreganillo, prontoalivio, quitadolor, venturosa, zorrillo **Habit:** Herb, Shrub **Use:** ME, SU **Origin:** Native **Region:** Andean **Dept.:** ANT, BOG, CAU, CUN, MAG, NAR, SAN, VAC **Elev.:** 570-2680 msnm **IUCN Habitat code:** 1,2,3,14 **Conservation status: (IUCN)** NE **(CPSG)** Potential LC

5141 *Lantana caracasana* Turcz. (Verbenaceae)
IPNI: 863239-1
Common name: venturosa, zorrillo **Use:** ME **Dept.:** MAG **Conservation status: (IUCN)** NE

5142 *Lantana cujabensis* Schauer (Verbenaceae)
IPNI: 324302-2
Common name: carrasquillo, gavilana negra, venturosa **Habit:** Shrub, Subshrub **Use:** HF, ME **Origin:** Native **Region:** Amazonia, Andean, Caribbean, Pacific **Dept.:** AMA, ANT, CHO, CUN, MET, PUT, VAC, VAU, VID **Elev.:** 170–1400 m a.s.l. **IUCN Habitat code:** 1,3 **Conservation status: (IUCN)** NE

5143 *Lantana fucata* Lindl. (Verbenaceae)
IPNI: 324303-2
Common name: maizorro, maíz tostao, venturosa **Habit:** Shrub, Subshrub **Use:** HF, ME **Origin:** Native **Region:** Andean **Dept.:** ANT, BOY, CAU, CUN, HUI, SAN, TOL, VAC **Elev.:** 400-2824 msnm **IUCN Habitat code:** 1,2,3,4,14 **Conservation status: (IUCN)** NE **(CPSG)** Potential LC

5144 *Lantana hirsuta* M.Martens & Galeotti (Verbenaceae)
IPNI: 30032194-2
Use: ME **Origin:** Native **IUCN Habitat code:** 2,14 **Conservation status: (IUCN)** NE

5145 *Lantana hirta* Graham (Verbenaceae)
IPNI: 30105602-2
Use: HF **Origin:** Native **IUCN Habitat code:** 1,3,14 **Conservation status: (IUCN)** NE

5146 *Lantana horrida* Kunth (Verbenaceae)
IPNI: 863305-1
Habit: Shrub, Subshrub **Use:** ME **Origin:** Native **Region:** Andean **Dept.:** BOY, CUN, NAR, SAN, TOL, VAC **Elev.:** 1100–2130 m a.s.l. **IUCN Habitat code:** 1,2,3,4,14 **Conservation status: (IUCN)** NE

5147 *Lantana involucrata* L. (Verbenaceae)
IPNI: 135206-2
Use: HF, MA, ME **Origin:** Native **Dept.:** BOG **IUCN Habitat code:** 3,14 **Conservation status: (IUCN)** NE

5148 *Lantana lopez-palacii* Moldenke (Verbenaceae)
IPNI: 287900-2
Common name: chicharrón, venturosa **Habit:** Shrub, Subshrub **Use:** ME **Origin:** Native **Region:** Andean **Dept.:** ANT, BOG, CAU, CUN, HUI, NAR, NSA, SAN, TOL, VAC **Elev.:** 1200–2800 m a.s.l. **IUCN Habitat code:** 1,2,3,14 **Conservation status: (IUCN)** NE

5149 *Lantana montevidensis* (Spreng.) Briq. (Verbenaceae)
IPNI: 135260-2
Use: EU, ME, PO **Region:** Andean **Dept.:** BOG **IUCN Habitat code:** 1,14 **Conservation status: (IUCN)** NE

5150 *Lantana rugulosa* Kunth (Verbenaceae)
IPNI: 863431-1
Common name: sanguinaria, venturosa **Habit:** Shrub **Use:** HF **Origin:** Native **Region:** Andean **Dept.:** ANT, BOG, CAU, CUN, MAG, NAR, PUT, TOL, VAC **Elev.:** 650-2800 msnm **IUCN Habitat code:** 1,2,3,4,5,14 **Conservation status: (IUCN)** NE **(CPSG)** Potential LC

5151 *Lantana trifolia* L. (Verbenaceae)
IPNI: 135333-2
Common name: barba de chivo, cariaquito, cariaquito morado, carrasposa, macheco, venturosa, verbena, yerba de conejo **Habit:** Herb, Shrub, Subshrub **Use:** AF, EU, FU, GS, HF, MA, ME, SU **Origin:** Native **Region:** Andean, Caribbean, Orinoquia, Pacific **Dept.:** ANT, BOY, CAL, CAU, CHO, COR, CUN, HUI, MAG, MET, NAR, NSA, PUT, SAP, SAN, SUC, TOL, VAC **Elev.:** 20-2080 msnm **IUCN Habitat code:** 1,2,3,5,14 **Conservation status: (IUCN)** NE **(CPSG)** Potential LC

5152 *Lippia alba* (Mill.) N.E.Br. ex Britton & P.Wilson (Verbenaceae)
IPNI: 141020-2
Common name: albahaca de páramo, aliviadolor, chiva, curalotodo, curayá, maiztostado, martinica, menta americana, oreganito, oreganito de La Sierra, orégano calentano, orégano de cerro, pronto alivio, prontoalivio, quitadolor, toronjil americano, verbena **Habit:** Herb, Shrub **Use:** EU, HF, MA, ME **Origin:** Native **Region:** Amazonia, Andean, Caribbean, Orinoquia **Dept.:** AMA, ANT, BOG, CUN,

LAG, MAG, MET, SAP, SAN **Elev.:** 400-2120 msnm **IUCN Habitat code:** 1,2,3,4,5,14 **Conservation status: (IUCN)** NE **(CPSG)** Potential LC

5153 *Lippia americana* L. (Verbenaceae) **IPNI:** 863509-1 **Common name:** orégano, varita de lazo, velita, yuco **Habit:** Shrub **Use:** ME **Origin:** Native **Region:** Andean, Caribbean, Orinoquia **Dept.:** ANT, ATL, BOL, BOY, CES, LAG, MAG, MET, SAN, SUC **Elev.:** 10-1300 msnm **IUCN Habitat code:** 1,2,3,4,14 **Conservation status: (IUCN)** LC **(CPSG)** Potential LC

5154 *Lippia dulcis* Trevir. (Verbenaceae) **IPNI:** 141103-2 **Common name:** herb dulce, menta, oro azul, orosul, orosú, orozul, orozuz, orozú, orozús, prontoalivio, salvia santa **Habit:** Herb **Use:** HF, ME **Origin:** Native **Region:** Andean, Pacific **Dept.:** BOG, CAU, NAR, VAC **Elev.:** 215-1000 msnm **IUCN Habitat code:** 1,3,14 **Conservation status: (IUCN)** NE **(CPSG)** Potential LC

5155 *Lippia hirsuta* L.f. (Verbenaceae) **IPNI:** 863659-1 **Common name:** bunquín, gallinazo, sacaojo, sacaojo blanco, salvia blanca **Habit:** Shrub, Tree **Use:** ME **Origin:** Native **Region:** Andean **Dept.:** ANT, BOG, BOY, CUN, HUI, QUI **Elev.:** 520-2850 msnm **IUCN Habitat code:** 1,3,4,5,14 **Conservation status: (IUCN)** NE **(CPSG)** Potential LC

5156 *Lippia micromera* Schauer (Verbenaceae) **IPNI:** 30135678-2 **Common name:** orégano **Use:** HF, ME **Origin:** Native **IUCN Habitat code:** 2,14 **Conservation status: (IUCN)** NE

5157 *Lippia origanoides* Kunth (Verbenaceae) **IPNI:** 863752-1 **Common name:** orégano, orégano ancho, orégano cimarrón, orégano de cerro, orégano de monte **Habit:** Shrub **Use:** HF, MA, ME, PO **Origin:** Native **Region:** Andean, Caribbean **Dept.:** BOG, CAU, CUN, LAG, MAG, NAR, NSA, SAN **Elev.:** 30–2350 m a.s.l. **IUCN Habitat code:** 1,2,3,4,14 **Conservation status: (IUCN)** NE

5158 *Petrea blanchetiana* Schauer (Verbenaceae) **IPNI:** 300267-2 **Habit:** Liana **Use:** MA **Origin:** Native **Region:** Amazonia, Pacific **Dept.:** AMA, CHO, GUV, VAU **Elev.:** 100–300 m a.s.l. **IUCN Habitat code:** 1 **Conservation status: (IUCN)** NE

5159 *Petrea rugosa* Kunth (Verbenaceae) **IPNI:** 300390-2 **Common name:** azulina, azulito, chaparrillo, chaparrito, chaparro, chicharrón, copo morado, florazul, mayadena, pluma de reina, plumito **Habit:** Shrub, Small tree, Tree **Use:** EU, ME **Origin:** Native **Region:** Andean **Dept.:** ANT, BOL, CAL, CAU, COR, CUN, HUI, MAG, RIS, TOL, VAC **Elev.:** 80-1400 msnm **IUCN Habitat code:** 1,3,14 **Conservation status: (IUCN)** LC **(CPSG)** Potential LC

5160 *Petrea volubilis* L. (Verbenaceae) **IPNI:** 330591-2 **Common name:** chaparrito, chicharrón, jazmín azul, pluma de reina **Habit:** Shrub, Liana, Climbing **Use:** EU, MA, ME **Origin:** Native **Region:** Amazonia, Andean, Caribbean, Orinoquia, Pacific **Dept.:** AMA, ANT, ATL, BOL, CAL, CAQ, CES, CHO, MAG, MET, NAR, PUT, SAN, SUC, VAC **Elev.:** 0-2100 msnm **IUCN Habitat code:** 1,2,3,4,5,14 **Conservation status: (IUCN)** NE **(CPSG)** Potential LC

5161 *Phyla nodiflora* (L.) Greene (Verbenaceae) **IPNI:** 194567-2 **Common name:** yerba'emaco **Habit:** Herb **Use:** EU, GS, HF, MA, ME, SU **Origin:** Native **Region:** Andean, Caribbean, Pacific **Dept.:** ANT, ATL, BOL, CAU, CHO, COR, LAG, MAG, SAP, SAN, VAC **Elev.:** 0-900 msnm **IUCN Habitat code:** 1,2,3,5,14 **Conservation status: (IUCN)** LC **(CPSG)** Potential LC

5162 *Priva lappulacea* (L.) Pers. (Verbenaceae) **IPNI:** 1194372-2 **Common name:** albahaca brasilera, cadillo, cadillo bola, cadillo de bolsa, cadillo pegajoso, florella, marrubio, sanalotodo **Habit:** Herb **Use:** ME **Origin:** Native **Region:** Andean, Caribbean **Dept.:** ANT, BOL, COR, CUN, HUI, MAG, SAP, SAN, SUC, TOL, VAC **Elev.:** 20-1200 msnm **IUCN Habitat code:** 1,2,3,4,5,14 **Conservation status: (IUCN)** NE **(CPSG)** Potential LC

5163 *Stachytarpheta cayennensis* (Rich.) Vahl (Verbenaceae) **IPNI:** 1126344-2 **Common name:** caraqueña, cola de armadillo, escobo macho, mata de quincha, rabo de zorra, rabo de zorro, rinchao, sacha verbena, verbena, verbena blanca, verbena macha, verbena macho, verbena morada, verbena negra, verbenilla, virginia **Habit:** Herb, Shrub, Subshrub **Use:** EU, HF, MA, ME, SU **Origin:** Native **Region:** Amazonia, Andean, Caribbean, Orinoquia, Pacific **Dept.:** AMA, ANT, BOY, CAL, CAU, CHO, CUN, HUI, MAG, MET, NAR, NSA, QUI, RIS, SAN, TOL, VAC **Elev.:** 0-1800 msnm **IUCN Habitat code:** 1,2,3,4,5,14 **Conservation status: (IUCN)** NE **(CPSG)** Potential LC

5164 *Stachytarpheta indica* (L.) Vahl (Verbenaceae) **IPNI:** 864749-1 **Habit:** Herb **Use:** AF, EU, GS, HF, MA, ME, SU **Origin:** Native **Region:** Caribbean, Orinoquia **Dept.:** COR, VID **Elev.:** 120–120 m a.s.l. **IUCN Habitat code:** 2,3,14 **Conservation status: (IUCN)** LC

5165 *Stachytarpheta jamaicensis* (L.) Vahl (Verbenaceae) **IPNI:** 1092351-2 **Common name:** divina verbena, verbena negra **Habit:** Herb **Use:** AF, EU, GS, HF, MA, ME, SU **Origin:** Native **Region:** Caribbean, Pacific **Dept.:** CHO, SAP **Elev.:** 100-605 msnm **IUCN Habitat code:** 1,3,5,14 **Conservation status: (IUCN)** NE **(CPSG)** Potential LC

5166 *Stachytarpheta mutabilis* (Jacq.) Vahl (Verbenaceae) **IPNI:** 243385-2 **Common name:** petunia de monte, rabo de alacrán, rabo de zorro **Habit:** Herb, Shrub **Use:** EU, HF, MA, ME, PO **Origin:** Native **Region:** Andean, Orinoquia **Dept.:** ANT, BOL, CAS, MAG, NAR, NSA, SAN **Elev.:** 300–1800 m a.s.l. **IUCN Habitat code:** 1,2,3,14 **Conservation status: (IUCN)** NE

5167 *Stachytarpheta orubica* (L.) Vahl (Verbenaceae) **IPNI:** 864784-1 **Habit:** Herb, Shrub **Use:** ME **Origin:** Native **Region:** Andean, Caribbean **Dept.:** BOL, CES, LAG, MAG **Elev.:** 0–1130 m a.s.l. **IUCN Habitat code:** 1,2,14 **Conservation status: (IUCN)** NE

5168 *Stachytarpheta sprucei* Moldenke (Verbenaceae) **IPNI:** 243419-2 **Habit:** Shrub **Use:** ME **Origin:** Native **Region:** Orinoquia **Dept.:** CAS **Elev.:** 500–550 m a.s.l. **IUCN Habitat code:** 1,3,4,14 **Conservation status: (IUCN)** NE

5169 *Stachytarpheta straminea* Moldenke (Verbenaceae) **IPNI:** 243421-2 **Common name:** amor de madre, verbena, verbena negra **Habit:** Subshrub **Use:** ME **Origin:** Native **Region:** Andean, Pacific **Dept.:** ANT, CHO, CUN **Elev.:** 10–1800 m a.s.l. **IUCN Habitat code:** 1,3,14 **Conservation status: (IUCN)** NE

5170 *Tamonea spicata* Aubl. (Verbenaceae) **IPNI:** 864913-1 **Habit:** Subshrub **Use:** HF, ME, PO **Origin:** Native **Region:** Amazonia, Orinoquia **Conservation status: (IUCN)** NE

5171 *Verbena × hybrida* Groenland & Rümpler (Verbenaceae) **IPNI:** 330557-2 **Common name:** verbena **Habit:** Herb **Use:** EU **Origin:** Endemic, Cultivated **Region:** Andean **Dept.:** ANT, CAU, SAN **Elev.:** 2040–2650 m a.s.l. **Conservation status: (IUCN)** NE

5172 *Verbena elegans* Kunth (Verbenaceae) **IPNI:** 263032-2 **Habit:** Herb, Subshrub **Use:** EU **Origin:** Cultivated **Region:** Andean **Dept.:** BOY, CUN **Elev.:** 2670–3000 m a.s.l. **Conservation status: (IUCN)** NE

5173 *Verbena glabrata* Kunth (Verbenaceae) **IPNI:** 865182-1 **Habit:** Herb **Use:** ME **Origin:** Native **Region:** Andean **Dept.:** CAU, NAR, PUT **Elev.:** 1100–4000 m a.s.l. **Conservation status: (IUCN)** NE

5174 *Verbena hispida* Ruiz & Pav. (Verbenaceae) **IPNI:** 865215-1 **Common name:** espina de raya, verbena, verbena chiquita, verbena del campo **Habit:** Herb **Use:** EU, HF, ME **Origin:** Native **Region:** Andean **Dept.:** ANT, BOG, BOY, CAU, CUN, HUI, VAC **Elev.:** 1700–3015 m a.s.l. **IUCN Habitat code:** 1,3,4,14 **Conservation status: (IUCN)** NE

5175 *Verbena litoralis* Kunth (Verbenaceae) **IPNI:** 30007120-2 **Common name:** catano, valbuena, verbena, verbena blanca, verbena negra, verbena verde **Habit:** Herb, Shrub, Subshrub **Use:** HF, ME **Origin:** Native **Region:** Amazonia, Andean, Caribbean, Orinoquia, Pacific **Dept.:** ANT, BOG, BOY, CAU, CHO, CUN, HUI, MAG, MET, NAR, NSA, PUT, QUI, RIS, SAN, TOL, VAC **Elev.:** 341-2880 msnm **IUCN Habitat code:** 1,2,3,4,5,14 **Conservation status: (IUCN)** NE **(CPSG)** Potential LC

5176 *Verbena officinalis* L. (Verbenaceae) **IPNI:** 330554-2 **Habit:** Herb **Use:** EU, FU, HF, MA, ME, PO, SU **Origin:** Cultivated **Region:** Andean **Dept.:** ANT, BOG

Elev.: 1300–1400 m a.s.l. **IUCN Habitat code:** 1,3,14 **Conservation status: (IUCN)** NE

5177 *Verbena valerianoides* Kunth (Verbenaceae) **IPNI:** 865542-1 **Common name:** verbena, verbena negra **Habit:** Herb **Use:** ME **Origin:** Endemic **Region:** Andean **Dept.:** BOG, BOY, CUN **Elev.:** 2650–3250 m a.s.l. **IUCN Habitat code:** 4,14 **Conservation status: (IUCN)** NE

Malpighiales

5178 *Carpotroche amazonica* Mart. ex Eichler (Achariaceae) **IPNI:** 365030-1 **Common name:** achotillo crespo, cocaito, mataconga, palo de raya, tablón, upvonia **Habit:** Small tree **Use:** HF, ME, PO **Origin:** Native **Region:** Amazonia **Dept.:** AMA, CAQ, PUT, VAU **Elev.:** 240–240 m a.s.l. **IUCN Habitat code:** 1,5 **Conservation status: (IUCN)** NE

5179 *Carpotroche grandiflora* Spruce ex Eichler (Achariaceae) **IPNI:** 111509-1 **Common name:** achiote de venado, cacao, guaco, huevo de gato, huevo de morrocoyo, manzana de venado, morrocoy, puercoespín, táparo baudoseño **Habit:** Shrub, Small tree, Tree **Use:** AF, HF, ME **Origin:** Native **Region:** Amazonia, Guiana Shield **Dept.:** GUA, VAU **Elev.:** 150–300 m a.s.l. **IUCN Habitat code:** 1,3 **Conservation status: (IUCN)** NE

5180 *Carpotroche integrifolia* Kuhlm. (Achariaceae) **IPNI:** 47696-2 **Common name:** peinemono, tabaco de serpiente **Habit:** Shrub, Small tree **Use:** HF **Origin:** Native **Region:** Amazonia **Dept.:** AMA **Elev.:** 240–240 m a.s.l. **IUCN Habitat code:** 1,5 **Conservation status: (IUCN)** NE

5181 *Carpotroche longifolia* (Poepp.) Benth. (Achariaceae) **IPNI:** 111511-1 **Common name:** ape-takii, cabello de ángel, cacahuillo, cacao, cacao blanco, cacao de monte, cacaíto, coirdeimiento, cumala, grasoso, herb del cáncer, huira guayo cacar-blanco, pelotillo, yerba de cáncer **Habit:** Shrub, Small tree, Tree **Use:** HF, MA, ME, PO **Origin:** Native **Region:** Amazonia, Caribbean, Orinoquia, Pacific **Dept.:** AMA, ANT, CAQ, CHO, PUT, VAU **Elev.:** 0–1200 m a.s.l. **IUCN Habitat code:** 1,3,4,5,14 **Conservation status: (IUCN)** LC

5182 *Carpotroche pacifica* (Cuatrec.) Cuatrec. (Achariaceae) **IPNI:** 47701-2 **Common name:** pacó **Habit:** Shrub, Small tree, Tree **Use:** HF, MA, ME **Origin:** Endemic **Region:** Amazonia, Pacific **Dept.:** AMA, CAU, CHO, VAC **Elev.:** 0–350 m a.s.l. **IUCN Habitat code:** 1,5,14 **Conservation status: (IUCN)** NE

5183 *Carpotroche platyptera* Pittier (Achariaceae) **IPNI:** 47704-2 **Common name:** grape point **Habit:** Shrub, Small tree **Use:** HF, ME **Origin:** Native **Region:** Pacific **Dept.:** CHO, VAC **Elev.:** 70–70 m a.s.l. **IUCN Habitat code:** 1 **Conservation status: (IUCN)** LC

5184 *Carpotroche ramosii* (Cuatrec.) Cuatrec. (Achariaceae) **IPNI:** 47705-2 **Habit:** Small tree, Tree **Use:** ME **Origin:** Native **Region:** Pacific **Dept.:** CAU, NAR **Elev.:** 5–50 m a.s.l. **IUCN Habitat code:** 14 **Conservation status: (IUCN)** NE

5185 *Carpotroche surinamensis* Uittien (Achariaceae) **IPNI:** 47706-2 **Habit:** Small tree **Use:** HF **Origin:** Native **Region:** Amazonia, Guiana Shield **Dept.:** AMA **Elev.:** 100–100 m a.s.l. **Conservation status: (IUCN)** LC

5186 *Hydnocarpus castaneus* Hook.f. & Thomson (Achariaceae) **IPNI:** 111741-1 **Habit:** Shrub, Small tree, Tree **Use:** AF, EU, FU, HF, MA, ME **Origin:** Cultivated **Dept.:** CUN **Elev.:** 350–350 m a.s.l. **IUCN Habitat code:** 3 **Conservation status: (IUCN)** NE

5187 *Hydnocarpus heterophyllus* Blume (Achariaceae) **IPNI:** 111747-1 **Use:** AF, HF, ME **Conservation status: (IUCN)** NE

5188 *Hydnocarpus macrocarpus* (Bedd.) Warb. (Achariaceae) **IPNI:** 111756-1 **Use:** HF, ME **Conservation status: (IUCN)** NE

5189 *Lindackeria laurina* C.Presl (Achariaceae) **IPNI:** 111811-1 **Common name:** carbonero, erizo, guayabo amarillo, huesito, mataganado, platero **Habit:** Small tree, Tree **Use:** FU, MA, ME **Origin:** Native **Region:** Amazonia, Caribbean, Orinoquia, Pacific **Dept.:** AMA, ANT, BOL, BOY, CAQ, CHO, COR, MAG, MET, SAN **Elev.:** 0–1000 m a.s.l. **IUCN Habitat code:** 1,2,3,5,14 **Conservation status: (IUCN)** LC

5190 *Lindackeria paludosa* (Benth.) Gilg (Achariaceae) **IPNI:** 140493-2 **Common name:** achiotillo, higuerillo, huacapú, huesillo, huichu caspi **Habit:** Shrub, Small tree, Tree **Use:** HF, ME **Origin:** Native **Region:** Amazonia, Guiana Shield, Orinoquia **Dept.:** AMA, CAQ, GUV, MET, PUT, VAU, VID **Elev.:** 100–500 m a.s.l. **IUCN Habitat code:** 1,3,4,5,14 **Conservation status: (IUCN)** LC

5191 *Mayna grandifolia* (H.Karst. & Triana) Warb. (Achariaceae) **IPNI:** 155434-2 **Common name:** manzana de venado **Habit:** Shrub, Small tree, Tree **Use:** AF, FU, HF, MA, ME **Origin:** Native **Region:** Amazonia, Guiana Shield, Caribbean, Pacific **Dept.:** AMA, ANT, ATL, BOL, CHO, COR, HUI, MAG, PUT, SAN, SUC **Elev.:** 0–1950 m a.s.l. **IUCN Habitat code:** 1,2,3,14 **Conservation status: (IUCN)** LC

5192 *Mayna odorata* Aubl. (Achariaceae) **IPNI:** 111855-1 **Common name:** borojó de monte, eriza café, erizo, erizo café, papaya de gallinazo **Habit:** Shrub, Small tree, Tree **Use:** FU, HF, ME **Origin:** Native **Region:** Amazonia, Guiana Shield, Caribbean, Orinoquia, Pacific **Dept.:** AMA, ANT, CAQ, CHO, CUN, MAG, MET, PUT, SAN, VAU, VID **Elev.:** 0–1550 m a.s.l. **IUCN Habitat code:** 1,2,3,4,5,14 **Conservation status: (IUCN)** LC

5193 *Mayna pubescens* (H.Karst. & Triana) Warb. (Achariaceae) **IPNI:** 155451-2 **Habit:** Shrub, Small tree **Use:** ME **Origin:** Endemic **Region:** Andean, Pacific **Dept.:** CHO, CUN, VAC **Elev.:** 1000-2000 msnm **IUCN Habitat code:** 14 **Conservation status: (IUCN)** EN **(CPSG)** EN

5194 *Mayna suaveolens* (H.Karst. & Triana) Warb. (Achariaceae) **IPNI:** 155453-2 **Common name:** madroñito **Habit:** Tree **Use:** ME **Origin:** Native **Region:** Pacific **Dept.:** ANT, CHO, CUN, TOL **Elev.:** 300-1000 msnm **IUCN Habitat code:** 1,3 **Conservation status: (IUCN)** EN **(CPSG)** EN

5195 *Bonnetia paniculata* Spruce ex Benth. (Bonnetiaceae) **IPNI:** 828487-1 **Habit:** Shrub, Small tree **Use:** ME **Origin:** Native **Region:** Amazonia, Guiana Shield **Dept.:** CAQ **Elev.:** 250–300 m a.s.l. **IUCN Habitat code:** 1,3,14 **Conservation status: (IUCN)** LC

5196 *Calophyllum brasiliense* Cambess. (Calophyllaceae) **IPNI:** 427115-1 **Common name:** aceite, aceite de María, aceite de maría, aceite maría, barcino, bario, bálsamo de María, cachicamo, calaba, caucho, cerezo, chagualo, charrado, chicharrón, costillo, galba, lagarto, mario, maría, marío, palo de María, palo de aceite, palo maría, árbol de **Habit:** Shrub, Small tree, Tree **Use:** AF, FU, HF, MA, ME **Origin:** Native **Region:** Amazonia, Andean, Guiana Shield, Caribbean, Orinoquia, Pacific **Dept.:** AMA, ANT, CAQ, CAS, CES, GUA, GUV, MET, NAR, NSA, QUI, RIS, SAP, TOL, VAC, VAU, VID **Elev.:** 0-2300 msnm **IUCN Habitat code:** 1,2,3,4,5,14 **Conservation status: (IUCN)** LC **(CPSG)** Potential LC

5197 *Calophyllum calaba* L. (Calophyllaceae) **IPNI:** 427119-1 **Common name:** maría **Use:** HF, MA, ME, PO **IUCN Habitat code:** 1 **Conservation status: (IUCN)** NE

5198 *Calophyllum inophyllum* L. (Calophyllaceae) **IPNI:** 427190-1 **Habit:** Tree **Use:** AF, EU, HF, IF, MA, ME, PO, SU **Origin:** Cultivated **Region:** Andean, Caribbean, Pacific **Dept.:** ANT, SAP **Elev.:** 0–1550 m a.s.l. **IUCN Habitat code:** 1,3,5,14 **Conservation status: (IUCN)** LC

5199 *Calophyllum longifolium* Willd. (Calophyllaceae) **IPNI:** 427212-1 **Common name:** aceite, aceite maría, cuen, marío, yacaeruba **Habit:** Shrub, Tree **Use:** MA, ME **Origin:** Native **Region:** Amazonia, Guiana Shield, Pacific **Dept.:** AMA, CHO, GUA, VAC **Elev.:** 0–220 m a.s.l. **IUCN Habitat code:** 1,3,5,14 **Conservation status: (IUCN)** LC

5200 *Caraipa densifolia* Mart. (Calophyllaceae) **IPNI:** 828563-1 **Habit:** Tree **Use:** ME **Origin:** Native **Region:** Amazonia, Guiana Shield **Dept.:** AMA, CAQ, GUA, GUV, MET **Elev.:** 220–500 m a.s.l. **IUCN Habitat code:** 1,3,4,5,14 **Conservation status: (IUCN)** LC

5201 *Caraipa grandifolia* Mart. (Calophyllaceae) **IPNI:** 828566-1 **Habit:** Tree **Use:** ME **Origin:** Native **Region:** Amazonia, Guiana Shield **Dept.:** AMA, CAQ, GUA, VAU **Elev.:** 100–100 m a.s.l. **IUCN Habitat code:** 1 **Conservation status: (IUCN)** LC

5202 *Caraipa llanorum* Cuatrec. (Calophyllaceae) **IPNI:** 45033-2 **Common name:** alcornoque, saladillo, saladillo colorado, saladillo colorao, saladillo rojo, árbol de pilón **Habit:** Shrub, Small tree, Tree **Use:** MA **Origin:** Native **Region:** Amazonia, Guiana Shield, Orinoquia **Dept.:** ARA, BOY, CAQ, CAS, GUA, MET, VID **Elev.:** 90–300 m a.s.l. **IUCN Habitat code:** 1,2,3,4,5,14 **Conservation status: (IUCN)** NE

5203 *Caraipa parvielliptica* Cuatrec. (Calophyllaceae) **IPNI:** 45044-2 **Habit:** Tree **Use:** ME **Origin:** Native **Region:** Guiana Shield **Dept.:** CAQ, GUV, VAU **Elev.:** 200–700 m a.s.l. **IUCN Habitat code:** 1,4,14 **Conservation status: (IUCN)** NE

5204 *Haploclathra paniculata* Benth. (Calophyllaceae) **IPNI:** 828939-1 **Use:** MA **Origin:** Native **IUCN Habitat code:** 1 **Conservation status: (IUCN)** NE

5205 *Mammea americana* L. (Calophyllaceae) **IPNI:** 428743-1 **Common name:** albaricoque, lechero amarillo, lucumo, mamei, mamey, mamey colorado, mamey de cartagena, zapote **Habit:** Tree **Use:** AF, EU, GS, HF, MA, ME, PO, SU **Origin:** Cultivated **Region:** Andean, Caribbean, Pacific **Dept.:** ANT, BOL, CAL, CUN, LAG, SUC, VAC **Elev.:** 100–1670 m a.s.l. **IUCN Habitat code:** 1,2,3,14 **Conservation status: (IUCN)** NE

5206 *Marila tomentosa* Poepp. (Calophyllaceae) **IPNI:** 829109-1 **Common name:** lacre **Habit:** Tree **Use:** ME **Origin:** Native **Region:** Amazonia, Andean **Dept.:** AMA, ANT, CAQ, GUV, MET, PUT **Elev.:** 109–1300 m a.s.l. **IUCN Habitat code:** 1,3,14 **Conservation status: (IUCN)** LC

5207 *Anthodiscus amazonicus* Gleason & A.C.Sm. (Caryocaraceae) **IPNI:** 15001-2 **Common name:** acapú macho, barbasco, castaño **Habit:** Tree **Use:** ME **Origin:** Native **Region:** Amazonia, Guiana Shield **Dept.:** AMA, GUA, VAU **Elev.:** 100–100 m a.s.l. **IUCN Habitat code:** 1 **Conservation status: (IUCN)** LC

5208 *Anthodiscus mazarunensis* Gilly (Caryocaraceae) **IPNI:** 15006-2 **Habit:** Small tree **Use:** HF **Origin:** Native **Region:** Guiana Shield **Dept.:** CAQ **Elev.:** 100–100 m a.s.l. **IUCN Habitat code:** 1,3 **Conservation status: (IUCN)** NE

5209 *Anthodiscus pilosus* Ducke (Caryocaraceae) **IPNI:** 15009-2 **Habit:** Tree **Use:** ME **Origin:** Native **Region:** Amazonia, Guiana Shield **Dept.:** AMA, CAQ, VAU **Elev.:** 100–100 m a.s.l. **IUCN Habitat code:** 1,3,4,14 **Conservation status: (IUCN)** NE

5210 *Caryocar amygdaliferum* Mutis (Caryocaraceae) **IPNI:** 47856-2 **Common name:** achiotillo, achiotillo mani chalmugra, achotillo, almendrillo, almendrón, almendrón de mariquita, cagüi, cagüí, caqui, chalmagra, chalmugra, genene, genené, jenené, jenine, maní **Habit:** Tree **Use:** HF, MA, ME **Origin:** Endemic **Region:** Pacific **Dept.:** ANT, BOL, CHO, SAN **Elev.:** 50-900 msnm **IUCN Habitat code:** 1,3,5,14 **Conservation status: (IUCN)** NE **(CPSG)** VU

5211 *Caryocar glabrum* (Aubl.) Pers. (Caryocaraceae) **IPNI:** 828599-1 **Common name:** almendra, almendro, almendrón, barbasco, barbasco de monte, barbasco de pepa grande, cagüi, cahuí, caimo buñuelo, castaña, castaña de mico, castaña espinosa, castañito, castaño, palo de trueno venenoso, palo de veneno, peine de abuela **Habit:** Tree **Use:** AF, FU, HF, MA, ME, PO **Origin:** Native **Region:** Amazonia **Dept.:** AMA, ANT, VAU **Elev.:** 100–1000 m a.s.l. **IUCN Habitat code:** 1,3,5,14 **Conservation status: (IUCN)** LC

5212 *Caryocar gracile* Wittm. (Caryocaraceae) **IPNI:** 47871-2 **Common name:** almendra, barbasco, barbasco mediano, palo barbasco **Habit:** Small tree **Use:** HF, ME **Origin:** Native **Region:** Amazonia **Dept.:** AMA, GUA, VAU **Elev.:** 200–700 m a.s.l. **IUCN Habitat code:** 1,3,14 **Conservation status: (IUCN)** LC

5213 *Caryocar harlingii* Prance & Encarn. (Caryocaraceae) **IPNI:** 274157-2 **Habit:** Small tree **Use:** HF **Origin:** Native **Region:** Amazonia **Dept.:** AMA, VAU **Elev.:** 100–100 m a.s.l. **IUCN Habitat code:** 1 **Conservation status: (IUCN)** NE

5214 *Caryocar microcarpum* Ducke (Caryocaraceae) **IPNI:** 47874-2 **Common name:** barbasco del río, castaña **Habit:** Tree **Use:** HF, MA, ME **Origin:** Native **Region:** Amazonia, Guiana Shield, Orinoquia **Dept.:** AMA, CAQ, GUV, VAU, VID **Elev.:** 100–250 m a.s.l. **IUCN Habitat code:** 1,3,4,5,14 **Conservation status: (IUCN)** LC

5215 *Caryocar nuciferum* L. (Caryocaraceae) **IPNI:** 828602-1 **Common name:** barbasco **Habit:** Tree **Use:** HF, MA, ME **Origin:** Native **Region:** Pacific **Dept.:** ANT, CHO **Elev.:** 60–700 m a.s.l. **IUCN Habitat code:** 1 **Conservation status: (IUCN)** DD

5216 *Caryocar pallidum* A.C.Sm. (Caryocaraceae) **IPNI:** 47876-2 **Common name:** barbasco **Habit:** Tree **Use:** HF **Origin:** Native **Region:** Amazonia, Guiana Shield **Dept.:** AMA, CAQ, VAU **Elev.:** 100–200 m a.s.l. **IUCN Habitat code:** 1 **Conservation status: (IUCN)** NE

5217 *Acioa dolichopoda* (Prance) Sothers & Prance (Chrysobalanaceae) **IPNI:** 77141743-1 **Common name:** aguire, agüire, algarrobo, ayuirai, ayurai, cenizo, huevo de danta, huevos de danta **Habit:** Tree **Use:** HF **Origin:** Native **Region:** Amazonia **Dept.:** AMA, CAQ **Elev.:** 100-300 msnm **IUCN Habitat code:** 1,5 **Conservation status: (IUCN)** NE **(CPSG)** NT

5218 *Acioa longipendula* (Pilg.) Sothers & Prance (Chrysobalanaceae) **IPNI:** 77141742-1 **Use:** HF, MA, ME **IUCN Habitat code:** 1 **Conservation status: (IUCN)** NE

5219 *Chrysobalanus icaco* L. (Chrysobalanaceae) **IPNI:** 722181-1 **Common name:** ciruela de algodón, cocoa plum, hicaco, icaco **Habit:** Shrub, Tree **Use:** AF, EU, FU, GS, HF, IF, MA, ME **Origin:** Native **Region:** Caribbean, Pacific **Dept.:** ANT, ATL, BOL, CHO, COR, CUN, HUI, MAG, NAR, SAP, SAN, SUC, TOL, VAC **Elev.:** 0-550 msnm **IUCN Habitat code:** 1,2,3,4,5,14 **Conservation status: (IUCN)** LC **(CPSG)** Potential LC

5220 *Couepia bracteosa* Benth. (Chrysobalanaceae) **IPNI:** 722743-1 **Habit:** Tree **Use:** HF **Origin:** Native **Region:** Guiana Shield **Dept.:** VAU **Elev.:** 200-200 msnm **IUCN Habitat code:** 1,3 **Conservation status: (IUCN)** NE **(CPSG)** NT

5221 *Couepia chrysocalyx* (Poepp.) Benth. ex Hook.f. (Chrysobalanaceae) **IPNI:** 722748-1 **Common name:** corderillo, culefierro, guaco **Habit:** Tree **Use:** HF **Origin:** Native **Region:** Amazonia, Guiana Shield **Dept.:** AMA, CAQ, CAU, GUA, PUT, VAU **Elev.:** 110-420 msnm **IUCN Habitat code:** 1,5,14 **Conservation status: (IUCN)** LC **(CPSG)** LC

5222 *Couepia guianensis* Aubl. (Chrysobalanaceae) **IPNI:** 67154-2 **Common name:** guacure de chimbe, guaro, salao **Habit:** Shrub, Tree **Use:** HF, MA **Origin:** Native **Region:** Amazonia, Guiana Shield, Orinoquia, Pacific **Dept.:** AMA, ANT, CAQ, CHO, GUA, MET, SAN, VAU **Elev.:** 20-600 msnm **IUCN Habitat code:** 1,3,4,5,14 **Conservation status: (IUCN)** LC **(CPSG)** LC

5223 *Couepia macrophylla* Spruce ex Hook.f. (Chrysobalanaceae) **IPNI:** 722774-1 **Habit:** Shrub, Small tree, Tree **Use:** HF **Origin:** Native **Region:** Amazonia **Dept.:** AMA **Elev.:** 100–120 m a.s.l. **IUCN Habitat code:** 1 **Conservation status: (IUCN)** LC

5224 *Couepia obovata* Ducke (Chrysobalanaceae) **IPNI:** 67170-2 **Common name:** curiza,vacùcù **Habit:** Tree **Use:** HF **Origin:** Native **Region:** Guiana Shield **Dept.:** VAU **Elev.:** 200-200 msnm **IUCN Habitat code:** 1,3 **Conservation status: (IUCN)** NE **(CPSG)** NT

5225 *Couepia paraensis* (Mart. & Zucc.) Benth. (Chrysobalanaceae) **IPNI:** 722782-1 **Habit:** Tree **Use:** HF, MA **Origin:** Native **Region:** Orinoquia **Dept.:** ARA, GUA, VID **Elev.:** 80-110 msnm **IUCN Habitat code:** 1,2,3,4,5,14 **Conservation status: (IUCN)** LC **(CPSG)** LC

5226 *Couepia subcordata* Benth. ex Hook.f. (Chrysobalanaceae) **IPNI:** 722799-1 **Common name:** mariramba, marirana, umarí **Habit:** Tree **Use:** EU, HF, ME **Origin:** Native, Cultivated **Region:** Amazonia **Dept.:** AMA **Elev.:** 90–90 m a.s.l. **IUCN Habitat code:** 1 **Conservation status: (IUCN)** NE

5227 *Couepia ulei* Pilg. (Chrysobalanaceae) **IPNI:** 67194-2 **Common name:** palo de grulla **Habit:** Tree **Use:** HF **Origin:** Native **Region:** Amazonia **Dept.:** AMA **Elev.:** 200-200 msnm **IUCN Habitat code:** 1 **Conservation status: (IUCN)** LC **(CPSG)** LC

5228 *Hirtella americana* L. (Chrysobalanaceae) **IPNI:** 725498-1 **Common name:** carita de negro, garrapato, garrapo, guamo mestizo, moradito, pasito, querrico **Habit:** Tree **Use:** EU, HF, MA **Origin:** Native **Region:** Andean, Guiana Shield, Caribbean, Orinoquia **Dept.:** BOL, BOY, CAL, CES, COR, CUN, GUV, HUI, LAG, MAG, MET, NSA, TOL **Elev.:** 50-1450 msnm **IUCN Habitat code:** 1,2,3,4,5,14 **Conservation status: (IUCN)** LC **(CPSG)** Potential LC

5229 *Hirtella bicornis* Mart. & Zucc. (Chrysobalanaceae) **IPNI:** 725506-1 **Habit:** Tree **Use:** HF **Origin:** Native **Region:** Amazonia, Andean, Guiana Shield, Pacific **Dept.:** AMA, ANT, CAQ, CUN, GUA, SAN, VAC **Elev.:** 80-950 msnm **IUCN Habitat code:** 1,3,5,14 **Conservation status: (IUCN)** LC **(CPSG)** LC

5230 *Hirtella bullata* Benth. (Chrysobalanaceae) **IPNI:** 725510-1 **Habit:** Shrub, Tree **Use:** HF **Origin:** Native **Region:** Guiana Shield **Dept.:** AMA, GUA, VAU **Elev.:** 100-150 msnm **IUCN Habitat code:** 1,3,4,14 **Conservation status: (IUCN)** LC **(CPSG)** LC

5231 *Hirtella carbonaria* Little (Chrysobalanaceae) **IPNI:** 123102-2 **Common name:** carboncillo, carbonero, garrapato **Habit:** Shrub, Tree **Use:** HF, MA **Origin:** Native **Region:** Pacific **Dept.:** CAU, CHO, NAR, VAC **Elev.:** 0-100 msnm **IUCN Habitat code:** 1,4,5,14 **Conservation status: (IUCN)** NE **(CPSG)** LC

5232 *Hirtella pilosissima* Mart. & Zucc. (Chrysobalanaceae) **IPNI:** 725595-1 **Common name:** palo de cemento **Habit:** Shrub, Tree **Use:** HF **Origin:** Native **Region:** Amazonia, Guiana Shield, Pacific **Dept.:** AMA, COR, PUT, VAU **Elev.:** 100-200 msnm **IUCN Habitat code:** 1,3,14 **Conservation status: (IUCN)** LC **(CPSG)** LC

5233 *Hirtella racemosa* Lam. (Chrysobalanaceae) **IPNI:** 725607-1 **Common name:** arrecheche, garrapato **Habit:** Shrub, Subshrub, Small tree, Tree **Use:** HF **Origin:** Native **Region:** Amazonia, Andean, Guiana Shield, Orinoquia, Pacific **Dept.:** AMA, ANT, ARA, BOL, BOY, CAQ, CAS, CAU, CHO, GUV, MAG, MET, NAR, SAN, TOL, VAC, VAU, VID **Elev.:** 0-1400 msnm **IUCN Habitat code:** 1,2,3,4,5,14 **Conservation status: (IUCN)** LC **(CPSG)** Potential LC

5234 *Hirtella triandra* Sw. (Chrysobalanaceae) **IPNI:** 725631-1 **Common name:** carita de negro, garrapato, pasito, tobo **Habit:** Tree **Use:** HF, MA **Origin:** Native **Region:** Amazonia, Andean, Guiana Shield, Caribbean, Orinoquia, Pacific **Dept.:** AMA, ANT, CES, CHO, LAG, MAG, MET, VAC **Elev.:** 50-1060 msnm **Conservation status: (IUCN)** NE **(CPSG)** LC

5235 *Hymenopus arachnoideus* (Fanshawe & Maguire) Sothers & Prance (Chrysobalanaceae) **IPNI:** 77156954-1 **Habit:** Tree **Use:** HF **Origin:** Native **Region:** Amazonia **Dept.:** AMA **Elev.:** 200-300 msnm **IUCN Habitat code:** 1,4 **Conservation status: (IUCN)** NE **(CPSG)** LC

5236 *Hymenopus glabriflorus* (Prance) Sothers & Prance (Chrysobalanaceae) **IPNI:** 77156959-1 **Habit:** Small tree, Tree **Use:** HF **Origin:** Native **Region:** Guiana Shield **Dept.:** VAU **Elev.:** 200-200 msnm **IUCN Habitat code:** 1 **Conservation status: (IUCN)** NE **(CPSG)** NT

5237 *Hymenopus heteromorphus* (Benth.) Sothers & Prance (Chrysobalanaceae) **IPNI:** 77156960-1 **Common name:** apacharama, capinurí, cucama, guacuri danta, guacurí de danta, merecure, palo de manteca, umarí de chimbe **Habit:** Shrub, Small tree, Tree **Use:** HF, MA, ME **Origin:** Native **Region:** Amazonia, Guiana Shield, Orinoquia **Dept.:** AMA, CAQ, GUA, GUV, MET, VAU, VID **Elev.:** 80-900 msnm **IUCN Habitat code:** 1,2,3,4,5,14 **Conservation status: (IUCN)** LC **(CPSG)** LC

5238 *Hymenopus intrapetiolaris* (Spreng. ex Hook.f.) Sothers & Prance (Chrysobalanaceae) **IPNI:** 77156976-1 **Habit:** Tree **Use:** HF **Origin:** Native **Region:** Amazonia, Guiana Shield **Dept.:** AMA, GUA, VAU **Elev.:** 115-925 msnm **IUCN Habitat code:** 1,3,4,14 **Conservation status: (IUCN)** NE **(CPSG)** LC

5239 *Leptobalanus apetalus* (E.Mey.) Sothers & Prance (Chrysobalanaceae) **IPNI:** 60472826-2 **Common name:** cagüí, caue, cemento, joropenaque, mezcla, pacharama, palo de cemento, palo de manteca **Habit:** Shrub, Small tree, Tree **Use:** HF, MA, ME **Origin:** Native **Region:** Amazonia, Guiana Shield, Caribbean, Orinoquia **Dept.:** AMA, ANT, BOL, COR, GUA, MET, PUT, SAN, VAU, VID **Elev.:** 80-1000 msnm **IUCN Habitat code:** 1,2,3,4,5,14 **Conservation status: (IUCN)** NE **(CPSG)** LC

5240 *Leptobalanus longistylus* (Hook.f.) Sothers & Prance (Chrysobalanaceae) **IPNI:** 60472856-2 **Common name:** palo de cemento **Habit:** Small tree, Tree **Use:** HF **Origin:** Native **Region:** Amazonia, Orinoquia, Pacific **Dept.:** AMA, CAQ, VAC, VAU, VID **Elev.:** 50-220 msnm **IUCN Habitat code:** 1,3,5 **Conservation status: (IUCN)** NE **(CPSG)** LC

5241 *Leptobalanus octandrus* (Hoffmanns. ex Schult.) Sothers & Prance (Chrysobalanaceae) **IPNI:** 60472870-2 **Common name:** apacharama, apacharana, palo de cemento **Habit:** Small tree, Tree **Use:** MA **Origin:** Native **Region:** Amazonia, Guiana Shield, Pacific **Dept.:** AMA, ANT, CAQ, COR, MET **Elev.:** 200-780 msnm **IUCN Habitat code:** 1,3 **Conservation status: (IUCN)** NE **(CPSG)** LC

5242 *Leptobalanus sclerophyllus* (Mart. ex Hook.f.) Sothers & Prance (Chrysobalanaceae) **IPNI:** 60472875-2 **Habit:** Small tree, Tree **Use:** HF **Origin:** Native **Region:** Amazonia **Dept.:** AMA **Elev.:** 455-455 msnm **IUCN Habitat code:** 1,14 **Conservation status: (IUCN)** NE **(CPSG)** NT

5243 *Licania canescens* Benoist (Chrysobalanaceae) **IPNI:** 139732-2 **Habit:** Shrub, Tree **Use:** HF **Origin:** Native **Region:** Amazonia, Guiana Shield, Orinoquia **Dept.:** AMA, GUA, VAU, VID **Elev.:** 100–120 m a.s.l. **IUCN Habitat code:** 1,3 **Conservation status: (IUCN)** LC

5244 *Licania hebantha* Mart. ex Hook.f. (Chrysobalanaceae) **IPNI:** 726046-1 **Habit:** Tree **Use:** ME **Origin:** Native **Region:** Guiana Shield **Dept.:** AMA, CAQ, VAU **Elev.:** 110-550 msnm **Conservation status: (IUCN)** NE **(CPSG)** NT

5245 *Licania hypoleuca* Benth. (Chrysobalanaceae) **IPNI:** 726057-1 **Common name:** palo de piedra, palo piedra **Habit:** Tree **Use:** HF, MA, ME **Origin:** Native **Region:** Guiana Shield, Orinoquia, Pacific **Dept.:** ANT, CHO, MET, VAC, VAU, VID **Elev.:** 0-450 msnm **IUCN Habitat code:** 1,3,4,5,14 **Conservation status: (IUCN)** LC **(CPSG)** LC

5246 *Licania kunthiana* Hook.f. (Chrysobalanaceae) **IPNI:** 315580-2 **Habit:** Tree **Use:** HF **Origin:** Native **Region:** Amazonia, Guiana Shield, Orinoquia **Dept.:** AMA, GUA, VAU, VID **Elev.:** 100–200 m a.s.l. **IUCN Habitat code:** 1,3 **Conservation status: (IUCN)** LC

5247 *Licania micrantha* Miq. (Chrysobalanaceae) **IPNI:** 726088-1 **Common name:** apacharana, beijoropenaci, carbonero, copay, lamparina, pavito, reventillo **Habit:** Tree **Use:** HF **Origin:** Native **Region:** Amazonia, Guiana Shield, Pacific **Dept.:** AMA, CAQ, CHO, COR, GUV, VAC, VAU, VID **Elev.:** 0-300 msnm **IUCN Habitat code:** 1,3,4,5,14 **Conservation status: (IUCN)** LC **(CPSG)** LC

5248 *Licania parviflora* Benth. (Chrysobalanaceae) **IPNI:** 726118-1 **Habit:** Tree **Use:** HF, ME **Origin:** Native **Region:** Amazonia, Guiana Shield, Orinoquia **Dept.:** AMA, NSA, VAU, VID **Elev.:** 50-400 msnm **IUCN Habitat code:** 1,4,5,14 **Conservation status: (IUCN)** LC **(CPSG)** LC

5249 *Licania polita* Spruce ex Hook.f. (Chrysobalanaceae) **IPNI:** 726129-1 **Habit:** Tree **Use:** MA **Origin:** Native **Region:** Guiana Shield **Dept.:** GUA, VAU **Elev.:** 150–200 m a.s.l. **Conservation status: (IUCN)** LC

5250 *Licania steyermarkii* Maguire (Chrysobalanaceae) **IPNI:** 139891-2 **Habit:** Tree **Use:** HF **Origin:** Native **Region:** Guiana Shield **Dept.:** VAU **Elev.:** 250-250 msnm **IUCN Habitat code:** 1,3,4,14 **Conservation status: (IUCN)** NE **(CPSG)** NT

5251 *Licania triandra* Mart. ex Hook.f. (Chrysobalanaceae) **IPNI:** 726172-1 **Common name:** palo de manteca **Habit:** Tree **Use:** HF **Origin:** Native **Region:** Amazonia, Guiana Shield **Dept.:** AMA, VAU **Elev.:** 200-230 msnm **IUCN Habitat** code: 1,5,14 **Conservation status:** (IUCN) NE (CPSG) LC

5252 *Microdesmia arborea* (Seem.) Sothers & Prance (Chrysobalanaceae) **IPNI:** 77156125-1 **Common name:** cañagria, garcero, tapaliso **Habit:** Tree **Use:** HF, MA, ME **Origin:** Native **Region:** Amazonia, Caribbean, Pacific **Dept.:** ANT, BOL, CES, CHO, MAG, PUT **Elev.:** 20-230 msnm **IUCN Habitat code:** 1,2,3,5,14 **Conservation status:** (IUCN) LC (CPSG) EN

5253 *Moquilea egleri* (Prance) Sothers & Prance (Chrysobalanaceae) **IPNI:** 60472909-2 **Common name:** castaña de picure, cuyubí de rebalse, guacurí de danta, guayacán polvillo **Habit:** Small tree, Tree **Use:** HF **Origin:** Native **Region:** Amazonia, Guiana Shield **Dept.:** AMA, CAQ, PUT, VAU **Elev.:** 150-200 msnm **IUCN Habitat code:** 1 **Conservation status:** (IUCN) NE (CPSG) NT

5254 *Moquilea magnifructa* Sothers & Prance (Chrysobalanaceae) **IPNI:** 77200625-1 **Common name:** apacharama, caimito, carbonera, carbonero, carbonero hojicenizo, parinari **Habit:** Tree **Use:** HF, MA, ME **Origin:** Native **Region:** Amazonia, Pacific **Dept.:** AMA, CHO, VAC **Elev.:** 0-280 msnm **Conservation status:** (IUCN) NE (CPSG) LC

5255 *Moquilea maritima* (Prance) Sothers & Prance (Chrysobalanaceae) **IPNI:** 60472931-2 **Habit:** Tree **Use:** HF **Origin:** Endemic **Region:** Pacific **Dept.:** CHO **Elev.:** 0-100 msnm **Conservation status:** (IUCN) LC (CPSG) EN

5256 *Moquilea platypus* Hemsl. (Chrysobalanaceae) **IPNI:** 726774-1 **Common name:** carbonero, chanu, chupa, mango de monte **Habit:** Tree **Use:** HF, ME **Origin:** Native **Region:** Caribbean **Dept.:** BOL, COR **Elev.:** 0-100 msnm **IUCN Habitat code:** 1 **Conservation status:** (IUCN) LC (CPSG) EN

5257 *Moquilea pyrifolia* (Griseb.) R.O.Williams (Chrysobalanaceae) **IPNI:** 164227-2 **Common name:** merecure **Habit:** Small tree, Tree **Use:** HF, ME **Origin:** Native **Region:** Orinoquia **Dept.:** ARA, CAS, MET **Elev.:** 150-350 msnm **IUCN Habitat code:** 2,4,14 **Conservation status:** (IUCN) NE (CPSG) DD

5258 *Moquilea tomentosa* Benth. (Chrysobalanaceae) **IPNI:** 726782-1 **Common name:** aití, oití, árbol de Cúcuta **Habit:** Small tree, Tree **Use:** EU, HF, MA **Origin:** Cultivated **Region:** Andean, Caribbean **Dept.:** ARA, ATL, CES, HUI, NSA, QUI, SAN **Elev.:** 0–1500 m a.s.l. **IUCN Habitat code:** 1,2,3,4,5,14 **Conservation status:** (IUCN) NE

5259 *Parinari campestris* Aubl. (Chrysobalanaceae) **IPNI:** 727083-1 **Habit:** Tree **Use:** HF, MA, ME **Origin:** Native **Region:** Amazonia, Guiana Shield **Dept.:** AMA, GUA **Elev.:** 150-730 msnm **IUCN Habitat code:** 1,2 **Conservation status:** (IUCN) LC (CPSG) NT

5260 *Parinari excelsa* Sabine (Chrysobalanaceae) **IPNI:** 727101-1 **Common name:** guaraí **Habit:** Tree **Use:** AF, EU, FU, GS, HF, IF, MA, ME, SU **Origin:** Native **Region:** Amazonia, Guiana Shield, Orinoquia **Dept.:** AMA, CAQ, GUV, VID **Elev.:** 90-250 msnm **IUCN Habitat code:** 1,3,4,5,14 **Conservation status:** (IUCN) LC (CPSG) LC

5261 *Parinari klugii* Prance (Chrysobalanaceae) **IPNI:** 183235-2 **Common name:** palo de manteca **Habit:** Tree **Use:** HF **Origin:** Native **Region:** Amazonia **Dept.:** AMA **Elev.:** 200-200 msnm **IUCN Habitat code:** 1,14 **Conservation status:** (IUCN) NE (CPSG) NT

5262 *Parinari montana* Aubl. (Chrysobalanaceae) **IPNI:** 727140-1 **Common name:** palo tintín, toñeca, toñeca propia **Habit:** Tree **Use:** HF, MA **Origin:** Native **Region:** Amazonia **Dept.:** CAQ **Elev.:** 200-300 msnm **IUCN Habitat code:** 1,5 **Conservation status:** (IUCN) NE (CPSG) NT

5263 *Parinari pachyphylla* Rusby (Chrysobalanaceae) **IPNI:** 183245-2 **Common name:** garrapato, peregüetero, peregüétano, perehuétano, pergüétano **Habit:** Tree **Use:** HF **Origin:** Native **Region:** Andean, Caribbean, Orinoquia, Pacific **Dept.:** ANT, BOL, CES, LAG, MAG, VID **Elev.:** 20-1300 msnm **IUCN Habitat code:** 1,2,3,5,14 **Conservation status:** (IUCN) NE (CPSG) EN

5264 *Parinari parilis* J.F.Macbr. (Chrysobalanaceae) **IPNI:** 183247-2 **Common name:** palo de manteca **Habit:** Small tree, Tree **Use:** HF **Origin:** Native **Region:** Amazonia **Dept.:** AMA **Elev.:** 100-150 msnm **IUCN Habitat code:** 1,5 **Conservation status:** (IUCN) NE (CPSG) LC

5265 *Parinari sprucei* Hook.f. (Chrysobalanaceae) **IPNI:** 727170-1 **Habit:** Tree **Use:** HF **Origin:** Native **Region:** Guiana Shield **Dept.:** GUA, VAU, VID **Elev.:** 110-200 msnm **IUCN Habitat code:** 1,4 **Conservation status:** (IUCN) LC (CPSG) LC

5266 *Parinariopsis licaniiflora* (Sagot) Sothers & Prance (Chrysobalanaceae) **IPNI:** 77156118-1 **Common name:** guaré, jinisico, palo de cemento **Habit:** Shrub, Small tree, Tree **Use:** HF **Origin:** Native **Region:** Amazonia **Dept.:** AMA, CAQ **Elev.:** 100-100 msnm **Conservation status:** (IUCN) NE (CPSG) NT

5267 *Chrysochlamys colombiana* (Cuatrec.) Cuatrec. (Clusiaceae) **IPNI:** 57606-2 **Common name:** ajo de loma, cacao, chagualo, crucero, cucharo, masato, rapabarbo **Habit:** Shrub, Small tree, Tree **Use:** AF, EU, FU, MA **Origin:** Native **Region:** Andean **Dept.:** ANT, BOY, CAL, CAU, CES, CHO, CUN, HUI, MAG, MET, NAR, NSA, PUT, QUI, RIS, SAN, TOL, VAC **Elev.:** 500-2900 msnm **IUCN Habitat code:** 1,3,14 **Conservation status:** (IUCN) NE (CPSG) LC

5268 *Chrysochlamys dependens* Planch. & Triana (Clusiaceae) **IPNI:** 427393-1 **Common name:** chagualo, coloradito, impamo, negrito, rapabarba, rapabarbo, salcillo, zanca de araña, zanco de araña **Habit:** Shrub, Small tree, Tree **Use:** HF, MA, ME **Origin:** Native **Region:** Amazonia, Andean, Guiana Shield **Dept.:** ANT, CAL, CHO, CUN, NAR, NSA, PUT, QUI, SAN, TOL, VAC, VAU **Elev.:** 80-2250 msnm **IUCN Habitat code:** 1,2,3,14 **Conservation status:** (IUCN) LC (CPSG) LC

5269 *Chrysochlamys tenuifolia* Cuatrec. (Clusiaceae) **IPNI:** 57626-2 **Common name:** caraña **Habit:** Shrub, Small tree, Tree **Use:** HF **Origin:** Native **Region:** Andean **Dept.:** ANT, CAQ, CAU, PUT **Elev.:** 300–1100 m a.s.l. **IUCN Habitat code:** 1,3,14 **Conservation status:** (IUCN) LC

5270 *Chrysochlamys weberbaueri* Engl. (Clusiaceae) **IPNI:** 57628-2 **Common name:** icainé **Habit:** Shrub, Small tree, Tree **Use:** HF **Origin:** Native **Region:** Amazonia, Andean, Guiana Shield **Dept.:** AMA, ANT, CAQ, PUT, VAU **Elev.:** 100–2000 m a.s.l. **IUCN Habitat code:** 1,3,5,14 **Conservation status:** (IUCN) LC

5271 *Clusia alata* Planch. & Triana (Clusiaceae) **IPNI:** 427426-1 **Common name:** cape, cape grande, chagualo, chagualón, copey, copé, cucharo, gaque **Habit:** Shrub, Tree, Epiphyte **Use:** AF, EU, MA, ME, SU **Origin:** Native **Region:** Andean **Dept.:** ANT, BOG, BOL, BOY, CAU, CES, CUN, HUI, LAG, MAG, NAR, NSA, QUI, SAN, TOL, VAC **Elev.:** 1100-3500 msnm **IUCN Habitat code:** 1,3,4,14 **Conservation status:** (IUCN) LC (CPSG) LC

5272 *Clusia amazonica* Planch. & Triana (Clusiaceae) **IPNI:** 427431-1 **Common name:** capey, copé, ka-hee-wa-ka, matapalo, pap-ká, ree - ka-ne'-to-ma, rica-netoma **Habit:** Shrub, Tree, Epiphyte, Liana, Climbing **Use:** ME **Origin:** Native **Region:** Amazonia, Andean, Guiana Shield, Orinoquia, Pacific **Dept.:** AMA, ANT, BOY, CAQ, CAU, CHO, HUI, MET, PUT, VAC, VAU **Elev.:** 30–1300 m a.s.l. **IUCN Habitat code:** 1,3,5,14 **Conservation status:** (IUCN) LC

5273 *Clusia chiribiquetensis* Maguire (Clusiaceae) **IPNI:** 61404-2 **Common name:** copé **Habit:** Shrub, Small tree, Tree **Use:** ME **Origin:** Native **Region:** Guiana Shield **Dept.:** CAQ, VAU **Elev.:** 100-645 msnm **IUCN Habitat code:** 1,3,4,14 **Conservation status:** (IUCN) LC (CPSG) LC

5274 *Clusia columnaris* Engl. (Clusiaceae) **IPNI:** 61412-2 **Common name:** boriche, chagualo, cucharo, dicadá, gaque, matapalo, palo blanco, ricaré-tumisi-nama **Habit:** Shrub, Small tree, Tree, Epiphyte, Liana, Climbing **Use:** ME **Origin:** Native **Region:** Amazonia, Andean, Guiana Shield, Orinoquia **Dept.:** AMA, ANT, ARA, BOY, CAQ, CAU, CUN, GUA, GUV, MET, SAN, TOL, VAC, VAU, VID **Elev.:** 0–1100 m a.s.l. **IUCN Habitat code:** 1,2,3,4,5,14 **Conservation status:** (IUCN) LC

5275 *Clusia dixonii* Little (Clusiaceae) **IPNI:** 61440-2 **Common name:** mandurillo blanco **Habit:** Shrub, Tree, Epiphyte **Use:** HF **Origin:** Native **Region:** Andean **Dept.:** ANT, CHO, NAR, PUT **Elev.:** 500-2200 msnm **IUCN Habitat code:** 1,3,14 **Conservation status: (IUCN)** NE **(CPSG)** LC

5276 *Clusia elliptica* Kunth (Clusiaceae) **IPNI:** 427499-1 **Common name:** gaque **Habit:** Small tree, Tree, Epiphyte **Use:** AF, EU, FU, MA **Origin:** Native **Region:** Andean **Dept.:** ANT, BOG, BOY, CUN, HUI, SAN **Elev.:** 2550-3600 msnm **IUCN Habitat code:** 1,3,4,14 **Conservation status: (IUCN)** LC **(CPSG)** LC

5277 *Clusia ellipticifolia* Cuatrec. (Clusiaceae) **IPNI:** 304204-2 **Common name:** cape, copé, gaque, incienso **Habit:** Shrub, Small tree, Tree **Use:** ME **Origin:** Endemic **Region:** Andean **Dept.:** ANT, BOG, CAL, CAU, CES, CHO, CUN, HUI, QUI, SAN, TOL, VAC **Elev.:** 1000–3200 m a.s.l. **IUCN Habitat code:** 1,3,14 **Conservation status: (IUCN)** NE

5278 *Clusia eugenioides* Planch. & Linden (Clusiaceae) **IPNI:** 427503-1 **Habit:** Shrub, Small tree, Tree **Use:** ME **Origin:** Native **Region:** Andean **Dept.:** BOY, LAG, MAG, SAN, TOL **Elev.:** 750-1900 msnm **IUCN Habitat code:** 1,2,3,14 **Conservation status: (IUCN)** LC **(CPSG)** LC

5279 *Clusia flavida* (Benth.) Pipoly (Clusiaceae) **IPNI:** 1016004-2 **Common name:** fruto de lora, impa **Habit:** Shrub, Tree, Epiphyte, Hemiepiphyte **Use:** ME **Origin:** Native **Region:** Amazonia, Andean, Guiana Shield, Orinoquia, Pacific **Dept.:** AMA, ANT, BOY, CAQ, CHO, CUN, GUA, MET, PUT, VAC, VAU, VID **Elev.:** 10–1480 m a.s.l. **IUCN Habitat code:** 1,3,4,5,14 **Conservation status: (IUCN)** LC

5280 *Clusia grandiflora* Splitg. (Clusiaceae) **IPNI:** 427524-1 **Common name:** cape, cucharo, gaque, gaque rosado, guaque, madroño, sombrero **Habit:** Tree, Epiphyte, Climbing **Use:** HF, ME **Origin:** Native **Region:** Amazonia, Guiana Shield **Dept.:** AMA, CAQ, GUA, GUV, VAU **Elev.:** 120–291 m a.s.l. **IUCN Habitat code:** 1,3,4,5,14 **Conservation status: (IUCN)** LC

5281 *Clusia hammeliana* Pipoly (Clusiaceae) **IPNI:** 1195925-2 **Common name:** matapalo **Habit:** Shrub, Tree, Epiphyte, Liana, Climbing **Use:** ME **Origin:** Native **Region:** Amazonia, Andean, Guiana Shield, Pacific **Dept.:** AMA, ANT, CAQ, CHO, GUA, MET, PUT, SAN, VAC, VAU **Elev.:** 0–2000 m a.s.l. **IUCN Habitat code:** 1,3,5,14 **Conservation status: (IUCN)** LC

5282 *Clusia huberi* Pipoly (Clusiaceae) **IPNI:** 1190784-2 **Habit:** Shrub, Small tree, Tree **Use:** ME **Origin:** Native **Region:** Guiana Shield **Dept.:** CAQ, GUA, VAU **Elev.:** 122-760 msnm **IUCN Habitat code:** 1,3,14 **Conservation status: (IUCN)** NE **(CPSG)** LC

5283 *Clusia inesiana* Cuatrec. (Clusiaceae) **IPNI:** 61480-2 **Common name:** gaque, gaque de monte **Habit:** Tree **Use:** ME **Origin:** Endemic **Region:** Andean **Dept.:** ANT, BOY, NSA **Elev.:** 1600-2820 msnm **IUCN Habitat code:** 1,3,4,14 **Conservation status: (IUCN)** LC **(CPSG)** LC

5284 *Clusia insignis* Mart. (Clusiaceae) **IPNI:** 427542-1 **Common name:** gaque, matapalo **Habit:** Shrub, Small tree, Tree **Use:** MA, ME **Origin:** Native **Region:** Andean, Guiana Shield, Pacific **Dept.:** ANT, CAQ, CAS, CAU, CHO, GUA, GUV, HUI, MET, PUT, VAU **Elev.:** 95–1090 m a.s.l. **IUCN Habitat code:** 1,3,4,14 **Conservation status: (IUCN)** LC

5285 *Clusia lineata* (Benth.) Planch. & Triana (Clusiaceae) **IPNI:** 427560-1 **Common name:** madroño **Habit:** Subshrub, Tree **Use:** HF, ME **Origin:** Native **Region:** Amazonia, Andean, Guiana Shield, Orinoquia, Pacific **Dept.:** AMA, ANT, CAQ, CHO, CUN, GUA, MET, NAR, PUT, QUI, RIS, TOL, VAC, VAU, VID **Elev.:** 5–2300 m a.s.l. **IUCN Habitat code:** 1,3,4,5,14 **Conservation status: (IUCN)** LC

5286 *Clusia lopezii* Maguire (Clusiaceae) **IPNI:** 61497-2 **Common name:** ma, ma-mê-tê-tê', mê, tê, tê' **Habit:** Small tree, Tree **Use:** ME **Origin:** Native **Region:** Amazonia, Guiana Shield **Dept.:** AMA, GUA, VAU **Elev.:** 120–380 m a.s.l. **IUCN Habitat code:** 1 **Conservation status: (IUCN)** NE

5287 *Clusia microstemon* Planch. & Triana (Clusiaceae) **IPNI:** 427586-1 **Habit:** Tree, Climbing **Use:** ME **Origin:** Native **Region:** Amazonia, Guiana Shield **Dept.:** AMA, GUA, VAU **Elev.:** 150–250 m a.s.l. **IUCN Habitat code:** 1,5,14 **Conservation status: (IUCN)** LC

5288 *Clusia minor* L. (Clusiaceae) **IPNI:** 315124-2 **Common name:** bateíto, capecape, cucharo, gaque, matapalo, rampacho colorado, sangre de doncella, sietelenguas, tapas **Habit:** Shrub, Small tree, Tree **Use:** MA, ME **Origin:** Native **Region:** Amazonia, Andean, Orinoquia **Dept.:** ANT, ARA, CAS, CUN, MET, NSA, PUT, QUI, RIS, SAP, TOL, VAC **Elev.:** 0–1800 m a.s.l. **IUCN Habitat code:** 1,2,3,4,5,14 **Conservation status: (IUCN)** LC

5289 *Clusia multiflora* Kunth (Clusiaceae) **IPNI:** 427594-1 **Common name:** cape, capé, caucho, caucho gaque, chagualo, chagualón, copey cucharo, cucharo, gaque, imparmo, incienso, manduro, moque, rapancho, sape, tampaco, tapas **Habit:** Shrub, Small tree, Tree, Epiphyte, Climbing **Use:** AF, EU, MA, ME **Origin:** Native **Region:** Andean, Guiana Shield, Pacific **Dept.:** ANT, BOG, BOY, CAL, CAQ, CAU, CES, CHO, CUN, HUI, LAG, MAG, MET, NAR, NSA, PUT, QUI, RIS, SAN, TOL, VAC **Elev.:** 80-3500 msnm **IUCN Habitat code:** 1,2,3,4,5,14 **Conservation status: (IUCN)** LC **(CPSG)** Potential LC

5290 *Clusia obovata* (Spruce ex Planch. & Triana) Pipoly (Clusiaceae) **IPNI:** 1082351-2 **Common name:** copé **Habit:** Small tree, Tree, Climbing **Use:** ME **Origin:** Native **Region:** Amazonia, Guiana Shield **Dept.:** AMA, VAU, VID **Elev.:** 100–300 m a.s.l. **IUCN Habitat code:** 1,3,5,14 **Conservation status: (IUCN)** LC

5291 *Clusia octandra* (Poepp.) Pipoly (Clusiaceae) **IPNI:** 1174011-2 **Common name:** matapalo **Habit:** Shrub, Tree, Epiphyte, Hemiepiphyte, Climbing **Use:** ME **Origin:** Native **Region:** Amazonia, Andean, Guiana Shield, Pacific **Dept.:** AMA, ANT, CAQ, CAU, CHO, CUN, NAR, PUT, SAN, VAC, VAU **Elev.:** 50–2060 m a.s.l. **IUCN Habitat code:** 1,3,5,14 **Conservation status: (IUCN)** LC

5292 *Clusia opaca* Maguire (Clusiaceae) **IPNI:** 61533-2 **Common name:** matapalo **Habit:** Small tree, Tree **Use:** ME **Origin:** Native **Region:** Amazonia, Guiana Shield **Dept.:** CAQ, GUA, VAU **Elev.:** 360-830 msnm **IUCN Habitat code:** 1,4 **Conservation status: (IUCN)** NE **(CPSG)** LC

5293 *Clusia orthoneura* Standl. (Clusiaceae) **IPNI:** 61535-2 **Common name:** cape, caucho, tapas **Habit:** Shrub, Subshrub, Tree, Hemiepiphyte **Use:** EU, ME **Origin:** Endemic **Region:** Andean **Dept.:** ANT, BOG, BOY, CAU, CUN, HUI, SAN **Elev.:** 1200–2300 m a.s.l. **IUCN Habitat code:** 1,2,3,14 **Conservation status: (IUCN)** NE

5294 *Clusia palmicida* Rich. ex Planch. & Triana (Clusiaceae) **IPNI:** 427616-1 **Common name:** copé **Habit:** Shrub, Tree, Hemiepiphyte, Liana **Use:** ME **Origin:** Native **Region:** Amazonia, Andean, Guiana Shield, Pacific **Dept.:** AMA, ANT, CAQ, CHO, MET, PUT **Elev.:** 0–1250 m a.s.l. **IUCN Habitat code:** 1,2,3,5,14 **Conservation status: (IUCN)** LC

5295 *Clusia penduliflora* Engl. (Clusiaceae) **IPNI:** 61546-2 **Common name:** matapalo **Habit:** Tree, Hemiepiphyte, Liana, Climbing **Use:** ME **Origin:** Native **Region:** Amazonia, Andean, Guiana Shield, Pacific **Dept.:** AMA, ANT, CHO, VAU **Elev.:** 0–1300 m a.s.l. **IUCN Habitat code:** 1,3 **Conservation status: (IUCN)** LC

5296 *Clusia planchoniana* Engl. (Clusiaceae) **IPNI:** 61556-2 **Use:** ME **IUCN Habitat code:** 1,2,3,4,5,14 **Conservation status: (IUCN)** NE

5297 *Clusia pseudomangle* Planch. & Triana (Clusiaceae) **IPNI:** 427650-1 **Common name:** cape **Use:** ME **Dept.:** CUN **Conservation status: (IUCN)** LC

5298 *Clusia renggerioides* Planch. & Triana (Clusiaceae) **IPNI:** 427663-1 **Common name:** boco, boco-á, á **Habit:** Small tree, Tree **Use:** ME **Origin:** Native **Region:** Amazonia, Guiana Shield, Orinoquia **Dept.:** AMA, CAQ, CAU, MET, VAU **Elev.:** 180–640 m a.s.l. **IUCN Habitat code:** 1,3,5,14 **Conservation status: (IUCN)** LC

5299 *Clusia rosea* Jacq. (Clusiaceae) **IPNI:** 61583-2 **Common name:** cape, cape grande, cerillo, chagualo, chuchi, chuchul, copei, copey, cucharo, cupay, cupey, gaque, isfuque, quiripiti, sombrero, tampeque **Habit:** Small tree, Tree **Use:** EU, MA, ME **Origin:** Native **Region:** Guiana Shield, Caribbean **Dept.:** LAG, MAG, MET, SAP, VAU **Elev.:** 5–450 m a.s.l. **IUCN Habitat code:** 1,3,4,14 **Conservation status: (IUCN)** LC

5300 *Clusia schomburgkiana* (Planch. & Triana) Benth. ex Engl. (Clusiaceae) **IPNI:** 61594-2 **Common name:** caucho, chagualo, copé, gaque **Habit:** Shrub, Small tree, Tree, Hemiepiphyte, Climbing **Use:** ME **Origin:** Native **Region:** Amazonia, Andean, Guiana Shield, Orinoquia **Dept.:** AMA, ANT, BOG, CHO, CUN, GUA, MET, SAN, TOL, VAC, VAU, VID **Elev.:** 50–1990 m a.s.l. **Conservation status: (IUCN)** LC

5301 *Clusia schultesii* Maguire (Clusiaceae) **IPNI:** 61597-2 **Habit:** Shrub, Tree **Use:** ME **Origin:** Native **Region:** Amazonia, Guiana Shield **Dept.:** AMA, CAQ, VAU **Elev.:** 250–700 m a.s.l. **IUCN Habitat code:** 1,14 **Conservation status: (IUCN)** NE

5302 *Clusia spathulifolia* Engl. (Clusiaceae) **IPNI:** 300766-2 **Common name:** matapalo **Habit:** Shrub, Small tree, Tree, Hemiepiphyte, Liana **Use:** ME **Origin:** Native **Region:** Amazonia, Guiana Shield **Dept.:** AMA, CAQ, GUA, VAU **Elev.:** 150–500 m a.s.l. **IUCN Habitat code:** 1,3,5,14 **Conservation status: (IUCN)** NE

5303 *Clusia sphaerocarpa* Planch. & Triana (Clusiaceae) **IPNI:** 427694-1 **Habit:** Tree **Use:** HF **Origin:** Native **Region:** Andean **Dept.:** PUT **Elev.:** 1000–2500 m a.s.l. **IUCN Habitat code:** 1 **Conservation status: (IUCN)** LC

5304 *Garcinia benthamiana* (Planch. & Triana) Pipoly (Clusiaceae) **IPNI:** 77100471-1 **Common name:** madroño, mamey **Habit:** Tree **Use:** HF, MA, ME **Origin:** Native **Region:** Andean, Caribbean **Dept.:** BOL, MET, SAN **Elev.:** 100–1700 m a.s.l. **Conservation status: (IUCN)** NE

5305 *Garcinia brasiliensis* Mart. (Clusiaceae) **IPNI:** 427844-1 **Use:** HF, ME **IUCN Habitat code:** 1,3,5 **Conservation status: (IUCN)** LC

5306 *Garcinia gardneriana* (Planch. & Triana) Zappi (Clusiaceae) **IPNI:** 972610-1 **Use:** HF **Origin:** Native **IUCN Habitat code:** 1 **Conservation status: (IUCN)** LC

5307 *Garcinia intermedia* (Pittier) Hammel (Clusiaceae) **IPNI:** 277320-2 **Common name:** aceite, madroño, madroño de monte, martín doble, tometo **Use:** HF **IUCN Habitat code:** 1,3,5,14 **Conservation status: (IUCN)** LC

5308 *Garcinia macrophylla* Mart. (Clusiaceae) **IPNI:** 428065-1 **Common name:** chuchuhaso, macuripí, madroño, madroño de cáscara lisa, madroño liso **Habit:** Tree **Use:** HF, MA, ME **Origin:** Native **Region:** Amazonia, Andean, Guiana Shield, Orinoquia **Dept.:** AMA, ANT, CAQ, CAS, GUA, GUV, MET, PUT **Elev.:** 100–1500 m a.s.l. **IUCN Habitat code:** 1,3,5,14 **Conservation status: (IUCN)** LC

5309 *Garcinia madruno* (Kunth) Hammel (Clusiaceae) **IPNI:** 277321-2 **Common name:** aceite, algarrobo, bacuri, bacurí, bebasojo, charichuelo, chuchuhasa de tierra baja, cozoiba, fruta de mono, kamururu, machari, madrona, madrono, madroña, madroño, madroño rugoso, madruño, mamey de monte, naranjuelo, pada **Habit:** Small tree, Tree **Use:** AF, EU, FU, HF, MA, ME **Origin:** Native, Cultivated **Region:** Amazonia, Andean, Guiana Shield, Orinoquia, Pacific **Dept.:** AMA, ANT, ARA, BOY, CAQ, CAS, CES, CHO, CUN, MET, PUT, QUI, RIS, SAN, TOL, VAC **Elev.:** 0–1700 m a.s.l. **IUCN Habitat code:** 1,2,3,4,5,14 **Conservation status: (IUCN)** LC

5310 *Garcinia magnifolia* (Pittier) Hammel (Clusiaceae) **IPNI:** 277318-2 **Common name:** aceituno, madroño, madruño **Habit:** Shrub, Tree **Use:** HF **Origin:** Native **Region:** Andean, Pacific **Dept.:** ANT, CHO, VAC **Elev.:** 0–1420 m a.s.l. **IUCN Habitat code:** 1,3,5,14 **Conservation status: (IUCN)** LC

5311 *Garcinia mangostana* L. (Clusiaceae) **IPNI:** 428073-1 **Common name:** mangostino, mangostán, mangostín, mangostín de Mariquita **Habit:** Tree **Use:** HF, MA, ME **Origin:** Cultivated **Region:** Orinoquia **Dept.:** MET, TOL **Elev.:** 600–600 m a.s.l. **IUCN Habitat code:** 1,3,14 **Conservation status: (IUCN)** NE

5312 *Lorostemon bombaciflorus* Ducke (Clusiaceae) **IPNI:** 143833-2 **Habit:** Tree **Use:** HF **Origin:** Native **Region:** Guiana Shield **Dept.:** AMA, CAQ, VAU **Elev.:** 100–100 m a.s.l. **IUCN Habitat code:** 1 **Conservation status: (IUCN)** NE

5313 *Lorostemon colombianus* Maguire (Clusiaceae) **IPNI:** 143834-2 **Habit:** Small tree **Use:** HF **Origin:** Native **Region:** Amazonia, Guiana Shield **Dept.:** AMA, CAQ, VAU **Elev.:** 213–213 m a.s.l. **Conservation status: (IUCN)** NE

5314 *Moronobea coccinea* Aubl. (Clusiaceae) **IPNI:** 428883-1 **Common name:** breo, charechuelo, charichuelo, palo breo **Habit:** Tree **Use:** HF, MA, ME **Origin:** Native **Region:** Amazonia, Guiana Shield **Dept.:** AMA, CAQ, VAU **Elev.:** 100–250 m a.s.l. **IUCN Habitat code:** 1,3 **Conservation status: (IUCN)** LC

5315 *Platonia insignis* Mart. (Clusiaceae) **IPNI:** 326389-2 **Common name:** madroño, madroño silvestre, naranjillo **Habit:** Tree **Use:** FU, HF, MA, ME **Origin:** Native **Region:** Guiana Shield **Dept.:** GUA, VAU **Elev.:** 120–600 m a.s.l. **IUCN Habitat code:** 1 **Conservation status: (IUCN)** NE

5316 *Symphonia globulifera* L.f. (Clusiaceae) **IPNI:** 429281-1 **Common name:** aceitillo, azufre, brea, breo, breo para calafatear, charechuelo, chucho, inse, jifeo, la brea, labrea, machare, machare de loma, machare de mangual, machare hojiancho, machare hojimenudo, machare teu, macharre, madroño, mangle de montaña, manie, manil, m **Habit:** Tree **Use:** EU, FU, HF, MA, ME **Origin:** Native **Region:** Amazonia, Orinoquia, Pacific **Dept.:** AMA, ANT, CAQ, CAU, CHO, NAR, PUT, VAC, VAU, VID **Elev.:** 0–1000 m a.s.l. **IUCN Habitat code:** 1,3,4,5,14 **Conservation status: (IUCN)** LC

5317 *Tovomita brevistaminea* Engl. (Clusiaceae) **IPNI:** 255606-2 **Habit:** Shrub, Tree **Use:** HF **Origin:** Native **Region:** Amazonia **Dept.:** AMA, CAQ, MET **Elev.:** 100–300 m a.s.l. **IUCN Habitat code:** 1,3,4 **Conservation status: (IUCN)** LC

5318 *Tovomita fructipendula* (Ruiz & Pav.) Cambess. (Clusiaceae) **IPNI:** 429349-1 **Use:** ME **Origin:** Native **IUCN Habitat code:** 1 **Conservation status: (IUCN)** LC

5319 *Tovomita laurina* Planch. & Triana (Clusiaceae) **IPNI:** 429368-1 **Habit:** Small tree **Use:** ME **Origin:** Native **Region:** Amazonia **Dept.:** AMA, CAQ, VAU **Elev.:** 100–400 m a.s.l. **IUCN Habitat code:** 1 **Conservation status: (IUCN)** LC

5320 *Tovomita schomburgkii* Planch. & Triana (Clusiaceae) **IPNI:** 429403-1 **Habit:** Tree **Use:** HF **Origin:** Native **Region:** Amazonia **Dept.:** AMA **Elev.:** 100–240 m a.s.l. **IUCN Habitat code:** 1 **Conservation status: (IUCN)** LC

5321 *Tovomita stylosa* Hemsl. (Clusiaceae) **IPNI:** 429411-1 **Common name:** alacoque **Habit:** Shrub, Small tree, Tree **Use:** ME **Origin:** Native **Region:** Andean, Pacific **Dept.:** ANT, BOY, CHO, COR, NAR, PUT, VAC **Elev.:** 0–1800 m a.s.l. **IUCN Habitat code:** 1,3,14 **Conservation status: (IUCN)** LC

5322 *Stephanopodium peruvianum* Poepp. (Dichapetalaceae) **IPNI:** 163297-1 **Common name:** bijo, costillo, naranjo de monte **Habit:** Small tree, Tree **Use:** HF, ME **Origin:** Native **Region:** Amazonia, Guiana Shield **Dept.:** AMA, CAQ, GUV **Elev.:** 150-450 msnm **IUCN Habitat code:** 1,3,5,14 **Conservation status: (IUCN)** NE **(CPSG)** LC

5323 *Tapura amazonica* Poepp. (Dichapetalaceae) **IPNI:** 163302-1 **Common name:** conga **Habit:** Small tree, Tree **Use:** HF **Origin:** Native **Region:** Amazonia **Dept.:** AMA, CAQ, GUV, VAU **Elev.:** 100-150 msnm **IUCN Habitat code:** 1,3,5,14 **Conservation status: (IUCN)** LC **(CPSG)** LC

5324 *Tapura coriacea* J.F.Macbr. (Dichapetalaceae) **IPNI:** 249421-2 **Habit:** Shrub, Small tree, Tree **Use:** HF **Origin:** Native **Region:** Amazonia **Dept.:** AMA **Elev.:** 100–220 m a.s.l.

IUCN Habitat code: 1,14 **Conservation status: (IUCN)** NE

5325 *Tapura juruana* (Ule) Rizzini (Dichapetalaceae) **IPNI:** 249430-2 **Habit:** Tree **Use:** HF **Origin:** Native **Region:** Amazonia **Dept.:** AMA, GUV **Elev.:** 110-218 msnm **IUCN Habitat code:** 1,3,14 **Conservation status: (IUCN)** LC **(CPSG)** LC

5326 *Tapura peruviana* K.Krause (Dichapetalaceae) **IPNI:** 249439-2 **Common name:** coca de antiguo **Habit:** Small tree **Use:** HF, ME **Origin:** Native **Region:** Amazonia **Dept.:** PUT **Elev.:** 600-600 msnm **IUCN Habitat code:** 1,14 **Conservation status: (IUCN)** NE **(CPSG)** NT

5327 *Erythroxylum acrobeles* W.A.Gentner (Erythroxylaceae) **IPNI:** 96866-2 **Use:** ME **Origin:** Endemic **Region:** Pacific **Dept.:** VAC **Elev.:** 130–200 m a.s.l. **Conservation status: (IUCN)** NE

5328 *Erythroxylum acutum* W.A.Gentner (Erythroxylaceae) **IPNI:** 96867-2 **Habit:** Shrub **Use:** ME **Origin:** Endemic **Region:** Pacific **Dept.:** CAU **Elev.:** 0–20 m a.s.l. **IUCN Habitat code:** 1 **Conservation status: (IUCN)** NE

5329 *Erythroxylum amazonicum* Peyr. (Erythroxylaceae) **IPNI:** 276833-2 **Common name:** cancho, coca de monte, comino, palacón ajicito **Habit:** Shrub, Small tree, Tree **Use:** MA, ME **Origin:** Native **Region:** Amazonia, Andean, Guiana Shield, Pacific **Dept.:** AMA, ANT, CAQ, CHO, GUA, MET, PUT, SAN, TOL **Elev.:** 60–2000 m a.s.l. **IUCN Habitat code:** 1,3,4,14 **Conservation status: (IUCN)** LC

5330 *Erythroxylum carthagenense* Jacq. (Erythroxylaceae) **IPNI:** 544013-1 **Common name:** ají de pájaro, coca, jayo, pintauñas **Habit:** Shrub, Small tree, Tree **Use:** ME **Origin:** Native **Region:** Caribbean **Dept.:** ANT, ATL, BOL, CES, LAG, MAG, SUC **Elev.:** 0–125 m a.s.l. **IUCN Habitat code:** 1,2,5 **Conservation status: (IUCN)** NE

5331 *Erythroxylum cataractarum* Spruce ex Peyr. (Erythroxylaceae) **IPNI:** 276842-2 **Common name:** coca de pescado **Habit:** Shrub, Small tree, Tree **Use:** MA, ME **Origin:** Native **Region:** Amazonia, Guiana Shield, Caribbean **Dept.:** ANT, BOL, CAS, GUA, MET, SAN, VAU, VID **Elev.:** 20–600 m a.s.l. **IUCN Habitat code:** 1,3,4,5,14 **Conservation status: (IUCN)** NE

5332 *Erythroxylum citrifolium* A.St.-Hil. (Erythroxylaceae) **IPNI:** 330131-2 **Common name:** cafecito, coca cimarrona, coca de monte, comino, jayito **Habit:** Shrub, Tree **Use:** AF, EU, MA, ME **Origin:** Native **Region:** Amazonia, Andean, Guiana Shield, Caribbean, Orinoquia, Pacific **Dept.:** AMA, ANT, BOY, CAL, CAS, CAU, CES, CHO, CUN, GUV, LAG, MET, NSA, QUI, RIS, SAN, TOL, VAC **Elev.:** 10-2000 msnm **IUCN Habitat code:** 1,2,3,4,14 **Conservation status: (IUCN)** LC **(CPSG)** Potential LC

5333 *Erythroxylum coca* Lam. (Erythroxylaceae) **IPNI:** 330132-2 **Common name:** amarga, coca, coca de danta, coca del sur, coca peruana, coca propia, coca suave, coca tinga, coca zambico, coca-á, epadu, hayo, hayuelo, huangana coca, ipadá, ipatú, javo, jayuelo, ka-heé, pa-too, patú, peruana, pringamaría, pussachpan, tinga **Habit:** Shrub **Use:** EU, HF, MA, ME, SU **Origin:** Native, Cultivated **Region:** Amazonia, Andean, Guiana Shield, Orinoquia **Dept.:** AMA, ANT, ARA, BOY, CAQ, CAS, CES, COR, CUN, GUA, GUV, HUI, MAG, MET, NAR, NSA, PUT, SAN, TOL, VAU, VID **Elev.:** 120-500 msnm **IUCN Habitat code:** 1,3,4,14 **Conservation status: (IUCN)** NE **(CPSG)** Potential LC

5334 *Erythroxylum densum* Rusby (Erythroxylaceae) **IPNI:** 96936-2 **Habit:** Shrub, Small tree **Use:** MA, ME **Origin:** Native **Region:** Andean, Caribbean **Dept.:** MAG, TOL **Elev.:** 30–700 m a.s.l. **IUCN Habitat code:** 2,14 **Conservation status: (IUCN)** NE

5335 *Erythroxylum fimbriatum* Peyr. (Erythroxylaceae) **IPNI:** 330146-2 **Habit:** Small tree, Tree **Use:** ME **Origin:** Native **Region:** Amazonia, Andean **Dept.:** ANT, CUN, PUT **Elev.:** 100–850 m a.s.l. **IUCN Habitat code:** 1,3 **Conservation status: (IUCN)** LC

5336 *Erythroxylum glaucum* O.E.Schulz (Erythroxylaceae) **IPNI:** 96953-2 **Use:** ME **Conservation status: (IUCN)** NE

5337 *Erythroxylum gracilipes* Peyr. (Erythroxylaceae) **IPNI:** 276848-2 **Common name:** coca, coca de danta, coca de monte **Habit:** Shrub, Small tree, Tree **Use:** HF, ME **Origin:** Native **Region:** Amazonia, Andean, Caribbean, Orinoquia, Pacific **Dept.:** AMA, ANT, CAQ, CHO, COR, CUN, GUA, GUV, MAG, MET, NAR, NSA, PUT, VAC, VID **Elev.:** 0-770 msnm **IUCN Habitat code:** 1,2,3,4,5,14 **Conservation status: (IUCN)** LC **(CPSG)** Potential LC

5338 *Erythroxylum haughtii* W.A.Gentner (Erythroxylaceae) **IPNI:** 96968-2 **Habit:** Shrub, Small tree, Tree **Use:** MA, ME **Origin:** Endemic **Region:** Andean **Dept.:** CAU, HUI, NAR, VAC **Elev.:** 468-1260 msnm **IUCN Habitat code:** 1,14 **Conservation status: (IUCN)** LC **(CPSG)** LC

5339 *Erythroxylum havanense* Jacq. (Erythroxylaceae) **IPNI:** 330154-2 **Common name:** coya, palo blanco **Habit:** Shrub **Use:** FU, HF, MA, ME, PO **Origin:** Native **Region:** Caribbean **Dept.:** ATL, BOL, CES, CUN, MAG, SUC **Elev.:** 5–450 m a.s.l. **IUCN Habitat code:** 1,2,3,4,14 **Conservation status: (IUCN)** LC

5340 *Erythroxylum hondense* Kunth (Erythroxylaceae) **IPNI:** 330155-2 **Common name:** coca, palo blanco **Habit:** Shrub, Small tree, Tree **Use:** AF,ME, SU **Origin:** Native **Region:** Andean, Caribbean, Orinoquia **Dept.:** ANT, ARA, CAU, CES, CUN, HUI, MAG, TOL, VAC **Elev.:** 150–1300 m a.s.l. **IUCN Habitat code:** 1,2,3,14 **Conservation status: (IUCN)** NE

5341 *Erythroxylum impressum* O.E.Schulz (Erythroxylaceae) **IPNI:** 276851-2 **Habit:** Shrub **Use:** ME **Origin:** Native **Region:** Guiana Shield **Dept.:** VID **Elev.:** 110–110 m a.s.l. **IUCN Habitat code:** 1,2,3,4,5 **Conservation status: (IUCN)** LC

5342 *Erythroxylum ligustrinum* DC. (Erythroxylaceae) **IPNI:** 330159-2 **Habit:** Small tree, Tree **Use:** MA **Origin:** Native **Region:** Orinoquia **Dept.:** ARA, VID **Elev.:** 150–150 m a.s.l. **IUCN Habitat code:** 5 **Conservation status: (IUCN)** LC

5343 *Erythroxylum lindemanii* Plowman (Erythroxylaceae) **IPNI:** 276859-2 **Habit:** Shrub **Use:** MA **Origin:** Native **Region:** Guiana Shield **Dept.:** GUA **Elev.:** 90–90 m a.s.l. **IUCN Habitat code:** 1 **Conservation status: (IUCN)** NE

5344 *Erythroxylum macrophyllum* Cav. (Erythroxylaceae) **IPNI:** 330167-2 **Common name:** ajicillo, ají, coca, coca de monte, pata de torcaza **Habit:** Shrub, Small tree, Tree **Use:** AF, EU, MA, ME **Origin:** Native **Region:** Amazonia, Andean, Orinoquia, Pacific **Dept.:** AMA, ANT, ARA, BOL, BOY, CAL, CAQ, CAS, CAU, CES, CHO, CUN, GUA, GUV, MAG, MET, NAR, NSA, PUT, SAN, TOL, VAC, VAU, VID **Elev.:** 10-1800 msnm **IUCN Habitat code:** 1,2,3,4,5,14 **Conservation status: (IUCN)** LC **(CPSG)** Potential LC

5345 *Erythroxylum novogranatense* (D.Morris) Hieron. (Erythroxylaceae) **IPNI:** 97004-2 **Common name:** coca, coca de trujillo, coca montana, coca rayo, colombian coca, pajarita, trujillo coca **Habit:** Shrub, Small tree **Use:** EU, HF, MA, ME, PO, SU **Origin:** Native, Cultivated **Region:** Amazonia, Andean, Caribbean **Dept.:** ANT, CAQ, CES, COR, CUN, HUI, LAG, MAG, NSA, PUT, QUI, TOL, VAC, VID **Elev.:** 265-2000 msnm **IUCN Habitat code:** 1,2,3,14 **Conservation status: (IUCN)** NE **(CPSG)** Potential LC

5346 *Erythroxylum orinocense* Kunth (Erythroxylaceae) **IPNI:** 97014-2 **Habit:** Shrub, Subshrub **Use:** ME **Origin:** Native **Region:** Orinoquia **Dept.:** CAS, VID **Elev.:** 259–259 m a.s.l. **IUCN Habitat code:** 1,3,4 **Conservation status: (IUCN)** NE

5347 *Erythroxylum oxycarpum* O.E.Schulz (Erythroxylaceae) **IPNI:** 97017-2 **Habit:** Shrub, Tree **Use:** MA **Origin:** Native **Region:** Andean, Caribbean **Dept.:** ATL, CES, LAG, MAG, SAN **Elev.:** 50–750 m a.s.l. **IUCN Habitat code:** 1,2,3,14 **Conservation status: (IUCN)** LC

5348 *Erythroxylum rufum* Cav. (Erythroxylaceae) **IPNI:** 330191-2 **Habit:** Tree **Use:** MA **Origin:** Native **Region:** Amazonia **Dept.:** CAQ **Elev.:** 270–300 m a.s.l. **IUCN Habitat code:** 3,4,14 **Conservation status: (IUCN)** LC

5349 *Erythroxylum spruceanum* Peyr. (Erythroxylaceae) **IPNI:** 276883-2 **Habit:** Tree **Use:** ME **Origin:** Native **Region:** Orinoquia **Dept.:** GUA, VAU **Elev.:** 65–65 m a.s.l. **Conservation status: (IUCN)** NE

5350 *Erythroxylum ulei* O.E.Schulz (Erythroxylaceae) **IPNI:** 276893-2 **Common name:** coca, coca mono, cocamono **Habit:** Shrub **Use:** ME **Origin:** Native **Region:** Andean **Dept.:** ANT, CAL, CAU, CUN, TOL, VAC **Elev.:** 200–1350 m a.s.l. **IUCN Habitat code:** 1,3,14 **Conservation status: (IUCN)** LC

5351 *Erythroxylum williamsii* Standl. ex Plowman (Erythroxylaceae) **IPNI:** 97075-2 **Habit:** Shrub **Use:** MA **Origin:** Native **Region:** Guiana Shield **Dept.:** VID **Elev.:** 90–130 m a.s.l. **IUCN Habitat code:** 1,3,4,5,14 **Conservation status: (IUCN)** NE

5352 *Acalypha alopecuroidea* Jacq. (Euphorbiaceae) **IPNI:** 337150-1 **Common name:** ajito con pelo, cadillo, cadillo de mazorca, churrusco, churrusquillo, churrusquillos, degonflé, gusanillo, rabo de ardilla, rabo de gato **Habit:** Herb **Use:** ME **Origin:** Native **Region:** Andean, Caribbean, Orinoquia, Pacific **Dept.:** ATL, BOL, CAU, COR, MET, NAR, SAP, SAN, SUC, TOL **Elev.:** 0-1000 msnm **IUCN Habitat code:** 1,2,3,4,14 **Conservation status: (IUCN)** NE **(CPSG)** Potential LC

5353 *Acalypha amentacea* Roxb. (Euphorbiaceae) **IPNI:** 337154-1 **Habit:** Shrub **Use:** AF, EU, FU, HF, MA, ME, SU **Origin:** Cultivated **Region:** Andean, Caribbean, Pacific **Dept.:** ANT, ATL, BOL, CAL, CHO, CUN, NAR, PUT, QUI, SAP, SAN, VAC **Elev.:** 0–1600 m a.s.l. **IUCN Habitat code:** 1,3,14 **Conservation status: (IUCN)** NE

5354 *Acalypha arvensis* Poepp. (Euphorbiaceae) **IPNI:** 1115-2 **Common name:** coirdeimiento, gusanillo, hierba del cáncer, piojito, piojo, plumilla, yerba de cáncer **Habit:** Herb **Use:** ME **Origin:** Native **Region:** Andean, Caribbean **Dept.:** ANT, ATL, MAG, QUI, SAP, SUC, VAC **Elev.:** 0-1800 msnm **IUCN Habitat code:** 1,2,3,14 **Conservation status: (IUCN)** NE **(CPSG)** Potential LC

5355 *Acalypha chamaedrifolia* (Lam.) Müll.Arg. (Euphorbiaceae) **IPNI:** 1012504-2 **Habit:** Herb **Use:** EU **Origin:** Cultivated **Region:** Andean **Dept.:** BOG, CAL, NSA, RIS **Elev.:** 1600–1950 m a.s.l. **Conservation status: (IUCN)** NE

5356 *Acalypha diversifolia* Jacq. (Euphorbiaceae) **IPNI:** 1228-2 **Common name:** borlas, coloraíto, cordoncillo, coya, dorotiango, juanajuana, juanita, juanjuán, madrejuano, maiztostao, marroquina, odorate, planchido, planchira, sangre de toro, trencilla, vara de alcalde, vara negra, zanca de mula, zanca de mula blanca, zanca de mula **Habit:** Shrub **Use:** AF,ME **Origin:** Native **Region:** Amazonia, Andean, Guiana Shield, Caribbean, Orinoquia, Pacific **Dept.:** ANT, BOL, BOY, CAL, CAU, CES, CHO, COR, CUN, GUV, HUI, MAG, MET, NAR, PUT, QUI, RIS, SAN, TOL, VAC **Elev.:** 0-2000 msnm **IUCN Habitat code:** 1,2,3,4,5,14 **Conservation status: (IUCN)** LC **(CPSG)** Potential LC

5357 *Acalypha hispida* Burm.f. (Euphorbiaceae) **IPNI:** 337478-1 **Common name:** amaranto de arbolito, cola de gato, combo, cordón de obispo, cresta de gallo, gusanillo, gusano, navidad, rabo de mico, suspiro, torbellino **Habit:** Shrub **Use:** EU, HF, MA, ME, SU **Origin:** Cultivated **Region:** Andean, Caribbean, Pacific **Dept.:** ANT, ATL, CAL, CHO, CUN, MAG, QUI, SAP, VAC **Elev.:** 100–1600 m a.s.l. **IUCN Habitat code:** 1,2,3,14 **Conservation status: (IUCN)** NE

5358 *Acalypha macrostachya* Jacq. (Euphorbiaceae) **IPNI:** 1362-2 **Common name:** barbas, barbas de guasco, bollo de puerco rojo, churrusco, cordoncillo, coroncorito, jaulo, mulato, ortiguillo, querida, querido, salado, zanca de araña, zancaemula, zancamula, zanco de mula, zanquemula, zábila de monte **Habit:** Tree **Use:** EU, MA, ME, PO **Origin:** Native **Region:** Amazonia, Andean, Caribbean, Orinoquia, Pacific **Dept.:** ANT, ATL, BOG, BOL, BOY, CAU, CES, CHO, COR, CUN, HUI, MAG, MET, NAR, NSA, PUT, QUI, SAN, TOL, VAC **Elev.:** 10-2000 msnm **IUCN Habitat code:** 1,2,3,4,5,14 **Conservation status: (IUCN)** LC **(CPSG)** Potential LC

5359 *Acalypha platyphylla* Müll.Arg. (Euphorbiaceae) **IPNI:** 1444-2 **Common name:** zanca de mula blanca **Habit:** Shrub **Use:** ME **Origin:** Native **Region:** Andean, Pacific **Dept.:** ANT, CAL, CAU, CHO, HUI, MAG, NAR, PUT, QUI, RIS, SAN, VAC **Elev.:** 30-2500 msnm **IUCN Habitat code:** 1,2,3,4,5,14 **Conservation status: (IUCN)** LC **(CPSG)** Potential LC

5360 *Acalypha plicata* Müll.Arg. (Euphorbiaceae) **IPNI:** 337706-1 **Habit:** Shrub **Use:** HF **Origin:** Native **Region:** Andean **Dept.:** HUI, NSA, SAN **Elev.:** 700–1300 m a.s.l. **IUCN Habitat code:** 1,2,3,14 **Conservation status: (IUCN)** NE

5361 *Acalypha setosa* A.Rich. (Euphorbiaceae) **IPNI:** 1520-2 **Habit:** Herb **Use:** ME **Origin:** Native **Dept.:** ANT, CUN, QUI, TOL, VAC **Elev.:** 300–1000 m a.s.l. **IUCN Habitat code:** 1,3,5,14 **Conservation status: (IUCN)** NE

5362 *Actinostemon concolor* (Spreng.) Müll.Arg. (Euphorbiaceae) **IPNI:** 1187218-2 **Habit:** Small tree **Use:** MA **Origin:** Native **Region:** Caribbean **Dept.:** LAG **Elev.:** 650–650 m a.s.l. **IUCN Habitat code:** 1 **Conservation status: (IUCN)** LC

5363 *Alchornea castaneifolia* (Humb. & Bonpl. ex Willd.) A.Juss. (Euphorbiaceae) **IPNI:** 338372-1 **Common name:** jarito, maçoya guafa, mangle, pájaro árbol, serensere **Habit:** Shrub **Use:** ME **Origin:** Native **Region:** Amazonia, Orinoquia **Dept.:** AMA, ARA, BOL, GUA, MET, VAU, VID **Elev.:** 20–600 m a.s.l. **IUCN Habitat code:** 1,2,3,4,5 **Conservation status: (IUCN)** LC

5364 *Alchornea costaricensis* Pax & K.Hoffm. (Euphorbiaceae) **IPNI:** 8183-2 **Common name:** quiebrapatas **Habit:** Tree **Use:** MA **Origin:** Native **Region:** Andean, Pacific **Dept.:** ANT, CAL, CHO **Elev.:** 20–750 m a.s.l. **IUCN Habitat code:** 1,3,4,14 **Conservation status: (IUCN)** LC

5365 *Alchornea discolor* Poepp. (Euphorbiaceae) **IPNI:** 338389-1 **Habit:** Tree **Use:** MA **Origin:** Native **Region:** Amazonia, Andean, Guiana Shield, Caribbean, Orinoquia **Dept.:** AMA, BOL, BOY, CAQ, COR, GUV, MET, VAU, VID **Elev.:** 140–2300 m a.s.l. **IUCN Habitat code:** 1,2,3,4,5,14 **Conservation status: (IUCN)** LC

5366 *Alchornea glandulosa* Poepp. (Euphorbiaceae) **IPNI:** 30043708-2 **Common name:** algodoncillo, algodón, carnegallina, clavo pasado, fresno, montefrío, sangregao, uyamo **Habit:** Tree **Use:** AF, EU, FU, MA, ME, PO **Origin:** Native **Region:** Andean, Guiana Shield, Caribbean, Pacific **Dept.:** ANT, BOY, CAL, CAQ, CAU, CES, CHO, CUN, HUI, MAG, MET, NAR, NSA, PUT, QUI, RIS, SAN, TOL, VAC **Elev.:** 20-2450 msnm **IUCN Habitat code:** 1,2,3,4,5,14 **Conservation status: (IUCN)** LC **(CPSG)** Potential LC

5367 *Alchornea grandiflora* Müll.Arg. (Euphorbiaceae) **IPNI:** 8192-2 **Common name:** algodoncillo, arenoso, clavo pasado, hojancho, lombricero, montefrío, ortiguillo, palo rosa, requesón **Habit:** Tree **Use:** AF, EU, FU, MA **Origin:** Native **Region:** Andean, Caribbean, Orinoquia **Dept.:** ANT, BOG, BOY, CAU, CES, CUN, HUI, MAG, MET, NAR, NSA, QUI, RIS, SAN, TOL, VAC **Elev.:** 1300-2900 msnm **IUCN Habitat code:** 1,3,4,5,14 **Conservation status: (IUCN)** LC **(CPSG)** Potential LC

5368 *Alchornea latifolia* Sw. (Euphorbiaceae) **IPNI:** 338417-1 **Common name:** algodoncillo, carcomo, escobo, fresno, gargantilla, gargantillo, lombricero, masato, mazamorro, tambor, álamo, álamo de monte **Habit:** Tree **Use:** AF, EU, FU, MA **Origin:** Native **Region:** Amazonia, Andean, Caribbean, Orinoquia, Pacific **Dept.:** AMA, ANT, BOY, CAQ, CAU, CES, CUN, HUI, MET, NAR, PUT, QUI, RIS, SAN, TOL, VAC **Elev.:** 200-2800 msnm **IUCN Habitat code:** 1,2,3,4,5,14 **Conservation status: (IUCN)** LC **(CPSG)** Potential LC

5369 *Alchornea triplinervia* (Spreng.) Müll.Arg. (Euphorbiaceae) **IPNI:** 30014592-2 **Common name:** algodoncillo, algodón, carcomo, cáscara de yuca, hojancho, jadebano, magameai, palo rosa, parta, reventillo **Habit:** Tree **Use:** HF, ME **Origin:** Native **Region:** Amazonia, Andean, Orinoquia **Dept.:** AMA, ANT, CAQ, CAU, CHO, MET, SAN, VAU **Elev.:** 120–1000 m a.s.l. **IUCN Habitat code:** 1,3,4,14 **Conservation status: (IUCN)** LC

5370 *Alchorneopsis floribunda* (Benth.) Müll.Arg. (Euphorbiaceae) **IPNI:** 338476-1 **Common name:** camaroncillo **Habit:** Tree **Use:** MA **Origin:** Native **Region:** Amazonia, Andean, Orinoquia, Pacific **Dept.:** AMA, ANT, CAQ, CHO, CUN, NAR, SAN, VAC, VID **Elev.:** 20–1200 m a.s.l. **IUCN Habitat code:** 1,3,4,5,14 **Conservation status: (IUCN)** LC

5371 *Aleurites moluccanus* (L.) Willd. (Euphorbiaceae) **IPNI:** 338514-1 **Habit:** Tree **Use:** AF, EU, FU, GS, HF, IF, MA, ME, PO, SU **Origin:** Cultivated **Region:** Andean, Caribbean, Pacific **Dept.:** CHO, CUN, SAP, VAC **IUCN Habitat code:** 5,14 **Conservation status: (IUCN)** LC

5372 *Aparisthmium cordatum* (A.Juss.) Baill. (Euphorbiaceae) **IPNI:** 339164-1 **Common name:** algodoncillo, guacamayo, palo de garrapata **Habit:** Shrub, Small tree **Use:** HF, MA **Origin:** Native **Region:** Amazonia, Andean, Guiana Shield, Caribbean, Orinoquia, Pacific **Dept.:** AMA, ANT, BOL, CAQ, CHO, CUN, GUA, MET, PUT, SAN, VAU **Elev.:** 200–1400 m a.s.l. **Conservation status: (IUCN)** LC

5373 *Astraea lobata* (L.) Klotzsch (Euphorbiaceae) **IPNI:** 339515-1 **Common name:** frailecillo cimarrón **Habit:** Herb **Use:** AF, EU, HF, MA, ME, PO, SU **Origin:** Native **Region:** Andean, Caribbean **Dept.:** ATL, BOL, COR, HUI, LAG, MAG, SAN, VAC **Elev.:** 0–1000 m a.s.l. **IUCN Habitat code:** 1,2,3,14 **Conservation status: (IUCN)** NE

5374 *Caperonia castaneifolia* (L.) A.St.-Hil. (Euphorbiaceae) **IPNI:** 44529-2 **Habit:** Herb **Use:** ME **Origin:** Native **Region:** Amazonia, Andean, Caribbean, Orinoquia, Pacific **Dept.:** AMA, BOL, CHO, COR, MAG, MET, TOL, VAC **Elev.:** 30–1400 m a.s.l. **IUCN Habitat code:** 1,2,3,4,5,14 **Conservation status: (IUCN)** NE

5375 *Caperonia palustris* (L.) A.St.-Hil. (Euphorbiaceae) **IPNI:** 44564-2 **Habit:** Herb **Use:** HF, MA, ME **Origin:** Native **Region:** Andean, Caribbean, Orinoquia, Pacific **Dept.:** ANT, ATL, BOL, BOY, CAS, CAU, CHO, COR, CUN, MAG, MET, NSA, PUT, QUI, TOL, VAC **Elev.:** 0-1100 msnm **IUCN Habitat code:** 1,2,3,4,14 **Conservation status: (IUCN)** NE **(CPSG)** Potential LC

5376 *Caryodendron orinocense* H.Karst. (Euphorbiaceae) **IPNI:** 340439-1 **Common name:** abay, almendro, arrayana, cacay, cahay, casepache, castaño, cumaná, guayabo de león, hambi, inche, inche hembra, inchi, iracana, kakarí, maní, nambi, nogal, nuez, nuez de barinas, nuez de barquisimeto, nuez llanera, palo de nuez, palo maní, tacay, taque, **Habit:** Tree **Use:** HF, MA, ME **Origin:** Native **Region:** Amazonia, Andean, Orinoquia **Dept.:** ANT, CAQ, CUN, MET, PUT **Elev.:** 350–1300 m a.s.l. **IUCN Habitat code:** 1,3,4,14 **Conservation status: (IUCN)** LC

5377 *Cnidoscolus aconitifolius* (Mill.) I.M.Johnst. (Euphorbiaceae) **IPNI:** 61821-2 **Common name:** azuceno, caquezá, manolo, panameño, panamá, papayote, papayuelo, pringamoza extranjera, árnica **Habit:** Shrub **Use:** EU, GS, HF, MA, ME, PO **Origin:** Naturalised **Region:** Andean **Dept.:** ANT **Elev.:** 1190–1190 m a.s.l. **IUCN Habitat code:** 1,2,3,14 **Conservation status: (IUCN)** LC

5378 *Cnidoscolus longipes* (Pax) I.M.Johnst. (Euphorbiaceae) **IPNI:** 61855-2 **Common name:** árnica **Habit:** Shrub, Small tree **Use:** ME **Origin:** Endemic **Region:** Andean, Caribbean **Dept.:** CES, CUN, HUI, MAG, TOL **Elev.:** 50-1300 msnm **IUCN Habitat code:** 1,2,3,14 **Conservation status: (IUCN)** NE **(CPSG)** LC

5379 *Cnidoscolus tubulosus* (Müll.Arg.) I.M.Johnst. (Euphorbiaceae) **IPNI:** 61897-2 **Common name:** ortiga, ortiga blanca, pringamoza **Habit:** Shrub **Use:** HF, ME **Origin:** Native **Region:** Andean, Caribbean, Pacific **Dept.:** ANT, BOL, HUI, MAG, TOL, VAC **Elev.:** 30–1350 m a.s.l. **IUCN Habitat code:** 1,2,3,14 **Conservation status: (IUCN)** LC

5380 *Cnidoscolus urens* (L.) Arthur (Euphorbiaceae) **IPNI:** 61899-2 **Common name:** guaritoto, ortiga, pringamosa, pringamoza, yawa **Habit:** Shrub **Use:** AF, EU, HF, MA, ME, PO **Origin:** Native **Region:** Andean, Caribbean, Orinoquia **Dept.:** ANT, ATL, BOL, BOY, CAL, CES, COR, CUN, HUI, LAG, MAG, MET, NAR, SAN, SUC, TOL, VAC **Elev.:** 0-2100 msnm **IUCN Habitat code:** 1,2,3,5,14 **Conservation status: (IUCN)** LC **(CPSG)** Potential LC

5381 *Codiaeum variegatum* (L.) Rumph. ex A.Juss. (Euphorbiaceae) **IPNI:** 85073-3 **Common name:** ancla, caramelo, cinta, cobalonga extranjera, corona de príncipe, croto, croton, crotos, granito de oro, grano de oro, guerra santa, lluviadeoro, manguito, manguito extranjero, naranjito, remiendoepobre, turquete **Habit:** Shrub **Use:** EU, FU, HF, MA, ME, SU **Origin:** Cultivated **Region:** Amazonia, Andean, Caribbean, Orinoquia, Pacific **Dept.:** AMA, ANT, BOG, CHO, CUN, MAG, MET, SAP, VAC **Elev.:** 880–1800 m a.s.l. **IUCN Habitat code:** 1,3,5,14 **Conservation status: (IUCN)** LC

5382 *Conceveiba guianensis* Aubl. (Euphorbiaceae) **IPNI:** 320623-2 **Common name:** parta, varasana **Habit:** Tree **Use:** HF, MA **Origin:** Native **Region:** Amazonia **Dept.:** AMA, CAQ **Elev.:** 200–500 m a.s.l. **IUCN Habitat code:** 1,3,4,5,14 **Conservation status: (IUCN)** NE

5383 *Conceveiba martiana* Baill. (Euphorbiaceae) **IPNI:** 341924-1 **Common name:** mantecoso, palo tábano **Habit:** Tree **Use:** HF **Origin:** Native **Region:** Amazonia **Dept.:** AMA, CAQ, GUA, VAU **Elev.:** 100–1000 m a.s.l. **IUCN Habitat code:** 1,4 **Conservation status: (IUCN)** LC

5384 *Croton argenteus* L. (Euphorbiaceae) **IPNI:** 69894-2 **Common name:** cotorrera, hierba de cotorra **Habit:** Herb **Use:** ME **Origin:** Native **Region:** Andean, Caribbean **Dept.:** ANT, ATL, BOL, BOY, COR, MAG **Elev.:** 0–200 m a.s.l. **IUCN Habitat code:** 1,2,3,5,14 **Conservation status: (IUCN)** NE

5385 *Croton conduplicatus* Kunth (Euphorbiaceae) **IPNI:** 342329-1 **Common name:** aloukemasu, amargosito, guayabito **Habit:** Shrub **Use:** ME **Origin:** Native **Region:** Andean, Caribbean **Dept.:** BOY, LAG, SAP, SAN **Elev.:** 0–3000 m a.s.l. **IUCN Habitat code:** 2 **Conservation status: (IUCN)** NE

5386 *Croton cuneatus* Klotzsch (Euphorbiaceae) **IPNI:** 342381-1 **Common name:** balso, cafetana, cafeterana, caimo, drago, reventillo, varablanca **Habit:** Tree **Use:** HF, ME **Origin:** Native **Region:** Amazonia, Orinoquia **Dept.:** AMA, MET, VAU **Elev.:** 100–300 m a.s.l. **IUCN Habitat code:** 1,2,3,4,5,14 **Conservation status: (IUCN)** LC

5387 *Croton cupreatus* Croizat (Euphorbiaceae) **IPNI:** 70101-2 **Common name:** candelero, candelero real, danto, guacamayo, mopo **Habit:** Tree **Use:** ME **Origin:** Endemic **Region:** Andean **Dept.:** CAQ **Elev.:** 2000-2250 msnm **IUCN Habitat code:** 1,3,14 **Conservation status: (IUCN)** VU **(CPSG)** VU

5388 *Croton flavens* L. (Euphorbiaceae) **IPNI:** 342537-1 **Use:** ME, PO **Origin:** Native **Region:** Caribbean **Dept.:** LAG, SAP **Elev.:** 0–50 m a.s.l. **Conservation status: (IUCN)** LC

5389 *Croton fragilis* Kunth (Euphorbiaceae) **IPNI:** 342550-1 **Habit:** Shrub **Use:** ME **Origin:** Native **Region:** Andean, Caribbean **Dept.:** LAG, MAG **Elev.:** 30–1500 m a.s.l. **IUCN Habitat code:** 1,2,14 **Conservation status: (IUCN)** NE

5390 *Croton fragrans* Kunth (Euphorbiaceae) **IPNI:** 342551-1 **Common name:** algodoncillo, algodón, cascarillero, colpachi, cordoncillo, mosquero **Habit:** Shrub **Use:** AF, FU, ME **Origin:** Native **Region:** Andean, Caribbean, Pacific **Dept.:** ANT, BOL, CES, CHO, CUN, MAG, SUC, TOL **Elev.:** 0–1000 m a.s.l. **IUCN Habitat code:** 1,2,3,14 **Conservation status: (IUCN)** LC

5391 *Croton glabellus* L. (Euphorbiaceae) **IPNI:** 1029203-2 **Common name:** algayubo, almizclillo, caguaneje, caguanejo, carcanapire, pulvide, quina blanca **Use:** MA, ME **IUCN Habitat code:** 1,2,3,5,14 **Conservation status: (IUCN)** LC

5392 *Croton glandulosus* L. (Euphorbiaceae) **IPNI:** 70233-2 **Use:** ME **Origin:** Native **IUCN Habitat code:** 1,2,3,14 **Conservation status: (IUCN)** NE

5393 *Croton gossypiifolius* Vahl (Euphorbiaceae) **IPNI:** 70274-2 **Common name:** barbasco, candelero, cascarilla, chicharrón, drago, guacamayo, sangre de drago, sangregao, sangregrado, sangregrago, sangro, sangrón, árbol de nuche **Habit:** Shrub, Tree **Use:** MA, ME **Origin:** Native **Region:** Andean, Caribbean **Dept.:** CAU, CES, CUN, MAG, NSA, SAN **Elev.:** 700–2300 m a.s.l. **IUCN Habitat code:** 1,2,3,14 **Conservation status: (IUCN)** LC

5394 *Croton hibiscifolius* Kunth ex Spreng. (Euphorbiaceae) **IPNI:** 342669-1 **Common name:** drago, puntalanza, sangre de cristo, sangre de drago, sangre de dragón, sangregadillo, sangregado, sangregao, sangregrago **Habit:** Tree **Use:** AF, EU, ME **Origin:** Native **Region:** Andean, Pacific **Dept.:** ANT, BOG, BOY, CAU, CUN, HUI, TOL, VAC **Elev.:** 300–2700 m a.s.l. **IUCN Habitat code:** 1,2,3,14 **Conservation status: (IUCN)** NE

5395 *Croton hircinus* Vent. (Euphorbiaceae) **IPNI:** 325164-2 **Habit:** Shrub **Use:** MA **Origin:** Native **Region:** Caribbean **Dept.:** ATL, LAG

Elev.: 10–860 m a.s.l. **IUCN Habitat code:** 1,2,3 **Conservation status: (IUCN)** LC

5396 *Croton hirtus* L'Hér. (Euphorbiaceae) **IPNI:** 70337-2 **Common name:** drago, guacamayo **Habit:** Herb **Use:** ME **Origin:** Native **Region:** Andean, Caribbean, Orinoquia, Pacific **Dept.:** ANT, ATL, BOL, CAL, CAU, COR, CUN, LAG, MAG, MET, NAR, QUI, SAN, TOL, VAC **Elev.:** 0-1800 msnm **IUCN Habitat code:** 1,2,3,4,14 **Conservation status: (IUCN)** NE **(CPSG)** Potential LC

5397 *Croton lechleri* Müll.Arg. (Euphorbiaceae) **IPNI:** 342842-1 **Common name:** balsa macho, drago, drago sangregado, palo de drago, sangre de drago, sangre de grado, sangregado, sangregrado, sangro, topa roja **Habit:** Tree **Use:** MA, ME **Origin:** Native **Region:** Pacific **Dept.:** PUT **Elev.:** 370–700 m a.s.l. **IUCN Habitat code:** 1,3,4,14 **Conservation status: (IUCN)** NE

5398 *Croton leptostachyus* Kunth (Euphorbiaceae) **IPNI:** 342854-1 **Common name:** drago, mosquerillo, mosquerito, mosquero, sangregao, árbol de nuche **Habit:** Shrub **Use:** ME **Origin:** Native **Region:** Andean **Dept.:** CAL, CUN, HUI, SAN, TOL, VAC **Elev.:** 175–1600 m a.s.l. **IUCN Habitat code:** 1,2,3,5,14 **Conservation status: (IUCN)** NE

5399 *Croton malambo* H.Karst. (Euphorbiaceae) **IPNI:** 342933-1 **Common name:** canela falsa, cáscara de lombrices, malambo, palomatías **Habit:** Small tree **Use:** MA, ME, PO **Origin:** Native **Region:** Caribbean **Dept.:** ATL, BOL, LAG, MAG **Elev.:** 0–200 m a.s.l. **IUCN Habitat code:** 1,2,3,5,14 **Conservation status: (IUCN)** NE

5400 *Croton matourensis* Aubl. (Euphorbiaceae) **IPNI:** 70513-2 **Common name:** drago, lengua de vaca, madera blanca, ojo de sardina, sardina, tabaquillo, tablero **Habit:** Tree **Use:** MA, ME **Origin:** Native **Region:** Amazonia, Andean, Pacific **Dept.:** AMA, ANT, CAU, CHO, GUV, VAU **Elev.:** 70–1100 m a.s.l. **IUCN Habitat code:** 1,3,5,14 **Conservation status: (IUCN)** LC

5401 *Croton menthodorus* Benth. (Euphorbiaceae) **IPNI:** 342979-1 **Common name:** chala, cucharilla, mosquero, purga **Use:** ME **Origin:** Native **Conservation status: (IUCN)** NT

5402 *Croton micans* Sw. (Euphorbiaceae) **IPNI:** 70536-2 **Common name:** amargoso, guayabito **Habit:** Shrub **Use:** ME **Origin:** Native **Region:** Andean **Dept.:** CUN **Elev.:** 500–500 m a.s.l. **IUCN Habitat code:** 1,2,3 **Conservation status: (IUCN)** NE

5403 *Croton morifolius* Willd. (Euphorbiaceae) **IPNI:** 70562-2 **Use:** ME **Origin:** Native **Conservation status: (IUCN)** NE

5404 *Croton mutisianus* Kunth (Euphorbiaceae) **IPNI:** 343058-1 **Common name:** drago, grado, grago, grao, guacamayo, sangregao **Habit:** Tree **Use:** ME **Origin:** Native **Region:** Andean **Dept.:** ANT, BOG, CAU, CUN, HUI, MAG, MET, QUI, RIS, VAC **Elev.:** 1080-2900 msnm **IUCN Habitat code:** 1,2,3,4,14 **Conservation status: (IUCN)** NE **(CPSG)** Potential LC

5405 *Croton nitens* Sw. (Euphorbiaceae) **IPNI:** 343083-1 **Habit:** Shrub **Use:** ME **Origin:** Native **Region:** Caribbean **Dept.:** SAP **Elev.:** 0–100 m a.s.l. **Conservation status: (IUCN)** NE

5406 *Croton niveus* Jacq. (Euphorbiaceae) **IPNI:** 70599-2 **Common name:** copachí, copalchi, plateado, plateao, platero, platiado, platiao, platino **Habit:** Shrub **Use:** HF, ME **Origin:** Native **Region:** Caribbean **Dept.:** ATL, BOL, CES, COR, LAG, MAG **Elev.:** 0–400 m a.s.l. **IUCN Habitat code:** 1,2,4,14 **Conservation status: (IUCN)** LC

5407 *Croton ovalifolius* Vahl (Euphorbiaceae) **IPNI:** 325185-2 **Common name:** aloukemasu, amargosito **Habit:** Shrub **Use:** ME **Origin:** Native **Region:** Andean, Caribbean **Dept.:** LAG, MAG, NSA **Elev.:** 50–650 m a.s.l. **IUCN Habitat code:** 1,2,14 **Conservation status: (IUCN)** NE

5408 *Croton palanostigma* Klotzsch (Euphorbiaceae) **IPNI:** 343150-1 **Common name:** sangregao **Habit:** Tree **Use:** MA, ME **Origin:** Native **Region:** Amazonia, Orinoquia **Dept.:** AMA, CAQ, MET, VAU, VID **Elev.:** 250–300 m a.s.l. **IUCN Habitat code:** 1,3,14 **Conservation status: (IUCN)** LC

5409 *Croton schiedeanus* Schltdl. (Euphorbiaceae) **IPNI:** 70846-2 **Common name:** sabaleto **Habit:** Tree **Use:** ME **Origin:** Native **Region:** Andean, Caribbean, Pacific **Dept.:** ANT, CAL, CHO, CUN, HUI, NAR, SAP, SAN, TOL **Elev.:** 0–900 m a.s.l. **IUCN Habitat code:** 1,2,3,14 **Conservation status: (IUCN)** LC

5410 *Croton smithianus* Croizat (Euphorbiaceae) **IPNI:** 70906-2 **Common name:** candelero, drago, guacamayo, sangregado, sangregao, sangrerillo, sangrero **Habit:** Tree **Use:** ME **Origin:** Native **Region:** Andean, Pacific **Dept.:** ANT, CAL, CHO, CUN, QUI, RIS, SAN, VAC **Elev.:** 200–2100 m a.s.l. **IUCN Habitat code:** 1,2,3,14 **Conservation status: (IUCN)** LC

5411 *Croton tessmannii* Mansf. (Euphorbiaceae) **IPNI:** 70990-2 **Habit:** Tree **Use:** HF **Origin:** Native **Region:** Amazonia **Dept.:** CAQ **Elev.:** 100–200 m a.s.l. **IUCN Habitat code:** 3 **Conservation status: (IUCN)** NE

5412 *Croton tricolor* Klotzsch ex Baill. (Euphorbiaceae) **IPNI:** 343663-1 **Common name:** amargoso **Habit:** Shrub, Small tree **Use:** ME **Origin:** Native **Region:** Andean **Dept.:** CUN, TOL **Elev.:** 500–1000 m a.s.l. **IUCN Habitat code:** 14 **Conservation status: (IUCN)** LC

5413 *Croton trinitatis* Millsp. (Euphorbiaceae) **IPNI:** 71013-2 **Common name:** pata de paloma **Habit:** Subshrub **Use:** HF, MA, ME **Origin:** Native **Region:** Amazonia, Andean, Caribbean, Orinoquia, Pacific **Dept.:** AMA, ANT, ATL, BOL, BOY, CAL, CAQ, CAS, CHO, COR, CUN, GUA, MAG, MET, SAN, TOL, VAC, VAU, VID **Elev.:** 40-1050 msnm **IUCN Habitat code:** 1,2,3,4,5,14 **Conservation status: (IUCN)** NE **(CPSG)** Potential LC

5414 *Dalechampia cissifolia* Poepp. (Euphorbiaceae) **IPNI:** 344031-1 **Habit:** Climbing **Use:** ME **Origin:** Native **Region:** Andean **Dept.:** CUN, QUI **Elev.:** 670–1650 m a.s.l. **IUCN Habitat code:** 1,3 **Conservation status: (IUCN)** NE

5415 *Dalechampia dioscoreifolia* Poepp. (Euphorbiaceae) **IPNI:** 314158-2 **Common name:** fríjol morado **Habit:** Climbing **Use:** ME **Origin:** Native **Region:** Andean, Pacific **Dept.:** ANT, CAQ, CHO, CUN, NAR, QUI, SAN, VAC **Elev.:** 50–1100 m a.s.l. **IUCN Habitat code:** 1,3,14 **Conservation status: (IUCN)** NE

5416 *Dalechampia scandens* L. (Euphorbiaceae) **IPNI:** 319177-2 **Common name:** conejillo, picapica, vidita mía **Habit:** Climbing **Use:** AF, MA, ME, PO **Origin:** Native **Region:** Andean, Caribbean, Pacific **Dept.:** ANT, ATL, BOL, CUN, HUI, MAG, NAR, NSA, SAP, SAN, SUC, TOL, VAC **Elev.:** 30-1300 msnm **IUCN Habitat code:** 1,2,3,14 **Conservation status: (IUCN)** LC **(CPSG)** Potential LC

5417 *Dalechampia tiliifolia* Lam. (Euphorbiaceae) **IPNI:** 344174-1 **Habit:** Climbing **Use:** ME **Origin:** Native **Region:** Andean, Caribbean, Orinoquia, Pacific **Dept.:** ANT, ATL, BOL, CAL, CHO, COR, CUN, MAG, MET, TOL **Elev.:** 20–700 m a.s.l. **IUCN Habitat code:** 1,2,3,4,5,14 **Conservation status: (IUCN)** NE

5418 *Ditaxis argothamnoides* (Bertol. ex Spreng.) Radcl.-Sm. & Govaerts (Euphorbiaceae) **IPNI:** 995988-1 **Common name:** kasirusa outka, maravilla **Habit:** Herb **Use:** ME **Origin:** Native **Region:** Caribbean **Dept.:** ATL, BOL, CES, GUA, MAG **Elev.:** 0–200 m a.s.l. **IUCN Habitat code:** 1,2 **Conservation status: (IUCN)** NE

5419 *Euphorbia comosa* Vell. (Euphorbiaceae) **IPNI:** 323896-2 **Habit:** Herb **Use:** ME **Origin:** Native **Region:** Caribbean **Dept.:** MAG **Elev.:** 200–200 m a.s.l. **IUCN Habitat code:** 2,3 **Conservation status: (IUCN)** LC

5420 *Euphorbia cotinifolia* L. (Euphorbiaceae) **IPNI:** 346143-1 **Common name:** arbolito de navidad, barbasco, caucho, caucho lechoso, caucho rojo, curacarate, estoraque, lechero, lechero colorao, lechero rojo, lechero verde, lechoso, lechudo colorao, liberal, lorencillo, manzanillo, manzanillo caucano, manzanillo de cerca, manzanill **Habit:** Small tree **Use:** EU, ME, PO **Origin:** Native **Region:** Andean, Caribbean, Pacific **Dept.:** ANT, CAL, CAU, CES, CHO, CUN, HUI, LAG, MAG, NSA, QUI,

SAN, TOL, VAC **Elev.:** 250-2500 msnm **IUCN Habitat code:** 1,2,3,4,5,14 **Conservation status: (IUCN)** LC **(CPSG)** Potential LC

5421 *Euphorbia cyparissias* L. (Euphorbiaceae) **IPNI:** 346205-1 **Habit:** Herb **Use:** EU, HF, ME, PO **Origin:** Naturalised **Region:** Andean **Dept.:** CUN **Elev.:** 1900–2400 m a.s.l. **Conservation status: (IUCN)** NE

5422 *Euphorbia densiflora* (Klotzsch) Klotzsch (Euphorbiaceae) **IPNI:** 1100936-2 **Habit:** Herb **Use:** ME **Origin:** Native **IUCN Habitat code:** 2,14 **Conservation status: (IUCN)** NE

5423 *Euphorbia fischeriana* Steud. (Euphorbiaceae) **IPNI:** 346510-1 **Use:** FU, MA, ME, PO **Conservation status: (IUCN)** NE

5424 *Euphorbia heterophylla* L. (Euphorbiaceae) **IPNI:** 30062970-2 **Common name:** clavo de cristo, golondrina, gota de sangre, viernes santo, yerba lechera **Habit:** Herb **Use:** AF, EU, GS, HF, MA, ME, PO, SU **Origin:** Native **Region:** Andean, Caribbean, Pacific **Dept.:** ANT, ATL, BOL, CUN, HUI, MAG, NAR, QUI, SAP, SAN, TOL, VAC **Elev.:** 80–2600 m a.s.l. **IUCN Habitat code:** 1,2,3,5,14 **Conservation status: (IUCN)** NE

5425 *Euphorbia heyneana* Spreng. (Euphorbiaceae) **IPNI:** 346789-1 **Use:** HF, ME **Conservation status: (IUCN)** NE

5426 *Euphorbia hirta* L. (Euphorbiaceae) **IPNI:** 101651-2 **Common name:** atu rui, fiafianae, hierba de pollo, hierba golondrina, pimpinela, tripa de pollo, tripaepollo, vua, yerba de pollo, yerba de sapo **Habit:** Herb **Use:** AF, HF, MA, ME, PO **Origin:** Native **Region:** Andean, Caribbean, Orinoquia, Pacific **Dept.:** ANT, ATL, BOL, BOY, CAL, CAS, CAU, CHO, COR, CUN, HUI, LAG, MAG, MET, NAR, NSA, QUI, SAP, SAN, SUC, TOL, VAC, VID **Elev.:** 0-2800 msnm **IUCN Habitat code:** 1,2,3,4,5,14 **Conservation status: (IUCN)** NE **(CPSG)** Potential LC

5427 *Euphorbia hypericifolia* L. (Euphorbiaceae) **IPNI:** 101671-2 **Common name:** canchalagua, chickenweed, palomitas, pimpinela blanca **Habit:** Herb **Use:** ME **Origin:** Native **Region:** Andean, Caribbean **Dept.:** ANT, ATL, BOL, BOY, CAL, COR, CUN, HUI, MAG, NAR, QUI, RIS, SAP, SAN, TOL, VAC **Elev.:** 0-2700 msnm **IUCN Habitat code:** 1,2,3,4,5,14 **Conservation status: (IUCN)** NE **(CPSG)** Potential LC

5428 *Euphorbia hyssopifolia* L. (Euphorbiaceae) **IPNI:** 101676-2 **Common name:** canchalagua, celedonia, herb de pollo, lechosa, verruguero **Habit:** Herb **Use:** ME **Origin:** Native **Region:** Andean, Caribbean, Orinoquia, Pacific **Dept.:** ANT, ATL, BOL, BOY, CAL, CAS, CAU, CHO, COR, CUN, HUI, MAG, MET, NAR, QUI, SAP, SAN, SUC, TOL, VAC **Elev.:** 0-2700 msnm **IUCN Habitat code:** 1,2,3,5,14 **Conservation status: (IUCN)** NE **(CPSG)** Potential LC

5429 *Euphorbia lactea* Haw. (Euphorbiaceae) **IPNI:** 347035-1 **Common name:** carató, karatón, lechero espinoso **Habit:** Subshrub **Use:** EU, MA, ME, PO **Origin:** Cultivated **Region:** Andean **Dept.:** SAN **Elev.:** 1100–1100 m a.s.l. **IUCN Habitat code:** 2,14 **Conservation status: (IUCN)** NE

5430 *Euphorbia lancifolia* Schltdl. (Euphorbiaceae) **IPNI:** 347056-1 **Common name:** ixbut, lechero hojas de lanza **Habit:** Herb **Use:** ME **Origin:** Native **Region:** Andean **Dept.:** VAC **Elev.:** 1000–1000 m a.s.l. **IUCN Habitat code:** 1 **Conservation status: (IUCN)** NE

5431 *Euphorbia lasiocarpa* Klotzsch (Euphorbiaceae) **IPNI:** 347064-1 **Common name:** canchalagua **Habit:** Herb **Use:** ME **Origin:** Native **Region:** Andean, Caribbean, Orinoquia **Dept.:** ANT, BOG, BOY, CAL, CUN, HUI, LAG, MAG, MET, NAR, NSA, RIS, SAN, SUC, TOL, VAC **Elev.:** 40-2660 msnm **IUCN Habitat code:** 1,2,3,5,14 **Conservation status: (IUCN)** NE **(CPSG)** Potential LC

5432 *Euphorbia lathyris* L. (Euphorbiaceae) **IPNI:** 347072-1 **Common name:** estrella, simetría, tártaro, álamo **Habit:** Herb **Use:** FU, HF, MA, ME, PO **Origin:** Cultivated **Region:** Andean **Dept.:** BOG, BOY, CUN **Elev.:** 2500–2647 m a.s.l. **Conservation status: (IUCN)** NE

5433 *Euphorbia laurifolia* Juss. ex Lam. (Euphorbiaceae) **IPNI:** 347079-1 **Common name:** caucho, caucho macho, lechero, lechero de cerca, nacedera, piglio, pinillo, pinllo **Habit:** Herb, Shrub, Small tree **Use:** ME **Origin:** Native **Region:** Andean **Dept.:** ANT, BOG, CAL, CAU, CHO, CUN, HUI, NAR, QUI, TOL, VAC **Elev.:** 1050–2850 m a.s.l. **IUCN Habitat code:** 1,3,14 **Conservation status: (IUCN)** LC

5434 *Euphorbia leucocephala* Lotsy (Euphorbiaceae) **IPNI:** 101612-2 **Common name:** garza, granito de arroz, lluvia de estrellas, palomitas **Habit:** Herb, Shrub **Use:** EU **Origin:** Native **Region:** Andean **Dept.:** ANT, CUN, QUI **Elev.:** 950–2000 m a.s.l. **IUCN Habitat code:** 1,14 **Conservation status: (IUCN)** NE

5435 *Euphorbia maculata* L. (Euphorbiaceae) **IPNI:** 347236-1 **Habit:** Herb **Use:** ME, PO **Origin:** Naturalised **Region:** Caribbean **Dept.:** BOL **Elev.:** 0–200 m a.s.l. **Conservation status: (IUCN)** NE

5436 *Euphorbia mauritanica* L. (Euphorbiaceae) **IPNI:** 347292-1 **Common name:** elástica, yellow milkbush **Use:** EU, ME, PO **IUCN Habitat code:** 14 **Conservation status: (IUCN)** NE

5437 *Euphorbia mesembryanthemifolia* Jacq. (Euphorbiaceae) **IPNI:** 347331-1 **Habit:** Herb **Use:** ME **Origin:** Native **Region:** Caribbean **Dept.:** BOL, LAG, MAG, SAP **Elev.:** 0–360 m a.s.l. **IUCN Habitat code:** 1,5 **Conservation status: (IUCN)** NE

5438 *Euphorbia milii* Des Moul. (Euphorbiaceae) **IPNI:** 347359-1 **Common name:** clavos de Cristo, corona de Cristo, corona de cristo, corona de espinas, disciplinas, espina de Cristo, lengua del diablo, llaga de Cristo, martirios, sangrecristo, tuiyó, tuyo, tú y yo **Habit:** Subshrub **Use:** EU, ME, PO **Origin:** Cultivated **Region:** Andean, Caribbean **Dept.:** BOG, CAU, CUN, QUI, SAP, VAC **Elev.:** 1000–1700 m a.s.l. **IUCN Habitat code:** 1,3,14 **Conservation status: (IUCN)** LC

5439 *Euphorbia neriifolia* L. (Euphorbiaceae) **IPNI:** 347489-1 **Common name:** díctamo real, leche de cerca, lechero de cercas, lechosa, penca de cercas **Habit:** Shrub, Subshrub **Use:** EU, HF, MA, ME, PO **Origin:** Cultivated **Dept.:** CUN **Elev.:** 500–500 m a.s.l. **IUCN Habitat code:** 3 **Conservation status: (IUCN)** LC

5440 *Euphorbia nutans* Lag. (Euphorbiaceae) **IPNI:** 101777-2 **Habit:** Herb **Use:** ME, PO **Origin:** Native **Conservation status: (IUCN)** NE

5441 *Euphorbia ophthalmica* Pers. (Euphorbiaceae) **IPNI:** 101793-2 **Habit:** Herb **Use:** ME **Origin:** Native **Conservation status: (IUCN)** NE

5442 *Euphorbia orbiculata* Kunth (Euphorbiaceae) **IPNI:** 347590-1 **Common name:** canchalagua, canchalagüe, lecherito, picantina, taforita, teología, yerba de pollo **Habit:** Herb **Use:** HF, ME **Origin:** Native **Region:** Andean **Dept.:** BOG, BOY, CUN **Elev.:** 2100–4000 m a.s.l. **IUCN Habitat code:** 1,3,4,5,14 **Conservation status: (IUCN)** NE

5443 *Euphorbia peplus* L. (Euphorbiaceae) **IPNI:** 347724-1 **Common name:** hierba de leche, leche eterna, lechero pequeño, picantico, picantina, pituya, teología, yerbaeleche **Habit:** Herb **Use:** AF, HF, ME, PO **Origin:** Naturalised **Region:** Andean **Dept.:** ANT, BOG, CAU, CUN, NAR, QUI **Elev.:** 1700–3100 m a.s.l. **IUCN Habitat code:** 1,3,14 **Conservation status: (IUCN)** NE

5444 *Euphorbia prostrata* Aiton (Euphorbiaceae) **IPNI:** 101913-2 **Common name:** tripita de pollo **Habit:** Herb **Use:** HF, MA, ME, PO, SU **Origin:** Native **Region:** Andean, Caribbean **Dept.:** ATL, CUN, HUI, MAG **Elev.:** 0–700 m a.s.l. **IUCN Habitat code:** 1,3,5,14 **Conservation status: (IUCN)** NE

5445 *Euphorbia pulcherrima* Willd. ex Klotzsch (Euphorbiaceae) **IPNI:** 30176564-2 **Common name:** alfonsolópez, arbolito de navidad, bandera de Pizarro, bandera de guerra, barbegallo, blas del mar, cardenal, carnaval, cayeno, corbata de Gaitán, corbata de Uribe, corona de oliva, crestegallo, cuiche rojo, flor de navidad, liberal, liberalito, moño, nav **Habit:** Shrub **Use:** EU, GS, HF, IF, MA, ME, PO, SU **Origin:** Cultivated **Region:** Andean, Caribbean **Dept.:** BOG, BOL, CUN, HUI, MAG, NAR,

PUT, QUI **Elev.:** 470–2650 m a.s.l. **IUCN Habitat code:** 1,2,3,14 **Conservation status: (IUCN)** LC

5446 *Euphorbia seguieriana* Neck. (Euphorbiaceae)
IPNI: 348221-1
Habit: Herb **Use:** ME **Origin:** Cultivated **Region:** Andean **Dept.:** CUN, MET **Elev.:** 1900–3200 m a.s.l. **IUCN Habitat code:** 14 **Conservation status: (IUCN)** NE

5447 *Euphorbia serpens* Kunth (Euphorbiaceae)
IPNI: 101976-2
Habit: Herb **Use:** ME **Origin:** Native **Region:** Caribbean **Dept.:** MAG **Elev.:** 75–75 m a.s.l. **IUCN Habitat code:** 5,14 **Conservation status: (IUCN)** NE

5448 *Euphorbia thymifolia* L. (Euphorbiaceae)
IPNI: 102034-2
Common name: bejuco para nube de ojo, hierba lechera, lechera, lágrima de madre, quebradura, yerba lechera **Habit:** Herb **Use:** HF, MA, ME, PO **Origin:** Native **Region:** Amazonia, Andean, Caribbean, Orinoquia, Pacific **Dept.:** AMA, ANT, ATL, BOL, CAQ, CHO, COR, GUA, MAG, MET, NAR, NSA, PUT, SAP, TOL, VAC, VID **Elev.:** 0-1200 msnm **IUCN Habitat code:** 1,2,3,4,5,14 **Conservation status: (IUCN)** NE **(CPSG)** Potential LC

5449 *Euphorbia tirucalli* L. (Euphorbiaceae)
IPNI: 348517-1
Common name: cactus, quebradura **Habit:** Shrub **Use:** AF, EU, FU, HF, IF, MA, ME, PO, SU **Origin:** Cultivated **Region:** Andean, Caribbean **Dept.:** MAG, SAP, SAN, TOL **Elev.:** 100–670 m a.s.l. **IUCN Habitat code:** 1,2,14 **Conservation status: (IUCN)** LC

5450 *Euphorbia tithymaloides* L. (Euphorbiaceae)
IPNI: 102036-2
Common name: bitamorrial, clavitos, clavo de cristo, clavo de huerto, clavos de Cristo, dictamo real, ditamo real, díctamo, dítamo real, esperma, gallito colorado, ipeca, ipecacuana, itamo real, itamorreal, itamorrial, lechero, lechero de palito, lechiterna, pajarito, **Habit:** Herb **Use:** EU, MA, ME, PO **Origin:** Native **Region:** Andean, Caribbean, Orinoquia, Pacific **Dept.:** ATL, BOL, CHO, COR, CUN, HUI, LAG, MAG, MET, SAP **Elev.:** 0–900 m a.s.l. **IUCN Habitat code:** 1,2,3,5,14 **Conservation status: (IUCN)** LC

5451 *Euphorbia trigona* Mill. (Euphorbiaceae)
IPNI: 348565-1
Common name: cactus **Habit:** Shrub **Use:** EU, HF, ME, PO **Origin:** Cultivated **Region:** Andean, Caribbean **Dept.:** CUN, MAG **Elev.:** 1100–1100 m a.s.l. **IUCN Habitat code:** 1,2,3 **Conservation status: (IUCN)** NE

5452 *Euphorbia xylophylloides* Brongn. ex Lem. (Euphorbiaceae)
IPNI: 348758-1
Habit: Shrub **Use:** EU **Origin:** Cultivated **Region:** Andean **Dept.:** SAN **Elev.:** 780–780 m a.s.l. **Conservation status: (IUCN)** VU

5453 *Garcia nutans* Vahl ex Rohr (Euphorbiaceae)
IPNI: 349162-1
Common name: avellana, avellano, avellano del Nare, cedrón, naranjito, pacualito, pascualito **Habit:** Tree **Use:** AF, EU, MA, ME **Origin:** Native **Region:** Caribbean **Dept.:** BOL, MAG, SUC **Elev.:** 100–300 m a.s.l. **IUCN Habitat code:** 1,2,3,14 **Conservation status: (IUCN)** EN

5454 *Glycydendron amazonicum* Ducke (Euphorbiaceae)
IPNI: 111068-2
Habit: Tree **Use:** HF **Origin:** Native **Region:** Amazonia **Dept.:** CAQ **Elev.:** 300–300 m a.s.l. **IUCN Habitat code:** 1,3,14 **Conservation status: (IUCN)** LC

5455 *Hevea benthamiana* Müll.Arg. (Euphorbiaceae)
IPNI: 349911-1
Common name: caucho, cucho, lechero, siringa, siringa hoja de ratón **Habit:** Tree **Use:** HF **Origin:** Native **Region:** Amazonia **Dept.:** AMA, CAQ, PUT, VAU **Elev.:** 100–480 m a.s.l. **IUCN Habitat code:** 1,4,5 **Conservation status: (IUCN)** LC

5456 *Hevea brasiliensis* (Willd. ex A.Juss.) Müll.Arg. (Euphorbiaceae)
IPNI: 349913-1
Common name: caucho, caucho amazónico, caucho de Pará, caucho de pará, cuacho amazónico, hule, lechoso, siringa **Habit:** Tree **Use:** AF, EU, FU, GS, HF, MA, ME **Origin:** Native **Region:** Amazonia, Pacific **Dept.:** AMA, ANT, BOY, CAQ, GUA, NAR, VAU **Elev.:** 100–400 m a.s.l. **IUCN Habitat code:** 1,3,4,5,14 **Conservation status: (IUCN)** LC

5457 *Hevea guianensis* Aubl. (Euphorbiaceae)
IPNI: 121724-2
Common name: bizcocho, buyecacu, caucho, siringa, siringa cauchosa, siringa de hoja larga, siringa propia, siringalo, siringo **Habit:** Tree **Use:** HF, MA, ME **Origin:** Native **Region:** Amazonia **Dept.:** AMA, CAQ, PUT, VAU **Elev.:** 80–650 m a.s.l. **IUCN Habitat code:** 1,3,4,5,14 **Conservation status: (IUCN)** LC

5458 *Hevea microphylla* Ule (Euphorbiaceae)
IPNI: 121748-2
Habit: Tree **Use:** MA **Origin:** Native **Region:** Amazonia **Dept.:** VAU **Elev.:** 850–900 m a.s.l. **IUCN Habitat code:** 1 **Conservation status: (IUCN)** NE

5459 *Hevea nitida* Mart. ex Müll.Arg. (Euphorbiaceae)
IPNI: 349938-1
Common name: buyecacu, caucho, lechoso, pegajoso, siringa, siringa de panguana, siringa de sabana, siringa pegajosa, siringarana **Habit:** Tree **Use:** HF, MA **Origin:** Native **Region:** Amazonia **Dept.:** AMA, CAQ, GUV, VAU **Elev.:** 250–450 m a.s.l. **IUCN Habitat code:** 1,3,4,5,14 **Conservation status: (IUCN)** LC

5460 *Hevea pauciflora* (Spruce ex Benth.) Müll.Arg. (Euphorbiaceae)
IPNI: 349941-1
Common name: caucho, siringa **Habit:** Tree **Use:** HF, MA **Origin:** Native **Region:** Amazonia **Dept.:** AMA, VAU, VID **Elev.:** 100–300 m a.s.l. **IUCN Habitat code:** 1,3,4,5 **Conservation status: (IUCN)** LC

5461 *Hevea rigidifolia* (Spruce ex Benth.) Müll.Arg. (Euphorbiaceae)
IPNI: 349944-1
Habit: Tree **Use:** HF **Origin:** Native **Region:** Amazonia **Dept.:** AMA, GUA, VAU **Elev.:** 320–320 m a.s.l. **IUCN Habitat code:** 1,3 **Conservation status: (IUCN)** NE

5462 *Hevea spruceana* (Benth.) Müll.Arg. (Euphorbiaceae)
IPNI: 349947-1
Use: HF, MA, ME **IUCN Habitat code:** 1 **Conservation status: (IUCN)** LC

5463 *Hippomane mancinella* L. (Euphorbiaceae)
IPNI: 349995-1
Common name: manzanilla, manzanillo, manzanillo de playa **Habit:** Tree **Use:** EU, MA, ME, PO **Origin:** Native **Region:** Caribbean, Pacific **Dept.:** ATL, BOL, CAU, CHO, MAG, SAP **Elev.:** 0–2 m a.s.l. **IUCN Habitat code:** 1,2,5 **Conservation status: (IUCN)** LC

5464 *Hura crepitans* L. (Euphorbiaceae)
IPNI: 350060-1
Common name: acuapa, acuapar, arenillero, arenillo, barbasco, castañeto, catagua, catahua, ceiba, ceiba amarilla, ceiba blanca, ceiba brava, ceiba bruja, ceiba de leche, ceiba lechera, ceiba lechosa, ceiba lechuda, ceiba pacua, ceiba tarro, ceiba tronadora, ceibo, cei **Habit:** Tree **Use:** AF, EU, FU, HF, MA, ME, PO, SU **Origin:** Native **Region:** Amazonia, Andean, Caribbean, Orinoquia, Pacific **Dept.:** AMA, ANT, ATL, BOL, BOY, CES, CHO, COR, CUN, LAG, MAG, MET, NSA, SAN **Elev.:** 0-1550 msnm **IUCN Habitat code:** 1,2,3,4,5,14 **Conservation status: (IUCN)** NE **(CPSG)** Potential LC

5465 *Jatropha curcas* L. (Euphorbiaceae)
IPNI: 131462-2
Common name: coquillo, cumataci, frailecillo, higo de infierno, jaquillo, mañolo, medsiyen, papayito, piñón, piñón blanco, piñón botija, piñón de indias, piñón de la India, piñón de purga, purga, purga de fraile, tree santo, tua-tua, tártago, túa-túa **Habit:** Tree **Use:** AF, EU, FU, GS, HF, MA, ME, PO **Origin:** Native **Region:** Andean, Orinoquia, Pacific **Dept.:** AMA, ANT, BOL, BOY, CAS, CES, CHO, COR, CUN, GUV, HUI, NAR, SAN, VID **Elev.:** 0-1850 msnm **IUCN Habitat code:** 1,2,3,4,14 **Conservation status: (IUCN)** LC **(CPSG)** Potential LC

5466 *Jatropha gossypiifolia* L. (Euphorbiaceae)
IPNI: 323930-2
Common name: aichwua, arracachillo, brailejón, carretillo, frailecillo, frailejón calentano, higuerilla, jaquillo, piñón negro, pringamoza macho, purga de fraile, túa tua, túa túa, túa-túa, túa-túa morada **Habit:** Shrub **Use:** EU, FU, HF, MA, ME, PO, SU **Origin:** Native **Region:** Andean, Caribbean, Orinoquia, Pacific **Dept.:** AMA, ANT, ATL, BOL, BOY, CAS, CAU, CUN, HUI, LAG, MAG, NAR, SAP, SAN, SUC, TOL, VAC, VID **Elev.:** 0-2100 msnm **IUCN Habitat code:** 1,2,3,4,5,14 **Conservation status: (IUCN)** LC **(CPSG)** Potential LC

5467 *Jatropha integerrima* Jacq. (Euphorbiaceae)
IPNI: 350290-1
Common name: amor de madre, beso de madre, robertico, todoelaño, veinte de julio **Habit:** Shrub **Use:** EU, ME **Origin:** Naturalised **Region:** Andean, Caribbean, Pacific **Dept.:** ANT, BOL, CHO, CUN, MAG, QUI **Elev.:** 0–2550 m a.s.l. **IUCN Habitat code:** 1,2,3,14 **Conservation status: (IUCN)** NE

5468 *Jatropha multifida* L. (Euphorbiaceae)
IPNI: 314308-2
Common name: avellano purgante, coralplant, emético, flor de coral, tártago, tártago emético, yuca de la India **Habit:** Shrub **Use:** EU, FU, HF, MA, ME, PO, SU

Origin: Naturalised **Region:** Caribbean **Dept.:** CES, QUI **Elev.:** 500–1500 m a.s.l. **IUCN Habitat code:** 1,2,3,14 **Conservation status: (IUCN)** NE

5469 *Jatropha podagrica* Hook. (Euphorbiaceae) **IPNI:** 350415-1 **Common name:** corales, guatemalan rhubard, novio, patihinchado, white rhubard, yuca, yuco **Habit:** Shrub **Use:** EU, ME, PO **Origin:** Naturalised **Region:** Andean, Caribbean, Pacific **Dept.:** BOL, CHO, CUN **Elev.:** 100–300 m a.s.l. **IUCN Habitat code:** 1,2,3,4,5,14 **Conservation status: (IUCN)** NE

5470 *Mabea nitida* Spruce ex Benth. (Euphorbiaceae) **IPNI:** 280156-2 **Common name:** reventilla, reventillo **Habit:** Tree **Use:** ME **Origin:** Native **Region:** Amazonia, Orinoquia **Dept.:** AMA, ARA, CAQ, CAS, GUA, GUV, MET, PUT, VAU, VID **Elev.:** 100–500 m a.s.l. **IUCN Habitat code:** 1,2,3,4,5,14 **Conservation status: (IUCN)** LC

5471 *Mabea occidentalis* Benth. (Euphorbiaceae) **IPNI:** 280157-2 **Common name:** cenizo, mare blanco **Habit:** Shrub, Tree **Use:** MA **Origin:** Native **Region:** Amazonia, Andean, Guiana Shield, Caribbean, Pacific **Dept.:** ANT, BOL, CAQ, CAU, CHO, COR, MAG, MET, NAR, SAN **Elev.:** 50–1400 m a.s.l. **IUCN Habitat code:** 1,2,3,4,5,14 **Conservation status: (IUCN)** NE

5472 *Mabea piriri* Aubl. (Euphorbiaceae) **IPNI:** 350987-1 **Habit:** Small tree, Tree **Use:** MA, ME **Origin:** Native **Region:** Amazonia, Andean, Orinoquia **Dept.:** AMA, ANT, CAQ, CUN, GUV, MET, PUT, VAU **Elev.:** 100–950 m a.s.l. **IUCN Habitat code:** 1,3,5,14 **Conservation status: (IUCN)** LC

5473 *Mabea subsessilis* Pax & K.Hoffm. (Euphorbiaceae) **IPNI:** 148081-2 **Common name:** palo de batidor **Habit:** Shrub, Small tree **Use:** HF **Origin:** Native **Region:** Amazonia **Dept.:** AMA, CAQ, VAU **Elev.:** 100–600 m a.s.l. **IUCN Habitat code:** 1,3,14 **Conservation status: (IUCN)** LC

5474 *Mabea trianae* Pax (Euphorbiaceae) **IPNI:** 148215-2 **Common name:** tsakibeiwa **Habit:** Shrub, Tree **Use:** ME **Origin:** Native **Region:** Andean, Guiana Shield, Orinoquia **Dept.:** ARA, BOY, CAS, GUV, MET, VAU, VID **Elev.:** 100–450 m a.s.l. **IUCN Habitat code:** 1,2,3,4,5,14 **Conservation status: (IUCN)** NE

5475 *Manihot brachyloba* Müll.Arg. (Euphorbiaceae) **IPNI:** 351747-1 **Common name:** yuca nativa, yuca silvestre **Habit:** Shrub **Use:** GS **Origin:** Native **Region:** Amazonia, Andean, Caribbean, Orinoquia, Pacific **Dept.:** AMA, ANT, BOL, CHO, GUV, MAG, MET, NSA, RIS, SAN, VAC **Elev.:** 0–900 m a.s.l. **IUCN Habitat code:** 1,2,3,5,14 **Conservation status: (IUCN)** LC

5476 *Manihot carthagenensis* (Jacq.) Müll.Arg. (Euphorbiaceae) **IPNI:** 1100922-2 **Common name:** matapalaa, yuca brava, yuca de monte, yuca escorzonera, yuca guajira, yuco **Habit:** Shrub **Use:** AF, EU, GS, HF, MA, ME, PO **Origin:** Native **Region:** Caribbean **Dept.:** ATL, BOL, CUN, LAG, MAG **Elev.:** 0–350 m a.s.l. **IUCN Habitat code:** 1,2,3,5,14 **Conservation status: (IUCN)** LC

5477 *Manihot esculenta* Crantz (Euphorbiaceae) **IPNI:** 351790-1 **Common name:** botón, buii, buñuelo, casanareña, caá, cui, hoja de canangucha, lengüevenado, loqueta, mandioca, manicuera, manyok, mejorada, monablanca, monita, newajü, newajüsinae, newaüibü, nolí, palo de yuca, pari, patetórtola, quii, quivá, riogrande, sardinaprieta, **Habit:** Shrub **Use:** AF, EU, FU, GS, HF, IF, MA, ME, PO, SU **Origin:** Native **Region:** Amazonia, Andean, Guiana Shield, Caribbean, Orinoquia, Pacific **Dept.:** AMA, ANT, ARA, CAL, CAQ, CAS, CHO, CUN, GUA, MAG, MET, QUI, SAN, TOL, VAC, VAU, VID **Elev.:** 50-1550 msnm **IUCN Habitat code:** 1,2,3,4,5,14 **Conservation status: (IUCN)** NE **(CPSG)** Potential LC

5478 *Manihot leptophylla* Pax & K.Hoffm. (Euphorbiaceae) **IPNI:** 152449-2 **Habit:** Shrub **Use:** GS, HF **Origin:** Native **Region:** Amazonia **Dept.:** CAQ **Elev.:** 270–300 m a.s.l. **IUCN Habitat code:** 1,3,14 **Conservation status: (IUCN)** LC

5479 *Maprounea guianensis* Aubl. (Euphorbiaceae) **IPNI:** 352054-1 **Common name:** arrayán blanco, coca macho, kuipirai, morado, nibagaigaraoau, usiyab-ai **Habit:** Tree **Use:** MA, ME **Origin:** Native **Region:** Amazonia, Andean, Guiana Shield, Caribbean, Orinoquia **Dept.:** AMA, ANT, BOY, CAQ, COR, GUA, MET, NSA, SAN, TOL, VAU, VID **Elev.:** 50-850 msnm **IUCN Habitat code:** 1,2,3,4,5,14 **Conservation status: (IUCN)** LC **(CPSG)** Potential LC

5480 *Micrandra minor* Benth. (Euphorbiaceae) **IPNI:** 352256-1 **Habit:** Tree **Use:** HF **Origin:** Native **Region:** Amazonia **Dept.:** AMA, VAU **Elev.:** 250–250 m a.s.l. **IUCN Habitat code:** 1,3 **Conservation status: (IUCN)** NE

5481 *Micrandra rossiana* R.E.Schult. (Euphorbiaceae) **IPNI:** 159987-2 **Habit:** Tree **Use:** HF **Origin:** Native **Region:** Amazonia **Dept.:** AMA, VAU, VID **Elev.:** 200–300 m a.s.l. **IUCN Habitat code:** 1,5,14 **Conservation status: (IUCN)** NE

5482 *Micrandra siphonioides* Benth. (Euphorbiaceae) **IPNI:** 352260-1 **Common name:** caimo, madroña, sirina de sabana, siringa blanca, siringa de panguana, siringa de pescado, siringa de sabana, tree dulce, árbol dulce **Habit:** Tree **Use:** HF **Origin:** Native **Region:** Amazonia **Dept.:** AMA, GUA, VAU **Elev.:** 100–300 m a.s.l. **IUCN Habitat code:** 1,3,4 **Conservation status: (IUCN)** LC

5483 *Micrandra spruceana* (Baill.) R.E.Schult. (Euphorbiaceae) **IPNI:** 159993-2 **Common name:** mortecino, reventillo, siringa, siringa blanca **Habit:** Tree **Use:** HF, MA **Origin:** Native **Region:** Amazonia **Dept.:** AMA, CAQ, VAU **Elev.:** 100–650 m a.s.l. **IUCN Habitat code:** 1,3,4,5,14 **Conservation status: (IUCN)** LC

5484 *Micrandra sprucei* (Müll.Arg.) R.E.Schult. (Euphorbiaceae) **IPNI:** 159994-2 **Habit:** Tree **Use:** HF **Origin:** Native **Region:** Amazonia **Dept.:** AMA, GUA, VAU **Elev.:** 250–450 m a.s.l. **IUCN Habitat code:** 1,3 **Conservation status: (IUCN)** NE

5485 *Nealchornea yapurensis* Huber (Euphorbiaceae) **IPNI:** 167726-2 **Common name:** bizcocho, coral, fariña seca, mantecoso, marupá, tamara, yucutare **Habit:** Tree **Use:** HF **Origin:** Native **Region:** Amazonia **Dept.:** AMA, CAQ, VAU **Elev.:** 100–300 m a.s.l. **IUCN Habitat code:** 1,5,14 **Conservation status: (IUCN)** LC

5486 *Omphalea diandra* L. (Euphorbiaceae) **IPNI:** 352565-1 **Common name:** mantecona, zaragoza **Habit:** Liana **Use:** HF, MA, ME **Origin:** Native **Region:** Amazonia, Caribbean, Orinoquia, Pacific **Dept.:** AMA, ANT, BOL, CHO, MET **Elev.:** 0–300 m a.s.l. **IUCN Habitat code:** 1,3,5,14 **Conservation status: (IUCN)** NE

5487 *Plukenetia polyadenia* Müll.Arg. (Euphorbiaceae) **IPNI:** 203935-2 **Habit:** Climbing **Use:** HF, ME **Origin:** Native **Region:** Orinoquia **Dept.:** MET **Elev.:** 450–2140 m a.s.l. **IUCN Habitat code:** 1,14 **Conservation status: (IUCN)** NE

5488 *Plukenetia volubilis* L. (Euphorbiaceae) **IPNI:** 354970-1 **Common name:** chocolate corona, maní estrella, sacha inchi **Habit:** Climbing **Use:** HF, MA, ME **Origin:** Native **Region:** Amazonia, Pacific **Dept.:** CHO, PUT, QUI **Elev.:** 100–1500 m a.s.l. **IUCN Habitat code:** 1,2,3,14 **Conservation status: (IUCN)** NE

5489 *Rhodothyrsus macrophyllus* (Ducke) Esser (Euphorbiaceae) **IPNI:** 1008822-1 **Common name:** bizcocho **Habit:** Tree **Use:** HF **Origin:** Native **Region:** Amazonia **Dept.:** AMA, CAQ, VAU **Elev.:** 150–200 m a.s.l. **IUCN Habitat code:** 1,3,14 **Conservation status: (IUCN)** LC

5490 *Ricinus communis* L. (Euphorbiaceae) **IPNI:** 355498-1 **Common name:** carapate, castor, cawapat, higuera, higuereta, higuereto, higuerilla, higuerillo, higuerillo blanco, higuerillo rojo, higuerito, jieratapi, masketi, palm kristi, palmacristi, palmchristis, popo, rejalgar, relajar, ricin, ricino, tirraya, tártago **Habit:** Shrub, Small tree **Use:** AF, EU, FU, HF, MA, ME, PO, SU **Origin:** Cultivated, Naturalised **Region:** Andean, Caribbean, Pacific **Dept.:** ANT, BOG, BOL, BOY, CHO, COR, CUN, HUI, LAG, MAG, PUT, QUI, SAP, SUC, VAC **Elev.:** 50–2650 m a.s.l. **IUCN Habitat code:** 1,2,3,5,14 **Conservation status: (IUCN)** NE

5491 *Sagotia racemosa* Baill. (Euphorbiaceae) **IPNI:** 355722-1 **Habit:** Tree **Use:** MA, ME **Origin:** Native **Region:** Amazonia, Pacific **Dept.:** AMA, CHO, GUA, VAU **Elev.:** 100–300 m a.s.l. **IUCN Habitat code:** 1,2,3,14 **Conservation status: (IUCN)** LC

5492 *Sapium glandulosum* (L.) Morong (Euphorbiaceae) **IPNI:** 227503-2 **Common name:** azucenillo, cauchillo, caucho, caucho balato, caucho blanco, caucho negro, caucho piñique, ceiba, clavo, higuerón, hule, lechero, lechito, liga, manzanillo, nacedero, palo de leche, peñico, peñique, piñi, piñico, piñique, piñiquepiñique, tinguitingui, ñip **Habit:** Tree **Use:** HF, MA, ME, PO **Origin:** Native **Region:** Amazonia, Andean, Caribbean, Orinoquia, Pacific **Dept.:** AMA, ANT, ATL, BOG, BOL, BOY, CAQ, CHO, COR, CUN, LAG, MAG, MET, NAR, SUC **Elev.:** 0-2300 msnm **IUCN Habitat code:** 1,2,3,4,5,14 **Conservation status: (IUCN)** LC **(CPSG)** Potential LC

5493 *Sapium marmieri* Huber (Euphorbiaceae) **IPNI:** 355874-1 **Common name:** caucho, cauchorana, mantequillo, palo de tabaco, siringarana **Habit:** Tree **Use:** MA, ME **Origin:** Native **Region:** Amazonia, Orinoquia **Dept.:** AMA, CAQ, GUV, MET, PUT **Elev.:** 100–600 m a.s.l. **IUCN Habitat code:** 1,2,3,14 **Conservation status: (IUCN)** NE

5494 *Sapium stylare* Müll.Arg. (Euphorbiaceae) **IPNI:** 355950-1 **Common name:** cauchillo serrano, caucho, caucho blanco, cebo, chilco, lechero, lechero del blanco, lechudo, lengua de vaca, lengua de vaca cebo, mantequilla, mantequillo, murillo, palo de leche **Habit:** Tree **Use:** MA **Origin:** Native **Region:** Andean, Orinoquia **Dept.:** ANT, BOY, CAL, CAS, CAU, CUN, HUI, MET, NAR, PUT, QUI, RIS, TOL, VAC, VID **Elev.:** 100-3150 msnm **IUCN Habitat code:** 1,3,4,5,14 **Conservation status: (IUCN)** NE **(CPSG)** Potential LC

5495 *Sebastiania brasiliensis* Spreng. (Euphorbiaceae) **IPNI:** 356183-1 **Habit:** Shrub **Use:** ME **Origin:** Native **Region:** Andean **Dept.:** NSA **Elev.:** 1060–1060 m a.s.l. **IUCN Habitat code:** 14 **Conservation status: (IUCN)** LC

5496 *Tetrorchidium andinum* Müll.Arg. (Euphorbiaceae) **IPNI:** 356853-1 **Common name:** costillo, palo de tunda, palo tumba, tunda **Habit:** Shrub, Tree **Use:** ME **Origin:** Native **Region:** Andean, Pacific **Dept.:** ANT, CAU, NAR, VAC **Elev.:** 30–3150 m a.s.l. **IUCN Habitat code:** 1,3,5,14 **Conservation status: (IUCN)** LC

5497 *Tetrorchidium rubrivenium* Poepp. (Euphorbiaceae) **IPNI:** 356873-1 **Common name:** arenillo, copachí, granizo, hojarasco, manteco blanco **Habit:** Small tree **Use:** MA **Origin:** Native **Region:** Amazonia, Andean, Guiana Shield **Dept.:** ANT, BOY, CUN, HUI, MAG, MET, NSA, PUT, QUI, SAN, TOL, VAC **Elev.:** 45-2800 msnm **IUCN Habitat code:** 1,3,5,14 **Conservation status: (IUCN)** LC **(CPSG)** Potential LC

5498 *Tragia volubilis* L. (Euphorbiaceae) **IPNI:** 315307-2 **Common name:** conejillo, ortiguilla **Habit:** Climbing **Use:** ME, PO **Origin:** Native **Region:** Andean, Caribbean, Orinoquia, Pacific **Dept.:** ANT, BOL, CES, CHO, CUN, MET, NAR, SUC, VAC **Elev.:** 550–2000 m a.s.l. **IUCN Habitat code:** 1,2,3,14 **Conservation status: (IUCN)** NE

5499 *Vaupesia cataractarum* R.E.Schult. (Euphorbiaceae) **IPNI:** 262733-2 **Common name:** arirai, doopoje, jufigü yeneckke, jufigüna, yeneckke **Habit:** Tree **Use:** HF, ME **Origin:** Native **Region:** Amazonia **Dept.:** AMA, CAQ, VAU **Elev.:** 100–300 m a.s.l. **IUCN Habitat code:** 1,14 **Conservation status: (IUCN)** NE

5500 *Vernicia fordii* (Hemsl.) Airy Shaw (Euphorbiaceae) **IPNI:** 358198-1 **Habit:** Tree **Use:** HF, MA, ME, PO **Origin:** Cultivated **Region:** Andean **Dept.:** ANT **Elev.:** 400–1500 m a.s.l. **IUCN Habitat code:** 14 **Conservation status: (IUCN)** LC

5501 *Goupia glabra* Aubl. (Goupiaceae) **IPNI:** 161264-1 **Common name:** barbasco, buchesapo, chaquiro, chipro, cuerosapo, jotduna, látigo, macano, parature, pecoje, pesá, piande, piaunde, poraquero, saino, saino negro, sajino, sajino negro, saíno, saíno negro, taracana, todikai **Habit:** Tree **Use:** HF, MA, ME **Origin:** Native **Region:** Amazonia, Guiana Shield, Pacific **Dept.:** AMA, ANT, BOL, CAQ, CAU, CHO, NAR, NSA, SAN, VAC, VAU, VID **Elev.:** 0–900 m a.s.l. **IUCN Habitat code:** 1,3,4,5,14 **Conservation status: (IUCN)** LC

5502 *Humiria balsamifera* Aubl. (Humiriaceae) **IPNI:** 431431-1 **Common name:** capirón,chíchicù, chichique, chilco, oloroso, tree de los lamentos, árbol de los lamentos **Habit:** Shrub, Small tree, Tree **Use:** HF **Origin:** Native **Region:** Amazonia, Guiana Shield **Dept.:** AMA, CAQ, GUA, GUV, MET, NAR, PUT, VAC, VAU **Elev.:** 0–1400 m a.s.l. **IUCN Habitat code:** 1,3,4,5,14 **Conservation status: (IUCN)** LC

5503 *Humiriastrum colombianum* (Cuatrec.) Cuatrec. (Humiriaceae) **IPNI:** 124794-2 **Common name:** aceituno, aceituno rojo, achote de monte, bateo, milpesos de cimbe **Habit:** Tree **Use:** HF, MA **Origin:** Native **Dept.:** ANT, SAN **Elev.:** 100-700 msnm **IUCN Habitat code:** 1,2,3,4 **Conservation status: (IUCN)** LC **(CPSG)** LC

5504 *Humiriastrum diguense* (Cuatrec.) Cuatrec. (Humiriaceae) **IPNI:** 124798-2 **Common name:** ají **Habit:** Small tree, Tree **Use:** HF, MA **Origin:** Native **Region:** Andean, Pacific **Dept.:** CAU, VAC **Elev.:** 50–1180 m a.s.l. **IUCN Habitat code:** 1,3 **Conservation status: (IUCN)** LC

5505 *Humiriastrum piraparanense* Cuatrec. (Humiriaceae) **IPNI:** 124806-2 **Habit:** Small tree, Tree **Use:** HF **Origin:** Native **Region:** Amazonia, Guiana Shield **Dept.:** AMA, CAQ, VAU **Elev.:** 250–660 m a.s.l. **IUCN Habitat code:** 1,3,4,14 **Conservation status: (IUCN)** LC

5506 *Humiriastrum procerum* (Little) Cuatrec. (Humiriaceae) **IPNI:** 124807-2 **Common name:** chanul, chanó, chanú, chilco **Habit:** Tree **Use:** HF, MA **Origin:** Native **Region:** Pacific **Dept.:** VAC **Elev.:** 5-100 msnm **IUCN Habitat code:** 1,5 **Conservation status: (IUCN)** VU **(CPSG)** CR

5507 *Sacoglottis amazonica* Mart. (Humiriaceae) **IPNI:** 431482-1 **Habit:** Small tree, Tree **Use:** HF **Origin:** Native **Region:** Amazonia **Dept.:** AMA **Elev.:** 100–220 m a.s.l. **IUCN Habitat code:** 1,3,4,5,14 **Conservation status: (IUCN)** LC

5508 *Sacoglottis ceratocarpa* Ducke (Humiriaceae) **IPNI:** 225095-2 **Habit:** Small tree, Tree **Use:** HF **Origin:** Native **Region:** Amazonia, Guiana Shield **Dept.:** AMA, GUA **Elev.:** 100–100 m a.s.l. **IUCN Habitat code:** 1 **Conservation status: (IUCN)** LC

5509 *Sacoglottis guianensis* Benth. (Humiriaceae) **IPNI:** 431492-1 **Common name:** aceituno rojo, cabeza de picure, fierrito, palo picure **Habit:** Small tree, Tree **Use:** HF, MA, ME **Origin:** Native **Region:** Guiana Shield, Orinoquia **Dept.:** GUA, VID **Elev.:** 80–150 m a.s.l. **Conservation status: (IUCN)** LC

5510 *Sacoglottis mattogrossensis* Malme (Humiriaceae) **IPNI:** 225114-2 **Habit:** Small tree, Tree **Use:** MA **Origin:** Native **Region:** Amazonia, Orinoquia **Dept.:** AMA, VID **Elev.:** 200–220 m a.s.l. **IUCN Habitat code:** 1,14 **Conservation status: (IUCN)** LC

5511 *Schistostemon retusus* (Ducke) Cuatrec. (Humiriaceae) **IPNI:** 229304-2 **Habit:** Shrub, Small tree, Tree **Use:** HF **Origin:** Native **Region:** Amazonia, Guiana Shield **Dept.:** VAU **Elev.:** 180–400 m a.s.l. **Conservation status: (IUCN)** LC

5512 *Vantanea magdalenensis* Cuatrec. (Humiriaceae) **IPNI:** 262606-2 **Common name:** coral, marfil, masaba, masábalo, sangretoro **Habit:** Tree **Use:** MA **Origin:** Endemic **Dept.:** ANT, SAN **Elev.:** 600-1000 msnm **IUCN Habitat code:** 1,2,3 **Conservation status: (IUCN)** EN **(CPSG)** EN

5513 *Vantanea parviflora* Lam. (Humiriaceae) **IPNI:** 431534-1 **Habit:** Tree **Use:** HF **Origin:** Native **Region:** Guiana Shield **Dept.:** GUA, VAU **Elev.:** 120–120 m a.s.l. **IUCN Habitat code:** 1,3,14 **Conservation status: (IUCN)** LC

5514 *Hypericum androsaemum* L. (Hypericaceae) **IPNI:** 433187-1 **Common name:** hipérico, impérico **Habit:** Subshrub **Use:** EU, ME **Origin:** Cultivated **Region:** Andean **Dept.:** CUN **Elev.:** 2600–2600 m a.s.l. **Conservation status: (IUCN)** NE

5515 *Hypericum calycinum* L. (Hypericaceae) **IPNI:** 433287-1 **Use:** EU, GS, ME **Region:** Andean **Dept.:** BOG **IUCN Habitat code:** 1 **Conservation status: (IUCN)** NE

5516 *Hypericum juniperinum* Kunth (Hypericaceae) **IPNI:** 433524-1 **Common name:** chite, escobo, guardarrocío **Habit:** Shrub, Subshrub **Use:** AF, EU, FU, MA, ME **Origin:** Native **Region:** Andean **Dept.:** ANT, BOG, BOY, CAL, CAU, CHO, CUN, HUI, MAG, MET, NSA, PUT, SAN, TOL **Elev.:** 1990-3810 msnm **IUCN Habitat code:** 1,3,4,8,14 **Conservation status: (IUCN)** NE **(CPSG)** Potential LC

5517 *Hypericum laricifolium* Juss. (Hypericaceae) **IPNI:** 433547-1 **Common name:** chite, guardarrocío, velillo **Habit:** Shrub, Small tree **Use:** ME **Origin:** Native **Region:** Andean **Dept.:** ANT, BOG, BOY, CAL, CAU, CES, CHO, CUN, LAG, MAG, NAR, NSA, PUT, QUI, RIS, SAN, TOL, VAC **Elev.:** 2500-4300 msnm **IUCN Habitat code:** 1,2,3,4,8,14 **Conservation status: (IUCN)** LC **(CPSG)** Potential LC

5518 *Hypericum mexicanum* L. (Hypericaceae) **IPNI:** 428535-1 **Common name:** chite, guardarrocío, lunaria **Habit:** Shrub, Subshrub **Use:** ME **Origin:** Native **Region:** Andean **Dept.:** ANT, BOG, BOY, CUN, MET, NSA, SAN, VAC **Elev.:** 1700-4200 msnm **IUCN Habitat code:** 1,3,4,8,14 **Conservation status: (IUCN)** NE **(CPSG)** LC

5519 *Hypericum mutilum* L. (Hypericaceae) **IPNI:** 126289-2 **Habit:** Herb **Use:** ME **Origin:** Naturalised **Region:** Andean **Dept.:** CUN, HUI, NAR, VAC **Elev.:** 1500–2000 m a.s.l. **IUCN Habitat code:** 1,3,14 **Conservation status: (IUCN)** NE

5520 *Hypericum patulum* Thunb. (Hypericaceae) **IPNI:** 927423-1 **Habit:** Shrub **Use:** EU, ME **Origin:** Cultivated **Region:** Andean **Dept.:** BOG, CUN **Elev.:** 2620–2800 m a.s.l. **IUCN Habitat code:** 14 **Conservation status: (IUCN)** NE

5521 *Hypericum perforatum* L. (Hypericaceae) **IPNI:** 433719-1 **Common name:** alfalfa argenti herb amarilla, alfalfa argentina, corazoncillo, herb amarilla, hierba de San Juan, hierba de san juan, hipérico, hypericón **Use:** HF, MA, ME, PO **Conservation status: (IUCN)** NE

5522 *Hypericum silenoides* Juss. (Hypericaceae) **IPNI:** 317526-2 **Common name:** chinchinali **Habit:** Herb **Use:** ME **Origin:** Native **Region:** Andean **Dept.:** ANT, CAQ, CAU, NAR, VAC **Elev.:** 700–2850 m a.s.l. **IUCN Habitat code:** 1,3,14 **Conservation status: (IUCN)** NE

5523 *Hypericum xylosteifolium* (Spach) N.Robson (Hypericaceae) **IPNI:** 433966-1 **Use:** EU **Region:** Andean **Dept.:** BOG **IUCN Habitat code:** 3 **Conservation status: (IUCN)** NE

5524 *Vismia baccifera* (L.) Planch. & Triana (Hypericaceae) **IPNI:** 434100-1 **Common name:** achiotillo, achotillo, aguacatillo, carate, carate blanco, carate rojo, caratillo, cuero de gallina, hacillo, lacre, lancillo, lanzo, mancha mancha, manchador, manchamancha, mancharropa, mandur, morochillo colorado, papamo, punta de lanza, punta lanza, pu **Habit:** Shrub, Small tree, Tree **Use:** EU, FU, MA, ME **Origin:** Native **Region:** Amazonia, Andean, Orinoquia, Pacific **Dept.:** AMA, ANT, BOG, BOY, CAL, CAU, CES, CHO, COR, CUN, GUA, HUI, MAG, MET, NAR, NSA, PUT, SAN, TOL, VAC **Elev.:** 0-2880 msnm **IUCN Habitat code:** 1,2,3,4,14 **Conservation status: (IUCN)** LC **(CPSG)** Potential LC

5525 *Vismia cayennensis* (Jacq.) Pers. (Hypericaceae) **IPNI:** 434107-1 **Common name:** carate, manchador **Habit:** Tree **Use:** EU, FU, HF, MA, ME **Origin:** Native **Region:** Andean **Dept.:** ANT **Elev.:** 760–1380 m a.s.l. **IUCN Habitat code:** 1,3,4,5,14 **Conservation status: (IUCN)** LC

5526 *Vismia ferruginea* Kunth (Hypericaceae) **IPNI:** 434114-1 **Habit:** Tree **Use:** ME **Origin:** Native **Region:** Amazonia, Guiana Shield **Dept.:** AMA, GUA, VAU, VID **Elev.:** 30–600 m a.s.l. **IUCN Habitat code:** 1,3,4,14 **Conservation status: (IUCN)** NE

5527 *Vismia guianensis* (Aubl.) Pers. (Hypericaceae) **IPNI:** 30051774-2 **Common name:** San Francisco, caimito, caparrosa, carate, carate blanco, carate negro, carrasposa, lacre, lance, manchado, manchador, mandur, mandús, matamatá ripero, muñeco, onotillo, papama, papamo, punta de lanza, puntaelanza, puntelanza, puntelanzo, sangre, sangre g **Habit:** Tree **Use:** MA, ME **Origin:** Native **Region:** Orinoquia **Dept.:** VID **Elev.:** 100–100 m a.s.l. **IUCN Habitat code:** 1,3,4,5,14 **Conservation status: (IUCN)** LC

5528 *Vismia jefensis* N.Robson (Hypericaceae) **IPNI:** 267303-2 **Habit:** Shrub, Small tree **Use:** ME **Origin:** Native **Region:** Andean **Dept.:** ANT **Elev.:** 1030–2020 m a.s.l. **IUCN Habitat code:** 1,3 **Conservation status: (IUCN)** LC

5529 *Vismia lateriflora* Ducke (Hypericaceae) **IPNI:** 267306-2 **Use:** ME **Origin:** Native **IUCN Habitat code:** 1 **Conservation status: (IUCN)** NE

5530 *Vismia lauriformis* (Lam.) Choisy (Hypericaceae) **IPNI:** 434138-1 **Common name:** achiotillo, carate, carate rojo, hierro, hierrolanzo, lacre, lancillo, manchado, manchador, manchará, manchari, onotillo, punta de lanza, puntaelanza, sangre **Habit:** Tree **Use:** MA, ME **Origin:** Native **Region:** Amazonia, Andean, Orinoquia, Pacific **Dept.:** AMA, ANT, CAU, CHO, CUN, HUI, NAR, NSA, PUT, TOL, VAC **Elev.:** 5–2350 m a.s.l. **IUCN Habitat code:** 1,3,14 **Conservation status: (IUCN)** LC

5531 *Vismia macrophylla* Kunth (Hypericaceae) **IPNI:** 434144-1 **Common name:** aguacacho, carate, fierro lanzo, hierrolanzo, lacre, lacre hojiancho, lanzo, manchador, mancharra, palo de chicharra, pichirina, puntelanza, sangregallo, siete cueros, sietecuero, sietecueros, yodo **Habit:** Shrub, Small tree, Tree **Use:** EU, FU, HF, MA, ME **Origin:** Native **Region:** Amazonia, Andean, Orinoquia **Dept.:** AMA, ANT, BOL, CAQ, CAU, CHO, COR, CUN, MET, NAR, PUT, SAN, TOL, VAC, VID **Elev.:** 5-1400 msnm **IUCN Habitat code:** 1,2,3,4,5,14 **Conservation status: (IUCN)** LC **(CPSG)** Potential LC

5532 *Vismia minutiflora* Ewan (Hypericaceae) **IPNI:** 267320-2 **Habit:** Shrub, Tree **Use:** ME **Origin:** Native **Region:** Amazonia, Orinoquia **Dept.:** AMA, GUV, MET, VID **Elev.:** 100–300 m a.s.l. **IUCN Habitat code:** 1,14 **Conservation status: (IUCN)** LC

5533 *Vismia sandwithii* Ewan (Hypericaceae) **IPNI:** 267332-2 **Common name:** palo de chicharra **Habit:** Shrub, Tree **Use:** ME **Origin:** Native **Region:** Amazonia **Dept.:** CAQ, PUT **Elev.:** 200–850 m a.s.l. **IUCN Habitat code:** 1 **Conservation status: (IUCN)** LC

5534 *Vismia sessilifolia* (Aubl.) Choisy (Hypericaceae) **IPNI:** 434162-1 **Use:** ME **Origin:** Native **IUCN Habitat code:** 1 **Conservation status: (IUCN)** LC

5535 *Lacistema aggregatum* (P.J.Bergius) Rusby (Lacistemataceae) **IPNI:** 134457-2 **Common name:** amancayo colorado, cafecillo, café de monte, coral, coralito, laurel rosado, palo de monte, palo de perezoso, porrejao **Habit:** Shrub, Tree **Use:** HF, ME **Origin:** Native **Region:** Amazonia, Andean, Guiana Shield, Orinoquia, Pacific **Dept.:** AMA, ANT, BOL, BOY, CAL, CAQ, CAU, CHO, CUN, GUA, HUI, LAG, MAG, MET, PUT, QUI, SAN, TOL, VAC, VAU, VID **Elev.:** 0-2100 msnm **IUCN Habitat code:** 1,2,3,4,5,14 **Conservation status: (IUCN)** LC **(CPSG)** Potential LC

5536 *Lacistema nena* J.F.Macbr. (Lacistemataceae) **IPNI:** 134470-2 **Common name:** llanchamilla **Habit:** Shrub, Tree **Use:** HF **Origin:** Native **Region:** Amazonia, Orinoquia **Dept.:** AMA, ANT, VID **Elev.:** 100–220 m a.s.l. **IUCN Habitat code:** 1,3,4,14 **Conservation status: (IUCN)** NE

5537 *Linum usitatissimum* L. (Linaceae) **IPNI:** 544772-1 **Common name:** linaza, lino **Habit:** Herb **Use:** AF, FU, HF, MA, ME, PO **Origin:** Cultivated **Region:** Andean **Dept.:** ANT, BOG, BOY, CUN, NAR **Elev.:** 1000–3000 m a.s.l. **IUCN Habitat code:** 1,2,3,14 **Conservation status: (IUCN)** NE

5538 *Roucheria calophylla* Planch. (Linaceae) **IPNI:** 544878-1 **Habit:** Shrub, Tree **Use:** ME **Origin:** Native **Region:** Amazonia **Dept.:** AMA, GUA, VAU **Elev.:** 100–300 m a.s.l. **IUCN Habitat code:** 1,3,4,14 **Conservation status: (IUCN)** NE

5539 *Alicia macrodisca* (Triana & Planch.) Griseb. (Malpighiaceae) **IPNI:** 60442801-2 **Habit:** Liana **Use:** HF **Origin:** Native **Region:** Amazonia, Andean, Guiana Shield, Orinoquia **Dept.:** AMA, BOY, GUV, MET, PUT, VAU **Elev.:** 240–980 m a.s.l. **IUCN Habitat code:** 1,3 **Conservation status: (IUCN)** NE

5540 *Banisteriopsis caapi* (Spruce ex Griseb.) Morton (Malpighiaceae) **IPNI:** 29180-2 **Common name:** ayahuasca, bejuco bravo, bejuco de oro, bejuco dulce, caapi, caji, caji-idirecaii, capi, capi o caapi, changropanga, cordón umbilical, dapa, kapi, liana del napo, natema, oofa, pejí, pildé, pinde, yaco borrachero, yagé, yagé de cacería, yagé de monte, yaj **Habit:** Liana **Use:** ME, SU **Origin:** Native, Cultivated **Region:** Amazonia, Andean, Orinoquia, Pacific **Dept.:** AMA, ANT, CAQ, CAU, CHO, CUN, GUA, NAR, PUT, VAC, VAU, VID **Elev.:** 100–2200 m a.s.l. **IUCN Habitat code:** 1,3,4,5,14 **Conservation status: (IUCN)** NE

5541 *Banisteriopsis martiniana* (A.Juss.) Cuatrec. (Malpighiaceae) **IPNI:** 29229-2 **Common name:** yagé, yajé **Habit:** Liana **Use:** MA, ME **Origin:** Native **Region:** Amazonia **Dept.:** AMA, CAQ, VAU **Elev.:** 100–550 m a.s.l. **IUCN Habitat code:** 1,3,5,14 **Conservation status: (IUCN)** NE

5542 *Banisteriopsis muricata* (Cav.) Cuatrec. (Malpighiaceae) **IPNI:** 29242-2 **Common name:** alegría, ayahuasca, ayahuasca de los brujos, ayahuasca negra, ayahuasca rosada, muérdago, yagé **Habit:** Liana, Climbing **Use:** ME **Origin:** Native **Region:** Andean, Caribbean, Orinoquia **Dept.:** ANT, ARA, ATL, BOG, BOL, BOY, CES, COR, CUN, HUI, LAG, MAG, MET, NSA, SAN, SUC, TOL **Elev.:** 0–1900 m a.s.l. **IUCN Habitat code:** 1,2,3,5,14 **Conservation status: (IUCN)** NE

5543 *Bronwenia cornifolia* (Kunth) W.R.Anderson & C.Davis (Malpighiaceae) **IPNI:** 60447688-2 **Habit:** Shrub, Liana, Climbing **Use:** ME **Origin:** Native **Region:** Andean, Caribbean, Pacific **Dept.:** ANT, BOG, CUN, HUI, LAG, TOL **Elev.:** 50–3000 m a.s.l. **IUCN Habitat code:** 1,2,3,14 **Conservation status: (IUCN)** NE

5544 *Bunchosia argentea* (Jacq.) DC. (Malpighiaceae) **IPNI:** 555883-1 **Common name:** ciruelo **Habit:** Shrub, Tree **Use:** HF, ME **Origin:** Native **Region:** Amazonia, Andean, Caribbean, Pacific **Dept.:** ANT, BOL, CAU, CES, CHO, HUI, LAG, MAG, PUT, QUI, VAC **Elev.:** 0–2055 m a.s.l. **IUCN Habitat code:** 1,3,5,14 **Conservation status: (IUCN)** LC

5545 *Bunchosia armeniaca* (Cav.) DC. (Malpighiaceae) **IPNI:** 555884-1 **Common name:** café de monte, ciruela, ciruela de fraile, ciruelillo, ciruelo, ciruelo de fraile, diomate, guáimaro, mamey, mamey blanco de tierra fría, mamey de tierra fria, mamey de tierra fría, manzano de monte, marmelo, míspero **Habit:** Shrub, Small tree, Tree **Use:** EU, FU, HF **Origin:** Native **Region:** Amazonia, Andean, Orinoquia, Pacific **Dept.:** AMA, ANT, BOG, BOY, CAL, CAQ, CAS, CAU, CUN, GUV, HUI, MET, NAR, PUT, QUI, RIS, VAC **Elev.:** 100–2800 m a.s.l. **IUCN Habitat code:** 1,2,3,14 **Conservation status: (IUCN)** NE

5546 *Bunchosia deflexa* Triana & Planch. (Malpighiaceae) **IPNI:** 555898-1 **Use:** HF **Origin:** Native **Conservation status: (IUCN)** NE

5547 *Bunchosia glandulifera* (Jacq.) Kunth (Malpighiaceae) **IPNI:** 555908-1 **Common name:** cerezo, ciruela, ciruela silvestre **Habit:** Shrub, Small tree, Tree **Use:** EU, HF, ME **Origin:** Native, Cultivated **Region:** Amazonia, Andean, Guiana Shield, Caribbean, Orinoquia **Dept.:** AMA, ANT, BOL, CAQ, CAS, CUN, GUV, MET, PUT, SAN, TOL, VAU, VID **Elev.:** 70–2250 m a.s.l. **IUCN Habitat code:** 1,3,4,14 **Conservation status: (IUCN)** LC

5548 *Bunchosia lindeniana* A.Juss. (Malpighiaceae) **IPNI:** 303745-2 **Use:** HF, ME **Origin:** Native **Conservation status: (IUCN)** LC

5549 *Bunchosia nitida* (Jacq.) A.Rich. (Malpighiaceae) **IPNI:** 1122895-2 **Common name:** ciruelo cimarrón, ciruelo de perro, mamey de tierra fría, manzano de monte **Habit:** Shrub, Small tree, Tree **Use:** HF, ME **Origin:** Native, Cultivated **Region:** Andean, Caribbean, Orinoquia, Pacific **Dept.:** ANT, BOL, BOY, CAS, CAU, CHO, CUN, HUI, QUI, RIS, TOL, VAC **Elev.:** 0–1830 m a.s.l. **IUCN Habitat code:** 1,2,3,5,14 **Conservation status: (IUCN)** NE

5550 *Bunchosia odorata* (Jacq.) Juss. (Malpighiaceae) **IPNI:** 555944-1 **Common name:** vara de piedra **Habit:** Shrub, Small tree, Tree **Use:** HF, ME **Origin:** Native **Region:** Andean, Caribbean **Dept.:** ATL, BOL, CUN, LAG, MAG, NSA, SUC **Elev.:** 5–1300 m a.s.l. **IUCN Habitat code:** 1,2,3,4,14 **Conservation status: (IUCN)** LC

5551 *Bunchosia pseudonitida* Cuatrec. (Malpighiaceae) **IPNI:** 38339-2 **Common name:** cerezo rojo, ciruelo cimarrón, ciruelo de loma, huesito, pateperro **Habit:** Shrub, Small tree, Tree **Use:** HF, PO **Origin:** Native **Region:** Andean, Caribbean **Dept.:** ANT, CAU, CUN, HUI, NSA, QUI, RIS, SUC, TOL, VAC **Elev.:** 50–1600 m a.s.l. **IUCN Habitat code:** 1,2,3,14 **Conservation status: (IUCN)** NE

5552 *Bunchosia swartziana* Griseb. (Malpighiaceae) **IPNI:** 555969-1 **Habit:** Shrub, Small tree, Tree **Use:** ME **Origin:** Native **Region:** Caribbean **Dept.:** AMA, SAP **Elev.:** 50–50 m a.s.l. **Conservation status: (IUCN)** LC

5553 *Byrsonima aerugo* Sagot (Malpighiaceae) **IPNI:** 555981-1 **Habit:** Tree **Use:** HF, MA **Origin:** Native **Region:** Andean, Pacific **Dept.:** ANT, NAR **Elev.:** 112–1060 m a.s.l. **IUCN Habitat code:** 1 **Conservation status: (IUCN)** LC

5554 *Byrsonima arthropoda* A.Juss. (Malpighiaceae) **IPNI:** 555988-1 **Habit:** Tree **Use:** HF **Origin:** Native **Region:** Amazonia, Andean, Guiana Shield, Orinoquia, Pacific **Dept.:** AMA, ANT, CAU, MET, PUT, SAN, VAC, VAU **Elev.:** 5–1100 m a.s.l. **IUCN Habitat code:** 1,3,14 **Conservation status: (IUCN)** LC

5555 *Byrsonima chrysophylla* Kunth (Malpighiaceae) **IPNI:** 556014-1 **Common name:** chaparro **Habit:** Shrub, Tree **Use:** MA **Origin:** Native **Region:** Amazonia, Orinoquia **Dept.:** AMA, GUA, VAU, VID **Elev.:** 65–350 m a.s.l. **IUCN Habitat code:** 1,4,5,14 **Conservation status: (IUCN)** NE

5556 *Byrsonima coccolobifolia* Kunth (Malpighiaceae) **IPNI:** 556020-1 **Use:** HF, ME **Origin:** Native **Dept.:** VID **Conservation status: (IUCN)** LC

5557 *Byrsonima crassifolia* (L.) Kunth (Malpighiaceae) **IPNI:** 556033-1 **Common name:** alcornoco, barbasco, candelero, chaparro, chaparro manteco, chaparromanteco, chaviaco, encenillo, itsonae, maache, manero, manteco, manteco peralejo, mantequera, mantequero, mora, nanache, nance, nero, noro, noro lanoso, noro rubio, oreja de burro, pajari **Habit:** Shrub, Small tree, Tree **Use:** AF, EU, FU, HF, IF, MA, ME **Origin:** Native **Region:** Amazonia, Andean, Guiana Shield, Caribbean, Orinoquia, Pacific **Dept.:** ANT, ARA, BOL, BOY, CAQ, CAS, CAU, CES, CHO, COR, CUN, GUA, GUV, HUI, LAG, MAG, MET, NSA, SAP, SAN, SUC, TOL, VAU, VID **Elev.:** 0–2500 m a.s.l. **IUCN Habitat code:** 1,2,3,4,5,14 **Conservation status: (IUCN)** LC

5558 *Byrsonima crispa* A.Juss. (Malpighiaceae) **IPNI:** 556037-1 **Common name:** canastillo, noro, peralejo, peralejo de montaña **Habit:** Tree **Use:** HF **Origin:** Native **Region:** Amazonia, Andean, Guiana Shield, Orinoquia, Pacific **Dept.:** CAQ, GUV, MET, NAR **Elev.:** 60–1400 m a.s.l. **IUCN Habitat code:** 1,3,4,14 **Conservation status: (IUCN)** LC

5559 *Byrsonima fernandezii* Cuatrec. (Malpighiaceae) **IPNI:** 38852-2 **Habit:** Tree **Use:** HF **Origin:** Native **Region:** Amazonia, Guiana Shield **Dept.:** CAQ, GUA, GUV **Elev.:** 230–230 m a.s.l. **IUCN Habitat code:** 1,3,4 **Conservation status: (IUCN)** NE

5560 *Byrsonima hypoleuca* Turcz. (Malpighiaceae) **IPNI:** 556088-1 **Common name:** quina **Habit:** Tree **Use:** HF **Origin:** Native **Region:** Andean, Pacific **Dept.:** ANT, BOY, CAS, NSA, SAN **Elev.:** 112–1780 m a.s.l. **IUCN Habitat code:** 1,2,3,14 **Conservation status: (IUCN)** NE

5561 *Byrsonima japurensis* A.Juss. (Malpighiaceae) **IPNI:** 556093-1 **Common name:** alcornoco, coropo, diablorrojo **Habit:** Shrub, Small tree, Tree **Use:** HF **Origin:** Native **Region:** Amazonia, Guiana Shield, Orinoquia **Dept.:** AMA, CAQ, CAS, GUV, MET, PUT, VAU, VID **Elev.:** 100–450 m a.s.l. **IUCN Habitat code:** 1,2,3,4,5 **Conservation status: (IUCN)** LC

5562 *Byrsonima putumayensis* Cuatrec. (Malpighiaceae) **IPNI:** 38924-2 **Habit:** Tree **Use:** HF **Origin:** Native **Region:** Andean **Dept.:** PUT **Elev.:** 1350–1400 m a.s.l. **IUCN Habitat code:** 3 **Conservation status: (IUCN)** NE

5563 *Byrsonima roigii* Urb. (Malpighiaceae) **IPNI:** 38926-2 **Habit:** Shrub **Use:** HF **Origin:** Native **Region:** Caribbean **Dept.:** SAP **Elev.:** 0–85 m a.s.l. **Conservation status: (IUCN)** NE

5564 *Byrsonima spicata* (Cav.) Rich. ex Kunth (Malpighiaceae) **IPNI:** 556192-1 **Common name:** noro, peralejo, peralejo montañero **Habit:** Shrub, Tree **Use:** HF, ME **Origin:** Native **Region:** Amazonia, Andean, Guiana Shield, Caribbean, Orinoquia, Pacific **Dept.:** ANT, BOL, BOY, CAL, CAQ, CAS, CHO, COR, CUN, GUA, GUV, LAG, MET, NSA, SAN,

TOL, VID **Elev.:** 5–1156 m a.s.l. **IUCN Habitat code:** 1,2,3,4,5,14 **Conservation status: (IUCN)** LC

5565 *Byrsonima verbascifolia* (L.) DC. (Malpighiaceae) **IPNI:** 556209-1 **Common name:** manteco orejón **Habit:** Subshrub **Use:** HF, ME **Origin:** Native **Region:** Amazonia, Andean, Guiana Shield, Caribbean, Orinoquia **Dept.:** BOY, CAQ, CAS, GUV, LAG, MET, SAN, VID **Elev.:** 100–1700 m a.s.l. **Conservation status: (IUCN)** LC

5566 *Callaeum antifebrile* (Ruiz ex Griseb.) D.M.Johnson (Malpighiaceae) **IPNI:** 273852-2 **Habit:** Liana **Use:** ME, SU **Origin:** Native **Region:** Amazonia **Dept.:** AMA **Elev.:** 200–200 m a.s.l. **Conservation status: (IUCN)** NE

5567 *Diplopterys cabrerana* (Cuatrec.) B.Gates (Malpighiaceae) **IPNI:** 81736-2 **Common name:** chagro panga, chaguarpanga, yagé **Habit:** Liana **Use:** MA, ME, SU **Origin:** Native, Cultivated **Region:** Amazonia, Andean **Dept.:** AMA, PUT, VAU **Elev.:** 100–720 m a.s.l. **IUCN Habitat code:** 1 **Conservation status: (IUCN)** NE

5568 *Diplopterys longialata* (Nied.) W.R.Anderson & C.Davis (Malpighiaceae) **IPNI:** 60444585-2 **Common name:** chagrupanga, ocoó, ocoó-yagé, yagé **Habit:** Liana **Use:** ME, SU **Origin:** Native **Region:** Amazonia **Dept.:** CAQ, PUT **Elev.:** 400–600 m a.s.l. **IUCN Habitat code:** 1,14 **Conservation status: (IUCN)** NE

5569 *Galphimia gracilis* Bartl. (Malpighiaceae) **IPNI:** 556349-1 **Habit:** Shrub **Use:** EU, ME **Origin:** Native, Cultivated **Region:** Andean, Caribbean **Dept.:** ANT, ATL, BOL, CAU, CUN, HUI, MET, TOL, VAC **Elev.:** 30–2350 m a.s.l. **IUCN Habitat code:** 1,3,14 **Conservation status: (IUCN)** NE

5570 *Galphimia paniculata* Bartl. (Malpighiaceae) **IPNI:** 556362-1 **Habit:** Shrub **Use:** ME **Origin:** Cultivated **Region:** Andean **Dept.:** NAR **Elev.:** 1700–1700 m a.s.l. **IUCN Habitat code:** 1 **Conservation status: (IUCN)** NE

5571 *Heteropterys laurifolia* (L.) A.Juss. (Malpighiaceae) **IPNI:** 556563-1 **Habit:** Shrub, Small tree, Liana **Use:** ME **Origin:** Native **Region:** Andean, Caribbean **Dept.:** CUN, MAG, SAN, SUC, TOL **Elev.:** 20–1290 m a.s.l. **IUCN Habitat code:** 1,3,14 **Conservation status: (IUCN)** LC

5572 *Heteropterys leona* (Cav.) Exell (Malpighiaceae) **IPNI:** 1079574-2 **Habit:** Shrub, Liana **Use:** MA, ME **Origin:** Native **Region:** Pacific **Dept.:** CHO, VAC **Elev.:** 5–190 m a.s.l. **Conservation status: (IUCN)** LC

5573 *Heteropterys macrostachya* A.Juss. (Malpighiaceae) **IPNI:** 556578-1 **Habit:** Shrub, Liana **Use:** MA **Origin:** Native **Region:** Amazonia, Andean, Guiana Shield, Caribbean, Orinoquia **Dept.:** AMA, ANT, CAQ, CUN, GUV, MAG, MET, NSA, SAN, TOL, VAU **Elev.:** 60–1830 m a.s.l. **IUCN Habitat code:** 1,3,14 **Conservation status: (IUCN)** NE

5574 *Heteropterys riparia* Cuatrec. (Malpighiaceae) **IPNI:** 121328-2 **Common name:** mi, mi-sigó, sigó **Habit:** Tree, Liana **Use:** ME **Origin:** Native **Region:** Amazonia, Orinoquia **Dept.:** AMA, MET, PUT, VAU **Elev.:** 100–300 m a.s.l. **IUCN Habitat code:** 1 **Conservation status: (IUCN)** NE

5575 *Hiraea reclinata* Jacq. (Malpighiaceae) **IPNI:** 556914-1 **Habit:** Liana, Climbing **Use:** HF, ME **Origin:** Native **Region:** Caribbean, Pacific **Dept.:** ANT, ATL, BOL, CES, CHO, COR, MAG, SUC **Elev.:** 0–590 m a.s.l. **IUCN Habitat code:** 1,2,3,14 **Conservation status: (IUCN)** LC

5576 *Malpighia coccigera* L. (Malpighiaceae) **IPNI:** 323776-2 **Common name:** arbolito de navidad, holliday, holly brillante, holly piracanta, joli **Habit:** Subshrub **Use:** EU, HF, ME **Origin:** Cultivated **Region:** Amazonia, Guiana Shield, Pacific **Dept.:** CAL, CHO, GUV, SAN, VAC **Elev.:** 70–1000 m a.s.l. **IUCN Habitat code:** 14 **Conservation status: (IUCN)** NE

5577 *Malpighia emarginata* DC. (Malpighiaceae) **IPNI:** 323670-2 **Common name:** acerola, cerezo **Habit:** Shrub, Small tree **Use:** EU, GS, HF, MA, ME **Origin:** Native, Cultivated **Region:** Andean, Caribbean, Orinoquia **Dept.:** ANT, ARA, ATL, BOL, COR, HUI, LAG, MAG, NSA, SUC, TOL, VAC **Elev.:** 10–1600 m a.s.l. **IUCN Habitat code:** 1,2,3,14 **Conservation status: (IUCN)** NE

5578 *Malpighia glabra* L. (Malpighiaceae) **IPNI:** 557048-1 **Common name:** acerola, arrayancito, arrayán, arrayán colorado, cemeruco, cerecilla, cerecillo, cerecito, cereza, cereza montañera, cerezo, cerezo silvestre, chandeperro, ciruela de perro, ciruelo, ciruelo de perro, grosello, guayaba de perro, guayabo de perro, guindill **Habit:** Shrub, Small tree, Tree **Use:** AF, EU, HF, MA, ME, SU **Origin:** Native, Cultivated **Region:** Andean, Caribbean, Pacific **Dept.:** ANT, ATL, BOL, CAU, CES, CHO, CUN, HUI, LAG, MAG, NAR, SAN, SUC, TOL, VAC **Elev.:** 0–2100 m a.s.l. **IUCN Habitat code:** 1,2,3,14 **Conservation status: (IUCN)** LC

5579 *Mascagnia ovatifolia* (Kunth) Griseb. (Malpighiaceae) **IPNI:** 30237926-2 **Common name:** botones **Habit:** Liana, Climbing **Use:** ME **Origin:** Native **Region:** Caribbean, Orinoquia, Pacific **Dept.:** ANT, ARA, ATL, BOL, CES, LAG, MAG, MET, SAN **Elev.:** 0–1000 m a.s.l. **IUCN Habitat code:** 1,2,3,4,5,14 **Conservation status: (IUCN)** NE

5580 *Niedenzuella stannea* (Griseb.) W.R.Anderson (Malpighiaceae) **IPNI:** 60443063-2 **Habit:** Liana **Use:** ME **Origin:** Native **Region:** Amazonia **Dept.:** AMA, SAN **Elev.:** 150–500 m a.s.l. **IUCN Habitat code:** 3,14 **Conservation status: (IUCN)** NE

5581 *Spachea elegans* (G.Mey.) A.Juss. (Malpighiaceae) **IPNI:** 557422-1 **Use:** MA **Origin:** Native **IUCN Habitat code:** 1,3,14 **Conservation status: (IUCN)** LC

5582 *Stigmaphyllon ciliatum* (Lam.) A.Juss. (Malpighiaceae) **IPNI:** 557487-1 **Habit:** Climbing **Use:** EU **Origin:** Native **Region:** Pacific **Dept.:** ANT **Elev.:** 0–0 m a.s.l. **IUCN Habitat code:** 1,3,5,14 **Conservation status: (IUCN)** NE

5583 *Stigmaphyllon columbicum* Nied. (Malpighiaceae) **IPNI:** 557610-1 **Common name:** bejuco San Juan, bejuco san juan **Habit:** Climbing **Use:** ME **Origin:** Native **Region:** Andean, Caribbean, Pacific **Dept.:** ANT, BOL, BOY, CAL, CES, CHO, CUN, MAG, SAN, SUC, TOL **Elev.:** 30–3200 m a.s.l. **IUCN Habitat code:** 1,2,3,4,14 **Conservation status: (IUCN)** NE

5584 *Stigmaphyllon dichotomum* (L.) Griseb. (Malpighiaceae) **IPNI:** 557494-1 **Common name:** bejuco de San Juan **Habit:** Herb, Liana **Use:** ME **Origin:** Native **Region:** Andean, Caribbean, Orinoquia, Pacific **Dept.:** ANT, ATL, BOL, BOY, CAL, CES, CHO, CUN, MAG, MET, NSA, SAN, TOL **Elev.:** 0–1300 m a.s.l. **IUCN Habitat code:** 1,2,3,5,14 **Conservation status: (IUCN)** NE

5585 *Stigmaphyllon ellipticum* (Kunth) A.Juss. (Malpighiaceae) **IPNI:** 30241255-2 **Habit:** Liana, Climbing **Use:** ME **Origin:** Native **Region:** Andean, Caribbean, Pacific **Dept.:** ANT, ATL, CAU, CES, CHO, LAG, MAG, NAR, VAC **Elev.:** 0–1480 m a.s.l. **Conservation status: (IUCN)** NE

5586 *Stigmaphyllon sinuatum* (DC.) A.Juss. (Malpighiaceae) **IPNI:** 557592-1 **Common name:** batatilla, batatillo **Habit:** Liana, Climbing **Use:** ME **Origin:** Native **Region:** Amazonia, Guiana Shield, Orinoquia **Dept.:** AMA, ANT, BOY, CAQ, GUA, GUV, MET, NSA, SAN, VAU, VID **Elev.:** 40–800 m a.s.l. **IUCN Habitat code:** 1,3,4,5,14 **Conservation status: (IUCN)** NE

5587 *Tetrapterys crispa* A.Juss. (Malpighiaceae) **IPNI:** 557660-1 **Habit:** Shrub, Liana **Use:** HF **Origin:** Native **Region:** Andean, Caribbean, Orinoquia **Dept.:** ATL, BOL, CES, LAG, MAG, MET, NSA, SAN, TOL **Elev.:** 10–1300 m a.s.l. **IUCN Habitat code:** 1,2,3,4,14 **Conservation status: (IUCN)** NE

5588 *Tetrapterys discolor* (G.Mey.) DC. (Malpighiaceae) **IPNI:** 557670-1 **Habit:** Liana **Use:** ME **Origin:** Native **Region:** Amazonia, Andean, Orinoquia **Dept.:** ANT, CAL, CUN, GUV, MET, PUT, SAN, VAC, VID **Elev.:** 100–1900 m a.s.l. **IUCN Habitat code:** 1,2,3,4,14 **Conservation status: (IUCN)** NE

5589 *Tetrapterys styloptera* A.Juss. (Malpighiaceae) **IPNI:** 557785-1 **Common name:** caapi **Habit:** Shrub, Liana, Climbing **Use:** ME **Origin:** Native **Region:** Amazonia, Caribbean, Orinoquia, Pacific **Dept.:** AMA, ANT, CES,

CHO, COR, GUA, LAG, MAG, MET, NSA, SAN, SUC, VAU **Elev.:** 60–1300 m a.s.l. **IUCN Habitat code:** 1,2,3,5,14 **Conservation status: (IUCN)** LC

5590 *Cespedesia spathulata* (Ruiz & Pav.) Planch. (Ochnaceae) **IPNI:** 605994-1 **Common name:** alejandro, casaco, copaseca, falso cocoro, flor de mayo, lengua de vaca, lenguadevaca, lengüevaca, lengüevaco, manzano, matatoro, muguel colorado, pacora, pacó, pacó casaco, pacó de monte, pacó guitarro, pedro tomín **Habit:** Tree **Use:** EU, FU, MA, ME **Origin:** Native, Cultivated **Region:** Amazonia, Andean, Guiana Shield, Caribbean, Pacific **Dept.:** AMA, ANT, BOL, BOY, CAL, CAQ, CAU, CHO, COR, CUN, MET, NAR, NSA, PUT, SAN, TOL, VAC **Elev.:** 100–1200 m a.s.l. **IUCN Habitat code:** 1,3,5,14 **Conservation status: (IUCN)** LC

5591 *Godoya antioquiensis* Planch. (Ochnaceae) **IPNI:** 606173-1 **Common name:** caunce, floramarillo **Habit:** Tree **Use:** MA **Origin:** Native **Region:** Andean **Dept.:** ANT, CAU, TOL **Elev.:** 1000-2500 msnm **IUCN Habitat code:** 1,3,14 **Conservation status: (IUCN)** LC **(CPSG)** LC

5592 *Lacunaria crenata* (Tul.) A.C.Sm. (Ochnaceae) **IPNI:** 134573-2 **Habit:** Tree **Use:** HF **Origin:** Native **Region:** Amazonia, Andean, Pacific **Dept.:** ANT, BOY, CAS, CHO, MET, SAN **Elev.:** 50–1400 m a.s.l. **IUCN Habitat code:** 1,3,14 **Conservation status: (IUCN)** LC

5593 *Lacunaria jenmanii* (Oliv.) Ducke (Ochnaceae) **IPNI:** 134577-2 **Common name:** palo de venado **Habit:** Tree **Use:** HF, MA, ME **Origin:** Native **Region:** Amazonia, Guiana Shield, Pacific **Dept.:** AMA, ANT, BOL, CAL, CAQ, CHO, MET, SAN, VAU **Elev.:** 70–1400 m a.s.l. **IUCN Habitat code:** 1,3,4,14 **Conservation status: (IUCN)** NE

5594 *Lacunaria macrostachya* (Tul.) A.C.Sm. (Ochnaceae) **IPNI:** 134578-2 **Common name:** ají **Habit:** Small tree, Tree **Use:** HF **Origin:** Native **Region:** Amazonia, Guiana Shield **Dept.:** AMA, GUA, PUT **Elev.:** 100–300 m a.s.l. **IUCN Habitat code:** 1 **Conservation status: (IUCN)** NE

5595 *Ochna thomasiana* Engl. & Gilg (Ochnaceae) **IPNI:** 606804-1 **Habit:** Shrub **Use:** EU, MA **Origin:** Cultivated **Region:** Andean **Dept.:** ANT, QUI **Elev.:** 1200–1600 m a.s.l. **IUCN Habitat code:** 14 **Conservation status: (IUCN)** LC

5596 *Ouratea kananariensis* Sastre (Ochnaceae) **IPNI:** 281844-2 **Habit:** Tree **Use:** HF **Origin:** Endemic **Region:** Guiana Shield **Dept.:** AMA, VAU **Elev.:** 700-700 msnm **IUCN Habitat code:** 1,3,4 **Conservation status: (IUCN)** LC **(CPSG)** LC

5597 *Ouratea lucens* (Kunth) Engl. (Ochnaceae) **IPNI:** 607063-1 **Common name:** carbonero blanco, marmolejo **Habit:** Shrub, Small tree, Tree **Use:** ME **Origin:** Native **Region:** Andean, Caribbean, Pacific **Dept.:** ANT, ATL, CHO, CUN, MET **Elev.:** 0-1150 msnm **IUCN Habitat code:** 1,2,3,5,14 **Conservation status: (IUCN)** LC **(CPSG)** LC

5598 *Ouratea nitida* (Sw.) Engl. (Ochnaceae) **IPNI:** 607105-1 **Habit:** Shrub, Small tree **Use:** HF **Origin:** Native **Region:** Caribbean **Dept.:** LAG, SAP **Elev.:** 0–800 m a.s.l. **IUCN Habitat code:** 1 **Conservation status: (IUCN)** LC

5599 *Quiina amazonica* A.C.Sm. (Ochnaceae) **IPNI:** 216643-2 **Common name:** ají de monte, golondrina **Habit:** Shrub, Small tree **Use:** ME **Origin:** Native **Region:** Amazonia **Dept.:** AMA, PUT **Elev.:** 0–700 m a.s.l. **IUCN Habitat code:** 1,4,5,14 **Conservation status: (IUCN)** LC

5600 *Quiina cruegeriana* Griseb. (Ochnaceae) **IPNI:** 429118-1 **Habit:** Tree **Use:** MA **Origin:** Native **Region:** Amazonia, Pacific **Dept.:** AMA, ANT, CHO, VAC **Elev.:** 5–925 m a.s.l. **IUCN Habitat code:** 1,3,14 **Conservation status: (IUCN)** LC

5601 *Quiina florida* Tul. (Ochnaceae) **IPNI:** 429121-1 **Habit:** Shrub, Tree **Use:** HF **Origin:** Native **Region:** Amazonia, Orinoquia, Pacific **Dept.:** AMA, ANT, VAC, VID **Elev.:** 100–800 m a.s.l. **IUCN Habitat code:** 1,3,4,14 **Conservation status: (IUCN)** LC

5602 *Quiina integrifolia* Pulle (Ochnaceae) **IPNI:** 216655-2 **Habit:** Shrub, Tree **Use:** HF **Origin:** Native **Region:** Amazonia, Guiana Shield **Dept.:** VAU **Elev.:** 200–200 m a.s.l. **IUCN Habitat code:** 1 **Conservation status: (IUCN)** NE

5603 *Quiina leptoclada* Tul. (Ochnaceae) **IPNI:** 429126-1 **Habit:** Small tree **Use:** ME **Origin:** Native **Region:** Amazonia, Guiana Shield **Dept.:** VAU **Elev.:** 230–230 m a.s.l. **IUCN Habitat code:** 1 **Conservation status: (IUCN)** NE

5604 *Quiina longifolia* Spruce ex Planch. & Triana (Ochnaceae) **IPNI:** 429127-1 **Habit:** Shrub, Tree **Use:** ME **Origin:** Native **Region:** Guiana Shield, Orinoquia **Dept.:** GUA, VAU **Elev.:** 75–120 m a.s.l. **IUCN Habitat code:** 1,3,4,5 **Conservation status: (IUCN)** LC

5605 *Quiina macrophylla* Tul. (Ochnaceae) **IPNI:** 429129-1 **Common name:** juma?dacoci **Habit:** Shrub, Tree **Use:** HF, ME **Origin:** Native **Region:** Amazonia, Guiana Shield, Orinoquia **Dept.:** AMA, ANT, CAQ, CAS, MET, VAU, VID **Elev.:** 100–800 m a.s.l. **IUCN Habitat code:** 1,3,4,5,14 **Conservation status: (IUCN)** LC

5606 *Quiina obovata* Tul. (Ochnaceae) **IPNI:** 429133-1 **Habit:** Tree **Use:** HF **Origin:** Native **Region:** Amazonia, Guiana Shield **Dept.:** CAQ, VAU **Elev.:** 200–200 m a.s.l. **IUCN Habitat code:** 1,5 **Conservation status: (IUCN)** LC

5607 *Quiina rhytidopus* Tul. (Ochnaceae) **IPNI:** 429136-1 **Habit:** Shrub, Small tree **Use:** HF **Origin:** Native **Region:** Amazonia, Guiana Shield **Dept.:** GUA, VAU **Elev.:** 80–80 m a.s.l. **IUCN Habitat code:** 1 **Conservation status: (IUCN)** NE

5608 *Sauvagesia erecta* L. (Ochnaceae) **IPNI:** 867452-1 **Habit:** Herb **Use:** HF, MA, ME **Origin:** Native **Region:** Amazonia, Andean, Guiana Shield, Caribbean, Orinoquia, Pacific **Dept.:** AMA, ANT, BOG, BOY, CAL, CAQ, CAU, CHO, CUN, GUA, GUV, HUI, MAG, MET, NAR, NSA, PUT, SAN, TOL, VAC, VAU, VID **Elev.:** 0-2000 msnm **IUCN Habitat code:** 1,2,3,4,5,14 **Conservation status: (IUCN)** LC **(CPSG)** Potential LC

5609 *Sauvagesia rubiginosa* A.St.-Hil. (Ochnaceae) **IPNI:** 867469-1 **Habit:** Herb, Subshrub **Use:** ME **Origin:** Native **Region:** Amazonia, Pacific **Dept.:** AMA, CHO **Elev.:** 100–300 m a.s.l. **IUCN Habitat code:** 3,4,5,14 **Conservation status: (IUCN)** NE

5610 *Touroulia guianensis* Aubl. (Ochnaceae) **IPNI:** 429315-1 **Habit:** Small tree, Tree **Use:** HF, MA **Origin:** Native **Region:** Amazonia **Dept.:** AMA, CAQ **Elev.:** 100–200 m a.s.l. **IUCN Habitat code:** 1,3,14 **Conservation status: (IUCN)** LC

5611 *Wallacea insignis* Spruce ex Benth. & Hook.f. (Ochnaceae) **IPNI:** 607558-1 **Habit:** Shrub, Tree **Use:** PO **Origin:** Native **Region:** Guiana Shield, Orinoquia **Dept.:** AMA, CAQ, GUA, VAU, VID **Elev.:** 100–300 m a.s.l. **IUCN Habitat code:** 1,3,4,14 **Conservation status: (IUCN)** LC

5612 *Dilkea retusa* Mast. (Passifloraceae) **IPNI:** 674426-1 **Common name:** canilla de tente, cocorillo **Habit:** Shrub, Liana **Use:** HF **Origin:** Native **Region:** Amazonia **Dept.:** AMA, CAQ, PUT, VAC, VAU **Elev.:** 100-500 msnm **IUCN Habitat code:** 1,3,5,14 **Conservation status: (IUCN)** NE **(CPSG)** LC

5613 *Passiflora adenopoda* DC. (Passifloraceae) **IPNI:** 674754-1 **Habit:** Climbing **Use:** HF, ME, PO **Origin:** Native **Region:** Andean, Caribbean **Dept.:** ANT, BOY, CAL, CES, CUN, MAG, QUI, RIS, TOL, VAC **Elev.:** 100-1600 msnm **IUCN Habitat code:** 1,2,3,14 **Conservation status: (IUCN)** NE **(CPSG)** LC

5614 *Passiflora alata* Curtis (Passifloraceae) **IPNI:** 321312-2 **Common name:** maracuyá dulce **Habit:** Climbing **Use:** EU, HF, ME **Origin:** Cultivated **Region:** Andean **Dept.:** QUI, VAC **Elev.:** 1000–1200 m a.s.l. **IUCN Habitat code:** 1,3,5,14 **Conservation status: (IUCN)** NE

5615 *Passiflora alnifolia* Kunth (Passifloraceae) **IPNI:** 674768-1 **Common name:** granadilla, granadilla silvestre, maracuyá de monte, tausilla **Habit:** Climbing **Use:**

HF, ME **Origin:** Native **Region:** Andean **Dept.:** ANT, CAL, CAU, CES, CUN, HUI, MAG, NAR, QUI, RIS, SAN, TOL, VAC **Elev.:** 1500-2500 msnm **IUCN Habitat code:** 1,2,3,4,5,14 **Conservation status: (IUCN)** NE **(CPSG)** LC

5616 *Passiflora ambigua* Hemsl. (Passifloraceae) **IPNI:** 184791-2 **Common name:** chulupa de mico, chulupo, granadilla silvestre **Habit:** Climbing **Use:** HF **Origin:** Native **Region:** Andean, Pacific **Dept.:** ANT, CAL, CHO, HUI, MAG, MET, SAN, VAC **Elev.:** 0-2000 msnm **IUCN Habitat code:** 1,3,5,14 **Conservation status: (IUCN)** NE **(CPSG)** LC

5617 *Passiflora antioquiensis* H.Karst. (Passifloraceae) **IPNI:** 674787-1 **Common name:** curuba, curuba antioqueña, curuba quiteña, curuba silvestre, curubo antioqueño, curubo quiteño, gulupa **Habit:** Climbing **Use:** AF, EU, HF, ME **Origin:** Endemic, Cultivated **Region:** Andean **Dept.:** ANT, BOG, BOY, CAL, CUN, PUT, QUI, RIS, VAC **Elev.:** 1800–2700 m a.s.l. **IUCN Habitat code:** 1,3,14 **Conservation status: (IUCN)** NE

5618 *Passiflora arborea* Spreng. (Passifloraceae) **IPNI:** 674793-1 **Common name:** capafraile, chirimoyitos parchas, curuba de monte, curubo de árbol, granadilla de árbol, granadillo arbóreo, granadillo de árbol **Habit:** Tree **Use:** HF **Origin:** Endemic **Region:** Andean **Dept.:** ANT, BOY, CAL, CHO, CUN, HUI, MAG, NSA, QUI, RIS, VAC **Elev.:** 1000-2000 msnm **IUCN Habitat code:** 1,2,3,4,14 **Conservation status: (IUCN)** NE **(CPSG)** LC

5619 *Passiflora auriculata* Kunth (Passifloraceae) **IPNI:** 674805-1 **Common name:** bejuco de rejo, dentón, granadilla chica negra, rejito **Habit:** Climbing **Use:** HF **Origin:** Native **Region:** Amazonia, Andean, Caribbean, Orinoquia, Pacific **Dept.:** AMA, ANT, BOL, BOY, CAL, CAQ, CAS, CAU, CHO, COR, CUN, GUA, GUV, MAG, MET, NAR, NSA, PUT, QUI, SAN, TOL, VAC, VAU, VID **Elev.:** 0-1500 msnm **IUCN Habitat code:** 1,2,3,4,5,14 **Conservation status: (IUCN)** NE **(CPSG)** LC

5620 *Passiflora bicornis* Mill. (Passifloraceae) **IPNI:** 674821-1 **Habit:** Climbing **Use:** HF **Origin:** Native **Region:** Caribbean **Dept.:** ANT, ATL, BOL, LAG, MAG **Elev.:** 0-500 msnm **IUCN Habitat code:** 1,2,3,4,5,14 **Conservation status: (IUCN)** NE **(CPSG)** LC

5621 *Passiflora biflora* Lam. (Passifloraceae) **IPNI:** 674824-1 **Common name:** cachito de venado **Habit:** Climbing **Use:** HF, ME **Origin:** Native **Region:** Andean, Guiana Shield, Caribbean, Pacific **Dept.:** ANT, ATL, BOL, CAL, CAU, CES, CHO, COR, CUN, GUA, HUI, MAG, NAR, NSA, RIS, SAP, SAN, TOL, VAC **Elev.:** 0-1600 msnm **IUCN Habitat code:** 1,2,3,5,14 **Conservation status: (IUCN)** NE **(CPSG)** LC

5622 *Passiflora bogotensis* Benth. (Passifloraceae) **IPNI:** 674832-1 **Common name:** granadilla, granadilla silvestre, gulupa **Habit:** Climbing **Use:** HF **Origin:** Native **Region:** Andean **Dept.:** BOG, BOY, CAL, CUN, LAG, MAG, SAN **Elev.:** 2000-3000 msnm **IUCN Habitat code:** 1,3,4,6,8,14 **Conservation status: (IUCN)** NE **(CPSG)** LC

5623 *Passiflora bryonioides* Kunth (Passifloraceae) **IPNI:** 674847-1 **Habit:** Climbing **Use:** HF **Origin:** Native **Region:** Andean **Dept.:** BOG, CUN **Conservation status: (IUCN)** NE

5624 *Passiflora caerulea* L. (Passifloraceae) **IPNI:** 321949-2 **Habit:** Climbing **Use:** EU, GS, HF, MA, ME **Origin:** Cultivated **Region:** Andean **Dept.:** BOG, CAL, CUN, QUI **Elev.:** 1000–2700 m a.s.l. **IUCN Habitat code:** 1 **Conservation status: (IUCN)** NE

5625 *Passiflora candollei* Triana & Planch. (Passifloraceae) **IPNI:** 77063117-1 **Common name:** rejito **Habit:** Climbing **Use:** HF **Origin:** Native **Region:** Amazonia, Andean **Dept.:** AMA, CAQ, CUN, MET **Elev.:** 300–1500 m a.s.l. **IUCN Habitat code:** 1,3,14 **Conservation status: (IUCN)** NE

5626 *Passiflora capsularis* L. (Passifloraceae) **IPNI:** 321950-2 **Habit:** Climbing **Use:** HF, ME, PO **Origin:** Native **Region:** Andean, Caribbean **Dept.:** ANT, BOY, CAL, CHO, CUN, HUI, LAG, MAG, QUI, RIS, SAN, TOL, VAC **Elev.:** 0-1600 msnm **IUCN Habitat code:** 1,2,3,5,14 **Conservation status: (IUCN)** NE **(CPSG)** LC

5627 *Passiflora ciliata* Aiton (Passifloraceae) **IPNI:** 184857-2 **Use:** HF **Origin:** Native **IUCN Habitat code:** 2,5,14 **Conservation status: (IUCN)** NE

5628 *Passiflora cincinnata* Mast. (Passifloraceae) **IPNI:** 321963-2 **Habit:** Climbing **Use:** HF, MA, ME **Origin:** Native **Region:** Andean **Dept.:** ANT, SAN **Elev.:** 900–1300 m a.s.l. **IUCN Habitat code:** 14 **Conservation status: (IUCN)** NE

5629 *Passiflora coccinea* Aubl. (Passifloraceae) **IPNI:** 674909-1 **Common name:** granadilla roja **Habit:** Climbing **Use:** HF, MA, ME **Origin:** Native **Region:** Amazonia, Andean, Guiana Shield **Dept.:** AMA, GUA, GUV, MET, SAN, VAU, VID **Elev.:** 0-1500 msnm **IUCN Habitat code:** 1,3,4,5,14 **Conservation status: (IUCN)** NE **(CPSG)** LC

5630 *Passiflora coriacea* Juss. (Passifloraceae) **IPNI:** 674920-1 **Common name:** ala de murciélago, bejuco chimbilá, desgarretadera, granadilla, granadilla de monte, granadilla silvestre, hoja de murciélago, murciélago **Habit:** Climbing **Use:** EU, HF, ME **Origin:** Native **Region:** Andean, Caribbean **Dept.:** ANT, BOG, BOL, CAL, CHO, CUN, HUI, MAG, QUI, RIS, SAN, TOL, VAC **Elev.:** 0-1600 msnm **IUCN Habitat code:** 1,2,3,4,5,14 **Conservation status: (IUCN)** NE **(CPSG)** LC

5631 *Passiflora cumbalensis* (H.Karst.) Harms (Passifloraceae) **IPNI:** 184886-2 **Common name:** curuba, curubo, tausa, tauso **Habit:** Climbing **Use:** GS, HF, ME **Origin:** Native **Region:** Andean **Dept.:** ANT, BOG, BOL, BOY, CAL, CAQ, CAU, CUN, HUI, MAG, NAR, PUT, QUI, RIS, SAN, TOL, VAC **Elev.:** 2000-3500 msnm **IUCN Habitat code:** 1,2,3,4,14 **Conservation status: (IUCN)** NE **(CPSG)** LC

5632 *Passiflora cuneata* Willd. (Passifloraceae) **IPNI:** 674933-1 **Habit:** Climbing **Use:** HF **Origin:** Native **Region:** Andean **Dept.:** BOY, CHO, CUN, MAG, MET, NSA, SAN, VAC **Elev.:** 1000-2000 msnm **IUCN Habitat code:** 1,3,4,14 **Conservation status: (IUCN)** NE **(CPSG)** LC

5633 *Passiflora edulis* Sims (Passifloraceae) **IPNI:** 321964-2 **Common name:** curubo, maracuyá, maracuyá de monte, maracuyá silvestre, pachita amarilla, pasiflora **Habit:** Climbing **Use:** AF, EU, GS, HF, IF, MA, ME, PO **Origin:** Cultivated, Naturalised **Region:** Amazonia, Andean, Caribbean, Pacific **Dept.:** ANT, BOG, CAU, CHO, CUN, GUV, LAG, QUI, SAN, SUC, VAC, VID **Elev.:** 0–2000 m a.s.l. **IUCN Habitat code:** 1,2,3,4,5,14 **Conservation status: (IUCN)** NE

5634 *Passiflora emarginata* Bonpl. (Passifloraceae) **IPNI:** 674975-1 **Habit:** Climbing **Use:** HF **Origin:** Endemic **Region:** Andean **Dept.:** CAU, CHO, NAR, TOL, VAC **Elev.:** 1500–2000 m a.s.l. **IUCN Habitat code:** 1,3,14 **Conservation status: (IUCN)** NE

5635 *Passiflora filipes* Benth. (Passifloraceae) **IPNI:** 674993-1 **Habit:** Climbing **Use:** HF, ME **Origin:** Native **Region:** Andean **Dept.:** RIS **Elev.:** 1100-1200 msnm **Conservation status: (IUCN)** NE **(CPSG)** LC

5636 *Passiflora flexipes* Triana & Planch. (Passifloraceae) **IPNI:** 674996-1 **Habit:** Climbing **Use:** HF **Origin:** Endemic **Region:** Andean **Dept.:** ANT, CAL, QUI, RIS, TOL **Elev.:** 2500–3000 m a.s.l. **IUCN Habitat code:** 1,3,14 **Conservation status: (IUCN)** NE

5637 *Passiflora foetida* L. (Passifloraceae) **IPNI:** 321966-2 **Common name:** bejuco canastilla, bejuco canastillo, bejuco siete llagas, chulupa de loma, cinco llagas, cincollagas, cocorilla, cocorillo, curubo, esponjilla, flor de la pasión, gulupo, mashula, pasionaria, patillita, sietellagas, tapallito **Habit:** Climbing **Use:** AF, EU, GS, HF, MA, ME, PO, SU **Origin:** Native **Region:** Andean, Caribbean, Orinoquia, Pacific **Dept.:** AMA, ANT, ARA, ATL, BOG, BOL, BOY, CAL, CAU, CES, CHO, COR, CUN, GUA, HUI, LAG, MAG, MET, NAR, NSA, QUI, SAN, SUC, TOL, VAC **Elev.:** 0-1500 msnm **IUCN Habitat code:** 1,2,3,4,5,14 **Conservation status: (IUCN)** NE **(CPSG)** LC

5638 *Passiflora garckei* Mast. (Passifloraceae) **IPNI:** 675010-1 **Habit:** Climbing **Use:** HF, MA **Origin:** Native **Region:** Orinoquia **Dept.:** MET **Elev.:** 480–480 m a.s.l. **Conservation status: (IUCN)** NE

5639 *Passiflora glandulosa* Cav. (Passifloraceae) **IPNI:** 675022-1 **Habit:** Climbing **Use:** HF **Origin:** Native **Region:** Amazonia **Dept.:** VAU **Elev.:** 0-500 msnm **IUCN Habitat code:** 1 **Conservation status: (IUCN)** NE **(CPSG)** LC

5640 *Passiflora gracilis* J.Jacq. ex Link (Passifloraceae) **IPNI:** 60461856-2 **Habit:** Climbing **Use:** HF **Origin:** Native **Region:** Andean **IUCN Habitat code:** 3 **Conservation status: (IUCN)** NE

5641 *Passiflora guazumifolia* Jacq. (Passifloraceae) **IPNI:** 675041-1 **Habit:** Climbing **Use:** HF **Origin:** Native **Region:** Caribbean **Dept.:** BOL, COR, MAG, MET, SAN, TOL **Elev.:** 0-500 msnm **IUCN Habitat code:** 1,2 **Conservation status: (IUCN)** NE **(CPSG)** LC

5642 *Passiflora holtii* Killip (Passifloraceae) **IPNI:** 185001-2 **Habit:** Liana **Use:** HF **Origin:** Native **Region:** Amazonia, Guiana Shield **Dept.:** AMA, GUA, PUT, VAU **Elev.:** 0-500 msnm **IUCN Habitat code:** 1 **Conservation status: (IUCN)** NE **(CPSG)** LC

5643 *Passiflora involucrata* (Mast.) A.H.Gentry (Passifloraceae) **IPNI:** 185013-2 **Common name:** granadilla, granadillo, granadillo de conga, maracuyá de monte **Habit:** Climbing **Use:** HF **Origin:** Native **Region:** Amazonia **Dept.:** AMA, CAQ, VAU **Elev.:** 0-500 msnm **IUCN Habitat code:** 1,3,5 **Conservation status: (IUCN)** NE **(CPSG)** LC

5644 *Passiflora killipiana* Cuatrec. (Passifloraceae) **IPNI:** 185020-2 **Habit:** Climbing **Use:** ME **Origin:** Endemic **Region:** Amazonia **Dept.:** AMA **Elev.:** 100–100 m a.s.l. **IUCN Habitat code:** 1,5 **Conservation status: (IUCN)** NE

5645 *Passiflora laurifolia* L. (Passifloraceae) **IPNI:** 675132-1 **Habit:** Climbing **Use:** EU, GS, HF, MA, ME, PO **Origin:** Native **Region:** Amazonia, Andean, Guiana Shield, Orinoquia, Pacific **Dept.:** AMA, ANT, CAQ, MET, VAU, VID **Elev.:** 0-1500 msnm **IUCN Habitat code:** 1,3,4,5,14 **Conservation status: (IUCN)** NE **(CPSG)** LC

5646 *Passiflora leptomischa* Harms (Passifloraceae) **IPNI:** 185027-2 **Habit:** Climbing **Use:** HF **Origin:** Endemic **Region:** Andean **Dept.:** ANT, CAU, VAC **Elev.:** 2000–2800 m a.s.l. **IUCN Habitat code:** 1,3,14 **Conservation status: (IUCN)** NE

5647 *Passiflora ligularis* Juss. (Passifloraceae) **IPNI:** 675142-1 **Common name:** desjarretadera, granadilla, granadillo, guampana, parchita amarilla **Habit:** Climbing **Use:** EU, GS, HF, IF, ME **Origin:** Native, Cultivated **Region:** Andean **Dept.:** ANT, BOG, BOY, CAU, CUN, HUI, MET, NAR, NSA, QUI, RIS, SAN, TOL, VAC **Elev.:** 1000-3000 msnm **IUCN Habitat code:** 1,2,3,5,14 **Conservation status: (IUCN)** NE **(CPSG)** LC

5648 *Passiflora macrophylla* Spruce ex Mast. (Passifloraceae) **IPNI:** 675177-1 **Common name:** chirimoyitos parchas, granadilla de árbol **Habit:** Shrub, Small tree **Use:** HF, MA **Origin:** Native **Region:** Amazonia, Pacific **Dept.:** ANT, CHO, NAR, PUT **Elev.:** 60-600 msnm **IUCN Habitat code:** 1,3,14 **Conservation status: (IUCN)** NE **(CPSG)** LC

5649 *Passiflora magnifica* L.K.Escobar (Passifloraceae) **IPNI:** 282204-2 **Habit:** Climbing **Use:** HF **Origin:** Endemic **Region:** Andean **Dept.:** ANT **Elev.:** 1250–1750 m a.s.l. **IUCN Habitat code:** 1 **Conservation status: (IUCN)** NE

5650 *Passiflora maliformis* L. (Passifloraceae) **IPNI:** 324148-2 **Common name:** cholupa, cholupo, chulupa, colopa, colupa, culupa, culupo, curubo, cóngolo, golupo, granadilla, granadilla de monte, granadilla de piedra, granadillo, gulupa, tripona **Habit:** Climbing **Use:** AF, HF, MA, ME **Origin:** Native **Region:** Andean **Dept.:** ANT, BOY, CAL, CUN, HUI, MAG, NAR, QUI, RIS, SAN, TOL, VAC **Elev.:** 0-2000 msnm **IUCN Habitat code:** 1,2,3,5,14 **Conservation status: (IUCN)** NE **(CPSG)** LC

5651 *Passiflora manicata* (Juss.) Pers. (Passifloraceae) **IPNI:** 675190-1 **Habit:** Climbing **Use:** EU, HF, ME **Origin:** Native **Region:** Andean **Dept.:** ANT, BOY, CAL, CAU, CUN, NAR, NSA, QUI, SAN, TOL, VAC **Elev.:** 1500-3000 msnm **IUCN Habitat code:** 1,3,4,14 **Conservation status: (IUCN)** NE **(CPSG)** LC

5652 *Passiflora menispermifolia* Kunth (Passifloraceae) **IPNI:** 675211-1 **Common name:** chulupe **Habit:** Climbing **Use:** HF **Origin:** Native **Region:** Andean, Orinoquia, Pacific **Dept.:** ANT, BOG, BOL, BOY, CAQ, CHO, CUN, MET, SAN, TOL, VAC **Elev.:** 0-1600 msnm **IUCN Habitat code:** 1,3,5,14 **Conservation status: (IUCN)** NE **(CPSG)** LC

5653 *Passiflora miniata* Vanderpl. (Passifloraceae) **IPNI:** 77078146-1 **Use:** EU, HF **Origin:** Native **IUCN Habitat code:** 4 **Conservation status: (IUCN)** NE

5654 *Passiflora misera* Kunth (Passifloraceae) **IPNI:** 321967-2 **Habit:** Climbing **Use:** HF, ME **Origin:** Native **Region:** Amazonia, Andean, Caribbean, Orinoquia, Pacific **Dept.:** AMA, ANT, ATL, BOL, BOY, CAL, CAS, CAU, CHO, COR, CUN, LAG, MAG, MET, NSA, VAC **Elev.:** 0-500 msnm **IUCN Habitat code:** 1,2,3,4,5,14 **Conservation status: (IUCN)** NE **(CPSG)** LC

5655 *Passiflora mixta* L.f. (Passifloraceae) **IPNI:** 675224-1 **Common name:** curuba, curuba de indio, curuba silvestre, curubito, curubito de indio, curubo, curubo de indio, curubo de monte, palchiguaca **Habit:** Climbing **Use:** EU, GS, HF, ME **Origin:** Native, Cultivated **Region:** Andean **Dept.:** ANT, BOG, BOY, CAL, CAU, CUN, LAG, MAG, NAR, NSA, PUT, QUI, RIS, SAN, TOL, VAC **Elev.:** 1300–3700 m a.s.l. **IUCN Habitat code:** 1,2,3,4,5,6,8,14 **Conservation status: (IUCN)** NE

5656 *Passiflora mollis* Kunth (Passifloraceae) **IPNI:** 675225-1 **Common name:** granadilla de monte **Habit:** Climbing **Use:** EU **Origin:** Native, Cultivated **Region:** Andean **Dept.:** BOY, CAL, CUN, MAG, QUI, TOL, VAC **Elev.:** 400–2300 m a.s.l. **Conservation status: (IUCN)** NE

5657 *Passiflora mollissima* (Kunth) L.H.Bailey (Passifloraceae) **IPNI:** 185076-2 **Common name:** curuba, curubo, curubo de castilla, kama kama, kita jampaway, parcha, tacso, tintin, trompos, tumbis, tumbo, tumpu, yopokolo **Use:** EU, GS, HF, IF, MA, ME, PO **Origin:** Native **IUCN Habitat code:** 1,3,4,5,8,14 **Conservation status: (IUCN)** NE

5658 *Passiflora morifolia* Mast. (Passifloraceae) **IPNI:** 321969-2 **Habit:** Climbing **Use:** EU, HF **Origin:** Native **Region:** Andean **Dept.:** NAR **Elev.:** 0-500 msnm **IUCN Habitat code:** 14 **Conservation status: (IUCN)** NE **(CPSG)** LC

5659 *Passiflora multiformis* Jacq. (Passifloraceae) **IPNI:** 675238-1 **Habit:** Climbing **Use:** HF **Origin:** Native **Region:** Andean, Caribbean **Dept.:** LAG, MAG **Elev.:** 0-1000 msnm **IUCN Habitat code:** 1,4,14 **Conservation status: (IUCN)** NE **(CPSG)** LC

5660 *Passiflora nitida* Kunth (Passifloraceae) **IPNI:** 675263-1 **Common name:** burucuña, granadilla, granadilla babosa, granadilla de monte, granadilla silvestre, granadillo, tuchica **Habit:** Climbing **Use:** HF, ME **Origin:** Native **Region:** Amazonia, Andean, Guiana Shield, Orinoquia, Pacific **Dept.:** AMA, BOG, CAQ, CHO, CUN, GUA, GUV, MET, NAR, PUT, TOL, VAC, VAU **Elev.:** 0-650 msnm **IUCN Habitat code:** 1,3,4,5,14 **Conservation status: (IUCN)** NE **(CPSG)** LC

5661 *Passiflora palenquensis* Holm-Niels. & Lawesson (Passifloraceae) **IPNI:** 282212-2 **Habit:** Climbing **Use:** HF **Origin:** Native **Region:** Andean, Pacific **Dept.:** ANT, CAU, CHO, NAR, VAC **Elev.:** 0-1200 msnm **IUCN Habitat code:** 1,5,14 **Conservation status: (IUCN)** NE **(CPSG)** LC

5662 *Passiflora pallida* L. (Passifloraceae) **IPNI:** 185104-2 **Habit:** Climbing **Use:** ME **Origin:** Native **Region:** Caribbean **Dept.:** ATL, BOL, MAG, SAP, SUC **Elev.:** 0–300 m a.s.l. **Conservation status: (IUCN)** NE

5663 *Passiflora parritae* (Mast.) L.H.Bailey (Passifloraceae) **IPNI:** 185110-2 **Habit:** Climbing **Use:** HF **Origin:** Endemic **Region:** Andean **Dept.:** CAL, RIS, TOL **Elev.:** 1900–2800 m a.s.l. **IUCN Habitat code:** 1,3,4,14 **Conservation status: (IUCN)** NE

5664 *Passiflora picturata* Ker Gawl. (Passifloraceae) **IPNI:** 675340-1 **Habit:** Climbing **Use:** HF **Origin:** Native **Region:** Orinoquia **Dept.:** MET **Elev.:** 450–450 m a.s.l. **Conservation status: (IUCN)** NE

5665 *Passiflora pinnatistipula* Cav. (Passifloraceae) **IPNI:** 675348-1 **Common name:** curuba, gulupa, palchas **Habit:** Climbing **Use:** HF, ME **Origin:** Cultivated, Naturalised **Region:** Andean **Dept.:** ANT, BOG, BOY, CUN, NAR, NSA **Elev.:** 2500–3800 m a.s.l. **IUCN Habitat code:** 1,3,4,14 **Conservation status: (IUCN)** NE

5666 *Passiflora platyloba* Killip (Passifloraceae) **IPNI:** 185126-2 **Habit:** Climbing **Use:** HF, ME **Origin:** Native **Region:** Pacific **Dept.:** ANT, CHO **Elev.:** 0-100 msnm **IUCN Habitat code:** 1,3,14 **Conservation status: (IUCN)** NE **(CPSG)** LC

5667 *Passiflora popenovii* Killip (Passifloraceae) **IPNI:** 185132-2 **Common name:** granadilla de quijos **Habit:** Climbing **Use:** HF, ME **Origin:** Native **Region:** Andean **Dept.:** ANT, BOG, CAU, NAR, VAC **Elev.:** 1200–2050 m a.s.l. **IUCN Habitat code:** 1 **Conservation status: (IUCN)** NE

5668 *Passiflora punctata* L. (Passifloraceae) **IPNI:** 675382-1 **Habit:** Climbing **Use:** HF **Origin:** Native **Region:** Andean, Pacific **Dept.:** CAU, NAR, VAC **Elev.:** 200–1500 m a.s.l. **IUCN Habitat code:** 1,14 **Conservation status: (IUCN)** NE

5669 *Passiflora putumayensis* Killip (Passifloraceae) **IPNI:** 185144-2 **Habit:** Shrub **Use:** HF **Origin:** Native **Region:** Andean **Dept.:** CAQ, PUT **Elev.:** 1800-2500 msnm **IUCN Habitat code:** 1,14 **Conservation status: (IUCN)** NE **(CPSG)** LC

5670 *Passiflora quadrangularis* L. (Passifloraceae) **IPNI:** 675390-1 **Common name:** badea, bejuco badeo, bejuco de granadillo, corvejo, granadilla, granadilla real, granadillo, granadillo grande, maracuyá **Habit:** Climbing **Use:** EU, GS, HF, MA, ME, PO **Origin:** Native, Cultivated **Region:** Amazonia, Andean, Guiana Shield, Pacific **Dept.:** AMA, ANT, BOY, CAQ, CAU, CHO, COR, CUN, GUA, HUI, MAG, MET, NSA, QUI, SAN, VAC, VAU **Elev.:** 0–1200 m a.s.l. **IUCN Habitat code:** 1,2,3,4,5,14 **Conservation status: (IUCN)** NE

5671 *Passiflora quadriglandulosa* Rodschied (Passifloraceae) **IPNI:** 675392-1 **Habit:** Climbing **Use:** HF, ME **Origin:** Native **Region:** Amazonia **Dept.:** AMA, GUV **Elev.:** 0-500 msnm **IUCN Habitat code:** 1,5,14 **Conservation status: (IUCN)** NE **(CPSG)** LC

5672 *Passiflora resticulata* Mast. & André (Passifloraceae) **IPNI:** 675404-1 **Habit:** Climbing **Use:** HF **Origin:** Native **Region:** Amazonia, Andean, Caribbean **Dept.:** BOL, CAU, QUI, TOL, VAC **Elev.:** 70-2100 msnm **IUCN Habitat code:** 1,3,14 **Conservation status: (IUCN)** NE **(CPSG)** LC

5673 *Passiflora riparia* Mart. ex Mast. (Passifloraceae) **IPNI:** 675414-1 **Habit:** Climbing **Use:** HF **Origin:** Native **Region:** Amazonia, Andean, Guiana Shield **Dept.:** CAQ, GUA, MET, VAC, VAU **Elev.:** 0-1050 msnm **IUCN Habitat code:** 3,4,14 **Conservation status: (IUCN)** NE **(CPSG)** LC

5674 *Passiflora rubra* L. (Passifloraceae) **IPNI:** 675423-1 **Use:** ME **Conservation status: (IUCN)** NE

5675 *Passiflora schlimiana* Triana & Planch. (Passifloraceae) **IPNI:** 675442-1 **Common name:** curuba **Habit:** Climbing **Use:** HF **Origin:** Endemic **Region:** Andean **Dept.:** CES, LAG, MAG **Elev.:** 2400–3100 m a.s.l. **IUCN Habitat code:** 1,3,4,14 **Conservation status: (IUCN)** NE

5676 *Passiflora seemannii* Griseb. (Passifloraceae) **IPNI:** 675446-1 **Habit:** Climbing **Use:** HF, ME **Origin:** Native **Region:** Andean, Orinoquia, Pacific **Dept.:** ANT, CHO, CUN, GUA, MET, SAN, VAC, VAU, VID **Elev.:** 0-1050 msnm **IUCN Habitat code:** 1,3,5,14 **Conservation status: (IUCN)** NE **(CPSG)** LC

5677 *Passiflora serratodigitata* L. (Passifloraceae) **IPNI:** 675456-1 **Common name:** cocorilla, granadilla **Habit:** Climbing **Use:** HF **Origin:** Native **Region:** Amazonia, Pacific **Dept.:** AMA, ANT, CHO **Elev.:** 0-100 msnm **IUCN Habitat code:** 1,4,5,14 **Conservation status: (IUCN)** NE **(CPSG)** LC

5678 *Passiflora serrulata* Jacq. (Passifloraceae) **IPNI:** 675457-1 **Habit:** Climbing **Use:** HF **Origin:** Native **Region:** Caribbean **Dept.:** ANT, ATL, BOL, CES, LAG, MAG **Elev.:** 0-500 msnm **IUCN Habitat code:** 1,2,3,14 **Conservation status: (IUCN)** NE **(CPSG)** LC

5679 *Passiflora suberosa* L. (Passifloraceae) **IPNI:** 321973-2 **Habit:** Climbing **Use:** EU, HF, ME **Origin:** Native **Region:** Andean **Dept.:** ANT, BOG, CAL, CAQ, CAU, CUN, MAG, NSA, QUI, SAN, TOL, VAC **Elev.:** 350-2000 msnm **IUCN Habitat code:** 1,2,3,4,14 **Conservation status: (IUCN)** NE **(CPSG)** LC

5680 *Passiflora subpeltata* Ortega (Passifloraceae) **IPNI:** 675494-1 **Common name:** cocorilla, granadillo **Habit:** Climbing **Use:** EU, HF, ME, PO **Origin:** Native **Region:** Andean, Caribbean, Orinoquia **Dept.:** ANT, BOL, CES, COR, CUN, MAG, TOL, VAC **Elev.:** 0-1600 msnm **IUCN Habitat code:** 1,2,3,4,14 **Conservation status: (IUCN)** NE **(CPSG)** LC

5681 *Passiflora tarminiana* Coppens & V.E.Barney (Passifloraceae) **IPNI:** 321062-2 **Habit:** Climbing **Use:** HF **Origin:** Native, Cultivated **Region:** Andean **Dept.:** ANT, BOG, BOY, NAR, VAC **Elev.:** 1850–3000 m a.s.l. **IUCN Habitat code:** 1,2,3,14 **Conservation status: (IUCN)** NE

5682 *Passiflora tenerifensis* L.K.Escobar (Passifloraceae) **IPNI:** 282223-2 **Habit:** Climbing **Use:** HF **Origin:** Endemic **Region:** Andean **Dept.:** VAC **Elev.:** 2800–3100 m a.s.l. **IUCN Habitat code:** 1,3,14 **Conservation status: (IUCN)** NE

5683 *Passiflora tica* Gómez-Laur. & L.D.Gómez (Passifloraceae) **IPNI:** 898371-1 **Common name:** San José, san josé **Habit:** Tree **Use:** HF **Origin:** Native **Region:** Andean, Pacific **Dept.:** ANT, CHO **Elev.:** 450-1500 msnm **IUCN Habitat code:** 1 **Conservation status: (IUCN)** NE **(CPSG)** LC

5684 *Passiflora tiliifolia* L. (Passifloraceae) **IPNI:** 675522-1 **Common name:** granadillo de monte **Habit:** Climbing **Use:** HF **Origin:** Native **Region:** Andean **Dept.:** ANT, CAL, CAU, CHO, QUI, RIS, TOL, VAC **Elev.:** 1500–2800 m a.s.l. **IUCN Habitat code:** 1,14 **Conservation status: (IUCN)** NE

5685 *Passiflora tolimana* Harms (Passifloraceae) **IPNI:** 185210-2 **Habit:** Climbing **Use:** HF **Origin:** Endemic **Region:** Andean **Dept.:** ANT, CHO, TOL, VAC **Elev.:** 1500–2000 m a.s.l. **IUCN Habitat code:** 1,3 **Conservation status: (IUCN)** NE

5686 *Passiflora tripartita* (Juss.) Poir. (Passifloraceae) **IPNI:** 675547-1 **Habit:** Climbing **Use:** GS, HF, ME **Origin:** Native, Cultivated **Region:** Andean **Dept.:** BOG, BOY, CUN, SAN **Elev.:** 2500–3500 m a.s.l. **IUCN Habitat code:** 1,2,3,4,5,8,14 **Conservation status: (IUCN)** NE

5687 *Passiflora vesicaria* L. (Passifloraceae) **IPNI:** 675582-1 **Use:** ME **Origin:** Native **Conservation status: (IUCN)** NE

5688 *Passiflora vespertilio* L. (Passifloraceae) **IPNI:** 675584-1 **Habit:** Climbing **Use:** EU, HF, ME **Origin:** Native **Region:** Amazonia, Orinoquia **Dept.:** AMA, CAQ, CUN, GUV, MET, NAR, VAU **Elev.:** 0-500 msnm **IUCN Habitat code:** 1,2,3,4,5,14 **Conservation status: (IUCN)** NE **(CPSG)** LC

5689 *Passiflora vestita* Killip (Passifloraceae) **IPNI:** 185238-2 **Habit:** Climbing **Use:** HF **Origin:** Native **Region:** Amazonia **Dept.:** PUT **Elev.:** 0-140 msnm **IUCN Habitat code:** 3 **Conservation status: (IUCN)** NE **(CPSG)** LC

5690 *Passiflora vitifolia* Kunth (Passifloraceae) **IPNI:** 30044451-2 **Common name:** chalupo, chulupa de mico, chulupo, cocorilla de monte, flor de la pasión roja, granadilla, granadilla de monte, granadilla de murciélago, granadilla silvestre, granadillo, granadillo silvestre, gulupa, gulupo, maracuyá de monte, vinagrera **Habit:** Liana **Use:** EU, HF, ME **Origin:** Native **Region:** Amazonia, Andean, Pacific **Dept.:** AMA, ANT, BOY, CAL, CAQ, CAU, CHO, COR, CUN, GUV, LAG, MAG, MET, NSA, PUT, SAN, TOL, VAC, VAU **Elev.:** 0-1000 msnm **IUCN Habitat code:** 1,2,3,4,5,14 **Conservation status: (IUCN)** NE **(CPSG)** LC

5691 *Piriqueta cistoides* (L.) Griseb. (Passifloraceae) **IPNI:** 30068867-2 **Habit:** Herb **Use:** ME **Origin:** Native **Region:** Guiana Shield, Caribbean **Dept.:** CAS, CES, GUA, GUV, MET, TOL, VAC, VAU, VID **Elev.:** 150–500 m a.s.l. **IUCN Habitat code:** 2,3,4,5,14 **Conservation status: (IUCN)** NE

5692 *Piriqueta racemosa* (Jacq.) Sweet (Passifloraceae)
IPNI: 836478-1
Habit: Herb **Use:** ME **Origin:** Native **Region:** Caribbean **Dept.:** MAG **Elev.:** 80–160 m a.s.l. **Conservation status: (IUCN)** NE

5693 *Turnera acuta* Willd. ex Schult. (Passifloraceae)
IPNI: 259040-2
Habit: Shrub **Use:** ME **Origin:** Native **Region:** Amazonia, Guiana Shield, Orinoquia **Dept.:** AMA, CAQ, GUA, GUV, PUT, VAU, VID **Elev.:** 100–120 m a.s.l. **IUCN Habitat code:** 1,3,5,14 **Conservation status: (IUCN)** NE

5694 *Turnera diffusa* Willd. ex Schult. (Passifloraceae)
IPNI: 836578-1
Common name: damiana, herb de la pastora, herb del venado, pastorcita, rompe camisa macho, yerba damia damia herb de la pastora, yerba damiana **Use:** EU, HF, MA, ME, PO **IUCN Habitat code:** 14 **Conservation status: (IUCN)** NE

5695 *Turnera melochia* Triana & Planch. (Passifloraceae)
IPNI: 836629-1
Habit: Herb **Use:** ME **Origin:** Native **Region:** Amazonia, Orinoquia **Dept.:** BOY, CAQ, CAS, CUN, MET, SAN, VID **Elev.:** 100–250 m a.s.l. **IUCN Habitat code:** 1,3,4,14 **Conservation status: (IUCN)** NE

5696 *Turnera pumilea* L. (Passifloraceae)
IPNI: 836658-1
Habit: Herb **Use:** ME **Origin:** Native **Region:** Caribbean, Orinoquia **Dept.:** CES, MAG, VID **Elev.:** 100–100 m a.s.l. **IUCN Habitat code:** 2,4 **Conservation status: (IUCN)** NE

5697 *Turnera scabra* Millsp. (Passifloraceae)
IPNI: 259112-2
Habit: Herb, Shrub, Subshrub **Use:** ME **Origin:** Native **Region:** Andean, Caribbean, Orinoquia **Dept.:** ANT, BOY, CAS, CAU, CES, CUN, HUI, MAG, NSA, TOL, VAC **Elev.:** 50-1300 msnm **IUCN Habitat code:** 1,2,3,4,14 **Conservation status: (IUCN)** NE **(CPSG)** LC

5698 *Turnera subulata* Sm. (Passifloraceae)
IPNI: 836689-1
Common name: bella hasta las once, bellaalasnueve, brillalasonce, matrimonio **Habit:** Herb, Shrub, Subshrub **Use:** EU, ME, PO **Origin:** Native, Cultivated **Region:** Andean, Caribbean **Dept.:** ANT, ATL, BOL, CAL, CAU, CES, CUN, GUA, HUI, LAG, MAG, MET, NSA, SAN, TOL, VAC **Elev.:** 30–1225 m a.s.l. **IUCN Habitat code:** 1,3,4,14 **Conservation status: (IUCN)** NE

5699 *Turnera ulmifolia* L. (Passifloraceae)
IPNI: 836702-1
Common name: amapola, chanana, clavel de oro, cumaná, damiana, granizo, herb de la pastorcita, hierba damiana, malva, manzanilla de los campos, marilope, oreja de coyote, panchecito, santalucía, vidita mía, yerba damiana **Habit:** Shrub, Subshrub **Use:** AF, EU, HF, ME **Origin:** Native **Region:** Caribbean **Dept.:** SAP **Elev.:** 100–100 m a.s.l. **IUCN Habitat code:** 1,2,3,14 **Conservation status: (IUCN)** NE

5700 *Paulownia tomentosa* (Thunb.) Steud. (Paulowniaceae)
IPNI: 185784-2
Use: EU, MA, ME **Region:** Andean **Dept.:** BOG **IUCN Habitat code:** 14 **Conservation status: (IUCN)** NE

5701 *Chaetocarpus schomburgkianus* (Kuntze) Pax & K.Hoffm. (Peraceae)
IPNI: 54072-2
Habit: Tree **Use:** MA **Origin:** Native **Region:** Amazonia **Dept.:** AMA **Elev.:** 100–500 m a.s.l. **IUCN Habitat code:** 1 **Conservation status: (IUCN)** LC

5702 *Pera glabrata* (Schott) Poepp. ex Baill. (Peraceae)
IPNI: 353360-1
Habit: Tree **Use:** MA **Origin:** Native **Region:** Andean, Pacific **Dept.:** CHO, CUN **Elev.:** 100–650 m a.s.l. **IUCN Habitat code:** 1,14 **Conservation status: (IUCN)** LC

5703 *Pogonophora schomburgkiana* Miers ex Benth. (Peraceae)
IPNI: 354987-1
Common name: cura, ñoño, ñuñucurá **Habit:** Tree **Use:** MA **Origin:** Native **Region:** Amazonia, Orinoquia **Dept.:** AMA, VID **Elev.:** 100–100 m a.s.l. **IUCN Habitat code:** 1,5 **Conservation status: (IUCN)** LC

5704 *Amanoa guianensis* Aubl. (Phyllanthaceae)
IPNI: 338575-1
Habit: Tree **Use:** MA, ME **Origin:** Native **Region:** Amazonia, Pacific **Dept.:** AMA, CAQ, CHO, GUA, VID **Elev.:** 0–200 m a.s.l. **IUCN Habitat code:** 1,3,4 **Conservation status: (IUCN)** LC

5705 *Amanoa oblongifolia* Müll.Arg. (Phyllanthaceae)
IPNI: 338582-1
Habit: Tree **Use:** ME **Origin:** Native **Region:** Amazonia **Dept.:** AMA, VAU **Elev.:** 90–600 m a.s.l. **IUCN Habitat code:** 1,3,5,14 **Conservation status: (IUCN)** LC

5706 *Bischofia javanica* Blume (Phyllanthaceae)
IPNI: 339921-1
Habit: Tree **Use:** AF, EU, FU, GS, HF, IF, MA, ME, PO, SU **Origin:** Cultivated **Region:** Andean **Dept.:** ANT **Elev.:** 1000–2200 m a.s.l. **IUCN Habitat code:** 14 **Conservation status: (IUCN)** LC

5707 *Breynia disticha* J.R.Forst. & G.Forst. (Phyllanthaceae)
IPNI: 340073-1
Common name: mimosa **Habit:** Shrub **Use:** EU, ME **Origin:** Cultivated **Region:** Andean, Caribbean **Dept.:** ANT, BOL, MAG, VAC **Elev.:** 570–1100 m a.s.l. **IUCN Habitat code:** 3,14 **Conservation status: (IUCN)** NE

5708 *Hieronyma alchorneoides* Allemão (Phyllanthaceae)
IPNI: 350066-1
Common name: arenillo, candelillo, candelo, canelo, cargamanto, cargamento, cedro mondey, chaparro, chaparro de monte, chaparro montañero, chuchuguaca, chuguacá, chuhuacá, hojiancho negro, mascarey, mondey, motilón, motilón rosado, muñequito, pantanillo, pantano, pedr **Habit:** Tree **Use:** AF, HF, MA, ME **Origin:** Native **Region:** Amazonia, Andean, Caribbean, Orinoquia, Pacific **Dept.:** AMA, ANT, BOL, CAQ, CES, CHO, COR, CUN, LAG, MAG, MET, NAR, PUT, RIS, SAN, TOL, VAC, VAU **Elev.:** 0-1580 msnm **IUCN Habitat code:** 1,2,3,4,5,14 **Conservation status: (IUCN)** LC **(CPSG)** Potential LC

5709 *Hieronyma asperifolia* Pax & K.Hoffm. (Phyllanthaceae)
IPNI: 122555-2
Common name: colorado clavel
Habit: Tree **Use:** HF **Origin:** Native **Region:** Andean **Dept.:** BOG, CUN, NAR **Elev.:** 2100–3270 m a.s.l. **IUCN Habitat code:** 1,3,14 **Conservation status: (IUCN)** NE

5710 *Hieronyma duquei* Cuatrec. (Phyllanthaceae)
IPNI: 122566-2
Common name: candelo, chuguacá, coacha, colorado, granadillo, motilón, motilón silvestre, ortegón **Habit:** Tree **Use:** HF **Origin:** Native **Region:** Andean **Dept.:** BOY, CAS, CAU, CES, HUI, PUT, QUI, VAC **Elev.:** 1000–2600 m a.s.l. **IUCN Habitat code:** 1,3,14 **Conservation status: (IUCN)** LC

5711 *Hieronyma fendleri* Briq. (Phyllanthaceae)
IPNI: 122567-2
Habit: Tree **Use:** HF **Origin:** Native **Region:** Andean **Dept.:** ANT, BOY, CES, CUN, HUI, MAG, SAN, TOL **Elev.:** 1200–2700 m a.s.l. **IUCN Habitat code:** 1,3,14 **Conservation status: (IUCN)** LC

5712 *Hieronyma macrocarpa* Müll.Arg. (Phyllanthaceae)
IPNI: 350075-1
Common name: arenillo, candelo, chuguacá, chuguacá hojiancho, colorado, granadillo, motilón, motilón colorado, motilón dulce, mulato, mulatón, pantano, pibal **Habit:** Tree **Use:** HF **Origin:** Native **Region:** Andean **Dept.:** ANT, BOG, BOY, CAL, CAU, CHO, CUN, HUI, NAR, PUT, QUI, SAN, TOL, VAC **Elev.:** 1200-3200 msnm **IUCN Habitat code:** 1,3,4,8,14 **Conservation status: (IUCN)** NE **(CPSG)** Potential LC

5713 *Hieronyma oblonga* (Tul.) Müll.Arg. (Phyllanthaceae)
IPNI: 350078-1
Common name: candelo, candelo blanco, candelo rojo, chaparro de vega, motilón **Habit:** Tree **Use:** HF, MA **Origin:** Native **Region:** Amazonia, Andean, Orinoquia, Pacific **Dept.:** AMA, ANT, BOY, CAQ, CAU, CES, CHO, GUV, MET, QUI, RIS, VAC **Elev.:** 0-2780 msnm **IUCN Habitat code:** 1,2,3,5,14 **Conservation status: (IUCN)** LC **(CPSG)** Potential LC

5714 *Margaritaria nobilis* L.f. (Phyllanthaceae)
IPNI: 352077-1
Common name: azulejo, curubo
Habit: Tree **Use:** HF, MA, ME **Origin:** Native **Region:** Amazonia, Andean, Caribbean, Orinoquia, Pacific **Dept.:** AMA, ANT, ARA, BOL, BOY, CAQ, CES, CHO, COR, CUN, MAG, PUT, SAN, VID **Elev.:** 0-1300 msnm **IUCN Habitat code:** 1,2,3,4,5,14 **Conservation status: (IUCN)** NE **(CPSG)** Potential LC

5715 *Meineckia neogranatensis* (Müll.Arg.) G.L.Webster (Phyllanthaceae)
IPNI: 156150-2
Common name: barbasco, patacón, porita **Habit:** Shrub **Use:** ME **Origin:** Native **Dept.:** CUN **Elev.:** 350–650 m a.s.l. **IUCN Habitat code:** 1,14 **Conservation status: (IUCN)** NE

5716 *Phyllanthus acidus* (L.) Skeels (Phyllanthaceae)
IPNI: 194610-2
Common name: cereza, cilantrillo, ciruela, grosella, grosellero, grosello, guinda, manzana de estrella **Habit:**

Shrub **Use:** EU, FU, GS, HF, MA, ME, PO **Origin:** Cultivated **Region:** Andean **Dept.:** VAC **Elev.:** 1088–1088 m a.s.l. **IUCN Habitat code:** 1,2,3,14 **Conservation status: (IUCN)** NE

5717 *Phyllanthus acuminatus* Vahl (Phyllanthaceae) **IPNI:** 194611-2 **Common name:** barbasco, barbasquillo, chirrinche, chirrinchoga, enuraci, maduraplátano, matasardina **Habit:** Shrub, Small tree **Use:** HF, MA, ME, PO **Origin:** Native **Region:** Amazonia, Andean, Guiana Shield, Caribbean, Pacific **Dept.:** AMA, ANT, BOL, BOY, CAL, CAQ, CAU, CHO, CUN, HUI, LAG, MAG, NSA, PUT, RIS, SAN, TOL, VAC **Elev.:** 50-1750 msnm **IUCN Habitat code:** 1,2,3,4,14 **Conservation status: (IUCN)** NE **(CPSG)** Potential LC

5718 *Phyllanthus amarus* Schumach. & Thonn. (Phyllanthaceae) **IPNI:** 326408-2 **Habit:** Herb **Use:** EU, HF, ME **Origin:** Native **Region:** Andean, Caribbean, Orinoquia, Pacific **Dept.:** ANT, ATL, BOL, CHO, CUN, MAG, MET, SAP, SUC, TOL, VAC **Elev.:** 0–1600 m a.s.l. **IUCN Habitat code:** 1,2,3,5,14 **Conservation status: (IUCN)** NE

5719 *Phyllanthus brasiliensis* (Aubl.) Poir. (Phyllanthaceae) **IPNI:** 353613-1 **Common name:** barbascajo, barbasco, barbasco de hoja, barbasco de árbol, barbasco patagón, barbasco tinta, barbasquillo, cedrillo, chirrinchao, chirrinchinchao, granadillo, hoja morada, maduraplátano, patacón, tinta barbasco **Habit:** Shrub **Use:** ME, PO **Origin:** Native **Region:** Amazonia **Dept.:** BOY, CAQ, PUT, VAU **Elev.:** 200–200 m a.s.l. **IUCN Habitat code:** 1,4 **Conservation status: (IUCN)** NE

5720 *Phyllanthus caroliniensis* Walter (Phyllanthaceae) **IPNI:** 353652-1 **Common name:** cocohui **Habit:** Herb **Use:** ME **Origin:** Native **Region:** Amazonia, Caribbean, Orinoquia, Pacific **Dept.:** AMA, ANT, CAQ, CAU, CHO, MAG, MET, NAR, VAC **Elev.:** 0-300 msnm **IUCN Habitat code:** 1,2,3,4 **Conservation status: (IUCN)** NE **(CPSG)** Potential LC

5721 *Phyllanthus fluitans* Benth. ex Müll.Arg. (Phyllanthaceae) **IPNI:** 353898-1 **Habit:** Herb **Use:** EU **Origin:** Native **Region:** Amazonia, Caribbean **Dept.:** AMA, BOL **Elev.:** 0–100 m a.s.l. **IUCN Habitat code:** 3,5,14 **Conservation status: (IUCN)** NE

5722 *Phyllanthus myrsinites* Kunth (Phyllanthaceae) **IPNI:** 354314-1 **Habit:** Shrub **Use:** ME **Origin:** Native **Region:** Amazonia, Guiana Shield **Dept.:** AMA, CAQ, VAU **Elev.:** 100–700 m a.s.l. **IUCN Habitat code:** 1,4,14 **Conservation status: (IUCN)** LC

5723 *Phyllanthus niruri* L. (Phyllanthaceae) **IPNI:** 194900-2 **Common name:** balsilla, barbasco, barbasquillo, barbasquito, chanca piedra, flor escondida, guaranguilla, hierba del caspe, huevo abajo, paracelsa, quiebrapiedra, viernes santo **Habit:** Herb **Use:** AF, EU, HF, MA, ME, PO **Origin:** Native **Region:** Andean, Caribbean, Orinoquia, Pacific **Dept.:** ANT, BOY, CAL, CAU, CUN, HUI, MAG, MET, NAR, NSA, PUT, QUI, RIS, TOL, VAC **Elev.:** 0-2470 msnm **IUCN Habitat code:** 1,2,3,4,5,14 **Conservation status: (IUCN)** NE **(CPSG)** Potential LC

5724 *Phyllanthus orbiculatus* Rich. (Phyllanthaceae) **IPNI:** 194933-2 **Habit:** Herb **Use:** ME **Origin:** Native **Region:** Caribbean **Dept.:** MAG **Elev.:** 75–75 m a.s.l. **IUCN Habitat code:** 1,4,14 **Conservation status: (IUCN)** NE

5725 *Phyllanthus salviifolius* Kunth (Phyllanthaceae) **IPNI:** 354635-1 **Common name:** barbasco, barbasquillo, camándulo, cedrillo, cedrito, cedrito de fucha, cedro, chiraco, chiriaco, chirriador, chirrinchao, chirrincho, maduraplátano, maduro, mute, plátano, teñidero, teñidor, yuco, yuco serrano **Habit:** Small tree **Use:** HF, MA, PO **Origin:** Native **Region:** Andean **Dept.:** ANT, BOG, BOY, CAU, CUN, MAG, NAR, QUI, TOL **Elev.:** 1700–3400 m a.s.l. **IUCN Habitat code:** 1,3,4,14 **Conservation status: (IUCN)** LC

5726 *Phyllanthus stipulatus* (Raf.) G.L.Webster (Phyllanthaceae) **IPNI:** 195027-2 **Common name:** balsilla, barbasquillo, celedonia, diente de culebra, mancidibo, piñoncito de sabana, viernes santo **Habit:** Herb, Shrub **Use:** ME **Origin:** Native **Region:** Amazonia, Andean, Caribbean, Orinoquia, Pacific **Dept.:** AMA, CAQ, CAS, CAU, CHO, CUN, GUV, MAG, MET, NSA, PUT, RIS, SUC, VAC, VAU **Elev.:** 0-1900 msnm **IUCN Habitat code:** 1,2,3,4,5,14 **Conservation status: (IUCN)** NE **(CPSG)** Potential LC

5727 *Phyllanthus urinaria* L. (Phyllanthaceae) **IPNI:** 354815-1 **Common name:** viernes santo **Habit:** Herb **Use:** HF, MA, ME, PO **Origin:** Naturalised **Region:** Andean, Orinoquia, Pacific **Dept.:** ANT, CAL, CAU, CHO, MET, VAC **Elev.:** 0–550 m a.s.l. **IUCN Habitat code:** 1,2,3,5,14 **Conservation status: (IUCN)** NE

5728 *Phyllanthus valleanus* Croizat (Phyllanthaceae) **IPNI:** 195049-2 **Common name:** barbasco, yawati **Habit:** Shrub **Use:** MA, ME, SU **Origin:** Endemic **Region:** Pacific **Dept.:** CHO, VAC **Elev.:** 0–100 m a.s.l. **IUCN Habitat code:** 1,3,14 **Conservation status: (IUCN)** NE

5729 *Richeria grandis* Vahl (Phyllanthaceae) **IPNI:** 355200-1 **Common name:** alcafeto negro, trapiche **Habit:** Tree **Use:** ME **Origin:** Native **Region:** Amazonia, Andean, Orinoquia, Pacific **Dept.:** AMA, ANT, CAQ, CHO, COR, GUA, MET, NAR, SAN, VAC, VAU, VID **Elev.:** 100-1450 msnm **IUCN Habitat code:** 1,3,4,5,14 **Conservation status: (IUCN)** NE **(CPSG)** Potential LC

5730 *Savia sessiliflora* (Sw.) Willd. (Phyllanthaceae) **IPNI:** 356125-1 **Habit:** Small tree **Use:** HF **Origin:** Native **Region:** Caribbean **Dept.:** LAG **Elev.:** 170–230 m a.s.l. **Conservation status: (IUCN)** NE

5731 *Marathrum foeniculaceum* Bonpl. (Podostemaceae) **IPNI:** 688390-1 **Habit:** Herb, Aquatic **Use:** HF **Origin:** Native **Region:** Andean, Caribbean **Dept.:** ANT, CUN, HUI, MAG, SAN **Elev.:** 100–2200 m a.s.l. **IUCN Habitat code:** 1,3,14 **Conservation status: (IUCN)** NE

5732 *Drypetes variabilis* Uittien (Putranjivaceae) **IPNI:** 85512-2 **Common name:** aceitillo, chirateté, cucharo, hoja de sal, naranjuelo, palo de sal **Habit:** Tree **Use:** HF **Origin:** Native **Region:** Andean **Dept.:** CES **Elev.:** 1820–1820 m a.s.l. **IUCN Habitat code:** 1 **Conservation status: (IUCN)** LC

5733 *Cassipourea guianensis* Aubl. (Rhizophoraceae) **IPNI:** 719636-1 **Common name:** ajo, ajo de monte **Habit:** Shrub, Small tree, Tree **Use:** MA **Origin:** Native **Region:** Amazonia, Andean, Pacific **Dept.:** AMA, ANT, CAQ, CAU, CHO, COR, CUN, NAR, SAN, VAC **Elev.:** 2–1650 m a.s.l. **IUCN Habitat code:** 1,3,5,14 **Conservation status: (IUCN)** LC

5734 *Rhizophora × harrisonii* Leechm. (Rhizophoraceae) **IPNI:** 219395-2 **Common name:** mangle, mangle colorado, mangle rojo **Use:** ME **Origin:** Native **IUCN Habitat code:** 1,5,14 **Conservation status: (IUCN)** NE

5735 *Rhizophora mangle* L. (Rhizophoraceae) **IPNI:** 30061148-2 **Common name:** kino, mangle, mangle cativo, mangle colorado, mangle piñón, mangle rojo, mangle rosado, mangle salado, pargua **Habit:** Tree **Use:** FU, GS, HF, MA, ME, PO **Origin:** Native **Region:** Caribbean, Pacific **Dept.:** ANT, ATL, BOL, CAU, CHO, COR, LAG, MAG, NAR, SAP, SUC, VAC **Elev.:** 0-6 msnm **IUCN Habitat code:** 1,2,3,5,14 **Conservation status: (IUCN)** LC **(CPSG)** Potential LC

5736 *Rhizophora racemosa* G.Mey. (Rhizophoraceae) **IPNI:** 719838-1 **Habit:** Tree **Use:** FU, HF, MA, ME **Origin:** Native **Region:** Pacific **Dept.:** NAR, VAC **Elev.:** 0–6 m a.s.l. **Conservation status: (IUCN)** LC

5737 *Abatia parviflora* Ruiz & Pav. (Salicaceae) **IPNI:** 779362-1 **Common name:** chirlobirlo, duraznillo, sacaojos colorado **Habit:** Shrub **Use:** MA **Origin:** Native **Region:** Andean **Dept.:** ANT, BOG, BOY, CAL, CUN, NAR, PUT, QUI, RIS, TOL, VAC **Elev.:** 1990–3382 m a.s.l. **IUCN Habitat code:** 1,3,4,5,8,14 **Conservation status: (IUCN)** LC

5738 *Azara microphylla* Hook.f. (Salicaceae) **IPNI:** 111471-1 **Habit:** Shrub, Small tree **Use:** HF, MA, ME **Origin:** Cultivated **Region:** Andean **Dept.:** BOG, CUN **Elev.:** 2700–2700 m a.s.l. **Conservation status: (IUCN)** NE

5739 *Banara arguta* Briq. (Salicaceae) **IPNI:** 28885-2 **Common name:** varablanca **Habit:** Shrub, Small tree, Tree **Use:** ME **Origin:** Native **Region:** Orinoquia **Dept.:** MET **Elev.:** 240–240 m a.s.l. **IUCN Habitat code:** 1,3,4,14 **Conservation status: (IUCN)** LC

5740 *Banara guianensis* Aubl. (Salicaceae) **IPNI:** 779396-1 **Common name:** chirilla, garlucho, gualpite, guasquilla, varablanca **Habit:** Shrub, Small tree, Tree **Use:** AF, FU, MA, ME **Origin:** Native **Region:** Amazonia,

Andean, Guiana Shield, Caribbean, Orinoquia, Pacific **Dept.:** ANT, BOG, BOL, BOY, CAQ, CAU, CHO, CUN, HUI, MAG, MET, NAR, PUT, QUI, RIS, TOL, VAC **Elev.:** 20-2090 msnm **IUCN Habitat code:** 1,2,3,4,5,14 **Conservation status: (IUCN)** LC **(CPSG)** Potential LC

5741 *Banara nitida* Spruce ex Benth. (Salicaceae)
IPNI: 779403-1
Habit: Tree **Use:** HF, MA **Origin:** Native **Region:** Amazonia, Guiana Shield **Dept.:** AMA, VAU **Elev.:** 100–150 m a.s.l. **IUCN Habitat code:** 1,3 **Conservation status: (IUCN)** LC

5742 *Casearia aculeata* Jacq. (Salicaceae)
IPNI: 47939-2
Common name: chocolatillo, huesito, limoncillo **Habit:** Shrub, Small tree, Tree **Use:** HF, ME **Origin:** Native **Region:** Amazonia, Andean, Caribbean, Pacific **Dept.:** AMA, ANT, ATL, BOL, CAQ, CAU, CHO, COR, CUN, MAG, MET, SAP, SAN, SUC, TOL, VAC **Elev.:** 0-1040 msnm **IUCN Habitat code:** 1,2,3,4,5,14 **Conservation status: (IUCN)** LC **(CPSG)** Potential LC

5743 *Casearia americana* (L.) T.Samar. & M.H.Alford (Salicaceae)
IPNI: 77197250-1
Common name: dorado, manteco, pino **Habit:** Shrub, Small tree, Tree **Use:** HF, MA **Origin:** Native **Region:** Andean, Caribbean **Dept.:** ATL, BOL, CAL, CAU, COR, CUN, MAG, SUC, VAC **Elev.:** 6–1100 m a.s.l. **IUCN Habitat code:** 1,2,3,14 **Conservation status: (IUCN)** LC

5744 *Casearia arborea* (Rich.) Urb. (Salicaceae)
IPNI: 47945-2
Common name: escobo, espadero, granadillo, huesillo, nigüito, peralejo montañero, varablanca, árbol de coca **Habit:** Shrub, Small tree, Tree **Use:** HF **Origin:** Native **Region:** Amazonia, Andean, Guiana Shield, Caribbean, Orinoquia, Pacific **Dept.:** AMA, ANT, BOL, BOY, CAQ, CAS, CHO, CUN, GUV, MAG, MET, NAR, PUT, SAN, VAC, VAU **Elev.:** 5-1850 msnm **IUCN Habitat code:** 1,2,3,4,5,14 **Conservation status: (IUCN)** LC **(CPSG)** Potential LC

5745 *Casearia arguta* Kunth (Salicaceae)
IPNI: 779498-1
Common name: huesito, perlito, tábano **Habit:** Shrub, Small tree, Tree **Use:** HF **Origin:** Native **Region:** Caribbean, Pacific **Dept.:** AMA, ANT, BOL, CHO, COR, LAG, MAG, SAN, TOL **Elev.:** 0–700 m a.s.l. **IUCN Habitat code:** 1,2,3,5,14 **Conservation status: (IUCN)** LC

5746 *Casearia bicolor* Urb. (Salicaceae)
IPNI: 47957-2
Common name: barbasquillo, chula, curubo, fariña seca, lagunero, magaleto, manteco, marcelo, mojarra, niguo, paloetigre, sajino blanco, saíno blanco, varablanca, varecasa **Habit:** Small tree, Tree **Use:** HF, MA **Origin:** Native **Region:** Amazonia, Andean, Guiana Shield, Pacific **Dept.:** AMA, ANT, CAQ, CHO, CUN, GUA, MAG, MET, NAR, SAN, VAC **Elev.:** 50–1450 m a.s.l. **IUCN Habitat code:** 1,2,3,4,14 **Conservation status: (IUCN)** LC

5747 *Casearia cajambrensis* Cuatrec. (Salicaceae)
IPNI: 47964-2
Common name: casero **Habit:** Small tree, Tree **Use:** HF **Origin:** Native **Region:** Andean, Pacific **Dept.:** ANT, CHO, NAR, VAC **Elev.:** 0–2050 m a.s.l. **IUCN Habitat code:** 1,3,14 **Conservation status: (IUCN)** NE

5748 *Casearia combaymensis* Tul. (Salicaceae)
IPNI: 779528-1
Habit: Shrub, Small tree, Tree **Use:** HF **Origin:** Native **Region:** Amazonia, Andean, Pacific **Dept.:** AMA, ANT, CHO, CUN, NSA, TOL, VAC **Elev.:** 110–1000 m a.s.l. **IUCN Habitat code:** 1,2,14 **Conservation status: (IUCN)** LC

5749 *Casearia corymbosa* Kunth (Salicaceae)
IPNI: 47977-2
Common name: café del diablo, cerezo, domingo feliz, dondequiera, endiquero, ondequera, ondequero, sauco, tacaloa, tica, undequiera, vara de piedra, vara de piedra de humo, varablanca, varasanta, varazón **Habit:** Shrub, Small tree, Tree **Use:** HF, MA, ME **Origin:** Native **Region:** Andean, Caribbean **Dept.:** ANT, ATL, BOL, CAL, CES, CUN, HUI, LAG, MAG, MET, SUC, TOL, VAC **Elev.:** 3-1200 msnm **IUCN Habitat code:** 1,2,3,4,5,14 **Conservation status: (IUCN)** LC **(CPSG)** Potential LC

5750 *Casearia decandra* Jacq. (Salicaceae)
IPNI: 779541-1
Common name: huesillo **Habit:** Shrub, Small tree, Tree **Use:** HF **Origin:** Native **Region:** Amazonia, Andean, Guiana Shield, Caribbean **Dept.:** AMA, ANT, MAG, NAR, SUC **Elev.:** 100–1750 m a.s.l. **IUCN Habitat code:** 1,2,14 **Conservation status: (IUCN)** NE

5751 *Casearia fasciculata* (Ruiz & Pav.) Sleumer (Salicaceae)
IPNI: 47989-2
Habit: Shrub, Small tree, Tree **Use:** HF **Origin:** Native **Region:** Amazonia, Pacific **Dept.:** VAC, VAU **Elev.:** 10–800 m a.s.l. **IUCN Habitat code:** 1,3,14 **Conservation status: (IUCN)** LC

5752 *Casearia laetioides* (A.Rich.) Warb. (Salicaceae)
IPNI: 48022-2
Common name: bagatá **Habit:** Shrub, Small tree, Tree **Use:** HF, MA **Origin:** Native **Region:** Caribbean **Dept.:** ANT, SAP **Elev.:** 220–220 m a.s.l. **IUCN Habitat code:** 1,3,4,5,14 **Conservation status: (IUCN)** NE

5753 *Casearia mariquitensis* Kunth (Salicaceae)
IPNI: 779611-1
Common name: dondequiera, oso blanco, varoblanco **Habit:** Shrub, Small tree, Tree **Use:** HF **Origin:** Native **Region:** Amazonia, Andean, Guiana Shield **Dept.:** AMA, ANT, BOY, CUN, MET, NAR, RIS, SAN, TOL **Elev.:** 210–2150 m a.s.l. **IUCN Habitat code:** 1,2,3,4,5,14 **Conservation status: (IUCN)** LC

5754 *Casearia negrensis* Eichler (Salicaceae)
IPNI: 779627-1
Habit: Shrub, Small tree, Tree **Use:** ME **Origin:** Native **Region:** Amazonia **Dept.:** AMA, CAQ **Elev.:** 100–200 m a.s.l. **IUCN Habitat code:** 1,3,14 **Conservation status: (IUCN)** LC

5755 *Casearia nigricans* Sleumer (Salicaceae)
IPNI: 48049-2
Habit: Shrub, Small tree **Use:** HF **Origin:** Native **Region:** Andean **Dept.:** PUT **Elev.:** 1500–1670 m a.s.l. **IUCN Habitat code:** 1,3,14 **Conservation status: (IUCN)** NE

5756 *Casearia obovalis* Poepp. ex Griseb. (Salicaceae)
IPNI: 779633-1
Habit: Shrub, Small tree, Tree **Use:** HF **Origin:** Native **Region:** Amazonia **Dept.:** AMA, ANT **Elev.:** 20–220 m a.s.l. **IUCN Habitat code:** 1 **Conservation status: (IUCN)** LC

5757 *Casearia pitumba* Sleumer (Salicaceae)
IPNI: 297979-2
Habit: Shrub, Small tree, Tree **Use:** HF **Origin:** Native **Region:** Amazonia, Andean, Guiana Shield, Pacific **Dept.:** AMA, ANT, CAQ, CHO, MET, NAR, PUT, VAU **Elev.:** 10–1500 m a.s.l. **IUCN Habitat code:** 1,3,14 **Conservation status: (IUCN)** LC

5758 *Casearia praecox* Griseb. (Salicaceae)
IPNI: 48069-2
Common name: limoncillo, naranjito, varaepiedra, zapatero **Habit:** Shrub, Small tree, Tree **Use:** HF, MA, ME **Origin:** Native **Region:** Caribbean **Dept.:** ANT, BOL, MAG, SAN, SUC **Elev.:** 50–650 m a.s.l. **IUCN Habitat code:** 1,2,3,14 **Conservation status: (IUCN)** LC

5759 *Casearia prunifolia* Kunth (Salicaceae)
IPNI: 779661-1
Common name: naranjito **Habit:** Shrub, Small tree **Use:** HF **Origin:** Native **Region:** Amazonia **Dept.:** AMA, PUT **Elev.:** 100–100 m a.s.l. **IUCN Habitat code:** 1 **Conservation status: (IUCN)** LC

5760 *Casearia sylvestris* Sw. (Salicaceae)
IPNI: 274165-2
Common name: cucaracho, dondequiera, elemento, huesito, laurel, limoncillo, perico, piloncillo, plomo, ratón **Habit:** Shrub, Small tree, Tree **Use:** HF, MA, ME, PO **Origin:** Native **Region:** Amazonia, Andean, Guiana Shield, Caribbean, Orinoquia, Pacific **Dept.:** AMA, ANT, ARA, ATL, BOL, CAL, CAQ, CAU, CES, CHO, CUN, GUV, HUI, LAG, MAG, MET, NAR, PUT, SAP, SAN, TOL, VAC, VID **Elev.:** 0-2100 msnm **IUCN Habitat code:** 1,2,3,4,5,14 **Conservation status: (IUCN)** LC **(CPSG)** Potential LC

5761 *Casearia tachirensis* Steyerm. (Salicaceae)
IPNI: 48107-2
Common name: casero **Habit:** Shrub, Small tree, Tree **Use:** HF **Origin:** Native **Region:** Andean **Dept.:** RIS, VAC **Elev.:** 1860–2620 m a.s.l. **IUCN Habitat code:** 1,14 **Conservation status: (IUCN)** NE

5762 *Casearia uleana* Sleumer (Salicaceae)
IPNI: 48111-2
Habit: Small tree, Tree **Use:** HF **Origin:** Native **Region:** Amazonia **Dept.:** AMA **Elev.:** 100–125 m a.s.l. **IUCN Habitat code:** 1 **Conservation status: (IUCN)** LC

5763 *Casearia ulmifolia* Vahl ex Vent. (Salicaceae)
IPNI: 779712-1
Common name: varablanca **Habit:** Shrub, Small tree, Tree **Use:** HF, MA **Origin:** Native **Region:** Amazonia, Andean, Guiana Shield, Caribbean, Orinoquia, Pacific **Dept.:** AMA, ANT, BOY, CAS, MET **Elev.:** 100–900 m a.s.l. **IUCN Habitat code:** 1,2,3,4,5,14 **Conservation status: (IUCN)** LC

5764 *Casearia zahlbruckneri* Szyszyl. (Salicaceae)
IPNI: 365270-1
Habit: Shrub, Small tree, Tree **Use:** HF **Origin:** Native **Region:** Andean **Dept.:** ANT, CAU, NAR, VAC **Elev.:** 1750–2050 m a.s.l. **IUCN Habitat code:** 1,3,14 **Conservation status: (IUCN)** NE

5765 *Casearia zizyphoides* Kunth (Salicaceae)
IPNI: 779723-1
Habit: Shrub, Small tree, Tree **Use:** HF **Origin:** Native **Region:** Caribbean **Dept.:** CES, LAG, MAG **Elev.:** 60–150 m a.s.l. **IUCN Habitat code:** 1,2,14 **Conservation status: (IUCN)** LC

5766 *Dovyalis hebecarpa* (Gardner) Warb. (Salicaceae)
IPNI: 111561-1
Habit: Shrub, Small tree, Tree **Use:** GS, HF, ME **Origin:** Cultivated **Region:** Andean **Dept.:** ANT **Elev.:** 1480–1480 m a.s.l. **IUCN Habitat code:** 3,14 **Conservation status: (IUCN)** NE

5767 *Flacourtia indica* (Burm.f.) Merr. (Salicaceae)
IPNI: 365348-1
Common name: cerezo del gobernador, ciruela del gobernador **Habit:** Tree **Use:** AF, EU, FU, GS, HF, IF, MA, ME, SU **Origin:** Cultivated **Dept.:** ANT **Elev.:** 0–1500 m a.s.l. **IUCN Habitat code:** 1,14 **Conservation status: (IUCN)** LC

5768 *Hasseltia floribunda* Kunth (Salicaceae)
IPNI: 834658-1
Common name: copal, coronillo, cuero de marrano, huesito, hueso, lameardita, laurel, manquetillo macho, marfil, mestizo, pataegallina, pechuga de gallina, pepero, pie de paloma **Habit:** Shrub, Small tree, Tree **Use:** AF, EU, FU, HF, MA **Origin:** Native **Region:** Amazonia, Andean, Guiana Shield, Caribbean, Orinoquia, Pacific **Dept.:** AMA, ANT, BOL, BOY, CAL, CAQ, CAS, CAU, CHO, CUN, MAG, MET, PUT, QUI, SAN, SUC, VAC **Elev.:** 5-2400 msnm **IUCN Habitat code:** 1,3,5,14 **Conservation status: (IUCN)** LC **(CPSG)** Potential LC

5769 *Homalium guianense* (Aubl.) Oken (Salicaceae)
IPNI: 365491-1
Common name: carrapo, marfil **Habit:** Shrub, Small tree, Tree **Use:** ME **Origin:** Native **Region:** Amazonia, Guiana Shield, Pacific **Dept.:** AMA, ANT, CHO, GUV **Elev.:** 5–270 m a.s.l. **IUCN Habitat code:** 1,2,3,4,5,14 **Conservation status: (IUCN)** LC

5770 *Homalium racemosum* Jacq. (Salicaceae)
IPNI: 779849-1
Common name: angelino, caramacate, chamizo, iguano, vara de piedra **Habit:** Shrub, Small tree, Tree **Use:** MA, ME **Origin:** Native **Region:** Amazonia, Andean, Guiana Shield, Caribbean, Orinoquia, Pacific **Dept.:** AMA, BOL, BOY, CHO, MET, SAN, VAU **Elev.:** 5–800 m a.s.l. **IUCN Habitat code:** 1,2,3,4,5,14 **Conservation status: (IUCN)** LC

5771 *Lunania parviflora* Spruce ex Benth. (Salicaceae)
IPNI: 779887-1
Common name: hormigo, hormigón, yumbé **Habit:** Shrub, Small tree, Tree **Use:** HF, MA, ME **Origin:** Native **Region:** Amazonia, Caribbean, Pacific **Dept.:** AMA, ANT, BOL, CHO, MAG, NAR, SAN **Elev.:** 50–700 m a.s.l. **IUCN Habitat code:** 1,3,14 **Conservation status: (IUCN)** LC

5772 *Piparea dentata* Aubl. (Salicaceae)
IPNI: 779914-1
Habit: Shrub, Small tree, Tree **Use:** ME **Origin:** Native **Region:** Amazonia, Guiana Shield, Caribbean, Orinoquia, Pacific **Dept.:** ANT, CAQ, CHO, GUA, PUT, SAP, VAC, VID **Elev.:** 0–1000 m a.s.l. **IUCN Habitat code:** 1 **Conservation status: (IUCN)** NE

5773 *Piparea spruceana* (Benth. ex Eichler) T.Samar. & M.H.Alford (Salicaceae)
IPNI: 77197258-1
Habit: Shrub, Small tree **Use:** ME **Origin:** Native **Region:** Guiana Shield **Dept.:** GUA **Elev.:** 85–85 m a.s.l. **IUCN Habitat code:** 1,5 **Conservation status: (IUCN)** NE

5774 *Pleuranthodendron lindenii* (Turcz.) Sleumer (Salicaceae)
IPNI: 202149-2
Common name: antioqueño, hormigo **Habit:** Shrub, Small tree, Tree **Use:** HF **Origin:** Native **Region:** Amazonia, Pacific **Dept.:** AMA, ANT, BOY, CHO, PUT, VAU **Elev.:** 20–914 m a.s.l. **IUCN Habitat code:** 1,3,5,14 **Conservation status: (IUCN)** LC

5775 *Populus alba* L. (Salicaceae)
IPNI: 776573-1
Use: EU, FU, GS, HF, MA, ME **Region:** Andean **Dept.:** BOG **Conservation status: (IUCN)** LC

5776 *Populus deltoides* W.Bartram ex Marshall (Salicaceae)
IPNI: 30102841-2
Habit: Tree **Use:** AF, EU, FU, HF, MA, ME **Origin:** Cultivated **Region:** Andean **Dept.:** ANT **Elev.:** 1200–1500 m a.s.l. **IUCN Habitat code:** 14 **Conservation status: (IUCN)** LC

5777 *Populus nigra* L. (Salicaceae)
IPNI: 776790-1
Common name: chopo, álamo, álamo negro **Habit:** Tree **Use:** EU, HF, MA, ME **Origin:** Cultivated **Region:** Andean **Dept.:** BOG **Elev.:** 2560–2560 m a.s.l. **IUCN Habitat code:** 14 **Conservation status: (IUCN)** DD

5778 *Prockia crucis* P.Browne ex L. (Salicaceae)
IPNI: 321266-2
Common name: huesito **Habit:** Shrub, Small tree **Use:** HF, MA **Origin:** Native **Region:** Amazonia, Andean, Caribbean, Orinoquia **Dept.:** AMA, BOY, NSA, SAP, SAN **Elev.:** 0–1300 m a.s.l. **IUCN Habitat code:** 1,14 **Conservation status: (IUCN)** NE

5779 *Ryania angustifolia* (Turcz.) Monach. (Salicaceae)
IPNI: 224736-2
Common name: cri, cri-jé-ric, jé, ric **Habit:** Shrub, Small tree **Use:** MA, ME, PO **Origin:** Native **Region:** Amazonia, Guiana Shield **Dept.:** AMA, GUA, VAU **Elev.:** 80–250 m a.s.l. **IUCN Habitat code:** 1,3,4,5 **Conservation status: (IUCN)** LC

5780 *Ryania dentata* Miq. (Salicaceae)
IPNI: 112045-1
Common name: borrachero, guaricamo, matacucarachas **Habit:** Shrub, Small tree **Use:** ME **Origin:** Native **Region:** Guiana Shield, Orinoquia **Dept.:** BOY, MET, VID **Elev.:** 100–250 m a.s.l. **IUCN Habitat code:** 1,2,4,5 **Conservation status: (IUCN)** NE

5781 *Ryania pyrifera* (Rich.) Uittien & Sleumer (Salicaceae)
IPNI: 365862-1
Common name: berraquillo, cucaracho, jagüillo, marcelo, matacucarachas **Habit:** Shrub, Small tree **Use:** ME **Origin:** Native **Region:** Guiana Shield **Dept.:** ANT, CAQ, VAU **Elev.:** 0–325 m a.s.l. **Conservation status: (IUCN)** LC

5782 *Ryania speciosa* Vahl (Salicaceae)
IPNI: 112052-1
Common name: anón de monte, berraquillo, guayabilla, marcelo, matacucaracha, zaino **Habit:** Shrub, Small tree, Tree **Use:** HF, MA, ME, PO **Origin:** Native **Region:** Amazonia, Andean, Guiana Shield, Caribbean, Orinoquia, Pacific **Dept.:** AMA, ANT, BOL, CAQ, CAU, CHO, COR, CUN, GUA, HUI, MET, NAR, SAN, TOL, VAC, VAU, VID **Elev.:** 0-1440 msnm **IUCN Habitat code:** 1,3,4,5,14 **Conservation status: (IUCN)** LC **(CPSG)** Potential LC

5783 *Salix babylonica* L. (Salicaceae)
IPNI: 777133-1
Common name: sauce llorón **Habit:** Tree **Use:** EU, HF, MA, ME **Origin:** Cultivated **Region:** Andean **Dept.:** ANT, BOG **Elev.:** 1486–1486 m a.s.l. **IUCN Habitat code:** 1,3,14 **Conservation status: (IUCN)** NE

5784 *Salix chilensis* Molina (Salicaceae)
IPNI: 777304-1
Common name: pajarito, pájaro bobo, sauce, sauce colorado, sauce llorón, sauce macho, sausa **Habit:** Small tree, Tree **Use:** EU, MA, ME **Origin:** Native, Cultivated **Region:** Amazonia, Andean, Caribbean **Dept.:** AMA, ANT, CAU, CUN, MAG, PUT, SAN, VAC **Elev.:** 100–2600 m a.s.l. **IUCN Habitat code:** 1,3,5,14 **Conservation status: (IUCN)** NE

5785 *Salix humboldtiana* Willd. (Salicaceae)
IPNI: 777818-1
Common name: sauce, sauce amargo, sauce macho, sausa **Use:** EU, MA, ME **Region:** Andean **Dept.:** BOG **Conservation status: (IUCN)** LC

5786 *Salix martiana* Leyb. (Salicaceae)
IPNI: 778154-1
Use: MA **Origin:** Endemic **Conservation status: (IUCN)** NE

5787 *Salix purpurea* L. (Salicaceae)
IPNI: 301321-2
Habit: Shrub, Small tree **Use:** EU, HF, MA, ME **Origin:** Cultivated **Region:** Andean **Dept.:** BOG, CUN **Elev.:** 2700–2700 m a.s.l. **Conservation status: (IUCN)** LC

5788 *Salix viminalis* L. (Salicaceae)
IPNI: 779152-1
Use: EU, FU, MA, ME, SU **Region:** Andean **Dept.:** BOG **IUCN Habitat code:** 14 **Conservation status: (IUCN)** LC

5789 *Tetrathylacium macrophyllum* Poepp. (Salicaceae)
IPNI: 867537-1
Common name: lomo de machete, marcelo, mestizo **Habit:** Small tree, Tree **Use:** HF, MA **Origin:** Native

Region: Amazonia, Andean, Guiana Shield, Caribbean, Pacific **Dept.:** AMA, ANT, BOL, BOY, CAQ, CHO, COR, GUV, MET, NAR, PUT, SAN, VAC **Elev.:** 0-1350 msnm **IUCN Habitat code:** 1,3,5,14 **Conservation status: (IUCN)** LC **(CPSG)** Potential LC

5790 *Xylosma flexuosa* (Kunth) Hemsl. (Salicaceae)
IPNI: 112138-1
Habit: Shrub, Small tree **Use:** HF, ME **Origin:** Native **Region:** Andean, Caribbean **Dept.:** MAG **Elev.:** 0–1800 m a.s.l. **Conservation status: (IUCN)** LC

5791 *Xylosma intermedia* (Seem.) Triana & Planch. (Salicaceae)
IPNI: 112146-1
Common name: bulletigre, carita de santo **Habit:** Shrub, Small tree, Tree **Use:** AF, MA **Origin:** Native **Region:** Amazonia, Andean, Guiana Shield, Caribbean, Orinoquia **Dept.:** ANT, ATL, BOL, CAL, CAU, CUN, MAG, MET, SUC, TOL, VAC **Elev.:** 200–2000 m a.s.l. **IUCN Habitat code:** 1,2,3,14 **Conservation status: (IUCN)** LC

5792 *Xylosma spiculifera* (Tul.) Triana & Planch. (Salicaceae)
IPNI: 112178-1
Common name: aguja de arria, cacho de venado, corona, corona de espinas, corona de espino, corono, corono espina, corono espino, corono espinoso, espino, espino cabro, espino de cabra, espino de cabras, pino, puyón, tachuelo **Habit:** Small tree **Use:** ME **Origin:** Native **Region:** Andean **Dept.:** BOG, BOY, CAU, CUN, NSA **Elev.:** 1750–2900 m a.s.l. **Conservation status: (IUCN)** NE

5793 *Anchietea frangulifolia* (Kunth) Melch. (Violaceae)
IPNI: 12522-2
Habit: Shrub, Liana, Climbing **Use:** ME **Origin:** Native **Region:** Andean **Dept.:** ANT, BOG, BOY, CUN, NAR, PUT, QUI, RIS, VAC **Elev.:** 1100–3000 m a.s.l. **Conservation status: (IUCN)** NE

5794 *Calyptrion arboreum* (L.) Paula-Souza (Violaceae)
IPNI: 77144891-1
Habit: Climbing **Use:** ME **Origin:** Native **Region:** Amazonia, Caribbean, Orinoquia, Pacific **Dept.:** AMA, ANT, BOL, CAS, CHO, SAN **Elev.:** 0–300 m a.s.l. **IUCN Habitat code:** 1,2,3,4,5,14 **Conservation status: (IUCN)** NE

5795 *Leonia crassa* L.B.Sm. & A.Fernández (Violaceae)
IPNI: 137351-2
Common name: fono cocoro **Habit:** Tree **Use:** HF **Origin:** Native **Region:** Amazonia, Guiana Shield, Orinoquia **Dept.:** AMA, CAQ, GUV, MET, PUT, VAU **Elev.:** 100–1000 m a.s.l. **IUCN Habitat code:** 1,3,14 **Conservation status: (IUCN)** LC

5796 *Leonia cymosa* Mart. (Violaceae)
IPNI: 866955-1
Common name: caimo silvestre, caricapuñoruaci, yema de huevo **Habit:** Small tree **Use:** HF **Origin:** Native **Region:** Amazonia, Guiana Shield, Orinoquia **Dept.:** AMA, CAQ, GUA, GUV, PUT, VAU **Elev.:** 80–600 m a.s.l. **IUCN Habitat code:** 1,3,5,14 **Conservation status: (IUCN)** LC

5797 *Leonia equatoriensis* (Hekking) Byng & Christenh. (Violaceae)
IPNI: 77184219-1
Habit: Shrub, Tree **Use:** HF **Origin:** Native **Region:** Amazonia, Andean, Pacific **Dept.:** AMA, CHO, NAR, PUT **Elev.:** 80–1200 m a.s.l. **Conservation status: (IUCN)** NE

5798 *Leonia glycycarpa* Ruiz & Pav. (Violaceae)
IPNI: 866956-1
Common name: bastimento, palo de sapo **Habit:** Tree **Use:** HF **Origin:** Native **Region:** Amazonia, Guiana Shield, Orinoquia, Pacific **Dept.:** AMA, ANT, CAQ, CAU, GUV, MET, NAR, PUT, VAC, VAU **Elev.:** 30–900 m a.s.l. **IUCN Habitat code:** 1,3,5,14 **Conservation status: (IUCN)** LC

5799 *Leonia occidentalis* Cuatrec. (Violaceae)
IPNI: 866958-1
Habit: Tree **Use:** HF **Origin:** Native **Region:** Andean, Pacific **Dept.:** ANT, CHO, NAR, VAC **Elev.:** 90–750 m a.s.l. **IUCN Habitat code:** 1 **Conservation status: (IUCN)** NE

5800 *Leonia racemosa* Mart. (Violaceae)
IPNI: 866960-1
Habit: Tree **Use:** HF **Origin:** Native **Region:** Amazonia, Guiana Shield, Orinoquia **Dept.:** AMA, ANT, CAQ, GUV, MET, PUT, SAN **Elev.:** 50–450 m a.s.l. **IUCN Habitat code:** 1,3,14 **Conservation status: (IUCN)** LC

5801 *Leonia sphaerocarpa* (Triana & Planch.) Byng & Christenh. (Violaceae)
IPNI: 77184111-1
Common name: chilco blanco, huesito, pepito **Habit:** Shrub, Tree **Use:** HF, ME **Origin:** Native **Region:** Amazonia, Andean, Guiana Shield, Orinoquia, Pacific **Dept.:** AMA, ANT, CAL, CAQ, CHO, MAG, MET, PUT, SAN **Elev.:** 20–1600 m a.s.l. **Conservation status: (IUCN)** NE

5802 *Leonia triandra* Cuatrec. (Violaceae)
IPNI: 137356-2
Common name: azuceno, popó de mula, sapotillo, tocino, veneno, yema de huevo, yemaehuevo **Habit:** Tree **Use:** HF **Origin:** Native **Region:** Andean, Pacific **Dept.:** ANT, BOL, CAL, CHO, COR, NAR, RIS, SAN, VAC **Elev.:** 30–900 m a.s.l. **IUCN Habitat code:** 1,2,3,5,14 **Conservation status: (IUCN)** NE

5803 *Paypayrola guianensis* Aubl. (Violaceae)
IPNI: 867031-1
Habit: Shrub, Small tree **Use:** MA **Origin:** Native **Region:** Amazonia, Guiana Shield **Dept.:** AMA, CAQ **Elev.:** 200–300 m a.s.l. **IUCN Habitat code:** 1 **Conservation status: (IUCN)** LC

5804 *Pombalia calceolaria* (L.) Paula-Souza (Violaceae)
IPNI: 77143549-1
Habit: Herb, Subshrub **Use:** ME **Origin:** Native **Region:** Amazonia, Caribbean **Dept.:** LAG, SUC, VAU **Elev.:** 2–200 m a.s.l. **IUCN Habitat code:** 2 **Conservation status: (IUCN)** NE

5805 *Pombalia glabra* (Dowell) Paula-Souza (Violaceae)
IPNI: 77143553-1
Habit: Herb **Use:** ME **Origin:** Native **Region:** Andean, Caribbean, Orinoquia **Dept.:** ANT, BOL, CAL, CAU, CUN, MAG, NAR, NSA, SAN, TOL, VAC **Elev.:** 0–1560 m a.s.l. **IUCN Habitat code:** 2 **Conservation status: (IUCN)** NE

5806 *Pombalia oppositifolia* (L.) Paula-Souza (Violaceae)
IPNI: 77143566-1
Habit: Herb **Use:** ME **Origin:** Native **Region:** Amazonia, Caribbean, Orinoquia **Dept.:** AMA, ATL, BOL, CUN, MET, TOL **Elev.:** 0–450 m a.s.l. **IUCN Habitat code:** 1,2 **Conservation status: (IUCN)** NE

5807 *Pombalia parviflora* (L.f.) Paula-Souza (Violaceae)
IPNI: 77143568-1
Common name: chuchunchullo, teatina **Habit:** Herb **Use:** ME **Origin:** Native **Region:** Andean **Dept.:** BOG, BOY, CUN, NAR **Elev.:** 2300–3600 m a.s.l. **IUCN Habitat code:** 1,3,4,8,14 **Conservation status: (IUCN)** NE

5808 *Pombalia prunifolia* (Humb. & Bonpl. ex Willd.) Paula-Souza (Violaceae)
IPNI: 77143573-1
Common name: cartage chuchunchullo, cartagena, chuchunchullo **Habit:** Shrub **Use:** ME **Origin:** Native **Region:** Amazonia, Guiana Shield, Caribbean, Orinoquia, Pacific **Dept.:** ANT, ARA, BOG, BOL, CAS, CES, CHO, COR, GUV, LAG, MAG, MET, PUT, SAN, SUC **Elev.:** 20–1000 m a.s.l. **IUCN Habitat code:** 1,2,3,4,5,14 **Conservation status: (IUCN)** LC

5809 *Rinorea flavescens* (Aubl.) Kuntze (Violaceae)
IPNI: 221409-2
Common name: carbonero cáscara, hormigo, patevenado, patevenao **Habit:** Shrub, Small tree, Tree **Use:** ME **Origin:** Native **Region:** Amazonia, Orinoquia, Pacific **Dept.:** AMA, CAQ, CHO, VAC, VAU, VID **Elev.:** 10–500 m a.s.l. **IUCN Habitat code:** 1,3,4,5,14 **Conservation status: (IUCN)** LC

5810 *Rinorea guianensis* Aubl. (Violaceae)
IPNI: 867224-1
Habit: Tree **Use:** MA **Origin:** Native **Region:** Amazonia **Dept.:** CAQ, SAN **Elev.:** 150–200 m a.s.l. **IUCN Habitat code:** 1 **Conservation status: (IUCN)** NE

5811 *Rinorea racemosa* (Mart.) Kuntze (Violaceae)
IPNI: 221443-2
Common name: coca de águila, coca del creador **Habit:** Shrub, Small tree, Tree **Use:** MA **Origin:** Native **Region:** Amazonia, Andean, Guiana Shield **Dept.:** AMA, CAQ, GUV, PUT, VAU **Elev.:** 100–1300 m a.s.l. **IUCN Habitat code:** 1,3,5,14 **Conservation status: (IUCN)** NE

5812 *Rinorea viridifolia* Rusby (Violaceae)
IPNI: 284322-2
Common name: trabuclo **Habit:** Small tree **Use:** HF **Origin:** Native **Region:** Amazonia, Andean, Orinoquia, Pacific **Dept.:** AMA, ANT, BOL, CHO, MET, PUT, SAN, VAU **Elev.:** 10–750 m a.s.l. **IUCN Habitat code:** 1,3,14 **Conservation status: (IUCN)** LC

5813 *Viola arguta* Humb. & Bonpl. ex Schult. (Violaceae)
IPNI: 867656-1
Habit: Subshrub, Climbing **Use:** ME **Origin:** Native **Region:** Andean **Dept.:** CES, MAG **Elev.:** 2000–3250 m a.s.l. **IUCN Habitat code:** 1,14 **Conservation status: (IUCN)** NE

5814 *Viola arvensis* Murray (Violaceae)
IPNI: 291320-2
Habit: Herb **Use:** HF, ME **Origin:** Naturalised **Region:** Andean **Dept.:** ANT, BOG, CAU, CUN **Elev.:** 2600–3100 m a.s.l. **Conservation status: (IUCN)** NE

5815 *Viola cuatrecasasii* L.B.Sm. & A.Fernández (Violaceae)
IPNI: 265843-2
Common name: pascuala **Habit:** Herb **Use:** ME **Origin:** Native **Region:** Andean **Dept.:** NAR, VAC **Elev.:** 2400–3300 m a.s.l. **IUCN Habitat code:** 1,3,4,8,14 **Conservation status: (IUCN)** NE

5816 *Viola dombeyana* DC. ex Ging. (Violaceae)
IPNI: 868059-1
Common name: cerrilla, violeta silvestre **Habit:** Herb **Use:** ME **Origin:** Native **Region:** Andean **Dept.:** CAU, HUI, NAR, NSA, VAC **Elev.:** 2100–3900 m a.s.l. **IUCN Habitat code:** 1,3,4,8,14 **Conservation status: (IUCN)** NE

5817 *Viola hirta* L. (Violaceae)
IPNI: 868309-1
Use: ME **Region:** Andean **Dept.:** BOG **Conservation status: (IUCN)** NE

5818 *Viola humilis* Kunth (Violaceae)
IPNI: 868344-1
Common name: violeta **Habit:** Herb **Use:** ME **Origin:** Native **Region:** Andean **Dept.:** ANT, BOG, BOY, CUN, PUT, SAN **Elev.:** 2500–3645 m a.s.l. **IUCN Habitat code:** 3,4,14 **Conservation status: (IUCN)** NE

5819 *Viola odorata* L. (Violaceae)
IPNI: 868828-1
Common name: violeta, violeta morada **Habit:** Herb **Use:** EU, HF, MA, ME **Origin:** Cultivated, Naturalised **Region:** Andean **Dept.:** ANT, BOG, BOY, CAL, CAU, CHO, CUN, HUI, PUT, VAC **Elev.:** 900–2700 m a.s.l. **IUCN Habitat code:** 1,3,14 **Conservation status: (IUCN)** NE

5820 *Viola scandens* Humb. & Bonpl. ex Schult. (Violaceae)
IPNI: 869203-1
Common name: violeta, violeta de páramo, violeta silvestre **Habit:** Herb **Use:** ME **Origin:** Native **Region:** Andean **Dept.:** ANT, BOG, BOY, CAU, CUN, NAR, NSA, PUT, QUI, RIS, SAN, TOL, VAC **Elev.:** 2400-3750 m.a.s.l. **IUCN Habitat code:** 1,3,4,5,8,14 **Conservation status: (IUCN)** NE **(CPSG)** Potential LC

5821 *Viola stipularis* Sw. (Violaceae)
IPNI: 266985-2
Common name: violeta, violeta silvestre **Habit:** Herb **Use:** ME **Origin:** Native **Region:** Andean **Dept.:** ANT, BOG, BOY, CAL, CAQ, CAU, CHO, CUN, HUI, MAG, MET, NAR, NSA, PUT, RIS, SAN, VAC **Elev.:** 1900–3240 m a.s.l. **IUCN Habitat code:** 1,3,4,5,14 **Conservation status: (IUCN)** NE

5822 *Viola tricolor* L. (Violaceae)
IPNI: 869428-1
Common name: pensamiento **Habit:** Herb **Use:** EU, HF, ME **Origin:** Naturalised **Region:** Andean **Dept.:** ANT, BOG, CAU, CUN, NAR, NSA, PUT **Elev.:** 2440–3000 m a.s.l. **IUCN Habitat code:** 1,2,3,4,14 **Conservation status: (IUCN)** NE

5823 *Viola veronicifolia* Planch. & Linden (Violaceae)
IPNI: 869498-1
Common name: violeta de páramo **Habit:** Herb **Use:** ME **Origin:** Native **Region:** Andean **Dept.:** BOG, CAU, CUN, HUI, NSA, SAN **Elev.:** 2500–3500 m a.s.l. **IUCN Habitat code:** 4,14 **Conservation status: (IUCN)** NE

Malvales

5824 *Bixa orellana* L. (Bixaceae)
IPNI: 33335-2
Common name: abujo, abujumujariyocù, achiote, achiote amarillo, achiote de bija, achiote de cholo, achiote de monte, achiote pepa roja, achiote rojo, achiote verde, achiotillo, achoote, achote, achote amarillo, achote casero, achote colorado, achote de bija, achote d **Habit:** Shrub, Small tree, Tree **Use:** AF, EU, FU, HF, MA, ME, PO, SU **Origin:** Native, Cultivated **Region:** Amazonia, Andean, Guiana Shield, Caribbean, Orinoquia, Pacific **Dept.:** AMA, ANT, ARA, ATL, BOL, BOY, CAL, CAQ, CAS, CAU, CES, CHO, COR, CUN, GUV, HUI, MAG, MET, NAR, NSA, PUT, QUI, RIS, SAP, SAN, SUC, TOL, VAC, VAU, VID **Elev.:** 0–2500 m a.s.l. **IUCN Habitat code:** 1,2,3,4,5,14 **Conservation status: (IUCN)** LC

5825 *Bixa platycarpa* Ruiz & Pav. ex G.Don (Bixaceae)
IPNI: 111499-1
Use: HF, ME **IUCN Habitat code:** 1,3 **Conservation status: (IUCN)** LC

5826 *Bixa urucurana* Willd. (Bixaceae)
IPNI: 111504-1
Common name: achiote, achiote cimarrón, achiote de monte, achote, achote cimarrón, achote montañero, bija, onoto, urucurú, urucú **Habit:** Tree **Use:** EU, GS, HF, MA, ME **Origin:** Native **Region:** Amazonia, Guiana Shield, Orinoquia, Pacific **Dept.:** AMA, ANT, ARA, BOY, CAQ, CAS, CES, CHO, GUV, MET, PUT, VAC, VID **Elev.:** 200–600 m a.s.l. **IUCN Habitat code:** 1,2,3,4,14 **Conservation status: (IUCN)** LC

5827 *Cochlospermum orinocense* (Kunth) Steud. (Bixaceae)
IPNI: 111532-1
Common name: bototo, botuto, ceibo, compadre bototo, flechero, floramarillo, poroporo, yuco **Habit:** Tree **Use:** HF, MA, ME, SU **Origin:** Native **Region:** Amazonia, Guiana Shield, Orinoquia, Pacific **Dept.:** AMA, ANT, ARA, BOY, CAL, CAQ, CAS, COR, GUA, GUV, LAG, NSA, SAN, VID **Elev.:** 20–770 m a.s.l. **IUCN Habitat code:** 1,2,3,4,5,14 **Conservation status: (IUCN)** LC

5828 *Cochlospermum palmatifidum* (DC.) Byng & Christenh. (Bixaceae)
IPNI: 77183799-1
Habit: Herb **Use:** HF **Origin:** Native **Dept.:** HUI, TOL **Elev.:** 400–800 m a.s.l. **IUCN Habitat code:** 14 **Conservation status: (IUCN)** NE

5829 *Cochlospermum vitifolium* (Willd.) Spreng. (Bixaceae)
IPNI: 111539-1
Common name: batabana, bototo, botuto, calastoliendra, carneasado, carnestolenda, carnestolendas, carnestolendo, carrestoliendo, carretosendro, ceiba, ceiba arracacha, ceiba yuca, ceibo, ceibo yuco, compadre bototo, flechero, papayote, papayuelo, poroporo, santocristo **Habit:** Tree **Use:** AF, EU, FU, HF, MA, ME **Origin:** Native **Region:** Andean, Caribbean, Pacific **Dept.:** ANT, BOL, CAS, CES, CHO, COR, GUV, LAG, MAG, MET, SUC, VID **Elev.:** 0–1490 m a.s.l. **IUCN Habitat code:** 1,2,3,4,5,14 **Conservation status: (IUCN)** LC

5830 *Abelmoschus esculentus* (L.) Moench (Malvaceae)
IPNI: 558006-1
Common name: algalia, almizclillo, angelonia, argalia, candia, chimbombó, gicombo, gombo, guicombo, lagarto, malvisco, monito, morito, ocra, ocro, quimbombó **Habit:** Herb **Use:** AF, EU, FU, GS, HF, MA, ME, PO, SU **Origin:** Cultivated **Region:** Amazonia, Andean, Caribbean, Pacific **Dept.:** ANT, CAU, PUT, SAP **Elev.:** 20–1500 m a.s.l. **IUCN Habitat code:** 1,3,5,14 **Conservation status: (IUCN)** NE

5831 *Abelmoschus moschatus* Medik. (Malvaceae)
IPNI: 558023-1
Common name: abuichillo, algalia, algodón, almizclillo, ambarina, ambra, argalia, candia, lagarto, majagüito de playa, ojo de gato, toullí **Habit:** Herb **Use:** EU, GS, HF, IF, MA, ME, PO **Origin:** Cultivated **Region:** Amazonia, Andean, Caribbean, Pacific **Dept.:** AMA, ANT, BOL, CHO, MAG, NAR, NSA, PUT **Elev.:** 0–1400 m a.s.l. **IUCN Habitat code:** 1,3,5,14 **Conservation status: (IUCN)** NE

5832 *Abutilon abutiloides* (Jacq.) Garcke ex Hochr. (Malvaceae)
IPNI: 1055043-2
Habit: Subshrub **Use:** MA **Origin:** Native **Conservation status: (IUCN)** NE

5833 *Abutilon grandifolium* (Willd.) Sweet (Malvaceae)
IPNI: 1026852-2
Use: EU, MA, ME **Region:** Andean **Dept.:** BOG **Conservation status: (IUCN)** NE

5834 *Abutilon hirtum* (Lam.) Sweet (Malvaceae)
IPNI: 238-2
Habit: Herb, Subshrub **Use:** EU, GS, HF, MA, ME **Origin:** Naturalised **Region:** Caribbean **Dept.:** SAP **Elev.:** 0–300 m a.s.l. **Conservation status: (IUCN)** NE

5835 *Abutilon ibarrense* Kunth (Malvaceae)
IPNI: 323515-2
Common name: panecito **Habit:** Shrub, Small tree **Use:** ME **Origin:** Native **Region:** Andean **Dept.:** CUN, HUI, SAN, VAC **Elev.:** 350–2500 m a.s.l. **IUCN Habitat code:** 1,2,14 **Conservation status: (IUCN)** NE

5836 *Abutilon indicum* (L.) Sweet (Malvaceae)
IPNI: 1101632-2
Common name: malva, yute de China **Habit:** Subshrub **Use:** AF, HF, MA, ME, PO **Origin:** Naturalised **Dept.:** BOG **Conservation status: (IUCN)** NE **(CPSG)** Potential LC

5837 *Abutilon insigne* Planch. (Malvaceae)
IPNI: 558295-1
Common name: abutilón, abutilón rojo, campana, campanita, gueva de indio, güevo de indio, yema de huevo **Habit:** Shrub **Use:** EU, ME **Origin:** Native, Cultivated **Region:** Andean **Dept.:** CAL, CUN **Elev.:** 2600–2700 m a.s.l. **IUCN Habitat code:** 1,3,4,14 **Conservation status: (IUCN)** NE

5838 *Abutilon viscosum* (L.) Dorr (Malvaceae)
IPNI: 77146518-1
Common name: algodoncillo, escoba hedionda, escobilla, escobilla blanca, escobo, hoja de sebo, kashushira, manteco **Habit:** Shrub, Subshrub **Use:** ME **Origin:** Native **Region:** Caribbean **Dept.:** ANT, ATL, BOL, CUN, HUI, LAG, MAG, NSA, SAN, SUC **Elev.:** 0–1000 m a.s.l. **IUCN Habitat code:** 1,2,3,5,14 **Conservation status: (IUCN)** NE

5839 *Alcea rosea* L. (Malvaceae)
IPNI: 558713-1
Common name: amapola, malvaloca, malvarrosa, malvavisco, malvavisco de flor **Habit:** Shrub, Subshrub **Use:** EU, HF, MA, ME **Origin:** Cultivated **Region:** Andean **Dept.:** ANT, BOG, CUN **Elev.:** 2620–2620 m a.s.l. **IUCN Habitat code:** 1,14 **Conservation status: (IUCN)** NE

5840 *Allosidastrum pyramidatum* (Desp. ex Cav.) Krapov., Fryxell & Bates (Malvaceae)
IPNI: 945985-1
Habit: Shrub **Use:** ME **Origin:** Native **Region:** Andean, Caribbean **Dept.:** ATL, CUN, MAG **Elev.:** 5–1700 m a.s.l. **IUCN Habitat code:** 2,3,14 **Conservation status: (IUCN)** NE

5841 *Althaea officinalis* L. (Malvaceae)
IPNI: 558872-1
Common name: malva, malvavisco **Use:** EU, HF, MA, ME **Dept.:** BOG **IUCN Habitat code:** 1,14 **Conservation status: (IUCN)** NE

5842 *Anoda cristata* (L.) Schltdl. (Malvaceae)
IPNI: 1218005-2
Common name: cadillo azul, campanilla morada, escobo, estrella **Habit:** Herb **Use:** EU, HF, ME **Origin:** Native **Region:** Andean, Caribbean **Dept.:** ANT, BOL, BOY, CAQ, CAU, COR, CUN, HUI, LAG, MAG, NAR, NSA, QUI, SAP, SAN, VAC **Elev.:** 20-2700 m.a.s.l. **IUCN Habitat code:** 1,2,3,5,14 **Conservation status: (IUCN)** NE **(CPSG)** Potential LC

5843 *Apeiba glabra* Aubl. (Malvaceae)
IPNI: 833257-1
Common name: balso, corcho, peine de mono, peine mico, peinemono **Habit:** Tree **Use:** MA, ME **Origin:** Native **Region:** Amazonia, Andean, Pacific **Dept.:** AMA, ANT, CAL, CAQ, CAU, CHO, CUN, GUV, MET, NAR, SAN, VAC **Elev.:** 0–1000 m a.s.l. **IUCN Habitat code:** 1,3,4,5,14 **Conservation status: (IUCN)** LC

5844 *Apeiba membranacea* Spruce ex Benth. (Malvaceae)
IPNI: 833266-1
Common name: balso, cadillo, cadillo grande, corcho, erizo, esponjilla, galleta, guásimo baba, guásimo blanco, malagano, maradague, melá, peine de mico, peine de mono, peinemico, peinemono, pcludo, topa, árbol baboso **Habit:** Small tree, Tree **Use:** HF, MA, ME **Origin:** Native **Region:** Amazonia, Andean, Pacific **Dept.:** AMA, ANT, BOL, CAL, CAQ, CAU, CHO, CUN, GUV, MET, NAR, SAN, VAC **Elev.:** 0–1120 m a.s.l. **IUCN Habitat code:** 1,2,3,4,5,14 **Conservation status: (IUCN)** LC

5845 *Apeiba petoumo* Aubl. (Malvaceae)
IPNI: 833267-1
Use: MA **Origin:** Native **Conservation status: (IUCN)** LC

5846 *Apeiba tibourbou* Aubl. (Malvaceae)
IPNI: 833270-1
Common name: ajínae, cadillo, corcho, corcho blanco, cortezo, cortizo, erizo, esponjilla, guásimo colorado, malagano, malgano, mapaso, matay, melá, mico, palo jobo, peine, peine de mico, peine de mono, peinemono, penemono, wetsuenae **Habit:** Shrub, Tree **Use:** EU, FU, HF, MA, ME **Origin:** Native **Region:** Amazonia, Caribbean, Orinoquia, Pacific **Dept.:** AMA, ANT, BOL, CAS, CHO, COR, GUV, MAG, SAN, SUC, VID **Elev.:** 0–800 m a.s.l. **IUCN Habitat code:** 1,2,3,4,5,14 **Conservation status: (IUCN)** LC

5847 *Ayenia aculeata* (Jacq.) Christenh. & Byng (Malvaceae)
IPNI: 77183732-1
Common name: bejuco espinoso, zarza hueca **Habit:** Climbing **Use:** HF, ME **Origin:** Native **Region:** Andean, Guiana Shield, Caribbean, Orinoquia **Dept.:** ANT, ATL, BOL, CAL, CHO, COR, CUN, MAG, MET, NSA, SAN, SUC, TOL, VAC **Elev.:** 10-1320 m.a.s.l. **Conservation status: (IUCN)** NE **(CPSG)** Potential LC

5848 *Ayenia catalpifolia* (Jacq.) Christenh. & Byng (Malvaceae)
IPNI: 77184121-1
Common name: zarza **Habit:** Climbing **Use:** HF, ME, PO **Origin:** Native **Region:** Caribbean, Orinoquia, Pacific **Dept.:** ANT, ATL, BOL, LAG, MAG, MET **Elev.:** 100–350 m a.s.l. **Conservation status: (IUCN)** NE

5849 *Ayenia magna* L. (Malvaceae)
IPNI: 822190-1
Common name: escobillita **Habit:** Shrub **Use:** ME **Origin:** Native **Region:** Andean, Caribbean **Dept.:** ANT, ATL, BOL, CUN, HUI, MAG, SUC, TOL **Elev.:** 25–710 m a.s.l. **IUCN Habitat code:** 1,2,3,14 **Conservation status: (IUCN)** NE

5850 *Ayenia scabra* (L.) Christenh. & Byng (Malvaceae)
IPNI: 77183967-1
Common name: puya, raya **Habit:** Shrub **Use:** ME **Origin:** Native **Region:** Orinoquia **Dept.:** CUN, MET, VID **Elev.:** 350–450 m a.s.l. **Conservation status: (IUCN)** NE

5851 *Ayenia uaupensis* (Spruce ex K.Schum.) Christenh. & Byng (Malvaceae)
IPNI: 77184160-1
Habit: Liana **Use:** MA **Origin:** Native **Region:** Guiana Shield **Dept.:** VAU **Elev.:** 200–200 m a.s.l. **Conservation status: (IUCN)** NE

5852 *Bombax ceiba* L. (Malvaceae)
IPNI: 559120-1
Common name: carrecillo, carretillo, cedro, cedro espino, cedro espinoso, cedro macho, cedro tolúa, ceiba, ceiba bruja, ceiba colorada, ceiba colorá, ceiba espinosa, ceiba roja, ceiba tolú, ceiba tolúa, saquisaqui, tolú, tolúa **Habit:** Tree **Use:** AF, EU, FU, GS, HF, IF, MA, ME, SU **Origin:** Native **Region:** Amazonia, Andean, Caribbean, Orinoquia, Pacific **Dept.:** ATL, BOL, CAS, CES, CHO, COR, GUV, MAG, MET, SUC **Elev.:** 0-500 m.a.s.l. **Conservation status: (IUCN)** LC **(CPSG)** VU

5853 *Brachychiton populneus* (Schott & Endl.) R.Br. (Malvaceae)
IPNI: 822268-1
Habit: Tree **Use:** EU, HF, MA **Origin:** Cultivated **Region:** Andean **Dept.:** ANT **Elev.:** 1500–1500 m a.s.l. **Conservation status: (IUCN)** NE

5854 *Briquetia spicata* (Kunth) Fryxell (Malvaceae)
IPNI: 36556-2
Habit: Herb, Subshrub **Use:** ME **Origin:** Native **Region:** Caribbean **Dept.:** ANT, ATL, BOL, MAG **Elev.:** 80–850 m a.s.l. **IUCN Habitat code:** 1,2,3,14 **Conservation status: (IUCN)** NE

5855 *Callianthe picta* (Gillies ex Hook. & Arn.) Donnell (Malvaceae)
IPNI: 77130215-1
Use: EU, MA, ME **Region:** Andean **Dept.:** BOG **IUCN Habitat code:** 1,3,4,14 **Conservation status: (IUCN)** NE

5856 *Callianthe striata* (G.F.Dicks. ex Lindl.) Donnell (Malvaceae)
IPNI: 77130224-1
Habit: Shrub **Use:** ME **Origin:** Cultivated, Naturalised **Region:** Andean **Dept.:** ANT, BOG, CUN **Elev.:** 1485–2900 m a.s.l. **Conservation status: (IUCN)** NE

5857 *Callianthe vexillaria* (É.Morren) Donnell (Malvaceae)
IPNI: 77130227-1
Habit: Shrub **Use:** EU **Origin:** Native, Cultivated **Region:** Andean **Dept.:** ANT, CUN **Elev.:** 2200–2700 m a.s.l. **Conservation status: (IUCN)** NE

5858 *Catostemma commune* Sandwith (Malvaceae)
IPNI: 50159-2
Habit: Tree **Use:** HF **Origin:** Native **Region:** Amazonia **Dept.:** AMA, CAQ, GUA, VAU **Elev.:** 100–700 m a.s.l. **IUCN Habitat code:** 1,5 **Conservation status: (IUCN)** LC

5859 *Cavanillesia platanifolia* (Bonpl.) Kunth (Malvaceae)
IPNI: 559273-1
Common name: balso, barrigón, barril, bonga, bongo, ceiba bruja, ceibo, cuipo, maconda, macondo, malambo, pericoaguao, pretino, quipo, volado, volandero, volao, volaó **Habit:** Tree **Use:** EU, HF, MA, ME **Origin:** Native **Region:** Caribbean, Pacific **Dept.:** ANT, BOL, CES, CHO, COR, MAG, SAP, SAN **Elev.:** 0–300 m a.s.l. **IUCN Habitat code:** 1,2,5,14 **Conservation status: (IUCN)** NT

5860 *Cavanillesia umbellata* Ruiz & Pav. (Malvaceae)
IPNI: 559275-1
Common name: lupuna **Habit:** Tree **Use:** HF **Origin:** Native **Region:** Amazonia **Dept.:** AMA, CAQ **Elev.:** 100–200 m a.s.l. **IUCN Habitat code:** 1 **Conservation status: (IUCN)** NE

5861 *Ceiba pentandra* (L.) Gaertn. (Malvaceae)
IPNI: 1166232-2
Common name: balso, bonga, bonga bruja, bongo, ceiba, ceiba Cartagena, ceiba blanca, ceiba bonga, ceiba bongo, ceiba bruja, ceiba colorada, ceiba de agua, ceiba de lana, ceibaebruja, ceibo, ceibo macho, chivecha, cumaca, jabillo, lano, lupuna, toborachi, yuque **Habit:** Tree **Use:** AF, EU, FU, GS, HF, IF, MA, ME, PO, SU **Origin:** Native, Cultivated **Region:** Amazonia, Andean, Guiana Shield, Caribbean, Orinoquia, Pacific **Dept.:** AMA, ANT, BOL, BOY, CAL, CAQ, CES, CHO, COR, CUN, GUV, HUI, MAG, MET, QUI, RIS, SAN, SUC, TOL, VAC **Elev.:** 0-2000 m.a.s.l. **IUCN Habitat code:** 1,2,3,4,5,14 **Conservation status: (IUCN)** LC **(CPSG)** Potential LC

5862 *Ceiba samauma* (Mart.) K.Schum. (Malvaceae) **IPNI:** 51083-2 **Habit:** Tree **Use:** MA, ME **Origin:** Native, Cultivated **Region:** Amazonia, Andean **Dept.:** AMA, CAQ, PUT, VAC **Elev.:** 500–1000 m a.s.l. **IUCN Habitat code:** 5 **Conservation status: (IUCN)** NE

5863 *Ceiba speciosa* (A.St.-Hil., A.Juss. & Cambess.) Ravenna (Malvaceae) **IPNI:** 1005661-1 **Habit:** Tree **Use:** EU, HF, MA, ME **Origin:** Cultivated **Region:** Andean **Dept.:** ANT, CAL, TOL, VAC **Elev.:** 400–1600 m a.s.l. **IUCN Habitat code:** 1,14 **Conservation status: (IUCN)** NE

5864 *Corchorus aestuans* L. (Malvaceae) **IPNI:** 320794-2 **Common name:** traquitraqui **Habit:** Herb **Use:** GS, HF, MA, ME **Origin:** Native **Region:** Caribbean **Dept.:** ATL, BOL, COR, MAG **Elev.:** 10–70 m a.s.l. **IUCN Habitat code:** 1,2,14 **Conservation status: (IUCN)** NE

5865 *Corchorus hirtus* L. (Malvaceae) **IPNI:** 320795-2 **Habit:** Herb **Use:** ME **Origin:** Native **Region:** Amazonia, Andean, Guiana Shield, Orinoquia **Dept.:** ANT, BOY, CAS, CAU, CUN, GUV, MAG, MET, NSA, TOL **Elev.:** 830–1320 m a.s.l. **IUCN Habitat code:** 1,2,3,4,14 **Conservation status: (IUCN)** NE

5866 *Corchorus orinocensis* Kunth (Malvaceae) **IPNI:** 65167-2 **Common name:** escobilla negra, espadilla **Habit:** Herb **Use:** ME **Origin:** Native **Region:** Andean, Caribbean, Orinoquia **Dept.:** ANT, ATL, BOL, BOY, CAL, CES, CUN, HUI, LAG, MAG, MET, TOL, VAC **Elev.:** 10–1050 m a.s.l. **IUCN Habitat code:** 1,2,3,4,14 **Conservation status: (IUCN)** NE

5867 *Corchorus siliquosus* L. (Malvaceae) **IPNI:** 833664-1 **Common name:** té, yute **Habit:** Shrub **Use:** HF, MA, ME **Origin:** Native **Region:** Pacific **Dept.:** ANT, BOL, CHO, COR, CUN, PUT, SUC **Elev.:** 0–736 m a.s.l. **IUCN Habitat code:** 1,2,3,14 **Conservation status: (IUCN)** NE

5868 *Dombeya × cayeuxii* André (Malvaceae) **IPNI:** 822968-1 **Use:** ME **Origin:** Endemic **Conservation status: (IUCN)** NE

5869 *Dombeya burgessiae* Gerrard ex Harv. (Malvaceae) **IPNI:** 822960-1 **Habit:** Tree **Use:** FU, GS, HF, MA, ME **Origin:** Cultivated **Region:** Andean **Dept.:** ANT **Elev.:** 1500–1500 m a.s.l. **IUCN Habitat code:** 14 **Conservation status: (IUCN)** LC

5870 *Dombeya wallichii* (Lindl.) K.Schum. (Malvaceae) **IPNI:** 823281-1 **Use:** EU, ME **Region:** Andean **Dept.:** BOG **IUCN Habitat code:** 3,14 **Conservation status: (IUCN)** NE

5871 *Eriotheca globosa* (Aubl.) A.Robyns (Malvaceae) **IPNI:** 95971-2 **Common name:** algodón, algodón de tierra firme **Habit:** Tree **Use:** ME **Origin:** Native **Region:** Amazonia, Andean, Orinoquia **Dept.:** AMA, ANT, CAQ, COR, GUV, MET, PUT **Elev.:** 100–1200 m a.s.l. **IUCN Habitat code:** 1,3,14 **Conservation status: (IUCN)** LC

5872 *Eriotheca macrophylla* (K.Schum.) A.Robyns (Malvaceae) **IPNI:** 95976-2 **Common name:** duende caspi **Habit:** Tree **Use:** HF **Origin:** Native **Region:** Amazonia, Guiana Shield **Dept.:** AMA, CAQ, GUV, PUT, VAU, VID **Elev.:** 100–800 m a.s.l. **IUCN Habitat code:** 1,3,4 **Conservation status: (IUCN)** LC

5873 *Eriotheca roseorum* (Cuatrec.) A.Robyns (Malvaceae) **IPNI:** 95985-2 **Habit:** Tree **Use:** HF **Origin:** Native **Dept.:** ANT **Elev.:** 400–600 m a.s.l. **Conservation status: (IUCN)** NE

5874 *Fuertesimalva limensis* (L.) Fryxell (Malvaceae) **IPNI:** 993881-1 **Common name:** malva, malva blanca, malvisco **Habit:** Herb, Subshrub **Use:** ME **Origin:** Native **Region:** Andean **Dept.:** BOG, BOY, CES, CUN, NAR **Elev.:** 1400–3500 m a.s.l. **IUCN Habitat code:** 1,3,14 **Conservation status: (IUCN)** NE

5875 *Fuertesimalva peruviana* (L.) Fryxell (Malvaceae) **IPNI:** 993885-1 **Common name:** malva, malva blanca, malva grande, malvavisco, malvisco **Use:** ME **Dept.:** BOG **IUCN Habitat code:** 1,3,14 **Conservation status: (IUCN)** NE

5876 *Gossypium arboreum* L. (Malvaceae) **IPNI:** 559664-1 **Common name:** algodonero, algodonero pajarito, algodón, algodón pajarito **Use:** AF, EU, GS, HF, IF, MA, ME, PO **IUCN Habitat code:** 1,2,3,5,14 **Conservation status: (IUCN)** NE

5877 *Gossypium barbadense* L. (Malvaceae) **IPNI:** 559677-1 **Common name:** algodonero, algodón, algodón macho, cuitote, hoja de algodón **Habit:** Shrub **Use:** EU, FU, GS, HF, MA, ME, PO, SU **Origin:** Native, Cultivated **Region:** Amazonia, Andean, Caribbean, Pacific **Dept.:** AMA, CAL, CAQ, CAU, CHO, CUN, NSA, PUT, SAN, VAC **Elev.:** 100–1800 m a.s.l. **Conservation status: (IUCN)** LC

5878 *Gossypium herbaceum* L. (Malvaceae) **IPNI:** 559730-1 **Common name:** algodonero, algodón, algodón de árbol, algodón silvestre, maawui **Use:** AF, EU, GS, HF, MA, ME, PO **IUCN Habitat code:** 1,3,14 **Conservation status: (IUCN)** DD

5879 *Gossypium hirsutum* L. (Malvaceae) **IPNI:** 30042843-2 **Common name:** algodonero, algodón, algodón morado **Habit:** Shrub **Use:** AF, EU, FU, GS, HF, MA, ME, PO **Origin:** Native, Cultivated **Region:** Andean, Guiana Shield, Caribbean, Orinoquia **Dept.:** ANT, ATL, BOL, CAU, HUI, MAG, NSA, SAN, TOL, VAU, VID **Elev.:** 0–1500 m a.s.l. **IUCN Habitat code:** 1,2,3,4,5,14 **Conservation status: (IUCN)** VU

5880 *Grewia caffra* Meisn. (Malvaceae) **IPNI:** 834119-1 **Use:** EU, GS, HF, MA, ME **Region:** Andean **Dept.:** BOG **Conservation status: (IUCN)** NE

5881 *Guazuma ulmifolia* Lam. (Malvaceae) **IPNI:** 823393-1 **Common name:** bolaina, burŭcunae, bwa dom, cabeza de negrito, cara de mico, caulote, chicarrón, guácima, guácimo, guácimo colorado, guásima, guásimo, guásimo blanco, guásimo colorado, guásimo macho, guásimo negro, guásimo rojo, huacimo,ja̰odedimicù, mielquemada, na **Habit:** Shrub, Tree **Use:** AF, EU, FU, GS, HF, IF, MA, ME, SU **Origin:** Native **Region:** Amazonia, Andean, Caribbean, Orinoquia, Pacific **Dept.:** AMA, ANT, ATL, BOL, BOY, CAL, CAS, CAU, CHO, COR, CUN, HUI, LAG, MAG, MET, NSA, SAN, SUC, TOL, VAC, VAU, VID **Elev.:** 0-1800 m.a.s.l. **IUCN Habitat code:** 1,2,3,4,5,14 **Conservation status: (IUCN)** LC **(CPSG)** Potential LC

5882 *Hampea thespesioides* Triana & Planch. (Malvaceae) **IPNI:** 559872-1 **Common name:** achote cimarrón, balso negro, maiztostado, sapotillo **Habit:** Tree **Use:** HF **Origin:** Endemic **Region:** Andean **Dept.:** ANT, BOG, BOY, CUN **Elev.:** 1000-2000 m.a.s.l. **IUCN Habitat code:** 1,3,14 **Conservation status: (IUCN)** LC **(CPSG)** LC

5883 *Helicteres baruensis* Jacq. (Malvaceae) **IPNI:** 321205-2 **Common name:** alfandoquillo, alfandoquitos, guacimo, guasimillo, guásimo, majaguo de playa, mosquero, sacatrapos, sacatropis, tuercemadre **Habit:** Shrub **Use:** ME **Origin:** Native **Region:** Andean, Caribbean **Dept.:** BOL, CUN, HUI, LAG, MAG, SAN, TOL **Elev.:** 50–1900 m a.s.l. **IUCN Habitat code:** 1,2,3,4,5,14 **Conservation status: (IUCN)** LC

5884 *Helicteres carthagenensis* Jacq. (Malvaceae) **IPNI:** 321203-2 **Habit:** Subshrub, Small tree **Use:** ME **Origin:** Native **Region:** Caribbean **Dept.:** ATL, BOL, CES, MAG **Elev.:** 10–350 m a.s.l. **IUCN Habitat code:** 1,2 **Conservation status: (IUCN)** NE

5885 *Helicteres guazumifolia* Kunth (Malvaceae) **IPNI:** 320786-2 **Common name:** guásimo torcido, lavaplatos, manelito, rabo de puerco, vara blanca **Habit:** Shrub **Use:** AF,ME **Origin:** Native **Region:** Andean, Caribbean, Orinoquia, Pacific **Dept.:** ANT, ARA, BOL, BOY, CAS, CHO, COR, GUV, MAG, MET, NSA, SAN, VID **Elev.:** 50-1850 m.a.s.l. **Conservation status: (IUCN)** NE **(CPSG)** Potential LC

5886 *Helicteres jamaicensis* Jacq. (Malvaceae) **IPNI:** 823472-1 **Habit:** Shrub **Use:** ME **Origin:** Native **Region:** Caribbean **Dept.:** MAG **Elev.:** 95–95 m a.s.l. **Conservation status: (IUCN)** LC

5887 *Heliocarpus americanus* L. (Malvaceae) **IPNI:** 834663-1 **Common name:** algodón, baba india, babo, baho blanco, balso,

balso baboso, balso blanco, balso cadillo, balso menudito, cadillo, guasca dulce, majagua, majagua melada, majaguo, majaguo melao, majao, majuo, melao, moho, ortiguillo, palobobo, panecito, pestaña de mula, p **Habit:** Tree **Use:** HF, MA, ME **Origin:** Native **Region:** Andean, Pacific **Dept.:** ANT, BOG, BOY, CAL, CAQ, CAS, CAU, CES, CHO, CUN, HUI, MAG, MET, NAR, NSA, PUT, QUI, RIS, SAN, TOL, VAC **Elev.:** 30–2500 m a.s.l. **IUCN Habitat code:** 1,2,3,5,14 **Conservation status: (IUCN)** LC

5888 *Herissantia crispa* (L.) Brizicky (Malvaceae) **IPNI:** 120630-2 **Habit:** Subshrub **Use:** EU, MA, ME **Origin:** Native **Region:** Caribbean **Dept.:** ANT, ATL, BOL, LAG, MAG, TOL **Elev.:** 0–857 m a.s.l. **IUCN Habitat code:** 1,2,3,14 **Conservation status: (IUCN)** NE

5889 *Herrania albiflora* Goudot (Malvaceae) **IPNI:** 824010-1 **Common name:** cacao cimarrón, cacao de monte, cacaíto, chocolatillo **Habit:** Shrub, Tree **Use:** HF, ME **Origin:** Native **Region:** Andean **Dept.:** ANT, BOL, BOY, NSA, SAN **Elev.:** 330–1000 m a.s.l. **IUCN Habitat code:** 1,3,14 **Conservation status: (IUCN)** NE

5890 *Herrania balaensis* H.Preuss (Malvaceae) **IPNI:** 824014-1 **Habit:** Small tree **Use:** HF **Origin:** Native **Region:** Pacific **Dept.:** ANT, CHO, VAC **Elev.:** 0–170 m a.s.l. **IUCN Habitat code:** 1,3 **Conservation status: (IUCN)** EN

5891 *Herrania breviligulata* R.E.Schult. (Malvaceae) **IPNI:** 120753-2 **Common name:** cacao de monte, cacao del monte **Habit:** Small tree **Use:** HF, ME **Origin:** Native **Region:** Amazonia, Andean **Dept.:** PUT **Elev.:** 370–850 m a.s.l. **IUCN Habitat code:** 1 **Conservation status: (IUCN)** NE

5892 *Herrania camargoana* R.E.Schult. (Malvaceae) **IPNI:** 120754-2 **Habit:** Tree **Use:** HF, ME **Origin:** Native **Dept.:** AMA, VAU **Elev.:** 50–450 m a.s.l. **IUCN Habitat code:** 1 **Conservation status: (IUCN)** NE

5893 *Herrania cuatrecasasiana* García-Barr. (Malvaceae) **IPNI:** 120755-2 **Common name:** cacao de monte **Habit:** Tree **Use:** HF, ME **Origin:** Native **Region:** Amazonia, Andean, Pacific **Dept.:** CHO, PUT, VAC **Elev.:** 0–1500 m a.s.l. **IUCN Habitat code:** 1 **Conservation status: (IUCN)** NE

5894 *Herrania dugandii* García-Barr. (Malvaceae) **IPNI:** 120756-2 **Common name:** cacao silvestre **Habit:** Small tree **Use:** ME **Origin:** Native **Region:** Amazonia **Dept.:** PUT **Elev.:** 230–250 m a.s.l. **IUCN Habitat code:** 1 **Conservation status: (IUCN)** NE

5895 *Herrania laciniifolia* Goudot ex Triana & Planch. (Malvaceae) **IPNI:** 824022-1 **Common name:** cacao de monte, cacao silvestre **Habit:** Shrub, Small tree **Use:** ME **Origin:** Endemic **Region:** Andean, Pacific **Dept.:** ANT, CHO, TOL **Elev.:** 300-1120 m.a.s.l. **IUCN Habitat code:** 1,3,14 **Conservation status: (IUCN)** LC **(CPSG)** LC

5896 *Herrania lemniscata* (M.R.Schomb.) R.E.Schult. (Malvaceae) **IPNI:** 824024-1 **Habit:** Shrub, Tree **Use:** HF **Origin:** Native **Dept.:** BOL, SAN **Elev.:** 130–200 m a.s.l. **IUCN Habitat code:** 1 **Conservation status: (IUCN)** NE

5897 *Herrania mariae* (Mart.) Goudot (Malvaceae) **IPNI:** 824025-1 **Common name:** cacao silvestre, mu-se-ge-ke, mu-se-na **Habit:** Small tree **Use:** HF, ME **Origin:** Native **Region:** Amazonia, Andean, Pacific **Dept.:** AMA, ANT, PUT **Elev.:** 30–700 m a.s.l. **IUCN Habitat code:** 1 **Conservation status: (IUCN)** LC

5898 *Herrania nitida* (Poepp.) R.E.Schult. (Malvaceae) **IPNI:** 824026-1 **Common name:** beñoó, cacao, cacao amarillo, cacao de altura, cacao de boruga, cacao de monte, cacao higuillo, cacao maní, cacao silvestre, cacay, cacaíto de monte, maduraverde, maraca de boruga, maraca de ratón, otonasaré, otonosare, pepi, puila, puilá **Habit:** Shrub, Tree **Use:** HF, ME **Origin:** Native **Region:** Amazonia, Andean, Pacific **Dept.:** AMA, ANT, CAQ, CUN, MET, PUT, VAU **Elev.:** 20–800 m a.s.l. **IUCN Habitat code:** 1,3,4,5,14 **Conservation status: (IUCN)** NE

5899 *Herrania nycterodendron* R.E.Schult. (Malvaceae) **IPNI:** 824027-1 **Common name:** cacao de monte, cacao de muerciélago, cacao de murciélago, cacao monte, chocolate de monte, palo de chimbe, palo de murciélago, trompa de cusumbo **Habit:** Tree **Use:** HF, ME **Origin:** Native **Region:** Amazonia **Dept.:** AMA, CAQ, PUT **Elev.:** 100–180 m a.s.l. **IUCN Habitat code:** 1,3,4,14 **Conservation status: (IUCN)** LC

5900 *Herrania pulcherrima* Goudot (Malvaceae) **IPNI:** 824028-1 **Common name:** cacahuillo, cacao cahuí, cacao cuadrado, cacao de monte, cacaíto de monte **Habit:** Shrub, Tree **Use:** HF, ME **Origin:** Native **Region:** Andean, Orinoquia, Pacific **Dept.:** BOY, CHO, CUN, MET, NAR, VAC **Elev.:** 0–1300 m a.s.l. **IUCN Habitat code:** 1,3,5,14 **Conservation status: (IUCN)** LC

5901 *Herrania purpurea* (Pittier) R.E.Schult. (Malvaceae) **IPNI:** 120768-2 **Common name:** cacao de monte, cacao montañero, cacauillo **Habit:** Small tree, Tree **Use:** HF, MA **Origin:** Native **Region:** Andean, Pacific **Dept.:** ANT, CHO **Elev.:** 20–700 m a.s.l. **IUCN Habitat code:** 1,3,14 **Conservation status: (IUCN)** LC

5902 *Herrania umbratica* R.E.Schult. (Malvaceae) **IPNI:** 120770-2 **Common name:** cacao de monte **Habit:** Small tree **Use:** ME **Origin:** Endemic **Region:** Andean **Dept.:** SAN **Elev.:** 695-695 m.a.s.l. **Conservation status: (IUCN)** CR **(CPSG)** CR

5903 *Hibiscus acetosella* Welw. ex Hiern (Malvaceae) **IPNI:** 559896-1 **Common name:** malvarrosa, malvavisco, malvisco **Habit:** Subshrub **Use:** EU, GS, HF, ME **Origin:** Cultivated **Region:** Andean, Orinoquia, Pacific **Dept.:** CAL, CHO, MET, VAC **Elev.:** 0–1500 m a.s.l. **IUCN Habitat code:** 1,14 **Conservation status: (IUCN)** NE

5904 *Hibiscus bifurcatus* Cav. (Malvaceae) **IPNI:** 560013-1 **Use:** HF, ME **Origin:** Native **IUCN Habitat code:** 1,2,4,5,14 **Conservation status: (IUCN)** NE

5905 *Hibiscus cannabinus* L. (Malvaceae) **IPNI:** 560068-1 **Common name:** rosella, yute de Java **Habit:** Herb, Subshrub **Use:** AF, EU, FU, GS, HF, MA, ME, SU **Origin:** Cultivated **Region:** Caribbean **Dept.:** MAG, TOL, VAC **Elev.:** 420–1088 m a.s.l. **IUCN Habitat code:** 3,14 **Conservation status: (IUCN)** NE

5906 *Hibiscus grandiflorus* Michx. (Malvaceae) **IPNI:** 1188401-2 **Use:** EU **IUCN Habitat code:** 1,2,14 **Conservation status: (IUCN)** NE

5907 *Hibiscus maculatus* Lam. (Malvaceae) **IPNI:** 560486-1 **Habit:** Shrub **Use:** HF **Origin:** Native **Region:** Andean **Dept.:** ANT, CAU, CHO, VAC **Elev.:** 1000–1700 m a.s.l. **Conservation status: (IUCN)** NE

5908 *Hibiscus mutabilis* L. (Malvaceae) **IPNI:** 560549-1 **Common name:** algodón, amistad del día, amor al uso, cayeno, flor del día, malva, piñón, rojo, rosa de engaño, variedad **Habit:** Shrub **Use:** EU, HF, MA, ME **Origin:** Cultivated **Region:** Amazonia, Andean, Caribbean, Orinoquia **Dept.:** AMA, BOL, CUN, MET, SAN **Elev.:** 87–2551 m a.s.l. **IUCN Habitat code:** 1,3,14 **Conservation status: (IUCN)** NE

5909 *Hibiscus radiatus* Cav. (Malvaceae) **IPNI:** 560723-1 **Common name:** malvavisco, malvisco **Habit:** Shrub **Use:** EU, HF, MA, ME **Origin:** Cultivated, Naturalised **Region:** Andean **Dept.:** ANT, CAL, CUN, PUT **Elev.:** 500–1500 m a.s.l. **IUCN Habitat code:** 1 **Conservation status: (IUCN)** NE

5910 *Hibiscus rosa-sinensis* L. (Malvaceae) **IPNI:** 560756-1 **Common name:** alsaciana, arrebatamacho, arrematamacho, astromelia, astromelio, aurora, avispa, bailarina, barbegallo, bombillo, bonche, boncheaí, cajeto, canasta, canastilla, canastillo, cayena, cayeno, cañeno, choublak, chupeta, clavel japonés, cupón, escandalosa, esc **Habit:** Shrub **Use:** AF, EU, GS, HF, MA, ME, SU **Origin:** Cultivated **Region:** Amazonia, Andean, Caribbean, Orinoquia, Pacific **Dept.:** AMA, ANT, ATL, BOG, BOL, BOY, CAL, CAQ, CAS, CAU, CES, CHO, COR, CUN, HUI, LAG, MAG, MET, NAR, NSA, PUT, QUI, RIS, SAP, SAN, SUC, TOL, VAC **Elev.:** 0–2750 m a.s.l. **IUCN Habitat code:** 1,2,3,5,14 **Conservation status: (IUCN)** NE

5911 *Hibiscus sabdariffa* L. (Malvaceae) **IPNI:** 326388-2 **Common name:** abutilón, flor de Jamaica, rosella, saril, sorrel, vino **Habit:** Herb, Shrub, Subshrub **Use:** AF, EU, GS, HF, MA, ME, PO, SU **Origin:** Cultivated, Naturalised **Region:** Andean, Caribbean, Pacific **Dept.:** ANT, ATL, BOG, CHO, QUI, SAN **Elev.:** 0–1700 m a.s.l.

IUCN Habitat code: 1,3,14 **Conservation status: (IUCN)** NE

5912 *Hibiscus schizopetalus* (Mast.) Hook.f. (Malvaceae) **IPNI:** 560792-1 **Habit:** Shrub **Use:** EU, HF, ME **Origin:** Cultivated **Region:** Amazonia, Andean, Pacific **Dept.:** AMA, ANT, CHO, CUN, VAC **Elev.:** 0–1500 m a.s.l. **IUCN Habitat code:** 1,3,14 **Conservation status: (IUCN)** NE

5913 *Hibiscus striatus* Cav. (Malvaceae) **IPNI:** 560861-1 **Habit:** Shrub **Use:** HF, ME **Origin:** Native **Region:** Andean **Dept.:** ANT, TOL **Elev.:** 260–1500 m a.s.l. **IUCN Habitat code:** 2,3,4,5 **Conservation status: (IUCN)** NE

5914 *Hibiscus syriacus* L. (Malvaceae) **IPNI:** 560890-1 **Common name:** resucitado **Habit:** Shrub **Use:** EU, HF, MA, ME **Origin:** Cultivated **Region:** Andean **Dept.:** BOG **Elev.:** 2600–2600 m a.s.l. **IUCN Habitat code:** 14 **Conservation status: (IUCN)** NE

5915 *Hibiscus tiliaceus* L. (Malvaceae) **IPNI:** 30143354-2 **Common name:** amajagua, chitú, flor de salmón, flor variante, majagua, majagua de playa, majaguo, majagüito, majagüito de playa, sanjoaquín **Habit:** Tree **Use:** AF, EU, FU, GS, HF, MA, ME, SU **Origin:** Native, Cultivated **Region:** Caribbean, Pacific **Dept.:** ANT, BOL, CAU, CHO, MAG, NAR, SAP, VAC **Elev.:** 0–50 m a.s.l. **IUCN Habitat code:** 1,3,5,14 **Conservation status: (IUCN)** LC

5916 *Hibiscus verbasciformis* Klotzsch ex Hochr. (Malvaceae) **IPNI:** 60443335-2 **Common name:** abutilón, malvarrosa **Habit:** Herb, Subshrub **Use:** ME **Origin:** Native **Region:** Amazonia, Andean, Pacific **Dept.:** ANT, CAQ, CUN, MET, NAR, VAC **Elev.:** 0–1250 m a.s.l. **IUCN Habitat code:** 1,3,14 **Conservation status: (IUCN)** LC

5917 *Huberodendron patinoi* Cuatrec. (Malvaceae) **IPNI:** 124373-2 **Common name:** carrá, carrás, ceiba sama, ceiba samá, coco volador, masábalo, naguare, nogal, volador **Habit:** Tree **Use:** MA **Origin:** Native **Region:** Andean, Pacific **Dept.:** ANT, BOY, CAU, CHO, COR, NAR, RIS, SAN, VAC **Elev.:** 0-1000 m.a.s.l. **IUCN Habitat code:** 1,3,5,14 **Conservation status: (IUCN)** VU **(CPSG)** VU

5918 *Kearnemalvastrum lacteum* (Aiton) D.M.Bates (Malvaceae) **IPNI:** 133329-2 **Habit:** Subshrub **Use:** ME **Origin:** Native **Region:** Andean **Dept.:** CAU **Elev.:** 1750–1850 m a.s.l. **IUCN Habitat code:** 14 **Conservation status: (IUCN)** NE

5919 *Kosteletzkya depressa* (L.) O.J.Blanch., Fryxell & D.M.Bates (Malvaceae) **IPNI:** 133833-2 **Habit:** Herb, Subshrub **Use:** ME **Origin:** Native **Region:** Caribbean **Dept.:** BOL, COR, MAG **Elev.:** 0–20 m a.s.l. **IUCN Habitat code:** 1,2,14 **Conservation status: (IUCN)** NE

5920 *Luehea candida* (Moc. & Sessé ex DC.) Mart. (Malvaceae) **IPNI:** 834727-1 **Common name:** algodoncillo, algodón de monte **Habit:** Tree **Use:** HF **Origin:** Native **Region:** Caribbean **Dept.:** BOL, CES, MAG **Elev.:** 200–250 m a.s.l. **IUCN Habitat code:** 1,2,14 **Conservation status: (IUCN)** LC

5921 *Luehea seemannii* Triana & Planch. (Malvaceae) **IPNI:** 834758-1 **Common name:** algodoncillo, bagazo, colorado, guácimo, guácimo colorado, guácimo colorao, guácimo coloro, guásimo, guásimo colorado, guásimo colorao, guásimo serrano, malagano, mata de jumo, peinemono **Habit:** Tree **Use:** ME **Origin:** Native **Region:** Amazonia, Caribbean, Orinoquia, Pacific **Dept.:** AMA, ANT, BOY, CAL, CAQ, CAS, CHO, MAG, MET, SAN, TOL, VAC **Elev.:** 5–1000 m a.s.l. **IUCN Habitat code:** 1,2,3,4,5,14 **Conservation status: (IUCN)** LC

5922 *Luehea speciosa* Willd. (Malvaceae) **IPNI:** 834759-1 **Habit:** Tree **Use:** HF **Origin:** Native **Region:** Andean, Caribbean **Dept.:** CES, LAG, MAG **Elev.:** 150–1000 m a.s.l. **IUCN Habitat code:** 1,3 **Conservation status: (IUCN)** LC

5923 *Malachra alceifolia* Jacq. (Malvaceae) **IPNI:** 561421-1 **Common name:** malauya, malva, malvón **Habit:** Herb, Subshrub **Use:** HF, MA, ME **Origin:** Native **Region:** Andean, Caribbean, Pacific **Dept.:** ANT, ATL, BOL, BOY, CAL, CHO, COR, CUN, LAG, MAG, MET, NSA, SAP, SUC, TOL, VAC **Elev.:** 0-1800 m.a.s.l. **IUCN Habitat code:** 1,2,3,5,14 **Conservation status: (IUCN)** NE **(CPSG)** Potential LC

5924 *Malachra capitata* (L.) L. (Malvaceae) **IPNI:** 561424-1 **Common name:** malva, malvisco **Habit:** Herb, Subshrub **Use:** EU, MA, ME **Origin:** Native **Region:** Andean, Caribbean, Pacific **Dept.:** AMA, ANT, BOL, BOY, CAU, NAR **Elev.:** 200–1480 m a.s.l. **IUCN Habitat code:** 1,3,14 **Conservation status: (IUCN)** NE

5925 *Malachra fasciata* Jacq. (Malvaceae) **IPNI:** 561430-1 **Habit:** Herb, Subshrub **Use:** MA **Origin:** Native **Region:** Caribbean, Orinoquia **Dept.:** BOL, MET, TOL **Elev.:** 30–830 m a.s.l. **IUCN Habitat code:** 2,3 **Conservation status: (IUCN)** NE

5926 *Malachra radiata* (L.) L. (Malvaceae) **IPNI:** 561446-1 **Habit:** Shrub **Use:** MA, ME **Origin:** Native **Region:** Amazonia, Caribbean **Dept.:** AMA, COR, MAG **Elev.:** 0–200 m a.s.l. **IUCN Habitat code:** 2 **Conservation status: (IUCN)** NE

5927 *Malachra rudis* Benth. (Malvaceae) **IPNI:** 561450-1 **Common name:** albahaca, malva, malva brusca, malva cimarrona, malva peluda, malva pintada, malva áspera, malvavisco, malvisco, malvón **Habit:** Subshrub **Use:** ME **Origin:** Native **Dept.:** AMA, ANT, BOL, CAQ, CAU, CHO, COR, CUN, HUI, NAR, PUT, RIS, SAN, SUC, TOL, VAC **Elev.:** 0-2330 m.a.s.l. **IUCN Habitat code:** 1,2,3,5,14 **Conservation status: (IUCN)** NE **(CPSG)** Potential LC

5928 *Malva arborea* (L.) Webb & Berthel. (Malvaceae) **IPNI:** 561541-1 **Habit:** Subshrub **Use:** EU, MA, ME **Origin:** Cultivated **Region:** Andean **Dept.:** ANT, BOG, CUN, PUT **Elev.:** 2000–2650 m a.s.l. **IUCN Habitat code:** 1,4,14 **Conservation status: (IUCN)** NE

5929 *Malva assurgentiflora* (Kellogg) M.F.Ray (Malvaceae) **IPNI:** 1003822-1 **Habit:** Shrub **Use:** EU, ME **Origin:** Cultivated **Region:** Andean **Dept.:** BOG, BOY, CUN **Elev.:** 2620–2620 m a.s.l. **IUCN Habitat code:** 14 **Conservation status: (IUCN)** NE

5930 *Malva nicaeensis* All. (Malvaceae) **IPNI:** 561798-1 **Common name:** malva **Use:** EU, HF, ME **IUCN Habitat code:** 3,14 **Conservation status: (IUCN)** NE

5931 *Malva parviflora* L. (Malvaceae) **IPNI:** 561830-1 **Common name:** malva, malva tendida **Habit:** Herb **Use:** AF, EU, HF, ME, PO **Origin:** Cultivated, Naturalised **Region:** Andean **Dept.:** ANT, BOG, BOY, CAU, CUN **Elev.:** 1800–2600 m a.s.l. **IUCN Habitat code:** 1,3,14 **Conservation status: (IUCN)** NE

5932 *Malva pusilla* Sm. (Malvaceae) **IPNI:** 561868-1 **Use:** ME **Region:** Andean **Dept.:** BOG **Conservation status: (IUCN)** NE

5933 *Malva sylvestris* L. (Malvaceae) **IPNI:** 561932-1 **Common name:** malva, malva blanca, malva común, malva morada, malva pequeña, malva silvestre, malvarrosa **Habit:** Shrub **Use:** EU, HF, MA, ME, PO **Origin:** Cultivated **Region:** Andean **Dept.:** ANT, BOG, CUN **Elev.:** 1600–2600 m a.s.l. **IUCN Habitat code:** 1,2,3,14 **Conservation status: (IUCN)** NE

5934 *Malvastrum americanum* (L.) Torr. (Malvaceae) **IPNI:** 561999-1 **Common name:** babosita, bretónica, escoba, escoba real, escobilla, escobilla amarilla, escobo, estancadera, malvilla, melosa **Habit:** Herb, Shrub, Subshrub **Use:** ME **Origin:** Native **Region:** Andean, Guiana Shield, Caribbean, Orinoquia **Dept.:** ANT, ARA, ATL, BOL, BOY, CAS, CAU, COR, CUN, HUI, MAG, MET, NAR, NSA, QUI, SAN, TOL, VAC, VAU, VID **Elev.:** 0-2900 m.a.s.l. **IUCN Habitat code:** 1,2,3,4,5,14 **Conservation status: (IUCN)** NE **(CPSG)** Potential LC

5935 *Malvastrum coromandelianum* (L.) Garcke (Malvaceae) **IPNI:** 1063417-2 **Common name:** escobilla, escobillo, escobo, escubilla, malva, malva blanca **Habit:** Herb, Subshrub **Use:** AF, HF, MA, ME **Origin:** Native **Region:** Andean, Caribbean, Orinoquia **Dept.:** ANT, BOL, BOY, CAL, CUN, MAG, MET, SAP, SAN, VAC **Elev.:** 340–2780 m a.s.l. **IUCN Habitat code:** 1,2,3,5,14 **Conservation status: (IUCN)** NE

5936 *Malvaviscus arboreus* Dill. ex Cav. (Malvaceae) **IPNI:** 30041958-2 **Common name:** ardilla de montaña, arito, astromelio, auroro, calzoncillo del diablo, flor de santos, malvavisco, manzanillo, monacillo, obelisco de la sierra, quesito, resucitado de monte, tripa de buey, tulipancillo **Habit:** Shrub **Use:** AF, EU, HF, MA, ME, SU **Origin:** Native, Cultivated **Region:** Caribbean **Dept.:** BOG, CES, SAP, SUC **Elev.:** 0-100 m.a.s.l. **IUCN Habitat code:** 1,2,3,14 **Conservation status: (IUCN)** LC **(CPSG)** Potential LC

5937 *Malvaviscus penduliflorus* Moc. & Sessé ex DC. (Malvaceae) **IPNI:** 150478-2 **Common name:** crespón, quesito, resucitado de monte, resucito **Habit:** Shrub **Use:** EU, ME **Origin:** Native, Cultivated **Region:** Andean **Dept.:** ANT, BOG, NAR, SAN, VAC **Elev.:** 240–1600 m a.s.l. **IUCN Habitat code:** 1,2,14 **Conservation status: (IUCN)** NE

5938 *Matisia alata* Little (Malvaceae) **IPNI:** 154666-2 **Common name:** castaño **Habit:** Tree **Use:** HF **Origin:** Native **Region:** Pacific **Dept.:** ANT, NAR **Elev.:** 0–500 m a.s.l. **Conservation status: (IUCN)** EN

5939 *Matisia bicolor* Ducke (Malvaceae) **IPNI:** 154672-2 **Habit:** Tree **Use:** HF **Origin:** Native **Region:** Amazonia **Dept.:** AMA, CAQ, CAU **Elev.:** 100–500 m a.s.l. **IUCN Habitat code:** 1 **Conservation status: (IUCN)** NE

5940 *Matisia bracteolosa* Ducke (Malvaceae) **IPNI:** 154675-2 **Common name:** abuelo de cacao, arracacho, lúa **Habit:** Tree **Use:** HF **Origin:** Native **Region:** Amazonia **Dept.:** AMA, PUT **Elev.:** 0–600 m a.s.l. **IUCN Habitat code:** 1,3,5,14 **Conservation status: (IUCN)** LC

5941 *Matisia calimana* Cuatrec. (Malvaceae) **IPNI:** 154676-2 **Common name:** palo de oso caballuno **Habit:** Tree **Use:** HF **Origin:** Native **Region:** Andean, Pacific **Dept.:** CHO, NAR, VAC **Elev.:** 0–1100 m a.s.l. **IUCN Habitat code:** 1,3 **Conservation status: (IUCN)** NE

5942 *Matisia castano* H.Karst. & Triana (Malvaceae) **IPNI:** 112441-1 **Common name:** bacaíto, bonifacio, castaño, molinillo, sare, vaina, zapotillo **Habit:** Tree **Use:** HF **Origin:** Native **Region:** Andean, Pacific **Dept.:** ANT, CAU, CHO, NAR, RIS, VAC **Elev.:** 0–1400 m a.s.l. **IUCN Habitat code:** 1,3,5,14 **Conservation status: (IUCN)** NE

5943 *Matisia cordata* Bonpl. (Malvaceae) **IPNI:** 562392-1 **Common name:** chupachupa, sapote, zapote, zapote chupachupa **Habit:** Tree **Use:** HF, ME **Origin:** Native, Cultivated **Region:** Amazonia, Andean, Pacific **Dept.:** AMA, ANT, BOY, CAL, CAU, CHO, CUN, HUI, NAR, PUT, QUI, RIS, SAN, TOL, VAC **Elev.:** 0-1500 m.a.s.l. **IUCN Habitat code:** 1,3,5,14 **Conservation status: (IUCN)** LC **(CPSG)** Potential LC

5944 *Matisia dowdingii* Sprague (Malvaceae) **IPNI:** 154683-2 **Habit:** Tree **Use:** MA **Origin:** Endemic **Region:** Amazonia, Andean **Dept.:** PUT **Elev.:** 300-800 m.a.s.l. **IUCN Habitat code:** 14 **Conservation status: (IUCN)** CR **(CPSG)** CR

5945 *Matisia gentryi* Fern.Alonso (Malvaceae) **IPNI:** 324613-2 **Habit:** Tree **Use:** HF **Origin:** Endemic **Region:** Pacific **Dept.:** CHO **Elev.:** 50-300 m.a.s.l. **IUCN Habitat code:** 1,14 **Conservation status: (IUCN)** LC **(CPSG)** LC

5946 *Matisia giacomettoi* Romero (Malvaceae) **IPNI:** 154684-2 **Common name:** sapotillo **Habit:** Tree **Use:** HF **Origin:** Endemic **Region:** Andean, Pacific **Dept.:** NAR **Elev.:** 0–700 m a.s.l. **Conservation status: (IUCN)** NE

5947 *Matisia idroboi* Cuatrec. (Malvaceae) **IPNI:** 154689-2 **Common name:** almirajó macho, molinillo, vaina **Habit:** Tree **Use:** HF **Origin:** Endemic **Region:** Amazonia, Andean, Pacific **Dept.:** AMA, CAQ, CAU, CHO, MET, NAR, VAC **Elev.:** 100–2000 m a.s.l. **IUCN Habitat code:** 1,3,14 **Conservation status: (IUCN)** LC

5948 *Matisia lasiocalyx* K.Schum. (Malvaceae) **IPNI:** 154692-2 **Common name:** abuelo de cacao, majaguo, maraca silvestre **Habit:** Tree **Use:** HF **Origin:** Native **Region:** Amazonia, Orinoquia **Dept.:** AMA, CAQ, CAS, CAU, MET, PUT, VID **Elev.:** 100–600 m a.s.l. **IUCN Habitat code:** 1,3,4,5,14 **Conservation status: (IUCN)** LC

5949 *Matisia lomensis* (Cuatrec.) Cuatrec. (Malvaceae) **IPNI:** 154695-2 **Habit:** Tree **Use:** HF **Origin:** Native **Region:** Amazonia, Andean **Dept.:** AMA, CAU, NAR, PUT **Elev.:** 0–1500 m a.s.l. **IUCN Habitat code:** 1,3,14 **Conservation status: (IUCN)** NE

5950 *Matisia longiflora* Gleason (Malvaceae) **IPNI:** 154696-2 **Habit:** Tree **Use:** HF **Origin:** Native **Region:** Andean **Dept.:** ANT, BOY, COR, CUN **Elev.:** 200-1300 m.a.s.l. **IUCN Habitat code:** 1,3,14 **Conservation status: (IUCN)** LC **(CPSG)** LC

5951 *Matisia longipes* Little (Malvaceae) **IPNI:** 280582-2 **Habit:** Tree **Use:** HF **Origin:** Native **Region:** Pacific **Dept.:** CAU, NAR **Elev.:** 0–800 m a.s.l. **IUCN Habitat code:** 1 **Conservation status: (IUCN)** NE

5952 *Matisia lozanoi* Fern.Alonso (Malvaceae) **IPNI:** 322421-2 **Habit:** Tree **Use:** HF **Origin:** Endemic **Region:** Andean **Dept.:** BOY, CUN **Elev.:** 1800-2700 m.a.s.l. **IUCN Habitat code:** 1,3 **Conservation status: (IUCN)** LC **(CPSG)** LC

5953 *Matisia malacocalyx* (A.Robyns & S.Nilsson) W.S.Alverson (Malvaceae) **IPNI:** 946934-1 **Common name:** sapote de monte **Habit:** Tree **Use:** HF **Origin:** Native **Region:** Amazonia **Dept.:** AMA, CAQ, CAU, PUT **Elev.:** 50–500 m a.s.l. **IUCN Habitat code:** 1,3,5,14 **Conservation status: (IUCN)** LC

5954 *Matisia obliquifolia* Standl. (Malvaceae) **IPNI:** 154700-2 **Habit:** Tree **Use:** HF **Origin:** Native **Region:** Amazonia, Andean **Dept.:** AMA, ANT, BOY, CAQ, CAU, COR, PUT, QUI, RIS, VAC **Elev.:** 0–2200 m a.s.l. **IUCN Habitat code:** 1,3,5,14 **Conservation status: (IUCN)** LC

5955 *Matisia ochrocalyx* K.Schum. (Malvaceae) **IPNI:** 154702-2 **Common name:** abuelo de cacao, cacao **Habit:** Tree **Use:** HF **Origin:** Native **Region:** Amazonia, Andean, Orinoquia **Dept.:** AMA, ANT, BOY, CAQ, CUN, GUV, MET, SAN, VAU **Elev.:** 0–1900 m a.s.l. **IUCN Habitat code:** 1,3,5,14 **Conservation status: (IUCN)** LC

5956 *Matisia sclerophylla* Cuatrec. (Malvaceae) **IPNI:** 154709-2 **Common name:** molinillo, vaina, zapote de monte, zapotillo **Habit:** Tree **Use:** HF **Origin:** Endemic **Region:** Pacific **Dept.:** ANT, CAU, VAC **Elev.:** 0-400 m.a.s.l. **IUCN Habitat code:** 1,3 **Conservation status: (IUCN)** LC **(CPSG)** LC

5957 *Matisia soegengii* Cuatrec. (Malvaceae) **IPNI:** 154710-2 **Habit:** Tree **Use:** HF **Origin:** Native **Region:** Andean, Pacific **Dept.:** CHO, NAR, VAC **Elev.:** 0–800 m a.s.l. **IUCN Habitat code:** 1 **Conservation status: (IUCN)** LC

5958 *Matisia spathacea* Fern.Alonso (Malvaceae) **IPNI:** 322423-2 **Common name:** zapotillo **Habit:** Tree **Use:** HF **Origin:** Endemic **Region:** Pacific **Dept.:** ANT, CHO **Elev.:** 0-800 m.a.s.l. **IUCN Habitat code:** 1,14 **Conservation status: (IUCN)** LC **(CPSG)** LC

5959 *Melochia caracasana* Jacq. (Malvaceae) **IPNI:** 824432-1 **Common name:** bretónica, escoba babosa **Habit:** Shrub, Tree **Use:** ME **Origin:** Native **Region:** Caribbean **Dept.:** CAL, CUN, MAG **Elev.:** 0–1800 m a.s.l. **IUCN Habitat code:** 1,3 **Conservation status: (IUCN)** NE

5960 *Melochia crenata* Vahl (Malvaceae) **IPNI:** 824451-1 **Common name:** ratania **Habit:** Herb **Use:** ME **Origin:** Native **Region:** Caribbean **Dept.:** LAG, MAG **Elev.:** 0–100 m a.s.l. **IUCN Habitat code:** 1,2,3,14 **Conservation status: (IUCN)** NE

5961 *Melochia melissifolia* Benth. (Malvaceae) **IPNI:** 824503-1 **Habit:** Shrub **Use:** HF, MA, ME **Origin:** Native **Dept.:** ANT **Elev.:** 400–700 m a.s.l. **IUCN Habitat code:** 1 **Conservation status: (IUCN)** NE

5962 *Melochia nodiflora* Sw. (Malvaceae) **IPNI:** 824514-1 **Habit:** Herb, Shrub, Subshrub **Use:** ME **Origin:** Native **Region:** Caribbean **Dept.:** ATL, BOL, LAG, MAG **Elev.:** 20–120 m a.s.l. **IUCN Habitat code:** 2,4,14 **Conservation status: (IUCN)** NE

5963 *Melochia pilosa* (Mill.) Fawc. & Rendle (Malvaceae)
IPNI: 157002-2
Common name: babosa amarilla, cola de oveja, escobo rosado **Habit:** Herb, Climbing **Use:** ME **Origin:** Native **Region:** Andean, Caribbean, Orinoquia **Dept.:** ANT, CAS, CUN, MAG, MET, SAN **Elev.:** 250–1280 m a.s.l. **IUCN Habitat code:** 1,2,3,4,14 **Conservation status: (IUCN)** NE

5964 *Melochia pyramidata* L. (Malvaceae)
IPNI: 157004-2
Common name: escoba, escoba babosa, escoba blanca, escobilla, escobilla blanca **Habit:** Herb, Shrub, Tree **Use:** HF, ME **Origin:** Native **Region:** Andean, Guiana Shield, Caribbean, Orinoquia **Dept.:** ANT, ATL, BOL, CAS, COR, CUN, HUI, MAG, MET, NSA, SAN, TOL, VAC, VAU **Elev.:** 5-2000 m.a.s.l. **IUCN Habitat code:** 1,2,3,4,5,14 **Conservation status: (IUCN)** LC **(CPSG)** Potential LC

5965 *Melochia tomentosa* L. (Malvaceae)
IPNI: 157023-2
Common name: bretónica morada, escoba, escobilla morada, jawaapia **Habit:** Shrub, Subshrub **Use:** AF, EU, ME, PO **Origin:** Native **Region:** Caribbean **Dept.:** ATL, BOL, LAG, MAG, NSA **Elev.:** 10–400 m a.s.l. **Conservation status: (IUCN)** NE

5966 *Melochia villosa* (Mill.) Fawc. & Rendle (Malvaceae)
IPNI: 157043-2
Use: ME **Origin:** Native **IUCN Habitat code:** 1,2,3,4,14 **Conservation status: (IUCN)** NE

5967 *Modiola caroliniana* (L.) G.Don (Malvaceae)
IPNI: 30017389-2
Common name: pata de chulo, pata de gallinazo, patechula **Habit:** Herb **Use:** HF, ME, PO **Origin:** Native **Region:** Andean **Dept.:** BOG, BOY, CUN **Elev.:** 1800–2950 m a.s.l. **IUCN Habitat code:** 1,3,14 **Conservation status: (IUCN)** NE

5968 *Mollia lepidota* Spruce ex Benth. (Malvaceae)
IPNI: 834881-1
Common name: palo achiote, palo de achote, palo de almizcle **Habit:** Tree **Use:** HF **Origin:** Native **Region:** Amazonia, Guiana Shield **Dept.:** AMA, CAQ, GUA, VAU **Elev.:** 120–400 m a.s.l. **IUCN Habitat code:** 1,5 **Conservation status: (IUCN)** LC

5969 *Ochroma pyramidale* (Cav. ex Lam.) Urb. (Malvaceae)
IPNI: 171458-2
Common name: balsa, balsillo, balso, balso algodón, balso blanco, balso colorado, balso negro, balso real, balso tambor, carguero, lana, lano, majagua, majao, sama, samo, tacarigua, tambor, tocumo, topa, tucumo, tucuno **Habit:** Tree **Use:** EU, HF, IF, MA, ME, SU **Origin:** Native **Region:** Amazonia, Andean, Guiana Shield, Caribbean, Orinoquia, Pacific **Dept.:** AMA, ANT, ATL, CAL, CAU, CES, CHO, CUN, HUI, MAG, MET, NAR, PUT, QUI, RIS, SAN, TOL, VAC, VAU **Elev.:** 0-1800 m.a.s.l. **IUCN Habitat code:** 1,2,3,5,14 **Conservation status: (IUCN)** LC **(CPSG)** Potential LC

5970 *Pachira aquatica* Aubl. (Malvaceae)
IPNI: 179762-2
Common name: algodón, bonguito, cabecetigre, cabeza de tigre, cacagüillo, cacao de agua, cacao de monte, castaño, ceiba, ceibo, dedimicù, malambo, pandeaño, salero, sapotolongo, sapotorongo, sapotón, zapote de agua, zapotolongo **Habit:** Tree **Use:** EU, GS, HF, MA, ME, SU **Origin:** Native **Region:** Andean, Pacific **Dept.:** AMA, ANT, CAU, CHO, NAR, VAC **Elev.:** 0–1200 m a.s.l. **IUCN Habitat code:** 1,2,3,5,14 **Conservation status: (IUCN)** LC

5971 *Pachira glabra* Pasq. (Malvaceae)
IPNI: 562664-1
Common name: cabeza de tigre, yaigue **Use:** EU, GS, HF, MA, ME **IUCN Habitat code:** 3 **Conservation status: (IUCN)** NE

5972 *Pachira insignis* (Sw.) Savigny (Malvaceae)
IPNI: 562668-1
Common name: algodón, palo de yuca **Habit:** Tree **Use:** GS, HF, MA, ME **Origin:** Native **Region:** Amazonia, Guiana Shield, Orinoquia **Dept.:** AMA, ARA, CAQ, NSA, PUT, VAU, VID **Elev.:** 0–500 m a.s.l. **IUCN Habitat code:** 1,3,5,14 **Conservation status: (IUCN)** NE

5973 *Pachira morae* Fern.Alonso (Malvaceae)
IPNI: 327590-2
Habit: Tree **Use:** HF **Origin:** Endemic **Region:** Amazonia, Guiana Shield **Dept.:** AMA, GUA, VAU **Elev.:** 0–500 m a.s.l. **IUCN Habitat code:** 1,5 **Conservation status: (IUCN)** NE

5974 *Pachira patinoi* (Dugand & A.Robyns) Fern.Alonso (Malvaceae)
IPNI: 1002081-1
Common name: guascayuca, guayaba, guayabilla, guayabilla agria, piscande, piscandé **Habit:** Tree **Use:** HF, MA, ME **Origin:** Native **Region:** Andean, Pacific **Dept.:** NAR **Elev.:** 200–1500 m a.s.l. **IUCN Habitat code:** 1,3 **Conservation status: (IUCN)** NE

5975 *Pachira sessilis* Benth. (Malvaceae)
IPNI: 281912-2
Common name: huimba **Habit:** Tree **Use:** HF, MA **Origin:** Native **Region:** Caribbean, Pacific **Dept.:** ANT, BOY, CES, CHO, SAN **Elev.:** 100–600 m a.s.l. **IUCN Habitat code:** 1,3,14 **Conservation status: (IUCN)** LC

5976 *Pachira speciosa* Triana & Planch. (Malvaceae)
IPNI: 281913-2
Common name: cacao, cacao de monte, ceibo de agua, mauricio **Habit:** Tree **Use:** ME **Origin:** Endemic, Cultivated **Region:** Andean, Orinoquia **Dept.:** ANT, BOY, CAL, CAS, CAU, CUN, HUI, MET, NAR, QUI, SAN, TOL, VAC **Elev.:** 400-1700 m.a.s.l. **IUCN Habitat code:** 1,2,3,14 **Conservation status: (IUCN)** LC **(CPSG)** LC

5977 *Patinoa almirajo* Cuatrec. (Malvaceae)
IPNI: 185556-2
Common name: almirajo, almirajó **Habit:** Tree **Use:** HF, ME **Origin:** Native **Region:** Andean, Pacific **Dept.:** ANT, CHO, RIS, VAC **Elev.:** 0–1100 m a.s.l. **IUCN Habitat code:** 1,3,14 **Conservation status: (IUCN)** NE

5978 *Patinoa paraensis* (Huber) Cuatrec. (Malvaceae)
IPNI: 185557-2
Common name: laurel bracilargo **Habit:** Tree **Use:** HF, MA **Origin:** Native **Region:** Amazonia **Dept.:** AMA, CAQ, PUT **Elev.:** 50–700 m a.s.l. **IUCN Habitat code:** 1 **Conservation status: (IUCN)** NE

5979 *Pavonia fruticosa* (Mill.) Fawc. & Rendle (Malvaceae)
IPNI: 185868-2
Common name: abrojo, cadillo, cadillo de bestia, imbiana, imbiande, jacintica, piojo de marrano **Habit:** Shrub **Use:** MA, ME **Origin:** Native **Region:** Amazonia, Andean, Caribbean, Orinoquia, Pacific **Dept.:** AMA, ANT, ARA, BOL, BOY, CAL, CAQ, CAU, CHO, CUN, MAG, MET, NAR, NSA, SAN, TOL, VAC **Elev.:** 20-2780 m.a.s.l. **IUCN Habitat code:** 1,2,3,4,5,14 **Conservation status: (IUCN)** NE **(CPSG)** Potential LC

5980 *Pavonia mutisii* Kunth (Malvaceae)
IPNI: 562981-1
Common name: algodoncillo, balsilla, balsillo, quesillo **Habit:** Subshrub **Use:** ME **Origin:** Native **Region:** Andean **Dept.:** ANT, BOY, NSA, VAC **Elev.:** 80–1650 m a.s.l. **IUCN Habitat code:** 1,3,14 **Conservation status: (IUCN)** NE

5981 *Pavonia schiedeana* Steud. (Malvaceae)
IPNI: 563076-1
Habit: Shrub, Subshrub **Use:** ME **Origin:** Native **Region:** Andean, Pacific **Dept.:** ANT, CHO, MAG **Elev.:** 0–1800 m a.s.l. **IUCN Habitat code:** 1,3,14 **Conservation status: (IUCN)** NE

5982 *Pavonia sepium* A.St.-Hil. (Malvaceae)
IPNI: 186058-2
Common name: cadillo, embiandé, escoba, escobilla, llausa, malva, melosita **Use:** ME **Origin:** Native **Dept.:** BOG **IUCN Habitat code:** 1,3,14 **Conservation status: (IUCN)** NE

5983 *Pavonia spinifex* (L.) Cav. (Malvaceae)
IPNI: 186083-2
Common name: cadillo, escobo, escobo de jardín, guazumilla, pajarito amarillo **Use:** EU, MA, ME **Dept.:** BOG **IUCN Habitat code:** 1,3,14 **Conservation status: (IUCN)** LC

5984 *Pentapetes phoenicea* L. (Malvaceae)
IPNI: 824676-1
Habit: Herb, Subshrub **Use:** EU, HF, ME **Origin:** Cultivated **Dept.:** ANT, BOL, COR, CUN **Elev.:** 120–400 m a.s.l. **Conservation status: (IUCN)** NE

5985 *Phragmotheca mammosa* W.S.Alverson (Malvaceae)
IPNI: 282736-2
Common name: molinillo, pacocito de monte, punula, vaina **Habit:** Tree **Use:** HF **Origin:** Native **Region:** Andean, Pacific **Dept.:** ANT, CHO **Elev.:** 0–900 m a.s.l. **IUCN Habitat code:** 1,5 **Conservation status: (IUCN)** LC

5986 *Phymosia rosea* (DC.) Kearney (Malvaceae)
IPNI: 563228-1
Use: EU **Region:** Andean **Dept.:** BOG **Conservation status: (IUCN)** LC

5987 *Pseudobombax grandiflorum* (Cav.) A.Robyns (Malvaceae)
IPNI: 211230-2
Use: MA **Conservation status: (IUCN)** NE

5988 *Pseudobombax munguba* (Mart.) Dugand (Malvaceae) **IPNI:** 211238-2 **Common name:** algodón, algodón de árbol, ceiba acuática, higuillo, imbira, ivira, palo de algodón, tree de algodón, árbol de algodón **Habit:** Tree **Use:** HF **Origin:** Native **Region:** Amazonia, Orinoquia **Dept.:** AMA, CAQ, MET, PUT, VAU **Elev.:** 0–500 m a.s.l. **IUCN Habitat code:** 1,5 **Conservation status: (IUCN)** LC

5989 *Pseudobombax septenatum* (Jacq.) Dugand (Malvaceae) **IPNI:** 211241-2 **Common name:** barrigón, bonga, bonga colorada, cartagena, cartageno, ceiba, ceiba barrigona, ceiba barrigón, ceiba bomba, ceiba de majagua, ceiba majagua, ceiba menché, ceiba vaquera, ceiba verde, ceibo, ceibo barrigón, chitua, guarabilla, guarabillo, majagua, majagua **Habit:** Tree **Use:** HF, MA, ME, SU **Origin:** Native **Region:** Amazonia, Andean, Caribbean, Orinoquia, Pacific **Dept.:** AMA, ANT, ATL, BOL, BOY, CAL, CAS, CES, CHO, COR, CUN, HUI, MAG, NSA, SAN, SUC, TOL, VAC **Elev.:** 0-1800 m.a.s.l. **IUCN Habitat code:** 1,2,3,5,14 **Conservation status: (IUCN)** LC **(CPSG)** Potential LC

5990 *Quararibea amazonica* Ulbr. (Malvaceae) **IPNI:** 215358-2 **Habit:** Tree **Use:** HF **Origin:** Native **Region:** Amazonia **Dept.:** AMA, CAQ **Elev.:** 50–300 m a.s.l. **IUCN Habitat code:** 1,14 **Conservation status: (IUCN)** LC

5991 *Quararibea asterolepis* Pittier (Malvaceae) **IPNI:** 215362-2 **Common name:** bolinillo, molinillo, zapatillo **Habit:** Tree **Use:** HF **Origin:** Native **Region:** Caribbean, Pacific **Dept.:** ANT, CHO, COR **Elev.:** 0–300 m a.s.l. **IUCN Habitat code:** 1,2,3,14 **Conservation status: (IUCN)** LC

5992 *Quararibea pterocalyx* Hemsl. (Malvaceae) **IPNI:** 563304-1 **Common name:** araño, botón, molinillo, zapotillo **Habit:** Tree **Use:** HF **Origin:** Native **Region:** Andean, Pacific **Dept.:** ANT, BOL, CHO **Elev.:** 0–800 m a.s.l. **IUCN Habitat code:** 1,3,5,14 **Conservation status: (IUCN)** VU

5993 *Quararibea wittii* K.Schum. & Ulbr. (Malvaceae) **IPNI:** 112602-1 **Habit:** Tree **Use:** HF **Origin:** Native **Region:** Amazonia, Guiana Shield **Dept.:** AMA, CAQ, CAU, MET, PUT **Elev.:** 100–500 m a.s.l. **IUCN Habitat code:** 1,14 **Conservation status: (IUCN)** LC

5994 *Scleronema micranthum* (Ducke) Ducke (Malvaceae) **IPNI:** 230850-2 **Common name:** abuelo de cacao, abuelo del cacao, baboso, boque, carguero yolombillo, castaña, castaño, yolombo **Habit:** Tree **Use:** HF **Origin:** Native **Region:** Amazonia **Dept.:** AMA, CAQ, VAU **Elev.:** 100–300 m a.s.l. **IUCN Habitat code:** 1,3,5,14 **Conservation status: (IUCN)** LC

5995 *Scleronema praecox* (Ducke) Ducke (Malvaceae) **IPNI:** 230851-2 **Common name:** abuelo de cacao, bogüe, castaña de paca, castaño de paca, castaño del rojo, palo cebo, yolombo **Habit:** Tree **Use:** ME **Origin:** Native **Region:** Amazonia **Dept.:** AMA, VAU **Elev.:** 100–500 m a.s.l. **IUCN Habitat code:** 1 **Conservation status: (IUCN)** NE

5996 *Sida abutifolia* Mill. (Malvaceae) **IPNI:** 563387-1 **Common name:** escoba babosa, escubilla macha **Habit:** Herb **Use:** ME **Origin:** Native **Region:** Andean, Caribbean **Dept.:** ATL, BOG, BOL, BOY, CAU, CUN, HUI, LAG, MAG, NAR, NSA, SAN, TOL, VAC **Elev.:** 0-2130 m.a.s.l. **IUCN Habitat code:** 1,2,3,14 **Conservation status: (IUCN)** NE **(CPSG)** Potential LC

5997 *Sida acuta* Burm.f. (Malvaceae) **IPNI:** 235518-2 **Common name:** babosa, babosilla, escoba, escoba babosa, escoba blanca, escoba de babosa, escoba negra, escobababosa, escobabosa, escobadura, escobilla, escobilla amarilla, escobilla babosa, escobilla blanca, escobilla negra, escobo, escubilla, esparto, huinar, malva co **Use:** AF, HF, MA, ME, PO **IUCN Habitat code:** 1,2,3,4,5,14 **Conservation status: (IUCN)** NE

5998 *Sida ciliaris* L. (Malvaceae) **IPNI:** 313702-2 **Common name:** babosa, escoba peluda, escoba pestañosa, escobilla negra, huinar, kotoloinsu **Habit:** Herb, Subshrub **Use:** ME **Origin:** Native **Region:** Andean, Caribbean **Dept.:** ATL, BOL, BOY, CUN, HUI, LAG, MAG, NSA, SAN **Elev.:** 60-2130 m.a.s.l. **IUCN Habitat code:** 1,2,14 **Conservation status: (IUCN)** NE **(CPSG)** Potential LC

5999 *Sida cordifolia* L. (Malvaceae) **IPNI:** 563600-1 **Common name:** brocas, escobadura calentana, escobilla, escobo, malva de Castilla, malva de castilla, mazamorrera, quitamanteca, yerba amarga **Habit:** Herb, Subshrub **Use:** AF, HF, MA, ME, PO **Origin:** Native **Dept.:** ANT, ATL, BOY, CAU, CUN, HUI, LAG, MAG, NSA, SAN, TOL, VAC **Elev.:** 10–2230 m a.s.l. **IUCN Habitat code:** 1,2,3,4,14 **Conservation status: (IUCN)** NE

6000 *Sida dictyocarpa* F.Muell. ex Benth. (Malvaceae) **IPNI:** 563645-1 **Common name:** escoba **Use:** ME **Origin:** Endemic **IUCN Habitat code:** 1,2,14 **Conservation status: (IUCN)** NE

6001 *Sida glabra* Mill. (Malvaceae) **IPNI:** 329579-2 **Common name:** escobo **Habit:** Herb, Subshrub **Use:** HF, MA, ME **Origin:** Native **Region:** Andean, Caribbean **Dept.:** ANT, ATL, BOL, CUN, HUI, LAG, MAG, NAR, SAN, TOL **Elev.:** 30-800 m.a.s.l. **IUCN Habitat code:** 1,2,5,14 **Conservation status: (IUCN)** NE **(CPSG)** Potential LC

6002 *Sida glomerata* Cav. (Malvaceae) **IPNI:** 329545-2 **Common name:** escoba, escobilla, pichanquilla **Habit:** Subshrub **Use:** ME **Origin:** Native **Region:** Amazonia, Andean, Caribbean, Orinoquia, Pacific **Dept.:** AMA, ANT, ATL, BOL, BOY, CAS, CAU, CES, CHO, CUN, HUI, LAG, MAG, MET, NAR, PUT, SAN, TOL, VAC, VID **Elev.:** 0-2240 m.a.s.l. **IUCN Habitat code:** 1,2,3,4,14 **Conservation status: (IUCN)** NE **(CPSG)** Potential LC

6003 *Sida glutinosa* Comm. ex Cav. (Malvaceae) **IPNI:** 563741-1 **Habit:** Herb, Subshrub **Use:** MA, ME **Origin:** Native **Region:** Andean **Dept.:** CAU, CUN, MAG, TOL, VAC **Elev.:** 400–1500 m a.s.l. **IUCN Habitat code:** 5,14 **Conservation status: (IUCN)** NE

6004 *Sida hyssopifolia* C.Presl (Malvaceae) **IPNI:** 563820-1 **Habit:** Subshrub **Use:** AF,ME **Origin:** Native **Region:** Andean, Orinoquia **Dept.:** ANT, MAG, MET, VAC **Elev.:** 300–1800 m a.s.l. **IUCN Habitat code:** 1,2 **Conservation status: (IUCN)** NE

6005 *Sida linifolia* Juss. ex Cav. (Malvaceae) **IPNI:** 563888-1 **Common name:** escobo macho **Habit:** Herb **Use:** MA, ME **Origin:** Native **Region:** Andean, Caribbean, Orinoquia **Dept.:** ANT, CAS, CAU, CUN, MAG, MET, SAN, VAC **Elev.:** 300–1700 m a.s.l. **IUCN Habitat code:** 1,2,3,4,14 **Conservation status: (IUCN)** NE

6006 *Sida poeppigiana* (K.Schum.) Fryxell (Malvaceae) **IPNI:** 969172-1 **Common name:** escoba, escoba babosa, escoba negra, escobadura, escobilla, escobo, salvia **Habit:** Subshrub **Use:** ME **Origin:** Native **Region:** Andean, Caribbean, Pacific **Dept.:** ANT, BOG, BOY, CAL, CAU, CES, CHO, CUN, LAG, MAG, NAR, NSA, PUT, QUI, SAN, TOL, VAC **Elev.:** 40-2800 m.a.s.l. **IUCN Habitat code:** 1,2,3,4,14 **Conservation status: (IUCN)** NE **(CPSG)** Potential LC

6007 *Sida rhombifolia* L. (Malvaceae) **IPNI:** 235798-2 **Common name:** escoba, escoba amarilla, escoba babosa, escoba cimarrona, escoba negra, escobabosa, escobadura, escobilla, escobilla babosa, escobo, escobo amarillo, escubilla, escubilla babosa **Habit:** Herb, Subshrub **Use:** AF, HF, MA, ME, PO, SU **Origin:** Naturalised **Region:** Amazonia, Andean, Caribbean, Orinoquia, Pacific **Dept.:** AMA, ANT, ARA, ATL, BOG, BOL, BOY, CAL, CAQ, CAS, CAU, CES, CHO, COR, CUN, HUI, LAG, MAG, MET, NSA, RIS, SAP, SAN, TOL, VAC, VID **Elev.:** 0-1800 m.a.s.l. **IUCN Habitat code:** 1,2,3,4,5,14 **Conservation status: (IUCN)** NE **(CPSG)** Potential LC

6008 *Sida santaremensis* Monteiro (Malvaceae) **IPNI:** 313705-2 **Habit:** Herb **Use:** ME **Origin:** Native **Region:** Orinoquia **Dept.:** MET **Elev.:** 500–600 m a.s.l. **IUCN Habitat code:** 14 **Conservation status: (IUCN)** NE

6009 *Sida setosa* Mart. ex Colla (Malvaceae) **IPNI:** 564227-1 **Common name:** pichanga **Habit:** Herb, Subshrub **Use:** ME **Origin:** Native **Region:** Amazonia, Andean, Guiana Shield, Caribbean, Orinoquia, Pacific **Dept.:** AMA, ANT, BOG, BOL, CAQ, CAU, CHO, CUN, GUV, LAG, MAG, MET, NAR, PUT, VAC, VAU, VID **Elev.:** 0-1800 m.a.s.l. **IUCN Habitat code:** 1,3,4,5,14 **Conservation status: (IUCN)** NE **(CPSG)** Potential LC

6010 *Sida spinosa* L. (Malvaceae) **IPNI:** 235838-2 **Habit:** Herb, Subshrub **Use:** AF, GS, HF, MA, ME **Origin:** Native **Region:** Andean, Caribbean **Dept.:** ANT, CAU, CES, CUN, HUI, MAG, NSA, SAN, TOL, VAC **Elev.:** 500–1700 m a.s.l. **IUCN Habitat code:** 1,2,3,14 **Conservation status: (IUCN)** NE

6011 *Sida ulmifolia* Mill. (Malvaceae)
IPNI: 329574-2
Habit: Subshrub **Use:** ME **Origin:** Native **Dept.:** CUN **Conservation status: (IUCN)** NE

6012 *Sida urens* L. (Malvaceae)
IPNI: 564343-1
Common name: escobo **Habit:** Herb, Subshrub, Climbing **Use:** MA, ME, PO **Origin:** Native **Region:** Andean, Caribbean, Orinoquia **Dept.:** ANT, BOY, CAL, CUN, LAG, MAG, MET, RIS, SAN, TOL, VAC **Elev.:** 400–1600 m a.s.l. **IUCN Habitat code:** 1,14 **Conservation status: (IUCN)** NE

6013 *Sidastrum micranthum* (A.St.-Hil.) Fryxell (Malvaceae)
IPNI: 235977-2
Habit: Subshrub **Use:** MA, ME **Origin:** Native **Region:** Andean, Orinoquia **Dept.:** SAN **Elev.:** 0–1000 m a.s.l. **IUCN Habitat code:** 3,4 **Conservation status: (IUCN)** NE

6014 *Sidastrum paniculatum* (L.) Fryxell (Malvaceae)
IPNI: 235979-2
Common name: lavaplatos **Habit:** Shrub **Use:** MA **Origin:** Native **Region:** Andean, Caribbean **Dept.:** CAU, CUN, HUI, MAG, SAP, TOL, VAC **Elev.:** 0–1600 m a.s.l. **IUCN Habitat code:** 1,2,3,14 **Conservation status: (IUCN)** NE

6015 *Sparrmannia africana* L.f. (Malvaceae)
IPNI: 835200-1
Habit: Shrub **Use:** EU, ME **Origin:** Cultivated **Region:** Andean **Dept.:** BOG, CUN **Elev.:** 2600–2600 m a.s.l. **IUCN Habitat code:** 1,3,14 **Conservation status: (IUCN)** LC

6016 *Spirotheca rosea* (Seem.) P.E.Gibbs & W.S.Alverson (Malvaceae)
IPNI: 60446802-2
Common name: ceiba, palosanto **Habit:** Tree **Use:** EU, FU **Origin:** Native **Region:** Andean, Pacific **Dept.:** ANT, CHO **Elev.:** 0–1500 m a.s.l. **IUCN Habitat code:** 1,3,14 **Conservation status: (IUCN)** LC

6017 *Sterculia apetala* (Jacq.) H.Karst. (Malvaceae)
IPNI: 245150-2
Common name: almendro, cacagüillo, cacahuito, cacaíto, camajonduro, camajorú, camajurú, camajón, camoruco, campano común, camuruco, castaño, castaño de Indias, chichí, cumaruco, guarauno, majao, mano de león, panamá, pata de danta, piñones, piñón, piñón de puerco, sun **Habit:** Tree **Use:** AF, EU, HF, MA, ME, PO **Origin:** Native **Region:** Caribbean, Pacific **Dept.:** ANT, ATL, COR, CUN, LAG, MAG, SAN, SUC, VAC **Elev.:** 0–1000 m a.s.l. **IUCN Habitat code:** 1,2,3,4,5,14 **Conservation status: (IUCN)** LC

6018 *Sterculia colombiana* Sprague (Malvaceae)
IPNI: 245153-2
Common name: babasaíno, bobena, cabeza de churuco, castañete, gomo blanco, sapote, sihuaruma montañera, zapote **Habit:** Tree **Use:** HF, MA **Origin:** Native **Region:** Amazonia, Andean, Pacific **Dept.:** AMA, MET, NAR, PUT, VAC **Elev.:** 0–1180 m a.s.l. **IUCN Habitat code:** 1,2,3,4,14 **Conservation status: (IUCN)** LC

6019 *Sterculia pruriens* (Aubl.) K.Schum. (Malvaceae)
IPNI: 825412-1
Habit: Tree **Use:** MA **Origin:** Native **Region:** Amazonia, Caribbean **Dept.:** AMA, BOL, COR **Elev.:** 100–150 m a.s.l. **IUCN Habitat code:** 1,14 **Conservation status: (IUCN)** LC

6020 *Sterculia rugosa* R.Br. (Malvaceae)
IPNI: 927898-1
Common name: camajorú, castaño, majao, piñón, zapote, zapote de monte, zapoterana, zapotillo **Habit:** Tree **Use:** HF **Origin:** Native **Dept.:** ANT, SAN **Elev.:** 200–900 m a.s.l. **IUCN Habitat code:** 1,3 **Conservation status: (IUCN)** LC

6021 *Theobroma bernoullii* Pittier (Malvaceae)
IPNI: 252439-2
Common name: cacao de monte **Habit:** Tree **Use:** GS, HF, ME **Origin:** Native **Region:** Pacific **Dept.:** ANT, COR, NAR, VAC **Elev.:** 5–910 m a.s.l. **IUCN Habitat code:** 1,3,14 **Conservation status: (IUCN)** LC

6022 *Theobroma bicolor* Bonpl. (Malvaceae)
IPNI: 825565-1
Common name: ao, bacao, bacao de monte, cacao, cacao cimarrón, cacao de monte, cacao maracos, cacao marraco, cacao'emonte, chocolatillo, heé-a, há-ha, macambo, macambu, macao, maraca, maraca de armadillo, maraca de bejuco, maraca de boruga, maraca de cucarrón, maraca **Habit:** Tree **Use:** GS, HF, MA, ME, SU **Origin:** Native, Cultivated **Region:** Amazonia, Guiana Shield, Pacific **Dept.:** AMA, ANT, CAQ, CHO, GUA, PUT, VAC, VAU **Elev.:** 20–1000 m a.s.l. **IUCN Habitat code:** 1,2,3,4,5,14 **Conservation status: (IUCN)** LC

6023 *Theobroma cacao* L. (Malvaceae)
IPNI: 320783-2
Common name: abacaraá, aso-ya-ee, bacao, baiuc, cabecera, cabecerá, cacahua caspi, cacao, cacao caturro, cacao criollo, cacao de iguana, cacao de monte, cacao de semilla, cacao dulce, cacao hartón, cacao manavén, cacao morado, cacao silvestre, cacaotero, cacau, capuí, **Habit:** Shrub, Small tree, Tree **Use:** AF, EU, GS, HF, MA, ME, PO, SU **Origin:** Native, Cultivated **Region:** Amazonia, Andean, Guiana Shield, Pacific **Dept.:** AMA, ANT, CAQ, CHO, GUV, HUI, LAG, MAG, MET, NSA, PUT, SAN, TOL, VAC, VAU **Elev.:** 0-1540 m.a.s.l. **IUCN Habitat code:** 1,2,3,5,14 **Conservation status: (IUCN)** NE **(CPSG)** Potential LC

6024 *Theobroma cirmolinae* Cuatrec. (Malvaceae)
IPNI: 252453-2
Common name: bacao, cacao de indio, cacao de monte, cacao indio **Habit:** Tree **Use:** GS, HF, ME **Origin:** Endemic **Region:** Andean **Dept.:** ANT, VAC **Elev.:** 700-1250 m.a.s.l. **IUCN Habitat code:** 1,3 **Conservation status: (IUCN)** LC **(CPSG)** LC

6025 *Theobroma gileri* Cuatrec. (Malvaceae)
IPNI: 252454-2
Habit: Tree **Use:** GS, HF **Origin:** Native **Region:** Pacific **Dept.:** ANT **Elev.:** 95–250 m a.s.l. **IUCN Habitat code:** 1 **Conservation status: (IUCN)** VU

6026 *Theobroma glaucum* H.Karst. (Malvaceae)
IPNI: 825579-1
Common name: bicco, cacao, cacao cimarrón, cacao de monte, cacao maduraverde, cacao silvestre, cacaíto, chucú, copoazú, maraca de búho, maraca de danta, mortecino **Habit:** Small tree, Tree **Use:** GS, HF, ME **Origin:** Native **Region:** Amazonia, Andean, Guiana Shield, Pacific **Dept.:** AMA, ANT, CAQ, CHO, GUV, MET, PUT, VAU **Elev.:** 10–1000 m a.s.l. **IUCN Habitat code:** 1,3,14 **Conservation status: (IUCN)** LC

6027 *Theobroma grandiflorum* (Willd. ex Spreng.) K.Schum. (Malvaceae)
IPNI: 825580-1
Common name: cacao de monte, copoazú, copoazú, cupuassú, cupuazú, maraca silvestre, taraira, usper, ñee-au **Habit:** Small tree, Tree **Use:** GS, HF, MA, ME **Origin:** Cultivated **Region:** Amazonia, Andean **Dept.:** AMA, CAQ, QUI, VAC, VAU **Elev.:** 120–1200 m a.s.l. **IUCN Habitat code:** 1,3,14 **Conservation status: (IUCN)** LC

6028 *Theobroma hylaeum* Cuatrec. (Malvaceae)
IPNI: 252455-2
Common name: cacao de monte **Habit:** Tree **Use:** GS, ME **Origin:** Native **Region:** Pacific **Dept.:** ANT, CHO **Elev.:** 40–280 m a.s.l. **IUCN Habitat code:** 1 **Conservation status: (IUCN)** LC

6029 *Theobroma microcarpum* Mart. (Malvaceae)
IPNI: 825595-1
Common name: cacao silvestre, metró-reemooeé **Habit:** Tree **Use:** GS, HF, ME **Origin:** Native **Region:** Amazonia **Dept.:** AMA **Elev.:** 100–120 m a.s.l. **IUCN Habitat code:** 1,14 **Conservation status: (IUCN)** NE

6030 *Theobroma nemorale* Cuatrec. (Malvaceae)
IPNI: 252464-2
Common name: bacao, chocolate de monte, chocolatillo **Habit:** Tree **Use:** GS, HF **Origin:** Endemic **Region:** Pacific **Dept.:** CHO, VAC **Elev.:** 0-50 m.a.s.l. **IUCN Habitat code:** 1,3 **Conservation status: (IUCN)** LC **(CPSG)** LC

6031 *Theobroma obovatum* Klotzsch ex Bernoulli (Malvaceae)
IPNI: 825599-1
Common name: cacao, cacao de mico, cacao de monte, cacao de rama, cacao silvestre, cacaíto, copoazú, maoo-hee-rée, maraca de lorito, maraca de monte, pan y cacao, win-cheék **Habit:** Small tree, Tree **Use:** GS, HF, ME **Origin:** Native **Region:** Amazonia **Dept.:** AMA, CAQ, PUT **Elev.:** 100–220 m a.s.l. **IUCN Habitat code:** 1,5,14 **Conservation status: (IUCN)** LC

6032 *Theobroma simiarum* Donn.Sm. (Malvaceae)
IPNI: 825610-1
Common name: chocolate **Habit:** Tree **Use:** GS, HF, ME **Origin:** Native **Region:** Amazonia, Pacific **Dept.:** AMA, ANT, CHO, NAR, VAC **Elev.:** 5–940 m a.s.l. **IUCN Habitat code:** 1,3,14 **Conservation status: (IUCN)** LC

6033 *Theobroma subincanum* Mart. (Malvaceae)
IPNI: 825619-1
Common name: abekará, bauk, cacao, cacao cojón, cacao de mico, cacao de monte, copoazú, copoazú de altura, copoazú de monte, copoazú silvestre, maraca, maraca de búho, maraca de iguana, maraca de monte, maraca silvestre **Habit:** Shrub, Small tree, Tree **Use:** GS, HF, ME **Origin:** Native **Region:** Amazonia **Dept.:** AMA, MET, PUT, VAU **Elev.:** 100–600 m a.s.l. **IUCN Habitat code:** 1,3,4,5,14 **Conservation status: (IUCN)** LC

6034 *Thespesia populnea* (L.) Sol. ex Corrêa (Malvaceae)
IPNI: 564713-1
Common name: algodoncillo, algodón de monte, algodón silvestre, almendro, cacambal, clemón, majagua, milo **Habit:** Tree **Use:** AF, EU, FU, GS, HF, IF, MA, ME, PO, SU **Origin:** Cultivated, Naturalised **Region:** Caribbean **Dept.:** ANT, ATL, BOL, CUN, HUI, LAG, MAG, SAP, SUC **Elev.:** 0–50 m a.s.l. **IUCN Habitat code:** 1,2,3,5,14 **Conservation status:** (IUCN) LC

6035 *Trichospermum lessertianum* (Hochr.) Dorr (Malvaceae)
IPNI: 77109528-1
Use: MA **Origin:** Native **Conservation status:** (IUCN) NE

6036 *Trichospermum mexicanum* (DC.) Baill. (Malvaceae)
IPNI: 1213322-2
Common name: chillalde, melao **Habit:** Small tree, Tree **Use:** MA **Origin:** Native **Region:** Andean, Pacific **Dept.:** ANT, CHO, NAR, NSA, SAN, VAC **Elev.:** 0–1200 m a.s.l. **IUCN Habitat code:** 1,3,5,14 **Conservation status:** (IUCN) NE

6037 *Triumfetta althaeoides* Lam. (Malvaceae)
IPNI: 835726-1
Habit: Shrub **Use:** MA, ME **Origin:** Native **Region:** Amazonia, Guiana Shield **Dept.:** AMA, MET **Elev.:** 100–250 m a.s.l. **IUCN Habitat code:** 1,3,14 **Conservation status:** (IUCN) NE

6038 *Triumfetta bogotensis* DC. (Malvaceae)
IPNI: 320790-2
Common name: cadillo, cadillo común, cadillo de bestia, cadillo de mula, cadillo de perro **Habit:** Herb, Shrub, Subshrub **Use:** ME **Origin:** Native **Region:** Andean, Caribbean **Dept.:** ANT, ATL, BOG, BOL, BOY, CAL, CAU, CUN, LAG, MAG, NAR, NSA, SAN, VAC **Elev.:** 5-2260 m.a.s.l. **IUCN Habitat code:** 1,2,3,14 **Conservation status:** (IUCN) LC (CPSG) Potential LC

6039 *Triumfetta grandiflora* Vahl (Malvaceae)
IPNI: 835819-1
Habit: Shrub, Tree **Use:** MA **Origin:** Native **Region:** Andean, Pacific **Dept.:** ANT, BOY, NAR, SAN, VAC **Elev.:** 325–1800 m a.s.l. **IUCN Habitat code:** 1,3,14 **Conservation status:** (IUCN) LC

6040 *Triumfetta lappula* L. (Malvaceae)
IPNI: 835864-1
Common name: cadilla, cadillo, cadillo cimarrón, cadillo de bestia, cadillo mulato, carretón, escobilla, pegaparuma, pegapega, vejigón negro **Habit:** Shrub **Use:** HF, MA, ME **Origin:** Native **Region:** Amazonia, Andean, Caribbean, Orinoquia, Pacific **Dept.:** ANT, ATL, CAU, CHO, GUV, MAG, MET, NSA, SAP, SUC, TOL, VAC, VID **Elev.:** 50–1300 m a.s.l. **IUCN Habitat code:** 1,2,3,5,14 **Conservation status:** (IUCN) LC

6041 *Triumfetta mollissima* Kunth (Malvaceae)
IPNI: 835901-1
Common name: cadillo, cadillo blanco, cadillo de bestia, cadillo lanudo, cadillo mulato, canotillo, pegapega **Habit:** Herb **Use:** ME **Origin:** Native **Region:** Andean, Caribbean **Dept.:** ANT, CAU, CES, CUN, HUI, LAG, MAG **Elev.:** 350–2600 m a.s.l. **IUCN Habitat code:** 1,2,3,5,14 **Conservation status:** (IUCN) NE

6042 *Triumfetta rhomboidea* Jacq. (Malvaceae)
IPNI: 320787-2
Habit: Shrub **Use:** AF, EU, GS, HF, MA, ME, PO, SU **Origin:** Naturalised **Region:** Andean, Caribbean **Dept.:** MAG **Elev.:** 600-600 m.a.s.l. **Conservation status:** (IUCN) NE (CPSG) Potential LC

6043 *Triumfetta semitriloba* Jacq. (Malvaceae)
IPNI: 320793-2
Habit: Shrub **Use:** MA, ME **Origin:** Native **Region:** Andean **Dept.:** ANT, CAU, MAG **Elev.:** 800–1490 m a.s.l. **IUCN Habitat code:** 1,2,3,14 **Conservation status:** (IUCN) LC

6044 *Urena lobata* L. (Malvaceae)
IPNI: 564786-1
Common name: aramina, cadillo, malva blanca, yute congo **Habit:** Subshrub **Use:** AF, EU, FU, GS, HF, MA, ME, SU **Origin:** Native **Region:** Amazonia, Andean, Orinoquia, Pacific **Dept.:** AMA, ANT, CAQ, CHO, CUN, HUI, MET, NAR **Elev.:** 50–2300 m a.s.l. **IUCN Habitat code:** 1,2,3,4,5,14 **Conservation status:** (IUCN) LC

6045 *Waltheria berteroi* (Spreng.) J.G.Saunders (Malvaceae)
IPNI: 60439289-2
Common name: mallorquín **Habit:** Shrub **Use:** ME, PO **Origin:** Native **Region:** Andean, Caribbean, Orinoquia **Dept.:** ATL, BOL, CES, COR, LAG, MAG, SAN, VID **Elev.:** 100–1600 m a.s.l. **IUCN Habitat code:** 1,2 **Conservation status:** (IUCN) NE

6046 *Waltheria indica* L. (Malvaceae)
IPNI: 320781-2
Common name: babosa amarilla, babosa hojimenuda, cadillo, escoba de puerco, escoba real, escobilla **Habit:** Herb, Shrub, Subshrub **Use:** AF, HF, MA, ME, SU **Origin:** Native **Region:** Andean, Caribbean **Dept.:** ANT, ATL, BOL, CAL, CAU, COR, CUN, HUI, LAG, MAG, NAR, NSA, SAP, SAN, TOL, VAC **Elev.:** 50-1450 m.a.s.l. **IUCN Habitat code:** 1,2,3,4,14 **Conservation status:** (IUCN) NE (CPSG) Potential LC

6047 *Waltheria viscosissima* A.St.-Hil. (Malvaceae)
IPNI: 825848-1
Habit: Herb **Use:** MA, ME **Origin:** Native **Region:** Caribbean, Orinoquia **Dept.:** MAG, VID **Elev.:** 0–100 m a.s.l. **IUCN Habitat code:** 1 **Conservation status:** (IUCN) NE

6048 *Wissadula amplissima* (L.) R.E.Fr. (Malvaceae)
IPNI: 268896-2
Use: AF, EU, MA, ME **Origin:** Native **IUCN Habitat code:** 2,3,14 **Conservation status:** (IUCN) NE

6049 *Wissadula contracta* (Link) R.E.Fr. (Malvaceae)
IPNI: 268909-2
Habit: Shrub **Use:** MA, ME **Origin:** Native **Conservation status:** (IUCN) NE

6050 *Wissadula periplocifolia* (L.) Thwaites (Malvaceae)
IPNI: 1213890-2
Common name: escoba, salvia blanca **Habit:** Shrub, Subshrub **Use:** MA, ME **Origin:** Native **Dept.:** ATL, BOL, CES, CUN, LAG, NSA, SAN, TOL **Elev.:** 1750–1750 m a.s.l. **IUCN Habitat code:** 1,2,3,14 **Conservation status:** (IUCN) NE

6051 *Dicraspidia donnell-smithii* Standl. (Muntingiaceae)
IPNI: 79965-2
Habit: Shrub, Small tree **Use:** MA **Origin:** Native **Region:** Amazonia, Andean, Pacific **Dept.:** ANT, CAL, CAQ **Elev.:** 10–1200 m a.s.l. **IUCN Habitat code:** 1,3 **Conservation status:** (IUCN) LC

6052 *Muntingia calabura* L. (Muntingiaceae)
IPNI: 320779-2
Common name: acaruco, acurruco, acuruco, acurucu, acurucú, burrillo, capulín, chicató, chicható, chirriador, chitato, chitató, comida paloma, guayuyu, itató, majagüillo, majagüito, majaquillo, manirito, nigua, nigüito, pasito, periquito, tapabotija, tapabotijo, uvito, **Habit:** Shrub, Small tree, Tree **Use:** AF, EU, FU, HF, MA, ME **Origin:** Native **Region:** Amazonia, Andean, Caribbean, Pacific **Dept.:** AMA, ANT, ATL, BOL, CAL, CHO, CUN, HUI, LAG, MAG, MET, SAN, VAC **Elev.:** 0-1100 m.a.s.l. **IUCN Habitat code:** 1,2,3,4,5,14 **Conservation status:** (IUCN) NE (CPSG) Potential LC

6053 *Daphnopsis americana* (Mill.) J.R.Johnst. (Thymelaeaceae)
IPNI: 76687-2
Common name: ajicillo, ají, barbasco, barbasquillo, majagüillo, majagüito, mata de mora, pelamano, pelamanos, pelamán **Habit:** Shrub, Tree **Use:** ME **Origin:** Native **Region:** Caribbean **Dept.:** ANT, MAG, TOL **Elev.:** 40–500 m a.s.l. **IUCN Habitat code:** 1,2,3,14 **Conservation status:** (IUCN) NE

6054 *Schoenobiblus daphnoides* Mart. (Thymelaeaceae)
IPNI: 832744-1
Habit: Shrub, Small tree **Use:** ME **Origin:** Native **Region:** Amazonia, Andean **Dept.:** AMA, ANT, NAR **Elev.:** 200–1800 m a.s.l. **IUCN Habitat code:** 1,3,14 **Conservation status:** (IUCN) LC

6055 *Schoenobiblus peruviana* Standl. (Thymelaeaceae)
IPNI: 229780-2
Common name: porson, rayo caspi, yanchama de pelazón **Habit:** Subshrub, Small tree, Tree **Use:** AF, MA, ME, PO **Origin:** Native **Region:** Amazonia, Andean, Guiana Shield **Dept.:** AMA, ANT, CAQ, CUN, GUV, QUI, VAU **Elev.:** 100–1350 m a.s.l. **Conservation status:** (IUCN) NE

Metteniusales

6056 *Metteniusa edulis* H.Karst. (Metteniusaceae)
IPNI: 608042-1
Common name: canjí, canyí, kandjí **Habit:** Tree **Use:** HF **Origin:** Native **Region:** Andean **Dept.:** LAG **Elev.:** 910–2000 m a.s.l. **IUCN Habitat code:** 1,14 **Conservation status:** (IUCN) VU

6057 *Metteniusa nucifera* (Pittier) Sleumer (Metteniusaceae)
IPNI: 158626-2
Habit: Tree **Use:** HF, ME **Origin:** Native **Elev.:** 1000–1200 m a.s.l. **Conservation status:** (IUCN) NE

Myrtales

6058 *Combretum cacoucia* Exell ex Sandwith (Combretaceae)
IPNI: 63781-2
Habit: Shrub, Small tree, Liana **Use:** ME **Origin:** Native **Region:** Caribbean, Pacific **Dept.:** ANT, CHO **Elev.:** 0–50 m a.s.l. **IUCN Habitat code:** 1,5 **Conservation status:** (IUCN) LC

6059 *Combretum decandrum* Jacq. (Combretaceae) **IPNI:** 170030-1 **Habit:** Shrub, Small tree, Liana **Use:** ME **Origin:** Native **Region:** Caribbean **Dept.:** ATL, BOL, LAG, MAG, SUC **Elev.:** 5–500 m a.s.l. **IUCN Habitat code:** 1,2,14 **Conservation status: (IUCN)** NE

6060 *Combretum fruticosum* (Loefl.) Stuntz (Combretaceae) **IPNI:** 63788-2 **Common name:** chupachupa, mangle blanco, mangle bobo, mangle salado, peinecillo, peineto, peinicillo **Habit:** Shrub, Small tree, Tree, Liana **Use:** AF, IF, MA, ME **Origin:** Native **Region:** Amazonia, Guiana Shield, Caribbean, Orinoquia, Pacific **Dept.:** AMA, ANT, ATL, BOL, BOY, CES, CHO, COR, CUN, HUI, LAG, MAG, MET, NSA, SUC, TOL, VAU **Elev.:** 30-1130 m.a.s.l. **IUCN Habitat code:** 1,2,3,4,5,14 **Conservation status: (IUCN)** NE **(CPSG)** Potential LC

6061 *Combretum indicum* (L.) DeFilipps (Combretaceae) **IPNI:** 77101543-1 **Common name:** albicia, amor al día, carácter, carácter de hombre, carácter de los hombres, carácter del hombre, genio, variedad, variedades **Habit:** Liana **Use:** EU, HF, MA, ME, PO **Origin:** Cultivated **Region:** Caribbean, Orinoquia **Dept.:** ANT, ATL, BOL, BOY, CAL, CES, CUN, HUI, MAG, NSA, SAP, TOL, VAC **Elev.:** 250–1000 m a.s.l. **IUCN Habitat code:** 1,2,3,14 **Conservation status: (IUCN)** NE

6062 *Combretum rotundifolium* Rich. (Combretaceae) **IPNI:** 170434-1 **Common name:** cresta de gallo **Habit:** Shrub, Small tree, Tree, Liana **Use:** ME **Origin:** Native **Region:** Amazonia, Guiana Shield, Orinoquia **Dept.:** AMA, GUV, VAU **Elev.:** 250-300 m.a.s.l. **IUCN Habitat code:** 1,2,3,4,5,14 **Conservation status: (IUCN)** LC **(CPSG)** Potential LC

6063 *Conocarpus erectus* L. (Combretaceae) **IPNI:** 319151-2 **Common name:** botoncillo, feliz torcido, mangle, mangle botoncillo, mangle botón, mangle garbancillo, mangle jelí, mangle negro, mangle zaragoza, zaragoza **Habit:** Shrub, Small tree, Tree **Use:** EU, FU, HF, MA, ME **Origin:** Native **Region:** Caribbean, Pacific **Dept.:** ANT, ATL, BOL, CAU, CHO, COR, MAG, NAR, SAP, SUC, VAC **Elev.:** 0-200 m.a.s.l. **IUCN Habitat code:** 1,2,3,5,14 **Conservation status: (IUCN)** LC **(CPSG)** Potential LC

6064 *Laguncularia racemosa* (L.) C.F.Gaertn. (Combretaceae) **IPNI:** 170672-1 **Common name:** comedero, feliz blanco, iguanero, mangle, mangle amarillo, mangle blanco, mangle bobo, mangle salado, mangle zaragoza, mereicillo **Habit:** Shrub, Tree **Use:** AF, EU, FU, GS, HF, MA, ME **Origin:** Native **Region:** Caribbean, Pacific **Dept.:** ANT, ATL, BOL, CAU, CHO, COR, LAG, MAG, NAR, SAP, SUC, VAC **Elev.:** 0-200 m.a.s.l. **IUCN Habitat code:** 1,2,3,4,5,8,14 **Conservation status: (IUCN)** LC **(CPSG)** Potential LC

6065 *Terminalia amazonia* (J.F.Gmel.) Exell (Combretaceae) **IPNI:** 250537-2 **Common name:** amarillo, amarillón, caneletillo, capacho, chipó, curibano amararillo, curichí, granadillo, guasco blanco, guayabillo, guayabo de hierro, guayabo de león, guayabo león, guayacán amarillo, macano, macano amarillo, mosco, muchilero, roble, vara de león **Habit:** Tree **Use:** MA, ME **Origin:** Native **Region:** Amazonia, Andean, Caribbean, Orinoquia, Pacific **Dept.:** AMA, ANT, BOL, CAU, CHO, CUN, LAG, MAG, MET, NAR, NSA, SAN, SUC, VAC, VID **Elev.:** 100-2000 m.a.s.l. **IUCN Habitat code:** 1,2,3,4,5,14 **Conservation status: (IUCN)** LC **(CPSG)** Potential LC

6066 *Terminalia buceras* (L.) C.Wright (Combretaceae) **IPNI:** 171012-1 **Habit:** Shrub, Tree **Use:** MA, ME **Origin:** Native **Region:** Caribbean **Dept.:** SAP **Elev.:** 0–50 m a.s.l. **IUCN Habitat code:** 14 **Conservation status: (IUCN)** NE

6067 *Terminalia catappa* L. (Combretaceae) **IPNI:** 171034-1 **Common name:** almendra, almendrillo, almendro, almendro de Indias, almendro de la India, almendro de la costa, almendro de las Indias, almendro de playa, almendrón, almondo, guayabillo **Habit:** Tree **Use:** AF, EU, FU, GS, HF, MA, ME, PO, SU **Origin:** Cultivated **Region:** Andean, Caribbean, Orinoquia, Pacific **Dept.:** AMA, ANT, ARA, ATL, BOL, BOY, CAU, CES, CHO, COR, CUN, HUI, MAG, MET, NSA, SAP, SAN, SUC, TOL, VAC **Elev.:** 0–1500 m a.s.l. **IUCN Habitat code:** 1,2,3,5,14 **Conservation status: (IUCN)** LC

6068 *Terminalia grandis* (Ducke) Gere & Boatwr. (Combretaceae) **IPNI:** 77164566-1 **Common name:** bocarique, tanimboca, tanimbuco **Habit:** Tree **Use:** HF **Origin:** Native **Region:** Amazonia **Dept.:** AMA, VAU **Elev.:** 100–250 m a.s.l. **IUCN Habitat code:** 1 **Conservation status: (IUCN)** NE

6069 *Terminalia ivorensis* A.Chev. (Combretaceae) **IPNI:** 171174-1 **Habit:** Tree **Use:** EU, FU, MA, ME **Origin:** Cultivated **Region:** Andean **Dept.:** ANT, SAN **Elev.:** 1020–1500 m a.s.l. **IUCN Habitat code:** 1,14 **Conservation status: (IUCN)** VU

6070 *Terminalia lucida* Hoffmanns. ex Mart. (Combretaceae) **IPNI:** 171207-1 **Habit:** Tree **Use:** MA **Origin:** Native **Region:** Caribbean **Dept.:** ANT **Elev.:** 0–0 m a.s.l. **IUCN Habitat code:** 1,5,14 **Conservation status: (IUCN)** LC

6071 *Terminalia macrophylla* (Spruce ex Eichler) Gere & Boatwr. (Combretaceae) **IPNI:** 77164571-1 **Common name:** cumé de rebalse, palo de Brasil, palo de brazil **Habit:** Tree **Use:** HF **Origin:** Native **Region:** Amazonia, Pacific **Dept.:** AMA, CAQ, VAC, VAU **Elev.:** 50–400 m a.s.l. **IUCN Habitat code:** 1,3 **Conservation status: (IUCN)** LC

6072 *Terminalia oblonga* (Ruiz & Pav.) Steud. (Combretaceae) **IPNI:** 171280-1 **Common name:** guayabillo, guayabito, guayabo, vara de león **Habit:** Tree **Use:** MA **Origin:** Native **Region:** Amazonia, Caribbean, Pacific **Dept.:** AMA, ANT, BOL, BOY, CAQ, CHO, COR, LAG, MAG, SAN, VAC, VAU **Elev.:** 30-900 m.a.s.l. **IUCN Habitat code:** 1,3,5,14 **Conservation status: (IUCN)** LC **(CPSG)** Potential LC

6073 *Terminalia oxycarpa* Mart. (Combretaceae) **IPNI:** 171300-1 **Habit:** Tree **Use:** HF **Origin:** Native **Region:** Amazonia, Guiana Shield **Dept.:** AMA, GUA **Elev.:** 100–100 m a.s.l. **IUCN Habitat code:** 1 **Conservation status: (IUCN)** NE

6074 *Terminalia parvifolia* (Ducke) Gere & Boatwr. (Combretaceae) **IPNI:** 77164576-1 **Habit:** Shrub, Tree **Use:** HF **Origin:** Native **Region:** Amazonia **Dept.:** AMA, CAQ, VAU **Elev.:** 100–550 m a.s.l. **IUCN Habitat code:** 1,3,4,14 **Conservation status: (IUCN)** LC

6075 *Terminalia tetraphylla* (Aubl.) Gere & Boatwr. (Combretaceae) **IPNI:** 77164595-1 **Common name:** almendro, amarillón serrano, balauruma, capacho, granadillo, guayabillo negro, guayacán mochilero, macano, macano negro, mulí, olivo negro, parasiempre, piedro amarillo, tiricio, vacaricuara, vacaricuara amarillo **Habit:** Tree **Use:** MA, ME **Origin:** Native **Region:** Amazonia, Andean, Caribbean **Dept.:** AMA, ANT, CUN, LAG, MET, SAN, VAU **Elev.:** 100–1420 m a.s.l. **IUCN Habitat code:** 1,3,4,5,14 **Conservation status: (IUCN)** NE

6076 *Adenaria floribunda* Kunth (Lythraceae) **IPNI:** 30023859-2 **Common name:** amor de zumba, arrayán, chaparral, chaparro, colorado, colorado chilacó, coralito, guayabillo, guayabito, guayabo, guayabo colorao, lumbí **Habit:** Shrub, Small tree, Tree **Use:** MA, ME **Origin:** Native **Region:** Amazonia, Andean, Pacific **Dept.:** AMA, ANT, BOY, CHO, CUN, HUI, NSA, SAN, TOL, VAC **Elev.:** 50–2800 m a.s.l. **IUCN Habitat code:** 1,2,3,5,14 **Conservation status: (IUCN)** LC

6077 *Ammannia auriculata* Willd. (Lythraceae) **IPNI:** 30030870-2 **Habit:** Herb **Use:** ME **Origin:** Native **Region:** Caribbean **Dept.:** BOL, CAL, CUN, LAG, MAG, TOL **Elev.:** 7–350 m a.s.l. **IUCN Habitat code:** 14 **Conservation status: (IUCN)** LC

6078 *Cuphea calophylla* Cham. & Schltdl. (Lythraceae) **IPNI:** 552922-1 **Habit:** Herb **Use:** ME **Origin:** Native **Region:** Pacific **Dept.:** CHO **Elev.:** 60–250 m a.s.l. **IUCN Habitat code:** 1,3 **Conservation status: (IUCN)** NE

6079 *Cuphea carthagenensis* (Jacq.) J.F.Macbr. (Lythraceae) **IPNI:** 72326-2 **Common name:** San Antonio, colombian waxweed, moradita, sanantonio, siete sangrías **Habit:** Herb, Subshrub **Use:** FU, ME **Origin:** Native **Region:** Andean, Caribbean, Orinoquia, Pacific **Dept.:** ANT, ATL, BOL, BOY, CAU, CHO, CUN, HUI, LAG, MET, NAR, RIS, SAN, SUC, TOL, VAC **Elev.:** 0-2600 m.a.s.l. **IUCN Habitat code:** 1,2,3,5,14 **Conservation status: (IUCN)** NE **(CPSG)** Potential LC

6080 *Cuphea ciliata* Ruiz & Pav. (Lythraceae) **IPNI:** 552931-1 **Common name:** herbbuenilla, moradita, sanalotodo, yerbabuenilla **Habit:** Herb **Use:** PO **Origin:** Native **Dept.:** ANT, BOG, BOY, CUN, NSA, SAN **Elev.:** 1100–3500 m a.s.l. **IUCN Habitat code:** 1,3,4,14 **Conservation status: (IUCN)** NE

6081 *Cuphea decandra* W.T.Aiton (Lythraceae) **IPNI:** 552959-1 **Habit:** Subshrub **Use:** ME **Origin:** Native **Region:** Andean **Dept.:** CUN,

HUI, TOL **Elev.:** 900–1600 m a.s.l. **IUCN Habitat code:** 2,3,14 **Conservation status: (IUCN)** NE

6082 *Cuphea dipetala* (L.f.) Koehne (Lythraceae)
IPNI: 552967-1
Common name: chinchimaní, moradita, zarcillejo **Habit:** Shrub **Use:** ME **Origin:** Native **Region:** Andean **Dept.:** BOG, BOY, CUN, NSA, SAN **Elev.:** 1400–3215 m a.s.l. **IUCN Habitat code:** 1,3,4,5,14 **Conservation status: (IUCN)** NE

6083 *Cuphea hyssopifolia* Kunth (Lythraceae)
IPNI: 553043-1
Common name: arbolito de navidad, carpintero, cupea, diosma, escoba, escubilla, escudilla, frailejón, gatico, gato, moradita, morita de alambre, navidad, patepalomo, pichindecillo, púa, sabañón, sanalotodo, verbenilla, yerbabuenilla **Habit:** Herb, Shrub **Use:** EU, HF **Origin:** Cultivated, Naturalised **Region:** Andean, Pacific **Dept.:** ANT, BOG, BOY, CHO, SAN, VAC **Elev.:** 340–2300 m a.s.l. **IUCN Habitat code:** 1,2,3,4,14 **Conservation status: (IUCN)** NE

6084 *Cuphea ignea* A.DC. (Lythraceae)
IPNI: 553048-1
Common name: cigarrillitos **Habit:** Herb, Subshrub **Use:** EU, ME **Origin:** Cultivated, Naturalised **Region:** Andean **Dept.:** ANT, BOG, CHO, CUN **Elev.:** 2000–2400 m a.s.l. **IUCN Habitat code:** 1,2,3,14 **Conservation status: (IUCN)** NE

6085 *Cuphea kubeorum* Lourteig (Lythraceae)
IPNI: 72437-2
Habit: Herb, Subshrub **Use:** ME **Origin:** Endemic **Region:** Amazonia, Guiana Shield **Dept.:** AMA, CAQ, GUV, MET, VAU **Elev.:** 200–1300 m a.s.l. **IUCN Habitat code:** 1,3,4,14 **Conservation status: (IUCN)** NE

6086 *Cuphea micrantha* Kunth (Lythraceae)
IPNI: 553109-1
Common name: hierbabuenilla, moradita, rosadita, sanalotodo, verbenilla **Habit:** Subshrub **Use:** ME **Origin:** Native **Region:** Caribbean, Orinoquia **Dept.:** BOG, CES, MAG, MET, SAP **Elev.:** 10–300 m a.s.l. **IUCN Habitat code:** 1,2,3,14 **Conservation status: (IUCN)** NE

6087 *Cuphea micropetala* Kunth (Lythraceae)
IPNI: 553111-1
Use: EU, ME **Region:** Andean **Dept.:** BOG **Conservation status: (IUCN)** NE

6088 *Cuphea racemosa* (L.f.) Spreng. (Lythraceae)
IPNI: 290454-2
Common name: amoraús, escancerillo, escobilla, fraile, hierbabuenilla, moradita, rosadita, sanalotodo, sietoluchas, verbenilla, yerbabuena **Habit:** Herb, Subshrub **Use:** ME **Origin:** Native **Region:** Amazonia, Andean, Caribbean, Orinoquia, Pacific **Dept.:** AMA, ANT, BOG, BOY, CAU, CHO, CUN, HUI, LAG, MAG, MET, NAR, PUT, RIS, SAN, VAC **Elev.:** 40-3000 m.a.s.l. **IUCN Habitat code:** 1,2,3,14 **Conservation status: (IUCN)** NE **(CPSG)** Potential LC

6089 *Cuphea wrightii* A.Gray (Lythraceae)
IPNI: 553315-1
Habit: Herb **Use:** EU **Origin:** Cultivated **Region:** Andean **Dept.:** NAR **Elev.:** 2500–2500 m a.s.l. **Conservation status: (IUCN)** NE

6090 *Lafoensia acuminata* (Ruiz & Pav.) DC. (Lythraceae)
IPNI: 1054383-2
Common name: guayabo manizaleño, guayacán, guayacán amarillo, guayacán de Manizales, mirto, roble amarillo **Habit:** Tree **Use:** ME **Origin:** Native, Cultivated **Region:** Andean **Dept.:** ANT, BOG, BOY, CAL, CAU, CUN, HUI, NAR, VAC **Elev.:** 1200–3350 m a.s.l. **IUCN Habitat code:** 1,3,14 **Conservation status: (IUCN)** LC

6091 *Lafoensia punicifolia* DC. (Lythraceae)
IPNI: 553519-1
Common name: arrayán escobo, guayacán, guayacán de Manizales, guayacán del Cauca, pelincú **Habit:** Tree **Use:** EU, MA **Origin:** Native, Cultivated **Region:** Caribbean, Pacific **Dept.:** BOG, BOL, CES, CHO, MAG, SUC **Elev.:** 80–1130 m a.s.l. **IUCN Habitat code:** 1,2,3,14 **Conservation status: (IUCN)** LC

6092 *Lagerstroemia indica* L. (Lythraceae)
IPNI: 553576-1
Common name: astromelia, astromelia púrpura, astromelia roja, astromelio, astromelio rosado, estromelia, gólgota, júpiter **Habit:** Tree **Use:** AF, EU, FU, HF, MA, ME **Origin:** Cultivated **Region:** Andean **Dept.:** ANT, BOY, CUN **Elev.:** 600–1800 m a.s.l. **IUCN Habitat code:** 1,3,14 **Conservation status: (IUCN)** LC

6093 *Lagerstroemia speciosa* (L.) Pers. (Lythraceae)
IPNI: 553618-1
Common name: astromelia, embrujo de la India, flor de la reina, flor de reina, orgullo de la India, reina de las flores, zaral **Habit:** Tree **Use:** EU, HF, MA, ME, PO, SU **Origin:** Cultivated **Region:** Andean, Pacific **Dept.:** ANT, CHO **Elev.:** 0–1500 m a.s.l. **IUCN Habitat code:** 1,2,3,14 **Conservation status: (IUCN)** NE

6094 *Lawsonia inermis* L. (Lythraceae)
IPNI: 553638-1
Common name: jazmín, mielito, resedad, resedo, resedá **Habit:** Shrub, Tree **Use:** AF, EU, FU, HF, IF, MA, ME, PO, SU **Origin:** Cultivated **Region:** Caribbean, Pacific **Dept.:** ANT, BOL, CES, CHO, CUN, MAG **Elev.:** 20–450 m a.s.l. **IUCN Habitat code:** 1,2,8,14 **Conservation status: (IUCN)** LC

6095 *Lythrum maritimum* Kunth (Lythraceae)
IPNI: 553756-1
Habit: Herb **Use:** ME **Origin:** Native **Region:** Andean **Dept.:** BOG, BOY, CUN **Elev.:** 1990–2700 m a.s.l. **IUCN Habitat code:** 3,14 **Conservation status: (IUCN)** NE

6096 *Punica granatum* L. (Lythraceae)
IPNI: 554129-1
Common name: granada, granado **Habit:** Shrub **Use:** EU, FU, GS, HF, MA, ME, PO, SU **Origin:** Cultivated **Region:** Andean **Dept.:** ANT, BOY, VAC **Elev.:** 1000–2775 m a.s.l. **IUCN Habitat code:** 1,2,3,4,14 **Conservation status: (IUCN)** LC

6097 *Aciotis acuminifolia* (Mart. ex DC.) Triana (Melastomataceae)
IPNI: 1025557-2
Common name: cañagria **Habit:** Herb, Shrub, Aquatic **Use:** ME **Origin:** Native **Region:** Amazonia, Andean, Orinoquia, Pacific **Dept.:** AMA, ANT, BOY, CAQ, CAS, CHO, GUV, MET, NAR, NSA, SAN, VAC, VAU, VID **Elev.:** 0-1300 m.a.s.l. **IUCN Habitat code:** 1,2,3,4,5,14 **Conservation status: (IUCN)** NE **(CPSG)** Potential LC

6098 *Aciotis purpurascens* (Aubl.) Triana (Melastomataceae)
IPNI: 1031600-2
Common name: cañagria, chulpa, doradilla, fruncida, mora, morilla, pulmonaria **Habit:** Herb, Shrub, Subshrub **Use:** HF, ME **Origin:** Native **Region:** Amazonia, Andean, Guiana Shield, Orinoquia, Pacific **Dept.:** AMA, ANT, BOY, CAL, CAQ, CAS, CAU, CHO, COR, CUN, GUV, MAG, MET, NAR, NSA, PUT, SAN, TOL, VAC, VAU, VID **Elev.:** 0-2700 m.a.s.l. **IUCN Habitat code:** 1,2,3,4,5,14 **Conservation status: (IUCN)** NE **(CPSG)** Potential LC

6099 *Aciotis rubricaulis* (Mart. ex DC.) Triana (Melastomataceae)
IPNI: 1101289-2
Common name: caña agria chiquita, chulco **Habit:** Herb, Shrub, Subshrub **Use:** ME **Origin:** Native **Region:** Amazonia, Andean, Caribbean, Orinoquia, Pacific **Dept.:** AMA, ANT, CAU, CHO, COR, MAG, MET, NAR, PUT, RIS, SAN, VAC **Elev.:** 0–2000 m a.s.l. **IUCN Habitat code:** 1,3,14 **Conservation status: (IUCN)** NE

6100 *Adelobotrys adscendens* Triana (Melastomataceae)
IPNI: 566051-1
Common name: hoja de solimán, hoja dulce, joquetaramo, yerba dulce **Habit:** Shrub, Epiphyte, Hemiepiphyte, Liana, Climbing **Use:** ME, SU **Origin:** Native **Region:** Amazonia, Andean, Caribbean, Pacific **Dept.:** ANT, BOL, CAQ, CAU, CHO, CUN, MAG, NAR, NSA, PUT, RIS, SAN, VAC **Elev.:** 0-1370 m.a.s.l. **IUCN Habitat code:** 1,3,5,14 **Conservation status: (IUCN)** NE **(CPSG)** Potential LC

6101 *Andesanthus lepidotus* (Humb. & Bonpl.) P.J.F.Guim. & Michelang. (Melastomataceae)
IPNI: 77205900-1
Common name: lengüebuey, mayo, pucasacha, sanjuanito, sietecueros, sietecueros real **Habit:** Shrub, Small tree, Tree **Use:** MA, ME **Origin:** Endemic **Region:** Andean **Dept.:** ANT, BOY, CAL, CAQ, CAU, CHO, CUN, HUI, MET, NAR, NSA, PUT, QUI, RIS, SAN, TOL, VAC **Elev.:** 400-3990 m.a.s.l. **IUCN Habitat code:** 1,3,4 **Conservation status: (IUCN)** NE **(CPSG)** Potential LC

6102 *Arthrostemma ciliatum* Pav. ex D.Don (Melastomataceae)
IPNI: 60465985-2
Common name: cañagri, cañagria, cañuela, chulco, cuadrada, pilpe, yerbaesiote, yuyo **Habit:** Herb, Shrub, Subshrub, Small tree, Climbing **Use:** HF, ME **Origin:** Native **Region:** Amazonia, Andean, Caribbean, Orinoquia, Pacific **Dept.:** ANT, BOY, CAL, CAQ, CAU, CHO, CUN, LAG, MAG, MET, NAR, NSA, PUT, QUI, RIS, SAN, TOL, VAC **Elev.:** 15-2260 m.a.s.l. **IUCN Habitat code:** 1,2,3,14 **Conservation status: (IUCN)** NE **(CPSG)** Potential LC

6103 *Bellucia acutata* Pilg. (Melastomataceae)
IPNI: 273207-2
Habit: Shrub, Small tree **Use:** HF **Origin:** Native **Region:** Guiana Shield **Dept.:** GUA **Elev.:** 100–100 m a.s.l. **IUCN Habitat code:** 1 **Conservation status: (IUCN)** NE

6104 *Bellucia arborescens* Baill. (Melastomataceae)
IPNI: 566597-1
Habit: Small tree, Tree **Use:** HF **Origin:** Native **Region:** Amazonia, Guiana Shield **Dept.:** AMA, CAQ, VAU,

VID **Elev.:** 140–259 m a.s.l. **IUCN Habitat code:** 1,4,5 **Conservation status: (IUCN)** NE

6105 *Bellucia egensis* (DC.) Penneys, Michelang., Judd & Almeda (Melastomataceae) **IPNI:** 77112284-1 **Common name:** tuno **Habit:** Shrub, Small tree, Tree, Liana **Use:** HF, ME **Origin:** Native **Region:** Amazonia, Guiana Shield **Dept.:** AMA, GUA, MET, VAU **Elev.:** 240–427 m a.s.l. **IUCN Habitat code:** 1,3,4,14 **Conservation status: (IUCN)** LC

6106 *Bellucia grossularioides* (L.) Triana (Melastomataceae) **IPNI:** 1157663-2 **Common name:** anacao, chompa, coronilla, coronillo, grosella, guayaba de danta, guayaba de pava, guayabilla, guayabo, guayabo cimarrón, guayabo de mico, guayabo de mono, guayabo de monte, guayabo de pava, guayabo silvestre, manzanilla, míspero, níspero, pomarrosa, poma **Habit:** Shrub, Small tree, Tree **Use:** EU, HF, MA, ME **Origin:** Native **Region:** Amazonia, Andean, Guiana Shield, Orinoquia **Dept.:** AMA, ANT, BOL, BOY, CAQ, CAS, CAU, CUN, GUV, MET, NAR, PUT, SAN, VAU, VID **Elev.:** 0-2000 m.a.s.l. **IUCN Habitat code:** 1,2,3,4,5,14 **Conservation status: (IUCN)** LC **(CPSG)** Potential LC

6107 *Bellucia mespiloides* (Miq.) J.F.Macbr. (Melastomataceae) **IPNI:** 31112-2 **Habit:** Shrub, Tree **Use:** HF, MA **Origin:** Native **Region:** Amazonia, Pacific **Dept.:** ANT, CHO, PUT **Elev.:** 380–600 m a.s.l. **IUCN Habitat code:** 1,3,14 **Conservation status: (IUCN)** LC

6108 *Bellucia pentamera* Naudin (Melastomataceae) **IPNI:** 323200-2 **Common name:** anselmo, bispera, coronilla, coronillo, cueruda, dakali, dakalinae, decalinae, guayaba de coronilla, guayaba de monte, guayaba de pava, guayabillo, guayabita peruana, guayabo coronillo, guayabo de mico, guayabo de monte, guayabo de pava, guayabo mico, gua **Habit:** Shrub, Small tree, Tree **Use:** EU, GS, HF, MA, ME **Origin:** Native **Region:** Amazonia, Andean, Guiana Shield, Pacific **Dept.:** AMA, ANT, BOY, CAL, CAQ, CAS, CAU, CHO, CUN, GUV, LAG, MET, NAR, NSA, PUT, RIS, SAN, TOL, VAC **Elev.:** 0-1600 m.a.s.l. **IUCN Habitat code:** 1,3,4,5,14 **Conservation status: (IUCN)** LC **(CPSG)** Potential LC

6109 *Bellucia spruceana* (Triana) J.F.Macbr. (Melastomataceae) **IPNI:** 31113-2 **Habit:** Small tree, Tree **Use:** HF **Origin:** Native **Region:** Andean **Dept.:** ANT, LAG **Elev.:** 330–800 m a.s.l. **IUCN Habitat code:** 1,3,14 **Conservation status: (IUCN)** LC

6110 *Blakea alternifolia* Gleason (Melastomataceae) **IPNI:** 33356-2 **Common name:** quereme menudito **Habit:** Shrub, Small tree, Hemiepiphyte, Climbing **Use:** ME **Origin:** Native **Region:** Andean, Pacific **Dept.:** CHO, NAR, VAC **Elev.:** 5–2080 m a.s.l. **IUCN Habitat code:** 1,3,14 **Conservation status: (IUCN)** NE

6111 *Blakea bracteata* Gleason (Melastomataceae) **IPNI:** 33365-2 **Habit:** Shrub, Small tree, Epiphyte, Climbing **Use:** HF **Origin:** Native **Region:** Amazonia, Andean **Dept.:** CAQ, CAU, PUT **Elev.:** 230–1460 m a.s.l. **IUCN Habitat code:** 1,3,14 **Conservation status: (IUCN)** NE

6112 *Blakea brasiliensis* Cogn. (Melastomataceae) **IPNI:** 33367-2 **Habit:** Shrub, Tree, Epiphyte **Use:** HF **Origin:** Native **Region:** Amazonia **Dept.:** AMA, CAQ, MET, NAR, VAU **Elev.:** 150–250 m a.s.l. **IUCN Habitat code:** 1,14 **Conservation status: (IUCN)** NE

6113 *Blakea quadrangularis* Triana (Melastomataceae) **IPNI:** 566748-1 **Common name:** miona, mioncita, mión, nigüito, tiraagua, tiramiaos **Habit:** Shrub, Small tree, Tree, Liana, Climbing **Use:** HF **Origin:** Endemic **Region:** Andean **Dept.:** ANT, SAN **Elev.:** 400-2900 m.a.s.l. **IUCN Habitat code:** 1,3,14 **Conservation status: (IUCN)** LC **(CPSG)** LC

6114 *Blakea rosea* D.Don (Melastomataceae) **IPNI:** 323201-2 **Common name:** morochillo **Habit:** Shrub, Epiphyte, Hemiepiphyte, Liana, Climbing **Use:** HF **Origin:** Native **Region:** Amazonia, Orinoquia **Dept.:** AMA, ANT, BOY, CAQ, CUN, HUI, MET, NSA, PUT, SAN, VAU **Elev.:** 100–850 m a.s.l. **IUCN Habitat code:** 1,3,5,14 **Conservation status: (IUCN)** LC

6115 *Blakea subconnata* O.Berg ex Triana (Melastomataceae) **IPNI:** 566768-1 **Common name:** matapalo **Habit:** Shrub, Small tree, Tree, Epiphyte, Climbing **Use:** HF **Origin:** Native **Region:** Andean, Pacific **Dept.:** ANT, CAU, CHO, NAR, RIS, VAC **Elev.:** 5–2200 m a.s.l. **IUCN Habitat code:** 1,3,14 **Conservation status: (IUCN)** LC

6116 *Brachyotum ledifolium* (Desr.) Triana (Melastomataceae) **IPNI:** 566862-1 **Common name:** almorrana, pichanga, zarcillejo **Habit:** Shrub **Use:** HF **Origin:** Native **Region:** Andean **Dept.:** ANT, BOG, BOY, CAU, CUN, HUI, MET, NAR, PUT, TOL, VAC **Elev.:** 2700–4500 m a.s.l. **IUCN Habitat code:** 1,3,4,8,14 **Conservation status: (IUCN)** NE

6117 *Brachyotum strigosum* (L.f.) Triana (Melastomataceae) **IPNI:** 566891-1 **Common name:** almorrana, moradito, tuno, zarcillejo, zarcillejo morado, zarcillo **Habit:** Shrub, Climbing **Use:** EU, ME **Origin:** Native **Region:** Andean **Dept.:** ANT, BOG, BOY, CUN, MET, QUI **Elev.:** 2500-3900 m.a.s.l. **IUCN Habitat code:** 1,3,4,5,6,8,14 **Conservation status: (IUCN)** NE **(CPSG)** LC

6118 *Bucquetia glutinosa* DC. (Melastomataceae) **IPNI:** 566936-1 **Common name:** angelito, ayuela, canelo, charne, charne saltón, chispeador, chíspero, guayabo, quebraolla, quebrollo, quebroollo, quiebraollo, saltón, sietecueros, tuno **Habit:** Shrub, Small tree, Tree **Use:** EU, MA **Origin:** Endemic **Region:** Andean **Dept.:** ANT, BOG, BOY, CUN, MET, SAN **Elev.:** 2100-4104 m.a.s.l. **IUCN Habitat code:** 1,3,4,5,8,14 **Conservation status: (IUCN)** LC **(CPSG)** LC

6119 *Chaetolepis lindeniana* Triana (Melastomataceae) **IPNI:** 567327-1 **Common name:** chinchimaní, doradita **Habit:** Shrub **Use:** ME **Origin:** Native **Region:** Andean **Dept.:** NSA, SAN **Elev.:** 2500–3800 m a.s.l. **IUCN Habitat code:** 1,4,14 **Conservation status: (IUCN)** NE

6120 *Chaetolepis microphylla* Miq. (Melastomataceae) **IPNI:** 567329-1 **Common name:** chimachimalí, chinchimalí, chinchimalía, doradilla, doradillo, flor de muerto, injerta de tierra, rema, venadillo **Habit:** Herb, Shrub, Subshrub **Use:** ME **Origin:** Native **Region:** Andean **Dept.:** ANT, BOG, BOY, CAU, CUN, NSA, SAN **Elev.:** 2000–3700 m a.s.l. **IUCN Habitat code:** 1,3,4,8,14 **Conservation status: (IUCN)** NE

6121 *Chalybea corymbifera* Naudin (Melastomataceae) **IPNI:** 567365-1 **Habit:** Small tree, Tree **Use:** HF **Origin:** Endemic **Region:** Andean **Dept.:** NSA, SAN **Elev.:** 2170-3500 m.a.s.l. **IUCN Habitat code:** 1 **Conservation status: (IUCN)** EN **(CPSG)** EN

6122 *Chalybea macrocarpa* (L.Uribe) M.E.Morales & Penneys (Melastomataceae) **IPNI:** 77146610-1 **Common name:** tuno **Habit:** Tree **Use:** HF, ME **Origin:** Endemic **Region:** Andean **Dept.:** BOY **Elev.:** 2350-2850 m.a.s.l. **IUCN Habitat code:** 1,3,4,14 **Conservation status: (IUCN)** EN **(CPSG)** EN

6123 *Desmoscelis villosa* (Aubl.) Naudin (Melastomataceae) **IPNI:** 1168615-2 **Habit:** Herb, Shrub, Subshrub **Use:** ME **Origin:** Native **Region:** Amazonia, Andean, Guiana Shield, Orinoquia **Dept.:** AMA, ANT, BOY, CAQ, CAS, GUA, MET, NSA, PUT, SAN, VAU, VID **Elev.:** 65–2250 m a.s.l. **IUCN Habitat code:** 1,2,3,4,5,14 **Conservation status: (IUCN)** LC

6124 *Graffenrieda galeottii* (Naudin) L.O.Williams (Melastomataceae) **IPNI:** 112942-2 **Habit:** Shrub, Small tree, Tree **Use:** HF **Origin:** Native **Region:** Andean, Guiana Shield **Dept.:** ANT, MET, RIS, TOL, VAC **Elev.:** 200–1650 m a.s.l. **IUCN Habitat code:** 1,3,5,14 **Conservation status: (IUCN)** LC

6125 *Graffenrieda rupestris* Ducke (Melastomataceae) **IPNI:** 112967-2 **Habit:** Tree **Use:** ME **Origin:** Native **Region:** Guiana Shield **Dept.:** AMA, CAQ, GUA, GUV, VAU **Elev.:** 110–800 m a.s.l. **IUCN Habitat code:** 1,4,14 **Conservation status: (IUCN)** LC

6126 *Henriettea ramiflora* (Sw.) DC. (Melastomataceae) **IPNI:** 569004-1 **Habit:** Tree **Use:** HF, MA **Origin:** Native **Region:** Caribbean **Dept.:** NSA **Elev.:** 330–330 m a.s.l. **IUCN Habitat code:** 1,14 **Conservation status: (IUCN)** NE

6127 *Henriettea stellaris* O.Berg ex Triana (Melastomataceae) **IPNI:** 569013-1 **Habit:** Small tree, Tree, Climbing **Use:** HF **Origin:** Native **Region:** Amazonia, Guiana Shield **Dept.:** AMA, CAQ, PUT, VAU, VID **Elev.:** 100–250 m a.s.l. **IUCN Habitat code:** 1,4,5 **Conservation status: (IUCN)** LC

6128 *Henriettea succosa* DC. (Melastomataceae) **IPNI:** 569017-1 **Habit:** Shrub, Tree **Use:** HF, MA **Origin:** Native **Region:** Andean **Dept.:** LAG **Elev.:** 500–1000 m a.s.l. **IUCN Habitat code:** 1 **Conservation status: (IUCN)** LC

6129 *Heterocentron elegans* (Schltdl.) Kuntze (Melastomataceae) **IPNI:** 121047-2 **Habit:** Herb **Use:** EU **Origin:** Cultivated, Naturalised **Region:** Andean **Dept.:** BOG, CUN **Elev.:** 2600–2600 m a.s.l. **Conservation status: (IUCN)** NE

6130 *Heterocentron glandulosum* Schrenk (Melastomataceae) **IPNI:** 569094-1 **Habit:** Shrub **Use:** EU **Origin:** Native, Cultivated **Region:** Andean **Dept.:** ANT **Elev.:** 1500–1500 m a.s.l. **Conservation status: (IUCN)** NE

6131 *Heterocentron subtriplinervium* (Link & Otto) A.Braun & C.D.Bouché (Melastomataceae) **IPNI:** 569110-1 **Habit:** Shrub, Tree **Use:** EU **Origin:** Cultivated, Naturalised **Region:** Andean **Dept.:** ANT **Elev.:** 1500–1500 m a.s.l. **IUCN Habitat code:** 3,14 **Conservation status: (IUCN)** NE

6132 *Heterotis rotundifolia* (Sm.) Jacq.-Fél. (Melastomataceae) **IPNI:** 908858-1 **Habit:** Herb **Use:** EU, GS, HF, ME, SU **Origin:** Cultivated, Naturalised **IUCN Habitat code:** 1,14 **Conservation status: (IUCN)** LC

6133 *Maieta poeppigii* Mart. ex Cogn. (Melastomataceae) **IPNI:** 77178142-1 **Habit:** Shrub, Subshrub **Use:** HF **Origin:** Endemic **Region:** Amazonia **Dept.:** AMA, ANT, CAQ, VAU **Elev.:** 100–700 m a.s.l. **IUCN Habitat code:** 1,3,5,14 **Conservation status: (IUCN)** NE

6134 *Miconia abbreviata* Markgr. (Melastomataceae) **IPNI:** 158668-2 **Common name:** miguel colorado **Habit:** Shrub, Subshrub, Small tree **Use:** HF **Origin:** Native **Region:** Amazonia **Dept.:** CAQ, PUT **Elev.:** 150–300 m a.s.l. **IUCN Habitat code:** 1,3,14 **Conservation status: (IUCN)** NT

6135 *Miconia acostae* (Wurdack) Michelang. (Melastomataceae) **IPNI:** 77207432-1 **Habit:** Shrub **Use:** HF **Origin:** Endemic **Region:** Pacific **Dept.:** CHO **Elev.:** 300–900 m a.s.l. **IUCN Habitat code:** 1 **Conservation status: (IUCN)** VU

6136 *Miconia aeruginosa* Naudin (Melastomataceae) **IPNI:** 158684-2 **Common name:** mortiño, nigüito **Habit:** Shrub, Small tree, Tree **Use:** HF **Origin:** Native **Region:** Andean, Caribbean, Orinoquia **Dept.:** ANT, BOL, BOY, CAL, CAQ, CAS, CAU, CES, CUN, HUI, LAG, MAG, NAR, NSA, QUI, RIS, SAN, TOL, VAC **Elev.:** 40-2600 m.a.s.l. **IUCN Habitat code:** 1,2,3,4,14 **Conservation status: (IUCN)** LC **(CPSG)** Potential LC

6137 *Miconia affinis* DC. (Melastomataceae) **IPNI:** 158685-2 **Common name:** migüito, morochillo, tresrayas, zafiro **Habit:** Shrub, Small tree, Tree, Climbing **Use:** HF, ME **Origin:** Native **Region:** Andean, Orinoquia, Pacific **Dept.:** AMA, ANT, CAL, CAQ, CAS, CHO, CUN, GUV, HUI, MET, NAR, NSA, PUT, RIS, VAC, VID **Elev.:** 20-2200 m.a.s.l. **IUCN Habitat code:** 1,3,4,5,14 **Conservation status: (IUCN)** LC **(CPSG)** Potential LC

6138 *Miconia albicans* (Sw.) Steud. (Melastomataceae) **IPNI:** 1125568-2 **Common name:** cenizo, chilco, esmeralda, mortiño, tuno, tuno sabanero **Habit:** Herb, Shrub, Subshrub, Small tree **Use:** HF, MA, ME **Origin:** Native **Region:** Andean, Caribbean, Orinoquia **Dept.:** ANT, ARA, BOL, BOY, CAQ, CAS, CAU, CES, COR, CUN, GUV, HUI, MAG, MET, NAR, NSA, SAN, TOL, VAC, VID **Elev.:** 25-2500 m.a.s.l. **IUCN Habitat code:** 1,2,3,4,5,14 **Conservation status: (IUCN)** LC **(CPSG)** Potential LC

6139 *Miconia allardii* (Wurdack) Michelang. (Melastomataceae) **IPNI:** 77207407-1 **Habit:** Shrub, Subshrub **Use:** HF **Origin:** Endemic **Region:** Amazonia **Dept.:** AMA, CAQ, CAU, PUT **Elev.:** 100–820 m a.s.l. **IUCN Habitat code:** 1,14 **Conservation status: (IUCN)** LC

6140 *Miconia amazonica* Triana (Melastomataceae) **IPNI:** 572075-1 **Habit:** Small tree, Tree **Use:** HF **Origin:** Native **Region:** Amazonia **Dept.:** AMA, CAQ **Elev.:** 100–120 m a.s.l. **IUCN Habitat code:** 1,3,14 **Conservation status: (IUCN)** LC

6141 *Miconia ampla* Triana (Melastomataceae) **IPNI:** 572080-1 **Common name:** morochillo **Habit:** Shrub, Tree **Use:** MA **Origin:** Native **Region:** Andean, Guiana Shield, Pacific **Dept.:** ANT, CAQ, CAU, CHO, MET, PUT **Elev.:** 50–1020 m a.s.l. **IUCN Habitat code:** 1,3,14 **Conservation status: (IUCN)** LC

6142 *Miconia aplostachya* (Humb. & Bonpl.) DC. (Melastomataceae) **IPNI:** 1039241-2 **Common name:** tunito **Habit:** Herb, Shrub, Subshrub, Small tree, Tree **Use:** MA **Origin:** Native **Region:** Amazonia, Orinoquia **Dept.:** AMA, CAQ, GUA, GUV, MET, VAU, VID **Elev.:** 80–500 m a.s.l. **IUCN Habitat code:** 1,2,3,4,5,14 **Conservation status: (IUCN)** LC

6143 *Miconia argentea* DC. (Melastomataceae) **IPNI:** 572109-1 **Habit:** Small tree, Tree **Use:** HF, MA **Origin:** Native **Region:** Caribbean, Orinoquia **Dept.:** MET, SAP **Elev.:** 200–200 m a.s.l. **IUCN Habitat code:** 1,14 **Conservation status: (IUCN)** LC

6144 *Miconia argyrophylla* DC. (Melastomataceae) **IPNI:** 158734-2 **Habit:** Shrub, Small tree **Use:** HF **Origin:** Native **Region:** Guiana Shield **Dept.:** MET, VAU **Elev.:** 350–1000 m a.s.l. **IUCN Habitat code:** 1,3 **Conservation status: (IUCN)** LC

6145 *Miconia aristigera* (Naudin) R.Goldenb. & Michelang. (Melastomataceae) **IPNI:** 77207374-1 **Habit:** Herb, Shrub, Small tree **Use:** HF **Origin:** Endemic **Region:** Amazonia, Andean **Dept.:** AMA, ANT, CAQ, NAR, PUT, VAU **Elev.:** 100–935 m a.s.l. **IUCN Habitat code:** 1,4 **Conservation status: (IUCN)** NE

6146 *Miconia asclepiadea* Triana (Melastomataceae) **IPNI:** 572118-1 **Habit:** Herb, Shrub, Subshrub, Climbing **Use:** HF **Origin:** Native **Region:** Andean **Dept.:** ANT, CAL, CAQ, HUI, NAR, PUT **Elev.:** 1200–2600 m a.s.l. **IUCN Habitat code:** 1,3,14 **Conservation status: (IUCN)** NE

6147 *Miconia asperrima* Triana (Melastomataceae) **IPNI:** 572120-1 **Habit:** Shrub, Small tree, Tree **Use:** HF **Origin:** Native **Region:** Andean **Dept.:** ANT, BOY, CAQ, CAU, HUI, MET, PUT, RIS, SAN, TOL, VAC **Elev.:** 1200–3000 m a.s.l. **IUCN Habitat code:** 1,3,14 **Conservation status: (IUCN)** NE

6148 *Miconia astroplocama* Donn.Sm. (Melastomataceae) **IPNI:** 158746-2 **Habit:** Shrub, Small tree **Use:** HF **Origin:** Native **Region:** Andean **Dept.:** SAN **Elev.:** 1260–1260 m a.s.l. **IUCN Habitat code:** 1,14 **Conservation status: (IUCN)** LC

6149 *Miconia biappendiculata* (Naudin) L.Uribe (Melastomataceae) **IPNI:** 158784-2 **Common name:** sietecueros **Habit:** Tree **Use:** MA **Origin:** Endemic **Region:** Andean **Dept.:** BOG, BOY, CAL, CAU, CUN, HUI, MAG, PUT, VAC **Elev.:** 2500-3650 m.a.s.l. **IUCN Habitat code:** 1,2,3,4,8,14 **Conservation status: (IUCN)** LC **(CPSG)** LC

6150 *Miconia biglandulosa* Gleason (Melastomataceae) **IPNI:** 158788-2 **Habit:** Shrub, Small tree, Tree **Use:** HF **Origin:** Native **Region:** Amazonia, Guiana Shield **Dept.:** AMA, CAQ, PUT **Elev.:** 100–300 m a.s.l. **IUCN Habitat code:** 1,3,5,14 **Conservation status: (IUCN)** NE

6151 *Miconia bracteolata* DC. (Melastomataceae) **IPNI:** 572199-1 **Habit:** Tree **Use:** HF **Origin:** Native **Region:** Andean **Dept.:** CAL, NAR, QUI, TOL **Elev.:** 2800–3600 m a.s.l. **IUCN Habitat code:** 1,3,4,8,14 **Conservation status: (IUCN)** NE

6152 *Miconia bubalina* Naudin (Melastomataceae) **IPNI:** 572209-1 **Habit:** Shrub, Small tree, Tree **Use:** HF **Origin:** Native **Region:** Amazonia, Andean, Orinoquia, Pacific **Dept.:** AMA, CAL, CAU, CHO, MET, PUT, RIS **Elev.:** 100–600 m a.s.l. **IUCN Habitat code:** 1,3,5,14 **Conservation status: (IUCN)** LC

6153 *Miconia bullosa* (DC.) Michelang. (Melastomataceae) **IPNI:** 77207390-1 **Habit:** Herb, Shrub **Use:** MA **Origin:** Endemic **Region:** Amazonia, Andean, Guiana Shield **Dept.:** AMA, ANT, CAQ, SAN, VAU **Elev.:** 200–1700 m a.s.l. **IUCN Habitat code:** 1,3,5,14 **Conservation status: (IUCN)** NE

6154 *Miconia calvescens* DC. (Melastomataceae) **IPNI:** 326247-2 **Common name:** tuno **Habit:** Shrub, Small tree, Tree **Use:** EU, HF, MA, ME **Origin:** Native **Region:** Amazonia **Dept.:** AMA, CAQ, CHO, GUV, NSA, PUT, RIS **Elev.:** 50–920 m a.s.l. **IUCN Habitat code:** 1,3,5,14 **Conservation status: (IUCN)** LC

6155 *Miconia caquetensis* (Gleason) Ocampo & Almeda (Melastomataceae) **IPNI:** 77207690-1

Habit: Shrub **Use:** HF **Origin:** Endemic **Region:** Amazonia, Andean **Dept.:** CAQ, PUT **Elev.:** 325–800 m a.s.l. **IUCN Habitat code:** 1 **Conservation status: (IUCN)** NE

6156 *Miconia caudata* DC. (Melastomataceae) **IPNI:** 572256-1 **Common name:** cacho de venado, cenizo, cerizo, chilco, lanzo, macanillo, morochillo, mortiño, nigüito, panocho, puntelanzo, sabanero, suaso, tunchín, tuno, tuno blanco, tuno coludo, zanca de mulo, zanco de mula **Habit:** Shrub, Small tree, Tree **Use:** FU, ME **Origin:** Native **Region:** Andean, Pacific **Dept.:** ANT, BOY, CAL, CAU, CES, CHO, CUN, HUI, MAG, NAR, QUI, RIS, SAN, TOL, VAC **Elev.:** 150-3200 m.a.s.l. **IUCN Habitat code:** 1,3,8,14 **Conservation status: (IUCN)** LC **(CPSG)** Potential LC

6157 *Miconia centrodesma* Naudin (Melastomataceae) **IPNI:** 158869-2 **Habit:** Shrub, Tree **Use:** HF **Origin:** Native **Region:** Amazonia, Andean, Guiana Shield, Pacific **Dept.:** AMA, ANT, CAQ, CAU, CHO, CUN, GUA, GUV, HUI, MET, NAR, NSA, PUT, VAC, VAU **Elev.:** 5-2100 m.a.s.l. **IUCN Habitat code:** 1,3,14 **Conservation status: (IUCN)** LC **(CPSG)** Potential LC

6158 *Miconia chamissois* Naudin (Melastomataceae) **IPNI:** 158886-2 **Habit:** Subshrub **Use:** HF, MA **Origin:** Native **Region:** Andean, Guiana Shield **Dept.:** MAG, MET, SAN **Elev.:** 260–1530 m a.s.l. **IUCN Habitat code:** 1,14 **Conservation status: (IUCN)** LC

6159 *Miconia ciliata* (Rich.) DC. (Melastomataceae) **IPNI:** 1108920-2 **Habit:** Shrub, Tree **Use:** HF **Origin:** Native **Region:** Amazonia, Andean, Guiana Shield, Caribbean, Orinoquia **Dept.:** AMA, ANT, CAQ, CAU, CHO, CUN, GUV, MET, NSA, SAP, SAN, TOL, VAU, VID **Elev.:** 100-1250 m.a.s.l. **IUCN Habitat code:** 1,3,4,5,14 **Conservation status: (IUCN)** LC **(CPSG)** Potential LC

6160 *Miconia cinnamomea* Beurl. (Melastomataceae) **IPNI:** 158912-2 **Habit:** Shrub, Small tree **Use:** HF **Origin:** Endemic **Region:** Andean, Pacific **Dept.:** ANT, BOY, CHO, MAG, NAR **Elev.:** 50–1500 m a.s.l. **IUCN Habitat code:** 1,3 **Conservation status: (IUCN)** NE

6161 *Miconia crenata* (Vahl) Michelang. (Melastomataceae) **IPNI:** 77164247-1 **Common name:** amaine, caiuia, camasey, canot-macaque, choroticos, curse, kosters curse, mortiño, mucurita, mucuritos, múcura, sietecueros, soapbush, tuno, árbol de hormiga **Habit:** Herb, Shrub, Tree **Use:** EU, GS, HF, ME **Origin:** Native **Region:** Amazonia, Andean, Guiana Shield, Caribbean, Orinoquia, Pacific **Dept.:** AMA, ANT, BOL, BOY, CAQ, CAS, CAU, CHO, COR, CUN, GUA, GUV, HUI, LAG, MAG, MET, NAR, NSA, PUT, RIS, SAN, TOL, VAC, VAU **Elev.:** 0–2300 m a.s.l. **IUCN Habitat code:** 1,3,14 **Conservation status: (IUCN)** NE

6162 *Miconia crenulata* (Gleason) Michelang. (Melastomataceae) **IPNI:** 77207399-1 **Common name:** morita de palo, morita espumosa **Habit:** Herb, Shrub **Use:** ME **Origin:** Endemic **Region:** Andean, Orinoquia, Pacific **Dept.:** ANT, CAQ, CAU, CHO, CUN, MET, VAC **Elev.:** 0–1650 m a.s.l. **IUCN Habitat code:** 1,3,14 **Conservation status: (IUCN)** NE

6163 *Miconia crocea* Naudin (Melastomataceae) **IPNI:** 572353-1 **Habit:** Shrub **Use:** HF **Origin:** Native **Region:** Andean **Dept.:** CAU, HUI, NAR, PUT **Elev.:** 2600–3380 m a.s.l. **IUCN Habitat code:** 1,4,14 **Conservation status: (IUCN)** NE

6164 *Miconia cursoris* (Wurdack) Michelang. (Melastomataceae) **IPNI:** 77207381-1 **Habit:** Shrub **Use:** HF **Origin:** Endemic **Region:** Andean **Dept.:** ANT, NSA, SAN, TOL **Elev.:** 1460–2050 m a.s.l. **IUCN Habitat code:** 1,3,14 **Conservation status: (IUCN)** NE

6165 *Miconia dentata* (D.Don) Michelang. (Melastomataceae) **IPNI:** 77207400-1 **Common name:** desvanecedora, mortiño **Habit:** Herb, Shrub, Subshrub, Tree **Use:** HF **Origin:** Endemic **Region:** Andean, Guiana Shield, Orinoquia, Pacific **Dept.:** ANT, CAQ, CAU, CHO, CUN, LAG, MAG, MET, NAR, NSA, PUT, RIS, SAN, VAC **Elev.:** 0-1400 m.a.s.l. **IUCN Habitat code:** 1,2,3,5,14 **Conservation status: (IUCN)** LC **(CPSG)** Potential LC

6166 *Miconia dependens* (Pav. ex D.Don) Judd & Majure (Melastomataceae) **IPNI:** 77196165-1 **Common name:** guahibo, mora, mora blanca, mortiño, nigüito, popuyo **Habit:** Herb, Shrub, Subshrub, Tree **Use:** HF, ME **Origin:** Native **Region:** Amazonia, Andean, Guiana Shield, Orinoquia, Pacific **Dept.:** AMA, ANT, ARA, BOL, BOY, CAL, CAQ, CAS, CAU, CHO, CUN, GUA, GUV, HUI, MAG, MET, NAR, NSA, PUT, QUI, SAN, TOL, VAC, VAU, VID **Elev.:** 0–2500 m a.s.l. **IUCN Habitat code:** 1,2,3,4,5,14 **Conservation status: (IUCN)** NE

6167 *Miconia desmantha* Benth. (Melastomataceae) **IPNI:** 572388-1 **Common name:** mortiño **Habit:** Shrub, Small tree **Use:** HF **Origin:** Native **Region:** Andean **Dept.:** CAU, CUN, HUI, NSA **Elev.:** 1600–2080 m a.s.l. **IUCN Habitat code:** 3,14 **Conservation status: (IUCN)** LC

6168 *Miconia dimorphica* (J.F.Macbr.) Michelang. (Melastomataceae) **IPNI:** 77207368-1 **Habit:** Herb, Shrub, Subshrub, Small tree, Tree **Use:** HF **Origin:** Endemic **Region:** Amazonia, Andean **Dept.:** AMA, CAQ, PUT **Elev.:** 80–700 m a.s.l. **IUCN Habitat code:** 1,14 **Conservation status: (IUCN)** LC

6169 *Miconia dodecandra* Cogn. (Melastomataceae) **IPNI:** 572411-1 **Common name:** tuno **Habit:** Small tree, Tree **Use:** HF **Origin:** Native **Region:** Amazonia, Andean, Guiana Shield **Dept.:** AMA, ANT, BOL, BOY, CAQ, CAU, CES, CHO, CUN, HUI, MAG, MET, NSA, PUT, SAN, VAC **Elev.:** 775-1950 m.a.s.l. **IUCN Habitat code:** 1,2,3,4,14 **Conservation status: (IUCN)** LC **(CPSG)** Potential LC

6170 *Miconia domociliata* Michelang. (Melastomataceae) **IPNI:** 77207714-1 **Common name:** esmeraldo, mortiño, pulmonaria **Habit:** Herb, Shrub, Subshrub, Small tree **Use:** HF **Origin:** Endemic **Region:** Andean, Orinoquia **Dept.:** AMA, ANT, BOG, BOY, CAL, CAQ, CAS, CAU, CES, CHO, CUN, MAG, MET, NAR, NSA, RIS, SAN, TOL, VAC **Elev.:** 100-2260 m.a.s.l. **IUCN Habitat code:** 1,2,3,14 **Conservation status: (IUCN)** NE **(CPSG)** Potential LC

6171 *Miconia elata* (Sw.) DC. (Melastomataceae) **IPNI:** 1092872-2 **Common name:** amainae, chilco negro, chirco de montaña, guayabilla, macanillo, macanillo negro, morochillo, mortiño negro, tuno, wanalienae **Habit:** Shrub, Tree **Use:** FU, MA, ME **Origin:** Native **Region:** Amazonia, Andean, Orinoquia, Pacific **Dept.:** ANT, BOY, CAQ, CAS, GUV, MET, NAR, NSA, PUT **Elev.:** 60–1500 m a.s.l. **IUCN Habitat code:** 1,3,4,5,8,14 **Conservation status: (IUCN)** NE

6172 *Miconia erioclada* Triana (Melastomataceae) **IPNI:** 159046-2 **Habit:** Tree **Use:** HF **Origin:** Native **Region:** Amazonia, Orinoquia **Dept.:** MET, PUT, SAN **Elev.:** 300–855 m a.s.l. **IUCN Habitat code:** 1,3,14 **Conservation status: (IUCN)** NE

6173 *Miconia ferruginea* (Desr.) DC. (Melastomataceae) **IPNI:** 572470-1 **Common name:** amarillo, quillo **Use:** ME **Dept.:** BOG **IUCN Habitat code:** 1,3,4,14 **Conservation status: (IUCN)** NE

6174 *Miconia formicoheterophylla* Michelang. (Melastomataceae) **IPNI:** 77207843-1 **Habit:** Shrub, Subshrub **Use:** HF **Origin:** Endemic **Region:** Amazonia, Andean **Dept.:** AMA, ANT, CAQ, CAU, MET, PUT **Elev.:** 150–1420 m a.s.l. **IUCN Habitat code:** 1,3,14 **Conservation status: (IUCN)** NE

6175 *Miconia formicosetosa* Michelang. (Melastomataceae) **IPNI:** 77207860-1 **Habit:** Shrub, Small tree, Tree **Use:** HF **Origin:** Endemic **Region:** Andean, Pacific **Dept.:** CHO, NAR, RIS **Elev.:** 0–1150 m a.s.l. **IUCN Habitat code:** 1 **Conservation status: (IUCN)** NE

6176 *Miconia gleasoniana* Wurdack (Melastomataceae) **IPNI:** 159113-2 **Common name:** nigüito, tuno **Habit:** Shrub, Small tree, Tree **Use:** ME **Origin:** Endemic **Region:** Andean **Dept.:** ANT, BOG, BOY, CAL, CAU, CUN, HUI, NAR, PUT, QUI, RIS, TOL, VAC **Elev.:** 1500-3950 m.a.s.l. **IUCN Habitat code:** 1,3,4,8,14 **Conservation status: (IUCN)** LC **(CPSG)** LC

6177 *Miconia granatensis* (Gleason) Ocampo & Almeda (Melastomataceae) **IPNI:** 77207722-1 **Common name:** mora **Habit:** Herb, Shrub, Tree **Use:** HF **Origin:** Endemic **Region:** Andean, Pacific **Dept.:** ANT, CAU, CHO, NAR, NSA, SAN, VAC **Elev.:** 0–1000 m a.s.l. **IUCN Habitat code:** 1,3,5,14 **Conservation status: (IUCN)** LC

6178 *Miconia heteroclita* (Naudin) Michelang. (Melastomataceae) **IPNI:** 77207474-1 **Common name:** ara de la altura **Habit:** Shrub, Subshrub **Use:** HF **Origin:** Endemic **Region:** Amazonia, Orinoquia **Dept.:** AMA, ANT, CAQ, GUA, GUV, MET, VAU **Elev.:** 65–700 m a.s.l. **IUCN Habitat code:** 1,3,4,5,14 **Conservation status: (IUCN)** NE

6179 *Miconia holosericea* (L.) DC. (Melastomataceae)
IPNI: 1024004-2
Habit: Shrub, Small tree **Use:** HF, MA **Origin:** Native **Region:** Amazonia, Guiana Shield, Orinoquia, Pacific **Dept.:** AMA, CAQ, CHO, MET, NSA, VID **Elev.:** 100–410 m a.s.l. **IUCN Habitat code:** 1,2,3,4,5,14 **Conservation status: (IUCN)** LC

6180 *Miconia hookeriana* Triana (Melastomataceae)
IPNI: 572575-1
Habit: Shrub **Use:** EU, ME **Origin:** Native **Region:** Amazonia **Dept.:** PUT **Elev.:** 325–325 m a.s.l. **IUCN Habitat code:** 1 **Conservation status: (IUCN)** NE

6181 *Miconia ibaguensis* Triana (Melastomataceae)
IPNI: 572590-1
Common name: mortiño **Habit:** Herb, Shrub **Use:** HF **Origin:** Native **Region:** Andean, Orinoquia **Dept.:** ANT, BOL, CAL, CAS, CAU, CES, CUN, MAG, MET, NAR, NSA, RIS, SAN, TOL **Elev.:** 240-2740 m.a.s.l. **IUCN Habitat code:** 1,2,3,4,14 **Conservation status: (IUCN)** NE **(CPSG)** Potential LC

6182 *Miconia impetiolaris* D.Don (Melastomataceae)
IPNI: 572600-1
Common name: danto **Habit:** Shrub, Tree **Use:** ME **Origin:** Native **Region:** Guiana Shield, Caribbean, Pacific **Dept.:** ANT, BOG, BOL, CAL, CES, CHO, CUN, MAG, SAN, TOL, VAC, VAU **Elev.:** 60–1000 m a.s.l. **IUCN Habitat code:** 1,2,3,4,5,14 **Conservation status: (IUCN)** LC

6183 *Miconia laevigata* (L.) DC. (Melastomataceae)
IPNI: 572641-1
Habit: Tree **Use:** ME **Origin:** Native **Region:** Caribbean, Orinoquia **Dept.:** ANT, CUN, LAG, MAG, MET, SAP **Elev.:** 250–730 m a.s.l. **Conservation status: (IUCN)** NE

6184 *Miconia latifolia* Naudin (Melastomataceae)
IPNI: 572660-1
Habit: Shrub **Use:** HF **Origin:** Native **Region:** Andean **Dept.:** BOG, BOY, CAL, CAU, CHO, CUN, MAG, MET, NAR, RIS, TOL **Elev.:** 2700–4040 m a.s.l. **IUCN Habitat code:** 1,3,4,6,14 **Conservation status: (IUCN)** LC

6185 *Miconia laxivenula* (Wurdack) Gamba & Almeda (Melastomataceae)
IPNI: 77144073-1
Habit: Shrub, Small tree **Use:** HF **Origin:** Native **Region:** Amazonia **Dept.:** CAU, NAR, PUT **Elev.:** 300-400 m.a.s.l. **IUCN Habitat code:** 1,3 **Conservation status: (IUCN)** LC **(CPSG)** Potential LC

6186 *Miconia lepidota* DC. (Melastomataceae)
IPNI: 572671-1
Common name: achiotillo, mora, morochillo **Habit:** Shrub, Tree **Use:** MA **Origin:** Native **Region:** Andean, Caribbean, Pacific **Dept.:** AMA, ANT, CAQ, CAU, CES, CHO, HUI, NAR, PUT, VAC **Elev.:** 0–1380 m a.s.l. **IUCN Habitat code:** 1,3,4,5,14 **Conservation status: (IUCN)** LC

6187 *Miconia ligustrina* Triana (Melastomataceae)
IPNI: 572682-1
Common name: esmeraldo, mortiño, tuno **Habit:** Shrub, Small tree, Tree **Use:** HF **Origin:** Native **Region:** Andean **Dept.:** ANT, BOG, BOY, CAU, CUN, HUI, MAG, MET, NAR, NSA, PUT, RIS, SAN, VAC **Elev.:** 2050–3800 m a.s.l. **IUCN Habitat code:** 1,3,4,5,8,14 **Conservation status: (IUCN)** NE

6188 *Miconia longifolia* (Aubl.) DC. (Melastomataceae)
IPNI: 1220088-2
Common name: aguanoso, morochillo, nigüito, tuno esmeraldo **Habit:** Herb, Shrub, Small tree, Tree **Use:** HF **Origin:** Native **Region:** Amazonia, Andean, Guiana Shield, Orinoquia, Pacific **Dept.:** AMA, ANT, BOY, CAQ, CAS, CAU, CHO, CUN, GUV, HUI, MAG, MET, PUT, SAN, VAC, VID **Elev.:** 5-2900 m.a.s.l. **IUCN Habitat code:** 1,3,5,14 **Conservation status: (IUCN)** LC **(CPSG)** Potential LC

6189 *Miconia longispicata* Triana (Melastomataceae)
IPNI: 159286-2
Habit: Shrub, Small tree **Use:** HF **Origin:** Native **Region:** Amazonia, Guiana Shield **Dept.:** AMA, CAQ, GUA, VAU **Elev.:** 250–900 m a.s.l. **IUCN Habitat code:** 1,5,14 **Conservation status: (IUCN)** LC

6190 *Miconia macrothyrsa* Benth. (Melastomataceae)
IPNI: 159311-2
Habit: Herb, Shrub **Use:** HF **Origin:** Native **Region:** Amazonia, Orinoquia **Dept.:** CAQ, CAS, GUV, MET, VID **Elev.:** 80–500 m a.s.l. **IUCN Habitat code:** 1,3,4,14 **Conservation status: (IUCN)** NE

6191 *Miconia magnifolia* Gamba & Almeda (Melastomataceae)
IPNI: 77144075-1
Habit: Herb, Shrub, Subshrub, Small tree, Tree **Use:** HF **Origin:** Native **Region:** Amazonia, Andean, Guiana Shield, Pacific **Dept.:** ANT, CAU, CHO, MET, NAR, PUT, VAC **Elev.:** 0-1400 m.a.s.l. **IUCN Habitat code:** 1,14 **Conservation status: (IUCN)** NE **(CPSG)** Potential LC

6192 *Miconia mayeta* (D.Don) Michelang. (Melastomataceae)
IPNI: 77164249-1
Common name: churijuique, labios de negra, palo de majiña **Habit:** Herb, Shrub, Subshrub **Use:** HF **Origin:** Native **Region:** Amazonia **Dept.:** AMA, ANT, CAQ, CAU, GUA, GUV, MET, NAR, PUT, SAN, VAU **Elev.:** 80–890 m a.s.l. **IUCN Habitat code:** 1,3,4,5,14 **Conservation status: (IUCN)** NE

6193 *Miconia minutiflora* (Bonpl.) DC. (Melastomataceae)
IPNI: 1098452-2
Common name: arrayán, hermosito, lengüevenado, mortiño, mutsibulibulinae, nigua, nigüito, paramuerto, teñidor, tuno, tuno blanco, tuno negro **Habit:** Shrub, Small tree, Tree **Use:** FU, MA, ME **Origin:** Native **Region:** Amazonia, Andean, Guiana Shield, Orinoquia, Pacific **Dept.:** AMA, ANT, BOY, CAL, CAQ, CAS, CHO, COR, CUN, GUA, GUV, MAG, MET, PUT, SAN, TOL, VAC, VAU, VID **Elev.:** 5-2500 m.a.s.l. **IUCN Habitat code:** 1,2,3,4,5,14 **Conservation status: (IUCN)** LC **(CPSG)** Potential LC

6194 *Miconia neoepiphyitca* Michelang. (Melastomataceae)
IPNI: 77207810-1
Habit: Herb, Tree, Epiphyte, Liana, Climbing **Use:** ME **Origin:** Endemic **Region:** Amazonia, Andean, Pacific **Dept.:** AMA, ANT, CAQ, CAU, CHO, CUN, NAR, NSA, PUT, RIS, SAN, VAC, VAU **Elev.:** 0–2120 m a.s.l. **IUCN Habitat code:** 1,3,4,5,14 **Conservation status: (IUCN)** LC

6195 *Miconia neopilosa* Michelang. (Melastomataceae)
IPNI: 77207921-1
Common name: mortiño, puntelanza **Habit:** Shrub, Small tree **Use:** HF **Origin:** Endemic **Region:** Andean, Pacific **Dept.:** ANT, CAU, RIS, VAC **Elev.:** 20–1700 m a.s.l. **IUCN Habitat code:** 1,3,5,14 **Conservation status: (IUCN)** NE

6196 *Miconia neosecunda* Ocampo & Almeda (Melastomataceae)
IPNI: 77207935-1
Habit: Herb, Shrub, Subshrub **Use:** HF **Origin:** Endemic **Region:** Amazonia, Andean, Orinoquia **Dept.:** AMA, CAQ, CHO, CUN, MET, PUT, VAU **Elev.:** 100–1650 m a.s.l. **IUCN Habitat code:** 1,3,4 **Conservation status: (IUCN)** NE

6197 *Miconia nervosa* Triana (Melastomataceae)
IPNI: 572835-1
Habit: Herb, Shrub, Small tree, Tree **Use:** HF, ME **Origin:** Native **Region:** Amazonia, Andean, Guiana Shield, Orinoquia, Pacific **Dept.:** AMA, ANT, BOL, CAQ, CAU, CHO, CUN, GUA, GUV, MET, NAR, PUT, QUI, RIS, SAN, VAC, VAU **Elev.:** 0-1450 m.a.s.l. **IUCN Habitat code:** 1,3,4,5,14 **Conservation status: (IUCN)** LC **(CPSG)** Potential LC

6198 *Miconia ochracea* Triana (Melastomataceae)
IPNI: 572860-1
Habit: Tree **Use:** HF **Origin:** Native **Region:** Andean **Dept.:** CAL, CAU, NAR, PUT, RIS, VAC **Elev.:** 1800–3300 m a.s.l. **IUCN Habitat code:** 1,3,4,5,14 **Conservation status: (IUCN)** NE

6199 *Miconia octona* (Bonpl.) Judd & Majure (Melastomataceae)
IPNI: 77196163-1
Common name: chilco colorado, mortiño, noorijujuique **Habit:** Shrub, Subshrub, Tree **Use:** HF **Origin:** Native **Region:** Amazonia, Andean, Guiana Shield, Orinoquia, Pacific **Dept.:** ANT, BOY, CAL, CAQ, CAS, CAU, CES, CHO, COR, CUN, GUV, MAG, MET, NAR, PUT, RIS, SAN, TOL, VAC, VAU **Elev.:** 30–2300 m a.s.l. **IUCN Habitat code:** 1,2,3,5,14 **Conservation status: (IUCN)** NE

6200 *Miconia paleacea* Cogn. (Melastomataceae)
IPNI: 159460-2
Habit: Shrub, Tree **Use:** ME **Origin:** Native **Region:** Amazonia **Dept.:** PUT **Elev.:** 0–350 m a.s.l. **IUCN Habitat code:** 1 **Conservation status: (IUCN)** LC

6201 *Miconia phaeochaeta* Wurdack (Melastomataceae)
IPNI: 159506-2
Habit: Shrub, Small tree **Use:** HF **Origin:** Native **Region:** Andean **Dept.:** PUT **Elev.:** 600–600 m a.s.l. **IUCN Habitat code:** 14 **Conservation status: (IUCN)** LC

6202 *Miconia phanerostila* Pilg. (Melastomataceae)
IPNI: 159509-2
Habit: Shrub, Small tree **Use:** HF **Origin:** Native **Region:** Amazonia **Dept.:** AMA, CAQ, PUT **Elev.:** 100–160 m a.s.l. **IUCN Habitat code:** 1,3,5,14 **Conservation status: (IUCN)** NE

6203 *Miconia poeppigii* Triana (Melastomataceae) **IPNI:** 159532-2 **Common name:** quesito **Habit:** Tree **Use:** HF, MA **Origin:** Native **Region:** Amazonia, Guiana Shield, Pacific **Dept.:** AMA, ANT, CAQ, CHO, MET, VAU **Elev.:** 60–700 m a.s.l. **IUCN Habitat code:** 1,2,3,4,14 **Conservation status: (IUCN)** LC

6204 *Miconia prasina* (Sw.) DC. (Melastomataceae) **IPNI:** 1069702-2 **Common name:** chiquito, esmeralda, guayabo de mono, macanillo, mortiño **Habit:** Shrub, Subshrub, Small tree, Tree **Use:** HF, MA **Origin:** Native **Region:** Amazonia, Andean, Caribbean, Orinoquia, Pacific **Dept.:** AMA, ANT, ARA, BOG, BOY, CAL, CAQ, CAU, CHO, COR, CUN, GUV, HUI, MAG, MET, NAR, NSA, PUT, QUI, RIS, SAP, SAN, TOL, VAC, VID **Elev.:** 40-2700 m.a.s.l. **IUCN Habitat code:** 1,2,3,4,5,14 **Conservation status: (IUCN)** LC **(CPSG)** Potential LC

6205 *Miconia punctata* (Desr.) D.Don (Melastomataceae) **IPNI:** 1069703-2 **Common name:** marfil, nigüito, tonca **Habit:** Shrub, Small tree, Tree **Use:** HF **Origin:** Native **Region:** Amazonia, Andean, Guiana Shield, Orinoquia, Pacific **Dept.:** AMA, CAQ, CAU, CHO, CUN, GUA, GUV, MAG, MET, NSA, PUT, SAN, VAC, VID **Elev.:** 0–1700 m a.s.l. **IUCN Habitat code:** 1,3,4,14 **Conservation status: (IUCN)** LC

6206 *Miconia quadrisulca* (Naudin) Ocampo & Almeda (Melastomataceae) **IPNI:** 77207567-1 **Habit:** Shrub, Subshrub **Use:** HF **Origin:** Endemic **Region:** Andean **Dept.:** AMA, ANT **Elev.:** 350–1150 m a.s.l. **IUCN Habitat code:** 1,3,14 **Conservation status: (IUCN)** NE

6207 *Miconia quinquenervia* (Mill.) Gamba & Almeda (Melastomataceae) **IPNI:** 77144083-1 **Habit:** Shrub, Subshrub **Use:** HF **Origin:** Native **Region:** Andean, Pacific **Dept.:** ANT, CAL, CHO, COR, CUN, MAG, MET, QUI, TOL **Elev.:** 0-1200 m.a.s.l. **IUCN Habitat code:** 1,2,3,5,14 **Conservation status: (IUCN)** NE **(CPSG)** VU

6208 *Miconia racemosa* (Aubl.) DC. (Melastomataceae) **IPNI:** 1158780-2 **Habit:** Shrub, Tree **Use:** HF, MA **Origin:** Native **Region:** Amazonia, Pacific **Dept.:** CHO, GUA, VAC, VAU **Elev.:** 0–400 m a.s.l. **Conservation status: (IUCN)** LC

6209 *Miconia richardsprucei* Michelang. (Melastomataceae) **IPNI:** 77207920-1 **Habit:** Shrub **Use:** HF **Origin:** Endemic **Region:** Amazonia **Dept.:** PUT **Elev.:** 350–350 m a.s.l. **IUCN Habitat code:** 1,14 **Conservation status: (IUCN)** NE

6210 *Miconia rivetii* Danguy & Cherm. (Melastomataceae) **IPNI:** 159648-2 **Habit:** Tree **Use:** HF **Origin:** Native **Region:** Andean **Dept.:** NAR **Elev.:** 1900–1900 m a.s.l. **IUCN Habitat code:** 1 **Conservation status: (IUCN)** LC

6211 *Miconia rubiginosa* (Bonpl.) DC. (Melastomataceae) **IPNI:** 1155116-2 **Common name:** chirco de sabana, mortiño, mortiño rojo, tuno, yuco **Habit:** Shrub, Small tree, Tree **Use:** HF, MA **Origin:** Native **Region:** Andean, Guiana Shield, Orinoquia **Dept.:** ANT, BOY, CAS, CAU, CES, CUN, GUV, HUI, LAG, MET, NSA, SAN, TOL, VAC, VID **Elev.:** 80-3250 m.a.s.l. **IUCN Habitat code:** 1,2,3,4,14 **Conservation status: (IUCN)** LC **(CPSG)** Potential LC

6212 *Miconia rubra* (Aubl.) Mabb. (Melastomataceae) **IPNI:** 77164250-1 **Habit:** Shrub, Tree **Use:** HF **Origin:** Endemic **Region:** Amazonia, Andean, Caribbean, Orinoquia, Pacific **Dept.:** AMA, ANT, BOY, CAQ, CAS, CAU, CES, CHO, COR, CUN, GUA, GUV, MET, NSA, SAP, SAN, TOL, VAC, VAU, VID **Elev.:** 40-2400 m.a.s.l. **IUCN Habitat code:** 1,2,3,4,5,14 **Conservation status: (IUCN)** LC **(CPSG)** Potential LC

6213 *Miconia rufescens* (Aubl.) DC. (Melastomataceae) **IPNI:** 1018450-2 **Common name:** cascajero, chirco sabanero, mortiño, pematatsabanaewa **Habit:** Shrub, Tree **Use:** ME **Origin:** Native **Region:** Andean, Caribbean, Orinoquia **Dept.:** AMA, ANT, ARA, CAQ, CAS, CAU, CUN, GUV, HUI, LAG, MAG, MET, NSA, SAN, TOL, VAC, VID **Elev.:** 100–1850 m a.s.l. **IUCN Habitat code:** 1,2,3,4,5,14 **Conservation status: (IUCN)** LC

6214 *Miconia ruficalyx* Gleason (Melastomataceae) **IPNI:** 289925-2 **Common name:** casposo, mora **Habit:** Shrub **Use:** HF **Origin:** Native **Region:** Pacific **Dept.:** CAU **Elev.:** 0–200 m a.s.l. **IUCN Habitat code:** 1,14 **Conservation status: (IUCN)** LC

6215 *Miconia rupicola* Gleason (Melastomataceae) **IPNI:** 159672-2 **Common name:** mora **Habit:** Shrub, Small tree, Tree **Use:** HF **Origin:** Endemic **Region:** Andean, Pacific **Dept.:** CAU, CHO, NAR, VAC **Elev.:** 0–1200 m a.s.l. **IUCN Habitat code:** 1,3,5,14 **Conservation status: (IUCN)** NE

6216 *Miconia salicifolia* Naudin (Melastomataceae) **IPNI:** 573124-1 **Common name:** cherte, chirte, coneja, esmeraldo, tuno, vara negra, velillo **Habit:** Shrub, Subshrub, Small tree, Tree **Use:** HF, ME **Origin:** Native **Region:** Andean **Dept.:** ANT, BOG, BOY, CAL, CAU, CHO, CUN, MET, NAR, NSA, QUI, RIS, SAN, TOL **Elev.:** 2850-4200 m.a.s.l. **IUCN Habitat code:** 1,2,3,4,8,14 **Conservation status: (IUCN)** NE **(CPSG)** Potential LC

6217 *Miconia secuncaquetana* Ocampo & Almeda (Melastomataceae) **IPNI:** 77207884-1 **Habit:** Shrub **Use:** HF **Origin:** Endemic **Region:** Amazonia **Dept.:** CAQ, PUT **Elev.:** 300–305 m a.s.l. **Conservation status: (IUCN)** NE

6218 *Miconia secundivaricata* Ocampo & Almeda (Melastomataceae) **IPNI:** 77207868-1 **Habit:** Subshrub **Use:** HF **Origin:** Endemic **Region:** Amazonia **Dept.:** AMA **Elev.:** 100–100 m a.s.l. **IUCN Habitat code:** 1 **Conservation status: (IUCN)** NE

6219 *Miconia septuplinervia* (Cogn.) Judd & Ionta (Melastomataceae) **IPNI:** 77196168-1 **Common name:** arachigue, mora **Habit:** Herb, Shrub, Subshrub, Epiphyte **Use:** HF **Origin:** Native **Region:** Amazonia, Guiana Shield, Orinoquia, Pacific **Dept.:** AMA, ANT, CAQ, CAU, CHO, GUV, MAG, MET, NAR, PUT, SAN, VAC **Elev.:** 0–1800 m a.s.l. **IUCN Habitat code:** 1,3,5,14 **Conservation status: (IUCN)** NE

6220 *Miconia sericea* (D.Don) Michelang. (Melastomataceae) **IPNI:** 77207578-1 **Common name:** azulejo, curruco, mora, moradillo, morado, mortiño, mortiño de alambre, tuno, tuno morado **Habit:** Herb, Shrub, Subshrub **Use:** HF **Origin:** Endemic **Region:** Andean, Guiana Shield, Orinoquia, Pacific **Dept.:** AMA, ANT, CAL, CAQ, CAS, CAU, CHO, CUN, GUA, GUV, HUI, MAG, MET, NSA, SAN, TOL, VAC, VID **Elev.:** 20-2500 m.a.s.l. **IUCN Habitat code:** 1,2,3,4,5,14 **Conservation status: (IUCN)** NE **(CPSG)** Potential LC

6221 *Miconia serrulata* (DC.) Naudin (Melastomataceae) **IPNI:** 1000103-2 **Common name:** dueño de pajarito, macanillo, morochillo guayaba, mortiño, tuno, tuno blanco, tuno rosado **Habit:** Herb, Shrub, Small tree, Tree **Use:** HF **Origin:** Native **Region:** Amazonia, Andean, Guiana Shield, Caribbean, Orinoquia, Pacific **Dept.:** AMA, ANT, BOL, BOY, CAL, CAQ, CAS, CAU, CES, CHO, CUN, GUV, MAG, MET, NAR, NSA, PUT, SAN, VAU **Elev.:** 5–2640 m a.s.l. **IUCN Habitat code:** 1,3,4,5,14 **Conservation status: (IUCN)** LC

6222 *Miconia splendens* (Sw.) Griseb. (Melastomataceae) **IPNI:** 1143367-2 **Common name:** ara **Habit:** Shrub, Small tree, Tree **Use:** HF **Origin:** Native **Region:** Amazonia, Andean, Orinoquia, Pacific **Dept.:** AMA, ANT, CAQ, CAU, CHO, CUN, GUA, GUV, MET, PUT, VAU **Elev.:** 5–1650 m a.s.l. **IUCN Habitat code:** 1,3,4,5,14 **Conservation status: (IUCN)** LC

6223 *Miconia squamulosa* Triana (Melastomataceae) **IPNI:** 573207-1 **Common name:** esmeraldo, niguito, nigüito, tuno, tuno esmeralda, tuno esmeraldo **Habit:** Herb, Shrub, Small tree, Tree **Use:** AF, EU, FU, HF, MA, ME **Origin:** Native **Region:** Andean **Dept.:** BOG, BOY, CAU, CUN, NSA, SAN **Elev.:** 2100–3600 m a.s.l. **IUCN Habitat code:** 1,3,4,5,14 **Conservation status: (IUCN)** NE

6224 *Miconia stenostachya* DC. (Melastomataceae) **IPNI:** 159763-2 **Common name:** gallinaza, nigüito, punteblanco, sabanero, tuno **Habit:** Shrub, Subshrub, Small tree, Tree **Use:** HF, MA **Origin:** Native **Region:** Amazonia, Andean, Orinoquia **Dept.:** AMA, ANT, ARA, BOY, CAS, CES, COR, CUN, GUV, HUI, MAG, MET, NAR, NSA, SAN, TOL, VAC, VID **Elev.:** 100-2700 m.a.s.l. **IUCN Habitat code:** 1,2,3,4,14 **Conservation status: (IUCN)** LC **(CPSG)** Potential LC

6225 *Miconia strigillosa* (Sw.) Judd & Ionta (Melastomataceae) **IPNI:** 77196164-1 **Common name:** chorote **Habit:** Shrub **Use:** HF **Origin:** Native **Region:** Amazonia, Andean, Guiana Shield, Orinoquia, Pacific **Dept.:** AMA, ANT, CAQ, CAS, CHO, CUN, GUA, HUI, MET, PUT, SAN, TOL, VAC, VAU **Elev.:**

30–1700 m a.s.l. **IUCN Habitat code:** 1,2,3,4,5,14 **Conservation status: (IUCN)** NE

6226 *Miconia subhirsuta* (DC.) M.Gómez (Melastomataceae) **IPNI:** 159783-2 **Habit:** Shrub, Tree **Use:** HF **Origin:** Endemic **Region:** Andean, Pacific **Dept.:** ANT, LAG, MAG **Elev.:** 140–1450 m a.s.l. **IUCN Habitat code:** 1,2,3,14 **Conservation status: (IUCN)** NE

6227 *Miconia subseriata* (Naudin) R.Goldenb. & Michelang. (Melastomataceae) **IPNI:** 77207614-1 **Common name:** mortiño de monte, mortiño lanudo, munchirero, nigüito, nigüitos **Habit:** Shrub, Small tree, Tree **Use:** HF **Origin:** Endemic **Region:** Andean **Dept.:** ANT, BOG, CAL, CAU, CUN, HUI, NAR, PUT, QUI, SAN, TOL, VAC **Elev.:** 1300–2910 m a.s.l. **IUCN Habitat code:** 1,3,4,14 **Conservation status: (IUCN)** LC

6228 *Miconia sulcicaulis* (Poepp. ex Naudin) Ocampo & Almeda (Melastomataceae) **IPNI:** 77207668-1 **Common name:** hierbamora **Habit:** Shrub, Subshrub **Use:** HF, ME **Origin:** Endemic **Region:** Andean, Pacific **Dept.:** AMA, ANT, BOY, CHO, CUN, HUI, MET, NAR, NSA, QUI, RIS, SAN, VAC **Elev.:** 0-2000 m.a.s.l. **IUCN Habitat code:** 1,3,5,14 **Conservation status: (IUCN)** NE **(CPSG)** Potential LC

6229 *Miconia theizans* (Bonpl.) Cogn. (Melastomataceae) **IPNI:** 1110588-2 **Common name:** amarillo, cántaro, danto, morochillo, mortiño, nigüita, nigüito, nigüito chamicero, palo amarillo, sabanerito, tuno, tuno negro, uvo **Habit:** Shrub, Subshrub, Small tree, Tree **Use:** HF **Origin:** Native **Region:** Andean, Guiana Shield, Pacific **Dept.:** ANT, BOG, BOY, CAL, CAQ, CAU, CES, CHO, CUN, HUI, LAG, MAG, MET, NAR, NSA, PUT, QUI, RIS, SAN, TOL, VAC **Elev.:** 230-3900 m.a.s.l. **IUCN Habitat code:** 1,3,4,5,14 **Conservation status: (IUCN)** NE **(CPSG)** Potential LC

6230 *Miconia tococa* (Desr.) Michelang. (Melastomataceae) **IPNI:** 77207761-1 **Common name:** ara, currucay, hormiguero, mora, moracotura, tuno, tuno vejigoso, árbol de chupaflor **Habit:** Herb, Shrub, Subshrub, Small tree, Tree, Liana **Use:** EU, HF **Origin:** Endemic **Region:** Amazonia, Andean, Guiana Shield, Orinoquia, Pacific **Dept.:** AMA, ANT,BOG, BOY, CAL, CAQ, CAS, CAU, CHO, COR, CUN, GUA, GUV, MET, NAR, NSA, PUT, SAN, VAC, VAU, VID **Elev.:** 0-1600 m.a.s.l. **IUCN Habitat code:** 1,2,3,4,5,8,14 **Conservation status: (IUCN)** NE **(CPSG)** Potential LC

6231 *Miconia tococoidea* (DC.) Michelang. (Melastomataceae) **IPNI:** 77207700-1 **Habit:** Shrub, Subshrub **Use:** HF **Origin:** Endemic **Region:** Amazonia, Andean, Guiana Shield **Dept.:** AMA, CAQ, GUA, GUV, NAR, SAN, VAC, VAU **Elev.:** 30–1530 m a.s.l. **IUCN Habitat code:** 1,2,3,14 **Conservation status: (IUCN)** NE

6232 *Miconia tococoronata* Michelang. (Melastomataceae) **IPNI:** 77207926-1 **Common name:** mora **Habit:** Shrub, Tree **Use:** HF **Origin:** Endemic **Region:** Amazonia, Guiana Shield **Dept.:** AMA, CAQ, GUA, GUV, MET, PUT, VAU, VID **Elev.:** 100–230 m a.s.l. **IUCN Habitat code:** 1,3,4,5,14 **Conservation status: (IUCN)** LC

6233 *Miconia tomentosa* (Rich.) D.Don (Melastomataceae) **IPNI:** 1088633-2 **Common name:** pupuyo, sietecueros **Habit:** Shrub, Small tree, Tree **Use:** HF, ME **Origin:** Native **Region:** Amazonia, Andean, Orinoquia, Pacific **Dept.:** AMA, ANT, BOY, CAQ, CAU, CHO, CUN, GUA, GUV, MET, NSA, PUT, SAN, VAU, VID **Elev.:** 20-1250 m.a.s.l. **IUCN Habitat code:** 1,3,4,5,14 **Conservation status: (IUCN)** LC **(CPSG)** Potential LC

6234 *Miconia triplinervis* Ruiz & Pav. (Melastomataceae) **IPNI:** 159880-2 **Habit:** Shrub, Small tree **Use:** HF **Origin:** Native **Region:** Amazonia, Andean, Pacific **Dept.:** ANT, CAU, CHO, MAG, PUT, QUI, RIS **Elev.:** 50–1150 m a.s.l. **IUCN Habitat code:** 1,3,5,14 **Conservation status: (IUCN)** LC

6235 *Miconia variabilis* Gamba & Almeda (Melastomataceae) **IPNI:** 77144090-1 **Habit:** Shrub, Small tree **Use:** HF **Origin:** Native **Region:** Amazonia, Andean, Pacific **Dept.:** ANT, CAU, CHO, NAR, PUT, RIS, VAC **Elev.:** 0-1800 m.a.s.l. **IUCN Habitat code:** 1,3,14 **Conservation status: (IUCN)** LC **(CPSG)** Potential LC

6236 *Miconia versicolor* Naudin (Melastomataceae) **IPNI:** 573368-1 **Common name:** chilco blanco, esmeraldo, mortiño, nigüito, pandaré, tuno **Habit:** Shrub, Subshrub, Tree **Use:** HF **Origin:** Native **Region:** Andean **Dept.:** BOY, CAU, CUN, HUI, NAR, TOL, VAC **Elev.:** 1760–3200 m a.s.l. **IUCN Habitat code:** 1,3,14 **Conservation status: (IUCN)** NE

6237 *Miconia xalapensis* (Bonpl.) M.Gómez (Melastomataceae) **IPNI:** 159943-2 **Common name:** tuno rosado, tuno roso **Habit:** Shrub, Small tree, Tree **Use:** HF **Origin:** Endemic **Region:** Andean **Dept.:** ANT, CAL, CES, CUN, MAG, TOL **Elev.:** 0–1400 m a.s.l. **IUCN Habitat code:** 1,2,3,14 **Conservation status: (IUCN)** NE

6238 *Monochaetum myrtoideum* (Bonpl.) Naudin (Melastomataceae) **IPNI:** 573702-1 **Common name:** angelito, charne, mortiño, saltón, tortolito **Habit:** Shrub, Subshrub **Use:** EU **Origin:** Native **Region:** Andean **Dept.:** BOG, BOY, CAS, CUN, MET, NAR, NSA, PUT, QUI, SAN, TOL **Elev.:** 1900–3900 m a.s.l. **IUCN Habitat code:** 1,3,4,5,8,14 **Conservation status: (IUCN)** NE

6239 *Monolena primuliflora* Hook.f. (Melastomataceae) **IPNI:** 323262-2 **Common name:** chulco, churco, hoja limón, muela de vaca, muelavaca **Habit:** Herb, Epiphyte **Use:** ME **Origin:** Native **Region:** Andean, Orinoquia, Pacific **Dept.:** ANT, CAQ, CAU, CHO, CUN, MET, NAR, PUT, RIS, VAC **Elev.:** 0–2600 m a.s.l. **IUCN Habitat code:** 1,3,5,14 **Conservation status: (IUCN)** NE

6240 *Mouriri acutiflora* Naudin (Melastomataceae) **IPNI:** 573777-1 **Habit:** Shrub, Tree **Use:** HF **Origin:** Native **Region:** Amazonia, Orinoquia **Dept.:** AMA, CAQ, MET, VAU **Elev.:** 100–250 m a.s.l. **IUCN Habitat code:** 1,3,5 **Conservation status: (IUCN)** LC

6241 *Mouriri cauliflora* DC. (Melastomataceae) **IPNI:** 573790-1 **Common name:** guayaba de monte, guayabo de monte, guayacán llanero **Habit:** Shrub, Small tree, Tree **Use:** HF **Origin:** Native **Region:** Guiana Shield, Orinoquia **Dept.:** AMA, CAQ, GUA, GUV, VAU **Elev.:** 180–300 m a.s.l. **IUCN Habitat code:** 1,3,5,14 **Conservation status: (IUCN)** LC

6242 *Mouriri collocarpa* Ducke (Melastomataceae) **IPNI:** 164540-2 **Habit:** Tree **Use:** HF **Origin:** Native **Region:** Orinoquia **Dept.:** VID **Elev.:** 250–250 m a.s.l. **IUCN Habitat code:** 1,3,4,5 **Conservation status: (IUCN)** LC

6243 *Mouriri completens* Burret (Melastomataceae) **IPNI:** 164541-2 **Common name:** arrecheche, gaspadillo, gasparillo **Habit:** Shrub, Tree **Use:** MA **Origin:** Native **Region:** Pacific **Dept.:** ANT, CHO, VAC **Elev.:** 0–100 m a.s.l. **IUCN Habitat code:** 1,3,5,14 **Conservation status: (IUCN)** EN

6244 *Mouriri crassifolia* Sagot (Melastomataceae) **IPNI:** 573797-1 **Habit:** Tree **Use:** HF, MA **Origin:** Native **Region:** Amazonia **Dept.:** AMA **Elev.:** 65–65 m a.s.l. **IUCN Habitat code:** 1 **Conservation status: (IUCN)** NE

6245 *Mouriri densifoliata* Ducke (Melastomataceae) **IPNI:** 164543-2 **Habit:** Tree **Use:** HF **Origin:** Native **Region:** Amazonia **Dept.:** VAU **Elev.:** 200–200 m a.s.l. **IUCN Habitat code:** 3 **Conservation status: (IUCN)** NE

6246 *Mouriri ficoides* Morley (Melastomataceae) **IPNI:** 164554-2 **Habit:** Tree **Use:** HF **Origin:** Native **Region:** Amazonia **Dept.:** GUA **Elev.:** 100–600 m a.s.l. **IUCN Habitat code:** 1 **Conservation status: (IUCN)** NE

6247 *Mouriri grandiflora* DC. (Melastomataceae) **IPNI:** 573813-1 **Common name:** jujucapeci **Habit:** Small tree, Tree **Use:** HF **Origin:** Native **Region:** Amazonia, Orinoquia, Pacific **Dept.:** AMA, ANT, CAQ, CHO, GUA, PUT, VAU, VID **Elev.:** 10–280 m a.s.l. **IUCN Habitat code:** 1,3,14 **Conservation status: (IUCN)** LC

6248 *Mouriri guianensis* Aubl. (Melastomataceae) **IPNI:** 306748-2 **Common name:** comiture, corocillo **Habit:** Tree **Use:** HF **Origin:** Native **Region:** Orinoquia **Elev.:** 100–300 m a.s.l. **IUCN Habitat code:** 1,4 **Conservation status: (IUCN)** LC

6249 *Mouriri myrtifolia* Spruce ex Triana (Melastomataceae) **IPNI:** 573830-1 **Habit:** Shrub, Tree **Use:** HF **Origin:** Native **Region:** Amazonia **Dept.:** AMA, CAQ **Elev.:** 200–200 m a.s.l. **IUCN Habitat code:** 1,2,3 **Conservation status: (IUCN)** LC

6250 *Mouriri myrtilloides* Poir. (Melastomataceae) **IPNI:** 573831-1 **Common name:** arraiján, guayacán, raiján **Habit:** Shrub, Small tree, Tree **Use:** HF **Origin:** Native **Region:** Amazonia, Andean, Caribbean, Orinoquia, Pacific **Dept.:** ANT, CES, GUA, HUI, LAG, MAG, MET, SAP, VID **Elev.:** 0–1200 m a.s.l. **IUCN Habitat code:** 1,3,5,14 **Conservation status: (IUCN)** LC

6251 *Mouriri nigra* (DC.) Morley & Morley (Melastomataceae) **IPNI:** 164583-2 **Common name:** caimito, moo-há-hê-tê **Habit:** Small tree, Tree **Use:** HF, MA **Origin:** Native **Region:** Amazonia, Guiana Shield, Caribbean, Orinoquia **Dept.:** AMA, ANT, CAQ, CAS, COR, GUA, GUV, PUT, VAU, VID **Elev.:** 200–925 m a.s.l. **IUCN Habitat code:** 1,3,4,5 **Conservation status: (IUCN)** LC

6252 *Mouriri oligantha* Pilg. (Melastomataceae) **IPNI:** 311816-2 **Habit:** Small tree **Use:** HF **Origin:** Native **Region:** Amazonia **Dept.:** AMA **Elev.:** 100–100 m a.s.l. **IUCN Habitat code:** 1 **Conservation status: (IUCN)** LC

6253 *Mouriri vernicosa* Naudin (Melastomataceae) **IPNI:** 573863-1 **Habit:** Tree **Use:** HF **Origin:** Native **Region:** Amazonia **Dept.:** AMA, VAU **Elev.:** 200–243 m a.s.l. **IUCN Habitat code:** 1 **Conservation status: (IUCN)** LC

6254 *Nepsera aquatica* Naudin (Melastomataceae) **IPNI:** 573916-1 **Habit:** Herb, Shrub, Subshrub **Use:** ME **Origin:** Native **Region:** Pacific **Dept.:** CAU, CHO, NAR, VAC **Elev.:** 0–700 m a.s.l. **IUCN Habitat code:** 1,3,14 **Conservation status: (IUCN)** NE

6255 *Pleroma urvilleanum* (DC.) P.J.F.Guim. & Michelang. (Melastomataceae) **IPNI:** 77206410-1 **Common name:** gamuzón, nazareno, sietecueros, sietecueros brasilero, sietecueros nazareno **Habit:** Shrub, Small tree, Tree **Use:** EU **Origin:** Cultivated, Naturalised **Region:** Andean **Dept.:** ANT, BOG, CAL, CAU, CUN, MET **Elev.:** 1600–2600 m a.s.l. **IUCN Habitat code:** 1,2,3,14 **Conservation status: (IUCN)** NE

6256 *Pterolepis trichotoma* (Rottb.) Cogn. (Melastomataceae) **IPNI:** 1131052-2 **Habit:** Herb **Use:** ME **Origin:** Native **Region:** Andean, Caribbean, Orinoquia **Dept.:** ANT, ARA, CAS, CES, CUN, GUV, LAG, MAG, MET, TOL, VAC, VID **Elev.:** 175–1150 m a.s.l. **IUCN Habitat code:** 1,2,3,4,5,14 **Conservation status: (IUCN)** NE

6257 *Rhynchanthera grandiflora* DC. (Melastomataceae) **IPNI:** 575460-1 **Habit:** Herb, Shrub, Subshrub **Use:** ME **Origin:** Native **Region:** Andean, Caribbean, Orinoquia **Dept.:** ARA, BOY, CAS, CES, GUA, GUV, MAG, MET, NSA, VID **Elev.:** 70–700 m a.s.l. **IUCN Habitat code:** 1,2,3,4,5,14 **Conservation status: (IUCN)** NE

6258 *Triolena spicata* (Triana) L.O.Williams (Melastomataceae) **IPNI:** 258514-2 **Common name:** rabichucha **Habit:** Herb, Shrub, Subshrub, Epiphyte, Liana, Climbing **Use:** ME **Origin:** Native **Region:** Andean, Pacific **Dept.:** ANT, CAU, CHO, NAR, VAC **Elev.:** 0–950 m a.s.l. **IUCN Habitat code:** 1,3,14 **Conservation status: (IUCN)** NE

6259 *Calycolpus calophyllus* (Kunth) O.Berg (Myrtaceae) **IPNI:** 1061290-2 **Habit:** Shrub **Use:** HF **Origin:** Native **Region:** Amazonia, Guiana Shield **Dept.:** GUA, MET, VAU, VID **Elev.:** 200–400 m a.s.l. **IUCN Habitat code:** 1,3,4,14 **Conservation status: (IUCN)** LC

6260 *Calycolpus goetheanus* (Mart. ex DC.) O.Berg (Myrtaceae) **IPNI:** 591482-1 **Habit:** Shrub, Tree **Use:** HF **Origin:** Native **Dept.:** TOL **Elev.:** 300–300 m a.s.l. **IUCN Habitat code:** 1,14 **Conservation status: (IUCN)** LC

6261 *Calycolpus moritzianus* (O.Berg) Burret (Myrtaceae) **IPNI:** 43033-2 **Common name:** arrayana, arrayán, arrayán guayabo, cinaro, guayabito, guayabo, guayabo liso, gumán, jumangue, keiyá, palo blanco, sinaro **Habit:** Small tree, Tree **Use:** HF, ME **Origin:** Native **Region:** Andean **Dept.:** ANT, BOG, BOY, CES, CUN, HUI, MAG, MET, NSA, SAN **Elev.:** 700–3500 m a.s.l. **IUCN Habitat code:** 1,2,3,4,14 **Conservation status: (IUCN)** LC

6262 *Campomanesia lineatifolia* Ruiz & Pav. (Myrtaceae) **IPNI:** 592044-1 **Common name:** ansermo, bichinche, chamba, champa, champe, champito, champo, chiga, guamo chigo, guayaba champa, guayaba de leche, guayabito, guayabo anselmo, guayabo brasilero, guayabo champo, guayabo cimarrón, guayabo de Anselmo, guayabo de leche, guayabo de mono, mic **Habit:** Tree **Use:** AF, EU, FU, HF, MA, ME, PO **Origin:** Native, Cultivated **Region:** Amazonia, Andean, Orinoquia, Pacific **Dept.:** AMA, ANT, BOY, CAU, CHO, CUN, GUV, NAR, SAN, VAU **Elev.:** 20–1850 m a.s.l. **IUCN Habitat code:** 1,2,3,14 **Conservation status: (IUCN)** LC

6263 *Corymbia calophylla* (Lindl.) K.D.Hill & L.A.S.Johnson (Myrtaceae) **IPNI:** 986246-1 **Habit:** Tree **Use:** MA, ME **Origin:** Cultivated **Region:** Andean **Dept.:** BOG, CUN **Elev.:** 2600–2600 m a.s.l. **IUCN Habitat code:** 14 **Conservation status: (IUCN)** NT

6264 *Corymbia citriodora* (Hook.) K.D.Hill & L.A.S.Johnson (Myrtaceae) **IPNI:** 986336-1 **Common name:** eucalipto, eucalipto aromático, eucalipto azul, ocalipto **Habit:** Tree **Use:** EU, FU, GS, HF, MA, ME **Origin:** Cultivated **Region:** Andean **Dept.:** BOG, CUN **Elev.:** 2600–2600 m a.s.l. **IUCN Habitat code:** 1,3,14 **Conservation status: (IUCN)** LC

6265 *Corymbia ficifolia* (F.Muell.) K.D.Hill & L.A.S.Johnson (Myrtaceae) **IPNI:** 986249-1 **Common name:** eucalipto, eucalipto pomarroso **Habit:** Tree **Use:** EU, MA, ME **Origin:** Cultivated **Region:** Andean **Dept.:** BOG, CAL, CUN **Elev.:** 2000–2600 m a.s.l. **IUCN Habitat code:** 3,14 **Conservation status: (IUCN)** NT

6266 *Corymbia tessellaris* (F.Muell.) K.D.Hill & L.A.S.Johnson (Myrtaceae) **IPNI:** 986340-1 **Use:** MA **Region:** Andean **Dept.:** BOG **Conservation status: (IUCN)** NT

6267 *Eucalyptus camaldulensis* Dehnh. (Myrtaceae) **IPNI:** 592777-1 **Common name:** eucalipto, eucaliptus **Habit:** Tree **Use:** AF, EU, FU, HF, IF, MA, ME **Origin:** Cultivated **Region:** Andean **Dept.:** ANT, BOG, CUN **Elev.:** 1600–2600 m a.s.l. **IUCN Habitat code:** 1,2,3,5,14 **Conservation status: (IUCN)** NT

6268 *Eucalyptus cinerea* F.Muell. ex Benth. (Myrtaceae) **IPNI:** 592798-1 **Common name:** eucalipto, eucalipto plateado, ocalita **Habit:** Tree **Use:** EU, MA, ME, SU **Origin:** Cultivated **Region:** Andean **Dept.:** BOG, CUN **Elev.:** 2600–2600 m a.s.l. **IUCN Habitat code:** 1,3,14 **Conservation status: (IUCN)** NT

6269 *Eucalyptus deanei* Maiden (Myrtaceae) **IPNI:** 592863-1 **Use:** MA **Region:** Andean **Dept.:** BOG **IUCN Habitat code:** 1,3,14 **Conservation status: (IUCN)** LC

6270 *Eucalyptus elata* Dehnh. (Myrtaceae) **IPNI:** 592904-1 **Use:** EU, MA **Region:** Andean **Dept.:** BOG **Conservation status: (IUCN)** LC

6271 *Eucalyptus globulus* Labill. (Myrtaceae) **IPNI:** 592965-1 **Common name:** caucapito, eucalipto, eucalipto común, eucaliptus, gomero azul, ocal, ocalito, ucal **Habit:** Tree **Use:** AF, EU, FU, HF, IF, MA, ME, PO, SU **Origin:** Cultivated **Region:** Andean **Dept.:** ANT, BOG, CUN **Elev.:** 1500–2900 m a.s.l. **IUCN Habitat code:** 1,3,5,14 **Conservation status: (IUCN)** LC

6272 *Eucalyptus grandis* W.Hill ex Maiden (Myrtaceae) **IPNI:** 592976-1 **Common name:** eucalipto, eucaliptus, flooded gum, rose gum, saligna gum **Habit:** Tree **Use:** EU, FU, HF, IF, MA, ME **Origin:** Cultivated, Naturalised **Region:** Andean **Dept.:** ANT, CUN **Elev.:** 1600–1600 m a.s.l. **IUCN Habitat code:** 1,3,14 **Conservation status: (IUCN)** NT

6273 *Eucalyptus polyanthemos* Schauer (Myrtaceae) **IPNI:** 593261-1 **Use:** FU, MA, ME **Region:** Andean **Dept.:** BOG **Conservation status: (IUCN)** NT

6274 *Eucalyptus pulverulenta* Sims (Myrtaceae) **IPNI:** 593283-1 **Common name:** baby blue, eucalipto, eucalipto plateado **Habit:** Tree **Use:** EU, HF, MA, ME **Origin:** Cultivated **Region:** Andean **Dept.:** ANT, BOG, CUN **Elev.:** 1600–2600 m a.s.l. **IUCN Habitat code:** 3,14 **Conservation status: (IUCN)** LC

6275 *Eucalyptus resinifera* J.White (Myrtaceae)
IPNI: 593305-1
Common name: eucalipto, kino gum, ocal, red mahogany **Habit:** Tree **Use:** FU, HF, MA, ME **Origin:** Cultivated **Region:** Andean **Dept.:** ANT, BOG, CUN **Elev.:** 1400–2600 m a.s.l. **IUCN Habitat code:** 4,14 **Conservation status: (IUCN)** LC

6276 *Eucalyptus robusta* Sm. (Myrtaceae)
IPNI: 593318-1
Use: AF, EU, FU, HF, MA, ME **Region:** Andean **Dept.:** BOG **IUCN Habitat code:** 1,3,14 **Conservation status: (IUCN)** NT

6277 *Eucalyptus saligna* Sm. (Myrtaceae)
IPNI: 593334-1
Use: EU, MA, ME **Region:** Andean **Dept.:** BOG **IUCN Habitat code:** 14 **Conservation status: (IUCN)** LC

6278 *Eucalyptus tereticornis* Sm. (Myrtaceae)
IPNI: 593412-1
Common name: eucalipto **Habit:** Tree **Use:** AF, EU, FU, HF, IF, MA, ME, PO **Origin:** Cultivated **Dept.:** ANT **Elev.:** 1200–1200 m a.s.l. **IUCN Habitat code:** 1,2,3,14 **Conservation status: (IUCN)** LC

6279 *Eugenia acapulcensis* Steud. (Myrtaceae)
IPNI: 593490-1
Common name: arrayán, café macho, escobillo **Habit:** Shrub **Use:** AF, HF, MA, ME **Origin:** Native **Region:** Caribbean **Dept.:** ATL, SAP **Elev.:** 5–15 m a.s.l. **IUCN Habitat code:** 1,2,3,14 **Conservation status: (IUCN)** LC

6280 *Eugenia anastomosans* DC. (Myrtaceae)
IPNI: 593585-1
Common name: pisuilse **Habit:** Tree **Use:** HF **Origin:** Native **Region:** Amazonia **Dept.:** AMA, VAU **Elev.:** 100–200 m a.s.l. **IUCN Habitat code:** 1,3,5,14 **Conservation status: (IUCN)** LC

6281 *Eugenia axillaris* (Sw.) Willd. (Myrtaceae)
IPNI: 593690-1
Use: HF, MA, ME **Conservation status: (IUCN)** LC

6282 *Eugenia biflora* (L.) DC. (Myrtaceae)
IPNI: 593768-1
Common name: arrayán, tsitsinae **Habit:** Shrub **Use:** HF, MA, ME **Origin:** Native **Region:** Amazonia, Andean, Guiana Shield, Caribbean, Orinoquia, Pacific **Dept.:** ANT, CAL, CES, CHO, CUN, HUI, LAG, MET, NSA, RIS, SAN, TOL, VAC, VID **Elev.:** 50-1850 m.a.s.l. **IUCN Habitat code:** 1,2,3,4,5,14 **Conservation status: (IUCN)** LC **(CPSG)** Potential LC

6283 *Eugenia cotinifolia* Jacq. (Myrtaceae)
IPNI: 594126-1
Use: HF, MA **Origin:** Endemic **IUCN Habitat code:** 1,2,3,4,14 **Conservation status: (IUCN)** NE

6284 *Eugenia cribrata* McVaugh (Myrtaceae)
IPNI: 98260-2
Habit: Shrub **Use:** HF **Origin:** Native **Region:** Orinoquia **Dept.:** MET **Elev.:** 400–500 m a.s.l. **IUCN Habitat code:** 1,2,3,14 **Conservation status: (IUCN)** NE

6285 *Eugenia egensis* DC. (Myrtaceae)
IPNI: 326111-2
Habit: Shrub, Small tree **Use:** HF **Origin:** Native **Region:** Amazonia **Dept.:** AMA, ANT **Elev.:** 100–925 m a.s.l. **IUCN Habitat code:** 1,3,5,14 **Conservation status: (IUCN)** LC

6286 *Eugenia florida* DC. (Myrtaceae)
IPNI: 594452-1
Common name: arrayán, guayabillo, guayabo, palo de guayabilla, payo **Habit:** Shrub, Small tree, Tree **Use:** HF, ME **Origin:** Native **Region:** Amazonia, Andean, Orinoquia, Pacific **Dept.:** AMA, ANT, BOY, CAQ, CHO, CUN, GUV, MET, NSA, VAC, VAU **Elev.:** 5–1720 m a.s.l. **IUCN Habitat code:** 1,2,3,4,5,14 **Conservation status: (IUCN)** LC

6287 *Eugenia galalonensis* (C.Wright ex Griseb.) Krug & Urb. (Myrtaceae)
IPNI: 594499-1
Habit: Shrub, Small tree **Use:** HF **Origin:** Native **Region:** Caribbean **Dept.:** MAG, SAP **Elev.:** 40–40 m a.s.l. **IUCN Habitat code:** 1 **Conservation status: (IUCN)** NE

6288 *Eugenia inundata* DC. (Myrtaceae)
IPNI: 594718-1
Use: HF **Origin:** Native **Conservation status: (IUCN)** LC

6289 *Eugenia lambertiana* DC. (Myrtaceae)
IPNI: 594830-1
Common name: arrayán **Habit:** Shrub, Small tree **Use:** HF **Origin:** Native **Region:** Amazonia, Orinoquia **Dept.:** ANT, MET, VAU **Elev.:** 300–300 m a.s.l. **IUCN Habitat code:** 1,3,4,5,14 **Conservation status: (IUCN)** LC

6290 *Eugenia monticola* (Sw.) DC. (Myrtaceae)
IPNI: 595185-1
Habit: Shrub, Tree **Use:** HF, MA **Origin:** Native **Dept.:** ANT, TOL, VAC **Elev.:** 350–800 m a.s.l. **IUCN Habitat code:** 1,3,14 **Conservation status: (IUCN)** LC

6291 *Eugenia multirimosa* McVaugh (Myrtaceae)
IPNI: 98702-2
Habit: Tree **Use:** HF **Origin:** Native **Region:** Pacific **Dept.:** CHO **Elev.:** 0–100 m a.s.l. **IUCN Habitat code:** 1 **Conservation status: (IUCN)** LC

6292 *Eugenia oerstediana* O.Berg (Myrtaceae)
IPNI: 595343-1
Habit: Small tree **Use:** HF **Origin:** Native **Region:** Caribbean **Dept.:** BOL **Elev.:** 200–200 m a.s.l. **IUCN Habitat code:** 1,2,3 **Conservation status: (IUCN)** LC

6293 *Eugenia patrisii* Vahl (Myrtaceae)
IPNI: 595457-1
Habit: Tree **Use:** HF, MA **Origin:** Native **Region:** Amazonia, Guiana Shield **Dept.:** AMA, CAQ, GUA, VAU **Elev.:** 80–300 m a.s.l. **IUCN Habitat code:** 1,3,4,5 **Conservation status: (IUCN)** LC

6294 *Eugenia polystachya* Rich. (Myrtaceae)
IPNI: 595593-1
Habit: Tree **Use:** HF **Origin:** Native **Region:** Amazonia **Dept.:** AMA **Elev.:** 100–220 m a.s.l. **Conservation status: (IUCN)** LC

6295 *Eugenia procera* (Sw.) Poir. (Myrtaceae)
IPNI: 595618-1
Common name: arrayán, arrayán chizo, guayabo de monte, guayabo rojo **Habit:** Shrub **Use:** HF, MA **Origin:** Native **Dept.:** VAC **Elev.:** 800–800 m a.s.l. **IUCN Habitat code:** 1,2,3,14 **Conservation status: (IUCN)** NE

6296 *Eugenia punicifolia* (Kunth) DC. (Myrtaceae)
IPNI: 1046564-2
Habit: Shrub, Subshrub **Use:** HF, MA **Origin:** Native **Region:** Amazonia, Andean, Guiana Shield, Orinoquia **Dept.:** CAQ, GUA, GUV, MET, SAN, VAU, VID **Elev.:** 90–1550 m a.s.l. **IUCN Habitat code:** 1,2,3,4,5,14 **Conservation status: (IUCN)** LC

6297 *Eugenia rhombea* (O.Berg) Krug & Urb. (Myrtaceae)
IPNI: 98934-2
Use: HF, MA **IUCN Habitat code:** 1,2 **Conservation status: (IUCN)** LC

6298 *Eugenia stipitata* McVaugh (Myrtaceae)
IPNI: 99059-2
Common name: aracá-boi, araza, arazá, arrayán, guayaba brasileña, mulchis, paraguaycito, pichi, rupiña caspi **Habit:** Shrub, Tree **Use:** HF, MA, ME **Origin:** Native, Cultivated **Region:** Amazonia, Orinoquia **Dept.:** ANT, CAQ, MET, PUT **Elev.:** 50–500 m a.s.l. **IUCN Habitat code:** 1,2,3,14 **Conservation status: (IUCN)** LC

6299 *Eugenia subterminalis* DC. (Myrtaceae)
IPNI: 326119-2
Habit: Small tree **Use:** HF **Origin:** Native **Region:** Amazonia **Dept.:** AMA **Elev.:** 100–100 m a.s.l. **IUCN Habitat code:** 1 **Conservation status: (IUCN)** NE

6300 *Eugenia umbellulifera* (Kunth) Krug & Urb. (Myrtaceae)
IPNI: 596253-1
Common name: fruto de morrocón, marira **Habit:** Shrub **Use:** HF, ME **Origin:** Endemic **Region:** Caribbean **Dept.:** LAG **Elev.:** 200–200 m a.s.l. **Conservation status: (IUCN)** NE

6301 *Eugenia uniflora* L. (Myrtaceae)
IPNI: 596263-1
Common name: cereza, cereza cuadrada, cereza de Cayena, cereza de Surinam, cereza de morrocoy, cerezo, cerezo de morrocoy, grosella, grosella roja, mirto, pitanga **Habit:** Shrub **Use:** EU, GS, HF, IF, MA, ME **Origin:** Cultivated **Region:** Andean, Caribbean **Dept.:** ANT, CAU, CUN, SAP, VAC **Elev.:** 1000–1850 m a.s.l. **IUCN Habitat code:** 1,2,3,14 **Conservation status: (IUCN)** LC

6302 *Eugenia venezuelensis* O.Berg (Myrtaceae)
IPNI: 596302-1
Habit: Shrub **Use:** HF, IF **Origin:** Native **Region:** Andean, Caribbean **Dept.:** ANT, SAP, TOL **Elev.:** 50–1300 m a.s.l. **IUCN Habitat code:** 1,2,3,14 **Conservation status: (IUCN)** LC

6303 *Eugenia victoriana* Cuatrec. (Myrtaceae) **IPNI:** 99187-2 **Common name:** guayabillo **Habit:** Tree **Use:** HF, ME **Origin:** Endemic **Region:** Pacific **Dept.:** CHO, NAR **Elev.:** 50-50 m.a.s.l. **IUCN Habitat code:** 1,3,14 **Conservation status: (IUCN)** LC **(CPSG)** LC

6304 *Feijoa sellowiana* (O.Berg) O.Berg (Myrtaceae) **IPNI:** 596467-1 **Common name:** feijoa, fraijoa, freijoa, guayaba feijoa **Habit:** Shrub **Use:** EU, HF, MA, ME **Origin:** Cultivated **Region:** Andean **Dept.:** ANT, CAL, CUN **Elev.:** 1500–2900 m a.s.l. **IUCN Habitat code:** 1,3,14 **Conservation status: (IUCN)** NE

6305 *Leptospermum scoparium* J.R.Forst. & G.Forst. (Myrtaceae) **IPNI:** 597505-1 **Common name:** eletroespermo, leptospermon, letoespermo **Habit:** Shrub **Use:** EU, HF, MA, ME **Origin:** Cultivated **Region:** Andean **Dept.:** BOG, CUN **Elev.:** 2600–2600 m a.s.l. **IUCN Habitat code:** 14 **Conservation status: (IUCN)** LC

6306 *Melaleuca armillaris* (Sol. ex Gaertn.) Sm. (Myrtaceae) **IPNI:** 597845-1 **Habit:** Shrub **Use:** EU, FU, MA **Origin:** Cultivated **Region:** Andean **Dept.:** CUN **Elev.:** 2800–2800 m a.s.l. **Conservation status: (IUCN)** LC

6307 *Melaleuca citrina* (Curtis) Dum.Cours. (Myrtaceae) **IPNI:** 77108602-1 **Habit:** Shrub **Use:** EU, FU, GS, HF, MA, ME, SU **Origin:** Cultivated **Region:** Andean **Dept.:** BOG, CUN **Elev.:** 2800–2800 m a.s.l. **IUCN Habitat code:** 3,14 **Conservation status: (IUCN)** NE

6308 *Melaleuca decussata* R.Br. (Myrtaceae) **IPNI:** 597912-1 **Habit:** Shrub **Use:** MA **Origin:** Cultivated **Region:** Andean **Dept.:** CUN **Elev.:** 2600–2600 m a.s.l. **Conservation status: (IUCN)** LC

6309 *Melaleuca quinquenervia* (Cav.) S.T.Blake (Myrtaceae) **IPNI:** 598073-1 **Common name:** calistemo blanco, corcho **Habit:** Shrub **Use:** EU, HF, IF, MA, ME **Origin:** Cultivated **Region:** Andean **Dept.:** ANT **Elev.:** 1470–1470 m a.s.l. **IUCN Habitat code:** 14 **Conservation status: (IUCN)** LC

6310 *Melaleuca viminalis* (Sol. ex Gaertn.) Byrnes (Myrtaceae) **IPNI:** 914221-1 **Habit:** Shrub **Use:** EU, FU, IF, MA **Origin:** Cultivated **Dept.:** BOG, VAC **Elev.:** 1100–1100 m a.s.l. **IUCN Habitat code:** 3,14 **Conservation status: (IUCN)** LC

6311 *Metrosideros excelsa* Sol. ex Gaertn. (Myrtaceae) **IPNI:** 598207-1 **Habit:** Small tree **Use:** EU, HF, MA, ME **Origin:** Cultivated **Region:** Andean **Dept.:** BOG, CUN **Elev.:** 2600–2600 m a.s.l. **IUCN Habitat code:** 14 **Conservation status: (IUCN)** NE

6312 *Myrcia albidotomentosa* (Amshoff) McVaugh (Myrtaceae) **IPNI:** 165538-2 **Use:** HF **Origin:** Native **IUCN Habitat code:** 1 **Conservation status: (IUCN)** LC

6313 *Myrcia argentigemma* E.Lucas & C.E.Wilson (Myrtaceae) **IPNI:** 77160821-1 **Common name:** arajuimuci, arrayán **Habit:** Tree **Use:** HF, ME **Origin:** Native **Region:** Amazonia, Guiana Shield **Dept.:** AMA, VAU **Elev.:** 250–700 m a.s.l. **IUCN Habitat code:** 1,4 **Conservation status: (IUCN)** LC

6314 *Myrcia bracteata* (Rich.) DC. (Myrtaceae) **IPNI:** 598646-1 **Common name:** arrayán **Habit:** Shrub **Use:** HF, MA, ME, SU **Origin:** Native **Region:** Amazonia, Guiana Shield, Orinoquia **Dept.:** AMA, ANT, CAQ, CAS, CUN, GUA, MET, SAN, VAU **Elev.:** 80–1000 m a.s.l. **IUCN Habitat code:** 1,2,3,4,5,14 **Conservation status: (IUCN)** LC

6315 *Myrcia celaenensis* E.Lucas & K.Samra (Myrtaceae) **IPNI:** 77199415-1 **Habit:** Tree **Use:** HF **Origin:** Native **Region:** Amazonia, Guiana Shield **Dept.:** GUA **Elev.:** 100–900 m a.s.l. **IUCN Habitat code:** 1 **Conservation status: (IUCN)** NE

6316 *Myrcia chytraculia* (L.) A.R.Lourenço & E.Lucas (Myrtaceae) **IPNI:** 77191371-1 **Habit:** Tree **Use:** MA, ME **Origin:** Endemic **Region:** Caribbean **Dept.:** MAG **Elev.:** 100–100 m a.s.l. **Conservation status: (IUCN)** LC

6317 *Myrcia clusiifolia* (Kunth) DC. (Myrtaceae) **IPNI:** 1005985-2 **Common name:** guayabo agrio **Habit:** Shrub, Small tree **Use:** HF, MA **Origin:** Native **Region:** Guiana Shield **Dept.:** GUA **Elev.:** 150–150 m a.s.l. **IUCN Habitat code:** 1 **Conservation status: (IUCN)** LC

6318 *Myrcia cuspidata* (Mart. ex DC.) A.R.Lourenço & E.Lucas (Myrtaceae) **IPNI:** 77191379-1 **Habit:** Tree **Use:** MA **Origin:** Native **Region:** Amazonia **Dept.:** AMA **Elev.:** 100–200 m a.s.l. **IUCN Habitat code:** 1,14 **Conservation status: (IUCN)** NE

6319 *Myrcia deflexa* (Poir.) DC. (Myrtaceae) **IPNI:** 598743-1 **Habit:** Tree **Use:** MA **Origin:** Native **Region:** Amazonia, Guiana Shield **Dept.:** CAQ, GUA **Elev.:** 320–320 m a.s.l. **IUCN Habitat code:** 1 **Conservation status: (IUCN)** LC

6320 *Myrcia fenzliana* O.Berg (Myrtaceae) **IPNI:** 328930-2 **Habit:** Tree **Use:** HF **Origin:** Native **Region:** Andean **Dept.:** HUI, SAN **Elev.:** 1500–1500 m a.s.l. **Conservation status: (IUCN)** NE

6321 *Myrcia guianensis* (Aubl.) DC. (Myrtaceae) **IPNI:** 598851-1 **Habit:** Shrub, Small tree **Use:** HF, MA **Origin:** Native **Region:** Amazonia, Guiana Shield, Orinoquia **Dept.:** AMA, CAQ, GUV, MET, VAU, VID **Elev.:** 80–630 m a.s.l. **Conservation status: (IUCN)** LC

6322 *Myrcia hylobates* (Standl. ex Amshoff) E.Lucas & K.Samra (Myrtaceae) **IPNI:** 77199448-1 **Common name:** bajé-guayaba, guayaba, guayabo de monte, molleja de paujil, nimekogore, nimé kogore **Habit:** Tree **Use:** HF, MA, ME **Origin:** Native **Region:** Pacific **Dept.:** CHO **Elev.:** 150–200 m a.s.l. **Conservation status: (IUCN)** LC

6323 *Myrcia inaequiloba* (DC.) Lemée (Myrtaceae) **IPNI:** 598883-1 **Common name:** guacure, guayabo hojiancho, guayabo hojimenudo, huesito **Habit:** Small tree, Tree **Use:** HF, MA **Origin:** Native **Region:** Amazonia, Orinoquia, Pacific **Dept.:** AMA, CAQ, CHO, VID **Elev.:** 10–520 m a.s.l. **IUCN Habitat code:** 1,3,4,5,14 **Conservation status: (IUCN)** LC

6324 *Myrcia megapaniculata* A.R.Lourenço & Parra-Os. (Myrtaceae) **IPNI:** 77191465-1 **Habit:** Tree **Use:** HF **Origin:** Native **Region:** Amazonia **Dept.:** AMA, CAQ **Elev.:** 250–250 m a.s.l. **IUCN Habitat code:** 1 **Conservation status: (IUCN)** NE

6325 *Myrcia mollis* (Kunth) DC. (Myrtaceae) **IPNI:** 598986-1 **Common name:** arrayán, arrayán blanco, guayabo **Habit:** Small tree, Tree **Use:** HF **Origin:** Native **Region:** Andean **Dept.:** ANT, CUN **Elev.:** 1000–2150 m a.s.l. **IUCN Habitat code:** 1,2,3,14 **Conservation status: (IUCN)** LC

6326 *Myrcia neoforsteri* A.R.Lourenço & E.Lucas (Myrtaceae) **IPNI:** 77191393-1 **Habit:** Tree **Use:** MA **Origin:** Native **Region:** Amazonia **Dept.:** AMA **Elev.:** 100–100 m a.s.l. **IUCN Habitat code:** 1 **Conservation status: (IUCN)** LC

6327 *Myrcia neolucida* A.R.Lourenço & E.Lucas (Myrtaceae) **IPNI:** 77191459-1 **Common name:** guayabilla **Habit:** Tree **Use:** HF, MA **Origin:** Native **Region:** Amazonia, Andean **Dept.:** AMA, ANT **Elev.:** 240–1700 m a.s.l. **IUCN Habitat code:** 1,5 **Conservation status: (IUCN)** NE

6328 *Myrcia neomontana* E.Lucas & C.E.Wilson (Myrtaceae) **IPNI:** 77160848-1 **Use:** MA **Origin:** Native **IUCN Habitat code:** 1 **Conservation status: (IUCN)** LC

6329 *Myrcia neoschomburgkiana* E.Lucas & C.E.Wilson (Myrtaceae) **IPNI:** 77160852-1 **Use:** MA **Origin:** Native **IUCN Habitat code:** 1,3 **Conservation status: (IUCN)** LC

6330 *Myrcia neospeciosa* A.R.Lourenço & E.Lucas (Myrtaceae) **IPNI:** 77191461-1 **Common name:** berraquillo, chontaduro blanco, poro **Habit:** Tree **Use:** HF, MA **Origin:** Native **Region:** Amazonia, Guiana Shield, Orinoquia, Pacific **Dept.:** AMA, ANT, CAQ, CHO, CUN, GUV, PUT, VAU, VID **Elev.:**

30–1400 m a.s.l. **IUCN Habitat code:** 1,3,5,14 **Conservation status: (IUCN)** LC

6331 *Myrcia paivae* O.Berg (Myrtaceae)
IPNI: 599026-1
Common name: arrayancito, arrayán **Habit:** Shrub, Small tree, Tree **Use:** HF, ME **Origin:** Native **Region:** Amazonia, Andean, Guiana Shield, Pacific **Dept.:** AMA, ANT, CAQ, CHO, GUV, MET, VAU **Elev.:** 100–1700 m a.s.l. **IUCN Habitat code:** 1,3,4,5,14 **Conservation status: (IUCN)** LC

6332 *Myrcia popayanensis* Hieron. (Myrtaceae)
IPNI: 165914-2
Common name: arrayancito, arrayán, guayabo blanco, árbol de cera **Habit:** Small tree **Use:** HF, ME **Origin:** Endemic **Region:** Andean **Dept.:** ANT, BOG, CUN, HUI **Elev.:** 1280-2350 m.a.s.l. **IUCN Habitat code:** 1,3,5,14 **Conservation status: (IUCN)** LC **(CPSG)** LC

6333 *Myrcia pulchella* (DC.) A.R.Lourenço & E.Lucas (Myrtaceae)
IPNI: 77191477-1
Common name: guayabilla **Habit:** Small tree, Tree **Use:** MA **Origin:** Native **Region:** Amazonia, Guiana Shield, Orinoquia **Dept.:** AMA, CAQ, GUV, MET, VAU **Elev.:** 220–700 m a.s.l. **IUCN Habitat code:** 1,3,14 **Conservation status: (IUCN)** LC

6334 *Myrcia pyrifolia* (Desv.) Nied. (Myrtaceae)
IPNI: 165926-2
Habit: Tree **Use:** MA **Origin:** Native **Region:** Amazonia **Dept.:** AMA **Elev.:** 270–270 m a.s.l. **IUCN Habitat code:** 1 **Conservation status: (IUCN)** LC

6335 *Myrcia salicifolia* DC. (Myrtaceae)
IPNI: 599124-1
Common name: guayabilla **Habit:** Shrub, Small tree **Use:** HF **Origin:** Native **Region:** Amazonia **Dept.:** AMA, CAQ **Elev.:** 200–250 m a.s.l. **IUCN Habitat code:** 1,4 **Conservation status: (IUCN)** LC

6336 *Myrcia splendens* (Sw.) DC. (Myrtaceae)
IPNI: 599170-1
Common name: arrayancito, arrayán, arrayán de hoja pequeña, arrayán, arrayán guayabo, chiso sabanero, cinaro, payo, tambe, usmillo, usmillo de corteza **Habit:** Shrub, Small tree, Tree **Use:** HF, MA, ME **Origin:** Native **Region:** Amazonia, Andean, Guiana Shield, Caribbean, Orinoquia, Pacific **Dept.:** AMA, ANT, BOL, BOY, CAQ, CAU, CHO, COR, CUN, GUV, HUI, MAG, MET, PUT, SAN, VAC, VAU, VID **Elev.:** 0-2800 m.a.s.l. **IUCN Habitat code:** 1,2,3,4,5,14 **Conservation status: (IUCN)** LC **(CPSG)** Potential LC

6337 *Myrcia subsessilis* O.Berg (Myrtaceae)
IPNI: 599191-1
Common name: arrayán **Habit:** Shrub, Small tree, Tree **Use:** HF, ME **Origin:** Native **Region:** Amazonia, Andean, Guiana Shield, Orinoquia **Dept.:** AMA, ANT, BOY, CAQ, CAS, GUV, MET, VAU, VID **Elev.:** 100–2300 m a.s.l. **IUCN Habitat code:** 1,2,3,4,5,14 **Conservation status: (IUCN)** LC

6338 *Myrcia sylvatica* (G.Mey.) DC. (Myrtaceae)
IPNI: 599199-1
Common name: arrayan, espinazo de picure **Habit:** Shrub, Tree **Use:** HF, MA, ME **Origin:** Native **Region:** Amazonia, Guiana Shield **Dept.:** AMA, CAQ, GUA, MET, VAU **Elev.:** 120–850 m a.s.l. **IUCN Habitat code:** 1,3,5,14 **Conservation status: (IUCN)** LC

6339 *Myrcia tomentosa* (Aubl.) DC. (Myrtaceae)
IPNI: 599215-1
Habit: Shrub, Small tree **Use:** HF, MA **Origin:** Native **Region:** Andean **Dept.:** ANT, CAU, NSA **Elev.:** 650–1610 m a.s.l. **IUCN Habitat code:** 1,2,3,14 **Conservation status: (IUCN)** LC

6340 *Myrcia umbraticola* (Kunth) E.Lucas & C.E.Wilson (Myrtaceae)
IPNI: 77160867-1
Habit: Tree **Use:** HF, MA **Origin:** Native **Region:** Amazonia **Dept.:** AMA **Elev.:** 98–98 m a.s.l. **IUCN Habitat code:** 1 **Conservation status: (IUCN)** LC

6341 *Myrcianthes alaternifolia* (Benth.) Grifo ex B.Holst & M.L.Kawas. (Myrtaceae)
IPNI: 77198138-1
Habit: Small tree **Use:** HF **Origin:** Native **Region:** Andean **Dept.:** SAN **Elev.:** 2800–2900 m a.s.l. **IUCN Habitat code:** 14 **Conservation status: (IUCN)** NE

6342 *Myrcianthes borealis* McVaugh (Myrtaceae)
IPNI: 166070-2
Common name: arrayán **Habit:** Tree **Use:** ME **Origin:** Native **Region:** Andean **Dept.:** NSA **Elev.:** 2500–2800 m a.s.l. **IUCN Habitat code:** 1,14 **Conservation status: (IUCN)** NE

6343 *Myrcianthes discolor* (Kunth) McVaugh (Myrtaceae)
IPNI: 166084-2
Common name: arrayán **Habit:** Shrub, Small tree **Use:** HF **Origin:** Native **Region:** Andean **Dept.:** CAU **Elev.:** 2900–3000 m a.s.l. **IUCN Habitat code:** 1,14 **Conservation status: (IUCN)** NE

6344 *Myrcianthes fragrans* (Sw.) McVaugh (Myrtaceae)
IPNI: 166091-2
Habit: Shrub, Small tree **Use:** HF, MA, ME **Origin:** Native **Region:** Andean, Caribbean **Dept.:** ANT, BOG, BOY, CAL, CUN, HUI, LAG, MAG, NSA, SAP, VAC **Elev.:** 50–2560 m a.s.l. **IUCN Habitat code:** 1,3,4,14 **Conservation status: (IUCN)** LC

6345 *Myrcianthes hallii* (O.Berg) McVaugh (Myrtaceae)
IPNI: 166097-2
Habit: Shrub, Tree **Use:** HF **Origin:** Native **Region:** Andean **Dept.:** CAU **Elev.:** 2600–2600 m a.s.l. **IUCN Habitat code:** 14 **Conservation status: (IUCN)** NE

6346 *Myrcianthes leucoxyla* (Ortega) McVaugh (Myrtaceae)
IPNI: 166103-2
Common name: arrayán, arrayán grande, guayabito, guayabito liso, guayabo de Castilla, guayabo liso, guayabón, levaduro, mirto, mortiño, palo blanco **Habit:** Shrub, Small tree **Use:** AF, EU, HF, MA, ME, PO **Origin:** Native **Region:** Andean **Dept.:** BOG, BOY, CUN **Elev.:** 2650–3200 m a.s.l. **IUCN Habitat code:** 1,3,4,5,8,14 **Conservation status: (IUCN)** LC

6347 *Myrcianthes myrsinoides* (Kunth) Grifo (Myrtaceae)
IPNI: 974565-1
Common name: tulá **Habit:** Shrub **Use:** ME **Origin:** Native **Region:** Andean **Dept.:** ANT, CES, CHO, LAG, MAG, NSA, SAN **Elev.:** 2000–3500 m a.s.l. **IUCN Habitat code:** 1,3,4,14 **Conservation status: (IUCN)** LC

6348 *Myrcianthes orthostemon* (O.Berg) Grifo (Myrtaceae)
IPNI: 974566-1
Common name: arrayán negro **Habit:** Small tree **Use:** HF, ME **Origin:** Native **Region:** Andean **Dept.:** ANT, BOG, BOY, CAU, CUN, NAR, NSA, RIS, SAN **Elev.:** 2000–2600 m a.s.l. **IUCN Habitat code:** 1,2,3,4,5,14 **Conservation status: (IUCN)** LC

6349 *Myrcianthes rhopaloides* (Kunth) McVaugh (Myrtaceae)
IPNI: 166115-2
Common name: arrayán, arrayán guayabo, arrayán negro, chisa, chiso, garrapato, guayabo cascarillo, guayabo de monte, paio, riso **Habit:** Small tree **Use:** HF, MA, ME **Origin:** Native **Region:** Andean **Dept.:** ANT, BOG, BOY, CAU, CES, CHO, CUN, HUI, NAR, PUT, QUI, SAN, TOL, VAC **Elev.:** 1800-3200 m.a.s.l. **IUCN Habitat code:** 1,3,4,5,14 **Conservation status: (IUCN)** LC **(CPSG)** Potential LC

6350 *Myrciaria dubia* (Kunth) McVaugh (Myrtaceae)
IPNI: 166146-2
Common name: camucamu, guayabo, guayabo arrayán, minuake **Habit:** Shrub **Use:** HF, MA, ME **Origin:** Native, Cultivated **Region:** Amazonia, Guiana Shield **Dept.:** AMA, GUA **Elev.:** 100–150 m a.s.l. **IUCN Habitat code:** 1,4 **Conservation status: (IUCN)** LC

6351 *Myrciaria floribunda* (H.West ex Willd.) O.Berg (Myrtaceae)
IPNI: 599327-1
Common name: caimito canelo, guayaba, guayabilla, guayabito **Habit:** Small tree **Use:** HF, MA **Origin:** Native **Region:** Amazonia, Caribbean, Orinoquia, Pacific **Dept.:** ANT, BOL, CHO, MET, SAN, VAU **Elev.:** 50–1800 m a.s.l. **IUCN Habitat code:** 1,2,3,4,5,14 **Conservation status: (IUCN)** LC

6352 *Myrciaria vismeifolia* (Benth.) O.Berg (Myrtaceae)
IPNI: 1052354-2
Use: HF **Origin:** Native **Conservation status: (IUCN)** NE

6353 *Myrrhinium atropurpureum* Schott (Myrtaceae)
IPNI: 599403-1
Habit: Shrub **Use:** HF, MA, ME **Origin:** Native **Region:** Andean **Dept.:** VAC **Elev.:** 2150–2150 m a.s.l. **Conservation status: (IUCN)** LC

6354 *Myrteola nummularia* (Poir.) O.Berg (Myrtaceae)
IPNI: 599434-1
Habit: Subshrub **Use:** EU, HF, ME **Origin:** Native **Region:** Andean **Dept.:** ANT, BOG, BOY, CAU, CHO, CUN, HUI, NAR, PUT, QUI, RIS, TOL, VAC **Elev.:** 2380-3450 m.a.s.l. **IUCN Habitat code:** 1,3,4,8,14 **Conservation status: (IUCN)** LC **(CPSG)** Potential LC

6355 *Myrtus communis* L. (Myrtaceae)
IPNI: 599719-1
Common name: arrayán, mirto **Habit:** Subshrub **Use:** EU, FU, GS, HF, IF, MA, ME **Origin:** Cultivated **Region:** Andean **Dept.:** CUN **Elev.:** 2600–2600 m a.s.l. **IUCN Habitat code:** 3,14 **Conservation status: (IUCN)** LC

6356 *Pimenta dioica* (L.) Merr. (Myrtaceae) **IPNI:** 196799-2 **Common name:** clavo, malaqueta, mint, pimienta de olor, pimienta dulce, pimiento, promento **Habit:** Tree **Use:** AF, GS, HF, MA, ME **Origin:** Naturalised **Region:** Caribbean **Dept.:** SAP **Elev.:** 40–40 m a.s.l. **IUCN Habitat code:** 1 **Conservation status: (IUCN)** LC

6357 *Pimenta racemosa* (Mill.) J.W.Moore (Myrtaceae) **IPNI:** 196820-2 **Common name:** bay run, bay run bwaden, bayrum, berrón, malagueta **Habit:** Tree **Use:** EU, GS, HF, MA, ME **Origin:** Cultivated **Region:** Andean **Dept.:** ANT **Elev.:** 1550–1550 m a.s.l. **IUCN Habitat code:** 14 **Conservation status: (IUCN)** NE

6358 *Plinia cauliflora* (Mart.) Kausel (Myrtaceae) **IPNI:** 203847-2 **Common name:** ibapurú, jaboticaba **Use:** HF, ME **IUCN Habitat code:** 3,14 **Conservation status: (IUCN)** LC

6359 *Plinia duplipilosa* McVaugh (Myrtaceae) **IPNI:** 203854-2 **Common name:** ñumeroca **Habit:** Small tree **Use:** HF **Origin:** Native **Region:** Amazonia **Dept.:** AMA **Elev.:** 240–240 m a.s.l. **IUCN Habitat code:** 1,5 **Conservation status: (IUCN)** NE

6360 *Plinia pinnata* L. (Myrtaceae) **IPNI:** 203871-2 **Use:** HF **IUCN Habitat code:** 1 **Conservation status: (IUCN)** NE

6361 *Psidium acutangulum* DC. (Myrtaceae) **IPNI:** 600701-1 **Common name:** guayabilla, guayabo rebalsero **Habit:** Tree **Use:** HF, MA, ME **Origin:** Native **Region:** Amazonia, Orinoquia **Dept.:** AMA, CUN, MET **Elev.:** 100–850 m a.s.l. **IUCN Habitat code:** 1,2,3,4 **Conservation status: (IUCN)** LC

6362 *Psidium cattleyanum* Sabine (Myrtaceae) **IPNI:** 600760-1 **Habit:** Shrub **Use:** EU, GS, HF, MA, ME **Origin:** Cultivated, Naturalised **Region:** Andean **Dept.:** ANT, CAU, SAN, VAC **Elev.:** 1000–2300 m a.s.l. **Conservation status: (IUCN)** NE

6363 *Psidium densicomum* Mart. ex DC. (Myrtaceae) **IPNI:** 600795-1 **Common name:** guayabilla **Habit:** Tree **Use:** HF **Origin:** Native **Region:** Amazonia **Dept.:** AMA **Elev.:** 100–100 m a.s.l. **IUCN Habitat code:** 1,2,3,5 **Conservation status: (IUCN)** LC

6364 *Psidium friedrichsthalianum* (O.Berg) Nied. (Myrtaceae) **IPNI:** 211742-2 **Common name:** cas, guayaba agria, guayaba de choco, guayabo de Castilla **Habit:** Tree **Use:** EU, GS, HF, MA **Origin:** Native, Cultivated **Region:** Andean, Caribbean **Dept.:** ANT, ATL, CAL, CES, CUN, SAN, VAC **Elev.:** 300–1560 m a.s.l. **IUCN Habitat code:** 1,3,14 **Conservation status: (IUCN)** LC

6365 *Psidium guajava* L. (Myrtaceae) **IPNI:** 600841-1 **Common name:** guayaba, guayaba común, guayaba dulce, guayaba manzana, guayaba pera, guayabo, guayabo colorado, guayabo dulce, guayabo pera, jujuli, jujulinae, vayavacù **Habit:** Small tree **Use:** AF, EU, FU, GS, HF, IF, MA, ME, PO **Origin:** Native, Cultivated **Region:** Amazonia, Andean, Caribbean, Orinoquia, Pacific **Dept.:** AMA, ANT, BOG, BOY, CAL, CAQ, CAU, CUN, HUI, MAG, MET, NAR, PUT, SAN, TOL, VAC **Elev.:** 50–2040 m a.s.l. **IUCN Habitat code:** 1,2,3,5,14 **Conservation status: (IUCN)** LC

6366 *Psidium guineense* Sw. (Myrtaceae) **IPNI:** 211886-2 **Common name:** arrayán, balazo, choba, chobo, chocolate, guayaba, guayaba agria, guayaba choba, guayaba sabanera, guayabilla, guayabillo, guayabita de conejo, guayabita sabanera, guayabito, guayabito agrio, guayabito de cerro, guayabo, guayabo agrio, guayabo chobo, guay **Habit:** Small tree **Use:** EU, HF, MA, ME **Origin:** Native, Cultivated **Region:** Amazonia, Andean, Caribbean, Orinoquia **Dept.:** AMA, ANT, BOY, CAL, CAQ, CAS, CAU, CES, CUN, HUI, LAG, MAG, MET, NAR, NSA, QUI, RIS, SAN, VAC, VAU, VID **Elev.:** 80-2300 m.a.s.l. **IUCN Habitat code:** 1,2,3,4,5,14 **Conservation status: (IUCN)** LC **(CPSG)** Potential LC

6367 *Psidium guyanense* Pers. (Myrtaceae) **IPNI:** 600845-1 **Use:** HF **IUCN Habitat code:** 1,2,3,4,14 **Conservation status: (IUCN)** LC

6368 *Psidium humile* Vell. (Myrtaceae) **IPNI:** 600857-1 **Common name:** guayabo del brazil **Use:** HF, ME **IUCN Habitat code:** 1,3,5,14 **Conservation status: (IUCN)** NE

6369 *Psidium laruotteanum* Cambess. (Myrtaceae) **IPNI:** 600884-1 **Habit:** Shrub **Use:** HF **Origin:** Native **Region:** Orinoquia **Dept.:** VID **Elev.:** 150–150 m a.s.l. **IUCN Habitat code:** 4 **Conservation status: (IUCN)** NE

6370 *Psidium oligospermum* Mart. ex DC. (Myrtaceae) **IPNI:** 600952-1 **Habit:** Shrub, Tree **Use:** EU, HF, MA, ME **Origin:** Native **Region:** Caribbean **Dept.:** MAG, SAP **Elev.:** 300–300 m a.s.l. **IUCN Habitat code:** 1,3,14 **Conservation status: (IUCN)** LC

6371 *Psidium salutare* (Kunth) O.Berg (Myrtaceae) **IPNI:** 1115374-2 **Habit:** Shrub **Use:** HF, ME **Origin:** Native **Region:** Andean, Orinoquia **Dept.:** BOY, CAU, VID **Elev.:** 150–1750 m a.s.l. **IUCN Habitat code:** 1,3,4,14 **Conservation status: (IUCN)** LC

6372 *Syzygium cumini* (L.) Skeels (Myrtaceae) **IPNI:** 601603-1 **Common name:** aceituno, aceituno dulce, jambolán, pomarroso, uva, uvito, yambolán **Habit:** Tree **Use:** EU, FU, GS, HF, IF, MA, ME, SU **Origin:** Cultivated **Region:** Andean, Caribbean **Dept.:** CES, CUN, VAC **Elev.:** 150–2430 m a.s.l. **IUCN Habitat code:** 1,2,3,5,14 **Conservation status: (IUCN)** LC

6373 *Syzygium jambos* (L.) Alston (Myrtaceae) **IPNI:** 601802-1 **Common name:** manzanita de rosa, poma, pomarrosa, pomarrosa común, pomarroso, pomo, pomorroso, pomorroso de Castilla **Habit:** Tree **Use:** EU, FU, GS, HF, MA, ME, PO **Origin:** Cultivated **Region:** Amazonia, Andean, Caribbean, Orinoquia, Pacific **Dept.:** ANT, BOG, BOY, CAL, CAU, CHO, CUN, HUI, LAG, MAG, MET, NSA, PUT, QUI, SAN, TOL, VAC **Elev.:** 30–2350 m a.s.l. **IUCN Habitat code:** 1,2,3,4,5,14 **Conservation status: (IUCN)** LC

6374 *Syzygium maingayi* Chantar. & J.Parn. (Myrtaceae) **IPNI:** 973178-1 **Use:** ME **IUCN Habitat code:** 1 **Conservation status: (IUCN)** NE

6375 *Syzygium malaccense* (L.) Merr. & L.M.Perry (Myrtaceae) **IPNI:** 601907-1 **Common name:** manzana de agua, manzano chino, marañón, pera, pera de Malasia, pera de agua, perita, perito, pero de agua, pomarrosa, pomarrosa del Brasil, pomarroso, pomarroso brasilero, pomarroso brasileño, pomo, pomorroso, pomorroso brasilero, ñambo **Habit:** Tree **Use:** AF, EU, FU, HF, MA, ME **Origin:** Cultivated **Region:** Amazonia, Andean, Orinoquia, Pacific **Dept.:** AMA, ANT, ARA, CAQ, CHO, CUN, GUA, GUV, QUI, TOL, VAC **Elev.:** 20–1200 m a.s.l. **IUCN Habitat code:** 1,2,3,4,5,14 **Conservation status: (IUCN)** LC

6376 *Syzygium paniculatum* Gaertn. (Myrtaceae) **IPNI:** 602052-1 **Common name:** eugenia, eugenio **Habit:** Small tree **Use:** EU, HF **Origin:** Cultivated **Region:** Andean **Dept.:** ANT, BOG, CUN **Elev.:** 1500–2600 m a.s.l. **IUCN Habitat code:** 1,3,5,14 **Conservation status: (IUCN)** NE

6377 *Ugni myricoides* (Kunth) O.Berg (Myrtaceae) **IPNI:** 602600-1 **Common name:** arrayancillo **Habit:** Shrub, Small tree **Use:** HF, ME **Origin:** Native **Region:** Andean **Dept.:** ANT, BOG, BOY, CAL, CAU, CHO, CUN, HUI, NAR, NSA, QUI, RIS, SAN, VAC **Elev.:** 2190-3500 m.a.s.l. **IUCN Habitat code:** 1,3,4,8,14 **Conservation status: (IUCN)** LC **(CPSG)** Potential LC

6378 *Epilobium denticulatum* Ruiz & Pav. (Onagraceae) **IPNI:** 611783-1 **Habit:** Herb **Use:** ME **Origin:** Native **Region:** Andean **Dept.:** ANT, BOG, BOY, CAL, CAU, CUN, HUI, MAG, MET, NAR, NSA, PUT, QUI, SAN, TOL, VAC **Elev.:** 2000–4160 m a.s.l. **IUCN Habitat code:** 1,3,4,5,6,14 **Conservation status: (IUCN)** NE

6379 *Fuchsia × hybrida* Voss (Onagraceae) **IPNI:** 612545-1 **Common name:** bailarina, fucsia, zarcillejo **Habit:** Shrub, Subshrub **Use:** EU, ME **Origin:** Endemic, Cultivated **Dept.:** ANT, BOG, CAL, CUN, RIS, VAC **Elev.:** 1025–3000 m a.s.l. **IUCN Habitat code:** 3,4,8,14 **Conservation status: (IUCN)** NE

6380 *Fuchsia ampliata* Benth. (Onagraceae) **IPNI:** 612463-1 **Habit:** Shrub **Use:** HF **Origin:** Native **Region:** Andean **Dept.:** PUT **Elev.:**

3000–3500 m a.s.l. **IUCN Habitat code:** 14 **Conservation status: (IUCN)** NE

6381 *Fuchsia arborescens* Sims (Onagraceae) **IPNI:** 106136-2 **Habit:** Shrub **Use:** EU, HF **Origin:** Cultivated **Dept.:** BOG, CUN **Elev.:** 2620–2750 m a.s.l. **IUCN Habitat code:** 1,3,5,14 **Conservation status: (IUCN)** NE

6382 *Fuchsia boliviana* Carrière (Onagraceae) **IPNI:** 106150-2 **Common name:** cigarrillito, fucsia, fucsia arbustiva, platanito, zarcillero **Habit:** Shrub **Use:** EU, HF, ME **Origin:** Naturalised **Region:** Andean **Dept.:** ANT, BOG, CUN, QUI, SAN, TOL **Elev.:** 1790–2700 m a.s.l. **IUCN Habitat code:** 1,3,4,6,8,14 **Conservation status: (IUCN)** LC

6383 *Fuchsia canescens* Benth. (Onagraceae) **IPNI:** 612484-1 **Common name:** bailarina, fucsia **Habit:** Shrub **Use:** HF, ME **Origin:** Endemic **Region:** Andean **Dept.:** CAU, HUI, NAR **Elev.:** 2800–3400 m a.s.l. **IUCN Habitat code:** 1,3,4,14 **Conservation status: (IUCN)** NE

6384 *Fuchsia corymbiflora* Ruiz & Pav. (Onagraceae) **IPNI:** 612501-1 **Use:** EU, HF **Dept.:** BOG **IUCN Habitat code:** 1,3,14 **Conservation status: (IUCN)** NE

6385 *Fuchsia hirtella* Kunth (Onagraceae) **IPNI:** 612542-1 **Habit:** Shrub **Use:** AF, EU **Origin:** Native **Region:** Andean **Dept.:** ANT, BOG, BOY, CUN, MET, QUI, VAC **Elev.:** 1900–3200 m a.s.l. **IUCN Habitat code:** 1,3,4,14 **Conservation status: (IUCN)** NE

6386 *Fuchsia magdalenae* Munz (Onagraceae) **IPNI:** 106221-2 **Common name:** sunshulá **Habit:** Shrub, Climbing **Use:** ME **Origin:** Endemic **Region:** Andean **Dept.:** MAG **Elev.:** 2670–3350 m a.s.l. **IUCN Habitat code:** 1,4 **Conservation status: (IUCN)** NE

6387 *Fuchsia magellanica* Lam. (Onagraceae) **IPNI:** 612572-1 **Habit:** Shrub **Use:** EU, GS, HF, MA, ME **Origin:** Naturalised **Region:** Andean **Dept.:** BOG, BOY, CUN **Elev.:** 2300–2800 m a.s.l. **IUCN Habitat code:** 1,3,4,14 **Conservation status: (IUCN)** NE

6388 *Fuchsia paniculata* Lindl. (Onagraceae) **IPNI:** 106252-2 **Habit:** Shrub **Use:** HF **Origin:** Cultivated, Naturalised **Dept.:** BOG, CUN **Elev.:** 2600–2600 m a.s.l. **Conservation status: (IUCN)** LC

6389 *Fuchsia regia* (Vand. ex Vell.) Munz (Onagraceae) **IPNI:** 106266-2 **Use:** MA **Region:** Andean **Dept.:** BOG **Conservation status: (IUCN)** NE

6390 *Fuchsia triphylla* L. (Onagraceae) **IPNI:** 106297-2 **Common name:** fucsia **Habit:** Shrub **Use:** HF **Origin:** Cultivated **Dept.:** BOG, CUN **Elev.:** 2950–2950 m a.s.l. **IUCN Habitat code:** 14 **Conservation status: (IUCN)** NE

6391 *Fuchsia vulcanica* André (Onagraceae) **IPNI:** 612680-1 **Habit:** Shrub **Use:** HF **Origin:** Native **Region:** Andean **Dept.:** NAR **Elev.:** 2800–3700 m a.s.l. **IUCN Habitat code:** 1,3,4,14 **Conservation status: (IUCN)** NE

6392 *Ludwigia adscendens* (L.) H.Hara (Onagraceae) **IPNI:** 144324-2 **Common name:** clavito, clavo de laguna **Use:** HF, ME **IUCN Habitat code:** 14 **Conservation status: (IUCN)** LC

6393 *Ludwigia affinis* (DC.) H.Hara (Onagraceae) **IPNI:** 144328-2 **Habit:** Herb, Shrub, Subshrub **Use:** ME **Origin:** Native **Region:** Amazonia, Andean, Caribbean, Orinoquia, Pacific **Dept.:** AMA, ANT, ARA, CAQ, CAS, CHO, CUN, MET, NAR, TOL, VAC, VAU **Elev.:** 0-1700 m.a.s.l. **IUCN Habitat code:** 1,2,3,4,5,14 **Conservation status: (IUCN)** NE **(CPSG)** Potential LC

6394 *Ludwigia decurrens* Walter (Onagraceae) **IPNI:** 613333-1 **Common name:** clavillo rojo, gallito de laguna **Habit:** Herb **Use:** HF, ME **Origin:** Native **Region:** Amazonia, Andean, Caribbean, Orinoquia, Pacific **Dept.:** AMA, ANT, ARA, ATL, CAL, CAQ, CAS, CHO, CUN, MAG, MET, PUT, SAP, TOL, VAC, VID **Elev.:** 5-1380 m.a.s.l. **IUCN Habitat code:** 1,2,3,4,5,14 **Conservation status: (IUCN)** NE **(CPSG)** Potential LC

6395 *Ludwigia erecta* (L.) H.Hara (Onagraceae) **IPNI:** 144351-2 **Common name:** bejuco campanita, clavito, clavos, hartón, mimbre **Habit:** Herb, Subshrub **Use:** AF, GS, HF, ME, SU **Origin:** Native **Region:** Amazonia, Andean, Caribbean, Pacific **Dept.:** AMA, ANT, ATL, BOL, CAL, CAQ, CHO, COR, CUN, HUI, LAG, MAG, SAN, TOL, VAC **Elev.:** 5-1500 m.a.s.l. **IUCN Habitat code:** 1,2,3,4,5,14 **Conservation status: (IUCN)** NE **(CPSG)** Potential LC

6396 *Ludwigia hyssopifolia* (G.Don) Exell (Onagraceae) **IPNI:** 613360-1 **Common name:** clavito de agua, clavo de agua **Habit:** Herb, Shrub **Use:** AF, EU, HF, MA, ME **Origin:** Native **Region:** Amazonia, Andean, Caribbean, Orinoquia, Pacific **Dept.:** AMA, ATL, BOL, CAQ, CAS, CAU, CHO, GUV, MET, NAR, PUT, VAC, VAU, VID **Elev.:** 5–700 m a.s.l. **IUCN Habitat code:** 1,2,3,4,5,14 **Conservation status: (IUCN)** LC

6397 *Ludwigia inclinata* (L.f.) M.Gómez (Onagraceae) **IPNI:** 613361-1 **Habit:** Herb, Aquatic **Use:** ME **Origin:** Native **Region:** Amazonia, Orinoquia, Pacific **Dept.:** AMA, ANT, CAL, CAQ, CAS, CHO, MET, PUT, TOL, VAC, VID **Elev.:** 100–700 m a.s.l. **IUCN Habitat code:** 1,2,3,4,5,14 **Conservation status: (IUCN)** NE

6398 *Ludwigia leptocarpa* (Nutt.) H.Hara (Onagraceae) **IPNI:** 144366-2 **Common name:** clavito, clavito de pantano, clavo de agua, guayabito de agua, palo de agua **Habit:** Herb, Shrub, Subshrub **Use:** EU, HF, MA, ME, PO **Origin:** Native **Region:** Amazonia, Guiana Shield, Caribbean, Pacific **Dept.:** AMA, ANT, ATL, BOL, CHO, COR, VAU, VID **Elev.:** 0–1500 m a.s.l. **IUCN Habitat code:** 1,2,3,4,5,14 **Conservation status: (IUCN)** LC

6399 *Ludwigia nervosa* (Poir.) H.Hara (Onagraceae) **IPNI:** 144384-2 **Habit:** Shrub **Use:** SU **Origin:** Native **Region:** Amazonia, Orinoquia, Pacific **Dept.:** ANT, CAS, CUN, GUA, GUV, MET, SAN, TOL, VAC, VID **Elev.:** 10-700 m.a.s.l. **IUCN Habitat code:** 1,2,3,4,5,14 **Conservation status: (IUCN)** LC **(CPSG)** Potential LC

6400 *Ludwigia octovalvis* (Jacq.) P.H.Raven (Onagraceae) **IPNI:** 144385-2 **Common name:** clavito, clavito de agua, clavo de laguna **Habit:** Subshrub, Aquatic **Use:** AF, HF, MA, ME, SU **Origin:** Native **Region:** Amazonia, Andean, Guiana Shield, Caribbean, Orinoquia, Pacific **Dept.:** AMA, ANT, ARA, CAQ, CAU, CHO, COR, CUN, GUA, HUI, MAG, MET, NAR, NSA, PUT, SAP, SAN, TOL, VAC, VAU, VID **Elev.:** 0-1900 m.a.s.l. **IUCN Habitat code:** 1,2,3,4,5,14 **Conservation status: (IUCN)** LC **(CPSG)** Potential LC

6401 *Ludwigia palustris* (L.) Elliott (Onagraceae) **IPNI:** 30031106-2 **Habit:** Herb, Aquatic **Use:** EU, HF **Origin:** Native **Region:** Andean **Dept.:** BOG, CUN **Elev.:** 1800–2560 m a.s.l. **IUCN Habitat code:** 1,3,14 **Conservation status: (IUCN)** LC

6402 *Ludwigia peploides* (Kunth) P.H.Raven (Onagraceae) **IPNI:** 613407-1 **Common name:** clavito, clavo de laguna, floating seedbox, guayabillo, hierba de clavo, tripa de pollo **Habit:** Herb, Aquatic **Use:** ME **Origin:** Native **Region:** Amazonia, Andean, Caribbean, Orinoquia **Dept.:** ANT, BOG, BOY, CAQ, CAS, CAU, CUN, MET, PUT, RIS, SAN, VAU **Elev.:** 0–3015 m a.s.l. **IUCN Habitat code:** 1,2,3,4,5,14 **Conservation status: (IUCN)** NE

6403 *Ludwigia peruviana* (L.) H.Hara (Onagraceae) **IPNI:** 144406-2 **Common name:** clavito, clavo, clavo amarillo, clavo de laguna, yerba de rejo **Habit:** Herb, Shrub, Subshrub **Use:** EU, ME, SU **Origin:** Native **Region:** Andean, Caribbean, Pacific **Dept.:** ANT, BOG, BOY, CAL, CAQ, CAU, CES, CHO, CUN, HUI, LAG, MAG, MET, NSA, PUT, RIS, SAN, VAC **Elev.:** 40–2600 m a.s.l. **IUCN Habitat code:** 1,2,3,5,14 **Conservation status: (IUCN)** NE

6404 *Oenothera biennis* L. (Onagraceae) **IPNI:** 172755-2 **Common name:** evening primrose, german rampion, onagra **Use:** EU, HF, ME **Dept.:** BOG **IUCN Habitat code:** 14 **Conservation status: (IUCN)** NE

6405 *Oenothera glazioviana* Micheli (Onagraceae) **IPNI:** 317809-2 **Habit:** Herb **Use:** EU, HF, ME **Origin:** Naturalised **Region:** Andean **Dept.:** BOG, CUN **Elev.:** 2100–2100 m a.s.l. **IUCN Habitat code:** 14 **Conservation status: (IUCN)** NE

6406 *Oenothera laciniata* Hill (Onagraceae)
IPNI: 613817-1
Habit: Herb **Use:** ME **Origin:** Naturalised **Region:** Andean **Dept.:** BOG, NAR **Elev.:** 2510–2900 m a.s.l. **Conservation status: (IUCN)** NE

6407 *Oenothera macrocarpa* Nutt. (Onagraceae)
IPNI: 173157-2
Use: EU **Region:** Andean **Dept.:** BOG
Conservation status: (IUCN) NE

6408 *Oenothera multicaulis* Ruiz & Pav. (Onagraceae)
IPNI: 173202-2
Common name: antañawi, apio, flor de muerto, gahuar-chuchuncca, injerta, injerto de tierra, sarya, saya, saya saya, saya-saya, shullu amarillo, yahuar-shonka, yerba de las tres flores **Habit:** Herb, Subshrub **Use:** ME **Origin:** Native **Region:** Andean **Dept.:** BOG, BOY, CAU, CUN, MET, NAR, QUI, SAN **Elev.:** 1990–4040 m a.s.l. **IUCN Habitat code:** 1,3,4,14 **Conservation status: (IUCN)** NE

6409 *Oenothera rosea* L'Hér. ex Aiton (Onagraceae)
IPNI: 173348-2
Habit: Herb **Use:** ME **Origin:** Native **Region:** Andean **Dept.:** ANT **Elev.:** 2000–2300 m a.s.l. **IUCN Habitat code:** 1,3,5,14 **Conservation status: (IUCN)** NE

6410 *Oenothera tetraptera* Cav. (Onagraceae)
IPNI: 614141-1
Common name: buenastardes, crepuscularia, flor de cinco, platanilla, platanillo **Habit:** Herb **Use:** ME **Origin:** Native **Region:** Andean **Dept.:** BOG, BOY, CAU, CUN, NSA, SAN **Elev.:** 2210–2210 m a.s.l. **IUCN Habitat code:** 14 **Conservation status: (IUCN)** NE

6411 *Erisma bicolor* Ducke (Vochysiaceae)
IPNI: 95991-2
Common name: brazanegra, flormorado, lengua de lombriz, oreja de chimbe, palo de chimbe, palo largo, palo santo, pepa de nigua, salado **Habit:** Tree **Use:** HF **Origin:** Native **Region:** Amazonia, Guiana Shield **Dept.:** AMA, CAQ **Elev.:** 200–200 m a.s.l. **IUCN Habitat code:** 1 **Conservation status: (IUCN)** NE

6412 *Erisma calcaratum* (Link) Warm. (Vochysiaceae)
IPNI: 870864-1
Habit: Shrub, Tree **Use:** HF **Origin:** Native **Region:** Amazonia **Dept.:** AMA **Elev.:** 100–100 m a.s.l. **IUCN Habitat code:** 1 **Conservation status: (IUCN)** LC

6413 *Erisma japura* Spruce ex Warm. (Vochysiaceae)
IPNI: 870870-1
Common name: flormorado, kinobeay, oreja de chimbe, piña de morroco, quiquiijej, yapurá **Habit:** Tree **Use:** HF, MA, ME **Origin:** Native **Region:** Guiana Shield **Dept.:** GUA, VAU **Elev.:** 200–400 m a.s.l. **IUCN Habitat code:** 1,5 **Conservation status: (IUCN)** NE

6414 *Erisma splendens* Stafleu (Vochysiaceae)
IPNI: 96007-2
Common name: arenillo de paloma, bachizomo, badizomo **Habit:** Tree **Use:** HF **Origin:** Native **Region:** Guiana Shield **Dept.:** GUA **Elev.:** 120–120 m a.s.l. **IUCN Habitat code:** 1,5 **Conservation status: (IUCN)** NE

6415 *Erisma uncinatum* Warm. (Vochysiaceae)
IPNI: 318964-2
Common name: arenillo, flormorado, flormorao, milpo, oreja de chimbe, palo perico **Habit:** Tree **Use:** HF, MA **Origin:** Native **Region:** Amazonia, Andean, Orinoquia **Dept.:** CAQ, PUT, VID **Elev.:** 110–600 m a.s.l. **IUCN Habitat code:** 1,3,4,5,14 **Conservation status: (IUCN)** LC

6416 *Qualea acuminata* Spruce ex Warm. (Vochysiaceae)
IPNI: 318965-2
Common name: arenillo, arenillo blanco, gomo, gomo blanco **Habit:** Tree **Use:** HF, ME **Origin:** Native **Region:** Amazonia, Guiana Shield **Dept.:** AMA, CAQ, VAU **Elev.:** 150–250 m a.s.l. **IUCN Habitat code:** 1,3,4,5,14 **Conservation status: (IUCN)** LC

6417 *Qualea paraensis* Ducke (Vochysiaceae)
IPNI: 215272-2
Common name: arenillo de paloma, azulejo, cintura de mono, guará, palo de picón, palo mono, vainillo **Habit:** Tree **Use:** HF, ME **Origin:** Native **Region:** Amazonia, Guiana Shield, Orinoquia **Dept.:** AMA, GUA, VAU, VID **Elev.:** 52–105 m a.s.l. **IUCN Habitat code:** 1,3,4,5,14 **Conservation status: (IUCN)** LC

6418 *Vochysia ferruginea* Mart. (Vochysiaceae)
IPNI: 266306-2
Common name: arracacho, bayo blanco, botagajo, cenascuras, cenascuro, cenescudo, cenoscuro, dormilón, goma, palo santillo, senescudo, sonoscuro, soragá, soroga **Habit:** Tree **Use:** EU, MA, ME, PO **Origin:** Native **Region:** Amazonia, Andean, Guiana Shield, Pacific **Dept.:** AMA, ANT, CHO, CUN, GUA, MET, NAR, NSA, SAN, TOL, VAC, VAU, VID **Elev.:** 0–800 m a.s.l. **IUCN Habitat code:** 1,3,4,5,14 **Conservation status: (IUCN)** LC

6419 *Vochysia guatemalensis* Donn.Sm. (Vochysiaceae)
IPNI: 266313-2
Common name: laguno **Habit:** Tree **Use:** MA **Origin:** Native **Region:** Pacific **Dept.:** CHO **Elev.:** 56-60 m.a.s.l. **Conservation status: (IUCN)** LC **(CPSG)** LC

6420 *Vochysia laxiflora* Stafleu (Vochysiaceae)
IPNI: 266328-2
Common name: palo santo **Habit:** Tree **Use:** ME **Origin:** Native **Region:** Guiana Shield **Dept.:** VAU **Elev.:** 250-250 m.a.s.l. **IUCN Habitat code:** 1,4 **Conservation status: (IUCN)** LC **(CPSG)** LC

6421 *Vochysia lehmannii* Hieron. (Vochysiaceae)
IPNI: 266331-2
Common name: barroso, casquillo, centello, cucharo, manteco, palo amarillo, saladillo, saladillo blanco, árbol de zancudo **Habit:** Tree **Use:** HF **Origin:** Native **Region:** Andean, Orinoquia **Dept.:** CAS, CUN, MET, NSA, SAN, TOL **Elev.:** 140–900 m a.s.l. **IUCN Habitat code:** 1,3,4,5,14 **Conservation status: (IUCN)** LC

6422 *Vochysia lomatophylla* Standl. (Vochysiaceae)
IPNI: 266333-2
Common name: guasicaspi blanco, palo seco **Habit:** Tree **Use:** HF **Origin:** Native **Region:** Amazonia **Dept.:** AMA, CAQ **Elev.:** 200–250 m a.s.l. **IUCN Habitat code:** 1 **Conservation status: (IUCN)** LC

6423 *Vochysia obscura* Warm. (Vochysiaceae)
IPNI: 871092-1
Habit: Tree **Use:** MA **Origin:** Native **Region:** Guiana Shield **Dept.:** CAQ, GUA, MET, VAU **Elev.:** 180–850 m a.s.l. **IUCN Habitat code:** 1,3,14 **Conservation status: (IUCN)** LC

6424 *Vochysia vismiifolia* Spruce ex Warm. (Vochysiaceae)
IPNI: 871122-1
Common name: almidón, arenillo, bizcochito, cacho, cacho de venado, cachovenado, gomo, labio de niña, saladillo **Habit:** Tree **Use:** HF, MA **Origin:** Native **Region:** Guiana Shield, Pacific **Dept.:** CHO, GUA, VID **Elev.:** 80–105 m a.s.l. **IUCN Habitat code:** 1,3,4 **Conservation status: (IUCN)** LC

Oxalidales

6425 *Brunellia comocladiifolia* Bonpl. (Brunelliaceae)
IPNI: 813633-1
Common name: cedrillo, riñón, yuco **Habit:** Tree **Use:** ME **Origin:** Native **Region:** Andean **Dept.:** ANT, BOY, CAU, CES, CUN, HUI, MAG, NAR, RIS, SAN, TOL, VAC **Elev.:** 800–3100 m a.s.l. **Conservation status: (IUCN)** NE

6426 *Brunellia propinqua* Kunth (Brunelliaceae)
IPNI: 813640-1
Habit: Small tree, Tree **Use:** ME **Origin:** Endemic **Region:** Andean **Dept.:** ANT, ARA, BOG, BOY, CHO, CUN, HUI, NSA, SAN, TOL **Elev.:** 1800-3350 m.a.s.l. **IUCN Habitat code:** 1,3,4,14 **Conservation status: (IUCN)** LC **(CPSG)** LC

6427 *Brunellia sibundoya* Cuatrec. (Brunelliaceae)
IPNI: 37499-2
Common name: cedrillo, riñón, yuco **Habit:** Tree **Use:** AF, EU, FU, ME **Origin:** Native **Region:** Andean **Dept.:** ANT, BOG, BOY, CAL, CAU, CUN, PUT, QUI, SAN, TOL, VAC **Elev.:** 1500–3400 m a.s.l. **IUCN Habitat code:** 1,3,4,14 **Conservation status: (IUCN)** NE

6428 *Brunellia stuebelii* Hieron. (Brunelliaceae)
IPNI: 37504-2
Common name: matapuercos, yuco riñón **Habit:** Shrub, Tree **Use:** ME **Origin:** Endemic **Region:** Andean **Dept.:** CAU **Elev.:** 2400-3100 m.a.s.l. **IUCN Habitat code:** 1 **Conservation status: (IUCN)** EN **(CPSG)** EN

6429 *Brunellia trigyna* Cuatrec. (Brunelliaceae)
IPNI: 37508-2
Habit: Tree **Use:** AF, FU **Origin:** Native **Region:** Andean **Dept.:** BOY, NSA, SAN **Elev.:** 1200–3300 m a.s.l. **IUCN Habitat code:** 1,4,14 **Conservation status: (IUCN)** NE

6430 *Connarus coriaceus* G.Schellenb. (Connaraceae)
IPNI: 64164-2
Habit: Liana **Use:** PO **Origin:** Native **Region:** Orinoquia **Dept.:** GUA **Elev.:** 100–100 m a.s.l. **Conservation status: (IUCN)** NE

6431 *Connarus lambertii* (DC.) Britton (Connaraceae) **IPNI:** 64181-2 **Habit:** Shrub, Liana **Use:** MA, ME **Origin:** Native **Region:** Orinoquia, Pacific **Dept.:** BOY, GUV, MET, VAC, VAU, VID **Elev.:** 5–300 m a.s.l. **IUCN Habitat code:** 1,2,3,4,5,14 **Conservation status: (IUCN)** NE

6432 *Connarus ruber* (Poepp.) Planch. (Connaraceae) **IPNI:** 264496-1 **Common name:** barbasco de ají **Habit:** Shrub, Liana **Use:** HF, MA, PO **Origin:** Native **Region:** Amazonia, Guiana Shield, Orinoquia **Dept.:** AMA, GUA, MET, VAU, VID **Elev.:** 75–700 m a.s.l. **IUCN Habitat code:** 1,2,4,5,14 **Conservation status: (IUCN)** NE

6433 *Pseudoconnarus macrophyllus* (Poepp.) Radlk. (Connaraceae) **IPNI:** 211312-2 **Common name:** apamu, bejuco de venado **Habit:** Liana **Use:** ME **Origin:** Native **Region:** Amazonia, Guiana Shield, Orinoquia, Pacific **Dept.:** AMA, CAQ, CHO, GUA, SAN, VAC, VAU, VID **Elev.:** 50–500 m a.s.l. **IUCN Habitat code:** 1,4,5,14 **Conservation status: (IUCN)** NE

6434 *Rourea glabra* Kunth (Connaraceae) **IPNI:** 264767-1 **Common name:** bejuco de garrote, bella maria, cedrillo, chalde, granada de monte, labiorin, mata negro, matanegro **Habit:** Liana **Use:** HF, ME, PO **Origin:** Native **Region:** Andean, Caribbean, Orinoquia, Pacific **Dept.:** ANT, ATL, CAS, LAG, MAG, MET, RIS, SAP, VAU **Elev.:** 0–770 m a.s.l. **IUCN Habitat code:** 1,2,3,4,5,14 **Conservation status: (IUCN)** NE

6435 *Rourea puberula* Baker (Connaraceae) **IPNI:** 264853-1 **Habit:** Liana **Use:** ME **Origin:** Native **Region:** Amazonia **Dept.:** CAQ **Elev.:** 230–230 m a.s.l. **IUCN Habitat code:** 1,3,14 **Conservation status: (IUCN)** NE

6436 *Weinmannia auriculata* D.Don (Cunoniaceae) **IPNI:** 794947-1 **Common name:** encenillo **Habit:** Tree **Use:** ME **Origin:** Native **Region:** Andean **Dept.:** ANT, BOG, CAU, CUN, NAR, RIS, TOL **Elev.:** 2500–3820 m a.s.l. **IUCN Habitat code:** 1,3,14 **Conservation status: (IUCN)** NE

6437 *Weinmannia balbisiana* Kunth (Cunoniaceae) **IPNI:** 319003-2 **Common name:** encenillo, encino **Habit:** Tree **Use:** ME **Origin:** Native **Region:** Andean **Dept.:** ANT, BOG, CAQ, CAU, CHO, CUN, HUI, MET, NAR, NSA, PUT, RIS, SAN, VAC **Elev.:** 710–3050 m a.s.l. **IUCN Habitat code:** 1,3,4,8,14 **Conservation status: (IUCN)** LC

6438 *Weinmannia elliptica* Kunth (Cunoniaceae) **IPNI:** 268187-2 **Common name:** encenillo **Habit:** Tree **Use:** ME **Origin:** Native **Region:** Andean **Dept.:** ANT, CAQ, HUI, PUT, VAC **Elev.:** 2300–3600 m a.s.l. **IUCN Habitat code:** 1,3,14 **Conservation status: (IUCN)** NE

6439 *Weinmannia fagaroides* Kunth (Cunoniaceae) **IPNI:** 268190-2 **Common name:** encenillo, encenillo crespo **Habit:** Tree **Use:** ME **Origin:** Native **Region:** Andean **Dept.:** ANT, BOG, BOY, CUN, HUI, LAG, MAG, NSA, SAN **Elev.:** 2550–4000 m a.s.l. **IUCN Habitat code:** 1,3,4,14 **Conservation status: (IUCN)** LC

6440 *Weinmannia glabra* L.f. (Cunoniaceae) **IPNI:** 268195-2 **Common name:** cuicuillá **Use:** ME **IUCN Habitat code:** 1,2,3,4,8,14 **Conservation status: (IUCN)** NE

6441 *Weinmannia mariquitae* Szyszyl. (Cunoniaceae) **IPNI:** 268241-2 **Common name:** encenillo **Habit:** Tree **Use:** ME **Origin:** Native **Region:** Andean **Dept.:** BOG, CAU, CUN, HUI, NAR, RIS, TOL **Elev.:** 2900–3700 m a.s.l. **IUCN Habitat code:** 1,3,4,8,14 **Conservation status: (IUCN)** NE

6442 *Weinmannia pinnata* L. (Cunoniaceae) **IPNI:** 268273-2 **Common name:** encenillo **Habit:** Tree **Use:** ME **Origin:** Native **Region:** Andean **Dept.:** ANT, BOG, BOY, CAU, CES, CHO, CUN, HUI, LAG, MAG, MET, NAR, NSA, RIS, SAN, TOL, VAC **Elev.:** 920–3230 m a.s.l. **Conservation status: (IUCN)** LC

6443 *Weinmannia pubescens* Kunth (Cunoniaceae) **IPNI:** 268284-2 **Common name:** carbonero, encenillo, encenillo crespo, encino **Habit:** Shrub, Small tree, Tree **Use:** ME **Origin:** Native **Region:** Andean **Dept.:** ANT, BOG, BOY, CAL, CAU, CUN, NAR, QUI, RIS, SAN, TOL, VAC **Elev.:** 1500–3400 m a.s.l. **IUCN Habitat code:** 1,3,4,14 **Conservation status: (IUCN)** NE

6444 *Weinmannia tomentosa* L.f. (Cunoniaceae) **IPNI:** 268326-2 **Common name:** encenillo, encenillo negro, encenillo peludo, encino **Habit:** Shrub, Small tree, Tree **Use:** EU, MA, ME **Origin:** Native **Region:** Andean **Dept.:** BOG, BOY, CUN, NSA, SAN **Elev.:** 1700–3724 m a.s.l. **IUCN Habitat code:** 1,3,4,5,8,14 **Conservation status: (IUCN)** NE

6445 *Sloanea eichleri* K.Schum. (Elaeocarpaceae) **IPNI:** 237404-2 **Habit:** Small tree, Tree **Use:** HF, MA **Origin:** Native **Region:** Amazonia, Orinoquia **Dept.:** AMA, VID **Elev.:** 0-500 m.a.s.l. **IUCN Habitat code:** 1,3,4,14 **Conservation status: (IUCN)** LC **(CPSG)** CR

6446 *Sloanea guianensis* (Aubl.) Benth. (Elaeocarpaceae) **IPNI:** 835145-1 **Common name:** palo de achiote, táparo **Habit:** Tree **Use:** MA **Origin:** Native **Region:** Amazonia, Andean, Guiana Shield, Orinoquia, Pacific **Dept.:** AMA, ANT, CHO, COR, MET, VAC, VID **Elev.:** 0-925 m.a.s.l. **IUCN Habitat code:** 1,3,4,5,14 **Conservation status: (IUCN)** LC **(CPSG)** Potential LC

6447 *Sloanea hirsuta* (Schott) Planch. ex Benth. (Elaeocarpaceae) **IPNI:** 835147-1 **Habit:** Tree **Use:** MA **Origin:** Native **Region:** Amazonia, Orinoquia **Dept.:** AMA, CAQ, CAS **Elev.:** 100–130 m a.s.l. **Conservation status: (IUCN)** NE

6448 *Sloanea macrophylla* Benth. ex Turcz. (Elaeocarpaceae) **IPNI:** 835157-1 **Habit:** Tree **Use:** MA **Origin:** Native **Region:** Andean, Pacific **Dept.:** CUN, VAC **Elev.:** 0–1000 m a.s.l. **IUCN Habitat code:** 1,3,5,14 **Conservation status: (IUCN)** NE

6449 *Sloanea medusula* K.Schum. & Pittier (Elaeocarpaceae) **IPNI:** 324537-1 **Habit:** Tree **Use:** HF **Origin:** Native **Region:** Pacific **Dept.:** CHO, VAC **Elev.:** 0–100 m a.s.l. **IUCN Habitat code:** 1,14 **Conservation status: (IUCN)** LC

6450 *Sloanea meianthera* Donn.Sm. (Elaeocarpaceae) **IPNI:** 237445-2 **Habit:** Tree **Use:** ME **Origin:** Native **Region:** Pacific **Dept.:** VAC **Elev.:** 50–50 m a.s.l. **IUCN Habitat code:** 1 **Conservation status: (IUCN)** LC

6451 *Sloanea obtusifolia* (Moric.) K.Schum. (Elaeocarpaceae) **IPNI:** 237456-2 **Habit:** Tree **Use:** MA **Origin:** Native **Conservation status: (IUCN)** NE

6452 *Sloanea rufa* Planch. ex Benth. (Elaeocarpaceae) **IPNI:** 835186-1 **Habit:** Small tree, Tree **Use:** HF **Origin:** Native **Region:** Andean, Pacific **Dept.:** ANT, LAG, MAG, VAC **Elev.:** 50-1720 m.a.s.l. **IUCN Habitat code:** 1,3,14 **Conservation status: (IUCN)** LC **(CPSG)** CR

6453 *Sloanea synandra* Spruce ex Benth. (Elaeocarpaceae) **IPNI:** 835194-1 **Habit:** Tree **Use:** HF **Origin:** Native **Region:** Amazonia **Dept.:** AMA **Elev.:** 100–110 m a.s.l. **IUCN Habitat code:** 1 **Conservation status: (IUCN)** NE

6454 *Sloanea terniflora* (Moc. & Sessé ex DC.) Standl. (Elaeocarpaceae) **IPNI:** 324587-1 **Habit:** Small tree, Tree **Use:** ME **Origin:** Native **Region:** Amazonia, Andean, Caribbean, Orinoquia **Dept.:** AMA, ARA, BOL, MAG, PUT **Elev.:** 80–1160 m a.s.l. **IUCN Habitat code:** 1,2,4,5,14 **Conservation status: (IUCN)** LC

6455 *Sloanea zuliaensis* Pittier (Elaeocarpaceae) **IPNI:** 237492-2 **Habit:** Tree **Use:** ME **Origin:** Native **Region:** Pacific **Dept.:** ANT, CHO **Elev.:** 0-600 m.a.s.l. **IUCN Habitat code:** 1,3,5,14 **Conservation status: (IUCN)** LC **(CPSG)** VU

6456 *Vallea stipularis* L.f. (Elaeocarpaceae) **IPNI:** 836044-1 **Common name:** campanito, campano, chaque, chaquén, chirriador, gache, gaco, gaque, majua, manzano, palo roso, paloroso, raco, raque, rojo, rosadito, roso, sanjuanito **Habit:** Shrub, Small tree, Tree **Use:** AF, EU, FU, HF, MA, ME, PO **Origin:** Native **Region:** Andean **Dept.:** ANT, BOG, BOY, CAU, CES, CUN, LAG, MAG, NAR, NSA, QUI, SAN, TOL **Elev.:** 1990-4300 m.a.s.l. **IUCN Habitat code:** 1,2,3,4,5,6,8,14 **Conservation status: (IUCN)** LC **(CPSG)** Potential LC

6457 *Averrhoa bilimbi* L. (Oxalidaceae) **IPNI:** 371869-1 **Common name:** curuba **Habit:** Small tree **Use:** EU, GS, HF, MA, ME **Origin:** Cultivated **Region:** Caribbean **Dept.:** COR, SAN **Elev.:** 0–766 m a.s.l. **Conservation status: (IUCN)** NE

6458 *Averrhoa carambola* L. (Oxalidaceae) **IPNI:** 371870-1 **Common name:** calambolo, carambola, carambolas, carambolo, tamarinda, tamarindo, tamarindo llanero, torombolo, árbol de carambolas **Habit:** Shrub, Small tree **Use:** AF, EU, GS, HF, MA, ME **Origin:** Cultivated **Region:** Amazonia, Andean, Orinoquia, Pacific **Dept.:** AMA, ANT, CHO, CUN, MET, QUI, TOL, VAC **Elev.:** 0–1460 m a.s.l. **IUCN Habitat code:** 1,2,3,14 **Conservation status: (IUCN)** NE

6459 *Biophytum dendroides* (Kunth) DC. (Oxalidaceae) **IPNI:** 371953-1 **Habit:** Herb **Use:** ME **Origin:** Native **Region:** Amazonia, Andean, Pacific **Dept.:** AMA, CAU, CHO, CUN, HUI, QUI, TOL, VAC **Elev.:** 250–1600 m a.s.l. **IUCN Habitat code:** 1,2,3,14 **Conservation status: (IUCN)** NE

6460 *Oxalis acetosella* L. (Oxalidaceae) **IPNI:** 30012601-2 **Use:** HF, ME, PO **Region:** Andean **Dept.:** BOG **IUCN Habitat code:** 1,3,14 **Conservation status: (IUCN)** NE

6461 *Oxalis articulata* Savigny (Oxalidaceae) **IPNI:** 374673-1 **Habit:** Herb **Use:** EU, HF, ME **Origin:** Cultivated **IUCN Habitat code:** 14 **Conservation status: (IUCN)** NE

6462 *Oxalis aureoflava* Steud. (Oxalidaceae) **IPNI:** 177771-2 **Common name:** acedera **Use:** ME **IUCN Habitat code:** 14 **Conservation status: (IUCN)** NE

6463 *Oxalis barrelieri* L. (Oxalidaceae) **IPNI:** 177777-2 **Common name:** doradilla, flor de mar **Habit:** Herb **Use:** HF, ME, PO **Origin:** Native **Region:** Amazonia, Caribbean, Orinoquia, Pacific **Dept.:** ANT, CAQ, CAS, MAG, MET, NSA, TOL, VAC **Elev.:** 30–1500 m a.s.l. **IUCN Habitat code:** 1,3,4,14 **Conservation status: (IUCN)** NE

6464 *Oxalis conorrhiza* Jacq. (Oxalidaceae) **IPNI:** 177883-2 **Use:** ME **Region:** Andean **Dept.:** BOG **IUCN Habitat code:** 14 **Conservation status: (IUCN)** NE

6465 *Oxalis corniculata* L. (Oxalidaceae) **IPNI:** 177893-2 **Common name:** acedera, acederilla, acederita, cachudita, chulco, chulquillo, chulquito macho, chupachupa, platanillo, platanito, platanitos, trébol, trébol amarillo **Habit:** Herb **Use:** AF, EU, GS, HF, IF, MA, ME, PO, SU **Origin:** Native **Region:** Andean **Dept.:** ANT, BOG, BOY, CAL, CAU, CHO, CUN, HUI, MAG, NAR, PUT, SAN, TOL, VAC **Elev.:** 800–3200 m a.s.l. **IUCN Habitat code:** 1,2,3,4,14 **Conservation status: (IUCN)** NE

6466 *Oxalis debilis* Kunth (Oxalidaceae) **IPNI:** 374833-1 **Habit:** Herb **Use:** EU, HF, ME, PO **Origin:** Native **Region:** Andean **Dept.:** ANT, BOG, BOY, CAU, CUN, NSA, VAC **Elev.:** 900-2660 m.a.s.l. **IUCN Habitat code:** 1,2,3,4,14 **Conservation status: (IUCN)** NE **(CPSG)** Potential LC

6467 *Oxalis drummondii* A.Gray (Oxalidaceae) **IPNI:** 374872-1 **Use:** ME **Region:** Andean **Dept.:** BOG **Conservation status: (IUCN)** NE

6468 *Oxalis filiformis* Kunth (Oxalidaceae) **IPNI:** 178000-2 **Common name:** acedera, cedera, chulco, tache **Habit:** Herb **Use:** ME **Origin:** Native **Region:** Andean **Dept.:** BOG, BOY, CUN, NAR, SAN **Elev.:** 1700-3900 m.a.s.l. **IUCN Habitat code:** 1,3,4,14 **Conservation status: (IUCN)** NE **(CPSG)** Potential LC

6469 *Oxalis frutescens* L. (Oxalidaceae) **IPNI:** 374932-1 **Habit:** Herb, Shrub, Subshrub **Use:** HF, ME **Origin:** Native **Region:** Caribbean, Orinoquia **Dept.:** LAG, MAG, SAP, VID **Elev.:** 100–450 m a.s.l. **IUCN Habitat code:** 2,4 **Conservation status: (IUCN)** NE

6470 *Oxalis hedysaroides* Kunth (Oxalidaceae) **IPNI:** 374990-1 **Common name:** acedera, cartagena, moradilla, moradita **Habit:** Shrub, Subshrub **Use:** MA, ME, SU **Origin:** Native **Region:** Andean **Dept.:** ANT, BOY, CAL, CAU, CUN, HUI, NAR, QUI, RIS, TOL, VAC **Elev.:** 870-2300 m.a.s.l. **IUCN Habitat code:** 1,3,14 **Conservation status: (IUCN)** NE **(CPSG)** Potential LC

6471 *Oxalis latifolia* Kunth (Oxalidaceae) **IPNI:** 375051-1 **Habit:** Herb **Use:** EU, HF, ME, PO **Origin:** Native **Region:** Andean **Dept.:** ANT, BOG, BOY, CAU, CUN, NAR, QUI, SAN **Elev.:** 1200-3700 m.a.s.l. **IUCN Habitat code:** 1,3,4,8,14 **Conservation status: (IUCN)** NE **(CPSG)** Potential LC

6472 *Oxalis lotoides* Kunth (Oxalidaceae) **IPNI:** 178186-2 **Common name:** vinagrera **Habit:** Herb **Use:** ME **Origin:** Native **Region:** Andean **Dept.:** BOG, BOY, CAL, CAU, CUN, HUI, NAR, PUT, QUI, RIS, TOL, VAC **Elev.:** 1600-3700 m.a.s.l. **IUCN Habitat code:** 1,3,4,14 **Conservation status: (IUCN)** NE **(CPSG)** Potential LC

6473 *Oxalis medicaginea* Kunth (Oxalidaceae) **IPNI:** 178220-2 **Common name:** acedera, chulco **Habit:** Herb **Use:** ME **Origin:** Native **Region:** Andean **Dept.:** BOG, BOY, CAL, CAU, CUN, HUI, MET, NAR, PUT, TOL, VAC **Elev.:** 1400-4000 m.a.s.l. **IUCN Habitat code:** 1,2,3,4,14 **Conservation status: (IUCN)** NE **(CPSG)** Potential LC

6474 *Oxalis melilotoides* Zucc. (Oxalidaceae) **IPNI:** 178225-2 **Habit:** Herb **Use:** ME **Origin:** Native **Region:** Andean **Dept.:** CAU **Elev.:** 3480–3480 m a.s.l. **Conservation status: (IUCN)** NE

6475 *Oxalis mollis* Kunth (Oxalidaceae) **IPNI:** 178245-2 **Common name:** yuyo **Habit:** Herb **Use:** ME **Origin:** Native **Region:** Andean **Dept.:** ANT, CAL, CAU, CUN, NAR, PUT, QUI, RIS, SAN, TOL **Elev.:** 1900–3500 m a.s.l. **IUCN Habitat code:** 1,3,4,14 **Conservation status: (IUCN)** NE

6476 *Oxalis peduncularis* Kunth (Oxalidaceae) **IPNI:** 178322-2 **Common name:** chulco, oquilla **Habit:** Herb **Use:** HF, ME **Origin:** Native **Region:** Andean **Dept.:** NAR **Elev.:** 2050–2050 m a.s.l. **Conservation status: (IUCN)** NE

6477 *Oxalis scandens* Kunth (Oxalidaceae) **IPNI:** 178449-2 **Common name:** acedera, acederita, chulco, platanitos, yuyo **Habit:** Herb **Use:** ME **Origin:** Native **Region:** Andean **Dept.:** BOG, CUN, QUI, TOL **Elev.:** 2000–2800 m a.s.l. **IUCN Habitat code:** 1,3,4,14 **Conservation status: (IUCN)** NE

6478 *Oxalis spiralis* Ruiz & Pav. ex G.Don (Oxalidaceae) **IPNI:** 178480-2 **Common name:** acedera, chulco **Habit:** Herb **Use:** HF, ME **Origin:** Native **Region:** Andean **Dept.:** BOG, BOY, CAL, CAU, CES, CUN, MAG, NAR, NSA, PUT, QUI, SAN, TOL, VAC **Elev.:** 1000-3750 m.a.s.l. **IUCN Habitat code:** 1,3,4,5,8,14 **Conservation status: (IUCN)** NE **(CPSG)** Potential LC

6479 *Oxalis tabaconasensis* R.Knuth (Oxalidaceae) **IPNI:** 178524-2 **Habit:** Herb **Use:** HF **Origin:** Native **Region:** Andean **Dept.:** ANT, BOY, CUN, NSA, SAN **Elev.:** 1400–3700 m a.s.l. **IUCN Habitat code:** 1,3,14 **Conservation status: (IUCN)** NE

6480 *Oxalis tetraphylla* Cav. (Oxalidaceae) **IPNI:** 375466-1 **Common name:** trébol, trébol de cuatro hojas **Use:** EU, HF, ME, PO, SU **IUCN Habitat code:** 3,14 **Conservation status: (IUCN)** NE

6481 *Oxalis triangularis* A.St.-Hil. (Oxalidaceae) **IPNI:** 375478-1 **Common name:** tinto, trébol, trébol morado **Habit:** Herb **Use:** HF **Origin:** Cultivated **Region:** Orinoquia **Dept.:** BOG, MET **Elev.:** 470–470 m a.s.l. **IUCN Habitat code:** 14 **Conservation status: (IUCN)** NE

6482 *Oxalis tuberosa* Molina (Oxalidaceae) **IPNI:** 178566-2 **Common name:** apilja, apilla, apio blanco, apiña, cubio, cuiba, hibia, ibia, ibias, kawi, lamaki, mafa, mishi-polé, oca, oca de chile, ocas, ocra, okka, papaíbia, piga-mishi, ruba, timbo, uca **Habit:** Herb **Use:** HF, ME **Origin:** Native, Cultivated **Region:** Andean **Dept.:** BOG, CAU, CUN, NAR **Elev.:** 1780–3000 m a.s.l. **IUCN Habitat code:** 1,3,8,14 **Conservation status: (IUCN)** NE

Picramniales

6483 *Nothotalisia peruviana* (Standl.) W.W.Thomas (Picramniaceae) **IPNI:** 77115312-1 **Habit:** Shrub, Small tree **Use:** HF **Origin:** Native **Region:** Amazonia **Dept.:** ANT, PUT **Elev.:** 250–410 m a.s.l. **IUCN Habitat code:** 1,3 **Conservation status: (IUCN)** LC

6484 *Picramnia antidesma* Sw. (Picramniaceae) **IPNI:** 813869-1 **Common name:** engajado, fresno amargo **Habit:** Shrub, Tree **Use:** MA, ME **Origin:** Native **Region:** Andean **Dept.:** ANT **Elev.:** 325–1400 m a.s.l. **IUCN Habitat code:** 1,3,14 **Conservation status: (IUCN)** NE

6485 *Picramnia gracilis* Tul. (Picramniaceae) **IPNI:** 813896-1 **Common name:** arrayán **Habit:** Shrub, Small tree, Tree **Use:** HF **Origin:** Native **Region:** Andean, Pacific **Dept.:** ANT, CHO, MAG, PUT, RIS, TOL **Elev.:** 45–2440 m a.s.l. **IUCN Habitat code:** 1,3,14 **Conservation status: (IUCN)** LC

6486 *Picramnia latifolia* Tul. (Picramniaceae) **IPNI:** 813902-1 **Common name:** chocanari, coralito, indígena, quina, quina indígena **Habit:** Shrub, Small tree, Tree **Use:** HF, ME **Origin:** Native **Region:** Amazonia, Guiana Shield, Pacific **Dept.:** AMA, ANT, BOL, CAL, COR, HUI, MET, TOL **Elev.:** 0–540 m a.s.l. **IUCN Habitat code:** 1,2,3,4,5,14 **Conservation status: (IUCN)** LC

6487 *Picramnia pentandra* Sw. (Picramniaceae) **IPNI:** 813923-1 **Use:** HF, ME **Origin:** Native **Conservation status: (IUCN)** LC

6488 *Picramnia sellowii* Planch. (Picramniaceae) **IPNI:** 301399-2 **Common name:** carabiri, chocanari, morado, palo de bostezo **Habit:** Shrub, Tree **Use:** HF **Origin:** Native **Dept.:** AMA, ANT **Elev.:** 100–1440 m a.s.l. **IUCN Habitat code:** 1,3,5,14 **Conservation status: (IUCN)** NE

6489 *Picramnia sphaerocarpa* Planch. (Picramniaceae) **IPNI:** 813935-1 **Common name:** corales, cuentecita, cuentecito, indiecito, lágrimas de San Pedro, manubo, árbol de coral, árbol de cuentas **Habit:** Shrub, Tree **Use:** ME **Origin:** Native **Region:** Andean, Pacific **Dept.:** BOY, CAL, CHO, CUN, NAR, QUI, RIS, SAN **Elev.:** 100–2700 m a.s.l. **IUCN Habitat code:** 1,2,3,14 **Conservation status: (IUCN)** LC

6490 *Picramnia teapensis* Tul. (Picramniaceae) **IPNI:** 813938-1 **Habit:** Shrub **Use:** HF **Origin:** Native **Region:** Andean **Dept.:** ANT **Elev.:** 1450–1450 m a.s.l. **Conservation status: (IUCN)** LC

Proteales

6491 *Nelumbo lutea* (Willd.) Pers. (Nelumbonaceae) **IPNI:** 605419-1 **Habit:** Herb, Aquatic **Use:** HF, ME **Origin:** Native **Region:** Andean **Dept.:** ANT **Elev.:** 2000–2000 m a.s.l. **IUCN Habitat code:** 1,3,5 **Conservation status: (IUCN)** LC

6492 *Euplassa saxicola* (R.E.Schult.) Steyerm. (Proteaceae) **IPNI:** 102211-2 **Habit:** Tree **Use:** MA **Origin:** Endemic **Region:** Amazonia, Guiana Shield, Orinoquia **Dept.:** GUV, MET, VAU, VID **Elev.:** 100-700 m.a.s.l. **IUCN Habitat code:** 1,3,4,14 **Conservation status: (IUCN)** LC **(CPSG)** LC

6493 *Grevillea robusta* A.Cunn. ex R.Br. (Proteaceae) **IPNI:** 50798-3 **Common name:** rabo de pavo, roble australiano, árbol de seda **Habit:** Tree **Use:** AF, EU, FU, GS, HF, IF, MA, ME, PO **Origin:** Cultivated **Region:** Andean **Dept.:** ANT, BOG, TOL **Elev.:** 1500–1500 m a.s.l. **IUCN Habitat code:** 1,3,14 **Conservation status: (IUCN)** LC

6494 *Leucadendron salignum* P.J.Bergius (Proteaceae) **IPNI:** 704928-1 **Common name:** protea **Habit:** Shrub **Use:** EU **Origin:** Cultivated **Region:** Andean **Dept.:** CUN **Elev.:** 2600–2600 m a.s.l. **Conservation status: (IUCN)** LC

6495 *Macadamia integrifolia* Maiden & Betche (Proteaceae) **IPNI:** 705192-1 **Habit:** Tree **Use:** AF, EU, GS, HF, MA, ME **Origin:** Cultivated **Region:** Andean **Dept.:** CAL, QUI **Elev.:** 1200–1500 m a.s.l. **IUCN Habitat code:** 1,3,14 **Conservation status: (IUCN)** VU

6496 *Panopsis costaricensis* Standl. (Proteaceae) **IPNI:** 182715-2 **Use:** HF **Origin:** Native **Conservation status: (IUCN)** LC

6497 *Panopsis hernandezii* L.E.Gut. (Proteaceae) **IPNI:** 304026-2 **Habit:** Tree **Use:** HF **Origin:** Endemic **Region:** Andean **Dept.:** CAU **Elev.:** 3000–3000 m a.s.l. **Conservation status: (IUCN)** CR

6498 *Panopsis lozanoi* L.E.Gut. (Proteaceae) **IPNI:** 304937-2 **Habit:** Tree **Use:** HF **Origin:** Native **Region:** Andean **Dept.:** SAN **Elev.:** 2300–2800 m a.s.l. **IUCN Habitat code:** 1,14 **Conservation status: (IUCN)** LC

6499 *Panopsis polystachya* (Kunth) Kuntze (Proteaceae) **IPNI:** 1076423-2 **Common name:** pepo, yolombo **Habit:** Shrub, Tree **Use:** HF **Origin:** Native **Region:** Andean **Dept.:** ANT, CAU, NAR, QUI, VAC **Elev.:** 2100–2900 m a.s.l. **IUCN Habitat code:** 1,3,14 **Conservation status: (IUCN)** NE

6500 *Panopsis rubescens* (Pohl) Pittier (Proteaceae) **IPNI:** 182723-2 **Habit:** Shrub, Small tree, Tree **Use:** HF, ME **Origin:** Native **Region:** Amazonia, Guiana Shield, Orinoquia **Dept.:** AMA, ARA, CAQ, CAS, GUA, GUV, MET, VAU, VID **Elev.:** 100–1200 m a.s.l. **IUCN Habitat code:** 1,2,3,4,5 **Conservation status: (IUCN)** LC

6501 *Panopsis sessilifolia* (Rich.) Sandwith (Proteaceae) **IPNI:** 182728-2 **Habit:** Tree **Use:** MA, ME **Origin:** Native **Region:** Amazonia, Guiana Shield, Orinoquia **Dept.:** AMA, CAQ, CAS, GUV, PUT, VAU **Elev.:** 125–300 m a.s.l. **IUCN Habitat code:** 1,4,5 **Conservation status: (IUCN)** LC

6502 *Panopsis suaveolens* (Klotzsch) Pittier (Proteaceae) **IPNI:** 182731-2 **Common name:** malla, uyamo, yolombo **Habit:** Tree **Use:** HF **Origin:** Native **Region:** Andean **Dept.:** ANT, BOG, BOY, CAU, CES, CUN, MAG, QUI, SAN **Elev.:** 1900–3150 m a.s.l. **IUCN Habitat code:** 1,3,4,14 **Conservation status: (IUCN)** NE

6503 *Panopsis yolombo* (Posada-Ar.) Killip (Proteaceae) **IPNI:** 182733-2 **Common name:** carnefiambre, cedro hembra, colombo, yolombo, yolombó, zambo **Habit:** Tree **Use:** HF, ME **Origin:** Native **Region:** Andean **Dept.:** ANT, CAL, CAU **Elev.:** 1100–2600 m a.s.l. **IUCN Habitat code:** 1,3,4,14 **Conservation status: (IUCN)** LC

6504 *Roupala monosperma* (Ruiz & Pav.) I.M.Johnst. (Proteaceae) **IPNI:** 222667-2 **Common name:** azufre, carafiambre, carne de fiambre, carneasao, carnefiambre, fiambre, jigracaspi, laurel tuno, salchicho, verraco **Habit:** Shrub, Tree **Use:** ME **Origin:** Native **Region:** Andean **Dept.:** ANT, BOY, CAL, CAU, CUN, HUI, NAR, RIS, SAN, TOL, VAC **Elev.:** 1400–3100 m a.s.l. **IUCN Habitat code:** 1,2,3,14 **Conservation status: (IUCN)** LC

6505 *Roupala montana* Aubl. (Proteaceae) **IPNI:** 706342-1 **Common name:** guaná, holguín, mapurito **Habit:** Shrub, Tree **Use:** MA, ME **Origin:** Native **Region:** Amazonia, Andean, Guiana Shield, Caribbean, Orinoquia **Dept.:** AMA, ANT, BOL, CAU, CUN, GUV, HUI, LAG, MAG, MET, NSA, SAN, VAC, VID **Elev.:** 100–2000 m a.s.l. **IUCN Habitat code:** 1,2,3,4,5,14 **Conservation status: (IUCN)** LC

6506 *Meliosma frondosa* Cuatrec. & Idrobo (Sabiaceae) **IPNI:** 156682-2 **Common name:** babillo, baboso, calabazo **Habit:** Tree **Use:** HF **Origin:** Native **Region:** Andean **Dept.:** ANT, BOG, CAU, CUN, MAG, QUI, TOL, VAC **Elev.:** 1240–3300 m a.s.l. **IUCN Habitat code:** 1,3,4,14 **Conservation status: (IUCN)** LC

Ranunculales

6507 *Berberis aurahuacensis* Lem. (Berberidaceae) **IPNI:** 106441-1 **Use:** ME **Origin:** Endemic **Region:** Andean **Dept.:** MAG **Elev.:** 2100–2700 m a.s.l. **Conservation status: (IUCN)** NE

6508 *Berberis glauca* Kunth (Berberidaceae) **IPNI:** 106685-1 **Common name:** doncel, espino de oro, capuelo, guilache, tachuelo, tachuelo de tierra fría, uña de gato **Habit:** Small tree **Use:** EU, HF, ME **Origin:** Native **Region:** Andean **Dept.:** BOG, BOY, CES, CUN, TOL **Elev.:** 2400–3200 m a.s.l. **IUCN Habitat code:** 1,3,4,8,14 **Conservation status: (IUCN)** NE

6509 *Berberis goudotii* Triana & Planch. (Berberidaceae) **IPNI:** 106695-1 **Common name:** adorote, espino, espuelo, tachuelo, uña de gato **Habit:** Shrub **Use:** ME **Origin:** Native **Region:** Andean **Dept.:** ANT, BOG, BOY, CUN, MET **Elev.:** 1900-3960 m.a.s.l. **IUCN Habitat code:** 1,3,4,5,14 **Conservation status: (IUCN)** NE **(CPSG)** LC

6510 *Berberis loxensis* Benth. (Berberidaceae) **IPNI:** 106858-1 **Use:** ME **Conservation status: (IUCN)** NE

6511 *Berberis lutea* Ruiz & Pav. (Berberidaceae) **IPNI:** 106862-1 **Use:** ME **Origin:** Native **Region:** Andean **Dept.:** CAU, NAR **Elev.:** 3000–4000 m a.s.l. **Conservation status: (IUCN)** LC

6512 *Berberis pichinchensis* Turcz. (Berberidaceae) **IPNI:** 106985-1 **Use:** ME **Origin:** Native **Region:** Andean **Dept.:** NAR **Elev.:** 3000–3700 m a.s.l. **IUCN Habitat code:** 4,14 **Conservation status: (IUCN)** NE

6513 *Berberis psilopoda* Turcz. (Berberidaceae) **IPNI:** 107016-1 **Use:** ME **Origin:** Native **Region:** Andean **Dept.:** ANT, RIS, TOL **Elev.:** 2600–4260 m a.s.l. **IUCN Habitat code:** 1,3,4 **Conservation status: (IUCN)** NE

6514 *Berberis quinduensis* Kunth (Berberidaceae) **IPNI:** 107024-1 **Common name:** espino **Habit:** Shrub **Use:** ME **Origin:** Endemic **Region:** Andean **Dept.:** BOY, CUN, QUI, TOL **Elev.:** 2780–3640 m a.s.l. **IUCN Habitat code:** 1,3,4,14 **Conservation status: (IUCN)** NE

6515 *Berberis rigidifolia* Kunth (Berberidaceae) **IPNI:** 107043-1 **Common name:** arrasquillo, casha, chinia, espino, espino amarillo, espino de oro, espuela, espuelo, quilache, tachuelo, uña de gato **Habit:** Shrub **Use:** EU, ME **Origin:** Native **Region:** Andean **Dept.:** BOG, BOY, CUN, NSA, SAN **Elev.:** 1750-4200 m.a.s.l. **IUCN Habitat code:** 1,3,4,8,14 **Conservation status: (IUCN)** NE **(CPSG)** LC

6516 *Abuta grandifolia* (Mart.) Sandwith (Menispermaceae) **IPNI:** 147-2 **Common name:** abuta, ancabesux, aralén de monte, buuyutoci, calentuno, castaño, chilicaspi, chin caspi, macana, mirsimarika, pico de paujil, taque-puraque, taquecurauqe, tresrayas, trompetero, veneno, vibuajeiria **Habit:** Shrub, Subshrub, Small tree **Use:** HF, MA, ME **Origin:** Native **Region:** Amazonia, Guiana Shield, Orinoquia **Dept.:** AMA, CAQ, CAU, GUA, GUV, MET, PUT **Elev.:** 75–550 m a.s.l. **IUCN Habitat code:** 1,3,4,5,14 **Conservation status: (IUCN)** LC

6517 *Abuta grisebachii* Triana & Planch. (Menispermaceae) **IPNI:** 580056-1 **Common name:** cipó **Habit:** Liana **Use:** ME **Origin:** Native **Region:** Amazonia **Dept.:** AMA, VAU **Elev.:** 0–200 m a.s.l. **IUCN Habitat code:** 1 **Conservation status: (IUCN)** LC

6518 *Abuta imene* (Mart.) Eichler (Menispermaceae) **IPNI:** 580059-1 **Common name:** curare **Habit:** Liana **Use:** ME **Origin:** Native **Region:** Amazonia **Dept.:** AMA, VAU **Elev.:** 120–250 m a.s.l. **IUCN Habitat code:** 1,3,5 **Conservation status: (IUCN)** LC

6519 *Abuta obovata* Diels (Menispermaceae) **IPNI:** 154-2 **Habit:** Liana **Use:** HF **Origin:** Native **Region:** Amazonia **Dept.:** AMA, VAU **Elev.:** 200–300 m a.s.l. **IUCN Habitat code:** 1 **Conservation status: (IUCN)** NE

6520 *Abuta racemosa* Triana & Planch. (Menispermaceae) **IPNI:** 580077-1 **Common name:** kuyawisi unuboto, yawatinae **Habit:** Liana **Use:** ME **Origin:** Native **Dept.:** ANT, BOL, COR, SAN, TOL, VAC **Elev.:** 80–1000 m a.s.l. **IUCN Habitat code:** 1,3,5,14 **Conservation status: (IUCN)** NE

6521 *Abuta rufescens* Aubl. (Menispermaceae) **IPNI:** 580080-1 **Common name:** curare, sapepá **Habit:** Liana **Use:** MA, ME **Origin:** Native **Region:** Amazonia **Dept.:** AMA, CAU, PUT **Elev.:** 100–600 m a.s.l. **IUCN Habitat code:** 1,5 **Conservation status: (IUCN)** LC

6522 *Abuta solimoesensis* Krukoff & Barneby (Menispermaceae) **IPNI:** 159-2 **Habit:** Liana **Use:** HF **Origin:** Native **Region:** Amazonia **Dept.:** AMA **Elev.:** 200–300 m a.s.l. **IUCN Habitat code:** 1,4 **Conservation status: (IUCN)** NE

6523 *Abuta vaupesensis* Krukoff & Barneby (Menispermaceae) **IPNI:** 164-2 **Habit:** Liana **Use:** ME **Origin:** Endemic **Region:** Amazonia **Dept.:** VAU **Elev.:** 150–150 m a.s.l. **IUCN Habitat code:** 1 **Conservation status: (IUCN)** NE

6524 *Anomospermum chloranthum* Diels (Menispermaceae) **IPNI:** 14455-2 **Common name:** max soeo, veneno de churuco **Habit:** Liana, Climbing **Use:** HF, ME **Origin:** Native **Region:** Amazonia, Andean, Pacific **Dept.:** AMA, ANT, CAL, CHO, NAR, RIS, VAC **Elev.:** 5–1900 m a.s.l. **IUCN Habitat code:** 1,5,14 **Conservation status: (IUCN)** NE

6525 *Anomospermum reticulatum* (Mart.) Eichler (Menispermaceae) **IPNI:** 580176-1 **Common name:** huevo de gallo **Habit:** Small tree, Liana, Climbing **Use:** ME **Origin:** Native **Region:** Amazonia, Andean, Guiana Shield, Pacific **Dept.:** AMA, ANT, CHO, MET, NSA **Elev.:** 0–2300 m a.s.l. **IUCN Habitat code:** 1,2,14 **Conservation status: (IUCN)** NE

6526 *Borismene japurensis* (Mart.) Barneby (Menispermaceae) **IPNI:** 34845-2 **Habit:** Climbing **Use:** ME **Origin:** Native **Region:** Pacific **Dept.:** CAU, CHO, VAC **Elev.:** 5–250 m a.s.l. **IUCN Habitat code:** 1,5,14 **Conservation status: (IUCN)** NE

6527 *Chondrodendron tomentosum* Ruiz & Pav. (Menispermaceae) **IPNI:** 580301-1 **Common name:** nasacatá **Habit:** Liana **Use:** HF, ME, PO **Origin:** Native **Region:** Amazonia, Guiana Shield, Pacific **Dept.:** AMA, ANT, BOY, CAQ, CHO, MET, SAN, VAC **Elev.:** 20–430 m a.s.l. **IUCN Habitat code:** 1,3,5,14 **Conservation status: (IUCN)** NE

6528 *Cissampelos andromorpha* DC. (Menispermaceae) **IPNI:** 580308-1 **Habit:** Liana, Climbing **Use:** ME **Origin:** Native **Region:** Amazonia, Andean **Dept.:** AMA, ANT, BOY, CAQ, CAU, CHO, GUA, MET, NAR, PUT, TOL, VAC, VAU **Elev.:** 150-1800 m.a.s.l. **IUCN Habitat code:** 1,3,14 **Conservation status: (IUCN)** NE **(CPSG)** Potential LC

6529 *Cissampelos fasciculata* Benth. (Menispermaceae) **IPNI:** 580358-1 **Habit:** Liana, Climbing **Use:** HF, ME **Origin:** Native **Region:** Andean **Dept.:** ANT, BOG, CAQ, CAU, CUN, HUI, NAR, PUT, QUI, SAN, VAC **Elev.:** 700–2860 m a.s.l. **IUCN Habitat code:** 1,2,3,14 **Conservation status: (IUCN)** NE

6530 *Cissampelos glaberrima* A.St.-Hil. (Menispermaceae) **IPNI:** 580364-1 **Use:** ME **Origin:** Native **IUCN Habitat code:** 14 **Conservation status: (IUCN)** NE

6531 *Cissampelos grandifolia* Triana & Planch. (Menispermaceae) **IPNI:** 580371-1 **Common name:** bejuco, capitana blanca **Habit:** Herb, Climbing **Use:** ME **Origin:** Native **Region:** Andean **Dept.:** ANT, BOY, CAL, CAU, CUN, HUI, NAR, PUT, QUI, RIS, TOL **Elev.:** 325-2610 m.a.s.l. **IUCN Habitat code:** 1,2,3,14 **Conservation status: (IUCN)** NE **(CPSG)** Potential LC

6532 *Cissampelos ovalifolia* DC. (Menispermaceae) **IPNI:** 580420-1 **Habit:** Herb, Subshrub **Use:** ME **Origin:** Native **Region:** Guiana Shield, Orinoquia **Dept.:** CAQ, CAS, CUN, MAG, MET, VID **Elev.:** 150–850 m a.s.l. **IUCN Habitat code:** 1,2,3,4,14 **Conservation status: (IUCN)** NE

6533 *Cissampelos pareira* L. (Menispermaceae) **IPNI:** 59219-2 **Common name:** amargoso, barbasco, batatilla, batato, batato de monte, bejuco bravo, bejuco de cerca, bejuco de salud, bejuco guayacán, bejuco huavo, bejuco pitilla, bejuco ratón, bejuco zaragozo, guaco, guayacán, masquiaunsabe, oreja de burro, oreja de ratón, pareira, **Habit:** Climbing **Use:** AF, EU, HF, MA, ME, PO **Origin:** Native **Region:** Amazonia, Andean, Guiana Shield, Caribbean, Pacific **Dept.:** AMA, ANT, ATL, BOG, BOL, CAL, CAQ, CAU, CES, CHO, COR, CUN, GUV, HUI, MAG, MET, NAR, NSA, PUT, QUI, RIS, SAP, TOL, VAC **Elev.:** 0-2800 m.a.s.l. **IUCN Habitat code:** 1,2,3,5,14 **Conservation status: (IUCN)** NE **(CPSG)** Potential LC

6534 *Cissampelos tropaeolifolia* DC. (Menispermaceae) **IPNI:** 580466-1 **Common name:** abrazadera, gallito, pareira **Habit:** Climbing **Use:** HF, ME, SU **Origin:** Native **Region:** Andean, Pacific **Dept.:** ANT, BOY, CAL, CAU, CHO, COR, CUN, MET, QUI, RIS, SAN **Elev.:** 100–1800 m a.s.l. **IUCN**

Habitat code: 1,2,3,4,5,14 **Conservation status: (IUCN)** NE

6535 *Curarea cuatrecasasii* Barneby & Krukoff (Menispermaceae) **IPNI:** 72707-2 **Habit:** Liana, Climbing **Use:** HF **Origin:** Native **Region:** Pacific **Dept.:** ANT, BOL, CHO **Elev.:** 10–400 m a.s.l. **IUCN Habitat code:** 1,3,14 **Conservation status: (IUCN)** NE

6536 *Curarea tecunarum* Barneby & Krukoff (Menispermaceae) **IPNI:** 72708-2 **Common name:** curare, veneno **Habit:** Liana, Climbing **Use:** HF, ME **Origin:** Native **Region:** Amazonia **Dept.:** AMA **Elev.:** 100–300 m a.s.l. **IUCN Habitat code:** 1,3,14 **Conservation status: (IUCN)** NE

6537 *Curarea toxicofera* (Wedd.) Barneby & Krukoff (Menispermaceae) **IPNI:** 72709-2 **Common name:** bejuco bravo, curare, dioimujiriñame, veneno, yamitá **Habit:** Liana, Climbing **Use:** HF, ME **Origin:** Native **Region:** Amazonia **Dept.:** AMA, GUV **Elev.:** 100–100 m a.s.l. **IUCN Habitat code:** 1,3,5,14 **Conservation status: (IUCN)** NE

6538 *Disciphania ernstii* Eichler (Menispermaceae) **IPNI:** 580772-1 **Habit:** Climbing **Use:** HF, ME **Origin:** Native **Region:** Andean, Caribbean, Pacific **Dept.:** ANT, BOL, BOY, CHO, CUN, MAG, MET, VAC **Elev.:** 0–1000 m a.s.l. **IUCN Habitat code:** 1,3,14 **Conservation status: (IUCN)** NE

6539 *Elissarrhena grandifolia* (Eichler) Diels (Menispermaceae) **IPNI:** 89958-2 **Habit:** Liana, Climbing **Use:** ME **Origin:** Native **Region:** Pacific **Dept.:** BOL, CHO, SAN **Elev.:** 80–300 m a.s.l. **IUCN Habitat code:** 1,14 **Conservation status: (IUCN)** NE

6540 *Hyperbaena domingensis* (DC.) Benth. (Menispermaceae) **IPNI:** 580887-1 **Habit:** Climbing **Use:** HF, MA, ME **Origin:** Native **Region:** Caribbean **Dept.:** LAG, MAG, VAC **Elev.:** 950–1000 m a.s.l. **IUCN Habitat code:** 1,14 **Conservation status: (IUCN)** NE

6541 *Odontocarya emarginata* Barneby (Menispermaceae) **IPNI:** 172169-2 **Habit:** Climbing **Use:** ME **Origin:** Native **Region:** Amazonia **Dept.:** AMA, VAU **Elev.:** 250–250 m a.s.l. **IUCN Habitat code:** 1,5 **Conservation status: (IUCN)** NE

6542 *Odontocarya macarenae* Barneby (Menispermaceae) **IPNI:** 172176-2 **Habit:** Liana **Use:** ME **Origin:** Endemic **Region:** Guiana Shield **Dept.:** MET **Elev.:** 310–650 m a.s.l. **IUCN Habitat code:** 1,3 **Conservation status: (IUCN)** NE

6543 *Odontocarya micrantha* (Diels) Barneby (Menispermaceae) **IPNI:** 172179-2 **Habit:** Liana, Climbing **Use:** HF **Origin:** Native **Region:** Amazonia **Dept.:** AMA **Elev.:** 100–100 m a.s.l. **IUCN Habitat code:** 1 **Conservation status: (IUCN)** NE

6544 *Odontocarya tamoides* (DC.) Miers (Menispermaceae) **IPNI:** 581190-1 **Common name:** bejuco rascarrasca, nigua de puerco, rascarrasca **Habit:** Liana, Climbing **Use:** ME **Origin:** Native **Region:** Caribbean, Pacific **Dept.:** ANT, ATL, BOL, BOY, CHO, LAG, MAG, SUC, TOL **Elev.:** 0–400 m a.s.l. **IUCN Habitat code:** 1,2,3,5,14 **Conservation status: (IUCN)** NE

6545 *Odontocarya tenacissima* Diels (Menispermaceae) **IPNI:** 172195-2 **Common name:** amargoso **Habit:** Liana, Climbing **Use:** ME **Origin:** Endemic **Region:** Andean **Dept.:** ANT **Elev.:** 1500–2250 m a.s.l. **IUCN Habitat code:** 3,14 **Conservation status: (IUCN)** NE

6546 *Odontocarya tripetala* Diels (Menispermaceae) **IPNI:** 172196-2 **Common name:** bejuco murcia **Habit:** Climbing **Use:** HF, ME **Origin:** Native **Region:** Andean, Pacific **Dept.:** AMA, ANT, BOY, CHO, NAR **Elev.:** 10–1050 m a.s.l. **IUCN Habitat code:** 1,5,14 **Conservation status: (IUCN)** NE

6547 *Orthomene schomburgkii* (Miers) Barneby & Krukoff (Menispermaceae) **IPNI:** 176827-2 **Common name:** bejuco de culebra **Habit:** Liana, Climbing **Use:** HF, ME **Origin:** Native **Region:** Amazonia, Andean, Guiana Shield, Pacific **Dept.:** AMA, ANT, CAQ, CHO, GUA, PUT, RIS, VAU **Elev.:** 5–1040 m a.s.l. **IUCN Habitat code:** 1,3,5,14 **Conservation status: (IUCN)** NE

6548 *Orthomene verruculosa* (Krukoff & Barneby) Barneby & Krukoff (Menispermaceae) **IPNI:** 176828-2 **Habit:** Climbing **Use:** ME **Origin:** Native **Region:** Guiana Shield **Dept.:** VAU **Elev.:** 200–200 m a.s.l. **IUCN Habitat code:** 1 **Conservation status: (IUCN)** NE

6549 *Sciadotenia nitida* (L.Riley) Krukoff & Barneby (Menispermaceae) **IPNI:** 230118-2 **Habit:** Climbing **Use:** ME **Origin:** Native **Region:** Pacific **Dept.:** ANT, SAN **Elev.:** 50–200 m a.s.l. **IUCN Habitat code:** 1 **Conservation status: (IUCN)** NE

6550 *Sciadotenia sprucei* Diels (Menispermaceae) **IPNI:** 230126-2 **Habit:** Liana **Use:** ME **Origin:** Native **Region:** Amazonia **Dept.:** AMA **Elev.:** 100–100 m a.s.l. **Conservation status: (IUCN)** NE

6551 *Sciadotenia toxifera* Krukoff & A.C.Sm. (Menispermaceae) **IPNI:** 581368-1 **Habit:** Liana, Climbing **Use:** ME **Origin:** Native **Region:** Amazonia **Dept.:** AMA, GUV, MET, PUT **Elev.:** 120–240 m a.s.l. **IUCN Habitat code:** 1,3,14 **Conservation status: (IUCN)** NE

6552 *Telitoxicum krukovii* Moldenke (Menispermaceae) **IPNI:** 250112-2 **Common name:** picón **Habit:** Liana, Climbing **Use:** HF **Origin:** Native **Region:** Amazonia **Dept.:** AMA, CAQ, GUA, PUT, VAU **Elev.:** 100–300 m a.s.l. **IUCN Habitat code:** 1,3 **Conservation status: (IUCN)** NE

6553 *Argemone mexicana* L. (Papaveraceae) **IPNI:** 18816-2 **Common name:** cardosanto, tissly **Habit:** Herb **Use:** AF, EU, FU, HF, MA, ME, PO, SU **Origin:** Naturalised **Region:** Andean **Dept.:** ANT, BOL, BOY, CUN, MAG, NSA, SAN, TOL, VAC **Elev.:** 75–2210 m a.s.l. **IUCN Habitat code:** 1,2,3,14 **Conservation status: (IUCN)** NE

6554 *Bocconia frutescens* L. (Papaveraceae) **IPNI:** 316358-2 **Common name:** albarracín, albarrecín, curaarador, curador, curarador, palo amarillo, sarno, trompeto, trompeto curador, tuno, yemehuevo **Habit:** Shrub, Small tree **Use:** AF, EU, MA, ME **Origin:** Native **Region:** Andean **Dept.:** ANT, BOG, BOY, CAL, CAU, CHO, CUN, HUI, LAG, MAG, NAR, PUT, QUI, RIS, VAC **Elev.:** 1190–3200 m a.s.l. **IUCN Habitat code:** 1,2,3,4,5,8,14 **Conservation status: (IUCN)** LC

6555 *Bocconia integrifolia* Bonpl. (Papaveraceae) **IPNI:** 671757-1 **Common name:** albarracín, albarrecín, salvarrecín, sarno, trompeto **Habit:** Small tree **Use:** ME **Origin:** Native **Region:** Andean **Dept.:** BOG, CUN **Elev.:** 2100–3560 m a.s.l. **IUCN Habitat code:** 1,2,3,4,14 **Conservation status: (IUCN)** NE

6556 *Eschscholzia californica* Cham. (Papaveraceae) **IPNI:** 30043364-2 **Common name:** campanilla, rasete, raso **Habit:** Herb **Use:** EU, HF, ME, PO **Origin:** Cultivated **Region:** Andean **Dept.:** BOG, CUN **Elev.:** 2600–2800 m a.s.l. **IUCN Habitat code:** 3,14 **Conservation status: (IUCN)** NE

6557 *Fumaria capreolata* L. (Papaveraceae) **IPNI:** 672920-1 **Common name:** palomitas, pollitos **Habit:** Herb **Use:** ME **Origin:** Naturalised **Region:** Andean **Dept.:** BOG, CUN **Elev.:** 2570–2620 m a.s.l. **IUCN Habitat code:** 14 **Conservation status: (IUCN)** NE

6558 *Fumaria officinalis* L. (Papaveraceae) **IPNI:** 30017594-2 **Common name:** fumaria, palomilla, plumaria **Habit:** Herb **Use:** HF, MA, ME **Origin:** Naturalised **Region:** Andean **Dept.:** BOG, BOY, CUN **Elev.:** 2580–2800 m a.s.l. **IUCN Habitat code:** 14 **Conservation status: (IUCN)** NE

6559 *Papaver orientale* L. (Papaveraceae) **IPNI:** 673641-1 **Use:** EU, ME **Region:** Andean **Dept.:** BOG **Conservation status: (IUCN)** NE

6560 *Papaver rhoeas* L. (Papaveraceae) **IPNI:** 306058-2 **Common name:** ababol, adormidera, amapola, amapola común **Use:** EU, HF, ME, PO **Dept.:** BOG **IUCN Habitat code:** 14 **Conservation status: (IUCN)** NE

6561 *Papaver somniferum* L. (Papaveraceae) **IPNI:** 673724-1 **Common name:** adormidera, amapola **Habit:** Herb **Use:** EU, GS, HF, MA, ME, PO, SU **Origin:** Cultivated **Region:** Andean **Dept.:** ANT, BOG, BOY,

CUN, HUI, NAR, QUI, TOL, VAC **Elev.:** 2000–3500 m a.s.l. **IUCN Habitat code:** 1,3,4,14 **Conservation status: (IUCN)** NE

6562 *Aconitum napellus* L. (Ranunculaceae) **IPNI:** 707615-1 **Use:** EU, HF, MA, ME, PO **Conservation status: (IUCN)** LC

6563 *Anemone coronaria* L. (Ranunculaceae) **IPNI:** 708314-1 **Common name:** anémona **Habit:** Herb **Use:** EU, ME **Origin:** Cultivated **Region:** Andean **Dept.:** CUN **Elev.:** 2600–2600 m a.s.l. **Conservation status: (IUCN)** NE

6564 *Aquilegia canadensis* L. (Ranunculaceae) **IPNI:** 30028287-2 **Habit:** Herb **Use:** EU, HF, ME **Origin:** Cultivated **Region:** Andean **Dept.:** BOG, CUN **Elev.:** 2600–2600 m a.s.l. **Conservation status: (IUCN)** NE

6565 *Aquilegia pubescens* Coville (Ranunculaceae) **IPNI:** 16871-2 **Use:** ME **Region:** Andean **Dept.:** BOG **Conservation status: (IUCN)** NE

6566 *Aquilegia vulgaris* L. (Ranunculaceae) **IPNI:** 306409-2 **Common name:** aguileña, colombina, golondrina, viuda **Use:** EU, GS, HF, ME, PO **Dept.:** BOG **IUCN Habitat code:** 14 **Conservation status: (IUCN)** NE

6567 *Clematis haenkeana* C.Presl (Ranunculaceae) **IPNI:** 709750-1 **Common name:** bejuco blanco, cascarillo, juanlanas, pelo de ángel, zanca de guacharaca **Habit:** Liana, Climbing **Use:** ME **Origin:** Native **Region:** Andean **Dept.:** ANT, BOG, BOY, CAL, CAU, CES, CUN, HUI, LAG, MAG, NAR, NSA, PUT, QUI, RIS, SAN, TOL, VAC **Elev.:** 750-3500 m.a.s.l. **IUCN Habitat code:** 1,3,4,14 **Conservation status: (IUCN)** NE **(CPSG)** Potential LC

6568 *Delphinium ajacis* L. (Ranunculaceae) **IPNI:** 77254-2 **Common name:** espuela de caballero, espuela de galán, pajarito **Habit:** Herb **Use:** EU, HF, ME **Origin:** Cultivated **Region:** Andean, Orinoquia **Dept.:** BOY, CAS, CUN, TOL **Elev.:** 200–2600 m a.s.l. **IUCN Habitat code:** 3 **Conservation status: (IUCN)** NE

6569 *Delphinium elatum* L. (Ranunculaceae) **IPNI:** 710585-1 **Common name:** belladona, delfinio **Habit:** Herb **Use:** EU, GS, ME **Origin:** Cultivated **Region:** Andean **Dept.:** CUN **Elev.:** 2600–2600 m a.s.l. **Conservation status: (IUCN)** NE

6570 *Eriocapitella hupehensis* (É.Lemoine) Christenh. & Byng (Ranunculaceae) **IPNI:** 77184471-1 **Common name:** anémone **Habit:** Herb **Use:** HF, ME **Origin:** Cultivated **Region:** Andean **Dept.:** CUN **Elev.:** 2620–2620 m a.s.l. **Conservation status: (IUCN)** NE

6571 *Halerpestes cymbalaria* (Pursh) Greene (Ranunculaceae) **IPNI:** 711250-1 **Habit:** Herb **Use:** HF, ME, PO **Origin:** Native **Region:** Andean **Dept.:** CAL **Elev.:** 3700–3700 m a.s.l. **Conservation status: (IUCN)** NE

6572 *Nigella damascena* L. (Ranunculaceae) **IPNI:** 711645-1 **Common name:** araña, arañas, azulejos, pasionaria **Habit:** Herb **Use:** EU, HF, ME **Origin:** Cultivated **Region:** Andean **Dept.:** BOG, CUN **Elev.:** 2600–2600 m a.s.l. **Conservation status: (IUCN)** NE

6573 *Ranunculus limoselloides* Turcz. (Ranunculaceae) **IPNI:** 713095-1 **Common name:** ñijuto **Habit:** Herb, Aquatic **Use:** ME **Origin:** Native **Region:** Andean **Dept.:** BOG, BOY, CUN, MET, NAR, NSA, SAN, TOL **Elev.:** 3300–4300 m a.s.l. **IUCN Habitat code:** 1,3,4,8,14 **Conservation status: (IUCN)** NE

6574 *Ranunculus nubigenus* Humb., Bonpl. & Kunth ex DC. (Ranunculaceae) **IPNI:** 713340-1 **Common name:** centello, oreja de ratón, rejoncito **Habit:** Herb, Aquatic **Use:** ME **Origin:** Native **Region:** Andean **Dept.:** ANT, BOG, BOY, CAU, CES, CUN, LAG, MAG, MET, NAR, NSA, PUT, QUI, SAN, VAC **Elev.:** 2090-3900 m.a.s.l. **IUCN Habitat code:** 1,3,4,5,8,14 **Conservation status: (IUCN)** NE **(CPSG)** Potential LC

6575 *Ranunculus petiolaris* Humb., Bonpl. & Kunth ex DC. (Ranunculaceae) **IPNI:** 713491-1 **Habit:** Herb **Use:** ME **Origin:** Native **Region:** Andean **Dept.:** ANT, BOG, BOY, CAU, NSA, QUI, SAN, TOL **Elev.:** 1300–2800 m a.s.l. **IUCN Habitat code:** 3,4 **Conservation status: (IUCN)** NE

6576 *Ranunculus praemorsus* Humb., Bonpl. & Kunth ex DC. (Ranunculaceae) **IPNI:** 713577-1 **Common name:** arracachuela, centella, centello, rejoncito **Habit:** Herb **Use:** EU, ME **Origin:** Native **Region:** Andean **Dept.:** ANT, BOG, BOY, CAL, CAU, CUN, MAG, MET, NAR, NSA, QUI, RIS, SAN, TOL, VAC **Elev.:** 1600-4300 m.a.s.l. **IUCN Habitat code:** 3,4,14 **Conservation status: (IUCN)** NE **(CPSG)** Potential LC

Rosales

6577 *Cannabis sativa* L. (Cannabaceae) **IPNI:** 306087-2 **Common name:** bareta, cáñamo, cáñamo de la India, cáñamo índico, maracachafa, mariguana, marihuana, marimba, mona **Habit:** Herb **Use:** AF, FU, HF, MA, ME, PO, SU **Origin:** Cultivated, Naturalised **Region:** Andean **Dept.:** ANT, BOG, CAQ, CAU, CHO, CUN, MAG, QUI, TOL, VAC **Elev.:** 1500–2600 m a.s.l. **IUCN Habitat code:** 1,2,3,14 **Conservation status: (IUCN)** NE

6578 *Celtis iguanaea* (Jacq.) Sarg. (Cannabaceae) **IPNI:** 1213898-2 **Common name:** doñaelena, maiztostado, maiztostao, pomarroso, tabardillo, uña de gato, uña de águila, uñegato, zarza **Habit:** Shrub, Small tree, Tree, Climbing **Use:** HF, MA, ME **Origin:** Native **Region:** Amazonia, Andean, Caribbean, Orinoquia, Pacific **Dept.:** AMA, ANT, ATL, BOL, CES, CHO, CUN, GUV, HUI, LAG, MAG, MET, SAN, SUC, TOL, VAC, VID **Elev.:** 5–1280 m a.s.l. **IUCN Habitat code:** 1,2,3,4,5,14 **Conservation status: (IUCN)** LC

6579 *Celtis schippii* Standl. (Cannabaceae) **IPNI:** 51219-2 **Common name:** cajuche, cuero de marrano **Habit:** Tree **Use:** MA **Origin:** Native **Region:** Amazonia, Caribbean **Dept.:** AMA, ANT, GUV, MET, SAP **Elev.:** 100–700 m a.s.l. **IUCN Habitat code:** 1,3,5,14 **Conservation status: (IUCN)** LC

6580 *Celtis trinervia* Lam. (Cannabaceae) **IPNI:** 851195-1 **Common name:** cupiagua, raspador, verraco, zurrumbo **Habit:** Tree **Use:** HF, MA, ME **Origin:** Native **Region:** Caribbean **Dept.:** BOL, COR, SUC **Elev.:** 40–300 m a.s.l. **IUCN Habitat code:** 1,2,3,14 **Conservation status: (IUCN)** DD

6581 *Humulus lupulus* L. (Cannabaceae) **IPNI:** 303502-2 **Common name:** lúpulo **Use:** EU, GS, HF, MA, ME **Conservation status: (IUCN)** NE

6582 *Trema micranthum* (L.) Blume (Cannabaceae) **IPNI:** 1018059-2 **Common name:** aliso, atadijo, balso, berraquillo, cagalero, capulín, cargadero, guacharaco, guayuyo, guayuyu, huitoto, isinde, jalapo, jordanillo, kaka, látigo, majagua, majagua colorada, majagüillo, majagüito, palorrejo, peñaemico, raspador, surundé, toropate, vara de **Habit:** Shrub, Tree **Use:** AF, EU, FU, HF, MA, ME, PO, SU **Origin:** Native **Region:** Amazonia, Andean, Caribbean, Orinoquia, Pacific **Dept.:** AMA, ANT, ARA, BOL, BOY, CAL, CAQ, CAU, CHO, CUN, GUA, GUV, HUI, LAG, MAG, MET, NAR, PUT, QUI, RIS, SAP, SAN, SUC, TOL, VAC, VID **Elev.:** 0–2420 m a.s.l. **Conservation status: (IUCN)** LC

6583 *Artocarpus altilis* (Parkinson) Fosberg (Moraceae) **IPNI:** 582598-1 **Common name:** castaña, fruta de pan, fruta del pan, frutapán, frutepan, fruto del pan, guaipán, marure, masa de pan, norte, pan, pan de pobre, pan de todo el año, pan de árbol, pan de ñame, pan del norte, pan del árbol, pandiaño, pepaepán, pepapán, tanta, árbol de pan, **Habit:** Tree **Use:** AF, EU, FU, GS, HF, IF, MA, ME, PO, SU **Origin:** Cultivated **Region:** Andean, Caribbean, Pacific **Dept.:** ANT, CAU, CHO, CUN, NAR, QUI, VAC **Elev.:** 80–1200 m a.s.l. **IUCN Habitat code:** 1,2,3,14 **Conservation status: (IUCN)** NE

6584 *Artocarpus heterophyllus* Lam. (Moraceae) **IPNI:** 850389-1 **Common name:** yaca, yaco, árbol del pan **Habit:** Tree **Use:** AF, EU, GS, HF, MA, ME, SU **Origin:** Cultivated **Region:** Andean **Dept.:** ANT, CUN, SAN **Elev.:** 1450–1500 m a.s.l. **IUCN Habitat code:** 1,2,3,14 **Conservation status: (IUCN)** NE

6585 *Artocarpus integer* (Thunb.) Merr. (Moraceae) **IPNI:** 582622-1 **Use:** AF, EU, GS, HF, IF, MA, ME, SU **IUCN Habitat code:** 1,14 **Conservation status: (IUCN)** NE

6586 *Batocarpus amazonicus* (Ducke) Fosberg (Moraceae) **IPNI:** 29952-2 **Common name:** guáimaro, pan de tree, pan de árbol, tree del pan propio, árbol del pan propio **Habit:** Tree **Use:** HF **Origin:** Native **Region:** Amazonia **Dept.:** AMA, GUA **Elev.:** 100–100 m a.s.l. **IUCN Habitat code:** 1,5,14 **Conservation status: (IUCN)** LC

6587 *Batocarpus costaricensis* Standl. & L.O.Williams (Moraceae) **IPNI:** 29953-2 **Common name:** bardon, bardón, lechera, lechero, mare **Habit:** Tree **Use:** HF, MA **Origin:** Native **Region:** Caribbean, Pacific **Dept.:** ANT, CAL, CHO **Elev.:** 20–200 m a.s.l. **IUCN Habitat code:** 1,2,3,5,14 **Conservation status: (IUCN)** LC

6588 *Batocarpus orinocensis* H.Karst. (Moraceae) **IPNI:** 850488-1 **Common name:** mano de tigre, piernaepisco, árbol de pan, árbol del pan de danta, árbol del pan de tigre, árbol del pan de zambico, árbol del pan propio **Habit:** Tree **Use:** HF **Origin:** Native **Region:** Amazonia, Andean, Orinoquia **Dept.:** AMA, MET, PUT, VAC **Elev.:** 150–1860 m a.s.l. **IUCN Habitat code:** 1,3,14 **Conservation status: (IUCN)** LC

6589 *Brosimum acutifolium* Huber (Moraceae) **IPNI:** 37254-2 **Common name:** tamamuri **Habit:** Tree **Use:** HF, ME **Origin:** Native **Region:** Amazonia, Orinoquia **Dept.:** AMA, MET, VAU **Elev.:** 100–350 m a.s.l. **IUCN Habitat code:** 1,3,14 **Conservation status: (IUCN)** LC

6590 *Brosimum alicastrum* Sw. (Moraceae) **IPNI:** 850824-1 **Common name:** caucho, charo, guaimarón, guácimo, guáimaro, lechero, mestizo, negro en cuero, numa cucú, palanco, sande, yumbá **Habit:** Small tree, Tree **Use:** AF, HF, MA, ME **Origin:** Native **Region:** Andean, Caribbean, Pacific **Dept.:** ANT, BOL, CES, CHO, COR, LAG, MAG, RIS, SUC, TOL, VAC **Elev.:** 15–1250 m a.s.l. **IUCN Habitat code:** 1,2,3,5,14 **Conservation status: (IUCN)** NE

6591 *Brosimum guianense* (Aubl.) Huber ex Ducke (Moraceae) **IPNI:** 582693-1 **Common name:** azulito, bordón, coca, fruta de pava, guáimaro, lechero, limoncillo, memecucú, mestizo, muí, oquendo, palo brasil, palo brazil **Habit:** Tree **Use:** AF, HF, MA, ME **Origin:** Native **Region:** Amazonia, Caribbean, Pacific **Dept.:** AMA, ANT, BOL, CAQ, CHO, COR, GUA, GUV, NAR, SAN, SUC, VAC, VAU, VID **Elev.:** 50–800 m a.s.l. **IUCN Habitat code:** 1,2,3,4,5,14 **Conservation status: (IUCN)** LC

6592 *Brosimum lactescens* (S.Moore) C.C.Berg (Moraceae) **IPNI:** 37286-2 **Common name:** cajeto, chichirimico, costillo, cuero de sapo, granadillo rojo, guáimaro, leche de chiva, lecheperra, lechero, pepa de pava, árbol vaca **Habit:** Small tree, Tree **Use:** HF, MA, ME **Origin:** Native **Region:** Amazonia, Andean, Orinoquia, Pacific **Dept.:** AMA, CAQ, CAS, CHO, GUV, MET, VAC, VID **Elev.:** 90–1600 m a.s.l. **IUCN Habitat code:** 1,2,3,4,5,14 **Conservation status: (IUCN)** LC

6593 *Brosimum longifolium* Ducke (Moraceae) **IPNI:** 37291-2 **Habit:** Tree **Use:** HF **Origin:** Native **Region:** Amazonia **Dept.:** AMA, VAU **Elev.:** 100–200 m a.s.l. **IUCN Habitat code:** 1 **Conservation status: (IUCN)** NE

6594 *Brosimum parinarioides* Ducke (Moraceae) **IPNI:** 37302-2 **Common name:** vaco, árbol vaco **Habit:** Tree **Use:** HF, MA, ME **Origin:** Native **Region:** Amazonia **Dept.:** AMA, VAU **Elev.:** 100–100 m a.s.l. **IUCN Habitat code:** 1,5 **Conservation status: (IUCN)** LC

6595 *Brosimum potabile* Ducke (Moraceae) **IPNI:** 37305-2 **Common name:** lechero amarillo, vaco **Habit:** Tree **Use:** HF, MA, ME **Origin:** Native **Region:** Amazonia **Dept.:** VAU **Elev.:** 200–200 m a.s.l. **IUCN Habitat code:** 1 **Conservation status: (IUCN)** NE

6596 *Brosimum rubescens* Taub. (Moraceae) **IPNI:** 37310-2 **Common name:** chimico, cococu joanicu, granadillo, granadillo rosado, guariuba, guáimaro, mare, mirapiranga, palo brasil, palo de sangre, palosangre, vaco **Habit:** Tree **Use:** HF, MA, ME **Origin:** Native **Region:** Amazonia, Caribbean, Pacific **Dept.:** AMA, BOL, CAQ, COR, VAC, VAU **Elev.:** 50-250 m.a.s.l. **IUCN Habitat code:** 1,3,4,5,14 **Conservation status: (IUCN)** NE **(CPSG)** NT

6597 *Brosimum utile* (Kunth) Oken (Moraceae) **IPNI:** 582724-1 **Common name:** avichure, avichuri, avichurí, bibosi, caucho, charo, frutipan de monte, granadillo, guáimaro, jidonuma, lechero, lechoso, lujuji, mare, marimari, marimá, marina, mestizo, palo de vaca, pan de árbol, pasai, perillo, pucre, pío, sande, sandé, tururi, tururí **Habit:** Tree **Use:** HF, MA, ME, SU **Origin:** Native **Region:** Amazonia, Andean, Caribbean, Orinoquia, Pacific **Dept.:** AMA, ANT, BOL, BOY, CAQ, CAU, CHO, COR, CUN, GUV, MET, NAR, PUT, QUI, SAN, VAC **Elev.:** 0–2200 m a.s.l. **IUCN Habitat code:** 1,3,4,5,14 **Conservation status: (IUCN)** LC

6598 *Castilla elastica* Cerv. (Moraceae) **IPNI:** 850895-1 **Common name:** cauchillo, caucho, caucho blanco, caucho de Pará, caucho hembra, caucho negro, iparo, árbol de caucho, árbol de caucho negro, árbol del caucho **Habit:** Tree **Use:** AF, EU, HF, MA, ME **Origin:** Native **Region:** Andean, Caribbean, Orinoquia, Pacific **Dept.:** ANT, CHO, COR, MET, NAR, VAC **Elev.:** 120–1850 m a.s.l. **IUCN Habitat code:** 1,3,5,14 **Conservation status: (IUCN)** LC

6599 *Castilla tunu* Hemsl. (Moraceae) **IPNI:** 49375-2 **Common name:** castilla, cauchillo, cauchillo de rastrojo, caucho, caucho negro **Habit:** Tree **Use:** HF, MA, ME **Origin:** Native **Region:** Pacific **Dept.:** ANT, CHO, NAR **Elev.:** 0–300 m a.s.l. **IUCN Habitat code:** 1,3,5,14 **Conservation status: (IUCN)** LC

6600 *Castilla ulei* Warb. (Moraceae) **IPNI:** 49376-2 **Habit:** Tree **Use:** HF, MA **Origin:** Native **Region:** Amazonia **Dept.:** AMA, VAU **Elev.:** 120–240 m a.s.l. **IUCN Habitat code:** 1,14 **Conservation status: (IUCN)** LC

6601 *Clarisia biflora* Ruiz & Pav. (Moraceae) **IPNI:** 851228-1 **Common name:** cauchillo, cauchillo colorao, cauchillo de loma, dinde, dinde mora, dinde negro, guayacán, lecheperra, lechero blanco, lechero colorado, lechero plomo, lechoso, mare dinde, mora, oquendo, perillo, sande, yambó, yumbá **Habit:** Small tree, Tree **Use:** HF, ME **Origin:** Native **Region:** Andean, Caribbean, Orinoquia, Pacific **Dept.:** ANT, CHO, CUN, HUI, LAG, MAG, NAR, PUT, QUI, RIS, VAC **Elev.:** 100–2000 m a.s.l. **IUCN Habitat code:** 1,2,3,14 **Conservation status: (IUCN)** LC

6602 *Clarisia ilicifolia* (Spreng.) Lanj. & Rossbach (Moraceae) **IPNI:** 59683-2 **Habit:** Shrub, Tree **Use:** HF **Origin:** Native **Region:** Amazonia, Guiana Shield, Orinoquia **Dept.:** GUV, VAU, VID **Elev.:** 100–200 m a.s.l. **IUCN Habitat code:** 1 **Conservation status: (IUCN)** LC

6603 *Clarisia racemosa* Ruiz & Pav. (Moraceae) **IPNI:** 851229-1 **Common name:** amaji, arracacha, arracacho, piyuque, bagtzue, basee, chingongo, dinde, gemena, guariuba, guayabero, guayacán, juansoco de la sabana, juansoco de sabana, justarrazón, maloine, mora, palo amarillo, pelacar, pelacara, piyu, sande mora, tolú, verruga **Habit:** Tree **Use:** AF, HF, MA, ME **Origin:** Native **Region:** Amazonia, Caribbean, Orinoquia, Pacific **Dept.:** AMA, ANT, BOL, CAQ, MET, VAC, VAU, VID **Elev.:** 100–325 m a.s.l. **IUCN Habitat code:** 1,3,4,14 **Conservation status: (IUCN)** LC

6604 *Dorstenia brasiliensis* Lam. (Moraceae) **IPNI:** 851513-1 **Habit:** Herb **Use:** ME **Origin:** Native **Region:** Orinoquia **Dept.:** MET **Elev.:** 300–300 m a.s.l. **IUCN Habitat code:** 1,2,3,14 **Conservation status: (IUCN)** NE

6605 *Dorstenia contrajerva* L. (Moraceae) **IPNI:** 319180-2 **Common name:** contrahierba, contrayerba, cresta de gallo, raíz de resfrío, tabardillo **Habit:** Herb **Use:** EU, HF, ME **Origin:** Native **Region:** Caribbean, Orinoquia, Pacific **Dept.:** ANT, BOL, CAL, CES, CHO, CUN, MAG, MET, SAN, TOL **Elev.:** 20–750 m a.s.l. **IUCN Habitat code:** 1,2,3,5,14 **Conservation status: (IUCN)** NE

6606 *Ficus altissima* Blume (Moraceae) **IPNI:** 852369-1 **Habit:** Tree **Use:** AF, EU, FU, HF, MA, ME **Origin:** Cultivated **Region:** Caribbean **Dept.:** BOL **Elev.:** 0–50 m a.s.l. **IUCN Habitat code:** 2,5 **Conservation status: (IUCN)** LC

6607 *Ficus amazonica* (Miq.) André (Moraceae) **IPNI:** 852372-1 **Common name:** matapalo **Use:** HF **Origin:** Native **IUCN Habitat code:** 1,2,3,5,14 **Conservation status: (IUCN)** LC

6608 *Ficus americana* Aubl. (Moraceae) **IPNI:** 316336-2 **Common name:** higueroncillo, higuerón, matapalo, pepero, sueldo

Habit: Shrub, Tree, Epiphyte, Hemiepiphyte **Use:** HF, ME **Origin:** Native **Region:** Amazonia, Andean, Orinoquia, Pacific **Dept.:** AMA, ANT, BOG, BOY, CAL, CAQ, CAS, CAU, CHO, CUN, GUA, HUI, LAG, MAG, MET, NAR, NSA, PUT, QUI, SAN, TOL, VAC, VAU, VID **Elev.:** 0–3000 m a.s.l. **IUCN Habitat code:** 1,2,3,4,5,14 **Conservation status: (IUCN)** LC

6609 *Ficus benghalensis* L. (Moraceae)
IPNI: 852482-1
Common name: caucho, caucho cartagenero **Habit:** Tree **Use:** AF, EU, HF, MA, ME, SU **Origin:** Cultivated **Region:** Caribbean **Dept.:** MAG **Elev.:** 5–5 m a.s.l. **IUCN Habitat code:** 3,14 **Conservation status: (IUCN)** NE

6610 *Ficus benjamina* L. (Moraceae)
IPNI: 852486-1
Common name: benjamín, caucho, caucho benjamina, falso laurel, fico, ficus, laurel, laurel de la India, pivijay **Habit:** Tree **Use:** AF, EU, FU, HF, MA, ME, PO, SU **Origin:** Cultivated **Region:** Andean **Dept.:** ANT, BOG, MAG, VAC **Elev.:** 0–1550 m a.s.l. **IUCN Habitat code:** 1,2,3,14 **Conservation status: (IUCN)** LC

6611 *Ficus caballina* Standl. (Moraceae)
IPNI: 583293-1
Habit: Hemiepiphyte **Use:** HF, ME **Origin:** Native **Region:** Amazonia, Guiana Shield **Dept.:** AMA, CAQ, GUA, VAU **Elev.:** 100–400 m a.s.l. **IUCN Habitat code:** 1,3,14 **Conservation status: (IUCN)** LC

6612 *Ficus caldasiana* Dugand (Moraceae)
IPNI: 104473-2
Common name: caucho, mediacara **Habit:** Small tree, Tree **Use:** HF **Origin:** Native **Region:** Andean **Dept.:** ANT, CAL, CUN, NAR, VAC **Elev.:** 1200–1900 m a.s.l. **IUCN Habitat code:** 1,3,14 **Conservation status: (IUCN)** NE

6613 *Ficus carica* L. (Moraceae)
IPNI: 852556-1
Common name: breva, brevo, higo, higuera **Habit:** Shrub **Use:** AF, EU, GS, HF, MA, ME, PO, SU **Origin:** Cultivated **Region:** Andean **Dept.:** ANT, BOG, CUN, TOL, VAC **Elev.:** 1450–2800 m a.s.l. **IUCN Habitat code:** 1,3,14 **Conservation status: (IUCN)** LC

6614 *Ficus casapiensis* (Miq.) Miq. (Moraceae)
IPNI: 852560-1
Habit: Tree **Use:** ME **Origin:** Native **Region:** Amazonia, Pacific **Dept.:** ANT, VAU **Elev.:** 0–200 m a.s.l. **IUCN Habitat code:** 1,14 **Conservation status: (IUCN)** NE

6615 *Ficus citrifolia* Mill. (Moraceae)
IPNI: 316340-2
Common name: abrazapalo, caucho, copé, higuerón, higuito, matapalo, palo de leche **Habit:** Small tree, Tree, Hemiepiphyte **Use:** HF, MA, ME **Origin:** Native **Region:** Amazonia, Andean, Caribbean, Pacific **Dept.:** AMA, ANT, ATL, BOL, CAL, CAQ, CAU, CHO, CUN, HUI, LAG, MAG, NSA, PUT, SAN, SUC, TOL, VAC, VAU **Elev.:** 0–2400 m a.s.l. **IUCN Habitat code:** 1,2,3,5,14 **Conservation status: (IUCN)** LC

6616 *Ficus costaricana* (Liebm.) Miq. (Moraceae)
IPNI: 852672-1
Habit: Tree **Use:** EU **Origin:** Native **Region:** Pacific **Dept.:** ANT, CHO **Elev.:** 50–450 m a.s.l. **IUCN Habitat code:** 1,2,14 **Conservation status: (IUCN)** LC

6617 *Ficus crocata* (Miq.) Mart. ex Miq. (Moraceae)
IPNI: 852689-1
Common name: caucho, higuerón, lechero, matapalo, tapás **Habit:** Hemiepiphyte **Use:** HF **Origin:** Native **Region:** Amazonia, Andean, Caribbean **Dept.:** ANT, ATL, BOL, BOY, CAQ, CAS, CES, COR, CUN, HUI, MAG, NSA, TOL, VAC **Elev.:** 10–1800 m a.s.l. **IUCN Habitat code:** 1,2,3,14 **Conservation status: (IUCN)** LC

6618 *Ficus cuatrecasasiana* Dugand (Moraceae)
IPNI: 104497-2
Habit: Tree, Hemiepiphyte **Use:** ME **Origin:** Native **Region:** Andean, Pacific **Dept.:** ANT, CHO, CUN, HUI, MAG, MET, NAR, NSA, PUT, QUI, SAN, TOL, VAC **Elev.:** 150–3200 m a.s.l. **IUCN Habitat code:** 1,3,14 **Conservation status: (IUCN)** LC

6619 *Ficus donnell-smithii* Standl. (Moraceae)
IPNI: 104505-2
Common name: lechoso, mapanae, matapalo **Habit:** Hemiepiphyte **Use:** ME **Origin:** Native **Region:** Amazonia, Andean, Caribbean, Orinoquia **Dept.:** AMA, ANT, BOL, CAQ, CUN, HUI, MAG, MET, SAN, SUC, TOL, VAC **Elev.:** 50–2300 m a.s.l. **IUCN Habitat code:** 1,2,3,5,14 **Conservation status: (IUCN)** LC

6620 *Ficus dulciaria* Dugand (Moraceae)
IPNI: 104508-2
Common name: breva de monte, guayabilla, guayabo, higuerón, higuerón guayabo **Habit:** Tree **Use:** HF, ME **Origin:** Native **Region:** Andean **Dept.:** ANT, PUT, RIS **Elev.:** 2000–2800 m a.s.l. **IUCN Habitat code:** 1,3,14 **Conservation status: (IUCN)** NE

6621 *Ficus elastica* Roxb. ex Hornem. (Moraceae)
IPNI: 60458499-2
Common name: caucho, caucho de la India, falso caucho, gomero, lechón **Habit:** Tree **Use:** EU, FU, GS, HF, MA, ME, SU **Origin:** Cultivated **Region:** Andean **Dept.:** ANT, BOG, CUN, VAC **Elev.:** 1000–2600 m a.s.l. **IUCN Habitat code:** 1,2,3,14 **Conservation status: (IUCN)** NE

6622 *Ficus eliadis* Standl. (Moraceae)
IPNI: 104514-2
Common name: copé, matapalo **Habit:** Shrub, Tree **Use:** EU, MA **Origin:** Endemic **Region:** Caribbean **Dept.:** ATL, MAG, SUC **Elev.:** 50-350 m.a.s.l. **IUCN Habitat code:** 1,2,14 **Conservation status: (IUCN)** EN **(CPSG)** EN

6623 *Ficus eximia* Schott (Moraceae)
IPNI: 852825-1
Habit: Tree **Use:** HF, ME **Origin:** Native **Region:** Amazonia, Andean, Orinoquia **Dept.:** AMA, CAQ, MET, PUT **Elev.:** 600–1850 m a.s.l. **IUCN Habitat code:** 1,3,14 **Conservation status: (IUCN)** LC

6624 *Ficus garcia-barrigae* Dugand (Moraceae)
IPNI: 104537-2
Common name: higuerón **Use:** ME **Origin:** Native **IUCN Habitat code:** 1,14 **Conservation status: (IUCN)** NE

6625 *Ficus gigantosyce* Dugand (Moraceae)
IPNI: 104541-2
Common name: higuerón, quiveche, quivechi **Habit:** Tree **Use:** HF, ME **Origin:** Native **Region:** Andean **Dept.:** ANT, BOY, CUN, HUI, QUI, RIS, TOL **Elev.:** 1950–2900 m a.s.l. **IUCN Habitat code:** 1,3,14 **Conservation status: (IUCN)** LC

6626 *Ficus hartwegii* Miq. (Moraceae)
IPNI: 852958-1
Common name: carretillo, cauchito, caucho, caucho higuerón, caucho rosado, conservo, lechero, matapalo, topaz **Habit:** Shrub, Hemiepiphyte **Use:** ME **Origin:** Native **Region:** Andean, Caribbean, Pacific **Dept.:** ANT, BOL, CAU, CHO, CUN, NAR, QUI, TOL, VAC **Elev.:** 0–2250 m a.s.l. **IUCN Habitat code:** 1,2,3,5,14 **Conservation status: (IUCN)** LC

6627 *Ficus heterophylla* L.f. (Moraceae)
IPNI: 852970-1
Use: ME **Region:** Andean **Dept.:** BOG **IUCN Habitat code:** 14 **Conservation status: (IUCN)** NE

6628 *Ficus insipida* Willd. (Moraceae)
IPNI: 853024-1
Common name: bamba, bibosi, bibrán, cachinguba, caucho, caucho menudito, chanchama, chanchinguba, chibechi, chibechí, cocoba, cope higuerón, copei, copey, copé, cumacanae, damagua, ficus, gomelero, higueroncillo, higuerón, ira, leche de casingua, leche de higuerón men **Habit:** Tree **Use:** AF, EU, FU, HF, MA, ME, PO, SU **Origin:** Native **Region:** Amazonia, Andean, Caribbean, Orinoquia, Pacific **Dept.:** AMA, ANT, BOL, BOY, CAQ, CAS, CAU, CHO, CUN, GUV, HUI, LAG, MAG, MET, NAR, NSA, PUT, RIS, SAN, TOL, VAC **Elev.:** 0–2850 m a.s.l. **IUCN Habitat code:** 1,2,3,4,5,14 **Conservation status: (IUCN)** LC

6629 *Ficus lyrata* Warb. (Moraceae)
IPNI: 853158-1
Common name: caucho lirata, lirata, pandurata **Habit:** Tree **Use:** EU, HF, MA, ME, SU **Origin:** Cultivated **Region:** Andean **Dept.:** ANT, BOG, VAC **Elev.:** 100–1550 m a.s.l. **IUCN Habitat code:** 3,14 **Conservation status: (IUCN)** NE

6630 *Ficus macbridei* Standl. (Moraceae)
IPNI: 104612-2
Habit: Tree **Use:** ME **Origin:** Native **Region:** Andean, Pacific **Dept.:** ANT, CHO, NAR, PUT, SAN, VAC **Elev.:** 0–1900 m a.s.l. **IUCN Habitat code:** 1,3,5,14 **Conservation status: (IUCN)** LC

6631 *Ficus macrosyce* Pittier (Moraceae)
IPNI: 104614-2
Use: ME **Origin:** Native **IUCN Habitat code:** 1,2,3,5,14 **Conservation status: (IUCN)** NE

6632 *Ficus mathewsii* (Miq.) Miq. (Moraceae)
IPNI: 853198-1
Common name: caucho, juansoco de alacrán **Use:** HF **Origin:** Native **IUCN Habitat code:** 1,3,4,5,14 **Conservation status: (IUCN)** LC

6633 *Ficus maxima* Mill. (Moraceae)
IPNI: 853200-1
Common name: abuelo de higuerón, bujurqui, caucho, higuerón, lechero, lechudo, tapás, yanchama, yanchama blanca, yanchama bujurqui **Habit:** Small tree, Tree **Use:** AF, EU, FU, HF, MA, ME, PO **Origin:** Native **Region:** Amazonia, Andean, Caribbean, Orinoquia, Pacific **Dept.:** AMA, ANT, ATL, BOL, BOY, CAL, CAQ, CAU, CHO, CUN, HUI, LAG, MAG, MET, NAR, PUT, SAN, TOL, VAC **Elev.:** 0–2100 m a.s.l. **IUCN Habitat code:** 1,2,3,5,14 **Conservation status: (IUCN)** LC

6634 *Ficus membranacea* C.Wright (Moraceae)
IPNI: 583812-1
Habit: Tree **Use:** HF **Origin:** Native **Region:** Caribbean **Dept.:** MAG **Elev.:** 50–100 m a.s.l. **IUCN Habitat code:** 1,3 **Conservation status: (IUCN)** NE

6635 *Ficus mexicana* (Miq.) Miq. (Moraceae)
IPNI: 853212-1
Use: ME **Origin:** Native **Conservation status: (IUCN)** NE

6636 *Ficus microcarpa* L.f. (Moraceae)
IPNI: 853217-1
Common name: caucho, matapalo **Habit:** Tree **Use:** AF, EU, FU, HF, IF, MA, ME, PO, SU **Origin:** Cultivated **Region:** Amazonia, Caribbean **Dept.:** AMA, ATL, LAG, VAC **Elev.:** 100–1000 m a.s.l. **IUCN Habitat code:** 1,14 **Conservation status: (IUCN)** LC

6637 *Ficus mutisii* Dugand (Moraceae)
IPNI: 104651-2
Common name: chanchama, higuerón, yanchama **Habit:** Shrub, Small tree, Tree **Use:** MA **Origin:** Native **Region:** Andean **Dept.:** ANT, CUN, HUI, MAG, MET, NAR, PUT, VAC **Elev.:** 100–3000 m a.s.l. **IUCN Habitat code:** 1,3,14 **Conservation status: (IUCN)** LC

6638 *Ficus nymphaeifolia* Mill. (Moraceae)
IPNI: 853301-1
Common name: arepero, caucha, caucho, caucho blanco, cope negro, copey, copé, copé negro, higuerona, higuerón, hojenegro, laurel abrazapalo, matapalo **Habit:** Small tree, Tree, Hemiepiphyte **Use:** MA, ME, SU **Origin:** Native **Region:** Andean, Caribbean, Pacific **Dept.:** ANT, ATL, BOL, CAS, CHO, COR, CUN, HUI, MAG, MET, NAR, NSA, SAN, TOL, VAC **Elev.:** 10–1700 m a.s.l. **IUCN Habitat code:** 1,2,3,4,14 **Conservation status: (IUCN)** LC

6639 *Ficus obtusifolia* Kunth (Moraceae)
IPNI: 853317-1
Common name: caucho, herrerún, higuerón, lechero, matapalo, mechino, renaco, vetamo **Habit:** Tree, Hemiepiphyte **Use:** HF, MA, ME, PO **Origin:** Native **Region:** Amazonia, Andean, Orinoquia, Pacific **Dept.:** AMA, ANT, BOY, CAL, CAS, CAU, CHO, CUN, HUI, LAG, MAG, MET, NAR, NSA, PUT, TOL, VAC, VAU **Elev.:** 80–1800 m a.s.l. **IUCN Habitat code:** 1,2,3,5,14 **Conservation status: (IUCN)** LC

6640 *Ficus pallida* Vahl (Moraceae)
IPNI: 853355-1
Common name: benjamín, caucho, caucho menudito, caucho menudo, cope, copé, higo, higo hembra, higuerón, higuito, matapalo, pivijay **Habit:** Tree **Use:** HF, MA, ME **Origin:** Native **Region:** Amazonia, Andean, Caribbean, Orinoquia **Dept.:** AMA, ANT, ATL, BOL, CES, COR, CUN, HUI, LAG, MAG, MET, NSA, PUT, SAN, TOL **Elev.:** 0–3000 m a.s.l. **IUCN Habitat code:** 1,2,3,5,14 **Conservation status: (IUCN)** LC

6641 *Ficus paraensis* (Miq.) Miq. (Moraceae)
IPNI: 1004295-2
Common name: caucho, cauchos, higuerón, matapalo, renaco **Habit:** Shrub, Small tree, Tree, Hemiepiphyte **Use:** ME **Origin:** Native **Region:** Amazonia, Andean, Orinoquia **Dept.:** AMA, ANT, CAQ, CAU, CUN, LAG, MET, NAR, PUT, SAN, VAC **Elev.:** 0–1800 m a.s.l. **IUCN Habitat code:** 1,2,3,5,14 **Conservation status: (IUCN)** LC

6642 *Ficus pertusa* L.f. (Moraceae)
IPNI: 853405-1
Common name: caucho, caucho común, caucho menudito, higuerón, lechudo, matapalo, matapalo de agua **Habit:** Tree, Hemiepiphyte **Use:** EU, HF, MA, ME **Origin:** Native **Region:** Amazonia, Andean, Caribbean, Orinoquia, Pacific **Dept.:** AMA, ANT, ATL, BOL, BOY, CAQ, CAU, CHO, COR, CUN, HUI, MET, NSA, RIS, SAN, TOL, VAC **Elev.:** 5–1940 m a.s.l. **IUCN Habitat code:** 1,2,3,4,5,14 **Conservation status: (IUCN)** LC

6643 *Ficus pumila* L. (Moraceae)
IPNI: 853513-1
Common name: uña de gato, uñita, yedro **Habit:** Climbing **Use:** EU, GS, HF, MA, ME, PO **Origin:** Cultivated **Region:** Andean **Dept.:** ANT, BOG, CUN, QUI **Elev.:** 1500–2600 m a.s.l. **IUCN Habitat code:** 1,3,14 **Conservation status: (IUCN)** NE

6644 *Ficus religiosa* L. (Moraceae)
IPNI: 853563-1
Common name: caucho de la India **Habit:** Tree **Use:** AF, EU, FU, GS, HF, MA, ME, SU **Origin:** Cultivated **Region:** Caribbean **Dept.:** BOL **IUCN Habitat code:** 1,14 **Conservation status: (IUCN)** NE

6645 *Ficus schippii* Standl. (Moraceae)
IPNI: 104716-2
Common name: daruja, higuerón, matapalo, renaco, wabama **Habit:** Tree, Hemiepiphyte **Use:** FU **Origin:** Native **Region:** Amazonia, Andean, Pacific **Dept.:** AMA, ANT, CAQ, CHO, NAR, NSA, PUT, TOL, VAC, VAU **Elev.:** 5–1800 m a.s.l. **IUCN Habitat code:** 1,2,3,5,14 **Conservation status: (IUCN)** LC

6646 *Ficus tonduzii* Standl. (Moraceae)
IPNI: 104743-2
Common name: ariza, caucho, higueron, higueroncillo, higuerón, higuerón grande, higuerón pequeño, lechero, lechero de hoja ancha, matapalo **Habit:** Tree **Use:** MA, ME, SU **Origin:** Native **Region:** Andean, Pacific **Dept.:** ANT, CAL, CAU, CHO, CUN, HUI, MAG, NAR, NSA, PUT, QUI, SAN, TOL, VAC **Elev.:** 100–2400 m a.s.l. **Conservation status: (IUCN)** LC

6647 *Ficus trigona* L.f. (Moraceae)
IPNI: 853876-1
Common name: caucho, caucho matapalo, matapalos, renaco **Habit:** Small tree, Tree, Hemiepiphyte **Use:** HF, ME **Origin:** Native **Region:** Amazonia, Andean, Orinoquia **Dept.:** AMA, CAQ, MET, PUT, VAU **Elev.:** 30–1500 m a.s.l. **IUCN Habitat code:** 1,2,3,14 **Conservation status: (IUCN)** LC

6648 *Ficus trigonata* L. (Moraceae)
IPNI: 316338-2
Use: HF **IUCN Habitat code:** 1,3,14 **Conservation status: (IUCN)** LC

6649 *Ficus velutina* Humb. & Bonpl. ex Willd. (Moraceae)
IPNI: 853933-1
Common name: caucho, copey de tierra fría, herrerún, higuerón, higuerón raspador, machimbi, machimbí, matamorro, oticón, pobo, raspador **Habit:** Tree **Use:** AF, EU, FU, HF, MA, ME **Origin:** Native **Region:** Andean **Dept.:** ANT, BOG, BOY, CAL, CAU, CUN, HUI, NAR, NSA, QUI, RIS, SAN, TOL, VAC **Elev.:** 350–3100 m a.s.l. **IUCN Habitat code:** 1,3,14 **Conservation status: (IUCN)** LC

6650 *Ficus yoponensis* Desv. (Moraceae)
IPNI: 853989-1
Habit: Tree **Use:** ME **Origin:** Native **Region:** Amazonia, Andean, Pacific **Dept.:** ANT, CHO, CUN, VAC, VAU **Elev.:** 0–1810 m a.s.l. **IUCN Habitat code:** 1,3,5,14 **Conservation status: (IUCN)** LC

6651 *Ficus ypsilophlebia* Dugand (Moraceae)
IPNI: 104771-2
Common name: caucho, ceroto, copé, matapalo **Habit:** Tree **Use:** EU, ME, SU **Origin:** Native **Region:** Caribbean **Dept.:** ATL, BOL, MAG **Elev.:** 10–200 m a.s.l. **IUCN Habitat code:** 1,2,14 **Conservation status: (IUCN)** NE

6652 *Helianthostylis sprucei* Baill. (Moraceae)
IPNI: 317010-2
Common name: castaño, guaimaro, guáimaro, leche de perra, leche perra, lechudo, palanco, piyu, quematatabro, suerpo, yema de huevo **Habit:** Shrub, Small tree, Tree **Use:** FU, HF, MA, ME **Origin:** Native **Region:** Andean, Guiana Shield, Caribbean, Orinoquia, Pacific **Dept.:** ANT, BOL, CAL, CHO, COR, CUN, GUA, LAG, SAN **Elev.:** 10–1400 m a.s.l. **IUCN Habitat code:** 1,2,3,5,14 **Conservation status: (IUCN)** LC

6653 *Helicostylis elegans* (J.F.Macbr.) C.C.Berg (Moraceae)
IPNI: 119532-2
Habit: Tree **Use:** HF **Origin:** Native **Region:** Amazonia **Dept.:** CAQ, VAU **Elev.:** 400–500 m a.s.l. **IUCN Habitat code:** 1,3,14 **Conservation status: (IUCN)** LC

6654 *Helicostylis heterotricha* Ducke (Moraceae)
IPNI: 119533-2
Common name: leche de cabra **Habit:** Small tree, Tree **Use:** HF **Origin:** Native **Region:** Amazonia **Dept.:** AMA, CAQ **Elev.:** 100–100 m a.s.l. **IUCN Habitat code:** 1,3,14 **Conservation status: (IUCN)** EN

6655 *Helicostylis scabra* (J.F.Macbr.) C.C.Berg (Moraceae)
IPNI: 119540-2
Common name: chimigua, ojo de culebra, raiberana **Habit:** Small tree, Tree **Use:** HF **Origin:** Native **Region:** Amazonia **Dept.:** AMA, VAU **Elev.:** 80–250 m a.s.l. **IUCN Habitat code:** 1,3,4,5,14 **Conservation status: (IUCN)** LC

6656 *Helicostylis tomentosa* (Poepp. & Endl.) J.F.Macbr. (Moraceae) **IPNI:** 584303-1 **Common name:** guáimaro, leche de perra, lecheperra, mare casaca, potacaci **Habit:** Small tree, Tree **Use:** HF, MA **Origin:** Native **Region:** Amazonia, Andean, Caribbean, Orinoquia, Pacific **Dept.:** AMA, ANT, BOL, CHO, COR, MET, NAR, SAN, VAC, VID **Elev.:** 20–1200 m a.s.l. **IUCN Habitat code:** 1,2,3,4,5,14 **Conservation status: (IUCN)** LC

6657 *Helicostylis tovarensis* (Klotzsch & H.Karst.) C.C.Berg (Moraceae) **IPNI:** 119542-2 **Common name:** cauchillo, jumbá, lechero **Habit:** Tree **Use:** HF **Origin:** Native **Region:** Andean **Dept.:** ANT, BOY, HUI, NSA, VAC **Elev.:** 1000–2500 m a.s.l. **IUCN Habitat code:** 1,3,14 **Conservation status: (IUCN)** LC

6658 *Maclura tinctoria* (L.) D.Don ex Steud. (Moraceae) **IPNI:** 1151750-2 **Common name:** abrojo, ají mora, avinje, charo, chimane, dinde, espino mora, fustete, insira, iron odd, laurel mora, mora, morajil, moral, moral fustete, morita, morito, morón, muche, palo am, palo amarillo, palo de mora, palo mora, palo moro, rabito, rabito de mora, ta **Habit:** Shrub, Tree **Use:** AF, EU, FU, HF, MA, ME **Origin:** Native **Region:** Amazonia, Andean, Caribbean **Dept.:** AMA, ANT, ATL, BOL, BOY, CAL, CAQ, CES, CUN, HUI, LAG, MAG, NSA, SAP, SAN, SUC, TOL, VAC **Elev.:** 0-1500 m.a.s.l. **IUCN Habitat code:** 1,2,3,4,14 **Conservation status: (IUCN)** LC **(CPSG)** Potential LC

6659 *Maquira calophylla* (Poepp. & Endl.) C.C.Berg (Moraceae) **IPNI:** 152821-2 **Common name:** leche chivo **Habit:** Small tree, Tree **Use:** HF, ME **Origin:** Native **Region:** Amazonia **Dept.:** AMA **Elev.:** 100–220 m a.s.l. **IUCN Habitat code:** 1,3,4,14 **Conservation status: (IUCN)** LC

6660 *Maquira coriacea* (H.Karst.) C.C.Berg (Moraceae) **IPNI:** 152822-2 **Common name:** lechechiva, lechero **Habit:** Tree **Use:** HF, ME **Origin:** Native **Region:** Amazonia, Orinoquia **Dept.:** AMA, MET, VID **Elev.:** 110–300 m a.s.l. **IUCN Habitat code:** 1,2,3,4,5,14 **Conservation status: (IUCN)** LC

6661 *Maquira guianensis* Aubl. (Moraceae) **IPNI:** 854491-1 **Common name:** lechero, marecasaca, oquendo, quematatabro **Habit:** Small tree, Tree **Use:** HF, MA **Origin:** Native **Region:** Amazonia, Andean, Pacific **Dept.:** AMA, ANT, CAQ, CAU, CHO, NAR, VAC **Elev.:** 100–980 m a.s.l. **IUCN Habitat code:** 1,2,3,5,14 **Conservation status: (IUCN)** LC

6662 *Morus alba* L. (Moraceae) **IPNI:** 30051955-2 **Common name:** morera **Habit:** Small tree **Use:** AF, EU, FU, GS, HF, MA, ME, PO **Origin:** Cultivated, Naturalised **Region:** Andean **Dept.:** ANT, BOG, CAL, CAU, QUI **Elev.:** 1200–1500 m a.s.l. **IUCN Habitat code:** 1,3,14 **Conservation status: (IUCN)** NE

6663 *Morus celtidifolia* Kunth (Moraceae) **IPNI:** 854634-1 **Use:** HF, MA, ME **Origin:** Native **IUCN Habitat code:** 14 **Conservation status: (IUCN)** LC

6664 *Morus insignis* Bureau (Moraceae) **IPNI:** 316991-2 **Common name:** aliso, caucho, huesito, justarrazón, lechero, queso fresco, salvio **Habit:** Small tree, Tree **Use:** HF, ME **Origin:** Native **Region:** Andean **Dept.:** ANT, BOG, BOY, CUN, HUI, LAG, NAR, QUI, RIS, VAC **Elev.:** 1450–2610 m a.s.l. **IUCN Habitat code:** 1,3,14 **Conservation status: (IUCN)** LC

6665 *Morus nigra* L. (Moraceae) **IPNI:** 854688-1 **Common name:** morera **Use:** AF, EU, FU, HF, MA, ME, SU **Conservation status: (IUCN)** NE

6666 *Naucleopsis concinna* (Standl.) C.C.Berg (Moraceae) **IPNI:** 167394-2 **Habit:** Tree **Use:** HF **Origin:** Native **Region:** Amazonia **Dept.:** AMA, VAU **Elev.:** 200–200 m a.s.l. **IUCN Habitat code:** 1 **Conservation status: (IUCN)** LC

6667 *Naucleopsis glabra* Spruce ex Pittier (Moraceae) **IPNI:** 167395-2 **Common name:** anón de monte, anón lechoso, anón montañero, caimo chicle, cerezo, lechero, mamoncillo, mamoncillo silvestre, pacurú-niaara, pata de morrocoy, sande, semilla peluda **Habit:** Small tree, Tree **Use:** HF, ME **Origin:** Native **Region:** Amazonia, Caribbean, Pacific **Dept.:** AMA, ANT, BOL, CAQ, CHO, COR, PUT, SAN **Elev.:** 50–900 m a.s.l. **IUCN Habitat code:** 1,3,4,5,14 **Conservation status: (IUCN)** LC

6668 *Naucleopsis herrerensis* C.C.Berg (Moraceae) **IPNI:** 304351-2 **Habit:** Tree **Use:** HF **Origin:** Native **Region:** Amazonia **Dept.:** CAQ, PUT **Elev.:** 300–300 m a.s.l. **Conservation status: (IUCN)** NE

6669 *Naucleopsis imitans* (Ducke) C.C.Berg (Moraceae) **IPNI:** 167397-2 **Habit:** Tree **Use:** HF **Origin:** Native **Region:** Amazonia **Dept.:** AMA, CAQ, VAU **Elev.:** 200–200 m a.s.l. **IUCN Habitat code:** 1,5,14 **Conservation status: (IUCN)** NE

6670 *Naucleopsis krukovii* (Standl.) C.C.Berg (Moraceae) **IPNI:** 167401-2 **Common name:** anón lechoso, madroño **Habit:** Small tree, Tree **Use:** HF **Origin:** Native **Region:** Amazonia **Dept.:** AMA, GUV, VAU **Elev.:** 100–200 m a.s.l. **IUCN Habitat code:** 1,3,14 **Conservation status: (IUCN)** LC

6671 *Naucleopsis naga* Pittier (Moraceae) **IPNI:** 167403-2 **Common name:** peinemono **Habit:** Tree **Use:** HF **Origin:** Native **Region:** Andean, Caribbean, Pacific **Dept.:** ANT, BOL, CHO, NAR, VAC **Elev.:** 50–1950 m a.s.l. **IUCN Habitat code:** 1,14 **Conservation status: (IUCN)** NE

6672 *Naucleopsis oblongifolia* (Kuhlm.) Carauta (Moraceae) **IPNI:** 1182949-2 **Common name:** amaji **Habit:** Small tree, Tree **Use:** HF, ME **Origin:** Native **Region:** Amazonia **Dept.:** AMA, CAQ **Elev.:** 100–200 m a.s.l. **IUCN Habitat code:** 1,3,14 **Conservation status: (IUCN)** VU

6673 *Naucleopsis pseudonaga* (Mildbr.) C.C.Berg (Moraceae) **IPNI:** 167405-2 **Habit:** Tree **Use:** HF **Origin:** Native **Region:** Amazonia **Dept.:** VAU **Elev.:** 200–200 m a.s.l. **IUCN Habitat code:** 1 **Conservation status: (IUCN)** NE

6674 *Naucleopsis ternstroemiiflora* (Mildbr.) C.C.Berg (Moraceae) **IPNI:** 584418-1 **Common name:** niaara, pacurú, pacurú-niaara **Habit:** Tree **Use:** HF, ME **Origin:** Native **Region:** Amazonia **Dept.:** AMA, GUV **Elev.:** 100–250 m a.s.l. **IUCN Habitat code:** 1 **Conservation status: (IUCN)** LC

6675 *Naucleopsis ulei* (Warb.) Ducke (Moraceae) **IPNI:** 167409-2 **Common name:** almendra, amacá, capinurí, cauchillo, madroño, mano de tigre, pama, patemorroco, patemorrocoy **Habit:** Small tree, Tree **Use:** HF, ME **Origin:** Native **Region:** Amazonia, Andean, Caribbean, Pacific **Dept.:** AMA, ANT, CHO, COR, NAR, PUT, VAC **Elev.:** 100–1520 m a.s.l. **IUCN Habitat code:** 1,14 **Conservation status: (IUCN)** LC

6676 *Perebea angustifolia* (Poepp. & Endl.) C.C.Berg (Moraceae) **IPNI:** 190063-2 **Common name:** cauchillo de monte **Habit:** Small tree, Tree **Use:** HF **Origin:** Native **Region:** Amazonia, Andean, Pacific **Dept.:** AMA, CHO, PUT, VAC **Elev.:** 50–1940 m a.s.l. **IUCN Habitat code:** 1,3,5,14 **Conservation status: (IUCN)** LC

6677 *Perebea guianensis* Aubl. (Moraceae) **IPNI:** 855103-1 **Common name:** caucho, echaduy, lechero, lechudo, palo diablo, pamá, varasanta **Habit:** Small tree, Tree **Use:** HF, ME **Origin:** Native **Region:** Amazonia, Andean, Pacific **Dept.:** AMA, ANT, CAQ, CHO, MET, PUT, VAU **Elev.:** 30–1850 m a.s.l. **IUCN Habitat code:** 1,3,14 **Conservation status: (IUCN)** LC

6678 *Perebea longepedunculata* C.C.Berg (Moraceae) **IPNI:** 190080-2 **Use:** HF **Origin:** Native **IUCN Habitat code:** 1 **Conservation status: (IUCN)** NE

6679 *Perebea mollis* (Poepp. & Endl.) Huber (Moraceae) **IPNI:** 1008109-1 **Habit:** Tree **Use:** HF **Origin:** Native **Region:** Amazonia, Guiana Shield **Dept.:** AMA, GUV, MET, VAU **Elev.:** 100–300 m a.s.l. **IUCN Habitat code:** 1,3,14 **Conservation status: (IUCN)** LC

6680 *Perebea xanthochyma* H.Karst. (Moraceae) **IPNI:** 855107-1 **Common name:** cauchillo, cauchillo de loma, cerezo, gure, jidadiapique, lechoso, pampanillo **Habit:** Shrub, Small tree, Tree, Hemiepiphyte **Use:** HF, ME **Origin:** Native **Region:** Amazonia, Andean, Caribbean, Orinoquia, Pacific **Dept.:** AMA, ANT, BOL, CAQ, CAU,

CHO, GUA, GUV, MET, NAR, PUT, SAN, VAC, VAU **Elev.:** 0–1800 m a.s.l. **IUCN Habitat code:** 1,3,4,5,14 **Conservation status: (IUCN)** LC

6681 *Poulsenia armata* (Miq.) Standl. (Moraceae)
IPNI: 1059330-2
Common name: algodoncillo, amarillo espino, carapacha, carbón, caucho, cocua, cocuá, corbón, cucuá, damagua, demaguo, fichacha tasa, guáimaro, higuerón tunoso, hueporo, lechero, majagua, maragua, mare casaca, namagua, pacua, punte, yanchama, yanchama colorada, yancham **Habit:** Tree **Use:** AF, HF, MA, ME, SU **Origin:** Native **Region:** Amazonia, Andean, Orinoquia, Pacific **Dept.:** AMA, ANT, CHO, CUN, LAG, MAG, MET, NAR, PUT, QUI, TOL, VAC **Elev.:** 0–2350 m a.s.l. **IUCN Habitat code:** 1,3,5,14 **Conservation status: (IUCN)** LC

6682 *Pseudolmedia laevigata* Trécul (Moraceae)
IPNI: 856393-1
Common name: caimito leche de perra, cajeto, calostro, capinurí, cauchillo, cauchillo de monte, cenizo, cervera, charo, chinicua, corazón de envuelto, guayabo, guáimaro, leche de chiva, leche de perra, lechechiva, lecheperra, lechesapo, mirapiranga **Habit:** Small tree, Tree **Use:** HF, MA, ME **Origin:** Native **Region:** Amazonia, Andean, Caribbean, Pacific **Dept.:** AMA, ANT, BOL, CAQ, CHO, COR, GUA, SAN, VAC, VAU **Elev.:** 10–1100 m a.s.l. **IUCN Habitat code:** 1,2,3,4,5,14 **Conservation status: (IUCN)** LC

6683 *Pseudolmedia laevis* (Ruiz & Pav.) J.F.Macbr. (Moraceae)
IPNI: 211411-2
Common name: ajicillo, amarillo peludo, bamneé, caimo, capinurí, guayacán, hojarasco, leche, leche chiva, leche perra, lechechiva, lecheperra, pica pica, picapica, pomo, sarrapio **Habit:** Shrub, Small tree, Tree **Use:** HF, ME **Origin:** Native **Region:** Amazonia, Andean, Orinoquia **Dept.:** AMA, CAQ, GUV, MET, PUT, VAC, VAU **Elev.:** 100–1400 m a.s.l. **IUCN Habitat code:** 1,3,4,14 **Conservation status: (IUCN)** LC

6684 *Pseudolmedia macrophylla* Trécul (Moraceae)
IPNI: 856394-1
Habit: Tree **Use:** HF, ME **Origin:** Native **Region:** Amazonia **Dept.:** AMA, CAQ, PUT, VAU **IUCN Habitat code:** 1,3,14 **Conservation status: (IUCN)** NE

6685 *Pseudolmedia rigida* (Klotzsch & H.Karst.) Cuatrec. (Moraceae)
IPNI: 211419-2
Common name: caimo, caimo fierro, caimo lechoso, candelo, cauchillo, cristal blanco, guaimarillo, guáimaro, lecheperra, lechero, lechero plomo, torcacero, yamba, yumbá, yumbá colorado **Habit:** Small tree, Tree **Use:** HF, MA **Origin:** Native **Region:** Amazonia, Andean, Pacific **Dept.:** AMA, ANT, CHO, CUN, LAG, MAG, MET, QUI, VAC **Elev.:** 10–2600 m a.s.l. **IUCN Habitat code:** 1,3,14 **Conservation status: (IUCN)** LC

6686 *Sorocea affinis* Hemsl. (Moraceae)
IPNI: 856562-1
Habit: Shrub, Tree **Use:** ME **Origin:** Native **Region:** Andean, Pacific **Dept.:** ANT, CHO, CUN **Elev.:** 20–1465 m a.s.l. **IUCN Habitat code:** 1,3,14 **Conservation status: (IUCN)** LC

6687 *Sorocea guilleminiana* Gaudich. (Moraceae)
IPNI: 856566-1
Common name: lechero **Habit:** Tree **Use:** HF **Origin:** Native **Region:** Amazonia **Dept.:** AMA, CAQ **Elev.:** 100–100 m a.s.l. **IUCN Habitat code:** 1,14 **Conservation status: (IUCN)** VU

6688 *Sorocea muriculata* Miq. (Moraceae)
IPNI: 316993-2
Common name: caimo, leche de chiva, leche de chivo **Habit:** Shrub, Small tree, Tree **Use:** HF **Origin:** Native **Region:** Amazonia, Andean, Guiana Shield, Orinoquia **Dept.:** AMA, CAQ, GUA, GUV, MET, PUT, VAU, VID **Elev.:** 100–1000 m a.s.l. **IUCN Habitat code:** 1,3,4,5,14 **Conservation status: (IUCN)** LC

6689 *Sorocea pubivena* Hemsl. (Moraceae)
IPNI: 856575-1
Common name: cresta de guacamayo, lechero, mare casaca, palo de lombriz **Habit:** Shrub, Small tree, Tree **Use:** HF, ME **Origin:** Native **Region:** Amazonia, Andean, Caribbean, Orinoquia **Dept.:** AMA, ANT, BOL, BOY, CAQ, COR, MET, NAR, PUT, VAC **Elev.:** 20–1670 m a.s.l. **IUCN Habitat code:** 1,2,3,4,5,14 **Conservation status: (IUCN)** LC

6690 *Sorocea sprucei* (Baill.) J.F.Macbr. (Moraceae)
IPNI: 240971-2
Common name: guáimaro, lechero, mamón leche, pimpinillo **Habit:** Small tree, Tree **Use:** AF, HF, MA **Origin:** Native **Region:** Andean, Caribbean, Orinoquia **Dept.:** ATL, BOL, CAS, HUI, LAG, MAG, NSA, QUI, RIS, SAN, SUC, TOL, VAC **Elev.:** 10–1070 m a.s.l. **IUCN Habitat code:** 1,2,3,14 **Conservation status: (IUCN)** LC

6691 *Sorocea steinbachii* C.C.Berg (Moraceae)
IPNI: 285398-2
Habit: Shrub, Small tree, Tree **Use:** HF **Origin:** Native **Region:** Amazonia, Orinoquia **Dept.:** AMA, CAQ, MET, PUT, VID **Elev.:** 100–350 m a.s.l. **IUCN Habitat code:** 1,3,4,5,14 **Conservation status: (IUCN)** LC

6692 *Sorocea trophoides* W.C.Burger (Moraceae)
IPNI: 240975-2
Common name: quematatabro **Habit:** Shrub, Tree **Use:** HF, ME **Origin:** Native **Region:** Andean, Caribbean **Dept.:** ANT, BOL, CUN, QUI, RIS, TOL, VAC **Elev.:** 300–1600 m a.s.l. **IUCN Habitat code:** 1,2,3,5,14 **Conservation status: (IUCN)** LC

6693 *Trophis caucana* (Pittier) C.C.Berg (Moraceae)
IPNI: 945890-1
Common name: cauchillo, guáimaro, lechero, pacurú-niaara **Habit:** Shrub, Small tree, Tree **Use:** HF, MA, ME **Origin:** Native **Region:** Andean, Pacific **Dept.:** ANT, BOL, CAL, CAU, CHO, CUN, HUI, PUT, QUI, RIS, TOL, VAC **Elev.:** 0–3200 m a.s.l. **IUCN Habitat code:** 1,2,3,5,14 **Conservation status: (IUCN)** LC

6694 *Trophis racemosa* (L.) Urb. (Moraceae)
IPNI: 260203-2
Common name: barbarito, chicharrón, gusano, guáimaro, lechero colorado, oquendo **Habit:** Shrub, Small tree, Tree **Use:** AF, HF, MA, ME **Origin:** Native **Region:** Andean, Caribbean, Orinoquia, Pacific **Dept.:** AMA, ANT, BOL, CAL, CAQ, CAS, CHO, CUN, GUV, LAG, MAG, MET, PUT, SAP, SAN, VAC **Elev.:** 0–1000 m a.s.l. **IUCN Habitat code:** 1,2,3,4,5,14 **Conservation status: (IUCN)** LC

6695 *Trymatococcus amazonicus* Poepp. & Endl. (Moraceae)
IPNI: 317009-2
Common name: comida de danta **Habit:** Shrub **Use:** HF, MA **Origin:** Native **Region:** Amazonia, Guiana Shield **Dept.:** AMA, CAQ, GUA, VAU **Elev.:** 80–100 m a.s.l. **IUCN Habitat code:** 1,3,4,5,14 **Conservation status: (IUCN)** LC

6696 *Ampelozizyphus amazonicus* Ducke (Rhamnaceae)
IPNI: 11511-2
Common name: bejuco de jabón, hierba de umarí **Habit:** Climbing **Use:** ME **Origin:** Native **Region:** Amazonia, Pacific **Dept.:** AMA, CAQ, VAC **Elev.:** 50–200 m a.s.l. **IUCN Habitat code:** 1,4,5 **Conservation status: (IUCN)** LC

6697 *Colubrina arborescens* (Mill.) Sarg. (Rhamnaceae)
IPNI: 63423-2
Habit: Shrub, Small tree **Use:** AF, EU, FU, HF, MA, ME **Origin:** Native **Region:** Andean **Dept.:** CAL, CHO, VAC **Elev.:** 1000–1100 m a.s.l. **IUCN Habitat code:** 1,3 **Conservation status: (IUCN)** LC

6698 *Colubrina elliptica* (Sw.) Brizicky & W.L.Stern (Rhamnaceae)
IPNI: 63435-2
Common name: bois mabi, mabi **Habit:** Tree **Use:** HF, MA, ME **Origin:** Native **Region:** Caribbean **Dept.:** MAG, VAC **Elev.:** 50–50 m a.s.l. **IUCN Habitat code:** 14 **Conservation status: (IUCN)** LC

6699 *Colubrina glandulosa* G.Perkins (Rhamnaceae)
IPNI: 63438-2
Habit: Tree **Use:** ME **Origin:** Native **Region:** Amazonia **Dept.:** CAQ, GUV, HUI, MET **Elev.:** 200–900 m a.s.l. **IUCN Habitat code:** 1,3,14 **Conservation status: (IUCN)** LC

6700 *Frangula goudotiana* (Triana & Planch.) Grubov (Rhamnaceae)
IPNI: 105605-2
Common name: cabo de hacha, naranjo, naranjuelo, palo amarillo **Habit:** Shrub, Small tree **Use:** ME **Origin:** Native **Region:** Andean **Dept.:** ANT, BOY, CUN, LAG, MAG, NAR, NSA, SAN **Elev.:** 2000–3700 m a.s.l. **IUCN Habitat code:** 1,2,3,4,14 **Conservation status: (IUCN)** NE

6701 *Frangula granulosa* (Ruiz & Pav.) Grubov (Rhamnaceae)
IPNI: 105606-2
Habit: Shrub, Tree **Use:** HF, ME **Origin:** Native **Region:** Andean **Dept.:** ANT, BOL, CAQ, CAU, CHO, HUI, NAR, VAC **Elev.:** 300–2700 m a.s.l. **IUCN Habitat code:** 1,3,14 **Conservation status: (IUCN)** NE

6702 *Frangula purshiana* (DC.) A.Gray ex J.G.Cooper (Rhamnaceae)
IPNI: 30426867-2
Common name: cáscara, cáscara sagrada **Use:** HF, MA, ME, PO **Conservation status: (IUCN)** LC

6703 *Gouania lupuloides* (L.) Urb. (Rhamnaceae)
IPNI: 112858-2
Common name: bejuco de reuma, jaboncillo, limpiadientes, rabo de mono **Habit:** Liana **Use:** HF, IF, MA, ME **Origin:** Native **Region:** Amazonia, Andean, Guiana Shield, Caribbean, Pacific **Dept.:** ANT, BOL, CAQ, CHO, CUN, HUI, MAG, VAU **Elev.:** 0–1650 m a.s.l. **IUCN**

Habitat code: 1,2,3,5,14 **Conservation status: (IUCN)** NE

6704 *Gouania polygama* (Jacq.) Urb. (Rhamnaceae) **IPNI:** 112865-2 **Common name:** agraz, bejuco de Urabá, bejuco de rema, canillo, corono, cueroevieja, gancho, jaboncilla, limpiadientes, mascapalo, sueldo, verdinegro, zarzaparrilla **Habit:** Shrub, Liana **Use:** HF, IF, MA, ME **Origin:** Native **Region:** Amazonia, Andean, Guiana Shield, Caribbean, Pacific **Dept.:** AMA, ANT, CHO, COR, HUI, LAG, MAG, MET, QUI, SAN, SUC, TOL, VAC, VAU **Elev.:** 5-1700 m.a.s.l. **IUCN Habitat code:** 1,2,3,4,5,14 **Conservation status: (IUCN)** NE **(CPSG)** Potential LC

6705 *Karwinskia humboldtiana* (Willd. ex Schult.) Zucc. (Rhamnaceae) **IPNI:** 717579-1 **Common name:** cacachilla, cogotillo, frángula, tullidora **Use:** ME, PO **Dept.:** CUN **Conservation status: (IUCN)** LC

6706 *Krugiodendron ferreum* (Vahl) Urb. (Rhamnaceae) **IPNI:** 133956-2 **Habit:** Tree **Use:** HF, MA **Origin:** Native **Region:** Caribbean **Dept.:** SAP **Elev.:** 10–120 m a.s.l. **Conservation status: (IUCN)** LC

6707 *Rhamnus infectoria* L. (Rhamnaceae) **IPNI:** 718405-1 **Use:** ME **Dept.:** BOG **IUCN Habitat code:** 14 **Conservation status: (IUCN)** NE

6708 *Sageretia elegans* (Kunth) Brongn. (Rhamnaceae) **IPNI:** 1013652-1 **Common name:** adorote, corono, gancho, patudo, teñidor, uña de gato, uña de águila **Habit:** Climbing **Use:** HF **Origin:** Native, Cultivated **Dept.:** ANT, CAU, CUN, RIS, SAN, VAC **Elev.:** 200–2000 m a.s.l. **IUCN Habitat code:** 1,2,3,14 **Conservation status: (IUCN)** NE

6709 *Sarcomphalus cinnamomum* (Triana & Planch.) Hauenschild (Rhamnaceae) **IPNI:** 77154236-1 **Habit:** Shrub, Small tree **Use:** HF **Origin:** Native **Region:** Amazonia, Pacific **Dept.:** AMA, ANT **Elev.:** 30–325 m a.s.l. **IUCN Habitat code:** 1,3,14 **Conservation status: (IUCN)** LC

6710 *Sarcomphalus cyclocardius* (S.F.Blake) Hauenschild (Rhamnaceae) **IPNI:** 77154237-1 **Habit:** Tree **Use:** HF **Origin:** Native **Region:** Caribbean **Dept.:** LAG, MAG **Elev.:** 20–20 m a.s.l. **IUCN Habitat code:** 1 **Conservation status: (IUCN)** NE

6711 *Sarcomphalus saeri* (Pittier) Hauenschild (Rhamnaceae) **IPNI:** 77154251-1 **Habit:** Tree **Use:** HF **Origin:** Native **Region:** Caribbean **Dept.:** ATL, BOL, LAG **Elev.:** 5–100 m a.s.l. **IUCN Habitat code:** 2,14 **Conservation status: (IUCN)** NE

6712 *Ziziphus jujuba* Mill. (Rhamnaceae) **IPNI:** 719213-1 **Common name:** guinda, manzanita, manzanito **Habit:** Shrub **Use:** AF, EU, FU, GS, HF, IF, MA, ME, PO, SU **Origin:** Cultivated **Region:** Caribbean **Dept.:** BOL, CAL, COR, VAC **Elev.:** 0–100 m a.s.l. **IUCN Habitat code:** 1,2,3,14 **Conservation status: (IUCN)** LC

6713 *Ziziphus mauritiana* Lam. (Rhamnaceae) **IPNI:** 719349-1 **Common name:** guinda, guindaguinda, manzanita costeña, pistacho **Habit:** Shrub, Small tree **Use:** AF, FU, HF, MA, ME **Origin:** Cultivated, Naturalised **Region:** Caribbean **Dept.:** ANT, ATL, BOL, BOY, CAL, LAG, MAG, SAP, VAC **Elev.:** 10–900 m a.s.l. **Conservation status: (IUCN)** LC

6714 *Acaena cylindristachya* Ruiz & Pav. (Rosaceae) **IPNI:** 719970-1 **Common name:** abrojo, abrotano, amorseco, cadillo, cadillo blanco, cadulo, cagalungasé, cunde, piojo, quiebrabarrigo, quin, quin felpa, yerba del sol **Habit:** Herb **Use:** EU, ME **Origin:** Native **Region:** Andean **Dept.:** ANT, BOG, BOY, CAU, CES, CUN, LAG, MAG, MET, NSA, SAN **Elev.:** 2500–4300 m a.s.l. **IUCN Habitat code:** 1,3,4,8,14 **Conservation status: (IUCN)** NE

6715 *Acaena elongata* L. (Rosaceae) **IPNI:** 719987-1 **Common name:** acaena, cadillo, cadillo blanco, cardillo **Habit:** Herb, Subshrub **Use:** ME **Origin:** Native **Region:** Andean **Dept.:** BOG, BOY, CAL, CAU, CES, CHO, CUN, MAG, MET, NAR, NSA, PUT, QUI, SAN, TOL **Elev.:** 2500-4000 m.a.s.l. **IUCN Habitat code:** 1,2,3,4,5,8,14 **Conservation status: (IUCN)** NE **(CPSG)** Potential LC

6716 *Acaena ovalifolia* Ruiz & Pav. (Rosaceae) **IPNI:** 720073-1 **Common name:** cadillo **Habit:** Herb, Subshrub **Use:** ME **Origin:** Native **Region:** Andean **Dept.:** BOG, BOY, CAL, CUN, NAR, QUI, RIS, TOL, VAC **Elev.:** 2000–3700 m a.s.l. **IUCN Habitat code:** 1,4,6,14 **Conservation status: (IUCN)** NE

6717 *Alchemilla aphanoides* Mutis ex L.f. (Rosaceae) **IPNI:** 720378-1 **Common name:** caripacunga, orejuela, plegadera **Habit:** Herb **Use:** ME **Origin:** Native **Region:** Andean **Dept.:** ANT, BOY, CAU, CUN, HUI, MAG, NSA, QUI, RIS, SAN, TOL **Elev.:** 2100–4200 m a.s.l. **IUCN Habitat code:** 4,14 **Conservation status: (IUCN)** NE

6718 *Alchemilla fulvescens* (L.M.Perry) Rothm. (Rosaceae) **IPNI:** 8079-2 **Common name:** orejuelita **Habit:** Herb **Use:** ME **Origin:** Native **Region:** Andean **Dept.:** ANT, BOY, CAL, CAU, CUN, MET, NAR, PUT, QUI, RIS, TOL, VAC **Elev.:** 2300-3800 m.a.s.l. **IUCN Habitat code:** 14 **Conservation status: (IUCN)** NE **(CPSG)** Potential LC

6719 *Alchemilla galioides* Benth. (Rosaceae) **IPNI:** 720586-1 **Habit:** Herb **Use:** ME **Origin:** Native **Region:** Andean **Dept.:** CAL, CAU, CUN, MET, SAN **Elev.:** 3000–4200 m a.s.l. **IUCN Habitat code:** 4 **Conservation status: (IUCN)** NE

6720 *Alchemilla hirta* (L.M.Perry) Rothm. (Rosaceae) **IPNI:** 8090-2 **Habit:** Herb **Use:** ME **Origin:** Native **Region:** Andean **Dept.:** BOY, CAL, CAU, CUN, TOL **Elev.:** 2900–4100 m a.s.l. **Conservation status: (IUCN)** NE

6721 *Alchemilla hispidula* L.M.Perry (Rosaceae) **IPNI:** 8091-2 **Habit:** Herb **Use:** ME **Origin:** Native **Region:** Andean **Dept.:** ANT, BOY, CAL, CAU, CUN, MET, NAR, NSA, QUI, RIS, SAN, TOL, VAC **Elev.:** 3100–4700 m a.s.l. **IUCN Habitat code:** 6,14 **Conservation status: (IUCN)** NE

6722 *Alchemilla holosericea* L.M.Perry (Rosaceae) **IPNI:** 8093-2 **Habit:** Herb **Use:** ME **Origin:** Native **Region:** Andean **Dept.:** CAL, CAU, CUN, MET, RIS, TOL **Elev.:** 2900–4300 m a.s.l. **Conservation status: (IUCN)** NE

6723 *Alchemilla killipii* Rothm. (Rosaceae) **IPNI:** 8097-2 **Habit:** Herb **Use:** ME **Origin:** Endemic **Region:** Andean **Dept.:** BOY, CAL, CAU, CES, CUN, NAR, NSA, RIS, SAN **Elev.:** 2900–4400 m a.s.l. **Conservation status: (IUCN)** NE

6724 *Alchemilla moritziana* (Dammer) L.M.Perry (Rosaceae) **IPNI:** 8105-2 **Habit:** Herb **Use:** ME **Origin:** Native **Region:** Andean **Dept.:** BOY, CAU, CUN, SAN **Elev.:** 2700–3500 m a.s.l. **IUCN Habitat code:** 1,14 **Conservation status: (IUCN)** NE

6725 *Alchemilla mutisii* Rothm. (Rosaceae) **IPNI:** 8106-2 **Common name:** orejuela, plegadera **Habit:** Herb **Use:** ME **Origin:** Endemic **Region:** Andean **Dept.:** ANT, BOY, CUN **Elev.:** 2300–3200 m a.s.l. **IUCN Habitat code:** 1,14 **Conservation status: (IUCN)** NE

6726 *Alchemilla nivalis* Kunth (Rosaceae) **IPNI:** 720798-1 **Habit:** Herb **Use:** ME **Origin:** Native **Region:** Andean **Dept.:** BOY, CAL, CAU, CES, CUN, HUI, NAR, QUI, TOL **Elev.:** 3000–4500 m a.s.l. **IUCN Habitat code:** 1,4,14 **Conservation status: (IUCN)** NE

6727 *Alchemilla orbiculata* Ruiz & Pav. (Rosaceae) **IPNI:** 720816-1 **Common name:** cargarrocío, caripacunga, oreja de ratón, orejuela, oreja de ratón, orijuela, plegadera **Habit:** Herb **Use:** EU, ME **Origin:** Native **Region:** Andean **Dept.:** ANT, BOY, CAL, CAU, CUN, HUI, NAR, NSA, QUI, RIS, SAN, TOL, VAC **Elev.:** 1500-4000 m.a.s.l. **IUCN Habitat code:** 1,3,4,14 **Conservation status: (IUCN)** NE **(CPSG)** Potential LC

6728 *Alchemilla paludicola* Rothm. (Rosaceae) **IPNI:** 8117-2 **Habit:** Herb **Use:** ME **Origin:** Native **Region:** Andean **Dept.:** ANT, BOY, CAL, CHO, QUI **Elev.:** 3400–4200 m a.s.l. **Conservation status: (IUCN)** NE

6729 *Alchemilla pectinata* Kunth (Rosaceae) **IPNI:** 720846-1 **Common name:** orejuela, plegadera **Habit:** Herb **Use:** ME **Origin:** Native **Region:** Andean **Dept.:** ANT, CUN, SAN **Elev.:** 1800–3400 m a.s.l. **IUCN Habitat code:** 14 **Conservation status: (IUCN)** NE

6730 *Alchemilla pinnata* Ruiz & Pav. (Rosaceae)
IPNI: 720867-1
Use: ME **Origin:** Native **Region:** Andean **Dept.:** ARA, BOY, SAN **Elev.:** 3500–4300 m a.s.l. **Conservation status: (IUCN)** NE

6731 *Alchemilla polylepis* Wedd. (Rosaceae)
IPNI: 720877-1
Common name: caripacunga **Habit:** Shrub, Subshrub **Use:** ME **Origin:** Native **Region:** Andean **Dept.:** BOY, CES, MAG, NSA, SAN **Elev.:** 3300–4200 m a.s.l. **IUCN Habitat code:** 4,14 **Conservation status: (IUCN)** NE

6732 *Alchemilla procumbens* Rose (Rosaceae)
IPNI: 8135-2
Common name: serenilla **Use:** ME **IUCN Habitat code:** 1 **Conservation status: (IUCN)** NE

6733 *Cotoneaster coriaceus* Franch. (Rosaceae)
IPNI: 722459-1
Use: ME **Region:** Andean **Dept.:** BOG **Conservation status: (IUCN)** NE

6734 *Cotoneaster frigidus* Wall. ex Lindl. (Rosaceae)
IPNI: 722495-1
Use: EU, ME **Region:** Andean **Dept.:** BOG **IUCN Habitat code:** 3 **Conservation status: (IUCN)** NE

6735 *Cotoneaster multiflorus* C.A.Mey. (Rosaceae)
IPNI: 722592-1
Use: ME **Region:** Andean **Dept.:** BOG **IUCN Habitat code:** 14 **Conservation status: (IUCN)** NE

6736 *Cotoneaster pannosus* Franch. (Rosaceae)
IPNI: 722623-1
Habit: Shrub **Use:** EU **Origin:** Cultivated **Region:** Andean **Dept.:** ANT, BOG, BOY, CUN, NAR **Elev.:** 2500–3000 m a.s.l. **IUCN Habitat code:** 3,14 **Conservation status: (IUCN)** NE

6737 *Cotoneaster racemiflorus* (Desf.) Schltdl. (Rosaceae)
IPNI: 77120298-1
Use: AF, EU, ME, PO **Region:** Andean **Dept.:** BOG **Conservation status: (IUCN)** NE

6738 *Cydonia oblonga* Mill. (Rosaceae)
IPNI: 724549-1
Common name: membrillero **Habit:** Tree **Use:** GS, HF, MA, ME **Origin:** Cultivated **Region:** Andean **Dept.:** ANT, CUN, NAR **Elev.:** 1500–3000 m a.s.l. **IUCN Habitat code:** 14 **Conservation status: (IUCN)** NE

6739 *Fragaria chiloensis* (L.) Mill. (Rosaceae)
IPNI: 724988-1
Common name: frambuesa, fresa, fresa de Chile, fresal de Chile, fresón, morisca **Habit:** Herb **Use:** GS, HF, ME **Origin:** Naturalised **Region:** Andean **Dept.:** BOG, CUN, NAR **Elev.:** 2500–3000 m a.s.l. **IUCN Habitat code:** 14 **Conservation status: (IUCN)** NE

6740 *Fragaria vesca* L. (Rosaceae)
IPNI: 30074127-2
Common name: fresa, frutilla, morisca **Habit:** Herb **Use:** EU, GS, HF, MA, ME **Origin:** Cultivated **Region:** Andean **Dept.:** ANT, BOG, BOY, CAL, CAU, CUN, NAR, QUI, RIS, SAN, TOL, VAC **Elev.:** 1500–3500 m a.s.l. **IUCN Habitat code:** 1,3,4,14 **Conservation status: (IUCN)** NE

6741 *Fragaria virginiana* Mill. (Rosaceae)
IPNI: 318974-2
Habit: Herb **Use:** GS, HF, ME **Origin:** Naturalised **Region:** Andean **Dept.:** ANT, CUN **Elev.:** 2500–3000 m a.s.l. **Conservation status: (IUCN)** NE

6742 *Geum peruvianum* Focke (Rosaceae)
IPNI: 109852-2
Common name: valerian, valeriana **Habit:** Herb **Use:** ME **Origin:** Native **Region:** Andean **Dept.:** CAU, NAR **Elev.:** 2500–4000 m a.s.l. **IUCN Habitat code:** 1,8,14 **Conservation status: (IUCN)** NE

6743 *Geum quellyon* Sweet (Rosaceae)
IPNI: 725342-1
Use: ME **Region:** Andean **Dept.:** BOG **Conservation status: (IUCN)** NE

6744 *Hesperomeles ferruginea* (Juss. ex Pers.) Benth. (Rosaceae)
IPNI: 725473-1
Common name: cerote, guayabo de páramo, manzano, mortiño, noro **Habit:** Shrub, Small tree **Use:** HF, ME **Origin:** Native **Region:** Andean **Dept.:** ANT, BOG, BOY, CAL, CAU, CES, CUN, HUI, MAG, NAR, NSA, SAN, VAC **Elev.:** 2900-3800 m.a.s.l. **IUCN Habitat code:** 1,2,3,4,6,8,14 **Conservation status: (IUCN)** LC **(CPSG)** Potential LC

6745 *Hesperomeles goudotiana* (Decne.) Killip (Rosaceae)
IPNI: 120960-2
Common name: cerote, mortiño, mortiño verdadero, mote mote, motemote **Habit:** Shrub, Small tree **Use:** AF, EU, FU, HF, MA, ME **Origin:** Endemic **Region:** Andean **Dept.:** BOG, BOY, CES, CUN, SAN **Elev.:** 2600–3700 m a.s.l. **IUCN Habitat code:** 1,3,4,8,14 **Conservation status: (IUCN)** NE

6746 *Hesperomeles nitida* Killip (Rosaceae)
IPNI: 120963-2
Habit: Shrub **Use:** HF **Origin:** Endemic **Region:** Andean **Dept.:** BOG, SAN **Elev.:** 3200–3800 m a.s.l. **IUCN Habitat code:** 1,4,14 **Conservation status: (IUCN)** NE

6747 *Hesperomeles obtusifolia* (Pers.) Lindl. (Rosaceae)
IPNI: 725484-1
Common name: amarillo, mortillo, mortiño, mote, mote mote, noro **Habit:** Shrub, Small tree **Use:** HF, ME **Origin:** Native **Region:** Andean **Dept.:** ANT, BOG, BOY, CAL, CAU, CUN, HUI, MET, NAR, NSA, QUI, RIS, SAN, TOL, VAC **Elev.:** 2200-3800 m.a.s.l. **IUCN Habitat code:** 1,2,3,4,5,8,14 **Conservation status: (IUCN)** LC **(CPSG)** Potential LC

6748 *Malus domestica* (Suckow) Borkh. (Rosaceae)
IPNI: 726282-1
Common name: manzana, manzana apí, manzana camuera, manzana común, manzana criolla, manzana de Saint-Valery, manzana reineta **Habit:** Shrub, Tree **Use:** EU, FU, GS, HF, MA, ME, PO, SU **Origin:** Cultivated **Region:** Andean **Dept.:** ANT, BOY, CAL, CAU, CUN, NAR, VAC **Elev.:** 2000–3000 m a.s.l. **IUCN Habitat code:** 1,3,14 **Conservation status: (IUCN)** NE

6749 *Margyricarpus pinnatus* (Lam.) Kuntze (Rosaceae)
IPNI: 153124-2
Common name: cansacerro, churapanga, nigua, nigüita, nigüitas, piquimullo, yerba de nigua **Habit:** Herb, Subshrub **Use:** EU, HF, ME **Origin:** Native **Region:** Andean **Dept.:** BOG, BOY, CUN, NAR **Elev.:** 2500–3600 m a.s.l. **IUCN Habitat code:** 14 **Conservation status: (IUCN)** NE

6750 *Polylepis quadrijuga* Bitter (Rosaceae)
IPNI: 206511-2
Common name: colorado, sietecapas, sietecueros **Habit:** Shrub, Small tree **Use:** EU, MA **Origin:** Endemic **Region:** Andean **Dept.:** ANT, BOG, BOY, CUN, SAN **Elev.:** 2800–4000 m a.s.l. **IUCN Habitat code:** 1,2,3,4,8,14 **Conservation status: (IUCN)** NE

6751 *Potentilla indica* (Andrews) Th.Wolf (Rosaceae)
IPNI: 728289-1
Common name: fresa, fresa de monte, fresa silvestre **Habit:** Herb **Use:** HF, ME **Origin:** Naturalised **Region:** Andean **Dept.:** ANT, BOG, BOY, CAL, CAQ, CAU, CUN, HUI, NAR, QUI, RIS, SAN, TOL, VAC **Elev.:** 1500–3200 m a.s.l. **IUCN Habitat code:** 1,3,8,14 **Conservation status: (IUCN)** NE

6752 *Prunus amplifolia* Pilg. (Rosaceae)
IPNI: 210724-2
Habit: Tree **Use:** ME **Origin:** Native **Region:** Amazonia, Andean **Dept.:** CAQ, MET, PUT **Elev.:** 500–1500 m a.s.l. **Conservation status: (IUCN)** LC

6753 *Prunus debilis* Koehne (Rosaceae)
IPNI: 210756-2
Habit: Tree **Use:** HF **Origin:** Native **Region:** Amazonia, Andean, Orinoquia **Dept.:** BOY, MET, PUT **Elev.:** 250-1570 m.a.s.l. **IUCN Habitat code:** 1,3,14 **Conservation status: (IUCN)** LC **(CPSG)** DD

6754 *Prunus domestica* L. (Rosaceae)
IPNI: 30221374-2
Common name: almendro, ciruela, ciruela Claudia, ciruela chilena, ciruela europea, reina Claudia **Habit:** Shrub, Tree **Use:** GS, HF, MA, ME, PO **Origin:** Cultivated **Region:** Andean **Dept.:** ANT, BOG, BOY, CAU, CUN, NAR, QUI, VAC **Elev.:** 2500–2900 m a.s.l. **IUCN Habitat code:** 14 **Conservation status: (IUCN)** DD

6755 *Prunus opaca* (Benth.) Walp. (Rosaceae)
IPNI: 730059-1
Common name: botumbo, cerezo, crespo, duraznillo, naranjuelo, truco **Habit:** Tree **Use:** HF **Origin:** Native **Region:** Andean **Dept.:** ANT, BOG, BOY, CAL, CAU, CUN, HUI, NAR, PUT, SAN **Elev.:** 1860–3300 m a.s.l. **IUCN Habitat code:** 1,3,4,14 **Conservation status: (IUCN)** LC

6756 *Prunus persica* (L.) Batsch (Rosaceae) **IPNI:** 1212858-2 **Common name:** albérchigo, durazno, durazno camueso, durazno común, durazno de carne amarilla, durazno de carne blanca, durazno melocotón, melocotón **Habit:** Shrub, Tree **Use:** EU, FU, GS, HF, IF, MA, ME, PO **Origin:** Cultivated **Region:** Andean **Dept.:** ANT, BOG, BOY, CUN, PUT, VAC **Elev.:** 1500–2680 m a.s.l. **IUCN Habitat code:** 1,3,14 **Conservation status: (IUCN)** NE

6757 *Prunus serotina* Ehrh. (Rosaceae) **IPNI:** 730262-1 **Common name:** capulí, cereza, cerezo, cerezo criollo, guinda **Habit:** Tree **Use:** AF, EU, GS, HF, MA, ME, PO **Origin:** Naturalised **Region:** Andean **Dept.:** ANT, BOG, BOY, CUN, HUI, NAR **Elev.:** 1700–3500 m a.s.l. **IUCN Habitat code:** 1,3,4,14 **Conservation status: (IUCN)** LC

6758 *Prunus subcorymbosa* Ruiz ex Koehne (Rosaceae) **IPNI:** 210915-2 **Common name:** batundo, botundo, cerezo, duraznillo **Habit:** Tree **Use:** FU, MA **Origin:** Native **Region:** Andean, Pacific **Dept.:** ANT, BOG, CAU, CUN, HUI, QUI, RIS, VAC **Elev.:** 870-2300 m.a.s.l. **IUCN Habitat code:** 1,3,14 **Conservation status: (IUCN)** LC **(CPSG)** NT

6759 *Pyracantha angustifolia* (Franch.) C.K.Schneid. (Rosaceae) **IPNI:** 730608-1 **Habit:** Small tree **Use:** EU, HF **Origin:** Cultivated **Region:** Andean **Dept.:** ANT, BOG, BOY, CUN **Elev.:** 2300–2800 m a.s.l. **IUCN Habitat code:** 14 **Conservation status: (IUCN)** LC

6760 *Pyracantha coccinea* M.Roem. (Rosaceae) **IPNI:** 730611-1 **Use:** EU, MA, ME **Region:** Andean **Dept.:** BOG **IUCN Habitat code:** 1,3,5,14 **Conservation status: (IUCN)** NE

6761 *Pyracantha crenulata* (D.Don) M.Roem. (Rosaceae) **IPNI:** 730613-1 **Use:** EU, MA, ME **Region:** Andean **Dept.:** BOG **Conservation status: (IUCN)** NE

6762 *Pyrus communis* L. (Rosaceae) **IPNI:** 30065762-2 **Common name:** pera, pera de Sogamoso, peral **Habit:** Shrub, Tree **Use:** GS, HF, MA, ME, PO **Origin:** Cultivated **Region:** Andean **Dept.:** ANT, BOG, BOY, CUN **Elev.:** 2000–2800 m a.s.l. **IUCN Habitat code:** 3,14 **Conservation status: (IUCN)** LC

6763 *Rhaphiolepis bibas* (Lour.) Galasso & Banfi (Rosaceae) **IPNI:** 77209608-1 **Common name:** ciruelo, durazno de Israel, míspero, nisperito, níspero, níspero del Japón, níspero del japón, níspero japonés, oropel **Habit:** Shrub **Use:** AF, EU, FU, GS, HF, MA, ME, PO, SU **Origin:** Cultivated **Region:** Andean **Dept.:** ANT, BOG, BOY, CAL, CUN, HUI, NAR, QUI, RIS, SAN, TOL, VAC **Elev.:** 1000–3000 m a.s.l. **IUCN Habitat code:** 1,2,3,14 **Conservation status: (IUCN)** NE

6764 *Rhaphiolepis indica* (L.) Lindl. (Rosaceae) **IPNI:** 731389-1 **Use:** EU, ME **Region:** Andean **Dept.:** BOG **IUCN Habitat code:** 14 **Conservation status: (IUCN)** LC

6765 *Rosa chinensis* Jacq. (Rosaceae) **IPNI:** 732029-1 **Common name:** rosa china **Habit:** Shrub **Use:** EU, GS, HF, ME **Origin:** Cultivated **Region:** Andean, Caribbean **Dept.:** ANT, BOG, BOY, CAL, CUN, NAR, NSA, SAP, SAN, TOL, VAC **Elev.:** 1000–3000 m a.s.l. **IUCN Habitat code:** 3,14 **Conservation status: (IUCN)** NE

6766 *Rosa multiflora* Thunb. (Rosaceae) **IPNI:** 927374-1 **Common name:** rosa bebé, rosa japonesa, rosal de enredadera, sietehermanas **Habit:** Subshrub **Use:** EU, GS, HF, MA, ME **Origin:** Cultivated **Dept.:** CUN **Elev.:** 1500–2000 m a.s.l. **IUCN Habitat code:** 1 **Conservation status: (IUCN)** NE

6767 *Rosa odorata* (Andrews) Sweet (Rosaceae) **IPNI:** 733521-1 **Common name:** rosa té **Use:** ME **IUCN Habitat code:** 2,3,14 **Conservation status: (IUCN)** NE

6768 *Rubus acanthophyllos* Focke (Rosaceae) **IPNI:** 734809-1 **Common name:** mora, zarzamora **Habit:** Herb, Subshrub **Use:** HF, ME **Origin:** Native **Region:** Andean **Dept.:** ARA, BOG, BOY, CES, CUN, NAR, NSA, PUT, SAN, TOL **Elev.:** 2800–4000 m a.s.l. **IUCN Habitat code:** 1,3,4,8,14 **Conservation status: (IUCN)** NE

6769 *Rubus adenotrichos* Schltdl. (Rosaceae) **IPNI:** 734894-1 **Common name:** mora, mora común, mora silvestre, zarzamora **Habit:** Subshrub **Use:** HF, MA, ME **Origin:** Native **Region:** Andean **Dept.:** BOG, BOY, CAL, CAU, CUN, MAG, PUT, QUI, VAC **Elev.:** 1600–3200 m a.s.l. **IUCN Habitat code:** 1,3,4,5,8,14 **Conservation status: (IUCN)** NE

6770 *Rubus bogotensis* Kunth (Rosaceae) **IPNI:** 735457-1 **Common name:** mora, mora borrachera, mora de cerro, zarza blanca, zarza de mora, zarzamora **Habit:** Subshrub **Use:** HF, ME **Origin:** Native **Region:** Andean **Dept.:** ANT, BOG, BOY, CAU, CES, CUN, NAR, NSA, QUI, SAN, TOL **Elev.:** 1990–4100 m a.s.l. **IUCN Habitat code:** 1,3,4,8,14 **Conservation status: (IUCN)** NE

6771 *Rubus boliviensis* Focke (Rosaceae) **IPNI:** 735458-1 **Habit:** Subshrub **Use:** HF **Origin:** Native **Region:** Andean **Dept.:** ANT, CAU, CHO, CUN, MAG, NAR, PUT, RIS, SAN, TOL, VAC **Elev.:** 1300–3200 m a.s.l. **IUCN Habitat code:** 1,3,14 **Conservation status: (IUCN)** NE

6772 *Rubus choachiensis* A.Berger (Rosaceae) **IPNI:** 223008-2 **Common name:** mora, mora de páramo **Habit:** Subshrub **Use:** HF, ME **Origin:** Endemic **Region:** Andean **Dept.:** BOG, CUN, MET **Elev.:** 2900–4000 m a.s.l. **IUCN Habitat code:** 1,4,8,14 **Conservation status: (IUCN)** NE

6773 *Rubus compactus* Benth. (Rosaceae) **IPNI:** 735930-1 **Common name:** mora **Habit:** Herb, Subshrub **Use:** HF, ME **Origin:** Native **Region:** Andean **Dept.:** ANT, CAU, CUN, NAR, TOL **Elev.:** 2500–3600 m a.s.l. **IUCN Habitat code:** 1,3,4,14 **Conservation status: (IUCN)** NE

6774 *Rubus coriaceus* Poir. (Rosaceae) **IPNI:** 736037-1 **Common name:** mora de páramo **Habit:** Subshrub **Use:** HF **Origin:** Native **Region:** Andean **Dept.:** BOG, BOY, CAU, CUN, HUI, MAG, NAR, PUT, SAN, TOL **Elev.:** 2800–4000 m a.s.l. **IUCN Habitat code:** 1,3,4,8,14 **Conservation status: (IUCN)** NE

6775 *Rubus eriocarpus* Liebm. (Rosaceae) **IPNI:** 60464737-2 **Common name:** mora, mora común, mora de Castilla, mora de castilla, moral, moral de Castilla, morrocota, zarzamora azul **Use:** HF, ME **Dept.:** BOG **IUCN Habitat code:** 1,3,14 **Conservation status: (IUCN)** NE

6776 *Rubus floribundus* Kunth (Rosaceae) **IPNI:** 736846-1 **Common name:** mora, mora silvestre, moral, moral de tierra fría, zarzamora **Habit:** Subshrub **Use:** AF, EU, HF, ME **Origin:** Native **Region:** Andean **Dept.:** ANT, BOY, CAL, CAU, CES, CUN, HUI, MAG, MET, NAR, NSA, PUT, QUI, RIS, SAN, TOL, VAC **Elev.:** 1200-3600 m.a.s.l. **IUCN Habitat code:** 1,3,4,14 **Conservation status: (IUCN)** NE **(CPSG)** Potential LC

6777 *Rubus glabratus* Kunth (Rosaceae) **IPNI:** 737034-1 **Common name:** mora, mora de la virgen, moral, moral de tierra fría **Habit:** Herb, Subshrub **Use:** HF, ME **Origin:** Native **Region:** Andean **Dept.:** ANT, BOY, CAU, NAR **Elev.:** 3000–3800 m a.s.l. **IUCN Habitat code:** 1,3,4,8,14 **Conservation status: (IUCN)** NE

6778 *Rubus glaucus* Benth. (Rosaceae) **IPNI:** 737067-1 **Common name:** mora, mora andi mora blanca, mora andina, mora blanca, mora de castilla, mora de los andes, zarzamora azul **Habit:** Shrub, Subshrub **Use:** HF, MA, ME **Origin:** Native, Cultivated **Region:** Andean **Dept.:** ANT, BOG, CAL, CAU, CHO, CUN, HUI, MAG, NSA, PUT, QUI, SAN, TOL, VAC **Elev.:** 1800-3200 m.a.s.l. **IUCN Habitat code:** 1,3,4,8,14 **Conservation status: (IUCN)** NE **(CPSG)** Potential LC

6779 *Rubus guyanensis* Focke (Rosaceae) **IPNI:** 737191-1 **Common name:** mora, morita **Habit:** Subshrub **Use:** ME **Origin:** Native **Region:** Andean **Dept.:** ANT, BOG, BOY, CAU, CUN, HUI, MAG, PUT **Elev.:** 2000–3000 m a.s.l. **IUCN Habitat code:** 1,3,4,14 **Conservation status: (IUCN)** NE

6780 *Rubus idaeus* L. (Rosaceae) **IPNI:** 298089-2 **Common name:** frambuesa **Habit:** Subshrub **Use:** GS, HF, ME **Origin:** Cultivated **Region:** Andean **Dept.:** BOG, CUN, NAR **Elev.:** 1500–2500 m a.s.l. **IUCN Habitat code:** 1,3,14 **Conservation status:** **(IUCN)** NE

6781 *Rubus megalococcus* Focke (Rosaceae)
IPNI: 738332-1
Common name: mora **Habit:** Subshrub **Use:** HF, ME, PO **Origin:** Native **Region:** Andean **Dept.:** ANT, BOG, CAU **Elev.:** 2000–3020 m a.s.l. **IUCN Habitat code:** 1,4,5,14 **Conservation status:** **(IUCN)** NE

6782 *Rubus nubigenus* Kunth (Rosaceae)
IPNI: 738683-1
Common name: mora, mora de Castilla, moral, morón **Habit:** Shrub, Subshrub **Use:** AF, EU, HF, ME **Origin:** Native **Region:** Andean **Dept.:** ANT, BOG, BOY, CAL, CAQ, CAU, CUN, HUI, MAG, MET, NAR, PUT, QUI, RIS, SAN, TOL, VAC **Elev.:** 2500-3600 m.a.s.l. **IUCN Habitat code:** 1,2,3,4,8,14 **Conservation status:** **(IUCN)** NE **(CPSG)** Potential LC

6783 *Rubus peruvianus* Fritsch (Rosaceae)
IPNI: 739108-1
Habit: Subshrub **Use:** HF **Origin:** Native **Region:** Andean **Dept.:** BOY, CES, CUN **Elev.:** 2200–3500 m a.s.l. **Conservation status:** **(IUCN)** NE

6784 *Rubus porphyromallos* Focke (Rosaceae)
IPNI: 223459-2
Habit: Subshrub **Use:** HF **Origin:** Endemic **Region:** Andean **Dept.:** ANT, CAU, CES, HUI, MAG, NSA, QUI, SAN **Elev.:** 1800–3000 m a.s.l. **IUCN Habitat code:** 1 **Conservation status:** **(IUCN)** NE

6785 *Rubus roseus* Poir. (Rosaceae)
IPNI: 739815-1
Common name: mora, mora silvestre **Habit:** Shrub, Subshrub **Use:** HF, ME **Origin:** Native **Region:** Andean **Dept.:** ANT, CAU, CUN, HUI, NAR, PUT **Elev.:** 2600–3510 m a.s.l. **IUCN Habitat code:** 1,4,14 **Conservation status:** **(IUCN)** NE

6786 *Rubus rosifolius* Sm. (Rosaceae)
IPNI: 739800-1
Common name: frambuesa, fresa de monte, la mora, mora silvestre **Habit:** Herb, Subshrub **Use:** AF, EU, HF, MA, ME, SU **Origin:** Naturalised **Region:** Andean **Dept.:** ANT, CAU, CUN, SAN **Elev.:** 1500–2500 m a.s.l. **IUCN Habitat code:** 1,2,3,4,14 **Conservation status:** **(IUCN)** NE

6787 *Rubus urticifolius* Poir. (Rosaceae)
IPNI: 741023-1
Common name: mora, mora blanca, mora de Castilla, mora de monte, mora silvestre, moral, morita, morita de cercas, zarzamora, zarzaparrilla **Habit:** Shrub, Subshrub **Use:** HF, ME, PO **Origin:** Native **Region:** Andean **Dept.:** ANT, BOG, BOY, CAL, CAU, CES, COR, CUN, HUI, LAG, MAG, MET, NAR, NSA, QUI, RIS, SAN, TOL, VAC **Elev.:** 900-2640 m.a.s.l. **IUCN Habitat code:** 1,3,4,5,8,14 **Conservation status:** **(IUCN)** NE **(CPSG)** Potential LC

6788 *Rubus wardii* Merr. (Rosaceae)
IPNI: 741240-1
Common name: mora **Use:** ME **Dept.:** CUN **Conservation status:** **(IUCN)** NE

6789 *Rubus weberbaueri* Focke (Rosaceae)
IPNI: 223690-2
Common name: mora **Use:** ME **Dept.:** CUN **Conservation status:** **(IUCN)** NE

6790 *Sanguisorba minor* Scop. (Rosaceae)
IPNI: 741394-1
Common name: amorseco, cadillo blanco, fodder burnet, pimpinela, piojito, yerba del sol **Habit:** Herb **Use:** EU, HF, ME **Origin:** Cultivated **Region:** Andean **Dept.:** BOG, BOY, CUN, NSA **Elev.:** 2500–3600 m a.s.l. **Conservation status:** **(IUCN)** NE

6791 *Sanguisorba officinalis* L. (Rosaceae)
IPNI: 741402-1
Common name: pimpinela, pimpinella mayor, sanguisorba **Use:** EU, HF, MA, ME **Dept.:** CUN **Conservation status:** **(IUCN)** LC

6792 *Spiraea cantoniensis* Lour. (Rosaceae)
IPNI: 742160-1
Common name: mirto **Habit:** Shrub **Use:** EU, HF **Origin:** Cultivated **Region:** Andean **Dept.:** ANT, BOG, BOY, CAU, CUN, RIS, VAC **Elev.:** 1500–2600 m a.s.l. **IUCN Habitat code:** 14 **Conservation status:** **(IUCN)** LC

6793 *Ampelocera albertiae* Todzia (Ulmaceae)
IPNI: 272079-2
Common name: costillo **Habit:** Small tree **Use:** HF, MA **Origin:** Endemic **Region:** Andean **Dept.:** QUI, RIS, VAC **Elev.:** 1750-1900 m.a.s.l. **IUCN Habitat code:** 1,3,14 **Conservation status:** **(IUCN)** EN **(CPSG)** EN

6794 *Ampelocera edentula* Kuhlm. (Ulmaceae)
IPNI: 11466-2
Common name: cucharo, verraquillo **Habit:** Shrub, Tree **Use:** MA **Origin:** Native **Region:** Amazonia, Pacific **Dept.:** AMA, CAQ, CHO, GUV, MET, VAU **Elev.:** 0–600 m a.s.l. **IUCN Habitat code:** 1,3,4,14 **Conservation status:** **(IUCN)** LC

6795 *Ampelocera hottlei* (Standl.) Standl. (Ulmaceae)
IPNI: 11469-2
Common name: verraco **Habit:** Tree **Use:** HF, ME **Origin:** Native **Region:** Caribbean **Dept.:** ATL, SUC **Elev.:** 340–340 m a.s.l. **Conservation status:** **(IUCN)** LC

6796 *Ampelocera longissima* Todzia (Ulmaceae)
IPNI: 272080-2
Habit: Tree **Use:** HF **Origin:** Native **Region:** Pacific **Dept.:** ANT **Elev.:** 100–100 m a.s.l. **IUCN Habitat code:** 1,2,3,14 **Conservation status:** **(IUCN)** NT

6797 *Ampelocera macrocarpa* Ferero & Gentry (Ulmaceae)
IPNI: 11471-2
Common name: zapotillo **Habit:** Tree **Use:** MA **Origin:** Native **Region:** Andean, Caribbean, Pacific **Dept.:** ANT, BOL, BOY, CHO, VAC **Elev.:** 50–775 m a.s.l. **IUCN Habitat code:** 1,3,5,14 **Conservation status:** **(IUCN)** LC

6798 *Phyllostylon rhamnoides* (J.Poiss.) Taub. (Ulmaceae)
IPNI: 855199-1
Common name: membrillo, sabonero, varo blanco, varoblanco **Habit:** Tree **Use:** HF, MA, ME **Origin:** Native **Region:** Caribbean **Dept.:** CES, LAG, MAG **Elev.:** 20–200 m a.s.l. **IUCN Habitat code:** 1,2,14 **Conservation status:** **(IUCN)** LC

6799 *Boehmeria caudata* Sw. (Urticaceae)
IPNI: 850541-1
Common name: aguanoso, baboso, chaguanoso, codillo, comblilla, cortés, japahue, ortiguillo, parietaria, rabo de gato, zanca de mula **Habit:** Herb, Shrub, Subshrub, Small tree, Tree **Use:** HF, MA, ME **Origin:** Native **Region:** Andean **Dept.:** ANT, CAL, CAU, CES, CHO, CUN, HUI, MAG, MET, NAR, NSA, PUT, QUI, RIS, SAN, TOL, VAC **Elev.:** 630-3150 m.a.s.l. **IUCN Habitat code:** 1,2,3,4,5,14 **Conservation status:** **(IUCN)** LC **(CPSG)** Potential LC

6800 *Boehmeria nivea* (L.) Gaudich. (Urticaceae)
IPNI: 55676-3
Common name: aguanoso, combia, cortezo, ortiga mansa, palo bobo, ramio **Habit:** Herb, Shrub, Small tree **Use:** AF, EU, HF, MA, ME, PO **Origin:** Cultivated, Naturalised **Region:** Andean **Dept.:** ANT, BOY, CUN, HUI, MAG, PUT, QUI, RIS, TOL, VAC **Elev.:** 420–1410 m a.s.l. **IUCN Habitat code:** 1,2,3,14 **Conservation status:** **(IUCN)** NE

6801 *Boehmeria ramiflora* Jacq. (Urticaceae)
IPNI: 34156-2
Habit: Shrub, Subshrub **Use:** ME **Origin:** Native **Region:** Andean, Caribbean **Dept.:** ATL, BOY, CUN, NSA, VAC **Elev.:** 350–2800 m a.s.l. **IUCN Habitat code:** 1,3 **Conservation status:** **(IUCN)** LC

6802 *Cecropia angustifolia* Trécul (Urticaceae)
IPNI: 850916-1
Common name: cosedera negra, guarumo negro, yarumo, yarumo negro **Habit:** Tree **Use:** HF **Origin:** Native **Region:** Amazonia, Andean, Guiana Shield, Orinoquia, Pacific **Dept.:** ANT, BOY, CAL, CAQ, CAS, CAU, CHO, CUN, HUI, MAG, MET, NAR, NSA, PUT, QUI, RIS, SAN, TOL, VAC **Elev.:** 950-2800 m.a.s.l. **IUCN Habitat code:** 1,2,3,5,14 **Conservation status:** **(IUCN)** NE **(CPSG)** Potential LC

6803 *Cecropia distachya* Huber (Urticaceae)
IPNI: 50853-2
Common name: guarumo **Habit:** Tree **Use:** HF **Origin:** Native **Region:** Amazonia, Guiana Shield **Dept.:** AMA, CAQ, GUA, GUV, MET, VAU **Elev.:** 80–550 m a.s.l. **IUCN Habitat code:** 1,3,4,5,14 **Conservation status:** **(IUCN)** LC

6804 *Cecropia ficifolia* Warb. ex Snethl. (Urticaceae)
IPNI: 582776-1
Common name: guarumo, yarumo, yarumo blanco, yarumo hembra **Habit:** Tree **Use:** HF, ME **Origin:** Native **Region:** Amazonia, Andean, Guiana Shield, Orinoquia **Dept.:** AMA, CAQ, GUA, GUV, MET, PUT, VAU, VID **Elev.:** 100-1000 m.a.s.l. **IUCN Habitat code:** 1,3,4,5,14 **Conservation status:** **(IUCN)** LC **(CPSG)** Potential LC

6805 *Cecropia garciae* Standl. (Urticaceae)
IPNI: 50860-2
Common name: burriadá, yarumo, yarumo colorado **Habit:** Tree **Use:** ME **Origin:** Native **Region:** Andean, Pacific **Dept.:** CAU, NAR, VAC **Elev.:** 5–1050 m a.s.l. **IUCN Habitat code:** 1,3,14 **Conservation status:** **(IUCN)** LC

6806 *Cecropia goudotiana* Trécul (Urticaceae)
IPNI: 850932-1
Common name: yarumo **Habit:** Tree **Use:** ME **Origin:** Endemic **Dept.:** TOL **Elev.:** 400-400 m.a.s.l. **IUCN Habitat code:** 3,14 **Conservation status:** **(IUCN)** EN **(CPSG)** EN

6807 *Cecropia herthae* Diels (Urticaceae)
IPNI: 50864-2
Habit: Tree **Use:** HF **Origin:** Native **Region:** Amazonia **Dept.:** AMA, PUT **Elev.:** 300–350 m a.s.l. **IUCN Habitat code:** 1,3 **Conservation status:** **(IUCN)** NE

6808 *Cecropia insignis* Liebm. (Urticaceae)
IPNI: 850935-1
Common name: yarumo, yarumo rojo **Habit:** Tree **Use:** HF **Origin:** Native **Region:** Andean, Pacific **Dept.:** ANT, BOY, CHO, TOL, VAC **Elev.:** 20–1200 m a.s.l. **IUCN Habitat code:** 1,3,14 **Conservation status:** **(IUCN)** LC

6809 *Cecropia latiloba* Miq. (Urticaceae)
IPNI: 850938-1
Habit: Tree **Use:** HF **Origin:** Native **Region:** Amazonia, Guiana Shield **Dept.:** AMA, CAQ, PUT, VAU **Elev.:** 100–500 m a.s.l. **IUCN Habitat code:** 1,5 **Conservation status:** **(IUCN)** LC

6810 *Cecropia membranacea* Trécul (Urticaceae)
IPNI: 850943-1
Common name: cetico, guarumo, yarumo, yarumo blanco, yarumo macho **Habit:** Small tree, Tree **Use:** HF, ME **Origin:** Native **Region:** Amazonia, Guiana Shield, Orinoquia, Pacific **Dept.:** AMA, ANT, BOL, CAQ, CAU, CES, CHO, MET, NSA, PUT, SAN, VAC, VAU **Elev.:** 0-1000 m.a.s.l. **IUCN Habitat code:** 1,3,5,14 **Conservation status:** **(IUCN)** LC **(CPSG)** Potential LC

6811 *Cecropia montana* Warb. ex Snethl. (Urticaceae)
IPNI: 582813-1
Habit: Tree **Use:** HF **Origin:** Native **Region:** Amazonia, Andean **Dept.:** PUT **Elev.:** 570–1700 m a.s.l. **IUCN Habitat code:** 3,5,14 **Conservation status:** **(IUCN)** NE

6812 *Cecropia mutisiana* Mildbr. (Urticaceae)
IPNI: 50901-2
Common name: yarumo **Habit:** Small tree, Tree **Use:** ME **Origin:** Endemic **Region:** Andean **Dept.:** CAL, CAU, CUN, HUI, QUI, RIS, TOL, VAC **Elev.:** 500-1800 m.a.s.l. **IUCN Habitat code:** 1,3,14 **Conservation status:** **(IUCN)** LC **(CPSG)** LC

6813 *Cecropia obtusifolia* Bertol. (Urticaceae)
IPNI: 850947-1
Common name: burriadá, burriará, cetico, guarumo, orumo, yarumo **Habit:** Small tree, Tree **Use:** AF, FU, HF, MA, ME, PO **Origin:** Native **Region:** Andean, Pacific **Dept.:** ANT, CAU, CHO, COR, NAR, RIS, SAN, VAC **Elev.:** 0-1900 m.a.s.l. **IUCN Habitat code:** 1,3,5,14 **Conservation status:** **(IUCN)** LC **(CPSG)** Potential LC

6814 *Cecropia peltata* L. (Urticaceae)
IPNI: 850956-1
Common name: guarumo, tukulipainae, yagrumo, yarumo, yarumo blanco **Habit:** Small tree, Tree **Use:** FU, HF, MA, ME **Origin:** Native **Region:** Andean, Caribbean, Pacific **Dept.:** ANT, ATL, BOL, BOY, CAL, CES, CHO, COR, CUN, HUI, LAG, MAG, NSA, SAP, SAN, SUC, TOL, VAC **Elev.:** 0-1800 m.a.s.l. **IUCN Habitat code:** 1,2,3,4,5,14 **Conservation status:** **(IUCN)** LC **(CPSG)** Potential LC

6815 *Cecropia schreberiana* Miq. (Urticaceae)
IPNI: 50937-2
Use: FU, HF, MA, ME, SU **Dept.:** SAP **Conservation status:** **(IUCN)** NE

6816 *Cecropia sciadophylla* Mart. (Urticaceae)
IPNI: 850970-1
Common name: guarumo, guarumo blanco, yarumo, yarumo macho, yarumo negro **Habit:** Tree **Use:** HF, MA, ME **Origin:** Native **Region:** Amazonia, Andean, Guiana Shield, Orinoquia **Dept.:** AMA, CAQ, GUV, MET, PUT, VAU **Elev.:** 120-1300 m.a.s.l. **IUCN Habitat code:** 1,3,4,5,14 **Conservation status:** **(IUCN)** LC **(CPSG)** Potential LC

6817 *Cecropia telealba* Cuatrec. (Urticaceae)
IPNI: 50959-2
Common name: guarumo, yarumo, yarumo blanco, yarumo plateado **Habit:** Tree **Use:** ME **Origin:** Native **Region:** Andean **Dept.:** CAL, CUN, QUI, RIS, VAC **Elev.:** 1650-2600 m.a.s.l. **IUCN Habitat code:** 1,3,5,14 **Conservation status:** **(IUCN)** LC **(CPSG)** LC

6818 *Cecropia telenitida* Cuatrec. (Urticaceae)
IPNI: 50964-2
Common name: agrumo, guarumo, yarumo, yarumo blanco, yarumo plateado **Habit:** Tree **Use:** HF, MA, ME **Origin:** Native **Region:** Andean **Dept.:** ANT, BOY, CAQ, CAS, CAU, CES, CUN, HUI, NSA, PUT, QUI, SAN, TOL **Elev.:** 1200-2620 m.a.s.l. **IUCN Habitat code:** 1,3,14 **Conservation status:** **(IUCN)** LC **(CPSG)** Potential LC

6819 *Coussapoa asperifolia* Trécul (Urticaceae)
IPNI: 851268-1
Common name: matapalo, patudo **Habit:** Tree **Use:** HF **Origin:** Native **Region:** Amazonia, Andean, Guiana Shield, Pacific **Dept.:** AMA, ANT, BOY, CAQ, CHO, CUN, VAC, VAU **Elev.:** 0–1300 m a.s.l. **IUCN Habitat code:** 1,3,5,14 **Conservation status:** **(IUCN)** LC

6820 *Coussapoa brevipes* Pittier (Urticaceae)
IPNI: 582910-1
Use: HF **IUCN Habitat code:** 3 **Conservation status:** **(IUCN)** LC

6821 *Coussapoa duquei* Standl. (Urticaceae)
IPNI: 67344-2
Common name: patudo, suelda de roca **Habit:** Tree **Use:** HF **Origin:** Native **Region:** Andean **Dept.:** ANT, CAL, NSA **Elev.:** 1500–1850 m a.s.l. **IUCN Habitat code:** 1,3,14 **Conservation status:** **(IUCN)** NE

6822 *Coussapoa nitida* Miq. (Urticaceae)
IPNI: 851283-1
Habit: Tree **Use:** HF, ME **Origin:** Native **Region:** Amazonia, Guiana Shield **Dept.:** AMA, VAU **Elev.:** 200–200 m a.s.l. **IUCN Habitat code:** 1 **Conservation status:** **(IUCN)** LC

6823 *Coussapoa orthoneura* Standl. (Urticaceae)
IPNI: 67367-2
Common name: matapalo **Habit:** Hemiepiphyte **Use:** HF **Origin:** Native **Region:** Amazonia, Andean, Guiana Shield **Dept.:** AMA, GUA, MET, SAN, VAU **Elev.:** 70–1000 m a.s.l. **IUCN Habitat code:** 1,3,14 **Conservation status:** **(IUCN)** LC

6824 *Coussapoa villosa* Poepp. & Endl. (Urticaceae)
IPNI: 851294-1
Common name: caucho, cirpo, matapalo, matapalos, patudo, sapá, uva de monte **Habit:** Shrub, Tree, Hemiepiphyte **Use:** HF **Origin:** Native **Region:** Amazonia, Andean, Guiana Shield, Pacific **Dept.:** AMA, ANT, BOL, BOY, CAL, CAQ, CHO, CUN, MET, QUI, VAU **Elev.:** 0-2200 m.a.s.l. **IUCN Habitat code:** 1,2,3,4,5,14 **Conservation status:** **(IUCN)** LC **(CPSG)** Potential LC

6825 *Laportea aestuans* (L.) Chew (Urticaceae)
IPNI: 135476-2
Common name: ichanga, ortiga blanca, pringamoza **Habit:** Herb, Shrub, Subshrub, Small tree **Use:** AF, GS, HF, MA, ME, PO, SU **Origin:** Native **Region:** Amazonia, Andean, Pacific **Dept.:** AMA, ANT, CHO, VAC **Elev.:** 100–1480 m a.s.l. **IUCN Habitat code:** 1,2,3,5,14 **Conservation status:** **(IUCN)** NE

6826 *Myriocarpa longipes* Liebm. (Urticaceae)
IPNI: 854756-1
Habit: Shrub, Small tree **Use:** HF, ME **Origin:** Native **Region:** Andean, Caribbean, Pacific **Dept.:** ANT, CHO, LAG, VAC **Elev.:** 0–830 m a.s.l. **IUCN Habitat code:** 1,3,5,14 **Conservation status:** **(IUCN)** LC

6827 *Myriocarpa stipitata* Benth. (Urticaceae)
IPNI: 854765-1
Common name: aguanchinoso, aguanosa, aguanoso, ortigo macho **Habit:** Herb, Shrub, Small tree, Tree **Use:** ME **Origin:** Native **Region:** Andean, Caribbean, Pacific **Dept.:** ANT, CAL, CHO, CUN, HUI, LAG, MAG, NAR, QUI, RIS, VAC **Elev.:** 0–2780 m a.s.l. **IUCN Habitat code:** 1,2,3,5,14 **Conservation status:** **(IUCN)** NE

6828 *Parietaria debilis* G.Forst. (Urticaceae)
IPNI: 854908-1
Common name: espaletaria, espaletario, paletaria, palitaria, parietaria **Habit:** Herb **Use:** AF, HF, ME **Origin:** Native **Region:** Andean **Dept.:** BOY, CUN, MAG, MET, NSA, QUI, SAN **Elev.:** 1900–3750 m a.s.l. **IUCN Habitat code:** 1,3,4,14 **Conservation status:** **(IUCN)** NE

6829 *Parietaria officinalis* L. (Urticaceae)
IPNI: 854958-1
Common name: desempreñadora, parietaria **Habit:** Herb **Use:** HF, ME **Origin:** Cultivated, Naturalised **Region:** Andean **Dept.:** CUN **Elev.:** 2600–2600 m a.s.l. **IUCN Habitat code:** 4 **Conservation status:** **(IUCN)** NE

6830 *Phenax rugosus* (Poir.) Wedd. (Urticaceae)
IPNI: 855180-1
Common name: palo de puerco, parietaria, rumia **Habit:** Herb, Shrub, Subshrub, Climbing **Use:** ME **Origin:** Native **Region:** Andean, Pacific **Dept.:** ANT, BOG, BOY, CAL, CAU, CHO, CUN, MET, RIS, VAC **Elev.:** 910–2850 m a.s.l. **IUCN Habitat code:** 1,3,14 **Conservation status:** **(IUCN)** NE

6831 *Phenax uliginosus* Wedd. (Urticaceae) **IPNI:** 855189-1 **Common name:** ortigo, parietaria **Habit:** Herb, Shrub, Subshrub **Use:** HF **Origin:** Native **Region:** Andean, Caribbean **Dept.:** ANT, ATL, CAL, LAG, MAG, QUI, RIS, SAN **Elev.:** 70–2680 m a.s.l. **IUCN Habitat code:** 1,3,14 **Conservation status: (IUCN)** NE

6832 *Pilea cadierei* Gagnep. & Guillaumin (Urticaceae) **IPNI:** 855275-1 **Common name:** congonilla **Habit:** Herb **Use:** EU, HF, ME, SU **Origin:** Cultivated **Region:** Andean **Dept.:** ANT, BOG, BOL, RIS, VAC **Elev.:** 100–1500 m a.s.l. **IUCN Habitat code:** 3,14 **Conservation status: (IUCN)** NE

6833 *Pilea hyalina* Fenzl (Urticaceae) **IPNI:** 855481-1 **Common name:** espalitaria, tripepollo **Habit:** Herb **Use:** ME **Origin:** Native **Region:** Andean **Dept.:** ANT, BOY, CUN, MAG, QUI, VAC **Elev.:** 400–2000 m a.s.l. **IUCN Habitat code:** 1,2,3,4,5,14 **Conservation status: (IUCN)** NE

6834 *Pilea imparifolia* Wedd. (Urticaceae) **IPNI:** 321319-2 **Common name:** miao de arco pequeño, yerba de ojo **Habit:** Herb, Epiphyte, Climbing **Use:** ME **Origin:** Native **Region:** Amazonia, Andean, Pacific **Dept.:** AMA, ANT, CHO, PUT, VAC **Elev.:** 5–1000 m a.s.l. **IUCN Habitat code:** 1,5,14 **Conservation status: (IUCN)** NE

6835 *Pilea involucrata* (Sims) C.H.Wright & Dewar (Urticaceae) **IPNI:** 77091436-1 **Common name:** carianabo, corianobo, sígueme caleño **Habit:** Herb **Use:** ME, SU **Origin:** Native **Region:** Andean, Caribbean **Dept.:** CUN, HUI, LAG, MAG, PUT, SAN, VAC **Elev.:** 300–1000 m a.s.l. **IUCN Habitat code:** 1,3,5,14 **Conservation status: (IUCN)** NE

6836 *Pilea melastomoides* (Poir.) Wedd. (Urticaceae) **IPNI:** 855573-1 **Use:** HF, ME **Region:** Andean **Dept.:** BOG **Conservation status: (IUCN)** NE

6837 *Pilea microphylla* (L.) Liebm. (Urticaceae) **IPNI:** 1162156-2 **Common name:** colchón de niño, encaje, falso helecho de arroz, granizo, lombricera, lágrimas de niño, paletaria, tristeza **Habit:** Herb **Use:** EU, HF, ME **Origin:** Native, Cultivated **Region:** Andean, Caribbean, Pacific **Dept.:** ANT, ATL, BOG, BOL, CAU, CES, CHO, CUN, HUI, LAG, MAG, MET, NAR, NSA, QUI, RIS, SAP, SAN, TOL, VAC **Elev.:** 0-2600 m.a.s.l. **IUCN Habitat code:** 1,2,3,4,5,14 **Conservation status: (IUCN)** NE **(CPSG)** Potential LC

6838 *Pilea parietaria* (L.) Blume (Urticaceae) **IPNI:** 30064315-2 **Habit:** Herb, Shrub **Use:** ME **Origin:** Native **Region:** Andean **Dept.:** BOG, CUN, MAG, NSA, SAN **Elev.:** 1900–2940 m a.s.l. **IUCN Habitat code:** 1,3,14 **Conservation status: (IUCN)** NE

6839 *Pilea pubescens* Liebm. (Urticaceae) **IPNI:** 855698-1 **Common name:** ichanga **Habit:** Herb, Epiphyte **Use:** ME **Origin:** Native **Region:** Andean, Pacific **Dept.:** ANT, BOG, BOY, CAU, CES, CUN, VAC **Elev.:** 300–3150 m a.s.l. **IUCN Habitat code:** 1,14 **Conservation status: (IUCN)** NE

6840 *Pourouma acuminata* Mart. ex Miq. (Urticaceae) **IPNI:** 856074-1 **Common name:** caimarón de mono, uva de monte **Habit:** Tree **Use:** HF **Origin:** Native **Region:** Amazonia **Dept.:** AMA **Elev.:** 100–100 m a.s.l. **IUCN Habitat code:** 1,5 **Conservation status: (IUCN)** NE

6841 *Pourouma bicolor* Mart. (Urticaceae) **IPNI:** 856077-1 **Common name:** caimarón, caimarón de mico, caimarón de mono, chichicuvo, cirpe, cirpe macho, cirpo, cirpo macho, corme, fruta de indio, guagaripo, guagay, guarumo uva, lija, sirpo, tuno negro, upujicabue, uva, uva de monte, uva silvestre, uvilla lija, uvillo, uvillo de **Habit:** Small tree, Tree **Use:** HF, ME **Origin:** Native **Region:** Amazonia, Andean, Caribbean, Orinoquia, Pacific **Dept.:** AMA, ANT, BOL, BOY, CHO, COR, CUN, GUV, NAR, RIS, SAN, VAC, VID **Elev.:** 50-1550 m.a.s.l. **IUCN Habitat code:** 1,3,4,5,14 **Conservation status: (IUCN)** LC **(CPSG)** Potential LC

6842 *Pourouma cecropiifolia* Mart. (Urticaceae) **IPNI:** 2856079-1 **Common name:** amabaibillo, ambaiba uva, bacoe, birajo, bóchoa tsáha, caima, caimaron, caimaron silvestre, caimarona, caimarón, caimarón de montaña, caimarón de monte, caimarón silvestre, camarón silvestre, chiah, cirpe, cucupa, cucura, cucuva, curura, cusoiritu, guarum **Habit:** Tree **Use:** HF, MA, ME **Origin:** Native, Cultivated **Region:** Amazonia, Andean, Caribbean, Orinoquia **Dept.:** AMA, ANT, CAQ, CHO, LAG, MET, PUT, VAU **Elev.:** 100–1100 m a.s.l. **IUCN Habitat code:** 1,2,3,5,14 **Conservation status: (IUCN)** LC

6843 *Pourouma cucura* Standl. & Cuatrec. (Urticaceae) **IPNI:** 209636-2 **Common name:** uvilla, uvo **Habit:** Small tree **Use:** HF, ME **Origin:** Native **Region:** Amazonia **Dept.:** AMA, CAQ, VAU **Elev.:** 200–250 m a.s.l. **IUCN Habitat code:** 1,3,5 **Conservation status: (IUCN)** LC

6844 *Pourouma cuspidata* Mildbr. (Urticaceae) **IPNI:** 283688-2 **Common name:** uvo **Habit:** Shrub, Tree **Use:** HF **Origin:** Native **Region:** Amazonia **Dept.:** CAQ **Elev.:** 100–220 m a.s.l. **IUCN Habitat code:** 1,3,5,14 **Conservation status: (IUCN)** NE

6845 *Pourouma ferruginea* Standl. (Urticaceae) **IPNI:** 209639-2 **Common name:** uvilla **Habit:** Small tree, Tree **Use:** HF **Origin:** Native **Region:** Amazonia **Dept.:** AMA **Elev.:** 100–180 m a.s.l. **IUCN Habitat code:** 1,3 **Conservation status: (IUCN)** NE

6846 *Pourouma floccosa* C.C.Berg (Urticaceae) **IPNI:** 306888-2 **Habit:** Tree **Use:** HF **Origin:** Native **Region:** Amazonia **Dept.:** AMA **Elev.:** 120–120 m a.s.l. **IUCN Habitat code:** 1 **Conservation status: (IUCN)** NE

6847 *Pourouma guianensis* Aubl. (Urticaceae) **IPNI:** 856084-1 **Common name:** uva, uvo **Habit:** Small tree, Tree **Use:** HF, MA, ME **Origin:** Native **Region:** Amazonia, Orinoquia **Dept.:** AMA, MET **Elev.:** 80–1000 m a.s.l. **IUCN Habitat code:** 1,3,4,5,14 **Conservation status: (IUCN)** LC

6848 *Pourouma hirsutipetiolata* Mildbr. (Urticaceae) **IPNI:** 209643-2 **Common name:** cirpe, cirpo **Habit:** Tree **Use:** HF **Origin:** Endemic **Region:** Andean **Dept.:** ANT, SAN, VAC **Elev.:** 200–1240 m a.s.l. **IUCN Habitat code:** 1,3,14 **Conservation status: (IUCN)** NE

6849 *Pourouma melinonii* Benoist (Urticaceae) **IPNI:** 209652-2 **Common name:** mocaiyeibu, uva de orilla **Habit:** Tree **Use:** HF **Origin:** Native **Region:** Amazonia, Guiana Shield, Pacific **Dept.:** AMA, ANT, CAQ, CAU, VAU **Elev.:** 0–700 m a.s.l. **IUCN Habitat code:** 1,3,5,14 **Conservation status: (IUCN)** LC

6850 *Pourouma minor* Benoist (Urticaceae) **IPNI:** 283697-2 **Common name:** blanquecino, blanquísimo, bobotomobu, caimarón, caimarón de monte, uvilla, uvo, uvo silvestre **Habit:** Tree **Use:** HF **Origin:** Native **Region:** Amazonia, Andean, Guiana Shield, Pacific **Dept.:** AMA, ANT, CHO, MET, VAC, VAU **Elev.:** 100–1300 m a.s.l. **IUCN Habitat code:** 1,3,14 **Conservation status: (IUCN)** LC

6851 *Pourouma mollis* Trécul (Urticaceae) **IPNI:** 856087-1 **Common name:** caimarón de mico, uvo caimarón **Habit:** Tree **Use:** HF **Origin:** Native **Region:** Guiana Shield **Dept.:** MET **Elev.:** 600–600 m a.s.l. **IUCN Habitat code:** 1,14 **Conservation status: (IUCN)** LC

6852 *Pourouma myrmecophila* Ducke (Urticaceae) **IPNI:** 283699-2 **Common name:** uva de monte **Habit:** Shrub, Tree **Use:** HF **Origin:** Native **Region:** Amazonia **Dept.:** AMA **Elev.:** 100–100 m a.s.l. **IUCN Habitat code:** 1,3,5 **Conservation status: (IUCN)** LC

6853 *Pourouma ovata* Trécul (Urticaceae) **IPNI:** 856089-1 **Habit:** Tree **Use:** HF **Origin:** Native **Region:** Amazonia **Dept.:** AMA, VAU **Elev.:** 200–220 m a.s.l. **IUCN Habitat code:** 1,14 **Conservation status: (IUCN)** NE

6854 *Pourouma tomentosa* Mart. ex Miq. (Urticaceae) **IPNI:** 283703-2 **Common name:** uva, uva de monte, uvilla **Habit:** Tree **Use:** HF **Origin:** Native **Region:** Amazonia, Guiana Shield **Dept.:** AMA, CAQ **Elev.:** 100–200 m a.s.l. **IUCN Habitat code:** 1,3,5,14 **Conservation status: (IUCN)** LC

6855 *Pouzolzia parasitica* (Forssk.) Schweinf. (Urticaceae)
IPNI: 856206-1
Use: HF, ME **Origin:** Native **Dept.:** BOY **Conservation status: (IUCN)** NE

6856 *Soleirolia soleirolii* (Req.) Dandy (Urticaceae)
IPNI: 856545-1
Common name: centavito, lágrimas de bebé **Habit:** Herb **Use:** EU, SU **Origin:** Cultivated, Naturalised **Region:** Andean **Dept.:** ANT, BOG, CUN **Elev.:** 1500–2600 m a.s.l. **IUCN Habitat code:** 1,14 **Conservation status: (IUCN)** NE

6857 *Urera aurantiaca* Wedd. (Urticaceae)
IPNI: 856972-1
Habit: Shrub **Use:** HF, ME **Origin:** Native **Region:** Amazonia **Dept.:** AMA **IUCN Habitat code:** 1,3,14 **Conservation status: (IUCN)** NE

6858 *Urera baccifera* (L.) Gaudich. ex Wedd. (Urticaceae)
IPNI: 1044269-2
Common name: bosopa, miø juji?, ofaebü wonotakirinae, ortiga, ortiga blanca, ortiga morada, ortiga negra, ortigo, ortigo blanco, pringamosa, pringamoza, pringamoza arbórea, pringamoza blanca, pringamoza de caballo, pringamoza dientona, pringamoza negra, tsemane, yuyo **Habit:** Shrub, Subshrub, Small tree, Tree **Use:** EU, HF, ME, PO, SU **Origin:** Native **Region:** Amazonia, Andean, Guiana Shield, Orinoquia, Pacific **Dept.:** AMA, ANT, BOY, CAL, CAQ, CES, CHO, CUN, GUV, HUI, MAG, MET, NAR, NSA, PUT, QUI, RIS, SAN, SUC, VAC **Elev.:** 15-3000 m.a.s.l. **IUCN Habitat code:** 1,2,3,4,5,14 **Conservation status: (IUCN)** LC **(CPSG)** Potential LC

6859 *Urera caracasana* (Jacq.) Gaudich. ex Griseb. (Urticaceae)
IPNI: 856982-1
Common name: acquai, hortiga, ichanga, juji?, katalawanae, lozanelia, origa de buho, ortiga, ortiga de búho, ortiga de flor, ortiga lisa, ortiga roja, ortigo, ortigo negro, ortiguillo rojo, pringamosa, pringamosa de flores, pringamosa morada, pringamoso, pringamoza, p **Habit:** Herb, Shrub, Small tree, Tree **Use:** AF, HF, MA, ME **Origin:** Native **Region:** Amazonia, Andean, Guiana Shield, Caribbean, Orinoquia, Pacific **Dept.:** AMA, ANT, BOY, CAQ, CAU, CES, CHO, CUN, HUI, LAG, MAG, MET, NAR, NSA, PUT, QUI, SAN, SUC, TOL, VAC, VAU **Elev.:** 20-2290 m.a.s.l. **IUCN Habitat code:** 1,2,3,5,14 **Conservation status: (IUCN)** LC **(CPSG)** Potential LC

6860 *Urera elata* (Sw.) Griseb. (Urticaceae)
IPNI: 856998-1
Common name: ortiga **Habit:** Herb, Shrub, Climbing **Use:** HF **Origin:** Native **Region:** Amazonia, Andean, Orinoquia, Pacific **Dept.:** ANT, CHO, CUN, GUV, HUI, VAC **Elev.:** 60–2100 m a.s.l. **IUCN Habitat code:** 1,14 **Conservation status: (IUCN)** LC

6861 *Urera laciniata* Wedd. (Urticaceae)
IPNI: 857027-1
Common name: ortiga, ortiga macho, ortigo, pringamosa, pringamoza **Habit:** Herb, Shrub, Subshrub **Use:** ME **Origin:** Native **Region:** Amazonia, Pacific **Dept.:** ANT, CAQ, CHO, NAR, PUT, VAC **Elev.:** 15–975 m a.s.l. **IUCN Habitat code:** 1,3,14 **Conservation status: (IUCN)** LC

6862 *Urera verrucosa* (Liebm.) V.W.Steinm. (Urticaceae)
IPNI: 60439275-2
Habit: Shrub, Tree **Use:** HF **Origin:** Native **Dept.:** ANT **Elev.:** 350–350 m a.s.l. **IUCN Habitat code:** 1,3 **Conservation status: (IUCN)** LC

6863 *Urtica ballotifolia* Wedd. (Urticaceae)
IPNI: 857434-1
Common name: lengua de buey, ortiga blanca **Habit:** Herb, Subshrub **Use:** ME **Origin:** Native **Region:** Andean **Dept.:** CAL, CAQ, CAU, CUN, NAR, NSA, PUT, QUI, SAN, VAC **Elev.:** 2400–3830 m a.s.l. **IUCN Habitat code:** 1,3,4,14 **Conservation status: (IUCN)** NE

6864 *Urtica dioica* L. (Urticaceae)
IPNI: 260630-2
Common name: ortiga, ortiga blanca, ortiga mayor, ortiga negra **Habit:** Herb **Use:** AF, GS, HF, MA, ME, PO, SU **Origin:** Naturalised **Region:** Andean **Dept.:** ANT, BOG, BOY, CUN, RIS, VAC **Elev.:** 2440–2600 m a.s.l. **IUCN Habitat code:** 1,3,4,14 **Conservation status: (IUCN)** LC

6865 *Urtica echinata* Benth. (Urticaceae)
IPNI: 857552-1
Habit: Herb **Use:** ME **Origin:** Native **Region:** Andean **Dept.:** BOY **Elev.:** 4300–4300 m a.s.l. **Conservation status: (IUCN)** NE

6866 *Urtica leptophylla* Kunth (Urticaceae)
IPNI: 857681-1
Habit: Herb, Shrub, Climbing **Use:** ME **Origin:** Native **Region:** Andean **Dept.:** ANT, BOG, BOY, CAU, CUN, HUI, NAR, NSA, SAN **Elev.:** 1800–4300 m a.s.l. **Conservation status: (IUCN)** NE

6867 *Urtica urens* L. (Urticaceae)
IPNI: 857987-1
Common name: ortiga, ortiga blanca, ortiga menor, ortiga pequeña, pringamoza **Habit:** Herb **Use:** HF, ME **Origin:** Cultivated, Naturalised **Region:** Andean **Dept.:** BOG, BOY, CUN, NSA **Elev.:** 500–2900 m a.s.l. **IUCN Habitat code:** 1,3,14 **Conservation status: (IUCN)** NE

Santalales

6868 *Corynaea crassa* Hook.f. (Balanophoraceae)
IPNI: 103322-1
Common name: mazorquitas **Habit:** Parasitic **Use:** ME **Origin:** Native **Region:** Andean **Dept.:** ANT, BOG, BOY, CAU, CHO, CUN, HUI, MET, NAR, RIS, VAC **Elev.:** 1300–3000 m a.s.l. **Conservation status: (IUCN)** NE

6869 *Helosis cayennensis* (Sw.) Spreng. (Balanophoraceae)
IPNI: 103346-1
Common name: velacho **Habit:** Parasitic, Saprophyte **Use:** ME **Origin:** Native **Region:** Amazonia, Andean, Pacific **Dept.:** AMA, ANT, BOL, CAQ, CHO, MET, NAR, SAN, TOL, VAC, VAU **Elev.:** 100–2700 m a.s.l. **IUCN Habitat code:** 1,3,4,5,14 **Conservation status: (IUCN)** NE

6870 *Langsdorffia hypogaea* Mart. (Balanophoraceae)
IPNI: 315558-2
Habit: Parasitic **Use:** ME **Origin:** Native **Region:** Andean **Dept.:** ANT, BOG, BOY, CAQ, CAU, CHO, CUN, HUI, MET, NAR, PUT, VAC **Elev.:** 1100–2740 m a.s.l. **IUCN Habitat code:** 1,3,14 **Conservation status: (IUCN)** NE

6871 *Gaiadendron punctatum* (Ruiz & Pav.) G.Don (Loranthaceae)
IPNI: 548708-1
Common name: azarito, cabrito, dontagua, limoncillo, matapalo, matapalos, mayo, pajarito, platero, tagua, taguapiche, taguo, tábano **Habit:** Shrub, Subshrub, Tree, Hemiparasitic **Use:** AF, EU, FU, MA, ME, SU **Origin:** Native **Region:** Andean **Dept.:** ANT, ARA, BOG, BOY, CAL, CAU, CES, CHO, CUN, HUI, LAG, MAG, MET, NAR, NSA, PUT, QUI, RIS, SAN, TOL, VAC **Elev.:** 1330-3950 m.a.s.l. **IUCN Habitat code:** 1,3,4,5,6,8,14 **Conservation status: (IUCN)** LC **(CPSG)** Potential LC

6872 *Oryctanthus alveolatus* (Kunth) Kuijt (Loranthaceae)
IPNI: 176899-2
Common name: golondrina, matapalo, pajarito, rubiarbo, suelda redonda, sueldaconsuelda **Habit:** Shrub, Subshrub, Hemiparasitic, Hyperparasitic **Use:** ME, SU **Origin:** Native **Region:** Amazonia, Andean, Guiana Shield, Caribbean, Orinoquia, Pacific **Dept.:** AMA, ANT, ATL, BOL, BOY, CAL, CAQ, CAU, CHO, COR, CUN, GUV, HUI, LAG, MAG, MET, NAR, PUT, QUI, RIS, SAN, TOL, VAC, VAU, VID **Elev.:** 0-2700 m.a.s.l. **IUCN Habitat code:** 1,2,3,4,5,14 **Conservation status: (IUCN)** NE **(CPSG)** Potential LC

6873 *Oryctanthus occidentalis* (L.) Eichler (Loranthaceae)
IPNI: 550656-1
Common name: pajarito **Habit:** Shrub, Hemiparasitic **Use:** ME **Origin:** Native **Region:** Andean, Pacific **Dept.:** ANT, CAL, CAU, CHO, CUN, TOL, VAC **Elev.:** 0–2050 m a.s.l. **IUCN Habitat code:** 1,2,3,5,14 **Conservation status: (IUCN)** NE

6874 *Passovia pedunculata* (Jacq.) Kuijt (Loranthaceae)
IPNI: 77147531-1
Common name: cagá de pájaro, consuelda, golondrina, golondrino, pajarillo, pajarito, pulijpanaa **Habit:** Shrub, Subshrub, Liana, Climbing, Hemiparasitic **Use:** AF, MA, ME **Origin:** Native **Region:** Amazonia, Andean, Guiana Shield, Caribbean, Orinoquia, Pacific **Dept.:** AMA, ANT, ATL, BOL, BOY, CAL, CAQ, CAS, CAU, CES, CHO, COR, CUN, GUA, GUV, HUI, LAG, MAG, MET, NAR, NSA, PUT, QUI, SAN, SUC, TOL, VAC, VAU, VID **Elev.:** 0-3250 m.a.s.l. **IUCN Habitat code:** 1,2,3 **Conservation status: (IUCN)** NE **(CPSG)** Potential LC

6875 *Passovia pyrifolia* (Kunth) Tiegh. (Loranthaceae)
IPNI: 185262-2
Common name: hiedra, pajarito, pajarito abejucado **Habit:** Shrub, Subshrub, Liana, Hemiparasitic **Use:** HF, ME, SU **Origin:** Native **Region:** Amazonia, Andean, Guiana Shield, Caribbean, Orinoquia, Pacific **Dept.:** AMA, ANT, ATL, BOL, BOY, CAL, CAQ, CAU, CHO, CUN, GUA, GUV, MET, NAR, NSA, PUT, SAN, SUC, TOL, VAC, VAU **Elev.:** 0-2400 m.a.s.l. **IUCN Habitat code:** 1,2,3,4,5,14 **Conservation status: (IUCN)** NE **(CPSG)** Potential LC

6876 *Psittacanthus calyculatus* (DC.) G.Don (Loranthaceae)
IPNI: 551347-1
Common name: claro de lu muérdago, claro de luna, injerto, mal de ojo, matapalo, media lu mal de ojo, media luna, muérdago, picadora, visco **Habit:** Hemiparasitic **Use:** ME **Origin:** Native **Dept.:** COR **Elev.:** 50–50 m a.s.l. **IUCN Habitat code:** 1,14 **Conservation status: (IUCN)** NE

6877 *Psittacanthus cucullaris* (Lam.) G.Don (Loranthaceae)
IPNI: 551367-1
Common name: pajarito **Habit:** Herb, Shrub, Subshrub, Hemiparasitic **Use:** ME **Origin:** Native **Region:** Amazonia, Andean, Guiana Shield, Orinoquia **Dept.:** AMA, ARA, CAQ, CAS, CAU, MET, NAR, PUT, SAN, VAU, VID **Elev.:** 100–2989 m a.s.l. **IUCN Habitat code:** 1,2,3,4,5,14 **Conservation status: (IUCN)** NE

6878 *Struthanthus marginatus* (Desr.) G.Don (Loranthaceae)
IPNI: 551718-1
Habit: Liana, Hemiparasitic **Use:** ME **Origin:** Native **Region:** Andean **Dept.:** CHO **Elev.:** 1500–1600 m a.s.l. **IUCN Habitat code:** 2 **Conservation status: (IUCN)** NE

6879 *Struthanthus orbicularis* (Kunth) Eichler (Loranthaceae)
IPNI: 551737-1
Common name: golondrino, pie de pajarito **Habit:** Shrub, Liana, Climbing, Hemiparasitic **Use:** ME **Origin:** Native **Region:** Amazonia, Andean, Guiana Shield, Caribbean, Pacific **Dept.:** AMA, ANT, BOL, CAQ, CAU, CES, CHO, COR, CUN, LAG, NAR, SAN, SUC, TOL, VAU **Elev.:** 0-1800 m.a.s.l. **IUCN Habitat code:** 1,2,3,4,5,14 **Conservation status: (IUCN)** NE **(CPSG)** Potential LC

6880 *Tripodanthus acutifolius* (Ruiz & Pav.) Tiegh. (Loranthaceae)
IPNI: 258636-2
Habit: Liana, Hemiparasitic **Use:** ME **Origin:** Native **Region:** Amazonia **Dept.:** PUT **IUCN Habitat code:** 1 **Conservation status: (IUCN)** LC

6881 *Aptandra tubicina* (Poepp.) Benth. ex Miers (Olacaceae)
IPNI: 607627-1
Common name: hueso, turmo **Habit:** Shrub, Tree **Use:** HF **Origin:** Native **Region:** Amazonia **Dept.:** AMA, ANT, CAQ **Elev.:** 100–1000 m a.s.l. **IUCN Habitat code:** 1,3,14 **Conservation status: (IUCN)** LC

6882 *Cathedra acuminata* (Benth.) Miers (Olacaceae)
IPNI: 607671-1
Habit: Shrub, Tree **Use:** HF **Origin:** Native **Region:** Amazonia, Guiana Shield, Orinoquia **Dept.:** AMA, CAQ, GUA, VAU, VID **Elev.:** 90–120 m a.s.l. **IUCN Habitat code:** 1,2,3,4,14 **Conservation status: (IUCN)** LC

6883 *Chaunochiton kappleri* (Sagot ex Engl.) Ducke (Olacaceae)
IPNI: 55672-2
Habit: Tree **Use:** MA **Origin:** Native **Region:** Amazonia, Orinoquia **Dept.:** AMA, VID **Elev.:** 100–100 m a.s.l. **IUCN Habitat code:** 1 **Conservation status: (IUCN)** LC

6884 *Chaunochiton loranthoides* Benth. (Olacaceae)
IPNI: 607695-1
Common name: palo sal **Habit:** Tree **Use:** HF **Origin:** Native **Region:** Amazonia, Guiana Shield **Dept.:** AMA, GUA **Elev.:** 100–100 m a.s.l. **IUCN Habitat code:** 2,3,5 **Conservation status: (IUCN)** NE

6885 *Curupira tefeensis* G.A.Black (Olacaceae)
IPNI: 72734-2
Common name: castaño espinoso, coco **Habit:** Tree **Use:** HF **Origin:** Native **Region:** Amazonia, Guiana Shield **Dept.:** CAQ **Elev.:** 100–200 m a.s.l. **IUCN Habitat code:** 1 **Conservation status: (IUCN)** NE

6886 *Heisteria acuminata* (Bonpl.) Engl. (Olacaccac)
IPNI: 70029599-1
Common name: cabo de hacha, cafecillo, granicillo, murciélago, ojo de venado, quiebracha **Habit:** Shrub, Small tree, Tree **Use:** HF, ME **Origin:** Native **Region:** Amazonia, Andean, Guiana Shield, Orinoquia, Pacific **Dept.:** AMA, ANT, BOL, CAQ, CAU, CHO, CUN, MET, NAR, NSA, PUT, QUI, SAN, TOL, VAC, VAU, VID **Elev.:** 0-1930 m.a.s.l. **IUCN Habitat code:** 1,3,4,5,14 **Conservation status: (IUCN)** LC **(CPSG)** Potential LC

6887 *Heisteria latifolia* Standl. (Olacaceae)
IPNI: 118612-2
Habit: Shrub, Tree **Use:** HF **Origin:** Native **Region:** Amazonia, Andean, Pacific **Dept.:** AMA, ANT, BOY, VAC **Elev.:** 100–1100 m a.s.l. **IUCN Habitat code:** 1,3,5 **Conservation status: (IUCN)** LC

6888 *Minquartia guianensis* Aubl. (Olacaceae)
IPNI: 110121-1
Common name: acapú, acapú negro, acaricuara, aceituno, acupú, ahumado, ahumado caqueteño, ahumado de hoja pequeña, ahumado pajarito, ahumado pedro, ahumao, arrayán, barbasco, barbasco ahumado, barbasco negro, curichichi, cuyubí, diapechimuci, guacapurana, guacarí, gua **Habit:** Tree **Use:** HF, MA, ME **Origin:** Native **Region:** Amazonia, Andean, Guiana Shield, Pacific **Dept.:** AMA, CAQ, CAU, CHO, GUV, NAR, PUT, VAC, VAU **Elev.:** 28–1200 m a.s.l. **IUCN Habitat code:** 1,3,4,5,14 **Conservation status: (IUCN)** NT

6889 *Olax inopiflora* (Miers) Christenh. & Byng (Olacaceae)
IPNI: 77183948-1
Habit: Shrub, Tree **Use:** ME **Origin:** Native **Region:** Amazonia, Guiana Shield **Dept.:** AMA, CAQ, GUA, VAU **Elev.:** 120–240 m a.s.l. **Conservation status: (IUCN)** NE

6890 *Olax macrophylla* Benth. (Olacaceae)
IPNI: 608118-1
Common name: barbasco, maní de perico **Habit:** Tree **Use:** HF **Origin:** Native **Region:** Amazonia, Guiana Shield **Dept.:** AMA, CAQ, GUA, VAU **Elev.:** 120–500 m a.s.l. **Conservation status: (IUCN)** NE

6891 *Ximenia americana* L. (Olacaceae)
IPNI: 316341-2
Common name: caimito de monte, caimito demonte, ciruelo, espino de brujo, huevo de morrocoyo, irima?ucoro, limoncillo, limoncillo de playa, limoncito, limoracho **Habit:** Shrub, Tree **Use:** AF, EU, FU, GS, HF, MA, ME, PO, SU **Origin:** Native **Region:** Caribbean, Pacific **Dept.:** ANT, BOL, CHO, MAG, SUC **Elev.:** 0–100 m a.s.l. **IUCN Habitat code:** 1,2,5,14 **Conservation status: (IUCN)** LC

6892 *Agonandra brasiliensis* Benth. & Hook.f. (Opiliaceae)
IPNI: 301518-2
Common name: caimancillo, conchito, frutenevado, frutevenado **Habit:** Tree **Use:** AF, FU, HF, MA, ME, PO **Origin:** Native **Region:** Amazonia, Andean, Caribbean, Orinoquia **Dept.:** ATL, BOL, LAG, MAG, PUT, VID **Elev.:** 0–610 m a.s.l. **IUCN Habitat code:** 1,2,3,14 **Conservation status: (IUCN)** LC

6893 *Agonandra silvatica* Ducke (Opiliaceae)
IPNI: 308951-2
Common name: gomo **Habit:** Tree, Hemiepiphyte **Use:** HF **Origin:** Native **Region:** Amazonia **Dept.:** AMA, CAQ, PUT, VAU **Elev.:** 100–300 m a.s.l. **IUCN Habitat code:** 1 **Conservation status: (IUCN)** LC

6894 *Acanthosyris annonagustata* C.Ulloa & P.M.Jørg. (Santalaceae)
IPNI: 315517-2
Habit: Tree **Use:** HF **Origin:** Native **Region:** Amazonia **Dept.:** PUT **Elev.:** 230–300 m a.s.l. **Conservation status: (IUCN)** NT

6895 *Acanthosyris glabrata* (Stapf) Stauffer ex Govaerts (Santalaceae)
IPNI: 1712-2
Common name: yaya de sangre **Habit:** Tree **Use:** AF, HF, MA **Origin:** Native **Region:** Caribbean **Dept.:** ATL, BOL **Elev.:** 30–300 m a.s.l. **IUCN Habitat code:** 3 **Conservation status: (IUCN)** NE

6896 *Cervantesia tomentosa* Ruiz & Pav. (Santalaceae)
IPNI: 780054-1
Habit: Tree **Use:** HF **Origin:** Native **Region:** Andean **Dept.:** CAU, QUI **Elev.:** 2500–3420 m a.s.l. **IUCN Habitat code:** 1,3,4,14 **Conservation status: (IUCN)** NE

6897 *Dendrophthora chrysostachya* (C.Presl) Urb. (Santalaceae)
IPNI: 77735-2
Habit: Shrub, Hemiparasitic **Use:** HF **Origin:** Native **Region:** Andean **Dept.:** CAU, NAR **Elev.:** 2600–3350 m a.s.l. **IUCN Habitat code:** 1,3,4,14 **Conservation status: (IUCN)** NE

6898 *Dendrophthora clavata* (Benth.) Urb. (Santalaceae)
IPNI: 77736-2
Common name: injerto, matapalo, matapalos, pajarito, suelda, sueldaconsuelda **Habit:** Shrub, Hemiparasitic **Use:** ME **Origin:** Native **Region:** Andean **Dept.:** ANT, BOG, BOY, CAL, CAU, CES, CUN, MAG, MET, NAR, NSA, QUI, RIS, SAN, TOL, VAC **Elev.:** 1940-3900 m.a.s.l. **IUCN Habitat code:** 1,3,4,8,14 **Conservation status: (IUCN)** NE **(CPSG)** Potential LC

6899 *Dendrophthora lindeniana* Tiegh. (Santalaceae)
IPNI: 77769-2
Common name: sanahuesos, suelda **Habit:** Shrub, Hemiparasitic **Use:** ME **Origin:** Native **Region:** Andean **Dept.:** ANT, BOG, BOY, CUN, LAG, MAG, NSA, PUT, QUI, SAN **Elev.:** 1500–3350 m a.s.l. **IUCN Habitat code:** 1,3,4,14 **Conservation status: (IUCN)** NE

6900 *Phoradendron chrysocladon* A.Gray (Santalaceae)
IPNI: 329775-2
Common name: matapalos, suelda **Habit:** Herb, Shrub, Hemiparasitic **Use:** ME **Origin:** Native **Region:** Amazonia, Andean, Pacific **Dept.:** ANT, CAQ, CAU, CHO, CUN, MET, NAR, NSA, PUT, QUI, SAN, TOL, VAC, VAU **Elev.:** 5-2480 m.a.s.l. **IUCN Habitat code:** 1,3,5,14 **Conservation status: (IUCN)** NE **(CPSG)** Potential LC

6901 *Phoradendron crassifolium* (Pohl ex DC.) Eichler (Santalaceae)
IPNI: 550798-1
Common name: pajarillo **Habit:** Herb, Shrub, Hemiparasitic **Use:** ME **Origin:** Native **Region:** Amazonia,

Andean, Guiana Shield, Orinoquia, Pacific **Dept.:** AMA, ANT, BOY, CAQ, CAU, CHO, CUN, GUA, MAG, MET, NAR, NSA, SAN, VAC, VAU, VID **Elev.:** 0-1600 m.a.s.l. **IUCN Habitat code:** 1,2,3,4,5,14 **Conservation status: (IUCN)** NE **(CPSG)** Potential LC

6902 *Phoradendron madisonii* Kuijt (Santalaceae) **IPNI:** 282719-2 **Common name:** golondrina **Habit:** Shrub, Hemiparasitic **Use:** ME **Origin:** Native **Region:** Pacific **Dept.:** NAR **Elev.:** 15–15 m a.s.l. **IUCN Habitat code:** 14 **Conservation status: (IUCN)** EN

6903 *Phoradendron mucronatum* (DC.) Krug & Urb. (Santalaceae) **IPNI:** 194156-2 **Common name:** consuelda **Habit:** Shrub, Hemiparasitic **Use:** ME **Origin:** Native **Region:** Caribbean **Dept.:** ANT, ATL, BOL, CES, CUN, LAG, MAG, NSA, TOL **Elev.:** 5–700 m a.s.l. **IUCN Habitat code:** 1,2,4,5,14 **Conservation status: (IUCN)** NE

6904 *Phoradendron nervosum* Oliv. (Santalaceae) **IPNI:** 550970-1 **Habit:** Shrub, Hemiparasitic **Use:** HF **Origin:** Native **Region:** Andean **Dept.:** ANT, BOG, BOY, CAL, CAU, CUN, MAG, MET, NAR, NSA, RIS, SAN, VAC **Elev.:** 400-2650 m.a.s.l. **IUCN Habitat code:** 1,3,4,14 **Conservation status: (IUCN)** NE **(CPSG)** Potential LC

6905 *Phoradendron piperoides* (Kunth) Trel. (Santalaceae) **IPNI:** 194192-2 **Common name:** golondrina, injerto, matapalo, pajarito **Habit:** Shrub, Subshrub, Hemiparasitic **Use:** ME **Origin:** Native **Region:** Amazonia, Andean, Guiana Shield, Caribbean, Orinoquia, Pacific **Dept.:** AMA, ANT, BOG, BOL, BOY, CAL, CAQ, CAU, CHO, CUN, GUA, GUV, LAG, MAG, MET, NAR, NSA, PUT, QUI, SAN, VAC, VAU **Elev.:** 10-3200 m.a.s.l. **IUCN Habitat code:** 1,2,3,4,5,14 **Conservation status: (IUCN)** NE **(CPSG)** Potential LC

6906 *Phoradendron quadrangulare* (Kunth) Griseb. (Santalaceae) **IPNI:** 60443757-2 **Common name:** cagada de pajarito, matapalos, pajarito, suelda, sueldaconsuelda **Habit:** Herb, Shrub, Hemiparasitic **Use:** HF, ME **Origin:** Native **Region:** Amazonia, Andean, Caribbean, Orinoquia, Pacific **Dept.:** ANT, ATL, BOL, CAU, CHO, COR, CUN, GUV, HUI, LAG, MAG, MET, NAR, NSA, SAN, SUC, TOL, VAC, VID **Elev.:** 5-1225 m.a.s.l. **IUCN Habitat code:** 1,2,3,5,14 **Conservation status: (IUCN)** NE **(CPSG)** Potential LC

6907 *Phoradendron trinervium* (Lam.) Griseb. (Santalaceae) **IPNI:** 1075239-2 **Habit:** Herb, Hemiparasitic **Use:** ME **Origin:** Native **Region:** Amazonia, Caribbean **Dept.:** GUA, LAG, MAG **Elev.:** 65–500 m a.s.l. **IUCN Habitat code:** 1,3 **Conservation status: (IUCN)** NE

6908 *Phoradendron undulatum* (Pohl ex DC.) Eichler (Santalaceae) **IPNI:** 1006438-2 **Common name:** suelda **Habit:** Shrub, Hemiparasitic **Use:** ME **Origin:** Native **Region:** Andean, Guiana Shield **Dept.:** ANT, CAU, CUN, HUI, LAG, MAG, MET, NSA, PUT, QUI, SAN, VAC **Elev.:** 100-1800 m.a.s.l. **IUCN Habitat code:** 1,3,4,14 **Conservation status: (IUCN)** NE **(CPSG)** Potential LC

6909 *Schoepfia schreberi* J.F.Gmel. (Schoepfiaceae) **IPNI:** 608391-1 **Habit:** Shrub, Tree **Use:** ME **Origin:** Native **Region:** Caribbean **Dept.:** MAG, TOL **Elev.:** 50–50 m a.s.l. **Conservation status: (IUCN)** LC

Sapindales

6910 *Anacardium excelsum* (Bertero & Balb. ex Kunth) Skeels (Anacardiaceae) **IPNI:** 12224-2 **Common name:** aspave, aspavé, assu, cahu, caracol, caracolí, caracolí amarillo, caracolí blanco, cayuvana, cojueiro de campo, coyo de campo, espavé, merey, mija, mijagua, mijao **Habit:** Tree **Use:** AF, EU, HF, IF, MA, ME **Origin:** Native **Region:** Andean, Caribbean, Orinoquia, Pacific **Dept.:** ANT, ATL, BOL, BOY, CAL, CES, CHO, CUN, HUI, LAG, MAG, QUI, SAN, TOL, VAC **Elev.:** 0-1100 m.a.s.l. **IUCN Habitat code:** 1,2,3,5,14 **Conservation status: (IUCN)** NE **(CPSG)** NT

6911 *Anacardium giganteum* Hancock ex Engl. (Anacardiaceae) **IPNI:** 69177-1 **Common name:** cayuasú, marañón de monte, marañón gigante, marañón grande, marañón monte, oreja de burro **Habit:** Tree **Use:** HF, MA, ME **Origin:** Native **Region:** Amazonia, Guiana Shield, Pacific **Dept.:** AMA, CAQ, GUA, SAN, VAC **Elev.:** 200–200 m a.s.l. **IUCN Habitat code:** 1,14 **Conservation status: (IUCN)** NE

6912 *Anacardium microsepalum* Loes. (Anacardiaceae) **IPNI:** 12228-2 **Use:** HF **Conservation status: (IUCN)** NE

6913 *Anacardium negrense* Pires & Fróes (Anacardiaceae) **IPNI:** 12229-2 **Habit:** Tree **Use:** HF **Origin:** Native **Region:** Amazonia **Dept.:** AMA, CAQ **Elev.:** 100–200 m a.s.l. **Conservation status: (IUCN)** NE

6914 *Anacardium occidentale* L. (Anacardiaceae) **IPNI:** 319068-2 **Common name:** amarillo, anacardo, arbol de jobo, cajuil, cajú, cashew, casú, caujil, caujir, caují, cayú, chura, churá, marañón, marañón amarillo, marañón rojo, marey, mercy, merey, pauji, paují **Habit:** Small tree, Tree **Use:** AF, EU, FU, GS, HF, IF, MA, ME, PO, SU **Origin:** Native **Region:** Amazonia, Andean, Guiana Shield, Caribbean, Orinoquia, Pacific **Dept.:** AMA, ANT, BOL, CAL, CAQ, CAS, GUA, GUV, HUI, MAG, MET, NSA, PUT, SAP, SAN, TOL, VAC, VAU **Elev.:** 0–900 m a.s.l. **IUCN Habitat code:** 1,2,3,4,5,14 **Conservation status: (IUCN)** NE

6915 *Anacardium parvifolium* Ducke (Anacardiaceae) **IPNI:** 12232-2 **Common name:** marañón de breo, marañón de monte **Habit:** Tree **Use:** HF **Origin:** Native **Region:** Amazonia **Dept.:** AMA, PUT **Elev.:** 200–600 m a.s.l. **IUCN Habitat code:** 1,5 **Conservation status: (IUCN)** NE

6916 *Anacardium spruceanum* Benth. ex Engl. (Anacardiaceae) **IPNI:** 272107-2 **Use:** HF **Origin:** Native **Dept.:** ANT, GUA **Conservation status: (IUCN)** NE

6917 *Antrocaryon amazonicum* (Ducke) B.L.Burtt & A.W.Hill (Anacardiaceae) **IPNI:** 16050-2 **Habit:** Tree **Use:** HF **Origin:** Native **Region:** Guiana Shield **Dept.:** MET **Elev.:** 150–150 m a.s.l. **IUCN Habitat code:** 1 **Conservation status: (IUCN)** NE

6918 *Astronium fraxinifolium* Schott (Anacardiaceae) **IPNI:** 69236-1 **Habit:** Tree **Use:** FU, MA, ME **Origin:** Native **Region:** Caribbean **Dept.:** BOL, CES, LAG, MAG **Elev.:** 5–500 m a.s.l. **IUCN Habitat code:** 1,2,4,14 **Conservation status: (IUCN)** NE

6919 *Astronium graveolens* Jacq. (Anacardiaceae) **IPNI:** 69240-1 **Common name:** abejón, carnefresca, diamante, diomante, diomate, diomato, gateado, gatiao, gusanera, gusanero, gusanero morado, palo gateado, quebracho, quiebrahacha, santacruz, santacrú, tibigaro, tibligaro, tigrillo, yomate **Habit:** Tree **Use:** AF, MA, ME **Origin:** Native **Region:** Caribbean, Orinoquia, Pacific **Dept.:** BOL, CAS, CHO, LAG, MAG, SUC **Elev.:** 5–1000 m a.s.l. **IUCN Habitat code:** 1,2,3,4,5,14 **Conservation status: (IUCN)** NE

6920 *Astronium lecointei* Ducke (Anacardiaceae) **IPNI:** 25916-2 **Habit:** Tree **Use:** MA **Origin:** Native **IUCN Habitat code:** 1,3,4,14 **Conservation status: (IUCN)** NE

6921 *Astronium obliquum* Griseb. (Anacardiaceae) **IPNI:** 69247-1 **Habit:** Tree **Use:** MA **Origin:** Native **Region:** Amazonia, Orinoquia **Dept.:** CAQ, CAS, VAU **Elev.:** 350–750 m a.s.l. **IUCN Habitat code:** 1 **Conservation status: (IUCN)** NE

6922 *Campnosperma panamense* Standl. (Anacardiaceae) **IPNI:** 44024-2 **Common name:** sajo, sajo conchecaimán, vaguerá **Habit:** Tree **Use:** HF, MA, ME **Origin:** Native **Region:** Pacific **Dept.:** CAU, CHO, NAR, VAC **Elev.:** 5-100 m.a.s.l. **IUCN Habitat code:** 1,4,5,14 **Conservation status: (IUCN)** LC **(CPSG)** NT

6923 *Cyrtocarpa velutinifolia* (R.S.Cowan) J.D.Mitch. & Daly (Anacardiaceae) **IPNI:** 275630-2 **Habit:** Tree **Use:** HF **Origin:** Native **Region:** Andean **Dept.:** MAG **Conservation status: (IUCN)** LC

6924 *Mangifera indica* L. (Anacardiaceae) **IPNI:** 69913-1 **Common name:** manga, manga albania, manga de Mariquita, manga mariquiteña, manga tomi, mango, mango canime, mango chancleto, mango de azúcar, mango de chupa, mango de masa, mango de rosa, mango farchil, mango filipino, mango tommy **Habit:** Tree **Use:** AF, EU, FU, GS, HF, MA, ME, PO, SU **Origin:** Cultivated, Naturalised **Region:** Amazonia, Andean, Caribbean, Pacific **Dept.:** AMA, ANT, ARA, ATL, BOL, BOY, CAL, CAQ, CAS, CAU, CES, CHO, COR, CUN, GUA, GUV, HUI, LAG, MAG, MET, NAR, NSA, PUT, QUI, RIS, SAP, SAN, SUC, TOL, VAC, VAU, VID **Elev.:** 0–2850 m a.s.l. **IUCN Habitat code:** 1,2,3,5,14 **Conservation status: (IUCN)** DD

6925 *Mauria heterophylla* Kunth (Anacardiaceae) **IPNI:** 70000-1 **Common name:** caspi, caspicolorado, chiraco, cuasio, manguito, mantequilla, mantequillo, pacua, palo amargo, palo de sarno, palodulce, quebracho, rascador, sarno **Habit:** Shrub, Tree **Use:** ME **Origin:** Native **Region:** Andean, Guiana Shield, Caribbean, Pacific **Dept.:** ANT, BOY, CAU, CES, CUN, HUI, LAG, MAG, NSA, PUT, RIS, SAN, TOL, VAC **Elev.:** 325-3300 m.a.s.l. **IUCN Habitat code:** 1,2,3,14 **Conservation status: (IUCN)** LC **(CPSG)** Potential LC

6926 *Mauria simplicifolia* Kunth (Anacardiaceae) **IPNI:** 70012-1 **Common name:** almanegra, birringo, caspi, caspicaracha, mantequillo, palo de incienso, pedrohernández, sarno **Habit:** Tree **Use:** ME **Origin:** Native **Region:** Andean **Dept.:** ANT, BOY, CAU, CUN, HUI, NAR, QUI, TOL, VAC **Elev.:** 2000–3200 m a.s.l. **IUCN Habitat code:** 1,3,4,14 **Conservation status: (IUCN)** NE

6927 *Metopium brownei* (Jacq.) Urb. (Anacardiaceae) **IPNI:** 158605-2 **Common name:** hog doctor **Habit:** Tree **Use:** MA, ME, PO **Origin:** Native **Region:** Caribbean **Dept.:** SAP **Elev.:** 220–220 m a.s.l. **IUCN Habitat code:** 1 **Conservation status: (IUCN)** LC

6928 *Mosquitoxylum jamaicense* Krug & Urb. (Anacardiaceae) **IPNI:** 164514-2 **Habit:** Tree **Use:** MA, ME **Origin:** Native **Conservation status: (IUCN)** LC

6929 *Schinus areira* L. (Anacardiaceae) **IPNI:** 71006-1 **Habit:** Tree **Use:** HF, ME **Origin:** Naturalised **Region:** Andean **Dept.:** ANT, BOY, CUN, VAC **Elev.:** 1025–2620 m a.s.l. **IUCN Habitat code:** 3,14 **Conservation status: (IUCN)** NE

6930 *Schinus molle* L. (Anacardiaceae) **IPNI:** 71044-1 **Common name:** aroeira, bálsamo sanalotodo, falso pimiento, molle, muelle, mulli, pimenteira, pimentera, pimiento, piru, terebinto, tree of la vida **Use:** AF, EU, FU, GS, HF, IF, MA, ME, PO, SU **Origin:** Naturalised **Dept.:** BOG **IUCN Habitat code:** 2,3,14 **Conservation status: (IUCN)** LC

6931 *Schinus terebinthifolia* Raddi (Anacardiaceae) **IPNI:** 71069-1 **Habit:** Tree **Use:** EU, GS, HF, IF, MA, ME, PO **Origin:** Naturalised **Region:** Andean, Caribbean **Dept.:** ANT, ATL, BOG, VAC **Elev.:** 10–1500 m a.s.l. **IUCN Habitat code:** 1,14 **Conservation status: (IUCN)** NE

6932 *Spondias dulcis* Parkinson (Anacardiaceae) **IPNI:** 71455-1 **Common name:** cancharama, cancharana, canllarana, canyarana, ciruelo dulce **Habit:** Tree **Use:** AF, EU, GS, HF, MA, ME, SU **Origin:** Cultivated, Naturalised **Region:** Amazonia **Dept.:** AMA **Elev.:** 100–100 m a.s.l. **IUCN Habitat code:** 1,5,14 **Conservation status: (IUCN)** NE

6933 *Spondias mombin* L. (Anacardiaceae) **IPNI:** 71480-1 **Common name:** arisco, camalitenae, cancharana, canyarana, ciruela, ciruelo, ciruelo calentano, ciruelo de castilla, ciruelo hobo, cocote, hobo, hobo arisco, hobo blanco, hobo colorado, hog plum, joba, jobo, jobo arisco, jobo blanco, jobo colorado, jobo de Castilla, job **Habit:** Tree **Use:** AF, EU, FU, GS, HF, IF, MA, ME, PO, SU **Origin:** Native, Cultivated **Region:** Amazonia, Andean, Guiana Shield, Caribbean, Orinoquia, Pacific **Dept.:** AMA, ANT, ARA, BOL, CAL, CAQ, CHO, COR, CUN, GUV, HUI, LAG, MAG, MET, NSA, SAP, SAN, SUC, TOL, VAC, VID **Elev.:** 0-1960 m.a.s.l. **IUCN Habitat code:** 1,2,3,4,5,14 **Conservation status: (IUCN)** LC **(CPSG)** Potential LC

6934 *Spondias purpurea* L. (Anacardiaceae) **IPNI:** 71494-1 **Common name:** cirgüelo, ciruela, ciruela calentana, ciruelo, ciruelo blanco, ciruelo calentano, ciruelo de Castilla, ciruelo hobo, ciruelo rojo, cocota, cocoto, hobo, hobo colorado, hobo manso, jam plum, jobito, jobo colorado **Habit:** Tree **Use:** AF, EU, GS, HF, MA, ME, PO **Origin:** Native, Cultivated **Region:** Andean, Caribbean, Pacific **Dept.:** ANT, ATL, CAL, CAU, CHO, CUN, MAG, SAP, SUC, TOL **Elev.:** 0–1470 m a.s.l. **IUCN Habitat code:** 1,2,3,14 **Conservation status: (IUCN)** LC

6935 *Spondias radlkoferi* Donn.Sm. (Anacardiaceae) **IPNI:** 242911-2 **Habit:** Tree **Use:** HF, ME **Origin:** Native **Region:** Andean, Caribbean, Pacific **Dept.:** ANT, CAL, CHO, COR, SAN, SUC **Elev.:** 0–700 m a.s.l. **IUCN Habitat code:** 1,2,3,5,14 **Conservation status: (IUCN)** LC

6936 *Tapirira guianensis* Aubl. (Anacardiaceae) **IPNI:** 71559-1 **Common name:** algodoncillo, borkai, cariquiay, cedrillo, cedro macho, fresno, garata, garotá, guarupayo, hojarasco, maduraverde, manteco, palo balsudo, palo de gusano, palo tigre, pechuga, pechugo, poogu, quincedías, tapaculo, uvilla, árbol de borugo **Habit:** Shrub, Small tree, Tree **Use:** FU, HF, MA, ME **Origin:** Native **Region:** Amazonia, Andean, Guiana Shield, Caribbean, Orinoquia, Pacific **Dept.:** AMA, ANT, BOL, BOY, CAQ, CHO, COR, GUA, GUV, HUI, MAG, MET, NSA, PUT, SAN, TOL, VAC, VAU, VID **Elev.:** 0-1960 m.a.s.l. **IUCN Habitat code:** 1,2,3,4,5,14 **Conservation status: (IUCN)** LC **(CPSG)** Potential LC

6937 *Tapirira obtusa* (Benth.) J.D.Mitch. (Anacardiaceae) **IPNI:** 970707-1 **Common name:** añame, cedrillo, papelilla **Habit:** Tree **Use:** HF, MA **Origin:** Native **Region:** Amazonia **Dept.:** AMA **Elev.:** 150–200 m a.s.l. **IUCN Habitat code:** 1,3,5 **Conservation status: (IUCN)** LC

6938 *Tapirira retusa* Ducke (Anacardiaceae) **IPNI:** 249404-2 **Common name:** balsudo, palo balsudo **Habit:** Tree **Use:** HF **Origin:** Native **Region:** Amazonia **Dept.:** AMA, CAQ **Elev.:** 100–100 m a.s.l. **IUCN Habitat code:** 1,5,14 **Conservation status: (IUCN)** LC

6939 *Thyrsodium spruceanum* Benth. (Anacardiaceae) **IPNI:** 71652-1 **Habit:** Small tree, Tree **Use:** HF **Origin:** Native **Region:** Guiana Shield **Dept.:** MET **Elev.:** 600–1300 m a.s.l. **IUCN Habitat code:** 1,3,4,14 **Conservation status: (IUCN)** LC

6940 *Toxicodendron striatum* (Ruiz & Pav.) Kuntze (Anacardiaceae) **IPNI:** 255790-2 **Common name:** ajicito, biringo, birringo, buenamigo, caratero, caspe, caspi, caspicaracho, cedromacho, chiraco, chunche, compadre, compadre chiraco, fresno, fumo, güeraca, juampintao, manzanillo, palo de sarno, pedrofernández, pedrohernández, pedrojucho, pedrote, pedru **Habit:** Shrub, Small tree, Tree **Use:** AF,ME, PO **Origin:** Native **Region:** Andean, Caribbean, Pacific **Dept.:** ANT, BOY, CAL, CAU, CHO, CUN, HUI, LAG, MAG, MET, NAR, NSA, RIS, SAN, VAC **Elev.:** 56-3000 m.a.s.l. **IUCN Habitat code:** 1,2,3,5,14 **Conservation status: (IUCN)** LC **(CPSG)** Potential LC

6941 *Bursera glabra* (Jacq.) Triana & Planch. (Burseraceae) **IPNI:** 127130-1 **Common name:** bija, bijo **Habit:** Tree **Use:** EU **Origin:** Native, Cultivated **Region:** Caribbean **Dept.:** ATL, BOL, LAG, MAG **Elev.:** 5–5 m a.s.l. **Conservation status: (IUCN)** NT

6942 *Bursera graveolens* (Kunth) Triana & Planch. (Burseraceae) **IPNI:** 127138-1 **Common name:** ali a, bijá, caraño, chachique, crespino, crespín, crispín, palo santo, sasafrán hembra, sasafrás, tamagaco, taspín, tatamaco **Habit:** Tree **Use:** HF, MA, ME, SU **Origin:** Native **Region:** Andean, Caribbean **Dept.:** ANT, ATL, BOL, BOY, CAU, CUN, HUI, LAG, MAG, TOL, VAC **Elev.:** 25-1300 m.a.s.l. **IUCN Habitat code:** 1,2,3,4,5,14 **Conservation status: (IUCN)** LC **(CPSG)** NT

6943 *Bursera simaruba* (L.) Sarg. (Burseraceae) **IPNI:** 127217-1 **Common name:** almácigo, birch, carate, caratero, caratoso, caraña, chibú, cuero de vieja, faca-naca o aría, garatero, guácimo, indio, indio desnudo, indio en cuero, indiodesnudo, indioencuero, indioencueros, kute ena, palo chino, palo colorado, palo mulato, resbala mon **Habit:** Small tree, Tree **Use:** AF, EU, FU, HF, IF, MA, ME, PO **Origin:** Native **Region:** Guiana Shield, Caribbean, Orinoquia, Pacific **Dept.:** ANT, BOL, CAS, CHO, COR, CUN, HUI, LAG, MAG, SAP, SAN, SUC, VID **Elev.:** 0-920 m.a.s.l. **IUCN Habitat code:** 1,2,3,4,5,14 **Conservation status: (IUCN)** LC **(CPSG)** Potential LC

6944 *Bursera tomentosa* (Jacq.) Triana & Planch. (Burseraceae) **IPNI:** 127234-1 **Common name:** bija amarilla, bijo, caraña, caricarito, maluua, sasafrás, tacamaca, tacamaco, tatamaco **Habit:** Shrub, Tree **Use:** ME **Origin:** Native **Region:** Andean, Caribbean **Dept.:** ANT, CES, CUN, HUI, LAG, NAR, TOL **Elev.:** 50–2500 m a.s.l. **IUCN Habitat code:** 1,2,3,4,5,14 **Conservation status: (IUCN)** LC

6945 *Dacryodes belemensis* Cuatrec. (Burseraceae) **IPNI:** 75717-2 **Common name:** copal **Habit:** Tree **Use:** HF, ME **Origin:** Native **Region:** Amazonia, Guiana Shield **Dept.:** AMA, VAU **Elev.:** 100–200 m a.s.l. **IUCN Habitat code:** 1,3,4,5 **Conservation status: (IUCN)** NE

6946 *Dacryodes chimantensis* Steyerm. & Maguire (Burseraceae) **IPNI:** 75718-2 **Common name:** ibapichuna, ibapichuna de pepa pequeña, ibapichuna pequeño, laurel, laurel de monte, laurel de pepa pequeña, laurel de perico, laurel perico **Habit:** Tree **Use:** HF **Origin:** Native

Region: Amazonia, Guiana Shield **Dept.:** AMA, CAQ, GUA, GUV, VAU **Elev.:** 50–200 m a.s.l. **IUCN Habitat code:** 1,3,4,5,14 **Conservation status: (IUCN)** LC

6947 *Dacryodes cupularis* Cuatrec. (Burseraceae)
IPNI: 75721-2
Habit: Tree **Use:** HF **Origin:** Native **Region:** Andean, Pacific **Dept.:** CAU, VAC **Elev.:** 600–600 m a.s.l. **IUCN Habitat code:** 1 **Conservation status: (IUCN)** NE

6948 *Dacryodes granatensis* Cuatrec. (Burseraceae)
IPNI: 75723-2
Common name: laurel **Habit:** Tree **Use:** HF **Origin:** Endemic **Region:** Amazonia, Andean, Pacific **Dept.:** AMA, CAQ, CAU, VAC **Elev.:** 50-1000 m.a.s.l. **IUCN Habitat code:** 1,3,5 **Conservation status: (IUCN)** NT **(CPSG)** NT

6949 *Dacryodes negrensis* Daly & M.C.Martínez (Burseraceae)
IPNI: 326810-2
Common name: laurel **Habit:** Tree **Use:** HF **Origin:** Native **Region:** Amazonia, Guiana Shield **Dept.:** AMA, CAQ, GUV, MET, VAU **Elev.:** 100–200 m a.s.l. **IUCN Habitat code:** 1,3,4,5 **Conservation status: (IUCN)** LC

6950 *Dacryodes nitens* Cuatrec. (Burseraceae)
IPNI: 75728-2
Common name: anime, laurel, palo de laurel **Habit:** Small tree, Tree **Use:** HF **Origin:** Native **Region:** Amazonia **Dept.:** CAQ, VAU **Elev.:** 0–830 m a.s.l. **IUCN Habitat code:** 1,3,14 **Conservation status: (IUCN)** LC

6951 *Dacryodes olivifera* Cuatrec. (Burseraceae)
IPNI: 75730-2
Common name: anime **Habit:** Tree **Use:** HF **Origin:** Native **Region:** Andean **Dept.:** VAC **Elev.:** 900–1400 m a.s.l. **IUCN Habitat code:** 1,14 **Conservation status: (IUCN)** NE

6952 *Dacryodes peruviana* (Loes.) H.J.Lam (Burseraceae)
IPNI: 75732-2
Common name: caraño, ibapichuna, ibapichuna grande, incienso blanco, laurel, laurel propio, laurel silvestre **Habit:** Tree **Use:** HF **Origin:** Native **Region:** Amazonia **Dept.:** AMA, CAQ, VAU **Elev.:** 50–900 m a.s.l. **IUCN Habitat code:** 1,3,14 **Conservation status: (IUCN)** LC

6953 *Dacryodes roraimensis* Cuatrec. (Burseraceae)
IPNI: 75734-2
Common name: laurel, laurel de mierda de perico, laurel de perico **Habit:** Tree **Use:** HF **Origin:** Native **Region:** Amazonia **Dept.:** AMA, CAQ, GUA **Elev.:** 120–120 m a.s.l. **IUCN Habitat code:** 1,3,4 **Conservation status: (IUCN)** NE

6954 *Dacryodes sclerophylla* Cuatrec. (Burseraceae)
IPNI: 75735-2
Habit: Tree **Use:** HF **Origin:** Native **Region:** Amazonia **Dept.:** AMA **Elev.:** 200–300 m a.s.l. **IUCN Habitat code:** 1 **Conservation status: (IUCN)** NE

6955 *Protium altsonii* Sandwith (Burseraceae)
IPNI: 210488-2
Common name: acarana, guamo de sapo, incienso, palo de breo **Habit:** Tree **Use:** MA **Origin:** Native **Region:** Amazonia, Guiana Shield **Dept.:** AMA, CAQ, GUA, VAU **Elev.:** 100–200 m a.s.l. **IUCN Habitat code:** 1,3,5 **Conservation status: (IUCN)** NE

6956 *Protium amazonicum* (Cuatrec.) Daly (Burseraceae)
IPNI: 1063489-2
Common name: guapichuna, laurel silvestre, marapiranga, tubaci, uguray **Habit:** Tree **Use:** HF, ME **Origin:** Native **Region:** Amazonia, Guiana Shield **Dept.:** AMA, CAQ, CHO, GUV, MET, NAR, PUT, VAU **Elev.:** 100–450 m a.s.l. **IUCN Habitat code:** 1,3,14 **Conservation status: (IUCN)** LC

6957 *Protium apiculatum* Swart (Burseraceae)
IPNI: 210491-2
Common name: anime, anime rojo, breo, laurel de perico **Habit:** Tree **Use:** ME **Origin:** Native **Region:** Amazonia, Pacific **Dept.:** AMA, ANT, CAQ, CHO **Elev.:** 400–800 m a.s.l. **IUCN Habitat code:** 1,3,5,14 **Conservation status: (IUCN)** LC

6958 *Protium aracouchini* (Aubl.) Marchand (Burseraceae)
IPNI: 128278-1
Common name: anime, anime alcanfor, anime toroó, carne de vaca, copal, incienso, pepa de loro, pepeloro **Habit:** Shrub, Small tree, Tree **Use:** HF, MA, ME **Origin:** Native **Region:** Amazonia, Guiana Shield, Pacific **Dept.:** AMA, ANT, CAU, GUA, SAN, VAU **Elev.:** 50–800 m a.s.l. **IUCN Habitat code:** 1,3,4,5,14 **Conservation status: (IUCN)** LC

6959 *Protium calanense* Cuatrec. (Burseraceae)
IPNI: 210507-2
Common name: anime, towanae **Habit:** Small tree, Tree **Use:** HF, ME, SU **Origin:** Native **Region:** Amazonia, Guiana Shield, Orinoquia **Dept.:** AMA, CAQ, GUV, MET, VAU, VID **Elev.:** 100–1100 m a.s.l. **IUCN Habitat code:** 1,3,4,5,14 **Conservation status: (IUCN)** NE

6960 *Protium calendulinum* Daly (Burseraceae)
IPNI: 60446820-2
Habit: Tree **Use:** HF **Origin:** Native **Region:** Amazonia **Dept.:** AMA, CAQ, VAU **Elev.:** 300–300 m a.s.l. **IUCN Habitat code:** 1 **Conservation status: (IUCN)** LC

6961 *Protium carana* (Humb.) Marchand (Burseraceae)
IPNI: 128295-1
Common name: caraña, crispín, elemí **Use:** ME **Conservation status: (IUCN)** DD

6962 *Protium crassipetalum* Cuatrec. (Burseraceae)
IPNI: 210520-2
Common name: anime, caraño, tree de borugo, árbol de borugo **Habit:** Tree **Use:** HF **Origin:** Native **Region:** Amazonia, Guiana Shield, Orinoquia **Dept.:** AMA, CAQ, GUA, VAU, VID **Elev.:** 110–300 m a.s.l. **IUCN Habitat code:** 1,3,4,5,14 **Conservation status: (IUCN)** NE

6963 *Protium decandrum* (Aubl.) Marchand (Burseraceae)
IPNI: 128316-1
Common name: breo blanco, laurel **Habit:** Tree **Use:** HF, MA, ME **Origin:** Native **Region:** Amazonia, Guiana Shield **Dept.:** AMA, CAQ **Elev.:** 190–220 m a.s.l. **IUCN Habitat code:** 1,14 **Conservation status: (IUCN)** LC

6964 *Protium divaricatum* Engl. (Burseraceae)
IPNI: 128318-1
Common name: insince **Habit:** Tree **Use:** ME **Origin:** Native **Region:** Amazonia **Dept.:** AMA **Elev.:** 100–280 m a.s.l. **IUCN Habitat code:** 1,14 **Conservation status: (IUCN)** LC

6965 *Protium ecuadorense* Benoist (Burseraceae)
IPNI: 210535-2
Habit: Small tree, Tree **Use:** HF **Origin:** Native **Region:** Andean **Dept.:** ANT, NAR **Elev.:** 1200–1500 m a.s.l. **IUCN Habitat code:** 1,3 **Conservation status: (IUCN)** NE

6966 *Protium elegans* Engl. (Burseraceae)
IPNI: 128322-1
Common name: breo **Habit:** Small tree **Use:** HF **Origin:** Native **Region:** Amazonia, Guiana Shield **Dept.:** AMA, CAQ **Elev.:** 200–300 m a.s.l. **IUCN Habitat code:** 1 **Conservation status: (IUCN)** NE

6967 *Protium excelsior* Byng & Christenh. (Burseraceae)
IPNI: 77183601-1
Common name: laurel **Habit:** Tree **Use:** HF, MA, ME **Origin:** Native **Region:** Amazonia **Dept.:** AMA **Elev.:** 100–100 m a.s.l. **Conservation status: (IUCN)** NE

6968 *Protium glabrescens* Swart (Burseraceae)
IPNI: 210542-2
Common name: anime, copal **Habit:** Tree **Use:** HF, ME **Origin:** Native **Region:** Amazonia, Guiana Shield, Orinoquia **Dept.:** AMA, CAQ, GUV, MET, PUT **Elev.:** 100–600 m a.s.l. **IUCN Habitat code:** 1,3,4,5,14 **Conservation status: (IUCN)** LC

6969 *Protium goudotianum* (Tul.) Byng & Christenh. (Burseraceae)
IPNI: 77183718-1
Common name: anime, pepa de loro, yataro **Habit:** Tree **Use:** HF, ME **Origin:** Native **Region:** Amazonia, Guiana Shield **Dept.:** AMA, GUV, MET, PUT, VAU **Elev.:** 200–600 m a.s.l. **Conservation status: (IUCN)** NE

6970 *Protium grandifolium* Engl. (Burseraceae)
IPNI: 128335-1
Common name: breo **Habit:** Tree **Use:** HF **Origin:** Native **Region:** Amazonia, Orinoquia **Dept.:** AMA, GUA, GUV, VID **Elev.:** 200–800 m a.s.l. **IUCN Habitat code:** 1,3,5,14 **Conservation status: (IUCN)** LC

6971 *Protium guacayanum* Cuatrec. (Burseraceae)
IPNI: 210549-2
Habit: Tree **Use:** HF **Origin:** Native **Region:** Amazonia **Dept.:** AMA **Elev.:** 213–600 m a.s.l. **IUCN Habitat code:** 1 **Conservation status: (IUCN)** NE

6972 *Protium guianense* (Aubl.) Marchand (Burseraceae)
IPNI: 128337-1
Common name: anime, anime blanco, caraño **Habit:** Shrub, Small tree, Tree **Use:** HF, MA, ME **Origin:** Native **Region:** Amazonia, Pacific **Dept.:** CAQ, CHO **Elev.:** 5–500 m a.s.l. **IUCN Habitat code:** 1,3,4,14 **Conservation status: (IUCN)** LC

6973 *Protium heptaphyllum* (Aubl.) Marchand (Burseraceae) **IPNI:** 128338-1 **Common name:** anime, anime blanco, breo de cucaracha, caraño, currucay, currusay, elemí, incienso, matupa, o-mo-ta', sasafrás, tacamahaco, tacamahaco colombiano, ysi **Habit:** Shrub, Small tree, Tree **Use:** FU, HF, MA, ME, PO, SU **Origin:** Native **Region:** Amazonia, Orinoquia **Dept.:** ANT, ARA, BOL, BOY, CAS, CES, COR, CUN, MAG, MET, NSA, SAN, VID **Elev.:** 20-600 m.a.s.l. **IUCN Habitat code:** 1,2,3,4,5,14 **Conservation status:** **(IUCN)** LC **(CPSG)** Potential LC

6974 *Protium llanorum* Cuatrec. (Burseraceae) **IPNI:** 210570-2 **Common name:** acaracay, anime, caraño, hugacal, hugucal, incienso **Habit:** Shrub, Tree **Use:** AF, EU, ME **Origin:** Native **Region:** Amazonia, Guiana Shield, Orinoquia **Dept.:** AMA, CAQ, GUV, MET, VID **Elev.:** 110–500 m a.s.l. **IUCN Habitat code:** 1,2,3,4,5,14 **Conservation status:** **(IUCN)** NE

6975 *Protium macrocarpum* Cuatrec. (Burseraceae) **IPNI:** 210573-2 **Habit:** Tree **Use:** HF **Origin:** Native **Region:** Amazonia, Guiana Shield **Dept.:** AMA, CAQ, VAU **Elev.:** 200–300 m a.s.l. **IUCN Habitat code:** 1,5 **Conservation status:** **(IUCN)** NE

6976 *Protium macrophyllum* (Kunth) Engl. (Burseraceae) **IPNI:** 128356-1 **Common name:** anime, anime blanco, guacamayo, guacharaco hediondo, incienso **Habit:** Tree **Use:** HF, ME **Origin:** Endemic **Region:** Andean **Dept.:** ANT, CAL, RIS, SAN, TOL **Elev.:** 100-1290 m.a.s.l. **IUCN Habitat code:** 1,3,5,14 **Conservation status:** **(IUCN)** LC **(CPSG)** LC

6977 *Protium mucronatum* Rusby (Burseraceae) **IPNI:** 210584-2 **Common name:** copal **Habit:** Tree **Use:** HF **Origin:** Native **Region:** Caribbean **Dept.:** ANT, CES, MAG **Elev.:** 100–600 m a.s.l. **Conservation status:** **(IUCN)** NE

6978 *Protium nodulosum* Swart (Burseraceae) **IPNI:** 210595-2 **Common name:** anime, cariaño, copal, dídao, incienso, jurmmejeguio, laurel, laurel ordinario, mastoko, mee, mejeguío, mégeguio, néejogua, peri, tsica tabütsujunae, tumoho-screbano, wejerinae **Habit:** Small tree, Tree **Use:** HF, MA, ME **Origin:** Native **Region:** Amazonia, Guiana Shield **Dept.:** AMA, CAQ, PUT, VAU **Elev.:** 100–750 m a.s.l. **IUCN Habitat code:** 1,3,4,5,14 **Conservation status:** **(IUCN)** LC

6979 *Protium panamense* (Rose) I.M.Johnst. (Burseraceae) **IPNI:** 210603-2 **Habit:** Tree **Use:** HF, ME **Origin:** Native **Region:** Pacific **Dept.:** CHO **Elev.:** 0–250 m a.s.l. **IUCN Habitat code:** 1,3,5 **Conservation status:** **(IUCN)** LC

6980 *Protium pittieri* (Rose) Engl. (Burseraceae) **IPNI:** 210615-2 **Habit:** Tree **Use:** HF **Origin:** Native **Region:** Pacific **Dept.:** ANT **Elev.:** 200–200 m a.s.l. **Conservation status:** **(IUCN)** VU

6981 *Protium polybotryum* Engl. (Burseraceae) **IPNI:** 128400-1 **Common name:** caraño, guapichuna, laurel **Habit:** Tree **Use:** HF **Origin:** Native **Region:** Amazonia **Dept.:** AMA **Elev.:** 140–300 m a.s.l. **IUCN Habitat code:** 1,5 **Conservation status:** **(IUCN)** LC

6982 *Protium prancei* (Daly) Byng & Christenh. (Burseraceae) **IPNI:** 77183907-1 **Habit:** Tree **Use:** HF **Origin:** Native **Region:** Amazonia, Guiana Shield **Dept.:** AMA, CAQ **Elev.:** 50–300 m a.s.l. **Conservation status:** **(IUCN)** NE

6983 *Protium puncticulatum* J.F.Macbr. (Burseraceae) **IPNI:** 210625-2 **Habit:** Shrub, Tree **Use:** HF **Origin:** Native **Region:** Amazonia **Dept.:** ANT, MET **Elev.:** 15–350 m a.s.l. **IUCN Habitat code:** 1,3,14 **Conservation status:** **(IUCN)** NE

6984 *Protium rhoifolium* (Benth.) Byng & Christenh. (Burseraceae) **IPNI:** 77184241-1 **Common name:** anime, anime blanco, anime macho, anime ruda, coquindillo, laurel, palo blanco, palo calavera, palo de breo, yataro **Habit:** Shrub, Small tree, Tree **Use:** HF **Origin:** Native **Region:** Amazonia, Guiana Shield, Orinoquia, Pacific **Dept.:** AMA, ANT, BOL, CAQ, CAU, CHO, COR, CUN, GUV, MET, PUT, VAC, VAU, VID **Elev.:** 50–900 m a.s.l. **Conservation status:** **(IUCN)** NE

6985 *Protium robustum* (Swart) D.M.Porter (Burseraceae) **IPNI:** 210629-2 **Common name:** copal **Habit:** Tree **Use:** HF, MA **Origin:** Native **Region:** Amazonia, Guiana Shield **Dept.:** AMA, CAQ, GUV, MET **Elev.:** 200–310 m a.s.l. **IUCN Habitat code:** 1,3,14 **Conservation status:** **(IUCN)** LC

6986 *Protium sagotianum* Marchand (Burseraceae) **IPNI:** 128413-1 **Common name:** anime, caraño, copal, gupediaque, pepemono **Habit:** Tree **Use:** HF, MA **Origin:** Native **Region:** Amazonia, Guiana Shield, Pacific **Dept.:** AMA, ANT, BOY, CHO, COR, GUV, MAG, MET, NAR, VAC **Elev.:** 100–1200 m a.s.l. **IUCN Habitat code:** 1,2,3,14 **Conservation status:** **(IUCN)** LC

6987 *Protium spruceanum* (Benth.) Engl. (Burseraceae) **IPNI:** 128420-1 **Common name:** laurel **Habit:** Tree **Use:** HF, MA, ME **Origin:** Native **Region:** Amazonia, Orinoquia, Pacific **Dept.:** AMA, VAC **Elev.:** 50–50 m a.s.l. **IUCN Habitat code:** 1 **Conservation status:** **(IUCN)** LC

6988 *Protium stevensonii* (Standl.) Daly (Burseraceae) **IPNI:** 77202519-1 **Common name:** anime, caraño, fruta de loro, guapichuna, guira **Habit:** Small tree, Tree **Use:** FU, HF, MA, ME **Origin:** Native **Region:** Amazonia, Andean, Orinoquia, Pacific **Dept.:** AMA, ANT, ARA, BOY, CAQ, CAS, CES, CHO, GUV, MET, PUT, SAN, VID **Elev.:** 55-1690 m.a.s.l. **IUCN Habitat code:** 1,3,14 **Conservation status:** **(IUCN)** LC **(CPSG)** Potential LC

6989 *Protium subserratum* (Engl.) Engl. (Burseraceae) **IPNI:** 128422-1 **Common name:** caraño, copal, laurel de búho, laurel de pajarito **Habit:** Small tree, Tree **Use:** HF, MA, ME **Origin:** Native **Region:** Amazonia, Guiana Shield **Dept.:** AMA, ANT, CAQ, CAU, GUA, PUT, VAU **Elev.:** 200–900 m a.s.l. **IUCN Habitat code:** 1,3,5,14 **Conservation status:** **(IUCN)** LC

6990 *Protium tenuifolium* (Engl.) Engl. (Burseraceae) **IPNI:** 128424-1 **Common name:** anime, noucda **Habit:** Tree **Use:** HF, ME **Origin:** Native **Region:** Amazonia, Andean **Dept.:** CHO, COR, GUV, LAG, MAG **Elev.:** 175–1000 m a.s.l. **IUCN Habitat code:** 1,3,5,14 **Conservation status:** **(IUCN)** LC

6991 *Protium trifoliolatum* Engl. (Burseraceae) **IPNI:** 128429-1 **Common name:** palo de cuchara **Habit:** Small tree, Tree **Use:** HF **Origin:** Native **Region:** Amazonia, Orinoquia **Dept.:** AMA, CAQ, GUV, VAU, VID **Elev.:** 100–250 m a.s.l. **IUCN Habitat code:** 1,3,4,5,14 **Conservation status:** **(IUCN)** LC

6992 *Protium unifoliolatum* Engl. (Burseraceae) **IPNI:** 128430-1 **Habit:** Small tree, Tree **Use:** HF, ME **Origin:** Native **Region:** Amazonia, Guiana Shield **Dept.:** AMA, CAQ, CUN, GUV, MET, PUT, VAU **Elev.:** 90–550 m a.s.l. **IUCN Habitat code:** 1,3,5,14 **Conservation status:** **(IUCN)** LC

6993 *Trattinnickia aspera* (Standl.) Swart (Burseraceae) **IPNI:** 256487-2 **Common name:** ampó, caraña, caraño, cariaño **Habit:** Tree **Use:** MA, ME **Origin:** Native **Region:** Pacific **Dept.:** ANT, CHO, NAR **Elev.:** 18–1000 m a.s.l. **IUCN Habitat code:** 1,3,4,14 **Conservation status:** **(IUCN)** LC

6994 *Trattinnickia burserifolia* Mart. (Burseraceae) **IPNI:** 128598-1 **Common name:** caraño, incienso, palo de incienso **Habit:** Shrub, Small tree, Tree **Use:** HF, ME **Origin:** Native **Region:** Amazonia, Guiana Shield **Dept.:** AMA, CAQ, GUA, VAU **Elev.:** 100–850 m a.s.l. **IUCN Habitat code:** 1,3,4,5,14 **Conservation status:** **(IUCN)** LC

6995 *Trattinnickia glaziovii* Swart (Burseraceae) **IPNI:** 256497-2 **Common name:** breo, incienso hembra, palo de incienso **Habit:** Tree **Use:** HF **Origin:** Native **Region:** Amazonia, Orinoquia **Dept.:** AMA, CAQ, MET, VID **Elev.:** 200–600 m a.s.l. **IUCN Habitat code:** 1,3,4,14 **Conservation status:** **(IUCN)** LC

6996 *Trattinnickia lawrancei* Standl. ex Swart (Burseraceae) **IPNI:** 256498-2 **Common name:** caraño, cariaño, trementino **Habit:** Tree **Use:** ME **Origin:** Native **Region:** Andean **Dept.:** CAQ, CUN, PUT **Elev.:** 200–1160 m a.s.l. **IUCN Habitat code:** 1,3,4,5,14 **Conservation status:** **(IUCN)** NE

6997 *Trattinnickia peruviana* Loes. (Burseraceae) **IPNI:** 256503-2 **Common name:** breo blanco, caraño, chipeci **Habit:** Tree **Use:** ME **Origin:** Native **Region:** Amazonia, Pacific **Dept.:** AMA, CAQ, CHO, MET, PUT, VAU **Elev.:** 100–600 m a.s.l. **IUCN Habitat code:** 1,3,5,14 **Conservation status:** **(IUCN)** LC

6998 *Trattinnickia rhoifolia* Willd. (Burseraceae) **IPNI:** 128602-1 **Common name:** anime, caraño, falso caraño **Habit:** Small tree, Tree **Use:** MA **Origin:** Native **Region:** Amazonia, Orinoquia, Pacific **Dept.:** ARA, CAQ, GUV, MET, VAC, VID **Elev.:** 250–500 m a.s.l. **IUCN Habitat code:** 1,2,3,4,14 **Conservation status: (IUCN)** LC

6999 *Azadirachta indica* A.Juss. (Meliaceae) **IPNI:** 1213180-2 **Common name:** barbasco, barbato, min, neem, nim, árbol de la vida **Habit:** Shrub, Small tree **Use:** AF, EU, FU, GS, HF, MA, ME, PO, SU **Origin:** Cultivated **Region:** Amazonia, Andean, Caribbean **Dept.:** ANT, ATL, CAQ, COR, CUN, QUI, SUC, TOL **Elev.:** 200–1300 m a.s.l. **IUCN Habitat code:** 1,2,3,4,5,14 **Conservation status: (IUCN)** LC

7000 *Cabralea canjerana* (Vell.) Mart. (Meliaceae) **IPNI:** 1021145-2 **Habit:** Tree **Use:** HF, MA, ME **Origin:** Native **Region:** Amazonia **Dept.:** AMA, MET **Elev.:** 100–120 m a.s.l. **IUCN Habitat code:** 1 **Conservation status: (IUCN)** LC

7001 *Carapa guianensis* Aubl. (Meliaceae) **IPNI:** 45059-2 **Common name:** andiroba, carapa, carapo, cedro bateo, cedro güino, cedro macho, chocó, figueroa, guina, guino, güina, güino, güino rojo, houiri, huina, huino, jandiroba, jimo, karapa, krapa, maco, masábalo, masábolo, mazábalo, nandiroba, pialde, tangare, tangare pipa, t **Habit:** Tree **Use:** EU, HF, MA, ME, PO **Origin:** Native **Region:** Amazonia, Andean, Pacific **Dept.:** AMA, ANT, CAU, CHO, NAR, SAN, VAC **Elev.:** 0–870 m a.s.l. **IUCN Habitat code:** 1,3,5,14 **Conservation status: (IUCN)** LC

7002 *Carapa procera* DC. (Meliaceae) **IPNI:** 577711-1 **Use:** EU, FU, GS, HF, MA, ME, PO **IUCN Habitat code:** 1,3 **Conservation status: (IUCN)** LC

7003 *Cedrela angustifolia* DC. (Meliaceae) **IPNI:** 577723-1 **Common name:** Cedro bayo, Cedro coya, Cedro de altura, Cedro del cerro, Cedro peludo, Cedro salteno, Cedro vírgen **Habit:** Tree **Use:** HF, MA, ME **Origin:** Native **Region:** Andean, Amazonia, Caribbean, Pacific, Magdalena valley **Dept.:** AMA, ANT, BOL, CAL, CAQ, CHO, CUN, HUI, MAG, MET, NAR, QUI, RIS, SAP, SAN, TOL, VAC **IUCN Habitat code:** 1,2,3,5,14 **Conservation status: (IUCN)** NE

7004 *Cedrela fissilis* Vell. (Meliaceae) **IPNI:** 577744-1 **Common name:** astillo, bastardo, cebolla, cedro, cedro blanco, cedro hembra, cedro jaspeado, cedro oloroso, monde **Habit:** Tree **Use:** HF, MA, ME **Origin:** Native **Region:** Amazonia, Andean, Caribbean, Pacific **Dept.:** AMA, ANT, CHO, HUI, LAG, MAG, VAC **Elev.:** 0–1650 m a.s.l. **IUCN Habitat code:** 1,14 **Conservation status: (IUCN)** VU

7005 *Cedrela montana* Moritz ex Turcz. (Meliaceae) **IPNI:** 577770-1 **Common name:** cedro, cedro cebolla, cedro cebollo, cedro clavel, cedro colorado, cedro de montaña, cedro de tierra fría, cedro dulce, cedro monde, cedro mondé, cedro oloroso, cedro rosado, monde, monde bogotano, munde, serrano **Habit:** Tree **Use:** AF, MA, ME **Origin:** Native **Region:** Andean **Dept.:** ANT, BOG, BOY, CAL, CAU, CUN, HUI, LAG, PUT, QUI, RIS, SAN, VAC **Elev.:** 1350-3000 m.a.s.l. **IUCN Habitat code:** 1,2,3,4,14 **Conservation status: (IUCN)** NE **(CPSG)** NT

7006 *Cedrela odorata* L. (Meliaceae) **IPNI:** 51010-2 **Common name:** asna cedro, bastardo, caoba de interior, cedro, cedro amargo, cedro bastardo, cedro blanco, cedro caoba, cedro cebollino, cedro cebollo, cedro cebollín, cedro clavel, cedro colorado, cedro hembra, cedro macho, cedro oloroso, cedro rojo, cedro **Habit:** Tree **Use:** EU, FU, HF, IF, MA, ME **Origin:** Native **Region:** Amazonia, Andean, Caribbean, Pacific **Dept.:** AMA, ANT, BOL, CAL, CAQ, CHO, CUN, HUI, MAG, MET, NAR, QUI, RIS, SAP, SAN, TOL, VAC **Elev.:** 10-2700 m.a.s.l. **IUCN Habitat code:** 1 **Conservation status: (IUCN)** VU **(CPSG)** VU

7007 *Guarea fistulosa* W.Palacios (Meliaceae) **IPNI:** 77091142-1 **Use:** HF **Origin:** Native **IUCN Habitat code:** 1 **Conservation status: (IUCN)** LC

7008 *Guarea glabra* Vahl (Meliaceae) **IPNI:** 578602-1 **Common name:** bilibil, cananco, cedrillo, cedro macho, güino cedro, trompillo **Habit:** Shrub, Small tree, Tree **Use:** HF, MA, ME **Origin:** Native **Region:** Amazonia, Andean, Caribbean, Pacific **Dept.:** AMA, ANT, BOL, CAQ, CHO, CUN, LAG, MAG, NAR, QUI, SAN, VAC **Elev.:** 0-2050 m.a.s.l. **IUCN Habitat code:** 1,2,3,4,5,14 **Conservation status: (IUCN)** LC **(CPSG)** Potential LC

7009 *Guarea gomma* Pulle (Meliaceae) **IPNI:** 114055-2 **Habit:** Shrub, Small tree, Tree **Use:** MA **Origin:** Native **Region:** Amazonia **Dept.:** AMA, CAQ, PUT, VAU **Elev.:** 100–300 m a.s.l. **IUCN Habitat code:** 1,3,5,14 **Conservation status: (IUCN)** LC

7010 *Guarea guidonia* (L.) Sleumer (Meliaceae) **IPNI:** 114060-2 **Common name:** bailador, bailador trompillo, bilibil, bilibile, bilibili, cacao, cacao de monte, cacao silvestre, cananeo, cartagueño, cartagüeño, cascarillo, cedrillo, cedro, cedro blanco, cedro guamo, cedro macho, cedro tigre, cedro trompillo, cedro zambo, cedrotigre, **Habit:** Small tree, Tree **Use:** AF, EU, FU, HF, MA, ME, PO, SU **Origin:** Native **Region:** Amazonia, Andean, Guiana Shield, Caribbean, Pacific **Dept.:** AMA, ANT, ARA, ATL, BOL, BOY, CAL, CAQ, CAS, CAU, CES, CHO, CUN, GUV, HUI, LAG, MAG, MET, NSA, PUT, QUI, SAN, SUC, TOL, VAC, VAU **Elev.:** 0-1400 m.a.s.l. **IUCN Habitat code:** 1,2,3,4,5,14 **Conservation status: (IUCN)** LC **(CPSG)** Potential LC

7011 *Guarea kunthiana* A.Juss. (Meliaceae) **IPNI:** 578634-1 **Common name:** bailador, bailador trompillo, bilibil, bilibile, bilibili, bitibiti, bombo, breo, cacao, cacao de monte, cacao de montero, cacao montañero, café, cananeo, carguero, cascarillo, cedrillo, cedro guamo, cedro macho, cedro pataegallo, cedro tigre, cedro tromp **Habit:** Small tree **Use:** HF, MA, ME **Origin:** Native **Region:** Amazonia, Andean, Guiana Shield **Dept.:** AMA, ANT, ARA, BOL, CAL, CAQ, CAS, CAU, CHO, CUN, HUI, LAG, MAG, MET, NAR, NSA, PUT, QUI, RIS, VAC, VAU **Elev.:** 10-2850 m.a.s.l. **IUCN Habitat code:** 1,2,3,5,14 **Conservation status: (IUCN)** LC **(CPSG)** Potential LC

7012 *Guarea macrophylla* Vahl (Meliaceae) **IPNI:** 578653-1 **Common name:** bilibil, ijebioici **Habit:** Shrub, Small tree, Tree **Use:** MA, ME **Origin:** Native **Region:** Amazonia, Andean, Caribbean, Pacific **Dept.:** AMA, ANT, CAQ, CAS, CHO, CUN, GUA, LAG, MAG, NAR, PUT, VAU **Elev.:** 20-950 m.a.s.l. **IUCN Habitat code:** 1,3,4,5,14 **Conservation status: (IUCN)** LC **(CPSG)** Potential LC

7013 *Guarea megantha* A.Juss. (Meliaceae) **IPNI:** 578660-1 **Use:** MA **Origin:** Native **IUCN Habitat code:** 1 **Conservation status: (IUCN)** NE

7014 *Guarea pubescens* (Rich.) A.Juss. (Meliaceae) **IPNI:** 578711-1 **Common name:** bilibil **Habit:** Small tree, Tree **Use:** ME **Origin:** Native **Region:** Amazonia, Andean, Pacific **Dept.:** AMA, ANT, BOL, BOY, CAQ, CHO, COR, GUA, GUV, MET, VAC, VAU **Elev.:** 50-1900 m.a.s.l. **IUCN Habitat code:** 1,3,4,5,14 **Conservation status: (IUCN)** LC **(CPSG)** Potential LC

7015 *Guarea purusana* C.DC. (Meliaceae) **IPNI:** 114114-2 **Common name:** bilibil, cedrillo, cedro macho, guacuray, guaira caspi, latape, latape altura **Habit:** Shrub, Small tree, Tree **Use:** HF **Origin:** Native **Region:** Amazonia **Dept.:** AMA, CAQ, PUT **Elev.:** 100–220 m a.s.l. **IUCN Habitat code:** 1,5,14 **Conservation status: (IUCN)** LC

7016 *Guarea silvatica* C.DC. (Meliaceae) **IPNI:** 578741-1 **Common name:** coquindo cimarrón, vara blanca **Habit:** Small tree, Tree **Use:** HF, MA **Origin:** Native **Region:** Amazonia **Dept.:** AMA, ANT, CAQ, PUT **Elev.:** 110–400 m a.s.l. **IUCN Habitat code:** 1,3,14 **Conservation status: (IUCN)** LC

7017 *Melia azedarach* L. (Meliaceae) **IPNI:** 578949-1 **Common name:** alelí, alhelí, chirrinchao, jardín, jazmín, jazmín de España, jazmín de caballo, jazmín de luto, nido de paloma, paraíso, pestaña, árbol del paraíso **Habit:** Shrub, Small tree **Use:** AF, EU, FU, GS, HF, IF, MA, ME, PO, SU **Origin:** Cultivated **Region:** Andean, Caribbean, Orinoquia, Pacific **Dept.:** ANT, ATL, BOL, BOY, CAU, CHO, COR, CUN, HUI, MAG, MET, NAR, NSA, PUT, SAN, TOL, VAC **Elev.:** 0–2000 m a.s.l. **IUCN Habitat code:** 1,2,3,4,14 **Conservation status: (IUCN)** LC

7018 *Schmardaea microphylla* (Hook.) H.Karst. ex Müll.Berol. (Meliaceae) **IPNI:** 579221-1 **Common name:** curito **Habit:** Shrub, Small tree, Tree **Use:** MA **Origin:** Native **Region:** Caribbean **Dept.:** ANT, LAG, MAG **Elev.:** 76–590 m a.s.l. **IUCN Habitat code:** 1,14 **Conservation status: (IUCN)** VU

7019 *Swietenia macrophylla* King (Meliaceae) **IPNI:** 246905-2 **Common name:** caoba, caoba americana, caobo, cedro caoba, palo santo **Habit:** Tree **Use:** AF, EU, HF, IF, MA, ME **Origin:** Native, Cultivated **Region:** Pacific **Dept.:** BOL, BOY, CHO, CUN, MAG **Elev.:** 200–500 m a.s.l. **IUCN Habitat code:** 1,2,3,5,14 **Conservation status: (IUCN)** VU

7020 *Trichilia acuminata* (Humb. & Bonpl. ex Schult.) C.DC. (Meliaceae) **IPNI:** 579322-1 **Common name:** coya colorada, jayo colorado, mangle blanco, mangle colorado, mangle rojo, manglecito, negrito, yayo blanco, yayo colorado **Habit:** Small tree, Tree **Use:** AF, EU, HF, MA **Origin:** Native **Region:** Caribbean **Dept.:** ANT, ATL, BOL, CUN, SUC, TOL **Elev.:** 50-600 m.a.s.l. **IUCN Habitat code:** 1,2,3,14 **Conservation status: (IUCN)** VU **(CPSG)** EN

7021 *Trichilia adolfi* Harms (Meliaceae) **IPNI:** 256778-2 **Habit:** Tree **Use:** HF **Origin:** Native **Region:** Amazonia, Andean **Dept.:** AMA, ANT **Elev.:** 120–750 m a.s.l. **Conservation status: (IUCN)** LC

7022 *Trichilia appendiculata* (Triana & Planch.) C.DC. (Meliaceae) **IPNI:** 579340-1 **Common name:** bilibil, cedro trompillo, coya blanco, mangle blanco, manglecita, manglecito, yaya, yayo, yayo blanco **Habit:** Shrub, Tree **Use:** MA **Origin:** Native **Region:** Caribbean **Dept.:** ATL, BOL, CES, CHO, MAG, SUC, TOL, VAC **Elev.:** 30-1040 m.a.s.l. **IUCN Habitat code:** 1,2,3,14 **Conservation status: (IUCN)** LC **(CPSG)** EN

7023 *Trichilia cipo* (A.Juss.) C.DC. (Meliaceae) **IPNI:** 579382-1 **Habit:** Small tree **Use:** ME **Origin:** Native **Region:** Amazonia **Dept.:** AMA, MET **Elev.:** 150–350 m a.s.l. **IUCN Habitat code:** 1 **Conservation status: (IUCN)** LC

7024 *Trichilia havanensis* Jacq. (Meliaceae) **IPNI:** 322349-2 **Common name:** cacique, cedrillo, cedro **Habit:** Shrub, Tree **Use:** HF, ME **Origin:** Native **Region:** Andean, Caribbean **Dept.:** ANT, BOY, CUN, HUI, LAG, MAG, TOL **Elev.:** 450–2130 m a.s.l. **IUCN Habitat code:** 1,2,3,14 **Conservation status: (IUCN)** LC

7025 *Trichilia hirta* L. (Meliaceae) **IPNI:** 322348-2 **Common name:** bilibil, jobo, jobo macho, jobo verde, lobo, mantequilla, meretiro, monben bata, perico, tautano, trompillo, tuétano **Habit:** Shrub, Tree **Use:** AF, HF, MA, ME, PO **Origin:** Native **Region:** Caribbean, Orinoquia, Pacific **Dept.:** ANT, ATL, BOL, BOY, CHO, COR, CUN, HUI, LAG, MAG, MET, SAP, SUC, TOL, VAC **Elev.:** 0-700 m.a.s.l. **IUCN Habitat code:** 1,2,3,5,14 **Conservation status: (IUCN)** LC **(CPSG)** Potential LC

7026 *Trichilia martiana* C.DC. (Meliaceae) **IPNI:** 579520-1 **Common name:** cedrillo, tuétano, vara de piedra **Habit:** Shrub, Tree **Use:** AF, MA **Origin:** Native **Region:** Amazonia, Andean, Caribbean, Pacific **Dept.:** ANT, BOL, CAQ, CHO, COR, CUN, HUI, LAG, MAG, MET, NAR, SAP, TOL, VAC **Elev.:** 20-1020 m.a.s.l. **IUCN Habitat code:** 1,2,3,5,14 **Conservation status: (IUCN)** LC **(CPSG)** Potential LC

7027 *Trichilia micrantha* Benth. (Meliaceae) **IPNI:** 579528-1 **Common name:** breo, palo balsudo **Habit:** Small tree, Tree **Use:** HF, MA **Origin:** Native **Region:** Amazonia, Guiana Shield **Dept.:** AMA, CAQ, GUV, SAN, VAU **Elev.:** 100–220 m a.s.l. **IUCN Habitat code:** 1,3,4,5,14 **Conservation status: (IUCN)** LC

7028 *Trichilia moschata* Sw. (Meliaceae) **IPNI:** 579549-1 **Habit:** Tree **Use:** MA **Origin:** Native **Region:** Pacific **Dept.:** ANT, HUI **Elev.:** 325–600 m a.s.l. **IUCN Habitat code:** 1,2,3,14 **Conservation status: (IUCN)** LC

7029 *Trichilia pallida* Sw. (Meliaceae) **IPNI:** 579573-1 **Common name:** arrayán, bilibil, cafetillo, caico, caratejo, cedrillo, guacharaca, injerto, lobo, mote, naranjuelo, rejudo, trompillo, yema de huevo, zapatillo **Habit:** Shrub, Small tree, Tree **Use:** HF, MA, ME, PO **Origin:** Native **Region:** Amazonia, Andean, Pacific **Dept.:** AMA, ANT, BOL, BOY, CAL, CAQ, CHO, CUN, HUI, MET, NAR, PUT, QUI, RIS, SAN, TOL, VAC, VAU **Elev.:** 2-2500 m.a.s.l. **IUCN Habitat code:** 1,2,3,4,5,14 **Conservation status: (IUCN)** NE **(CPSG)** Potential LC

7030 *Trichilia pleeana* (A.Juss.) C.DC. (Meliaceae) **IPNI:** 579591-1 **Habit:** Shrub, Small tree, Tree **Use:** ME **Origin:** Native **Region:** Amazonia, Andean, Guiana Shield, Caribbean, Pacific **Dept.:** AMA, ANT, BOL, CAQ, CHO, GUV, LAG, MAG, MET, VAU **Elev.:** 0–1900 m a.s.l. **IUCN Habitat code:** 1,2,3,5,14 **Conservation status: (IUCN)** LC

7031 *Trichilia quadrijuga* Kunth (Meliaceae) **IPNI:** 579615-1 **Common name:** cedro macho, chagará, oso, tostado, yagará **Habit:** Small tree, Tree **Use:** HF, MA **Origin:** Native **Region:** Amazonia, Pacific **Dept.:** AMA, ANT, BOL, CHO, GUV, MAG **Elev.:** 10–520 m a.s.l. **IUCN Habitat code:** 1,3,5,14 **Conservation status: (IUCN)** LC

7032 *Trichilia rubra* C.DC. (Meliaceae) **IPNI:** 579636-1 **Habit:** Tree **Use:** HF, ME **Origin:** Native **Region:** Amazonia, Guiana Shield, Orinoquia **Dept.:** AMA, CAQ, GUA, GUV, MET, VAU, VID **Elev.:** 100–220 m a.s.l. **IUCN Habitat code:** 1,3,4 **Conservation status: (IUCN)** LC

7033 *Trichilia schomburgkii* C.DC. (Meliaceae) **IPNI:** 579645-1 **Common name:** cedrillo, oropo, pintado **Habit:** Shrub, Tree **Use:** MA **Origin:** Native **Region:** Amazonia, Guiana Shield, Orinoquia, Pacific **Dept.:** AMA, CAS, CHO, MET **Elev.:** 500–1300 m a.s.l. **IUCN Habitat code:** 1,3,14 **Conservation status: (IUCN)** LC

7034 *Trichilia septentrionalis* C.DC. (Meliaceae) **IPNI:** 579652-1 **Common name:** fono negro **Habit:** Shrub, Small tree, Tree **Use:** ME **Origin:** Native **Region:** Amazonia, Andean, Orinoquia, Pacific **Dept.:** AMA, ANT, CAQ, CAS, CHO, COR, PUT, VAC, VAU **Elev.:** 50–1420 m a.s.l. **IUCN Habitat code:** 1,3,14 **Conservation status: (IUCN)** LC

7035 *Trichilia singularis* C.DC. (Meliaceae) **IPNI:** 579659-1 **Habit:** Shrub, Tree **Use:** ME **Origin:** Native **Region:** Amazonia, Orinoquia **Dept.:** AMA, MET, SAN **Elev.:** 100–200 m a.s.l. **IUCN Habitat code:** 1,4 **Conservation status: (IUCN)** LC

7036 *Trichilia tuberculata* (Triana & Planch.) C.DC. (Meliaceae) **IPNI:** 579709-1 **Common name:** catagua, cedrón, maiztostao **Habit:** Tree **Use:** HF **Origin:** Native **Region:** Amazonia **Dept.:** AMA, CAQ, GUV, MET, PUT **Elev.:** 110–500 m a.s.l. **IUCN Habitat code:** 1,3,14 **Conservation status: (IUCN)** LC

7037 *Amyris balsamifera* L. (Rutaceae) **IPNI:** 12146-2 **Common name:** candil, guayabo ulanda, naranjito quigua, ulanda **Habit:** Tree **Use:** MA, ME **Origin:** Native **Region:** Caribbean **Dept.:** BOL **Elev.:** 200–250 m a.s.l. **IUCN Habitat code:** 1,2,14 **Conservation status: (IUCN)** LC

7038 *Amyris elemifera* L. (Rutaceae) **IPNI:** 126885-1 **Common name:** peraco **Habit:** Shrub **Use:** HF, MA, ME **Origin:** Native **Region:** Andean, Caribbean **Dept.:** BOL, CUN, LAG, MAG, NSA, SAP, SAN **Elev.:** 10–1275 m a.s.l. **Conservation status: (IUCN)** LC

7039 *Amyris pinnata* Kunth (Rutaceae) **IPNI:** 126928-1 **Common name:** bilanda, chaquiro, chaquiro amargo, chaquiro marfil, marfil, ruda **Habit:** Small tree, Tree **Use:** HF **Origin:** Native **Region:** Andean **Dept.:** ANT, CUN, HUI, RIS, VAC **Elev.:** 500–1200 m a.s.l. **IUCN Habitat code:** 1,2,3,14 **Conservation status: (IUCN)** LC

7040 *Amyris plumieri* DC. (Rutaceae) **IPNI:** 126929-1 **Use:** ME **Origin:** Native **Dept.:** BOL, CUN, LAG, MAG, NSA, SAP, SAN **Conservation status: (IUCN)** LC

7041 *Angostura trifoliata* (Willd.) T.S.Elias (Rutaceae) **IPNI:** 771231-1 **Common name:** angostura, cascarilla, cuspare, palo amarillo, quina blanca, quina orinoco **Use:** HF, MA, ME **Origin:** Native **Conservation status: (IUCN)** NE

7042 *Casimiroa edulis* La Llave (Rutaceae) **IPNI:** 771774-1 **Common name:** mangosán, matasano, zapote, zapote blanco **Habit:** Tree **Use:** EU, FU, GS, HF, MA, ME **Origin:** Cultivated **Region:** Andean **Dept.:** ANT, BOG, CUN, RIS, TOL **Elev.:** 200–2560 m a.s.l. **IUCN Habitat code:** 1,3,14 **Conservation status: (IUCN)** LC

7043 *Citrus × aurantiifolia* (Christm.) Swingle (Rutaceae) **IPNI:** 59599-2 **Common name:** citron vert, lima, lima agria, lime, limonero, limón, limón criollo, sitron **Use:** GS, HF, MA, ME, PO, SU **IUCN Habitat code:** 1,2,3,14 **Conservation status: (IUCN)** NE

7044 *Citrus × aurantium* L. (Rutaceae)
IPNI: 59600-2
Common name: mandarina, mandarina arrayana, mandarina cleopatra, mandarina común, mandarina israelita, mandarina oneco, mandarina reina, mandarina satsuma, mandarina tangolea, mandarino, naranja, naranja agria, naranja agrio, naranja china, naranja común, naranja crio **Habit:** Small tree **Use:** AF, EU, GS, HF, MA, ME, PO, SU **Origin:** Cultivated **Region:** Amazonia, Andean, Caribbean, Pacific **Dept.:** AMA, ANT, BOY, CUN, HUI, MET, NAR, PUT, QUI, RIS, SAN, TOL, VAC, VAU **Elev.:** 200–2500 m a.s.l. **IUCN Habitat code:** 1,2,3,14 **Conservation status: (IUCN)** NE

7045 *Citrus × limon* (L.) Osbeck (Rutaceae)
IPNI: 60454758-2
Common name: lima, lima dulce, limón dulce **Habit:** Small tree **Use:** AF, EU, FU, GS, HF, MA, ME **Origin:** Cultivated, Naturalised **Region:** Amazonia, Andean, Caribbean, Pacific **Dept.:** AMA, BOY, CUN, NAR, PUT, VAC **Elev.:** 80–2620 m a.s.l. **IUCN Habitat code:** 1,2,3,14 **Conservation status: (IUCN)** NE

7046 *Citrus maxima* (Burm.) Merr. (Rutaceae)
IPNI: 30075266-2
Common name: lima, lima dulce, lima ácida, pamplemusa, pomelo, toronja, toronjo, árbol de lima **Habit:** Small tree, Tree **Use:** AF, EU, GS, HF, MA, ME **Origin:** Cultivated **Region:** Amazonia, Andean, Caribbean, Pacific **Dept.:** AMA, ANT, ATL, CAQ, CAU, CES, CHO, CUN, MAG, MET, NAR, NSA, PUT, TOL **Elev.:** 0–2950 m a.s.l. **IUCN Habitat code:** 1,3,14 **Conservation status: (IUCN)** LC

7047 *Citrus medica* L. (Rutaceae)
IPNI: 59603-2
Common name: cidra, cidrero, cidro, greifruta **Habit:** Small tree **Use:** AF, EU, GS, HF, MA, ME, PO, SU **Origin:** Cultivated **Region:** Amazonia, Guiana Shield, Pacific **Dept.:** AMA, MET, NAR, PUT **Elev.:** 15–300 m a.s.l. **IUCN Habitat code:** 1,3,4,14 **Conservation status: (IUCN)** NE

7048 *Citrus volkameriana* Pasq. (Rutaceae)
IPNI: 1567547
Common name: mandarina volkameriana **Habit:** Small tree **Use:** HF **Origin:** Endemic, Cultivated **Region:** Orinoquia **Dept.:** MET **Elev.:** 400–400 m a.s.l. **Conservation status: (IUCN)** NE

7049 *Coleonema album* (Thunb.) Bartl. & H.L.Wendl. (Rutaceae)
IPNI: 772187-1
Common name: diosme, falso buchú **Habit:** Subshrub **Use:** ME **Origin:** Cultivated **Region:** Andean **Dept.:** BOG, CUN **Elev.:** 2600–2700 m a.s.l. **IUCN Habitat code:** 3,14 **Conservation status: (IUCN)** NE

7050 *Conchocarpus toxicarius* (Spruce ex Engl.) Kallunki & Pirani (Rutaceae)
IPNI: 1001963-1
Habit: Shrub **Use:** ME **Origin:** Native **Dept.:** SAN **Elev.:** 100–700 m a.s.l. **IUCN Habitat code:** 3 **Conservation status: (IUCN)** LC

7051 *Ertela trifolia* (L.) Kuntze (Rutaceae)
IPNI: 96244-2
Common name: barbasco **Habit:** Herb, Subshrub **Use:** ME **Origin:** Native **Region:** Amazonia, Guiana Shield, Caribbean, Orinoquia **Dept.:** AMA, CAQ, CES, GUA, GUV, MAG, MET, VAU **Elev.:** 200–350 m a.s.l. **IUCN Habitat code:** 1,2,3,14 **Conservation status: (IUCN)** NE

7052 *Erythrochiton brasiliensis* Nees & Mart. (Rutaceae)
IPNI: 772802-1
Habit: Shrub **Use:** EU, ME **Origin:** Native **Region:** Guiana Shield **Dept.:** MET **Elev.:** 500–600 m a.s.l. **IUCN Habitat code:** 1,3 **Conservation status: (IUCN)** LC

7053 *Esenbeckia panamensis* T.S.Elias (Rutaceae)
IPNI: 97592-2
Common name: duaná **Habit:** Shrub, Tree **Use:** ME **Origin:** Native **Region:** Andean, Pacific **Dept.:** ANT, CHO **Elev.:** 10–1000 m a.s.l. **IUCN Habitat code:** 1,3,14 **Conservation status: (IUCN)** LC

7054 *Hortia brasiliana* Vand. ex DC. (Rutaceae)
IPNI: 774036-1
Habit: Tree **Use:** MA, ME **Origin:** Native **Region:** Andean, Pacific **Dept.:** ANT, BOY, CHO, VAC **Elev.:** 50–1500 m a.s.l. **IUCN Habitat code:** 1,3,5,14 **Conservation status: (IUCN)** LC

7055 *Murraya exotica* L. (Rutaceae)
IPNI: 774426-1
Common name: azahar de la india **Use:** ME **IUCN Habitat code:** 1,2,3,14 **Conservation status: (IUCN)** NE

7056 *Murraya paniculata* (L.) Jack (Rutaceae)
IPNI: 774441-1
Common name: azahar, azahar de la India, boj de la China, boj de tierra caliente, buenasuerte, jazmín de la India, mirto, mirto blanco, mirto de Malabar, mirto de la India **Habit:** Shrub, Tree **Use:** EU, GS, HF, IF, MA, ME, SU **Origin:** Cultivated **Region:** Amazonia, Andean, Caribbean, Pacific **Dept.:** AMA, ANT, ATL, BOL, CES, CHO, CUN, LAG, MAG, SAN, TOL, VAC **Elev.:** 20–2350 m a.s.l. **Conservation status: (IUCN)** NE

7057 *Pilocarpus goudotianus* Tul. (Rutaceae)
IPNI: 774816-1
Common name: jaborandi **Habit:** Shrub, Tree **Use:** ME **Origin:** Native **Region:** Caribbean **Dept.:** BOL, HUI, LAG, MAG, SUC, TOL **Elev.:** 80–460 m a.s.l. **IUCN Habitat code:** 1,2,14 **Conservation status: (IUCN)** NE

7058 *Pilocarpus racemosus* Vahl (Rutaceae)
IPNI: 774840-1
Use: ME **Origin:** Native **Conservation status: (IUCN)** NE

7059 *Ruta chalepensis* L. (Rutaceae)
IPNI: 775070-1
Common name: ruda, ruda de arao **Habit:** Herb, Subshrub **Use:** EU, HF, MA, ME, PO, SU **Origin:** Cultivated **Region:** Andean **Dept.:** ANT, BOG, CUN, VAC **Elev.:** 1000–3000 m a.s.l. **IUCN Habitat code:** 1,2,3,4,14 **Conservation status: (IUCN)** NE

7060 *Ruta graveolens* L. (Rutaceae)
IPNI: 775099-1
Common name: ruda, ruda común, ruda de Castilla, ruda de guala, ruda de huerta, ruda silvestre **Habit:** Herb, Shrub **Use:** EU, GS, HF, MA, ME, PO, SU **Origin:** Cultivated **Region:** Andean **Dept.:** ANT, BOG, BOY, CUN, PUT, SAN, VAC **Elev.:** 600–3000 m a.s.l. **Conservation status: (IUCN)** NE

7061 *Swinglea glutinosa* (Blanco) Merr. (Rutaceae)
IPNI: 775281-1
Common name: limoncillo, limón suingla, limón suingle, limón swinglia, naranja swinglia, suingla, suingle, swinglia **Habit:** Shrub **Use:** HF, MA, ME **Origin:** Cultivated **Region:** Amazonia, Andean, Caribbean, Orinoquia **Dept.:** AMA, ANT, BOL, CAL, CAU, CES, COR, CUN, MET, PUT, QUI, RIS, SUC, TOL, VAC **Elev.:** 5–2000 m a.s.l. **IUCN Habitat code:** 1,2,3,14 **Conservation status: (IUCN)** NE

7062 *Triphasia trifolia* (Burm.f.) P.Wilson (Rutaceae)
IPNI: 258565-2
Use: EU, FU, GS, HF, MA, ME **Origin:** Naturalised **IUCN Habitat code:** 1,2,5 **Conservation status: (IUCN)** NE

7063 *Zanthoxylum acuminatum* (Sw.) Sw. (Rutaceae)
IPNI: 775552-1
Habit: Tree **Use:** MA, ME **Origin:** Nativc **Region:** Andean, Caribbean, Pacific **Dept.:** CHO, LAG, TOL **Elev.:** 100–1200 m a.s.l. **IUCN Habitat code:** 1,3,14 **Conservation status: (IUCN)** LC

7064 *Zanthoxylum caribaeum* Lam. (Rutaceae)
IPNI: 775634-1
Common name: amamor, mapurito, mapurito prieto, palo carey, tachuelo **Habit:** Tree **Use:** MA, ME **Origin:** Native **Dept.:** ANT, ATL, BOL, CUN, NSA, VAC **Elev.:** 500–1100 m a.s.l. **IUCN Habitat code:** 1,2,3,14 **Conservation status: (IUCN)** LC

7065 *Zanthoxylum fagara* (L.) Sarg. (Rutaceae)
IPNI: 270659-2
Common name: palo carey, ruda, tachuelo, uña de gato, uña de gavilán, uñaegato, uñegato **Habit:** Shrub, Small tree **Use:** HF, MA, ME **Origin:** Native **Region:** Andean, Caribbean, Orinoquia **Dept.:** ANT, ARA, BOY, CAU, CUN, HUI, MAG, NSA, PUT, SAP, VAC **Elev.:** 0–1600 m a.s.l. **IUCN Habitat code:** 1,2,3,5,14 **Conservation status: (IUCN)** LC

7066 *Zanthoxylum martinicense* (Lam.) DC. (Rutaceae)
IPNI: 775850-1
Common name: espino, justarrazón, snissn wood, tachuelo **Habit:** Tree **Use:** HF, MA, ME **Origin:** Native **Region:** Andean **Dept.:** ANT, CAL, CAU, LAG, VAC **Elev.:** 210–1050 m a.s.l. **IUCN Habitat code:** 1,3,5,14 **Conservation status: (IUCN)** NE

7067 *Zanthoxylum mollissimum* (Engl.) P.Wilson (Rutaceae)
IPNI: 270717-2
Common name: matijón **Habit:** Small tree **Use:** ME **Origin:** Native **Region:** Caribbean **Dept.:** ATL, LAG **Elev.:** 100–200 m a.s.l. **Conservation status: (IUCN)** LC

7068 *Zanthoxylum rhoifolium* Lam. (Rutaceae)
IPNI: 775991-1
Common name: cilantrón, doncel, tacho, tachuelo **Habit:** Small tree, Tree **Use:** HF, MA, ME **Origin:** Native **Region:** Amazonia, Andean, Caribbean **Dept.:** ANT, CAQ, COR, LAG, RIS, VAC **Elev.:** 0–1500 m a.s.l. **IUCN Habitat code:** 1,2,3,4,14 **Conservation status: (IUCN)** LC

7069 *Zanthoxylum rigidum* Humb. & Bonpl. ex Willd. (Rutaceae) **IPNI:** 775996-1 **Common name:** abataque, cedrito, chipuelo, chipuelo aterciopelado, chipuelo hediondo, cilanda, cuero de sapo, doncel, jeta de perro, justarrazón, mapurito, matahijo, matahijón, matijón, moló, moló blanco, muela, tachuelo, tachuelo blanco, ulando, uña de gato, uña de pe **Habit:** Small tree, Tree **Use:** ME **Origin:** Native **Region:** Andean **Dept.:** CAU, CUN, TOL, VAC **Elev.:** 500–1600 m a.s.l. **IUCN Habitat code:** 1,2,3,14 **Conservation status: (IUCN)** LC

7070 *Zanthoxylum schreberi* (J.F.Gmel.) Reynel ex C.Nelson (Rutaceae) **IPNI:** 77146708-1 **Common name:** justarrazón, tachuelo **Habit:** Small tree, Tree **Use:** MA, ME **Origin:** Native **Region:** Caribbean **Dept.:** CUN, MAG, TOL, VAC **Elev.:** 0–1050 m a.s.l. **IUCN Habitat code:** 1,2,3,4,14 **Conservation status: (IUCN)** LC

7071 *Acer negundo* L. (Sapindaceae) **IPNI:** 781412-1 **Use:** EU, FU, MA, ME **Region:** Andean **Dept.:** BOG **Conservation status: (IUCN)** LC

7072 *Aesculus hippocastanum* L. (Sapindaceae) **IPNI:** 781594-1 **Use:** EU, HF, MA, ME, PO **Conservation status: (IUCN)** VU

7073 *Allophylus incanus* Radlk. (Sapindaceae) **IPNI:** 781797-1 **Habit:** Small tree **Use:** HF **Origin:** Native **Region:** Pacific **Dept.:** VAC **Elev.:** 50–50 m a.s.l. **IUCN Habitat code:** 1 **Conservation status: (IUCN)** NE

7074 *Allophylus mollis* (Kunth) Radlk. (Sapindaceae) **IPNI:** 781858-1 **Common name:** pata de loro **Habit:** Subshrub, Small tree, Tree **Use:** HF **Origin:** Endemic **Region:** Andean **Dept.:** CAL, CAU, CUN, PUT, QUI, RIS, TOL **Elev.:** 1350–3200 m a.s.l. **IUCN Habitat code:** 1,2,3,14 **Conservation status: (IUCN)** NE

7075 *Allophylus myrianthus* Radlk. (Sapindaceae) **IPNI:** 297212-2 **Habit:** Tree **Use:** HF **Origin:** Native **Region:** Andean **Dept.:** BOG, NAR, QUI **Elev.:** 300–1800 m a.s.l. **Conservation status: (IUCN)** NE

7076 *Allophylus pilosus* (J.F.Macbr.) A.H.Gentry (Sapindaceae) **IPNI:** 969448-1 **Common name:** mote **Habit:** Tree **Use:** HF **Origin:** Native **Region:** Amazonia **Dept.:** AMA, CAQ **Elev.:** 150–500 m a.s.l. **IUCN Habitat code:** 1,3,14 **Conservation status: (IUCN)** LC

7077 *Allophylus psilospermus* Radlk. (Sapindaceae) **IPNI:** 781899-1 **Habit:** Shrub, Tree **Use:** HF **Origin:** Native **Region:** Caribbean, Pacific **Dept.:** ANT, SAP **Elev.:** 80–500 m a.s.l. **IUCN Habitat code:** 1,2,3,14 **Conservation status: (IUCN)** LC

7078 *Allophylus racemosus* Sw. (Sapindaceae) **IPNI:** 781904-1 **Habit:** Shrub, Small tree **Use:** HF, MA, ME **Origin:** Native **Region:** Amazonia, Caribbean, Orinoquia **Dept.:** BOL, CAS, MAG **Elev.:** 200–700 m a.s.l. **IUCN Habitat code:** 1,2,3,14 **Conservation status: (IUCN)** LC

7079 *Billia rosea* (Planch. & Linden) C.Ulloa & P.M.Jørg. (Sapindaceae) **IPNI:** 1021491-1 **Common name:** achotillo, capulí, cariseco, curapia, duro, fruta de indio, manzanillo, manzano, manzano de monte, maru, mutundó, pechugo, rayo, tauquin, treshojas **Habit:** Tree **Use:** AF, EU, MA **Origin:** Native **Region:** Andean **Dept.:** ANT, BOG, BOY, CAL, CAQ, CAU, CHO, CUN, HUI, LAG, MAG, MET, NAR, PUT, QUI, RIS, SAN, TOL, VAC **Elev.:** 600–3150 m a.s.l. **IUCN Habitat code:** 1,2,3,4,8,14 **Conservation status: (IUCN)** LC

7080 *Blighia sapida* K.D.Koenig (Sapindaceae) **IPNI:** 782152-1 **Common name:** aquí, bienmesabe, biernesabe, huevo vegetal, pan, pan y quesito, quesito, seso vegetal **Habit:** Tree **Use:** EU, FU, GS, HF, MA, ME, PO **Origin:** Cultivated **Region:** Andean, Caribbean **Dept.:** ANT, ATL, BOL, HUI, MAG, VAC **Elev.:** 400–970 m a.s.l. **IUCN Habitat code:** 1,2,3,5,14 **Conservation status: (IUCN)** LC

7081 *Cardiospermum corindum* L. (Sapindaceae) **IPNI:** 782195-1 **Common name:** farolito, globito, topotopo, topotoropo, vejiga **Habit:** Herb, Climbing **Use:** ME **Origin:** Native **Region:** Andean, Caribbean **Dept.:** ANT, ATL, LAG, MAG, SAN, VAC **Elev.:** 0–1400 m a.s.l. **IUCN Habitat code:** 1,2,3,5,14 **Conservation status: (IUCN)** NE

7082 *Cardiospermum grandiflorum* Sw. (Sapindaceae) **IPNI:** 782205-1 **Common name:** bejuco de globo, farolito **Habit:** Climbing **Use:** EU, HF, MA, ME, SU **Origin:** Native **Region:** Amazonia, Andean, Caribbean, Pacific **Dept.:** ANT, ATL, BOL, CAL, CAQ, CAS, CAU, CUN, HUI, MAG, MET, NAR, NSA, PUT, SAN, SUC, TOL, VAC **Elev.:** 76-1800 m.a.s.l. **IUCN Habitat code:** 1,2,3,4,14 **Conservation status: (IUCN)** NE **(CPSG)** Potential LC

7083 *Cardiospermum halicacabum* L. (Sapindaceae) **IPNI:** 782206-1 **Common name:** farolito, farolitos, globitos, opio, shuru wina, topo toropo, ubasoshi, yeyomo **Habit:** Herb, Climbing **Use:** AF, EU, HF, IF, MA, ME, PO, SU **Origin:** Native **Region:** Amazonia, Andean, Caribbean **Dept.:** AMA, ANT, ATL, COR, GUA, HUI, LAG, MAG, MET, NAR, PUT, SAN, VAC, VAU, VID **Elev.:** 20-1800 m.a.s.l. **IUCN Habitat code:** 1,2,3,4,5,14 **Conservation status: (IUCN)** LC **(CPSG)** Potential LC

7084 *Cardiospermum microcarpum* Kunth (Sapindaceae) **IPNI:** 782221-1 **Use:** HF, ME **Origin:** Native **IUCN Habitat code:** 1 **Conservation status: (IUCN)** NE

7085 *Cupania americana* L. (Sapindaceae) **IPNI:** 319175-2 **Common name:** arevalo, arévalo, cacho de venado, carapo, cariseco, chircagüero, coropo, guacamayo, guacharaco, guara, guara blanca, mestizo **Habit:** Small tree, Tree **Use:** HF, MA, ME **Origin:** Native **Region:** Andean, Caribbean, Pacific **Dept.:** ANT, CAS, CHO, CUN, GUA, HUI, LAG, MAG, MET, NSA, QUI, SAN, VAC **Elev.:** 80-2340 m.a.s.l. **IUCN Habitat code:** 1,2,3,4,14 **Conservation status: (IUCN)** NE **(CPSG)** Potential LC

7086 *Cupania hirsuta* Radlk. (Sapindaceae) **IPNI:** 782448-1 **Habit:** Tree **Use:** ME **Origin:** Native **Region:** Pacific **Dept.:** ANT **Elev.:** 150–150 m a.s.l. **IUCN Habitat code:** 1,2,14 **Conservation status: (IUCN)** LC

7087 *Cupania livida* (Radlk.) Croat (Sapindaceae) **IPNI:** 72250-2 **Habit:** Small tree, Tree **Use:** HF **Origin:** Native **Region:** Amazonia, Andean, Pacific **Dept.:** AMA, ANT, BOL, CHO, VAC **Elev.:** 50–2500 m a.s.l. **IUCN Habitat code:** 1,3,5,14 **Conservation status: (IUCN)** LC

7088 *Cupania rufescens* Triana & Planch. (Sapindaceae) **IPNI:** 782543-1 **Common name:** guacharaco colorado **Habit:** Tree **Use:** ME **Origin:** Native **Region:** Andean **Dept.:** HUI, MAG **Elev.:** 200–800 m a.s.l. **IUCN Habitat code:** 1 **Conservation status: (IUCN)** LC

7089 *Cupania scrobiculata* Rich. (Sapindaceae) **IPNI:** 782556-1 **Common name:** mestizo, temutorabueaci **Habit:** Shrub, Small tree, Tree **Use:** AF, FU, HF, MA, ME, SU **Origin:** Native **Region:** Amazonia, Andean, Guiana Shield, Caribbean, Pacific **Dept.:** AMA, ANT, BOL, BOY, CAQ, CHO, GUV, MAG, MET, SAN, VAC, VAU **Elev.:** 100-2350 m.a.s.l. **IUCN Habitat code:** 1,2,3,4,5,14 **Conservation status: (IUCN)** LC **(CPSG)** Potential LC

7090 *Dilodendron costaricense* (Radlk.) A.H.Gentry & Steyerm. (Sapindaceae) **IPNI:** 944887-1 **Common name:** carbonero, chicharrón, cilantrón, harino, iguano, loro **Habit:** Small tree **Use:** EU, HF, MA, SU **Origin:** Native **Region:** Caribbean, Pacific **Dept.:** ANT, BOL, CAL, CHO, SUC **Elev.:** 10–500 m a.s.l. **IUCN Habitat code:** 1,2,3,14 **Conservation status: (IUCN)** LC

7091 *Dodonaea viscosa* Jacq. (Sapindaceae) **IPNI:** 30058367-2 **Common name:** ayullo, ayuno, chamana, chamano, chamarro, coca silvestre, escobo, florida hopbush, hayo, hayuelo, hopshrup, jayo **Habit:** Shrub **Use:** AF, EU, FU, HF, IF, MA, ME, PO, SU **Origin:** Native **Region:** Andean, Caribbean **Dept.:** ANT, BOG, BOL, BOY, CES, CUN, HUI, LAG, MAG, MET, NAR, NSA, SAP, SAN **Elev.:** 250-3900 m.a.s.l. **IUCN Habitat code:** 1,2,3,4,5,14 **Conservation status: (IUCN)** LC **(CPSG)** Potential LC

7092 *Exothea paniculata* (Juss.) Radlk. (Sapindaceae) **IPNI:** 102890-2 **Habit:** Tree **Use:** MA **Origin:** Native **Dept.:** ANT **Elev.:** 325–325 m a.s.l. **IUCN Habitat code:** 1,3 **Conservation status: (IUCN)** LC

7093 *Matayba camptoneura* Radlk. (Sapindaceae) **IPNI:** 783591-1 **Common name:** pata de paujil, pataepajuil, quiebramachete, varablanca **Habit:** Tree **Use:** HF

Origin: Native **Region:** Andean, Caribbean, Orinoquia **Dept.:** CAS **Elev.:** 100–1400 m a.s.l. **IUCN Habitat code:** 1,2,3,14 **Conservation status:** **(IUCN)** LC

7094 *Matayba elegans* Radlk. (Sapindaceae) **IPNI:** 783598-1 **Common name:** bálsamo, cachicarnero, guacharaco, joropo, palo tigre, patepaujil **Habit:** Tree **Use:** HF **Origin:** Native **Region:** Amazonia, Andean, Guiana Shield, Orinoquia, Pacific **Dept.:** ANT, BOY, MET, PUT, SAN, VAC, VAU, VID **Elev.:** 50–2450 m a.s.l. **IUCN Habitat code:** 1,2,3,4,5,14 **Conservation status:** **(IUCN)** LC

7095 *Matayba guianensis* Aubl. (Sapindaceae) **IPNI:** 783604-1 **Common name:** paloerrayo **Habit:** Shrub, Small tree, Tree **Use:** MA, ME **Origin:** Native **Region:** Amazonia, Andean, Orinoquia, Pacific **Dept.:** AMA, ANT, CAS, CHO, MET, TOL, VAU, VID **Elev.:** 100–1000 m a.s.l. **IUCN Habitat code:** 1,2,3,4,5,14 **Conservation status:** **(IUCN)** LC

7096 *Matayba inelegans* Radlk. (Sapindaceae) **IPNI:** 783606-1 **Common name:** jubuicacurameaci **Habit:** Shrub, Small tree **Use:** HF **Origin:** Native **Region:** Amazonia, Guiana Shield **Dept.:** AMA, CAQ, GUA, VAU **Elev.:** 15–550 m a.s.l. **IUCN Habitat code:** 1,3,4,5,14 **Conservation status:** **(IUCN)** LC

7097 *Matayba opaca* Radlk. (Sapindaceae) **IPNI:** 783625-1 **Use:** MA **Origin:** Native **IUCN Habitat code:** 1,3 **Conservation status:** **(IUCN)** LC

7098 *Matayba oppositifolia* (A.Rich.) Britton (Sapindaceae) **IPNI:** 154460-2 **Use:** ME **IUCN Habitat code:** 1 **Conservation status:** **(IUCN)** LC

7099 *Matayba purgans* Radlk. (Sapindaceae) **IPNI:** 783633-1 **Common name:** belanae, bibisinae, coquindo cimarrón, guayoba silvestre, huesito, hueso, mamoncillo, mebaganoó, molinillo, oobeoba, podai ijie juici, tiirai, trestablas **Habit:** Shrub, Small tree, Tree **Use:** HF, MA, ME, SU **Origin:** Native **Region:** Andean, Guiana Shield, Orinoquia, Pacific **Dept.:** AMA, ANT, CAQ, CAS, GUV, MET, PUT, VAU, VID **Elev.:** 50–1200 m a.s.l. **IUCN Habitat code:** 1,3,4,5,14 **Conservation status:** **(IUCN)** LC

7100 *Matayba scrobiculata* Radlk. (Sapindaceae) **IPNI:** 783637-1 **Common name:** asaó, guacharaco, mangle arroyero, árbol de indio **Habit:** Tree **Use:** AF, MA **Origin:** Native **Region:** Andean, Caribbean **Dept.:** ANT, ATL, BOY, CES, LAG, MAG **Elev.:** 10–925 m a.s.l. **Conservation status:** **(IUCN)** LC

7101 *Melicoccus antioquensis* Acev.-Rodr. (Sapindaceae) **IPNI:** 50426541-2 **Habit:** Tree **Use:** HF **Origin:** Endemic **Dept.:** ANT **Elev.:** 325-425 m.a.s.l. **IUCN Habitat code:** 3 **Conservation status:** **(IUCN)** CR **(CPSG)** CR

7102 *Melicoccus bijugatus* Jacq. (Sapindaceae) **IPNI:** 783682-1 **Common name:** casero, macao, maco, mamoncillo, mamoncillo de monte, mamoncillo macho, mamón, mamón común, muco, quenepa **Habit:** Tree **Use:** AF, EU, FU, GS, HF, MA, ME **Origin:** Native, Cultivated **Region:** Andean, Caribbean, Pacific **Dept.:** ANT, BOL, CES, CHO, COR, CUN, HUI, LAG, MAG, QUI, SAP, SUC, TOL, VAC **Elev.:** 20-1500 m.a.s.l. **IUCN Habitat code:** 1,2,3,4,14 **Conservation status:** **(IUCN)** LC **(CPSG)** Potential LC

7103 *Melicoccus oliviformis* Kunth (Sapindaceae) **IPNI:** 783676-1 **Common name:** cotoperi, cotoplí, cotoplís, cotoprí, cotoprís, cutuplí, guacharaco, mamoncillo silvestre, mamón, mamón cotopli, mamón cotoprix, mamón cotoprí, mamón cutuplí, mamón de María, mamón de mico, mamón de puerco **Habit:** Tree **Use:** AF, EU, HF, MA, ME **Origin:** Native **Region:** Caribbean **Dept.:** ANT, ATL, BOL, CES, LAG, MAG, SUC **Elev.:** 0–800 m a.s.l. **IUCN Habitat code:** 1,2,3,5,6,14 **Conservation status:** **(IUCN)** NE

7104 *Paullinia alata* (Ruiz & Pav.) G.Don (Sapindaceae) **IPNI:** 784051-1 **Common name:** bejuco cruceta, bejuco tripadanta, bejuco tripedanta, yoco colorado, yoco cusumbe **Habit:** Liana, Climbing **Use:** HF, ME, PO **Origin:** Native **Region:** Amazonia, Andean, Caribbean, Orinoquia, Pacific **Dept.:** AMA, ANT, ARA, BOL, BOY, CAQ, CAS, CHO, COR, CUN, HUI, MET, PUT, QUI, TOL **Elev.:** 2-1625 m.a.s.l. **IUCN Habitat code:** 1,2,3,4,5,14 **Conservation status:** **(IUCN)** NE **(CPSG)** Potential LC

7105 *Paullinia anodonta* Radlk. (Sapindaceae) **IPNI:** 185586-2 **Habit:** Climbing **Use:** ME **Origin:** Native **Region:** Pacific **Dept.:** CHO **Elev.:** 40–40 m a.s.l. **IUCN Habitat code:** 1 **Conservation status:** **(IUCN)** NE

7106 *Paullinia bracteosa* Radlk. (Sapindaceae) **IPNI:** 185601-2 **Habit:** Liana, Climbing **Use:** HF **Origin:** Native **Region:** Amazonia, Andean, Orinoquia, Pacific **Dept.:** AMA, ANT, CHO, CUN, MET, QUI, SAN **Elev.:** 0–1420 m a.s.l. **IUCN Habitat code:** 1,2,3,14 **Conservation status:** **(IUCN)** NE

7107 *Paullinia clavigera* Schltdl. (Sapindaceae) **IPNI:** 784100-1 **Common name:** tsukuiiniibo **Habit:** Liana, Climbing **Use:** ME **Origin:** Native **Region:** Amazonia, Andean, Caribbean, Pacific **Dept.:** AMA, ANT, BOL, CHO, VAC **Elev.:** 50–1070 m a.s.l. **IUCN Habitat code:** 1,4,14 **Conservation status:** **(IUCN)** NE

7108 *Paullinia costata* Schltdl. & Cham. (Sapindaceae) **IPNI:** 784108-1 **Habit:** Climbing **Use:** ME **Origin:** Native **Region:** Andean **Dept.:** CHO **Elev.:** 800–1200 m a.s.l. **Conservation status:** **(IUCN)** NE

7109 *Paullinia cupana* Kunth (Sapindaceae) **IPNI:** 784110-1 **Common name:** auranazeiro, cupana, guaranazeiro, guaraná, rabo de iguana, yocoo **Habit:** Liana, Climbing **Use:** HF, MA, ME, SU **Origin:** Native **Region:** Amazonia, Guiana Shield, Orinoquia **Dept.:** AMA, MET, VAU **Elev.:** 100–100 m a.s.l. **IUCN Habitat code:** 1,2 **Conservation status:** **(IUCN)** NE

7110 *Paullinia cururu* L. (Sapindaceae) **IPNI:** 784115-1 **Common name:** azucarerito, barbasco, bejuco de San Pedro, bejuco mulato, bollo mohoso, cupani, cupania, ojito de nene, ojo de perdiz, perita de monte, yocoó **Habit:** Liana, Climbing **Use:** AF,ME **Origin:** Native **Region:** Caribbean **Dept.:** ANT, ATL, BOL, BOY, MAG, SAP, SAN, SUC **Elev.:** 0–900 m a.s.l. **IUCN Habitat code:** 1,2,3,5,14 **Conservation status:** **(IUCN)** NE

7111 *Paullinia elegans* Cambess. (Sapindaceae) **IPNI:** 784130-1 **Habit:** Climbing **Use:** HF, ME **Origin:** Native **Region:** Amazonia **Dept.:** AMA, PUT **Elev.:** 100–250 m a.s.l. **IUCN Habitat code:** 1,2,3 **Conservation status:** **(IUCN)** NE

7112 *Paullinia emetica* R.E.Schult. (Sapindaceae) **IPNI:** 185640-2 **Habit:** Climbing **Use:** ME **Origin:** Endemic **Region:** Amazonia **Dept.:** VAU **Elev.:** 350–350 m a.s.l. **Conservation status:** **(IUCN)** NE

7113 *Paullinia eriocarpa* Triana & Planch. (Sapindaceae) **IPNI:** 784136-1 **Habit:** Liana, Climbing **Use:** HF **Origin:** Native **Region:** Amazonia, Andean, Caribbean, Pacific **Dept.:** AMA, ANT, BOL, CHO, CUN, MET, PUT **Elev.:** 0–800 m a.s.l. **IUCN Habitat code:** 1,2,3,4,14 **Conservation status:** **(IUCN)** NE

7114 *Paullinia fuscescens* Kunth (Sapindaceae) **IPNI:** 784152-1 **Common name:** bollolimpio, ojo de perdiz **Habit:** Liana, Climbing **Use:** HF, ME **Origin:** Native **Region:** Andean, Caribbean, Orinoquia **Dept.:** ANT, ATL, BOL, CAQ, CAS, CAU, COR, HUI, MAG, MET, NSA, SAP, SAN, TOL, VAC **Elev.:** 0-2950 m.a.s.l. **IUCN Habitat code:** 1,2,3,4,5,14 **Conservation status:** **(IUCN)** NE **(CPSG)** Potential LC

7115 *Paullinia leiocarpa* Griseb. (Sapindaceae) **IPNI:** 784185-1 **Common name:** bejuco costillo, bejuco cruceta **Habit:** Liana, Climbing **Use:** MA, PO **Origin:** Native **Region:** Andean, Caribbean **Dept.:** ANT, BOL, LAG, MAG, SAN, SUC **Elev.:** 10–1420 m a.s.l. **IUCN Habitat code:** 1,2,3,14 **Conservation status:** **(IUCN)** NE

7116 *Paullinia macrophylla* Kunth (Sapindaceae) **IPNI:** 784191-1 **Common name:** barbasco, bejuco costillo, bejuco prieto, ojito de nene, ojo de nene **Habit:** Liana, Climbing **Use:** HF **Origin:** Endemic **Region:** Caribbean **Dept.:** ANT, ATL, MAG, NSA **Elev.:** 100–1260 m a.s.l. **IUCN Habitat code:** 1,2,3,14 **Conservation status:** **(IUCN)** NE

7117 *Paullinia obovata* (Ruiz & Pav.) Pers. (Sapindaceae) **IPNI:** 784226-1 **Habit:** Liana, Climbing **Use:** HF **Origin:** Native **Region:** Andean **Dept.:** ANT, MET, SAN, VAC **Elev.:** 325–

1930 m a.s.l. **IUCN Habitat code:** 1,3,5,14 **Conservation status: (IUCN)** NE

7118 *Paullinia pachycarpa* Benth. (Sapindaceae) **IPNI:** 784232-1 **Habit:** Liana, Climbing **Use:** HF **Origin:** Native **Region:** Amazonia, Andean, Orinoquia **Dept.:** AMA, BOL, CAQ, MET, NSA, PUT **Elev.:** 80–1440 m a.s.l. **IUCN Habitat code:** 1,3,4,5,14 **Conservation status: (IUCN)** NE

7119 *Paullinia pinnata* L. (Sapindaceae) **IPNI:** 185718-2 **Common name:** barbasco, carneasada **Habit:** Liana, Climbing **Use:** AF, GS, HF, MA, ME, PO **Origin:** Native **Region:** Andean, Caribbean, Pacific **Dept.:** ANT, BOL, CHO, MAG, NSA, TOL **Elev.:** 0–1450 m a.s.l. **IUCN Habitat code:** 1,3,4,5,14 **Conservation status: (IUCN)** NE

7120 *Paullinia pterophylla* Triana & Planch. (Sapindaceae) **IPNI:** 784253-1 **Common name:** yoco **Habit:** Liana, Climbing **Use:** ME **Origin:** Native **Region:** Amazonia, Caribbean, Orinoquia, Pacific **Dept.:** CAQ, CHO, MET **Elev.:** 20–450 m a.s.l. **IUCN Habitat code:** 1,3,4,14 **Conservation status: (IUCN)** NE

7121 *Paullinia rubiginosa* Cambess. (Sapindaceae) **IPNI:** 784268-1 **Habit:** Climbing **Use:** HF **Origin:** Native **Region:** Andean, Pacific **Dept.:** ANT **Elev.:** 20–925 m a.s.l. **IUCN Habitat code:** 1,3,14 **Conservation status: (IUCN)** NE

7122 *Paullinia sphaerocarpa* Rich. ex Juss. (Sapindaceae) **IPNI:** 784289-1 **Habit:** Climbing **Use:** HF, MA **Origin:** Native **Region:** Orinoquia **Dept.:** CAS **Elev.:** 300–300 m a.s.l. **IUCN Habitat code:** 1,14 **Conservation status: (IUCN)** NE

7123 *Paullinia spicata* Benth. (Sapindaceae) **IPNI:** 784290-1 **Habit:** Liana, Climbing **Use:** HF, PO **Origin:** Native **Region:** Orinoquia, Pacific **Dept.:** ANT, CAL, CHO, MET, SAN **Elev.:** 10–240 m a.s.l. **IUCN Habitat code:** 1,14 **Conservation status: (IUCN)** NE

7124 *Paullinia yoco* R.E.Schult. & Killip (Sapindaceae) **IPNI:** 185782-2 **Common name:** turuca yoco, yagé-yoco, yo'co, yoco, yocó **Habit:** Liana, Climbing **Use:** HF, ME, SU **Origin:** Native **Region:** Amazonia, Andean **Dept.:** PUT **Elev.:** 325–1150 m a.s.l. **IUCN Habitat code:** 1,3,4,14 **Conservation status: (IUCN)** NE

7125 *Sapindus saponaria* L. (Sapindaceae) **IPNI:** 30035906-2 **Common name:** chambimba, chambimbe, chambimbo, chocho, chumbimbe, chumbimbo, chumbimbomichú, cuchimbo, cueresapo, guamo, jabo, jaboncillo, jabón, michu, michú, parapara, pepo, pipo, siminduñe, siminuñe, árbol de jabón **Habit:** Small tree, Tree **Use:** AF, EU, FU, HF, MA, ME, PO, SU **Origin:** Native **Region:** Andean, Caribbean, Pacific **Dept.:** ANT, ATL, BOL, BOY, CES, CHO, COR, CUN, HUI, LAG, MAG, MET, NSA, SUC, VAC **Elev.:** 0–2150 m a.s.l. **IUCN Habitat code:** 1,2,3,4,14 **Conservation status: (IUCN)** LC

7126 *Sapindus trifoliatus* Turcz. (Sapindaceae) **IPNI:** 784698-1 **Use:** EU, GS, HF, IF, MA, ME, PO **Origin:** Endemic **Conservation status: (IUCN)** NE

7127 *Serjania caracasana* (Jacq.) Willd. (Sapindaceae) **IPNI:** 784925-1 **Habit:** Liana **Use:** ME **Origin:** Native **Dept.:** ANT **Elev.:** 500–500 m a.s.l. **Conservation status: (IUCN)** NE

7128 *Serjania clematidea* Triana & Planch. (Sapindaceae) **IPNI:** 784933-1 **Common name:** bejuco gonorrea, enredadera, mutsualiunubo **Habit:** Climbing **Use:** ME **Origin:** Native **Region:** Amazonia, Andean, Orinoquia **Dept.:** ANT, CAL, CAQ, CAS, CUN, MET, NAR, TOL, VID **Elev.:** 140–1820 m a.s.l. **IUCN Habitat code:** 1,2,3,4,14 **Conservation status: (IUCN)** NE

7129 *Serjania communis* Cambess. (Sapindaceae) **IPNI:** 784938-1 **Habit:** Climbing **Use:** ME **Origin:** Native **Region:** Amazonia, Andean, Guiana Shield, Caribbean **Dept.:** AMA, CAQ, CUN, MAG, MET, PUT, TOL **Elev.:** 80–1600 m a.s.l. **IUCN Habitat code:** 1,3,5,14 **Conservation status: (IUCN)** NE

7130 *Serjania curassavica* (L.) Radlk. (Sapindaceae) **IPNI:** 784949-1 **Habit:** Climbing **Use:** ME **Origin:** Native **Region:** Caribbean **Dept.:** ATL, BOL **Elev.:** 50–100 m a.s.l. **IUCN Habitat code:** 1,2 **Conservation status: (IUCN)** NE

7131 *Serjania glabrata* Kunth (Sapindaceae) **IPNI:** 784992-1 **Habit:** Liana, Climbing **Use:** ME, PO **Origin:** Native **Region:** Amazonia **Dept.:** CAQ **Elev.:** 450–450 m a.s.l. **IUCN Habitat code:** 1,3,14 **Conservation status: (IUCN)** NE

7132 *Serjania membranacea* Splitg. (Sapindaceae) **IPNI:** 785059-1 **Common name:** rabo de iguana, tocá **Habit:** Liana, Climbing **Use:** ME, PO **Origin:** Native **Region:** Andean, Pacific **Dept.:** ANT, BOY, CHO, HUI, MET, VAC **Elev.:** 0–1300 m a.s.l. **Conservation status: (IUCN)** NE

7133 *Serjania mexicana* (L.) Willd. (Sapindaceae) **IPNI:** 785062-1 **Common name:** bejuco buendía, bejuco esquinero macho, bejuco güendía, bejuco pegahueso, bejuco sierra, ojo de nene, rabo de iguana, raya, sardinato **Habit:** Liana, Climbing **Use:** MA, ME, PO **Origin:** Native **Region:** Andean, Caribbean **Dept.:** ANT, ATL, BOL, CAL, CUN, HUI, LAG, MAG, SUC, TOL **Elev.:** 10–1460 m a.s.l. **IUCN Habitat code:** 1,2,3,4,5,14 **Conservation status: (IUCN)** NE

7134 *Serjania paucidentata* DC. (Sapindaceae) **IPNI:** 234951-2 **Habit:** Climbing **Use:** MA, PO **Origin:** Native **Region:** Amazonia, Caribbean **Dept.:** BOL, CAQ **Elev.:** 150–450 m a.s.l. **IUCN Habitat code:** 2,3,14 **Conservation status: (IUCN)** NE

7135 *Talisia cerasina* (Benth.) Radlk. (Sapindaceae) **IPNI:** 785288-1 **Habit:** Shrub, Small tree, Tree **Use:** HF, ME **Origin:** Native **Region:** Amazonia, Guiana Shield, Pacific **Dept.:** ANT, CHO, GUV, MET, PUT **Elev.:** 200–855 m a.s.l. **IUCN Habitat code:** 1,2,3,4,14 **Conservation status: (IUCN)** LC

7136 *Talisia clathrata* Radlk. (Sapindaceae) **IPNI:** 785289-1 **Habit:** Shrub, Small tree **Use:** HF **Origin:** Native **Region:** Amazonia **Dept.:** AMA **Elev.:** 200–200 m a.s.l. **IUCN Habitat code:** 1 **Conservation status: (IUCN)** LC

7137 *Talisia croatii* Acev.-Rodr. (Sapindaceae) **IPNI:** 50426551-2 **Habit:** Shrub, Tree **Use:** HF **Origin:** Native **Region:** Andean, Caribbean, Pacific **Dept.:** ANT, BOL, CHO, VAC **Elev.:** 30–1570 m a.s.l. **IUCN Habitat code:** 1,2,5,14 **Conservation status: (IUCN)** LC

7138 *Talisia esculenta* (A.St.-Hil., A.Juss. & Cambess.) Radlk. (Sapindaceae) **IPNI:** 785297-1 **Use:** HF, MA, ME, PO **IUCN Habitat code:** 2 **Conservation status: (IUCN)** NE

7139 *Talisia firma* Radlk. (Sapindaceae) **IPNI:** 785299-1 **Habit:** Small tree, Tree **Use:** HF **Origin:** Native **Region:** Amazonia, Guiana Shield **Dept.:** AMA, GUA, VAU **Elev.:** 75–260 m a.s.l. **IUCN Habitat code:** 1,4,5,14 **Conservation status: (IUCN)** LC

7140 *Talisia hexaphylla* Vahl (Sapindaceae) **IPNI:** 785306-1 **Habit:** Tree **Use:** HF **Origin:** Native **Region:** Andean, Pacific **Dept.:** ANT, CAL, CHO, SAN **Elev.:** 10–770 m a.s.l. **IUCN Habitat code:** 1,3,14 **Conservation status: (IUCN)** LC

7141 *Talisia japurensis* (C.DC.) Radlk. (Sapindaceae) **IPNI:** 249229-2 **Habit:** Shrub **Use:** HF **Origin:** Native **Region:** Amazonia **Dept.:** AMA **Elev.:** 80–100 m a.s.l. **IUCN Habitat code:** 1 **Conservation status: (IUCN)** NE

7142 *Talisia macrophylla* (Mart.) Radlk. (Sapindaceae) **IPNI:** 785313-1 **Common name:** mestizo **Habit:** Shrub, Small tree, Tree **Use:** HF **Origin:** Native **Region:** Amazonia, Andean, Guiana Shield, Pacific **Dept.:** AMA, ANT, CAQ, CHO, GUA, VAC **Elev.:** 10–300 m a.s.l. **IUCN Habitat code:** 1,3,5,14 **Conservation status: (IUCN)** LC

7143 *Talisia nervosa* Radlk. (Sapindaceae) **IPNI:** 249235-2 **Habit:** Small tree, Tree **Use:** HF **Origin:** Native **Region:** Amazonia, Andean, Pacific **Dept.:** ANT, BOL, CAQ, CHO, VAC **Elev.:** 5–940 m a.s.l. **IUCN Habitat code:** 1,5,14 **Conservation status: (IUCN)** LC

7144 *Talisia setigera* Radlk. (Sapindaceae)
IPNI: 249244-2
Habit: Small tree **Use:** HF **Origin:** Native **Region:** Andean **Dept.:** ANT **Elev.:** 600–600 m a.s.l. **IUCN Habitat code:** 1,2 **Conservation status: (IUCN)** EN

7145 *Talisia stricta* (H.Karst. & Triana) Radlk. (Sapindaceae)
IPNI: 785342-1
Common name: matapuerco **Habit:** Shrub, Small tree, Tree **Use:** ME **Origin:** Native **Dept.:** CUN, TOL **Elev.:** 265-616 m.a.s.l. **IUCN Habitat code:** 1,3,14 **Conservation status: (IUCN)** VU **(CPSG)** VU

7146 *Toulicia guianensis* Aubl. (Sapindaceae)
IPNI: 785498-1
Habit: Tree **Use:** MA **Origin:** Native **Region:** Orinoquia **Dept.:** VID **Elev.:** 85–85 m a.s.l. **Conservation status: (IUCN)** LC

7147 *Urvillea ulmacea* Kunth (Sapindaceae)
IPNI: 785593-1
Habit: Liana, Climbing **Use:** ME **Origin:** Native **Region:** Andean, Caribbean, Orinoquia **Dept.:** ANT, ATL, BOL, CAL, CES, CUN, HUI, LAG, MET, SAN, SUC, TOL, VAC, VID **Elev.:** 0-1420 m.a.s.l. **IUCN Habitat code:** 1,2,3,14 **Conservation status: (IUCN)** NE **(CPSG)** Potential LC

7148 *Vouarana guianensis* Aubl. (Sapindaceae)
IPNI: 785604-1
Habit: Tree **Use:** MA **Origin:** Native **Region:** Amazonia, Guiana Shield **Dept.:** AMA, GUA, MET, PUT **Elev.:** 100–450 m a.s.l. **IUCN Habitat code:** 1,3,14 **Conservation status: (IUCN)** LC

7149 *Castela erecta* Turpin (Simaroubaceae)
IPNI: 813657-1
Common name: chaparro amargoso, juluwa, jurva, mancamulo, retama, revienta puerco, revientapuerco **Habit:** Shrub **Use:** ME **Origin:** Native **Region:** Caribbean **Dept.:** LAG **Elev.:** 10–50 m a.s.l. **IUCN Habitat code:** 1,2,3,14 **Conservation status: (IUCN)** LC

7150 *Castela tortuosa* Liebm. (Simaroubaceae)
IPNI: 813672-1
Use: ME **Origin:** Native **Conservation status: (IUCN)** DD

7151 *Homalolepis cedron* (Planch.) Devecchi & Pirani (Simaroubaceae)
IPNI: 77190108-1
Common name: cedrón, mapurito **Habit:** Shrub, Small tree, Tree **Use:** HF, MA, ME **Origin:** Native, Cultivated **Region:** Amazonia, Orinoquia, Pacific **Dept.:** AMA, ANT, BOL, BOY, CHO, COR, GUV, SAN, VID **Elev.:** 0–700 m a.s.l. **IUCN Habitat code:** 1,2,3,4,5,14 **Conservation status: (IUCN)** LC

7152 *Homalolepis ferruginea* (A.St.-Hil.) Devecchi & Pirani (Simaroubaceae)
IPNI: 77190094-1
Habit: Tree **Use:** MA **Origin:** Native **Region:** Caribbean **Dept.:** LAG **Elev.:** 10–10 m a.s.l. **Conservation status: (IUCN)** LC

7153 *Picrasma excelsa* (Sw.) Planch. (Simaroubaceae)
IPNI: 813955-1
Common name: coralito, cuasio, cáscara amarga, fresno, fresno amargo **Habit:** Tree **Use:** HF, MA, ME, PO **Origin:** Native **Region:** Andean, Orinoquia **Dept.:** MET, RIS, VAC **Elev.:** 450–2100 m a.s.l. **IUCN Habitat code:** 1,3,14 **Conservation status: (IUCN)** VU

7154 *Picrolemma huberi* Ducke (Simaroubaceae)
IPNI: 813977-1
Common name: barbasco, cedro rojo, cedrón, cedrón rojo, madibia, valdivia **Habit:** Shrub, Small tree, Tree **Use:** ME **Origin:** Native **Region:** Andean **Dept.:** ANT, CUN, VAC **Elev.:** 500–1500 m a.s.l. **IUCN Habitat code:** 1,3,14 **Conservation status: (IUCN)** NE

7155 *Picrolemma sprucei* Benth. & Hook.f. (Simaroubaceae)
IPNI: 813979-1
Common name: amargo, barbasco **Habit:** Tree **Use:** ME **Origin:** Native **Region:** Amazonia, Guiana Shield **Dept.:** AMA, CAQ, PUT **Elev.:** 200–300 m a.s.l. **IUCN Habitat code:** 1,5 **Conservation status: (IUCN)** LC

7156 *Quassia amara* L. (Simaroubaceae)
IPNI: 814006-1
Common name: aceituno, acuasia, aguasio, cedrillo, cimarrú, contracruceta, contracruceto, creceto morado, cruceta, crucete, cruceto, cruceto blanco, cruceto morado, cuasia, cuasia amarga, guabito, guavito amargo, hombre grande, hombregrande, machaco, palo amarillo, pa **Habit:** Shrub, Small tree **Use:** EU, HF, MA, ME, PO **Origin:** Native, Cultivated **Region:** Caribbean, Pacific **Dept.:** ANT, BOL, CHO, MAG, SUC **Elev.:** 10–400 m a.s.l. **IUCN Habitat code:** 1,2,3,5,14 **Conservation status: (IUCN)** LC

7157 *Simaba guianensis* Aubl. (Simaroubaceae)
IPNI: 814098-1
Common name: ají de pescado **Habit:** Tree **Use:** HF, MA **Origin:** Native **Region:** Amazonia, Guiana Shield **Dept.:** AMA, CAQ, PUT, VAU **Elev.:** 200–220 m a.s.l. **IUCN Habitat code:** 1,4,5 **Conservation status: (IUCN)** LC

7158 *Simaba orinocensis* Kunth (Simaroubaceae)
IPNI: 814110-1
Habit: Shrub **Use:** HF, MA **Origin:** Native **Region:** Amazonia, Guiana Shield **Dept.:** AMA, GUA, GUV **Elev.:** 120–120 m a.s.l. **IUCN Habitat code:** 1,2,4,14 **Conservation status: (IUCN)** LC

7159 *Simaba polyphylla* (Cavalcante) W.W.Thomas (Simaroubaceae)
IPNI: 285037-2
Common name: laurel **Habit:** Tree **Use:** HF **Origin:** Native **Region:** Amazonia, Guiana Shield **Dept.:** AMA, CAQ **Elev.:** 200–220 m a.s.l. **IUCN Habitat code:** 1,5 **Conservation status: (IUCN)** LC

7160 *Simarouba amara* Aubl. (Simaroubaceae)
IPNI: 814124-1
Common name: aceituno, aliso, amargo, arenillo blanco, caratero, cedrillo, chapul, cimarrú, garza, luna, machaco, manciribo, marfil, marfil otara, marupá, nogal, olivo silvestre, palo blanco, papelillo, pavito, pavo, perico, proporó, ruda, tara **Habit:** Small tree, Tree **Use:** AF, EU, FU, HF, MA, ME, PO **Origin:** Native **Region:** Amazonia, Andean, Guiana Shield, Caribbean, Orinoquia, Pacific **Dept.:** AMA, ANT, ARA, BOL, CAQ, CHO, COR, GUA, GUV, MET, PUT, SAP, VAC, VID **Elev.:** 0–1850 m a.s.l. **IUCN Habitat code:** 1,2,3,4,5,14 **Conservation status: (IUCN)** LC

Saxifragales

7161 *Liquidambar styraciflua* L. (Altingiaceae)
IPNI: 430718-1
Common name: alcornoque, liquidambar **Habit:** Tree **Use:** EU, HF, MA, ME **Origin:** Cultivated **Region:** Andean **Dept.:** BOG, CUN **Elev.:** 2600–2600 m a.s.l. **IUCN Habitat code:** 14 **Conservation status: (IUCN)** LC

7162 *Aeonium arboreum* (L.) Webb & Berthel. (Crassulaceae)
IPNI: 272194-1
Use: EU, ME **Region:** Andean **Dept.:** BOG **IUCN Habitat code:** 14 **Conservation status: (IUCN)** NE

7163 *Aeonium canariense* (L.) Webb & Berthel. (Crassulaceae)
IPNI: 272205-1
Common name: chupahuevo, piña, yema **Habit:** Herb **Use:** EU **Origin:** Cultivated **Region:** Andean **Dept.:** BOG, CUN **Elev.:** 2600–2600 m a.s.l. **IUCN Habitat code:** 4 **Conservation status: (IUCN)** NE

7164 *Aeonium tabuliforme* (Haw.) Webb & Berthel. (Crassulaceae)
IPNI: 272263-1
Common name: flor de Jericó, rosa de Jericó **Habit:** Herb **Use:** EU **Origin:** Cultivated **Region:** Andean **Dept.:** CUN **Elev.:** 2600–2600 m a.s.l. **Conservation status: (IUCN)** NE

7165 *Crassula multicava* Lem. (Crassulaceae)
IPNI: 273315-1
Use: ME **Region:** Andean **Dept.:** BOG **IUCN Habitat code:** 1,14 **Conservation status: (IUCN)** NE

7166 *Crassula ovata* (Mill.) Druce (Crassulaceae)
IPNI: 273350-1
Use: EU, HF, ME, SU **Dept.:** BOG **IUCN Habitat code:** 14 **Conservation status: (IUCN)** NE

7167 *Echeveria bicolor* (Kunth) E.Walther (Crassulaceae)
IPNI: 86883-2
Common name: chupahuevo, chupahuevos, chupahuevos de tejados, conchita **Habit:** Herb **Use:** EU, ME **Origin:** Native **Region:** Andean **Dept.:** BOG, BOY, CAU, CUN, LAG, NAR **Elev.:** 2200–4000 m a.s.l. **IUCN Habitat code:** 1,3,4,5,14 **Conservation status: (IUCN)** NE

7168 *Echeveria harmsii* J.F.Macbr. (Crassulaceae)
IPNI: 298512-2
Use: EU **Region:** Andean **Dept.:** BOG **Conservation status: (IUCN)** NE

7169 *Echeveria quitensis* (Kunth) Lindl. (Crassulaceae)
IPNI: 274010-1
Common name: sugiscala **Habit:** Herb **Use:** ME **Origin:** Native **Region:** Andean **Dept.:** BOY, CAU, CES, NAR, NSA, SAN, TOL **Elev.:** 2700–4100 m a.s.l. **IUCN Habitat code:** 1,3,4,14 **Conservation status: (IUCN)** NE

7170 *Kalanchoe × poincarei* Raym.-Hamet & H.Perrier (Crassulaceae) **IPNI:** 274440-1 **Use:** EU **Origin:** Cultivated **Region:** Caribbean **Dept.:** BOL **Elev.:** 100–100 m a.s.l. **Conservation status: (IUCN)** NE

7171 *Kalanchoe beharensis* Drake (Crassulaceae) **IPNI:** 274270-1 **Common name:** imán **Habit:** Shrub **Use:** EU **Origin:** Cultivated **Dept.:** BOG, CUN, QUI, SAN **Elev.:** 1100–1500 m a.s.l. **Conservation status: (IUCN)** VU

7172 *Kalanchoe blossfeldiana* Poelln. (Crassulaceae) **IPNI:** 274279-1 **Habit:** Herb **Use:** EU, ME **Origin:** Cultivated **Region:** Andean **Dept.:** ANT, BOG **Elev.:** 1500–2500 m a.s.l. **IUCN Habitat code:** 3,5,14 **Conservation status: (IUCN)** NE

7173 *Kalanchoe crenata* (Andrews) Haw. (Crassulaceae) **IPNI:** 274311-1 **Use:** EU, ME, SU **Region:** Andean **Dept.:** BOG **IUCN Habitat code:** 14 **Conservation status: (IUCN)** NE

7174 *Kalanchoe daigremontiana* Raym.-Hamet & H.Perrier (Crassulaceae) **IPNI:** 274315-1 **Common name:** hojarancín, lenguaesuegra **Habit:** Herb, Subshrub **Use:** EU,PO, SU **Origin:** Cultivated, Naturalised **Region:** Andean **Dept.:** ANT, BOG **Elev.:** 1600–1600 m a.s.l. **IUCN Habitat code:** 1,2,3,5,14 **Conservation status: (IUCN)** EN

7175 *Kalanchoe delagoensis* Eckl. & Zeyh. (Crassulaceae) **IPNI:** 274320-1 **Habit:** Herb **Use:** EU,PO **Origin:** Naturalised **Dept.:** ANT, BOG, BOL, BOY, CUN, QUI, SAN **IUCN Habitat code:** 2,14 **Conservation status: (IUCN)** NE

7176 *Kalanchoe densiflora* Rolfe (Crassulaceae) **IPNI:** 274321-1 **Use:** ME **Region:** Andean **Dept.:** BOG **IUCN Habitat code:** 1,3,14 **Conservation status: (IUCN)** NE

7177 *Kalanchoe gastonis-bonnieri* Raym.-Hamet & H.Perrier (Crassulaceae) **IPNI:** 274343-1 **Habit:** Herb **Use:** EU **Origin:** Cultivated **Region:** Andean **Dept.:** BOG, CUN, QUI, SAN **Elev.:** 1200–2600 m a.s.l. **Conservation status: (IUCN)** NE

7178 *Kalanchoe grandiflora* Wight & Arn. (Crassulaceae) **IPNI:** 274356-1 **Habit:** Herb **Use:** EU **Origin:** Cultivated **Region:** Andean **Dept.:** CUN **Elev.:** 1500–1500 m a.s.l. **Conservation status: (IUCN)** NE

7179 *Kalanchoe laciniata* (L.) DC. (Crassulaceae) **IPNI:** 274383-1 **Habit:** Herb **Use:** EU, MA, ME **Origin:** Cultivated **Region:** Andean **Dept.:** BOY **Elev.:** 2120–2120 m a.s.l. **Conservation status: (IUCN)** NE

7180 *Kalanchoe manginii* Raym.-Hamet & H.Perrier (Crassulaceae) **IPNI:** 274404-1 **Habit:** Herb **Use:** EU **Origin:** Cultivated **Region:** Andean **Dept.:** CUN **Elev.:** 1500–1500 m a.s.l. **Conservation status: (IUCN)** NE

7181 *Kalanchoe marmorata* Baker (Crassulaceae) **IPNI:** 274406-1 **Use:** ME **Region:** Andean **Dept.:** BOG **IUCN Habitat code:** 14 **Conservation status: (IUCN)** NE

7182 *Kalanchoe millotii* Raym.-Hamet & H.Perrier (Crassulaceae) **IPNI:** 274409-1 **Habit:** Subshrub **Use:** EU **Origin:** Cultivated **Region:** Andean **Dept.:** CUN **Elev.:** 1500–1500 m a.s.l. **Conservation status: (IUCN)** NE

7183 *Kalanchoe orgyalis* Baker (Crassulaceae) **IPNI:** 274427-1 **Habit:** Shrub **Use:** EU **Origin:** Cultivated **Region:** Andean **Dept.:** CUN **Elev.:** 1500–1500 m a.s.l. **Conservation status: (IUCN)** NE

7184 *Kalanchoe pinnata* (Lam.) Pers. (Crassulaceae) **IPNI:** 274438-1 **Common name:** air plant, bruja, centaura, chupahuevo, colombiana, espiritusanto, espíritu santo, fey choche, hierbasanta, hoja de la fortuna, hoja del aire, hoja del soldado, hoja inmortal, hoja santa, hojasanta, leaf of live, lou gawou, malamadre, patriota, pereirana **Habit:** Herb **Use:** AF, EU, ME, PO, SU **Origin:** Naturalised **Region:** Amazonia **Dept.:** AMA, ANT, BOG, BOL, BOY, CAQ, CAU, CUN, HUI, MET, PUT, QUI, SAP, SAN, VAU **Elev.:** 150–2700 m a.s.l. **IUCN Habitat code:** 1,2,3,14 **Conservation status: (IUCN)** NE

7185 *Kalanchoe scapigera* Welw. ex Britten (Crassulaceae) **IPNI:** 274469-1 **Habit:** Herb **Use:** EU **Origin:** Cultivated **Region:** Andean **Dept.:** CUN **Elev.:** 1500–1500 m a.s.l. **Conservation status: (IUCN)** NE

7186 *Kalanchoe tomentosa* Baker (Crassulaceae) **IPNI:** 274502-1 **Use:** EU **Region:** Andean **Dept.:** BOG **IUCN Habitat code:** 14 **Conservation status: (IUCN)** NE

7187 *Rosularia sempervivum* (M.Bieb.) A.Berger (Crassulaceae) **IPNI:** 274966-1 **Common name:** chupahuevos, conchitas, estrellitas **Use:** EU, ME **Dept.:** BOG **IUCN Habitat code:** 14 **Conservation status: (IUCN)** NE

7188 *Sedum moranense* Kunth (Crassulaceae) **IPNI:** 275718-1 **Habit:** Herb **Use:** EU **Origin:** Cultivated, Naturalised **Dept.:** BOG, BOY, CUN **Elev.:** 2600–3000 m a.s.l. **Conservation status: (IUCN)** NE

7189 *Sedum morganianum* E.Walther (Crassulaceae) **IPNI:** 231789-2 **Common name:** cordón de novia, guinea, moño, platanito, platanitos, porcelana, racimo de San José, trencita **Habit:** Herb **Use:** EU **Origin:** Cultivated **Region:** Andean **Dept.:** BOG, BOY, CAL, QUI, RIS, VAC **Elev.:** 1000–2700 m a.s.l. **Conservation status: (IUCN)** NE

7190 *Sedum pachyphyllum* Rose (Crassulaceae) **IPNI:** 231817-2 **Common name:** arrocito, chupahuevos, platanitos **Habit:** Herb **Use:** EU **Origin:** Cultivated **Region:** Andean **Dept.:** CUN, RIS **Elev.:** 1400–2600 m a.s.l. **IUCN Habitat code:** 1 **Conservation status: (IUCN)** NE

7191 *Sedum praealtum* A.DC. (Crassulaceae) **IPNI:** 231837-2 **Habit:** Shrub, Subshrub **Use:** ME **Origin:** Cultivated **Region:** Andean **Dept.:** BOG, CUN **Elev.:** 2600–2600 m a.s.l. **IUCN Habitat code:** 1,14 **Conservation status: (IUCN)** NE

7192 *Sedum spathulifolium* Hook. (Crassulaceae) **IPNI:** 284830-2 **Habit:** Herb **Use:** EU, HF, ME **Origin:** Cultivated **Region:** Andean **Dept.:** ANT **Elev.:** 2150–2300 m a.s.l. **IUCN Habitat code:** 3,14 **Conservation status: (IUCN)** NE

7193 *Ribes andicola* Jancz. (Grossulariaceae) **IPNI:** 792564-1 **Common name:** niubajalé **Habit:** Shrub **Use:** HF, ME **Origin:** Native **Region:** Andean **Dept.:** BOG, BOY, CUN, MET, NAR, NSA, SAN **Elev.:** 2800–3720 m a.s.l. **IUCN Habitat code:** 1,3,4,14 **Conservation status: (IUCN)** NE

7194 *Myriophyllum aquaticum* (Vell.) Verdc. (Haloragaceae) **IPNI:** 166319-2 **Habit:** Herb, Aquatic **Use:** EU, HF, ME **Origin:** Native, Cultivated **Region:** Andean **Dept.:** ANT, BOG, BOY, CUN, NAR **Elev.:** 2100–3300 m a.s.l. **IUCN Habitat code:** 1,3,4,5,14 **Conservation status: (IUCN)** NE

7195 *Hamamelis virginiana* L. (Hamamelidaceae) **IPNI:** 430697-1 **Common name:** hamamelis **Use:** EU, GS, HF, MA, ME **Conservation status: (IUCN)** LC

7196 *Bergenia crassifolia* (L.) Fritsch (Saxifragaceae) **IPNI:** 790361-1 **Use:** EU, MA, ME **Region:** Andean **Dept.:** BOG **IUCN Habitat code:** 14 **Conservation status: (IUCN)** NE

7197 *Bergenia stracheyi* (Hook.f. & Thomson) Engl. (Saxifragaceae) **IPNI:** 790385-1 **Use:** ME **Region:** Andean **Dept.:** BOG **Conservation status: (IUCN)** NE

7198 *Saxifraga stolonifera* Curtis (Saxifragaceae) **IPNI:** 794584-1 **Use:** EU, HF, ME **Origin:** Naturalised **Dept.:** BOG **IUCN Habitat code:** 1,3,14 **Conservation status: (IUCN)** NE

7199 *Tolmiea menziesii* (Pursh) Torr. & A.Gray (Saxifragaceae) **IPNI:** 794907-1 **Common name:** milhoja **Habit:** Herb **Use:** HF, ME **Origin:** Cultivated **Region:** Andean **Dept.:** BOG, CUN

Elev.: 2600–2600 m a.s.l. **Conservation status: (IUCN)** NE

Solanales

7200 *Aniseia martinicensis* (Jacq.) Choisy (Convolvulaceae) **IPNI:** 30070865-2 **Habit:** Climbing **Use:** GS, HF, ME **Origin:** Native **Region:** Amazonia, Caribbean, Orinoquia, Pacific **Dept.:** AMA, ATL, CHO, COR, MET, TOL, VAU **Elev.:** 15–235 m a.s.l. **IUCN Habitat code:** 1,2,14 **Conservation status: (IUCN)** LC

7201 *Camonea umbellata* (L.) A.R.Simões & Staples (Convolvulaceae) **IPNI:** 77163301-1 **Common name:** batatilla, batatilla amarilla, batatilla morada, batatillo, bejuco barbero, bejuco batatero **Habit:** Climbing **Use:** EU, HF, MA, ME **Origin:** Native **Region:** Andean, Caribbean, Orinoquia, Pacific **Dept.:** ANT, ARA, ATL, BOL, CAL, CAS, CAU, CES, CHO, COR, CUN, HUI, MAG, MET, NSA, PUT, SAP, SAN, TOL, VAC **Elev.:** 0-2000 m.a.s.l. **IUCN Habitat code:** 1,2,3,4,5,14 **Conservation status: (IUCN)** NE **(CPSG)** Potential LC

7202 *Cuscuta americana* L. (Convolvulaceae) **IPNI:** 267279-1 **Common name:** barbasco, bejuco de bruja, bejuquillo, cuerda de violín, cuerdas de violín, cuerdilla, cuerdillas, cuscuta, enredacotorra, fideos, hilo de oro, matalimón, rejilla, rejillo, rema **Habit:** Hemiparasitic **Use:** MA, ME **Origin:** Native **Region:** Andean, Caribbean **Dept.:** CUN, HUI, MAG, TOL **Elev.:** 500–1090 m a.s.l. **IUCN Habitat code:** 1,14 **Conservation status: (IUCN)** NE

7203 *Cuscuta campestris* Yunck. (Convolvulaceae) **IPNI:** 72786-2 **Use:** ME **Origin:** Native **Conservation status: (IUCN)** NE

7204 *Cuscuta colombiana* Yunck. (Convolvulaceae) **IPNI:** 72792-2 **Habit:** Hemiparasitic **Use:** ME **Origin:** Native **Region:** Caribbean **Dept.:** LAG, MAG **Elev.:** 30–300 m a.s.l. **Conservation status: (IUCN)** NE

7205 *Cuscuta corymbosa* Ruiz & Pav. (Convolvulaceae) **IPNI:** 267380-1 **Habit:** Hemiparasitic **Use:** ME **Origin:** Native **Region:** Andean **Dept.:** CAU, NSA, SAN **Elev.:** 1500–2100 m a.s.l. **Conservation status: (IUCN)** NE

7206 *Cuscuta grandiflora* Kunth (Convolvulaceae) **IPNI:** 72853-2 **Common name:** bejuco de tinto, rejillo **Habit:** Hemiepiphyte **Use:** ME **Origin:** Native **Region:** Andean **Dept.:** BOG, BOY, CUN, NAR **Elev.:** 500–3200 m a.s.l. **IUCN Habitat code:** 1,4 **Conservation status: (IUCN)** NE

7207 *Cuscuta indecora* Choisy (Convolvulaceae) **IPNI:** 72866-2 **Habit:** Hemiparasitic **Use:** ME **Origin:** Native, Cultivated **Region:** Andean **Dept.:** BOG, VAC **Elev.:** 1000–1000 m a.s.l. **IUCN Habitat code:** 3,14 **Conservation status: (IUCN)** NE

7208 *Cuscuta obtusiflora* Kunth (Convolvulaceae) **IPNI:** 72898-2 **Habit:** Hemiparasitic **Use:** ME **Origin:** Native **Region:** Andean **Dept.:** ANT, CUN **Elev.:** 1500–1949 m a.s.l. **Conservation status: (IUCN)** NE

7209 *Cuscuta partita* Choisy (Convolvulaceae) **IPNI:** 294964-2 **Common name:** hilo de oro **Habit:** Hemiparasitic **Use:** ME **Origin:** Native **Region:** Caribbean **Dept.:** CES, LAG **Elev.:** 10–50 m a.s.l. **IUCN Habitat code:** 1,14 **Conservation status: (IUCN)** NE

7210 *Cuscuta pentagona* Engelm. (Convolvulaceae) **IPNI:** 72922-2 **Use:** ME **Origin:** Naturalised **Conservation status: (IUCN)** NE

7211 *Cuscuta umbellata* Kunth (Convolvulaceae) **IPNI:** 72987-2 **Use:** HF, ME **Origin:** Native **Conservation status: (IUCN)** NE

7212 *Dichondra argentea* Willd. (Convolvulaceae) **IPNI:** 267747-1 **Habit:** Herb **Use:** ME **Origin:** Native **Region:** Andean **Dept.:** CUN, TOL **Elev.:** 2600–2600 m a.s.l. **IUCN Habitat code:** 14 **Conservation status: (IUCN)** NE

7213 *Dichondra micrantha* Urb. (Convolvulaceae) **IPNI:** 79457-2 **Use:** EU, HF, ME, PO **Origin:** Naturalised **Dept.:** VAC **IUCN Habitat code:** 1,3,5,14 **Conservation status: (IUCN)** LC

7214 *Dichondra repens* J.R.Forst. & G.Forst. (Convolvulaceae) **IPNI:** 267765-1 **Habit:** Herb **Use:** HF, ME **Origin:** Native **Dept.:** BOG **Conservation status: (IUCN)** NE

7215 *Dicranostyles ampla* Ducke (Convolvulaceae) **IPNI:** 79947-2 **Habit:** Liana **Use:** HF **Origin:** Native **Region:** Amazonia **Dept.:** AMA **Elev.:** 80–200 m a.s.l. **IUCN Habitat code:** 1,5 **Conservation status: (IUCN)** NE

7216 *Distimake aegyptius* (L.) A.R.Simões & Staples (Convolvulaceae) **IPNI:** 77163214-1 **Common name:** batatilla blanca, campanita blanca **Habit:** Climbing **Use:** AF, HF, MA, ME **Origin:** Native **Region:** Andean, Caribbean **Dept.:** ATL, BOL, CAU, CES, COR, LAG, MAG, NSA **Elev.:** 10–575 m a.s.l. **IUCN Habitat code:** 1,2,3,5,14 **Conservation status: (IUCN)** NE

7217 *Distimake cissoides* (Lam.) A.R.Simões & Staples (Convolvulaceae) **IPNI:** 77163224-1 **Habit:** Climbing **Use:** EU, ME **Origin:** Native **Region:** Andean **Dept.:** CAL, CUN, TOL, VAC **Elev.:** 300–1200 m a.s.l. **IUCN Habitat code:** 1,3,14 **Conservation status: (IUCN)** NE

7218 *Distimake dissectus* (Jacq.) A.R.Simões & Staples (Convolvulaceae) **IPNI:** 77163242-1 **Common name:** batatilla **Habit:** Climbing **Use:** EU, HF, MA, ME, PO **Origin:** Native **Region:** Caribbean **Dept.:** ANT, BOL, CUN, HUI, SAP, TOL, VAC **Elev.:** 0–980 m a.s.l. **IUCN Habitat code:** 2,3,5,14 **Conservation status: (IUCN)** NE

7219 *Distimake quinquefolius* (L.) A.R.Simões & Staples (Convolvulaceae) **IPNI:** 77163286-1 **Habit:** Liana **Use:** HF, MA, ME **Origin:** Native **Region:** Andean, Caribbean, Pacific **Dept.:** ANT, ATL, BOL, CAL, CAQ, CHO, COR, CUN, HUI, LAG, MAG, VAC **Elev.:** 0-1050 m.a.s.l. **IUCN Habitat code:** 1,3,4,5,14 **Conservation status: (IUCN)** NE **(CPSG)** Potential LC

7220 *Distimake tuberosus* (L.) A.R.Simões & Staples (Convolvulaceae) **IPNI:** 77163293-1 **Common name:** flor de madera **Habit:** Liana **Use:** EU, HF, ME, PO **Origin:** Native, Cultivated **Region:** Andean, Caribbean, Pacific **Dept.:** ANT, BOL, BOY, CHO, CUN, NAR, NSA, QUI, RIS, SAN, VAC **Elev.:** 200–1600 m a.s.l. **IUCN Habitat code:** 1,2,3,14 **Conservation status: (IUCN)** NE

7221 *Evolvulus alsinoides* (L.) L. (Convolvulaceae) **IPNI:** 267957-1 **Common name:** cielito, hierba de lagartija, lagartija, lluviadeoro **Habit:** Herb **Use:** AF, EU, HF, MA, ME, SU **Origin:** Native **Region:** Andean, Caribbean **Dept.:** BOG, CAU, CUN, MAG, NSA, SAN, TOL, VAC **Elev.:** 20–1200 m a.s.l. **IUCN Habitat code:** 1,2,3,14 **Conservation status: (IUCN)** NE

7222 *Evolvulus convolvuloides* (Willd. ex Schult.) Stearn (Convolvulaceae) **IPNI:** 268002-1 **Habit:** Herb **Use:** ME **Origin:** Native **Region:** Andean, Caribbean **Dept.:** ATL, BOL, LAG, MAG **Elev.:** 0–1800 m a.s.l. **IUCN Habitat code:** 2,14 **Conservation status: (IUCN)** NE

7223 *Evolvulus nummularius* (L.) L. (Convolvulaceae) **IPNI:** 102674-2 **Common name:** hierba del cuartillo **Habit:** Herb **Use:** HF, ME **Origin:** Native **Region:** Andean, Caribbean **Dept.:** ANT, ATL, BOL, CAU, NSA, TOL, VAC **Elev.:** 10–1100 m a.s.l. **IUCN Habitat code:** 1,2,3,14 **Conservation status: (IUCN)** NE

7224 *Evolvulus sericeus* Sw. (Convolvulaceae) **IPNI:** 268156-1 **Common name:** vira, vira-vira **Habit:** Herb **Use:** ME **Origin:** Native **Region:** Andean, Caribbean **Dept.:** CES, CUN, MAG, TOL **Elev.:** 0–1000 m a.s.l. **IUCN Habitat code:** 1,2,3,14 **Conservation status: (IUCN)** NE

7225 *Ipomoea alba* L. (Convolvulaceae) **IPNI:** 129128-2 **Common name:** batata, batatilla, dama de noche **Habit:** Climbing **Use:** EU, HF, MA, ME, PO **Origin:** Native, Cultivated **Region:** Amazonia, Andean, Guiana Shield, Caribbean, Pacific **Dept.:** AMA, ANT, BOL, CES, CHO, CUN, HUI, MAG, MET, PUT, QUI, SAN, VAC **Elev.:** 0-1780 m.a.s.l. **IUCN Habitat code:** 1,2,3,4,5,8,14 **Conservation status: (IUCN)** LC **(CPSG)** Potential LC

7226 *Ipomoea aquatica* Forssk. (Convolvulaceae) **IPNI:** 268410-1 **Common name:** batatilla de puerco, bejuco batatilla, bejuco de agua **Habit:** Aquatic **Use:** AF, EU, FU, GS, HF, MA, ME, PO, SU **Origin:** Naturalised **Region:** Amazonia, Caribbean **Dept.:** AMA, ATL, BOL, COR, MAG **Elev.:** 6–30 m a.s.l. **IUCN Habitat code:** 1,2,5,14 **Conservation status: (IUCN)** LC

7227 *Ipomoea argentea* Meisn. (Convolvulaceae) **IPNI:** 268419-1 **Habit:** Herb **Use:** ME **Origin:** Native **Region:** Amazonia, Guiana Shield, Orinoquia **Dept.:** CAQ, CAS, MET, VAU **Elev.:** 200–480 m a.s.l. **IUCN Habitat code:** 1,2,3,4,14 **Conservation status: (IUCN)** NE

7228 *Ipomoea asarifolia* (Desr.) Roem. & Schult. (Convolvulaceae) **IPNI:** 268434-1 **Habit:** Climbing **Use:** AF, EU, FU, HF, MA, ME, PO, SU **Origin:** Native **Region:** Caribbean **Dept.:** COR **Elev.:** 12–12 m a.s.l. **IUCN Habitat code:** 1,2 **Conservation status: (IUCN)** NE

7229 *Ipomoea batatas* (L.) Lam. (Convolvulaceae) **IPNI:** 1101088-2 **Common name:** apichu, batata, batata amarilla, batata blanca, batata morada, batata roja, batatilla, batatilla morada, batatillo, bejuquillo, camote, camote blanco, camote morado, chaco, cumara, papa dulce, yapima **Habit:** Climbing **Use:** AF, EU, GS, HF, MA, ME, PO, SU **Origin:** Cultivated, Naturalised **Region:** Amazonia, Andean, Guiana Shield, Caribbean, Orinoquia, Pacific **Dept.:** AMA, ANT, ARA, BOY, CAL, CAQ, CAS, CAU, CES, CHO, CUN, GUA, HUI, MAG, MET, NAR, NSA, QUI, RIS, SAN, TOL, VAC **Elev.:** 0-2200 m.a.s.l. **Conservation status: (IUCN)** DD **(CPSG)** Potential LC

7230 *Ipomoea batatoides* Choisy (Convolvulaceae) **IPNI:** 268480-1 **Common name:** campanilla, elamoroso **Habit:** Liana, Climbing **Use:** ME **Origin:** Native **Region:** Andean, Caribbean, Orinoquia, Pacific **Dept.:** ANT, CES, MAG, MET, PUT **Elev.:** 80–1570 m a.s.l. **IUCN Habitat code:** 1,3,14 **Conservation status: (IUCN)** NE

7231 *Ipomoea cairica* (L.) Sweet (Convolvulaceae) **IPNI:** 268564-1 **Common name:** bejuco estrellita **Habit:** Climbing **Use:** AF, EU, HF, MA, ME, PO **Origin:** Cultivated, Naturalised **Region:** Caribbean **Dept.:** ANT, COR, SAN **Elev.:** 6–900 m a.s.l. **IUCN Habitat code:** 1,2,14 **Conservation status: (IUCN)** LC

7232 *Ipomoea capillacea* (Kunth) G.Don (Convolvulaceae) **IPNI:** 268591-1 **Habit:** Herb **Use:** ME **Origin:** Native **Region:** Andean, Pacific **Dept.:** CAU **Elev.:** 400–1800 m a.s.l. **Conservation status: (IUCN)** NE

7233 *Ipomoea carnea* Jacq. (Convolvulaceae) **IPNI:** 30299665-2 **Common name:** bambadu, batatilla, batatilla morada, bejuco campana, bejuco pupú, bejuco tapabotijo, bulto, campana, casisé, hija de sapo, kasisu, ojo de sapo, pitanga, pitongo, tapabotija, tapabotijo **Habit:** Shrub **Use:** EU, FU, HF, MA, ME **Origin:** Native, Cultivated **Region:** Amazonia, Andean, Caribbean, Orinoquia, Pacific **Dept.:** AMA, ANT, ARA, ATL, BOL, BOY, CAS, CES, CHO, COR, CUN, HUI, LAG, MAG, NAR, SAP, SAN, SUC, TOL, VAC **Elev.:** 0-1750 m.a.s.l. **IUCN Habitat code:** 1,2,3,5,14 **Conservation status: (IUCN)** NE **(CPSG)** Potential LC

7234 *Ipomoea cholulensis* Kunth (Convolvulaceae) **IPNI:** 268637-1 **Habit:** Climbing **Use:** ME **Origin:** Native **Region:** Andean **Dept.:** ANT, CAU, HUI, NSA, SAN **Elev.:** 1200–2300 m a.s.l. **IUCN Habitat code:** 3,14 **Conservation status: (IUCN)** NE

7235 *Ipomoea clavata* (G.Don) Ooststr. ex J.F.Macbr. (Convolvulaceae) **IPNI:** 129258-2 **Common name:** batatilla, batatilla rosada **Habit:** Climbing **Use:** ME **Origin:** Native **Region:** Andean **Dept.:** CUN **Elev.:** 460–1800 m a.s.l. **IUCN Habitat code:** 1,3,14 **Conservation status: (IUCN)** NE

7236 *Ipomoea coccinea* L. (Convolvulaceae) **IPNI:** 268659-1 **Common name:** convólvulo rojo **Use:** EU, ME **Dept.:** BOG **IUCN Habitat code:** 1,2,3,4,5,14 **Conservation status: (IUCN)** NE

7237 *Ipomoea cordatotriloba* Dennst. (Convolvulaceae) **IPNI:** 30231094-2 **Habit:** Climbing **Use:** GS, ME **Origin:** Native **Region:** Andean **Dept.:** HUI, TOL, VAC **Elev.:** 450–1100 m a.s.l. **IUCN Habitat code:** 14 **Conservation status: (IUCN)** LC

7238 *Ipomoea corymbosa* (L.) Roth (Convolvulaceae) **IPNI:** 60458008-2 **Common name:** campanita blanca **Habit:** Liana **Use:** EU, MA, ME, SU **Origin:** Native **Region:** Caribbean **Dept.:** ATL, BOL, MAG **Elev.:** 30–250 m a.s.l. **IUCN Habitat code:** 5 **Conservation status: (IUCN)** NE

7239 *Ipomoea dumetorum* (Kunth) Willd. (Convolvulaceae) **IPNI:** 30061510-2 **Habit:** Climbing **Use:** HF, ME **Origin:** Native **Region:** Andean **Dept.:** BOG, MAG **Elev.:** 2400–2400 m a.s.l. **IUCN Habitat code:** 2,14 **Conservation status: (IUCN)** NE

7240 *Ipomoea dumosa* (Benth.) L.O.Williams (Convolvulaceae) **IPNI:** 129313-2 **Use:** HF, ME **IUCN Habitat code:** 1 **Conservation status: (IUCN)** NE

7241 *Ipomoea hederacea* Jacq. (Convolvulaceae) **IPNI:** 268998-1 **Common name:** batatilla **Use:** EU, ME **Origin:** Naturalised **IUCN Habitat code:** 1,2,3,14 **Conservation status: (IUCN)** NE

7242 *Ipomoea hederifolia* L. (Convolvulaceae) **IPNI:** 268999-1 **Common name:** batatilla, batatilla roja, batatilla rosada, cundamor, cundeamor, trompetica roja **Habit:** Climbing **Use:** EU, HF, MA, ME, PO **Origin:** Native **Region:** Andean, Caribbean, Orinoquia **Dept.:** ANT, ATL, BOL, CAL, CAU, CES, CUN, HUI, LAG, MAG, MET, NAR, QUI, SAN, SUC, VAC **Elev.:** 40-1610 m.a.s.l. **Conservation status: (IUCN)** NE **(CPSG)** Potential LC

7243 *Ipomoea imperati* (Vahl) Griseb. (Convolvulaceae) **IPNI:** 60442184-2 **Common name:** batata **Habit:** Climbing **Use:** EU, ME **Origin:** Native **Region:** Pacific **Dept.:** CAU, CHO, NAR **Elev.:** 0–70 m a.s.l. **IUCN Habitat code:** 1,3,5,14 **Conservation status: (IUCN)** NE

7244 *Ipomoea incarnata* (Vahl) Choisy (Convolvulaceae) **IPNI:** 269080-1 **Common name:** batatilla, batatilla lila, campanita **Habit:** Climbing **Use:** ME **Origin:** Native **Region:** Caribbean **Dept.:** BOL, CES, MAG, NSA **Elev.:** 0–370 m a.s.l. **IUCN Habitat code:** 1,2,14 **Conservation status: (IUCN)** NE

7245 *Ipomoea indica* (Burm.) Merr. (Convolvulaceae) **IPNI:** 269086-1 **Common name:** batata, batatilla, neguilla **Habit:** Climbing **Use:** AF, EU, HF, ME, PO, SU **Origin:** Native, Cultivated **Region:** Andean, Caribbean, Pacific **Dept.:** ANT, CHO, CUN, HUI, SAN, TOL, VAC **Elev.:** 20–1900 m a.s.l. **IUCN Habitat code:** 1,2,3,14 **Conservation status: (IUCN)** NE

7246 *Ipomoea involucrata* P.Beauv. (Convolvulaceae) **IPNI:** 269099-1 **Use:** HF, ME **Origin:** Naturalised **Conservation status: (IUCN)** NE

7247 *Ipomoea lacunosa* L. (Convolvulaceae) **IPNI:** 269155-1 **Habit:** Climbing **Use:** GS **Origin:** Naturalised **Region:** Pacific **Dept.:** CHO **IUCN Habitat code:** 3 **Conservation status: (IUCN)** LC

7248 *Ipomoea lobata* (Cerv.) Thell. (Convolvulaceae) **IPNI:** 129456-2 **Habit:** Climbing **Use:** EU, ME **Origin:** Cultivated, Naturalised **Region:** Andean **Dept.:** CUN **Elev.:** 1550–1800 m a.s.l. **Conservation status: (IUCN)** NE

7249 *Ipomoea mairetii* Choisy (Convolvulaceae) **IPNI:** 269266-1 **Common name:** neguilla **Use:** ME **Dept.:** CUN **Conservation status: (IUCN)** NE

7250 *Ipomoea mauritiana* Jacq. (Convolvulaceae) **IPNI:** 269292-1 **Common name:** batata de monte **Habit:** Climbing **Use:** AF, EU, HF, MA, ME **Origin:** Native **Region:** Amazonia, Orinoquia, Pacific **Dept.:** AMA, ANT, CAS **Elev.:** 0–100 m a.s.l. **IUCN Habitat code:** 1,3 **Conservation status: (IUCN)** NE

7251 *Ipomoea muricata* (L.) Jacq. (Convolvulaceae) **IPNI:** 269359-1 **Habit:** Climbing **Use:** EU, HF, MA, ME, PO **Origin:** Native **IUCN Habitat code:** 1,2 **Conservation status: (IUCN)** NE

7252 *Ipomoea nil* (L.) Roth (Convolvulaceae) **IPNI:** 1071575-2 **Common name:** batatilla, campanilla, campanita, coralito **Habit:** Climbing **Use:** AF, EU, HF, MA, ME, PO, SU **Origin:** Native, Cultivated **Region:** Andean, Caribbean **Dept.:** ANT, ATL, BOL, CUN, HUI, MAG, MET, NAR, NSA, SUC, TOL, VAC **Elev.:** 25–2560 m a.s.l. **IUCN Habitat code:** 1,2,3,14 **Conservation status: (IUCN)** NE

7253 *Ipomoea parasitica* (Kunth) G.Don (Convolvulaceae) **IPNI:** 269470-1 **Habit:** Climbing **Use:** MA **Origin:** Native **Dept.:** VAC **Elev.:** 1000–1000 m a.s.l. **Conservation status: (IUCN)** NE

7254 *Ipomoea pauciflora* M.Martens & Galeotti (Convolvulaceae) **IPNI:** 269479-1 **Habit:** Herb, Climbing **Use:** ME **Origin:** Native **Region:** Andean **Dept.:** BOY **Elev.:** 2050–2050 m a.s.l. **Conservation status: (IUCN)** LC

7255 *Ipomoea pes-caprae* (L.) R.Br. (Convolvulaceae) **IPNI:** 30040623-2 **Common name:** batata, batatilla, batatilla de playa, flor de playa **Habit:** Climbing **Use:** AF, EU, IF, MA, ME, PO, SU **Origin:** Native **Region:** Caribbean, Pacific **Dept.:** ANT, ATL, BOL, CAU, CHO, MAG, NAR, SAP, SUC **Elev.:** 0–100 m a.s.l. **IUCN Habitat code:** 1,2,5,14 **Conservation status: (IUCN)** NE

7256 *Ipomoea philomega* (Vell.) House (Convolvulaceae) **IPNI:** 129579-2 **Common name:** batatilla **Habit:** Liana **Use:** ME **Origin:** Native **Region:** Amazonia, Andean, Caribbean, Orinoquia, Pacific **Dept.:** AMA, ANT, BOL, BOY, CAQ, CHO, CUN, MAG, MET, NAR, PUT, RIS, SAN, TOL **Elev.:** 10-1480 m.a.s.l. **IUCN Habitat code:** 1,3,5,14 **Conservation status: (IUCN)** NE **(CPSG)** Potential LC

7257 *Ipomoea pubescens* Lam. (Convolvulaceae) **IPNI:** 269612-1 **Habit:** Climbing **Use:** HF, ME **Origin:** Native **Region:** Andean **Dept.:** CUN **Elev.:** 2500–2600 m a.s.l. **IUCN Habitat code:** 14 **Conservation status: (IUCN)** NE

7258 *Ipomoea purpurea* (L.) Roth (Convolvulaceae) **IPNI:** 30043043-2 **Common name:** batatilla **Habit:** Climbing **Use:** EU, GS, HF, IF, MA, ME, PO **Origin:** Native, Cultivated **Region:** Andean, Caribbean **Dept.:** ANT, BOG, BOY, CAL, CAU, CUN, HUI, LAG, MAG, PUT, SAN, TOL, VAC **Elev.:** 0–2560 m a.s.l. **Conservation status: (IUCN)** NE

7259 *Ipomoea quamoclit* L. (Convolvulaceae) **IPNI:** 326361-2 **Common name:** batatilla, cundeamor, sombra **Habit:** Climbing **Use:** AF, EU, HF, MA, ME, PO, SU **Origin:** Naturalised **Region:** Amazonia, Andean, Guiana Shield, Caribbean, Orinoquia, Pacific **Dept.:** AMA, ANT, CAL, CAS, CAU, CHO, CUN, GUA, GUV, MAG, MET, NAR, PUT, QUI, SAN, TOL, VAC **Elev.:** 0-1500 m.a.s.l. **IUCN Habitat code:** 1,2,3,4,5,14 **Conservation status: (IUCN)** NE **(CPSG)** Potential LC

7260 *Ipomoea ramosissima* (Poir.) Choisy (Convolvulaceae) **IPNI:** 269655-1 **Habit:** Climbing **Use:** GS **Origin:** Native **Region:** Amazonia **Dept.:** AMA, PUT **Elev.:** 260–260 m a.s.l. **IUCN Habitat code:** 1,2,3,14 **Conservation status: (IUCN)** LC

7261 *Ipomoea rubens* Choisy (Convolvulaceae) **IPNI:** 269697-1 **Habit:** Climbing **Use:** AF **Origin:** Naturalised **Region:** Andean **Dept.:** CAU **Elev.:** 1300–1300 m a.s.l. **IUCN Habitat code:** 1,2,14 **Conservation status: (IUCN)** LC

7262 *Ipomoea setifera* Poir. (Convolvulaceae) **IPNI:** 269793-1 **Common name:** batatilla **Habit:** Climbing **Use:** EU, ME **Origin:** Native **Region:** Amazonia, Andean, Orinoquia, Pacific **Dept.:** ANT, CAS, CHO, GUA, MET **Elev.:** 75–550 m a.s.l. **IUCN Habitat code:** 1,3,4,5,14 **Conservation status: (IUCN)** NE

7263 *Ipomoea tiliacea* (Willd.) Choisy (Convolvulaceae) **IPNI:** 269949-1 **Common name:** batatilla **Habit:** Climbing **Use:** GS, HF, ME **Origin:** Native **Region:** Pacific **Dept.:** CHO **Elev.:** 0–100 m a.s.l. **IUCN Habitat code:** 1,2,3,4,5,14 **Conservation status: (IUCN)** LC

7264 *Ipomoea tricolor* Cav. (Convolvulaceae) **IPNI:** 269966-1 **Common name:** batatilla, bella de día **Habit:** Climbing **Use:** EU, MA, ME, PO, SU **Origin:** Cultivated, Naturalised **Dept.:** VAC **Elev.:** 1000–1000 m a.s.l. **IUCN Habitat code:** 1,3,14 **Conservation status: (IUCN)** NE

7265 *Ipomoea trifida* (Kunth) G.Don (Convolvulaceae) **IPNI:** 269968-1 **Common name:** batata, batatilla, bejuco campanita, campanilla, campanita, campanita morada **Habit:** Climbing **Use:** GS **Origin:** Native **Region:** Caribbean **Dept.:** ATL, BOL, CES, LAG, MAG, SUC **Elev.:** 10–200 m a.s.l. **IUCN Habitat code:** 1,2,3,4,14 **Conservation status: (IUCN)** LC

7266 *Ipomoea triloba* L. (Convolvulaceae) **IPNI:** 129763-2 **Common name:** bejuco batatillo, trompetica roja **Habit:** Climbing **Use:** GS, HF, IF, ME **Origin:** Native **Region:** Caribbean, Orinoquia, Pacific **Dept.:** CHO, COR, MAG, MET, VAC **Elev.:** 40–990 m a.s.l. **IUCN Habitat code:** 1,2,3,4,14 **Conservation status: (IUCN)** LC

7267 *Ipomoea violacea* L. (Convolvulaceae) **IPNI:** 270063-1 **Habit:** Liana **Use:** EU, HF, ME **Origin:** Native **Region:** Caribbean **Dept.:** SAP **Elev.:** 0–0 m a.s.l. **IUCN Habitat code:** 1 **Conservation status: (IUCN)** NE

7268 *Jacquemontia pentanthos* (Jacq.) G.Don (Convolvulaceae) **IPNI:** 270229-1 **Habit:** Climbing **Use:** ME **Origin:** Native **Region:** Andean, Caribbean **Dept.:** ANT, ATL, BOL, CAL, CAU, CUN, GUA, HUI, LAG, MAG, NAR, NSA, SAN, TOL, VAC **Elev.:** 10–1320 m a.s.l. **Conservation status: (IUCN)** NE

7269 *Jacquemontia tamnifolia* (L.) Griseb. (Convolvulaceae) **IPNI:** 270265-1 **Habit:** Climbing **Use:** AF, EU, GS, HF, MA, ME, PO, SU **Origin:** Native **Region:** Amazonia, Caribbean, Orinoquia **Dept.:** BOL, BOY, GUA, HUI, MAG, MET, TOL, VAU **Elev.:** 20–600 m a.s.l. **IUCN Habitat code:** 1,2,3,4,5,14 **Conservation status: (IUCN)** NE

7270 *Maripa nicaraguensis* Hemsl. (Convolvulaceae) **IPNI:** 270455-1 **Common name:** mielquemada **Habit:** Liana **Use:** HF **Origin:** Native **Region:** Amazonia, Guiana Shield, Caribbean, Pacific **Dept.:** AMA, ANT, CHO, GUV, MAG, MET, NAR, QUI, SAN, VAC **Elev.:** 100–1885 m a.s.l. **IUCN Habitat code:** 1,3,14 **Conservation status: (IUCN)** NE

7271 *Maripa panamensis* Hemsl. (Convolvulaceae) **IPNI:** 270456-1 **Common name:** mielquemada **Habit:** Liana **Use:** HF, ME **Origin:** Native **Region:** Orinoquia, Pacific **Dept.:** ANT, BOY, CAL, CHO, CUN, MET, NSA, VAC **Elev.:** 0–1920 m a.s.l. **IUCN Habitat code:** 1,3,5,14 **Conservation status: (IUCN)** NE

7272 *Maripa scandens* Aubl. (Convolvulaceae) **IPNI:** 270465-1 **Habit:** Liana **Use:** HF **Origin:** Native **IUCN Habitat code:** 1,14 **Conservation status: (IUCN)** NE

7273 *Merremia hederacea* (Burm.f.) Hallier f. (Convolvulaceae) **IPNI:** 270531-1 **Use:** AF, HF, ME, SU **Origin:** Naturalised **IUCN Habitat code:** 2 **Conservation status: (IUCN)** NE

7274 *Operculina hamiltonii* (G.Don) D.F.Austin & Staples (Convolvulaceae) **IPNI:** 174541-2 **Habit:** Climbing **Use:** ME **Origin:** Native **Region:** Caribbean, Pacific **Dept.:** CHO, COR, MAG **Elev.:** 30–230 m a.s.l. **IUCN Habitat code:** 1,2,5,14 **Conservation status: (IUCN)** NE

7275 *Operculina pteripes* (G.Don) O'Donell (Convolvulaceae) **IPNI:** 174555-2 **Habit:** Climbing **Use:** ME **Origin:** Native **Region:** Caribbean **Dept.:** MAG **Elev.:** 300–300 m a.s.l. **IUCN Habitat code:** 14 **Conservation status: (IUCN)** NE

7276 *Poranopsis paniculata* (Roxb.) Roberty (Convolvulaceae) **IPNI:** 270975-1 **Habit:** Liana **Use:** EU, ME **Origin:** Cultivated **Region:** Andean **Dept.:** VAC **Elev.:** 1070–1070 m a.s.l. **IUCN Habitat code:** 1 **Conservation status: (IUCN)** NE

7277 *Hydrolea spinosa* L. (Hydroleaceae) **IPNI:** 432240-1 **Common name:** cáncer, cáncer de cocha, de la potra, flaco maluco, hierba de la potra, hierba del cáncer, yerba de la potra, yerba del cáncer **Habit:** Herb, Shrub **Use:** ME **Origin:** Native **Region:** Andean, Caribbean, Orinoquia **Dept.:** ARA, CAS, CAU, CES, LAG, MET, SAN, TOL, VAC **Elev.:** 100–1000 m a.s.l. **IUCN Habitat code:** 1,2,3,4,14 **Conservation status: (IUCN)** NE

7278 *Browallia americana* L. (Solanaceae)
IPNI: 814568-1
Common name: campesina, caucana, cielito, clavillo, flor de Santana, flor de nube, pensamiento de pobre, pimpinilla, sanalotodo, señorita, sombrilla morada, teresita, teresita azul, verónica, yerbamora, zulia **Habit:** Herb **Use:** EU, ME **Origin:** Native **Region:** Andean, Caribbean, Orinoquia **Dept.:** ANT, BOG, BOY, CAU, CHO, CUN, HUI, LAG, MAG, MET, NAR, PUT, QUI, RIS, SAN, TOL, VAC **Elev.:** 10-2620 m.a.s.l. **IUCN Habitat code:** 1,2,3,5,14 **Conservation status: (IUCN)** NE **(CPSG)** Potential LC

7279 *Browallia speciosa* Hook. (Solanaceae)
IPNI: 814588-1
Common name: alondra, caucana, coqueta **Habit:** Herb, Subshrub **Use:** EU **Origin:** Native **Region:** Andean **Dept.:** ANT, CHO, CUN, PUT, QUI, RIS, TOL, VAC **Elev.:** 600–2740 m a.s.l. **IUCN Habitat code:** 1,3,14 **Conservation status: (IUCN)** NE

7280 *Brugmansia × candida* Pers. (Solanaceae)
IPNI: 814597-1
Use: AF, EU, MA, ME, SU **Origin:** Native **Dept.:** BOG **IUCN Habitat code:** 1,3,14 **Conservation status: (IUCN)** NE

7281 *Brugmansia arborea* (L.) Sweet (Solanaceae)
IPNI: 77109574-1
Common name: borrachera, borrachero, borrachero blanco, cacao sabanero, chamico, corneta, floripondio, floripón, guamuco, guanto, tonca, tonga, yopa **Habit:** Shrub, Small tree **Use:** AF, EU, HF, MA, ME, SU **Origin:** Cultivated, Naturalised **Region:** Andean **Dept.:** NAR **Elev.:** 2700–2700 m a.s.l. **Conservation status: (IUCN)** EW

7282 *Brugmansia aurea* Lagerh. (Solanaceae)
IPNI: 37434-2
Common name: amarón, borrachera, borrachero, borrachero amarillo, borrachero blanco, cacao sabanero, culebra borrachera, culebra borrachero, culebra sabanera, munchira **Habit:** Shrub, Small tree **Use:** EU, ME, SU **Origin:** Native, Cultivated **Region:** Andean **Dept.:** ANT, BOG, BOY, CAL, CAQ, CAU, CES, CUN, HUI, MAG, MET, NAR, NSA, PUT, QUI, RIS, SAN, TOL, VAC **Elev.:** 600–3200 m a.s.l. **IUCN Habitat code:** 1,2,3,4,5,8,14 **Conservation status: (IUCN)** EW

7283 *Brugmansia insignis* (Barb.Rodr.) Lockwood ex R.E.Schult. (Solanaceae)
IPNI: 814604-1
Habit: Shrub **Use:** ME, SU **Origin:** Native, Cultivated **Region:** Amazonia, Andean, Pacific **Dept.:** AMA, ANT, CHO, HUI, MET, NAR, PUT **Elev.:** 80–2200 m a.s.l. **IUCN Habitat code:** 1,3,14 **Conservation status: (IUCN)** EW

7284 *Brugmansia sanguinea* (Ruiz & Pav.) D.Don (Solanaceae)
IPNI: 814612-1
Common name: borrachero, borrachero colorado, borrachero rojo, borrachero rosado, campanilla, campanilla encarnada, floripondio, floripondio colorado, floripondio encarna, floripondio encarnado, floripondio rojo, floripón, guamuca, guamuca borrachera, guamuco, guanto, **Habit:** Small tree, Tree **Use:** EU, MA, ME, PO, SU **Origin:** Native, Cultivated **Region:** Andean **Dept.:** ANT, BOG, BOY, CAU, CUN, NAR, PUT, SAN **Elev.:** 1100–3700 m a.s.l. **Conservation status: (IUCN)** EW

7285 *Brugmansia suaveolens* (Humb. & Bonpl. ex Willd.) Sweet (Solanaceae)
IPNI: 77109576-1
Common name: borrachera, borrachero, borrachero blanco, borrachero rojo, floripondio blanco, fut-avá-a, guanto, pimpinella, quinde borrachera **Use:** AF, EU, MA, ME, PO, SU **IUCN Habitat code:** 1,3,4,14 **Conservation status: (IUCN)** EW

7286 *Brugmansia versicolor* Lagerh. (Solanaceae)
IPNI: 37444-2
Common name: borrachero, corneto, tonga **Habit:** Shrub, Small tree **Use:** EU, ME **Origin:** Cultivated **Region:** Andean, Caribbean, Pacific **Dept.:** CHO, CUN, MAG, NAR, SAN **Elev.:** 0–2800 m a.s.l. **IUCN Habitat code:** 1,3,14 **Conservation status: (IUCN)** EW

7287 *Brugmansia vulcanicola* (A.S.Barclay) R.E.Schult. (Solanaceae)
IPNI: 37445-2
Common name: borrachero **Habit:** Shrub **Use:** ME **Origin:** Native, Cultivated **Region:** Andean **Dept.:** CAU, CUN, NAR, PUT **Elev.:** 2600–3500 m a.s.l. **IUCN Habitat code:** 3,4,14 **Conservation status: (IUCN)** EW

7288 *Brunfelsia americana* L. (Solanaceae)
IPNI: 319114-2
Habit: Shrub **Use:** EU, ME **Origin:** Cultivated **Region:** Caribbean **Dept.:** ATL **Elev.:** 0–10 m a.s.l. **IUCN Habitat code:** 2 **Conservation status: (IUCN)** NE

7289 *Brunfelsia australis* Benth. (Solanaceae)
IPNI: 37517-2
Common name: jazmín de Malabar **Habit:** Shrub **Use:** PO **Origin:** Cultivated **Region:** Andean **Dept.:** ANT, BOG, CAU **Elev.:** 1500–1800 m a.s.l. **IUCN Habitat code:** 1,3,14 **Conservation status: (IUCN)** LC

7290 *Brunfelsia chiricaspi* Plowman (Solanaceae)
IPNI: 37528-2
Common name: chiricaspi, zanango **Habit:** Shrub **Use:** ME, SU **Origin:** Native **Region:** Amazonia **Dept.:** CAQ, NAR, PUT **Elev.:** 200–600 m a.s.l. **IUCN Habitat code:** 1,3 **Conservation status: (IUCN)** NE

7291 *Brunfelsia chocoensis* Plowman (Solanaceae)
IPNI: 37529-2
Common name: carabajo **Habit:** Shrub **Use:** HF **Origin:** Native **Region:** Pacific **Dept.:** ANT, CHO **Elev.:** 50–200 m a.s.l. **IUCN Habitat code:** 1,14 **Conservation status: (IUCN)** NE

7292 *Brunfelsia grandiflora* D.Don (Solanaceae)
IPNI: 37537-2
Common name: chiricaspi, chiriguey, chirigüey, francesina, francesino, jardín de mayo, jazmín lavanda, juanramos, pino blanco, sietecueros, sirigüé **Habit:** Shrub, Small tree **Use:** EU, ME, SU **Origin:** Native, Cultivated **Region:** Amazonia, Guiana Shield, Orinoquia **Dept.:** AMA, CAQ, MET, PUT **Elev.:** 100–720 m a.s.l. **IUCN Habitat code:** 1,2,3,4,5,14 **Conservation status: (IUCN)** LC

7293 *Brunfelsia macrocarpa* Plowman (Solanaceae)
IPNI: 37546-2
Common name: mielquemada **Habit:** Shrub, Small tree **Use:** HF **Origin:** Native **Region:** Pacific **Dept.:** CAU, NAR, VAC **Elev.:** 0–50 m a.s.l. **IUCN Habitat code:** 1,5 **Conservation status: (IUCN)** NE

7294 *Brunfelsia pauciflora* (Cham. & Schltdl.) Benth. (Solanaceae)
IPNI: 814676-1
Common name: martín galve **Use:** EU, ME **IUCN Habitat code:** 2,3,14 **Conservation status: (IUCN)** LC

7295 *Capsicum annuum* L. (Solanaceae)
IPNI: 316944-2
Common name: aji charapilla, aji cheroso, aji piquiucho, ají, ají ajuja, ají amarillo, ají amarillo de culebra, ají bejuco, ají blanco, ají botón, ají bravo, ají camarón, ají carcarón de tiro, ají cascarón, ají catuto, ají cerbatano, ají chiche de perro, ají chichegat **Habit:** Herb, Shrub, Subshrub **Use:** EU, GS, HF, MA, ME, PO, SU **Origin:** Cultivated, Naturalised **Region:** Amazonia, Andean, Guiana Shield, Caribbean, Orinoquia, Pacific **Dept.:** AMA, ANT, BOG, BOL, BOY, CAU, CES, CHO, CUN, GUA, HUI, MAG, MET, NAR, NSA, QUI, RIS, TOL, VAC, VAU, VID **Elev.:** 5–1700 m a.s.l. **IUCN Habitat code:** 1,2,3,4,5,14 **Conservation status: (IUCN)** LC

7296 *Capsicum baccatum* L. (Solanaceae)
IPNI: 814733-1
Common name: ají, ají chirel, ají chivato, ají cimarrón, ají de babilla, ají de la capitanía, ají dulce, ají pajarito, ají pimienta, ají pimiento, ají pique, conguito, pimentón, pimiento **Habit:** Shrub, Subshrub **Use:** EU, GS, HF, ME **Origin:** Naturalised **Region:** Caribbean **Dept.:** MAG **Elev.:** 0-500 m.a.s.l. **IUCN Habitat code:** 1,2,3,4,5,14 **Conservation status: (IUCN)** NE **(CPSG)** Potential LC

7297 *Capsicum chinense* Jacq. (Solanaceae)
IPNI: 814752-1
Common name: ea, kumá, tot, tot-ea-kumá **Habit:** Herb, Subshrub **Use:** GS, HF, ME **Origin:** Cultivated **Region:** Amazonia, Guiana Shield, Caribbean **Dept.:** AMA, ANT, BOL, CAQ, CHO, NSA, SAP, SAN, VAU **Elev.:** 80–1000 m a.s.l. **IUCN Habitat code:** 1,3,4,5,14 **Conservation status: (IUCN)** NE

7298 *Capsicum dimorphum* (Miers) Kuntze (Solanaceae)
IPNI: 993722-1
Common name: ahuyamo **Habit:** Shrub, Subshrub **Use:** ME **Origin:** Native **Region:** Andean **Dept.:** ANT, BOY, CAL, CAQ, CAU, CUN, HUI, MET, NAR, QUI, RIS, SAN, TOL, VAC **Elev.:** 1200-2850 m.a.s.l. **IUCN Habitat code:** 1,3,4,14 **Conservation status: (IUCN)** NE **(CPSG)** Potential LC

7299 *Capsicum frutescens* L. (Solanaceae)
IPNI: 316959-2
Common name: ají, ají - pique, ají caribe, ají huevo de aragua ají, ají huevo de araguana, piman, pique **Habit:** Herb, Shrub, Subshrub **Use:** EU, HF, MA, ME, PO **Origin:** Cultivated, Naturalised **Region:** Amazonia, Andean, Guiana Shield **Dept.:** AMA, ARA, BOL, BOY, CAQ, CAU, CUN, GUA, GUV, MAG, MET, PUT, VAC, VAU, VID **Elev.:** 140–1700 m a.s.l. **Conservation status: (IUCN)** LC

7300 *Capsicum pubescens* Ruiz & Pav. (Solanaceae)
IPNI: 814839-1
Common name: ají, ají de guacamayo, ají morado, ají rocote, ají rocoto, ají vocato, rocote, rocoto **Habit:** Shrub, Small tree, Climbing **Use:** GS, HF, ME **Origin:** Cultivated, Naturalised **Region:** Andean **Dept.:**

BOG, HUI, NAR, PUT **Elev.:** 1600–3200 m a.s.l. **IUCN Habitat code:** 1,3,14 **Conservation status: (IUCN)** NE

7301 *Capsicum rhomboideum* (Humb. & Bonpl. ex Dunal) Kuntze (Solanaceae) **IPNI:** 993719-1 **Common name:** ají, ají silvestre **Habit:** Shrub, Small tree, Tree **Use:** HF, ME **Origin:** Native **Region:** Andean, Caribbean **Dept.:** ANT, ATL, CAL, CAU, CUN, HUI, LAG, MAG, NAR, NSA, SAN, TOL, VAC **Elev.:** 110-3000 m.a.s.l. **IUCN Habitat code:** 1,2,3,14 **Conservation status: (IUCN)** LC **(CPSG)** Potential LC

7302 *Cestrum affine* Kunth (Solanaceae) **IPNI:** 814887-1 **Habit:** Shrub **Use:** ME **Origin:** Native **Region:** Andean **Dept.:** ANT, CAU, NAR **Elev.:** 2350–3460 m a.s.l. **Conservation status: (IUCN)** NE

7303 *Cestrum alternifolium* (Jacq.) O.E.Schulz (Solanaceae) **IPNI:** 53464-2 **Common name:** pepa de culebra, tinto **Habit:** Shrub, Small tree, Tree **Use:** ME **Origin:** Native **Region:** Andean, Caribbean, Orinoquia **Dept.:** ATL, CES, CUN, HUI, LAG, MAG, MET, NSA, SAP, SAN, SUC, TOL **Elev.:** 0-750 m.a.s.l. **IUCN Habitat code:** 1,2,3,4,5,14 **Conservation status: (IUCN)** LC **(CPSG)** Potential LC

7304 *Cestrum buxifolium* Kunth (Solanaceae) **IPNI:** 814934-1 **Common name:** sauco, sauco cimarrón, sauco hediondo, tinto, tinto blanco, tinto negro, uvilla **Habit:** Herb, Shrub, Subshrub, Small tree **Use:** AF, EU, MA, ME **Origin:** Native **Region:** Andean **Dept.:** ANT, BOG, BOY, CAU, CUN, MET, NSA, SAN **Elev.:** 1870–4100 m a.s.l. **IUCN Habitat code:** 1,2,3,4,5,8,14 **Conservation status: (IUCN)** NE

7305 *Cestrum elegans* (Brongn. ex Neumann) Schltdl. (Solanaceae) **IPNI:** 815003-1 **Habit:** Shrub, Small tree **Use:** EU, ME **Origin:** Cultivated **Region:** Andean **Dept.:** BOG, CUN **Elev.:** 2600–2600 m a.s.l. **IUCN Habitat code:** 14 **Conservation status: (IUCN)** LC

7306 *Cestrum intermedium* Sendtn. (Solanaceae) **IPNI:** 815080-1 **Common name:** tinta pango **Use:** ME **IUCN Habitat code:** 1 **Conservation status: (IUCN)** LC

7307 *Cestrum latifolium* Lam. (Solanaceae) **IPNI:** 324587-2 **Common name:** venenito **Habit:** Shrub, Tree, Liana **Use:** HF, ME, PO **Origin:** Native **Region:** Andean, Caribbean, Orinoquia **Dept.:** ANT, ARA, CAS, CUN, MAG, MET, NSA **Elev.:** 200–1000 m a.s.l. **IUCN Habitat code:** 1,2,3,4,14 **Conservation status: (IUCN)** LC

7308 *Cestrum mariquitense* Kunth (Solanaceae) **IPNI:** 324588-2 **Common name:** jazmín de monte **Habit:** Shrub, Subshrub, Small tree **Use:** ME **Origin:** Native **Region:** Amazonia, Andean, Guiana Shield, Caribbean **Dept.:** AMA, ANT, BOG, BOY, CAL, CAS, CAU, CUN, HUI, MET, PUT, QUI, RIS, SAN, TOL, VAC **Elev.:** 20-2580 m.a.s.l. **IUCN Habitat code:** 1,2,3,14 **Conservation status: (IUCN)** LC **(CPSG)** Potential LC

7309 *Cestrum megalophyllum* Dunal (Solanaceae) **IPNI:** 316383-2 **Common name:** mazorco playero, palomaní, sauco negro, tabaco, uvillo **Habit:** Shrub, Small tree **Use:** HF **Origin:** Native **Region:** Andean, Pacific **Dept.:** ANT, CHO, CUN, HUI, NAR, QUI, VAC **Elev.:** 70–2780 m a.s.l. **IUCN Habitat code:** 1,2,3,5,14 **Conservation status: (IUCN)** NE

7310 *Cestrum microcalyx* Francey (Solanaceae) **IPNI:** 53665-2 **Common name:** chucho **Habit:** Shrub, Tree **Use:** ME **Origin:** Native **Region:** Amazonia, Andean, Orinoquia, Pacific **Dept.:** AMA, ANT, BOY, CAL, CAQ, CAU, CHO, GUV, MET, NAR, PUT, QUI, RIS, VAC **Elev.:** 80-1700 m.a.s.l. **IUCN Habitat code:** 1,3,5,14 **Conservation status: (IUCN)** LC **(CPSG)** Potential LC

7311 *Cestrum mutisii* Willd. ex Roem. & Schult. (Solanaceae) **IPNI:** 815171-1 **Common name:** tinto **Habit:** Shrub **Use:** ME **Origin:** Native **Region:** Andean **Dept.:** BOG, BOY, CUN **Elev.:** 1990–3000 m a.s.l. **IUCN Habitat code:** 1,2,3,14 **Conservation status: (IUCN)** NE

7312 *Cestrum nocturnum* L. (Solanaceae) **IPNI:** 316386-2 **Common name:** dama de noche, flautilla, galán de la noche, jazmín, jazmín de noche, jazmín nocturno **Habit:** Shrub, Subshrub, Small tree **Use:** EU, HF, IF, MA, ME, PO, SU **Origin:** Native **Region:** Amazonia, Andean, Pacific **Dept.:** AMA, ANT, BOG, BOY, CAL, CHO, CUN, NAR, NSA, QUI, RIS, SAN, TOL, VAC **Elev.:** 200–2600 m a.s.l. **IUCN Habitat code:** 1,2,3,14 **Conservation status: (IUCN)** LC

7313 *Cestrum ochraceum* Francey (Solanaceae) **IPNI:** 53678-2 **Common name:** higo, tabaquillo **Habit:** Shrub, Tree **Use:** ME **Origin:** Native **Region:** Andean **Dept.:** ANT, BOG, CAL, CAU, CHO, CUN, HUI, NAR, PUT, QUI, RIS, TOL, VAC **Elev.:** 1400-3400 m.a.s.l. **IUCN Habitat code:** 1,2,3,4,14 **Conservation status: (IUCN)** LC **(CPSG)** Potential LC

7314 *Cestrum racemosum* Ruiz & Pav. (Solanaceae) **IPNI:** 330718-2 **Common name:** galán de noche **Habit:** Small tree, Tree **Use:** HF, ME, SU **Origin:** Native **Region:** Andean, Caribbean **Dept.:** ANT, BOY, CAL, CAQ, CAS, CAU, CUN, MAG, MET, NAR, NSA, PUT, QUI, RIS, SAN, VAC **Elev.:** 50-2300 m.a.s.l. **IUCN Habitat code:** 1,3,4,14 **Conservation status: (IUCN)** LC **(CPSG)** Potential LC

7315 *Cestrum reflexum* Sendtn. (Solanaceae) **IPNI:** 316388-2 **Habit:** Shrub, Liana, Climbing **Use:** ME **Origin:** Native **Region:** Amazonia, Andean, Pacific **Dept.:** AMA, ANT, CAQ, CHO, MET, QUI, RIS, SAN, TOL, VAC **Elev.:** 10–1100 m a.s.l. **IUCN Habitat code:** 1,3,4,14 **Conservation status: (IUCN)** NE

7316 *Cestrum schlechtendalii* G.Don (Solanaceae) **IPNI:** 330712-2 **Habit:** Shrub, Subshrub, Small tree **Use:** HF, ME **Origin:** Native **Region:** Amazonia, Andean **Dept.:** AMA, ANT, BOY, CAQ, CAU, CHO, CUN, MAG, MET, NAR, NSA, PUT, RIS, SAN, TOL, VAC **Elev.:** 40–2220 m a.s.l. **IUCN Habitat code:** 1,2,3,5,8,14 **Conservation status: (IUCN)** LC

7317 *Cestrum strigilatum* Ruiz & Pav. (Solanaceae) **IPNI:** 330710-2 **Common name:** clavellino verde **Use:** ME **Origin:** Native **Conservation status: (IUCN)** LC

7318 *Cestrum strigillatum* Ruiz & Pav. (Solanaceae) **IPNI:** 815306-1 **Habit:** Shrub, Small tree, Liana, Climbing **Use:** ME **Origin:** Endemic **Region:** Andean **Dept.:** ANT, CUN, SAN **Elev.:** 300–1500 m a.s.l. **IUCN Habitat code:** 1,4,14 **Conservation status: (IUCN)** NE

7319 *Cestrum tomentosum* L.f. (Solanaceae) **IPNI:** 330708-2 **Common name:** hoja de mono, pasmo, tinto **Habit:** Shrub, Tree **Use:** MA, ME, PO **Origin:** Native **Region:** Andean **Dept.:** ANT, BOG, BOY, CAU, CUN, NSA, QUI, RIS, VAC **Elev.:** 1100–2845 m a.s.l. **IUCN Habitat code:** 1,2,3,4,14 **Conservation status: (IUCN)** LC

7320 *Datura candida* Saff. (Solanaceae) **IPNI:** 76934-2 **Common name:** amarón borrachero, borrachero, borrachero blanco, cacao sabanero, floripondio, guamsian borrachero, kinde borrachero, munchira, muscuai borrachera, salamán borrachero **Use:** ME **Origin:** Endemic **Conservation status: (IUCN)** NE

7321 *Datura ferox* L. (Solanaceae) **IPNI:** 815555-1 **Habit:** Herb **Use:** ME, PO, SU **Origin:** Naturalised **Region:** Andean **Dept.:** SAN **Elev.:** 1000–1000 m a.s.l. **IUCN Habitat code:** 1,2,3,5,14 **Conservation status: (IUCN)** NE

7322 *Datura innoxia* Mill. (Solanaceae) **IPNI:** 316945-2 **Common name:** chamico, malaatapui, mata de San Pedro, ñongué **Habit:** Herb, Shrub **Use:** EU, HF, MA, ME, PO, SU **Origin:** Naturalised **Region:** Caribbean **Dept.:** ATL, BOL, CES, CUN, LAG, MAG **Elev.:** 50–800 m a.s.l. **IUCN Habitat code:** 1,2,3,14 **Conservation status: (IUCN)** NE

7323 *Datura metel* L. (Solanaceae) **IPNI:** 314739-2 **Common name:** chamico, gigantón morado, guaco, higantón **Habit:** Herb, Shrub **Use:** EU, GS, HF, MA, ME, PO, SU **Origin:** Native **Region:** Caribbean **Dept.:** ANT, BOL, SAN **Elev.:** 0–200 m a.s.l. **IUCN Habitat code:** 2,3,14 **Conservation status: (IUCN)** NE

7324 *Datura stramonium* L. (Solanaceae) **IPNI:** 314738-2 **Common name:** carda, chamico, chamico morado, datira, estramonio, herb hedionda, manzano espinoso, nongué, pedro noche, ñongué, ñoque **Habit:** Herb, Shrub **Use:** EU, FU, HF, MA, ME, PO, SU **Origin:** Naturalised **Region:** Andean **Dept.:** ANT, BOG, BOY, CUN, HUI, NAR, NSA, TOL, VAC **Elev.:** 1000–2600 m a.s.l. **Conservation status: (IUCN)** NE

7325 *Hawkesiophyton ochraceum* (Cuatrec.) A.Orejuela & C.I.Orozco (Solanaceae) **IPNI:** 77164742-1 **Common name:** ayahuasca **Habit:** Shrub, Small tree, Tree, Epiphyte, Climbing **Use:** ME **Origin:** Native **Region:** Andean, Guiana Shield, Orinoquia **Dept.:** ANT, ARA, BOY, CAQ, CUN, MET, PUT, SAN **Elev.:** 300–1500 m a.s.l. **IUCN Habitat code:** 1,3,4,14 **Conservation status: (IUCN)** NE

7326 *Iochroma arborescens* (L.) J.M.H.Shaw (Solanaceae) **IPNI:** 77186334-1 **Common name:** borrachero, cojoco, frutogallino, mínigo, míniga, míniguo, nocua, tabalgue, tobo, tocua, tomatoquina, tomatoquín, toquín, uvito de gallina **Habit:** Shrub, Small tree, Tree **Use:** HF, MA, ME **Origin:** Native **Region:** Andean **Dept.:** ANT, BOY, CUN, HUI, MET, NSA, SAN **Elev.:** 1000–2400 m a.s.l. **Conservation status: (IUCN)** NE

7327 *Iochroma fuchsioides* (Bonpl.) Miers (Solanaceae) **IPNI:** 815977-1 **Common name:** corazón de pollo, gallino **Habit:** Shrub, Small tree **Use:** HF, ME, SU **Origin:** Native **Region:** Andean **Dept.:** CAU, CUN, NAR, PUT, QUI **Elev.:** 2000–3150 m a.s.l. **IUCN Habitat code:** 1,3,4,14 **Conservation status: (IUCN)** NE

7328 *Jaltomata procumbens* (Cav.) J.L.Gentry (Solanaceae) **IPNI:** 131311-2 **Common name:** uchuva negra, yerbamora **Habit:** Herb **Use:** HF, ME **Origin:** Native **Region:** Andean **Dept.:** ANT, MAG, NAR, PUT, QUI **Elev.:** 1200–2900 m a.s.l. **IUCN Habitat code:** 1,3,14 **Conservation status: (IUCN)** NE

7329 *Jaltomata repandidentata* (Dunal) Hunz. (Solanaceae) **IPNI:** 131312-2 **Habit:** Herb **Use:** HF **Origin:** Native **Region:** Andean **Dept.:** ANT, BOY, MAG, RIS **Elev.:** 750–1700 m a.s.l. **IUCN Habitat code:** 1,3 **Conservation status: (IUCN)** NE

7330 *Jaltomata sinuosa* (Miers) Mione (Solanaceae) **IPNI:** 1018958-1 **Common name:** tomatillo **Habit:** Herb, Shrub **Use:** HF **Origin:** Native **Region:** Andean **Dept.:** ANT, BOG, BOY, CUN, MAG **Elev.:** 1360–3420 m a.s.l. **IUCN Habitat code:** 1,3,14 **Conservation status: (IUCN)** NE

7331 *Jaltomata viridiflora* (Kunth) M.Nee & Mione (Solanaceae) **IPNI:** 1029212-2 **Common name:** belladona, guchuva de perro, uchuva blanca **Habit:** Herb, Shrub, Subshrub **Use:** HF **Origin:** Native **Region:** Andean **Dept.:** ANT, BOG, BOY, CAL, CAU, CUN, NAR, PUT, QUI, RIS, SAN **Elev.:** 2100–3700 m a.s.l. **IUCN Habitat code:** 1,3,4,8,14 **Conservation status: (IUCN)** NE

7332 *Lycianthes inaequilatera* (Rusby) Bitter (Solanaceae) **IPNI:** 146533-2 **Habit:** Herb, Shrub, Subshrub **Use:** HF **Origin:** Native **Region:** Andean **Dept.:** ANT, CHO, CUN, MAG, NSA, QUI, RIS **Elev.:** 280–1800 m a.s.l. **IUCN Habitat code:** 1,3,5,14 **Conservation status: (IUCN)** NE

7333 *Lycianthes lycioides* (L.) Hassl. (Solanaceae) **IPNI:** 146555-2 **Common name:** cobán, gurrubo, gurumo, sambo, sombo, tote, zumbito **Habit:** Herb, Shrub, Subshrub **Use:** HF, ME **Origin:** Native **Region:** Andean **Dept.:** BOG, BOY, CUN, VAC **Elev.:** 1200–3420 m a.s.l. **IUCN Habitat code:** 1,3,14 **Conservation status: (IUCN)** NE

7334 *Lycianthes medusocalyx* (Bitter) Bitter (Solanaceae) **IPNI:** 146567-2 **Habit:** Herb, Shrub, Subshrub **Use:** HF **Origin:** Native **Region:** Amazonia, Andean **Dept.:** AMA, ANT, CAU, PUT **Elev.:** 100–1600 m a.s.l. **IUCN Habitat code:** 1,3,14 **Conservation status: (IUCN)** NE

7335 *Lycianthes pauciflora* (Vahl) Bitter (Solanaceae) **IPNI:** 146582-2 **Habit:** Shrub, Subshrub, Liana, Climbing **Use:** HF **Origin:** Native **Region:** Amazonia, Andean, Pacific **Dept.:** AMA, ANT, CAU, CHO, CUN, HUI, MAG, MET, NAR, PUT, QUI, RIS, VAC **Elev.:** 0-1840 m.a.s.l. **IUCN Habitat code:** 1,2,3,4,5,14 **Conservation status: (IUCN)** NE **(CPSG)** Potential LC

7336 *Lycianthes synanthera* (Sendtn.) Bitter (Solanaceae) **IPNI:** 146628-2 **Habit:** Herb, Shrub, Subshrub, Climbing **Use:** HF **Origin:** Native **Region:** Amazonia, Andean, Guiana Shield, Caribbean, Pacific **Dept.:** ANT, CAQ, CAU, CHO, LAG, MAG, MET, RIS **Elev.:** 20–2600 m a.s.l. **IUCN Habitat code:** 1,3,14 **Conservation status: (IUCN)** LC

7337 *Lycium americanum* Jacq. (Solanaceae) **IPNI:** 816336-1 **Habit:** Shrub, Subshrub **Use:** HF **Origin:** Native **Region:** Caribbean **Dept.:** BOL, LAG, MAG **Elev.:** 30–50 m a.s.l. **Conservation status: (IUCN)** NE

7338 *Markea coccinea* Rich. (Solanaceae) **IPNI:** 816751-1 **Habit:** Shrub, Epiphyte, Hemiepiphyte, Liana **Use:** ME **Origin:** Native **Region:** Amazonia, Guiana Shield **Dept.:** AMA, CAQ, VAU **Elev.:** 200–500 m a.s.l. **IUCN Habitat code:** 1,14 **Conservation status: (IUCN)** NE

7339 *Nicandra physalodes* (L.) Gaertn. (Solanaceae) **IPNI:** 816832-1 **Common name:** chamico, mostaza extranjera, sacabuche, suspiro, tomatillo, uchuva de perro, uchuva de perros, uchuva del duende, uva, uvilla **Habit:** Herb, Shrub **Use:** EU, GS, HF, ME, PO **Origin:** Naturalised **Region:** Andean **Dept.:** ANT, BOG, BOY, CAU, CHO, CUN, NAR, QUI, TOL, VAC **Elev.:** 1000–2800 m a.s.l. **IUCN Habitat code:** 1,3,14 **Conservation status: (IUCN)** NE

7340 *Nicotiana alata* Link & Otto (Solanaceae) **IPNI:** 816845-1 **Common name:** tabaco, tabaquillo **Habit:** Herb **Use:** EU, GS, ME, SU **Origin:** Cultivated, Naturalised **Region:** Andean **Dept.:** BOG, CUN **Elev.:** 1700–2620 m a.s.l. **IUCN Habitat code:** 3,14 **Conservation status: (IUCN)** NE

7341 *Nicotiana glauca* Graham (Solanaceae) **IPNI:** 816941-1 **Habit:** Shrub, Small tree **Use:** EU, GS, HF, MA, ME, PO, SU **Origin:** Naturalised **Region:** Andean **Dept.:** ANT, BOY **Elev.:** 1450–2700 m a.s.l. **IUCN Habitat code:** 3,14 **Conservation status: (IUCN)** NE

7342 *Nicotiana rustica* L. (Solanaceae) **IPNI:** 817055-1 **Common name:** ambirá, tabaco, tabaco sabanero, tabaco yuá **Habit:** Subshrub **Use:** EU, GS, HF, MA, ME, PO, SU **Origin:** Naturalised **Region:** Andean **Dept.:** BOG, CUN, MAG, NAR **Elev.:** 1750–4000 m a.s.l. **IUCN Habitat code:** 14 **Conservation status: (IUCN)** NE

7343 *Nicotiana tabacum* L. (Solanaceae) **IPNI:** 817077-1 **Common name:** tabaco, tabaco calzao, tabaco cuba, tabaco de puntas, tabaco frenteloro, tabaco habano, tabaco negrolindo, tabaco puntalargo, tabak, tawaku **Habit:** Herb, Liana **Use:** EU, GS, HF, IF, MA, ME, PO, SU **Origin:** Cultivated, Naturalised **Region:** Amazonia, Andean, Orinoquia **Dept.:** AMA, ANT, BOG, CAU, CES, CUN, MAG, NAR, PUT, SAN, VAC, VAU **Elev.:** 100–2600 m a.s.l. **IUCN Habitat code:** 1,2,3,5,14 **Conservation status: (IUCN)** NE

7344 *Nierembergia repens* Ruiz & Pav. (Solanaceae) **IPNI:** 817157-1 **Habit:** Herb **Use:** EU **Origin:** Native **Region:** Andean **Dept.:** BOG, BOY, CUN **Elev.:** 2550–3300 m a.s.l. **IUCN Habitat code:** 8,14 **Conservation status: (IUCN)** NE

7345 *Petunia × atkinsiana* (Sweet) D.Don ex W.H.Baxter (Solanaceae) **IPNI:** 817227-1 **Common name:** conservadora, petunia **Habit:** Herb **Use:** EU, ME **Origin:** Endemic, Cultivated **Region:** Andean **Dept.:** ANT, CAL, CUN, MET, NSA **Elev.:** 2150–2620 m a.s.l. **IUCN Habitat code:** 1,3,14 **Conservation status: (IUCN)** NE

7346 *Physalis angulata* L. (Solanaceae) **IPNI:** 195334-2 **Common name:** ajenje, ají, huilla, mullaca, tofetorofe, topotopo, topotoropo, uchuva, uva, uvilla, uvillas, uvita de monte, vejigón **Habit:** Herb, Subshrub **Use:** GS, HF, MA, ME, SU **Origin:** Native **Region:** Amazonia, Andean, Caribbean, Orinoquia, Pacific **Dept.:** AMA, ANT, ARA, ATL, BOL, CAL, CAU, CHO, COR, CUN, LAG, MAG, MET, NAR, NSA, PUT, QUI, SAP, SAN, SUC, TOL, VAC **Elev.:** 0-1780 m.a.s.l. **IUCN Habitat code:** 1,2,3,5,14 **Conservation status: (IUCN)** LC **(CPSG)** Potential LC

7347 *Physalis cordata* Houst. ex Mill. (Solanaceae) **IPNI:** 817370-1 **Common name:** topotoropo **Habit:** Herb, Subshrub **Use:** HF, ME **Origin:** Native **Region:** Caribbean **Dept.:** LAG **Elev.:** 0–0 m a.s.l. **IUCN Habitat code:** 1,3 **Conservation status: (IUCN)** LC

7348 *Physalis lagascae* Roem. & Schult. (Solanaceae) **IPNI:** 817449-1 **Common name:** uchuvilla, vejigón **Habit:** Herb, Subshrub **Use:** GS, HF, ME **Origin:** Native **Dept.:** CUN, SAN, TOL **Elev.:** 100–438 m a.s.l. **Conservation status: (IUCN)** LC

7349 *Physalis peruviana* L. (Solanaceae)
IPNI: 195473-2
Common name: alquenque, bolsa de amor, buchuba, buchuva, capulí, chamico, guchero, guchuba, guchubo, guchuvo, herbbuena, hierbabuena, huevo de sapo, ochuva, tomate, topo-topo, uchuba, uchubo, uchuva, uchuvo, uvilla, uvilla buchona, vejigon, vejigón **Habit:** Herb, Subshrub **Use:** EU, GS, HF, MA, ME, PO **Origin:** Cultivated, Naturalised **Region:** Andean **Dept.:** ANT, BOG, BOY, CES, HUI, MAG, PUT, QUI, TOL **Elev.:** 1300–3700 m a.s.l. **IUCN Habitat code:** 1,3,4,8,14 **Conservation status: (IUCN)** NE

7350 *Physalis pubescens* L. (Solanaceae)
IPNI: 195489-2
Common name: uchuva, uvillo, vejigón **Habit:** Herb **Use:** AF, GS, HF, MA, ME **Origin:** Native **Region:** Andean, Pacific **Dept.:** ANT, CHO **Elev.:** 0–700 m a.s.l. **IUCN Habitat code:** 1,3,5,14 **Conservation status: (IUCN)** LC

7351 *Poortmannia speciosa* Drake (Solanaceae)
IPNI: 208218-2
Habit: Epiphyte, Hemiepiphyte, Liana **Use:** HF **Origin:** Native **Region:** Andean **Dept.:** CAU, HUI, MAG, NAR, SAN **Elev.:** 1300–2600 m a.s.l. **IUCN Habitat code:** 1,14 **Conservation status: (IUCN)** NE

7352 *Salpichroa diffusa* Walp. (Solanaceae)
IPNI: 817738-1
Common name: zarcillo de bruja **Use:** HF, ME **Conservation status: (IUCN)** NE

7353 *Salpichroa tristis* Walp. (Solanaceae)
IPNI: 817756-1
Common name: belladona, cherche, uvilla **Habit:** Herb, Shrub, Liana, Climbing **Use:** HF **Origin:** Native **Region:** Andean **Dept.:** BOG, BOY, CAL, CAU, CUN, HUI, MET, NAR, SAN **Elev.:** 460–3950 m a.s.l. **IUCN Habitat code:** 1,3,4,8,14 **Conservation status: (IUCN)** NE

7354 *Schwenckia americana* L. (Solanaceae)
IPNI: 817941-1
Habit: Herb **Use:** AF, HF, MA, ME **Origin:** Native **Region:** Guiana Shield, Caribbean, Orinoquia, Pacific **Dept.:** CAS, GUA, MAG, MET, VAC, VAU **Elev.:** 65–500 m a.s.l. **IUCN Habitat code:** 1,4,14 **Conservation status: (IUCN)** NE

7355 *Solandra grandiflora* Sw. (Solanaceae)
IPNI: 329815-2
Habit: Shrub, Liana **Use:** EU, HF, ME **Origin:** Cultivated **Region:** Andean, Caribbean **Dept.:** MAG, SAP, VAC **Elev.:** 0–1800 m a.s.l. **IUCN Habitat code:** 14 **Conservation status: (IUCN)** NE

7356 *Solandra longiflora* Tussac (Solanaceae)
IPNI: 818075-1
Habit: Shrub, Liana **Use:** EU **Origin:** Cultivated **Region:** Caribbean **Dept.:** MAG **Elev.:** 0–10 m a.s.l. **Conservation status: (IUCN)** NE

7357 *Solandra maxima* (Moc. & Sessé ex Dunal) P.S.Green (Solanaceae)
IPNI: 818078-1
Habit: Shrub, Small tree, Liana **Use:** EU, HF, MA, ME **Origin:** Native **Region:** Andean **Dept.:** ANT, BOG, CUN, SAN **Elev.:** 1300–2600 m a.s.l. **IUCN Habitat code:** 1,14 **Conservation status: (IUCN)** NE

7358 *Solanum agrarium* Sendtn. (Solanaceae)
IPNI: 818168-1
Common name: hoja de luna **Habit:** Herb **Use:** HF, ME **Origin:** Native **Region:** Andean, Caribbean **Dept.:** BOL, CUN, HUI, LAG, MAG, MET, QUI, RIS, SAN, TOL, VAC **Elev.:** 20–830 m a.s.l. **IUCN Habitat code:** 1,2,3,14 **Conservation status: (IUCN)** NE

7359 *Solanum albidum* Dunal (Solanaceae)
IPNI: 322192-2
Habit: Shrub **Use:** ME **Origin:** Naturalised **Region:** Andean **Dept.:** ANT, VAC **Elev.:** 1000–2300 m a.s.l. **IUCN Habitat code:** 14 **Conservation status: (IUCN)** LC

7360 *Solanum allophyllum* (Miers) Standl. (Solanaceae)
IPNI: 238268-2
Common name: bleo de gallinazo, bleo de golero, chucho fruteador **Habit:** Herb, Subshrub **Use:** HF, ME **Origin:** Native **Region:** Caribbean, Pacific **Dept.:** ANT, BOL, CES, CHO, LAG, MAG, NSA, SAN, SUC **Elev.:** 0–300 m a.s.l. **IUCN Habitat code:** 1,2,14 **Conservation status: (IUCN)** NE

7361 *Solanum americanum* Mill. (Solanaceae)
IPNI: 238279-2
Common name: charrul, chumbalo, hierbamora, hierbamora blanca, tomaticos, yerba mora, yerbamora **Habit:** Herb, Shrub, Subshrub **Use:** AF, EU, GS, HF, MA, ME, PO **Origin:** Native **Region:** Amazonia, Andean, Caribbean, Orinoquia, Pacific **Dept.:** AMA, ANT, ATL, BOG, BOL, BOY, CAS, CAU, CES, CHO, CUN, HUI, MAG, MET, NAR, NSA, RIS, SAP, SAN, TOL, VAC **Elev.:** 0-2600 m.a.s.l. **IUCN Habitat code:** 1,2,3,4,5,14 **Conservation status: (IUCN)** NE **(CPSG)** Potential LC

7362 *Solanum andreanum* Baker (Solanaceae)
IPNI: 818239-1
Habit: Herb **Use:** GS **Origin:** Native **Region:** Andean **Dept.:** CAU, NAR, PUT **Elev.:** 2200–3190 m a.s.l. **IUCN Habitat code:** 1,3,4,14 **Conservation status: (IUCN)** NE

7363 *Solanum apaporanum* R.E.Schult. (Solanaceae)
IPNI: 238333-2
Habit: Liana **Use:** ME **Origin:** Native **Region:** Amazonia, Orinoquia **Dept.:** GUV, MET, VAU **Elev.:** 100–640 m a.s.l. **IUCN Habitat code:** 1,3 **Conservation status: (IUCN)** NE

7364 *Solanum asperolanatum* Ruiz & Pav. (Solanaceae)
IPNI: 818330-1
Habit: Shrub, Small tree, Tree **Use:** EU, GS, MA, ME **Origin:** Native **Region:** Andean **Dept.:** ANT, BOG, CHO, CUN, HUI, NAR, QUI, RIS, SAN, TOL **Elev.:** 800–3100 m a.s.l. **IUCN Habitat code:** 1,3,4,8,14 **Conservation status: (IUCN)** LC

7365 *Solanum asperum* Rich. (Solanaceae)
IPNI: 818335-1
Common name: cucuo, mapuro **Habit:** Herb, Shrub, Small tree **Use:** HF, ME **Origin:** Native **Region:** Amazonia, Caribbean, Orinoquia, Pacific **Dept.:** ANT, CHO, COR, GUV, MET, SAN **Elev.:** 100–900 m a.s.l. **IUCN Habitat code:** 1,3,4,5,14 **Conservation status: (IUCN)** NE

7366 *Solanum atropurpureum* Schrank (Solanaceae)
IPNI: 330638-2
Common name: mancadera, mancamano, peloepuerco **Habit:** Herb, Shrub, Subshrub **Use:** HF **Origin:** Native **Region:** Andean **Dept.:** ANT, COR, QUI, VAC **Elev.:** 1500–2500 m a.s.l. **IUCN Habitat code:** 1,3,5,14 **Conservation status: (IUCN)** NE

7367 *Solanum barbeyanum* Huber (Solanaceae)
IPNI: 238405-2
Habit: Subshrub, Climbing **Use:** HF **Origin:** Native **Region:** Amazonia, Andean **Dept.:** AMA, PUT, VAU **Elev.:** 200–1350 m a.s.l. **IUCN Habitat code:** 1,3,14 **Conservation status: (IUCN)** LC

7368 *Solanum betaceum* Cav. (Solanaceae)
IPNI: 306169-2
Common name: pepino de tree, pepino de árbol, tamarillo, tomate, tomate de montaña, tomate de monte, tomate de tree, tomate de árbol, tomate guiteño, tomate quiteño, tomate silvestre **Habit:** Shrub, Small tree **Use:** AF, GS, HF, ME **Origin:** Native, Cultivated **Region:** Andean **Dept.:** ANT, BOG, BOY, CAL, CAU, CUN, MAG, NAR, NSA, PUT, QUI, RIS, VAC **Elev.:** 1000–3000 m a.s.l. **IUCN Habitat code:** 1,3,4,5,8,14 **Conservation status: (IUCN)** DD

7369 *Solanum brevifolium* Humb. & Bonpl. ex Dunal (Solanaceae)
IPNI: 818501-1
Common name: papa silvestre, riñonada **Habit:** Herb, Liana, Climbing **Use:** HF, ME **Origin:** Native **Region:** Andean **Dept.:** ANT, BOG, BOY, CAL, CAU, CUN, HUI, MAG, NAR, PUT, QUI, RIS, SAN, TOL, VAC **Elev.:** 1800-3950 m.a.s.l. **IUCN Habitat code:** 1,3,4,8,14 **Conservation status: (IUCN)** NE **(CPSG)** Potential LC

7370 *Solanum cajanumense* Kunth (Solanaceae)
IPNI: 301585-2
Habit: Small tree **Use:** GS, HF **Origin:** Native **Region:** Andean **Dept.:** TOL, VAC **Elev.:** 1950–2050 m a.s.l. **IUCN Habitat code:** 1 **Conservation status: (IUCN)** NT

7371 *Solanum calidum* Bohs (Solanaceae)
IPNI: 985913-1
Habit: Shrub, Subshrub **Use:** GS, HF **Origin:** Native **Region:** Amazonia, Andean **Dept.:** CAQ, PUT **Elev.:** 150–1700 m a.s.l. **IUCN Habitat code:** 1,3,5,14 **Conservation status: (IUCN)** NE

7372 *Solanum candidum* Lindl. (Solanaceae)
IPNI: 818589-1
Common name: lulo **Habit:** Herb, Shrub **Use:** HF **Origin:** Native, Cultivated **Region:** Andean, Pacific **Dept.:** CHO, VAC **Elev.:** 50–1000 m a.s.l. **IUCN Habitat code:** 14 **Conservation status: (IUCN)** NE

7373 *Solanum canense* Rydb. (Solanaceae)
IPNI: 238501-2
Common name: papa silvestre **Habit:** Herb, Climbing **Use:** HF **Origin:** Native **Region:** Caribbean, Pacific **Dept.:** ANT, BOL, CHO, MAG **Elev.:** 100–770 m a.s.l. **IUCN Habitat code:** 1,2,4,5 **Conservation status: (IUCN)** NE

7374 *Solanum capsicoides* All. (Solanaceae) **IPNI:** 330645-2 **Habit:** Herb, Shrub **Use:** EU, GS, HF, MA, ME, PO **Origin:** Native **Region:** Andean **Dept.:** ANT, CUN, SAN, VAC **Elev.:** 900–1800 m a.s.l. **IUCN Habitat code:** 1,3,14 **Conservation status: (IUCN)** NE

7375 *Solanum caripense* Humb. & Bonpl. ex Dunal (Solanaceae) **IPNI:** 818619-1 **Common name:** berriadores, chimbalo, chimbolo, dulonsoga, dulunsoga, llorones, llorón, mamoncillo, mamón de monte, pepino llorón **Habit:** Herb, Climbing **Use:** GS, HF, ME **Origin:** Native **Region:** Andean **Dept.:** ANT, BOG, CAU, CES, CUN, HUI, MAG, NAR, QUI, SAN, TOL, VAC **Elev.:** 800–3900 m a.s.l. **IUCN Habitat code:** 1,3,4,5,14 **Conservation status: (IUCN)** NE

7376 *Solanum colombianum* Dunal (Solanaceae) **IPNI:** 818757-1 **Common name:** llorones, papa, papa de duende, papa de indio, papa de la vieja, papa de monte, papa de vieja, papa silvestre, sachapapa **Habit:** Herb **Use:** GS **Origin:** Native **Region:** Andean **Dept.:** ANT, BOG, BOY, CAL, CAQ, CAU, CES, CUN, HUI, MAG, NAR, NSA, PUT, QUI, RIS, TOL, VAC **Elev.:** 2200-3900 m.a.s.l. **IUCN Habitat code:** 1,2,3,4,5,8,14 **Conservation status: (IUCN)** NE **(CPSG)** Potential LC

7377 *Solanum confine* Dunal (Solanaceae) **IPNI:** 818775-1 **Habit:** Shrub **Use:** MA **Origin:** Native **Region:** Amazonia **Dept.:** VAU **Elev.:** 200–200 m a.s.l. **IUCN Habitat code:** 4 **Conservation status: (IUCN)** LC

7378 *Solanum crinitipes* Dunal (Solanaceae) **IPNI:** 818842-1 **Common name:** cucubo **Habit:** Shrub, Subshrub, Small tree, Tree **Use:** ME **Origin:** Native **Region:** Andean **Dept.:** BOG, BOY, CUN, MAG **Elev.:** 2200–3600 m a.s.l. **IUCN Habitat code:** 1,14 **Conservation status: (IUCN)** NE

7379 *Solanum crinitum* Lam. (Solanaceae) **IPNI:** 818843-1 **Common name:** cocubo, lulo, lulo de monte, lulo de perro, lulo silvestre, tachuelo, tupiro **Habit:** Shrub, Tree **Use:** HF, ME **Origin:** Native **Region:** Amazonia, Guiana Shield, Orinoquia **Dept.:** AMA, ANT, BOL, CAQ, GUA, MET, SAN, VAU, VID **Elev.:** 100–1000 m a.s.l. **IUCN Habitat code:** 1,2,3,4,5,14 **Conservation status: (IUCN)** LC

7380 *Solanum diversifolium* Humb. & Bonpl. ex Dunal (Solanaceae) **IPNI:** 819037-1 **Habit:** Shrub, Subshrub, Small tree **Use:** GS, HF, MA, ME, PO **Origin:** Native **Region:** Andean, Pacific **Dept.:** ANT, CHO, HUI, MAG, NSA, SAN **Elev.:** 90–1850 m a.s.l. **IUCN Habitat code:** 1,2,3,14 **Conservation status: (IUCN)** LC

7381 *Solanum endopogon* (Bitter) Bohs (Solanaceae) **IPNI:** 985906-1 **Common name:** gallinazo panga **Habit:** Small tree **Use:** GS, HF, MA, ME **Origin:** Native **Region:** Amazonia, Pacific **Dept.:** AMA, GUV, PUT, VAC **Elev.:** 250–280 m a.s.l. **IUCN Habitat code:** 1 **Conservation status: (IUCN)** LC

7382 *Solanum erianthum* D.Don (Solanaceae) **IPNI:** 819125-1 **Habit:** Small tree **Use:** EU, GS, HF, MA, ME, PO **Origin:** Native **Dept.:** ANT **Elev.:** 400–700 m a.s.l. **IUCN Habitat code:** 1,3,14 **Conservation status: (IUCN)** NE

7383 *Solanum fallax* Bohs (Solanaceae) **IPNI:** 985910-1 **Habit:** Shrub, Small tree **Use:** GS **Origin:** Native **Dept.:** VAC **Elev.:** 900–1200 m a.s.l. **IUCN Habitat code:** 1 **Conservation status: (IUCN)** NT

7384 *Solanum galeatum* André (Solanaceae) **IPNI:** 819272-1 **Common name:** tomate de tree **Use:** ME **Origin:** Endemic **IUCN Habitat code:** 14 **Conservation status: (IUCN)** NE

7385 *Solanum garcia-barrigae* Ochoa (Solanaceae) **IPNI:** 238854-2 **Habit:** Herb **Use:** GS **Origin:** Endemic **Region:** Andean **Dept.:** MAG, NSA **Elev.:** 3000–3900 m a.s.l. **IUCN Habitat code:** 1,2,3,4,5 **Conservation status: (IUCN)** LC

7386 *Solanum gardneri* Sendtn. (Solanaceae) **IPNI:** 819278-1 **Common name:** cucubo, frutillo, frutillo macho **Habit:** Herb, Shrub, Subshrub **Use:** HF **Origin:** Native **Region:** Andean **Dept.:** BOY, HUI, SAN **Elev.:** 300–3400 m a.s.l. **IUCN Habitat code:** 1,2,3,14 **Conservation status: (IUCN)** NE

7387 *Solanum gonocladum* Dunal (Solanaceae) **IPNI:** 819340-1 **Use:** ME **Conservation status: (IUCN)** NE

7388 *Solanum grandiflorum* Ruiz & Pav. (Solanaceae) **IPNI:** 819358-1 **Common name:** cojaca, cucubo, cujaco, pepito, tachuelo **Habit:** Shrub, Small tree, Tree **Use:** EU, GS, HF, ME **Origin:** Native **Region:** Amazonia, Andean, Guiana Shield **Dept.:** AMA, ANT, GUA, QUI, VAC **Elev.:** 50–780 m a.s.l. **IUCN Habitat code:** 1,3,4,5,14 **Conservation status: (IUCN)** NE

7389 *Solanum hayesii* Fernald (Solanaceae) **IPNI:** 238943-2 **Common name:** sauco macho **Habit:** Shrub, Tree **Use:** MA **Origin:** Native **Region:** Andean, Pacific **Dept.:** ANT, CHO, VAC **Elev.:** 0–2000 m a.s.l. **IUCN Habitat code:** 1,3,14 **Conservation status: (IUCN)** LC

7390 *Solanum hazenii* Britton (Solanaceae) **IPNI:** 238944-2 **Common name:** blanquito, pepito, tabaco de monte, tabaquillo **Habit:** Shrub, Small tree, Tree **Use:** EU, MA, ME, PO, SU **Origin:** Native **Region:** Andean, Caribbean, Pacific **Dept.:** ANT, ATL, BOL, CAU, CES, CHO, COR, CUN, HUI, LAG, MAG, NAR, NSA, SAN, VAC **Elev.:** 0-2000 m.a.s.l. **IUCN Habitat code:** 1,2,3,4,14 **Conservation status: (IUCN)** LC **(CPSG)** Potential LC

7391 *Solanum hirtum* Vahl (Solanaceae) **IPNI:** 819477-1 **Common name:** bolaegato, bolegato, cabello de ángel, huevo de gato, huevoegato, lulo, tunito espinoso, turmas de gato **Habit:** Herb, Shrub, Subshrub, Tree **Use:** GS, HF, ME **Origin:** Native **Region:** Amazonia, Andean, Caribbean **Dept.:** AMA, ATL, BOL, CUN, LAG, MAG, NSA, SAN, TOL **Elev.:** 0–2200 m a.s.l. **IUCN Habitat code:** 1,2,3,4,5,14 **Conservation status: (IUCN)** LC

7392 *Solanum jamaicense* Mill. (Solanaceae) **IPNI:** 819638-1 **Common name:** arañaplato, ayainae, chichu, cocubo, cucubo, cucuo, cujaco, frega plato, frega platos, fregaplato, fregaplatos, friega plato, friegaplato, lola, matico, rajateta, rasgateta, tacomocho, uña de gato **Habit:** Herb, Subshrub, Liana **Use:** HF, ME **Origin:** Native **Region:** Amazonia, Andean, Guiana Shield, Orinoquia, Pacific **Dept.:** AMA, ANT, BOY, CAL, CAQ, CAS, CAU, CES, CHO, COR, CUN, GUV, HUI, MET, NAR, NSA, PUT, QUI, RIS, SAN, TOL, VAC, VAU **Elev.:** 0-2000 m.a.s.l. **IUCN Habitat code:** 1,2,3,4,5,14 **Conservation status: (IUCN)** LC **(CPSG)** Potential LC

7393 *Solanum juglandifolium* Humb. & Bonpl. ex Dunal (Solanaceae) **IPNI:** 315945-2 **Common name:** riñonada, riñón, tomate silvestre **Habit:** Liana, Climbing **Use:** GS **Origin:** Native **Region:** Andean **Dept.:** ANT, BOG, CAL, CAU, CHO, CUN, HUI, NAR, PUT, QUI, RIS, SAN, VAC **Elev.:** 1500-3100 m.a.s.l. **IUCN Habitat code:** 1,2,3,5,14 **Conservation status: (IUCN)** LC **(CPSG)** Potential LC

7394 *Solanum lanceifolium* Jacq. (Solanaceae) **IPNI:** 819733-1 **Common name:** arañagato **Habit:** Subshrub, Liana **Use:** HF, ME **Origin:** Native **Region:** Andean, Caribbean, Pacific **Dept.:** ANT, CAU, CES, CUN, GUV, LAG, MAG, NAR, NSA, QUI, SAP, TOL, VAC **Elev.:** 0-1600 m.a.s.l. **IUCN Habitat code:** 1,2,3,5,14 **Conservation status: (IUCN)** NE **(CPSG)** Potential LC

7395 *Solanum laxum* Spreng. (Solanaceae) **IPNI:** 330664-2 **Common name:** buqué de novia, manto de María, manto de la Virgen, manto de novia, ponqué, velo de novia **Habit:** Climbing **Use:** EU **Origin:** Cultivated, Naturalised **Region:** Andean **Dept.:** ANT, BOG, CUN, SAN **Elev.:** 850–2800 m a.s.l. **IUCN Habitat code:** 1,2,3,14 **Conservation status: (IUCN)** NE

7396 *Solanum lepidotum* Humb. & Bonpl. ex Dunal (Solanaceae) **IPNI:** 819795-1 **Common name:** frutillo, pepito, tomate de monte **Habit:** Shrub, Small tree, Tree **Use:** ME **Origin:** Native **Region:** Amazonia, Andean, Orinoquia, Pacific **Dept.:** ANT, BOY, CAQ, CHO, CUN, GUV, MET, NAR, PUT, QUI, RIS, SAN, VAC **Elev.:** 100-2400 m.a.s.l. **IUCN Habitat code:** 1,3,4,5,14 **Conservation status: (IUCN)** LC **(CPSG)** Potential LC

7397 *Solanum leucocarpon* Dunal (Solanaceae) **IPNI:** 819814-1 **Common name:** aguarrús **Habit:** Shrub, Subshrub, Small tree, Tree **Use:** HF **Origin:** Native **Region:** Amazonia, Andean, Guiana Shield **Dept.:** AMA, ANT, BOL, BOY, CAQ, CES, CHO, CUN, GUV, HUI, MET, NAR,

PUT, SAN, VAU **Elev.:** 150-3100 m.a.s.l. **IUCN Habitat code:** 1,3,4,14 **Conservation status: (IUCN)** LC **(CPSG)** Potential LC

7398 *Solanum leucopogon* Huber (Solanaceae) **IPNI:** 239141-2 **Habit:** Liana **Use:** HF **Origin:** Native **Region:** Amazonia, Andean **Dept.:** AMA, CAQ, CAU, HUI, PUT **Elev.:** 200–2150 m a.s.l. **IUCN Habitat code:** 1,3,14 **Conservation status: (IUCN)** NE

7399 *Solanum lobbianum* Bitter (Solanaceae) **IPNI:** 239156-2 **Habit:** Herb **Use:** GS **Origin:** Endemic **Region:** Andean **Dept.:** CAL **Elev.:** 3000–3000 m a.s.l. **IUCN Habitat code:** 1,3,4,8 **Conservation status: (IUCN)** DD

7400 *Solanum lycopersicum* L. (Solanaceae) **IPNI:** 316947-2 **Common name:** bovista, tomate, tomate común, tomate de riñón, tomate grande **Habit:** Herb, Climbing **Use:** AF, EU, GS, HF, IF, MA, ME, PO, SU **Origin:** Cultivated, Naturalised **Region:** Amazonia, Andean, Caribbean, Pacific **Dept.:** AMA, ANT, BOG, BOY, CAU, CUN, HUI, NSA, PUT, SAP, VAC **Elev.:** 100–2500 m a.s.l. **IUCN Habitat code:** 1,2,3,4,5,14 **Conservation status: (IUCN)** NE

7401 *Solanum macrotonum* Bitter (Solanaceae) **IPNI:** 239202-2 **Use:** ME **Origin:** Native **Dept.:** BOG **Conservation status: (IUCN)** NE

7402 *Solanum mammosum* L. (Solanaceae) **IPNI:** 285235-2 **Common name:** friegaplatos, kukuma, lulo de perro, matacucarachas, pepito, rejalgar, teticas, tetilla **Habit:** Herb, Shrub, Subshrub **Use:** EU, GS, HF, MA, ME, PO, SU **Origin:** Native **Region:** Amazonia, Andean, Guiana Shield, Caribbean, Orinoquia, Pacific **Dept.:** AMA, ANT, ATL, CAL, CAQ, CAU, CHO, MET, PUT, RIS, TOL, VAC **Elev.:** 150–2400 m a.s.l. **IUCN Habitat code:** 1,2,3,4,14 **Conservation status: (IUCN)** NE

7403 *Solanum marginatum* L.f. (Solanaceae) **IPNI:** 819994-1 **Common name:** lulo, lulo de perro, pedronoche, toronja, toronjilla **Habit:** Shrub, Subshrub **Use:** EU, GS, MA, ME **Origin:** Naturalised **Region:** Andean **Dept.:** BOG, BOY, CUN, NAR **Elev.:** 2200–4000 m a.s.l. **IUCN Habitat code:** 1,3,5,14 **Conservation status: (IUCN)** LC

7404 *Solanum mauritianum* Scop. (Solanaceae) **IPNI:** 330666-2 **Use:** EU, HF, MA, ME, PO, SU **IUCN Habitat code:** 1,2,3,4,5,14 **Conservation status: (IUCN)** NE

7405 *Solanum melongena* L. (Solanaceae) **IPNI:** 820053-1 **Common name:** berenjena, berenjena de monte, manzana de amor, pepino morado, ponedora **Habit:** Herb, Climbing **Use:** GS, HF, ME, PO **Origin:** Cultivated **Region:** Andean **Dept.:** ANT **Elev.:** 1500–1500 m a.s.l. **IUCN Habitat code:** 2,3,14 **Conservation status: (IUCN)** NE

7406 *Solanum mite* Ruiz & Pav. (Solanaceae) **IPNI:** 820119-1 **Habit:** Shrub, Subshrub **Use:** ME **Origin:** Native **Region:** Amazonia **Dept.:** AMA, PUT **Elev.:** 90–350 m a.s.l. **IUCN Habitat code:** 1,14 **Conservation status: (IUCN)** NE

7407 *Solanum muricatum* Aiton (Solanaceae) **IPNI:** 820187-1 **Common name:** pepinito redondo, pepino, pepino dulce, pepino dulce morado, pepino morado, pepino redondo, pepino redondo morado, pepino silvestre **Habit:** Herb, Subshrub **Use:** GS, HF, ME **Origin:** Native, Cultivated **Region:** Andean **Dept.:** ANT, BOG, BOY, CAU, CUN, NAR, PUT, QUI **Elev.:** 2500–2700 m a.s.l. **IUCN Habitat code:** 1,3,4,14 **Conservation status: (IUCN)** NE

7408 *Solanum nemorense* Dunal (Solanaceae) **IPNI:** 820224-1 **Habit:** Liana **Use:** HF **Origin:** Native **Region:** Amazonia **Dept.:** AMA, CAU **Elev.:** 200–400 m a.s.l. **IUCN Habitat code:** 1 **Conservation status: (IUCN)** NE

7409 *Solanum nigrescens* M.Martens & Galeotti (Solanaceae) **IPNI:** 239356-2 **Common name:** herb mora, yerba mora **Habit:** Herb, Shrub, Subshrub, Climbing **Use:** HF, ME **Origin:** Native **Region:** Andean, Guiana Shield **Dept.:** ANT, BOG, BOY, CES, CUN, HUI, MAG, MET, NAR, PUT, QUI, SAN **Elev.:** 1100–3300 m a.s.l. **IUCN Habitat code:** 1,2,3,4,14 **Conservation status: (IUCN)** NE

7410 *Solanum nigrum* L. (Solanaceae) **IPNI:** 30048260-2 **Common name:** campano, herbmora, hierbamora, macuy, mora, quielete, suspiros, yerba mora, yerbamora, zeb ame **Use:** HF, ME **IUCN Habitat code:** 1,2,3,8,14 **Conservation status: (IUCN)** NE

7411 *Solanum nudum* Humb. & Bonpl. ex Dunal (Solanaceae) **IPNI:** 820269-1 **Common name:** cauco, cucubo, frutillo, frutillo hembra, gediondito, pepito, sauco, sauco amargo, sauco comun, sauco común, toronjo, zapata, zapatico **Habit:** Herb, Shrub, Subshrub, Small tree, Tree **Use:** ME **Origin:** Native **Region:** Amazonia, Andean, Orinoquia, Pacific **Dept.:** ANT, BOY, CAL, CAQ, CAS, CAU, CES, CHO, CUN, HUI, MAG, MET, NAR, PUT, QUI, RIS, SAN, TOL, VAC **Elev.:** 0-3700 m.a.s.l. **IUCN Habitat code:** 1,2,3,4,5,14 **Conservation status: (IUCN)** LC **(CPSG)** Potential LC

7412 *Solanum nutans* Ruiz & Pav. (Solanaceae) **IPNI:** 820273-1 **Habit:** Shrub, Small tree, Tree, Climbing **Use:** ME **Origin:** Native **Region:** Andean **Dept.:** ANT, CAU, NAR, PUT **Elev.:** 1750–2500 m a.s.l. **IUCN Habitat code:** 1,3,14 **Conservation status: (IUCN)** LC

7413 *Solanum obliquum* Ruiz & Pav. (Solanaceae) **IPNI:** 300622-2 **Habit:** Shrub, Small tree, Tree **Use:** HF, ME **Origin:** Native **Region:** Amazonia **Dept.:** AMA **Elev.:** 100–120 m a.s.l. **IUCN Habitat code:** 1 **Conservation status: (IUCN)** LC

7414 *Solanum occultum* Bohs (Solanaceae) **IPNI:** 985919-1 **Common name:** gallinazo pajo **Habit:** Small tree, Tree **Use:** GS, HF **Origin:** Native **Region:** Amazonia **Dept.:** AMA, VAU **Elev.:** 100–250 m a.s.l. **IUCN Habitat code:** 1,14 **Conservation status: (IUCN)** LC

7415 *Solanum ochranthum* Humb. & Bonpl. ex Dunal (Solanaceae) **IPNI:** 315944-2 **Common name:** papa, papa silvestre, payaca **Habit:** Liana, Climbing **Use:** GS, HF **Origin:** Native **Region:** Andean **Dept.:** CAL, NAR, PUT, QUI, TOL, VAC **Elev.:** 2100–3300 m a.s.l. **IUCN Habitat code:** 1,3,4,14 **Conservation status: (IUCN)** LC

7416 *Solanum opacum* A.Braun & C.D.Bouché (Solanaceae) **IPNI:** 820333-1 **Use:** HF **Conservation status: (IUCN)** NE

7417 *Solanum ovalifolium* Humb. & Bonpl. ex Dunal (Solanaceae) **IPNI:** 820354-1 **Common name:** chucho, cocubo, cocuy, cucu, cucubo, cucubo espinoso, frutillo, fruto, pepito, pepo, tachuelo, tigrillo **Habit:** Shrub, Subshrub, Small tree, Tree **Use:** ME **Origin:** Native **Region:** Andean **Dept.:** ANT, BOG, BOL, BOY, CAL, CAQ, CAU, CUN, HUI, MAG, NAR, PUT, QUI, RIS, SAN, VAC **Elev.:** 1100-2600 m.a.s.l. **IUCN Habitat code:** 1,2,3,5,14 **Conservation status: (IUCN)** LC **(CPSG)** Potential LC

7418 *Solanum oxyphyllum* C.V.Morton (Solanaceae) **IPNI:** 239427-2 **Habit:** Shrub, Small tree **Use:** GS **Origin:** Native **Region:** Andean **Dept.:** NAR, PUT **Elev.:** 800–800 m a.s.l. **IUCN Habitat code:** 1,14 **Conservation status: (IUCN)** LC

7419 *Solanum pectinatum* Dunal (Solanaceae) **IPNI:** 820443-1 **Common name:** huevo de gato, naranjillo, naranjuelo, toronja **Habit:** Herb, Shrub **Use:** HF **Origin:** Native **Region:** Amazonia, Andean **Dept.:** ANT, CAQ, GUV, MET, PUT, SAN **Elev.:** 40–850 m a.s.l. **IUCN Habitat code:** 1,3,14 **Conservation status: (IUCN)** NE

7420 *Solanum pseudocapsicum* L. (Solanaceae) **IPNI:** 316948-2 **Common name:** grano de oro, mirto **Habit:** Shrub, Subshrub, Small tree **Use:** EU, HF, MA, ME, PO, SU **Origin:** Cultivated, Naturalised **Region:** Andean **Dept.:** BOG, CUN, MAG, NAR, PUT **Elev.:** 200–2800 m a.s.l. **IUCN Habitat code:** 1,2,3,14 **Conservation status: (IUCN)** NE

7421 *Solanum pseudolulo* Heiser (Solanaceae) **IPNI:** 239554-2 **Common name:** lulo, lulo de perro, lulo pequeño, luloeperro, toronja **Habit:** Shrub, Subshrub **Use:** HF **Origin:** Endemic, Cultivated **Region:** Andean, Pacific **Dept.:** ANT, BOY, CAU, CHO, CUN, HUI, RIS, SAN, TOL, VAC **Elev.:** 450–2300 m a.s.l. **IUCN Habitat code:** 1,2,3,14 **Conservation status: (IUCN)** LC

7422 *Solanum quitoense* Lam. (Solanaceae)
IPNI: 820696-1
Common name: bombolo, bómbolo, lulo, lulo de Castilla, lulo de perro, lulo morado, lulo pequeño, maschactie, naranjilla, naranjillo, naranjillo de pájaro, piño, pómpulo, toronja **Habit:** Shrub, Subshrub **Use:** AF, EU, GS, HF, MA, ME **Origin:** Native, Cultivated **Region:** Andean, Orinoquia, Pacific **Dept.:** ANT, BOG, BOY, CHO, NAR, NSA, SAN, TOL, VAC, VID **Elev.:** 100–2600 m a.s.l. **IUCN Habitat code:** 1,2,3,4,14 **Conservation status: (IUCN)** NE

7423 *Solanum rude-pannum* Dunal (Solanaceae)
IPNI: 820821-1
Common name: capiria, friegaplatos, frutillo, hoja de luna, limpiaplato, mancatigre, pepito, pepo, sacaestaca, sacalaestaca, yuca de espíritu **Habit:** Shrub, Small tree, Tree **Use:** HF, ME **Origin:** Native **Region:** Andean, Caribbean, Pacific **Dept.:** ANT, BOL, BOY, CAU, CHO, CUN, MAG, QUI, RIS, TOL, VAC **Elev.:** 0-1500 m.a.s.l. **Conservation status: (IUCN)** NE **(CPSG)** Potential LC

7424 *Solanum rugosum* Dunal (Solanaceae)
IPNI: 820831-1
Common name: pelotillo **Habit:** Shrub, Small tree **Use:** HF, MA **Origin:** Native **Region:** Amazonia, Andean, Guiana Shield, Orinoquia **Dept.:** AMA, ANT, CAQ, GUA, GUV, MET, PUT, VAC, VAU **Elev.:** 100–1200 m a.s.l. **Conservation status: (IUCN)** NE

7425 *Solanum scabrum* Mill. (Solanaceae)
IPNI: 820912-1
Common name: arañagato, huevo de gato, uña de gato **Use:** AF, EU, GS, HF, MA, ME **IUCN Habitat code:** 1,2,3,5,14 **Conservation status: (IUCN)** NE

7426 *Solanum seaforthianum* Andrews (Solanaceae)
IPNI: 316949-2
Common name: guatemala, guatemalteca, manto de la Virgen **Habit:** Liana, Climbing **Use:** EU, HF, MA, ME, PO **Origin:** Native, Cultivated **Region:** Andean, Caribbean **Dept.:** ANT, CUN, HUI, MAG, NSA, SAN, TOL, VAC **Elev.:** 100–1500 m a.s.l. **IUCN Habitat code:** 1,2,3,14 **Conservation status: (IUCN)** NE

7427 *Solanum sessile* Ruiz & Pav. (Solanaceae)
IPNI: 820998-1
Habit: Shrub, Tree **Use:** HF, ME **Origin:** Native **Region:** Amazonia, Andean **Dept.:** AMA, CAQ, CAU, CUN, GUV, PUT, RIS **Elev.:** 100–2950 m a.s.l. **IUCN Habitat code:** 1,5,14 **Conservation status: (IUCN)** LC

7428 *Solanum sessiliflorum* Dunal (Solanaceae)
IPNI: 927606-1
Common name: betaca, cocona, cubiyú, eto-a, lulo, lulo dulce, lulo grande, ma-ra, naranjilla, naranjillo, tonto grande, tupiro, tupiru, tupirú **Habit:** Herb, Shrub, Subshrub, Small tree, Climbing **Use:** HF, MA, ME, PO **Origin:** Native, Cultivated **Region:** Amazonia, Andean, Guiana Shield, Pacific **Dept.:** AMA, CAQ, CHO, GUA, MET, PUT, VAU **Elev.:** 100–1000 m a.s.l. **IUCN Habitat code:** 1,3,5,14 **Conservation status: (IUCN)** NE

7429 *Solanum sibundoyense* (Bohs) Bohs (Solanaceae)
IPNI: 985918-1
Common name: tomate, tomate salvaje, tomate silvestre **Habit:** Shrub, Small tree, Tree **Use:** GS, HF **Origin:** Endemic **Region:** Andean **Dept.:** HUI, NAR, PUT **Elev.:** 1400-2300 m.a.s.l. **IUCN Habitat code:** 1,3,14 **Conservation status: (IUCN)** LC **(CPSG)** LC

7430 *Solanum sisymbriifolium* Lam. (Solanaceae)
IPNI: 330681-2
Common name: guinda **Habit:** Herb, Shrub, Subshrub **Use:** EU, GS, HF, ME **Origin:** Native **Region:** Andean **Dept.:** ANT, BOY, CAU, CUN, HUI, NAR, SAN, TOL, VAC **Elev.:** 270-2900 m.a.s.l. **Conservation status: (IUCN)** NE **(CPSG)** Potential LC

7431 *Solanum splendens* (Dunal) Bohs (Solanaceae)
IPNI: 77178345-1
Common name: gallinazo pago, huevo de perro, huevo de toro, naranjilla, regalgar, rejalgar, reventadera, tomate de monte, tomate de tree silvestre, tomate de árbol silvestre, tomate silvestre, venenillo **Habit:** Shrub, Small tree, Tree **Use:** GS, HF, ME, PO **Origin:** Native **Region:** Andean, Guiana Shield, Caribbean, Pacific **Dept.:** ANT, BOL, BOY, CAL, CAQ, CES, CHO, COR, CUN, MET, NAR, PUT, RIS, SAN, TOL, VAC **Elev.:** 50-2200 m.a.s.l. **IUCN Habitat code:** 1,2,3,5,14 **Conservation status: (IUCN)** LC **(CPSG)** Potential LC

7432 *Solanum stellatiglandulosum* Bitter (Solanaceae)
IPNI: 239753-2
Common name: pepo **Habit:** Herb, Shrub, Subshrub, Small tree, Tree **Use:** HF **Origin:** Endemic **Region:** Andean **Dept.:** ANT, BOG, BOY, CES, CHO, CUN, HUI, MAG, NAR, NSA, PUT, QUI, SAN, VAC **Elev.:** 800-3700 m.a.s.l. **IUCN Habitat code:** 1,2,3,4,14 **Conservation status: (IUCN)** LC **(CPSG)** Potential LC

7433 *Solanum stramoniifolium* Jacq. (Solanaceae)
IPNI: 30162226-2
Common name: betacacù, chiri, cobuia, coconilla, cucuma, e-to-pa-a, etopaa, lulillo, lulo, lulo de charapa, lulo de monte, lulo pequeño, lulo silvestre, pipica, roya, sevilla, uvilla **Habit:** Shrub, Subshrub **Use:** HF, MA, ME **Origin:** Native, Cultivated **Region:** Amazonia, Andean **Dept.:** AMA, GUA, MET, NAR, PUT, VAU **Elev.:** 300–1000 m a.s.l. **Conservation status: (IUCN)** NE

7434 *Solanum suaveolens* Kunth & C.D.Bouché (Solanaceae)
IPNI: 821130-1
Common name: pepino llorón **Habit:** Herb, Climbing **Use:** HF **Origin:** Native **Region:** Andean **Dept.:** ANT, CAL, CUN, HUI, NAR, NSA, PUT, SAN, VAC **Elev.:** 200–2300 m a.s.l. **IUCN Habitat code:** 1,3,14 **Conservation status: (IUCN)** NE

7435 *Solanum subinerme* Jacq. (Solanaceae)
IPNI: 821146-1
Common name: cucunanay, cujaco, hoja de luna **Habit:** Shrub, Subshrub **Use:** MA, ME, PO **Origin:** Native **Region:** Amazonia, Andean, Guiana Shield, Caribbean, Orinoquia **Dept.:** AMA, ANT, ARA, ATL, BOL, BOY, CAS, COR, GUA, GUV, MAG, MET, NSA, PUT, SAN, SUC, VAU, VID **Elev.:** 0-1700 m.a.s.l. **IUCN Habitat code:** 1,2,3,4,5,14 **Conservation status: (IUCN)** LC **(CPSG)** Potential LC

7436 *Solanum tabanoense* Correll (Solanaceae)
IPNI: 239822-2
Common name: berriadores **Habit:** Herb, Climbing **Use:** GS, HF **Origin:** Native **Region:** Andean **Dept.:** ANT, CAU, NSA, PUT **Elev.:** 2200–3500 m a.s.l. **IUCN Habitat code:** 1,3,4,5,14 **Conservation status: (IUCN)** NE

7437 *Solanum tampicense* Dunal (Solanaceae)
IPNI: 821208-1
Use: EU, ME **Origin:** Native **IUCN Habitat code:** 2,5 **Conservation status: (IUCN)** NE

7438 *Solanum torvum* Sw. (Solanaceae)
IPNI: 821287-1
Common name: chucho hediondo, cujaca, cujaco, fregaplatos, friegaplato, friegaplatos, pepo cujaco, quiebraplatos **Habit:** Herb, Shrub, Small tree **Use:** EU, GS, HF, MA, ME, PO, SU **Origin:** Native **Region:** Andean, Caribbean, Pacific **Dept.:** ANT, CAL, CHO, MAG, NSA, QUI, RIS, SAP, VAU **Elev.:** 20–1700 m a.s.l. **IUCN Habitat code:** 5 **Conservation status: (IUCN)** NE

7439 *Solanum tuberosum* L. (Solanaceae)
IPNI: 821337-1
Common name: chauacha, chava, guata roja, paduana, papa, papa abondaria, papa alemana, papa americana, papa arbolona, papa argentina, papa azul, papa azulita, papa bambera, papa bandera, papa batata, papa blanca, papa bogotana, papa boliviana, papa brucha, papa caiced **Habit:** Herb **Use:** AF, EU, GS, HF, MA, ME, PO, SU **Origin:** Native, Cultivated **Region:** Andean **Dept.:** ANT, BOG, BOY, CAU, CUN, HUI, NAR, PUT **Elev.:** 2000–3600 m a.s.l. **IUCN Habitat code:** 1,2,3,4,5,8,14 **Conservation status: (IUCN)** NE

7440 *Solanum umbellatum* Mill. (Solanaceae)
IPNI: 821363-1
Common name: borrachero, chucho lanudo, pepo **Habit:** Herb, Shrub, Tree **Use:** HF **Origin:** Native **Region:** Andean **Dept.:** ANT, CAU, CUN, NAR, NSA, QUI, SAN, VAC **Elev.:** 850–2350 m a.s.l. **IUCN Habitat code:** 1,2,3,5,14 **Conservation status: (IUCN)** NE

7441 *Solanum uncinellum* Lindl. (Solanaceae)
IPNI: 821373-1
Habit: Subshrub, Liana **Use:** ME **Origin:** Native **Region:** Amazonia, Andean, Pacific **Dept.:** AMA, ANT, BOY, CHO, CUN, MET, PUT **Elev.:** 400–1830 m a.s.l. **IUCN Habitat code:** 1,3,5,14 **Conservation status: (IUCN)** NE

7442 *Solanum vestissimum* Dunal (Solanaceae)
IPNI: 821460-1
Common name: bobolo, espina, pelones, toronja **Habit:** Herb, Shrub, Tree, Liana **Use:** HF, ME **Origin:** Native, Cultivated **Region:** Andean **Dept.:** ANT, BOY, CES, CUN, MAG, NSA, SAN **Elev.:** 1400–3250 m a.s.l. **IUCN Habitat code:** 1,3,4,14 **Conservation status: (IUCN)** LC

7443 *Solanum viarum* Dunal (Solanaceae)
IPNI: 330693-2
Habit: Herb, Shrub **Use:** GS, HF, ME **Origin:** Native **Region:** Amazonia, Orinoquia **Dept.:** CAQ, MET **Elev.:** 300–600 m a.s.l. **IUCN Habitat code:** 3,5,14 **Conservation status: (IUCN)** LC

7444 *Solanum volubile* Sw. (Solanaceae)
IPNI: 821520-1
Habit: Herb, Subshrub, Liana **Use:** AF, HF **Origin:** Native **Region:**

Amazonia, Andean, Guiana Shield, Caribbean, Orinoquia, Pacific **Dept.:** AMA, ANT, ARA, ATL, BOL, BOY, CES, CHO, COR, MAG, MET, NSA, SAN, SUC **Elev.:** 0–1500 m a.s.l. **Conservation status: (IUCN)** NE

7445 *Solanum wendlandii* Hook.f. (Solanaceae)
IPNI: 240016-2
Common name: cecilia, guatemala, guatemalteca **Habit:** Liana **Use:** EU, HF **Origin:** Native, Cultivated **Region:** Andean **Dept.:** ANT, CAU, CUN, NSA **Elev.:** 1200–1900 m a.s.l. **IUCN Habitat code:** 1,14 **Conservation status: (IUCN)** NE

7446 *Streptosolen jamesonii* (Benth.) Miers (Solanaceae)
IPNI: 821609-1
Common name: mermelada, milflores, peruana, sandiego, tamarindo, tamaris **Habit:** Shrub **Use:** EU **Origin:** Cultivated, Naturalised **Region:** Andean **Dept.:** ANT, BOG, BOY, CUN, PUT, QUI **Elev.:** 1480–2900 m a.s.l. **IUCN Habitat code:** 1,2,3,5,14 **Conservation status: (IUCN)** NE

7447 *Trianaea nobilis* Planch. & Linden (Solanaceae)
IPNI: 821639-1
Habit: Epiphyte, Liana **Use:** HF **Origin:** Native **Region:** Andean **Dept.:** NAR, VAC **Elev.:** 1500–2800 m a.s.l. **IUCN Habitat code:** 1 **Conservation status: (IUCN)** NE

7448 *Witheringia solanacea* L'Hér. (Solanaceae)
IPNI: 821802-1
Common name: chamico, culebrilla, culebrillera, hierbamora, lumbi, lumbí, yerba amarga, yerba culebrillera, yerba mora, yerbamora **Habit:** Herb, Shrub, Small tree **Use:** HF, ME **Origin:** Native **Region:** Amazonia, Andean, Caribbean, Pacific **Dept.:** ANT, BOL, BOY, CAL, CAQ, CAU, CES, CHO, CUN, HUI, LAG, MAG, MET, NAR, PUT, QUI, RIS, SAN, VAC **Elev.:** 0-2800 m.a.s.l. **IUCN Habitat code:** 1,2,3,5,14 **Conservation status: (IUCN)** LC **(CPSG)** Potential LC

7449 *Sphenoclea zeylanica* Gaertn. (Sphenocleaceae)
IPNI: 145463-1
Habit: Herb **Use:** AF, HF, ME **Origin:** Naturalised **Region:** Amazonia, Guiana Shield, Caribbean, Orinoquia **Dept.:** AMA, ANT, ATL, BOL, CAU, CHO, COR, CUN, GUA, MAG, MET, SAN, SUC, VAC **Elev.:** 0-900 m.a.s.l. **IUCN Habitat code:** 1,2,3,4,5,14 **Conservation status: (IUCN)** LC **(CPSG)** Potential LC

Vitales

7450 *Cissus alata* Jacq. (Vitaceae)
IPNI: 311226-2
Habit: Liana **Use:** EU, HF, ME **Origin:** Native **Region:** Caribbean, Orinoquia, Pacific **Dept.:** ANT, BOL, CAQ, CES, CHO, MAG, MET, TOL, VID **Elev.:** 20–950 m a.s.l. **IUCN Habitat code:** 1,2,3,5,14 **Conservation status: (IUCN)** NE

7451 *Cissus erosa* Rich. (Vitaceae)
IPNI: 67646-1
Common name: maculculu champu, matano waü ünuboto, naecono **Habit:** Shrub, Liana, Climbing **Use:** EU, HF, MA, ME, SU **Origin:** Native **Region:** Amazonia, Andean, Guiana Shield, Caribbean, Pacific **Dept.:** AMA, ANT, BOL, BOY, CAL, CAQ, CAS, CAU, CHO, CUN, MAG, MET, NAR, NSA, PUT, RIS, SAN, VAC, VAU, VID **Elev.:** 5-1850 m.a.s.l. **IUCN Habitat code:** 1,2,3,4,5,14 **Conservation status: (IUCN)** NE **(CPSG)** Potential LC

7452 *Cissus gongylodes* (Baker) Burch. ex Baker (Vitaceae)
IPNI: 67686-1
Habit: Liana **Use:** HF **Origin:** Native **Region:** Amazonia **Dept.:** AMA **Elev.:** 0–100 m a.s.l. **Conservation status: (IUCN)** NE

7453 *Cissus gossypiifolia* Standl. (Vitaceae)
IPNI: 59292-2
Habit: Liana **Use:** ME **Origin:** Native **Region:** Guiana Shield **Dept.:** MET **Elev.:** 30–200 m a.s.l. **Conservation status: (IUCN)** NE

7454 *Cissus microcarpa* Vahl (Vitaceae)
IPNI: 316225-2
Common name: bejuco comemano **Habit:** Liana, Climbing **Use:** ME **Origin:** Native **Region:** Andean, Caribbean, Pacific **Dept.:** ANT, CAU, CES, CHO, MAG **Elev.:** 0–2000 m a.s.l. **IUCN Habitat code:** 1,3,4,14 **Conservation status: (IUCN)** NE

7455 *Cissus palmata* Poir. (Vitaceae)
IPNI: 67839-1
Habit: Liana, Climbing **Use:** GS **Origin:** Native **Region:** Amazonia, Pacific **Dept.:** AMA, CHO, MET **Elev.:** 70–120 m a.s.l. **IUCN Habitat code:** 1,5,14 **Conservation status: (IUCN)** NE

7456 *Cissus peruviana* Lombardi (Vitaceae)
IPNI: 303957-2
Habit: Liana **Use:** HF **Origin:** Native **Region:** Guiana Shield **Dept.:** MET **Elev.:** 300–300 m a.s.l. **IUCN Habitat code:** 1 **Conservation status: (IUCN)** NE

7457 *Cissus trifoliata* (L.) L. (Vitaceae)
IPNI: 68021-1
Habit: Climbing **Use:** ME **Origin:** Native **Region:** Caribbean **Dept.:** LAG, MAG **Elev.:** 3–50 m a.s.l. **IUCN Habitat code:** 1,2,14 **Conservation status: (IUCN)** NE

7458 *Cissus verticillata* (L.) Nicolson & C.E.Jarvis (Vitaceae)
IPNI: 980361-1
Common name: agraz, bejuco Castro, bejuco caro, bejuco castro, bejuco chirriador, bejuco comemano, bejuco de agua, bejuco de caro, bejuco ubí, chirriador, fuente, hierba de buey, lyann mol, uva de culebra, uva de gorro, uva de pajarito, uvilla, yedr, yedra **Habit:** Liana, Climbing **Use:** EU, HF, MA, ME **Origin:** Native **Region:** Amazonia, Andean, Guiana Shield, Caribbean, Pacific **Dept.:** AMA, ANT, ATL, BOL, BOY, CAL, CAQ, CAU, CES, CHO, COR, CUN, HUI, LAG, MAG, MET, NAR, NSA, PUT, QUI, RIS, SAN, SUC, TOL, VAC **Elev.:** 0-2100 m.a.s.l. **IUCN Habitat code:** 1,2,3,4,5,14 **Conservation status: (IUCN)** NE **(CPSG)** Potential LC

7459 *Vitis tiliifolia* Humb. & Bonpl. ex Schult. (Vitaceae)
IPNI: 69103-1
Common name: agraz, bejuco de agua, fuente, parra de monte, uva cimarrona, uva de monte, uva de playa, uva silvestre **Habit:** Climbing **Use:** GS, HF, MA, ME **Origin:** Native **Region:** Andean, Caribbean, Pacific **Dept.:** ANT, CAQ, CHO, HUI, MAG, PUT, SAN **Elev.:** 40–1900 m a.s.l. **IUCN Habitat code:** 1,2,3,5,14 **Conservation status: (IUCN)** NE

7460 *Vitis vinifera* L. (Vitaceae)
IPNI: 30478388-2
Common name: parra, uva **Habit:** Climbing **Use:** AF, FU, GS, HF, MA, ME, SU **Origin:** Cultivated **Region:** Andean **Dept.:** BOY, CUN, QUI, VAC **Elev.:** 450–2100 m a.s.l. **IUCN Habitat code:** 1,14 **Conservation status: (IUCN)** LC

Zygophyllales

7461 *Krameria ixine* L. (Krameriaceae)
IPNI: 690428-1
Common name: cardillo, carretón, estancadera, ratania de nueva granada, ratania de sabanilla, sabanilla **Habit:** Shrub, Subshrub **Use:** ME **Origin:** Native **Region:** Andean, Caribbean **Dept.:** LAG, MAG, SAN **Elev.:** 70–1005 m a.s.l. **IUCN Habitat code:** 1,2,3,14 **Conservation status: (IUCN)** NE

7462 *Krameria spartioides* Klotzsch ex O.Berg (Krameriaceae)
IPNI: 279266-2
Common name: estancadera **Habit:** Subshrub, Climbing **Use:** ME **Origin:** Native **Region:** Caribbean **Dept.:** CES, LAG, MAG **Elev.:** 40–40 m a.s.l. **Conservation status: (IUCN)** NE

7463 *Guaiacum officinale* L. (Zygophyllaceae)
IPNI: 873252-1
Common name: florazul, guayaco, guayacán, guayacán azul, guayacán chaparro, guayacán de playa, guayacán extranjero, guayacán flor azul, guayacán negro, guayacán verdadero, palo sano, palo santo, palosanto, polo o guayacán de playa, poloó, wira kupa **Habit:** Shrub, Small tree, Tree **Use:** EU, HF, IF, MA, ME **Origin:** Native **Region:** Caribbean **Dept.:** LAG, MAG, SUC, VAC **Elev.:** 30-1050 m.a.s.l. **IUCN Habitat code:** 2,3,5,14 **Conservation status: (IUCN)** EN **(CPSG)** EN

7464 *Kallstroemia maxima* (L.) Hook. & Arn. (Zygophyllaceae)
IPNI: 77076481-1
Common name: abrojo, coclí, hierba de pollo, verdolaga, yerba de pasmo **Habit:** Herb **Use:** HF, ME **Origin:** Native **Region:** Andean, Caribbean **Dept.:** ANT, ATL, BOL, CAU, COR, CUN, MAG, TOL, VAC **Elev.:** 20–1700 m a.s.l. **IUCN Habitat code:** 1,2,3,14 **Conservation status: (IUCN)** NE

7465 *Kallstroemia parviflora* Norton (Zygophyllaceae)
IPNI: 873291-1
Habit: Herb **Use:** PO **Origin:** Naturalised **Dept.:** CUN **Elev.:** 400–400 m a.s.l. **IUCN Habitat code:** 3 **Conservation status: (IUCN)** NE

7466 *Kallstroemia pubescens* (G.Don) Dandy (Zygophyllaceae)
IPNI: 873297-1
Common name: abrojo, atarraya, mancatigre, perrito **Habit:** Herb **Use:** ME **Origin:** Native **Region:** Andean, Caribbean **Dept.:** ANT, ATL, BOL, BOY, COR, CUN, HUI, MAG, SUC, VAC **Elev.:** 5–1400 m a.s.l. **IUCN Habitat code:** 1,2,3,14 **Conservation status: (IUCN)** NE

7467 *Plectrocarpa arborea* (Jacq.) Christenh. & Byng (Zygophyllaceae)
IPNI: 77184861-1
Common name: bola, guayacán, guayacán blanco, guayacán carrapo, guayacán chaparro, guayacán de bola, guayacán de piedra, guayacán polvillo, paraa, polo **Habit:** Tree **Use:** AF, FU, MA, ME **Origin:** Native, Cultivated **Region:** Caribbean, Pacific **Dept.:** ANT, ATL,

BOL, CES, COR, LAG, MAG, NAR, SAN, SUC, VAC **Elev.:** 0–1050 m a.s.l. **Conservation status: (IUCN)** NE

7468 *Plectrocarpa carrapo* (Killip & Dugand) Christenh. & Byng (Zygophyllaceae)
IPNI: 77184859-1
Common name: guayacán, guayacán blanco, guayacán carrapo, guayacán de bola **Habit:** Tree **Use:** ME **Origin:** Endemic **Region:** Andean **Dept.:** BOY, CUN, TOL **Elev.:** 340-1400 m.a.s.l. **Conservation status: (IUCN)** NE **(CPSG)** LC

7469 *Tribulus cistoides* L. (Zygophyllaceae)
IPNI: 873411-1
Common name: abrojo, hierba de pollo, mancatigre, manna, pecado mortal, perrito **Habit:** Herb **Use:** GS, HF, MA, ME **Origin:** Naturalised **Region:** Caribbean **Dept.:** ATL, BOL, BOY, LAG, MAG **Elev.:** 0–1000 m a.s.l. **IUCN Habitat code:** 1,2,5,14 **Conservation status: (IUCN)** NE

Rhodophyta Florideophyceae Gracilariales

7470 *Gracilaria cervicornis* (Turner) J.Agardh (Gracilariaceae)
IPNI: 679
Use: HF, ME **Origin:** Endemic **IUCN Habitat code:** 14 **Conservation status: (IUCN)** NE

Plocamiales

7471 *Plocamium cartilagineum* (Linnaeus) P.S.Dixon (Plocamiaceae)
IPNI: 31
Use: ME **Origin:** Endemic **Dept.:** MAG **Conservation status: (IUCN)** NE

Ochrophyta Phaeophyceae Fucales

7472 *Fucus vesiculosus* L. (Fucaceae)
IPNI: 87
Use: ME **Origin:** Endemic **Conservation status: (IUCN)** NE

Moringa oleifera Lam. (Moringaceae), had been recorded as presenting all categories of uses.
Laura Kor

Laminae

Selected species of Colombian plants

Alnus acuminata Mauricio Diazgranados

Alternanthera philoxeroides William Milliken

Astronium graveolens William Milliken

Azolla filiculoides Ori Fragman-Sapir

Bejaria resinosa Mauricio Diazgranados

Cassia moschata Mauricio Diazgranados

Castilleja integrifolia Mauricio Diazgranados

Castilleja fissifolia Mauricio Diazgranados

Ceiba pentandra **Mauricio Diazgranados**

Cestrum buxifolium **Mauricio Diazgranados**

Cissampelos pareira **Mauricio Diazgranados**

Cyperus eragrostis **William Milliken**

Disakisperma dubium **Doug Goldman**

Doliocarpus dentatus **Lisa Campbell**

Echinolaena inflexa **W.Milliken**

Eirmocephala brachiata **Mauricio Diazgranados**

Erythrina edulis **Carlos Magdalena**

Euterpe precatoria **Juan Carlos Copete**

Fraxinus uhdei **Mauricio Diazgranados**

Eryngium foetidum Mauricio Diazgranados

Gaultheria myrsinoides Mauricio Diazgranados

Gaultheria rigida Mauricio Diazgranados

Garcinia madruno Mauricio Diazgranados

Halenia asclepiadea Mauricio Diazgranados

Gaultheria anastomosans Mauricio Diazgranados

Heliconia marginata Denise Sasaki

Holcus lanatus Mauricio Diazgranados

Macroptilium lathyroides Oliver Whaley

Inga edulis Mauricio Diazgranados

Limnocharis flava Ori Fragman-Sapir

Mauritia flexuosa Mauricio Diazgranados

Ipomoea rubens Denise Sasaki

Ligustrum lucidum Mauricio Diazgranados

Macleania rupestris Mauricio Diazgranados

Miconia squamulosa Mauricio Diazgranados

Momordica charantia Rafaël Govaerts

Monticalia ledifolia Mauricio Diazgranados

Pistia stratiotes Lucinda Lay

Protium llanorum Mauricio Diazgranados

Puya trianae Mauricio Diazgranados

Quercus humboldtii Mauricio Diazgranados

Saurauia scabra David Granados

Setaria parviflora William Milliken

Sicyos edulis Mauricio Diazgranados

Solanum betaceum Mauricio Diazgranados

Solanum quitoense Mauricio Diazgranados

Spirodela polyrhiza Bing Liu

Stuckenia pectinata Ori Fragman-Sapir

Vachellia pennatula G. P. Lewis

Zinnia peruviana Sheila Gregory

Abutilon insigne Mauricio Diazgranados

Acaena cylindristachya Mauricio Diazgranados

Ageratina tinifolia Mauricio Diazgranados

Alchemilla orbiculata Mauricio Diazgranados

Barnadesia spinosa Mauricio Diazgranados

Anthurium crystallinum Mauricio Diazgranados

Araucaria angustifolia Mauricio Diazgranados

Brachyotum strigosum Mauricio Diazgranados

Brunfelsia pauciflora Mauricio Diazgranados

Carex pichinchensis Mauricio Diazgranados

Cattleya violacea Mauricio Diazgranados

Clusia orthoneura Mauricio Diazgranados

Cordyline australis Mauricio Diazgranados

Cotoneaster pannosus Mauricio Diazgranados

Cyathea caracasana Mauricio Diazgranados

Dimerocostus strobilaceus Mauricio Diazgranados

Dryopteris wallichiana Mauricio Diazgranados

Epidendrum × elongatum Mauricio Diazgranados

Episcia cupreata Mauricio Diazgranados

Espeletia arbelaezii **Mauricio Diazgranados**

Espeletia boyacensis **Mauricio Diazgranados**

Espeletia chocontana **Mauricio Diazgranados**

Espeletia garciae **Mauricio Diazgranados**

Espeletia grandiflora **Mauricio Diazgranados**

Espeletia incana **Mauricio Diazgranados**

Espeletia jimenez-quesadae **Mauricio Diazgranados**

Espeletia muiska **Mauricio Diazgranados**

Espeletia pycnophylla **Mauricio Diazgranados**

Espeletia tunjana **Mauricio Diazgranados**

Fuchsia arborescens **Mauricio Diazgranados**

Hypochaeris sessiliflora **Mauricio Diazgranados**

Jacaranda obtusifolia **Mauricio Diazgranados**

Hesperocyparis lusitanica **Mauricio Diazgranados**

Jatropha integerrima **Laura Kor**

Juglans neotropica **Mauricio Diazgranados**

Justicia carnea **Rafaël Govaerts**

Kohleria spicata Mauricio Diazgranados

Lafoensia acuminata Mauricio Diazgranados

Myrsine dependens Mauricio Diazgranados

Orthrosanthus chimboracensis Mauricio Diazgranados

Parajubaea cocoides Mauricio Diazgranados

Phyllostachys aurea Mauricio Diazgranados

Pittosporum undulatum Mauricio Diazgranados

Pontederia crassipes Ori Fragman-Sapir

Pleroma urvilleanum Mauricio Diazgranados

Retrophyllum rospigliosii Mauricio Diazgranados

Sisyrinchium convolutum Mauricio Diazgranados

Ranunculus praemorsus Mauricio Diazgranados

Streptosolen jamesonii Mauricio Diazgranados

Vinca major Mauricio Diazgranados

Apeiba tibourbou Daniela Zappi

Astrocaryum chambira William J. Baker

Attalea maripa Denise Sasaki

Austroeupatorium inulifolium Himesh Dilruwan Jayasinghe

Baccharis prunifolia Mauricio Diazgranados

Bauhinia variegata Mauricio Shlomit Heymann

Byrsonima crassifolia Mauricio Diazgranados

Calophyllum brasiliense Mauricio Diazgranados

Caryocar glabrum William Milliken

Cascabela thevetia Igor Sheremetyev

Cereus repandus Mauricio Diazgranados

Cespedesia spathulata Mauricio Diazgranados

Clethra fimbriata Mauricio Diazgranados

Crateva tapia Mauricio Diazgranados

Cupressus sempervirens Egon Krogsgaard

Delonix regia D. Du Puy

Erythrina crista-galli Ori Fragman-Sapir

Dodonaea viscosa Mauricio Diazgranados

Escallonia pendula Mauricio Diazgranados

Eucalyptus globulus Mauricio Diazgranados

Gaiadendron punctatum Mauricio Diazgranados

Guarea guidonia William Milliken

Gyrocarpus americanus Ken Behrens

Hesperomeles goudotiana Mauricio Diazgranados

Heterostemon conjugatus Dairon Cárdenas L.

Hura crepitans Pradeep Rajatewa

Inga punctata Denise Sasaki

Jacaranda copaia William Milliken

Hypericum juniperinum Mauricio Diazgranados

Jacaranda mimosifolia Igor Sheremetyev

Jatropha multifida Rafaël Govaerts

Lachesiodendron viridiflorum G. P. Lewis

Lantana trifolia Pradeep Rajatewa

Lycopodium clavatum Mauricio Diazgranados

Linochilus rosmarinifolius Mauricio Diazgranados

Maclura tinctoria William Milliken

Mayna odorata Dairon Cárdenas L.

Parkinsonia aculeata Oliver Whaley

Muntingia calabura Ángela Rodríguez-C.

Pinus patula Mauricio Diazgranados

Piper aduncum M. Reith

Prosopis juliflora D. Du Puy

Samanea saman Pradeep Rajatewa

Socratea exorrhiza Denise Sasaki

Symphonia globulifera William Milliken

Tabebuia rosea Himesh Dilruwan Jayasinghe

Typha domingensis Ori Fragman-Sapir

Ulex europaeus Mauricio Diazgranados

Vachellia macracantha Oliver Whaley

Xylopia aromatica Dairon Cárdenas L.

Achillea millefolium Ángela Rodríguez-C.

Alstroemeria aurea Mauricio Diazgranados

Aniseia martinicensis William Milliken

Begonia cucullata Mauricio Diazgranados

Bellucia pentamera Mauricio Diazgranados

Bertholletia excelsa Denise Sasaki

Bixa urucurana Mauricio Diazgranados

Cajanus cajan Oliver Whaley

Caladium bicolor Egon Krogsgaard

Camellia japonica Mauricio Diazgranados

Carludovica palmata Laura Kor

Ceratopteris thalictroides Nawendy

Cystopteris fragilis Bing Liu

Gliricidia sepium Mauricio Diazgranados

Cinchona pubescens Andrew McRobb

Fuchsia magellanica Egon Krogsgaard

Guazuma ulmifolia Mauricio Diazgranados

Cycas revoluta Igor Sheremetyev

Ficus elastica Mauricio Diazgranados

Heliconia bihai William Milliken

Hevea brasiliensis Mauricio Diazgranados

Hymenaea courbaril William Milliken

Ipomoea aquatica Himesh Dilruwan Jayasinghe

Jacaratia digitata William Milliken

Laguncularia racemosa Rafaël Govaerts

Lupinus bogotensis Mauricio Diazgranados

Lupinus mutabilis G. P. Lewis

Melaleuca citrina Mauricio Diazgranados

Melicoccus bijugatus Mauricio Diazgranados

Moringa oleifera Laura Kor

Oryza sativa Egon Krogsgaard

Pachira aquatica Ori Fragman-Sapir

Paepalanthus alpinus Mauricio Diazgranados

Passiflora cumbalensis David Granados

Pelargonium cucullatum Ori Fragman-Sapir

Persea americana Mauricio Diazgranados

Phoenix canariensis Mauricio Diazgranados

Phoenix roebelenii Mauricio Diazgranados

Physalis peruviana Michael Wolf

Phytolacca octandra Oliver Whaley

Pithecellobium dulce Egon Krogsgaard

Psidium guajava Mauricio Diazgranados

Rhizophora mangle Rafaël Govaerts

Sagittaria guayanensis Edson Guilherme

Sonchus oleraceus Himesh Dilruwan Jayasinghe

Theobroma cacao Mauricio Diazgranados

Trifolium repens Mauricio Diazgranados

Urena lobata Ori Fragman-Sapir

Vanilla planifolia Egon Krogsgaard

Vigna luteola Oliver Whaley

Anacardium occidentale Wolfgang Stuppy

Annona muricata Himesh Dilruwan Jayasinghe

Brosimum alicastrum Néstor Plata

Canna indica Mauricio Diazgranados

Carica papaya Andrew McRobb

Coccoloba ovata Mauricio Diazgranados

Costus arabicus Mauricio Diazgranados

Couepia bracteosa Dairon Cárdenas L.

Cynometra bauhiniifolia Mauricio Diazgranados

Daucus montanus **Mauricio Diazgranados**

Dipteryx rosea **Mauricio Diazgranados**

Disterigma empetrifolium **Mauricio Diazgranados**

Elaeagia pastoensis **Mauricio Diazgranados**

Duroia micrantha **Mauricio Diazgranados**

Escallonia myrtilloides **Mauricio Diazgranados**

Eschweilera parvifolia **Dairon Cárdenas L.**

Euterpe catinga **David Hammond**

Fuchsia boliviana **Mauricio Diazgranados**

Galium hypocarpium **Mauricio Diazgranados**

Gaylussacia buxifolia Mauricio Diazgranados

Hypochaeris radicata Mauricio Diazgranados

Macrolobium acaciifolium Mauricio Diazgranados

Hieronyma macrocarpa Mauricio Diazgranados

Lecythis ampla Laura Kor

Lycianthes lycioides Ángela Rodríguez-C.

Miconia astroplocama Mauricio Diazgranados

Hirtella racemosa Mauricio Diazgranados

Leopoldinia pulchra Mauricio Diazgranados

Manihot esculenta Vengolis

Miconia salicifolia Mauricio Diazgranados

Moquilea pyrifolia Mauricio Diazgranados

Ocotea caesariata Mauricio Diazgranados

Parahancornia oblonga Mauricio Diazgranados

Passiflora ambigua Mauricio Diazgranados

Passiflora tarminiana Ángela Rodríguez-C.

Pelliciera rhizophorae Laura Kor

Pontederia azurea Mauricio Diazgranados

Plantago australis Mauricio Diazgranados

Ribes andicola Mauricio Diazgranados

Rubus glabratus Mauricio Diazgranados

Sambucus peruviana Mauricio Diazgranados

Solanum betaceum (red fruit) Laura Kor

Syzygium paniculatum Mauricio Diazgranados

Theobroma bicolor William Milliken

Ugni myricoides Mauricio Diazgranados

Veronica serpyllifolia Mauricio Diazgranados

Vitex orinocensis Mauricio Diazgranados

Vochysia lehmannii Mauricio Diazgranados

Xanthosoma sagittifolium Mauricio Diazgranados

Zygia cataractae Mauricio Diazgranados

Agave americana var. *franzosini* Rafaël Govaerts

Ageratum conyzoides Himesh Dilruwan Jayasingh

Angelica archangelica Egon Krogsgaard

Areca catechu Himesh Dilruwan Jayasinghe

Asclepias curassavica Himesh Dilruwan Jayasinghe

Bidens bipinnata Bing Liu

Bidens pilosa William Milliken

Borago officinalis William Milliken

Bursera simaruba William Milliken

Cananga odorata Ori Fragman-Sapir

Cardiospermum halicacabum Wolfgang Stuppy

Carthamus tinctorius Dr Henry Oakeley

Casuarina cunninghamiana Igor Sheremetyev

Cedrela odorata Denise Sasaki

Centaurea cyanus Ori Fragman-Sapir

Chrysobalanus icaco Entofille

Cordia alliodora Richard Moore

Cocos nucifera A. McRobb

Cordyline fruticosa Rafaël Govaerts

Crescentia cujete Ori Fragman-Sapir

Cucumis melo Ori Fragman-Sapir

Enneapogon desvauxii Ori Fragman-Sapir

Dipsacus fullonum Rafaël Govaerts

Erigeron bonariensis Ori Fragman-Sapir

Erigeron sumatrensis Ori Fragman-Sapir

Guaiacum officinale Pradeep Rajatewa

Helianthus annuus Igor Sheremetyev

Ipomoea pes-caprae Himesh Dilruwan Jayasinghe

Ipomoea purpurea Igor Sheremetyev

Ipomoea triloba Daniela Zappi

Kigelia africana Wolfgang Stuppy

Nicoteba betonica Himesh Dilruwan Jayasinghe

Nymphoides indica Ori Fragman-Sapir

Ochroma pyramidale Egon Krogsgaard

Opuntia cochenillifera Dr Henry Oakeley

Origanum vulgare Mauricio Diazgranados

Passiflora edulis Egon Krogsgaard

Parinari excelsa Denise Sasaki

Phoenix reclinata A. McRobb

Psilotum nudum Rich Ackerman

Silybum marianum Ori Fragman-Sapir

Swietenia macrophylla Pradeep Rajatewa

Schinus terebinthifolia Ori Fragman-Sapir

Spondias mombin Denise Sasaki

Tecoma stans Himesh Dilruwan Jayasinghe

Taraxacum officinale Igor Sheremetyev

Tithonia diversifolia Himesh Dilruwan Jayasinghe

Tithonia rotundifolia Egon Krogsgaard

Varronia curassavica Himesh Dilruwan Jayasinghe

Verbesina encelioides Helen Pickering

Acacia melanoxylon **Mauricio Diazgranados**

Acrocomia aculeata **Mauricio Diazgranados**

Andesanthus lepidotus **Mauricio Diazgranados**

Aspidosperma desmanthum **Dairon Cárdenas L.**

Astrocaryum jauari **Mauricio Diazgranados**

Anthoxanthum odoratum **Mauricio Diazgranados**

Bactris brongniartii **Mauricio Diazgranados**

Bucquetia glutinosa **Mauricio Diazgranados**

Bactris guineensis **Mateo Fernández**

Bambusa vulgaris Mauricio Diazgranados

Calliandra trinervia Mauricio Diazgranados

Cecropia telenitida Mauricio Diazgranados

Cyclanthus bipartitus Mauricio Diazgranados

Bowdichia virgilioides Mauricio Diazgranados

Campsiandra comosa Mauricio Diazgranados

Coccoloba mollis Mauricio Diazgranados

Calamagrostis bogotensis Mauricio Diazgranados

Caraipa llanorum Mauricio Diazgranados

Curatella americana Mauricio Diazgranados

Dioscorea coriacea Mauricio Diazgranados

Espeletia lopezii Mauricio Diazgranados

Espeletia pleiochasia Mauricio Diazgranados

Feijoa sellowiana Egon Krogsgaard

Handroanthus barbatus Mauricio Diazgranados

Hevea nitida Dairon Cárdenas L.

Hippeastrum vittatum Mauricio Diazgranados

Hypericum laricifolium Mauricio Diazgranados

Jasminum officinale Mauricio Diazgranados

Hymenopus heteromorphus Mauricio Diazgranados

Liquidambar styraciflua **Mauricio Diazgranados**

Monnina aestuans **Mauricio Diazgranados**

Mucuna sloanei **Mauricio Diazgranados**

Magnolia grandiflora **Mauricio Diazgranados**

Monticalia pulchella **Mauricio Diazgranados**

Myrcianthes leucoxyla **Mauricio Diazgranados**

Oreopanax incisus **Mauricio Diazgranados**

Mespilodaphne cymbarum **Mauricio Diazgranados**

Neurolaena lobata **Mauricio Diazgranados**

Paramochloa effusa **Mauricio Diazgranados**

Phytelephas macrocarpa **Laura Kor**

Podocarpus oleifolius Mauricio Diazgranados

Polylepis quadrijuga Mauricio Diazgranados

Prunus serotina Mauricio Diazgranados

Psoralea pinnata Mauricio Diazgranados

Siparuna thecaphora Mauricio Diazgranados

Tabernaemontana grandiflora Néstor Plata

Trattinnickia aspera Mauricio Diazgranados

Valeriana arborea Mauricio Diazgranados

Weinmannia fagaroides Mauricio Diazgranados

Virola carinata Mauricio Diazgranados

Acaena elongata Mauricio Diazgranados

Achyrocline lehmannii Mauricio Diazgranados

Alchemilla nivalis Mauricio Diazgranados

Arcytophyllum nitidum Mauricio Diazgranados

Bidens andicola Mauricio Diazgranados

Bidens triplinervia Mauricio Diazgranados

Chamaecrista desvauxii Mauricio Diazgranados

Callianthe striata Mauricio Diazgranados

Cerastium arvense Mauricio Diazgranados

Chuquiraga jussieui Mauricio Diazgranados

Croton mutisianus Mauricio Diazgranados

Culcitium canescens Mauricio Diazgranados

Dieffenbachia parlatorei Mauricio Diazgranados

Draba litamo Mauricio Diazgranados

Eryngium humboldtii Mauricio Diazgranados

Espeletia congestiflora Mauricio Diazgranados

Espeletia argentea f. *phaneractis* Mauricio Diazgranados

Espeletia barclayana Mauricio Diazgranados

Espeletia caldasii Mauricio Diazgranados

Espeletia corymbosa Mauricio Diazgranados

Espeletia curialensis Mauricio Diazgranados

Espeletia glandulosa Mauricio Diazgranados

Espeletia nemekenei **Mauricio Diazgranados**

Espeletia jaramilloi **Mauricio Diazgranados**

Espeletia killipii var. *chisacana* **Mauricio Diazgranados**

Espeletia oswaldiana **Mauricio Diazgranados**

Espeletia occulta subsp. *oroquensis* **Mauricio Diazgranados**

Espeletia santanderensis **Mauricio Diazgranados**

Gamochaeta americana **Mauricio Diazgranados**

Gentiana sedifolia **Mauricio Diazgranados**

Gentianella corymbosa **Mauricio Diazgranados**

Loricaria complanata Mauricio Diazgranados

Lycopodium thyoides Mauricio Diazgranados

Pseudorhipsalis amazonica Mauricio Diazgranados

Hypericum mexicanum Mauricio Diazgranados

Monnina padifolia Mauricio Diazgranados

Pseudomalmea boyacana Mauricio Diazgranados

Lobelia tenera Mauricio Diazgranados

Monotropa uniflora Mauricio Diazgranados

Monticalia vaccinioides Mauricio Diazgranados

Napeanthus apodemus Mauricio Diazgranados

Schiekia orinocensis Mauricio Diazgranados

Senna macrophylla Mauricio Diazgranados

Simira rubescens Mauricio Diazgranados

Siphocampylus columnae Mauricio Diazgranados

Strychnos guianensis Mauricio Diazgranados

Valeriana rigida Mauricio Diazgranados

Valeriana triphylla Mauricio Diazgranados

Warszewiczia coccinea Mauricio Diazgranados

Zygia latifolia Mauricio Diazgranados

Acacia melanoxylon Mauricio Diazgranados

Allamanda cathartica Rafaël Govaerts

Allium cepa Sarath Rajapakshe

Amaranthus tricolor Himesh Dilruwan Jayasinghe

Anthurium andraeanum Egon Krogsgaard

Amaranthus spinosus Himesh Dilruwan Jayasinghe

Bixa orellana HEDS Digitisation Services

Caryota mitis Pradeep Rajatewa

Celosia argentea Bing Liu

Chamissoa altissima William Milliken

Chenopodium album Igor Sheremetycv

Conium maculatum Igor Sheremetyev

Connarus coriaceus Denise Sasaki

Copaifera officinalis Mauricio Diazgranados

Datura metel Laura Kor

Dysphania ambrosioides Himesh Dilruwan Jayasinghe

Erythrina fusca Pradeep Rajatewa

Ficus insipida Denise Sasaki

Galinsoga parviflora Igor Sheremetyev

Gloriosa superba Ori Fragman-Sapir

Hippobroma longiflora Himesh Dilruwan Jayasingh

Malouetia nitida Mauricio Diazgranados

Nerium oleander Igor Sheremetyev

Heliotropium curassavicum Himesh Dilruwan Jayasinghe

Hyptis brachiata Mauricio Diazgranados

Manihot carthagenensis Pradeep Rajatewa

Hippeastrum puniceum Himesh Dilruwan Jayasingh

Kalanchoe delagoensis Mauricio Diazgranados

Monstera deliciosa Mauricio Diazgranados

Oxalis barrelieri Christian Moliné

Palicourea crocea William Milliken

Piper peltatum Mauricio Diazgranados

Rumex acetosella Mauricio Diazgranados

Palicourea tomentosa Mauricio Diazgranados

Plumeria rubra Igor Sheremetyev

Ricinus communis Igor Sheremetyev

Senna occidentalis Mauricio Diazgranados

Philodendron hederaceum Thomas Bernard Croat

Schinus molle Mauricio Diazgranados

Spigelia anthelmia Pradeep Rajatewa

Spinacia oleracea Rafaël Govaerts

Sterculia apetala Mateo Fernández

Swartzia arborescens Denise Sasaki

Tabernaemontana divaricata Himesh Dilruwan Jayasinghe

Terminalia catappa Egon Krogsgaard

Thunbergia alata Mauricio Diazgranados

Trianthema portulacastrum Himesh Dilruwan Jayasinghe

Turnera subulata Vero Alpizar

Vallea stipularis Mauricio Diazgranados

Vallea stipularis Helen Pickering

Abarema jupunba William Milliken

Anadenanthera peregrina Daniela Zapp

Baccharis latifolia Mauricio Diazgranados

Banisteriopsis caapi Andrew McRobb

Calathea lutea Mauricio Diazgranados

Brownea grandiceps Egon Krogsgaard

Brugmansia insignis Mauricio Diazgranados

Calophyllum inophyllum Sarath Rajapakshe

Capsicum annuum Rafaël Govaerts

Cassytha filiformis Bing Liu

Cattleya trianae Mauricio Diazgranados

Ceroxylon quindiuense Mauricio Diazgranados

Citrus × limon Mauricio Diazgranados

Cobaea scandens Mauricio Diazgranados

Cochlospermum orinocense Mauricio Diazgranados

Coffea arabica Ori Fragman-Sapir

Cornutia pyramidata Alexey Yakovlev

Cymbopogon citratus Mauricio Diazgranados

Cyperus alternifolius Mauricio Diazgranados

Dianthera pectoralis Eve Lucas

Dichorisandra hexandra William Milliken

Dimorphotheca jucunda Mauricio Diazgranados

Dioclea guianensis William Milliken

Erythroxylum coca Egon Krogsgaard

Espeletia discoidea Mauricio Diazgranados

Espeletia uribei Mauricio Diazgranados

Euphorbia pulcherrima André Karwath

Genipa americana Mauricio Diazgranados

Hibiscus rosa-sinensis Mauricio Diazgranados

Jasminum multiflorum Pradeep Rajatewa

Lantana camara Mauricio Diazgranados

Jatropha gossypiifolia Mauricio Diazgranados

Leonotis leonurus Ori Fragman-Sapir

Lepidocaryum tenue Dairon Cárdenas L.

Malvaviscus arboreus Mauricio Diazgranados

Manilkara zapota Mauricio Diazgranados

Mangifera indica Mauricio Diazgranados

Mesosphaerum suaveolens Ori Fragman-Sapir

Nertera granadensis Mauricio Diazgranados

Mimosa pudica G. P. Lewis

Nicotiana alata Henry Oakeley

Nicotiana tabacum William Milliken

Piper arboreum Eve Lucas

Psychotria carthagenensis Lidyanne Aona

Salvia rosmarinus Igor Sheremetyev

Tamarindus indica G. P. Lewis

Sida rhombifolia Mauricio Diazgranados

Urtica dioica Wolfgang Stuppy

Vaccinium floribundum Mauricio Diazgranados

Xiphidium caeruleum Egon Krogsgaard

Tropical rain forest in the Guaviare department, Colombia.
Mauricio Diazgranados

Index of Families

A
Acanthaceae 55, 75, 97, 138, 168, 359–362
Achariaceae 168, 380
Achatocarpaceae 168, 280
Actinidiaceae 168, 297
Aizoaceae 168, 280
Alismataceae 168, 197, 198
Alstroemeriaceae 168, 226
Altingiaceae 168, 460
Amaranthaceae 72, 74, 97, 168, 281–283
Amaryllidaceae 85, 99, 168, 213, 214
Anacardiaceae 74, 97, 100, 168, 450, 451
Annonaceae 66, 95, 96, 99, 100, 138, 168, 186–190
Apiaceae 168, 251, 252, 253
Apocynaceae 88, 95, 96, 97, 98, 100, 168, 339–344
Aquifoliaceae 168, 254, 255
Araceae 85, 95, 96, 97, 99, 138, 168, 198–203
Araliaceae 55, 138, 168, 253, 254
Araucariaceae 68, 98, 168, 180
Arecaceae 66, 67, 68, 73, 74, 75, 85, 87, 95, 96, 97, 98, 99, 100, 144, 168, 203–213
Aristolochiaceae 98, 168, 192, 193
Asparagaceae 168, 214, 215
Asphodelaceae 168, 215, 216
Aspleniaceae 168, 174
Asteraceae 6, 12, 49, 55, 56, 66, 73, 79, 85, 88, 90, 95, 97, 116, 135, 138, 139, 140, 144, 146, 161, 168, 255–271

B
Balanophoraceae 168, 448
Balsaminaceae 138, 168, 297
Basellaceae 73, 168, 283
Bataceae 168
Begoniaceae 168, 292, 293
Berberidaceae 49, 55, 56, 88, 168, 433, 434
Betulaceae 168, 338
Bignoniaceae 97, 98, 100, 168, 362–365
Bixaceae 74, 96, 168, 406
Bonnetiaceae 168, 380
Boraginaceae 168, 273–276
Brassicaceae 55, 139, 168, 276–278
Bromeliaceae 16, 100, 138, 168, 227–229
Brunelliaceae 168, 430
Burseraceae 95, 97, 168, 451–454
Buxaceae 168, 280

C
Cabombaceae 168, 181
Cactaceae 168, 283–285
Calceolariaceae 168, 365
Calophyllaceae 98, 100, 168, 380, 381
Campanulaceae 55, 168, 272
Cannabaceae 168, 436
Cannaceae 74, 99, 168, 247
Capparaceae 168, 278
Caprifoliaceae 55, 168, 295, 296
Cardiopteridaceae 145
Caricaceae 67, 73, 99, 168, 279
Caryocaraceae 99, 100, 168, 381
Caryophyllaceae 168, 285, 286
Casuarinaceae 168, 338
Celastraceae 97, 168, 290, 291
Ceratophyllaceae 168, 251
Chloranthaceae 168, 181, 182
Chrysobalanaceae 100, 168, 381–383
Cleomaceae 168, 279, 280
Clethraceae 168, 297, 298
Clusiaceae 55, 92, 96, 97, 100, 168, 383–385
Colchicaceae 168, 226
Combretaceae 168, 415, 416
Commelinaceae 168, 223, 224
Compositae 6, 55, 56, 73, 79, 85, 87, 88, 90, 255–271
Connaraceae 168, 430, 431
Convolvulaceae 75, 139, 168, 461–464
Coriariaceae 168, 293
Costaceae 168, 247, 248
Crassulaceae 168, 460, 461
Cucurbitaceae 75, 97, 99, 139, 168, 293–295
Cunoniaceae 168, 431
Cupressaceae 98, 168, 180
Cyatheaceae 168, 172, 173
Cycadaceae 168, 179
Cyclanthaceae 168, 227
Cyperaceae 85, 97, 168, 229–233
Cyrillaceae 168, 298
Cystopteridaceae 168, 174

D
Davalliaceae 86
Dennstaedtiaceae 86, 168, 175
Dichapetalaceae 168, 385, 386
Dilleniaceae 168, 295
Dioscoreaceae 168, 226
Droseraceae 168, 286
Dryopteridaceae 86

E
Ebenaceae 168, 298
Elaeocarpaceae 168, 431
Elatinaceae 145
Equisetaceae 86, 168, 173
Ericaceae 72, 73, 168, 298–300
Eriocaulaceae 168, 233
Erythroxylaceae 55, 96, 98, 99, 168, 386, 387
Escalloniaceae 168, 309
Euphorbiaceae 85, 95, 97, 98, 99, 100, 138, 168, 387–393

F
Fabaceae 12, 55, 66, 68, 72, 73, 74, 75, 79, 85, 95, 97, 98, 99, 100, 135, 138, 139, 140, 144, 146, 168, 309–337
Fagaceae 168, 338
Fucaceae 168, 473

G
Gentianaceae 55, 97, 98, 100, 168, 344, 345
Geraniaceae 99, 168, 358
Gesneriaceae 55, 168, 365–368
Ginkgoaceae 168, 179
Gleicheniaceae 168, 173
Gnetaceae 168, 179, 180
Goupiaceae 96, 168, 393
Gracilariaceae 168, 473
Gramineae 85
Grossulariaceae 168, 461
Gunneraceae 168, 358, 359

H
Haemodoraceae 168, 224
Haloragaceae 168, 461
Hamamelidaceae 168, 461
Heliconiaceae 168, 248, 249
Hernandiaceae 168, 182
Humiriaceae 168, 393
Hydrangeaceae 168, 291
Hydrocharitaceae 168, 203
Hydroleaceae 168, 464
Hymenophyllaceae 168, 173
Hypericaceae 97, 168, 393, 394
Hypoxidaceae 168, 216

I
Icacinaceae 168, 359
Iridaceae 10, 168, 216, 217

J
Juglandaceae 72, 168, 339
Juncaceae 168, 233

K
Krameriaceae 168, 472

L
Lacistemataceae 168, 394
Lamiaceae 55, 85, 98, 99, 139, 168, 368–374
Lauraceae 49, 56, 89, 95, 96, 98, 99, 100, 168, 182–185
Lecythidaceae 95, 96, 97, 98, 99, 100, 168, 300–302
Leguminosae 72, 73, 74, 75, 85
Lejeuneaceae 138
Lentibulariaceae 85, 145, 168, 374
Liliaceae 168, 226
Linaceae 168, 394
Linderniaceae 168, 374
Loasaceae 168, 291, 292
Loganiaceae 98, 168, 345, 346
Loranthaceae 168, 448, 449
Lunulariaceae 168, 171
Lycopodiaceae 86, 168, 171
Lythraceae 168, 416, 417

M
Magnoliaceae 168, 190
Malpighiaceae 96, 98, 100, 168, 394–396
Malvaceae 55, 66, 75, 95, 97, 99, 100, 138, 168, 406–415
Marantaceae 75, 98, 168, 249, 250
Marcgraviaceae 168, 302
Marchantiaceae 168, 171
Marsileaceae 168, 179
Martyniaceae 168, 374
Melastomataceae 55, 66, 72, 95, 100, 116, 138, 140, 168, 417–424
Meliaceae 95, 97, 98, 99, 168, 454, 455
Menispermaceae 98, 100, 168, 434, 435
Menyanthaceae 168, 273
Metteniusaceae 100, 168, 359, 415
Microteaceae 168, 286
Molluginaceae 168, 286
Monimiaceae 168, 185
Montiaceae 168, 286
Moraceae 66, 74, 95, 97, 99, 168, 436–441
Moringaceae 168, 280, 474
Muntingiaceae 168, 415
Musaceae 99, 138, 168, 250
Myricaceae 168, 339
Myristicaceae 96, 98, 168, 190–192
Myrtaceae 66, 68, 72, 73, 75, 97, 168, 424–428

N
Nelumbonaceae 168, 433
Nyctaginaceae 72, 168, 286, 287
Nymphaeaceae 98, 168, 181

O
Ochnaceae 99, 100, 168, 397
Olacaceae 98, 168, 449
Oleaceae 168, 374, 375
Onagraceae 168, 428–430
Ophioglossaceae 168, 173
Opiliaceae 168, 449
Orchidaceae 95, 100, 138, 140, 145, 168, 217–223
Orobanchaceae 96, 168, 375
Osmundaceae 168
Oxalidaceae 73, 168, 432

P
Palmae 73, 74, 75, 85
Pandanaceae 168, 227
Papaveraceae 168, 435
Passifloraceae 14, 49, 56, 67, 73, 138, 168, 397–401
Paulowniaceae 168, 401
Pedaliaceae 168, 375
Pentaphylacaceae 168, 302, 303
Peraceae 168, 401
Petiveriaceae 168, 287
Phrymaceae 168, 375
Phyllanthaceae 97, 168, 401, 402
Phytolaccaceae 168, 287, 288
Picramniaceae 96, 100, 168, 433
Pinaceae 98, 168, 180
Piperaceae 55, 95, 116, 138, 168, 193–197
Pittosporaceae 168
Plantaginaceae 168, 375–377
Platanaceae 168, 377
Plocamiaceae 168, 473
Plumbaginaceae 168, 288
Poaceae 12, 66, 85, 97, 99, 135, 138, 139, 140, 144, 146, 168, 233–247
Podocarpaceae 99, 168, 180
Podostemaceae 168, 402
Polemoniaceae 168, 303
Polygalaceae 168, 337, 338
Polygonaceae 139, 168, 288–290
Polypodiaceae 138, 168, 175–177
Polytrichaceae 168, 171
Pontederiaceae 168, 225
Portulacaceae 168, 290
Potamogetonaceae 168, 203
Primulaceae 168, 303–305
Proteaceae 68, 168, 433
Psilotaceae 168, 179
Pteridaceae 168, 177, 178
Putranjivaceae 168, 402

R
Ranunculaceae 168, 436
Rapateaceae 98, 99, 100, 168, 247
Resedaceae 168, 280
Rhamnaceae 168, 441, 442
Rhizophoraceae 168, 402
Rosaceae 55, 72, 73, 168, 442–445
Rubiaceae 55, 66, 74, 95, 96, 97, 100, 138, 139, 144, 168, 346–358
Ruppiaceae 168, 203
Rutaceae 97, 99, 168, 455–457

S
Sabiaceae 145, 168, 433
Salicaceae 89, 168, 402–405
Salviniaceae 168, 179
Santalaceae 168, 449, 450
Sapindaceae 68, 138, 168, 457–460
Sapotaceae 75, 168, 305–308
Saxifragaceae 168, 461
Schisandraceae 168
Schizaeaceae 98, 168, 179
Schlegeliaceae 168, 377
Schoepfiaceae 168, 450
Scrophulariaceae 168, 377
Selaginellaceae 55, 168, 171, 172
Simaroubaceae 168, 460
Siparunaceae 168, 185, 186
Smilacaceae 168, 226, 227
Solanaceae 67, 73, 75, 79, 85, 95, 96, 98, 99, 138, 139, 144, 168, 465–472
Sphagnaceae 168, 171
Sphenocleaceae 138, 168, 472
Staphyleaceae 168, 292
Strelitziaceae 168, 250
Styracaceae 55, 168, 308, 309
Surianaceae 168, 338
Symplocaceae 55, 168, 309

T
Talinaceae 168, 290
Tetrameristaceae 168, 309
Theaceae 168, 309
Thymelaeaceae 168, 415
Triuridaceae 168, 227
Tropaeolaceae 73, 168, 280
Typhaceae 168, 247

U
Ulmaceae 168, 445
Urticaceae 49, 55, 56, 96, 168, 445–448

V
Velloziaceae 98, 100, 145
Verbenaceae 116, 118, 168, 377–380
Viburnaceae 168, 296, 297
Violaceae 168, 405, 406
Vitaceae 168, 472
Vochysiaceae 100, 168, 430

W
Winteraceae 168, 181

X
Xyridaceae 168, 247

Z
Zamiaceae 99, 168, 179
Zingiberaceae 85, 168, 250, 251
Zygophyllaceae 99, 168, 472, 473

Index of Genera

Page numbers in **bold** text correspond to photographs

A

Aa 217
Abarema 309, 310, **522**
Abatia 402
Abelmoschus 406
Abrus 310
Abuta 98, 434
Abutilon 406, 407, **481**
Acacia 144, 153, 160, 310, **507**, **517**
Acaciella 310
Acaena 442, **481**, **512**
Acalypha 387
Acanthella 100
Acanthocereus 283
Acanthospermum 255
Acanthosyris 449
Acanthus 359
Acer 457
Achatocarpus 280
Achillea 255, **492**
Achimenes 365
Achyranthes 281
Achyrocline 49, 50, 55, 56, 57, 59, 60, 255, **512**
Acineta 217
Acioa 381
Aciotis 417
Acmella 255, 256
Acoelorraphe 203
Aconitum 436
Acrocarpus 310
Acroceras 233
Acrocomia **507**
Acrostichum 177
Actinostemon 387
Adelobotrys 417
Adenanthera 310
Adenaria 416
Adenocalymma 362
Adenostemma 256
Adiantum 177
Adonidia 204
Aechmea 100, 227, 228
Aegiphila 368
Aeonium 460
Aequatorium 256
Aeschynomene 310, 311
Aesculus 457
Aetheolaena 256
Agalinis 375
Agapanthus 213
Agastache 368
Agave 214, **502**
Ageratina 79, 87, 89, 90, 256, **481**
Ageratum 90, 256, **502**
Aglaonema 198
Agonandra 449
Agouticarpa 346
Agrostis 233
Aiouea 182
Aiphanes 204
Aira 233
Ajuga 368
Albizia 311
Albuca 214
Alcea 407
Alchemilla 442, 443, **481**, **512**
Alchornea 387
Alchorneopsis 387
Aldina 311
Aleurites 387
Alibertia 346, 347
Alicia 394
Alisma 197
Allamanda 98, 339, **517**
Allantoma 300
Allionia 286
Allium 99, 213, **517**
Alloispermum 256
Allophylus 457
Allosidastrum 407
Alnus 338, **474**
Alocasia 198
Aloe 215
Alonsoa 377
Alopecurus 233
Aloysia 377
Alpinia 250
Alseis 347
Alsophila 172
Alstroemeria 226, **492**
Alternanthera 281, **474**
Althaea 407
Alysicarpus 311
Amaioua 347
Amanoa 401
Amaranthus 72, 74, 282, **517**
Ambelania 100, 339
Ambrosia 256, 257
Ammandra 204
Ammannia 416
Ammi 251
Ampelocera 445
Ampelozizyphus 441
Amphidasya 347
Amphilophium 362
Amphitecna 362
Amyris 455
Anacardium 74, 100, 140, 450, **497**
Anadenanthera 98, 311, **522**
Ananas 228
Anathallis 217, 257
Anaueria 100
Anaxagorea 186
Anchietea 405
Ancistrotropis 311
Andesanthus 417, **507**
Andira 311
Andropogon 233, 234
Anechites 339
Aneilema 223
Anemia 179
Anemone 436
Anemopaegma 362
Anethum 251
Angelica 251, **502**
Angelonia 375
Angostura 455
Anguloa 217
Aniba 96, 98, 99, 100, 107, **108**, 182
Anigozanthos 224
Aniseia 461, **492**
Annona 66, 67, 90, 96, 99, 186, 187, **497**
Anoda 407
Anomospermum 434
Anredera 283
Anthephora 234
Anthodiscus 100, 381
Anthopterus 298
Anthoxanthum 234, **507**
Anthurium 138, 144, 198, 199, **481**, **517**
Antigonon 288
Antirrhinum 375
Antonia 345
Antrocaryon 450
Aosa 291
Aparisthmium 388
Apeiba 407, **487**
Aphelandra 359
Apium 251
Appunia 347
Aptandra 449
Apuleia 311
Aquarius 197
Aquilegia 436
Arachis 99, 140, 311
Arachnothryx 347
Aralia 253
Araucaria 68, 98, 180, **481**
Arawakia 97
Archontophoenix 204
Arcytophyllum 347, **512**
Ardisia 303
Areca 204, **502**
Arenaria 285
Argemone 435
Argyranthemum 257
Aristeguietia 257
Aristida 234
Aristolochia 98, 192, 193
Armeria 288
Arracacia 251
Arrhenatherum 234
Artemisia 257

Arthrostemma 417
Artocarpus 74, 436
Arundinella 234
Arundo 234
Asclepias 339, **502**
Asemeia 337
Asparagus 214
Aspidogyne 217
Aspidosperma 97, 100, 339, **507**
Asplenium 174
Asplundia 227
Aster 257
Asterogyne 204
Astraea 388
Astrocaryum 97, 99, **110**, 140, 204, 205, **487**, **507**
Astronium 450, **474**
Asystasia 359
Athyrium 174
Atocion 285
Attalea 205, 206, **487**
Augusta 347
Aulonemia 234
Aureolaria 375
Austrocylindropuntia 283
Austroeupatorium 257, **487**
*Avena*234
Averrhoa 432
Avicennia 359
*Axonopus*234
Ayapana 257
Ayenia 407
Azadirachta 454
Azara 402
Azolla 179, **474**
Azorella 251

B
Baccharis 257, 258, **487**, **522**
Bacopa 375, 376
Bactris 57, 66, 67, 68, 74, 99, 206, 207, **507**
Badilloa 258
Balizia 311
Baltimora 258
Bambusa 139, 234, 235, **508**
Banara 402, 403
Banisteriopsis 96, 98, 100, 108, 394, 395, **522**
Barbarea 276
Barleria 359
Barnadesia 258, **481**
Basella 283
Batis 276
Batocarpus 437
Bauhinia 311, 312, **487**
Begonia 292, 293, **492**
Beilschmiedia 182
Bejaria 298, **474**
Bellis 258
Bellucia 417, 418, **492**
Berberis 49, 55, 56, 57, 59, 60, 79, 87, 88, 89, 90, 433, 434
Bergenia 461
Bertholletia 99, 301, **492**
Bertiera 347
Besleria 365, 366
Beta 282
Biancaea 312
Bidens 258, **502**, **512**
Bignonia **106**, 362
Billia 457
Biophytum 432
Bischofia 401
Bismarckia 98, 207
Bixa 74, 96, 122, 406, **492**, **517**
Blakea 418
Blechnum 174
Blepharodon 339
Bletia 217
Blighia 457
Blutaparon 282
Bocageopsis 187
Bocconia 435
Boehmeria 445
Boerhavia 286
Bolboschoenus 229
Bomarea 226
Bombax 407
Bonellia 303
Bonnaya 374
Bonnetia 380
Bontia 377
Borago 273, **502**
Borismene 434
Borojoa 74
Borzicactus 283
Bothriochloa 235
Botryarrhena 347
Botrychium 173
Bouchea 378
Bougainvillea 286
Bourreria 273
Bouteloua 235
Bouvardia 347
Bowdichia 312, **508**
Brachychiton 407
Brachyotum 55, 418, **481**
Brasiliopuntia 283
Brassavola 217
Brassica 276
Bravaisia 359
Bredemeyera 337
Breynia 401
Briquetia 407
Briza 235
Bromelia 228
Bromus 235
Bronwenia 395
Brosimum **20**, 97, 437, **497**
Browallia 465
Brownea 312, **522**
Browneopsis 312
Brugmansia 465, **522**
Brunellia 430
Brunfelsia 465, **481**
Buchnera 375
Bucquetia 418, **507**
Buddleja 377
Bulbostylis 229
Bulnesia 99, 107
Bunchosia 395
Bupleurum 252
Burmeistera 272
Bursera 451, **502**
Buxus 280
Byrsonima 395, 396, **487**

C
Caamembeca 337
Cabomba 181
Cabralea 454
Caesalpinia 312
Cajanus 74, 313, **492**
Cakile 276
Caladium 199, **492**
Calamagrostis **508**
Calandrinia 286
Calathea **522**
Calatola 359
Calceolaria 365
Calea 57, 258
Calendula 259
Callaeum 396
Calliandra 313, **508**
Callianthe 407, **512**
Callicarpa 368
Callichlamys 362
Callisia 223, 224
Callitriche 376
Calophyllum 98, 100, 380, **487**, **522**
Calopogonium 313
Calotropis 339
Calycolpus 424
Calycophyllum 347
Calyptrion 405
Camellia 105, 309, **492**
Camonea 461
Campanula 272
Campnosperma 450
Campomanesia 424
Campsiandra 313, **508**
Campuloclinium 259
Campyloneurum 175
Cananga 118, 187, **503**
Canavalia 74, 313
Canna 74, 99, 247, **497**
Cannabis 436
Canscora 344
Cantinoa 368
Cantua 303
Caperonia 388
Capirona 347
Capraria 377
Capsella 276
Capsicum 465, 466, **522**
Caraipa **108**, 380, 381, **508**
Carapa 97, 98, 454
Carapichea 107, 347
Cardamine 276
Cardionema 285
Cardiospermum 457, **503**
Carex 230, **481**
Carica 67, 99, 279, **497**
Cariniana 98, 99, 107, 301
Carissa 340
Carludovica 227, **492**
Carpobrotus 280
Carpotroche 380
Carthamus 259, **503**
Caryocar 99, 100, 381, **487**
Caryodaphnopsis 182
Caryodendron 74, 388
Caryota 207, **517**
Cascabela 340, **487**
Cascarilla 107
Casearia 67, 403, 404
Casimirella 359
Casimiroa 455
Cassia 313, **474**
Cassipourea 402
Cassytha 182, **523**
Castela 460
Castilla **106**, 107, 437
Castilleja 375, **474**
Casuarina 338, **503**
Catabrosa 235
Catalpa 362
Catasetum 217
Catharanthus 340
Cathedra 449
Catoferia 368
Catopsis 228
Catostemma 407
Cattleya **164**, 217, **482**, **523**
Catunaregam 348
Caularthron 218
Cavanillesia 407
Cavendishia 72, 298
Cayaponia 293

Cecropia 49, 50, 55, 56, 57, 59, 60, 445, 446, **508**
Cedrela 98, 99, 107, 153, 454, **503**
Cedrelinga 98, 314
Ceiba 407, 408, **475**
Celosia 282, **518**
Celtis 436
Cenchrus 139, 235, 236
Cenostigma 314
Centaurea 259, **503**
Centaurium 344
Centella 252
Centratherum 259
Centrolobium 314
Centropogon 272, 273
Centrosema 314
Cerastium 285, **512**
Ceratophyllum 251
Ceratopteris 178, **493**
Ceratosanthes 293
Ceratostema 299
Cereus 283, **488**
Ceroxylon 207, **523**
Cervantesia 449
Cespedesia 397, **488**
Cestrum 466, **475**
Chaerophyllum 252
Chaetocarpus 401
Chaetolepis 418
Chalybea 72, 418
Chamaecrista 314, **512**
Chamaecyparis 180
Chamaedorea 207
Chamaemelum 259
Chamaerops 208
Chamissoa 282, **518**
Chaptalia 259
Chascolytrum 236
Chaunochiton 449
Cheiloclinium 290
Chelonanthus 344
Chelyocarpus 208
Chenopodiastrum 282
Chenopodium 72, 97, 282, **518**
Chimarrhis 348
Chiococca 348
Chionanthus 374
Chione 348
Chloris 236
Chlorocardium 183
Chloroleucon 315
Chloropetalum 340
Chlorophytum 215
Chomelia 348
Chondrodendron 434
Chromolaena 49, 55, 56, 57, 59, 60, 79, 87, 88, 89, 90, 259
Chromolucuma 305
Chrysanthemum 259
Chryselium 259
Chrysobalanus 381, **503**
Chrysochlamys 383
Chrysojasminum 374
Chrysophyllum 305
Chrysopogon 236
Chrysothemis 366
Chuquiraga 260, **512**
Chusquea 236
Cicer 315
Cichorium 260
Ciliosemina 107, 348
Cinchona **105**, 107, 348, **493**
Cinnamomum 183
Cirsium 260
Cischweinfia 218
Cissampelopsis 260
Cissampelos 434, **475**
Cissus 472
Citharexylum 378
Citrullus 293
Citrus 97, 99, 455, 456, **523**
Cladium 230
Clarisia 437
Clathrotropis 100, 315
Clavija 303
Clematis 436
Cleome 279, 280
Clerodendrum 369
Clethra 297, 298, **488**
Clianthus 315
Clibadium 260
Clidemia 138, 140
Clinopodium 369
Clitoria 315
Clivia 213
Clusia 57, 60, 383–385, **482**
Cnidoscolus 388
Cobaea 303, **523**
Coccineorchis 218
Coccinia 293
Coccocypselum 348
Coccoloba 288, **497**, **508**
Coccothrinax 208
Cochliasanthus 315
Cochlospermum 406, **523**
Cocos 96, 140, 208, **503**
Codiaeum 388
Codonanthopsis 366
Coffea 96, 348, **523**
Coincya 276
Coix 97, 236
Cojoba 315
Coleonema 456
Coleus 369
Colocasia 99, 199
Colubrina 441
Columnea 366, 367
Colutea 315
Combretum 415, 416
Commelina 224
Commicarpus 286
Compsoneura 190
Conceveiba 388
Conchocarpus 456
Condaminea 348
Condea 369
Congea 369
Conium 252, **518**
Connarus 430, 431, **518**
Conobea 376
Conocarpus 416
Conocliniopsis 260
Copaifera 55, 97, 100, 315, **518**
Copernicia 208
Corchorus 408
Cordia 273, **503**
Cordiera 349
Cordyline 215, **482**, **504**
Coriandrum 252
Coriaria 293
Cornutia 369, **524**
Cortaderia 236
Coryanthes 218
Corymbia 424
Corynaea 448
Cosmibuena 349
Cosmos 260
Costus 247, 248, **497**
Cotoneaster 443, **482**
Couepia 381, 382, **497**
Coulteria 315
Couma 100, 105, 107, 340
Couratari 97, 301
Couroupita 301
Coursetia 315
Coussapoa 446
Coutoubea 344
Craniolaria 374
Crassula 460
Crateva 278, **488**
Cremastosperma 187
Crepis 260
Crescentia 97, 363, **504**
Crinum 213
Critonia 260
Critoniella 260
Critoniopsis 260
Crocosmia 216
Crossopetalum 290
Crotalaria 139, 315, 316
Croton 97, 100, 388, 389, **513**
Cruciata 349
Crudia 316
Cryosophila 208
Cryptomeria 180
Cryptostegia 340
Ctenodon 316
Cucumis 293, **504**
Cucurbita 99, 293, 294
Cuervea 290
Culcitium 260, **513**
Cuminum 252
Cupania 457
Cuphea 416, 417
Cupressus 180, **488**
Curarea 98, 100, 434, 435
Curatella 295, **508**
Curculigo 216
Curcuma 250
Curio 260
Curupira 449
Cuscuta 461
Cuspidaria 363
Cyanthillium 260
Cyathea 172, **482**
Cyathula 282
Cybianthus 304
Cycas 179, **493**
Cyclamen 304
Cyclanthera 294
Cyclanthus 227, **508**
Cyclopeltis 175
Cyclopogon 218
Cyclospermum 252
Cycnoches 218
Cydonia 443
Cylindropuntia 283
Cymbalaria 376
Cymbopetalum 187
Cymbopogon 118, 236, **524**
Cynanchum 340
Cynara 260
Cynodon 236
Cynoglossum 274
Cynometra 316, **497**
Cyperus 67, 139, 230, **475**, **524**
Cyrilla 298
Cyrtocarpa 450
Cyrtochilum 218
Cyrtocymura 261
Cyrtopodium 218
Cyrtostachys 208
Cystopteris 174, **493**

D

Dacryodes 451, 452
Dactylis 236
Dactyloctenium 236
Dahlia 261
Dalbergia 317
Dalea 317

Dalechampia 389
Damburneya 183
Daphnopsis 415
Datura 252, 466, **498**, **518**
Davallia 175
Davilla 295
Deguelia 317
Delairea 261
Delonix 98, 317, **488**
Delosperma 281
Delostoma 363
Delphinium 436
Dendrobangia 255
Dendrocalamus 237
Dendropanax 253
Dendrophorbium 261
Dendrophthora 449
Dennstaedtia 175
Desmanthus 317
Desmodium 317, 318
Desmoncus 208
Desmopsis 187
Desmoscelis 418
Dialium 318
Dianthera 359, **524**
Dianthus 285
Diastatea 273
Dichaea 218
Dichanthelium 237
Dichanthium 237
Dichapetalum 145
Dichondra 461
Dichorisandra 224, **524**
Dicksonia 172
Diclidanthera 338
Diclinanona 187
Dicliptera 359
Dicorynia 318
Dicranopteris 173
Dicranopygium 227
Dicranostyles 461
Dicraspidia 415
Dictyocaryum 208
Didymochlaena 175
Didymopanax 253
Diectomis 237
Dieffenbachia 199, **513**
Dietes 216
Digitalis 376
Digitaria 139, 140, 237, 238
Dilkea 397
Dillenia 295
Dilodendron 457
Dimerocostus 248, **482**
Dimorphotheca 261, **524**
Dioclea 318, **524**
Dioscorea 67, 226, **509**
Diospyros 66, 298
Diphysa 318
Diplachne 238
Diplazium 174
Diplopterys 396
Diplotropis 318
Dipogon 318
Dipsacus 295, **504**
Dipteryx 100, 318, 319, **498**
Disakisperma 238, **475**
Disciphania 435
Disterigma 299, **498**
Distichlis 238
Distimake 461
Ditaxis 389
Dodonaea 457, **488**
Dolichandra 363
Doliocarpus 295, **475**
Dombeya 408
Dorstenia 437
Dovyalis 404
Doyerea 294
Draba 55, 276, **513**
Dracaena 215
Dracontium 200
Dracula 218
Drimys 181
Drosera 286
Drymaria 285
Drymonia 367
Dryopteris 175, **482**
Drypetes 402
Duguetia 96, 187, 188
Duranta 378
Duroia 349, 350, **498**
Dussia 319
Dypsis 208
Dysphania 282, **518**

E

Ecclinusa 305
Echeveria 460
Echinochloa 238
Echinolaena 238, **475**
Echinopepon 294
Echites 340
Eclipta 261
Egletes 261
Eirmocephala 261, **475**
Elaeagia 350, **498**
Elaeis 97, 99, 145, 208
Elaeoluma 305
Elaphoglossum 175
Eleocharis 231, 232
Elephantopus 261
Elettaria 250
Eleusine 238
Eleutheranthera 261
Eleutherine 216
Elionurus 238
Elissarrhena 435
Elleanthus 218
Elodea 203
Elymus 238
Elytraria 360
Emilia 261
Emmeorhiza 350
Endlicheria 183
Enneapogon 238, **504**
Enneatypus 288
Ensete 250
Entada **102**, 319
Enterolobium 319
Enydra 261
Eperua 319
Ephedranthus 188
Epidendrum 100, 138, 144, 218, **482**
Epilobium 428
Epiphyllum 283
Epipremnum 200
Episcia 367, **482**
Equisetum 173
Eragrostis 67, 139, 140, 238, 239
Eranthemum 360
Erechtites 261
Eremocaulon 239
Erica 299
Erigeron 261, 262, **504**
Eriocapitella 436
Eriocaulon 233
Eriochloa 239
Eriope 369
Eriopsis 219
Eriosema 319
Eriotheca 408
Erisma 100, 430
Erithalis 350
Ernodea 350
Erodium 358
Ertela 456
Eruca 277
Erycina 219
Eryngium 252, **476**, **513**
Erysimum 277
Erythranthe 375
Erythrina 74, 319, 320, **475**, **488**, **518**
Erythrochiton 456
Erythroxylum 96, 98, 99, 386, 387, **524**
Escallonia 309, **488**, **498**
Eschscholzia 435
Eschweilera 96, 97, 100, 301, **498**
Escobedia 96, 375
Esenbeckia 456
Espeleptiosis 57
Espeletia **6**, **48**, 49, 55, 56, 57, 60, 79, 87, 88, 89, 138, 140, 145, 161, 262, 263, **483**, **484**, **509**, **513**, **514**, **524**
Espeletiopsis 49, 55, 56, 57, 60
Etlingera 251
Eucalyptus 144, 153, 160, 424, 425, **488**
Euclasta 239
Eugenia 68, 75, 425, 426
Eumachia 350
Euonymus 290
Euphorbia 99, 389–391, **525**
Euplassa 433
Euploca 274
Eustoma 344
Euterpe 12, 27, 96, 100, 120, 121, 122, 123, 208, **475**, **498**
Evodianthus 227
Evolvulus 461
Exostema 350
Exothea 457

F

Fagopyrum 289
Faramea 350
Fatsia 253
Feijoa 426, **509**
Ferdinandusa 350
Festuca 239
Fevillea 294
Ficus 66, 67, 97, 99, 138, 437–439, **493**, **518**
Fimbristylis 232
Fischeria 340
Fittonia 360
Flacourtia 404
Fleischmannia 263
Flemingia 320
Foeniculum 252
Forsteronia 340
Fragaria 443
Frangula 441
Fraxinus 374, **475**
Freesia 216
Freziera 302, 303
Fridericia **106**, 107, 363
Fuchsia 428, **484**, **493**, **498**
Fucus 473
Fuertesimalva 408
Fuirena 232
Fumaria 435
Funastrum 340
Furcraea **106**, 107, 215
Fusaea 188

G

Gaiadendron 448, **489**
Gaillardia 263
Galactia 320
Galianthe 350
Galinsoga 263, **518**
Galium 350, **498**
Galphimia 396
Gamochaeta 263, **514**
Garcia 391
Garcinia 66, 96, 100, 385, **476**
Gardenia 350
Gasteranthus 367
Gaultheria 299, **476**
Gaylussacia 72, 299, **499**
Gazania 263
Geissanthus 304
Genipa 96, 122, 351, **525**
Genista 320
Gentiana 344, **514**
Gentianella 344, 345, **514**
Geoffroea 320
Geogenanthus 224
Geonoma 209
Geophila 351
Geranium 358
Gerbera 263
Geum 443
Gibasis 224
Gigantochloa 239
Ginkgo 179
Gladiolus 216
Glebionis 263
Gleichenella 173
Glinus 286
Gliricidia 97, 153, 320, **493**
Gloriosa 226, **519**
Glossoloma 367
Gloxinia 367
Glyceria 239
Glycine 320
Glycydendron 391
Gmelina 98, 369
Gnetum 179, 180
Godmania 363
Godoya 397
Goeppertia 98, 249, 345
Gomesa 219
Gomphocarpus 340
Gomphrena 282
Gongora 219
Gonzalagunia 351
Gordonia 309
Gossypium 408
Gouania 441, 442
Goupia 96, 393
Gracilaria 473
Graffenrieda 418
Gratiola 376
Greigia 228
Grevillea 433
Grewia 408
Grias 96, 301
Grona 320
Gronovia 291
Guacamaya 98, 134
Guadua 239
Guaiacum 472, **504**
Guapira 286, 287
Guarea 454, **489**
Guatteria 96, 99, 100, 188, 189
Guazuma 153, 408, **493**
Guettarda 351
Guilandina 320
Guilleminea 282
Gunnera 358, 359
Gurania 294
Gustavia 301, 302
Guzmania 228, 229
Gymnopogon 240
Gynerium 108, 240
Gynura 263
Gypsophila 285
Gyrocarpus 182, **489**

H

Habenaria 219
Hackelochloa 240
Haematoxylum 320
Halenia 345, **476**
Halerpestes 436
Hamamelis 461
Hamelia 351
Hampea 408
Handroanthus 98, 100, 363, 364, **509**
Haploclathra 381
Hasseltia 404
Hawkesiophyton 467
Haworthiopsis 215
Hebeclinium 263
Hedera 253
Hedychium 251
Hedyosmum 181, 182
Heisteria 449
Helanthium 197
Helia 345
Helianthostylis 439
Helianthus 263, **504**
Helichrysum 264
Heliconia 96, 248, 249, **476**, **493**
Helicostylis 439, 440
Helicotropis 320
Helicteres 408
Heliocarpus 408
Heliopsis 264
Heliotropium 274, 275, **519**
Helosis 448
Hemarthria 240
Hemerocallis 215
Hemionitis 178
Henriettea 418
Heppiella 367
Heptapleurum 253
Herbertia 216
Herissantia 409
Hernandia 182
Herrania 409
Hesperocyparis 98, 180, **484**
Hesperomeles 57, 72, 140, 443, **489**
Heteranthera 225
Heterocentron 419
Heteropogon 240
Heteropsis 96, 97, 200
Heteropterys 396
Heterosperma 264
Heterostachys 283
Heterostemon 320, **489**
Heterotis 419
Hevea 100, 107, 391, **494**, **509**
Hexasepalum 351
Hibiscus 99, 409, 410, **525**
Hieronyma 97, 401, **499**
Hilleria 287
Hillia 351
Himatanthus 340
Hippeastrum 213, **509**, **519**
Hippobroma 273, **519**
Hippocratea 290
Hippomane 391
Hiraea 396
Hirtella 382, **499**
Hoffmannia 351
Holcus 240, **478**
Holmskioldia 369
Homalium 404
Homalolepis 460
Homalomena 145
Homolepis 240
Hordeum 240
Hortia 456
Howea 209
Hoya 341
Huberodendron 410
Humiria 393
Humiriastrum 393
Humulus 436
Huntleya 219
Huperzia 171
Hura 100, 391, **489**
Hydnocarpus 380
Hydrangea 291
Hydrilla 203
Hydrocharis 203
Hydrochorea 320
Hydrocleys 198
Hydrocotyle 253
Hydrolea 464
Hydrotriche 376
Hygrophila 360
Hylenaea 291
Hymenachne 240
Hymenaea 320, 321, **494**
Hymenasplenium 174
Hymenocallis 213
Hymenolobium 321
Hymenophyllum 173
Hymenopus 382, **509**
Hyophorbe 68, 209
Hyospathe 209
Hyparrhenia 97, 240
Hyperbaena 435
Hypericum 393, 394, **489**, **509**, **515**
Hyperthelia 240
Hypochaeris 264, **484**, **499**
Hypoestes 360
Hypolepis 175
Hypoxis 216
Hyptidendron 369
Hyptis 369, 370, **519**

I

Ibatia 341
Iberis 277
Ichnanthus 240
Ichthyothere 264
Ilex 254, 255
Illicium 181
Impatiens 297
Imperata 97, 240
Incarvillea 364
Indigofera 321
Inga 12, 66, 67, 68, 95, 96, 135, 138, 146, 321–324, **478**, **489**
Iochroma 467
Ipomoea 67, 75, 138, 139, 140, 461–464, **478**, **494**, **504**, **505**
Iresine 283
Iriartea 96
Iriartella 210
Iris 216
Irlbachia 345
Iryanthera 98, 190, 191
Ischaemum 240
Ischnosiphon 249
Isertia 351, 352
Isocarpha 264

Isolepis 232
Itaya 210
Ixora 352

J
Jacaranda 364, **484**, **489**
Jacaratia 279, **494**
Jacobaea 264
Jacquemontia 464
Jacquinia 304
Jaltomata 467
Jamesonia 178
Jasminum 374, 375, **509**, **525**
Jatropha 99, 391, 392, **484**, **489**, **525**
Juglans 72, 338, 339, **484**
Juncus 233
Jungia 264
Justicia 55, 360, **484**

K
Kalanchoe 461, **519**
Kallstroemia 472
Karwinskia 442
Kearnemalvastrum 410
Khaya 98
Kigelia 364, **505**
Klarobelia 189
Klopstockia 210
Kniphofia 216
Kohleria 367, **485**
Kosteletzkya 410
Krameria 472
Krugiodendron 442
Kutchubaea 352

L
Lablab 324
Lachesiodendron 325, **490**
Lacistema 394
Lacmellea 100, 341
Lactuca 264
Lacunaria 397
Ladenbergia 107, 352
Laelia 219
Laennecia 264
Lafoensia 417, **485**
Lagascea 264
Lagenaria 294
Lagerstroemia 417
Laguncularia 416, **494**
Lagurus 240
Lamarckia 240
Lamium 370
Lamourouxia 375
Lampranthus 281
Langsdorffia 448
Lantana 378, **490**, **525**
Laportea 446
Lapsana 264
Lasiacis 240, 241
Lasiocephalus 264
Lathyrus 68, 325
Launaea 264
Laurus 183
Lavandula 370
Lawsonia 417
Leandra 138, 140
Lecointea 325
Lecythis 302, **499**, **540**
Ledenbergia 287
Leersia 241
Lemna 200
Lennoa 275
Leonia 405
Leonotis 370, **525**
Leonurus 370
Leopoldinia 97, 210, **499**
Lepanthes 138, 144
Lepechinia 370, 371
Lepidagathis 360
Lepidaploa 265
Lepidium 277
Lepidocaryum 96, 210, **526**
Leptobalanus 382
Leptochloa 241
Leptolobium 325
Leptospermum 426
Leptospron 325
Leretia 359
Lessingianthus 265
Leucadendron 433
Leucaena 97, 325
Leucanthemum 265
Leucophyllum 377
Leuenbergeria 284
Liabum 265
Liatris 265
Libidibia 325
Licania 382
Licaria 183
Licuala 210
Ligustrum 375, **478**
Lilaeopsis 252
Lilium 226
Limnocharis 198, **478**
Limonium 288
Limosella 376
Lindackeria 380
Lindernia 374
Linochilus 265, **490**
Linum 394
Liparis 219
Lippia **117**, 118, 378, 379
Liquidambar 460, **510**
Lithospermum 275
Livistona 210
Lobelia 273, **515**
Lobularia 277
Lockhartia 219
Lolium 241
Lomariopsis 175
Lomelosia 296
Lonchocarpus 325
Lonicera 296
Lophosoria 173
Lophospermum 376
Loricaria 265, **515**
Lorostemon 385
Lotus 325
Louisiella 241
Lourteigia 79, 87, 88, 265
Ludovia 227
Ludwigia 429
Luehea 410
Luffa 294
Lunania 404
Lunaria 277
Lundia 364
Lunularia 171
Lupinus 72, 325, 326, **494**
Lutheria 229
Luziola 241
Lycianthes 467, **499**
Lycium 467
Lycopodiella 171
Lycopodium **490**, **515**
Lycoseris 265
Lygodium 179
Lysimachia 304
Lythrum 417

M
Mabea 392
Macadamia 433
Machaerium 326
Macleania 72, 299, **478**
Maclura 440, **490**
Macoubea 96, 341
Macrocnemum 352
Macrolobium 326, **499**
Macropsychanthus 326, 327
Macroptilium 327, **478**
Macrothelypteris 174
Magnolia 138, 140, 190, **510**
Maieta 419
Malachra 410
Malephora 281
Malouetia 341, **519**
Malpighia 396
Malus 443
Malva 410
Malvastrum 410, 411
Malvaviscus **526**
Mammea 381
Mammillaria 284
Mandevilla 79, 87, 88, 341, 342
Manettia 352
Mangifera 96, 450, **526**
Manicaria 210
Manihot 99, 392, **499**, **519**
Manilkara 305, 306, **526**
Mansoa 364
Maprounea 392
Maquira 440
Maranta 75, 249, 250
Marathrum 402
Marcgravia 302
Marcgraviastrum 302
Marchantia 171
Margaritaria 401
Margyricarpus 443
Marila 381
Maripa 464
Markea 467
Marrubium 371
Marsdenia 342
Marsilea 179
Marsypianthes 371
Martinella 364
Martynia 374
Mascagnia 396
Masdevallia 138, 219
Matayba 457, 458
Matelea 342
Matisia 67, 138, 140, 411
Matricaria 265
Matthiola 277
Maurandella 376
Maurandya 376
Mauria 451
Mauritia 75, 96, 210, **478**
Mauritiella 108, 210
Maxillaria 138, 220–222
Mayna 380, **490**
Maytenus 97, 291
Mecardonia 376
Medicago 327
Megaskepasma 360
Megathyrsus 241
Meineckia 401
Melaleuca 426, **494**
Melampodium 265
Melanthera 266
Melia 454
Melicoccus 68, 458, **495**
Melilotus 327
Melinis 241
Meliosma 145, 433
Melissa 371
Melocactus 284
Melocanna 241

Melochia 411, 412
Melothria 294
Mendoncia 360
Mentha 371
Mentzelia 291
Merremia 145, 464
Mesechites 342
Mesembryanthemum 281
Mesosphaerum 371, **526**
Mespilodaphne 90, 98, 183, **510**
Metopium 451
Metrosideros 426
Metteniusa 415
Mezilaurus 98, 183
Miconia 66, 67, 72, 95, 96, 138, 419–423, **478**, **499**, **500**
Micrandra 392
Microchloa 241
Microdesmia 383
Microgramma 175, 176
Micropholis 306
Microsorum 176
Microtea 286
Mikania 266
Milleria 266
Miltoniopsis 222
Mimosa 327, 328, **526**
Minquartia 98, 449
Minthostachys 371
Mirabilis 72, 287
Mitracarpus 352
Modiola 412
Mollia 412
Mollinedia 185
Mollugo 286
Molongum 100, 342
Moluccella 371
Momordica 97, 295, **479**
Monnina 338, **510**, **515**
Monochaetum 423
Monolena 423
Monopteryx 328
Monotropa 300, **515**
Monstera 200, **519**
Montanoa 266
Montia 286
Monticalia 266, 267, **479**, **510**, **515**
Montrichardia 200
Moquilea 383, **500**
Mora 328
Morinda 352
Moringa 280, **474**, **495**
Morisonia 278
Mormodes 222
Moronobea 385
Morus 440
Mosannona 189
Mosquitoxylum 451
Mouriri 423, 424
Moutabea 338
Mucoa 342
Mucuna 328, **510**
Muehlenbeckia 289
Muellera 328
Muhlenbergia 241
Munnozia 267
Muntingia 415, **490**
Murdannia 224
Murraya 456
Musa 99, 250
Mussaenda 352
Mutisia 267
Myosotis 275
Myrcia 426, 427
Myrcianthes 72, **510**
Myrciaria 75
Myrica 339
Myriocarpa 446
Myriophyllum 461
Myriopus 275
Myristica 191
Myrocarpus 97
Myrosma 250
Myrospermum 105, 328
Myroxylon 105, 328
Myrsine 304, **485**

N
Napeanthus 367, **515**
Nasa 292
Nassella 241
Nasturtium 277
Naucleopsis 440
Nautilocalyx 368
Navia 100
Nealchornea 392
Nectandra 183, 184
Neea 287
Nelsonia 360
Nelumbo 433
Neobartsia 375
Neocouma 342
Neomoorea 222
*Neonelsonia*252
Neoregelia 229
Nepeta 371
Nephrolepis 176
Nepsera 424
Neptunia 328
Nerium 342, **519**
Nertera 352, **526**
Neurolaena 267, **510**
Neustanthus 328
Nicandra 467
Nicoteba 360, **505**
Nicotiana 67, 89, 96, 98, 99, 467, **527**
Niedenzuella 396
Nierembergia 467
Nigella 436
Niphogeton 252
Nissolia 329
Nothoscordum 214
Nothotalisia 433
Notopleura 353
Nuttallanthus 376
Nymphaea 181
Nymphoides 273, **505**

O
Obtegomeria 371
Ocellochloa 241
Ochna 397
Ochroma 412, **505**
Ocimum 99, 371
Ocotea 49, 55, 56, 57, 59, 60, 79, 87, 89, 90, 96, 184, **500**
Odontadenia 342
Odontocarya 435
Odontonema 360
Oeceoclades 222
Oenocarpus 96, 100, 210, 211
Oenothera 429, 430
Olax 449
Oldenlandia 353
Olea 375
Oleandra 176
Oligactis 267
Olyra 241
Omphalea 100, 392
Oncidium 222
Onoseris 267
Operculina 464
Ophioglossum 173
Oplismenus 242
Opuntia 284, **505**
Oreopanax 138, 254, **510**
Origanum 372, **505**
Ormosia 97, 329
Ornithogalum 215
Orobanche 375
Orphanodendron 329
Orthomene 435
Orthopappus 267
Orthrosanthus **10**, 216, **485**
Oryctanthus 448
Oryza 242, **495**
Osmorhiza 252
Osmunda 173
Osmundastrum 173
Osteophloeum 96, 191
Otholobium 329
Otoba 191
Ouratea 99, 100, 397
Oxalis 67, 73, 432, **519**
Oxandra 96, 189
Oxyrhynchus 329

P
Pachira 55, 99, 100, 412, **495**
Pachyrhizus 75, 329
Pachystachys 360, 361
Pacouria 342
Pacourina 267
Paepalanthus 233, **495**
Pagamea 353
Palicourea 96, 138, 353, 354, **520**
Panax 254
Pandanus 227
Panicum 242
Panopsis 68, 433
Papaver 435
Pappophorum 242
Parahancornia 100, 342, **500**
Parajubaea 211, **485**
Paramochloa 242, **510**
Paraserianthes 329
Parietaria 446
Parinari 100, 383, **505**
Parinariopsis 383
Parkia 329
Parkinsonia 329, 330, **490**
Parmentiera 364
Paronychia 285
Parthenium 267
Paspalum 242, 243
Passiflora 12, **14**, 49, 55, 56, 57, 60, 66, 67, 68, 70, 73, 135, 138, 140, 144, 146, 397–400, **495**, **500**, **505**
Passovia 448
Patinoa 412
Paullinia 67, 458, 459
Paulownia 401
Pavonia 412
Paypayrola 405
Pectinopitys 180
Pectis 267
Pedersenia 283
Pelargonium 99, 118, 358, **495**
Pelliciera 309, **500**
Peltogyne 330
Peltophorum 330
Penstemon 376
Pentacalia 49, 55, 56, 57, 59, 60, 267
Pentaclethra 330
Pentagonia 355
Pentanema 267

Pentapetes 412
Pentas 355
Peperomia 55, 138, 193, 194, 195
Pera 401
Perebea 440
Pereskia 284
Perezia 268
Pericallis 268
Peristeria 222
Peritassa 291
Persea 99, 140, 185, **495**
Persicaria 289
Pescatoria 222
Petitia 372
Petiveria 287
Petrea 379
Petroselinum 252
Petunia 467
Peumus 185
Pfaffia 283
Phalaris 243
Pharus 243
Phaseolus 330
Phenakospermum 250
Phenax 446, 447
Philodendron 138, 200, 201, **520**
PhilodendronIpomea 144
Phlebodium 176
Phlox 303
Phoenix 211, **495**, **505**
Pholidostachys 211
Phoradendron 449, 450
Phormium 216
Phragmipedium 222
Phragmites 244
Phragmotheca 412
Phyla 379
Phyllanthus 401
Phyllostachys 244, **485**
Phyllostylon 445
Phymosia 412
Physalis 73, 467, 468, **496**
Phytelephas 211, **510**
Phytolacca 287, 288, **496**
Picramnia 96, 100, 433
Picrasma 460
Picrolemma 460
Pilea 447
Pilocarpus 456
Pilosocereus 284
Pimenta 428
Pimentelia 355
Pimpinella 252
Pinus 98, 144, 153, 160, 180, **490**
Pinzona 295
Piparea 404
Piper 67, 95, 96, 138, 195–197, **491**, **520**, **527**
Piptadenia 330
Piptocoma 97, 268
Piresia 244
Piriqueta 400, 401
Piscidia 330
Pisonia 287
Pistia 202, **479**
Pitcairnia 229
Pithecellobium 330, **496**
Pittosporum 254, **485**
Pityrogramma 178
Plagiocheilus 268
Plantago 376, **500**
Platanus 377
Platonia 385
Platycarpum 355
Platycerium 176
Platycladus 180
Platymiscium 97, 98, 330, 331
Platypodium 331
Plectranthus 372
Plectrocarpa 472, 473
Pleioblastus 244
Pleonotoma 364
Pleopeltis 176
Plerandra 254
Pleroma 424, **486**
Pleuranthodendron 404
Pleurothallis 138, 222
Pleurothyrium 185
Plinia 428
Plocamium 473
Pluchea 268
Plukenetia 75, 122, 392
Plumbago 288
Plumeria 342, **520**
Plutarchia 73, 300
Poa 244
Podachaenium 268
Podocarpus 99, 180, **511**
Podranea 364
Poeppigia 331
Pogonophora 401
Pogonopus 355
Pogostemon 118
Polyalthia 189
Polybotrya 176
Polygala 338
Polygonum 289
Polylepis 443, **511**
Polypodium 176
Polypogon 244
Polyscias 254
Polystachya 222
Polystichum 176
Polytaenium 178
Polytrias 244
Polytrichum 171
Pombalia 405
Pontederia 225, **486**, **500**
Poortmannia 468
Populus 404
Poranopsis 464
Poraqueiba 100, 359
Porcelia 189
Porophyllum 268
Portulaca 290
Posoqueria 355
Potalia 97, 98, 100, 345
Potamogeton 203
Potentilla 443
Poulsenia 441
Pourouma 447
Pouteria 66, 67, 68, 75, 138, 144, 306–308
Pouteris 96
Pouzolzia 448
Pradosia 308
Praecereus 284
Prestoea 73, 211
Prestonia 342
Primula 304
Prionostemma 291
Prioria 331
Pristimera 291
Pritchardia 211
Priva 379
Prockia 404
Prosopis 75, 331, **491**
Prosthechea 222
Protium 66, 67, 95, 96, 97, 452, 453, **479**
Prumnopitys 107
Prunella 372
Prunus 443, 444, **511**
Psammisia 300
Pseudechinolaena 244
Pseudelephantopus 268
Pseuderanthemum 361
Pseudobombax 97, 100, 412, 413
Pseudoconnarus 431
Pseudoconyza 268
Pseudognaphalium 268
Pseudogynoxys 269
Pseudolmedia 97, 441
Pseudomalmea 189, **515**
Pseudopiptadenia 331
Pseudorhipsalis 284, **515**
Pseudosamanea 331
Pseudosenefeldera 97
Pseudoxandra 189
Psidium 428, **496**
Psiguria 295
Psilotum 179, **505**
Psittacanthus 448, 449
Psoralea 331, **511**
Psychopsis 222
Psychotria 55, 355, 356, **527**
Pteridium 175
Pteris 178
Pterocarpus 331
Pterocaulon 269
Pterogyne 331
Pterolepis 424
Pterozonium 178
Ptychosperma 211
Pueraria 331
Pulchranthus 361
Punica 417
Puya **16**, 229, **479**
Pyracantha 444
Pyrostegia 365
Pyrus 444

Q
Qualea 430
Quararibea 413
Quassia 460
Quercus 153, 338, **479**
Quiina 397

R
Racinaea 229
Randia 356
Ranunculus 436, **486**
Rapatea 247
Raphanus 277
Raphia 211
Rapistrum 277
Rauvolfia 343
Ravenala 250
Reldia 368
Remijia 356
Renealmia 251
Reseda 280
Restrepia 222
Retrophyllum 180, 356, **486**
Rhabdadenia 343
Rhamnus 442
Rhaphidophora 202
Rhaphiolepis 444
Rheum 289
Rhigospira 100, 343
Rhipidocladum 244
Rhipsalis 284
Rhizophora 402, **496**
Rhodanthe 269
Rhodocactus 284
Rhododendron 300
Rhodospatha 202

Rhodostemonodaphne 185
Rhodothyrsus 392
Rhynchanthera 424
Rhynchosia 331
Rhynchospora 232
Rhytachne 244
Ribes 461, **501**
Richardia 356
Richeria 402
Ricinus 392, **520**
Rinorea 405
Rivina 287
Rodriguezia 222, 223
Rolandra 269
Ronabea 356
Rorippa 277
Rosa 444
Rosenbergiodendron 356
Roseodendron 365
Rosmarinus 118
Rosularia 461
Rottboellia 97, 244
Roucheria 394
Roupala 433
Rourea 431
Roystonea 98, 99, 212
Rubus 55, 73, 444, 445, **501**
Rudbeckia 269
Rudgea 356, 357
Ruehssia 343
Ruellia 361
Rugoloa 244
Ruizodendron 189
Rumex 289, 290, **520**
Rumohra 176
Ruppia 203
Ruscus 215
Russelia 376
Ruta 456
Ryania 79, 87, 89, 90, 404

S
Sabal 212
Sabicea 357
Saccharum 99, 244
Sacciolepis 244
Sacoglottis 393
Sageretia 442
Sagina 285
Sagittaria 198, **496**
Sagotia 392
Salacia 291
Salix 404
Salmea 269
Salpianthus 287
Salpichroa 468
Salvia 57, 60, 372, 373, **527**
Salvinia 179
Samanea 331, **491**
Sambucus 296, **501**
Sanchezia 361
Sanguisorba 445
Sanicula 253
Santolina 269
Sapindus 459
Sapium 392, 393
Sarcaulus 308
Sarcomphalus 442
Saribus 212
Satureja 373
Satyria 300
Saurauia 297, **479**
Sauvagesia 397
Savia 402
Saxifraga 461
Scabiosa 296
Scaphyglottis 223
Schaefferia 291
Schiekia 224, **516**
Schinus 451, **506**, **520**
Schistostemon 393
Schizachyrium 244
Schizaea 98, 179
Schizocalyx 357
Schizolobium 332
Schkuhria 269
Schlegelia 377
Schlumbergera 284
Schmardaea 454
Schnella 332
Schoenobiblus 415
Schoenocephalium **18**, 98, 99, 100
Schoenoplectus 232
Schoepfia 450
Schradera 357
Schultesia 345
Schwenckia 468
Sciadotenia 435
Sciaphila 227
Sciodaphyllum 254
Scleranthus 286
Scleria 232, 233
Sclerocarpus 269
Scleronema 413
Scoparia 376
Scutellaria 373
Sebastiania 393
Secale 244
Sechium 75
Securidaca 338
Sedum 461
Seguieria 287
Selaginella 55, 171, 172
Selenicereus 284
Senecio 49, 55, 56, 57, 60, 269
Senefeldera 98
Senegalia 332
Senna 67, 139, 332–334, **516**, **520**
Sequoia 180
Serenoa 212
Serjania 459
Serpocaulon 176, 177
Sesamum 375
Sesbania 334
Sesuvium 281
Setaria 139, 245, **480**
Sextonia 185
Sicana 295
Sicyos 295, **480**
Sida 413, 414, **527**
Sidastrum 414
Sideroxylon 308
Sigesbeckia 270
Silene 286
Silybum 270, **506**
Simaba 460
Simarouba 460
Simira 357, **516**
Sinapis 278
Sinningia 368
Sipanea 357
Siparuna 185, 186, **511**
Siphocampylus 273, **516**
Sisymbrium 278
Sisyrinchium 216, **486**
Sloanea 431
Smallanthus 73, 270
Smilax 226, 227
Sobralia 223
Socratea 212, **491**
Solandra 468
Solanum 12, 59, 66, 67, 73, 75, 89, 95, 96, 99, 135, 138, 144, 146, 468–472, **480**, **501**
Soleirolia 448
Solidago 270
Soliva 270
Sonchus 270, **496**
Sophora 334
Sorghastrum 245
Sorghum 245
Sorocea 441
Souroubea 302
Spachea 396
Sparattanthelium 182
Sparganium 247
Sparrmannia 414
Spartium 334
Spathiphyllum 202
Spathodea 98, 365
Spergula 286
Spergularia 286
Spermacoce 357
Sphaeradenia 227
Sphaeropteris 173
Sphagneticola 270
Sphagnum 171
Sphenoclea 472
Sphyrospermum 300
Spigelia 345, **520**
Spilanthes 270
Spinacia 283, **521**
Spiraea 445
Spirodela 202, **480**
Spirotheca 414
Spondias **109**, 140, 451, **506**, **988**
Spongiosperma 343
Sporobolus 245
Stachyarrhena 358
Stachys 373
Stachytarpheta 379
Stanhopea 223
Stapfochloa 245
Steinchisma 246
Stelis 138, 144
Stellaria 286
Stemodia 377
Stenandrium 361
Stenocereus 285
Stenorrhynchos 223
Stenospermation 202
Stenostephanus 361
Stenostomum 358
Stenotaphrum 246
Stephanopodium 385
Sterculia 414, **521**
Stevia 270
Sticherus 173
Stigmaphyllon 396
Stizophyllum 365
Strelitzia 250
Streptocarpus 368
Streptosolen 472, **486**
*Strobilanthes*361
Stromanthe 250
Strongylodon 334
Struchium 270
Struthanthus 449
Strychnos 98, 105, 345, 346, **516**
Stryphnodendron 334
Stuckenia 203, **480**
Styloceras 280
Stylogyne 304, 305
Stylosanthes 334
Styrax 308, 309
Sudamerlycaste 223
Sulla 334

Suriana 338
Swainsona 334
Swartzia 334, 335, **521**
Swietenia 99, 107, 454, **506**
Swinglea 456
Syagrus 99, 212
Symmeria 290
Symphonia **92**, 385, **491**
Symphyotrichum 270, 271
Symphytum 275
Symplocos 55, **105**, 138, 140, 309
Synechanthus 212
Synedrella 271
Syngonium 202
Syzygium 97, 428, **501**

T
Tabebuia 365, **491**
Tabernaemontana 343, 344, **511**, **521**
Tachigali 335
Tagetes 271
Talinum 290
Talisia 459, 460
Tamarindus 335, **527**
Tamonea 379
Tanacetum 271
Tanaecium 365
Tapeinochilos 248
Tapirira 153, 451
Tapura 385, 386
Tara 335
Taralea 335
Taraxacum 271, **506**
Tecoma 98, 365, **506**
Tecomaria 365
Tectaria 177
Tectona 373
Telipogon 223
Telitoxicum 435
Telmatoblechnum 174
Tephrosia 335
Teramnus 335
Terminalia 107, 416, **521**
Ternstroemia 303
Tessaria 271
Tetracera 295
Tetrameranthus 189
Tetramerium 361
*Tetrapanax*254
Tetrapterys 396
Tetrathylacium 404
Tetrorchidium 393
Teucrium 373
Thalia 250
Thaumatophyllum 202
Thelypteris 174
Themistoclesia 300
Theobroma 75, 99, 414, **496**, **501**
Thespesia 415
Thevetia 344
Thibaudia 300
Thlaspi 278
Thoracocarpus 227
Thrinax 212
Thuja 180
Thunbergia 361, 362, **521**
Thymus 118, 373
Thyrsodium 451
Tigridia 216, 217
Tillandsia 229
Tinantia 224
Tithonia 97, 271, **506**
Tocoyena 358
Tolmiea 461
Tontelea 291
Torenia 374
Toulicia 460
Tournefortia 275
Touroulia 397
Tovomita 385
Toxicodendron 451
Trachycarpus 212
Trachypogon 246
Tradescantia 224
Tragia 393
Tragopogon 271
Tragus 246
Trattinnickia 97, 453, 454, **511**
Trema 436
Trianaea 472
Trianthema 281, **521**
Tribulus 473
Trichanthera 75, 362
Trichilia 455
Trichocentrum 223
Trichocereus 285
Trichomanes 173
Trichopilia 223
Trichospermum 415
Trichostigma 287
Tridax 271
Trifolium 139, 335, 336, **496**
Triodanis 273
Triolena 424
Triphasia 456
Triplaris 290
Triplophyllum 177
Tripodanthus 449
Tripsacum 246
Trisetum 246
Triticum 246
Triumfetta 415
Trixis 272
Tropaeolum 73, 280
Trophis 441
Trymatococcus 441
Tulbaghia 214
Turnera 401, **521**
Turpinia 292
Tynanthus 365
Typha 247, **491**

U
Ugni 73, 428, **501**
Ulex 336, **491**
Ullucus 73, 283
Uncaria 97, 100, 358
Unonopsis 189
Unxia 272
Urceolina 214
Urena 415, **496**
Urera 96, 448
Uribea 336
Urochloa 97, 99, 139, 246
Urospatha 202
Urtica 448, **527**
Urvillea 460
Utricularia 145, 374

V
Vaccinium 73, 300, **527**
Vachellia 336, **480**, **491**
Valeriana 55, 296, **511**, **516**
Vallea 431, **521**
Vallesia 344
Vallisneria 203
Vandellia 374
Vanilla 118, 223, **496**
Vantanea 393
Varronia 275, **506**
Vasconcellea 55, 57, 60, 67, 68, 73, 279
Vatairea 336
Vataireopsis 336
Vaupesia 393
Vellozia 98, 100
Verbascum 377
Verbena 55, 379, 380
Verbesina 272, **506**
Vernicia 393
Vernonanthura 272
Veronica 377, **501**
Viburnum 57, 296, 297
Vicia 73, 336
Victoria 98, 181
Vigna 336, 337, **496**
Viguiera 272
Vinca 344, **486**
Viola 405, 406
Virola 96, 191, 192, **511**
Vismia 97, 394
Visnaga 253
Vitex 373, 374, **501**
Vitis 472
Vittaria 178
Vochysia 430, **501**
Volkameria 374
Vouarana 460
Voyria 345
Vriesea 145

W
Wallacea 397
Waltheria 415
Warczewiczella 223
Warszewiczia 358, **516**
Washingtonia 212
Watsonia 217
Wedelia 272
Weinmannia 431, **511**
Welfia 212
Werneria 272
Wettinia 212, 213
Wigandia 276
Wissadula 415
Witheringia 472
Wodyetia 213
Wolffia 202

X
Xanthium 272
Xanthosoma 202, 203, **501**
Xerochrysum 272
Xestaea 345
Ximenia 449
Xiphidium 224, **527**
Xylopia 96, 189, 190, **491**
Xylosma 405
Xyris 247

Y
Youngia 272
Yucca 215

Z
Zamia 99, 179
Zannichellia 203
Zantedeschia 203
Zanthoxylum 456, 457
Zapoteca 337
Zea 99, 140, 246, 247
Zephyranthes 214
Zingiber 251
Zinnia 272, **480**
Ziziphus 442
Zornia 337
Zoysia 247
Zuloagaea 247
Zygia 337, **501**, **516**

Index of Synonyms

A

Aa hartwegii IPNI=*302812-2* 217
Abarema racemosa IPNI=*996977-1* 337
Abasoloa taboada IPNI=*1186403-2* 261
Abatia borealis IPNI=*779362-1* 402
Abatia verbascifolia IPNI=*779362-1* 402
Abelmoschus abelmoschus IPNI=*558023-1* 406
Abelmoschus bammia IPNI=*558006-1* 406
Abelmoschus betulifolia IPNI=*558023-1* 406
Abelmoschus chinensis IPNI=*558023-1* 406
Abelmoschus ciliaris IPNI=*558023-1* 406
Abelmoschus cisplatinus IPNI=*560861-1* 410
Abelmoschus congener IPNI=*560068-1* 409
Abelmoschus cruentus IPNI=*326388-2* 409
Abelmoschus cubensis IPNI=*558023-1* 406
Abelmoschus cucurbitaceus IPNI=*558023-1* 406
Abelmoschus guineensis IPNI=*30143354-2* 410
Abelmoschus haenkeanus IPNI=*558023-1* 406
Abelmoschus lanyunatus IPNI=*558023-1* 406
Abelmoschus longifolius IPNI=*558006-1* 406
Abelmoschus magnificus IPNI=*558023-1* 406
Abelmoschus marianus IPNI=*558023-1* 406
Abelmoschus moschatus var. *betulifolius* IPNI=*558023-1* 406
Abelmoschus moschatus var. *lanyunatus* IPNI=*558023-1* 406
Abelmoschus moschatus var. *multiformis* IPNI=*558023-1* 406
Abelmoschus mutabilis IPNI=*560549-1* 409
Abelmoschus officinalis IPNI=*558006-1* 406
Abelmoschus palustris IPNI=*558023-1* 406
Abelmoschus praecox IPNI=*558006-1* 406
Abelmoschus roseus IPNI=*558023-1* 406
Abelmoschus sublobatus IPNI=*558023-1* 406
Abelmoschus tuberculatus IPNI=*558006-1* 406
Abelmoschus tuberculatus var. *deltoidefolius* IPNI=*558006-1* 406
Abelmoschus verrucosus IPNI=*560068-1* 409
Abena cayennensis IPNI=*1126344-2* 379
Abena jamaicensis IPNI=*1092351-2* 379
Aberemoa guianensis IPNI=*114255-2* 188
Aberemoa guianensis var. *glabrescens* IPNI=*114255-2* 188
Aberemoa hadrantha IPNI=*72868-1* 188
Aberemoa lepidota IPNI=*1077481-2* 188
Aberemoa longifolia IPNI=*106440-2* 188
Aberemoa odorata IPNI=*72891-1* 188
Aberemoa quitarensis IPNI=*85728-2* 188
Aberemoa spixiana IPNI=*309665-2* 188
Aberia gardneri IPNI=*111561-1* 404
Aberia hebecarpa IPNI=*111561-1* 404
Abies araucana IPNI=*261681-1* 180
Abies columbaria IPNI=*676669-1* 180
Abildgaardia capillaris IPNI=*1186511-2* 229
Abildgaardia compressa IPNI=*104972-2* 232
Abildgaardia hispidula IPNI=*912172-1* 229
Abildgaardia indica IPNI=*104972-2* 232
Abildgaardia javanensis IPNI=*104972-2* 232
Abildgaardia ovata IPNI=*104972-2* 232
Abildgaardia polystachya IPNI=*415708-1* 243
Ablania digitata IPNI=*835145-1* 431
Ablania guianensis IPNI=*835145-1* 431
Ablania laurifolia IPNI=*835145-1* 431
Abroma mariae IPNI=*824025-1* 409
Abroma nitidum IPNI=*824026-1* 409
Abrotanum foemina IPNI=*241927-1* 269
Abrus lusorius IPNI=*517469-1* 331
Abryanthemum acinaciforme IPNI=*360773-1* 280
Abryanthemum edule IPNI=*1200427-2* 280
Absinthium bipedale IPNI=*300106-2* 257
Absinthium cernuum IPNI=*307848-2* 257
Absinthium majus IPNI=*300106-2* 257
Absinthium officinale IPNI=*300106-2* 257
Absinthium vulgare IPNI=*20812-2* 257
Absinthium vulgare IPNI=*300106-2* 257
Abumon africanum IPNI=*527258-1* 213
Abuta concolor IPNI=*147-2* 434
Abuta convexa IPNI=*580080-1* 434
Abuta guianensis IPNI=*147-2* 434
Abuta heterophylla IPNI=*580080-1* 434
Abuta macrocarpa IPNI=*89958-2* 435
Abuta macrophylla IPNI=*580080-1* 434
Abuta rufescens IPNI=*580056-1* 434
Abuta scandens IPNI=*580080-1* 434
Abuta splendida IPNI=*580080-1* 434
Abuta toxifera IPNI=*72709-2* 435
Abuta trinervis IPNI=*580059-1* 434
Abuta verruculosa IPNI=*176828-2* 435
Abuta wilson-brownei IPNI=*580080-1* 434
Abutilodes carolinianum IPNI=*30017389-2* 412
Abutilon albidum IPNI=*1101632-2* 406
Abutilon album IPNI=*1101632-2* 406
Abutilon americanum IPNI=*1055043-2* 406
Abutilon amplissimum IPNI=*268896-2* 415
Abutilon arborescens IPNI=*1101632-2* 406
Abutilon arnottianum IPNI=*1026852-2* 406
Abutilon asiaticum IPNI=*1101632-2* 406
Abutilon asiaticum var. *subasperum* IPNI=*1101632-2* 406
Abutilon asiaticum var. *supraviride* IPNI=*1101632-2* 406
Abutilon badium IPNI=*1101632-2* 406
Abutilon cavaleriei IPNI=*1101632-2* 406
Abutilon chapelieri IPNI=*268896-2* 415
Abutilon circinatum IPNI=*77146518-1* 407
Abutilon contractum IPNI=*268909-2* 415
Abutilon crassifolium IPNI=*1055043-2* 406
Abutilon crassifolium var. *tomentosum* IPNI=*1055043-2* 406
Abutilon crinitum IPNI=*268896-2* 415
Abutilon crispum IPNI=*120630-2* 409
Abutilon crispum var. *imberbe* IPNI=*120630-2* 409
Abutilon crispum var. *svensonii* IPNI=*120630-2* 409
Abutilon croizatianum IPNI=*1101632-2* 406

Abutilon cunninghamii IPNI=*1101632-2* 406
Abutilon decarpum IPNI=*1055043-2* 406
Abutilon dentatum IPNI=*1055043-2* 406
Abutilon domingense IPNI=*1055043-2* 406
Abutilon elongatum IPNI=*1101632-2* 406
Abutilon foetidum IPNI=*77146518-1* 407
Abutilon frutescens IPNI=*1101632-2* 406
Abutilon graveolens IPNI=*238-2* 406
Abutilon graveolens var. *hirtum* IPNI=*238-2* 406
Abutilon graveolens var. *queenslandicum* IPNI=*238-2* 406
Abutilon hastatum IPNI=*1213890-2* 415
Abutilon heterotrichum IPNI=*238-2* 406
Abutilon hirsutissimum IPNI=*1101632-2* 406
Abutilon hirtum var. *heterotrichum* IPNI=*238-2* 406
Abutilon imberbe IPNI=*120630-2* 409
Abutilon indicum var. *albidum* IPNI=*1101632-2* 406
Abutilon indicum var. *asiaticum* IPNI=*1101632-2* 406
Abutilon indicum var. *hirtum* IPNI=*238-2* 406
Abutilon indicum var. *welwitschii* IPNI=*1101632-2* 406
Abutilon jacquinii IPNI=*1055043-2* 406
Abutilon kauaiense IPNI=*1026852-2* 406
Abutilon laxiflorum IPNI=*268896-2* 415
Abutilon leiospermum IPNI=*1101632-2* 406
Abutilon leschenaultianum IPNI=*268909-2* 415
Abutilon leucanthemum IPNI=*268909-2* 415
Abutilon lignosum IPNI=*1055043-2* 406
Abutilon lucianum IPNI=*268909-2* 415
Abutilon lugardii IPNI=*238-2* 406
Abutilon molle IPNI=*1026852-2* 406
Abutilon mollissimum var. *sandwicense* IPNI=*1026852-2* 406
Abutilon neurocarpum IPNI=*120630-2* 409
Abutilon periplocifolium IPNI=*1213890-2* 415
Abutilon periplocifolium var. *albicans* IPNI=*268909-2* 415
Abutilon periplocifolium var. *peruvianum* IPNI=*1213890-2* 415
Abutilon periplocifolium var. *zeylanicum* IPNI=*1213890-2* 415
Abutilon pictum IPNI=*77130215-1* 407
Abutilon pleniflorum IPNI=*77130224-1* 407
Abutilon polyanthon IPNI=*268896-2* 415
Abutilon populifolium IPNI=*1101632-2* 406
Abutilon pseudostriatum IPNI=*77130224-1* 407
Abutilon scabrum IPNI=*1055043-2* 406
Abutilon sessilifolium IPNI=*120630-2* 409
Abutilon sordidum IPNI=*1026852-2* 406
Abutilon spicatum IPNI=*36556-2* 407
Abutilon spinifex IPNI=*186083-2* 412
Abutilon striatum IPNI=*77130224-1* 407
Abutilon subpapyraceum IPNI=*1101632-2* 406
Abutilon thompsonii IPNI=*77130224-1* 407
Abutilon tortuosum IPNI=*1026852-2* 406
Abutilon trichodum IPNI=*120630-2* 409
Abutilon truncatum IPNI=*563600-1* 413
Abutilon verbascoides IPNI=*268909-2* 415
Abutilon vesicarium IPNI=*1101632-2* 406
Abutilon vexillarium IPNI=*77130227-1* 407
Acacia adenophora IPNI=*470138-1* 310
Acacia adhaerens IPNI=*77085269-1* 332
Acacia albescens IPNI=*60441241-2* 336
Acacia albida IPNI=*60441241-2* 336
Acacia amambayensis IPNI=*518289-1* 332
Acacia ampeloclada IPNI=*77178320-1* 325
Acacia angulata IPNI=*470138-1* 310
Acacia angustiloba IPNI=*12268-2* 311
Acacia angustisiliqua IPNI=*1125234-2* 317
Acacia angustissima IPNI=*712-2* 310
Acacia angustissima subsp. *typica* IPNI=*712-2* 310
Acacia arborea IPNI=*62882-2* 315
Acacia arcuata IPNI=*470873-1* 310
Acacia arenosa IPNI=*507659-1* 327
Acacia aspidioides IPNI=*188076-2* 330
Acacia baileyana var. *aurea* IPNI=*469798-1* 310
Acacia baileyana var. *purpurea* IPNI=*469798-1* 310
Acacia bracteata IPNI=*471383-1* 310
Acacia buchananiana IPNI=*234370-2* 332
Acacia calderonii IPNI=*762-2* 310
Acacia caracasana IPNI=*287164-2* 337
Acacia carbonaria IPNI=*762-2* 310
Acacia caudata IPNI=*1009172-2* 319
Acacia collinsii IPNI=*60441206-2* 336
Acacia colubrina IPNI=*12264-2* 311
Acacia costaricensis IPNI=*60441206-2* 336
Acacia cowellii IPNI=*60441223-2* 336
Acacia cumingii IPNI=*762-2* 310
Acacia curassavica IPNI=*77077786-1* 310
Acacia cyanophylla IPNI=*471383-1* 310
Acacia decurrens f. *normalis* IPNI=*470138-1* 310
Acacia decurrens var. *angulata* IPNI=*470138-1* 310
Acacia decurrens var. *dealbata* IPNI=*470130-1* 310
Acacia decurrens var. *normalis* IPNI=*470138-1* 310
Acacia decussata IPNI=*470782-1* 310
Acacia depauperata IPNI=*1125234-2* 317
Acacia farnesiana IPNI=*1099713-2* 336
Acacia fasciculata IPNI=*316281-2* 313
Acacia fisheri IPNI=*762-2* 310
Acacia flava IPNI=*77208976-1* 330
Acacia flexuosa IPNI=*77100888-1* 336
Acacia flexuosa var. *ambigua* IPNI=*60441223-2* 336
Acacia flexuosa var. *lasiocarpa* IPNI=*60441223-2* 336
Acacia fluviatilis IPNI=*518289-1* 332
Acacia foliosa IPNI=*470782-1* 310
Acacia formosa IPNI=*287168-2* 337
Acacia fraseri IPNI=*76522-3* 310
Acacia glauca IPNI=*77077786-1* 310
Acacia glomerosa IPNI=*518289-1* 332
Acacia glutea IPNI=*60441206-2* 336
Acacia guachapele IPNI=*211561-2* 331
Acacia hostilis IPNI=*508775-1* 328
Acacia humboldtii IPNI=*60441223-2* 336
Acacia juliflora IPNI=*1002432-2* 331
Acacia jupunba IPNI=*469538-1* 310
Acacia jurema IPNI=*508775-1* 328
Acacia lacustris IPNI=*509612-1* 328
Acacia langlassei IPNI=*518289-1* 332
Acacia latifolia IPNI=*470873-1* 310
Acacia laxa IPNI=*1161746-2* 313
Acacia lebbeck IPNI=*99109-3* 311
Acacia leptosperma IPNI=*1125234-2* 317
Acacia leucacantha IPNI=*60441241-2* 336
Acacia leucocephala IPNI=*138955-2* 325
Acacia lindleyi IPNI=*471383-1* 310
Acacia longifolia var. *angustata* IPNI=*470782-1* 310
Acacia longifolia var. *bylongensis* IPNI=*470782-1* 310
Acacia longifolia var. *lanceolata* IPNI=*470782-1* 310
Acacia longifolia var. *latifolia* IPNI=*470782-1* 310
Acacia longifolia var. *prostrata* IPNI=*470782-1* 310
Acacia longifolia var. *typica* IPNI=*470782-1* 310
Acacia lophantha IPNI=*912944-1* 329
Acacia lophanthoides IPNI=*762-2* 310
Acacia lutea IPNI=*60441223-2* 336
Acacia lycopodioides IPNI=*1056199-2* 328
Acacia macracantha IPNI=*60441223-2* 336
Acacia macracantha var. *glabra* IPNI=*60441223-2* 336
Acacia macracantha var. *glabrescens* IPNI=*60441223-2* 336
Acacia macracanthoides IPNI=*60441223-2* 336
Acacia macroloba IPNI=*188076-2* 330
Acacia macrophylla IPNI=*99109-3* 311
Acacia magdalenae IPNI=*1051245-2* 313

Acacia martiusiana IPNI=*77085269-1* 332
Acacia melanoxylon f. *frutescens* IPNI=*470873-1* 310
Acacia melanoxylon var. *arcuata* IPNI=*470873-1* 310
Acacia melanoxylon var. *obtusifolia* IPNI=*470873-1* 310
Acacia micradena IPNI=*77085269-1* 332
Acacia microcephala IPNI=*60441223-2* 336
Acacia microphylla IPNI=*12268-2* 311
Acacia mollissima var. *angulata* IPNI=*470138-1* 310
Acacia multiflora IPNI=*989448-1* 311
Acacia multipinnata IPNI=*234346-2* 332
Acacia nelsonii IPNI=*60441206-2* 336
Acacia nilotica IPNI=*77089275-1* 336
Acacia niopo IPNI=*12268-2* 311
Acacia obliquifolia IPNI=*1003508-2* 330
Acacia obtusa IPNI=*60441223-2* 336
Acacia oerstedii IPNI=*762-2* 310
Acacia pallida IPNI=*1175026-2* 331
Acacia panamensis IPNI=*60441206-2* 336
Acacia paniculata IPNI=*234346-2* 332
Acacia pectinata IPNI=*512223-1* 329
Acacia pellacantha IPNI=*60441223-2* 336
Acacia pennatula IPNI=*60441228-2* 336
Acacia peregrina IPNI=*12268-2* 311
Acacia pinnata IPNI=*234370-2* 332
Acacia platyacantha IPNI=*60441223-2* 336
Acacia podalyriifolia var. *typica* IPNI=*76522-3* 310
Acacia podalyriifolia var. *viridis* IPNI=*76522-3* 310
Acacia polyphylla IPNI=*518289-1* 332
Acacia polypodioides IPNI=*762-2* 310
Acacia portoricensis IPNI=*943976-1* 337
Acacia propinqua IPNI=*227126-2* 331
Acacia psilostachya IPNI=*962800-1* 331
Acacia pulcherrima IPNI=*519643-1* 334
Acacia punctata IPNI=*1056199-2* 328
Acacia punctata IPNI=*60441223-2* 336
Acacia riparia IPNI=*518289-1* 332
Acacia scandens IPNI=*493866-1* 319
Acacia secundiflora IPNI=*1009172-2* 319
Acacia seeressa IPNI=*99109-3* 311
Acacia seifriziana IPNI=*60441241-2* 336
Acacia spathulata IPNI=*470782-1* 310
Acacia speciosa IPNI=*99109-3* 311
Acacia spegazziniana IPNI=*77085269-1* 332
Acacia stenostachya IPNI=*962800-1* 331
Acacia subinermis IPNI=*60441223-2* 336
Acacia subpaniculata IPNI=*77085269-1* 332
Acacia subtilifolia IPNI=*77178320-1* 325
Acacia tamarindifolia IPNI=*234370-2* 332
Acacia tenuiflora IPNI=*508775-1* 328
Acacia thegonocarpa IPNI=*470782-1* 310
Acacia tortuosa IPNI=*60441241-2* 336
Acacia tortuosa var. *glabrior* IPNI=*60441241-2* 336
Acacia trichophylloides IPNI=*12268-2* 311
Acacia trigonocarpa IPNI=*470782-1* 310
Acacia valenzuelana IPNI=*762-2* 310
Acacia verticillata var. *angusta* IPNI=*471751-1* 310
Acacia villosa IPNI=*762-2* 310
Acacia virescens IPNI=*60441241-2* 336
Acacia virgata IPNI=*1125234-2* 317
Acacia viridiflora IPNI=*77178320-1* 325
Acacia yucatanensis IPNI=*60441206-2* 336
Acaciella bakeri IPNI=*762-2* 310
Acaciella calderonii IPNI=*762-2* 310
Acaciella carbonaria IPNI=*762-2* 310
Acaciella cumingii IPNI=*762-2* 310
Acaciella curassavica IPNI=*77077786-1* 310
Acaciella fisheri IPNI=*762-2* 310
Acaciella oerstedii IPNI=*762-2* 310
Acaciella polypodioides IPNI=*762-2* 310
Acaciella rekoi IPNI=*762-2* 310
Acaciella valenzuelana IPNI=*762-2* 310
Acaena agrimonioides IPNI=*719987-1* 442
Acaena cylindristachya subvar. *pusilla* IPNI=*719970-1* 442
Acaena cylindristachya var. *composita* IPNI=*719970-1* 442
Acaena cylindristachya var. *macrophyllidion* IPNI=*719970-1* 442
Acaena cylindristachya var. *nitidissima* IPNI=*719970-1* 442
Acaena cylindristachya var. *pilosinervis* IPNI=*719970-1* 442
Acaena elegans IPNI=*720073-1* 442
Acaena elongata subvar. *compacta* IPNI=*719987-1* 442
Acaena elongata subvar. *pergrandis* IPNI=*719987-1* 442
Acaena elongata var. *gracilis* IPNI=*719987-1* 442
Acaena elongata var. *incisa* IPNI=*719987-1* 442
Acaena elongata var. *lappacea* IPNI=*719987-1* 442
Acaena elongata var. *robusta* IPNI=*719987-1* 442
Acaena elongata var. *villosula* IPNI=*719987-1* 442
Acaena hirsuta IPNI=*720073-1* 442
Acaena krausei var. *hirsuta* IPNI=*720073-1* 442
Acaena lappacea IPNI=*719987-1* 442
Acaena micranthera IPNI=*720073-1* 442
Acaena ovalifolia subsp. *australis* IPNI=*720073-1* 442
Acaena ovalifolia subsp. *chamaephyllon* IPNI=*720073-1* 442
Acaena ovalifolia subsp. *elegans* IPNI=*720073-1* 442
Acaena ovalifolia subsp. *glabricaulis* IPNI=*720073-1* 442
Acaena ovalifolia var. *calophylla* IPNI=*720073-1* 442
Acaena ovalifolia var. *calvescenticupula* IPNI=*720073-1* 442
Acaena ovalifolia var. *chamanthera* IPNI=*720073-1* 442
Acaena ovalifolia var. *chubutensis* IPNI=*720073-1* 442
Acaena ovalifolia var. *dolichacantha* IPNI=*720073-1* 442
Acaena ovalifolia var. *elegans* IPNI=*720073-1* 442
Acaena ovalifolia var. *hirsuta* IPNI=*720073-1* 442
Acaena ovalifolia var. *incisiserrata* IPNI=*720073-1* 442
Acaena ovalifolia var. *insulae-exterioris* IPNI=*720073-1* 442
Acaena ovalifolia var. *nirihuaoensis* IPNI=*720073-1* 442
Acaena ovalifolia var. *serrata* IPNI=*720073-1* 442
Acaena ovalifolia var. *subglabrescens* IPNI=*720073-1* 442
Acaena ovalifolia var. *subserrata* IPNI=*720073-1* 442
Acaena ovalifolia var. *subsexjuga* IPNI=*720073-1* 442
Acaena ovalifolia var. *subtuspellita* IPNI=*720073-1* 442
Acaena ovalifolia var. *tenerrima* IPNI=*720073-1* 442
Acaena ovalis IPNI=*720073-1* 442
Acaena tenuipila IPNI=*720073-1* 442
Acajuba occidentalis IPNI=*319068-2* 450
Acalypha adscendens IPNI=*1012504-2* 387
Acalypha alchorneoides IPNI=*1228-2* 387
Acalypha alopecuroidea f. *glanduligera* IPNI=*337150-1* 387
Acalypha alopecuroidea f. *polycephala* IPNI=*337150-1* 387
Acalypha alopecuroidea lusus *glanduligera* IPNI=*337150-1* 387
Acalypha alopecuroidea lusus *hispida* IPNI=*337150-1* 387
Acalypha amplifolia IPNI=*1362-2* 387
Acalypha aristata IPNI=*337150-1* 387
Acalypha arvensis var. *belangeri* IPNI=*1115-2* 387
Acalypha arvensis var. *genuina* IPNI=*1115-2* 387
Acalypha arvensis var. *pavoniana* IPNI=*1115-2* 387
Acalypha betuloides IPNI=*1228-2* 387
Acalypha billbergiana IPNI=*1228-2* 387
Acalypha callosa IPNI=*1362-2* 387
Acalypha cancana IPNI=*1362-2* 387
Acalypha capitellata IPNI=*1115-2* 387
Acalypha caucana IPNI=*1362-2* 387
Acalypha caudata IPNI=*1362-2* 387
Acalypha chamaedrifolia var. *brevipes* IPNI=*1012504-2* 387
Acalypha chamaedrifolia var. *genuina* IPNI=*1012504-2* 387
Acalypha corchorifolia IPNI=*1012504-2* 387
Acalypha cordifolia IPNI=*337706-1* 387
Acalypha cucullata IPNI=*1362-2* 387
Acalypha densiflora IPNI=*337478-1* 387

Acalypha diversifolia var. *genuina* IPNI=*1228-2* 387
Acalypha diversifolia var. *leptostachya* IPNI=*1228-2* 387
Acalypha diversifolia var. *popayanensis* IPNI=*1228-2* 387
Acalypha diversifolia var. *squarrosa* IPNI=*1228-2* 387
Acalypha fertilis IPNI=*1362-2* 387
Acalypha flabellifera IPNI=*337706-1* 387
Acalypha foliosa IPNI=*1362-2* 387
Acalypha hartwegiana IPNI=*1228-2* 387
Acalypha heterodonta IPNI=*1362-2* 387
Acalypha heterodonta var. *hirsuta* IPNI=*1362-2* 387
Acalypha heterodonta var. *psiloclada* IPNI=*1362-2* 387
Acalypha heterodonta var. *trichoclada* IPNI=*1362-2* 387
Acalypha heteromorpha IPNI=*1362-2* 387
Acalypha hicksii IPNI=*1362-2* 387
Acalypha hirsutissima IPNI=*1362-2* 387
Acalypha hispaniolae IPNI=*1012504-2* 387
Acalypha hotteana IPNI=*1012504-2* 387
Acalypha hystrix IPNI=*1115-2* 387
Acalypha inaequalis IPNI=*1228-2* 387
Acalypha lehmanniana IPNI=*1362-2* 387
Acalypha leptostachya f. *diversifolia* IPNI=*1228-2* 387
Acalypha leptostachya IPNI=*1228-2* 387
Acalypha leptostachya var. *genuina* IPNI=*1228-2* 387
Acalypha leptostachya var. *popayanensis* IPNI=*1228-2* 387
Acalypha macrophylla IPNI=*1362-2* 387
Acalypha macrostachya f. *androgyna* IPNI=*1362-2* 387
Acalypha macrostachya f. *macrophylla* IPNI=*1362-2* 387
Acalypha macrostachya var. *genuina* IPNI=*1362-2* 387
Acalypha macrostachya var. *hirsutissima* IPNI=*1362-2* 387
Acalypha macrostachya var. *macrophylla* IPNI=*1362-2* 387
Acalypha macrostachya var. *puberula* IPNI=*1362-2* 387
Acalypha macrostachya var. *sidifolia* IPNI=*1362-2* 387
Acalypha macrostachya var. *tristis* IPNI=*1362-2* 387
Acalypha microgyna IPNI=*1228-2* 387
Acalypha neogranatensis IPNI=*1362-2* 387
Acalypha obtusifolia IPNI=*1362-2* 387
Acalypha panamensis IPNI=*1228-2* 387
Acalypha pavoniana IPNI=*1115-2* 387
Acalypha piperoides IPNI=*1362-2* 387
Acalypha popayanensis IPNI=*1228-2* 387
Acalypha reptans IPNI=*1012504-2* 387
Acalypha reptans var. *brevipes* IPNI=*1012504-2* 387
Acalypha reptans var. *genuina* IPNI=*1012504-2* 387
Acalypha salicoides IPNI=*1228-2* 387
Acalypha samydifolia IPNI=*1228-2* 387
Acalypha sanderi IPNI=*337478-1* 387
Acalypha seemannii IPNI=*1362-2* 387
Acalypha sidifolia IPNI=*1362-2* 387
Acalypha subandina IPNI=*1444-2* 387
Acalypha tabascensis IPNI=*1228-2* 387
Acalypha tarapotensis IPNI=*1362-2* 387
Acalypha tristis IPNI=*1362-2* 387
Acalypha ulmifolia IPNI=*1228-2* 387
Acalypha vermifera IPNI=*1228-2* 387
Acanthinophyllum ilicifolium IPNI=*59683-2* 437
Acanthinophyllum spruceanum IPNI=*59683-2* 437
Acanthinophyllum strepitans IPNI=*59683-2* 437
Acanthocereus baxaniensis IPNI=*1646-2* 283
Acanthocereus colombianus IPNI=*1646-2* 283
Acanthocereus floridanus IPNI=*1646-2* 283
Acanthocereus horribarbis IPNI=*130627-1* 283
Acanthocereus horridus IPNI=*1646-2* 283
Acanthocereus occidentalis IPNI=*1646-2* 283
Acanthocereus pentagonus IPNI=*1646-2* 283
Acanthocereus pitajaya IPNI=*1646-2* 283
Acanthocereus princeps IPNI=*1646-2* 283
Acanthocereus subinermis IPNI=*1646-2* 283
Acanthocereus tetragonus var. *micracanthus* IPNI=*1646-2* 283
Acanthocereus thalassinus IPNI=*130283-1* 283
Acanthodium spicatum IPNI=*52558-1* 360
Acanthonychia polycnemoides IPNI=*45444-2* 285
Acanthonychia ramosissima IPNI=*45444-2* 285
Acanthonychia ramosissima var. *andina* IPNI=*45444-2* 285
Acanthonychia ramosissima var. *rosetta* IPNI=*45444-2* 285
Acanthorhiza kalbreyeri IPNI=*666356-1* 208
Acanthosabal caespitosa IPNI=*663912-1* 203
Acanthospermum brasilium IPNI=*1692-2* 255
Acanthospermum hirsutum IPNI=*1692-2* 255
Acanthospermum humile var. *hispidum* IPNI=*173718-1* 255
Acanthospermum xanthioides IPNI=*1692-2* 255
Acanthospermum xanthioides var. *acutifolium* IPNI=*1692-2* 255
Acanthospermum xanthioides var. *obtusifolium* IPNI=*1692-2* 255
Acanthosphaera ulei IPNI=*167409-2* 440
Acanthosyris colombiana IPNI=*1712-2* 449
Acanthoxanthium ambrosioides IPNI=*260892-1* 272
Acanthus latifolius IPNI=*44892-1* 359
Acanthus longifolius IPNI=*44892-1* 359
Acanthus niger IPNI=*44892-1* 359
Acanthus platyphyllus IPNI=*44892-1* 359
Acanthus spinosissimus IPNI=*44892-1* 359
Acca sellowiana IPNI=*596467-1* 426
Acca sellowiana var. *rugosa* IPNI=*596467-1* 426
Acetosa acetosella IPNI=*224391-2* 289
Acetosa conglomerata IPNI=*697003-1* 289
Acetosa hastata IPNI=*224391-2* 289
Acetosa obtusifolia IPNI=*310739-2* 290
Acetosa parva IPNI=*224391-2* 289
Acetosa patientia IPNI=*697388-1* 290
Acetosa repens IPNI=*224391-2* 289
Acetosa sterilis IPNI=*224391-2* 289
Acetosella acetosella IPNI=*224391-2* 289
Acetosella alba IPNI=*30012601-2* 432
Acetosella amazonica IPNI=*177777-2* 432
Acetosella andicola IPNI=*177883-2* 432
Acetosella articulata IPNI=*374673-1* 432
Acetosella aureoflava IPNI=*177771-2* 432
Acetosella bahiensis IPNI=*374932-1* 432
Acetosella barrelieri IPNI=*177777-2* 432
Acetosella caespitosa IPNI=*177883-2* 432
Acetosella chrysantha IPNI=*177883-2* 432
Acetosella cineracea IPNI=*177883-2* 432
Acetosella commersonii IPNI=*177883-2* 432
Acetosella conorrhiza IPNI=*177883-2* 432
Acetosella corniculata IPNI=*177893-2* 432
Acetosella corniculata var. *subglabra* IPNI=*177893-2* 432
Acetosella corniculata var. *villosa* IPNI=*177893-2* 432
Acetosella crassicaulis IPNI=*178566-2* 432
Acetosella crenata IPNI=*178566-2* 432
Acetosella debilis IPNI=*374833-1* 432
Acetosella deppei IPNI=*375466-1* 432
Acetosella diffusa IPNI=*374932-1* 432
Acetosella distans IPNI=*374932-1* 432
Acetosella drummondii IPNI=*374872-1* 432
Acetosella fasciculata IPNI=*374932-1* 432
Acetosella filiformis IPNI=*178000-2* 432
Acetosella frutescens IPNI=*374932-1* 432
Acetosella grandifolia IPNI=*374833-1* 432
Acetosella hedysaroides IPNI=*374990-1* 432
Acetosella herpestica IPNI=*177893-2* 432
Acetosella lilacina IPNI=*375051-1* 432
Acetosella lotoides IPNI=*178186-2* 432
Acetosella martiana IPNI=*374833-1* 432
Acetosella medicaginea IPNI=*178220-2* 432
Acetosella megapotamica IPNI=*177883-2* 432
Acetosella melilotoides IPNI=*178225-2* 432
Acetosella mollis IPNI=*178245-2* 432

Acetosella multifida subsp. *tenuifolia* IPNI=*224391-2* 289
Acetosella multifida subsp. *vulgaris* IPNI=*224391-2* 289
Acetosella oxyptera IPNI=*375478-1* 432
Acetosella papilionacea IPNI=*375478-1* 432
Acetosella parvifolia IPNI=*178000-2* 432
Acetosella peduncularis IPNI=*178322-2* 432
Acetosella pentantha IPNI=*374932-1* 432
Acetosella pilosissima IPNI=*374932-1* 432
Acetosella platensis IPNI=*374673-1* 432
Acetosella psilotricha IPNI=*374932-1* 432
Acetosella pubescens IPNI=*178480-2* 432
Acetosella radiata IPNI=*177883-2* 432
Acetosella regnellii IPNI=*375478-1* 432
Acetosella rubra IPNI=*374673-1* 432
Acetosella scandens IPNI=*178449-2* 432
Acetosella schomburgkiana IPNI=*374932-1* 432
Acetosella simulans IPNI=*177893-2* 432
Acetosella tephrodes IPNI=*374932-1* 432
Acetosella tetraphylla IPNI=*375466-1* 432
Acetosella triangularis IPNI=*375478-1* 432
Acetosella tuberosa IPNI=*178566-2* 432
Acetosella violacea subsp. *latifolia* IPNI=*375051-1* 432
Acetosella violacea var. *albida* IPNI=*375051-1* 432
Acetosella vulgaris f. *integrifolia* IPNI=*224391-2* 289
Acetosella vulgaris IPNI=*224391-2* 289
Acetosella vulgaris subsp. *tenuifolia* IPNI=*224391-2* 289
Acetosella yapacaniensis IPNI=*375478-1* 432
Achaetogeron sophiifolius IPNI=*947840-1* 264
Achania ciliata IPNI=*30041958-2* 410
Achania coccinea IPNI=*30041958-2* 410
Achania cordata IPNI=*30041958-2* 410
Achania malvaviscus IPNI=*30041958-2* 410
Achania mollis IPNI=*30041958-2* 410
Achania pilosa IPNI=*30041958-2* 410
Achatocarpus mexicanus IPNI=*58636-1* 280
Achillea albicaulis IPNI=*2294-2* 255
Achillea albida IPNI=*2294-2* 255
Achillea alpicola IPNI=*2294-2* 255
Achillea ambigua IPNI=*2294-2* 255
Achillea anethifolia IPNI=*2294-2* 255
Achillea angustissima IPNI=*2294-2* 255
Achillea arenicola IPNI=*2294-2* 255
Achillea bicolor IPNI=*2294-2* 255
Achillea borealis f. *fusca* IPNI=*2294-2* 255
Achillea borealis IPNI=*2294-2* 255
Achillea borealis subsp. *arenicola* IPNI=*2294-2* 255
Achillea borealis subsp. *californica* IPNI=*2294-2* 255
Achillea borealis subsp. *typica* IPNI=*2294-2* 255
Achillea borealis var. *arenicola* IPNI=*2294-2* 255
Achillea borealis var. *californica* IPNI=*2294-2* 255
Achillea borealis var. *fusca* IPNI=*2294-2* 255
Achillea californica IPNI=*2294-2* 255
Achillea chamaecyparissus IPNI=*241927-1* 269
Achillea compacta IPNI=*2294-2* 255
Achillea coronopifolia IPNI=*2294-2* 255
Achillea crassifolia IPNI=*2294-2* 255
Achillea cristata IPNI=*2294-2* 255
Achillea cuspidata IPNI=*2294-2* 255
Achillea dentifera IPNI=*2294-2* 255
Achillea dracunculus IPNI=*307848-2* 257
Achillea eradiata IPNI=*2294-2* 255
Achillea fusca IPNI=*2294-2* 255
Achillea gigantea IPNI=*2294-2* 255
Achillea gracilis IPNI=*2294-2* 255
Achillea haenkeana IPNI=*2294-2* 255
Achillea lanata IPNI=*2294-2* 255
Achillea lanulosa f. *peroutkyi* IPNI=*2294-2* 255
Achillea lanulosa f. *rubicunda* IPNI=*2294-2* 255
Achillea lanulosa IPNI=*2294-2* 255
Achillea lanulosa subsp. *alpicola* IPNI=*2294-2* 255
Achillea lanulosa subsp. *megacephala* IPNI=*2294-2* 255
Achillea lanulosa subsp. *typica* IPNI=*2294-2* 255
Achillea lanulosa var. *alpicola* IPNI=*2294-2* 255
Achillea lanulosa var. *arachnoidea* IPNI=*2294-2* 255
Achillea lanulosa var. *eradiata* IPNI=*2294-2* 255
Achillea laxiflora IPNI=*2294-2* 255
Achillea magna IPNI=*2294-2* 255
Achillea marginata IPNI=*2294-2* 255
Achillea megacephala IPNI=*2294-2* 255
Achillea millefolium f. *albiflora* IPNI=*2294-2* 255
Achillea millefolium f. *californica* IPNI=*2294-2* 255
Achillea millefolium f. *discolor* IPNI=*2294-2* 255
Achillea millefolium f. *iserana* IPNI=*2294-2* 255
Achillea millefolium f. *pseudopannonica* IPNI=*2294-2* 255
Achillea millefolium f. *rhodantha* IPNI=*2294-2* 255
Achillea millefolium f. *roseiflora* IPNI=*2294-2* 255
Achillea millefolium f. *roseoides* IPNI=*2294-2* 255
Achillea millefolium f. *rubicunda* IPNI=*2294-2* 255
Achillea millefolium subsp. *atrotegula* IPNI=*2294-2* 255
Achillea millefolium subsp. *balearica* IPNI=*2294-2* 255
Achillea millefolium subsp. *borealis* IPNI=*2294-2* 255
Achillea millefolium subsp. *lanulosa* IPNI=*2294-2* 255
Achillea millefolium subsp. *occidentalis* IPNI=*2294-2* 255
Achillea millefolium subsp. *pallidotegula* IPNI=*2294-2* 255
Achillea millefolium subvar. *parviligulata* IPNI=*2294-2* 255
Achillea millefolium var. *alpicola* IPNI=*2294-2* 255
Achillea millefolium var. *arenicola* IPNI=*2294-2* 255
Achillea millefolium var. *borealis* IPNI=*2294-2* 255
Achillea millefolium var. *californica* IPNI=*2294-2* 255
Achillea millefolium var. *colliniformis* IPNI=*2294-2* 255
Achillea millefolium var. *densiloba* IPNI=*2294-2* 255
Achillea millefolium var. *dipetala* IPNI=*2294-2* 255
Achillea millefolium var. *dissecta* IPNI=*2294-2* 255
Achillea millefolium var. *fulva* IPNI=*2294-2* 255
Achillea millefolium var. *fusca* IPNI=*2294-2* 255
Achillea millefolium var. *gigantea* IPNI=*2294-2* 255
Achillea millefolium var. *gracilis* IPNI=*2294-2* 255
Achillea millefolium var. *iserana* IPNI=*2294-2* 255
Achillea millefolium var. *lanata* IPNI=*2294-2* 255
Achillea millefolium var. *lanulosa* IPNI=*2294-2* 255
Achillea millefolium var. *litoralis* IPNI=*2294-2* 255
Achillea millefolium var. *lobata* IPNI=*2294-2* 255
Achillea millefolium var. *maritima* IPNI=*2294-2* 255
Achillea millefolium var. *megacephala* IPNI=*2294-2* 255
Achillea millefolium var. *nigrescens* IPNI=*2294-2* 255
Achillea millefolium var. *occidentalis* IPNI=*2294-2* 255
Achillea millefolium var. *pacifica* IPNI=*2294-2* 255
Achillea millefolium var. *parviligula* IPNI=*2294-2* 255
Achillea millefolium var. *parvula* IPNI=*2294-2* 255
Achillea millefolium var. *puberula* IPNI=*2294-2* 255
Achillea millefolium var. *purpurea* IPNI=*2294-2* 255
Achillea millefolium var. *rosea* IPNI=*2294-2* 255
Achillea millefolium var. *russeolata* IPNI=*2294-2* 255
Achillea millefolium var. *sordida* IPNI=*2294-2* 255
Achillea millefolium var. *spathulata* IPNI=*2294-2* 255
Achillea millefolium var. *sylvatica* IPNI=*2294-2* 255
Achillea nabelekii IPNI=*2294-2* 255
Achillea nigrescens IPNI=*2294-2* 255
Achillea occidentalis IPNI=*2294-2* 255
Achillea ochroleuca IPNI=*2294-2* 255
Achillea ossica IPNI=*2294-2* 255
Achillea pacifica IPNI=*2294-2* 255
Achillea palmeri IPNI=*2294-2* 255
Achillea pecten-veneris IPNI=*2294-2* 255
Achillea pratensis IPNI=*2294-2* 255
Achillea pseudotanacetifolia IPNI=*2294-2* 255

Achillea puberula IPNI=*2294-2* 255
Achillea scabra IPNI=*2294-2* 255
Achillea setacea IPNI=*2294-2* 255
Achillea sordida IPNI=*2294-2* 255
Achillea subalpina IPNI=*2294-2* 255
Achillea subhirsuta IPNI=*2294-2* 255
Achillea submellifolium IPNI=*2294-2* 255
Achillea sylvatica IPNI=*2294-2* 255
Achillea tanacetifolia IPNI=*2294-2* 255
Achillea tanacetifolia var. *dentifera* IPNI=*2294-2* 255
Achillea tenuifolia IPNI=*2294-2* 255
Achillea tenuis IPNI=*2294-2* 255
Achillea tomentosa IPNI=*2294-2* 255
Achillea virgata IPNI=*2294-2* 255
Achillios millefoliatus IPNI=*2294-2* 255
Achimenes autumnalis IPNI=*2352-2* 365
Achimenes coccinea IPNI=*2352-2* 365
Achimenes coccinea var. *pyropaea* IPNI=*2352-2* 365
Achimenes coccinea var. *rosea* IPNI=*2352-2* 365
Achimenes cupreata IPNI=*380772-1* 367
Achimenes jaureguia IPNI=*377615-1* 365
Achimenes kleei IPNI=*377615-1* 365
Achimenes longiflora var. *alba* IPNI=*377615-1* 365
Achimenes longiflora var. *albida* IPNI=*377615-1* 365
Achimenes longiflora var. *ilicifolia* IPNI=*377615-1* 365
Achimenes longiflora var. *jauregiana* IPNI=*377615-1* 365
Achimenes longiflora var. *latifolia* IPNI=*377615-1* 365
Achimenes longiflora var. *margaritae* IPNI=*377615-1* 365
Achimenes longiflora var. *nobilis* IPNI=*377615-1* 365
Achimenes longiflora var. *rosea* IPNI=*377615-1* 365
Achimenes longiflora var. *sessilis* IPNI=*377615-1* 365
Achimenes margarita IPNI=*377615-1* 365
Achimenes pauciflora IPNI=*377615-1* 365
Achimenes picta IPNI=*133755-2* 367
Achimenes pulchella IPNI=*2352-2* 365
Achimenes pyropaea IPNI=*2352-2* 365
Achimenes rosea IPNI=*2352-2* 365
Achimenes tenella IPNI=*2352-2* 365
Achimenes tyrianthina IPNI=*377615-1* 365
Achradelpha mammosa IPNI=*152641-2* 306
Achras acana IPNI=*209824-2* 307
Achras breviloba IPNI=*152641-2* 306
Achras buchananiifolia IPNI=*943696-1* 307
Achras caimito IPNI=*788845-1* 306
Achras calcicola IPNI=*152591-2* 305
Achras calderonii IPNI=*152641-2* 306
Achras chicle IPNI=*152591-2* 305
Achras conzattii IPNI=*152641-2* 306
Achras coriacea IPNI=*152641-2* 306
Achras dactylina IPNI=*152641-2* 306
Achras gaumeri IPNI=*152641-2* 306
Achras glaucescens IPNI=*209808-2* 307
Achras guapeda IPNI=*788845-1* 306
Achras latiloba IPNI=*152641-2* 306
Achras lobulata IPNI=*152641-2* 306
Achras lucuma IPNI=*152641-2* 306
Achras lucuma IPNI=*209804-2* 307
Achras mammosa IPNI=*1110421-2* 307
Achras mammosa IPNI=*152641-2* 306
Achras mayana IPNI=*948088-1* 305
Achras meridionalis IPNI=*152641-2* 306
Achras occidentalis IPNI=*152641-2* 306
Achras paludosa IPNI=*152641-2* 306
Achras petenensis IPNI=*152641-2* 306
Achras rojasii IPNI=*152641-2* 306
Achras sapatilla IPNI=*152641-2* 306
Achras sapota IPNI=*152641-2* 306
Achras striata IPNI=*152641-2* 306
Achras tabogaensis IPNI=*152641-2* 306
Achras tainteriana IPNI=*152641-2* 306
Achras tchicomame IPNI=*152641-2* 306
Achras venosa IPNI=*209920-2* 308
Achras verrucosa IPNI=*152641-2* 306
Achras zapota IPNI=*152641-2* 306
Achras zapota var. *major* IPNI=*152641-2* 306
Achras zapota var. *pedicellaris* IPNI=*152641-2* 306
Achras zapota var. *zapotilla* IPNI=*152641-2* 306
Achras zapotilla IPNI=*152641-2* 306
Achrouteria pomifera IPNI=*948086-1* 305
Achyranthes altissima IPNI=*60077-1* 282
Achyranthes amabilis IPNI=*10291-2* 281
Achyranthes aphylla IPNI=*1024156-2* 281
Achyranthes aspera var. *australis* IPNI=*2468-2* 280
Achyranthes aspera var. *simplex* IPNI=*2468-2* 280
Achyranthes aspera var. *villosior* IPNI=*2468-2* 280
Achyranthes australis IPNI=*2468-2* 280
Achyranthes baccata IPNI=*60077-1* 282
Achyranthes bangii IPNI=*10447-2* 281
Achyranthes bettzickiana IPNI=*10291-2* 281
Achyranthes brasiliana IPNI=*10294-2* 281
Achyranthes calea IPNI=*60789-1* 283
Achyranthes canescens IPNI=*2468-2* 280
Achyranthes conferta IPNI=*1183159-2* 282
Achyranthes cyclophylla IPNI=*10291-2* 281
Achyranthes debilis IPNI=*60185-1* 282
Achyranthes diffusa IPNI=*60185-1* 282
Achyranthes ellipticifolia IPNI=*2468-2* 280
Achyranthes ficoidea IPNI=*1078972-2* 281
Achyranthes frutescens IPNI=*10362-2* 281
Achyranthes fruticosa IPNI=*2468-2* 280
Achyranthes geminata IPNI=*60139-1* 282
Achyranthes globosa IPNI=*60185-1* 282
Achyranthes grandifolia IPNI=*2468-2* 280
Achyranthes halimifolia IPNI=*10362-2* 281
Achyranthes hirtiflora IPNI=*60139-1* 282
Achyranthes incana IPNI=*10362-2* 281
Achyranthes laguroides IPNI=*10377-2* 281
Achyranthes lehmannii IPNI=*10378-2* 281
Achyranthes leiantha IPNI=*327342-2* 281
Achyranthes linkiana IPNI=*60077-1* 282
Achyranthes lorentzii IPNI=*327342-2* 281
Achyranthes martinicensis IPNI=*10362-2* 281
Achyranthes mexicana IPNI=*10378-2* 281
Achyranthes mollis IPNI=*60185-1* 282
Achyranthes mucronata IPNI=*327342-2* 281
Achyranthes nivea IPNI=*327342-2* 281
Achyranthes nodosa IPNI=*1002675-1* 283
Achyranthes obovata IPNI=*2468-2* 280
Achyranthes obovata IPNI=*327342-2* 281
Achyranthes obovatifolia IPNI=*2468-2* 280
Achyranthes okinawensis IPNI=*2468-2* 280
Achyranthes paludosa IPNI=*1024156-2* 281
Achyranthes paronychioides IPNI=*327343-2* 281
Achyranthes peploides IPNI=*327344-2* 281
Achyranthes philoxeroides IPNI=*1024156-2* 281
Achyranthes picta IPNI=*10291-2* 281
Achyranthes pilosa IPNI=*1002675-1* 283
Achyranthes polygonoides IPNI=*1069792-2* 281
Achyranthes porrigens IPNI=*10447-2* 281
Achyranthes portoricensis IPNI=*10362-2* 281
Achyranthes prostrata IPNI=*60185-1* 282
Achyranthes purpurea IPNI=*10447-2* 281
Achyranthes radicans IPNI=*327342-2* 281
Achyranthes repens IPNI=*1069792-2* 281
Achyranthes repens IPNI=*327342-2* 281
Achyranthes robusta IPNI=*2468-2* 280

Achyranthes sarmentosa IPNI=*60077-1* 282
Achyranthes sessilis IPNI=*1069792-2* 281
Achyranthes tenuifolia IPNI=*2468-2* 280
Achyranthes ternata IPNI=*112097-2* 282
Achyranthes triandra IPNI=*1069792-2* 281
Achyranthes uncinata IPNI=*60139-1* 282
Achyranthes vermicularis IPNI=*34033-2* 282
Achyranthes villosa IPNI=*1069792-2* 281
Achyranthes virgata IPNI=*60011-1* 282
Achyranthes williamsii IPNI=*10450-2* 281
Achyrocline argentina IPNI=*174368-1* 255
Achyrocline candicans IPNI=*174407-1* 255
Achyrocline flavescens IPNI=*174368-1* 255
Achyrocline madioides IPNI=*174368-1* 255
Achyrocline polycephala IPNI=*2571-2* 255
Achyrocline pterocaula IPNI=*174368-1* 255
Achyrocline rufescens IPNI=*174368-1* 255
Achyrocline saturejoides var. *candicans* IPNI=*174407-1* 255
Achyrocline saturejoides var. *longifolia* IPNI=*174407-1* 255
Achyrocline saturejoides var. *remotifolia* IPNI=*174407-1* 255
Achyrodes aureum IPNI=*406803-1* 240
Achyrophorus albiflorus IPNI=*317594-2* 264
Achyrophorus balbisii IPNI=*225504-1* 264
Achyrophorus barbatus IPNI=*317594-2* 264
Achyrophorus brasiliensis IPNI=*126428-2* 264
Achyrophorus chillensis IPNI=*126428-2* 264
Achyrophorus elatus IPNI=*225495-1* 264
Achyrophorus humboldtii IPNI=*317594-2* 264
Achyrophorus jussieui IPNI=*225575-1* 264
Achyrophorus microcephalus IPNI=*126475-2* 264
Achyrophorus minimus IPNI=*225504-1* 264
Achyrophorus quitensis IPNI=*317594-2* 264
Achyrophorus quitensis var. *sonchoides* IPNI=*317594-2* 264
Achyrophorus radicatus IPNI=*225575-1* 264
Achyrophorus sagittatus IPNI=*126428-2* 264
Achyrophorus selloi IPNI=*126428-2* 264
Achyrophorus sessiliflorus IPNI=*317594-2* 264
Achyrophorus sessiliflorus var. *barbatus* IPNI=*317594-2* 264
Achyrophorus sonchoides IPNI=*317594-2* 264
Achyrophorus stenocephala IPNI=*296211-2* 264
Achyrophorus taraxacifolius IPNI=*296211-2* 264
Achyrophorus taraxacoides IPNI=*296211-2* 264
Acia amara IPNI=*67154-2* 381
Acicarpa sacchariflora IPNI=*1050107-2* 237
Acilepis cirsiifolia IPNI=*236351-1* 267
Acineta fulva IPNI=*614722-1* 217
Acineta humboldtii IPNI=*614722-1* 217
Acineta schilleriana IPNI=*614722-1* 217
Acinodendron aeruginosum IPNI=*158684-2* 419
Acinodendron affine IPNI=*572641-1* 421
Acinodendron albicans IPNI=*1125568-2* 419
Acinodendron amazonicum IPNI=*572075-1* 419
Acinodendron ambiguum IPNI=*572641-1* 421
Acinodendron amplexans IPNI=*1088633-2* 423
Acinodendron amplum IPNI=*572080-1* 419
Acinodendron andinum IPNI=*572660-1* 421
Acinodendron aplostachyum IPNI=*1039241-2* 419
Acinodendron arboreum IPNI=*326247-2* 419
Acinodendron argenteum IPNI=*572109-1* 419
Acinodendron asclepiadeum IPNI=*572118-1* 419
Acinodendron asperrimum IPNI=*572120-1* 419
Acinodendron beurlingii IPNI=*158685-2* 419
Acinodendron boissierianum IPNI=*1000103-2* 422
Acinodendron bourgaeanum IPNI=*572388-1* 420
Acinodendron bracteolatum IPNI=*572199-1* 419
Acinodendron bubalinum IPNI=*572209-1* 419
Acinodendron carioanum IPNI=*572388-1* 420
Acinodendron caudatum IPNI=*572256-1* 420
Acinodendron cecidophorum IPNI=*158685-2* 419
Acinodendron centrodesmum IPNI=*158869-2* 420
Acinodendron chamissois IPNI=*158886-2* 420
Acinodendron ciliatum IPNI=*1108920-2* 420
Acinodendron croceum IPNI=*572353-1* 420
Acinodendron decussatum IPNI=*1108920-2* 420
Acinodendron desmanthum IPNI=*572388-1* 420
Acinodendron dodecandrum IPNI=*572411-1* 420
Acinodendron elatum IPNI=*1092872-2* 420
Acinodendron ellipticum IPNI=*1143367-2* 422
Acinodendron epiphyticum IPNI=*572660-1* 421
Acinodendron eriocladum IPNI=*159046-2* 420
Acinodendron ferrugineum IPNI=*572470-1* 420
Acinodendron granulosum IPNI=*158784-2* 419
Acinodendron holosericeum IPNI=*1024004-2* 421
Acinodendron holosericeum var. *pauciflorum* IPNI=*1024004-2* 421
Acinodendron holosericum IPNI=*1024004-2* 421
Acinodendron hookerianum IPNI=*572575-1* 421
Acinodendron ibaguense IPNI=*572590-1* 421
Acinodendron impetiolare IPNI=*572600-1* 421
Acinodendron laevigatum IPNI=*572641-1* 421
Acinodendron latifolium IPNI=*572660-1* 421
Acinodendron lepidotum IPNI=*572671-1* 421
Acinodendron ligustrinum IPNI=*572682-1* 421
Acinodendron lindenii IPNI=*158684-2* 419
Acinodendron longifolium IPNI=*1220088-2* 421
Acinodendron longipedunculatum IPNI=*159286-2* 421
Acinodendron longispicatum IPNI=*159286-2* 421
Acinodendron macrophyllum IPNI=*1000103-2* 422
Acinodendron macrostachyum IPNI=*1039241-2* 419
Acinodendron macrothyrsum IPNI=*159311-2* 421
Acinodendron microcarpum IPNI=*158685-2* 419
Acinodendron minutiflorum IPNI=*1098452-2* 421
Acinodendron mucronatum IPNI=*1024004-2* 421
Acinodendron multinervulosum IPNI=*1110588-2* 423
Acinodendron nervosum IPNI=*572835-1* 421
Acinodendron ochraceum IPNI=*572860-1* 421
Acinodendron paleaceum IPNI=*159460-2* 421
Acinodendron planinerve IPNI=*158685-2* 419
Acinodendron plantaginifolium IPNI=*1143367-2* 422
Acinodendron poeppigii IPNI=*159532-2* 422
Acinodendron prasinum IPNI=*1069702-2* 422
Acinodendron pteropodum IPNI=*1069702-2* 422
Acinodendron punctatum IPNI=*1069703-2* 422
Acinodendron racemosum IPNI=*1158780-2* 422
Acinodendron rubiginosum IPNI=*1155116-2* 422
Acinodendron rufescens IPNI=*1018450-2* 422
Acinodendron salicifolium IPNI=*573124-1* 422
Acinodendron setinode IPNI=*572353-1* 420
Acinodendron squamulosum IPNI=*573207-1* 422
Acinodendron stenostachyum IPNI=*159763-2* 422
Acinodendron strictum IPNI=*573368-1* 423
Acinodendron theaezans IPNI=*1110588-2* 423
Acinodendron tomentosum IPNI=*1088633-2* 423
Acinodendron triplinerve IPNI=*159880-2* 423
Acinodendron versicolor IPNI=*573368-1* 423
Acinolis elliptica IPNI=*573207-1* 422
Acinopetala wendlandiana IPNI=*154376-2* 220
Acioa amara IPNI=*67154-2* 381
Aciotis aequatorialis IPNI=*1025557-2* 417
Aciotis alata IPNI=*1101289-2* 417
Aciotis amazonica IPNI=*1025557-2* 417
Aciotis amazonica var. *radicans* IPNI=*1025557-2* 417
Aciotis aquatica IPNI=*573916-1* 424
Aciotis aristata IPNI=*1101289-2* 417
Aciotis aristellata IPNI=*1101289-2* 417
Aciotis asplundii IPNI=*1101289-2* 417
Aciotis caulialata IPNI=*1101289-2* 417

Aciotis circaeifolia var. *major* IPNI=*1031600-2* 417
Aciotis dichotoma IPNI=*1025557-2* 417
Aciotis dichotoma var. *anomala* IPNI=*1025557-2* 417
Aciotis dichotoma var. *longifolia* IPNI=*1025557-2* 417
Aciotis discolor IPNI=*1031600-2* 417
Aciotis discolor var. *martinicensis* IPNI=*1031600-2* 417
Aciotis fragilis IPNI=*1031600-2* 417
Aciotis fragilis var. *lancifolia* IPNI=*1031600-2* 417
Aciotis levyana IPNI=*1101289-2* 417
Aciotis longifolia var. *glabra* IPNI=*1031600-2* 417
Aciotis martinicensis IPNI=*1031600-2* 417
Aciotis pellucida IPNI=*1031600-2* 417
Aciotis purpurascens var. *alata* IPNI=*1101289-2* 417
Aciotis purpurascens var. *longifolia* IPNI=*1031600-2* 417
Aciotis purpurascens var. *martinensis* IPNI=*1101289-2* 417
Aciotis purpurascens var. *martinicensis* IPNI=*1031600-2* 417
Aciotis purpurascens var. *pellucida* IPNI=*1031600-2* 417
Aciotis sieberi IPNI=*573916-1* 424
Aciotis sileniflora IPNI=*1031600-2* 417
Aciotis sphaeranthera IPNI=*1025557-2* 417
Aciotis trinitensis IPNI=*1031600-2* 417
Aciotis uliginosa IPNI=*1031600-2* 417
Acmanthera radlkoferi IPNI=*303745-2* 395
Acmella brasiliensis IPNI=*1093589-2* 270
Acmella buphthalmoides IPNI=*213524-1* 264
Acmella debilis IPNI=*271548-2* 256
Acmella filipes var. *cayensis* IPNI=*271533-2* 255
Acmella filipes var. *parvifolia* IPNI=*271533-2* 255
Acmella fimbriata IPNI=*174534-1* 255
Acmella flavicaulis IPNI=*213534-1* 264
Acmella lanceolata IPNI=*1186403-2* 261
Acmella mutisii IPNI=*213524-1* 264
Acmella nudicaulis IPNI=*213534-1* 264
Acmella nuttaliana IPNI=*174552-1* 256
Acmella occidentalis IPNI=*174552-1* 256
Acmella occidentalis IPNI=*213524-1* 264
Acmella oppositifolia IPNI=*213524-1* 264
Acmella oppositifolia var. *repens* IPNI=*174552-1* 256
Acmella parvifolia IPNI=*213534-1* 264
Acmella radicans var. *debilis* IPNI=*271548-2* 256
Acmella spilanthoides IPNI=*1093589-2* 270
Acmella tenella IPNI=*271548-2* 256
Acnida alabamensis IPNI=*10611-2* 281
Acnida australis IPNI=*10611-2* 281
Acnida cannabina var. *australis* IPNI=*10611-2* 281
Acnida cannabina var. *cuspidata* IPNI=*10611-2* 281
Acnida cuspidata IPNI=*10611-2* 281
Acoelorraphe arborescens IPNI=*663912-1* 203
Acoelorraphe pinetorum IPNI=*663912-1* 203
Acoelorraphe wrightii f. *inermis* IPNI=*663912-1* 203
Acoelorraphe wrightii var. *novogeronensis* IPNI=*663912-1* 203
Aconisia grandis IPNI=*77200715-1* 240
Aconitum anglicum IPNI=*707615-1* 436
Acontias hastifolius IPNI=*269304-2* 202
Acontias helleborifolius IPNI=*89333-1* 203
Acontias plumieri IPNI=*89333-1* 203
Acontias striatipes IPNI=*909842-1* 203
Acontias variegatus IPNI=*89333-1* 203
Acorellus laevigatus IPNI=*304938-1* 231
Acosmium nitens IPNI=*502385-1* 325
Acosmium panamense IPNI=*77091925-1* 325
Acosmium tomentellum IPNI=*502385-1* 325
Acosmium trichonema IPNI=*316401-2* 320
Acosta aculeata IPNI=*690705-1* 338
Acridocarpus cavanillesii IPNI=*1079574-2* 396
Acrocarpidium brevipes IPNI=*189588-2* 194
Acrocarpidium exile IPNI=*678531-1* 193
Acrocarpidium guayaquillense IPNI=*189588-2* 194
Acrocarpidium guildingianum IPNI=*679841-1* 194
Acrocarpidium jamesonianum IPNI=*679597-1* 194
Acrocarpidium mexicanum IPNI=*189588-2* 194
Acrocarpidium nummulariifolium f. *pilosior* IPNI=*189588-2* 194
Acrocarpidium nummulariifolium IPNI=*189588-2* 194
Acrocarpidium nummulariifolium var. *obcordatum* IPNI=*189588-2* 194
Acrocarpidium pulicare IPNI=*679841-1* 194
Acrocarpidium repens f. *minus* IPNI=*679841-1* 194
Acrocarpidium repens IPNI=*679841-1* 194
Acrocarpidium rotundifolium IPNI=*189588-2* 194
Acrocarpidium scandens IPNI=*679841-1* 194
Acrocarpidium serpens IPNI=*679841-1* 194
Acrocarpus combretiflorus IPNI=*472105-1* 310
Acrocarpus fraxinifolius var. *guangxiensis* IPNI=*472105-1* 310
Acrocarpus grandis IPNI=*472105-1* 310
Acroceras oryzoides IPNI=*3538-2* 233
Acrocomia antiguana IPNI=*663914-1* 203
Acrocomia antioquiensis IPNI=*663914-1* 203
Acrocomia belizensis IPNI=*663914-1* 203
Acrocomia christopherensis IPNI=*663914-1* 203
Acrocomia chunta IPNI=*663914-1* 203
Acrocomia cubensis IPNI=*663914-1* 203
Acrocomia erioacantha IPNI=*663914-1* 203
Acrocomia fusiformis IPNI=*663914-1* 203
Acrocomia glaucophylla IPNI=*663914-1* 203
Acrocomia globosa IPNI=*663914-1* 203
Acrocomia grenadana IPNI=*663914-1* 203
Acrocomia guianensis IPNI=*663914-1* 203
Acrocomia horrida IPNI=*663914-1* 203
Acrocomia hospes IPNI=*663914-1* 203
Acrocomia ierensis IPNI=*663914-1* 203
Acrocomia karukerana IPNI=*663914-1* 203
Acrocomia lasiospatha IPNI=*663914-1* 203
Acrocomia mexicana IPNI=*663914-1* 203
Acrocomia microcarpa IPNI=*663914-1* 203
Acrocomia minor IPNI=*663914-1* 203
Acrocomia mokayayba IPNI=*663914-1* 203
Acrocomia odorata IPNI=*663914-1* 203
Acrocomia panamensis IPNI=*663914-1* 203
Acrocomia pilosa IPNI=*663914-1* 203
Acrocomia quisqueyana IPNI=*663914-1* 203
Acrocomia sclerocarpa IPNI=*663914-1* 203
Acrocomia sclerocarpa var. *wallaceana* IPNI=*663914-1* 203
Acrocomia sphaerocarpa IPNI=*663914-1* 203
Acrocomia spinosa IPNI=*663914-1* 203
Acrocomia subinermis IPNI=*663914-1* 203
Acrocomia tenuifrons IPNI=*663914-1* 203
Acrocomia ulei IPNI=*663914-1* 203
Acrocomia viegasii IPNI=*663914-1* 203
Acrocomia vinifera IPNI=*663914-1* 203
Acrocomia wallaceana IPNI=*663914-1* 203
Acrocomia zapotecis IPNI=*663914-1* 203
Acrodiclidium anacardioides IPNI=*158649-2* 183
Acrodiclidium appelii IPNI=*465275-1* 183
Acrodiclidium armeniacum IPNI=*465248-1* 183
Acrodiclidium aubletii IPNI=*465275-1* 183
Acrodiclidium benense IPNI=*465312-1* 183
Acrodiclidium cannella IPNI=*139921-2* 183
Acrodiclidium cayennense IPNI=*139921-2* 183
Acrodiclidium chrysophyllum IPNI=*465260-1* 183
Acrodiclidium cinnamomoides IPNI=*77198017-1* 183
Acrodiclidium ita-uba IPNI=*158649-2* 183
Acrodiclidium ita-uba var. *amarella* IPNI=*158649-2* 183
Acrodiclidium jamaicense IPNI=*465312-1* 183
Acrodiclidium limbatum IPNI=*465312-1* 183
Acrodiclidium limbosum IPNI=*465312-1* 183
Acrodiclidium macrophyllum IPNI=*465280-1* 183
Acrodiclidium parviflorum IPNI=*465248-1* 183

Acrodiclidium puchury-major IPNI=*13872-2* 182
Acrodiclidium rigidum IPNI=*465260-1* 183
Acrodiclidium triandrum IPNI=*465312-1* 183
Acrolasia squalida IPNI=*545374-1* 291
Acronia baudoensis IPNI=*316910-2* 222
Acropterygium pectinatum IPNI=*110733-2* 173
Acropterygium pectinatum var. *sublineare* IPNI=*110733-2* 173
Acrostichum albidulum IPNI=*77182941-1* 178
Acrostichum album IPNI=*17345640-1* 178
Acrostichum aureum f. *cristata* IPNI=*319061-2* 177
Acrostichum aureum var. *hirsutum* IPNI=*319061-2* 177
Acrostichum aureum var. *marginatum* IPNI=*319061-2* 177
Acrostichum botryoides IPNI=*17180780-1* 176
Acrostichum calomelanos f. *sulfurea* IPNI=*17345640-1* 178
Acrostichum calomelanos IPNI=*17345640-1* 178
Acrostichum cayennense IPNI=*319061-2* 177
Acrostichum conforme var. *alpinum* IPNI=*17104440-1* 175
Acrostichum crassifolium IPNI=*319061-2* 177
Acrostichum danaeifolium f. *glabrifolium* IPNI=*17002360-1* 177
Acrostichum danaeifolium f. *incanum* IPNI=*17002360-1* 177
Acrostichum danaeifolium f. *lobatum* IPNI=*17002360-1* 177
Acrostichum dissimile IPNI=*89270-2* 175
Acrostichum ebeneum IPNI=*200935-2* 178
Acrostichum elegans IPNI=*17220460-1* 179
Acrostichum emarginatum IPNI=*319061-2* 177
Acrostichum excelsum f. *lobatum* IPNI=*17002360-1* 177
Acrostichum excelsum IPNI=*17002360-1* 177
Acrostichum fasciculatum IPNI=*319061-2* 177
Acrostichum ferrugineum IPNI=*1039912-2* 176
Acrostichum ferruginosum IPNI=*1039912-2* 176
Acrostichum flavens IPNI=*77182941-1* 178
Acrostichum formosum IPNI=*319061-2* 177
Acrostichum gayanum IPNI=*17103530-1* 175
Acrostichum guineense IPNI=*319061-2* 177
Acrostichum hartii IPNI=*17181340-1* 176
Acrostichum inaequale IPNI=*319061-2* 177
Acrostichum japurense IPNI=*17442740-1* 175
Acrostichum juglandifolium IPNI=*319061-2* 177
Acrostichum karstenianum IPNI=*89270-2* 175
Acrostichum leptophyllum IPNI=*17103530-1* 175
Acrostichum lomarioides f. *lobatum* IPNI=*17002360-1* 177
Acrostichum lomarioides IPNI=*17002360-1* 177
Acrostichum marginatum IPNI=*313710-2* 178
Acrostichum marginatum IPNI=*319061-2* 177
Acrostichum minutum IPNI=*17104440-1* 175
Acrostichum neriifolium IPNI=*89270-2* 175
Acrostichum niveum IPNI=*77182941-1* 178
Acrostichum nudum IPNI=*17104440-1* 175
Acrostichum obliquum IPNI=*319061-2* 177
Acrostichum osmundaceum IPNI=*30017613-2* 176
Acrostichum osmundaceum var. *caudatum* IPNI=*30017613-2* 176
Acrostichum petiolatum IPNI=*89270-2* 175
Acrostichum phlebodes IPNI=*17442740-1* 175
Acrostichum plumieri IPNI=*89270-2* 175
Acrostichum polypodioides IPNI=*1039912-2* 176
Acrostichum punctatum IPNI=*17341250-1* 176
Acrostichum rigens IPNI=*319061-2* 177
Acrostichum scalpturatum IPNI=*319061-2* 177
Acrostichum schiedei IPNI=*89270-2* 175
Acrostichum schmitzii IPNI=*89270-2* 175
Acrostichum sorbifolium var. *japurense* IPNI=*17442740-1* 175
Acrostichum spectabile IPNI=*319061-2* 177
Acrostichum tartareum IPNI=*200935-2* 178
Acrostichum tereticaulon IPNI=*77182941-1* 178
Acrostichum thalictroides IPNI=*17064190-1* 178
Acrostichum unitum IPNI=*17104440-1* 175
Acrostichum viscosum IPNI=*89270-2* 175
Acrostichum viscosum var. *obtusum* IPNI=*89270-2* 175
Acrostichum viscosum var. *wrightii* IPNI=*89270-2* 175
Acrostichum wightianum IPNI=*319061-2* 177
Acrostigma aequale IPNI=*983907-1* 212
Actinochloa humilis IPNI=*328564-2* 235
Actinochloa procumbens IPNI=*328564-2* 235
Actinochloa prostrata IPNI=*328564-2* 235
Actinochloa simplex IPNI=*328564-2* 235
Actinochloa tenuis IPNI=*328564-2* 235
Actinophloeus bleeseri IPNI=*55245-3* 211
Actinophloeus capitis-yorki IPNI=*669378-1* 211
Actinophloeus hospitus IPNI=*55245-3* 211
Actinophloeus macarthurii IPNI=*55245-3* 211
Actinostemon acuminatus IPNI=*1187218-2* 387
Actinostemon angustifolius IPNI=*1187218-2* 387
Actinostemon australis IPNI=*1187218-2* 387
Actinostemon brasiliensis IPNI=*356183-1* 393
Actinostemon cantagallensis IPNI=*1187218-2* 387
Actinostemon concolor f. *latifolius* IPNI=*1187218-2* 387
Actinostemon concolor f. *sessilifolius* IPNI=*1187218-2* 387
Actinostemon concolor lusus biattenuatus IPNI=*1187218-2* 387
Actinostemon concolor var. *acuminatus* IPNI=*1187218-2* 387
Actinostemon concolor var. *acutissimus* IPNI=*1187218-2* 387
Actinostemon concolor var. *angustifolius* IPNI=*1187218-2* 387
Actinostemon concolor var. *bicolor* IPNI=*1187218-2* 387
Actinostemon concolor var. *ellipticus* IPNI=*1187218-2* 387
Actinostemon concolor var. *gardneri* IPNI=*1187218-2* 387
Actinostemon concolor var. *genuinus* IPNI=*1187218-2* 387
Actinostemon concolor var. *grandifolius* IPNI=*1187218-2* 387
Actinostemon concolor var. *intermedius* IPNI=*1187218-2* 387
Actinostemon concolor var. *longifolius* IPNI=*1187218-2* 387
Actinostemon concolor var. *microphyllus* IPNI=*1187218-2* 387
Actinostemon concolor var. *mucronatus* IPNI=*1187218-2* 387
Actinostemon concolor var. *obovatus* IPNI=*1187218-2* 387
Actinostemon concolor var. *riedelii* IPNI=*1187218-2* 387
Actinostemon concolor var. *sellowii* IPNI=*1187218-2* 387
Actinostemon concolor var. *variifolius* IPNI=*1187218-2* 387
Actinostemon grandifolius IPNI=*1187218-2* 387
Actinostemon marginatus IPNI=*1187218-2* 387
Actinostemon multiflorus IPNI=*1187218-2* 387
Actinostemon multiflorus var. *leptostachys* IPNI=*1187218-2* 387
Actinostemon oligandrus IPNI=*1187218-2* 387
Actinostemon polymorphus f. *angustifolius* IPNI=*1187218-2* 387
Actinostemon polymorphus f. *biattenuatus* IPNI=*1187218-2* 387
Actinostemon polymorphus f. *concolor* IPNI=*1187218-2* 387
Actinostemon polymorphus f. *cuspidatus* IPNI=*1187218-2* 387
Actinostemon polymorphus f. *latifolius* IPNI=*1187218-2* 387
Actinostemon polymorphus f. *microphyllus* IPNI=*1187218-2* 387
Actinostemon polymorphus f. *minor* IPNI=*1187218-2* 387
Actinostemon polymorphus IPNI=*1187218-2* 387
Actinostemon polymorphus var. *acutissimus* IPNI=*1187218-2* 387
Actinostemon polymorphus var. *angustatus* IPNI=*1187218-2* 387
Actinostemon polymorphus var. *bicolor* IPNI=*1187218-2* 387
Actinostemon polymorphus var. *ellipticus* IPNI=*1187218-2* 387
Actinostemon polymorphus var. *gardneri* IPNI=*1187218-2* 387
Actinostemon polymorphus var. *grandifolius* IPNI=*1187218-2* 387
Actinostemon polymorphus var. *intermedius* IPNI=*1187218-2* 387
Actinostemon polymorphus var. *longifolius* IPNI=*1187218-2* 387
Actinostemon polymorphus var. *mucronatus* IPNI=*1187218-2* 387
Actinostemon polymorphus var. *obovatus* IPNI=*1187218-2* 387
Actinostemon polymorphus var. *sellowii* IPNI=*1187218-2* 387
Actinostemon polymorphus var. *variifolius* IPNI=*1187218-2* 387
Actinostemon sessilifolius IPNI=*1187218-2* 387
Actinostemon tortuosus IPNI=*1187218-2* 387
Actinostigma speciosum IPNI=*429281-1* 385
Acuan arboreum IPNI=*1056199-2* 328
Acuan depressum IPNI=*1125234-2* 317
Acuan texanum IPNI=*1125234-2* 317
Acuan tracyi IPNI=*1125234-2* 317

Acuan virgatum IPNI=*1125234-2* 317
Acuan virgatum var. *albiflorum* IPNI=*1125234-2* 317
Acuan virgatum var. *flaviflorum* IPNI=*1125234-2* 317
Acunna lanceolata IPNI=*317630-2* 298
Acunna oblonga IPNI=*317630-2* 298
Acyntha guineensis IPNI=*77164234-1* 215
Acyntha thyrsiflora IPNI=*77164234-1* 215
Adambea glabra IPNI=*553618-1* 417
Adambea hirsuta IPNI=*553618-1* 417
Adelaster albivenis IPNI=*105033-2* 360
Adelobotrys guianensis IPNI=*566051-1* 417
Adelobotrys multiflorus IPNI=*566051-1* 417
Adenanthera bonplandiana IPNI=*1009172-2* 319
Adenanthera scandens IPNI=*493866-1* 319
Adenaria floribunda var. *grisleoides* IPNI=*30023859-2* 416
Adenaria floribunda var. *microphylla* IPNI=*30023859-2* 416
Adenaria floribunda var. *parvifolia* IPNI=*30023859-2* 416
Adenaria floribunda var. *purpurata* IPNI=*30023859-2* 416
Adenaria griseleoides IPNI=*30023859-2* 416
Adenaria lanceolata IPNI=*30023859-2* 416
Adenaria parvifolia IPNI=*30023859-2* 416
Adenaria purpurata IPNI=*30023859-2* 416
Adenaria purpurata var. *australis* IPNI=*30023859-2* 416
Adenobasium obtusifolium IPNI=*237456-2* 431
Adenobasium salicifolium IPNI=*835147-1* 431
Adenocalymma alboviolaceum IPNI=*152677-2* 364
Adenocalymma alliaceum IPNI=*152672-2* 364
Adenocalymma bilabiatum IPNI=*77100467-1* 365
Adenocalymma ciliolatum IPNI=*152677-2* 364
Adenocalymma fissum IPNI=*152688-2* 364
Adenocalymma flos-ardeae IPNI=*246385-2* 365
Adenocalymma helicocalyx IPNI=*152688-2* 364
Adenocalymma hosmeca IPNI=*152677-2* 364
Adenocalymma laevigatum IPNI=*152677-2* 364
Adenocalymma macrocarpum IPNI=*152677-2* 364
Adenocalymma micradenium IPNI=*152688-2* 364
Adenocalymma moritzii IPNI=*152688-2* 364
Adenocalymma obovatum IPNI=*152672-2* 364
Adenocalymma pachypus IPNI=*152672-2* 364
Adenocalymma paniculatum IPNI=*77141996-1* 363
Adenocalymma perezii IPNI=*152688-2* 364
Adenocalymma pohlianum IPNI=*152677-2* 364
Adenocalymma punctifolium IPNI=*246385-2* 365
Adenocalymma sagotii IPNI=*152672-2* 364
Adenocalymma seleri IPNI=*152688-2* 364
Adenocalymma verruciferum IPNI=*152688-2* 364
Adenocalyx racemosus IPNI=*67201-2* 315
Adenocyclus condensatus IPNI=*988439-1* 268
Adenola glauca IPNI=*144366-2* 429
Adenola longipes IPNI=*144385-2* 429
Adenopodia polystachya IPNI=*1009172-2* 319
Adenoropium gossypiifolium IPNI=*323930-2* 391
Adenoropium hastatum IPNI=*350290-1* 391
Adenoropium integerrimum IPNI=*350290-1* 391
Adenoropium multifidum IPNI=*314308-2* 391
Adenoropium pandurifolium IPNI=*350290-1* 391
Adenosma chenopodiifolia IPNI=*51951-1* 360
Adenostemma aquaticum IPNI=*175030-1* 256
Adenostemma dregei IPNI=*175030-1* 256
Adenostemma erectum IPNI=*175030-1* 256
Adenostemma fastigiatum IPNI=*175030-1* 256
Adenostemma glutinosum IPNI=*175030-1* 256
Adenostemma lavenia var. *typicum* IPNI=*175003-1* 256
Adenostemma lavenia var. *viscosum* IPNI=*175030-1* 256
Adenostemma leiocarpum IPNI=*175030-1* 256
Adenostemma natalense IPNI=*175030-1* 256
Adenostemma ovatum IPNI=*175030-1* 256
Adenostemma perrottetii IPNI=*175030-1* 256
Adenostemma rivale IPNI=*175030-1* 256
Adenostemma tinctorium IPNI=*175030-1* 256
Adesmia mimosoides IPNI=*77211890-1* 316
Adhatoda arenaria IPNI=*52612-1* 360
Adhatoda betonica IPNI=*52612-1* 360
Adhatoda carthaginensis IPNI=*50468-1* 360
Adhatoda carthaginensis var. *glabra* IPNI=*50468-1* 360
Adhatoda carthaginensis var. *minor* IPNI=*50468-1* 360
Adhatoda cheiranthifolia IPNI=*52612-1* 360
Adhatoda felisbertiana IPNI=*50468-1* 360
Adhatoda lupulina IPNI=*52612-1* 360
Adhatoda ramosissima IPNI=*52612-1* 360
Adhatoda scabra IPNI=*941389-1* 359
Adhatoda sericea IPNI=*1140399-2* 359
Adhatoda trinervia IPNI=*52612-1* 360
Adhatoda variegata IPNI=*52612-1* 360
Adiantopsis chlorophylla f. *siccanea* IPNI=*77182519-1* 178
Adiantopsis chlorophylla IPNI=*77182519-1* 178
Adiantopsis chlorophylla var. *contracta* IPNI=*77182519-1* 178
Adiantopsis chlorophylla var. *siccanea* IPNI=*77182519-1* 178
Adiantopsis spectabilis IPNI=*77182519-1* 178
Adiantum acuminatum IPNI=*315262-2* 177
Adiantum aemulum IPNI=*17012580-1* 177
Adiantum affine IPNI=*315077-2* 177
Adiantum affine IPNI=*325030-2* 177
Adiantum africanum IPNI=*325030-2* 177
Adiantum asperum IPNI=*17011270-1* 177
Adiantum assimile IPNI=*17013290-1* 177
Adiantum baenitzii IPNI=*17012580-1* 177
Adiantum brasiliense IPNI=*271633-2* 177
Adiantum capillaceum IPNI=*325030-2* 177
Adiantum capillus IPNI=*325030-2* 177
Adiantum capillus-veneris f. *cristatum* IPNI=*325030-2* 177
Adiantum capillus-veneris f. *fissum* IPNI=*325030-2* 177
Adiantum capillus-veneris f. *rimicola* IPNI=*325030-2* 177
Adiantum capillus-veneris var. *fissum* IPNI=*325030-2* 177
Adiantum capillus-veneris var. *grande* IPNI=*325030-2* 177
Adiantum capillus-veneris var. *modestum* IPNI=*325030-2* 177
Adiantum capillus-veneris var. *protrusum* IPNI=*325030-2* 177
Adiantum capillus-veneris var. *trifidum* IPNI=*325030-2* 177
Adiantum colpodes IPNI=*17013420-1* 177
Adiantum concinnum var. *integrum* IPNI=*17013420-1* 177
Adiantum concinnum var. *subscandens* IPNI=*315077-2* 177
Adiantum coriandrifolium IPNI=*325030-2* 177
Adiantum crenatum IPNI=*17012320-1* 177
Adiantum crenatum var. *glaucescens* IPNI=*17012320-1* 177
Adiantum cuneatum f. *elongatum* IPNI=*17012580-1* 177
Adiantum cuneatum f. *typicum* IPNI=*17012580-1* 177
Adiantum cuneatum IPNI=*17012580-1* 177
Adiantum cuneatum var. *majus* IPNI=*17012580-1* 177
Adiantum cuneatum var. *subintegrum* IPNI=*17012580-1* 177
Adiantum cuneatum var. *vastum* IPNI=*17012580-1* 177
Adiantum cuneifolium IPNI=*325030-2* 177
Adiantum cuneipinnulum IPNI=*17012580-1* 177
Adiantum cycloides IPNI=*17012320-1* 177
Adiantum cyclosorum IPNI=*17012580-1* 177
Adiantum decorum IPNI=*17012580-1* 177
Adiantum deflectens var. *subaristatum* IPNI=*17009750-1* 177
Adiantum deflectens var. *tremulum* IPNI=*17009750-1* 177
Adiantum denticulatum IPNI=*271633-2* 177
Adiantum denticulatum var. *humile* IPNI=*271633-2* 177
Adiantum denticulatum var. *obtusum* IPNI=*271633-2* 177
Adiantum diaphanum var. *affine* IPNI=*325030-2* 177
Adiantum discretodenticulatum IPNI=*17009750-1* 177
Adiantum dissimile IPNI=*17009750-1* 177
Adiantum dolabriforme IPNI=*17009750-1* 177
Adiantum elatum IPNI=*271633-2* 177
Adiantum emarginatum IPNI=*325030-2* 177

Adiantum extensum IPNI=*17013290-1* 177
Adiantum falcatum IPNI=*315261-2* 177
Adiantum falcatum IPNI=*315262-2* 177
Adiantum farleyense IPNI=*17013290-1* 177
Adiantum fergusonii IPNI=*17013290-1* 177
Adiantum flagellum var. *schizaeoides* IPNI=*17009750-1* 177
Adiantum fontanum IPNI=*325030-2* 177
Adiantum formosum IPNI=*325030-2* 177
Adiantum fovearum IPNI=*271633-2* 177
Adiantum fovearum var. *major* IPNI=*271633-2* 177
Adiantum fovearum var. *reductum* IPNI=*271633-2* 177
Adiantum fumarioides IPNI=*325030-2* 177
Adiantum ghiesbreghtii IPNI=*17013290-1* 177
Adiantum glaucophyllum IPNI=*17013290-1* 177
Adiantum glaziovii IPNI=*271633-2* 177
Adiantum gratum IPNI=*17012320-1* 177
Adiantum hendersonii IPNI=*315261-2* 177
Adiantum intermedium var. *medioximum* IPNI=*271633-2* 177
Adiantum intermedium var. *triangulatum* IPNI=*271633-2* 177
Adiantum kohautianum IPNI=*315261-2* 177
Adiantum lancea IPNI=*17011380-1* 177
Adiantum lancea IPNI=*315261-2* 177
Adiantum lanceolatum IPNI=*315262-2* 177
Adiantum latifolium var. *grisebachianum* IPNI=*271633-2* 177
Adiantum laxum IPNI=*315261-2* 177
Adiantum lingii IPNI=*325030-2* 177
Adiantum littorale IPNI=*17013290-1* 177
Adiantum lucidum var. *anomalum* IPNI=*17011270-1* 177
Adiantum lucidum var. *bipinnatum* IPNI=*271633-2* 177
Adiantum lucidum var. *majus* IPNI=*17011270-1* 177
Adiantum lunatum IPNI=*17009750-1* 177
Adiantum lunulatum IPNI=*17084790-1* 175
Adiantum lutescens IPNI=*315077-2* 177
Adiantum macrocarpum IPNI=*315261-2* 177
Adiantum macrophyllum var. *glaucum* IPNI=*17011380-1* 177
Adiantum maderense IPNI=*325030-2* 177
Adiantum marginatum IPNI=*325030-2* 177
Adiantum mexicanum IPNI=*17012580-1* 177
Adiantum michelii IPNI=*325030-2* 177
Adiantum microphyllum IPNI=*77183177-1* 178
Adiantum microphyllum var. *coriaceum* IPNI=*77183177-1* 178
Adiantum modestum IPNI=*325030-2* 177
Adiantum moritzianum IPNI=*325030-2* 177
Adiantum multiforme IPNI=*17013290-1* 177
Adiantum oblique-truncatum IPNI=*315262-2* 177
Adiantum pachysorum IPNI=*315261-2* 177
Adiantum paradiseae IPNI=*325030-2* 177
Adiantum pellucidum IPNI=*17012320-1* 177
Adiantum persimile IPNI=*17009750-1* 177
Adiantum poiretii var. *glaucescens* IPNI=*17012320-1* 177
Adiantum prionophyllum IPNI=*315261-2* 177
Adiantum prionophyllum var. *curtum* IPNI=*315261-2* 177
Adiantum prionophyllum var. *subcoriaceum* IPNI=*315261-2* 177
Adiantum pseudocapillus IPNI=*325030-2* 177
Adiantum pubescens IPNI=*77183177-1* 178
Adiantum ramyam IPNI=*315077-2* 177
Adiantum remyanum IPNI=*325030-2* 177
Adiantum repandum IPNI=*325030-2* 177
Adiantum rigidum IPNI=*315261-2* 177
Adiantum rimicola IPNI=*325030-2* 177
Adiantum ripidium IPNI=*17009750-1* 177
Adiantum rubellum IPNI=*17012580-1* 177
Adiantum schaffneri IPNI=*325030-2* 177
Adiantum sordidum IPNI=*271633-2* 177
Adiantum subaristatum IPNI=*17009750-1* 177
Adiantum subemarginatum IPNI=*325030-2* 177
Adiantum tenerum IPNI=*315077-2* 177
Adiantum tenerum var. *farleyense* IPNI=*17013290-1* 177
Adiantum tenerum var. *littorale* IPNI=*17013290-1* 177
Adiantum tenerum var. *majus* IPNI=*17013290-1* 177
Adiantum tenerum var. *minus* IPNI=*17013290-1* 177
Adiantum tenerum var. *obtusissimum* IPNI=*17013290-1* 177
Adiantum tenerum var. *rotundatum* IPNI=*17013290-1* 177
Adiantum ternatum IPNI=*315261-2* 177
Adiantum ternatum IPNI=*77182326-1* 178
Adiantum tetraphyllum f. *apiculatum* IPNI=*315261-2* 177
Adiantum tetraphyllum f. *genuina* IPNI=*315261-2* 177
Adiantum tetraphyllum f. *obtusum* IPNI=*315261-2* 177
Adiantum tetraphyllum var. *costaricense* IPNI=*315261-2* 177
Adiantum tetraphyllum var. *macropterum* IPNI=*315261-2* 177
Adiantum tetraphyllum var. *microphyllum* IPNI=*315261-2* 177
Adiantum tetraphyllum var. *obtusum* IPNI=*315261-2* 177
Adiantum tetraphyllum var. *paucijugum* IPNI=*315261-2* 177
Adiantum tetraphyllum var. *subsimplex* IPNI=*315261-2* 177
Adiantum thalictroides IPNI=*17012320-1* 177
Adiantum tinctum var. *wagneri* IPNI=*17013420-1* 177
Adiantum trapeziforme var. *trapezoides* IPNI=*17013290-1* 177
Adiantum trapezoides IPNI=*17013290-1* 177
Adiantum tremulum IPNI=*17009750-1* 177
Adiantum tremulum var. *deflectens* IPNI=*17009750-1* 177
Adiantum triangulatum IPNI=*271633-2* 177
Adiantum triangulatum var. *acuminatum* IPNI=*271633-2* 177
Adiantum trifidum IPNI=*325030-2* 177
Adiantum varium IPNI=*17011270-1* 177
Adiantum villosum var. *falcatum* IPNI=*315262-2* 177
Adiantum villosum var. *macrosorum* IPNI=*315262-2* 177
Adiantum visianii IPNI=*325030-2* 177
Adiantum wagneri IPNI=*17013420-1* 177
Adiantum weigandii IPNI=*17012580-1* 177
Adiantum werckleanum IPNI=*17012580-1* 177
Adiantum wilsonii var. *semicordatum* IPNI=*17013890-1* 177
Adicea chrysosplenioides IPNI=*77091436-1* 447
Adicea ciliaris IPNI=*30064315-2* 447
Adicea dendrophila IPNI=*321319-2* 447
Adicea integrifolia IPNI=*30064315-2* 447
Adicea involucrata IPNI=*855698-1* 447
Adicea lundii IPNI=*855481-1* 447
Adicea melastomodes IPNI=*855573-1* 447
Adicea microphylla IPNI=*1162156-2* 447
Adicea rubiifolia IPNI=*30064315-2* 447
Adicea scrobiculata IPNI=*855481-1* 447
Adipera bicapsularis IPNI=*518330-1* 332
Adipera laevigata IPNI=*234652-2* 333
Adipera tomentosa IPNI=*234543-2* 333
Adnaria caracasana IPNI=*858406-1* 299
Adnaria crenata IPNI=*261842-2* 300
Adnaria glandulosa IPNI=*858406-1* 299
Adoketon maritimum IPNI=*10362-2* 281
Adoketon saxatile IPNI=*10362-2* 281
Adventina ciliata IPNI=*208557-1* 263
Adventina parviflora IPNI=*208554-1* 263
Adyseton halimifolium IPNI=*1085072-2* 277
Adyseton maritimum IPNI=*1085072-2* 277
Adyseton orbiculare IPNI=*1085072-2* 277
Aechmea barleei IPNI=*121570-1* 227
Aechmea bernoulliana IPNI=*271722-2* 228
Aechmea bracteata var. *pacifica* IPNI=*121570-1* 227
Aechmea chiriquensis IPNI=*121847-1* 228
Aechmea colombiana var. *laxa* IPNI=*5475-2* 228
Aechmea friedrichsthalii IPNI=*5555-2* 228
Aechmea inermis IPNI=*5555-2* 228
Aechmea isabellina IPNI=*121570-1* 227
Aechmea kienastii IPNI=*121847-1* 228
Aechmea laxiflora IPNI=*121570-1* 227
Aechmea magdalenae var. *quadricolor* IPNI=*5516-2* 228
Aechmea mitis IPNI=*5405-2* 228

Aechmea regularis IPNI=*121570-1* 227
Aechmea schiedeana IPNI=*121570-1* 227
Aechmea schultesiana IPNI=*5555-2* 228
Aechmea tillandsioides var. *kienastii* IPNI=*121847-1* 228
Aechmea vrieseoides IPNI=*121847-1* 228
Aechmea xiphophylla IPNI=*121847-1* 228
Aegilops fluviatilis IPNI=*909653-1* 244
Aegilops saccharina IPNI=*397387-1* 236
Aeginetia longiflora IPNI=*745065-1* 347
Aegiphila acutangula IPNI=*319153-2* 369
Aegiphila amazonica IPNI=*860891-1* 368
Aegiphila arborea IPNI=*860891-1* 368
Aegiphila arborescens IPNI=*860891-1* 368
Aegiphila arborescens var. *breviflora* IPNI=*860891-1* 368
Aegiphila arborescens var. *longiflora* IPNI=*860891-1* 368
Aegiphila barbadensis IPNI=*860919-1* 368
Aegiphila bogotensis f. *ternata* IPNI=*860809-1* 368
Aegiphila bogotensis var. *aequinoctialis* IPNI=*860809-1* 368
Aegiphila congesta IPNI=*229565-2* 377
Aegiphila cowanii IPNI=*860891-1* 368
Aegiphila diffusa IPNI=*860919-1* 368
Aegiphila dumosa IPNI=*860919-1* 368
Aegiphila glandulifera IPNI=*5786-2* 368
Aegiphila glandulifera var. *paraensis* IPNI=*5786-2* 368
Aegiphila glandulifera var. *peruviana* IPNI=*5786-2* 368
Aegiphila glandulifera var. *pyramidata* IPNI=*5786-2* 368
Aegiphila grandifolia IPNI=*860919-1* 368
Aegiphila guianensis IPNI=*860891-1* 368
Aegiphila integrifolia var. *lopez-palacii* IPNI=*860891-1* 368
Aegiphila laxicupulis IPNI=*5786-2* 368
Aegiphila macrophylla IPNI=*770293-1* 358
Aegiphila magnifica IPNI=*5786-2* 368
Aegiphila magnifica var. *pubescens* IPNI=*5786-2* 368
Aegiphila magnifolia IPNI=*860919-1* 368
Aegiphila martinicensis f. *barbadensis* IPNI=*860919-1* 368
Aegiphila martinicensis var. *oligoneura* IPNI=*860919-1* 368
Aegiphila meridensis IPNI=*860809-1* 368
Aegiphila minima IPNI=*860919-1* 368
Aegiphila nigrescens IPNI=*5786-2* 368
Aegiphila oligoneura IPNI=*860919-1* 368
Aegiphila paniculata IPNI=*5786-2* 368
Aegiphila parviflora IPNI=*860891-1* 368
Aegiphila pendula IPNI=*5786-2* 368
Aegiphila punctata IPNI=*864000-1* 372
Aegiphila pyramidata IPNI=*5786-2* 368
Aegiphila straminea IPNI=*860919-1* 368
Aegiphila variabilis IPNI=*860919-1* 368
Aegopricum betulinum IPNI=*352054-1* 392
Aequatorium verrucosum IPNI=*970931-1* 256
Aerobion maculatum IPNI=*1073117-2* 222
Aeschynomene americana var. *villosa* IPNI=*6008-2* 311
Aeschynomene biflora IPNI=*77211517-1* 316
Aeschynomene brasiliana IPNI=*77211517-1* 316
Aeschynomene cassioides IPNI=*284993-2* 334
Aeschynomene elegans IPNI=*77211590-1* 316
Aeschynomene emerus IPNI=*284993-2* 334
Aeschynomene falcata IPNI=*77211592-1* 316
Aeschynomene falcata var. *elegans* IPNI=*77211590-1* 316
Aeschynomene falcata var. *paucijuga* IPNI=*77211592-1* 316
Aeschynomene fascicularis IPNI=*77211890-1* 316
Aeschynomene fistulosa IPNI=*472949-1* 311
Aeschynomene fruticosa IPNI=*77211890-1* 316
Aeschynomene grandiflora IPNI=*518476-1* 334
Aeschynomene herbacea IPNI=*509612-1* 328
Aeschynomene hispidula IPNI=*472933-1* 310
Aeschynomene honesta IPNI=*472949-1* 311
Aeschynomene incana IPNI=*1015219-2* 318
Aeschynomene natans IPNI=*472933-1* 310
Aeschynomene oligantha IPNI=*77211890-1* 316
Aeschynomene paucijuga IPNI=*77211517-1* 316
Aeschynomene paucijuga var. *subscabra* IPNI=*77211517-1* 316
Aeschynomene pubescens IPNI=*1015399-2* 334
Aeschynomene pumila IPNI=*509612-1* 328
Aeschynomene sensitiva f. *paucifoliolata* IPNI=*472949-1* 311
Aeschynomene sensitiva var. *amazonica* IPNI=*472949-1* 311
Aeschynomene sensitiva var. *hispidula* IPNI=*472933-1* 310
Aeschynomene spicata IPNI=*1015219-2* 318
Aeschynomene sulcata IPNI=*472949-1* 311
Aeschynomene triflora IPNI=*77187443-1* 320
Aeschynomene trisperma IPNI=*77211517-1* 316
Aetheilema alopecuroidea IPNI=*51951-1* 360
Aetheilema haenkei IPNI=*51951-1* 360
Aethonopogon praemorsus IPNI=*963150-1* 244
Aethusa ammi IPNI=*837486-1* 251
Aethusa leptophylla IPNI=*840899-1* 252
Agalinis levigata IPNI=*799460-1* 375
Agallostachys chrysantha IPNI=*122289-1* 228
Agallostachys fastuosa IPNI=*319113-2* 228
Agallostachys pinguin IPNI=*319113-2* 228
Aganippea dentata IPNI=*1186403-2* 261
Agapanthus medius IPNI=*527258-1* 213
Agapanthus minor IPNI=*527258-1* 213
Agapanthus tuberosus IPNI=*527258-1* 213
Agapanthus umbellatus IPNI=*527258-1* 213
Agapanthus variegatus IPNI=*527258-1* 213
Agarista anastomosans IPNI=*330459-1* 299
Agastache rugosa f. *alba* IPNI=*444504-1* 368
Agathophytum ambrosioides IPNI=*1198060-2* 282
Agathophytum anthelminticum IPNI=*1090820-2* 282
Agati grandiflora IPNI=*518476-1* 334
Agati grandiflora var. *albiflora* IPNI=*518476-1* 334
Agati sericea IPNI=*1015399-2* 334
Agave aspera IPNI=*105442-2* 215
Agave australis IPNI=*105442-2* 215
Agave cocui var. *cucutensis* IPNI=*6425-2* 214
Agave cocui var. *laguayrensis* IPNI=*6425-2* 214
Agave commelyni IPNI=*64429-1* 215
Agave cubensis IPNI=*105442-2* 215
Agave foetida IPNI=*64429-1* 215
Agave gigantea IPNI=*64429-1* 215
Agave hexapetala IPNI=*105442-2* 215
Agave madagascariensis IPNI=*64429-1* 215
Agave odorata IPNI=*105442-2* 215
Agave polianthes IPNI=*77153983-1* 214
Agave spectabilis IPNI=*319063-2* 214
Agave tuberosa IPNI=*77153983-1* 214
Agave vivipara IPNI=*105442-2* 215
Ageratina ageratoides IPNI=*6815-2* 256
Ageratina articulata IPNI=*6988-2* 256
Ageratina aschenborniana IPNI=*6988-2* 256
Ageratina bimatra IPNI=*105109-2* 263
Ageratina ciliata IPNI=*6988-2* 256
Ageratina humboldtii IPNI=*271815-2* 256
Ageratina pacifica IPNI=*6988-2* 256
Ageratina piurae IPNI=*6988-2* 256
Ageratina pomaderrifolia IPNI=*271815-2* 256
Ageratina vulcanica IPNI=*6988-2* 256
Ageratum album IPNI=*7086-2* 256
Ageratum altissimum IPNI=*6815-2* 256
Ageratum aquaticum IPNI=*175030-1* 256
Ageratum arsenei IPNI=*7086-2* 256
Ageratum brachystephanum IPNI=*7086-2* 256
Ageratum caeruleum IPNI=*1092582-2* 263
Ageratum ciliare IPNI=*7086-2* 256
Ageratum coeruleum IPNI=*1092582-2* 263
Ageratum coeruleum IPNI=*7086-2* 256

Ageratum conyzoides subsp. *houstonianum* IPNI=*175500-1* 256
Ageratum conyzoides subsp. *latifolium* IPNI=*7086-2* 256
Ageratum conyzoides var. *houstonianum* IPNI=*175500-1* 256
Ageratum conyzoides var. *inaequipaleaceum* IPNI=*7086-2* 256
Ageratum conyzoides var. *mexicanum* IPNI=*175500-1* 256
Ageratum cordifolium IPNI=*7086-2* 256
Ageratum guianense IPNI=*1092582-2* 263
Ageratum hirsutum IPNI=*7086-2* 256
Ageratum hirtum IPNI=*7086-2* 256
Ageratum houstonianum f. *isochroum* IPNI=*175500-1* 256
Ageratum houstonianum f. *luteum* IPNI=*175500-1* 256
Ageratum houstonianum f. *niveum* IPNI=*175500-1* 256
Ageratum houstonianum f. *normale* IPNI=*175500-1* 256
Ageratum houstonianum f. *versicolor* IPNI=*175500-1* 256
Ageratum houstonianum var. *angustatum* IPNI=*175500-1* 256
Ageratum houstonianum var. *muticescens* IPNI=*175500-1* 256
Ageratum houstonianum var. *typicum* IPNI=*175500-1* 256
Ageratum humile IPNI=*7086-2* 256
Ageratum koulianum IPNI=*7086-2* 256
Ageratum latifolium IPNI=*7086-2* 256
Ageratum latifolium var. *galapageium* IPNI=*7086-2* 256
Ageratum mexicanum IPNI=*175500-1* 256
Ageratum mexicanum var. *majus* IPNI=*175500-1* 256
Ageratum mexicanum var. *wendlandii* IPNI=*175500-1* 256
Ageratum muticum IPNI=*7086-2* 256
Ageratum nanum IPNI=*7086-2* 256
Ageratum obtusifolium IPNI=*7086-2* 256
Ageratum odoratum IPNI=*7086-2* 256
Ageratum oerstedii IPNI=*7086-2* 256
Ageratum oliveri IPNI=*7086-2* 256
Ageratum perplexans IPNI=*208557-1* 263
Ageratum pinetorum IPNI=*175500-1* 256
Ageratum punctatum IPNI=*251571-1* 270
Ageratum quadriflorum IPNI=*239804-1* 268
Ageratum sandwicense IPNI=*7086-2* 256
Ageratum striatum IPNI=*251571-1* 270
Ageratum strictum IPNI=*175030-1* 256
Ageratum suffruticosum IPNI=*7086-2* 256
Ageratum wendlandii IPNI=*175500-1* 256
Agerella arvensis IPNI=*811649-1* 377
Ageria acuminata IPNI=*127855-2* 255
Ageria retusa IPNI=*83691-1* 255
Ageria uniflora IPNI=*308592-2* 255
Aglaonema acutispathum IPNI=*84081-1* 198
Aglaonema costatum f. *foxii* IPNI=*84052-1* 198
Aglaonema laoticum IPNI=*84081-1* 198
Aglycia coarctata IPNI=*77164258-1* 246
Agnus-castus robusta IPNI=*865568-1* 373
Agnus-castus vulgaris IPNI=*865568-1* 373
Agrestis polymorpha IPNI=*30197865-2* 233
Agricolaea fragrans IPNI=*946989-1* 369
Agropyron attenuatum IPNI=*961772-1* 238
Agropyron boliviacum IPNI=*961772-1* 238
Agropyron geminatum IPNI=*74675-3* 238
Agropyron peruvianum IPNI=*82343-2* 238
Agrostemma coronaria IPNI=*60461121-2* 286
Agrostemma flos-jovis IPNI=*60461121-2* 286
Agrosticula muralis IPNI=*243044-2* 245
Agrostis aberrans IPNI=*1128639-2* 233
Agrostis adscendens IPNI=*386353-1* 233
Agrostis affinis IPNI=*423010-1* 245
Agrostis alba subsp. *decumbens* IPNI=*386353-1* 233
Agrostis alba subsp. *pallens* IPNI=*386353-1* 233
Agrostis alba var. *aristata* IPNI=*30197865-2* 233
Agrostis alba var. *colorata* IPNI=*386353-1* 233
Agrostis alba var. *decumbens* IPNI=*386353-1* 233
Agrostis alba var. *densiflora* IPNI=*386353-1* 233
Agrostis alba var. *minor* IPNI=*30197865-2* 233
Agrostis alba var. *narbonensis* IPNI=*386353-1* 233
Agrostis alba var. *stricta* IPNI=*30197865-2* 233
Agrostis albida IPNI=*386353-1* 233
Agrostis altissima IPNI=*1128639-2* 233
Agrostis altissima var. *laxa* IPNI=*1128639-2* 233
Agrostis ambigua IPNI=*386353-1* 233
Agrostis anomala IPNI=*1128639-2* 233
Agrostis aristulata IPNI=*386353-1* 233
Agrostis atropurpurea IPNI=*30197865-2* 233
Agrostis atroviolacea IPNI=*30197865-2* 233
Agrostis barbata IPNI=*423010-1* 245
Agrostis bermudiana IPNI=*397065-1* 236
Agrostis berteroana IPNI=*21052-2* 234
Agrostis bodaibensis IPNI=*386353-1* 233
Agrostis borbonica IPNI=*243044-2* 245
Agrostis borbonica IPNI=*30197865-2* 233
Agrostis brevis IPNI=*386353-1* 233
Agrostis bryoides IPNI=*386353-1* 233
Agrostis caespitosa IPNI=*386353-1* 233
Agrostis campyla IPNI=*1128639-2* 233
Agrostis candollei IPNI=*386353-1* 233
Agrostis canina f. *varians* IPNI=*30197865-2* 233
Agrostis canina subsp. *grandiflora* IPNI=*1128639-2* 233
Agrostis capillaris subsp. *oreophila* IPNI=*30197865-2* 233
Agrostis capillaris var. *alpigena* IPNI=*30197865-2* 233
Agrostis capillaris var. *aristata* IPNI=*30197865-2* 233
Agrostis capillaris var. *aristulata* IPNI=*30197865-2* 233
Agrostis capillaris var. *canina* IPNI=*30197865-2* 233
Agrostis capillaris var. *dubia* IPNI=*30197865-2* 233
Agrostis capillaris var. *glaucina* IPNI=*30197865-2* 233
Agrostis capillaris var. *hispida* IPNI=*30197865-2* 233
Agrostis capillaris var. *sambukii* IPNI=*30197865-2* 233
Agrostis caryophyllea IPNI=*386633-1* 233
Agrostis ceretana IPNI=*30197865-2* 233
Agrostis chinantlae IPNI=*1128639-2* 233
Agrostis chinensis IPNI=*407138-1* 241
Agrostis claudii IPNI=*30197865-2* 233
Agrostis colorata IPNI=*60445990-2* 241
Agrostis congener IPNI=*423010-1* 245
Agrostis crinita IPNI=*1048074-2* 244
Agrostis debilis IPNI=*409081-1* 241
Agrostis decipiens IPNI=*1021229-2* 239
Agrostis decumbens IPNI=*1128639-2* 233
Agrostis decumbens IPNI=*386353-1* 233
Agrostis densissima IPNI=*386353-1* 233
Agrostis depressa IPNI=*386353-1* 233
Agrostis difficilis IPNI=*386353-1* 233
Agrostis diluta IPNI=*386353-1* 233
Agrostis distichophylla IPNI=*423010-1* 245
Agrostis divaricata IPNI=*30197865-2* 233
Agrostis dubia IPNI=*30197865-2* 233
Agrostis dulcis IPNI=*386353-1* 233
Agrostis dura IPNI=*423010-1* 245
Agrostis elata IPNI=*1128639-2* 233
Agrostis elegans IPNI=*1128639-2* 233
Agrostis elongata IPNI=*1207790-2* 245
Agrostis ericetorum IPNI=*30197865-2* 233
Agrostis exarata var. *angustifolia* IPNI=*1128639-2* 233
Agrostis exarata var. *stolonifera* IPNI=*386353-1* 233
Agrostis exilis IPNI=*30197865-2* 233
Agrostis extensa IPNI=*422898-1* 245
Agrostis fasciculata IPNI=*1128639-2* 233
Agrostis filifolia IPNI=*386353-1* 233
Agrostis filiformis IPNI=*397065-1* 236
Agrostis flavidula IPNI=*1128639-2* 233
Agrostis glabrata IPNI=*60445990-2* 241
Agrostis glaucescens IPNI=*386353-1* 233
Agrostis glaucina IPNI=*30197865-2* 233

Agrostis glutinosa IPNI=*408232-1* 241
Agrostis helenae IPNI=*30197865-2* 233
Agrostis hispida IPNI=*30197865-2* 233
Agrostis hornungiana IPNI=*30197865-2* 233
Agrostis humboldtiana IPNI=*1128639-2* 233
Agrostis humifusa IPNI=*243026-2* 245
Agrostis hyemalis var. *elata* IPNI=*1128639-2* 233
Agrostis indica IPNI=*1207790-2* 245
Agrostis indica IPNI=*422898-1* 245
Agrostis insignita IPNI=*386353-1* 233
Agrostis intermedia IPNI=*1128639-2* 233
Agrostis intermedia IPNI=*386353-1* 233
Agrostis jacutica IPNI=*386353-1* 233
Agrostis juncea IPNI=*1103484-2* 245
Agrostis juncea IPNI=*423010-1* 245
Agrostis karsensis IPNI=*386353-1* 233
Agrostis kitaibelii IPNI=*386353-1* 233
Agrostis kufuim IPNI=*1128639-2* 233
Agrostis langei IPNI=*386353-1* 233
Agrostis lapponica IPNI=*386353-1* 233
Agrostis laxa IPNI=*30197865-2* 233
Agrostis lenta IPNI=*398933-1* 237
Agrostis linearis IPNI=*397065-1* 236
Agrostis littoralis IPNI=*423010-1* 245
Agrostis macrantha IPNI=*386353-1* 233
Agrostis maritima IPNI=*30197865-2* 233
Agrostis maritima IPNI=*386353-1* 233
Agrostis maritima var. *pseudopungens* IPNI=*386353-1* 233
Agrostis marysae-tortiae IPNI=*30197865-2* 233
Agrostis matrella IPNI=*1214917-2* 247
Agrostis michauxii IPNI=*1128639-2* 233
Agrostis microcarpa IPNI=*409081-1* 241
Agrostis microsperma IPNI=*409081-1* 241
Agrostis minutiflora IPNI=*243044-2* 245
Agrostis monostachya IPNI=*405429-1* 240
Agrostis multicaulis IPNI=*386353-1* 233
Agrostis mutabilis IPNI=*386353-1* 233
Agrostis myuros IPNI=*225088-2* 244
Agrostis navarroi IPNI=*30197865-2* 233
Agrostis neglecta IPNI=*386353-1* 233
Agrostis nemoralis IPNI=*386353-1* 233
Agrostis nigrescens IPNI=*1156911-2* 246
Agrostis nigricans IPNI=*422090-1* 245
Agrostis novae-angliae IPNI=*1128639-2* 233
Agrostis noveboracensis IPNI=*1128639-2* 233
Agrostis obliqua IPNI=*386353-1* 233
Agrostis oreophila IPNI=*1128639-2* 233
Agrostis orientalis IPNI=*1207790-2* 245
Agrostis owarensis IPNI=*422898-1* 245
Agrostis palustris IPNI=*386353-1* 233
Agrostis palustris var. *stricta* IPNI=*30197865-2* 233
Agrostis panicea IPNI=*1048074-2* 244
Agrostis patula IPNI=*386353-1* 233
Agrostis perennans f. *atherophora* IPNI=*1128639-2* 233
Agrostis perennans f. *chaetophora* IPNI=*1128639-2* 233
Agrostis perennans var. *aestivalis* IPNI=*1128639-2* 233
Agrostis perennans var. *elata* IPNI=*1128639-2* 233
Agrostis perennans var. *humilis* IPNI=*1128639-2* 233
Agrostis plicata IPNI=*421711-1* 245
Agrostis polymorpha IPNI=*30197865-2* 233
Agrostis polymorpha var. *palustris* IPNI=*386353-1* 233
Agrostis polypogon IPNI=*408232-1* 241
Agrostis procera IPNI=*402044-1* 239
Agrostis procumbens IPNI=*386353-1* 233
Agrostis prostrata IPNI=*386353-1* 233
Agrostis pseudoalba IPNI=*386353-1* 233
Agrostis pseudointermedia IPNI=*1128639-2* 233
Agrostis pseudopungens IPNI=*386353-1* 233
Agrostis pubescens IPNI=*386353-1* 233
Agrostis pulchella IPNI=*1128639-2* 233
Agrostis pumila IPNI=*30197865-2* 233
Agrostis punctata IPNI=*1021229-2* 239
Agrostis pungens IPNI=*423010-1* 245
Agrostis pyramidalis IPNI=*243026-2* 245
Agrostis pyramidata IPNI=*243026-2* 245
Agrostis quitensis IPNI=*60445990-2* 241
Agrostis ramosa IPNI=*402044-1* 239
Agrostis reptans IPNI=*386353-1* 233
Agrostis retroflexa IPNI=*30197865-2* 233
Agrostis salina IPNI=*386353-1* 233
Agrostis scabra IPNI=*1128639-2* 233
Agrostis scabrida IPNI=*386353-1* 233
Agrostis scabriglumis IPNI=*386353-1* 233
Agrostis schiedeana IPNI=*1128639-2* 233
Agrostis schweinitzii IPNI=*1128639-2* 233
Agrostis scribneriana IPNI=*1128639-2* 233
Agrostis setosa IPNI=*409081-1* 241
Agrostis sibirica IPNI=*386353-1* 233
Agrostis sibthorpii IPNI=*386353-1* 233
Agrostis sicula IPNI=*386353-1* 233
Agrostis signata IPNI=*386353-1* 233
Agrostis simulans IPNI=*30197865-2* 233
Agrostis sinaica IPNI=*386353-1* 233
Agrostis sokolovskajae IPNI=*386353-1* 233
Agrostis spicata IPNI=*386353-1* 233
Agrostis spicigera IPNI=*386353-1* 233
Agrostis stellata IPNI=*397065-1* 236
Agrostis stolonifera f. *amethystina* IPNI=*386353-1* 233
Agrostis stolonifera f. *caespitosa* IPNI=*386353-1* 233
Agrostis stolonifera f. *decumbens* IPNI=*386353-1* 233
Agrostis stolonifera f. *flava* IPNI=*386353-1* 233
Agrostis stolonifera f. *gracilis* IPNI=*386353-1* 233
Agrostis stolonifera f. *humilis* IPNI=*386353-1* 233
Agrostis stolonifera f. *lutescens* IPNI=*386353-1* 233
Agrostis stolonifera f. *natans* IPNI=*386353-1* 233
Agrostis stolonifera f. *oligantha* IPNI=*386353-1* 233
Agrostis stolonifera f. *patula* IPNI=*386353-1* 233
Agrostis stolonifera f. *pontica* IPNI=*386353-1* 233
Agrostis stolonifera f. *subchlorantha* IPNI=*386353-1* 233
Agrostis stolonifera f. *substerilis* IPNI=*386353-1* 233
Agrostis stolonifera subsp. *albida* IPNI=*386353-1* 233
Agrostis stolonifera subsp. *filifolia* IPNI=*386353-1* 233
Agrostis stolonifera subsp. *gaditana* IPNI=*386353-1* 233
Agrostis stolonifera subsp. *karsensis* IPNI=*386353-1* 233
Agrostis stolonifera subsp. *maritima* IPNI=*386353-1* 233
Agrostis stolonifera subsp. *palustris* IPNI=*386353-1* 233
Agrostis stolonifera subsp. *scabrida* IPNI=*386353-1* 233
Agrostis stolonifera subsp. *sokolovskajae* IPNI=*386353-1* 233
Agrostis stolonifera subsp. *stolonizans* IPNI=*386353-1* 233
Agrostis stolonifera subsp. *straminea* IPNI=*386353-1* 233
Agrostis stolonifera subsp. *transcaspica* IPNI=*386353-1* 233
Agrostis stolonifera subvar. *malcuitiana* IPNI=*386353-1* 233
Agrostis stolonifera var. *calcicola* IPNI=*386353-1* 233
Agrostis stolonifera var. *densiflora* IPNI=*386353-1* 233
Agrostis stolonifera var. *filifolia* IPNI=*386353-1* 233
Agrostis stolonifera var. *hispida* IPNI=*30197865-2* 233
Agrostis stolonifera var. *marina* IPNI=*386353-1* 233
Agrostis stolonifera var. *minor* IPNI=*30197865-2* 233
Agrostis stolonifera var. *narbonensis* IPNI=*386353-1* 233
Agrostis stolonifera var. *nievrensis* IPNI=*386353-1* 233
Agrostis stolonifera var. *palustris* IPNI=*386353-1* 233
Agrostis stolonifera var. *pseudopungens* IPNI=*386353-1* 233
Agrostis stolonizans IPNI=*386353-1* 233
Agrostis straminea IPNI=*386353-1* 233
Agrostis stricta IPNI=*30197865-2* 233
Agrostis sylvatica IPNI=*30197865-2* 233

Agrostis tarda IPNI=*30197865-2* 233
Agrostis tenacissima IPNI=*1207790-2* 245
Agrostis tenuifolia var. *fretensis* IPNI=*1128639-2* 233
Agrostis tenuis f. *aristata* IPNI=*30197865-2* 233
Agrostis tenuis f. *contracta* IPNI=*30197865-2* 233
Agrostis tenuis f. *montana* IPNI=*30197865-2* 233
Agrostis tenuis f. *parviflora* IPNI=*30197865-2* 233
Agrostis tenuis f. *umbrosa* IPNI=*30197865-2* 233
Agrostis tenuis IPNI=*30197865-2* 233
Agrostis tenuis var. *aristata* IPNI=*30197865-2* 233
Agrostis tenuis var. *hornungiana* IPNI=*30197865-2* 233
Agrostis tenuis var. *repens* IPNI=*30197865-2* 233
Agrostis tenuis var. *schurii* IPNI=*30197865-2* 233
Agrostis tenuissima IPNI=*1048074-2* 244
Agrostis thyrsoides IPNI=*1103484-2* 245
Agrostis tillieri IPNI=*395086-1* 235
Agrostis transcaspica IPNI=*386353-1* 233
Agrostis tremula IPNI=*423010-1* 245
Agrostis triaristata IPNI=*1048074-2* 244
Agrostis trichotoma IPNI=*386353-1* 233
Agrostis trinervata IPNI=*30197865-2* 233
Agrostis triracemosa IPNI=*398933-1* 237
Agrostis turiensis IPNI=*386353-1* 233
Agrostis umbrosa IPNI=*30197865-2* 233
Agrostis valentina IPNI=*386353-1* 233
Agrostis varia IPNI=*386353-1* 233
Agrostis varians IPNI=*30197865-2* 233
Agrostis velutina IPNI=*80280-2* 237
Agrostis vernalis IPNI=*30197865-2* 233
Agrostis versicolor IPNI=*30197865-2* 233
Agrostis verticillata IPNI=*396213-1* 236
Agrostis violacea IPNI=*1128639-2* 233
Agrostis violacea IPNI=*30197865-2* 233
Agrostis virgata IPNI=*407138-1* 241
Agrostis virginica IPNI=*423010-1* 245
Agrostis virletii IPNI=*386353-1* 233
Agrostis vivipara IPNI=*386353-1* 233
Agrostis vulgaris IPNI=*30197865-2* 233
Agrostis vulgaris var. *alpigena* IPNI=*30197865-2* 233
Agrostis vulgaris var. *aristata* IPNI=*30197865-2* 233
Agrostis weberbaueri IPNI=*1128639-2* 233
Agrostis welwitschii IPNI=*386353-1* 233
Agrostis zerovii IPNI=*386353-1* 233
Agrostomia barbata IPNI=*396063-1* 236
Aguava guianensis IPNI=*598851-1* 426
Aguava tomentosa IPNI=*599215-1* 427
Ahouai nitida IPNI=*82440-1* 344
Aidema albiella IPNI=*270910-2* 214
Aigiros nigra IPNI=*776790-1* 404
Aiouea costaricensis IPNI=*171697-2* 184
Aiouea guyanensis IPNI=*13878-2* 182
Aiouea lundelliana IPNI=*171697-2* 184
Aiouea talamancensis IPNI=*171697-2* 184
Aiouea vexatrix IPNI=*171697-2* 184
Aiphanes aculeata IPNI=*7909-2* 204
Aiphanes caryotifolia IPNI=*7909-2* 204
Aiphanes chocoensis IPNI=*7916-2* 204
Aiphanes echinocarpa IPNI=*7914-2* 204
Aiphanes elegans IPNI=*7909-2* 204
Aiphanes ernestii IPNI=*7909-2* 204
Aiphanes orinocensis IPNI=*7909-2* 204
Aiphanes praemorsa IPNI=*7909-2* 204
Aiphanes schultzeana IPNI=*7926-2* 204
Aiphanes stergiosii IPNI=*664018-1* 204
Aira aggregata IPNI=*386633-1* 233
Aira aquatica IPNI=*395086-1* 235
Aira armoricana IPNI=*386633-1* 233
Aira baetica IPNI=*386633-1* 233
Aira bengalensis IPNI=*390837-1* 234
Aira brasiliensis IPNI=*21052-2* 234
Aira canescens IPNI=*386633-1* 233
Aira caryophyllea subsp. *armoricana* IPNI=*386633-1* 233
Aira caryophyllea subsp. *reverchonii* IPNI=*386633-1* 233
Aira caryophyllea var. *biaristata* IPNI=*386633-1* 233
Aira curta IPNI=*386633-1* 233
Aira duriuscula IPNI=*386633-1* 233
Aira edouardii IPNI=*386633-1* 233
Aira filiformis IPNI=*407138-1* 241
Aira holcus-lanata IPNI=*405034-1* 240
Aira holcus-mollis IPNI=*123530-2* 240
Aira hostii IPNI=*386633-1* 233
Aira koenigii IPNI=*401189-1* 239
Aira latigluma IPNI=*386633-1* 233
Aira leersii IPNI=*386633-1* 233
Aira lensii IPNI=*386633-1* 233
Aira lenzei IPNI=*386633-1* 233
Aira mollis IPNI=*123530-2* 240
Aira multicaulis IPNI=*386633-1* 233
Aira multiculmis IPNI=*386633-1* 233
Aira nebrodensis IPNI=*386633-1* 233
Aira patulipes IPNI=*386633-1* 233
Aira plesiantha IPNI=*386633-1* 233
Aira pulchella var. *biaristata* IPNI=*386633-1* 233
Aira pumila IPNI=*320035-2* 244
Aira purpureoargentea IPNI=*386633-1* 233
Aira reverchonii IPNI=*386633-1* 233
Aira rigida IPNI=*386633-1* 233
Aira sabuletorum IPNI=*423010-1* 245
Aira sabulonum IPNI=*423010-1* 245
Aira scabra IPNI=*395086-1* 235
Aira spicata IPNI=*258855-2* 246
Aira subspicata IPNI=*258855-2* 246
Aira subulorum IPNI=*423010-1* 245
Aira tenera IPNI=*386633-1* 233
Aira tenorei IPNI=*30197865-2* 233
Aira tenuissima IPNI=*243044-2* 245
Aira todaroi IPNI=*386633-1* 233
Aira triglumis IPNI=*1103484-2* 245
Aira truncata IPNI=*386633-1* 233
Aira uniaristata var. *biaristata* IPNI=*386633-1* 233
Aira viridula IPNI=*395086-1* 235
Airella caryophyllea IPNI=*386633-1* 233
Airopsis caryophyllea IPNI=*386633-1* 233
Airopsis jubata IPNI=*401189-1* 239
Aizoon canariense IPNI=*235145-2* 280
Aizoon montevidense IPNI=*235145-2* 280
Ajuga alpestris IPNI=*444576-1* 368
Ajuga alpicola IPNI=*444576-1* 368
Ajuga alpina IPNI=*444576-1* 368
Ajuga alpina IPNI=*444664-1* 368
Ajuga barrelieri IPNI=*444664-1* 368
Ajuga breviproles IPNI=*444664-1* 368
Ajuga densiflora IPNI=*444664-1* 368
Ajuga foliosa IPNI=*444576-1* 368
Ajuga glabrifolia IPNI=*444576-1* 368
Ajuga interrupta IPNI=*444576-1* 368
Ajuga lanata IPNI=*444576-1* 368
Ajuga latifolia IPNI=*444576-1* 368
Ajuga montana IPNI=*444576-1* 368
Ajuga pyramidalis IPNI=*444576-1* 368
Ajuga pyramidalis IPNI=*444664-1* 368
Ajuga repens IPNI=*444664-1* 368
Ajuga reptans var. *lerezana* IPNI=*444664-1* 368
Ajuga reptans var. *pyramidata* IPNI=*444664-1* 368
Ajuga rugosa IPNI=*444576-1* 368
Ajuga stolonifera IPNI=*444664-1* 368

Ajuga vulgaris IPNI=*444664-1* 368
Aklema cotinifolia IPNI=*346143-1* 389
Alaticaulia schlimii IPNI=*643106-1* 220
Alaticaulia trochilus IPNI=*643178-1* 220
Albersia arenaria IPNI=*10619-2* 281
Albersia ascendens IPNI=*10619-2* 281
Albersia blitum IPNI=*10619-2* 281
Albersia caudata IPNI=*316349-2* 282
Albersia gracilis IPNI=*316349-2* 282
Albersia livida IPNI=*10619-2* 281
Albersia polygama IPNI=*327386-2* 282
Albersia polystachya IPNI=*316349-2* 282
Albertokuntzea americana IPNI=*676493-1* 287
Albertokuntzea floribunda IPNI=*676493-1* 287
Albertokuntzea foliosa IPNI=*676493-1* 287
Albertokuntzea longifolia IPNI=*676493-1* 287
Albertokuntzea macrophylla IPNI=*676510-1* 287
Albertokuntzea vauthieri IPNI=*676493-1* 287
Albizia corymbosa IPNI=*989341-1* 320
Albizia distachya IPNI=*912944-1* 329
Albizia dulcis IPNI=*1003508-2* 330
Albizia flavovirens IPNI=*227126-2* 331
Albizia guachapele IPNI=*211561-2* 331
Albizia latifolia IPNI=*99109-3* 311
Albizia longepedata IPNI=*211561-2* 331
Albizia longipes IPNI=*1037928-2* 319
Albizia lophantha IPNI=*912944-1* 329
Albizia malacocarpa IPNI=*8016-2* 311
Albizia pedicellaris IPNI=*989349-1* 311
Albizia saman IPNI=*227126-2* 331
Albizia speciosa IPNI=*99109-3* 311
Albizia sumatrana IPNI=*8016-2* 311
Alcanna spinosa IPNI=*553638-1* 417
Alcea biennis subsp. *cretica* IPNI=*558713-1* 407
Alcea cretica IPNI=*558713-1* 407
Alcea microchiton IPNI=*558713-1* 407
Alchemilla achilleifolia IPNI=*720867-1* 443
Alchemilla alata IPNI=*720867-1* 443
Alchemilla aphanoides var. *subalpestris* IPNI=*720378-1* 442
Alchemilla aphanoides var. *tripartita* IPNI=*720378-1* 442
Alchemilla calchaquina IPNI=*720867-1* 443
Alchemilla ciliata IPNI=*720816-1* 442
Alchemilla galioides var. *sericea* IPNI=*720586-1* 442
Alchemilla glandulosa IPNI=*720378-1* 442
Alchemilla hirsuta IPNI=*720378-1* 442
Alchemilla hirsuta var. *alpestris* IPNI=*720378-1* 442
Alchemilla hirsuta var. *calvescens* IPNI=*720378-1* 442
Alchemilla hultenii IPNI=*720378-1* 442
Alchemilla imbricata IPNI=*720798-1* 442
Alchemilla mandoniana f. *pilosa* IPNI=*8117-2* 442
Alchemilla orbiculata f. *villifera* IPNI=*720816-1* 442
Alchemilla orbiculata var. *villifera* IPNI=*720816-1* 442
Alchemilla paludicola f. *pilosa* IPNI=*8117-2* 442
Alchemilla pectinata var. *mexicana* IPNI=*720846-1* 442
Alchemilla pinnata f. *argentea* IPNI=*720867-1* 443
Alchemilla pinnata f. *robusta* IPNI=*720867-1* 443
Alchemilla pinnata f. *rosulata* IPNI=*720867-1* 443
Alchemilla pinnata var. *minima* IPNI=*720867-1* 443
Alchemilla procumbens f. *deformata* IPNI=*8135-2* 443
Alchemilla procumbens var. *hirta* IPNI=*8090-2* 442
Alchemilla radicans IPNI=*720586-1* 442
Alchemilla rotundifolia IPNI=*720816-1* 442
Alchemilla subalpestris IPNI=*720378-1* 442
Alchemilla tripartita IPNI=*720378-1* 442
Alchemilla venusta var. *fulvescens* IPNI=*8079-2* 442
Alchornea brachygyne IPNI=*338389-1* 387
Alchornea brevistyla IPNI=*30014592-2* 387
Alchornea castaneifolia var. *genuina* IPNI=*338372-1* 387
Alchornea castaneifolia var. *puberula* IPNI=*338372-1* 387
Alchornea cordata IPNI=*339164-1* 388
Alchornea costaricensis f. *longispicata* IPNI=*8183-2* 387
Alchornea cyclophylla IPNI=*338417-1* 387
Alchornea gardneri IPNI=*338389-1* 387
Alchornea glandulosa IPNI=*338417-1* 387
Alchornea glandulosa var. *floribunda* IPNI=*338476-1* 387
Alchornea glandulosa var. *genuina* IPNI=*30043708-2* 387
Alchornea glandulosa var. *parvifolia* IPNI=*30014592-2* 387
Alchornea haitiensis IPNI=*338417-1* 387
Alchornea intermedia IPNI=*30014592-2* 387
Alchornea janeirensis IPNI=*30014592-2* 387
Alchornea latifolia IPNI=*339164-1* 388
Alchornea latifolia var. *islaensis* IPNI=*338417-1* 387
Alchornea macrophylla IPNI=*339164-1* 388
Alchornea martiana IPNI=*341924-1* 388
Alchornea nemoralis IPNI=*30014592-2* 387
Alchornea nemoralis var. *lanceolata* IPNI=*30014592-2* 387
Alchornea obovata IPNI=*30014592-2* 387
Alchornea orinocensis IPNI=*339164-1* 388
Alchornea parvifolia IPNI=*30014592-2* 387
Alchornea passargei IPNI=*338372-1* 387
Alchornea platyphylla IPNI=*338417-1* 387
Alchornea polyantha IPNI=*338417-1* 387
Alchornea psilorhachis IPNI=*30014592-2* 387
Alchornea rotundifolia IPNI=*30014592-2* 387
Alchornea schomburgkii IPNI=*338389-1* 387
Alchornea similis IPNI=*338417-1* 387
Alchornea triplinervia f. *intermedia* IPNI=*30014592-2* 387
Alchornea triplinervia f. *psilorhachis* IPNI=*30014592-2* 387
Alchornea triplinervia var. *boliviana* IPNI=*30014592-2* 387
Alchornea triplinervia var. *crassifolia* IPNI=*30014592-2* 387
Alchornea triplinervia var. *genuina* IPNI=*30014592-2* 387
Alchornea triplinervia var. *iricuranoides* IPNI=*30014592-2* 387
Alchornea triplinervia var. *janeirensis* IPNI=*30014592-2* 387
Alchornea triplinervia var. *laevigata* IPNI=*30014592-2* 387
Alchornea triplinervia var. *lanceolata* IPNI=*30014592-2* 387
Alchornea triplinervia var. *meridensis* IPNI=*8192-2* 387
Alchornea triplinervia var. *nemoralis* IPNI=*30014592-2* 387
Alchornea triplinervia var. *parvifolia* IPNI=*30014592-2* 387
Alchornea triplinervia var. *tomentella* IPNI=*30014592-2* 387
Alchornea triplinervia var. *trinitatis* IPNI=*30014592-2* 387
Alchorneopsis floribunda var. *genuina* IPNI=*338476-1* 387
Alchorneopsis floribunda var. *sessiliflora* IPNI=*338476-1* 387
Alchorneopsis trimera IPNI=*338476-1* 387
Alcicornium angolense IPNI=*17177470-1* 176
Alcina minor IPNI=*156179-2* 265
Alcina ovatifolia IPNI=*156179-2* 265
Aldina amazonica IPNI=*473440-1* 311
Aldina latifolia var. *auyantepuiensis* IPNI=*473440-1* 311
Aldina latifolia var. *pubescens* IPNI=*473440-1* 311
Alectoroctonum cotinifolium IPNI=*346143-1* 389
Alegria candida IPNI=*834727-1* 410
Aletris chinensis IPNI=*533580-1* 215
Aletris fragrans IPNI=*534207-1* 215
Aletris guineensis IPNI=*77164234-1* 215
Aletris hyacinthoides IPNI=*77164234-1* 215
Aletris hyacinthoides var. *guineensis* IPNI=*77164234-1* 215
Aletris uvaria IPNI=*537112-1* 216
Aleurites ambinux IPNI=*338514-1* 387
Aleurites angustifolius IPNI=*338514-1* 387
Aleurites commutatus IPNI=*338514-1* 387
Aleurites cordifolius IPNI=*338514-1* 387
Aleurites fordii IPNI=*358198-1* 393
Aleurites integrifolius IPNI=*338514-1* 387
Aleurites lanceolatus IPNI=*338514-1* 387
Aleurites lobatus IPNI=*338514-1* 387
Aleurites moluccanus var. *floccosus* IPNI=*338514-1* 387

Aleurites moluccanus var. *katoi* IPNI=*338514-1* 387
Aleurites remyi IPNI=*338514-1* 387
Aleurites trilobus IPNI=*338514-1* 387
Aleuritopteris chihuahuaensis IPNI=*77182808-1* 178
Aleuritopteris farinosa IPNI=*77182808-1* 178
Aleuritopteris mexicana IPNI=*77182808-1* 178
Aleuritopteris peruviana IPNI=*77182808-1* 178
Alfonsia oleifera IPNI=*88982-2* 208
Algarobia juliflora IPNI=*1002432-2* 331
Alguelaguen jamesonii IPNI=*137859-2* 370
Alguelaguen lindenianum IPNI=*137860-2* 370
Alguelaguen sprucei IPNI=*137859-2* 370
Alguelagum alpinum IPNI=*449287-1* 370
Alguelagum bullatum IPNI=*137860-2* 370
Alguelagum confertum IPNI=*137869-2* 370
Alguelagum parviflorum IPNI=*137860-2* 370
Alguelagum salviifolium IPNI=*137894-2* 370
Alibertia granulosa IPNI=*8399-2* 347
Alibertia hadrantha IPNI=*77089908-1* 349
Alibertia isernii IPNI=*1095647-2* 346
Alibertia latifolia var. *pargueniana* IPNI=*8399-2* 347
Alibertia latifolia var. *parvifolia* IPNI=*8399-2* 347
Alibertia myrciifolia IPNI=*60436803-2* 349
Alibertia obidensis IPNI=*35222-2* 347
Alibertia oligantha IPNI=*8399-2* 347
Alicastrum guianense IPNI=*582693-1* 437
Alicastrum rubescens IPNI=*37310-2* 437
Alifana canescens IPNI=*566862-1* 418
Alifana striata IPNI=*566891-1* 418
Alisma bolivianum IPNI=*77089409-1* 197
Alisma cordifolium IPNI=*77182733-1* 197
Alisma flavum IPNI=*58404-1* 198
Alisma floribundum IPNI=*77183476-1* 197
Alisma pubescens var. *claussenii* IPNI=*77183476-1* 197
Alisma tenellum f. *latifolium* IPNI=*77089409-1* 197
Alisma tenellum IPNI=*118649-2* 197
Aliteria sagotii IPNI=*59683-2* 437
Alitubus millefolium IPNI=*2294-2* 255
Alkanna spinosa IPNI=*553638-1* 417
Allamanda aubletii IPNI=*319065-2* 339
Allamanda cathartica var. *grandiflora* IPNI=*319065-2* 339
Allamanda cathartica var. *hendersonii* IPNI=*319065-2* 339
Allamanda chelsonii IPNI=*319065-2* 339
Allamanda grandiflora IPNI=*319065-2* 339
Allamanda latifolia IPNI=*319065-2* 339
Allamanda linnaei IPNI=*319065-2* 339
Allamanda violacea IPNI=*76468-1* 339
Allamanda wardleyana IPNI=*319065-2* 339
Allantoma aulacocarpa IPNI=*590593-1* 300
Allantoma burchelliana IPNI=*590593-1* 300
Allantoma caudata IPNI=*590593-1* 300
Allantoma corbula IPNI=*590593-1* 300
Allantoma dictyocarpa IPNI=*590593-1* 300
Allantoma dilatata IPNI=*590593-1* 300
Allantoma impressa IPNI=*590593-1* 300
Allantoma macrocarpa IPNI=*590593-1* 300
Allantoma scutellata IPNI=*590593-1* 300
Allantoma torulosa IPNI=*590593-1* 300
Allardtia potockii IPNI=*254406-2* 229
Allionia craterimorpha IPNI=*8606-2* 286
Allionia cristata IPNI=*8606-2* 286
Allionia incarnata f. *cristata* IPNI=*8606-2* 286
Allionia incarnata f. *cylindrica* IPNI=*8606-2* 286
Allionia incarnata f. *glomerata* IPNI=*8606-2* 286
Allionia incarnata f. *grandiflora* IPNI=*8606-2* 286
Allionia incarnata f. *obtusilobata* IPNI=*8606-2* 286
Allionia incarnata f. *villosiceps* IPNI=*8606-2* 286
Allionia jarae IPNI=*8606-2* 286
Allionia mendocina IPNI=*8606-2* 286
Allionia puberula IPNI=*8606-2* 286
Allium acutum IPNI=*528823-1* 213
Allium adscendens IPNI=*332079-2* 213
Allium albescens IPNI=*332079-2* 213
Allium alpinum IPNI=*528823-1* 213
Allium ampeloprasum subsp. *polyanthum* IPNI=*332079-2* 213
Allium angolense IPNI=*527795-1* 213
Allium aobanum IPNI=*527795-1* 213
Allium arenarium IPNI=*528796-1* 213
Allium ascalonicum var. *condensum* IPNI=*527795-1* 213
Allium ascalonicum var. *fertile* IPNI=*527795-1* 213
Allium ascalonicum var. *sterile* IPNI=*527795-1* 213
Allium ascendens IPNI=*332079-2* 213
Allium babingtonii IPNI=*332079-2* 213
Allium bertolonii IPNI=*332079-2* 213
Allium bouddae IPNI=*528046-1* 213
Allium broteri IPNI=*528823-1* 213
Allium buhseanum IPNI=*528823-1* 213
Allium byzantinum IPNI=*332079-2* 213
Allium carneum IPNI=*528823-1* 213
Allium cepaeum IPNI=*527795-1* 213
Allium coloratum IPNI=*528823-1* 213
Allium commune IPNI=*527795-1* 213
Allium controversum IPNI=*528796-1* 213
Allium cumaria IPNI=*527795-1* 213
Allium duriaeanum IPNI=*332079-2* 213
Allium durieuanum IPNI=*332079-2* 213
Allium esculentum IPNI=*527795-1* 213
Allium foliosum IPNI=*528823-1* 213
Allium gasparrinii IPNI=*332079-2* 213
Allium glaucum IPNI=*528823-1* 213
Allium gracile IPNI=*130726-3* 214
Allium halleri IPNI=*332079-2* 213
Allium holmense IPNI=*332079-2* 213
Allium idzuense IPNI=*528823-1* 213
Allium kashgaricum IPNI=*528046-1* 213
Allium kurrat IPNI=*332079-2* 213
Allium laetum IPNI=*332079-2* 213
Allium leucanthum IPNI=*332079-2* 213
Allium leucanthum var. *tridentatum* IPNI=*332079-2* 213
Allium longicuspis IPNI=*528796-1* 213
Allium lusitanicum IPNI=*528823-1* 213
Allium luteum IPNI=*528506-1* 213
Allium montanum IPNI=*528823-1* 213
Allium multiflorum IPNI=*332079-2* 213
Allium napus IPNI=*527795-1* 213
Allium nigritanum IPNI=*527795-1* 213
Allium palustre IPNI=*528823-1* 213
Allium pauciflorum IPNI=*527795-1* 213
Allium pekinense IPNI=*528796-1* 213
Allium polyanthum IPNI=*332079-2* 213
Allium porraceum IPNI=*332079-2* 213
Allium porrum IPNI=*332079-2* 213
Allium porrum IPNI=*528506-1* 213
Allium porrum subsp. *polyanthum* IPNI=*332079-2* 213
Allium porrum var. *kurrat* IPNI=*332079-2* 213
Allium punctulatum IPNI=*528823-1* 213
Allium purpurascens IPNI=*528823-1* 213
Allium pylium IPNI=*332079-2* 213
Allium raddeanum IPNI=*528823-1* 213
Allium ramosum IPNI=*528506-1* 213
Allium reflexum IPNI=*528823-1* 213
Allium riparium IPNI=*528823-1* 213
Allium roseum IPNI=*528823-1* 213
Allium salota IPNI=*527795-1* 213
Allium sativum f. *sagittatum* IPNI=*528796-1* 213
Allium sativum f. *vulgare* IPNI=*528796-1* 213

Allium sativum subsp. *asiae-mediae* IPNI=*528796-1* 213
Allium schmitzii var. *duriminium* IPNI=*528823-1* 213
Allium schoenoprasum f. *albiflorum* IPNI=*528823-1* 213
Allium schoenoprasum f. *purpuratum* IPNI=*528823-1* 213
Allium schoenoprasum subsp. *gredense* IPNI=*528823-1* 213
Allium schoenoprasum subsp. *latiorifolium* IPNI=*528823-1* 213
Allium schoenoprasum subsp. *orosiae* IPNI=*528823-1* 213
Allium schoenoprasum var. *idzuense* IPNI=*528823-1* 213
Allium schoenoprasum var. *laurentianum* IPNI=*528823-1* 213
Allium schoenoprasum var. *litorale* IPNI=*528823-1* 213
Allium scopulicola IPNI=*332079-2* 213
Allium scorodoprasum var. *multibulbillosum* IPNI=*528796-1* 213
Allium sibiricum IPNI=*528823-1* 213
Allium spectabile IPNI=*332079-2* 213
Allium syriacum IPNI=*332079-2* 213
Allium tenuifolium IPNI=*528823-1* 213
Allium thessalum IPNI=*332079-2* 213
Allium ubinicum IPNI=*528823-1* 213
Allium udinicum IPNI=*528823-1* 213
Allocalyx microphylla IPNI=*1072674-2* 376
Allocarpus caracasanus IPNI=*9082-2* 256
Allocarpus sillaensis IPNI=*9082-2* 256
Allohemia pedunculata IPNI=*77147531-1* 448
Allohemia purpurea IPNI=*550656-1* 448
Alloispermum divaricatum IPNI=*9082-2* 256
Alloispermum liebmannii IPNI=*187778-1* 259
Alloplectus affinis IPNI=*378985-1* 367
Alloplectus aureonitens IPNI=*378985-1* 367
Alloplectus calotrichus IPNI=*63523-2* 366
Alloplectus circinnatus IPNI=*84424-2* 367
Alloplectus coccineus IPNI=*84424-2* 367
Alloplectus coccineus var. *fuscomaculatus* IPNI=*84424-2* 367
Alloplectus coriaceus IPNI=*84425-2* 367
Alloplectus cubensis IPNI=*378985-1* 367
Alloplectus dimidiatus IPNI=*63559-2* 366
Alloplectus discolor IPNI=*167431-2* 368
Alloplectus glaber IPNI=*380742-1* 367
Alloplectus ichthyoderma f. *rubescens* IPNI=*60441766-2* 367
Alloplectus ichthyoderma IPNI=*60441766-2* 367
Alloplectus ichthyoderma var. *pallidus* IPNI=*60441766-2* 367
Alloplectus lynchii IPNI=*167431-2* 368
Alloplectus meridensis IPNI=*60441766-2* 367
Alloplectus panamensis IPNI=*60441770-2* 367
Alloplectus patrisii IPNI=*84424-2* 367
Alloplectus ruacophilus IPNI=*60441766-2* 367
Alloplectus varelanus IPNI=*60441766-2* 367
Alloplectus ventricosus IPNI=*84425-2* 367
Alloplectus viridis IPNI=*378968-1* 367
Allosorus acutifolius IPNI=*17210060-1* 175
Allosorus aquilinus IPNI=*17210060-1* 175
Allosorus arachnoideus IPNI=*213798-2* 175
Allosorus caespitosus IPNI=*77182463-1* 178
Allosorus capensis IPNI=*17210060-1* 175
Allosorus caudatus IPNI=*213799-2* 175
Allosorus ciliatus IPNI=*77182463-1* 178
Allosorus ciliatus var. *laxus* IPNI=*77182463-1* 178
Allosorus ciliatus var. *minor* IPNI=*77182463-1* 178
Allosorus concolor IPNI=*77182373-1* 178
Allosorus farinosus IPNI=*77182808-1* 178
Allosorus flexuosus IPNI=*77183106-1* 178
Allosorus marginatus IPNI=*77182463-1* 178
Allosorus marginatus var. *brevilobus* IPNI=*77182463-1* 178
Allosorus marginatus var. *normalis* IPNI=*77182463-1* 178
Allosorus microphyllus IPNI=*77183177-1* 178
Allosorus microphyllus var. *elongata* IPNI=*77183177-1* 178
Allosorus myriophyllus IPNI=*77183191-1* 178
Allosorus myriophyllus var. *elegans* IPNI=*77183191-1* 178
Allosorus psittacinus IPNI=*213798-2* 175
Allosorus subverticillatus IPNI=*77182326-1* 178
Allosorus tauricus IPNI=*17210060-1* 175
Allosorus ternifolius IPNI=*77182326-1* 178
Allosorus villosus IPNI=*17210060-1* 175
Allosperma tuberosa IPNI=*302798-2* 224
Allostelites arvensis IPNI=*300073-2* 173
Allouya americana IPNI=*60458762-2* 345
Alnus jorullensis var. *acuminata* IPNI=*294882-1* 338
Alocasia cordifolia IPNI=*60444565-2* 198
Alocasia dussii IPNI=*1170772-2* 199
Alocasia grandis IPNI=*60444565-2* 198
Alocasia illustris IPNI=*1170772-2* 199
Alocasia marginata IPNI=*60444565-2* 198
Alocasia pallida IPNI=*60444565-2* 198
Alocasia plumbea IPNI=*60444565-2* 198
Alocasia rapiformis IPNI=*60444565-2* 198
Alocasia rugosa IPNI=*84151-1* 198
Alocasia talihan IPNI=*89373-1* 203
Alocasia undipes IPNI=*89384-1* 203
Aloe americana IPNI=*319063-2* 214
Aloe barbadensis IPNI=*530017-1* 215
Aloe chinensis IPNI=*530017-1* 215
Aloe elongata IPNI=*530017-1* 215
Aloe flava IPNI=*530017-1* 215
Aloe foetida IPNI=*64429-1* 215
Aloe fragrantissima IPNI=*534207-1* 215
Aloe guineensis IPNI=*77164234-1* 215
Aloe hyacinthoides IPNI=*77164234-1* 215
Aloe hyacinthoides var. *guineensis* IPNI=*77164234-1* 215
Aloe indica IPNI=*530017-1* 215
Aloe lanzae IPNI=*530017-1* 215
Aloe longifolia IPNI=*537112-1* 216
Aloe maculata IPNI=*530017-1* 215
Aloe perfoliata var. *arborescens* IPNI=*529214-1* 215
Aloe perfoliata var. *barbadensis* IPNI=*530017-1* 215
Aloe perfoliata var. *vera* IPNI=*530017-1* 215
Aloe rigida IPNI=*537112-1* 216
Aloe rubescens IPNI=*530017-1* 215
Aloe uvaria IPNI=*537112-1* 216
Aloe variegata IPNI=*530017-1* 215
Aloe vera var. *littoralis* IPNI=*530017-1* 215
Aloe vulgaris IPNI=*530017-1* 215
Alomia coelestina IPNI=*7086-2* 256
Alomia pinetorum IPNI=*175500-1* 256
Alonsoa albiflora IPNI=*9536-2* 377
Alonsoa hirsuta IPNI=*1072674-2* 376
Alonsoa incisifolia IPNI=*9536-2* 377
Alonsoa meridionalis var. *carnea* IPNI=*9536-2* 377
Alonsoa meridionalis var. *cinnabarina* IPNI=*9536-2* 377
Alonsoa meridionalis var. *crocea* IPNI=*9536-2* 377
Alonsoa meridionalis var. *flava* IPNI=*9536-2* 377
Alonsoa meridionalis var. *lactea* IPNI=*9536-2* 377
Alonsoa meridionalis var. *micans* IPNI=*9536-2* 377
Alonsoa meridionalis var. *rubra* IPNI=*9536-2* 377
Alonsoa mutisii IPNI=*9536-2* 377
Alonsoa myrtifolia IPNI=*9536-2* 377
Alonsoa parviflora IPNI=*9536-2* 377
Alonsoa peruviana IPNI=*9536-2* 377
Alonsoa procumbens IPNI=*9536-2* 377
Alonsoa quadrifolia IPNI=*9536-2* 377
Alonsoa urticifolia IPNI=*9536-2* 377
Alonsoa warscewiczii f. *coccinea* IPNI=*9536-2* 377
Alonsoa warscewiczii IPNI=*9536-2* 377
Alopecurus aequalis f. *fluitans* IPNI=*387005-1* 233
Alopecurus aequalis f. *foliosus* IPNI=*387005-1* 233
Alopecurus aequalis f. *nigricans* IPNI=*387005-1* 233
Alopecurus aequalis f. *violaceus* IPNI=*387005-1* 233
Alopecurus aequalis var. *aristulatus* IPNI=*387005-1* 233

Alopecurus aequalis var. *fluitans* IPNI=*387005-1* 233
Alopecurus aequalis var. *natans* IPNI=*387005-1* 233
Alopecurus aequalis var. *violaceus* IPNI=*387005-1* 233
Alopecurus amurensis IPNI=*387005-1* 233
Alopecurus aristatus IPNI=*1048074-2* 244
Alopecurus aristatus var. *monspeliensis* IPNI=*1048074-2* 244
Alopecurus aristulatus IPNI=*387005-1* 233
Alopecurus aristulatus var. *merriamii* IPNI=*387005-1* 233
Alopecurus aristulatus var. *natans* IPNI=*387005-1* 233
Alopecurus brachytrichus IPNI=*387005-1* 233
Alopecurus caesius IPNI=*387005-1* 233
Alopecurus caespitosus IPNI=*387005-1* 233
Alopecurus carolinianus IPNI=*1128639-2* 233
Alopecurus caudatus IPNI=*1207294-2* 245
Alopecurus diandrus IPNI=*387005-1* 233
Alopecurus fulvus f. *violaceus* IPNI=*387005-1* 233
Alopecurus fulvus IPNI=*387005-1* 233
Alopecurus geniculatus var. *aristulatus* IPNI=*387005-1* 233
Alopecurus geniculatus var. *fulvus* IPNI=*387005-1* 233
Alopecurus geniculatus var. *natans* IPNI=*387005-1* 233
Alopecurus geniculatus var. *robustus* IPNI=*387005-1* 233
Alopecurus howellii var. *merrimanii* IPNI=*387005-1* 233
Alopecurus konradii IPNI=*387005-1* 233
Alopecurus monspeliensis IPNI=*1048074-2* 244
Alopecurus otiporensis IPNI=*415632-1* 243
Alopecurus paludosus IPNI=*387005-1* 233
Alopecurus palustris subsp. *fulvus* IPNI=*387005-1* 233
Alopecurus paniceus IPNI=*1048074-2* 244
Alopecurus rubicundus IPNI=*943811-1* 245
Alopecurus typhoides IPNI=*77105978-1* 235
Alophia pulchella IPNI=*278134-2* 216
Aloysia sleumeri IPNI=*1015994-1* 377
Aloysia triphylla f. *serrulata* IPNI=*1015994-1* 377
Aloysia triphylla IPNI=*1015994-1* 377
Alpinia acrostachya IPNI=*942355-1* 251
Alpinia arabica IPNI=*796267-1* 247
Alpinia aromatica IPNI=*798014-1* 251
Alpinia bicalyculata IPNI=*218583-2* 251
Alpinia bisulcata IPNI=*798014-1* 251
Alpinia breviscapa IPNI=*798020-1* 251
Alpinia cardamomum IPNI=*796556-1* 250
Alpinia comosa IPNI=*796280-1* 247
Alpinia cristata IPNI=*872083-1* 250
Alpinia diracodes IPNI=*942355-1* 251
Alpinia elatior IPNI=*942355-1* 251
Alpinia exaltata IPNI=*218583-2* 251
Alpinia fimbriata IPNI=*872083-1* 250
Alpinia fluvitialis IPNI=*872083-1* 250
Alpinia grandis IPNI=*795383-1* 250
Alpinia javanica IPNI=*942355-1* 251
Alpinia macrantha IPNI=*218583-2* 251
Alpinia magnifica IPNI=*942355-1* 251
Alpinia multicaulis IPNI=*798014-1* 251
Alpinia occidentalis IPNI=*798014-1* 251
Alpinia penicillata IPNI=*872083-1* 250
Alpinia presliana IPNI=*218597-2* 251
Alpinia renealmia IPNI=*218583-2* 251
Alpinia schumanniana IPNI=*872083-1* 250
Alpinia speciosa IPNI=*942355-1* 251
Alpinia spicata IPNI=*218597-2* 251
Alpinia spicata IPNI=*329964-2* 248
Alpinia spiralis IPNI=*796387-1* 248
Alpinia strobilifera IPNI=*218597-2* 251
Alpinia thyrsoidea IPNI=*798093-1* 251
Alseodaphne amara IPNI=*77198011-1* 183
Alseodaphne cymbarum IPNI=*77198011-1* 183
Alseodaphne opifera IPNI=*171526-2* 184
Alsine alpina IPNI=*158983-1* 286
Alsine apetala IPNI=*159330-1* 286
Alsine arvensis IPNI=*158725-1* 286
Alsine arvensis IPNI=*329786-2* 285
Alsine avicularum IPNI=*159330-1* 286
Alsine barbata IPNI=*159330-1* 286
Alsine bipartita IPNI=*159330-1* 286
Alsine brachypetala IPNI=*159330-1* 286
Alsine cuspidata IPNI=*244185-2* 286
Alsine elongata IPNI=*159330-1* 286
Alsine glomerata IPNI=*152274-1* 285
Alsine gussonii IPNI=*159330-1* 286
Alsine latifolia IPNI=*159330-1* 286
Alsine marginata IPNI=*241336-2* 286
Alsine marina IPNI=*241336-2* 286
Alsine media IPNI=*159330-1* 286
Alsine media IPNI=*30172747-2* 285
Alsine media var. *transiens* IPNI=*159330-1* 286
Alsine procumbens IPNI=*302082-2* 285
Alsine radicans IPNI=*158983-1* 286
Alsine repens IPNI=*159330-1* 286
Alsine rotundifolia IPNI=*30172747-2* 285
Alsine rubra IPNI=*158983-1* 286
Alsine salina IPNI=*241336-2* 286
Alsine sparsiflora IPNI=*241336-2* 286
Alsine tenuis IPNI=*241336-2* 286
Alsine trivialis IPNI=*152274-1* 285
Alsine vulgaris IPNI=*159330-1* 286
Alsinella media IPNI=*241336-2* 286
Alsinella procumbens IPNI=*302082-2* 285
Alsodeia flavescens IPNI=*221409-2* 405
Alsodeia floribunda IPNI=*867224-1* 405
Alsodeia guianensis IPNI=*867224-1* 405
Alsodeia japurana IPNI=*221443-2* 405
Alsodeia piparea IPNI=*779914-1* 404
Alsodeia racemosa IPNI=*221443-2* 405
Alsophila aculeata IPNI=*73288-2* 172
Alsophila affinis IPNI=*143624-2* 173
Alsophila aterrima IPNI=*73117-2* 172
Alsophila bicrenata IPNI=*73125-2* 172
Alsophila bilineata IPNI=*143624-2* 173
Alsophila caracasana IPNI=*73146-2* 172
Alsophila caracasana var. *petiolularis* IPNI=*73146-2* 172
Alsophila costalis IPNI=*9976-2* 172
Alsophila deckeriana IPNI=*143624-2* 173
Alsophila delicatula IPNI=*17072730-1* 172
Alsophila ferox IPNI=*73288-2* 172
Alsophila fulva IPNI=*17072730-1* 172
Alsophila furcata IPNI=*9976-2* 172
Alsophila glauca IPNI=*143624-2* 173
Alsophila ichthyolepis IPNI=*73125-2* 172
Alsophila ichtyolepis IPNI=*73125-2* 172
Alsophila killipii IPNI=*73267-2* 172
Alsophila lasiosora IPNI=*73267-2* 172
Alsophila mapiriensis IPNI=*73267-2* 172
Alsophila microdonta IPNI=*73288-2* 172
Alsophila millefolium IPNI=*143624-2* 173
Alsophila monticola IPNI=*143624-2* 173
Alsophila nigra IPNI=*73267-2* 172
Alsophila ochroleuca IPNI=*73125-2* 172
Alsophila pansamalana IPNI=*73125-2* 172
Alsophila pruinata IPNI=*143624-2* 173
Alsophila pruinata var. *concolor* IPNI=*143624-2* 173
Alsophila pruinata var. *conspicua* IPNI=*143624-2* 173
Alsophila pruinata var. *tenuis* IPNI=*143624-2* 173
Alsophila quadripinnata IPNI=*143624-2* 173
Alsophila scabriuscula IPNI=*73125-2* 172
Alsophila scabriuscula var. *guatemalensis* IPNI=*73125-2* 172
Alsophila schaffneriana IPNI=*143624-2* 173

Alsophila scopulina IPNI=*73117-2* 172
Alsophila stipularis IPNI=*73125-2* 172
Alsophila tarapotensis IPNI=*73267-2* 172
Alsophila tenerifrons IPNI=*9976-2* 172
Alsophila trichiata IPNI=*73125-2* 172
Alsophila tryonorum IPNI=*275500-2* 172
Alstroemeria affinis IPNI=*63283-1* 226
Alstroemeria araucana IPNI=*62457-1* 226
Alstroemeria aurantiaca IPNI=*62457-1* 226
Alstroemeria caldasiana IPNI=*63359-1* 226
Alstroemeria caldasii IPNI=*63359-1* 226
Alstroemeria chiloensis IPNI=*62457-1* 226
Alstroemeria concolor IPNI=*62457-1* 226
Alstroemeria edulis IPNI=*63283-1* 226
Alstroemeria floribunda IPNI=*63359-1* 226
Alstroemeria gloriosa IPNI=*63283-1* 226
Alstroemeria grandifolia IPNI=*63283-1* 226
Alstroemeria hirsuta IPNI=*63318-1* 226
Alstroemeria hirtella IPNI=*63283-1* 226
Alstroemeria jacquesiana IPNI=*63283-1* 226
Alstroemeria miniata IPNI=*63283-1* 226
Alstroemeria multiflora IPNI=*63359-1* 226
Alstroemeria nivali IPNI=*62457-1* 226
Alstroemeria pauciflora IPNI=*63283-1* 226
Alstroemeria peruviana IPNI=*62457-1* 226
Alstroemeria pulchella IPNI=*62457-1* 226
Alstroemeria purpurea IPNI=*63359-1* 226
Alstroemeria salsilla IPNI=*63283-1* 226
Alstroemeria salsilloides IPNI=*63283-1* 226
Alstroemeria sepium IPNI=*63283-1* 226
Alstroemeria setacea IPNI=*63411-1* 226
Alstroemeria xanthina IPNI=*62457-1* 226
Altensteinia fragosa IPNI=*302812-2* 217
Alternanthera achyrantha IPNI=*327342-2* 281
Alternanthera achyrantha var. *dasyantha* IPNI=*327344-2* 281
Alternanthera achyrantha var. *echinata* IPNI=*327342-2* 281
Alternanthera achyrantha var. *leiantha* IPNI=*327342-2* 281
Alternanthera achyrantha var. *parvifolia* IPNI=*327344-2* 281
Alternanthera achyranthoides IPNI=*1069792-2* 281
Alternanthera amabilis IPNI=*10291-2* 281
Alternanthera amabilis var. *tricolor* IPNI=*10291-2* 281
Alternanthera amoena f. *amabilis* IPNI=*10291-2* 281
Alternanthera aphylla IPNI=*1024156-2* 281
Alternanthera argentata IPNI=*1002675-1* 283
Alternanthera argentata var. *amazonica* IPNI=*1002675-1* 283
Alternanthera aristata f. *glabra* IPNI=*10378-2* 281
Alternanthera aristata IPNI=*10378-2* 281
Alternanthera articulata f. *capitellata* IPNI=*10291-2* 281
Alternanthera articulata f. *capitulifera* IPNI=*10291-2* 281
Alternanthera articulata IPNI=*10291-2* 281
Alternanthera articulata var. *minuta* IPNI=*10291-2* 281
Alternanthera asterotricha IPNI=*10362-2* 281
Alternanthera aurata IPNI=*1002675-1* 283
Alternanthera bangii IPNI=*10447-2* 281
Alternanthera brachiata IPNI=*1078972-2* 281
Alternanthera brasiliana f. *angustifolia* IPNI=*10294-2* 281
Alternanthera brasiliana f. *bicolor* IPNI=*10294-2* 281
Alternanthera brasiliana f. *obtusifolia* IPNI=*10294-2* 281
Alternanthera brasiliana f. *rubicunda* IPNI=*10294-2* 281
Alternanthera brasiliana var. *glabriuscula* IPNI=*10294-2* 281
Alternanthera brasiliana var. *longiseta* IPNI=*10294-2* 281
Alternanthera brasiliana var. *sericea* IPNI=*10294-2* 281
Alternanthera brasiliana var. *straminea* IPNI=*10294-2* 281
Alternanthera canescens IPNI=*129900-2* 283
Alternanthera ciliata IPNI=*327342-2* 281
Alternanthera crucis IPNI=*10362-2* 281
Alternanthera culebrasensis IPNI=*10362-2* 281
Alternanthera cyclophylla IPNI=*10291-2* 281
Alternanthera dentata f. *pubescens* IPNI=*10294-2* 281
Alternanthera dentata f. *rubiginosa* IPNI=*10294-2* 281
Alternanthera dentata IPNI=*10294-2* 281
Alternanthera dolichocephala IPNI=*10362-2* 281
Alternanthera dubia IPNI=*1069792-2* 281
Alternanthera dubia IPNI=*129900-2* 283
Alternanthera echinata IPNI=*327342-2* 281
Alternanthera elongata var. *nigriceps* IPNI=*10447-2* 281
Alternanthera ficoidea var. *amoena* IPNI=*10291-2* 281
Alternanthera ficoidea var. *bettzickiana* IPNI=*10291-2* 281
Alternanthera ficoidea var. *brachiata* IPNI=*1078972-2* 281
Alternanthera ficoidea var. *diffusa* IPNI=*1078972-2* 281
Alternanthera ficoidea var. *erecta* IPNI=*1078972-2* 281
Alternanthera ficoidea var. *flavogrisea* IPNI=*10362-2* 281
Alternanthera ficoidea var. *halimifolia* IPNI=*10362-2* 281
Alternanthera ficoidea var. *normalis* IPNI=*327343-2* 281
Alternanthera ficoidea var. *radicans* IPNI=*1078972-2* 281
Alternanthera ficoidea var. *spathulata* IPNI=*10291-2* 281
Alternanthera flavogrisea IPNI=*10362-2* 281
Alternanthera frutescens IPNI=*10362-2* 281
Alternanthera glabra IPNI=*1069792-2* 281
Alternanthera gomphrenoides IPNI=*10447-2* 281
Alternanthera halimifolia subsp. *macrophylla* IPNI=*10362-2* 281
Alternanthera iresinoides IPNI=*61055-1* 283
Alternanthera junciflora IPNI=*10447-2* 281
Alternanthera kerberi IPNI=*10291-2* 281
Alternanthera latifolia IPNI=*60789-1* 283
Alternanthera lehmannii IPNI=*10378-2* 281
Alternanthera leiantha IPNI=*327342-2* 281
Alternanthera leucantha IPNI=*327343-2* 281
Alternanthera lorentzii IPNI=*327342-2* 281
Alternanthera macrorhiza IPNI=*327344-2* 281
Alternanthera major IPNI=*1069792-2* 281
Alternanthera manillensis IPNI=*10362-2* 281
Alternanthera martinicensis IPNI=*10362-2* 281
Alternanthera meridensis IPNI=*10378-2* 281
Alternanthera mexicana f. *lanuginosa* IPNI=*10378-2* 281
Alternanthera mexicana IPNI=*10378-2* 281
Alternanthera mexicana var. *divergens* IPNI=*10378-2* 281
Alternanthera mexicana var. *gracilis* IPNI=*10378-2* 281
Alternanthera mexicana var. *lehmannii* IPNI=*10378-2* 281
Alternanthera meyeriana IPNI=*10378-2* 281
Alternanthera microcephala IPNI=*10378-2* 281
Alternanthera moquinii var. *grandiceps* IPNI=*10294-2* 281
Alternanthera nana IPNI=*1069792-2* 281
Alternanthera nana var. *major* IPNI=*1069792-2* 281
Alternanthera nigriceps IPNI=*10447-2* 281
Alternanthera nigripes IPNI=*10447-2* 281
Alternanthera paniculata IPNI=*10447-2* 281
Alternanthera paronychioides var. *amazonica* IPNI=*327343-2* 281
Alternanthera paronychioides var. *bettzickiana* IPNI=*10291-2* 281
Alternanthera paronychioides var. *floribunda* IPNI=*327343-2* 281
Alternanthera parviflora IPNI=*1002675-1* 283
Alternanthera parvifolia IPNI=*327344-2* 281
Alternanthera peploides IPNI=*327344-2* 281
Alternanthera philoxerina IPNI=*1024156-2* 281
Alternanthera philoxeroides f. *acutifolia* IPNI=*1024156-2* 281
Alternanthera philoxeroides f. *angustifolia* IPNI=*1024156-2* 281
Alternanthera philoxeroides f. *phyllantha* IPNI=*1024156-2* 281
Alternanthera philoxeroides subf. *philoxerina* IPNI=*1024156-2* 281
Alternanthera philoxeroides var. *acutifolia* IPNI=*1024156-2* 281
Alternanthera philoxeroides var. *lancifolia* IPNI=*1024156-2* 281
Alternanthera philoxeroides var. *luxurians* IPNI=*1024156-2* 281
Alternanthera philoxeroides var. *obtusifolia* IPNI=*1024156-2* 281
Alternanthera philoxeroides var. *phyllantha* IPNI=*1024156-2* 281
Alternanthera polygonoides f. *halimioides* IPNI=*327343-2* 281
Alternanthera polygonoides IPNI=*1069792-2* 281
Alternanthera polygonoides var. *austrotrinitatis* IPNI=*327343-2* 281

Alternanthera polygonoides var. *diffusa* IPNI=*1078972-2* 281
Alternanthera polygonoides var. *ficoidea* IPNI=*1078972-2* 281
Alternanthera polygonoides var. *glabrescens* IPNI=*1078972-2* 281
Alternanthera polygonoides var. *radicans* IPNI=*327343-2* 281
Alternanthera portoricensis IPNI=*10362-2* 281
Alternanthera procumbens IPNI=*327342-2* 281
Alternanthera prostrata IPNI=*1069792-2* 281
Alternanthera pubiflora f. *purpurea* IPNI=*10450-2* 281
Alternanthera pulchella var. *ciliata* IPNI=*435017-1* 285
Alternanthera pungens f. *pauciflora* IPNI=*327342-2* 281
Alternanthera pungens var. *leiantha* IPNI=*327342-2* 281
Alternanthera pungens var. *monocephala* IPNI=*327342-2* 281
Alternanthera purpurea IPNI=*10291-2* 281
Alternanthera repens IPNI=*1069792-2* 281
Alternanthera repens IPNI=*327342-2* 281
Alternanthera sessilis var. *amoena* IPNI=*10291-2* 281
Alternanthera sibirica IPNI=*1069792-2* 281
Alternanthera spathulata IPNI=*10291-2* 281
Alternanthera straminea IPNI=*10294-2* 281
Alternanthera tenella IPNI=*1069792-2* 281
Alternanthera tenella IPNI=*1078972-2* 281
Alternanthera tenella subsp. *flavogrisea* IPNI=*10362-2* 281
Alternanthera tenella var. *bettzickiana* IPNI=*10291-2* 281
Alternanthera tenuissima IPNI=*1069792-2* 281
Alternanthera thomensis IPNI=*327342-2* 281
Alternanthera triandra IPNI=*1078972-2* 281
Alternanthera tucumana IPNI=*10378-2* 281
Alternanthera uliginosa IPNI=*1069792-2* 281
Alternanthera villiflora IPNI=*327344-2* 281
Alternanthera virgata IPNI=*327343-2* 281
Alternanthera williamsii f. *purpurea* IPNI=*10450-2* 281
Alternanthera williamsii IPNI=*10450-2* 281
Althaea arborea IPNI=*561541-1* 410
Althaea assurgentifolia IPNI=*1003822-1* 410
Althaea balearica IPNI=*558872-1* 407
Althaea borealis IPNI=*561868-1* 410
Althaea caribaea IPNI=*558713-1* 407
Althaea chinensis IPNI=*558713-1* 407
Althaea coromandeliana IPNI=*558713-1* 407
Althaea cretica IPNI=*558713-1* 407
Althaea flexuosa IPNI=*558713-1* 407
Althaea frutex IPNI=*560890-1* 410
Althaea godronii IPNI=*561932-1* 410
Althaea kragujevacensis IPNI=*558872-1* 407
Althaea mareotica IPNI=*561830-1* 410
Althaea mauritiana IPNI=*561932-1* 410
Althaea meonantha IPNI=*558713-1* 407
Althaea mexicana IPNI=*558713-1* 407
Althaea microcarpa IPNI=*561830-1* 410
Althaea microchiton IPNI=*558713-1* 407
Althaea multiflora IPNI=*558872-1* 407
Althaea nicaeensis IPNI=*561798-1* 410
Althaea parviflora IPNI=*561830-1* 410
Althaea pulchra IPNI=*558872-1* 407
Althaea rosea IPNI=*558713-1* 407
Althaea sinensis IPNI=*558713-1* 407
Althaea sublobata IPNI=*558872-1* 407
Althaea ulmifolia IPNI=*329545-2* 413
Althaea vulgaris IPNI=*561932-1* 410
Alvarezia parviflora IPNI=*224052-2* 361
Alycia coarctata IPNI=*77164258-1* 246
Alyssum halimifolium IPNI=*1085072-2* 277
Alyssum maritimum IPNI=*1085072-2* 277
Alyssum minimum IPNI=*1085072-2* 277
Alyssum odoratum IPNI=*1085072-2* 277
Alyssum strigulosum IPNI=*1085072-2* 277
Amaioua corymbosa IPNI=*77124830-1* 347
Amaioua edulis IPNI=*743001-1* 347
Amaioua fagifolia IPNI=*77124830-1* 347
Amaioua genipoides IPNI=*86133-2* 349
Amaioua guianensis var. *macrantha* IPNI=*743074-1* 347
Amaioua hirsuta IPNI=*86135-2* 349
Amaioua peruviana IPNI=*77124830-1* 347
Amaioua utilis IPNI=*743001-1* 347
Amalobotrya latifolia IPNI=*697710-1* 290
Amanoa bracteata IPNI=*338575-1* 401
Amanoa grandiflora IPNI=*338575-1* 401
Amanoa guianensis var. *genuina* IPNI=*338575-1* 401
Amanoa guianensis var. *grandiflora* IPNI=*338575-1* 401
Amanoa guianensis var. *poeppigii* IPNI=*338575-1* 401
Amanoa macrocarpa IPNI=*338575-1* 401
Amanoa potamophila IPNI=*338575-1* 401
Amaracus × majorica IPNI=*453305-1* 372
Amaracus majorana IPNI=*453303-1* 372
Amarantesia brasiliensis IPNI=*10291-2* 281
Amaranthoides decumbens IPNI=*112097-2* 282
Amaranthus abyssinicus IPNI=*316347-2* 281
Amaranthus albus IPNI=*10619-2* 281
Amaranthus alius IPNI=*10619-2* 281
Amaranthus anacardana IPNI=*10631-2* 281
Amaranthus arardhanus IPNI=*10631-2* 281
Amaranthus ascendens IPNI=*10619-2* 281
Amaranthus aureus IPNI=*10631-2* 281
Amaranthus aureus IPNI=*10648-2* 281
Amaranthus batalleri IPNI=*10648-2* 281
Amaranthus bellardii IPNI=*10648-2* 281
Amaranthus berchtoldii IPNI=*10619-2* 281
Amaranthus bernhardii IPNI=*10665-2* 281
Amaranthus bicolor IPNI=*327386-2* 282
Amaranthus blitonius IPNI=*10619-2* 281
Amaranthus blitum var. *ascendens* IPNI=*10619-2* 281
Amaranthus bouchonii IPNI=*10648-2* 281
Amaranthus brasiliensis IPNI=*10631-2* 281
Amaranthus caracasanus IPNI=*10711-2* 281
Amaranthus cararu IPNI=*316347-2* 281
Amaranthus carneus IPNI=*10631-2* 281
Amaranthus caudatus subsp. *saueri* IPNI=*316347-2* 281
Amaranthus chlorostachys convar. *erythrostachys* IPNI=*10665-2* 281
Amaranthus chlorostachys IPNI=*10648-2* 281
Amaranthus coracanus IPNI=*10711-2* 281
Amaranthus cristatus IPNI=*316350-2* 282
Amaranthus cruentus var. *albus* IPNI=*10631-2* 281
Amaranthus cruentus var. *patulus* IPNI=*10648-2* 281
Amaranthus cuspidatus IPNI=*327386-2* 282
Amaranthus diacanthus IPNI=*10711-2* 281
Amaranthus diffusus IPNI=*10619-2* 281
Amaranthus dubius IPNI=*59693-1* 281
Amaranthus dussii IPNI=*316347-2* 281
Amaranthus edulis IPNI=*316347-2* 281
Amaranthus edulis var. *spadiceus* IPNI=*316347-2* 281
Amaranthus edulis var. *typicus* IPNI=*316347-2* 281
Amaranthus eugenii IPNI=*10648-2* 281
Amaranthus farinaceus IPNI=*10631-2* 281
Amaranthus fasciatus IPNI=*316349-2* 282
Amaranthus flavescens IPNI=*10648-2* 281
Amaranthus flavus IPNI=*10631-2* 281
Amaranthus flexuosus IPNI=*327386-2* 282
Amaranthus frumentaceus IPNI=*10665-2* 281
Amaranthus gangeticus IPNI=*10619-2* 281
Amaranthus gangeticus IPNI=*327386-2* 282
Amaranthus gangeticus var. *angustior* IPNI=*327386-2* 282
Amaranthus graecizans var. *blitum* IPNI=*10619-2* 281
Amaranthus guadelupensis IPNI=*10631-2* 281
Amaranthus hecticus IPNI=*10648-2* 281
Amaranthus huttonii IPNI=*316350-2* 282
Amaranthus hybridus f. *aciculatus* IPNI=*10648-2* 281

Amaranthus hybridus IPNI=*10665-2* 281
Amaranthus hybridus subsp. *bouchonii* IPNI=*10648-2* 281
Amaranthus hybridus subsp. *caudatus* IPNI=*316347-2* 281
Amaranthus hybridus subsp. *cruentus* IPNI=*10631-2* 281
Amaranthus hybridus subsp. *hypochondriacus* IPNI=*10665-2* 281
Amaranthus hybridus subsp. *incurvatus* IPNI=*10631-2* 281
Amaranthus hybridus subsp. *patulus* IPNI=*10631-2* 281
Amaranthus hybridus subvar. *aciculatus* IPNI=*10648-2* 281
Amaranthus hybridus subvar. *aristulatus* IPNI=*10648-2* 281
Amaranthus hybridus subvar. *genuinus* IPNI=*10648-2* 281
Amaranthus hybridus subvar. *multispiculatus* IPNI=*10648-2* 281
Amaranthus hybridus var. *batalleri* IPNI=*10648-2* 281
Amaranthus hybridus var. *bellardii* IPNI=*10648-2* 281
Amaranthus hybridus var. *bouchonii* IPNI=*10648-2* 281
Amaranthus hybridus var. *densus* IPNI=*10648-2* 281
Amaranthus hybridus var. *erythrostachys* IPNI=*10665-2* 281
Amaranthus hybridus var. *hecticus* IPNI=*10648-2* 281
Amaranthus hybridus var. *hypochondriacus* IPNI=*10665-2* 281
Amaranthus hybridus var. *laetus* IPNI=*10648-2* 281
Amaranthus hybridus var. *leucocarpus* IPNI=*10665-2* 281
Amaranthus hybridus var. *paniculatus* IPNI=*10631-2* 281
Amaranthus hybridus var. *patulus* IPNI=*10631-2* 281
Amaranthus hybridus var. *prostratus* IPNI=*10648-2* 281
Amaranthus hybridus var. *rubricaulis* IPNI=*10648-2* 281
Amaranthus hybridus var. *sanguineus* IPNI=*10631-2* 281
Amaranthus hypochondriacus var. *macrostachys* IPNI=*10665-2* 281
Amaranthus hypochondriacus var. *monstrosus* IPNI=*10665-2* 281
Amaranthus hypochondriacus var. *racemosus* IPNI=*10665-2* 281
Amaranthus hypochondriacus var. *tortuosus* IPNI=*10665-2* 281
Amaranthus inamoenus IPNI=*327386-2* 282
Amaranthus incarnatus IPNI=*10631-2* 281
Amaranthus incomptus IPNI=*59693-1* 281
Amaranthus intermedius IPNI=*10648-2* 281
Amaranthus japonicus IPNI=*327386-2* 282
Amaranthus laetus IPNI=*10648-2* 281
Amaranthus lanceolatus IPNI=*327386-2* 282
Amaranthus lancifolius IPNI=*327386-2* 282
Amaranthus laxiflorus IPNI=*10648-2* 281
Amaranthus leucocarpus IPNI=*10665-2* 281
Amaranthus leucospermus IPNI=*10665-2* 281
Amaranthus lividus IPNI=*10619-2* 281
Amaranthus lividus IPNI=*327386-2* 282
Amaranthus lividus subsp. *ascendens* IPNI=*10619-2* 281
Amaranthus lividus var. *ascendens* IPNI=*10619-2* 281
Amaranthus macrostachyus IPNI=*10665-2* 281
Amaranthus mangostanus IPNI=*327386-2* 282
Amaranthus mantegazzianus IPNI=*316347-2* 281
Amaranthus maximus IPNI=*316347-2* 281
Amaranthus melancholicus IPNI=*327386-2* 282
Amaranthus minor IPNI=*10619-2* 281
Amaranthus montevidensis IPNI=*10631-2* 281
Amaranthus mucronatus IPNI=*10619-2* 281
Amaranthus mucronatus IPNI=*327386-2* 282
Amaranthus neglectus IPNI=*10648-2* 281
Amaranthus oleraceus IPNI=*327386-2* 282
Amaranthus paniculatus IPNI=*10631-2* 281
Amaranthus paniculatus var. *cruentus* IPNI=*10631-2* 281
Amaranthus paniculatus var. *sanguineus* IPNI=*10631-2* 281
Amaranthus paniculatus var. *sanguineus* IPNI=*10648-2* 281
Amaranthus paniculatus var. *speciosus* IPNI=*10631-2* 281
Amaranthus patulus IPNI=*10648-2* 281
Amaranthus pendulinus IPNI=*316347-2* 281
Amaranthus pendulus IPNI=*316347-2* 281
Amaranthus polygamus IPNI=*327386-2* 282
Amaranthus polystachyus IPNI=*316349-2* 282
Amaranthus powellii subsp. *bouchonii* IPNI=*10648-2* 281
Amaranthus prostratus IPNI=*10619-2* 281
Amaranthus pseudoretroflexus IPNI=*10648-2* 281
Amaranthus purgans IPNI=*10631-2* 281
Amaranthus pyramidalis IPNI=*316350-2* 282
Amaranthus rotundifolius IPNI=*327386-2* 282
Amaranthus roxburghianus var. *angustifolius* IPNI=*327386-2* 282
Amaranthus rubescens IPNI=*10631-2* 281
Amaranthus ruderalis IPNI=*10619-2* 281
Amaranthus salicifolius IPNI=*327386-2* 282
Amaranthus sanguineus IPNI=*10631-2* 281
Amaranthus sanguinolentus IPNI=*10631-2* 281
Amaranthus speciosus IPNI=*10631-2* 281
Amaranthus spicatus IPNI=*10631-2* 281
Amaranthus spicatus IPNI=*10648-2* 281
Amaranthus strictus IPNI=*10631-2* 281
Amaranthus tenuiflorus IPNI=*10619-2* 281
Amaranthus tenuifolius IPNI=*10619-2* 281
Amaranthus tenuifolius IPNI=*327386-2* 282
Amaranthus timeroyi IPNI=*10648-2* 281
Amaranthus tricolor var. *acutus* IPNI=*327386-2* 282
Amaranthus tristis IPNI=*327386-2* 282
Amaranthus tristis var. *leptostachys* IPNI=*10631-2* 281
Amaranthus trivialis IPNI=*10648-2* 281
Amaranthus violaceus IPNI=*10631-2* 281
Amaranthus viridis IPNI=*10619-2* 281
Amarella californica IPNI=*108639-2* 344
Amarella conferta IPNI=*108639-2* 344
Amarella copelandii IPNI=*108639-2* 344
Amarella distegia IPNI=*108639-2* 344
Amarella gracilis IPNI=*108639-2* 344
Amarella lembertii IPNI=*108639-2* 344
Amarella plebeia var. *holmii* IPNI=*108639-2* 344
Amarella quinquefida IPNI=*108639-2* 344
Amarella revoluta IPNI=*108639-2* 344
Amarella tenuis IPNI=*108639-2* 344
Amarella theiantha IPNI=*108639-2* 344
Amarella theiantha var. *lactea* IPNI=*108639-2* 344
Amarella theiantha var. *livida* IPNI=*108639-2* 344
Amarenus campestris IPNI=*522975-1* 335
Amaryllis araripina IPNI=*122772-2* 213
Amaryllis araripina subsp. *conspicua* IPNI=*122772-2* 213
Amaryllis araripina subsp. *rubrituba* IPNI=*122772-2* 213
Amaryllis araripina subsp. *striata* IPNI=*122772-2* 213
Amaryllis berteroi IPNI=*270986-2* 214
Amaryllis bulbisperma IPNI=*63828-1* 213
Amaryllis capensis IPNI=*63828-1* 213
Amaryllis carinata IPNI=*270933-2* 214
Amaryllis carnea IPNI=*67293-1* 214
Amaryllis concinna IPNI=*270933-2* 214
Amaryllis crociflora IPNI=*122815-2* 213
Amaryllis depauperata IPNI=*67314-1* 214
Amaryllis elegans IPNI=*122772-2* 213
Amaryllis elegans var. *ambigua* IPNI=*122772-2* 213
Amaryllis elegans var. *rubrituba* IPNI=*122772-2* 213
Amaryllis elegans var. *striata* IPNI=*122772-2* 213
Amaryllis elegans var. *vittata* IPNI=*122772-2* 213
Amaryllis equestris IPNI=*122815-2* 213
Amaryllis gigantea IPNI=*63934-1* 213
Amaryllis jagus IPNI=*63934-1* 213
Amaryllis latifolia IPNI=*63934-1* 213
Amaryllis lindleyana IPNI=*270933-2* 214
Amaryllis lineata IPNI=*65185-1* 213
Amaryllis longiflora IPNI=*122772-2* 213
Amaryllis procera IPNI=*30032850-2* 213
Amaryllis punicea IPNI=*122815-2* 213
Amaryllis riparia IPNI=*63828-1* 213
Amaryllis robusta IPNI=*270986-2* 214
Amaryllis rosea IPNI=*67293-1* 214
Amaryllis solandriflora IPNI=*122772-2* 213
Amaryllis solandriflora var. *vittata* IPNI=*122772-2* 213

Amaryllis tubispatha IPNI=*67314-1* 214
Amaryllis vittata IPNI=*65185-1* 213
Amaryllis vittata subsp. *guarapuavae* IPNI=*65185-1* 213
Amaryllis vittata var. *harrisoniae* IPNI=*65185-1* 213
Amaryllis vittata var. *major* IPNI=*65185-1* 213
Amaryllis vittata var. *tweedieana* IPNI=*65185-1* 213
Amaryllis zeylanica IPNI=*64125-1* 213
Ambaiba acutifolia IPNI=*850916-1* 445
Ambaiba commutata IPNI=*850947-1* 446
Ambaiba costaricensis IPNI=*850947-1* 446
Ambaiba digitata IPNI=*850916-1* 445
Ambaiba goudotiana IPNI=*850932-1* 446
Ambaiba hemsleyana IPNI=*850947-1* 446
Ambaiba humboldtiana IPNI=*850956-1* 446
Ambaiba latiloba IPNI=*850938-1* 446
Ambaiba membranacea IPNI=*850943-1* 446
Ambaiba mexicana IPNI=*850947-1* 446
Ambaiba obtusifolia IPNI=*850947-1* 446
Ambaiba panamensis IPNI=*850947-1* 446
Ambaiba peltata IPNI=*850956-1* 446
Ambaiba propinqua IPNI=*850956-1* 446
Ambaiba schiedeana IPNI=*850956-1* 446
Ambaiba schreberiana IPNI=*50937-2* 446
Ambaiba sciadophylla IPNI=*850970-1* 446
Ambaiba surinamensis IPNI=*850956-1* 446
Ambaiba tubulosa IPNI=*850916-1* 445
Ambelania cuneata IPNI=*945268-1* 342
Ambelania lopezii IPNI=*945265-1* 343
Ambelania lucida IPNI=*945268-1* 342
Ambelania macrophylla IPNI=*945265-1* 343
Ambelania markgrafiana IPNI=*281069-2* 342
Ambelania quadrangularis IPNI=*81716-1* 343
Ambelania ternstroemiacea IPNI=*168521-2* 342
Amblovenatum opulentum IPNI=*17460390-1* 174
Amblyanthera campestris IPNI=*151979-2* 341
Amblyanthera ciliata IPNI=*151979-2* 341
Amblyanthera claussenii IPNI=*151979-2* 341
Amblyanthera fluminensis IPNI=*151979-2* 341
Amblyanthera fluminensis var. *stadelmeyeri* IPNI=*151979-2* 341
Amblyanthera hirsuta IPNI=*151979-2* 341
Amblyanthera hispida IPNI=*151979-2* 341
Amblyanthera hispida var. *tomentosa* IPNI=*151979-2* 341
Amblyanthera lasiocarpa IPNI=*151979-2* 341
Amblyanthera lasiocarpa var. *pubescens* IPNI=*151979-2* 341
Amblyanthera lasiocarpa var. *tomentosa* IPNI=*151979-2* 341
Amblyanthera macrophylla IPNI=*151979-2* 341
Amblyanthera membranacea IPNI=*152067-2* 342
Amblyanthera microcalyx IPNI=*152067-2* 342
Amblyanthera microcalyx var. *glabra* IPNI=*152067-2* 342
Amblyanthera microcalyx var. *hirtiflora* IPNI=*152067-2* 342
Amblyanthera microcalyx var. *leiantha* IPNI=*152067-2* 342
Amblyanthera microcalyx var. *leiophylla* IPNI=*152067-2* 342
Amblyanthera microcalyx var. *trichantha* IPNI=*152067-2* 342
Amblyanthera ovata IPNI=*151979-2* 341
Amblyanthera palustris IPNI=*151979-2* 341
Amblyanthera palustris var. *almadensis* IPNI=*151979-2* 341
Amblyanthera spruceana IPNI=*151987-2* 341
Amblyanthera versicolor IPNI=*152052-2* 341
Amblyanthera versicolor var. *glabrata* IPNI=*152052-2* 341
Amblyanthera versicolor var. *intcrmcdia* IPNI=*152052 2* 341
Amblyanthera versicolor var. *olivacea* IPNI=*152052-2* 341
Amblyanthera versicolor var. *psilophylla* IPNI=*152052-2* 341
Amblyanthera versicolor var. *pubiflora* IPNI=*152052-2* 341
Amblyanthera versicolor var. *stenosolen* IPNI=*152052-2* 341
Amblygonum glabrum IPNI=*1005755-2* 289
Amblyopappus mendocinus IPNI=*242907-1* 269
Ambrina ambrosioides IPNI=*1198060-2* 282
Ambrina ambrosioides var. *anthelmintica* IPNI=*1090820-2* 282
Ambrina anthelmintica IPNI=*1090820-2* 282
Ambrina incisa IPNI=*1198060-2* 282
Ambrina parvula IPNI=*1198060-2* 282
Ambrina spathulata IPNI=*1198060-2* 282
Ambrosia artemisiifolia f. *gracilissima* IPNI=*176058-1* 256
Ambrosia artemisiifolia f. *villosa* IPNI=*176058-1* 256
Ambrosia artemisiifolia subsp. *diversifolia* IPNI=*176058-1* 256
Ambrosia artemisiifolia var. *elatior* IPNI=*176058-1* 256
Ambrosia artemisiifolia var. *jamaicensis* IPNI=*176058-1* 256
Ambrosia artemisiifolia var. *octocornis* IPNI=*176058-1* 256
Ambrosia artemisiifolia var. *paniculata* IPNI=*176058-1* 256
Ambrosia artemisiifolia var. *quadricornis* IPNI=*176058-1* 256
Ambrosia artemisiifolia var. *trinitensis* IPNI=*11053-2* 257
Ambrosia chilensis IPNI=*176058-1* 256
Ambrosia diversifolia IPNI=*176058-1* 256
Ambrosia elata IPNI=*176058-1* 256
Ambrosia elatior f. *aurea* IPNI=*176058-1* 256
Ambrosia elatior IPNI=*176058-1* 256
Ambrosia elatior var. *artemisiifolia* IPNI=*176058-1* 256
Ambrosia elatior var. *heterophylla* IPNI=*176058-1* 256
Ambrosia fruticosa IPNI=*176055-1* 256
Ambrosia fruticosa var. *intermedia* IPNI=*176059-1* 257
Ambrosia glandulosa IPNI=*176058-1* 256
Ambrosia monophylla IPNI=*176058-1* 256
Ambrosia orobanchifera IPNI=*11053-2* 257
Ambrosia paniculata f. *dissecta* IPNI=*11053-2* 257
Ambrosia paniculata f. *litoralis* IPNI=*11053-2* 257
Ambrosia paniculata IPNI=*176058-1* 256
Ambrosia paniculata var. *cumanensis* IPNI=*11053-2* 257
Ambrosia paniculata var. *peruviana* IPNI=*11053-2* 257
Ambrosia peruviana IPNI=*11053-2* 257
Ambrosia senegalensis IPNI=*176058-1* 256
Ambrosia tacorensis IPNI=*176059-1* 257
Ambrosia umbellata IPNI=*176058-1* 256
Ambulia micrantha IPNI=*230902-2* 376
Amellus asper f. *bicolor* IPNI=*156327-2* 266
Amellus asper IPNI=*156327-2* 266
Amellus asper var. *canescens* IPNI=*156327-2* 266
Amellus asper var. *glabriusculus* IPNI=*156327-2* 266
Amellus asper var. *normalis* IPNI=*156327-2* 266
Amellus carolinianus IPNI=*1186403-2* 261
Amellus discoideus IPNI=*156327-2* 266
Amellus echinocephalon IPNI=*156327-2* 266
Amellus floribundus IPNI=*230435-1* 265
Amellus lanceolatus IPNI=*156327-2* 266
Amellus microphyllus IPNI=*156327-2* 266
Amellus niveus IPNI=*156327-2* 266
Amellus oxylepis IPNI=*156327-2* 266
Amellus pedunculatus IPNI=*257735-2* 271
Amerimnon brownei IPNI=*75919-2* 317
Amerimnon ecastaphyllum IPNI=*75934-2* 317
Amerimnon foliaceum IPNI=*1006519-1* 317
Amerimnon frutescens IPNI=*75939-2* 317
Amerimnon inundatum IPNI=*75954-2* 317
Amerimnon latifolium IPNI=*502971-1* 325
Amerimnon malifolium IPNI=*75934-2* 317
Amerimnon pinnatum IPNI=*201767-2* 330
Amerimnon riparium IPNI=*490433-1* 317
Amerimnon sieberi IPNI=*75934-2* 317
Amerina tomentosa IPNI−*860809-1* 368
Ammandra dasyneura IPNI=*11322-2* 204
Ammannia arenaria IPNI=*30030870-2* 416
Ammannia auriculata f. *brasiliensis* IPNI=*30030870-2* 416
Ammannia auriculata f. *hyrcanica* IPNI=*30030870-2* 416
Ammannia auriculata f. *indica* IPNI=*30030870-2* 416
Ammannia auriculata var. *arenaria* IPNI=*30030870-2* 416
Ammannia auriculata var. *elata* IPNI=*30030870-2* 416
Ammannia hyrcanica IPNI=*30030870-2* 416

Ammannia longipes IPNI=*30030870-2* 416
Ammannia racemosa IPNI=*30030870-2* 416
Ammannia sanguinolenta IPNI=*30030870-2* 416
Ammannia senegalensis var. *brasiliensis* IPNI=*30030870-2* 416
Ammannia wrightii IPNI=*30030870-2* 416
Ammi boeberi IPNI=*837486-1* 251
Ammi broussonetii IPNI=*837486-1* 251
Ammi cicutifolium IPNI=*837486-1* 251
Ammi dilatatum IPNI=*849956-1* 253
Ammi elatum IPNI=*837486-1* 251
Ammi glaucifolium IPNI=*837486-1* 251
Ammi intermedium IPNI=*837486-1* 251
Ammi pauciradiatum IPNI=*837486-1* 251
Ammi petroselinoides IPNI=*60442790-2* 252
Ammi pumilum IPNI=*837486-1* 251
Ammi topalii IPNI=*837486-1* 251
Ammi visnaga IPNI=*849956-1* 253
Ammi visnaga var. *longibracteatum* IPNI=*849956-1* 253
Ammocallis rosea IPNI=*77880-1* 340
Ammoseris surinamensis IPNI=*229022-1* 264
Amomophyllum floribundum IPNI=*88963-1* 202
Amomum alpinia IPNI=*218583-2* 251
Amomum angustifolium IPNI=*798372-1* 251
Amomum cardamomum IPNI=*796556-1* 250
Amomum curcuma IPNI=*796451-1* 250
Amomum ensal IPNI=*796556-1* 250
Amomum filiforme IPNI=*796836-1* 251
Amomum magnificum IPNI=*942355-1* 251
Amomum nutans IPNI=*872083-1* 250
Amomum petiolatum IPNI=*329964-2* 248
Amomum racemosum IPNI=*796556-1* 250
Amomum renealmia IPNI=*218583-2* 251
Amomum repens IPNI=*796556-1* 250
Amomum spirale IPNI=*796387-1* 248
Amomum thyrsoideum IPNI=*798093-1* 251
Amomum tridentatum IPNI=*942355-1* 251
Amomum uncinatum IPNI=*796556-1* 250
Amomum zingiber IPNI=*798372-1* 251
Amomum zinziba IPNI=*798372-1* 251
Amoreuxia colombiana IPNI=*77183799-1* 406
Amoreuxia palmatifida IPNI=*77183799-1* 406
Amoreuxia schiedeana IPNI=*77183799-1* 406
Amoria hybrida IPNI=*30240399-2* 335
Amoria repens IPNI=*523626-1* 336
Amorpha glandulosa IPNI=*76085-2* 317
Amorphospermum schomburgkianum IPNI=*89010-2* 305
Ampelocera crenulata IPNI=*48069-2* 403
Ampelocera latifolia IPNI=*11466-2* 445
Ampelodaphne dasyantha IPNI=*90972-2* 183
Ampherephis aristata IPNI=*51409-2* 259
Ampherephis intermedia IPNI=*51409-2* 259
Ampherephis mutica IPNI=*51409-2* 259
Ampherephis pilosa IPNI=*51409-2* 259
Ampherephis pulchella IPNI=*51409-2* 259
Amphibecis violacea IPNI=*51409-2* 259
Amphiglottis lacera IPNI=*92159-2* 219
Amphiglottis nocturna IPNI=*30010607-2* 219
Amphiglottis secunda IPNI=*92159-2* 219
Amphilophis barbinodis IPNI=*35184-2* 235
Amphilophis emersa IPNI=*35184-2* 235
Amphilophis haenkei IPNI=*392701-1* 235
Amphilophis intermedia IPNI=*392701-1* 235
Amphilophis intermedia var. *acidula* IPNI=*392701-1* 235
Amphilophis ischaemum IPNI=*1191417-2* 235
Amphilophis leucopogon IPNI=*35184-2* 235
Amphilophis odorata IPNI=*392701-1* 235
Amphilophis palmeri IPNI=*35184-2* 235
Amphilophis perforata IPNI=*35184-2* 235
Amphilophis pertusa IPNI=*65406-3* 235
Amphilophis piptathera IPNI=*97916-2* 239
Amphilophis saccharoides IPNI=*35205-2* 235
Amphilophis schlumbergeri IPNI=*35184-2* 235
Amphilophium glanduliferum IPNI=*108202-1* 362
Amphilophium macrophyllum IPNI=*108202-1* 362
Amphilophium molle IPNI=*108202-1* 362
Amphilophium mollicomum IPNI=*108202-1* 362
Amphilophium mutisii IPNI=*108202-1* 362
Amphilophium paniculatum var. *imatacense* IPNI=*108202-1* 362
Amphilophium paniculatum var. *molle* IPNI=*108202-1* 362
Amphilophium paniculatum var. *mollicomum* IPNI=*108202-1* 362
Amphilophium purpureum IPNI=*108202-1* 362
Amphilophium xerophilum IPNI=*108202-1* 362
Amphineuron opulentum IPNI=*17460390-1* 174
Amphirrhox sprucei IPNI=*866955-1* 405
Amphiscopia ciliata IPNI=*47308-1* 359
Amphiscopia moricandiana IPNI=*47308-1* 359
Amphiscopia pohliana IPNI=*132826-2* 360
Amphiscopia pohliana var. *angustifolia* IPNI=*132826-2* 360
Amphiscopia retusa IPNI=*50468-1* 360
Amphitecna obovata IPNI=*108218-1* 362
Amphymenium floridum IPNI=*516561-1* 331
Amphymenium laxiflorum IPNI=*516561-1* 331
Amphymenium longifolium IPNI=*516568-1* 331
Amphymenium rohrii IPNI=*516561-1* 331
Amphymenium violaceum IPNI=*516561-1* 331
Amygdalus ferganensis IPNI=*1212858-2* 444
Amygdalus nucipersica IPNI=*1212858-2* 444
Amygdalus persica f. *pendula* IPNI=*1212858-2* 444
Amygdalus persica f. *stellata* IPNI=*1212858-2* 444
Amygdalus persica f. *versicolor* IPNI=*1212858-2* 444
Amygdalus persica IPNI=*1212858-2* 444
Amygdalus persica var. *aganonucipersica* IPNI=*1212858-2* 444
Amygdalus persica var. *compressa* IPNI=*1212858-2* 444
Amygdalus persica var. *nucipersica* IPNI=*1212858-2* 444
Amygdalus persica var. *scleronucipersica* IPNI=*1212858-2* 444
Amygdalus persica var. *sclеropersica* IPNI=*1212858-2* 444
Amygdalus persica var. *stellata* IPNI=*1212858-2* 444
Amylocarpus acanthocnemis IPNI=*664750-1* 207
Amylocarpus angustifolius IPNI=*664750-1* 207
Amylocarpus arenarius IPNI=*664750-1* 207
Amylocarpus hirtus IPNI=*664621-1* 206
Amylocarpus luetzelburgii IPNI=*664750-1* 207
Amylocarpus microspathus IPNI=*664750-1* 207
Amylocarpus obovatus IPNI=*664750-1* 207
Amylocarpus simplicifrons IPNI=*664750-1* 207
Amylocarpus tenuissimus IPNI=*664750-1* 207
Amylocarpus xanthocarpus IPNI=*664750-1* 207
Amyris altissima IPNI=*77183601-1* 452
Amyris ambrosiaca IPNI=*128337-1* 452
Amyris ambrosiaca IPNI=*128338-1* 453
Amyris brasiliensis IPNI=*128338-1* 453
Amyris carana IPNI=*128295-1* 452
Amyris caranifera IPNI=*127138-1* 451
Amyris decandra IPNI=*128316-1* 452
Amyris elaphrium IPNI=*127130-1* 451
Amyris enneandra IPNI=*128316-1* 452
Amyris graveolens IPNI=*127138-1* 451
Amyris guianensis IPNI=*128337-1* 452
Amyris heterophylla IPNI=*128278-1* 452
Amyris pubescens IPNI=*127138-1* 451
Amyris tomentosa IPNI=*127234-1* 451
Amyrsia compressa IPNI=*166091-2* 427
Amyrsia discolor IPNI=*166084-2* 427
Amyrsia foliosa IPNI=*166103-2* 427
Amyrsia hallii IPNI=*166097-2* 427
Amyrsia limbata IPNI=*166091-2* 427

Anacampta albescens IPNI=*82371-1* 344
Anacampta echinata IPNI=*20001045-1* 344
Anacampta hirtula IPNI=*20001045-1* 344
Anacampta longifolia IPNI=*20001045-1* 344
Anacampta macrocalyx IPNI=*285812-2* 343
Anacampta maxima IPNI=*82212-1* 344
Anacampta muelleriana IPNI=*285812-2* 343
Anacampta palustris IPNI=*248891-2* 344
Anacardium microcarpum IPNI=*319068-2* 450
Anacardium occidentale var. *gardneri* IPNI=*319068-2* 450
Anacardium rhinocarpus IPNI=*12224-2* 450
Anacharis densa IPNI=*431710-1* 203
Anacheilium cochleatum IPNI=*1001247-1* 222
Anacheilium fragrans IPNI=*1001253-1* 222
Anacyclus aureus IPNI=*192489-1* 259
Anacyclus nobilis IPNI=*192489-1* 259
Anacyclus pyrethraria IPNI=*914778-1* 255
Anadenanthera falcata IPNI=*12268-2* 311
Anadenanthera peregrina var. *falcata* IPNI=*12268-2* 311
Anagallidastrum exiguum IPNI=*77100404-1* 304
Anagallis arabica IPNI=*77100386-1* 304
Anagallis arvensis f. *purpurea* IPNI=*77100386-1* 304
Anagallis arvensis IPNI=*77100386-1* 304
Anagallis arvensis subsp. *phoenicea* IPNI=*77100386-1* 304
Anagallis arvensis var. *phoenicea* IPNI=*77100386-1* 304
Anagallis arvensis var. *purpurea* IPNI=*77100386-1* 304
Anagallis carnea IPNI=*77100386-1* 304
Anagallis centunculus IPNI=*77100404-1* 304
Anagallis hadidii IPNI=*77100386-1* 304
Anagallis indica IPNI=*77100386-1* 304
Anagallis jacquemontii IPNI=*77100386-1* 304
Anagallis latifolia IPNI=*77100386-1* 304
Anagallis mas IPNI=*77100386-1* 304
Anagallis minima IPNI=*77100404-1* 304
Anagallis monelli IPNI=*77100386-1* 304
Anagallis orientalis IPNI=*77100386-1* 304
Anagallis parviflora IPNI=*77100386-1* 304
Anagallis phoenicea IPNI=*77100386-1* 304
Anagallis platyphylla IPNI=*77100386-1* 304
Anagallis pulchella IPNI=*77100386-1* 304
Anagallis pumila var. *ruizii* IPNI=*77100404-1* 304
Anagallis punctifolia IPNI=*77100386-1* 304
Anagallis pusilla IPNI=*77100404-1* 304
Anagallis repens IPNI=*77100386-1* 304
Anagallis sessilis IPNI=*77100404-1* 304
Anamomis dichotoma IPNI=*166091-2* 427
Anamomis dicrana IPNI=*166091-2* 427
Anamomis esculenta IPNI=*596253-1* 425
Anamomis fragrans IPNI=*166091-2* 427
Anamomis fragrans var. *brachyrhiza* IPNI=*166091-2* 427
Anamomis fragrans var. *cuneata* IPNI=*166091-2* 427
Anamomis grandis IPNI=*166091-2* 427
Anamomis guayabillo IPNI=*166091-2* 427
Anamomis lucayana IPNI=*166091-2* 427
Anamomis punctata IPNI=*166091-2* 427
Anamomis punctata var. *dichotoma* IPNI=*166091-2* 427
Anamomis simpsonii IPNI=*166091-2* 427
Anamomis umbellulifera IPNI=*596253-1* 425
Ananas magdalenae IPNI=*5516-2* 228
Ananas pinguin IPNI=*319113-2* 228
Anapeltis lycopodioides IPNI=*160142-2* 175
Anapeltis nitida IPNI=*160142-2* 175
Anapeltis palmeri IPNI=*17444940-1* 175
Anapeltis stigmatica IPNI=*160142-2* 175
Anapeltis vacciniifolia IPNI=*17406250-1* 176
Anapeltis venosa IPNI=*160142-2* 175
Anartia flavescens IPNI=*79896-1* 341
Anartia meyeri IPNI=*82371-1* 344
Anastrophus compressus IPNI=*392087-1* 234
Anastrophus deludens IPNI=*27415-2* 234
Anastrophus paspalodes IPNI=*184275-2* 243
Anastrophus platicaulon IPNI=*392087-1* 234
Anathallis modesta IPNI=*1147281-2* 217
Anatherum berteroanum IPNI=*405817-1* 240
Anatherum bicorne IPNI=*12758-2* 233
Anatherum brasiliense IPNI=*12920-2* 234
Anatherum caudatum IPNI=*405817-1* 240
Anatherum domingense IPNI=*12920-2* 234
Anatherum inerme IPNI=*13112-2* 234
Anatherum megapotamicum IPNI=*90411-2* 238
Anatherum montanum IPNI=*392701-1* 235
Anatherum muricatum IPNI=*396213-1* 236
Anatherum pedunculosum IPNI=*420877-1* 244
Anatherum portoricense IPNI=*405817-1* 240
Anatherum spathiflorum IPNI=*13112-2* 234
Anatherum virgatum IPNI=*13112-2* 234
Anatherum virginicum subvar. *glomeratus* IPNI=*12850-2* 233
Anatherum virginicum subvar. *leucostachyum* IPNI=*12920-2* 234
Anatherum virginicum subvar. *reinoldii* IPNI=*12850-2* 233
Anatherum zizanioides IPNI=*396213-1* 236
Anaueria brasiliensis IPNI=*31042-2* 182
Anavinga parvifolia IPNI=*779541-1* 403
Anavinga samyda IPNI=*274165-2* 403
Anaxagorea krukoffii IPNI=*72027-1* 186
Anaxagorea megalophylla IPNI=*12508-2* 186
Anaxagorea minor IPNI=*72027-1* 186
Anaxagorea multiflora IPNI=*72023-1* 186
Anaxagorea mutica IPNI=*12508-2* 186
Anaxetum crassifolium IPNI=*77183236-1* 175
Ancistrum barbatum IPNI=*153124-2* 443
Ancistrum cylindristachyum IPNI=*719970-1* 442
Ancistrum elongatum IPNI=*719987-1* 442
Ancistrum lappaceum IPNI=*719987-1* 442
Ancistrum ovalifolium IPNI=*720073-1* 442
Ancistrum repens IPNI=*720073-1* 442
Ancylocalyx acuminata IPNI=*516568-1* 331
Ancylogyne macrocnemis IPNI=*1014801-1* 361
Ancylogyne peruviana IPNI=*54890-1* 361
Andinocleome anomala IPNI=*146919-1* 279
Andinocleome glandulosa IPNI=*147061-1* 279
Andinocleome pilosa IPNI=*147242-1* 279
Andira amazonum IPNI=*524446-1* 336
Andira bracteosa IPNI=*524446-1* 336
Andira jamaicensis IPNI=*1075415-2* 311
Andira oblonga IPNI=*12607-2* 311
Andira retusa IPNI=*12607-2* 311
Andira retusa var. *oblonga* IPNI=*12607-2* 311
Andira retusa var. *paraguariensis* IPNI=*12607-2* 311
Andira retusa var. *surinamensis* IPNI=*12607-2* 311
Andira surinamensis var. *ovatifoliolata* IPNI=*12607-2* 311
Andrachne trifoliata IPNI=*339921-1* 401
Andreoskia cardaminifolia IPNI=*286421-1* 277
Andriapetalum cayennense IPNI=*182728-2* 433
Andriapetalum polystachyum IPNI=*1076423-2* 433
Andriapetalum rubescens IPNI=*182723-2* 433
Andriapetalum rubescens var. *acuminatum* IPNI=*182723-2* 433
Andriapetalum sessilifolium IPNI=*182728-2* 433
Andriapetalum sprucei IPNI=*182723-2* 433
Andriapetalum suaveolens IPNI=*182731-2* 433
Andriapetalum yolombo IPNI=*182733-2* 433
Andrieuxia mexicana IPNI=*213524-1* 264
Androcentrum multiflorum IPNI=*36223-2* 359
Andromachia igniaria IPNI=*230435-1* 265
Andromachia sessiliflora IPNI=*173688-2* 267
Andromachia sessilifolia IPNI=*173688-2* 267
Andromeda anastomosans IPNI=*330459-1* 299

Andromeda glomerata IPNI=*107641-2* 299
Andromeda plumata IPNI=*319176-2* 298
Andromeda prostrata IPNI=*330618-1* 299
Andromeda purpurea IPNI=*330618-1* 299
Andromeda trinervis IPNI=*60458979-2* 298
Androphoranthus glandulosus IPNI=*44564-2* 388
Andropogon acicularis IPNI=*396110-1* 236
Andropogon aciculatus IPNI=*396110-1* 236
Andropogon aculeatus IPNI=*396110-1* 236
Andropogon acuminatus IPNI=*420951-1* 244
Andropogon adustus IPNI=*90411-2* 238
Andropogon allionii IPNI=*404772-1* 240
Andropogon altissimus IPNI=*404684-1* 240
Andropogon amaurus IPNI=*963150-1* 244
Andropogon amboinicus IPNI=*1156566-2* 238
Andropogon angustifolius IPNI=*1191417-2* 235
Andropogon angustifolius IPNI=*424391-1* 246
Andropogon annulatus IPNI=*398528-1* 237
Andropogon annulatus var. *subrepens* IPNI=*398532-1* 237
Andropogon aristatus IPNI=*398529-1* 237
Andropogon armillaris IPNI=*65406-3* 235
Andropogon arnottianus IPNI=*406281-1* 240
Andropogon articulatus IPNI=*1191417-2* 235
Andropogon arundinaceus IPNI=*422078-1* 245
Andropogon attenuatus IPNI=*12758-2* 233
Andropogon austrocaledonicus IPNI=*404772-1* 240
Andropogon avenaceus IPNI=*422139-1* 245
Andropogon barbatus IPNI=*60469006-2* 245
Andropogon barbinodis IPNI=*35184-2* 235
Andropogon barbinodis var. *perforatus* IPNI=*35184-2* 235
Andropogon bellardii IPNI=*404772-1* 240
Andropogon benthamianus IPNI=*420918-1* 244
Andropogon besseri IPNI=*422090-1* 245
Andropogon besukiensis IPNI=*404772-1* 240
Andropogon bicolor IPNI=*422090-1* 245
Andropogon bicornis IPNI=*396888-1* 236
Andropogon bicornis var. *absconditus* IPNI=*12758-2* 233
Andropogon bicornis var. *angustifolius* IPNI=*12758-2* 233
Andropogon bicornis var. *burchellii* IPNI=*12758-2* 233
Andropogon bicornis var. *gracillimus* IPNI=*12758-2* 233
Andropogon bicornis var. *hybridus* IPNI=*12758-2* 233
Andropogon bicornis var. *paranensis* IPNI=*12758-2* 233
Andropogon bicornis var. *virginicoides* IPNI=*12758-2* 233
Andropogon binatus IPNI=*398532-1* 237
Andropogon bipennatus IPNI=*240818-2* 245
Andropogon bladhii IPNI=*392701-1* 235
Andropogon brevifolius IPNI=*1126606-2* 244
Andropogon caesius IPNI=*396888-1* 236
Andropogon caespitosus IPNI=*420951-1* 244
Andropogon caespitosus IPNI=*90411-2* 238
Andropogon candidus IPNI=*90411-2* 238
Andropogon canescens IPNI=*424391-1* 246
Andropogon canus IPNI=*424391-1* 246
Andropogon caricosus IPNI=*398532-1* 237
Andropogon caucasicus IPNI=*392701-1* 235
Andropogon cerifer IPNI=*396896-1* 236
Andropogon chlorostachys IPNI=*90411-2* 238
Andropogon ciliaris IPNI=*400052-1* 238
Andropogon cirratus IPNI=*420874-1* 244
Andropogon citratus IPNI=*396896-1* 236
Andropogon citriodorus IPNI=*396896-1* 236
Andropogon citrosus IPNI=*396972-1* 236
Andropogon compactus IPNI=*422090-1* 245
Andropogon condensatus IPNI=*420877-1* 244
Andropogon condensatus subsp. *corymbosus* IPNI=*420877-1* 244
Andropogon condensatus subvar. *exserens* IPNI=*420877-1* 244
Andropogon condensatus subvar. *lactiflorus* IPNI=*420877-1* 244
Andropogon condensatus subvar. *latens* IPNI=*420877-1* 244
Andropogon condensatus var. *genuinus* IPNI=*420877-1* 244
Andropogon condensatus var. *paniculatus* IPNI=*420918-1* 244
Andropogon condylotrichus IPNI=*97916-2* 239
Andropogon condylotrichus var. *palmeri* IPNI=*97916-2* 239
Andropogon condylotrichus var. *piptatherus* IPNI=*97916-2* 239
Andropogon confertiflorus IPNI=*396972-1* 236
Andropogon connatus IPNI=*396888-1* 236
Andropogon contortus IPNI=*404772-1* 240
Andropogon contortus subvar. *hispidissimus* IPNI=*404772-1* 240
Andropogon contortus subvar. *secundus* IPNI=*404772-1* 240
Andropogon contortus var. *glaber* IPNI=*404772-1* 240
Andropogon contortus var. *macroglumis* IPNI=*404772-1* 240
Andropogon corymbosus IPNI=*12850-2* 233
Andropogon corymbosus var. *abbreviatus* IPNI=*12850-2* 233
Andropogon crupina IPNI=*422139-1* 245
Andropogon curtipendulus IPNI=*1198286-2* 235
Andropogon curvatus IPNI=*398532-1* 237
Andropogon dactyloides IPNI=*424391-1* 246
Andropogon debilis IPNI=*1126606-2* 244
Andropogon decolorans IPNI=*422139-1* 245
Andropogon densus IPNI=*12850-2* 233
Andropogon depressus IPNI=*398532-1* 237
Andropogon dianae IPNI=*12920-2* 234
Andropogon diatherus IPNI=*398608-1* 237
Andropogon digitatus IPNI=*1191417-2* 235
Andropogon digitatus IPNI=*424391-1* 246
Andropogon dissolutus IPNI=*424391-1* 246
Andropogon diversiflorus IPNI=*963150-1* 244
Andropogon doloensis IPNI=*97916-2* 239
Andropogon domingensis IPNI=*12920-2* 234
Andropogon domingensis IPNI=*420951-1* 244
Andropogon dubitatus IPNI=*422139-1* 245
Andropogon dubius IPNI=*90411-2* 238
Andropogon emersus IPNI=*35184-2* 235
Andropogon epiphyllus IPNI=*21052-2* 234
Andropogon excavatus IPNI=*396888-1* 236
Andropogon exsertus IPNI=*1126606-2* 244
Andropogon fabricii IPNI=*398888-1* 237
Andropogon fasciculatus IPNI=*404684-1* 240
Andropogon fastigiatus IPNI=*398608-1* 237
Andropogon feensis IPNI=*420951-1* 244
Andropogon filiformis IPNI=*398532-1* 237
Andropogon firmandus IPNI=*963150-1* 244
Andropogon firmus IPNI=*404772-1* 240
Andropogon flaccidus IPNI=*1126606-2* 244
Andropogon flavescens IPNI=*420951-1* 244
Andropogon floridus IPNI=*1126606-2* 244
Andropogon foliatus IPNI=*396888-1* 236
Andropogon fundaensis IPNI=*424391-1* 246
Andropogon galeottii IPNI=*240818-2* 245
Andropogon glaber IPNI=*404772-1* 240
Andropogon glomeratus subsp. *reinoldii* IPNI=*12850-2* 233
Andropogon glomeratus var. *abbreviatus* IPNI=*12850-2* 233
Andropogon glomeratus var. *corymbosus* IPNI=*12850-2* 233
Andropogon glomeratus var. *pumilus* IPNI=*12850-2* 233
Andropogon grandis IPNI=*396972-1* 236
Andropogon griffithsiae IPNI=*406281-1* 240
Andropogon haenkei IPNI=*392701-1* 235
Andropogon halepensis IPNI=*422139-1* 245
Andropogon halepensis subsp. *anatherus* IPNI=*422139-1* 245
Andropogon halepensis var. *muticus* IPNI=*422139-1* 245
Andropogon hamulatus IPNI=*396972-1* 236
Andropogon heteranthus IPNI=*12758-2* 233
Andropogon hirtiflorus IPNI=*420951-1* 244
Andropogon hirtiflorus var. *brevipedicellatus* IPNI=*420951-1* 244
Andropogon hirtiflorus var. *feensis* IPNI=*420951-1* 244
Andropogon hirtiflorus var. *genuinus* IPNI=*420951-1* 244
Andropogon hirtiflorus var. *oligostachyus* IPNI=*420951-1* 244

Andropogon hispidissimus IPNI=*404772-1* 240
Andropogon hispidus IPNI=*21052-2* 234
Andropogon hochstetteri IPNI=*398608-1* 237
Andropogon hohenackeri IPNI=*404772-1* 240
Andropogon humboldtianus IPNI=*240834-2* 245
Andropogon incompletus IPNI=*240818-2* 245
Andropogon incurvatus IPNI=*398529-1* 237
Andropogon inermis IPNI=*13112-2* 234
Andropogon insularis IPNI=*398888-1* 237
Andropogon integer IPNI=*1126606-2* 244
Andropogon intermedius IPNI=*392701-1* 235
Andropogon intermedius var. *acidula* IPNI=*392701-1* 235
Andropogon intortum IPNI=*1207790-2* 245
Andropogon inundatus IPNI=*392701-1* 235
Andropogon ischaemum IPNI=*1191417-2* 235
Andropogon javanicus IPNI=*396110-1* 236
Andropogon koenigii IPNI=*398529-1* 237
Andropogon lanuginosus IPNI=*12920-2* 234
Andropogon laxiflorus IPNI=*424391-1* 246
Andropogon lechleri IPNI=*420918-1* 244
Andropogon leptanthus IPNI=*392701-1* 235
Andropogon leptostachyus IPNI=*420951-1* 244
Andropogon leucopogon IPNI=*35184-2* 235
Andropogon leucostachyus subvar. *mas* IPNI=*12920-2* 234
Andropogon leucostachyus subvar. *subvillosus* IPNI=*12920-2* 234
Andropogon leucostachyus subvar. *typicus* IPNI=*12920-2* 234
Andropogon lhotskyi IPNI=*420918-1* 244
Andropogon ligularis IPNI=*424391-1* 246
Andropogon lolioides IPNI=*420874-1* 244
Andropogon lumeneensis IPNI=*420951-1* 244
Andropogon luteolus IPNI=*126405-2* 240
Andropogon macrolepis IPNI=*126405-2* 240
Andropogon macrourus IPNI=*12850-2* 233
Andropogon macrourus var. *abbreviatus* IPNI=*12850-2* 233
Andropogon macrourus var. *corymbosus* IPNI=*12850-2* 233
Andropogon macrourus var. *genuinus* IPNI=*12850-2* 233
Andropogon macrourus var. *pumilus* IPNI=*12850-2* 233
Andropogon megapotamicus IPNI=*424391-1* 246
Andropogon melanocarpus IPNI=*404802-1* 240
Andropogon messanensis IPNI=*404772-1* 240
Andropogon microstachyus IPNI=*420918-1* 244
Andropogon miliaceus IPNI=*422139-1* 245
Andropogon miliformis IPNI=*422139-1* 245
Andropogon minimus IPNI=*1126606-2* 244
Andropogon mollicomus IPNI=*398529-1* 237
Andropogon mollis IPNI=*424391-1* 246
Andropogon montanus IPNI=*392701-1* 235
Andropogon montufarii IPNI=*424391-1* 246
Andropogon muricatus IPNI=*396213-1* 236
Andropogon myosurus IPNI=*420951-1* 244
Andropogon nardus IPNI=*396972-1* 236
Andropogon nardus subsp. *cerifer* IPNI=*396896-1* 236
Andropogon nardus subsp. *nilagiricus* IPNI=*396972-1* 236
Andropogon nardus var. *luridus* IPNI=*396972-1* 236
Andropogon niger IPNI=*422090-1* 245
Andropogon nutans var. *incompletus* IPNI=*240818-2* 245
Andropogon nutans var. *stipoides* IPNI=*240834-2* 245
Andropogon obtusifolius IPNI=*1126606-2* 244
Andropogon odoratus IPNI=*392701-1* 235
Andropogon odoratus IPNI=*396213-1* 236
Andropogon oligostachyus IPNI=*420951-1* 244
Andropogon palmeri IPNI=*35184-2* 235
Andropogon paniculatus IPNI=*420918-1* 244
Andropogon panormitanus IPNI=*1191417-2* 235
Andropogon parviflorus IPNI=*1126606-2* 244
Andropogon patulus IPNI=*1191417-2* 235
Andropogon pennatus IPNI=*12920-2* 234
Andropogon perforatus IPNI=*35184-2* 235
Andropogon perfossus IPNI=*392701-1* 235
Andropogon pertusus IPNI=*65406-3* 235
Andropogon piptatherus IPNI=*97916-2* 239
Andropogon piptatherus var. *palmeri* IPNI=*97916-2* 239
Andropogon plumiger IPNI=*420918-1* 244
Andropogon plumosus IPNI=*424391-1* 246
Andropogon polydactylos IPNI=*60469006-2* 245
Andropogon polystachyos IPNI=*404772-1* 240
Andropogon polystictus IPNI=*404802-1* 240
Andropogon pseudograya IPNI=*420951-1* 244
Andropogon pseudohirtus IPNI=*396972-1* 236
Andropogon pubescens IPNI=*398529-1* 237
Andropogon punctatus IPNI=*392701-1* 235
Andropogon racemosus IPNI=*77105978-1* 235
Andropogon rectirhachis IPNI=*420877-1* 244
Andropogon reinoldii IPNI=*12850-2* 233
Andropogon riedelii IPNI=*420951-1* 244
Andropogon riedelii var. *genuinus* IPNI=*420951-1* 244
Andropogon riedelii var. *multirameus* IPNI=*420951-1* 244
Andropogon rostratus IPNI=*90411-2* 238
Andropogon roxburghii IPNI=*396896-1* 236
Andropogon rubens IPNI=*422090-1* 245
Andropogon rufus IPNI=*405583-1* 240
Andropogon rugosus IPNI=*406281-1* 240
Andropogon ruprechtii IPNI=*126405-2* 240
Andropogon saccharatus IPNI=*422090-1* 245
Andropogon saccharoides IPNI=*35205-2* 235
Andropogon saccharoides subsp. *leucopogon* IPNI=*35184-2* 235
Andropogon saccharoides subvar. *perforatus* IPNI=*35184-2* 235
Andropogon saccharoides var. *barbinodis* IPNI=*35184-2* 235
Andropogon sanguineus IPNI=*420951-1* 244
Andropogon sanguineus var. *brevipedicellatus* IPNI=*420951-1* 244
Andropogon schlumbergeri IPNI=*35184-2* 235
Andropogon schlumbergeri var. *andreae* IPNI=*35184-2* 235
Andropogon scoparius IPNI=*420918-1* 244
Andropogon scrobiculatus IPNI=*404802-1* 240
Andropogon secundus IPNI=*424391-1* 246
Andropogon segetus IPNI=*406281-1* 240
Andropogon semiberbis IPNI=*420951-1* 244
Andropogon semiberbis subvar. *pruinatus* IPNI=*420951-1* 244
Andropogon semiberbis var. *incertus* IPNI=*420951-1* 244
Andropogon semiglaber IPNI=*420951-1* 244
Andropogon serratus IPNI=*398532-1* 237
Andropogon sorghum IPNI=*422090-1* 245
Andropogon sorghum subsp. *abyssinicus* IPNI=*422078-1* 245
Andropogon sorghum subsp. *halepensis* IPNI=*422139-1* 245
Andropogon sorghum subsp. *sativus* IPNI=*422090-1* 245
Andropogon sorghum subsp. *verticilliflorus* IPNI=*422078-1* 245
Andropogon sorghum subsp. *vogelianus* IPNI=*422078-1* 245
Andropogon sorghum subvar. *aristatus* IPNI=*422078-1* 245
Andropogon sorghum subvar. *badius* IPNI=*422090-1* 245
Andropogon sorghum subvar. *breviaristatus* IPNI=*422078-1* 245
Andropogon sorghum subvar. *fragilis* IPNI=*422090-1* 245
Andropogon sorghum subvar. *japonicus* IPNI=*422090-1* 245
Andropogon sorghum subvar. *leiocladus* IPNI=*422139-1* 245
Andropogon sorghum subvar. *lividus* IPNI=*422090-1* 245
Andropogon sorghum subvar. *longiaristatus* IPNI=*422078-1* 245
Andropogon sorghum subvar. *muticus* IPNI=*422139-1* 245
Andropogon sorghum subvar. *niger* IPNI=*422090-1* 245
Andropogon sorghum subvar. *rubens* IPNI=*422090-1* 245
Andropogon sorghum subvar. *splendidus* IPNI=*422090-1* 245
Andropogon sorghum subvar. *submuticus* IPNI=*422078-1* 245
Andropogon sorghum subvar. *trachycladus* IPNI=*422139-1* 245
Andropogon sorghum var. *abyssinicus* IPNI=*422090-1* 245
Andropogon sorghum var. *aegyptiacus* IPNI=*422090-1* 245
Andropogon sorghum var. *aethiopicus* IPNI=*422078-1* 245
Andropogon sorghum var. *ankolib* IPNI=*422090-1* 245
Andropogon sorghum var. *bicarinatus* IPNI=*422090-1* 245

Andropogon sorghum var. *burmanicus* IPNI=*422090-1* 245
Andropogon sorghum var. *campanus* IPNI=*422090-1* 245
Andropogon sorghum var. *caudatus* IPNI=*422090-1* 245
Andropogon sorghum var. *corymbosus* IPNI=*422090-1* 245
Andropogon sorghum var. *durra* IPNI=*422090-1* 245
Andropogon sorghum var. *effusus* IPNI=*422078-1* 245
Andropogon sorghum var. *ehrenbergianus* IPNI=*422090-1* 245
Andropogon sorghum var. *exiguus* IPNI=*422139-1* 245
Andropogon sorghum var. *fulvus* IPNI=*422090-1* 245
Andropogon sorghum var. *glaberrimus* IPNI=*422090-1* 245
Andropogon sorghum var. *globosus* IPNI=*422090-1* 245
Andropogon sorghum var. *halepensis* IPNI=*422139-1* 245
Andropogon sorghum var. *hybridus* IPNI=*422090-1* 245
Andropogon sorghum var. *irungu* IPNI=*422090-1* 245
Andropogon sorghum var. *javanicus* IPNI=*422090-1* 245
Andropogon sorghum var. *lasiorhachis* IPNI=*422090-1* 245
Andropogon sorghum var. *leiostachyus* IPNI=*422090-1* 245
Andropogon sorghum var. *melanospermus* IPNI=*422090-1* 245
Andropogon sorghum var. *miliiformis* IPNI=*422090-1* 245
Andropogon sorghum var. *nervosus* IPNI=*422090-1* 245
Andropogon sorghum var. *obovatus* IPNI=*422090-1* 245
Andropogon sorghum var. *ovulifer* IPNI=*422090-1* 245
Andropogon sorghum var. *perennis* IPNI=*422139-1* 245
Andropogon sorghum var. *pulcher* IPNI=*422090-1* 245
Andropogon sorghum var. *roxburghii* IPNI=*422090-1* 245
Andropogon sorghum var. *rufescens* IPNI=*422090-1* 245
Andropogon sorghum var. *rugulosus* IPNI=*422090-1* 245
Andropogon sorghum var. *schimperi* IPNI=*422090-1* 245
Andropogon sorghum var. *subglabrescens* IPNI=*422090-1* 245
Andropogon sorghum var. *subglobosus* IPNI=*422090-1* 245
Andropogon sorghum var. *technicus* IPNI=*422090-1* 245
Andropogon sorghum var. *vulgaris* IPNI=*422090-1* 245
Andropogon sorghum var. *wightii* IPNI=*422090-1* 245
Andropogon spathaceus IPNI=*12850-2* 233
Andropogon spathiflorus IPNI=*13112-2* 234
Andropogon spathiflorus var. *inermis* IPNI=*13112-2* 234
Andropogon spicatus IPNI=*424391-1* 246
Andropogon stapfii IPNI=*422078-1* 245
Andropogon stenophyllus IPNI=*424391-1* 246
Andropogon stipoides IPNI=*240834-2* 245
Andropogon subglabrescens IPNI=*422090-1* 245
Andropogon subulatus IPNI=*396110-1* 236
Andropogon tenellus IPNI=*1126606-2* 244
Andropogon tenellus IPNI=*398532-1* 237
Andropogon tenuifolius IPNI=*90411-2* 238
Andropogon thwaitesii IPNI=*396972-1* 236
Andropogon thymiodorus IPNI=*90411-2* 238
Andropogon tong-dong IPNI=*406281-1* 240
Andropogon tortus IPNI=*424391-1* 246
Andropogon trichopus IPNI=*240834-2* 245
Andropogon trichospirus IPNI=*424391-1* 246
Andropogon truchmenorum IPNI=*422090-1* 245
Andropogon truncatus IPNI=*424391-1* 246
Andropogon tumbackianus IPNI=*422139-1* 245
Andropogon undatus IPNI=*409823-1* 242
Andropogon usorum IPNI=*422090-1* 245
Andropogon vachellii IPNI=*392701-1* 235
Andropogon vaginatus IPNI=*420951-1* 244
Andropogon velatus IPNI=*420951-1* 244
Andropogon verticilliflorus IPNI=*422078-1* 245
Andropogon vestitus IPNI=*424391-1* 246
Andropogon villosus IPNI=*1191417-2* 235
Andropogon virens IPNI=*21052-2* 234
Andropogon virginicus IPNI=*12920-2* 234
Andropogon virginicus subsp. *leucostachyus* IPNI=*12920-2* 234
Andropogon virginicus var. *abbreviatus* IPNI=*12850-2* 233
Andropogon virginicus var. *corymbosus* IPNI=*12850-2* 233
Andropogon vulgaris IPNI=*422090-1* 245
Andropogon zizanioides IPNI=*396213-1* 236
Androsaemum adenophyllum IPNI=*433966-1* 394
Androsaemum androsaemum IPNI=*433187-1* 393
Androsaemum calycinum IPNI=*433287-1* 393
Androsaemum officinale IPNI=*433187-1* 393
Androsaemum vulgare IPNI=*433187-1* 393
Androsaemum xylosteifolium IPNI=*433966-1* 394
Androstylanthus paraensis IPNI=*317010-2* 439
Andrzeiowskia cardaminifolia IPNI=*286421-1* 277
Anechites asperuginis IPNI=*13196-2* 339
Anechites lappulaceus IPNI=*13196-2* 339
Aneilema bracteolatum var. *majus* IPNI=*172873-1* 224
Aneilema bracteolatum var. *minus* IPNI=*172873-1* 224
Aneilema compressum IPNI=*172873-1* 224
Aneilema debile IPNI=*172873-1* 224
Aneilema diversifolium IPNI=*172873-1* 224
Aneilema ecuadoriense IPNI=*79497-2* 224
Aneilema floribundum IPNI=*77184055-1* 224
Aneilema foliosum IPNI=*172873-1* 224
Aneilema gracile IPNI=*42443-2* 223
Aneilema gracilis f. *bicolor* IPNI=*42443-2* 223
Aneilema junghunianum IPNI=*172873-1* 224
Aneilema lancifolium IPNI=*172873-1* 224
Aneilema malabaricum IPNI=*172873-1* 224
Aneilema minutum IPNI=*172873-1* 224
Aneilema nudicaule IPNI=*172873-1* 224
Aneilema nudiflorum IPNI=*172873-1* 224
Aneilema radicans IPNI=*172873-1* 224
Aneilema trichocoleum IPNI=*172873-1* 224
Anelasma concolor IPNI=*147-2* 434
Anelasma gardnerianum IPNI=*147-2* 434
Anelasma guianense IPNI=*147-2* 434
Anelasma jamaicense IPNI=*580887-1* 435
Anelasma laurifolium IPNI=*147-2* 434
Anelasma martianum IPNI=*147-2* 434
Anelasma minutiflora IPNI=*580887-1* 435
Anelasma minutiflorum IPNI=*580887-1* 435
Anelasma pallidum IPNI=*147-2* 434
Anelasma spruceanum IPNI=*147-2* 434
Anelytrum avenaceum IPNI=*391443-1* 234
Anemia adiantifolia var. *carvifolia* IPNI=*1031990-2* 179
Anemia adiantifolia var. *pumila* IPNI=*1031990-2* 179
Anemia adiantifolia var. *subaurita* IPNI=*1031990-2* 179
Anemia aspleniifolia IPNI=*1031990-2* 179
Anemia carvifolia IPNI=*1031990-2* 179
Anemia cordifolia IPNI=*1202412-2* 179
Anemia densa IPNI=*1202412-2* 179
Anemia filiculifolia IPNI=*30017613-2* 176
Anemia fraxinifolia IPNI=*1202412-2* 179
Anemia haenkei IPNI=*1202412-2* 179
Anemia hirta IPNI=*1202412-2* 179
Anemia laciniata IPNI=*1202412-2* 179
Anemia lanceolata IPNI=*1202412-2* 179
Anemia langsdorffiana IPNI=*1202412-2* 179
Anemia longifolia IPNI=*1202412-2* 179
Anemia longifolia var. *laevis* IPNI=*1202412-2* 179
Anemia longifolia var. *undulata* IPNI=*1202412-2* 179
Anemia obliqua IPNI=*1202412-2* 179
Anemia phyllitidis f. *auritolobata* IPNI=*1202412-2* 179
Anemia phyllitidis f. *minor* IPNI=*1202412-2* 179
Anemia phyllitidis f. *subtripinnatifida* IPNI=*1202412-2* 179
Anemia phyllitidis f. *transiens* IPNI=*1202412-2* 179
Anemia phyllitidis f. *transitoria* IPNI=*1202412-2* 179
Anemia phyllitidis var. *caryotidea* IPNI=*1202412-2* 179
Anemia phyllitidis var. *fraxinifolia* IPNI=*1202412-2* 179
Anemia phyllitidis var. *hirta* IPNI=*1202412-2* 179
Anemia phyllitidis var. *longifolia* IPNI=*1202412-2* 179
Anemia phyllitidis var. *obliqua* IPNI=*1202412-2* 179

Anemia phyllitidis var. *pygmaea* IPNI=*1202412-2* 179
Anemia sorbifolia IPNI=*1202412-2* 179
Anemidictyon densum IPNI=*1202412-2* 179
Anemidictyon fraxinifolium IPNI=*1202412-2* 179
Anemidictyon haenkei IPNI=*1202412-2* 179
Anemidictyon laciniatum IPNI=*1202412-2* 179
Anemidictyon obliquum IPNI=*1202412-2* 179
Anemidictyon phyllitidis IPNI=*1202412-2* 179
Anemidictyon phyllitidis var. *cordifolium* IPNI=*1202412-2* 179
Anemidictyon phyllitidis var. *fraxinifolia* IPNI=*1202412-2* 179
Anemidictyon phyllitidis var. *fraxinifolium* IPNI=*1202412-2* 179
Anemidictyon phyllitidis var. *laciniatum* IPNI=*1202412-2* 179
Anemidictyon phyllitidis var. *longifolium* IPNI=*1202412-2* 179
Anemirhiza adiantifolia IPNI=*1031990-2* 179
Anemone alba IPNI=*708314-1* 436
Anemone albiflora IPNI=*708314-1* 436
Anemone coccinea IPNI=*708314-1* 436
Anemone coronaria f. *albiflora* IPNI=*708314-1* 436
Anemone coronaria f. *parviflora* IPNI=*708314-1* 436
Anemone coronaria var. *albiflora* IPNI=*708314-1* 436
Anemone coronaria var. *cyanea* IPNI=*708314-1* 436
Anemone coronaria var. *mouansii* IPNI=*708314-1* 436
Anemone coronaria var. *parviflora* IPNI=*708314-1* 436
Anemone coronaria var. *phaenicea* IPNI=*708314-1* 436
Anemone coronaria var. *purpurea* IPNI=*708314-1* 436
Anemone coronaria var. *rissoana* IPNI=*708314-1* 436
Anemone coronaria var. *ventreana* IPNI=*708314-1* 436
Anemone cyanea IPNI=*708314-1* 436
Anemone grassensis IPNI=*708314-1* 436
Anemone hupehensis f. *alba* IPNI=*77184471-1* 436
Anemone hupehensis IPNI=*77184471-1* 436
Anemone japonica var. *hupehensis* IPNI=*77184471-1* 436
Anemone kusnetzowii IPNI=*708314-1* 436
Anemone matsudae IPNI=*77184471-1* 436
Anemone messarensis IPNI=*708314-1* 436
Anemone nobilis IPNI=*708314-1* 436
Anemone oenanthe IPNI=*708314-1* 436
Anemone pusilla IPNI=*708314-1* 436
Anemone rosea IPNI=*708314-1* 436
Anemone rosea var. *mouansii* IPNI=*708314-1* 436
Anemone variata IPNI=*708314-1* 436
Anemone ventreana IPNI=*708314-1* 436
Anemone versicolor IPNI=*708314-1* 436
Anemopaegma lehmannii IPNI=*13609-2* 362
Anemopaegma leptosiphon IPNI=*153710-2* 364
Anemopaegma pachypus IPNI=*152672-2* 364
Anemopaegma tonduzianum IPNI=*108550-1* 362
Anepsias moritzianus IPNI=*88631-1* 202
Anethum arvense IPNI=*837530-1* 251
Anethum dulce IPNI=*842680-1* 252
Anethum foeniculum IPNI=*842680-1* 252
Anethum graveolens subsp. *australe* IPNI=*837530-1* 251
Anethum graveolens subsp. *sowa* IPNI=*837530-1* 251
Anethum graveolens var. *anatolicum* IPNI=*837530-1* 251
Anethum graveolens var. *copiosum* IPNI=*837530-1* 251
Anethum graveolens var. *nanum* IPNI=*837530-1* 251
Anethum graveolens var. *parvifolium* IPNI=*837530-1* 251
Anethum graveolens var. *tenerifrons* IPNI=*837530-1* 251
Anethum minus IPNI=*842680-1* 252
Anethum panmorium IPNI=*842680-1* 252
Anethum pinnatum IPNI=*837486-1* 251
Anethum rupestre IPNI=*842680-1* 252
Anethum sowa IPNI=*837530-1* 251
Aneuriscus aubletii IPNI=*429281-1* 385
Aneuriscus exserens IPNI=*429281-1* 385
Angadenia almadensis IPNI=*151979-2* 341
Angadenia cururu IPNI=*291380-2* 342
Angadenia grandifolia IPNI=*291380-2* 342
Angadenia nitida IPNI=*1054185-2* 342
Angadenia reticulata IPNI=*152052-2* 341
Angadenia sprucei IPNI=*291380-2* 342
Angadenia sylvestris IPNI=*77191232-1* 342
Angelica archangelica subsp. *norwegica* IPNI=*837560-1* 251
Angelica commutata IPNI=*837560-1* 251
Angelica discocarpa IPNI=*837560-1* 251
Angelica graveolens IPNI=*837530-1* 251
Angelica intermedia IPNI=*837560-1* 251
Angelica major IPNI=*837560-1* 251
Angelica norwegica IPNI=*837560-1* 251
Angelica officinalis IPNI=*837560-1* 251
Angelica procera IPNI=*837560-1* 251
Angelica sativa IPNI=*837560-1* 251
Angelina divergentifolia IPNI=*582285-1* 185
Angelonia evitae IPNI=*798952-1* 375
Angelonia gardneri IPNI=*798952-1* 375
Angelonia hirta IPNI=*798952-1* 375
Angelonia leandrii IPNI=*798952-1* 375
Angraecum maculatum IPNI=*1073117-2* 222
Angraecum monophyllum IPNI=*1073117-2* 222
Anguillaria coriacea IPNI=*1028287-2* 304
Anguillaria icacorea IPNI=*18168-2* 303
Anguloa grandiflora IPNI=*243490-2* 223
Anguloa superba IPNI=*614722-1* 217
Anguloa uniflora var. *treyeranii* IPNI=*616523-1* 217
Anguria arrabidae IPNI=*1010655-1* 294
Anguria balfoureana IPNI=*114819-2* 294
Anguria bignoniacea IPNI=*114819-2* 294
Anguria boissieriana IPNI=*211990-2* 295
Anguria cissoides IPNI=*114819-2* 294
Anguria citrullus IPNI=*291938-1* 293
Anguria diversifolia IPNI=*211990-2* 295
Anguria dunlapii IPNI=*211990-2* 295
Anguria elliptica IPNI=*211990-2* 295
Anguria eriantha IPNI=*292789-1* 294
Anguria eriantha IPNI=*292814-1* 294
Anguria glomerata IPNI=*592558-1* 294
Anguria heterophylla IPNI=*114819-2* 294
Anguria inaequalis IPNI=*114819-2* 294
Anguria kegeliana IPNI=*1010655-1* 294
Anguria lanata IPNI=*292789-1* 294
Anguria lobata IPNI=*1010655-1* 294
Anguria longipedunculata IPNI=*211990-2* 295
Anguria longiracemosa IPNI=*211990-2* 295
Anguria makoyana IPNI=*292814-1* 294
Anguria mathewsii IPNI=*114819-2* 294
Anguria multiflora IPNI=*1010655-1* 294
Anguria oblongifolia IPNI=*211990-2* 295
Anguria pachyphylla IPNI=*211990-2* 295
Anguria pallida IPNI=*211990-2* 295
Anguria pedatisecta IPNI=*292512-1* 294
Anguria pedisecta IPNI=*292512-1* 294
Anguria pyrrhocephala IPNI=*292789-1* 294
Anguria schomburgkiana IPNI=*211990-2* 295
Anguria speciosa IPNI=*292789-1* 294
Anguria spinulosa IPNI=*1010655-1* 294
Anguria tabascensis IPNI=*211990-2* 295
Anguria tessmannii IPNI=*114819-2* 294
Anguria treslingiana IPNI=*211990-2* 295
Anguria trifoliata IPNI=*114819-2* 294
Anguria triphylla IPNI=*211990-2* 295
Anguria triphylla var. *acuminata* IPNI=*211990-2* 295
Anguria triphylla var. *pallida* IPNI=*211990-2* 295
Anguria ucayalina IPNI=*114819-2* 294
Anguria vogliana IPNI=*211990-2* 295
Anguria wageneriana IPNI=*1010655-1* 294
Anguriopsis margaritensis IPNI=*592558-1* 294

Aniba amazonica IPNI=*13872-2* 182
Aniba compacta IPNI=*13869-2* 182
Aniba duckei IPNI=*13880-2* 182
Aniba elliptica IPNI=*13816-2* 182
Aniba gigantifolia IPNI=*13841-2* 182
Aniba gonggrijpii IPNI=*13863-2* 182
Aniba mas IPNI=*13863-2* 182
Aniba megacarpa IPNI=*13816-2* 182
Aniba pittieri IPNI=*13841-2* 182
Aniba pseudocoto IPNI=*31062-2* 182
Aniba rosodora f. *typica* IPNI=*13880-2* 182
Aniba rosodora var. *amazonica* IPNI=*13880-2* 182
Aniba sericea IPNI=*464535-1* 183
Anigosia flavida IPNI=*429575-1* 224
Anigozanthos coccineus IPNI=*429575-1* 224
Anigozanthos flavescens IPNI=*429575-1* 224
Anigozanthos flavidus var. *bicolor* IPNI=*429575-1* 224
Anigozanthos grandiflorus IPNI=*429575-1* 224
Anigozanthos manglesii IPNI=*429575-1* 224
Anil hirsuta IPNI=*499456-1* 321
Anil tinctoria var. *brachycarpa* IPNI=*500063-1* 321
Anil tinctoria var. *vera* IPNI=*500063-1* 321
Anisactis segetalis IPNI=*60442790-2* 252
Anisanthera hastata IPNI=*489042-1* 316
Anisanthera versicolor IPNI=*489042-1* 316
Anisocalyx limnanthiflorus IPNI=*1072674-2* 376
Anisolobus cururu IPNI=*291380-2* 342
Anisolobus cururu var. *grandifolius* IPNI=*291380-2* 342
Anisolobus distinctus IPNI=*291380-2* 342
Anisolobus fockei IPNI=*291380-2* 342
Anisolobus hostmannii IPNI=*291380-2* 342
Anisolobus kappleri IPNI=*291380-2* 342
Anisolobus oblongus IPNI=*291380-2* 342
Anisolobus puncticulosus IPNI=*291380-2* 342
Anisolobus sprucei IPNI=*291380-2* 342
Anisomeris boliviana IPNI=*57096-2* 348
Anisomeris brachyloba IPNI=*746662-1* 348
Anisomeris paniculata IPNI=*57096-2* 348
Anisomeris peruviana IPNI=*57096-2* 348
Anisomeris purpusii IPNI=*746662-1* 348
Anisomeris spinosa IPNI=*746662-1* 348
Anisophyllum densiflorum IPNI=*1100936-2* 390
Anisophyllum emarginatum IPNI=*101976-2* 391
Anisophyllum flexuosum IPNI=*347331-1* 390
Anisophyllum glabratum IPNI=*347331-1* 390
Anisophyllum hypericifolium IPNI=*101671-2* 390
Anisophyllum hyssopifolium IPNI=*101676-2* 390
Anisophyllum lasiocarpum IPNI=*347064-1* 390
Anisophyllum maculatum IPNI=*347236-1* 390
Anisophyllum orbiculatum IPNI=*347590-1* 390
Anisophyllum prostratum IPNI=*101913-2* 390
Anisophyllum serpens IPNI=*101976-2* 391
Anisophyllum thymifolium IPNI=*102034-2* 391
Anisostichus crucigera IPNI=*77100444-1* 362
Anisum odoratum IPNI=*846658-1* 252
Anisum officinale IPNI=*846658-1* 252
Anisum officinarum IPNI=*846658-1* 252
Anisum vulgare IPNI=*846658-1* 252
Anneslea amazonica IPNI=*605776-1* 181
Anneslea brasiliana IPNI=*605776-1* 181
Anneslia caracasana IPNI=*287164-2* 337
Anneslia cumingii IPNI=*1161746-2* 313
Anneslia fasciculata IPNI=*316281-2* 313
Anneslia formosa IPNI=*287168-2* 337
Anneslia fulgens IPNI=*42154-2* 313
Anneslia guildingii IPNI=*42179-2* 313
Anneslia haematocephala IPNI=*42180-2* 313
Anneslia magdalenae IPNI=*1051245-2* 313
Anneslia pittieri IPNI=*42259-2* 313
Anneslia portoricensis IPNI=*943976-1* 337
Annona amazonica var. *lancifolia* IPNI=*14221-2* 186
Annona ambotay subsp. *occidentalis* IPNI=*72148-1* 186
Annona asiatica IPNI=*72319-1* 187
Annona australis IPNI=*72208-1* 186
Annona axilliflora IPNI=*912648-1* 188
Annona biflora IPNI=*72259-1* 187
Annona bonplandiana IPNI=*14308-2* 187
Annona cearaensis IPNI=*14308-2* 187
Annona cherimolioides var. *amplifolia* IPNI=*300526-2* 186
Annona chrysopetala IPNI=*912648-1* 188
Annona cinerea IPNI=*72319-1* 187
Annona distincta IPNI=*72319-1* 187
Annona excelsa IPNI=*72295-1* 187
Annona forsskalii IPNI=*72319-1* 187
Annona glabra IPNI=*72319-1* 187
Annona guaricensis IPNI=*14285-2* 186
Annona hostmannii IPNI=*73626-1* 189
Annona humboldtiana IPNI=*72295-1* 187
Annona humboldtii IPNI=*72295-1* 187
Annona involucrata IPNI=*72289-1* 187
Annona jenmanii IPNI=*72310-1* 187
Annona klainei IPNI=*72208-1* 186
Annona laevis IPNI=*72295-1* 187
Annona lanceolata IPNI=*73503-1* 188
Annona laurifolia IPNI=*72208-1* 186
Annona lepidota IPNI=*1077481-2* 188
Annona longifolia IPNI=*106440-2* 188
Annona longifolia IPNI=*72295-1* 187
Annona lutescens IPNI=*72295-1* 187
Annona manirote IPNI=*72289-1* 187
Annona marcgravii IPNI=*72257-1* 186
Annona montana f. *marcgravii* IPNI=*72257-1* 186
Annona muricata f. *mirabilis* IPNI=*14308-2* 187
Annona muscosa IPNI=*72259-1* 187
Annona nitida IPNI=*73503-1* 188
Annona obtusiflora IPNI=*72259-1* 187
Annona obtusifolia IPNI=*72259-1* 187
Annona palustris IPNI=*72208-1* 186
Annona palustris var. *grandifolia* IPNI=*72208-1* 186
Annona peduncularis IPNI=*259508-2* 189
Annona peruviana IPNI=*72208-1* 186
Annona pisonis IPNI=*72257-1* 186
Annona poeppigiana IPNI=*72324-1* 187
Annona prestoei IPNI=*72289-1* 187
Annona primigenia IPNI=*72295-1* 187
Annona pteropetala IPNI=*72259-1* 187
Annona pubescens IPNI=*927235-1* 186
Annona punctata IPNI=*912648-1* 188
Annona reticulata IPNI=*72259-1* 187
Annona reticulata var. *mucosa* IPNI=*72259-1* 187
Annona reticulata var. *primigenia* IPNI=*72295-1* 187
Annona rhombipetala IPNI=*106440-2* 188
Annona riparia IPNI=*72295-1* 187
Annona sphaerocarpa IPNI=*72257-1* 186
Annona squamosa f. *parvifolia* IPNI=*72319-1* 187
Annona squamosa IPNI=*72259-1* 187
Annona squamosa IPNI=*72319-1* 187
Annona tessmannii IPNI=*72224-1* 186
Annona trinitensis IPNI=*72310-1* 187
Annona tripetala IPNI=*927235-1* 186
Annona trunciflora IPNI=*72204-1* 186
Annona uliginosa IPNI=*72208-1* 186
Anoda acerifolia var. *minoriflora* IPNI=*1218005-2* 407
Anoda arizonica IPNI=*1218005-2* 407
Anoda arizonica var. *digitata* IPNI=*1218005-2* 407
Anoda brachyantha IPNI=*1218005-2* 407

Anoda cristata var. *brachyantha* IPNI=*1218005-2* 407
Anoda cristata var. *typica* IPNI=*1218005-2* 407
Anoda dilleniana IPNI=*1218005-2* 407
Anoda fernandeziana IPNI=*1218005-2* 407
Anoda hastata IPNI=*1218005-2* 407
Anoda incarnata IPNI=*1218005-2* 407
Anoda lavateroides IPNI=*1218005-2* 407
Anoda populifolia IPNI=*1218005-2* 407
Anoda strictiflora IPNI=*30017389-2* 412
Anoda triangularis IPNI=*1218005-2* 407
Anoda triangularis var. *digitata* IPNI=*1218005-2* 407
Anoda triloba IPNI=*1218005-2* 407
Anoectochilus querceticola IPNI=*77125705-1* 217
Anogramma bifida IPNI=*77110633-1* 178
Anoma moringa IPNI=*584736-1* 280
Anomalosicyos quinquelobatus IPNI=*294002-1* 295
Anomopanax cumingianus IPNI=*91749-1* 254
Anomospermum grandifolium IPNI=*89958-2* 435
Anomospermum hostmannii IPNI=*176827-2* 435
Anomospermum japurense IPNI=*34845-2* 434
Anomospermum lucidum IPNI=*176827-2* 435
Anomospermum oblongatum IPNI=*176827-2* 435
Anomospermum schomburgkii IPNI=*176827-2* 435
Anomospermum schomburgkii var. *lucidum* IPNI=*176827-2* 435
Anomospermum ulei IPNI=*89958-2* 435
Anonis americana IPNI=*247544-2* 334
Anonocarpus amazonicus IPNI=*29952-2* 437
Anonocarpus peruvianus IPNI=*850488-1* 437
Anoplophytum flexuosum var. *pallidum* IPNI=*254406-2* 229
Anotis conferta IPNI=*17992-2* 347
Anotis filiformis IPNI=*17992-2* 347
Anotis hypnoides IPNI=*17992-2* 347
Anotis juniperifolia IPNI=*18012-2* 347
Anotis laricifolia IPNI=*18012-2* 347
Anotis longiflora IPNI=*745065-1* 347
Anotis microphylla IPNI=*17992-2* 347
Anotis pilifera IPNI=*17992-2* 347
Anotis serpens IPNI=*17992-2* 347
Anotis thymifolia IPNI=*18012-2* 347
Anotis thymifolia var. *hyssopifolia* IPNI=*18012-2* 347
Anotis thymifolia var. *thesioides* IPNI=*18012-2* 347
Anredera americana IPNI=*14549-2* 283
Anredera cordifolia subsp. *gracilis* IPNI=*14549-2* 283
Anredera cumingii IPNI=*1153747-2* 283
Anredera leptostachys IPNI=*1153747-2* 283
Anredera scandens IPNI=*1153747-2* 283
Anredera spicata IPNI=*1153747-2* 283
Anredera vesiculosa IPNI=*1153747-2* 283
Antelaea azadirachta IPNI=*1213180-2* 454
Antelaea canescens IPNI=*1213180-2* 454
Antelaea javanica IPNI=*1213180-2* 454
Antennaria gnaphalioides IPNI=*77197189-1* 259
Antennaria margaritacea IPNI=*30173179-2* 257
Antennaria monoica IPNI=*77197189-1* 259
Anthactinia walkeri IPNI=*321973-2* 400
Anthadenia sesamoides IPNI=*675971-1* 375
Antheilema paniculata IPNI=*54389-1* 361
Anthema arborea IPNI=*561541-1* 410
Anthemis × artemisifolia IPNI=*77111328-1* 259
Anthemis × grandiflora IPNI=*77111328-1* 259
Anthemis × stipulacea IPNI=*77111328-1* 259
Anthemis abyssinica IPNI=*1186403-2* 261
Anthemis americana IPNI=*213524-1* 264
Anthemis apiifolia IPNI=*192489-1* 259
Anthemis aurea IPNI=*192489-1* 259
Anthemis buphthalmoides IPNI=*213524-1* 264
Anthemis chamomilla-romana IPNI=*192489-1* 259
Anthemis cotula IPNI=*1186403-2* 261
Anthemis cotula subsp. *lithuanica* IPNI=*1186403-2* 261
Anthemis cotula var. *hierosolymitana* IPNI=*1186403-2* 261
Anthemis cotula-foetida IPNI=*1186403-2* 261
Anthemis cotuloides IPNI=*1186403-2* 261
Anthemis frutescens IPNI=*178745-1* 257
Anthemis galilaea var. *hierosolymitana* IPNI=*1186403-2* 261
Anthemis hondensis IPNI=*30439090-2* 272
Anthemis nobilis IPNI=*192489-1* 259
Anthemis nobilis var. *aurea* IPNI=*192489-1* 259
Anthemis nobilis var. *discoidea* IPNI=*192489-1* 259
Anthemis occidentalis IPNI=*213524-1* 264
Anthemis odorata IPNI=*192489-1* 259
Anthemis oppositifolia IPNI=*213524-1* 264
Anthemis ovatifolia IPNI=*213524-1* 264
Anthemis parthenioides IPNI=*192489-1* 259
Anthemis repens IPNI=*174552-1* 256
Anthemis santolinoides IPNI=*192489-1* 259
Anthemis sulphurea IPNI=*1186403-2* 261
Anthemis trinervia IPNI=*174552-1* 256
Anthemis valentina IPNI=*208446-2* 268
Anthemis variabilis IPNI=*1186403-2* 261
Anthemis viridis IPNI=*1186403-2* 261
Anthemis vulgaris IPNI=*154715-2* 265
Anthenantia columbiensis IPNI=*27415-2* 234
Anthenantia gigantea IPNI=*27415-2* 234
Anthephora cuspidata IPNI=*14931-2* 234
Anthephora elegans IPNI=*14931-2* 234
Anthephora elegans var. *armata* IPNI=*14931-2* 234
Anthephora elegans var. *villosa* IPNI=*14931-2* 234
Anthephora villosa IPNI=*14931-2* 234
Anthericum comosum IPNI=*532810-1* 215
Anthericum ensiforme IPNI=*65385-1* 216
Anthericum longituberosum IPNI=*532810-1* 215
Anthericum sternbergianum IPNI=*532810-1* 215
Anthericum vallis-trappii IPNI=*532810-1* 215
Antherylium floribundum IPNI=*30023859-2* 416
Antherylium grisleoides IPNI=*30023859-2* 416
Anthodon apiculatum IPNI=*72123-2* 290
Anthodon ellipticum IPNI=*162581-1* 291
Anthodon glomeratum IPNI=*162581-1* 291
Anthodon grandiflorum IPNI=*225455-2* 291
Anthodon laevigatum IPNI=*190399-2* 291
Antholyza meriana IPNI=*89940-2* 216
Anthopterus bracteatus IPNI=*858211-1* 298
Anthoxanthum asperum IPNI=*389034-1* 234
Anthoxanthum maderense IPNI=*389034-1* 234
Anthoxanthum nebrodense IPNI=*389034-1* 234
Anthoxanthum odoratum f. *montanum* IPNI=*389034-1* 234
Anthoxanthum odoratum f. *tenerum* IPNI=*389034-1* 234
Anthoxanthum pauciflorum IPNI=*389034-1* 234
Anthoxanthum sommierianum IPNI=*389034-1* 234
Anthurium acrobates IPNI=*84611-1* 198
Anthurium acuminatum IPNI=*84849-1* 198
Anthurium aeranthe IPNI=*85321-1* 199
Anthurium alienatum IPNI=*85321-1* 199
Anthurium amazonicum IPNI=*85321-1* 199
Anthurium andersonii IPNI=*85170-1* 199
Anthurium belangeri IPNI=*84849-1* 198
Anthurium cardiophyllum IPNI=*85127-1* 199
Anthurium cochleatum IPNI=*85127-1* 199
Anthurium cordatosagittatum IPNI=*85321-1* 199
Anthurium crystallinum f. *peltifolium* IPNI=*84713-1* 198
Anthurium dechardii IPNI=*88955-1* 202
Anthurium elegans IPNI=*85170-1* 199
Anthurium fissum IPNI=*85170-1* 199
Anthurium flexuosum IPNI=*121108-2* 200
Anthurium floribundum IPNI=*88963-1* 202
Anthurium fraternum IPNI=*85127-1* 199

Anthurium fucatum IPNI=*85127-1* 199
Anthurium gloriosum IPNI=*87777-1* 201
Anthurium gracile subsp. *belangeri* IPNI=*84849-1* 198
Anthurium gracile var. *poiteanum* IPNI=*84849-1* 198
Anthurium huamaliesense IPNI=*84611-1* 198
Anthurium humboldtianum IPNI=*85321-1* 199
Anthurium igarapeum IPNI=*84611-1* 198
Anthurium inamoenum IPNI=*85127-1* 199
Anthurium inconditum IPNI=*84849-1* 198
Anthurium indecorum IPNI=*84611-1* 198
Anthurium killipianum IPNI=*84713-1* 198
Anthurium lindenianum IPNI=*85127-1* 199
Anthurium longipes IPNI=*84849-1* 198
Anthurium macilentum IPNI=*84849-1* 198
Anthurium nymphaeifolium var. *ornatum* IPNI=*85127-1* 199
Anthurium nymphaeifolium var. *ovatum* IPNI=*85127-1* 199
Anthurium nymphaeifolium var. *roezlii* IPNI=*85127-1* 199
Anthurium ornatum IPNI=*85127-1* 199
Anthurium poiteanum IPNI=*84849-1* 198
Anthurium poiteauanum IPNI=*84849-1* 198
Anthurium polyrrhizon IPNI=*85321-1* 199
Anthurium quindiuense IPNI=*85127-1* 199
Anthurium rubrinervium IPNI=*85321-1* 199
Anthurium rudgeanum IPNI=*84849-1* 198
Anthurium scolopendrinum IPNI=*84849-1* 198
Anthurium scolopendrinum var. *belangeri* IPNI=*84849-1* 198
Anthurium scolopendrinum var. *contractum* IPNI=*84849-1* 198
Anthurium sororium IPNI=*85321-1* 199
Anthurium terrestre IPNI=*85321-1* 199
Anthurium tridigitatum IPNI=*15067-2* 198
Anthurium trinerve var. *obtusum* IPNI=*1001003-1* 199
Anthurium veitchii var. *acuminatum* IPNI=*85473-1* 199
Anthurium venustum IPNI=*84549-1* 198
Antidesma triplinervium IPNI=*30014592-2* 387
Antigonon cordatum IPNI=*15860-2* 288
Antigonon grandiflorum IPNI=*693090-1* 288
Antigonon insigne IPNI=*693090-1* 288
Antigonon macrocarpum IPNI=*693090-1* 288
Antigonon platypus IPNI=*15860-2* 288
Antiphyla serrata IPNI=*320795-2* 408
Antirhea panamensis IPNI=*77107342-1* 358
Antirhea surinamensis IPNI=*77107342-1* 358
Antirrhinum acutangulum IPNI=*325773-2* 376
Antirrhinum antirrhiniflorum IPNI=*805502-1* 376
Antirrhinum canadense IPNI=*281417-2* 376
Antirrhinum cimbalaria IPNI=*325773-2* 376
Antirrhinum cymbalaria IPNI=*325773-2* 376
Antirrhinum grandiflorum IPNI=*799180-1* 375
Antirrhinum hederaceum IPNI=*325773-2* 376
Antirrhinum hederifolium IPNI=*325773-2* 376
Antirrhinum hendersonii IPNI=*799180-1* 375
Antirrhinum majus var. *hendersonii* IPNI=*799180-1* 375
Antirrhinum maurandioides IPNI=*805502-1* 376
Antirrhinum murale IPNI=*799180-1* 375
Antirrhinum muralis IPNI=*325773-2* 376
Antirrhinum pauciflorum IPNI=*281417-2* 376
Antirrhinum quinquelobum IPNI=*325773-2* 376
Antirrhinum stellatum IPNI=*281417-2* 376
Antirrhinum vulgare IPNI=*799180-1* 375
Antonia ovata var. *excelsa* IPNI=*545655-1* 345
Antonia ovata var. *glabra* IPNI=*545655-1* 345
Antonia ovata var. *pilosa* IPNI=*545655-1* 345
Antonia pilosa IPNI=*545655-1* 345
Antonia pubescens IPNI=*545655-1* 345
Antoniana laurifolia IPNI=*1187456-2* 350
Antrophyum guayanense IPNI=*17355560-1* 178
Antrophyum lineatum IPNI=*17209080-1* 178
Apalatoa amazonica IPNI=*489094-1* 316
Apargia chillensis IPNI=*126428-2* 264
Apargia serotina IPNI=*225575-1* 264
Aparine hispida IPNI=*30007294-2* 350
Aparine latifolia IPNI=*747959-1* 349
Aparine vulgaris IPNI=*30007294-2* 350
Aparisthmium spruceanum IPNI=*339164-1* 388
Apatelia scabra IPNI=*829326-1* 297
Apatelia tomentosa IPNI=*829436-1* 297
Apeiba aspera IPNI=*833257-1* 407
Apeiba burchellii IPNI=*833257-1* 407
Apeiba cimbalaria IPNI=*833270-1* 407
Apeiba echinata IPNI=*833267-1* 407
Apeiba hirsuta IPNI=*833270-1* 407
Apeiba hispida IPNI=*833267-1* 407
Apeiba hypoleuca IPNI=*833267-1* 407
Apeiba laevis IPNI=*833257-1* 407
Apeiba levis IPNI=*833257-1* 407
Apeiba ulmifolia IPNI=*834759-1* 410
Apera palustris IPNI=*386353-1* 233
Apera procumbens IPNI=*386353-1* 233
Aphanes alata IPNI=*720867-1* 443
Aphanes orbiculata IPNI=*720816-1* 442
Aphanes pinnata IPNI=*720867-1* 443
Aphanes tripartita IPNI=*720378-1* 442
Aphelandra acutifolia IPNI=*1053173-2* 359
Aphelandra aurantiaca var. *roezlii* IPNI=*1053173-2* 359
Aphelandra aurantiaca var. *stenophylla* IPNI=*1053173-2* 359
Aphelandra bullata IPNI=*1053173-2* 359
Aphelandra chartacea IPNI=*284442-2* 361
Aphelandra chrysops IPNI=*315544-2* 359
Aphelandra concinna IPNI=*315544-2* 359
Aphelandra cristata IPNI=*941389-1* 359
Aphelandra deppeana IPNI=*941389-1* 359
Aphelandra fascinator IPNI=*1053173-2* 359
Aphelandra fulgens IPNI=*941389-1* 359
Aphelandra ghiesbreghtiana IPNI=*133119-2* 360
Aphelandra glischrochlamys IPNI=*1053173-2* 359
Aphelandra haenkeana IPNI=*941389-1* 359
Aphelandra illustrata IPNI=*315544-2* 359
Aphelandra incarnata IPNI=*16313-2* 359
Aphelandra leiophylla IPNI=*1053173-2* 359
Aphelandra leopoldii IPNI=*315544-2* 359
Aphelandra lonchochlamys IPNI=*1053173-2* 359
Aphelandra nana IPNI=*1053173-2* 359
Aphelandra oostachya IPNI=*315544-2* 359
Aphelandra paraensis IPNI=*1053173-2* 359
Aphelandra pectinata IPNI=*941389-1* 359
Aphelandra pectinata var. *macra* IPNI=*941389-1* 359
Aphelandra phalacra IPNI=*1053173-2* 359
Aphelandra repanda IPNI=*1053173-2* 359
Aphelandra roezlei IPNI=*1053173-2* 359
Aphelandra scabra var. *angustifolia* IPNI=*941389-1* 359
Aphelandra sciophila IPNI=*1053173-2* 359
Aphelandra simplex IPNI=*1053173-2* 359
Aphelandra squarrosa var. *angustifolia* IPNI=*315544-2* 359
Aphelandra squarrosa var. *citrina* IPNI=*315544-2* 359
Aphelandra squarrosa var. *laxior* IPNI=*315544-2* 359
Aphelandra uribei IPNI=*1053173-2* 359
Aphelexis spectabilis IPNI=*962307-1* 272
Aphora blodgettii IPNI=*995988-1* 389
Aphragmia haenkei IPNI=*54193-1* 361
Aphragmia inundata IPNI=*54193-1* 361
Aphyllocalpa regalis IPNI=*305721-2* 173
Apicra fasciata IPNI=*77138009-1* 215
Apiospermum obcordatum IPNI=*317684-2* 202
Apium ammi f. *nanum* IPNI=*840899-1* 252
Apium ammi f. *pedunculatum* IPNI=*840899-1* 252
Apium ammi IPNI=*837486-1* 251

Apium ammi var. *caespitosum* IPNI=*840899-1* 252
Apium ammi var. *genuinum* IPNI=*840899-1* 252
Apium ammi var. *leptophyllum* IPNI=*840899-1* 252
Apium ammi-maius IPNI=*837486-1* 251
Apium andinum IPNI=*840899-1* 252
Apium anisum IPNI=*846658-1* 252
Apium australe var. *latisectum* IPNI=*838067-1* 251
Apium candollei IPNI=*837486-1* 251
Apium celleri IPNI=*838067-1* 251
Apium crispum IPNI=*60442790-2* 252
Apium decumbens IPNI=*838067-1* 251
Apium depressum IPNI=*840899-1* 252
Apium dulce IPNI=*838067-1* 251
Apium elatum IPNI=*837486-1* 251
Apium glaucescens IPNI=*170074-2* 252
Apium graveolens f. *lusitanicum* IPNI=*838067-1* 251
Apium graveolens subsp. *rapaceum* IPNI=*838067-1* 251
Apium graveolens var. *lusitanicum* IPNI=*838067-1* 251
Apium integrilobum IPNI=*838067-1* 251
Apium laetum IPNI=*60442790-2* 252
Apium latifolium IPNI=*60442790-2* 252
Apium leptophyllum IPNI=*840899-1* 252
Apium leptophyllum var. *dubium* IPNI=*840899-1* 252
Apium lobatum IPNI=*838067-1* 251
Apium lusitanicum IPNI=*838067-1* 251
Apium maritimum IPNI=*838067-1* 251
Apium montanum IPNI=*170084-2* 252
Apium montanum subsp. *ranunculifolium* IPNI=*170084-2* 252
Apium occidentale IPNI=*60442790-2* 252
Apium palustre IPNI=*838067-1* 251
Apium peregrinum IPNI=*60442790-2* 252
Apium petraeum IPNI=*837486-1* 251
Apium petroselinum IPNI=*60442790-2* 252
Apium petroselinum var. *angustifolium* IPNI=*60442790-2* 252
Apium petroselinum var. *variegatum* IPNI=*60442790-2* 252
Apium petroselinum var. *vulgare* IPNI=*60442790-2* 252
Apium pumilum IPNI=*837486-1* 251
Apium ranunculifolium IPNI=*170084-2* 252
Apium rapaceum IPNI=*838067-1* 251
Apium rivulare IPNI=*840899-1* 252
Apium romanum IPNI=*60442790-2* 252
Apium tenuifolium IPNI=*840899-1* 252
Apium ternatum IPNI=*170084-2* 252
Apium ternatum var. *genuinum* IPNI=*170084-2* 252
Apium ternatum var. *ranunculifolium* IPNI=*170084-2* 252
Apium tuberosum IPNI=*60442790-2* 252
Apium visnaga IPNI=*849956-1* 253
Apium vulgare IPNI=*60442790-2* 252
Apium vulgare IPNI=*838067-1* 251
Apium wolffianum IPNI=*840899-1* 252
Aplarina microphylla IPNI=*102034-2* 391
Aplarina prostrata IPNI=*101913-2* 390
Aplostemon triquetrus IPNI=*307702-1* 232
Apluda rugosa IPNI=*406281-1* 240
Apocopis pallidus IPNI=*398532-1* 237
Apocynum cordatum IPNI=*81684-1* 343
Apocynum nerium IPNI=*13196-2* 339
Apocynum nervosum IPNI=*81684-1* 343
Apocynum obliquum IPNI=*78821-1* 340
Apocynum proliferum IPNI=*106366-2* 340
Apocynum syriacum IPNI=*1004515-2* 339
Apodanthera congestiflora IPNI=*592558-1* 294
Apodanthera gracilis IPNI=*293293-1* 294
Apodanthera pedisecta IPNI=*292512-1* 294
Apodanthera scaberrima IPNI=*292141-1* 293
Apodanthera trifoliata IPNI=*592558-1* 294
Apodina cucullaris IPNI=*551367-1* 449
Apodina cupulifer IPNI=*551367-1* 449
Apodina mexicana IPNI=*551367-1* 449
Apoleya leiocarpa IPNI=*16766-2* 311
Apoleya molaris IPNI=*16766-2* 311
Apozia chamaedryoides IPNI=*61179-2* 369
Apozia serpyllacea IPNI=*61179-2* 369
Appunia longiloba IPNI=*1061341-2* 347
Appunia parviflora IPNI=*764729-1* 356
Appunia sparsiflora IPNI=*1061341-2* 347
Appunia tenuiflora var. *leiophylla* IPNI=*1061341-2* 347
Aptandra spruceana IPNI=*607627-1* 449
Aptenia cordifolia IPNI=*322566-2* 280
Apuleia grazielana IPNI=*16766-2* 311
Apuleia molaris IPNI=*16766-2* 311
Apuleia praecox IPNI=*16766-2* 311
Aquilegia alba IPNI=*306409-2* 436
Aquilegia arbascensis IPNI=*306409-2* 436
Aquilegia australis IPNI=*30028287-2* 436
Aquilegia canadensis f. *albiflora* IPNI=*30028287-2* 436
Aquilegia canadensis f. *ecalcarata* IPNI=*30028287-2* 436
Aquilegia canadensis f. *gartneri* IPNI=*30028287-2* 436
Aquilegia canadensis f. *phippenii* IPNI=*30028287-2* 436
Aquilegia canadensis subsp. *americana* IPNI=*30028287-2* 436
Aquilegia canadensis var. *australis* IPNI=*30028287-2* 436
Aquilegia canadensis var. *coccinea* IPNI=*30028287-2* 436
Aquilegia canadensis var. *eminens* IPNI=*30028287-2* 436
Aquilegia canadensis var. *flaviflora* IPNI=*30028287-2* 436
Aquilegia canadensis var. *hybrida* IPNI=*30028287-2* 436
Aquilegia canadensis var. *latiuscula* IPNI=*30028287-2* 436
Aquilegia canadensis var. *phippenii* IPNI=*30028287-2* 436
Aquilegia coccinea IPNI=*30028287-2* 436
Aquilegia coerulea f. *pubescens* IPNI=*16871-2* 436
Aquilegia collina IPNI=*306409-2* 436
Aquilegia corniculata IPNI=*306409-2* 436
Aquilegia cornuta IPNI=*306409-2* 436
Aquilegia cyclophylla IPNI=*306409-2* 436
Aquilegia ebneri IPNI=*306409-2* 436
Aquilegia ecalcarata IPNI=*306409-2* 436
Aquilegia elegans IPNI=*30028287-2* 436
Aquilegia eminens IPNI=*30028287-2* 436
Aquilegia eynensis IPNI=*306409-2* 436
Aquilegia flaviflora IPNI=*30028287-2* 436
Aquilegia glaucescens IPNI=*306409-2* 436
Aquilegia glaucophylla IPNI=*306409-2* 436
Aquilegia inversa IPNI=*306409-2* 436
Aquilegia latiuscula IPNI=*30028287-2* 436
Aquilegia mollis IPNI=*306409-2* 436
Aquilegia nemoralis IPNI=*306409-2* 436
Aquilegia phoenicantha IPNI=*30028287-2* 436
Aquilegia platysepala IPNI=*306409-2* 436
Aquilegia plena IPNI=*306409-2* 436
Aquilegia praecox IPNI=*306409-2* 436
Aquilegia recticornu IPNI=*306409-2* 436
Aquilegia ruscinonensis IPNI=*306409-2* 436
Aquilegia silvestris IPNI=*306409-2* 436
Aquilegia speciosa IPNI=*306409-2* 436
Aquilegia stellata IPNI=*306409-2* 436
Aquilegia subalpina IPNI=*306409-2* 436
Aquilegia sylvestris IPNI=*306409-2* 436
Aquilegia variegata IPNI=*30028287-2* 436
Aquilegia versicolor IPNI=*306409-2* 436
Aquilegia vulgaris subsp. *subalpina* IPNI=*306409-2* 436
Aquilegia winterbottomiana IPNI=*306409-2* 436
Aquilina vulgaris IPNI=*306409-2* 436
Arabis charbonnelii IPNI=*237940-2* 278
Arabis heterophylla IPNI=*280380-1* 276
Arabis nasturtium IPNI=*287534-1* 277
Arabis pinnata IPNI=*222187-2* 277
Arachidna hypogaea IPNI=*318562-2* 311

Arachis africana IPNI=*318562-2* 311
Arachis americana IPNI=*318562-2* 311
Arachis asiatica IPNI=*318562-2* 311
Arachis guaraniana IPNI=*318562-2* 311
Arachis hypogaea f. *nambyquarae* IPNI=*318562-2* 311
Arachis hypogaea f. *typica* IPNI=*318562-2* 311
Arachis hypogaea subsp. *fastigiata* IPNI=*318562-2* 311
Arachis hypogaea subsp. *nambyquarae* IPNI=*318562-2* 311
Arachis hypogaea subsp. *procumbens* IPNI=*318562-2* 311
Arachis hypogaea var. *aequatoriana* IPNI=*318562-2* 311
Arachis hypogaea var. *nambyquarae* IPNI=*318562-2* 311
Arachis hypogaea var. *peruviana* IPNI=*318562-2* 311
Arachis nambyquarae IPNI=*318562-2* 311
Arachis oleifera IPNI=*318562-2* 311
Arachis rasteiro IPNI=*318562-2* 311
Arachnothryx inconstans IPNI=*743570-1* 347
Arachnothryx reflexa var. *breviloba* IPNI=*743570-1* 347
Arachnothryx reflexa var. *calycina* IPNI=*743570-1* 347
Arachnothryx reflexa var. *glandulifera* IPNI=*743570-1* 347
Arachnothryx reflexa var. *inconstans* IPNI=*743570-1* 347
Arachnothryx reflexa var. *meridensis* IPNI=*743570-1* 347
Arachnothryx reflexa var. *trichotoma* IPNI=*743570-1* 347
Aralia arborea IPNI=*1133937-2* 253
Aralia capitata IPNI=*30050876-2* 254
Aralia catalpifolia IPNI=*91160-1* 254
Aralia cochleata IPNI=*91817-1* 254
Aralia elegantissima IPNI=*77131172-1* 254
Aralia floribunda IPNI=*91222-1* 254
Aralia fruticosa IPNI=*1133937-2* 253
Aralia fruticosa IPNI=*847351-1* 254
Aralia ginseng IPNI=*91472-1* 254
Aralia guilfoylei IPNI=*91773-1* 254
Aralia heterophylla IPNI=*91261-1* 254
Aralia incisa IPNI=*91222-1* 254
Aralia japonica IPNI=*90496-1* 253
Aralia latifolia IPNI=*91817-1* 254
Aralia longifolia IPNI=*30050876-2* 254
Aralia mairei IPNI=*86417-3* 254
Aralia micans IPNI=*90355-1* 253
Aralia monstrosa IPNI=*91773-1* 254
Aralia multiflora IPNI=*30050876-2* 254
Aralia mutisiana IPNI=*91261-1* 254
Aralia naumannii IPNI=*91749-1* 254
Aralia ovata IPNI=*30050876-2* 254
Aralia papyrifera IPNI=*86417-3* 254
Aralia septemnervia IPNI=*91160-1* 254
Aralia sieboldii IPNI=*90496-1* 253
Aralia tripinnata IPNI=*847351-1* 254
Aralia tuxtlensis IPNI=*1133937-2* 253
Araucaria angustifolia f. *catharinensis* IPNI=*17518-2* 180
Araucaria angustifolia var. *alba* IPNI=*17518-2* 180
Araucaria angustifolia var. *caiova* IPNI=*17518-2* 180
Araucaria angustifolia var. *caiuva* IPNI=*17518-2* 180
Araucaria angustifolia var. *dependens* IPNI=*17518-2* 180
Araucaria angustifolia var. *indehiscens* IPNI=*17518-2* 180
Araucaria angustifolia var. *monoica* IPNI=*17518-2* 180
Araucaria angustifolia var. *nigra* IPNI=*17518-2* 180
Araucaria angustifolia var. *sancti-josephi* IPNI=*17518-2* 180
Araucaria angustifolia var. *semialba* IPNI=*17518-2* 180
Araucaria angustifolia var. *stricta* IPNI=*17518-2* 180
Araucaria angustifolia var. *vinacea* IPNI=*17518-2* 180
Araucaria araucana subsp. *conguillioensis* IPNI=*261681-1* 180
Araucaria brasiliana IPNI=*17518-2* 180
Araucaria brasiliana var. *elegans* IPNI=*17518-2* 180
Araucaria chilensis IPNI=*261681-1* 180
Araucaria cookii IPNI=*261688-1* 180
Araucaria dioica IPNI=*17518-2* 180
Araucaria dombeyi IPNI=*261681-1* 180
Araucaria elegans IPNI=*17518-2* 180
Araucaria imbricata IPNI=*261681-1* 180
Araucaria intermedia IPNI=*261688-1* 180
Araucaria lavoixii IPNI=*261688-1* 180
Araucaria ridolfiana IPNI=*17518-2* 180
Araucaria saviana IPNI=*17518-2* 180
Arbutus pilosa IPNI=*330618-1* 299
Arbutus prostrata IPNI=*330618-1* 299
Archangelica archangelica IPNI=*837560-1* 251
Archangelica commutata IPNI=*837560-1* 251
Archangelica discocarpa IPNI=*837560-1* 251
Archangelica littoralis IPNI=*837560-1* 251
Archangelica norwegica IPNI=*837560-1* 251
Archangelica officinalis IPNI=*837560-1* 251
Archangelica officinalis monstr. phyllomana IPNI=*837560-1* 251
Archangelica officinalis subsp. *littoralis* IPNI=*837560-1* 251
Archangelica sativa IPNI=*837560-1* 251
Archangelica slavica IPNI=*837560-1* 251
Archangelica spuria IPNI=*837560-1* 251
Archontophoenix alexandrae var. *beatriceae* IPNI=*664083-1* 204
Archontophoenix alexandrae var. *schizanthera* IPNI=*664083-1* 204
Archontophoenix jardinei IPNI=*669378-1* 211
Archontophoenix veitchii IPNI=*669378-1* 211
Arctotis aspleniifolia IPNI=*941423-1* 263
Arctotis elegans IPNI=*193510-1* 259
Arcytophyllum bryoides IPNI=*17992-2* 347
Arcytophyllum caracasanum IPNI=*743608-1* 347
Arcytophyllum confertum IPNI=*17992-2* 347
Arcytophyllum confertum var. *bryoides* IPNI=*17992-2* 347
Arcytophyllum juniperifolium IPNI=*18012-2* 347
Arcytophyllum laricifolium IPNI=*18012-2* 347
Arcytophyllum nitidum f. *hispidulum* IPNI=*743608-1* 347
Arcytophyllum nitidum subsp. *caracasanum* IPNI=*743608-1* 347
Arcytophyllum nitidum var. *culmenicola* IPNI=*743608-1* 347
Arcytophyllum nitidum var. *glabratum* IPNI=*743608-1* 347
Arcytophyllum puberulentum IPNI=*743608-1* 347
Arcytophyllum thymifolium var. *scolnikii* IPNI=*18012-2* 347
Arcytophyllum vareschii IPNI=*743608-1* 347
Arcytophyllum weberbaueri IPNI=*18012-2* 347
Ardisia acuminata IPNI=*18168-2* 303
Ardisia acuminifolia IPNI=*18168-2* 303
Ardisia alstonii IPNI=*586863-1* 303
Ardisia amanuensis IPNI=*18168-2* 303
Ardisia angustifolia IPNI=*18168-2* 303
Ardisia aurantiaca IPNI=*586863-1* 303
Ardisia belizensis IPNI=*586863-1* 303
Ardisia brasiliensis IPNI=*18168-2* 303
Ardisia capollina IPNI=*586863-1* 303
Ardisia catharinensis IPNI=*18168-2* 303
Ardisia caudatifolia IPNI=*586863-1* 303
Ardisia chahalana IPNI=*586863-1* 303
Ardisia cookii IPNI=*18168-2* 303
Ardisia coriacea IPNI=*1028287-2* 304
Ardisia coriacea var. *berteroana* IPNI=*1028287-2* 304
Ardisia cuspidata IPNI=*586863-1* 303
Ardisia decipiens IPNI=*247441-2* 304
Ardisia deminuta IPNI=*18168-2* 303
Ardisia digitata IPNI=*586863-1* 303
Ardisia edwardsii IPNI=*586863-1* 303
Ardisia furfuracella subsp. *veraguasensis* IPNI=*586863-1* 303
Ardisia herbert-smithii IPNI=*18168-2* 303
Ardisia hostmannii IPNI=*247474-2* 304
Ardisia ibaguensis IPNI=*247490-2* 305
Ardisia irazuensis IPNI=*586863-1* 303
Ardisia ixcanensis IPNI=*586863-1* 303
Ardisia lewisii IPNI=*18048-2* 303
Ardisia longicaudata IPNI=*18168-2* 303
Ardisia longifolia IPNI=*247465-2* 304

Ardisia longistaminea IPNI=*975391-1* 304
Ardisia martinensis IPNI=*18168-2* 303
Ardisia metensis IPNI=*18168-2* 303
Ardisia micrantha IPNI=*247469-2* 304
Ardisia nicaraguensis IPNI=*586863-1* 303
Ardisia nigricans IPNI=*247472-2* 304
Ardisia oerstediana IPNI=*586863-1* 303
Ardisia orinocensis IPNI=*247474-2* 304
Ardisia panurensis IPNI=*247472-2* 304
Ardisia pergracilis IPNI=*18168-2* 303
Ardisia platyphylla IPNI=*247490-2* 305
Ardisia principis IPNI=*18168-2* 303
Ardisia reflexiflora IPNI=*586863-1* 303
Ardisia romeroi IPNI=*18048-2* 303
Ardisia rusbyana IPNI=*247490-2* 305
Ardisia salvadorensis IPNI=*586863-1* 303
Ardisia samalana IPNI=*586863-1* 303
Ardisia sapoana IPNI=*18048-2* 303
Ardisia schomburgkiana IPNI=*247474-2* 304
Ardisia semicrenata IPNI=*18168-2* 303
Ardisia spectabilis IPNI=*73559-2* 304
Ardisia surinamensis IPNI=*247474-2* 304
Ardisia tetrandra IPNI=*18168-2* 303
Ardisia tuirana IPNI=*18048-2* 303
Ardisia turbacensis IPNI=*247490-2* 305
Ardisia veraguasensis IPNI=*586863-1* 303
Ardisia whitei IPNI=*586863-1* 303
Arduina grandiflora IPNI=*77727-1* 340
Arduina macrocarpa IPNI=*77727-1* 340
Areca bacaba IPNI=*668509-1* 210
Areca cathechu IPNI=*664107-1* 204
Areca erythropoda IPNI=*666411-1* 208
Areca faufel IPNI=*664107-1* 204
Areca himalayana IPNI=*664107-1* 204
Areca hortensis IPNI=*664107-1* 204
Areca macrocarpa IPNI=*664107-1* 204
Areca nigra IPNI=*664107-1* 204
Areca oleracea IPNI=*222805-2* 212
Areca verschaffeltii IPNI=*667456-1* 209
Arecastrum romanzoffianum IPNI=*246966-2* 212
Arecastrum romanzoffianum subvar. *minus* IPNI=*246966-2* 212
Arecastrum romanzoffianum var. *australe* IPNI=*246966-2* 212
Arecastrum romanzoffianum var. *genuinum* IPNI=*246966-2* 212
Arecastrum romanzoffianum var. *micropindo* IPNI=*246966-2* 212
Aregelia carolinae IPNI=*169003-2* 229
Arenaria alsinoides IPNI=*30040285-2* 285
Arenaria alsinoides var. *ovatifolia* IPNI=*30040285-2* 285
Arenaria arvensis IPNI=*158725-1* 286
Arenaria berteroana IPNI=*159015-1* 286
Arenaria campestris IPNI=*158983-1* 286
Arenaria diffusa IPNI=*30040285-2* 285
Arenaria guatemalensis IPNI=*30040285-2* 285
Arenaria guatemalensis var. *ensifolia* IPNI=*30040285-2* 285
Arenaria heterosperma IPNI=*241336-2* 286
Arenaria lanuginosa subsp. *guatemalensis* IPNI=*30040285-2* 285
Arenaria lanuginosa var. *diffusa* IPNI=*30040285-2* 285
Arenaria lanuginosa var. *ensifolia* IPNI=*30040285-2* 285
Arenaria lanuginosa var. *longipedunculata* IPNI=*30040285-2* 285
Arenaria lanuginosa var. *megalantha* IPNI=*30040285-2* 285
Arenaria maritima IPNI=*241336-2* 286
Arenaria media IPNI=*241336-2* 286
Arenaria megalantha IPNI=*30040285-2* 285
Arenaria megalantha var. *ensifolia* IPNI=*30040285-2* 285
Arenaria megalantha var. *tucumanensis* IPNI=*30040285-2* 285
Arenaria membranacea IPNI=*158983-1* 286
Arenaria miquelonensis IPNI=*241336-2* 286
Arenaria nemorosa IPNI=*30040285-2* 285
Arenaria nemorosa var. *novogranatensis* IPNI=*30040285-2* 285
Arenaria nemorosa var. *quitensis* IPNI=*30040285-2* 285
Arenaria peruviana IPNI=*30040285-2* 285
Arenaria pulchella IPNI=*154453-1* 285
Arenaria radicans IPNI=*158983-1* 286
Arenaria rubra IPNI=*158983-1* 286
Arenaria rubra var. *campestris* IPNI=*158983-1* 286
Arenaria rubra var. *canadensis* IPNI=*158983-1* 286
Arenaria rubra var. *marina* IPNI=*241336-2* 286
Arenaria salina IPNI=*241336-2* 286
Arenaria vulgaris IPNI=*159330-1* 286
Arenbergia glauca IPNI=*367440-1* 344
Argemone alba IPNI=*18816-2* 435
Argemone alba var. *leiocarpa* IPNI=*18816-2* 435
Argemone leiocarpa IPNI=*18816-2* 435
Argemone leiocarpa var. *mexicanoides* IPNI=*18816-2* 435
Argemone leiocarpa var. *ochroleucoides* IPNI=*18816-2* 435
Argemone mexicana f. *leiocarpa* IPNI=*18816-2* 435
Argemone mexicana subvar. *inermis* IPNI=*18816-2* 435
Argemone mexicana var. *anacanthoidea* IPNI=*18816-2* 435
Argemone mexicana var. *parviflora* IPNI=*18816-2* 435
Argemone mexicana var. *typica* IPNI=*18816-2* 435
Argemone mucronata IPNI=*18816-2* 435
Argemone sexvalvis IPNI=*18816-2* 435
Argemone spinosa IPNI=*18816-2* 435
Argemone versicolor IPNI=*18816-2* 435
Argusia gnaphalodes IPNI=*120998-1* 275
Argyrochaeta bipinnatifida IPNI=*236510-1* 267
Argyrochosma flava IPNI=*77182941-1* 178
Argyrochosma nivea IPNI=*77182941-1* 178
Argyrochosma nivea var. *flava* IPNI=*77182941-1* 178
Argyrochosma nivea var. *tenera* IPNI=*77182941-1* 178
Argyrochosma tenera IPNI=*77182941-1* 178
Argyrocome manglesii IPNI=*241451-1* 269
Argythamnia argothamnoides IPNI=*995988-1* 389
Argythamnia blodgettii IPNI=*995988-1* 389
Argythamnia castaneifolia IPNI=*44529-2* 388
Argythamnia fendleri IPNI=*995988-1* 389
Argythamnia paludosa IPNI=*44529-2* 388
Argythamnia palustris IPNI=*44564-2* 388
Argythamnia rubricaulis IPNI=*995988-1* 389
Argythamnia savanillensis IPNI=*995988-1* 389
Ariaria superba IPNI=*481718-1* 312
Aridaria aurea IPNI=*362165-1* 280
Aristida abyssinica IPNI=*319759-2* 234
Aristida adscensionis subsp. *coerulescens* IPNI=*319759-2* 234
Aristida adscensionis subvar. *condensata* IPNI=*319759-2* 234
Aristida adscensionis var. *abortiva* IPNI=*319759-2* 234
Aristida adscensionis var. *breviseta* IPNI=*319759-2* 234
Aristida adscensionis var. *bromoides* IPNI=*319759-2* 234
Aristida adscensionis var. *coarctata* IPNI=*319759-2* 234
Aristida adscensionis var. *coerulescens* IPNI=*319759-2* 234
Aristida adscensionis var. *condensata* IPNI=*319759-2* 234
Aristida adscensionis var. *decolorata* IPNI=*319759-2* 234
Aristida adscensionis var. *glabricallis* IPNI=*319759-2* 234
Aristida adscensionis var. *humilis* IPNI=*319759-2* 234
Aristida adscensionis var. *interrupta* IPNI=*319759-2* 234
Aristida adscensionis var. *modesta* IPNI=*319759-2* 234
Aristida adscensionis var. *nigrescens* IPNI=*319759-2* 234
Aristida adscensionis var. *normalis* IPNI=*319759-2* 234
Aristida adscensionis var. *scabriflora* IPNI=*319759-2* 234
Aristida aethiopica IPNI=*319759-2* 234
Aristida americana var. *bromoides* IPNI=*319759-2* 234
Aristida arabica IPNI=*319759-2* 234
Aristida bromoides IPNI=*319759-2* 234
Aristida canariensis IPNI=*319759-2* 234
Aristida cardosoi IPNI=*319759-2* 234
Aristida chaetophylla IPNI=*319759-2* 234
Aristida coarctata IPNI=*319759-2* 234

Aristida coerulescens IPNI=*319759-2* 234
Aristida coerulescens var. *pumila* IPNI=*319759-2* 234
Aristida curvata IPNI=*319759-2* 234
Aristida debilis IPNI=*319759-2* 234
Aristida depressa IPNI=*319759-2* 234
Aristida dispersa IPNI=*319759-2* 234
Aristida dispersa var. *bromoides* IPNI=*319759-2* 234
Aristida dispersa var. *coarctata* IPNI=*319759-2* 234
Aristida dispersa var. *nigrescens* IPNI=*319759-2* 234
Aristida domingensis IPNI=*420951-1* 244
Aristida effusa IPNI=*319759-2* 234
Aristida ehrenbergii IPNI=*319759-2* 234
Aristida elatior IPNI=*319759-2* 234
Aristida fasciculata IPNI=*319759-2* 234
Aristida festucoides IPNI=*319759-2* 234
Aristida gigantea IPNI=*319759-2* 234
Aristida grisebachiana IPNI=*319759-2* 234
Aristida grisebachiana var. *decolorata* IPNI=*319759-2* 234
Aristida guineensis IPNI=*319759-2* 234
Aristida heymannii IPNI=*319759-2* 234
Aristida humilis IPNI=*319759-2* 234
Aristida interrupta IPNI=*319759-2* 234
Aristida luzoniensis IPNI=*319759-2* 234
Aristida macrochloa IPNI=*319759-2* 234
Aristida maritima IPNI=*319759-2* 234
Aristida mauritiana IPNI=*319759-2* 234
Aristida modatica IPNI=*319759-2* 234
Aristida nana IPNI=*319759-2* 234
Aristida nigrescens IPNI=*319759-2* 234
Aristida nutans IPNI=*319759-2* 234
Aristida paniculata IPNI=*319759-2* 234
Aristida peruviana IPNI=*319759-2* 234
Aristida pumila IPNI=*319759-2* 234
Aristida pusilla IPNI=*319759-2* 234
Aristida racemosa IPNI=*319759-2* 234
Aristida schaffneri IPNI=*319759-2* 234
Aristida setacea IPNI=*319759-2* 234
Aristida simplicissima IPNI=*319759-2* 234
Aristida spicigera IPNI=*319759-2* 234
Aristida submucronata IPNI=*319759-2* 234
Aristida swartziana IPNI=*319759-2* 234
Aristida teneriffae IPNI=*319759-2* 234
Aristida thonningii IPNI=*319759-2* 234
Aristida unilateralis IPNI=*1049662-2* 235
Aristida viciosorum IPNI=*319759-2* 234
Aristida vulgaris IPNI=*319759-2* 234
Aristida vulgaris var. *coerulescens* IPNI=*319759-2* 234
Aristida vulpioides IPNI=*319759-2* 234
Aristida waibeliana IPNI=*319759-2* 234
Aristoclesia esculenta IPNI=*326389-2* 385
Aristolochia appendiculata IPNI=*319092-2* 193
Aristolochia argentea IPNI=*19406-2* 192
Aristolochia argyroneura IPNI=*272659-2* 192
Aristolochia aurantiaca IPNI=*316342-2* 193
Aristolochia biflora IPNI=*316426-2* 193
Aristolochia caracasana IPNI=*319092-2* 193
Aristolochia chapmaniana IPNI=*19637-2* 193
Aristolochia claussenii IPNI=*93167-1* 193
Aristolochia claussenii subsp. *pyrenea* IPNI=*93167-1* 193
Aristolochia claveriana IPNI=*93231-1* 193
Aristolochia clypeata IPNI=*92858-1* 192
Aristolochia conferta IPNI=*93219-1* 193
Aristolochia cordata IPNI=*316342-2* 193
Aristolochia costaricensis IPNI=*93231-1* 193
Aristolochia costaricensis var. *zamorensis* IPNI=*93231-1* 193
Aristolochia dammeriana IPNI=*316426-2* 193
Aristolochia dodsonii IPNI=*19503-2* 192
Aristolochia duchartrei IPNI=*93298-1* 193
Aristolochia egensis IPNI=*272643-2* 192
Aristolochia eggersii IPNI=*93231-1* 193
Aristolochia elegans IPNI=*93074-1* 192
Aristolochia elegans var. *hassleriana* IPNI=*93074-1* 192
Aristolochia ernestulei IPNI=*93231-1* 193
Aristolochia esoterica IPNI=*92858-1* 192
Aristolochia exigua IPNI=*93167-1* 193
Aristolochia ferruginea IPNI=*93231-1* 193
Aristolochia foetens IPNI=*92743-1* 192
Aristolochia geminiflora IPNI=*316426-2* 193
Aristolochia gibbosa IPNI=*93027-1* 192
Aristolochia gigantea IPNI=*92978-1* 192
Aristolochia gigas IPNI=*92978-1* 192
Aristolochia gigas var. *sturtevantii* IPNI=*92978-1* 192
Aristolochia glaucescens IPNI=*92858-1* 192
Aristolochia glaziovii IPNI=*316342-2* 193
Aristolochia hassleriana IPNI=*93074-1* 192
Aristolochia hassleriana var. *guaranitica* IPNI=*316342-2* 193
Aristolochia hastata IPNI=*93219-1* 193
Aristolochia haughtiana IPNI=*93231-1* 193
Aristolochia loriflora IPNI=*316428-2* 192
Aristolochia macrophylla IPNI=*19503-2* 192
Aristolochia marshii IPNI=*93219-1* 193
Aristolochia martiniana IPNI=*316342-2* 193
Aristolochia maxima var. *angustifolia* IPNI=*316426-2* 193
Aristolochia maxima var. *cordata* IPNI=*19637-2* 193
Aristolochia maxima var. *geminiflora* IPNI=*316426-2* 193
Aristolochia medellinensis IPNI=*93231-1* 193
Aristolochia mexicana IPNI=*316426-2* 193
Aristolochia moschata IPNI=*316342-2* 193
Aristolochia oblongifolia IPNI=*316426-2* 193
Aristolochia odorata IPNI=*316342-2* 193
Aristolochia odoratissima f. *brasiliensis* IPNI=*316342-2* 193
Aristolochia odoratissima f. *paraguariensis* IPNI=*316342-2* 193
Aristolochia odoratissima IPNI=*93027-1* 192
Aristolochia odoratissima var. *genuina* IPNI=*316342-2* 193
Aristolochia odoratissima var. *glaziovii* IPNI=*316342-2* 193
Aristolochia odoratissima var. *grandiflora* IPNI=*316342-2* 193
Aristolochia odoratissima var. *guaranitica* IPNI=*316342-2* 193
Aristolochia odoratissima var. *hastata* IPNI=*316342-2* 193
Aristolochia odoratissima var. *pandurata* IPNI=*316342-2* 193
Aristolochia ottonis IPNI=*316342-2* 193
Aristolochia pandurata IPNI=*316342-2* 193
Aristolochia pandurata var. *warscewiczii* IPNI=*316342-2* 193
Aristolochia panduriformis IPNI=*316342-2* 193
Aristolochia pannosa IPNI=*93231-1* 193
Aristolochia pavoniana IPNI=*316428-2* 192
Aristolochia pentandra var. *hastata* IPNI=*93219-1* 193
Aristolochia picta IPNI=*316342-2* 193
Aristolochia pilgeriana IPNI=*93298-1* 193
Aristolochia pilosa var. *ligulifera* IPNI=*93231-1* 193
Aristolochia podocarpa IPNI=*93027-1* 192
Aristolochia pusilla IPNI=*93167-1* 193
Aristolochia pyrenea IPNI=*93167-1* 193
Aristolochia racemosa IPNI=*93219-1* 193
Aristolochia reticulata IPNI=*316426-2* 193
Aristolochia rimbachii IPNI=*316342-2* 193
Aristolochia rodriguesii IPNI=*19457-2* 192
Aristolochia schubertioides IPNI=*93231-1* 193
Aristolochia securidata IPNI=*92853-1* 192
Aristolochia sprucei IPNI=*92853-1* 192
Aristolochia sturtevantii IPNI=*92978-1* 192
Aristolochia sylvicola IPNI=*92858-1* 192
Aristolochia tenera IPNI=*93167-1* 193
Aristolochia torta IPNI=*93027-1* 192
Aristolochia translucida IPNI=*19637-2* 193
Aristolochia trifida IPNI=*319092-2* 193
Aristolochia triloba IPNI=*319092-2* 193

Aristolochia turbacensis IPNI=*93285-1* 193
Aristolochia veraguensis IPNI=*272659-2* 192
Aristolochia wageneriana IPNI=*316426-2* 193
Arivela viscosa IPNI=*320133-2* 280
Arkezostis glandulosa IPNI=*291801-1* 293
Arkezostis granatensis IPNI=*291805-1* 293
Arkezostis heterophylla IPNI=*291801-1* 293
Arkezostis latebrosa IPNI=*291845-1* 293
Arkezostis ottoniana IPNI=*291801-1* 293
Arkezostis podantha IPNI=*291841-1* 293
Arkezostis poeppigii IPNI=*291801-1* 293
Arkezostis racemosa IPNI=*291845-1* 293
Arkezostis tomentosa IPNI=*291801-1* 293
Arkopoda luteola IPNI=*715535-1* 280
Armeria armeria IPNI=*19793-2* 288
Armeria campestris IPNI=*19793-2* 288
Armeria drummondii IPNI=*193463-2* 303
Armeria duriuscula IPNI=*19793-2* 288
Armeria elongata var. *maritima* IPNI=*19793-2* 288
Armeria intermedia IPNI=*19793-2* 288
Armeria maritima var. *duriuscula* IPNI=*19793-2* 288
Armeria maritima var. *pubescens* IPNI=*19793-2* 288
Armeria maritima var. *scotica* IPNI=*19793-2* 288
Armeria miscella IPNI=*19793-2* 288
Armeria planifolia IPNI=*19793-2* 288
Armeria pubescens IPNI=*19793-2* 288
Armeria pubigera var. *scotica* IPNI=*19793-2* 288
Armeria vulgaris f. *purpurea* IPNI=*19793-2* 288
Armeria vulgaris IPNI=*19793-2* 288
Armeria vulgaris var. *alba* IPNI=*19793-2* 288
Armeria vulgaris var. *glabra* IPNI=*19793-2* 288
Armeria vulgaris var. *grandiflora* IPNI=*19793-2* 288
Armeria vulgaris var. *laucheana* IPNI=*19793-2* 288
Armeria vulgaris var. *longiinvolucrata* IPNI=*19793-2* 288
Armeria vulgaris var. *maritima* IPNI=*19793-2* 288
Armeria vulgaris var. *purpurea* IPNI=*19793-2* 288
Armeria vulgaris var. *splendens* IPNI=*19793-2* 288
Arnica coronopifolia IPNI=*941423-1* 263
Arnoglossum lanceolatum IPNI=*321285-2* 376
Arodes aethiopicum IPNI=*89403-1* 203
Aron colocasium IPNI=*1170772-2* 199
Arosma obtusifolia IPNI=*87769-1* 201
Arouna divaricatum IPNI=*78900-2* 318
Arouna guianensis IPNI=*78900-2* 318
Arrabidaea acutifolia IPNI=*77142115-1* 363
Arrabidaea amazonica IPNI=*77141996-1* 363
Arrabidaea bilabiata IPNI=*77100467-1* 365
Arrabidaea blanchetii IPNI=*77142059-1* 363
Arrabidaea blanchetii var. *desertorum* IPNI=*77142059-1* 363
Arrabidaea calodictyos IPNI=*77141989-1* 363
Arrabidaea candicans IPNI=*77141989-1* 363
Arrabidaea candicans var. *latifolia* IPNI=*77141989-1* 363
Arrabidaea carabobensis IPNI=*77142059-1* 363
Arrabidaea chica f. *cuprea* IPNI=*77142115-1* 363
Arrabidaea chica IPNI=*77142115-1* 363
Arrabidaea chica var. *acutifolia* IPNI=*77142115-1* 363
Arrabidaea chica var. *angustifolia* IPNI=*77142115-1* 363
Arrabidaea chica var. *cuprea* IPNI=*77142115-1* 363
Arrabidaea chica var. *thyrsoidea* IPNI=*77142115-1* 363
Arrabidaea conjugata IPNI=*77141996 1* 363
Arrabidaea cordifolia IPNI=*77142059-1* 363
Arrabidaea cuminaensis IPNI=*77100467-1* 365
Arrabidaea cuprea IPNI=*77142115-1* 363
Arrabidaea desertorum IPNI=*77142059-1* 363
Arrabidaea duckei IPNI=*153710-2* 364
Arrabidaea guatemalensis IPNI=*108550-1* 362
Arrabidaea inaequalis IPNI=*77100463-1* 363
Arrabidaea isthmica IPNI=*108550-1* 362
Arrabidaea japurensis IPNI=*77142016-1* 363
Arrabidaea kuhlmannii IPNI=*77100467-1* 365
Arrabidaea larensis IPNI=*77142115-1* 363
Arrabidaea lenticellosa IPNI=*77142016-1* 363
Arrabidaea littoralis IPNI=*77142038-1* 363
Arrabidaea lundellii IPNI=*77142059-1* 363
Arrabidaea magnifica IPNI=*108858-1* 362
Arrabidaea mollicoma IPNI=*77142038-1* 363
Arrabidaea mollissima IPNI=*77142038-1* 363
Arrabidaea orbignyana IPNI=*77141989-1* 363
Arrabidaea pachycalyx IPNI=*77141989-1* 363
Arrabidaea paniculata IPNI=*108202-1* 362
Arrabidaea pleei IPNI=*77142059-1* 363
Arrabidaea pseudochica IPNI=*108550-1* 362
Arrabidaea pubescens IPNI=*77142059-1* 363
Arrabidaea quinquenervia IPNI=*77141989-1* 363
Arrabidaea rhodothyrsus IPNI=*77141989-1* 363
Arrabidaea rosea IPNI=*77142115-1* 363
Arrabidaea sanctae-marthae IPNI=*77142038-1* 363
Arrabidaea seleriana IPNI=*77142038-1* 363
Arrabidaea sieberi IPNI=*77142059-1* 363
Arrabidaea subincana IPNI=*77141989-1* 363
Arrabidaea virescens IPNI=*77141996-1* 363
Arrabidaea virescens var. *farinosa* IPNI=*77141996-1* 363
Arrabidaea xanthophylla IPNI=*77142096-1* 365
Arrabidaea zuliaensis IPNI=*77141989-1* 363
Arracacia acuminata IPNI=*168905-2* 252
Arracacia andina IPNI=*838272-1* 251
Arracacia esculenta IPNI=*838272-1* 251
Arrhostoxylum ciliatiflorum IPNI=*53920-1* 361
Arrhostoxylum microphyllum IPNI=*54585-1* 361
Arrhostoxylum terminale IPNI=*1014800-1* 361
Artanthe adenophora IPNI=*681251-1* 196
Artanthe adenophylla IPNI=*681224-1* 196
Artanthe adunca f. *angustifolia* IPNI=*680296-1* 195
Artanthe adunca IPNI=*680296-1* 195
Artanthe aequalis IPNI=*680300-1* 195
Artanthe alaris IPNI=*682138-1* 196
Artanthe amazonica IPNI=*681027-1* 195
Artanthe andicola IPNI=*680559-1* 195
Artanthe anonifolia IPNI=*1003227-1* 195
Artanthe anonifolia var. *latifolia* IPNI=*1003227-1* 195
Artanthe apiculata IPNI=*1003227-1* 195
Artanthe aspera IPNI=*681674-1* 196
Artanthe asperifolia f. *surinamensis* IPNI=*681674-1* 196
Artanthe asperifolia IPNI=*681674-1* 196
Artanthe augusta IPNI=*680507-1* 195
Artanthe aurita IPNI=*680520-1* 195
Artanthe bahiensis IPNI=*680296-1* 195
Artanthe barbata IPNI=*680559-1* 195
Artanthe bartlingiana IPNI=*680574-1* 195
Artanthe beecheyana IPNI=*683806-1* 197
Artanthe benthamiana IPNI=*683316-1* 197
Artanthe brachypoda IPNI=*680665-1* 195
Artanthe bredemeyeri IPNI=*680680-1* 195
Artanthe caracasana IPNI=*682429-1* 196
Artanthe caripensis IPNI=*680300-1* 195
Artanthe catalpifolia IPNI=*682138-1* 196
Artanthe caudata IPNI=*682138-1* 196
Artanthe cearensis IPNI=*680296-1* 195
Artanthe celtidifolia IPNI=*680296-1* 195
Artanthe chavicoides IPNI=*680871-1* 195
Artanthe colubrina var. *crassa* IPNI=*681224-1* 196
Artanthe colubrina var. *crassifolia* IPNI=*681224-1* 196
Artanthe colubrina var. *glandulosa* IPNI=*681224-1* 196
Artanthe consanguinea IPNI=*680988-1* 195
Artanthe controversa IPNI=*681674-1* 196
Artanthe coruscans IPNI=*681027-1* 195

Artanthe corylifolia IPNI=*681193-1* 196
Artanthe crassispicata IPNI=*681559-1* 196
Artanthe decurrens IPNI=*683806-1* 197
Artanthe demerarana IPNI=*681157-1* 196
Artanthe dimidiata IPNI=*680435-1* 195
Artanthe elongata f. *brasiliensis* IPNI=*680296-1* 195
Artanthe elongata f. *glabrior* IPNI=*680296-1* 195
Artanthe enckeoides IPNI=*681041-1* 195
Artanthe exserens IPNI=*681041-1* 195
Artanthe flagellaris IPNI=*682429-1* 196
Artanthe flexuosa IPNI=*680680-1* 195
Artanthe galeottii IPNI=*680296-1* 195
Artanthe grandifolia IPNI=*681559-1* 196
Artanthe granulosa IPNI=*680296-1* 195
Artanthe guilleminiana IPNI=*681041 1* 195
Artanthe heterophylla f. *angustifolia* IPNI=*681651-1* 196
Artanthe heterophylla IPNI=*681651-1* 196
Artanthe hirsuta IPNI=*681674-1* 196
Artanthe hirtula IPNI=*681193-1* 196
Artanthe hispida IPNI=*681674-1* 196
Artanthe hostmanniana IPNI=*681692-1* 196
Artanthe insignis IPNI=*682429-1* 196
Artanthe lacunosa IPNI=*681864-1* 196
Artanthe lanceata IPNI=*680296-1* 195
Artanthe leprieurii IPNI=*680988-1* 195
Artanthe lessertiana IPNI=*680435-1* 195
Artanthe linguifolia IPNI=*680296-1* 195
Artanthe luschnathiana IPNI=*680435-1* 195
Artanthe luschnathiana var. *glabrata* IPNI=*680435-1* 195
Artanthe macrophylla f. *latifolia* IPNI=*680435-1* 195
Artanthe marginata IPNI=*682138-1* 196
Artanthe meyeri IPNI=*681224-1* 196
Artanthe moritziana IPNI=*682429-1* 196
Artanthe nubigena IPNI=*682399-1* 196
Artanthe obesa IPNI=*681674-1* 196
Artanthe obovata IPNI=*681224-1* 196
Artanthe olfersiana IPNI=*681674-1* 196
Artanthe opizii IPNI=*681674-1* 196
Artanthe plantaginea IPNI=*680359-1* 195
Artanthe pothomorpha IPNI=*683316-1* 197
Artanthe pseudochurumayu IPNI=*681027-1* 195
Artanthe punctata IPNI=*681651-1* 196
Artanthe radula IPNI=*680680-1* 195
Artanthe ramiflora IPNI=*681692-1* 196
Artanthe riparia IPNI=*681559-1* 196
Artanthe ruiziana IPNI=*198988-2* 197
Artanthe salviifolia IPNI=*680296-1* 195
Artanthe sancta IPNI=*680520-1* 195
Artanthe saururoides IPNI=*680359-1* 195
Artanthe scabra IPNI=*681674-1* 196
Artanthe scabra var. *hirsuta* IPNI=*681674-1* 196
Artanthe seemanniana IPNI=*680520-1* 195
Artanthe sellowiana IPNI=*681041-1* 195
Artanthe staminea IPNI=*680435-1* 195
Artanthe subaequalis IPNI=*681224-1* 196
Artanthe superba IPNI=*682429-1* 196
Artanthe tejucensis IPNI=*681224-1* 196
Artanthe tiliifolia IPNI=*683316-1* 197
Artanthe tuberculata IPNI=*683806-1* 197
Artanthe tuberculata var. *brevispicata* IPNI=*683806-1* 197
Artanthe ulmifolia IPNI=*681193-1* 196
Artanthe ulmifolia var. *brevipedunculata* IPNI=*681193-1* 196
Artanthe variegata IPNI=*679043-1* 194
Artanthe velloziana IPNI=*680296-1* 195
Artanthe verrucosa IPNI=*680435-1* 195
Artanthe warakabacoura IPNI=*680574-1* 195
Artanthe xestophylla var. *latifolia* IPNI=*680435-1* 195
Artanthe xylopioides IPNI=*680435-1* 195
Artemisia absinthia IPNI=*300106-2* 257
Artemisia absinthium var. *insipida* IPNI=*300106-2* 257
Artemisia albida IPNI=*300106-2* 257
Artemisia arborescens f. *rehan* IPNI=*300106-2* 257
Artemisia arborescens var. *cupaniana* IPNI=*300106-2* 257
Artemisia aromatica IPNI=*307848-2* 257
Artemisia baldaccii IPNI=*300106-2* 257
Artemisia cernua IPNI=*307848-2* 257
Artemisia coarctata IPNI=*20812-2* 257
Artemisia desertorum var. *macrocephala* IPNI=*307848-2* 257
Artemisia doonense IPNI=*300106-2* 257
Artemisia dracunculiformis IPNI=*307848-2* 257
Artemisia dracunculoides IPNI=*307848-2* 257
Artemisia dracunculoides var. *brevifolia* IPNI=*307848-2* 257
Artemisia dracunculoides var. *incana* IPNI=*307848-2* 257
Artemisia dracunculoides var. *tenuifolia* IPNI=*307848-2* 257
Artemisia dracunculoides var. *wolfii* IPNI=*307848-2* 257
Artemisia dracunculus f. *humilis* IPNI=*307848-2* 257
Artemisia dracunculus subsp. *typica* IPNI=*307848-2* 257
Artemisia dracunculus var. *humilis* IPNI=*307848-2* 257
Artemisia dracunculus var. *pratorum* IPNI=*307848-2* 257
Artemisia dracunculus var. *redowskyi* IPNI=*307848-2* 257
Artemisia dracunculus var. *wolfii* IPNI=*307848-2* 257
Artemisia glauca var. *cernua* IPNI=*307848-2* 257
Artemisia glauca var. *dracunculoides* IPNI=*307848-2* 257
Artemisia glauca var. *magacephala* IPNI=*307848-2* 257
Artemisia inodora IPNI=*300106-2* 257
Artemisia inodora IPNI=*307848-2* 257
Artemisia kulbadica IPNI=*300106-2* 257
Artemisia nutans IPNI=*307848-2* 257
Artemisia pendula IPNI=*300106-2* 257
Artemisia redowskyi IPNI=*307848-2* 257
Artemisia rehan IPNI=*300106-2* 257
Artemisia rhaetica IPNI=*300106-2* 257
Artemisia simplicifolia IPNI=*307848-2* 257
Artemisia viridis IPNI=*1186403-2* 261
Artemisia vulgaris subsp. *litoralis* IPNI=*20812-2* 257
Artemisia vulgaris subsp. *typica* IPNI=*20812-2* 257
Artemisia vulgaris var. *coarctata* IPNI=*20812-2* 257
Artemisia vulgaris var. *typica* IPNI=*20812-2* 257
Arthratherum adscensionis subsp. *heymannii* IPNI=*319759-2* 234
Arthraxon falcifrons IPNI=*551367-1* 449
Arthrolobus rugosus IPNI=*1061021-2* 277
Arthrophyllum pinnatum IPNI=*91749-1* 254
Arthrosamanea marginata IPNI=*989342-1* 320
Arthrosamanea multiflora IPNI=*989448-1* 311
Arthrostemma campanulare IPNI=*60465985-2* 417
Arthrostemma caulialatum IPNI=*1101289-2* 417
Arthrostemma exiguum IPNI=*1131052-2* 424
Arthrostemma fragile IPNI=*60465985-2* 417
Arthrostemma goudotianum IPNI=*1131052-2* 424
Arthrostemma grandiflorum IPNI=*60465985-2* 417
Arthrostemma ladanoides IPNI=*1131052-2* 424
Arthrostemma macrodesmum IPNI=*60465985-2* 417
Arthrostemma myrtoideum IPNI=*573702-1* 423
Arthrostemma pumilum IPNI=*1131052-2* 424
Arthrostemma pusillum IPNI=*1131052-2* 424
Arthrostemma subtriplinervium IPNI=*569110-1* 419
Arthrostemma volubile IPNI=*60465985-2* 417
Arthrostylidium queko IPNI=*26916-2* 234
Arthrostylidium racemiflorum IPNI=*219234-2* 244
Arthrothamnus tirucalli IPNI=*348517-1* 391
Artocarpus altilis var. *non-seminifer* IPNI=*582598-1* 436
Artocarpus altilis var. *seminifer* IPNI=*582598-1* 436
Artocarpus blancoi IPNI=*582598-1* 436
Artocarpus brasiliensis IPNI=*850389-1* 436
Artocarpus camansi IPNI=*582598-1* 436
Artocarpus champeden IPNI=*582622-1* 436

Artocarpus communis IPNI=*582598-1* 436
Artocarpus hirsutissimus IPNI=*582622-1* 436
Artocarpus horridus IPNI=*582598-1* 436
Artocarpus incisifolius IPNI=*582598-1* 436
Artocarpus incisus IPNI=*582598-1* 436
Artocarpus integrifolius IPNI=*582622-1* 436
Artocarpus laevis IPNI=*582598-1* 436
Artocarpus leeuwenii IPNI=*582598-1* 436
Artocarpus mariannensis IPNI=*582598-1* 436
Artocarpus maximus IPNI=*850389-1* 436
Artocarpus multifidus IPNI=*582598-1* 436
Artocarpus nanca IPNI=*850389-1* 436
Artocarpus papuanus IPNI=*582598-1* 436
Artocarpus philippensis IPNI=*850389-1* 436
Artocarpus pilosus IPNI=*582622-1* 436
Artocarpus pinnatisectus IPNI=*582598-1* 436
Artocarpus polyphemus IPNI=*582622-1* 436
Artocarpus rima IPNI=*582598-1* 436
Aruana silvestris IPNI=*586076-1* 191
Arum aculeatum IPNI=*87563-1* 200
Arum arborescens IPNI=*87563-1* 200
Arum bicolor IPNI=*30010088-2* 199
Arum colocasia IPNI=*1170772-2* 199
Arum colocasioides IPNI=*1170772-2* 199
Arum cordifolium IPNI=*60444565-2* 198
Arum crudele IPNI=*86911-1* 199
Arum cucullatum IPNI=*84151-1* 198
Arum esculentum IPNI=*1170772-2* 199
Arum hederaceum IPNI=*87797-1* 201
Arum helleborifolium IPNI=*89333-1* 203
Arum liniferum IPNI=*87565-1* 200
Arum lividum IPNI=*1170772-2* 199
Arum macrorrhizon IPNI=*60444565-2* 198
Arum mucronatum IPNI=*60444565-2* 198
Arum nigrum IPNI=*89373-1* 203
Arum nymphaeifolium IPNI=*1170772-2* 199
Arum pellucidum IPNI=*30010088-2* 199
Arum peltatum IPNI=*1170772-2* 199
Arum peregrinum IPNI=*60444565-2* 198
Arum pinnatifidum IPNI=*77178497-1* 202
Arum pulchrum IPNI=*30010088-2* 199
Arum rapiforme IPNI=*60444565-2* 198
Arum sagittifolium IPNI=*89296-1* 202
Arum sagittifolium IPNI=*89333-1* 203
Arum sagittifolium IPNI=*89373-1* 203
Arum seguine IPNI=*86911-1* 199
Arum seguinum IPNI=*86911-1* 199
Arum tripartitum IPNI=*88045-1* 201
Arum venosum IPNI=*89333-1* 203
Arum vermitoxicum IPNI=*30010088-2* 199
Arum xanthorrhizon IPNI=*89373-1* 203
Arundarbor amplexifolia IPNI=*113747-2* 239
Arundarbor angulata IPNI=*392553-1* 235
Arundarbor aurea IPNI=*1097548-2* 234
Arundarbor beecheyana IPNI=*392260-1* 234
Arundarbor bitung IPNI=*397862-1* 237
Arundarbor blancoi IPNI=*392574-1* 235
Arundarbor brevifolia IPNI=*392553-1* 235
Arundarbor fera IPNI=*392574-1* 235
Arundarbor fortunei IPNI=*416271-1* 244
Arundarbor guadua IPNI=*1074014-2* 239
Arundarbor mitis IPNI=*392574-1* 235
Arundarbor monogyna IPNI=*392574-1* 235
Arundarbor multiplex IPNI=*1097548-2* 234
Arundarbor nana IPNI=*1097548-2* 234
Arundarbor oldhamii IPNI=*392445-1* 234
Arundarbor pygmaea IPNI=*416271-1* 244
Arundarbor striata IPNI=*392574-1* 235
Arundinaria fortunei f. *glabra* IPNI=*416271-1* 244
Arundinaria fortunei IPNI=*416271-1* 244
Arundinaria glaucescens IPNI=*1097548-2* 234
Arundinaria hispida var. *glabrivaginata* IPNI=*21052-2* 234
Arundinaria multiflora IPNI=*26923-2* 234
Arundinaria queko IPNI=*26916-2* 234
Arundinaria racemiflora IPNI=*219234-2* 244
Arundinaria trianae IPNI=*26923-2* 234
Arundinaria variegata f. *glabra* IPNI=*416271-1* 244
Arundinaria variegata IPNI=*416271-1* 244
Arundinella brasiliensis IPNI=*21052-2* 234
Arundinella confinis IPNI=*21052-2* 234
Arundinella elata IPNI=*21052-2* 234
Arundinella martinicensis IPNI=*21052-2* 234
Arundinella mikanii IPNI=*21052-2* 234
Arundinella pallida IPNI=*21052-2* 234
Arundinella scoparia IPNI=*21052-2* 234
Arundo aegyptia IPNI=*390837-1* 234
Arundo aegyptiaca IPNI=*390837-1* 234
Arundo aspera IPNI=*397862-1* 237
Arundo australis IPNI=*30337627-2* 244
Arundo bambusifolia IPNI=*390837-1* 234
Arundo bengalensis IPNI=*390837-1* 234
Arundo bifaria IPNI=*390837-1* 234
Arundo calycina IPNI=*66306-2* 236
Arundo coleotricha IPNI=*390837-1* 234
Arundo colorata IPNI=*415571-1* 243
Arundo donax f. *versicolor* IPNI=*390837-1* 234
Arundo donax var. *angustifolia* IPNI=*390837-1* 234
Arundo donax var. *lanceolata* IPNI=*390837-1* 234
Arundo donax var. *versicolor* IPNI=*390837-1* 234
Arundo effusa IPNI=*77206801-1* 242
Arundo fera IPNI=*392574-1* 235
Arundo glauca IPNI=*390837-1* 234
Arundo kila IPNI=*66317-2* 236
Arundo latifolia IPNI=*390837-1* 234
Arundo maxima IPNI=*404020-1* 239
Arundo mitis IPNI=*392574-1* 235
Arundo multiplex IPNI=*1097548-2* 234
Arundo nitida IPNI=*66306-2* 236
Arundo piscatoria IPNI=*397862-1* 237
Arundo riparia IPNI=*415571-1* 243
Arundo saccharifera IPNI=*419977-1* 244
Arundo sagittata IPNI=*404486-1* 240
Arundo sativa IPNI=*390837-1* 234
Arundo scriptoria IPNI=*390837-1* 234
Arundo selloana IPNI=*66317-2* 236
Arundo triflora IPNI=*390837-1* 234
Arundo versicolor IPNI=*390837-1* 234
Arviela tubispatha IPNI=*67314-1* 214
Asarina antirrhiniflora IPNI=*805502-1* 376
Asarina barclayana IPNI=*280591-2* 376
Asarina erubescens IPNI=*279831-2* 376
Ascalonicum sativum IPNI=*527795-1* 213
Ascalonicum schoenoprasum IPNI=*528823-1* 213
Aschamia equestris IPNI=*122815-2* 213
Aschenbornia heteropoda IPNI=*187778-1* 259
Asclepias alba IPNI=*106366-2* 340
Asclepias altissima IPNI=*60478412-2* 343
Asclepias argentata IPNI=*95173-1* 339
Asclepias aurantiaca IPNI=*94213-1* 339
Asclepias bicolor IPNI=*94213-1* 339
Asclepias brasiliensis IPNI=*97670-1* 340
Asclepias carnosa IPNI=*107308-3* 341
Asclepias clausa IPNI=*106366-2* 340
Asclepias cordata IPNI=*74165-2* 340
Asclepias cubensis IPNI=*94213-1* 339
Asclepias curassavica f. *flaviflora* IPNI=*94213-1* 339

Asclepias curassavica f. *kentingensis* IPNI=*94213-1* 339
Asclepias curassavica var. *concolor* IPNI=*94213-1* 339
Asclepias cynanchoides IPNI=*106382-2* 340
Asclepias filiformis IPNI=*106366-2* 340
Asclepias fruticosa IPNI=*97610-1* 340
Asclepias gigantea IPNI=*1004515-2* 339
Asclepias gigantea IPNI=*95173-1* 339
Asclepias jacquiniana IPNI=*106366-2* 340
Asclepias linearifolia IPNI=*106366-2* 340
Asclepias macrophylla IPNI=*100911-1* 343
Asclepias margaritacea IPNI=*94213-1* 339
Asclepias maritima IPNI=*98854-1* 341
Asclepias nivea var. *curassavica* IPNI=*94213-1* 339
Asclepias patula IPNI=*1004515-2* 339
Asclepias physocarpa IPNI=*97670-1* 340
Asclepias procera IPNI=*1004515-2* 339
Asclepias scandens IPNI=*106366-2* 340
Asclepias vicinalis IPNI=*106366-2* 340
Asclepias viminalis IPNI=*106366-2* 340
Asclepias volubilis IPNI=*106366-2* 340
Ascyrum calycinum IPNI=*433287-1* 393
Ascyrum coriaceum IPNI=*433287-1* 393
Asimina arborea IPNI=*72208-1* 186
Asparagopsis aethiopica IPNI=*530994-1* 214
Asparagopsis densiflora IPNI=*531072-1* 214
Asparagopsis lancea IPNI=*530994-1* 214
Asparagopsis setacea IPNI=*531312-1* 214
Asparagus altilis IPNI=*531229-1* 214
Asparagus caspius IPNI=*531229-1* 214
Asparagus collinus IPNI=*531229-1* 214
Asparagus esculentus IPNI=*531229-1* 214
Asparagus fiori IPNI=*531229-1* 214
Asparagus hedecarpus IPNI=*531229-1* 214
Asparagus hortensis IPNI=*531229-1* 214
Asparagus laetus IPNI=*530994-1* 214
Asparagus lanceus IPNI=*530994-1* 214
Asparagus littoralis IPNI=*531229-1* 214
Asparagus lujae IPNI=*531312-1* 214
Asparagus myriocladus IPNI=*531072-1* 214
Asparagus oxycarpus IPNI=*531229-1* 214
Asparagus paragus IPNI=*531229-1* 214
Asparagus plumosus IPNI=*531312-1* 214
Asparagus polyphyllus IPNI=*531229-1* 214
Asparagus sativus IPNI=*531229-1* 214
Asparagus setiformis IPNI=*531229-1* 214
Asparagus setiformis var. *tenuior* IPNI=*531229-1* 214
Asparagus setiformis var. *typicus* IPNI=*531229-1* 214
Asparagus sprengeri IPNI=*531072-1* 214
Asparagus sylvaticus IPNI=*531381-1* 214
Asparagus tenuifolius IPNI=*531229-1* 214
Asparagus terminalis IPNI=*533580-1* 215
Asparagus vulgaris IPNI=*531229-1* 214
Asparagus zanzibaricus IPNI=*531312-1* 214
Asperula aparine IPNI=*30007294-2* 350
Asperula aparine var. *aparine* IPNI=*30007294-2* 350
Aspidium aculeatum var. *montevidense* IPNI=*208026-2* 176
Aspidium acuminatum IPNI=*17160710-1* 176
Aspidium acutilobum IPNI=*249914-2* 177
Aspidium acutum IPNI=*17160710-1* 176
Aspidium adiantiforme IPNI=*224543-2* 176
Aspidium amoa IPNI=*77069554-1* 176
Aspidium andinum IPNI=*17226640-1* 177
Aspidium aquapimense IPNI=*251308-2* 174
Aspidium articulatum IPNI=*1138771-2* 176
Aspidium articulatum IPNI=*17160710-1* 176
Aspidium asplenioides IPNI=*17054310-1* 174
Aspidium auriculatum IPNI=*17160730-1* 176
Aspidium bifidum IPNI=*17226640-1* 177
Aspidium biserratum IPNI=*17160710-1* 176
Aspidium biserratum var. *paraense* IPNI=*17160710-1* 176
Aspidium blechnoides IPNI=*1148359-2* 175
Aspidium caducum IPNI=*1148359-2* 175
Aspidium calyptratum IPNI=*224543-2* 176
Aspidium capense IPNI=*224543-2* 176
Aspidium cicutarium subsp. *latifolium* IPNI=*249914-2* 177
Aspidium cicutarium var. *angustius* IPNI=*249914-2* 177
Aspidium clypeolarium IPNI=*224543-2* 176
Aspidium coniifolium IPNI=*77182519-1* 178
Aspidium conionevron IPNI=*17460390-1* 174
Aspidium continuum IPNI=*17458930-1* 174
Aspidium cordifolium IPNI=*17160730-1* 176
Aspidium cordifolium var. *pectinatum* IPNI=*17161100-1* 176
Aspidium cultratum IPNI=*17084790-1* 175
Aspidium cumingianum IPNI=*224543-2* 176
Aspidium cunninghamianum IPNI=*224543-2* 176
Aspidium cunninghamii IPNI=*224543-2* 176
Aspidium discolor IPNI=*224543-2* 176
Aspidium dissectum IPNI=*17458930-1* 174
Aspidium donianum IPNI=*17392950-1* 175
Aspidium duriusculum IPNI=*224543-2* 176
Aspidium ecklonii IPNI=*17458930-1* 174
Aspidium edule IPNI=*17160730-1* 176
Aspidium emarginatum IPNI=*319061-2* 177
Aspidium ensifolium IPNI=*17160710-1* 176
Aspidium exaltatum IPNI=*17160830-1* 176
Aspidium exaltatum var. *longipinna* IPNI=*17160710-1* 176
Aspidium expansum IPNI=*17086820-1* 174
Aspidium expansum IPNI=*17226640-1* 177
Aspidium extensum IPNI=*17460390-1* 174
Aspidium ferrugineum IPNI=*224543-2* 176
Aspidium filix-femina IPNI=*30055316-2* 174
Aspidium floccigerum IPNI=*77069554-1* 176
Aspidium fragile IPNI=*30013429-2* 174
Aspidium fraxinifolium IPNI=*17226640-1* 177
Aspidium funestum IPNI=*286436-2* 177
Aspidium gaudichaudii IPNI=*1148359-2* 175
Aspidium gibbosum IPNI=*17160710-1* 176
Aspidium gongylodes IPNI=*251353-2* 174
Aspidium guineense IPNI=*17160710-1* 176
Aspidium hemicardion IPNI=*1148359-2* 175
Aspidium heracleifolium IPNI=*249899-2* 177
Aspidium hoffmannseggii IPNI=*17160710-1* 176
Aspidium hookeri IPNI=*17460390-1* 174
Aspidium imbricatum IPNI=*17161310-1* 176
Aspidium latifolium IPNI=*249914-2* 177
Aspidium leucopis IPNI=*1019960-2* 174
Aspidium longifolium IPNI=*17226640-1* 177
Aspidium lucens IPNI=*17458930-1* 174
Aspidium macrolepidum IPNI=*208026-2* 176
Aspidium macrophyllum var. *pittieri* IPNI=*17226640-1* 177
Aspidium malaccense IPNI=*17460390-1* 174
Aspidium martinicense IPNI=*17226640-1* 177
Aspidium martinicense var. *puberulum* IPNI=*17226640-1* 177
Aspidium mauritianum IPNI=*17160710-1* 176
Aspidium microcarpon IPNI=*17341250-1* 176
Aspidium molle IPNI=*251308-2* 174
Aspidium molle var. *anoticum* IPNI=*251308-2* 174
Aspidium molle var. *decurtatum* IPNI=*251308-2* 174
Aspidium molle var. *glabrindusiatum* IPNI=*251308-2* 174
Aspidium molle var. *guadalupense* IPNI=*251308-2* 174
Aspidium mollissimum IPNI=*1019960-2* 174
Aspidium montevidense f. *genuina* IPNI=*208026-2* 176
Aspidium montevidense f. *imbricata* IPNI=*208026-2* 176
Aspidium montevidense IPNI=*208026-2* 176
Aspidium montevidense var. *subtripinnatum* IPNI=*208026-2* 176
Aspidium multijugum IPNI=*17460390-1* 174

Aspidium muricatum var. *montevidense* IPNI=*208026-2* 176
Aspidium natalense IPNI=*251308-2* 174
Aspidium nephrodioides IPNI=*17460390-1* 174
Aspidium nigropunctatum IPNI=*17160710-1* 176
Aspidium nodosum IPNI=*1138771-2* 176
Aspidium nymphale IPNI=*251308-2* 174
Aspidium obtusatum IPNI=*17458930-1* 174
Aspidium oligophlebium IPNI=*1019960-2* 174
Aspidium opulentum IPNI=*17460390-1* 174
Aspidium orientale IPNI=*224543-2* 176
Aspidium paleaceum IPNI=*17392950-1* 175
Aspidium pallidivenium IPNI=*17458930-1* 174
Aspidium paludosum IPNI=*17458930-1* 174
Aspidium paraense IPNI=*17160710-1* 176
Aspidium pectinatum IPNI=*17161100-1* 176
Aspidium pendulum IPNI=*173662-2* 176
Aspidium plantianum IPNI=*17458930-1* 174
Aspidium pohlianum IPNI=*251353-2* 174
Aspidium politum IPNI=*224543-2* 176
Aspidium propinquum IPNI=*17458930-1* 174
Aspidium pteroides var. *terminans* IPNI=*17460390-1* 174
Aspidium puberulum IPNI=*17226640-1* 177
Aspidium punctulatum IPNI=*17160710-1* 176
Aspidium remotum IPNI=*224543-2* 176
Aspidium resiniferum IPNI=*17458930-1* 174
Aspidium rufescens IPNI=*17160710-1* 176
Aspidium schkuhrii IPNI=*77069554-1* 176
Aspidium semicordatum IPNI=*1148359-2* 175
Aspidium semicordatum var. *americanum* IPNI=*1148359-2* 175
Aspidium serra IPNI=*17458930-1* 174
Aspidium serratum IPNI=*17458930-1* 174
Aspidium splendens IPNI=*17160710-1* 176
Aspidium squamatum IPNI=*17084790-1* 175
Aspidium squamosum IPNI=*17084790-1* 175
Aspidium sublanosum IPNI=*17160730-1* 176
Aspidium tavoyanum IPNI=*17160730-1* 176
Aspidium tenericaule IPNI=*1019960-2* 174
Aspidium terminans IPNI=*17460390-1* 174
Aspidium trapezoides IPNI=*17161100-1* 176
Aspidium trifoliatum var. *heracleifolium* IPNI=*249899-2* 177
Aspidium truncatulum IPNI=*17084790-1* 175
Aspidium tuberosum IPNI=*17160730-1* 176
Aspidium uliginosum IPNI=*1019960-2* 174
Aspidium undulatum IPNI=*17161310-1* 176
Aspidium unitum var. *glabrum* IPNI=*17458930-1* 174
Aspidium unitum var. *hirsutum* IPNI=*17458930-1* 174
Aspidium unitum var. *propinquum* IPNI=*17458930-1* 174
Aspidium venulosum IPNI=*17458930-1* 174
Aspidium violascens IPNI=*251308-2* 174
Aspidium volubile var. *cavernicola* IPNI=*17160730-1* 176
Aspidium wakefieldii IPNI=*17460390-1* 174
Aspidium wallichianum IPNI=*17392950-1* 175
Aspidogyne maculata IPNI=*77125705-1* 217
Aspidogyne mayoriana IPNI=*77125705-1* 217
Aspidogyne sagrana IPNI=*77125705-1* 217
Aspidogyne vaginata IPNI=*77125705-1* 217
Aspidosperma acreanum IPNI=*22058-2* 339
Aspidosperma aquaticum IPNI=*77367-1* 339
Aspidosperma bicolor IPNI=*77347-1* 339
Aspidosperma chiapense f. *tenax* IPNI=*77353-1* 339
Aspidosperma chiapense IPNI=*77353-1* 339
Aspidosperma cruentum IPNI=*77353-1* 339
Aspidosperma decipiens IPNI=*77347-1* 339
Aspidosperma domingense IPNI=*77347-1* 339
Aspidosperma dugandii IPNI=*77433-1* 339
Aspidosperma elatum IPNI=*77411-1* 339
Aspidosperma ellipticum IPNI=*77347-1* 339
Aspidosperma igapoanum IPNI=*22069-2* 339
Aspidosperma ingratum IPNI=*315552-2* 339
Aspidosperma jaunechense IPNI=*22058-2* 339
Aspidosperma latisiliqua IPNI=*77325-1* 339
Aspidosperma laxiflorum IPNI=*22058-2* 339
Aspidosperma leucomelanum IPNI=*22069-2* 339
Aspidosperma lhotzkyanum IPNI=*77347-1* 339
Aspidosperma lhotzkyanum var. *hypolasium* IPNI=*77347-1* 339
Aspidosperma lucentivenium IPNI=*77347-1* 339
Aspidosperma marcgravianum IPNI=*77367-1* 339
Aspidosperma matudae IPNI=*77353-1* 339
Aspidosperma melanocalyx IPNI=*22069-2* 339
Aspidosperma nitidum IPNI=*77367-1* 339
Aspidosperma paniculatum IPNI=*22069-2* 339
Aspidosperma peroba IPNI=*77433-1* 339
Aspidosperma polyneuron var. *genuinum* IPNI=*77433-1* 339
Aspidosperma polyneuron var. *longifolium* IPNI=*77433-1* 339
Aspidosperma polyneuron var. *puberulum* IPNI=*77433-1* 339
Aspidosperma quadriovulatum IPNI=*217932-2* 343
Aspidosperma rawolfioides IPNI=*22058-2* 339
Aspidosperma sanguineum IPNI=*77353-1* 339
Aspidosperma sessiliflorum IPNI=*77347-1* 339
Aspidosperma subumbellatum IPNI=*22058-2* 339
Aspidosperma vargasii IPNI=*315552-2* 339
Aspidosperma venosum IPNI=*77433-1* 339
Aspidosperma verruculosum IPNI=*22069-2* 339
Aspidosperma verruculosum var. *laeve* IPNI=*22069-2* 339
Aspidosperma woodsonianum IPNI=*22069-2* 339
Aspidosperma woronowii IPNI=*77347-1* 339
Aspilia bolivarana IPNI=*993359-1* 271
Aspilia laxa IPNI=*259782-1* 272
Asplenium allantodioides IPNI=*17086820-1* 174
Asplenium alloeopteron IPNI=*22662-2* 174
Asplenium aquaticum IPNI=*60458652-2* 174
Asplenium aquilinum IPNI=*17210060-1* 175
Asplenium arifolium IPNI=*319061-2* 177
Asplenium athyrium IPNI=*17054310-1* 174
Asplenium auritum f. *angustisectum* IPNI=*22222-2* 174
Asplenium auritum f. *diversifolium* IPNI=*22222-2* 174
Asplenium auritum f. *genuina* IPNI=*22222-2* 174
Asplenium auritum f. *laxum* IPNI=*22222-2* 174
Asplenium auritum f. *pendens* IPNI=*22222-2* 174
Asplenium auritum IPNI=*17039970-1* 174
Asplenium auritum subvar. *pinnatifidum* IPNI=*22222-2* 174
Asplenium auritum var. *auriculatum* IPNI=*22222-2* 174
Asplenium auritum var. *bipinnatifidum* IPNI=*22222-2* 174
Asplenium auritum var. *dissectum* IPNI=*22222-2* 174
Asplenium auritum var. *longicaudatum* IPNI=*22222-2* 174
Asplenium auritum var. *macilentum* IPNI=*22222-2* 174
Asplenium auritum var. *moritziana* IPNI=*22222-2* 174
Asplenium auritum var. *serratum* IPNI=*22222-2* 174
Asplenium bidentatum IPNI=*22222-2* 174
Asplenium brasiliense IPNI=*22751-2* 174
Asplenium bulbosum IPNI=*17160730-1* 176
Asplenium crenatum IPNI=*328699-2* 174
Asplenium crenulatum IPNI=*22751-2* 174
Asplenium crenulatum var. *expansum* IPNI=*17086820-1* 174
Asplenium cuneatum var. *caripense* IPNI=*328699-2* 174
Asplenium cuneatum var. *obtusum* IPNI=*328699-2* 174
Asplenium curvatum IPNI=*22222-2* 174
Asplenium cyrtopteron IPNI=*22662-2* 174
Asplenium dareoides IPNI=*328699-2* 174
Asplenium dimidiatum f. *laceratum* IPNI=*328509-2* 174
Asplenium dimidiatum var. *laceratum* IPNI=*328509-2* 174
Asplenium distentum IPNI=*17086820-1* 174
Asplenium expansum IPNI=*17086820-1* 174
Asplenium expansum var. *glabriusculum* IPNI=*17086820-1* 174
Asplenium filix-femina IPNI=*30055316-2* 174
Asplenium filix-femina var. *exile* IPNI=*30055316-2* 174

Asplenium filix-femina var. *purpureum* IPNI=*30055316-2* 174
Asplenium flagelliferum IPNI=*22662-2* 174
Asplenium formosum var. *incultum* IPNI=*22417-2* 174
Asplenium fruticosum IPNI=*17084790-1* 175
Asplenium gracile IPNI=*328699-2* 174
Asplenium hartianum IPNI=*17086820-1* 174
Asplenium integrum IPNI=*22751-2* 174
Asplenium lanciforme IPNI=*22751-2* 174
Asplenium levyi IPNI=*22222-2* 174
Asplenium lineare IPNI=*17217170-1* 178
Asplenium lonchophyllum IPNI=*17088960-1* 174
Asplenium longifolium IPNI=*22751-2* 174
Asplenium macilentum IPNI=*22222-2* 174
Asplenium marinum var. *auriculatum* IPNI=*22222-2* 174
Asplenium martinicense IPNI=*328699-2* 174
Asplenium mastigophyllum IPNI=*22662-2* 174
Asplenium mocquerysii IPNI=*17039970-1* 174
Asplenium monanthemum IPNI=*22558-2* 174
Asplenium nanum IPNI=*22417-2* 174
Asplenium nitidulum IPNI=*328509-2* 174
Asplenium obtusifolium IPNI=*60458652-2* 174
Asplenium obtusifolium var. *aquatica* IPNI=*60458652-2* 174
Asplenium obtusilobum IPNI=*328699-2* 174
Asplenium odontophyllum IPNI=*22417-2* 174
Asplenium plumbeum IPNI=*22222-2* 174
Asplenium polyodon var. *squamulosum* IPNI=*328509-2* 174
Asplenium praemorsum var. *artemisiifolium* IPNI=*328509-2* 174
Asplenium praemorsum var. *laceratum* IPNI=*328509-2* 174
Asplenium prolixum IPNI=*22222-2* 174
Asplenium pygmaeum IPNI=*17039970-1* 174
Asplenium raddii IPNI=*22751-2* 174
Asplenium radicans f. *fusagasugensis* IPNI=*22662-2* 174
Asplenium radicans f. *genuina* IPNI=*22662-2* 174
Asplenium radicans var. *alloeopteron* IPNI=*22662-2* 174
Asplenium radicans var. *argutidentatum* IPNI=*22662-2* 174
Asplenium radicans var. *bipinnatifidum* IPNI=*22662-2* 174
Asplenium radicans var. *crenatum* IPNI=*22662-2* 174
Asplenium radicans var. *cyrtopteron* IPNI=*22662-2* 174
Asplenium radicans var. *expansum* IPNI=*17086820-1* 174
Asplenium radicans var. *longedecurrens* IPNI=*22662-2* 174
Asplenium radicans var. *pachoensis* IPNI=*22662-2* 174
Asplenium radicans var. *remotum* IPNI=*22662-2* 174
Asplenium ramosum IPNI=*17084790-1* 175
Asplenium rhizophorum IPNI=*22662-2* 174
Asplenium rhizophorum var. *alloeopteron* IPNI=*22662-2* 174
Asplenium rhizophorum var. *bipinnatum* IPNI=*22662-2* 174
Asplenium rhizophorum var. *cyrtopteron* IPNI=*22662-2* 174
Asplenium rhizophorum var. *pinnatopinnatifidum* IPNI=*22662-2* 174
Asplenium rhizophorum var. *radicans* IPNI=*22662-2* 174
Asplenium rhizophorum var. *supersum* IPNI=*22662-2* 174
Asplenium rhizophyllum var. *alloeoptera* IPNI=*22662-2* 174
Asplenium riparium var. *obtusifolium* IPNI=*60458652-2* 174
Asplenium schomburgkianum IPNI=*22751-2* 174
Asplenium schottii IPNI=*22222-2* 174
Asplenium serratum f. *incisum* IPNI=*22751-2* 174
Asplenium serratum var. *blanchetianum* IPNI=*22751-2* 174
Asplenium serratum var. *caudatum* IPNI=*22751-2* 174
Asplenium serratum var. *crenulatum* IPNI=*22751-2* 174
Asplenium serratum var. *lanciforme* IPNI=*22751-2* 174
Asplenium serratum var. *latius* IPNI=*22751-2* 174
Asplenium serratum var. *minor* IPNI=*22751-2* 174
Asplenium striatum IPNI=*17086820-1* 174
Asplenium striatum var. *expansum* IPNI=*17086820-1* 174
Asplenium subalatum IPNI=*22417-2* 174
Asplenium subserratum IPNI=*17088960-1* 174
Asplenium subsessile IPNI=*22751-2* 174
Asplenium sulcatum IPNI=*17039970-1* 174
Asplenium sulcatum var. *auritum* IPNI=*22222-2* 174
Asplenium sulcatum var. *rigidum* IPNI=*22222-2* 174
Asplenium umbrosum IPNI=*22222-2* 174
Asplenium weberbaueri IPNI=*22751-2* 174
Asplenium willdenowii IPNI=*17086820-1* 174
Asplundia costaricensis IPNI=*22967-2* 227
Asprella australis IPNI=*137026-2* 241
Asprella brasiliensis IPNI=*137026-2* 241
Asprella digitaria IPNI=*1038558-2* 237
Asprella hexandra IPNI=*137026-2* 241
Asprella mexicana IPNI=*137026-2* 241
Asprella purpurea IPNI=*137026-2* 241
Aspris caryophyllea IPNI=*386633-1* 233
Assonia × cayeuxii IPNI=*822968-1* 408
Assonia burgessiae IPNI=*822960-1* 408
Assonia calantha IPNI=*822960-1* 408
Assonia phoenicea IPNI=*824676-1* 412
Assonia sparmannioides IPNI=*822960-1* 408
Assonia wallichii IPNI=*823281-1* 408
Astephanus mucronatus IPNI=*94927-1* 339
Aster ambiguus IPNI=*204572-1* 262
Aster amellus IPNI=*181476-1* 257
Aster amplexicaulis IPNI=*247114-2* 270
Aster autranii IPNI=*247114-2* 270
Aster bellis IPNI=*184409-1* 258
Aster bupleurifolius IPNI=*247114-2* 270
Aster canadensis IPNI=*203653-1* 261
Aster conyzae IPNI=*60478388-2* 267
Aster cyaneus IPNI=*247114-2* 270
Aster decorus IPNI=*247114-2* 270
Aster exilis f. *subalpinus* IPNI=*981909-1* 271
Aster exilis var. *australis* IPNI=*981909-1* 271
Aster expansus IPNI=*247114-2* 270
Aster falcidens IPNI=*247114-2* 270
Aster flexicaulis IPNI=*981909-1* 271
Aster glaucescens IPNI=*247114-2* 270
Aster glaucus IPNI=*247114-2* 270
Aster laevigatus IPNI=*247114-2* 270
Aster laevis f. *amplifolius* IPNI=*247114-2* 270
Aster laevis f. *beckwithiae* IPNI=*247114-2* 270
Aster laevis f. *falcatus* IPNI=*247114-2* 270
Aster laevis f. *latifolius* IPNI=*247114-2* 270
Aster laevis f. *purpurascens* IPNI=*247114-2* 270
Aster laevis IPNI=*247114-2* 270
Aster laevis var. *abbreviatus* IPNI=*247114-2* 270
Aster laevis var. *amplifolius* IPNI=*247114-2* 270
Aster laevis var. *cyaneus* IPNI=*247114-2* 270
Aster laevis var. *falcatus* IPNI=*247114-2* 270
Aster laevis var. *latifolius* IPNI=*247114-2* 270
Aster laevis var. *panduratus* IPNI=*247114-2* 270
Aster laevis var. *simplex* IPNI=*247114-2* 270
Aster laevis var. *sourisensis* IPNI=*247114-2* 270
Aster laevis var. *strictiflorus* IPNI=*247114-2* 270
Aster laevis var. *thyrsoideus* IPNI=*247114-2* 270
Aster laevis var. *undulatifolius* IPNI=*247114-2* 270
Aster latissimifolius var. *serotinus* IPNI=*249575-1* 270
Aster linifolius IPNI=*981909-1* 271
Aster lucidus IPNI=*247114-2* 270
Aster pensilvanicus IPNI=*247114-2* 270
Aster pickettianus IPNI=*247114-2* 270
Aster politus IPNI=*247114-2* 270
Aster pseudamellus IPNI=*181476-1* 257
Aster rubicaulis IPNI=*247114-2* 270
Aster rubricaulis IPNI=*247114-2* 270
Aster steeleorum IPNI=*247114-2* 270
Aster subulatus IPNI=*981909-1* 271
Aster subulatus var. *australis* IPNI=*981909-1* 271
Aster subulatus var. *euroauster* IPNI=*981909-1* 271
Aster subulatus var. *obtusifolius* IPNI=*981909-1* 271

Asteriastigma macrocarpa IPNI=*111756-1* 380
Asterocephalus acutiflorus IPNI=*319717-1* 296
Asterocephalus ambiguus IPNI=*319717-1* 296
Asterocephalus atropurpureus IPNI=*319717-1* 296
Asterocephalus caucasicus IPNI=*915212-1* 296
Asterocephalus elegans IPNI=*915212-1* 296
Asterocephalus grandiflorus IPNI=*319717-1* 296
Asterochlaena cuspidata IPNI=*562981-1* 412
Asterogyne minor IPNI=*664278-1* 204
Asterolepidion elatum IPNI=*77582-2* 255
Asterophyllum aparine IPNI=*30007294-2* 350
Asterostoma rotundifolia IPNI=*908858-1* 419
Astraea palmata IPNI=*339515-1* 388
Astraea seemannii IPNI=*339515-1* 388
Astragalus spicatus IPNI=*499456-1* 321
Astrapaea penduliflora IPNI=*823281-1* 408
Astrapaea wallichii IPNI=*823281-1* 408
Astrephia chaerophylloides IPNI=*859680-1* 296
Astrephia chinensis IPNI=*860012-1* 296
Astrephia laciniata IPNI=*859680-1* 296
Astrocaryum acaule var. *angustifolium* IPNI=*664283-1* 204
Astrocaryum aculeatum IPNI=*664513-1* 206
Astrocaryum aureum IPNI=*326537-2* 205
Astrocaryum caudescens IPNI=*326537-2* 205
Astrocaryum cuatrecasasianum IPNI=*25891-2* 205
Astrocaryum dasychaetum IPNI=*326535-2* 204
Astrocaryum granatense IPNI=*326537-2* 205
Astrocaryum guara IPNI=*664323-1* 204
Astrocaryum gymnopus IPNI=*326535-2* 204
Astrocaryum gynacanthum IPNI=*326535-2* 204
Astrocaryum gynacanthum var. *dasychaetum* IPNI=*326535-2* 204
Astrocaryum huebneri IPNI=*664283-1* 204
Astrocaryum jucuma IPNI=*326537-2* 205
Astrocaryum luetzelburgii IPNI=*664283-1* 204
Astrocaryum macrocarpum IPNI=*326537-2* 205
Astrocaryum manaoense IPNI=*326537-2* 205
Astrocaryum minus var. *terrae-firmae* IPNI=*326535-2* 204
Astrocaryum munbaca IPNI=*326535-2* 204
Astrocaryum murumuru var. *ciliatum* IPNI=*984104-1* 204
Astrocaryum murumuru var. *ferrugineum* IPNI=*984103-1* 204
Astrocaryum murumuru var. *macrocalyx* IPNI=*25873-2* 204
Astrocaryum murumuru var. *urostachys* IPNI=*25891-2* 205
Astrocaryum plicatum IPNI=*25885-2* 205
Astrocaryum princeps IPNI=*326537-2* 205
Astrocaryum princeps var. *aurantiacum* IPNI=*326537-2* 205
Astrocaryum princeps var. *flavum* IPNI=*326537-2* 205
Astrocaryum princeps var. *sulphureum* IPNI=*326537-2* 205
Astrocaryum princeps var. *vitellinum* IPNI=*326537-2* 205
Astrocaryum sclerocarpum IPNI=*663914-1* 203
Astrocaryum standleyanum var. *calimense* IPNI=*25887-2* 205
Astrocaryum trachycarpum IPNI=*25887-2* 205
Astrocaryum yauaperyense IPNI=*664335-1* 205
Astronium conzattii IPNI=*69240-1* 450
Astronium conzattii var. *lundellii* IPNI=*69240-1* 450
Astronium conzattii var. *standleyi* IPNI=*69240-1* 450
Astronium fraxinifolium f. *mollissimum* IPNI=*69236-1* 450
Astronium fraxinifolium f. *pilosum* IPNI=*69236-1* 450
Astronium gracile f. *acuminatum* IPNI=*69240-1* 450
Astronium gracile f. *serratum* IPNI=*69240-1* 450
Astronium gracile IPNI=*69240-1* 450
Astronium gracile var. *acuminatum* IPNI=*69240-1* 450
Astronium graveolens var. *brasiliense* IPNI=*69236-1* 450
Astronium graveolens var. *dugandii* IPNI=*69240-1* 450
Astronium graveolens var. *inodorum* IPNI=*69240-1* 450
Astronium graveolens var. *planchonianum* IPNI=*69240-1* 450
Astronium lecointei f. *tomentosum* IPNI=*25916-2* 450
Astronium lecointei var. *tomentosum* IPNI=*25916-2* 450
Astronium planchonianum IPNI=*69240-1* 450
Astronium zongolicum IPNI=*69240-1* 450
Astrophea emarginata IPNI=*674975-1* 398
Astyposanthes guianensis IPNI=*1020451-2* 334
Astyposanthes humilis IPNI=*519681-1* 334
Astyposanthes viscosa IPNI=*1176701-2* 334
Asystasia coromandeliana IPNI=*45742-1* 359
Asystasia gangetica var. *mendeliana* IPNI=*45742-1* 359
Asystasia plumbaginea IPNI=*45742-1* 359
Asystasia quarterna IPNI=*45742-1* 359
Asystasia violacea IPNI=*45742-1* 359
Atamosco carinata IPNI=*270933-2* 214
Atamosco tubispatha IPNI=*67314-1* 214
Athanasia hastata IPNI=*156327-2* 266
Atheropogon aristidoides IPNI=*1049662-2* 235
Atheropogon curtipendulus IPNI=*1198286-2* 235
Atheropogon hirsutus IPNI=*1049662-2* 235
Atheropogon humilis IPNI=*328564-2* 235
Atheropogon procumbens IPNI=*328564-2* 235
Athyrium asplenioides f. *ellipticum* IPNI=*17054310-1* 174
Athyrium asplenioides f. *subtripinnatum* IPNI=*17054310-1* 174
Athyrium asplenioides var. *subtripinnatum* IPNI=*17054310-1* 174
Athyrium asplenioides var. *typicum* IPNI=*17054310-1* 174
Athyrium distentum IPNI=*17086820-1* 174
Athyrium expansum IPNI=*17086820-1* 174
Athyrium filix-femina f. *ellipticum* IPNI=*17054310-1* 174
Athyrium filix-femina f. *plano-rhoeticum* IPNI=*17054310-1* 174
Athyrium filix-femina f. *rectangulare* IPNI=*17054310-1* 174
Athyrium filix-femina f. *subtripinnatum* IPNI=*17054310-1* 174
Athyrium filix-femina subsp. *asplenioides* IPNI=*17054310-1* 174
Athyrium filix-femina var. *asplenioides* IPNI=*17054310-1* 174
Athyrium filix-femina var. *commune* IPNI=*17054310-1* 174
Athyrium filix-femina var. *incisa* IPNI=*17054310-1* 174
Athyrium fragile IPNI=*30013429-2* 174
Athyrium laetum IPNI=*30055316-2* 174
Athyrium subserratum IPNI=*17088960-1* 174
Athyrocarpus mathewsii IPNI=*63900-2* 224
Atimeta martii IPNI=*88634-1* 202
Atimeta videniana IPNI=*88634-1* 202
Atitara aerea IPNI=*666648-1* 208
Atitara caespitosa IPNI=*666648-1* 208
Atitara dubia IPNI=*666648-1* 208
Atitara horrida IPNI=*666606-1* 208
Atitara inermis IPNI=*666648-1* 208
Atitara lophacantha IPNI=*666637-1* 208
Atitara macroacantha IPNI=*666648-1* 208
Atitara macrodon IPNI=*666648-1* 208
Atitara mitis IPNI=*666632-1* 208
Atitara nemorosa IPNI=*666648-1* 208
Atitara oligacantha IPNI=*666648-1* 208
Atitara orthacantha IPNI=*666637-1* 208
Atitara oxyacantha IPNI=*666648-1* 208
Atitara paraensis IPNI=*666648-1* 208
Atitara phengophylla IPNI=*666648-1* 208
Atitara philippiana IPNI=*666648-1* 208
Atitara phoenicocarpa IPNI=*666648-1* 208
Atitara polyacantha IPNI=*666648-1* 208
Atitara pycnacantha IPNI=*666648-1* 208
Atitara riparia IPNI=*666648-1* 208
Atocion armeria var. *lituanicum* IPNI=*77093328-1* 285
Atocion armeria var. *sparsiflorum* IPNI=*77093328-1* 285
Atocion armeroides IPNI=*77093328-1* 285
Atocion lituanicum IPNI=*77093328-1* 285
Atractylis mexicana IPNI=*231341-1* 265
Atractylis purpurata IPNI=*77181848-1* 267
Atractylis purpurea IPNI=*77181848-1* 267
Atractylis triplinervia IPNI=*231347-1* 265
Atriplex alba IPNI=*55993-2* 282
Atriplex ambrosioides IPNI=*1198060-2* 282

Atriplex anthelmintica IPNI=*1090820-2* 282
Atriplex muralis IPNI=*77121007-1* 282
Atriplex viridis IPNI=*55993-2* 282
Attalea agrestis IPNI=*319686-2* 205
Attalea blepharopus IPNI=*1172851-2* 205
Attalea cryptanthera IPNI=*664452-1* 205
Attalea excelsa IPNI=*26580-2* 205
Attalea ferruginea IPNI=*664463-1* 205
Attalea goeldiana IPNI=*26571-2* 205
Attalea gomphococca IPNI=*1172851-2* 205
Attalea humboldtiana IPNI=*1172851-2* 205
Attalea macrocarpa IPNI=*1172851-2* 205
Attalea macropetala IPNI=*664452-1* 205
Attalea phalerata var. *concinna* IPNI=*26580-2* 205
Attalea polysticha IPNI=*319686-2* 205
Attalea pycnocarpa IPNI=*1172851-2* 205
Attalea regia IPNI=*664452-1* 205
Attalea rhynchocarpa IPNI=*664411-1* 205
Attalea sagotii IPNI=*319686-2* 205
Attalea uberrima IPNI=*664411-1* 205
Attalea victoriana IPNI=*664411-1* 205
Attalea wallisii IPNI=*1172851-2* 205
Aubletia aspera IPNI=*833267-1* 407
Aubletia laevis IPNI=*833257-1* 407
Aubletia petoumo IPNI=*833267-1* 407
Aubletia tibourbou IPNI=*833270-1* 407
Aubletia tiburbu IPNI=*833270-1* 407
Aucellia ruberrima IPNI=*155262-2* 221
Aucellia sophronitis IPNI=*155290-2* 221
Auganthus obconicus IPNI=*702340-1* 304
Auganthus praenitens IPNI=*702446-1* 304
Auganthus sinensis IPNI=*702446-1* 304
Augustinea major IPNI=*664668-1* 206
Aulacocarpus completens IPNI=*164541-2* 423
Aulacophyllum ortgiesii IPNI=*297256-1* 179
Aulacophyllum roezlii IPNI=*297372-1* 179
Auliphas ligustrina IPNI=*572682-1* 421
Auliza ciliaris IPNI=*631578-1* 218
Auliza nocturna IPNI=*30010607-2* 219
Aulomyrcia albidotomentosa IPNI=*165538-2* 426
Aulomyrcia alloiota IPNI=*599215-1* 427
Aulomyrcia alloiota var. *cuneata* IPNI=*599215-1* 427
Aulomyrcia alloiota var. *obovata* IPNI=*599215-1* 427
Aulomyrcia alloiota var. *ovalis* IPNI=*599215-1* 427
Aulomyrcia alloiota var. *pyramidalis* IPNI=*599215-1* 427
Aulomyrcia alloiota var. *subcordata* IPNI=*599215-1* 427
Aulomyrcia alternifolia IPNI=*598851-1* 426
Aulomyrcia amshoffiana IPNI=*165538-2* 426
Aulomyrcia androsaemoides IPNI=*598851-1* 426
Aulomyrcia angustifolia IPNI=*598851-1* 426
Aulomyrcia aureolanata IPNI=*599215-1* 427
Aulomyrcia biformis IPNI=*598851-1* 426
Aulomyrcia bimarginata IPNI=*598851-1* 426
Aulomyrcia botrys IPNI=*598851-1* 426
Aulomyrcia bracteata IPNI=*598851-1* 426
Aulomyrcia buxifolia var. *elliptica* IPNI=*598851-1* 426
Aulomyrcia buxifolia var. *ovalis* IPNI=*598851-1* 426
Aulomyrcia buxizans IPNI=*598851-1* 426
Aulomyrcia capivarhyensis IPNI=*599215-1* 427
Aulomyrcia caracasana IPNI=*599215-1* 427
Aulomyrcia cassinioides IPNI=*598851-1* 426
Aulomyrcia cassinioides var. *glabrata* IPNI=*598851-1* 426
Aulomyrcia cassinioides var. *velutina* IPNI=*598851-1* 426
Aulomyrcia clausseniana IPNI=*598851-1* 426
Aulomyrcia conduplicata IPNI=*598851-1* 426
Aulomyrcia confusa IPNI=*599215-1* 427
Aulomyrcia costata IPNI=*599170-1* 427
Aulomyrcia crassicaulis IPNI=*598851-1* 426
Aulomyrcia cuneata IPNI=*598851-1* 426
Aulomyrcia curatellifolia IPNI=*599215-1* 427
Aulomyrcia curatellifolia var. *australis* IPNI=*599215-1* 427
Aulomyrcia curatellifolia var. *grandifolia* IPNI=*599215-1* 427
Aulomyrcia curatellifolia var. *parvifolia* IPNI=*599215-1* 427
Aulomyrcia daphnoides IPNI=*598851-1* 426
Aulomyrcia decrescens IPNI=*598851-1* 426
Aulomyrcia desertorum IPNI=*598851-1* 426
Aulomyrcia dichroma IPNI=*598851-1* 426
Aulomyrcia elaeodendra IPNI=*598851-1* 426
Aulomyrcia emarginata IPNI=*598851-1* 426
Aulomyrcia exsucca IPNI=*598851-1* 426
Aulomyrcia fragilis IPNI=*598851-1* 426
Aulomyrcia gardneriana IPNI=*598851-1* 426
Aulomyrcia gardneriana var. *caerulescens* IPNI=*598851-1* 426
Aulomyrcia gardneriana var. *virescens* IPNI=*598851-1* 426
Aulomyrcia glandulosa IPNI=*598851-1* 426
Aulomyrcia glandulosa var. *elliptica* IPNI=*598851-1* 426
Aulomyrcia glandulosa var. *longifolia* IPNI=*598851-1* 426
Aulomyrcia glandulosa var. *obovata* IPNI=*598851-1* 426
Aulomyrcia hepatica IPNI=*598851-1* 426
Aulomyrcia inaequiloba IPNI=*598883-1* 426
Aulomyrcia inaequiloba var. *nitida* IPNI=*598883-1* 426
Aulomyrcia inaequiloba var. *paniculata* IPNI=*598883-1* 426
Aulomyrcia jequitinhonhensis var. *glauca* IPNI=*598851-1* 426
Aulomyrcia lancea IPNI=*599215-1* 427
Aulomyrcia lancifolia IPNI=*598883-1* 426
Aulomyrcia lanuginosa IPNI=*599215-1* 427
Aulomyrcia lanuginosa var. *pyramidata* IPNI=*599215-1* 427
Aulomyrcia lauriflora IPNI=*598851-1* 426
Aulomyrcia leucadendron IPNI=*598851-1* 426
Aulomyrcia leucantha IPNI=*599215-1* 427
Aulomyrcia lingua IPNI=*598851-1* 426
Aulomyrcia lingua var. *glabrata* IPNI=*598851-1* 426
Aulomyrcia lingua var. *rufa* IPNI=*598851-1* 426
Aulomyrcia longipes IPNI=*599215-1* 427
Aulomyrcia longipes var. *latifolia* IPNI=*599215-1* 427
Aulomyrcia longipes var. *obovata* IPNI=*599215-1* 427
Aulomyrcia longipes var. *spathulata* IPNI=*599215-1* 427
Aulomyrcia mansoni IPNI=*598851-1* 426
Aulomyrcia maritima IPNI=*598851-1* 426
Aulomyrcia micrantha IPNI=*599215-1* 427
Aulomyrcia nigrescens IPNI=*77160867-1* 427
Aulomyrcia nigrescens var. *angustifolia* IPNI=*77160867-1* 427
Aulomyrcia nigrescens var. *dives* IPNI=*77160867-1* 427
Aulomyrcia nigrescens var. *latifolia* IPNI=*77160867-1* 427
Aulomyrcia nigropunctata IPNI=*598851-1* 426
Aulomyrcia obscura IPNI=*598851-1* 426
Aulomyrcia obscura var. *longipes* IPNI=*598851-1* 426
Aulomyrcia obtecta IPNI=*598851-1* 426
Aulomyrcia obtusa IPNI=*598851-1* 426
Aulomyrcia obtusa var. *grandifolia* IPNI=*598851-1* 426
Aulomyrcia obtusa var. *longipes* IPNI=*598851-1* 426
Aulomyrcia obtusa var. *panicularis* IPNI=*598851-1* 426
Aulomyrcia obtusa var. *pauciflora* IPNI=*598851-1* 426
Aulomyrcia obtusa var. *surinamensis* IPNI=*598851-1* 426
Aulomyrcia obtusa var. *tenuifolia* IPNI=*598851-1* 426
Aulomyrcia ottonis IPNI=*599215-1* 427
Aulomyrcia pallens IPNI=*598851-1* 426
Aulomyrcia pallens var. *ovalis* IPNI=*598851-1* 426
Aulomyrcia pallens var. *ovata* IPNI=*598851-1* 426
Aulomyrcia pallens var. *subcordata* IPNI=*598851-1* 426
Aulomyrcia paniculata IPNI=*598883-1* 426
Aulomyrcia pirarensis IPNI=*598883-1* 426
Aulomyrcia plumbea IPNI=*598851-1* 426
Aulomyrcia poeppigiana IPNI=*598851-1* 426
Aulomyrcia pruinosa IPNI=*598851-1* 426
Aulomyrcia prunifolia IPNI=*599215-1* 427

Aulomyrcia prunifolia var. *brevipes* IPNI=*599215-1* 427
Aulomyrcia prunifolia var. *longipes* IPNI=*599215-1* 427
Aulomyrcia puberula IPNI=*599215-1* 427
Aulomyrcia pusilla IPNI=*598851-1* 426
Aulomyrcia pyrifolia var. *gracilis* IPNI=*165926-2* 427
Aulomyrcia pyrifolia var. *robusta* IPNI=*165926-2* 427
Aulomyrcia regeliana IPNI=*598851-1* 426
Aulomyrcia regeliana var. *oppositifolia* IPNI=*598851-1* 426
Aulomyrcia regeliana var. *sparsifolia* IPNI=*598851-1* 426
Aulomyrcia roraimae IPNI=*598851-1* 426
Aulomyrcia roraimensis IPNI=*598851-1* 426
Aulomyrcia rosulans IPNI=*599215-1* 427
Aulomyrcia rotundifolia IPNI=*598851-1* 426
Aulomyrcia salicifolia IPNI=*599124-1* 427
Aulomyrcia schomburgkiana IPNI=*598851-1* 426
Aulomyrcia schrankiana IPNI=*598851-1* 426
Aulomyrcia scrobiculata IPNI=*598851-1* 426
Aulomyrcia spathulata IPNI=*599215-1* 427
Aulomyrcia suaveolens IPNI=*598851-1* 426
Aulomyrcia suffruticosa IPNI=*598851-1* 426
Aulomyrcia surinamensis IPNI=*598851-1* 426
Aulomyrcia tomentosa IPNI=*599215-1* 427
Aulomyrcia vacciniifolia IPNI=*598851-1* 426
Aulomyrcia velhensis IPNI=*598851-1* 426
Aulomyrcia wullschlaegeliana IPNI=*599170-1* 427
Auriculardisia fimbrillifera IPNI=*18143-2* 303
Austrocylindropuntia intermedia IPNI=*128942-1* 283
Austroeupatorium entreriense IPNI=*27149-2* 257
Aveledoa nucifera IPNI=*158626-2* 415
Avena airoides IPNI=*258855-2* 246
Avena algeriensis IPNI=*391732-1* 234
Avena anglica IPNI=*391732-1* 234
Avena caryophyllea IPNI=*386633-1* 233
Avena cinerea IPNI=*391732-1* 234
Avena dispermis IPNI=*391732-1* 234
Avena distans IPNI=*391732-1* 234
Avena dura IPNI=*30010036-2* 239
Avena elata IPNI=*389923-1* 234
Avena elatior IPNI=*389923-1* 234
Avena fatua f. *subcontracta* IPNI=*391443-1* 234
Avena fatua subsp. *brevipilosa* IPNI=*391443-1* 234
Avena fatua subsp. *glabrata* IPNI=*391443-1* 234
Avena fatua subsp. *praegravis* IPNI=*391732-1* 234
Avena fatua subsp. *sativa* IPNI=*391732-1* 234
Avena fatua subvar. *naniformis* IPNI=*391443-1* 234
Avena fatua subvar. *pseudonana* IPNI=*391443-1* 234
Avena fatua subvar. *pumila* IPNI=*391443-1* 234
Avena fatua subvar. *zine* IPNI=*391443-1* 234
Avena fatua var. *acidophila* IPNI=*391443-1* 234
Avena fatua var. *alcaliphila* IPNI=*391443-1* 234
Avena fatua var. *alta* IPNI=*391443-1* 234
Avena fatua var. *altissima* IPNI=*391443-1* 234
Avena fatua var. *gravis* IPNI=*391443-1* 234
Avena fatua var. *hyugaensis* IPNI=*391443-1* 234
Avena fatua var. *levis* IPNI=*391443-1* 234
Avena fatua var. *nipponica* IPNI=*391443-1* 234
Avena fatua var. *pilosiformis* IPNI=*391443-1* 234
Avena fatua var. *transiens* IPNI=*391732-1* 234
Avena flava IPNI=*391732-1* 234
Avena fusciflora IPNI=*391732-1* 234
Avena geogica IPNI=*391732-1* 234
Avena georgiana IPNI=*391732-1* 234
Avena grandis IPNI=*391732-1* 234
Avena hungarica IPNI=*391732-1* 234
Avena intermedia IPNI=*391443-1* 234
Avena japonica IPNI=*391443-1* 234
Avena lanata IPNI=*405034-1* 240
Avena lanuginosa IPNI=*391443-1* 234
Avena macrantha IPNI=*391732-1* 234
Avena meridionalis IPNI=*391443-1* 234
Avena mollis IPNI=*123530-2* 240
Avena mollis IPNI=*258855-2* 246
Avena multiculmis IPNI=*386633-1* 233
Avena muralis IPNI=*403134-1* 239
Avena mutata IPNI=*391732-1* 234
Avena mutica IPNI=*391732-1* 234
Avena nigra IPNI=*391443-1* 234
Avena nodipilosa IPNI=*391732-1* 234
Avena occidentalis IPNI=*391443-1* 234
Avena orientalis IPNI=*391732-1* 234
Avena pallida IPNI=*405034-1* 240
Avena paupercula IPNI=*393559-1* 235
Avena pendula IPNI=*391732-1* 234
Avena pilosa IPNI=*391443-1* 234
Avena podolica IPNI=*391732-1* 234
Avena polonica IPNI=*391732-1* 234
Avena ponderosa IPNI=*391732-1* 234
Avena praecocioides IPNI=*391732-1* 234
Avena praecoqua IPNI=*391732-1* 234
Avena pseudosativa IPNI=*391732-1* 234
Avena racemosa IPNI=*391732-1* 234
Avena redolens IPNI=*389043-1* 234
Avena sativa convar. *nodipilosa* IPNI=*391732-1* 234
Avena sativa f. *arundinacea* IPNI=*391732-1* 234
Avena sativa subsp. *macrantha* IPNI=*391732-1* 234
Avena sativa var. *brachytricha* IPNI=*391732-1* 234
Avena sativa var. *flavescens* IPNI=*391732-1* 234
Avena sativa var. *fuscoatra* IPNI=*391732-1* 234
Avena sativa var. *macrotricha* IPNI=*391732-1* 234
Avena sativa var. *microtricha* IPNI=*391732-1* 234
Avena sativa var. *pilifera* IPNI=*391732-1* 234
Avena sativa var. *praegravis* IPNI=*391732-1* 234
Avena sativa var. *subuniflora* IPNI=*391732-1* 234
Avena secalina IPNI=*394095-1* 235
Avena septentrionalis IPNI=*391443-1* 234
Avena sexflora IPNI=*391732-1* 234
Avena shatilowiana IPNI=*391732-1* 234
Avena squarrosa IPNI=*258855-2* 246
Avena sterilis IPNI=*391443-1* 234
Avena subspicata IPNI=*258855-2* 246
Avena sylvatica IPNI=*123530-2* 240
Avena tatarica IPNI=*391732-1* 234
Avena thellungii IPNI=*391732-1* 234
Avena tolucensis IPNI=*258855-2* 246
Avena trabutiana IPNI=*391732-1* 234
Avena transiens IPNI=*391732-1* 234
Avena trisperma IPNI=*391732-1* 234
Avena unilateralis IPNI=*391732-1* 234
Avena vilis IPNI=*391443-1* 234
Avena virescens IPNI=*258855-2* 246
Avenastrum caryophylleum IPNI=*386633-1* 233
Avenastrum elatius IPNI=*389923-1* 234
Averrhoa acida IPNI=*194610-2* 401
Averrhoa acutangula IPNI=*371870-1* 432
Averrhoa obtusangula IPNI=*371869-1* 432
Averrhoa pentandra IPNI=*371870-1* 432
Avicennia africana IPNI=*861125-1* 359
Avicennia elliptica IPNI=*861125-1* 359
Avicennia elliptica var. *martii* IPNI=*861125-1* 359
Avicennia floridana IPNI=*861125-1* 359
Avicennia germinans f. *aberrans* IPNI=*861125-1* 359
Avicennia germinans f. *brasiliensis* IPNI=*861125-1* 359
Avicennia germinans f. *venezuelensis* IPNI=*861125-1* 359
Avicennia germinans var. *cumanensis* IPNI=*861125-1* 359
Avicennia germinans var. *guayaquilensis* IPNI=*861125-1* 359
Avicennia lamarckiana IPNI=*861125-1* 359

Avicennia meyeri IPNI=*861125-1* 359
Avicennia nitida IPNI=*861125-1* 359
Avicennia nitida var. *trinitensis* IPNI=*861125-1* 359
Avicennia oblongifolia IPNI=*861125-1* 359
Avicennia officinalis var. *lanceolata* IPNI=*861125-1* 359
Avicennia officinalis var. *nitida* IPNI=*861125-1* 359
Avicennia tomentosa IPNI=*861125-1* 359
Avicennia tomentosa var. *cumanensis* IPNI=*861125-1* 359
Avicennia tomentosa var. *guayaquilensis* IPNI=*861125-1* 359
Avicularia vulgaris IPNI=*695422-1* 289
Avornela canariensis IPNI=*495939-1* 320
Avornela candicans IPNI=*496189-1* 320
Axia cochinchinensis IPNI=*326899-2* 286
Axinanthera bellucia IPNI=*323200-2* 418
Axinanthera hostmannii IPNI=*1157663-2* 418
Axinanthera macrophylla IPNI=*323200-2* 418
Axonopus affinis IPNI=*27338-2* 234
Axonopus amplifolius IPNI=*392087-1* 234
Axonopus arsenei IPNI=*392087-1* 234
Axonopus ater IPNI=*27338-2* 234
Axonopus brevipedunculatus IPNI=*392087-1* 234
Axonopus burchellii IPNI=*321761-2* 234
Axonopus canescens IPNI=*321761-2* 234
Axonopus canescens var. *psilachne* IPNI=*321761-2* 234
Axonopus carinatovaginatus IPNI=*321761-2* 234
Axonopus carinatovaginatus var. *sprucei* IPNI=*321761-2* 234
Axonopus chrysites IPNI=*321761-2* 234
Axonopus chrysodactylus IPNI=*321761-2* 234
Axonopus chrysostachyus IPNI=*321761-2* 234
Axonopus compressus subsp. *brevipedunculatus* IPNI=*392087-1* 234
Axonopus compressus var. *affinis* IPNI=*27338-2* 234
Axonopus compressus var. *australis* IPNI=*392087-1* 234
Axonopus compressus var. *itirapinensis* IPNI=*392087-1* 234
Axonopus compressus var. *jesuiticus* IPNI=*392087-1* 234
Axonopus compressus var. *macropodius* IPNI=*392087-1* 234
Axonopus deludens IPNI=*27415-2* 234
Axonopus digitatus IPNI=*329018-2* 237
Axonopus elongatus IPNI=*392087-1* 234
Axonopus exasperatus IPNI=*321761-2* 234
Axonopus fissifolius var. *coronatus* IPNI=*27338-2* 234
Axonopus hirsutus IPNI=*27338-2* 234
Axonopus holochrysus IPNI=*321761-2* 234
Axonopus iridifolius IPNI=*27415-2* 234
Axonopus jesuiticus IPNI=*392087-1* 234
Axonopus mathewsii IPNI=*27415-2* 234
Axonopus mexicanus IPNI=*27415-2* 234
Axonopus micay IPNI=*27415-2* 234
Axonopus minutus IPNI=*321761-2* 234
Axonopus multipes IPNI=*392087-1* 234
Axonopus paucisetosus IPNI=*321761-2* 234
Axonopus poiretii IPNI=*1103484-2* 245
Axonopus pulcher IPNI=*321761-2* 234
Axonopus radiatus IPNI=*321761-2* 234
Axonopus ramosissimus IPNI=*321761-2* 234
Axonopus rosengurttii IPNI=*392087-1* 234
Axonopus sprucei IPNI=*321761-2* 234
Axonopus sprucei var. *glabratus* IPNI=*321761-2* 234
Axonopus stragulus IPNI=*27338-2* 234
Axonopus tenuis IPNI=*321761-2* 234
Aydendron affine IPNI=*13808-2* 182
Aydendron amazonicum IPNI=*13872-2* 182
Aydendron cannella IPNI=*139921-2* 183
Aydendron cayennense IPNI=*139921-2* 183
Aydendron cubense IPNI=*465312-1* 183
Aydendron hostmannianum IPNI=*13841-2* 182
Aydendron laurel IPNI=*1171177-2* 184
Aydendron pallidum IPNI=*466885-1* 184
Aydendron panurense IPNI=*13863-2* 182
Aydendron riparium IPNI=*13878-2* 182
Aydendron sericeum IPNI=*464535-1* 183
Ayenia cardiopetala IPNI=*822190-1* 407
Azadirachta indica subsp. *vartakii* IPNI=*1213180-2* 454
Azalea balsaminiflora IPNI=*1137397-2* 300
Azalea burgeri IPNI=*1137397-2* 300
Azalea calycina IPNI=*1137397-2* 300
Azalea crispiflora IPNI=*1137397-2* 300
Azalea danielsiana IPNI=*1137397-2* 300
Azalea decumbens IPNI=*1137397-2* 300
Azalea fimbriata IPNI=*333359-1* 300
Azalea indica IPNI=*1137397-2* 300
Azalea macrantha IPNI=*1137397-2* 300
Azalea obtusa IPNI=*1137397-2* 300
Azalea ramentacea IPNI=*1137397-2* 300
Azalea rosiflora IPNI=*1137397-2* 300
Azalea simsii IPNI=*333359-1* 300
Azalea vittata IPNI=*333359-1* 300
Azalea vittata-punctata IPNI=*333359-1* 300
Azara borealis IPNI=*111471-1* 402
Azara umbellata IPNI=*77197250-1* 403
Azara valdiviae IPNI=*111471-1* 402
Azolla bonariensis IPNI=*17056910-1* 179
Azolla densa IPNI=*17056910-1* 179
Azolla mexicana IPNI=*17057000-1* 179
Azolla portoricensis IPNI=*17056910-1* 179
Azorella biloba f. *subglabra* IPNI=*838552-1* 251
Azorella biloba var. *normalis* IPNI=*838552-1* 251
Azorella biloba var. *strigosa* IPNI=*838552-1* 251
Azorella cladorrhiza IPNI=*838571-1* 251
Azorella crenata var. *compacta* IPNI=*838571-1* 251
Azorella crenata var. *mutisiana* IPNI=*838571-1* 251
Azorella ecuadorensis IPNI=*838571-1* 251
Azorella lehmanii IPNI=*838552-1* 251
Azorella reniformis IPNI=*838571-1* 251
Azorella vareschii IPNI=*838571-1* 251
Azukia angularis IPNI=*525291-1* 336
Azukia umbellata IPNI=*525554-1* 337

B

Baccharis alata IPNI=*183309-1* 257
Baccharis brasiliana IPNI=*969180-1* 272
Baccharis braunii IPNI=*28074-2* 257
Baccharis capitata IPNI=*183793-1* 258
Baccharis capitata var. *entera* IPNI=*183793-1* 258
Baccharis chilquilla IPNI=*183704-1* 258
Baccharis chubutensis IPNI=*30159244-2* 258
Baccharis cinerea IPNI=*183821-1* 258
Baccharis cinnamomifolia IPNI=*28074-2* 257
Baccharis cotinifolia IPNI=*28074-2* 257
Baccharis decurrens IPNI=*183309-1* 257
Baccharis divergens IPNI=*183821-1* 258
Baccharis eggersii IPNI=*183821-1* 258
Baccharis erioptera IPNI=*240147-1* 269
Baccharis fallax IPNI=*183580-1* 257
Baccharis floribunda IPNI=*183399-1* 257
Baccharis foetida IPNI=*238826-1* 268
Baccharis genistelloides IPNI=*30159244-2* 258
Baccharis genistelloides var. *angusta* IPNI=*183309-1* 257
Baccharis genistelloides var. *resinosa* IPNI=*183309-1* 257
Baccharis genistelloides var. *typica* IPNI=*183309-1* 257
Baccharis genistelloides var. *venosa* IPNI=*183309-1* 257
Baccharis guascensis IPNI=*183631-1* 258
Baccharis ibaguensis IPNI=*28074-2* 257
Baccharis ilinissae IPNI=*143805-2* 265
Baccharis incarum f. *spathulata* IPNI=*183170-1* 257
Baccharis ivifolia IPNI=*204572-1* 262
Baccharis jelskii var. *cladotricha* IPNI=*27895-2* 257

Baccharis laxa IPNI=*183821-1* 258
Baccharis lepidota IPNI=*183511-1* 257
Baccharis loxensis IPNI=*183511-1* 257
Baccharis loxensis var. *oligocephala* IPNI=*183511-1* 257
Baccharis macroptera IPNI=*30159244-2* 258
Baccharis microphylla IPNI=*183814-1* 258
Baccharis microphylla subvar. *prostrata* IPNI=*183814-1* 258
Baccharis microphylla var. *pulverulenta* IPNI=*183814-1* 258
Baccharis moritziana var. *subcrenatoserrata* IPNI=*27895-2* 257
Baccharis nitida f. *angustifolia* IPNI=*183511-1* 257
Baccharis odorata var. *spathulata* IPNI=*183170-1* 257
Baccharis oronocensis IPNI=*183511-1* 257
Baccharis oronocensis var. *perua* IPNI=*183511-1* 257
Baccharis oxyphylla IPNI=*183821-1* 258
Baccharis pentlandii subsp. *sculpta* IPNI=*183580-1* 257
Baccharis polyantha IPNI=*183399-1* 257
Baccharis polyantha var. *macrophylla* IPNI=*183399-1* 257
Baccharis popayanensis IPNI=*183511-1* 257
Baccharis prinoides IPNI=*183511-1* 257
Baccharis prunifolia subsp. *sierrana* IPNI=*183631-1* 258
Baccharis prunifolia var. *sierrana* IPNI=*183631-1* 258
Baccharis racemosa var. *obtusifolia* IPNI=*183511-1* 257
Baccharis reticulata IPNI=*183309-1* 257
Baccharis rhexioides IPNI=*183821-1* 258
Baccharis rhexioides var. *parvifolia* IPNI=*183821-1* 258
Baccharis riparia IPNI=*183399-1* 257
Baccharis rubricaulis IPNI=*183580-1* 257
Baccharis rugosa IPNI=*969218-1* 272
Baccharis sagittalis var. *poeppigii* IPNI=*30159244-2* 258
Baccharis sculpta IPNI=*183580-1* 257
Baccharis speciosa IPNI=*28074-2* 257
Baccharis splendens IPNI=*28074-2* 257
Baccharis subpenninervis IPNI=*183399-1* 257
Baccharis thuyoides IPNI=*231180-1* 265
Baccharis tolimensis IPNI=*183814-1* 258
Baccharis trichoclada IPNI=*183821-1* 258
Baccharis tricuneata var. *antioquensis* IPNI=*183814-1* 258
Baccharis tricuneata var. *parvifolia* IPNI=*183814-1* 258
Baccharis tricuneata var. *pulverulenta* IPNI=*183814-1* 258
Baccharis tricuneata var. *tolimensis* IPNI=*183814-1* 258
Baccharis tricuneata var. *variifolia* IPNI=*183814-1* 258
Baccharis trinervis var. *cinerea* IPNI=*183821-1* 258
Baccharis trinervis var. *rhexioides* IPNI=*183821-1* 258
Baccharis triptera IPNI=*183309-1* 257
Baccharis tripterix IPNI=*30159244-2* 258
Baccharis variifolia IPNI=*183814-1* 258
Baccharis venosa IPNI=*183309-1* 257
Baccharis venosa var. *apiciflora* IPNI=*183309-1* 257
Baccharis venusta IPNI=*183821-1* 258
Baccharis weddelliana IPNI=*183680-1* 258
Baccharoides brachylepis IPNI=*51409-2* 259
Baccharoides holtonii IPNI=*51409-2* 259
Baccharoides mutica IPNI=*51409-2* 259
Baccharoides punctata IPNI=*51409-2* 259
Baccharoides violacea IPNI=*51409-2* 259
Bacopa chamaedryoides f. *intermedia* IPNI=*155661-2* 376
Bacopa chamaedryoides IPNI=*155661-2* 376
Bacopa dianthera IPNI=*155661-2* 376
Bacopa diffusa IPNI=*244449-2* 377
Bacopa divaricata IPNI=*155661-2* 376
Bacopa monnieria IPNI=*1072674-2* 376
Bacopa monnieria var. *cuneifolia* IPNI=*1072674-2* 376
Bacopa montevidensis IPNI=*155661-2* 376
Bacopa peduncularis IPNI=*155661-2* 376
Bacopa procumbens IPNI=*155661-2* 376
Bacopa procumbens var. *peduncularis* IPNI=*155661-2* 376
Bacopa procumbens var. *schottii* IPNI=*155661-2* 376
Bacopa procumbens var. *tenuis* IPNI=*155661-2* 376
Bacopa tenuis IPNI=*155661-2* 376
Bactris acanthocnemis IPNI=*664750-1* 207
Bactris acanthospatha IPNI=*664663-1* 206
Bactris amoena IPNI=*664750-1* 207
Bactris angustifolia IPNI=*664520-1* 206
Bactris arenaria IPNI=*664750-1* 207
Bactris aristata IPNI=*664596-1* 206
Bactris bergantina IPNI=*664746-1* 207
Bactris bifida IPNI=*664607-1* 206
Bactris bifida var. *humaitensis* IPNI=*664520-1* 206
Bactris bifida var. *puruensis* IPNI=*664520-1* 206
Bactris brevifolia IPNI=*664750-1* 207
Bactris burretii IPNI=*664527-1* 206
Bactris carolensis IPNI=*664750-1* 207
Bactris circularis IPNI=*664746-1* 207
Bactris concinna subsp. *depauperata* IPNI=*664555-1* 206
Bactris concinna var. *inundata* IPNI=*664555-1* 206
Bactris confluens var. *acanthospatha* IPNI=*664663-1* 206
Bactris corazillo IPNI=*664560-1* 206
Bactris cuesco IPNI=*664560-1* 206
Bactris cuspidata var. *tenuis* IPNI=*664750-1* 207
Bactris cuvaro IPNI=*664746-1* 207
Bactris cuyabaensis IPNI=*664527-1* 206
Bactris dakamana IPNI=*664750-1* 207
Bactris duidae IPNI=*664560-1* 206
Bactris duplex IPNI=*664560-1* 206
Bactris elegantissima IPNI=*664582-1* 206
Bactris essequiboensis IPNI=*664750-1* 207
Bactris falcata IPNI=*664746-1* 207
Bactris fissifrons var. *robusta* IPNI=*664596-1* 206
Bactris flavispina IPNI=*664527-1* 206
Bactris gleasonii IPNI=*664750-1* 207
Bactris globosa IPNI=*663914-1* 203
Bactris gracilis IPNI=*664750-1* 207
Bactris hirsuta IPNI=*664717-1* 207
Bactris hondurensis IPNI=*664697-1* 207
Bactris horrida IPNI=*28460-2* 206
Bactris huberiana IPNI=*664750-1* 207
Bactris inermis IPNI=*664750-1* 207
Bactris inundata IPNI=*664736-1* 207
Bactris juruensis var. *lissospatha* IPNI=*664750-1* 207
Bactris kalbreyeri IPNI=*664746-1* 207
Bactris kuhlmannii IPNI=*664750-1* 207
Bactris kuhlmannii var. *aculeata* IPNI=*664750-1* 207
Bactris littoralis IPNI=*664736-1* 207
Bactris longifrons IPNI=*664736-1* 207
Bactris luetzelburgii IPNI=*664750-1* 207
Bactris luetzelburgii var. *anacantha* IPNI=*664750-1* 207
Bactris maguirei IPNI=*664750-1* 207
Bactris maraya-acu IPNI=*664527-1* 206
Bactris microspatha IPNI=*664750-1* 207
Bactris minor IPNI=*28460-2* 206
Bactris minor IPNI=*663914-1* 203
Bactris mitis subsp. *inermis* IPNI=*664750-1* 207
Bactris mitis subsp. *tenuis* IPNI=*664750-1* 207
Bactris mitis subsp. *uaupensis* IPNI=*664750-1* 207
Bactris naevia IPNI=*664750-1* 207
Bactris negrensis IPNI=*664750-1* 207
Bactris negrensis var. *carolensis* IPNI=*664750-1* 207
Bactris negrensis var. *minor* IPNI=*664750-1* 207
Bactris nigrispina IPNI=*664519-1* 206
Bactris obovata IPNI=*664750-1* 207
Bactris oerstediana IPNI=*664607-1* 206
Bactris oraria IPNI=*28460-2* 206
Bactris ovata IPNI=*664668-1* 206
Bactris pallidispina IPNI=*664527-1* 206
Bactris palustris IPNI=*664519-1* 206
Bactris paucisecta IPNI=*664750-1* 207

Bactris paula IPNI=*664697-1* 207
Bactris pavoniana IPNI=*663914-1* 203
Bactris piritu IPNI=*28460-2* 206
Bactris piscatorum IPNI=*664527-1* 206
Bactris platyacantha IPNI=*664663-1* 206
Bactris porschiana IPNI=*28417-2* 206
Bactris praemorsa IPNI=*7909-2* 204
Bactris pubescens IPNI=*664697-1* 207
Bactris pulchra var. *inermis* IPNI=*664750-1* 207
Bactris rivularis IPNI=*664527-1* 206
Bactris rotunda IPNI=*28460-2* 206
Bactris sciophila IPNI=*25885-2* 205
Bactris simplex IPNI=*664750-1* 207
Bactris simplicifrons subvar. *minor* IPNI=*664750-1* 207
Bactris simplicifrons var. *acanthocnemis* IPNI=*664750-1* 207
Bactris simplicifrons var. *brevifolia* IPNI=*664750-1* 207
Bactris simplicifrons var. *carolensis* IPNI=*664750-1* 207
Bactris simplicifrons var. *negrensis* IPNI=*664750-1* 207
Bactris simplicifrons var. *subpinnata* IPNI=*664750-1* 207
Bactris sororopanae IPNI=*664750-1* 207
Bactris sphaerocarpa subsp. *pinnatisecta* IPNI=*664754-1* 207
Bactris sphaerocarpa var. *ensifolia* IPNI=*664754-1* 207
Bactris sphaerocarpa var. *minor* IPNI=*664754-1* 207
Bactris sphaerocarpa var. *pinnatisecta* IPNI=*664754-1* 207
Bactris sphaerocarpa var. *platyphylla* IPNI=*664754-1* 207
Bactris sphaerocarpa var. *schizophylla* IPNI=*664754-1* 207
Bactris stahelii IPNI=*664750-1* 207
Bactris standleyana IPNI=*664697-1* 207
Bactris stictacantha IPNI=*664527-1* 206
Bactris sworderiana IPNI=*664746-1* 207
Bactris tenera IPNI=*664527-1* 206
Bactris tenuis IPNI=*664750-1* 207
Bactris tenuis var. *inermis* IPNI=*664750-1* 207
Bactris tomentosa var. *sphaerocarpa* IPNI=*664754-1* 207
Bactris trinitensis IPNI=*664750-1* 207
Bactris uaupensis IPNI=*664750-1* 207
Bactris ulei IPNI=*664750-1* 207
Bactris venezuelensis IPNI=*664560-1* 206
Bactris villosa IPNI=*664697-1* 207
Bactris wendlandiana IPNI=*664697-1* 207
Bactris xanthocarpa IPNI=*664750-1* 207
Bactyrilobium bacillare IPNI=*234415-2* 332
Bactyrilobium fistula IPNI=*484507-1* 313
Bactyrilobium grande IPNI=*484607-1* 314
Bactyrilobium javanica IPNI=*484704-1* 314
Bactyrilobium molle IPNI=*484607-1* 314
Badiera acuminata IPNI=*77121649-1* 337
Badiera diversifolia IPNI=*231560-2* 338
Badula nigricans IPNI=*247472-2* 304
Badula schomburgkiana IPNI=*247474-2* 304
Badula schomburgkiana var. *brasiliensis* IPNI=*247474-2* 304
Baeumerta nasturtium IPNI=*287534-1* 277
Baillieria aspera IPNI=*319147-2* 260
Baillieria barbasco IPNI=*1050569-2* 260
Baillieria commelinoides IPNI=*127605-2* 264
Baillieria sylvestris IPNI=*1050569-2* 260
Baillieria terebinthacea IPNI=*1006328-2* 260
Balantium cordifolium IPNI=*727083-1* 383
Balantium karstenianum IPNI=*1190416-2* 172
Balantium sellowianum IPNI=*275823-2* 172
Balbisia canescens IPNI=*257735-2* 271
Balbisia divaricata IPNI=*257735-2* 271
Balbisia elongata IPNI=*257735-2* 271
Balbisia pedunculata IPNI=*257735-2* 271
Baldingera arundinacea IPNI=*415571-1* 243
Baldingera arundinacea var. *rotgesii* IPNI=*415571-1* 243
Baldingera colorata IPNI=*415571-1* 243
Baldwinia peltata IPNI=*321973-2* 400
Ballota disticha IPNI=*158242-2* 371
Ballota parviflora IPNI=*158242-2* 371
Ballota suaveolens IPNI=*158242-2* 371
Ballota suaveolens IPNI=*158297-2* 371
Balsamaria inophyllum IPNI=*427190-1* 380
Balsamina balsamina IPNI=*373978-1* 297
Balsamina coccinea IPNI=*373978-1* 297
Balsamina cornuta IPNI=*373978-1* 297
Balsamina foeminea IPNI=*373978-1* 297
Balsamina hortensis IPNI=*373978-1* 297
Balsamina lacca IPNI=*373978-1* 297
Balsamina lutea IPNI=*30208246-2* 297
Balsamina minutiflora IPNI=*373978-1* 297
Balsamina mollis IPNI=*373978-1* 297
Balsamina noli-tangere IPNI=*30208246-2* 297
Balsamina odorata IPNI=*373978-1* 297
Balsamina racemosa IPNI=*373978-1* 297
Balsamina salicifolia IPNI=*373978-1* 297
Balsamona pinto IPNI=*72326-2* 416
Baltimora ovata IPNI=*28847-2* 258
Bambos nana IPNI=*1097548-2* 234
Bambos verticillata IPNI=*404020-1* 239
Bambusa aculeata IPNI=*1074014-2* 239
Bambusa albifolia IPNI=*1097548-2* 234
Bambusa alphonse-karrii IPNI=*1097548-2* 234
Bambusa amplexifolia IPNI=*113747-2* 239
Bambusa angulata IPNI=*392553-1* 235
Bambusa annulata IPNI=*392542-1* 234
Bambusa argentea IPNI=*1097548-2* 234
Bambusa argentea var. *vittata* IPNI=*1097548-2* 234
Bambusa aspera IPNI=*397862-1* 237
Bambusa atrovirens IPNI=*392445-1* 234
Bambusa auriculata IPNI=*392574-1* 235
Bambusa baccifera IPNI=*408271-1* 241
Bambusa bitung IPNI=*397862-1* 237
Bambusa blancoi IPNI=*392574-1* 235
Bambusa blumeana IPNI=*392553-1* 235
Bambusa breviflora IPNI=*392553-1* 235
Bambusa breviflora var. *hainanensis* IPNI=*392553-1* 235
Bambusa caesia IPNI=*1097548-2* 234
Bambusa dolichomerithalla IPNI=*1097548-2* 234
Bambusa eutuldoides var. *viridivittata* IPNI=*392318-1* 234
Bambusa excelsa IPNI=*404020-1* 239
Bambusa fauriei IPNI=*392553-1* 235
Bambusa fera IPNI=*392574-1* 235
Bambusa flagellifera IPNI=*397862-1* 237
Bambusa flavonoda IPNI=*392553-1* 235
Bambusa fortunei IPNI=*416271-1* 244
Bambusa glaucescens f. *albostriata* IPNI=*1097548-2* 234
Bambusa glaucescens f. *albovariegata* IPNI=*1097548-2* 234
Bambusa glaucescens f. *alphonso-karrii* IPNI=*1097548-2* 234
Bambusa glaucescens f. *elegans* IPNI=*1097548-2* 234
Bambusa glaucescens f. *fu-komachi* IPNI=*1097548-2* 234
Bambusa glaucescens f. *gimmei* IPNI=*1097548-2* 234
Bambusa glaucescens f. *kimmei-suhou* IPNI=*1097548-2* 234
Bambusa glaucescens f. *midori* IPNI=*1097548-2* 234
Bambusa glaucescens f. *midori-beni* IPNI=*1097548-2* 234
Bambusa glaucescens f. *shirosuji* IPNI=*1097548-2* 234
Bambusa glaucescens f. *solida* IPNI=*1097548-2* 234
Bambusa glaucescens f. *tukushi-komachi* IPNI=*1097548-2* 234
Bambusa glaucescens f. *variegata* IPNI=*1097548-2* 234
Bambusa glaucescens f. *viridistriata* IPNI=*1097548-2* 234
Bambusa glaucescens IPNI=*1097548-2* 234
Bambusa glaucescens var. *annulata* IPNI=*392542-1* 234
Bambusa glaucescens var. *floribunda* IPNI=*1097548-2* 234
Bambusa glaucescens var. *lutea* IPNI=*1097548-2* 234
Bambusa glaucescens var. *pubivagina* IPNI=*1097548-2* 234
Bambusa glaucescens var. *riviereorum* IPNI=*1097548-2* 234

Bambusa glaucescens var. *shimadae* IPNI=*1097548-2* 234
Bambusa guadua IPNI=*1074014-2* 239
Bambusa humilis IPNI=*392574-1* 235
Bambusa inermis IPNI=*1074014-2* 239
Bambusa latiflora IPNI=*392574-1* 235
Bambusa liukiuensis IPNI=*1097548-2* 234
Bambusa longiflora IPNI=*392553-1* 235
Bambusa macala IPNI=*392551-1* 234
Bambusa madagascariensis IPNI=*392574-1* 235
Bambusa maximowiczii IPNI=*416271-1* 244
Bambusa minutiligulata IPNI=*392542-1* 234
Bambusa mitis IPNI=*392574-1* 235
Bambusa monogyna IPNI=*392574-1* 235
Bambusa multiplex var. *gracillima* IPNI=*1097548-2* 234
Bambusa multiplex var. *incana* IPNI=*1097548-2* 234
Bambusa multiplex var. *lutea* IPNI=*1097548-2* 234
Bambusa multiplex var. *pubivagina* IPNI=*1097548-2* 234
Bambusa multiplex var. *strigosa* IPNI=*1097548-2* 234
Bambusa nana IPNI=*1097548-2* 234
Bambusa nana var. *alphonse-karrii* IPNI=*1097548-2* 234
Bambusa nguyenii IPNI=*392574-1* 235
Bambusa oldhamii f. *revoluta* IPNI=*392445-1* 234
Bambusa pallescens IPNI=*392553-1* 235
Bambusa parvifolia IPNI=*392553-1* 235
Bambusa picta IPNI=*416271-1* 244
Bambusa pseudoarundinacea IPNI=*404020-1* 239
Bambusa pubivaginata IPNI=*1097548-2* 234
Bambusa pygmaea IPNI=*416271-1* 244
Bambusa revoluta IPNI=*392445-1* 234
Bambusa shimadae IPNI=*1097548-2* 234
Bambusa sieberi IPNI=*392574-1* 235
Bambusa sterilis IPNI=*1097548-2* 234
Bambusa striata IPNI=*392574-1* 235
Bambusa strigosa IPNI=*1097548-2* 234
Bambusa surinamensis IPNI=*392574-1* 235
Bambusa textilis f. *purpurascens* IPNI=*392542-1* 234
Bambusa textilis var. *pubescens* IPNI=*392542-1* 234
Bambusa textilis var. *purpurascens* IPNI=*392542-1* 234
Bambusa thouarsii IPNI=*392574-1* 235
Bambusa trigyna IPNI=*392551-1* 234
Bambusa tulda IPNI=*392553-1* 235
Bambusa variegata IPNI=*416271-1* 244
Bambusa varioaurita IPNI=*392542-1* 234
Bambusa verticillata IPNI=*392260-1* 234
Bambusa verticillata IPNI=*404020-1* 239
Bambusa viridivittata IPNI=*392318-1* 234
Banara amazonica IPNI=*28885-2* 402
Banara fagifolia IPNI=*779396-1* 402
Banara glabrata IPNI=*28885-2* 402
Banara glandulosa IPNI=*779396-1* 402
Banara guianensis var. *argentina* IPNI=*28885-2* 402
Banara guianensis var. *martiana* IPNI=*28885-2* 402
Banara guianensis var. *mollis* IPNI=*779396-1* 402
Banara guianensis var. *spruceana* IPNI=*779396-1* 402
Banara laxiflora IPNI=*834658-1* 404
Banara leucothyrsa IPNI=*202149-2* 404
Banara mexicana IPNI=*202149-2* 404
Banara mollis IPNI=*779396-1* 402
Banara pubescens IPNI=*779396-1* 402
Banara pyramidata IPNI=*779396-1* 402
Banara tessmannii IPNI=*28885-2* 402
Banara tulasnei IPNI=*779396-1* 402
Bancroftia decipiens IPNI=*838272-1* 251
Bancroftia xanthorrhiza IPNI=*838272-1* 251
Banisteria acanthocarpa IPNI=*29242-2* 395
Banisteria acanthocarpa var. *eglandulosa* IPNI=*29242-2* 395
Banisteria acanthocarpa var. *glandulifera* IPNI=*29242-2* 395
Banisteria antifebrilis IPNI=*273852-2* 396
Banisteria apiculata IPNI=*556578-1* 396
Banisteria argentea f. *eglandulosa* IPNI=*29242-2* 395
Banisteria argentea f. *glandulifera* IPNI=*29242-2* 395
Banisteria argentea var. *acuminata* IPNI=*29242-2* 395
Banisteria argentea var. *obtusiuscula* IPNI=*29242-2* 395
Banisteria argentea var. *transiens* IPNI=*29242-2* 395
Banisteria atrosanguinea IPNI=*29242-2* 395
Banisteria benthamiana IPNI=*29242-2* 395
Banisteria bicolor IPNI=*29242-2* 395
Banisteria billbergiana IPNI=*30241255-2* 396
Banisteria bopiana IPNI=*60444585-2* 396
Banisteria caapi IPNI=*29180-2* 394
Banisteria caerulea IPNI=*556563-1* 396
Banisteria ciliata IPNI=*557487-1* 396
Banisteria cornifolia IPNI=*60447688-2* 395
Banisteria cornifolia var. *discolor* IPNI=*60443063-2* 396
Banisteria cornifolia var. *seibertii* IPNI=*60447688-2* 395
Banisteria cornifolia var. *typica* IPNI=*60447688-2* 395
Banisteria dichotoma IPNI=*557494-1* 396
Banisteria discolor IPNI=*60443063-2* 396
Banisteria elliptica IPNI=*30241255-2* 396
Banisteria glauca IPNI=*557487-1* 396
Banisteria heterophylla IPNI=*557592-1* 396
Banisteria humboldtiana IPNI=*557494-1* 396
Banisteria illustris IPNI=*29242-2* 395
Banisteria inebrians IPNI=*29180-2* 394
Banisteria lancifolia IPNI=*556563-1* 396
Banisteria laurifolia IPNI=*556563-1* 396
Banisteria leona IPNI=*1079574-2* 396
Banisteria leptocarpa IPNI=*29229-2* 395
Banisteria longialata IPNI=*60444585-2* 396
Banisteria lupuloides IPNI=*112858-2* 441
Banisteria magnoliifolia IPNI=*556563-1* 396
Banisteria maracaybensis var. *cornifolia* IPNI=*60447688-2* 395
Banisteria martiniana IPNI=*29229-2* 395
Banisteria metallicolor f. *eglandulosa* IPNI=*29242-2* 395
Banisteria metallicolor f. *glandulifera* IPNI=*29242-2* 395
Banisteria metallicolor IPNI=*29242-2* 395
Banisteria metallicolor var. *aurea* IPNI=*29242-2* 395
Banisteria metallicolor var. *falcata* IPNI=*29242-2* 395
Banisteria metallicolor var. *pruinosa* IPNI=*29242-2* 395
Banisteria metallicolor var. *sericea* IPNI=*29242-2* 395
Banisteria metallicolor var. *subrotunda* IPNI=*29242-2* 395
Banisteria metallicolor var. *subsalicina* IPNI=*29242-2* 395
Banisteria metallicolor var. *typica* IPNI=*29242-2* 395
Banisteria mucronata IPNI=*30241255-2* 396
Banisteria multiflora IPNI=*1079574-2* 396
Banisteria muricata IPNI=*29242-2* 395
Banisteria muricata var. *atrosanguinea* IPNI=*29242-2* 395
Banisteria nitida IPNI=*557487-1* 396
Banisteria paniculata IPNI=*556563-1* 396
Banisteria pruinosa IPNI=*29242-2* 395
Banisteria pubiflora IPNI=*556563-1* 396
Banisteria quitensis IPNI=*29180-2* 394
Banisteria reticulata IPNI=*1079574-2* 396
Banisteria rusbyana IPNI=*60444585-2* 396
Banisteria sanguinea var. *benthamiana* IPNI=*29242-2* 395
Banisteria schlimii IPNI=*29242-2* 395
Banisteria schomburgkiana IPNI=*29242-2* 395
Banisteria sinuata IPNI=*557592-1* 396
Banisteria splendens IPNI=*557592-1* 396
Banisteria ternata IPNI=*30241255-2* 396
Banisteria tiliifolia IPNI=*557494-1* 396
Banisteria varia IPNI=*557494-1* 396
Banisteria variifolia IPNI=*557494-1* 396
Banisteria vitifolia IPNI=*557494-1* 396
Banisteria williamsii IPNI=*29242-2* 395
Banisteriopsis argentea IPNI=*29242-2* 395

Banisteriopsis cabrerana IPNI=*81736-2* 396
Banisteriopsis cornifolia IPNI=*60447688-2* 395
Banisteriopsis discolor IPNI=*60443063-2* 396
Banisteriopsis illustris IPNI=*29242-2* 395
Banisteriopsis inebrians IPNI=*29180-2* 394
Banisteriopsis leptocarpa IPNI=*29229-2* 395
Banisteriopsis longialata IPNI=*60444585-2* 396
Banisteriopsis metallicolor IPNI=*29242-2* 395
Banisteriopsis metallicolor var. *sericea* IPNI=*29242-2* 395
Banisteriopsis metallicolor var. *subrotunda* IPNI=*29242-2* 395
Banisteriopsis quitensis IPNI=*29180-2* 394
Banisteriopsis rusbyana IPNI=*60444585-2* 396
Banisteriopsis schomburgkiana IPNI=*29242-2* 395
Banisteriopsis williamsii IPNI=*29242-2* 395
Barbarea australis IPNI=*1109261-2* 276
Barbarea brevistyla IPNI=*1109261-2* 276
Barbarea derchiensis IPNI=*279485-1* 276
Barbarea erysimoides IPNI=*1109261-2* 276
Barbarea longisiliqua IPNI=*1109261-2* 276
Barbarea patula IPNI=*1109261-2* 276
Barhamia hispida IPNI=*70599-2* 389
Barhamia macrostachya IPNI=*70599-2* 389
Barhamia ovalifolia IPNI=*325185-2* 389
Barhamia panamensis IPNI=*325164-2* 388
Barleria afzelii IPNI=*46114-1* 359
Barleria alba IPNI=*45927-1* 359
Barleria anomala IPNI=*36223-2* 359
Barleria chegosa IPNI=*45927-1* 359
Barleria ciliata IPNI=*45927-1* 359
Barleria cristata var. *lactea* IPNI=*45927-1* 359
Barleria dichotoma IPNI=*45927-1* 359
Barleria discolor IPNI=*46114-1* 359
Barleria flava IPNI=*46114-1* 359
Barleria gentianoides IPNI=*46114-1* 359
Barleria indica IPNI=*45927-1* 359
Barleria laciniata IPNI=*45927-1* 359
Barleria lactea IPNI=*45927-1* 359
Barleria micans IPNI=*46114-1* 359
Barleria napalensis IPNI=*45927-1* 359
Barleria nuda IPNI=*45927-1* 359
Barleria prinoides IPNI=*45927-1* 359
Barleria pyramidalis IPNI=*224052-2* 361
Barleria pyramidata IPNI=*224052-2* 361
Barleria senegalensis IPNI=*46114-1* 359
Barleria venulosa IPNI=*45927-1* 359
Barleriopsis discolor IPNI=*46114-1* 359
Barleriopsis glandulosa IPNI=*46114-1* 359
Barleriopsis glandulosa var. *brevifolia* IPNI=*46114-1* 359
Barleriopsis micans IPNI=*46114-1* 359
Barleriopsis micans var. *brachystachya* IPNI=*46114-1* 359
Barnadesia arborea var. *vestita* IPNI=*273073-2* 258
Barnadesia media IPNI=*273073-2* 258
Barnadesia sodiroi IPNI=*273073-2* 258
Barnadesia spinosa IPNI=*273073-2* 258
Barnadesia spinosa var. *recurvata* IPNI=*184250-1* 258
Barnadesia vestita IPNI=*273073-2* 258
Barreria guianensis IPNI=*608256-1* 359
Barreria theobromifolia IPNI=*608256-1* 359
Bartramia angulata IPNI=*320787-2* 415
Bartramia crispifolia IPNI=*320787-2* 415
Bartramia glandulosa IPNI=*320787-2* 415
Bartramia indica IPNI=*320787-2* 415
Bartramia lappago IPNI=*320787-2* 415
Bartramia rhombifolia IPNI=*320787-2* 415
Bartsia santolinifolia IPNI=*77160774-1* 375
Bartsia trinervis IPNI=*800871-1* 375
Baryosma oppositifolium IPNI=*520179-1* 335
Baryosma tongo IPNI=*492769-1* 318
Baryxylum ferrugineum IPNI=*513021-1* 330
Baryxylum inerme IPNI=*513021-1* 330
Basanacantha armata IPNI=*1115268-2* 356
Basanacantha calycina IPNI=*764017-1* 356
Basanacantha dioica IPNI=*764082-1* 356
Basanacantha erythropoda IPNI=*764273-1* 356
Basanacantha hebecarpa IPNI=*764142-1* 356
Basanacantha humboldtiana IPNI=*764273-1* 356
Basanacantha macrocarpa IPNI=*222509-2* 356
Basanacantha mollis IPNI=*764273-1* 356
Basanacantha mucronata IPNI=*764273-1* 356
Basanacantha spinosa var. *nitida* IPNI=*764273-1* 356
Basanacantha spinosa var. *pubescens* IPNI=*764273-1* 356
Basella alba var. *cordifolia* IPNI=*164286-1* 283
Basella cananifolia IPNI=*164286-1* 283
Basella cordifolia IPNI=*164286-1* 283
Basella crassifolia IPNI=*164286-1* 283
Basella japonica IPNI=*164286-1* 283
Basella lucida IPNI=*164286-1* 283
Basella nigra IPNI=*164286-1* 283
Basella rubra IPNI=*164286-1* 283
Basella tuberosa IPNI=*167633-1* 283
Basella vesicaria IPNI=*1153747-2* 283
Basella volubilis IPNI=*164286-1* 283
Bassia jussaei IPNI=*1110421-2* 307
Bastardia crispa IPNI=*120630-2* 409
Bastardia cubensis IPNI=*77146518-1* 407
Bastardia foetida IPNI=*77146518-1* 407
Bastardia guayquilensis IPNI=*77146518-1* 407
Bastardia parvifolia IPNI=*77146518-1* 407
Bastardia viscosa f. *crenulata* IPNI=*77146518-1* 407
Bastardia viscosa f. *subtriloba* IPNI=*77146518-1* 407
Bastardia viscosa IPNI=*77146518-1* 407
Bastardia viscosa var. *abutiloides* IPNI=*77146518-1* 407
Bastardia viscosa var. *cubensis* IPNI=*77146518-1* 407
Bastardia viscosa var. *luteovirens* IPNI=*77146518-1* 407
Bastardia viscosa var. *macrantha* IPNI=*77146518-1* 407
Bastardia viscosa var. *novogranatensis* IPNI=*77146518-1* 407
Bastardia viscosa var. *parvifolia* IPNI=*77146518-1* 407
Bastardia viscosa var. *typica* IPNI=*77146518-1* 407
Batemannia lalindei IPNI=*60440679-2* 222
Batemannia meleagris IPNI=*639074-1* 219
Bathmium fraxinifolium IPNI=*17226640-1* 177
Bathmium heracleifolium IPNI=*249899-2* 177
Bathmium macrophyllum IPNI=*17226640-1* 177
Bathmium martinicense IPNI=*17226640-1* 177
Bathmium trifoliatum var. *heracleifolium* IPNI=*249899-2* 177
Bathysa bracteosa IPNI=*766156-1* 357
Bathysa difformis IPNI=*308173-2* 348
Batidaea idaea IPNI=*298089-2* 444
Batidaea vulgaris IPNI=*298089-2* 444
Batis americana IPNI=*319368-2* 276
Batis californica IPNI=*319368-2* 276
Batocarpus maranhensis IPNI=*29952-2* 437
Batschia nivea IPNI=*6815-2* 256
Batschia racemosa IPNI=*580077-1* 434
Bauhinia aculeata IPNI=*481315-1* 311
Bauhinia amplifolia IPNI=*481718-1* 312
Bauhinia armata IPNI=*481315-1* 311
Bauhinia baina IPNI=*481718-1* 312
Bauhinia bicuspidata IPNI=*30050541-2* 311
Bauhinia brachystachya IPNI=*229758-2* 332
Bauhinia brasiliensis IPNI=*481315-1* 311
Bauhinia breviloba IPNI=*481315-1* 311
Bauhinia castrata IPNI=*481605-1* 312
Bauhinia chrysophylla IPNI=*518025-1* 332
Bauhinia chrysophylla IPNI=*77111925-1* 332
Bauhinia columbiensis IPNI=*229758-2* 332

Bauhinia coromandeliana IPNI=*481605-1* 312
Bauhinia cumanensis IPNI=*229758-2* 332
Bauhinia divaricata IPNI=*481315-1* 311
Bauhinia eriophora IPNI=*229758-2* 332
Bauhinia forficata var. *latifolia* IPNI=*481315-1* 311
Bauhinia forficata var. *longiflora* IPNI=*481315-1* 311
Bauhinia geminata IPNI=*481456-1* 312
Bauhinia glabra IPNI=*229758-2* 332
Bauhinia guianensis IPNI=*77111925-1* 332
Bauhinia guianensis var. *kunthiana* IPNI=*77111927-1* 332
Bauhinia guianensis var. *splendens* IPNI=*518025-1* 332
Bauhinia herrerae IPNI=*229759-2* 332
Bauhinia heterophylla IPNI=*229758-2* 332
Bauhinia holophylla var. *paraensis* IPNI=*30050541-2* 311
Bauhinia hondurensis IPNI=*229758-2* 332
Bauhinia kalbreyeri IPNI=*481565-1* 312
Bauhinia kappleri IPNI=*30129-2* 312
Bauhinia krugii IPNI=*30129-2* 312
Bauhinia kunthiana IPNI=*77111927-1* 332
Bauhinia kurzii IPNI=*481605-1* 312
Bauhinia ligulata IPNI=*481565-1* 312
Bauhinia longicuspis var. *bicuspidata* IPNI=*30050541-2* 311
Bauhinia longicuspis var. *paraensis* IPNI=*30050541-2* 311
Bauhinia longiflora IPNI=*481315-1* 311
Bauhinia longifolia var. *acuminata* IPNI=*481456-1* 312
Bauhinia longipetala IPNI=*229758-2* 332
Bauhinia macrophylla IPNI=*77111925-1* 332
Bauhinia manca IPNI=*77111925-1* 332
Bauhinia microstachya IPNI=*229762-2* 332
Bauhinia obtusata IPNI=*481456-1* 312
Bauhinia porosa IPNI=*30129-2* 312
Bauhinia punctata IPNI=*229758-2* 332
Bauhinia punctiflora IPNI=*30129-2* 312
Bauhinia recurva IPNI=*481456-1* 312
Bauhinia rosea IPNI=*481605-1* 312
Bauhinia rosea IPNI=*77111927-1* 332
Bauhinia sericella IPNI=*77111925-1* 332
Bauhinia speciosa IPNI=*77111925-1* 332
Bauhinia splendens IPNI=*518025-1* 332
Bauhinia splendens var. *latifolia* IPNI=*518025-1* 332
Bauhinia standleyi IPNI=*229758-2* 332
Bauhinia stenocardia IPNI=*30050541-2* 311
Bauhinia storkii IPNI=*229758-2* 332
Bauhinia suaveolens IPNI=*229758-2* 332
Bauhinia suaveolens var. *loretana* IPNI=*229758-2* 332
Bauhinia superba IPNI=*77111925-1* 332
Bauhinia tarapotensis IPNI=*481718-1* 312
Bauhinia triandra IPNI=*481605-1* 312
Bauhinia variegata var. *purpurascens* IPNI=*481756-1* 312
Bayonia fissa IPNI=*152688-2* 364
Bayonia helicocalyx IPNI=*152688-2* 364
Baziasa microglossa IPNI=*208554-1* 263
Baziasa urticifolia IPNI=*208557-1* 263
Beatonia grandiflora IPNI=*441553-1* 216
Beatonia lutea IPNI=*441553-1* 216
Beauharnoisia fructipendula IPNI=*429349-1* 385
Beesha baccifera IPNI=*408271-1* 241
Beesha rheedii IPNI=*408271-1* 241
Beethovenia cerifera IPNI=*53436-2* 207
Begonia aculeata IPNI=*105130-1* 292
Begonia albidosetulosa IPNI=*104855-1* 292
Begonia alemanii IPNI=*30121655-2* 292
Begonia argentinensis IPNI=*104291-1* 292
Begonia argyrostigma IPNI=*105130-1* 292
Begonia bakeri IPNI=*104371-1* 292
Begonia biolleyi IPNI=*105718-1* 293
Begonia boliviensis var. *latitepala* IPNI=*104291-1* 292
Begonia boliviensis var. *volcanensis* IPNI=*104291-1* 292
Begonia bowerae var. *major* IPNI=*30424-2* 292
Begonia bowerae var. *roseflora* IPNI=*30424-2* 292
Begonia brasila IPNI=*104855-1* 292
Begonia brasiliana IPNI=*104855-1* 292
Begonia brasiliensis IPNI=*315565-2* 292
Begonia caudilimba IPNI=*273181-2* 292
Begonia charadrophila IPNI=*104848-1* 292
Begonia chepoensis IPNI=*104848-1* 292
Begonia ciliata IPNI=*104855-1* 292
Begonia ciliata var. *nana* IPNI=*104855-1* 292
Begonia ciliibracteola IPNI=*315565-2* 292
Begonia corallina IPNI=*105130-1* 292
Begonia dasypoda IPNI=*104855-1* 292
Begonia derycxiana IPNI=*273181-2* 292
Begonia diversifolia var. *nana* IPNI=*104855-1* 292
Begonia elata IPNI=*315565-2* 292
Begonia elliptica IPNI=*104759-1* 292
Begonia ermanii IPNI=*315565-2* 292
Begonia filipes IPNI=*104848-1* 292
Begonia fischeri var. *brasiliensis* IPNI=*315565-2* 292
Begonia fischeri var. *brevipilosa* IPNI=*315565-2* 292
Begonia fischeri var. *crenatoglabra* IPNI=*315565-2* 292
Begonia fischeri var. *crenulatoglabra* IPNI=*315565-2* 292
Begonia fischeri var. *elata* IPNI=*315565-2* 292
Begonia fischeri var. *ermanii* IPNI=*315565-2* 292
Begonia fischeri var. *klugii* IPNI=*315565-2* 292
Begonia fischeri var. *macroptera* IPNI=*315565-2* 292
Begonia fischeri var. *malvacea* IPNI=*315565-2* 292
Begonia fischeri var. *moritziana* IPNI=*315565-2* 292
Begonia fischeri var. *palustris* IPNI=*315565-2* 292
Begonia fischeri var. *tovarensis* IPNI=*315565-2* 292
Begonia flexuosa IPNI=*105712-1* 293
Begonia floribunda IPNI=*104715-1* 292
Begonia foliosa var. *miniata* IPNI=*104715-1* 292
Begonia fuchsioides f. *miniata* IPNI=*104715-1* 292
Begonia fuchsioides var. *miniata* IPNI=*104715-1* 292
Begonia fulgens IPNI=*104715-1* 292
Begonia glabra var. *amplifolia* IPNI=*104759-1* 292
Begonia grewiifolia IPNI=*105087-1* 292
Begonia guyanensis IPNI=*105712-1* 293
Begonia guyanensis var. *cearensis* IPNI=*30121655-2* 292
Begonia haematotricha IPNI=*30121655-2* 292
Begonia hassleri IPNI=*315565-2* 292
Begonia hederacea IPNI=*105168-1* 292
Begonia heracleifolia f. *punctata* IPNI=*104828-1* 292
Begonia heracleifolia var. *longipila* IPNI=*104828-1* 292
Begonia heracleifolia var. *nigricans* IPNI=*104828-1* 292
Begonia heracleifolia var. *sunderbruckii* IPNI=*104828-1* 292
Begonia hernandiifolia IPNI=*273181-2* 292
Begonia herteri IPNI=*104855-1* 292
Begonia heterodonta IPNI=*104848-1* 292
Begonia hirsuta IPNI=*30121655-2* 292
Begonia hirtella var. *nana* IPNI=*104855-1* 292
Begonia hoegeana IPNI=*104759-1* 292
Begonia humilis IPNI=*104848-1* 292
Begonia humilis var. *glabrata* IPNI=*105712-1* 293
Begonia humilis var. *porteriana* IPNI=*30121655-2* 292
Begonia hygrophila IPNI=*104848-1* 292
Begonia hygrophila var. *puberula* IPNI=*104848-1* 292
Begonia hypolipara IPNI=*105718-1* 293
Begonia inanis IPNI=*105280-1* 292
Begonia intercedens IPNI=*315565-2* 292
Begonia jatrophifolia IPNI=*104828-1* 292
Begonia kaietukensis IPNI=*315565-2* 292
Begonia lanuginosa IPNI=*105718-1* 293
Begonia lepidota IPNI=*105149-1* 292
Begonia leptopoda IPNI=*104848-1* 292
Begonia locellata IPNI=*104759-1* 292

Begonia lokobeensis IPNI=*30121655-2* 292
Begonia longiciliata IPNI=*105573-1* 292
Begonia longipila IPNI=*104828-1* 292
Begonia lucida IPNI=*104759-1* 292
Begonia lucida IPNI=*30121655-2* 292
Begonia macroptera IPNI=*315565-2* 292
Begonia macroptera var. *paludum* IPNI=*315565-2* 292
Begonia macroptera var. *pohliana* IPNI=*315565-2* 292
Begonia maculata f. *argentea* IPNI=*105130-1* 292
Begonia maculata var. *argentea* IPNI=*105130-1* 292
Begonia malvacea IPNI=*315565-2* 292
Begonia mameiana IPNI=*104848-1* 292
Begonia manicata var. *aureomaculata* IPNI=*105149-1* 292
Begonia manicata var. *manicativiridi* IPNI=*105149-1* 292
Begonia manicata var. *ocozocoautlaensis* IPNI=*105149-1* 292
Begonia manicata var. *peltata* IPNI=*105149-1* 292
Begonia meyeniana IPNI=*30121655-2* 292
Begonia miniata IPNI=*104715-1* 292
Begonia moritziana IPNI=*104759-1* 292
Begonia moritziana IPNI=*315565-2* 292
Begonia multiflora IPNI=*104715-1* 292
Begonia nicaraguensis IPNI=*105718-1* 293
Begonia nitida var. *pilosula* IPNI=*105214-1* 292
Begonia obliqua IPNI=*315565-2* 292
Begonia palustris IPNI=*315565-2* 292
Begonia parvifolia IPNI=*315565-2* 292
Begonia patula IPNI=*315565-2* 292
Begonia pauciflora IPNI=*315565-2* 292
Begonia pavoniana IPNI=*30121655-2* 292
Begonia peltata IPNI=*273181-2* 292
Begonia physalifolia IPNI=*104759-1* 292
Begonia pilifera IPNI=*105718-1* 293
Begonia pohliana IPNI=*315565-2* 292
Begonia populifolia IPNI=*104759-1* 292
Begonia populifolia IPNI=*315565-2* 292
Begonia porteriana IPNI=*30121655-2* 292
Begonia pulchra IPNI=*105214-1* 292
Begonia punctata IPNI=*104828-1* 292
Begonia punctata IPNI=*105130-1* 292
Begonia quetamensis IPNI=*105280-1* 292
Begonia radiata IPNI=*104828-1* 292
Begonia repens IPNI=*104759-1* 292
Begonia robustior IPNI=*105149-1* 292
Begonia roraimensis IPNI=*315565-2* 292
Begonia rosea IPNI=*105712-1* 293
Begonia scandens IPNI=*104759-1* 292
Begonia scandens var. *amplifolia* IPNI=*104759-1* 292
Begonia scandens var. *cordifolia* IPNI=*104759-1* 292
Begonia schizolepis IPNI=*105149-1* 292
Begonia setosa IPNI=*315565-2* 292
Begonia speciosa IPNI=*105214-1* 292
Begonia spruceana IPNI=*105712-1* 293
Begonia suaveolens IPNI=*105214-1* 292
Begonia subhumilis IPNI=*30121655-2* 292
Begonia tanacetifolia IPNI=*104828-1* 292
Begonia tovarensis IPNI=*315565-2* 292
Begonia tovarensis var. *ocanensis* IPNI=*315565-2* 292
Begonia tovarensis var. *palustris* IPNI=*315565-2* 292
Begonia trigonoptera IPNI=*104828-1* 292
Begonia ulei IPNI=*315565-2* 292
Begonia uliginosa IPNI=*315565-2* 292
Begonia uliginosa var. *ermanii* IPNI=*315565-2* 292
Begonia vellerea IPNI=*315565-2* 292
Begonia villosa IPNI=*104855-1* 292
Begonia villosa IPNI=*315565-2* 292
Beilschmiedia brenesii IPNI=*1057986-2* 182
Beilschmiedia cuneata IPNI=*190632-2* 185
Beilschmiedia lundelliana IPNI=*190632-2* 185
Beilschmiedia thomaea IPNI=*1057986-2* 182
Bejaria aestuans var. *macrantha* IPNI=*327015-1* 298
Bejaria albiflora IPNI=*327015-1* 298
Bejaria antioquiae IPNI=*317630-2* 298
Bejaria boliviensis IPNI=*317630-2* 298
Bejaria caxamarcensis IPNI=*327015-1* 298
Bejaria cinnamomea IPNI=*327015-1* 298
Bejaria coarctata IPNI=*317630-2* 298
Bejaria congesta IPNI=*327015-1* 298
Bejaria corymbosa IPNI=*327003-1* 298
Bejaria decora IPNI=*317630-2* 298
Bejaria densa IPNI=*317630-2* 298
Bejaria denticulata IPNI=*317630-2* 298
Bejaria discolor IPNI=*317630-2* 298
Bejaria dryanderae IPNI=*327015-1* 298
Bejaria drymifolia IPNI=*317630-2* 298
Bejaria floribunda IPNI=*317630-2* 298
Bejaria ghiesbreghtiana IPNI=*317630-2* 298
Bejaria glabra IPNI=*317630-2* 298
Bejaria glauca IPNI=*317630-2* 298
Bejaria glauca var. *coarctata* IPNI=*317630-2* 298
Bejaria glauca var. *glandulosa* IPNI=*317630-2* 298
Bejaria glauca var. *setosa* IPNI=*317630-2* 298
Bejaria glauca var. *tomentella* IPNI=*317630-2* 298
Bejaria glauca var. *typica* IPNI=*317630-2* 298
Bejaria grandiflora IPNI=*327015-1* 298
Bejaria grisea IPNI=*327015-1* 298
Bejaria guatemalensis IPNI=*317630-2* 298
Bejaria hintonii IPNI=*317630-2* 298
Bejaria hispida IPNI=*317630-2* 298
Bejaria laevis IPNI=*317630-2* 298
Bejaria lanceolata IPNI=*327015-1* 298
Bejaria lehmanniana IPNI=*327015-1* 298
Bejaria lindeniana IPNI=*317630-2* 298
Bejaria mathewsii f. *pulcherrima* IPNI=*327003-1* 298
Bejaria mexicana IPNI=*317630-2* 298
Bejaria myrtifolia IPNI=*327015-1* 298
Bejaria pallens IPNI=*317630-2* 298
Bejaria parviflora IPNI=*317630-2* 298
Bejaria parvifolia IPNI=*317630-2* 298
Bejaria pedunculata IPNI=*317630-2* 298
Bejaria peruviana IPNI=*327015-1* 298
Bejaria phillyreifolia IPNI=*327003-1* 298
Bejaria popayana IPNI=*327003-1* 298
Bejaria subserrata IPNI=*317630-2* 298
Bejaria tricolor IPNI=*327015-1* 298
Bejaria weberhaueriana IPNI=*327015-1* 298
Bejuco pendulus IPNI=*161653-1* 290
Belencita hagenii IPNI=*77183738-1* 278
Belencita nemorosa IPNI=*77183738-1* 278
Belicea hoffimannioides IPNI=*756446-1* 352
Beliceodendron tango IPNI=*136896-2* 325
Bellardia guianensis IPNI=*61955-2* 348
Bellardia mollis IPNI=*315609-2* 348
Bellardia repens IPNI=*61955-2* 348
Bellardia tontanea IPNI=*61955-2* 348
Bellis alpina IPNI=*184409-1* 258
Bellis armena IPNI=*184409-1* 258
Bellis croatica IPNI=*184409-1* 258
Bellis hortensis IPNI=*184409-1* 258
Bellis hybrida IPNI=*184409-1* 258
Bellis integrifolia IPNI=*184409-1* 258
Bellis minor IPNI=*184409-1* 258
Bellis perennis f. *disciformis* IPNI=*184409-1* 258
Bellis perennis f. *discoidea* IPNI=*184409-1* 258
Bellis pumila IPNI=*184409-1* 258
Bellis racemosa IPNI=*1186403-2* 261
Bellis ramosa IPNI=*1186403-2* 261

Bellis scaposa IPNI=184409-1 258
Bellis validula IPNI=184409-1 258
Bellota costaricensis IPNI=171697-2 184
Bellucia aricuaizensium IPNI=323200-2 418
Bellucia aubletii IPNI=1157663-2 418
Bellucia axinanthera IPNI=323200-2 418
Bellucia brasiliensis IPNI=1157663-2 418
Bellucia circumscissa IPNI=1157663-2 418
Bellucia costaricensis IPNI=323200-2 418
Bellucia hostmannii IPNI=1157663-2 418
Bellucia macrophylla IPNI=1157663-2 418
Bellucia multiflora IPNI=1157663-2 418
Bellucia nervosa IPNI=1157663-2 418
Bellucia pluricaulis IPNI=1157663-2 418
Bellucia quinquenervia IPNI=1157663-2 418
Bellucia riparia IPNI=1157663-2 418
Bellucia superba IPNI=1157663-2 418
Bellucia weberbaueri IPNI=323200-2 418
Beloperone carthaginensis IPNI=50468-1 360
Beloperone guttata IPNI=289220-2 360
Beloperone pulchella IPNI=50468-1 360
Beloperone violacea var. pulchella IPNI=50468-1 360
Belotia campbellii IPNI=77109528-1 415
Belotia caribaea IPNI=77109528-1 415
Belotia grewiifolia IPNI=1213322-2 415
Belotia insignis IPNI=1213322-2 415
Belotia lessertiana IPNI=77109528-1 415
Belotia mexicana IPNI=1213322-2 415
Belotia panamensis IPNI=1213322-2 415
Belotia reticulata IPNI=77109528-1 415
Belotia tabascana IPNI=77109528-1 415
Bemsetia paniculata IPNI=753834-1 352
Bennetia juncea IPNI=1103484-2 245
Benthamantha caribaea IPNI=275194-2 315
Bentinckia renda IPNI=666411-1 208
Benzoin glaucescens IPNI=465312-1 183
Berberis bergeriana IPNI=107016-1 434
Berberis carrikeri IPNI=106685-1 433
Berberis chocontana IPNI=107024-1 434
Berberis conferta IPNI=106862-1 434
Berberis conferta var. hypopyrrantha IPNI=106862-1 434
Berberis conferta var. karsteniana IPNI=106862-1 434
Berberis conferta var. typica IPNI=106862-1 434
Berberis cretata IPNI=106695-1 434
Berberis densa IPNI=107043-1 434
Berberis discolor IPNI=107043-1 434
Berberis globosa IPNI=107043-1 434
Berberis globosa var. typica IPNI=107043-1 434
Berberis medellinensis IPNI=107016-1 434
Berberis nitida IPNI=107024-1 434
Berberis paniculata IPNI=106685-1 433
Berberis pichinchensis var. rivetii IPNI=106985-1 434
Berberis retinervis IPNI=106985-1 434
Berberis rigidifolia var. densa IPNI=107043-1 434
Berberis rigidifolia var. globosa IPNI=107043-1 434
Berinia stricta IPNI=199753-1 260
Bermudiana bermudiana var. iridifolia IPNI=330260-2 216
Bermudiana bermudiana var. micrantha IPNI=330260-2 216
Bermudiana bogotensis IPNI=441310-1 216
Bermudiana bulbosa IPNI=89940-2 216
Bermudiana chilensis IPNI=330256-2 216
Bermudiana congesta IPNI=89940-2 216
Bermudiana convoluta IPNI=285111-2 216
Bermudiana iridifolia IPNI=330260-2 216
Bermudiana micrantha IPNI=330260-2 216
Bermudiana palmifolia IPNI=237727-2 216
Bermudiana tinctoria IPNI=441310-1 216
Bermudiana vaginata IPNI=237808-2 216
Bernhardia antillarum IPNI=212103-2 179
Bernhardia capensis IPNI=212103-2 179
Bernhardia complanata IPNI=114220-3 179
Bernhardia deppeana IPNI=212103-2 179
Bernhardia dichotoma IPNI=212103-2 179
Bernhardia novae-hollandiae IPNI=212103-2 179
Bernhardia oahuensis IPNI=212103-2 179
Bertholletia nobilis IPNI=591270-1 301
Bertiera diversiramea IPNI=744444-1 347
Bertiera ferrugrinea IPNI=329821-2 355
Bertiera guianensis var. leiophylla IPNI=744444-1 347
Bertiera mucronata IPNI=744444-1 347
Bertiera tenuis IPNI=744444-1 347
Bertolonia primuliflora IPNI=323262-2 423
Besleria acutifolia IPNI=378279-1 366
Besleria coccinea IPNI=84424-2 367
Besleria corallina IPNI=107520-2 367
Besleria corallinoides IPNI=107520-2 367
Besleria costaricensis IPNI=378279-1 366
Besleria crenata IPNI=107520-2 367
Besleria cristata IPNI=380742-1 367
Besleria drymonia IPNI=380742-1 367
Besleria ecuadorensis IPNI=107533-2 367
Besleria incarnata IPNI=237093-2 368
Besleria obliqua IPNI=378279-1 366
Besleria pansamalana IPNI=107533-2 367
Besleria pansamalana var. ecuadorensis IPNI=107533-2 367
Besleria pulchella IPNI=378695-1 366
Besleria reticulata var. macrophylla IPNI=32018-2 366
Besleria reticulata var. pubistyla IPNI=32018-2 366
Besleria sanguinea IPNI=378985-1 367
Besleria scandens IPNI=279831-2 376
Besleria serrulata IPNI=380742-1 367
Besleria solanoides var. dentata IPNI=378279-1 366
Besleria solanoides var. parvifolia IPNI=378279-1 366
Besleria solanoides var. tenera IPNI=378279-1 366
Besleria spectabilis IPNI=380742-1 367
Besleria splendens IPNI=380742-1 367
Besleria surinamensis IPNI=56505-1 362
Besleria sylvarum IPNI=107520-2 367
Besleria umbellata IPNI=378695-1 366
Besleria verrucosa IPNI=56505-1 362
Beta alba IPNI=164505-1 282
Beta altissima IPNI=164505-1 282
Beta atriplicifolia IPNI=164505-1 282
Beta bengalensis IPNI=164505-1 282
Beta brasiliensis IPNI=164505-1 282
Beta braziliensis IPNI=164505-1 282
Beta carnulosa IPNI=164505-1 282
Beta cicla IPNI=164505-1 282
Beta cicla var. argentea IPNI=164505-1 282
Beta cicla var. viridis IPNI=164505-1 282
Beta crispa IPNI=164505-1 282
Beta decumbens IPNI=164505-1 282
Beta esculenta IPNI=164505-1 282
Beta foliosa IPNI=164505-1 282
Beta hortensis IPNI=164505-1 282
Beta hybrida IPNI=164505-1 282
Beta incarnata IPNI=164505-1 282
Beta lutea IPNI=164505-1 282
Beta marina IPNI=164505-1 282
Beta maritima IPNI=164505-1 282
Beta maritima subsp. atriplicifolia IPNI=164505-1 282
Beta maritima subsp. marcosii IPNI=164505-1 282
Beta maritima subsp. orientalis IPNI=164505-1 282
Beta noeana IPNI=164505-1 282
Beta orientalis IPNI=164505-1 282
Beta purpurea IPNI=164505-1 282

Beta rapa IPNI=*164505-1* 282
Beta rapacea IPNI=*164505-1* 282
Beta rosea IPNI=*164505-1* 282
Beta sativa IPNI=*164505-1* 282
Beta stricta IPNI=*164505-1* 282
Beta sulcata IPNI=*164505-1* 282
Beta triflora IPNI=*164505-1* 282
Beta vulgaris var. *aurantia* IPNI=*164505-1* 282
Beta vulgaris var. *coniciformis* IPNI=*164505-1* 282
Beta vulgaris var. *grisea* IPNI=*164505-1* 282
Beta vulgaris var. *marcosii* IPNI=*164505-1* 282
Beta vulgaris var. *maritima* IPNI=*164505-1* 282
Beta vulgaris var. *mediasiatica* IPNI=*164505-1* 282
Beta vulgaris var. *ovaliformis* IPNI=*164505-1* 282
Beta vulgaris var. *rapacea* IPNI=*164505-1* 282
Beta vulgaris var. *rubidus* IPNI=*164505-1* 282
Beta vulgaris var. *rubra* IPNI=*164505-1* 282
Beta vulgaris var. *rubrifolia* IPNI=*164505-1* 282
Beta vulgaris var. *virescens* IPNI=*164505-1* 282
Beta vulgaris var. *viridifolia* IPNI=*164505-1* 282
Biancaea ferox IPNI=*481887-1* 312
Biancaea scandens IPNI=*481887-1* 312
Biancaea sepiaria IPNI=*481887-1* 312
Biauricula amara IPNI=*324660-2* 277
Biauricula panduriformis IPNI=*324660-2* 277
Biauricula resedifolia IPNI=*324660-2* 277
Bidens abadiae IPNI=*32564-2* 258
Bidens abadiae var. *pilosoides* IPNI=*32564-2* 258
Bidens abadiae var. *typica* IPNI=*32564-2* 258
Bidens abortiva IPNI=*184711-1* 258
Bidens acmelloides IPNI=*914778-1* 255
Bidens adhaerescens IPNI=*32564-2* 258
Bidens africana IPNI=*32564-2* 258
Bidens alausensis IPNI=*32564-2* 258
Bidens alba var. *radiata* IPNI=*184711-1* 258
Bidens andicola f. *dissecta* IPNI=*184720-1* 258
Bidens andicola f. *tarijensis* IPNI=*184720-1* 258
Bidens andicola var. *diversifolia* IPNI=*184720-1* 258
Bidens andicola var. *heterophylla* IPNI=*184720-1* 258
Bidens andicola var. *normalis* IPNI=*184720-1* 258
Bidens andicola var. *tarijensis* IPNI=*184720-1* 258
Bidens angustifolia IPNI=*250890-1* 270
Bidens angustissima IPNI=*184720-1* 258
Bidens antiguensis IPNI=*185190-1* 258
Bidens antiguensis var. *procumbens* IPNI=*185190-1* 258
Bidens antiguensis var. *salvadorensis* IPNI=*185190-1* 258
Bidens arenaria IPNI=*32564-2* 258
Bidens artemisiifolia f. *rubra* IPNI=*198324-1* 260
Bidens artemisiifolia IPNI=*185294-1* 258
Bidens artemisiifolia IPNI=*198368-1* 260
Bidens artemisiifolia var. *caudata* IPNI=*198324-1* 260
Bidens artemisiifolia var. *intermedia* IPNI=*198324-1* 260
Bidens artemisiifolia var. *rosea* IPNI=*198324-1* 260
Bidens artemisiifolia var. *rubra* IPNI=*198324-1* 260
Bidens attenuata IPNI=*185294-1* 258
Bidens aurantiaca IPNI=*32564-2* 258
Bidens bardanna IPNI=*193510-1* 259
Bidens barrancae IPNI=*32564-2* 258
Bidens berteroana IPNI=*198324-1* 260
Bidens bimucronata IPNI=*32564-2* 258
Bidens bipinnata IPNI=*198320-1* 260
Bidens bipinnata var. *biternatoides* IPNI=*184769-1* 258
Bidens bipinnata var. *cynapiifolia* IPNI=*32394-2* 258
Bidens bipinnata var. *minor* IPNI=*184769-1* 258
Bidens bipinnata var. *simplicior* IPNI=*184769-1* 258
Bidens bonplandii IPNI=*184711-1* 258
Bidens boquetiensis IPNI=*185190-1* 258
Bidens brachycarpa IPNI=*184711-1* 258
Bidens brittonii IPNI=*185190-1* 258
Bidens californica IPNI=*32564-2* 258
Bidens canescens IPNI=*185294-1* 258
Bidens caracasana IPNI=*32564-2* 258
Bidens carnea IPNI=*198324-1* 260
Bidens caucalidea IPNI=*184711-1* 258
Bidens caudata IPNI=*198324-1* 260
Bidens chilensis IPNI=*32564-2* 258
Bidens chilensis var. *apiifolia* IPNI=*32564-2* 258
Bidens chrysanthemoides IPNI=*32497-2* 258
Bidens chrysanthemoides var. *nashii* IPNI=*32497-2* 258
Bidens cicutifolia IPNI=*184769-1* 258
Bidens consolidifolia IPNI=*185294-1* 258
Bidens coreopsidis IPNI=*185190-1* 258
Bidens coreopsidis var. *procumbens* IPNI=*185190-1* 258
Bidens coreopsidis var. *reptans* IPNI=*185190-1* 258
Bidens coronata IPNI=*184711-1* 258
Bidens cosmanthus var. *diversifolia* IPNI=*184720-1* 258
Bidens crithmifolia IPNI=*185294-1* 258
Bidens cynapiifolia var. *portoricensis* IPNI=*184769-1* 258
Bidens cynapiifolia var. *tenuis* IPNI=*32394-2* 258
Bidens cynapiifolia var. *typica* IPNI=*32394-2* 258
Bidens daucifolia IPNI=*184711-1* 258
Bidens deamii IPNI=*184711-1* 258
Bidens decomposita IPNI=*184769-1* 258
Bidens decomposita var. *hirsutior* IPNI=*185294-1* 258
Bidens decumbens IPNI=*184711-1* 258
Bidens decussata IPNI=*32564-2* 258
Bidens delphiniifolia IPNI=*185294-1* 258
Bidens dissecta IPNI=*185190-1* 258
Bidens elegans IPNI=*32497-2* 258
Bidens elongata IPNI=*184769-1* 258
Bidens exaristata IPNI=*184711-1* 258
Bidens expansa IPNI=*32497-2* 258
Bidens fervida IPNI=*184769-1* 258
Bidens fervida IPNI=*914778-1* 255
Bidens fixa IPNI=*914778-1* 255
Bidens formosa IPNI=*198320-1* 260
Bidens formosa IPNI=*32497-2* 258
Bidens fruticulosa IPNI=*184720-1* 258
Bidens fusca IPNI=*914778-1* 255
Bidens geraniifolia IPNI=*185294-1* 258
Bidens glaberrima IPNI=*185294-1* 258
Bidens grandiflora var. *breviloba* IPNI=*185294-1* 258
Bidens grandiflora var. *humilis* IPNI=*185294-1* 258
Bidens helianthoides IPNI=*32497-2* 258
Bidens hirsuta IPNI=*32564-2* 258
Bidens hirta IPNI=*32564-2* 258
Bidens hirtella IPNI=*185294-1* 258
Bidens hispida IPNI=*32564-2* 258
Bidens humilis IPNI=*184720-1* 258
Bidens humilis IPNI=*185294-1* 258
Bidens humilis var. *macrantha* IPNI=*185294-1* 258
Bidens humilis var. *major* IPNI=*185294-1* 258
Bidens humilis var. *tenuifolia* IPNI=*185294-1* 258
Bidens inermis IPNI=*184711-1* 258
Bidens izabalensis IPNI=*185190-1* 258
Bidens leucantha IPNI=*184711-1* 258
Bidens leucanthema IPNI=*184711-1* 258
Bidens leucanthema var. *humilis* IPNI=*184711-1* 258
Bidens leucanthema var. *pilosa* IPNI=*32564-2* 258
Bidens lindleyi IPNI=*198320-1* 260
Bidens lugens IPNI=*32497-2* 258
Bidens mollis IPNI=*185294-1* 258
Bidens montaubani IPNI=*32564-2* 258
Bidens myrrhidifolia IPNI=*184769-1* 258
Bidens nashii IPNI=*32497-2* 258
Bidens nivea IPNI=*156327-2* 266

Bidens ocymifolia IPNI=*271528-2* 255
Bidens odorata IPNI=*32564-2* 258
Bidens odorata var. *calcicola* IPNI=*32564-2* 258
Bidens odorata var. *chilpancingensis* IPNI=*184711-1* 258
Bidens odorata var. *oaxacensis* IPNI=*184711-1* 258
Bidens odorata var. *rosea* IPNI=*184711-1* 258
Bidens oleracea IPNI=*914778-1* 255
Bidens orendainae IPNI=*32564-2* 258
Bidens oxyodonta IPNI=*184711-1* 258
Bidens paleacea IPNI=*32564-2* 258
Bidens parryi IPNI=*32497-2* 258
Bidens pectinata IPNI=*185294-1* 258
Bidens pedunculata IPNI=*185294-1* 258
Bidens persicifolia IPNI=*32497-2* 258
Bidens pilosa f. *alausensis* IPNI=*32564-2* 258
Bidens pilosa f. *bimucronata* IPNI=*32564-2* 258
Bidens pilosa f. *calcicola* IPNI=*32564-2* 258
Bidens pilosa f. *decumbens* IPNI=*184711-1* 258
Bidens pilosa f. *discoidea* IPNI=*32564-2* 258
Bidens pilosa f. *dissecta* IPNI=*32564-2* 258
Bidens pilosa f. *dondiifolia* IPNI=*32564-2* 258
Bidens pilosa f. *indivisa* IPNI=*32564-2* 258
Bidens pilosa f. *monophylla* IPNI=*32564-2* 258
Bidens pilosa f. *odorata* IPNI=*32564-2* 258
Bidens pilosa f. *pilosior* IPNI=*32564-2* 258
Bidens pilosa f. *pinnata* IPNI=*32564-2* 258
Bidens pilosa f. *radiata* IPNI=*184711-1* 258
Bidens pilosa f. *scandicina* IPNI=*32564-2* 258
Bidens pilosa f. *simplex* IPNI=*32564-2* 258
Bidens pilosa f. *subbiternata* IPNI=*32564-2* 258
Bidens pilosa f. *subsimplicifolia* IPNI=*32564-2* 258
Bidens pilosa f. *ternata* IPNI=*32564-2* 258
Bidens pilosa f. *triaristata* IPNI=*32564-2* 258
Bidens pilosa f. *umbrosa* IPNI=*32564-2* 258
Bidens pilosa subvar. *discoidea* IPNI=*32564-2* 258
Bidens pilosa subvar. *radiata* IPNI=*184711-1* 258
Bidens pilosa var. *alausesis* IPNI=*32564-2* 258
Bidens pilosa var. *albus* IPNI=*184711-1* 258
Bidens pilosa var. *apiifolia* IPNI=*32564-2* 258
Bidens pilosa var. *bimucronata* IPNI=*32564-2* 258
Bidens pilosa var. *bipinnata* IPNI=*184769-1* 258
Bidens pilosa var. *brachycarpa* IPNI=*184711-1* 258
Bidens pilosa var. *brevifoliata* IPNI=*32564-2* 258
Bidens pilosa var. *calcicola* IPNI=*32564-2* 258
Bidens pilosa var. *discoidea* IPNI=*32564-2* 258
Bidens pilosa var. *dubius* IPNI=*32564-2* 258
Bidens pilosa var. *leucanthema* IPNI=*184711-1* 258
Bidens pilosa var. *minor* IPNI=*32564-2* 258
Bidens pilosa var. *radiata* IPNI=*184711-1* 258
Bidens pilosa var. *subbiternata* IPNI=*32564-2* 258
Bidens pilosa var. *typica* IPNI=*32564-2* 258
Bidens pinnata IPNI=*32564-2* 258
Bidens platensis IPNI=*185264-1* 258
Bidens portoricensis IPNI=*184769-1* 258
Bidens procumbens IPNI=*185294-1* 258
Bidens pusilla IPNI=*184720-1* 258
Bidens quadrangularis IPNI=*185264-1* 258
Bidens quadriaristata IPNI=*32497-2* 258
Bidens ramosissima IPNI=*184711-1* 258
Bidens reflexa IPNI=*32564-2* 258
Bidens refracta IPNI=*185198-1* 258
Bidens reptans IPNI=*198320-1* 260
Bidens reptans var. *bipartita* IPNI=*185190-1* 258
Bidens reptans var. *brittonii* IPNI=*185190-1* 258
Bidens reptans var. *coreopsidis* IPNI=*185190-1* 258
Bidens reptans var. *dissecta* IPNI=*185190-1* 258
Bidens reptans var. *tomentosa* IPNI=*185190-1* 258
Bidens reptans var. *urbanii* IPNI=*185190-1* 258
Bidens riparia var. *refracta* IPNI=*185198-1* 258
Bidens riparia var. *typica* IPNI=*185198-1* 258
Bidens rosea IPNI=*184711-1* 258
Bidens rosea var. *calcicola* IPNI=*32564-2* 258
Bidens rubifolia var. *coreopsidis* IPNI=*185190-1* 258
Bidens scandens IPNI=*241903-1* 269
Bidens scandicina IPNI=*32564-2* 258
Bidens serrata IPNI=*185294-1* 258
Bidens serratula IPNI=*185294-1* 258
Bidens speciosa IPNI=*32497-2* 258
Bidens squarrosa IPNI=*185190-1* 258
Bidens squarrosa var. *atrostriata* IPNI=*185190-1* 258
Bidens squarrosa var. *hondurensis* IPNI=*185190-1* 258
Bidens squarrosa var. *indivisa* IPNI=*185190-1* 258
Bidens squarrosa var. *speciosa* IPNI=*185190-1* 258
Bidens squarrosa var. *tereticaulis* IPNI=*185190-1* 258
Bidens striata IPNI=*32564-2* 258
Bidens subalternans var. *typica* IPNI=*185264-1* 258
Bidens sulphurea IPNI=*198368-1* 260
Bidens sundaica IPNI=*32564-2* 258
Bidens sundaica var. *minor* IPNI=*32564-2* 258
Bidens taquetii IPNI=*32564-2* 258
Bidens tereticaulis IPNI=*185190-1* 258
Bidens tereticaulis var. *antiguensis* IPNI=*185190-1* 258
Bidens tereticaulis var. *indivisa* IPNI=*185190-1* 258
Bidens tereticaulis var. *sordida* IPNI=*185190-1* 258
Bidens triplinervia f. *exaristata* IPNI=*185294-1* 258
Bidens triplinervia f. *macrantha* IPNI=*185294-1* 258
Bidens triplinervia f. *octoradiata* IPNI=*185294-1* 258
Bidens triplinervia var. *boyocana* IPNI=*185294-1* 258
Bidens triplinervia var. *eurymera* IPNI=*185294-1* 258
Bidens triplinervia var. *hirtella* IPNI=*185294-1* 258
Bidens triplinervia var. *macrantha* IPNI=*185294-1* 258
Bidens triplinervia var. *mollis* IPNI=*185294-1* 258
Bidens triplinervia var. *nematoidea* IPNI=*185294-1* 258
Bidens triplinervia var. *typica* IPNI=*185294-1* 258
Bidens urbanii IPNI=*185190-1* 258
Bidens valparadisiaca IPNI=*32564-2* 258
Bidens variabilis IPNI=*30027297-2* 261
Bidens viciosoi IPNI=*32564-2* 258
Bidens wallichii IPNI=*184769-1* 258
Bifora loureiroi IPNI=*840760-1* 252
Bigelovia commutata IPNI=*77154345-1* 351
Bigelovia cornifolia IPNI=*317930-2* 357
Bigelovia distans IPNI=*317930-2* 357
Bigelovia gracilis IPNI=*766863-1* 357
Bigelovia laevicaulis IPNI=*766863-1* 357
Bigelovia linifolia IPNI=*241474-2* 357
Bigelovia ocymoides IPNI=*766863-1* 357
Bigelovia parviflora IPNI=*766863-1* 357
Bigelovia stricta IPNI=*766863-1* 357
Bigelovia tenera IPNI=*766863-1* 357
Bigelovia umbellata IPNI=*90501-2* 350
Bigelovia vaginata IPNI=*319724-2* 357
Bigelovia verticillata IPNI=*767031-1* 357
Bignonia acutistipula IPNI=*77100466-1* 363
Bignonia aequinoctialis f. *spectabilis* IPNI=*108550-1* 362
Bignonia aequinoctialis var. *hirtella* IPNI=*108550-1* 362
Bignonia aequinoctialis var. *spectabilis* IPNI=*108550-1* 362
Bignonia aesculifolia IPNI=*111572-2* 363
Bignonia africana IPNI=*109874-1* 364
Bignonia alliacea IPNI=*152672-2* 364
Bignonia araliacea IPNI=*77085096-1* 364
Bignonia barbata IPNI=*109687-1* 363
Bignonia blanchetii IPNI=*77142059-1* 363
Bignonia botryoides IPNI=*77100444-1* 362
Bignonia brevipes IPNI=*246385-2* 365
Bignonia californica IPNI=*77100466-1* 363

Bignonia candicans IPNI=*77141989-1* 363
Bignonia capensis IPNI=*111333-1* 365
Bignonia caryophyllea IPNI=*259274-2* 365
Bignonia catharinae IPNI=*77100444-1* 362
Bignonia catharinensis IPNI=*77100466-1* 363
Bignonia chica IPNI=*77142115-1* 363
Bignonia chrysantha IPNI=*77085066-1* 363
Bignonia ciliata IPNI=*77141996-1* 363
Bignonia citrifolia IPNI=*152672-2* 364
Bignonia coito IPNI=*60474996-2* 362
Bignonia colorata IPNI=*77141989-1* 363
Bignonia conjugata IPNI=*77141996-1* 363
Bignonia conspicua IPNI=*77085096-1* 364
Bignonia copaia IPNI=*109789-1* 364
Bignonia crucifera IPNI=*42392-2* 362
Bignonia crucigera IPNI=*77100444-1* 362
Bignonia cuprea IPNI=*77142115-1* 363
Bignonia cuprea var. *grandiflora* IPNI=*77142115-1* 363
Bignonia cuprea var. *parviflora* IPNI=*77142115-1* 363
Bignonia dasyonyx IPNI=*77100466-1* 363
Bignonia decomposita IPNI=*77100463-1* 363
Bignonia dombeyana IPNI=*109429-1* 363
Bignonia dura IPNI=*248703-2* 365
Bignonia echinata IPNI=*77100444-1* 362
Bignonia exoleta IPNI=*77100466-1* 363
Bignonia farinosa IPNI=*77141996-1* 363
Bignonia flavescens IPNI=*77085096-1* 364
Bignonia fluviatilis IPNI=*111027-1* 365
Bignonia fockeana IPNI=*153710-2* 364
Bignonia fraxinifolia IPNI=*77100464-1* 363
Bignonia gracilis IPNI=*77100466-1* 363
Bignonia hexagona IPNI=*77100444-1* 362
Bignonia hondensis IPNI=*42392-2* 362
Bignonia hostmannii IPNI=*108550-1* 362
Bignonia hymenaea IPNI=*152677-2* 364
Bignonia ignea IPNI=*110483-1* 365
Bignonia inaequalis IPNI=*77100463-1* 363
Bignonia incarnata var. *caribaea* IPNI=*108550-1* 362
Bignonia inflata IPNI=*77100466-1* 363
Bignonia jasminifolia IPNI=*110423-1* 364
Bignonia lanuginosa IPNI=*77100466-1* 363
Bignonia latifolia IPNI=*42392-2* 362
Bignonia latisiliqua IPNI=*77325-1* 339
Bignonia laxiflora IPNI=*259274-2* 365
Bignonia lepidota IPNI=*77141971-1* 362
Bignonia leucoxylon IPNI=*248689-2* 365
Bignonia litoralis IPNI=*77142038-1* 363
Bignonia longisiliqua IPNI=*153710-2* 364
Bignonia longissima IPNI=*77176826-1* 362
Bignonia lundii IPNI=*77100444-1* 362
Bignonia magnoliifolia IPNI=*77100449-1* 362
Bignonia martini IPNI=*153710-2* 364
Bignonia melioides IPNI=*202082-2* 364
Bignonia modesta IPNI=*246385-2* 365
Bignonia mollissima IPNI=*77142038-1* 363
Bignonia muricata IPNI=*77100444-1* 362
Bignonia nervosa IPNI=*109429-1* 363
Bignonia nitidissima IPNI=*108550-1* 362
Bignonia obovata IPNI=*153710-2* 364
Bignonia pachyptera IPNI=*77100511-1* 363
Bignonia paniculata IPNI=*108202-1* 362
Bignonia patrisiana IPNI=*77085096-1* 364
Bignonia pentaphylla IPNI=*248689-2* 365
Bignonia phaseoloides IPNI=*77100444-1* 362
Bignonia picta IPNI=*108550-1* 362
Bignonia pilosa IPNI=*108550-1* 362
Bignonia populifolia IPNI=*77100464-1* 363
Bignonia procera IPNI=*109789-1* 364
Bignonia pseudoquercus IPNI=*77176826-1* 362
Bignonia pseudounguis IPNI=*77100466-1* 363
Bignonia pubescens IPNI=*77142059-1* 363
Bignonia quadrivalvis IPNI=*77100464-1* 363
Bignonia quercus IPNI=*77176826-1* 362
Bignonia riparia IPNI=*246385-2* 365
Bignonia robusta IPNI=*13613-2* 362
Bignonia rodigasiana IPNI=*77100466-1* 363
Bignonia roseoalba IPNI=*248789-2* 365
Bignonia rubescens IPNI=*77142115-1* 363
Bignonia rufescens IPNI=*77142115-1* 363
Bignonia rufinervis IPNI=*42392-2* 362
Bignonia sarmentosa IPNI=*108550-1* 362
Bignonia sarmentosa var. *hirtella* IPNI=*108550-1* 362
Bignonia serratifolia IPNI=*77085096-1* 364
Bignonia sieberi IPNI=*77142059-1* 363
Bignonia simplicifolia IPNI=*109429-1* 363
Bignonia spectabilis IPNI=*108550-1* 362
Bignonia squalus IPNI=*77100444-1* 362
Bignonia stans IPNI=*111284-1* 365
Bignonia tecomiflora IPNI=*110483-1* 365
Bignonia tenuisiliqua IPNI=*77176826-1* 362
Bignonia tetragonocaulis IPNI=*110423-1* 364
Bignonia thyrsoidea IPNI=*77142115-1* 363
Bignonia tiliifolia IPNI=*77100444-1* 362
Bignonia tomentella IPNI=*77142059-1* 363
Bignonia triantha IPNI=*77100466-1* 363
Bignonia triflora IPNI=*67491-2* 349
Bignonia triphylla IPNI=*77142115-1* 363
Bignonia triternata IPNI=*77100463-1* 363
Bignonia tubulosa IPNI=*110483-1* 365
Bignonia tweedieana IPNI=*77100466-1* 363
Bignonia uncata IPNI=*77100511-1* 363
Bignonia uncinata IPNI=*77100511-1* 363
Bignonia unguis IPNI=*77100466-1* 363
Bignonia unguis-cati IPNI=*77100466-1* 363
Bignonia unguis-cati var. *exoleta* IPNI=*77100466-1* 363
Bignonia unguis-cati var. *guatemalensis* IPNI=*77100466-1* 363
Bignonia unguis-cati var. *radicans* IPNI=*77100466-1* 363
Bignonia unguis-cati var. *serrata* IPNI=*77100466-1* 363
Bignonia venusta IPNI=*110483-1* 365
Bignonia verrucifera IPNI=*152688-2* 364
Bignonia vespertilia IPNI=*77100466-1* 363
Bignonia villosa IPNI=*108550-1* 362
Bignonia vitalba IPNI=*77100444-1* 362
Bignonia vitalba var. *aequinoctialis* IPNI=*77100444-1* 362
Bignonia vitalba var. *extratropica* IPNI=*77100444-1* 362
Bihai acuminata IPNI=*796910-1* 248
Bihai bihai IPNI=*796925-1* 248
Bihai cannoidea IPNI=*796972-1* 248
Bihai curtispatha IPNI=*119344-2* 248
Bihai distans IPNI=*796925-1* 248
Bihai elegans IPNI=*796965-1* 248
Bihai elongata IPNI=*119524-2* 249
Bihai episcopalis IPNI=*119355-2* 248
Bihai ferdinando-coburgi IPNI=*119355-2* 248
Bihai harrisiana IPNI=*796953-1* 248
Bihai hirsuta IPNI=*796953-1* 248
Bihai humilis IPNI=*796972-1* 248
Bihai latispatha IPNI=*796959-1* 248
Bihai longa IPNI=*119402-2* 248
Bihai marginata IPNI=*119411-2* 248
Bihai mariae IPNI=*796965-1* 248
Bihai metallica IPNI=*796967-1* 248
Bihai poeppigiana IPNI=*796978-1* 249
Bihai psittacorum IPNI=*796972-1* 248
Bihai punicea IPNI=*796965-1* 248
Bihai purpurea IPNI=*796925-1* 248

Bihai rostrata IPNI=*796978-1* 249
Bihai rutila IPNI=*796925-1* 248
Bihai straminea IPNI=*796953-1* 248
Bihai stricta IPNI=*119505-2* 249
Bihai sylvestris IPNI=*796972-1* 248
Bihai wagneriana IPNI=*119524-2* 249
Billbergia chantinii IPNI=*5385-2* 228
Billbergia corymbosa IPNI=*5405-2* 228
Billbergia gracilis IPNI=*121847-1* 228
Billbergia meyendorffii IPNI=*169003-2* 229
Billbergia mitis IPNI=*5405-2* 228
Billbergia nudicaulis IPNI=*121752-1* 228
Billbergia paniculata IPNI=*5555-2* 228
Billbergia tetrantha IPNI=*970621-1* 229
Billbergia tillandsioides IPNI=*121847-1* 228
Binaria cumanensis IPNI=*229758-2* 332
Binaria herrerae IPNI=*229759-2* 332
Binaria hondurensis IPNI=*229758-2* 332
Binaria kunthiana IPNI=*77111927-1* 332
Binaria longipetala IPNI=*229758-2* 332
Binaria splendens IPNI=*518025-1* 332
Binaria splendens var. *latifolia* IPNI=*518025-1* 332
Biophytum ferrugineum IPNI=*371953-1* 432
Biophytum mimosoides IPNI=*371953-1* 432
Biota chengii IPNI=*677124-1* 180
Biota coraeana IPNI=*677124-1* 180
Biota dumosa IPNI=*677124-1* 180
Biota elegantissima IPNI=*677124-1* 180
Biota ericoides IPNI=*677124-1* 180
Biota excelsa IPNI=*677124-1* 180
Biota falcata IPNI=*677124-1* 180
Biota fortunei IPNI=*677124-1* 180
Biota freneloides IPNI=*677124-1* 180
Biota funiculata IPNI=*677124-1* 180
Biota glauca IPNI=*677124-1* 180
Biota gracilifolia IPNI=*677124-1* 180
Biota intermedia IPNI=*677124-1* 180
Biota japonica IPNI=*677124-1* 180
Biota meldensis IPNI=*677124-1* 180
Biota nepalensis IPNI=*677124-1* 180
Biota orientalis IPNI=*677124-1* 180
Biota pyramidalis IPNI=*677124-1* 180
Biota semperaurescens IPNI=*677124-1* 180
Biota stricta IPNI=*677124-1* 180
Biota tatarica IPNI=*677124-1* 180
Biota variegata IPNI=*677124-1* 180
Biota wareana IPNI=*677124-1* 180
Biota zuccarinii IPNI=*677124-1* 180
Bischofia leptopoda IPNI=*339921-1* 401
Bischofia oblongifolia IPNI=*339921-1* 401
Bischofia roeperiana IPNI=*339921-1* 401
Bischofia toui IPNI=*339921-1* 401
Bischofia trifoliata IPNI=*339921-1* 401
Biscutella apetala IPNI=*286148-1* 277
Bivonea urens IPNI=*61899-2* 388
Bixa acuminata IPNI=*33335-2* 406
Bixa americana IPNI=*33335-2* 406
Bixa katangensis IPNI=*33335-2* 406
Bixa orellana f. *leiocarpa* IPNI=*33335-2* 406
Bixa orellana var. *leiocarpa* IPNI=*33335-2* 406
Bixa orellana var. *platycarpa* IPNI=*111499-1* 406
Bixa orleana IPNI=*33335-2* 406
Bixa purpurea IPNI=*33335-2* 406
Bixa sphaerocarpa IPNI=*111504-1* 406
Bixa tinctaria IPNI=*33335-2* 406
Bixa upatensis IPNI=*33335-2* 406
Blackia egensis IPNI=*77112284-1* 418
Blairia lappulacea IPNI=*1194372-2* 379
Blairia nodiflora IPNI=*194567-2* 379
Blakea bracteata subsp. *ecuadorensis* IPNI=*33365-2* 418
Blakea calycanthus IPNI=*323201-2* 418
Blakea caudata IPNI=*323201-2* 418
Blakea macrophylla IPNI=*1157663-2* 418
Blakea quinquenervia IPNI=*1157663-2* 418
Blakea sphaerica IPNI=*566748-1* 418
Blakea triplinervia IPNI=*1157663-2* 418
Blattaria phoenicea IPNI=*824676-1* 412
Blattaria virgata IPNI=*770211-1* 377
Blechnopsis brasiliensis IPNI=*17058320-1* 174
Blechnopsis serrulata IPNI=*77142412-1* 174
Blechnum acuminatum IPNI=*33632-2* 174
Blechnum arborescens IPNI=*33533-2* 174
Blechnum arborescens var. *subspinuloso-serrulata* IPNI=*33533-2* 174
Blechnum brasiliense f. *multifida* IPNI=*17058320-1* 174
Blechnum brasiliense var. *corcovadense* IPNI=*17058320-1* 174
Blechnum campylotis IPNI=*33632-2* 174
Blechnum capense var. *ornifolium* IPNI=*33533-2* 174
Blechnum cartilagineum IPNI=*33632-2* 174
Blechnum chilense f. *imbricatum* IPNI=*33533-2* 174
Blechnum chilense IPNI=*33533-2* 174
Blechnum chilense var. *reedii* IPNI=*33533-2* 174
Blechnum ciliatum IPNI=*33632-2* 174
Blechnum cognatum IPNI=*33632-2* 174
Blechnum corcovadense IPNI=*17058320-1* 174
Blechnum cunninghamii IPNI=*33632-2* 174
Blechnum drepanophyllum IPNI=*33632-2* 174
Blechnum euraddianum IPNI=*33533-2* 174
Blechnum extensum IPNI=*33632-2* 174
Blechnum falcatum IPNI=*33632-2* 174
Blechnum falciculatum IPNI=*33632-2* 174
Blechnum fluminense IPNI=*17058320-1* 174
Blechnum galeottii IPNI=*33632-2* 174
Blechnum gilliesii IPNI=*33533-2* 174
Blechnum glandulosum IPNI=*33632-2* 174
Blechnum haenkeanum IPNI=*33632-2* 174
Blechnum helveolum IPNI=*33632-2* 174
Blechnum humile IPNI=*33632-2* 174
Blechnum itatiaiense IPNI=*33533-2* 174
Blechnum lomariaceum IPNI=*33632-2* 174
Blechnum lomarioides IPNI=*33632-2* 174
Blechnum macahense IPNI=*33533-2* 174
Blechnum mucronatum IPNI=*33632-2* 174
Blechnum nigro-squamatum IPNI=*17058320-1* 174
Blechnum occidentale f. *bifidum* IPNI=*33632-2* 174
Blechnum occidentale var. *campylote* IPNI=*33632-2* 174
Blechnum occidentale var. *cunninghamii* IPNI=*33632-2* 174
Blechnum occidentale var. *lacerata* IPNI=*33632-2* 174
Blechnum occidentale var. *lomariaceum* IPNI=*33632-2* 174
Blechnum occidentale var. *majus* IPNI=*33632-2* 174
Blechnum occidentale var. *minus* IPNI=*33632-2* 174
Blechnum occidentale var. *pallidum* IPNI=*33632-2* 174
Blechnum occidentale var. *pectinatum* IPNI=*33632-2* 174
Blechnum occidentale var. *solutum* IPNI=*33632-2* 174
Blechnum ornifolium IPNI=*33533-2* 174
Blechnum pectinatum IPNI=*33632-2* 174
Blechnum peruvianum IPNI=*33533-2* 174
Blechnum raddianum IPNI=*33533-2* 174
Blechnum regnellianum IPNI=*33533-2* 174
Blechnum regnellianum var. *elatius* IPNI=*33533-2* 174
Blechnum rugosum IPNI=*33632-2* 174
Blechnum scaberulum IPNI=*33632-2* 174
Blechnum serrulatum f. *nanum* IPNI=*77142412-1* 174
Blechnum serrulatum IPNI=*77142412-1* 174
Blechnum serrulatum var. *crespianum* IPNI=*77142412-1* 174
Blechnum serrulatum var. *distans* IPNI=*77142412-1* 174
Blechnum serrulatum var. *stierii* IPNI=*77142412-1* 174

Blechnum simile IPNI=*33533-2* 174
Blechnum sodiroi IPNI=*33632-2* 174
Blechnum squamulosum IPNI=*77186436-1* 174
Blechnum stierii IPNI=*77142412-1* 174
Blechnum suburbicum IPNI=*33632-2* 174
Blechnum triangulatum IPNI=*33632-2* 174
Blechnum validum IPNI=*17058320-1* 174
Blechum angustius IPNI=*56148-1* 361
Blechum blechum IPNI=*224052-2* 361
Blechum brownei f. *albiflora* IPNI=*224052-2* 361
Blechum brownei f. *coeruleum* IPNI=*224052-2* 361
Blechum brownei f. *puberulum* IPNI=*224052-2* 361
Blechum brownei IPNI=*224052-2* 361
Blechum brownei var. *laxum* IPNI=*224052-2* 361
Blechum brownei var. *subcordatum* IPNI=*224052-2* 361
Blechum cordatum IPNI=*77147688-1* 361
Blechum haenkei IPNI=*224052-2* 361
Blechum linnaei IPNI=*224052-2* 361
Blechum linnaei var. *laxum* IPNI=*224052-2* 361
Blechum linnaei var. *nanum* IPNI=*224052-2* 361
Blechum luzonium IPNI=*224052-2* 361
Blechum pyramidatum IPNI=*224052-2* 361
Blechum sternbergii IPNI=*56148-1* 361
Blechum trinitense IPNI=*224052-2* 361
Blepharis verticillata IPNI=*49672-1* 360
Blepharochloa ciliata IPNI=*137026-2* 241
Blepharodon anomalum IPNI=*94927-1* 339
Bletia acutipetala IPNI=*1218040-2* 217
Bletia anomala IPNI=*617639-1* 217
Bletia catenulata f. *alba* IPNI=*617644-1* 217
Bletia catenulata f. *caerulea* IPNI=*617644-1* 217
Bletia catenulata var. *alba* IPNI=*617644-1* 217
Bletia catenulata var. *caerulea* IPNI=*617644-1* 217
Bletia cucullata IPNI=*618026-1* 217
Bletia cucullata var. *cuspidata* IPNI=*618026-1* 217
Bletia ecuadorensis IPNI=*617644-1* 217
Bletia edwardsii IPNI=*617639-1* 217
Bletia expansa IPNI=*1218040-2* 217
Bletia gloriosa IPNI=*134675-2* 219
Bletia havannensis IPNI=*1218040-2* 217
Bletia landsbergii IPNI=*617639-1* 217
Bletia lankesteri IPNI=*617639-1* 217
Bletia mandonii IPNI=*617639-1* 217
Bletia martiana IPNI=*325518-2* 217
Bletia pallida IPNI=*1218040-2* 217
Bletia pottsii IPNI=*1218040-2* 217
Bletia purpurea var. *alba* IPNI=*1218040-2* 217
Bletia purpurea var. *pittieri* IPNI=*1218040-2* 217
Bletia repanda IPNI=*617644-1* 217
Bletia rodriguesii IPNI=*617644-1* 217
Bletia sanguinea IPNI=*617644-1* 217
Bletia verecunda IPNI=*1218040-2* 217
Bletia wageneri IPNI=*617639-1* 217
Bletia wageneri var. *cobra* IPNI=*617639-1* 217
Bletia wallisii IPNI=*134675-2* 219
Blitum ambrosioides IPNI=*1198060-2* 282
Blitum gangeticum IPNI=*327386-2* 282
Blitum maius IPNI=*10619-2* 281
Blitum melancholicum IPNI=*327386-2* 282
Blitum viride IPNI=*55993-2* 282
Blochmannia weigeltiana IPNI=*258621-2* 290
Blumea aurita IPNI=*211317-2* 268
Blumea bojeri IPNI=*211317-2* 268
Blumea chinensis IPNI=*962413-1* 260
Blumea esquirolii IPNI=*962413-1* 260
Blumea glutinosa IPNI=*211317-2* 268
Blumea guineensis IPNI=*211317-2* 268
Blumea lyrata IPNI=*211317-2* 268
Blumea microphylla IPNI=*211317-2* 268
Blumea obliqua var. *aurita* IPNI=*211317-2* 268
Blumea senegalensis IPNI=*211317-2* 268
Blumea viscosa IPNI=*211317-2* 268
Blumea viscosa var. *lyrata* IPNI=*211317-2* 268
Blumenbachia halepensis IPNI=*422139-1* 245
Blutaparon brevifolium IPNI=*34033-2* 282
Blutaparon portulacoides var. *commersonii* IPNI=*34030-2* 282
Blutaparon repens IPNI=*34033-2* 282
Blutaparon vermiculare var. *aggregatum* IPNI=*34033-2* 282
Blutaparon vermiculare var. *longispicatum* IPNI=*34033-2* 282
Bocagea aromatica IPNI=*1013110-1* 189
Bocagea asbeckii IPNI=*178617-2* 189
Bocagea espintana IPNI=*1013110-1* 189
Bocagea multiflora IPNI=*34051-2* 187
Bocageopsis multiflora var. *angustifolia* IPNI=*34051-2* 187
Bocconia frutescens f. *glaucescens* IPNI=*316358-2* 435
Bocconia frutescens f. *subtomentosa* IPNI=*316358-2* 435
Bocconia frutescens var. *cernua* IPNI=*316358-2* 435
Bocconia frutescens var. *integrifolia* IPNI=*671757-1* 435
Bocconia frutescens var. *normalis* IPNI=*316358-2* 435
Bocconia frutescens var. *subtomentosa* IPNI=*316358-2* 435
Bocconia glauca IPNI=*316358-2* 435
Bocconia integrifolia subf. *glaucescens* IPNI=*671757-1* 435
Bocconia integrifolia var. *peruviana* IPNI=*671757-1* 435
Bocconia integrifolia var. *seleri* IPNI=*671757-1* 435
Bocconia pearcei IPNI=*671757-1* 435
Bocconia quercifolia IPNI=*316358-2* 435
Bocconia sinuatifolia IPNI=*316358-2* 435
Bocconia subtomentosa IPNI=*316358-2* 435
Boehmeria ampullacea IPNI=*850541-1* 445
Boehmeria arborescens IPNI=*850541-1* 445
Boehmeria arguta IPNI=*850541-1* 445
Boehmeria ballotifolia IPNI=*855180-1* 446
Boehmeria candicans IPNI=*55676-3* 445
Boehmeria caudata var. *arborescens* IPNI=*850541-1* 445
Boehmeria caudata var. *arguta* IPNI=*850541-1* 445
Boehmeria caudata var. *petiolaris* IPNI=*850541-1* 445
Boehmeria cavaleriei IPNI=*855573-1* 447
Boehmeria compacta IPNI=*55676-3* 445
Boehmeria cuspidata IPNI=*34156-2* 445
Boehmeria ehrenbergiana IPNI=*34156-2* 445
Boehmeria flagelliformis IPNI=*850541-1* 445
Boehmeria guatemalensis IPNI=*850541-1* 445
Boehmeria jamaicensis IPNI=*34156-2* 445
Boehmeria juncea IPNI=*55676-3* 445
Boehmeria mollicoma IPNI=*55676-3* 445
Boehmeria niponoivea IPNI=*55676-3* 445
Boehmeria nivea f. *viridula* IPNI=*55676-3* 445
Boehmeria nivea var. *niponoivea* IPNI=*55676-3* 445
Boehmeria palmeri IPNI=*850541-1* 445
Boehmeria peruviana IPNI=*850541-1* 445
Boehmeria petiolaris IPNI=*850541-1* 445
Boehmeria procridioides IPNI=*856206-1* 448
Boehmeria ramiflora var. *cuspidata* IPNI=*34156-2* 445
Boehmeria ramiflora var. *vinacea* IPNI=*34156-2* 445
Boehmeria rhynchophylla IPNI=*34156-2* 445
Boehmeria rugosa IPNI=*855180-1* 446
Boehmeria scabrella IPNI=*850541-1* 445
Boehmeria sordida IPNI=*850541-1* 445
Boehmeria stipulata IPNI=*850541-1* 445
Boehmeria tenuistachys IPNI=*850541-1* 445
Boehmeria thailandica IPNI=*55676-3* 445
Boehmeria uliginosa IPNI=*855189-1* 447
Boehmeria utilis IPNI=*55676-3* 445
Boenninghausia vincentina IPNI=*77191263-1* 329
Boerhavia adscendens IPNI=*326899-2* 286
Boerhavia africana IPNI=*326899-2* 286

Boerhavia agglutinans IPNI=*326873-2* 286
Boerhavia atomaria IPNI=*319111-2* 286
Boerhavia bracteata IPNI=*326873-2* 286
Boerhavia caespitosa IPNI=*326899-2* 286
Boerhavia caribaea IPNI=*326873-2* 286
Boerhavia chaerophylloides IPNI=*859680-1* 296
Boerhavia coccinea f. *parcehirsuta* IPNI=*326873-2* 286
Boerhavia coccinea var. *leiocarpa* IPNI=*326899-2* 286
Boerhavia coccinea var. *paniculata* IPNI=*326873-2* 286
Boerhavia coccinea var. *viscosa* IPNI=*326873-2* 286
Boerhavia decumbens IPNI=*326899-2* 286
Boerhavia diffusa f. *crassifolia* IPNI=*326873-2* 286
Boerhavia diffusa IPNI=*326873-2* 286
Boerhavia diffusa var. *hirsuta* IPNI=*326873-2* 286
Boerhavia diffusa var. *leiocarpa* IPNI=*326899-2* 286
Boerhavia diffusa var. *obtusifolia* IPNI=*326873-2* 286
Boerhavia diffusa var. *pubescens* IPNI=*326899-2* 286
Boerhavia diffusa var. *viscosa* IPNI=*326873-2* 286
Boerhavia diffusa var. *xerophila* IPNI=*326873-2* 286
Boerhavia discolor IPNI=*319111-2* 286
Boerhavia elongata IPNI=*319111-2* 286
Boerhavia erecta f. *subepunctata* IPNI=*319111-2* 286
Boerhavia erecta var. *thornberi* IPNI=*319111-2* 286
Boerhavia friesii IPNI=*326899-2* 286
Boerhavia glandulosa IPNI=*326873-2* 286
Boerhavia glutinosa IPNI=*319111-2* 286
Boerhavia grahamii IPNI=*63937-2* 286
Boerhavia hirsuta IPNI=*326873-2* 286
Boerhavia ixodes IPNI=*326873-2* 286
Boerhavia lateriflora IPNI=*319111-2* 286
Boerhavia laxa IPNI=*326899-2* 286
Boerhavia marlothii IPNI=*326873-2* 286
Boerhavia nantocana IPNI=*326873-2* 286
Boerhavia obtusifolia IPNI=*326873-2* 286
Boerhavia obtusifolia IPNI=*326899-2* 286
Boerhavia paniculata f. *esetosa* IPNI=*326899-2* 286
Boerhavia paniculata IPNI=*326899-2* 286
Boerhavia paniculata var. *guaranitica* IPNI=*326899-2* 286
Boerhavia paniculata var. *leiocarpa* IPNI=*326899-2* 286
Boerhavia paniculata var. *subacuta* IPNI=*319111-2* 286
Boerhavia patula IPNI=*326873-2* 286
Boerhavia polymorpha IPNI=*326873-2* 286
Boerhavia ramulosa IPNI=*326873-2* 286
Boerhavia reboudiana IPNI=*326873-2* 286
Boerhavia repens var. *diffusa* IPNI=*326899-2* 286
Boerhavia rhomboidea IPNI=*605135-1* 287
Boerhavia rottleri IPNI=*326899-2* 286
Boerhavia rugosa IPNI=*326899-2* 286
Boerhavia sarmentosa IPNI=*63937-2* 286
Boerhavia scabrida IPNI=*319111-2* 286
Boerhavia scandens IPNI=*63937-2* 286
Boerhavia sonorae IPNI=*326873-2* 286
Boerhavia squamata IPNI=*326873-2* 286
Boerhavia thornberi IPNI=*319111-2* 286
Boerhavia virgata IPNI=*319111-2* 286
Boerhavia viscosa f. *oligadena* IPNI=*326873-2* 286
Boerhavia viscosa IPNI=*326873-2* 286
Boerhavia viscosa subsp. *apiculata* IPNI=*326873-2* 286
Boerhavia xerophila IPNI=*326899-2* 286
Bogenhardia crispa IPNI=*120630-2* 409
Bohadschia humifusa IPNI=*836578-1* 401
Bohadschia microphylla IPNI=*836578-1* 401
Bojeria magnifica IPNI=*942355-1* 251
Bolax cladorrhiza IPNI=*838571-1* 251
Bolax crenata IPNI=*838571-1* 251
Boldea boldus IPNI=*191376-2* 185
Boldea fragrans IPNI=*191376-2* 185
Boldoa ovatifolia IPNI=*605135-1* 287
Boldoa paniculata IPNI=*605135-1* 287
Boldu chilanum IPNI=*191376-2* 185
Boldus chilensis IPNI=*191376-2* 185
Bollea lalindei IPNI=*60440679-2* 222
Bollea patinii IPNI=*60440679-2* 222
Bomarea affinis IPNI=*63283-1* 226
Bomarea allenii IPNI=*63253-1* 226
Bomarea alpicola IPNI=*63318-1* 226
Bomarea ambigua IPNI=*63359-1* 226
Bomarea bakeriana IPNI=*63283-1* 226
Bomarea borjae IPNI=*63359-1* 226
Bomarea brauniana IPNI=*63283-1* 226
Bomarea caldasiana IPNI=*63359-1* 226
Bomarea caldasiana var. *concolor* IPNI=*63318-1* 226
Bomarea caldasiana var. *villosa* IPNI=*63318-1* 226
Bomarea caldasii var. *quitensis* IPNI=*63359-1* 226
Bomarea caraccensis IPNI=*63283-1* 226
Bomarea edulis var. *furcata* IPNI=*63283-1* 226
Bomarea edulis var. *grandis* IPNI=*63283-1* 226
Bomarea edulis var. *hirtula* IPNI=*63283-1* 226
Bomarea edulis var. *maranensis* IPNI=*63283-1* 226
Bomarea edulis var. *parvifolia* IPNI=*63283-1* 226
Bomarea floribunda IPNI=*63359-1* 226
Bomarea foliolosa IPNI=*63359-1* 226
Bomarea frondea IPNI=*63359-1* 226
Bomarea furcata IPNI=*63283-1* 226
Bomarea glomerata IPNI=*63411-1* 226
Bomarea gloriosa IPNI=*63283-1* 226
Bomarea grandifolia IPNI=*63283-1* 226
Bomarea guianensis IPNI=*63283-1* 226
Bomarea halliana IPNI=*63359-1* 226
Bomarea hirsuta var. *concolor* IPNI=*63318-1* 226
Bomarea hirta IPNI=*63283-1* 226
Bomarea hirtella IPNI=*63283-1* 226
Bomarea jacquesiana IPNI=*63283-1* 226
Bomarea janeirensis IPNI=*63283-1* 226
Bomarea kalbreyeri IPNI=*63318-1* 226
Bomarea lanata IPNI=*63318-1* 226
Bomarea maakiana IPNI=*63283-1* 226
Bomarea macranthera IPNI=*63411-1* 226
Bomarea maranensis IPNI=*63283-1* 226
Bomarea martiana IPNI=*63283-1* 226
Bomarea microcephala IPNI=*63359-1* 226
Bomarea miniata IPNI=*63283-1* 226
Bomarea oligantha IPNI=*63359-1* 226
Bomarea ovata var. *tatiana* IPNI=*63283-1* 226
Bomarea pachyphlebia IPNI=*63411-1* 226
Bomarea perlongipes IPNI=*63283-1* 226
Bomarea petiolata IPNI=*63283-1* 226
Bomarea rigidifolia IPNI=*63359-1* 226
Bomarea salsilla IPNI=*63283-1* 226
Bomarea salsilloides IPNI=*63283-1* 226
Bomarea salsilloides var. *pauciflora* IPNI=*63283-1* 226
Bomarea salsilloides var. *pubescens* IPNI=*63283-1* 226
Bomarea salsilloides var. *sepium* IPNI=*63283-1* 226
Bomarea sororia IPNI=*63283-1* 226
Bomarea spectabilis IPNI=*63283-1* 226
Bomarea spectabilis var. *parvifolia* IPNI=*63283-1* 226
Bomarea tatiana IPNI=*63283-1* 226
Bomarea tomentosa var. *pangoensis* IPNI=*63411-1* 226
Bomarea turneriana IPNI=*63359-1* 226
Bomarea vegasana IPNI=*63359-1* 226
Bomarea vestita IPNI=*63359-1* 226
Bomarea williamsiae IPNI=*63253-1* 226
Bombacopsis glabra IPNI=*562664-1* 412
Bombacopsis patinoi IPNI=*1002081-1* 412
Bombacopsis quinata IPNI=*559120-1* 407
Bombacopsis sessilis IPNI=*281912-2* 412

Bombacopsis speciosa IPNI=*281913-2* 412
Bombax aculeatum IPNI=*1005661-1* 408
Bombax aculeatum IPNI=*559120-1* 407
Bombax affine IPNI=*562668-1* 412
Bombax album IPNI=*211230-2* 412
Bombax angulatum IPNI=*171458-2* 412
Bombax anisophyllum IPNI=*562664-1* 412
Bombax aquaticum IPNI=*179762-2* 412
Bombax balanoides IPNI=*211241-2* 413
Bombax barrigon IPNI=*211241-2* 413
Bombax carabobense IPNI=*211241-2* 413
Bombax ceiba IPNI=*559120-1* 407
Bombax columellatum IPNI=*562664-1* 412
Bombax cumanense IPNI=*1166232-2* 407
Bombax cyathophorum IPNI=*211230-2* 412
Bombax decaphyllum IPNI=*211230-2* 412
Bombax glabrum IPNI=*562664-1* 412
Bombax globiferum IPNI=*95971-2* 408
Bombax globosum IPNI=*95971-2* 408
Bombax grandiflorum IPNI=*211230-2* 412
Bombax heptaphyllum IPNI=*211241-2* 413
Bombax heptaphyllum IPNI=*559120-1* 407
Bombax insigne IPNI=*562668-1* 412
Bombax kimuenzae IPNI=*562664-1* 412
Bombax macrocarpum IPNI=*179762-2* 412
Bombax macrophyllum IPNI=*95976-2* 408
Bombax malabaricum IPNI=*559120-1* 407
Bombax mompoxense IPNI=*1166232-2* 407
Bombax munguba IPNI=*211238-2* 413
Bombax occidentale IPNI=*1166232-2* 407
Bombax oleagineum IPNI=*562664-1* 412
Bombax orientale IPNI=*1166232-2* 407
Bombax orinocense IPNI=*111532-1* 406
Bombax pentandrum IPNI=*1166232-2* 407
Bombax plumosum IPNI=*1166232-2* 407
Bombax pyramidale IPNI=*171458-2* 412
Bombax pyramidatum IPNI=*171458-2* 412
Bombax rigidifolium IPNI=*179762-2* 412
Bombax roseorum IPNI=*95985-2* 408
Bombax samauma IPNI=*51083-2* 408
Bombax septenatum IPNI=*211241-2* 413
Bombax sessile IPNI=*281912-2* 412
Bombax spectabile IPNI=*562668-1* 412
Bombax spruceanum IPNI=*562668-1* 412
Bombax thorelii IPNI=*559120-1* 407
Bombax tussacii IPNI=*559120-1* 407
Bombax vitifolium IPNI=*111539-1* 406
Bonafousia columbiensis IPNI=*285800-2* 343
Bonafousia guyanensis IPNI=*20001045-1* 344
Bonafousia hirtula IPNI=*20001045-1* 344
Bonafousia juruana IPNI=*20001045-1* 344
Bonafousia killipii IPNI=*20001045-1* 344
Bonafousia longituba IPNI=*288916-2* 344
Bonafousia macrocalyx IPNI=*285812-2* 343
Bonafousia maxima IPNI=*82212-1* 344
Bonafousia morettii IPNI=*285812-2* 343
Bonafousia muelleriana IPNI=*285812-2* 343
Bonafousia obliqua IPNI=*82371-1* 344
Bonafousia palustris IPNI=*248891-2* 344
Bonafousia panamensis IPNI=*248892-2* 344
Bonafousia perrottetii IPNI=*82371-1* 344
Bonafousia rimulosa IPNI=*248891-2* 344
Bonafousia sananho IPNI=*82327-1* 344
Bonafousia siphilitica IPNI=*20001045-1* 344
Bonafousia siphilitica var. *hirtula* IPNI=*20001045-1* 344
Bonafousia siphilitica var. *juruana* IPNI=*20001045-1* 344
Bonafousia speciosa IPNI=*20001045-1* 344
Bonafousia tessmannii IPNI=*20001045-1* 344
Bonafousia tetrastachya IPNI=*20001045-1* 344
Bonafousia undulata IPNI=*82371-1* 344
Bonafousia undulata var. *ovalifolia* IPNI=*82371-1* 344
Bonannia officinalis IPNI=*288952-1* 278
Bonduc canadense IPNI=*114706-2* 320
Bonduc majus IPNI=*114718-2* 320
Bonduc minus IPNI=*114706-2* 320
Bonnaya alata IPNI=*810361-1* 374
Bonnaya bracteata IPNI=*799766-1* 374
Bonnaya cyanea IPNI=*799766-1* 374
Bonnaya grandiflora IPNI=*799766-1* 374
Bonnaya origanifolia IPNI=*262512-2* 374
Bonnaya penicillata IPNI=*799766-1* 374
Bonnaya pulegiifolia IPNI=*799766-1* 374
Bonnaya rothii IPNI=*799766-1* 374
Bonnaya rotundifolia IPNI=*1061345-2* 374
Bonnaya roxburghiana IPNI=*799766-1* 374
Bonnaya verbenifolia IPNI=*799766-1* 374
Bonnaya veronicifolia var. *angustifolia* IPNI=*799766-1* 374
Bonnetia dinizii IPNI=*828487-1* 380
Bonnetia paniculata var. *parviflora* IPNI=*828487-1* 380
Bonnetia paniculata var. *typica* IPNI=*828487-1* 380
Bonnetia parviflora IPNI=*828487-1* 380
Bontia daphnoides var. *minor* IPNI=*319112-2* 377
Bontia germinans IPNI=*861125-1* 359
Bontia minor IPNI=*319112-2* 377
Borago advena IPNI=*113618-1* 273
Borago aspera IPNI=*113618-1* 273
Borago hortensis IPNI=*113618-1* 273
Borassus pinnatifrons IPNI=*665826-1* 207
Borbonia globosa IPNI=*167822-2* 184
Borbonia membranacea IPNI=*466773-1* 184
Borbonia peruviana IPNI=*77198017-1* 183
Borojoa claviflora IPNI=*8381-2* 346
Borojoa duckei IPNI=*77175918-1* 346
Borojoa patinoi IPNI=*77116169-1* 347
Borojoa sorbilis IPNI=*8417-2* 347
Borojoa verticillata IPNI=*943695-1* 347
Borreria alata IPNI=*766585-1* 357
Borreria alternans IPNI=*766862-1* 357
Borreria anthospermoides IPNI=*999397-1* 350
Borreria aralioides IPNI=*90501-2* 350
Borreria arenicola IPNI=*945476-1* 357
Borreria asperula IPNI=*766862-1* 357
Borreria assurgens IPNI=*319724-2* 357
Borreria aturensis IPNI=*317930-2* 357
Borreria bartlingiana IPNI=*766800-1* 357
Borreria bogotensis IPNI=*999397-1* 350
Borreria capitata f. *ferruginea* IPNI=*317930-2* 357
Borreria capitata f. *glabra* IPNI=*317930-2* 357
Borreria capitata IPNI=*317930-2* 357
Borreria capitata subsp. *restingicola* IPNI=*317930-2* 357
Borreria capitellata IPNI=*241474-2* 357
Borreria commutata IPNI=*767031-1* 357
Borreria coriacea IPNI=*317930-2* 357
Borreria cornifolia IPNI=*317930-2* 357
Borreria densiflora IPNI=*945476-1* 357
Borreria densiflora var. *latifolia* IPNI=*945476-1* 357
Borreria densiflora var. *minima* IPNI=*945476-1* 357
Borreria densiflora var. *perennis* IPNI=*945476-1* 357
Borreria densiflora var. *pilosa* IPNI=*945476-1* 357
Borreria distans IPNI=*317930-2* 357
Borreria domingensis IPNI=*766900-1* 357
Borreria elongata IPNI=*317930-2* 357
Borreria eradii IPNI=*766800-1* 357
Borreria ferruginea IPNI=*317930-2* 357
Borreria ferruginea IPNI=*756228-1* 352
Borreria fockeana IPNI=*766800-1* 357

Borreria globularioides IPNI=*767031-1* 357
Borreria gracilis IPNI=*317930-2* 357
Borreria graminifolia IPNI=*767031-1* 357
Borreria herbert-smithii IPNI=*241474-2* 357
Borreria heterophylla IPNI=*766900-1* 357
Borreria kohautiana IPNI=*767031-1* 357
Borreria laevicaulis IPNI=*766863-1* 357
Borreria laevigata IPNI=*767031-1* 357
Borreria laevis IPNI=*241474-2* 357
Borreria latifolia f. *fockeana* IPNI=*766800-1* 357
Borreria latifolia f. *minor* IPNI=*766800-1* 357
Borreria latifolia IPNI=*766800-1* 357
Borreria latifolia var. *fockeana* IPNI=*766800-1* 357
Borreria latifolia var. *minor* IPNI=*766800-1* 357
Borreria latifolia var. *scabrida* IPNI=*766800-1* 357
Borreria latifolia var. *sideritis* IPNI=*766800-1* 357
Borreria linifolia IPNI=*319724-2* 357
Borreria lutescens IPNI=*317930-2* 357
Borreria malacophylla IPNI=*319724-2* 357
Borreria micrantha IPNI=*766900-1* 357
Borreria microphylla IPNI=*766900-1* 357
Borreria minima IPNI=*767031-1* 357
Borreria molleri IPNI=*767031-1* 357
Borreria oaxacana IPNI=*767031-1* 357
Borreria ocymifolia IPNI=*766862-1* 357
Borreria ocymoides f. *bisepala* IPNI=*766900-1* 357
Borreria ocymoides f. *tenuis* IPNI=*766900-1* 357
Borreria ocymoides IPNI=*766863-1* 357
Borreria ocymoides var. *bisepala* IPNI=*766900-1* 357
Borreria ocymoides var. *minima* IPNI=*766900-1* 357
Borreria ocymoides var. *thekkumalensis* IPNI=*766863-1* 357
Borreria oligodonta IPNI=*767031-1* 357
Borreria parviflora IPNI=*766900-1* 357
Borreria patula IPNI=*766900-1* 357
Borreria penicillata IPNI=*766800-1* 357
Borreria perrottetii IPNI=*766800-1* 357
Borreria platyphylla IPNI=*766800-1* 357
Borreria podocephala IPNI=*767031-1* 357
Borreria prostrata IPNI=*766900-1* 357
Borreria psyllioides IPNI=*317930-2* 357
Borreria radicosa IPNI=*945476-1* 357
Borreria ramisparsa IPNI=*766900-1* 357
Borreria remota IPNI=*319724-2* 357
Borreria remotifolia IPNI=*756228-1* 352
Borreria rubrostipulata IPNI=*317930-2* 357
Borreria saturejifolia IPNI=*317930-2* 357
Borreria scabrida IPNI=*766800-1* 357
Borreria sideritis IPNI=*766800-1* 357
Borreria sphaerica IPNI=*317930-2* 357
Borreria spinosa IPNI=*77154343-1* 351
Borreria spinosa var. *latifolia* IPNI=*945476-1* 357
Borreria spinosa var. *minima* IPNI=*945476-1* 357
Borreria splitgerheri IPNI=*766800-1* 357
Borreria stricta IPNI=*766863-1* 357
Borreria stricta IPNI=*767031-1* 357
Borreria strictissima IPNI=*317930-2* 357
Borreria tampicana IPNI=*766900-1* 357
Borreria tenella var. *caesia* IPNI=*317930-2* 357
Borreria tenella var. *coriacea* IPNI=*317930-2* 357
Borreria tenella var. *lavandulifolia* IPNI=*317930-2* 357
Borreria tenella var. *macrocephala* IPNI=*317930-2* 357
Borreria tenella var. *pubescens* IPNI=*317930-2* 357
Borreria tenella var. *suffruticosa* IPNI=*317930-2* 357
Borreria tenella var. *tenera* IPNI=*317930-2* 357
Borreria tetraptera IPNI=*766800-1* 357
Borreria trichantha IPNI=*766863-1* 357
Borreria umbellata IPNI=*90501-2* 350
Borreria vaginata IPNI=*319724-2* 357
Borreria verticillata IPNI=*767031-1* 357
Borreria verticillata var. *brevifolia* IPNI=*767031-1* 357
Borreria verticillata var. *caesia* IPNI=*767031-1* 357
Borreria verticillata var. *thymiformis* IPNI=*767031-1* 357
Borreria virgata IPNI=*766862-1* 357
Borreria wydleriana IPNI=*319724-2* 357
Borzicactus aequatorialis IPNI=*35149-2* 283
Borzicactus roezlii IPNI=*35149-2* 283
Borzicactus sepium var. *roezlii* IPNI=*35149-2* 283
Bothriochloa anamitica IPNI=*392701-1* 235
Bothriochloa barbinodis var. *palmeri* IPNI=*35184-2* 235
Bothriochloa barbinodis var. *perforata* IPNI=*35184-2* 235
Bothriochloa barbinodis var. *schlumbergeri* IPNI=*35184-2* 235
Bothriochloa bladhii var. *punctata* IPNI=*392701-1* 235
Bothriochloa caucasica IPNI=*392701-1* 235
Bothriochloa emersa IPNI=*35184-2* 235
Bothriochloa haenkei IPNI=*392701-1* 235
Bothriochloa insculpta subsp. *panormitana* IPNI=*1191417-2* 235
Bothriochloa intermedia IPNI=*392701-1* 235
Bothriochloa inundata IPNI=*392701-1* 235
Bothriochloa ischaemum f. *songarica* IPNI=*1191417-2* 235
Bothriochloa ischaemum var. *songarica* IPNI=*1191417-2* 235
Bothriochloa leucopogon IPNI=*35184-2* 235
Bothriochloa nana IPNI=*65406-3* 235
Bothriochloa odorata IPNI=*392701-1* 235
Bothriochloa palmeri IPNI=*35184-2* 235
Bothriochloa panormitana IPNI=*1191417-2* 235
Bothriochloa perforata IPNI=*35184-2* 235
Bothriochloa piptanthera IPNI=*97916-2* 239
Bothriochloa punctata IPNI=*392701-1* 235
Bothriochloa schlumbergeri IPNI=*35184-2* 235
Bothriochloa taiwanensis IPNI=*1191417-2* 235
Botrychium lunarioides var. *daedaleum* IPNI=*35373-2* 173
Botrychium pusillum IPNI=*35345-2* 173
Botrychium schaffneri var. *pusillum* IPNI=*35345-2* 173
Botrychium schaffneri var. *typicum* IPNI=*35345-2* 173
Botrychium virginianum subsp. *typicum* IPNI=*1106719-2* 173
Botrychium virginianum var. *americanum* IPNI=*1106719-2* 173
Botrychium virginianum var. *typicum* IPNI=*1106719-2* 173
Botryodendrum capitatum IPNI=*30050876-2* 254
Botryopsis spruceana IPNI=*580301-1* 434
Botrypus virginianus IPNI=*1106719-2* 173
Botrypus virginicus IPNI=*1106719-2* 173
Botrys albus IPNI=*55993-2* 282
Botrys albus var. *pauper* IPNI=*55993-2* 282
Botrys ambrosioides IPNI=*1198060-2* 282
Botrys anthelminticus IPNI=*1090820-2* 282
Botrys ferulatus IPNI=*55993-2* 282
Botrys pagana IPNI=*55993-2* 282
Bouchea boyacana IPNI=*35430-2* 378
Bouchea ehrenbergii IPNI=*35430-2* 378
Bouchea prismatica var. *longirostra* IPNI=*35430-2* 378
Bougainvillea arborea IPNI=*35444-2* 286
Bougainvillea brachycarpa IPNI=*35444-2* 286
Bougainvillea bracteata IPNI=*604358-1* 286
Bougainvillea brasiliensis IPNI=*604358-1* 286
Bougainvillea glabra var. *acutibracteata* IPNI=*35444-2* 286
Bougainvillea glabra var. *alba* IPNI=*35444-2* 286
Bougainvillea glabra var. *brachycarpa* IPNI=*35444-2* 286
Bougainvillea glabra var. *graciliflora* IPNI=*35444-2* 286
Bougainvillea glabra var. *obtusibracteata* IPNI=*35444-2* 286
Bougainvillea glabra var. *sanderiana* IPNI=*35444-2* 286
Bougainvillea glabra var. *typica* IPNI=*35444-2* 286
Bougainvillea peruviana IPNI=*604358-1* 286
Bougainvillea rubicunda IPNI=*35444-2* 286
Bougainvillea sanderiana IPNI=*35444-2* 286
Bougainvillea speciosa IPNI=*604358-1* 286
Bougainvillea spectabilis var. *glabra* IPNI=*35444-2* 286

Bougainvillea spectabilis var. *hirsutissima* IPNI=*604358-1* 286
Bougainvillea spectabilis var. *parviflora* IPNI=*604358-1* 286
Bougainvillea spectabilis var. *typica* IPNI=*604358-1* 286
Bougainvillea spectabilis var. *virescens* IPNI=*604358-1* 286
Bougainvillea virescens IPNI=*604358-1* 286
Bourreria baccata IPNI=*1088317-2* 273
Bourreria clariuscula IPNI=*1088317-2* 273
Bourreria cumanensis IPNI=*77069564-1* 275
Bourreria domingensis IPNI=*1088317-2* 273
Bourreria grandiflora IPNI=*1088317-2* 273
Bourreria ovata IPNI=*1088317-2* 273
Bourreria ovata var. *hirtella* IPNI=*1088317-2* 273
Bourreria oxyphylla IPNI=*35505-2* 273
Bourreria recurva IPNI=*1088317-2* 273
Bourreria revoluta IPNI=*1088317-2* 273
Bourreria rigida IPNI=*1088317-2* 273
Bourreria succulenta f. *umbrosa* IPNI=*1088317-2* 273
Bourreria succulenta var. *canescens* IPNI=*1088317-2* 273
Bourreria succulenta var. *revoluta* IPNI=*1088317-2* 273
Bourreria tomentosa var. *velutina* IPNI=*1088317-2* 273
Bourreria velutina IPNI=*1088317-2* 273
Bourreria velutina var. *venosa* IPNI=*1088317-2* 273
Bourreria venosa IPNI=*1088317-2* 273
Bourreria veracruziana IPNI=*35505-2* 273
Bourreria viridis IPNI=*275053-2* 274
Bourreria wagneri IPNI=*35505-2* 273
Bourreria wrightii IPNI=*1088317-2* 273
Boussingaultia cordata IPNI=*14549-2* 283
Boussingaultia cordifolia IPNI=*14549-2* 283
Boussingaultia gracilis f. *pseudobaselloides* IPNI=*14549-2* 283
Boussingaultia gracilis f. *typica* IPNI=*14549-2* 283
Boussingaultia gracilis IPNI=*14549-2* 283
Boussingaultia gracilis var. *pseudobaselloides* IPNI=*14549-2* 283
Boussingaultia leptostachya IPNI=*1153747-2* 283
Bouteloua aristidoides var. *arizonica* IPNI=*1049662-2* 235
Bouteloua brachyathera IPNI=*328564-2* 235
Bouteloua ciliata IPNI=*1049662-2* 235
Bouteloua gracilis IPNI=*1049662-2* 235
Bouteloua humilis IPNI=*328564-2* 235
Bouteloua procumbens IPNI=*328564-2* 235
Bouteloua prostrata IPNI=*328564-2* 235
Bouteloua pusilla IPNI=*328564-2* 235
Bouteloua rahmeri IPNI=*328564-2* 235
Bouteloua simplex var. *actinochloides* IPNI=*328564-2* 235
Bouteloua tenuis IPNI=*328564-2* 235
Bouteloua tenuis var. *humilis* IPNI=*328564-2* 235
Bouvardia havanensis IPNI=*755756-1* 352
Bouvardia humboldtii IPNI=*745065-1* 347
Bouvardia longiflora var. *latifolia* IPNI=*745065-1* 347
Bouvardia uniflora IPNI=*755756-1* 352
Bowdichia brachypetala IPNI=*59774-2* 315
Bowdichia brevipes IPNI=*482144-1* 312
Bowdichia densiflora IPNI=*482144-1* 312
Bowdichia ferruginea IPNI=*482144-1* 312
Bowdichia floribunda IPNI=*482144-1* 312
Bowdichia kuhlmannii IPNI=*482144-1* 312
Bowdichia major IPNI=*482144-1* 312
Bowdichia major var. *fruticosa* IPNI=*482144-1* 312
Bowdichia martiusii IPNI=*492746-1* 318
Bowdichia parvifolia IPNI=*482144-1* 312
Bowdichia pubescens IPNI=*482144-1* 312
Bowdichia sebipira IPNI=*482144-1* 312
Bowdichia virgilioides f. *villosa* IPNI=*482144-1* 312
Bowdichia virgilioides var. *ferruginea* IPNI=*482144-1* 312
Bowdichia virgilioides var. *glabrata* IPNI=*482144-1* 312
Bowdichia virgilioides var. *pubescens* IPNI=*482144-1* 312
Bowdichia virgilioides var. *tomentosa* IPNI=*482144-1* 312
Brachiaria arrecta IPNI=*972082-1* 246
Brachiaria balansae IPNI=*426280-1* 246
Brachiaria bequaertii IPNI=*974586-1* 246
Brachiaria brizantha IPNI=*943736-1* 246
Brachiaria brizantha var. *ciliata* IPNI=*943736-1* 246
Brachiaria decumbens IPNI=*974586-1* 246
Brachiaria decumbens var. *ruziziensis* IPNI=*974586-1* 246
Brachiaria dictyoneura IPNI=*992521-1* 246
Brachiaria dictyoneura subsp. *humidicola* IPNI=*992521-1* 246
Brachiaria distachya IPNI=*426250-1* 246
Brachiaria eminii IPNI=*974586-1* 246
Brachiaria fasciculata IPNI=*1156554-2* 246
Brachiaria fasciculata var. *carthaginensis* IPNI=*1156554-2* 246
Brachiaria fasciculata var. *reticulata* IPNI=*1156554-2* 246
Brachiaria gangangalaensis IPNI=*943736-1* 246
Brachiaria humidicola IPNI=*992521-1* 246
Brachiaria hybrida IPNI=*426250-1* 246
Brachiaria keniensis IPNI=*992521-1* 246
Brachiaria latifolia IPNI=*972082-1* 246
Brachiaria longifolia IPNI=*399655-1* 238
Brachiaria manzonzoensis IPNI=*943736-1* 246
Brachiaria miliiformis IPNI=*943744-1* 246
Brachiaria mollis IPNI=*972086-1* 246
Brachiaria mutica IPNI=*426266-1* 246
Brachiaria numidiana IPNI=*426266-1* 246
Brachiaria obvoluta IPNI=*992521-1* 246
Brachiaria plantaginea IPNI=*286530-2* 246
Brachiaria prostrata IPNI=*426280-1* 246
Brachiaria purpurascens IPNI=*426266-1* 246
Brachiaria radicans IPNI=*972082-1* 246
Brachiaria rautanenii IPNI=*992521-1* 246
Brachiaria reptans IPNI=*426280-1* 246
Brachiaria reptans var. *hispida* IPNI=*426280-1* 246
Brachiaria ruziziensis IPNI=*974586-1* 246
Brachiaria subquadripara IPNI=*943744-1* 246
Brachiaria subquadripara var. *miliiformis* IPNI=*943744-1* 246
Brachiolobos palustris IPNI=*288669-1* 277
Brachychiton diversifolius var. *muelleri* IPNI=*822268-1* 407
Brachyloma erianthum IPNI=*381572-1* 367
Brachyloma hirsutum IPNI=*381572-1* 367
Brachyloma incurvum IPNI=*381638-1* 367
Brachyloma karstenianum IPNI=*381572-1* 367
Brachyloma leucomallon IPNI=*381572-1* 367
Brachyloma longiflorum IPNI=*381572-1* 367
Brachyloma longifolium IPNI=*381628-1* 367
Brachyloma molle IPNI=*381572-1* 367
Brachyloma moritzianum IPNI=*381572-1* 367
Brachyloma pictum IPNI=*381638-1* 367
Brachyloma pilosum IPNI=*381638-1* 367
Brachyloma rhodomallon IPNI=*381572-1* 367
Brachyloma rhynchocarpum IPNI=*381638-1* 367
Brachyloma rubricaule IPNI=*381572-1* 367
Brachyloma strictum IPNI=*381638-1* 367
Brachyloma tubiflora IPNI=*381638-1* 367
Brachyloma ventricosum IPNI=*381572-1* 367
Brachyloma vestitum IPNI=*381572-1* 367
Brachyotum canescens IPNI=*566862-1* 418
Brachyramphus caribaeus IPNI=*229022-1* 264
Brachyramphus goraeensis IPNI=*229022-1* 264
Brachyramphus intybaceus IPNI=*229022-1* 264
Brachyramphus remotiflorus IPNI=*229022-1* 264
Brachyramphus schimperi IPNI=*229022-1* 264
Brachyramphus sonchifolius IPNI=*229022-1* 264
Brachystachys hirta IPNI=*70337-2* 389
Bracteantha bracteata IPNI=*962307-1* 272
Bradburya acutifolia IPNI=*485540-1* 314
Bradburya biflora IPNI=*51897-2* 314
Bradburya brasiliana IPNI=*485555-1* 314
Bradburya cologanioides IPNI=*60457232-2* 311

Bradburya conjugata IPNI=*51897-2* 314
Bradburya decumbens IPNI=*51897-2* 314
Bradburya dubia IPNI=*485618-1* 314
Bradburya erecta IPNI=*51897-2* 314
Bradburya ferruginea IPNI=*485611-1* 314
Bradburya insulana IPNI=*485555-1* 314
Bradburya macrocarpa IPNI=*485597-1* 314
Bradburya plumieri IPNI=*485609-1* 314
Bradburya pubescens IPNI=*485611-1* 314
Bradburya sagittata IPNI=*485618-1* 314
Bradburya schiedeana IPNI=*485611-1* 314
Bradburya virginiana IPNI=*51897-2* 314
Bradburya virginiana var. *angustifolia* IPNI=*51897-2* 314
Bradburya virginiana var. *flavida* IPNI=*51897-2* 314
Bradleia sinica IPNI=*352077-1* 401
Brahea dulcis IPNI=*267788-2* 212
Brahea filamentosa IPNI=*267788-2* 212
Brahea filifera IPNI=*267788-2* 212
Brahea psilocalyx IPNI=*663912-1* 203
Brahea serrulata IPNI=*941782-1* 212
Bramia indica IPNI=*1072674-2* 376
Bramia monnieri IPNI=*1072674-2* 376
Brandesia lanceolata IPNI=*10378-2* 281
Brandesia martiana IPNI=*1024156-2* 281
Brandesia mexicana IPNI=*10378-2* 281
Brandesia porrigens IPNI=*10447-2* 281
Brandesia pubiflora IPNI=*10450-2* 281
Brandesia pubiflora var. *glomerata* IPNI=*10450-2* 281
Brasilettia ferruginea IPNI=*513021-1* 330
Brasilettia mollis IPNI=*67201-2* 315
Brasiliopuntia argentina IPNI=*129102-1* 283
Brasiliopuntia bahiensis IPNI=*129102-1* 283
Brasiliopuntia brasiliensis subsp. *bahiensis* IPNI=*129102-1* 283
Brasiliopuntia brasiliensis subsp. *subacarpa* IPNI=*129102-1* 283
Brasiliopuntia neoargentina IPNI=*129102-1* 283
Brasiliopuntia schulzii IPNI=*129102-1* 283
Brasiliopuntia subacarpa IPNI=*129102-1* 283
Brasilium glabrum IPNI=*813869-1* 433
Brassaia actinophylla IPNI=*77213393-1* 253
Brassaia singaporensis IPNI=*77213393-1* 253
Brassaia trianae IPNI=*230167-2* 254
Brassavola cuspidata IPNI=*618026-1* 217
Brassavola duckeana IPNI=*325518-2* 217
Brassavola gardneri IPNI=*325518-2* 217
Brassavola martiana var. *multiflora* IPNI=*325518-2* 217
Brassavola multiflora IPNI=*325518-2* 217
Brassavola odoratissima IPNI=*618026-1* 217
Brassica alba IPNI=*279435-1* 276
Brassica alba IPNI=*288952-1* 278
Brassica alboglabra IPNI=*279435-1* 276
Brassica amblyorhyncha IPNI=*999093-1* 276
Brassica amplexicaulis IPNI=*279485-1* 276
Brassica antiquorum IPNI=*279485-1* 276
Brassica arborea IPNI=*279435-1* 276
Brassica argyi IPNI=*60442520-2* 276
Brassica arvensis IPNI=*279485-1* 276
Brassica arvensis var. *juncea* IPNI=*60442520-2* 276
Brassica asperifolia IPNI=*279485-1* 276
Brassica besseriana IPNI=*60442520-2* 276
Brassica brachycarpa IPNI=*279422-1* 276
Brassica bracteolata IPNI=*279422-1* 276
Brassica brassicata IPNI=*279485-1* 276
Brassica briggsii IPNI=*279485-1* 276
Brassica bullata IPNI=*279435-1* 276
Brassica campestris f. *auriculata* IPNI=*279485-1* 276
Brassica campestris f. *tenuis* IPNI=*279485-1* 276
Brassica campestris IPNI=*279485-1* 276
Brassica campestris subsp. *annua* IPNI=*279419-1* 276
Brassica campestris subsp. *biennis* IPNI=*279419-1* 276
Brassica campestris subsp. *napobrassica* IPNI=*279419-1* 276
Brassica campestris subsp. *napus* IPNI=*279419-1* 276
Brassica campestris subsp. *rapifera* IPNI=*279419-1* 276
Brassica campestris subsp. *vulgaris* IPNI=*279485-1* 276
Brassica campestris var. *sarson* IPNI=*279485-1* 276
Brassica capitala IPNI=*279435-1* 276
Brassica cauliflora IPNI=*279435-1* 276
Brassica caulorapa IPNI=*279435-1* 276
Brassica celerifolia IPNI=*279485-1* 276
Brassica cephala IPNI=*279435-1* 276
Brassica chenopodiifolia IPNI=*60442520-2* 276
Brassica chinensis f. *aichi* IPNI=*279485-1* 276
Brassica chinensis f. *chifu* IPNI=*279485-1* 276
Brassica chinensis f. *hiberna* IPNI=*279485-1* 276
Brassica chinensis f. *hiroshimana* IPNI=*279485-1* 276
Brassica chinensis f. *honsaitai* IPNI=*279485-1* 276
Brassica chinensis f. *indivisa* IPNI=*279485-1* 276
Brassica chinensis f. *japonica* IPNI=*60442520-2* 276
Brassica chinensis f. *komatsuna* IPNI=*279485-1* 276
Brassica chinensis f. *mana* IPNI=*279485-1* 276
Brassica chinensis f. *nigra* IPNI=*279485-1* 276
Brassica chinensis f. *nozawensis* IPNI=*279485-1* 276
Brassica chinensis f. *osakana* IPNI=*279485-1* 276
Brassica chinensis f. *piaoh-tsai* IPNI=*279485-1* 276
Brassica chinensis f. *taito* IPNI=*279485-1* 276
Brassica chinensis f. *youngtungtsai* IPNI=*279485-1* 276
Brassica chinensis IPNI=*279485-1* 276
Brassica chinensis subf. *peking* IPNI=*279485-1* 276
Brassica chinensis subf. *shigatsuna* IPNI=*279485-1* 276
Brassica chinensis var. *dentata* IPNI=*279485-1* 276
Brassica chinensis var. *musifolia* IPNI=*279485-1* 276
Brassica chinensis var. *narinosa* IPNI=*279485-1* 276
Brassica chinensis var. *rosularis* IPNI=*279485-1* 276
Brassica colza IPNI=*279485-1* 276
Brassica cyrenaica IPNI=*279485-1* 276
Brassica dubiosa IPNI=*279485-1* 276
Brassica eruca IPNI=*1137625-2* 277
Brassica erucoides IPNI=*1137625-2* 277
Brassica fimbriata IPNI=*279435-1* 276
Brassica gemmifera IPNI=*279435-1* 276
Brassica gongylodes IPNI=*279419-1* 276
Brassica harmsiana IPNI=*279419-1* 276
Brassica hirta IPNI=*288952-1* 278
Brassica hispida IPNI=*1137625-2* 277
Brassica integrifolia IPNI=*60442520-2* 276
Brassica japonica IPNI=*279485-1* 276
Brassica japonica IPNI=*60442520-2* 276
Brassica juncea subsp. *napiformis* IPNI=*60442520-2* 276
Brassica juncea subsp. *tsatsai* IPNI=*60442520-2* 276
Brassica juncea var. *japonica* IPNI=*60442520-2* 276
Brassica laciniata IPNI=*279435-1* 276
Brassica lanceolata IPNI=*60442520-2* 276
Brassica lativalvis IPNI=*1137625-2* 277
Brassica lutea IPNI=*279485-1* 276
Brassica macrorhlza IPNI=*279419-1* 276
Brassica maritima IPNI=*279435-1* 276
Brassica mesopotamica IPNI=*999093-1* 276
Brassica millecapitata IPNI=*279435-1* 276
Brassica muscovita IPNI=*279435-1* 276
Brassica musifolia IPNI=*279485-1* 276
Brassica napella IPNI=*279485-1* 276
Brassica napiformis IPNI=*60442520-2* 276
Brassica napobrassica IPNI=*279419-1* 276
Brassica napus convar. *pabularia* IPNI=*279419-1* 276
Brassica napus convar. *sibirica* IPNI=*279419-1* 276
Brassica napus subsp. *oleifera* IPNI=*279419-1* 276
Brassica napus subsp. *rapifera* IPNI=*279419-1* 276

Brassica napus var. *schebalinae* IPNI=*279419-1* 276
Brassica narinosa IPNI=*279485-1* 276
Brassica nigra subsp. *hispida* IPNI=*279422-1* 276
Brassica nigra var. *subglabra* IPNI=*279422-1* 276
Brassica nipposinica IPNI=*279485-1* 276
Brassica nipposinica var. *laciniifolia* IPNI=*279485-1* 276
Brassica odorata IPNI=*279435-1* 276
Brassica oleifera IPNI=*279419-1* 276
Brassica oleracea convar. *costata* IPNI=*279435-1* 276
Brassica oleracea convar. *gemmifera* IPNI=*279435-1* 276
Brassica oleracea subsp. *acephala* IPNI=*279435-1* 276
Brassica oleracea subsp. *bullata* IPNI=*279435-1* 276
Brassica oleracea subsp. *caulorapa* IPNI=*279435-1* 276
Brassica oleracea subsp. *fruticosa* IPNI=*279435-1* 276
Brassica oleracea subsp. *laevis* IPNI=*279435-1* 276
Brassica oleracea var. *capitata* IPNI=*279435-1* 276
Brassica oleracea var. *costata* IPNI=*279435-1* 276
Brassica oleracea var. *gemmifera* IPNI=*279435-1* 276
Brassica oleracea var. *gongylodes* IPNI=*279435-1* 276
Brassica oleracea var. *kashmiriana* IPNI=*279435-1* 276
Brassica oleracea var. *laciniata* IPNI=*279435-1* 276
Brassica oleracea var. *napobrassica* IPNI=*279419-1* 276
Brassica oleracea var. *palmifolia* IPNI=*279435-1* 276
Brassica oleracea var. *rubra* IPNI=*279435-1* 276
Brassica oleracea var. *sabauda* IPNI=*279435-1* 276
Brassica oleracea var. *sabellica* IPNI=*279435-1* 276
Brassica oleracea var. *viridis* IPNI=*279435-1* 276
Brassica oleronensis IPNI=*279485-1* 276
Brassica parachinensis IPNI=*279485-1* 276
Brassica pekinensis var. *cephalata* IPNI=*279485-1* 276
Brassica pekinensis var. *cylindrica* IPNI=*279485-1* 276
Brassica pekinensis var. *laxa* IPNI=*279485-1* 276
Brassica peregrina IPNI=*279435-1* 276
Brassica perfoliata IPNI=*279485-1* 276
Brassica persoonii IPNI=*279422-1* 276
Brassica perviridis IPNI=*279485-1* 276
Brassica polymorpha IPNI=*279485-1* 276
Brassica praecox IPNI=*279419-1* 276
Brassica pseudocolza IPNI=*279485-1* 276
Brassica quadrivalvis IPNI=*279419-1* 276
Brassica quercifolia IPNI=*279435-1* 276
Brassica rapa convar. *cephalata* IPNI=*279485-1* 276
Brassica rapa convar. *cylindrica* IPNI=*279485-1* 276
Brassica rapa convar. *europea* IPNI=*279485-1* 276
Brassica rapa convar. *laxa* IPNI=*279485-1* 276
Brassica rapa convar. *parachinensis* IPNI=*279485-1* 276
Brassica rapa convar. *rosularis* IPNI=*279485-1* 276
Brassica rapa f. *glabra* IPNI=*279485-1* 276
Brassica rapa subf. *depressa* IPNI=*279485-1* 276
Brassica rapa subf. *globosa* IPNI=*279485-1* 276
Brassica rapa subf. *oblonga* IPNI=*279485-1* 276
Brassica rapa subsp. *afghanica* IPNI=*279485-1* 276
Brassica rapa subsp. *campestris* IPNI=*279485-1* 276
Brassica rapa subsp. *chinensis* IPNI=*279485-1* 276
Brassica rapa subsp. *dichotoma* IPNI=*279485-1* 276
Brassica rapa subsp. *japonica* IPNI=*279485-1* 276
Brassica rapa subsp. *narinosa* IPNI=*279485-1* 276
Brassica rapa subsp. *nipposinica* IPNI=*279485-1* 276
Brassica rapa subsp. *oleifera* IPNI=*279485-1* 276
Brassica rapa subsp. *pekinensis* IPNI=*279485-1* 276
Brassica rapa subsp. *rapifera* IPNI=*279419-1* 276
Brassica rapa subsp. *sarson* IPNI=*279485-1* 276
Brassica rapa subsp. *sylvestris* IPNI=*279485-1* 276
Brassica rapa subsp. *trilocularis* IPNI=*279485-1* 276
Brassica rapa subvar. *pe-tsai* IPNI=*279485-1* 276
Brassica rapa var. *alborosea* IPNI=*279485-1* 276
Brassica rapa var. *amplexicaulis* IPNI=*279485-1* 276
Brassica rapa var. *cephalata* IPNI=*279485-1* 276
Brassica rapa var. *communis* IPNI=*279485-1* 276
Brassica rapa var. *cylindrica* IPNI=*279485-1* 276
Brassica rapa var. *dichotoma* IPNI=*279485-1* 276
Brassica rapa var. *dissecta* IPNI=*279485-1* 276
Brassica rapa var. *hybrida* IPNI=*279485-1* 276
Brassica rapa var. *intermedia* IPNI=*279485-1* 276
Brassica rapa var. *laxa* IPNI=*279485-1* 276
Brassica rapa var. *lorifolia* IPNI=*279485-1* 276
Brassica rapa var. *narinosa* IPNI=*279485-1* 276
Brassica rapa var. *nipposinica* IPNI=*279485-1* 276
Brassica rapa var. *parachinensis* IPNI=*279485-1* 276
Brassica rapa var. *pekinensis* IPNI=*279485-1* 276
Brassica rapa var. *perviridis* IPNI=*279485-1* 276
Brassica rapa var. *rosularis* IPNI=*279485-1* 276
Brassica rapa var. *rubra* IPNI=*279485-1* 276
Brassica rapa var. *sarson* IPNI=*279485-1* 276
Brassica rapa var. *septiceps* IPNI=*279485-1* 276
Brassica rapa var. *sylvestris* IPNI=*279485-1* 276
Brassica rapa var. *violascens* IPNI=*279485-1* 276
Brassica richeri IPNI=*60442520-2* 276
Brassica rubra IPNI=*279435-1* 276
Brassica rugosa IPNI=*60442520-2* 276
Brassica sabellica IPNI=*279435-1* 276
Brassica saruna IPNI=*279485-1* 276
Brassica sativa IPNI=*279485-1* 276
Brassica sativa subsp. *napus* IPNI=*279419-1* 276
Brassica sativa var. *campestris* IPNI=*279485-1* 276
Brassica sativa var. *rapa* IPNI=*279485-1* 276
Brassica septiceps IPNI=*279485-1* 276
Brassica sinapioides IPNI=*279422-1* 276
Brassica sinapis IPNI=*279422-1* 276
Brassica sisymbrioides IPNI=*999093-1* 276
Brassica sphaerorhiza IPNI=*279485-1* 276
Brassica stocksii IPNI=*999093-1* 276
Brassica stricta IPNI=*279419-1* 276
Brassica suttoniana IPNI=*279435-1* 276
Brassica sylvestris IPNI=*279435-1* 276
Brassica taquetii IPNI=*60442520-2* 276
Brassica tournefortii IPNI=*999093-1* 276
Brassica tournefortii var. *sisymbrioides* IPNI=*999093-1* 276
Brassica trilocularis IPNI=*279485-1* 276
Brassica trimestris IPNI=*279485-1* 276
Brassica tuberosa IPNI=*279485-1* 276
Brassica turgida IPNI=*1137625-2* 277
Brassica turgida IPNI=*279422-1* 276
Brassica uechtritziana IPNI=*1137625-2* 277
Brassica urbaniana IPNI=*60442520-2* 276
Brassica vesicaria IPNI=*1137625-2* 277
Brassica willdenovii IPNI=*60442520-2* 276
Brathys acerosa IPNI=*433547-1* 394
Brathys juniperina IPNI=*433524-1* 393
Brathys laricifolia IPNI=*433547-1* 394
Brathys mutisiana IPNI=*428535-1* 394
Bravaisia floribunda IPNI=*36223-2* 359
Bravaisia integerrima var. *pilosa* IPNI=*36223-2* 359
Brayulinea australis IPNI=*1183159-2* 282
Brayulinea densa IPNI=*1183159-2* 282
Bredemeyera floribunda f. *elliptica* IPNI=*690122-1* 337
Bredemeyera floribunda f. *subvestita* IPNI=*690122-1* 337
Bredemeyera floribunda var. *puberula* IPNI=*690122-1* 337
Bredemeyera lhotzkyana IPNI=*690122-1* 337
Breynia axillaris IPNI=*340073-1* 401
Breynia disticha f. *nivosa* IPNI=*340073-1* 401
Breynia indica IPNI=*77217338-1* 278
Breynia nivosa IPNI=*340073-1* 401
Brickellia microphylla IPNI=*183814-1* 258
Brignolia acuminata IPNI=*753641-1* 352
Brignolia pubigera IPNI=*753641-1* 352

Briquetiastrum spicatum IPNI=*36556-2* 407
Brittonamra caribaea IPNI=*275194-2* 315
Briza aspera IPNI=*393299-1* 235
Briza capensis IPNI=*393293-1* 235
Briza dalmatica IPNI=*393293-1* 235
Briza deltoidea IPNI=*393299-1* 235
Briza dura IPNI=*322530-2* 239
Briza elegans IPNI=*400813-1* 238
Briza eragrostis IPNI=*1104599-2* 238
Briza gracilescens IPNI=*393293-1* 235
Briza gracilis IPNI=*393299-1* 235
Briza grandis IPNI=*393293-1* 235
Briza lilloi IPNI=*77116853-1* 236
Briza major IPNI=*393293-1* 235
Briza mandoniana IPNI=*77116853-1* 236
Briza mandoniana var. *herzogiana* IPNI=*77116853-1* 236
Briza mandoniana var. *tuberculata* IPNI=*77116853-1* 236
Briza mandoniana var. *vallegrandensis* IPNI=*77116853-1* 236
Briza megastachya IPNI=*1104599-2* 238
Briza microclada IPNI=*393293-1* 235
Briza minima IPNI=*393299-1* 235
Briza minor f. *purpurascens* IPNI=*393299-1* 235
Briza monandra IPNI=*77116853-1* 236
Briza monandra var. *condensata* IPNI=*77116853-1* 236
Briza monspessulana IPNI=*393293-1* 235
Briza montana IPNI=*393293-1* 235
Briza oblonga IPNI=*1104599-2* 238
Briza portenschlagii IPNI=*393293-1* 235
Briza pulchella IPNI=*393293-1* 235
Briza purpurascens IPNI=*1104599-2* 238
Briza rubens IPNI=*393293-1* 235
Briza rubra IPNI=*393293-1* 235
Briza spicata IPNI=*82343-2* 238
Briza toluccensis IPNI=*258855-2* 246
Briza trichotoma IPNI=*393299-1* 235
Briza virens IPNI=*393299-1* 235
Brizopyrum americanum IPNI=*82343-2* 238
Brizopyrum boreale IPNI=*82343-2* 238
Brizopyrum ovatum IPNI=*82343-2* 238
Brizopyrum prostratum IPNI=*82343-2* 238
Brizopyrum prostratum var. *erectum* IPNI=*82343-2* 238
Brizopyrum prostratum var. *humile* IPNI=*82343-2* 238
Brizopyrum spicatum IPNI=*82343-2* 238
Bromelia acanga IPNI=*328361-2* 228
Bromelia acaulis IPNI=*328361-2* 228
Bromelia argentina IPNI=*36836-2* 228
Bromelia aurea IPNI=*122289-1* 228
Bromelia balansae var. *tricolor* IPNI=*36836-2* 228
Bromelia bracteata IPNI=*121570-1* 227
Bromelia caratas IPNI=*328361-2* 228
Bromelia carolinae IPNI=*169003-2* 229
Bromelia comosa IPNI=*12322-2* 228
Bromelia fastuosa IPNI=*319113-2* 228
Bromelia ignea IPNI=*319113-2* 228
Bromelia magdalenae IPNI=*5516-2* 228
Bromelia nudicaulis IPNI=*121752-1* 228
Bromelia plumieri IPNI=*328361-2* 228
Bromelia sepiaria IPNI=*319113-2* 228
Bromelia tricolor IPNI=*115178-2* 229
Bromus ambiguus IPNI=*402472-1* 239
Bromus angustatus IPNI=*37001-2* 235
Bromus arundinaceus IPNI=*973267-1* 241
Bromus badensis IPNI=*394095-1* 235
Bromus barbatoides IPNI=*393559-1* 235
Bromus barbatoides var. *sulcatus* IPNI=*393559-1* 235
Bromus bicuspis IPNI=*393559-1* 235
Bromus bohemicus IPNI=*403134-1* 239
Bromus bolivianus IPNI=*37001-2* 235
Bromus breviaristatus IPNI=*37001-2* 235
Bromus brevisetus IPNI=*394095-1* 235
Bromus brongniartii IPNI=*37001-2* 235
Bromus buchtienii IPNI=*393995-1* 235
Bromus carinatus var. *californicus* IPNI=*30065892-2* 235
Bromus carinatus var. *densus* IPNI=*30065892-2* 235
Bromus carinatus var. *hookerianus* IPNI=*30065892-2* 235
Bromus carinatus var. *linearis* IPNI=*30065892-2* 235
Bromus commutatus subsp. *decipiens* IPNI=*394095-1* 235
Bromus compressus IPNI=*30065892-2* 235
Bromus decipiens IPNI=*394095-1* 235
Bromus dertonensis IPNI=*402472-1* 239
Bromus ehrhartii IPNI=*394095-1* 235
Bromus elongatus IPNI=*394095-1* 235
Bromus glaber IPNI=*393995-1* 235
Bromus glomeratus IPNI=*397309-1* 236
Bromus haenkeanus IPNI=*37001-2* 235
Bromus hookeri var. *pendulinus* IPNI=*30065892-2* 235
Bromus hookeri var. *schaffneri* IPNI=*30065892-2* 235
Bromus hookeri var. *schlechtendalii* IPNI=*30065892-2* 235
Bromus hookerianus IPNI=*30065892-2* 235
Bromus hordeaceus IPNI=*394095-1* 235
Bromus hybridus IPNI=*394095-1* 235
Bromus laciniatus IPNI=*30065892-2* 235
Bromus lenis IPNI=*393995-1* 235
Bromus leyboldtii IPNI=*393559-1* 235
Bromus luzonensis IPNI=*30065892-2* 235
Bromus matthewsii IPNI=*37001-2* 235
Bromus mucronatus IPNI=*37001-2* 235
Bromus multiflorus IPNI=*394095-1* 235
Bromus mutabilis IPNI=*394095-1* 235
Bromus mutabilis var. *hirtus* IPNI=*394095-1* 235
Bromus pendulinus IPNI=*30065892-2* 235
Bromus pitensis var. *latifolius* IPNI=*393995-1* 235
Bromus polymorphus IPNI=*394095-1* 235
Bromus polystachios IPNI=*60458632-2* 238
Bromus preslii IPNI=*37001-2* 235
Bromus proximus var. *schlechtendalii* IPNI=*30065892-2* 235
Bromus schaffneri IPNI=*30065892-2* 235
Bromus schraderi IPNI=*37001-2* 235
Bromus secalinus f. *divergens* IPNI=*394095-1* 235
Bromus secalinus f. *hirtus* IPNI=*394095-1* 235
Bromus secalinus subsp. *decipiens* IPNI=*394095-1* 235
Bromus secalinus subsp. *infestus* IPNI=*394095-1* 235
Bromus secundus IPNI=*30010036-2* 239
Bromus sitchensis var. *carinatus* IPNI=*30065892-2* 235
Bromus strictus IPNI=*37001-2* 235
Bromus submuticus IPNI=*394095-1* 235
Bromus subvelutinus IPNI=*30065892-2* 235
Bromus tacna IPNI=*37001-2* 235
Bromus temulentus IPNI=*407516-1* 241
Bromus tenuis IPNI=*393995-1* 235
Bromus trinii IPNI=*393559-1* 235
Bromus trinii var. *effusus* IPNI=*393559-1* 235
Bromus trinii var. *excelsus* IPNI=*393559-1* 235
Bromus trinii var. *manicatus* IPNI=*393559-1* 235
Bromus trinii var. *micrantherus* IPNI=*393559-1* 235
Bromus trinii var. *pallidiflorus* IPNI=*393559-1* 235
Bromus trinii var. *strictus* IPNI=*393559-1* 235
Bromus unioloides f. *chasmogamus* IPNI=*37001-2* 235
Bromus unioloides f. *oliveus* IPNI=*37001-2* 235
Bromus unioloides IPNI=*37001-2* 235
Bromus unioloides subf. *achalensis* IPNI=*37001-2* 235
Bromus unioloides var. *haenkeanus* IPNI=*37001-2* 235
Bromus unioloides var. *humilis* IPNI=*37001-2* 235
Bromus unioloides var. *micranthus* IPNI=*37001-2* 235
Bromus unioloides var. *montanus* IPNI=*37001-2* 235
Bromus unioloides var. *multiflorus* IPNI=*37001-2* 235

Bromus unioloides var. *sanjuaninus* IPNI=*37001-2* 235
Bromus virens IPNI=*30065892-2* 235
Bromus vitiosus IPNI=*394095-1* 235
Bromus willdenowii IPNI=*37001-2* 235
Bromus willdenowii var. *haenkeanus* IPNI=*37001-2* 235
Bromus willdenowii var. *lasiophyllus* IPNI=*37001-2* 235
Brosimopsis acutifolia IPNI=*37254-2* 437
Brosimopsis amplifolia IPNI=*37286-2* 437
Brosimopsis diandra IPNI=*37286-2* 437
Brosimopsis lactescens IPNI=*37286-2* 437
Brosimopsis oblongifolia IPNI=*37286-2* 437
Brosimum angustifolium IPNI=*37310-2* 437
Brosimum aubletii IPNI=*582693-1* 437
Brosimum belizense IPNI=*37286-2* 437
Brosimum brevipedunculatum IPNI=*37310-2* 437
Brosimum caloxylon IPNI=*37310-2* 437
Brosimum conzattii IPNI=*850824-1* 437
Brosimum discolor IPNI=*582693-1* 437
Brosimum echinocarpum IPNI=*111855-1* 380
Brosimum foetidum IPNI=*37291-2* 437
Brosimum galactodendron IPNI=*582724-1* 437
Brosimum gentlei IPNI=*850824-1* 437
Brosimum humboldtii IPNI=*582724-1* 437
Brosimum lanciferum IPNI=*37310-2* 437
Brosimum lecointei IPNI=*582693-1* 437
Brosimum longistipulatum IPNI=*37310-2* 437
Brosimum mello-barretoi IPNI=*1182949-2* 440
Brosimum myristicoides IPNI=*37305-2* 437
Brosimum ojoche IPNI=*37286-2* 437
Brosimum palmarum IPNI=*582693-1* 437
Brosimum panamense IPNI=*582693-1* 437
Brosimum paraense IPNI=*37310-2* 437
Brosimum platyneurum IPNI=*37310-2* 437
Brosimum refractum IPNI=*582724-1* 437
Brosimum rotundatum IPNI=*582693-1* 437
Brosimum terrabanum IPNI=*850824-1* 437
Brosimum tessmannii IPNI=*582693-1* 437
Brosimum utile subsp. *longifolium* IPNI=*37291-2* 437
Brosimum velutinum IPNI=*582693-1* 437
Brossaea anastomosans IPNI=*330459-1* 299
Brossaea brachybotrys IPNI=*107641-2* 299
Brossaea bracteata var. *cordifolia* IPNI=*330527-1* 299
Brossaea bracteata var. *odorata* IPNI=*330527-1* 299
Brossaea conferta IPNI=*330459-1* 299
Brossaea erecta IPNI=*330527-1* 299
Brossaea foliolosa IPNI=*330535-1* 299
Brossaea formosa IPNI=*330527-1* 299
Brossaea hirtiflora IPNI=*330527-1* 299
Brossaea insipida IPNI=*330570-1* 299
Brossaea lanigera IPNI=*330584-1* 299
Brossaea loxensis IPNI=*330527-1* 299
Brossaea odorata IPNI=*330527-1* 299
Brossaea pichinchensis IPNI=*107641-2* 299
Brossaea ramosissima IPNI=*330459-1* 299
Brossaea rigida IPNI=*330676-1* 299
Brossaea roraimae IPNI=*330527-1* 299
Brossaea scabra IPNI=*330527-1* 299
Brossaea shallon IPNI=*317388-2* 299
Brossaea trichocalycina IPNI=*330527-1* 299
Brossaea vestita IPNI=*330527-1* 299
Brotera maritima IPNI=*834759-1* 410
Brotera persica IPNI=*158242-2* 371
Brotera phoenicea IPNI=*824676-1* 412
Brotobroma asperum IPNI=*824026-1* 409
Broughtonia alba IPNI=*643250-1* 220
Broughtonia grandiflora IPNI=*643478-1* 220
Broussonetia plumeri IPNI=*1151750-2* 440
Broussonetia tinctoria IPNI=*1151750-2* 440
Broussonetia zanthoxylon IPNI=*1151750-2* 440
Browallia alienata IPNI=*54389-1* 361
Brownea antioquensis IPNI=*482431-1* 312
Brownea ariza IPNI=*482418-1* 312
Brownea coccinea IPNI=*482440-1* 312
Brownea excelsa IPNI=*37425-2* 312
Brownea herthae IPNI=*37413-2* 312
Brownea holtonii IPNI=*482418-1* 312
Brownea loretensis IPNI=*482418-1* 312
Brownea macbrideana IPNI=*482431-1* 312
Brownea princeps IPNI=*482418-1* 312
Brownea rosa IPNI=*482440-1* 312
Brownea rosea IPNI=*482440-1* 312
Brownea speciosa IPNI=*482440-1* 312
Bruguiera decangulata IPNI=*30061148-2* 402
Bruinsmania isertioides IPNI=*753641-1* 352
Brunellia antioquensis IPNI=*37499-2* 430
Brunellia colombiana IPNI=*813640-1* 430
Brunellia coroicoana IPNI=*37499-2* 430
Brunellia sibundoya subsp. *antioquensis* IPNI=*37499-2* 430
Brunellia sibundoya subsp. *sebastopola* IPNI=*37499-2* 430
Brunyera biennis IPNI=*172755-2* 429
Bryodesma sellowii IPNI=*232292-2* 172
Bryonia acerifolia IPNI=*291972-1* 293
Bryonia alceifolia IPNI=*291972-1* 293
Bryonia barbata IPNI=*291972-1* 293
Bryonia cheirophylla IPNI=*293060-1* 294
Bryonia collosa IPNI=*292238-1* 293
Bryonia convolvulifolia IPNI=*293293-1* 294
Bryonia filiformis IPNI=*293293-1* 294
Bryonia glandulosa IPNI=*291801-1* 293
Bryonia grandis IPNI=*291972-1* 293
Bryonia guadalupensis IPNI=*293293-1* 294
Bryonia latebrosa IPNI=*291845-1* 293
Bryonia melothria IPNI=*293293-1* 294
Bryonia moimoi IPNI=*291972-1* 293
Bryonia punctata IPNI=*104422-2* 294
Bryonia racemosa IPNI=*291845-1* 293
Bryonia sinuosa IPNI=*291972-1* 293
Bubroma grandiflorum IPNI=*825580-1* 414
Bubroma guasuma IPNI=*823393-1* 408
Bubroma polybotryum IPNI=*823393-1* 408
Bubroma tomentosum IPNI=*823393-1* 408
Bubroma ulmifolia IPNI=*823393-1* 408
Buccaferrea crispata IPNI=*323116-2* 203
Buccaferrea maritima IPNI=*224565-2* 203
Buccaferrea pectinata IPNI=*77099639-1* 203
Bucephalon racemosum IPNI=*260203-2* 441
Buceras angustifolium IPNI=*250537-2* 416
Buceras bucida IPNI=*171012-1* 416
Buceras catappa IPNI=*171034-1* 416
Buchenavia capitata IPNI=*77164595-1* 416
Buchenavia grandis IPNI=*77164566-1* 416
Buchenavia huberi IPNI=*77164566-1* 416
Buchenavia macrophylla IPNI=*77164571-1* 416
Buchenavia oxycarpa IPNI=*171300-1* 416
Buchenavia parvifolia IPNI=*77164576-1* 416
Buchenavia ptariensis IPNI=*77164595-1* 416
Buchenavia stellae IPNI=*77164571-1* 416
Buchenavia tetraphylla IPNI=*77164595-1* 416
Buchenavia vaupesana IPNI=*77164595-1* 416
Buchnera capillaris IPNI=*1189750-2* 374
Buchnera coccinea IPNI=*2352-2* 365
Buchnera elongata f. *glabriuscula* IPNI=*800009-1* 375
Buchnera elongata f. *parviflora* IPNI=*800009-1* 375
Buchnera elongata f. *pilosior* IPNI=*800009-1* 375
Buchnera elongata f. *rigidior* IPNI=*800009-1* 375
Buchnera elongata var. *obtusa* IPNI=*800009-1* 375

Buchnera grandiflora IPNI=*97540-2* 375
Buchnera macrocarpa IPNI=*800009-1* 375
Buchnera tinctoria IPNI=*800009-1* 375
Bucholzia brachiata IPNI=*1078972-2* 281
Bucholzia ficoidea IPNI=*1078972-2* 281
Bucholzia frutescens IPNI=*10362-2* 281
Bucholzia philoxeroides IPNI=*1024156-2* 281
Bucholzia polygonoides IPNI=*1069792-2* 281
Bucholzia polygonoides var. *diffusa* IPNI=*327343-2* 281
Bucholzia polygonoides var. *erecta* IPNI=*1078972-2* 281
Bucholzia polygonoides var. *radicans* IPNI=*1078972-2* 281
Bucholzia spinescens IPNI=*327342-2* 281
Buchosia aquatica IPNI=*698080-1* 225
Bucida angustifolia IPNI=*250537-2* 416
Bucida angustifolia IPNI=*77164595-1* 416
Bucida buceras IPNI=*171012-1* 416
Bucida buceras var. *angustifolia* IPNI=*250537-2* 416
Bucida capitata IPNI=*77164595-1* 416
Bucida ophiticola IPNI=*171012-1* 416
Bucida palustris IPNI=*171012-1* 416
Bucida subinermis IPNI=*171012-1* 416
Bucquetia glutinosa var. *rosea* IPNI=*566936-1* 418
Buda borealis IPNI=*158983-1* 286
Buda campestris IPNI=*158983-1* 286
Buda marina IPNI=*241336-2* 286
Buda marina var. *minor* IPNI=*241336-2* 286
Buda media var. *marina* IPNI=*241336-2* 286
Buda rubra IPNI=*158983-1* 286
Buda rupestris IPNI=*159015-1* 286
Buddleja americana IPNI=*319122-2* 377
Buddleja americana var. *albiflora* IPNI=*319122-2* 377
Buddleja americana var. *rothschuhii* IPNI=*319122-2* 377
Buddleja callicarpoides IPNI=*319122-2* 377
Buddleja dentata IPNI=*319122-2* 377
Buddleja floribunda IPNI=*319122-2* 377
Buddleja incana var. *pannulosa* IPNI=*545779-1* 377
Buddleja occidentalis IPNI=*319122-2* 377
Buddleja rondeletiiflora IPNI=*545779-1* 377
Buddleja rugosa IPNI=*545779-1* 377
Buddleja shaanxiensis IPNI=*545734-1* 377
Buddleja shimidzuana IPNI=*545734-1* 377
Buddleja spicata IPNI=*319122-2* 377
Buddleja striata IPNI=*545734-1* 377
Buddleja striata var. *zhouquensis* IPNI=*545734-1* 377
Buddleja variabilis IPNI=*545734-1* 377
Buddleja verbasci IPNI=*319122-2* 377
Buddleja verbascifolia IPNI=*319122-2* 377
Buena carua IPNI=*754813-1* 352
Buena crassifolia IPNI=*1215741-2* 352
Buena hookeriana IPNI=*134610-2* 352
Buena latifolia IPNI=*66725-2* 349
Buena magnifolia IPNI=*1106418-2* 352
Buena muzonensis IPNI=*134610-2* 352
Buena nitida IPNI=*1106418-2* 352
Buena obtusifolia IPNI=*66725-2* 349
Buena panamensis IPNI=*30107991-2* 351
Buena skinneri IPNI=*66725-2* 349
Buena triflora IPNI=*66725-2* 349
Bufonia rotundifolia IPNI=*30172747-2* 285
Bugula alpina IPNI=*444576-1* 368
Bugula decumbens IPNI=*444664-1* 368
Bugula genevensis IPNI=*444576-1* 368
Bugula pyramidalis IPNI=*444657-1* 368
Bugula reptans IPNI=*444664-1* 368
Bugula tomentosa IPNI=*444576-1* 368
Bulbine uncinata IPNI=*319172-2* 213
Bulbostylis brasiliensis IPNI=*38049-2* 229
Bulbostylis capillaris var. *genuina* IPNI=*1186511-2* 229
Bulbostylis capitata IPNI=*1029981-2* 232
Bulbostylis elegans IPNI=*27442-2* 257
Bulbostylis geniculatus IPNI=*1029981-2* 232
Bulbostylis glandulosa IPNI=*27442-2* 257
Bulbostylis junciformis var. *congesta* IPNI=*38049-2* 229
Bulbostylis junciformis var. *fulvicoma* IPNI=*38049-2* 229
Bulbostylis junciformis var. *laxiuscula* IPNI=*38049-2* 229
Bulbostylis lanata var. *xyrioides* IPNI=*30060098-2* 229
Bulbostylis micrantha IPNI=*27442-2* 257
Bulbostylis microcephala IPNI=*27442-2* 257
Bulbostylis nepetifolia IPNI=*251582-1* 270
Bulbostylis tomentosa IPNI=*27442-2* 257
Bulga pyramidalis IPNI=*444657-1* 368
Bumelia affinis IPNI=*948032-1* 308
Bumelia amazonica IPNI=*209738-2* 306
Bumelia angustifolia IPNI=*948032-1* 308
Bumelia celastrina IPNI=*948032-1* 308
Bumelia celastrina var. *angustifolia* IPNI=*948032-1* 308
Bumelia egensis IPNI=*160374-2* 306
Bumelia eggersii IPNI=*948032-1* 308
Bumelia ferox IPNI=*948032-1* 308
Bumelia hayesii IPNI=*948032-1* 308
Bumelia megacocca IPNI=*948032-1* 308
Bumelia nervosa IPNI=*209808-2* 307
Bumelia obtusifolia IPNI=*1008737-2* 308
Bumelia obtusifolia subsp. *typica* IPNI=*1008737-2* 308
Bumelia parvifolia IPNI=*948032-1* 308
Bumelia persimilis IPNI=*948021-1* 308
Bumelia persimilis subsp. *typica* IPNI=*948021-1* 308
Bumelia reclinata IPNI=*948032-1* 308
Bumelia schottii IPNI=*948032-1* 308
Bumelia spiniflora IPNI=*948032-1* 308
Bumelia spinosa IPNI=*948032-1* 308
Bunchosia armeniaca f. *parvifolia* IPNI=*555884-1* 395
Bunchosia armeniaca f. *systyla* IPNI=*555884-1* 395
Bunchosia cornifolia IPNI=*1122895-2* 395
Bunchosia emarginata var. *martinicensis* IPNI=*555908-1* 395
Bunchosia gentlei IPNI=*303745-2* 395
Bunchosia glauca IPNI=*1122895-2* 395
Bunchosia hypoleuca IPNI=*555908-1* 395
Bunchosia lanceolata f. *leiocarpa* IPNI=*303745-2* 395
Bunchosia lanceolata IPNI=*303745-2* 395
Bunchosia leonis f. *rotundata* IPNI=*555969-1* 395
Bunchosia lindeniana var. *boliviensis* IPNI=*303745-2* 395
Bunchosia lindeniana var. *mexicana* IPNI=*303745-2* 395
Bunchosia macrophylla var. *chocoensis* IPNI=*555944-1* 395
Bunchosia martinicensis IPNI=*555908-1* 395
Bunchosia media f. *bicarpellaris* IPNI=*1122895-2* 395
Bunchosia mollis IPNI=*555944-1* 395
Bunchosia nitida var. *domingensis* IPNI=*1122895-2* 395
Bunchosia pilosa IPNI=*555944-1* 395
Bunchosia pilosa var. *retusa* IPNI=*555944-1* 395
Bunchosia pringlei IPNI=*303745-2* 395
Bunchosia retusa IPNI=*555944-1* 395
Bunchosia sessilifolia IPNI=*555944-1* 395
Bunchosia silvestris IPNI=*557670-1* 396
Bunchosia squarrosa IPNI=*557785-1* 396
Bunchosia swartziana var. *yucatanensis* IPNI=*555969-1* 395
Bunias perennis IPNI=*1061021-2* 277
Bunias raphanifolia IPNI=*1061021-2* 277
Bunias verrucosa IPNI=*1061021-2* 277
Bupariti altissima IPNI=*564713-1* 415
Bupariti populnea IPNI=*564713-1* 415
Buphthalmum diffusum IPNI=*1186403-2* 261
Buphthalmum helianthoides IPNI=*213534-1* 264
Buphthalmum heterophyllum IPNI=*1093589-2* 270
Buphthalmum oleraceum IPNI=*1013480-1* 263
Buphthalmum procumbens IPNI=*1093589-2* 270

Buphthalmum repens IPNI=*1093589-2* 270
Buphthalmum strigosum IPNI=*1093589-2* 270
Bupleurum perfoliatum IPNI=*38361-2* 252
Bupleurum perfoliatum var. *rotundifolium* IPNI=*38361-2* 252
Bupleurum petroselinoides IPNI=*60442790-2* 252
Burcardia villosa IPNI=*30068867-2* 400
Burghartia villosa IPNI=*30068867-2* 400
Burlingtonia granadensis IPNI=*655792-1* 222
Burmeistera macrocalyx IPNI=*38509-2* 272
Burmeistera pittieri IPNI=*38509-2* 272
Burmeistera pittieri var. *decorans* IPNI=*38509-2* 272
Bursa abscissa IPNI=*30092589-2* 276
Bursa alandica IPNI=*30092589-2* 276
Bursa anglica IPNI=*30092589-2* 276
Bursa astoris IPNI=*30092589-2* 276
Bursa batavorum IPNI=*30092589-2* 276
Bursa belgica IPNI=*30092589-2* 276
Bursa bremensis IPNI=*30092589-2* 276
Bursa brittonii IPNI=*30092589-2* 276
Bursa bursa-pastoris IPNI=*30092589-2* 276
Bursa bursa-pastoris var. *heteris* IPNI=*30092589-2* 276
Bursa bursa-pastoris var. *integrifolia* IPNI=*30092589-2* 276
Bursa bursa-pastoris var. *minor* IPNI=*30092589-2* 276
Bursa bursa-pastoris var. *rhomboidea* IPNI=*30092589-2* 276
Bursa bursa-pastoris var. *simplex* IPNI=*30092589-2* 276
Bursa bursa-pastoris var. *tenuis* IPNI=*30092589-2* 276
Bursa concava IPNI=*30092589-2* 276
Bursa divaricata IPNI=*30092589-2* 276
Bursa djurdjurae IPNI=*30092589-2* 276
Bursa druceana IPNI=*30092589-2* 276
Bursa gallica IPNI=*30092589-2* 276
Bursa germanica IPNI=*30092589-2* 276
Bursa grossa IPNI=*30092589-2* 276
Bursa heegeri IPNI=*30092589-2* 276
Bursa hiatula IPNI=*30092589-2* 276
Bursa integrella IPNI=*30092589-2* 276
Bursa laevigata IPNI=*30092589-2* 276
Bursa mediterranea IPNI=*30092589-2* 276
Bursa occidentalis subsp. *madeirae* IPNI=*30092589-2* 276
Bursa occidentalis var. *concava* IPNI=*30092589-2* 276
Bursa orientalis IPNI=*30092589-2* 276
Bursa origo IPNI=*30092589-2* 276
Bursa patagonica IPNI=*30092589-2* 276
Bursa penarthae IPNI=*30092589-2* 276
Bursa pergrossa IPNI=*30092589-2* 276
Bursa robusta IPNI=*30092589-2* 276
Bursa segetum IPNI=*30092589-2* 276
Bursa sinuosa IPNI=*30092589-2* 276
Bursa sublaevigata IPNI=*30092589-2* 276
Bursa trevirorum IPNI=*30092589-2* 276
Bursa turoniensis IPNI=*30092589-2* 276
Bursa viguieri IPNI=*30092589-2* 276
Bursa viminalis IPNI=*30092589-2* 276
Bursera altissima IPNI=*77183601-1* 452
Bursera anderssonii IPNI=*127138-1* 451
Bursera aracouchini IPNI=*128278-1* 452
Bursera arborea IPNI=*127217-1* 451
Bursera bahiensis IPNI=*71559-1* 451
Bursera bonairensis IPNI=*127217-1* 451
Bursera carana IPNI=*128295-1* 452
Bursera caudata IPNI=*128413-1* 453
Bursera decandra IPNI=*128316-1* 452
Bursera eracouchii IPNI=*128278-1* 452
Bursera graveolens var. *pubescens* IPNI=*127138-1* 451
Bursera graveolens var. *villosula* IPNI=*127138-1* 451
Bursera guianensis IPNI=*128337-1* 452
Bursera gummifera IPNI=*127217-1* 451
Bursera gummifera var. *glabrata* IPNI=*127217-1* 451
Bursera gummifera var. *polyphylla* IPNI=*127217-1* 451
Bursera howellii IPNI=*127130-1* 451
Bursera integerrima IPNI=*127217-1* 451
Bursera panamensis IPNI=*127234-1* 451
Bursera pubescens IPNI=*127138-1* 451
Bursera simaruba var. *yucatanensis* IPNI=*127217-1* 451
Bursera subpubescens IPNI=*127217-1* 451
Bursera tacahamaca IPNI=*128338-1* 453
Bursera tatamaco IPNI=*127138-1* 451
Bursera tomentosa var. *pubescens* IPNI=*127234-1* 451
Bursera velutinifolia IPNI=*275630-2* 450
Bustamenta ovata IPNI=*6988-2* 256
Butia argentea IPNI=*670066-1* 212
Buxus angustifolia IPNI=*340317-1* 280
Buxus arborescens IPNI=*340317-1* 280
Buxus argentea IPNI=*340317-1* 280
Buxus aurea IPNI=*340317-1* 280
Buxus caucasica IPNI=*340317-1* 280
Buxus colchica IPNI=*340317-1* 280
Buxus crispa IPNI=*340317-1* 280
Buxus cucullata IPNI=*340317-1* 280
Buxus elegantissima IPNI=*340317-1* 280
Buxus fortunei IPNI=*340303-1* 280
Buxus fruticosa IPNI=*340317-1* 280
Buxus handsworthii IPNI=*340317-1* 280
Buxus hyrcana IPNI=*340317-1* 280
Buxus japonica IPNI=*340303-1* 280
Buxus kitashimae IPNI=*340303-1* 280
Buxus marginata IPNI=*340317-1* 280
Buxus microphylla var. *kitashimae* IPNI=*340303-1* 280
Buxus mucronata IPNI=*340317-1* 280
Buxus myrtifolia IPNI=*340317-1* 280
Buxus obcordatavariegata IPNI=*340303-1* 280
Buxus ovalifolia IPNI=*340303-1* 280
Buxus rosmarinifolia IPNI=*340317-1* 280
Buxus rotundifolia IPNI=*340303-1* 280
Buxus salicifolia IPNI=*340317-1* 280
Buxus sempervirens subsp. *hyrcana* IPNI=*340317-1* 280
Buxus sempervirens var. *arborescens* IPNI=*340317-1* 280
Buxus sempervirens var. *suffruticosa* IPNI=*340317-1* 280
Buxus suffruticosa IPNI=*340317-1* 280
Buxus tenuifolia IPNI=*340317-1* 280
Buxus variegata IPNI=*340317-1* 280
Buxus virens IPNI=*340303-1* 280
Buxus vulgaris IPNI=*340317-1* 280
Byrsanthes humboldtiana IPNI=*145280-1* 273
Byrsanthes nivea IPNI=*145280-1* 273
Byrsella coriacea IPNI=*642756-1* 219
Byrsella fragrans IPNI=*642833-1* 219
Byrsella velifera IPNI=*643196-1* 220
Byrsonima adenophylla IPNI=*555988-1* 395
Byrsonima angustifolia IPNI=*556192-1* 395
Byrsonima berteroana IPNI=*556020-1* 395
Byrsonima biacuminata IPNI=*556192-1* 395
Byrsonima brachystachya IPNI=*556037-1* 395
Byrsonima bucidifolia IPNI=*38926-2* 395
Byrsonima carmeniana IPNI=*556037-1* 395
Byrsonima chrysophylla f. *glandulifera* IPNI=*556192-1* 395
Byrsonima chrysophylla f. *kunthiana* IPNI=*556192-1* 395
Byrsonima chrysophylla subf. *obtusa* IPNI=*556192-1* 395
Byrsonima cinerea IPNI=*556033-1* 395
Byrsonima coccolobifolia f. *parvifolia* IPNI=*556020-1* 395
Byrsonima coccolobifolia var. *latifolia* IPNI=*556020-1* 395
Byrsonima coccolobifolia var. *obtusifolia* IPNI=*556020-1* 395
Byrsonima coriacea f. *eglandulosa* IPNI=*556192-1* 395
Byrsonima coriacea f. *propinqua* IPNI=*556192-1* 395
Byrsonima coriacea f. *typica* IPNI=*556033-1* 395
Byrsonima coriacea IPNI=*556033-1* 395

Byrsonima coriacea var. *spicata* IPNI=*556192-1* 395
Byrsonima coriacea var. *swartziana* IPNI=*556033-1* 395
Byrsonima cotinifolia IPNI=*556033-1* 395
Byrsonima crassifolia f. *cubensis* IPNI=*556033-1* 395
Byrsonima crassifolia IPNI=*556033-1* 395
Byrsonima crassifolia IPNI=*556209-1* 396
Byrsonima crassifolia subsp. *insulata* IPNI=*556033-1* 395
Byrsonima crassifolia var. *cinerea* IPNI=*556033-1* 395
Byrsonima crassifolia var. *jamaicensis* IPNI=*556033-1* 395
Byrsonima crassifolia var. *lanceolata* IPNI=*556033-1* 395
Byrsonima crassifolia var. *moureila* IPNI=*556033-1* 395
Byrsonima crassifolia var. *peruviana* IPNI=*556033-1* 395
Byrsonima crassifolia var. *spruceana* IPNI=*556033-1* 395
Byrsonima crassifolia var. *typica* IPNI=*556033-1* 395
Byrsonima cubensis IPNI=*556033-1* 395
Byrsonima cumingiana IPNI=*556033-1* 395
Byrsonima elegans IPNI=*557422-1* 396
Byrsonima fagifolia IPNI=*556033-1* 395
Byrsonima fendleri IPNI=*556033-1* 395
Byrsonima ferruginea IPNI=*556033-1* 395
Byrsonima ferruginea var. *macrophylla* IPNI=*555981-1* 395
Byrsonima ferruginea var. *moureila* IPNI=*556033-1* 395
Byrsonima fluminensis IPNI=*556093-1* 395
Byrsonima guadalupensis IPNI=*556192-1* 395
Byrsonima hostmannii IPNI=*556192-1* 395
Byrsonima indorum IPNI=*555988-1* 395
Byrsonima inundata IPNI=*556093-1* 395
Byrsonima japurensis subsp. *fluminensis* IPNI=*556093-1* 395
Byrsonima japurensis subsp. *silvatica* IPNI=*556093-1* 395
Byrsonima karwinskiana IPNI=*556033-1* 395
Byrsonima lanceolata IPNI=*556033-1* 395
Byrsonima latifolia IPNI=*556209-1* 396
Byrsonima laurifolia IPNI=*556033-1* 395
Byrsonima laurifolia var. *guatemalensis* IPNI=*556033-1* 395
Byrsonima montana IPNI=*556033-1* 395
Byrsonima monticola IPNI=*556020-1* 395
Byrsonima moritziana IPNI=*556033-1* 395
Byrsonima pachyphylla IPNI=*556209-1* 396
Byrsonima panamensis IPNI=*556033-1* 395
Byrsonima peruviana IPNI=*556014-1* 395
Byrsonima peruviana var. *eglandulosa* IPNI=*556192-1* 395
Byrsonima propinqua IPNI=*556192-1* 395
Byrsonima pulchra IPNI=*556033-1* 395
Byrsonima reticulata IPNI=*1079574-2* 396
Byrsonima rhopalifolia IPNI=*556033-1* 395
Byrsonima rufescens IPNI=*556033-1* 395
Byrsonima schultesiana IPNI=*555988-1* 395
Byrsonima sessilifolia IPNI=*556020-1* 395
Byrsonima spathulifolia IPNI=*556209-1* 396
Byrsonima spicata f. *propinqua* IPNI=*556192-1* 395
Byrsonima stigmatophora IPNI=*556563-1* 396
Byrsonima verbascifolia f. *vulgaris* IPNI=*556209-1* 396
Byrsonima verbascifolia subsp. *discolor* IPNI=*556209-1* 396
Byrsonima verbascifolia subsp. *villosa* IPNI=*556209-1* 396
Byrsonima verbascifolia var. *angustifolia* IPNI=*556209-1* 396
Byrsonima verbascifolia var. *denudata* IPNI=*556209-1* 396
Byrsonima verbascifolia var. *intermedia* IPNI=*556209-1* 396
Byrsonima verbascifolia var. *latifolia* IPNI=*556209-1* 396
Byrsonima verbascifolia var. *leiocarpa* IPNI=*556209-1* 396
Byrsonima verbascifolia var. *paraguayensis* IPNI=*556209-1* 396
Byrsonima verbascifolia var. *petiolata* IPNI=*556209-1* 396
Byrsonima verbascifolia var. *spatulifolia* IPNI=*556209-1* 396
Byrsonima verbascifolia var. *villosa* IPNI=*556209-1* 396
Bystropogon coarctatus IPNI=*158242-2* 371
Bystropogon graveolens IPNI=*158297-2* 371
Bystropogon mollis IPNI=*451957-1* 371
Bystropogon pectinatus IPNI=*158242-2* 371
Bystropogon sidifolius IPNI=*77121103-1* 371
Bystropogon suaveolens IPNI=*158297-2* 371
Byttneria aculeata IPNI=*77183732-1* 407
Byttneria acuminata IPNI=*77183732-1* 407
Byttneria brevipes IPNI=*77183732-1* 407
Byttneria catalpifolia IPNI=*77184121-1* 407
Byttneria guatemalensis IPNI=*77183732-1* 407
Byttneria lanceolata IPNI=*77183732-1* 407
Byttneria lateralis IPNI=*77183732-1* 407
Byttneria longifolia IPNI=*77183967-1* 407
Byttneria macrocarpa IPNI=*77184121-1* 407
Byttneria oligacantha IPNI=*77183732-1* 407
Byttneria peruviana IPNI=*77183732-1* 407
Byttneria pohliana IPNI=*77183967-1* 407
Byttneria rubicaulis IPNI=*77183732-1* 407
Byttneria salicifolia IPNI=*77183967-1* 407
Byttneria scabra IPNI=*77183967-1* 407
Byttneria scabra var. *brasiliensis* IPNI=*77183967-1* 407
Byttneria scabra var. *dentata* IPNI=*77183967-1* 407
Byttneria scabra var. *hastata* IPNI=*77183967-1* 407
Byttneria scabra var. *latissima* IPNI=*77183967-1* 407
Byttneria scabra var. *pilosiuscula* IPNI=*77183967-1* 407
Byttneria scabra var. *serrata* IPNI=*77183967-1* 407
Byttneria scabra var. *typica* IPNI=*77183967-1* 407
Byttneria sulcata IPNI=*77183732-1* 407
Byttneria tahitensis IPNI=*77184121-1* 407
Byttneria tereticaulis IPNI=*77183732-1* 407
Byttneria tessmannii IPNI=*77183732-1* 407
Byttneria tiliifolia IPNI=*77183732-1* 407
Byttneria uaupensis IPNI=*77184160-1* 407
Byttneria virgata IPNI=*77183967-1* 407

C

Caballeria coriacea IPNI=*166565-2* 304
Caballeria dependens IPNI=*1119456-2* 304
Caballeria ferruginea IPNI=*1028287-2* 304
Caballeria latifolia IPNI=*1033305-2* 304
Caballeria myrtifolia IPNI=*1119456-2* 304
Cabanisia martiana IPNI=*698175-1* 225
Cabi paraensis IPNI=*273852-2* 396
Cabralea brachystachya IPNI=*1021145-2* 454
Cabralea burchellii IPNI=*1021145-2* 454
Cabralea cauliflora IPNI=*1021145-2* 454
Cabralea eichleriana IPNI=*1021145-2* 454
Cabralea eichleriana var. *macrantha* IPNI=*1021145-2* 454
Cabralea erismatica IPNI=*1021145-2* 454
Cabralea estrellensis IPNI=*1021145-2* 454
Cabralea gaudichaudii IPNI=*1021145-2* 454
Cabralea glaberrima IPNI=*1021145-2* 454
Cabralea glaziovii IPNI=*1021145-2* 454
Cabralea jussiaeana IPNI=*1021145-2* 454
Cacalia angulata IPNI=*232826-1* 266
Cacalia aschenborniana IPNI=*969218-1* 272
Cacalia asclepiadea IPNI=*271815-2* 256
Cacalia aurantiaca IPNI=*211219-1* 263
Cacalia baccharoides IPNI=*969218-1* 272
Cacalia brachiata IPNI=*944095-1* 261
Cacalia brachylepis IPNI=*51409-2* 259
Cacalia brasiliana IPNI=*969180-1* 272
Cacalia bullata IPNI=*948489-1* 265
Cacalia cancellata IPNI=*238826-1* 268
Cacalia canescens IPNI=*948489-1* 265
Cacalia cinerea IPNI=*962413-1* 260
Cacalia coccinea IPNI=*203038-1* 261
Cacalia cordata IPNI=*232826-1* 266
Cacalia cordifolia IPNI=*232826-1* 266
Cacalia decurrens IPNI=*183309-1* 257
Cacalia erigerodes IPNI=*962413-1* 260
Cacalia exilis IPNI=*962413-1* 260

Cacalia geminata IPNI=*948489-1* 265
Cacalia glabra IPNI=*203080-1* 261
Cacalia glabrata IPNI=*963618-1* 267
Cacalia glandulosa IPNI=*208446-2* 268
Cacalia haenkeana IPNI=*969218-1* 272
Cacalia intermedia IPNI=*962014-1* 265
Cacalia kroneana IPNI=*962413-1* 260
Cacalia lanceolaris IPNI=*969218-1* 272
Cacalia ledifolia IPNI=*963586-1* 267
Cacalia lehmannii IPNI=*948528-1* 265
Cacalia mattogrossensis IPNI=*969218-1* 272
Cacalia mentrasto IPNI=*175500-1* 256
Cacalia micradenia IPNI=*969218-1* 272
Cacalia mollis IPNI=*948489-1* 265
Cacalia nitida IPNI=*963597-1* 267
Cacalia patens IPNI=*6096-2* 256
Cacalia patens IPNI=*969218-1* 272
Cacalia pilosa IPNI=*232826-1* 266
Cacalia porophyllum IPNI=*208446-2* 268
Cacalia prenanthoides IPNI=*203080-1* 261
Cacalia pulchella IPNI=*963604-1* 267
Cacalia radicans IPNI=*1011411-1* 260
Cacalia rotundifolia IPNI=*962413-1* 260
Cacalia rubricaulis IPNI=*962014-1* 265
Cacalia ruderalis IPNI=*208446-2* 268
Cacalia scandens IPNI=*211361-2* 269
Cacalia scorpioides IPNI=*944993-1* 261
Cacalia scorpioides var. *glabriuscula* IPNI=*944993-1* 261
Cacalia scorpioides var. *subrepanda* IPNI=*944993-1* 261
Cacalia scorpioides var. *subtomentosa* IPNI=*944993-1* 261
Cacalia scorpioides var. *tomentosa* IPNI=*944993-1* 261
Cacalia sonchifolia IPNI=*203080-1* 261
Cacalia sonchifolia IPNI=*229022-1* 264
Cacalia tenuissima IPNI=*203038-1* 261
Cacalia tournefortioides IPNI=*944993-1* 261
Cacalia triangularis IPNI=*232826-1* 266
Cacalia vaccinioides IPNI=*963618-1* 267
Cacao bicolor IPNI=*825565-1* 414
Cacao guianensis IPNI=*825619-1* 414
Cacao minus IPNI=*320783-2* 414
Cacao sativa IPNI=*320783-2* 414
Cacao sylvestris IPNI=*825619-1* 414
Cacao theobroma IPNI=*320783-2* 414
Cacara tuberosa IPNI=*512027-1* 329
Caconapea axillaris IPNI=*28282-2* 375
Cacoucia coccinea IPNI=*63781-2* 415
Cactus arboreus IPNI=*129102-1* 283
Cactus bleo IPNI=*60461844-2* 284
Cactus brasiliensis IPNI=*129102-1* 283
Cactus caripensis IPNI=*219250-2* 284
Cactus cassythoides IPNI=*219250-2* 284
Cactus chinensis IPNI=*1151735-2* 284
Cactus cochenillifer IPNI=*1008816-2* 284
Cactus conicus IPNI=*134797-1* 284
Cactus coranarius IPNI=*128942-1* 283
Cactus cylindricus IPNI=*128942-1* 283
Cactus dillenii IPNI=*136534-1* 284
Cactus eburneus IPNI=*244634-2* 285
Cactus elatior IPNI=*136561-1* 284
Cactus epidendrum IPNI=*219250-2* 284
Cactus fasciculatus IPNI=*219250-2* 284
Cactus ficus IPNI=*1151735-2* 284
Cactus ficus-indica IPNI=*1151735-2* 284
Cactus grandiflorus IPNI=*138488-1* 285
Cactus grandifolius IPNI=*219420-2* 284
Cactus hexagonus IPNI=*130283-1* 283
Cactus horridus IPNI=*137125-1* 284
Cactus humilis IPNI=*136561-1* 284
Cactus lanuginosus IPNI=*196706-2* 284
Cactus lucidus IPNI=*137452-1* 284
Cactus mammillaris IPNI=*134797-1* 284
Cactus mammillaris var. *glaber* IPNI=*134797-1* 284
Cactus mammillaris var. *lanuginosus* IPNI=*134797-1* 284
Cactus mammillaris var. *prolifer* IPNI=*134797-1* 284
Cactus matanzanus IPNI=*156850-2* 284
Cactus maxonii IPNI=*305646-2* 284
Cactus micranthus IPNI=*138328-1* 284
Cactus microdasys IPNI=*136806-1* 284
Cactus microthele IPNI=*134797-1* 284
Cactus nigricans IPNI=*136561-1* 284
Cactus oaxacensis IPNI=*305646-2* 284
Cactus obtusipetalus IPNI=*305646-2* 284
Cactus obvallatus IPNI=*134797-1* 284
Cactus octogonus IPNI=*130283-1* 283
Cactus opuntia IPNI=*1151735-2* 284
Cactus opuntia var. *inermis* IPNI=*137078-1* 284
Cactus paradoxus IPNI=*129102-1* 283
Cactus parasiticus IPNI=*219250-2* 284
Cactus parvimamma IPNI=*134797-1* 284
Cactus pendulinus IPNI=*219250-2* 284
Cactus pendulus IPNI=*219250-2* 284
Cactus pentagonus IPNI=*1646-2* 283
Cactus pereskia IPNI=*137452-1* 284
Cactus peruvianus IPNI=*130283-1* 283
Cactus peruvianus IPNI=*130627-1* 283
Cactus phyllanthus IPNI=*132939-1* 283
Cactus polyanthos IPNI=*136561-1* 284
Cactus prismaticus IPNI=*1646-2* 283
Cactus prolifer IPNI=*134797-1* 284
Cactus quadrangularis IPNI=*219250-2* 284
Cactus regalis IPNI=*130283-1* 283
Cactus repandus IPNI=*130627-1* 283
Cactus reptans IPNI=*1646-2* 283
Cactus rosa IPNI=*219420-2* 284
Cactus ruestii IPNI=*305646-2* 284
Cactus salvador IPNI=*305646-2* 284
Cactus sepium IPNI=*35149-2* 283
Cactus strictus IPNI=*137078-1* 284
Cactus subinermis IPNI=*1008816-2* 284
Cactus tetragonus IPNI=*1646-2* 283
Cactus truncatus IPNI=*229597-2* 284
Cactus tuna IPNI=*137125-1* 284
Cactus tuna var. *elatior* IPNI=*136561-1* 284
Cactus tuna var. *flava* IPNI=*136561-1* 284
Cactus tuna var. *nigricans* IPNI=*136561-1* 284
Cactus tunicatus IPNI=*73905-2* 283
Cacuvallum altissimus IPNI=*1197939-2* 328
Cadelaria punctata IPNI=*2468-2* 280
Caelestina cordata IPNI=*1092582-2* 263
Caelestina latifolia IPNI=*7086-2* 256
Caelestina suffruticosa IPNI=*7086-2* 256
Caenopteris canariensis IPNI=*17078300-1* 175
Caenotus canadensis IPNI=*203653-1* 261
Caenotus pusillus IPNI=*203653-1* 261
Caesalpinia acutifolia IPNI=*67201-2* 315
Caesalpinia andreana IPNI=*482664-1* 312
Caesalpinia arborea IPNI=*513021-1* 330
Caesalpinia benguetensis IPNI=*481887-1* 312
Caesalpinia bicolor IPNI=*482664-1* 312
Caesalpinia bonduc IPNI=*114706-2* 320
Caesalpinia bonducella IPNI=*114706-2* 320
Caesalpinia bonducella var. *aequiaculeata* IPNI=*114706-2* 320
Caesalpinia bonducella var. *inaequiaculeata* IPNI=*114706-2* 320
Caesalpinia brea IPNI=*1006943-1* 330
Caesalpinia broomensis IPNI=*114718-2* 320
Caesalpinia cassioides var. *pardoana* IPNI=*482664-1* 312

Caesalpinia coriacea IPNI=*1010207-2* 325
Caesalpinia coriaria IPNI=*1010207-2* 325
Caesalpinia corymbosa IPNI=*77130395-1* 325
Caesalpinia decapetala IPNI=*481887-1* 312
Caesalpinia decapetala var. *japonica* IPNI=*481887-1* 312
Caesalpinia decapetala var. *pubescens* IPNI=*481887-1* 312
Caesalpinia ebano IPNI=*139695-2* 325
Caesalpinia ferox IPNI=*481887-1* 312
Caesalpinia ferruginea IPNI=*513021-1* 330
Caesalpinia glabra IPNI=*114706-2* 320
Caesalpinia glabrata IPNI=*77130395-1* 325
Caesalpinia glenieii IPNI=*513021-1* 330
Caesalpinia globulorum IPNI=*114718-2* 320
Caesalpinia granadillo IPNI=*139695-2* 325
Caesalpinia hispida IPNI=*1164658-2* 312
Caesalpinia horrida IPNI=*249443-2* 335
Caesalpinia horrida IPNI=*481887-1* 312
Caesalpinia inermis IPNI=*513021-1* 330
Caesalpinia insignis IPNI=*482664-1* 312
Caesalpinia japonica IPNI=*481887-1* 312
Caesalpinia jayabo IPNI=*114718-2* 320
Caesalpinia jayabo var. *cyanosperma* IPNI=*114706-2* 320
Caesalpinia jayabo var. *microphylla* IPNI=*114718-2* 320
Caesalpinia kilaroe IPNI=*513021-1* 330
Caesalpinia macklottii IPNI=*513021-1* 330
Caesalpinia major IPNI=*114718-2* 320
Caesalpinia mollis IPNI=*67201-2* 315
Caesalpinia mucronata IPNI=*249443-2* 335
Caesalpinia nitida IPNI=*114718-2* 320
Caesalpinia paipai IPNI=*77130395-1* 325
Caesalpinia paipai var. *pubens* IPNI=*77130395-1* 325
Caesalpinia parahyba IPNI=*517988-1* 332
Caesalpinia pardoana IPNI=*482664-1* 312
Caesalpinia paucijuga IPNI=*139695-2* 325
Caesalpinia pectinata IPNI=*249443-2* 335
Caesalpinia pluviosa IPNI=*77158081-1* 314
Caesalpinia praecox IPNI=*1006943-1* 330
Caesalpinia procera IPNI=*515473-1* 331
Caesalpinia pulcherrima f. *flava* IPNI=*1164658-2* 312
Caesalpinia pulcherrima var. *insignis* IPNI=*1164658-2* 312
Caesalpinia punctata IPNI=*139695-2* 325
Caesalpinia regia IPNI=*491231-1* 317
Caesalpinia roxburghii IPNI=*513021-1* 330
Caesalpinia sepiaria IPNI=*481887-1* 312
Caesalpinia sepiaria var. *pubescens* IPNI=*481887-1* 312
Caesalpinia sogerensis IPNI=*114706-2* 320
Caesalpinia spinosa IPNI=*249443-2* 335
Caesalpinia tara IPNI=*249443-2* 335
Caesalpinia thomaea IPNI=*1010207-2* 325
Caesalpinia tinctoria IPNI=*249443-2* 335
Caesalpinia tinctoria var. *horrida* IPNI=*249443-2* 335
Caesia comosa IPNI=*532810-1* 215
Caesulia scandens IPNI=*993359-1* 271
Cailliea macrostachya IPNI=*188076-2* 330
Cainito pomiferum IPNI=*304384-2* 305
Cajanum thora IPNI=*1152177-2* 313
Cajanus bicolor IPNI=*1152177-2* 313
Cajanus flavus IPNI=*1152177-2* 313
Cajanus indicus IPNI=*1152177-2* 313
Cajanus indicus var. *bicolor* IPNI=*1152177-2* 313
Cajanus indicus var. *flavus* IPNI=*1152177-2* 313
Cajanus indicus var. *maculatus* IPNI=*1152177-2* 313
Cajanus luteus IPNI=*1152177-2* 313
Cajanus obcordifolia IPNI=*1152177-2* 313
Cajanus pseudocajan IPNI=*1152177-2* 313
Cajanus striatus IPNI=*1152177-2* 313
Cakile aequalis IPNI=*118574-3* 276
Cakile americana var. *cubensis* IPNI=*118574-3* 276
Cakile clavata IPNI=*1061021-2* 277
Cakile cubensis IPNI=*118574-3* 276
Cakile domingensis IPNI=*118574-3* 276
Cakile lanceolata subsp. *domingensis* IPNI=*118574-3* 276
Cakile lanceolata var. *integrifolia* IPNI=*118574-3* 276
Cakile lanceolata var. *pinnatifida* IPNI=*118574-3* 276
Cakile maritima var. *aequalis* IPNI=*118574-3* 276
Cakile maritima var. *cubensis* IPNI=*118574-3* 276
Cakile rugosa IPNI=*1061021-2* 277
Cakile rugosa var. *stylosa* IPNI=*1061021-2* 277
Calacinum leptobotrys IPNI=*694785-1* 289
Calacinum peruvianum IPNI=*694785-1* 289
Calacinum platycladum IPNI=*694774-1* 289
Calacinum tamnifolium IPNI=*694785-1* 289
Calacinum volcanicum IPNI=*1022243-2* 289
Caladium acre IPNI=*1170772-2* 199
Caladium aculeatum IPNI=*87563-1* 200
Caladium acuminatum IPNI=*89310-1* 202
Caladium albopunctatissimum IPNI=*30010088-2* 199
Caladium amabile IPNI=*30010088-2* 199
Caladium amoenum IPNI=*30010088-2* 199
Caladium angustifolium IPNI=*909842-1* 203
Caladium angustilobum IPNI=*86379-1* 199
Caladium appunianum IPNI=*30010088-2* 199
Caladium arborescens IPNI=*87563-1* 200
Caladium arboreum IPNI=*87563-1* 200
Caladium argyrospilum IPNI=*30010088-2* 199
Caladium barraquinii IPNI=*30010088-2* 199
Caladium belophyllum IPNI=*89310-1* 202
Caladium bicolor f. *argyrospilum* IPNI=*30010088-2* 199
Caladium bicolor f. *barraquinii* IPNI=*30010088-2* 199
Caladium bicolor f. *brongniartii* IPNI=*30010088-2* 199
Caladium bicolor f. *chantinii* IPNI=*30010088-2* 199
Caladium bicolor f. *devosianum* IPNI=*30010088-2* 199
Caladium bicolor f. *haematostigma* IPNI=*30010088-2* 199
Caladium bicolor f. *macrophyllum* IPNI=*30010088-2* 199
Caladium bicolor f. *mirabile* IPNI=*30010088-2* 199
Caladium bicolor f. *neumannii* IPNI=*30010088-2* 199
Caladium bicolor f. *perrieri* IPNI=*30010088-2* 199
Caladium bicolor f. *regale* IPNI=*30010088-2* 199
Caladium bicolor f. *robustum* IPNI=*30010088-2* 199
Caladium bicolor f. *splendens* IPNI=*30010088-2* 199
Caladium bicolor f. *vellozianum* IPNI=*30010088-2* 199
Caladium bicolor f. *verschaffeltii* IPNI=*30010088-2* 199
Caladium bicolor f. *wightii* IPNI=*30010088-2* 199
Caladium bicolor subvar. *rosaceum* IPNI=*30010088-2* 199
Caladium bicolor var. *roseomaculatum* IPNI=*30010088-2* 199
Caladium bicolor var. *surinamense* IPNI=*30010088-2* 199
Caladium bicolor var. *vermitoxicum* IPNI=*30010088-2* 199
Caladium brasiliense IPNI=*269304-2* 202
Caladium brongniartii IPNI=*30010088-2* 199
Caladium caracasanum IPNI=*89310-1* 202
Caladium chantinii IPNI=*30010088-2* 199
Caladium colocasia IPNI=*84151-1* 198
Caladium colocasioides IPNI=*1170772-2* 199
Caladium concolor IPNI=*30010088-2* 199
Caladium connaertii IPNI=*30010088-2* 199
Caladium cucullatum IPNI=*84151-1* 198
Caladium curwadlii IPNI=*30010088-2* 199
Caladium devosianum IPNI=*30010088-2* 199
Caladium discolor IPNI=*30010088-2* 199
Caladium duchartrei IPNI=*30010088-2* 199
Caladium eckhartii IPNI=*30010088-2* 199
Caladium edule IPNI=*89373-1* 203
Caladium enkeanum IPNI=*30010088-2* 199
Caladium esculentum IPNI=*1170772-2* 199
Caladium firmulum IPNI=*30010088-2* 199
Caladium fragrantissimum IPNI=*87769-1* 201

Caladium gaerdtii IPNI=*30010088-2* 199
Caladium glycyrrhizum IPNI=*1170772-2* 199
Caladium griseoargenteum IPNI=*30010088-2* 199
Caladium haageanum IPNI=*30010088-2* 199
Caladium hastifolium IPNI=*89333-1* 203
Caladium helleborifolium IPNI=*89333-1* 203
Caladium hendersonii IPNI=*30010088-2* 199
Caladium heterotypicum IPNI=*909842-1* 203
Caladium holtonianum IPNI=*89347-1* 203
Caladium houbyanum IPNI=*30010088-2* 199
Caladium houlletii IPNI=*30010088-2* 199
Caladium indicum IPNI=*60444565-2* 198
Caladium jacquinii IPNI=*30010088-2* 199
Caladium ketteleri IPNI=*30010088-2* 199
Caladium kochii IPNI=*30010088-2* 199
Caladium kramerianum IPNI=*30010088-2* 199
Caladium laucheanum IPNI=*30010088-2* 199
Caladium leopoldii IPNI=*30010088-2* 199
Caladium lilliputiense IPNI=*30010088-2* 199
Caladium lindenii IPNI=*30010088-2* 199
Caladium lindenii var. *sylvestre* IPNI=*40313-2* 199
Caladium liniferum IPNI=*87565-1* 200
Caladium macrophyllum IPNI=*30010088-2* 199
Caladium macrorrhizon IPNI=*60444565-2* 198
Caladium maculatum IPNI=*86911-1* 199
Caladium mafaffa IPNI=*89373-1* 203
Caladium marginatum IPNI=*30010088-2* 199
Caladium marmoratum IPNI=*30010088-2* 199
Caladium martersteigianum IPNI=*30010088-2* 199
Caladium medioradiatum IPNI=*30010088-2* 199
Caladium metallicum IPNI=*60444565-2* 198
Caladium mirabile IPNI=*30010088-2* 199
Caladium mooreanum IPNI=*30010088-2* 199
Caladium neumannii IPNI=*30010088-2* 199
Caladium nymphaeifolium IPNI=*1170772-2* 199
Caladium ottonis IPNI=*30010088-2* 199
Caladium pallidinervium IPNI=*30010088-2* 199
Caladium pallidum IPNI=*30010088-2* 199
Caladium paradoxum IPNI=*77071638-1* 203
Caladium pedatum IPNI=*87928-1* 201
Caladium pellucidum IPNI=*30010088-2* 199
Caladium perrieri IPNI=*30010088-2* 199
Caladium pictum IPNI=*30010088-2* 199
Caladium pilosum IPNI=*89347-1* 203
Caladium plumbeum IPNI=*60444565-2* 198
Caladium puberulum IPNI=*89347-1* 203
Caladium punctatissimum IPNI=*30010088-2* 199
Caladium purdieanum IPNI=*30010088-2* 199
Caladium pusillum IPNI=*30010088-2* 199
Caladium regale IPNI=*30010088-2* 199
Caladium reichenbachianum IPNI=*30010088-2* 199
Caladium rougieri IPNI=*30010088-2* 199
Caladium rubellum IPNI=*30010088-2* 199
Caladium rubricaule IPNI=*30010088-2* 199
Caladium rugosum IPNI=*84151-1* 198
Caladium sagittifolium IPNI=*30010088-2* 199
Caladium sagittifolium IPNI=*89373-1* 203
Caladium seguine IPNI=*86911-1* 199
Caladium seguine var. *maculatum* IPNI=*86911-1* 199
Caladium sieboldii IPNI=*30010088-2* 199
Caladium sororium IPNI=*30010088-2* 199
Caladium splendens IPNI=*30010088-2* 199
Caladium spruceanum IPNI=*30010088-2* 199
Caladium stangeanum IPNI=*30010088-2* 199
Caladium striatipes IPNI=*909842-1* 203
Caladium striatipes var. *subhastata* IPNI=*909842-1* 203
Caladium surinamense IPNI=*30010088-2* 199
Caladium thelemannii IPNI=*30010088-2* 199
Caladium thripedestum IPNI=*30010088-2* 199
Caladium trifoliatum IPNI=*88045-1* 201
Caladium tripartitum IPNI=*88045-1* 201
Caladium utile IPNI=*89373-1* 203
Caladium variegatum IPNI=*89333-1* 203
Caladium vellozianum IPNI=*30010088-2* 199
Caladium verschaffeltii IPNI=*30010088-2* 199
Caladium violaceum IPNI=*1170772-2* 199
Caladium wagneri IPNI=*30010088-2* 199
Caladium wightii IPNI=*30010088-2* 199
Caladium xanthorrhizon IPNI=*89373-1* 203
Calamagrostis areantha IPNI=*77206801-1* 242
Calamagrostis colorata IPNI=*415571-1* 243
Calamagrostis effusa IPNI=*77206801-1* 242
Calamagrostis falklandiae IPNI=*258855-2* 246
Calamagrostis funckii IPNI=*77206801-1* 242
Calamagrostis piscatoria IPNI=*397862-1* 237
Calamagrostis subspicata IPNI=*258855-2* 246
Calamagrostis variegata IPNI=*415571-1* 243
Calamintha albiflora IPNI=*452333-1* 371
Calamintha arvensis IPNI=*30026217-2* 371
Calamintha caerulescens IPNI=*1001364-1* 371
Calamintha fenzlii IPNI=*451022-1* 371
Calandrinia andrewsii IPNI=*1182764-2* 290
Calandrinia andrewskii IPNI=*1007146-2* 290
Calandrinia arizonica IPNI=*698357-1* 286
Calandrinia axilliflora var. *foliosa* IPNI=*698357-1* 286
Calandrinia bonariensis IPNI=*698357-1* 286
Calandrinia caulescens IPNI=*698357-1* 286
Calandrinia caulescens subsp. *menziesii* IPNI=*698357-1* 286
Calandrinia caulescens var. *menziesii* IPNI=*698357-1* 286
Calandrinia cerrateae IPNI=*311155-2* 286
Calandrinia ciliata var. *menziesii* IPNI=*698357-1* 286
Calandrinia elegans IPNI=*698357-1* 286
Calandrinia feltonii IPNI=*698357-1* 286
Calandrinia lockhartii IPNI=*1182764-2* 290
Calandrinia megarhiza IPNI=*698305-1* 286
Calandrinia menziesii IPNI=*698357-1* 286
Calandrinia menziesii var. *alba* IPNI=*698357-1* 286
Calandrinia meyeniana IPNI=*698305-1* 286
Calandrinia micrantha IPNI=*698357-1* 286
Calandrinia muricata IPNI=*698357-1* 286
Calandrinia pachypoda IPNI=*1182764-2* 290
Calandrinia phacosperma IPNI=*698357-1* 286
Calandrinia pulchella IPNI=*698357-1* 286
Calandrinia pusilla IPNI=*311155-2* 286
Calandrinia saltensis IPNI=*698305-1* 286
Calandrinia speciosa IPNI=*698357-1* 286
Calandrinia stenophylla IPNI=*698357-1* 286
Calandrinia tenella IPNI=*698357-1* 286
Calanthea pulcherrima IPNI=*77183588-1* 278
Calappa acrocomioides IPNI=*246966-2* 212
Calappa australis IPNI=*246966-2* 212
Calappa datil IPNI=*246966-2* 212
Calappa elegantissima IPNI=*246979-2* 212
Calappa martiana IPNI=*246966-2* 212
Calappa nucifera IPNI=*666160-1* 208
Calappa orinocensis IPNI=*246957-2* 212
Calappa plumosa IPNI=*246966-2* 212
Calappa romanzoffiana IPNI=*246966-2* 212
Calappa sancona IPNI=*670066-1* 212
Calappa speciosa IPNI=*246946-2* 212
Calathea albicans IPNI=*60461053-2* 249
Calathea allouia IPNI=*60458762-2* 345
Calathea arrecta IPNI=*60461169-2* 249
Calathea bachemiana IPNI=*60458927-2* 249
Calathea binotii IPNI=*796810-1* 249
Calathea cachibou IPNI=*60458866-2* 249

Calathea cyclophora IPNI=*60459273-2* 249
Calathea discolor IPNI=*60458866-2* 249
Calathea gigas IPNI=*60461040-2* 249
Calathea glaziovii IPNI=*77174968-1* 249
Calathea illustris IPNI=*60461319-2* 249
Calathea imperialis IPNI=*60461040-2* 249
Calathea insignis IPNI=*273648-2* 249
Calathea juncea IPNI=*797103-1* 249
Calathea klugii IPNI=*60461053-2* 249
Calathea lanceolata IPNI=*60459373-2* 249
Calathea latifolia IPNI=*60459373-2* 249
Calathea lietzei IPNI=*77174968-1* 249
Calathea loeseneri IPNI=*77128944-1* 249
Calathea macrochlamys IPNI=*285700-2* 250
Calathea macrosepala IPNI=*60461039-2* 249
Calathea macrosepala var. *reflexa* IPNI=*60461039-2* 249
Calathea magnifica IPNI=*60458866-2* 249
Calathea majestica IPNI=*60461040-2* 249
Calathea micans IPNI=*60461053-2* 249
Calathea micans var. *amabilis* IPNI=*60461053-2* 249
Calathea micans var. *robustior* IPNI=*60461053-2* 249
Calathea myrosma IPNI=*797633-1* 250
Calathea ornata IPNI=*60461169-2* 249
Calathea ornata var. *albolineata* IPNI=*60461169-2* 249
Calathea ornata var. *roseolineata* IPNI=*60461169-2* 249
Calathea panamensis IPNI=*60461175-2* 249
Calathea paucifolia IPNI=*60461176-2* 249
Calathea princeps IPNI=*60461040-2* 249
Calathea propinqua IPNI=*60461303-2* 249
Calathea pulchella IPNI=*796810-1* 249
Calathea quadratispica IPNI=*273648-2* 249
Calathea roseopicta IPNI=*60461319-2* 249
Calathea roseopicta var. *illustris* IPNI=*60461319-2* 249
Calathea rotundifolia IPNI=*797124-1* 249
Calathea sclerobractea IPNI=*273648-2* 249
Calathea sprucei IPNI=*60461053-2* 249
Calathea trifasciata IPNI=*60461303-2* 249
Calathea zebrina IPNI=*796810-1* 249
Calatola columbiana IPNI=*41180-2* 359
Calatola sanquininensis IPNI=*41180-2* 359
Calatola venezuelana IPNI=*41180-2* 359
Calawaya insolita IPNI=*280607-2* 220
Calceolaria anagalloides IPNI=*800278-1* 365
Calceolaria angustata IPNI=*77143566-1* 405
Calceolaria arborea IPNI=*77144891-1* 405
Calceolaria bangii IPNI=*77143568-1* 405
Calceolaria biflora IPNI=*800278-1* 365
Calceolaria celendinensis IPNI=*41696-2* 365
Calceolaria chelidonioides var. *flaccida* IPNI=*41696-2* 365
Calceolaria chelidonioides var. *parvifolia* IPNI=*41696-2* 365
Calceolaria circaeodes IPNI=*77143566-1* 405
Calceolaria concinna IPNI=*41696-2* 365
Calceolaria corazonica IPNI=*41696-2* 365
Calceolaria glabra IPNI=*77143553-1* 405
Calceolaria gracilis IPNI=*41696-2* 365
Calceolaria gracilis IPNI=*800278-1* 365
Calceolaria heterophylla IPNI=*41696-2* 365
Calceolaria hirsuta IPNI=*77143549-1* 405
Calceolaria ipecacuanha IPNI=*77143549-1* 405
Calceolaria laciniata IPNI=*41487-2* 365
Calceolaria larecajensis IPNI=*41696-2* 365
Calceolaria longifolia IPNI=*77143566-1* 405
Calceolaria lysimachioides IPNI=*800278-1* 365
Calceolaria mandoniana IPNI=*41696-2* 365
Calceolaria meistantha IPNI=*800278-1* 365
Calceolaria mexicana var. *pichinchensis* IPNI=*41487-2* 365
Calceolaria micrantha IPNI=*41696-2* 365
Calceolaria microphylla IPNI=*77143568-1* 405
Calceolaria oppositifolia IPNI=*77143566-1* 405
Calceolaria ovata IPNI=*800278-1* 365
Calceolaria palustris IPNI=*41487-2* 365
Calceolaria palustris subsp. *hitchcockii* IPNI=*41487-2* 365
Calceolaria palustris subsp. *megacarpa* IPNI=*41487-2* 365
Calceolaria palustris subsp. *micrantha* IPNI=*41487-2* 365
Calceolaria palustris subsp. *pennatipartita* IPNI=*41487-2* 365
Calceolaria palustris subsp. *radiculoides* IPNI=*41487-2* 365
Calceolaria palustris subsp. *santanderensis* IPNI=*41487-2* 365
Calceolaria pinnata IPNI=*41696-2* 365
Calceolaria puru-puru IPNI=*41696-2* 365
Calceolaria radiculoides IPNI=*41487-2* 365
Calceolaria ranunculoides IPNI=*41696-2* 365
Calceolaria sarmentosa IPNI=*41696-2* 365
Calceolaria scabiosifolia IPNI=*41696-2* 365
Calceolaria soukupii IPNI=*41696-2* 365
Calceolaria tracheliifolia IPNI=*41487-2* 365
Calceolaria tripartita subsp. *cajamarcensis* IPNI=*41696-2* 365
Calceolaria tripartita subsp. *candelabris* IPNI=*41696-2* 365
Calceolaria tripartita subsp. *chumpitazii* IPNI=*41696-2* 365
Calceolaria tripartita subsp. *crenatifolia* IPNI=*41696-2* 365
Calceolaria tripartita subsp. *diversifolia* IPNI=*41696-2* 365
Calceolaria tripartita subsp. *inaequalis* IPNI=*41696-2* 365
Calceolaria tripartita subsp. *longipedunculata* IPNI=*41696-2* 365
Calceolaria tripartita subsp. *retroflexa* IPNI=*41696-2* 365
Calceolaria tripartita var. *typica* IPNI=*41696-2* 365
Calceolaria urticina IPNI=*41487-2* 365
Calceolaria velardei IPNI=*41696-2* 365
Calceolaria villosissima IPNI=*77143549-1* 405
Calchas acuminatus IPNI=*446165-1* 369
Calchas atropurpureus IPNI=*446165-1* 369
Calchas crispipilus IPNI=*446165-1* 369
Calchas scutellarioides IPNI=*446165-1* 369
Calchas scutellarioides var. *angustifolia* IPNI=*446165-1* 369
Calchas scutellarioides var. *crispipilus* IPNI=*446165-1* 369
Calchas scutellarioides var. *limnophila* IPNI=*446165-1* 369
Calcitrapa tinctoria IPNI=*324467-2* 259
Caldasia andicola IPNI=*77085100-1* 252
Caldasia brachycarpa IPNI=*77085100-1* 252
Caldasia brasiliensis IPNI=*103346-1* 448
Caldasia chaerophyllaea IPNI=*77085100-1* 252
Caldasia chaerophyllaea var. *glabriuscula* IPNI=*77085100-1* 252
Caldasia chaerophyllaea var. *hirsuta* IPNI=*77085100-1* 252
Caldasia chaerophylloides IPNI=*77085100-1* 252
Caldasia daucoides IPNI=*77085100-1* 252
Caldasia eriopus IPNI=*77085100-1* 252
Caldasia fungiformis IPNI=*103346-1* 448
Caldasia lasiopetala IPNI=*77085100-1* 252
Caldenbachia macrogranum IPNI=*55411-1* 361
Calea acuminata IPNI=*187778-1* 259
Calea acuminata var. *xanthactis* IPNI=*187778-1* 259
Calea albida IPNI=*187778-1* 259
Calea amellus IPNI=*241903-1* 269
Calea aspera IPNI=*156327-2* 266
Calea brevipes IPNI=*187759-1* 259
Calea caracasana f. *discoidea* IPNI=*9082-2* 256
Calea caracasana IPNI=*9082-2* 256
Calea caracasana var. *glabrior* IPNI=*9082-2* 256
Calea caracasana var. *pilosior* IPNI=*9082-2* 256
Calea chocoensis IPNI=*187732-1* 259
Calea collina IPNI=*187576-1* 258
Calea cordata IPNI=*962413-1* 260
Calea cordifolia IPNI=*1157019-2* 259
Calea dichotoma IPNI=*187778-1* 259
Calea glomerata IPNI=*187759-1* 259
Calea holtonii IPNI=*187576-1* 258
Calea hypoleuca IPNI=*187778-1* 259
Calea jamaicensis var. *parvifolia* IPNI=*1157019-2* 259

Calea leontophthalmum IPNI=*41956-2* 259
Calea leptocephala IPNI=*187778-1* 259
Calea liebmannii IPNI=*187778-1* 259
Calea lobata IPNI=*234386-1* 267
Calea oppositifolia IPNI=*1008406-1* 264
Calea ottonis IPNI=*187576-1* 258
Calea pennellii IPNI=*41956-2* 259
Calea pittieri IPNI=*187732-1* 259
Calea pringlei IPNI=*187778-1* 259
Calea pringlei var. *rubida* IPNI=*187778-1* 259
Calea rugosa IPNI=*187778-1* 259
Calea salmeifolia IPNI=*187778-1* 259
Calea savannarum IPNI=*271533-2* 255
Calea serrata IPNI=*1157019-2* 259
Calea sessiliflora IPNI=*236351-1* 267
Calea sillaensis IPNI=*9082-2* 256
Calea sororia IPNI=*187778-1* 259
Calea suriani IPNI=*234386-1* 267
Calea tejadae IPNI=*187778-1* 259
Calea ternifolia var. *calyculata* IPNI=*187778-1* 259
Calea ternifolia var. *hypoleuca* IPNI=*187778-1* 259
Calea trianae IPNI=*41956-2* 259
Calea trianae var. *tolimensis* IPNI=*41956-2* 259
Calea trifida IPNI=*1157019-2* 259
Calea zacatechichi IPNI=*187778-1* 259
Calea zacatechichi var. *calyculata* IPNI=*187778-1* 259
Calea zacatechichi var. *laevigata* IPNI=*187778-1* 259
Calea zacatechichi var. *macrophylla* IPNI=*187778-1* 259
Calea zacatechichi var. *rugosa* IPNI=*187778-1* 259
Calea zacatechichi var. *xanthina* IPNI=*187778-1* 259
Caleatia caimito IPNI=*788845-1* 306
Calendula aurantiaca IPNI=*187894-1* 259
Calendula eriocarpa IPNI=*187894-1* 259
Calendula hydruntina IPNI=*187894-1* 259
Calendula officinalis f. *pleniflora* IPNI=*187894-1* 259
Calendula officinalis var. *parviflora* IPNI=*187894-1* 259
Calendula prolifera IPNI=*187894-1* 259
Calendula ranunculodes IPNI=*187894-1* 259
Calendula santamariae IPNI=*187894-1* 259
Calendula sinuata IPNI=*187894-1* 259
Calendula sinuata var. *aurantiaca* IPNI=*187894-1* 259
Calinea bentegeati IPNI=*317113-1* 295
Caliphruria candida IPNI=*261572-2* 214
Caliphruria subedentata IPNI=*67017-1* 214
Calla aethiopica IPNI=*89403-1* 203
Calla ambigua IPNI=*89403-1* 203
Calla badian IPNI=*60444565-2* 198
Calla dracontium IPNI=*87464-1* 200
Calla gaby IPNI=*1170772-2* 199
Calla maxima IPNI=*60444565-2* 198
Calla moschata IPNI=*89403-1* 203
Calla pertusa IPNI=*87464-1* 200
Calliandra amblyphylla IPNI=*42154-2* 313
Calliandra angustidens IPNI=*316281-2* 313
Calliandra caracasana IPNI=*287164-2* 337
Calliandra clavellina IPNI=*42269-2* 313
Calliandra cumingii IPNI=*1161746-2* 313
Calliandra decrescens IPNI=*42179-2* 313
Calliandra formosa IPNI=*287168-2* 337
Calliandra fulgens IPNI=*42154-2* 313
Calliandra latifolia IPNI=*271432-2* 337
Calliandra marginata IPNI=*989342-1* 320
Calliandra panlosia IPNI=*1161746-2* 313
Calliandra portoricensis IPNI=*943976-1* 337
Calliandra saman IPNI=*227126-2* 331
Calliandra schultzei IPNI=*483259-1* 313
Calliandra schwackeana IPNI=*962821-1* 337
Calliandra serjanioides IPNI=*42154-2* 313
Calliandra sodiroi IPNI=*42066-2* 313
Calliandra stricta IPNI=*42066-2* 313
Calliandra subnervosa IPNI=*42066-2* 313
Calliandra tenuiflora IPNI=*316281-2* 313
Calliandra trijugata IPNI=*1161746-2* 313
Calliaspidia guttata IPNI=*289220-2* 360
Callicarpa cinerea IPNI=*864000-1* 372
Callicarpa discolor IPNI=*860891-1* 368
Callicarpa globiflora IPNI=*860891-1* 368
Callicarpa integrifolia IPNI=*860891-1* 368
Callicarpa reticulata IPNI=*864000-1* 372
Callichlamys garnieri IPNI=*42392-2* 362
Callichlamys riparia IPNI=*42392-2* 362
Callichlamys rubiginosa IPNI=*42392-2* 362
Callichlamys rufinervis IPNI=*42392-2* 362
Callichlamys splendida IPNI=*42392-2* 362
Callicocca alba IPNI=*60473597-2* 353
Callicocca elata IPNI=*77115600-1* 353
Callicocca ipecacuanha IPNI=*1214950-2* 347
Callicocca tomentosa IPNI=*77115635-1* 354
Callicysthus volubilis IPNI=*525428-1* 336
Callisemaea elegans IPNI=*515259-1* 331
Callisia ciliata IPNI=*77184055-1* 224
Callisia hexandra IPNI=*171826-1* 224
Callisia hexandra var. *caracasana* IPNI=*171826-1* 224
Callisia hexandra var. *mandonii* IPNI=*171826-1* 224
Callisia hexandra var. *salzmannii* IPNI=*171826-1* 224
Callisia parvula IPNI=*172873-1* 224
Callisia repens var. *ciliata* IPNI=*171826-1* 224
Callisia repens var. *mandonii* IPNI=*171826-1* 224
Callisia rosea IPNI=*77184055-1* 224
Callisia tripogandra IPNI=*77205602-1* 224
Callista flavescens IPNI=*654273-1* 222
Callistemon citrinus IPNI=*77108602-1* 426
Callistemon citrinus var. *splendens* IPNI=*77108602-1* 426
Callistemon cunninghamii IPNI=*77108602-1* 426
Callistemon flavescens IPNI=*77108602-1* 426
Callistemon laevis IPNI=*77108602-1* 426
Callistemon lanceolatus f. *semperflorens* IPNI=*77108602-1* 426
Callistemon lanceolatus IPNI=*77108602-1* 426
Callistemon lanceolatus var. *pendulus* IPNI=*77108602-1* 426
Callistemon lanceolatus var. *semperflorens* IPNI=*77108602-1* 426
Callistemon lanceolatus var. *sparsus* IPNI=*77108602-1* 426
Callistemon lilacinus f. *albus* IPNI=*77108602-1* 426
Callistemon lilacinus f. *carminus* IPNI=*77108602-1* 426
Callistemon lilacinus IPNI=*77108602-1* 426
Callistemon longifolius IPNI=*77108602-1* 426
Callistemon marginatus IPNI=*77108602-1* 426
Callistemon pendulus IPNI=*77108602-1* 426
Callistemon pendulus var. *confertus* IPNI=*77108602-1* 426
Callistemon pendulus var. *genuinus* IPNI=*77108602-1* 426
Callistemon pendulus var. *rigidus* IPNI=*77108602-1* 426
Callistemon semperflorens IPNI=*77108602-1* 426
Callistemon viminalis IPNI=*914221-1* 426
Callistylon arboreum IPNI=*944647-1* 315
Callitriche austinii IPNI=*430053-1* 376
Callitriche deflexa var. *austinii* IPNI=*430053-1* 376
Callitriche deflexa var. *subsessilis* IPNI=*430053-1* 376
Callitriche pedunculata IPNI=*430053-1* 376
Callitriche terrestris subsp. *subsessilis* IPNI=*430053-1* 376
Callitriche verna var. *terrestris* IPNI=*430053-1* 376
Callitropsis arizonica IPNI=*60451546-2* 180
Callitropsis lusitanica IPNI=*60451554-2* 180
Callitropsis macrocarpa IPNI=*60451558-2* 180
Calocarpum buchananiifolium IPNI=*943696-1* 307
Calocarpum huastecanum IPNI=*1110421-2* 307
Calocarpum mammosum IPNI=*152641-2* 306
Calocarpum mammosum var. *bonplandii* IPNI=*1110421-2* 307

Calocarpum mammosum var. *candollei* IPNI=*1110421-2* 307
Calocarpum mammosum var. *ovoideum* IPNI=*1110421-2* 307
Calocarpum procerum IPNI=*943696-1* 307
Calocarpum procerum var. *cuspidatum* IPNI=*943696-1* 307
Calocarpum procerum var. *oblongifolium* IPNI=*943696-1* 307
Calocarpum sapota IPNI=*1110421-2* 307
Calocarpum sclerocarpum IPNI=*209883-2* 308
Calochlamys capitata IPNI=*862610-1* 369
Calodium cochinchinensis IPNI=*463202-1* 182
Calodracon heliconiifolia IPNI=*533580-1* 215
Calodracon jacquinii IPNI=*533580-1* 215
Calodracon nobilis IPNI=*533580-1* 215
Calodracon sieberi IPNI=*533580-1* 215
Calodracon terminalis IPNI=*533580-1* 215
Calonnea pulcherrima IPNI=*329185-2* 263
Calophanes cubensis IPNI=*49672-1* 360
Calophyllum apetalum IPNI=*427190-1* 380
Calophyllum blumei IPNI=*427190-1* 380
Calophyllum brasiliense subsp. *longifolium* IPNI=*427212-1* 380
Calophyllum brasiliense subsp. *mariae* IPNI=*427115-1* 380
Calophyllum brasiliense subsp. *verum* IPNI=*427115-1* 380
Calophyllum brasiliense subsp. *wrightii* IPNI=*427115-1* 380
Calophyllum brasiliense var. *burchellii* IPNI=*427115-1* 380
Calophyllum brasiliense var. *elongatum* IPNI=*427115-1* 380
Calophyllum brasiliense var. *gardneri* IPNI=*427115-1* 380
Calophyllum brasiliense var. *genuina* IPNI=*427115-1* 380
Calophyllum brasiliense var. *spruceana* IPNI=*427115-1* 380
Calophyllum burmanni IPNI=*427119-1* 380
Calophyllum dubium IPNI=*427212-1* 380
Calophyllum edule IPNI=*277320-2* 385
Calophyllum ellipticum IPNI=*427115-1* 380
Calophyllum inophyllum var. *takamaka* IPNI=*427190-1* 380
Calophyllum inophyllum var. *wakamatsui* IPNI=*427190-1* 380
Calophyllum lucidum IPNI=*427115-1* 380
Calophyllum madruno IPNI=*277321-2* 385
Calophyllum mariae IPNI=*427115-1* 380
Calophyllum ovatifolium IPNI=*427190-1* 380
Calophyllum piaroanum IPNI=*427115-1* 380
Calophyllum revolutum IPNI=*427115-1* 380
Calophyllum spurium IPNI=*427190-1* 380
Calophyllum wakamatsui IPNI=*427190-1* 380
Calophysa ciliata IPNI=*77207714-1* 420
Calophysa dentata IPNI=*77207920-1* 422
Calophysa dispar IPNI=*77207368-1* 420
Calophysa heterophylla IPNI=*77207843-1* 420
Calophysa pilosa IPNI=*77207921-1* 421
Calophysa setosa IPNI=*77207860-1* 420
Calophysa testiculata IPNI=*77207921-1* 421
Calophysa tococoidea IPNI=*77207700-1* 423
Calophysa vesiculosa IPNI=*77207761-1* 423
Calopogonium brachycarpum IPNI=*483437-1* 313
Calopogonium caeruleum var. *glabrescens* IPNI=*60471356-2* 313
Calopogonium ferrugineum IPNI=*945254-1* 329
Calopogonium flavidum IPNI=*483437-1* 313
Calopogonium orthocarpum IPNI=*483437-1* 313
Calopogonium pedunculatum IPNI=*525419-1* 336
Calopogonium pumilum IPNI=*60471356-2* 313
Caloseris rupestris IPNI=*174455-2* 267
Calospermum mammosum IPNI=*152641-2* 306
Calospermum mammosum var. *bonplandii* IPNI=*1110421-2* 307
Calospermum mammosum var. *candollei* IPNI=*1110421-2* 307
Calospermum mammosum var. *ovoidea* IPNI=*1110421-2* 307
Calospermum parvum IPNI=*1110421-2* 307
Calostima aculeata IPNI=*1044269-2* 448
Calotheca purpurascens IPNI=*1104599-2* 238
Calotheca stricta var. *mandoniana* IPNI=*77116853-1* 236
Calotropis gigantea var. *procera* IPNI=*1004515-2* 339
Calotropis hamiltonii IPNI=*1004515-2* 339
Calotropis inflexa IPNI=*1004515-2* 339
Calotropis persica IPNI=*1004515-2* 339
Calotropis procera subsp. *hamiltonii* IPNI=*1004515-2* 339
Calotropis syriaca IPNI=*1004515-2* 339
Calotropis wallichii IPNI=*1004515-2* 339
Caltha officinalis IPNI=*187894-1* 259
Calusia emarginata IPNI=*509434-1* 328
Calycodaphne dysoxylon IPNI=*1146025-2* 184
Calycodaphne floribunda IPNI=*1146025-2* 184
Calycolpus angustifolius IPNI=*591482-1* 424
Calycolpus calophyllus var. *angustifolius* IPNI=*1061290-2* 424
Calycolpus campomanesioides IPNI=*1061290-2* 424
Calycolpus cordatus IPNI=*591482-1* 424
Calycolpus glaber IPNI=*591482-1* 424
Calycolpus glaber var. *angustilobus* IPNI=*591482-1* 424
Calycolpus ovalifolius IPNI=*591482-1* 424
Calycolpus schomburgkianus IPNI=*591482-1* 424
Calycolpus schomburgkianus var. *recurvatus* IPNI=*591482-1* 424
Calycolpus schomburgkianus var. *speciosus* IPNI=*591482-1* 424
Calycophyllum acreanum IPNI=*969466-1* 347
Calycophyllum coccineum IPNI=*770293-1* 358
Calycophyllum spruceanum f. *brasiliensis* IPNI=*745249-1* 347
Calycophyllum spruceanum f. *peruvianum* IPNI=*745249-1* 347
Calycophyllum tubulosum IPNI=*204802-2* 355
Calycorectes protractus IPNI=*600952-1* 428
Calydermos rugosus IPNI=*187778-1* 259
Calydermos salmeifolius IPNI=*187778-1* 259
Calyplectus acuminatus IPNI=*1054383-2* 417
Calyplectus dependens IPNI=*1054383-2* 417
Calyplectus punicifolius IPNI=*553519-1* 417
Calyplectus speciosus IPNI=*1054383-2* 417
Calyptracordia alba IPNI=*1018059-2* 436
Calyptracordia rotundifolia IPNI=*274993-2* 274
Calyptranthes affinis IPNI=*77191459-1* 426
Calyptranthes bipennis IPNI=*77199448-1* 426
Calyptranthes capitellata IPNI=*601603-1* 428
Calyptranthes caryophyllifolia IPNI=*601603-1* 428
Calyptranthes chytraculia IPNI=*77191371-1* 426
Calyptranthes chytraculia var. *genuina* IPNI=*77191371-1* 426
Calyptranthes cumini IPNI=*601603-1* 428
Calyptranthes cuminodora IPNI=*601603-1* 428
Calyptranthes cuspidata IPNI=*77191379-1* 426
Calyptranthes eugenioides IPNI=*600952-1* 428
Calyptranthes forsteri IPNI=*77191393-1* 426
Calyptranthes fragrans IPNI=*77191465-1* 426
Calyptranthes gigantifolia IPNI=*77191461-1* 426
Calyptranthes grandiflora IPNI=*77191459-1* 426
Calyptranthes hylobates IPNI=*77199448-1* 426
Calyptranthes jambolana IPNI=*601603-1* 428
Calyptranthes jambolifera IPNI=*601603-1* 428
Calyptranthes lucida f. *hilariana* IPNI=*77191459-1* 426
Calyptranthes lucida IPNI=*77191459-1* 426
Calyptranthes lucida var. *polyantha* IPNI=*77191459-1* 426
Calyptranthes luetzelburgii IPNI=*77191459-1* 426
Calyptranthes macrophylla IPNI=*77191461-1* 426
Calyptranthes nigrescens IPNI=*77199415-1* 426
Calyptranthes obtusa IPNI=*77160848-1* 426
Calyptranthes oneillii IPNI=*601603-1* 428
Calyptranthes paniculata IPNI=*77191465-1* 426
Calyptranthes pedunculata IPNI=*601603-1* 428
Calyptranthes pohliana IPNI=*77191459-1* 426
Calyptranthes polyantha IPNI=*77191459-1* 426
Calyptranthes pulchella IPNI=*77191477-1* 427
Calyptranthes pulchella var. *cuneata* IPNI=*77191477-1* 427
Calyptranthes pulchella var. *grandiflora* IPNI=*77191477-1* 427
Calyptranthes pulchella var. *latifolia* IPNI=*77191477-1* 427
Calyptranthes pulchella var. *parviflora* IPNI=*77191477-1* 427
Calyptranthes speciosa IPNI=*77191461-1* 426

Calyptranthes speciosa var. *gigantifolia* IPNI=*77191461-1* 426
Calyptranthes tenuis IPNI=*601603-1* 428
Calyptranthes tonduzii IPNI=*600952-1* 428
Calyptrella cycliophylla IPNI=*112942-2* 418
Calyptrella galeottii IPNI=*112942-2* 418
Calyptridium depressum IPNI=*698357-1* 286
Calyptrion aubletii IPNI=*77144891-1* 405
Calyptrion berteroi IPNI=*77144891-1* 405
Calyptrion citrifolium IPNI=*77144891-1* 405
Calyptrion excelsum IPNI=*77144891-1* 405
Calyptrion frangulifolium IPNI=*12522-2* 405
Calyptrion hybanthus IPNI=*77144891-1* 405
Calyptrion nitidum IPNI=*77144891-1* 405
Calyptrion orinocense IPNI=*77144891-1* 405
Calyptrogyne synanthera IPNI=*193857-2* 211
Calyptromyrcia elegans IPNI=*598851-1* 426
Calyptromyrcia spixiana IPNI=*598851-1* 426
Calyptronoma synanthera IPNI=*193857-2* 211
Calyptropsidium friedrichsthalianum IPNI=*211742-2* 428
Calyptropsidium sartorianum IPNI=*600952-1* 428
Calyptrostylis articulata IPNI=*220271-2* 232
Calyptrostylis divergens IPNI=*220271-2* 232
Calyptrostylis fascicularis IPNI=*220271-2* 232
Calyptrostylis florida IPNI=*220271-2* 232
Calyptrostylis gaudichaudii IPNI=*220271-2* 232
Calyptrostylis rudgei IPNI=*220271-2* 232
Calytriplex obovata IPNI=*1072674-2* 376
Camara alba IPNI=*141020-2* 378
Camara canescens IPNI=*863236-1* 378
Camara involucrata IPNI=*135206-2* 378
Camara lilacina f. *albiflora* IPNI=*324303-2* 378
Camara lilacina IPNI=*324303-2* 378
Camara lilacina var. *grandifolia* IPNI=*324303-2* 378
Camara lilacina var. *media* IPNI=*324303-2* 378
Camara lilacina var. *parvifolia* IPNI=*324303-2* 378
Camara lippioides IPNI=*863236-1* 378
Camara macropoda IPNI=*324301-2* 378
Camara microcephala IPNI=*863236-1* 378
Camara montevidensis IPNI=*135260-2* 378
Camara purpurea IPNI=*324301-2* 378
Camara rugulosa IPNI=*863431-1* 378
Camara salviifolia IPNI=*324303-2* 378
Camara sellowiana IPNI=*135260-2* 378
Camara trifolia f. *rosea* IPNI=*135333-2* 378
Camara trifolia IPNI=*135333-2* 378
Camara trifolia var. *grandifolia* IPNI=*135333-2* 378
Camara trifolia var. *indica* IPNI=*135333-2* 378
Camara vulgaris IPNI=*325686-2* 378
Camaridium affine IPNI=*643592-1* 221
Camaridium album IPNI=*643250-1* 220
Camaridium amazonicum IPNI=*643592-1* 221
Camaridium ampliflorum IPNI=*643344-1* 220
Camaridium carinulatum IPNI=*643344-1* 220
Camaridium cyrtopodanthum IPNI=*643592-1* 221
Camaridium grandiflorum IPNI=*643344-1* 220
Camaridium inauditum IPNI=*643519-1* 220
Camaridium insolitum IPNI=*280607-2* 220
Camaridium lutescens IPNI=*643592-1* 221
Camaridium ochroleucum IPNI=*643592-1* 221
Camaridium squamatum IPNI=*155331-2* 222
Camaridium uncatum IPNI=*155331-2* 222
Camarilla obliqua IPNI=*528506-1* 213
Camellia arborescens IPNI=*828548-1* 309
Camellia bohea IPNI=*828548-1* 309
Camellia chinensis IPNI=*828548-1* 309
Camellia florida IPNI=*828524-1* 309
Camellia formosensis IPNI=*828548-1* 309
Camellia hayaoi IPNI=*828524-1* 309
Camellia hozanensis IPNI=*828524-1* 309
Camellia kaempferia IPNI=*828524-1* 309
Camellia longlingensis IPNI=*828548-1* 309
Camellia mutabilis IPNI=*828524-1* 309
Camellia nakaii IPNI=*828524-1* 309
Camellia oleosa IPNI=*828548-1* 309
Camellia planipetala IPNI=*828524-1* 309
Camellia sinensis var. *waldenae* IPNI=*828548-1* 309
Camellia thea IPNI=*828548-1* 309
Camellia thea var. *bohea* IPNI=*828548-1* 309
Camellia thea var. *stricta* IPNI=*828548-1* 309
Camellia thea var. *viridis* IPNI=*828548-1* 309
Camellia tsubakki IPNI=*828524-1* 309
Camellia viridis IPNI=*828548-1* 309
Camellia wabiske IPNI=*828524-1* 309
Camellia waldenae IPNI=*828548-1* 309
Cameraria fontana IPNI=*311155-2* 286
Cameraria tamaquarina IPNI=*79926-1* 341
Camerium moluccanum IPNI=*338514-1* 387
Camirium cordifolium IPNI=*338514-1* 387
Camirium oleosum IPNI=*338514-1* 387
Camomilla deflexa IPNI=*154715-2* 265
Camomilla patens IPNI=*154715-2* 265
Campanula bourdiniana IPNI=*140670-1* 272
Campanula florida IPNI=*140670-1* 272
Campanula grandiflora IPNI=*140670-1* 272
Campanula perfoliata IPNI=*258447-2* 273
Campderia gracilis IPNI=*693795-1* 288
Campe praecox IPNI=*1109261-2* 276
Campe verna IPNI=*1109261-2* 276
Campelia bibracteata IPNI=*1191409-2* 224
Campelia boucheana IPNI=*1191409-2* 224
Campelia fastigiata IPNI=*1191409-2* 224
Campelia fendleri IPNI=*1191409-2* 224
Campelia glabrata IPNI=*1191409-2* 224
Campelia hoffmannii IPNI=*1191409-2* 224
Campelia mexicana IPNI=*1191409-2* 224
Campelia pseudozanonia IPNI=*1191409-2* 224
Campelia scandens IPNI=*1191409-2* 224
Campelia zanonia IPNI=*1191409-2* 224
Campelia zanonia var. *glabrata* IPNI=*1191409-2* 224
Campelia zanonia var. *sessilis* IPNI=*1191409-2* 224
Camphora camelliifolia IPNI=*463336-1* 183
Camphora camphora IPNI=*463336-1* 183
Camphora decaisnei IPNI=*463336-1* 183
Camphora decandollei IPNI=*463336-1* 183
Camphora humboldtii IPNI=*463336-1* 183
Camphora humboldtii var. *syringifolia* IPNI=*463336-1* 183
Camphora mauritiana IPNI=*463752-1* 183
Camphora neesii IPNI=*463336-1* 183
Camphora neesii var. *microphylla* IPNI=*463336-1* 183
Camphora neesii var. *pharbitifolia* IPNI=*463336-1* 183
Camphora officinarum var. *hahnemannii* IPNI=*463336-1* 183
Camphora officinarum var. *hippocratei* IPNI=*463336-1* 183
Camphora oldhamii IPNI=*463336-1* 183
Camphora procera IPNI=*463336-1* 183
Camphora rougieri IPNI=*463336-1* 183
Camphora sieboldii IPNI=*463336-1* 183
Camphora thouarsii IPNI=*463336-1* 183
Camphora thunbergii IPNI=*463336-1* 183
Camphora thwaitesii IPNI=*463336-1* 183
Camphora vera IPNI=*463336-1* 183
Camphora wrightii IPNI=*463336-1* 183
Camphora zollingeri IPNI=*463336-1* 183
Camphorina camphora IPNI=*463336-1* 183
Camphorina cinnamomum IPNI=*463752-1* 183
Camphoromoea surinamensis IPNI=*171727-2* 184
Campomanesia cornifolia IPNI=*592044-1* 424

Campomanesia glabra IPNI=*591482-1* 424
Campomanesia goetheana IPNI=*591482-1* 424
Campomanesia rivularis IPNI=*592044-1* 424
Campomanesia suffruticosa IPNI=*600884-1* 428
Campomanesia tomentosa IPNI=*211886-2* 428
Campsiandra comosa var. *laurifolia* IPNI=*44114-2* 313
Campteria biaurita IPNI=*213840-2* 178
Campteria pseudolonchitis IPNI=*17217170-1* 178
Campuloclinium ascendens IPNI=*64237-2* 260
Campuloclinium macrocephalum var. *strigosum* IPNI=*188196-1* 259
Campuloclinium pohlianum IPNI=*188196-1* 259
Campuloclinium strigosum IPNI=*188196-1* 259
Campylanthera samauma IPNI=*51083-2* 408
Campylia variegata IPNI=*377229-1* 358
Campyloneurum angustifolium var. *amphostemon* IPNI=*17063000-1* 175
Campyloneurum angustifolium var. *taeniosum* IPNI=*17063000-1* 175
Campyloneurum angustifolium var. *tenerum* IPNI=*17063000-1* 175
Campyloneurum decumanum IPNI=*17175400-1* 176
Campyloneurum dimorphum IPNI=*17063000-1* 175
Campyloneurum levigatum IPNI=*77075105-1* 177
Campyloneurum loreum IPNI=*17063000-1* 175
Campyloneurum moritzianum IPNI=*17063380-1* 175
Campyloneurum sieberianum IPNI=*17063380-1* 175
Campyloneurum taeniosum IPNI=*17063000-1* 175
Campyloneurum undulatum IPNI=*17063380-1* 175
Cananga schomburgkiana IPNI=*73626-1* 189
Cananga scortechinii IPNI=*72580-1* 187
Canangium mitrastigma IPNI=*72580-1* 187
Canangium odoratum IPNI=*72580-1* 187
Canangium scortechinii IPNI=*72580-1* 187
Canavalia altissima IPNI=*1197939-2* 328
Canavalia amazonica IPNI=*483623-1* 313
Canavalia anomala IPNI=*483623-1* 313
Canavalia apiculata IPNI=*30187778-2* 313
Canavalia arenicola IPNI=*30187778-2* 313
Canavalia baueriana IPNI=*30187778-2* 313
Canavalia campylocarpa IPNI=*483623-1* 313
Canavalia caribaea IPNI=*483623-1* 313
Canavalia dictyota IPNI=*483623-1* 313
Canavalia dictyota var. *ferruginea* IPNI=*483623-1* 313
Canavalia emarginata IPNI=*30187778-2* 313
Canavalia ensiformis var. *albida* IPNI=*1209500-2* 313
Canavalia ensiformis var. *coriacea* IPNI=*1209500-2* 313
Canavalia ensiformis var. *normalis* IPNI=*1209500-2* 313
Canavalia ensiformis var. *truncata* IPNI=*1209500-2* 313
Canavalia fendleri IPNI=*483623-1* 313
Canavalia leptophylla IPNI=*483623-1* 313
Canavalia maritima IPNI=*30187778-2* 313
Canavalia mexicana IPNI=*483623-1* 313
Canavalia miniata IPNI=*30187778-2* 313
Canavalia moneta IPNI=*30187778-2* 313
Canavalia obcordata IPNI=*30187778-2* 313
Canavalia obtusifolia IPNI=*30187778-2* 313
Canavalia panamensis IPNI=*483623-1* 313
Canavalia paraguayensis IPNI=*483623-1* 313
Canavalia podocarpa IPNI=*30187778-2* 313
Canavalia prolifica IPNI=*483623-1* 313
Canavalia rutilans IPNI=*30187778-2* 313
Canavalia stocksii IPNI=*1209500-2* 313
Candollea polypodioides IPNI=*1039912-2* 176
Canhamo braziliensis IPNI=*560723-1* 409
Canna achiras IPNI=*319130-2* 247
Canna altensteinii IPNI=*319130-2* 247
Canna amabilis IPNI=*319130-2* 247
Canna amambayensis IPNI=*315579-2* 247
Canna anahuacensis IPNI=*44441-2* 247
Canna angustifolia IPNI=*796016-1* 247
Canna annaei IPNI=*796016-1* 247
Canna ascendens IPNI=*319130-2* 247
Canna aurantiaca IPNI=*319130-2* 247
Canna aureovittata IPNI=*319130-2* 247
Canna barbadica IPNI=*319130-2* 247
Canna bidentata IPNI=*319130-2* 247
Canna bifida IPNI=*319130-2* 247
Canna bihorellii IPNI=*319130-2* 247
Canna brasiliensis IPNI=*319130-2* 247
Canna caledonis-peltata IPNI=*319130-2* 247
Canna carnea IPNI=*319130-2* 247
Canna cearensis IPNI=*319130-2* 247
Canna chinensis IPNI=*319130-2* 247
Canna cinnabarina IPNI=*319130-2* 247
Canna coccinea f. *flaviflora* IPNI=*319130-2* 247
Canna coccinea IPNI=*319130-2* 247
Canna coccinea var. *bicolor* IPNI=*319130-2* 247
Canna commutata IPNI=*319130-2* 247
Canna compacta IPNI=*319130-2* 247
Canna concinna IPNI=*319130-2* 247
Canna confusa IPNI=*315579-2* 247
Canna crocea IPNI=*319130-2* 247
Canna curviflora IPNI=*44441-2* 247
Canna densifolia IPNI=*319130-2* 247
Canna denudata IPNI=*315579-2* 247
Canna denudata var. *grandis* IPNI=*319130-2* 247
Canna discolor IPNI=*319130-2* 247
Canna discolor var. *rubripunctata* IPNI=*319130-2* 247
Canna discolor var. *viridifolia* IPNI=*319130-2* 247
Canna domingensis IPNI=*44425-2* 247
Canna edulis IPNI=*319130-2* 247
Canna ehrenbergii IPNI=*319130-2* 247
Canna elegans IPNI=*319130-2* 247
Canna ellipticifolia IPNI=*319130-2* 247
Canna excelsa IPNI=*315579-2* 247
Canna exigua IPNI=*319130-2* 247
Canna eximia IPNI=*319130-2* 247
Canna expansa-rubra IPNI=*319130-2* 247
Canna fintelmannii IPNI=*796016-1* 247
Canna flavescens IPNI=*319130-2* 247
Canna floribunda IPNI=*319130-2* 247
Canna formosa IPNI=*319130-2* 247
Canna fuchsina IPNI=*319130-2* 247
Canna fulgida IPNI=*319130-2* 247
Canna gaboniensis IPNI=*319130-2* 247
Canna gemella IPNI=*44441-2* 247
Canna gigantea IPNI=*44441-2* 247
Canna glauca var. *rubrolutea* IPNI=*796016-1* 247
Canna glauca var. *rufa* IPNI=*796016-1* 247
Canna glauca var. *siamensis* IPNI=*796016-1* 247
Canna hassleriana IPNI=*796016-1* 247
Canna heliconiifolia IPNI=*319130-2* 247
Canna heliconiifolia var. *xalapensis* IPNI=*319130-2* 247
Canna humilis IPNI=*319130-2* 247
Canna indica var. *edwardsii* IPNI=*319130-2* 247
Canna indica var. *limbata* IPNI=*319130-2* 247
Canna indica var. *patens* IPNI=*319130-2* 247
Canna indica var. *sanctae-rosae* IPNI=*319130-2* 247
Canna indica var. *warszewiczii* IPNI=*319130-2* 247
Canna insignis IPNI=*319130-2* 247
Canna iridiflora IPNI=*44441-2* 247
Canna jacobiniflora IPNI=*796016-1* 247
Canna jacquinii IPNI=*315579-2* 247
Canna juncea IPNI=*319130-2* 247
Canna kunzei IPNI=*315579-2* 247
Canna laeta IPNI=*319130-2* 247
Canna lagunensis IPNI=*319130-2* 247
Canna lambertii IPNI=*319130-2* 247

Canna lanuginosa IPNI=*319130-2* 247
Canna latifolia IPNI=*44441-2* 247
Canna leptochila IPNI=*319130-2* 247
Canna leucocarpa IPNI=*44425-2* 247
Canna limbata IPNI=*319130-2* 247
Canna linkii IPNI=*315579-2* 247
Canna liturata IPNI=*796016-1* 247
Canna longifolia IPNI=*796016-1* 247
Canna lutea IPNI=*319130-2* 247
Canna lutea var. *genuina* IPNI=*319130-2* 247
Canna macrophylla IPNI=*319130-2* 247
Canna maculata IPNI=*319130-2* 247
Canna maxima IPNI=*319130-2* 247
Canna meridensis IPNI=*315579-2* 247
Canna mexicana IPNI=*796016-1* 247
Canna miniata IPNI=*315579-2* 247
Canna montana IPNI=*319130-2* 247
Canna moritziana IPNI=*319130-2* 247
Canna musifolia IPNI=*319130-2* 247
Canna neglecta IPNI=*315579-2* 247
Canna neglecta IPNI=*44441-2* 247
Canna orientalis IPNI=*319130-2* 247
Canna ottonis IPNI=*315579-2* 247
Canna pallida IPNI=*319130-2* 247
Canna patens IPNI=*319130-2* 247
Canna pedicellata IPNI=*796016-1* 247
Canna pentaphylla IPNI=*319130-2* 247
Canna pertusa IPNI=*44425-2* 247
Canna peruviana-purpurea IPNI=*319130-2* 247
Canna peruviana-robusta IPNI=*319130-2* 247
Canna platyphylla IPNI=*319130-2* 247
Canna plurituberosa IPNI=*319130-2* 247
Canna poeppigii IPNI=*319130-2* 247
Canna polyclada IPNI=*319130-2* 247
Canna polymorpha IPNI=*319130-2* 247
Canna portoricensis IPNI=*319130-2* 247
Canna pruinosa IPNI=*319130-2* 247
Canna pulchra IPNI=*319130-2* 247
Canna purpurea-spectabilis IPNI=*319130-2* 247
Canna recurvata IPNI=*319130-2* 247
Canna rendatleri IPNI=*319130-2* 247
Canna roscoeana IPNI=*319130-2* 247
Canna rotundifolia IPNI=*319130-2* 247
Canna rubra IPNI=*319130-2* 247
Canna rubricaulis IPNI=*319130-2* 247
Canna sanctae-rosae IPNI=*319130-2* 247
Canna sanguinea IPNI=*319130-2* 247
Canna saturate-rubra IPNI=*319130-2* 247
Canna schlechtendaliana IPNI=*796016-1* 247
Canna schubertii IPNI=*319130-2* 247
Canna seleriana IPNI=*319130-2* 247
Canna sellowii IPNI=*319130-2* 247
Canna siamensis IPNI=*796016-1* 247
Canna speciosa IPNI=*319130-2* 247
Canna spectabilis IPNI=*319130-2* 247
Canna stenantha IPNI=*796016-1* 247
Canna stolonifera IPNI=*796016-1* 247
Canna stricta IPNI=*796016-1* 247
Canna sulphurea IPNI=*319130-2* 247
Canna surinamensis IPNI=*319130-2* 247
Canna sylvestris IPNI=*44441-2* 247
Canna tenuiflora IPNI=*319130-2* 247
Canna textoria IPNI=*319130-2* 247
Canna thyrsiflora IPNI=*319130-2* 247
Canna tubiflora IPNI=*315579-2* 247
Canna tulianensis IPNI=*44441-2* 247
Canna variabilis IPNI=*319130-2* 247
Canna variegata IPNI=*319130-2* 247
Canna variegatifolia IPNI=*319130-2* 247
Canna ventricosa IPNI=*319130-2* 247
Canna violacea IPNI=*44441-2* 247
Canna warszewiczii IPNI=*319130-2* 247
Canna warszewiczii var. *flameus* IPNI=*319130-2* 247
Canna xalapensis IPNI=*319130-2* 247
Cannabis americana IPNI=*306087-2* 436
Cannabis chinensis IPNI=*306087-2* 436
Cannabis erratica IPNI=*306087-2* 436
Cannabis foetens IPNI=*306087-2* 436
Cannabis generalis IPNI=*306087-2* 436
Cannabis gigantea IPNI=*306087-2* 436
Cannabis indica IPNI=*306087-2* 436
Cannabis intersita IPNI=*306087-2* 436
Cannabis kafiristanica IPNI=*306087-2* 436
Cannabis lupulus IPNI=*306087-2* 436
Cannabis macrosperma IPNI=*306087-2* 436
Cannabis ruderalis IPNI=*306087-2* 436
Cannabis sativa subsp. *indica* IPNI=*306087-2* 436
Cannabis sativa subsp. *intersita* IPNI=*306087-2* 436
Cannabis sativa var. *afghanica* IPNI=*306087-2* 436
Cannabis sativa var. *indica* IPNI=*306087-2* 436
Cannabis sativa var. *kafiristanica* IPNI=*306087-2* 436
Cannabis sativa var. *ruderalis* IPNI=*306087-2* 436
Cannacorus indicus IPNI=*319130-2* 247
Cannacorus ovatus IPNI=*319130-2* 247
Canscora decurrens IPNI=*366552-1* 344
Canscora diffusa var. *tetraptera* IPNI=*366552-1* 344
Canscora divaricata IPNI=*366552-1* 344
Canscora foliosa IPNI=*366552-1* 344
Canscora khandalensis IPNI=*366552-1* 344
Canscora kirkii IPNI=*366552-1* 344
Canscora lancifolia IPNI=*366552-1* 344
Canscora lauri IPNI=*366552-1* 344
Canscora lawii IPNI=*366552-1* 344
Canscora pusilla IPNI=*366552-1* 344
Canscora rubiflora IPNI=*366552-1* 344
Canscora sanjappae IPNI=*366552-1* 344
Canthium chinense IPNI=*745904-1* 348
Canthium coronatum IPNI=*745904-1* 348
Canthium spinosum IPNI=*745904-1* 348
Cantua dependens IPNI=*688721-1* 303
Caopia acuminata IPNI=*30051774-2* 394
Caopia baccifera IPNI=*434100-1* 394
Caopia cayennensis IPNI=*434107-1* 394
Caopia ferruginea IPNI=*434114-1* 394
Caopia guianensis IPNI=*30051774-2* 394
Caopia lauriformis IPNI=*434138-1* 394
Caopia macrophylla IPNI=*434144-1* 394
Caopia reichardtiana IPNI=*30051774-2* 394
Caopia sessilifolia IPNI=*434162-1* 394
Caopia sessilifolia var. *rufescens* IPNI=*434162-1* 394
Caoutchoua elastica IPNI=*121724-2* 391
Caoutchoua guianensis IPNI=*121724-2* 391
Caperonia angusta IPNI=*44529-2* 388
Caperonia liebmanniana IPNI=*44564-2* 388
Caperonia nervosa IPNI=*44529-2* 388
Caperonia paludosa IPNI=*44529-2* 388
Caperonia palustris var. *linearifolia* IPNI=*44564-2* 388
Caperonia panamensis IPNI=*44529-2* 388
Caperonia pubescens IPNI=*44564-2* 388
Caperonia stenomeres IPNI=*44529-2* 388
Capirona decorticans IPNI=*77210192-1* 347
Capirona duckei IPNI=*77210192-1* 347
Capirona huberiana IPNI=*77210192-1* 347
Capirona leiophloea IPNI=*77210192-1* 347
Capirona surinamensis IPNI=*77210192-1* 347
Capirona wurdackii IPNI=*77210192-1* 347

Capparicordis crotonoides IPNI=*77183834-1* 278
Capparidastrum elegans IPNI=*77183570-1* 278
Capparidastrum frondosum IPNI=*77183570-1* 278
Capparidastrum osmanthum IPNI=*77184046-1* 278
Capparidastrum pachaca IPNI=*77184114-1* 278
Capparidastrum sola IPNI=*77184186-1* 278
Capparis acutifolia IPNI=*77184186-1* 278
Capparis amplissima IPNI=*77184082-1* 278
Capparis amygdalifolia IPNI=*77217338-1* 278
Capparis amygdalina IPNI=*77217338-1* 278
Capparis arborescens IPNI=*77183588-1* 278
Capparis arrabidae IPNI=*77183570-1* 278
Capparis barcellonensis IPNI=*77217338-1* 278
Capparis blanchetii IPNI=*77183644-1* 278
Capparis brevipes IPNI=*77183644-1* 278
Capparis brevisiliqua IPNI=*316361-2* 278
Capparis breynia f. *parviflora* IPNI=*77217338-1* 278
Capparis breynia IPNI=*77184072-1* 278
Capparis breynia var. *uniflora* IPNI=*77217338-1* 278
Capparis cardiophylla IPNI=*77183680-1* 278
Capparis commutata IPNI=*77183570-1* 278
Capparis crotonantha IPNI=*77184198-1* 278
Capparis crotonoides IPNI=*77183834-1* 278
Capparis cuneata IPNI=*77183570-1* 278
Capparis cynophallophora f. *laetevirens* IPNI=*316361-2* 278
Capparis cynophallophora var. *pubescens* IPNI=*316361-2* 278
Capparis cynophallophora var. *saligna* IPNI=*316361-2* 278
Capparis detonsa IPNI=*77183701-1* 278
Capparis diversifolia IPNI=*316361-2* 278
Capparis elegans IPNI=*77183570-1* 278
Capparis eucalyptifolia IPNI=*316361-2* 278
Capparis eustachiana IPNI=*316361-2* 278
Capparis ferriginea IPNI=*77184072-1* 278
Capparis flexuosa IPNI=*316361-2* 278
Capparis flexuosa IPNI=*77183570-1* 278
Capparis flexuosa var. *cordifolia* IPNI=*316361-2* 278
Capparis flexuosa var. *saligna* IPNI=*316361-2* 278
Capparis frondosa IPNI=*77183570-1* 278
Capparis furfuracea IPNI=*77217338-1* 278
Capparis guaguaensis IPNI=*77184046-1* 278
Capparis guayaquilensis IPNI=*316361-2* 278
Capparis indica IPNI=*77217338-1* 278
Capparis intermedia IPNI=*77184072-1* 278
Capparis laeta IPNI=*77183644-1* 278
Capparis laetevirens IPNI=*316361-2* 278
Capparis laevigata IPNI=*316361-2* 278
Capparis lepidota IPNI=*77184072-1* 278
Capparis linearifolia IPNI=*77184255-1* 278
Capparis linearis IPNI=*77184255-1* 278
Capparis mollis IPNI=*316361-2* 278
Capparis morisonia IPNI=*316366-2* 278
Capparis nemorosa IPNI=*77183738-1* 278
Capparis nitida IPNI=*77184082-1* 278
Capparis noctiflora IPNI=*77184082-1* 278
Capparis odoratissima IPNI=*77184072-1* 278
Capparis osmantha IPNI=*77184046-1* 278
Capparis pachaca IPNI=*77184114-1* 278
Capparis palmeri IPNI=*77183644-1* 278
Capparis pendula IPNI=*77184082-1* 278
Capparis petiolaris IPNI=*77183570-1* 278
Capparis pittieri IPNI=*77184198-1* 278
Capparis pluvialis IPNI=*316361-2* 278
Capparis populifolia IPNI=*77183834-1* 278
Capparis portoricensis IPNI=*77184082-1* 278
Capparis pulcherrima IPNI=*77183588-1* 278
Capparis racemosa IPNI=*316366-2* 278
Capparis rugulosa IPNI=*77183644-1* 278
Capparis saligna IPNI=*316361-2* 278
Capparis schunkei IPNI=*77183701-1* 278
Capparis securidacea IPNI=*77183680-1* 278
Capparis sessilis IPNI=*77183680-1* 278
Capparis sessilis var. *genuina* IPNI=*77183680-1* 278
Capparis sessilis var. *securidacea* IPNI=*77183680-1* 278
Capparis sidifolia IPNI=*77183834-1* 278
Capparis sinclairii IPNI=*316361-2* 278
Capparis sola IPNI=*77184186-1* 278
Capparis sola var. *longiracemosa* IPNI=*77184186-1* 278
Capparis stenophylla IPNI=*77183570-1* 278
Capparis subbiloba IPNI=*316361-2* 278
Capparis tonduzii IPNI=*77217338-1* 278
Capparis torulosa IPNI=*77184072-1* 278
Capparis triflora IPNI=*77183570-1* 278
Capparis trifoliata IPNI=*316363-2* 278
Capparis trinitensis IPNI=*77183644-1* 278
Capparis usiacurina IPNI=*77183680-1* 278
Capparis verrucosa IPNI=*77183644-1* 278
Capraria biflora f. *hirta* IPNI=*800773-1* 377
Capraria biflora var. *pilosa* IPNI=*800773-1* 377
Capraria crustacea IPNI=*1189750-2* 374
Capraria dulcis IPNI=*230902-2* 376
Capraria dulcis var. *albiflora* IPNI=*230902-2* 376
Capraria dulcis var. *coerulea* IPNI=*230902-2* 376
Capraria durantifolia IPNI=*328499-2* 377
Capraria hirsuta IPNI=*800773-1* 377
Capraria humilis IPNI=*244449-2* 377
Capraria lanceolata IPNI=*800773-1* 377
Capraria monnieria IPNI=*1072674-2* 376
Capraria semiserrata IPNI=*800773-1* 377
Capraria semiserrata var. *berteroi* IPNI=*800773-1* 377
Caprificus gigantea IPNI=*316340-2* 438
Caprificus insectifera IPNI=*852556-1* 438
Caprificus leucocarpa IPNI=*852556-1* 438
Caprificus oblongata IPNI=*852556-1* 438
Caprificus pedunculata IPNI=*852556-1* 438
Caprificus rugosa IPNI=*852556-1* 438
Caprificus sphaerocarpa IPNI=*852556-1* 438
Caprifolium brachypodum IPNI=*148850-1* 296
Caprifolium japonicum IPNI=*148850-1* 296
Caprifolium roseum IPNI=*148850-1* 296
Capriola dactylon IPNI=*397065-1* 236
Capriola dactylon var. *maritima* IPNI=*397065-1* 236
Capsella abscissa IPNI=*30092589-2* 276
Capsella acutifolia IPNI=*30092589-2* 276
Capsella aestivalis IPNI=*30092589-2* 276
Capsella agrestis IPNI=*30092589-2* 276
Capsella agrestis-rubella IPNI=*30092589-2* 276
Capsella alandica IPNI=*30092589-2* 276
Capsella algida IPNI=*30092589-2* 276
Capsella alpestris IPNI=*30092589-2* 276
Capsella altissima IPNI=*30092589-2* 276
Capsella amblodes IPNI=*30092589-2* 276
Capsella angermannica IPNI=*30092589-2* 276
Capsella anglica IPNI=*30092589-2* 276
Capsella angustiloba IPNI=*30092589-2* 276
Capsella apetala IPNI=*30092589-2* 276
Capsella austriaca IPNI=*30092589-2* 276
Capsella autumnalis IPNI=*30092589-2* 276
Capsella batavorum IPNI=*30092589-2* 276
Capsella belgica IPNI=*30092589-2* 276
Capsella bicuspis IPNI=*30092589-2* 276
Capsella bifida IPNI=*30092589-2* 276
Capsella biformis IPNI=*30092589-2* 276
Capsella bottnica IPNI=*30092589-2* 276
Capsella bremensis IPNI=*30092589-2* 276
Capsella brevisiliqua IPNI=*30092589-2* 276
Capsella brittonii IPNI=*30092589-2* 276

Capsella bursa IPNI=*30092589-2* 276
Capsella bursa-pastoris f. *arachnoidea* IPNI=*30092589-2* 276
Capsella bursa-pastoris f. *attenuata* IPNI=*30092589-2* 276
Capsella bursa-pastoris f. *coronopifolia* IPNI=*30092589-2* 276
Capsella bursa-pastoris f. *orbicularis* IPNI=*30092589-2* 276
Capsella bursa-pastoris f. *pinnatifida* IPNI=*30092589-2* 276
Capsella bursa-pastoris f. *rhomboidea* IPNI=*30092589-2* 276
Capsella bursa-pastoris f. *setchelliana* IPNI=*30092589-2* 276
Capsella bursa-pastoris f. *simplex* IPNI=*30092589-2* 276
Capsella bursa-pastoris f. *treleaseana* IPNI=*30092589-2* 276
Capsella bursa-pastoris var. *bifida* IPNI=*30092589-2* 276
Capsella bursa-pastoris var. *integrifolia* IPNI=*30092589-2* 276
Capsella bursa-pastoris var. *minor* IPNI=*30092589-2* 276
Capsella bursa-pastoris var. *pinnata* IPNI=*30092589-2* 276
Capsella bursa-pastoris var. *rubriflora* IPNI=*30092589-2* 276
Capsella bursa-pastoris var. *stenocarpa* IPNI=*30092589-2* 276
Capsella calmariensis IPNI=*30092589-2* 276
Capsella caucasica IPNI=*30092589-2* 276
Capsella cavata IPNI=*30092589-2* 276
Capsella collina IPNI=*30092589-2* 276
Capsella compacta IPNI=*30092589-2* 276
Capsella convexiformis IPNI=*30092589-2* 276
Capsella cordata IPNI=*30092589-2* 276
Capsella coroncpus IPNI=*30092589-2* 276
Capsella cuneiformis IPNI=*30092589-2* 276
Capsella cuneolata IPNI=*30092589-2* 276
Capsella curvisiliqua IPNI=*30092589-2* 276
Capsella cuspidata IPNI=*30092589-2* 276
Capsella dentata IPNI=*30092589-2* 276
Capsella denticulata IPNI=*30092589-2* 276
Capsella difformis IPNI=*30092589-2* 276
Capsella diodonta IPNI=*30092589-2* 276
Capsella divaricata IPNI=*30092589-2* 276
Capsella druceana IPNI=*30092589-2* 276
Capsella elegans IPNI=*30092589-2* 276
Capsella emarginata IPNI=*30092589-2* 276
Capsella erecta IPNI=*30092589-2* 276
Capsella exotica IPNI=*30092589-2* 276
Capsella faucialis IPNI=*30092589-2* 276
Capsella foliosa IPNI=*30092589-2* 276
Capsella frigida IPNI=*30092589-2* 276
Capsella fucorum IPNI=*30092589-2* 276
Capsella furcata IPNI=*30092589-2* 276
Capsella galicica IPNI=*30092589-2* 276
Capsella gallica IPNI=*30092589-2* 276
Capsella germanica IPNI=*30092589-2* 276
Capsella gothica IPNI=*30092589-2* 276
Capsella gotlandica IPNI=*30092589-2* 276
Capsella gracilescens IPNI=*30092589-2* 276
Capsella grossa IPNI=*30092589-2* 276
Capsella grossotriangularis IPNI=*30092589-2* 276
Capsella hanseatica IPNI=*30092589-2* 276
Capsella heegeri IPNI=*30092589-2* 276
Capsella helsingica IPNI=*30092589-2* 276
Capsella herjedalica IPNI=*30092589-2* 276
Capsella hians IPNI=*30092589-2* 276
Capsella hiatula IPNI=*30092589-2* 276
Capsella hiemalis IPNI=*30092589-2* 276
Capsella hispida IPNI=*30092589-2* 276
Capsella humilis IPNI=*30092589-2* 276
Capsella hyrcana IPNI=*30092589-2* 276
Capsella incisura IPNI=*30092589-2* 276
Capsella integrella IPNI=*30092589-2* 276
Capsella integrifolia IPNI=*30092589-2* 276
Capsella jemtlandica IPNI=*30092589-2* 276
Capsella jeniseiensis IPNI=*30092589-2* 276
Capsella kyphosa IPNI=*30092589-2* 276
Capsella lacerata IPNI=*30092589-2* 276
Capsella laevigata IPNI=*30092589-2* 276
Capsella lanceolatocaspica IPNI=*30092589-2* 276
Capsella latisiliqua IPNI=*30092589-2* 276
Capsella lenaensis IPNI=*30092589-2* 276
Capsella leptoloba IPNI=*30092589-2* 276
Capsella linearis IPNI=*30092589-2* 276
Capsella lingulata IPNI=*30092589-2* 276
Capsella lobulata IPNI=*30092589-2* 276
Capsella longiracemosa IPNI=*30092589-2* 276
Capsella longirostris IPNI=*30092589-2* 276
Capsella longisiliqua IPNI=*30092589-2* 276
Capsella lulensis IPNI=*30092589-2* 276
Capsella lutetiana IPNI=*30092589-2* 276
Capsella macroclada IPNI=*30092589-2* 276
Capsella matura IPNI=*30092589-2* 276
Capsella mediterranea IPNI=*30092589-2* 276
Capsella microcarpa IPNI=*30092589-2* 276
Capsella monasterialis IPNI=*30092589-2* 276
Capsella nanella IPNI=*30092589-2* 276
Capsella nylandica IPNI=*30092589-2* 276
Capsella obovata IPNI=*30092589-2* 276
Capsella obtusa IPNI=*30092589-2* 276
Capsella odontophylla IPNI=*30092589-2* 276
Capsella oelandica IPNI=*30092589-2* 276
Capsella origo IPNI=*30092589-2* 276
Capsella oviculata IPNI=*30092589-2* 276
Capsella ovifera IPNI=*30092589-2* 276
Capsella pastoralis IPNI=*30092589-2* 276
Capsella pastoris IPNI=*30092589-2* 276
Capsella patagonica IPNI=*30092589-2* 276
Capsella penarthae IPNI=*30092589-2* 276
Capsella perconcava IPNI=*30092589-2* 276
Capsella perdentata IPNI=*30092589-2* 276
Capsella pergrossa IPNI=*30092589-2* 276
Capsella perhians IPNI=*30092589-2* 276
Capsella perversa IPNI=*30092589-2* 276
Capsella pinnata IPNI=*30092589-2* 276
Capsella pinnatofoliosa IPNI=*30092589-2* 276
Capsella poimenobalantion IPNI=*30092589-2* 276
Capsella polymorpha IPNI=*30092589-2* 276
Capsella praecox IPNI=*30092589-2* 276
Capsella praematura IPNI=*30092589-2* 276
Capsella prionophylla IPNI=*30092589-2* 276
Capsella provincialis IPNI=*30092589-2* 276
Capsella querceti IPNI=*30092589-2* 276
Capsella ramselensis IPNI=*30092589-2* 276
Capsella rhenana IPNI=*30092589-2* 276
Capsella rhombea IPNI=*30092589-2* 276
Capsella robusta IPNI=*30092589-2* 276
Capsella rubelliformis IPNI=*30092589-2* 276
Capsella rubelloagrestis IPNI=*30092589-2* 276
Capsella rubiginosa IPNI=*30092589-2* 276
Capsella ruderalis IPNI=*30092589-2* 276
Capsella salinula IPNI=*30092589-2* 276
Capsella savonica IPNI=*30092589-2* 276
Capsella scanica IPNI=*30092589-2* 276
Capsella scoliocaspica IPNI=*30092589-2* 276
Capsella segetum IPNI=*30092589-2* 276
Capsella seleniaca IPNI=*30092589-2* 276
Capsella semilobata IPNI=*30092589-2* 276
Capsella semirubella IPNI=*30092589-2* 276
Capsella sinuatolinearis IPNI=*30092589-2* 276
Capsella sinuosa IPNI=*30092589-2* 276
Capsella smolandica IPNI=*30092589-2* 276
Capsella spinosa IPNI=*286421-1* 277
Capsella stenocarpa IPNI=*30092589-2* 276
Capsella subalgida IPNI=*30092589-2* 276
Capsella subalpina IPNI=*30092589-2* 276

Capsella subarctica IPNI=*30092589-2* 276
Capsella subbergiana IPNI=*30092589-2* 276
Capsella subcanescens IPNI=*30092589-2* 276
Capsella subcavata IPNI=*30092589-2* 276
Capsella subseleniaca IPNI=*30092589-2* 276
Capsella tenerescens IPNI=*30092589-2* 276
Capsella tibelensis IPNI=*30092589-2* 276
Capsella tjustiana IPNI=*30092589-2* 276
Capsella trevirorum IPNI=*30092589-2* 276
Capsella triangularis IPNI=*30092589-2* 276
Capsella turoniensis IPNI=*30092589-2* 276
Capsella umensis IPNI=*30092589-2* 276
Capsella ursina IPNI=*30092589-2* 276
Capsella varia IPNI=*30092589-2* 276
Capsella vestrogothica IPNI=*30092589-2* 276
Capsella viguieri IPNI=*30092589-2* 276
Capsella viminalis IPNI=*30092589-2* 276
Capsella virgata IPNI=*30092589-2* 276
Capsella viridis IPNI=*30092589-2* 276
Caracalla pulcherrima IPNI=*60471746-2* 315
Caraguata lingulata IPNI=*115161-2* 228
Caraguata musaica IPNI=*115185-2* 229
Caraguata sanguinea IPNI=*115224-2* 229
Caraipa calophylla IPNI=*828566-1* 381
Caraipa excelsa IPNI=*828563-1* 380
Caraipa fasciculata IPNI=*828563-1* 380
Caraipa fasciculata var. *laxiflora* IPNI=*828563-1* 380
Caraipa fasciculata var. *typica* IPNI=*828563-1* 380
Caraipa glabrata IPNI=*828566-1* 381
Caraipa glabrata var. *typica* IPNI=*828566-1* 381
Caraipa insidiosa IPNI=*828563-1* 380
Caraipa laurifolia IPNI=*828563-1* 380
Caraipa laxiflora IPNI=*828563-1* 380
Caraipa melhemiana IPNI=*828563-1* 380
Caraipa myrciifolia IPNI=*828563-1* 380
Caraipa palustris IPNI=*828566-1* 381
Caraipa paniculata IPNI=*828939-1* 381
Caraipa sellowii IPNI=*828563-1* 380
Caraipa silvatica IPNI=*828566-1* 381
Caraipa variabilis IPNI=*828563-1* 380
Carapichea patrisii IPNI=*77115613-1* 354
Carara didyma IPNI=*286148-1* 277
Caraxeron brasiliense IPNI=*10294-2* 281
Caraxeron vermicularis IPNI=*34033-2* 282
Cardamindum majus IPNI=*310974-2* 280
Cardamine alsophila IPNI=*280189-1* 276
Cardamine alsophila var. *caespitosa* IPNI=*280189-1* 276
Cardamine alsophila var. *tridens* IPNI=*280189-1* 276
Cardamine andicola IPNI=*280189-1* 276
Cardamine angulata IPNI=*280380-1* 276
Cardamine aquatica IPNI=*287534-1* 277
Cardamine axillaris IPNI=*280189-1* 276
Cardamine bonariensis subsp. *minima* IPNI=*280189-1* 276
Cardamine borbonica IPNI=*280380-1* 276
Cardamine bracteata IPNI=*280189-1* 276
Cardamine caespitosa IPNI=*280189-1* 276
Cardamine demissa IPNI=*280189-1* 276
Cardamine drymeja IPNI=*45189-2* 276
Cardamine duraniensis IPNI=*45189-2* 276
Cardamine fagetina IPNI=*280380-1* 276
Cardamine flaccida f. *bracteata* IPNI=*280189-1* 276
Cardamine flaccida proles laxa IPNI=*280189-1* 276
Cardamine flaccida proles pusilla IPNI=*280189-1* 276
Cardamine flaccida subsp. *alsophila* IPNI=*280189-1* 276
Cardamine flaccida subsp. *bonariensis* IPNI=*280189-1* 276
Cardamine flaccida subsp. *ebracteata* IPNI=*280189-1* 276
Cardamine flaccida subsp. *minima* IPNI=*280189-1* 276
Cardamine flexuosa var. *gracilipes* IPNI=*45189-2* 276
Cardamine flexuosa var. *haleakalensis* IPNI=*45189-2* 276
Cardamine flexuosa var. *ovatifolia* IPNI=*45189-2* 276
Cardamine flexuosa var. *petiolulata* IPNI=*45189-2* 276
Cardamine fontana IPNI=*287534-1* 277
Cardamine glandulosa IPNI=*222144-2* 277
Cardamine hirsuta f. *pilosa* IPNI=*280380-1* 276
Cardamine hirsuta f. *unicaulis* IPNI=*280380-1* 276
Cardamine hirsuta proles lasiocarpa IPNI=*280380-1* 276
Cardamine hirsuta var. *apetala* IPNI=*280380-1* 276
Cardamine hirsuta var. *montana* IPNI=*280380-1* 276
Cardamine hirsuta var. *omeiensis* IPNI=*45189-2* 276
Cardamine hirsuta var. *petiolulata* IPNI=*280380-1* 276
Cardamine hirsuta var. *pilosa* IPNI=*280380-1* 276
Cardamine hirsuta var. *unicaulis* IPNI=*280380-1* 276
Cardamine humilis IPNI=*280380-1* 276
Cardamine ibaguense IPNI=*280543-1* 276
Cardamine indica subvar. *plebeia* IPNI=*222160-2* 277
Cardamine johnstonii IPNI=*280530-1* 276
Cardamine johnstonii proles punicea IPNI=*280530-1* 276
Cardamine johnstonii var. *superba* IPNI=*280530-1* 276
Cardamine killipii IPNI=*280189-1* 276
Cardamine konaensis IPNI=*45189-2* 276
Cardamine lamontii IPNI=*222144-2* 277
Cardamine laxa IPNI=*280189-1* 276
Cardamine lehmannii IPNI=*280543-1* 276
Cardamine micrantha IPNI=*280380-1* 276
Cardamine minima IPNI=*280189-1* 276
Cardamine multicaulis IPNI=*280380-1* 276
Cardamine muscosa IPNI=*45189-2* 276
Cardamine nana IPNI=*222164-2* 277
Cardamine nasturtium-aquaticum IPNI=*287534-1* 277
Cardamine nemophila IPNI=*280189-1* 276
Cardamine obliqua var. *stylosa* IPNI=*280530-1* 276
Cardamine ovata f. *lehmannii* IPNI=*280543-1* 276
Cardamine ovata subsp. *lehmannii* IPNI=*280543-1* 276
Cardamine ovata var. *bracteata* IPNI=*280543-1* 276
Cardamine ovata var. *corymbosa* IPNI=*280543-1* 276
Cardamine ovata var. *eriocarpa* IPNI=*280543-1* 276
Cardamine ovata var. *glabrata* IPNI=*280543-1* 276
Cardamine palustris IPNI=*288669-1* 277
Cardamine palustris subvar. *erecta* IPNI=*288669-1* 277
Cardamine palustris subvar. *lyratopinnatifida* IPNI=*288669-1* 277
Cardamine palustris subvar. *umbrosa* IPNI=*288669-1* 277
Cardamine palustris var. *intermedia* IPNI=*288669-1* 277
Cardamine palustris var. *mexicana* IPNI=*222160-2* 277
Cardamine palustris var. *pumila* IPNI=*288669-1* 277
Cardamine parviflora IPNI=*280380-1* 276
Cardamine praecox IPNI=*280380-1* 276
Cardamine punicea IPNI=*280530-1* 276
Cardamine pusilla IPNI=*280189-1* 276
Cardamine pusilla IPNI=*45189-2* 276
Cardamine repanda IPNI=*280189-1* 276
Cardamine scutata subsp. *flexuosa* IPNI=*45189-2* 276
Cardamine scutata var. *formosana* IPNI=*280380-1* 276
Cardamine setigera IPNI=*45189-2* 276
Cardamine simensis IPNI=*280380-1* 276
Cardamine sublyrata IPNI=*288735-1* 277
Cardamine sylvatica IPNI=*45189-2* 276
Cardamine tenella IPNI=*280380-1* 276
Cardamine tetrandra IPNI=*280380-1* 276
Cardamine tolimensis IPNI=*280530-1* 276
Cardamine tridens IPNI=*280189-1* 276
Cardamine virginica IPNI=*280380-1* 276
Cardaminum nasturtium IPNI=*287534-1* 277
Cardamomum elletari IPNI=*796556-1* 250
Cardamomum magnificum IPNI=*942355-1* 251
Cardamomum malabaricum IPNI=*796556-1* 250
Cardamomum officinale IPNI=*796556-1* 250

Cardamomum speciosum IPNI=*942355-1* 251
Cardamomum tridentatum IPNI=*942355-1* 251
Cardamomum verum IPNI=*796556-1* 250
Cardenasia setacea IPNI=*229758-2* 332
Cardia arvensis IPNI=*811649-1* 377
Cardia filiformis IPNI=*30198355-2* 377
Cardia multiflora IPNI=*812650-1* 377
Cardia officinalis IPNI=*812650-1* 377
Cardiaca arvensis IPNI=*459319-1* 373
Cardiochlaena ampla IPNI=*17226640-1* 177
Cardiochlaena macrophylla IPNI=*17226640-1* 177
Cardiochlaena macrophylla var. *acuminata* IPNI=*17226640-1* 177
Cardiochlaena macrophylla var. *decrescens* IPNI=*17226640-1* 177
Cardiochlaena macrophylla var. *discreta* IPNI=*17226640-1* 177
Cardiochlaena macrophylla var. *distans* IPNI=*17226640-1* 177
Cardiochlaena macrophylla var. *lata* IPNI=*17226640-1* 177
Cardiochlaena macrophylla var. *pinnata* IPNI=*17226640-1* 177
Cardiochlaena macrophylla var. *sinuata* IPNI=*17226640-1* 177
Cardiolepis obtusa IPNI=*30426867-2* 441
Cardionema multicaule IPNI=*45444-2* 285
Cardionema rosetta IPNI=*45444-2* 285
Carduus amplexicaulis IPNI=*250268-1* 270
Carduus cardunculus IPNI=*200876-1* 260
Carduus chinensis IPNI=*196191-1* 260
Carduus cynara IPNI=*200876-1* 260
Carduus cynaroides IPNI=*200876-1* 260
Carduus echinatus IPNI=*195292-1* 260
Carduus firmus IPNI=*196191-1* 260
Carduus lactifolius IPNI=*249211-1* 270
Carduus mariae IPNI=*249211-1* 270
Carduus marianus IPNI=*249211-1* 270
Carduus mexicanus IPNI=*195679-1* 260
Carduus scolymus IPNI=*200876-1* 260
Carduus spinosissimus IPNI=*196191-1* 260
Carduus tinctorius IPNI=*324467-2* 259
Carduus versicolor IPNI=*249211-1* 270
Carduus vulgaris IPNI=*196191-1* 260
Carelia brachystephana IPNI=*7086-2* 256
Carelia conyzoides IPNI=*7086-2* 256
Carelia houstoniana IPNI=*175500-1* 256
Carelia latifolia IPNI=*7086-2* 256
Carelia mutica IPNI=*7086-2* 256
Carex atrorufa IPNI=*301577-1* 230
Carex dura IPNI=*301577-1* 230
Carex flagellum IPNI=*230730-2* 233
Carex fuscoatia IPNI=*301577-1* 230
Carex hermaphrodita IPNI=*74768-2* 230
Carex hirtella IPNI=*329993-2* 232
Carex jamaicensis IPNI=*46311-2* 230
Carex margaritifera IPNI=*230730-2* 233
Carex myosuroides IPNI=*304256-1* 230
Carex opetiola IPNI=*304256-1* 230
Carex ovata IPNI=*104972-2* 232
Carex pichinchensis var. *simplex* IPNI=*301577-1* 230
Carex ruiziana IPNI=*301577-1* 230
Carex sachapata IPNI=*301577-1* 230
Carex tabina IPNI=*301577-1* 230
Carex uncinata IPNI=*46311-2* 230
Carica × *chrysopetala* IPNI=*77164259-1* 279
Carica × *fructifragrans* IPNI=*77164259-1* 279
Carica × *heilbornii* IPNI=*77164259-1* 279
Carica × *heilbornii* var. *pentagona* IPNI=*77164259-1* 279
Carica × *pentagona* IPNI=*77164259-1* 279
Carica boissieri IPNI=*675769-1* 279
Carica boliviana IPNI=*675779-1* 279
Carica bourgeaui IPNI=*675769-1* 279
Carica candamarcensis IPNI=*675782-1* 279
Carica cauliflora IPNI=*675769-1* 279
Carica cestriflora IPNI=*675782-1* 279
Carica chiriquensis IPNI=*675782-1* 279
Carica citriformis IPNI=*30011248-2* 279
Carica crassipetala IPNI=*1015560-1* 279
Carica cubensis IPNI=*30011248-2* 279
Carica cundinamarcensis IPNI=*675782-1* 279
Carica digitata IPNI=*130970-2* 279
Carica dodecaphylla IPNI=*30104966-2* 279
Carica erythrocarpa IPNI=*675779-1* 279
Carica goudotiana IPNI=*675773-1* 279
Carica hermaphrodita IPNI=*30011248-2* 279
Carica jamaicensis IPNI=*30011248-2* 279
Carica jimenezii IPNI=*30011248-2* 279
Carica longiflora IPNI=*1015565-1* 279
Carica mamaya IPNI=*30011248-2* 279
Carica manihot IPNI=*675778-1* 279
Carica microcarpa IPNI=*675778-1* 279
Carica monoica IPNI=*675779-1* 279
Carica papaya f. *correae* IPNI=*30011248-2* 279
Carica papaya f. *ernstii* IPNI=*30011248-2* 279
Carica papaya f. *mamaya* IPNI=*30011248-2* 279
Carica papaya f. *portoricensis* IPNI=*30011248-2* 279
Carica papaya var. *jimenezii* IPNI=*30011248-2* 279
Carica peltata IPNI=*30011248-2* 279
Carica pennata IPNI=*675769-1* 279
Carica pinnatifida IPNI=*30011248-2* 279
Carica portoricensis IPNI=*30011248-2* 279
Carica posopora IPNI=*30011248-2* 279
Carica pubescens IPNI=*675782-1* 279
Carica pulchra IPNI=*1015572-1* 279
Carica pyriformis IPNI=*30011248-2* 279
Carica quinqueloba IPNI=*675769-1* 279
Carica rochefortii IPNI=*30011248-2* 279
Carica sphaerocarpa IPNI=*1015573-1* 279
Carica spinosa IPNI=*30104966-2* 279
Cariniana clavata IPNI=*592127-1* 301
Cariniana exigua IPNI=*592127-1* 301
Cariniana multiflora IPNI=*77094248-1* 300
Cariniana paraensis IPNI=*592371-1* 301
Carinta repens IPNI=*109024-2* 351
Carissa africana IPNI=*77727-1* 340
Carissa carandas IPNI=*77727-1* 340
Carissa grandiflora IPNI=*77727-1* 340
Carissa praetermissa IPNI=*77727-1* 340
Carludovica bracteosa IPNI=*252718-2* 227
Carludovica costaricensis IPNI=*22967-2* 227
Carludovica funifera IPNI=*102548-2* 227
Carludovica gigantea IPNI=*297578-1* 227
Carludovica humilis IPNI=*297578-1* 227
Carludovica incisa IPNI=*297578-1* 227
Carludovica jamaicensis IPNI=*297578-1* 227
Carludovica kegeliana IPNI=*252718-2* 227
Carludovica macropoda IPNI=*22935-2* 227
Carludovica mattogrossensis IPNI=*252718-2* 227
Carludovica moritziana IPNI=*22935-2* 227
Carludovica palmata var. *humilis* IPNI=*297578-1* 227
Carludovica plicata IPNI=*22935-2* 227
Carludovica sarmentosa IPNI=*252718-2* 227
Carludovica speciosa IPNI=*297529-1* 227
Carludovica tabascana IPNI=*297529-1* 227
Carmenocania porphyrantha IPNI=*204802-2* 355
Carolinea affinis IPNI=*562668-1* 412
Carolinea alba IPNI=*211230-2* 412
Carolinea grandiflora IPNI=*179762-2* 412
Carolinea insignis IPNI=*562668-1* 412
Carolinea macrocarpa IPNI=*179762-2* 412
Carolinea princeps IPNI=*179762-2* 412
Carota sylvestris IPNI=*841063-1* 252

Carpanthus axillaris IPNI=*17056910-1* 179
Carpentaria bleeseri IPNI=*55245-3* 211
Carpiphea dentata IPNI=*927551-1* 274
Carpobrotus concavus IPNI=*360773-1* 280
Carpobrotus edulis var. *chrysophthalmus* IPNI=*1200427-2* 280
Carpobrotus laevigatus IPNI=*360773-1* 280
Carpobrotus rubrocinctus IPNI=*360773-1* 280
Carpobrotus vanzijliae IPNI=*360773-1* 280
Carpoceras longiflora IPNI=*675840-1* 374
Carpopogon pruriens IPNI=*509275-1* 328
Carpotroche angustifolia IPNI=*112597-2* 309
Carpotroche crassiramea IPNI=*47704-2* 380
Carpotroche glaucescens IPNI=*47704-2* 380
Carpotroche linguifolia IPNI=*111511-1* 380
Carpotroche longifolia var. *heliocarpa* IPNI=*111511-1* 380
Carpotroche longifolia var. *phasmatocarpa* IPNI=*111511-1* 380
Carpotroche mollis IPNI=*155434-2* 380
Carpotroche pacifica var. *brachycarpa* IPNI=*47701-2* 380
Carpotroche subintegra IPNI=*111855-1* 380
Carpotroche zuliana IPNI=*111509-1* 380
Carpunya lessertiana IPNI=*680810-1* 195
Carpunya peruviana IPNI=*680810-1* 195
Carpupica odorata IPNI=*680810-1* 195
Carthamus glaber IPNI=*324467-2* 259
Carthamus maculatus IPNI=*249211-1* 270
Carthamus tinctorius var. *spinosus* IPNI=*324467-2* 259
Caruelina filipes IPNI=*746662-1* 348
Caruelina spinosa IPNI=*746662-1* 348
Carum anisum IPNI=*846658-1* 252
Carum graveolens IPNI=*838067-1* 251
Carum majus IPNI=*837486-1* 251
Carum peregrinum IPNI=*60442790-2* 252
Carum petroselinum IPNI=*60442790-2* 252
Carum visnaga IPNI=*849956-1* 253
Carum vulgare IPNI=*60442790-2* 252
Caryocar glabrum var. *pilosum* IPNI=*47874-2* 381
Caryocar krukovii IPNI=*47871-2* 381
Caryocar riparium IPNI=*47874-2* 381
Caryocar tomentosum IPNI=*828602-1* 381
Caryocar tuberculosum IPNI=*828602-1* 381
Caryophyllea airoides IPNI=*386633-1* 233
Caryophyllus armerius IPNI=*310755-2* 285
Caryophyllus barbatus IPNI=*306674-2* 285
Caryophyllus corticosus IPNI=*601603-1* 428
Caryophyllus cotinifolius IPNI=*594126-1* 425
Caryophyllus fruticosus IPNI=*593768-1* 425
Caryophyllus jambos IPNI=*601603-1* 428
Caryophyllus malaccensis IPNI=*601907-1* 428
Caryophyllus pimenta IPNI=*196799-2* 428
Caryophyllus plumarius IPNI=*153717-1* 285
Caryophyllus racemosus IPNI=*196820-2* 428
Caryophyllus tunica IPNI=*153205-1* 285
Caryota furfuracea IPNI=*665616-1* 207
Caryota griffithii IPNI=*665616-1* 207
Caryota horrida IPNI=*7909-2* 204
Caryota javanica IPNI=*665616-1* 207
Caryota nana IPNI=*665616-1* 207
Caryota propinqua IPNI=*665616-1* 207
Caryota purpuracea IPNI=*665616-1* 207
Caryota speciosa IPNI=*665616-1* 207
Casarettoa diversifolia IPNI=*866038-1* 374
Cascabela peruviana IPNI=*47927-2* 340
Cascarilla caduciflora IPNI=*1106418-2* 352
Cascarilla calyptrata IPNI=*1215741-2* 352
Cascarilla carua IPNI=*754813-1* 352
Cascarilla crassifolia IPNI=*1215741-2* 352
Cascarilla gavanensis IPNI=*1106418-2* 352
Cascarilla hookeriana IPNI=*134610-2* 352
Cascarilla macrocarpa IPNI=*1215741-2* 352
Cascarilla magnifolia IPNI=*1106418-2* 352
Cascarilla magnifolia var. *caduciflora* IPNI=*1106418-2* 352
Cascarilla magnifolia var. *rostrata* IPNI=*1106418-2* 352
Cascarilla muzonensis IPNI=*134610-2* 352
Cascarilla nitida IPNI=*1106418-2* 352
Cascarilla oblongifolia IPNI=*1106418-2* 352
Cascarilla officinalis IPNI=*58723-2* 348
Cascarilla primatostylis IPNI=*1215741-2* 352
Cascarilla rostrata IPNI=*1106418-2* 352
Casearia aculeata var. *tafallana* IPNI=*47939-2* 403
Casearia adamantium IPNI=*779541-1* 403
Casearia adstringens IPNI=*779541-1* 403
Casearia affinis IPNI=*274165-2* 403
Casearia alba IPNI=*47939-2* 403
Casearia albicaulis IPNI=*779541-1* 403
Casearia attenuata IPNI=*274165-2* 403
Casearia avellana IPNI=*47939-2* 403
Casearia bangii IPNI=*47945-2* 403
Casearia banquitana IPNI=*47977-2* 403
Casearia banquitana var. *laevis* IPNI=*47977-2* 403
Casearia belizensis IPNI=*47957-2* 403
Casearia benthamiana IPNI=*274165-2* 403
Casearia berberoidea IPNI=*47939-2* 403
Casearia berteroana IPNI=*47939-2* 403
Casearia blanchetiana IPNI=*779712-1* 403
Casearia boliviana IPNI=*47939-2* 403
Casearia bracteifera IPNI=*779528-1* 403
Casearia brasiliensis IPNI=*47945-2* 403
Casearia brighamii IPNI=*779914-1* 404
Casearia cambessedesii IPNI=*779712-1* 403
Casearia cambessedesii var. *angustifolia* IPNI=*779712-1* 403
Casearia cambessedesii var. *parvifolia* IPNI=*779712-1* 403
Casearia camporum IPNI=*779611-1* 403
Casearia caudata IPNI=*274165-2* 403
Casearia celastroides IPNI=*779723-1* 404
Casearia celtidifolia IPNI=*779712-1* 403
Casearia chlorophoroidea IPNI=*274165-2* 403
Casearia commersoniana IPNI=*779914-1* 404
Casearia commutata IPNI=*779633-1* 403
Casearia congestiflora IPNI=*779914-1* 404
Casearia densiflora IPNI=*779914-1* 404
Casearia densiflora var. *parviflora* IPNI=*779914-1* 404
Casearia dentata IPNI=*779914-1* 404
Casearia dolichophylla IPNI=*47977-2* 403
Casearia ekmanii IPNI=*274165-2* 403
Casearia eriophora IPNI=*48069-2* 403
Casearia floribunda IPNI=*779541-1* 403
Casearia fockeana IPNI=*779611-1* 403
Casearia gentlei IPNI=*47977-2* 403
Casearia glaberrima IPNI=*47945-2* 403
Casearia glaziovii IPNI=*47945-2* 403
Casearia guantanamensis IPNI=*274165-2* 403
Casearia guianensis var. *rafflesioides* IPNI=*47939-2* 403
Casearia guidonia IPNI=*48022-2* 403
Casearia hassleri IPNI=*47939-2* 403
Casearia herbert-smithii IPNI=*274165-2* 403
Casearia hirsuta var. *glabrata* IPNI=*47939-2* 403
Casearia hirta IPNI=*47939-2* 403
Casearia hostmanniana IPNI=*47945-2* 403
Casearia icosandra IPNI=*48022-2* 403
Casearia integrifolia IPNI=*274165-2* 403
Casearia javitensis var. *myriantha* IPNI=*779914-1* 404
Casearia laevis IPNI=*47977-2* 403
Casearia lanceolata IPNI=*47945-2* 403
Casearia lasiosperma IPNI=*779914-1* 404
Casearia laurifolia IPNI=*779914-1* 404
Casearia lindeniana IPNI=*274165-2* 403

Casearia lindeniana IPNI=*47977-2* 403
Casearia macrophylla IPNI=*297979-2* 403
Casearia macrophylla var. *barbatula* IPNI=*297979-2* 403
Casearia maculata IPNI=*47989-2* 403
Casearia mathewsii IPNI=*779661-1* 403
Casearia maximiliani IPNI=*779914-1* 404
Casearia membranacea IPNI=*779528-1* 403
Casearia micrantha IPNI=*779541-1* 403
Casearia mollifolia IPNI=*47977-2* 403
Casearia multiflora IPNI=*553638-1* 417
Casearia myriantha IPNI=*779914-1* 404
Casearia nicoyensis IPNI=*47939-2* 403
Casearia nitida IPNI=*779541-1* 403
Casearia obovata IPNI=*47939-2* 403
Casearia obtusifolia IPNI=*47939-2* 403
Casearia odorata IPNI=*47939-2* 403
Casearia oligantha IPNI=*47945-2* 403
Casearia onacaensis IPNI=*274165-2* 403
Casearia orizabana IPNI=*47977-2* 403
Casearia ovoidea IPNI=*274165-2* 403
Casearia parviflora IPNI=*274165-2* 403
Casearia parviflora var. *microphylla* IPNI=*779541-1* 403
Casearia parvifolia IPNI=*779541-1* 403
Casearia parvifolia var. *genuina* IPNI=*779541-1* 403
Casearia parvifolia var. *microcarpa* IPNI=*274165-2* 403
Casearia parvifolia var. *paraguariensis* IPNI=*779541-1* 403
Casearia pavoniana IPNI=*779541-1* 403
Casearia petiolaris IPNI=*47989-2* 403
Casearia petraea IPNI=*779712-1* 403
Casearia phegocarpa IPNI=*47977-2* 403
Casearia piparea IPNI=*779914-1* 404
Casearia platyphylla IPNI=*779611-1* 403
Casearia poeppigii IPNI=*47945-2* 403
Casearia pringlei IPNI=*47977-2* 403
Casearia prunifolia IPNI=*47939-2* 403
Casearia pubiflora IPNI=*779611-1* 403
Casearia punctata IPNI=*274165-2* 403
Casearia reflexa IPNI=*779541-1* 403
Casearia reginae IPNI=*779712-1* 403
Casearia riparia IPNI=*47939-2* 403
Casearia rufidula IPNI=*47939-2* 403
Casearia salicifolia IPNI=*47977-2* 403
Casearia samyda IPNI=*274165-2* 403
Casearia schulziana IPNI=*274165-2* 403
Casearia sericea IPNI=*47939-2* 403
Casearia serrata IPNI=*47945-2* 403
Casearia serrulata IPNI=*47945-2* 403
Casearia serrulata IPNI=*779541-1* 403
Casearia spinosa IPNI=*47939-2* 403
Casearia spinosa var. *coriacifolia* IPNI=*47939-2* 403
Casearia spinosa var. *tafallana* IPNI=*47939-2* 403
Casearia spruceana IPNI=*77197258-1* 404
Casearia stipularis IPNI=*47945-2* 403
Casearia subopaca IPNI=*779528-1* 403
Casearia subsessiliflora IPNI=*274165-2* 403
Casearia sylvestris subvar. *campestris* IPNI=*274165-2* 403
Casearia sylvestris subvar. *pedicellaris* IPNI=*274165-2* 403
Casearia sylvestris var. *angustifolia* IPNI=*274165-2* 403
Casearia sylvestris var. *benthamiana* IPNI=*274165-2* 403
Casearia sylvestris var. *carpinifolia* IPNI=*274165-2* 403
Casearia sylvestris var. *chlorophoroidea* IPNI=*274165-2* 403
Casearia sylvestris var. *eichleri* IPNI=*274165-2* 403
Casearia sylvestris var. *paraensis* IPNI=*274165-2* 403
Casearia sylvestris var. *platyphylla* IPNI=*274165-2* 403
Casearia sylvestris var. *tomentella* IPNI=*274165-2* 403
Casearia sylvestris var. *wydleri* IPNI=*274165-2* 403
Casearia tarapotina IPNI=*779611-1* 403
Casearia timbuchi IPNI=*297979-2* 403
Casearia tremifolia IPNI=*779712-1* 403
Casearia ulmifolia IPNI=*779712-1* 403
Casearia umbellifera IPNI=*47945-2* 403
Casearia urbaniana IPNI=*47939-2* 403
Casimiroa tomentosa IPNI=*780054-1* 449
Casinga procera IPNI=*47957-2* 403
Caspareopsis monandra IPNI=*30129-2* 312
Caspareopsis purpurea IPNI=*481605-1* 312
Casparia castrata IPNI=*481605-1* 312
Casparya grewiifolia IPNI=*105087-1* 292
Casparya grewiifolia var. *pavoniana* IPNI=*105087-1* 292
Casparya longirostris IPNI=*105087-1* 292
Cassebeera farinosa IPNI=*77182808-1* 178
Cassebeera marginata IPNI=*77182463-1* 178
Cassebeera microphylla IPNI=*77183177-1* 178
Cassebeera ternifolia IPNI=*77182326-1* 178
Cassia × *floribunda* var. *pubescens* IPNI=*234652-2* 333
Cassia absus IPNI=*54310-2* 314
Cassia acutifolia IPNI=*518323-1* 332
Cassia aegyptiaca IPNI=*518364-1* 333
Cassia alata IPNI=*1192167-2* 332
Cassia albida IPNI=*234543-2* 333
Cassia alexandrina IPNI=*518323-1* 332
Cassia annunciata IPNI=*234636-2* 333
Cassia arayatensis IPNI=*1192167-2* 332
Cassia arborea IPNI=*911410-1* 333
Cassia arborescens IPNI=*234410-2* 332
Cassia atomaria IPNI=*234410-2* 332
Cassia atroviridis IPNI=*518364-1* 333
Cassia aurata IPNI=*234652-2* 333
Cassia axillaris IPNI=*234475-2* 332
Cassia bacillaris IPNI=*234415-2* 332
Cassia berlandieri IPNI=*234475-2* 332
Cassia bicapsularis IPNI=*518330-1* 332
Cassia biflora IPNI=*77211517-1* 316
Cassia bimarginata IPNI=*234686-2* 333
Cassia bonplandiana IPNI=*484507-1* 313
Cassia borneensis IPNI=*518369-1* 333
Cassia bracteata IPNI=*1192167-2* 332
Cassia brasiliana IPNI=*484607-1* 314
Cassia brittoniana IPNI=*911344-1* 332
Cassia cana IPNI=*234543-2* 333
Cassia canca IPNI=*518364-1* 333
Cassia caroliniana IPNI=*1117728-2* 333
Cassia chinensis IPNI=*518364-1* 333
Cassia chrysophylla IPNI=*234410-2* 332
Cassia ciliaris IPNI=*54568-2* 314
Cassia ciliata IPNI=*1117728-2* 333
Cassia ciliata IPNI=*234687-2* 334
Cassia contorta IPNI=*518369-1* 333
Cassia coromandeliana IPNI=*518364-1* 333
Cassia coymbosa IPNI=*234652-2* 333
Cassia densiflora IPNI=*234475-2* 332
Cassia desvauxii IPNI=*54460-2* 314
Cassia diphylla IPNI=*54479-2* 314
Cassia dumetorum IPNI=*234636-2* 333
Cassia dysophylla IPNI=*234691-2* 334
Cassia elegans IPNI=*234652-2* 333
Cassia elliptica IPNI=*234410-2* 332
Cassia emarginata IPNI=*234410-2* 332
Cassia emarginata var. *subunijuga* IPNI=*234410-2* 332
Cassia esculenta IPNI=*518364-1* 333
Cassia falcata IPNI=*1117728-2* 333
Cassia fastigiata IPNI=*911423-1* 333
Cassia fistuloides IPNI=*484507-1* 313
Cassia flexuosa IPNI=*54508-2* 314
Cassia florida IPNI=*911410-1* 333
Cassia foetida IPNI=*1117728-2* 333

Cassia foetida IPNI=*518369-1* 333
Cassia frutescens IPNI=*518364-1* 333
Cassia fruticosa IPNI=*234475-2* 332
Cassia gallinaria IPNI=*518369-1* 333
Cassia geminiflora IPNI=*518364-1* 333
Cassia gigantea IPNI=*911410-1* 333
Cassia glaucescens IPNI=*1117728-2* 333
Cassia grandiflora IPNI=*234652-2* 333
Cassia graveolens IPNI=*234410-2* 332
Cassia grisea IPNI=*234410-2* 332
Cassia hayesiana IPNI=*911344-1* 332
Cassia herpetica IPNI=*1192167-2* 332
Cassia hirsuta IPNI=*234491-2* 332
Cassia hoffmannseggii IPNI=*911354-1* 333
Cassia hornemannii IPNI=*911354-1* 333
Cassia houstoniana IPNI=*77211517-1* 316
Cassia humilis IPNI=*518369-1* 333
Cassia humilis IPNI=*54568-2* 314
Cassia inaequilatera IPNI=*911344-1* 332
Cassia indica IPNI=*518364-1* 333
Cassia italica IPNI=*518341-1* 332
Cassia ketschta IPNI=*72208-1* 186
Cassia kunthiana IPNI=*54568-2* 314
Cassia laevigata IPNI=*234652-2* 333
Cassia lanceolata IPNI=*518364-1* 333
Cassia latifolia IPNI=*911354-1* 333
Cassia latifolia var. *falcistipula* IPNI=*911354-1* 333
Cassia ligustrina IPNI=*518364-1* 333
Cassia linearis IPNI=*518364-1* 333
Cassia longisiliqua IPNI=*234410-2* 332
Cassia lorentzii IPNI=*234410-2* 332
Cassia lutescens IPNI=*234543-2* 333
Cassia macradenia IPNI=*1117728-2* 333
Cassia macranthera IPNI=*911359-1* 333
Cassia macrophylla IPNI=*234523-2* 333
Cassia marimari IPNI=*77211517-1* 316
Cassia maxonii IPNI=*911344-1* 332
Cassia michoacanensis IPNI=*234410-2* 332
Cassia mollis IPNI=*484607-1* 314
Cassia monantha IPNI=*234687-2* 334
Cassia moschata IPNI=*48715-2* 314
Cassia mucronulosa IPNI=*234687-2* 334
Cassia multiglandulosa IPNI=*234543-2* 333
Cassia multijuga IPNI=*234544-2* 333
Cassia nelsonii IPNI=*911412-1* 333
Cassia nictitans IPNI=*30485908-2* 314
Cassia obliquifolia IPNI=*1117728-2* 333
Cassia obovata IPNI=*518341-1* 332
Cassia obtusifolia IPNI=*234560-2* 333
Cassia obtusifolia var. *humilis* IPNI=*518369-1* 333
Cassia occidentalis IPNI=*1117728-2* 333
Cassia occidentalis var. *glabra* IPNI=*1117728-2* 333
Cassia ornata IPNI=*911354-1* 333
Cassia ornithopodioides IPNI=*234687-2* 334
Cassia ornithopoides IPNI=*234687-2* 334
Cassia oxyphylla var. *dariensis* IPNI=*911330-1* 332
Cassia pachycarpa IPNI=*484607-1* 314
Cassia pallida IPNI=*234569-2* 333
Cassia papillosa IPNI=*234589-2* 333
Cassia papulosa IPNI=*1117728-2* 333
Cassia parahyba IPNI=*517988-1* 332
Cassia paramariboensis IPNI=*472949-1* 311
Cassia pendula IPNI=*911386-1* 333
Cassia pilifera IPNI=*911389-1* 333
Cassia planisiliqua IPNI=*1117728-2* 333
Cassia plumieri IPNI=*1117728-2* 333
Cassia proboscidea IPNI=*518364-1* 333
Cassia pubescens IPNI=*234543-2* 333
Cassia purpurea IPNI=*518364-1* 333
Cassia quadrangularis IPNI=*234652-2* 333
Cassia racemosa IPNI=*911344-1* 332
Cassia reticulata IPNI=*234636-2* 333
Cassia rhombifolia IPNI=*484507-1* 313
Cassia robinioides IPNI=*518364-1* 333
Cassia rogeonii IPNI=*234560-2* 333
Cassia rotundifolia IPNI=*54723-2* 314
Cassia ruiziana IPNI=*911404-1* 333
Cassia ruscifolia IPNI=*518364-1* 333
Cassia sagera IPNI=*518369-1* 333
Cassia sagotiana IPNI=*485255-1* 314
Cassia sclerocarpa IPNI=*911354-1* 333
Cassia senna IPNI=*518323-1* 332
Cassia sensitiva IPNI=*234687-2* 334
Cassia septemtrionalis IPNI=*234652-2* 333
Cassia septentrionalis IPNI=*234652-2* 333
Cassia sericea IPNI=*234687-2* 334
Cassia siamea IPNI=*911410-1* 333
Cassia silvestris IPNI=*234653-2* 333
Cassia skinneri IPNI=*911412-1* 333
Cassia socotrana IPNI=*518364-1* 333
Cassia sophera IPNI=*518364-1* 333
Cassia sophera subsp. *purpurea* IPNI=*518364-1* 333
Cassia sophera var. *ligustrinoides* IPNI=*518364-1* 333
Cassia sophera var. *penghuana* IPNI=*518364-1* 333
Cassia sophera var. *purpurea* IPNI=*518364-1* 333
Cassia sophera var. *torulosa* IPNI=*518364-1* 333
Cassia sopheroides IPNI=*518364-1* 333
Cassia spectabilis IPNI=*234663-2* 333
Cassia spinigera var. *suffruticosa* IPNI=*911423-1* 333
Cassia standleyi IPNI=*911344-1* 332
Cassia strobilacea IPNI=*234636-2* 333
Cassia suffruticosa IPNI=*911423-1* 333
Cassia sumatrana IPNI=*911410-1* 333
Cassia sunsub IPNI=*518369-1* 333
Cassia surattensis IPNI=*911423-1* 333
Cassia surattensis subsp. *suffruticosa* IPNI=*911423-1* 333
Cassia surattensis var. *suffruticosa* IPNI=*911423-1* 333
Cassia tagera IPNI=*518369-1* 333
Cassia tala IPNI=*518369-1* 333
Cassia tarantan IPNI=*234636-2* 333
Cassia tetraphylla IPNI=*54460-2* 314
Cassia tomentosa IPNI=*234543-2* 333
Cassia tora f. *hirsuta* IPNI=*518369-1* 333
Cassia tora IPNI=*518369-1* 333
Cassia tora var. *obtusifolia* IPNI=*234560-2* 333
Cassia torosa IPNI=*518364-1* 333
Cassia torulosa IPNI=*518364-1* 333
Cassia trichoneura IPNI=*911412-1* 333
Cassia triflora IPNI=*234410-2* 332
Cassia tula IPNI=*518369-1* 333
Cassia undulata IPNI=*234686-2* 333
Cassia uniflora IPNI=*234687-2* 334
Cassia velutina IPNI=*234691-2* 334
Cassia velutina IPNI=*234694-2* 334
Cassia vernicosa IPNI=*234652-2* 333
Cassia viarum IPNI=*234694-2* 334
Cassia wightiana IPNI=*234543-2* 333
Cassine tomentosa IPNI=*780054-1* 449
Cassipourea broadwayi IPNI=*719636-1* 402
Cassipourea elliptica var. *latifolia* IPNI=*719636-1* 402
Cassipourea guianensis var. *serrata* IPNI=*719636-1* 402
Cassipourea guianensis var. *trichopoda* IPNI=*719636-1* 402
Cassipourea latifolia IPNI=*719636-1* 402
Cassipourea macrophylla IPNI=*719636-1* 402
Cassipourea quadrilocularis IPNI=*719636-1* 402
Cassipourea serrata IPNI=*719636-1* 402

Cassupa alba IPNI=*278892-2* 351
Cassupa juruana IPNI=*278892-2* 351
Cassupa laevis IPNI=*278892-2* 351
Cassupa panamensis IPNI=*278892-2* 351
Cassupa pittieri IPNI=*130336-2* 352
Cassupa scarlatina IPNI=*753635-1* 351
Cassuvium pomiferum IPNI=*319068-2* 450
Cassuvium reniforme IPNI=*319068-2* 450
Cassuvium solitarium IPNI=*319068-2* 450
Cassytha americana IPNI=*463202-1* 182
Cassytha americana var. *brachystachya* IPNI=*463202-1* 182
Cassytha americana var. *brasiliensis* IPNI=*463202-1* 182
Cassytha americana var. *puberula* IPNI=*463202-1* 182
Cassytha aphylla IPNI=*463202-1* 182
Cassytha archboldiana IPNI=*463202-1* 182
Cassytha brasiliensis IPNI=*463202-1* 182
Cassytha corniculata IPNI=*463202-1* 182
Cassytha dissitiflora IPNI=*463202-1* 182
Cassytha filiformis f. *pycnantha* IPNI=*463202-1* 182
Cassytha filiformis var. *pseudopubescens* IPNI=*463202-1* 182
Cassytha guineensis IPNI=*463202-1* 182
Cassytha macrocarpa IPNI=*463202-1* 182
Cassytha novoguineensis IPNI=*463202-1* 182
Cassytha paradoxae IPNI=*463202-1* 182
Cassytha polysperma IPNI=*219250-2* 284
Cassytha senegalensis IPNI=*463202-1* 182
Cassytha timoriensis IPNI=*463202-1* 182
Cassytha zeylanica IPNI=*463202-1* 182
Castalia amazonum IPNI=*605491-1* 181
Castalia ampla IPNI=*605494-1* 181
Castalia edulis IPNI=*605604-1* 181
Castalia elegans IPNI=*171178-2* 181
Castalia gilbertii IPNI=*605588-1* 181
Castalia jamesoniana IPNI=*605588-1* 181
Castalia lotus IPNI=*605604-1* 181
Castalia mystica IPNI=*605604-1* 181
Castalia pulchella IPNI=*605675-1* 181
Castalia rudgeana IPNI=*281442-2* 181
Castalia sacra IPNI=*605604-1* 181
Castilla australis IPNI=*855103-1* 440
Castilla elastica var. *liga* IPNI=*850895-1* 437
Castilla fallax IPNI=*49375-2* 437
Castilla guatemalensis IPNI=*850895-1* 437
Castilla lactiflua IPNI=*850895-1* 437
Castilla markhamiana IPNI=*850895-1* 437
Castilla markhamiana IPNI=*855107-1* 440
Castilla nicoyensis IPNI=*850895-1* 437
Castilleja agrestis IPNI=*800871-1* 375
Castilleja breviflora IPNI=*800957-1* 375
Castilleja communis f. *johnstoniae* IPNI=*800871-1* 375
Castilleja communis IPNI=*800871-1* 375
Castilleja coronopifolia IPNI=*800957-1* 375
Castilleja divaricata IPNI=*800957-1* 375
Castilleja fissifolia subvar. *coronopifolia* IPNI=*800957-1* 375
Castilleja fissifolia subvar. *hispidior* IPNI=*800957-1* 375
Castilleja fissifolia subvar. *longifolia* IPNI=*800957-1* 375
Castilleja fissifolia subvar. *stricta* IPNI=*800957-1* 375
Castilleja fissifolia subvar. *tenuifolia* IPNI=*800957-1* 375
Castilleja fissifolia var. *angustifolia* IPNI=*800957-1* 375
Castilleja fissifolia var. *breviflora* IPNI=*800957-1* 375
Castilleja fissifolia var. *divaricata* IPNI=*800957-1* 375
Castilleja fissifolia var. *integrifolia* IPNI=*801003-1* 375
Castilleja fissifolia var. *longifolia* IPNI=*800957-1* 375
Castilleja haitiensis IPNI=*800871-1* 375
Castilleja hyssopifolia IPNI=*800871-1* 375
Castilleja integrifolia var. *alpigena* IPNI=*801003-1* 375
Castilleja stricta IPNI=*800957-1* 375
Casuarina equisetifolia var. *typica* IPNI=*159856-1* 338
Catabrosa airoides IPNI=*395086-1* 235
Catabrosa angusta IPNI=*395086-1* 235
Catabrosa aquatica f. *latifolia* IPNI=*395086-1* 235
Catabrosa aquatica f. *schurii* IPNI=*395086-1* 235
Catabrosa aquatica f. *uniflora* IPNI=*395086-1* 235
Catabrosa aquatica f. *zapalowiczii* IPNI=*395086-1* 235
Catabrosa aquatica var. *atrata* IPNI=*395086-1* 235
Catabrosa aquatica var. *laurentiana* IPNI=*395086-1* 235
Catabrosa aquatica var. *minor* IPNI=*395086-1* 235
Catabrosa aquatica var. *uniflora* IPNI=*395086-1* 235
Catabrosa atrata IPNI=*395086-1* 235
Catabrosa capusii IPNI=*395086-1* 235
Catabrosa elbursensis IPNI=*395086-1* 235
Catabrosa kneuckeri IPNI=*395086-1* 235
Catabrosa micrantha IPNI=*401189-1* 239
Catabrosa minor IPNI=*395086-1* 235
Catabrosa ochroleuca IPNI=*395086-1* 235
Catabrosa pseudairoides IPNI=*395086-1* 235
Catabrosa pumila IPNI=*320035-2* 244
Catabrosa scabra IPNI=*395086-1* 235
Catabrosa sikkimensis IPNI=*395086-1* 235
Catabrosa verticillata IPNI=*1180620-2* 239
Catabrosa viridula IPNI=*395086-1* 235
Catalpa longisiliqua IPNI=*77176826-1* 362
Catapodium aquaticum IPNI=*395086-1* 235
Cataputia major IPNI=*355498-1* 392
Cataputia minor IPNI=*355498-1* 392
Cataria tomentosa IPNI=*452333-1* 371
Cataria vulgaris IPNI=*452333-1* 371
Catasetum baraquinianum IPNI=*621924-1* 217
Catasetum brachybulbon IPNI=*1092236-2* 217
Catasetum buchtienii IPNI=*1092236-2* 217
Catasetum christyanum IPNI=*621924-1* 217
Catasetum christyanum var. *chlorops* IPNI=*621924-1* 217
Catasetum christyanum var. *obscurum* IPNI=*621924-1* 217
Catasetum ciliatum IPNI=*621811-1* 217
Catasetum claesianum IPNI=*621811-1* 217
Catasetum claveringii IPNI=*325522-2* 217
Catasetum comosum IPNI=*1092236-2* 217
Catasetum cruciatum IPNI=*621924-1* 217
Catasetum discolor f. *genuinum* IPNI=*621811-1* 217
Catasetum discolor var. *claesianum* IPNI=*621811-1* 217
Catasetum discolor var. *vinosum* IPNI=*621811-1* 217
Catasetum discolor var. *viridiflorum* IPNI=*621811-1* 217
Catasetum floribundum IPNI=*325522-2* 217
Catasetum macrocarpum var. *bellum* IPNI=*325522-2* 217
Catasetum macrocarpum var. *carnosissimum* IPNI=*325522-2* 217
Catasetum macrocarpum var. *chrysanthum* IPNI=*325522-2* 217
Catasetum polydactylon IPNI=*1092236-2* 217
Catasetum proboscideum IPNI=*1092236-2* 217
Catasetum saccatum var. *christyanum* IPNI=*621924-1* 217
Catasetum saccatum var. *eusaccatum* IPNI=*621924-1* 217
Catasetum saccatum var. *pliciferum* IPNI=*621924-1* 217
Catasetum saccatum var. *typum* IPNI=*621924-1* 217
Catasetum spinosum IPNI=*1092236-2* 217
Catasetum tridentatum IPNI=*325522-2* 217
Catasetum tridentatum var. *viridiflorum* IPNI=*325522-2* 217
Catasetum variabile IPNI=*1092236-2* 217
Catevala arborescens IPNI=*529214-1* 215
Catevala fasciata IPNI=*77138009-1* 215
Cathartocarpus bacillaris IPNI=*234415-2* 332
Cathartocarpus bicapsularis IPNI=*518330-1* 332
Cathartocarpus brasilianus IPNI=*484607-1* 314
Cathartocarpus erubescens IPNI=*484607-1* 314
Cathartocarpus excelsus IPNI=*484507-1* 313
Cathartocarpus fistula IPNI=*484507-1* 313
Cathartocarpus fistuloides IPNI=*484507-1* 313
Cathartocarpus grandis IPNI=*484607-1* 314

Cathartocarpus javanicus IPNI=*484704-1* 314
Cathartocarpus moschatus IPNI=*484879-1* 314
Cathartocarpus rhombifolius IPNI=*484507-1* 313
Cathedra aestuaria IPNI=*607671-1* 449
Cathedra caurensis IPNI=*607671-1* 449
Cathedra crassifolia IPNI=*607671-1* 449
Cathedra guianensis IPNI=*607671-1* 449
Cathedra inaequilatera IPNI=*607671-1* 449
Cathedra oblonga IPNI=*607671-1* 449
Cathormion corymbosum IPNI=*989341-1* 320
Cathormion mangense IPNI=*56985-2* 315
Cathormion multiflorum IPNI=*989448-1* 311
Catimbium exaltatum IPNI=*218583-2* 251
Catimbium speciosum IPNI=*872083-1* 250
Catis martiana IPNI=*666941-1* 209
Catoblastus aequalis IPNI=*983907-1* 212
Catoblastus distichus IPNI=*983909-1* 212
Catoblastus drudei IPNI=*981485-1* 212
Catoblastus dryanderae IPNI=*1169616-2* 213
Catoblastus inconstans IPNI=*983910-1* 213
Catoblastus kalbreyeri IPNI=*983910-1* 213
Catoblastus maynensis IPNI=*670297-1* 213
Catoblastus megalocarpus IPNI=*983910-1* 213
Catoblastus microcaryus IPNI=*983910-1* 213
Catoblastus radiatus IPNI=*1169616-2* 213
Catoblastus sphaerocarpus IPNI=*983910-1* 213
Catoblastus velutinus IPNI=*983907-1* 212
Catocoma floribunda IPNI=*690122-1* 337
Catopsis mosenii IPNI=*50112-2* 228
Catostemma micranthum IPNI=*230850-2* 413
Catostemma praecox IPNI=*230851-2* 413
Catostigma aequale IPNI=*983907-1* 212
Catostigma drudei IPNI=*981485-1* 212
Catostigma dryanderae IPNI=*1169616-2* 213
Catostigma inconstans IPNI=*983910-1* 213
Catostigma kalbreyeri IPNI=*983910-1* 213
Catostigma megalocarpum IPNI=*983910-1* 213
Catostigma microcaryum IPNI=*983910-1* 213
Catostigma radiatum IPNI=*1169616-2* 213
Catostigma sphaerocarpum IPNI=*983910-1* 213
Catostigma sphaerocarpum var. *microcaryum* IPNI=*983910-1* 213
Cattleya bogotensis IPNI=*622257-1* 217
Cattleya caucaensis IPNI=*622208-1* 217
Cattleya chlorantha IPNI=*657915-1* 223
Cattleya chocoensis IPNI=*622208-1* 217
Cattleya decora IPNI=*657877-1* 223
Cattleya gigas IPNI=*622282-1* 217
Cattleya gloriosa IPNI=*622282-1* 217
Cattleya imperialis IPNI=*622282-1* 217
Cattleya kimballiana IPNI=*622257-1* 217
Cattleya labiata var. *bella* IPNI=*70028975-1* 217
Cattleya labiata var. *dowiana* IPNI=*622063-1* 217
Cattleya labiata var. *percivaliana* IPNI=*622189-1* 217
Cattleya labiata var. *trianae* IPNI=*622257-1* 217
Cattleya malouana IPNI=*622161-1* 217
Cattleya maxima f. *alba* IPNI=*622161-1* 217
Cattleya maxima f. *aphlebia* IPNI=*622161-1* 217
Cattleya maxima var. *aphlebia* IPNI=*622161-1* 217
Cattleya maxima var. *backhousii* IPNI=*622161-1* 217
Cattleya maxima var. *hrubyana* IPNI=*622161-1* 217
Cattleya maxima var. *marchettiana* IPNI=*622161-1* 217
Cattleya mendelii f. *morganiae* IPNI=*70028975-1* 217
Cattleya mendelii var. *bella* IPNI=*70028975-1* 217
Cattleya mendelii var. *grandiflora* IPNI=*70028975-1* 217
Cattleya mendelii var. *morganiae* IPNI=*70028975-1* 217
Cattleya morganiae IPNI=*70028975-1* 217
Cattleya odoratissima IPNI=*622273-1* 217
Cattleya percivaliana f. *alba* IPNI=*622189-1* 217
Cattleya percivaliana f. *bicolor* IPNI=*622189-1* 217
Cattleya percivaliana var. *alba* IPNI=*622189-1* 217
Cattleya percivaliana var. *bicolor* IPNI=*622189-1* 217
Cattleya ruckeri IPNI=*622208-1* 217
Cattleya sanderiana IPNI=*622282-1* 217
Cattleya schroederae f. *alba* IPNI=*50291-2* 217
Cattleya schroederae var. *alba* IPNI=*50291-2* 217
Cattleya superba IPNI=*622273-1* 217
Cattleya superba var. *alba* IPNI=*622273-1* 217
Cattleya superba var. *splendens* IPNI=*622273-1* 217
Cattleya trianae f. *alba* IPNI=*622257-1* 217
Cattleya trianae var. *alba* IPNI=*622257-1* 217
Cattleya trianae var. *formosa* IPNI=*622257-1* 217
Cattleya trianae var. *hooleana* IPNI=*622257-1* 217
Cattleya trianae var. *reginae* IPNI=*622257-1* 217
Cattleya trianae var. *splendidissima* IPNI=*622257-1* 217
Cattleya trianae var. *vanneriana* IPNI=*622257-1* 217
Cattleya violacea f. *alba* IPNI=*622273-1* 217
Cattleya violacea var. *alba* IPNI=*622273-1* 217
Cattleya violacea var. *ashworthii* IPNI=*622273-1* 217
Cattleya violacea var. *huebneri* IPNI=*622273-1* 217
Cattleya violacea var. *splendens* IPNI=*622273-1* 217
Cattleya violacea var. *wellsiana* IPNI=*622273-1* 217
Cattleya warscewiczii f. *franconvillensis* IPNI=*622282-1* 217
Cattleya warscewiczii f. *rochellensis* IPNI=*622282-1* 217
Cattleya warscewiczii var. *burfordiensis* IPNI=*622282-1* 217
Cattleya warscewiczii var. *franconvillensis* IPNI=*622282-1* 217
Cattleya warscewiczii var. *lageriana* IPNI=*622282-1* 217
Cattleya warscewiczii var. *rochellensis* IPNI=*622282-1* 217
Cattleya warscewiczii var. *sanderiana* IPNI=*622282-1* 217
Catunaregam brandisii IPNI=*745904-1* 348
Caturus spiciflorus IPNI=*337478-1* 387
Caucalis carnosa IPNI=*841063-1* 252
Caucalis carota IPNI=*841063-1* 252
Caucalis daucus IPNI=*841063-1* 252
Caulanthus sulfureus IPNI=*279485-1* 276
Caulotretus eriophorus IPNI=*229758-2* 332
Caulotretus heterophyllus IPNI=*229758-2* 332
Cavanilla phoenicea IPNI=*824676-1* 412
Cavanillea hastata IPNI=*1218005-2* 407
Cavanillea mabolo IPNI=*322328-1* 298
Cavanillea philippensis IPNI=*322328-1* 298
Cavanillesia cordata IPNI=*559275-1* 407
Cavaraea elegans IPNI=*520167-1* 335
Cavendishia acuminata IPNI=*60458979-2* 298
Cavendishia beckmanniana IPNI=*60458979-2* 298
Cavendishia campii IPNI=*60458979-2* 298
Cavendishia chiapensis IPNI=*60458979-2* 298
Cavendishia cordifolia IPNI=*60458979-2* 298
Cavendishia crassifolia IPNI=*60458979-2* 298
Cavendishia dichroa IPNI=*60458979-2* 298
Cavendishia dugandiana IPNI=*60458979-2* 298
Cavendishia durifolia IPNI=*60458979-2* 298
Cavendishia gracilis IPNI=*60458979-2* 298
Cavendishia guascensis IPNI=*204021-2* 300
Cavendishia guatemalensis IPNI=*60458979-2* 298
Cavendishia guatemalensis var. *chiapensis* IPNI=*60458979-2* 298
Cavendishia hendersonii IPNI=*60458979-2* 298
Cavendishia hoffmannii IPNI=*60458979-2* 298
Cavendishia kalbreyeri IPNI=*858240-1* 298
Cavendishia killipii IPNI=*60458979-2* 298
Cavendishia latifolia IPNI=*60458979-2* 298
Cavendishia lehmannii IPNI=*60458979-2* 298
Cavendishia melastomoides IPNI=*60458979-2* 298
Cavendishia mexiae IPNI=*858240-1* 298
Cavendishia miconioides IPNI=*60458979-2* 298
Cavendishia montana IPNI=*60458979-2* 298
Cavendishia obtusa IPNI=*50474-2* 298

Cavendishia peruviana IPNI=*60458979-2* 298
Cavendishia pilgeriana IPNI=*60458979-2* 298
Cavendishia pubescens var. *boliviensis* IPNI=*858236-1* 298
Cavendishia pubescens var. *microphylla* IPNI=*60458979-2* 298
Cavendishia pubescens var. *typica* IPNI=*858236-1* 298
Cavendishia rigidifolia IPNI=*60458979-2* 298
Cavendishia sandemanii IPNI=*60458979-2* 298
Cavendishia secundiflora IPNI=*60458979-2* 298
Cavendishia sillarensis IPNI=*60458979-2* 298
Cavendishia smithii IPNI=*60458979-2* 298
Cavendishia strobilifera IPNI=*60458979-2* 298
Cavendishia tolimensis IPNI=*60458979-2* 298
Cavendishia ulbrichiana IPNI=*60458979-2* 298
Cavendishia ulei IPNI=*60458979-2* 298
Cavendishia veraguensis IPNI=*60458979-2* 298
Cavendishia warscewiczii IPNI=*60458979-2* 298
Cavendishia weberbaueri IPNI=*858240-1* 298
Cayaponia heterophylla IPNI=*291801-1* 293
Cayaponia latebrosa IPNI=*291845-1* 293
Cayaponia leucosticta IPNI=*50589-2* 293
Cayaponia metensis IPNI=*291841-1* 293
Cayaponia ottoniana IPNI=*291801-1* 293
Cayaponia poeppigii IPNI=*291801-1* 293
Cayaponia racemosa var. *acutiloba* IPNI=*291845-1* 293
Cayaponia racemosa var. *laevis* IPNI=*291845-1* 293
Cayaponia racemosa var. *microcarpa* IPNI=*291845-1* 293
Cayaponia racemosa var. *palmatipartita* IPNI=*291845-1* 293
Cayaponia racemosa var. *plumieri* IPNI=*291845-1* 293
Cayaponia racemosa var. *scaberrima* IPNI=*291845-1* 293
Cayaponia racemosa var. *subintegrifolia* IPNI=*291845-1* 293
Cayaponia tomentosa IPNI=*291801-1* 293
Ceanothus arborescens IPNI=*63423-2* 441
Ceanothus colubrinus IPNI=*63423-2* 441
Ceanothus ferreus IPNI=*133956-2* 442
Ceanothus granulosus IPNI=*105606-2* 441
Ceanothus reclinatus IPNI=*63435-2* 441
Cecropia acutifolia IPNI=*850916-1* 445
Cecropia alborugosa IPNI=*50959-2* 446
Cecropia alexandrina IPNI=*582813-1* 446
Cecropia alvarezii IPNI=*850947-1* 446
Cecropia amphichlora IPNI=*850947-1* 446
Cecropia angelica IPNI=*50964-2* 446
Cecropia arachnoidea IPNI=*850956-1* 446
Cecropia arbelaezii IPNI=*850932-1* 446
Cecropia argentea IPNI=*850956-1* 446
Cecropia asperrima IPNI=*850956-1* 446
Cecropia bifurcata IPNI=*850943-1* 446
Cecropia burriada IPNI=*850947-1* 446
Cecropia caucana IPNI=*850916-1* 445
Cecropia commutata IPNI=*850947-1* 446
Cecropia concolor IPNI=*850947-1* 446
Cecropia congesta IPNI=*50864-2* 446
Cecropia coriacea IPNI=*850916-1* 445
Cecropia dabeibana IPNI=*850947-1* 446
Cecropia danielis IPNI=*850916-1* 445
Cecropia dielsiana IPNI=*850956-1* 446
Cecropia digitata IPNI=*850916-1* 445
Cecropia discolor IPNI=*582776-1* 445
Cecropia eximia IPNI=*850935-1* 446
Cecropia ferreyrae IPNI=*582776-1* 445
Cecropia goodspeedii IPNI=*850956-1* 446
Cecropia hachensis IPNI=*850916-1* 445
Cecropia hondurensis IPNI=*850956-1* 446
Cecropia hormigiana IPNI=*582776-1* 445
Cecropia humboldtiana IPNI=*850956-1* 446
Cecropia ibaguensis IPNI=*50901-2* 446
Cecropia inchuensis IPNI=*850970-1* 446
Cecropia juranyiana IPNI=*850970-1* 446
Cecropia laetevirens IPNI=*850943-1* 446
Cecropia lanciloba IPNI=*582813-1* 446
Cecropia levyana IPNI=*850947-1* 446
Cecropia magnifolia IPNI=*582776-1* 445
Cecropia maxonii IPNI=*850947-1* 446
Cecropia mexicana f. *glabrescens* IPNI=*850947-1* 446
Cecropia mexicana IPNI=*850947-1* 446
Cecropia mexicana var. *macrostachya* IPNI=*850947-1* 446
Cecropia mituana IPNI=*582776-1* 445
Cecropia mocoana IPNI=*582776-1* 445
Cecropia moniquirana IPNI=*850916-1* 445
Cecropia obtusifolia subsp. *burriada* IPNI=*850947-1* 446
Cecropia occidentalis IPNI=*850943-1* 446
Cecropia orinocensis IPNI=*850938-1* 446
Cecropia palmatisecta IPNI=*850916-1* 445
Cecropia panamensis IPNI=*850947-1* 446
Cecropia paraensis IPNI=*850938-1* 446
Cecropia philipsonii IPNI=*850916-1* 445
Cecropia polyandrophora IPNI=*850935-1* 446
Cecropia polyphlebia IPNI=*850916-1* 445
Cecropia porvenirensis IPNI=*582776-1* 445
Cecropia propinqua IPNI=*850956-1* 446
Cecropia radlkoferiana IPNI=*850947-1* 446
Cecropia richardii IPNI=*50853-2* 445
Cecropia riparia IPNI=*50853-2* 445
Cecropia robusta IPNI=*850943-1* 446
Cecropia sandersoniana IPNI=*850935-1* 446
Cecropia santanderensis IPNI=*50964-2* 446
Cecropia scabra IPNI=*856084-1* 447
Cecropia scabrifolia IPNI=*850956-1* 446
Cecropia schiedeana IPNI=*850956-1* 446
Cecropia sciadophylla var. *decurrens* IPNI=*850970-1* 446
Cecropia sciadophylla var. *guamuesensis* IPNI=*850970-1* 446
Cecropia sciadophylla var. *juranyiana* IPNI=*850970-1* 446
Cecropia sciadophylla var. *pedroa* IPNI=*850970-1* 446
Cecropia sciadophylla var. *subsessilis* IPNI=*850970-1* 446
Cecropia sciadophylla var. *typica* IPNI=*850970-1* 446
Cecropia setico IPNI=*850943-1* 446
Cecropia standleyi IPNI=*582776-1* 445
Cecropia strigilosa IPNI=*850916-1* 445
Cecropia surinamensis IPNI=*850956-1* 446
Cecropia sylvicola IPNI=*850916-1* 445
Cecropia telealba var. *hirsuta* IPNI=*50959-2* 446
Cecropia telealbida IPNI=*50964-2* 446
Cecropia teleargentea IPNI=*50959-2* 446
Cecropia teleincana IPNI=*50964-2* 446
Cecropia telenivea IPNI=*50964-2* 446
Cecropia tessmannii IPNI=*850943-1* 446
Cecropia tolimensis IPNI=*50901-2* 446
Cecropia tubulosa IPNI=*850916-1* 445
Cecropia urbaniana IPNI=*50937-2* 446
Cecropia vageleri IPNI=*850943-1* 446
Cecropia villosa IPNI=*850916-1* 445
Cedrela alternifolia IPNI=*834759-1* 410
Cedronella japonica IPNI=*444504-1* 368
Ceiba allenii IPNI=*60446802-2* 414
Ceiba burchellii IPNI=*51083-2* 408
Ceiba caribaea IPNI=*1166232-2* 407
Ceiba casearia IPNI=*1166232-2* 407
Ceiba guineensis IPNI=*1166232-2* 407
Ceiba occidentalis IPNI=*1166232-2* 407
Ceiba pentandra var. *caribaea* IPNI=*1166232-2* 407
Ceiba phaeosantha IPNI=*51083-2* 408
Ceiba rosea IPNI=*60446802-2* 414
Ceiba thonningii IPNI=*1166232-2* 407
Celastrus macrocarpus IPNI=*155542-2* 291
Celastrus ovatus IPNI=*63423-2* 441
Celastrus verticillatus IPNI=*162118-1* 291

Celeri graveolens IPNI=*838067-1* 251
Celosia amphibia IPNI=*1024156-2* 281
Celosia argentea f. *cristata* IPNI=*316350-2* 282
Celosia argentea var. *cristata* IPNI=*316350-2* 282
Celosia argentea var. *margaritacea* IPNI=*316350-2* 282
Celosia argentea var. *mumbaiana* IPNI=*316350-2* 282
Celosia aurea IPNI=*316350-2* 282
Celosia castrensis IPNI=*316350-2* 282
Celosia cernua IPNI=*316350-2* 282
Celosia coccinea IPNI=*316350-2* 282
Celosia comosa IPNI=*316350-2* 282
Celosia cristata IPNI=*316350-2* 282
Celosia cristata var. *castrensis* IPNI=*316350-2* 282
Celosia debilis IPNI=*316350-2* 282
Celosia decumbens IPNI=*112097-2* 282
Celosia echinata IPNI=*327342-2* 281
Celosia elongata IPNI=*10447-2* 281
Celosia humifusa IPNI=*327344-2* 281
Celosia huttonii IPNI=*316350-2* 282
Celosia japonica IPNI=*316350-2* 282
Celosia linearis IPNI=*316350-2* 282
Celosia major IPNI=*60077-1* 282
Celosia margaritacea IPNI=*316350-2* 282
Celosia maritima IPNI=*34033-2* 282
Celosia marylandica IPNI=*316350-2* 282
Celosia pallida IPNI=*316350-2* 282
Celosia paniculata IPNI=*129956-2* 283
Celosia peruviana IPNI=*10362-2* 281
Celosia plumosa IPNI=*316350-2* 282
Celosia procumbens IPNI=*112097-2* 282
Celosia prostrata IPNI=*112097-2* 282
Celosia purpurea IPNI=*316350-2* 282
Celosia pyramidalis IPNI=*316350-2* 282
Celosia sparsa IPNI=*60077-1* 282
Celosia splendens IPNI=*316350-2* 282
Celosia stricta IPNI=*316350-2* 282
Celosia swinhoei IPNI=*316350-2* 282
Celosia virgata var. *paniculata* IPNI=*60011-1* 282
Celsia urticifolia IPNI=*9536-2* 377
Celsia viscosa IPNI=*770211-1* 377
Celtidopsis citrifolia IPNI=*1213898-2* 436
Celtis aculeata IPNI=*1213898-2* 436
Celtis aculeata var. *laevigata* IPNI=*1213898-2* 436
Celtis aculeata var. *pubescens* IPNI=*1213898-2* 436
Celtis aculeata var. *serrata* IPNI=*1213898-2* 436
Celtis albicans IPNI=*1018059-2* 436
Celtis alnifolia IPNI=*1213898-2* 436
Celtis anfractuosa IPNI=*1213898-2* 436
Celtis asperula IPNI=*1213898-2* 436
Celtis biflora IPNI=*1213898-2* 436
Celtis brevifolia IPNI=*1213898-2* 436
Celtis canescens IPNI=*1018059-2* 436
Celtis chichilea IPNI=*1018059-2* 436
Celtis curiandiuba IPNI=*1018059-2* 436
Celtis dichotoma IPNI=*1213898-2* 436
Celtis diffusa IPNI=*1213898-2* 436
Celtis ehrenbergiana IPNI=*1213898-2* 436
Celtis epiphylladena IPNI=*1213898-2* 436
Celtis ferarum IPNI=*51219-2* 436
Celtis gardneri IPNI=*1213898-2* 436
Celtis glabrata IPNI=*1213898-2* 436
Celtis glycycarpa IPNI=*1213898-2* 436
Celtis goudotii IPNI=*1213898-2* 436
Celtis hilariana IPNI=*1213898-2* 436
Celtis hottlei IPNI=*11469-2* 445
Celtis laevigata IPNI=*1213898-2* 436
Celtis lima IPNI=*1018059-2* 436
Celtis macrophylla IPNI=*1018059-2* 436
Celtis membranacea IPNI=*1213898-2* 436
Celtis micrantha IPNI=*1018059-2* 436
Celtis microcarpa IPNI=*1018059-2* 436
Celtis mollis IPNI=*1018059-2* 436
Celtis morifolia IPNI=*1213898-2* 436
Celtis pavonii IPNI=*1213898-2* 436
Celtis platycaulis IPNI=*1213898-2* 436
Celtis pubescens IPNI=*1213898-2* 436
Celtis rhamnoides IPNI=*1213898-2* 436
Celtis riparia IPNI=*1018059-2* 436
Celtis rufescens IPNI=*1018059-2* 436
Celtis rugosa IPNI=*1018059-2* 436
Celtis scabra IPNI=*1018059-2* 436
Celtis schiedeana IPNI=*1018059-2* 436
Celtis spinosa IPNI=*1213898-2* 436
Celtis spinosissima IPNI=*1213898-2* 436
Celtis swartzii IPNI=*851195-1* 436
Celtis triflora IPNI=*1213898-2* 436
Celtis velutina IPNI=*1213898-2* 436
Celtis williamsii IPNI=*1213898-2* 436
Celtis zizyphoides IPNI=*1213898-2* 436
Cenchropsis ekmaniana IPNI=*51261-2* 235
Cenchropsis myosuroides IPNI=*51261-2* 235
Cenchropsis myosuroides var. *longiseta* IPNI=*51261-2* 235
Cenchrus aegyptius IPNI=*397387-1* 236
Cenchrus albertsonii IPNI=*395325-1* 236
Cenchrus alopecuroides IPNI=*51261-2* 235
Cenchrus anjania IPNI=*395214-1* 235
Cenchrus bulbosus IPNI=*395214-1* 235
Cenchrus carolinianus IPNI=*395325-1* 236
Cenchrus cavanillesii IPNI=*316518-2* 235
Cenchrus ciliaris subsp. *ibrahimii* IPNI=*395214-1* 235
Cenchrus crinitus IPNI=*316518-2* 235
Cenchrus dactylolepis IPNI=*395204-1* 235
Cenchrus echinatus IPNI=*395204-1* 235
Cenchrus echinatus var. *brevisetus* IPNI=*316518-2* 235
Cenchrus echinatus var. *morisonii* IPNI=*316518-2* 235
Cenchrus echinatus var. *viridis* IPNI=*395204-1* 235
Cenchrus ekmanianus IPNI=*51261-2* 235
Cenchrus elliotii IPNI=*51261-2* 235
Cenchrus glaucus IPNI=*395214-1* 235
Cenchrus granularis IPNI=*116848-2* 240
Cenchrus hexaflorus IPNI=*395204-1* 235
Cenchrus hillebrandianus IPNI=*316518-2* 235
Cenchrus humilis IPNI=*395325-1* 236
Cenchrus incertus IPNI=*395325-1* 236
Cenchrus inflexus IPNI=*395204-1* 235
Cenchrus inflexus IPNI=*88264-2* 238
Cenchrus insularis IPNI=*316518-2* 235
Cenchrus laevigatus IPNI=*14931-2* 234
Cenchrus lechleri IPNI=*316518-2* 235
Cenchrus longifolius IPNI=*395214-1* 235
Cenchrus macrocarpus IPNI=*316518-2* 235
Cenchrus marginalis IPNI=*88264-2* 238
Cenchrus melanostachyus IPNI=*395214-1* 235
Cenchrus microcephalus IPNI=*395325-1* 236
Cenchrus mucronatus IPNI=*397387-1* 236
Cenchrus muricatus IPNI=*395325-1* 236
Cenchrus mutabilis IPNI=*395214-1* 235
Cenchrus myosuroides var. *longisetus* IPNI=*51261-2* 235
Cenchrus paniceus IPNI=*77105978-1* 235
Cenchrus parviceps IPNI=*395325-1* 236
Cenchrus parviflorus IPNI=*943811-1* 245
Cenchrus pauciflorus IPNI=*395325-1* 236
Cenchrus polystachios IPNI=*943811-1* 245
Cenchrus pungens IPNI=*316518-2* 235
Cenchrus pycnostachyus IPNI=*77105978-1* 235
Cenchrus rigidus IPNI=*395204-1* 235

Cenchrus rufescens IPNI=*395214-1* 235
Cenchrus scabridus IPNI=*51261-2* 235
Cenchrus setoides IPNI=*51261-2* 235
Cenchrus spicatus IPNI=*77105978-1* 235
Cenchrus spinosus IPNI=*395325-1* 236
Cenchrus strictus IPNI=*395325-1* 236
Cenchrus tripsacoides IPNI=*14931-2* 234
Cenchrus villosus IPNI=*14931-2* 234
Cenchrus villosus IPNI=*51255-2* 235
Cenchrus viridis IPNI=*395204-1* 235
Centaurea carthamus IPNI=*324467-2* 259
Centaurea cyaneum IPNI=*190310-1* 259
Centaurea cyanocephala IPNI=*190310-1* 259
Centaurea cyanus subsp. *coa* IPNI=*190310-1* 259
Centaurea cyanus var. *denudata* IPNI=*190310-1* 259
Centaurea dalmatica IPNI=*249211-1* 270
Centaurea hoffmanniana IPNI=*190310-1* 259
Centaurea hortorum IPNI=*190310-1* 259
Centaurea lanata IPNI=*190310-1* 259
Centaurea pulcherrima IPNI=*190310-1* 259
Centaurea pulchra IPNI=*190310-1* 259
Centaurea rhizocephala IPNI=*190310-1* 259
Centaurea riparia IPNI=*51409-2* 259
Centaurea segetalis IPNI=*190310-1* 259
Centaurea umbrosa IPNI=*190310-1* 259
Centaurella dichotoma IPNI=*326619-2* 344
Centaurium capitatum IPNI=*326619-2* 344
Centaurium corymbosum IPNI=*326619-2* 344
Centaurium erythraea var. *capitatum* IPNI=*326619-2* 344
Centaurium erythraea var. *fasciculare* IPNI=*326619-2* 344
Centaurium erythraea var. *latifolium* IPNI=*326619-2* 344
Centaurium erythraea var. *laxum* IPNI=*326619-2* 344
Centaurium erythraea var. *masclansii* IPNI=*326619-2* 344
Centaurium erythraea var. *subcapitatum* IPNI=*326619-2* 344
Centaurium erythraea var. *sublitorale* IPNI=*326619-2* 344
Centaurium latifolium IPNI=*326619-2* 344
Centaurium lomae IPNI=*326619-2* 344
Centaurium minus IPNI=*326619-2* 344
Centaurium minus var. *austriacum* IPNI=*326619-2* 344
Centaurium minus var. *transiens* IPNI=*326619-2* 344
Centaurium umbellatum f. *album* IPNI=*326619-2* 344
Centaurium umbellatum var. *fasciculare* IPNI=*326619-2* 344
Centaurium vulgare IPNI=*326619-2* 344
Centella asiatica var. *floridana* IPNI=*51383-2* 252
Centella biflora IPNI=*51383-2* 252
Centella boninensis IPNI=*1197718-2* 252
Centella coriacea IPNI=*1197718-2* 252
Centella dusenii IPNI=*51383-2* 252
Centella floridana IPNI=*51383-2* 252
Centella glochidiata IPNI=*1197718-2* 252
Centella hirtella IPNI=*1197718-2* 252
Centella homalocarpa IPNI=*843759-1* 253
Centella repanda IPNI=*51383-2* 252
Centella repanda var. *floridana* IPNI=*51383-2* 252
Centella triflora IPNI=*51383-2* 252
Centella tussilaginifolia IPNI=*1197718-2* 252
Centella ulugurensis IPNI=*1197718-2* 252
Centella uniflora IPNI=*1197718-2* 252
Centella verticillata IPNI=*843759-1* 253
Centinodia agrestina IPNI=*695422-1* 289
Centinodia avicularis IPNI=*695422-1* 289
Centinodia humifusa IPNI=*695422-1* 289
Centinodium aviculare IPNI=*695422-1* 289
Centranthus latifolius IPNI=*860141-1* 296
Centranthus marinus IPNI=*860141-1* 296
Centranthus maritimus IPNI=*860141-1* 296
Centranthus ruber IPNI=*860141-1* 296
Centratherum aristatum IPNI=*51409-2* 259
Centratherum australianum IPNI=*51409-2* 259
Centratherum brachylepis IPNI=*51409-2* 259
Centratherum brevispinum IPNI=*51409-2* 259
Centratherum camporum IPNI=*51409-2* 259
Centratherum camporum var. *albicans* IPNI=*51409-2* 259
Centratherum chinense IPNI=*962413-1* 260
Centratherum confertum IPNI=*51409-2* 259
Centratherum fruticosum IPNI=*51409-2* 259
Centratherum holtonii IPNI=*51409-2* 259
Centratherum intermedium IPNI=*51409-2* 259
Centratherum longispinum IPNI=*51409-2* 259
Centratherum muticum IPNI=*51409-2* 259
Centratherum pulchellum IPNI=*51409-2* 259
Centratherum punctatum f. *brachyphyllum* IPNI=*51409-2* 259
Centratherum punctatum f. *foliosum* IPNI=*51409-2* 259
Centratherum punctatum subsp. *australianum* IPNI=*51409-2* 259
Centratherum punctatum subsp. *camporum* IPNI=*51409-2* 259
Centratherum punctatum subsp. *fruticosum* IPNI=*51409-2* 259
Centratherum punctatum var. *albicans* IPNI=*51409-2* 259
Centratherum punctatum var. *foliosum* IPNI=*51409-2* 259
Centratherum punctatum var. *longipes* IPNI=*51409-2* 259
Centratherum punctatum var. *parviflorum* IPNI=*51409-2* 259
Centratherum punctatum var. *viscossissimum* IPNI=*51409-2* 259
Centratherum riparium IPNI=*51409-2* 259
Centratherum violaceum IPNI=*51409-2* 259
Centrolobium orinocense IPNI=*485534-1* 314
Centrolobium paraense var. *orenocense* IPNI=*485534-1* 314
Centrolobium patinense IPNI=*485534-1* 314
Centrophorum chinense IPNI=*396110-1* 236
Centropogon andropogon IPNI=*51553-2* 272
Centropogon austin-smithii IPNI=*141634-1* 273
Centropogon bonplandianus f. *glabrescens* IPNI=*51553-2* 272
Centropogon bonplandianus IPNI=*51553-2* 272
Centropogon bonplandianus var. *intermedius* IPNI=*51553-2* 272
Centropogon cornutus f. *leucanthus* IPNI=*51553-2* 272
Centropogon cornutus f. *leucostomus* IPNI=*51553-2* 272
Centropogon cornutus f. *vellozoanus* IPNI=*51553-2* 272
Centropogon cornutus f. *ynesae* IPNI=*51553-2* 272
Centropogon cornutus var. *angustifolius* IPNI=*51553-2* 272
Centropogon cornutus var. *laevigatus* IPNI=*51553-2* 272
Centropogon discolor IPNI=*141634-1* 273
Centropogon edulis IPNI=*51553-2* 272
Centropogon erianthus var. *brachysepalus* IPNI=*141468-1* 272
Centropogon fastuosus IPNI=*51553-2* 272
Centropogon flexuosus IPNI=*141634-1* 273
Centropogon intermedius IPNI=*51553-2* 272
Centropogon laevigatus IPNI=*51553-2* 272
Centropogon longipes IPNI=*141634-1* 273
Centropogon montanus f. *subpubescens* IPNI=*141634-1* 273
Centropogon montanus IPNI=*141634-1* 273
Centropogon nubicola IPNI=*141634-1* 273
Centropogon oblongus IPNI=*51553-2* 272
Centropogon ovalifolius IPNI=*141634-1* 273
Centropogon ovalifolius var. *glabristamineus* IPNI=*141634-1* 273
Centropogon pedicellaris IPNI=*141634-1* 273
Centropogon pedicellaris var. *breviflorus* IPNI=*141634-1* 273
Centropogon pedicellaris var. *gallerensis* IPNI=*141634-1* 273
Centropogon planchonis IPNI=*141634-1* 273
Centropogon prostratus IPNI=*141634-1* 273
Centropogon psilandrus IPNI=*141634-1* 273
Centropogon puerilis IPNI=*51553-2* 272
Centropogon radicans IPNI=*77083518-1* 272
Centropogon riparius IPNI=*141634-1* 273
Centropogon semperflorens IPNI=*141634-1* 273
Centropogon solanifolius var. *inconstans* IPNI=*141634-1* 273
Centropogon solanifolius var. *speciosus* IPNI=*141634-1* 273
Centropogon speciosus IPNI=*141634-1* 273
Centropogon subfalcatus IPNI=*141634-1* 273

Centropogon surinamensis IPNI=*51553-2* 272
Centropogon surinamensis var. *angustifolius* IPNI=*51553-2* 272
Centropogon surinamensis var. *vestitus* IPNI=*51553-2* 272
Centropogon xestus IPNI=*141634-1* 273
Centrosema biflorum IPNI=*51897-2* 314
Centrosema brasilianum var. *angustifolium* IPNI=*485555-1* 314
Centrosema conjugatum IPNI=*51897-2* 314
Centrosema decumbens IPNI=*51897-2* 314
Centrosema dubium IPNI=*485618-1* 314
Centrosema ferrugineum IPNI=*485611-1* 314
Centrosema galeottii IPNI=*485611-1* 314
Centrosema grandiflorum IPNI=*485611-1* 314
Centrosema hastatum IPNI=*485618-1* 314
Centrosema insulanum IPNI=*485555-1* 314
Centrosema intermedium IPNI=*485611-1* 314
Centrosema lisboae IPNI=*485597-1* 314
Centrosema macrocarpum var. *andinum* IPNI=*485597-1* 314
Centrosema magnificum IPNI=*485597-1* 314
Centrosema salzmannii IPNI=*485611-1* 314
Centrosema scabriusculum IPNI=*51897-2* 314
Centrosema seymourianum IPNI=*485597-1* 314
Centrosema simulans IPNI=*485540-1* 314
Centrosema virginianum subsp. *angustifolium* IPNI=*51897-2* 314
Centrosema virginianum var. *angustifolium* IPNI=*51897-2* 314
Centrosema virginianum var. *ellipticum* IPNI=*51897-2* 314
Centrosema virginianum var. *genuinum* IPNI=*51897-2* 314
Centrosolenia glabra IPNI=*77139499-1* 366
Centrospermum xanthioides IPNI=*1692-2* 255
Centrostachys aspera IPNI=*2468-2* 280
Centrostachys australis IPNI=*2468-2* 280
Centrostachys canescens IPNI=*2468-2* 280
Centrostachys grandifolia IPNI=*2468-2* 280
Centunculus angustifolius IPNI=*329786-2* 285
Centunculus arvensis IPNI=*329786-2* 285
Centunculus crassulus IPNI=*77100404-1* 304
Centunculus lanceolatus IPNI=*77100404-1* 304
Centunculus mexicanus IPNI=*77100404-1* 304
Centunculus minimus IPNI=*77100404-1* 304
Centunculus rigidus IPNI=*329786-2* 285
Centunculus simplex IPNI=*77100404-1* 304
Cepa alba IPNI=*527795-1* 213
Cepa esculenta IPNI=*527795-1* 213
Cepa fissilis IPNI=*528046-1* 213
Cepa fistulosa IPNI=*528046-1* 213
Cepa obliqua IPNI=*528506-1* 213
Cepa pallens IPNI=*527795-1* 213
Cepa rubra IPNI=*527795-1* 213
Cepa schoenoprasa IPNI=*528823-1* 213
Cepa tenuifolia IPNI=*528823-1* 213
Cepa ventricosa IPNI=*528046-1* 213
Cepa vulgaris IPNI=*527795-1* 213
Cephaelis acuminata IPNI=*1214950-2* 347
Cephaelis alba IPNI=*60473597-2* 353
Cephaelis amoena IPNI=*77161054-1* 353
Cephaelis barcellana IPNI=*77115635-1* 354
Cephaelis blepharophora IPNI=*77161948-1* 353
Cephaelis bracteocardia IPNI=*60473604-2* 353
Cephaelis colorata IPNI=*77161054-1* 353
Cephaelis costaricensis IPNI=*77115600-1* 353
Cephaelis cyanocarpa IPNI=*77115635-1* 354
Cephaelis cymosa IPNI=*77161077-1* 354
Cephaelis dichotoma IPNI=*77115613-1* 354
Cephaelis elata f. *lutea* IPNI=*77115600-1* 353
Cephaelis elata IPNI=*77115600-1* 353
Cephaelis emetica IPNI=*764727-1* 356
Cephaelis furcata IPNI=*77115613-1* 354
Cephaelis glabrescens IPNI=*77161054-1* 353
Cephaelis glomerulata IPNI=*77115607-1* 353
Cephaelis hirsuta IPNI=*77115635-1* 354
Cephaelis hoffmannseggiana IPNI=*77115613-1* 354
Cephaelis horridula IPNI=*77161076-1* 354
Cephaelis humboldtiana IPNI=*77161073-1* 354
Cephaelis humboldtiana var. *ornata* IPNI=*77161073-1* 354
Cephaelis iodotricha IPNI=*77161076-1* 354
Cephaelis ipecacuanha IPNI=*1214950-2* 347
Cephaelis justiciifolia IPNI=*77161077-1* 354
Cephaelis microcephala IPNI=*77161098-1* 354
Cephaelis paraensis IPNI=*60473597-2* 353
Cephaelis patrisii IPNI=*77115613-1* 354
Cephaelis peruviana IPNI=*759372-1* 354
Cephaelis phoenicia IPNI=*77115600-1* 353
Cephaelis psychotriifolia IPNI=*213293-2* 356
Cephaelis pubescens IPNI=*60473604-2* 353
Cephaelis punicea IPNI=*77115600-1* 353
Cephaelis reniformis IPNI=*109024-2* 351
Cephaelis rosea IPNI=*77161054-1* 353
Cephaelis rubra IPNI=*77115613-1* 354
Cephaelis sagotii IPNI=*77115635-1* 354
Cephaelis sylvatica IPNI=*60473604-2* 353
Cephaelis tenuis IPNI=*77161098-1* 354
Cephaelis tomentosa IPNI=*77115635-1* 354
Cephaelis tricholoba IPNI=*77161076-1* 354
Cephaelis triplocephala IPNI=*170751-2* 353
Cephaelis violifolia IPNI=*1080914-2* 351
Cephaelis vultusmimi IPNI=*77115635-1* 354
Cephaelis williamsii IPNI=*77153569-1* 354
Cephalandra grandis IPNI=*291972-1* 293
Cephalandra indica IPNI=*291972-1* 293
Cephalandra moghadd IPNI=*291972-1* 293
Cephalandra schimperi IPNI=*291972-1* 293
Cephalocereus atroviridis IPNI=*130627-1* 283
Cephalocereus fricii IPNI=*52982-2* 283
Cephalocereus lanuginosus IPNI=*196706-2* 284
Cephalocereus perlucens IPNI=*130283-1* 283
Cephalocereus remolinensis IPNI=*130627-1* 283
Cephalocereus russellianus IPNI=*52982-2* 283
Cephalomanes elegans IPNI=*17229030-1* 173
Cephalophilon capitatum IPNI=*60433515-2* 289
Cephalophilon nepalense IPNI=*60433514-2* 289
Cephaloschoenus articulatus IPNI=*220271-2* 232
Cephaloschoenus divergens IPNI=*220271-2* 232
Ceramanthus bonariensis IPNI=*106366-2* 340
Ceramocarpium cubense IPNI=*77173628-1* 182
Ceranthus corniculatus IPNI=*301087-2* 374
Ceranthus schreberi IPNI=*301087-2* 374
Cerastium acutatum IPNI=*152274-1* 285
Cerastium acutifolium IPNI=*329786-2* 285
Cerastium alpestre IPNI=*152312-1* 285
Cerastium alpicola IPNI=*329786-2* 285
Cerastium alpigenum IPNI=*152312-1* 285
Cerastium alpinum IPNI=*152274-1* 285
Cerastium alpinum IPNI=*329786-2* 285
Cerastium ambiguum IPNI=*329786-2* 285
Cerastium andinum IPNI=*152312-1* 285
Cerastium angustifolium IPNI=*329786-2* 285
Cerastium apetalum IPNI=*152274-1* 285
Cerastium arenosum IPNI=*152274-1* 285
Cerastium arvense var. *alpicola* IPNI=*329786-2* 285
Cerastium arvense var. *eglandulosum* IPNI=*329786-2* 285
Cerastium arvense var. *latifolium* IPNI=*329786-2* 285
Cerastium arvense var. *taimyrense* IPNI=*329786-2* 285
Cerastium ballsii IPNI=*329786-2* 285
Cerastium brachycarpum IPNI=*152274-1* 285
Cerastium brachycarpum IPNI=*329786-2* 285
Cerastium busambarense IPNI=*329786-2* 285
Cerastium caespitosum IPNI=*152274-1* 285

Cerastium caespitosum IPNI=*329786-2* 285
Cerastium caespitosum subsp. *alpinum* IPNI=*152255-1* 285
Cerastium caespitosum var. *montevidense* IPNI=*152312-1* 285
Cerastium caespitosum var. *pseudoalpestre* IPNI=*152312-1* 285
Cerastium calcicola IPNI=*329786-2* 285
Cerastium camphoratifolium IPNI=*329786-2* 285
Cerastium carnosulum IPNI=*329786-2* 285
Cerastium collinum IPNI=*329786-2* 285
Cerastium connatum IPNI=*152312-1* 285
Cerastium constantinopolitanum IPNI=*152274-1* 285
Cerastium cordatum IPNI=*30172747-2* 285
Cerastium cordifolium IPNI=*30172747-2* 285
Cerastium coronense IPNI=*329786-2* 285
Cerastium corsicum IPNI=*329786-2* 285
Cerastium fasciculatum IPNI=*329786-2* 285
Cerastium fontanum f. *glutinosum* IPNI=*152312-1* 285
Cerastium fontanum f. *gracile* IPNI=*152312-1* 285
Cerastium fontanum f. *holosteoides* IPNI=*152312-1* 285
Cerastium fontanum subsp. *holosteoides* IPNI=*152312-1* 285
Cerastium fontanum subsp. *scandicum* IPNI=*152312-1* 285
Cerastium fontanum subsp. *triviale* IPNI=*152274-1* 285
Cerastium fontanum subsp. *vulgare* IPNI=*152312-1* 285
Cerastium fontanum var. *eglandulosum* IPNI=*152312-1* 285
Cerastium fontanum var. *glandulosum* IPNI=*152312-1* 285
Cerastium fontanum var. *lucens* IPNI=*152312-1* 285
Cerastium fontanum var. *tibeticum* IPNI=*152312-1* 285
Cerastium fontanum var. *vulgare* IPNI=*152312-1* 285
Cerastium fulvum IPNI=*152274-1* 285
Cerastium glandulosum IPNI=*152312-1* 285
Cerastium glomeratum var. *brachycarpum* IPNI=*152274-1* 285
Cerastium glomeratum var. *robustum* IPNI=*152274-1* 285
Cerastium grandiflorum IPNI=*329786-2* 285
Cerastium heterophyllum IPNI=*329786-2* 285
Cerastium hirsutum IPNI=*152274-1* 285
Cerastium holosteoides subsp. *glandulosum* IPNI=*152312-1* 285
Cerastium holosteoides subsp. *serpentini* IPNI=*152312-1* 285
Cerastium holosteoides subsp. *vulgare* IPNI=*152312-1* 285
Cerastium holosteoides var. *glabrescens* IPNI=*152312-1* 285
Cerastium holosteoides var. *vulgare* IPNI=*152312-1* 285
Cerastium incanum IPNI=*329786-2* 285
Cerastium insubricum IPNI=*329786-2* 285
Cerastium longirostre IPNI=*152255-1* 285
Cerastium macrocarpum IPNI=*152312-1* 285
Cerastium malachiforme IPNI=*152312-1* 285
Cerastium mauritianum IPNI=*152274-1* 285
Cerastium membranaceum IPNI=*152274-1* 285
Cerastium minus IPNI=*152312-1* 285
Cerastium minutulum IPNI=*152274-1* 285
Cerastium mutabile IPNI=*329786-2* 285
Cerastium ovale IPNI=*152274-1* 285
Cerastium ovatum IPNI=*329786-2* 285
Cerastium petrosum IPNI=*329786-2* 285
Cerastium poelianum IPNI=*329786-2* 285
Cerastium polyphyllum IPNI=*329786-2* 285
Cerastium pseudosylvaticum IPNI=*152312-1* 285
Cerastium pseudoviscosum IPNI=*152274-1* 285
Cerastium pumilum IPNI=*152274-1* 285
Cerastium rotundifolium IPNI=*152274-1* 285
Cerastium scandicum IPNI=*152312-1* 285
Cerastium serpyllifolium IPNI=*329786-2* 285
Cerastium stevenii IPNI=*152274-1* 285
Cerastium sylvaticum IPNI=*152274-1* 285
Cerastium sylvaticum IPNI=*152312-1* 285
Cerastium tenellum IPNI=*152274-1* 285
Cerastium tomentosum IPNI=*152274-1* 285
Cerastium triviale IPNI=*152274-1* 285
Cerastium triviale var. *alpinum* IPNI=*152255-1* 285
Cerastium uliginosum IPNI=*152312-1* 285
Cerastium uralense IPNI=*329786-2* 285
Cerastium vidalii IPNI=*329786-2* 285
Cerastium villosum IPNI=*152274-1* 285
Cerastium viscidum IPNI=*152312-1* 285
Cerastium viscosioides IPNI=*152274-1* 285
Cerastium viscosum IPNI=*152312-1* 285
Cerastium viscosum var. *elongatum* IPNI=*152312-1* 285
Cerastium viscosum var. *glabrescens* IPNI=*152312-1* 285
Cerastium viscosum var. *glandulosum* IPNI=*152312-1* 285
Cerastium vulgare IPNI=*152312-1* 285
Cerastium vulgare var. *holosteoides* IPNI=*152312-1* 285
Cerastium vulgatum f. *lucens* IPNI=*152312-1* 285
Cerastium vulgatum IPNI=*152274-1* 285
Cerastium vulgatum subsp. *alpinum* IPNI=*152255-1* 285
Cerastium vulgatum var. *alpestre* IPNI=*152255-1* 285
Cerastium vulgatum var. *americanum* IPNI=*152312-1* 285
Cerastium vulgatum var. *andinum* IPNI=*152312-1* 285
Cerastium vulgatum var. *anomalum* IPNI=*152255-1* 285
Cerastium vulgatum var. *filiforme* IPNI=*152255-1* 285
Cerastium vulgatum var. *montevidense* IPNI=*152312-1* 285
Cerastium vulgatum var. *peruvianum* IPNI=*152312-1* 285
Cerastium vulgatum var. *viscidum* IPNI=*152255-1* 285
Cerasus asplenifolia IPNI=*730262-1* 444
Cerasus opaca IPNI=*730059-1* 443
Cerasus serotina IPNI=*730262-1* 444
Cerasus serotina var. *montana* IPNI=*730262-1* 444
Ceratocephalus acmella var. *depauperata* IPNI=*174556-1* 256
Ceratocephalus acmella var. *uliginosa* IPNI=*174556-1* 256
Ceratocephalus americanus IPNI=*213524-1* 264
Ceratocephalus beccabunga IPNI=*174552-1* 256
Ceratocephalus bicolor IPNI=*250890-1* 270
Ceratocephalus caespitosus IPNI=*174530-1* 255
Ceratocephalus ciliatus IPNI=*174534-1* 255
Ceratocephalus commutatus IPNI=*250890-1* 270
Ceratocephalus debilis IPNI=*271548-2* 256
Ceratocephalus diffusus IPNI=*213524-1* 264
Ceratocephalus exasperatus IPNI=*271548-2* 256
Ceratocephalus fimbriatus IPNI=*174534-1* 255
Ceratocephalus nitidus IPNI=*241903-1* 269
Ceratocephalus parvifolius IPNI=*271533-2* 255
Ceratocephalus pilosus IPNI=*32564-2* 258
Ceratocephalus poeppigii IPNI=*174534-1* 255
Ceratocephalus repens IPNI=*174552-1* 256
Ceratocephalus salivaria IPNI=*271528-2* 255
Ceratocephalus subhirsutus IPNI=*174552-1* 256
Ceratocephalus tenellus IPNI=*271548-2* 256
Ceratocephalus urens IPNI=*250890-1* 270
Ceratocephalus wedeliodes IPNI=*1186403-2* 261
Ceratochloa australis IPNI=*37001-2* 235
Ceratochloa boryana IPNI=*37001-2* 235
Ceratochloa breviaristata IPNI=*37001-2* 235
Ceratochloa carinata IPNI=*30065892-2* 235
Ceratochloa carinata var. *hookeriana* IPNI=*30065892-2* 235
Ceratochloa cathartica IPNI=*37001-2* 235
Ceratochloa festucoides IPNI=*37001-2* 235
Ceratochloa grandiflora IPNI=*30065892-2* 235
Ceratochloa haenkeana IPNI=*37001-2* 235
Ceratochloa laciniata IPNI=*30065892-2* 235
Ceratochloa pendula IPNI=*37001-2* 235
Ceratochloa pitensis IPNI=*393995-1* 235
Ceratochloa secunda IPNI=*37001-2* 235
Ceratochloa unioloides IPNI=*37001-2* 235
Ceratochloa willdenowii IPNI=*37001-2* 235
Ceratophyllum demersum var. *commune* IPNI=*30006047-2* 251
Ceratopteris lockhartii IPNI=*17242590-1* 178
Ceratopteris parkeri IPNI=*17242590-1* 178
Ceratosanthes corniculata IPNI=*52543-2* 293
Ceratosanthes gracilis IPNI=*52543-2* 293

Ceratosanthes latiloba IPNI=*52543-2* 293
Ceratosanthes tuberosa IPNI=*52543-2* 293
Ceratosepalum micranthum IPNI=*674754-1* 397
Ceratosepalum parviflorum IPNI=*674754-1* 397
Ceratostanthus ajacis IPNI=*77254-2* 436
Ceratostema colombianum IPNI=*204021-2* 300
Ceratostema odoratissimum IPNI=*1064712-2* 298
Ceratostema parvifolium IPNI=*252657-2* 300
Ceratozamia muricata IPNI=*270552-2* 179
Cerbera ahouai IPNI=*82440-1* 344
Cerbera linearifolia IPNI=*47927-2* 340
Cerbera nitida IPNI=*82440-1* 344
Cerbera peruviana IPNI=*47927-2* 340
Cerbera thevetia IPNI=*47927-2* 340
Cercidium goldmanii IPNI=*1006943-1* 330
Cercidium plurifoliolatum IPNI=*1006943-1* 330
Cercidium praecox IPNI=*1006943-1* 330
Cercidium spinosum IPNI=*1006943-1* 330
Cercidium unijugum IPNI=*1006943-1* 330
Cercidium viride IPNI=*1006943-1* 330
Cercophora anomala IPNI=*597250-1* 302
Cercouratea lucens IPNI=*607063-1* 397
Cerdana alliodora IPNI=*113943-1* 273
Cerdana cujabensis IPNI=*113943-1* 273
Cerdana gerascanthus IPNI=*65321-2* 274
Ceresia fluitans IPNI=*30475806-2* 243
Cereus alticostatus IPNI=*1001185-1* 284
Cereus antoinii IPNI=*138488-1* 285
Cereus atroviridis IPNI=*130627-1* 283
Cereus baccifer IPNI=*219250-2* 284
Cereus baxaniensis f. *pellucidus* IPNI=*1646-2* 283
Cereus baxaniensis IPNI=*1646-2* 283
Cereus brevispinulus IPNI=*138488-1* 285
Cereus campinensis IPNI=*1001185-1* 284
Cereus caripensis IPNI=*219250-2* 284
Cereus chotaensis IPNI=*35149-2* 283
Cereus clavatus IPNI=*244634-2* 285
Cereus colombianus IPNI=*196706-2* 284
Cereus coniflorus IPNI=*138488-1* 285
Cereus costaricensis IPNI=*77177683-1* 284
Cereus cylindricus IPNI=*128942-1* 283
Cereus deficiens IPNI=*244634-2* 285
Cereus dussii IPNI=*1646-2* 283
Cereus eburneus IPNI=*244634-2* 285
Cereus eburneus var. *polygonus* IPNI=*244634-2* 285
Cereus euchlorus IPNI=*1001185-1* 284
Cereus euchlorus subsp. *alticostatus* IPNI=*1001185-1* 284
Cereus euchlorus subsp. *leucanthus* IPNI=*1001185-1* 284
Cereus gracilis IPNI=*130627-1* 283
Cereus grandiflorus IPNI=*138488-1* 285
Cereus grandiflorus var. *armatus* IPNI=*138488-1* 285
Cereus grandiflorus var. *uranos* IPNI=*138488-1* 285
Cereus grenadensis IPNI=*130627-1* 283
Cereus griseus IPNI=*244634-2* 285
Cereus haitiensis IPNI=*138488-1* 285
Cereus hallensis IPNI=*138488-1* 285
Cereus hookeri IPNI=*132926-1* 283
Cereus horribarbis IPNI=*130627-1* 283
Cereus horribilis IPNI=*1646-2* 283
Cereus horridus IPNI=*130283-1* 283
Cereus horridus var. *alatosquamatus* IPNI=*130283-1* 283
Cereus jajoianus IPNI=*35149-2* 283
Cereus jalapaensis IPNI=*138488-1* 285
Cereus karstenii IPNI=*130283-1* 283
Cereus lanuginosus IPNI=*130627-1* 283
Cereus lanuginosus IPNI=*196706-2* 284
Cereus lauterbachii IPNI=*1001185-1* 284
Cereus lemairei IPNI=*77163180-1* 285
Cereus longiflorus IPNI=*130283-1* 283
Cereus macrogonus IPNI=*257107-2* 285
Cereus margaritensis IPNI=*130627-1* 283
Cereus margaritensis var. *micracanthus* IPNI=*52982-2* 283
Cereus marginatus IPNI=*132926-1* 283
Cereus megalanthus IPNI=*232386-2* 285
Cereus monacanthus IPNI=*77163180-1* 285
Cereus nitidus IPNI=*1646-2* 283
Cereus northumberlandianus IPNI=*130283-1* 283
Cereus nycticalus IPNI=*138488-1* 285
Cereus nycticalus var. *armatus* IPNI=*138488-1* 285
Cereus obtusus IPNI=*138488-1* 285
Cereus ocamponis IPNI=*77163181-1* 285
Cereus paradisiacus IPNI=*138488-1* 285
Cereus paradoxus IPNI=*129102-1* 283
Cereus parasiticus IPNI=*219250-2* 284
Cereus pentagonus IPNI=*1646-2* 283
Cereus perlucens IPNI=*130283-1* 283
Cereus peruvianus IPNI=*130627-1* 283
Cereus peruvianus IPNI=*244634-2* 285
Cereus phyllanthus IPNI=*132939-1* 283
Cereus pitajaya IPNI=*1646-2* 283
Cereus polygonatus IPNI=*244634-2* 285
Cereus polyrhizus IPNI=*77163180-1* 285
Cereus princeps IPNI=*1646-2* 283
Cereus prismaticus IPNI=*1646-2* 283
Cereus quadrangularis IPNI=*219250-2* 284
Cereus ramulosus IPNI=*963240-1* 284
Cereus regalis IPNI=*130283-1* 283
Cereus remolinensis IPNI=*130627-1* 283
Cereus reptans IPNI=*1646-2* 283
Cereus ritteri subsp. *parapetiensis* IPNI=*1001185-1* 284
Cereus roezlii IPNI=*35149-2* 283
Cereus rosaceus IPNI=*138488-1* 285
Cereus roseanus IPNI=*138488-1* 285
Cereus rosei IPNI=*257107-2* 285
Cereus royenii IPNI=*196706-2* 284
Cereus scandens IPNI=*77163180-1* 285
Cereus schomburgkii IPNI=*77163180-1* 285
Cereus sepium IPNI=*35149-2* 283
Cereus tetragonus IPNI=*1646-2* 283
Cereus tetragonus var. *minor* IPNI=*1646-2* 283
Cereus thalassinus IPNI=*130283-1* 283
Cereus tricostatus IPNI=*77163184-1* 285
Cereus trigonus var. *costaricensis* IPNI=*77177683-1* 284
Cereus trinitatensis IPNI=*77163180-1* 285
Cereus truncatus IPNI=*229597-2* 284
Cereus undatus IPNI=*77163184-1* 285
Cereus undulatus IPNI=*1646-2* 283
Cereus undulatus IPNI=*77163184-1* 285
Cereus uranus IPNI=*138488-1* 285
Cereus urbanianus IPNI=*138488-1* 285
Cereus variabilis IPNI=*1646-2* 283
Cereus vasmerii IPNI=*1646-2* 283
Cereus venezuelensis IPNI=*77163180-1* 285
Ceriscus malabaricus IPNI=*745904-1* 348
Cerophora angustifolia IPNI=*166220-2* 339
Cerophora inodora IPNI=*166220-2* 339
Cerophora lanceolata IPNI=*166220-2* 339
Cerophora spicans IPNI=*166220-2* 339
Ceropteris calomelanos IPNI=*17345640-1* 178
Ceropteris calomelanos var. *gracilis* IPNI=*17345640-1* 178
Ceropteris martensii IPNI=*17345640-1* 178
Ceropteris massonii IPNI=*17345640-1* 178
Ceropteris pallescens IPNI=*200935-2* 178
Ceropteris peruviana IPNI=*200935-2* 178
Ceropteris plicata IPNI=*200935-2* 178
Ceropteris serrata IPNI=*17345640-1* 178

Ceropteris tartarea IPNI=*200935-2* 178
Cerothamnus cerifer IPNI=*166220-2* 339
Cerothamnus pumilus IPNI=*166220-2* 339
Ceroxylon andicola IPNI=*30119294-2* 207
Ceroxylon andicola var. *occidentale* IPNI=*30119294-2* 207
Ceroxylon beethovenia IPNI=*53436-2* 207
Ceroxylon coarctatum IPNI=*665703-1* 207
Ceroxylon crispum IPNI=*665703-1* 207
Ceroxylon flexuosum IPNI=*665703-1* 207
Ceroxylon floccosum IPNI=*665697-1* 207
Ceroxylon hexandrum IPNI=*665703-1* 207
Ceroxylon interruptum IPNI=*53436-2* 207
Ceroxylon klopstockia IPNI=*53436-2* 207
Ceroxylon latisectum IPNI=*665695-1* 207
Ceroxylon mooreanum IPNI=*665695-1* 207
Ceroxylon schultzei IPNI=*53436-2* 207
Ceroxylon sclerophyllum IPNI=*665703-1* 207
Ceroxylon utile IPNI=*667716-1* 210
Ceroxylon verruculosum IPNI=*665703-1* 207
Cervantesia colombiana IPNI=*1712-2* 449
Cervantesia glabrata IPNI=*1712-2* 449
Cervantesia kunthiana IPNI=*780054-1* 449
Cervantesia macrocarpa IPNI=*780054-1* 449
Cervantesia tomentosa IPNI=*780054-1* 449
Cervispina infectoria IPNI=*718405-1* 442
Cespedesia amazonica IPNI=*605994-1* 397
Cespedesia bonplandii IPNI=*605994-1* 397
Cespedesia brasiliana IPNI=*605994-1* 397
Cespedesia excelsa IPNI=*605994-1* 397
Cespedesia macrophylla IPNI=*605994-1* 397
Cespedesia repanda IPNI=*605994-1* 397
Cespedesia repanda var. *lanceolata* IPNI=*605994-1* 397
Cespedesia scandens IPNI=*605994-1* 397
Cespedesia sprucei IPNI=*605994-1* 397
Chaenopleura longifolia IPNI=*1220088-2* 421
Chaetaea aculeata IPNI=*77183732-1* 407
Chaetanthera multiflora IPNI=*237026-1* 268
Chaetaria adscensionis IPNI=*319759-2* 234
Chaetaria bromoides IPNI=*319759-2* 234
Chaetaria canariensis IPNI=*319759-2* 234
Chaetaria coarctata IPNI=*319759-2* 234
Chaetaria coerulescens IPNI=*319759-2* 234
Chaetaria curvata IPNI=*319759-2* 234
Chaetaria depressa IPNI=*319759-2* 234
Chaetaria elatior IPNI=*319759-2* 234
Chaetaria fasciculata IPNI=*319759-2* 234
Chaetaria festucoides IPNI=*319759-2* 234
Chaetaria forskalii IPNI=*319759-2* 234
Chaetaria gigantea IPNI=*319759-2* 234
Chaetaria humilis IPNI=*319759-2* 234
Chaetaria interrupta IPNI=*319759-2* 234
Chaetaria luzoniensis IPNI=*319759-2* 234
Chaetaria mauritiana IPNI=*319759-2* 234
Chaetaria nana IPNI=*319759-2* 234
Chaetaria teneriffae IPNI=*319759-2* 234
Chaetocalyx fissus IPNI=*77191263-1* 329
Chaetocalyx magniflorus IPNI=*77191263-1* 329
Chaetocalyx nigrescens IPNI=*77191263-1* 329
Chaetocalyx parviflorus IPNI=*77191263-1* 329
Chaetocalyx paucifolius IPNI=*77191263-1* 329
Chaetocalyx perglandulosus IPNI=*77191263-1* 329
Chaetocalyx pubescens IPNI=*77191263-1* 329
Chaetocalyx retusus IPNI=*77191263-1* 329
Chaetocalyx scandens IPNI=*77191263-1* 329
Chaetocalyx scandens var. *fissus* IPNI=*77191263-1* 329
Chaetocalyx scandens var. *pubescens* IPNI=*77191263-1* 329
Chaetocalyx tenuipedicellatus IPNI=*77191263-1* 329
Chaetocalyx vestitus IPNI=*77191263-1* 329
Chaetocalyx vincentinus IPNI=*77191263-1* 329
Chaetocarpus myrsinites var. *stipularis* IPNI=*54072-2* 401
Chaetocarpus pouteria IPNI=*788972-1* 307
Chaetocarpus stipularis IPNI=*54072-2* 401
Chaetocarpus williamsii IPNI=*54072-2* 401
Chaetochloa ambigua IPNI=*421857-1* 245
Chaetochloa amplifolia IPNI=*1182687-2* 245
Chaetochloa barbata IPNI=*1172068-2* 245
Chaetochloa brevispica IPNI=*421428-1* 245
Chaetochloa composita IPNI=*1182687-2* 245
Chaetochloa corrugata var. *parviflora* IPNI=*943811-1* 245
Chaetochloa costata IPNI=*1172068-2* 245
Chaetochloa crus-ardeae IPNI=*421828-1* 245
Chaetochloa effusa IPNI=*421711-1* 245
Chaetochloa flava IPNI=*943811-1* 245
Chaetochloa geniculata IPNI=*943811-1* 245
Chaetochloa geniculata var. *perennis* IPNI=*943811-1* 245
Chaetochloa geniculata var. *purpurascens* IPNI=*943811-1* 245
Chaetochloa germanica IPNI=*1207294-2* 245
Chaetochloa gibbosa IPNI=*235234-2* 245
Chaetochloa glauca IPNI=*77105978-1* 235
Chaetochloa glauca var. *purpurea* IPNI=*421764-1* 245
Chaetochloa gracilis IPNI=*943811-1* 245
Chaetochloa grisebachii IPNI=*319034-2* 245
Chaetochloa grisebachii var. *ampla* IPNI=*319034-2* 245
Chaetochloa grisebachii var. *mexicana* IPNI=*319034-2* 245
Chaetochloa imberbis IPNI=*943811-1* 245
Chaetochloa imberbis var. *geniculata* IPNI=*943811-1* 245
Chaetochloa imberbis var. *penicillata* IPNI=*943811-1* 245
Chaetochloa imberbis var. *perennis* IPNI=*943811-1* 245
Chaetochloa imberbis var. *streptobotrys* IPNI=*943811-1* 245
Chaetochloa imberbis var. *versicolor* IPNI=*943811-1* 245
Chaetochloa italica IPNI=*1207294-2* 245
Chaetochloa italica var. *germanica* IPNI=*1207294-2* 245
Chaetochloa liebmannii IPNI=*319037-2* 245
Chaetochloa liebmannii var. *pauciflora* IPNI=*319037-2* 245
Chaetochloa lutescens IPNI=*77105978-1* 235
Chaetochloa macrostachya IPNI=*235234-2* 245
Chaetochloa membranifolia IPNI=*319034-2* 245
Chaetochloa occidentalis IPNI=*943811-1* 245
Chaetochloa palmifolia IPNI=*421711-1* 245
Chaetochloa penicillata IPNI=*943811-1* 245
Chaetochloa perennis IPNI=*943811-1* 245
Chaetochloa purpurascens IPNI=*943811-1* 245
Chaetochloa sulcata IPNI=*421711-1* 245
Chaetochloa trichorhachis IPNI=*1182687-2* 245
Chaetochloa ventenatii IPNI=*943811-1* 245
Chaetochloa versicolor IPNI=*943811-1* 245
Chaetochloa verticillata IPNI=*421857-1* 245
Chaetochloa verticillata var. *breviseta* IPNI=*421857-1* 245
Chaetochloa vulpiseta IPNI=*1182687-2* 245
Chaetocrater capitatus IPNI=*274165-2* 403
Chaetocrater fasciculatus IPNI=*47989-2* 403
Chaetocrater reflexus IPNI=*779541-1* 403
Chaetocyperus acicularis IPNI=*1169169-2* 231
Chaetogastra bonplandiana IPNI=*566862-1* 418
Chaetogastra canescens IPNI=*566862-1* 418
Chaetogastra goudotii IPNI=*566891-1* 418
Chaetogastra hypericoides IPNI=*1168615-2* 418
Chaetogastra lepidota IPNI=*77205900-1* 417
Chaetogastra lychnitoides IPNI=*1168615-2* 418
Chaetogastra stachyoides IPNI=*1168615-2* 418
Chaetogastra stricta IPNI=*566891-1* 418
Chaetogastra strigosa IPNI=*566891-1* 418
Chaetogastra sulphurea IPNI=*566862-1* 418
Chaetolepis alpestris IPNI=*567327-1* 418
Chaetospira funckii IPNI=*211400-2* 268
Chaetospira spiralis IPNI=*211400-2* 268

Chaetospora aurea IPNI=*220271-2* 232
Chamaecassia grandiflora IPNI=*234652-2* 333
Chamaecassia laevigata IPNI=*234652-2* 333
Chamaecnide microphylla IPNI=*1162156-2* 447
Chamaecrista tagera IPNI=*518369-1* 333
Chamaecrista tetraphylla IPNI=*54460-2* 314
Chamaecyparis boursieri f. *viridis* IPNI=*261839-1* 180
Chamaecyparis boursieri IPNI=*263956-1* 180
Chamaecyparis decussata IPNI=*677124-1* 180
Chamaecyparis glauca IPNI=*677124-1* 180
Chamaecyparis lawsoniana var. *stewartii* IPNI=*261839-1* 180
Chamaecyparis thujiformis IPNI=*261974-1* 180
Chamaedorea aguilariana IPNI=*665826-1* 207
Chamaedorea amazonica IPNI=*54928-2* 207
Chamaedorea anomospadix IPNI=*54964-2* 208
Chamaedorea bartlingiana IPNI=*665826-1* 207
Chamaedorea bifurcata IPNI=*665826-1* 207
Chamaedorea boliviensis IPNI=*665826-1* 207
Chamaedorea bracteata IPNI=*665826-1* 207
Chamaedorea brevifrons IPNI=*665826-1* 207
Chamaedorea casperiana IPNI=*54964-2* 208
Chamaedorea columbica IPNI=*54964-2* 208
Chamaedorea concinna IPNI=*665826-1* 207
Chamaedorea concolor IPNI=*665826-1* 207
Chamaedorea conocarpa IPNI=*665826-1* 207
Chamaedorea corallina IPNI=*54901-2* 207
Chamaedorea depauperata IPNI=*665826-1* 207
Chamaedorea deppeana IPNI=*54865-2* 207
Chamaedorea donnell-smithii IPNI=*54954-2* 208
Chamaedorea dryanderae IPNI=*665826-1* 207
Chamaedorea eburnea IPNI=*54901-2* 207
Chamaedorea elegans var. *angustifolia* IPNI=*54865-2* 207
Chamaedorea erumpens IPNI=*54954-2* 208
Chamaedorea exorrhiza IPNI=*54964-2* 208
Chamaedorea flavovirens IPNI=*665826-1* 207
Chamaedorea formosa IPNI=*54901-2* 207
Chamaedorea geonomoides IPNI=*665826-1* 207
Chamaedorea gracilis IPNI=*665826-1* 207
Chamaedorea heilbornii IPNI=*665826-1* 207
Chamaedorea helleriana IPNI=*54865-2* 207
Chamaedorea herrerae IPNI=*665826-1* 207
Chamaedorea holmgrenii IPNI=*665826-1* 207
Chamaedorea hoppii IPNI=*665826-1* 207
Chamaedorea integrifolia IPNI=*54928-2* 207
Chamaedorea kalbreyeriana IPNI=*665826-1* 207
Chamaedorea lanceolata IPNI=*665826-1* 207
Chamaedorea lanceolata var. *littoralis* IPNI=*665826-1* 207
Chamaedorea latisecta IPNI=*54901-2* 207
Chamaedorea lechleriana IPNI=*54928-2* 207
Chamaedorea macroloba IPNI=*665826-1* 207
Chamaedorea megaphylla IPNI=*54901-2* 207
Chamaedorea membranacea IPNI=*665826-1* 207
Chamaedorea micrantha IPNI=*665826-1* 207
Chamaedorea minor IPNI=*665826-1* 207
Chamaedorea oerstedii IPNI=*665826-1* 207
Chamaedorea pacaya IPNI=*665826-1* 207
Chamaedorea poeppigiana IPNI=*54901-2* 207
Chamaedorea polyclada IPNI=*54901-2* 207
Chamaedorea pulchella IPNI=*54865-2* 207
Chamaedorea rhombea IPNI=*665826-1* 207
Chamaedorea simplicifrons IPNI=*54867-2* 207
Chamaedorea sphaerocarpa IPNI=*54964-2* 208
Chamaedorea wendlandiana IPNI=*54964-2* 208
Chamaedrys latifolia IPNI=*460416-1* 373
Chamaefilix erosa var. *serrata* IPNI=*22222-2* 174
Chamaefilix formosa IPNI=*22417-2* 174
Chamaefilix monanthes IPNI=*22558-2* 174
Chamaefilix obtusifolia IPNI=*60458652-2* 174
Chamaefilix rhizophora IPNI=*22662-2* 174
Chamaefilix serrata IPNI=*22751-2* 174
Chamaefilix serrata var. *crenulata* IPNI=*22751-2* 174
Chamaefistula acuminata IPNI=*911344-1* 332
Chamaefistula anconis IPNI=*911344-1* 332
Chamaefistula apiculata IPNI=*911344-1* 332
Chamaefistula bacillaris IPNI=*234415-2* 332
Chamaefistula berlandieri IPNI=*234475-2* 332
Chamaefistula chiapensis IPNI=*911344-1* 332
Chamaefistula chinensis IPNI=*518364-1* 333
Chamaefistula contorta IPNI=*518369-1* 333
Chamaefistula coromandelina IPNI=*518364-1* 333
Chamaefistula dariensis IPNI=*911330-1* 332
Chamaefistula deficiens IPNI=*911344-1* 332
Chamaefistula densiflora IPNI=*234475-2* 332
Chamaefistula elegans IPNI=*234652-2* 333
Chamaefistula fruticosa IPNI=*234475-2* 332
Chamaefistula gigantea IPNI=*911410-1* 333
Chamaefistula hayesiana IPNI=*911344-1* 332
Chamaefistula ignota IPNI=*234475-2* 332
Chamaefistula laevigata IPNI=*234652-2* 333
Chamaefistula lanata IPNI=*911344-1* 332
Chamaefistula latifolia IPNI=*911354-1* 333
Chamaefistula macrophylla IPNI=*234523-2* 333
Chamaefistula maxonii IPNI=*911344-1* 332
Chamaefistula membranacea IPNI=*911344-1* 332
Chamaefistula papillosa IPNI=*234589-2* 333
Chamaefistula pendula IPNI=*911386-1* 333
Chamaefistula ruiziana IPNI=*911404-1* 333
Chamaefistula sinaloana IPNI=*234475-2* 332
Chamaefistula sophera IPNI=*518364-1* 333
Chamaefistula standleyi IPNI=*911344-1* 332
Chamaefistula subcoriacea IPNI=*911344-1* 332
Chamaefistula subpilosa IPNI=*911344-1* 332
Chamaefistula torosa IPNI=*518364-1* 333
Chamaefistula undulata IPNI=*234686-2* 333
Chamaefistula valerioi IPNI=*911344-1* 332
Chamaefistula williamsii IPNI=*911344-1* 332
Chamaemelum chamomilla IPNI=*154715-2* 265
Chamaemelum coronarium IPNI=*1013480-1* 263
Chamaemelum foetidum IPNI=*1186403-2* 261
Chamaemelum millefolium IPNI=*2294-2* 255
Chamaemelum nobile f. *discoideum* IPNI=*192489-1* 259
Chamaemelum nobile f. *duplicatum* IPNI=*192489-1* 259
Chamaemelum nobile f. *nudum* IPNI=*192489-1* 259
Chamaemelum nobile var. *aureum* IPNI=*192489-1* 259
Chamaemelum nobile var. *discoideum* IPNI=*192489-1* 259
Chamaemelum nobile var. *duplicatum* IPNI=*192489-1* 259
Chamaemelum nobile var. *nudum* IPNI=*192489-1* 259
Chamaemelum parthenium IPNI=*252460-1* 271
Chamaemelum romanum IPNI=*192489-1* 259
Chamaemelum segetum IPNI=*209007-1* 263
Chamaemelum suaveolens IPNI=*154715-2* 265
Chamaemelum tanacetifolium IPNI=*2294-2* 255
Chamaemelum vulgare IPNI=*154715-2* 265
Chamaepeuce cynaroides IPNI=*200876-1* 260
Chamaeplium officinale IPNI=*289589-1* 278
Chamaeraphis brachiariiformis IPNI=*399655-1* 238
Chamaeraphis crus-ardeae IPNI=*421828-1* 245
Chamaeraphis effusa IPNI=*421711-1* 245
Chamaeraphis glauca IPNI=*77105978-1* 235
Chamaeraphis glauca var. *geniculata* IPNI=*943811-1* 245
Chamaeraphis glauca var. *imberbis* IPNI=*943811-1* 245
Chamaeraphis glauca var. *laevigata* IPNI=*943811-1* 245
Chamaeraphis glauca var. *penicillata* IPNI=*943811-1* 245
Chamaeraphis glauca var. *perennis* IPNI=*943811-1* 245
Chamaeraphis gracilis IPNI=*943811-1* 245
Chamaeraphis imberbis IPNI=*943811-1* 245

Chamaeraphis italica IPNI=*1207294-2* 245
Chamaeraphis italica var. *ambigua* IPNI=*421857-1* 245
Chamaeraphis muricata IPNI=*396213-1* 236
Chamaeraphis nepalensis IPNI=*421711-1* 245
Chamaeraphis neurodes IPNI=*421711-1* 245
Chamaeraphis palmifolia IPNI=*421711-1* 245
Chamaeraphis paniculifera IPNI=*421711-1* 245
Chamaeraphis penicillata IPNI=*943811-1* 245
Chamaeraphis rhachitricha IPNI=*1172068-2* 245
Chamaeraphis setosa var. *macrostachya* IPNI=*235234-2* 245
Chamaeraphis setosa var. *vulpiseta* IPNI=*1182687-2* 245
Chamaeraphis sulcata IPNI=*421711-1* 245
Chamaeraphis sulcata IPNI=*421828-1* 245
Chamaeraphis ventenatii IPNI=*943811-1* 245
Chamaeraphis verticillata IPNI=*421857-1* 245
Chamaeraphis viatica IPNI=*1172068-2* 245
Chamaerops excelsa IPNI=*669577-1* 211
Chamaerops fortunei IPNI=*670182-1* 212
Chamaerops kwanwortsik IPNI=*669577-1* 211
Chamaerops serrulata IPNI=*941782-1* 212
Chamaesenna inaequilatera IPNI=*911344-1* 332
Chamaesenna laevigata IPNI=*234652-2* 333
Chamaesenna multijuga IPNI=*234544-2* 333
Chamaesenna reticulata IPNI=*234636-2* 333
Chamaesenna velutina IPNI=*234694-2* 334
Chamaesyce biramensis IPNI=*101976-2* 391
Chamaesyce boliviana IPNI=*101671-2* 390
Chamaesyce brasiliensis IPNI=*101676-2* 390
Chamaesyce buxifolia IPNI=*347331-1* 390
Chamaesyce densiflora IPNI=*1100936-2* 390
Chamaesyce emarginata IPNI=*101976-2* 391
Chamaesyce gemella IPNI=*101651-2* 390
Chamaesyce glomerifera IPNI=*101671-2* 390
Chamaesyce heyneana IPNI=*346789-1* 390
Chamaesyce hirta IPNI=*101651-2* 390
Chamaesyce hirta subsp. *procumbens* IPNI=*101793-2* 390
Chamaesyce hirta var. *laeticincta* IPNI=*101651-2* 390
Chamaesyce hirta var. *procumbens* IPNI=*101793-2* 390
Chamaesyce hypericifolia IPNI=*101671-2* 390
Chamaesyce hyssopifolia IPNI=*101676-2* 390
Chamaesyce jonesii IPNI=*101676-2* 390
Chamaesyce jovetii IPNI=*347236-1* 390
Chamaesyce karwinskyi IPNI=*101651-2* 390
Chamaesyce lansingii IPNI=*101777-2* 390
Chamaesyce lasiocarpa IPNI=*347064-1* 390
Chamaesyce maculata IPNI=*347236-1* 390
Chamaesyce malaca IPNI=*101913-2* 390
Chamaesyce mangletii IPNI=*101976-2* 391
Chamaesyce mauritiana IPNI=*102034-2* 391
Chamaesyce mesembryanthemifolia IPNI=*347331-1* 390
Chamaesyce microclada IPNI=*101976-2* 391
Chamaesyce microphylla IPNI=*102034-2* 391
Chamaesyce nirurioides IPNI=*101676-2* 390
Chamaesyce nutans IPNI=*101777-2* 390
Chamaesyce ophthalmica IPNI=*101793-2* 390
Chamaesyce orbiculata IPNI=*347590-1* 390
Chamaesyce pileoides IPNI=*101976-2* 391
Chamaesyce pilulifera var. *procumbens* IPNI=*101793-2* 390
Chamaesyce preslii IPNI=*101777-2* 390
Chamaesyce prostrata IPNI=*101913-2* 390
Chamaesyce pseudonutans IPNI=*347236-1* 390
Chamaesyce radicans IPNI=*101976-2* 391
Chamaesyce rosei IPNI=*101651-2* 390
Chamaesyce rubrosperma IPNI=*102034-2* 391
Chamaesyce serpens IPNI=*101976-2* 391
Chamaesyce serpens var. *montevidensis* IPNI=*101976-2* 391
Chamaesyce supina IPNI=*347236-1* 390
Chamaesyce thymifolia IPNI=*102034-2* 391
Chamaesyce tracyi IPNI=*347236-1* 390
Chamaesyce villosior IPNI=*101913-2* 390
Chamaesyce yayalesia IPNI=*347331-1* 390
Chamissoa altissima f. *semispicata* IPNI=*60077-1* 282
Chamissoa altissima subsp. *albogrisea* IPNI=*60077-1* 282
Chamissoa altissima var. *densiflora* IPNI=*60077-1* 282
Chamissoa altissima var. *densipaniculata* IPNI=*60077-1* 282
Chamissoa altissima var. *glabrata* IPNI=*60077-1* 282
Chamissoa altissima var. *grandispicata* IPNI=*60077-1* 282
Chamissoa altissima var. *laxiflora* IPNI=*60077-1* 282
Chamissoa altissima var. *rubella* IPNI=*60077-1* 282
Chamissoa frondosa IPNI=*60077-1* 282
Chamissoa macrocarpa IPNI=*60077-1* 282
Chamissoa margaritacea IPNI=*316350-2* 282
Chamissoa stricta IPNI=*316350-2* 282
Chamomilla chamomilla IPNI=*154715-2* 265
Chamomilla courrantiana IPNI=*154715-2* 265
Chamomilla meridionalis IPNI=*154715-2* 265
Chamomilla nobilis IPNI=*192489-1* 259
Chamomilla officinalis IPNI=*154715-2* 265
Chamomilla recutita IPNI=*154715-2* 265
Chamomilla recutita var. *bayeri* IPNI=*154715-2* 265
Chamomilla unilateralis IPNI=*154715-2* 265
Chamomilla vulgaris IPNI=*154715-2* 265
Chaptalia diversifolia IPNI=*192653-1* 259
Chaptalia ebracteata IPNI=*192653-1* 259
Chaptalia erosa IPNI=*192653-1* 259
Chaptalia majuscula IPNI=*192653-1* 259
Chaptalia subcordata IPNI=*192653-1* 259
Charlwoodia australis IPNI=*533544-1* 215
Charlwoodia rubra IPNI=*97540-3* 215
Chasea prolifera IPNI=*410542-1* 242
Chatiakella platyglossa IPNI=*993359-1* 271
Chatiakella stenoglossa IPNI=*993359-1* 271
Chatinia calyculata IPNI=*551347-1* 448
Chaunochiton breviflorum IPNI=*55672-2* 449
Chaunochiton mouririoides IPNI=*607627-1* 449
Chaunochiton purpurascens IPNI=*55672-2* 449
Chaunochiton tubicinum IPNI=*607627-1* 449
Chavica grandis IPNI=*681559-1* 196
Chayota edulis IPNI=*293946-1* 295
Cheilanthes aculeata IPNI=*1031307-2* 175
Cheilanthes balansae IPNI=*77182303-1* 178
Cheilanthes brasiliensis IPNI=*77182519-1* 178
Cheilanthes caespitosa IPNI=*77182463-1* 178
Cheilanthes chaerophylloides IPNI=*77182463-1* 178
Cheilanthes chihuahuaensis IPNI=*77182808-1* 178
Cheilanthes chlorophylla IPNI=*77182519-1* 178
Cheilanthes ciliata IPNI=*77182463-1* 178
Cheilanthes concolor IPNI=*77182373-1* 178
Cheilanthes elegans IPNI=*77183191-1* 178
Cheilanthes elongata IPNI=*77183177-1* 178
Cheilanthes farinosa IPNI=*77182808-1* 178
Cheilanthes fournieri IPNI=*77183191-1* 178
Cheilanthes glaberrima IPNI=*77182519-1* 178
Cheilanthes heterotricha IPNI=*77183177-1* 178
Cheilanthes hostilis IPNI=*17126480-1* 175
Cheilanthes intermedia IPNI=*77183191-1* 178
Cheilanthes marginata f. *typica* IPNI=*77182463-1* 178
Cheilanthes marginata IPNI=*77182463-1* 178
Cheilanthes marginata var. *ovalis* IPNI=*77182463-1* 178
Cheilanthes marginata var. *reflexa* IPNI=*77182463-1* 178
Cheilanthes mexicana IPNI=*77182808-1* 178
Cheilanthes microphylla f. *elongata* IPNI=*77183177-1* 178
Cheilanthes microphylla IPNI=*77183177-1* 178
Cheilanthes microphylla var. *elongata* IPNI=*77183177-1* 178
Cheilanthes microphylla var. *minor* IPNI=*77183177-1* 178
Cheilanthes myriophylla IPNI=*77183191-1* 178

Cheilanthes myriophylla var. *elegans* IPNI=*77183191-1* 178
Cheilanthes obducta IPNI=*77182303-1* 178
Cheilanthes paleacea IPNI=*77183191-1* 178
Cheilanthes pohliana IPNI=*77182373-1* 178
Cheilanthes pulveracea IPNI=*77182808-1* 178
Cheilanthes pygmaea IPNI=*77183177-1* 178
Cheilanthes repens IPNI=*1031307-2* 175
Cheilanthes rigidula IPNI=*77182808-1* 178
Cheilanthes spectabilis IPNI=*77182519-1* 178
Cheilanthes ternifolia IPNI=*77182326-1* 178
Cheilanthes ternifolia var. *lanuginosa* IPNI=*77182326-1* 178
Cheilanthes weddelliana IPNI=*77182326-1* 178
Cheiloclinium glaziovii IPNI=*430947-1* 290
Cheiloclinium gleasonianum IPNI=*55849-2* 290
Cheiloclinium krukovii IPNI=*55849-2* 290
Cheiloclinium lineolatum IPNI=*55844-2* 290
Cheiloclinium minutiflorum IPNI=*55849-2* 290
Cheiranthus × *cheiri* f. *glaucescens* IPNI=*283703-1* 277
Cheiranthus × *cheiri* IPNI=*283703-1* 277
Cheiranthus × *fruticulosus* IPNI=*283703-1* 277
Cheiranthus × *helveticus* IPNI=*283703-1* 277
Cheiranthus × *intermedius* IPNI=*283703-1* 277
Cheiranthus × *luteus* IPNI=*283703-1* 277
Cheiranthus × *muralis* IPNI=*283703-1* 277
Cheiranthus × *silenifolius* IPNI=*283703-1* 277
Cheiranthus albus IPNI=*286910-1* 277
Cheiranthus annuus IPNI=*286910-1* 277
Cheiranthus arborescens IPNI=*286910-1* 277
Cheiranthus coccineus IPNI=*286910-1* 277
Cheiranthus fenestralis IPNI=*286910-1* 277
Cheiranthus glaber IPNI=*286910-1* 277
Cheiranthus glaberrimus IPNI=*286910-1* 277
Cheiranthus graecus IPNI=*286910-1* 277
Cheiranthus hortensis IPNI=*286910-1* 277
Cheiranthus incanus IPNI=*286910-1* 277
Cheiranthus viridis IPNI=*286910-1* 277
Cheiri × *montanum* IPNI=*283703-1* 277
Cheiri × *vulgare* IPNI=*283703-1* 277
Cheiri murale IPNI=*283703-1* 277
Cheirinia × *suffruticosa* IPNI=*283703-1* 277
Chelidonium multifidum IPNI=*30043364-2* 435
Chelonanthus alatus IPNI=*118811-2* 345
Chelonanthus cardonae IPNI=*130185-2* 345
Chelonanthus spruceanus IPNI=*1002418-1* 344
Chelonanthus uliginosus IPNI=*1002418-1* 344
Chenopodium × *borbasioides* f. *hircinifolium* IPNI=*55993-2* 282
Chenopodium × *zobelii* f. *hircinifolium* IPNI=*55993-2* 282
Chenopodium × *zobelii* f. *multidentatum* IPNI=*55993-2* 282
Chenopodium acutifolium IPNI=*55993-2* 282
Chenopodium agreste IPNI=*55993-2* 282
Chenopodium album f. *cymigerum* IPNI=*55993-2* 282
Chenopodium album f. *glomerulosum* IPNI=*55993-2* 282
Chenopodium album f. *lanceolatum* IPNI=*55993-2* 282
Chenopodium album f. *leiospermum* IPNI=*55993-2* 282
Chenopodium album f. *opuliforme* IPNI=*55993-2* 282
Chenopodium album f. *ovalifolium* IPNI=*55993-2* 282
Chenopodium album f. *paucidentatum* IPNI=*55993-2* 282
Chenopodium album f. *pseudozschackei* IPNI=*55993-2* 282
Chenopodium album f. *spicatum* IPNI=*55993-2* 282
Chenopodium album f. *subspontaneum* IPNI=*165175-1* 282
Chenopodium album f. *viride* IPNI=*55993-2* 282
Chenopodium album subf. *corymbosodentatum* IPNI=*55993-2* 282
Chenopodium album subf. *corymbosolanceolatum* IPNI=*55993-2* 282
Chenopodium album subf. *medium* IPNI=*55993-2* 282
Chenopodium album subf. *ovalifolium* IPNI=*55993-2* 282
Chenopodium album subf. *paucidens* IPNI=*55993-2* 282
Chenopodium album subf. *thyrsoideum* IPNI=*55993-2* 282
Chenopodium album subsp. *collinsii* IPNI=*55993-2* 282
Chenopodium album subsp. *densifoliatum* IPNI=*55993-2* 282
Chenopodium album subsp. *quinoa* IPNI=*165175-1* 282
Chenopodium album subvar. *leucospermum* IPNI=*165175-1* 282
Chenopodium album var. *coronatum* IPNI=*55993-2* 282
Chenopodium album var. *cymigerum* IPNI=*55993-2* 282
Chenopodium album var. *dacoticum* IPNI=*55993-2* 282
Chenopodium album var. *desertorum* IPNI=*55993-2* 282
Chenopodium album var. *polymorphum* IPNI=*55993-2* 282
Chenopodium album var. *purpureum* IPNI=*55993-2* 282
Chenopodium album var. *spicatum* IPNI=*55993-2* 282
Chenopodium album var. *stevensii* IPNI=*55993-2* 282
Chenopodium album var. *striatiforme* IPNI=*55993-2* 282
Chenopodium album var. *subaphyllum* IPNI=*55993-2* 282
Chenopodium amboanum IPNI=*1198060-2* 282
Chenopodium ambrosioides f. *angustifolium* IPNI=*1198060-2* 282
Chenopodium ambrosioides f. *dentatum* IPNI=*1198060-2* 282
Chenopodium ambrosioides f. *genuinum* IPNI=*1198060-2* 282
Chenopodium ambrosioides f. *integrifolium* IPNI=*1198060-2* 282
Chenopodium ambrosioides f. *minus* IPNI=*1198060-2* 282
Chenopodium ambrosioides f. *normale* IPNI=*1198060-2* 282
Chenopodium ambrosioides f. *pinnatifidum* IPNI=*1198060-2* 282
Chenopodium ambrosioides f. *rotundatum* IPNI=*1198060-2* 282
Chenopodium ambrosioides f. *spathulatum* IPNI=*1198060-2* 282
Chenopodium ambrosioides IPNI=*1198060-2* 282
Chenopodium ambrosioides subsp. *euambrosioides* IPNI=*1198060-2* 282
Chenopodium ambrosioides subsp. *pinnatifidum* IPNI=*1198060-2* 282
Chenopodium ambrosioides subsp. *suffruticosum* IPNI=*1198060-2* 282
Chenopodium ambrosioides var. *angustifolium* IPNI=*1198060-2* 282
Chenopodium ambrosioides var. *anthelminticum* IPNI=*1090820-2* 282
Chenopodium ambrosioides var. *costei* IPNI=*1198060-2* 282
Chenopodium ambrosioides var. *dentatum* IPNI=*1198060-2* 282
Chenopodium ambrosioides var. *genuinum* IPNI=*1198060-2* 282
Chenopodium ambrosioides var. *incisum* IPNI=*1198060-2* 282
Chenopodium ambrosioides var. *integrifolium* IPNI=*1198060-2* 282
Chenopodium ambrosioides var. *pinnatifidum* IPNI=*1198060-2* 282
Chenopodium ambrosioides var. *querciforme* IPNI=*1198060-2* 282
Chenopodium ambrosioides var. *suffruticosum* IPNI=*1198060-2* 282
Chenopodium ambrosioides var. *typicum* IPNI=*1198060-2* 282
Chenopodium angustifolium IPNI=*1198060-2* 282
Chenopodium anthelminticum IPNI=*1090820-2* 282
Chenopodium anthelminticum var. *glabratum* IPNI=*1198060-2* 282
Chenopodium anthelminticum var. *hastatum* IPNI=*1198060-2* 282
Chenopodium anthelminticum var. *subhirsutum* IPNI=*1198060-2* 282
Chenopodium atriplicis IPNI=*165175-1* 282
Chenopodium bicolor IPNI=*55993-2* 282
Chenopodium biforme IPNI=*77121007-1* 282
Chenopodium bolivianum IPNI=*165126-1* 282
Chenopodium bonariense IPNI=*164974-1* 282
Chenopodium browneanum IPNI=*55993-2* 282
Chenopodium candicans IPNI=*55993-2* 282
Chenopodium canihua IPNI=*56201-2* 282
Chenopodium catenulatum IPNI=*55993-2* 282
Chenopodium caudatum IPNI=*316349-2* 282
Chenopodium ccoyto IPNI=*165175-1* 282
Chenopodium ccuchi-huila IPNI=*165175-1* 282
Chenopodium chamrium IPNI=*77121007-1* 282
Chenopodium chilense IPNI=*165175-1* 282
Chenopodium chilense var. *incisa* IPNI=*1198060-2* 282
Chenopodium collae IPNI=*165126-1* 282
Chenopodium concatenatum IPNI=*55993-2* 282
Chenopodium congestum IPNI=*77121007-1* 282
Chenopodium cuneifolium IPNI=*1198060-2* 282
Chenopodium densifoliatum IPNI=*55993-2* 282
Chenopodium divaricatum IPNI=*55993-2* 282
Chenopodium diversifolium var. *montuosum* IPNI=*55993-2* 282
Chenopodium elatum IPNI=*55993-2* 282

Chenopodium ferulatum IPNI=*55993-2* 282
Chenopodium flavum IPNI=*77121007-1* 282
Chenopodium gandhium IPNI=*77121007-1* 282
Chenopodium glomerulosum IPNI=*55993-2* 282
Chenopodium guineense IPNI=*77121007-1* 282
Chenopodium guinoa IPNI=*165175-1* 282
Chenopodium hastatum IPNI=*165126-1* 282
Chenopodium hircinum f. *acutatum* IPNI=*164974-1* 282
Chenopodium hircinum f. *angustifolium* IPNI=*164974-1* 282
Chenopodium hircinum f. *deminutum* IPNI=*164974-1* 282
Chenopodium hircinum f. *genuinum* IPNI=*164974-1* 282
Chenopodium hircinum f. *laciniatum* IPNI=*165175-1* 282
Chenopodium hircinum f. *microphyllum* IPNI=*164974-1* 282
Chenopodium hircinum f. *multidentatum* IPNI=*164974-1* 282
Chenopodium hircinum f. *purpureum* IPNI=*164974-1* 282
Chenopodium hircinum f. *rubescens* IPNI=*165175-1* 282
Chenopodium hircinum f. *subtrilobum* IPNI=*164974-1* 282
Chenopodium hircinum f. *viridescens* IPNI=*165175-1* 282
Chenopodium hircinum subsp. *euhircinum* IPNI=*164974-1* 282
Chenopodium hircinum subsp. *milleanum* IPNI=*165175-1* 282
Chenopodium hircinum var. *andinum* IPNI=*165175-1* 282
Chenopodium hircinum var. *crataeginum* IPNI=*164974-1* 282
Chenopodium hircinum var. *densum* IPNI=*164974-1* 282
Chenopodium hircinum var. *erectum* IPNI=*164974-1* 282
Chenopodium hircinum var. *quinoa* IPNI=*165175-1* 282
Chenopodium hircinum var. *rhombicum* IPNI=*165175-1* 282
Chenopodium hircinum var. *subtrilobum* IPNI=*164974-1* 282
Chenopodium hircinum var. *typicum* IPNI=*164974-1* 282
Chenopodium ilicifolium IPNI=*77121007-1* 282
Chenopodium integrifolium IPNI=*1198060-2* 282
Chenopodium laciniatum IPNI=*55993-2* 282
Chenopodium lanceolatum f. *opizii* IPNI=*55993-2* 282
Chenopodium lanceolatum f. *sessiliflorum* IPNI=*55993-2* 282
Chenopodium lanceolatum IPNI=*55993-2* 282
Chenopodium lanceolatum var. *antiquitum* IPNI=*55993-2* 282
Chenopodium leiospermum IPNI=*55993-2* 282
Chenopodium leucospermum IPNI=*165175-1* 282
Chenopodium lobatum IPNI=*55993-2* 282
Chenopodium longidjawense IPNI=*77121007-1* 282
Chenopodium lucidum IPNI=*77121007-1* 282
Chenopodium maroccanum IPNI=*77121007-1* 282
Chenopodium missouriense var. *bushianum* IPNI=*55993-2* 282
Chenopodium murale IPNI=*77121007-1* 282
Chenopodium murale var. *angustatum* IPNI=*77121007-1* 282
Chenopodium murale var. *biforme* IPNI=*77121007-1* 282
Chenopodium murale var. *latifolium* IPNI=*77121007-1* 282
Chenopodium neglectum IPNI=*55993-2* 282
Chenopodium neoalbum IPNI=*55993-2* 282
Chenopodium oleraceum IPNI=*167384-1* 283
Chenopodium opulaceum IPNI=*55993-2* 282
Chenopodium ovalifolium IPNI=*55993-2* 282
Chenopodium paganum IPNI=*55993-2* 282
Chenopodium pallidicaule f. *melanospermum* IPNI=*56201-2* 282
Chenopodium pallidicaule f. *purpureum* IPNI=*56201-2* 282
Chenopodium pallidicaule f. *typicum* IPNI=*56201-2* 282
Chenopodium paniculatum IPNI=*165126-1* 282
Chenopodium paniculatum var. *reniforme* IPNI=*165126-1* 282
Chenopodium paucidentatum IPNI=*55993-2* 282
Chenopodium pedunculare IPNI=*55993-2* 282
Chenopodium petiolare f. *hastatum* IPNI=*165126-1* 282
Chenopodium petiolare f. *hookeri* IPNI=*165126-1* 282
Chenopodium petiolare f. *incanum* IPNI=*165126-1* 282
Chenopodium petiolare f. *reniforme* IPNI=*165126-1* 282
Chenopodium petiolare f. *scutatum* IPNI=*165126-1* 282
Chenopodium petiolare f. *trilobum* IPNI=*165126-1* 282
Chenopodium petiolare IPNI=*605135-1* 287
Chenopodium probstii f. *lanceolatum* IPNI=*55993-2* 282
Chenopodium probstii f. *parvoangustifolium* IPNI=*55993-2* 282
Chenopodium probstii IPNI=*55993-2* 282
Chenopodium probstii var. *glaucescens* IPNI=*55993-2* 282
Chenopodium pseudoborbasii f. *aellenii* IPNI=*55993-2* 282
Chenopodium pseudoborbasii f. *albiforme* IPNI=*55993-2* 282
Chenopodium pseudoborbasii f. *borbasiiforme* IPNI=*55993-2* 282
Chenopodium pseudoborbasii f. *longipedicellatum* IPNI=*55993-2* 282
Chenopodium pseudoborbasii f. *ramosum* IPNI=*55993-2* 282
Chenopodium pseudoborbasii var. *austromoravicum* IPNI=*55993-2* 282
Chenopodium punctulatum IPNI=*165175-1* 282
Chenopodium purpurascens IPNI=*165175-1* 282
Chenopodium querciforme IPNI=*1198060-2* 282
Chenopodium querciforme var. *minus* IPNI=*1198060-2* 282
Chenopodium quinoa f. *purpureum* IPNI=*165175-1* 282
Chenopodium quinoa subsp. *milleanum* IPNI=*165175-1* 282
Chenopodium quinoa var. *laciniatum* IPNI=*165175-1* 282
Chenopodium quinoa var. *lutescens* IPNI=*165175-1* 282
Chenopodium quinoa var. *melanospermum* IPNI=*165175-1* 282
Chenopodium quinoa var. *orbicans* IPNI=*165175-1* 282
Chenopodium quinoa var. *rubescens* IPNI=*165175-1* 282
Chenopodium quinoa var. *viridescens* IPNI=*165175-1* 282
Chenopodium riparium IPNI=*55993-2* 282
Chenopodium rugosum IPNI=*55993-2* 282
Chenopodium sancta-maria IPNI=*1198060-2* 282
Chenopodium sparsiflorum IPNI=*165126-1* 282
Chenopodium spathulatum IPNI=*1198060-2* 282
Chenopodium spathulatum var. *platyphyllum* IPNI=*1198060-2* 282
Chenopodium strictum IPNI=*55993-2* 282
Chenopodium subaphyllum IPNI=*55993-2* 282
Chenopodium suffruticosum IPNI=*1198060-2* 282
Chenopodium suffruticosum subsp. *remotum* IPNI=*1198060-2* 282
Chenopodium superalbum f. *kuehnii* IPNI=*55993-2* 282
Chenopodium superalbum IPNI=*55993-2* 282
Chenopodium variegatum IPNI=*1198060-2* 282
Chenopodium viride IPNI=*55993-2* 282
Chenopodium viridescens IPNI=*55993-2* 282
Chenopodium vulgare IPNI=*55993-2* 282
Chenopodium vulpinum IPNI=*55993-2* 282
Chetocrater hirta IPNI=*47939-2* 403
Chevaliera magdalenae IPNI=*5516-2* 228
Chevaliera magdalenae var. *quadricolor* IPNI=*5516-2* 228
Chevaliera rubiginosa IPNI=*271755-2* 228
Chichaea acerifolia IPNI=*245150-2* 414
Chichicaste grandis IPNI=*77188059-1* 291
Chiloporus andinus IPNI=*572660-1* 421
Chimarrhis duckei IPNI=*308173-2* 348
Chimarrhis williamsii IPNI=*56522-2* 348
Chimonobambusa angulata IPNI=*392553-1* 235
Chiococca alba f. *pilosa* IPNI=*56535-2* 348
Chiococca alba subsp. *parvifolia* IPNI=*56535-2* 348
Chiococca alba var. *micrantha* IPNI=*56535-2* 348
Chiococca alba var. *parvifolia* IPNI=*56535-2* 348
Chiococca anguifuga IPNI=*56535-2* 348
Chiococca anguifuga var. *pubescens* IPNI=*56535-2* 348
Chiococca bermudiana IPNI=*56535-2* 348
Chiococca brachiata IPNI=*56535-2* 348
Chiococca brachiata var. *acuminata* IPNI=*56535-2* 348
Chiococca brachiata var. *acutifolia* IPNI=*56535-2* 348
Chiococca brachiata var. *biformis* IPNI=*56535-2* 348
Chiococca brachiata var. *conjungens* IPNI=*56535-2* 348
Chiococca brachiata var. *densifolia* IPNI=*56535-2* 348
Chiococca brachiata var. *diplomorpha* IPNI=*56535-2* 348
Chiococca brachiata var. *genuina* IPNI=*56535-2* 348
Chiococca brachiata var. *grandifolia* IPNI=*56535-2* 348
Chiococca brachiata var. *intercedens* IPNI=*56535-2* 348
Chiococca brachiata var. *intermedia* IPNI=*56535-2* 348
Chiococca brachiata var. *lanceolata* IPNI=*56535-2* 348
Chiococca brachiata var. *microphylla* IPNI=*56535-2* 348

Chiococca brachiata var. *petiolaris* IPNI=*56535-2* 348
Chiococca brachiata var. *rigidula* IPNI=*56535-2* 348
Chiococca brachiata var. *subrhombea* IPNI=*56535-2* 348
Chiococca brachiata var. *tenuifolia* IPNI=*56535-2* 348
Chiococca brachiata var. *valida* IPNI=*56535-2* 348
Chiococca capitata IPNI=*56535-2* 348
Chiococca densifolia IPNI=*56535-2* 348
Chiococca densifolia var. *cubensis* IPNI=*56535-2* 348
Chiococca floridana IPNI=*56535-2* 348
Chiococca latifolia IPNI=*56535-2* 348
Chiococca macrocarpa IPNI=*56535-2* 348
Chiococca micrantha IPNI=*56535-2* 348
Chiococca paniculata IPNI=*56535-2* 348
Chiococca parvifolia IPNI=*56535-2* 348
Chiococca petensis IPNI=*56535-2* 348
Chiococca pinetorum IPNI=*56535-2* 348
Chiococca pubescens IPNI=*56535-2* 348
Chiococca pubescens var. *peninsularis* IPNI=*56535-2* 348
Chiococca racemosa var. *cubensis* IPNI=*56535-2* 348
Chiococca racemosa var. *floridana* IPNI=*56535-2* 348
Chiococca racemosa var. *jacquiniana* IPNI=*56535-2* 348
Chiococca racemosa var. *laxiflora* IPNI=*56535-2* 348
Chiococca racemosa var. *longifolia* IPNI=*56535-2* 348
Chiococca racemosa var. *yucatana* IPNI=*56535-2* 348
Chiococca trisperma IPNI=*56535-2* 348
Chiococca trisperma var. *angustifolia* IPNI=*56535-2* 348
Chionanthus angustifolius IPNI=*319141-2* 374
Chionanthus caribaeus IPNI=*301087-2* 374
Chionanthus cotinifolius IPNI=*319141-2* 374
Chionanthus fragrans IPNI=*319141-2* 374
Chionanthus gorgonae IPNI=*301087-2* 374
Chionanthus heterophylus IPNI=*319141-2* 374
Chionanthus latifolius IPNI=*319141-2* 374
Chionanthus longifolius IPNI=*319141-2* 374
Chionanthus luteus IPNI=*319141-2* 374
Chionanthus maritimus IPNI=*319141-2* 374
Chionanthus montanus IPNI=*319141-2* 374
Chionanthus obovatus IPNI=*319141-2* 374
Chionanthus roseus IPNI=*319141-2* 374
Chionanthus trifidus IPNI=*319141-2* 374
Chionanthus triflorus IPNI=*319141-2* 374
Chionanthus vernalis IPNI=*319141-2* 374
Chionanthus vernus IPNI=*319141-2* 374
Chionanthus virginicus subsp. *maritimus* IPNI=*319141-2* 374
Chionanthus virginicus var. *angustifolius* IPNI=*319141-2* 374
Chionanthus virginicus var. *latifolius* IPNI=*319141-2* 374
Chionanthus virginicus var. *maritimus* IPNI=*319141-2* 374
Chionanthus virginicus var. *montanus* IPNI=*319141-2* 374
Chionanthus zeylanicus IPNI=*319141-2* 374
Chirocalyx candolleanus IPNI=*494614-1* 320
Chirocalyx divaricatus IPNI=*494614-1* 320
Chirocalyx indicus IPNI=*494614-1* 320
Chirocalyx pictus IPNI=*494614-1* 320
Chirocalyx rubrinervius IPNI=*494567-1* 319
Chirocalyx velutinus IPNI=*494616-1* 320
Chironia centaurium IPNI=*326619-2* 344
Chironia centaurium var. *fascicularis* IPNI=*326619-2* 344
Chironia erythraea IPNI=*326619-2* 344
Chitonia bubalina IPNI=*572209-1* 419
Chitonia macrophylla IPNI=*1000103-2* 422
Chlaenobolus alopecuroides IPNI=*240147-1* 269
Chlaenobolus spicatus IPNI=*30022662-2* 269
Chlaenobolus virgatus IPNI=*30022662-2* 269
Chlamydia tenacissima IPNI=*539926-1* 216
Chlora exaltata IPNI=*367440-1* 344
Chloris abyssinica IPNI=*395891-1* 236
Chloris alba IPNI=*396063-1* 236
Chloris alba var. *aristulata* IPNI=*396063-1* 236
Chloris alberti IPNI=*396063-1* 236
Chloris anomala IPNI=*395997-1* 236
Chloris arundinacea IPNI=*60469006-2* 245
Chloris barbata IPNI=*60469006-2* 245
Chloris barbata var. *decora* IPNI=*396063-1* 236
Chloris beyreichiana IPNI=*395997-1* 236
Chloris beyrichiana IPNI=*395997-1* 236
Chloris brachystachys IPNI=*396063-1* 236
Chloris breviaristata IPNI=*60469006-2* 245
Chloris compressa IPNI=*396063-1* 236
Chloris consanguinea IPNI=*60469006-2* 245
Chloris curtipendula IPNI=*1198286-2* 235
Chloris cynodon IPNI=*397065-1* 236
Chloris dandyana IPNI=*60469006-2* 245
Chloris dandyana var. *breviaristata* IPNI=*60469006-2* 245
Chloris dandyana var. *stolonifera* IPNI=*60469006-2* 245
Chloris decora IPNI=*396063-1* 236
Chloris dubia IPNI=*60458636-2* 238
Chloris elata IPNI=*60469006-2* 245
Chloris elegans IPNI=*396063-1* 236
Chloris filiformis IPNI=*328564-2* 235
Chloris gabrielae IPNI=*396063-1* 236
Chloris glabrata IPNI=*395891-1* 236
Chloris guineensis IPNI=*397387-1* 236
Chloris humilis IPNI=*395997-1* 236
Chloris inflata IPNI=*1168070-2* 236
Chloris intermedia IPNI=*395997-1* 236
Chloris leptostachya IPNI=*395997-1* 236
Chloris longifolia IPNI=*1168070-2* 236
Chloris madagascariensis IPNI=*396063-1* 236
Chloris maritima IPNI=*397065-1* 236
Chloris meccana IPNI=*396063-1* 236
Chloris mucronata IPNI=*397387-1* 236
Chloris multiradiata IPNI=*396063-1* 236
Chloris notocoma IPNI=*396063-1* 236
Chloris obtusifolia IPNI=*395997-1* 236
Chloris paraguaiensis IPNI=*1168070-2* 236
Chloris paytensis IPNI=*397065-1* 236
Chloris penicillata IPNI=*396063-1* 236
Chloris pluriflora IPNI=*60458641-2* 241
Chloris polydactyla f. *stolonifera* IPNI=*60469006-2* 245
Chloris polydactyla IPNI=*396063-1* 236
Chloris polydactyla IPNI=*60469006-2* 245
Chloris polydactyla var. *breviaristata* IPNI=*60469006-2* 245
Chloris procumbens IPNI=*328564-2* 235
Chloris prostrata IPNI=*397387-1* 236
Chloris pubescens IPNI=*396063-1* 236
Chloris pulchra IPNI=*396063-1* 236
Chloris repens IPNI=*395891-1* 236
Chloris repens IPNI=*74675-3* 238
Chloris rogeonii IPNI=*396063-1* 236
Chloris rufescens IPNI=*1168070-2* 236
Chloris salzmannii IPNI=*395997-1* 236
Chloris tenuis IPNI=*328564-2* 235
Chloris tetrastachys IPNI=*396063-1* 236
Chloris tibestica IPNI=*396063-1* 236
Chlorocharis capitata IPNI=*1029981-2* 232
Chlorocharis geniculata IPNI=*1029981-2* 232
Chlorocyperus abyssinicus IPNI=*74698-2* 230
Chlorocyperus articulatus IPNI=*303822-1* 230
Chlorocyperus aureus IPNI=*300668-2* 230
Chlorocyperus cimicinus IPNI=*74915-2* 231
Chlorocyperus cordobensis IPNI=*303822-1* 230
Chlorocyperus eragrostis IPNI=*304426-1* 230
Chlorocyperus flavescens IPNI=*74698-2* 230
Chlorocyperus giganteus IPNI=*304595-1* 230
Chlorocyperus inflexus IPNI=*324890-2* 231
Chlorocyperus intermedius IPNI=*74698-2* 230

Chlorocyperus iria IPNI=*304844-1* 230
Chlorocyperus laevigatus IPNI=*304938-1* 231
Chlorocyperus melanostachyus IPNI=*74915-2* 231
Chlorocyperus papyrus IPNI=*330934-2* 231
Chlorocyperus phymatodes IPNI=*300668-2* 230
Chlorocyperus polystachyus IPNI=*75004-2* 231
Chlorocyperus prolixus IPNI=*305627-1* 231
Chlorocyperus rotundus IPNI=*305797-1* 231
Chlorocyperus salaamensis IPNI=*305797-1* 231
Chloromyron verticillatum IPNI=*277321-2* 385
Chlorophora mora IPNI=*1151750-2* 440
Chlorophora reticulata IPNI=*1151750-2* 440
Chlorophora tinctoria f. *glabrescens* IPNI=*1151750-2* 440
Chlorophora tinctoria f. *miqueliana* IPNI=*1151750-2* 440
Chlorophora tinctoria f. *ovata* IPNI=*1151750-2* 440
Chlorophora tinctoria f. *polyneura* IPNI=*1151750-2* 440
Chlorophora tinctoria f. *tataiba* IPNI=*1151750-2* 440
Chlorophora tinctoria IPNI=*1151750-2* 440
Chlorophora tinctoria subsp. *eutinctoria* IPNI=*1151750-2* 440
Chlorophora tinctoria subsp. *mora* IPNI=*1151750-2* 440
Chlorophora tinctoria subsp. *zanthoxylon* IPNI=*1151750-2* 440
Chlorophora tinctoria var. *affinis* IPNI=*1151750-2* 440
Chlorophora tinctoria var. *mora* IPNI=*1151750-2* 440
Chlorophora tinctoria var. *ovata* IPNI=*1151750-2* 440
Chlorophytum brevipes IPNI=*532810-1* 215
Chlorophytum bukobense IPNI=*532810-1* 215
Chlorophytum burchellii IPNI=*532810-1* 215
Chlorophytum delagoense IPNI=*532810-1* 215
Chlorophytum elatulum IPNI=*532810-1* 215
Chlorophytum elgonense IPNI=*532810-1* 215
Chlorophytum gazense IPNI=*532810-1* 215
Chlorophytum glaucidulum IPNI=*532810-1* 215
Chlorophytum inopinum IPNI=*532810-1* 215
Chlorophytum kirkii IPNI=*532810-1* 215
Chlorophytum limurense IPNI=*532810-1* 215
Chlorophytum longum IPNI=*532810-1* 215
Chlorophytum magnum IPNI=*532810-1* 215
Chlorophytum miserum IPNI=*532810-1* 215
Chlorophytum nemorosum IPNI=*532810-1* 215
Chlorophytum paludicola IPNI=*532810-1* 215
Chlorophytum ramiferum IPNI=*532810-1* 215
Chlorophytum rugosum IPNI=*532810-1* 215
Chlorophytum sternbergianum IPNI=*532810-1* 215
Chlorophytum turritum IPNI=*532810-1* 215
Chlorophytum usambarense IPNI=*532810-1* 215
Chloropsis pluriflora IPNI=*60458641-2* 241
Chodanthus puberulus IPNI=*13609-2* 362
Chomelia boliviana IPNI=*57096-2* 348
Chomelia brachyloba IPNI=*746662-1* 348
Chomelia chambersii IPNI=*77107342-1* 358
Chomelia dimorpha IPNI=*57096-2* 348
Chomelia longicaudata IPNI=*746662-1* 348
Chomelia multiflora IPNI=*57096-2* 348
Chomelia panamensis IPNI=*77107342-1* 358
Chomelia peruviana IPNI=*57096-2* 348
Chomelia purpusii IPNI=*746662-1* 348
Chonais vittata IPNI=*65185-1* 213
Chondrilla japonica IPNI=*1124129-2* 272
Chondrilla multiflora IPNI=*1124129-2* 272
Chondrilla runcinata IPNI=*1124129-2* 272
Chondrocarpus asiaticus IPNI=*1197718-2* 252
Chondrocarpus repandus IPNI=*51383-2* 252
Chondrocarpus triflorus IPNI=*1197718-2* 252
Chondrodendron hypoleucum IPNI=*580301-1* 434
Chondrodendron polyanthum IPNI=*72709-2* 435
Chondrodendron toxicoferum IPNI=*72709-2* 435
Chondrophylla americana IPNI=*108504-2* 344
Chondrophylla prostrata IPNI=*108504-2* 344
Chondrorhyncha marginata IPNI=*661384-1* 223
Chondrosum humile IPNI=*328564-2* 235
Chondrosum prostratum IPNI=*328564-2* 235
Chondrosum simplex IPNI=*328564-2* 235
Chorisia rosea IPNI=*60446802-2* 414
Chorisia speciosa IPNI=*1005661-1* 408
Chorisia speciosa var. *minor* IPNI=*1005661-1* 408
Chorisia speciosa var. *paraguariensis* IPNI=*1005661-1* 408
Christella dentata IPNI=*251308-2* 174
Christella dentata var. *caespitosa* IPNI=*251308-2* 174
Christella dentata var. *himalayensis* IPNI=*251308-2* 174
Christella malabariensis IPNI=*251308-2* 174
Christella normalis IPNI=*251400-2* 174
Christella nymphalis IPNI=*251308-2* 174
Christella taprobanica IPNI=*251308-2* 174
Christensonella squamata IPNI=*155331-2* 222
Christensonella uncata IPNI=*155331-2* 222
Chromolaena barranquillensis IPNI=*57443-2* 259
Chromolaena clematitis IPNI=*57443-2* 259
Chromolaena furcata IPNI=*57410-2* 259
Chromolaena laevigata var. *arenaria* IPNI=*57414-2* 259
Chromolaena laevigata var. *submembranacea* IPNI=*57414-2* 259
Chromolaena luquensis IPNI=*57410-2* 259
Chromolaena odorata f. *squarrosa* IPNI=*57443-2* 259
Chromolaena pratensis IPNI=*188196-1* 259
Chrysalidocarpus glaucescens IPNI=*985360-1* 208
Chrysalidocarpus lucubensis IPNI=*77211330-1* 208
Chrysalidocarpus lutescens IPNI=*985360-1* 208
Chrysalidocarpus madagascariensis IPNI=*77211330-1* 208
Chrysalidocarpus oleraceus IPNI=*77211330-1* 208
Chrysallidosperma smithii IPNI=*246971-2* 212
Chrysanthemum × *hortorum* IPNI=*77111328-1* 259
Chrysanthemum × *indicum* IPNI=*77111328-1* 259
Chrysanthemum × *maximoviczianum* IPNI=*77111328-1* 259
Chrysanthemum × *morifolium* IPNI=*77111328-1* 259
Chrysanthemum × *morifolium* var. *gracile* IPNI=*77111328-1* 259
Chrysanthemum × *morifolium* var. *sinense* IPNI=*77111328-1* 259
Chrysanthemum × *sabinii* IPNI=*193510-1* 259
Chrysanthemum × *sinense* IPNI=*77111328-1* 259
Chrysanthemum × *sinense* var. *spontaneum* IPNI=*77111328-1* 259
Chrysanthemum × *stipulaceum* IPNI=*77111328-1* 259
Chrysanthemum breviradiatum IPNI=*1013480-1* 263
Chrysanthemum burbankii IPNI=*230047-1* 265
Chrysanthemum carnosulum IPNI=*178745-1* 257
Chrysanthemum chamomilla IPNI=*154715-2* 265
Chrysanthemum cinerariifolium IPNI=*252275-1* 271
Chrysanthemum coronarium IPNI=*1013480-1* 263
Chrysanthemum coronarium var. *discolor* IPNI=*1013480-1* 263
Chrysanthemum frutescens IPNI=*178745-1* 257
Chrysanthemum grandiflorum IPNI=*230047-1* 265
Chrysanthemum holophyllum IPNI=*209007-1* 263
Chrysanthemum indicum var. *acutum* IPNI=*193510-1* 259
Chrysanthemum indicum var. *coreanum* IPNI=*193510-1* 259
Chrysanthemum indicum var. *edule* IPNI=*193510-1* 259
Chrysanthemum indicum var. *hibernum* IPNI=*193510-1* 259
Chrysanthemum indicum var. *litorale* IPNI=*193510-1* 259
Chrysanthemum indicum var. *lushanense* IPNI=*193510-1* 259
Chrysanthemum indicum var. *procumbens* IPNI=*193510-1* 259
Chrysanthemum indicum var. *purpureum* IPNI=*77111328-1* 259
Chrysanthemum japonicola IPNI=*193510-1* 259
Chrysanthemum koraiense IPNI=*193510-1* 259
Chrysanthemum laciniatum IPNI=*209007-1* 263
Chrysanthemum lavandulifolium var. *acutum* IPNI=*193510-1* 259
Chrysanthemum leucanthemum IPNI=*30296198-2* 265
Chrysanthemum littorale IPNI=*193510-1* 259
Chrysanthemum lushanense IPNI=*193510-1* 259
Chrysanthemum matricariodes IPNI=*1013480-1* 263
Chrysanthemum maximum IPNI=*230047-1* 265

Chrysanthemum nankingense IPNI=*193510-1* 259
Chrysanthemum partheniifolium f. *sanguineum* IPNI=*252460-1* 271
Chrysanthemum partheniifolium IPNI=*252460-1* 271
Chrysanthemum partheniodes IPNI=*192489-1* 259
Chrysanthemum parthenium IPNI=*252460-1* 271
Chrysanthemum persicum IPNI=*252460-1* 271
Chrysanthemum praealtum IPNI=*252460-1* 271
Chrysanthemum procumbens IPNI=*193510-1* 259
Chrysanthemum procumbens IPNI=*257735-2* 271
Chrysanthemum purpureum IPNI=*193510-1* 259
Chrysanthemum rigidum IPNI=*252275-1* 271
Chrysanthemum roxburghii IPNI=*1013480-1* 263
Chrysanthemum segetale IPNI=*209007-1* 263
Chrysanthemum segetum IPNI=*209007-1* 263
Chrysanthemum senecioides IPNI=*1013480-1* 263
Chrysanthemum speciosum IPNI=*1013480-1* 263
Chrysanthemum suaveolens IPNI=*154715-2* 265
Chrysanthemum tripartitum IPNI=*193510-1* 259
Chrysanthemum turreanum IPNI=*252275-1* 271
Chrysanthemum umbrosum IPNI=*209007-1* 263
Chrysanthemum vulgare IPNI=*252568-1* 271
Chrysanthemum welwitschii IPNI=*209007-1* 263
Chrysanthemum yamazutanum IPNI=*193510-1* 259
Chrysaspis campestris IPNI=*522975-1* 335
Chrysaspis dubia IPNI=*523099-1* 335
Chryseis caespitosa IPNI=*30043364-2* 435
Chryseis californica IPNI=*30043364-2* 435
Chryseis compacta IPNI=*30043364-2* 435
Chryseis crocea IPNI=*30043364-2* 435
Chryseis douglasii IPNI=*30043364-2* 435
Chryseis hypecoides IPNI=*30043364-2* 435
Chryseis tenuifolia IPNI=*30043364-2* 435
Chrysobalanus americanus IPNI=*725498-1* 382
Chrysobalanus mollicomus IPNI=*725498-1* 382
Chrysobalanus montanus IPNI=*727140-1* 383
Chrysobalanus triandrus IPNI=*725631-1* 382
Chrysochlamys dependens var. *sararensis* IPNI=*427393-1* 383
Chrysocoma linifolia IPNI=*981909-1* 271
Chrysocoma oppositifolia IPNI=*27442-2* 257
Chrysocoma phosphorica IPNI=*1038787-2* 272
Chrysocoma punctata IPNI=*57414-2* 259
Chrysocoma punctulata IPNI=*57414-2* 259
Chrysocoma purpurascens IPNI=*27442-2* 257
Chrysocoma repanda IPNI=*944993-1* 261
Chrysocoma violacea IPNI=*962413-1* 260
Chrysocoma volubilis IPNI=*57443-2* 259
Chrysodium aureum IPNI=*319061-2* 177
Chrysodium cayennense IPNI=*319061-2* 177
Chrysodium danaeifolium IPNI=*17002360-1* 177
Chrysodium fasciculatum IPNI=*319061-2* 177
Chrysodium hirsutum IPNI=*319061-2* 177
Chrysodium inaequale IPNI=*319061-2* 177
Chrysodium lomarioides IPNI=*17002360-1* 177
Chrysodium lomarioides var. *hastatum* IPNI=*17002360-1* 177
Chrysodium scalpturatum IPNI=*319061-2* 177
Chrysodium speciosum IPNI=*319061-2* 177
Chrysodium vulgare IPNI=*319061-2* 177
Chrysogonum peruvianum IPNI=*261363-1* 272
Chrysomelon pomiferum IPNI=*71455-1* 451
Chrysophyllum abbreviatum IPNI=*786784-1* 305
Chrysophyllum alnifolium IPNI=*291747-2* 307
Chrysophyllum bonplandii IPNI=*304384-2* 305
Chrysophyllum brasiliense IPNI=*227688-2* 308
Chrysophyllum caeruleum IPNI=*304384-2* 305
Chrysophyllum cainito var. *caeruleum* IPNI=*304384-2* 305
Chrysophyllum cainito var. *pomiferum* IPNI=*304384-2* 305
Chrysophyllum cainito var. *portoricense* IPNI=*304384-2* 305
Chrysophyllum cainito var. *typicum* IPNI=*304384-2* 305
Chrysophyllum caracasanum IPNI=*948100-1* 308
Chrysophyllum cayennense IPNI=*209723-2* 306
Chrysophyllum cochlearium IPNI=*948096-1* 308
Chrysophyllum collinum IPNI=*1048092-2* 306
Chrysophyllum cyanogenum IPNI=*57816-2* 305
Chrysophyllum excelsum IPNI=*948088-1* 305
Chrysophyllum glaucescens IPNI=*209758-2* 306
Chrysophyllum guianense IPNI=*86806-2* 305
Chrysophyllum jamaicense IPNI=*304384-2* 305
Chrysophyllum leptocarpum IPNI=*209758-2* 306
Chrysophyllum macrophyllum IPNI=*209808-2* 307
Chrysophyllum maliforme IPNI=*304384-2* 305
Chrysophyllum minutiflorum IPNI=*209732-2* 306
Chrysophyllum monopyrenum IPNI=*304384-2* 305
Chrysophyllum oleifolium IPNI=*89010-2* 305
Chrysophyllum ottonis IPNI=*304384-2* 305
Chrysophyllum ramiflorum IPNI=*786784-1* 305
Chrysophyllum reticulatum IPNI=*209864-2* 307
Chrysophyllum schomburgkianum IPNI=*209950-2* 308
Chrysophyllum sericeum IPNI=*304384-2* 305
Chrysophyllum sessiliflorum IPNI=*209723-2* 306
Chrysophyllum wurdackii IPNI=*209738-2* 306
Chrysopogon acicularis IPNI=*396110-1* 236
Chrysopogon aciculatus var. *elatior* IPNI=*396110-1* 236
Chrysopogon stipoides IPNI=*240834-2* 245
Chrysopogon subulatus IPNI=*396110-1* 236
Chrysopogon trivialis IPNI=*396110-1* 236
Chrysopteris araneosa IPNI=*17175380-1* 176
Chrysopteris areolata IPNI=*17175380-1* 176
Chrysopteris aurea IPNI=*17175390-1* 176
Chrysopteris decumana IPNI=*17175400-1* 176
Chrysopteris dictyocallis IPNI=*17175400-1* 176
Chrysopteris glauca IPNI=*17175380-1* 176
Chrysopteris glaucina IPNI=*17175380-1* 176
Chrysopteris grandis IPNI=*17175390-1* 176
Chrysopteris lanosa IPNI=*17175380-1* 176
Chrysopteris martinicensis IPNI=*17175390-1* 176
Chrysopteris microdictya IPNI=*17175390-1* 176
Chrysopteris multiserialis IPNI=*17175400-1* 176
Chrysopteris pulvinata IPNI=*17175390-1* 176
Chrysopteris sporadocarpa IPNI=*17175380-1* 176
Chrysopteris trilobata IPNI=*17175380-1* 176
Chrysothemis aurantiaca IPNI=*378695-1* 366
Chrysothemis semiclausa IPNI=*378695-1* 366
Chrysothemis villosa IPNI=*378695-1* 366
Chrysoxylon speciosum IPNI=*204802-2* 355
Chrysurus aureus IPNI=*406803-1* 240
Chrysurus cynosuroides IPNI=*406803-1* 240
Chthonia elongata IPNI=*236598-1* 267
Chthonia prostrata IPNI=*236681-1* 267
Chulusium acuminatum IPNI=*694896-1* 289
Chuncoa amazonia IPNI=*250537-2* 416
Chuncoa diptera IPNI=*171280-1* 416
Chuncoa oblonga IPNI=*171280-1* 416
Chuncoa obovata IPNI=*250537-2* 416
Chupalon acuminatum IPNI=*60458979-2* 298
Chupalon benthamianum IPNI=*60458979-2* 298
Chupalon bracteatum IPNI=*60458979-2* 298
Chupalon cordatum IPNI=*60458979-2* 298
Chupalon cordifolium IPNI=*60458979-2* 298
Chupalon crassifolium IPNI=*60458979-2* 298
Chupalon hartwegianum IPNI=*60458979-2* 298
Chupalon latifolium IPNI=*60458979-2* 298
Chupalon laurifolium IPNI=*60458979-2* 298
Chupalon pseudopubescens IPNI=*60458979-2* 298
Chupalon pubescens IPNI=*858236-1* 298
Chupalon quereme IPNI=*1064712-2* 298
Chupalon scabriusculum IPNI=*60458979-2* 298

Chupalon splendens IPNI=*60458979-2* 298
Chupalon strobiliferum IPNI=*60458979-2* 298
Chupalon tarapotanum IPNI=*858240-1* 298
Chupalon veraguense IPNI=*60458979-2* 298
Chupalon warscewiczii IPNI=*60458979-2* 298
Chuquiraga fruticosa IPNI=*194359-1* 260
Chuquiraga insignis IPNI=*194359-1* 260
Chuquiraga insignis var. *genuina* IPNI=*194359-1* 260
Chuquiraga insignis var. *lancifolia* IPNI=*194359-1* 260
Chuquiraga insignis var. *microphylla* IPNI=*194359-1* 260
Chuquiraga jussieui var. *lancifolia* IPNI=*194359-1* 260
Chuquiraga lancifolia IPNI=*194359-1* 260
Chuquiraga microphylla IPNI=*194359-1* 260
Chuquiraga peruviana IPNI=*194359-1* 260
Chuquiraga pseudoruscifolia IPNI=*194359-1* 260
Chusquea jamesonii IPNI=*1135523-2* 236
Chusquea meyeriana var. *patentissima* IPNI=*1135523-2* 236
Chusquea quitensis IPNI=*1135523-2* 236
Chusquea quitensis var. *patentissima* IPNI=*1135523-2* 236
Chylodia sarmentosa IPNI=*993359-1* 271
Chymocarpus tuberosus IPNI=*310353-2* 280
Chytraculia arborea IPNI=*77191371-1* 426
Chytraculia bipennis IPNI=*77199448-1* 426
Chytraculia chytraculia IPNI=*77191371-1* 426
Chytraculia chytraculia var. *genuina* IPNI=*77191371-1* 426
Chytraculia cuspidata IPNI=*77191379-1* 426
Chytraculia eugenioides IPNI=*600952-1* 428
Chytraculia forsteri IPNI=*77191393-1* 426
Chytraculia gardneriana IPNI=*600952-1* 428
Chytraculia lucida IPNI=*77191459-1* 426
Chytraculia macrophylla IPNI=*77191461-1* 426
Chytraculia paniculata IPNI=*77191465-1* 426
Chytraculia pohliana IPNI=*77191459-1* 426
Chytraculia polyantha IPNI=*77191459-1* 426
Chytraculia pulchella IPNI=*77191477-1* 427
Chytraculia sartoriana IPNI=*600952-1* 428
Chytroma bipartita IPNI=*597290-1* 302
Chytroma chartacea IPNI=*597250-1* 302
Chytroma cincturata IPNI=*947902-1* 301
Chytroma cistella IPNI=*597250-1* 302
Chytroma cupularis IPNI=*597250-1* 302
Chytroma glossiniformis IPNI=*136980-2* 302
Chytroma grandifolia IPNI=*947902-1* 301
Chytroma laevicula IPNI=*597250-1* 302
Chytroma langsdorffii IPNI=*597250-1* 302
Chytroma marawynensis IPNI=*597250-1* 302
Chytroma monosperma IPNI=*597250-1* 302
Chytroma perspicua IPNI=*276907-2* 301
Chytroma pilacarpa IPNI=*597250-1* 302
Chytroma rorida IPNI=*597250-1* 302
Chytroma spruceana IPNI=*597250-1* 302
Chytroma urceolata IPNI=*597250-1* 302
Chytroma valida IPNI=*597290-1* 302
Cicca acida IPNI=*194610-2* 401
Cicca acidissima IPNI=*194610-2* 401
Cicca antillana IPNI=*352077-1* 401
Cicca antillana var. *pedicellaris* IPNI=*352077-1* 401
Cicca brasiliensis IPNI=*353613-1* 402
Cicca chinensis IPNI=*352077-1* 401
Cicca disticha IPNI=*194610-2* 401
Cicca nodiflora IPNI=*194610-2* 401
Cicca pavoniana IPNI=*352077-1* 401
Cicca racemosa IPNI=*194610-2* 401
Cicca sinica IPNI=*352077-1* 401
Cicca surinamensis IPNI=*352077-1* 401
Cicer grossum IPNI=*486336-1* 315
Cicer lens IPNI=*524872-1* 336
Cicer physodes IPNI=*486336-1* 315
Cicer sativum IPNI=*486336-1* 315
Cicerbita intybacea IPNI=*229022-1* 264
Cicercula alata IPNI=*501990-1* 325
Cicercula alba IPNI=*501990-1* 325
Cicercula caerulea IPNI=*501990-1* 325
Cicercula sativa IPNI=*501990-1* 325
Cichorium balearicum IPNI=*194533-1* 260
Cichorium byzanthinum IPNI=*194533-1* 260
Cichorium caeruleum IPNI=*194533-1* 260
Cichorium callosum IPNI=*194533-1* 260
Cichorium casnia IPNI=*194533-1* 260
Cichorium cicorea IPNI=*194533-1* 260
Cichorium commune IPNI=*194533-1* 260
Cichorium cosnia IPNI=*194533-1* 260
Cichorium divaricatum IPNI=*194533-1* 260
Cichorium glabratum IPNI=*194533-1* 260
Cichorium glaucum IPNI=*194533-1* 260
Cichorium hirsutum IPNI=*194533-1* 260
Cichorium intybus convar. *foliosum* IPNI=*194533-1* 260
Cichorium intybus convar. *radicosum* IPNI=*194533-1* 260
Cichorium intybus f. *album* IPNI=*194533-1* 260
Cichorium intybus f. *crispum* IPNI=*194533-1* 260
Cichorium intybus f. *rubicunda* IPNI=*194533-1* 260
Cichorium intybus f. *sativum* IPNI=*194533-1* 260
Cichorium intybus f. *sylvestre* IPNI=*194533-1* 260
Cichorium intybus subsp. *balearica* IPNI=*194533-1* 260
Cichorium intybus subsp. *glabratum* IPNI=*194533-1* 260
Cichorium intybus subsp. *glaucum* IPNI=*194533-1* 260
Cichorium intybus subsp. *sativum* IPNI=*194533-1* 260
Cichorium intybus var. *callosum* IPNI=*194533-1* 260
Cichorium intybus var. *khatoonae* IPNI=*194533-1* 260
Cichorium officinale IPNI=*194533-1* 260
Cichorium perenne IPNI=*194533-1* 260
Cichorium rigidum IPNI=*194533-1* 260
Cichorium sylvestre IPNI=*194533-1* 260
Ciconium × coccineum IPNI=*376335-1* 358
Ciconium × hybridum IPNI=*376335-1* 358
Ciconium × oxyphyllum IPNI=*376335-1* 358
Ciconium clarum IPNI=*377229-1* 358
Ciconium cocciniflorum IPNI=*377229-1* 358
Ciconium densiflorum IPNI=*377229-1* 358
Ciconium leptopetalum IPNI=*377229-1* 358
Ciconium leucanthum IPNI=*377229-1* 358
Ciconium marginatum IPNI=*377229-1* 358
Ciconium placidum IPNI=*377229-1* 358
Ciconium rosula IPNI=*377229-1* 358
Ciconium siccifolium IPNI=*377229-1* 358
Ciconium stenopetalum IPNI=*377229-1* 358
Ciconium zonale IPNI=*377229-1* 358
Cicuta major IPNI=*840668-1* 252
Cicuta officinalis IPNI=*840668-1* 252
Cieca angustifolia IPNI=*321973-2* 400
Cieca appendiculata IPNI=*674805-1* 398
Cieca argentea IPNI=*69894-2* 388
Cieca auriculata IPNI=*674805-1* 398
Cieca bauhiniifolia IPNI=*674768-1* 397
Cieca cinerea IPNI=*674805-1* 398
Cieca coriacea IPNI=*674920-1* 398
Cieca difformis IPNI=*674920-1* 398
Cieca discolor IPNI=*321967-2* 399
Cieca glabrata IPNI=*674824-1* 398
Cieca gracilis IPNI=*60461856-2* 399
Cieca hederacea IPNI=*321973-2* 400
Cieca heterophylla IPNI=*321973-2* 400
Cieca maculata IPNI=*321973-2* 400
Cieca minima IPNI=*185104-2* 399
Cieca misera IPNI=*321967-2* 399
Cieca montevidensis IPNI=*69894-2* 388

Cieca nigra IPNI=*185104-2* 399
Cieca pallida IPNI=*185104-2* 399
Cieca peltata IPNI=*321973-2* 400
Cieca pubescens IPNI=*321950-2* 398
Cieca suberosa IPNI=*321973-2* 400
Cieca viridis IPNI=*185104-2* 399
Cieca warei IPNI=*185104-2* 399
Ciminalis alpina IPNI=*1167766-2* 345
Ciminalis angustifolia IPNI=*1167766-2* 345
Ciminalis grandiflora IPNI=*1167766-2* 345
Ciminalis prostrata IPNI=*108504-2* 344
Cinchona amygdalifolia IPNI=*746718-1* 348
Cinchona angustifolia IPNI=*746806-1* 348
Cinchona australis IPNI=*746718-1* 348
Cinchona caduciflora IPNI=*1106418-2* 352
Cinchona calisaya var. *josephiana* IPNI=*58723-2* 348
Cinchona caloptera IPNI=*746872-1* 348
Cinchona carabayensis IPNI=*746718-1* 348
Cinchona caribaea IPNI=*1125915-2* 350
Cinchona caribbeana IPNI=*1125915-2* 350
Cinchona carua IPNI=*754813-1* 352
Cinchona china IPNI=*66725-2* 349
Cinchona chomeliana IPNI=*746872-1* 348
Cinchona condaminea IPNI=*58723-2* 348
Cinchona coronulata IPNI=*746872-1* 348
Cinchona corymbosa IPNI=*746868-1* 348
Cinchona crassifolia IPNI=*1215741-2* 352
Cinchona crispa IPNI=*58723-2* 348
Cinchona cuatrecasasii IPNI=*1106418-2* 352
Cinchona cucumifolia IPNI=*58723-2* 348
Cinchona decurrentifolia IPNI=*746872-1* 348
Cinchona delondriana IPNI=*746718-1* 348
Cinchona elliptica IPNI=*746872-1* 348
Cinchona euneura IPNI=*746718-1* 348
Cinchona fusca IPNI=*755513-1* 352
Cinchona gammiana IPNI=*746718-1* 348
Cinchona gironensis IPNI=*746718-1* 348
Cinchona globifera IPNI=*768322-1* 358
Cinchona goudotiana IPNI=*746872-1* 348
Cinchona govana IPNI=*746872-1* 348
Cinchona grandiflora IPNI=*66725-2* 349
Cinchona grandifolia IPNI=*1106418-2* 352
Cinchona hasskarliana IPNI=*746718-1* 348
Cinchona henleana IPNI=*134610-2* 352
Cinchona henleana var. *trichophylla* IPNI=*134610-2* 352
Cinchona herbacea IPNI=*1125915-2* 350
Cinchona heterocarpa IPNI=*1106418-2* 352
Cinchona hookeriana IPNI=*134610-2* 352
Cinchona howardiana IPNI=*746872-1* 348
Cinchona jamaicensis IPNI=*1125915-2* 350
Cinchona josephiana IPNI=*58723-2* 348
Cinchona laccifera IPNI=*236591-2* 357
Cinchona lancifolia var. *angustifolia* IPNI=*746806-1* 348
Cinchona lechleriana IPNI=*746872-1* 348
Cinchona ledgeriana IPNI=*746718-1* 348
Cinchona legitima IPNI=*58723-2* 348
Cinchona longiflora IPNI=*66725-2* 349
Cinchona lutea IPNI=*746872-1* 348
Cinchona lutescens IPNI=*1106418-2* 352
Cinchona macrocalyx var. *obtusifolia* IPNI=*58723-2* 348
Cinchona macrocalyx var. *uritusinga* IPNI=*58723-2* 348
Cinchona macrocarpa IPNI=*1215741-2* 352
Cinchona macrophylla IPNI=*764577-1* 356
Cinchona magniflora IPNI=*1215741-2* 352
Cinchona magnifolia IPNI=*1106418-2* 352
Cinchona morado IPNI=*746872-1* 348
Cinchona muzonensis IPNI=*134610-2* 352
Cinchona myrtifolia IPNI=*1125915-2* 350
Cinchona oblongifolia IPNI=*1106418-2* 352
Cinchona obtusifolia IPNI=*58723-2* 348
Cinchona obtusifolia IPNI=*66725-2* 349
Cinchona officinalis var. *josephiana* IPNI=*58723-2* 348
Cinchona officinalis var. *vera* IPNI=*58723-2* 348
Cinchona ovalifolia IPNI=*1215741-2* 352
Cinchona ovata IPNI=*746872-1* 348
Cinchona pahudiana IPNI=*746718-1* 348
Cinchona palescens IPNI=*746872-1* 348
Cinchona palton IPNI=*58723-2* 348
Cinchona pedunculata IPNI=*77065717-1* 348
Cinchona pelalba IPNI=*746872-1* 348
Cinchona pelletieriana IPNI=*746872-1* 348
Cinchona peruviana IPNI=*58723-2* 348
Cinchona peruviana IPNI=*746718-1* 348
Cinchona platyphylla IPNI=*746872-1* 348
Cinchona prismatostylis IPNI=*1215741-2* 352
Cinchona pubescens var. *cordata* IPNI=*746872-1* 348
Cinchona pubescens var. *ovata* IPNI=*746872-1* 348
Cinchona purdieana IPNI=*77065718-1* 348
Cinchona purpurascens IPNI=*746872-1* 348
Cinchona purpurea IPNI=*746872-1* 348
Cinchona quina IPNI=*66725-2* 349
Cinchona racemosa IPNI=*1125915-2* 350
Cinchona rosea IPNI=*755513-1* 352
Cinchona rosulenta IPNI=*746872-1* 348
Cinchona rotundifolia IPNI=*746872-1* 348
Cinchona rubicunda IPNI=*746872-1* 348
Cinchona rufinervis IPNI=*746872-1* 348
Cinchona rugosa IPNI=*746872-1* 348
Cinchona souzana IPNI=*67491-2* 349
Cinchona suberosa IPNI=*58723-2* 348
Cinchona subsessilis IPNI=*746872-1* 348
Cinchona succirubra IPNI=*746872-1* 348
Cinchona thwaitesii IPNI=*746718-1* 348
Cinchona trianae IPNI=*746868-1* 348
Cinchona tucujensis IPNI=*746872-1* 348
Cinchona tunita IPNI=*746806-1* 348
Cinchona uritusinga IPNI=*58723-2* 348
Cinchona viridiflora IPNI=*760466-1* 355
Cinchona vritusino IPNI=*58723-2* 348
Cinchona weddelliana IPNI=*746718-1* 348
Cincinalis aquilina IPNI=*17210060-1* 175
Cincinalis arachnoidea IPNI=*213798-2* 175
Cincinalis caudata IPNI=*213799-2* 175
Cincinalis chrysophylla IPNI=*77182941-1* 178
Cincinalis flavens IPNI=*77182941-1* 178
Cincinalis gleditschii IPNI=*17210060-1* 175
Cincinalis lanuginosa IPNI=*17210060-1* 175
Cincinalis nivea IPNI=*77182941-1* 178
Cincinalis tarapacana IPNI=*77182941-1* 178
Cincinalis tenera IPNI=*77182941-1* 178
Cineraria × hybrida IPNI=*1164364-2* 268
Cineraria canadensis IPNI=*1152557-2* 261
Cineraria cruenta IPNI=*237108-1* 268
Cineraria glaberrima IPNI=*962413-1* 260
Cineraria lactea IPNI=*237108-1* 268
Cineraria lanata IPNI=*237108-1* 268
Cineraria maritima IPNI=*77075194-1* 264
Cinna glomerata IPNI=*12850-2* 233
Cinnamomum alexii IPNI=*463752-1* 183
Cinnamomum antillanum IPNI=*77173628-1* 182
Cinnamomum areolatocostae IPNI=*282532-2* 185
Cinnamomum aromaticum IPNI=*463752-1* 183
Cinnamomum australe IPNI=*77173628-1* 182
Cinnamomum brasiliense IPNI=*77173628-1* 182
Cinnamomum camphora f. *cyclophyllum* IPNI=*463336-1* 183
Cinnamomum camphora var. *cyclophyllum* IPNI=*463336-1* 183

Cinnamomum camphora var. *hosyo* IPNI=*463336-1* 183
Cinnamomum camphoriferum IPNI=*463336-1* 183
Cinnamomum camphoroides IPNI=*463336-1* 183
Cinnamomum chana IPNI=*77173628-1* 182
Cinnamomum cinnamomifolium IPNI=*77173628-1* 182
Cinnamomum cinnamomum IPNI=*463752-1* 183
Cinnamomum corzoanum IPNI=*326508-2* 184
Cinnamomum cubense IPNI=*77173628-1* 182
Cinnamomum elongatum IPNI=*77173628-1* 182
Cinnamomum filamentosum IPNI=*77173628-1* 182
Cinnamomum fruticosum IPNI=*77173628-1* 182
Cinnamomum helicterifolium IPNI=*326508-2* 184
Cinnamomum henricii IPNI=*463336-1* 183
Cinnamomum heterotepalum IPNI=*77173628-1* 182
Cinnamomum iners IPNI=*463752-1* 183
Cinnamomum johnstonii IPNI=*77173628-1* 182
Cinnamomum karrouwa IPNI=*463752-1* 183
Cinnamomum leptopus IPNI=*463752-1* 183
Cinnamomum maheanum IPNI=*463752-1* 183
Cinnamomum mathewianum IPNI=*463752-1* 183
Cinnamomum mayanum IPNI=*171797-2* 184
Cinnamomum maynense IPNI=*77173628-1* 182
Cinnamomum mexicanum IPNI=*77173628-1* 182
Cinnamomum montanum IPNI=*77173628-1* 182
Cinnamomum nominale IPNI=*463336-1* 183
Cinnamomum obtusatum IPNI=*326508-2* 184
Cinnamomum officinarum IPNI=*463336-1* 183
Cinnamomum paraguariense IPNI=*77173628-1* 182
Cinnamomum peruvianum IPNI=*77173628-1* 182
Cinnamomum phoebe IPNI=*77173628-1* 182
Cinnamomum pichisense IPNI=*77173628-1* 182
Cinnamomum pickellii IPNI=*77173628-1* 182
Cinnamomum portosecurianum IPNI=*77173628-1* 182
Cinnamomum riedelii IPNI=*463752-1* 183
Cinnamomum simondii IPNI=*463336-1* 183
Cinnamomum taquetii IPNI=*463336-1* 183
Cinnamomum triplinerve IPNI=*77173628-1* 182
Cinnamomum xinguense IPNI=*77173628-1* 182
Cinnamomum zeylanicum IPNI=*463752-1* 183
Cinnamomum zeylanicum var. *antillanum* IPNI=*463752-1* 183
Cinnamomum zeylanicum var. *barthii* IPNI=*463752-1* 183
Cinnamomum zeylanicum var. *bengalense* IPNI=*463752-1* 183
Cinnamomum zeylanicum var. *biafranum* IPNI=*463752-1* 183
Cinnamomum zeylanicum var. *bonplandii* IPNI=*463752-1* 183
Cinnamomum zeylanicum var. *boutonii* IPNI=*463752-1* 183
Cinnamomum zeylanicum var. *capense* IPNI=*463752-1* 183
Cinnamomum zeylanicum var. *cayennense* IPNI=*463752-1* 183
Cinnamomum zeylanicum var. *commersonii* IPNI=*463752-1* 183
Cinnamomum zeylanicum var. *cordifolium* IPNI=*463752-1* 183
Cinnamomum zeylanicum var. *decandollei* IPNI=*463752-1* 183
Cinnamomum zeylanicum var. *delessertii* IPNI=*463752-1* 183
Cinnamomum zeylanicum var. *ellipticum* IPNI=*463752-1* 183
Cinnamomum zeylanicum var. *erectum* IPNI=*463752-1* 183
Cinnamomum zeylanicum var. *humboldtii* IPNI=*463752-1* 183
Cinnamomum zeylanicum var. *leschenaultii* IPNI=*463752-1* 183
Cinnamomum zeylanicum var. *madrassicum* IPNI=*463752-1* 183
Cinnamomum zeylanicum var. *mauritianum* IPNI=*463752-1* 183
Cinnamomum zeylanicum var. *meissneri* IPNI=*463752-1* 183
Cinnamomum zeylanicum var. *ovatum* IPNI=*463752-1* 183
Cinnamomum zeylanicum var. *pallasii* IPNI=*463752-1* 183
Cinnamomum zeylanicum var. *pleei* IPNI=*463752-1* 183
Cinnamomum zeylanicum var. *pourretii* IPNI=*463752-1* 183
Cinnamomum zeylanicum var. *regelii* IPNI=*463752-1* 183
Cinnamomum zeylanicum var. *roxburghii* IPNI=*463752-1* 183
Cinnamomum zeylanicum var. *sieberi* IPNI=*463752-1* 183
Cinnamomum zeylanicum var. *sonneratii* IPNI=*463752-1* 183
Cinnamomum zeylanicum var. *vaillantii* IPNI=*463752-1* 183
Cinnamomum zeylanicum var. *variabile* IPNI=*463752-1* 183
Cinnamomum zeylanicum var. *wolkensteinii* IPNI=*463752-1* 183
Cinnamomum zeylanicum var. *zollingeri* IPNI=*463752-1* 183
Cionandra cuspidata IPNI=*291845-1* 293
Cionandra graciliflora IPNI=*291845-1* 293
Cionandra racemosa IPNI=*291845-1* 293
Cipura longifolia IPNI=*77182480-1* 217
Cipura martinicensis IPNI=*441642-1* 217
Cipura plicata IPNI=*89940-2* 216
Cipura semiaperta IPNI=*441642-1* 217
Cirsium costaricense IPNI=*195679-1* 260
Cirsium echinatum var. *willkommianum* IPNI=*195292-1* 260
Cirsium maculatum IPNI=*249211-1* 270
Cirsium mexicanum var. *bracteatum* IPNI=*195679-1* 260
Cirsium portoricense IPNI=*195679-1* 260
Cirsium willkommianum IPNI=*195292-1* 260
Cissampelos acuminata IPNI=*59219-2* 434
Cissampelos acuta IPNI=*580371-1* 434
Cissampelos amazonica IPNI=*580420-1* 434
Cissampelos argentea IPNI=*59219-2* 434
Cissampelos assimilis IPNI=*580420-1* 434
Cissampelos auriculata IPNI=*59219-2* 434
Cissampelos australis IPNI=*59219-2* 434
Cissampelos benthamiana IPNI=*59219-2* 434
Cissampelos boivinii IPNI=*59219-2* 434
Cissampelos bojeriana IPNI=*59219-2* 434
Cissampelos caapeba IPNI=*59219-2* 434
Cissampelos canescens IPNI=*59219-2* 434
Cissampelos ciliata IPNI=*580466-1* 434
Cissampelos clematidea IPNI=*580364-1* 434
Cissampelos cocculus IPNI=*59219-2* 434
Cissampelos communis IPNI=*580420-1* 434
Cissampelos consociata IPNI=*59219-2* 434
Cissampelos convexa IPNI=*580080-1* 434
Cissampelos convolvulacea IPNI=*59219-2* 434
Cissampelos cordifolia IPNI=*59219-2* 434
Cissampelos coriacea IPNI=*580358-1* 434
Cissampelos crenata IPNI=*580420-1* 434
Cissampelos cumingiana IPNI=*59219-2* 434
Cissampelos delicatula IPNI=*59219-2* 434
Cissampelos denudata IPNI=*580308-1* 434
Cissampelos diffusa IPNI=*59219-2* 434
Cissampelos discolor IPNI=*59219-2* 434
Cissampelos diversa IPNI=*59219-2* 434
Cissampelos ebracteata IPNI=*580420-1* 434
Cissampelos elata IPNI=*59219-2* 434
Cissampelos ellenbeckii IPNI=*59219-2* 434
Cissampelos eriantha IPNI=*59219-2* 434
Cissampelos eriocarpa IPNI=*59219-2* 434
Cissampelos errabunda IPNI=*580364-1* 434
Cissampelos floribunda IPNI=*580358-1* 434
Cissampelos fluminensis IPNI=*580466-1* 434
Cissampelos galapagensis IPNI=*580364-1* 434
Cissampelos glaberrima var. *orbicularis* IPNI=*580364-1* 434
Cissampelos glaucescens IPNI=*59219-2* 434
Cissampelos gracilis IPNI=*59219-2* 434
Cissampelos grallatoria IPNI=*59219-2* 434
Cissampelos guayaquilensis IPNI=*59219-2* 434
Cissampelos haenkeana IPNI=*59219-2* 434
Cissampelos hederacea IPNI=*59219-2* 434
Cissampelos heterophylla IPNI=*59219-2* 434
Cissampelos hirsutissima IPNI=*59219-2* 434
Cissampelos kohautiana IPNI=*59219-2* 434
Cissampelos limbata IPNI=*59219-2* 434
Cissampelos littoralis IPNI=*59219-2* 434
Cissampelos littoralis var. *minutiflora* IPNI=*59219-2* 434
Cissampelos longipes IPNI=*59219-2* 434
Cissampelos madagascariensis IPNI=*59219-2* 434
Cissampelos mallophylla IPNI=*580420-1* 434

Cissampelos mauritiana IPNI=59219-2 434
Cissampelos microcarpa IPNI=59219-2 434
Cissampelos monoica IPNI=59219-2 434
Cissampelos myriocarpa IPNI=59219-2 434
Cissampelos nepalensis IPNI=59219-2 434
Cissampelos nephrophylla IPNI=59219-2 434
Cissampelos obtecta IPNI=59219-2 434
Cissampelos orinocensis IPNI=59219-2 434
Cissampelos ovalifolia f. latifolia IPNI=580420-1 434
Cissampelos ovalifolia f. longipetiolata IPNI=580420-1 434
Cissampelos ovalifolia f. ovatomucronata IPNI=580420-1 434
Cissampelos ovalifolia f. reniformis IPNI=580420-1 434
Cissampelos ovalifolia var. cinereoviridis IPNI=580420-1 434
Cissampelos ovalifolia var. cinerescens IPNI=580420-1 434
Cissampelos ovalifolia var. rufescens IPNI=580420-1 434
Cissampelos pannosa IPNI=59219-2 434
Cissampelos pareira f. emarginatomucronata IPNI=59219-2 434
Cissampelos pareira f. reniformis IPNI=59219-2 434
Cissampelos pareira var. australis IPNI=59219-2 434
Cissampelos pareira var. caapeba IPNI=59219-2 434
Cissampelos pareira var. crenata IPNI=580420-1 434
Cissampelos pareira var. gardneri IPNI=59219-2 434
Cissampelos pareira var. haenkeana IPNI=59219-2 434
Cissampelos pareira var. laevis IPNI=59219-2 434
Cissampelos pareira var. monoica IPNI=59219-2 434
Cissampelos pareira var. racemiflora IPNI=59219-2 434
Cissampelos pareira var. tamoides IPNI=59219-2 434
Cissampelos pareira var. typica IPNI=59219-2 434
Cissampelos pareira var. vestita IPNI=580420-1 434
Cissampelos pareira var. wildei IPNI=59219-2 434
Cissampelos pareiroides IPNI=59219-2 434
Cissampelos parmata IPNI=580364-1 434
Cissampelos parriera IPNI=580364-1 434
Cissampelos pata IPNI=59219-2 434
Cissampelos peltata IPNI=580466-1 434
Cissampelos perrieri IPNI=59219-2 434
Cissampelos pilgeri IPNI=59219-2 434
Cissampelos poilanei IPNI=59219-2 434
Cissampelos ramiflora IPNI=580308-1 434
Cissampelos reticulata IPNI=59219-2 434
Cissampelos rotundata IPNI=580420-1 434
Cissampelos salzmannii IPNI=59219-2 434
Cissampelos scutigera IPNI=580358-1 434
Cissampelos suborbicularis IPNI=580420-1 434
Cissampelos suborbicularis var. sublanata IPNI=580420-1 434
Cissampelos subpeltata IPNI=59219-2 434
Cissampelos subreniformis IPNI=59219-2 434
Cissampelos subtriangularis IPNI=580420-1 434
Cissampelos sympodialis var. grandifolia IPNI=580466-1 434
Cissampelos tamoides IPNI=580358-1 434
Cissampelos tamoides IPNI=59219-2 434
Cissampelos testudinum IPNI=59219-2 434
Cissampelos tetrandra IPNI=59219-2 434
Cissampelos tomentosa IPNI=580080-1 434
Cissampelos tomentosa IPNI=59219-2 434
Cissampelos tropaeolifolia var. fluminensis IPNI=580466-1 434
Cissampelos velutina IPNI=580420-1 434
Cissampelos vestita IPNI=580420-1 434
Cissampelos violifolia IPNI=59219-2 434
Cistus tricuspis IPNI=30143354-2 410
Citharexylum album IPNI=330600-2 378
Citharexylum amazonicum IPNI=861796-1 378
Citharexylum berteroi IPNI=330600-2 378
Citharexylum caudatum f. angustifolium IPNI=330600-2 378
Citharexylum caudatum f. lindenii IPNI=330600-2 378
Citharexylum caudatum f. parvifolium IPNI=330600-2 378
Citharexylum erectum IPNI=330600-2 378
Citharexylum lindenii IPNI=330600-2 378
Citharexylum melanocardium IPNI=864000-1 372
Citharexylum mucronatum IPNI=330600-2 378
Citharexylum paniculatum IPNI=864000-1 372
Citharexylum poeppigii f. anomalum IPNI=861827-1 378
Citharexylum poeppigii var. calvescens IPNI=861827-1 378
Citharexylum poeppigii var. margaritaceum IPNI=861827-1 378
Citharexylum ramizii IPNI=229565-2 377
Citharexylum ternatum IPNI=330600-2 378
Citrosma andina IPNI=582392-1 186
Citrosma apicifera IPNI=582257-1 185
Citrosma buddleiifolia IPNI=582257-1 185
Citrosma camporum IPNI=582285-1 185
Citrosma chiridota IPNI=582392-1 186
Citrosma cuspidata IPNI=582248-1 185
Citrosma decipiens IPNI=582253-1 185
Citrosma dentata IPNI=582223-1 185
Citrosma dimidiata IPNI=90974-2 183
Citrosma discolor IPNI=582285-1 185
Citrosma dubia IPNI=582373-1 186
Citrosma echinata IPNI=582257-1 185
Citrosma guianensis IPNI=582285-1 185
Citrosma guianensis var. divergentifolia IPNI=582285-1 185
Citrosma guianensis var. nuda IPNI=582285-1 185
Citrosma lagopus IPNI=582392-1 186
Citrosma limoniodora IPNI=582223-1 185
Citrosma macrophylla IPNI=582223-1 185
Citrosma mollicoma IPNI=582373-1 186
Citrosma mollis IPNI=582373-1 186
Citrosma oblongifolia IPNI=582354-1 186
Citrosma oligocarpa IPNI=582285-1 185
Citrosma paniculata IPNI=90974-2 183
Citrosma pauciflora IPNI=582342-1 185
Citrosma petiolaris IPNI=582347-1 185
Citrosma poeppigii IPNI=582354-1 186
Citrosma pyricarpa IPNI=582223-1 185
Citrosma radiata IPNI=582373-1 186
Citrosma riparia IPNI=582392-1 186
Citrosma sessiliflora IPNI=582373-1 186
Citrosma thecaphora IPNI=582392-1 186
Citrullus anguria IPNI=291938-1 293
Citrullus aquosus IPNI=291938-1 293
Citrullus battich IPNI=291938-1 293
Citrullus caffrorum IPNI=291938-1 293
Citrullus chodospermus IPNI=291938-1 293
Citrullus citrullus IPNI=291938-1 293
Citrullus edulis IPNI=291938-1 293
Citrullus lanatus f. nigroseminius IPNI=291938-1 293
Citrullus lanatus var. albidus IPNI=291938-1 293
Citrullus lanatus var. caffrorum IPNI=291938-1 293
Citrullus lanatus var. cordophanus IPNI=291938-1 293
Citrullus lanatus var. minor IPNI=291938-1 293
Citrullus lanatus var. oblongus IPNI=291938-1 293
Citrullus lanatus var. pulcherrimus IPNI=291938-1 293
Citrullus lanatus var. pumilus IPNI=291938-1 293
Citrullus lanatus var. rotundus IPNI=291938-1 293
Citrullus lanatus var. senegalicus IPNI=291938-1 293
Citrullus lanatus var. shami IPNI=291938-1 293
Citrullus lanatus var. variegatus IPNI=291938-1 293
Citrullus lanatus var. virgatus IPNI=291938-1 293
Citrullus lanatus var. viridis IPNI=291938-1 293
Citrullus mucosospermus var. senegalicus IPNI=291938-1 293
Citrullus pasteca IPNI=291938-1 293
Citrullus variegatus IPNI=292416-1 294
Citrullus vulgaris IPNI=291938-1 293
Cladocolea biflora IPNI=608391-1 450
Cladosicyos edulis IPNI=77110571-1 294
Clandestinaria indica IPNI=222144-2 277
Claotrachelus rupestris IPNI=962413-1 260

Clarisia colombiana IPNI=*851228-1* 437
Clarisia ilicifolia var. *cuyunensis* IPNI=*59683-2* 437
Clarisia ilicifolia var. *guianensis* IPNI=*59683-2* 437
Clarisia ilicifolia var. *micranthera* IPNI=*59683-2* 437
Clarisia ilicifolia var. *paraensis* IPNI=*59683-2* 437
Clarisia mattogrossensis IPNI=*851228-1* 437
Clarisia mollis IPNI=*856575-1* 441
Clarisia nitida IPNI=*851229-1* 437
Clarisia panamensis IPNI=*851228-1* 437
Clarisia spruceana IPNI=*59683-2* 437
Clarisia strepitans IPNI=*59683-2* 437
Clarisia strepitans var. *cuyunensis* IPNI=*59683-2* 437
Clarisia strepitans var. *guianensis* IPNI=*59683-2* 437
Clarisia strepitans var. *micranthera* IPNI=*59683-2* 437
Clarisia strepitans var. *paraensis* IPNI=*59683-2* 437
Clarisia urophylla IPNI=*260203-2* 441
Clarisia volubilis IPNI=*1153747-2* 283
Clavapetalum elatum IPNI=*77582-2* 255
Clavapetalum surinamense IPNI=*77582-2* 255
Clavija antioquensis IPNI=*587939-1* 303
Clavija brachystachys IPNI=*587931-1* 303
Clavija caliensis IPNI=*311309-2* 303
Clavija densilineata IPNI=*309153-2* 303
Clavija longifolia IPNI=*587939-1* 303
Clavija magdalenae IPNI=*587931-1* 303
Clavija ornata var. *coriacea* IPNI=*587939-1* 303
Clavija reflexiflora IPNI=*308892-2* 303
Clavija sparsifolia IPNI=*866956-1* 405
Clavija spathulata IPNI=*311911-2* 303
Clavula acicularis IPNI=*1169169-2* 231
Claytonia alba var. *caulescens* IPNI=*698357-1* 286
Claytonia caulescens IPNI=*698357-1* 286
Claytonia ciliata IPNI=*698357-1* 286
Claytonia fontana IPNI=*311155-2* 286
Claytonia hallii IPNI=*311155-2* 286
Claytonia humboldtiana IPNI=*698305-1* 286
Claytonia megarhiza IPNI=*698305-1* 286
Claytonia menziesii IPNI=*698357-1* 286
Claytonia meyeniana IPNI=*698305-1* 286
Claytonia nemorosa IPNI=*130185-2* 345
Claytonia patens IPNI=*1007146-2* 290
Claytonia peruviana IPNI=*698357-1* 286
Claytonia pusilla IPNI=*311155-2* 286
Claytonia reflexa IPNI=*1007146-2* 290
Claytonia sarmentosa IPNI=*1007146-2* 290
Claytonia triandria IPNI=*698357-1* 286
Claytonia triangularis IPNI=*1182764-2* 290
Claytonia triangularis var. *crassifolia* IPNI=*1182764-2* 290
Cleianthus coccineus IPNI=*862301-1* 369
Cleistocactus chotaensis IPNI=*35149-2* 283
Cleistocactus roezlii IPNI=*35149-2* 283
Cleistocactus sepium IPNI=*35149-2* 283
Clematis dioica subsp. *sericea* IPNI=*709750-1* 436
Clematis grossa var. *striatula* IPNI=*709750-1* 436
Clematis sericea IPNI=*709750-1* 436
Clematis sericea var. *costata* IPNI=*709750-1* 436
Clematis sericea var. *striatula* IPNI=*709750-1* 436
Clematis sericea var. *tereticaulis* IPNI=*709750-1* 436
Clematis thalictroides IPNI=*709750-1* 436
Clematis virginiana subsp. *sericea* IPNI=*709750-1* 436
Cleome aculeata var. *normalis* IPNI=*316362-2* 279
Cleome acuta IPNI=*60518-2* 279
Cleome acutifolia IPNI=*320133-2* 280
Cleome affinis IPNI=*60518-2* 279
Cleome alliacea IPNI=*60518-2* 279
Cleome alliodora IPNI=*60518-2* 279
Cleome amazonica IPNI=*147057-1* 279
Cleome applanata IPNI=*147242-1* 279
Cleome arborea IPNI=*316363-2* 278
Cleome blumeana IPNI=*60518-2* 279
Cleome brachypoda IPNI=*147226-1* 279
Cleome brasiliensis IPNI=*147226-1* 279
Cleome bungei IPNI=*60518-2* 279
Cleome candelabrum IPNI=*60518-2* 279
Cleome capparoides IPNI=*146919-1* 279
Cleome cardinalis IPNI=*147320-1* 279
Cleome christii IPNI=*147242-1* 279
Cleome cremoloba IPNI=*147242-1* 279
Cleome denticulata IPNI=*60518-2* 279
Cleome eckloniana IPNI=*60518-2* 279
Cleome erucago IPNI=*60589-2* 279
Cleome filisepala IPNI=*147057-1* 279
Cleome friesii IPNI=*147057-1* 279
Cleome gigantea IPNI=*147320-1* 279
Cleome gilletii IPNI=*316362-2* 279
Cleome hamiltoniana IPNI=*316362-2* 279
Cleome hassleriana IPNI=*147105-1* 279
Cleome heptaphylla IPNI=*60589-2* 279
Cleome heterotricha IPNI=*60518-2* 279
Cleome hirsuticaulis IPNI=*147057-1* 279
Cleome hullettii IPNI=*316362-2* 279
Cleome humboldtii IPNI=*147226-1* 279
Cleome icosandra IPNI=*320133-2* 280
Cleome kerberi IPNI=*147057-1* 279
Cleome longifolia IPNI=*147320-1* 279
Cleome macrantha IPNI=*147320-1* 279
Cleome malhadensis IPNI=*316362-2* 279
Cleome margaritae IPNI=*316362-2* 279
Cleome micrantha IPNI=*147226-1* 279
Cleome monochroma IPNI=*147057-1* 279
Cleome muricata IPNI=*60518-2* 279
Cleome oleracea IPNI=*60518-2* 279
Cleome pachystigma IPNI=*147320-1* 279
Cleome palustris IPNI=*147226-1* 279
Cleome panamensis IPNI=*147226-1* 279
Cleome pedunculata IPNI=*147057-1* 279
Cleome pentaphylla IPNI=*60518-2* 279
Cleome pentaphylla var. *glabra* IPNI=*60518-2* 279
Cleome pilosa var. *costaricensis* IPNI=*147242-1* 279
Cleome pittieri IPNI=*147242-1* 279
Cleome pubescens IPNI=*60518-2* 279
Cleome pubescens IPNI=*60589-2* 279
Cleome pungens IPNI=*60589-2* 279
Cleome pungens var. *swartziana* IPNI=*60589-2* 279
Cleome purpurea IPNI=*147320-1* 279
Cleome rosea IPNI=*60518-2* 279
Cleome rubiginosa IPNI=*146919-1* 279
Cleome sandwicensis IPNI=*60589-2* 279
Cleome sessilifolia IPNI=*60518-2* 279
Cleome speciosa IPNI=*147320-1* 279
Cleome speciosissima IPNI=*147320-1* 279
Cleome spinosa f. *pungens* IPNI=*60589-2* 279
Cleome spinosa IPNI=*60589-2* 279
Cleome spinosa var. *subinermis* IPNI=*60589-2* 279
Cleome surinamensis IPNI=*316362-2* 279
Cleome tetraphylla IPNI=*147226-1* 279
Cleome tonduzii IPNI=*60589-2* 279
Cleome triphylla IPNI=*60518-2* 279
Cleome villosa IPNI=*147226-1* 279
Cleome viridiflora IPNI=*147057-1* 279
Cleoserrata speciosa IPNI=*147320-1* 279
Clercia tweedieana IPNI=*210308-2* 291
Clerodendrum aculeatum f. *acutifolium* IPNI=*866088-1* 374
Clerodendrum aculeatum f. *lanceolatum* IPNI=*866088-1* 374
Clerodendrum aculeatum f. *mucronatum* IPNI=*866088-1* 374
Clerodendrum aculeatum f. *orientale* IPNI=*866088-1* 374

Clerodendrum aculeatum f. *rotundatum* IPNI=*866088-1* 374
Clerodendrum aculeatum IPNI=*866088-1* 374
Clerodendrum aculeatum var. *gracile* IPNI=*866088-1* 374
Clerodendrum aculeatum var. *grandifolium* IPNI=*866088-1* 374
Clerodendrum aculeatum var. *parvifolium* IPNI=*866088-1* 374
Clerodendrum angustifolium IPNI=*866088-1* 374
Clerodendrum balfourii IPNI=*862437-1* 369
Clerodendrum blumeanum IPNI=*50426038-2* 369
Clerodendrum chinense var. *hamrense* IPNI=*946989-1* 369
Clerodendrum chinense var. *parviflorum* IPNI=*946989-1* 369
Clerodendrum chinense var. *plenum* IPNI=*946989-1* 369
Clerodendrum chinense var. *simplex* IPNI=*946989-1* 369
Clerodendrum citrinum IPNI=*862301-1* 369
Clerodendrum congense IPNI=*862461-1* 369
Clerodendrum cordifolium IPNI=*862461-1* 369
Clerodendrum diversifolium IPNI=*862301-1* 369
Clerodendrum duckei IPNI=*866088-1* 374
Clerodendrum fallax IPNI=*50426038-2* 369
Clerodendrum fragrans f. *pleniflorum* IPNI=*946989-1* 369
Clerodendrum fragrans IPNI=*946989-1* 369
Clerodendrum fragrans var. *pleniflorum* IPNI=*946989-1* 369
Clerodendrum greyi IPNI=*50426038-2* 369
Clerodendrum hirsutum IPNI=*862461-1* 369
Clerodendrum horsfieldii IPNI=*50426038-2* 369
Clerodendrum japonicum IPNI=*946989-1* 369
Clerodendrum lasiocephalum IPNI=*946989-1* 369
Clerodendrum macradenium IPNI=*946989-1* 369
Clerodendrum papuanum IPNI=*50426038-2* 369
Clerodendrum philippinum f. *multiplex* IPNI=*946989-1* 369
Clerodendrum philippinum f. *pleniflorum* IPNI=*946989-1* 369
Clerodendrum philippinum f. *subfertile* IPNI=*946989-1* 369
Clerodendrum philippinum IPNI=*946989-1* 369
Clerodendrum philippinum var. *simplex* IPNI=*946989-1* 369
Clerodendrum pulchrum IPNI=*50426038-2* 369
Clerodendrum pyramidale IPNI=*862301-1* 369
Clerodendrum riedelii IPNI=*946989-1* 369
Clerodendrum roseum IPNI=*946989-1* 369
Clerodendrum scandens IPNI=*862461-1* 369
Clerodendrum simplex IPNI=*862461-1* 369
Clerodendrum speciosum IPNI=*50426038-2* 369
Clerodendrum ternifolium IPNI=*866088-1* 374
Clerodendrum ternifolium var. *serratifolium* IPNI=*866088-1* 374
Clerodendrum ternifolium var. *velutinosum* IPNI=*866088-1* 374
Clerodendrum thomsoniae var. *balfourii* IPNI=*862437-1* 369
Clerodendrum umbellatum f. *scandens* IPNI=*862461-1* 369
Clethra bimatris IPNI=*327610-1* 298
Clethra brevifolia IPNI=*327593-1* 298
Clethra integerrima IPNI=*327610-1* 298
Clethra integrifolia IPNI=*327610-1* 298
Clethra kowalewskii IPNI=*327610-1* 298
Clethra michoacana IPNI=*327610-1* 298
Clethra ovalifolia IPNI=*327610-1* 298
Clethra punctata IPNI=*327593-1* 298
Clianthus dampieri IPNI=*948198-1* 334
Clianthus dampieri var. *marginatus* IPNI=*948198-1* 334
Clianthus dampieri var. *tricolor* IPNI=*948198-1* 334
Clianthus formosus IPNI=*948198-1* 334
Clianthus oxleyi IPNI=*948198-1* 334
Clianthus speciosus IPNI=*948198-1* 334
Clianthus speciosus subsp. *marginatus* IPNI=*948198-1* 334
Clianthus speciosus var. *oxleyi* IPNI=*948198-1* 334
Clianthus speciosus var. *typicus* IPNI=*948198-1* 334
Clibadium alexandri IPNI=*1006328-2* 260
Clibadium appressipilum IPNI=*1050569-2* 260
Clibadium asperum IPNI=*319147-2* 260
Clibadium badieri IPNI=*1050569-2* 260
Clibadium barbasco IPNI=*1050569-2* 260
Clibadium caracasanum IPNI=*319147-2* 260
Clibadium caudatum IPNI=*1050569-2* 260
Clibadium chocoense IPNI=*60769-2* 260
Clibadium commelinoides IPNI=*127605-2* 264
Clibadium foetidum IPNI=*319147-2* 260
Clibadium havanense IPNI=*1050569-2* 260
Clibadium lanceolatum IPNI=*319147-2* 260
Clibadium latifolium IPNI=*1050569-2* 260
Clibadium pallidum IPNI=*319147-2* 260
Clibadium pittieri f. *phrixium* IPNI=*60769-2* 260
Clibadium pittieri IPNI=*60769-2* 260
Clibadium polygynum IPNI=*60769-2* 260
Clibadium propinquum IPNI=*60769-2* 260
Clibadium schomburgkii IPNI=*1050569-2* 260
Clibadium strigillosum IPNI=*1050569-2* 260
Clibadium surinamense var. *asperum* IPNI=*319147-2* 260
Clibadium surinamense var. *macrophyllum* IPNI=*319147-2* 260
Clibadium terebinthinaceum var. *badieri* IPNI=*1050569-2* 260
Clibadium terebinthinaceum var. *pittieri* IPNI=*60769-2* 260
Clibadium trinitatis IPNI=*319147-2* 260
Clibadium vargasianum IPNI=*319147-2* 260
Clibadium vargasii IPNI=*1050569-2* 260
Clibadium villosum IPNI=*319147-2* 260
Clidemia acostae IPNI=*77207432-1* 419
Clidemia allardii IPNI=*77207407-1* 419
Clidemia allardii subsp. *maranonensis* IPNI=*77207407-1* 419
Clidemia amblyandra IPNI=*77207614-1* 423
Clidemia aristigera IPNI=*77207374-1* 419
Clidemia benthamiana IPNI=*77164247-1* 420
Clidemia brachystephana IPNI=*77207400-1* 420
Clidemia brachystephana var. *longidentata* IPNI=*77207400-1* 420
Clidemia bullosa IPNI=*77207390-1* 419
Clidemia candolleana IPNI=*77207390-1* 419
Clidemia candolleana var. *schwackeana* IPNI=*77207390-1* 419
Clidemia capitellata IPNI=*77196165-1* 420
Clidemia capitellata var. *dependens* IPNI=*77196165-1* 420
Clidemia capitellata var. *levelii* IPNI=*77196165-1* 420
Clidemia capitellata var. *neglecta* IPNI=*77196165-1* 420
Clidemia chamissois IPNI=*572590-1* 421
Clidemia ciliata IPNI=*77207714-1* 420
Clidemia ciliata var. *elata* IPNI=*77207714-1* 420
Clidemia ciliata var. *grandifolia* IPNI=*77207714-1* 420
Clidemia ciliata var. *testiculata* IPNI=*77207921-1* 421
Clidemia cognata IPNI=*77164247-1* 420
Clidemia crenata IPNI=*77164247-1* 420
Clidemia crenulata IPNI=*77207399-1* 420
Clidemia cursoris IPNI=*77207381-1* 420
Clidemia cursoris subsp. *angustifolia* IPNI=*77207381-1* 420
Clidemia cymosa IPNI=*77207921-1* 421
Clidemia decurrens IPNI=*77144083-1* 422
Clidemia dentata IPNI=*77207400-1* 420
Clidemia dependens IPNI=*77196165-1* 420
Clidemia dichotoma IPNI=*77207668-1* 423
Clidemia dimorphica IPNI=*77207368-1* 420
Clidemia dispar IPNI=*77207368-1* 420
Clidemia divaricata IPNI=*77207868-1* 422
Clidemia diversifolia IPNI=*77144083-1* 422
Clidemia elata IPNI=*77207714-1* 420
Clidemia elegans IPNI=*77164247-1* 420
Clidemia epiphytica IPNI=*77207810-1* 421
Clidemia epiphytica var. *trichocalyx* IPNI=*77207810-1* 421
Clidemia fenestrata IPNI=*77196164-1* 422
Clidemia glabrata IPNI=*572590-1* 421
Clidemia glabriflora IPNI=*1108920-2* 420
Clidemia heterobasis IPNI=*77207474-1* 420
Clidemia heteroclita IPNI=*77207474-1* 420
Clidemia heteromalla IPNI=*77164250-1* 422
Clidemia heterophylla IPNI=*77207843-1* 420
Clidemia hirta IPNI=*77164247-1* 420

Clidemia hirta var. *chrysantha* IPNI=*77164247-1* 420
Clidemia hirta var. *tiliifolia* IPNI=*77164247-1* 420
Clidemia impetiolaris IPNI=*77207921-1* 421
Clidemia japurensis IPNI=*77207474-1* 420
Clidemia japurensis var. *heterobasis* IPNI=*77207474-1* 420
Clidemia kuhlmannii IPNI=*77196165-1* 420
Clidemia leptoclades IPNI=*77207400-1* 420
Clidemia longisetosa IPNI=*77196165-1* 420
Clidemia macrophylla IPNI=*77144075-1* 421
Clidemia maculata IPNI=*572590-1* 421
Clidemia manoacensis IPNI=*77207390-1* 419
Clidemia martii IPNI=*158869-2* 420
Clidemia miconioides IPNI=*572590-1* 421
Clidemia naevula IPNI=*77207474-1* 420
Clidemia neglecta IPNI=*77196165-1* 420
Clidemia neglecta var. *angustifolia* IPNI=*77196165-1* 420
Clidemia neglecta var. *squarrosa* IPNI=*77196165-1* 420
Clidemia novemnervia var. *depauperata* IPNI=*77207390-1* 419
Clidemia octona IPNI=*77196163-1* 421
Clidemia octona subsp. *guayanensis* IPNI=*77196163-1* 421
Clidemia panamensis IPNI=*77207810-1* 421
Clidemia pauciflora IPNI=*77164247-1* 420
Clidemia pilosa IPNI=*77207921-1* 421
Clidemia pilosissima IPNI=*77196163-1* 421
Clidemia pilosissima var. *glabrata* IPNI=*77196163-1* 421
Clidemia platyphylla IPNI=*77164250-1* 422
Clidemia pleiostemon IPNI=*77207390-1* 419
Clidemia quadrisulca IPNI=*77207567-1* 422
Clidemia quinquenervia IPNI=*77144083-1* 422
Clidemia radicans IPNI=*77207810-1* 421
Clidemia rariflora IPNI=*77207390-1* 419
Clidemia rhamnifolia var. *macrodon* IPNI=*77207868-1* 422
Clidemia rubra IPNI=*77164250-1* 422
Clidemia rubra var. *brevifolia* IPNI=*77164250-1* 422
Clidemia rubra var. *cordifolia* IPNI=*77164250-1* 422
Clidemia rubra var. *intermedia* IPNI=*77164250-1* 422
Clidemia rubra var. *macrophylla* IPNI=*77164250-1* 422
Clidemia rubra var. *ursina* IPNI=*77164250-1* 422
Clidemia secunda IPNI=*77207935-1* 421
Clidemia septuplinervia IPNI=*77196168-1* 422
Clidemia sericea IPNI=*77207578-1* 422
Clidemia setosa IPNI=*77207860-1* 420
Clidemia spicata IPNI=*572835-1* 421
Clidemia spicata IPNI=*77196165-1* 420
Clidemia spiciformis IPNI=*572835-1* 421
Clidemia spondylantha IPNI=*572835-1* 421
Clidemia sprucei IPNI=*77207920-1* 422
Clidemia strigillosa IPNI=*77196164-1* 422
Clidemia subseriata IPNI=*77207614-1* 423
Clidemia subspicata IPNI=*77196165-1* 420
Clidemia sulcicaulis IPNI=*77207668-1* 423
Clidemia testiculata IPNI=*77207921-1* 421
Clidemia tiliifolia IPNI=*77164247-1* 420
Clidemia tococoidea IPNI=*77207700-1* 423
Clidemia tovarensis IPNI=*77207714-1* 420
Clidemia trichocalyx IPNI=*77207810-1* 421
Clidemia trichogona IPNI=*77207860-1* 420
Clidemia tristis IPNI=*77196164-1* 422
Clidemia umbonata IPNI=*77196164-1* 422
Clidemia valenzuelana IPNI=*77196163-1* 421
Clidemia virgata IPNI=*572590-1* 421
Cliffordiochloa parvispicula IPNI=*1156911-2* 246
Clinopodium caerulescens IPNI=*1001364-1* 371
Clinopodium capitatum IPNI=*447996-1* 370
Clinopodium chamaedrys IPNI=*449997-1* 371
Clinopodium imbricatum IPNI=*158242-2* 371
Clinopodium montanum IPNI=*457736-1* 373
Clinopodium stoloniferum IPNI=*61179-2* 369
Clinopodium xalapense IPNI=*61179-2* 369
Clipteria dichotoma IPNI=*1186403-2* 261
Clitoria alabamensis IPNI=*51897-2* 314
Clitoria albiflora IPNI=*486606-1* 315
Clitoria amoena IPNI=*486481-1* 315
Clitoria amoena IPNI=*51897-2* 314
Clitoria bracteata IPNI=*486606-1* 315
Clitoria brasiliana IPNI=*485555-1* 314
Clitoria calcarata IPNI=*485609-1* 314
Clitoria calcarigera IPNI=*51897-2* 314
Clitoria fluminensis IPNI=*485609-1* 314
Clitoria formosa IPNI=*485555-1* 314
Clitoria gemina IPNI=*51897-2* 314
Clitoria gladiata IPNI=*51897-2* 314
Clitoria grandiflora IPNI=*485611-1* 314
Clitoria insulana IPNI=*485555-1* 314
Clitoria mearnsii IPNI=*486606-1* 315
Clitoria micrantha IPNI=*51897-2* 314
Clitoria oblonga IPNI=*51897-2* 314
Clitoria parviflora IPNI=*486606-1* 315
Clitoria philippensis IPNI=*486606-1* 315
Clitoria plumieri IPNI=*485609-1* 314
Clitoria poitaei IPNI=*486481-1* 315
Clitoria pubescens var. *amphigona* IPNI=*485611-1* 314
Clitoria pudica IPNI=*485609-1* 314
Clitoria racemosa IPNI=*291409-2* 315
Clitoria racemosa IPNI=*525491-1* 336
Clitoria rotundifolia IPNI=*30187778-2* 313
Clitoria schiedeana IPNI=*485611-1* 314
Clitoria spectabilis IPNI=*486606-1* 315
Clitoria tanganicensis IPNI=*486606-1* 315
Clitoria ternatea f. *albiflora* IPNI=*486606-1* 315
Clitoria ternatea f. *fasciculata* IPNI=*486606-1* 315
Clitoria ternatea f. *leucopetala* IPNI=*486606-1* 315
Clitoria ternatea f. *subpolyadelpha* IPNI=*486606-1* 315
Clitoria ternatea var. *major* IPNI=*486606-1* 315
Clitoria ternatea var. *pleniflora* IPNI=*486606-1* 315
Clitoria ternatensium IPNI=*486606-1* 315
Clitoria virginiana IPNI=*51897-2* 314
Clitoria virginiana var. *angustifolia* IPNI=*51897-2* 314
Clitoria virginiana var. *elliptica* IPNI=*51897-2* 314
Clitoria virginiana var. *ovata* IPNI=*51897-2* 314
Clitoria zanzibarensis IPNI=*486606-1* 315
Clitoria zoophthalmum IPNI=*1197939-2* 328
Clivia sulphurea IPNI=*994513-1* 213
Clompanus apetala IPNI=*245150-2* 414
Clompanus chicha IPNI=*245150-2* 414
Clompanus diversifolia IPNI=*822268-1* 407
Clompanus haenkeana IPNI=*245150-2* 414
Clompanus pruriens IPNI=*825412-1* 414
Clompanus punctata IPNI=*245150-2* 414
Clompanus rugosa IPNI=*927898-1* 414
Clonodia sessilis IPNI=*60443063-2* 396
Cluacena myrsinoides IPNI=*974565-1* 427
Cluacena vaccinioides IPNI=*599434-1* 427
Clusia acuminata IPNI=*427431-1* 383
Clusia amazonica var. *parvifolia* IPNI=*427431-1* 383
Clusia cartilaginea IPNI=*315124-2* 384
Clusia columnaris var. *magdalenensis* IPNI=*61412-2* 383
Clusia couleti IPNI=*315124-2* 384
Clusia galactodendron IPNI=*315124-2* 384
Clusia maxima IPNI=*427524-1* 384
Clusia oedematopoidea IPNI=*427431-1* 383
Clusia parvicapsula IPNI=*315124-2* 384
Clusia peruviana IPNI=*427499-1* 384
Clusia planchoniana var. *oblongifolia* IPNI=*61594-2* 385
Clusia reducta IPNI=*61533-2* 384
Clusia retusa IPNI=*61583-2* 385

Clusia rosea var. *colombiana* IPNI=*427616-1* 384
Clusia schomburgkiana var. *lakei* IPNI=*61594-2* 385
Clusia silvicola IPNI=*61583-2* 385
Clusia utilis IPNI=*315124-2* 384
Clusia venosa IPNI=*315124-2* 384
Clusia williamsii IPNI=*61412-2* 383
Clusiophyllum sprucei IPNI=*159994-2* 392
Clypeola caroliniana IPNI=*138164-2* 277
Clypeola halimifolia IPNI=*1085072-2* 277
Clypeola maritima IPNI=*1085072-2* 277
Clytostoma cuneatum IPNI=*77100457-1* 362
Cnicus costaricensis IPNI=*195679-1* 260
Cnicus echinatus IPNI=*195292-1* 260
Cnicus firmus IPNI=*196191-1* 260
Cnicus mexicanus IPNI=*195679-1* 260
Cnicus misilmerensis IPNI=*196191-1* 260
Cnicus portoricensis IPNI=*195679-1* 260
Cnidium petroselinum IPNI=*60442790-2* 252
Cnidium tenuifolium IPNI=*840899-1* 252
Cnidium tolucense IPNI=*838268-1* 251
Cnidoscolus adenophilus IPNI=*61899-2* 388
Cnidoscolus cordifolius IPNI=*61897-2* 388
Cnidoscolus herbaceus IPNI=*61899-2* 388
Cnidoscolus hernandezii IPNI=*61897-2* 388
Cnidoscolus jurgensenii IPNI=*61897-2* 388
Cnidoscolus marcgravii IPNI=*61899-2* 388
Cnidoscolus mexicanus IPNI=*61899-2* 388
Cnidoscolus osteocarpus IPNI=*61899-2* 388
Cnidoscolus tomentosus IPNI=*61897-2* 388
Cnidoscolus urens subsp. *adenophilus* IPNI=*61899-2* 388
Cnidoscolus vitifolius var. *repandus* IPNI=*61899-2* 388
Cobaea lasseri IPNI=*61922-2* 303
Cobamba dichotoma IPNI=*366552-1* 344
Coccineorchis corymbosa IPNI=*61935-2* 218
Coccinia grandis var. *wightiana* IPNI=*291972-1* 293
Coccinia helenae IPNI=*291972-1* 293
Coccinia indica IPNI=*291972-1* 293
Coccinia loureiroana IPNI=*291972-1* 293
Coccinia moghadd IPNI=*291972-1* 293
Coccinia moimoi IPNI=*291972-1* 293
Coccinia palmatisecta IPNI=*291972-1* 293
Coccinia schimperi IPNI=*291972-1* 293
Coccinia wightiana IPNI=*291972-1* 293
Coccocypselum apurense IPNI=*746945-1* 348
Coccocypselum aureum var. *capitatum* IPNI=*746936-1* 348
Coccocypselum brittonii IPNI=*746945-1* 348
Coccocypselum canescens IPNI=*315609-2* 348
Coccocypselum ciliatum IPNI=*746963-1* 348
Coccocypselum condalia var. *caaguazuense* IPNI=*746945-1* 348
Coccocypselum condalia var. *genuinum* IPNI=*746945-1* 348
Coccocypselum crassifolium IPNI=*315609-2* 348
Coccocypselum croatii IPNI=*746945-1* 348
Coccocypselum decumbens IPNI=*746945-1* 348
Coccocypselum dichroolasium IPNI=*315609-2* 348
Coccocypselum discolor IPNI=*746963-1* 348
Coccocypselum glabrum IPNI=*61955-2* 348
Coccocypselum glabrum IPNI=*746945-1* 348
Coccocypselum guianense var. *glabratum* IPNI=*61955-2* 348
Coccocypselum guianense var. *patens* IPNI=*61955-2* 348
Coccocypselum hirsutum var. *glabrum* IPNI=*61955-2* 348
Coccocypselum hirsutum var. *uniflorum* IPNI=*746963-1* 348
Coccocypselum huberi IPNI=*746945-1* 348
Coccocypselum macropodum IPNI=*1080914-2* 351
Coccocypselum tontanea IPNI=*61955-2* 348
Coccocypselum trinitense IPNI=*746945-1* 348
Coccocypselum umbellatum IPNI=*746945-1* 348
Coccoloba acuminata var. *glabra* IPNI=*693560-1* 288
Coccoloba acuminata var. *pubescens* IPNI=*693560-1* 288
Coccoloba alagoensis IPNI=*693788-1* 288
Coccoloba alagoensis var. *major* IPNI=*693788-1* 288
Coccoloba barbadensis IPNI=*693597-1* 288
Coccoloba barbeyana IPNI=*693638-1* 288
Coccoloba bejuco IPNI=*693572-1* 288
Coccoloba billbergii IPNI=*693788-1* 288
Coccoloba bracteolosa IPNI=*693654-1* 288
Coccoloba caracasana f. *glabra* IPNI=*693597-1* 288
Coccoloba caribaea IPNI=*693617-1* 288
Coccoloba carinata IPNI=*694785-1* 289
Coccoloba coriacea IPNI=*693787-1* 288
Coccoloba cyclophylla IPNI=*693597-1* 288
Coccoloba duckei IPNI=*693795-1* 288
Coccoloba dugandiana IPNI=*693754-1* 288
Coccoloba excelsa var. *glabra* IPNI=*693654-1* 288
Coccoloba fagifolia IPNI=*693617-1* 288
Coccoloba goudotiana IPNI=*693787-1* 288
Coccoloba grandis IPNI=*693712-1* 288
Coccoloba guianensis IPNI=*693777-1* 288
Coccoloba guianensis var. *angustifolia* IPNI=*693777-1* 288
Coccoloba guianensis var. *macrostachya* IPNI=*693777-1* 288
Coccoloba guianensis var. *major* IPNI=*693777-1* 288
Coccoloba japurana IPNI=*221443-2* 405
Coccoloba latifolia IPNI=*693754-1* 288
Coccoloba marginata IPNI=*693777-1* 288
Coccoloba microneura IPNI=*693777-1* 288
Coccoloba micropunctata IPNI=*693654-1* 288
Coccoloba monoica IPNI=*694785-1* 289
Coccoloba moritziana IPNI=*693795-1* 288
Coccoloba moritzii IPNI=*693795-1* 288
Coccoloba moritzii var. *lucida* IPNI=*693795-1* 288
Coccoloba moritzii var. *opaca* IPNI=*693795-1* 288
Coccoloba nigra IPNI=*693795-1* 288
Coccoloba nitida IPNI=*693617-1* 288
Coccoloba nitida IPNI=*693795-1* 288
Coccoloba nitida var. *cordata* IPNI=*693795-1* 288
Coccoloba nitida var. *rotundata* IPNI=*693795-1* 288
Coccoloba novogranatensis IPNI=*693617-1* 288
Coccoloba ovata var. *lanceolata* IPNI=*693795-1* 288
Coccoloba ovata var. *major* IPNI=*693795-1* 288
Coccoloba ovata var. *minor* IPNI=*693795-1* 288
Coccoloba paniculata IPNI=*693754-1* 288
Coccoloba parimensis IPNI=*693654-1* 288
Coccoloba parimensis var. *hostmannii* IPNI=*693654-1* 288
Coccoloba parimensis var. *schomburgkii* IPNI=*693654-1* 288
Coccoloba pendula IPNI=*693777-1* 288
Coccoloba pichuna IPNI=*693638-1* 288
Coccoloba platyclada IPNI=*694774-1* 289
Coccoloba polystachya IPNI=*693754-1* 288
Coccoloba polystachya var. *glabra* IPNI=*693754-1* 288
Coccoloba polystachya var. *mollis* IPNI=*693754-1* 288
Coccoloba polystachya var. *pubescens* IPNI=*693754-1* 288
Coccoloba quadrifida IPNI=*693572-1* 288
Coccoloba recurva IPNI=*693777-1* 288
Coccoloba rheifolia IPNI=*693712-1* 288
Coccoloba riparia IPNI=*693787-1* 288
Coccoloba scandens IPNI=*693654-1* 288
Coccoloba sparsifolia IPNI=*693572-1* 288
Coccoloba standleyana IPNI=*693754-1* 288
Coccoloba strobilulifera IPNI=*693560-1* 288
Coccoloba trinitatis IPNI=*693777-1* 288
Coccoloba virens IPNI=*693617-1* 288
Coccoloba waittii IPNI=*693617-1* 288
Coccolobis uvifera IPNI=*693918-1* 288
Cocculus abuta IPNI=*580080-1* 434
Cocculus amazonum IPNI=*580059-1* 434
Cocculus domingensis IPNI=*580887-1* 435
Cocculus grandifolius IPNI=*147-2* 434

Cocculus imene IPNI=*580059-1* 434
Cocculus japurensis IPNI=*34845-2* 434
Cocculus laevigatus IPNI=*147-2* 434
Cocculus membranaceus IPNI=*59219-2* 434
Cocculus paniculiger IPNI=*580887-1* 435
Cocculus reticulatus IPNI=*580176-1* 434
Cocculus tamoides IPNI=*581190-1* 435
Cocculus tomentosus IPNI=*580080-1* 434
Cocculus toxicofer IPNI=*72709-2* 435
Cochleanthes marginata IPNI=*661384-1* 223
Cochlearia humifusa IPNI=*286148-1* 277
Cochlearia rugosa IPNI=*1061021-2* 277
Cochlospermum codinae IPNI=*111539-1* 406
Cochlospermum hibiscoides IPNI=*111539-1* 406
Cochlospermum hibiscoides var. *dasycarpum* IPNI=*111539-1* 406
Cochlospermum hibiscoides var. *gymnocarpum* IPNI=*111539-1* 406
Cochlospermum luetzeiburgii IPNI=*111539-1* 406
Cochlospermum parkeri IPNI=*111532-1* 406
Cochlospermum serratifolium IPNI=*111539-1* 406
Cochlospermum triphyllum IPNI=*111539-1* 406
Cochlospermum wentii IPNI=*111532-1* 406
Cochlospermum williamsii IPNI=*111532-1* 406
Cochranella pilosa IPNI=*147242-1* 279
Cocos acicularis IPNI=*28460-2* 206
Cocos acrocomioides IPNI=*246966-2* 212
Cocos aculeata IPNI=*663914-1* 203
Cocos arechavaletana IPNI=*246966-2* 212
Cocos argentea IPNI=*670066-1* 212
Cocos australis IPNI=*246966-2* 212
Cocos butyracea IPNI=*1172851-2* 205
Cocos chavesiana IPNI=*246946-2* 212
Cocos chiragua IPNI=*670066-1* 212
Cocos chloroleuca IPNI=*664513-1* 206
Cocos datil IPNI=*246966-2* 212
Cocos equatorialis IPNI=*246946-2* 212
Cocos fusiformis IPNI=*663914-1* 203
Cocos geriba IPNI=*246966-2* 212
Cocos guineensis IPNI=*28460-2* 206
Cocos inajai IPNI=*246946-2* 212
Cocos indica IPNI=*666160-1* 208
Cocos mamillaris IPNI=*666160-1* 208
Cocos martiana IPNI=*246966-2* 212
Cocos nana IPNI=*666160-1* 208
Cocos nucifera f. *palmyrensis* IPNI=*666160-1* 208
Cocos nucifera var. *palmyrensis* IPNI=*666160-1* 208
Cocos nucifera var. *synphyllica* IPNI=*666160-1* 208
Cocos orinocensis IPNI=*246957-2* 212
Cocos plumosa IPNI=*246966-2* 212
Cocos romanzoffiana IPNI=*246966-2* 212
Cocos sancona IPNI=*670066-1* 212
Cocos speciosa IPNI=*246946-2* 212
Cocos weddelliana IPNI=*246979-2* 212
Cocos weddelliana var. *pynaertii* IPNI=*246979-2* 212
Codazzia nervosa IPNI=*109429-1* 363
Codazzia rosea IPNI=*109429-1* 363
Codazzia speciosa IPNI=*109429-1* 363
Codiaeum chrysosticton IPNI=*85073-3* 388
Codonanthe confusa IPNI=*77139499-1* 366
Codonanthe crassifolia IPNI=*77139499-1* 366
Codonanthe stenantha IPNI=*77139499-1* 366
Codonemma calycinum IPNI=*285812-2* 343
Codonemma macrocalyx IPNI=*285812-2* 343
Codonium arborescens IPNI=*608391-1* 450
Codonium arborescens var. *berterianum* IPNI=*608391-1* 450
Codonium chrysophylloides IPNI=*608391-1* 450
Coelocline lucida IPNI=*76067-1* 189
Coelorachis loricata IPNI=*325559-2* 244
Coffea bourbonica IPNI=*747038-1* 348
Coffea corymbulosa IPNI=*747038-1* 348
Coffea laurifolia IPNI=*747038-1* 348
Coffea moka IPNI=*747038-1* 348
Coffea occidentalis IPNI=*1187456-2* 350
Coffea racemosa IPNI=*1114920-2* 357
Coffea spicata IPNI=*77202779-1* 353
Coffea subsessilis IPNI=*764729-1* 356
Coffea sundana IPNI=*747038-1* 348
Coffea tenuiflora IPNI=*1061341-2* 347
Coffea umbellata IPNI=*749186-1* 350
Coffea verticillata IPNI=*1114920-2* 357
Coffea viburnoides IPNI=*1139608-2* 357
Coffea vulgaris IPNI=*747038-1* 348
Cohiba urens IPNI=*114242-3* 276
Coilanthera rotundifolia IPNI=*274993-2* 274
Coilostylis ciliaris IPNI=*631578-1* 218
Coilostylis cuspidata IPNI=*631578-1* 218
Coilostylis emarginata IPNI=*631578-1* 218
Coilotapalus peltata IPNI=*850956-1* 446
Coix angulata IPNI=*424990-1* 246
Coix dactyloides IPNI=*424990-1* 246
Coix lacryma IPNI=*30100521-2* 236
Coix ovata IPNI=*30100521-2* 236
Coix pendula IPNI=*30100521-2* 236
Colax parkeri IPNI=*643684-1* 221
Coldenia succulenta IPNI=*116784-1* 275
Coleogeton pectinatus IPNI=*77099639-1* 203
Coleosanthus tiliifolius IPNI=*1092582-2* 263
Coleus acuminatus IPNI=*446165-1* 369
Coleus aromaticus IPNI=*445902-1* 369
Coleus atropurpureus IPNI=*446165-1* 369
Coleus berkeleyi IPNI=*446165-1* 369
Coleus blancoi IPNI=*446165-1* 369
Coleus blumei IPNI=*446165-1* 369
Coleus carnosus IPNI=*445902-1* 369
Coleus crassifolius IPNI=*445902-1* 369
Coleus crispipilus IPNI=*446165-1* 369
Coleus dixii IPNI=*446165-1* 369
Coleus formosanus IPNI=*446165-1* 369
Coleus gaudichaudii IPNI=*446165-1* 369
Coleus gibbsiae IPNI=*446165-1* 369
Coleus gibsonii IPNI=*446165-1* 369
Coleus glabratus IPNI=*446113-1* 369
Coleus grandifolius IPNI=*446165-1* 369
Coleus hybridus IPNI=*446165-1* 369
Coleus igolotorum IPNI=*446165-1* 369
Coleus ingratus IPNI=*446165-1* 369
Coleus integrifolius IPNI=*446165-1* 369
Coleus laciniatus IPNI=*446165-1* 369
Coleus macranthus IPNI=*446165-1* 369
Coleus marshallii IPNI=*446165-1* 369
Coleus multiflorus IPNI=*446165-1* 369
Coleus pubescens IPNI=*446165-1* 369
Coleus pumilus IPNI=*446165-1* 369
Coleus rehneltianus IPNI=*446165-1* 369
Coleus ruckeri IPNI=*446165-1* 369
Coleus saundersii IPNI=*446165-1* 369
Coleus savannicola IPNI=*446165-1* 369
Coleus scutellarioides var. *angustifolius* IPNI=*446165-1* 369
Coleus scutellarioides var. *laxus* IPNI=*446165-1* 369
Coleus scutellarioides var. *limnophilus* IPNI=*446165-1* 369
Coleus secundiflorus IPNI=*446165-1* 369
Coleus subfrutectosus IPNI=*445902-1* 369
Coleus suborbicularis IPNI=*445902-1* 369
Coleus suganda IPNI=*445902-1* 369
Coleus vaalae IPNI=*445902-1* 369
Coleus veitchii IPNI=*446165-1* 369
Coleus verschaffeltii IPNI=*446165-1* 369

Coleus wightii IPNI=*446113-1* 369
Coleus zschokkei IPNI=*446165-1* 369
Colicodendron breynia IPNI=*77217338-1* 278
Colicodendron crotonoides IPNI=*77183834-1* 278
Colicodendron lepidotum IPNI=*77184072-1* 278
Colicodendron obliquifolium IPNI=*316363-2* 278
Colicodendron pulchellum IPNI=*77183588-1* 278
Colicodendron subbilobum IPNI=*316361-2* 278
Colinil cinereum IPNI=*1071691-2* 335
Colladoa distachya IPNI=*406281-1* 240
Colladoa monostachya IPNI=*14931-2* 234
Collandra atronitens IPNI=*378985-1* 367
Collandra aureonitens IPNI=*378985-1* 367
Collandra phoenicea IPNI=*378985-1* 367
Collandra picta IPNI=*378985-1* 367
Collandra pilosa IPNI=*378985-1* 367
Collandra sanguinea IPNI=*378985-1* 367
Collania macranthera IPNI=*63411-1* 226
Colletia iguanaea IPNI=*1213898-2* 436
Collococcus glaber IPNI=*1088317-2* 273
Collophora utilis IPNI=*78177-1* 340
Colocasia acris IPNI=*1170772-2* 199
Colocasia aegyptiaca IPNI=*1170772-2* 199
Colocasia aethiopica IPNI=*89403-1* 203
Colocasia antiquorum IPNI=*1170772-2* 199
Colocasia boryi IPNI=*60444565-2* 198
Colocasia cochleata IPNI=*84151-1* 198
Colocasia colocasia IPNI=*1170772-2* 199
Colocasia cucullata IPNI=*84151-1* 198
Colocasia esculenta var. *rupicola* IPNI=*1170772-2* 199
Colocasia esculenta var. *stolonifera* IPNI=*1170772-2* 199
Colocasia euchlora IPNI=*1170772-2* 199
Colocasia formosana IPNI=*1170772-2* 199
Colocasia gracilis IPNI=*1170772-2* 199
Colocasia himalensis IPNI=*1170772-2* 199
Colocasia konishii IPNI=*1170772-2* 199
Colocasia macrorrhizos IPNI=*60444565-2* 198
Colocasia mucronata IPNI=*60444565-2* 198
Colocasia neocaledonica IPNI=*1170772-2* 199
Colocasia nymphaeifolia IPNI=*1170772-2* 199
Colocasia peltata IPNI=*1170772-2* 199
Colocasia peregrina IPNI=*60444565-2* 198
Colocasia rapiformis IPNI=*60444565-2* 198
Colocasia rugosa IPNI=*84151-1* 198
Colocasia tonoimo IPNI=*1170772-2* 199
Colocasia vulgaris IPNI=*1170772-2* 199
Colocynthis amarissima IPNI=*291938-1* 293
Colocynthis citrullus IPNI=*291938-1* 293
Colpodium aquaticum IPNI=*395086-1* 235
Colpodium elbursense IPNI=*395086-1* 235
Colpodium junceum IPNI=*1103484-2* 245
Colubrina americana IPNI=*63423-2* 441
Colubrina colubrina IPNI=*63423-2* 441
Colubrina fermenta IPNI=*63435-2* 441
Colubrina ferruginosa IPNI=*63423-2* 441
Colubrina granulosa IPNI=*105606-2* 441
Colubrina hondurensis IPNI=*63435-2* 441
Colubrina obtusata IPNI=*63423-2* 441
Colubrina pittieriana IPNI=*63435-2* 441
Colubrina reclinata IPNI=*63435-2* 441
Colubrina rufa var. *glandulosa* IPNI=*63438-2* 441
Colubrina venezuelensis IPNI=*63435-2* 441
Columbea angustifolia IPNI=*17518-2* 180
Columbea excelsa IPNI=*261688-1* 180
Columbea imbricata IPNI=*261681-1* 180
Columbea quadrifaria IPNI=*261681-1* 180
Columnea acuminata IPNI=*63508-2* 366
Columnea affinis IPNI=*378985-1* 367
Columnea aurantiaca IPNI=*379021-1* 367
Columnea aureonitens IPNI=*378985-1* 367
Columnea calotricha var. *breviflora* IPNI=*63523-2* 366
Columnea campanulata IPNI=*379021-1* 367
Columnea cinnabarina IPNI=*378968-1* 367
Columnea circinnata IPNI=*84424-2* 367
Columnea citrina IPNI=*305407-2* 366
Columnea coccinea IPNI=*84424-2* 367
Columnea conferta IPNI=*305407-2* 366
Columnea consimilis IPNI=*63510-2* 366
Columnea coriacea IPNI=*84425-2* 367
Columnea cubensis IPNI=*378985-1* 367
Columnea densibracteata IPNI=*999467-1* 366
Columnea erecta IPNI=*2352-2* 365
Columnea eriantha IPNI=*381572-1* 367
Columnea gloriosa IPNI=*378924-1* 366
Columnea grandifolia IPNI=*378985-1* 367
Columnea humilis IPNI=*2352-2* 365
Columnea ichthyoderma IPNI=*60441766-2* 367
Columnea kalbreyeri IPNI=*305407-2* 366
Columnea kucyniakii IPNI=*379021-1* 367
Columnea localis IPNI=*378924-1* 366
Columnea lutea IPNI=*378924-1* 366
Columnea macrantha IPNI=*379021-1* 367
Columnea meridensis IPNI=*60441766-2* 367
Columnea microcalyx var. *macrophylla* IPNI=*378924-1* 366
Columnea mocinoana IPNI=*380742-1* 367
Columnea patrisii IPNI=*84424-2* 367
Columnea pichinchensis IPNI=*379021-1* 367
Columnea picta IPNI=*378985-1* 367
Columnea salmonea IPNI=*378924-1* 366
Columnea sanguinea var. *cubensis* IPNI=*378985-1* 367
Columnea sanguinea var. *trinitensis* IPNI=*378985-1* 367
Columnea sanmartensis IPNI=*63508-2* 366
Columnea silvatica IPNI=*328486-2* 366
Columnea tuerckheimii IPNI=*378924-1* 366
Columnea vestita IPNI=*63508-2* 366
Columnea viridis IPNI=*378968-1* 367
Columnea warscewicziana IPNI=*63508-2* 366
Colutea americana IPNI=*948330-1* 318
Colutea florida IPNI=*486805-1* 315
Colutea novae-hollandiae IPNI=*948198-1* 334
Comatocroton ovalifolia IPNI=*325185-2* 389
Combretum alternifolium IPNI=*170030-1* 416
Combretum aubletii IPNI=*170434-1* 416
Combretum aurantiacum IPNI=*63788-2* 416
Combretum basslerianum IPNI=*170434-1* 416
Combretum benthamianum IPNI=*63788-2* 416
Combretum bracteatum IPNI=*170030-1* 416
Combretum coccineum IPNI=*63781-2* 415
Combretum elegans IPNI=*170434-1* 416
Combretum farinosum var. *phaenopetalum* IPNI=*63788-2* 416
Combretum formosum IPNI=*63788-2* 416
Combretum gloriosum IPNI=*63788-2* 416
Combretum guianense IPNI=*170434-1* 416
Combretum latipaniculatum IPNI=*170030-1* 416
Combretum laxum IPNI=*170434-1* 416
Combretum lepidopetalum IPNI=*63788-2* 416
Combretum loeflingii IPNI=*63788-2* 416
Combretum loeflingii subsp. *ornithophilum* IPNI=*63788-2* 416
Combretum madolata IPNI=*170030-1* 416
Combretum magnificum IPNI=*170434-1* 416
Combretum micropetalum IPNI=*63788-2* 416
Combretum multidiscum IPNI=*63788-2* 416
Combretum nicoyanum IPNI=*170030-1* 416
Combretum occidentale IPNI=*63788-2* 416
Combretum oxypetalum IPNI=*63788-2* 416
Combretum palmeri IPNI=*170030-1* 416

Combretum phaenopetalum IPNI=*63788-2* 416
Combretum punctatum IPNI=*170434-1* 416
Combretum reticulatum IPNI=*63788-2* 416
Combretum secundum IPNI=*170434-1* 416
Combretum secundum IPNI=*63788-2* 416
Combretum superbum IPNI=*63788-2* 416
Combretum trinitense IPNI=*63788-2* 416
Combretum wandurraganum IPNI=*63788-2* 416
Comesperma floribundum IPNI=*690122-1* 337
Commelina acuminata IPNI=*302798-2* 224
Commelina acuminata IPNI=*332090-2* 224
Commelina alpestris IPNI=*302798-2* 224
Commelina bambusifolioides IPNI=*63900-2* 224
Commelina bicolor IPNI=*42443-2* 223
Commelina canescens IPNI=*332090-2* 224
Commelina cavaleriei IPNI=*332090-2* 224
Commelina chinensis IPNI=*172873-1* 224
Commelina clandestina IPNI=*302798-2* 224
Commelina clavata var. *hohenackeri* IPNI=*171972-1* 224
Commelina coelestis f. *albipetala* IPNI=*302798-2* 224
Commelina coelestis IPNI=*302798-2* 224
Commelina communis IPNI=*63900-2* 224
Commelina cucullata IPNI=*332090-2* 224
Commelina delicatula IPNI=*332090-2* 224
Commelina diandra IPNI=*172873-1* 224
Commelina diffusa IPNI=*171972-1* 224
Commelina dubia IPNI=*302798-2* 224
Commelina erecta var. *obliqua* IPNI=*63900-2* 224
Commelina erecta var. *typica* IPNI=*172048-1* 224
Commelina exilis IPNI=*172873-1* 224
Commelina fasciculata IPNI=*302798-2* 224
Commelina fasciculata subsp. *chacoensis* IPNI=*302798-2* 224
Commelina floribunda IPNI=*77184055-1* 224
Commelina glabra IPNI=*302798-2* 224
Commelina gracilis IPNI=*302798-2* 224
Commelina graminifolia IPNI=*302798-2* 224
Commelina graminifolia var. *clandestina* IPNI=*302798-2* 224
Commelina hexandra IPNI=*79497-2* 224
Commelina hirsuta IPNI=*332090-2* 224
Commelina humilis IPNI=*172043-1* 224
Commelina intermedia IPNI=*302798-2* 224
Commelina karwinskii IPNI=*172043-1* 224
Commelina kilimandscharica IPNI=*332090-2* 224
Commelina mathewsii IPNI=*63900-2* 224
Commelina mexicana IPNI=*77184055-1* 224
Commelina minuta IPNI=*172873-1* 224
Commelina mollis IPNI=*332090-2* 224
Commelina monticola IPNI=*63900-2* 224
Commelina monticola var. *vestita* IPNI=*63900-2* 224
Commelina nervosa IPNI=*302798-2* 224
Commelina nervosa IPNI=*332090-2* 224
Commelina nilagirica IPNI=*302798-2* 224
Commelina nudicaulis IPNI=*172873-1* 224
Commelina nudiflora f. *angustifolia* IPNI=*172873-1* 224
Commelina nudiflora IPNI=*172873-1* 224
Commelina obscura IPNI=*332090-2* 224
Commelina orchioides IPNI=*172043-1* 224
Commelina pallida var. *parviflora* IPNI=*302798-2* 224
Commelina parviflora IPNI=*302798-2* 224
Commelina polyspatha IPNI=*63900-2* 224
Commelina procurrens IPNI=*332090-2* 224
Commelina prostrata IPNI=*332090-2* 224
Commelina pyrrhoblepharis IPNI=*332090-2* 224
Commelina quitensis var. *mandonii* IPNI=*302798-2* 224
Commelina radicans IPNI=*172873-1* 224
Commelina radiciflora IPNI=*332090-2* 224
Commelina rhizocarpa IPNI=*332090-2* 224
Commelina robusta f. *vestita* IPNI=*63900-2* 224
Commelina robusta IPNI=*63900-2* 224
Commelina rosea IPNI=*173146-1* 224
Commelina rufociliata IPNI=*332090-2* 224
Commelina salicifolia IPNI=*171972-1* 224
Commelina saltiana IPNI=*332090-2* 224
Commelina scabrata IPNI=*63900-2* 224
Commelina senegalensis IPNI=*332090-2* 224
Commelina serrulata IPNI=*77184055-1* 224
Commelina thwaitesii IPNI=*171972-1* 224
Commelina tuberosa var. *coelestis* IPNI=*302798-2* 224
Commelina tuberosa var. *inflata* IPNI=*302798-2* 224
Commelina tuberosa var. *nana* IPNI=*302798-2* 224
Commelina turbinata IPNI=*332090-2* 224
Commelina umbrosa IPNI=*171758-1* 223
Commelina uncata IPNI=*332090-2* 224
Commelina variabilis IPNI=*172043-1* 224
Commelina vestita IPNI=*63900-2* 224
Commelina vilavelhensis IPNI=*63900-2* 224
Commelina villosiuscula IPNI=*332090-2* 224
Commelina zanonia IPNI=*1191409-2* 224
Commianthus schomburgkii IPNI=*764631-1* 356
Commicarpus africanus IPNI=*326899-2* 286
Comocladia loxensis IPNI=*813935-1* 433
Comocladia tapaculo IPNI=*71559-1* 451
Comomyrsine schlimii IPNI=*73559-2* 304
Comomyrsine sprucei IPNI=*73569-2* 304
Comostemum prolixum IPNI=*305627-1* 231
Comostemum schottii IPNI=*305627-1* 231
Complaya trilobata IPNI=*1093589-2* 270
Compsoneura tessmannii IPNI=*64004-2* 190
Conami brasiliensis IPNI=*353613-1* 402
Conami conami IPNI=*353613-1* 402
Conceveiba cordata IPNI=*339164-1* 388
Conceveiba macropylla IPNI=*339164-1* 388
Conceveiba megalophylla IPNI=*341924-1* 388
Conceveiba ovata IPNI=*320623-2* 388
Conceveiba simulata IPNI=*320623-2* 388
Conceveiba trigonocarpa IPNI=*320623-2* 388
Condalia ferrea IPNI=*133956-2* 442
Condalia lanceolata IPNI=*315609-2* 348
Condalia repens IPNI=*746945-1* 348
Condaminea angustifolia IPNI=*1010049-2* 348
Condaminea breviflora IPNI=*1010049-2* 348
Condaminea corymbosa var. *pubescens* IPNI=*1010049-2* 348
Condaminea corymbosa var. *subsessilis* IPNI=*1010049-2* 348
Condaminea macrophylla IPNI=*77210192-1* 347
Condaminea tinctoria IPNI=*236591-2* 357
Condaminea utilis IPNI=*748737-1* 350
Condylocarpus sempervirens IPNI=*263741-1* 180
Congea azurea IPNI=*862610-1* 369
Congea oblonga IPNI=*862610-1* 369
Congea petelotii IPNI=*862610-1* 369
Congea villosa IPNI=*862610-1* 369
Conium arracacia IPNI=*838272-1* 251
Conium ceretanum IPNI=*840668-1* 252
Conium cicuta IPNI=*840668-1* 252
Conium croaticum IPNI=*840668-1* 252
Conium maculatum subsp. *viride* IPNI=*840668-1* 252
Conium maculatum var. *barceloi* IPNI=*840668-1* 252
Conium maculosum IPNI=*840668-1* 252
Conium nodosum IPNI=*840668-1* 252
Conium pyrenaicum IPNI=*840668-1* 252
Conium sibiricum IPNI=*840668-1* 252
Conium strictum IPNI=*840668-1* 252
Conium tenuifolium IPNI=*840668-1* 252
Connaropsis philippica IPNI=*371870-1* 432
Connarus africanus IPNI=*64181-2* 431
Connarus fecundus IPNI=*211312-2* 431

Connarus gilgianus IPNI=*264853-1* 431
Connarus glaber IPNI=*264767-1* 431
Connarus guianensis IPNI=*64181-2* 431
Connarus lonchotus IPNI=*64181-2* 431
Connarus macrophyllus IPNI=*211312-2* 431
Connarus merizaldinua IPNI=*64181-2* 431
Connarus neglectus IPNI=*64164-2* 430
Connarus negrensis IPNI=*264496-1* 431
Connarus oblongifolius IPNI=*264767-1* 431
Connarus opacus IPNI=*64164-2* 430
Connarus rigidus IPNI=*64164-2* 430
Connarus ruber var. *sprucei* IPNI=*264496-1* 431
Connarus schomburgkii IPNI=*64181-2* 431
Connarus sprucei IPNI=*264496-1* 431
Conobea ovata IPNI=*244449-2* 377
Conobea pumila IPNI=*244449-2* 377
Conobea verticillaris IPNI=*328499-2* 377
Conobea viscosa IPNI=*328499-2* 377
Conocarpus erectus f. *sericeus* IPNI=*319151-2* 416
Conocarpus erectus var. *arboreus* IPNI=*319151-2* 416
Conocarpus erectus var. *argenteus* IPNI=*319151-2* 416
Conocarpus erectus var. *procumbens* IPNI=*319151-2* 416
Conocarpus erectus var. *sericeus* IPNI=*319151-2* 416
Conocarpus procumbens IPNI=*319151-2* 416
Conocarpus pubescens IPNI=*319151-2* 416
Conocarpus racemosus IPNI=*170672-1* 416
Conocarpus sericeus IPNI=*319151-2* 416
Conocarpus supina IPNI=*319151-2* 416
Conoclinium ballotifolium IPNI=*144056-2* 265
Conoclinium humile IPNI=*144058-2* 265
Conoclinium prasiifolium IPNI=*64237-2* 260
Conohoria aubletii IPNI=*779914-1* 404
Conohoria castaneifolia IPNI=*867224-1* 405
Conohoria dentata IPNI=*867224-1* 405
Conohoria flavescens IPNI=*221409-2* 405
Conohoria passoura IPNI=*221409-2* 405
Conohoria racemosa IPNI=*221443-2* 405
Conomorpha ampla IPNI=*73449-2* 304
Conomorpha candolleana IPNI=*73568-2* 304
Conomorpha glaucorubens IPNI=*73568-2* 304
Conomorpha gracilis IPNI=*73568-2* 304
Conomorpha grandifora IPNI=*73568-2* 304
Conomorpha guyanensis IPNI=*1166848-2* 304
Conomorpha heterantha IPNI=*1166848-2* 304
Conomorpha latifolia IPNI=*73568-2* 304
Conomorpha laxiflora IPNI=*73568-2* 304
Conomorpha laxiflora var. *latifolia* IPNI=*73568-2* 304
Conomorpha laxiflora var. *longifolia* IPNI=*73568-2* 304
Conomorpha macrophylla IPNI=*73449-2* 304
Conomorpha madeirensis IPNI=*73568-2* 304
Conomorpha spicata IPNI=*73568-2* 304
Conomorpha utiarityi IPNI=*73449-2* 304
Conoria cuspa IPNI=*77347-1* 339
Conostegia acutidentata IPNI=*159943-2* 423
Conostegia bernoulliana IPNI=*159783-2* 423
Conostegia cinnamomea IPNI=*158912-2* 420
Conostegia haughtii IPNI=*158912-2* 420
Conostegia holosericea IPNI=*159943-2* 423
Conostegia icosandra IPNI=*159783-2* 423
Conostegia icosandra subsp. *crenata* IPNI=*159783-2* 423
Conostegia icosandra var. *crenata* IPNI=*159783-2* 423
Conostegia lanceolata IPNI=*159943-2* 423
Conostegia lanceolata var. *subtrinervia* IPNI=*159943-2* 423
Conostegia lundellii IPNI=*159783-2* 423
Conostegia mexicana IPNI=*159783-2* 423
Conostegia micromeris IPNI=*158912-2* 420
Conostegia minutiflora IPNI=*159943-2* 423
Conostegia parviflora IPNI=*1069702-2* 422
Conostegia polyandra IPNI=*159672-2* 422
Conostegia quadrangularis IPNI=*159783-2* 423
Conostegia sphaerica IPNI=*159783-2* 423
Conostegia subhirsuta IPNI=*159783-2* 423
Conostegia viridis IPNI=*159943-2* 423
Conostegia xalapensis f. *canescens* IPNI=*159943-2* 423
Conostegia xalapensis IPNI=*159943-2* 423
Consolea bahamana IPNI=*137078-1* 284
Consolida ajacis IPNI=*77254-2* 436
Consolida major IPNI=*120815-1* 275
Consolida orientalis IPNI=*77254-2* 436
Convallaria diffusa IPNI=*79497-2* 224
Convallaria fruticosa IPNI=*533580-1* 215
Conyza abbreviata IPNI=*962413-1* 260
Conyza albida IPNI=*204572-1* 262
Conyza alopecuroides IPNI=*240147-1* 269
Conyza altissima IPNI=*204572-1* 262
Conyza ambigua IPNI=*204572-1* 262
Conyza amplexicaulis IPNI=*238826-1* 268
Conyza angustifolia IPNI=*1099254-2* 268
Conyza argentea IPNI=*211317-2* 268
Conyza aurita IPNI=*211317-2* 268
Conyza bellidifolia IPNI=*962413-1* 260
Conyza bifrons IPNI=*238826-1* 268
Conyza bilbaoana IPNI=*60473415-2* 261
Conyza bonariensis f. *subleiotheca* IPNI=*204572-1* 262
Conyza bonariensis IPNI=*93603-2* 261
Conyza bonariensis var. *angustifolia* IPNI=*93603-2* 261
Conyza bonariensis var. *leiotheca* IPNI=*60473415-2* 261
Conyza bonariensis var. *microcephala* IPNI=*204572-1* 262
Conyza brasiliensis IPNI=*203916-2* 268
Conyza canadensis IPNI=*203653-1* 261
Conyza canadensis var. *glabrata* IPNI=*203653-1* 261
Conyza canadensis var. *incisa* IPNI=*203653-1* 261
Conyza canadensis var. *obovoidea* IPNI=*203653-1* 261
Conyza canadensis var. *pusilla* IPNI=*203653-1* 261
Conyza canadensis var. *robusta* IPNI=*203653-1* 261
Conyza canadensis var. *simplex* IPNI=*203653-1* 261
Conyza carolinensis IPNI=*238803-1* 268
Conyza cayennensis IPNI=*1092582-2* 263
Conyza chenopodioides IPNI=*93603-2* 261
Conyza chiapensis IPNI=*211317-2* 268
Conyza chinensis IPNI=*962413-1* 260
Conyza cinerascens IPNI=*962413-1* 260
Conyza cinerea IPNI=*962413-1* 260
Conyza clethrifolia IPNI=*969180-1* 272
Conyza cortesii IPNI=*238803-1* 268
Conyza coulteri var. *tenuisecta* IPNI=*947840-1* 264
Conyza crispa IPNI=*93603-2* 261
Conyza elata IPNI=*60473415-2* 261
Conyza elegantula IPNI=*962413-1* 260
Conyza erigeroides IPNI=*204572-1* 262
Conyza evacioides IPNI=*228376-1* 264
Conyza filaginoides IPNI=*228375-1* 264
Conyza flahaultiana IPNI=*204572-1* 262
Conyza floribunda IPNI=*60473415-2* 261
Conyza floribunda var. *laciniata* IPNI=*60473415-2* 261
Conyza floribunda var. *linearifolia* IPNI=*60473415-2* 261
Conyza floribunda var. *subleiotheca* IPNI=*204572-1* 262
Conyza foetida IPNI=*238826-1* 268
Conyza genistelloides IPNI=*183309-1* 257
Conyza gnaphalioides IPNI=*228376-1* 264
Conyza groegeri IPNI=*204572-1* 262
Conyza guineensis IPNI=*211317-2* 268
Conyza heterophylla IPNI=*962413-1* 260
Conyza hispida IPNI=*93603-2* 261
Conyza incana IPNI=*962413-1* 260
Conyza ivifolia IPNI=*962413-1* 260

Conyza leucodasys IPNI=*93603-2* 261
Conyza linearis IPNI=*93603-2* 261
Conyza lobata IPNI=*234386-1* 267
Conyza lyrata IPNI=*211317-2* 268
Conyza lyrata var. *pilosa* IPNI=*211317-2* 268
Conyza marilandica IPNI=*1099254-2* 268
Conyza martii IPNI=*969180-1* 272
Conyza mollis IPNI=*962413-1* 260
Conyza myriocephala IPNI=*60473415-2* 261
Conyza naudinii IPNI=*204572-1* 262
Conyza odorata IPNI=*1099254-2* 268
Conyza parva IPNI=*203653-1* 261
Conyza pedunculata IPNI=*28074-2* 257
Conyza plebeia IPNI=*93603-2* 261
Conyza prolifera IPNI=*962413-1* 260
Conyza pulchella IPNI=*947840-1* 264
Conyza pulcherrima IPNI=*228376-1* 264
Conyza purpurascens IPNI=*1099254-2* 268
Conyza quitoc IPNI=*203916-2* 268
Conyza riparia IPNI=*30283369-2* 271
Conyza rudis IPNI=*211317-2* 268
Conyza sagittalis IPNI=*203916-2* 268
Conyza scorpioides IPNI=*944993-1* 261
Conyza senegalensis IPNI=*211317-2* 268
Conyza serpentaria IPNI=*947840-1* 264
Conyza sinensis IPNI=*962413-1* 260
Conyza sinuata IPNI=*93603-2* 261
Conyza sophiifolia IPNI=*947840-1* 264
Conyza sordescens IPNI=*93603-2* 261
Conyza spicata IPNI=*30022662-2* 269
Conyza squarrosa IPNI=*60478388-2* 267
Conyza sumatrensis IPNI=*204572-1* 262
Conyza sumatrensis var. *floribunda* IPNI=*60473415-2* 261
Conyza sumatrensis var. *leiotheca* IPNI=*60473415-2* 261
Conyza symphytifolia IPNI=*234386-1* 267
Conyza thuyoides IPNI=*231180-1* 265
Conyza tomentosa IPNI=*969218-1* 272
Conyza tricuneata IPNI=*183814-1* 258
Conyza trihecatactis IPNI=*94203-2* 262
Conyza trinervia IPNI=*183821-1* 258
Conyza trinervis IPNI=*183821-1* 258
Conyza uliginosa IPNI=*204638-1* 262
Conyza uliginosa IPNI=*238826-1* 268
Conyza uliginosa var. *columbiana* IPNI=*204638-1* 262
Conyza uliginosa var. *hirsuta* IPNI=*204638-1* 262
Conyza venosa IPNI=*27442-2* 257
Conyza villosa IPNI=*211317-2* 268
Conyza virgata IPNI=*30022662-2* 269
Conyza viscosa IPNI=*211317-2* 268
Conyza vulgaris IPNI=*60478388-2* 267
Conyzella canadensis IPNI=*203653-1* 261
Conyzella linifolia IPNI=*93603-2* 261
Copaiba bijuga IPNI=*487006-1* 315
Copaiba guianensis IPNI=*487006-1* 315
Copaiba officinalis IPNI=*64940-2* 315
Copaifera beyrichii IPNI=*487006-1* 315
Copaifera bijuga IPNI=*487006-1* 315
Copaifera jacquiniana IPNI=*64940-2* 315
Copaifera jacquinii IPNI=*64940-2* 315
Copaifera martii var. *pubiflora* IPNI=*487038-1* 315
Copaifera reticulata var. *peruviana* IPNI=*64945-2* 315
Copaiva guianensis IPNI=*487006-1* 315
Copaiva officinalis IPNI=*64940-2* 315
Copedesma nitens IPNI=*289925-2* 422
Copernicia sanctae-martae IPNI=*666287-1* 208
Copernicia wrightii IPNI=*663912-1* 203
Coprosma granadensis IPNI=*757486-1* 352
Corallocarpus angosturensis IPNI=*592558-1* 294
Corallocarpus emetocatharticus IPNI=*592558-1* 294
Corallocarpus glomeratus IPNI=*592558-1* 294
Corallocarpus glomeratus var. *gracilis* IPNI=*592558-1* 294
Corallocarpus guatemalensis IPNI=*592558-1* 294
Corallodendron aculeatum IPNI=*316421-2* 319
Corallodendron australe IPNI=*316421-2* 319
Corallodendron crista-galli IPNI=*494400-1* 319
Corallodendron divaricatum IPNI=*494614-1* 320
Corallodendron fuscum IPNI=*494431-1* 319
Corallodendron glaucum IPNI=*494431-1* 319
Corallodendron lobulatum IPNI=*494614-1* 320
Corallodendron occidentale IPNI=*316421-2* 319
Corallodendron orientale IPNI=*494614-1* 320
Corallodendron ovalifolium IPNI=*494431-1* 319
Corallodendron patens IPNI=*494431-1* 319
Corallodendron rubrinerve IPNI=*494567-1* 319
Corallodendron spathaceum IPNI=*494614-1* 320
Corallodendron velutinum IPNI=*494616-1* 320
Corallophyllum caeruleum IPNI=*526306-1* 275
Corchorus acutangulus IPNI=*320794-2* 408
Corchorus acutangulus var. *brachycarpus* IPNI=*320794-2* 408
Corchorus aesticans IPNI=*320794-2* 408
Corchorus argutus IPNI=*65167-2* 408
Corchorus argutus var. *australis* IPNI=*65167-2* 408
Corchorus argutus var. *benthamii* IPNI=*65167-2* 408
Corchorus argutus var. *blanchetii* IPNI=*65167-2* 408
Corchorus argutus var. *longicarpus* IPNI=*65167-2* 408
Corchorus argutus var. *prismatocarpa* IPNI=*65167-2* 408
Corchorus argutus var. *typica* IPNI=*65167-2* 408
Corchorus biflorus IPNI=*320795-2* 408
Corchorus bullatus IPNI=*320795-2* 408
Corchorus campestris IPNI=*320794-2* 408
Corchorus caryophylloides IPNI=*833664-1* 408
Corchorus coreta IPNI=*833664-1* 408
Corchorus furcatus IPNI=*320794-2* 408
Corchorus fuscus IPNI=*320794-2* 408
Corchorus hirtus f. *adscendens* IPNI=*320795-2* 408
Corchorus hirtus f. *hilarianus* IPNI=*320795-2* 408
Corchorus hirtus f. *humilis* IPNI=*320795-2* 408
Corchorus hirtus f. *longepetiolatus* IPNI=*320795-2* 408
Corchorus hirtus var. *argentinensis* IPNI=*320795-2* 408
Corchorus hirtus var. *brasiliensis* IPNI=*320795-2* 408
Corchorus hirtus var. *cuyabensis* IPNI=*320795-2* 408
Corchorus hirtus var. *glabellus* IPNI=*320795-2* 408
Corchorus hirtus var. *martii* IPNI=*320795-2* 408
Corchorus hirtus var. *orinocensis* IPNI=*65167-2* 408
Corchorus hirtus var. *pilosus* IPNI=*320795-2* 408
Corchorus hirtus var. *villosissima* IPNI=*320795-2* 408
Corchorus humilis IPNI=*320795-2* 408
Corchorus lasiolobus IPNI=*320795-2* 408
Corchorus linearis IPNI=*833664-1* 408
Corchorus mompoxensis IPNI=*320795-2* 408
Corchorus oppositiflorus IPNI=*320794-2* 408
Corchorus pilolobus IPNI=*320795-2* 408
Corchorus pilosus IPNI=*320795-2* 408
Corchorus prismatocarpus IPNI=*65167-2* 408
Corchorus secundiflorus IPNI=*833664-1* 408
Corchorus spinosus IPNI=*186083-2* 412
Corchorus tetrangulus IPNI=*320795-2* 408
Corchorus tortipes IPNI=*320795-2* 408
Corchorus villosissimus IPNI=*320795-2* 408
Corculum leptopus IPNI=*15860-2* 288
Corculum leptopus var. *album* IPNI=*15860-2* 288
Cordia acuta IPNI=*286671-2* 275
Cordia adnata IPNI=*262677-2* 276
Cordia alba IPNI=*1018059-2* 436
Cordia alliodora f. *albotomentosa* IPNI=*113943-1* 273
Cordia alliodora var. *boliviana* IPNI=*113943-1* 273

Cordia alliodora var. *glabra* IPNI=*113943-1* 273
Cordia alliodora var. *tomentosa* IPNI=*113943-1* 273
Cordia andina IPNI=*113943-1* 273
Cordia angustifolia IPNI=*262640-2* 275
Cordia araripensis IPNI=*274895-2* 274
Cordia asperrima IPNI=*318022-2* 275
Cordia avicularis IPNI=*114394-1* 274
Cordia belizensis IPNI=*274895-2* 274
Cordia beurreria IPNI=*1088317-2* 273
Cordia bifurcata IPNI=*121424-1* 276
Cordia boliviana IPNI=*262677-2* 276
Cordia bourreria IPNI=*1088317-2* 273
Cordia bracteata IPNI=*65321-2* 274
Cordia brevispicata IPNI=*262640-2* 275
Cordia brevispicata var. *hypomalaca* IPNI=*262640-2* 275
Cordia bullata IPNI=*318022-2* 275
Cordia bullata var. *angustata* IPNI=*318022-2* 275
Cordia calyptrata IPNI=*927551-1* 274
Cordia cana IPNI=*946177-1* 276
Cordia canescens IPNI=*262640-2* 275
Cordia caracasana IPNI=*262641-2* 276
Cordia caracasana var. *velutina* IPNI=*262641-2* 276
Cordia carnosa IPNI=*274895-2* 274
Cordia cerdana IPNI=*113943-1* 273
Cordia chacoensis IPNI=*262640-2* 275
Cordia chamissoniana var. *complicata* IPNI=*113943-1* 273
Cordia chepensis IPNI=*262640-2* 275
Cordia collococa IPNI=*114394-1* 274
Cordia coloradiphila IPNI=*946226-1* 276
Cordia cordifolia IPNI=*275071-2* 274
Cordia coriacea IPNI=*275053-2* 274
Cordia corylifolia IPNI=*927551-1* 274
Cordia corymbosa f. *detonsa* IPNI=*121424-1* 276
Cordia corymbosa f. *hirsuta* IPNI=*262677-2* 276
Cordia corymbosa f. *intonsa* IPNI=*262677-2* 276
Cordia corymbosa f. *typica* IPNI=*262677-2* 276
Cordia corymbosa IPNI=*262677-2* 276
Cordia costaricensis IPNI=*946226-1* 276
Cordia crenulata IPNI=*946226-1* 276
Cordia cujabensis IPNI=*113943-1* 273
Cordia cuneiformis IPNI=*262640-2* 275
Cordia curassavica IPNI=*262640-2* 275
Cordia cylindrostachya IPNI=*262641-2* 276
Cordia cylindrostachya var. *floribunda* IPNI=*262641-2* 276
Cordia cylindrostachya var. *interrupta* IPNI=*262640-2* 275
Cordia dichotoma IPNI=*121424-1* 276
Cordia divaricata IPNI=*262640-2* 275
Cordia ehretioides IPNI=*114058-1* 274
Cordia exaltata var. *melanoneura* IPNI=*114135-1* 274
Cordia expansa IPNI=*275076-2* 274
Cordia ferruginea IPNI=*946226-1* 276
Cordia flava IPNI=*274993-2* 274
Cordia floribunda IPNI=*262641-2* 276
Cordia formicarum IPNI=*114394-1* 274
Cordia gerascanthus f. *genuina* IPNI=*65321-2* 274
Cordia gerascanthus f. *martinicensis* IPNI=*113943-1* 273
Cordia gerascanthus f. *micrantha* IPNI=*65321-2* 274
Cordia gerascanthus var. *subcanescens* IPNI=*113943-1* 273
Cordia geraschanthoides IPNI=*65321-2* 274
Cordia glabra IPNI=*1088317-2* 273
Cordia goudotii IPNI=*113943-1* 273
Cordia grandiflora IPNI=*121434-1* 276
Cordia graveolens IPNI=*262640-2* 275
Cordia guazumifolia IPNI=*262677-2* 276
Cordia guianensis IPNI=*262640-2* 275
Cordia hartwissiana IPNI=*113943-1* 273
Cordia hebecarpa IPNI=*275076-2* 274
Cordia hebeclada IPNI=*275016-2* 274
Cordia hermanniifolia IPNI=*262677-2* 276
Cordia hermanniifolia var. *calycina* IPNI=*262677-2* 276
Cordia heterophylla IPNI=*275076-2* 274
Cordia hirsuta IPNI=*114394-1* 274
Cordia hirsuta IPNI=*262640-2* 275
Cordia hispida IPNI=*262640-2* 275
Cordia hispidissima IPNI=*114394-1* 274
Cordia humboldtii IPNI=*262641-2* 276
Cordia ierensis IPNI=*275053-2* 274
Cordia imparilis IPNI=*262640-2* 275
Cordia inermis IPNI=*946177-1* 276
Cordia insularis IPNI=*946177-1* 276
Cordia interrupta IPNI=*262640-2* 275
Cordia intonsa IPNI=*262640-2* 275
Cordia johnstonii IPNI=*274931-2* 274
Cordia juglandifolia IPNI=*65551-2* 274
Cordia laevis IPNI=*65551-2* 274
Cordia lanata IPNI=*262641-2* 276
Cordia langlassei IPNI=*65321-2* 274
Cordia lapensis IPNI=*262677-2* 276
Cordia laxiflora IPNI=*946226-1* 276
Cordia leptopoda IPNI=*927551-1* 274
Cordia linearicalycina IPNI=*262641-2* 276
Cordia linearis IPNI=*262640-2* 275
Cordia lineata IPNI=*77084349-1* 276
Cordia linnaei IPNI=*77084349-1* 276
Cordia littoralis IPNI=*262640-2* 275
Cordia llanense IPNI=*262641-2* 276
Cordia macrostachya IPNI=*262640-2* 275
Cordia macrostachya IPNI=*262641-2* 276
Cordia macuirensis IPNI=*262640-2* 275
Cordia marchionica IPNI=*274993-2* 274
Cordia mariquitensis IPNI=*262677-2* 276
Cordia martinicensis IPNI=*262640-2* 275
Cordia martinicensis IPNI=*262641-2* 276
Cordia melanoneura IPNI=*114135-1* 274
Cordia micrantha IPNI=*114058-1* 274
Cordia miranda IPNI=*114394-1* 274
Cordia mollis IPNI=*262640-2* 275
Cordia monosperma IPNI=*262677-2* 276
Cordia mucronata IPNI=*114135-1* 274
Cordia muneco IPNI=*275071-2* 274
Cordia nodosa f. *glabrior* IPNI=*114394-1* 274
Cordia nodosa var. *angustifolia* IPNI=*114394-1* 274
Cordia nodosa var. *hispidissima* IPNI=*114394-1* 274
Cordia opaca IPNI=*275053-2* 274
Cordia ovata IPNI=*927551-1* 274
Cordia oxyphylla IPNI=*262640-2* 275
Cordia palmeri IPNI=*262640-2* 275
Cordia paniculata IPNI=*274931-2* 274
Cordia paraguariensis IPNI=*262677-2* 276
Cordia parviflora IPNI=*121424-1* 276
Cordia patens f. *polycephala* IPNI=*262677-2* 276
Cordia patens IPNI=*262677-2* 276
Cordia patens var. *angustifolia* IPNI=*262677-2* 276
Cordia pauciflora IPNI=*946226-1* 276
Cordia pellita IPNI=*77069564-1* 275
Cordia peruviana var. *mexicana* IPNI=*262640-2* 275
Cordia petenensis IPNI=*274931-2* 274
Cordia platystachya IPNI=*946226-1* 276
Cordia poeppigii IPNI=*946226-1* 276
Cordia polycephala IPNI=*262677-2* 276
Cordia pubescens IPNI=*275076-2* 274
Cordia pyrifolia IPNI=*946226-1* 276
Cordia rariflora IPNI=*114058-1* 274
Cordia rhombifolia IPNI=*946226-1* 276
Cordia riparia IPNI=*946226-1* 276
Cordia rothschuhii IPNI=*65321-2* 274

Cordia rotundifolia IPNI=*274993-2* 274
Cordia rubescens IPNI=*262641-2* 276
Cordia rufa IPNI=*121434-1* 276
Cordia rugosa IPNI=*262640-2* 275
Cordia rusbyi IPNI=*113943-1* 273
Cordia salicina IPNI=*262640-2* 275
Cordia salviifolia IPNI=*262641-2* 276
Cordia salzmannii IPNI=*262677-2* 276
Cordia salzmannii var. *albiflora* IPNI=*262677-2* 276
Cordia salzmannii var. *lanceolata* IPNI=*262677-2* 276
Cordia scabrida IPNI=*114135-1* 274
Cordia sebestena var. *rubra* IPNI=*65551-2* 274
Cordia socorrensis IPNI=*262640-2* 275
Cordia speciosa IPNI=*65551-2* 274
Cordia spicata IPNI=*262640-2* 275
Cordia spinescens IPNI=*946226-1* 276
Cordia stenophylla IPNI=*262640-2* 275
Cordia subserrata IPNI=*262641-2* 276
Cordia sulfurata IPNI=*262677-2* 276
Cordia tarmensis IPNI=*262641-2* 276
Cordia tenuifolia IPNI=*927551-1* 274
Cordia tetraphylla IPNI=*77164595-1* 416
Cordia thibaudiana IPNI=*946226-1* 276
Cordia trichostyla IPNI=*274895-2* 274
Cordia ulmifolia IPNI=*262677-2* 276
Cordia ulmifolia var. *lineata* IPNI=*77084349-1* 276
Cordia ulmifolia var. *ovalis* IPNI=*262677-2* 276
Cordia ulmifolia var. *ovata* IPNI=*262677-2* 276
Cordia umbraculifera IPNI=*275071-2* 274
Cordia urticifolia IPNI=*262677-2* 276
Cordia velutina IPNI=*113943-1* 273
Cordia verbenacea IPNI=*262640-2* 275
Cordia viridis IPNI=*275053-2* 274
Cordia volubilis IPNI=*114394-1* 274
Cordia wickstroemii IPNI=*262677-2* 276
Cordiera claviflora IPNI=*8381-2* 346
Cordiera edulis IPNI=*743001-1* 347
Cordiera latifolia IPNI=*8399-2* 347
Cordiera oligantha IPNI=*8399-2* 347
Cordiopsis monosperma IPNI=*262677-2* 276
Cordyline albicans IPNI=*533580-1* 215
Cordyline alborosea IPNI=*533580-1* 215
Cordyline amabilis IPNI=*533580-1* 215
Cordyline amboynensis IPNI=*533580-1* 215
Cordyline angusta IPNI=*533580-1* 215
Cordyline argenteostriata IPNI=*533580-1* 215
Cordyline balmoreana IPNI=*533580-1* 215
Cordyline baptistii IPNI=*533580-1* 215
Cordyline bartelii IPNI=*533580-1* 215
Cordyline bellula IPNI=*533580-1* 215
Cordyline calocoma IPNI=*533544-1* 215
Cordyline cheesemanii IPNI=*533580-1* 215
Cordyline chelsonii IPNI=*533580-1* 215
Cordyline compacta IPNI=*533580-1* 215
Cordyline dennisonii IPNI=*533580-1* 215
Cordyline densicoma IPNI=*533580-1* 215
Cordyline duffii IPNI=*533580-1* 215
Cordyline excellens IPNI=*533580-1* 215
Cordyline excelsa IPNI=*533580-1* 215
Cordyline ferrea IPNI=*533580-1* 215
Cordyline forsteri IPNI=*533544-1* 215
Cordyline fragrans IPNI=*534207-1* 215
Cordyline fraseri IPNI=*533580-1* 215
Cordyline fruticosa IPNI=*534103-1* 215
Cordyline gloriosa IPNI=*533580-1* 215
Cordyline grandis IPNI=*533580-1* 215
Cordyline guilfoylei IPNI=*533580-1* 215
Cordyline guineensis IPNI=*77164234-1* 215
Cordyline hedychioides IPNI=*533580-1* 215
Cordyline heliconiifolia IPNI=*533580-1* 215
Cordyline hendersonii IPNI=*533580-1* 215
Cordyline hyacinthoides IPNI=*77164234-1* 215
Cordyline imperialis IPNI=*533580-1* 215
Cordyline indivisa IPNI=*533544-1* 215
Cordyline inscripta IPNI=*533580-1* 215
Cordyline jacquinii IPNI=*533580-1* 215
Cordyline javanica IPNI=*533580-1* 215
Cordyline kewensis IPNI=*533580-1* 215
Cordyline laingii IPNI=*533580-1* 215
Cordyline lentiginosa IPNI=*533544-1* 215
Cordyline lutescens-striata IPNI=*533580-1* 215
Cordyline macarthurii IPNI=*533580-1* 215
Cordyline madagascariensis IPNI=*533580-1* 215
Cordyline magnifica IPNI=*533580-1* 215
Cordyline metallica IPNI=*533580-1* 215
Cordyline mirabilis IPNI=*533580-1* 215
Cordyline mooreana IPNI=*533580-1* 215
Cordyline nigrorubra IPNI=*533580-1* 215
Cordyline nobilis IPNI=*533580-1* 215
Cordyline norwoodiensis IPNI=*533580-1* 215
Cordyline ornata IPNI=*533580-1* 215
Cordyline picturata IPNI=*533580-1* 215
Cordyline placida IPNI=*533580-1* 215
Cordyline porphyrophylla IPNI=*533580-1* 215
Cordyline pulchella IPNI=*533580-1* 215
Cordyline reali IPNI=*533580-1* 215
Cordyline reflexa IPNI=*534338-1* 215
Cordyline regina IPNI=*533580-1* 215
Cordyline rex IPNI=*533580-1* 215
Cordyline robinsoniana IPNI=*533580-1* 215
Cordyline rosacea IPNI=*533580-1* 215
Cordyline rubella IPNI=*533580-1* 215
Cordyline rumphii IPNI=*534103-1* 215
Cordyline sepiaria IPNI=*533580-1* 215
Cordyline sieberi IPNI=*533580-1* 215
Cordyline splendens IPNI=*533580-1* 215
Cordyline sturmii IPNI=*533544-1* 215
Cordyline sulcata IPNI=*533580-1* 215
Cordyline superbiens IPNI=*533544-1* 215
Cordyline terminalis IPNI=*533580-1* 215
Cordyline terminalis var. *baileyi* IPNI=*533580-1* 215
Cordyline terminalis var. *hedychioides* IPNI=*533580-1* 215
Cordyline thomsonii IPNI=*533580-1* 215
Cordyline timorensis IPNI=*533580-1* 215
Cordyline triumphans IPNI=*533580-1* 215
Cordyline veitchii IPNI=*533544-1* 215
Cordyline venosa IPNI=*533580-1* 215
Cordyline violascens IPNI=*97540-3* 215
Cordyline vivipara IPNI=*532810-1* 215
Cordyline williamsii IPNI=*533580-1* 215
Cordyline youngii IPNI=*533580-1* 215
Cordylocarpus glaber IPNI=*1061021-2* 277
Coreopsis acmella var. *uliginosa* IPNI=*174556-1* 256
Coreopsis alba IPNI=*184711-1* 258
Coreopsis artemisiifolia IPNI=*198368-1* 260
Coreopsis baccata IPNI=*993359-1* 271
Coreopsis bidens IPNI=*32497-2* 258
Coreopsis chrysantha IPNI=*185190-1* 258
Coreopsis coronata IPNI=*198324-1* 260
Coreopsis corymbifolia IPNI=*32564-2* 258
Coreopsis crassifolia IPNI=*30027297-2* 261
Coreopsis flammula IPNI=*32497-2* 258
Coreopsis formosa IPNI=*198320-1* 260
Coreopsis georgina IPNI=*30027297-2* 261
Coreopsis gracilis IPNI=*198324-1* 260
Coreopsis leucantha IPNI=*32564-2* 258

Coreopsis leucanthema IPNI=*184711-1* 258
Coreopsis multifida var. *mutica* IPNI=*32564-2* 258
Coreopsis odorata IPNI=*32564-2* 258
Coreopsis odoratissima IPNI=*184711-1* 258
Coreopsis patula IPNI=*184720-1* 258
Coreopsis perfoliata IPNI=*32497-2* 258
Coreopsis radiata IPNI=*32497-2* 258
Coreopsis reptans IPNI=*185190-1* 258
Coreopsis scandens IPNI=*185190-1* 258
Coreopsis trifoliata IPNI=*185190-1* 258
Coreopsis variifolia IPNI=*185190-1* 258
Coriandropsis syriaca IPNI=*840760-1* 252
Coriandrum cicuta IPNI=*840668-1* 252
Coriandrum diversifolium IPNI=*840760-1* 252
Coriandrum globosum IPNI=*840760-1* 252
Coriandrum maculatum IPNI=*840668-1* 252
Coriandrum majus IPNI=*840760-1* 252
Coriandrum melphitense IPNI=*840760-1* 252
Coriandrum sativum var. *afghanicum* IPNI=*840760-1* 252
Coriandrum sativum var. *africanum* IPNI=*840760-1* 252
Coriandrum sativum var. *anatolicum* IPNI=*840760-1* 252
Coriandrum sativum var. *arabicum* IPNI=*840760-1* 252
Coriandrum sativum var. *microcarpum* IPNI=*840760-1* 252
Coriandrum sativum var. *pygmaeum* IPNI=*840760-1* 252
Corion halophilum IPNI=*241336-2* 286
Corion marinum IPNI=*241336-2* 286
Corion marinum var. *leiospermum* IPNI=*241336-2* 286
Corion rubrum IPNI=*158983-1* 286
Cormophyllum arboreum IPNI=*17071820-1* 172
Cornucopiae altissima IPNI=*1128639-2* 233
Cornucopiae altissimum IPNI=*1128639-2* 233
Cornucopiae perennans IPNI=*1128639-2* 233
Cornuella venezuelanensis IPNI=*948088-1* 305
Cornutia crenata IPNI=*319153-2* 369
Cornutia divaricata IPNI=*319153-2* 369
Cornutia grandifolia f. *quadrangularis* IPNI=*319153-2* 369
Cornutia grandifolia IPNI=*319153-2* 369
Cornutia grandifolia var. *intermedia* IPNI=*319153-2* 369
Cornutia grandifolia var. *normalis* IPNI=*319153-2* 369
Cornutia grandifolia var. *purpusii* IPNI=*319153-2* 369
Cornutia grandifolia var. *quadrangularis* IPNI=*319153-2* 369
Cornutia grandifolia var. *storkii* IPNI=*319153-2* 369
Cornutia graveolens IPNI=*862634-1* 369
Cornutia latifolia f. *alba* IPNI=*319153-2* 369
Cornutia latifolia IPNI=*319153-2* 369
Cornutia lilacina IPNI=*319153-2* 369
Cornutia lilacina var. *velutina* IPNI=*319153-2* 369
Cornutia longifolia IPNI=*319153-2* 369
Cornutia microcalycina IPNI=*319153-2* 369
Cornutia microcalycina var. *anomala* IPNI=*319153-2* 369
Cornutia microcalycina var. *pulverulenta* IPNI=*319153-2* 369
Cornutia odorata var. *calvescens* IPNI=*862634-1* 369
Cornutia odorata var. *colombiana* IPNI=*862634-1* 369
Cornutia pyramidata f. *albida* IPNI=*319153-2* 369
Cornutia pyramidata f. *serrata* IPNI=*319153-2* 369
Cornutia pyramidata var. *dentata* IPNI=*319153-2* 369
Cornutia pyramidata var. *isthmica* IPNI=*319153-2* 369
Cornutia pyramidata var. *longifolia* IPNI=*319153-2* 369
Cornutia pyramidata var. *normalis* IPNI=*319153-2* 369
Cornutia pyramidata var. *punctata* IPNI=*319153-2* 369
Coronaria coriacea IPNI=*60461121-2* 286
Coronaria coronaria IPNI=*60461121-2* 286
Coronaria tomentosa IPNI=*60461121-2* 286
Corone agrestina IPNI=*332106-2* 286
Corone gallica IPNI=*332106-2* 286
Corone quinquevulnera IPNI=*332106-2* 286
Coronilla grandiflora IPNI=*518476-1* 334
Coronilla monilis IPNI=*77118402-1* 328
Coronilla occidentalis IPNI=*284993-2* 334
Coronilla pubescens IPNI=*1015399-2* 334
Coronilla scandens IPNI=*77191263-1* 329
Coronilla sericea IPNI=*1015399-2* 334
Coronopus didymus f. *pectinatus* IPNI=*286148-1* 277
Coronopus didymus IPNI=*286148-1* 277
Coronopus didymus subsp. *eudidymus* IPNI=*286148-1* 277
Coronopus didymus var. *incisus* IPNI=*286148-1* 277
Coronopus didymus var. *macrocarpus* IPNI=*286148-1* 277
Coronopus didymus var. *rosulatus* IPNI=*286148-1* 277
Coronopus incisus IPNI=*286148-1* 277
Coronopus leptocarpus IPNI=*286148-1* 277
Coronopus leptocarpus var. *microcarpus* IPNI=*286148-1* 277
Coronopus pectinatus IPNI=*286148-1* 277
Coronopus pinnatifidus IPNI=*286148-1* 277
Coronopus pinnatus IPNI=*286148-1* 277
Corozo oleifera IPNI=*88982-2* 208
Correlliana schlimii IPNI=*73559-2* 304
Correlliana spectabilis IPNI=*73559-2* 304
Cortaderia argentea IPNI=*66317-2* 236
Cortaderia selloana subsp. *jubata* IPNI=*66302-2* 236
Cortaderia sodiroana IPNI=*66306-2* 236
Cortusina odoratissima IPNI=*376654-1* 358
Coryanthes macrantha IPNI=*624302-1* 218
Coryanthes wolfii IPNI=*624302-1* 218
Corymbia citriodora subsp. *variegata* IPNI=*986336-1* 424
Corymbia variegata IPNI=*986336-1* 424
Corynaea purdiei IPNI=*103322-1* 448
Corynaea sphaerica IPNI=*103322-1* 448
Corynandra viscosa IPNI=*320133-2* 280
Corynosicyos edulis IPNI=*77110571-1* 294
Corynostigma jussiaeoides IPNI=*144384-2* 429
Corynostylis albiflora IPNI=*77144891-1* 405
Corynostylis arborea IPNI=*77144891-1* 405
Corynostylis aubletii IPNI=*77144891-1* 405
Corynostylis benthamii IPNI=*77144891-1* 405
Corynostylis berteroi IPNI=*77144891-1* 405
Corynostylis guayanensis IPNI=*77144891-1* 405
Corynostylis hybanthus IPNI=*77144891-1* 405
Corynostylis hybanthus var. *albiflora* IPNI=*77144891-1* 405
Corynostylis loeflingii IPNI=*77144891-1* 405
Corynostylis palustris IPNI=*77144891-1* 405
Corypha laevis IPNI=*895770-1* 227
Corypha obliqua IPNI=*941782-1* 212
Corypha repens IPNI=*941782-1* 212
Corypha rotundifolia IPNI=*669832-1* 212
Corypha tectorum IPNI=*666287-1* 208
Corytholoma caracasanum IPNI=*237093-2* 368
Corytholoma chelonoides IPNI=*237093-2* 368
Corytholoma ellipticum IPNI=*237093-2* 368
Corytholoma gracile IPNI=*237093-2* 368
Corytholoma rutilum IPNI=*237093-2* 368
Corytholoma tubiflorum IPNI=*381638-1* 367
Corytholoma warscewiczii IPNI=*237093-2* 368
Cosmea bipinnata IPNI=*198320-1* 260
Cosmea caudata IPNI=*198324-1* 260
Cosmea pilosa IPNI=*32564-2* 258
Cosmea sulphurea IPNI=*198368-1* 260
Cosmea tenuifolia IPNI=*198320-1* 260
Cosmia caulescens IPNI=*698357-1* 286
Cosmia montana IPNI=*698357-1* 286
Cosmibuena arborea IPNI=*66725-2* 349
Cosmibuena gardenioides IPNI=*66725-2* 349
Cosmibuena latifolia IPNI=*66725-2* 349
Cosmibuena obtusifolia IPNI=*66725-2* 349
Cosmibuena obtusifolia var. *latifolia* IPNI=*66725-2* 349
Cosmibuena ovalis IPNI=*66725-2* 349
Cosmibuena quinqueflora IPNI=*66725-2* 349

Cosmibuena skinneri IPNI=*66725-2* 349
Cosmibuena triflora IPNI=*66725-2* 349
Cosmophyllum cacaliifolium IPNI=*238923-1* 268
Cosmos artemisiifolius IPNI=*198368-1* 260
Cosmos aurantiacus IPNI=*198368-1* 260
Cosmos bipinnatus var. *albiflorus* IPNI=*198320-1* 260
Cosmos bipinnatus var. *exaristatus* IPNI=*198320-1* 260
Cosmos bipinnatus var. *typicus* IPNI=*198320-1* 260
Cosmos gracilis IPNI=*198368-1* 260
Cosmos hybridus IPNI=*198320-1* 260
Cosmos pacificus var. *chiapensis* IPNI=*198324-1* 260
Cosmos pilosus IPNI=*184711-1* 258
Cosmos pinnatus IPNI=*32564-2* 258
Cosmos spectabilis IPNI=*198320-1* 260
Cosmos sulphureus var. *exaristatus* IPNI=*198368-1* 260
Cosmos sulphureus var. *hirsuticaulis* IPNI=*198368-1* 260
Cosmos sulphureus var. *typicus* IPNI=*198368-1* 260
Cosmos tenellus IPNI=*184711-1* 258
Cosmos tenuifolius IPNI=*198320-1* 260
Costus anachiri IPNI=*796376-1* 248
Costus ananassae IPNI=*873025-1* 248
Costus arrabidae IPNI=*796267-1* 247
Costus brasiliensis IPNI=*796267-1* 247
Costus cernuus IPNI=*218597-2* 251
Costus ciliatus IPNI=*796376-1* 248
Costus conicus IPNI=*329964-2* 248
Costus cylindricus IPNI=*329964-2* 248
Costus cylindricus var. *anachiri* IPNI=*796376-1* 248
Costus cylindricus var. *ciliatus* IPNI=*796376-1* 248
Costus cylindricus var. *pulcherrimus* IPNI=*796315-1* 247
Costus discolor IPNI=*796267-1* 247
Costus formosus IPNI=*796364-1* 248
Costus friedrichsenii IPNI=*796399-1* 248
Costus giganteus IPNI=*796315-1* 247
Costus glabratus IPNI=*796267-1* 247
Costus gracilis IPNI=*796267-1* 247
Costus hirsutus IPNI=*796399-1* 248
Costus kuntzei IPNI=*796315-1* 247
Costus laxus IPNI=*796364-1* 248
Costus malortieanus var. *amazonicus* IPNI=*66817-2* 247
Costus maritimus IPNI=*796280-1* 247
Costus maximus IPNI=*796315-1* 247
Costus micranthus IPNI=*329964-2* 248
Costus niveopurpureus IPNI=*796267-1* 247
Costus niveus IPNI=*796267-1* 247
Costus nutans IPNI=*796376-1* 248
Costus pilgeri IPNI=*796267-1* 247
Costus podocephalus IPNI=*218597-2* 251
Costus pubescens f. *fibrillosus* IPNI=*796267-1* 247
Costus pubescens IPNI=*796267-1* 247
Costus puchucupango IPNI=*796376-1* 248
Costus pulcherrimus IPNI=*796315-1* 247
Costus pungens IPNI=*873025-1* 248
Costus quartus IPNI=*329964-2* 248
Costus quintus IPNI=*796376-1* 248
Costus ramosus IPNI=*796267-1* 247
Costus ruber IPNI=*796364-1* 248
Costus sanguineus IPNI=*796364-1* 248
Costus septimus IPNI=*796399-1* 248
Costus sextus IPNI=*796267-1* 247
Costus sinningiiflorus IPNI=*796308-1* 247
Costus skutchii IPNI=*796315-1* 247
Costus spiralis var. *hirsutus* IPNI=*796267-1* 247
Costus splendens IPNI=*796315-1* 247
Costus tatei IPNI=*796376-1* 248
Costus validus IPNI=*796267-1* 247
Costus verschaffeltianus IPNI=*796267-1* 247
Costus weberbaueri IPNI=*796315-1* 247
Costus zerumbet IPNI=*872083-1* 250
Cotinus metopium IPNI=*158605-2* 451
Cotoneaster amoenus IPNI=*722623-1* 443
Cotoneaster angustifolius IPNI=*730608-1* 444
Cotoneaster borealichinensis IPNI=*722592-1* 443
Cotoneaster crenulatus IPNI=*730613-1* 444
Cotoneaster fortunei IPNI=*730613-1* 444
Cotoneaster gamblei IPNI=*722495-1* 443
Cotoneaster lacteus IPNI=*722459-1* 443
Cotoneaster lidjiangensis IPNI=*722623-1* 443
Cotoneaster magnificus IPNI=*722592-1* 443
Cotoneaster multiflorus var. *borealichinensis* IPNI=*722592-1* 443
Cotoneaster oligocarpus IPNI=*722459-1* 443
Cotoneaster pyracantha IPNI=*730611-1* 444
Cotoneaster reflexus IPNI=*722592-1* 443
Cotoneaster smithii IPNI=*722459-1* 443
Cotoneaster vernae IPNI=*722623-1* 443
Cotoneaster wilsonii IPNI=*722592-1* 443
Cottea sarmentosa IPNI=*400626-1* 238
Cotula alba IPNI=*1186403-2* 261
Cotula debilis IPNI=*1186403-2* 261
Cotula dichotoma IPNI=*914778-1* 255
Cotula oederi IPNI=*1186403-2* 261
Cotula prostrata IPNI=*1186403-2* 261
Cotula pyretharia IPNI=*914778-1* 255
Cotula spilanthes IPNI=*250890-1* 270
Cotula viscosa IPNI=*1037395-2* 261
Coublandia frutescens IPNI=*77118402-1* 328
Couepia amazonica IPNI=*722799-1* 381
Couepia bracteosa f. *typica* IPNI=*722743-1* 381
Couepia bracteosa var. *grandifolia* IPNI=*722743-1* 381
Couepia bracteosa var. *minor* IPNI=*722743-1* 381
Couepia dolichopoda IPNI=*77141743-1* 381
Couepia longipendula IPNI=*77141742-1* 381
Couepia speciosa IPNI=*722774-1* 381
Coulteria chilensis IPNI=*249443-2* 335
Coulteria horrida IPNI=*249443-2* 335
Coulteria tinctoria IPNI=*249443-2* 335
Couma capiron IPNI=*67217-2* 340
Couma caurensis IPNI=*67217-2* 340
Couma dulcis IPNI=*78177-1* 340
Couma fasciculata IPNI=*182946-2* 342
Couma guatemalensis IPNI=*67217-2* 340
Couma mocuge IPNI=*78175-1* 340
Couma multinervis IPNI=*78177-1* 340
Couma oblonga IPNI=*182949-2* 342
Couma sapida IPNI=*67217-2* 340
Coumarouna micrantha IPNI=*492767-1* 318
Coumarouna odorata IPNI=*492769-1* 318
Coumarouna oleifera IPNI=*492770-1* 318
Coumarouna oppositifolia IPNI=*520179-1* 335
Coumarouna panamensis IPNI=*492770-1* 318
Coumarouna polyphylla IPNI=*82092-2* 318
Coumarouna punctata IPNI=*492778-1* 318
Coumarouna rosea IPNI=*492781-1* 319
Coumarouna trifoliolata IPNI=*492778-1* 318
Coupoui micrantha IPNI=*276143-2* 349
Couralia rosea IPNI=*111027-1* 365
Couralia toxophora IPNI=*109687-1* 363
Couratari aulacocarpa IPNI=*590593-1* 300
Couratari bragancae IPNI=*592371-1* 301
Couratari dictyocarpa IPNI=*590593-1* 300
Couratari lineata IPNI=*590593-1* 300
Couratari macrocarpa IPNI=*590593-1* 300
Couratari panamensis IPNI=*592371-1* 301
Couratari paraensis IPNI=*592371-1* 301
Couratari pedicellaris IPNI=*592385-1* 301
Couratari pulchra IPNI=*592371-1* 301

Couratari reticulata IPNI=*67276-2* 301
Couroupita acreensis IPNI=*592385-1* 301
Couroupita antillana IPNI=*592385-1* 301
Couroupita crenulata IPNI=*597315-1* 302
Couroupita cutteri IPNI=*592389-1* 301
Couroupita darienensis IPNI=*592389-1* 301
Couroupita froesii IPNI=*592385-1* 301
Couroupita guianensis var. *surinamensis* IPNI=*592385-1* 301
Couroupita idolica IPNI=*592385-1* 301
Couroupita lentula IPNI=*597315-1* 302
Couroupita magnifica IPNI=*592389-1* 301
Couroupita membranacea IPNI=*592385-1* 301
Couroupita odoratissima IPNI=*592389-1* 301
Couroupita parviflora IPNI=*592389-1* 301
Couroupita peruviana IPNI=*592385-1* 301
Couroupita surinamensis IPNI=*592385-1* 301
Couroupita venezuelensis IPNI=*592385-1* 301
Courrantia chamomilloides IPNI=*154715-2* 265
Coursetia arborea IPNI=*944647-1* 315
Coursetia caracasana IPNI=*944647-1* 315
Coussapoa apoda IPNI=*67344-2* 446
Coussapoa araneosa IPNI=*851294-1* 446
Coussapoa boliviana IPNI=*851294-1* 446
Coussapoa cayennensis IPNI=*851268-1* 446
Coussapoa danielis IPNI=*851294-1* 446
Coussapoa donnell-smithii IPNI=*851294-1* 446
Coussapoa eggersii IPNI=*851294-1* 446
Coussapoa embirana IPNI=*851294-1* 446
Coussapoa ficina IPNI=*851268-1* 446
Coussapoa intermedia IPNI=*851283-1* 446
Coussapoa laevigata IPNI=*851294-1* 446
Coussapoa lawrancei IPNI=*851294-1* 446
Coussapoa lehmannii IPNI=*851294-1* 446
Coussapoa macarenensis IPNI=*851294-1* 446
Coussapoa macarenensis var. *antioquiensis* IPNI=*851294-1* 446
Coussapoa martiana IPNI=*851294-1* 446
Coussapoa mutisii IPNI=*851294-1* 446
Coussapoa panamensis IPNI=*851294-1* 446
Coussapoa planitiensis IPNI=*851294-1* 446
Coussapoa prancei IPNI=*67367-2* 446
Coussapoa rekoi IPNI=*1059330-2* 441
Coussapoa schunkei IPNI=*851283-1* 446
Coussapoa standleyi IPNI=*851294-1* 446
Coussapoa subcrenata IPNI=*851294-1* 446
Coussapoa subincana IPNI=*851294-1* 446
Coussapoa vellerea IPNI=*851294-1* 446
Coussapoa williamsii IPNI=*67367-2* 446
Coussarea americana IPNI=*1187456-2* 350
Coussarea multiflora IPNI=*749186-1* 350
Coussarea neei IPNI=*67412-2* 349
Coussarea odoratissima IPNI=*1187456-2* 350
Coussarea scalaris IPNI=*749277-1* 350
Coussarea urophylla IPNI=*749277-1* 350
Coutarea campanilla IPNI=*67491-2* 349
Coutarea corymbosa IPNI=*67491-2* 349
Coutarea flavescens IPNI=*67491-2* 349
Coutarea hexandra f. *albiflora* IPNI=*67491-2* 349
Coutarea hexandra f. *grandiflora* IPNI=*67491-2* 349
Coutarea hexandra f. *pubescens* IPNI=*67491-2* 349
Coutarea hexandra f. *roseiflora* IPNI=*67491-2* 349
Coutarea hexandra f. *tarapotensis* IPNI=*67491-2* 349
Coutarea hexandra var. *amazonica* IPNI=*67491-2* 349
Coutarea hexandra var. *calycina* IPNI=*67491-2* 349
Coutarea hexandra var. *fluminensis* IPNI=*67491-2* 349
Coutarea hexandra var. *pubescens* IPNI=*67491-2* 349
Coutarea hexandra var. *speciosa* IPNI=*67491-2* 349
Coutarea hexandra var. *typica* IPNI=*67491-2* 349
Coutarea lindeniana IPNI=*67491-2* 349
Coutarea portlandia IPNI=*67491-2* 349
Coutarea pubescens IPNI=*67491-2* 349
Coutarea scherffiana IPNI=*67491-2* 349
Coutarea speciosa IPNI=*67491-2* 349
Coutoubea alba IPNI=*367036-1* 344
Coutoubea arenaria IPNI=*367033-1* 344
Coutoubea densiflora IPNI=*367036-1* 344
Coutoubea lutea IPNI=*367036-1* 344
Coutoubea purpurea IPNI=*367033-1* 344
Coutoubea racemosa IPNI=*367033-1* 344
Coutoubea ramosa f. *racemosa* IPNI=*367033-1* 344
Coutoubea ramosa f. *vulgaris* IPNI=*367033-1* 344
Coutoubea ramosa var. *latifolia* IPNI=*367033-1* 344
Coutoubea ramosa var. *longifolia* IPNI=*367033-1* 344
Coutoubea ramosa var. *racemosa* IPNI=*367033-1* 344
Coutoubea ramosa var. *vulgaris* IPNI=*367033-1* 344
Coutoubea ternifolia IPNI=*367036-1* 344
Cracca adunca IPNI=*520366-1* 335
Cracca benghalensis IPNI=*524572-1* 336
Cracca candida IPNI=*520444-1* 335
Cracca cinerea IPNI=*1071691-2* 335
Cracca littoralis IPNI=*1071691-2* 335
Cracca pauciflora IPNI=*524865-1* 336
Cracca purpurea IPNI=*520851-1* 335
Cracca schiedeana IPNI=*250423-2* 335
Cracca toxicaria IPNI=*250423-2* 335
Cracca villosa IPNI=*525259-1* 336
Cracca villosa var. *cinerea* IPNI=*1071691-2* 335
Cracca vogelii IPNI=*521002-1* 335
Craepalia temulenta IPNI=*407516-1* 241
Craepaloprumnon heterophyllum IPNI=*112178-1* 405
Craepaloprumnon prunifolium IPNI=*112146-1* 405
Craepaloprumnon racemosum IPNI=*112146-1* 405
Cranichis luteola IPNI=*654273-1* 222
Crantzia circinnata IPNI=*84424-2* 367
Crantzia coccinea IPNI=*84424-2* 367
Crantzia coriacea IPNI=*84425-2* 367
Crantzia lineata var. *subulata* IPNI=*140089-2* 252
Crantzia macloviana IPNI=*140089-2* 252
Crantzia ovata IPNI=*51409-2* 259
Crantzia varelana IPNI=*60441766-2* 367
Craspedaria cordifolia IPNI=*17406250-1* 176
Craspedaria surinamensis IPNI=*160142-2* 175
Craspedaria vacciniifolia IPNI=*17406250-1* 176
Crassina elegans IPNI=*261331-1* 272
Crassina intermedia IPNI=*261363-1* 272
Crassina leptopoda IPNI=*261363-1* 272
Crassina linearis IPNI=*261331-1* 272
Crassina multiflora IPNI=*261363-1* 272
Crassina peruviana IPNI=*261363-1* 272
Crassina peruviana var. *flava* IPNI=*261363-1* 272
Crassina tenuiflora IPNI=*261363-1* 272
Crassina verticillata IPNI=*261363-1* 272
Crassocephalum aurantiacum IPNI=*211219-1* 263
Crassocephalum flatmense IPNI=*962413-1* 260
Crassocephalum sonchifolium IPNI=*203080-1* 261
Crassocephalum valerianifolium IPNI=*203365-1* 261
Crassula scutellaria IPNI=*91817-1* 254
Crataegus bibas IPNI=*77209608-1* 444
Crataegus chinensis IPNI=*731389-1* 444
Crataegus crenulata IPNI=*730613-1* 444
Crataegus dumosa IPNI=*730611-1* 444
Crataegus escalloniifolia IPNI=*725484-1* 443
Crataegus excelsa IPNI=*30065762-2* 444
Crataegus ferruginea IPNI=*725473-1* 443
Crataegus indica IPNI=*731389-1* 444
Crataegus obtusifolia IPNI=*725484-1* 443
Crataegus pauciflora IPNI=*730611-1* 444

Crataegus pyracantha IPNI=*730611-1* 444
Crataegus rubra IPNI=*731389-1* 444
Crataegus serotina IPNI=*724549-1* 443
Crataegus sieboldii IPNI=*731389-1* 444
Crataegus sinensis IPNI=*731389-1* 444
Crateria capitata IPNI=*274165-2* 403
Crateria fasciculata IPNI=*47989-2* 403
Craterotecoma parviflora IPNI=*109921-1* 364
Crateva acuminata IPNI=*316363-2* 278
Crateva arborea IPNI=*316363-2* 278
Crateva bahiana IPNI=*316363-2* 278
Crateva benthamii IPNI=*316363-2* 278
Crateva benthamii var. *leptopetala* IPNI=*316363-2* 278
Crateva coriacea IPNI=*316363-2* 278
Crateva glauca IPNI=*316363-2* 278
Crateva gynandra IPNI=*316363-2* 278
Crateva radiatiflora IPNI=*316363-2* 278
Crateva tapia var. *glauca* IPNI=*316363-2* 278
Crateva tapioides IPNI=*316363-2* 278
Creatantha peruviana IPNI=*278892-2* 351
Cremanium ambiguum IPNI=*1110588-2* 423
Cremanium caraccasanum IPNI=*1155116-2* 422
Cremanium compressum IPNI=*572109-1* 419
Cremanium croceum IPNI=*572353-1* 420
Cremanium glaberrimum IPNI=*1110588-2* 423
Cremanium granulosum IPNI=*158784-2* 419
Cremanium latifolium IPNI=*572660-1* 421
Cremanium milleflorum IPNI=*1110588-2* 423
Cremanium minutiflorum IPNI=*1110588-2* 423
Cremanium nitidum IPNI=*572660-1* 421
Cremanium paludosum IPNI=*1110588-2* 423
Cremanium paniculatum IPNI=*1098452-2* 421
Cremanium rubens IPNI=*1098452-2* 421
Cremanium theezans IPNI=*1110588-2* 423
Cremanium trinerve IPNI=*159880-2* 423
Cremanium trinitatis IPNI=*1098452-2* 421
Cremastosperma anomalum IPNI=*1003565-1* 189
Crematomia attenuata IPNI=*1088317-2* 273
Crematomia elongata IPNI=*1088317-2* 273
Crematomia molliuscula IPNI=*1088317-2* 273
Crematomia revoluta IPNI=*1088317-2* 273
Crematomia velutina IPNI=*1088317-2* 273
Crematomia venosa IPNI=*1088317-2* 273
Cremosperma minutiflorum IPNI=*284180-2* 368
Crepidospermum goudotianum IPNI=*77183718-1* 452
Crepidospermum guyanense IPNI=*77184241-1* 453
Crepidospermum prancei IPNI=*77183907-1* 453
Crepidospermum rhoifolium IPNI=*77184241-1* 453
Crepidospermum sprucei IPNI=*77183718-1* 452
Crepinodendron crotonoides IPNI=*160368-2* 306
Crepis agrestis IPNI=*199753-1* 260
Crepis bauhiniana IPNI=*199753-1* 260
Crepis candollei IPNI=*199753-1* 260
Crepis capillaris f. *bischoffii* IPNI=*199753-1* 260
Crepis capillaris f. *integra* IPNI=*199753-1* 260
Crepis capillaris subsp. *arvensis* IPNI=*199753-1* 260
Crepis cooperi IPNI=*199753-1* 260
Crepis diffusa IPNI=*199753-1* 260
Crepis dioscoridis IPNI=*199753-1* 260
Crepis fastigiata IPNI=*1124129-2* 272
Crepis formosana IPNI=*1124129-2* 272
Crepis gaditana IPNI=*199753-1* 260
Crepis japonica IPNI=*1124129-2* 272
Crepis lanceolata IPNI=*199753-1* 260
Crepis leiosperma IPNI=*199753-1* 260
Crepis linifolia IPNI=*199753-1* 260
Crepis longe-pinnatifida IPNI=*199753-1* 260
Crepis lusitanica IPNI=*199753-1* 260
Crepis molokaiensis IPNI=*225575-1* 264
Crepis neglecta IPNI=*199753-1* 260
Crepis nemorum IPNI=*199753-1* 260
Crepis parviflora IPNI=*199753-1* 260
Crepis pinnatifida IPNI=*199753-1* 260
Crepis polymorpha IPNI=*199753-1* 260
Crepis pulchra IPNI=*199753-1* 260
Crepis sagittata IPNI=*199753-1* 260
Crepis taquetii IPNI=*1124129-2* 272
Crepis tectorum IPNI=*199753-1* 260
Crepis umbellata IPNI=*199753-1* 260
Crepis uniflora IPNI=*199753-1* 260
Crepis variabilis IPNI=*199753-1* 260
Crepis virens IPNI=*199753-1* 260
Crescentia aculeata IPNI=*110241-1* 364
Crescentia acuminata IPNI=*319161-2* 363
Crescentia angustifolia IPNI=*319161-2* 363
Crescentia arborea IPNI=*319161-2* 363
Crescentia coriacea IPNI=*108218-1* 362
Crescentia cucurbitifera IPNI=*108218-1* 362
Crescentia cucurbitina IPNI=*108218-1* 362
Crescentia cucurbitina var. *heterophylla* IPNI=*108218-1* 362
Crescentia cujete var. *puberula* IPNI=*319161-2* 363
Crescentia cuneifolia IPNI=*319161-2* 363
Crescentia cuspidata IPNI=*108218-1* 362
Crescentia edulis IPNI=*110241-1* 364
Crescentia elongata IPNI=*108218-1* 362
Crescentia fasciculata IPNI=*319161-2* 363
Crescentia latifolia IPNI=*108218-1* 362
Crescentia latifolia IPNI=*319161-2* 363
Crescentia lethifera IPNI=*108218-1* 362
Crescentia obovata IPNI=*108218-1* 362
Crescentia ovata IPNI=*108218-1* 362
Crescentia palustris IPNI=*108218-1* 362
Crescentia plectantha IPNI=*319161-2* 363
Crescentia pumila IPNI=*319161-2* 363
Crescentia spathulata IPNI=*319161-2* 363
Crescentia toxicaria IPNI=*108218-1* 362
Crinum africanum IPNI=*527258-1* 213
Crinum americanum subsp. *robustum* IPNI=*319172-2* 213
Crinum americanum var. *traubii* IPNI=*319172-2* 213
Crinum angustifolium IPNI=*77153983-1* 214
Crinum attenuatum IPNI=*30032850-2* 213
Crinum bequaertii IPNI=*63934-1* 213
Crinum capense IPNI=*63828-1* 213
Crinum caribaeum IPNI=*319172-2* 213
Crinum ceruleum IPNI=*319172-2* 213
Crinum commelinianum IPNI=*30032850-2* 213
Crinum commelyni IPNI=*30032850-2* 213
Crinum commelyni IPNI=*319172-2* 213
Crinum congolense IPNI=*63934-1* 213
Crinum conicum IPNI=*319172-2* 213
Crinum corantinum IPNI=*30032850-2* 213
Crinum corantynum IPNI=*30032850-2* 213
Crinum crucibulum IPNI=*30032850-2* 213
Crinum cruentum IPNI=*30032850-2* 213
Crinum dieterichii IPNI=*30032850-2* 213
Crinum erubescens var. *braziliense* IPNI=*30032850-2* 213
Crinum erubescens var. *caraccense* IPNI=*30032850-2* 213
Crinum erubescens var. *corantinum* IPNI=*30032850-2* 213
Crinum erubescens var. *octoflorum* IPNI=*30032850-2* 213
Crinum erubescens var. *rubrilimbum* IPNI=*30032850-2* 213
Crinum floridanum IPNI=*319172-2* 213
Crinum floridum IPNI=*527258-1* 213
Crinum giganteum IPNI=*63934-1* 213
Crinum gordonianum IPNI=*30032850-2* 213
Crinum govenium IPNI=*63828-1* 213
Crinum guianense IPNI=*30032850-2* 213

Crinum herbertianum IPNI=*319172-2* 213
Crinum inodorum IPNI=*30032850-2* 213
Crinum lanceanum IPNI=*30032850-2* 213
Crinum lancei IPNI=*30032850-2* 213
Crinum lindleyanum IPNI=*30032850-2* 213
Crinum loddigesianum IPNI=*30032850-2* 213
Crinum longiflorum IPNI=*319172-2* 213
Crinum octoflorum IPNI=*30032850-2* 213
Crinum odoratissimum IPNI=*30032850-2* 213
Crinum petiolatum IPNI=*63934-1* 213
Crinum podophyllum IPNI=*63934-1* 213
Crinum pumilum IPNI=*30032850-2* 213
Crinum revolutum IPNI=*30032850-2* 213
Crinum riparium IPNI=*63828-1* 213
Crinum roozenianum IPNI=*319172-2* 213
Crinum rubrilimbum IPNI=*30032850-2* 213
Crinum soboliferum IPNI=*30032850-2* 213
Crinum spectabile IPNI=*63934-1* 213
Crinum spofforthianum IPNI=*63828-1* 213
Crinum strictum IPNI=*319172-2* 213
Crinum strictum var. *traubii* IPNI=*319172-2* 213
Crinum suaveolens IPNI=*63934-1* 213
Crinum texanum IPNI=*319172-2* 213
Crinum vanillodorum IPNI=*63934-1* 213
Crinum viridifolium IPNI=*30032850-2* 213
Critesion muticum IPNI=*278265-2* 240
Critonia chrysocephala IPNI=*234386-1* 267
Critonia thyrsigera IPNI=*69558-2* 260
Critonia thyrsoidea IPNI=*69565-2* 260
Critoniopsis calerana IPNI=*200446-1* 260
Crocanthus croceus IPNI=*1011059-2* 280
Crocosmia × latifolia IPNI=*436427-1* 216
Crodisperma aspera IPNI=*993359-1* 271
Crossandra fascicularis IPNI=*55411-1* 361
Crossopetalum integrifolium IPNI=*133956-2* 442
Crossopetalum pallens IPNI=*60458018-2* 290
Crossopetalum rotundatum IPNI=*60458018-2* 290
Crotalaria acuminata IPNI=*489042-1* 316
Crotalaria alatipes IPNI=*322612-2* 316
Crotalaria albertiana IPNI=*487901-1* 316
Crotalaria altissima IPNI=*322616-2* 316
Crotalaria anagyroides IPNI=*322606-2* 316
Crotalaria anagyroides var. *angustifolia* IPNI=*322606-2* 316
Crotalaria anagyroides var. *minor* IPNI=*322606-2* 316
Crotalaria angulosa IPNI=*489042-1* 316
Crotalaria belizensis IPNI=*318914-2* 316
Crotalaria benghalensis IPNI=*322601-2* 316
Crotalaria bialata IPNI=*318914-2* 316
Crotalaria brachystachya IPNI=*322606-2* 316
Crotalaria bracteosa IPNI=*488624-1* 316
Crotalaria brasila IPNI=*322616-2* 316
Crotalaria caerulea IPNI=*489042-1* 316
Crotalaria coerulea IPNI=*489042-1* 316
Crotalaria dombeyana IPNI=*322606-2* 316
Crotalaria erecta IPNI=*488624-1* 316
Crotalaria espadilla IPNI=*488910-1* 316
Crotalaria exstipulata IPNI=*488559-1* 316
Crotalaria fenestrata IPNI=*322601-2* 316
Crotalaria flexuosa IPNI=*489042-1* 316
Crotalaria fruticosa IPNI=*318914-2* 316
Crotalaria fruticosa IPNI=*488660-1* 316
Crotalaria genistella IPNI=*488660-1* 316
Crotalaria hastata IPNI=*489042-1* 316
Crotalaria hexaptera IPNI=*488660-1* 316
Crotalaria intermedia IPNI=*487901-1* 316
Crotalaria leschenaultii IPNI=*322612-2* 316
Crotalaria lunulata IPNI=*318914-2* 316
Crotalaria lupiniflora IPNI=*322612-2* 316
Crotalaria lupinifolia IPNI=*322612-2* 316
Crotalaria macrophylla IPNI=*322612-2* 316
Crotalaria matthewsiana IPNI=*318914-2* 316
Crotalaria mollis IPNI=*489042-1* 316
Crotalaria nitens var. *trichina* IPNI=*488559-1* 316
Crotalaria nitidula IPNI=*488559-1* 316
Crotalaria paramariboensis IPNI=*489042-1* 316
Crotalaria parviflora IPNI=*318914-2* 316
Crotalaria parviflora var. *glabella* IPNI=*318914-2* 316
Crotalaria parviflora var. *hirsutissima* IPNI=*318914-2* 316
Crotalaria pilosa IPNI=*318914-2* 316
Crotalaria pilosa var. *skutchii* IPNI=*488660-1* 316
Crotalaria pilosa var. *typica* IPNI=*488660-1* 316
Crotalaria platycarpa IPNI=*318914-2* 316
Crotalaria poeppigii IPNI=*322616-2* 316
Crotalaria pohliana f. *intermedia* IPNI=*488910-1* 316
Crotalaria pohliana var. *prostrata* IPNI=*488910-1* 316
Crotalaria porrecta IPNI=*322601-2* 316
Crotalaria pringlei IPNI=*318914-2* 316
Crotalaria pterocaula IPNI=*488660-1* 316
Crotalaria pterocaula var. *genistella* IPNI=*488660-1* 316
Crotalaria purpureolineata IPNI=*487901-1* 316
Crotalaria racemosa IPNI=*488624-1* 316
Crotalaria retzii IPNI=*322612-2* 316
Crotalaria sagittalis var. *blumeriana* IPNI=*318914-2* 316
Crotalaria sagittalis var. *espadilla* IPNI=*488910-1* 316
Crotalaria sagittalis var. *fruticosa* IPNI=*318914-2* 316
Crotalaria sagittalis var. *oblonga* IPNI=*318914-2* 316
Crotalaria sagittalis var. *typica* IPNI=*318914-2* 316
Crotalaria sericea IPNI=*322601-2* 316
Crotalaria sericea IPNI=*322612-2* 316
Crotalaria serpyllifolia IPNI=*488910-1* 316
Crotalaria spectabilis subsp. *parvibracteata* IPNI=*322612-2* 316
Crotalaria stipularia f. *elliptica* IPNI=*488910-1* 316
Crotalaria stipularia f. *intermedia* IPNI=*488910-1* 316
Crotalaria stipularia f. *oblongifolia* IPNI=*488910-1* 316
Crotalaria stipularia var. *genuina* IPNI=*488910-1* 316
Crotalaria stipularia var. *grandifolia* IPNI=*488910-1* 316
Crotalaria stipularia var. *longepedunculata* IPNI=*488910-1* 316
Crotalaria stipularia var. *oblongata* IPNI=*488910-1* 316
Crotalaria stipularia var. *prostrata* IPNI=*488910-1* 316
Crotalaria stipularia var. *sericea* IPNI=*488910-1* 316
Crotalaria stipularia var. *serpyllifolia* IPNI=*488910-1* 316
Crotalaria stipularia var. *typica* IPNI=*488910-1* 316
Crotalaria stipulata IPNI=*322606-2* 316
Crotalaria tenuifolia IPNI=*322601-2* 316
Crotalaria triphylla IPNI=*322606-2* 316
Crotalaria tuerckheimii IPNI=*318914-2* 316
Crotalaria tuerckheimii var. *macrantha* IPNI=*318914-2* 316
Crotalaria undulata IPNI=*318914-2* 316
Crotalaria urbaniana IPNI=*488624-1* 316
Crotalaria virgata IPNI=*488624-1* 316
Crotalaria vitellina var. *glabrata* IPNI=*322616-2* 316
Crotalaria vitellina var. *minor* IPNI=*322616-2* 316
Crotalaria vitellina var. *oblongifolia* IPNI=*322616-2* 316
Crotalaria vitellina var. *schippii* IPNI=*322616-2* 316
Crotalaria vitellina var. *typica* IPNI=*322616-2* 316
Croton albidus IPNI=*342537-1* 388
Croton allenii IPNI=*325164-2* 388
Croton althaeifolius IPNI=*342537-1* 388
Croton argothamnoides IPNI=*995988-1* 389
Croton argyrophylloides IPNI=*343663-1* 389
Croton astroites IPNI=*342537-1* 388
Croton atwoodianus IPNI=*69894-2* 388
Croton balsamifer IPNI=*342537-1* 388
Croton benthamianus IPNI=*343150-1* 389
Croton benthamianus IPNI=*70513-2* 389
Croton bilocularis IPNI=*342381-1* 388

Croton bixoides IPNI=*70536-2* 389
Croton buchtienii IPNI=*71013-2* 389
Croton caryophyllus IPNI=*70513-2* 389
Croton castaneifolius IPNI=*44529-2* 388
Croton chamaedrifolius IPNI=*1012504-2* 387
Croton chamaedrifolius IPNI=*71013-2* 389
Croton choristolepis IPNI=*70536-2* 389
Croton cienagensis IPNI=*342550-1* 388
Croton citrifolius IPNI=*343083-1* 389
Croton comes IPNI=*70337-2* 389
Croton confusus IPNI=*70274-2* 388
Croton courtetii IPNI=*339515-1* 388
Croton curassavicus IPNI=*325164-2* 388
Croton cyathodenus IPNI=*70513-2* 389
Croton decorsei IPNI=*339515-1* 388
Croton dentatus IPNI=*70599-2* 389
Croton deppeanus IPNI=*70562-2* 389
Croton diffusus IPNI=*325185-2* 389
Croton digitatus IPNI=*70274-2* 388
Croton dolichostachyus IPNI=*342854-1* 389
Croton draco var. *cordatus* IPNI=*342842-1* 389
Croton draco var. *hibisciformis* IPNI=*342669-1* 388
Croton draco var. *rhombifolius* IPNI=*70274-2* 388
Croton eluteria IPNI=*343083-1* 389
Croton escathos IPNI=*325185-2* 389
Croton flabellifolius IPNI=*70274-2* 388
Croton flavens var. *balsamifer* IPNI=*342537-1* 388
Croton flavens var. *genuinus* IPNI=*342537-1* 388
Croton flavens var. *mucronatus* IPNI=*342537-1* 388
Croton flavens var. *pallidus* IPNI=*342537-1* 388
Croton flavens var. *rigidus* IPNI=*342537-1* 388
Croton flavescens IPNI=*70562-2* 389
Croton flavescens var. *brandegeeanus* IPNI=*70562-2* 389
Croton flocculosus IPNI=*342537-1* 388
Croton foetens IPNI=*70536-2* 389
Croton fragilis IPNI=*342550-1* 388
Croton fragilis var. *genuinus* IPNI=*342550-1* 388
Croton fragilis var. *sericeus* IPNI=*70562-2* 389
Croton funckianus IPNI=*342669-1* 388
Croton funckianus var. *hibisciformis* IPNI=*342669-1* 388
Croton funckianus var. *rhombifolius* IPNI=*70274-2* 388
Croton glandulosus subsp. *hirtus* IPNI=*70337-2* 389
Croton glandulosus var. *cordovensis* IPNI=*70337-2* 389
Croton glandulosus var. *genuinus* IPNI=*70233-2* 388
Croton glandulosus var. *hirtus* IPNI=*70337-2* 389
Croton glandulosus var. *martii* IPNI=*70337-2* 389
Croton glandulosus var. *sagotii* IPNI=*70337-2* 389
Croton glandulosus var. *schomburgkianus* IPNI=*70337-2* 389
Croton glandulosus var. *sieberi* IPNI=*70337-2* 389
Croton glandulosus var. *tenellus* IPNI=*70337-2* 389
Croton gossypiifolius var. *genuinus* IPNI=*70274-2* 388
Croton gossypiifolius var. *heterophyllus* IPNI=*70274-2* 388
Croton gossypiifolius var. *hibiscifolius* IPNI=*342669-1* 388
Croton gossypiifolius var. *stipularis* IPNI=*70274-2* 388
Croton helicoideus IPNI=*70536-2* 389
Croton heterophyllus IPNI=*70274-2* 388
Croton hispidus IPNI=*325164-2* 388
Croton impetiginosus IPNI=*70513-2* 389
Croton integer IPNI=*69894-2* 388
Croton isertii IPNI=*325164-2* 388
Croton juigalpensis IPNI=*70562-2* 389
Croton kaieteuri IPNI=*342381-1* 388
Croton kavanayensis IPNI=*70513-2* 389
Croton lamarckianus IPNI=*342537-1* 388
Croton lanjouwensis IPNI=*70513-2* 389
Croton leiocarpus IPNI=*339515-1* 388
Croton leprosus IPNI=*342537-1* 388
Croton leptostachyus var. *genuinus* IPNI=*342854-1* 389
Croton leptostachyus var. *malacophyllus* IPNI=*342854-1* 389
Croton leptostachyus var. *spathulatus* IPNI=*342854-1* 389
Croton leptostachyus var. *truxillanus* IPNI=*342854-1* 389
Croton lobatus IPNI=*339515-1* 388
Croton lobatus var. *genuinus* IPNI=*339515-1* 388
Croton lobatus var. *intermedius* IPNI=*339515-1* 388
Croton lobatus var. *palmatus* IPNI=*339515-1* 388
Croton lobatus var. *seemannii* IPNI=*339515-1* 388
Croton lobatus var. *sericeus* IPNI=*339515-1* 388
Croton magdalenensis IPNI=*343058-1* 389
Croton malacophyllus IPNI=*342854-1* 389
Croton margaritensis IPNI=*325164-2* 388
Croton marginatus IPNI=*325185-2* 389
Croton martii IPNI=*342381-1* 388
Croton martii var. *latifolius* IPNI=*342381-1* 388
Croton martii var. *longifolius* IPNI=*342381-1* 388
Croton matourensis var. *benthamianus* IPNI=*70513-2* 389
Croton matourensis var. *genuinus* IPNI=*70513-2* 389
Croton matourensis var. *poeppigianus* IPNI=*70513-2* 389
Croton matourensis var. *sericeus* IPNI=*70513-2* 389
Croton mauralis IPNI=*342537-1* 388
Croton menthodorus var. *riveti* IPNI=*342979-1* 389
Croton meridensis IPNI=*342329-1* 388
Croton micans var. *genuinus* IPNI=*70536-2* 389
Croton mimeticus IPNI=*342381-1* 388
Croton miquelensis IPNI=*71013-2* 389
Croton mocinoi IPNI=*339515-1* 388
Croton mollii IPNI=*70599-2* 389
Croton monachinoensis IPNI=*342381-1* 388
Croton morifolius var. *brandegeeanus* IPNI=*70562-2* 389
Croton morifolius var. *genuinus* IPNI=*70562-2* 389
Croton morifolius var. *lanatus* IPNI=*70562-2* 389
Croton morifolius var. *sphaerocarpus* IPNI=*70562-2* 389
Croton mucronatus IPNI=*342537-1* 388
Croton nervosus IPNI=*44529-2* 388
Croton niveus IPNI=*70536-2* 389
Croton obtusus IPNI=*342329-1* 388
Croton ochromifolius IPNI=*70274-2* 388
Croton odoratus IPNI=*325164-2* 388
Croton ovalifolius var. *genuinus* IPNI=*325185-2* 389
Croton padifolius IPNI=*342537-1* 388
Croton pakaraimae IPNI=*70513-2* 389
Croton palustris IPNI=*44529-2* 388
Croton palustris IPNI=*44564-2* 388
Croton perobtusus IPNI=*70846-2* 389
Croton polycarpus IPNI=*343058-1* 389
Croton polystachyus IPNI=*70513-2* 389
Croton populifolius IPNI=*325164-2* 388
Croton populifolius IPNI=*70599-2* 389
Croton populifolius var. *genuinus* IPNI=*70599-2* 389
Croton portoricensis IPNI=*342537-1* 388
Croton prunifolius IPNI=*343083-1* 389
Croton pseudochina IPNI=*70599-2* 389
Croton pseudochina var. *minor* IPNI=*70599-2* 389
Croton rhamnifolius IPNI=*342329-1* 388
Croton rhamnifolius var. *antillanus* IPNI=*342537-1* 388
Croton rhamnifolius var. *caudatus* IPNI=*70562-2* 389
Croton rhamnifolius var. *conduplicatus* IPNI=*342329-1* 388
Croton rhamnifolius var. *genuinus* IPNI=*342329-1* 388
Croton rhamnifolius var. *salviifolius* IPNI=*342329-1* 388
Croton richardii IPNI=*342537-1* 388
Croton rigidus IPNI=*342537-1* 388
Croton rufoargenteus IPNI=*343663-1* 389
Croton salviifolius IPNI=*342329-1* 388
Croton sanguifluus IPNI=*70274-2* 388
Croton scandens IPNI=*315307-2* 393
Croton septemnervius IPNI=*70599-2* 389
Croton sericeus IPNI=*70513-2* 389

Croton sericeus IPNI=*70562-2* 389
Croton sessiliflorus IPNI=*356125-1* 402
Croton sloanei IPNI=*343083-1* 389
Croton sphaerocarpus IPNI=*70562-2* 389
Croton spinosus IPNI=*355498-1* 392
Croton squamulosus IPNI=*343083-1* 389
Croton strigosus IPNI=*325185-2* 389
Croton subcoriaceus IPNI=*342381-1* 388
Croton surinamensis IPNI=*342381-1* 388
Croton syringifolius IPNI=*70599-2* 389
Croton tomentosus IPNI=*342537-1* 388
Croton tragioides IPNI=*71013-2* 389
Croton tricolor IPNI=*343663-1* 389
Croton truxillanus IPNI=*342854-1* 389
Croton variegatus IPNI=*85073-3* 388
Croton venezuelensis IPNI=*70274-2* 388
Crozophyla variegata IPNI=*85073-3* 388
Crucianella purpurea IPNI=*30007294-2* 350
Cruciata ciliata IPNI=*747959-1* 349
Cruciata hirsuta IPNI=*747959-1* 349
Cruciata laevipes var. *brutia* IPNI=*747959-1* 349
Crucifera × cheiri IPNI=*283703-1* 277
Crucifera altissima IPNI=*289144-1* 278
Crucifera brassica IPNI=*279435-1* 276
Crucifera capsella IPNI=*30092589-2* 276
Crucifera cardamine IPNI=*280380-1* 276
Crucifera erratica IPNI=*1061021-2* 277
Crucifera eruca IPNI=*1137625-2* 277
Crucifera fontana IPNI=*287534-1* 277
Crucifera iberis IPNI=*324660-2* 277
Crucifera irio IPNI=*237940-2* 278
Crucifera juncea IPNI=*60442520-2* 276
Crucifera koniga IPNI=*1085072-2* 277
Crucifera lampsana IPNI=*288952-1* 278
Crucifera lunaria IPNI=*286647-1* 277
Crucifera napus IPNI=*279419-1* 276
Crucifera palustris IPNI=*288669-1* 277
Crucifera praecox IPNI=*1109261-2* 276
Crucifera rapa IPNI=*279485-1* 276
Crucifera raphanistrum IPNI=*217826-2* 277
Crucifera rugosa IPNI=*1061021-2* 277
Crucifera senebiera IPNI=*286148-1* 277
Crucifera sinapis IPNI=*279422-1* 276
Crucifera sisymbrium IPNI=*289589-1* 278
Crucifera thlaspi IPNI=*290477-1* 278
Crucifera umbellata IPNI=*324671-2* 277
Crucifera virginica IPNI=*138164-2* 277
Crucita hispanica IPNI=*129900-2* 283
Cruminium giganteum IPNI=*485609-1* 314
Cruminium virginianum IPNI=*51897-2* 314
Cruzeta americana IPNI=*129900-2* 283
Cruzeta celosioides IPNI=*129956-2* 283
Cruzeta crassifolia IPNI=*34033-2* 282
Cruzeta gossypiantha IPNI=*129900-2* 283
Cruzeta hispanica IPNI=*129900-2* 283
Cruzeta vermicularis IPNI=*34033-2* 282
Cryphiacanthus angustifolius IPNI=*54585-1* 361
Cryphiacanthus barbadensis IPNI=*54671-1* 361
Cryphiospermum repens IPNI=*203207-1* 261
Crypsis juncea IPNI=*1103484-2* 245
Crypsis maritima IPNI=*423010-1* 245
Crypsis virginica IPNI=*423010-1* 245
Cryptanthus chinensis IPNI=*946989-1* 369
Cryptocarpus globosus IPNI=*605135-1* 287
Cryptocarpus paniculatus IPNI=*605135-1* 287
Cryptocarpus purpurascens IPNI=*605135-1* 287
Cryptocarpus rhomboideus IPNI=*605135-1* 287
Cryptocarya canelilla IPNI=*13816-2* 182
Cryptocarya hirsuta IPNI=*90974-2* 183
Cryptocarya kostermansiana IPNI=*1057986-2* 182
Cryptocarya pretiosa IPNI=*13816-2* 182
Cryptoloma anonymum IPNI=*381638-1* 367
Cryptoloma coloratum IPNI=*381638-1* 367
Cryptoloma cordifolium IPNI=*381638-1* 367
Cryptoloma pictum IPNI=*381638-1* 367
Cryptoloma pilosum IPNI=*381638-1* 367
Cryptoloma rhynchocarpum IPNI=*381638-1* 367
Cryptoloma strictum IPNI=*381638-1* 367
Cryptomeria compacta IPNI=*261870-1* 180
Cryptomeria generalis IPNI=*261870-1* 180
Cryptomeria japonica subsp. *sinensis* IPNI=*261870-1* 180
Cryptomeria kawaii IPNI=*261870-1* 180
Cryptomeria lobbiana IPNI=*261870-1* 180
Cryptomeria lobbii IPNI=*261870-1* 180
Cryptomeria mairei IPNI=*261870-1* 180
Cryptomeria mucronata IPNI=*261870-1* 180
Cryptomeria nigricans IPNI=*261870-1* 180
Cryptomeria pungens IPNI=*261870-1* 180
Cryptomeria variegata IPNI=*261870-1* 180
Cryptomeria viridis IPNI=*261870-1* 180
Cryptostegia glaberrima IPNI=*95963-1* 340
Cryptostegia madagascariensis var. *glaberrima* IPNI=*95963-1* 340
Cryptostegia madagascariensis var. *septentrionalis* IPNI=*95963-1* 340
Cryptostomum coriaceum IPNI=*690710-1* 338
Cryptostomum guianense IPNI=*690710-1* 338
Cryptostomum laurifolium IPNI=*690710-1* 338
Ctenitis funesta IPNI=*286436-2* 177
Ctenitis funestra IPNI=*286436-2* 177
Ctenitis protensa var. *funesta* IPNI=*286436-2* 177
Ctenopteris vulgaris IPNI=*300720-2* 176
Cuatrecasea spruceana IPNI=*667622-1* 210
Cuatrecasea vaupesana IPNI=*667622-1* 210
Cuba paniculata IPNI=*520138-1* 335
Cuba trigona IPNI=*520138-1* 335
Cubelium parviflorum IPNI=*77143568-1* 405
Cubospermum palustre IPNI=*144324-2* 429
Cucubalus fasciculatus IPNI=*77093328-1* 285
Cucubalus glaucus IPNI=*77093328-1* 285
Cucubalus sylvestris IPNI=*332106-2* 286
Cucubalus variegatus IPNI=*332106-2* 286
Cucullaria excelsa IPNI=*266306-2* 430
Cucullaria ferruginea IPNI=*266306-2* 430
Cucumeropsis edulis IPNI=*77110571-1* 294
Cucumeropsis mackennii IPNI=*77110571-1* 294
Cucumeropsis mannii IPNI=*77110571-1* 294
Cucumis acidus IPNI=*292238-1* 293
Cucumis acutangulus IPNI=*105280-3* 294
Cucumis agrestis IPNI=*292238-1* 293
Cucumis alba IPNI=*292238-1* 293
Cucumis amarissimus IPNI=*291938-1* 293
Cucumis ambiguus IPNI=*292238-1* 293
Cucumis anguria subsp. *longipes* IPNI=*292141-1* 293
Cucumis anguriοides IPNI=*292141-1* 293
Cucumis argyi IPNI=*293413-1* 295
Cucumis aromaticus IPNI=*292238-1* 293
Cucumis bardanus IPNI=*292238-1* 293
Cucumis bicirrha IPNI=*134809-2* 294
Cucumis bisexualis IPNI=*292238-1* 293
Cucumis campechianus IPNI=*292238-1* 293
Cucumis cantalupensis IPNI=*292238-1* 293
Cucumis cantalupo IPNI=*292238-1* 293
Cucumis chate IPNI=*292238-1* 293
Cucumis chinensis IPNI=*292238-1* 293
Cucumis chito IPNI=*292238-1* 293
Cucumis cicatrisatus IPNI=*292238-1* 293
Cucumis citrullus IPNI=*291938-1* 293

Cucumis cognata IPNI=*292238-1* 293
Cucumis collosus IPNI=*292238-1* 293
Cucumis conomon IPNI=*292238-1* 293
Cucumis crassiflorus IPNI=*293902-1* 295
Cucumis cubensis IPNI=*292238-1* 293
Cucumis cubensis var. *costatus* IPNI=*292238-1* 293
Cucumis deliciosus IPNI=*292238-1* 293
Cucumis dissectus IPNI=*291938-1* 293
Cucumis dudaim IPNI=*292238-1* 293
Cucumis echinatus IPNI=*292141-1* 293
Cucumis edulis IPNI=*291938-1* 293
Cucumis erinaceus IPNI=*292141-1* 293
Cucumis eriocarpus IPNI=*292238-1* 293
Cucumis erivanicus IPNI=*292238-1* 293
Cucumis esculentus IPNI=*292296-1* 293
Cucumis flexuosus IPNI=*292238-1* 293
Cucumis fricatorius IPNI=*293097-1* 294
Cucumis glaber IPNI=*293293-1* 294
Cucumis hardwickii IPNI=*292296-1* 293
Cucumis intermedius IPNI=*293413-1* 295
Cucumis jamaicensis IPNI=*292238-1* 293
Cucumis jucunda IPNI=*292238-1* 293
Cucumis laciniosus IPNI=*291938-1* 293
Cucumis laevigatus IPNI=*292238-1* 293
Cucumis lagenaria IPNI=*134809-2* 294
Cucumis lineatus IPNI=*105280-3* 294
Cucumis longipes IPNI=*292141-1* 293
Cucumis maculatus IPNI=*292238-1* 293
Cucumis mairei IPNI=*134809-2* 294
Cucumis megacarpus IPNI=*105280-3* 294
Cucumis melo f. *agrestis* IPNI=*292238-1* 293
Cucumis melo subsp. *agrestis* IPNI=*292238-1* 293
Cucumis melo var. *agrestis* IPNI=*292238-1* 293
Cucumis melo var. *conomon* IPNI=*292238-1* 293
Cucumis melo var. *cultus* IPNI=*292238-1* 293
Cucumis microcarpus IPNI=*292238-1* 293
Cucumis microspermus IPNI=*292238-1* 293
Cucumis momordica IPNI=*292238-1* 293
Cucumis moschatus IPNI=*292238-1* 293
Cucumis muricatus IPNI=*292296-1* 293
Cucumis odoratissimus IPNI=*292238-1* 293
Cucumis odoratissimus IPNI=*293902-1* 295
Cucumis officinarum-melo IPNI=*292238-1* 293
Cucumis operculatus IPNI=*105280-3* 294
Cucumis pancherianus IPNI=*292238-1* 293
Cucumis parviflorus IPNI=*292141-1* 293
Cucumis pavel IPNI=*291972-1* 293
Cucumis pedatifidus IPNI=*292238-1* 293
Cucumis pentandrus IPNI=*293060-1* 294
Cucumis pepo IPNI=*292416-1* 294
Cucumis persicodorus IPNI=*292238-1* 293
Cucumis persicus IPNI=*292238-1* 293
Cucumis pictus IPNI=*292238-1* 293
Cucumis princeps IPNI=*292238-1* 293
Cucumis pseudocolocynthis IPNI=*292238-1* 293
Cucumis pubescens IPNI=*292238-1* 293
Cucumis purpureus IPNI=*291938-1* 293
Cucumis pyriformis IPNI=*292238-1* 293
Cucumis rapallito IPNI=*30006538-2* 293
Cucumis reflexus IPNI=*292238-1* 293
Cucumis reginae IPNI=*292238-1* 293
Cucumis rumphii IPNI=*292296-1* 293
Cucumis sativus f. *albus* IPNI=*292296-1* 293
Cucumis sativus f. *hardwickii* IPNI=*292296-1* 293
Cucumis sativus subsp. *hardwickii* IPNI=*292296-1* 293
Cucumis sativus var. *xishuangbannanensis* IPNI=*292296-1* 293
Cucumis schraderianus IPNI=*292238-1* 293
Cucumis sepium IPNI=*1076251-2* 294
Cucumis serotinus IPNI=*292238-1* 293
Cucumis trigonus IPNI=*292238-1* 293
Cucumis turbinatus IPNI=*292238-1* 293
Cucumis umbilicatus IPNI=*292238-1* 293
Cucumis utilissimus IPNI=*292238-1* 293
Cucumis villosus IPNI=*292238-1* 293
Cucumis vulgaris IPNI=*291938-1* 293
Cucumis zapallo IPNI=*292416-1* 294
Cucurbita acutangula IPNI=*105280-3* 294
Cucurbita anguria IPNI=*291938-1* 293
Cucurbita aspera IPNI=*292238-1* 293
Cucurbita aurantia IPNI=*292416-1* 294
Cucurbita caffra IPNI=*291938-1* 293
Cucurbita campanulata IPNI=*105280-3* 294
Cucurbita ceratoceras IPNI=*292416-1* 294
Cucurbita ciceraria IPNI=*134809-2* 294
Cucurbita citrullus IPNI=*291938-1* 293
Cucurbita clodiensis IPNI=*292416-1* 294
Cucurbita colombiana IPNI=*320034-2* 293
Cucurbita courgero IPNI=*292416-1* 294
Cucurbita dioica IPNI=*291972-1* 293
Cucurbita elongata IPNI=*292416-1* 294
Cucurbita esculenta IPNI=*292416-1* 294
Cucurbita euodicarpa IPNI=*293902-1* 295
Cucurbita farinae IPNI=*30006538-2* 293
Cucurbita fastuosa IPNI=*292416-1* 294
Cucurbita ficifolia f. *melanosperma* IPNI=*292369-1* 293
Cucurbita gigantea IPNI=*291938-1* 293
Cucurbita hippopera IPNI=*320034-2* 293
Cucurbita hybrida IPNI=*292416-1* 294
Cucurbita idololatrica IPNI=*134809-2* 294
Cucurbita lagenaria IPNI=*134809-2* 294
Cucurbita leucantha IPNI=*134809-2* 294
Cucurbita lignosa IPNI=*292416-1* 294
Cucurbita longa IPNI=*134809-2* 294
Cucurbita macrocarpa IPNI=*320034-2* 293
Cucurbita mammeata IPNI=*292416-1* 294
Cucurbita mammosa IPNI=*292416-1* 294
Cucurbita marsupiiformis IPNI=*292416-1* 294
Cucurbita maxima var. *triloba* IPNI=*30006538-2* 293
Cucurbita maxima var. *turgida* IPNI=*30006538-2* 293
Cucurbita maxima var. *zapallito* IPNI=*30006538-2* 293
Cucurbita maxima var. *zipinka* IPNI=*30006538-2* 293
Cucurbita meloniformis IPNI=*320034-2* 293
Cucurbita mexicana IPNI=*292369-1* 293
Cucurbita moschata var. *claviformis* IPNI=*320034-2* 293
Cucurbita moschata var. *colombiana* IPNI=*320034-2* 293
Cucurbita moschata var. *depressa* IPNI=*320034-2* 293
Cucurbita moschata var. *magna* IPNI=*320034-2* 293
Cucurbita moschata var. *meloniformis* IPNI=*320034-2* 293
Cucurbita oblonga IPNI=*292416-1* 294
Cucurbita odorifera IPNI=*293902-1* 295
Cucurbita pepo subsp. *gumala* IPNI=*292416-1* 294
Cucurbita pepo subvar. *alba* IPNI=*292416-1* 294
Cucurbita pepo subvar. *nigra* IPNI=*292416-1* 294
Cucurbita pepo var. *americana* IPNI=*292416-1* 294
Cucurbita pepo var. *bicolor* IPNI=*292416-1* 294
Cucurbita pepo var. *elongata* IPNI=*292416-1* 294
Cucurbita pepo var. *flogra* IPNI=*292416-1* 294
Cucurbita pepo var. *georgica* IPNI=*292416-1* 294
Cucurbita pepo var. *praecox* IPNI=*292416-1* 294
Cucurbita pepo var. *vulgaris* IPNI=*292416-1* 294
Cucurbita pileiformis IPNI=*30006538-2* 293
Cucurbita pinnatifida IPNI=*291938-1* 293
Cucurbita polymorpha IPNI=*292416-1* 294
Cucurbita pomiformis IPNI=*292416-1* 294
Cucurbita pyridaris IPNI=*292416-1* 294
Cucurbita pyriformis IPNI=*134809-2* 294

Cucurbita pyxidaris IPNI=*292416-1* 294
Cucurbita rapallito IPNI=*30006538-2* 293
Cucurbita schimperiana IPNI=*291972-1* 293
Cucurbita siceraria IPNI=*134809-2* 294
Cucurbita succado IPNI=*292416-1* 294
Cucurbita succedo IPNI=*292416-1* 294
Cucurbita sulcata IPNI=*30006538-2* 293
Cucurbita tuberculosa IPNI=*292416-1* 294
Cucurbita turbaniformis IPNI=*30006538-2* 293
Cucurbita umbellata IPNI=*105280-3* 294
Cucurbita urnigera IPNI=*292416-1* 294
Cucurbita variegata IPNI=*292416-1* 294
Cucurbita venosa IPNI=*292416-1* 294
Cucurbita vittata IPNI=*134809-2* 294
Cucurbitella ecuadorensis IPNI=*77110571-1* 294
Cuervea granatensis IPNI=*72123-2* 290
Cuervea latifolia IPNI=*72123-2* 290
Cuervea megacarpa IPNI=*72123-2* 290
Cuervea mitchellae IPNI=*72123-2* 290
Culcitium boyacense IPNI=*200642-1* 260
Culcitium canescens var. *monocephalum* IPNI=*200642-1* 260
Culcitium nitidum IPNI=*963597-1* 267
Culcitium ovatum IPNI=*228818-1* 264
Culcitium reflexum IPNI=*228818-1* 264
Culcitium uniflorum IPNI=*228818-1* 264
Cumetea divaricata IPNI=*599170-1* 427
Cumetea fragrans IPNI=*166091-2* 427
Cumetea mini IPNI=*593768-1* 425
Cumetea montana IPNI=*77160848-1* 426
Cumetea tomentosa IPNI=*599215-1* 427
Cuminum aegyptiacum IPNI=*840882-1* 252
Cuminum aethiopicum IPNI=*837486-1* 251
Cuminum hispanicum IPNI=*840882-1* 252
Cuminum odorum IPNI=*840882-1* 252
Cuminum officinale IPNI=*840882-1* 252
Cuminum regium IPNI=*837486-1* 251
Cuminum sativum IPNI=*840882-1* 252
Cunuria australis IPNI=*159993-2* 392
Cunuria bracteosa IPNI=*159993-2* 392
Cunuria crassipes IPNI=*159994-2* 392
Cunuria glabra IPNI=*159993-2* 392
Cunuria spruceana IPNI=*159993-2* 392
Cunuria spruceana var. *bracteosa* IPNI=*159993-2* 392
Cupamenis chamedrifolia IPNI=*1012504-2* 387
Cupamenis reptans IPNI=*1012504-2* 387
Cuparilla sophorina IPNI=*470782-1* 310
Cuphea anisophylla IPNI=*552922-1* 416
Cuphea antisyphilitica IPNI=*552922-1* 416
Cuphea apetala IPNI=*553111-1* 417
Cuphea balsamona IPNI=*72326-2* 416
Cuphea beneradicata IPNI=*72437-2* 417
Cuphea calophylla f. *breuningii* IPNI=*552922-1* 416
Cuphea calophylla f. *deformis* IPNI=*552922-1* 416
Cuphea calophylla f. *plumbaginea* IPNI=*552922-1* 416
Cuphea calophylla var. *microstyla* IPNI=*552922-1* 416
Cuphea calophylla var. *orthodisca* IPNI=*552922-1* 416
Cuphea ciliata f. *vestita* IPNI=*552931-1* 416
Cuphea corymbifera IPNI=*290454-2* 417
Cuphea decandra var. *purpusii* IPNI=*552959-1* 416
Cuphea divaricata IPNI=*72326-2* 416
Cuphea eminens IPNI=*553111-1* 417
Cuphea euphorbioides IPNI=*553109-1* 417
Cuphea fruticosa IPNI=*290454-2* 417
Cuphea fruticulosa IPNI=*290454-2* 417
Cuphea hirta IPNI=*553109-1* 417
Cuphea hirtella IPNI=*552959-1* 416
Cuphea hyssopifolia f. *subrevoluta* IPNI=*553043-1* 417
Cuphea hyssopifolia var. *brachyphylla* IPNI=*553043-1* 417
Cuphea liebmannii IPNI=*553048-1* 417
Cuphea loxensis IPNI=*552931-1* 416
Cuphea micrantha var. *longiciliata* IPNI=*553109-1* 417
Cuphea microphylla IPNI=*552931-1* 416
Cuphea microstyla IPNI=*552922-1* 416
Cuphea nudiflora IPNI=*290454-2* 417
Cuphea obtusifolia IPNI=*290454-2* 417
Cuphea origanifolia IPNI=*290454-2* 417
Cuphea origanifolia var. *minor* IPNI=*290454-2* 417
Cuphea origanifolia var. *sublanata* IPNI=*290454-2* 417
Cuphea orthodisca IPNI=*552922-1* 416
Cuphea peplidioides IPNI=*72326-2* 416
Cuphea platycentra IPNI=*553048-1* 417
Cuphea plumbaginea IPNI=*552922-1* 416
Cuphea prunellifolia IPNI=*72326-2* 416
Cuphea purpusii IPNI=*552959-1* 416
Cuphea racemosa f. *corymbifera* IPNI=*290454-2* 417
Cuphea racemosa f. *divergens* IPNI=*290454-2* 417
Cuphea racemosa f. *grandifolia* IPNI=*290454-2* 417
Cuphea racemosa f. *nudiflora* IPNI=*290454-2* 417
Cuphea racemosa f. *parvifolia* IPNI=*290454-2* 417
Cuphea racemosa f. *polysperma* IPNI=*290454-2* 417
Cuphea racemosa f. *robusta* IPNI=*290454-2* 417
Cuphea racemosa var. *discolor* IPNI=*290454-2* 417
Cuphea racemosa var. *extratropica* IPNI=*290454-2* 417
Cuphea racemosa var. *tropica* IPNI=*290454-2* 417
Cuphea rapunculoides IPNI=*553109-1* 417
Cuphea rivularis IPNI=*553043-1* 417
Cuphea serpyllifolia IPNI=*552931-1* 416
Cuphea serpyllifolia var. *tachirensis* IPNI=*552931-1* 416
Cuphea spicata f. *corymbifera* IPNI=*290454-2* 417
Cuphea spicata f. *extratropica* IPNI=*290454-2* 417
Cuphea spicata f. *grandifolia* IPNI=*290454-2* 417
Cuphea spicata f. *nudiflora* IPNI=*290454-2* 417
Cuphea spicata f. *parvifolia* IPNI=*290454-2* 417
Cuphea spicata f. *polysperma* IPNI=*290454-2* 417
Cuphea spicata f. *racemosa* IPNI=*290454-2* 417
Cuphea spicata f. *robusta* IPNI=*290454-2* 417
Cuphea spicata f. *tropica* IPNI=*290454-2* 417
Cuphea spicata IPNI=*290454-2* 417
Cuphea spicata var. *extratropica* IPNI=*290454-2* 417
Cuphea spicata var. *molliuscula* IPNI=*290454-2* 417
Cuphea spicata var. *tropica* IPNI=*290454-2* 417
Cuphea strumosa IPNI=*553111-1* 417
Cuphea sunubana IPNI=*553043-1* 417
Cuphea tubiflora IPNI=*553048-1* 417
Cuphea valerii IPNI=*553109-1* 417
Cuphea verticillata IPNI=*552967-1* 417
Cuphea watsonii IPNI=*553048-1* 417
Cuphea wrightii var. *compacta* IPNI=*553315-1* 417
Cuphea wrightii var. *nematopetala* IPNI=*553315-1* 417
Cuphea wrightii var. *typica* IPNI=*553315-1* 417
Cupressus arborvitae IPNI=*263956-1* 180
Cupressus arizonica f. *glomerata* IPNI=*60451546-2* 180
Cupressus arizonica f. *minor* IPNI=*60451546-2* 180
Cupressus arizonica f. *typica* IPNI=*60451546-2* 180
Cupressus arizonica IPNI=*60451546-2* 180
Cupressus arizonica var. *bonita* IPNI=*60451546-2* 180
Cupressus assamica IPNI=*60451554-2* 180
Cupressus benthamii var. *arizonica* IPNI=*60451546-2* 180
Cupressus columnaris IPNI=*261688-1* 180
Cupressus conoidea IPNI=*261974-1* 180
Cupressus coulteri IPNI=*60451554-2* 180
Cupressus elongata IPNI=*261974-1* 180
Cupressus excelsa IPNI=*60451554-2* 180
Cupressus expansa IPNI=*261974-1* 180
Cupressus fastigiata IPNI=*261974-1* 180
Cupressus filiformis IPNI=*677124-1* 180

Cupressus foemina IPNI=*261974-1* 180
Cupressus fragrans IPNI=*261839-1* 180
Cupressus glauca IPNI=*60451554-2* 180
Cupressus globulifera IPNI=*261974-1* 180
Cupressus hartwegii IPNI=*60451558-2* 180
Cupressus horizontalis IPNI=*261974-1* 180
Cupressus japonica IPNI=*261870-1* 180
Cupressus karwinskiana IPNI=*60451554-2* 180
Cupressus lambertiana IPNI=*60451558-2* 180
Cupressus lawsoniana f. *argentea* IPNI=*261839-1* 180
Cupressus lawsoniana f. *lutea* IPNI=*261839-1* 180
Cupressus lawsoniana f. *stewartii* IPNI=*261839-1* 180
Cupressus lawsoniana f. *viridis* IPNI=*261839-1* 180
Cupressus lawsoniana IPNI=*261839-1* 180
Cupressus lawsoniana lutea IPNI=*261839-1* 180
Cupressus lawsoniana var. *aluminii* IPNI=*261839-1* 180
Cupressus lawsoniana var. *argentea* IPNI=*261839-1* 180
Cupressus lindleyi IPNI=*60451554-2* 180
Cupressus lindleyi subsp. *hondurensis* IPNI=*60451554-2* 180
Cupressus lindleyi var. *hondurensis* IPNI=*60451554-2* 180
Cupressus lugubris IPNI=*261974-1* 180
Cupressus lusitanica f. *vulgaris* IPNI=*60451554-2* 180
Cupressus lusitanica IPNI=*60451554-2* 180
Cupressus lusitanica subsp. *arizonica* IPNI=*60451546-2* 180
Cupressus lusitanica subsp. *genuina* IPNI=*60451554-2* 180
Cupressus lusitanica var. *chlorocarpa* IPNI=*60451554-2* 180
Cupressus lusitanica var. *communis* IPNI=*60451554-2* 180
Cupressus lusitanica var. *epruinosa* IPNI=*60451554-2* 180
Cupressus lusitanica var. *hondurensis* IPNI=*60451554-2* 180
Cupressus lusitanica var. *typica* IPNI=*60451554-2* 180
Cupressus macrocarpa IPNI=*60451558-2* 180
Cupressus macrocarpa subsp. *lobosensis* IPNI=*60451558-2* 180
Cupressus macrocarpa var. *angulata* IPNI=*60451558-2* 180
Cupressus macrocarpa var. *farallonensis* IPNI=*60451558-2* 180
Cupressus mairei IPNI=*261870-1* 180
Cupressus mariae IPNI=*261974-1* 180
Cupressus mas IPNI=*261974-1* 180
Cupressus nobleana IPNI=*263956-1* 180
Cupressus nutkanus IPNI=*261839-1* 180
Cupressus orientalis IPNI=*261974-1* 180
Cupressus patula IPNI=*261974-1* 180
Cupressus pendula IPNI=*60451554-2* 180
Cupressus pyramidalis IPNI=*261974-1* 180
Cupressus reinwardtii IPNI=*60451558-2* 180
Cupressus roylei IPNI=*261974-1* 180
Cupressus sempervirens subsp. *indica* IPNI=*261974-1* 180
Cupressus sempervirens var. *indica* IPNI=*261974-1* 180
Cupressus sinensis IPNI=*60451554-2* 180
Cupressus sphaerocarpa IPNI=*261974-1* 180
Cupressus stricta IPNI=*261974-1* 180
Cupressus thuja IPNI=*677124-1* 180
Cupressus thujifolia IPNI=*261974-1* 180
Cupressus thujiformis IPNI=*261974-1* 180
Cupressus thurifera IPNI=*60451554-2* 180
Cupressus uhdeana IPNI=*60451554-2* 180
Cupressus umbilicata IPNI=*261974-1* 180
Cupressus whitleyana IPNI=*261974-1* 180
Curatella americana var. *pentagyna* IPNI=*321019-2* 295
Curatella cambaiba IPNI=*321019-2* 295
Curatella coriacea IPNI=*317113-1* 295
Curatella glabra IPNI=*321019-2* 295
Curatella glaziovii IPNI=*82916-2* 295
Curatella grisebachiana IPNI=*321019-2* 295
Curcas adansonii IPNI=*131462-2* 391
Curcas curcas IPNI=*131462-2* 391
Curcas drastica IPNI=*131462-2* 391
Curcas indica IPNI=*131462-2* 391
Curcas lobata IPNI=*131462-2* 391
Curcas purgans IPNI=*131462-2* 391
Curculigo glabra IPNI=*64141-1* 216
Curculigo recurvata IPNI=*64141-1* 216
Curculigo strobiliformis IPNI=*64141-1* 216
Curculigo sumatrana IPNI=*64141-1* 216
Curcuma americana IPNI=*60458762-2* 345
Curcuma brog IPNI=*796451-1* 250
Curcuma domestica IPNI=*796451-1* 250
Curcuma longa var. *vanaharidra* IPNI=*796451-1* 250
Curcuma ochrorhiza IPNI=*796451-1* 250
Curcuma soloensis IPNI=*796451-1* 250
Cuspidaria hymenaea IPNI=*152677-2* 364
Cyamus flavicomus IPNI=*605419-1* 433
Cyamus luteus IPNI=*605419-1* 433
Cyamus mysticus IPNI=*605419-1* 433
Cyamus nelumbo IPNI=*605419-1* 433
Cyamus pentapetalus IPNI=*605419-1* 433
Cyamus reniformis IPNI=*605419-1* 433
Cyanococcus amoenus IPNI=*261823-2* 300
Cyanococcus atrococcus IPNI=*261823-2* 300
Cyanococcus corymbosus IPNI=*261823-2* 300
Cyanococcus cuthbertii IPNI=*261823-2* 300
Cyanococcus elliottii IPNI=*261823-2* 300
Cyanococcus fuscatus IPNI=*261823-2* 300
Cyanococcus holophyllus IPNI=*261823-2* 300
Cyanococcus simulatus IPNI=*261823-2* 300
Cyanococcus virgatus IPNI=*261823-2* 300
Cyanophyllum magnificum IPNI=*326247-2* 419
Cyanophyllum marmoreum IPNI=*326247-2* 419
Cyanophyllum spectandum IPNI=*326247-2* 419
Cyanopis erigeroides IPNI=*962413-1* 260
Cyanopsis decurrens IPNI=*962413-1* 260
Cyanotis gueinzii IPNI=*172873-1* 224
Cyanotis zebrina IPNI=*30049896-2* 224
Cyanthillium chinense IPNI=*962413-1* 260
Cyanthillium cinereum var. *ovatum* IPNI=*962413-1* 260
Cyanus arvensis IPNI=*190310-1* 259
Cyanus cyanus IPNI=*190310-1* 259
Cyanus dentato-folius IPNI=*190310-1* 259
Cyanus segetum IPNI=*190310-1* 259
Cyanus vulgaris IPNI=*190310-1* 259
Cyathea aphlebioides IPNI=*73125-2* 172
Cyathea arborea var. *pusilla* IPNI=*17071820-1* 172
Cyathea arborea var. *regnellii* IPNI=*17071820-1* 172
Cyathea arida IPNI=*9976-2* 172
Cyathea articulata IPNI=*9976-2* 172
Cyathea asperula IPNI=*17072730-1* 172
Cyathea aurea IPNI=*17072730-1* 172
Cyathea aureonitens IPNI=*9971-2* 172
Cyathea bilineata IPNI=*143624-2* 173
Cyathea bonapartii IPNI=*241965-2* 173
Cyathea conspersa IPNI=*17072730-1* 172
Cyathea costalis IPNI=*9976-2* 172
Cyathea crassipes IPNI=*241965-2* 173
Cyathea crenata IPNI=*73146-2* 172
Cyathea cuspidata IPNI=*9954-2* 172
Cyathea cuspidata var. *rigida* IPNI=*9971-2* 172
Cyathea decora IPNI=*143624-2* 173
Cyathea delicatula IPNI=*17072730-1* 172
Cyathea denudans IPNI=*9976-2* 172
Cyathea discolor IPNI=*143624-2* 173
Cyathea erinacea IPNI=*9971-2* 172
Cyathea filix-femina IPNI=*30055316-2* 174
Cyathea firma IPNI=*9976-2* 172
Cyathea firmula IPNI=*9976-2* 172
Cyathea fragilis IPNI=*30013429-2* 174
Cyathea fulva IPNI=*73146-2* 172
Cyathea fulva var. *minor* IPNI=*17072730-1* 172

Cyathea furfuracea IPNI=*17072730-1* 172
Cyathea gemmifera IPNI=*9976-2* 172
Cyathea guadalupensis IPNI=*17071820-1* 172
Cyathea hassleriana IPNI=*9954-2* 172
Cyathea hassleriana var. *rojasii* IPNI=*9954-2* 172
Cyathea ichtyolepis IPNI=*73125-2* 172
Cyathea mapiriensis IPNI=*73267-2* 172
Cyathea membranulosa IPNI=*17072730-1* 172
Cyathea mettenii IPNI=*73146-2* 172
Cyathea mettenii var. *caucana* IPNI=*73146-2* 172
Cyathea mexicana IPNI=*9976-2* 172
Cyathea mollis IPNI=*17072730-1* 172
Cyathea molliuscula IPNI=*17072730-1* 172
Cyathea monticola IPNI=*143624-2* 173
Cyathea ocanensis IPNI=*73146-2* 172
Cyathea onusta IPNI=*17072730-1* 172
Cyathea oxyacantha IPNI=*9954-2* 172
Cyathea oyapoka IPNI=*9954-2* 172
Cyathea pansamalana IPNI=*73125-2* 172
Cyathea papyracea IPNI=*17072730-1* 172
Cyathea patellaris IPNI=*9976-2* 172
Cyathea portoricensis IPNI=*9971-2* 172
Cyathea primaeva IPNI=*73267-2* 172
Cyathea producta IPNI=*73146-2* 172
Cyathea punctifera IPNI=*9954-2* 172
Cyathea purpurascens IPNI=*9971-2* 172
Cyathea quadripinnata IPNI=*143624-2* 173
Cyathea quindiuensis IPNI=*241965-2* 173
Cyathea scabriuscula IPNI=*73125-2* 172
Cyathea schaffneriana IPNI=*143624-2* 173
Cyathea serra IPNI=*17071820-1* 172
Cyathea sherringii IPNI=*73146-2* 172
Cyathea sodiroi IPNI=*73146-2* 172
Cyathea stipularis IPNI=*73125-2* 172
Cyathea subochroleuca IPNI=*73125-2* 172
Cyathea tenerifrons IPNI=*9976-2* 172
Cyathea trejoi IPNI=*9976-2* 172
Cyathea trichiata IPNI=*73125-2* 172
Cyathea underwoodii IPNI=*17072730-1* 172
Cyathea yungensis IPNI=*241965-2* 173
Cyathoglottis macrantha IPNI=*657915-1* 223
Cyathophora heterophyla IPNI=*30062970-2* 390
Cyathula achyranthoides var. *densiflora* IPNI=*60139-1* 282
Cyathula achyranthoides var. *glabrescens* IPNI=*60139-1* 282
Cyathula alternifolia IPNI=*60185-1* 282
Cyathula geminata IPNI=*60139-1* 282
Cyathula geniculata IPNI=*60185-1* 282
Cyathula globosa IPNI=*60185-1* 282
Cyathula prostrata var. *achyranthoides* IPNI=*60139-1* 282
Cyathula repens IPNI=*60185-1* 282
Cybianthus albiflorus IPNI=*73545-2* 304
Cybianthus foliosus IPNI=*247469-2* 304
Cybianthus gentryi IPNI=*73545-2* 304
Cybianthus glaucus IPNI=*247441-2* 304
Cybianthus humboldtii IPNI=*18168-2* 303
Cybianthus panamensis IPNI=*73569-2* 304
Cybianthus principis IPNI=*18168-2* 303
Cybianthus spectabilis IPNI=*73559-2* 304
Cybistax chrysea IPNI=*222517-2* 365
Cybistax donnell-smithii IPNI=*222518-2* 365
Cybistax macrocarpa IPNI=*111572-2* 363
Cybistax millsii IPNI=*222518-2* 365
Cycas celebica IPNI=*326820-2* 179
Cycas duivenbodei IPNI=*326820-2* 179
Cycas inermis IPNI=*328823-2* 179
Cycas miquelii IPNI=*328823-2* 179
Cycas speciosa IPNI=*326820-2* 179
Cyclamen albidum IPNI=*700765-1* 304
Cyclamen aleppicum IPNI=*700765-1* 304
Cyclamen antiochium IPNI=*700765-1* 304
Cyclamen hederaceum IPNI=*700765-1* 304
Cyclamen latifolium IPNI=*700765-1* 304
Cyclamen persicum f. *albidum* IPNI=*700765-1* 304
Cyclamen punicum IPNI=*700765-1* 304
Cyclamen pyrolifolium IPNI=*700765-1* 304
Cyclamen tunetanum IPNI=*700765-1* 304
Cyclamen utopicum IPNI=*700765-1* 304
Cyclamen vernale IPNI=*700765-1* 304
Cyclamen vulgare IPNI=*700765-1* 304
Cyclaminus persica IPNI=*700765-1* 304
Cyclanthera achocchilla IPNI=*292472-1* 294
Cyclanthera bourgaeana IPNI=*292473-1* 294
Cyclanthera cordifolia var. *angustifolia* IPNI=*292478-1* 294
Cyclanthera costaricensis IPNI=*292473-1* 294
Cyclanthera digitata IPNI=*292512-1* 294
Cyclanthera edulis IPNI=*292512-1* 294
Cyclanthera explodens IPNI=*292473-1* 294
Cyclanthera filiformis IPNI=*77110356-1* 294
Cyclanthera glauca IPNI=*292473-1* 294
Cyclanthera montana IPNI=*292478-1* 294
Cyclanthera pedata f. *edulis* IPNI=*292512-1* 294
Cyclanthera pedata var. *edulis* IPNI=*292512-1* 294
Cyclanthera quadrifida IPNI=*77110356-1* 294
Cyclanthera quinqueloba IPNI=*77110356-1* 294
Cyclanthera rusbyi IPNI=*292478-1* 294
Cyclanthera siemiradzkii IPNI=*292478-1* 294
Cyclanthera subinermis IPNI=*292478-1* 294
Cyclanthera tomentosa IPNI=*292472-1* 294
Cyclanthus bifolius IPNI=*30336798-2* 227
Cyclanthus bipartitus var. *gracilis* IPNI=*30336798-2* 227
Cyclanthus cristatus IPNI=*30336798-2* 227
Cyclanthus godseffianus IPNI=*30336798-2* 227
Cyclanthus plumieri IPNI=*30336798-2* 227
Cyclea madagascariensis IPNI=*59219-2* 434
Cyclopeltis semicordata f. *crenata* IPNI=*1148359-2* 175
Cyclopteris fragilis IPNI=*30013429-2* 174
Cyclosorus caespitosus IPNI=*251308-2* 174
Cyclosorus dentatus IPNI=*251308-2* 174
Cyclosorus dentatus var. *violascens* IPNI=*251308-2* 174
Cyclosorus dicarpus IPNI=*251308-2* 174
Cyclosorus extensus IPNI=*17460390-1* 174
Cyclosorus fengii IPNI=*251308-2* 174
Cyclosorus gongylodes IPNI=*251353-2* 174
Cyclosorus impressus IPNI=*17460390-1* 174
Cyclosorus incertus IPNI=*17460390-1* 174
Cyclosorus interruptus IPNI=*17458930-1* 174
Cyclosorus jiulungshanensis IPNI=*251308-2* 174
Cyclosorus kunthii IPNI=*251400-2* 174
Cyclosorus lunganensis IPNI=*251308-2* 174
Cyclosorus mekongensis IPNI=*251308-2* 174
Cyclosorus mianningensis IPNI=*251308-2* 174
Cyclosorus mindanaensis IPNI=*251308-2* 174
Cyclosorus mollis IPNI=*251308-2* 174
Cyclosorus nymphalis IPNI=*251308-2* 174
Cyclosorus opulentus IPNI=*17460390-1* 174
Cyclosorus paradentatus IPNI=*251308-2* 174
Cyclosorus pingshanensis IPNI=*251308-2* 174
Cyclosorus proximus IPNI=*251308-2* 174
Cyclosorus sapingbaensis IPNI=*251308-2* 174
Cyclosorus stenopes IPNI=*251308-2* 174
Cyclosorus sulfureus IPNI=*17460390-1* 174
Cyclosorus taprobanicus IPNI=*251308-2* 174
Cyclosorus tottus IPNI=*17458930-1* 174
Cyclosorus wakefieldii IPNI=*17460390-1* 174
Cyclosorus wangii IPNI=*251308-2* 174
Cyclospermum ammi IPNI=*840899-1* 252

Cyclospermum tenuifolium IPNI=*840899-1* 252
Cyclostigma hibiscifolium IPNI=*342669-1* 388
Cycnoches ventricosum var. *chlorochilon* IPNI=*73806-2* 218
Cycnoseris australis IPNI=*225504-1* 264
Cydista aequinoctialis IPNI=*108550-1* 362
Cydista aequinoctialis var. *hirtella* IPNI=*108550-1* 362
Cydista aequinoctialis var. *sarmentosa* IPNI=*108550-1* 362
Cydista amoena IPNI=*108550-1* 362
Cydista heterophylla IPNI=*77141971-1* 362
Cydista picta IPNI=*108550-1* 362
Cydista pubescens IPNI=*108550-1* 362
Cydista sarmentosa IPNI=*108550-1* 362
Cydista seemannii IPNI=*108550-1* 362
Cydista spectabilis IPNI=*108550-1* 362
Cydonia communis IPNI=*724549-1* 443
Cydonia cydonia IPNI=*724549-1* 443
Cydonia europaea IPNI=*724549-1* 443
Cydonia lusitanica IPNI=*724549-1* 443
Cydonia maliformis IPNI=*724549-1* 443
Cydonia oblonga subsp. *integerrima* IPNI=*724549-1* 443
Cydonia oblonga var. *integerrimosepala* IPNI=*724549-1* 443
Cydonia oblonga var. *obpyricarpa* IPNI=*724549-1* 443
Cydonia oblonga var. *obpyriformis* IPNI=*724549-1* 443
Cydonia oblonga var. *orbiculatocomplanata* IPNI=*724549-1* 443
Cydonia oblonga var. *ovalicarpa* IPNI=*724549-1* 443
Cydonia oblonga var. *ovalis* IPNI=*724549-1* 443
Cydonia oblonga var. *plano-cyclocarpa* IPNI=*724549-1* 443
Cydonia oblonga var. *pomiformis* IPNI=*724549-1* 443
Cydonia oblonga var. *urceolata* IPNI=*724549-1* 443
Cydonia silvestris IPNI=*724549-1* 443
Cydonia sumboshia IPNI=*724549-1* 443
Cylicadenia harrisii IPNI=*77191232-1* 342
Cylichnanthus barbatus IPNI=*306674-2* 285
Cylichnanthus maculatus IPNI=*310755-2* 285
Cylichnanthus plumarius IPNI=*153717-1* 285
Cylindropuntia cylindrica IPNI=*128942-1* 283
Cylindropuntia intermedia IPNI=*128942-1* 283
Cylindropuntia metuenda IPNI=*73852-2* 283
Cylindropuntia subulata IPNI=*27133-2* 283
Cylindropuntia tunicata var. *aricensis* IPNI=*73905-2* 283
Cylindropuntia tunicata var. *chilensis* IPNI=*73905-2* 283
Cymatochloa fluitans IPNI=*30475806-2* 243
Cymatochloa repens IPNI=*30475806-2* 243
Cymbachne fastigiata IPNI=*398608-1* 237
Cymbalaria cymbalaria IPNI=*325773-2* 376
Cymbalaria flabellifer IPNI=*325773-2* 376
Cymbalaria gerbaultii IPNI=*325773-2* 376
Cymbalaria glechomifolia IPNI=*325773-2* 376
Cymbalaria globosa IPNI=*325773-2* 376
Cymbalaria hederacea IPNI=*325773-2* 376
Cymbalaria toutonii IPNI=*325773-2* 376
Cymbidium altum IPNI=*1218040-2* 217
Cymbidium andersonii IPNI=*1173554-2* 218
Cymbidium candidum IPNI=*622208-1* 217
Cymbidium cucullatum IPNI=*618026-1* 217
Cymbidium muricatum IPNI=*79276-2* 218
Cymbidium nervosum IPNI=*641061-1* 219
Cymbidium ochroleucum IPNI=*643592-1* 221
Cymbidium pusillum IPNI=*1021256-1* 219
Cymbidium trifidum IPNI=*1218040-2* 217
Cymbidium trinerve IPNI=*1147853-2* 218
Cymbidium verecundum IPNI=*1218040-2* 217
Cymbidium violaceum IPNI=*622273-1* 217
Cymbopetalum brasiliense f. *latifolia* IPNI=*72706-1* 187
Cymbopetalum odoratissimum IPNI=*72706-1* 187
Cymbopogon afronardus IPNI=*396972-1* 236
Cymbopogon claessensii IPNI=*396972-1* 236
Cymbopogon condensatus IPNI=*420877-1* 244
Cymbopogon confertiflorus IPNI=*396972-1* 236
Cymbopogon connatus IPNI=*396888-1* 236
Cymbopogon excavatus IPNI=*396888-1* 236
Cymbopogon melanocarpus IPNI=*404802-1* 240
Cymbopogon nardus var. *confertiflorus* IPNI=*396972-1* 236
Cymbopogon nardus var. *luridus* IPNI=*396972-1* 236
Cymbopogon odoratus IPNI=*392701-1* 235
Cymbopogon plicatus IPNI=*396888-1* 236
Cymbopogon prolixus IPNI=*396972-1* 236
Cymbopogon rufus IPNI=*405583-1* 240
Cymbopogon ruprechtii IPNI=*126405-2* 240
Cymbopogon suaveolens IPNI=*396888-1* 236
Cymbopogon thwaitesii IPNI=*396972-1* 236
Cymbopogon validus IPNI=*396972-1* 236
Cymbopogon virgatus IPNI=*396972-1* 236
Cymbosepalum baronii IPNI=*316401-2* 320
Cymburus mutabilis IPNI=*243385-2* 379
Cymburus urticifolius IPNI=*1092351-2* 379
Cyminon longeinvolucellatum IPNI=*840882-1* 252
Cynanchum album IPNI=*106366-2* 340
Cynanchum altissimum IPNI=*60478412-2* 343
Cynanchum carnosum IPNI=*107308-3* 341
Cynanchum ciliatum IPNI=*97427-1* 340
Cynanchum clausum IPNI=*106366-2* 340
Cynanchum cordatum IPNI=*74165-2* 340
Cynanchum denticulatum IPNI=*77206440-1* 340
Cynanchum discolor IPNI=*154570-2* 342
Cynanchum filiforme IPNI=*106366-2* 340
Cynanchum filliforme IPNI=*106366-2* 340
Cynanchum jamaicense IPNI=*74165-2* 340
Cynanchum leucanthum IPNI=*106366-2* 340
Cynanchum magdalenicum IPNI=*74165-2* 340
Cynanchum maritimum IPNI=*98854-1* 341
Cynanchum mexicanum IPNI=*106366-2* 340
Cynanchum obliquum IPNI=*154570-2* 342
Cynanchum stellatum IPNI=*97427-1* 340
Cynanchum tamifolium IPNI=*74165-2* 340
Cynanchum tamifolium var. *major* IPNI=*74165-2* 340
Cynanchum tamifolium var. *minor* IPNI=*74165-2* 340
Cynanchum verticillare IPNI=*106366-2* 340
Cynanchum virginicum IPNI=*154570-2* 342
Cynanchum viride IPNI=*77206440-1* 340
Cynanchum viridiflorum IPNI=*77206440-1* 340
Cynara carduncellus IPNI=*200876-1* 260
Cynara cardunculus subsp. *scolymus* IPNI=*200876-1* 260
Cynara cardunculus var. *altilis* IPNI=*200876-1* 260
Cynara cardunculus var. *ferocissima* IPNI=*200876-1* 260
Cynara cardunculus var. *sativa* IPNI=*200876-1* 260
Cynara cardunculus var. *scolymus* IPNI=*200876-1* 260
Cynara cardunculus var. *sylvestris* IPNI=*200876-1* 260
Cynara communis IPNI=*200876-1* 260
Cynara corsica IPNI=*200876-1* 260
Cynara esculenta IPNI=*200876-1* 260
Cynara ferox IPNI=*200876-1* 260
Cynara horrida IPNI=*200876-1* 260
Cynara hortensis IPNI=*200876-1* 260
Cynara humilis IPNI=*200876-1* 260
Cynara pygmaea IPNI=*200876-1* 260
Cynara scolymus IPNI=*200876-1* 260
Cynara spinosissima IPNI=*200876-1* 260
Cynara sylvestris IPNI=*200876-1* 260
Cynaropsis gomerensis IPNI=*200876-1* 260
Cynocardamum virginicum IPNI=*138164-2* 277
Cynodendron argenteum IPNI=*311591-2* 305
Cynodon affinis IPNI=*397065-1* 236
Cynodon amabilis IPNI=*401703-1* 239
Cynodon aristiglumis IPNI=*397065-1* 236
Cynodon aristulatus IPNI=*397065-1* 236

Cynodon ciliaris IPNI=*1156566-2* 238
Cynodon curtipendulus IPNI=*1198286-2* 235
Cynodon dactylon f. *glabrescens* IPNI=*397065-1* 236
Cynodon dactylon f. *major* IPNI=*397065-1* 236
Cynodon dactylon f. *viviparus* IPNI=*397065-1* 236
Cynodon dactylon subsp. *nipponicus* IPNI=*397065-1* 236
Cynodon dactylon var. *affinis* IPNI=*397065-1* 236
Cynodon dactylon var. *longiglumis* IPNI=*397065-1* 236
Cynodon dactylon var. *maritimus* IPNI=*397065-1* 236
Cynodon dactylon var. *parviglumis* IPNI=*397065-1* 236
Cynodon dactylon var. *pilosus* IPNI=*397065-1* 236
Cynodon dactylon var. *pulchellus* IPNI=*397065-1* 236
Cynodon decipiens IPNI=*397065-1* 236
Cynodon distichloides IPNI=*397065-1* 236
Cynodon donax IPNI=*390837-1* 234
Cynodon erectus IPNI=*397065-1* 236
Cynodon filiformis IPNI=*407138-1* 241
Cynodon glabratus IPNI=*397065-1* 236
Cynodon grandispiculus IPNI=*397065-1* 236
Cynodon hirsutissimus IPNI=*397065-1* 236
Cynodon indicus IPNI=*74675-3* 238
Cynodon iraquensis IPNI=*397065-1* 236
Cynodon laeviglumis IPNI=*397065-1* 236
Cynodon linearis IPNI=*397065-1* 236
Cynodon maritimus IPNI=*397065-1* 236
Cynodon maritimus var. *breviglumis* IPNI=*397065-1* 236
Cynodon maritimus var. *grandispiculus* IPNI=*397065-1* 236
Cynodon maritimus var. *vaginiflorus* IPNI=*397065-1* 236
Cynodon mucronatus IPNI=*397065-1* 236
Cynodon nitidus IPNI=*397065-1* 236
Cynodon occidentalis IPNI=*397065-1* 236
Cynodon parviglumis IPNI=*397065-1* 236
Cynodon pascuus IPNI=*397065-1* 236
Cynodon pedicellatus IPNI=*397065-1* 236
Cynodon polevansii IPNI=*397065-1* 236
Cynodon portoricensis IPNI=*397065-1* 236
Cynodon praecox IPNI=*399115-1* 238
Cynodon procumbens IPNI=*328564-2* 235
Cynodon repens IPNI=*397065-1* 236
Cynodon sarmentosus IPNI=*397065-1* 236
Cynodon scabrifolius IPNI=*397065-1* 236
Cynodon stellatus IPNI=*397065-1* 236
Cynodon tenuis IPNI=*397065-1* 236
Cynodon umbellatus IPNI=*397065-1* 236
Cynoglossum amabile f. *leucanthum* IPNI=*115031-1* 274
Cynoglossum amabile f. *ruberum* IPNI=*115031-1* 274
Cynometra martiana IPNI=*498789-1* 321
Cynophalla amplissima IPNI=*77184082-1* 278
Cynophalla flexuosa IPNI=*316361-2* 278
Cynophalla guayaquilensis IPNI=*316361-2* 278
Cynophalla laeta IPNI=*77183644-1* 278
Cynophalla linearis IPNI=*77184255-1* 278
Cynophalla mollis IPNI=*316361-2* 278
Cynophalla saligna IPNI=*316361-2* 278
Cynophalla sessilis IPNI=*77183680-1* 278
Cynophalla verrucosa IPNI=*77183644-1* 278
Cynosurus aegyptiacus IPNI=*397387-1* 236
Cynosurus aegyptius IPNI=*397387-1* 236
Cynosurus ara IPNI=*74675-3* 238
Cynosurus aureus IPNI=*406803-1* 240
Cynosurus caplllaris IPNI=*407138-1* 241
Cynosurus carolinianus IPNI=*397387-1* 236
Cynosurus ciliaris IPNI=*397387-1* 236
Cynosurus distachyos IPNI=*397387-1* 236
Cynosurus indicus IPNI=*74675-3* 238
Cynosurus macara IPNI=*397387-1* 236
Cynosurus paniceus IPNI=*1048074-2* 244
Cynosurus pectinatus IPNI=*74675-3* 238
Cynosurus penicillatus IPNI=*396063-1* 236
Cynosurus tenerrimus IPNI=*407138-1* 241
Cynosurus tristachyos IPNI=*30127333-2* 238
Cynosurus uniflorus IPNI=*397065-1* 236
Cypella longifolia IPNI=*77182480-1* 217
Cypella lutea IPNI=*77182480-1* 217
Cyperus abyssinicus IPNI=*74698-2* 230
Cyperus acicularis IPNI=*1169169-2* 231
Cyperus acutus IPNI=*305060-1* 231
Cyperus adenophorus IPNI=*304701-1* 230
Cyperus adenophorus var. *aphyllus* IPNI=*304701-1* 230
Cyperus adustus IPNI=*74915-2* 231
Cyperus agrestis IPNI=*305797-1* 231
Cyperus alpinus IPNI=*74768-2* 230
Cyperus amabilis var. *macrostachyus* IPNI=*303735-1* 230
Cyperus amentaceus IPNI=*1111946-2* 232
Cyperus americanus IPNI=*304701-1* 230
Cyperus amphekes IPNI=*75184-2* 231
Cyperus amplissimus IPNI=*305627-1* 231
Cyperus anabaptistus IPNI=*304793-1* 230
Cyperus anceps IPNI=*75184-2* 231
Cyperus andersonianus IPNI=*304256-1* 230
Cyperus anisostachyos IPNI=*303735-1* 230
Cyperus aphyllus IPNI=*1009998-1* 231
Cyperus aphyllus IPNI=*304701-1* 230
Cyperus aphyllus IPNI=*74934-2* 231
Cyperus arabicus IPNI=*305797-1* 231
Cyperus aristatus f. *falciculosus* IPNI=*324890-2* 231
Cyperus aristatus f. *inflexus* IPNI=*324890-2* 231
Cyperus aristatus IPNI=*324890-2* 231
Cyperus aristatus var. *inflexus* IPNI=*324890-2* 231
Cyperus aristatus var. *runyonii* IPNI=*324890-2* 231
Cyperus arsitatus var. *perennis* IPNI=*324890-2* 231
Cyperus articulatus f. *longispiculosus* IPNI=*303822-1* 230
Cyperus articulatus IPNI=*303822-1* 230
Cyperus articulatus var. *multiflorus* IPNI=*303822-1* 230
Cyperus articulatus var. *nodosus* IPNI=*303822-1* 230
Cyperus aurantiacus IPNI=*303735-1* 230
Cyperus aureus IPNI=*300668-2* 230
Cyperus aureus IPNI=*303735-1* 230
Cyperus aureus IPNI=*324890-2* 231
Cyperus aureus var. *aurantiacus* IPNI=*303735-1* 230
Cyperus aureus var. *macrostachyus* IPNI=*303735-1* 230
Cyperus aureus var. *oligostachyus* IPNI=*303735-1* 230
Cyperus autumnalis IPNI=*303822-1* 230
Cyperus autumnalis IPNI=*304701-1* 230
Cyperus bahiensis IPNI=*300668-2* 230
Cyperus balbisii IPNI=*305967-1* 231
Cyperus balfourii IPNI=*305041-1* 231
Cyperus bangianus IPNI=*74851-2* 231
Cyperus berteroi IPNI=*304595-1* 230
Cyperus biceps IPNI=*304793-1* 230
Cyperus bicolor IPNI=*305797-1* 231
Cyperus bifax IPNI=*305797-1* 231
Cyperus biradiatus IPNI=*74915-2* 231
Cyperus bisumbellatus IPNI=*305627-1* 231
Cyperus borbonicus IPNI=*303822-1* 230
Cyperus brachiatus IPNI=*304155-1* 230
Cyperus brachyphyllus IPNI=*303735-1* 230
Cyperus breviradiatus IPNI=*75178-2* 231
Cyperus brownii IPNI=*324890-2* 231
Cyperus buchananii IPNI=*300668-2* 230
Cyperus bufonius IPNI=*30079279-2* 233
Cyperus bulamensis IPNI=*300668-2* 230
Cyperus bulbosostolonifer IPNI=*305797-1* 231
Cyperus caespitosus IPNI=*304701-1* 230
Cyperus caffer IPNI=*304155-1* 230
Cyperus callophorus IPNI=*305019-1* 231

Cyperus campestris IPNI=*304793-1* 230
Cyperus camphoratus var. *breviradiatus* IPNI=*75178-2* 231
Cyperus caracasanus IPNI=*75178-2* 231
Cyperus caribaeus IPNI=*104972-2* 232
Cyperus cayennensis IPNI=*304701-1* 230
Cyperus chromatolepis IPNI=*305967-1* 231
Cyperus chrysomelinus IPNI=*304844-1* 230
Cyperus chrysostachys IPNI=*300668-2* 230
Cyperus cimicinus IPNI=*74915-2* 231
Cyperus comosus IPNI=*305797-1* 231
Cyperus complanatus IPNI=*1111946-2* 232
Cyperus compressus f. *depauperatus* IPNI=*304155-1* 230
Cyperus compressus subsp. *micranthus* IPNI=*304155-1* 230
Cyperus compressus var. *capillaceus* IPNI=*304155-1* 230
Cyperus compressus var. *compositus* IPNI=*304155-1* 230
Cyperus compressus var. *simplex* IPNI=*304155-1* 230
Cyperus confertiflorus IPNI=*304793-1* 230
Cyperus confusus IPNI=*305041-1* 231
Cyperus conoideus IPNI=*74851-2* 231
Cyperus conspicuus IPNI=*304595-1* 230
Cyperus cordobensis IPNI=*303822-1* 230
Cyperus coriaceus IPNI=*305019-1* 231
Cyperus corymbosus var. *subnodosus* IPNI=*303822-1* 230
Cyperus curvispiculosus IPNI=*77140538-1* 231
Cyperus cylindrostachys IPNI=*304256-1* 230
Cyperus cyperinus f. *curvatus* IPNI=*304256-1* 230
Cyperus cyperinus subsp. *laxatus* IPNI=*304256-1* 230
Cyperus cyperinus var. *scabriculmis* IPNI=*304256-1* 230
Cyperus damiettensis IPNI=*300668-2* 230
Cyperus declinatus IPNI=*304426-1* 230
Cyperus declinatus var. *limbatus* IPNI=*304426-1* 230
Cyperus dense-spicatus IPNI=*304793-1* 230
Cyperus densicaespitosus IPNI=*77144505-1* 230
Cyperus densiflorus IPNI=*304595-1* 230
Cyperus densiflorus IPNI=*304793-1* 230
Cyperus densispicatus IPNI=*304793-1* 230
Cyperus diandrus var. *capitatus* IPNI=*74915-2* 231
Cyperus diandrus var. *castaneus* IPNI=*74915-2* 231
Cyperus diaphaniria IPNI=*304844-1* 230
Cyperus diffusus subsp. *chalaranthus* IPNI=*304087-1* 230
Cyperus diffusus subsp. *tolucensis* IPNI=*304973-1* 231
Cyperus diffusus var. *chalaranthus* IPNI=*304087-1* 230
Cyperus diffusus var. *tolucensis* IPNI=*304973-1* 231
Cyperus diffusus var. *umbrosus* IPNI=*304973-1* 231
Cyperus digitatus IPNI=*304793-1* 230
Cyperus disciger IPNI=*75178-2* 231
Cyperus disruptus IPNI=*305797-1* 231
Cyperus distans f. *pallidior* IPNI=*301880-2* 230
Cyperus distans subsp. *longibracteatus* IPNI=*301880-2* 230
Cyperus distans var. *densiflorus* IPNI=*301880-2* 230
Cyperus distans var. *rubrotinctus* IPNI=*301880-2* 230
Cyperus duchaisingii IPNI=*305967-1* 231
Cyperus durandii IPNI=*74698-2* 230
Cyperus dussianus IPNI=*75178-2* 231
Cyperus dussii IPNI=*75178-2* 231
Cyperus efoliatus IPNI=*304701-1* 230
Cyperus elongatus IPNI=*305797-1* 231
Cyperus entrerianus f. *abbreviatus* IPNI=*74636-2* 230
Cyperus entrerianus var. *parvicapitulatus* IPNI=*74636-2* 230
Cyperus eragrostis f. *latifrons* IPNI=*304426-1* 230
Cyperus eragrostis f. *tener* IPNI=*304426-1* 230
Cyperus eragrostis IPNI=*74698-2* 230
Cyperus eragrostis var. *compactus* IPNI=*304426-1* 230
Cyperus eragrostis var. *spectabilis* IPNI=*304426-1* 230
Cyperus esculentus f. *angustispicatus* IPNI=*300668-2* 230
Cyperus esculentus f. *evolutus* IPNI=*300668-2* 230
Cyperus esculentus f. *macrostachyus* IPNI=*300668-2* 230
Cyperus esculentus f. *princeps* IPNI=*300668-2* 230
Cyperus esculentus var. *angustispicatus* IPNI=*300668-2* 230
Cyperus esculentus var. *helodes* IPNI=*300668-2* 230
Cyperus esculentus var. *leptostachyus* IPNI=*300668-2* 230
Cyperus esculentus var. *lutescens* IPNI=*300668-2* 230
Cyperus esculentus var. *macrostachyus* IPNI=*300668-2* 230
Cyperus esculentus var. *phymatodes* IPNI=*300668-2* 230
Cyperus esculentus var. *sprucei* IPNI=*300668-2* 230
Cyperus eurystachys IPNI=*75178-2* 231
Cyperus falciculosus IPNI=*324890-2* 231
Cyperus fallaciosus IPNI=*74698-2* 230
Cyperus ferax IPNI=*305060-1* 231
Cyperus fistulosus IPNI=*303822-1* 230
Cyperus flavens IPNI=*74698-2* 230
Cyperus flavescens subsp. *fallaciosus* IPNI=*74698-2* 230
Cyperus flavescens subsp. *intermedius* IPNI=*74698-2* 230
Cyperus flavescens subsp. *microglumis* IPNI=*74698-2* 230
Cyperus flavescens subsp. *tanaensis* IPNI=*74698-2* 230
Cyperus flavescens subvar. *capitatus* IPNI=*74915-2* 231
Cyperus flavescens subvar. *mohrii* IPNI=*74698-2* 230
Cyperus flavescens var. *castaneus* IPNI=*74698-2* 230
Cyperus flavescens var. *paraensis* IPNI=*74698-2* 230
Cyperus flavescens var. *piceus* IPNI=*74698-2* 230
Cyperus flavescens var. *poiformis* IPNI=*74698-2* 230
Cyperus flavescens var. *rehmannianus* IPNI=*74698-2* 230
Cyperus flavus var. *aggregatus* IPNI=*303704-1* 230
Cyperus flexibilis IPNI=*75178-2* 231
Cyperus flexifolius IPNI=*304793-1* 230
Cyperus fontanesii IPNI=*74698-2* 230
Cyperus frondosus IPNI=*303729-1* 230
Cyperus fulvescens IPNI=*300668-2* 230
Cyperus giraudyi IPNI=*304155-1* 230
Cyperus glandulosus IPNI=*305019-1* 231
Cyperus glareosus IPNI=*303735-1* 230
Cyperus glaucoviridis IPNI=*305019-1* 231
Cyperus globulifer IPNI=*74851-2* 231
Cyperus globulosus IPNI=*74851-2* 231
Cyperus glomulifer IPNI=*74851-2* 231
Cyperus goeringii IPNI=*304311-1* 230
Cyperus gracilescens IPNI=*300668-2* 230
Cyperus gracilis IPNI=*300668-2* 230
Cyperus gracilis IPNI=*304426-1* 230
Cyperus gracilis IPNI=*304701-1* 230
Cyperus graeffei IPNI=*304256-1* 230
Cyperus gramineus IPNI=*304701-1* 230
Cyperus graminicola IPNI=*301880-2* 230
Cyperus graminifolius IPNI=*304701-1* 230
Cyperus guadalajaranus IPNI=*303735-1* 230
Cyperus guatemalensis IPNI=*74851-2* 231
Cyperus gymnos IPNI=*303822-1* 230
Cyperus haematodes IPNI=*305060-1* 231
Cyperus hartii IPNI=*75178-2* 231
Cyperus haspan f. *coarctatus* IPNI=*304701-1* 230
Cyperus haspan f. *major* IPNI=*304701-1* 230
Cyperus haspan subsp. *juncoides* IPNI=*304701-1* 230
Cyperus haspan var. *adenophorus* IPNI=*304701-1* 230
Cyperus haspan var. *americanus* IPNI=*304701-1* 230
Cyperus haspan var. *amplissimus* IPNI=*304701-1* 230
Cyperus haspan var. *coarctatus* IPNI=*304701-1* 230
Cyperus haspan var. *dietrichiae* IPNI=*304701-1* 230
Cyperus haspan var. *firmicaulis* IPNI=*304701-1* 230
Cyperus haspan var. *riparius* IPNI=*304701-1* 230
Cyperus heermannii IPNI=*300668-2* 230
Cyperus helodes IPNI=*300668-2* 230
Cyperus herbicavus IPNI=*305797-1* 231
Cyperus hermaphroditus var. *angustior* IPNI=*74768-2* 230
Cyperus hermaphroditus var. *condensatus* IPNI=*74768-2* 230
Cyperus hermaphroditus var. *longiradiatus* IPNI=*74768-2* 230
Cyperus hermaphroditus var. *longispicatus* IPNI=*74768-2* 230

Cyperus hexastachyos IPNI=*305797-1* 231
Cyperus hexastachyos IPNI=*305967-1* 231
Cyperus hildra IPNI=*305797-1* 231
Cyperus homoiiria IPNI=*304973-1* 231
Cyperus hostmannii IPNI=*74851-2* 231
Cyperus hydra IPNI=*300668-2* 230
Cyperus hydra IPNI=*305797-1* 231
Cyperus hyemalis IPNI=*304701-1* 230
Cyperus imbricatus subsp. *elongatus* IPNI=*304793-1* 230
Cyperus imbricatus var. *capitatus* IPNI=*304793-1* 230
Cyperus imbricatus var. *elongatus* IPNI=*304793-1* 230
Cyperus imbricatus var. *minor* IPNI=*304793-1* 230
Cyperus imbricatus var. *viridispicatus* IPNI=*304793-1* 230
Cyperus incompletus IPNI=*74768-2* 230
Cyperus inconspicuus IPNI=*305797-1* 231
Cyperus inflexus IPNI=*324890-2* 231
Cyperus inflexus var. *acaulis* IPNI=*324890-2* 231
Cyperus inflexus var. *elongatus* IPNI=*324890-2* 231
Cyperus interceptus IPNI=*303822-1* 230
Cyperus intermedius IPNI=*74698-2* 230
Cyperus intricatoramosus IPNI=*74915-2* 231
Cyperus iria f. *chrysomelinus* IPNI=*304844-1* 230
Cyperus iria f. *flavescens* IPNI=*304844-1* 230
Cyperus iria f. *multiflora* IPNI=*304844-1* 230
Cyperus iria var. *flavescens* IPNI=*304844-1* 230
Cyperus iria var. *paniciformis* IPNI=*304844-1* 230
Cyperus iria var. *santonici* IPNI=*304844-1* 230
Cyperus iria var. *typicus* IPNI=*304844-1* 230
Cyperus ischnostachys IPNI=*304256-1* 230
Cyperus junceus IPNI=*304701-1* 230
Cyperus juncoides IPNI=*304701-1* 230
Cyperus keniensis IPNI=*301880-2* 230
Cyperus kurrii IPNI=*301880-2* 230
Cyperus kyllingia IPNI=*311841-1* 232
Cyperus laevissimus IPNI=*305797-1* 231
Cyperus lateriflorus IPNI=*304311-1* 230
Cyperus laticulmis IPNI=*304701-1* 230
Cyperus laxus IPNI=*301880-2* 230
Cyperus laxus IPNI=*304973-1* 231
Cyperus lepidus IPNI=*303735-1* 230
Cyperus leptos IPNI=*304701-1* 230
Cyperus leptostachyus IPNI=*305797-1* 231
Cyperus leucocephalus IPNI=*77140538-1* 231
Cyperus ligusticus IPNI=*74698-2* 230
Cyperus longespicatus IPNI=*74768-2* 230
Cyperus longibracteatus IPNI=*301880-2* 230
Cyperus longibracteatus var. *niger* IPNI=*301880-2* 230
Cyperus longicruris IPNI=*74915-2* 231
Cyperus longus IPNI=*305797-1* 231
Cyperus lorentzianus IPNI=*74915-2* 231
Cyperus lorentzianus var. *compressus* IPNI=*74915-2* 231
Cyperus lorentzianus var. *elatus* IPNI=*74915-2* 231
Cyperus lucidus var. *acutus* IPNI=*305060-1* 231
Cyperus lucidus var. *grandiusculus* IPNI=*305060-1* 231
Cyperus lucidus var. *sanguineofuscus* IPNI=*305060-1* 231
Cyperus lutescens IPNI=*300668-2* 230
Cyperus lutescens IPNI=*304426-1* 230
Cyperus luzulae f. *pallidiflorens* IPNI=*74851-2* 231
Cyperus luzulae IPNI=*304426-1* 230
Cyperus luzulae var. *entrerianus* IPNI=*74636-2* 230
Cyperus luzulae var. *minor* IPNI=*74851-2* 231
Cyperus luzulae var. *umbellulatus* IPNI=*74851-2* 231
Cyperus macrosciadion IPNI=*304793-1* 230
Cyperus macrostachyus IPNI=*304973-1* 231
Cyperus mediorubescens IPNI=*304793-1* 230
Cyperus melanorhizus IPNI=*300668-2* 230
Cyperus melanostachyus IPNI=*74698-2* 230
Cyperus melanostachyus IPNI=*74915-2* 231
Cyperus melanostachyus var. *caespitosus* IPNI=*74915-2* 231
Cyperus melanostachyus var. *gracilis* IPNI=*74915-2* 231
Cyperus melanostachyus var. *robustus* IPNI=*74915-2* 231
Cyperus merkeri IPNI=*305797-1* 231
Cyperus meyenii IPNI=*304155-1* 230
Cyperus micreilema IPNI=*305797-1* 231
Cyperus microcarpus IPNI=*304701-1* 230
Cyperus microlepis IPNI=*304844-1* 230
Cyperus microstachyus IPNI=*303735-1* 230
Cyperus monandrus IPNI=*304426-1* 230
Cyperus monostachyos IPNI=*104972-2* 232
Cyperus muehlenbergii IPNI=*304701-1* 230
Cyperus muelleri IPNI=*303735-1* 230
Cyperus nangtciangensis IPNI=*304844-1* 230
Cyperus nemorosus IPNI=*305967-1* 231
Cyperus nervosus IPNI=*300668-2* 230
Cyperus nieuwlandii IPNI=*74698-2* 230
Cyperus niger var. *capitatus* IPNI=*74915-2* 231
Cyperus niger var. *castaneus* IPNI=*74915-2* 231
Cyperus niger var. *intricatoramosus* IPNI=*74915-2* 231
Cyperus niger var. *lorentzianus* IPNI=*74915-2* 231
Cyperus niger var. *pseudoelegantulus* IPNI=*74915-2* 231
Cyperus niger var. *robustus* IPNI=*74915-2* 231
Cyperus niloticus IPNI=*303822-1* 230
Cyperus nodosus IPNI=*303822-1* 230
Cyperus nodosus var. *aphyllus* IPNI=*303822-1* 230
Cyperus nubicus IPNI=*305797-1* 231
Cyperus nudus IPNI=*304701-1* 230
Cyperus nudus var. *firmior* IPNI=*304701-1* 230
Cyperus nudus var. *flaccidior* IPNI=*304701-1* 230
Cyperus nutans IPNI=*301880-2* 230
Cyperus obscurus IPNI=*304793-1* 230
Cyperus obtusatus var. *cylindrostachyus* IPNI=*74934-2* 231
Cyperus obtusatus var. *tenuis* IPNI=*74934-2* 231
Cyperus obtusiusculus IPNI=*74698-2* 230
Cyperus ochreoides IPNI=*305797-1* 231
Cyperus ochrocephalus IPNI=*304426-1* 230
Cyperus octophyllus IPNI=*304973-1* 231
Cyperus officinalis IPNI=*300668-2* 230
Cyperus officinalis IPNI=*305797-1* 231
Cyperus oliganthus IPNI=*305797-1* 231
Cyperus oligostachyus IPNI=*303735-1* 230
Cyperus olivaris IPNI=*305797-1* 231
Cyperus olivaris var. *brevibracteatus* IPNI=*305797-1* 231
Cyperus oryzetorum IPNI=*304311-1* 230
Cyperus pallens IPNI=*75184-2* 231
Cyperus pallescens IPNI=*305797-1* 231
Cyperus pallidus IPNI=*300668-2* 230
Cyperus pallidus IPNI=*74698-2* 230
Cyperus paniciformis IPNI=*304844-1* 230
Cyperus panicoides IPNI=*304844-1* 230
Cyperus paniculatus IPNI=*1088685-2* 239
Cyperus paraensis IPNI=*74698-2* 230
Cyperus patulus IPNI=*305797-1* 231
Cyperus pectinatus IPNI=*304155-1* 230
Cyperus pectiniformis IPNI=*304155-1* 230
Cyperus pedatus IPNI=*304701-1* 230
Cyperus pergracilis IPNI=*74915-2* 231
Cyperus peruvianus IPNI=*74934-2* 231
Cyperus peruvianus var. *foliatus* IPNI=*1009998-1* 231
Cyperus phaeorhizus IPNI=*304701-1* 230
Cyperus phymatodes IPNI=*300668-2* 230
Cyperus phymatodes var. *heermannii* IPNI=*300668-2* 230
Cyperus piceus f. *densior* IPNI=*74915-2* 231
Cyperus piceus f. *potosiensis* IPNI=*74698-2* 230
Cyperus piceus f. *squalidus* IPNI=*74698-2* 230
Cyperus piceus IPNI=*74698-2* 230
Cyperus pittieri IPNI=*75184-2* 231

Cyperus planiculmis IPNI=*304701-1* 230
Cyperus platyculmis IPNI=*304701-1* 230
Cyperus platystachys IPNI=*305797-1* 231
Cyperus platystachyus IPNI=*75178-2* 231
Cyperus pluristamineus IPNI=*30313989-2* 230
Cyperus poiformis IPNI=*74698-2* 230
Cyperus polycephalus IPNI=*74851-2* 231
Cyperus polystachyus IPNI=*74698-2* 230
Cyperus princeae IPNI=*304701-1* 230
Cyperus princeps IPNI=*304595-1* 230
Cyperus prionotropis IPNI=*304426-1* 230
Cyperus procerulus IPNI=*305797-1* 231
Cyperus prolixus var. *vialis* IPNI=*305627-1* 231
Cyperus proteinolepis IPNI=*305797-1* 231
Cyperus protractus IPNI=*304311-1* 230
Cyperus pseudosphacelatus IPNI=*75184-2* 231
Cyperus pseudosurinamensis IPNI=*74851-2* 231
Cyperus psilostachys IPNI=*301880-2* 230
Cyperus punctatifolius IPNI=*305019-1* 231
Cyperus purpureovaginatus IPNI=*301880-2* 230
Cyperus purpureovariegatus IPNI=*305797-1* 231
Cyperus purshii IPNI=*324890-2* 231
Cyperus pygmaeus IPNI=*324890-2* 231
Cyperus quitensis IPNI=*303735-1* 230
Cyperus radiatus IPNI=*304793-1* 230
Cyperus radiatus var. *capitatus* IPNI=*304793-1* 230
Cyperus radiatus var. *elongatus* IPNI=*304793-1* 230
Cyperus radiatus var. *minor* IPNI=*304793-1* 230
Cyperus radicosus IPNI=*305797-1* 231
Cyperus regiomontanus var. *pallens* IPNI=*75184-2* 231
Cyperus regiomontanus var. *pittieri* IPNI=*75184-2* 231
Cyperus rehmannianus IPNI=*74698-2* 230
Cyperus repens IPNI=*300668-2* 230
Cyperus rionensis IPNI=*305019-1* 231
Cyperus riparius IPNI=*304701-1* 230
Cyperus rotundus f. *depallescens* IPNI=*305797-1* 231
Cyperus rotundus subsp. *brevibracteatus* IPNI=*305797-1* 231
Cyperus rotundus subsp. *comosus* IPNI=*305797-1* 231
Cyperus rotundus subsp. *divaricatus* IPNI=*305797-1* 231
Cyperus rotundus subsp. *retzii* IPNI=*305797-1* 231
Cyperus rotundus subsp. *tuberosus* IPNI=*305797-1* 231
Cyperus rotundus var. *amaliae* IPNI=*305797-1* 231
Cyperus rotundus var. *brevibracteatus* IPNI=*305797-1* 231
Cyperus rotundus var. *carinalis* IPNI=*305797-1* 231
Cyperus rotundus var. *carinatus* IPNI=*305797-1* 231
Cyperus rotundus var. *centiflorus* IPNI=*305797-1* 231
Cyperus rotundus var. *disruptus* IPNI=*305797-1* 231
Cyperus rotundus var. *hydra* IPNI=*305797-1* 231
Cyperus rotundus var. *nubicus* IPNI=*305797-1* 231
Cyperus rotundus var. *pallidus* IPNI=*305797-1* 231
Cyperus rotundus var. *salsolus* IPNI=*305797-1* 231
Cyperus roylei IPNI=*304793-1* 230
Cyperus rubicundus IPNI=*305797-1* 231
Cyperus rubrotinctus IPNI=*301880-2* 230
Cyperus rudioi IPNI=*305797-1* 231
Cyperus rudioi var. *minor* IPNI=*305797-1* 231
Cyperus ruficomus IPNI=*300668-2* 230
Cyperus rufus IPNI=*305019-1* 231
Cyperus ruizianus IPNI=*74768-2* 230
Cyperus samoensis IPNI=*304256-1* 230
Cyperus sanguineofuscus IPNI=*305060-1* 231
Cyperus santonici IPNI=*304844-1* 230
Cyperus schottii IPNI=*305627-1* 231
Cyperus schraderianus IPNI=*301880-2* 230
Cyperus semidives IPNI=*304793-1* 230
Cyperus serrulatus IPNI=*304426-1* 230
Cyperus sertularinus IPNI=*74768-2* 230
Cyperus sertularinus var. *disciger* IPNI=*75178-2* 231
Cyperus sesquiflorus f. *latifolius* IPNI=*77140538-1* 231
Cyperus setaceus IPNI=*74698-2* 230
Cyperus sieberianus IPNI=*300668-2* 230
Cyperus sintenisii IPNI=*305019-1* 231
Cyperus sphaerostachys IPNI=*74851-2* 231
Cyperus spicatocapitatus IPNI=*305019-1* 231
Cyperus spicatus IPNI=*304793-1* 230
Cyperus squalidus IPNI=*74698-2* 230
Cyperus squamulatus IPNI=*301880-2* 230
Cyperus squarrosus var. *congestus* IPNI=*324890-2* 231
Cyperus squarrosus var. *cylindraceus* IPNI=*324890-2* 231
Cyperus squarrosus var. *runyonii* IPNI=*324890-2* 231
Cyperus stellatus IPNI=*304701-1* 230
Cyperus stramineus IPNI=*74698-2* 230
Cyperus subambiguus IPNI=*74768-2* 230
Cyperus subarticulatus IPNI=*303822-1* 230
Cyperus subintermedius IPNI=*74698-2* 230
Cyperus subnodosus IPNI=*303822-1* 230
Cyperus subrotundus IPNI=*304311-1* 230
Cyperus surinamensis var. *strictus* IPNI=*74636-2* 230
Cyperus tabernaemontani IPNI=*1019164-1* 232
Cyperus tanaensis IPNI=*74698-2* 230
Cyperus taylorii IPNI=*305797-1* 231
Cyperus tenoreanus IPNI=*300668-2* 230
Cyperus tenorei IPNI=*300668-2* 230
Cyperus tenuifolius IPNI=*305797-1* 231
Cyperus tenuis var. *brevior* IPNI=*75178-2* 231
Cyperus tenuis var. *grandiceps* IPNI=*75178-2* 231
Cyperus tetragonus IPNI=*74768-2* 230
Cyperus tetrastachyos IPNI=*305797-1* 231
Cyperus thyrsiflorus IPNI=*305019-1* 231
Cyperus thyrsiflorus var. *anceps* IPNI=*75184-2* 231
Cyperus thyrsiflorus var. *macrior* IPNI=*75184-2* 231
Cyperus thyrsiflorus var. *vegetior* IPNI=*75184-2* 231
Cyperus tolucensis IPNI=*304973-1* 231
Cyperus tonduzianus IPNI=*74698-2* 230
Cyperus tribrachiatus IPNI=*75184-2* 231
Cyperus tribrachiatus var. *tenuis* IPNI=*75184-2* 231
Cyperus trigonus IPNI=*305019-1* 231
Cyperus trinitatis IPNI=*74851-2* 231
Cyperus triqueter IPNI=*304701-1* 230
Cyperus tristachyus IPNI=*74698-2* 230
Cyperus tristachyus var. *elongatus* IPNI=*74698-2* 230
Cyperus tristachyus var. *minor* IPNI=*74698-2* 230
Cyperus tuberosus IPNI=*300668-2* 230
Cyperus tuberosus IPNI=*305797-1* 231
Cyperus tucumanensis IPNI=*74636-2* 230
Cyperus tunicatus IPNI=*304701-1* 230
Cyperus ubertus IPNI=*305019-1* 231
Cyperus umbellatus var. *laxatus* IPNI=*304256-1* 230
Cyperus umbrosus IPNI=*304973-1* 231
Cyperus uncinatus IPNI=*324890-2* 231
Cyperus vaginatus IPNI=*304701-1* 230
Cyperus variabilis IPNI=*300668-2* 230
Cyperus variegatus IPNI=*74915-2* 231
Cyperus vegetus IPNI=*304426-1* 230
Cyperus vegetus var. *compactus* IPNI=*304426-1* 230
Cyperus vegetus var. *obtusangulus* IPNI=*304426-1* 230
Cyperus vegetus var. *triangularis* IPNI=*304426-1* 230
Cyperus vegetus var. *trigonus* IPNI=*304426-1* 230
Cyperus venustus IPNI=*304256-1* 230
Cyperus versicolor IPNI=*324890-2* 231
Cyperus verticillatus IPNI=*304793-1* 230
Cyperus vialis IPNI=*305627-1* 231
Cyperus viridis IPNI=*304311-1* 230
Cyperus viridis IPNI=*305797-1* 231
Cyperus viridispicatus IPNI=*304793-1* 230
Cyperus vulnerans IPNI=*305019-1* 231

Cyperus weinlandii IPNI=*305797-1* 231
Cyperus xanthinus IPNI=*74698-2* 230
Cyperus yoshinagae IPNI=*305797-1* 231
Cypripedium caudatum IPNI=*194355-2* 222
Cypripedium cothurnum IPNI=*325522-2* 217
Cypripedium lindenii IPNI=*1119571-2* 222
Cyprolepis denudata IPNI=*74934-2* 231
Cyrilla antillana IPNI=*319176-2* 298
Cyrilla antillana var. *perrottetii* IPNI=*319176-2* 298
Cyrilla arida IPNI=*319176-2* 298
Cyrilla brevifolia IPNI=*319176-2* 298
Cyrilla caroliniana IPNI=*319176-2* 298
Cyrilla fuscata IPNI=*319176-2* 298
Cyrilla parvifolia IPNI=*319176-2* 298
Cyrilla perrottetii IPNI=*319176-2* 298
Cyrilla polystachia IPNI=*319176-2* 298
Cyrilla pulchella IPNI=*2352-2* 365
Cyrilla racemiflora subsp. *parvifolia* IPNI=*319176-2* 298
Cyrilla racemiflora var. *brevifolia* IPNI=*319176-2* 298
Cyrilla racemiflora var. *parvifolia* IPNI=*319176-2* 298
Cyrilla racemiflora var. *subglobosa* IPNI=*319176-2* 298
Cyrtacanthus corymbosus IPNI=*947317-1* 361
Cyrtandra capsularis IPNI=*799766-1* 374
Cyrtandra glaberrima IPNI=*799766-1* 374
Cyrtandra serrata IPNI=*799766-1* 374
Cyrtandromoea minor IPNI=*378695-1* 366
Cyrtanthera carnea IPNI=*132826-2* 360
Cyrtanthera ghiesbreghtiana IPNI=*133119-2* 360
Cyrtanthera longiflora IPNI=*132826-2* 360
Cyrtanthera magnifica IPNI=*132826-2* 360
Cyrtanthera magnifica var. *minor* IPNI=*132826-2* 360
Cyrtanthera pohliana IPNI=*132826-2* 360
Cyrtanthera pohliana var. *obtusior* IPNI=*132826-2* 360
Cyrtanthera pohliana var. *velutina* IPNI=*132826-2* 360
Cyrtochilum ramosissimum f. *albomaculatum* IPNI=*1021207-1* 218
Cyrtochilum ramosissimum f. *xanthinum* IPNI=*1021207-1* 218
Cyrtochilum superbiens IPNI=*75479-2* 218
Cyrtodeira chontalensis IPNI=*380794-1* 367
Cyrtodeira cupreata f. *metallica* IPNI=*380772-1* 367
Cyrtodeira cupreata IPNI=*380772-1* 367
Cyrtodeira fulgida IPNI=*380813-1* 367
Cyrtodeira trianae IPNI=*380772-1* 367
Cyrtoglottis gracilipes IPNI=*77155011-1* 221
Cyrtophlebium angustifolium IPNI=*17063000-1* 175
Cyrtophlebium phyllitidis IPNI=*17063380-1* 175
Cyrtopodium andersonii var. *flavescens* IPNI=*1173554-2* 218
Cyrtopodium flavescens IPNI=*1173554-2* 218
Cyrtopodium roraimense IPNI=*1173554-2* 218
Cyrtopodium speciosissimum IPNI=*1147853-2* 218
Cyrtorhyncha cymbalaria IPNI=*711250-1* 436
Cyrtorhyncha cymbalaria subsp. *alpina* IPNI=*711250-1* 436
Cyrtospadix bicolor IPNI=*30010088-2* 199
Cyrtospadix striatipes IPNI=*909842-1* 203
Cyrtosperma spruceanum IPNI=*991397-1* 200
Cyrtostachys lakka IPNI=*666411-1* 208
Cyrtostemma atropurpureum IPNI=*319717-1* 296
Cystea fragilis IPNI=*30013429-2* 174
Cystopteris filix-femina IPNI=*30055316-2* 174
Cystopteris filix-fragilis IPNI=*30013429-2* 174
Cystopteris fragilis subsp. *eufragilis* IPNI=*30013429-2* 174
Cytheraea dulcis IPNI=*71455-1* 451
Cytisus cajan IPNI=*1152177-2* 313
Cytisus canariensis IPNI=*495939-1* 320
Cytisus candicans IPNI=*496189-1* 320
Cytisus guineensis IPNI=*1152177-2* 313
Cytisus hillebrandtii IPNI=*495939-1* 320
Cytisus junceus IPNI=*519057-1* 334
Cytisus kunzeanus IPNI=*496189-1* 320
Cytisus major IPNI=*143380-2* 325
Cytisus membranaceus IPNI=*502971-1* 325
Cytisus mollis IPNI=*1152177-2* 313
Cytisus monspessulanus IPNI=*496189-1* 320
Cytisus paniculatus IPNI=*495939-1* 320
Cytisus philippii IPNI=*495939-1* 320
Cytisus pseudocajan IPNI=*1152177-2* 313
Cytisus pubescens IPNI=*496189-1* 320
Cytisus ramosissimus IPNI=*495939-1* 320
Cytisus rhodopnoa IPNI=*495939-1* 320
Cytisus syriacus IPNI=*496189-1* 320
Cytisus tetragonus IPNI=*284993-2* 334
Czerniaevia rupestris IPNI=*400626-1* 238

D

Dacryodes trinitensis IPNI=*75717-2* 451
Dactilon officinale IPNI=*397065-1* 236
Dactilon sanguinale IPNI=*399115-1* 238
Dactyloctenium aegyptiacum IPNI=*397387-1* 236
Dactyloctenium aegyptium f. *viviparum* IPNI=*397387-1* 236
Dactyloctenium aegyptium var. *mucronatum* IPNI=*397387-1* 236
Dactyloctenium ciliare IPNI=*397387-1* 236
Dactyloctenium figarei IPNI=*397387-1* 236
Dactyloctenium meridionale IPNI=*397387-1* 236
Dactyloctenium mpuetense IPNI=*397387-1* 236
Dactyloctenium mucronatum IPNI=*397387-1* 236
Dactyloctenium mucronatum var. *erectum* IPNI=*397387-1* 236
Dactyloctenium prostratum IPNI=*397387-1* 236
Dactylodes angulatum IPNI=*424990-1* 246
Dactylodes dactyloides IPNI=*424990-1* 246
Dactylophyllum fragaria IPNI=*30074127-2* 443
Dactylostegium assurgens IPNI=*1114607-2* 359
Dactylostegium portoricense IPNI=*1114607-2* 359
Dactylostemon australis IPNI=*1187218-2* 387
Dactylostemon oligandrus IPNI=*1187218-2* 387
Dactylostemon oligandrus var. *parvifolia* IPNI=*1187218-2* 387
Dactylus officinalis IPNI=*397065-1* 236
Daedalacanthus nervosus IPNI=*48879-1* 360
Daedalacanthus scaber IPNI=*48879-1* 360
Dahlgrenia ptariana IPNI=*79985-2* 208
Dahlia astrantiiflora IPNI=*30027297-2* 261
Dahlia dumicola IPNI=*200959-1* 261
Dahlia lehmannii IPNI=*200959-1* 261
Dahlia lehmannii var. *leucantha* IPNI=*200959-1* 261
Dahlia maximiliana IPNI=*200959-1* 261
Dahlia maxonii IPNI=*200959-1* 261
Dahlia mexicana IPNI=*30027297-2* 261
Dahlia nana IPNI=*30027297-2* 261
Dahlia pinnata var. *nana* IPNI=*30027297-2* 261
Dahlia pinnata var. *variabilis* IPNI=*30027297-2* 261
Dahlia purpurea IPNI=*30027297-2* 261
Dahlia purpurea var. *flavescens* IPNI=*30027297-2* 261
Dahlia purpurea var. *lilacina* IPNI=*30027297-2* 261
Dahlia purpurea var. *pallida* IPNI=*30027297-2* 261
Dahlia purpurea var. *rubra* IPNI=*30027297-2* 261
Dahlia pusilla IPNI=*30027297-2* 261
Dahlia rosea IPNI=*30027297-2* 261
Dahlia royleana IPNI=*30027297-2* 261
Dahlia sambucifolia IPNI=*30027297-2* 261
Dahlia sphondyliifolia IPNI=*30027297-2* 261
Dahlia superflua IPNI=*30027297-2* 261
Dahlia variabilis IPNI=*30027297-2* 261
Dahlia zimapani IPNI=*198320-1* 260
Dalbergaria aureonitens IPNI=*378985-1* 367
Dalbergaria cubensis IPNI=*378985-1* 367
Dalbergaria dimidiata IPNI=*63559-2* 366
Dalbergaria eburnea IPNI=*969729-1* 366
Dalbergaria ericae IPNI=*378844-1* 366

Dalbergaria evolvens IPNI=*63584-2* 366
Dalbergaria guttata IPNI=*63584-2* 366
Dalbergaria kahlbreyeriana IPNI=*305407-2* 366
Dalbergaria madisonii IPNI=*63584-2* 366
Dalbergaria medicinalis IPNI=*999467-1* 366
Dalbergaria phoenicea IPNI=*378985-1* 367
Dalbergaria picta IPNI=*63643-2* 366
Dalbergaria pulcherrima IPNI=*63653-2* 367
Dalbergaria puyana IPNI=*63584-2* 366
Dalbergaria sanguinea IPNI=*378985-1* 367
Dalbergia amerimnum IPNI=*75919-2* 317
Dalbergia aturensis IPNI=*75954-2* 317
Dalbergia cucullata IPNI=*16766-2* 311
Dalbergia domingensis IPNI=*143380-2* 325
Dalbergia foliacea IPNI=*1006519-1* 317
Dalbergia guineensis IPNI=*143380-2* 325
Dalbergia heptaphylla IPNI=*502971-1* 325
Dalbergia laevigata IPNI=*77091925-1* 325
Dalbergia macrophylla IPNI=*143380-2* 325
Dalbergia malifolia IPNI=*75934-2* 317
Dalbergia monetaria IPNI=*77117114-1* 317
Dalbergia monetaria var. *brachyphylla* IPNI=*77117114-1* 317
Dalbergia pentaphylla IPNI=*502971-1* 325
Dalbergia pseudoarborea IPNI=*490433-1* 317
Dalbergia punctata IPNI=*503074-1* 325
Dalbergia rimosa var. *foliacea* IPNI=*1006519-1* 317
Dalbergia tomentosa IPNI=*75939-2* 317
Dalbergia variabilis IPNI=*75939-2* 317
Dalbergia variabilis var. *intermedia* IPNI=*75939-2* 317
Dalbergia variabilis var. *tomentosa* IPNI=*75939-2* 317
Dalbergia violacea IPNI=*503131-1* 325
Dalbergia volubilis IPNI=*77117114-1* 317
Dalea acutifolia IPNI=*76085-2* 317
Dalea angustifolia IPNI=*76085-2* 317
Dalea annua IPNI=*76085-2* 317
Dalea annua var. *cliffortiana* IPNI=*76085-2* 317
Dalea annua var. *ebracteata* IPNI=*76085-2* 317
Dalea annua var. *willdenowii* IPNI=*76085-2* 317
Dalea dalea IPNI=*76085-2* 317
Dalea glandulosa IPNI=*76085-2* 317
Dalea mutisii IPNI=*76086-2* 317
Dalea nigra IPNI=*76085-2* 317
Dalea pumila IPNI=*76085-2* 317
Dalea robusta IPNI=*76085-2* 317
Dalea tephrosioides IPNI=*499658-1* 321
Dalea virgata IPNI=*76085-2* 317
Dalechampia cissifolia subsp. *panamensis* IPNI=*344031-1* 389
Dalechampia dioscoreifolia var. *genuina* IPNI=*314158-2* 389
Dalechampia dioscoreifolia var. *pubescens* IPNI=*314158-2* 389
Dalechampia duplicata IPNI=*344031-1* 389
Dalechampia guatemalensis IPNI=*344031-1* 389
Dalechampia heteromorpha IPNI=*344031-1* 389
Dalechampia heterophylla IPNI=*344174-1* 389
Dalechampia magnobracteata IPNI=*344174-1* 389
Dalechampia molliuscula IPNI=*344031-1* 389
Dalechampia panamensis IPNI=*344031-1* 389
Dalechampia peruviana IPNI=*344174-1* 389
Dalechampia pruriens IPNI=*344174-1* 389
Dalechampia ruiziana IPNI=*314158-2* 389
Dalechampia scandens var. *genuina* IPNI=*319177-2* 389
Dalechampia scandens var. *trisecta* IPNI=*344031-1* 389
Dalechampia serrulata IPNI=*344174-1* 389
Dalechampia tiliifolia var. *heterophylla* IPNI=*344174-1* 389
Dalechampia tiliifolia var. *peruviana* IPNI=*344174-1* 389
Dalechampia triphylla var. *cissifolia* IPNI=*344031-1* 389
Dalechampia triphylla var. *mexicana* IPNI=*344031-1* 389
Dallwatsonia pilosa IPNI=*77144198-1* 244
Dalrympelea domingensis IPNI=*785554-1* 292
Damasonium flavum IPNI=*58404-1* 198
Damasonium maximum IPNI=*58404-1* 198
Dancera hirta IPNI=*77164247-1* 420
Danthonia pseudospicata IPNI=*393559-1* 235
Dantia palustris IPNI=*30031106-2* 429
Daphne americana IPNI=*76687-2* 415
Daphne bonplandiana IPNI=*76687-2* 415
Daphne lagetto IPNI=*76687-2* 415
Daphne obovata IPNI=*76687-2* 415
Daphnopsis lindenii IPNI=*76687-2* 415
Darluca prostrata IPNI=*1187456-2* 350
Darwinia exaltata IPNI=*284993-2* 334
Dasistoma integrifolia IPNI=*799460-1* 375
Dasistoma levigata IPNI=*799460-1* 375
Dasyanthus conglobatus IPNI=*212994-1* 264
Dasycarpus quadrivalvis IPNI=*324587-1* 431
Dasynema alnifolium IPNI=*835145-1* 431
Dasynema cuneifolium IPNI=*835145-1* 431
Dasynema hirsutum IPNI=*835147-1* 431
Dasynema obtusum IPNI=*237456-2* 431
Dasynema ochrocarpum IPNI=*835147-1* 431
Dasynema pubescens IPNI=*835145-1* 431
Dasynema riparium IPNI=*835147-1* 431
Dasynema salicifolium IPNI=*835147-1* 431
Daucus agrestis IPNI=*841063-1* 252
Daucus alatus IPNI=*841063-1* 252
Daucus allionii IPNI=*841063-1* 252
Daucus australis IPNI=*841063-1* 252
Daucus blanchei IPNI=*841063-1* 252
Daucus brevicaulis IPNI=*841063-1* 252
Daucus carota f. *epurpuratus* IPNI=*841063-1* 252
Daucus carota f. *fischeri* IPNI=*841063-1* 252
Daucus carota f. *goodmanii* IPNI=*841063-1* 252
Daucus carota f. *roseus* IPNI=*841063-1* 252
Daucus carota subsp. *boissieri* IPNI=*841063-1* 252
Daucus carota var. *brachycaulos* IPNI=*841063-1* 252
Daucus carota var. *linearis* IPNI=*841063-1* 252
Daucus carota var. *pseudocarota* IPNI=*841063-1* 252
Daucus communis IPNI=*841063-1* 252
Daucus communis var. *pseudocarota* IPNI=*841063-1* 252
Daucus dentatus IPNI=*841063-1* 252
Daucus esculentus IPNI=*841063-1* 252
Daucus exiguus IPNI=*841063-1* 252
Daucus foliosus IPNI=*841063-1* 252
Daucus gibbosus IPNI=*841063-1* 252
Daucus gingidium IPNI=*841063-1* 252
Daucus glaber IPNI=*837486-1* 251
Daucus heterophylus IPNI=*841063-1* 252
Daucus kotovii IPNI=*841063-1* 252
Daucus laevis IPNI=*849956-1* 253
Daucus levis IPNI=*841063-1* 252
Daucus marcidus IPNI=*841063-1* 252
Daucus maritimus IPNI=*841063-1* 252
Daucus martellii IPNI=*841063-1* 252
Daucus montanus IPNI=*841063-1* 252
Daucus neglectus IPNI=*841063-1* 252
Daucus nudicaulis IPNI=*841063-1* 252
Daucus officinalis IPNI=*841063-1* 252
Daucus parsae IPNI=*837486-1* 251
Daucus polygamus IPNI=*841063-1* 252
Daucus scariosus IPNI=*841063-1* 252
Daucus sciadophylus IPNI=*841063-1* 252
Daucus strigosus IPNI=*841063-1* 252
Daucus sylvestris IPNI=*841063-1* 252
Daucus toriloides IPNI=*841167-1* 252
Daucus visnaga IPNI=*849956-1* 253
Daucus vulgaris IPNI=*841063-1* 252
Davallia bornmulleri IPNI=*17078300-1* 175

Davallia caudata IPNI=*17080730-1* 175
Davallia elegans IPNI=*17080730-1* 175
Davallia elmeri IPNI=*17080730-1* 175
Davallia falcata IPNI=*17160730-1* 176
Davallia lindleyi IPNI=*17080730-1* 175
Davallia magellanica IPNI=*17080730-1* 175
Davallia ornata IPNI=*17080730-1* 175
Davallia plumosa IPNI=*17080730-1* 175
Davallia procera IPNI=*17080730-1* 175
Davallia robinsonii IPNI=*17080730-1* 175
Davallia solida var. *caudata* IPNI=*17080730-1* 175
Davallia solida var. *lindleyi* IPNI=*17080730-1* 175
Davallia solida var. *ornata* IPNI=*17080730-1* 175
Davallia splendens IPNI=*17080730-1* 175
Davallia subsolida IPNI=*17080730-1* 175
Davallia tahitensis IPNI=*17080730-1* 175
Davalliopsis elegans IPNI=*17229030-1* 173
Davilla aspera var. *matudae* IPNI=*77025-2* 295
Davilla aspera var. *tenuis* IPNI=*316577-1* 295
Davilla asperrima IPNI=*316577-1* 295
Davilla brasiliana IPNI=*316577-1* 295
Davilla brasiliana IPNI=*328431-2* 295
Davilla ciliata IPNI=*77025-2* 295
Davilla densa IPNI=*316577-1* 295
Davilla densiflora IPNI=*77025-2* 295
Davilla itaparicensis IPNI=*77025-2* 295
Davilla lacunosa var. *minor* IPNI=*77025-2* 295
Davilla lucida IPNI=*316577-1* 295
Davilla lucida var. *tenuis* IPNI=*316577-1* 295
Davilla macrophylla IPNI=*77025-2* 295
Davilla matudae IPNI=*77025-2* 295
Davilla multiflora IPNI=*77025-2* 295
Davilla multiflora var. *acutangula* IPNI=*77025-2* 295
Davilla ovata IPNI=*316577-1* 295
Davilla piauhiensis IPNI=*77025-2* 295
Davilla pilosa IPNI=*328431-2* 295
Davilla pseudorugosa IPNI=*316577-1* 295
Davilla rugosa var. *capitata* IPNI=*328431-2* 295
Davilla rugosa var. *kunthii* IPNI=*328431-2* 295
Davilla rugosa var. *luschnathii* IPNI=*328431-2* 295
Davilla rugosa var. *martii* IPNI=*328431-2* 295
Davilla rugosa var. *sellowii* IPNI=*328431-2* 295
Davilla rugosa var. *willdenowii* IPNI=*328431-2* 295
Davilla sagrana IPNI=*77025-2* 295
Davilla suaveolens IPNI=*77025-2* 295
Davilla surinamensis IPNI=*316577-1* 295
Davilla tenuis IPNI=*316577-1* 295
Davya adscendens IPNI=*566051-1* 417
Davya guianensis IPNI=*566051-1* 417
Davya peruviana IPNI=*566051-1* 417
Debraea violacea IPNI=*870864-1* 430
Decaloba alnifolia IPNI=*674768-1* 397
Decaloba biflora IPNI=*674824-1* 398
Decaloba bilobata IPNI=*321950-2* 398
Decaloba bogotensis IPNI=*674832-1* 398
Decaloba bryonioides IPNI=*674847-1* 398
Decaloba capsularis IPNI=*321950-2* 398
Decaloba cuneata IPNI=*674933-1* 398
Decaloba cyathophora IPNI=*674805-1* 398
Decaloba filipes IPNI=*674993-1* 398
Decaloba geminiflora IPNI=*675584-1* 400
Decaloba hemicycla IPNI=*675584-1* 400
Decaloba jacquinii IPNI=*674821-1* 398
Decaloba lunata IPNI=*77063117-1* 398
Decaloba mollis IPNI=*675225-1* 399
Decaloba obscura IPNI=*321966-2* 398
Decaloba piligera IPNI=*321950-2* 398
Decaloba punctata IPNI=*675382-1* 400
Decaloba retusa IPNI=*321967-2* 399
Decaloba rohrii IPNI=*674805-1* 398
Decaloba rubra IPNI=*675423-1* 400
Decaloba semilunaris IPNI=*321950-2* 398
Decaloba smithii IPNI=*321950-2* 398
Decaloba surinamensis IPNI=*675584-1* 400
Decaloba vespertilio IPNI=*675584-1* 400
Decandolia pumila IPNI=*30197865-2* 233
Decandolia stolonifera IPNI=*386353-1* 233
Decandolia vulgaris IPNI=*30197865-2* 233
Decaraphe hostmannii IPNI=*1000103-2* 422
Decarinium glandulosum IPNI=*70233-2* 388
Deckeria corneto IPNI=*667594-1* 210
Deckeria lamarckiana IPNI=*666671-1* 208
Deckeria phaeocarpa IPNI=*667594-1* 210
Deckeria ventricosa IPNI=*667594-1* 210
Declieuxia roraimensis IPNI=*759372-1* 354
Decussocarpus rospigliosii IPNI=*946471-1* 180
Deguelia alata IPNI=*1005531-1* 317
Deguelia arborescens IPNI=*519836-1* 334
Deguelia javanica IPNI=*486606-1* 315
Deianira divaricata IPNI=*371484-1* 345
Delima castaneifolia IPNI=*332108-2* 295
Delima dasyphylla f. *angustifolia* IPNI=*82916-2* 295
Delima dasyphylla IPNI=*82916-2* 295
Delima guianensis IPNI=*30327947-2* 295
Delima mexicana IPNI=*317470-1* 295
Delima nitida IPNI=*77025-2* 295
Delima tomentosa IPNI=*82916-2* 295
Delonix regia var. *flavida* IPNI=*491231-1* 317
Delostoma dombeyanum IPNI=*109429-1* 363
Delostoma hookeri IPNI=*109429-1* 363
Delostoma latifolium IPNI=*42392-2* 362
Delostoma loxense IPNI=*109429-1* 363
Delostoma nervosum IPNI=*109429-1* 363
Delostoma weberbauerianum IPNI=*109429-1* 363
Delphinastrum elatum IPNI=*710585-1* 436
Delphinastrum hybridum IPNI=*710585-1* 436
Delphinastrum urceolatum IPNI=*710585-1* 436
Delphinium addendum IPNI=*77254-2* 436
Delphinium ajacis f. *alba* IPNI=*77254-2* 436
Delphinium alpinum var. *hebecarpum* IPNI=*710585-1* 436
Delphinium ambiguum IPNI=*77254-2* 436
Delphinium anomalum IPNI=*710585-1* 436
Delphinium atropurpureum IPNI=*710585-1* 436
Delphinium azureum IPNI=*77254-2* 436
Delphinium candelabrum IPNI=*77254-2* 436
Delphinium clusianum IPNI=*710585-1* 436
Delphinium cryophilum IPNI=*710585-1* 436
Delphinium davuricum IPNI=*710585-1* 436
Delphinium elatum f. *grandiflorum* IPNI=*710585-1* 436
Delphinium elatum f. *parviflorum* IPNI=*710585-1* 436
Delphinium elatum f. *productum* IPNI=*710585-1* 436
Delphinium elatum subsp. *bosniacum* IPNI=*710585-1* 436
Delphinium elatum subsp. *cryophilum* IPNI=*710585-1* 436
Delphinium elatum subsp. *macrotepalum* IPNI=*710585-1* 436
Delphinium elatum subsp. *pubicaule* IPNI=*710585-1* 436
Delphinium elatum subsp. *tiroliense* IPNI=*710585-1* 436
Delphinium elatum var. *alpinum* IPNI=*710585-1* 436
Delphinium elatum var. *giganteum* IPNI=*710585-1* 436
Delphinium elatum var. *glabrum* IPNI=*710585-1* 436
Delphinium elatum var. *gracile* IPNI=*710585-1* 436
Delphinium elatum var. *grandifolium* IPNI=*710585-1* 436
Delphinium elisabethae var. *eglandulosum* IPNI=*710585-1* 436
Delphinium gratum IPNI=*710585-1* 436
Delphinium hirsutum IPNI=*710585-1* 436
Delphinium hispidum IPNI=*710585-1* 436
Delphinium hybridum IPNI=*710585-1* 436

Delphinium kioviense IPNI=*710585-1* 436
Delphinium macrotepalum IPNI=*710585-1* 436
Delphinium mixtum IPNI=*710585-1* 436
Delphinium napellus IPNI=*707615-1* 436
Delphinium orientale IPNI=*77254-2* 436
Delphinium ornatum IPNI=*77254-2* 436
Delphinium palmatifidum IPNI=*710585-1* 436
Delphinium pauciflorum IPNI=*77254-2* 436
Delphinium simplex IPNI=*77254-2* 436
Delphinium speciosum IPNI=*710585-1* 436
Delphinium tauricum IPNI=*710585-1* 436
Delphinium tiroliense IPNI=*710585-1* 436
Delphinium urceolatum IPNI=*710585-1* 436
Dendranthema × *grandiflorum* IPNI=*77111328-1* 259
Dendranthema × *morifolium* IPNI=*77111328-1* 259
Dendranthema × *sinense* IPNI=*77111328-1* 259
Dendranthema coreanum subsp. *littorale* IPNI=*193510-1* 259
Dendranthema coronarium IPNI=*1013480-1* 263
Dendranthema indicum IPNI=*193510-1* 259
Dendranthema indicum var. *acutum* IPNI=*193510-1* 259
Dendranthema indicum var. *edule* IPNI=*193510-1* 259
Dendranthema indicum var. *huludaoense* IPNI=*193510-1* 259
Dendranthema indicum var. *iyoense* IPNI=*193510-1* 259
Dendranthema indicum var. *maruyamanum* IPNI=*193510-1* 259
Dendranthema indicum var. *procumbens* IPNI=*193510-1* 259
Dendranthema indicum var. *tsurugisanense* IPNI=*193510-1* 259
Dendranthema littorale IPNI=*193510-1* 259
Dendranthema littorale subsp. *coreana* IPNI=*193510-1* 259
Dendranthema nankingense IPNI=*193510-1* 259
Dendranthema parthenioides IPNI=*192489-1* 259
Dendranthema parthenium IPNI=*252460-1* 271
Dendrobangia tenuis IPNI=*77582-2* 255
Dendrobium album IPNI=*643250-1* 220
Dendrobium flavescens IPNI=*654273-1* 222
Dendrobium grandiflorum IPNI=*643478-1* 220
Dendrobium longipetalum IPNI=*155147-2* 221
Dendrobium parvum IPNI=*654273-1* 222
Dendrobium polystachyum IPNI=*654273-1* 222
Dendrocalamopsis atrovirens IPNI=*392445-1* 234
Dendrocalamopsis beecheyana IPNI=*392260-1* 234
Dendrocalamopsis beecheyana var. *pubescens* IPNI=*392260-1* 234
Dendrocalamopsis oldhamii f. *revoluta* IPNI=*392445-1* 234
Dendrocalamopsis oldhamii IPNI=*392445-1* 234
Dendrocalamus flagellifer IPNI=*397862-1* 237
Dendrocalamus tulda IPNI=*392551-1* 234
Dendropanax alaris IPNI=*1133937-2* 253
Dendropanax brachypodus IPNI=*1133937-2* 253
Dendropanax capillaris IPNI=*1133937-2* 253
Dendropanax compactus IPNI=*1133937-2* 253
Dendropanax concinnus IPNI=*1133937-2* 253
Dendropanax insularis IPNI=*1133937-2* 253
Dendropanax juergensenii IPNI=*1133937-2* 253
Dendropanax matudae IPNI=*1133937-2* 253
Dendropanax monticola IPNI=*1133937-2* 253
Dendropanax pavonii IPNI=*1133937-2* 253
Dendropanax samydifolius IPNI=*1133937-2* 253
Dendropanax schippii IPNI=*1133937-2* 253
Dendropemon domingensis IPNI=*1075239-2* 450
Dendrophthoe heyneanus IPNI=*551367-1* 449
Dendrophthora crassuloides IPNI=*77736-2* 449
Dendrophthora myrtilloides IPNI=*1075239-2* 450
Dendrophthora portulacoides IPNI=*77736-2* 449
Dendropogon usneoides IPNI=*1169665-2* 229
Dendrorkis extinctoria IPNI=*654273-1* 222
Dendrorkis jussieuana IPNI=*654273-1* 222
Dendrorkis minuta IPNI=*654273-1* 222
Dendrorkis polystachyon IPNI=*654273-1* 222
Dendrorkis purpurea IPNI=*654273-1* 222
Dendrorkis rufinula IPNI=*654273-1* 222
Dendrorkis tessellata IPNI=*654273-1* 222
Dendrorkis zollingeri IPNI=*654273-1* 222
Dendrosicus latifolius IPNI=*108218-1* 362
Dendrosicus saxatilis IPNI=*108218-1* 362
Dendrostylis apeibifolia IPNI=*111855-1* 380
Dendrostylis denticulata IPNI=*111855-1* 380
Dendrostylis echinata IPNI=*111855-1* 380
Dendrostylis grandifolia IPNI=*155434-2* 380
Dendrostylis microphylla IPNI=*111855-1* 380
Dendrostylis pubescens IPNI=*155451-2* 380
Dendrostylis suaveolens IPNI=*155453-2* 380
Denea forsteriana IPNI=*667434-1* 209
Dennstaedtia cicutaria var. *erosa* IPNI=*17081410-1* 175
Dennstaedtia decomposita IPNI=*17081410-1* 175
Dennstaedtia rubiginosa IPNI=*17081410-1* 175
Dermatocalyx parviflorus IPNI=*229565-2* 377
Derris amazonica IPNI=*77178-2* 317
Derris floribunda IPNI=*77143238-1* 317
Derris glabrescens IPNI=*60453072-2* 317
Derris guianensis IPNI=*491205-1* 317
Derris javanica IPNI=*486606-1* 315
Derris latifolia IPNI=*502971-1* 325
Derris moniliformis IPNI=*77118402-1* 328
Derris nicou IPNI=*502971-1* 325
Derris nitidula IPNI=*77143238-1* 317
Derris pterocarpus IPNI=*491205-1* 317
Derris rariflora IPNI=*77116882-1* 317
Derris scandens IPNI=*491205-1* 317
Derris sericea IPNI=*143380-2* 325
Derris utilis IPNI=*1005531-1* 317
Descantaria balbisii IPNI=*77184055-1* 224
Descantaria cumanensis IPNI=*77184055-1* 224
Descantaria diuretica IPNI=*77184055-1* 224
Descantaria elongata IPNI=*77184055-1* 224
Descantaria multiflora IPNI=*77205602-1* 224
Descantaria procumbens IPNI=*77205602-1* 224
Descantaria schlechtendalii IPNI=*77184055-1* 224
Deschampsia baetica IPNI=*386633-1* 233
Descurainia irio IPNI=*237940-2* 278
Desmanthus adenanthera IPNI=*1056199-2* 328
Desmanthus caudatus IPNI=*1009172-2* 319
Desmanthus comosus IPNI=*1056199-2* 328
Desmanthus depressus IPNI=*1125234-2* 317
Desmanthus lacustris IPNI=*509612-1* 328
Desmanthus natans IPNI=*509612-1* 328
Desmanthus plenus IPNI=*1056199-2* 328
Desmanthus polyphyllus IPNI=*1056199-2* 328
Desmanthus pratorum IPNI=*1125234-2* 317
Desmanthus punctatus IPNI=*1056199-2* 328
Desmanthus stolonifer IPNI=*509612-1* 328
Desmanthus tenellus IPNI=*1125234-2* 317
Desmanthus virgatus var. *depressus* IPNI=*1125234-2* 317
Desmochaeta achyranthoides IPNI=*60139-1* 282
Desmochaeta densiflora IPNI=*60139-1* 282
Desmochaeta globosa IPNI=*60185-1* 282
Desmochaeta micrantha IPNI=*60185-1* 282
Desmochaeta prostrata IPNI=*60185-1* 282
Desmochaeta repens IPNI=*60185-1* 282
Desmochaeta sordida IPNI=*327342-2* 281
Desmochaeta uncinata IPNI=*60139-1* 282
Desmodium adhaesivum IPNI=*134286-3* 318
Desmodium adscendens f. *glabrescens* IPNI=*77187388-1* 320
Desmodium adscendens IPNI=*77187388-1* 320
Desmodium adscendens var. *caeruleum* IPNI=*1124646-2* 318
Desmodium adscendens var. *robustum* IPNI=*77187388-1* 320
Desmodium akoense IPNI=*135026-3* 318
Desmodium albiflorum f. *grandifolia* IPNI=*78342-2* 317

Desmodium albiflorum IPNI=*77187443-1* 320
Desmodium albiflorum IPNI=*78342-2* 317
Desmodium ancistrocarpum IPNI=*1015219-2* 318
Desmodium angustifolium var. *gramineum* IPNI=*78351-2* 317
Desmodium angustifolium var. *typicum* IPNI=*78351-2* 317
Desmodium aparines IPNI=*134286-3* 318
Desmodium arenarium IPNI=*135026-3* 318
Desmodium arinense IPNI=*77187388-1* 320
Desmodium asperum IPNI=*78415-2* 317
Desmodium axillare f. *robustius* IPNI=*491595-1* 317
Desmodium axillare var. *acutifolium* IPNI=*491595-1* 317
Desmodium axillare var. *angustatum* IPNI=*491595-1* 317
Desmodium axillare var. *genuinum* IPNI=*491595-1* 317
Desmodium axillare var. *obtusifoliolum* IPNI=*491595-1* 317
Desmodium axillare var. *sintenisii* IPNI=*491595-1* 317
Desmodium axillare var. *stoloniferum* IPNI=*491595-1* 317
Desmodium barbatum IPNI=*77187394-1* 320
Desmodium bullamense IPNI=*77187443-1* 320
Desmodium caespitosum IPNI=*77187388-1* 320
Desmodium campestre IPNI=*1130757-2* 317
Desmodium canum IPNI=*1015219-2* 318
Desmodium canum var. *angustifolium* IPNI=*78351-2* 317
Desmodium cinereum IPNI=*135026-3* 318
Desmodium coeruleum IPNI=*77187388-1* 320
Desmodium distortum var. *pabulare* IPNI=*78415-2* 317
Desmodium diversifolium IPNI=*1015219-2* 318
Desmodium elatum IPNI=*78415-2* 317
Desmodium ellipticum IPNI=*77187388-1* 320
Desmodium emarginatum IPNI=*1130757-2* 317
Desmodium frutescens IPNI=*1015219-2* 318
Desmodium frutescens var. *amblyophyllum* IPNI=*78351-2* 317
Desmodium frutescens var. *angustifolium* IPNI=*78351-2* 317
Desmodium glaucescens IPNI=*77187388-1* 320
Desmodium gramineum IPNI=*78351-2* 317
Desmodium granulatum IPNI=*77187443-1* 320
Desmodium heterocarpon IPNI=*77187416-1* 320
Desmodium hjalmarsonii IPNI=*1109494-2* 318
Desmodium incanum f. *minor* IPNI=*1015219-2* 318
Desmodium incanum f. *obovatifolium* IPNI=*1015219-2* 318
Desmodium incanum var. *angustifolium* IPNI=*78351-2* 317
Desmodium incanum var. *supinum* IPNI=*1015219-2* 318
Desmodium intortum var. *apiculatum* IPNI=*134286-3* 318
Desmodium lindleyi IPNI=*1015219-2* 318
Desmodium linearifolium IPNI=*78351-2* 317
Desmodium mauritianum IPNI=*1015219-2* 318
Desmodium mexicanum IPNI=*1124646-2* 318
Desmodium molle IPNI=*1130757-2* 317
Desmodium multicaule IPNI=*135026-3* 318
Desmodium nantouensis IPNI=*134286-3* 318
Desmodium oblongifolium IPNI=*491595-1* 317
Desmodium obovatum IPNI=*77187388-1* 320
Desmodium ovalifolium IPNI=*77187388-1* 320
Desmodium oxalidifolium IPNI=*77187388-1* 320
Desmodium pabulare IPNI=*78415-2* 317
Desmodium parviflorum IPNI=*135026-3* 318
Desmodium perrottetii IPNI=*78415-2* 317
Desmodium physocarpos IPNI=*1028008-2* 318
Desmodium pilosiusculum IPNI=*1109494-2* 318
Desmodium portoricense IPNI=*1015219-2* 318
Desmodium procumbens var. *typicum* IPNI=*78541-2* 318
Desmodium pulcherrimum IPNI=*1028008-2* 318
Desmodium purpureum IPNI=*1028008-2* 318
Desmodium racemiferum IPNI=*77187388-1* 320
Desmodium radicans IPNI=*491595-1* 317
Desmodium reptans IPNI=*491595-1* 317
Desmodium rubiginosum IPNI=*78415-2* 317
Desmodium sandwicense IPNI=*1109494-2* 318
Desmodium simplex IPNI=*77187388-1* 320
Desmodium sinclairii IPNI=*1109494-2* 318
Desmodium sonorae IPNI=*134286-3* 318
Desmodium sparsiflorum IPNI=*1015219-2* 318
Desmodium spectabile IPNI=*78415-2* 317
Desmodium spirale IPNI=*78541-2* 318
Desmodium spirale var. *stoloniferum* IPNI=*491595-1* 317
Desmodium stipulaceum IPNI=*1028008-2* 318
Desmodium stipulaceum IPNI=*77187443-1* 320
Desmodium stoloniferum IPNI=*491595-1* 317
Desmodium subspicatum IPNI=*78351-2* 317
Desmodium supinum IPNI=*1015219-2* 318
Desmodium supinum var. *amblyophyllum* IPNI=*78351-2* 317
Desmodium supinum var. *angustifolium* IPNI=*78351-2* 317
Desmodium tenue IPNI=*1015219-2* 318
Desmodium terminale IPNI=*1130757-2* 317
Desmodium thwaitesii IPNI=*77187388-1* 320
Desmodium tortuosum var. *hirtellum* IPNI=*1028008-2* 318
Desmodium triflorum IPNI=*77187443-1* 320
Desmodium triflorum var. *pygmaeum* IPNI=*77187443-1* 320
Desmodium trifoliastrum IPNI=*77187388-1* 320
Desmodium trigonum IPNI=*134286-3* 318
Desmodium uncinatum var. *gracile* IPNI=*1109494-2* 318
Desmodium variifolium IPNI=*1015219-2* 318
Desmodium violaceum IPNI=*78415-2* 317
Desmodium virgatum IPNI=*135026-3* 318
Desmodium vogelii IPNI=*77187388-1* 320
Desmoncus aereus IPNI=*666648-1* 208
Desmoncus ataxacanthus IPNI=*666637-1* 208
Desmoncus brevisectus IPNI=*666648-1* 208
Desmoncus caespitosus IPNI=*666648-1* 208
Desmoncus campylacanthus IPNI=*666648-1* 208
Desmoncus dasyacanthus IPNI=*666648-1* 208
Desmoncus duidensis IPNI=*666648-1* 208
Desmoncus inermis IPNI=*666648-1* 208
Desmoncus longisectus IPNI=*666648-1* 208
Desmoncus lophacanthos IPNI=*666637-1* 208
Desmoncus macroacanthos IPNI=*666648-1* 208
Desmoncus macrodon IPNI=*666648-1* 208
Desmoncus maguirei IPNI=*666648-1* 208
Desmoncus minor IPNI=*666648-1* 208
Desmoncus mirandanus IPNI=*666648-1* 208
Desmoncus nemorosus IPNI=*666648-1* 208
Desmoncus oligacanthus IPNI=*666648-1* 208
Desmoncus orthacanthos var. *mitis* IPNI=*666637-1* 208
Desmoncus oxyacanthos IPNI=*666648-1* 208
Desmoncus paraensis IPNI=*666648-1* 208
Desmoncus peraltus IPNI=*666648-1* 208
Desmoncus phengophyllus IPNI=*666648-1* 208
Desmoncus philippianus IPNI=*666648-1* 208
Desmoncus phoenicocarpus IPNI=*666648-1* 208
Desmoncus polyacanthos var. *cuspidatus* IPNI=*666648-1* 208
Desmoncus polyacanthos var. *oxyacanthos* IPNI=*666648-1* 208
Desmoncus prestoei IPNI=*666648-1* 208
Desmoncus pycnacanthos IPNI=*666648-1* 208
Desmoncus pycnacanthos var. *sarmentosus* IPNI=*666648-1* 208
Desmoncus riparius IPNI=*666648-1* 208
Desmoncus ulei IPNI=*666648-1* 208
Desmoncus werdermannii IPNI=*666637-1* 208
Desmonema hirta IPNI=*101651-2* 390
Desmoscelis lychnitoides IPNI=*1168615-2* 418
Desmoscelis mollis IPNI=*1168615-2* 418
Desmoscelis stachyoides IPNI=*1168615-2* 418
Desmoscelis villosa var. *fuscescens* IPNI=*1168615-2* 418
Desmoscelis villosa var. *gracillima* IPNI=*1168615-2* 418
Desmoscelis villosa var. *hypericoides* IPNI=*1168615-2* 418
Desmoscelis villosa var. *lychnitoides* IPNI=*1168615-2* 418
Desmoscelis villosa var. *ovata* IPNI=*1168615-2* 418
Desmoscelis villosa var. *purpureoviolacea* IPNI=*1168615-2* 418

Desmoscelis villosa var. *stachyoides* IPNI=*1168615-2* 418
Deyeuxia araeantha IPNI=*77206801-1* 242
Deyeuxia effusa IPNI=*77206801-1* 242
Deyeuxia funckii IPNI=*77206801-1* 242
Deyeuxia spathacea IPNI=*77206801-1* 242
Deyeuxia subspicata IPNI=*258855-2* 246
Diacantha arborea IPNI=*273073-2* 258
Diacrium bicornutum IPNI=*622297-1* 218
Diadesma rhombifolia IPNI=*235798-2* 413
Dialesta discolor IPNI=*988437-1* 268
Dialesta discolor var. *polychaeta* IPNI=*988437-1* 268
Dialium divaricatum IPNI=*78900-2* 318
Diallobus bicapsularis IPNI=*518330-1* 332
Diallobus falcatus IPNI=*1117728-2* 333
Diallobus subuniflorus IPNI=*234687-2* 334
Diallobus tora IPNI=*518369-1* 333
Diallobus uniflorus IPNI=*234687-2* 334
Dialyanthera acuminata IPNI=*586382-1* 191
Dialyanthera gracilipes IPNI=*586384-1* 191
Dialyanthera otoba IPNI=*177341-2* 191
Dialyanthera parvifolia IPNI=*586389-1* 191
Diandrochloa diarrhena IPNI=*401189-1* 239
Diandrochloa diplachnoides IPNI=*401189-1* 239
Diandrochloa glomerata IPNI=*401189-1* 239
Diandrochloa japonica IPNI=*401189-1* 239
Diandrochloa namaquensis IPNI=*401189-1* 239
Dianella cubensis IPNI=*533580-1* 215
Dianthera angularis IPNI=*1114607-2* 359
Dianthera celebica IPNI=*52558-1* 360
Dianthera decumbens IPNI=*78920-2* 359
Dianthera elongata IPNI=*172440-2* 360
Dianthera eustachiana IPNI=*224052-2* 361
Dianthera humifusa IPNI=*78920-2* 359
Dianthera lucida IPNI=*47308-1* 359
Dianthera mexicana IPNI=*1140399-2* 359
Dianthera secunda var. *geniculata* IPNI=*47308-1* 359
Dianthera secunda var. *holtonii* IPNI=*47308-1* 359
Dianthera secunda var. *lucida* IPNI=*47308-1* 359
Dianthera sexangularis IPNI=*1114607-2* 359
Dianthera sonorae IPNI=*56148-1* 361
Dianthera tomentosa IPNI=*52558-1* 360
Dianthera violacea IPNI=*50468-1* 360
Dianthus acinifolius IPNI=*153205-1* 285
Dianthus aggregatus IPNI=*306674-2* 285
Dianthus altaicus IPNI=*153215-1* 285
Dianthus amurensis IPNI=*153215-1* 285
Dianthus arbuscula IPNI=*153205-1* 285
Dianthus armeria f. *albiviridis* IPNI=*310755-2* 285
Dianthus armeria f. *glabrissimus* IPNI=*310755-2* 285
Dianthus armeria var. *laevis* IPNI=*310755-2* 285
Dianthus arrectus IPNI=*153205-1* 285
Dianthus barbatus var. *latifolius* IPNI=*306674-2* 285
Dianthus barbatus var. *paniculatus* IPNI=*306674-2* 285
Dianthus barbatus var. *pedunculosus* IPNI=*306674-2* 285
Dianthus binatus IPNI=*153205-1* 285
Dianthus blandus IPNI=*153717-1* 285
Dianthus carolinianus IPNI=*310755-2* 285
Dianthus caryophyllus f. *intermedius* IPNI=*153205-1* 285
Dianthus caryophyllus var. *carduinus* IPNI=*153205-1* 285
Dianthus chinensis f. *albiflora* IPNI=*153215-1* 285
Dianthus chinensis subsp. *versicolor* IPNI=*153215-1* 285
Dianthus chinensis var. *amurensis* IPNI=*153215-1* 285
Dianthus chinensis var. *jingpoensis* IPNI=*153215-1* 285
Dianthus chinensis var. *liaotungensis* IPNI=*153215-1* 285
Dianthus chinensis var. *morii* IPNI=*153215-1* 285
Dianthus chinensis var. *serpens* IPNI=*153215-1* 285
Dianthus chinensis var. *shandongensis* IPNI=*153215-1* 285
Dianthus chinensis var. *subulifolius* IPNI=*153215-1* 285
Dianthus chinensis var. *versicolor* IPNI=*153215-1* 285
Dianthus coronarius IPNI=*153205-1* 285
Dianthus corsicus IPNI=*153205-1* 285
Dianthus corymbosus IPNI=*306674-2* 285
Dianthus dentosus IPNI=*153215-1* 285
Dianthus dubius IPNI=*153717-1* 285
Dianthus epirotus IPNI=*310755-2* 285
Dianthus fischeri IPNI=*153215-1* 285
Dianthus flaccidus IPNI=*153408-1* 285
Dianthus girardinii IPNI=*306674-2* 285
Dianthus hirsutus IPNI=*310755-2* 285
Dianthus hortensis IPNI=*153717-1* 285
Dianthus hybridus IPNI=*310755-2* 285
Dianthus ibericus IPNI=*153215-1* 285
Dianthus jeniseensis IPNI=*153215-1* 285
Dianthus kayserianus IPNI=*153205-1* 285
Dianthus laciniatus IPNI=*153215-1* 285
Dianthus latifolius IPNI=*306674-2* 285
Dianthus longicaulis IPNI=*153205-1* 285
Dianthus miniatus IPNI=*153205-1* 285
Dianthus morii IPNI=*153215-1* 285
Dianthus morrisii IPNI=*153205-1* 285
Dianthus moschatus IPNI=*153205-1* 285
Dianthus multinervis IPNI=*153205-1* 285
Dianthus ochroleucus IPNI=*153215-1* 285
Dianthus odoratus IPNI=*153717-1* 285
Dianthus patens IPNI=*153215-1* 285
Dianthus pineticola IPNI=*153215-1* 285
Dianthus plumarius var. *portensis* IPNI=*153717-1* 285
Dianthus portensis IPNI=*153717-1* 285
Dianthus pseudocorymbosus IPNI=*310755-2* 285
Dianthus pulchellus IPNI=*153408-1* 285
Dianthus pulcher IPNI=*153215-1* 285
Dianthus pulcherrimus IPNI=*306674-2* 285
Dianthus ruthenicus IPNI=*153215-1* 285
Dianthus scaber IPNI=*153215-1* 285
Dianthus schraderi IPNI=*153215-1* 285
Dianthus sinensis IPNI=*153215-1* 285
Dianthus splendidissimus IPNI=*306674-2* 285
Dianthus suavis IPNI=*153408-1* 285
Dianthus subulifolius IPNI=*153215-1* 285
Dianthus tataricus IPNI=*153215-1* 285
Dianthus umbellatus IPNI=*153215-1* 285
Dianthus versicolor f. *leucopetalus* IPNI=*153215-1* 285
Dianthus versicolor IPNI=*153215-1* 285
Dianthus versicolor subsp. *ninelli* IPNI=*153215-1* 285
Dianthus versicolor var. *ninelli* IPNI=*153215-1* 285
Dianthus versicolor var. *subulifolius* IPNI=*153215-1* 285
Dianthus villosus IPNI=*310755-2* 285
Dianthus vivariensis IPNI=*310755-2* 285
Dianthus willdenowii IPNI=*153215-1* 285
Diapedium assurgens IPNI=*1114607-2* 359
Diapedium sexangulare IPNI=*1114607-2* 359
Diaphoranthema recurvata IPNI=*1194855-2* 229
Diaphoranthema uniflora IPNI=*1194855-2* 229
Diarrhena aquatica IPNI=*395086-1* 235
Diasperus acidissimus IPNI=*194610-2* 401
Diasperus acuminatus IPNI=*194611-2* 402
Diasperus antillanus IPNI=*352077-1* 401
Diasperus brasiliensis IPNI=*353613-1* 402
Diasperus caroliniensis IPNI=*353652-1* 402
Diasperus diffusus IPNI=*195027-2* 402
Diasperus diffusus var. *palustris* IPNI=*195027-2* 402
Diasperus fluitans IPNI=*353898-1* 402
Diasperus myrsinites IPNI=*354314-1* 402
Diasperus nanus IPNI=*326408-2* 402
Diasperus neogranatensis IPNI=*156150-2* 401
Diasperus niruri IPNI=*194900-2* 402

Diasperus orbiculatus IPNI=*194933-2* 402
Diasperus salviifolius IPNI=*354635-1* 402
Diasperus urinaria IPNI=*354815-1* 402
Diastatea maximiliana IPNI=*79151-2* 273
Diastatea micrantha var. *longibracteata* IPNI=*79151-2* 273
Diastatea micrantha var. *ovatifolia* IPNI=*79151-2* 273
Diastatea serrata IPNI=*79151-2* 273
Diastema pictum IPNI=*133755-2* 367
Diastemanthe platystachys IPNI=*423051-1* 246
Diatropa rotundifolia IPNI=*38361-2* 252
Diazeuxis intermedia IPNI=*231347-1* 265
Diazeuxis latifolia IPNI=*231347-1* 265
Diazeuxis mutisiana IPNI=*231341-1* 265
Dibrachion riparium IPNI=*492746-1* 318
Dibrachya clypeata IPNI=*376728-1* 358
Dibrachya peltata IPNI=*376728-1* 358
Dibrachya scutata IPNI=*376728-1* 358
Dicalymma fragrans IPNI=*238923-1* 268
Dichaea bradeorum IPNI=*79276-2* 218
Dichaea muricata IPNI=*79276-2* 218
Dichaea ovatipetala IPNI=*79276-2* 218
Dichaea robusta IPNI=*79276-2* 218
Dichaea verrucosa IPNI=*79276-2* 218
Dichaeopsis morrisii IPNI=*79276-2* 218
Dichanthelium acuminatum var. *thurowii* IPNI=*79328-2* 237
Dichanthelium auburne IPNI=*79328-2* 237
Dichanthelium leucoblepharis IPNI=*79395-2* 237
Dichanthelium leucoblepharis var. *glabrescens* IPNI=*79395-2* 237
Dichanthelium leucoblepharis var. *pubescens* IPNI=*79395-2* 237
Dichanthelium strigosum subsp. *glabrescens* IPNI=*79395-2* 237
Dichanthelium strigosum subsp. *leucoblepharis* IPNI=*79395-2* 237
Dichanthelium strigosum var. *glabrescens* IPNI=*79395-2* 237
Dichanthelium strigosum var. *leucoblepharis* IPNI=*79395-2* 237
Dichanthium bladhii IPNI=*392701-1* 235
Dichanthium caucasicum IPNI=*392701-1* 235
Dichanthium condylotrichum IPNI=*97916-2* 239
Dichanthium glabrum var. *punctatum* IPNI=*392701-1* 235
Dichanthium intermedium IPNI=*392701-1* 235
Dichanthium ischaemum IPNI=*1191417-2* 235
Dichanthium ischaemum subvar. *intermedium* IPNI=*392701-1* 235
Dichanthium nodosum IPNI=*398528-1* 237
Dichanthium odoratum IPNI=*392701-1* 235
Dichanthium pallidum IPNI=*398532-1* 237
Dichanthium pertusum IPNI=*65406-3* 235
Dichanthium saccharoides IPNI=*35205-2* 235
Dichanthium theinlwinii IPNI=*398532-1* 237
Dichorisandra affinis IPNI=*79497-2* 224
Dichorisandra angustifolia IPNI=*172628-1* 224
Dichorisandra aubletiana IPNI=*79497-2* 224
Dichorisandra aubletiana var. *affinis* IPNI=*79497-2* 224
Dichorisandra aubletiana var. *intermedia* IPNI=*79497-2* 224
Dichorisandra aubletiana var. *ovata* IPNI=*79497-2* 224
Dichorisandra aubletiana var. *persicariifolia* IPNI=*79497-2* 224
Dichorisandra gracilis IPNI=*79497-2* 224
Dichorisandra hexandra var. *persicariifolia* IPNI=*79497-2* 224
Dichorisandra inaequalis IPNI=*79497-2* 224
Dichorisandra intermedia IPNI=*79497-2* 224
Dichorisandra longifolia IPNI=*79534-2* 224
Dichorisandra mexicana IPNI=*79497-2* 224
Dichorisandra ovalifolia IPNI=*79497-2* 224
Dichorisandra ovata IPNI=*79497-2* 224
Dichorisandra reginae IPNI=*172628-1* 224
Dichorisandra scandens IPNI=*79497-2* 224
Dichorisandra schomburgkiana IPNI=*79497-2* 224
Dichorisandra tenuior IPNI=*79497-2* 224
Dichostylis aristata IPNI=*324890-2* 231
Dichostylis radiata IPNI=*304793-1* 230
Dichostylis squarrosa IPNI=*324890-2* 231
Dichotophyllum demersum IPNI=*30006047-2* 251
Dichromena atrosanguinea IPNI=*229810-2* 232
Dichromena blepharophora IPNI=*220187-2* 232
Dichromena ciliata IPNI=*220187-2* 232
Dichromena ciliata IPNI=*312402-1* 232
Dichromena ciliata var. *angustifolia* IPNI=*312402-1* 232
Dichromena ciliata var. *cinnamomea* IPNI=*312402-1* 232
Dichromena ciliata var. *vahliana* IPNI=*312330-1* 232
Dichromena colorata IPNI=*311841-1* 232
Dichromena corymbosa IPNI=*220271-2* 232
Dichromena drummondiana IPNI=*311841-1* 232
Dichromena globosa IPNI=*312402-1* 232
Dichromena leucocephala IPNI=*311841-1* 232
Dichromena leucophylla IPNI=*312402-1* 232
Dichromena nervosa IPNI=*312330-1* 232
Dichromena obtusiflora IPNI=*312402-1* 232
Dichromena obtusifolia IPNI=*312402-1* 232
Dichromena persooniana IPNI=*311841-1* 232
Dichromena pilosa IPNI=*312330-1* 232
Dichromena pittieri IPNI=*220187-2* 232
Dichromena polystachys IPNI=*220575-2* 232
Dichromena preslii IPNI=*220187-2* 232
Dichromena pulchella IPNI=*220187-2* 232
Dichromena pura IPNI=*312402-1* 232
Dichromena schoenus IPNI=*312067-1* 232
Dichromena tenuifolia IPNI=*312330-1* 232
Dichrostachys macrostachya IPNI=*188076-2* 330
Dicksonia aculeata IPNI=*1031307-2* 175
Dicksonia angustidens IPNI=*17081410-1* 175
Dicksonia cicutaria IPNI=*17081410-1* 175
Dicksonia cicutaria var. *remota* IPNI=*17081410-1* 175
Dicksonia decomposita IPNI=*17081410-1* 175
Dicksonia ghiesbreghtii IPNI=*1190416-2* 172
Dicksonia gigantea IPNI=*1190416-2* 172
Dicksonia lobulata IPNI=*1190416-2* 172
Dicksonia navarrensis IPNI=*1190416-2* 172
Dicksonia organensis IPNI=*275823-2* 172
Dicksonia rubiginosa IPNI=*17081410-1* 175
Dicksonia sellowiana var. *karsteniana* IPNI=*1190416-2* 172
Dicksonia siifolia IPNI=*17081410-1* 175
Dicksonia umbrosa IPNI=*17081410-1* 175
Diclidanthera penduliflora var. *penaeantha* IPNI=*826364-1* 338
Diclidium odoratum IPNI=*74939-2* 231
Diclidostigma melothrioides IPNI=*293293-1* 294
Dicliptera assurgens IPNI=*1114607-2* 359
Dicliptera assurgens var. *conferta* IPNI=*1114607-2* 359
Dicliptera assurgens var. *vahliana* IPNI=*1114607-2* 359
Dicliptera lupulina IPNI=*52612-1* 360
Dicliptera mollis IPNI=*1114607-2* 359
Dicliptera ramentacea IPNI=*1167021-2* 360
Dicliptera retusa IPNI=*50468-1* 360
Dicliptera trinervia IPNI=*52612-1* 360
Dicliptera vahliana IPNI=*1114607-2* 359
Dicranopteris bifida IPNI=*245596-2* 173
Dicranopteris dichotoma var. *pubescens* IPNI=*245596-2* 173
Dicranopteris flexuosa f. *monstrosa* IPNI=*79839-2* 173
Dicranopteris linearis var. *lherminieri* IPNI=*110733-2* 173
Dicranopteris nitida IPNI=*110733-2* 173
Dicranopteris nuda IPNI=*245632-2* 173
Dicranopteris pectinata IPNI=*110733-2* 173
Dicranopteris pubescens IPNI=*245596-2* 173
Dicranopteris pumila IPNI=*79839-2* 173
Dicranopteris rigida IPNI=*79839-2* 173
Dicranopteris rubiginosa IPNI=*245647-2* 173
Dicranopteris scalpturata IPNI=*79839-2* 173
Dicrypta baueri IPNI=*643387-1* 220
Dicrypta crassifolia IPNI=*643387-1* 220
Dicrypta discolor IPNI=*643414-1* 220

Dicrypta irisphyta IPNI=*155338-2* 222
Dicrypta longifolia IPNI=*643845-1* 221
Dicrypta villosa IPNI=*155347-2* 222
Dictyocaryum globiferum IPNI=*666671-1* 208
Dictyocaryum platysepalum IPNI=*666671-1* 208
Dictyocaryum schultzei IPNI=*666671-1* 208
Dictyocaryum superbum IPNI=*666671-1* 208
Didymandra purpurea IPNI=*134457-2* 394
Didymocarpus blancoi IPNI=*799766-1* 374
Didymochlaena dimidiata IPNI=*17084790-1* 175
Didymochlaena lunulata IPNI=*17084790-1* 175
Didymochlaena lunulata var. *minor* IPNI=*17084790-1* 175
Didymochlaena pulcherrima IPNI=*17084790-1* 175
Didymochlaena sinuosa IPNI=*17084790-1* 175
Didymochlaena sinuosa var. *microtheca* IPNI=*17084790-1* 175
Didymochlaena squamata IPNI=*17084790-1* 175
Didymopanax chrysophyllus IPNI=*90355-1* 253
Didymopanax micans IPNI=*90355-1* 253
Didymopanax morototoni var. *poeppigii* IPNI=*90355-1* 253
Didymopanax morototoni var. *sessiliflorus* IPNI=*90355-1* 253
Didymopanax papyrifer IPNI=*86417-3* 254
Didymopanax poeppigii IPNI=*90355-1* 253
Didymopanax speciosus IPNI=*90355-1* 253
Didymopanax splendidus IPNI=*90355-1* 253
Diectomis fasciculata IPNI=*398608-1* 237
Dieffenbachia amazonica IPNI=*86911-1* 199
Dieffenbachia barraquiniana IPNI=*86911-1* 199
Dieffenbachia baumannii IPNI=*86846-1* 199
Dieffenbachia brasiliensis IPNI=*86911-1* 199
Dieffenbachia chelsonii IPNI=*86911-1* 199
Dieffenbachia cognata IPNI=*86911-1* 199
Dieffenbachia consobrina IPNI=*86911-1* 199
Dieffenbachia conspurcata IPNI=*86911-1* 199
Dieffenbachia decora IPNI=*86911-1* 199
Dieffenbachia gollmeriana IPNI=*86911-1* 199
Dieffenbachia irrorata IPNI=*86911-1* 199
Dieffenbachia jenmanii IPNI=*86911-1* 199
Dieffenbachia latimaculata IPNI=*86846-1* 199
Dieffenbachia lineata IPNI=*86911-1* 199
Dieffenbachia lingulata IPNI=*86911-1* 199
Dieffenbachia liturata IPNI=*86911-1* 199
Dieffenbachia maculata IPNI=*86911-1* 199
Dieffenbachia magnifica IPNI=*86911-1* 199
Dieffenbachia membranulifera IPNI=*86911-1* 199
Dieffenbachia mirabilis IPNI=*86911-1* 199
Dieffenbachia neglecta IPNI=*86911-1* 199
Dieffenbachia picta f. *brasiliensis* IPNI=*86911-1* 199
Dieffenbachia picta IPNI=*86911-1* 199
Dieffenbachia picta subvar. *angustifolia* IPNI=*86911-1* 199
Dieffenbachia picta subvar. *gigantea* IPNI=*86911-1* 199
Dieffenbachia picta subvar. *jenmanii* IPNI=*86911-1* 199
Dieffenbachia picta subvar. *magnifica* IPNI=*86911-1* 199
Dieffenbachia picta subvar. *memoria* IPNI=*86911-1* 199
Dieffenbachia picta subvar. *picturata* IPNI=*86911-1* 199
Dieffenbachia picta var. *angustior* IPNI=*86911-1* 199
Dieffenbachia picta var. *barraquiniana* IPNI=*86911-1* 199
Dieffenbachia picta var. *latior* IPNI=*86911-1* 199
Dieffenbachia picta var. *typica* IPNI=*86911-1* 199
Dieffenbachia picturata IPNI=*86911-1* 199
Dieffenbachia plumieri IPNI=*86911-1* 199
Dieffenbachia poeppigii IPNI=*86911-1* 199
Dieffenbachia regina IPNI=*86911-1* 199
Dieffenbachia robusta IPNI=*86911-1* 199
Dieffenbachia seguine f. *conspurcata* IPNI=*86911-1* 199
Dieffenbachia seguine f. *irrorata* IPNI=*86911-1* 199
Dieffenbachia seguine f. *lineata* IPNI=*86911-1* 199
Dieffenbachia seguine f. *liturata* IPNI=*86911-1* 199
Dieffenbachia seguine f. *nobilis* IPNI=*86911-1* 199
Dieffenbachia seguine f. *viridis* IPNI=*86911-1* 199
Dieffenbachia seguine var. *lingulata* IPNI=*86911-1* 199
Dieffenbachia seguine var. *minor* IPNI=*86911-1* 199
Dieffenbachia seguine var. *robusta* IPNI=*86911-1* 199
Dieffenbachia variegata IPNI=*86911-1* 199
Dieffenbachia ventenatiana IPNI=*86911-1* 199
Dieffenbachia wallisii IPNI=*86911-1* 199
Digitalis alba IPNI=*802077-1* 376
Digitalis epiglottidea IPNI=*802028-1* 376
Digitalis eriostachya IPNI=*802028-1* 376
Digitalis gloxinioides IPNI=*802077-1* 376
Digitalis gyspergerae IPNI=*802077-1* 376
Digitalis intermedia IPNI=*802077-1* 376
Digitalis libertiana IPNI=*802077-1* 376
Digitalis longiflora IPNI=*802077-1* 376
Digitalis mauretanica IPNI=*802077-1* 376
Digitalis media IPNI=*802077-1* 376
Digitalis miniana IPNI=*802077-1* 376
Digitalis nevadensis IPNI=*802077-1* 376
Digitalis nova IPNI=*802028-1* 376
Digitalis orientalis IPNI=*802028-1* 376
Digitalis purpurascens IPNI=*802077-1* 376
Digitalis purpurea subsp. *bocquetii* IPNI=*802077-1* 376
Digitalis purpurea subsp. *mauretanica* IPNI=*802077-1* 376
Digitalis speciosa IPNI=*802077-1* 376
Digitalis tomentosa IPNI=*802077-1* 376
Digitalis winterli IPNI=*802028-1* 376
Digitaria abortiva IPNI=*1038558-2* 237
Digitaria adpressa IPNI=*979871-1* 245
Digitaria adscendens IPNI=*1038558-2* 237
Digitaria adscendens subsp. *chrysoblephara* IPNI=*1038558-2* 237
Digitaria adscendens subsp. *marginata* IPNI=*1038558-2* 237
Digitaria adscendens var. *rhachiseta* IPNI=*329018-2* 237
Digitaria aegyptiaca IPNI=*399115-1* 238
Digitaria aegyptiaca subsp. *caucasica* IPNI=*399115-1* 238
Digitaria affinis IPNI=*979871-1* 245
Digitaria appressa IPNI=*979871-1* 245
Digitaria argillacea f. *asetosa* IPNI=*398643-1* 237
Digitaria argillacea var. *nulliseta* IPNI=*398643-1* 237
Digitaria aurea IPNI=*321761-2* 234
Digitaria australis IPNI=*399115-1* 238
Digitaria barbata IPNI=*1218147-2* 237
Digitaria barbulata IPNI=*1218147-2* 237
Digitaria bechuanica IPNI=*398792-1* 237
Digitaria biformis IPNI=*1218147-2* 237
Digitaria bogoriensis IPNI=*399226-1* 238
Digitaria borbonica IPNI=*329018-2* 237
Digitaria brevifolia IPNI=*1038558-2* 237
Digitaria caespitosa IPNI=*399226-1* 238
Digitaria californica var. *villosissima* IPNI=*80220-2* 237
Digitaria caucasica IPNI=*399115-1* 238
Digitaria chinensis IPNI=*1038558-2* 237
Digitaria chinensis IPNI=*399226-1* 238
Digitaria chrysoblephara IPNI=*1038558-2* 237
Digitaria ciliaris f. *albovariegata* IPNI=*1038558-2* 237
Digitaria ciliaris subsp. *chrysoblephara* IPNI=*1038558-2* 237
Digitaria ciliaris subsp. *marginata* IPNI=*1038558-2* 237
Digitaria ciliaris subsp. *nubica* IPNI=*1038558-2* 237
Digitaria ciliaris var. *chrysoblephara* IPNI=*1038558-2* 237
Digitaria ciliaris var. *criniformis* IPNI=*1038558-2* 237
Digitaria ciliaris var. *fimbriata* IPNI=*1038558-2* 237
Digitaria ciliaris var. *pes-avis* IPNI=*1038558-2* 237
Digitaria ciliaris var. *rhachiseta* IPNI=*329018-2* 237
Digitaria ciliaris var. *sericea* IPNI=*1038558-2* 237
Digitaria ciliata IPNI=*1038558-2* 237
Digitaria collina IPNI=*1191417-2* 235
Digitaria commutata IPNI=*1218147-2* 237
Digitaria composita IPNI=*409823-1* 242

Digitaria conjugata IPNI=*30011315-2* 242
Digitaria corradii IPNI=*398933-1* 237
Digitaria curvipes IPNI=*398933-1* 237
Digitaria cuspidata IPNI=*399655-1* 238
Digitaria dactylon IPNI=*397065-1* 236
Digitaria decumbens IPNI=*398792-1* 237
Digitaria diamesa IPNI=*329018-2* 237
Digitaria digitata IPNI=*329018-2* 237
Digitaria digitata IPNI=*399226-1* 238
Digitaria dilatata IPNI=*414070-1* 243
Digitaria dinteri IPNI=*398792-1* 237
Digitaria distachyos IPNI=*426250-1* 246
Digitaria disticha IPNI=*184275-2* 243
Digitaria diversiflora IPNI=*1218147-2* 237
Digitaria domingensis IPNI=*392087-1* 234
Digitaria effusa IPNI=*1209084-2* 237
Digitaria eichingeri IPNI=*1209084-2* 237
Digitaria elatior IPNI=*409823-1* 242
Digitaria eriantha subsp. *pentzii* IPNI=*398792-1* 237
Digitaria eriantha subsp. *stolonifera* IPNI=*398792-1* 237
Digitaria eriantha subsp. *transvaalensis* IPNI=*398792-1* 237
Digitaria eriogona IPNI=*399115-1* 238
Digitaria eriolepis IPNI=*398933-1* 237
Digitaria filiculmis IPNI=*398933-1* 237
Digitaria fimbriata IPNI=*1038558-2* 237
Digitaria flexilis IPNI=*398933-1* 237
Digitaria foliosa IPNI=*184742-2* 243
Digitaria friesii IPNI=*398933-1* 237
Digitaria fusca IPNI=*399226-1* 238
Digitaria geniculata IPNI=*398792-1* 237
Digitaria glauca IPNI=*398792-1* 237
Digitaria gracilis IPNI=*399115-1* 238
Digitaria hackelii IPNI=*1209084-2* 237
Digitaria hatusimae IPNI=*398933-1* 237
Digitaria henryi IPNI=*1038558-2* 237
Digitaria hentyi IPNI=*1209084-2* 237
Digitaria hiascens IPNI=*398792-1* 237
Digitaria hirtigluma IPNI=*398643-1* 237
Digitaria ischaemum var. *violascens* IPNI=*399226-1* 238
Digitaria jamaicensis IPNI=*329018-2* 237
Digitaria lecardii IPNI=*398643-1* 237
Digitaria lehmanniana var. *dasyantha* IPNI=*80276-2* 237
Digitaria lemeeana IPNI=*1209084-2* 237
Digitaria leucites var. *glabella* IPNI=*80280-2* 237
Digitaria leucophaea IPNI=*398888-1* 237
Digitaria linearis IPNI=*397065-1* 236
Digitaria littoralis IPNI=*397065-1* 236
Digitaria livida IPNI=*398792-1* 237
Digitaria malesiae IPNI=*398933-1* 237
Digitaria marginata IPNI=*1038558-2* 237
Digitaria marginata var. *fimbriata* IPNI=*1038558-2* 237
Digitaria maritima IPNI=*397065-1* 236
Digitaria megapotamica IPNI=*77147487-1* 241
Digitaria merkeri IPNI=*1209084-2* 237
Digitaria miliacea IPNI=*1156554-2* 246
Digitaria mollissima IPNI=*80280-2* 237
Digitaria mutica IPNI=*1209084-2* 237
Digitaria nealleyi IPNI=*399115-1* 238
Digitaria nemoralis IPNI=*398792-1* 237
Digitaria nervalis IPNI=*80276-2* 237
Digitaria nervosa IPNI=*399115-1* 238
Digitaria oblongo-ovata IPNI=*398933-1* 237
Digitaria panicea IPNI=*399115-1* 238
Digitaria paspalodes IPNI=*184275-2* 243
Digitaria pectiniformis IPNI=*399115-1* 238
Digitaria pentzii IPNI=*398792-1* 237
Digitaria pertenuis IPNI=*399226-1* 238
Digitaria pes-avis IPNI=*1038558-2* 237
Digitaria platicaulon IPNI=*392087-1* 234
Digitaria plebeia IPNI=*399115-1* 238
Digitaria praecox IPNI=*399115-1* 238
Digitaria preslii IPNI=*398933-1* 237
Digitaria propinqua IPNI=*398933-1* 237
Digitaria pseudodurva IPNI=*399226-1* 238
Digitaria pseudoischaemum IPNI=*80254-2* 237
Digitaria queenslandica IPNI=*1218147-2* 237
Digitaria recta IPNI=*399226-1* 238
Digitaria reflexa IPNI=*329014-2* 237
Digitaria rhachitricha IPNI=*80253-2* 237
Digitaria rottleri IPNI=*1218147-2* 237
Digitaria roxburghii IPNI=*398933-1* 237
Digitaria sabulosa IPNI=*399115-1* 238
Digitaria sanguinalis f. *chlorantha* IPNI=*399115-1* 238
Digitaria sanguinalis f. *composita* IPNI=*399115-1* 238
Digitaria sanguinalis f. *illinoensis* IPNI=*399115-1* 238
Digitaria sanguinalis f. *intercedens* IPNI=*399115-1* 238
Digitaria sanguinalis subsp. *pakistanica* IPNI=*399115-1* 238
Digitaria sanguinalis subsp. *pectiniformis* IPNI=*399115-1* 238
Digitaria sanguinalis subsp. *sabulosa* IPNI=*399115-1* 238
Digitaria sanguinalis subsp. *vulgaris* IPNI=*399115-1* 238
Digitaria sanguinalis var. *marginata* IPNI=*1038558-2* 237
Digitaria sanguinalis var. *rhachiseta* IPNI=*329018-2* 237
Digitaria sanguinalis var. *stewartii* IPNI=*399115-1* 238
Digitaria sanguinaria IPNI=*399115-1* 238
Digitaria sanguinea IPNI=*399115-1* 238
Digitaria sanguinolenta IPNI=*399115-1* 238
Digitaria sasakii IPNI=*1038558-2* 237
Digitaria scalarum IPNI=*1209084-2* 237
Digitaria sericea IPNI=*1038558-2* 237
Digitaria setosa IPNI=*329014-2* 237
Digitaria sienitica IPNI=*399115-1* 238
Digitaria similis IPNI=*80323-2* 238
Digitaria smutsii IPNI=*398792-1* 237
Digitaria somalensis IPNI=*1209084-2* 237
Digitaria speciosa IPNI=*398933-1* 237
Digitaria stentiana IPNI=*398792-1* 237
Digitaria stolonifera IPNI=*397065-1* 236
Digitaria stricta IPNI=*399115-1* 238
Digitaria tangaensis IPNI=*1209084-2* 237
Digitaria tarapacana IPNI=*1038558-2* 237
Digitaria tenuiflora IPNI=*398933-1* 237
Digitaria tristachya IPNI=*184742-2* 243
Digitaria tristachya IPNI=*399115-1* 238
Digitaria umbraticola IPNI=*329014-2* 237
Digitaria umbrosa IPNI=*329014-2* 237
Digitaria umbrosa IPNI=*426280-1* 246
Digitaria umfolozi IPNI=*398792-1* 237
Digitaria undulata IPNI=*329014-2* 237
Digitaria uniflora IPNI=*392087-1* 234
Digitaria vaginata IPNI=*184742-2* 243
Digitaria valida IPNI=*398792-1* 237
Digitaria velutina IPNI=*80280-2* 237
Digitaria vestita IPNI=*1209084-2* 237
Diglossus variabilis IPNI=*252097-1* 271
Digraphis arundinacea IPNI=*415571-1* 243
Dileptium diffusum IPNI=*138164-2* 277
Dileptium precox IPNI=*138164-2* 277
Dileptium virginicum IPNI=*138164-2* 277
Dilkea johannesii IPNI=*674426-1* 397
Dilkea magnifica IPNI=*674426-1* 397
Dilkea parviflora IPNI=*674426-1* 397
Dilkea ulei IPNI=*674426-1* 397
Dilkea wallisii IPNI=*674426-1* 397
Dillenia elongata IPNI=*316684-1* 295
Dillenia speciosa IPNI=*316684-1* 295
Dimorpha falcata IPNI=*493934-1* 319

Dimorphandra megacarpa IPNI=*512214-1* 329
Dimorphandra megistosperma IPNI=*164265-2* 328
Dimorphandra oleifera IPNI=*164265-2* 328
Dimorphanthes ambigua IPNI=*204572-1* 262
Dimorphanthes angustifolia IPNI=*93603-2* 261
Dimorphanthes crispa IPNI=*93603-2* 261
Dimorphanthes floribunda IPNI=*204572-1* 262
Dimorphanthes hispida IPNI=*93603-2* 261
Dimorphanthes linifolia IPNI=*93603-2* 261
Dimorphanthes pulchella IPNI=*947840-1* 264
Dimorphanthes sophiifolia IPNI=*947840-1* 264
Dimorphanthus japonicus IPNI=*90496-1* 253
Dimorphostachys monostachya IPNI=*77144198-1* 244
Dimorphostachys oaxacensis IPNI=*184275-2* 243
Dinebra aristidoides IPNI=*1049662-2* 235
Dinebra hirsuta IPNI=*1049662-2* 235
Dinebra panicea IPNI=*407138-1* 241
Dioclea altissima IPNI=*77212334-1* 326
Dioclea comosa IPNI=*77212307-1* 326
Dioclea comosa var. *panamensis* IPNI=*492636-1* 318
Dioclea densiflora IPNI=*77212324-1* 326
Dioclea guianensis var. *velutina* IPNI=*492636-1* 318
Dioclea guianensis var. *villosior* IPNI=*492636-1* 318
Dioclea macrocarpa IPNI=*77212340-1* 326
Dioclea megacarpa IPNI=*77212324-1* 326
Dioclea panamensis IPNI=*492636-1* 318
Dioclea paraguariensis IPNI=*77212334-1* 326
Dioclea pilifera IPNI=*77212334-1* 326
Dioclea reflexa IPNI=*77212307-1* 326
Dioclea reflexa var. *grandiflora* IPNI=*77212324-1* 326
Dioclea trinervia IPNI=*179469-2* 329
Dioclea ucayalina IPNI=*77212332-1* 326
Dioclea violacea IPNI=*77212334-1* 326
Dioclea wilsonii IPNI=*77212335-1* 327
Dioctis maculatum IPNI=*77069480-1* 289
Dioctis vernum IPNI=*77069480-1* 289
Diodella sarmentosa IPNI=*77154343-1* 351
Diodella serrulata IPNI=*77154345-1* 351
Diodia auriculosa IPNI=*77154343-1* 351
Diodia breviseta IPNI=*77154343-1* 351
Diodia littoralis IPNI=*77154343-1* 351
Diodia maritima IPNI=*77154345-1* 351
Diodia maritima var. *commutata* IPNI=*77154345-1* 351
Diodia ocymifolia IPNI=*766862-1* 357
Diodia pilosa IPNI=*77154343-1* 351
Diodia radicans IPNI=*77154345-1* 351
Diodia riparia IPNI=*77154343-1* 351
Diodia sarmentosa IPNI=*77154343-1* 351
Diodia sarmentosa var. *bisepala* IPNI=*77154343-1* 351
Diodia serrulata IPNI=*77154345-1* 351
Diodia villosa IPNI=*756228-1* 352
Diolena spicata IPNI=*258514-2* 424
Dioscorea aculeata IPNI=*317946-1* 225
Dioscorea affinis IPNI=*323927-2* 226
Dioscorea ainensis IPNI=*81088-2* 226
Dioscorea altissima IPNI=*323921-2* 226
Dioscorea amaranthoides var. *denuda* IPNI=*323907-2* 225
Dioscorea amaranthoides var. *elegantula* IPNI=*323907-2* 225
Dioscorea amaranthoides var. *glauca* IPNI=*323907-2* 225
Dioscorea amaranthoides var. *metallica* IPNI=*323907-2* 225
Dioscorea amaranthoides var. *paniculata* IPNI=*323907-2* 225
Dioscorea ancistrocarpa IPNI=*323920-2* 226
Dioscorea anconensis IPNI=*318262-1* 225
Dioscorea angustifolia IPNI=*323927-2* 226
Dioscorea anoectochilus IPNI=*323907-2* 225
Dioscorea anthropophagorum IPNI=*323908-2* 225
Dioscorea apaensis IPNI=*323907-2* 225
Dioscorea apurimacensis IPNI=*80658-2* 225
Dioscorea arcuata IPNI=*323920-2* 226
Dioscorea articulata IPNI=*323927-2* 226
Dioscorea astrostigma IPNI=*318262-1* 225
Dioscorea atropurpurea IPNI=*317625-1* 225
Dioscorea bangii IPNI=*323912-2* 225
Dioscorea bilbergiana IPNI=*318262-1* 225
Dioscorea brasiliensis IPNI=*323927-2* 226
Dioscorea bulbifera var. *albotuberosa* IPNI=*323908-2* 225
Dioscorea bulbifera var. *brachybotryum* IPNI=*323908-2* 225
Dioscorea bulbifera var. *elongata* IPNI=*323908-2* 225
Dioscorea caracasana IPNI=*323921-2* 226
Dioscorea cliffortiana IPNI=*318778-1* 226
Dioscorea colocasiifolia IPNI=*317625-1* 225
Dioscorea conferta IPNI=*323920-2* 226
Dioscorea corumbensis IPNI=*80658-2* 225
Dioscorea costaricensis IPNI=*318417-1* 226
Dioscorea crispata IPNI=*323908-2* 225
Dioscorea crumenigera IPNI=*323907-2* 225
Dioscorea deamii IPNI=*318262-1* 225
Dioscorea deppei IPNI=*318262-1* 225
Dioscorea dodecandra IPNI=*323912-2* 225
Dioscorea dodecaneura var. *maronensis* IPNI=*323912-2* 225
Dioscorea dodecaneura var. *villosa* IPNI=*323912-2* 225
Dioscorea duchassaingii IPNI=*318417-1* 226
Dioscorea eburina IPNI=*317625-1* 225
Dioscorea eburnea IPNI=*317625-1* 225
Dioscorea esculenta var. *fulvidotomentosa* IPNI=*317946-1* 225
Dioscorea esculenta var. *tiliifolia* IPNI=*317946-1* 225
Dioscorea fasciculata IPNI=*317946-1* 225
Dioscorea ferruginicaulis IPNI=*81088-2* 226
Dioscorea frutescens IPNI=*324018-2* 225
Dioscorea galipanensis IPNI=*324037-2* 226
Dioscorea glauca IPNI=*318778-1* 226
Dioscorea globosa IPNI=*317625-1* 225
Dioscorea gouanioides IPNI=*80658-2* 225
Dioscorea goyazensis IPNI=*323927-2* 226
Dioscorea guanaiensis IPNI=*80658-2* 225
Dioscorea hebantha IPNI=*323912-2* 225
Dioscorea helmiicarpa IPNI=*323920-2* 226
Dioscorea heterophylla IPNI=*323908-2* 225
Dioscorea hexaphylla IPNI=*318778-1* 226
Dioscorea hirticaulis IPNI=*318778-1* 226
Dioscorea hoffa IPNI=*323908-2* 225
Dioscorea huallagensis IPNI=*323912-2* 225
Dioscorea illustrata IPNI=*323912-2* 225
Dioscorea javanica IPNI=*317625-1* 225
Dioscorea kegeliana IPNI=*323921-2* 226
Dioscorea kita IPNI=*323912-2* 225
Dioscorea korrorensis IPNI=*323908-2* 225
Dioscorea lanosa IPNI=*318417-1* 226
Dioscorea larecajensis IPNI=*323920-2* 226
Dioscorea lasseriana IPNI=*318417-1* 226
Dioscorea latifolia IPNI=*323908-2* 225
Dioscorea leiboldiana IPNI=*318262-1* 225
Dioscorea lindmanii IPNI=*318417-1* 226
Dioscorea lloydiana IPNI=*318778-1* 226
Dioscorea longifolia IPNI=*318778-1* 226
Dioscorea longipetiolata IPNI=*323908-2* 225
Dioscorea lutea IPNI=*323921-2* 226
Dioscorea macrophylla IPNI=*318262-1* 225
Dioscorea macrostachya IPNI=*318262-1* 225
Dioscorea macrostachya var. *sessiliflora* IPNI=*318262-1* 225
Dioscorea mandonii IPNI=*323920-2* 226
Dioscorea martinicensis IPNI=*323921-2* 226
Dioscorea maynensis IPNI=*323920-2* 226
Dioscorea megaptera IPNI=*318778-1* 226
Dioscorea mexicana var. *sessiliflora* IPNI=*318262-1* 225
Dioscorea multicolor var. *chrysophylla* IPNI=*323907-2* 225

Dioscorea multicolor var. *melanoleuca* IPNI=*323907-2* 225
Dioscorea multicolor var. *metallica* IPNI=*323907-2* 225
Dioscorea multicolor var. *sagittaria* IPNI=*323907-2* 225
Dioscorea multiflora IPNI=*323921-2* 226
Dioscorea multiflora var. *gouanioides* IPNI=*80658-2* 225
Dioscorea narinensis IPNI=*80658-2* 225
Dioscorea nobilis IPNI=*323907-2* 225
Dioscorea palmata IPNI=*323927-2* 226
Dioscorea paniculata IPNI=*318778-1* 226
Dioscorea paniculata var. *glabrifolia* IPNI=*318778-1* 226
Dioscorea papillaris IPNI=*317946-1* 225
Dioscorea papuana IPNI=*317946-1* 225
Dioscorea paraguayensis IPNI=*80658-2* 225
Dioscorea pennellii var. *pilosula* IPNI=*324018-2* 225
Dioscorea perrieri IPNI=*323908-2* 225
Dioscorea pilosiuscula var. *panamensis* IPNI=*318417-1* 226
Dioscorea piperifolia IPNI=*323921-2* 226
Dioscorea piperifolia var. *apiculata* IPNI=*323920-2* 226
Dioscorea piperifolia var. *glandulosa* IPNI=*324020-2* 225
Dioscorea piperifolia var. *obtusifolia* IPNI=*323920-2* 226
Dioscorea polygonoides IPNI=*318367-1* 225
Dioscorea polygonoides var. *aperta* IPNI=*323921-2* 226
Dioscorea polygonoides var. *martinicensis* IPNI=*323921-2* 226
Dioscorea polygonoides var. *scorpioidea* IPNI=*323921-2* 226
Dioscorea polygonoides var. *sieberi* IPNI=*323921-2* 226
Dioscorea pozucoensis IPNI=*80658-2* 225
Dioscorea propinqua IPNI=*318262-1* 225
Dioscorea pruinosa IPNI=*318778-1* 226
Dioscorea pulchella IPNI=*323908-2* 225
Dioscorea purpurea IPNI=*317625-1* 225
Dioscorea quaternata var. *glauca* IPNI=*318778-1* 226
Dioscorea quinquelobata IPNI=*323927-2* 226
Dioscorea racemosa IPNI=*323912-2* 225
Dioscorea repanda IPNI=*318778-1* 226
Dioscorea rogersii IPNI=*323908-2* 225
Dioscorea rubella IPNI=*317625-1* 225
Dioscorea ruiziana IPNI=*323927-2* 226
Dioscorea sapindioides IPNI=*318417-1* 226
Dioscorea sapinii IPNI=*317625-1* 225
Dioscorea sativa IPNI=*318778-1* 226
Dioscorea sativa var. *elongata* IPNI=*323908-2* 225
Dioscorea sativa var. *rotunda* IPNI=*323908-2* 225
Dioscorea sieberi IPNI=*323921-2* 226
Dioscorea silvestris IPNI=*323907-2* 225
Dioscorea spinosa IPNI=*317946-1* 225
Dioscorea tamifolia IPNI=*323908-2* 225
Dioscorea tenuiflora IPNI=*323908-2* 225
Dioscorea tequendamensis IPNI=*80894-2* 225
Dioscorea tiliifolia IPNI=*317946-1* 225
Dioscorea trifoliata var. *amazonica* IPNI=*324037-2* 226
Dioscorea trifoliata var. *galipanensis* IPNI=*324037-2* 226
Dioscorea triloba IPNI=*323927-2* 226
Dioscorea triloba IPNI=*324037-2* 226
Dioscorea tuerckheimii IPNI=*318262-1* 225
Dioscorea tugui IPNI=*317946-1* 225
Dioscorea villosa f. *glabrifolia* IPNI=*318778-1* 226
Dioscorea villosa subsp. *glabrifolia* IPNI=*318778-1* 226
Dioscorea villosa var. *glabrifolia* IPNI=*318778-1* 226
Dioscorea villosa var. *hirticaulis* IPNI=*318778-1* 226
Dioscorea villosa var. *laeviuscula* IPNI=*318778-1* 226
Dioscorea violacea IPNI=*323908-2* 225
Dioscorea vittata IPNI=*323912-2* 225
Dioscorea vulgaris IPNI=*317625-1* 225
Dioscorea waltheri IPNI=*318778-1* 226
Dioscorea yucatanensis IPNI=*80927-2* 225
Diospyros anzoateguiensis IPNI=*322526-1* 298
Diospyros berteroi IPNI=*322526-1* 298
Diospyros blancoi IPNI=*322328-1* 298
Diospyros cuneata IPNI=*323099-1* 298
Diospyros digyna IPNI=*322764-1* 298
Diospyros durionoides IPNI=*322328-1* 298
Diospyros edulis IPNI=*322764-1* 298
Diospyros mabolo IPNI=*322328-1* 298
Diospyros malacapai IPNI=*322328-1* 298
Diospyros melinonii IPNI=*322202-1* 298
Diospyros merrillii IPNI=*322328-1* 298
Diospyros myristicoides IPNI=*322748-1* 298
Diospyros obovata IPNI=*323099-1* 298
Diospyros obtusifolia IPNI=*322526-1* 298
Diospyros obtusifolia IPNI=*322764-1* 298
Diospyros opacifolia IPNI=*322748-1* 298
Diospyros pauciflora IPNI=*322764-1* 298
Diospyros pearcei IPNI=*322093-1* 298
Diospyros peruviana IPNI=*322093-1* 298
Diospyros peruviana var. *ocanensis* IPNI=*322093-1* 298
Diospyros peruviana var. *sprucei* IPNI=*322093-1* 298
Diospyros philippensis IPNI=*322328-1* 298
Diospyros polyandra IPNI=*322093-1* 298
Diospyros praetermissa IPNI=*323004-1* 298
Diospyros sapota IPNI=*322764-1* 298
Diospyros sapotanigera IPNI=*322764-1* 298
Diospyros schippii IPNI=*323099-1* 298
Diospyros tliltzapotl IPNI=*322764-1* 298
Diospyros utilis IPNI=*322328-1* 298
Diospyros vestita IPNI=*323004-1* 298
Dipetalon speciosum IPNI=*552967-1* 417
Diphasiastrum complanatum f. *montellii* IPNI=*147262-2* 171
Diphasiastrum complanatum IPNI=*147262-2* 171
Diphasiastrum complanatum subsp. *hastulatum* IPNI=*147262-2* 171
Diphasiastrum complanatum subsp. *montellii* IPNI=*147262-2* 171
Diphasiastrum complanatum var. *dilatatum* IPNI=*147262-2* 171
Diphasiastrum hastulatum IPNI=*147262-2* 171
Diphasiastrum montellii IPNI=*147262-2* 171
Diphasiastrum thyoides IPNI=*147665-2* 171
Diphasiastrum wilceae IPNI=*147262-2* 171
Diphasium complanatum IPNI=*147262-2* 171
Diphasium complanatum subsp. *montellii* IPNI=*147262-2* 171
Diphasium hastulatum IPNI=*147262-2* 171
Diphasium heterophyllum IPNI=*147420-2* 171
Diphasium holtonii IPNI=*147420-2* 171
Diphasium jussiaei IPNI=*147420-2* 171
Diphasium lindseaceum IPNI=*147420-2* 171
Diphasium wallrothii IPNI=*147262-2* 171
Diphysa robinioides IPNI=*948330-1* 318
Diplachne dubia IPNI=*60458636-2* 238
Diplachne dubia var. *humboldtiana* IPNI=*60458636-2* 238
Diplachne dubia var. *kurtziana* IPNI=*60458636-2* 238
Diplachne dubia var. *pringleana* IPNI=*60458636-2* 238
Diplachne mendocina IPNI=*60458636-2* 238
Diplachne patens IPNI=*60458636-2* 238
Diplachne polystachia IPNI=*60458632-2* 238
Diplactis cyanea IPNI=*247114-2* 270
Dipladenia bella IPNI=*151935-2* 341
Dipladenia billbergii IPNI=*81684-1* 343
Dipladenia boliviensis IPNI=*151935-2* 341
Dipladenia harrisii IPNI=*77191232-1* 342
Dipladenia sanderi IPNI=*152049-2* 341
Diplandra compressa IPNI=*613333-1* 429
Diplandra decurrens IPNI=*613333-1* 429
Diplandra heterophyla IPNI=*613333-1* 429
Diplandra montana IPNI=*613333-1* 429
Diplandra pumila IPNI=*613333-1* 429
Diplasanthum lanosum IPNI=*398529-1* 237
Diplazium athyrioides IPNI=*17086820-1* 174
Diplazium bakerianum IPNI=*22222-2* 174
Diplazium bipartitum IPNI=*17039970-1* 174

Diplazium bonapartii IPNI=*17086820-1* 174
Diplazium brachysorum IPNI=*17086820-1* 174
Diplazium crenatum IPNI=*328699-2* 174
Diplazium decipiens IPNI=*17088960-1* 174
Diplazium distentum IPNI=*17086820-1* 174
Diplazium expansum var. *spruceana* IPNI=*17086820-1* 174
Diplazium haenkeanum IPNI=*17086820-1* 174
Diplazium lonchophyllum IPNI=*17088960-1* 174
Diplazium marattiifolium IPNI=*17086820-1* 174
Diplazium pulcherrimum IPNI=*17084790-1* 175
Diplazium sundense IPNI=*17088960-1* 174
Diplazium tenerifrons IPNI=*17086820-1* 174
Diplima viminalis IPNI=*779152-1* 404
Diplisca elliptica IPNI=*63435-2* 441
Diplocalyx chrysophylloides IPNI=*608391-1* 450
Diplochita bracteata IPNI=*1024004-2* 421
Diplochita bubalina IPNI=*572209-1* 419
Diplochita leucocephala IPNI=*1000103-2* 422
Diplochita macrophylla IPNI=*1000103-2* 422
Diplochita mucronata IPNI=*1024004-2* 421
Diplochita serrulata IPNI=*1000103-2* 422
Diplochita serrulata var. *latifolia* IPNI=*1000103-2* 422
Diplochita swartziana IPNI=*1024004-2* 421
Diplochita swartziana IPNI=*572411-1* 420
Diplochita tomentosa IPNI=*1088633-2* 423
Diplocrater acuminatus IPNI=*607671-1* 449
Diplodonta setosa IPNI=*77196163-1* 421
Diplopenta leptocarpa IPNI=*185868-2* 412
Diplostephium anactinotum IPNI=*77203458-1* 265
Diplostephium bacchariideum IPNI=*230916-1* 265
Diplostephium costaricense IPNI=*230916-1* 265
Diplostephium cyparissias IPNI=*77203471-1* 265
Diplostephium igniarium IPNI=*230435-1* 265
Diplostephium nevadense IPNI=*77203490-1* 265
Diplostephium revolutum f. *macrocephalum* IPNI=*77203500-1* 265
Diplostephium revolutum IPNI=*77203500-1* 265
Diplostephium revolutum var. *macrocephalum* IPNI=*77203500-1* 265
Diplostephium revolutum var. *rubrum* IPNI=*77203500-1* 265
Diplostephium rosmarinifolium IPNI=*230916-1* 265
Diplostephium sessiliflorum IPNI=*173688-2* 267
Diplostephium sessiliflorum IPNI=*77203515-1* 265
Diplostephium weddellii IPNI=*77203515-1* 265
Diplotax arborescens IPNI=*234410-2* 332
Diplothemium henryanum IPNI=*666160-1* 208
Diplotropis brachypetala IPNI=*59774-2* 315
Dipogon avenaceus IPNI=*240834-2* 245
Dipogon glycinoides IPNI=*492756-1* 318
Dipsacus arcimusci IPNI=*319260-1* 295
Dipsacus botterii IPNI=*319260-1* 295
Dipsacus carminatorius IPNI=*319260-1* 295
Dipsacus connatofolius IPNI=*319260-1* 295
Dipsacus divaricatus IPNI=*319260-1* 295
Dipsacus fullonum f. *ternatus* IPNI=*319260-1* 295
Dipsacus fullonum subsp. *sativus* IPNI=*319292-1* 295
Dipsacus fullonum var. *sativus* IPNI=*319292-1* 295
Dipsacus horridus IPNI=*319260-1* 295
Dipsacus meyeri IPNI=*319260-1* 295
Dipsacus mirabilis IPNI=*319260-1* 295
Dipsacus morisonii IPNI=*319260-1* 295
Dipsacus orsini IPNI=*319260-1* 295
Dipsacus palustris IPNI=*319260-1* 295
Dipsacus purpurascens IPNI=*319260-1* 295
Dipsacus silvester IPNI=*319260-1* 295
Dipsacus sinuatus IPNI=*319260-1* 295
Dipsacus sylvestris f. *albidus* IPNI=*319260-1* 295
Dipsacus sylvestris IPNI=*319260-1* 295
Dipsacus vulgaris IPNI=*319260-1* 295
Dipteracanthus brasiliensis IPNI=*49672-1* 360
Dipteracanthus clandestinus IPNI=*54671-1* 361
Dipteracanthus haenkei IPNI=*54193-1* 361
Dipteracanthus paniculatus IPNI=*54389-1* 361
Dipteracanthus turbacensis IPNI=*54389-1* 361
Dipteris crassifolia IPNI=*77183236-1* 175
Dipterocalyx scaberrimus IPNI=*863659-1* 379
Dipteryx applanata IPNI=*520179-1* 335
Dipteryx cordata IPNI=*249447-2* 335
Dipteryx oppositifolia IPNI=*520179-1* 335
Dipteryx oppositifolia var. *parviflora* IPNI=*520179-1* 335
Dipteryx panamensis IPNI=*492770-1* 318
Dipteryx parviflora IPNI=*520179-1* 335
Dipteryx trifoliolata IPNI=*492778-1* 318
Diracodes javanica IPNI=*942355-1* 251
Disakisperma mexicanum IPNI=*60458636-2* 238
Discanthus odoratus IPNI=*30336798-2* 227
Disciphania appendiculata IPNI=*580772-1* 435
Disciphania micrantha IPNI=*172179-2* 435
Discipiper reticulatum IPNI=*198988-2* 197
Discocarpus mazarunensis IPNI=*54072-2* 401
Discolenta lapathifolia IPNI=*77069480-1* 289
Discolenta punctata IPNI=*190855-2* 289
Discolenta scabra IPNI=*77069480-1* 289
Disocactus amazonicus IPNI=*60426912-2* 284
Disocactus ramulosus IPNI=*963240-1* 284
Disocactus ramulosus var. *angustissimus* IPNI=*963240-1* 284
Disphenia arborea IPNI=*17071820-1* 172
Dissopetalum mauritianum IPNI=*59219-2* 434
Dissotis plumosa IPNI=*908858-1* 419
Dissotis rotundifolia IPNI=*908858-1* 419
Disteira angulosa IPNI=*675840-1* 374
Distemon brasiliensis IPNI=*319130-2* 247
Distemon grandis IPNI=*319130-2* 247
Distemon jacquinii IPNI=*315579-2* 247
Distemon kunzei IPNI=*315579-2* 247
Distemon linkii IPNI=*315579-2* 247
Distemon miniatus IPNI=*315579-2* 247
Distemon ottonis IPNI=*315579-2* 247
Distemon roscoeanus IPNI=*315579-2* 247
Distephana fockeana IPNI=*675022-1* 398
Distephana glandulosa IPNI=*675022-1* 398
Distephana glandulosa var. *canaliculata* IPNI=*675022-1* 398
Distephana pubescens IPNI=*675392-1* 400
Distephana quadridentata IPNI=*675392-1* 400
Distephana quadriglandulosa IPNI=*675392-1* 400
Distephana quitensis IPNI=*675224-1* 399
Distephana rohriana IPNI=*675022-1* 398
Distephana stoupyana IPNI=*675022-1* 398
Disterigma alaternoides var. *parvifolium* IPNI=*82288-2* 299
Disterigma dendrophilum IPNI=*82286-2* 299
Disterigma elassanthum IPNI=*82286-2* 299
Disterigma epacridifolium IPNI=*327994-1* 299
Disterigma leiopodandrum IPNI=*327994-1* 299
Disterigma margaricoccum IPNI=*327994-1* 299
Disterigma mayanum IPNI=*82301-2* 299
Disterigma microphyllum IPNI=*82286-2* 299
Disterigma pachyphyllum IPNI=*82301-2* 299
Disterigma penaeoides IPNI=*327994-1* 299
Disterigma popenoi IPNI=*82288-2* 299
Distichlis araucana IPNI=*82343-2* 238
Distichlis dentata IPNI=*82343-2* 238
Distichlis hirta IPNI=*82343-2* 238
Distichlis marginata IPNI=*82343-2* 238
Distichlis maritima var. *laxa* IPNI=*82343-2* 238
Distichlis maritima var. *stricta* IPNI=*82343-2* 238
Distichlis mendocina IPNI=*82343-2* 238
Distichlis prostrata IPNI=*82343-2* 238
Distichlis spicata f. *hirta* IPNI=*82343-2* 238

Distichlis spicata subsp. *stricta* IPNI=*82343-2* 238
Distichlis spicata var. *andina* IPNI=*82343-2* 238
Distichlis spicata var. *borealis* IPNI=*82343-2* 238
Distichlis spicata var. *distichophylla* IPNI=*82343-2* 238
Distichlis spicata var. *divaricata* IPNI=*82343-2* 238
Distichlis spicata var. *laxa* IPNI=*82343-2* 238
Distichlis spicata var. *marginata* IPNI=*82343-2* 238
Distichlis spicata var. *mendocina* IPNI=*82343-2* 238
Distichlis spicata var. *mexicana* IPNI=*82343-2* 238
Distichlis spicata var. *nana* IPNI=*82343-2* 238
Distichlis spicata var. *stolonifera* IPNI=*82343-2* 238
Distichlis spicata var. *stricta* IPNI=*82343-2* 238
Distichlis spicata var. *thalassica* IPNI=*82343-2* 238
Distichlis stricta IPNI=*82343-2* 238
Distichlis stricta var. *dentata* IPNI=*82343-2* 238
Distichlis stricta var. *laxa* IPNI=*82343-2* 238
Distichlis tenuifolia IPNI=*82343-2* 238
Distichlis thalassica IPNI=*82343-2* 238
Distichlis thalassica var. *mendocina* IPNI=*82343-2* 238
Distichlis thalassica var. *pectinata* IPNI=*82343-2* 238
Distichlis viridis IPNI=*82343-2* 238
Distictella angustifolia IPNI=*77100449-1* 362
Distictella broadwayana IPNI=*77100449-1* 362
Distictella kochii IPNI=*77100449-1* 362
Distictella lutescens IPNI=*77100449-1* 362
Distictella magnoliifolia IPNI=*77100449-1* 362
Distictella negrensis IPNI=*77100449-1* 362
Distictis angustifolia IPNI=*77100449-1* 362
Distictis kochii IPNI=*77100449-1* 362
Distimus flavescens IPNI=*74698-2* 230
Distreptus angustifolius IPNI=*176829-2* 267
Distreptus crispus IPNI=*176829-2* 267
Distreptus nudiflorus IPNI=*176829-2* 267
Distreptus replicatus IPNI=*239804-1* 268
Distreptus spicatus IPNI=*239804-1* 268
Distreptus spicatus var. *interruptus* IPNI=*239804-1* 268
Distreptus spicatus var. *nicaraguensis* IPNI=*239804-1* 268
Distreptus spiralis IPNI=*211400-2* 268
Ditaxis blodgettii IPNI=*995988-1* 389
Ditaxis castaneifolia IPNI=*44529-2* 388
Ditaxis fendleri IPNI=*995988-1* 389
Ditaxis rubricaulis IPNI=*995988-1* 389
Diteilis nervosa IPNI=*641061-1* 219
Ditelesia nudiflora IPNI=*172873-1* 224
Ditmaria violacea IPNI=*870864-1* 430
Ditremexa caroliniana IPNI=*1117728-2* 333
Ditremexa fetida IPNI=*1117728-2* 333
Ditremexa hirsuta IPNI=*234491-2* 332
Ditremexa occidentalis IPNI=*1117728-2* 333
Ditremexa sophera IPNI=*518364-1* 333
Diuroglossum rufescens IPNI=*823393-1* 408
Dizygotheca elegantissima IPNI=*77131172-1* 254
Dizygotheca faguetii IPNI=*77131172-1* 254
Docanthe alba IPNI=*665826-1* 207
Dolichandra fenzliana IPNI=*77100511-1* 363
Dolichandra kohautiana IPNI=*77100466-1* 363
Dolicholus apoloensis IPNI=*517240-1* 331
Dolicholus ixodes IPNI=*517240-1* 331
Dolicholus melanostictus IPNI=*517240-1* 331
Dolicholus minimus IPNI=*1197938-2* 331
Dolicholus nigropunctatus IPNI=*517240-1* 331
Dolicholus phaseoloides IPNI=*1006990-2* 331
Dolicholus precatorius IPNI=*517450-1* 331
Dolicholus pyramidalis IPNI=*517469-1* 331
Dolicholus vailiae IPNI=*517450-1* 331
Dolichos abyssinicus IPNI=*525419-1* 336
Dolichos acinaciformis IPNI=*1209500-2* 313
Dolichos altissimus IPNI=*1197939-2* 328
Dolichos altissimus IPNI=*77212334-1* 326
Dolichos angularis IPNI=*525291-1* 336
Dolichos angulatus IPNI=*525291-1* 336
Dolichos arborescens IPNI=*518476-1* 334
Dolichos arboreus IPNI=*518476-1* 334
Dolichos aristatus IPNI=*51897-2* 314
Dolichos benthamii IPNI=*492756-1* 318
Dolichos cienskowskii IPNI=*1209500-2* 313
Dolichos comosus IPNI=*77212307-1* 326
Dolichos cultratus IPNI=*1209500-2* 313
Dolichos curtisii IPNI=*492756-1* 318
Dolichos emarginatus IPNI=*30187778-2* 313
Dolichos ensiformis IPNI=*1209500-2* 313
Dolichos gangeticus IPNI=*525419-1* 336
Dolichos gibbosus IPNI=*492756-1* 318
Dolichos glycinoides IPNI=*525419-1* 336
Dolichos helicopus IPNI=*525419-1* 336
Dolichos lignosus IPNI=*492756-1* 318
Dolichos littoralis IPNI=*30187778-2* 313
Dolichos luteolus IPNI=*525419-1* 336
Dolichos luteus IPNI=*525428-1* 336
Dolichos maritimus IPNI=*30187778-2* 313
Dolichos mexicanus IPNI=*525419-1* 336
Dolichos miniatus IPNI=*30187778-2* 313
Dolichos minimus IPNI=*1197938-2* 331
Dolichos montanus IPNI=*516708-1* 331
Dolichos mutabilis IPNI=*1209500-2* 313
Dolichos niloticus IPNI=*525419-1* 336
Dolichos obcordatus IPNI=*30187778-2* 313
Dolichos obovatus IPNI=*30187778-2* 313
Dolichos obtusifolius IPNI=*30187778-2* 313
Dolichos oleraceus IPNI=*60457245-2* 325
Dolichos ovalifolius IPNI=*30187778-2* 313
Dolichos phaseoloides IPNI=*509661-1* 328
Dolichos pruriens IPNI=*509275-1* 328
Dolichos pugioniformis IPNI=*1209500-2* 313
Dolichos purpureus IPNI=*501411-1* 324
Dolichos pyramidalis IPNI=*517469-1* 331
Dolichos repens IPNI=*525419-1* 336
Dolichos retusus IPNI=*525428-1* 336
Dolichos rhynchosioides IPNI=*492756-1* 318
Dolichos rhynchosioides IPNI=*57174-3* 327
Dolichos roseus IPNI=*30187778-2* 313
Dolichos rotundifolius IPNI=*30187778-2* 313
Dolichos savii IPNI=*525428-1* 336
Dolichos sinensis IPNI=*525419-1* 336
Dolichos spurius IPNI=*486481-1* 315
Dolichos tonkinensis IPNI=*513918-1* 330
Dolichos trilobus IPNI=*525554-1* 337
Dolichos tuberosus IPNI=*512027-1* 329
Dolichos umbellatus IPNI=*525554-1* 337
Dolichos uncinatus IPNI=*521046-1* 335
Dolichos unguiculatus IPNI=*1127257-2* 337
Dolichos unguiculatus IPNI=*525554-1* 337
Dolichos urens IPNI=*1197939-2* 328
Dolichos vexillatus IPNI=*525561-1* 337
Dolichos virgatus IPNI=*492690-1* 318
Dolichos viridis IPNI=*509661-1* 328
Doliocarpus aptus IPNI=*316807-1* 295
Doliocarpus aptus var. *curtus* IPNI=*316807-1* 295
Doliocarpus belizensis IPNI=*317113-1* 295
Doliocarpus calineoides IPNI=*317113-1* 295
Doliocarpus congestiflorus IPNI=*82916-2* 295
Doliocarpus coriacea IPNI=*317113-1* 295
Doliocarpus curtipes IPNI=*30327947-2* 295
Doliocarpus dentosus IPNI=*30327947-2* 295
Doliocarpus nicaraguensis IPNI=*317113-1* 295
Doliocarpus oaxacanus IPNI=*82916-2* 295

Doliocarpus piauhiensis IPNI=*30327947-2* 295
Doliocarpus platystigma IPNI=*82916-2* 295
Doliocarpus pubens IPNI=*82916-2* 295
Doliocarpus punctatus IPNI=*30327947-2* 295
Doliocarpus rolandri IPNI=*82916-2* 295
Doliocarpus rolandri var. *parcus* IPNI=*30327947-2* 295
Doliocarpus scandens IPNI=*30327947-2* 295
Doliocarpus semidentatus IPNI=*82916-2* 295
Doliocarpus spinulifer IPNI=*30327947-2* 295
Doliocarpus strictus IPNI=*30327947-2* 295
Doliocarpus virgatus IPNI=*316807-1* 295
Dombeya angulata IPNI=*822960-1* 408
Dombeya antunesii IPNI=*822960-1* 408
Dombeya auriculata IPNI=*822960-1* 408
Dombeya burttii IPNI=*822960-1* 408
Dombeya calantha IPNI=*822960-1* 408
Dombeya chilensis IPNI=*261681-1* 180
Dombeya concinna IPNI=*822960-1* 408
Dombeya dawei IPNI=*822960-1* 408
Dombeya endlichii IPNI=*822960-1* 408
Dombeya excelsa IPNI=*261688-1* 180
Dombeya gamwelliae IPNI=*822960-1* 408
Dombeya globiflora IPNI=*822960-1* 408
Dombeya greenwayi IPNI=*822960-1* 408
Dombeya johnstonii IPNI=*822960-1* 408
Dombeya kindtiana IPNI=*822960-1* 408
Dombeya lasiostylis IPNI=*822960-1* 408
Dombeya mastersii IPNI=*822960-1* 408
Dombeya nairobensis IPNI=*822960-1* 408
Dombeya nyasica IPNI=*822960-1* 408
Dombeya parvifolia IPNI=*822960-1* 408
Dombeya penduliflora IPNI=*823281-1* 408
Dombeya phoenicea IPNI=*824676-1* 412
Dombeya platypoda IPNI=*822960-1* 408
Dombeya rosea IPNI=*822960-1* 408
Dombeya sparmannioides IPNI=*822960-1* 408
Dombeya speciosa IPNI=*823281-1* 408
Dombeya sphaerantha IPNI=*822960-1* 408
Dombeya tanganyikensis IPNI=*822960-1* 408
Dombeya trichoclada IPNI=*822960-1* 408
Dombeya velutina IPNI=*822960-1* 408
Donax arundinaceus IPNI=*390837-1* 234
Donax bengalensis IPNI=*390837-1* 234
Donax donax IPNI=*390837-1* 234
Donax sativus IPNI=*390837-1* 234
Donax versicolor IPNI=*390837-1* 234
Donia formosa IPNI=*948198-1* 334
Donia punicea IPNI=*486457-1* 315
Donia speciosa IPNI=*948198-1* 334
Donkelaaria dichotoma IPNI=*114528-2* 351
Doria dumerorum IPNI=*249575-1* 270
Doria pitcheri IPNI=*249575-1* 270
Doronicum cruentum IPNI=*237108-1* 268
Dorstenia alexiteria IPNI=*319180-2* 437
Dorstenia amazonica IPNI=*851513-1* 437
Dorstenia brasiliensis f. *balansae* IPNI=*851513-1* 437
Dorstenia brasiliensis var. *guaranitica* IPNI=*851513-1* 437
Dorstenia brasiliensis var. *major* IPNI=*851513-1* 437
Dorstenia brasiliensis var. *palustris* IPNI=*851513-1* 437
Dorstenia brasiliensis var. *tomentosa* IPNI=*851513-1* 437
Dorstenia brasiliensis var. *tubicina* IPNI=*851513-1* 437
Dorstenia brasiliensis var. *typica* IPNI=*851513-1* 437
Dorstenia contrajerva subsp. *tenuiloba* IPNI=*319180-2* 437
Dorstenia contrajerva var. *maculata* IPNI=*319180-2* 437
Dorstenia contrajerva var. *tenuiloba* IPNI=*319180-2* 437
Dorstenia heringeri IPNI=*851513-1* 437
Dorstenia houstonii IPNI=*319180-2* 437
Dorstenia infundibuliformis IPNI=*851513-1* 437
Dorstenia montana IPNI=*851513-1* 437
Dorstenia montevidensis IPNI=*851513-1* 437
Dorstenia pernambucana IPNI=*851513-1* 437
Dorstenia placentoides IPNI=*851513-1* 437
Dorstenia quadrangularis IPNI=*319180-2* 437
Dorstenia sabanensis IPNI=*851513-1* 437
Dorstenia schulzii IPNI=*851513-1* 437
Dorstenia tomentosa IPNI=*851513-1* 437
Dorstenia tubicina f. *major* IPNI=*851513-1* 437
Dorstenia tubicina f. *subexcentrica* IPNI=*851513-1* 437
Dorstenia tubicina f. *typica* IPNI=*851513-1* 437
Dorstenia tubicina IPNI=*851513-1* 437
Dorstenia tubicina var. *genuina* IPNI=*851513-1* 437
Dortmanna acuminata IPNI=*142908-1* 273
Dortmanna cardinalis IPNI=*142115-2* 273
Dortmanna cordigera IPNI=*142115-2* 273
Dortmanna cymbalaria IPNI=*143584-1* 273
Dortmanna diversifolia IPNI=*143931-1* 273
Dortmanna engelmanniana IPNI=*142115-2* 273
Dortmanna erinoides IPNI=*143240-1* 273
Dortmanna erinus IPNI=*143240-1* 273
Dortmanna flexuosa IPNI=*143240-1* 273
Dortmanna fulgens IPNI=*142115-2* 273
Dortmanna gracilis IPNI=*1020602-1* 273
Dortmanna graminea IPNI=*142115-2* 273
Dortmanna lavandulacea IPNI=*143240-1* 273
Dortmanna laxiflora IPNI=*142260-2* 273
Dortmanna longifolia IPNI=*142115-2* 273
Dortmanna micrantha IPNI=*79151-2* 273
Dortmanna minutiflora IPNI=*79151-2* 273
Dortmanna nana IPNI=*143584-1* 273
Dortmanna parviflora IPNI=*79151-2* 273
Dortmanna phyllostachya IPNI=*142115-2* 273
Dortmanna retrorsa IPNI=*145335-1* 273
Dortmanna rupestris IPNI=*143931-1* 273
Dortmanna senegalensis IPNI=*143240-1* 273
Dortmanna splendens IPNI=*142115-2* 273
Dortmanna tenera IPNI=*143931-1* 273
Doryopteris concolor IPNI=*77182373-1* 178
Doryopteris concolor var. *typica* IPNI=*77182373-1* 178
Doxantha acutistipula IPNI=*77100466-1* 363
Doxantha adunca IPNI=*77100466-1* 363
Doxantha chelephora IPNI=*77100466-1* 363
Doxantha dasyonyx IPNI=*77100466-1* 363
Doxantha exoleta IPNI=*77100466-1* 363
Doxantha lanuginosa IPNI=*77100466-1* 363
Doxantha longisiliqua IPNI=*153710-2* 364
Doxantha mexicana IPNI=*77100466-1* 363
Doxantha praesignis IPNI=*77100466-1* 363
Doxantha radicans IPNI=*77100466-1* 363
Doxantha tenuicula IPNI=*77100466-1* 363
Doxantha torquata IPNI=*77100466-1* 363
Doxantha tweedieana IPNI=*77100466-1* 363
Doxantha uncata IPNI=*77100511-1* 363
Doxantha unguis var. *microphylla* IPNI=*77100466-1* 363
Doxantha unguis-cati var. *dasyonyx* IPNI=*77100466-1* 363
Doxantha unguis-cati var. *exoleta* IPNI=*77100466-1* 363
Doyerea angosturensis IPNI=*592558-1* 294
Draba maritima IPNI=*1085072-2* 277
Dracaena × bartetii IPNI=*533580-1* 215
Dracaena angustifolia var. *angustior* IPNI=*534103-1* 215
Dracaena angustifolia var. *honoriae* IPNI=*534103-1* 215
Dracaena argenteostriata IPNI=*533580-1* 215
Dracaena aurora IPNI=*533580-1* 215
Dracaena australasica IPNI=*534103-1* 215
Dracaena australis IPNI=*533544-1* 215
Dracaena bellula IPNI=*533580-1* 215
Dracaena brasiliensis IPNI=*533580-1* 215

Dracaena butayei IPNI=*534207-1* 215
Dracaena casanovae IPNI=*533580-1* 215
Dracaena chelsonii IPNI=*533580-1* 215
Dracaena cuprea IPNI=*533580-1* 215
Dracaena deisteliana IPNI=*534207-1* 215
Dracaena deremensis IPNI=*534207-1* 215
Dracaena ensiformis IPNI=*534103-1* 215
Dracaena esculenta IPNI=*533580-1* 215
Dracaena ferrea IPNI=*533580-1* 215
Dracaena flemingii IPNI=*533580-1* 215
Dracaena fruticosa IPNI=*534103-1* 215
Dracaena gibsonii IPNI=*533580-1* 215
Dracaena hendersonii IPNI=*533580-1* 215
Dracaena inscripta IPNI=*533580-1* 215
Dracaena kewensis IPNI=*533580-1* 215
Dracaena lennei IPNI=*543700-1* 215
Dracaena lindenii IPNI=*534207-1* 215
Dracaena litoralis IPNI=*534129-1* 215
Dracaena lutescens IPNI=*533580-1* 215
Dracaena menglaensis IPNI=*534103-1* 215
Dracaena metallica IPNI=*533580-1* 215
Dracaena neocaledonica IPNI=*533580-1* 215
Dracaena nobilis IPNI=*533580-1* 215
Dracaena parrei IPNI=*533544-1* 215
Dracaena picturata IPNI=*533580-1* 215
Dracaena placida IPNI=*533580-1* 215
Dracaena porteana IPNI=*533580-1* 215
Dracaena pulcherrima IPNI=*533580-1* 215
Dracaena quitensis IPNI=*534103-1* 215
Dracaena reali IPNI=*533580-1* 215
Dracaena regalis IPNI=*533580-1* 215
Dracaena regis IPNI=*533580-1* 215
Dracaena rothiana IPNI=*533580-1* 215
Dracaena rubra IPNI=*97540-3* 215
Dracaena salviati IPNI=*533580-1* 215
Dracaena sepiaria IPNI=*533580-1* 215
Dracaena siamensis IPNI=*533580-1* 215
Dracaena smithii IPNI=*534207-1* 215
Dracaena stenophylla IPNI=*533580-1* 215
Dracaena terminalis IPNI=*533580-1* 215
Dracaena thomsonii IPNI=*533580-1* 215
Dracaena troubetzkoi IPNI=*533580-1* 215
Dracaena ugandensis IPNI=*534207-1* 215
Dracaena utilis IPNI=*533580-1* 215
Dracaena venosa IPNI=*533580-1* 215
Dracaena warocquei IPNI=*533580-1* 215
Dracaena yuccoides IPNI=*543700-1* 215
Dracaena zeelandica IPNI=*97540-3* 215
Dracaenopsis australis IPNI=*533544-1* 215
Dracaenopsis calocoma IPNI=*533544-1* 215
Draco angustifolia IPNI=*534103-1* 215
Draco fragrans IPNI=*534207-1* 215
Draco fruticosa IPNI=*534103-1* 215
Draco reflexa IPNI=*534338-1* 215
Draco stenophylla IPNI=*533580-1* 215
Draconia dracunculiformis IPNI=*307848-2* 257
Draconia dracunculus IPNI=*307848-2* 257
Dracontium bissectum IPNI=*252718-2* 227
Dracontium carderi IPNI=*991397-1* 200
Dracontium costaricense IPNI=*991397-1* 200
Dracontium integerrimum IPNI=*87123-1* 200
Dracontium laciniatum IPNI=*87928-1* 201
Dracontium lanceifolium IPNI=*88983-1* 202
Dracontium loretense IPNI=*991397-1* 200
Dracontium ornatum IPNI=*991397-1* 200
Dracontium pentaphyllum IPNI=*85201-1* 199
Dracontium pertusum IPNI=*87464-1* 200
Dracontium sagittifolium IPNI=*89296-1* 202
Dracontium scandens IPNI=*85333-1* 199
Dracontium trianae IPNI=*991397-1* 200
Dracula senilis IPNI=*83967-2* 218
Dracunculus cannifolius IPNI=*88955-1* 202
Drejera willdenowiana IPNI=*133119-2* 360
Drejerella guttata IPNI=*289220-2* 360
Drepachenia lancifolia IPNI=*58516-1* 198
Drepanocarpus africanus IPNI=*148561-2* 326
Drepanocarpus discolor IPNI=*148542-2* 326
Drepanocarpus inundatus IPNI=*148542-2* 326
Drepanocarpus isadelphus IPNI=*148544-2* 326
Drepanocarpus lunatus IPNI=*148561-2* 326
Drepanocarpus microphyllus IPNI=*148544-2* 326
Drepanocarpus paludicola IPNI=*75954-2* 317
Drimys aromatica IPNI=*330748-2* 181
Drimys granadensis var. *grandiflora* IPNI=*84100-2* 181
Drimys punctata IPNI=*330748-2* 181
Drimys winterana IPNI=*330748-2* 181
Drimys winteri f. *granadensis* IPNI=*84100-2* 181
Drimys winteri f. *magellanica* IPNI=*330748-2* 181
Drimys winteri var. *morenonis* IPNI=*330748-2* 181
Drimys winteri var. *punctata* IPNI=*330748-2* 181
Drimys winteri var. *semiglobulosa* IPNI=*330748-2* 181
Drosera communis var. *breviscapa* IPNI=*321807-1* 286
Drosera communis var. *cubensis* IPNI=*321807-1* 286
Drosera minor IPNI=*321807-1* 286
Druparia prunus IPNI=*30221374-2* 443
Druparia racemosa IPNI=*291801-1* 293
Dryandra oleifera IPNI=*338514-1* 387
Drymaria adenophora IPNI=*30172747-2* 285
Drymaria adiantoides IPNI=*154201-1* 285
Drymaria barrancae IPNI=*154201-1* 285
Drymaria blasiana IPNI=*154139-1* 285
Drymaria ciliaris IPNI=*154201-1* 285
Drymaria ciliata IPNI=*154201-1* 285
Drymaria cordata var. *diandra* IPNI=*30172747-2* 285
Drymaria cordata var. *pacifica* IPNI=*30172747-2* 285
Drymaria cordata var. *pilosa* IPNI=*154201-1* 285
Drymaria cordata var. *puberula* IPNI=*30172747-2* 285
Drymaria cordata var. *villosa* IPNI=*154201-1* 285
Drymaria cubensis IPNI=*154201-1* 285
Drymaria fendleri IPNI=*154139-1* 285
Drymaria fendleri var. *perennis* IPNI=*154139-1* 285
Drymaria glandulosa var. *fendleri* IPNI=*154139-1* 285
Drymaria glandulosa var. *perennis* IPNI=*154139-1* 285
Drymaria hirsuta IPNI=*154201-1* 285
Drymaria idiopoda IPNI=*154201-1* 285
Drymaria leptoclados IPNI=*154139-1* 285
Drymaria leptoclados var. *peruviana* IPNI=*154139-1* 285
Drymaria palmeri IPNI=*159379-1* 286
Drymaria procumbens IPNI=*30172747-2* 285
Drymaria ramosissima IPNI=*154139-1* 285
Drymaria sessilifolia IPNI=*30172747-2* 285
Drymaria stylosa IPNI=*154201-1* 285
Drymaria tepicana IPNI=*154201-1* 285
Drymaria villosa f. *tepicana* IPNI=*154201-1* 285
Drymonia bicolor IPNI=*380742-1* 367
Drymonia buscalionii IPNI=*380742-1* 367
Drymonia calcarata IPNI=*380742-1* 367
Drymonia campbellii IPNI=*380742-1* 367
Drymonia chiapensis IPNI=*380742-1* 367
Drymonia coerulea IPNI=*380742-1* 367
Drymonia cristata IPNI=*380742-1* 367
Drymonia glabra IPNI=*380742-1* 367
Drymonia jacquinii IPNI=*380742-1* 367
Drymonia lindmaniana IPNI=*380742-1* 367
Drymonia maculata IPNI=*380742-1* 367
Drymonia mollis IPNI=*380742-1* 367

Drymonia ovata IPNI=*380742-1* 367
Drymonia parvifolia IPNI=*380742-1* 367
Drymonia sarmentosula IPNI=*380742-1* 367
Drymonia spectabilis IPNI=*380742-1* 367
Drymonia splendens IPNI=*380742-1* 367
Drymonia ulei IPNI=*380742-1* 367
Drymophloeus zippellii IPNI=*665616-1* 207
Drynaria araneosa IPNI=*17175380-1* 176
Drynaria compacta IPNI=*17448650-1* 176
Drynaria crassifolia IPNI=*77183236-1* 175
Drynaria fulva IPNI=*17175380-1* 176
Drynaria glaucina IPNI=*17175380-1* 176
Drynaria irioides IPNI=*17341250-1* 176
Drynaria lepidota IPNI=*17448650-1* 176
Drynaria lycopodioides IPNI=*160142-2* 175
Drynaria macrocarpa IPNI=*17448650-1* 176
Drynaria obtusata IPNI=*17341250-1* 176
Drynaria percussa IPNI=*17546710-1* 175
Drynaria polycarpa IPNI=*17341250-1* 176
Drynaria pulverulenta IPNI=*17341250-1* 176
Dryopteris adiantiformis IPNI=*224543-2* 176
Dryopteris alsophiloides IPNI=*1019960-2* 174
Dryopteris auriculata IPNI=*17160730-1* 176
Dryopteris capensis IPNI=*224543-2* 176
Dryopteris dentata IPNI=*251308-2* 174
Dryopteris dentata var. *urens* IPNI=*251308-2* 174
Dryopteris dicarpa IPNI=*251308-2* 174
Dryopteris discophora IPNI=*224543-2* 176
Dryopteris doniana IPNI=*17392950-1* 175
Dryopteris elegans IPNI=*1019960-2* 174
Dryopteris extensa IPNI=*17460390-1* 174
Dryopteris funesta IPNI=*286436-2* 177
Dryopteris gongylodes IPNI=*251353-2* 174
Dryopteris gongylodes var. *glabra* IPNI=*17458930-1* 174
Dryopteris gongylodes var. *hirsuta* IPNI=*17458930-1* 174
Dryopteris gongylodes var. *longipinna* IPNI=*251353-2* 174
Dryopteris impressa IPNI=*17460390-1* 174
Dryopteris incerta IPNI=*17460390-1* 174
Dryopteris interrupta IPNI=*17458930-1* 174
Dryopteris lasiocarpa IPNI=*1019960-2* 174
Dryopteris martinicensis IPNI=*17226640-1* 177
Dryopteris mediterranea IPNI=*17392950-1* 175
Dryopteris mindanaensis IPNI=*251308-2* 174
Dryopteris mollis IPNI=*251308-2* 174
Dryopteris mollis var. *glabrindusiata* IPNI=*251308-2* 174
Dryopteris mollissima IPNI=*1019960-2* 174
Dryopteris nephrodioides IPNI=*17460390-1* 174
Dryopteris nephrodioides var. *glandulosa* IPNI=*17460390-1* 174
Dryopteris nicaraguensis IPNI=*252198-2* 174
Dryopteris nobilis IPNI=*77205539-1* 175
Dryopteris normalis f. *austrobrasiliensis* IPNI=*251400-2* 174
Dryopteris normalis IPNI=*251400-2* 174
Dryopteris nymphalis IPNI=*251308-2* 174
Dryopteris obtusata IPNI=*17458930-1* 174
Dryopteris oligophlebia IPNI=*1019960-2* 174
Dryopteris orbicularis IPNI=*17460390-1* 174
Dryopteris paleacea IPNI=*17392950-1* 175
Dryopteris parasitica var. *decurtata* IPNI=*251308-2* 174
Dryopteris propinqua IPNI=*17458930-1* 174
Dryopteris protensa var. *funesta* IPNI=*286436-2* 177
Dryopteris punctata IPNI=*17460390-1* 174
Dryopteris radicans IPNI=*22662-2* 174
Dryopteris reducta IPNI=*251308-2* 174
Dryopteris saxatilis IPNI=*251400-2* 174
Dryopteris semicordata IPNI=*1148359-2* 175
Dryopteris serra var. *incisa* IPNI=*17226640-1* 177
Dryopteris submollis IPNI=*251308-2* 174
Dryopteris sulfurea IPNI=*17460390-1* 174
Dryopteris tenericaulis IPNI=*1019960-2* 174
Dryopteris thomassetii IPNI=*224543-2* 176
Dryopteris trichodes IPNI=*1019960-2* 174
Dryopteris uliginosa IPNI=*1019960-2* 174
Dryopteris unca IPNI=*251400-2* 174
Dryopteris wakefieldii IPNI=*17460390-1* 174
Drypetes maguireana IPNI=*54072-2* 401
Drypetes spruceana IPNI=*54072-2* 401
Dubrueilia ciliaris IPNI=*30064315-2* 447
Dubrueilia melastomoides IPNI=*855573-1* 447
Dubrueilia microphylla IPNI=*1162156-2* 447
Dubrueilia parietaria IPNI=*30064315-2* 447
Duchassaingia glauca IPNI=*494431-1* 319
Duchassaingia ovalifolia IPNI=*494431-1* 319
Duchesnea indica f. *albocaput* IPNI=*728289-1* 443
Duchesnea indica IPNI=*728289-1* 443
Duchesnea indica var. *albicarpa* IPNI=*728289-1* 443
Duchesnea indica var. *major* IPNI=*728289-1* 443
Duchesnea major IPNI=*728289-1* 443
Ducoudraea capensis IPNI=*111333-1* 365
Dugandiodendron calimaense IPNI=*991406-1* 190
Dugandiodendron urraoense IPNI=*991441-1* 190
Duggena hayesii IPNI=*30107991-2* 351
Duggena leptantha IPNI=*30107991-2* 351
Duggena ovatifolia IPNI=*112521-2* 351
Duggena panamensis IPNI=*30107991-2* 351
Duggena rudis IPNI=*112531-2* 351
Dugortia nucamentosa IPNI=*727083-1* 383
Duguetia candollei IPNI=*72920-1* 188
Duguetia caudata IPNI=*85740-2* 188
Duguetia glabra IPNI=*167395-2* 440
Duguetia heteroclada IPNI=*85700-2* 188
Duguetia ibonensis IPNI=*85728-2* 188
Duguetia insculpta IPNI=*85731-2* 188
Duguetia longifolia IPNI=*106440-2* 188
Duguetia spruceana IPNI=*85715-2* 188
Duguetia tessmannii IPNI=*85728-2* 188
Duguetia trichostemon IPNI=*85700-2* 188
Duhamelia coccinea IPNI=*30281446-2* 351
Duhamelia glabra IPNI=*743074-1* 347
Duhamelia odorata IPNI=*752194-1* 351
Duhamelia patens IPNI=*752194-1* 351
Duhamelia sphaerocarpa IPNI=*752194-1* 351
Dulacia inopiflora IPNI=*77183948-1* 449
Dulacia macrophylla IPNI=*608118-1* 449
Dulacia ovata IPNI=*77183948-1* 449
Dumerilia frutescens IPNI=*256729-1* 272
Dupatya alpina IPNI=*336004-1* 233
Dupatya dendroides IPNI=*336162-1* 233
Dupratzia scoparia IPNI=*367440-1* 344
Durandea unilocularis IPNI=*217826-2* 277
Duranta angustifolia IPNI=*330606-2* 378
Duranta baumgartenii IPNI=*330606-2* 378
Duranta brachypoda IPNI=*330606-2* 378
Duranta dentata IPNI=*330606-2* 378
Duranta ellisia IPNI=*330606-2* 378
Duranta erecta var. *grandiflora* IPNI=*330606-2* 378
Duranta hitchcockii IPNI=*862775-1* 378
Duranta inermis IPNI=*330606-2* 378
Duranta integrifolia IPNI=*330606-2* 378
Duranta latifolia IPNI=*330606-2* 378
Duranta macrocarpa IPNI=*862753-1* 378
Duranta macrodonta IPNI=*330606-2* 378
Duranta microphylla IPNI=*330606-2* 378
Duranta mutisiana IPNI=*862753-1* 378
Duranta parietariifolia IPNI=*862753-1* 378
Duranta parviflora IPNI=*1194372-2* 379
Duranta phytolaccifolia IPNI=*862753-1* 378

Duranta plumieri f. *variegata* IPNI=*330606-2* 378
Duranta plumieri IPNI=*330606-2* 378
Duranta plumieri var. *alba* IPNI=*330606-2* 378
Duranta plumieri var. *ellisia* IPNI=*330606-2* 378
Duranta plumieri var. *normalis* IPNI=*330606-2* 378
Duranta plumieri var. *strigillosa* IPNI=*330606-2* 378
Duranta racemosa IPNI=*330606-2* 378
Duranta recurvistachys IPNI=*862753-1* 378
Duranta repens f. *acuta* IPNI=*330606-2* 378
Duranta repens f. *grandiflora* IPNI=*330606-2* 378
Duranta repens f. *integrifolia* IPNI=*330606-2* 378
Duranta repens f. *microphylla* IPNI=*330606-2* 378
Duranta repens f. *obtusifolia* IPNI=*330606-2* 378
Duranta repens f. *serrata* IPNI=*330606-2* 378
Duranta repens f. *variegata* IPNI=*330606-2* 378
Duranta repens f. *vestita* IPNI=*330606-2* 378
Duranta repens IPNI=*330606-2* 378
Duranta repens var. *ellisia* IPNI=*330606-2* 378
Duranta repens var. *grandiflora* IPNI=*330606-2* 378
Duranta repens var. *microphylla* IPNI=*330606-2* 378
Duranta repens var. *multidentata* IPNI=*330606-2* 378
Duranta repens var. *mutisii* IPNI=*330606-2* 378
Duranta repens var. *normalis* IPNI=*330606-2* 378
Duranta repens var. *paucidentata* IPNI=*330606-2* 378
Duranta repens var. *serrata* IPNI=*330606-2* 378
Duranta repens var. *variegata* IPNI=*330606-2* 378
Duranta serratifolia var. *variegata* IPNI=*330606-2* 378
Duranta spinosa IPNI=*330606-2* 378
Duranta sprucei var. *breviracemosa* IPNI=*862753-1* 378
Duranta turbinata IPNI=*330606-2* 378
Duroia spraguei IPNI=*86135-2* 349
Duroia sprucei IPNI=*276143-2* 349
Duvaua pleuropogon IPNI=*70012-1* 451
Dypsis madagascariensis IPNI=*77211330-1* 208
Dysemone integrifolia IPNI=*430163-1* 358
Dysmicodon perfoliatum IPNI=*258447-2* 273
Dysodium divaricatum IPNI=*156179-2* 265
Dysodium radiatum IPNI=*156179-2* 265
Dysosmia acerifolia IPNI=*674754-1* 397
Dysosmia ciliata IPNI=*184857-2* 398
Dysosmia fluminensis IPNI=*321966-2* 398
Dysosmia foetida IPNI=*321966-2* 398
Dysosmia gossypiifolia IPNI=*321966-2* 398
Dysosmia hastata IPNI=*184857-2* 398
Dysosmia polyadena IPNI=*321966-2* 398
Dysphania ambrosioides var. *anthelmintica* IPNI=*1090820-2* 282
Dysphania ambrosioides var. *integrifolia* IPNI=*1198060-2* 282

E
Ebenus inconstans IPNI=*322526-1* 298
Ebenus melinonii IPNI=*322202-1* 298
Ebenus myrmecocarpa IPNI=*322748-1* 298
Ebenus sericea IPNI=*323004-1* 298
Eberlea glomerata IPNI=*49672-1* 360
Eburnax pudica IPNI=*313242-2* 327
Ecastaphyllum benthamianum IPNI=*77117114-1* 317
Ecastaphyllum brownei IPNI=*75934-2* 317
Ecastaphyllum brownei var. *glabrum* IPNI=*77117114-1* 317
Ecastaphyllum ecastaphyllum IPNI=*75934-2* 317
Ecastaphyllum microphyllum IPNI=*75954-2* 317
Ecastaphyllum molle IPNI=*77117114-1* 317
Ecastaphyllum monetaria IPNI=*77117114-1* 317
Ecastaphyllum plumieri IPNI=*77117114-1* 317
Ecastaphyllum richardi IPNI=*77117114-1* 317
Ecastaphyllum sieberi IPNI=*77117114-1* 317
Ecballium lambertianum IPNI=*292238-1* 293
Ecbolium betonica IPNI=*52612-1* 360
Ecbolium candelariae IPNI=*1140399-2* 359
Ecbolium carthaginense IPNI=*50468-1* 360
Ecbolium chapadanum IPNI=*941389-1* 359
Ecbolium comatum IPNI=*78920-2* 359
Ecbolium felisbertianum IPNI=*50468-1* 360
Ecbolium geniculatum IPNI=*47308-1* 359
Ecbolium havanense IPNI=*78920-2* 359
Ecbolium heterantherum IPNI=*50468-1* 360
Ecbolium lithospermifolium IPNI=*49672-1* 360
Ecbolium lucidum IPNI=*47308-1* 359
Ecbolium mexicanum IPNI=*1140399-2* 359
Ecbolium moricandianum IPNI=*47308-1* 359
Ecbolium orthotactus IPNI=*47308-1* 359
Ecbolium pectorale IPNI=*47290-1* 359
Ecbolium polygonoides IPNI=*51367-1* 360
Ecbolium retusum IPNI=*50468-1* 360
Ecbolium rostratum IPNI=*941389-1* 359
Ecbolium scabrum IPNI=*941389-1* 359
Ecbolium secundum IPNI=*47308-1* 359
Ecbolium sellinii IPNI=*132826-2* 360
Ecbolium trinervium IPNI=*52612-1* 360
Ecbolium willdenowii IPNI=*47290-1* 359
Ecclinusa abbreviata IPNI=*786784-1* 305
Ecclinusa bacuri IPNI=*86806-2* 305
Ecclinusa cochlearia IPNI=*948096-1* 308
Ecclinusa colombiana IPNI=*77176719-1* 308
Ecclinusa costata IPNI=*786784-1* 305
Ecclinusa cyanogena IPNI=*57816-2* 305
Ecclinusa prieurii IPNI=*57816-2* 305
Ecclinusa sanguinolenta IPNI=*57826-2* 305
Echetrosis pentasperma IPNI=*236510-1* 267
Echidnium spruceanum IPNI=*991397-1* 200
Echinacanthus dichotomus IPNI=*947317-1* 361
Echinocephalum discoideum IPNI=*156327-2* 266
Echinochloa aristata IPNI=*30149942-2* 238
Echinochloa colonum f. *vivipara* IPNI=*399655-1* 238
Echinochloa colonum f. *zonalis* IPNI=*399655-1* 238
Echinochloa compressa IPNI=*392087-1* 234
Echinochloa crus-galli f. *sabulicola* IPNI=*30149942-2* 238
Echinochloa crus-galli subsp. *colonum* IPNI=*399655-1* 238
Echinochloa crus-galli var. *crus-pavonis* IPNI=*30149942-2* 238
Echinochloa crus-galli var. *macera* IPNI=*30149942-2* 238
Echinochloa crus-pavonis var. *macera* IPNI=*30149942-2* 238
Echinochloa cubensis IPNI=*1004002-1* 242
Echinochloa disticha IPNI=*399660-1* 238
Echinochloa divaricata IPNI=*399655-1* 238
Echinochloa echinata IPNI=*211125-2* 244
Echinochloa elephantipes IPNI=*77147487-1* 241
Echinochloa equitans IPNI=*399655-1* 238
Echinochloa erythrosperma IPNI=*1207294-2* 245
Echinochloa geminata IPNI=*979871-1* 245
Echinochloa intermedia IPNI=*1207294-2* 245
Echinochloa kimayalaensis IPNI=*30149942-2* 238
Echinochloa lanceolata IPNI=*409823-1* 242
Echinochloa polystachya IPNI=*211125-2* 244
Echinochloa polystachya var. *spectabilis* IPNI=*88097-2* 238
Echinochloa reptans IPNI=*426280-1* 246
Echinochloa sabulicola IPNI=*30149942-2* 238
Echinochloa subcordata IPNI=*426280-1* 246
Echinochloa subverticillata IPNI=*399655-1* 238
Echinochloa zelayensis var. *macera* IPNI=*30149942-2* 238
Echinochloa zelayensis var. *subaristata* IPNI=*30149942-2* 238
Echinochloa zizanioides IPNI=*3538-2* 233
Echinochloa zonalis IPNI=*399655-1* 238
Echinocystis araneosa IPNI=*88306-2* 294
Echinocystis lanata IPNI=*88306-2* 294
Echinocystis macrocarpa IPNI=*88306-2* 294
Echinocystis muricata IPNI=*88306-2* 294
Echinocystis polycarpa IPNI=*88306-2* 294

Echinocystis racemosa IPNI=*88306-2* 294
Echinodorus amazonicus IPNI=*77183004-1* 197
Echinodorus amphibius IPNI=*77183004-1* 197
Echinodorus angustifolius IPNI=*77089409-1* 197
Echinodorus australis IPNI=*77089409-1* 197
Echinodorus austroamericanus IPNI=*77089409-1* 197
Echinodorus bleherae IPNI=*77183004-1* 197
Echinodorus bolivianus IPNI=*77089409-1* 197
Echinodorus botanicorum IPNI=*58564-1* 198
Echinodorus brevipedicellatus IPNI=*58564-1* 198
Echinodorus cordifolius IPNI=*77182733-1* 197
Echinodorus cordifolius subsp. *fluitans* IPNI=*77182733-1* 197
Echinodorus eglandulosus IPNI=*77183004-1* 197
Echinodorus floribundus IPNI=*77183476-1* 197
Echinodorus fluitans IPNI=*77182733-1* 197
Echinodorus gracilis IPNI=*77183004-1* 197
Echinodorus grandiflorus subsp. *aureus* IPNI=*77183476-1* 197
Echinodorus grandiflorus var. *aureus* IPNI=*77183476-1* 197
Echinodorus grandiflorus var. *claussenii* IPNI=*77183476-1* 197
Echinodorus grandiflorus var. *floribundus* IPNI=*77183476-1* 197
Echinodorus grisebachii IPNI=*77183004-1* 197
Echinodorus guayanensis IPNI=*58497-1* 198
Echinodorus horizontalis IPNI=*77183523-1* 197
Echinodorus isthmicus IPNI=*77089409-1* 197
Echinodorus latifolius IPNI=*77089409-1* 197
Echinodorus macrophyllus var. *muricatus* IPNI=*77183476-1* 197
Echinodorus magdalenensis IPNI=*77089409-1* 197
Echinodorus muricatus IPNI=*77183476-1* 197
Echinodorus ovalis IPNI=*77182733-1* 197
Echinodorus paniculatus f. *latifolius* IPNI=*77182731-1* 197
Echinodorus paniculatus IPNI=*77182731-1* 197
Echinodorus paniculatus var. *brevifolius* IPNI=*77182731-1* 197
Echinodorus paniculatus var. *dubius* IPNI=*77182731-1* 197
Echinodorus parviflorus IPNI=*77183004-1* 197
Echinodorus parvulus IPNI=*118649-2* 197
Echinodorus punctatus IPNI=*77183476-1* 197
Echinodorus quadricostatus IPNI=*77089409-1* 197
Echinodorus quadricostatus var. *magdalenensis* IPNI=*77089409-1* 197
Echinodorus tenellus f. *apanecae* IPNI=*77089409-1* 197
Echinodorus tenellus f. *randii* IPNI=*118649-2* 197
Echinodorus tenellus IPNI=*118649-2* 197
Echinodorus tenellus var. *ecostatus* IPNI=*118649-2* 197
Echinodorus tenellus var. *latifolius* IPNI=*77089409-1* 197
Echinodorus tenellus var. *parvulus* IPNI=*118649-2* 197
Echinolaena hirta IPNI=*88264-2* 238
Echinolaena polystachya IPNI=*211125-2* 244
Echinolaena scabra IPNI=*88264-2* 238
Echinolaena trinii IPNI=*211125-2* 244
Echinolobium coronarium IPNI=*519725-1* 334
Echinopanax japonicus IPNI=*90496-1* 253
Echinopanax papyrifer IPNI=*86417-3* 254
Echinopepon horridus IPNI=*88306-2* 294
Echinopepon lanatus IPNI=*88306-2* 294
Echinops fruticosus IPNI=*221674-2* 269
Echinops nodiflorus IPNI=*221674-2* 269
Echinopsis macrogona IPNI=*257107-2* 285
Echinopsis peruviana IPNI=*257107-2* 285
Echinopsis peruviana subsp. *puquiensis* IPNI=*257107-2* 285
Echinopsis puquiensis IPNI=*257107-2* 285
Echinopsis trichosa IPNI=*257107-2* 285
Echites acuminatus IPNI=*152067-2* 342
Echites almadensis IPNI=*151979-2* 341
Echites asperuginis IPNI=*13196-2* 339
Echites asperus IPNI=*152067-2* 342
Echites auriculatus IPNI=*151979-2* 341
Echites bangii IPNI=*81405-1* 342
Echites bicolor IPNI=*152052-2* 341
Echites bicorniculatus IPNI=*30132318-2* 342
Echites biflorus IPNI=*81684-1* 343
Echites bignoniiflorus IPNI=*285810-2* 343
Echites bilbergii IPNI=*81684-1* 343
Echites brasiliensis IPNI=*82371-1* 344
Echites campestris IPNI=*151979-2* 341
Echites canescens IPNI=*152052-2* 341
Echites chloranthus IPNI=*30132318-2* 342
Echites ciliatus IPNI=*151979-2* 341
Echites crassipes IPNI=*78821-1* 340
Echites cururu IPNI=*291380-2* 342
Echites cururu var. *angustifolius* IPNI=*291380-2* 342
Echites cururu var. *grandifolius* IPNI=*291380-2* 342
Echites cuspidatus IPNI=*30132318-2* 342
Echites cuspidifer IPNI=*152067-2* 342
Echites disadenus IPNI=*30132318-2* 342
Echites ehrenbergii IPNI=*81684-1* 343
Echites fluminensis IPNI=*151979-2* 341
Echites fluminensis var. *claussenii* IPNI=*151979-2* 341
Echites grandiflorus IPNI=*77191232-1* 342
Echites guayaquilensis IPNI=*152067-2* 342
Echites hirsutus IPNI=*151979-2* 341
Echites hirsutus var. *angustifolius* IPNI=*151979-2* 341
Echites hirsutus var. *latifolius* IPNI=*151979-2* 341
Echites hirtellus IPNI=*152067-2* 342
Echites hirtiflorus IPNI=*152067-2* 342
Echites hispidus IPNI=*151979-2* 341
Echites hoffmannseggianus IPNI=*77191232-1* 342
Echites hulkianus IPNI=*81405-1* 342
Echites insignis IPNI=*77191232-1* 342
Echites japurensis IPNI=*30132318-2* 342
Echites jasminiflorus IPNI=*152067-2* 342
Echites javitensis IPNI=*151987-2* 341
Echites lanatus IPNI=*152067-2* 342
Echites lappulaceus IPNI=*13196-2* 339
Echites lappulaceus var. *asperuginis* IPNI=*13196-2* 339
Echites lasiocarpus IPNI=*151979-2* 341
Echites lasiocarpus var. *lobbianus* IPNI=*151979-2* 341
Echites laurentiae-discus IPNI=*81405-1* 342
Echites leptolobus IPNI=*81405-1* 342
Echites littoreus IPNI=*78821-1* 340
Echites longiflorus IPNI=*78821-1* 340
Echites lucidus IPNI=*1054185-2* 342
Echites macranthus IPNI=*77191232-1* 342
Echites macrophyllus IPNI=*151979-2* 341
Echites mapirensis IPNI=*30132318-2* 342
Echites maranhamensis IPNI=*152052-2* 341
Echites membranaceus IPNI=*152067-2* 342
Echites meyerianus IPNI=*77191232-1* 342
Echites microcalyx IPNI=*152067-2* 342
Echites mollissimus IPNI=*152016-2* 341
Echites mucronatus IPNI=*152067-2* 342
Echites muricatus IPNI=*152067-2* 342
Echites nitidus IPNI=*1054185-2* 342
Echites nutans IPNI=*81405-1* 342
Echites obliquus IPNI=*78821-1* 340
Echites ovatus IPNI=*78821-1* 340
Echites pallidus IPNI=*30132318-2* 342
Echites paludosus IPNI=*81684-1* 343
Echites palustris IPNI=*151979-2* 341
Echites pubiflorus IPNI=*152052-2* 341
Echites puncticulosus IPNI=*291380-2* 342
Echites quinquangularis IPNI=*81405-1* 342
Echites richardii IPNI=*151979-2* 341
Echites rigidus IPNI=*30132318-2* 342
Echites rubrovenosus IPNI=*81405-1* 342
Echites sagittatus IPNI=*152067-2* 342
Echites sanctae-crucis IPNI=*30132318-2* 342

Echites scaber IPNI=*152052-2* 341
Echites secundiflorus IPNI=*152067-2* 342
Echites secundus IPNI=*152067-2* 342
Echites semidigynus IPNI=*77191232-1* 342
Echites siphiliticus IPNI=*20001045-1* 344
Echites stadelmeyeri IPNI=*151979-2* 341
Echites subcarnosus IPNI=*80081-1* 342
Echites subsagittatus IPNI=*152067-2* 342
Echites surinamensis IPNI=*30132318-2* 342
Echites sylvestris IPNI=*77191232-1* 342
Echites tenuicaulis IPNI=*152052-2* 341
Echites tenuicaulis var. *angustifolius* IPNI=*152052-2* 341
Echites tenuicaulis var. *latifolius* IPNI=*152052-2* 341
Echites tomentosus IPNI=*151979-2* 341
Echites tomentosus var. *laticordatus* IPNI=*151979-2* 341
Echites trifidus f. *puberulus* IPNI=*30132318-2* 342
Echites trifidus IPNI=*30132318-2* 342
Echites trifidus var. *sanctae-crucis* IPNI=*30132318-2* 342
Echites tubulosus IPNI=*30132318-2* 342
Echites umbellatus subsp. *crassipes* IPNI=*78821-1* 340
Echites umbellatus var. *crassipes* IPNI=*78821-1* 340
Echites umbellatus var. *typicus* IPNI=*78821-1* 340
Echites versicolor IPNI=*152052-2* 341
Echites verticillatus IPNI=*319065-2* 339
Echthronema convoluta IPNI=*285111-2* 216
Echtrus mexicanus IPNI=*18816-2* 435
Echtrus trivialis IPNI=*18816-2* 435
Eclipta adpressa IPNI=*1186403-2* 261
Eclipta alba f. *erecta* IPNI=*1186403-2* 261
Eclipta alba f. *longifolia* IPNI=*1186403-2* 261
Eclipta alba f. *prostrata* IPNI=*1186403-2* 261
Eclipta alba f. *zippeliana* IPNI=*1186403-2* 261
Eclipta alba IPNI=*1186403-2* 261
Eclipta alba var. *erecta* IPNI=*1186403-2* 261
Eclipta alba var. *longifolia* IPNI=*1186403-2* 261
Eclipta alba var. *parviflora* IPNI=*1186403-2* 261
Eclipta alba var. *prostrata* IPNI=*1186403-2* 261
Eclipta alba var. *zippeliana* IPNI=*1186403-2* 261
Eclipta angustata IPNI=*1186403-2* 261
Eclipta angustifolia IPNI=*1186403-2* 261
Eclipta arabica IPNI=*1186403-2* 261
Eclipta brachypoda IPNI=*1186403-2* 261
Eclipta ciliata IPNI=*1186403-2* 261
Eclipta dentata IPNI=*1186403-2* 261
Eclipta dichotoma IPNI=*1186403-2* 261
Eclipta dubia IPNI=*1186403-2* 261
Eclipta erecta IPNI=*1186403-2* 261
Eclipta erecta var. *brachypoda* IPNI=*1186403-2* 261
Eclipta erecta var. *diffusa* IPNI=*1186403-2* 261
Eclipta erecta var. *latifolia* IPNI=*1186403-2* 261
Eclipta erecta var. *prostrata* IPNI=*1186403-2* 261
Eclipta flexuosa IPNI=*1186403-2* 261
Eclipta heterophylla IPNI=*1186403-2* 261
Eclipta hirsuta IPNI=*1186403-2* 261
Eclipta linearis IPNI=*1186403-2* 261
Eclipta longifolia IPNI=*1186403-2* 261
Eclipta longifolia var. *linearis* IPNI=*1186403-2* 261
Eclipta marginata IPNI=*1186403-2* 261
Eclipta nutans IPNI=*1186403-2* 261
Eclipta oederi IPNI=*1186403-2* 261
Eclipta palustris IPNI=*1186403-2* 261
Eclipta parviflora IPNI=*1186403-2* 261
Eclipta patula IPNI=*1186403-2* 261
Eclipta philippinensis IPNI=*1186403-2* 261
Eclipta procumbens IPNI=*1186403-2* 261
Eclipta procumbens var. *brachypoda* IPNI=*1186403-2* 261
Eclipta procumbens var. *patula* IPNI=*1186403-2* 261
Eclipta prostrata f. *aureoreticulata* IPNI=*1186403-2* 261
Eclipta prostrata var. *dixitii* IPNI=*1186403-2* 261
Eclipta prostrata var. *undulata* IPNI=*202984-1* 261
Eclipta prostrata var. *zippeliana* IPNI=*1186403-2* 261
Eclipta pumila IPNI=*1186403-2* 261
Eclipta punctata IPNI=*1186403-2* 261
Eclipta pusilla IPNI=*1186403-2* 261
Eclipta simplex IPNI=*1186403-2* 261
Eclipta spicata IPNI=*1186403-2* 261
Eclipta strumosa IPNI=*1186403-2* 261
Eclipta sulcata IPNI=*1186403-2* 261
Eclipta thermalis IPNI=*1186403-2* 261
Eclipta tinctoria IPNI=*1186403-2* 261
Eclipta undulata IPNI=*202984-1* 261
Eclipta zippeliana IPNI=*1186403-2* 261
Ecliptica alba IPNI=*1186403-2* 261
Ecliptica alba var. *erecta* IPNI=*1186403-2* 261
Ecliptica alba var. *parviflora* IPNI=*1186403-2* 261
Ecliptica alba var. *prostrata* IPNI=*1186403-2* 261
Ecliptica alba var. *zippeliana* IPNI=*1186403-2* 261
Ectemis lutea IPNI=*274993-2* 274
Edanthe veraepacis IPNI=*54964-2* 208
Edechia inermis IPNI=*752006-1* 351
Edmonstonia pacifica IPNI=*867537-1* 404
Eganthus poeppigii IPNI=*110121-1* 449
Egena emirnensis IPNI=*862052-1* 369
Egeria densa IPNI=*431710-1* 203
Egletes viscosa f. *bipinnatifida* IPNI=*1037395-2* 261
Egletes viscosa var. *dissecta* IPNI=*1037395-2* 261
Egletes viscosa var. *sprucei* IPNI=*1037395-2* 261
Ehretia beurreria IPNI=*1088317-2* 273
Ehretia bogotensis IPNI=*860809-1* 368
Ehretia bourreria IPNI=*1088317-2* 273
Ehretia bourreria var. *paracalophylla* IPNI=*1088317-2* 273
Ehretia cumanensis IPNI=*77069564-1* 275
Ehretia cymosa IPNI=*77069564-1* 275
Ehretia domingensis IPNI=*1088317-2* 273
Ehretia dubia IPNI=*114058-1* 274
Ehretia exsucca IPNI=*1088317-2* 273
Ehretia grandiflora IPNI=*1088317-2* 273
Ehretia revoluta IPNI=*1088317-2* 273
Ehretia rupestris IPNI=*1088317-2* 273
Ehretia velutina IPNI=*1088317-2* 273
Eichhornia aquatica IPNI=*698138-1* 225
Eichhornia azurea IPNI=*698138-1* 225
Eichhornia azurea var. *rhizantha* IPNI=*698138-1* 225
Eichhornia cordifolia IPNI=*310928-2* 225
Eichhornia cordifolia IPNI=*698175-1* 225
Eichhornia crassicaulis IPNI=*310928-2* 225
Eichhornia crassipes IPNI=*310928-2* 225
Eichhornia diversifolia IPNI=*77188085-1* 225
Eichhornia martiana IPNI=*698175-1* 225
Eichhornia martiusiana IPNI=*698175-1* 225
Eichhornia paniculata IPNI=*698175-1* 225
Eichhornia speciosa IPNI=*310928-2* 225
Eichhornia tricolor IPNI=*698175-1* 225
Eichlerina orbicularis IPNI=*551737-1* 449
Eichlerina vulgaris IPNI=*551718-1* 449
Eichlerodendron calophyllum IPNI=*112146-1* 405
Eichlerodendron intermedium IPNI=*112146-1* 405
Eichlerodendron mexicanum IPNI=*112146-1* 405
Einomeia bracteata IPNI=*93219-1* 193
Einomeia hastata IPNI=*93219-1* 193
Einomeia pentandra IPNI=*93219-1* 193
Einsteinia sericantha IPNI=*133858-2* 352
Einsteinia speciosa IPNI=*133858-2* 352
Elaeagia brasiliensis IPNI=*308173-2* 348
Elaeagia karstenii var. *hispidula* IPNI=*88958-2* 350
Elaeagia pastoensis f. *acuminata* IPNI=*748734-1* 350

Elaeis dybowskii IPNI=*666802-1* 208
Elaeis macrophylla IPNI=*666802-1* 208
Elaeis madagascariensis IPNI=*666802-1* 208
Elaeis melanococca IPNI=*666802-1* 208
Elaeodendron javanicum IPNI=*160944-1* 290
Elaeodendron macrophyllum IPNI=*55844-2* 290
Elaeodendron tomentosum IPNI=*780054-1* 449
Elaeophora abutifolia IPNI=*203935-2* 392
Elaeophora polyadenia IPNI=*203935-2* 392
Elaeophorbia neriifolia IPNI=*347489-1* 390
Elaphoglossum dissimile IPNI=*89270-2* 175
Elaphoglossum karstenianum IPNI=*89270-2* 175
Elaphoglossum leptophyllum IPNI=*17103530-1* 175
Elaphoglossum petiolatum var. *minus* IPNI=*89270-2* 175
Elaphoglossum schiedeanum IPNI=*89270-2* 175
Elaphoglossum schiedei IPNI=*89270 2* 175
Elaphoglossum unduaviense var. *leptophylloides* IPNI=*17103530-1* 175
Elaphoglossum viscosum IPNI=*89270-2* 175
Elaphrium arboreum IPNI=*127217-1* 451
Elaphrium glabrum IPNI=*127130-1* 451
Elaphrium graveolens IPNI=*127138-1* 451
Elaphrium integerrimum IPNI=*127217-1* 451
Elaphrium jacquinianum IPNI=*127234-1* 451
Elaphrium octandrum IPNI=*127234-1* 451
Elaphrium pubescens IPNI=*127138-1* 451
Elaphrium simaruba IPNI=*127217-1* 451
Elaphrium subpubescens IPNI=*127217-1* 451
Elaphrium tatamaco IPNI=*127138-1* 451
Elaphrium tomentosum IPNI=*127234-1* 451
Elaterium amazonicum IPNI=*77110356-1* 294
Elaterium carthagenense IPNI=*77110356-1* 294
Elaterium carthagenense var. *cordatum* IPNI=*77110356-1* 294
Elaterium ciliatum IPNI=*77110356-1* 294
Elaterium cordatum IPNI=*77110356-1* 294
Elaterium filiforme IPNI=*77110356-1* 294
Elaterium gracile IPNI=*77110356-1* 294
Elaterium gracile var. *ottonianum* IPNI=*77110356-1* 294
Elaterium gracile var. *trilobum* IPNI=*77110356-1* 294
Elaterium heterophyllum IPNI=*77110356-1* 294
Elaterium macrophyllum IPNI=*77110356-1* 294
Elaterium pauciflorum IPNI=*77110356-1* 294
Elaterium quadrifidum IPNI=*77110356-1* 294
Elaterium quinquefidum IPNI=*293097-1* 294
Elaterium quinquelobum IPNI=*77110356-1* 294
Elaterium saepicola IPNI=*77110356-1* 294
Elaterium trilobatum IPNI=*77110356-1* 294
Elatine cymbalaria IPNI=*325773-2* 376
Elemifera decandra IPNI=*128316-1* 452
Eleocharis acicularis var. *typica* IPNI=*1169169-2* 231
Eleocharis articulata IPNI=*60444117-2* 232
Eleocharis brizantha IPNI=*1029981-2* 232
Eleocharis capitata IPNI=*1029981-2* 232
Eleocharis capitata var. *dispar* IPNI=*1029981-2* 232
Eleocharis capitata var. *typica* IPNI=*1029981-2* 232
Eleocharis caribaea IPNI=*1029981-2* 232
Eleocharis caribaea var. *achlamydea* IPNI=*1029981-2* 232
Eleocharis caribaea var. *dispar* IPNI=*1029981-2* 232
Eleocharis cognata IPNI=*60444117-2* 232
Eleocharis constricta IPNI=*306803-1* 231
Eleocharis cyperina IPNI=*74768-2* 230
Eleocharis densisquamata IPNI=*1029981-2* 232
Eleocharis dichotoma IPNI=*307853-1* 232
Eleocharis dispar IPNI=*1029981-2* 232
Eleocharis geniculata f. *brunnea* IPNI=*1029981-2* 232
Eleocharis geniculata var. *minor* IPNI=*1029981-2* 232
Eleocharis kirkii IPNI=*1029981-2* 232
Eleocharis melanosperma IPNI=*1029981-2* 232
Eleocharis mexicana IPNI=*1029981-2* 232
Eleocharis mexicana IPNI=*306803-1* 231
Eleocharis obsoleta IPNI=*60444117-2* 232
Eleocharis plantaginea IPNI=*60444117-2* 232
Eleocharis riparia IPNI=*1029981-2* 232
Eleocharis septata IPNI=*60444117-2* 232
Eleocharis setacea IPNI=*1029981-2* 232
Eleocharis singularis IPNI=*1029981-2* 232
Eleocharis sintenisii IPNI=*1029981-2* 232
Eleocharis valida IPNI=*1029981-2* 232
Eleocharis viridis IPNI=*306803-1* 231
Eleogenus capitatus IPNI=*1029981-2* 232
Eleogenus capitatus var. *major* IPNI=*1029981-2* 232
Eleogenus capitatus var. *minor* IPNI=*1029981-2* 232
Eleogiton cernua IPNI=*309238-1* 232
Eleogiton spadicea IPNI=*308274-1* 232
Eleogiton tabernaemontani IPNI=*1019164-1* 232
Elephantodon eburnea IPNI=*317625-1* 225
Elephantopus angustifolius IPNI=*176829-2* 267
Elephantopus bodinieri IPNI=*202967-1* 261
Elephantopus carolinensis IPNI=*202942-1* 261
Elephantopus carolinianus IPNI=*202942-1* 261
Elephantopus carolinianus var. *simplex* IPNI=*202967-1* 261
Elephantopus cernuus IPNI=*202942-1* 261
Elephantopus cervinus IPNI=*202942-1* 261
Elephantopus colimensis IPNI=*239804-1* 268
Elephantopus crispus f. *hirsuta* IPNI=*211400-2* 268
Elephantopus crispus IPNI=*176829-2* 267
Elephantopus cuneifolius IPNI=*156327-2* 266
Elephantopus dilatatus IPNI=*202942-1* 261
Elephantopus dubius IPNI=*239804-1* 268
Elephantopus glaber IPNI=*239804-1* 268
Elephantopus hypomalacus IPNI=*202942-1* 261
Elephantopus liatroides IPNI=*176829-2* 267
Elephantopus littoralis IPNI=*239804-1* 268
Elephantopus martii IPNI=*202942-1* 261
Elephantopus mollis var. *bracteosus* IPNI=*202942-1* 261
Elephantopus mollis var. *capitulatus* IPNI=*202942-1* 261
Elephantopus nudicaulis IPNI=*202967-1* 261
Elephantopus nudicaulis var. *major* IPNI=*202967-1* 261
Elephantopus nudiflorus IPNI=*176829-2* 267
Elephantopus pilosus IPNI=*202942-1* 261
Elephantopus quadriflora IPNI=*176829-2* 267
Elephantopus scaber var. *albiflorus* IPNI=*202942-1* 261
Elephantopus sericeus IPNI=*202942-1* 261
Elephantopus serratus IPNI=*202942-1* 261
Elephantopus sordidus IPNI=*202942-1* 261
Elephantopus spicatus IPNI=*239804-1* 268
Elephantopus spicatus var. *densiflorus* IPNI=*239804-1* 268
Elephantopus spicatus var. *flavidus* IPNI=*239804-1* 268
Elephantopus spicatus var. *laxiflorus* IPNI=*239804-1* 268
Elephantopus spicatus var. *lilacinus* IPNI=*239804-1* 268
Elephantopus spicatus var. *roseus* IPNI=*239804-1* 268
Elephantopus spiralis IPNI=*211400-2* 268
Elephantopus strigosus IPNI=*239804-1* 268
Elephantopus tomentosus f. *rotundatus* IPNI=*202967-1* 261
Elephantopus tomentosus var. *nudicaulis* IPNI=*202967-1* 261
Elephantopus virgatus IPNI=*202967-1* 261
Elephantosis angustifolia IPNI=*176829-2* 267
Elephantosis liatroides IPNI=*176829-2* 267
Elephantosis quadrifolia IPNI=*176829-2* 267
Elephantusia macrocarpa IPNI=*668996-1* 211
Elephantusia microcarpa IPNI=*668996-1* 211
Elettaria cardamomum var. *minor* IPNI=*796556-1* 250
Elettaria repens IPNI=*796556-1* 250
Elettaria speciosa IPNI=*942355-1* 251
Eleusine aegyptia IPNI=*397387-1* 236
Eleusine barcinonensis IPNI=*30127333-2* 238

Eleusine ciliata IPNI=*397387-1* 236
Eleusine cruciata IPNI=*397387-1* 236
Eleusine distachya IPNI=*74675-3* 238
Eleusine distans IPNI=*74675-3* 238
Eleusine domingensis IPNI=*74675-3* 238
Eleusine erythrosperma IPNI=*30127333-2* 238
Eleusine geminata IPNI=*30127333-2* 238
Eleusine glabra IPNI=*74675-3* 238
Eleusine gonantha IPNI=*74675-3* 238
Eleusine gouinii IPNI=*74675-3* 238
Eleusine inaequalis IPNI=*74675-3* 238
Eleusine indica var. *brachystachya* IPNI=*30127333-2* 238
Eleusine indica var. *condensata* IPNI=*30127333-2* 238
Eleusine indica var. *major* IPNI=*74675-3* 238
Eleusine indica var. *monostachya* IPNI=*74675-3* 238
Eleusine italica IPNI=*30127333-2* 238
Eleusine japonica IPNI=*74675-3* 238
Eleusine leucosperma IPNI=*30127333-2* 238
Eleusine macrosperma IPNI=*74675-3* 238
Eleusine marginata IPNI=*74675-3* 238
Eleusine oligostachya IPNI=*30127333-2* 238
Eleusine pectinata IPNI=*397387-1* 236
Eleusine pectinata IPNI=*74675-3* 238
Eleusine penicillata IPNI=*396063-1* 236
Eleusine polydactyla IPNI=*74675-3* 238
Eleusine prostrata IPNI=*397387-1* 236
Eleusine rigidifolia IPNI=*74675-3* 238
Eleusine scabra IPNI=*74675-3* 238
Eleusine tenerrima IPNI=*407138-1* 241
Eleusine textilis IPNI=*74675-3* 238
Eleutheranthera areolata IPNI=*202984-1* 261
Eleutheranthera discoidea IPNI=*202984-1* 261
Eleutheranthera divaricata IPNI=*156179-2* 265
Eleutheranthera ovata IPNI=*202984-1* 261
Eleutheranthera prostrata IPNI=*1186403-2* 261
Eleutherine americana IPNI=*89940-2* 216
Eleutherine anomala IPNI=*89940-2* 216
Eleutherine longifolia IPNI=*89940-2* 216
Eleutherine palmifolia IPNI=*237727-2* 216
Eleutherine plicata IPNI=*89940-2* 216
Eleutherine subaphylla IPNI=*89940-2* 216
Eleutheropetalum ernesti-augustii IPNI=*54867-2* 207
Eliokarmos caudatum IPNI=*77097737-1* 214
Eliokarmos ceresianus IPNI=*539581-1* 215
Eliokarmos coarctatum IPNI=*539581-1* 215
Eliokarmos thyrsoides IPNI=*539581-1* 215
Elionurus adustus IPNI=*90411-2* 238
Elionurus adustus var. *calvescens* IPNI=*90411-2* 238
Elionurus argenteus IPNI=*90411-2* 238
Elionurus candidus IPNI=*90411-2* 238
Elionurus candidus var. *bisetosus* IPNI=*90411-2* 238
Elionurus candidus var. *latiflorus* IPNI=*90411-2* 238
Elionurus chevalieri IPNI=*90411-2* 238
Elionurus ciliaris IPNI=*90411-2* 238
Elionurus glaber IPNI=*90411-2* 238
Elionurus gobariensis IPNI=*90411-2* 238
Elionurus latiflorus IPNI=*90411-2* 238
Elionurus latiflorus var. *adustus* IPNI=*90411-2* 238
Elionurus latiflorus var. *calvescens* IPNI=*90411-2* 238
Elionurus latiflorus var. *genuinus* IPNI=*90411-2* 238
Elionurus latiflorus var. *gracilescens* IPNI=*90411-2* 238
Elionurus latiflorus var. *pectinatus* IPNI=*90411-2* 238
Elionurus marunguensis IPNI=*90411-2* 238
Elionurus megapotamicus IPNI=*90411-2* 238
Elionurus muticus var. *calvescens* IPNI=*90411-2* 238
Elionurus muticus var. *gracilescens* IPNI=*90411-2* 238
Elionurus pertusus IPNI=*65406-3* 235
Elionurus pobeguinii IPNI=*400052-1* 238
Elionurus pretoriensis IPNI=*90411-2* 238
Elionurus prostratus IPNI=*90411-2* 238
Elionurus rostratus IPNI=*90411-2* 238
Elionurus tenax IPNI=*400052-1* 238
Elionurus thymiodorus IPNI=*90411-2* 238
Elionurus tripsacoides subvar. *ciliaris* IPNI=*400052-1* 238
Elionurus tripsacoides subvar. *rostratus* IPNI=*90411-2* 238
Elionurus tripsacoides var. *ciliaris* IPNI=*400052-1* 238
Eliopia riparia IPNI=*116938-1* 275
Eliopia serrata IPNI=*116938-1* 275
Elissarrhena longipes IPNI=*89958-2* 435
Elleanthus bractescens IPNI=*630704-1* 218
Elleanthus cajamarcae IPNI=*630704-1* 218
Elleanthus gallipanensis IPNI=*630704-1* 218
Elleanthus hallii IPNI=*630704-1* 218
Elleanthus hoppii IPNI=*630704-1* 218
Elleanthus lindenii IPNI=*630704-1* 218
Elleanthus lindenii var. *cymbibracteus* IPNI=*630704-1* 218
Elleanthus pallidiflavus IPNI=*630704-1* 218
Ellisia acuta IPNI=*330606-2* 378
Elodea verticillata IPNI=*114224-3* 203
Elsholtzia monostachys IPNI=*444504-1* 368
Elsota diversifolia IPNI=*231560-2* 338
Elsota lamarckii IPNI=*231560-2* 338
Elsota schlechtendaliana IPNI=*231560-2* 338
Elsota volubilis IPNI=*77117114-1* 317
Elwertia retusa IPNI=*61583-2* 385
Elymus attenuatus IPNI=*961772-1* 238
Elymus boliviacus IPNI=*961772-1* 238
Elytraria amara IPNI=*1167021-2* 360
Elytraria apargiifolia IPNI=*1167021-2* 360
Elytraria caulescens IPNI=*1167021-2* 360
Elytraria fasciculata IPNI=*1167021-2* 360
Elytraria frondosa IPNI=*1167021-2* 360
Elytraria microstachya IPNI=*1167021-2* 360
Elytraria pachystachya IPNI=*1167021-2* 360
Elytraria ramosa IPNI=*1167021-2* 360
Elytraria scorpioides IPNI=*1167021-2* 360
Elytraria squamosa IPNI=*1167021-2* 360
Elytraria tridentata IPNI=*1167021-2* 360
Elytraria tridentata var. *caulescens* IPNI=*1167021-2* 360
Elytraria tridentata var. *wrightii* IPNI=*1167021-2* 360
Elytrigia attenuata IPNI=*961772-1* 238
Elytrospermum californicum IPNI=*229810-2* 232
Embothrium chaparro IPNI=*706342-1* 433
Embothrium curvatum IPNI=*706342-1* 433
Embothrium monospermum IPNI=*222667-2* 433
Embothrium polystachyum IPNI=*1076423-2* 433
Embryopteris discolor IPNI=*322328-1* 298
Emelista mucronulosa IPNI=*234687-2* 334
Emelista obtusifolia IPNI=*234560-2* 333
Emelista pilifera IPNI=*911389-1* 333
Emelista tora IPNI=*518369-1* 333
Emerus grandiflorus IPNI=*518476-1* 334
Emerus herbaceus IPNI=*284993-2* 334
Emerus pubescens IPNI=*1015399-2* 334
Emilia caespitosa IPNI=*203038-1* 261
Emilia flammea IPNI=*203038-1* 261
Emilia gracilis IPNI=*203038-1* 261
Emilia humbertii IPNI=*203038-1* 261
Emilia macaulayae IPNI=*203038-1* 261
Emilia mucronata IPNI=*203080-1* 261
Emilia purpurea IPNI=*203080-1* 261
Emilia rigidula IPNI=*203080-1* 261
Emilia sagittata IPNI=*203038-1* 261
Emilia scabra IPNI=*203080-1* 261
Emilia sinica IPNI=*203080-1* 261
Emilia sonchifolia var. *typica* IPNI=*203080-1* 261

Emmeorhiza brasiliensis IPNI=*90501-2* 350
Emmeorhiza pohliana IPNI=*90501-2* 350
Emmeorhiza umbellata subsp. *septentrionalis* IPNI=*90501-2* 350
Emmeorhiza umbellata var. *pubens* IPNI=*90501-2* 350
Emmeorhiza umbellata var. *tomentosa* IPNI=*90501-2* 350
Empetrum pinnatum IPNI=*153124-2* 443
Emurtia micrantha IPNI=*595185-1* 425
Emurtia punicifolia IPNI=*1046564-2* 425
Enalcida foeniculifolia IPNI=*252097-1* 271
Enalcida pilifera IPNI=*252097-1* 271
Enallagma cucurbitina IPNI=*108218-1* 362
Enallagma latifolia IPNI=*108218-1* 362
Enallagma obovata IPNI=*108218-1* 362
Encelia albescens IPNI=*30147294-2* 272
Encelia montana IPNI=*259782-1* 272
Encephalartos latifolius IPNI=*270552-2* 179
Encephalartos pungens IPNI=*668912-1* 211
Enchysia erinoides IPNI=*143240-1* 273
Enckea amalago IPNI=*680359-1* 195
Enckea amalago var. *hirtella* IPNI=*680359-1* 195
Enckea laevigata IPNI=*681876-1* 196
Enckea lata IPNI=*198988-2* 197
Enckea martiana IPNI=*681876-1* 196
Enckea plantaginea IPNI=*680359-1* 195
Enckea reticulata IPNI=*198988-2* 197
Enckea sieberi IPNI=*680359-1* 195
Enckea smilacifolia IPNI=*198988-2* 197
Enckea vernicea IPNI=*680359-1* 195
Encyclia ciliaris IPNI=*631578-1* 218
Encyclia cochleata IPNI=*1001247-1* 222
Encyclia fragrans IPNI=*1001253-1* 222
Endiandra ita-uba IPNI=*158649-2* 183
Endiandra jamaicensis IPNI=*465312-1* 183
Endiandra viridis IPNI=*465312-1* 183
Endlichera brasiliensis IPNI=*90501-2* 350
Endlichera umbellata IPNI=*90501-2* 350
Endlicheria boliviensis IPNI=*90974-2* 183
Endlicheria endlicheriopsis IPNI=*307011-2* 183
Endlicheria grandis IPNI=*284250-2* 185
Endlicheria guadalupensis IPNI=*464535-1* 183
Endlicheria hirsuta IPNI=*90974-2* 183
Endlicheria hirsuta var. *glabrata* IPNI=*90974-2* 183
Endlicheria hirsuta var. *robusta* IPNI=*90974-2* 183
Endlicheria longifolia IPNI=*90974-2* 183
Endlicheria poeppigii IPNI=*90974-2* 183
Endlicheria racemosa IPNI=*90974-2* 183
Endlicheria villosa IPNI=*90972-2* 183
Endocodon zebrina IPNI=*796810-1* 249
Endoloma purpurea IPNI=*108202-1* 362
Endomelas alata IPNI=*56177-1* 361
Endusa punctata IPNI=*110121-1* 449
Englerella macrocarpa IPNI=*209892-2* 308
Englerella speciosa IPNI=*209892-2* 308
Englerodoxa alata IPNI=*52551-2* 299
Englerophoenix caribaeum IPNI=*664452-1* 205
Englerophoenix insignis IPNI=*26571-2* 205
Englerophoenix longirostrata IPNI=*664452-1* 205
Englerophoenix maripa IPNI=*664452-1* 205
Englerophoenix regia IPNI=*664452-1* 205
Enkianthus biflorus IPNI=*1137397-2* 300
Enneapogon borealis IPNI=*400626-1* 238
Enneapogon brachystachyus IPNI=*400626-1* 238
Enneapogon phleioides IPNI=*400626-1* 238
Enneapogon pusillus IPNI=*400626-1* 238
Enneapogon wrightii IPNI=*400626-1* 238
Ensete arnoldianum IPNI=*584810-1* 250
Ensete bagshawei IPNI=*584810-1* 250
Ensete buchananii IPNI=*584810-1* 250
Ensete davyae IPNI=*584810-1* 250
Ensete edule IPNI=*584810-1* 250
Ensete fecundum IPNI=*584810-1* 250
Ensete holstii IPNI=*584810-1* 250
Ensete laurentii IPNI=*584810-1* 250
Ensete proboscideum IPNI=*584810-1* 250
Ensete ruandense IPNI=*584810-1* 250
Ensete rubronervatum IPNI=*584810-1* 250
Ensete schweinfurthii IPNI=*584810-1* 250
Ensete ulugurense IPNI=*584810-1* 250
Ensete ventricosum var. *montbeliardii* IPNI=*584810-1* 250
Enslenia jamaicensis IPNI=*74165-2* 340
Entada acaciifolia IPNI=*1009172-2* 319
Entada adenanthera IPNI=*493866-1* 319
Entada chiliantha IPNI=*1009172-2* 319
Entada gandu IPNI=*493866-1* 319
Entada gigalobium IPNI=*493866-1* 319
Entada juliflora IPNI=*1002432-2* 331
Entada paranaguana IPNI=*91067-2* 319
Entada parrana IPNI=*493866-1* 319
Entada planoseminata IPNI=*1078643-2* 319
Entada plumeri IPNI=*1009172-2* 319
Entada polystachya var. *polyphylla* IPNI=*91067-2* 319
Entada rumphii IPNI=*493866-1* 319
Entada scandens IPNI=*493866-1* 319
Entada scandens var. *aequilatera* IPNI=*493866-1* 319
Entada umbonata IPNI=*1078643-2* 319
Entada werbaena IPNI=*188076-2* 330
Entadopsis polyphylla IPNI=*91067-2* 319
Entadopsis polystachya IPNI=*1009172-2* 319
Enterolobium mangense IPNI=*56985-2* 315
Enterolobium saman IPNI=*227126-2* 331
Enydra anagallis IPNI=*203207-1* 261
Enydra heloncha IPNI=*203207-1* 261
Enydra hingcha IPNI=*203207-1* 261
Enydra longifolia IPNI=*203207-1* 261
Enydra paludosa IPNI=*203207-1* 261
Enydra woollsii IPNI=*203207-1* 261
Ephemerum bicolor IPNI=*173471-1* 224
Ephemerum racemosum IPNI=*173146-1* 224
Epicycas miquelii IPNI=*328823-2* 179
Epidendrum × *elongatum* subsp. *rubrum* IPNI=*631770-1* 218
Epidendrum × *rubroticum* IPNI=*631770-1* 218
Epidendrum alternans IPNI=*631826-1* 218
Epidendrum altum IPNI=*1218040-2* 217
Epidendrum ansiferum IPNI=*92159-2* 219
Epidendrum antioquiense IPNI=*92159-2* 219
Epidendrum aromaticum IPNI=*661225-1* 223
Epidendrum bahiense IPNI=*30010607-2* 219
Epidendrum bicornutum IPNI=*622297-1* 218
Epidendrum bituberculatum IPNI=*633050-1* 218
Epidendrum brachyphyllum IPNI=*92159-2* 219
Epidendrum brachythyrsus IPNI=*632433-1* 219
Epidendrum bulkeleyi IPNI=*92159-2* 219
Epidendrum carolinianum IPNI=*30010607-2* 219
Epidendrum chrysostomum IPNI=*633050-1* 218
Epidendrum ciliare var. *cuspidatum* IPNI=*631578-1* 218
Epidendrum ciliare var. *typicum* IPNI=*631578-1* 218
Epidendrum ciliare var. *viscidum* IPNI=*631578-1* 218
Epidendrum cochleatum IPNI=*1001247-1* 222
Epidendrum cochleatum var. *grandiflorum* IPNI=*1001247-1* 222
Epidendrum concretum IPNI=*654273-1* 222
Epidendrum connivens IPNI=*1073117-2* 222
Epidendrum coroicoense IPNI=*92159-2* 219
Epidendrum corymbosum var. *latifolium* IPNI=*92159-2* 219
Epidendrum crassifolium var. *albescens* IPNI=*92159-2* 219
Epidendrum cucullatum IPNI=*618026-1* 217
Epidendrum cuspidatum IPNI=*631578-1* 218

Epidendrum cuzcoense IPNI=*92159-2* 219
Epidendrum decipiens IPNI=*633050-1* 218
Epidendrum discolor IPNI=*30010607-2* 219
Epidendrum dolichopus IPNI=*92159-2* 219
Epidendrum evectum IPNI=*632050-1* 218
Epidendrum fimbria IPNI=*633050-1* 218
Epidendrum fimbriatum IPNI=*134675-2* 219
Epidendrum fimbriatum var. *rhomboglossum* IPNI=*631826-1* 218
Epidendrum floribundum IPNI=*632433-1* 219
Epidendrum floribundum var. *convexum* IPNI=*632433-1* 219
Epidendrum fragrans IPNI=*1001253-1* 222
Epidendrum fragrans var. *pachypus* IPNI=*1001253-1* 222
Epidendrum giroudianum IPNI=*92159-2* 219
Epidendrum gracilicaule IPNI=*92159-2* 219
Epidendrum grandiflorum IPNI=*243490-2* 223
Epidendrum herzogii IPNI=*92159-2* 219
Epidendrum holmnielsenii IPNI=*91528-2* 218
Epidendrum huacapistanae IPNI=*631856-1* 218
Epidendrum incisum IPNI=*92159-2* 219
Epidendrum integrilabium IPNI=*631826-1* 218
Epidendrum labiatum var. *trianae* IPNI=*622257-1* 217
Epidendrum labiatum var. *warscewiczii* IPNI=*622282-1* 217
Epidendrum lacerum IPNI=*92159-2* 219
Epidendrum laetum IPNI=*633050-1* 218
Epidendrum leucarachne IPNI=*30010607-2* 219
Epidendrum lineatum IPNI=*1001253-1* 222
Epidendrum loddigesii IPNI=*622144-1* 217
Epidendrum longicrure IPNI=*632433-1* 219
Epidendrum longihastatum IPNI=*92159-2* 219
Epidendrum maximum IPNI=*622161-1* 217
Epidendrum mayzifolium IPNI=*1147853-2* 218
Epidendrum miguelii IPNI=*633050-1* 218
Epidendrum minutum IPNI=*654273-1* 222
Epidendrum miquelii IPNI=*633050-1* 218
Epidendrum mosenii IPNI=*92214-2* 219
Epidendrum muricatum IPNI=*79276-2* 218
Epidendrum nervosum IPNI=*641061-1* 219
Epidendrum nocturnum var. *angustifolium* IPNI=*30010607-2* 219
Epidendrum nocturnum var. *minor* IPNI=*30010607-2* 219
Epidendrum novogranatense IPNI=*92159-2* 219
Epidendrum obtusifolium IPNI=*30010607-2* 219
Epidendrum octandrum IPNI=*617644-1* 217
Epidendrum ornatum IPNI=*632433-1* 219
Epidendrum palmarum IPNI=*661237-1* 223
Epidendrum paniculatum var. *cuspidatum* IPNI=*632433-1* 219
Epidendrum planiceps IPNI=*633050-1* 218
Epidendrum polyschistum IPNI=*92159-2* 219
Epidendrum pratense IPNI=*632612-1* 219
Epidendrum pseudowallisii IPNI=*633025-1* 219
Epidendrum punctatum IPNI=*1147853-2* 218
Epidendrum pusillum IPNI=*1021256-1* 219
Epidendrum radicans var. *chiriquense* IPNI=*632612-1* 219
Epidendrum ramosum IPNI=*92214-2* 219
Epidendrum reflexum IPNI=*632433-1* 219
Epidendrum rhomboglossum IPNI=*631826-1* 218
Epidendrum rodriguesii IPNI=*92214-2* 219
Epidendrum rubrum IPNI=*262578-2* 223
Epidendrum sanctalucianum IPNI=*631578-1* 218
Epidendrum secundum f. *albescens* IPNI=*92159-2* 219
Epidendrum secundum var. *albescens* IPNI=*92159-2* 219
Epidendrum sessile IPNI=*643387-1* 220
Epidendrum sororium IPNI=*633050-1* 218
Epidendrum stamfordianum var. *leeanum* IPNI=*77100813-1* 219
Epidendrum sulfuratorium IPNI=*92159-2* 219
Epidendrum superbum IPNI=*622273-1* 217
Epidendrum syringiiflorum IPNI=*632433-1* 219
Epidendrum tarmense IPNI=*92159-2* 219
Epidendrum tricallosum IPNI=*92159-2* 219
Epidendrum vaginatum IPNI=*1001253-1* 222
Epidendrum vanilla IPNI=*661225-1* 223
Epidendrum ventilabrum IPNI=*1021256-1* 219
Epidendrum verecundum IPNI=*92214-2* 219
Epidendrum vermifugum IPNI=*661231-1* 223
Epidendrum versicolor IPNI=*92159-2* 219
Epidendrum violaceum IPNI=*622144-1* 217
Epidendrum violaceum IPNI=*622273-1* 217
Epidendrum viscidum IPNI=*631578-1* 218
Epigynanthus blumei IPNI=*114224-3* 203
Epigynium floribundum IPNI=*858930-1* 300
Epilobium aequinoctiale IPNI=*611783-1* 428
Epilobium andicola IPNI=*611783-1* 428
Epilobium asplundii IPNI=*611783-1* 428
Epilobium assurgens IPNI=*611783-1* 428
Epilobium bolivianum IPNI=*611783-1* 428
Epilobium bonplandianum IPNI=*611783-1* 428
Epilobium caesium IPNI=*611783-1* 428
Epilobium denticulatum var. *aberrans* IPNI=*611783-1* 428
Epilobium denticulatum var. *confertum* IPNI=*611783-1* 428
Epilobium denticulatum var. *macropetalum* IPNI=*611783-1* 428
Epilobium euphrasioides IPNI=*611783-1* 428
Epilobium fruticosum IPNI=*144385-2* 429
Epilobium glabellum var. *junceum* IPNI=*611783-1* 428
Epilobium helodes IPNI=*611783-1* 428
Epilobium hirtum IPNI=*611783-1* 428
Epilobium junceum IPNI=*611783-1* 428
Epilobium meridense IPNI=*611783-1* 428
Epilobium meridense var. *condensatum* IPNI=*611783-1* 428
Epilobium meridense var. *helodes* IPNI=*611783-1* 428
Epilobium repandum IPNI=*611783-1* 428
Epilobium tetragonum IPNI=*144385-2* 429
Epiphyllum altensteinii IPNI=*229597-2* 284
Epiphyllum ciliare IPNI=*963240-1* 284
Epiphyllum ciliatum IPNI=*963240-1* 284
Epiphyllum delicatum IPNI=*229597-2* 284
Epiphyllum elegans IPNI=*229597-2* 284
Epiphyllum gaillardae IPNI=*132939-1* 283
Epiphyllum guedneyrii IPNI=*229597-2* 284
Epiphyllum phyllanthus subsp. *cleistogamus* IPNI=*132939-1* 283
Epiphyllum phyllanthus subsp. *hookeri* IPNI=*132926-1* 283
Epiphyllum phyllanthus var. *boliviense* IPNI=*132939-1* 283
Epiphyllum phyllanthus var. *hookeri* IPNI=*132926-1* 283
Epiphyllum phyllanthus var. *paraguayense* IPNI=*132939-1* 283
Epiphyllum phyllanthus var. *schnetteri* IPNI=*132939-1* 283
Epiphyllum purpurascens IPNI=*229597-2* 284
Epiphyllum ramulosum IPNI=*963240-1* 284
Epiphyllum ruckeri IPNI=*229597-2* 284
Epiphyllum ruckerianum IPNI=*229597-2* 284
Epiphyllum ruestii IPNI=*132926-1* 283
Epiphyllum salmoneum IPNI=*229597-2* 284
Epiphyllum smithianum IPNI=*229597-2* 284
Epiphyllum stenopetalum IPNI=*132926-1* 283
Epiphyllum strictum IPNI=*132926-1* 283
Epiphyllum strictum var. *ruestii* IPNI=*132926-1* 283
Epiphyllum truncatum IPNI=*229597-2* 284
Epiphyllum truncatum var. *wagneri* IPNI=*229597-2* 284
Epiphyllum violaceum IPNI=*229597-2* 284
Epipremnum angustilobum IPNI=*87046-1* 200
Epipremnum crassifolium IPNI=*87046-1* 200
Epipremnum elegans IPNI=*87046-1* 200
Epipremnum formosanum IPNI=*87046-1* 200
Epipremnum merrillii IPNI=*87046-1* 200
Epipremnum mirabile IPNI=*87046-1* 200
Epipremnum robinsonii IPNI=*87046-1* 200
Episcia acaulis IPNI=*380794-1* 367
Episcia aurea IPNI=*107533-2* 367
Episcia chontalensis IPNI=*380794-1* 367

Episcia fendleriana IPNI=*380794-1* 367
Episcia fulgida IPNI=*380813-1* 367
Episcia hookeri IPNI=*77139499-1* 366
Episcia lynchii IPNI=*167431-2* 368
Episcia splendens IPNI=*380772-1* 367
Episcopia pendula IPNI=*252411-2* 300
Epistephium herzogianum IPNI=*617644-1* 217
Episyzygium oahuense IPNI=*600760-1* 428
Epithecia morrisii IPNI=*79276-2* 218
Epitrachys vulgaris IPNI=*196191-1* 260
Epurga lathyris IPNI=*347072-1* 390
Equisetum alpestre IPNI=*300073-2* 173
Equisetum araucanum IPNI=*320302-2* 173
Equisetum arvense f. *alpestre* IPNI=*300073-2* 173
Equisetum arvense f. *arcticum* IPNI=*300073-2* 173
Equisetum arvense f. *atratidens* IPNI=*300073-2* 173
Equisetum arvense f. *caespitosum* IPNI=*300073-2* 173
Equisetum arvense f. *decumbens* IPNI=*300073-2* 173
Equisetum arvense f. *diffusum* IPNI=*300073-2* 173
Equisetum arvense f. *microstachyum* IPNI=*300073-2* 173
Equisetum arvense f. *nanum* IPNI=*300073-2* 173
Equisetum arvense f. *nemorosum* IPNI=*300073-2* 173
Equisetum arvense f. *polystachyon* IPNI=*300073-2* 173
Equisetum arvense f. *proliferum* IPNI=*300073-2* 173
Equisetum arvense f. *pseudoalpestre* IPNI=*300073-2* 173
Equisetum arvense f. *pseudonemorosum* IPNI=*300073-2* 173
Equisetum arvense f. *pseudovarium* IPNI=*300073-2* 173
Equisetum arvense f. *ramulosum* IPNI=*300073-2* 173
Equisetum arvense monstr. proliferum IPNI=*300073-2* 173
Equisetum arvense subsp. *alpestre* IPNI=*300073-2* 173
Equisetum arvense subsp. *arcticum* IPNI=*300073-2* 173
Equisetum arvense subsp. *boreale* IPNI=*300073-2* 173
Equisetum arvense subsp. *ramulosum* IPNI=*300073-2* 173
Equisetum arvense subvar. *decumbens* IPNI=*300073-2* 173
Equisetum arvense var. *alpestre* IPNI=*300073-2* 173
Equisetum arvense var. *campestre* IPNI=*300073-2* 173
Equisetum arvense var. *decumbens* IPNI=*300073-2* 173
Equisetum arvense var. *diffusum* IPNI=*300073-2* 173
Equisetum arvense var. *frondescens* IPNI=*300073-2* 173
Equisetum arvense var. *granulatum* IPNI=*300073-2* 173
Equisetum arvense var. *nanum* IPNI=*300073-2* 173
Equisetum arvense var. *nemorosum* IPNI=*300073-2* 173
Equisetum arvense var. *nudum* IPNI=*300073-2* 173
Equisetum arvense var. *polystachyon* IPNI=*300073-2* 173
Equisetum arvense var. *proliferum* IPNI=*300073-2* 173
Equisetum arvense var. *ramulosum* IPNI=*300073-2* 173
Equisetum arvense var. *riparium* IPNI=*300073-2* 173
Equisetum arvense var. *triangulare* IPNI=*300073-2* 173
Equisetum bogotense var. *flagelliforme* IPNI=*329489-2* 173
Equisetum bogotense var. *nudum* IPNI=*329489-2* 173
Equisetum bolivianum IPNI=*320302-2* 173
Equisetum boreale IPNI=*300073-2* 173
Equisetum calderi IPNI=*300073-2* 173
Equisetum caracasanum IPNI=*320302-2* 173
Equisetum elongatum var. *scaberrimum* IPNI=*320302-2* 173
Equisetum giganteum var. *caracasanum* IPNI=*320302-2* 173
Equisetum giganteum var. *digitaliferum* IPNI=*320302-2* 173
Equisetum hyemale f. *geminatum* IPNI=*92939-2* 173
Equisetum hyemale f. *hondae* IPNI=*92939-2* 173
Equisetum martii var. *minus* IPNI=*320302-2* 173
Equisetum mexicanum IPNI=*77060052-1* 173
Equisetum philippii IPNI=*320302-2* 173
Equisetum ramosissimum IPNI=*320302-2* 173
Equisetum rinihuense IPNI=*329489-2* 173
Equisetum riparium IPNI=*300073-2* 173
Equisetum saxicola IPNI=*300073-2* 173
Equisetum tarapacanum IPNI=*320302-2* 173
Eragrostis acuminata IPNI=*1147585-2* 239
Eragrostis afghanica IPNI=*1180620-2* 239
Eragrostis alba IPNI=*1141212-2* 239
Eragrostis albensis IPNI=*1180620-2* 239
Eragrostis albescens IPNI=*1075146-2* 239
Eragrostis amabilis IPNI=*401703-1* 239
Eragrostis amabilis var. *insularis* IPNI=*1088685-2* 239
Eragrostis amabilis var. *peramangalamensis* IPNI=*1088685-2* 239
Eragrostis amboinensis IPNI=*1156566-2* 238
Eragrostis amboinica IPNI=*1156566-2* 238
Eragrostis amoena IPNI=*1147585-2* 239
Eragrostis arabica IPNI=*1156566-2* 238
Eragrostis argentina IPNI=*1104599-2* 238
Eragrostis articulata IPNI=*1104599-2* 238
Eragrostis atroviridis IPNI=*400813-1* 238
Eragrostis aturensis IPNI=*401189-1* 239
Eragrostis aurea IPNI=*401189-1* 239
Eragrostis bagdadensis IPNI=*1180620-2* 239
Eragrostis baguirmiensis IPNI=*1180620-2* 239
Eragrostis bahiensis f. *riparia* IPNI=*93057-2* 238
Eragrostis bahiensis var. *contracta* IPNI=*93057-2* 238
Eragrostis bahiensis var. *laxiuscula* IPNI=*93057-2* 238
Eragrostis barrelieri subsp. *ambigua* IPNI=*322547-2* 238
Eragrostis barrelieri subsp. *pygmaea* IPNI=*322547-2* 238
Eragrostis barrelieri subvar. *pygmaea* IPNI=*322547-2* 238
Eragrostis barrelieri var. *ambigua* IPNI=*322547-2* 238
Eragrostis barrelieri var. *pygmaea* IPNI=*322547-2* 238
Eragrostis beyrichii IPNI=*322572-2* 239
Eragrostis bicolor IPNI=*1180620-2* 239
Eragrostis biformis IPNI=*400813-1* 238
Eragrostis bithynica IPNI=*1180620-2* 239
Eragrostis blepharophylla IPNI=*93057-2* 238
Eragrostis boryana IPNI=*1156566-2* 238
Eragrostis brasiliana IPNI=*401189-1* 239
Eragrostis breviculmis IPNI=*1088685-2* 239
Eragrostis brizoides IPNI=*1015280-2* 239
Eragrostis bromoides IPNI=*400813-1* 238
Eragrostis cambessediana IPNI=*401103-1* 239
Eragrostis capillifolia IPNI=*1169955-2* 239
Eragrostis caroliniana IPNI=*1015280-2* 239
Eragrostis caudata IPNI=*401189-1* 239
Eragrostis chariis IPNI=*400813-1* 238
Eragrostis chloromelas IPNI=*1169955-2* 239
Eragrostis cilianensis subsp. *starosselskyi* IPNI=*1104599-2* 238
Eragrostis cilianensis subsp. *thyrsiflora* IPNI=*1104599-2* 238
Eragrostis cilianensis var. *starosselskyi* IPNI=*1104599-2* 238
Eragrostis cilianensis var. *subbiloba* IPNI=*1104599-2* 238
Eragrostis cilianensis var. *thyrsiflora* IPNI=*1104599-2* 238
Eragrostis ciliaris subsp. *brachystachya* IPNI=*1156566-2* 238
Eragrostis ciliaris var. *patens* IPNI=*1088685-2* 239
Eragrostis cognata IPNI=*1015280-2* 239
Eragrostis compacta IPNI=*322572-2* 239
Eragrostis compta IPNI=*1156566-2* 238
Eragrostis conferta IPNI=*401189-1* 239
Eragrostis confinis IPNI=*1088685-2* 239
Eragrostis costata IPNI=*1104599-2* 238
Eragrostis curvula var. *valida* IPNI=*1169955-2* 239
Eragrostis dakarensis IPNI=*401103-1* 239
Eragrostis damiensiana var. *laxior* IPNI=*1180620-2* 239
Eragrostis dayanandanii IPNI=*401103-1* 239
Eragrostis depauperata IPNI=*401189-1* 239
Eragrostis despiciens IPNI=*1088685-2* 239
Eragrostis diarrhena IPNI=*401189-1* 239
Eragrostis diffusa IPNI=*1015280-2* 239
Eragrostis diplachnoides IPNI=*401189-1* 239
Eragrostis diversiflora IPNI=*1075146-2* 239
Eragrostis domingensis IPNI=*1075146-2* 239
Eragrostis dumasiana IPNI=*401103-1* 239
Eragrostis elatior IPNI=*93057-2* 238

Eragrostis elegans IPNI=*401189-1* 239
Eragrostis elegans var. *laxiflora* IPNI=*401189-1* 239
Eragrostis elegantula IPNI=*400813-1* 238
Eragrostis elegantula IPNI=*401103-1* 239
Eragrostis elytroblephara IPNI=*1088685-2* 239
Eragrostis eragrostis IPNI=*1104599-2* 238
Eragrostis eragrostis subvar. *leersioides* IPNI=*1104599-2* 238
Eragrostis eragrostis var. *megastachya* IPNI=*1104599-2* 238
Eragrostis excelsa IPNI=*1075146-2* 239
Eragrostis expansa IPNI=*93057-2* 238
Eragrostis fascicularis IPNI=*1075146-2* 239
Eragrostis filiformis IPNI=*1169955-2* 239
Eragrostis filiformis IPNI=*1180620-2* 239
Eragrostis firma IPNI=*93057-2* 238
Eragrostis flamignii IPNI=*401103-1* 239
Eragrostis flexuosa IPNI=*1104599-2* 238
Eragrostis fracta IPNI=*400813-1* 238
Eragrostis gigantea IPNI=*1075146-2* 239
Eragrostis glandulosa IPNI=*1141212-2* 239
Eragrostis glomerata IPNI=*401189-1* 239
Eragrostis gracilis IPNI=*1180620-2* 239
Eragrostis gracillima IPNI=*1180620-2* 239
Eragrostis guineensis IPNI=*1075146-2* 239
Eragrostis hackelii IPNI=*93057-2* 238
Eragrostis haenkei IPNI=*1075146-2* 239
Eragrostis hagerupii IPNI=*1075146-2* 239
Eragrostis hapalantha IPNI=*401189-1* 239
Eragrostis hirsutissima IPNI=*401703-1* 239
Eragrostis hochstetteri IPNI=*401189-1* 239
Eragrostis huillensis IPNI=*1169955-2* 239
Eragrostis imberbis IPNI=*1180620-2* 239
Eragrostis indica IPNI=*1180620-2* 239
Eragrostis insulatlantica IPNI=*322547-2* 238
Eragrostis intermedia var. *appressa* IPNI=*93115-2* 239
Eragrostis interrupta IPNI=*322572-2* 239
Eragrostis interrupta var. *diplachnoides* IPNI=*401189-1* 239
Eragrostis interrupta var. *laxiflora* IPNI=*401189-1* 239
Eragrostis interrupta var. *parviflora* IPNI=*401189-1* 239
Eragrostis interrupta var. *tenuissima* IPNI=*401189-1* 239
Eragrostis jeffreysii IPNI=*1169955-2* 239
Eragrostis jeholensis IPNI=*1180620-2* 239
Eragrostis kalinaensis IPNI=*401103-1* 239
Eragrostis koenigii IPNI=*401189-1* 239
Eragrostis kossinskyi IPNI=*401189-1* 239
Eragrostis lasseri IPNI=*1156566-2* 238
Eragrostis laxa IPNI=*1088685-2* 239
Eragrostis leprieurii IPNI=*401189-1* 239
Eragrostis limbata f. *densiuscula* IPNI=*1141212-2* 239
Eragrostis limbata IPNI=*1141212-2* 239
Eragrostis limbata var. *major* IPNI=*1141212-2* 239
Eragrostis lindeniana IPNI=*1147585-2* 239
Eragrostis linearis IPNI=*1075146-2* 239
Eragrostis linkii IPNI=*1180620-2* 239
Eragrostis lobata IPNI=*1156566-2* 238
Eragrostis longipes IPNI=*1180620-2* 239
Eragrostis longispicula IPNI=*400813-1* 238
Eragrostis lugens var. *major* IPNI=*93115-2* 239
Eragrostis luzoniensis IPNI=*400813-1* 238
Eragrostis macra IPNI=*93057-2* 238
Eragrostis major IPNI=*1104599-2* 238
Eragrostis mangalorica IPNI=*401703-1* 239
Eragrostis maritima IPNI=*401189-1* 239
Eragrostis mauritii IPNI=*1088685-2* 239
Eragrostis maypurensis var. *densiuscula* IPNI=*1147585-2* 239
Eragrostis maypurensis var. *meratiana* IPNI=*1147585-2* 239
Eragrostis megastachya IPNI=*1104599-2* 238
Eragrostis megastachya var. *thyrsiflora* IPNI=*1104599-2* 238
Eragrostis mendocina IPNI=*60458636-2* 238
Eragrostis meratiana IPNI=*1147585-2* 239
Eragrostis mexicana var. *pilosior* IPNI=*1141212-2* 239
Eragrostis mexicana var. *pygmaea* IPNI=*1141212-2* 239
Eragrostis microstachya IPNI=*93057-2* 238
Eragrostis milleflora IPNI=*401189-1* 239
Eragrostis minutiflora IPNI=*401189-1* 239
Eragrostis monodii IPNI=*1104599-2* 238
Eragrostis moritzii IPNI=*322530-2* 239
Eragrostis mossulensis IPNI=*401189-1* 239
Eragrostis multiglandulosa IPNI=*1104599-2* 238
Eragrostis multispicula IPNI=*1180620-2* 239
Eragrostis namaquensis IPNI=*401189-1* 239
Eragrostis namaquensis var. *diplachnoides* IPNI=*401189-1* 239
Eragrostis nigricans var. *punensis* IPNI=*1141212-2* 239
Eragrostis nuttalliana IPNI=*1015280-2* 239
Eragrostis orcuttiana IPNI=*1141212-2* 239
Eragrostis oxylepis IPNI=*322572-2* 239
Eragrostis oxylepis var. *beyrichii* IPNI=*322572-2* 239
Eragrostis pallescens IPNI=*1075146-2* 239
Eragrostis pallida IPNI=*401189-1* 239
Eragrostis panamensis IPNI=*1147585-2* 239
Eragrostis panormitana IPNI=*322547-2* 238
Eragrostis pappii IPNI=*1104599-2* 238
Eragrostis parviglumis IPNI=*1064189-2* 239
Eragrostis pauciflora IPNI=*1141212-2* 239
Eragrostis pectinacea subsp. *felzinesii* IPNI=*1015280-2* 239
Eragrostis pensylvanica IPNI=*1015280-2* 239
Eragrostis petersii IPNI=*1180620-2* 239
Eragrostis pilosa subsp. *felzinesii* IPNI=*1015280-2* 239
Eragrostis pilosa subsp. *neglecta* IPNI=*1180620-2* 239
Eragrostis pilosa subsp. *subspontanea* IPNI=*1180620-2* 239
Eragrostis pilosa var. *bahiensis* IPNI=*93057-2* 238
Eragrostis pilosa var. *caroliniana* IPNI=*1015280-2* 239
Eragrostis plumosa IPNI=*1088685-2* 239
Eragrostis poa IPNI=*1169955-2* 239
Eragrostis polyadenia IPNI=*1104599-2* 238
Eragrostis polymorpha IPNI=*1104599-2* 238
Eragrostis polysperma IPNI=*1104599-2* 238
Eragrostis procerior IPNI=*1169955-2* 239
Eragrostis psammodes IPNI=*93057-2* 238
Eragrostis psammodes var. *microstachya* IPNI=*93057-2* 238
Eragrostis pubiculmis IPNI=*1169955-2* 239
Eragrostis pulchella IPNI=*1156566-2* 238
Eragrostis pumila IPNI=*401103-1* 239
Eragrostis punctata IPNI=*1180620-2* 239
Eragrostis purshii var. *major* IPNI=*1141212-2* 239
Eragrostis retinorrhoea IPNI=*401703-1* 239
Eragrostis robusta IPNI=*1169955-2* 239
Eragrostis rothii IPNI=*401189-1* 239
Eragrostis salzmannii IPNI=*1075146-2* 239
Eragrostis schweinfurthiana IPNI=*1104599-2* 238
Eragrostis secundiflora subsp. *oxylepis* IPNI=*322572-2* 239
Eragrostis secundiflora var. *capitata* IPNI=*322572-2* 239
Eragrostis sporoboloides IPNI=*401189-1* 239
Eragrostis squarrosa IPNI=*322572-2* 239
Eragrostis starosselskyi IPNI=*1104599-2* 238
Eragrostis stenoclada IPNI=*401103-1* 239
Eragrostis stenophylla IPNI=*401103-1* 239
Eragrostis stricta IPNI=*401189-1* 239
Eragrostis strigosa IPNI=*401703-1* 239
Eragrostis subulata IPNI=*1169955-2* 239
Eragrostis sudanica IPNI=*400813-1* 238
Eragrostis taffzagra IPNI=*1088685-2* 239
Eragrostis tenuiflora IPNI=*1180620-2* 239
Eragrostis tenuissima IPNI=*401189-1* 239
Eragrostis terebinthacea IPNI=*401703-1* 239
Eragrostis thalassica IPNI=*82343-2* 238
Eragrostis thunbergiana IPNI=*1169955-2* 239

Eragrostis thyrsiflora IPNI=*1104599-2* 238
Eragrostis transvaalensis IPNI=*401703-1* 239
Eragrostis triticea IPNI=*1104599-2* 238
Eragrostis unionis IPNI=*1015280-2* 239
Eragrostis vahlii IPNI=*1147585-2* 239
Eragrostis vahlii var. *polyantha* IPNI=*1147585-2* 239
Eragrostis vahlii var. *sejuncta* IPNI=*322572-2* 239
Eragrostis vahlii var. *subfasciculata* IPNI=*322572-2* 239
Eragrostis valida IPNI=*1169955-2* 239
Eragrostis verticillata IPNI=*1180620-2* 239
Eragrostis villosa IPNI=*1156566-2* 238
Eragrostis vinicolor IPNI=*401103-1* 239
Eragrostis virescens subsp. *verloovei* IPNI=*1141212-2* 239
Eragrostis volgensis IPNI=*401189-1* 239
Eragrostis vulgaris var. *megastachya* IPNI=*1104599-2* 238
Eragrostis warmingii IPNI=*401703-1* 239
Eragrostis willdenowii IPNI=*401103-1* 239
Eragrostis yucatana IPNI=*322572-2* 239
Eragrostis zeylanica IPNI=*93057-2* 238
Eranthemum asperum IPNI=*48879-1* 360
Eranthemum atropurpureum IPNI=*77158889-1* 361
Eranthemum carruthersii IPNI=*77158889-1* 361
Eranthemum cristatum IPNI=*46114-1* 359
Eranthemum edgeworthianum IPNI=*48879-1* 360
Eranthemum flavum IPNI=*46114-1* 359
Eranthemum grandiflorum IPNI=*48879-1* 360
Eranthemum moorei IPNI=*77158889-1* 361
Eranthemum nervosum IPNI=*48879-1* 360
Eranthemum nigrescens IPNI=*77158889-1* 361
Eranthemum nigrum IPNI=*77158889-1* 361
Eranthemum pulchellum var. *varians* IPNI=*48879-1* 360
Eranthemum reticulatum IPNI=*77158889-1* 361
Eranthemum rubrovenium IPNI=*105033-2* 360
Eranthemum sanguinolentum IPNI=*49962-1* 360
Eranthemum scabrum IPNI=*48879-1* 360
Eranthemum tricolor IPNI=*77158889-1* 361
Eranthemum varians IPNI=*48879-1* 360
Eranthemum versicolor IPNI=*77158889-1* 361
Eranthemum whartonianum IPNI=*77158889-1* 361
Erechtites agrestis IPNI=*1152557-2* 261
Erechtites ambiguus IPNI=*203365-1* 261
Erechtites cacalioides IPNI=*1152557-2* 261
Erechtites carduifolius IPNI=*1152557-2* 261
Erechtites elongatus IPNI=*1152557-2* 261
Erechtites gardnerianus IPNI=*203365-1* 261
Erechtites hieraciifolius var. *cacalioides* IPNI=*1152557-2* 261
Erechtites hieraciifolius var. *carduifolius* IPNI=*1152557-2* 261
Erechtites hieraciifolius var. *glabrescens* IPNI=*1152557-2* 261
Erechtites hieraciifolius var. *intermedius* IPNI=*1152557-2* 261
Erechtites hieraciifolius var. *megalocarpus* IPNI=*1152557-2* 261
Erechtites hieraciifolius var. *praealtus* IPNI=*1152557-2* 261
Erechtites hieraciifolius var. *typicus* IPNI=*1152557-2* 261
Erechtites megalocarpus IPNI=*1152557-2* 261
Erechtites organensis IPNI=*203365-1* 261
Erechtites praealtus IPNI=*1152557-2* 261
Erechtites sulcatus IPNI=*1152557-2* 261
Erechtites valerianifolius f. *organensis* IPNI=*203365-1* 261
Erechtites valerianifolius f. *reductus* IPNI=*203365-1* 261
Erechtites valerianifolius var. *organensis* IPNI=*203365-1* 261
Ereicoctis caracasana IPNI=*743608-1* 347
Ereicoctis conferta IPNI=*17992-2* 347
Ereicoctis filiformis IPNI=*17992-2* 347
Ereicoctis hypnoides IPNI=*17992-2* 347
Ereicoctis hyssopifolia IPNI=*18012-2* 347
Ereicoctis juniperifolia IPNI=*18012-2* 347
Ereicoctis laricifolia IPNI=*18012-2* 347
Ereicoctis nitida IPNI=*743608-1* 347
Ereicoctis thymifolia IPNI=*18012-2* 347
Eremanthe calycina IPNI=*433287-1* 393
Eremanthe patula IPNI=*927423-1* 394
Eremanthe venosa IPNI=*433287-1* 393
Eriachne rigida IPNI=*80220-2* 237
Erianthus angustifolius IPNI=*419853-1* 244
Erianthus balansae IPNI=*420046-1* 244
Erianthus biaristatus IPNI=*419853-1* 244
Erianthus fischerianus IPNI=*419853-1* 244
Erianthus glabrinodis IPNI=*420046-1* 244
Erianthus purpureus IPNI=*420046-1* 244
Erianthus repens IPNI=*2945817-1* 241
Erianthus saccharoides IPNI=*35205-2* 235
Erianthus saccharoides subsp. *angustifolius* IPNI=*419853-1* 244
Erianthus saccharoides subsp. *biaristatus* IPNI=*419853-1* 244
Erianthus saccharoides subvar. *glabrinodis* IPNI=*420046-1* 244
Erianthus saccharoides subvar. *villosus* IPNI=*420046-1* 244
Erianthus saccharoides var. *angustifolius* IPNI=*419853-1* 244
Erianthus saccharoides var. *biaristatus* IPNI=*419853-1* 244
Erianthus saccharoides var. *neesii* IPNI=*419853-1* 244
Erianthus saccharoides var. *trinii* IPNI=*420046-1* 244
Erianthus trinii IPNI=*420046-1* 244
Erianthus trinii var. *glabrinodis* IPNI=*420046-1* 244
Erianthus trinii var. *villosus* IPNI=*420046-1* 244
Erica arborea IPNI=*329044-1* 299
Erica codonodes IPNI=*329044-1* 299
Erica lusitanica subsp. *cantabrica* IPNI=*329044-1* 299
Erica polytrichifolia IPNI=*329044-1* 299
Ericoides polytrichofolium IPNI=*329044-1* 299
Ericoila prostrata IPNI=*108504-2* 344
Erigeron albidus IPNI=*204572-1* 262
Erigeron ambiguus IPNI=*204572-1* 262
Erigeron bilbaoanus IPNI=*60473415-2* 261
Erigeron bonariensis f. *grisea* IPNI=*93603-2* 261
Erigeron bonariensis var. *angustifolius* IPNI=*93603-2* 261
Erigeron bonariensis var. *leiothecus* IPNI=*60473415-2* 261
Erigeron bonariensis var. *major* IPNI=*93603-2* 261
Erigeron bonariensis var. *microcephalus* IPNI=*204572-1* 262
Erigeron canadensis f. *coloratus* IPNI=*203653-1* 261
Erigeron canadensis var. *glabratus* IPNI=*203653-1* 261
Erigeron canadensis var. *grandiflorus* IPNI=*203653-1* 261
Erigeron canadensis var. *pusillus* IPNI=*203653-1* 261
Erigeron canadensis var. *strictus* IPNI=*203653-1* 261
Erigeron chinensis IPNI=*211317-2* 268
Erigeron crispus IPNI=*93603-2* 261
Erigeron crispus subsp. *naudinii* IPNI=*204572-1* 262
Erigeron flahaultianus IPNI=*204572-1* 262
Erigeron gusalakensis IPNI=*93603-2* 261
Erigeron karvinskianus var. *mucronatus* IPNI=*204028-1* 262
Erigeron linearifolius IPNI=*93603-2* 261
Erigeron linifolius IPNI=*93603-2* 261
Erigeron linifolius IPNI=*981909-1* 271
Erigeron looseri IPNI=*93603-2* 261
Erigeron lyratus IPNI=*211317-2* 268
Erigeron mucronatus IPNI=*204028-1* 262
Erigeron musashensis IPNI=*204572-1* 262
Erigeron myriocephalus IPNI=*203653-1* 261
Erigeron myriocephalus IPNI=*60473415-2* 261
Erigeron naudinii IPNI=*204572-1* 262
Erigeron niveus IPNI=*228376-1* 264
Erigeron oreophilus IPNI=*93603-2* 261
Erigeron paniculatus IPNI=*203653-1* 261
Erigeron perennis IPNI=*184409-1* 258
Erigeron pisonis IPNI=*962413-1* 260
Erigeron pusillus IPNI=*203653-1* 261
Erigeron ruderalis IPNI=*203653-1* 261
Erigeron setiferus IPNI=*203653-1* 261
Erigeron sordidus IPNI=*93603-2* 261
Erigeron stipulatus IPNI=*211317-2* 268

Erigeron strictus IPNI=*203653-1* 261
Erigeron strictus IPNI=*228375-1* 264
Erigeron sulcatus var. *columbianus* IPNI=*204638-1* 262
Erigeron transsilvanicus IPNI=*93603-2* 261
Erigeron tricuneatus IPNI=*183814-1* 258
Erigeron trilobus IPNI=*204028-1* 262
Erigeron uliginosus var. *hirsutus* IPNI=*204638-1* 262
Erigone govenica IPNI=*63828-1* 213
Erinus procumbens IPNI=*155661-2* 376
Erinus verticillatus IPNI=*244449-2* 377
Eriobotrya cordata IPNI=*725473-1* 443
Eriobotrya glabrata IPNI=*725484-1* 443
Eriobotrya heterophylla IPNI=*725484-1* 443
Eriobotrya japonica IPNI=*77209608-1* 444
Eriobotrya oblonga IPNI=*725473-1* 443
Eriobotrya obtusifolia IPNI=*725484-1* 443
Eriocaulon aquaticum IPNI=*335722-1* 233
Eriocaulon bifistulosum IPNI=*335722-1* 233
Eriocaulon capillus-naiadis IPNI=*335722-1* 233
Eriocaulon cauliferum IPNI=*335722-1* 233
Eriocaulon dendroides IPNI=*336162-1* 233
Eriocaulon equisetoides IPNI=*335722-1* 233
Eriocaulon fluitans IPNI=*335722-1* 233
Eriocaulon intermedium IPNI=*335722-1* 233
Eriocaulon intermedium var. *brevicaule* IPNI=*335722-1* 233
Eriocaulon intermedium var. *glabrum* IPNI=*335722-1* 233
Eriocaulon limosum IPNI=*335722-1* 233
Eriocaulon schweinfurthii IPNI=*335722-1* 233
Eriocaulon setaceum var. *capillus-naiadis* IPNI=*335722-1* 233
Eriocaulon usterianum IPNI=*335722-1* 233
Eriochloa annulata IPNI=*402044-1* 239
Eriochloa hackelii IPNI=*402044-1* 239
Eriochloa kunthii IPNI=*1021229-2* 239
Eriochloa polystachya IPNI=*77164258-1* 246
Eriochloa polystachya var. *annulata* IPNI=*402044-1* 239
Eriochloa polystachya var. *punctata* IPNI=*1021229-2* 239
Eriochloa punctata f. *intermedia* IPNI=*1021229-2* 239
Eriochloa punctata IPNI=*77164258-1* 246
Eriochloa punctata var. *aristata* IPNI=*94589-2* 239
Eriochloa punctata var. *subglabra* IPNI=*77164258-1* 246
Eriochloa ramosa IPNI=*402044-1* 239
Eriochloa subglabra IPNI=*77164258-1* 246
Eriochloa sundaica IPNI=*402044-1* 239
Eriochloa texana IPNI=*1021229-2* 239
Eriocoma lehmannii IPNI=*233605-1* 266
Eriocoma moritziana IPNI=*233605-1* 266
Eriodendron anfractuosum IPNI=*1166232-2* 407
Eriodendron anfractuosum var. *caribaeum* IPNI=*1166232-2* 407
Eriodendron anfractuosum var. *guianense* IPNI=*1166232-2* 407
Eriodendron caribaeum IPNI=*1166232-2* 407
Eriodendron occidentale IPNI=*1166232-2* 407
Eriodendron orientale IPNI=*1166232-2* 407
Eriodendron pentandrum IPNI=*1166232-2* 407
Eriodendron phaeosanthum IPNI=*51083-2* 408
Eriodendron samauma IPNI=*51083-2* 408
Eriolepis echinata IPNI=*195292-1* 260
Eriopsis colombiana IPNI=*634223-1* 219
Eriopsis wercklei IPNI=*634223-1* 219
Eriorhaphe phoenicea IPNI=*824676-1* 412
Eriorhaphe punicea IPNI=*824676-1* 412
Eriosema edule IPNI=*517240-1* 331
Eriosema volubile IPNI=*517240-1* 331
Eriosorus flexuosus IPNI=*77110633-1* 178
Eriosorus villosulus IPNI=*77110633-1* 178
Eriostomum lanatum IPNI=*459393-1* 373
Erisma petiolatum IPNI=*95991-2* 430
Erisma violaceum IPNI=*870864-1* 430
Erithalis elliptica IPNI=*315444-2* 350
Erithalis inodora IPNI=*315444-2* 350
Erithalis parviflora IPNI=*315444-2* 350
Erithalis pentagonia IPNI=*1133937-2* 253
Erithalis procumbens IPNI=*315444-2* 350
Erithalis revoluta IPNI=*315444-2* 350
Ernodea uninervis IPNI=*748814-1* 350
Ernstamra caracasana IPNI=*114242-3* 276
Ernstamra macrophylla IPNI=*114242-3* 276
Ernstamra scorpioides IPNI=*114242-3* 276
Ernstamra urens IPNI=*114242-3* 276
Ernstia lyrata IPNI=*211317-2* 268
Erobathos coarctatum IPNI=*711645-1* 436
Erobathos damascenum IPNI=*711645-1* 436
Erodium allotrichum IPNI=*372354-1* 358
Erodium alsiniflorum IPNI=*372219-1* 358
Erodium arenarium IPNI=*372219-1* 358
Erodium atomarium IPNI=*372219-1* 358
Erodium australe IPNI=*372354-1* 358
Erodium boraeanum IPNI=*372219-1* 358
Erodium carneum IPNI=*372219-1* 358
Erodium cicutarium f. *chaerophyllum* IPNI=*372219-1* 358
Erodium cicutarium subsp. *ontigolanum* IPNI=*372219-1* 358
Erodium cicutarium subsp. *zairae* IPNI=*372219-1* 358
Erodium cicutarium var. *arenicola* IPNI=*372219-1* 358
Erodium cicutarium var. *triviale* IPNI=*372219-1* 358
Erodium cicutifolium IPNI=*372219-1* 358
Erodium commixtum IPNI=*372219-1* 358
Erodium danicum IPNI=*372219-1* 358
Erodium dentatum IPNI=*372354-1* 358
Erodium dissectum IPNI=*372219-1* 358
Erodium filicinum IPNI=*372219-1* 358
Erodium glutinosum IPNI=*372219-1* 358
Erodium himalayanum IPNI=*372219-1* 358
Erodium hirsutum IPNI=*372219-1* 358
Erodium immaculatum IPNI=*372219-1* 358
Erodium melanostigma IPNI=*372219-1* 358
Erodium millefolium IPNI=*372219-1* 358
Erodium minutiflorum IPNI=*372219-1* 358
Erodium moranense IPNI=*372219-1* 358
Erodium pallidiflorum IPNI=*372219-1* 358
Erodium parviflorum IPNI=*372219-1* 358
Erodium pentalobum IPNI=*372219-1* 358
Erodium petroselinum IPNI=*372219-1* 358
Erodium pimpinellifolium IPNI=*372219-1* 358
Erodium praetermissum IPNI=*372219-1* 358
Erodium robertianum IPNI=*372354-1* 358
Erodium sabulicola IPNI=*372219-1* 358
Erodium subalbidum IPNI=*372219-1* 358
Erodium tenuisectum IPNI=*372219-1* 358
Erodium triviale IPNI=*372219-1* 358
Erodium verbenifolium IPNI=*372219-1* 358
Erosion ciliare IPNI=*1156566-2* 238
Eroteum calophyllum IPNI=*828818-1* 302
Eroteum dombeyanum IPNI=*106032-2* 303
Eroteum sericeum IPNI=*106032-2* 303
Eroteum tomentosum IPNI=*106032-2* 303
Eruca alba IPNI=*288952-1* 278
Eruca cappadocica IPNI=*1137625-2* 277
Eruca deserti IPNI=*1137625-2* 277
Eruca erecta IPNI=*999093-1* 276
Eruca eruca IPNI=*1137625-2* 277
Eruca foetida IPNI=*1137625-2* 277
Eruca glabrescens IPNI=*1137625-2* 277
Eruca grandiflora IPNI=*1137625-2* 277
Eruca hispida IPNI=*1137625-2* 277
Eruca lanceolata IPNI=*1137625-2* 277
Eruca lativalvis IPNI=*1137625-2* 277
Eruca longirostris IPNI=*1137625-2* 277

Eruca longistyla IPNI=*1137625-2* 277
Eruca oleracea IPNI=*1137625-2* 277
Eruca orthosepala IPNI=*1137625-2* 277
Eruca permixta IPNI=*1137625-2* 277
Eruca ruchetta IPNI=*1137625-2* 277
Eruca sativa IPNI=*1137625-2* 277
Eruca sativa subsp. *lativalvis* IPNI=*1137625-2* 277
Eruca stenocarpa IPNI=*1137625-2* 277
Eruca subbipinnata IPNI=*1137625-2* 277
Eruca sylvestris IPNI=*1137625-2* 277
Eruca vesicaria subsp. *longirostris* IPNI=*1137625-2* 277
Eruca vesicaria subvar. *hispida* IPNI=*1137625-2* 277
Eruca vesicaria subvar. *oleracea* IPNI=*1137625-2* 277
Eruca vesicaria var. *lanceolata* IPNI=*1137625-2* 277
Erucaria lutescens IPNI=*328564-2* 235
Erucaria monostachia IPNI=*328564-2* 235
Erucastrum minutiflorum IPNI=*999093-1* 276
Erucastrum tournefortii IPNI=*999093-1* 276
Ervatamia coronaria IPNI=*82097-1* 343
Ervatamia divaricata IPNI=*82097-1* 343
Ervatamia divaricata var. *plena* IPNI=*82097-1* 343
Ervatamia flabelliformis IPNI=*82097-1* 343
Ervatamia jasminoides IPNI=*81996-1* 343
Ervatamia recurva IPNI=*82097-1* 343
Ervatamia siamensis IPNI=*82097-1* 343
Ervum gramineum IPNI=*524789-1* 336
Ervum lathyroides IPNI=*524865-1* 336
Ervum lens IPNI=*524872-1* 336
Ervum soloniense IPNI=*524865-1* 336
Ervum villosum IPNI=*525259-1* 336
Erxlebia fusiformis IPNI=*302798-2* 224
Erycina allemanii IPNI=*1021256-1* 219
Eryngium antihystericum IPNI=*841793-1* 252
Eryngium caricinum IPNI=*841769-1* 252
Eryngium ebracteatum var. *poterioides* IPNI=*841769-1* 252
Eryngium ebracteatum var. *typicum* IPNI=*841769-1* 252
Eryngium humboldtianum IPNI=*841840-1* 252
Eryngium humile subvar. *acaule* IPNI=*841842-1* 252
Eryngium humile var. *gigantophyllum* IPNI=*841842-1* 252
Eryngium humile var. *latipes* IPNI=*841842-1* 252
Eryngium humile var. *multiradiatum* IPNI=*841842-1* 252
Eryngium humile var. *typicum* IPNI=*841842-1* 252
Eryngium moritzii IPNI=*841842-1* 252
Eryngium nudiflorum IPNI=*841769-1* 252
Eryngium paniculatum f. *junior* IPNI=*841951-1* 252
Eryngium paniculatum var. *chilense* IPNI=*841951-1* 252
Eryngium paniculatum var. *litorale* IPNI=*841951-1* 252
Eryngium poterioides IPNI=*841769-1* 252
Eryngium pulchellum IPNI=*841842-1* 252
Eryngium ranunculoides IPNI=*841842-1* 252
Eryngium stellatum IPNI=*841842-1* 252
Eryngium subulatum IPNI=*841951-1* 252
Erysimum × cheiri subsp. *inexpectans* IPNI=*283703-1* 277
Erysimum × helveticum IPNI=*283703-1* 277
Erysimum × murale IPNI=*283703-1* 277
Erysimum × suffruticosum IPNI=*283703-1* 277
Erysimum barbarea IPNI=*288669-1* 277
Erysimum glabrum IPNI=*279422-1* 276
Erysimum irio IPNI=*237940-2* 278
Erysimum officinale IPNI=*289589-1* 278
Erysimum officinale var. *leiocarpum* IPNI=*289589-1* 278
Erysimum praecox IPNI=*1109261-2* 276
Erysimum runcinatum IPNI=*289589-1* 278
Erysimum tenuifolium IPNI=*1109261-2* 276
Erysimum vernum IPNI=*1109261-2* 276
Erysimum vulgare IPNI=*289589-1* 278
Erythraea capitata IPNI=*326619-2* 344
Erythraea centaurium f. *itatiaiaensis* IPNI=*326619-2* 344
Erythraea centaurium IPNI=*326619-2* 344
Erythraea corymbosa IPNI=*326619-2* 344
Erythraea germanica IPNI=*326619-2* 344
Erythraea jorulensis IPNI=*230042-2* 345
Erythraea latifolia IPNI=*326619-2* 344
Erythraea lomae IPNI=*326619-2* 344
Erythraea plumieri IPNI=*367440-1* 344
Erythraea vulgaris IPNI=*326619-2* 344
Erythranthe andicola IPNI=*77120251-1* 375
Erythranthe parviflora IPNI=*77120251-1* 375
Erythranthe pilosiuscula IPNI=*77120251-1* 375
Erythrina aculeatissima IPNI=*494616-1* 320
Erythrina alba IPNI=*494614-1* 320
Erythrina amasisa IPNI=*96637-2* 319
Erythrina argentea IPNI=*494431-1* 319
Erythrina atrosanguinea IPNI=*494431-1* 319
Erythrina aurantiaca IPNI=*494616-1* 320
Erythrina boninensis IPNI=*494614-1* 320
Erythrina caffra IPNI=*494431-1* 319
Erythrina colombiana IPNI=*96578-2* 319
Erythrina compacta IPNI=*494400-1* 319
Erythrina corallifera IPNI=*316421-2* 319
Erythrina corallodendron IPNI=*494614-1* 320
Erythrina corallodendron var. *bicolor* IPNI=*316421-2* 319
Erythrina corallodendron var. *connata* IPNI=*316421-2* 319
Erythrina corallodendron var. *occidentalis* IPNI=*316421-2* 319
Erythrina costaricensis var. *panamensis* IPNI=*96578-2* 319
Erythrina crista-galli var. *corallina* IPNI=*494400-1* 319
Erythrina crista-galli var. *longiflora* IPNI=*494400-1* 319
Erythrina divaricata IPNI=*494614-1* 320
Erythrina esculenta IPNI=*96586-2* 319
Erythrina fasciculata IPNI=*494400-1* 319
Erythrina glauca IPNI=*494431-1* 319
Erythrina graefferi IPNI=*494400-1* 319
Erythrina indica IPNI=*494614-1* 320
Erythrina indica var. *phlebocarpa* IPNI=*494614-1* 320
Erythrina inermis IPNI=*316421-2* 319
Erythrina laurifolia IPNI=*494400-1* 319
Erythrina lobulata IPNI=*494614-1* 320
Erythrina lorenoi IPNI=*96586-2* 319
Erythrina loureiroi IPNI=*494614-1* 320
Erythrina megistophylla IPNI=*96586-2* 319
Erythrina micropteryx IPNI=*96637-2* 319
Erythrina moelebei IPNI=*494431-1* 319
Erythrina neglecta IPNI=*96559-2* 319
Erythrina orientalis IPNI=*494614-1* 320
Erythrina ovalifolia IPNI=*494431-1* 319
Erythrina panamensis IPNI=*96578-2* 319
Erythrina panduriformis IPNI=*494614-1* 320
Erythrina patens IPNI=*494431-1* 319
Erythrina phlebocarpa IPNI=*494614-1* 320
Erythrina picta IPNI=*494614-1* 320
Erythrina pisamo IPNI=*96637-2* 319
Erythrina poeppigiana f. *redmondii* IPNI=*96637-2* 319
Erythrina pulcherrima IPNI=*494400-1* 319
Erythrina resupinata IPNI=*494614-1* 320
Erythrina rostrata IPNI=*494614-1* 320
Erythrina spathacea IPNI=*494614-1* 320
Erythrina speciosa IPNI=*494400-1* 319
Erythrina spinosa IPNI=*316421-2* 319
Erythrina splendida IPNI=*494616-1* 320
Erythrina variegata var. *alba* IPNI=*494614-1* 320
Erythrina velutina f. *aurantiaca* IPNI=*494616-1* 320
Erythrina xinguensis IPNI=*96661-2* 319
Erythrobalanus humboldtii IPNI=*296268-1* 338
Erythrobalanus lindenii IPNI=*296268-1* 338
Erythrobalanus tolimensis IPNI=*296268-1* 338
Erythrodes debilis IPNI=*77125687-1* 217

Erythrodes jamaicensis IPNI=*77125705-1* 217
Erythrodes maculata IPNI=*77125705-1* 217
Erythrodes mayoriana IPNI=*77125705-1* 217
Erythrodes querceticola IPNI=*77125705-1* 217
Erythrodes sagrana IPNI=*77125705-1* 217
Erythrodes trinitatis IPNI=*77125705-1* 217
Erythrodes vaginata IPNI=*77125705-1* 217
Erythroxylum acutifolium IPNI=*330131-2* 386
Erythroxylum areolatum IPNI=*330154-2* 386
Erythroxylum bolivianum IPNI=*330132-2* 386
Erythroxylum chiapense IPNI=*330154-2* 386
Erythroxylum chilpei IPNI=*330132-2* 386
Erythroxylum cienaguense IPNI=*96953-2* 386
Erythroxylum citrifolium var. *latifolium* IPNI=*330131-2* 386
Erythroxylum citrifolium var. *minus* IPNI=*330131-2* 386
Erythroxylum coca var. *novogranatense* IPNI=*97004-2* 386
Erythroxylum coca var. *spruceanum* IPNI=*97004-2* 386
Erythroxylum costaricense IPNI=*330167-2* 386
Erythroxylum cuatrecasasii IPNI=*276848-2* 386
Erythroxylum cumanense IPNI=*330154-2* 386
Erythroxylum duckei IPNI=*330131-2* 386
Erythroxylum durum IPNI=*330167-2* 386
Erythroxylum filipes IPNI=*330167-2* 386
Erythroxylum floribundum IPNI=*330167-2* 386
Erythroxylum fraternum IPNI=*330131-2* 386
Erythroxylum glaucum var. *patiacum* IPNI=*96968-2* 386
Erythroxylum gracilipes var. *exareolatum* IPNI=*276848-2* 386
Erythroxylum havanense IPNI=*330154-2* 386
Erythroxylum havanense var. *continentis* IPNI=*330154-2* 386
Erythroxylum hostmannianum IPNI=*330131-2* 386
Erythroxylum laurinum IPNI=*330167-2* 386
Erythroxylum lineolatum var. *sanctae-marthae* IPNI=*96936-2* 386
Erythroxylum lucidum IPNI=*330167-2* 386
Erythroxylum lucidum var. *costaricense* IPNI=*330167-2* 386
Erythroxylum manglilla IPNI=*330131-2* 386
Erythroxylum micranthum IPNI=*330131-2* 386
Erythroxylum mucronatum var. *majus* IPNI=*276833-2* 386
Erythroxylum multiflorum IPNI=*330167-2* 386
Erythroxylum obtusum IPNI=*330154-2* 386
Erythroxylum ovatum IPNI=*330154-2* 386
Erythroxylum ovatum var. *splendens* IPNI=*330154-2* 386
Erythroxylum paraense IPNI=*330131-2* 386
Erythroxylum recurrens IPNI=*276848-2* 386
Erythroxylum rigidulum IPNI=*544013-1* 386
Erythroxylum rigidulum var. *exareolatum* IPNI=*544013-1* 386
Erythroxylum rubrinerve IPNI=*330167-2* 386
Erythroxylum skutchii IPNI=*330167-2* 386
Erythroxylum squamatum var. *orinocense* IPNI=*97014-2* 386
Erythroxylum tabascense IPNI=*330167-2* 386
Erythroxylum tessmannii IPNI=*276893-2* 387
Erythroxylum uniflorum IPNI=*97017-2* 386
Erythroxylum venosum IPNI=*276893-2* 387
Escallonia adscendens IPNI=*791067-1* 309
Escallonia atahuallpae IPNI=*791075-1* 309
Escallonia caracasana IPNI=*791071-1* 309
Escallonia commutata IPNI=*791101-1* 309
Escallonia concinna IPNI=*791101-1* 309
Escallonia corymbosa IPNI=*791067-1* 309
Escallonia duplicatoserrata IPNI=*791101-1* 309
Escallonia floribunda IPNI=*791071-1* 309
Escallonia floribunda var. *caracasana* IPNI=*791071-1* 309
Escallonia hirta IPNI=*791015-1* 309
Escallonia hypsophila IPNI=*791067-1* 309
Escallonia littoralis IPNI=*791101-1* 309
Escallonia littoralis var. *concinna* IPNI=*791101-1* 309
Escallonia macrantha var. *duplicatoserrata* IPNI=*791101-1* 309
Escallonia paniculata var. *floribunda* IPNI=*791071-1* 309
Escallonia pilgeriana IPNI=*791071-1* 309
Escallonia poeppigiana IPNI=*791101-1* 309
Escallonia punctata IPNI=*791101-1* 309
Escallonia rahmeri IPNI=*791101-1* 309
Escallonia rubra var. *glabriuscula* IPNI=*791101-1* 309
Escallonia rubra var. *multiflora* IPNI=*791101-1* 309
Escallonia rubra var. *pubescens* IPNI=*791101-1* 309
Escallonia rubra var. *punctata* IPNI=*791101-1* 309
Escallonia rubra var. *uniflora* IPNI=*791101-1* 309
Escallonia tubar IPNI=*791015-1* 309
Eschenbachia ambigua IPNI=*204572-1* 262
Eschenbachia lyrata IPNI=*211317-2* 268
Eschenbachia tenuisecta IPNI=*947840-1* 264
Escheria gloxiniifolia IPNI=*381282-1* 367
Eschscholzia absinthiifolia IPNI=*30043364-2* 435
Eschscholzia ambigua IPNI=*30043364-2* 435
Eschscholzia angularis IPNI=*30043364-2* 435
Eschscholzia apiculata IPNI=*30043364-2* 435
Eschscholzia arizonica IPNI=*30043364-2* 435
Eschscholzia australis IPNI=*30043364-2* 435
Eschscholzia benedicta IPNI=*30043364-2* 435
Eschscholzia bernardina IPNI=*30043364-2* 435
Eschscholzia bernardina var. *coarctata* IPNI=*30043364-2* 435
Eschscholzia bicornuta IPNI=*30043364-2* 435
Eschscholzia biolettii IPNI=*30043364-2* 435
Eschscholzia brandegeei IPNI=*30043364-2* 435
Eschscholzia californica var. *ambigua* IPNI=*30043364-2* 435
Eschscholzia californica var. *compacta* IPNI=*30043364-2* 435
Eschscholzia californica var. *crocea* IPNI=*30043364-2* 435
Eschscholzia californica var. *douglasii* IPNI=*30043364-2* 435
Eschscholzia californica var. *luxurians* IPNI=*30043364-2* 435
Eschscholzia californica var. *maritima* IPNI=*30043364-2* 435
Eschscholzia californica var. *peninsularis* IPNI=*30043364-2* 435
Eschscholzia californica var. *stricta* IPNI=*30043364-2* 435
Eschscholzia calosperma IPNI=*30043364-2* 435
Eschscholzia chartacea IPNI=*30043364-2* 435
Eschscholzia clevelandi IPNI=*30043364-2* 435
Eschscholzia cognata IPNI=*30043364-2* 435
Eschscholzia columbiana IPNI=*30043364-2* 435
Eschscholzia compacta IPNI=*30043364-2* 435
Eschscholzia confinis IPNI=*30043364-2* 435
Eschscholzia crassula IPNI=*30043364-2* 435
Eschscholzia crocea IPNI=*30043364-2* 435
Eschscholzia crocea var. *apiifolia* IPNI=*30043364-2* 435
Eschscholzia crocea var. *longissima* IPNI=*30043364-2* 435
Eschscholzia crocea var. *rosea* IPNI=*30043364-2* 435
Eschscholzia crocea var. *sanctarum* IPNI=*30043364-2* 435
Eschscholzia cucullata IPNI=*30043364-2* 435
Eschscholzia cyathifera IPNI=*30043364-2* 435
Eschscholzia debilis IPNI=*30043364-2* 435
Eschscholzia diversiloba IPNI=*30043364-2* 435
Eschscholzia douglasii IPNI=*30043364-2* 435
Eschscholzia eastwoodiae IPNI=*30043364-2* 435
Eschscholzia exilis IPNI=*30043364-2* 435
Eschscholzia floribunda IPNI=*30043364-2* 435
Eschscholzia floribunda var. *gorgonica* IPNI=*30043364-2* 435
Eschscholzia floribunda var. *gracillima* IPNI=*30043364-2* 435
Eschscholzia foeniculacea IPNI=*30043364-2* 435
Eschscholzia gigas IPNI=*30043364-2* 435
Eschscholzia glauca IPNI=*30043364-2* 435
Eschscholzia granulata IPNI=*30043364-2* 435
Eschscholzia granulata var. *minuscula* IPNI=*30043364-2* 435
Eschscholzia helleriana IPNI=*30043364-2* 435
Eschscholzia helleriana var. *tilingii* IPNI=*30043364-2* 435
Eschscholzia humilis IPNI=*30043364-2* 435
Eschscholzia inflata IPNI=*30043364-2* 435
Eschscholzia isostigma IPNI=*30043364-2* 435
Eschscholzia juncea IPNI=*30043364-2* 435
Eschscholzia lacera IPNI=*30043364-2* 435

Eschscholzia leptandra IPNI=*30043364-2* 435
Eschscholzia leptomitra IPNI=*30043364-2* 435
Eschscholzia leucosticta IPNI=*30043364-2* 435
Eschscholzia ludens IPNI=*30043364-2* 435
Eschscholzia marcida IPNI=*30043364-2* 435
Eschscholzia marcida var. *monticola* IPNI=*30043364-2* 435
Eschscholzia maritima IPNI=*30043364-2* 435
Eschscholzia menziesiana IPNI=*30043364-2* 435
Eschscholzia menziesiana var. *anemophila* IPNI=*30043364-2* 435
Eschscholzia menziesiana var. *coarctata* IPNI=*30043364-2* 435
Eschscholzia menziesiana var. *nesiaca* IPNI=*30043364-2* 435
Eschscholzia menziesiana var. *recedens* IPNI=*30043364-2* 435
Eschscholzia microloba IPNI=*30043364-2* 435
Eschscholzia nitrophila IPNI=*30043364-2* 435
Eschscholzia oregana IPNI=*30043364-2* 435
Eschscholzia peninsularis IPNI=*30043364-2* 435
Eschscholzia physodes IPNI=*30043364-2* 435
Eschscholzia picta IPNI=*30043364-2* 435
Eschscholzia pseudoinflata IPNI=*30043364-2* 435
Eschscholzia pseudopraecox IPNI=*30043364-2* 435
Eschscholzia ptarmicoides IPNI=*30043364-2* 435
Eschscholzia quadrangularis IPNI=*30043364-2* 435
Eschscholzia recta IPNI=*30043364-2* 435
Eschscholzia revoluta IPNI=*30043364-2* 435
Eschscholzia revoluta var. *caudatocalyx* IPNI=*30043364-2* 435
Eschscholzia rigida IPNI=*30043364-2* 435
Eschscholzia robusta IPNI=*30043364-2* 435
Eschscholzia rosea IPNI=*30043364-2* 435
Eschscholzia sanctarum IPNI=*30043364-2* 435
Eschscholzia scapifera IPNI=*30043364-2* 435
Eschscholzia scariosa IPNI=*30043364-2* 435
Eschscholzia scariosa var. *dichasiophora* IPNI=*30043364-2* 435
Eschscholzia setchellii IPNI=*30043364-2* 435
Eschscholzia shastensis IPNI=*30043364-2* 435
Eschscholzia straminea IPNI=*30043364-2* 435
Eschscholzia stricta IPNI=*30043364-2* 435
Eschscholzia tenuisecta IPNI=*30043364-2* 435
Eschscholzia thermophila IPNI=*30043364-2* 435
Eschscholzia tristis IPNI=*30043364-2* 435
Eschscholzia vaccarum IPNI=*30043364-2* 435
Eschscholzia vernalis IPNI=*30043364-2* 435
Eschscholzia xylorrhiza IPNI=*30043364-2* 435
Eschscholzia yainacensis IPNI=*30043364-2* 435
Eschscholzia yainacensis var. *modocensis* IPNI=*30043364-2* 435
Eschweilera acuminatissima IPNI=*947902-1* 301
Eschweilera bolivarensis IPNI=*597290-1* 302
Eschweilera brancoensis IPNI=*597250-1* 302
Eschweilera carrii IPNI=*596743-1* 302
Eschweilera cauliflora IPNI=*97378-2* 301
Eschweilera chaffanjonii IPNI=*592624-1* 301
Eschweilera chartacea IPNI=*597250-1* 302
Eschweilera cincta IPNI=*114963-2* 302
Eschweilera cordata IPNI=*77176926-1* 302
Eschweilera elegans IPNI=*592624-1* 301
Eschweilera eymaana IPNI=*947902-1* 301
Eschweilera flaccida IPNI=*276907-2* 301
Eschweilera fractiflexa IPNI=*947902-1* 301
Eschweilera grandifolia IPNI=*947902-1* 301
Eschweilera idatimonoides IPNI=*276907-2* 301
Eschweilera iquitosensis IPNI=*592624-1* 301
Eschweilera krukovii IPNI=*592624-1* 301
Eschweilera lancifolia IPNI=*592624-1* 301
Eschweilera longipes f. *genuina* IPNI=*276907-2* 301
Eschweilera longipes f. *platycarpa* IPNI=*276907-2* 301
Eschweilera longipes IPNI=*276907-2* 301
Eschweilera macrophylla IPNI=*276907-2* 301
Eschweilera marawynensis IPNI=*597250-1* 302
Eschweilera matamata IPNI=*947902-1* 301
Eschweilera odora IPNI=*947902-1* 301
Eschweilera pallida IPNI=*947902-1* 301
Eschweilera pilosa IPNI=*276907-2* 301
Eschweilera pittieri IPNI=*97452-2* 301
Eschweilera platycarpa IPNI=*276907-2* 301
Eschweilera retroflexa IPNI=*947902-1* 301
Eschweilera rigida IPNI=*597250-1* 302
Eschweilera rorida IPNI=*597250-1* 302
Eschweilera spruceana IPNI=*597250-1* 302
Eschweilera timbuchensis IPNI=*592624-1* 301
Eschweilera vageleri IPNI=*947902-1* 301
Eschweilera valida IPNI=*597290-1* 302
Eschweilera verruculosa IPNI=*97452-2* 301
Escobedia brevipes IPNI=*97540-2* 375
Escobedia curialis IPNI=*97540-2* 375
Escobedia foliolosa IPNI=*97540-2* 375
Escobedia longiflora IPNI=*97540-2* 375
Escobedia obtusifolia IPNI=*97540-2* 375
Escobedia parimensis IPNI=*97540-2* 375
Escobedia parvifolia IPNI=*97540-2* 375
Escobedia reticulata IPNI=*97540-2* 375
Escobedia scabrifolia IPNI=*97540-2* 375
Escobedia scabrifolia var. *laevigata* IPNI=*97540-2* 375
Escobedia silvia IPNI=*97540-2* 375
Espeletia × *almorzana* f. *latifolia* IPNI=*289835-2* 262
Espeletia corymbosa subsp. *zipaquirana* IPNI=*97655-2* 262
Espeletia corymbosa var. *foliosa* IPNI=*97655-2* 262
Espeletia glandulosa var. *scaberrima* IPNI=*97673-2* 262
Espeletia grandiflora f. *longiligulata* IPNI=*205109-1* 262
Espeletia grandiflora f. *multiflora* IPNI=*205109-1* 262
Espeletia grandiflora f. *reducta* IPNI=*205109-1* 262
Espeletia hartwegiana IPNI=*205109-1* 262
Espeletia nivea IPNI=*205061-1* 262
Espeletia phaneractis subsp. *boyacensis* IPNI=*97631-2* 262
Espeletia platylepis IPNI=*97655-2* 262
Espeletia rigida IPNI=*97655-2* 262
Espeletiopsis × *bogotensis* IPNI=*97630-2* 262
Espeletiopsis caldasii IPNI=*97641-2* 262
Espeletiopsis corymbosa IPNI=*97655-2* 262
Espeletiopsis garciae IPNI=*97671-2* 262
Espeletiopsis glandulosa IPNI=*97673-2* 262
Espeletiopsis jimenez-quesadae IPNI=*97703-2* 262
Espeletiopsis muiska IPNI=*97726-2* 262
Espeletiopsis pleiochasia IPNI=*97751-2* 263
Espeletiopsis santanderensis IPNI=*97768-2* 263
Esquirolia sinensis IPNI=*610017-1* 375
Esula cupressina IPNI=*346205-1* 390
Esula cyparissias IPNI=*346205-1* 390
Esula minor IPNI=*346205-1* 390
Esula peplus IPNI=*347724-1* 390
Ethanium aromaticum IPNI=*798014-1* 251
Ethanium bracteosum IPNI=*218583-2* 251
Ethanium breviscapum IPNI=*798020-1* 251
Ethanium exaltatum IPNI=*218583-2* 251
Ethanium macranthum IPNI=*218583-2* 251
Ethanium mexicanum IPNI=*218639-2* 251
Ethanium pacoseroca IPNI=*218583-2* 251
Ethanium racemosum IPNI=*798020-1* 251
Ethanium thyrsodeum IPNI=*798093-1* 251
Ethesia carnea IPNI=*132826-2* 360
Ethnora maripa IPNI=*664452-1* 205
Ethulia aquatica IPNI=*175030-1* 256
Ethulia sparganophora IPNI=*1059151-2* 270
Ethulia struchium IPNI=*1059151-2* 270
Etlingera elatior var. *pileng* IPNI=*942355-1* 251
Eucalyptus andreana IPNI=*592904-1* 424
Eucalyptus andreana var. *stenophylla* IPNI=*592904-1* 424
Eucalyptus brunnea IPNI=*592863-1* 424

Eucalyptus calophylla IPNI=*986246-1* 424
Eucalyptus calophylla var. *maideniana* IPNI=*986246-1* 424
Eucalyptus calophylla var. *multiflora* IPNI=*986246-1* 424
Eucalyptus calophylla var. *parviflora* IPNI=*986246-1* 424
Eucalyptus calophylla var. *rosea* IPNI=*986246-1* 424
Eucalyptus calophylla var. *rubra* IPNI=*986246-1* 424
Eucalyptus citriodora IPNI=*986336-1* 424
Eucalyptus cordata IPNI=*593283-1* 424
Eucalyptus ficifolia IPNI=*986249-1* 424
Eucalyptus ficifolia var. *alba* IPNI=*986249-1* 424
Eucalyptus ficifolia var. *bakeri* IPNI=*986249-1* 424
Eucalyptus ficifolia var. *carmina* IPNI=*986249-1* 424
Eucalyptus ficifolia var. *rosea* IPNI=*986249-1* 424
Eucalyptus glaucophylla IPNI=*986246-1* 424
Eucalyptus hemilampra IPNI=*593305-1* 425
Eucalyptus hookeri IPNI=*986340-1* 424
Eucalyptus lindleyana IPNI=*592904-1* 424
Eucalyptus lindleyana var. *stenophylla* IPNI=*592904-1* 424
Eucalyptus maculata var. *citriodora* IPNI=*986336-1* 424
Eucalyptus melissiodora IPNI=*986336-1* 424
Eucalyptus multiflora IPNI=*593318-1* 425
Eucalyptus ovalifolia IPNI=*593261-1* 424
Eucalyptus ovalifolia var. *lanceolata* IPNI=*593261-1* 424
Eucalyptus pulverulenta var. *ovatifolia* IPNI=*593283-1* 424
Eucalyptus pulvigera IPNI=*593283-1* 424
Eucalyptus resinifera subsp. *hemilampra* IPNI=*593305-1* 425
Eucalyptus resinifera var. *grandiflora* IPNI=*593305-1* 425
Eucalyptus resinifera var. *hemilampra* IPNI=*593305-1* 425
Eucalyptus robusta var. *bivalvis* IPNI=*593318-1* 425
Eucalyptus robusta var. *rostrata* IPNI=*593318-1* 425
Eucalyptus rostrata IPNI=*593318-1* 425
Eucalyptus saligna var. *pallidivalvis* IPNI=*592976-1* 424
Eucalyptus saligna var. *parviflora* IPNI=*592863-1* 424
Eucalyptus saligna var. *protrusa* IPNI=*593334-1* 425
Eucalyptus splachnicarpa IPNI=*986246-1* 424
Eucalyptus tessellaris IPNI=*986340-1* 424
Eucalyptus variegata IPNI=*986336-1* 424
Eucalyptus viminalis IPNI=*986340-1* 424
Eucharis × grandiflora IPNI=*261579-2* 214
Eucharis × lowii IPNI=*261579-2* 214
Eucharis × mastersii IPNI=*261579-2* 214
Eucharis amazonica IPNI=*77182693-1* 214
Eucharis candida IPNI=*261572-2* 214
Eucharis formosa IPNI=*77183240-1* 214
Eucharis fosteri IPNI=*67017-1* 214
Eucharis paradoxa IPNI=*67017-1* 214
Eucharis sanderi IPNI=*261595-2* 214
Eucharis sanderi var. *multiflora* IPNI=*261595-2* 214
Eucharis subedentata IPNI=*67017-1* 214
Euchlaena luxurians IPNI=*270861-2* 247
Euchlaena mexicana var. *luxurians* IPNI=*270861-2* 247
Euclasta glumacea IPNI=*97916-2* 239
Eucolum crassifolium IPNI=*381282-1* 367
Eucyperus alternifolius IPNI=*303729-1* 230
Eucyperus longifolius IPNI=*305041-1* 231
Eucyperus luzulae IPNI=*74851-2* 231
Eudistemon humifusum IPNI=*286148-1* 277
Eugenia acuminatissima IPNI=*593768-1* 425
Eugenia acutiloba IPNI=*593768-1* 425
Eugenia alaternifolia IPNI=*77198138-1* 427
Eugenia alexandri IPNI=*593768-1* 425
Eugenia alfaroana IPNI=*593768-1* 425
Eugenia amanuensis IPNI=*593768-1* 425
Eugenia ambigua IPNI=*1046564-2* 425
Eugenia anthera IPNI=*593690-1* 425
Eugenia antiquae IPNI=*593490-1* 425
Eugenia arbutifolia IPNI=*1046564-2* 425
Eugenia arbutifolia var. *obscura* IPNI=*1046564-2* 425
Eugenia arbutifolia var. *pallida* IPNI=*1046564-2* 425
Eugenia arctostaphyloides IPNI=*1046564-2* 425
Eugenia arctostaphyloides var. *ovalis* IPNI=*1046564-2* 425
Eugenia arechavaletae IPNI=*596263-1* 425
Eugenia argyrea IPNI=*594499-1* 425
Eugenia aromatica IPNI=*166091-2* 427
Eugenia atropunctata IPNI=*594452-1* 425
Eugenia atropunctata var. *gracilis* IPNI=*594452-1* 425
Eugenia atropunctata var. *robusta* IPNI=*594452-1* 425
Eugenia axillaris var. *cozumelensis* IPNI=*593690-1* 425
Eugenia axillaris var. *microcarpa* IPNI=*593690-1* 425
Eugenia balbisiana IPNI=*166091-2* 427
Eugenia balbisiana var. *dives* IPNI=*166091-2* 427
Eugenia balbisiana var. *triflora* IPNI=*166091-2* 427
Eugenia balsamica IPNI=*166091-2* 427
Eugenia banghamii IPNI=*596302-1* 425
Eugenia bartlettiana IPNI=*593490-1* 425
Eugenia baruensis IPNI=*593690-1* 425
Eugenia baruensis IPNI=*595457-1* 425
Eugenia benthamii IPNI=*1046564-2* 425
Eugenia berlynensis IPNI=*595457-1* 425
Eugenia biflora var. *lancea* IPNI=*593768-1* 425
Eugenia biflora var. *ludibunda* IPNI=*593768-1* 425
Eugenia biflora var. *mini* IPNI=*593768-1* 425
Eugenia biflora var. *pallens* IPNI=*593768-1* 425
Eugenia biflora var. *virgultosa* IPNI=*593768-1* 425
Eugenia boliviensis IPNI=*1046564-2* 425
Eugenia boliviensis IPNI=*598986-1* 426
Eugenia bonplandiana IPNI=*593490-1* 425
Eugenia brachiata IPNI=*601603-1* 428
Eugenia brachythrix IPNI=*593768-1* 425
Eugenia bracteata IPNI=*598646-1* 426
Eugenia bracteolosa IPNI=*593490-1* 425
Eugenia cabanisiana IPNI=*593690-1* 425
Eugenia calycolpoides IPNI=*1046564-2* 425
Eugenia calyptrata IPNI=*601603-1* 428
Eugenia campechiana IPNI=*593490-1* 425
Eugenia canescens IPNI=*595185-1* 425
Eugenia carranzae IPNI=*593490-1* 425
Eugenia carthagenensis IPNI=*593490-1* 425
Eugenia cartilaginea IPNI=*974565-1* 427
Eugenia caryophyllifolia IPNI=*601603-1* 428
Eugenia cauliflora IPNI=*203847-2* 428
Eugenia caurensis IPNI=*593768-1* 425
Eugenia chacteana IPNI=*594499-1* 425
Eugenia ciarensis IPNI=*1046564-2* 425
Eugenia ciliolata IPNI=*599327-1* 427
Eugenia clinocarpa IPNI=*1046564-2* 425
Eugenia coarensis IPNI=*1046564-2* 425
Eugenia cobanensis IPNI=*596302-1* 425
Eugenia coccolobifolia IPNI=*166115-2* 427
Eugenia cocquericotensis IPNI=*595343-1* 425
Eugenia coloradoensis IPNI=*594452-1* 425
Eugenia comitanensis IPNI=*593490-1* 425
Eugenia conglobata IPNI=*596302-1* 425
Eugenia conzattii IPNI=*595343-1* 425
Eugenia correae IPNI=*594830-1* 425
Eugenia costata IPNI=*596263-1* 425
Eugenia cozumelensis IPNI=*593690-1* 425
Eugenia cumini IPNI=*601603-1* 428
Eugenia dasyantha IPNI=*1046564-2* 425
Eugenia dasyblasta IPNI=*596263-1* 425
Eugenia decidua IPNI=*596263-1* 425
Eugenia decora IPNI=*601802-1* 428
Eugenia decorticans IPNI=*1046564-2* 425
Eugenia deflexa IPNI=*598743-1* 426
Eugenia deltoidea IPNI=*593490-1* 425
Eugenia diantha IPNI=*1046564-2* 425

Eugenia diantha var. *ciliata* IPNI=*1046564-2* 425
Eugenia diantha var. *glabra* IPNI=*1046564-2* 425
Eugenia dichotoma IPNI=*166091-2* 427
Eugenia dichotoma var. *fragrans* IPNI=*166091-2* 427
Eugenia dicrana IPNI=*166091-2* 427
Eugenia dipoda f. *grandifolia* IPNI=*1046564-2* 425
Eugenia dipoda IPNI=*1046564-2* 425
Eugenia dipoda var. *brachypoda* IPNI=*1046564-2* 425
Eugenia discolor IPNI=*166084-2* 427
Eugenia disticha var. *galalonensis* IPNI=*594499-1* 425
Eugenia divaricata IPNI=*166146-2* 427
Eugenia divaricata IPNI=*599170-1* 427
Eugenia diversiflora IPNI=*1046564-2* 425
Eugenia djouat IPNI=*601603-1* 428
Eugenia domestica IPNI=*601907-1* 428
Eugenia dugandii IPNI=*166091-2* 427
Eugenia egensis var. *angustifolia* IPNI=*326111-2* 425
Eugenia egensis var. *bimarginata* IPNI=*326111-2* 425
Eugenia egensis var. *grandifolia* IPNI=*326111-2* 425
Eugenia egensis var. *latifolia* IPNI=*326111-2* 425
Eugenia egensis var. *parvifolia* IPNI=*326111-2* 425
Eugenia erythroxyloides IPNI=*1046564-2* 425
Eugenia escuintlensis IPNI=*593490-1* 425
Eugenia esculenta IPNI=*596253-1* 425
Eugenia eutenuipes IPNI=*595343-1* 425
Eugenia fallax IPNI=*599170-1* 427
Eugenia ferruginea IPNI=*600760-1* 428
Eugenia fieldingii IPNI=*593768-1* 425
Eugenia fiscalensis IPNI=*98934-2* 425
Eugenia flava IPNI=*1046564-2* 425
Eugenia flavonigra IPNI=*594830-1* 425
Eugenia flavonigra var. *guadalupensis* IPNI=*594830-1* 425
Eugenia flavonigra var. *martinicensis* IPNI=*594830-1* 425
Eugenia flavovirens IPNI=*595185-1* 425
Eugenia flavovirens var. *obscura* IPNI=*595185-1* 425
Eugenia flavovirens var. *pallida* IPNI=*595185-1* 425
Eugenia foetida var. *maleolens* IPNI=*595185-1* 425
Eugenia foetida var. *parvifolia* IPNI=*595185-1* 425
Eugenia foetida var. *rhombea* IPNI=*98934-2* 425
Eugenia foliosa IPNI=*166103-2* 427
Eugenia forsteri IPNI=*595593-1* 425
Eugenia fragrans IPNI=*166091-2* 427
Eugenia fragrans var. *brachyrhiza* IPNI=*166091-2* 427
Eugenia fragrans var. *obovata* IPNI=*166091-2* 427
Eugenia freireana IPNI=*593768-1* 425
Eugenia freireana var. *angustifolia* IPNI=*593768-1* 425
Eugenia freireana var. *latifolia* IPNI=*593768-1* 425
Eugenia friedrichsthalii IPNI=*602600-1* 428
Eugenia fruticosa IPNI=*601603-1* 428
Eugenia fruticulosa IPNI=*1046564-2* 425
Eugenia gardneriana IPNI=*594452-1* 425
Eugenia gardneriana var. *depauperata* IPNI=*594452-1* 425
Eugenia gardneriana var. *dives* IPNI=*594452-1* 425
Eugenia gardneriana var. *ovata* IPNI=*594452-1* 425
Eugenia gardneriana var. *rigida* IPNI=*594452-1* 425
Eugenia glareosa IPNI=*1046564-2* 425
Eugenia grandiglandulosa IPNI=*166146-2* 427
Eugenia granulata IPNI=*166091-2* 427
Eugenia guadalupensis IPNI=*593690-1* 425
Eugenia guapurium IPNI=*203847-2* 428
Eugenia guayabillo IPNI=*166091-2* 427
Eugenia guianensis IPNI=*598851-1* 426
Eugenia guttata IPNI=*593690-1* 425
Eugenia hallii IPNI=*166097-2* 427
Eugenia hartii IPNI=*593768-1* 425
Eugenia hauthalii IPNI=*211886-2* 428
Eugenia heteroclita IPNI=*166091-2* 427
Eugenia hirsuta IPNI=*598646-1* 426
Eugenia hoffmannseggii IPNI=*593768-1* 425
Eugenia hoffmannseggii var. *grandifolia* IPNI=*593768-1* 425
Eugenia hoffmannseggii var. *parvifolia* IPNI=*593768-1* 425
Eugenia humifusa IPNI=*599434-1* 427
Eugenia inaequiloba IPNI=*598883-1* 426
Eugenia inocarpa IPNI=*595457-1* 425
Eugenia insipida IPNI=*1046564-2* 425
Eugenia insularis IPNI=*595185-1* 425
Eugenia inundata var. *acutifolia* IPNI=*593768-1* 425
Eugenia inundata var. *coriacea* IPNI=*594718-1* 425
Eugenia inundata var. *membranacea* IPNI=*594718-1* 425
Eugenia itzana IPNI=*593690-1* 425
Eugenia jaboticaba IPNI=*203847-2* 428
Eugenia jamaicana IPNI=*593768-1* 425
Eugenia jamaicensis IPNI=*593768-1* 425
Eugenia jamboides IPNI=*601802-1* 428
Eugenia jambolana IPNI=*601603-1* 428
Eugenia jambolana var. *caryophyllifolia* IPNI=*601603-1* 428
Eugenia jambolifera IPNI=*601603-1* 428
Eugenia jambos IPNI=*601802-1* 428
Eugenia jambosa IPNI=*601802-1* 428
Eugenia kochiana IPNI=*1046564-2* 425
Eugenia kunthiana IPNI=*1046564-2* 425
Eugenia kunthiana var. *opaca* IPNI=*1046564-2* 425
Eugenia kunthiana var. *pellucida* IPNI=*1046564-2* 425
Eugenia lambertiana var. *hispidula* IPNI=*594830-1* 425
Eugenia lancea IPNI=*593768-1* 425
Eugenia laxiflora IPNI=*599170-1* 427
Eugenia leptantha IPNI=*594718-1* 425
Eugenia leptopa IPNI=*98934-2* 425
Eugenia leptophlebia IPNI=*593768-1* 425
Eugenia leucomyrtillus IPNI=*599434-1* 427
Eugenia leucophloea IPNI=*599327-1* 427
Eugenia leucophloea var. *warmingiana* IPNI=*599327-1* 427
Eugenia limbata IPNI=*166091-2* 427
Eugenia linearifolia IPNI=*1046564-2* 425
Eugenia linearifolia var. *oppositifolia* IPNI=*1046564-2* 425
Eugenia linearifolia var. *ternifolia* IPNI=*1046564-2* 425
Eugenia lopeziana IPNI=*166091-2* 427
Eugenia loretensis IPNI=*593768-1* 425
Eugenia lucayana IPNI=*166091-2* 427
Eugenia ludibunda IPNI=*593768-1* 425
Eugenia macarensis IPNI=*593768-1* 425
Eugenia macroclada IPNI=*1046564-2* 425
Eugenia macrophylla IPNI=*601907-1* 428
Eugenia maculata IPNI=*326111-2* 425
Eugenia malaccensis f. *cericarpa* IPNI=*601802-1* 428
Eugenia malaccensis IPNI=*601802-1* 428
Eugenia malaccensis IPNI=*601907-1* 428
Eugenia maleolens IPNI=*595185-1* 425
Eugenia maranhensis IPNI=*599327-1* 427
Eugenia mariquitensis IPNI=*166115-2* 427
Eugenia matanzensis IPNI=*593690-1* 425
Eugenia maximiliana IPNI=*1046564-2* 425
Eugenia maximiliana IPNI=*599327-1* 427
Eugenia melanosticta IPNI=*594452-1* 425
Eugenia membranacea IPNI=*594452-1* 425
Eugenia meyeriana IPNI=*593768-1* 425
Eugenia meyeriana var. *depauperata* IPNI=*593768-1* 425
Eugenia meyeriana var. *dives* IPNI=*593768-1* 425
Eugenia michelii IPNI=*596263-1* 425
Eugenia micrantha IPNI=*196799-2* 428
Eugenia micrantha IPNI=*595185-1* 425
Eugenia microcarpos IPNI=*593768-1* 425
Eugenia mikaniana IPNI=*599170-1* 427
Eugenia mini IPNI=*593768-1* 425
Eugenia minimiflora IPNI=*593690-1* 425
Eugenia modesta var. *jamaicensis* IPNI=*593768-1* 425

Eugenia mollis IPNI=*599170-1* 427
Eugenia monantha IPNI=*601802-1* 428
Eugenia montana IPNI=*77160848-1* 426
Eugenia monticola var. *flavovirens* IPNI=*595185-1* 425
Eugenia monticola var. *latifolia* IPNI=*595185-1* 425
Eugenia moraviana var. *gardneriana* IPNI=*594452-1* 425
Eugenia mosquitensis IPNI=*593490-1* 425
Eugenia mugiensis IPNI=*1046564-2* 425
Eugenia myriostigma IPNI=*593768-1* 425
Eugenia myrsinoides IPNI=*974565-1* 427
Eugenia myrtifolia IPNI=*596263-1* 425
Eugenia myrtillifolia IPNI=*1046564-2* 425
Eugenia nhanica IPNI=*1046564-2* 425
Eugenia nicholsii IPNI=*593768-1* 425
Eugenia nitida IPNI=*598883-1* 426
Eugenia oblongifolia IPNI=*596263-1* 425
Eugenia oblongifolia IPNI=*973178-1* 428
Eugenia obovata IPNI=*601603-1* 428
Eugenia obtusata IPNI=*595185-1* 425
Eugenia obtusifolia IPNI=*1046564-2* 425
Eugenia obtusifolia IPNI=*601603-1* 428
Eugenia obversa IPNI=*1046564-2* 425
Eugenia odorata IPNI=*601603-1* 428
Eugenia oleifolia IPNI=*1046564-2* 425
Eugenia oligoneura IPNI=*594452-1* 425
Eugenia oligophylla IPNI=*594830-1* 425
Eugenia oneillii IPNI=*599327-1* 427
Eugenia origanoides IPNI=*596302-1* 425
Eugenia orthostemon IPNI=*974566-1* 427
Eugenia ovalifolia IPNI=*1046564-2* 425
Eugenia ovatifolia IPNI=*593490-1* 425
Eugenia oxygona IPNI=*600760-1* 428
Eugenia pallens IPNI=*593768-1* 425
Eugenia paniculata IPNI=*598743-1* 426
Eugenia paniculata IPNI=*602052-1* 428
Eugenia paniculiflora IPNI=*599170-1* 427
Eugenia parkeriana IPNI=*595618-1* 425
Eugenia parodiana IPNI=*326111-2* 425
Eugenia patens IPNI=*595593-1* 425
Eugenia patula IPNI=*594452-1* 425
Eugenia perforata IPNI=*326111-2* 425
Eugenia periplocifolia IPNI=*599170-1* 427
Eugenia perorebi IPNI=*594452-1* 425
Eugenia petenensis IPNI=*595343-1* 425
Eugenia phillyraeoides IPNI=*1046564-2* 425
Eugenia pimenta IPNI=*196799-2* 428
Eugenia pimenta var. *longifolia* IPNI=*196799-2* 428
Eugenia pimenta var. *ovalifolia* IPNI=*196799-2* 428
Eugenia pinnata IPNI=*203871-2* 428
Eugenia platyclada IPNI=*1046564-2* 425
Eugenia platyclada var. *cuneata* IPNI=*1046564-2* 425
Eugenia platyclada var. *microphylla* IPNI=*1046564-2* 425
Eugenia platyclada var. *multiflora* IPNI=*1046564-2* 425
Eugenia platyclada var. *ovalis* IPNI=*1046564-2* 425
Eugenia plumieri IPNI=*203871-2* 428
Eugenia polyantha IPNI=*598883-1* 426
Eugenia polyneura IPNI=*599327-1* 427
Eugenia polyphylla IPNI=*1046564-2* 425
Eugenia polyphylla var. *obovata* IPNI=*1046564-2* 425
Eugenia porphyroclada IPNI=*166115-2* 427
Eugenia prominens IPNI=*1046564-2* 425
Eugenia protracta IPNI=*599327-1* 427
Eugenia psammophila IPNI=*1046564-2* 425
Eugenia pseudomalaccensis IPNI=*601907-1* 428
Eugenia pseudovenosa IPNI=*600760-1* 428
Eugenia psychotrioides IPNI=*1046564-2* 425
Eugenia ptariensis IPNI=*593585-1* 425
Eugenia punctata IPNI=*166091-2* 427
Eugenia pungens var. *subalterna* IPNI=*1046564-2* 425
Eugenia punicifolia var. *brachypoda* IPNI=*1046564-2* 425
Eugenia punicifolia var. *subalterna* IPNI=*1046564-2* 425
Eugenia purpurascens IPNI=*601907-1* 428
Eugenia purpurea IPNI=*601907-1* 428
Eugenia pusilana IPNI=*98934-2* 425
Eugenia pycnantha IPNI=*77198138-1* 427
Eugenia pycnoneura IPNI=*599327-1* 427
Eugenia pyramidalis IPNI=*1046564-2* 425
Eugenia pyramidalis var. *angustifolia* IPNI=*1046564-2* 425
Eugenia pyrifolia IPNI=*165926-2* 427
Eugenia pyrroclada IPNI=*1046564-2* 425
Eugenia pyrroclada var. *columbiensis* IPNI=*1046564-2* 425
Eugenia pyrroclada var. *guianensis* IPNI=*1046564-2* 425
Eugenia racemifera IPNI=*594452-1* 425
Eugenia racemosa IPNI=*593768-1* 425
Eugenia rheedioides IPNI=*602052-1* 428
Eugenia rhombocarpa IPNI=*1046564-2* 425
Eugenia rhopaloides IPNI=*166115-2* 427
Eugenia richardiana IPNI=*593768-1* 425
Eugenia romana IPNI=*1046564-2* 425
Eugenia romana var. *pickeli* IPNI=*1046564-2* 425
Eugenia rondonensis IPNI=*166091-2* 427
Eugenia rubrescens IPNI=*1046564-2* 425
Eugenia rufoflavescens IPNI=*1046564-2* 425
Eugenia ruscifolia IPNI=*595618-1* 425
Eugenia sabulosa IPNI=*1046564-2* 425
Eugenia salicifolia IPNI=*593768-1* 425
Eugenia salzmannii IPNI=*599327-1* 427
Eugenia sancta IPNI=*1046564-2* 425
Eugenia schlechtendaliana IPNI=*595593-1* 425
Eugenia schomburgkii IPNI=*594830-1* 425
Eugenia seriatoracemosa IPNI=*594452-1* 425
Eugenia sericiflora IPNI=*593768-1* 425
Eugenia sibunensis IPNI=*593490-1* 425
Eugenia simpsonii IPNI=*166091-2* 427
Eugenia smaragdina IPNI=*594830-1* 425
Eugenia smaragdina var. *angustifolia* IPNI=*594830-1* 425
Eugenia smaragdina var. *brevipes* IPNI=*594830-1* 425
Eugenia smaragdina var. *rigida* IPNI=*594830-1* 425
Eugenia sodiroi IPNI=*77198138-1* 427
Eugenia spathophylla IPNI=*1046564-2* 425
Eugenia spathulata IPNI=*1046564-2* 425
Eugenia sphaerosperma IPNI=*326111-2* 425
Eugenia squamata IPNI=*77160867-1* 427
Eugenia steyermarkii IPNI=*166091-2* 427
Eugenia strigosa IPNI=*596263-1* 425
Eugenia subalterna IPNI=*1046564-2* 425
Eugenia subcorymbosa IPNI=*1046564-2* 425
Eugenia suffruticosa IPNI=*1046564-2* 425
Eugenia suffruticosa var. *laeta* IPNI=*1046564-2* 425
Eugenia suffruticosa var. *latifolia* IPNI=*1046564-2* 425
Eugenia suffruticosa var. *opaca* IPNI=*1046564-2* 425
Eugenia surinamensis IPNI=*1046564-2* 425
Eugenia surinamensis var. *impunctata* IPNI=*1046564-2* 425
Eugenia surinamensis var. *punctata* IPNI=*1046564-2* 425
Eugenia sylvatica IPNI=*594452-1* 425
Eugenia teffensis IPNI=*595457-1* 425
Eugenia teffensis var. *subcordata* IPNI=*595457-1* 425
Eugenia teffensis var. *truncata* IPNI=*595457-1* 425
Eugenia tenuis IPNI=*601603-1* 428
Eugenia ternifolia IPNI=*974565-1* 427
Eugenia tinge-lingua IPNI=*594452-1* 425
Eugenia tomentosa IPNI=*599215-1* 427
Eugenia triflora IPNI=*166091-2* 427
Eugenia triphylla IPNI=*1046564-2* 425
Eugenia triquetra IPNI=*974565-1* 427
Eugenia triquetra var. *aequatorialis* IPNI=*974565-1* 427

Eugenia tsoi IPNI=*601603-1* 428
Eugenia turumiquirensis IPNI=*166091-2* 427
Eugenia uniflora var. *atropurpurea* IPNI=*596263-1* 425
Eugenia urceolata IPNI=*600760-1* 428
Eugenia vaga IPNI=*1046564-2* 425
Eugenia vaga var. *brachypoda* IPNI=*1046564-2* 425
Eugenia vaga var. *brasiliensis* IPNI=*1046564-2* 425
Eugenia vaga var. *dipoda* IPNI=*1046564-2* 425
Eugenia vaga var. *pumila* IPNI=*1046564-2* 425
Eugenia vaga var. *punicifolia* IPNI=*1046564-2* 425
Eugenia vaga var. *rigida* IPNI=*1046564-2* 425
Eugenia vallis IPNI=*599215-1* 427
Eugenia vellozoi IPNI=*595457-1* 425
Eugenia verrucosa IPNI=*593690-1* 425
Eugenia verruculosa IPNI=*595593-1* 425
Eugenia vincentina IPNI=*595343-1* 425
Eugenia virgata IPNI=*593768-1* 425
Eugenia virgultosa IPNI=*593768-1* 425
Eugenia virgultosa var. *jamaicensis* IPNI=*593768-1* 425
Eugenia vismeifolia IPNI=*1052354-2* 427
Eugenia vulgaris IPNI=*601802-1* 428
Eugenia warszewiczii IPNI=*602600-1* 428
Eugenia weberbaueri IPNI=*974565-1* 427
Eugenia willdenowii IPNI=*596263-1* 425
Eugenia xylopifolia IPNI=*593768-1* 425
Eugenia xylopifolia var. *brevipes* IPNI=*593768-1* 425
Eugenia yumuryensis IPNI=*593690-1* 425
Eugenia zeylanica IPNI=*596263-1* 425
Eugone superba IPNI=*535953-1* 226
Eukylista spruceana IPNI=*745249-1* 347
Eulalia amaura IPNI=*963150-1* 244
Eulalia mexicana IPNI=*35184-2* 235
Eulalia nana IPNI=*963150-1* 244
Eulalia praemorsa IPNI=*963150-1* 244
Eulophia maculata IPNI=*1073117-2* 222
Eulophia monophylla IPNI=*1073117-2* 222
Eulophidium mackenii IPNI=*1073117-2* 222
Eulophidium maculatum IPNI=*1073117-2* 222
Eulophidium maculatum var. *pterocarpum* IPNI=*1073117-2* 222
Eulophidium monophyllum IPNI=*1073117-2* 222
Eulophidium nyassanum IPNI=*1073117-2* 222
Eulophidium warneckeanum IPNI=*1073117-2* 222
Euodia divaricata IPNI=*128318-1* 452
Euonymus angustifolius IPNI=*160876-1* 290
Euonymus bulgaricus IPNI=*160876-1* 290
Euonymus carrierei IPNI=*160944-1* 290
Euonymus czernjaevii IPNI=*160876-1* 290
Euonymus europaeus f. *angustifolius* IPNI=*160876-1* 290
Euonymus europaeus f. *bulgaricus* IPNI=*160876-1* 290
Euonymus europaeus f. *microphyllus* IPNI=*160876-1* 290
Euonymus europaeus f. *scaberulus* IPNI=*160876-1* 290
Euonymus europaeus subsp. *moldavicus* IPNI=*160876-1* 290
Euonymus europaeus subsp. *subvelutinus* IPNI=*160876-1* 290
Euonymus europaeus var. *suberosus* IPNI=*160876-1* 290
Euonymus floribundus IPNI=*160876-1* 290
Euonymus japonicus var. *litoralis* IPNI=*160944-1* 290
Euonymus latifolius IPNI=*160876-1* 290
Euonymus medirossicus IPNI=*160876-1* 290
Euonymus medius IPNI=*160876-1* 290
Euonymus moldavicus IPNI=*160876-1* 290
Euonymus odessanus IPNI=*160876-1* 290
Euonymus pubescens IPNI=*160876-1* 290
Euonymus pulchellus IPNI=*160944-1* 290
Euonymus quadrigonus IPNI=*160876-1* 290
Euonymus suberosus IPNI=*160876-1* 290
Euonymus tobira IPNI=*684716-1* 254
Euonymus vulgaris IPNI=*160876-1* 290
Euonymus vulgaris var. *microphyllus* IPNI=*160876-1* 290
Euonymus yoshinagae IPNI=*160944-1* 290
Eupatoriophalacron album IPNI=*1186403-2* 261
Eupatorium aboriginum IPNI=*6815-2* 256
Eupatorium acuminatum IPNI=*69573-2* 260
Eupatorium affine IPNI=*57410-2* 259
Eupatorium ageratoides IPNI=*6815-2* 256
Eupatorium albertinae IPNI=*188196-1* 259
Eupatorium alternans IPNI=*57414-2* 259
Eupatorium altissimum IPNI=*6815-2* 256
Eupatorium amarum IPNI=*233110-1* 266
Eupatorium amygdalinum f. *elegans* IPNI=*27442-2* 257
Eupatorium amygdalinum f. *heterophyllum* IPNI=*27442-2* 257
Eupatorium amygdalinum f. *latifolium* IPNI=*27442-2* 257
Eupatorium amygdalinum f. *loniceroides* IPNI=*27442-2* 257
Eupatorium amygdalinum f. *serratum* IPNI=*27442-2* 257
Eupatorium amygdalinum f. *subobtusum* IPNI=*27442-2* 257
Eupatorium amygdalinum IPNI=*27442-2* 257
Eupatorium amygdalinum var. *dissitiflorum* IPNI=*27442-2* 257
Eupatorium amygdalinum var. *elegans* IPNI=*27442-2* 257
Eupatorium amygdalinum var. *glandulosum* IPNI=*27442-2* 257
Eupatorium amygdalinum var. *heterophyllum* IPNI=*27442-2* 257
Eupatorium amygdalinum var. *oxychlaenum* IPNI=*27442-2* 257
Eupatorium amygdalinum var. *serratum* IPNI=*27442-2* 257
Eupatorium amygdalinum var. *subglabrum* IPNI=*27442-2* 257
Eupatorium angosturae IPNI=*203365-1* 261
Eupatorium angustifolium IPNI=*6821-2* 256
Eupatorium arboreum IPNI=*962413-1* 260
Eupatorium articulatum IPNI=*6988-2* 256
Eupatorium aschenbornianum IPNI=*6988-2* 256
Eupatorium atriplicifolium IPNI=*57443-2* 259
Eupatorium australe IPNI=*57414-2* 259
Eupatorium baccharoides IPNI=*6833-2* 256
Eupatorium baccharoides var. *araneotectum* IPNI=*6833-2* 256
Eupatorium baccharoides var. *fratris* IPNI=*6833-2* 256
Eupatorium baccharoides var. *typicum* IPNI=*6833-2* 256
Eupatorium ballotifolium IPNI=*144056-2* 265
Eupatorium ballotifolium var. *caucense* IPNI=*144056-2* 265
Eupatorium ballotifolium var. *typicum* IPNI=*144056-2* 265
Eupatorium barclayanum IPNI=*27442-2* 257
Eupatorium barranquillense IPNI=*57443-2* 259
Eupatorium berteroanum IPNI=*105109-2* 263
Eupatorium bimatrum IPNI=*105109-2* 263
Eupatorium bogotense IPNI=*6833-2* 256
Eupatorium boreale IPNI=*6815-2* 256
Eupatorium brachiatum IPNI=*57443-2* 259
Eupatorium braunii IPNI=*28074-2* 257
Eupatorium bullatum IPNI=*57366-2* 259
Eupatorium calocephalum IPNI=*57410-2* 259
Eupatorium chrysocephalum IPNI=*234386-1* 267
Eupatorium ciliatum IPNI=*6988-2* 256
Eupatorium cinereum IPNI=*27149-2* 257
Eupatorium clematitis IPNI=*57443-2* 259
Eupatorium concinnum IPNI=*57410-2* 259
Eupatorium condensatum IPNI=*988439-1* 268
Eupatorium confertifolium IPNI=*7058-2* 256
Eupatorium conyzoides f. *angustiflorum* IPNI=*57443-2* 259
Eupatorium conyzoides f. *glabratum* IPNI=*57443-2* 259
Eupatorium conyzoides IPNI=*57443-2* 259
Eupatorium conyzoides IPNI=*7086-2* 256
Eupatorium conyzoides var. *affine* IPNI=*57410-2* 259
Eupatorium conyzoides var. *extensum* IPNI=*57443-2* 259
Eupatorium conyzoides var. *floribundum* IPNI=*57443-2* 259
Eupatorium conyzoides var. *glabrescens* IPNI=*57443-2* 259
Eupatorium conyzoides var. *heterolepis* IPNI=*57443-2* 259
Eupatorium conyzoides var. *incanum* IPNI=*57443-2* 259
Eupatorium conyzoides var. *maximiliani* IPNI=*57443-2* 259
Eupatorium conyzoides var. *pauciflorum* IPNI=*57443-2* 259
Eupatorium conyzoides var. *phyllocephalum* IPNI=*57443-2* 259

Eupatorium conyzoides var. *scaberulum* IPNI=*57443-2* 259
Eupatorium conyzoides var. *tambillense* IPNI=*57443-2* 259
Eupatorium coperense IPNI=*271815-2* 256
Eupatorium cordatum var. *fraseri* IPNI=*6815-2* 256
Eupatorium cornifolium IPNI=*988439-1* 268
Eupatorium cotacachense IPNI=*6999-2* 256
Eupatorium cotinifolium IPNI=*28074-2* 257
Eupatorium cowleyanum IPNI=*232826-1* 266
Eupatorium crenatum IPNI=*232826-1* 266
Eupatorium crenatum IPNI=*57489-2* 259
Eupatorium critonioides IPNI=*69558-2* 260
Eupatorium cujabense IPNI=*27442-2* 257
Eupatorium cuspidatum IPNI=*988437-1* 268
Eupatorium denticulatum IPNI=*161051-2* 266
Eupatorium denudatum IPNI=*188196-1* 259
Eupatorium dichotomum IPNI=*57489-2* 259
Eupatorium divergens IPNI=*57443-2* 259
Eupatorium dodonaeifolium IPNI=*27442-2* 257
Eupatorium donianum IPNI=*188196-1* 259
Eupatorium donnell-smithii var. *parvifolium* IPNI=*6988-2* 256
Eupatorium dryadeum IPNI=*1092582-2* 263
Eupatorium duodecimiflorum IPNI=*27149-2* 257
Eupatorium entreriense IPNI=*27149-2* 257
Eupatorium eurybiifolium IPNI=*6815-2* 256
Eupatorium exacoides IPNI=*127605-2* 264
Eupatorium extrorsum IPNI=*57410-2* 259
Eupatorium fasciculare IPNI=*57410-2* 259
Eupatorium floribundum IPNI=*57443-2* 259
Eupatorium fraseri IPNI=*6815-2* 256
Eupatorium frasieri IPNI=*6815-2* 256
Eupatorium fraternum IPNI=*27442-2* 257
Eupatorium furcatum IPNI=*57410-2* 259
Eupatorium glandulosissimum IPNI=*27442-2* 257
Eupatorium glandulosum IPNI=*27442-2* 257
Eupatorium goyazense IPNI=*27442-2* 257
Eupatorium graciliflorum IPNI=*57443-2* 259
Eupatorium grossedentatum IPNI=*144056-2* 265
Eupatorium guadalupense IPNI=*105109-2* 263
Eupatorium guadalupense var. *gardneri* IPNI=*105109-2* 263
Eupatorium guadalupense var. *wydlerianum* IPNI=*105109-2* 263
Eupatorium guanaiense IPNI=*57410-2* 259
Eupatorium hasslerianum IPNI=*27442-2* 257
Eupatorium hirtum IPNI=*232743-1* 266
Eupatorium horsfieldii IPNI=*27149-2* 257
Eupatorium houstonis IPNI=*241903-1* 269
Eupatorium humboldtii IPNI=*271815-2* 256
Eupatorium humile f. *caeruleum* IPNI=*144058-2* 265
Eupatorium humile f. *erubescens* IPNI=*144058-2* 265
Eupatorium humile f. *typicum* IPNI=*144058-2* 265
Eupatorium humile IPNI=*144058-2* 265
Eupatorium incisum IPNI=*57443-2* 259
Eupatorium inulifolium f. *lasiophlebium* IPNI=*27149-2* 257
Eupatorium inulifolium f. *suaveolens* IPNI=*27149-2* 257
Eupatorium inulifolium f. *typicum* IPNI=*27149-2* 257
Eupatorium inulifolium IPNI=*27149-2* 257
Eupatorium ivifolium IPNI=*57410-2* 259
Eupatorium ivifolium var. *asperum* IPNI=*57410-2* 259
Eupatorium ivifolium var. *extrorsum* IPNI=*57410-2* 259
Eupatorium ivifolium var. *foliosum* IPNI=*57410-2* 259
Eupatorium ivifolium var. *genuinum* IPNI=*57410-2* 259
Eupatorium ivifolium var. *hirsutum* IPNI=*57410-2* 259
Eupatorium ivifolium var. *laxiflorum* IPNI=*57410-2* 259
Eupatorium ivifolium var. *ludovicianum* IPNI=*57410-2* 259
Eupatorium ivifolium var. *perturbinatum* IPNI=*57410-2* 259
Eupatorium ivifolium var. *pilcomayense* IPNI=*57410-2* 259
Eupatorium ivifolium var. *serratum* IPNI=*57410-2* 259
Eupatorium ixodes IPNI=*27442-2* 257
Eupatorium klattii IPNI=*57443-2* 259
Eupatorium kunthianum IPNI=*57366-2* 259
Eupatorium laevigatum f. *albiflorum* IPNI=*57414-2* 259
Eupatorium laevigatum f. *crenatifolium* IPNI=*57414-2* 259
Eupatorium laevigatum f. *flavidum* IPNI=*57414-2* 259
Eupatorium laevigatum f. *lilacinum* IPNI=*57414-2* 259
Eupatorium laevigatum f. *serratifolium* IPNI=*57414-2* 259
Eupatorium laevigatum f. *subcapillatum* IPNI=*57414-2* 259
Eupatorium laevigatum IPNI=*57414-2* 259
Eupatorium laevigatum var. *arenarium* IPNI=*57414-2* 259
Eupatorium laevigatum var. *claussenii* IPNI=*57414-2* 259
Eupatorium laevigatum var. *genuinum* IPNI=*57414-2* 259
Eupatorium laevigatum var. *longepetiolatum* IPNI=*57414-2* 259
Eupatorium laevigatum var. *microcephalum* IPNI=*57414-2* 259
Eupatorium laevigatum var. *psidiifolium* IPNI=*57414-2* 259
Eupatorium laevigatum var. *squamulosum* IPNI=*57414-2* 259
Eupatorium laevigatum var. *submembranacea* IPNI=*57414-2* 259
Eupatorium laevigatum var. *tomentosum* IPNI=*57414-2* 259
Eupatorium lamiifolium IPNI=*19109-2* 257
Eupatorium lehmannianum IPNI=*7050-2* 256
Eupatorium loniceroides IPNI=*27442-2* 257
Eupatorium lozanoi IPNI=*144056-2* 265
Eupatorium ludovicianum IPNI=*232826-1* 266
Eupatorium luquense IPNI=*57410-2* 259
Eupatorium luquense var. *corymbosum* IPNI=*57410-2* 259
Eupatorium lyratum IPNI=*211317-2* 268
Eupatorium macrocephalum IPNI=*188196-1* 259
Eupatorium macrocephalum var. *angustifolium* IPNI=*188196-1* 259
Eupatorium macrocephalum var. *stigmatosum* IPNI=*188196-1* 259
Eupatorium macrocephalum var. *strigosum* IPNI=*188196-1* 259
Eupatorium macrophyllum IPNI=*1092582-2* 263
Eupatorium marquezianum IPNI=*232826-1* 266
Eupatorium martii IPNI=*57489-2* 259
Eupatorium martiusii IPNI=*57489-2* 259
Eupatorium megaphyllum IPNI=*69565-2* 260
Eupatorium micranthon IPNI=*233110-1* 266
Eupatorium microstemon IPNI=*105109-2* 263
Eupatorium microstemon var. *albiflorum* IPNI=*105109-2* 263
Eupatorium microstemon var. *lilacinum* IPNI=*105109-2* 263
Eupatorium microstemon var. *perennans* IPNI=*105109-2* 263
Eupatorium molle IPNI=*1092582-2* 263
Eupatorium molle IPNI=*27149-2* 257
Eupatorium morifolium IPNI=*69558-2* 260
Eupatorium myosotifolium IPNI=*962413-1* 260
Eupatorium nepetoides IPNI=*64237-2* 260
Eupatorium neurolepis IPNI=*57410-2* 259
Eupatorium obovatum IPNI=*969180-1* 272
Eupatorium obscurum IPNI=*57410-2* 259
Eupatorium odoratum f. *scandens* IPNI=*57443-2* 259
Eupatorium odoratum IPNI=*57443-2* 259
Eupatorium odoratum IPNI=*6815-2* 256
Eupatorium odoratum var. *brachiatum* IPNI=*57443-2* 259
Eupatorium odoratum var. *cubense* IPNI=*57443-2* 259
Eupatorium odoratum var. *mallotophyllum* IPNI=*57443-2* 259
Eupatorium odoratum var. *pauciflorum* IPNI=*57443-2* 259
Eupatorium opiferum IPNI=*232826-1* 266
Eupatorium orgyale IPNI=*27149-2* 257
Eupatorium orinocense IPNI=*161051-2* 266
Eupatorium orinocense var. *batatifolium* IPNI=*161051-2* 266
Eupatorium orinocense var. *tamoides* IPNI=*161051-2* 266
Eupatorium osseanum IPNI=*57414-2* 259
Eupatorium oxychlaenum f. *hasslerianum* IPNI=*27442-2* 257
Eupatorium oxychlaenum IPNI=*27442-2* 257
Eupatorium oxychlaenum var. *glandulosissimum* IPNI=*27442-2* 257
Eupatorium oxyhedrum IPNI=*188196-1* 259
Eupatorium pacificum IPNI=*6988-2* 256
Eupatorium paleaceum IPNI=*7086-2* 256
Eupatorium pallescens IPNI=*27149-2* 257
Eupatorium pallescens var. *bonariense* IPNI=*27149-2* 257

Eupatorium pallescens var. *hirsutum* IPNI=*27149-2* 257
Eupatorium pallidum IPNI=*27149-2* 257
Eupatorium paniculatum IPNI=*105109-2* 263
Eupatorium paranense IPNI=*27149-2* 257
Eupatorium parviflorum IPNI=*233110-1* 266
Eupatorium paulinum IPNI=*57489-2* 259
Eupatorium pauperatum IPNI=*144056-2* 265
Eupatorium pellucidum IPNI=*69573-2* 260
Eupatorium perglabrum IPNI=*57461-2* 259
Eupatorium pichinchense f. *typicum* IPNI=*6988-2* 256
Eupatorium pichinchense IPNI=*6988-2* 256
Eupatorium piurae IPNI=*6988-2* 256
Eupatorium polystachyum IPNI=*948489-1* 265
Eupatorium pomaderrifolium IPNI=*271815-2* 256
Eupatorium populifolium IPNI=*1092582-2* 263
Eupatorium populifolium IPNI=*69558-2* 260
Eupatorium propinquum IPNI=*57414-2* 259
Eupatorium pseudochilca IPNI=*6999-2* 256
Eupatorium pseudochilca var. *cotacachense* IPNI=*6999-2* 256
Eupatorium psiadiifolium IPNI=*57414-2* 259
Eupatorium psiadiifolium var. *latifolium* IPNI=*57414-2* 259
Eupatorium psiadiifolium var. *tereticaule* IPNI=*57414-2* 259
Eupatorium ptaretepuiense IPNI=*6988-2* 256
Eupatorium quadrangulare IPNI=*69565-2* 260
Eupatorium ramosissimum IPNI=*57489-2* 259
Eupatorium rebaudianum IPNI=*245468-2* 270
Eupatorium resinosum IPNI=*57414-2* 259
Eupatorium rhodanthum IPNI=*27442-2* 257
Eupatorium rugosum f. *villicaule* IPNI=*6815-2* 256
Eupatorium rugosum IPNI=*6815-2* 256
Eupatorium rugosum var. *chlorolepis* IPNI=*6815-2* 256
Eupatorium rugosum var. *tomentellum* IPNI=*6815-2* 256
Eupatorium rugosum var. *villicaule* IPNI=*6815-2* 256
Eupatorium sabeanum IPNI=*57443-2* 259
Eupatorium salzmannianum IPNI=*27442-2* 257
Eupatorium scabrum IPNI=*57486-2* 259
Eupatorium sitiense IPNI=*57489-2* 259
Eupatorium sladenianum IPNI=*188196-1* 259
Eupatorium squalidum IPNI=*57489-2* 259
Eupatorium squalidum var. *martiusii* IPNI=*57489-2* 259
Eupatorium squalidum var. *rusbyanum* IPNI=*57489-2* 259
Eupatorium squalidum var. *subvelutinum* IPNI=*57489-2* 259
Eupatorium squalidum var. *typicum* IPNI=*57489-2* 259
Eupatorium steetzii IPNI=*28627-2* 258
Eupatorium steetzii var. *columbiense* IPNI=*28627-2* 258
Eupatorium steetzii var. *macrocephalum* IPNI=*28627-2* 258
Eupatorium steetzii var. *typicum* IPNI=*28627-2* 258
Eupatorium stigmatosum IPNI=*188196-1* 259
Eupatorium stigmatosum IPNI=*57443-2* 259
Eupatorium stigmatosum var. *subcalvatum* IPNI=*188196-1* 259
Eupatorium stigmatosum var. *violaceum* IPNI=*188196-1* 259
Eupatorium stoechadifolium IPNI=*144063-2* 265
Eupatorium stoechadifolium var. *pamplonense* IPNI=*144063-2* 265
Eupatorium stoechadifolium var. *typicum* IPNI=*144063-2* 265
Eupatorium suaveolens IPNI=*27149-2* 257
Eupatorium subobtusum IPNI=*27442-2* 257
Eupatorium subvelutinum IPNI=*57489-2* 259
Eupatorium surinamense IPNI=*27442-2* 257
Eupatorium tacotanum f. *apodum* IPNI=*57497-2* 259
Eupatorium tacotanum f. *normale* IPNI=*57497-2* 259
Eupatorium tacotanum IPNI=*57497-2* 259
Eupatorium tacotanum var. *trineurolepis* IPNI=*57497-2* 259
Eupatorium tacotanum var. *typicum* IPNI=*57497-2* 259
Eupatorium tequendamense var. *glabratum* IPNI=*6988-2* 256
Eupatorium tequendamense var. *glandulosopubescens* IPNI=*6988-2* 256
Eupatorium thyrsigerum IPNI=*69558-2* 260
Eupatorium thyrsoideum IPNI=*69565-2* 260
Eupatorium thyrsoideum var. *puberum* IPNI=*69565-2* 260
Eupatorium tinifolium IPNI=*7050-2* 256
Eupatorium tolimense IPNI=*69573-2* 260
Eupatorium umbrosum IPNI=*6999-2* 256
Eupatorium urticifolium IPNI=*105109-2* 263
Eupatorium urticifolium IPNI=*6815-2* 256
Eupatorium urticifolium var. *tomentellum* IPNI=*6815-2* 256
Eupatorium urticifolium var. *villicaule* IPNI=*6815-2* 256
Eupatorium vacciniifolium IPNI=*7058-2* 256
Eupatorium valverdeanum IPNI=*234386-1* 267
Eupatorium venosum IPNI=*57489-2* 259
Eupatorium vincifolium IPNI=*233110-1* 266
Eupatorium violaceum IPNI=*105109-2* 263
Eupatorium vulcanicum IPNI=*6988-2* 256
Eupetalum punctatum IPNI=*104828-1* 292
Euphorbia afzelii IPNI=*102034-2* 391
Euphorbia androsaemifolia IPNI=*101777-2* 390
Euphorbia ascendens IPNI=*347590-1* 390
Euphorbia bancana IPNI=*101651-2* 390
Euphorbia begoniifolia IPNI=*101976-2* 391
Euphorbia bicapitata IPNI=*101793-2* 390
Euphorbia biramensis IPNI=*101976-2* 391
Euphorbia boliviana IPNI=*101671-2* 390
Euphorbia botryoides IPNI=*102034-2* 391
Euphorbia brasiliensis f. *angustior* IPNI=*101676-2* 390
Euphorbia brasiliensis f. *latior* IPNI=*101676-2* 390
Euphorbia brasiliensis f. *major* IPNI=*101676-2* 390
Euphorbia brasiliensis f. *pruinosa* IPNI=*101676-2* 390
Euphorbia brasiliensis f. *subsessilis* IPNI=*101676-2* 390
Euphorbia brasiliensis IPNI=*101676-2* 390
Euphorbia brasiliensis var. *blanchetii* IPNI=*101676-2* 390
Euphorbia brasiliensis var. *genuina* IPNI=*101676-2* 390
Euphorbia brasiliensis var. *hyssopifolia* IPNI=*101676-2* 390
Euphorbia brasiliensis var. *paraguayensis* IPNI=*101676-2* 390
Euphorbia brasiliensis var. *pruinosa* IPNI=*101676-2* 390
Euphorbia brasiliensis var. *pulchella* IPNI=*101676-2* 390
Euphorbia brasiliensis var. *uniflora* IPNI=*101676-2* 390
Euphorbia breonii IPNI=*347359-1* 390
Euphorbia buxifolia f. *florida* IPNI=*347331-1* 390
Euphorbia buxifolia f. *reclinata* IPNI=*347331-1* 390
Euphorbia buxifolia f. *seminuda* IPNI=*347331-1* 390
Euphorbia buxifolia IPNI=*347331-1* 390
Euphorbia buxifolia var. *flexuosa* IPNI=*347331-1* 390
Euphorbia buxifolia var. *ramosissima* IPNI=*347331-1* 390
Euphorbia callitrichoides IPNI=*101913-2* 390
Euphorbia capitata IPNI=*101651-2* 390
Euphorbia cestrifolia IPNI=*347079-1* 390
Euphorbia chrysochaeta IPNI=*101651-2* 390
Euphorbia coccinea IPNI=*30176564-2* 390
Euphorbia corallothamnus IPNI=*347292-1* 390
Euphorbia decussata IPNI=*347072-1* 390
Euphorbia depressa IPNI=*347236-1* 390
Euphorbia domingensis IPNI=*101676-2* 390
Euphorbia edulis IPNI=*347489-1* 390
Euphorbia emarginata IPNI=*101976-2* 391
Euphorbia enterophora IPNI=*348758-1* 391
Euphorbia erithrophylla IPNI=*30176564-2* 390
Euphorbia esuloides IPNI=*346205-1* 390
Euphorbia fastuosa IPNI=*30176564-2* 390
Euphorbia fischeriana f. *glaberrima* IPNI=*346510-1* 390
Euphorbia flexicaulis IPNI=*101976-2* 391
Euphorbia flexuosa IPNI=*347331-1* 390
Euphorbia foliata IPNI=*102034-2* 391
Euphorbia geayi IPNI=*348517-1* 391
Euphorbia gemella IPNI=*101651-2* 390
Euphorbia gibraltarica IPNI=*101777-2* 390
Euphorbia glabrata IPNI=*347331-1* 390
Euphorbia globulifera IPNI=*101651-2* 390

Euphorbia glomerifera IPNI=*101671-2* 390
Euphorbia goudotii IPNI=*347079-1* 390
Euphorbia hermentiana IPNI=*348565-1* 391
Euphorbia herniarioides IPNI=*101976-2* 391
Euphorbia heterophylla var. *genuina* IPNI=*30062970-2* 390
Euphorbia hirta var. *destituta* IPNI=*101651-2* 390
Euphorbia hirta var. *ophthalmica* IPNI=*101793-2* 390
Euphorbia hirta var. *typica* IPNI=*101651-2* 390
Euphorbia hypericifolia var. *communis* IPNI=*101777-2* 390
Euphorbia hypericifolia var. *maculata* IPNI=*101671-2* 390
Euphorbia hyssopifolia var. *blanchetii* IPNI=*101676-2* 390
Euphorbia hyssopifolia var. *paraguayensis* IPNI=*101676-2* 390
Euphorbia hyssopifolia var. *pruinosa* IPNI=*101676-2* 390
Euphorbia hyssopifolia var. *pulchella* IPNI=*101676-2* 390
Euphorbia hyssopifolia var. *uniflora* IPNI=*101676-2* 390
Euphorbia inflexa IPNI=*101976-2* 391
Euphorbia jonesii IPNI=*101676-2* 390
Euphorbia jovetii IPNI=*347236-1* 390
Euphorbia karwinskyi IPNI=*101651-2* 390
Euphorbia klotzschiana IPNI=*101676-2* 390
Euphorbia lancifolia var. *villicaulis* IPNI=*347056-1* 390
Euphorbia laro IPNI=*348517-1* 391
Euphorbia lasiocarpa var. *subprostrata* IPNI=*347064-1* 390
Euphorbia latazi IPNI=*347079-1* 390
Euphorbia latazi var. *glabra* IPNI=*347079-1* 390
Euphorbia lathyris IPNI=*346510-1* 390
Euphorbia lathyris var. *minor* IPNI=*347072-1* 390
Euphorbia lehmanniana IPNI=*347079-1* 390
Euphorbia ligularia IPNI=*347489-1* 390
Euphorbia litoralis IPNI=*347331-1* 390
Euphorbia lupulina IPNI=*323896-2* 389
Euphorbia lutea IPNI=*30176564-2* 390
Euphorbia maculata var. *detonsa* IPNI=*347236-1* 390
Euphorbia maculata var. *parvula* IPNI=*347236-1* 390
Euphorbia malaca IPNI=*101913-2* 390
Euphorbia mangletii IPNI=*101976-2* 391
Euphorbia media IPNI=*348517-1* 391
Euphorbia mesembryanthemifolia f. *florida* IPNI=*347331-1* 390
Euphorbia mesembryanthemifolia f. *reclinata* IPNI=*347331-1* 390
Euphorbia mesembryanthemifolia f. *seminuda* IPNI=*347331-1* 390
Euphorbia mesembryanthemifolia var. *flexuosa* IPNI=*347331-1* 390
Euphorbia mesembryanthemifolia var. *ramosissima* IPNI=*347331-1* 390
Euphorbia microclada IPNI=*101976-2* 391
Euphorbia microphylla IPNI=*102034-2* 391
Euphorbia microphylla IPNI=*346789-1* 390
Euphorbia minutiflora IPNI=*101976-2* 391
Euphorbia myrtilloides IPNI=*347331-1* 390
Euphorbia nirurioides IPNI=*101676-2* 390
Euphorbia nodiflora IPNI=*101651-2* 390
Euphorbia nutans var. *glaberrima* IPNI=*101777-2* 390
Euphorbia obliterata IPNI=*101651-2* 390
Euphorbia pallasii f. *glaberrima* IPNI=*346510-1* 390
Euphorbia papilligera IPNI=*101671-2* 390
Euphorbia paxiana IPNI=*347292-1* 390
Euphorbia pentagona IPNI=*347489-1* 390
Euphorbia perforata IPNI=*101913-2* 390
Euphorbia pileoides IPNI=*101976-2* 391
Euphorbia pilulifera f. *humifusa* IPNI=*101651-2* 390
Euphorbia pilulifera f. *rubromaculata* IPNI=*101651-2* 390
Euphorbia pilulifera f. *viridis* IPNI=*101651-2* 390
Euphorbia pilulifera var. *arechavaletae* IPNI=*101651-2* 390
Euphorbia pilulifera var. *discolor* IPNI=*101651-2* 390
Euphorbia pilulifera var. *glabrescens* IPNI=*101651-2* 390
Euphorbia pilulifera var. *guaranitica* IPNI=*101651-2* 390
Euphorbia pilulifera var. *hirta* IPNI=*101651-2* 390
Euphorbia pilulifera var. *obliterata* IPNI=*101651-2* 390
Euphorbia pilulifera var. *procumbens* IPNI=*101793-2* 390
Euphorbia poinsettiana IPNI=*30176564-2* 390
Euphorbia poinsettii IPNI=*30176564-2* 390
Euphorbia potosina var. *lamasis* IPNI=*101777-2* 390
Euphorbia preslii IPNI=*101777-2* 390
Euphorbia preslii var. *andicola* IPNI=*101777-2* 390
Euphorbia preslii var. *glaberrima* IPNI=*101777-2* 390
Euphorbia procumbens IPNI=*101793-2* 390
Euphorbia prostrata var. *caudirhiza* IPNI=*101913-2* 390
Euphorbia prostrata var. *vestita* IPNI=*101913-2* 390
Euphorbia pseudonutans IPNI=*101777-2* 390
Euphorbia pulchella IPNI=*101676-2* 390
Euphorbia pulcherrima f. *lutea* IPNI=*30176564-2* 390
Euphorbia punctata IPNI=*346205-1* 390
Euphorbia radicans IPNI=*101976-2* 391
Euphorbia ramosa var. *villosior* IPNI=*101913-2* 390
Euphorbia refracta IPNI=*101777-2* 390
Euphorbia rhipsaloides IPNI=*348517-1* 391
Euphorbia rubicunda IPNI=*102034-2* 391
Euphorbia rubrosperma IPNI=*102034-2* 391
Euphorbia salina IPNI=*347331-1* 390
Euphorbia sarcostemmatoides IPNI=*347292-1* 390
Euphorbia scoparia IPNI=*348517-1* 391
Euphorbia serpens f. *psilocyathia* IPNI=*101976-2* 391
Euphorbia serpens subsp. *fissistipula* IPNI=*101976-2* 391
Euphorbia serpens var. *fissistipula* IPNI=*101976-2* 391
Euphorbia serpens var. *flexicaulis* IPNI=*101976-2* 391
Euphorbia serpens var. *imbricata* IPNI=*101976-2* 391
Euphorbia serpens var. *microphylla* IPNI=*101976-2* 391
Euphorbia serpens var. *radicans* IPNI=*101976-2* 391
Euphorbia serrulata IPNI=*101676-2* 390
Euphorbia spongiosa IPNI=*347072-1* 390
Euphorbia stenomeres IPNI=*101676-2* 390
Euphorbia suareziana IPNI=*348517-1* 391
Euphorbia supina IPNI=*347236-1* 390
Euphorbia tenella IPNI=*101913-2* 390
Euphorbia thymifolia f. *laxifoliata* IPNI=*102034-2* 391
Euphorbia trichogona IPNI=*101913-2* 390
Euphorbia trinervis IPNI=*101777-2* 390
Euphorbia tyraica IPNI=*346205-1* 390
Euphorbia verticillata IPNI=*101651-2* 390
Euphorbia verticillata IPNI=*346510-1* 390
Euphorbia viminalis IPNI=*348517-1* 391
Euphorbia yayalesia IPNI=*347331-1* 390
Euphorbiodendron cestrifolium IPNI=*347079-1* 390
Euphorbiodendron goudotii IPNI=*347079-1* 390
Euphorbiodendron latazi IPNI=*347079-1* 390
Euphorbiodendron laurifolium IPNI=*347079-1* 390
Euphorbion lathyrum IPNI=*347072-1* 390
Euphorbion peplum IPNI=*347724-1* 390
Euphrasia santolinifolia IPNI=*77160774-1* 375
Eupritchardia pacifica IPNI=*669288-1* 211
Eupteris aquilina IPNI=*17210060-1* 175
Eurya calophylla IPNI=*828818-1* 302
Eurya dombeyana IPNI=*106032-2* 303
Eurya sericea IPNI=*106032-2* 303
Eurya verrucosa IPNI=*106037-2* 303
Euryale amazonica IPNI=*605776-1* 181
Euryale brasiliana IPNI=*605776-1* 181
Euryanthe schiedeana IPNI=*77183799-1* 406
Eurychaenia ferruginea IPNI=*572470-1* 420
Eurychaenia punctata var. *latifolia* IPNI=*1069703-2* 422
Eurygania parvifolia IPNI=*252657-2* 300
Euryspermum frondosum IPNI=*704928-1* 433
Euryspermum humifusum IPNI=*704928-1* 433
Euryspermum salignum IPNI=*704928-1* 433
Eustachys gayana IPNI=*395891-1* 236
Eustoma chironioides IPNI=*367440-1* 344
Eustoma exaltatum f. *albiflorum* IPNI=*367440-1* 344

Eustoma silenifolium IPNI=*367440-1* 344
Eutacta cookii IPNI=*261688-1* 180
Eutacta excelsa IPNI=*261688-1* 180
Eutacta humilis IPNI=*261688-1* 180
Eutacta minor IPNI=*261688-1* 180
Eutassa columnaris IPNI=*261688-1* 180
Eutassa heterophylla IPNI=*676669-1* 180
Euterpe aculeata IPNI=*7909-2* 204
Euterpe acuminata IPNI=*210057-2* 211
Euterpe badiocarpa IPNI=*666941-1* 209
Euterpe beardii IPNI=*666941-1* 209
Euterpe brasiliana IPNI=*666941-1* 209
Euterpe caribaea IPNI=*222805-2* 212
Euterpe catinga var. *aurantiaca* IPNI=*317294-2* 208
Euterpe decurrens IPNI=*210062-2* 211
Euterpe ensiformis IPNI=*210063-2* 211
Euterpe jenmanii IPNI=*222809-2* 212
Euterpe longepetiolata IPNI=*669239-1* 211
Euterpe mollissima IPNI=*317294-2* 208
Euterpe schultzeana IPNI=*210069-2* 211
Euterpe ventricosa IPNI=*222809-2* 212
Euthales fructipendula IPNI=*429349-1* 385
Euthales peruviana IPNI=*429349-1* 385
Eutriana aristidoides IPNI=*1049662-2* 235
Eutriana curtipendula IPNI=*1198286-2* 235
Eutriana hirsuta IPNI=*1049662-2* 235
Eutriana humilis IPNI=*328564-2* 235
Eutriana tenuis IPNI=*328564-2* 235
Euxenia grata IPNI=*202984-1* 261
Euxenia radiata IPNI=*993359-1* 271
Euxolus alius IPNI=*10619-2* 281
Euxolus arvensis IPNI=*316347-2* 281
Euxolus ascendens IPNI=*10619-2* 281
Euxolus blitum IPNI=*10619-2* 281
Euxolus caudatus IPNI=*316349-2* 282
Euxolus caudatus var. *gracilis* IPNI=*316349-2* 282
Euxolus caudatus var. *maximus* IPNI=*316349-2* 282
Euxolus crassipes IPNI=*327385-2* 281
Euxolus lividus var. *oblongifolius* IPNI=*10619-2* 281
Euxolus polygamus IPNI=*327386-2* 282
Euxolus polystachyus IPNI=*316349-2* 282
Euxolus viridis IPNI=*316349-2* 282
Euxolus viridis var. *ascendens* IPNI=*10619-2* 281
Euzomum hispidum IPNI=*1137625-2* 277
Euzomum sativum IPNI=*1137625-2* 277
Euzomum vesicarium IPNI=*1137625-2* 277
Evanesca crassifolia IPNI=*196799-2* 428
Evansia chinensis IPNI=*438741-1* 216
Evansia fimbriata IPNI=*438741-1* 216
Evansia japonica IPNI=*438741-1* 216
Evea elata IPNI=*77115600-1* 353
Evea emetica IPNI=*764727-1* 356
Evea ipecacuanha IPNI=*1214950-2* 347
Evea tomentosa IPNI=*77115635-1* 354
Evelyna aurantiaca IPNI=*630704-1* 218
Evelyna bractescens IPNI=*630704-1* 218
Evelyna hallii IPNI=*630704-1* 218
Evelyna kermesina IPNI=*630759-1* 218
Evelyna lindenii var. *cymbibractea* IPNI=*630704-1* 218
Evia acida IPNI=*71455-1* 451
Evia dulcis IPNI=*71455-1* 451
Evonymodaphne armeniaca IPNI=*465248-1* 183
Exacum aphyllum IPNI=*1167766-2* 345
Exacum arenarium IPNI=*367033-1* 344
Exacum cubense IPNI=*230042-2* 345
Exacum diffusum IPNI=*366552-1* 344
Exacum erectum IPNI=*366552-1* 344
Exacum guianense IPNI=*230042-2* 345
Exacum minus IPNI=*367036-1* 344
Exacum purpureum IPNI=*230042-2* 345
Exacum racemosum IPNI=*367033-1* 344
Exacum ramosum IPNI=*367033-1* 344
Exacum spicatum IPNI=*367036-1* 344
Exacum strictum IPNI=*367036-1* 344
Exacum tenellum IPNI=*366552-1* 344
Exacum ternifolium IPNI=*367036-1* 344
Excoecaria aerea IPNI=*227503-2* 392
Excoecaria biglandulosa IPNI=*227503-2* 392
Excoecaria biglandulosa var. *aubletianum* IPNI=*227503-2* 392
Excoecaria biglandulosa var. *aucuparia* IPNI=*227503-2* 392
Excoecaria biglandulosa var. *clausseniana* IPNI=*227503-2* 392
Excoecaria biglandulosa var. *cuneata* IPNI=*227503-2* 392
Excoecaria biglandulosa var. *dracunculoides* IPNI=*227503-2* 392
Excoecaria biglandulosa var. *glandulata* IPNI=*227503-2* 392
Excoecaria biglandulosa var. *grandifolia* IPNI=*227503-2* 392
Excoecaria biglandulosa var. *hamata* IPNI=*227503-2* 392
Excoecaria biglandulosa var. *hippomane* IPNI=*227503-2* 392
Excoecaria biglandulosa var. *intercedens* IPNI=*227503-2* 392
Excoecaria biglandulosa var. *klotzschiana* IPNI=*227503-2* 392
Excoecaria biglandulosa var. *lanceolata* IPNI=*227503-2* 392
Excoecaria biglandulosa var. *leptadenia* IPNI=*227503-2* 392
Excoecaria biglandulosa var. *montevidensis* IPNI=*227503-2* 392
Excoecaria biglandulosa var. *moritziana* IPNI=*227503-2* 392
Excoecaria biglandulosa var. *pavoniana* IPNI=*227503-2* 392
Excoecaria biglandulosa var. *petiolaris* IPNI=*227503-2* 392
Excoecaria biglandulosa var. *prunifolia* IPNI=*227503-2* 392
Excoecaria biglandulosa var. *salicifolia* IPNI=*227503-2* 392
Excoecaria biglandulosa var. *serrata* IPNI=*227503-2* 392
Excoecaria brasiliensis IPNI=*356183-1* 393
Excoecaria concolor IPNI=*1187218-2* 387
Excoecaria guianensis IPNI=*352054-1* 392
Excoecaria marginata IPNI=*227503-2* 392
Excoecaria marginata var. *conjungens* IPNI=*227503-2* 392
Excoecaria marginata var. *grandifolia* IPNI=*227503-2* 392
Excoecaria marginata var. *lanceolata* IPNI=*227503-2* 392
Excoecaria marginata var. *longifolia* IPNI=*227503-2* 392
Excoecaria marginata var. *puberula* IPNI=*227503-2* 392
Excoecaria marginata var. *spathulata* IPNI=*227503-2* 392
Excoecaria obtusiloba IPNI=*227503-2* 392
Excoecaria occidentalis IPNI=*227503-2* 392
Excoecaria stylaris IPNI=*355950-1* 393
Excoecaria suberosa IPNI=*227503-2* 392
Exocarpos ceramicus IPNI=*694774-1* 289
Exostema caribaeum var. *pubescens* IPNI=*1125915-2* 350
Exostema caribaeum var. *velutinum* IPNI=*1125915-2* 350
Exostema longicuspe IPNI=*1125915-2* 350
Exostema souzanum IPNI=*67491-2* 349
Exostema veraensis IPNI=*1125915-2* 350
Exothostemon javitensis IPNI=*151987-2* 341
Exothostemon macrophyllus IPNI=*151979-2* 341
Exothostemon mollissimus IPNI=*152016-2* 341
Exothostemon paludosus IPNI=*81684-1* 343
Eyrythalia amarella IPNI=*108639-2* 344
Eyrythalia conferta IPNI=*108639-2* 344
Eyrythalia lancifolia IPNI=*108639-2* 344
Eyrythalia pratensis IPNI=*108639-2* 344
Eyrythalia pulchella IPNI=*108639-2* 344
Eyrythalia uliginosa IPNI=*108639-2* 344
Ezehlsia palma IPNI=*533580-1* 215

F

Faba bona subsp. *equina* IPNI=*524737-1* 336
Faba equina IPNI=*524737-1* 336
Faba faba IPNI=*524737-1* 336
Faba major IPNI=*524737-1* 336
Faba minor IPNI=*524737-1* 336

Faba sativa IPNI=*524737-1* 336
Faba vulgaris IPNI=*524737-1* 336
Faba vulgaris subsp. *minor* IPNI=*524737-1* 336
Fagara octandra IPNI=*127234-1* 451
Fagelia crenatiloba IPNI=*41487-2* 365
Fagelia gracilis IPNI=*41696-2* 365
Fagelia mexicana IPNI=*41487-2* 365
Fagelia micrantha IPNI=*41487-2* 365
Fagelia ovata IPNI=*800278-1* 365
Fagelia radiculoides IPNI=*41487-2* 365
Fagelia scabiosifolia IPNI=*41696-2* 365
Fagelia scalaris IPNI=*41696-2* 365
Fagelia tripartita IPNI=*41696-2* 365
Fagopyrum cereale IPNI=*694526-1* 289
Fagopyrum dryandrii IPNI=*694526-1* 289
Fagopyrum emarginatum IPNI=*694526-1* 289
Fagopyrum esculentum subsp. *ancestralis* IPNI=*694526-1* 289
Fagopyrum fagopyrum IPNI=*694526-1* 289
Fagopyrum polygonum IPNI=*694526-1* 289
Fagopyrum sagittatum IPNI=*694526-1* 289
Fagopyrum sarracenicum IPNI=*694526-1* 289
Fagopyrum vulgare IPNI=*694526-1* 289
Faramea amazonica IPNI=*749186-1* 350
Faramea americana IPNI=*1187456-2* 350
Faramea angustifolia IPNI=*749186-1* 350
Faramea benensis IPNI=*749186-1* 350
Faramea breviflora IPNI=*749070-1* 350
Faramea cestroides IPNI=*1039243-2* 350
Faramea cyanea IPNI=*1187456-2* 350
Faramea egensis IPNI=*749186-1* 350
Faramea hondurae IPNI=*749120-1* 350
Faramea laxula IPNI=*749186-1* 350
Faramea leptoloba IPNI=*747692-1* 349
Faramea marginata IPNI=*1039243-2* 350
Faramea maynensis IPNI=*749186-1* 350
Faramea multiflora var. *amazonica* IPNI=*749186-1* 350
Faramea multiflora var. *benensis* IPNI=*749186-1* 350
Faramea multiflora var. *epedunculata* IPNI=*749186-1* 350
Faramea multiflora var. *maynensis* IPNI=*749186-1* 350
Faramea multiflora var. *salicifolia* IPNI=*749186-1* 350
Faramea occidentalis var. *brachycalyx* IPNI=*1187456-2* 350
Faramea occidentalis var. *meridionalis* IPNI=*1187456-2* 350
Faramea odoratissima IPNI=*1187456-2* 350
Faramea pachydictyon IPNI=*749277-1* 350
Faramea paludicola IPNI=*749277-1* 350
Faramea parviflora IPNI=*749186-1* 350
Faramea phyllonomoides IPNI=*749186-1* 350
Faramea salicifolia f. *paniculata* IPNI=*749186-1* 350
Faramea salicifolia f. *subumbellata* IPNI=*749186-1* 350
Faramea salicifolia IPNI=*749186-1* 350
Faramea stenura IPNI=*749120-1* 350
Faramea talamancarum f. *acutifolia* IPNI=*749186-1* 350
Faramea talamancarum IPNI=*749186-1* 350
Faramea tenuifolia IPNI=*1187456-2* 350
Faramea warmingiana IPNI=*1187456-2* 350
Faramea zetekii IPNI=*1187456-2* 350
Fasciculus ruber IPNI=*158983-1* 286
Fatsia japonica var. *liukiuensis* IPNI=*90496-1* 253
Fatsia papyrifera IPNI=*86417-3* 254
Faucibarba officinalis IPNI=*450084-1* 371
Faulia verrucosa IPNI=*610017-1* 375
Fedia chaerophylloides IPNI=*859680-1* 296
Feea boryi IPNI=*17228970-1* 173
Feea diversifrons IPNI=*17228970-1* 173
Feijoa obovata IPNI=*596467-1* 426
Feijoa schenckiana IPNI=*596467-1* 426
Feijoa sellowiana f. *elongata* IPNI=*596467-1* 426
Feijoa sellowiana var. *rugosa* IPNI=*596467-1* 426
Fenelonia bracteata IPNI=*77097737-1* 214
Fenixanthes splendens IPNI=*60466516-2* 373
Ferdinanda eminens IPNI=*238923-1* 268
Ferdinandusa andina IPNI=*103549-2* 350
Ferdinandusa paxii IPNI=*103549-2* 350
Fernandezia acuta IPNI=*641704-1* 219
Fernandezia longifolia IPNI=*142885-2* 219
Ferolia amazonica IPNI=*727101-1* 383
Ferolia campestris IPNI=*727083-1* 383
Ferolia excelsa IPNI=*727101-1* 383
Ferolia guianensis IPNI=*37310-2* 437
Ferolia montana IPNI=*727140-1* 383
Ferolia sprucei IPNI=*727170-1* 383
Ferraria lurida IPNI=*441642-1* 217
Ferraria parviflora IPNI=*89940-2* 216
Ferraria pavonia IPNI=*441553-1* 216
Ferraria tigridia IPNI=*441553-1* 216
Ferula graveolens IPNI=*837530-1* 251
Ferula marathrophylla IPNI=*837530-1* 251
Festuca alexeevii IPNI=*30010036-2* 239
Festuca alexeevii var. *villispiculata* IPNI=*30010036-2* 239
Festuca alpestris IPNI=*30010036-2* 239
Festuca alpigena IPNI=*30010036-2* 239
Festuca ammobia IPNI=*30010036-2* 239
Festuca arenaria IPNI=*30010036-2* 239
Festuca arundinacea IPNI=*30010036-2* 239
Festuca arundinacea IPNI=*973267-1* 241
Festuca askelofiana IPNI=*30010036-2* 239
Festuca atlantigena IPNI=*30010036-2* 239
Festuca aucta IPNI=*30010036-2* 239
Festuca aurasiaca IPNI=*30010036-2* 239
Festuca baicalensis IPNI=*30010036-2* 239
Festuca baicalensis var. *tjulinae* IPNI=*30010036-2* 239
Festuca balearica IPNI=*30010036-2* 239
Festuca baltica IPNI=*30010036-2* 239
Festuca barbata IPNI=*30010036-2* 239
Festuca brachypoda IPNI=*402472-1* 239
Festuca briquetii IPNI=*30010036-2* 239
Festuca bromoides var. *tenella* IPNI=*402472-1* 239
Festuca calcarea IPNI=*20008857-1* 239
Festuca clarkei IPNI=*30010036-2* 239
Festuca communis IPNI=*30010036-2* 239
Festuca commutata IPNI=*403134-1* 239
Festuca daghestanica IPNI=*30010036-2* 239
Festuca densiuscula IPNI=*30010036-2* 239
Festuca dertonensis IPNI=*402472-1* 239
Festuca distichophylla IPNI=*82343-2* 238
Festuca dumetorum IPNI=*30010036-2* 239
Festuca duriuscula IPNI=*30010036-2* 239
Festuca egena IPNI=*30010036-2* 239
Festuca elatior var. *arundinacea* IPNI=*973267-1* 241
Festuca eragrostoides IPNI=*20008857-1* 239
Festuca eriantha IPNI=*30010036-2* 239
Festuca fluitans IPNI=*404100-1* 239
Festuca fusca IPNI=*60458632-2* 238
Festuca geniculata var. *monandra* IPNI=*402472-1* 239
Festuca glabra IPNI=*30010036-2* 239
Festuca glauca IPNI=*30010036-2* 239
Festuca glaucescens IPNI=*30010036-2* 239
Festuca glomerata IPNI=*397309-1* 236
Festuca halleri IPNI=*30010036-2* 239
Festuca helgolandica IPNI=*30010036-2* 239
Festuca heterophylla var. *subheterophylla* IPNI=*30010036-2* 239
Festuca hondoensis IPNI=*30010036-2* 239
Festuca hordeiformis IPNI=*402472-1* 239
Festuca iberica subsp. *atlantigena* IPNI=*30010036-2* 239
Festuca inopoda IPNI=*30010036-2* 239
Festuca kirelowii IPNI=*30010036-2* 239

Festuca kitaibeliana IPNI=*30010036-2* 239
Festuca lanuginosa IPNI=*30010036-2* 239
Festuca limosa IPNI=*30010036-2* 239
Festuca linearis IPNI=*403134-1* 239
Festuca longifolia IPNI=*30010036-2* 239
Festuca megalura IPNI=*403134-1* 239
Festuca michiganica IPNI=*30010036-2* 239
Festuca musashiensis IPNI=*30010036-2* 239
Festuca nankotaizanensis IPNI=*30010036-2* 239
Festuca niitakensis IPNI=*30010036-2* 239
Festuca obtusiflora IPNI=*60458636-2* 238
Festuca onurus IPNI=*402472-1* 239
Festuca oregona IPNI=*30010036-2* 239
Festuca ovina subsp. *major* IPNI=*30010036-2* 239
Festuca pallidiuscula IPNI=*30010036-2* 239
Festuca panormitana IPNI=*402472-1* 239
Festuca paucispiculata IPNI=*30010036-2* 239
Festuca pendulina IPNI=*30065892-2* 235
Festuca perennis IPNI=*407493-1* 241
Festuca planifolia IPNI=*30010036-2* 239
Festuca pluriflora IPNI=*30010036-2* 239
Festuca polonica IPNI=*30010036-2* 239
Festuca polyantha IPNI=*30010036-2* 239
Festuca pratensis IPNI=*30010036-2* 239
Festuca pruinosa IPNI=*30010036-2* 239
Festuca pseudomyuros IPNI=*403134-1* 239
Festuca pseudonemorum IPNI=*30010036-2* 239
Festuca pseudorubra IPNI=*30010036-2* 239
Festuca pubescens IPNI=*30010036-2* 239
Festuca quitensis IPNI=*37001-2* 235
Festuca repens IPNI=*30010036-2* 239
Festuca rivularis subsp. *pseudorivularis* IPNI=*30010036-2* 239
Festuca rubra f. *caespitosa* IPNI=*30010036-2* 239
Festuca rubra f. *dasyphylla* IPNI=*30010036-2* 239
Festuca rubra f. *glabra* IPNI=*30010036-2* 239
Festuca rubra f. *grandiflora* IPNI=*30010036-2* 239
Festuca rubra f. *littoralis* IPNI=*30010036-2* 239
Festuca rubra f. *pilosa* IPNI=*30010036-2* 239
Festuca rubra f. *psilolemma* IPNI=*30010036-2* 239
Festuca rubra f. *subheterophylla* IPNI=*30010036-2* 239
Festuca rubra subsp. *arenaria* IPNI=*30010036-2* 239
Festuca rubra subsp. *arenicola* IPNI=*30010036-2* 239
Festuca rubra subsp. *aucta* IPNI=*30010036-2* 239
Festuca rubra subsp. *baicalensis* IPNI=*30010036-2* 239
Festuca rubra subsp. *commutata* IPNI=*30010036-2* 239
Festuca rubra subsp. *daghestanica* IPNI=*30010036-2* 239
Festuca rubra subsp. *densiuscula* IPNI=*30010036-2* 239
Festuca rubra subsp. *eriantha* IPNI=*30010036-2* 239
Festuca rubra subsp. *fraterculae* IPNI=*30010036-2* 239
Festuca rubra subsp. *glabra* IPNI=*30010036-2* 239
Festuca rubra subsp. *glaucodea* IPNI=*30010036-2* 239
Festuca rubra subsp. *kirelowii* IPNI=*30010036-2* 239
Festuca rubra subsp. *kitaibeliana* IPNI=*30010036-2* 239
Festuca rubra subsp. *lanuginosa* IPNI=*30010036-2* 239
Festuca rubra subsp. *limosa* IPNI=*30010036-2* 239
Festuca rubra subsp. *mediana* IPNI=*30010036-2* 239
Festuca rubra subsp. *pluriflora* IPNI=*30010036-2* 239
Festuca rubra subsp. *pruinosa* IPNI=*30010036-2* 239
Festuca rubra subsp. *pseudorivularis* IPNI=*30010036-2* 239
Festuca rubra subsp. *scotica* IPNI=*30010036-2* 239
Festuca rubra subsp. *secunda* IPNI=*30010036-2* 239
Festuca rubra subsp. *sochondinensis* IPNI=*30010036-2* 239
Festuca rubra subsp. *thessalica* IPNI=*30010036-2* 239
Festuca rubra subsp. *vallicola* IPNI=*30010036-2* 239
Festuca rubra subsp. *villosa* IPNI=*30010036-2* 239
Festuca rubra subvar. *barbata* IPNI=*30010036-2* 239
Festuca rubra subvar. *dasyphylla* IPNI=*30010036-2* 239
Festuca rubra subvar. *densiuscula* IPNI=*30010036-2* 239
Festuca rubra subvar. *glaucescens* IPNI=*30010036-2* 239
Festuca rubra subvar. *grandiflora* IPNI=*30010036-2* 239
Festuca rubra subvar. *pruinosa* IPNI=*30010036-2* 239
Festuca rubra var. *altaica* IPNI=*30010036-2* 239
Festuca rubra var. *arenaria* IPNI=*30010036-2* 239
Festuca rubra var. *atlantigena* IPNI=*30010036-2* 239
Festuca rubra var. *atrata* IPNI=*30010036-2* 239
Festuca rubra var. *caulispicula* IPNI=*30010036-2* 239
Festuca rubra var. *clarkei* IPNI=*30010036-2* 239
Festuca rubra var. *glabra* IPNI=*30010036-2* 239
Festuca rubra var. *glabrata* IPNI=*30010036-2* 239
Festuca rubra var. *lanuginosa* IPNI=*30010036-2* 239
Festuca rubra var. *magnelii* IPNI=*30010036-2* 239
Festuca rubra var. *mediana* IPNI=*30010036-2* 239
Festuca rubra var. *pauciflora* IPNI=*30010036-2* 239
Festuca rubra var. *radnensis* IPNI=*30010036-2* 239
Festuca rubra var. *secunda* IPNI=*30010036-2* 239
Festuca salina IPNI=*30010036-2* 239
Festuca sciuroides IPNI=*402472-1* 239
Festuca septiflora IPNI=*30010036-2* 239
Festuca siracusana IPNI=*30010036-2* 239
Festuca steineri IPNI=*30010036-2* 239
Festuca stricta IPNI=*37001-2* 235
Festuca strictifolia IPNI=*20008857-1* 239
Festuca subcontorta IPNI=*30010036-2* 239
Festuca subglaucescens IPNI=*30010036-2* 239
Festuca temulenta IPNI=*407516-1* 241
Festuca tenuiculmis IPNI=*320035-2* 244
Festuca trichodes IPNI=*1147585-2* 239
Festuca triticea IPNI=*82343-2* 238
Festuca triticoides IPNI=*82343-2* 238
Festuca unifaria IPNI=*30010036-2* 239
Festuca unifaria var. *izhorica* IPNI=*30010036-2* 239
Festuca unioloides IPNI=*37001-2* 235
Festuca vallicola IPNI=*30010036-2* 239
Festuca veneris IPNI=*20008857-1* 239
Festuca villiflora IPNI=*30010036-2* 239
Festuca villosa IPNI=*30010036-2* 239
Festuca viridis IPNI=*30010036-2* 239
Festuca waldsteinii IPNI=*30010036-2* 239
Festuca willemetii IPNI=*402472-1* 239
Festucaria heisteri IPNI=*404100-1* 239
Feuilleea acrocephala IPNI=*128354-2* 321
Feuilleea acuminata IPNI=*313763-2* 321
Feuilleea affinis IPNI=*128358-2* 321
Feuilleea aggregata IPNI=*1129244-2* 321
Feuilleea alba IPNI=*1129244-2* 321
Feuilleea angustifolia IPNI=*42066-2* 313
Feuilleea auriculata IPNI=*989358-1* 309
Feuilleea bahiensis IPNI=*1131764-2* 322
Feuilleea bertolonii IPNI=*1003508-2* 330
Feuilleea bonplandiana IPNI=*128732-2* 324
Feuilleea bourgoni IPNI=*500286-1* 321
Feuilleea brachyptera IPNI=*128596-2* 323
Feuilleea bracteosa IPNI=*128596-2* 323
Feuilleea brevispicata IPNI=*128764-2* 324
Feuilleea caracasana IPNI=*287164-2* 337
Feuilleea cataractae IPNI=*962821-1* 337
Feuilleea catharinae IPNI=*500844-1* 324
Feuilleea cayennensis IPNI=*128427-2* 322
Feuilleea chartacea IPNI=*128432-2* 322
Feuilleea ciliata IPNI=*128440-2* 322
Feuilleea cinnamomea IPNI=*128442-2* 322
Feuilleea coccinea IPNI=*962823-1* 337
Feuilleea conferta IPNI=*128478-2* 322
Feuilleea coruscans IPNI=*128458-2* 322
Feuilleea corymbosa IPNI=*989341-1* 320
Feuilleea cumingii IPNI=*1161746-2* 313

Feuilleea cyclocarpa IPNI=*1037928-2* 319
Feuilleea cylindrica IPNI=*1090709-2* 322
Feuilleea densiflora IPNI=*128466-2* 322
Feuilleea distachya IPNI=*912944-1* 329
Feuilleea divaricata IPNI=*962821-1* 337
Feuilleea dubia IPNI=*989342-1* 320
Feuilleea dulcis IPNI=*1003508-2* 330
Feuilleea edulis IPNI=*128478-2* 322
Feuilleea endlicheri IPNI=*128647-2* 323
Feuilleea fagifolia IPNI=*1199545-2* 323
Feuilleea falcata IPNI=*42154-2* 313
Feuilleea fasciculata IPNI=*316281-2* 313
Feuilleea fastuosa IPNI=*1040744-2* 322
Feuilleea fendleriana IPNI=*500432-1* 322
Feuilleea feuilleei IPNI=*128507-2* 322
Feuilleea formosa IPNI=*287168-2* 337
Feuilleea guildingii IPNI=*42179-2* 313
Feuilleea haematocephala IPNI=*42180-2* 313
Feuilleea hayesii IPNI=*128532-2* 322
Feuilleea heterophylla IPNI=*128535-2* 322
Feuilleea heteroptera IPNI=*1122303-2* 324
Feuilleea hymenaeifolia IPNI=*1183336-2* 330
Feuilleea ignota IPNI=*128586-2* 323
Feuilleea inaequalis IPNI=*271431-2* 337
Feuilleea inga IPNI=*1006087-2* 324
Feuilleea ingoides IPNI=*1131764-2* 322
Feuilleea insignis IPNI=*128550-2* 322
Feuilleea jupunba IPNI=*469538-1* 310
Feuilleea laeta IPNI=*989371-1* 310
Feuilleea lallensis IPNI=*128564-2* 323
Feuilleea lateriflora IPNI=*128568-2* 323
Feuilleea latifolia IPNI=*271432-2* 337
Feuilleea laurina IPNI=*1199545-2* 323
Feuilleea laxa IPNI=*1161746-2* 313
Feuilleea lebbeck IPNI=*99109-3* 311
Feuilleea leiocalycina IPNI=*500531-1* 322
Feuilleea leptoloba IPNI=*128693-2* 323
Feuilleea leucophylla IPNI=*989368-1* 310
Feuilleea ligustrina IPNI=*514978-1* 330
Feuilleea lindeniana IPNI=*128732-2* 324
Feuilleea longifolia IPNI=*271434-2* 337
Feuilleea macrostachys IPNI=*1183336-2* 330
Feuilleea magdalenae IPNI=*1051245-2* 313
Feuilleea mangensis IPNI=*56985-2* 315
Feuilleea marginata IPNI=*128599-2* 323
Feuilleea multiflora IPNI=*989448-1* 311
Feuilleea multijuga IPNI=*128628-2* 323
Feuilleea myriantha IPNI=*1035636-2* 324
Feuilleea niopoides IPNI=*8032-2* 311
Feuilleea nobilis IPNI=*128633-2* 323
Feuilleea obtusata IPNI=*128644-2* 323
Feuilleea oerstediana IPNI=*128647-2* 323
Feuilleea panamensis IPNI=*128732-2* 324
Feuilleea pavoniana IPNI=*128732-2* 324
Feuilleea pedicellaris IPNI=*989349-1* 311
Feuilleea pezizifera IPNI=*128673-2* 323
Feuilleea pilosula IPNI=*128675-2* 323
Feuilleea plumifera IPNI=*128680-2* 323
Feuilleea portoricensis IPNI=*943976-1* 337
Feuilleea pruriens IPNI=*128687-2* 323
Feuilleea pubescens IPNI=*996931-1* 330
Feuilleea punctata IPNI=*128693-2* 323
Feuilleea purdiei IPNI=*42269-2* 313
Feuilleea quassiifolia IPNI=*128675-2* 323
Feuilleea rubiginosa IPNI=*1068738-2* 323
Feuilleea rufescens IPNI=*62892-2* 315
Feuilleea rufinervis IPNI=*128693-2* 323
Feuilleea ruiziana IPNI=*128721-2* 324
Feuilleea salzmanniana IPNI=*500844-1* 324
Feuilleea saman IPNI=*227126-2* 331
Feuilleea sapinoides IPNI=*128732-2* 324
Feuilleea scabriuscula IPNI=*128478-2* 322
Feuilleea schomburgkii IPNI=*493928-1* 319
Feuilleea sciadion IPNI=*1035636-2* 324
Feuilleea sessiliflora IPNI=*128427-2* 322
Feuilleea sesuya IPNI=*1131764-2* 322
Feuilleea setifera IPNI=*128675-2* 323
Feuilleea setosa IPNI=*128746-2* 324
Feuilleea spectabilis IPNI=*1122303-2* 324
Feuilleea splendens IPNI=*128749-2* 324
Feuilleea stenoptera IPNI=*128764-2* 324
Feuilleea stipularis IPNI=*500843-1* 324
Feuilleea strigillosa IPNI=*128693-2* 323
Feuilleea subnervosa IPNI=*42066-2* 313
Feuilleea tenuiflora IPNI=*316281-2* 313
Feuilleea tenuifolia IPNI=*1090709-2* 322
Feuilleea thibaudiana IPNI=*128789-2* 324
Feuilleea tomentosa IPNI=*128792-2* 324
Feuilleea trinervia IPNI=*42330-2* 313
Feuilleea umbellifera IPNI=*1035636-2* 324
Feuilleea umbratica IPNI=*500893-1* 324
Feuilleea unguis-cati IPNI=*515137-1* 330
Feuilleea unguis-cati var. *latifolia* IPNI=*515137-1* 330
Feuilleea velutina IPNI=*128810-2* 324
Feuilleea venosa IPNI=*1040744-2* 322
Feuilleea villosissima IPNI=*128820-2* 324
Feuilleea vismiifolia IPNI=*500919-1* 324
Fevillea cordifolia var. *hederacea* IPNI=*104422-2* 294
Fevillea cordifolia var. *typica* IPNI=*104422-2* 294
Fevillea hederacea IPNI=*104422-2* 294
Fevillea javilla IPNI=*104422-2* 294
Fevillea karstenii IPNI=*104422-2* 294
Fevillea moorei IPNI=*104422-2* 294
Fevillea punctata IPNI=*104422-2* 294
Fevillea scandens IPNI=*104422-2* 294
Fevillea triloba IPNI=*104422-2* 294
Fevillea trilobata IPNI=*104422-2* 294
Fevillea uncipetala IPNI=*104422-2* 294
Fibichia dactylon IPNI=*397065-1* 236
Fibichia umbellata IPNI=*397065-1* 236
Ficus acutiloba IPNI=*852970-1* 438
Ficus aguaraguensis IPNI=*104497-2* 438
Ficus ajajuensis IPNI=*853876-1* 439
Ficus albescens IPNI=*852556-1* 438
Ficus albotomentosa IPNI=*852689-1* 438
Ficus altissima f. *laccifera* IPNI=*852369-1* 437
Ficus altissima var. *laccifera* IPNI=*852369-1* 437
Ficus alvareziana IPNI=*853876-1* 439
Ficus amblyphylla IPNI=*853217-1* 439
Ficus anguina IPNI=*853301-1* 439
Ficus angustifolia IPNI=*852372-1* 437
Ficus anthelminthica IPNI=*853024-1* 438
Ficus aquatica IPNI=*852970-1* 438
Ficus arbutifolia IPNI=*316336-2* 437
Ficus arbutifolia IPNI=*853405-1* 439
Ficus arpazusa IPNI=*853405-1* 439
Ficus arukensis IPNI=*1004295-2* 439
Ficus baccata IPNI=*853405-1* 439
Ficus banyana IPNI=*852482-1* 438
Ficus benjamina subsp. *comosa* IPNI=*852486-1* 438
Ficus benjamina var. *nuda* IPNI=*852486-1* 438
Ficus biglandula IPNI=*852970-1* 438
Ficus boapoy IPNI=*852825-1* 438
Ficus bonplandiana IPNI=*853317-1* 439
Ficus bopiana IPNI=*853200-1* 439
Ficus botryapioides IPNI=*316340-2* 438

Ficus brevifolia IPNI=*316340-2* 438
Ficus burdigalensis IPNI=*852556-1* 438
Ficus cairnsii IPNI=*853217-1* 439
Ficus campbellii IPNI=*852689-1* 438
Ficus cannabina IPNI=*852970-1* 438
Ficus caprificus IPNI=*852556-1* 438
Ficus caribaea IPNI=*316340-2* 438
Ficus carica var. *caprificus* IPNI=*852556-1* 438
Ficus catesbaei IPNI=*316340-2* 438
Ficus caucana IPNI=*316340-2* 438
Ficus caudata IPNI=*853563-1* 439
Ficus cerasifolia IPNI=*853405-1* 439
Ficus chaconiana IPNI=*853200-1* 439
Ficus chiapensis IPNI=*853317-1* 439
Ficus chiriquiana IPNI=*316336-2* 437
Ficus ciliolosa IPNI=*853405-1* 439
Ficus citrifolia IPNI=*853200-1* 439
Ficus citrifolia subsp. *brevifolia* IPNI=*316340-2* 438
Ficus citrifolia var. *brevifolia* IPNI=*316340-2* 438
Ficus clethrifolia IPNI=*853933-1* 439
Ficus clusiifolia IPNI=*60458499-2* 438
Ficus clusiifolia IPNI=*853317-1* 439
Ficus colchica IPNI=*852556-1* 438
Ficus colombra IPNI=*852556-1* 438
Ficus communis IPNI=*852556-1* 438
Ficus comosa IPNI=*852486-1* 438
Ficus complicata IPNI=*853405-1* 439
Ficus condaravia IPNI=*853217-1* 439
Ficus consanguinea IPNI=*316336-2* 437
Ficus cordata IPNI=*60458499-2* 438
Ficus coronata IPNI=*852689-1* 438
Ficus corpulenta IPNI=*853198-1* 438
Ficus cotoneaster IPNI=*853876-1* 439
Ficus cotoneifolia IPNI=*852482-1* 438
Ficus cotonifolia IPNI=*852482-1* 438
Ficus coybana IPNI=*853200-1* 439
Ficus crassinervia IPNI=*852482-1* 438
Ficus cuspidatocaudata IPNI=*852486-1* 438
Ficus cystopoda IPNI=*856394-1* 441
Ficus dahlii IPNI=*853217-1* 439
Ficus danielis IPNI=*853933-1* 439
Ficus daphniphylla IPNI=*853405-1* 439
Ficus deliciosa IPNI=*852556-1* 438
Ficus denticulata IPNI=*852970-1* 438
Ficus dictyophylla IPNI=*852486-1* 438
Ficus dilatata IPNI=*853217-1* 439
Ficus dottata IPNI=*852556-1* 438
Ficus dulciaria var. *antioquiensis* IPNI=*104508-2* 438
Ficus duquei IPNI=*853301-1* 439
Ficus duquei var. *obtusiloba* IPNI=*853301-1* 439
Ficus dyctiophleba IPNI=*853217-1* 439
Ficus elastica var. *rubrinervis* IPNI=*60458499-2* 438
Ficus elliptica IPNI=*853405-1* 439
Ficus elongata IPNI=*852970-1* 438
Ficus ernstiana IPNI=*852689-1* 438
Ficus erythrosticta IPNI=*853405-1* 439
Ficus eugeniifolia IPNI=*316336-2* 437
Ficus euomphala IPNI=*853876-1* 439
Ficus eximia f. *paraguariensis* IPNI=*316340-2* 438
Ficus expansa IPNI=*852825-1* 438
Ficus fadyenii IPNI=*316336-2* 437
Ficus fagifolia IPNI=*853876-1* 439
Ficus faginea IPNI=*850824-1* 437
Ficus fanshawei IPNI=*853876-1* 439
Ficus fasciculata IPNI=*853405-1* 439
Ficus finlayana IPNI=*853989-1* 439
Ficus florenciana IPNI=*104505-2* 438
Ficus floresina IPNI=*853317-1* 439
Ficus fontanesii IPNI=*316340-2* 438
Ficus foveata IPNI=*316340-2* 438
Ficus foveolata IPNI=*316340-2* 438
Ficus frigida IPNI=*853933-1* 439
Ficus galeata IPNI=*852689-1* 438
Ficus garcesii IPNI=*853405-1* 439
Ficus garcia-barrigae var. *ovoidea* IPNI=*104497-2* 438
Ficus gardeniifolia IPNI=*853405-1* 439
Ficus gardneriana IPNI=*853317-1* 439
Ficus gemina IPNI=*853405-1* 439
Ficus gentlei IPNI=*316340-2* 438
Ficus georgii IPNI=*316336-2* 437
Ficus gigantea IPNI=*316340-2* 438
Ficus glabra IPNI=*852825-1* 438
Ficus glabrata IPNI=*853024-1* 438
Ficus glabrata var. *obtusula* IPNI=*853024-1* 438
Ficus glabrata var. *typica* IPNI=*853024-1* 438
Ficus glandulosa IPNI=*852825-1* 438
Ficus glaucescens IPNI=*853200-1* 439
Ficus gleasonii IPNI=*853198-1* 438
Ficus globosa IPNI=*852556-1* 438
Ficus glycicarpa IPNI=*853933-1* 439
Ficus gramalotensis IPNI=*852689-1* 438
Ficus grandaeva IPNI=*853200-1* 439
Ficus guadalajarana IPNI=*853200-1* 439
Ficus guajavoides IPNI=*104614-2* 438
Ficus guanarensis IPNI=*316340-2* 438
Ficus gummifera IPNI=*850895-1* 437
Ficus haematocarpa IPNI=*852486-1* 438
Ficus halliana IPNI=*853405-1* 439
Ficus hanceana IPNI=*853513-1* 439
Ficus harrisii IPNI=*583812-1* 439
Ficus hartii IPNI=*852689-1* 438
Ficus hartii var. *tumacana* IPNI=*852689-1* 438
Ficus haughtii IPNI=*1004295-2* 439
Ficus havanensis IPNI=*852689-1* 438
Ficus helminthagoga IPNI=*853024-1* 438
Ficus hernandezii IPNI=*853200-1* 439
Ficus huilensis IPNI=*104508-2* 438
Ficus hydrophila IPNI=*1004295-2* 439
Ficus hypoleuca IPNI=*852556-1* 438
Ficus hyrcana IPNI=*852556-1* 438
Ficus ierensis IPNI=*853301-1* 439
Ficus indica IPNI=*852482-1* 438
Ficus insipida subsp. *radulina* IPNI=*853212-1* 439
Ficus insipida subsp. *scabra* IPNI=*853024-1* 438
Ficus interfector IPNI=*852689-1* 438
Ficus involuta IPNI=*853317-1* 439
Ficus involuta var. *urbaniana* IPNI=*853301-1* 439
Ficus iturbei IPNI=*852689-1* 438
Ficus jacquinifolia IPNI=*316336-2* 437
Ficus jonesii IPNI=*583812-1* 439
Ficus kanukuensis IPNI=*853405-1* 439
Ficus karet IPNI=*852482-1* 438
Ficus katherinae IPNI=*853198-1* 438
Ficus kellermanii IPNI=*852672-1* 438
Ficus kopetdagensis IPNI=*852556-1* 438
Ficus laccifera IPNI=*852369-1* 437
Ficus laevigata IPNI=*316340-2* 438
Ficus laevigata subvar. *cerasicarpa* IPNI=*316340-2* 438
Ficus laevigata subvar. *crassipetiolata* IPNI=*316340-2* 438
Ficus laevigata subvar. *obliquicuspis* IPNI=*316340-2* 438
Ficus laevigata subvar. *portoricensis* IPNI=*316340-2* 438
Ficus laevigata subvar. *reticulosa* IPNI=*316340-2* 438
Ficus laevigata subvar. *subcordata* IPNI=*316340-2* 438
Ficus laevigata subvar. *subcuspidata* IPNI=*316340-2* 438
Ficus laevigata subvar. *subdepressa* IPNI=*316340-2* 438
Ficus laevigata subvar. *subpyriformis* IPNI=*316340-2* 438

Ficus laevigata var. *brevifolia* IPNI=*316340-2* 438
Ficus laevigata var. *genuina* IPNI=*316340-2* 438
Ficus laevigata var. *lentiginosa* IPNI=*316340-2* 438
Ficus lancifolia IPNI=*852482-1* 438
Ficus lancifolia IPNI=*853405-1* 439
Ficus lasiophylla IPNI=*852482-1* 438
Ficus latifolia IPNI=*852369-1* 437
Ficus latifolia IPNI=*852556-1* 438
Ficus leavensii IPNI=*852689-1* 438
Ficus leavensii var. *thinophila* IPNI=*852689-1* 438
Ficus lentiginosa IPNI=*316340-2* 438
Ficus lentiginosa var. *imrayana* IPNI=*316340-2* 438
Ficus lentiginosa var. *subcuspidata* IPNI=*316340-2* 438
Ficus leucocarpa IPNI=*852556-1* 438
Ficus liebmanniana IPNI=*316336-2* 437
Ficus ligustrina IPNI=*853355-1* 439
Ficus littoralis IPNI=*853217-1* 439
Ficus llanensis IPNI=*853876-1* 439
Ficus llewelynii IPNI=*853876-1* 439
Ficus longipedicellata IPNI=*853513-1* 439
Ficus longistipula IPNI=*853024-1* 438
Ficus macrocarpa IPNI=*852556-1* 438
Ficus maroana IPNI=*853198-1* 438
Ficus mattogrossensis IPNI=*853317-1* 439
Ficus maximiliana IPNI=*852689-1* 438
Ficus meizonochlamys IPNI=*583812-1* 439
Ficus metensis IPNI=*853198-1* 438
Ficus microcarpa var. *crassifolia* IPNI=*853217-1* 439
Ficus microcarpa var. *fuyuensis* IPNI=*853217-1* 439
Ficus microcarpa var. *latifolia* IPNI=*853217-1* 439
Ficus microcarpa var. *nitida* IPNI=*853217-1* 439
Ficus microcarpa var. *oluangpiensis* IPNI=*853217-1* 439
Ficus microcarpa var. *pusillifolia* IPNI=*853217-1* 439
Ficus microchlamys IPNI=*853933-1* 439
Ficus microclada IPNI=*104716-2* 439
Ficus mitrophora IPNI=*852689-1* 438
Ficus morazaniana IPNI=*852689-1* 438
Ficus murilloi IPNI=*853200-1* 439
Ficus murilloi var. *cajambrensis* IPNI=*853200-1* 439
Ficus murilloi var. *typica* IPNI=*853200-1* 439
Ficus myrmecophila IPNI=*1004295-2* 439
Ficus myrtifolia IPNI=*853405-1* 439
Ficus naumannii IPNI=*853217-1* 439
Ficus neapolitana IPNI=*852556-1* 438
Ficus neglecta IPNI=*852486-1* 438
Ficus nepalensis IPNI=*852486-1* 438
Ficus nitida IPNI=*852486-1* 438
Ficus notobor IPNI=*852486-1* 438
Ficus nuda IPNI=*852486-1* 438
Ficus nymphoides IPNI=*853301-1* 439
Ficus oblongata IPNI=*316340-2* 438
Ficus obovata IPNI=*104716-2* 439
Ficus ochroleuca IPNI=*853405-1* 439
Ficus oerstediana IPNI=*316336-2* 437
Ficus omphalophora IPNI=*316336-2* 437
Ficus orinocensis IPNI=*1004295-2* 439
Ficus pachycarpa IPNI=*852556-1* 438
Ficus padifolia IPNI=*853405-1* 439
Ficus palmicida IPNI=*853405-1* 439
Ficus palmirana IPNI=*853024-1* 438
Ficus panamensis IPNI=*1004295-2* 439
Ficus pandurata IPNI=*853158-1* 438
Ficus panduriformis IPNI=*852970-1* 438
Ficus papyrifera IPNI=*852486-1* 438
Ficus parkeri IPNI=*853200-1* 439
Ficus parvifolia IPNI=*852486-1* 438
Ficus pascuorum IPNI=*853317-1* 439
Ficus pedunculata IPNI=*852556-1* 438
Ficus peepul IPNI=*853563-1* 439
Ficus pendula IPNI=*852486-1* 438
Ficus perez-arbelaezii IPNI=*852689-1* 438
Ficus perforata IPNI=*316336-2* 437
Ficus periplocifolia IPNI=*316336-2* 437
Ficus peroblonga IPNI=*1004295-2* 439
Ficus peruviana IPNI=*853405-1* 439
Ficus picardae IPNI=*853200-1* 439
Ficus pittieri IPNI=*852689-1* 438
Ficus planicostata IPNI=*316340-2* 438
Ficus plicato-ostiolata IPNI=*853876-1* 439
Ficus plumeri IPNI=*853200-1* 439
Ficus politoria IPNI=*852970-1* 438
Ficus polymorpha IPNI=*852556-1* 438
Ficus polypus IPNI=*853405-1* 439
Ficus populnea IPNI=*316340-2* 438
Ficus populnea subvar. *cerasicarpa* IPNI=*316340-2* 438
Ficus populnea subvar. *crassipetiolata* IPNI=*316340-2* 438
Ficus populnea subvar. *cubana* IPNI=*316340-2* 438
Ficus populnea subvar. *diversifolia* IPNI=*316340-2* 438
Ficus populnea subvar. *flavicosta* IPNI=*316340-2* 438
Ficus populnea subvar. *floridana* IPNI=*316340-2* 438
Ficus populnea subvar. *hispaniolae* IPNI=*316340-2* 438
Ficus populnea subvar. *longipetiolata* IPNI=*316340-2* 438
Ficus populnea subvar. *obliquicuspis* IPNI=*316340-2* 438
Ficus populnea subvar. *pedunculata* IPNI=*316340-2* 438
Ficus populnea subvar. *portoricensis* IPNI=*316340-2* 438
Ficus populnea subvar. *reticulosa* IPNI=*316340-2* 438
Ficus populnea subvar. *sancti-crucis* IPNI=*316340-2* 438
Ficus populnea subvar. *sancti-eustachii* IPNI=*316340-2* 438
Ficus populnea subvar. *sancti-joannis* IPNI=*316340-2* 438
Ficus populnea subvar. *subcordata* IPNI=*316340-2* 438
Ficus populnea subvar. *subcuspidata* IPNI=*316340-2* 438
Ficus populnea subvar. *subdepressa* IPNI=*316340-2* 438
Ficus populnea subvar. *subpyriformis* IPNI=*316340-2* 438
Ficus populnea subvar. *thomaea* IPNI=*316340-2* 438
Ficus populnea subvar. *typica* IPNI=*316340-2* 438
Ficus populnea var. *bahamensis* IPNI=*316340-2* 438
Ficus populnea var. *brevifolia* IPNI=*316340-2* 438
Ficus populnea var. *hispaniolae* IPNI=*316340-2* 438
Ficus populnea var. *laevigata* IPNI=*316340-2* 438
Ficus populnea var. *lentiginosa* IPNI=*316340-2* 438
Ficus populoides f. *syringifolia* IPNI=*316340-2* 438
Ficus populoides f. *umbrifera* IPNI=*316340-2* 438
Ficus populoides IPNI=*316340-2* 438
Ficus populoides var. *dilatata* IPNI=*316340-2* 438
Ficus populoides var. *elongata* IPNI=*316340-2* 438
Ficus populoides var. *maculosa* IPNI=*316340-2* 438
Ficus portoricensis IPNI=*316340-2* 438
Ficus praecox IPNI=*852556-1* 438
Ficus prinoides IPNI=*853355-1* 439
Ficus prinoides var. *genuina* IPNI=*853355-1* 439
Ficus prinoides var. *subtriplinervia* IPNI=*853405-1* 439
Ficus procera IPNI=*852482-1* 438
Ficus proctor-cooperi IPNI=*853317-1* 439
Ficus protensa IPNI=*853200-1* 439
Ficus pseudoradula IPNI=*853200-1* 439
Ficus putumayonis IPNI=*1004295-2* 439
Ficus pyrifolia IPNI=*316340-2* 438
Ficus pyrifolia IPNI=*852486-1* 438
Ficus quijosana IPNI=*104651-2* 439
Ficus radicans IPNI=*853405-1* 439
Ficus radula IPNI=*853200-1* 439
Ficus radulina IPNI=*853212-1* 439
Ficus ramiflora IPNI=*583293-1* 438
Ficus reclinata IPNI=*852486-1* 438
Ficus regnans IPNI=*853217-1* 439
Ficus regularis IPNI=*853876-1* 439

Ficus repens IPNI=*852970-1* 438
Ficus retusa var. *nitida* IPNI=*852486-1* 438
Ficus retusiformis IPNI=*853217-1* 439
Ficus rhynchophylla IPNI=*853563-1* 439
Ficus rolandri IPNI=*853405-1* 439
Ficus rubra IPNI=*853217-1* 439
Ficus rubricosta IPNI=*853200-1* 439
Ficus rubrinervis IPNI=*316340-2* 438
Ficus rufescens IPNI=*852970-1* 438
Ficus rugosa IPNI=*852556-1* 438
Ficus ruiziana IPNI=*316340-2* 438
Ficus salix IPNI=*777133-1* 404
Ficus sancti-crucis IPNI=*316340-2* 438
Ficus sanguinosa IPNI=*852689-1* 438
Ficus sapida IPNI=*853405-1* 439
Ficus scabrella IPNI=*852970-1* 438
Ficus scandens IPNI=*853513-1* 439
Ficus sibundoya IPNI=*104497-2* 438
Ficus silvestris IPNI=*852556-1* 438
Ficus sintenisii IPNI=*316336-2* 437
Ficus skytinodermis IPNI=*60458499-2* 438
Ficus sonorae IPNI=*853405-1* 439
Ficus sprucei IPNI=*853198-1* 438
Ficus squamulosa IPNI=*852689-1* 438
Ficus stipulata IPNI=*853513-1* 439
Ficus striata IPNI=*852486-1* 438
Ficus subandina IPNI=*316340-2* 438
Ficus subpanduriformis IPNI=*852970-1* 438
Ficus subscabrida IPNI=*853200-1* 439
Ficus subtriplinervia IPNI=*853405-1* 439
Ficus suffocans IPNI=*853200-1* 439
Ficus sulcipes IPNI=*853405-1* 439
Ficus superstitiosa IPNI=*853563-1* 439
Ficus surinamensis IPNI=*852372-1* 437
Ficus syringifolia IPNI=*316340-2* 438
Ficus taeda IPNI=*60458499-2* 438
Ficus tamatamae IPNI=*583293-1* 438
Ficus tapajozensis IPNI=*853405-1* 439
Ficus thelephora IPNI=*1004295-2* 439
Ficus thomaea IPNI=*316340-2* 438
Ficus thynneana IPNI=*853217-1* 439
Ficus thynneana var. *minor* IPNI=*853217-1* 439
Ficus thynneana var. *typica* IPNI=*853217-1* 439
Ficus tobagensis IPNI=*853989-1* 439
Ficus tomentella IPNI=*852689-1* 438
Ficus torteana IPNI=*852970-1* 438
Ficus truncata IPNI=*852970-1* 438
Ficus turbinata IPNI=*316340-2* 438
Ficus turbinata IPNI=*853405-1* 439
Ficus uberrima IPNI=*1004295-2* 439
Ficus umbrina IPNI=*852486-1* 438
Ficus umbrosa IPNI=*852482-1* 438
Ficus urbaniana IPNI=*853301-1* 439
Ficus usiacurina IPNI=*852689-1* 438
Ficus vaupesana IPNI=*853198-1* 438
Ficus venosissima IPNI=*852689-1* 438
Ficus vestita IPNI=*853513-1* 439
Ficus vicencionis IPNI=*853200-1* 439
Ficus vulpina IPNI=*853876-1* 439
Ficus weberbaueri IPNI=*853876-1* 439
Ficus werckleana IPNI=*853024-1* 438
Ficus whitei IPNI=*853024-1* 438
Ficus williamsii IPNI=*1004295-2* 439
Ficus williamsii IPNI=*852689-1* 438
Ficus wilsonii IPNI=*316336-2* 437
Ficus wrightii IPNI=*316338-2* 439
Ficus wuiana IPNI=*853876-1* 439
Ficus xavieri IPNI=*852486-1* 438
Ficus yucatanensis IPNI=*852689-1* 438
Ficus zarzalensis IPNI=*853405-1* 439
Filaginella luteoalba IPNI=*212994-1* 264
Filicula fragilis IPNI=*30013429-2* 174
Filix aquilina IPNI=*17210060-1* 175
Filix fragilis f. *magnasora* IPNI=*30013429-2* 174
Filix fragilis IPNI=*30013429-2* 174
Filix goggilodes IPNI=*17458930-1* 174
Filix-foemina aquilina IPNI=*17210060-1* 175
Filix-foemina aquilina var. *arachnoidea* IPNI=*213798-2* 175
Filix-foemina aquilina var. *caudata* IPNI=*213799-2* 175
Filix-foemina vulgaris IPNI=*17210060-1* 175
Filix-mas augescens var. *normalis* IPNI=*251400-2* 174
Fimbristylis amblyphylla IPNI=*1111946-2* 232
Fimbristylis anceps IPNI=*1111946-2* 232
Fimbristylis autumnalis var. *complanata* IPNI=*1111946-2* 232
Fimbristylis bequaertii IPNI=*1111946-2* 232
Fimbristylis boeckeleri IPNI=*1111946-2* 232
Fimbristylis capillaris IPNI=*1186511-2* 229
Fimbristylis castanea IPNI=*308274-1* 232
Fimbristylis complanata f. *exaltata* IPNI=*1111946-2* 232
Fimbristylis complanata subsp. *keniaeensis* IPNI=*1111946-2* 232
Fimbristylis complanata var. *exaltata* IPNI=*1111946-2* 232
Fimbristylis complanata var. *keniaeensis* IPNI=*1111946-2* 232
Fimbristylis complanata var. *macrocarya* IPNI=*1111946-2* 232
Fimbristylis compressa IPNI=*104972-2* 232
Fimbristylis conferta IPNI=*1111946-2* 232
Fimbristylis cylindrica IPNI=*308274-1* 232
Fimbristylis cylindrostachya IPNI=*1111946-2* 232
Fimbristylis exaltata IPNI=*1111946-2* 232
Fimbristylis hispidula IPNI=*912172-1* 229
Fimbristylis junciformis IPNI=*38049-2* 229
Fimbristylis keniaeensis IPNI=*1111946-2* 232
Fimbristylis kielmaieri IPNI=*1111946-2* 232
Fimbristylis monandra IPNI=*307702-1* 232
Fimbristylis monostachya IPNI=*104972-2* 232
Fimbristylis obscura IPNI=*1111946-2* 232
Fimbristylis pallidula IPNI=*308274-1* 232
Fimbristylis riehleana IPNI=*308274-1* 232
Fimbristylis rigida IPNI=*229810-2* 232
Fimbristylis rudgeana IPNI=*1111946-2* 232
Fimbristylis scabra IPNI=*1111946-2* 232
Fimbristylis spadicea f. *domingensis* IPNI=*308274-1* 232
Fimbristylis spadicea var. *castanea* IPNI=*308274-1* 232
Fimbristylis spadicea var. *longestigmata* IPNI=*308274-1* 232
Fimbristylis speciosa IPNI=*308274-1* 232
Fimbristylis sterilis IPNI=*308274-1* 232
Fimbristylis umbellata IPNI=*308274-1* 232
Fimbrolina incarnata IPNI=*237093-2* 368
Findlaya alba IPNI=*687109-1* 288
Fingalia hexagona IPNI=*202984-1* 261
Firensia fusca IPNI=*114058-1* 274
Firensia hirsuta IPNI=*114394-1* 274
Firensia lutea IPNI=*77164595-1* 416
Firkea rosea IPNI=*61583-2* 385
Firkea venosa IPNI=*315124-2* 384
Fischeria acuminata IPNI=*97427-1* 340
Fischeria boliviana IPNI=*97427-1* 340
Fischeria calycina IPNI=*97427-1* 340
Fischeria hilariana IPNI=*97427-1* 340
Fischeria martiana IPNI=*97427-1* 340
Fischeria multiflora IPNI=*97427-1* 340
Fischeria peruviana IPNI=*97427-1* 340
Fischeria propinqua IPNI=*97427-1* 340
Fischeria riedelii IPNI=*97427-1* 340
Fischeria rotundifolia IPNI=*97427-1* 340
Fischeria setosa IPNI=*97427-1* 340
Fischeria subaequalis IPNI=*97427-1* 340

Fischeria warmingii IPNI=*97427-1* 340
Fissendocarpa linifolia IPNI=*613360-1* 429
Fittonia argyroneura IPNI=*105033-2* 360
Fittonia pearcei IPNI=*105033-2* 360
Fittonia verschaffeltii IPNI=*105033-2* 360
Fitzgeraldia mitrastigma IPNI=*72580-1* 187
Flacourtia afra IPNI=*365348-1* 404
Flacourtia balansae IPNI=*365348-1* 404
Flacourtia calophylla IPNI=*112146-1* 405
Flacourtia celastrina IPNI=*112138-1* 405
Flacourtia elliptica IPNI=*365348-1* 404
Flacourtia flexuosa IPNI=*112138-1* 405
Flacourtia gambecola IPNI=*365348-1* 404
Flacourtia heterophylla IPNI=*365348-1* 404
Flacourtia hilsenbergii IPNI=*365348-1* 404
Flacourtia hirtiuscula IPNI=*365348-1* 404
Flacourtia indica var. *innocua* IPNI=*365348-1* 404
Flacourtia lenis IPNI=*365348-1* 404
Flacourtia lucida IPNI=*365348-1* 404
Flacourtia obcordata IPNI=*365348-1* 404
Flacourtia parvifolia IPNI=*365348-1* 404
Flacourtia perrottetiana IPNI=*365348-1* 404
Flacourtia prunifolia IPNI=*112146-1* 405
Flacourtia racemosa IPNI=*112146-1* 405
Flacourtia ramontchi var. *renvoizei* IPNI=*365348-1* 404
Flacourtia rotundifolia IPNI=*365348-1* 404
Flacourtia sapida IPNI=*365348-1* 404
Flacourtia sepiaria IPNI=*365348-1* 404
Flacourtia spiculifera IPNI=*112178-1* 405
Flacourtia thorelii IPNI=*365348-1* 404
Flemingia affinis IPNI=*495307-1* 320
Flemingia grandiflora IPNI=*56270-1* 362
Fleurya aestuans IPNI=*135476-2* 446
Fleurya aestuans var. *racemosa* IPNI=*135476-2* 446
Fleurya aestuans var. *tuberculata* IPNI=*135476-2* 446
Fleurya caraoellana IPNI=*135476-2* 446
Fleurya cordata IPNI=*135476-2* 446
Fleurya corylifolia IPNI=*135476-2* 446
Fleurya cymosa IPNI=*135476-2* 446
Fleurya glandulosa IPNI=*135476-2* 446
Fleurya ingrata IPNI=*135476-2* 446
Fleurya lurida IPNI=*135476-2* 446
Fleurya perrieri IPNI=*135476-2* 446
Fleurya petiolata IPNI=*135476-2* 446
Fleurya racemosa IPNI=*135476-2* 446
Fleuryopsis petiolata IPNI=*135476-2* 446
Flueckigera peruviana IPNI=*136986-2* 287
Foeniculum azoricum IPNI=*842680-1* 252
Foeniculum capillaceum IPNI=*842680-1* 252
Foeniculum divaricatum IPNI=*842680-1* 252
Foeniculum dulce IPNI=*842680-1* 252
Foeniculum foeniculum IPNI=*842680-1* 252
Foeniculum giganteum IPNI=*842680-1* 252
Foeniculum officinale IPNI=*842680-1* 252
Foeniculum panmorium IPNI=*842680-1* 252
Foeniculum piperitum IPNI=*842680-1* 252
Foeniculum rigidum IPNI=*842680-1* 252
Foeniculum sanguineum IPNI=*842680-1* 252
Foeniculum vulgare convar. *azoricum* IPNI=*842680-1* 252
Foeniculum vulgare subsp. *sativum* IPNI=*842680-1* 252
Foeniculum vulgare var. *sanguincum* IPNI=*842680-1* 252
Forasaccus breviaristatus IPNI=*37001-2* 235
Forasaccus secalinus IPNI=*394095-1* 235
Forgetina guianensis IPNI=*835145-1* 431
Fornicaria scandens IPNI=*241903-1* 269
Fournieria scandens IPNI=*605994-1* 397
Foveolaria cordata IPNI=*826890-1* 308
Foveolaria ovata IPNI=*826890-1* 308
Fragaria abnormis IPNI=*30074127-2* 443
Fragaria aliena IPNI=*30074127-2* 443
Fragaria alpina IPNI=*30074127-2* 443
Fragaria australis IPNI=*318974-2* 443
Fragaria bonariensis IPNI=*724988-1* 443
Fragaria botryformis IPNI=*30074127-2* 443
Fragaria canadensis IPNI=*318974-2* 443
Fragaria chilensis IPNI=*724988-1* 443
Fragaria chiloensis f. *patagonica* IPNI=*724988-1* 443
Fragaria efflagellis IPNI=*30074127-2* 443
Fragaria elatior IPNI=*30074127-2* 443
Fragaria florentina IPNI=*30074127-2* 443
Fragaria gillmanii IPNI=*30074127-2* 443
Fragaria grandiflora IPNI=*318974-2* 443
Fragaria grossa IPNI=*724988-1* 443
Fragaria hortensis IPNI=*30074127-2* 443
Fragaria indica IPNI=*728289-1* 443
Fragaria insularis IPNI=*30074127-2* 443
Fragaria malayana IPNI=*728289-1* 443
Fragaria michauxiana IPNI=*318974-2* 443
Fragaria minor IPNI=*30074127-2* 443
Fragaria multicipita IPNI=*318974-2* 443
Fragaria multiplex IPNI=*30074127-2* 443
Fragaria muricata IPNI=*30074127-2* 443
Fragaria nemoralis IPNI=*30074127-2* 443
Fragaria nilagirica IPNI=*728289-1* 443
Fragaria nuda IPNI=*30074127-2* 443
Fragaria odora IPNI=*318974-2* 443
Fragaria ovalis var. *quinata* IPNI=*318974-2* 443
Fragaria pauciflora IPNI=*318974-2* 443
Fragaria portentosa IPNI=*30074127-2* 443
Fragaria prolifica IPNI=*318974-2* 443
Fragaria pumila IPNI=*318974-2* 443
Fragaria retrorsa IPNI=*30074127-2* 443
Fragaria roseiflora IPNI=*30074127-2* 443
Fragaria roxburghii IPNI=*728289-1* 443
Fragaria sericea IPNI=*724988-1* 443
Fragaria serotina IPNI=*318974-2* 443
Fragaria sibbaldifolia IPNI=*318974-2* 443
Fragaria succulenta IPNI=*30074127-2* 443
Fragaria suksdorfii IPNI=*318974-2* 443
Fragaria sylvestris IPNI=*30074127-2* 443
Fragaria truncata IPNI=*318974-2* 443
Fragaria unifolia IPNI=*30074127-2* 443
Fragaria vesca f. *albida* IPNI=*30074127-2* 443
Fragaria vesca f. *landonii* IPNI=*30074127-2* 443
Fragaria vesca var. *alba* IPNI=*30074127-2* 443
Fragaria vesca var. *chiloensis* IPNI=*724988-1* 443
Fragaria vesca var. *sativa* IPNI=*30074127-2* 443
Fragaria virginiana f. *multicipita* IPNI=*318974-2* 443
Fragaria virginiana IPNI=*318974-2* 443
Fragaria virginiana var. *canadensis* IPNI=*318974-2* 443
Fragaria virginiana var. *sibbaldifolia* IPNI=*318974-2* 443
Fragaria vulgaris IPNI=*30074127-2* 443
Fragaria vulgaris var. *alba* IPNI=*30074127-2* 443
Fragaria yukonensis IPNI=*318974-2* 443
Fragariopsis paxii IPNI=*354970-1* 392
Fragosa biloba IPNI=*838552-1* 251
Fragosa cladorrhiza IPNI=*838571-1* 251
Fragosa crenata IPNI=*838571-1* 251
Fragosa reniformis IPNI=*838571-1* 251
Franchetella anibifolia IPNI=*209864-2* 307
Franchetella crassifolia IPNI=*209758-2* 306
Franchetella gongrijpii IPNI=*209784-2* 307
Franchetella neblinaensis IPNI=*209912-2* 308
Franchetella pubescens IPNI=*948067-1* 307
Franchetella reticulata IPNI=*209864-2* 307
Franchetella unilocularis IPNI=*209864-2* 307

Frangula atroviridis IPNI=*105605-2* 441
Frangula chimantensis IPNI=*105605-2* 441
Franseria ambrosioides IPNI=*176055-1* 256
Franseria arborescens IPNI=*176055-1* 256
Franseria artemisioides IPNI=*176055-1* 256
Franseria conwayi IPNI=*176055-1* 256
Franseria fruticosa IPNI=*176059-1* 257
Franseria meyeniana IPNI=*176059-1* 257
Franseria recurva IPNI=*176059-1* 257
Franseria tacorensis IPNI=*176059-1* 257
Fraxinus americana var. *uhdei* IPNI=*105922-2* 374
Fraxinus borealis IPNI=*609077-1* 374
Fraxinus caudata IPNI=*609077-1* 374
Fraxinus cavekiana IPNI=*105922-2* 374
Fraxinus chiapensis IPNI=*105922-2* 374
Fraxinus hondurensis IPNI=*105922-2* 374
Fraxinus kantoensis IPNI=*609077-1* 374
Fraxinus koehneana IPNI=*609077-1* 374
Fraxinus montana IPNI=*609128-1* 374
Fraxinus obovata IPNI=*609077-1* 374
Fraxinus ovalifolia IPNI=*105922-2* 374
Fraxinus pubinervis IPNI=*609077-1* 374
Fraxinus satsumana IPNI=*609077-1* 374
Fraxinus uhdei var. *pseudoperiptera* IPNI=*105922-2* 374
Fraxinus uhdei var. *typica* IPNI=*105922-2* 374
Fraxinus yamatense IPNI=*609077-1* 374
Freesia hurlingii IPNI=*437030-1* 216
Freesia middlemostii IPNI=*437022-1* 216
Freesia muirii IPNI=*437022-1* 216
Freirea australis IPNI=*854908-1* 446
Freirea debilis IPNI=*854908-1* 446
Freirea erecta IPNI=*854908-1* 446
Freirea gracilis IPNI=*854908-1* 446
Freirea humifusa IPNI=*854908-1* 446
Freirea laxiflora IPNI=*854908-1* 446
Freirea micrantha IPNI=*854908-1* 446
Freirea ruwenzoriensis IPNI=*854908-1* 446
Freirea squalida IPNI=*854908-1* 446
Freirea steudelii IPNI=*854908-1* 446
Freziera dioica IPNI=*429121-1* 397
Freziera dombeyana IPNI=*106032-2* 303
Freziera esmeraldina IPNI=*828818-1* 302
Freziera florida IPNI=*429121-1* 397
Freziera hirsuta IPNI=*106032-2* 303
Freziera sericea IPNI=*106032-2* 303
Freziera sericea var. *typica* IPNI=*106032-2* 303
Frumentum hordeum IPNI=*405391-1* 240
Frumentum sativum IPNI=*405391-1* 240
Frumentum triticum IPNI=*332110-2* 246
Fuchsia amoena IPNI=*106136-2* 429
Fuchsia araucana IPNI=*612572-1* 429
Fuchsia arborea IPNI=*106136-2* 429
Fuchsia arborescens f. *parva* IPNI=*106252-2* 429
Fuchsia arborescens f. *tenuis* IPNI=*106252-2* 429
Fuchsia arborescens f. *typica* IPNI=*106136-2* 429
Fuchsia arborescens var. *megalantha* IPNI=*106252-2* 429
Fuchsia arborescens var. *syringiflora* IPNI=*106252-2* 429
Fuchsia boliviana f. *puberulenta* IPNI=*106150-2* 429
Fuchsia boliviana var. *luxurians* IPNI=*106150-2* 429
Fuchsia boliviana var. *typica* IPNI=*106150-2* 429
Fuchsia caracasana IPNI=*612542-1* 429
Fuchsia chonotica IPNI=*612572-1* 429
Fuchsia coccinea var. *chonotica* IPNI=*612572-1* 429
Fuchsia coccinea var. *macrostemma* IPNI=*612572-1* 429
Fuchsia coccinea var. *robustior* IPNI=*612572-1* 429
Fuchsia conica IPNI=*612572-1* 429
Fuchsia cuspidata IPNI=*106150-2* 429
Fuchsia discolor IPNI=*612572-1* 429
Fuchsia globosa IPNI=*612572-1* 429
Fuchsia gracilis IPNI=*612572-1* 429
Fuchsia gracilis var. *macrostemma* IPNI=*612572-1* 429
Fuchsia gracilis var. *multiflora* IPNI=*612572-1* 429
Fuchsia gracilis var. *tenella* IPNI=*612572-1* 429
Fuchsia hitchcockii IPNI=*612680-1* 429
Fuchsia integrifolia IPNI=*106266-2* 429
Fuchsia lampadaria IPNI=*106221-2* 429
Fuchsia lenneana IPNI=*106150-2* 429
Fuchsia liebmannii IPNI=*106252-2* 429
Fuchsia macrostemma IPNI=*612572-1* 429
Fuchsia macrostemma var. *grandiflora* IPNI=*612572-1* 429
Fuchsia macrostemma var. *parviflora* IPNI=*612572-1* 429
Fuchsia macrostemma var. *tenella* IPNI=*612572-1* 429
Fuchsia magellanica var. *conica* IPNI=*612572-1* 429
Fuchsia magellanica var. *discolor* IPNI=*612572-1* 429
Fuchsia magellanica var. *eburnea* IPNI=*612572-1* 429
Fuchsia magellanica var. *globosa* IPNI=*612572-1* 429
Fuchsia magellanica var. *gracilis* IPNI=*612572-1* 429
Fuchsia magellanica var. *macrostemma* IPNI=*612572-1* 429
Fuchsia magellanica var. *molinae* IPNI=*612572-1* 429
Fuchsia magellanica var. *riccartonii* IPNI=*612572-1* 429
Fuchsia magellanica var. *typica* IPNI=*612572-1* 429
Fuchsia miniata IPNI=*612542-1* 429
Fuchsia multiflora IPNI=*612572-1* 429
Fuchsia munzii IPNI=*612501-1* 429
Fuchsia myrtifolia IPNI=*612572-1* 429
Fuchsia pumila IPNI=*612572-1* 429
Fuchsia pyrifolia IPNI=*106266-2* 429
Fuchsia racemosa IPNI=*106297-2* 429
Fuchsia recurvata IPNI=*612572-1* 429
Fuchsia regia var. *typica* IPNI=*106266-2* 429
Fuchsia ricoartonii IPNI=*612572-1* 429
Fuchsia syringiflora IPNI=*106136-2* 429
Fuchsia tenella IPNI=*612572-1* 429
Fuchsia thompsonii IPNI=*612572-1* 429
Fuchsia velutina IPNI=*612501-1* 429
Fuchsia virgata IPNI=*612572-1* 429
Fuirena appendiculata IPNI=*308552-1* 232
Fuirena brachylepis IPNI=*308552-1* 232
Fuirena brasiliensis IPNI=*308552-1* 232
Fuirena canescens IPNI=*308552-1* 232
Fuirena capitata IPNI=*308552-1* 232
Fuirena complotricha IPNI=*308552-1* 232
Fuirena glomerata IPNI=*308552-1* 232
Fuirena guianensis IPNI=*308552-1* 232
Fuirena haemaglottis IPNI=*308552-1* 232
Fuirena hildebrandtii IPNI=*308552-1* 232
Fuirena mahouxii IPNI=*308552-1* 232
Fuirena multiflora IPNI=*308552-1* 232
Fuirena paniculata IPNI=*308552-1* 232
Fuirena pentagona IPNI=*308552-1* 232
Fuirena philippinensis IPNI=*308552-1* 232
Fuirena quinguangularis IPNI=*308552-1* 232
Fuirena rivularis IPNI=*308552-1* 232
Fuirena seriata IPNI=*308552-1* 232
Fuirena tereticulmis IPNI=*308552-1* 232
Fuirena thouarsiana IPNI=*308552-1* 232
Fuirena umbellata var. *americana* IPNI=*308552-1* 232
Fuirena umbellata var. *pilosa* IPNI=*308552-1* 232
Fuirena umbellata var. *typica* IPNI=*308552-1* 232
Fuirena umbellata var. *unguiculata* IPNI=*308552-1* 232
Fulchironia senegalensis IPNI=*668943-1* 211
Fumaria angustifolia IPNI=*30017594-2* 435
Fumaria capreolata f. *sciophila* IPNI=*672920-1* 435
Fumaria capreolata f. *speciosa* IPNI=*672920-1* 435
Fumaria diffusa IPNI=*30017594-2* 435
Fumaria elegans IPNI=*672920-1* 435

Fumaria gasparinii IPNI=*30017594-2* 435
Fumaria leonardii IPNI=*672920-1* 435
Fumaria media IPNI=*30017594-2* 435
Fumaria media IPNI=*672920-1* 435
Fumaria neapolitana IPNI=*672920-1* 435
Fumaria nemorosa IPNI=*672920-1* 435
Fumaria officinalis var. *capreolata* IPNI=*672920-1* 435
Fumaria officinarum IPNI=*30017594-2* 435
Fumaria pallidiflora IPNI=*672920-1* 435
Fumaria petteri IPNI=*30017594-2* 435
Fumaria platycalyx IPNI=*672920-1* 435
Fumaria prehensilis IPNI=*672920-1* 435
Fumaria pulchella IPNI=*30017594-2* 435
Fumaria queri IPNI=*672920-1* 435
Fumaria speciosa IPNI=*672920-1* 435
Funastrum apiculatum IPNI=*106366-2* 340
Funastrum barbatum IPNI=*106366-2* 340
Funastrum bonariense IPNI=*106366-2* 340
Funastrum crassifolium IPNI=*106366-2* 340
Funastrum cumanense IPNI=*106366-2* 340
Funastrum cuspidatum IPNI=*106366-2* 340
Funastrum dombeyanum IPNI=*106366-2* 340
Funastrum filiforme IPNI=*106366-2* 340
Funastrum fragile IPNI=*106366-2* 340
Funastrum gardneri IPNI=*106366-2* 340
Funastrum glaziovii IPNI=*106366-2* 340
Funastrum lanceolatum IPNI=*106366-2* 340
Funastrum lasianthum IPNI=*106366-2* 340
Funastrum ovalifolium IPNI=*106382-2* 340
Funastrum pallidum IPNI=*106366-2* 340
Funastrum palmeri IPNI=*106366-2* 340
Funastrum pedunculatum IPNI=*106366-2* 340
Funastrum pubescens IPNI=*106366-2* 340
Funastrum riparium IPNI=*106366-2* 340
Funastrum schottii IPNI=*106366-2* 340
Funastrum seibertii IPNI=*106366-2* 340
Funium piliferum IPNI=*64429-1* 215
Furcaria cannabina IPNI=*560068-1* 409
Furcaria cavanillesii IPNI=*560068-1* 409
Furcaria sabdariffa IPNI=*326388-2* 409
Furcaria thalictroides IPNI=*17064190-1* 178
Furcraea agavephylla IPNI=*105442-2* 215
Furcraea aspera IPNI=*105442-2* 215
Furcraea barillettii IPNI=*64429-1* 215
Furcraea cahum IPNI=*105442-2* 215
Furcraea commelyni IPNI=*64429-1* 215
Furcraea cubensis IPNI=*105442-2* 215
Furcraea cubensis var. *inermis* IPNI=*105442-2* 215
Furcraea deledevantii IPNI=*106421-2* 215
Furcraea gigantea IPNI=*64429-1* 215
Furcraea lindenii IPNI=*64452-1* 215
Furcraea macrophylla IPNI=*105442-2* 215
Furcraea madagascariensis IPNI=*64429-1* 215
Furcraea valleculata IPNI=*105442-2* 215
Furcraea viridis IPNI=*64429-1* 215
Fusaea rhombipetala IPNI=*106440-2* 188
Fussia caryophyllea IPNI=*386633-1* 233
Fusticus glabra IPNI=*1151750-2* 440
Fusticus tataiba IPNI=*1151750-2* 440
Fusticus tinctoria IPNI=*1151750-2* 440
Fusticus vera IPNI=*1151750-2* 440
Fusticus zanthoxylon IPNI=*1151750-2* 440

G

Gaedawakka schomburgkiana IPNI=*54072-2* 401
Gaerdtia argentea IPNI=*105130-1* 292
Gaerdtia maculata IPNI=*105130-1* 292
Gaertnera hongkongensis IPNI=*753834-1* 352
Gaertneria artemisiodes IPNI=*176055-1* 256
Gaertneria fruticosa IPNI=*176055-1* 256
Gaertneria meyeniana IPNI=*176055-1* 256
Gaga marginata IPNI=*77182463-1* 178
Gaiadendron breviflorum IPNI=*548708-1* 448
Gaiadendron ellipticum IPNI=*548708-1* 448
Gaiadendron eugenioides IPNI=*258636-2* 449
Gaiadendron lanceolatum IPNI=*548708-1* 448
Gaiadendron laurifolium IPNI=*548708-1* 448
Gaiadendron macranthum IPNI=*548708-1* 448
Gaiadendron nitidum IPNI=*548708-1* 448
Gaiadendron poasense IPNI=*548708-1* 448
Gaiadendron punctatum var. *puracense* IPNI=*548708-1* 448
Gaiadendron puracense IPNI=*548708-1* 448
Gaiadendron revolutum IPNI=*548708-1* 448
Gaiadendron tagua IPNI=*548708-1* 448
Gaiadendron tagua var. *reductum* IPNI=*548708-1* 448
Gaiadendron tagua var. *revolutum* IPNI=*548708-1* 448
Gaillardia bicolor integrifolia IPNI=*329185-2* 263
Gaillardia bicolor IPNI=*329185-2* 263
Gaillardia bicolor var. *drummondii* IPNI=*329185-2* 263
Gaillardia drummondii IPNI=*329185-2* 263
Gaillardia lobata IPNI=*329185-2* 263
Gaillardia neomexicana IPNI=*329185-2* 263
Gaillardia picta IPNI=*329185-2* 263
Gaillardia pulchella f. *flaviflora* IPNI=*329185-2* 263
Gaillardia pulchella f. *lorenziana* IPNI=*329185-2* 263
Gaillardia pulchella f. *picta* IPNI=*329185-2* 263
Gaillardia pulchella var. *albiflora* IPNI=*329185-2* 263
Gaillardia pulchella var. *australis* IPNI=*329185-2* 263
Gaillardia pulchella var. *drummondii* IPNI=*329185-2* 263
Gaillardia pulchella var. *picta* IPNI=*329185-2* 263
Gaillardia pulchella var. *simplex* IPNI=*329185-2* 263
Gaillardia pulchella var. *tubulosa* IPNI=*329185-2* 263
Gaillardia scabrosa IPNI=*329185-2* 263
Gaillardia villosa IPNI=*329185-2* 263
Galactodendrum utile IPNI=*582724-1* 437
Galarhoeus cyparissias IPNI=*346205-1* 390
Galarhoeus decussatus IPNI=*347072-1* 390
Galarhoeus fischerianus IPNI=*346510-1* 390
Galarhoeus lathyris IPNI=*347072-1* 390
Galarhoeus pallasii IPNI=*346510-1* 390
Galarhoeus peplus IPNI=*347724-1* 390
Galarhoeus seguierianus IPNI=*348221-1* 391
Galatea americana IPNI=*89940-2* 216
Galatea bulbosa IPNI=*89940-2* 216
Galatea vespertina IPNI=*89940-2* 216
Galatella tenuifolia IPNI=*198320-1* 260
Galega caribaea IPNI=*275194-2* 315
Galega cinerea IPNI=*1071691-2* 335
Galega coerulea IPNI=*76086-2* 317
Galega littoralis IPNI=*1071691-2* 335
Galega purpurea IPNI=*520851-1* 335
Galega sinapou IPNI=*250423-2* 335
Galega toxicaria IPNI=*250423-2* 335
Galeobdolon amplexicaule IPNI=*448797-1* 370
Galeobdolon galeobdolon IPNI=*60452701-2* 370
Galeopsis galeobdolon IPNI=*60452701-2* 370
Galinsoga acmella IPNI=*208554-1* 263
Galinsoga allaeocarpa IPNI=*208557-1* 263
Galinsoga aristulata IPNI=*208557-1* 263
Galinsoga bicolorata IPNI=*208557-1* 263
Galinsoga brachystephana IPNI=*208557-1* 263
Galinsoga calva IPNI=*208554-1* 263
Galinsoga caracasana IPNI=*208557-1* 263
Galinsoga ciliata f. *vargasiana* IPNI=*208557-1* 263
Galinsoga ciliata IPNI=*208557-1* 263
Galinsoga eligulata IPNI=*208557-1* 263

Galinsoga hirsuta IPNI=*208554-1* 263
Galinsoga hispida IPNI=*208557-1* 263
Galinsoga humboldtii IPNI=*208557-1* 263
Galinsoga laciniata IPNI=*208554-1* 263
Galinsoga oblongifolia IPNI=*1186403-2* 261
Galinsoga parviflora f. *parceglandulosa* IPNI=*208554-1* 263
Galinsoga parviflora f. *subeglandulosa* IPNI=*208554-1* 263
Galinsoga parviflora subsp. *hispida* IPNI=*208557-1* 263
Galinsoga parviflora var. *adenophora* IPNI=*208554-1* 263
Galinsoga parviflora var. *caracasana* IPNI=*208557-1* 263
Galinsoga parviflora var. *hispida* IPNI=*208557-1* 263
Galinsoga parviflora var. *semicalva* IPNI=*208554-1* 263
Galinsoga plikeri IPNI=*208557-1* 263
Galinsoga purpurea IPNI=*208557-1* 263
Galinsoga quadriradiata f. *albiflora* IPNI=*208557-1* 263
Galinsoga quadriradiata f. *vargasiana* IPNI=*208557-1* 263
Galinsoga quadriradiata var. *hispida* IPNI=*208557-1* 263
Galinsoga quinqueradiata IPNI=*208554-1* 263
Galinsoga semicalva IPNI=*208554-1* 263
Galinsoga semicalva var. *percalva* IPNI=*208554-1* 263
Galinsoga urticifolia IPNI=*208557-1* 263
Galion aparinum IPNI=*30007294-2* 350
Galium aculeatissimum IPNI=*30007294-2* 350
Galium adhaerens IPNI=*30007294-2* 350
Galium aparine var. *intermedium* IPNI=*30007294-2* 350
Galium aparine var. *pseudoaparine* IPNI=*30007294-2* 350
Galium apiculatum IPNI=*749770-1* 350
Galium asperum IPNI=*30007294-2* 350
Galium benthamianum IPNI=*749770-1* 350
Galium charoides IPNI=*30007294-2* 350
Galium chilense IPNI=*30007294-2* 350
Galium chonosense IPNI=*30007294-2* 350
Galium ciliatum IPNI=*749770-1* 350
Galium ciliatum var. *normale* IPNI=*749770-1* 350
Galium ciliatum var. *weddellianum* IPNI=*749770-1* 350
Galium cruciata IPNI=*747959-1* 349
Galium cruciata var. *brutium* IPNI=*747959-1* 349
Galium hirsutum IPNI=*749770-1* 350
Galium hispidum IPNI=*30007294-2* 350
Galium horridum IPNI=*30007294-2* 350
Galium involucratum IPNI=*749770-1* 350
Galium lappaceum IPNI=*30007294-2* 350
Galium larecajense IPNI=*30007294-2* 350
Galium leucocarpum IPNI=*749770-1* 350
Galium luteocruciatum IPNI=*747959-1* 349
Galium monanthon IPNI=*749770-1* 350
Galium montanum IPNI=*749770-1* 350
Galium mucronatum IPNI=*749770-1* 350
Galium parviflorum IPNI=*30007294-2* 350
Galium pseudoaparine IPNI=*30007294-2* 350
Galium segetum IPNI=*30007294-2* 350
Galium tarmense IPNI=*749770-1* 350
Galium tenerrimum IPNI=*30007294-2* 350
Galium uliginosum IPNI=*30007294-2* 350
Galium uncinatum IPNI=*30007294-2* 350
Galium valantia IPNI=*747959-1* 349
Galliaria adscendens IPNI=*316349-2* 282
Galliaria hybrida IPNI=*10648-2* 281
Galliaria patula IPNI=*10648-2* 281
Galliaria spitosa IPNI=*10711-2* 281
Gallitrichum anglicum IPNI=*457495-1* 373
Gallitrichum arvale IPNI=*457495-1* 373
Gallitrichum candollei IPNI=*457495-1* 373
Gallitrichum clandestinum IPNI=*457495-1* 373
Gallitrichum dichroanthum IPNI=*457495-1* 373
Gallitrichum horminioides IPNI=*457495-1* 373
Gallitrichum maculatum IPNI=*457495-1* 373
Gallitrichum pallidiflorum IPNI=*457495-1* 373
Gallitrichum ptychophyllum IPNI=*457495-1* 373
Gallitrichum rosulatum IPNI=*457495-1* 373
Gallitrichum rubellum IPNI=*457495-1* 373
Gallitrichum stereocaulon IPNI=*457495-1* 373
Gallitrichum verbenacum IPNI=*457495-1* 373
Gallitrichum virgatum IPNI=*457495-1* 373
Galoglychia martinicensis IPNI=*853024-1* 438
Galordia alternifolia IPNI=*329185-2* 263
Galphimia chrysophylla IPNI=*556014-1* 395
Galphimia glandulosa IPNI=*556362-1* 396
Galphimia glauca IPNI=*556349-1* 396
Galphimia gracilis var. *gracillima* IPNI=*556349-1* 396
Galphimia humboldtiana IPNI=*556362-1* 396
Galphimia latifolia IPNI=*556362-1* 396
Galphimia paniculata var. *glandulosa* IPNI=*556362-1* 396
Galphimia paniculata var. *humboldtiana* IPNI=*556362-1* 396
Galvania vandellii IPNI=*759061-1* 353
Galvania vellosii IPNI=*759061-1* 353
Gambeya excelsa IPNI=*948088-1* 305
Gambeya mammosa IPNI=*152641-2* 306
Gamochaeta americana subvar. *interrupta* IPNI=*1213296-2* 263
Gamochaeta americana var. *alpina* IPNI=*1213296-2* 263
Gamochaeta americana var. *vulgaris* IPNI=*1213296-2* 263
Gamochaeta guatemalensis IPNI=*1213296-2* 263
Gamochaeta irazuensis IPNI=*1213296-2* 263
Gamochaeta paramora IPNI=*77154313-1* 268
Gamochaeta rosacea IPNI=*107379-2* 263
Gamochaeta spicata IPNI=*943810-1* 263
Gamotopea alba IPNI=*60473597-2* 353
Gandasulium angustifolium IPNI=*796834-1* 251
Gandasulium coccineum IPNI=*796834-1* 251
Gandasulium coronarium IPNI=*796836-1* 251
Gandasulium lingulatum IPNI=*796836-1* 251
Gandola nigra IPNI=*164286-1* 283
Garcia mayana IPNI=*349162-1* 391
Garcinia brasiliensis var. *parvifolia* IPNI=*427844-1* 385
Garcinia megaphylla IPNI=*428065-1* 385
Gardenia angustifolia IPNI=*751073-1* 350
Gardenia armata IPNI=*1115268-2* 356
Gardenia augusta f. *shanpinensis* IPNI=*751073-1* 350
Gardenia augusta IPNI=*751073-1* 350
Gardenia brasiliensis IPNI=*751331-1* 351
Gardenia dumetorum IPNI=*745904-1* 348
Gardenia dumosa IPNI=*745904-1* 348
Gardenia edulis IPNI=*743001-1* 347
Gardenia fagifolia IPNI=*764273-1* 356
Gardenia floribunda IPNI=*745904-1* 348
Gardenia florida IPNI=*751073-1* 350
Gardenia genipa IPNI=*751331-1* 351
Gardenia glabra IPNI=*745904-1* 348
Gardenia grandiflora IPNI=*751073-1* 350
Gardenia hexagona IPNI=*67491-2* 349
Gardenia humboldtiana IPNI=*764273-1* 356
Gardenia jasminoides f. *kueishanensis* IPNI=*751073-1* 350
Gardenia jasminoides f. *longicarpa* IPNI=*751073-1* 350
Gardenia latifolia IPNI=*745904-1* 348
Gardenia longiflora IPNI=*222509-2* 356
Gardenia longisepala IPNI=*751073-1* 350
Gardenia mussaenda IPNI=*222508-2* 356
Gardenia oblongifolia IPNI=*751331-1* 351
Gardenia pictorum IPNI=*751073-1* 350
Gardenia radicans IPNI=*751073-1* 350
Gardenia randia IPNI=*763958-1* 356
Gardenia spinosa IPNI=*745904-1* 348
Gardenia stipularis IPNI=*745904-1* 348
Gardoquia glabrata IPNI=*1009966-1* 369
Gardoquia origanoides IPNI=*30135678-2* 379
Gardoquia taxifolia IPNI=*1009966-1* 369

Gardoquia taxifolia var. *loxensis* IPNI=*1009966-1* 369
Garuga gigantea IPNI=*71652-1* 451
Garuga schomburgkiana IPNI=*71652-1* 451
Garuga schomburgkiana var. *salzmanniana* IPNI=*71652-1* 451
Garuga spruceana IPNI=*71652-1* 451
Gasteranthus corallinoides IPNI=*107520-2* 367
Gasteranthus crenatus IPNI=*107520-2* 367
Gasteranthus ecuadorensis IPNI=*107533-2* 367
Gasteranthus magentatus IPNI=*107533-2* 367
Gasteranthus sylvarum IPNI=*107520-2* 367
Gaultheria alpina IPNI=*330618-1* 299
Gaultheria antioquiensis IPNI=*330527-1* 299
Gaultheria apiculata IPNI=*330527-1* 299
Gaultheria bolivari IPNI=*330535-1* 299
Gaultheria brachybotrys IPNI=*107641-2* 299
Gaultheria ciliata IPNI=*330618-1* 299
Gaultheria coccinea IPNI=*330527-1* 299
Gaultheria conferta IPNI=*330459-1* 299
Gaultheria congesta IPNI=*330535-1* 299
Gaultheria conzattii IPNI=*330527-1* 299
Gaultheria conzattii var. *mijorum* IPNI=*330527-1* 299
Gaultheria cordata IPNI=*330527-1* 299
Gaultheria cordifolia IPNI=*330527-1* 299
Gaultheria cordifrons IPNI=*330527-1* 299
Gaultheria costaricensis IPNI=*330527-1* 299
Gaultheria cummingii IPNI=*330527-1* 299
Gaultheria donnellii IPNI=*330527-1* 299
Gaultheria epigeoides IPNI=*107641-2* 299
Gaultheria fendleri IPNI=*330527-1* 299
Gaultheria formosa IPNI=*330527-1* 299
Gaultheria glandulifera IPNI=*330527-1* 299
Gaultheria glandulosissima IPNI=*330527-1* 299
Gaultheria hartwegiana IPNI=*330527-1* 299
Gaultheria hidalgensis IPNI=*330527-1* 299
Gaultheria hirsuta IPNI=*330618-1* 299
Gaultheria hirtiflora IPNI=*330527-1* 299
Gaultheria jelskii IPNI=*330527-1* 299
Gaultheria lancifolia IPNI=*330527-1* 299
Gaultheria lancifolia var. *dulcis* IPNI=*330527-1* 299
Gaultheria lepida IPNI=*330527-1* 299
Gaultheria longiflora IPNI=*330527-1* 299
Gaultheria longipes IPNI=*330527-1* 299
Gaultheria loxensis IPNI=*330527-1* 299
Gaultheria meridensis IPNI=*330527-1* 299
Gaultheria microdonta IPNI=*330459-1* 299
Gaultheria montana IPNI=*330527-1* 299
Gaultheria mutisiana IPNI=*107644-2* 299
Gaultheria odorata IPNI=*330527-1* 299
Gaultheria odorata var. *costaricensis* IPNI=*330527-1* 299
Gaultheria odorata var. *leiocalyx* IPNI=*330527-1* 299
Gaultheria odorata var. *mexicana* IPNI=*330527-1* 299
Gaultheria opaca IPNI=*330527-1* 299
Gaultheria ornata IPNI=*330527-1* 299
Gaultheria parvifolia IPNI=*330527-1* 299
Gaultheria pennellii IPNI=*330527-1* 299
Gaultheria petraca IPNI=*107641-2* 299
Gaultheria pichinchensis IPNI=*107641-2* 299
Gaultheria pilosa IPNI=*330527-1* 299
Gaultheria poasana IPNI=*330527-1* 299
Gaultheria psilantha IPNI=*330676-1* 299
Gaultheria pubiflora IPNI=*330459-1* 299
Gaultheria ramosissima IPNI=*330459-1* 299
Gaultheria regia IPNI=*330527-1* 299
Gaultheria roraimae IPNI=*330527-1* 299
Gaultheria saltensis IPNI=*330527-1* 299
Gaultheria sanmartansis IPNI=*330618-1* 299
Gaultheria scabra IPNI=*330527-1* 299
Gaultheria schiedeana IPNI=*330527-1* 299
Gaultheria stuebeliana IPNI=*107644-2* 299
Gaultheria subrotunda IPNI=*330527-1* 299
Gaultheria tacanensis IPNI=*330527-1* 299
Gaultheria tatei IPNI=*330527-1* 299
Gaultheria tetriches IPNI=*330527-1* 299
Gaultheria trichocalycina IPNI=*330527-1* 299
Gaultheria vegasana IPNI=*330527-1* 299
Gaultheria vestita IPNI=*330527-1* 299
Gaultheria weberbaueriana IPNI=*330527-1* 299
Gaura epilobia IPNI=*173348-2* 430
Gaura fruticosa IPNI=*63788-2* 416
Gaura laxa IPNI=*63788-2* 416
Gaylussacia cacuminis IPNI=*858406-1* 299
Gaylussacia caracasana IPNI=*858406-1* 299
Gaylussacia microphylla IPNI=*82286-2* 299
Gayoides crispum IPNI=*120630-2* 409
Gazania coronopifolia IPNI=*941423-1* 263
Gazania grandis IPNI=*941423-1* 263
Gazania heterophylla IPNI=*208720-1* 263
Gazania kraussii IPNI=*208700-1* 263
Gazania longiscapa IPNI=*208700-1* 263
Gazania multijuga IPNI=*208700-1* 263
Gazania pectinata IPNI=*941423-1* 263
Gazania pinnata IPNI=*941423-1* 263
Gazania speciosa IPNI=*941423-1* 263
Gazania stenophylla IPNI=*208700-1* 263
Gazania subulata IPNI=*208700-1* 263
Gazaniopsis stenophylla IPNI=*208700-1* 263
Geboscon obliquum IPNI=*528506-1* 213
Geiseleria chamaedrifolia IPNI=*71013-2* 389
Geiseleria chamaedryfolia IPNI=*1012504-2* 387
Geiseleria corchorifolia IPNI=*71013-2* 389
Geiseleria glandulosa IPNI=*70233-2* 388
Geissospermum myristicifolium IPNI=*77411-1* 339
Gelseminum araliaceum IPNI=*77085096-1* 364
Gelseminum avellanedae IPNI=*117330-2* 363
Gelseminum capense IPNI=*111333-1* 365
Gelseminum chrysotrichum IPNI=*109691-1* 363
Gelseminum insigne IPNI=*248703-2* 365
Gelseminum stans IPNI=*111284-1* 365
Gendarussa arenaria IPNI=*52612-1* 360
Gendarussa betonica IPNI=*52612-1* 360
Gendarussa ramosissima IPNI=*52612-1* 360
Gendarussa trinervia IPNI=*52612-1* 360
Geniosporum tenuiflorum IPNI=*453130-1* 371
Genipa aculeata IPNI=*763958-1* 356
Genipa americana f. *grandifolia* IPNI=*751331-1* 351
Genipa americana f. *jorgensenii* IPNI=*751331-1* 351
Genipa americana f. *parvifolia* IPNI=*751331-1* 351
Genipa americana var. *caruto* IPNI=*751331-1* 351
Genipa americana var. *riobranquensis* IPNI=*751331-1* 351
Genipa barbata IPNI=*751331-1* 351
Genipa brasiliana IPNI=*751331-1* 351
Genipa brasiliensis IPNI=*751331-1* 351
Genipa caruto IPNI=*751331-1* 351
Genipa codonocalyx IPNI=*751331-1* 351
Genipa curviflora IPNI=*1011811-2* 346
Genipa dumetorum IPNI=*745904-1* 348
Genipa edulis IPNI=*743001-1* 347
Genipa excelsa IPNI=*751331-1* 351
Genipa florida IPNI=*751073-1* 350
Genipa grandiflora IPNI=*751073-1* 350
Genipa grandifolia IPNI=*751331-1* 351
Genipa humilis IPNI=*751331-1* 351
Genipa maxonii IPNI=*236581-2* 357
Genipa mussaendae IPNI=*222508-2* 356
Genipa oblongifolia IPNI=*751331-1* 351
Genipa oleosa IPNI=*751331-1* 351

Genipa pubescens IPNI=*751331-1* 351
Genipa radicans IPNI=*751073-1* 350
Genipa venosa IPNI=*751331-1* 351
Genista acutifolia IPNI=*519057-1* 334
Genista americana IPNI=*519057-1* 334
Genista candicans IPNI=*496189-1* 320
Genista discolor IPNI=*495939-1* 320
Genista eriocarpa IPNI=*496189-1* 320
Genista europaea IPNI=*524298-1* 336
Genista hillebrandii IPNI=*495939-1* 320
Genista hispanica IPNI=*519057-1* 334
Genista juncea IPNI=*519057-1* 334
Genista monspessulana var. *colmeiroi* IPNI=*496189-1* 320
Genista odorata IPNI=*519057-1* 334
Genista odoratissima IPNI=*519057-1* 334
Genista paniculata IPNI=*495939-1* 320
Genista philippii IPNI=*495939-1* 320
Genista ramosissima IPNI=*495939-1* 320
Genista rhodopnoa IPNI=*495939-1* 320
Genista syriaca IPNI=*496189-1* 320
Gentiana aggregata IPNI=*108639-2* 344
Gentiana ajanensis IPNI=*108639-2* 344
Gentiana amarella IPNI=*108639-2* 344
Gentiana amarella var. *lingulata* IPNI=*108639-2* 344
Gentiana amarella var. *stricta* IPNI=*108639-2* 344
Gentiana amarella var. *tenuis* IPNI=*108639-2* 344
Gentiana aphylla IPNI=*1167766-2* 345
Gentiana arbelaezii IPNI=*108639-2* 344
Gentiana autumnalis IPNI=*108639-2* 344
Gentiana axillaris IPNI=*108639-2* 344
Gentiana californica IPNI=*108639-2* 344
Gentiana campanulata IPNI=*108639-2* 344
Gentiana campestris IPNI=*108639-2* 344
Gentiana casapaltensis IPNI=*317620-2* 344
Gentiana centaurium IPNI=*326619-2* 344
Gentiana cerastioides IPNI=*108656-2* 344
Gentiana cerastioides var. *chimborazensis* IPNI=*108656-2* 344
Gentiana cerastioides var. *emarginata* IPNI=*108656-2* 344
Gentiana cerastioides var. *humilis* IPNI=*108656-2* 344
Gentiana cerastioides var. *minor* IPNI=*108656-2* 344
Gentiana cespitosa IPNI=*317620-2* 344
Gentiana chimboracensis IPNI=*317620-2* 344
Gentiana cistiflora IPNI=*108656-2* 344
Gentiana cocuyana IPNI=*108661-2* 344
Gentiana conferta IPNI=*108639-2* 344
Gentiana congesta IPNI=*108661-2* 344
Gentiana connata IPNI=*367440-1* 344
Gentiana connectens IPNI=*108639-2* 344
Gentiana copelandii IPNI=*108639-2* 344
Gentiana corymbosa IPNI=*108661-2* 344
Gentiana dacrydioides IPNI=*108667-2* 344
Gentiana dasyantha IPNI=*108732-2* 345
Gentiana diffusa IPNI=*366552-1* 344
Gentiana distegia IPNI=*108639-2* 344
Gentiana engleri IPNI=*108667-2* 344
Gentiana exaltata IPNI=*367440-1* 344
Gentiana eximia IPNI=*108639-2* 344
Gentiana flava IPNI=*108639-2* 344
Gentiana flavescens IPNI=*108639-2* 344
Gentiana gayi IPNI=*317620-2* 344
Gentiana gayi subsp. *magellanica* IPNI=*317620-2* 344
Gentiana gerardii IPNI=*326619-2* 344
Gentiana gracilis IPNI=*108639-2* 344
Gentiana hybrida IPNI=*108639-2* 344
Gentiana labradorica IPNI=*108639-2* 344
Gentiana laevicalyx IPNI=*108639-2* 344
Gentiana lancifolia IPNI=*108639-2* 344
Gentiana lingulata IPNI=*108639-2* 344
Gentiana microphylla IPNI=*317620-2* 344
Gentiana minima IPNI=*108504-2* 344
Gentiana nevadensis IPNI=*108708-2* 345
Gentiana nigrescens IPNI=*108639-2* 344
Gentiana obtusifolia IPNI=*108639-2* 344
Gentiana palustris IPNI=*326619-2* 344
Gentiana plebeia var. *holmii* IPNI=*108639-2* 344
Gentiana podocarpa IPNI=*317620-2* 344
Gentiana polyantha IPNI=*108639-2* 344
Gentiana pratensis IPNI=*108639-2* 344
Gentiana prostrata f. *magellanica* IPNI=*317620-2* 344
Gentiana prostrata subsp. *americana* IPNI=*108504-2* 344
Gentiana prostrata subsp. *podocarpa* IPNI=*317620-2* 344
Gentiana prostrata var. *americana* IPNI=*108504-2* 344
Gentiana prostrata var. *podocarpa* IPNI=*317620-2* 344
Gentiana prostrata var. *typica* IPNI=*108504-2* 344
Gentiana pulchella IPNI=*108639-2* 344
Gentiana purpurea IPNI=*108639-2* 344
Gentiana pyramidalis IPNI=*108639-2* 344
Gentiana ramosissima IPNI=*317620-2* 344
Gentiana remota IPNI=*108639-2* 344
Gentiana rupicola IPNI=*108656-2* 344
Gentiana sedifolia f. *acaulis* IPNI=*317620-2* 344
Gentiana sedifolia subvar. *imbricata* IPNI=*317620-2* 344
Gentiana sedifolia subvar. *splendens* IPNI=*317620-2* 344
Gentiana sedifolia var. *casapaltensis* IPNI=*317620-2* 344
Gentiana sedifolia var. *compacta* IPNI=*317620-2* 344
Gentiana sedifolia var. *elongata* IPNI=*317620-2* 344
Gentiana sedifolia var. *grandiflora* IPNI=*317620-2* 344
Gentiana sedifolia var. *imbricata* IPNI=*317620-2* 344
Gentiana sedifolia var. *lineata* IPNI=*317620-2* 344
Gentiana sedifolia var. *micrantha* IPNI=*317620-2* 344
Gentiana sedifolia var. *microphylla* IPNI=*317620-2* 344
Gentiana sedifolia var. *nana* IPNI=*317620-2* 344
Gentiana selaginifolia IPNI=*108732-2* 345
Gentiana seminuda IPNI=*108639-2* 344
Gentiana stricta IPNI=*108639-2* 344
Gentiana sylvestris IPNI=*108639-2* 344
Gentiana tatei IPNI=*108639-2* 344
Gentiana tenuiflora IPNI=*108639-2* 344
Gentiana tenuis IPNI=*108639-2* 344
Gentiana tetragona IPNI=*108639-2* 344
Gentianella amarella subsp. *euamarella* IPNI=*108639-2* 344
Gentianella amarella var. *cornubiensis* IPNI=*108639-2* 344
Gentianella amarella var. *lingulata* IPNI=*108639-2* 344
Gentianella anglica subsp. *cornubiensis* IPNI=*108639-2* 344
Gentianella axillaris IPNI=*108639-2* 344
Gentianella dasyantha IPNI=*108732-2* 345
Gentianella davidiana IPNI=*108639-2* 344
Gentianella hispanica IPNI=*108639-2* 344
Gentianella lingulata IPNI=*108639-2* 344
Gentianusa amarella IPNI=*108639-2* 344
Gentianusa flava IPNI=*108504-2* 344
Gentianusa lancifolia IPNI=*108639-2* 344
Geocardia picta IPNI=*60473597-2* 353
Geocardia repens IPNI=*109024-2* 351
Geocardia violacea IPNI=*759372-1* 354
Geocardia violifolia IPNI=*1080914-2* 351
Geoffroea acutifolia IPNI=*1075415-2* 311
Geoffroea bredemeyeri IPNI=*496501-1* 320
Geoffroea inermis IPNI=*1075415-2* 311
Geoffroea obtusifolia IPNI=*12607-2* 311
Geoffroea pubescens IPNI=*12607-2* 311
Geoffroea striata IPNI=*496501-1* 320
Geoffroea superba IPNI=*496501-1* 320
Geoffroea surinamensis IPNI=*12607-2* 311
Geonoma acaulis IPNI=*667158-1* 209
Geonoma acaulis subsp. *tapajotensis* IPNI=*667158-1* 209

Geonoma atrovirens IPNI=*667158-1* 209
Geonoma corallifera IPNI=*54867-2* 207
Geonoma ecuadoriensis IPNI=*667158-1* 209
Geonoma lagesiana IPNI=*667031-1* 209
Geonoma latifrons IPNI=*54867-2* 207
Geonoma macrostachys var. *acaulis* IPNI=*667158-1* 209
Geonoma martiana IPNI=*664278-1* 204
Geonoma oligoclada IPNI=*667216-1* 209
Geonoma polyandra IPNI=*108938-2* 209
Geonoma saramaccana IPNI=*667140-1* 209
Geonoma supracostata IPNI=*667158-1* 209
Geonoma synanthera IPNI=*193857-2* 211
Geonoma tamandua IPNI=*667158-1* 209
Geonoma trifurcata IPNI=*664278-1* 204
Geonoma woronowii IPNI=*667158-1* 209
Geophila herbacea var. *violifolia* IPNI=*1080914-2* 351
Geophila paraensis IPNI=*60473597-2* 353
Geophila picta IPNI=*60473597-2* 353
Geophila reniformis var. *americana* IPNI=*109024-2* 351
Geophila repens var. *violifolia* IPNI=*1080914-2* 351
Geophila violacea IPNI=*109024-2* 351
Geophila violifolia IPNI=*1080914-2* 351
Georgia bipinnata IPNI=*198320-1* 260
Georgia variabilis IPNI=*30027297-2* 261
Georgina astrantiiflora IPNI=*30027297-2* 261
Georgina purpurea IPNI=*30027297-2* 261
Georgina rosea IPNI=*30027297-2* 261
Georgina superflua IPNI=*30027297-2* 261
Georgina superflua var. *flavescens* IPNI=*30027297-2* 261
Georgina superflua var. *lilacina* IPNI=*30027297-2* 261
Georgina superflua var. *pallida* IPNI=*30027297-2* 261
Georgina superflua var. *purpurea* IPNI=*30027297-2* 261
Georgina superflua var. *rubra* IPNI=*30027297-2* 261
Georgina variabilis IPNI=*30027297-2* 261
Georgina variabilis var. *lilacina* IPNI=*30027297-2* 261
Georgina variabilis var. *pallida* IPNI=*30027297-2* 261
Georgina variabilis var. *purpurea* IPNI=*30027297-2* 261
Geraniospermum angulosum IPNI=*164626-3* 358
Geraniospermum capitatum IPNI=*375818-1* 358
Geraniospermum clavatum IPNI=*375818-1* 358
Geraniospermum cucculatum IPNI=*164626-3* 358
Geraniospermum odoratissimum IPNI=*376654-1* 358
Geraniospermum peltatum IPNI=*376728-1* 358
Geraniospermum terebintaceum IPNI=*376259-1* 358
Geraniospermum zonale IPNI=*377229-1* 358
Geranium × *crenatum* IPNI=*376335-1* 358
Geranium × *hybridum* IPNI=*376335-1* 358
Geranium × *miniatum* IPNI=*376335-1* 358
Geranium aequatoriale IPNI=*109179-2* 358
Geranium africanum IPNI=*376654-1* 358
Geranium angulosum IPNI=*164626-3* 358
Geranium arenicola IPNI=*372219-1* 358
Geranium aristatum IPNI=*373129-1* 358
Geranium aristisepalum IPNI=*373129-1* 358
Geranium aristisepalum var. *michoacanum* IPNI=*373129-1* 358
Geranium aristisepalum var. *typicum* IPNI=*373129-1* 358
Geranium asperum IPNI=*376259-1* 358
Geranium bicknellii var. *longipes* IPNI=*109083-2* 358
Geranium capitatum IPNI=*375818-1* 358
Geranium carolinianum var. *longipes* IPNI=*109083-2* 358
Geranium chaerophyllum IPNI=*372219-1* 358
Geranium chamaense IPNI=*372978-1* 358
Geranium chimborazense IPNI=*372978-1* 358
Geranium cicutarium IPNI=*372219-1* 358
Geranium cicutarium var. *moschatum* IPNI=*372354-1* 358
Geranium citriodorum IPNI=*164626-3* 358
Geranium columbianum IPNI=*372978-1* 358
Geranium cucullatum IPNI=*164626-3* 358
Geranium diffusum var. *sprucei* IPNI=*372978-1* 358
Geranium diffusum var. *subsericeum* IPNI=*372978-1* 358
Geranium elongatum IPNI=*109179-2* 358
Geranium fimbriatum IPNI=*164626-3* 358
Geranium grandiflorum IPNI=*373138-1* 358
Geranium harmsii IPNI=*372978-1* 358
Geranium hernandesii var. *michoacanum* IPNI=*373129-1* 358
Geranium hirtum IPNI=*373129-1* 358
Geranium holm-nielsenii IPNI=*372978-1* 358
Geranium imbaburae IPNI=*372978-1* 358
Geranium irbyanum IPNI=*164626-3* 358
Geranium knuthianum IPNI=*109179-2* 358
Geranium longipes IPNI=*109083-2* 358
Geranium marginatum IPNI=*377229-1* 358
Geranium moschatum IPNI=*372354-1* 358
Geranium nemorale IPNI=*109083-2* 358
Geranium nemorale var. *bicknellii* IPNI=*109083-2* 358
Geranium odoratissimum IPNI=*376654-1* 358
Geranium odoriferum IPNI=*376654-1* 358
Geranium peltatum IPNI=*376728-1* 358
Geranium pentandrum IPNI=*372219-1* 358
Geranium petroselinum IPNI=*372219-1* 358
Geranium pichinchense IPNI=*372978-1* 358
Geranium pimpinellifolium IPNI=*372219-1* 358
Geranium plicatum IPNI=*164626-3* 358
Geranium quinquelobum IPNI=*372978-1* 358
Geranium radula IPNI=*376259-1* 358
Geranium regelii IPNI=*373138-1* 358
Geranium robustipes IPNI=*372978-1* 358
Geranium rosa IPNI=*375818-1* 358
Geranium sodiroanum IPNI=*372978-1* 358
Geranium sonderianum IPNI=*327766-2* 358
Geranium stenopetalum IPNI=*377229-1* 358
Geranium subscandens IPNI=*109179-2* 358
Geranium terebinthinaceum IPNI=*376259-1* 358
Geranium trianae IPNI=*109179-2* 358
Geranium venezuelae IPNI=*109083-2* 358
Geranium zonale IPNI=*377229-1* 358
Gerardia dulcis IPNI=*55411-1* 361
Gerardia integrifolia IPNI=*799460-1* 375
Gerardia lanceolata IPNI=*6185-2* 375
Gerardia lanceolata var. *angustifolia* IPNI=*6185-2* 375
Gerardia lanceolata var. *parvifolia* IPNI=*6185-2* 375
Gerardia lanceolata var. *revoluta* IPNI=*6185-2* 375
Gerardia levigata IPNI=*799460-1* 375
Gerardia stemodiifolia IPNI=*9536-2* 377
Gerascanthus albus IPNI=*1018059-2* 436
Gerascanthus alliodorus IPNI=*113943-1* 273
Gerascanthus bicolor IPNI=*274895-2* 274
Gerascanthus bifurcatus IPNI=*121424-1* 276
Gerascanthus bogotensis IPNI=*274898-2* 274
Gerascanthus collococcus IPNI=*114058-1* 274
Gerascanthus cordifolius IPNI=*275071-2* 274
Gerascanthus cujabensis IPNI=*113943-1* 273
Gerascanthus cymosus IPNI=*65279-2* 274
Gerascanthus diversifolius IPNI=*274931-2* 274
Gerascanthus exaltatus IPNI=*114135-1* 274
Gerascanthus gerascanthoides IPNI=*65321-2* 274
Gerascanthus hebecladus IPNI=*275016-2* 274
Gerascanthus heterophyllus IPNI=*275076-2* 274
Gerascanthus lanceolatus IPNI=*65321-2* 274
Gerascanthus lasiocalyx IPNI=*274981-2* 274
Gerascanthus lucidulus IPNI=*274992-2* 274
Gerascanthus luteus IPNI=*274993-2* 274
Gerascanthus muneco IPNI=*275071-2* 274
Gerascanthus opacus IPNI=*275053-2* 274
Gerascanthus panamensis IPNI=*275016-2* 274
Gerascanthus petenensis IPNI=*274931-2* 274

Gerascanthus protractus IPNI=*275031-2* 274
Gerascanthus pubescens IPNI=*275076-2* 274
Gerascanthus rotundifolius IPNI=*274993-2* 274
Gerascanthus scabridus IPNI=*114135-1* 274
Gerascanthus sericicalyx IPNI=*275053-2* 274
Gerascanthus tetrandrus IPNI=*275071-2* 274
Gerascanthus tetraphyllus IPNI=*77164595-1* 416
Gerascanthus toqueve IPNI=*275076-2* 274
Gerascanthus velutinus IPNI=*113943-1* 273
Gerascanthus viridis IPNI=*275053-2* 274
Gerascanthus vulgaris IPNI=*65321-2* 274
Gerbera nutans IPNI=*192653-1* 259
Gerbera nutans var. *leiocarpa* IPNI=*192653-1* 259
Germanea nudiflora IPNI=*446165-1* 369
Gerontogea corymbosa IPNI=*757791-1* 353
Gesnera eggersii IPNI=*237093-2* 368
Gesnera lehmannii IPNI=*237093-2* 368
Gesneria acaulis IPNI=*2352-2* 365
Gesneria aurantiaca IPNI=*237093-2* 368
Gesneria breviflora IPNI=*381628-1* 367
Gesneria caracasana IPNI=*237093-2* 368
Gesneria chelonioides IPNI=*237093-2* 368
Gesneria coccinea IPNI=*237093-2* 368
Gesneria elliptica IPNI=*237093-2* 368
Gesneria elliptica var. *lutea* IPNI=*237093-2* 368
Gesneria eriantha IPNI=*381572-1* 367
Gesneria erubescens IPNI=*237093-2* 368
Gesneria flavescens IPNI=*237093-2* 368
Gesneria fragilis IPNI=*237093-2* 368
Gesneria gollmeriana IPNI=*237093-2* 368
Gesneria gracilis IPNI=*237093-2* 368
Gesneria guazumifolia IPNI=*381388-1* 367
Gesneria guianensis IPNI=*237093-2* 368
Gesneria hirsuta IPNI=*381572-1* 367
Gesneria ignorata IPNI=*381628-1* 367
Gesneria incurva IPNI=*381638-1* 367
Gesneria lindeniana IPNI=*237093-2* 368
Gesneria linkiana IPNI=*381628-1* 367
Gesneria longiflora IPNI=*377615-1* 365
Gesneria longiflora IPNI=*381572-1* 367
Gesneria longifolia IPNI=*381628-1* 367
Gesneria mollis IPNI=*381572-1* 367
Gesneria oblongata IPNI=*381572-1* 367
Gesneria petiolaris IPNI=*381628-1* 367
Gesneria picta IPNI=*381638-1* 367
Gesneria pilosa IPNI=*378985-1* 367
Gesneria pulchella IPNI=*2352-2* 365
Gesneria rhynchocarpa IPNI=*381638-1* 367
Gesneria rubricaulis IPNI=*381572-1* 367
Gesneria rutila IPNI=*237093-2* 368
Gesneria sceptroides IPNI=*237093-2* 368
Gesneria schiedeana IPNI=*381628-1* 367
Gesneria schomburgkiana IPNI=*237093-2* 368
Gesneria seemannii IPNI=*381628-1* 367
Gesneria spicata IPNI=*381628-1* 367
Gesneria spicata var. *schiedeana* IPNI=*381628-1* 367
Gesneria stachydifolia IPNI=*237093-2* 368
Gesneria tubiflora IPNI=*381638-1* 367
Gesneria ulmifolia IPNI=*381388-1* 367
Gesneria vargasii IPNI=*237093-2* 368
Gesneria ventricosa IPNI=*381572-1* 367
Gesneria vestita IPNI=*381572-1* 367
Gesneria warszewiczii IPNI=*237093-2* 368
Gethyra tubulata IPNI=*218583-2* 251
Geum chilense IPNI=*725342-1* 443
Geum chilense var. *grandiflorum* IPNI=*725342-1* 443
Geum grandiflorum IPNI=*725342-1* 443
Ghinia hirsuta IPNI=*280380-1* 276
Ghinia spicata IPNI=*864913-1* 379
Ghinia sylvatica IPNI=*280380-1* 276
Ghinia tamonea IPNI=*864913-1* 379
Gibasis schiedeana IPNI=*109899-2* 224
Giesleria picta IPNI=*381638-1* 367
Gigalobium scandens IPNI=*493866-1* 319
Gigantabies taxifolia IPNI=*263741-1* 180
Gigantochloa aspera IPNI=*397862-1* 237
Gigantochloa auriculata IPNI=*392574-1* 235
Gigantochloa pseudoarundinacea IPNI=*404020-1* 239
Gilibertia alaris IPNI=*1133937-2* 253
Gilibertia brachypoda IPNI=*1133937-2* 253
Gilibertia concinna IPNI=*1133937-2* 253
Gilibertia eurycarpa IPNI=*1133937-2* 253
Gilibertia insularis IPNI=*1133937-2* 253
Gilibertia juergensenii IPNI=*1133937-2* 253
Gilibertia langeana IPNI=*1133937-2* 253
Gilibertia matudae IPNI=*1133937-2* 253
Gilibertia pavonii IPNI=*1133937-2* 253
Gilibertia rothschuhii IPNI=*1133937-2* 253
Gilibertia samydifolia IPNI=*1133937-2* 253
Gilibertia schippii IPNI=*1133937-2* 253
Gilibertia smithiana IPNI=*1133937-2* 253
Gilibertia stenocarpa IPNI=*1133937-2* 253
Gimbernatia oblonga IPNI=*171280-1* 416
Gimbernatia obovata IPNI=*250537-2* 416
Ginannia lanata IPNI=*405034-1* 240
Ginannia pubescens IPNI=*405034-1* 240
Ginkgo biloba f. *microsperma* IPNI=*262125-1* 179
Ginkgo biloba f. *parvifolia* IPNI=*262125-1* 179
Ginkgo macrophylla IPNI=*262125-1* 179
Gireoudia cardiocarpa IPNI=*104371-1* 292
Gireoudia crassicaulis IPNI=*104473-1* 292
Gireoudia fibrillosa IPNI=*105718-1* 293
Gireoudia heracleifolia IPNI=*104828-1* 292
Gireoudia heracleifolia var. *punctata* IPNI=*104828-1* 292
Gireoudia heracleifolia var. *viridis* IPNI=*104828-1* 292
Gireoudia manicata IPNI=*105149-1* 292
Gireoudia nelumbiifolia IPNI=*273181-2* 292
Gireoudia pilifera IPNI=*105718-1* 293
Gireoudia punctata IPNI=*104828-1* 292
Gireoudia schizolepis IPNI=*105149-1* 292
Gireoudia sericoneura IPNI=*105718-1* 293
Girtanneria jujuba IPNI=*719213-1* 442
Girtanneria micrantha IPNI=*1018059-2* 436
Gissanthe spiralis IPNI=*796387-1* 248
Gladiolus collinus IPNI=*437372-1* 216
Gladiolus communis var. *montserratii* IPNI=*437372-1* 216
Gladiolus gussonei IPNI=*437372-1* 216
Gladiolus lannesii IPNI=*437372-1* 216
Gladiolus notarisii IPNI=*437372-1* 216
Gladiolus numidicus IPNI=*437372-1* 216
Gladiolus pottsii IPNI=*436437-1* 216
Gladiolus refractus IPNI=*437030-1* 216
Gladiolus resupinatus IPNI=*437030-1* 216
Gladiolus reuteri IPNI=*437372-1* 216
Glandularia × hybrida IPNI=*330557-2* 379
Glandularia elegans IPNI=*263032-2* 379
Glandularia elegans var. *asperata* IPNI=*263032-2* 379
Glaziova martiana IPNI=*246979-2* 212
Glebionis coronaria var. *discolor* IPNI=*1013480-1* 263
Glebionis discolor IPNI=*1013480-1* 263
Glebionis roxburghii IPNI=*1013480-1* 263
Glechoma arvensis IPNI=*459319-1* 373
Glechoma belgica IPNI=*459319-1* 373
Glechoma cataria IPNI=*452333-1* 371
Glechoma macrura IPNI=*452333-1* 371
Glechoma marrubiastrum IPNI=*459319-1* 373

Gleichenia bifida IPNI=*245596-2* 173
Gleichenia brasiliana IPNI=*110733-2* 173
Gleichenia crespiana IPNI=*245647-2* 173
Gleichenia flexuosa f. *monstrosa* IPNI=*79839-2* 173
Gleichenia flexuosa f. *zuccarinii* IPNI=*79839-2* 173
Gleichenia flexuosa IPNI=*79839-2* 173
Gleichenia glaucescens IPNI=*110733-2* 173
Gleichenia hermannii IPNI=*110733-2* 173
Gleichenia immersa IPNI=*245596-2* 173
Gleichenia linearis var. *depauperata* IPNI=*110733-2* 173
Gleichenia nitida IPNI=*110733-2* 173
Gleichenia nuda IPNI=*245632-2* 173
Gleichenia pectinata IPNI=*110733-2* 173
Gleichenia pectinata var. *sublinearis* IPNI=*110733-2* 173
Gleichenia pubescens IPNI=*245596-2* 173
Gleichenia pubescens var. *glabra* IPNI=*245596-2* 173
Gleichenia rigida IPNI=*79839-2* 173
Gleichenia rubiginosa f. *virescens* IPNI=*245647-2* 173
Gleichenia rubiginosa IPNI=*245647-2* 173
Gleichenia salesiana IPNI=*245647-2* 173
Glinus cambessedesii IPNI=*1008757-2* 286
Glinus glaber IPNI=*1008757-2* 286
Gliricidia lambii IPNI=*496639-1* 320
Gliricidia maculata var. *multijuga* IPNI=*496639-1* 320
Gloeospermum equatoriense IPNI=*77184219-1* 405
Gloeospermum melinonianum IPNI=*866956-1* 405
Gloeospermum sphaerocarpum IPNI=*77184111-1* 405
Gloeospermum sphaerocarpum var. *latifolium* IPNI=*77184111-1* 405
Gloeospermum sprucei IPNI=*866955-1* 405
Glomeraria bicolor IPNI=*327386-2* 282
Glomeraria blitum IPNI=*10619-2* 281
Glomeraria tricolor IPNI=*327386-2* 282
Glomeraria viridis IPNI=*316349-2* 282
Gloriosa angulata IPNI=*535953-1* 226
Gloriosa cirrhifolia IPNI=*535953-1* 226
Gloriosa doniana IPNI=*535953-1* 226
Gloriosa nepalensis IPNI=*535953-1* 226
Gloriosa rockefelleriana IPNI=*535953-1* 226
Gloriosa rothschildiana IPNI=*535953-1* 226
Gloriosa verschuurii IPNI=*535953-1* 226
Glossocentrum collinum IPNI=*1098452-2* 421
Glossogyne chinensis IPNI=*32564-2* 258
Glossopetalum glabrum IPNI=*161264-1* 393
Glossopetalum tomentosum IPNI=*161264-1* 393
Gloxinia bicolor IPNI=*381282-1* 367
Gloxinia caulescens IPNI=*382786-1* 368
Gloxinia digitaliflora IPNI=*382786-1* 368
Gloxinia discolor IPNI=*382786-1* 368
Gloxinia diversiflora IPNI=*382786-1* 368
Gloxinia floribunda IPNI=*382786-1* 368
Gloxinia fyfiana IPNI=*382786-1* 368
Gloxinia heterophylla IPNI=*381282-1* 367
Gloxinia maxima IPNI=*382786-1* 368
Gloxinia menziesiana IPNI=*382786-1* 368
Gloxinia merkii IPNI=*382786-1* 368
Gloxinia pallidiflora IPNI=*381282-1* 367
Gloxinia passinghamii IPNI=*382786-1* 368
Gloxinia rubra IPNI=*382786-1* 368
Gloxinia speciosa var. *albiflora* IPNI=*382786-1* 368
Gloxinia speciosa var. *macrophylla* IPNI=*382786-1* 368
Gloxinia speciosa var. *menziesii* IPNI=*382786-1* 368
Gloxinia suaveolens IPNI=*381282-1* 367
Gloxinia teuchleri IPNI=*382786-1* 368
Gloxinia trichantha IPNI=*381282-1* 367
Gloxinia trichotoma IPNI=*381282-1* 367
Gloxinia variabilis IPNI=*382786-1* 368
Glumosia palmifolia IPNI=*237727-2* 216
Glyce maritima IPNI=*1085072-2* 277
Glyceria acuminata IPNI=*404100-1* 239
Glyceria airoides IPNI=*395086-1* 235
Glyceria aquatica IPNI=*395086-1* 235
Glyceria asiatica IPNI=*1197718-2* 252
Glyceria denticulata IPNI=*404100-1* 239
Glyceria dulcis IPNI=*395086-1* 235
Glyceria fluitans var. *islandica* IPNI=*404100-1* 239
Glyceria integra IPNI=*404100-1* 239
Glyceria micrantha IPNI=*401189-1* 239
Glyceria ochroleuca IPNI=*395086-1* 235
Glyceria repanda IPNI=*51383-2* 252
Glyceria triflora IPNI=*1197718-2* 252
Glyceria viridula IPNI=*395086-1* 235
Glycine corallina IPNI=*1006990-2* 331
Glycine diffusa IPNI=*494021-1* 319
Glycine phaseoloides IPNI=*1006990-2* 331
Glycine precatoria IPNI=*517450-1* 331
Glycine procumbens IPNI=*494021-1* 319
Glycine pugiunculus IPNI=*51897-2* 314
Glycine rosea IPNI=*30187778-2* 313
Glycine sagittata IPNI=*485618-1* 314
Glycine striata IPNI=*106712-2* 320
Glycine uncinata IPNI=*521046-1* 335
Glycine vincentina IPNI=*77191263-1* 329
Glycyrrhiza aculeata IPNI=*114706-2* 320
Glycyrrhiza kansoo IPNI=*518369-1* 333
Gmelina oblongifolia IPNI=*77159191-1* 369
Gmelina rheedei IPNI=*77159191-1* 369
Gmelina sinuata IPNI=*77159191-1* 369
Gnaphalion dysodum IPNI=*909620-1* 268
Gnaphalium acuminatum IPNI=*962307-1* 272
Gnaphalium acutifolium IPNI=*909620-1* 268
Gnaphalium africanum IPNI=*213129-1* 264
Gnaphalium alatum IPNI=*174368-1* 255
Gnaphalium alausense IPNI=*174368-1* 255
Gnaphalium alboluteum IPNI=*212994-1* 264
Gnaphalium americanum IPNI=*1213296-2* 263
Gnaphalium antennarioides IPNI=*77197189-1* 259
Gnaphalium araucanum IPNI=*909620-1* 268
Gnaphalium banksii IPNI=*962307-1* 272
Gnaphalium bogotense IPNI=*174373-1* 255
Gnaphalium candicans IPNI=*174407-1* 255
Gnaphalium cheiranthemifolium IPNI=*909620-1* 268
Gnaphalium cheiranthifolium f. *citrinum* IPNI=*909620-1* 268
Gnaphalium cheiranthifolium IPNI=*283846-2* 268
Gnaphalium cheiranthifolium IPNI=*909620-1* 268
Gnaphalium cheiranthifolium var. *gaudichaudianum* IPNI=*283850-2* 268
Gnaphalium cheiranthifolium var. *graveolens* IPNI=*77142354-1* 268
Gnaphalium cheiranthifolium var. *multiflorum* IPNI=*909620-1* 268
Gnaphalium cheiranthifolium var. *paniculatum* IPNI=*909620-1* 268
Gnaphalium cheiranthifolium var. *riedelianum* IPNI=*909620-1* 268
Gnaphalium cheiranthifolium var. *subrufescens* IPNI=*909620-1* 268
Gnaphalium cheiranthifolium var. *typicum* IPNI=*909620-1* 268
Gnaphalium chrysanthum IPNI=*962307-1* 272
Gnaphalium citrinum IPNI=*909620-1* 268
Gnaphalium coarctatum IPNI=*943810-1* 263
Gnaphalium columbianum IPNI=*107379-2* 263
Gnaphalium conglobatum IPNI=*212994-1* 264
Gnaphalium conoideum IPNI=*909626-1* 268
Gnaphalium consanguineum IPNI=*1213296-2* 263
Gnaphalium crenatum IPNI=*283895-2* 269
Gnaphalium cymatoides IPNI=*283846-2* 268
Gnaphalium cymatoides var. *glabrum* IPNI=*283846-2* 268
Gnaphalium decurrens var. *glandulosum* IPNI=*283895-2* 269
Gnaphalium depressum IPNI=*212994-1* 264
Gnaphalium diffusum IPNI=*212994-1* 264
Gnaphalium dillenii IPNI=*107379-2* 263

Gnaphalium dombeyanum IPNI=*283847-2* 268
Gnaphalium dysodes IPNI=*77142354-1* 268
Gnaphalium elegans IPNI=*1039426-2* 268
Gnaphalium elegans IPNI=*213129-1* 264
Gnaphalium elichrysoides IPNI=*77197189-1* 259
Gnaphalium elongatum IPNI=*213129-1* 264
Gnaphalium ferrugineum IPNI=*174368-1* 255
Gnaphalium gaudichaudianum IPNI=*283850-2* 268
Gnaphalium gaudichaudianum var. *subrufescens* IPNI=*909620-1* 268
Gnaphalium glabratum IPNI=*962307-1* 272
Gnaphalium gracile IPNI=*212994-1* 264
Gnaphalium gracile IPNI=*283895-2* 269
Gnaphalium graveolens IPNI=*213129-1* 264
Gnaphalium graveolens IPNI=*77142354-1* 268
Gnaphalium guatemalense IPNI=*1213296-2* 263
Gnaphalium helichrysoides IPNI=*283847-2* 268
Gnaphalium hirtum IPNI=*283895-2* 269
Gnaphalium hortense IPNI=*213129-1* 264
Gnaphalium humillimum IPNI=*283847-2* 268
Gnaphalium hyemale IPNI=*107379-2* 263
Gnaphalium imbaburense IPNI=*283895-2* 269
Gnaphalium incanum IPNI=*174368-1* 255
Gnaphalium ivesii IPNI=*283895-2* 269
Gnaphalium javanum IPNI=*212994-1* 264
Gnaphalium jelskii IPNI=*283847-2* 268
Gnaphalium kunthianum IPNI=*174368-1* 255
Gnaphalium lanuginosum IPNI=*963030-1* 268
Gnaphalium leontopodinum IPNI=*212994-1* 264
Gnaphalium leptophyllum IPNI=*283895-2* 269
Gnaphalium littorale IPNI=*107379-2* 263
Gnaphalium liui IPNI=*943810-1* 263
Gnaphalium luteoalbum IPNI=*212994-1* 264
Gnaphalium luteoalbum var. *glandulosum* IPNI=*212994-1* 264
Gnaphalium luteoalbum var. *incanum* IPNI=*212994-1* 264
Gnaphalium luteoalbum var. *occidentale* IPNI=*212994-1* 264
Gnaphalium luteolum IPNI=*212994-1* 264
Gnaphalium macranthum IPNI=*962307-1* 272
Gnaphalium macrocephalum IPNI=*962307-1* 272
Gnaphalium mandonii IPNI=*283850-2* 268
Gnaphalium margaritaceum IPNI=*30173179-2* 257
Gnaphalium martabanicum IPNI=*212994-1* 264
Gnaphalium mendocinum IPNI=*283850-2* 268
Gnaphalium meridanum IPNI=*283867-2* 268
Gnaphalium molle IPNI=*212994-1* 264
Gnaphalium multicapitatum IPNI=*107382-2* 263
Gnaphalium nanum IPNI=*283847-2* 268
Gnaphalium obtusifolium IPNI=*107379-2* 263
Gnaphalium obtusifolium IPNI=*909626-1* 268
Gnaphalium obtusifolium var. *praecox* IPNI=*909626-1* 268
Gnaphalium obtusifolium var. *saxicola* IPNI=*909626-1* 268
Gnaphalium orientale IPNI=*213129-1* 264
Gnaphalium orixense IPNI=*212994-1* 264
Gnaphalium pallidum IPNI=*212994-1* 264
Gnaphalium paniculatum IPNI=*909620-1* 268
Gnaphalium papillosum IPNI=*962307-1* 272
Gnaphalium paramorum IPNI=*77154313-1* 268
Gnaphalium pellitum IPNI=*963040-1* 269
Gnaphalium petiolatum IPNI=*213170-1* 264
Gnaphalium piravira IPNI=*283893-2* 269
Gnaphalium poeppigianum IPNI=*1039426-2* 268
Gnaphalium polycephalum IPNI=*909626-1* 268
Gnaphalium pompejanum IPNI=*212994-1* 264
Gnaphalium purpureum IPNI=*107379-2* 263
Gnaphalium purpureum var. *americanum* IPNI=*1213296-2* 263
Gnaphalium purpureum var. *macrophyllum* IPNI=*1213296-2* 263
Gnaphalium purpureum var. *normale* IPNI=*107379-2* 263
Gnaphalium purpureum var. *simplicicaule* IPNI=*107382-2* 263
Gnaphalium purpureum var. *spicatum* IPNI=*943810-1* 263
Gnaphalium radians var. *spicatum* IPNI=*943810-1* 263
Gnaphalium regnellii IPNI=*909620-1* 268
Gnaphalium reinwardtianum IPNI=*212994-1* 264
Gnaphalium resedifolium IPNI=*283846-2* 268
Gnaphalium retusum IPNI=*107379-2* 263
Gnaphalium riedelianum IPNI=*909620-1* 268
Gnaphalium rosaceum IPNI=*107379-2* 263
Gnaphalium roseorufescens IPNI=*107379-2* 263
Gnaphalium rufescens IPNI=*174368-1* 255
Gnaphalium rufum IPNI=*174407-1* 255
Gnaphalium saturejifolium IPNI=*174407-1* 255
Gnaphalium saturejoides f. *genuina* IPNI=*174407-1* 255
Gnaphalium saturejoides IPNI=*174407-1* 255
Gnaphalium saturejoides var. *angustifolia* IPNI=*174407-1* 255
Gnaphalium saturejoides var. *candicans* IPNI=*174407-1* 255
Gnaphalium saturejoides var. *citrinum* IPNI=*174407-1* 255
Gnaphalium saturejoides var. *flavescens* IPNI=*174407-1* 255
Gnaphalium saturejoides var. *microcalyx* IPNI=*174407-1* 255
Gnaphalium saturejoides var. *normale* IPNI=*174407-1* 255
Gnaphalium saturejoides var. *patagonica* IPNI=*174407-1* 255
Gnaphalium saturejoides var. *vattuonei* IPNI=*174407-1* 255
Gnaphalium saxicola IPNI=*909626-1* 268
Gnaphalium simplicicaule IPNI=*107382-2* 263
Gnaphalium sphaericum IPNI=*212994-1* 264
Gnaphalium spicatum f. *consanguineum* IPNI=*943810-1* 263
Gnaphalium spicatum f. *purpurascens* IPNI=*107379-2* 263
Gnaphalium spicatum IPNI=*30022662-2* 269
Gnaphalium spicatum IPNI=*943810-1* 263
Gnaphalium spicatum var. *alpinum* IPNI=*943810-1* 263
Gnaphalium spicatum var. *chonoticum* IPNI=*943810-1* 263
Gnaphalium spicatum var. *consanguineum* IPNI=*943810-1* 263
Gnaphalium spicatum var. *hololeucum* IPNI=*943810-1* 263
Gnaphalium spicatum var. *interruptum* IPNI=*943810-1* 263
Gnaphalium spicatum var. *sabulosum* IPNI=*943810-1* 263
Gnaphalium splendens IPNI=*283895-2* 269
Gnaphalium suaveolens IPNI=*203916-2* 268
Gnaphalium tenue f. *latifolium* IPNI=*283895-2* 269
Gnaphalium tenue IPNI=*283895-2* 269
Gnaphalium trifidum IPNI=*212994-1* 264
Gnaphalium ulophyllum IPNI=*283846-2* 268
Gnaphalium uniflorum IPNI=*228818-1* 264
Gnaphalium valdivianum IPNI=*909620-1* 268
Gnaphalium valdiviense IPNI=*909620-1* 268
Gnaphalium vauthierianum IPNI=*174368-1* 255
Gnaphalium viravira IPNI=*283893-2* 269
Gnaphalium virgatum IPNI=*30022662-2* 269
Gnaphalium viscosum IPNI=*283895-2* 269
Gnaphalium viscosum IPNI=*909620-1* 268
Gnemon amazonica IPNI=*111373-2* 179
Gnemon leyboldii IPNI=*111368-2* 179
Gnemon microstachya IPNI=*111375-2* 179
Gnemon nigra IPNI=*111373-2* 179
Gnemon nodiflora IPNI=*111373-2* 179
Gnemon paniculata IPNI=*111375-2* 179
Gnemon urens IPNI=*383578-1* 180
Gnetum amazonicum IPNI=*111373-2* 179
Gnetum cruzianum IPNI=*111373-2* 179
Gnetum dioicum IPNI=*111368-2* 179
Gnetum leyboldii var. *woodsonianum* IPNI=*383578-1* 180
Gnetum melinonii IPNI=*383578-1* 180
Gnetum microstachyum IPNI=*111375-2* 179
Gnetum nigrum IPNI=*111373-2* 179
Gnetum oblongifolium IPNI=*111373-2* 179
Gnetum paraense IPNI=*111368-2* 179
Gnetum thoa IPNI=*383578-1* 180
Gnoteris cordata IPNI=*158297-2* 371
Gnoteris villosa IPNI=*158297-2* 371
Godetia heucki IPNI=*173348-2* 430

Godmania macrocarpa IPNI=*111572-2* 363
Godmania uleana IPNI=*111572-2* 363
Godoya repanda IPNI=*605994-1* 397
Godoya spathulata IPNI=*605994-1* 397
Goeldinia ovatifolia IPNI=*590593-1* 300
Goeldinia riparia IPNI=*590593-1* 300
Goeppertia anomala IPNI=*464481-1* 183
Goeppertia cantagallana IPNI=*90974-2* 183
Goeppertia hirsuta IPNI=*90974-2* 183
Goeppertia hirsuta var. *cantagallana* IPNI=*90974-2* 183
Goeppertia hirsuta var. *coriacea* IPNI=*90974-2* 183
Goeppertia hirsuta var. *hirsutior* IPNI=*90974-2* 183
Goeppertia hirsuta var. *latifolia* IPNI=*90974-2* 183
Goeppertia longifolia IPNI=*90974-2* 183
Goeppertia multiflora IPNI=*90972-2* 183
Goeppertia panicularis IPNI=*90974-2* 183
Goeppertia polyantha IPNI=*464481-1* 183
Goeppertia sericea IPNI=*464535-1* 183
Goeppertia sericea var. *opaca* IPNI=*464535-1* 183
Goeppertia sprucei IPNI=*60461053-2* 249
Goldbachia mikanii IPNI=*21052-2* 234
Goldfussia colorata IPNI=*77147688-1* 361
Gomidesia lindeniana IPNI=*328930-2* 426
Gomozia granadensis IPNI=*757486-1* 352
Gomphia guatemalensis IPNI=*607063-1* 397
Gomphia lucens IPNI=*607063-1* 397
Gomphia nitida IPNI=*607105-1* 397
Gomphiluma gomphiifolia IPNI=*788967-1* 307
Gomphiluma martiana IPNI=*788967-1* 307
Gomphocarpus brasiliensis IPNI=*97670-1* 340
Gomphocarpus fruticosus f. *brasiliensis* IPNI=*97670-1* 340
Gomphosia chlorantha IPNI=*103549-2* 350
Gomphosia laxiflora IPNI=*103549-2* 350
Gomphosia longiflora var. *laxiflora* IPNI=*103549-2* 350
Gomphrena aggregata IPNI=*34033-2* 282
Gomphrena albiflora IPNI=*34033-2* 282
Gomphrena arborescens IPNI=*1002675-1* 283
Gomphrena argentea IPNI=*112097-2* 282
Gomphrena bicolor IPNI=*112097-2* 282
Gomphrena brasiliana IPNI=*10294-2* 281
Gomphrena brasiliensis IPNI=*10294-2* 281
Gomphrena brasiliensis IPNI=*1069792-2* 281
Gomphrena crassifolia IPNI=*34033-2* 282
Gomphrena crucis IPNI=*10362-2* 281
Gomphrena decumbens f. *lanceolata* IPNI=*112097-2* 282
Gomphrena decumbens f. *magnifolia* IPNI=*112097-2* 282
Gomphrena decumbens IPNI=*112097-2* 282
Gomphrena decumbens var. *pringlei* IPNI=*112097-2* 282
Gomphrena dentata IPNI=*10294-2* 281
Gomphrena diffusa IPNI=*112097-2* 282
Gomphrena dispersa IPNI=*112097-2* 282
Gomphrena dunaliana IPNI=*191400-2* 283
Gomphrena eriopoda IPNI=*327333-2* 282
Gomphrena ficoidea IPNI=*10362-2* 281
Gomphrena ficoidea IPNI=*1078972-2* 281
Gomphrena flava IPNI=*112097-2* 282
Gomphrena fruticosa IPNI=*10447-2* 281
Gomphrena fulgens IPNI=*10447-2* 281
Gomphrena glauca IPNI=*191400-2* 283
Gomphrena iresinoides IPNI=*61055-1* 283
Gomphrena iresinoides var. *martiana* IPNI=*61055-1* 283
Gomphrena iresinoides var. *riedeliana* IPNI=*61055-1* 283
Gomphrena ixiamensis IPNI=*112097-2* 282
Gomphrena lancifolia IPNI=*112097-2* 282
Gomphrena latifolia IPNI=*60789-1* 283
Gomphrena luzuliflora IPNI=*191400-2* 283
Gomphrena perennis var. *divergens* IPNI=*10450-2* 281
Gomphrena polygonoides IPNI=*1069792-2* 281
Gomphrena prostrata IPNI=*112097-2* 282
Gomphrena purpurea IPNI=*10447-2* 281
Gomphrena rubra IPNI=*327333-2* 282
Gomphrena sericea IPNI=*1219488-2* 283
Gomphrena sessilis IPNI=*1069792-2* 281
Gomphrena stenophylla IPNI=*191400-2* 283
Gomphrena trichotoma IPNI=*10447-2* 281
Gomphrena trispicata IPNI=*112097-2* 282
Gomphrena tuberosa IPNI=*1219488-2* 283
Gomphrena tuberosa var. *glabrescens* IPNI=*1219488-2* 283
Gomphrena tuberosa var. *sericea* IPNI=*1219488-2* 283
Gomphrena tuberosa var. *vestita* IPNI=*1219488-2* 283
Gomphrena tumida IPNI=*327333-2* 282
Gomphrena vermicularis IPNI=*34033-2* 282
Gonatandra tradescantioides IPNI=*1191409-2* 224
Gongora ecornuta IPNI=*636033-1* 219
Gongora portentosa var. *rosea* IPNI=*636033-1* 219
Gongora rosea IPNI=*636033-1* 219
Goniophlebium albopunctatum IPNI=*77075127-1* 177
Goniophlebium ampliatum IPNI=*77075127-1* 177
Goniophlebium angustifolium IPNI=*17063000-1* 175
Goniophlebium areolatum IPNI=*17175380-1* 176
Goniophlebium attenuatum IPNI=*77075087-1* 176
Goniophlebium brasiliense IPNI=*77075127-1* 177
Goniophlebium brasiliense var. *preslianum* IPNI=*77075127-1* 177
Goniophlebium cardiophyllum IPNI=*17406250-1* 176
Goniophlebium ceteraccinum IPNI=*1039912-2* 176
Goniophlebium elatius IPNI=*77075127-1* 177
Goniophlebium ensifolium IPNI=*17063000-1* 175
Goniophlebium fraternum IPNI=*17189260-1* 176
Goniophlebium gladiatum IPNI=*77075127-1* 177
Goniophlebium glaucophyllum IPNI=*77075105-1* 177
Goniophlebium haenkei IPNI=*77075127-1* 177
Goniophlebium incanum IPNI=*1039912-2* 176
Goniophlebium intermedium IPNI=*77075127-1* 177
Goniophlebium kuhnii IPNI=*77075087-1* 176
Goniophlebium levigatum IPNI=*77075105-1* 177
Goniophlebium ligustrifolium IPNI=*160142-2* 175
Goniophlebium myrtillifolium IPNI=*160142-2* 175
Goniophlebium neriifolium IPNI=*77075127-1* 177
Goniophlebium oleandrifolium IPNI=*77075087-1* 176
Goniophlebium revertens IPNI=*17189260-1* 176
Goniophlebium serratum IPNI=*17189260-1* 176
Goniophlebium triseriale IPNI=*77075127-1* 177
Goniophlebium vacciniifolium IPNI=*17406250-1* 176
Goniophlebium xiphophorum IPNI=*77075127-1* 177
Goniopteris cheilocarpa IPNI=*17458930-1* 174
Goniopteris nicaraguensis IPNI=*252198-2* 174
Goniopteris pennigera IPNI=*17458930-1* 174
Goniopteris radicans IPNI=*22662-2* 174
Goniostachyum citrosum IPNI=*863236-1* 378
Goniostachyum graveolens IPNI=*863752-1* 379
Gonipia linearis IPNI=*326619-2* 344
Gonipia tenuiflora IPNI=*108639-2* 344
Gonocarpus jacquinii IPNI=*170030-1* 416
Gonolobus albus IPNI=*106366-2* 340
Gonolobus altissimus IPNI=*60478412-2* 343
Gonolobus ciliatus IPNI=*77206440-1* 340
Gonolobus cundurango IPNI=*99126-1* 342
Gonolobus denticulatus IPNI=*77206440-1* 340
Gonolobus discolor IPNI=*154570-2* 342
Gonolobus filiformis IPNI=*106366-2* 340
Gonolobus floccosus IPNI=*98854-1* 341
Gonolobus grenadensis IPNI=*77206440-1* 340
Gonolobus guianensis IPNI=*77206440-1* 340
Gonolobus hirsutus IPNI=*154570-2* 342
Gonolobus macrophyllus IPNI=*154570-2* 342
Gonolobus maritimus IPNI=*98854-1* 341

Gonolobus martianus IPNI=*97427-1* 340
Gonolobus obliquus IPNI=*154570-2* 342
Gonolobus obliquus var. *shortii* IPNI=*154570-2* 342
Gonolobus setosus IPNI=*97427-1* 340
Gonolobus shortii IPNI=*154570-2* 342
Gonolobus stelliflorus IPNI=*77206440-1* 340
Gonolobus suberosus IPNI=*98854-1* 341
Gonolobus viridiflorus IPNI=*77206440-1* 340
Gonolobus viridis IPNI=*77206440-1* 340
Gonufas paniculata IPNI=*129956-2* 283
Gonypetalum juruanum IPNI=*249430-2* 386
Gonzalagunia acutifolia IPNI=*112531-2* 351
Gonzalagunia dicocca subsp. *venezuelensis* IPNI=*751530-1* 351
Gonzalagunia flexuosa IPNI=*112495-2* 351
Gonzalagunia hirsuta var. *dicocca* IPNI=*751530-1* 351
Gonzalagunia leptantha IPNI=*30107991-2* 351
Gonzalagunia panamensis var. *cornifolia* IPNI=*112495-2* 351
Gonzalagunia panamensis var. *hirsuta* IPNI=*30107991-2* 351
Gonzalagunia panamensis var. *longistipula* IPNI=*112495-2* 351
Gonzalagunia villosa IPNI=*112490-2* 351
Gonzalea cornifolia IPNI=*112495-2* 351
Gonzalea dicocca IPNI=*751530-1* 351
Gonzalea hayesii IPNI=*30107991-2* 351
Gonzalea leptantha IPNI=*30107991-2* 351
Gonzalea ovatifolia IPNI=*112521-2* 351
Gonzalea panamensis IPNI=*30107991-2* 351
Gordonia humboldtii IPNI=*112597-2* 309
Gordonia obovata IPNI=*112597-2* 309
Gordonia planchonii IPNI=*112597-2* 309
Gordonia semiserrata IPNI=*112597-2* 309
Gorgasia oleracea IPNI=*222805-2* 212
Gorteria heterophylla IPNI=*208720-1* 263
Gorteria incisa IPNI=*941423-1* 263
Gorteria linearis IPNI=*208700-1* 263
Gorteria lyrato-pinnatifida IPNI=*941423-1* 263
Gorteria pectinata IPNI=*941423-1* 263
Gorteria personata IPNI=*208720-1* 263
Gorteria pinnata IPNI=*941423-1* 263
Gorteria rigens IPNI=*208720-1* 263
Gorteria rigida IPNI=*208720-1* 263
Gorteria speciosa IPNI=*941423-1* 263
Gorteria spectabilis IPNI=*208720-1* 263
Gossampinus alba IPNI=*1166232-2* 407
Gossampinus heptaphylla IPNI=*211241-2* 413
Gossampinus malabarica IPNI=*559120-1* 407
Gossampinus rubra IPNI=*559120-1* 407
Gossampinus rumphii IPNI=*1166232-2* 407
Gossampinus thorelii IPNI=*559120-1* 407
Gossypianthus australis IPNI=*1183159-2* 282
Gossypianthus mollis IPNI=*60789-1* 283
Gossypiospermum eriophorum IPNI=*48069-2* 403
Gossypiospermum praecox IPNI=*48069-2* 403
Gossypium acuminatum IPNI=*559677-1* 408
Gossypium albescens IPNI=*559730-1* 408
Gossypium albiflorum IPNI=*559664-1* 408
Gossypium album IPNI=*559730-1* 408
Gossypium amblospermum IPNI=*559730-1* 408
Gossypium anomalum IPNI=*559664-1* 408
Gossypium arboreum var. *wightianum* IPNI=*559664-1* 408
Gossypium asiaticum IPNI=*559664-1* 408
Gossypium aureum IPNI=*559730-1* 408
Gossypium auritum IPNI=*559677-1* 408
Gossypium bani IPNI=*559664-1* 408
Gossypium barbadense f. *casimiranum* IPNI=*559677-1* 408
Gossypium barbadense f. *eggersii* IPNI=*559677-1* 408
Gossypium barbadense subsp. *peruvianum* IPNI=*559677-1* 408
Gossypium barbadense subsp. *racemosum* IPNI=*559677-1* 408
Gossypium barbadense subsp. *sprucei* IPNI=*559677-1* 408
Gossypium barbadense subsp. *typicum* IPNI=*559677-1* 408
Gossypium barbadense subsp. *vitifolium* IPNI=*559677-1* 408
Gossypium barbadense var. *apospermum* IPNI=*559677-1* 408
Gossypium barbadense var. *brasiliense* IPNI=*559677-1* 408
Gossypium barbadense var. *caravonicum* IPNI=*559677-1* 408
Gossypium barbadense var. *cayanense* IPNI=*559677-1* 408
Gossypium barbadense var. *eubarbadense* IPNI=*559677-1* 408
Gossypium barbadense var. *hirsutum* IPNI=*30042843-2* 408
Gossypium barbadense var. *integrum* IPNI=*559677-1* 408
Gossypium barbadense var. *marie-galante* IPNI=*30042843-2* 408
Gossypium barbadense var. *maritimum* IPNI=*559677-1* 408
Gossypium barbadense var. *pedatum* IPNI=*559677-1* 408
Gossypium barbadense var. *peruvianum* IPNI=*559677-1* 408
Gossypium barbadense var. *rufum* IPNI=*559677-1* 408
Gossypium bicolor IPNI=*559730-1* 408
Gossypium birkinshawii IPNI=*30042843-2* 408
Gossypium brasiliense IPNI=*559677-1* 408
Gossypium brasiliense var. *apospermum* IPNI=*559677-1* 408
Gossypium caespitosum IPNI=*30042843-2* 408
Gossypium calycotum IPNI=*559677-1* 408
Gossypium cambayense IPNI=*559677-1* 408
Gossypium cavanillesianum IPNI=*30042843-2* 408
Gossypium cernuum IPNI=*559664-1* 408
Gossypium chinense IPNI=*559730-1* 408
Gossypium cinereum IPNI=*559730-1* 408
Gossypium comesii IPNI=*559664-1* 408
Gossypium convexum IPNI=*559730-1* 408
Gossypium croceum IPNI=*559730-1* 408
Gossypium decurrens IPNI=*559730-1* 408
Gossypium divaricatum IPNI=*30042843-2* 408
Gossypium eglandulosum IPNI=*559730-1* 408
Gossypium elatum IPNI=*559730-1* 408
Gossypium evertum IPNI=*559677-1* 408
Gossypium figarei IPNI=*559664-1* 408
Gossypium frutescens IPNI=*559730-1* 408
Gossypium fruticulosum IPNI=*559677-1* 408
Gossypium fuscum IPNI=*559677-1* 408
Gossypium fuscum IPNI=*559730-1* 408
Gossypium glabratum IPNI=*559664-1* 408
Gossypium glabrum IPNI=*559677-1* 408
Gossypium gracile IPNI=*559664-1* 408
Gossypium guyanense IPNI=*559677-1* 408
Gossypium guyanense var. *brasiliense* IPNI=*559677-1* 408
Gossypium harrisii IPNI=*30042843-2* 408
Gossypium herbaceum var. *wightianum* IPNI=*559664-1* 408
Gossypium hirsutum f. *atrocoloratum* IPNI=*30042843-2* 408
Gossypium hirsutum f. *berlandieri* IPNI=*30042843-2* 408
Gossypium hirsutum f. *cavelsenii* IPNI=*30042843-2* 408
Gossypium hirsutum f. *cearense* IPNI=*30042843-2* 408
Gossypium hirsutum f. *gaumeri* IPNI=*30042843-2* 408
Gossypium hirsutum f. *guatemaltequense* IPNI=*30042843-2* 408
Gossypium hirsutum f. *pollardii* IPNI=*30042843-2* 408
Gossypium hirsutum subsp. *archetypicum* IPNI=*30042843-2* 408
Gossypium hirsutum subsp. *caespitosum* IPNI=*30042843-2* 408
Gossypium hirsutum subsp. *mustelinum* IPNI=*30042843-2* 408
Gossypium hirsutum subsp. *prostratum* IPNI=*30042843-2* 408
Gossypium hirsutum var. *album* IPNI=*30042843-2* 408
Gossypium hirsutum var. *chacoense* IPNI=*30042843-2* 408
Gossypium hirsutum var. *colombianum* IPNI=*30042843-2* 408
Gossypium hirsutum var. *macrocarpa* IPNI=*30042843-2* 408
Gossypium hirsutum var. *marie-galante* IPNI=*30042843-2* 408
Gossypium hirsutum var. *mexicanum* IPNI=*30042843-2* 408
Gossypium hirsutum var. *nervosum* IPNI=*30042843-2* 408
Gossypium hirsutum var. *pellitum* IPNI=*30042843-2* 408
Gossypium hirsutum var. *uliginosum* IPNI=*30042843-2* 408
Gossypium hirsutum var. *volubile* IPNI=*30042843-2* 408
Gossypium hopi IPNI=*30042843-2* 408
Gossypium indicum IPNI=*559664-1* 408

Gossypium intermedium IPNI=*559664-1* 408
Gossypium isabelum IPNI=*559677-1* 408
Gossypium jamaicense IPNI=*30042843-2* 408
Gossypium janiphifolium IPNI=*30042843-2* 408
Gossypium javanicum IPNI=*559677-1* 408
Gossypium jumelianum IPNI=*559677-1* 408
Gossypium lanceolatum IPNI=*30042843-2* 408
Gossypium latifolium f. *cearense* IPNI=*30042843-2* 408
Gossypium latifolium f. *pseudoracemosum* IPNI=*30042843-2* 408
Gossypium latifolium IPNI=*30042843-2* 408
Gossypium latifolium var. *archetypicum* IPNI=*30042843-2* 408
Gossypium latifolium var. *paniculatum* IPNI=*559730-1* 408
Gossypium latifolium var. *prostratum* IPNI=*30042843-2* 408
Gossypium latifolium var. *pseudovolubile* IPNI=*30042843-2* 408
Gossypium latifolium var. *tricuspidatum* IPNI=*559730-1* 408
Gossypium leoninum IPNI=*559730-1* 408
Gossypium macedonicum IPNI=*559730-1* 408
Gossypium macranthum IPNI=*559664-1* 408
Gossypium macrospermum IPNI=*559730-1* 408
Gossypium marie-galante IPNI=*30042843-2* 408
Gossypium maritimum IPNI=*559677-1* 408
Gossypium mexicanum IPNI=*30042843-2* 408
Gossypium micranthum IPNI=*559730-1* 408
Gossypium microcarpum IPNI=*559677-1* 408
Gossypium molle IPNI=*559730-1* 408
Gossypium multiglandulosum IPNI=*559677-1* 408
Gossypium nankin IPNI=*559677-1* 408
Gossypium nanking IPNI=*559664-1* 408
Gossypium neglectum IPNI=*559664-1* 408
Gossypium nervosum IPNI=*30042843-2* 408
Gossypium nicaraguense IPNI=*30042843-2* 408
Gossypium nigrum IPNI=*559677-1* 408
Gossypium niveum IPNI=*559677-1* 408
Gossypium obtusifolium IPNI=*559664-1* 408
Gossypium obtusifolium var. *wightiana* IPNI=*559664-1* 408
Gossypium oligospermum IPNI=*30042843-2* 408
Gossypium pallens IPNI=*559677-1* 408
Gossypium palmeri IPNI=*30042843-2* 408
Gossypium paniculatum IPNI=*559730-1* 408
Gossypium pedatum IPNI=*559677-1* 408
Gossypium perennans IPNI=*559664-1* 408
Gossypium perenne IPNI=*559677-1* 408
Gossypium perrieri IPNI=*559730-1* 408
Gossypium peruvianum IPNI=*559677-1* 408
Gossypium peruvianum subsp. *typicum* IPNI=*559677-1* 408
Gossypium peruvianum var. *apospermum* IPNI=*559677-1* 408
Gossypium peruvianum var. *brasiliense* IPNI=*559677-1* 408
Gossypium peruvianum var. *braziliense* IPNI=*559677-1* 408
Gossypium peruvianum var. *glabrum* IPNI=*559677-1* 408
Gossypium prostratum IPNI=*30042843-2* 408
Gossypium pubescens IPNI=*559677-1* 408
Gossypium punctatum IPNI=*30042843-2* 408
Gossypium punctatum var. *jamaica* IPNI=*30042843-2* 408
Gossypium puniceum IPNI=*559664-1* 408
Gossypium purpurascens IPNI=*30042843-2* 408
Gossypium purpureum IPNI=*559730-1* 408
Gossypium quinacre IPNI=*559677-1* 408
Gossypium racemosum IPNI=*559677-1* 408
Gossypium religiosum IPNI=*30042843-2* 408
Gossypium religiosum IPNI=*559677-1* 408
Gossypium rhorii IPNI=*30042843-2* 408
Gossypium rohrianum IPNI=*559677-1* 408
Gossypium roseum IPNI=*559664-1* 408
Gossypium roxburghii IPNI=*559664-1* 408
Gossypium royleanum IPNI=*559664-1* 408
Gossypium rubicundum IPNI=*559664-1* 408
Gossypium rubrum IPNI=*559664-1* 408
Gossypium rufum IPNI=*30042843-2* 408
Gossypium rupestre IPNI=*559677-1* 408
Gossypium sanguineum IPNI=*559664-1* 408
Gossypium sarmentosum IPNI=*559677-1* 408
Gossypium schottii IPNI=*30042843-2* 408
Gossypium sericatum IPNI=*30042843-2* 408
Gossypium siamense IPNI=*559730-1* 408
Gossypium simpsonii IPNI=*559730-1* 408
Gossypium soudanense IPNI=*559664-1* 408
Gossypium speciosum IPNI=*559677-1* 408
Gossypium strictum IPNI=*559730-1* 408
Gossypium suffruticosum IPNI=*559677-1* 408
Gossypium teleium IPNI=*559677-1* 408
Gossypium tenax IPNI=*559677-1* 408
Gossypium trichospermum IPNI=*559677-1* 408
Gossypium tricuspidatum IPNI=*559730-1* 408
Gossypium tridens IPNI=*30042843-2* 408
Gossypium vaupelii IPNI=*559664-1* 408
Gossypium versicolor IPNI=*559677-1* 408
Gossypium virens IPNI=*559677-1* 408
Gossypium virgatum IPNI=*559677-1* 408
Gossypium vitifolium IPNI=*559677-1* 408
Gossypium vitifolium IPNI=*559730-1* 408
Gossypium volubile IPNI=*30042843-2* 408
Gossypium wattianum IPNI=*559664-1* 408
Gossypium wightianum IPNI=*559664-1* 408
Gossypium zaitzevii IPNI=*559730-1* 408
Gouania crenata IPNI=*112865-2* 442
Gouania domingensis IPNI=*112858-2* 441
Gouania glabra IPNI=*112858-2* 441
Gouania glabriuscula IPNI=*112858-2* 441
Gouania lehmannii IPNI=*1013652-1* 442
Gouania lupuloides var. *aptera* IPNI=*112858-2* 441
Gouania lupuloides var. *parvifolia* IPNI=*112858-2* 441
Gouania martinicensis IPNI=*112865-2* 442
Gouania paniculata IPNI=*112858-2* 441
Gouania pubescens IPNI=*112865-2* 442
Gouania tomentosa IPNI=*112865-2* 442
Gouania ulmifolia IPNI=*112865-2* 442
Gouania ursinicarpa IPNI=*112865-2* 442
Goupia paraensis IPNI=*161264-1* 393
Goupia tomentosa IPNI=*161264-1* 393
Gramen rubrum IPNI=*30010036-2* 239
Grammatotheca erinoides IPNI=*143240-1* 273
Grammitis angustifolia IPNI=*17063000-1* 175
Grammitis araucana IPNI=*17448650-1* 176
Grammitis flexuosa IPNI=*77110633-1* 178
Grammitis robusta IPNI=*17448650-1* 176
Granadilla caerulea IPNI=*321949-2* 398
Granadilla capsularis IPNI=*321950-2* 398
Granadilla foetida IPNI=*321966-2* 398
Granadilla laurifolia IPNI=*675132-1* 399
Granadilla quadrangularis IPNI=*675390-1* 400
Granadilla rubra IPNI=*675423-1* 400
Granadilla suberosa IPNI=*321973-2* 400
Granadilla vespertilio IPNI=*675584-1* 400
Granatum punicum IPNI=*554129-1* 417
Grangea domingensis var. *viscosa* IPNI=*1037395-2* 261
Grangea lanceolata IPNI=*1186403-2* 261
Grantia globosa IPNI=*1135607-2* 202
Graphardisia bartlettii IPNI=*18048-2* 303
Graphardisia lewisii IPNI=*18048-2* 303
Graphardisia romeroi IPNI=*18048-2* 303
Graphardisia sapoana IPNI=*18048-2* 303
Graphardisia tuirana IPNI=*18048-2* 303
Graphephorum shearii IPNI=*258855-2* 246
Graphorkis maculata IPNI=*1073117-2* 222
Graptophyllum caudifolium IPNI=*77158889-1* 361
Gratiola alata IPNI=*810361-1* 374

Gratiola aspera IPNI=*1189750-2* 374
Gratiola debilis IPNI=*1061345-2* 374
Gratiola glabra IPNI=*803515-1* 376
Gratiola grandiflora IPNI=*799766-1* 374
Gratiola integrifolia IPNI=*1061345-2* 374
Gratiola latifolia IPNI=*803515-1* 376
Gratiola lucida IPNI=*1189750-2* 374
Gratiola micrantha IPNI=*230902-2* 376
Gratiola monnieri IPNI=*1072674-2* 376
Gratiola monnieria IPNI=*1072674-2* 376
Gratiola origanifolia IPNI=*262512-2* 374
Gratiola peruviana var. *pusilla* IPNI=*803515-1* 376
Gratiola peruviana var. *uliginosa* IPNI=*803515-1* 376
Gratiola portulacacea IPNI=*1072674-2* 376
Gratiola pulegiifolia IPNI=*799766-1* 374
Gratiola racemosa IPNI=*799766-1* 374
Gratiola rotundifolia IPNI=*1061345-2* 374
Gratiola roxburghiana IPNI=*799766-1* 374
Gratiola tetrandra IPNI=*1072674-2* 376
Gratiola verbenifolia IPNI=*799766-1* 374
Greeneina affinis IPNI=*584303-1* 440
Greeneina poeppigiana IPNI=*584303-1* 440
Greenmania boladorensis IPNI=*257035-1* 272
Greenmania ulei IPNI=*257035-1* 272
Grevillea robusta var. *compacta* IPNI=*50798-3* 433
Grevillea robusta var. *forsteri* IPNI=*50798-3* 433
Grevillea umbratica IPNI=*50798-3* 433
Grevillea venusta IPNI=*50798-3* 433
Grewia fruticetorum IPNI=*834119-1* 408
Grewia mexicana IPNI=*1213322-2* 415
Grias aubletiana IPNI=*114915-2* 301
Grias dariensis IPNI=*596618-1* 301
Grias dukei IPNI=*596618-1* 301
Grias fendleri IPNI=*596618-1* 301
Grias foetidissima IPNI=*113443-2* 301
Grias gentlei IPNI=*596618-1* 301
Grias grandifolia IPNI=*113444-2* 301
Grias integrifolia IPNI=*596618-1* 301
Grias loretensis IPNI=*113443-2* 301
Grias maranonensis IPNI=*113444-2* 301
Grias pittieri IPNI=*596618-1* 301
Grias sternii IPNI=*596618-1* 301
Grias tessmannii IPNI=*113444-2* 301
Grias tetrapetala IPNI=*114915-2* 301
Grimaldia absus IPNI=*54310-2* 314
Grisebachia forsteriana IPNI=*667434-1* 209
Griseocactus griseus IPNI=*244634-2* 285
Griseocereus griseus IPNI=*244634-2* 285
Gromovia pulchella IPNI=*50468-1* 360
Gromovia violacea IPNI=*50468-1* 360
Grona adscendens var. *robusta* IPNI=*77187388-1* 320
Gronovia humboldtiana IPNI=*545139-1* 291
Gronovia jacquiniana IPNI=*545139-1* 291
Grusonia caribaea IPNI=*73852-2* 283
Grusonia tunicata IPNI=*73905-2* 283
Guachamaca toxicaria IPNI=*81405-1* 342
Guadua angustifolia var. *bicolor* IPNI=*1074014-2* 239
Guadua angustifolia var. *nigra* IPNI=*1074014-2* 239
Guadua intermedia IPNI=*1074014-2* 239
Guadua pallescens IPNI=*392553-1* 235
Guagnebina ignita IPNI=*755652-1* 352
Guaiabara uvifera IPNI=*693918-1* 288
Guajava acutangula IPNI=*600701-1* 428
Guajava albida IPNI=*211886-2* 428
Guajava arborea IPNI=*600760-1* 428
Guajava basantha IPNI=*600884-1* 428
Guajava benthamiana IPNI=*211886-2* 428
Guajava buxifolia IPNI=*600760-1* 428
Guajava campomanesiodes IPNI=*1061290-2* 424
Guajava cattleyana IPNI=*600760-1* 428
Guajava costaricensis IPNI=*211886-2* 428
Guajava densicoma IPNI=*600795-1* 428
Guajava fluviatilis IPNI=*600845-1* 428
Guajava glaucescens IPNI=*600884-1* 428
Guajava guineensis IPNI=*211886-2* 428
Guajava hians IPNI=*211886-2* 428
Guajava humilis IPNI=*600857-1* 428
Guajava laruotteana IPNI=*600884-1* 428
Guajava laurifolia IPNI=*211886-2* 428
Guajava mollis IPNI=*211886-2* 428
Guajava multiflora IPNI=*211886-2* 428
Guajava obovata IPNI=*600760-1* 428
Guajava oligosperma IPNI=*600952-1* 428
Guajava ooidea IPNI=*211886-2* 428
Guajava ovatifolia IPNI=*600795-1* 428
Guajava polycarpa IPNI=*211886-2* 428
Guajava pumila IPNI=*600841-1* 428
Guajava pyrifera IPNI=*600841-1* 428
Guajava richardiana IPNI=*600845-1* 428
Guajava salutaris IPNI=*1115374-2* 428
Guajava schiedeana IPNI=*211886-2* 428
Guajava sprucei IPNI=*600845-1* 428
Guajava umbrosa IPNI=*600845-1* 428
Guajava ypanemensis IPNI=*211886-2* 428
Guapeba caimito IPNI=*788845-1* 306
Guapeba coriacea IPNI=*209732-2* 306
Guapeba crassifolia IPNI=*209758-2* 306
Guapeba glomerata IPNI=*788965-1* 307
Guapeba gomphiifolia IPNI=*788967-1* 307
Guapeba laevigata IPNI=*789009-1* 307
Guapeba lasiocarpa IPNI=*788845-1* 306
Guapeba laurifolia IPNI=*788845-1* 306
Guapeba torta IPNI=*789233-1* 308
Guapebeira brasiliensis IPNI=*788845-1* 306
Guapebeira laurifolia IPNI=*788845-1* 306
Guapira ayacuchae IPNI=*113926-2* 286
Guapira coriifolia IPNI=*113934-2* 287
Guapira dussii IPNI=*113934-2* 287
Guapira harrisiana IPNI=*113934-2* 287
Guapira schomburgkiana IPNI=*113926-2* 286
Guapurium peruvianum IPNI=*203847-2* 428
Guatteria aberemoa IPNI=*114255-2* 188
Guatteria aberemoa var. *microcarpa* IPNI=*114255-2* 188
Guatteria acutissima IPNI=*912648-1* 188
Guatteria alutacea IPNI=*73441-1* 188
Guatteria alutacea var. *angustifolia* IPNI=*73441-1* 188
Guatteria alutacea var. *steinbachii* IPNI=*73441-1* 188
Guatteria asplundiana IPNI=*912648-1* 188
Guatteria atra IPNI=*912648-1* 188
Guatteria augusti IPNI=*73441-1* 188
Guatteria axilliflora IPNI=*912648-1* 188
Guatteria boliviana IPNI=*114422-2* 189
Guatteria boyacana IPNI=*1003592-1* 189
Guatteria brevipedicellata IPNI=*73441-1* 188
Guatteria buchtienii IPNI=*912648-1* 188
Guatteria calimensis IPNI=*912648-1* 188
Guatteria calliantha IPNI=*912648-1* 188
Guatteria calophylla IPNI=*114255-2* 188
Guatteria calva IPNI=*73503-1* 188
Guatteria caniflora IPNI=*912648-1* 188
Guatteria caniflora var. *angustifolia* IPNI=*912648-1* 188
Guatteria caniflora var. *latifolia* IPNI=*912648-1* 188
Guatteria cardoniana IPNI=*114422-2* 189
Guatteria cargadero IPNI=*912648-1* 188
Guatteria cestrifolia IPNI=*73441-1* 188
Guatteria chocoensis IPNI=*73441-1* 188

Guatteria chrysopetala IPNI=*912648-1* 188
Guatteria chrysopetala var. *major* IPNI=*912648-1* 188
Guatteria chrysopetala var. *tenuipes* IPNI=*912648-1* 188
Guatteria coeloneura IPNI=*912648-1* 188
Guatteria collina IPNI=*912648-1* 188
Guatteria curvipetala IPNI=*73441-1* 188
Guatteria cuspidata IPNI=*240971-2* 441
Guatteria diospyroides IPNI=*73306-1* 188
Guatteria diospyroides subsp. *hondurensis* IPNI=*73306-1* 188
Guatteria ecuadorensis IPNI=*73441-1* 188
Guatteria elliptica IPNI=*912648-1* 188
Guatteria excellens IPNI=*114255-2* 188
Guatteria excelsa IPNI=*73441-1* 188
Guatteria flavovirens IPNI=*73626-1* 189
Guatteria gamosepala IPNI=*912648-1* 188
Guatteria glauca IPNI=*912648-1* 188
Guatteria gracilipes IPNI=*912648-1* 188
Guatteria guentheri IPNI=*912648-1* 188
Guatteria guentheriana IPNI=*73441-1* 188
Guatteria inuncta IPNI=*73306-1* 188
Guatteria inuncta var. *caudata* IPNI=*73306-1* 188
Guatteria inuncta var. *minor* IPNI=*73306-1* 188
Guatteria jamundensis IPNI=*73441-1* 188
Guatteria juninensis IPNI=*912648-1* 188
Guatteria jurgensenii IPNI=*73306-1* 188
Guatteria juruensis IPNI=*73441-1* 188
Guatteria klugii IPNI=*73441-1* 188
Guatteria kuhlmannii IPNI=*114231-2* 188
Guatteria lanceolata IPNI=*912648-1* 188
Guatteria lasiocalyx IPNI=*912648-1* 188
Guatteria latipetala IPNI=*912648-1* 188
Guatteria latisepala IPNI=*73441-1* 188
Guatteria laurina IPNI=*73441-1* 188
Guatteria lawrancei IPNI=*114422-2* 189
Guatteria leiocarpa IPNI=*912648-1* 188
Guatteria longepetiolata IPNI=*114422-2* 189
Guatteria longestipitata IPNI=*912648-1* 188
Guatteria longipes IPNI=*73441-1* 188
Guatteria macropetala IPNI=*114422-2* 189
Guatteria maypurensis var. *attenuata* IPNI=*73503-1* 188
Guatteria maypurensis var. *pulchra* IPNI=*73503-1* 188
Guatteria megalophylla var. *deminuta* IPNI=*114300-2* 188
Guatteria melinii IPNI=*73441-1* 188
Guatteria melosma IPNI=*114300-2* 188
Guatteria metensis IPNI=*73441-1* 188
Guatteria multivenia IPNI=*114255-2* 188
Guatteria obliqua IPNI=*912648-1* 188
Guatteria occidentalis IPNI=*912648-1* 188
Guatteria olivacea IPNI=*912648-1* 188
Guatteria ovalifolia IPNI=*912648-1* 188
Guatteria ovalis IPNI=*224360-2* 189
Guatteria parviflora IPNI=*912648-1* 188
Guatteria peduncularis IPNI=*259508-2* 189
Guatteria persicifolia IPNI=*73441-1* 188
Guatteria persicifolia IPNI=*74692-1* 189
Guatteria platypetala IPNI=*73306-1* 188
Guatteria platyphylla IPNI=*912648-1* 188
Guatteria pleiocarpa IPNI=*912648-1* 188
Guatteria poeppigiana IPNI=*912648-1* 188
Guatteria pteropus IPNI=*912648-1* 188
Guatteria pteropus var. *angustior* IPNI=*912648-1* 188
Guatteria pteropus var. *cinerea* IPNI=*912648-1* 188
Guatteria recurvisepala IPNI=*114422-2* 189
Guatteria rhamnoides IPNI=*912648-1* 188
Guatteria robusta IPNI=*114255-2* 188
Guatteria rugosa IPNI=*114219-2* 188
Guatteria sagotiana IPNI=*912648-1* 188
Guatteria sagotiana var. *gracilior* IPNI=*912648-1* 188
Guatteria sandwithii IPNI=*73626-1* 189
Guatteria schomburgkiana var. *angustifolia* IPNI=*73626-1* 189
Guatteria schomburgkiana var. *holosericea* IPNI=*73626-1* 189
Guatteria schomburgkiana var. *latifolia* IPNI=*73626-1* 189
Guatteria schunkevigoi IPNI=*114422-2* 189
Guatteria sessilis IPNI=*73626-1* 189
Guatteria speciosa IPNI=*114422-2* 189
Guatteria spruceana IPNI=*73626-1* 189
Guatteria sylvicola IPNI=*912648-1* 188
Guatteria umbonata IPNI=*912648-1* 188
Guatteria velezii IPNI=*73503-1* 188
Guatteria veneficiorum IPNI=*259534-2* 189
Guatteria vestita IPNI=*73626-1* 189
Guatteria wessels-boeri IPNI=*912648-1* 188
Guazuma blumei IPNI=*823393-1* 408
Guazuma burbroma IPNI=*823393-1* 408
Guazuma coriacea IPNI=*823393-1* 408
Guazuma grandiflora IPNI=*825580-1* 414
Guazuma guazuma IPNI=*823393-1* 408
Guazuma guazuma var. *tomentosa* IPNI=*823393-1* 408
Guazuma guazuma var. *ulmifolia* IPNI=*823393-1* 408
Guazuma parvifolia IPNI=*823393-1* 408
Guazuma polybotrya IPNI=*823393-1* 408
Guazuma tomentosa IPNI=*823393-1* 408
Guazuma tomentosa var. *parvifolia* IPNI=*823393-1* 408
Guazuma ulmifolia var. *glabra* IPNI=*823393-1* 408
Guazuma ulmifolia var. *tomentosa* IPNI=*823393-1* 408
Guazuma ulmifolia var. *trianae* IPNI=*823393-1* 408
Guazuma ulmifolia var. *velutina* IPNI=*823393-1* 408
Guazuma utilis IPNI=*823393-1* 408
Guettarda acreana IPNI=*77107342-1* 358
Guettarda ambigua IPNI=*30058289-2* 351
Guettarda armata IPNI=*746662-1* 348
Guettarda coccinea IPNI=*30281446-2* 351
Guettarda costaricensis IPNI=*746662-1* 348
Guettarda dichotoma IPNI=*114528-2* 351
Guettarda fuertesii IPNI=*114528-2* 351
Guettarda havanensis IPNI=*30058289-2* 351
Guettarda leiantha IPNI=*77107342-1* 358
Guettarda lindeniana var. *reticulata* IPNI=*114528-2* 351
Guettarda mathiola IPNI=*30058289-2* 351
Guettarda paniculata IPNI=*57096-2* 348
Guettarda parviflora IPNI=*752006-1* 351
Guettarda parvifolia IPNI=*752006-1* 351
Guettarda reticulata IPNI=*114528-2* 351
Guettarda rugosa IPNI=*30058289-2* 351
Guettarda tetrandra IPNI=*114528-2* 351
Guettarda xylosteoides IPNI=*114524-2* 351
Guidonia adstringens IPNI=*779541-1* 403
Guidonia alba IPNI=*47939-2* 403
Guidonia hirsuta var. *glabrata* IPNI=*47939-2* 403
Guidonia hirta IPNI=*47939-2* 403
Guidonia hirta var. *glabriora* IPNI=*47939-2* 403
Guidonia laetioides IPNI=*48022-2* 403
Guidonia parvifolia IPNI=*779541-1* 403
Guidonia rosauriana IPNI=*48069-2* 403
Guidonia stipularis IPNI=*47945-2* 403
Guidonia stipularis var. *serrulata* IPNI=*47945-2* 403
Guidonia sylvestris IPNI=*274165-2* 403
Guidonia sylvestris var. *platyphylla* IPNI=*274165-2* 403
Guidonia ulmifolia IPNI=*779712-1* 403
Guilandina aculeata IPNI=*114706-2* 320
Guilandina bonduc var. *major* IPNI=*114706-2* 320
Guilandina bonducella IPNI=*114706-2* 320
Guilandina glabra IPNI=*114706-2* 320
Guilandina microphylla IPNI=*114718-2* 320
Guilandina moringa IPNI=*584736-1* 280
Guilandina socorroensis IPNI=*114706-2* 320

Guilandina viridiflora IPNI=*114718-2* 320
Guilielma gasipaes IPNI=*28453-2* 206
Guilielma granatensis IPNI=*664717-1* 207
Guilielma piritu IPNI=*28460-2* 206
Guilielma tenera IPNI=*664527-1* 206
Guillainia novoebudica IPNI=*795383-1* 250
Guillainia purpurata IPNI=*795383-1* 250
Guilleminea australis IPNI=*1183159-2* 282
Guilleminea illecebroides IPNI=*1183159-2* 282
Guilleminea illecebrum IPNI=*1183159-2* 282
Guilleminea procumbens IPNI=*327342-2* 281
Gumifera nilotica IPNI=*77089275-1* 336
Gunnera falklandica IPNI=*430163-1* 358
Gunnera integrifolia IPNI=*430163-1* 358
Gunnera plicata IPNI=*430163-1* 358
Gunnera relchei IPNI=*430163-1* 358
Gunnera reniformis IPNI=*430163-1* 358
Gunnera scabra IPNI=*430194-1* 359
Gurania andreana IPNI=*292789-1* 294
Gurania angustiflora IPNI=*1010655-1* 294
Gurania annulata IPNI=*1010655-1* 294
Gurania arabidae IPNI=*1010655-1* 294
Gurania autranii IPNI=*1010655-1* 294
Gurania balfoureana IPNI=*114819-2* 294
Gurania boliviana IPNI=*1010655-1* 294
Gurania breviflora IPNI=*114819-2* 294
Gurania breviflora var. *subintegrifolia* IPNI=*114819-2* 294
Gurania brevisepala IPNI=*1010655-1* 294
Gurania castroi IPNI=*114819-2* 294
Gurania cissoides IPNI=*114819-2* 294
Gurania coccinea IPNI=*114819-2* 294
Gurania costaricensis IPNI=*114819-2* 294
Gurania donnell-smithii IPNI=*292814-1* 294
Gurania dugandii IPNI=*1010655-1* 294
Gurania durandii IPNI=*1010655-1* 294
Gurania guianensis IPNI=*114819-2* 294
Gurania heteromorpha IPNI=*114819-2* 294
Gurania hirsuta IPNI=*292789-1* 294
Gurania inaequalis IPNI=*114819-2* 294
Gurania kegeliana IPNI=*1010655-1* 294
Gurania lanata IPNI=*292789-1* 294
Gurania levyana IPNI=*292814-1* 294
Gurania linkiana IPNI=*114819-2* 294
Gurania macrantha IPNI=*1010655-1* 294
Gurania macrophylla IPNI=*1010655-1* 294
Gurania macrotricha IPNI=*292789-1* 294
Gurania martiana IPNI=*292789-1* 294
Gurania multiflora IPNI=*1010655-1* 294
Gurania multiflora var. *longipedunculata* IPNI=*1010655-1* 294
Gurania nervosa IPNI=*1010655-1* 294
Gurania parviflora IPNI=*114819-2* 294
Gurania phanerosiphon IPNI=*292789-1* 294
Gurania pittieri IPNI=*292814-1* 294
Gurania plumosa IPNI=*292789-1* 294
Gurania pyrrhocephala IPNI=*292789-1* 294
Gurania racemifera IPNI=*114819-2* 294
Gurania seemanniana IPNI=*292814-1* 294
Gurania speciosa IPNI=*292789-1* 294
Gurania spinulosa IPNI=*1010655-1* 294
Gurania spinulosa var. *glabrata* IPNI=*1010655-1* 294
Gurania steinbachii IPNI=*292789-1* 294
Gurania sylvatica IPNI=*292789-1* 294
Gurania tessmannii IPNI=*114819-2* 294
Gurania tonduziana IPNI=*292814-1* 294
Gurania ucayalina IPNI=*114819-2* 294
Gurania variabilis IPNI=*114819-2* 294
Gurania wageneriana IPNI=*1010655-1* 294
Gussonia australis IPNI=*1187218-2* 387
Gussonia concolor IPNI=*1187218-2* 387
Gustavia angusta IPNI=*114915-2* 301
Gustavia antillana IPNI=*114915-2* 301
Gustavia augusta var. *brasiliensis* IPNI=*114915-2* 301
Gustavia augusta var. *calycaris* IPNI=*114915-2* 301
Gustavia augusta var. *guianensis* IPNI=*114915-2* 301
Gustavia augusta var. *verrucosa* IPNI=*114915-2* 301
Gustavia brasiliana IPNI=*596743-1* 302
Gustavia brasiliana var. *minor* IPNI=*596743-1* 302
Gustavia brasiliensis IPNI=*114915-2* 301
Gustavia caballoensis IPNI=*114966-2* 302
Gustavia calycaris IPNI=*114915-2* 301
Gustavia cincta IPNI=*114963-2* 302
Gustavia corymbosa IPNI=*114913-2* 301
Gustavia duckei IPNI=*114952-2* 302
Gustavia dumitiana IPNI=*114952-2* 302
Gustavia eximia IPNI=*596743-1* 302
Gustavia fastuosa IPNI=*596743-1* 302
Gustavia fastuosa var. *angustisepala* IPNI=*596743-1* 302
Gustavia fastuosa var. *latisepala* IPNI=*596743-1* 302
Gustavia foliosa IPNI=*114963-2* 302
Gustavia fustis-mortui IPNI=*596743-1* 302
Gustavia insignis IPNI=*114915-2* 301
Gustavia integrifolia IPNI=*596618-1* 301
Gustavia iquitosensis IPNI=*114952-2* 302
Gustavia laciniosa IPNI=*114915-2* 301
Gustavia magna IPNI=*114952-2* 302
Gustavia mangua IPNI=*114952-2* 302
Gustavia marcgraaviana IPNI=*114915-2* 301
Gustavia meizocarpa IPNI=*114915-2* 301
Gustavia mexiana IPNI=*114915-2* 301
Gustavia microcarpa IPNI=*596743-1* 302
Gustavia pleurocarpa IPNI=*596736-1* 302
Gustavia poeppigiana var. *nitens* IPNI=*114966-2* 302
Gustavia poeppigiana var. *opaca* IPNI=*114966-2* 302
Gustavia poeppigiana var. *rigida* IPNI=*114915-2* 301
Gustavia pterocarpa IPNI=*596743-1* 302
Gustavia ruiziana IPNI=*114913-2* 301
Gustavia speciosa subsp. *occidentalis* IPNI=*114963-2* 302
Gustavia spruceana IPNI=*114952-2* 302
Gustavia tessmannii IPNI=*114952-2* 302
Gustavia tetrapetala IPNI=*114915-2* 301
Gustavia theophrasta IPNI=*114915-2* 301
Gustavia ulei IPNI=*114966-2* 302
Gustavia urceolata IPNI=*114915-2* 301
Gustavia yaracuyensis IPNI=*114966-2* 302
Guzmania clavata IPNI=*115178-2* 229
Guzmania grandis IPNI=*115178-2* 229
Guzmania laxa IPNI=*115178-2* 229
Guzmania maculata IPNI=*115178-2* 229
Guzmania monostachia var. *alba* IPNI=*115178-2* 229
Guzmania musaica var. *concolor* IPNI=*115185-2* 229
Guzmania musaica var. *discolor* IPNI=*115185-2* 229
Guzmania musaica var. *rosea* IPNI=*115185-2* 229
Guzmania musaica var. *zebrina* IPNI=*115185-2* 229
Guzmania platysepala IPNI=*115178-2* 229
Guzmania sympaganthera IPNI=*115178-2* 229
Guzmania tricolor IPNI=*115178-2* 229
Guzmania tricolor var. *cuspidata* IPNI=*115178-2* 229
Guzmania tricolor var. *variegata* IPNI=*115178-2* 229
Gyas verecunda IPNI=*1218040-2* 217
Gymnandropogon annulatus IPNI=*398528-1* 237
Gymnanthelia connata IPNI=*396888-1* 236
Gymnanthes brasiliensis IPNI=*356183-1* 393
Gymnanthes brasiliensis var. *genuina* IPNI=*356183-1* 393
Gymnanthes concolor IPNI=*1187218-2* 387
Gymnobalanus guianensis IPNI=*467336-1* 184
Gymnobalanus latifolia IPNI=*77158867-1* 183

Gymnobalanus perseoides IPNI=*1032261-2* 185
Gymnogramma arifolia IPNI=*319061-2* 177
Gymnogramma bidentata IPNI=*17345640-1* 178
Gymnogramma bogotensis IPNI=*17127250-1* 178
Gymnogramma calomelanos f. *albo-farinosa* IPNI=*17345640-1* 178
Gymnogramma calomelanos f. *epruinata* IPNI=*17345640-1* 178
Gymnogramma calomelanos f. *flavofarinosa* IPNI=*17345640-1* 178
Gymnogramma calomelanos f. *peruviana* IPNI=*200935-2* 178
Gymnogramma calomelanos f. *tartarea* IPNI=*200935-2* 178
Gymnogramma calomelanos IPNI=*17345640-1* 178
Gymnogramma calomelanos var. *denudata* IPNI=*17345640-1* 178
Gymnogramma calomelanos var. *martensii* IPNI=*17345640-1* 178
Gymnogramma calomelanos var. *normalis* IPNI=*17345640-1* 178
Gymnogramma calomelanos var. *peruviana* IPNI=*200935-2* 178
Gymnogramma calomelanos var. *tartarea* IPNI=*200935-2* 178
Gymnogramma calomelas IPNI=*17345640-1* 178
Gymnogramma flavens IPNI=*77182941-1* 178
Gymnogramma flexuosa IPNI=*77110633-1* 178
Gymnogramma flexuosa var. *hirsuta* IPNI=*77110633-1* 178
Gymnogramma flexuosa var. *peruviana* IPNI=*77110633-1* 178
Gymnogramma haematodes IPNI=*77110633-1* 178
Gymnogramma hybrida IPNI=*17345640-1* 178
Gymnogramma kunzei IPNI=*77110633-1* 178
Gymnogramma martensii IPNI=*17345640-1* 178
Gymnogramma massonii IPNI=*17345640-1* 178
Gymnogramma mayi IPNI=*200935-2* 178
Gymnogramma nivea IPNI=*77182941-1* 178
Gymnogramma pallescens IPNI=*200935-2* 178
Gymnogramma peruviana IPNI=*200935-2* 178
Gymnogramma peruviana var. *argyrophylla* IPNI=*200935-2* 178
Gymnogramma peruviana var. *major* IPNI=*200935-2* 178
Gymnogramma plicata IPNI=*200935-2* 178
Gymnogramma recurvata IPNI=*77110633-1* 178
Gymnogramma reniformis IPNI=*17217830-1* 178
Gymnogramma serrata IPNI=*17345640-1* 178
Gymnogramma tartarea IPNI=*200935-2* 178
Gymnogramma tartarea var. *aureopallida* IPNI=*200935-2* 178
Gymnogramma tenera IPNI=*77182941-1* 178
Gymnogramma villosula IPNI=*77110633-1* 178
Gymnolomia cruciata IPNI=*993359-1* 271
Gymnolomia maculata IPNI=*993359-1* 271
Gymnolomia sylvatica IPNI=*213524-1* 264
Gymnoluma glabrescens IPNI=*89008-2* 305
Gymnoluma ucuqui IPNI=*209914-2* 308
Gymnopetalum calyculatum IPNI=*320034-2* 293
Gymnopogon beyrichianus IPNI=*395997-1* 236
Gymnopogon biflorus IPNI=*115792-2* 240
Gymnopogon filiformis IPNI=*115792-2* 240
Gymnopogon haumanii IPNI=*395997-1* 236
Gymnopogon laevis IPNI=*115792-2* 240
Gymnopogon laevis var. *pluriflorus* IPNI=*115792-2* 240
Gymnopogon spicatus f. *longiaristatus* IPNI=*115792-2* 240
Gymnopogon spicatus var. *longiaristatus* IPNI=*115792-2* 240
Gymnopsis divaricata IPNI=*242973-1* 269
Gymnopsis euxenioides IPNI=*993359-1* 271
Gymnopsis microcephala IPNI=*202984-1* 261
Gymnopyrenium pyracantha IPNI=*730611-1* 444
Gymnostachyum argyroneurum IPNI=*105033-2* 360
Gymnostachyum pearcei IPNI=*105033-2* 360
Gymnostachyum verschaffeltii IPNI=*105033-2* 360
Gymnostyles anthemifolia IPNI=*249968-1* 270
Gymnostyles marilandica IPNI=*1099254-2* 268
Gymnostyles mutisii IPNI=*249968-1* 270
Gymnotrix nitens IPNI=*77106033-1* 235
Gynandropsis adenocarpa IPNI=*147061-1* 279
Gynandropsis affinis IPNI=*60518-2* 279
Gynandropsis candelabrum IPNI=*60518-2* 279
Gynandropsis denticulata IPNI=*60518-2* 279
Gynandropsis glandulosa IPNI=*60518-2* 279
Gynandropsis grandiflora IPNI=*147320-1* 279
Gynandropsis gynandra IPNI=*60518-2* 279
Gynandropsis heterotricha IPNI=*60518-2* 279
Gynandropsis macrantha IPNI=*147320-1* 279
Gynandropsis muricata IPNI=*60518-2* 279
Gynandropsis ophitocarpa IPNI=*60518-2* 279
Gynandropsis palmipes IPNI=*60518-2* 279
Gynandropsis pentaphylla IPNI=*60518-2* 279
Gynandropsis sessilifolia IPNI=*60518-2* 279
Gynandropsis sinica IPNI=*60518-2* 279
Gynandropsis speciosa IPNI=*147320-1* 279
Gynandropsis tracyi IPNI=*147320-1* 279
Gynandropsis triphylla IPNI=*60518-2* 279
Gynandropsis viscida IPNI=*60518-2* 279
Gynerium argenteum IPNI=*66317-2* 236
Gynerium dioicum IPNI=*66317-2* 236
Gynerium jubatum IPNI=*66302-2* 236
Gynerium neesii IPNI=*66302-2* 236
Gynerium nitidum IPNI=*66306-2* 236
Gynerium purpureum IPNI=*66317-2* 236
Gynerium pygmaeum IPNI=*66302-2* 236
Gynestum deversum IPNI=*667064-1* 209
Gynestum maximum IPNI=*667168-1* 209
Gynestum strictum IPNI=*667262-1* 209
Gynheteria dentata IPNI=*30283369-2* 271
Gynheteria incana IPNI=*30283369-2* 271
Gynheteria salicifolia IPNI=*30283369-2* 271
Gynoxys auriculata IPNI=*6096-2* 256
Gynoxys berlandieri IPNI=*211361-2* 269
Gynoxys berlandieri var. *cordifolia* IPNI=*211361-2* 269
Gynoxys berlandieri var. *cuneata* IPNI=*211361-2* 269
Gynoxys cordifolia IPNI=*211361-2* 269
Gynoxys cordifolia var. *neaei* IPNI=*211361-2* 269
Gynoxys cummingii IPNI=*211365-2* 269
Gynoxys prenanthifolia IPNI=*6096-2* 256
Gynoxys verrucosa IPNI=*970931-1* 256
Gynoxys verrucosa var. *grandifolia* IPNI=*970931-1* 256
Gynoxys verrucosa var. *magna* IPNI=*970931-1* 256
Gynura aspera IPNI=*1152557-2* 261
Gynura ecalyculata IPNI=*203080-1* 261
Gynura malasica IPNI=*1152557-2* 261
Gynura rosea IPNI=*203365-1* 261
Gypsophila ceballosi IPNI=*154453-1* 285
Gypsophila producta IPNI=*154453-1* 285
Gyrocarpus acuminatus IPNI=*170637-1* 182
Gyrocarpus asiaticus IPNI=*170637-1* 182
Gyrocarpus jacquinii IPNI=*170637-1* 182
Gyrocarpus lobatus IPNI=*170637-1* 182
Gyrocarpus rugosus IPNI=*170637-1* 182
Gyrostachys haenkeana IPNI=*73688-2* 218
Gyrostachys minor IPNI=*73688-2* 218
Gyrostachys speciosa IPNI=*659300-1* 223

H

Habenaria autumnalis IPNI=*637263-1* 219
Habenaria garberi IPNI=*637263-1* 219
Habenaria herzogii IPNI=*637263-1* 219
Habenaria odontopetala f. *heatonii* IPNI=*637263-1* 219
Habenaria odontopetala IPNI=*637263-1* 219
Habenaria purdiei IPNI=*637263-1* 219
Habenaria selerorum IPNI=*637263-1* 219
Habenaria strictissima var. *odontopetala* IPNI=*637263-1* 219
Habenaria tetrodon IPNI=*637263-1* 219
Habenella floribunda IPNI=*637263-1* 219
Habenella garberi IPNI=*637263-1* 219
Habenella herzogii IPNI=*637263-1* 219
Habenella odontopetala IPNI=*637263-1* 219

Habershamia cuneifolia IPNI=*1072674-2* 376
Habranthus andersonii var. *texanus* IPNI=*67314-1* 214
Habranthus berteroi IPNI=*270986-2* 214
Habranthus parvulus IPNI=*67314-1* 214
Habranthus quilmesianus IPNI=*270986-2* 214
Habranthus robustus IPNI=*270986-2* 214
Habranthus texanus IPNI=*67314-1* 214
Habranthus tubispathus f. *bicolor* IPNI=*67314-1* 214
Habranthus tubispathus f. *roseus* IPNI=*67314-1* 214
Habranthus tubispathus IPNI=*67314-1* 214
Habranthus tubispathus subsp. *macranthus* IPNI=*67314-1* 214
Habranthus tubispathus subsp. *variabilis* IPNI=*67314-1* 214
Habranthus tubispathus var. *bicolor* IPNI=*67314-1* 214
Habranthus tubispathus var. *roseus* IPNI=*67314-1* 214
Habranthus variabilis IPNI=*67314-1* 214
Habzelia aromatica IPNI=*76067-1* 189
Habzelia cubensis IPNI=*76067-1* 189
Habzelia discreta IPNI=*76119-1* 189
Haemadictyon acutifolium IPNI=*81405-1* 342
Haemadictyon caliginosum IPNI=*81405-1* 342
Haemadictyon glabratum IPNI=*81387-1* 342
Haemadictyon grandiflorum IPNI=*77191232-1* 342
Haemadictyon molle IPNI=*81387-1* 342
Haemadictyon nutans IPNI=*81405-1* 342
Haemadictyon pallidum IPNI=*81387-1* 342
Haemadictyon tomentellum IPNI=*81387-1* 342
Haemadictyon trifidum IPNI=*210147-2* 342
Haemadictyon venosum IPNI=*81405-1* 342
Haematoxylum boreale IPNI=*497227-1* 320
Haemocharis camelliodes IPNI=*112597-2* 309
Haemocharis caracasana IPNI=*112597-2* 309
Haemocharis intermedia IPNI=*112597-2* 309
Haemocharis obovata IPNI=*112597-2* 309
Haemocharis parviflora IPNI=*112597-2* 309
Haemocharis praemorsa IPNI=*112597-2* 309
Haemocharis pubescens IPNI=*112597-2* 309
Haemocharis quinoderma IPNI=*112597-2* 309
Haemocharis semiserrata IPNI=*112597-2* 309
Haemocharis speciosa IPNI=*112597-2* 309
Haemocharis symplocodes IPNI=*112597-2* 309
Haenkea multiflora IPNI=*155542-2* 291
Hakea robusta IPNI=*50798-3* 433
Halenia antigonorrhoica IPNI=*117020-2* 345
Halenia campanulata IPNI=*369463-1* 345
Halenia cuatrecasasii IPNI=*369443-1* 345
Halenia dasyantha IPNI=*369463-1* 345
Halenia foliosa IPNI=*369463-1* 345
Halenia hoppii IPNI=*117020-2* 345
Halenia hygrophila IPNI=*369463-1* 345
Halenia hypericoides IPNI=*117020-2* 345
Halenia meyeri-johannis IPNI=*117020-2* 345
Halenia occulta IPNI=*369443-1* 345
Halenia phyllophora IPNI=*117020-2* 345
Halenia tolimae IPNI=*369463-1* 345
Halenia verticillata IPNI=*369463-1* 345
Halerpestes cymbalaria f. *hebecaulis* IPNI=*711250-1* 436
Halerpestes cymbalaria var. *saximontanus* IPNI=*711250-1* 436
Halerpestes tridentata IPNI=*711250-1* 436
Halimus maritima IPNI=*235145-2* 280
Halimus portulacastrum IPNI=*235145-2* 280
Hallia vaginalis IPNI=*473569-1* 311
Halocnemum americanum IPNI=*165886-1* 283
Halocnemum ritterianum IPNI=*165886-1* 283
Halopeplis gilliesii IPNI=*165886-1* 283
Halopeplis ritteriana IPNI=*165886-1* 283
Halostachys ritteriana IPNI=*165886-1* 283
Hamelia appendiculata IPNI=*117176-2* 351
Hamelia axillaris var. *appendiculata* IPNI=*117176-2* 351
Hamelia axillaris var. *breviflora* IPNI=*117176-2* 351
Hamelia brachystemon IPNI=*752194-1* 351
Hamelia breviflora IPNI=*117176-2* 351
Hamelia brittoniana IPNI=*752194-1* 351
Hamelia chrysantha IPNI=*117176-2* 351
Hamelia coccinea IPNI=*752194-1* 351
Hamelia corymbosa IPNI=*752194-1* 351
Hamelia declinata IPNI=*117176-2* 351
Hamelia erecta IPNI=*752194-1* 351
Hamelia glabra IPNI=*743074-1* 347
Hamelia intermedia IPNI=*752194-1* 351
Hamelia lanuginosa IPNI=*752194-1* 351
Hamelia latifolia IPNI=*752194-1* 351
Hamelia lucida IPNI=*117176-2* 351
Hamelia lutea IPNI=*117176-2* 351
Hamelia nodosa IPNI=*752194-1* 351
Hamelia ovata IPNI=*752194-1* 351
Hamelia patens var. *axillaroides* IPNI=*752194-1* 351
Hamelia patens var. *glabra* IPNI=*752194-1* 351
Hamelia patens var. *quinifolia* IPNI=*752194-1* 351
Hamelia pedicellata IPNI=*752194-1* 351
Hamelia sessiliflora IPNI=*743074-1* 347
Hamelia sphaerocarpa IPNI=*752194-1* 351
Hamelia suaveolens IPNI=*752194-1* 351
Hamelia tubiflora IPNI=*752194-1* 351
Hamelia verticillata IPNI=*752194-1* 351
Hamelia viridifolia IPNI=*752194-1* 351
Hancornia amapa IPNI=*182946-2* 342
Hancornia arborescens IPNI=*79531-1* 341
Hancornia floribunda IPNI=*79552-1* 341
Hancornia gracilis IPNI=*79536-1* 341
Hancornia macrophylla IPNI=*81716-1* 343
Hancornia monosperma IPNI=*79531-1* 341
Hancornia ramosissima IPNI=*79547-1* 341
Handroanthus araliaceus IPNI=*77085096-1* 364
Handroanthus atractocarpus IPNI=*77085096-1* 364
Handroanthus avellanedae IPNI=*117330-2* 363
Handroanthus durus IPNI=*248703-2* 365
Handroanthus flavescens IPNI=*77085096-1* 364
Handroanthus odontodiscus IPNI=*248789-2* 365
Handroanthus pentaphyllus IPNI=*248689-2* 365
Handroanthus piutinga IPNI=*248789-2* 365
Handroanthus roseoalbus IPNI=*248789-2* 365
Hapalanthus repens IPNI=*171826-1* 224
Haplodesmium lindenianum IPNI=*567327-1* 418
Haplopteris lineata IPNI=*17234040-1* 178
Hargasseria mexicana IPNI=*76687-2* 415
Hargasseria schiedeana IPNI=*76687-2* 415
Hariota cassytha IPNI=*219250-2* 284
Hariota coriacea IPNI=*963240-1* 284
Hariota fasciculata IPNI=*219250-2* 284
Hariota macrocarpa IPNI=*132939-1* 283
Hariota micrantha IPNI=*138328-1* 284
Hariota parasitica IPNI=*219250-2* 284
Hariota ramulosa IPNI=*963240-1* 284
Harrachia repanda IPNI=*1053173-2* 359
Harrisia undata IPNI=*77163184-1* 285
Hartigia barbigera IPNI=*1108920-2* 420
Hartigia milleflora IPNI=*1110588-2* 423
Hartigia oblongifolia IPNI=*1158780-2* 422
Hartigia spectabilis IPNI=*1158780-2* 422
Hartmannia affinis IPNI=*173348-2* 430
Hartmannia boliviana IPNI=*173202-2* 430
Hartmannia gauroides IPNI=*173348-2* 430
Hartmannia latiflora IPNI=*614141-1* 430
Hartmannia macrantha IPNI=*614141-1* 430
Hartmannia rosea IPNI=*173348-2* 430
Hartmannia rosea var. *parvifolia* IPNI=*173348-2* 430

Hartmannia tetraptera IPNI=*614141-1* 430
Hartmannia virgata IPNI=*173348-2* 430
Hartwegia comosa IPNI=*532810-1* 215
Hasseltia mexicana IPNI=*202149-2* 404
Hasseltia micrantha IPNI=*834658-1* 404
Hasseltia peruviana IPNI=*834658-1* 404
Hasseltia pubescens IPNI=*834658-1* 404
Hasseltia pyramidalis IPNI=*202149-2* 404
Hasseltia quinquenervia IPNI=*202149-2* 404
Hasseltia rigida IPNI=*834658-1* 404
Hasseltia tomentulosa IPNI=*834658-1* 404
Hasseltiopsis albomicans IPNI=*202149-2* 404
Hasseltiopsis leucothyrsa IPNI=*202149-2* 404
Hasseltiopsis mucronata IPNI=*202149-2* 404
Hastingia coccinea IPNI=*863073-1* 369
Hastingia scandens IPNI=*863073-1* 369
Havetia flavida IPNI=*1016004-2* 384
Havetia flexilis IPNI=*1016004-2* 384
Havetia obovata IPNI=*1082351-2* 384
Havetia octandra IPNI=*1174011-2* 384
Havetiopsis flavida IPNI=*1016004-2* 384
Havetiopsis flexilis IPNI=*1016004-2* 384
Havetiopsis flexilis var. *flavida* IPNI=*1016004-2* 384
Haworthia fasciata IPNI=*77138009-1* 215
Haworthia pumila subsp. *fasciata* IPNI=*77138009-1* 215
Haynea edulis IPNI=*236351-1* 267
Hebanthe aurata IPNI=*1002675-1* 283
Hebanthe mollis IPNI=*60789-1* 283
Hebanthe parviflora IPNI=*1002675-1* 283
Hebe anomala IPNI=*812400-1* 377
Hebe buxifolia IPNI=*812400-1* 377
Hebe elliptica IPNI=*811951-1* 377
Hebe fonkii IPNI=*812616-1* 377
Hebe haustrata IPNI=*812400-1* 377
Hebe imperialis IPNI=*812675-1* 377
Hebe magellanica IPNI=*811951-1* 377
Hebe menziesii IPNI=*811951-1* 377
Hebe odora IPNI=*812400-1* 377
Hebe salicifolia IPNI=*812616-1* 377
Hebe speciosa IPNI=*812675-1* 377
Hebeandra crassifolia IPNI=*690504-1* 338
Hebeandra fastigiata IPNI=*690538-1* 338
Hebeandra hirta IPNI=*163646-2* 338
Hebeandra padifolia IPNI=*77143255-1* 338
Hebeandra phillyreoides IPNI=*690654-1* 338
Hebeandra revoluta IPNI=*690504-1* 338
Hebeclinium macrophyllum var. *rivale* IPNI=*1092582-2* 263
Hebecocca panamensis IPNI=*352565-1* 392
Hebepetalum calophyllum IPNI=*544878-1* 394
Hebepetalum parviflorum IPNI=*544878-1* 394
Heckeria peltata IPNI=*316334-2* 197
Heckeria scutata IPNI=*316334-2* 197
Heckeria sidifolia IPNI=*316339-2* 197
Heckeria sidifolia var. *subglabrata* IPNI=*316339-2* 197
Heckeria speciosa IPNI=*316334-2* 197
Heckeria subpeltata IPNI=*316339-2* 197
Heckeria umbellata IPNI=*316339-2* 197
Hedeoma caerulescens IPNI=*1001364-1* 371
Hedeoma gracillima IPNI=*61179-2* 369
Hedera alaris IPNI=*1133937-2* 253
Hedera arborea IPNI=*1133937-2* 253
Hedera canariensis var. *arborescens* IPNI=*90665-1* 253
Hedera capitata IPNI=*30050876-2* 254
Hedera catalpifolia IPNI=*91160-1* 254
Hedera cochleata IPNI=*91817-1* 254
Hedera communis IPNI=*90723-1* 253
Hedera floribunda IPNI=*91222-1* 254
Hedera frondosa IPNI=*30050876-2* 254
Hedera grandifolia IPNI=*90665-1* 253
Hedera latifolia IPNI=*91817-1* 254
Hedera multiflora IPNI=*30050876-2* 254
Hedera mutisiana IPNI=*91261-1* 254
Hedera poetica IPNI=*90723-1* 253
Hedera septemnervia IPNI=*91160-1* 254
Hedwigia balsamifera IPNI=*77202519-1* 453
Hedwigia panamensis IPNI=*77202519-1* 453
Hedwigia rhoifolia IPNI=*77184241-1* 453
Hedychium angustifolium IPNI=*796834-1* 251
Hedychium aurantiacum IPNI=*796834-1* 251
Hedychium chrysoleucum IPNI=*796836-1* 251
Hedychium coronarium var. *maximum* IPNI=*796836-1* 251
Hedychium flavescens var. *chrysoleucum* IPNI=*796836-1* 251
Hedychium gandasulium IPNI=*796836-1* 251
Hedychium lingulatum IPNI=*796836-1* 251
Hedychium longifolium IPNI=*796834-1* 251
Hedychium maximum IPNI=*796836-1* 251
Hedychium prophetae IPNI=*796836-1* 251
Hedychium spicatum IPNI=*796836-1* 251
Hedychium squarrosum IPNI=*796834-1* 251
Hedychium sulphureum IPNI=*796836-1* 251
Hedychloe fragrans IPNI=*77144505-1* 230
Hedyosmum arborescens IPNI=*277998-2* 182
Hedyosmum artocarpus IPNI=*167876-1* 181
Hedyosmum callososerratum IPNI=*277969-2* 181
Hedyosmum crassifolium IPNI=*297228-2* 181
Hedyosmum flocculosum IPNI=*277998-2* 182
Hedyosmum glabratum IPNI=*167884-1* 181
Hedyosmum glabratum IPNI=*277980-2* 181
Hedyosmum glaucum IPNI=*167876-1* 181
Hedyosmum glaucum IPNI=*167884-1* 181
Hedyosmum granizo IPNI=*167850-1* 181
Hedyosmum hirsutum IPNI=*167887-1* 182
Hedyosmum huilense IPNI=*167884-1* 181
Hedyosmum laciniatum IPNI=*167832-1* 181
Hedyosmum mandonii IPNI=*167887-1* 182
Hedyosmum montanum IPNI=*277980-2* 181
Hedyosmum pavonii IPNI=*167832-1* 181
Hedyosmum scabrum var. *pavonii* IPNI=*167832-1* 181
Hedyosmum toxicum IPNI=*167850-1* 181
Hedyotis caracasana IPNI=*743608-1* 347
Hedyotis coarctata IPNI=*18012-2* 347
Hedyotis conferta IPNI=*17992-2* 347
Hedyotis conferta var. *pusilla* IPNI=*17992-2* 347
Hedyotis cordata IPNI=*746945-1* 348
Hedyotis corymbosa IPNI=*757791-1* 353
Hedyotis filiformis IPNI=*17992-2* 347
Hedyotis floribunda IPNI=*743608-1* 347
Hedyotis herbacea IPNI=*757892-1* 353
Hedyotis humboldtiana IPNI=*743608-1* 347
Hedyotis hypnoides IPNI=*17992-2* 347
Hedyotis hyssopifolia IPNI=*18012-2* 347
Hedyotis juniperifolia IPNI=*18012-2* 347
Hedyotis lancifolia IPNI=*757942-1* 353
Hedyotis laricifolia IPNI=*18012-2* 347
Hedyotis linearis IPNI=*757892-1* 353
Hedyotis micrantha IPNI=*757892-1* 353
Hedyotis microphylla IPNI=*17992-2* 347
Hedyotis nitida IPNI=*743608-1* 347
Hedyotis serpens IPNI=*17992-2* 347
Hedyotis thesioides IPNI=*18012-2* 347
Hedyotis thujoides IPNI=*18012-2* 347
Hedyotis thymifolia IPNI=*18012-2* 347
Hedyotis triflora IPNI=*757892-1* 353
Hedysarum adhaerens IPNI=*1109494-2* 318
Hedysarum adscendens IPNI=*77187388-1* 320
Hedysarum adscendens var. *caeruleum* IPNI=*1124646-2* 318

Hedysarum ancistrocarpum IPNI=*1015219-2* 318
Hedysarum angustifolium IPNI=*78351-2* 317
Hedysarum aparines IPNI=*134286-3* 318
Hedysarum ascendens IPNI=*77187388-1* 320
Hedysarum asperum IPNI=*78415-2* 317
Hedysarum axillare IPNI=*491595-1* 317
Hedysarum barbatum IPNI=*77187394-1* 320
Hedysarum biflorum IPNI=*77187443-1* 320
Hedysarum bifolium IPNI=*271325-2* 337
Hedysarum brasilianum IPNI=*77211517-1* 316
Hedysarum caespitosum IPNI=*77187388-1* 320
Hedysarum cajanifolium IPNI=*1108524-2* 317
Hedysarum canescens IPNI=*78415-2* 317
Hedysarum canum IPNI=*1015219-2* 318
Hedysarum celtifolium IPNI=*495307-1* 320
Hedysarum clandestinum IPNI=*526091-1* 337
Hedysarum coccineum IPNI=*499456-1* 321
Hedysarum conjugatum IPNI=*30026493-2* 337
Hedysarum conjunctum IPNI=*1015219-2* 318
Hedysarum coronarium IPNI=*519725-1* 334
Hedysarum diphyllum IPNI=*30026493-2* 337
Hedysarum distortum IPNI=*78415-2* 317
Hedysarum diversifolium IPNI=*1015219-2* 318
Hedysarum ecastaphyllum IPNI=*75934-2* 317
Hedysarum emarginatum IPNI=*1130757-2* 317
Hedysarum falcatum IPNI=*77211592-1* 316
Hedysarum gemellum IPNI=*30156577-2* 337
Hedysarum giganteum IPNI=*78415-2* 317
Hedysarum glabrum IPNI=*1130757-2* 317
Hedysarum granulatum IPNI=*77187443-1* 320
Hedysarum granuliferum IPNI=*77187443-1* 320
Hedysarum hamatum IPNI=*247544-2* 334
Hedysarum hamatum var. *viscosum* IPNI=*1176701-2* 334
Hedysarum heterocarpon IPNI=*77187416-1* 320
Hedysarum hirtum IPNI=*77211517-1* 316
Hedysarum incanum IPNI=*1015219-2* 318
Hedysarum intortum IPNI=*134286-3* 318
Hedysarum lindleyi IPNI=*77187388-1* 320
Hedysarum madagascariense IPNI=*1015219-2* 318
Hedysarum mexicanum IPNI=*1109494-2* 318
Hedysarum mexicanum IPNI=*78351-2* 317
Hedysarum molle IPNI=*1130757-2* 317
Hedysarum molliculum IPNI=*1124646-2* 318
Hedysarum mucronatum IPNI=*1108524-2* 317
Hedysarum oblongifolium IPNI=*491595-1* 317
Hedysarum portoricense IPNI=*1015219-2* 318
Hedysarum procumbens IPNI=*78541-2* 318
Hedysarum prostratum IPNI=*519725-1* 334
Hedysarum purpureum IPNI=*1028008-2* 318
Hedysarum racemosum IPNI=*1015219-2* 318
Hedysarum ramosum IPNI=*135026-3* 318
Hedysarum reptans IPNI=*491595-1* 317
Hedysarum scorpiurus IPNI=*135026-3* 318
Hedysarum spirale IPNI=*78541-2* 318
Hedysarum stipulaceum IPNI=*1028008-2* 318
Hedysarum stipulaceum IPNI=*77187443-1* 320
Hedysarum stoloniferum IPNI=*491595-1* 317
Hedysarum strobiliferum IPNI=*495307-1* 320
Hedysarum supinum IPNI=*1015219-2* 318
Hedysarum terminale IPNI=*1130757-2* 317
Hedysarum tortuosum IPNI=*1028008-2* 318
Hedysarum triflorum IPNI=*77187443-1* 320
Hedysarum trigonum IPNI=*134286-3* 318
Hedysarum uncinatum IPNI=*1109494-2* 318
Hedysarum vaginale IPNI=*473569-1* 311
Hedysarum violaceum IPNI=*491595-1* 317
Hedysarum virgatum IPNI=*1109494-2* 318
Heeria axillaris IPNI=*569110-1* 419
Heeria elegans IPNI=*121047-2* 419
Heeria macrostachya IPNI=*569110-1* 419
Heeria procumbens IPNI=*121047-2* 419
Heeria rosea IPNI=*569110-1* 419
Heeria subtriplinervia IPNI=*569110-1* 419
Heeria undulata IPNI=*569110-1* 419
Heinzia peregrina IPNI=*492769-1* 318
Heisteria burchellii IPNI=*70029599-1* 449
Heisteria celastrinea IPNI=*70029599-1* 449
Heisteria cyanocarpa IPNI=*70029599-1* 449
Heisteria cyanocarpa subsp. *occidentalis* IPNI=*70029599-1* 449
Heisteria ixiamensis IPNI=*70029599-1* 449
Heisteria kappleri IPNI=*55672-2* 449
Heisteria longipes IPNI=*70029599-1* 449
Heisteria megalophylla IPNI=*118612-2* 449
Heisteria olivae IPNI=*118612-2* 449
Heisteria pallida IPNI=*70029599-1* 449
Heisteria rhaptostylum IPNI=*70029599-1* 449
Heisteria tubicina IPNI=*607627-1* 449
Hekaterosachne elatior IPNI=*1004002-1* 242
Helanthium parvulum IPNI=*118649-2* 197
Helemonium canescens IPNI=*213524-1* 264
Helenium vulgare IPNI=*60478388-2* 267
Helenomoium canescens IPNI=*213524-1* 264
Heleochloa juncea IPNI=*1103484-2* 245
Heleogiton tabernaemontani IPNI=*1019164-1* 232
Heleophylax tabernaemontani IPNI=*1019164-1* 232
Helepta angustifolia IPNI=*213534-1* 264
Helepta grandiflora IPNI=*213534-1* 264
Helepta parviflora IPNI=*213534-1* 264
Helia amoena IPNI=*1002418-1* 344
Helia campanulacea IPNI=*118811-2* 345
Helia gracilis IPNI=*1002418-1* 344
Helia nemorosa IPNI=*130185-2* 345
Helia paniculata IPNI=*130185-2* 345
Helia purpurascens IPNI=*1002418-1* 344
Helia spruceana IPNI=*1002418-1* 344
Helia uliginosa IPNI=*1002418-1* 344
Heliacme scorzonerifolia IPNI=*64173-1* 216
Helianthella latifolia IPNI=*259782-1* 272
Helianthemoides patens IPNI=*1007146-2* 290
Helianthostylis guggenheimii IPNI=*317010-2* 439
Helianthostylis paraensis IPNI=*317010-2* 439
Helianthostylis salzedoi IPNI=*317010-2* 439
Helianthostylis schultesii IPNI=*317010-2* 439
Helianthostylis suerpo IPNI=*317010-2* 439
Helianthus annuus f. *fallax* IPNI=*119003-2* 263
Helianthus annuus f. *lenticularis* IPNI=*119003-2* 263
Helianthus annuus f. *silvester* IPNI=*119003-2* 263
Helianthus annuus lenticularis IPNI=*119003-2* 263
Helianthus annuus mut. inornatus IPNI=*119003-2* 263
Helianthus annuus subsp. *jaegeri* IPNI=*119003-2* 263
Helianthus annuus subsp. *lenticularis* IPNI=*119003-2* 263
Helianthus annuus subsp. *texanus* IPNI=*119003-2* 263
Helianthus annuus var. *apicalis* IPNI=*119003-2* 263
Helianthus annuus var. *aridus* IPNI=*119003-2* 263
Helianthus annuus var. *basalis* IPNI=*119003-2* 263
Helianthus annuus var. *chrysanthemoides* IPNI=*119003-2* 263
Helianthus annuus var. *convolutus* IPNI=*119003-2* 263
Helianthus annuus var. *coronatus* IPNI=*119003-2* 263
Helianthus annuus var. *dilutus* IPNI=*119003-2* 263
Helianthus annuus var. *flavobasis* IPNI=*119003-2* 263
Helianthus annuus var. *indicus* IPNI=*119003-2* 263
Helianthus annuus var. *latibasis* IPNI=*119003-2* 263
Helianthus annuus var. *lenticularis* IPNI=*119003-2* 263
Helianthus annuus var. *macrocarpus* IPNI=*119003-2* 263
Helianthus annuus var. *niger* IPNI=*119003-2* 263
Helianthus annuus var. *pallescens* IPNI=*119003-2* 263

Helianthus annuus var. *passiflora* IPNI=*119003-2* 263
Helianthus annuus var. *primulinus* IPNI=*119003-2* 263
Helianthus annuus var. *pumilus* IPNI=*119003-2* 263
Helianthus annuus var. *reversus* IPNI=*119003-2* 263
Helianthus annuus var. *revolutus* IPNI=*119003-2* 263
Helianthus annuus var. *ruberrimus* IPNI=*119003-2* 263
Helianthus annuus var. *selene* IPNI=*119003-2* 263
Helianthus annuus var. *semivinosus* IPNI=*119003-2* 263
Helianthus annuus var. *texanus* IPNI=*119003-2* 263
Helianthus annuus var. *tortuosus* IPNI=*119003-2* 263
Helianthus annuus var. *trizonatus* IPNI=*119003-2* 263
Helianthus annuus var. *vinosissimus* IPNI=*119003-2* 263
Helianthus annuus var. *vinosus* IPNI=*119003-2* 263
Helianthus aridus IPNI=*119003-2* 263
Helianthus aridus var. *citrinus* IPNI=*119003-2* 263
Helianthus baillonianus IPNI=*259782-1* 272
Helianthus buphthalmoides IPNI=*213524-1* 264
Helianthus cordatus IPNI=*213524-1* 264
Helianthus dentatus IPNI=*259782-1* 272
Helianthus erythrocarpus IPNI=*119003-2* 263
Helianthus grandiflorus IPNI=*119003-2* 263
Helianthus grosseserratus IPNI=*993359-1* 271
Helianthus indicus IPNI=*119003-2* 263
Helianthus jaegeri IPNI=*119003-2* 263
Helianthus laciniatus IPNI=*241716-1* 269
Helianthus laevis IPNI=*32497-2* 258
Helianthus lenticularis IPNI=*119003-2* 263
Helianthus lenticularis var. *angustus* IPNI=*119003-2* 263
Helianthus lenticularis var. *aridus* IPNI=*119003-2* 263
Helianthus lenticularis var. *primulinus* IPNI=*119003-2* 263
Helianthus lindheimerianus IPNI=*119003-2* 263
Helianthus macrocarpus IPNI=*119003-2* 263
Helianthus membranifolius IPNI=*993359-1* 271
Helianthus microclinus IPNI=*259782-1* 272
Helianthus ovatus IPNI=*119003-2* 263
Helianthus petiolaris IPNI=*119003-2* 263
Helianthus platycephalus IPNI=*119003-2* 263
Helianthus pumilus IPNI=*119003-2* 263
Helianthus quinquelobus IPNI=*255747-1* 271
Helianthus sarmentosus IPNI=*993359-1* 271
Helianthus scaber IPNI=*993359-1* 271
Helianthus speciosus IPNI=*254724-2* 271
Helianthus triqueter IPNI=*259782-1* 272
Helianthus ultiflorus IPNI=*119003-2* 263
Helichrysum acuminatum IPNI=*962307-1* 272
Helichrysum banksii IPNI=*962307-1* 272
Helichrysum bracteatum f. *incurvum* IPNI=*962307-1* 272
Helichrysum bracteatum IPNI=*962307-1* 272
Helichrysum bracteatum var. *albidum* IPNI=*962307-1* 272
Helichrysum bracteatum var. *angustifolium* IPNI=*962307-1* 272
Helichrysum bracteatum var. *atrococcineum* IPNI=*962307-1* 272
Helichrysum bracteatum var. *atrosanguineum* IPNI=*962307-1* 272
Helichrysum bracteatum var. *chrysanthemum* IPNI=*962307-1* 272
Helichrysum bracteatum var. *chrysanthum* IPNI=*962307-1* 272
Helichrysum bracteatum var. *eriopodum* IPNI=*962307-1* 272
Helichrysum bracteatum var. *normale* IPNI=*962307-1* 272
Helichrysum chrysanthum IPNI=*962307-1* 272
Helichrysum compositum IPNI=*962307-1* 272
Helichrysum conglomeratum IPNI=*212994-1* 264
Helichrysum elongatum IPNI=*213129-1* 264
Helichrysum glabratum IPNI=*962307-1* 272
Helichrysum gnaphalioides IPNI=*77197189-1* 259
Helichrysum lucidum IPNI=*962307-1* 272
Helichrysum lucidum var. *normalis* IPNI=*962307-1* 272
Helichrysum macrocephalum IPNI=*962307-1* 272
Helichrysum manglesii IPNI=*241451-1* 269
Helichrysum margaritaceum IPNI=*30173179-2* 257
Helichrysum monstruosum IPNI=*962307-1* 272
Helichrysum niveum IPNI=*962307-1* 272
Helichrysum petiolatum IPNI=*213170-1* 264
Helichrysum pichleri IPNI=*213129-1* 264
Heliconia aequatoriensis IPNI=*796959-1* 248
Heliconia andrewsii IPNI=*796972-1* 248
Heliconia bahiensis IPNI=*796972-1* 248
Heliconia ballia IPNI=*796972-1* 248
Heliconia bicolor IPNI=*796953-1* 248
Heliconia bihai IPNI=*798194-1* 250
Heliconia brasiliensis var. *concolor* IPNI=*796972-1* 248
Heliconia burle-marxii IPNI=*796953-1* 248
Heliconia cannoidea IPNI=*796972-1* 248
Heliconia cannoidea var. *villosa* IPNI=*796953-1* 248
Heliconia cararensis IPNI=*796953-1* 248
Heliconia cardenasii IPNI=*796953-1* 248
Heliconia costanensis IPNI=*796953-1* 248
Heliconia distans IPNI=*796925-1* 248
Heliconia elegans IPNI=*796965-1* 248
Heliconia elongata IPNI=*119524-2* 249
Heliconia ferdinando-coburgi IPNI=*119355-2* 248
Heliconia goiasensis IPNI=*796972-1* 248
Heliconia harrisiana IPNI=*796953-1* 248
Heliconia hirsuta var. *rubriflora* IPNI=*796953-1* 248
Heliconia hirsuta var. *villosula* IPNI=*796953-1* 248
Heliconia humilis IPNI=*796972-1* 248
Heliconia lennartiana IPNI=*119524-2* 249
Heliconia longissima IPNI=*119402-2* 248
Heliconia marantifolia IPNI=*796972-1* 248
Heliconia marginata f. *lutea* IPNI=*119411-2* 248
Heliconia nana IPNI=*796967-1* 248
Heliconia nigrescens IPNI=*796925-1* 248
Heliconia osaensis var. *rubescens* IPNI=*796967-1* 248
Heliconia poeppigiana IPNI=*796978-1* 249
Heliconia psittacorum var. *rhizomatosa* IPNI=*796972-1* 248
Heliconia punicea IPNI=*796965-1* 248
Heliconia purpurea IPNI=*796925-1* 248
Heliconia ravenala IPNI=*60466138-2* 250
Heliconia rutila IPNI=*796925-1* 248
Heliconia schaeferiana IPNI=*796925-1* 248
Heliconia schomburgkiana IPNI=*796972-1* 248
Heliconia straminea IPNI=*796953-1* 248
Heliconia swartziana IPNI=*796972-1* 248
Heliconia sylvestris IPNI=*796972-1* 248
Heliconia tricolor IPNI=*119505-2* 249
Heliconia variegata IPNI=*796925-1* 248
Helicostylis affinis IPNI=*584303-1* 440
Helicostylis asperifolia IPNI=*119540-2* 439
Helicostylis duckei IPNI=*584303-1* 440
Helicostylis lancifolia IPNI=*119532-2* 439
Helicostylis podogyne IPNI=*584303-1* 440
Helicostylis poeppigiana IPNI=*584303-1* 440
Helicostylis urophylla IPNI=*119542-2* 440
Helicotropis stenoloba IPNI=*60457240-2* 320
Helicteres althaeifolia IPNI=*321205-2* 408
Helicteres althaeifolia IPNI=*823472-1* 408
Helicteres apetala IPNI=*245150-2* 414
Helicteres barbadensis IPNI=*823472-1* 408
Helicteres baruensis IPNI=*320786-2* 408
Helicteres biflora IPNI=*320786-2* 408
Helicteres brevior IPNI=*823472-1* 408
Helicteres carpinifolia IPNI=*320786-2* 408
Helicteres dumosa IPNI=*823472-1* 408
Helicteres fetida IPNI=*321203-2* 408
Helicteres guazumifolia var. *parvifolia* IPNI=*320786-2* 408
Helicteres mexicana IPNI=*320786-2* 408
Helicteres mollis IPNI=*321205-2* 408
Helicteres retinophylla IPNI=*320786-2* 408
Helicteres spiralis IPNI=*823472-1* 408

Heliocarpus americanus var. *popayanensis* IPNI=*834663-1* 408
Heliocarpus americanus var. *schumannii* IPNI=*834663-1* 408
Heliocarpus australis IPNI=*834663-1* 408
Heliocarpus boliviensis IPNI=*834663-1* 408
Heliocarpus costaricensis IPNI=*834663-1* 408
Heliocarpus cuspidatus IPNI=*834663-1* 408
Heliocarpus diclinus IPNI=*834663-1* 408
Heliocarpus floribundus IPNI=*834663-1* 408
Heliocarpus gentlei IPNI=*834663-1* 408
Heliocarpus hirtus IPNI=*320793-2* 415
Heliocarpus horridus IPNI=*834663-1* 408
Heliocarpus nelsonii IPNI=*834663-1* 408
Heliocarpus oblongifolius IPNI=*834663-1* 408
Heliocarpus popayanensis IPNI=*834663-1* 408
Heliocarpus popayanensis var. *grandifolius* IPNI=*834663-1* 408
Heliocarpus popayanensis var. *purdiei* IPNI=*834663-1* 408
Heliocarpus popayanensis var. *trichopodus* IPNI=*834663-1* 408
Heliocarpus rosei IPNI=*834663-1* 408
Heliocarpus rudis IPNI=*834663-1* 408
Heliocarpus stipulatus IPNI=*834663-1* 408
Heliocarpus subtrilobus IPNI=*834663-1* 408
Heliocarpus tomentosus IPNI=*834663-1* 408
Heliocarpus trichopodus IPNI=*834663-1* 408
Heliocarpus viridis IPNI=*834663-1* 408
Heliophytum foetidum IPNI=*116938-1* 275
Heliophytum indicum IPNI=*116938-1* 275
Heliophytum parviflorum IPNI=*116662-1* 274
Heliophytum parviflorum var. *subellipticum* IPNI=*116662-1* 274
Heliophytum portoricensis IPNI=*116662-1* 274
Heliophytum velutinum IPNI=*116938-1* 275
Heliopsis canescens IPNI=*213524-1* 264
Heliopsis dubia IPNI=*213524-1* 264
Heliopsis helianthoides var. *fischeri* IPNI=*213534-1* 264
Heliopsis helianthoides var. *solidaginoides* IPNI=*213534-1* 264
Heliopsis laevis IPNI=*32497-2* 258
Heliopsis oppositifolia IPNI=*213524-1* 264
Heliopsis oppositifolia IPNI=*213534-1* 264
Heliopsis pulchra IPNI=*213524-1* 264
Heliopsis rugosa IPNI=*213524-1* 264
Heliotropium africanum IPNI=*116938-1* 275
Heliotropium album IPNI=*116833-1* 275
Heliotropium album IPNI=*77069564-1* 275
Heliotropium americanum IPNI=*60430582-2* 274
Heliotropium angustifolium IPNI=*116784-1* 275
Heliotropium anisophyllum IPNI=*116938-1* 275
Heliotropium arborescens var. *genuinum* IPNI=*116670-1* 275
Heliotropium arborescens var. *grisellum* IPNI=*116670-1* 275
Heliotropium brasilianum IPNI=*60430582-2* 274
Heliotropium bridgesii IPNI=*60430582-2* 274
Heliotropium canescens IPNI=*116833-1* 275
Heliotropium canescens IPNI=*60430582-2* 274
Heliotropium carduchorum IPNI=*116833-1* 275
Heliotropium chenopodioides IPNI=*116784-1* 275
Heliotropium chilense IPNI=*116784-1* 275
Heliotropium cinereum IPNI=*60430582-2* 274
Heliotropium commutatum IPNI=*116833-1* 275
Heliotropium cordifolium IPNI=*116938-1* 275
Heliotropium corymbosum IPNI=*116670-1* 275
Heliotropium curassavicum var. *violaceum* IPNI=*116784-1* 275
Heliotropium curassavicum var. *xerophilum* IPNI=*116784-1* 275
Heliotropium decumbens IPNI=*60430582-2* 274
Heliotropium dioscoridis IPNI=*116833-1* 275
Heliotropium ellipticum IPNI=*60430582-2* 274
Heliotropium erectum IPNI=*116833-1* 275
Heliotropium foetidum IPNI=*116662-1* 274
Heliotropium foetidum IPNI=*116938-1* 275
Heliotropium funkiae IPNI=*77069564-1* 275
Heliotropium glandulosum IPNI=*116833-1* 275
Heliotropium glaucophyllum IPNI=*116784-1* 275
Heliotropium glaucum IPNI=*116784-1* 275
Heliotropium gnaphalodes IPNI=*120998-1* 275
Heliotropium gracile var. *depressum* IPNI=*60430582-2* 274
Heliotropium grandiflorum IPNI=*116670-1* 275
Heliotropium gymnocarpum IPNI=*116833-1* 275
Heliotropium horminifolium IPNI=*116938-1* 275
Heliotropium houstonii IPNI=*60430582-2* 274
Heliotropium humboldtianum IPNI=*60430582-2* 274
Heliotropium humile IPNI=*116833-1* 275
Heliotropium inundatum f. *ellipticum* IPNI=*60430582-2* 274
Heliotropium inundatum IPNI=*60430582-2* 274
Heliotropium inundatum subf. *pusillum* IPNI=*60430582-2* 274
Heliotropium inundatum var. *chacoense* IPNI=*60430582-2* 274
Heliotropium inundatum var. *cubense* IPNI=*60430582-2* 274
Heliotropium lacunarium IPNI=*116833-1* 275
Heliotropium lanceolatum IPNI=*116938-1* 275
Heliotropium lancifolium IPNI=*116662-1* 274
Heliotropium latifolium IPNI=*116662-1* 274
Heliotropium lehmannianum IPNI=*116784-1* 275
Heliotropium leptostachyum IPNI=*60430582-2* 274
Heliotropium longepetiolatum IPNI=*60430582-2* 274
Heliotropium majus IPNI=*116833-1* 275
Heliotropium oblongifolium IPNI=*116662-1* 274
Heliotropium oblongifolium IPNI=*116833-1* 275
Heliotropium odoratissimum IPNI=*116670-1* 275
Heliotropium odoratum IPNI=*116670-1* 275
Heliotropium odorum IPNI=*116670-1* 275
Heliotropium oppositifolium IPNI=*121103-1* 275
Heliotropium ovalifolium var. *depressum* IPNI=*60430582-2* 274
Heliotropium parviflorum IPNI=*116662-1* 274
Heliotropium parviflorum IPNI=*116938-1* 275
Heliotropium patebilcense IPNI=*116662-1* 274
Heliotropium peruvianum IPNI=*116670-1* 275
Heliotropium peruvianum var. *voltaireanum* IPNI=*116670-1* 275
Heliotropium procumbens IPNI=*60430582-2* 274
Heliotropium procumbens var. *depressum* IPNI=*60430582-2* 274
Heliotropium rigidulum IPNI=*60430582-2* 274
Heliotropium scorpioides IPNI=*116662-1* 274
Heliotropium simplex IPNI=*60430582-2* 274
Heliotropium subcanescens IPNI=*116833-1* 275
Heliotropium synzystachyum IPNI=*116662-1* 274
Heliotropium tenuiflorum IPNI=*116833-1* 275
Heliotropium virens IPNI=*116784-1* 275
Heliotropium voltaireanum IPNI=*116670-1* 275
Heliotropium vulgare IPNI=*116833-1* 275
Heliotropium willdenowii IPNI=*60430582-2* 274
Heliotropium xerophilum IPNI=*116784-1* 275
Helipterum cryptanthum IPNI=*241451-1* 269
Helipterum manglesii IPNI=*241451-1* 269
Helipterum manglesii var. *maculatum* IPNI=*241451-1* 269
Helleborine americana IPNI=*1218040-2* 217
Helmia bulbifera IPNI=*323908-2* 225
Helmia coriacea IPNI=*324018-2* 225
Helmia galipanensis IPNI=*324037-2* 226
Helmia pilosiuscula IPNI=*318417-1* 226
Helmia schomburgkiana IPNI=*318417-1* 226
Helmia trifoliata IPNI=*324037-2* 226
Helopus annulatus IPNI=*402044-1* 239
Helopus cognatus IPNI=*1021229-2* 239
Helopus javanicus IPNI=*402044-1* 239
Helopus kunthii IPNI=*1021229-2* 239
Helopus laevis IPNI=*402044-1* 239
Helopus pilosus IPNI=*402044-1* 239
Helopus polystachyus IPNI=*77164258-1* 246
Helopus punctatus IPNI=*1021229-2* 239
Helopus punctatus var. *cognatus* IPNI=*1021229-2* 239
Helopus punctatus var. *grandispiculus* IPNI=*1021229-2* 239

Helosciadium graveolens IPNI=*838067-1* 251
Helosciadium heterophyllum IPNI=*840899-1* 252
Helosciadium lateriflorum IPNI=*837486-1* 251
Helosciadium lateriflorum IPNI=*840899-1* 252
Helosciadium leptophyllum IPNI=*840899-1* 252
Helosciadium leptophyllum var. *latifolium* IPNI=*840899-1* 252
Helosciadium oppositifolium IPNI=*60442790-2* 252
Helosciadium ranunculifolium IPNI=*840899-1* 252
Helosciadium ruta IPNI=*838067-1* 251
Helosciadium rutaceum IPNI=*838067-1* 251
Helosis brasiliensis IPNI=*103346-1* 448
Helosis guyannensis f. *brasilensis* IPNI=*103346-1* 448
Helosis guyannensis var. *andicola* IPNI=*103346-1* 448
Helosis mexicana var. *andicola* IPNI=*103346-1* 448
Helxine dioscoridis IPNI=*854958-1* 446
Helxine fagopyrum IPNI=*694526-1* 289
Helxine soleirolii IPNI=*856545-1* 448
Hemarthria capensis IPNI=*404684-1* 240
Hemarthria caudiculata IPNI=*404684-1* 240
Hemarthria compressa var. *fasciculata* IPNI=*404684-1* 240
Hemarthria fasciculata IPNI=*404684-1* 240
Hemarthria guyanensis IPNI=*404684-1* 240
Hematophyla villosa IPNI=*378985-1* 367
Hemerocallis crocea IPNI=*536335-1* 215
Hemerocallis flava IPNI=*536346-1* 215
Hemerocallis lilioasphodelus f. *aurantiaca* IPNI=*536346-1* 215
Hemerocallis lutea IPNI=*536346-1* 215
Hemestheum molle IPNI=*251308-2* 174
Hemicardion macrosorum IPNI=*1148359-2* 175
Hemicardion nephrolepis IPNI=*1148359-2* 175
Hemicrepidospermum rhoifolium IPNI=*77184241-1* 453
Hemidesmas diffusus IPNI=*1056199-2* 328
Hemidesmas plenus IPNI=*1056199-2* 328
Hemidiodia ocymifolia IPNI=*766862-1* 357
Hemigraphis alternata IPNI=*77147688-1* 361
Hemigraphis colorata IPNI=*77147688-1* 361
Hemimeris hirsuta IPNI=*1072674-2* 376
Hemimeris incisifolia IPNI=*9536-2* 377
Hemimeris intermedia IPNI=*9536-2* 377
Hemimeris mutisii IPNI=*9536-2* 377
Hemimeris parviflora IPNI=*9536-2* 377
Hemimeris procumbens IPNI=*9536-2* 377
Hemimeris urticifolia IPNI=*9536-2* 377
Heminema multiflora IPNI=*77205602-1* 224
Hemionitis arifolia IPNI=*319061-2* 177
Hemionitis chihuahuaensis IPNI=*77182808-1* 178
Hemionitis dealbata IPNI=*200935-2* 178
Hemionitis glaberrima IPNI=*77182519-1* 178
Hemionitis lineata IPNI=*17209080-1* 178
Hemisandra aurantiaca IPNI=*1053173-2* 359
Hemiscola aculeata IPNI=*316362-2* 279
Hemisodon leonurus IPNI=*449133-1* 370
Hemitelia arborea IPNI=*17071820-1* 172
Hemitelia crenata IPNI=*73146-2* 172
Hemitelia firma IPNI=*9976-2* 172
Hemitelia serra IPNI=*17071820-1* 172
Hemitelia sherringii IPNI=*73146-2* 172
Hemitomus urticifolius IPNI=*9536-2* 377
Hemolepis canescens IPNI=*213524-1* 264
Henckelia alata IPNI=*810361-1* 374
Henckelia grandiflora IPNI=*799766-1* 374
Henlea muzonensis IPNI=*134610-2* 352
Henlea rosea IPNI=*134610-2* 352
Henriettea brasiliensis IPNI=*569017-1* 418
Henriettea brunnescens IPNI=*31112-2* 418
Henriettea granularis IPNI=*569004-1* 418
Henriettea macrocalyx IPNI=*569017-1* 418
Henriettea sagotiana IPNI=*31112-2* 418
Henriettea succosa var. *guianensis* IPNI=*569004-1* 418
Henriettea surinamensis IPNI=*569004-1* 418
Henriettea trinervia IPNI=*569004-1* 418
Henriettella granularis IPNI=*569004-1* 418
Henriettella macrocalyx IPNI=*569017-1* 418
Hepetis andreana IPNI=*123813-1* 229
Hepetis heterophylla IPNI=*123989-1* 229
Hepetis heterophylla var. *exscapa* IPNI=*123989-1* 229
Hepetis pungens IPNI=*124143-1* 229
Heppiella ampla IPNI=*381388-1* 367
Heppiella grandifolia IPNI=*381388-1* 367
Heppiella guazumifolia IPNI=*381388-1* 367
Heppiella karsteniana IPNI=*381388-1* 367
Heppiella pauciflora IPNI=*381388-1* 367
Heppiella rosea IPNI=*381388-1* 367
Heppiella scandens IPNI=*381388-1* 367
Heppiella stenocalyx IPNI=*381388-1* 367
Heppiella trianae IPNI=*381388-1* 367
Heppiella warszewiczii IPNI=*381388-1* 367
Herbertia caerulea IPNI=*278134-2* 216
Herbertia lineata IPNI=*278134-2* 216
Herissantia trichoda IPNI=*120630-2* 409
Hermesia castaneifolia IPNI=*338372-1* 387
Hermesia salicifolia IPNI=*338372-1* 387
Hermesias ariza IPNI=*482406-1* 312
Hermesias coccinea IPNI=*482413-1* 312
Hermesias grandiceps IPNI=*482418-1* 312
Hermesias neglecta IPNI=*482418-1* 312
Hermesias negrensis IPNI=*482433-1* 312
Hermesias speciosa IPNI=*482440-1* 312
Hernandezia sonora IPNI=*316357-2* 182
Hernandia peltata IPNI=*316357-2* 182
Hernandia sonora var. *guadeloupensis* IPNI=*316357-2* 182
Herpestis africana IPNI=*1072674-2* 376
Herpestis axillaris IPNI=*28282-2* 375
Herpestis brownei IPNI=*1072674-2* 376
Herpestis calytriplex IPNI=*1072674-2* 376
Herpestis caprarioides IPNI=*155661-2* 376
Herpestis chamaedryoides IPNI=*155661-2* 376
Herpestis chamaedryoides var. *peduncularis* IPNI=*155661-2* 376
Herpestis chrysantha IPNI=*155661-2* 376
Herpestis colubrina IPNI=*155661-2* 376
Herpestis cuneifolia IPNI=*1072674-2* 376
Herpestis divaricata IPNI=*155661-2* 376
Herpestis macrantha IPNI=*155661-2* 376
Herpestis monnieri IPNI=*1072674-2* 376
Herpestis monnieria IPNI=*1072674-2* 376
Herpestis montevidensis IPNI=*155661-2* 376
Herpestis moranensis IPNI=*1072674-2* 376
Herpestis peduncularis IPNI=*155661-2* 376
Herpestis pedunculosa IPNI=*1072674-2* 376
Herpestis procumbens IPNI=*1072674-2* 376
Herpestis procumbens IPNI=*155661-2* 376
Herpestis spathulata IPNI=*1072674-2* 376
Herpetica alata IPNI=*1192167-2* 332
Herpetica rumphiana IPNI=*1192167-2* 332
Herrania albiflora f. *titanica* IPNI=*824010-1* 409
Herrania aspera IPNI=*824026-1* 409
Herrania atrorubens IPNI=*824026-1* 409
Herrania nitida var. *aspera* IPNI=*824026-1* 409
Herrania pulcherrima var. *pacifica* IPNI=*824014-1* 409
Hesperis aestiva IPNI=*286910-1* 277
Hesperis altissima IPNI=*289144-1* 278
Hesperis fasciculata IPNI=*286910-1* 277
Hesperis fenestralis IPNI=*286910-1* 277
Hesperis glabra IPNI=*286910-1* 277
Hesperis incana IPNI=*286910-1* 277
Hesperis iriodes IPNI=*237940-2* 278

Hesperis mandonii IPNI=*284416-2* 277
Hesperis maxima IPNI=*237940-2* 278
Hesperis niagarensis IPNI=*289589-1* 278
Hesperis officinalis IPNI=*289589-1* 278
Hesperis rigidula IPNI=*289144-1* 278
Hesperis ruderalis IPNI=*289589-1* 278
Hesperis septulata IPNI=*289144-1* 278
Hesperis subulata IPNI=*237940-2* 278
Hesperis violaria IPNI=*286910-1* 277
Hesperocyparis lindleyi IPNI=*60451554-2* 180
Hesperocyparis lindleyi subsp. *hondurensis* IPNI=*60451554-2* 180
Hesperocyparis lusitanica var. *lindleyi* IPNI=*60451554-2* 180
Hesperomeles chiriquensis IPNI=*725484-1* 443
Hesperomeles cordata IPNI=*725473-1* 443
Hesperomeles escalloniifolia IPNI=*725484-1* 443
Hesperomeles fieldii IPNI=*725484-1* 443
Hesperomeles glabrata IPNI=*725484-1* 443
Hesperomeles heterophylla IPNI=*725484-1* 443
Hesperomeles heterophylla var. *vacciniifolia* IPNI=*725484-1* 443
Hesperomeles latifolia IPNI=*725473-1* 443
Hesperomeles oblonga IPNI=*725473-1* 443
Hesperomeles obovata IPNI=*725484-1* 443
Hesperomeles pernettyoides IPNI=*725484-1* 443
Hesperomeles pernettyoides var. *glaucophylla* IPNI=*725484-1* 443
Heterachthia gaudichaudiana IPNI=*77184055-1* 224
Heterandra reniformis IPNI=*698080-1* 225
Heteranthera acuta IPNI=*698080-1* 225
Heteranthera cordata IPNI=*77188085-1* 225
Heteranthera diversifolia IPNI=*77188085-1* 225
Heteranthera formosa IPNI=*310928-2* 225
Heteranthera grandiflora IPNI=*77188085-1* 225
Heteranthera pubescens IPNI=*698080-1* 225
Heteranthera reniformis var. *conjungens* IPNI=*698080-1* 225
Heteranthera virginicus IPNI=*698080-1* 225
Heterocarpus obliquus IPNI=*63900-2* 224
Heterocentron alpestre IPNI=*569110-1* 419
Heterocentron axillare IPNI=*569110-1* 419
Heterocentron floribundum IPNI=*569110-1* 419
Heterocentron hondurense IPNI=*569110-1* 419
Heterocentron hondurensis IPNI=*569110-1* 419
Heterocentron macrostachyum IPNI=*569110-1* 419
Heterocentron mexicana IPNI=*569110-1* 419
Heterocentron roseum IPNI=*569110-1* 419
Heterocentron salvadoranum IPNI=*569094-1* 419
Heterocentron sessile IPNI=*121047-2* 419
Heterocentron sessilis IPNI=*121047-2* 419
Heterocentron sinuatum IPNI=*569094-1* 419
Heterocentron undulatum IPNI=*569110-1* 419
Heterochaeta stricta IPNI=*228375-1* 264
Heterochlamys quinquenervia IPNI=*69894-2* 388
Heteroloma lanatum IPNI=*1124646-2* 318
Heteroloma supinum IPNI=*1015219-2* 318
Heteronoma campanulare IPNI=*60465985-2* 417
Heteronoma ciliatum IPNI=*60465985-2* 417
Heteronoma cubense IPNI=*60465985-2* 417
Heteronoma diversifolium IPNI=*60465985-2* 417
Heteronoma diversifolium var. *rubijinosa* IPNI=*60465985-2* 417
Heteronoma galeottianum IPNI=*60465985-2* 417
Heteronoma goudotianum IPNI=*60465985-2* 417
Heteronoma mexicanum IPNI=*569110-1* 419
Heteronoma subtriplinervium IPNI=*569110-1* 419
Heteronoma volubile IPNI=*60465985-2* 417
Heteropogon acuminatus IPNI=*404802-1* 240
Heteropogon allionii IPNI=*404772-1* 240
Heteropogon besukiensis IPNI=*404772-1* 240
Heteropogon betafensis IPNI=*404802-1* 240
Heteropogon concinnus IPNI=*398532-1* 237
Heteropogon contortus subvar. *secundus* IPNI=*404772-1* 240
Heteropogon contortus var. *glaber* IPNI=*404772-1* 240
Heteropogon contortus var. *hirtus* IPNI=*404772-1* 240
Heteropogon contortus var. *hispidissimus* IPNI=*404772-1* 240
Heteropogon contortus var. *polystachyos* IPNI=*404772-1* 240
Heteropogon fertilis IPNI=*404772-1* 240
Heteropogon firmus IPNI=*404772-1* 240
Heteropogon glaber IPNI=*404772-1* 240
Heteropogon hirsutus IPNI=*404772-1* 240
Heteropogon hirtus IPNI=*404772-1* 240
Heteropogon hispidissimus IPNI=*404772-1* 240
Heteropogon hochstetteri IPNI=*398608-1* 237
Heteropogon hohenackeri IPNI=*404772-1* 240
Heteropogon megapotamicus IPNI=*424391-1* 246
Heteropogon messanensis IPNI=*404772-1* 240
Heteropogon polystictus IPNI=*404802-1* 240
Heteropogon roxburghii IPNI=*404772-1* 240
Heteropogon secundus IPNI=*424391-1* 246
Heteropogon stipoides IPNI=*424391-1* 246
Heteropogon tenellus IPNI=*398532-1* 237
Heteropogon truncatus IPNI=*424391-1* 246
Heteropsis integerrima IPNI=*87123-1* 200
Heteropsis rimbachii IPNI=*121107-2* 200
Heteropterys apiculata IPNI=*556578-1* 396
Heteropterys argentea IPNI=*29242-2* 395
Heteropterys belizensis IPNI=*556578-1* 396
Heteropterys bopiana IPNI=*60444585-2* 396
Heteropterys caerulea IPNI=*556563-1* 396
Heteropterys caerulea var. *angustifolia* IPNI=*556563-1* 396
Heteropterys caerulea var. *diversifolia* IPNI=*556563-1* 396
Heteropterys cornifolia IPNI=*60447688-2* 395
Heteropterys floribunda IPNI=*556563-1* 396
Heteropterys grisebachiana IPNI=*1079574-2* 396
Heteropterys jussieui IPNI=*1079574-2* 396
Heteropterys laurifolia subvar. *floribunda* IPNI=*556563-1* 396
Heteropterys laurifolia subvar. *longifolia* IPNI=*556563-1* 396
Heteropterys laurifolia var. *antillana* IPNI=*556563-1* 396
Heteropterys laurifolia var. *mexicana* IPNI=*556563-1* 396
Heteropterys lessertiana IPNI=*556563-1* 396
Heteropterys longifolia IPNI=*556563-1* 396
Heteropterys longifolia var. *borealis* IPNI=*556563-1* 396
Heteropterys longifolia var. *martinicensis* IPNI=*556563-1* 396
Heteropterys macrostachya f. *oblonga* IPNI=*556578-1* 396
Heteropterys macrostachya f. *ovata* IPNI=*556578-1* 396
Heteropterys macrostachya f. *transiens* IPNI=*556578-1* 396
Heteropterys macrostachya var. *rotundifolia* IPNI=*556578-1* 396
Heteropterys multiflora IPNI=*1079574-2* 396
Heteropterys pirayuensis IPNI=*29242-2* 395
Heteropterys pubiflora IPNI=*556563-1* 396
Heteropterys stannea IPNI=*60443063-2* 396
Heterostega curtipendula IPNI=*1198286-2* 235
Heterotaxis crassifolia IPNI=*643387-1* 220
Heterotaxis discolor IPNI=*643414-1* 220
Heterotaxis gatunensis IPNI=*643387-1* 220
Heterotaxis longifolia IPNI=*643845-1* 221
Heterotaxis maleolens IPNI=*155157-2* 221
Heterotaxis sessilis IPNI=*643387-1* 220
Heterotaxis superflua IPNI=*643845-1* 221
Heterotaxis valenzuelana IPNI=*155338-2* 222
Heterotaxis villosa IPNI=*155347-2* 222
Heterotaxis violaceopunctata IPNI=*643899-1* 222
Heterothalamus trinervis IPNI=*183821-1* 258
Heterotheca deltoidea IPNI=*164998-2* 267
Heterotis plumosa IPNI=*908858-1* 419
Heterotrichum octonum IPNI=*77196163-1* 421
Heterotrichum octonum var. *brasiliensis* IPNI=*77196163-1* 421
Hevea benthamiana f. *caudata* IPNI=*349911-1* 391
Hevea benthamiana f. *huberiana* IPNI=*349911-1* 391
Hevea benthamiana f. *obtusiloba* IPNI=*349911-1* 391

Hevea benthamiana f. *subglabrifolia* IPNI=*349911-1* 391
Hevea benthamiana var. *caudata* IPNI=*349911-1* 391
Hevea benthamiana var. *obtusiloba* IPNI=*349911-1* 391
Hevea benthamiana var. *subglabrifolia* IPNI=*349911-1* 391
Hevea benthamiana var. *typica* IPNI=*349911-1* 391
Hevea brasiliensis f. *acreana* IPNI=*349913-1* 391
Hevea brasiliensis f. *randiana* IPNI=*349913-1* 391
Hevea brasiliensis var. *acreana* IPNI=*349913-1* 391
Hevea brasiliensis var. *janeirensis* IPNI=*349913-1* 391
Hevea brasiliensis var. *randiana* IPNI=*349913-1* 391
Hevea brasiliensis var. *stylosa* IPNI=*349913-1* 391
Hevea discolor IPNI=*349911-1* 391
Hevea discolor IPNI=*349947-1* 391
Hevea duckei IPNI=*349911-1* 391
Hevea granthamii IPNI=*349913-1* 391
Hevea janeirensis IPNI=*349913-1* 391
Hevea microphylla var. *major* IPNI=*121748-2* 391
Hevea microphylla var. *typica* IPNI=*121748-2* 391
Hevea paraensis IPNI=*349947-1* 391
Hevea randiana IPNI=*349913-1* 391
Hevea sieberi IPNI=*349913-1* 391
Hevea similis IPNI=*349947-1* 391
Hevea spruceana var. *discolor* IPNI=*349947-1* 391
Hevea spruceana var. *similis* IPNI=*349947-1* 391
Hevea spruceana var. *tridentata* IPNI=*349947-1* 391
Hewardia wilsonii IPNI=*17013890-1* 177
Hexacentris mysorensis IPNI=*60447289-2* 362
Hexepta racemosa IPNI=*1114920-2* 357
Hexopetion alatum IPNI=*25851-2* 204
Heymassoli inermis IPNI=*316341-2* 449
Heymassoli spinosa IPNI=*316341-2* 449
Hibiscus abelmoschus IPNI=*558023-1* 406
Hibiscus abelmoschus var. *betulifolius* IPNI=*558023-1* 406
Hibiscus abelmoschus var. *genuinus* IPNI=*558023-1* 406
Hibiscus abelmoschus var. *multiformis* IPNI=*558023-1* 406
Hibiscus acerifolius IPNI=*560890-1* 410
Hibiscus acetosus IPNI=*326388-2* 409
Hibiscus acuminatus IPNI=*562981-1* 412
Hibiscus aestuans IPNI=*560549-1* 409
Hibiscus albiflorus IPNI=*559664-1* 408
Hibiscus americanus IPNI=*564786-1* 415
Hibiscus amoenus IPNI=*560861-1* 410
Hibiscus androphoropetaloides IPNI=*560756-1* 409
Hibiscus angustifolius IPNI=*560861-1* 410
Hibiscus arborescens IPNI=*560890-1* 410
Hibiscus argentinus IPNI=*560861-1* 410
Hibiscus aristatus IPNI=*186083-2* 412
Hibiscus arnottii IPNI=*560756-1* 409
Hibiscus asper IPNI=*560068-1* 409
Hibiscus baccifer IPNI=*564713-1* 415
Hibiscus bammia IPNI=*558006-1* 406
Hibiscus bancroftianus IPNI=*30041958-2* 410
Hibiscus barbadensis f. *fruticulosus* IPNI=*559677-1* 408
Hibiscus barbadensis f. *oligospermus* IPNI=*559677-1* 408
Hibiscus barbadensis IPNI=*559677-1* 408
Hibiscus barbadensis var. *lanceolatum* IPNI=*559677-1* 408
Hibiscus bicornis IPNI=*560013-1* 409
Hibiscus bifurcatus f. *glaber* IPNI=*560013-1* 409
Hibiscus bifurcatus f. *pilosus* IPNI=*560013-1* 409
Hibiscus bifurcatus var. *bicornis* IPNI=*560013-1* 409
Hibiscus blumei IPNI=*564713-1* 415
Hibiscus boninensis IPNI=*30143354-2* 410
Hibiscus cannabinus var. *unidens* IPNI=*560723-1* 409
Hibiscus capitatus IPNI=*561424-1* 410
Hibiscus cardiophyllus IPNI=*558023-1* 406
Hibiscus cernuus IPNI=*559664-1* 408
Hibiscus chinensis IPNI=*560890-1* 410
Hibiscus ciliaris IPNI=*558023-1* 406
Hibiscus circinnatus IPNI=*30143354-2* 410
Hibiscus cisplatinus IPNI=*560861-1* 410
Hibiscus coccineus var. *integrifolius* IPNI=*30041958-2* 410
Hibiscus congener IPNI=*560068-1* 409
Hibiscus cordatus IPNI=*133833-2* 410
Hibiscus cordifolius IPNI=*562981-1* 412
Hibiscus cordofanus IPNI=*560068-1* 409
Hibiscus corrugatus IPNI=*30143354-2* 410
Hibiscus cruentus IPNI=*326388-2* 409
Hibiscus deciduus IPNI=*30042843-2* 408
Hibiscus digitatus IPNI=*326388-2* 409
Hibiscus digitatus var. *kerrianus* IPNI=*326388-2* 409
Hibiscus diversifolius var. *granatensis* IPNI=*560486-1* 409
Hibiscus eetveldeanus IPNI=*559896-1* 409
Hibiscus esculentus IPNI=*558006-1* 406
Hibiscus fauriei IPNI=*560792-1* 410
Hibiscus festalis IPNI=*560756-1* 409
Hibiscus ficifolius IPNI=*558006-1* 406
Hibiscus flavescens IPNI=*558023-1* 406
Hibiscus floridus IPNI=*560890-1* 410
Hibiscus fluminensis IPNI=*560013-1* 409
Hibiscus fraternus IPNI=*326388-2* 409
Hibiscus fruticulosus IPNI=*559677-1* 408
Hibiscus fulgens IPNI=*560756-1* 409
Hibiscus gossypifolius IPNI=*326388-2* 409
Hibiscus guianensis IPNI=*185868-2* 412
Hibiscus guineensis IPNI=*30143354-2* 410
Hibiscus haenkeanus IPNI=*558023-1* 406
Hibiscus hastatus IPNI=*133833-2* 410
Hibiscus hastatus IPNI=*30143354-2* 410
Hibiscus heptaphyllus IPNI=*560723-1* 409
Hibiscus hexaphyllus IPNI=*186058-2* 412
Hibiscus hispidissimus IPNI=*558006-1* 406
Hibiscus hispidus IPNI=*133833-2* 410
Hibiscus integrifolius IPNI=*30041958-2* 410
Hibiscus javanicus IPNI=*560756-1* 409
Hibiscus jerroldianus IPNI=*30041958-2* 410
Hibiscus lambertianus var. *angustifolius* IPNI=*560861-1* 410
Hibiscus lindleyi IPNI=*560723-1* 409
Hibiscus lindmanii IPNI=*560861-1* 410
Hibiscus linearis IPNI=*560861-1* 410
Hibiscus litoreus IPNI=*564713-1* 415
Hibiscus longifolius IPNI=*558006-1* 406
Hibiscus macrophyllus IPNI=*564713-1* 415
Hibiscus malangensis IPNI=*560068-1* 409
Hibiscus malvarosa IPNI=*560549-1* 409
Hibiscus malvaviscus IPNI=*30041958-2* 410
Hibiscus masuianus IPNI=*326388-2* 409
Hibiscus metallicus IPNI=*560756-1* 409
Hibiscus moschatus IPNI=*558023-1* 406
Hibiscus multiformis IPNI=*558023-1* 406
Hibiscus nangking IPNI=*559664-1* 408
Hibiscus nutans IPNI=*30041958-2* 410
Hibiscus obtusatus IPNI=*560068-1* 409
Hibiscus oligospermus IPNI=*30042843-2* 408
Hibiscus palmatilobus IPNI=*326388-2* 409
Hibiscus pentaspermus IPNI=*133833-2* 410
Hibiscus pilosus IPNI=*30041958-2* 410
Hibiscus populifolius IPNI=*564713-1* 415
Hibiscus populneus IPNI=*564713-1* 415
Hibiscus porophyllus IPNI=*30143354-2* 410
Hibiscus praecox IPNI=*558006-1* 406
Hibiscus preslii IPNI=*133833-2* 410
Hibiscus pseudabelmoschus IPNI=*558023-1* 406
Hibiscus purpurascens IPNI=*30042843-2* 408
Hibiscus racemosus IPNI=*559677-1* 408
Hibiscus religiosus IPNI=*30042843-2* 408
Hibiscus rhombifolius IPNI=*560890-1* 410

Hibiscus ricinifolius IPNI=*558023-1* 406
Hibiscus rosa-sinensis f. *schizopetalus* IPNI=*560792-1* 410
Hibiscus rosa-sinensis var. *schizopetalus* IPNI=*560792-1* 410
Hibiscus rosiflorus IPNI=*560756-1* 409
Hibiscus roxburghii IPNI=*558023-1* 406
Hibiscus rudis IPNI=*561450-1* 410
Hibiscus sabdariffa subsp. *cannabinus* IPNI=*560068-1* 409
Hibiscus sagittatus IPNI=*133833-2* 410
Hibiscus sagranus IPNI=*560861-1* 410
Hibiscus salicifolius IPNI=*185868-2* 412
Hibiscus sanguineus IPNI=*326388-2* 409
Hibiscus sinensis IPNI=*560549-1* 409
Hibiscus spathulatus IPNI=*60443335-2* 410
Hibiscus spathulatus var. *froesianus* IPNI=*60443335-2* 410
Hibiscus spinifera IPNI=*186083-2* 412
Hibiscus spinifex IPNI=*186083-2* 412
Hibiscus subdariffa IPNI=*326388-2* 409
Hibiscus sublobatus IPNI=*558023-1* 406
Hibiscus submaritimus IPNI=*560861-1* 410
Hibiscus syriacus var. *micranthus* IPNI=*560890-1* 410
Hibiscus tiliaceus f. *albiflorus* IPNI=*30143354-2* 410
Hibiscus tiliaceus var. *genuinus* IPNI=*30143354-2* 410
Hibiscus tiliifolius IPNI=*30143354-2* 410
Hibiscus tortuosus IPNI=*30143354-2* 410
Hibiscus tricolor IPNI=*560756-1* 409
Hibiscus tricuspis IPNI=*30143354-2* 410
Hibiscus tripartitus IPNI=*560068-1* 409
Hibiscus umbellatus IPNI=*563076-1* 412
Hibiscus unidens IPNI=*560723-1* 409
Hibiscus urbanii IPNI=*1188401-2* 409
Hibiscus vanderystii IPNI=*560068-1* 409
Hibiscus verrucosus IPNI=*560068-1* 409
Hibiscus vitifolius IPNI=*560068-1* 409
Hibiscus wightianus IPNI=*560068-1* 409
Hieracium asperum IPNI=*1003017-2* 270
Hieracium cichorium IPNI=*194533-1* 260
Hieracium glabrum IPNI=*225504-1* 264
Hieracium pulchrum IPNI=*199753-1* 260
Hieracium sonchus IPNI=*1003017-2* 270
Hieracium virens IPNI=*199753-1* 260
Hierochloe banksiana IPNI=*389043-1* 234
Hierochloe redolens IPNI=*389043-1* 234
Hierochloe redolens var. *typica* IPNI=*389043-1* 234
Hieronia scabra IPNI=*316577-1* 295
Hieronyma andina IPNI=*350078-1* 401
Hieronyma blanchetiana IPNI=*350078-1* 401
Hieronyma buchtienii IPNI=*122567-2* 401
Hieronyma colombiana IPNI=*350075-1* 401
Hieronyma croizatii IPNI=*350075-1* 401
Hieronyma dichrophylla IPNI=*350078-1* 401
Hieronyma gentlei IPNI=*350078-1* 401
Hieronyma guatemalensis IPNI=*350078-1* 401
Hieronyma macrocarpa var. *moritziana* IPNI=*122567-2* 401
Hieronyma macrocarpa var. *spruceana* IPNI=*350075-1* 401
Hieronyma moritziana IPNI=*122567-2* 401
Hieronyma moritziana var. *fendleri* IPNI=*122567-2* 401
Hieronyma moritziana var. *yungasensis* IPNI=*122567-2* 401
Hieronyma oblonga f. *glabra* IPNI=*350078-1* 401
Hieronyma oblonga var. *benthamii* IPNI=*350078-1* 401
Hieronyma oblonga var. *blanchetiana* IPNI=*350078-1* 401
Hieronyma oblonga var. *crassifolia* IPNI=*350078-1* 401
Hieronyma oblonga var. *genuina* IPNI=*350078-1* 401
Hieronyma oblonga var. *nervata* IPNI=*350078-1* 401
Hieronyma oblonga var. *obtusa* IPNI=*350078-1* 401
Hieronyma poasana IPNI=*350078-1* 401
Hieronyma sararita IPNI=*122555-2* 401
Hilairanthus nitidus IPNI=*861125-1* 359
Hilairanthus tomentosus IPNI=*861125-1* 359
Hilairia paniculata IPNI=*174455-2* 267
Hildaea pallens IPNI=*405749-1* 240
Hilleria elastica IPNI=*122647-2* 287
Hillia marcano-bertii IPNI=*122668-2* 351
Himatanthus lancifolius IPNI=*122690-2* 340
Himatanthus rigidus IPNI=*122688-2* 340
Himatanthus speciosus IPNI=*122690-2* 340
Himatanthus sucuuba IPNI=*122688-2* 340
Hingtsha repens IPNI=*203207-1* 261
Hippagrostis burmanni IPNI=*30453766-2* 242
Hippagrostis composita IPNI=*409823-1* 242
Hippagrostis hirtella IPNI=*1004002-1* 242
Hippagrostis loliacea IPNI=*1004002-1* 242
Hippagrostis setaria IPNI=*409823-1* 242
Hippeastrum albiellum IPNI=*270910-2* 214
Hippeastrum ambiguum var. *longiflorum* IPNI=*122772-2* 213
Hippeastrum ambiguum var. *tweedieanum* IPNI=*65185-1* 213
Hippeastrum andersonii IPNI=*67314-1* 214
Hippeastrum araripinum IPNI=*122772-2* 213
Hippeastrum berteroi IPNI=*270986-2* 214
Hippeastrum carinatum IPNI=*270933-2* 214
Hippeastrum carneum IPNI=*67293-1* 214
Hippeastrum crociflorum IPNI=*122815-2* 213
Hippeastrum equestre IPNI=*122815-2* 213
Hippeastrum puertoricense IPNI=*270983-2* 214
Hippeastrum purpureum IPNI=*122815-2* 213
Hippeastrum quilmesianum IPNI=*270986-2* 214
Hippeastrum solandriflorum IPNI=*122772-2* 213
Hippeastrum solandriflorum var. *conspicuum* IPNI=*122772-2* 213
Hippeastrum solandriflorum var. *rubritubum* IPNI=*122772-2* 213
Hippeastrum solandriflorum var. *striatum* IPNI=*122772-2* 213
Hippeastrum texanum IPNI=*67314-1* 214
Hippeastrum tubispathum IPNI=*67314-1* 214
Hippeastrum vittatum var. *latifolium* IPNI=*65185-1* 213
Hippia bogotensis IPNI=*238661-1* 268
Hippia minuta IPNI=*249968-1* 270
Hippion aphyllum IPNI=*1167766-2* 345
Hippion axillare IPNI=*108639-2* 344
Hippion gentianella IPNI=*108639-2* 344
Hippion imbricatum IPNI=*108504-2* 344
Hippion lancifolium IPNI=*108639-2* 344
Hippion prostratum IPNI=*108504-2* 344
Hippocentaurea centaurium IPNI=*326619-2* 344
Hippochaete gigantea IPNI=*320302-2* 173
Hippochaete hyemalis IPNI=*92939-2* 173
Hippochaete myriochaeta IPNI=*77060052-1* 173
Hippocratea acapulcensis IPNI=*210308-2* 291
Hippocratea acuminata IPNI=*162403-1* 291
Hippocratea acutiflora IPNI=*161653-1* 290
Hippocratea affinis IPNI=*161653-1* 290
Hippocratea aggregata IPNI=*72123-2* 290
Hippocratea anafensis IPNI=*161653-1* 290
Hippocratea andina IPNI=*210308-2* 291
Hippocratea aspera IPNI=*162403-1* 291
Hippocratea aubletiana IPNI=*161653-1* 290
Hippocratea bonplandiana IPNI=*210308-2* 291
Hippocratea caribaea IPNI=*210308-2* 291
Hippocratea cearensis IPNI=*161653-1* 290
Hippocratea celastroides IPNI=*210308-2* 291
Hippocratea comosa IPNI=*161663-1* 291
Hippocratea comosa IPNI=*210308-2* 291
Hippocratea copiosiflora IPNI=*161653-1* 290
Hippocratea crinita IPNI=*161663-1* 291
Hippocratea diffusa IPNI=*161653-1* 290
Hippocratea discolor IPNI=*161653-1* 290
Hippocratea divaricata IPNI=*161653-1* 290
Hippocratea divaricata IPNI=*210308-2* 291
Hippocratea domingensis IPNI=*210308-2* 291

Hippocratea elliptica IPNI=*161653-1* 290
Hippocratea flaccida IPNI=*210308-2* 291
Hippocratea flava IPNI=*72123-2* 290
Hippocratea foliosa IPNI=*162403-1* 291
Hippocratea granatensis IPNI=*72123-2* 290
Hippocratea grandiflora IPNI=*162403-1* 291
Hippocratea grisebachii IPNI=*210308-2* 291
Hippocratea grisebachii var. *parvifolia* IPNI=*210308-2* 291
Hippocratea grisebachii var. *typica* IPNI=*210308-2* 291
Hippocratea hilariana IPNI=*161653-1* 290
Hippocratea infuscata IPNI=*161653-1* 290
Hippocratea inundata IPNI=*72123-2* 290
Hippocratea kappleriana IPNI=*72123-2* 290
Hippocratea lancifolia IPNI=*72123-2* 290
Hippocratea lonchophylla IPNI=*161653-1* 290
Hippocratea lutea IPNI=*72123-2* 290
Hippocratea malpighiifolia IPNI=*162403-1* 291
Hippocratea megacarpa IPNI=*72123-2* 290
Hippocratea meizantha IPNI=*161653-1* 290
Hippocratea miersii IPNI=*72123-2* 290
Hippocratea mitchellae IPNI=*72123-2* 290
Hippocratea obcordata IPNI=*161653-1* 290
Hippocratea oblongata IPNI=*72123-2* 290
Hippocratea ovalifolia IPNI=*161653-1* 290
Hippocratea ovata IPNI=*161653-1* 290
Hippocratea ovata var. *crassifolia* IPNI=*161653-1* 290
Hippocratea ovata var. *grandiflora* IPNI=*161653-1* 290
Hippocratea ovata var. *oblongifolia* IPNI=*161653-1* 290
Hippocratea ovata var. *parviflora* IPNI=*161653-1* 290
Hippocratea pallidula IPNI=*161653-1* 290
Hippocratea pauciflora IPNI=*210308-2* 291
Hippocratea perspicua IPNI=*161653-1* 290
Hippocratea plumieri IPNI=*161653-1* 290
Hippocratea riedeliana IPNI=*210308-2* 291
Hippocratea rovirosae IPNI=*161653-1* 290
Hippocratea scandens IPNI=*161653-1* 290
Hippocratea scandens IPNI=*162403-1* 291
Hippocratea scutellata IPNI=*162403-1* 291
Hippocratea serrulata IPNI=*161653-1* 290
Hippocratea swartziana IPNI=*161653-1* 290
Hippocratea tabascensis IPNI=*210308-2* 291
Hippocratea tenuiflora var. *grisebachii* IPNI=*210308-2* 291
Hippocratea utilis IPNI=*161653-1* 290
Hippocratea vahliana IPNI=*161653-1* 290
Hippocratea vahliana IPNI=*162403-1* 291
Hippocratea verrucosa IPNI=*210308-2* 291
Hippocratea versicolor IPNI=*161653-1* 290
Hippocratea viridis IPNI=*161653-1* 290
Hippocratea volubilis IPNI=*161653-1* 290
Hippocratea volubilis var. *caribaea* IPNI=*210308-2* 291
Hippocratea volubilis var. *circumalata* IPNI=*161653-1* 290
Hippocratea warmingii IPNI=*161653-1* 290
Hippocrepis humilis IPNI=*77187443-1* 320
Hippomane aucuparia IPNI=*227503-2* 392
Hippomane biglandulosa IPNI=*227503-2* 392
Hippomane dioica IPNI=*349995-1* 391
Hippomane fruticosa IPNI=*227503-2* 392
Hippomane glandulosa IPNI=*227503-2* 392
Hippomane zeocca IPNI=*227503-2* 392
Hiraea borealis var. *eglandulosa* IPNI=*556914-1* 396
Hiraea borealis var. *glandulifera* IPNI=*556914-1* 396
Hiraea crispa IPNI=*557660-1* 396
Hiraea macrodisca IPNI=*60442801-2* 394
Hiraea obovata f. *eglandulosa* IPNI=*556914-1* 396
Hiraea obovata f. *glandulifera* IPNI=*556914-1* 396
Hiraea obovata var. *angustifolia* IPNI=*556914-1* 396
Hiraea obovata var. *latifolia* IPNI=*556914-1* 396
Hiraea ovatifolia IPNI=*30237926-2* 396
Hiraea purpusii IPNI=*556914-1* 396
Hirschfeldia hispanica IPNI=*1061021-2* 277
Hirtella amplexicaulis IPNI=*725510-1* 382
Hirtella apetala IPNI=*60472826-2* 382
Hirtella aureohirsuta IPNI=*725510-1* 382
Hirtella mollicoma IPNI=*725498-1* 382
Hirtella mollissima IPNI=*725498-1* 382
Hirtella octandra IPNI=*60472870-2* 382
Hirtella purusana IPNI=*725595-1* 382
Hirtella stipitadenia IPNI=*725510-1* 382
Hisingera celastrina IPNI=*112138-1* 405
Hisingera flexuosa IPNI=*112138-1* 405
Hisingera intermedia IPNI=*112146-1* 405
Hisingera mexicana IPNI=*112146-1* 405
Hisingera nitida IPNI=*112138-1* 405
Hisingera obovata IPNI=*112138-1* 405
Hisingera paliurus IPNI=*112138-1* 405
Hisingera prunifolia IPNI=*112146-1* 405
Hisingera spiculifera IPNI=*112178-1* 405
Hoarea × coccinea IPNI=*376335-1* 358
Hoarea amoena IPNI=*377229-1* 358
Hoarea capitata IPNI=*375818-1* 358
Hoarea intermixta IPNI=*376259-1* 358
Hoarea marginata IPNI=*377229-1* 358
Hoffmannia aphylla IPNI=*212103-2* 179
Hoffmannia ostaurea IPNI=*123345-2* 351
Hoffmannia pallidiflora IPNI=*123345-2* 351
Hoffmannia rexmontis IPNI=*123345-2* 351
Hohenbergia bracteata IPNI=*121570-1* 227
Hohenbergia laxiflora IPNI=*121570-1* 227
Hohenbergia nudicaulis IPNI=*121752-1* 228
Hoiriri bromeliifolia IPNI=*121574-1* 227
Holcus aciculatus IPNI=*396110-1* 236
Holcus aestivalis IPNI=*405034-1* 240
Holcus albus IPNI=*422090-1* 245
Holcus arduinoi IPNI=*422090-1* 245
Holcus argenteus IPNI=*405034-1* 240
Holcus avenaceus IPNI=*389923-1* 234
Holcus bicolor IPNI=*422090-1* 245
Holcus cafer IPNI=*422090-1* 245
Holcus caffrorum IPNI=*422090-1* 245
Holcus cernuus IPNI=*422090-1* 245
Holcus compactus IPNI=*422090-1* 245
Holcus decolorans IPNI=*422139-1* 245
Holcus densus IPNI=*123530-2* 240
Holcus dochna IPNI=*422090-1* 245
Holcus dora IPNI=*422090-1* 245
Holcus duna IPNI=*422090-1* 245
Holcus durra IPNI=*422090-1* 245
Holcus elatior IPNI=*389923-1* 234
Holcus exiguus IPNI=*422139-1* 245
Holcus ferrugineus IPNI=*422090-1* 245
Holcus glaucus IPNI=*405034-1* 240
Holcus halepensis IPNI=*422139-1* 245
Holcus intermedius IPNI=*405034-1* 240
Holcus lanatus f. *viviparus* IPNI=*405034-1* 240
Holcus lanatus subsp. *tuberosus* IPNI=*405034-1* 240
Holcus lanatus subsp. *vaginatus* IPNI=*405034-1* 240
Holcus lanatus var. *sobolifer* IPNI=*405034-1* 240
Holcus lanatus var. *tuberosus* IPNI=*405034-1* 240
Holcus lanatus var. *vaginatus* IPNI=*405034-1* 240
Holcus longiaristatus IPNI=*123530-2* 240
Holcus mollis subsp. *reuteri* IPNI=*123530-2* 240
Holcus niger IPNI=*422090-1* 245
Holcus nigerrimus IPNI=*422090-1* 245
Holcus nigricans IPNI=*422090-1* 245
Holcus oriolis IPNI=*405034-1* 240
Holcus paniciformis IPNI=*77105978-1* 235

Holcus pertusus IPNI=*65406-3* 235
Holcus pyramidalis IPNI=*422090-1* 245
Holcus racemosus IPNI=*77105978-1* 235
Holcus reuteri IPNI=*123530-2* 240
Holcus rubens IPNI=*422090-1* 245
Holcus saccharatus IPNI=*422090-1* 245
Holcus saccharoides IPNI=*35205-2* 235
Holcus sorghum IPNI=*422090-1* 245
Holcus spicatus IPNI=*77105978-1* 235
Holcus triflorus IPNI=*123530-2* 240
Holcus tuberosus IPNI=*405034-1* 240
Hollia comosa IPNI=*532810-1* 215
Holmskioldia rubra IPNI=*863073-1* 369
Holmskioldia sanguinea f. *aurantiaca* IPNI=*863073-1* 369
Holmskioldia scandens IPNI=*863073-1* 369
Holosteum alsine IPNI=*159330-1* 286
Holosteum cordatum IPNI=*30172747-2* 285
Holosteum diandrum IPNI=*30172747-2* 285
Holosteum sperguloides IPNI=*241336-2* 286
Homalium anzoateguiense IPNI=*779849-1* 404
Homalium chocoense IPNI=*365491-1* 404
Homalium columbianum IPNI=*779849-1* 404
Homalium cuneifolium IPNI=*779849-1* 404
Homalium densiflorum IPNI=*365491-1* 404
Homalium eleutherostylum IPNI=*779849-1* 404
Homalium eurypetalum IPNI=*779849-1* 404
Homalium hemisystylum IPNI=*779849-1* 404
Homalium hondurense IPNI=*779849-1* 404
Homalium integrifolium IPNI=*779849-1* 404
Homalium leiogynum IPNI=*779849-1* 404
Homalium mattogrossense IPNI=*365491-1* 404
Homalium mituense IPNI=*779849-1* 404
Homalium napimoga IPNI=*365491-1* 404
Homalium nicaraguense IPNI=*779849-1* 404
Homalium obtusatum IPNI=*779849-1* 404
Homalium pedicellatum IPNI=*779849-1* 404
Homalium pittieri IPNI=*779849-1* 404
Homalium pleiandrum IPNI=*779849-1* 404
Homalium puberulum IPNI=*365491-1* 404
Homalium racemosum subsp. *barbellatum* IPNI=*779849-1* 404
Homalium racoubea IPNI=*365491-1* 404
Homalium riparium IPNI=*779849-1* 404
Homalium schippii IPNI=*779849-1* 404
Homalium spicatum IPNI=*365491-1* 404
Homalium stenosepalum IPNI=*779849-1* 404
Homalium surinamensis IPNI=*365491-1* 404
Homalium trichocladum IPNI=*779849-1* 404
Homalocenchrus gouinii IPNI=*137026-2* 241
Homalocenchrus hexander IPNI=*137026-2* 241
Homalocladium platycladum IPNI=*694774-1* 289
Homanthis multiflorus IPNI=*237026-1* 268
Homoianthus multiflorus IPNI=*237026-1* 268
Homonoma aridum IPNI=*573916-1* 424
Hopkirkia anthemoidea IPNI=*242907-1* 269
Hopkirkia eupatoria IPNI=*241903-1* 269
Hopkirkia scandens IPNI=*241903-1* 269
Hoplophytum bracteatum IPNI=*121570-1* 227
Hoplophytum grande IPNI=*271722-2* 228
Hoplophytum nudicaule IPNI=*121752-1* 228
Horau racemosus IPNI=*170672-1* 416
Hordeum aestivum IPNI=*405391-1* 240
Hordeum americanum IPNI=*405391-1* 240
Hordeum andicola IPNI=*278265-2* 240
Hordeum bifarium IPNI=*405391-1* 240
Hordeum brachyatherum IPNI=*405391-1* 240
Hordeum caspicum IPNI=*405391-1* 240
Hordeum chilense var. *muticum* IPNI=*278265-2* 240
Hordeum compressum IPNI=*407493-1* 241
Hordeum daghestanicum IPNI=*405391-1* 240
Hordeum defectoides IPNI=*405391-1* 240
Hordeum durum IPNI=*405391-1* 240
Hordeum elongatum IPNI=*405391-1* 240
Hordeum gymnodistichum IPNI=*405391-1* 240
Hordeum heterostychon IPNI=*405391-1* 240
Hordeum hexastichon IPNI=*405391-1* 240
Hordeum hibernaculum IPNI=*405391-1* 240
Hordeum hibernans IPNI=*405391-1* 240
Hordeum hirtiusculum IPNI=*405391-1* 240
Hordeum horsfordianum IPNI=*405391-1* 240
Hordeum ircutianum IPNI=*405391-1* 240
Hordeum jarenskianum IPNI=*405391-1* 240
Hordeum juliae IPNI=*405391-1* 240
Hordeum kalugense IPNI=*405391-1* 240
Hordeum karzinianum IPNI=*405391-1* 240
Hordeum kiarchanum IPNI=*405391-1* 240
Hordeum laevipaleatum IPNI=*405391-1* 240
Hordeum lapponicum IPNI=*405391-1* 240
Hordeum leptostachys IPNI=*405391-1* 240
Hordeum mandshuricum IPNI=*405391-1* 240
Hordeum mandshuroides IPNI=*405391-1* 240
Hordeum michalkowii IPNI=*405391-1* 240
Hordeum muticum var. *andicola* IPNI=*278265-2* 240
Hordeum nekludowii IPNI=*405391-1* 240
Hordeum nigrum IPNI=*405391-1* 240
Hordeum pamiricum IPNI=*405391-1* 240
Hordeum parvum IPNI=*405391-1* 240
Hordeum pensanum IPNI=*405391-1* 240
Hordeum perversum IPNI=*405391-1* 240
Hordeum polystichon IPNI=*405391-1* 240
Hordeum pyramidatum IPNI=*405391-1* 240
Hordeum sativum IPNI=*405391-1* 240
Hordeum scabriusculum IPNI=*405391-1* 240
Hordeum secalinum var. *andicola* IPNI=*278265-2* 240
Hordeum septentrionale IPNI=*405391-1* 240
Hordeum stassewitschii IPNI=*405391-1* 240
Hordeum strobelense IPNI=*405391-1* 240
Hordeum taganrocense IPNI=*405391-1* 240
Hordeum tanaiticum IPNI=*405391-1* 240
Hordeum tetrastichum IPNI=*405391-1* 240
Hordeum transcaucasicum IPNI=*405391-1* 240
Hordeum violaceum IPNI=*405391-1* 240
Hordeum vulgare convar. *revelatum* IPNI=*405391-1* 240
Hordeum vulgare f. *hexastichon* IPNI=*405391-1* 240
Hordeum vulgare var. *abdulbasirovii* IPNI=*405391-1* 240
Hordeum vulgare var. *addis-abebae* IPNI=*405391-1* 240
Hordeum vulgare var. *africanum* IPNI=*405391-1* 240
Hordeum vulgare var. *ancoberense* IPNI=*405391-1* 240
Hordeum vulgare var. *atratum* IPNI=*405391-1* 240
Hordeum vulgare var. *bachteevii* IPNI=*405391-1* 240
Hordeum vulgare var. *breviaristatum* IPNI=*405391-1* 240
Hordeum vulgare var. *brevispicatum* IPNI=*405391-1* 240
Hordeum vulgare var. *brunneinudum* IPNI=*405391-1* 240
Hordeum vulgare var. *chalunicum* IPNI=*405391-1* 240
Hordeum vulgare var. *chinense* IPNI=*405391-1* 240
Hordeum vulgare var. *colonicum* IPNI=*405391-1* 240
Hordeum vulgare var. *daghestanicum* IPNI=*405391-1* 240
Hordeum vulgare var. *euryhypatherum* IPNI=*405391-1* 240
Hordeum vulgare var. *glabribrevisetum* IPNI=*405391-1* 240
Hordeum vulgare var. *glabrideficiens* IPNI=*405391-1* 240
Hordeum vulgare var. *glabrierectum* IPNI=*405391-1* 240
Hordeum vulgare var. *glabrigracilius* IPNI=*405391-1* 240
Hordeum vulgare var. *glabrispicatum* IPNI=*405391-1* 240
Hordeum vulgare var. *glabriviride* IPNI=*405391-1* 240
Hordeum vulgare var. *gobicum* IPNI=*405391-1* 240
Hordeum vulgare var. *griseinigrum* IPNI=*405391-1* 240
Hordeum vulgare var. *griseinudum* IPNI=*405391-1* 240

Hordeum vulgare var. *hadaka* IPNI=*405391-1* 240
Hordeum vulgare var. *hangaicum* IPNI=*405391-1* 240
Hordeum vulgare var. *harlanii* IPNI=*405391-1* 240
Hordeum vulgare var. *hexastichon* IPNI=*405391-1* 240
Hordeum vulgare var. *hypatherum* IPNI=*405391-1* 240
Hordeum vulgare var. *ibericum* IPNI=*405391-1* 240
Hordeum vulgare var. *insularum* IPNI=*405391-1* 240
Hordeum vulgare var. *interpallidum* IPNI=*405391-1* 240
Hordeum vulgare var. *ismailii* IPNI=*405391-1* 240
Hordeum vulgare var. *japonicum* IPNI=*405391-1* 240
Hordeum vulgare var. *kobdicum* IPNI=*405391-1* 240
Hordeum vulgare var. *latibrevisetum* IPNI=*405391-1* 240
Hordeum vulgare var. *latinudipyramidatum* IPNI=*405391-1* 240
Hordeum vulgare var. *latirevelatum* IPNI=*405391-1* 240
Hordeum vulgare var. *leioheterolepis* IPNI=*405391-1* 240
Hordeum vulgare var. *lukyanovae* IPNI=*405391-1* 240
Hordeum vulgare var. *lvovii* IPNI=*405391-1* 240
Hordeum vulgare var. *micrurum* IPNI=*405391-1* 240
Hordeum vulgare var. *mongolicum* IPNI=*405391-1* 240
Hordeum vulgare var. *multispiculum* IPNI=*405391-1* 240
Hordeum vulgare var. *nanum* IPNI=*405391-1* 240
Hordeum vulgare var. *nigrinudum* IPNI=*405391-1* 240
Hordeum vulgare var. *nipponicum* IPNI=*405391-1* 240
Hordeum vulgare var. *nudijaponicum* IPNI=*405391-1* 240
Hordeum vulgare var. *nudinipponicum* IPNI=*405391-1* 240
Hordeum vulgare var. *palestinicum* IPNI=*405391-1* 240
Hordeum vulgare var. *patimatae* IPNI=*405391-1* 240
Hordeum vulgare var. *rarum* IPNI=*405391-1* 240
Hordeum vulgare var. *saidii* IPNI=*405391-1* 240
Hordeum vulgare var. *sessilifurcatum* IPNI=*405391-1* 240
Hordeum vulgare var. *sinicum* IPNI=*405391-1* 240
Hordeum vulgare var. *sinojaponicum* IPNI=*405391-1* 240
Hordeum vulgare var. *solitarum* IPNI=*405391-1* 240
Hordeum vulgare var. *suberectum* IPNI=*405391-1* 240
Hordeum vulgare var. *subneogenes* IPNI=*405391-1* 240
Hordeum vulgare var. *subnudipiramidatum* IPNI=*405391-1* 240
Hordeum vulgare var. *subnudum* IPNI=*405391-1* 240
Hordeum vulgare var. *subnutans* IPNI=*405391-1* 240
Hordeum vulgare var. *subpyramidatum* IPNI=*405391-1* 240
Hordeum vulgare var. *subzeocrithon* IPNI=*405391-1* 240
Hordeum vulgare var. *syriacum* IPNI=*405391-1* 240
Hordeum vulgare var. *tetranutans* IPNI=*405391-1* 240
Hordeum vulgare var. *tetrapallidum* IPNI=*405391-1* 240
Hordeum vulgare var. *tibetanum* IPNI=*405391-1* 240
Hordeum vulgare var. *trofimovskajae* IPNI=*405391-1* 240
Hordeum vulgare var. *uljassutaicum* IPNI=*405391-1* 240
Hordeum vulgare var. *urgaicum* IPNI=*405391-1* 240
Hordeum vulgare var. *valentinae* IPNI=*405391-1* 240
Hordeum vulgare var. *viride* IPNI=*405391-1* 240
Hordeum vulgare var. *zeocrithideficiens* IPNI=*405391-1* 240
Hordeum vulgare var. *zuleichatae* IPNI=*405391-1* 240
Hordeum walpersii IPNI=*405391-1* 240
Horminum coccineum IPNI=*455978-1* 372
Horminum occidentale IPNI=*456822-1* 372
Horminum parviflorum IPNI=*457495-1* 373
Horminum splendens IPNI=*60466516-2* 373
Horminum sylvestre IPNI=*457495-1* 373
Horminum verbenacum IPNI=*457495-1* 373
Hornemannia ovata IPNI=*1189750-2* 374
Hornera altissima IPNI=*1197939-2* 328
Hornera pruriens IPNI=*509275-1* 328
Hornera urens IPNI=*1197939-2* 328
Hornstedtia imperialis IPNI=*942355-1* 251
Hortensia japonica IPNI=*791637-1* 291
Hortensia macrophylla f. *isobue* IPNI=*791637-1* 291
Hortensia macrophylla f. *isonotaki* IPNI=*791637-1* 291
Hortensia macrophylla f. *normalis* IPNI=*791637-1* 291
Hortensia macrophylla IPNI=*791637-1* 291
Hortensia mutabilis IPNI=*791637-1* 291
Hortensia nigra IPNI=*791637-1* 291
Hortensia opuloides IPNI=*791637-1* 291
Hortensia rosea IPNI=*791637-1* 291
Hortensia speciosa IPNI=*791637-1* 291
Hosta grandiflora IPNI=*319153-2* 369
Hosta grandifolia IPNI=*319153-2* 369
Hosta latifolia IPNI=*319153-2* 369
Hosta longifolia IPNI=*319153-2* 369
Hosta odorata IPNI=*862634-1* 369
Hosta pyramidata IPNI=*319153-2* 369
Houmiri balsamifera IPNI=*431431-1* 393
Houstonia bryoides IPNI=*17992-2* 347
Houstonia longiflora IPNI=*745065-1* 347
Howardia anguicida IPNI=*316428-2* 192
Howardia benthamii IPNI=*93027-1* 192
Howardia caracasensis IPNI=*204802-2* 355
Howardia costaricensis IPNI=*93231-1* 193
Howardia cymbifera IPNI=*93285-1* 193
Howardia foetens IPNI=*92978-1* 192
Howardia galeata IPNI=*93285-1* 193
Howardia geminiflora IPNI=*316426-2* 193
Howardia glaucescens IPNI=*92858-1* 192
Howardia golmerii IPNI=*316426-2* 193
Howardia grandiflora IPNI=*204802-2* 355
Howardia grandiflora IPNI=*92978-1* 192
Howardia hoffmannii IPNI=*316426-2* 193
Howardia inflata IPNI=*93027-1* 192
Howardia maxima IPNI=*316426-2* 193
Howardia pandurata IPNI=*316342-2* 193
Howardia pilosa IPNI=*93231-1* 193
Howardia ringens IPNI=*93285-1* 193
Howardia ruiziana IPNI=*93298-1* 193
Howardia trifida IPNI=*319092-2* 193
Howardia trilobata IPNI=*319092-2* 193
Howardia veraguensis IPNI=*272659-2* 192
Howardia warszewiczii IPNI=*316342-2* 193
Hoya carnosa var. *gushanica* IPNI=*107308-3* 341
Hoya chinensis IPNI=*107308-3* 341
Hoya compacta IPNI=*107308-3* 341
Hoya crassifolia IPNI=*107308-3* 341
Hoya intermedia IPNI=*107308-3* 341
Hoya motoskei IPNI=*107308-3* 341
Hoya rotundifolia IPNI=*107308-3* 341
Hoya skinneriana IPNI=*107308-3* 341
Hoya variegata IPNI=*107308-3* 341
Huberia argentea IPNI=*573207-1* 422
Hudsonia arborea IPNI=*77164595-1* 416
Hufelandia pendula IPNI=*1057986-2* 182
Hufelandia sulcata IPNI=*31062-2* 182
Hufelandia thomaea IPNI=*1057986-2* 182
Hufelandia tovariensis IPNI=*1116029-2* 182
Huilaea macrocarpa IPNI=*77146610-1* 418
Huilaea multiflora IPNI=*567365-1* 418
Humata solida IPNI=*17080730-1* 175
Humboldtiella arborea IPNI=*944647-1* 315
Humboldtiella ferruginea IPNI=*944647-1* 315
Humboltia brevipes IPNI=*1147281-2* 217
Humiria amplexicaulis IPNI=*431431-1* 393
Humiria balsamifera f. *acuminata* IPNI=*431431-1* 393
Humiria montana IPNI=*431431-1* 393
Humiria procera IPNI=*124807-2* 393
Humirianthera ampla IPNI=*963946-1* 359
Humirianthera duckei IPNI=*963946-1* 359
Humiriastrum diguense subsp. *costaricense* IPNI=*124798-2* 393
Humulus lupulus var. *fengxianensis* IPNI=*303502-2* 436
Humulus volubilis IPNI=*303502-2* 436
Humulus vulgaris IPNI=*303502-2* 436

Huntleya albidofulva IPNI=*639074-1* 219
Huntleya marginata IPNI=*661384-1* 223
Huperzia erythraea IPNI=*124844-2* 171
Huperzia lindeneri IPNI=*1090207-2* 171
Huperzia mortonii IPNI=*1090207-2* 171
Huperzia pilgeriana IPNI=*124844-2* 171
Huperzia sanctae-barbarae IPNI=*60447820-2* 171
Huperzia schlechtendalii IPNI=*1090207-2* 171
Huperzia springii IPNI=*124844-2* 171
Hura brasiliensis IPNI=*350060-1* 391
Hura crepitans var. *genuina* IPNI=*350060-1* 391
Hura crepitans var. *membranacea* IPNI=*350060-1* 391
Hura crepitans var. *oblongifolia* IPNI=*350060-1* 391
Hura crepitans var. *orbicularis* IPNI=*350060-1* 391
Hura crepitans var. *ovata* IPNI=*350060-1* 391
Hura sencgalensis IPNI=*350060-1* 391
Hura strepens IPNI=*350060-1* 391
Huxleya linifolia IPNI=*866088-1* 374
Hybanthus angustifolius IPNI=*77143566-1* 405
Hybanthus anomalus IPNI=*77143573-1* 405
Hybanthus calceolaria IPNI=*77143549-1* 405
Hybanthus chiapensis IPNI=*77143553-1* 405
Hybanthus circaeoides IPNI=*77143566-1* 405
Hybanthus glaber IPNI=*77143553-1* 405
Hybanthus glutinosus IPNI=*77143568-1* 405
Hybanthus glutinosus var. *paraguayensis* IPNI=*77143568-1* 405
Hybanthus grandiflorus IPNI=*77143573-1* 405
Hybanthus hieronymi IPNI=*77143566-1* 405
Hybanthus indecorus IPNI=*77143549-1* 405
Hybanthus ipecacuanha IPNI=*77143549-1* 405
Hybanthus longifolius IPNI=*77143566-1* 405
Hybanthus maytensillo IPNI=*77143568-1* 405
Hybanthus microphyllus IPNI=*77143568-1* 405
Hybanthus oppositifolius IPNI=*77143566-1* 405
Hybanthus oppositifolius var. *genuinus* IPNI=*77143566-1* 405
Hybanthus parietariifolius IPNI=*77143566-1* 405
Hybanthus parviflorus f. *membranaceus* IPNI=*77143568-1* 405
Hybanthus parviflorus f. *microphyllus* IPNI=*77143568-1* 405
Hybanthus parviflorus f. *paraguayensis* IPNI=*77143568-1* 405
Hybanthus parviflorus IPNI=*77143568-1* 405
Hybanthus parviflorus var. *angustifolius* IPNI=*77143568-1* 405
Hybanthus parviflorus var. *argentinensis* IPNI=*77143568-1* 405
Hybanthus parviflorus var. *bangii* IPNI=*77143568-1* 405
Hybanthus parviflorus var. *chamaedrifolius* IPNI=*77143568-1* 405
Hybanthus parviflorus var. *glutinosus* IPNI=*77143568-1* 405
Hybanthus parviflorus var. *latifolius* IPNI=*77143568-1* 405
Hybanthus procumbens IPNI=*77143549-1* 405
Hybanthus prunifolius IPNI=*77143573-1* 405
Hybanthus scandens IPNI=*77144891-1* 405
Hybanthus supinus IPNI=*77143549-1* 405
Hybanthus teucriifolius IPNI=*77143566-1* 405
Hybanthus urticifolius IPNI=*77143566-1* 405
Hybanthus villosissimus IPNI=*77143549-1* 405
Hydnocarpus anthelminthicus IPNI=*111741-1* 380
Hydora lithuanica IPNI=*114224-3* 203
Hydrangea azisai IPNI=*791637-1* 291
Hydrangea florida IPNI=*791637-1* 291
Hydrangea hortensia IPNI=*791637-1* 291
Hydrangea hortensis IPNI=*791637-1* 291
Hydrangea macrocephala IPNI=*791637-1* 291
Hydrangea macrophylla f. *isobue* IPNI=*791637-1* 291
Hydrangea macrophylla f. *isonotaki* IPNI=*791637-1* 291
Hydrangea macrophylla f. *nuda* IPNI=*791637-1* 291
Hydrangea macrophylla var. *hattoriana* IPNI=*791637-1* 291
Hydrangea macrophylla var. *normalis* IPNI=*791637-1* 291
Hydrangea mandshurica IPNI=*791637-1* 291
Hydrangea mutabilis IPNI=*791637-1* 291
Hydrangea nigra IPNI=*791637-1* 291
Hydrangea opuloides IPNI=*791637-1* 291
Hydrangea otaksa IPNI=*791637-1* 291
Hydrangea sieboldii IPNI=*791637-1* 291
Hydrangea sinensis IPNI=*791637-1* 291
Hydrilla angustifolia IPNI=*114224-3* 203
Hydrilla dentata IPNI=*114224-3* 203
Hydrilla lithuanica IPNI=*114224-3* 203
Hydrilla najadifolia IPNI=*114224-3* 203
Hydrilla ovalifolia IPNI=*114224-3* 203
Hydrilla polysperma IPNI=*114224-3* 203
Hydrilla roxburghii IPNI=*114224-3* 203
Hydrilla subulata IPNI=*114224-3* 203
Hydrilla verticillata var. *crispa* IPNI=*114224-3* 203
Hydrilla verticillata var. *roxburghii* IPNI=*114224-3* 203
Hydrilla wightii IPNI=*114224-3* 203
Hydrocharis cordifolia IPNI=*431804-1* 203
Hydrochloa airoides IPNI=*395086-1* 235
Hydrochloa fluitans IPNI=*404100-1* 239
Hydrocleys azurea IPNI=*58399-1* 198
Hydrocleys commersonii IPNI=*58399-1* 198
Hydrocleys humboldtii IPNI=*58399-1* 198
Hydrocotyle abbreviata IPNI=*1197718-2* 252
Hydrocotyle abyssinica IPNI=*1197718-2* 252
Hydrocotyle adoensis IPNI=*843671-1* 253
Hydrocotyle ambigua IPNI=*843759-1* 253
Hydrocotyle artensis IPNI=*1197718-2* 252
Hydrocotyle asiatica IPNI=*1197718-2* 252
Hydrocotyle asiatica var. *floridana* IPNI=*51383-2* 252
Hydrocotyle asiatica var. *monantha* IPNI=*1197718-2* 252
Hydrocotyle batrachioides IPNI=*843671-1* 253
Hydrocotyle biflora IPNI=*51383-2* 252
Hydrocotyle bonariensis var. *multiflora* IPNI=*125117-2* 253
Hydrocotyle bonariensis var. *texana* IPNI=*125117-2* 253
Hydrocotyle bonariensis var. *tribotrys* IPNI=*125117-2* 253
Hydrocotyle bonplandii var. *glabra* IPNI=*843403-1* 253
Hydrocotyle bonplandii var. *hirtipes* IPNI=*843403-1* 253
Hydrocotyle brasiliensis IPNI=*1197718-2* 252
Hydrocotyle brevipedata IPNI=*1197718-2* 252
Hydrocotyle brevipes IPNI=*1197718-2* 252
Hydrocotyle caffra IPNI=*843746-1* 253
Hydrocotyle chilensis IPNI=*843403-1* 253
Hydrocotyle communis IPNI=*843759-1* 253
Hydrocotyle cuneata IPNI=*843759-1* 253
Hydrocotyle cymbalariifolia IPNI=*843671-1* 253
Hydrocotyle dentata IPNI=*1197718-2* 252
Hydrocotyle dux IPNI=*843574-1* 253
Hydrocotyle emarginata IPNI=*843574-1* 253
Hydrocotyle erecta IPNI=*51383-2* 252
Hydrocotyle ficarifolia IPNI=*1197718-2* 252
Hydrocotyle ficarioides IPNI=*1197718-2* 252
Hydrocotyle fluitans IPNI=*843746-1* 253
Hydrocotyle glochidiata IPNI=*1197718-2* 252
Hydrocotyle grossulariifolia var. *buchtienii* IPNI=*125145-2* 253
Hydrocotyle grossulariifolia var. *jamesonii* IPNI=*125145-2* 253
Hydrocotyle hazenii IPNI=*843574-1* 253
Hydrocotyle hebecarpa IPNI=*1197718-2* 252
Hydrocotyle humboldtii IPNI=*843403-1* 253
Hydrocotyle humboldtii var. *isoloba* IPNI=*843540-1* 253
Hydrocotyle humboldtii var. *macrophylla* IPNI=*843540-1* 253
Hydrocotyle humboldtii var. *pubescens* IPNI=*843540-1* 253
Hydrocotyle inaequipes IPNI=*1197718-2* 252
Hydrocotyle incrassata IPNI=*843746-1* 253
Hydrocotyle indivisa IPNI=*1197718-2* 252
Hydrocotyle interrupta IPNI=*843759-1* 253
Hydrocotyle interrupta var. *tuberosa* IPNI=*843759-1* 253
Hydrocotyle isoloba IPNI=*843540-1* 253
Hydrocotyle leptostachys IPNI=*1197718-2* 252
Hydrocotyle leucocephala f. *peltata* IPNI=*843574-1* 253

Hydrocotyle leucocephala var. *intermedia* IPNI=*843574-1* 253
Hydrocotyle leucocephala var. *minuta* IPNI=*843574-1* 253
Hydrocotyle leucocephala var. *obtusiloba* IPNI=*843574-1* 253
Hydrocotyle leucocephala var. *truncatiloba* IPNI=*843574-1* 253
Hydrocotyle leucocephala var. *villosa* IPNI=*843574-1* 253
Hydrocotyle lunata IPNI=*1197718-2* 252
Hydrocotyle lurida IPNI=*1197718-2* 252
Hydrocotyle maxonii IPNI=*843574-1* 253
Hydrocotyle multicaulis IPNI=*843574-1* 253
Hydrocotyle multiflora IPNI=*125117-2* 253
Hydrocotyle natans IPNI=*1197718-2* 252
Hydrocotyle natans IPNI=*843671-1* 253
Hydrocotyle natans IPNI=*843759-1* 253
Hydrocotyle nummularioides IPNI=*1197718-2* 252
Hydrocotyle pallida IPNI=*1197718-2* 252
Hydrocotyle pelviformis IPNI=*125117-2* 253
Hydrocotyle petiolaris IPNI=*125117-2* 253
Hydrocotyle polyrhiza IPNI=*843574-1* 253
Hydrocotyle polystachya IPNI=*843746-1* 253
Hydrocotyle polystachya var. *triradiata* IPNI=*125117-2* 253
Hydrocotyle prolifera IPNI=*843671-1* 253
Hydrocotyle quinqueloba f. *subglabra* IPNI=*843540-1* 253
Hydrocotyle quinqueradiata IPNI=*843746-1* 253
Hydrocotyle racemosa IPNI=*843759-1* 253
Hydrocotyle ranunculoides f. *minima* IPNI=*843671-1* 253
Hydrocotyle ranunculoides var. *brasiliensis* IPNI=*843671-1* 253
Hydrocotyle ranunculoides var. *genuina* IPNI=*843671-1* 253
Hydrocotyle ranunculoides var. *incisocrenata* IPNI=*843671-1* 253
Hydrocotyle ranunculoides var. *lobata* IPNI=*843671-1* 253
Hydrocotyle ranunculoides var. *natans* IPNI=*843671-1* 253
Hydrocotyle reniformis IPNI=*1197718-2* 252
Hydrocotyle repanda IPNI=*51383-2* 252
Hydrocotyle sarmentosa IPNI=*1197718-2* 252
Hydrocotyle scaposa IPNI=*843746-1* 253
Hydrocotyle superposita IPNI=*843759-1* 253
Hydrocotyle sylvicola IPNI=*1197718-2* 252
Hydrocotyle tenerrima var. *puberula* IPNI=*125218-2* 253
Hydrocotyle thunbergiana IPNI=*1197718-2* 252
Hydrocotyle tribotrys IPNI=*125117-2* 253
Hydrocotyle triflora IPNI=*51383-2* 252
Hydrocotyle trilobulata IPNI=*843759-1* 253
Hydrocotyle tussilaginifolia IPNI=*1197718-2* 252
Hydrocotyle ulugurensis IPNI=*1197718-2* 252
Hydrocotyle umbellata var. *ambigua* IPNI=*843759-1* 253
Hydrocotyle umbellata var. *bonariensis* IPNI=*125117-2* 253
Hydrocotyle umbellata var. *intermedia* IPNI=*843746-1* 253
Hydrocotyle umbellata var. *microphylla* IPNI=*843746-1* 253
Hydrocotyle umbellata var. *scaposa* IPNI=*843746-1* 253
Hydrocotyle umbellata var. *umbellulata* IPNI=*843746-1* 253
Hydrocotyle umbellulata IPNI=*843746-1* 253
Hydrocotyle uniflora IPNI=*1197718-2* 252
Hydrocotyle verticillata var. *cuneata* IPNI=*843759-1* 253
Hydrocotyle verticillata var. *longipedunculata* IPNI=*843759-1* 253
Hydrocotyle verticillata var. *pluriradiata* IPNI=*843759-1* 253
Hydrocotyle verticillata var. *racemosa* IPNI=*843759-1* 253
Hydrocotyle verticillata var. *tenella* IPNI=*843759-1* 253
Hydrocotyle verticillata var. *triradiata* IPNI=*125117-2* 253
Hydrocotyle volkmannii IPNI=*843759-1* 253
Hydrocotyle vulgaris IPNI=*843759-1* 253
Hydrocotyle vulgaris var. *communis* IPNI=*843759-1* 253
Hydrocotyle vulgaris var. *verticillata* IPNI=*843759-1* 253
Hydrocotyle wightiana IPNI=*1197718-2* 252
Hydrocotyle yucatanensis IPNI=*125117-2* 253
Hydroglossum expansum IPNI=*17143520-1* 179
Hydroglossum hastatum IPNI=*17143510-1* 179
Hydroglossum heptaphyllum IPNI=*17143520-1* 179
Hydroglossum hirsutum IPNI=*17143510-1* 179
Hydroglossum volubile IPNI=*17143520-1* 179
Hydrolea auriculata IPNI=*114242-3* 276
Hydrolea caracassana IPNI=*114242-3* 276
Hydrolea mollis IPNI=*114242-3* 276
Hydrolea scorpioides IPNI=*114242-3* 276
Hydrolea urens IPNI=*114242-3* 276
Hydrophace minor IPNI=*526194-1* 200
Hydrospondylus submersus IPNI=*114224-3* 203
Hydrotrida beccabunga IPNI=*1072674-2* 376
Hygrophila angusta IPNI=*30085162-2* 360
Hygrophila atricheta IPNI=*49672-1* 360
Hygrophila brasiliensis IPNI=*49672-1* 360
Hygrophila conferta IPNI=*30085162-2* 360
Hygrophila conferta var. *rubricaulis* IPNI=*30085162-2* 360
Hygrophila costata var. *strobilantha* IPNI=*49672-1* 360
Hygrophila guianensis var. *angustior* IPNI=*49672-1* 360
Hygrophila helodes IPNI=*49672-1* 360
Hygrophila helodes var. *multiflora* IPNI=*49672-1* 360
Hygrophila hispida IPNI=*49672-1* 360
Hygrophila lacustris IPNI=*49672-1* 360
Hygrophila latifolia IPNI=*49672-1* 360
Hygrophila portoricensis IPNI=*49672-1* 360
Hygrophila pubescens IPNI=*49672-1* 360
Hygrophila rivularis IPNI=*49672-1* 360
Hygroryza ciliata IPNI=*137026-2* 241
Hylenaea capilliflora IPNI=*161663-1* 291
Hylenaea jamaicensis IPNI=*72123-2* 290
Hylocereus bronxensis IPNI=*77163181-1* 285
Hylocereus costaricensis IPNI=*77177683-1* 284
Hylocereus estebanensis IPNI=*77163180-1* 285
Hylocereus lemairei IPNI=*77163180-1* 285
Hylocereus megalanthus IPNI=*232386-2* 285
Hylocereus microcladus IPNI=*77177683-1* 284
Hylocereus monacanthus IPNI=*77163180-1* 285
Hylocereus ocamponis IPNI=*77163181-1* 285
Hylocereus peruvianus IPNI=*77163180-1* 285
Hylocereus polyrhizus IPNI=*77163180-1* 285
Hylocereus scandens IPNI=*77163180-1* 285
Hylocereus schomburgkii IPNI=*77163180-1* 285
Hylocereus tricostatus IPNI=*77163184-1* 285
Hylocereus trinitatensis IPNI=*77163180-1* 285
Hylocereus undatus IPNI=*77163184-1* 285
Hylocereus undatus subsp. *luteocarpus* IPNI=*77163184-1* 285
Hylocereus venezuelensis IPNI=*77163180-1* 285
Hymenachne acutigluma IPNI=*405429-1* 240
Hymenachne auriculata IPNI=*125410-2* 240
Hymenachne calamitosa IPNI=*405429-1* 240
Hymenachne cordata IPNI=*125410-2* 240
Hymenachne frondescens IPNI=*77105897-1* 241
Hymenachne gouinii IPNI=*405429-1* 240
Hymenachne myosurus IPNI=*225088-2* 244
Hymenachne myuros IPNI=*225088-2* 244
Hymenachne palustris IPNI=*125410-2* 240
Hymenachne patula IPNI=*125410-2* 240
Hymenachne pseudointerrupta IPNI=*405429-1* 240
Hymenaea adenotricha IPNI=*125429-2* 320
Hymenaea animifera IPNI=*125420-2* 320
Hymenaea candolleana IPNI=*125420-2* 320
Hymenaea chrysophylla IPNI=*518025-1* 332
Hymenaea confertifolia IPNI=*125420-2* 320
Hymenaea courbaril var. *obtusifolia* IPNI=*125420-2* 320
Hymenaea courbaril var. *stilbocarpa* IPNI=*125420-2* 320
Hymenaea courbaril var. *subsessilis* IPNI=*125420-2* 320
Hymenaea courbaril var. *villosa* IPNI=*125420-2* 320
Hymenaea intermedia var. *adenotricha* IPNI=*125429-2* 320
Hymenaea microphylla IPNI=*498789-1* 321
Hymenaea multiflora IPNI=*125420-2* 320
Hymenaea pubescens IPNI=*498789-1* 321
Hymenaea resinifera IPNI=*125420-2* 320

Hymenaea retusa IPNI=*125420-2* 320
Hymenaea sellowiana IPNI=*498789-1* 321
Hymenaea splendida IPNI=*125420-2* 320
Hymenaea stilbocarpa IPNI=*125420-2* 320
Hymenaea venosa IPNI=*513006-1* 330
Hymenesthes nitida IPNI=*1088317-2* 273
Hymenocallis adnata var. *disticha* IPNI=*30048918-2* 213
Hymenocallis adnata var. *driandrina* IPNI=*30048918-2* 213
Hymenocallis adnata var. *staplesiana* IPNI=*30048918-2* 213
Hymenocallis americana f. *disticha* IPNI=*30048918-2* 213
Hymenocallis americana f. *dryanderi* IPNI=*30048918-2* 213
Hymenocallis americana f. *staplesiana* IPNI=*30048918-2* 213
Hymenocallis americana IPNI=*30048918-2* 213
Hymenocallis amoena IPNI=*65212-1* 213
Hymenocallis amoena var. *princeps* IPNI=*65212-1* 213
Hymenocallis angusta IPNI=*65212-1* 213
Hymenocallis arenaria IPNI=*30048918-2* 213
Hymenocallis biflora IPNI=*30048918-2* 213
Hymenocallis caribaea f. *angustifolia* IPNI=*65212-1* 213
Hymenocallis caribaea var. *cinerascens* IPNI=*65212-1* 213
Hymenocallis caribaea var. *patens* IPNI=*65212-1* 213
Hymenocallis cinerascens IPNI=*65212-1* 213
Hymenocallis disticha IPNI=*30048918-2* 213
Hymenocallis dryanderi IPNI=*30048918-2* 213
Hymenocallis formosa IPNI=*65212-1* 213
Hymenocallis insignis IPNI=*30048918-2* 213
Hymenocallis littoralis var. *longituba* IPNI=*30048918-2* 213
Hymenocallis niederleinii IPNI=*30048918-2* 213
Hymenocallis obtusata IPNI=*65212-1* 213
Hymenocallis panamensis IPNI=*30048918-2* 213
Hymenocallis patens IPNI=*65212-1* 213
Hymenocallis pedalis IPNI=*30048918-2* 213
Hymenocallis peruviana IPNI=*30048918-2* 213
Hymenocallis senegambica IPNI=*30048918-2* 213
Hymenocallis sessilis IPNI=*65212-1* 213
Hymenocallis sloanei IPNI=*65212-1* 213
Hymenocallis speciosa var. *fragrans* IPNI=*65212-1* 213
Hymenocallis staplesiana IPNI=*30048918-2* 213
Hymenocallis staplesii IPNI=*30048918-2* 213
Hymenocallis tenuiflora IPNI=*30048918-2* 213
Hymenochaeta tabernaemontani IPNI=*1019164-1* 232
Hymenocharis arouma IPNI=*797103-1* 249
Hymenocharis rotundifolia IPNI=*797124-1* 249
Hymenocyclus croceus IPNI=*1011059-2* 280
Hymenolobium elatum IPNI=*125608-2* 321
Hymenophyllum atrovirens IPNI=*1009149-2* 173
Hymenophyllum blepharodes IPNI=*1009149-2* 173
Hymenophyllum calodictyon IPNI=*1009149-2* 173
Hymenophyllum fucoides var. *atrovirens* IPNI=*1009149-2* 173
Hymenophyllum fucoides var. *calodictyon* IPNI=*1009149-2* 173
Hymenophyllum fucoides var. *chachapoyense* IPNI=*1009149-2* 173
Hymenophyllum fucoides var. *integrum* IPNI=*1009149-2* 173
Hymenophyllum intercalatum IPNI=*125766-2* 173
Hymenophyllum peruvianum IPNI=*1009149-2* 173
Hymenophyllum podocarpon IPNI=*1009149-2* 173
Hymenophyllum raddianum IPNI=*125766-2* 173
Hymenophyllum spinulosum IPNI=*1009149-2* 173
Hymenophyllum torquescens IPNI=*1009149-2* 173
Hymenostachys diversifrons IPNI=*17228970-1* 173
Hymenostachys elegans IPNI=*17228970-1* 173
Hyoseris pusilla IPNI=*225504-1* 264
Hyoseris tenella IPNI=*225504-1* 264
Hyospathe montana IPNI=*665826-1* 207
Hyparrhenia dissoluta IPNI=*126405-2* 240
Hyparrhenia macrolepis IPNI=*126405-2* 240
Hyparrhenia ruprechtii IPNI=*126405-2* 240
Hypelate geniculata IPNI=*128278-1* 452
Hyperanthera decandra IPNI=*584736-1* 280
Hyperanthera moringa IPNI=*584736-1* 280
Hyperanthera pterygosperma IPNI=*584736-1* 280
Hyperbaena cuatrecasasii IPNI=*34845-2* 434
Hyperbaena graciliflora IPNI=*580887-1* 435
Hyperbaena hostmannii IPNI=*580887-1* 435
Hyperbaena moricandii IPNI=*580887-1* 435
Hyperbaena polyantha IPNI=*72709-2* 435
Hyperbaena reticulata IPNI=*580887-1* 435
Hyperbaena retinervis IPNI=*580887-1* 435
Hyperbaena rotundiuscula IPNI=*580887-1* 435
Hyperbaena trinervis IPNI=*580059-1* 434
Hyperbaena tweediei IPNI=*580887-1* 435
Hypericum acerosum IPNI=*433547-1* 394
Hypericum acuminatum IPNI=*30051774-2* 394
Hypericum adenophyllum IPNI=*433966-1* 394
Hypericum arboreum IPNI=*434138-1* 394
Hypericum argyi IPNI=*927423-1* 394
Hypericum ascyron IPNI=*433287-1* 393
Hypericum bacciferum IPNI=*433187-1* 393
Hypericum bacciferum IPNI=*434100-1* 394
Hypericum bacciforme IPNI=*433187-1* 393
Hypericum brathys IPNI=*433524-1* 393
Hypericum brathys var. *juniperinum* IPNI=*433524-1* 393
Hypericum calycinum f. *luteum* IPNI=*433287-1* 393
Hypericum cayennense IPNI=*434107-1* 394
Hypericum cuspidatum IPNI=*30051774-2* 394
Hypericum eugeniifolium IPNI=*30051774-2* 394
Hypericum floridum IPNI=*433187-1* 393
Hypericum grandiflorum IPNI=*433287-1* 393
Hypericum guianense IPNI=*30051774-2* 394
Hypericum holtonii IPNI=*433524-1* 393
Hypericum hookerianum var. *dentatum* IPNI=*927423-1* 394
Hypericum jahnii IPNI=*433524-1* 393
Hypericum laricifolium var. *acerosum* IPNI=*433547-1* 394
Hypericum laricoides IPNI=*433547-1* 394
Hypericum laurifolium IPNI=*434138-1* 394
Hypericum lauriforme IPNI=*434138-1* 394
Hypericum mexicanum var. *intermedium* IPNI=*428535-1* 394
Hypericum mutisianum IPNI=*428535-1* 394
Hypericum officinale IPNI=*433719-1* 394
Hypericum officinarum IPNI=*433719-1* 394
Hypericum platypetalum IPNI=*433547-1* 394
Hypericum pseudobrathys IPNI=*433524-1* 393
Hypericum racemulosum IPNI=*433547-1* 394
Hypericum ramosissimum IPNI=*433966-1* 394
Hypericum rariflorum IPNI=*433966-1* 394
Hypericum resinosum IPNI=*428535-1* 394
Hypericum reticulatum IPNI=*434144-1* 394
Hypericum rufescens IPNI=*434162-1* 394
Hypericum sessilifolium IPNI=*434162-1* 394
Hypericum strictum var. *reductum* IPNI=*433524-1* 393
Hypericum tinctorum IPNI=*433547-1* 394
Hypericum venosum IPNI=*433287-1* 393
Hypericum vulgare IPNI=*433719-1* 394
Hyperthelia macrolepis IPNI=*126405-2* 240
Hyphipus bracteata IPNI=*551367-1* 449
Hyphipus trigona IPNI=*551347-1* 448
Hypochaeris adscendens IPNI=*225504-1* 264
Hypochaeris arachnites IPNI=*225504-1* 264
Hypochaeris arachnoidea IPNI=*225504-1* 264
Hypochaeris ascendens IPNI=*225504-1* 264
Hypochaeris barbata IPNI=*317594-2* 264
Hypochaeris bartholdii IPNI=*225504-1* 264
Hypochaeris brasiliensis IPNI=*126428-2* 264
Hypochaeris brasiliensis var. *chacoensis* IPNI=*126428-2* 264
Hypochaeris brasiliensis var. *microcephala* IPNI=*126475-2* 264
Hypochaeris brasiliensis var. *sulfurea* IPNI=*126428-2* 264
Hypochaeris brasiliensis var. *tweediei* IPNI=*126428-2* 264

Hypochaeris candollei IPNI=*225504-1* 264
Hypochaeris capensis IPNI=*225504-1* 264
Hypochaeris contexta IPNI=*225504-1* 264
Hypochaeris dimorpha IPNI=*225504-1* 264
Hypochaeris heterosperma IPNI=*225504-1* 264
Hypochaeris hirsuta IPNI=*225504-1* 264
Hypochaeris hispida IPNI=*225504-1* 264
Hypochaeris infesta IPNI=*225575-1* 264
Hypochaeris intermedia IPNI=*225504-1* 264
Hypochaeris intertexta IPNI=*225504-1* 264
Hypochaeris jussieui IPNI=*225575-1* 264
Hypochaeris lasiophylla IPNI=*225575-1* 264
Hypochaeris minima IPNI=*225504-1* 264
Hypochaeris montana IPNI=*225504-1* 264
Hypochaeris neapolitana IPNI=*225575-1* 264
Hypochaeris pumila IPNI=*225504-1* 264
Hypochaeris pusilla IPNI=*225504-1* 264
Hypochaeris radiata IPNI=*225575-1* 264
Hypochaeris radicata subsp. *glabra* IPNI=*225504-1* 264
Hypochaeris runcinata IPNI=*225575-1* 264
Hypochaeris sessiliflora f. *caulescens* IPNI=*317594-2* 264
Hypochaeris sessiliflora var. *albiflora* IPNI=*317594-2* 264
Hypochaeris simplex IPNI=*225504-1* 264
Hypochaeris sonchoides IPNI=*317594-2* 264
Hypochaeris sonchoides var. *fibrillosa* IPNI=*317594-2* 264
Hypochaeris stellata IPNI=*225504-1* 264
Hypochaeris stenocephala IPNI=*296211-2* 264
Hypochaeris stenocephala var. *integrifolia* IPNI=*296211-2* 264
Hypochaeris stenocephala var. *taraxacoides* IPNI=*296211-2* 264
Hypochaeris stuebelii IPNI=*317594-2* 264
Hypochaeris subalpina IPNI=*225575-1* 264
Hypochaeris tenorei IPNI=*225575-1* 264
Hypochaeris thracica IPNI=*225504-1* 264
Hypochaeris tweediei IPNI=*126428-2* 264
Hypocyrta crassifolia IPNI=*77139499-1* 366
Hypocyrta leucostoma IPNI=*378201-1* 366
Hypoestes sanguinolenta IPNI=*49962-1* 360
Hypogynium spathiflorum IPNI=*13112-2* 234
Hypogynium virgatum IPNI=*13112-2* 234
Hypolepis aculeata IPNI=*1031307-2* 175
Hypolepis coniifolia IPNI=*77182519-1* 178
Hypolepis delicatula IPNI=*1031307-2* 175
Hypolepis dicksonioides IPNI=*1031307-2* 175
Hypolepis hostilis var. *major* IPNI=*17126480-1* 175
Hypolepis mexicana IPNI=*1031307-2* 175
Hypolepis parviloba IPNI=*17126480-1* 175
Hypolepis repens var. *hostilis* IPNI=*17126480-1* 175
Hypolepis repens var. *selloana* IPNI=*1031307-2* 175
Hypolepis spectabilis IPNI=*77182519-1* 178
Hypolepis trichodes IPNI=*1019960-2* 174
Hypopeltis amygdalina IPNI=*17160710-1* 176
Hypopeltis articulata IPNI=*1138771-2* 176
Hypopeltis biserrata IPNI=*17160710-1* 176
Hypopeltis coriacea IPNI=*224543-2* 176
Hypopeltis exaltata IPNI=*17160830-1* 176
Hypopeltis marginifera IPNI=*17458930-1* 174
Hypopeltis palmoides IPNI=*17160710-1* 176
Hypopeltis propinquoides IPNI=*17458930-1* 174
Hypopitys uniflora IPNI=*281016-2* 300
Hypoporum hirtellum IPNI=*329993-2* 232
Hypoporum interruptum IPNI=*329993-2* 232
Hypostate elegans IPNI=*121047-2* 419
Hypoxis breviscapa IPNI=*65385-1* 216
Hypoxis caricifolia IPNI=*65385-1* 216
Hypoxis decumbens IPNI=*64173-1* 216
Hypoxis decumbens var. *dolichocarpa* IPNI=*65385-1* 216
Hypoxis decumbens var. *major* IPNI=*65385-1* 216
Hypoxis elongata IPNI=*65385-1* 216
Hypoxis gracilis IPNI=*65385-1* 216
Hypoxis luzulifolia IPNI=*64173-1* 216
Hypoxis pusilla IPNI=*65385-1* 216
Hypoxis racemosa IPNI=*65385-1* 216
Hypoxis scorzonerifolia IPNI=*64173-1* 216
Hyptis acuta IPNI=*447982-1* 370
Hyptis affinis IPNI=*448211-1* 370
Hyptis americana IPNI=*77120747-1* 368
Hyptis anomala IPNI=*448211-1* 370
Hyptis arborea IPNI=*278655-2* 369
Hyptis arborea subsp. *bracteosa* IPNI=*278655-2* 369
Hyptis arvensis IPNI=*77120760-1* 368
Hyptis aspera IPNI=*77120760-1* 368
Hyptis atrorubens var. *villosissima* IPNI=*447964-1* 369
Hyptis axillaris IPNI=*77120819-1* 369
Hyptis barbata IPNI=*77120760-1* 368
Hyptis belizensis IPNI=*448022-1* 370
Hyptis brevipes var. *glabrior* IPNI=*447982-1* 370
Hyptis brevipes var. *remotidens* IPNI=*447982-1* 370
Hyptis brevipes var. *robusta* IPNI=*447982-1* 370
Hyptis brevipes var. *serrata* IPNI=*447982-1* 370
Hyptis brevipes var. *vulgaris* IPNI=*447982-1* 370
Hyptis canaminensis IPNI=*77120760-1* 368
Hyptis canescens IPNI=*77120760-1* 368
Hyptis capitata f. *pilosa* IPNI=*448405-1* 370
Hyptis capitata var. *mexicana* IPNI=*447996-1* 370
Hyptis capitata var. *pilosa* IPNI=*447996-1* 370
Hyptis capitata var. *vulgaris* IPNI=*447996-1* 370
Hyptis celebica IPNI=*447996-1* 370
Hyptis chamaedrys IPNI=*449997-1* 371
Hyptis colombiana IPNI=*77120750-1* 368
Hyptis compacta IPNI=*448211-1* 370
Hyptis conferta var. *angustata* IPNI=*448022-1* 370
Hyptis conferta var. *excelsa* IPNI=*448022-1* 370
Hyptis congesta IPNI=*158297-2* 371
Hyptis constricta IPNI=*448022-1* 370
Hyptis decurrens IPNI=*447996-1* 370
Hyptis ebracteata IPNI=*158297-2* 371
Hyptis excelsa IPNI=*448022-1* 370
Hyptis ferruginea IPNI=*448022-1* 370
Hyptis fulva IPNI=*126717-2* 369
Hyptis glandulosa IPNI=*456822-1* 372
Hyptis globifera IPNI=*448173-1* 370
Hyptis gonocephala IPNI=*77120747-1* 368
Hyptis graveolens IPNI=*158297-2* 371
Hyptis hirsuta var. *villosior* IPNI=*448128-1* 370
Hyptis hispida IPNI=*77120747-1* 368
Hyptis inflata IPNI=*449997-1* 371
Hyptis jurgensenii IPNI=*448022-1* 370
Hyptis jurgensenii var. *angustata* IPNI=*448022-1* 370
Hyptis kerberi IPNI=*77120760-1* 368
Hyptis lanceifolia IPNI=*448173-1* 370
Hyptis lanicephala IPNI=*448057-1* 370
Hyptis lantanifolia var. *costaricensis* IPNI=*448177-1* 370
Hyptis lantanifolia var. *glabra* IPNI=*448177-1* 370
Hyptis lantanifolia var. *glabrescens* IPNI=*448177-1* 370
Hyptis lophantha IPNI=*77120747-1* 368
Hyptis lundii IPNI=*448022-1* 370
Hyptis lurida IPNI=*449997-1* 371
Hyptis macrochila IPNI=*447996-1* 370
Hyptis madagascariensis IPNI=*77120747-1* 368
Hyptis madrensis IPNI=*448177-1* 370
Hyptis mariannarum IPNI=*447996-1* 370
Hyptis marrubiastra IPNI=*448405-1* 370
Hyptis melanosticta IPNI=*447982-1* 370
Hyptis menthodora IPNI=*77120747-1* 368
Hyptis micrantha IPNI=*77120760-1* 368
Hyptis microcephala IPNI=*448358-1* 370

Hyptis mikanii IPNI=*77121103-1* 371
Hyptis mutabilis IPNI=*77120760-1* 368
Hyptis mutabilis var. *bromfieldii* IPNI=*77120760-1* 368
Hyptis mutabilis var. *canescens* IPNI=*77120760-1* 368
Hyptis mutabilis var. *cuneata* IPNI=*77120760-1* 368
Hyptis mutabilis var. *micrantha* IPNI=*77120760-1* 368
Hyptis mutabilis var. *pavoniana* IPNI=*77120760-1* 368
Hyptis mutabilis var. *polystachya* IPNI=*77120760-1* 368
Hyptis mutabilis var. *rostrata* IPNI=*77120760-1* 368
Hyptis mutabilis var. *spicata* IPNI=*77120760-1* 368
Hyptis nepetoides IPNI=*158242-2* 371
Hyptis parviflora IPNI=*77120819-1* 369
Hyptis pascuarum IPNI=*447964-1* 369
Hyptis pectinata IPNI=*158242-2* 371
Hyptis persica IPNI=*158242-2* 371
Hyptis pittieri IPNI−*447996-1* 370
Hyptis plumieri IPNI=*158297-2* 371
Hyptis pohliana IPNI=*77120747-1* 368
Hyptis polyantha IPNI=*77121103-1* 371
Hyptis polystachya IPNI=*77120760-1* 368
Hyptis polystachya var. *longiflora* IPNI=*77121103-1* 371
Hyptis pringlei IPNI=*77120819-1* 369
Hyptis procumbens IPNI=*447964-1* 369
Hyptis pseudochamaedrys IPNI=*449997-1* 371
Hyptis racemosa IPNI=*158242-2* 371
Hyptis racemosa IPNI=*77120760-1* 368
Hyptis radiata IPNI=*447982-1* 370
Hyptis radula IPNI=*448022-1* 370
Hyptis recurvata var. *grandifolia* IPNI=*448358-1* 370
Hyptis recurvata var. *hirsutior* IPNI=*448358-1* 370
Hyptis recurvata var. *megacephala* IPNI=*448358-1* 370
Hyptis rhomboidea IPNI=*447996-1* 370
Hyptis rostrata IPNI=*77120760-1* 368
Hyptis rufula IPNI=*448211-1* 370
Hyptis siderotricha IPNI=*448128-1* 370
Hyptis sidifolia IPNI=*77121103-1* 371
Hyptis silvestris IPNI=*77121103-1* 371
Hyptis singularis IPNI=*77120760-1* 368
Hyptis sinuata var. *borucensis* IPNI=*448405-1* 370
Hyptis sinuata var. *genuina* IPNI=*448405-1* 370
Hyptis sinuata var. *hispida* IPNI=*448405-1* 370
Hyptis sinuata var. *subglabra* IPNI=*448405-1* 370
Hyptis spicata IPNI=*77120760-1* 368
Hyptis spicata var. *bromfieldii* IPNI=*77120760-1* 368
Hyptis spicata var. *micrantha* IPNI=*77120760-1* 368
Hyptis spicata var. *rostrata* IPNI=*77120760-1* 368
Hyptis spicigera IPNI=*77120747-1* 368
Hyptis suaveolens IPNI=*158297-2* 371
Hyptis subverticillata IPNI=*77120747-1* 368
Hyptis tenella IPNI=*447964-1* 369
Hyptis tenuiflora IPNI=*77120760-1* 368
Hyptis trachychroa IPNI=*448211-1* 370
Hyptis trichocalyx IPNI=*77120760-1* 368
Hyptis umbrosa IPNI=*77121103-1* 371
Hyptis umbrosa var. *mikanii* IPNI=*77121103-1* 371
Hyptis verticillata IPNI=*77120819-1* 369

I
Ibatia muricata IPNI=*98854-1* 341
Iberis affinis IPNI=*324660-2* 277
Iberis amoena IPNI=*324671-2* 277
Iberis apricorum IPNI=*324660-2* 277
Iberis arvatica IPNI=*324660-2* 277
Iberis bicolor IPNI=*324660-2* 277
Iberis bicorymbifera IPNI=*324660-2* 277
Iberis bursa-pastoris IPNI=*30092589-2* 276
Iberis ciliata subsp. *vinetorum* IPNI=*324660-2* 277
Iberis contracta subsp. *vinetorum* IPNI=*324660-2* 277
Iberis coronaria IPNI=*324671-2* 277
Iberis corymbosa IPNI=*324671-2* 277
Iberis crenata IPNI=*324660-2* 277
Iberis decipiens IPNI=*324660-2* 277
Iberis forestieri IPNI=*324660-2* 277
Iberis hortensis IPNI=*324671-2* 277
Iberis linifolia var. *vinetorum* IPNI=*324660-2* 277
Iberis liviensis IPNI=*324660-2* 277
Iberis lusitanica IPNI=*324660-2* 277
Iberis martini IPNI=*324660-2* 277
Iberis montolivensis IPNI=*324660-2* 277
Iberis panduriformis IPNI=*324660-2* 277
Iberis pinetorum IPNI=*324660-2* 277
Iberis platyptera IPNI=*324671-2* 277
Iberis pulchra IPNI=*324671-2* 277
Iberis roseopurpurea IPNI=*324671-2* 277
Iberis sabauda IPNI=*324660-2* 277
Iberis serotina IPNI=*324660-2* 277
Iberis umbellata var. *brachyptera* IPNI=*324671-2* 277
Iberis umbellata var. *platyptera* IPNI=*324671-2* 277
Iberis vinetorum IPNI=*324660-2* 277
Iberis virginica IPNI=*138164-2* 277
Ibervillea guatemalensis IPNI=*592558-1* 294
Ibidium coloratum IPNI=*659300-1* 223
Ibidium elatum IPNI=*73688-2* 218
Ibidium speciosum IPNI=*659300-1* 223
Icacina poeppigiana IPNI=*607989-1* 359
Icacorea acuminata IPNI=*18168-2* 303
Icacorea acuminifolia IPNI=*18168-2* 303
Icacorea alstonii IPNI=*18168-2* 303
Icacorea amanuensis IPNI=*18168-2* 303
Icacorea angustifolia IPNI=*18168-2* 303
Icacorea aurantiaca IPNI=*586863-1* 303
Icacorea belizensis IPNI=*586863-1* 303
Icacorea brasiliensis IPNI=*18168-2* 303
Icacorea capollina IPNI=*586863-1* 303
Icacorea catharinensis IPNI=*18168-2* 303
Icacorea caudatifolia IPNI=*586863-1* 303
Icacorea chahalana IPNI=*586863-1* 303
Icacorea compressa IPNI=*586863-1* 303
Icacorea cookii IPNI=*18168-2* 303
Icacorea cuspidata IPNI=*586863-1* 303
Icacorea deminuta IPNI=*18168-2* 303
Icacorea digitata IPNI=*586863-1* 303
Icacorea edwardsii IPNI=*586863-1* 303
Icacorea granatensis IPNI=*247490-2* 305
Icacorea guianensis IPNI=*18168-2* 303
Icacorea herbert-smithii IPNI=*18168-2* 303
Icacorea humboldtii IPNI=*18168-2* 303
Icacorea ibaguensis IPNI=*247490-2* 305
Icacorea irazuensis IPNI=*586863-1* 303
Icacorea ixcanensis IPNI=*586863-1* 303
Icacorea longicaudata IPNI=*18168-2* 303
Icacorea manglillo IPNI=*18218-2* 303
Icacorea martinensis IPNI=*18168-2* 303
Icacorea metensis IPNI=*18168-2* 303
Icacorea nicaraguensis IPNI=*586863-1* 303
Icacorea oerstediana IPNI=*586863-1* 303
Icacorea panurensis IPNI=*247472-2* 304
Icacorea pergracilis IPNI=*18168-2* 303
Icacorea platyphylla IPNI=*247490-2* 305
Icacorea principis IPNI=*18168-2* 303
Icacorea reflexiflora IPNI=*586863-1* 303
Icacorea rusbyana IPNI=*247490-2* 305
Icacorea salvadorensis IPNI=*586863-1* 303
Icacorea samalana IPNI=*586863-1* 303
Icacorea semicrenata IPNI=*18168-2* 303
Icacorea veraguasensis IPNI=*586863-1* 303

Icacorea whitei IPNI=*586863-1* 303
Ichthyomethia acuminata IPNI=*316422-2* 330
Ichthyomethia americana IPNI=*316422-2* 330
Ichthyomethia piscipula var. *acuminata* IPNI=*316422-2* 330
Ichthyophora leptocarpa IPNI=*209758-2* 306
Ichthyothere cearaensis IPNI=*127605-2* 264
Ichthyothere curvifolia IPNI=*127605-2* 264
Ichthyothere latifolia IPNI=*127605-2* 264
Ichthyothere suffruticosa var. *pilosa* IPNI=*127605-2* 264
Icica altissima IPNI=*77183601-1* 452
Icica ambrosiaca IPNI=*128338-1* 453
Icica aracouchini IPNI=*128278-1* 452
Icica carana IPNI=*128295-1* 452
Icica cuspidata IPNI=*77183601-1* 452
Icica decandra IPNI=*128316-1* 452
Icica decandra IPNI=*128413-1* 453
Icica enneandra IPNI=*128316-1* 452
Icica goudotiana IPNI=*77183718-1* 452
Icica guianensis IPNI=*128337-1* 452
Icica guianensis IPNI=*128338-1* 453
Icica heptaphylla IPNI=*128337-1* 452
Icica heptaphylla IPNI=*128338-1* 453
Icica heterophylla IPNI=*128278-1* 452
Icica hostmannii IPNI=*128337-1* 452
Icica insignis IPNI=*128413-1* 453
Icica lucida IPNI=*70000-1* 451
Icica macrophylla IPNI=*128356-1* 453
Icica panamensis IPNI=*210603-2* 453
Icica parvifolia IPNI=*128278-1* 452
Icica pentandra IPNI=*128316-1* 452
Icica pittieri IPNI=*210615-2* 453
Icica polybotrya IPNI=*128400-1* 453
Icica pubescens IPNI=*128430-1* 453
Icica salzmannii IPNI=*128338-1* 453
Icica simplicifolia IPNI=*128430-1* 453
Icica spruceana IPNI=*128420-1* 453
Icica surinamensis IPNI=*128338-1* 453
Icica tacamahaca IPNI=*128338-1* 453
Icica trianensis IPNI=*128278-1* 452
Icica unifoliolata IPNI=*128430-1* 453
Icica viridiflora IPNI=*128337-1* 452
Icicariba simaruba IPNI=*127217-1* 451
Icicopsis caudata IPNI=*128413-1* 453
Icicopsis insignis IPNI=*128413-1* 453
Icicopsis subserrata IPNI=*128422-1* 453
Icicopsis tenuifolia IPNI=*128424-1* 453
Icicopsis tenuifolia var. *brevicalyx* IPNI=*128424-1* 453
Icicopsis tenuifolia var. *multijuga* IPNI=*128424-1* 453
Ida fulvescens IPNI=*77096738-1* 223
Ida fulvescens var. *pallida* IPNI=*77096738-1* 223
Ida gigantea IPNI=*77096741-1* 223
Ida labelloviridis IPNI=*77096741-1* 223
Iebine nervosa IPNI=*641061-1* 219
Ilex acuminata IPNI=*127855-2* 255
Ilex caliana IPNI=*83435-1* 255
Ilex caniensis IPNI=*308592-2* 255
Ilex celastroides IPNI=*127855-2* 255
Ilex cumaneuris IPNI=*127855-2* 255
Ilex fructiclipeata IPNI=*83432-1* 255
Ilex gentlei IPNI=*127855-2* 255
Ilex guianensis var. *cuencensis* IPNI=*127855-2* 255
Ilex guianensis var. *macoucoua* IPNI=*127855-2* 255
Ilex humbertii IPNI=*83432-1* 255
Ilex kunthiana f. *funckii* IPNI=*30054025-2* 255
Ilex kunthiana f. *genuina* IPNI=*30054025-2* 255
Ilex laureola var. *genuina* IPNI=*83432-1* 255
Ilex laureola var. *neglecta* IPNI=*83432-1* 255
Ilex lucumifolia IPNI=*948065-1* 307
Ilex macoucoua IPNI=*127855-2* 255
Ilex macrolaurus IPNI=*83432-1* 255
Ilex myricoides f. *euryterophylla* IPNI=*83545-1* 255
Ilex myricoides f. *genuina* IPNI=*83545-1* 255
Ilex myricoides var. *meridensis* IPNI=*83545-1* 255
Ilex myricoides var. *polyphylla* IPNI=*83545-1* 255
Ilex myricoides var. *trianiana* IPNI=*83545-1* 255
Ilex naiguatina IPNI=*83545-1* 255
Ilex panamensis IPNI=*127855-2* 255
Ilex parviflora IPNI=*306411-2* 255
Ilex polyphylla IPNI=*83545-1* 255
Ilex pseudomacoucoua IPNI=*127855-2* 255
Ilex retusa f. *brunnescens* IPNI=*83691-1* 255
Ilex retusa f. *genuina* IPNI=*83691-1* 255
Ilex retusa f. *glabra* IPNI=*83691-1* 255
Ilex retusa f. *glaucescens* IPNI=*83691-1* 255
Ilex retusa f. *major* IPNI=*83691-1* 255
Ilex stenophylla IPNI=*83691-1* 255
Ilex uniflora f. *pastoensis* IPNI=*308592-2* 255
Ilex uniflora f. *pitayensis* IPNI=*308592-2* 255
Ilex uniflora var. *paramensis* IPNI=*308592-2* 255
Ilex utilis IPNI=*127854-2* 255
Illecebrum achyrantha IPNI=*327342-2* 281
Illecebrum alsinefolium IPNI=*10362-2* 281
Illecebrum alsinifolium IPNI=*1183159-2* 282
Illecebrum densum IPNI=*1183159-2* 282
Illecebrum denticulatum IPNI=*1069792-2* 281
Illecebrum dubium IPNI=*129900-2* 283
Illecebrum ficoideum IPNI=*1078972-2* 281
Illecebrum frutescens IPNI=*10362-2* 281
Illecebrum glabrum IPNI=*1069792-2* 281
Illecebrum gomphrenoides IPNI=*10447-2* 281
Illecebrum incanum IPNI=*10362-2* 281
Illecebrum indicum IPNI=*1069792-2* 281
Illecebrum iresinoides IPNI=*61055-1* 283
Illecebrum limense IPNI=*10362-2* 281
Illecebrum nanum IPNI=*1069792-2* 281
Illecebrum obliquum IPNI=*327342-2* 281
Illecebrum paniculatum IPNI=*10447-2* 281
Illecebrum peploides IPNI=*327344-2* 281
Illecebrum polygonoides IPNI=*1069792-2* 281
Illecebrum pungens IPNI=*327342-2* 281
Illecebrum sessile IPNI=*1069792-2* 281
Illecebrum sibiricum IPNI=*1069792-2* 281
Illecebrum sinense IPNI=*327342-2* 281
Illecebrum tenellum IPNI=*1078972-2* 281
Illecebrum triandrum IPNI=*1069792-2* 281
Illecebrum vermiculatum IPNI=*34033-2* 282
Illicium stellatum IPNI=*554553-1* 181
Ilysanthes antipoda IPNI=*799766-1* 374
Ilysanthes hypericifolia IPNI=*1061345-2* 374
Ilysanthes madagascariensis IPNI=*1061345-2* 374
Ilysanthes oblongifolia IPNI=*1061345-2* 374
Ilysanthes rotundifolia IPNI=*1061345-2* 374
Ilysanthes verbenifolia IPNI=*799766-1* 374
Imantophyllum miniatum IPNI=*994513-1* 213
Imatophyllum atrosanguineum IPNI=*994513-1* 213
Impatiens angustifolia IPNI=*373978-1* 297
Impatiens arcuata IPNI=*373978-1* 297
Impatiens balsamina var. *arcuata* IPNI=*373978-1* 297
Impatiens balsamina var. *coccinea* IPNI=*373978-1* 297
Impatiens balsamina var. *longifolia* IPNI=*373978-1* 297
Impatiens balsamina var. *vulgaris* IPNI=*373978-1* 297
Impatiens borealis IPNI=*30208246-2* 297
Impatiens caucasica IPNI=*30208246-2* 297
Impatiens coccinea IPNI=*373978-1* 297
Impatiens cornuta IPNI=*373978-1* 297
Impatiens cucullata IPNI=*30208246-2* 297

Impatiens dahlii IPNI=*373978-1* 297
Impatiens elgonensis IPNI=*374351-1* 297
Impatiens episcopi IPNI=*374419-1* 297
Impatiens eriocarpa IPNI=*373978-1* 297
Impatiens giorgii IPNI=*373978-1* 297
Impatiens holstii IPNI=*374419-1* 297
Impatiens komarovii IPNI=*30208246-2* 297
Impatiens longifolia IPNI=*373978-1* 297
Impatiens lujae IPNI=*374419-1* 297
Impatiens lutea IPNI=*30208246-2* 297
Impatiens magnifica IPNI=*374351-1* 297
Impatiens malayensis IPNI=*373978-1* 297
Impatiens noli-me-tangere IPNI=*30208246-2* 297
Impatiens noli-tangere f. *hemileuca* IPNI=*30208246-2* 297
Impatiens noli-tangere f. *pallida* IPNI=*30208246-2* 297
Impatiens noli-tangere var. *pallida* IPNI=*30208246-2* 297
Impatiens occidentalis IPNI=*30208246-2* 297
Impatiens oliveri IPNI=*374351-1* 297
Impatiens palustris IPNI=*30208246-2* 297
Impatiens penduliflora IPNI=*30208246-2* 297
Impatiens petersiana IPNI=*374419-1* 297
Impatiens salicifolia IPNI=*373978-1* 297
Impatiens sinensis IPNI=*373978-1* 297
Impatiens stapfiana IPNI=*373978-1* 297
Impatiens sultani IPNI=*374419-1* 297
Impatiens tamsiana IPNI=*373978-1* 297
Impatiens thomsonii IPNI=*374351-1* 297
Impatiens uguenensis IPNI=*374351-1* 297
Imperata arundinacea var. *americana* IPNI=*405808-1* 240
Imperata caudata IPNI=*405817-1* 240
Imperata cylindrica var. *mexicana* IPNI=*405808-1* 240
Imperata exaltata var. *caudata* IPNI=*405817-1* 240
Imperata flexuosa IPNI=*405817-1* 240
Imperata longifolia IPNI=*405817-1* 240
Imperata ovata IPNI=*406799-1* 240
Imperata sape IPNI=*405817-1* 240
Incarvillea chinensis IPNI=*109760-1* 364
Incarvillea przewalskii IPNI=*109760-1* 364
Incarvillea sinensis var. *przewalskii* IPNI=*109760-1* 364
Incarvillea variabilis IPNI=*109760-1* 364
Indigofera acasonicae IPNI=*499569-1* 321
Indigofera angolensis IPNI=*500063-1* 321
Indigofera angustifolia IPNI=*499456-1* 321
Indigofera anil IPNI=*500063-1* 321
Indigofera anil var. *polyphylla* IPNI=*500063-1* 321
Indigofera argentea IPNI=*500063-1* 321
Indigofera argentea var. *caerulea* IPNI=*500063-1* 321
Indigofera barbata IPNI=*499456-1* 321
Indigofera berteroana IPNI=*1071691-2* 335
Indigofera boiviniana IPNI=*499658-1* 321
Indigofera cornezuelo IPNI=*500063-1* 321
Indigofera divaricata IPNI=*500063-1* 321
Indigofera domingensis IPNI=*499658-1* 321
Indigofera ferruginea IPNI=*499456-1* 321
Indigofera fusca IPNI=*499456-1* 321
Indigofera grisea IPNI=*499569-1* 321
Indigofera guatimala IPNI=*500063-1* 321
Indigofera guatimalensis IPNI=*500063-1* 321
Indigofera hirta IPNI=*499456-1* 321
Indigofera humboldtiana IPNI=*499569-1* 321
Indigofera indica IPNI=*499456-1* 321
Indigofera lespedezioides var. *acutifolia* IPNI=*499569-1* 321
Indigofera lespedezioides var. *pascuorum* IPNI=*499569-1* 321
Indigofera lespedezioides var. *typica* IPNI=*499569-1* 321
Indigofera mexicana IPNI=*281771-2* 329
Indigofera micrantha IPNI=*500063-1* 321
Indigofera mucronata IPNI=*499569-1* 321
Indigofera mutisii IPNI=*281771-2* 329
Indigofera pascuorum IPNI=*499569-1* 321
Indigofera pohliana IPNI=*499569-1* 321
Indigofera porrigens IPNI=*499658-1* 321
Indigofera punctata IPNI=*499658-1* 321
Indigofera rhodosantha IPNI=*499456-1* 321
Indigofera richardiana IPNI=*499658-1* 321
Indigofera sabulicola IPNI=*499658-1* 321
Indigofera suffruticosa f. *obtusifolia* IPNI=*500063-1* 321
Indigofera suffruticosa subsp. *guatimalensis* IPNI=*500063-1* 321
Indigofera suffruticosa var. *canescens* IPNI=*500063-1* 321
Indigofera tinctoria var. *brachycarpa* IPNI=*500063-1* 321
Indigofera uncinata IPNI=*500063-1* 321
Indigofera volubilis IPNI=*517450-1* 331
Inga acutifolia IPNI=*128358-2* 321
Inga affinis IPNI=*128675-2* 323
Inga aggregata IPNI=*1129244-2* 321
Inga alatocarpa IPNI=*128596-2* 323
Inga alatopetiola IPNI=*128732-2* 324
Inga albicans IPNI=*128414-2* 321
Inga albicoria IPNI=*1090709-2* 322
Inga altissima IPNI=*1129244-2* 321
Inga angustifolia IPNI=*313763-2* 321
Inga antioquensis IPNI=*128732-2* 324
Inga apta IPNI=*500286-1* 321
Inga archeri IPNI=*128564-2* 323
Inga aria IPNI=*128427-2* 322
Inga arinensis IPNI=*128358-2* 321
Inga arrabidae IPNI=*128358-2* 321
Inga assimilis IPNI=*500286-1* 321
Inga bahiensis IPNI=*1131764-2* 322
Inga benthamiana IPNI=*128478-2* 322
Inga biolleyana IPNI=*128732-2* 324
Inga bolivariana IPNI=*128432-2* 322
Inga bonplandiana IPNI=*128732-2* 324
Inga borbonica IPNI=*99109-3* 311
Inga brachyptera IPNI=*128596-2* 323
Inga brachystachya IPNI=*288863-2* 321
Inga bracteosa IPNI=*128596-2* 323
Inga brevialata IPNI=*500893-1* 324
Inga brevipedicellata IPNI=*128354-2* 321
Inga brunnescens IPNI=*128446-2* 322
Inga caldasiana IPNI=*128458-2* 322
Inga callicarpa IPNI=*128812-2* 324
Inga calocephala IPNI=*128596-2* 323
Inga calophylla IPNI=*128432-2* 322
Inga camatchili IPNI=*1003508-2* 330
Inga camuriensis IPNI=*128732-2* 324
Inga canaminensis IPNI=*500844-1* 324
Inga capitata IPNI=*989371-1* 310
Inga capitata var. *brevicalyx* IPNI=*128414-2* 321
Inga capitata var. *latifolia* IPNI=*128414-2* 321
Inga capitata var. *tenuior* IPNI=*128414-2* 321
Inga capsellata IPNI=*128732-2* 324
Inga capuchoi IPNI=*128414-2* 321
Inga caracasana IPNI=*128732-2* 324
Inga carachensis IPNI=*1129244-2* 321
Inga cataractae IPNI=*962821-1* 337
Inga catharinae IPNI=*500844-1* 324
Inga cayennensis var. *sessiliflora* IPNI=*128427-2* 322
Inga chardonii IPNI=*128732-2* 324
Inga chartana IPNI=*128647-2* 323
Inga chiriquensis IPNI=*128647-2* 323
Inga chorrerana IPNI=*128596-2* 323
Inga chrysotricha IPNI=*128746-2* 324
Inga cinerea IPNI=*227126-2* 331
Inga clavigera IPNI=*128647-2* 323
Inga cobanensis IPNI=*128647-2* 323
Inga coccinea IPNI=*962823-1* 337

Inga cocleensis subsp. *megantha* IPNI=*128446-2* 322
Inga cocleensis var. *megantha* IPNI=*128446-2* 322
Inga codonantha IPNI=*500662-1* 323
Inga colonchensis IPNI=*128432-2* 322
Inga comewynensis IPNI=*500844-1* 324
Inga complanata IPNI=*128478-2* 322
Inga conferta IPNI=*128478-2* 322
Inga confusa IPNI=*128721-2* 324
Inga crista IPNI=*128599-2* 323
Inga culagana IPNI=*128647-2* 323
Inga cumingiana IPNI=*128507-2* 322
Inga cycladenia IPNI=*128693-2* 323
Inga cyclocarpa IPNI=*1037928-2* 319
Inga cyclocarpa IPNI=*128721-2* 324
Inga davidsoniae IPNI=*500443-1* 322
Inga divaricata IPNI=*962821-1* 337
Inga dolichantha IPNI=*500662-1* 323
Inga donaeana IPNI=*128732-2* 324
Inga dulcis IPNI=*1003508-2* 330
Inga dulcis IPNI=*128358-2* 321
Inga dumosa IPNI=*128764-2* 324
Inga dysantha IPNI=*128427-2* 322
Inga edulis f. *typica* IPNI=*128478-2* 322
Inga edulis var. *grenadensis* IPNI=*128647-2* 323
Inga edulis var. *minutula* IPNI=*128647-2* 323
Inga edulis var. *parviflora* IPNI=*128478-2* 322
Inga eggersii IPNI=*128732-2* 324
Inga eglandulosa IPNI=*500662-1* 323
Inga ellsworthiana IPNI=*500844-1* 324
Inga endlicheri IPNI=*128647-2* 323
Inga eriorhachis IPNI=*128647-2* 323
Inga excelsa IPNI=*128599-2* 323
Inga fagifolia f. *genuina* IPNI=*1199545-2* 323
Inga fagifolia f. *pedicellaris* IPNI=*128599-2* 323
Inga fagifolia IPNI=*1199545-2* 323
Inga fagifolia IPNI=*128721-2* 324
Inga fagifolia var. *belemnensis* IPNI=*1199545-2* 323
Inga fagifolia var. *intermedia* IPNI=*128599-2* 323
Inga fagifolia var. *marginata* IPNI=*128599-2* 323
Inga fagifolia var. *pedicellaris* IPNI=*1199545-2* 323
Inga fagifolia var. *typica* IPNI=*1199545-2* 323
Inga falciformis IPNI=*271434-2* 337
Inga falcistipula IPNI=*128414-2* 321
Inga fasciculata IPNI=*128647-2* 323
Inga fasciculata IPNI=*316281-2* 313
Inga felina IPNI=*515137-1* 330
Inga floribunda IPNI=*128749-2* 324
Inga foliosa IPNI=*128721-2* 324
Inga forfex IPNI=*996931-1* 330
Inga forficata IPNI=*481315-1* 311
Inga fraxinea IPNI=*1129244-2* 321
Inga fredoniana IPNI=*500662-1* 323
Inga fulgens IPNI=*1122303-2* 324
Inga galibica IPNI=*1131764-2* 322
Inga glomerata IPNI=*962821-1* 337
Inga gracilipes IPNI=*1035636-2* 324
Inga grandifolia IPNI=*128732-2* 324
Inga guadalupensis IPNI=*515137-1* 330
Inga guaremalensis IPNI=*1040744-2* 322
Inga guayaquilensis IPNI=*128599-2* 323
Inga guentheri IPNI=*128784-2* 324
Inga hartii IPNI=*128732-2* 324
Inga heinei IPNI=*128466-2* 322
Inga heteroptera IPNI=*1122303-2* 324
Inga hintonii IPNI=*500443-1* 322
Inga hirsutissima IPNI=*128746-2* 324
Inga holtonii IPNI=*500662-1* 323
Inga hostmannii IPNI=*128749-2* 324
Inga hymenaeifolia IPNI=*1183336-2* 330
Inga icrensis IPNI=*128693-2* 323
Inga inaequalis IPNI=*271431-2* 337
Inga inga IPNI=*1006087-2* 324
Inga java IPNI=*128466-2* 322
Inga javana IPNI=*1003508-2* 330
Inga jimeneziana IPNI=*128732-2* 324
Inga klugii IPNI=*128458-2* 322
Inga laeta IPNI=*989371-1* 310
Inga lanceolata IPNI=*1003508-2* 330
Inga lanceolata IPNI=*514978-1* 330
Inga langlassei IPNI=*128466-2* 322
Inga lateriflora var. *latior* IPNI=*128568-2* 323
Inga latifolia IPNI=*271432-2* 337
Inga laurifolia IPNI=*44114-2* 313
Inga lawranceana IPNI=*1035636-2* 324
Inga laxiflora IPNI=*1131764-2* 322
Inga leiocalycina IPNI=*500531-1* 322
Inga leptoloba IPNI=*128693-2* 323
Inga leptostachya IPNI=*128599-2* 323
Inga leucantha IPNI=*1003508-2* 330
Inga leucoxylon IPNI=*99109-3* 311
Inga ligustrina IPNI=*514978-1* 330
Inga lindeniana IPNI=*128732-2* 324
Inga longifolia IPNI=*271434-2* 337
Inga longipedunculata IPNI=*500531-1* 322
Inga longituba IPNI=*1040744-2* 322
Inga lucida IPNI=*1122303-2* 324
Inga macrophylla IPNI=*128675-2* 323
Inga macrophylla var. *stenoptera* IPNI=*128596-2* 323
Inga macrostachys IPNI=*1183336-2* 330
Inga mapiriensis IPNI=*128535-2* 322
Inga marginata var. *itayensis* IPNI=*128599-2* 323
Inga medellinensis IPNI=*128732-2* 324
Inga megacarpa IPNI=*125420-2* 320
Inga melinonis IPNI=*128721-2* 324
Inga merianae IPNI=*1131764-2* 322
Inga meridensis IPNI=*128647-2* 323
Inga micheliana IPNI=*500443-1* 322
Inga microdonta IPNI=*128466-2* 322
Inga microgyna IPNI=*128732-2* 324
Inga microphylla IPNI=*515137-1* 330
Inga microstachya IPNI=*128673-2* 323
Inga minutula IPNI=*128647-2* 323
Inga mitaraka IPNI=*1129244-2* 321
Inga mollifoliola IPNI=*128466-2* 322
Inga montealegrei IPNI=*128466-2* 322
Inga monticola IPNI=*128466-2* 322
Inga multiflora IPNI=*500531-1* 322
Inga myriantha IPNI=*1035636-2* 324
Inga myriocephala IPNI=*128353-2* 321
Inga niopo IPNI=*12268-2* 311
Inga nitens IPNI=*1003508-2* 330
Inga nitida IPNI=*128675-2* 323
Inga nubigena IPNI=*500443-1* 322
Inga obscura IPNI=*128721-2* 324
Inga obtusa IPNI=*128644-2* 323
Inga ochroclada IPNI=*128732-2* 324
Inga odorata IPNI=*128599-2* 323
Inga odoratissima IPNI=*128591-2* 323
Inga olivacea IPNI=*128521-2* 322
Inga ornifolia IPNI=*128550-2* 322
Inga ouraphylla IPNI=*128596-2* 323
Inga pacai IPNI=*128507-2* 322
Inga pachycarpa IPNI=*128550-2* 322
Inga pamplonae IPNI=*128647-2* 323
Inga panamensis IPNI=*128732-2* 324
Inga panamensis var. *clavata* IPNI=*128732-2* 324

Inga panamensis var. *pittieri* IPNI=*128732-2* 324
Inga panamensis var. *rodrigueziana* IPNI=*128732-2* 324
Inga paniculata IPNI=*996931-1* 330
Inga pavoniana IPNI=*128732-2* 324
Inga pectinata IPNI=*512223-1* 329
Inga pedicellaris IPNI=*989349-1* 311
Inga pennatula IPNI=*60441228-2* 336
Inga perrottetii IPNI=*500844-1* 324
Inga pilosa IPNI=*128820-2* 324
Inga pilosiuscula IPNI=*128675-2* 323
Inga pilosiuscula var. *panurensis* IPNI=*128675-2* 323
Inga pisana IPNI=*128647-2* 323
Inga pittieri IPNI=*128732-2* 324
Inga platycarpa IPNI=*128675-2* 323
Inga polystachya IPNI=*1090709-2* 322
Inga popayanensis IPNI=*128693 2* 323
Inga preussii IPNI=*128732-2* 324
Inga prieurii IPNI=*500844-1* 324
Inga pringlei IPNI=*500443-1* 322
Inga protracta IPNI=*128535-2* 322
Inga pterocarpa IPNI=*513021-1* 330
Inga puberula IPNI=*128599-2* 323
Inga pubescens IPNI=*996931-1* 330
Inga pulchriflora IPNI=*500919-1* 324
Inga punctata subsp. *chagrensis* IPNI=*128693-2* 323
Inga punctata var. *elongata* IPNI=*128693-2* 323
Inga punctata var. *panamensis* IPNI=*128693-2* 323
Inga pungens IPNI=*1003508-2* 330
Inga purpusii IPNI=*128732-2* 324
Inga pycnostachya IPNI=*128599-2* 323
Inga quadrangularis IPNI=*128596-2* 323
Inga quassiifolia IPNI=*128675-2* 323
Inga racemiflora IPNI=*128800-2* 324
Inga radiata IPNI=*500893-1* 324
Inga ramiflora IPNI=*962821-1* 337
Inga rensonii IPNI=*128732-2* 324
Inga reticulata IPNI=*128507-2* 322
Inga riopalenquensis IPNI=*128673-2* 323
Inga rodrigueziana IPNI=*128732-2* 324
Inga rosea IPNI=*996931-1* 330
Inga rosea var. *subinermis* IPNI=*996931-1* 330
Inga rufescens IPNI=*62892-2* 315
Inga rufinervis IPNI=*128693-2* 323
Inga rugosa IPNI=*500432-1* 322
Inga rutilans IPNI=*1035636-2* 324
Inga salutaris IPNI=*227126-2* 331
Inga salvadorensis IPNI=*128732-2* 324
Inga salzmanniana IPNI=*500844-1* 324
Inga saman IPNI=*227126-2* 331
Inga sanctae-annae IPNI=*128675-2* 323
Inga sapida IPNI=*128599-2* 323
Inga sarmentosa IPNI=*128414-2* 321
Inga scabriuscula IPNI=*128478-2* 322
Inga scabriuscula var. *villosior* IPNI=*128478-2* 322
Inga schiedeana IPNI=*500443-1* 322
Inga schimpffii IPNI=*1122303-2* 324
Inga sciadion IPNI=*1035636-2* 324
Inga semialata IPNI=*128599-2* 323
Inga semialata var. *latifolia* IPNI=*128599-2* 323
Inga semiglabra IPNI=*500531-1* 322
Inga setifera IPNI=*128675-2* 323
Inga similis IPNI=*128732-2* 324
Inga sinemariensis IPNI=*1199545-2* 323
Inga smithii IPNI=*1122303-2* 324
Inga sodiroi IPNI=*128564-2* 323
Inga sordida IPNI=*128466-2* 322
Inga spectabilis var. *schimpffii* IPNI=*1122303-2* 324
Inga splendens IPNI=*128764-2* 324
Inga splendens var. *hostmannii* IPNI=*128749-2* 324
Inga splendens var. *superba* IPNI=*128749-2* 324
Inga spruceana IPNI=*1129244-2* 321
Inga spuria var. *sordida* IPNI=*128358-2* 321
Inga stenocarpa IPNI=*128535-2* 322
Inga stenostachya IPNI=*962800-1* 331
Inga strigillosa IPNI=*128693-2* 323
Inga subsericantha IPNI=*128673-2* 323
Inga superba IPNI=*128749-2* 324
Inga suturalis IPNI=*500893-1* 324
Inga tapajozensis IPNI=*500286-1* 321
Inga tenella IPNI=*500443-1* 322
Inga tenuifolia IPNI=*1090709-2* 322
Inga tenuirama IPNI=*500531-1* 322
Inga tessmannii var. *harmsii* IPNI=*128442-2* 322
Inga tetraphylla IPNI=*128599-2* 323
Inga tetraphylla var. *parvifolia* IPNI=*128599-2* 323
Inga thyrsoidea IPNI=*1129244-2* 321
Inga titiribiana IPNI=*128466-2* 322
Inga trapeziformis IPNI=*989341-1* 320
Inga tubaeformis IPNI=*128586-2* 323
Inga tuerckheimii IPNI=*128647-2* 323
Inga tysonii IPNI=*128599-2* 323
Inga umbellata IPNI=*128535-2* 322
Inga uncinata IPNI=*128478-2* 322
Inga unguis-cati IPNI=*515137-1* 330
Inga urabensis IPNI=*128812-2* 324
Inga urnifera IPNI=*128673-2* 323
Inga vahliana IPNI=*271434-2* 337
Inga vallensis IPNI=*500662-1* 323
Inga velloziana IPNI=*128358-2* 321
Inga venosa IPNI=*1040744-2* 322
Inga vera IPNI=*128478-2* 322
Inga vera subsp. *affinis* IPNI=*128358-2* 321
Inga verrucosa IPNI=*1199545-2* 323
Inga versicolor IPNI=*128675-2* 323
Inga vouapifolia IPNI=*128535-2* 322
Inga weberbaueri IPNI=*128687-2* 323
Inga ynga IPNI=*128478-2* 322
Inga yunckeri IPNI=*500531-1* 322
Inophloeum armatum IPNI=*1059330-2* 441
Inti bicallosa IPNI=*154952-2* 220
Intrusaria bicolor IPNI=*45742-1* 359
Intutis amygdalina IPNI=*77184072-1* 278
Intutis ferruginea IPNI=*77184072-1* 278
Inula canadensis IPNI=*203653-1* 261
Inula conyza IPNI=*60478388-2* 267
Inula conyzae IPNI=*60478388-2* 267
Inula foetida IPNI=*60478388-2* 267
Inula squarrosa IPNI=*60478388-2* 267
Inula trixis IPNI=*256729-1* 272
Inula vulgaris IPNI=*60478388-2* 267
Iocaulon lunulata IPNI=*318914-2* 316
Iocaulon pilosa IPNI=*318914-2* 316
Iocaulon sagittalis IPNI=*318914-2* 316
Ion tricolor IPNI=*869428-1* 406
Ionidium angustifolium IPNI=*77143566-1* 405
Ionidium anomalum IPNI=*77143573-1* 405
Ionidium aubletii IPNI=*77144891-1* 405
Ionidium barzelonense IPNI=*77143566-1* 405
Ionidium calceolaria IPNI=*77143549-1* 405
Ionidium chamaedrifolium IPNI=*77143568-1* 405
Ionidium circaeoides IPNI=*77143566-1* 405
Ionidium glaziovii IPNI=*77143566-1* 405
Ionidium glutinosum IPNI=*77143568-1* 405
Ionidium glutinosum var. *angustifolium* IPNI=*77143568-1* 405
Ionidium glutinosum var. *latifolium* IPNI=*77143568-1* 405
Ionidium glutinosum var. *paraguayense* IPNI=*77143568-1* 405

Ionidium glutinosum var. *parviflorum* IPNI=*77143568-1* 405
Ionidium grandiflorum IPNI=*77143573-1* 405
Ionidium hieronymi IPNI=*77143566-1* 405
Ionidium hybanthus IPNI=*77144891-1* 405
Ionidium indecorum IPNI=*77143549-1* 405
Ionidium ipecacuanha f. *grandiflora* IPNI=*77143549-1* 405
Ionidium ipecacuanha IPNI=*77143549-1* 405
Ionidium ipecacuanha var. *glabrum* IPNI=*77143549-1* 405
Ionidium ipecacuanha var. *pubescens* IPNI=*77143549-1* 405
Ionidium itoubou IPNI=*77143549-1* 405
Ionidium itubu IPNI=*77143549-1* 405
Ionidium loefflingianum IPNI=*77143549-1* 405
Ionidium longifolium IPNI=*77143566-1* 405
Ionidium microphyllum IPNI=*77143568-1* 405
Ionidium oppositifolium IPNI=*77143566-1* 405
Ionidium parviflorum IPNI=*77143568-1* 405
Ionidium stipulare IPNI=*266985-2* 406
Ionidium teucriifolium IPNI=*77143566-1* 405
Ionidium urticifolium IPNI=*77143566-1* 405
Ionidium villosissimum IPNI=*77143549-1* 405
Ionidium viscidulum IPNI=*77143566-1* 405
Ionoxalis amplifolia IPNI=*374872-1* 432
Ionoxalis attenuata IPNI=*375051-1* 432
Ionoxalis bipartita IPNI=*375051-1* 432
Ionoxalis buchtienii IPNI=*375051-1* 432
Ionoxalis calcaria IPNI=*375051-1* 432
Ionoxalis canaminensis IPNI=*374833-1* 432
Ionoxalis cuernavacana IPNI=*375466-1* 432
Ionoxalis deppei IPNI=*375466-1* 432
Ionoxalis divaricata IPNI=*375466-1* 432
Ionoxalis drummondii IPNI=*374872-1* 432
Ionoxalis intermedia IPNI=*375051-1* 432
Ionoxalis latifolia IPNI=*375051-1* 432
Ionoxalis martiana IPNI=*374833-1* 432
Ionoxalis scopulorum IPNI=*375466-1* 432
Ionoxalis stipitata IPNI=*375051-1* 432
Ionoxalis tenuiloba IPNI=*375051-1* 432
Ionoxalis tetraphylla IPNI=*375466-1* 432
Ionoxalis vespertilionis IPNI=*374872-1* 432
Ioxylon mora IPNI=*1151750-2* 440
Ipecacuanha fusca IPNI=*1214950-2* 347
Ipecacuanha officinalis IPNI=*1214950-2* 347
Ipnum mendocinum IPNI=*60458636-2* 238
Iposues obovata IPNI=*1137397-2* 300
Iresine aggregata IPNI=*34033-2* 282
Iresine argentata IPNI=*1002675-1* 283
Iresine argentata var. *colombiana* IPNI=*1002675-1* 283
Iresine argentata var. *latifolia* IPNI=*1002675 1* 283
Iresine aurata IPNI=*1002675-1* 283
Iresine aurata var. *amazonica* IPNI=*1002675-1* 283
Iresine bonariensis IPNI=*129900-2* 283
Iresine calea IPNI=*60789-1* 283
Iresine canescens IPNI=*1002675-1* 283
Iresine canescens IPNI=*129900-2* 283
Iresine celosia f. *ciliolata* IPNI=*129900-2* 283
Iresine celosia grex lutescens IPNI=*129900-2* 283
Iresine celosia IPNI=*129956-2* 283
Iresine celosia subsp. *tomentosa* IPNI=*129900-2* 283
Iresine celosia var. *alopecuroidea* IPNI=*129900-2* 283
Iresine celosia var. *diffusa* IPNI=*129900-2* 283
Iresine celosia var. *pubescens* IPNI=*129900-2* 283
Iresine celosia var. *verticillata* IPNI=*129900-2* 283
Iresine celosioides f. *albonervosa* IPNI=*129900-2* 283
Iresine celosioides f. *glabra* IPNI=*129900-2* 283
Iresine celosioides IPNI=*129956-2* 283
Iresine celosioides var. *eriophylla* IPNI=*129900-2* 283
Iresine celosioides var. *obtusifolia* IPNI=*129900-2* 283
Iresine celosioides var. *pubescens* IPNI=*129900-2* 283
Iresine celosioides var. *verticillata* IPNI=*129900-2* 283
Iresine crassifolia IPNI=*34033-2* 282
Iresine dysdicta IPNI=*1002675-1* 283
Iresine elatior IPNI=*1002675-1* 283
Iresine elongata IPNI=*129900-2* 283
Iresine endlichii IPNI=*129900-2* 283
Iresine eriophora IPNI=*129900-2* 283
Iresine eriophylla IPNI=*129900-2* 283
Iresine floribunda IPNI=*129900-2* 283
Iresine frutescens IPNI=*129900-2* 283
Iresine glomerata IPNI=*191400-2* 283
Iresine gossypiantha IPNI=*129900-2* 283
Iresine gracilis IPNI=*129900-2* 283
Iresine havanensis IPNI=*129900-2* 283
Iresine herbacea IPNI=*129900-2* 283
Iresine latifolia var. *velutina* IPNI=*60789-1* 283
Iresine laxa IPNI=*60789-1* 283
Iresine luzuliflora IPNI=*61055-1* 283
Iresine macbridei IPNI=*129900-2* 283
Iresine mexicana IPNI=*10378-2* 281
Iresine mutisii IPNI=*129900-2* 283
Iresine paniculata IPNI=*129956-2* 283
Iresine paniculata var. *floridana* IPNI=*129900-2* 283
Iresine papantlana IPNI=*129900-2* 283
Iresine parvifolia IPNI=*129900-2* 283
Iresine pilgeri IPNI=*129900-2* 283
Iresine polymorpha IPNI=*129900-2* 283
Iresine polymorpha var. *alopecuroidea* IPNI=*129900-2* 283
Iresine polymorpha var. *diffusa* IPNI=*129900-2* 283
Iresine polymorpha var. *effusa* IPNI=*129900-2* 283
Iresine polymorpha var. *verticillata* IPNI=*129900-2* 283
Iresine portulacoides IPNI=*34030-2* 282
Iresine portulacoides var. *commersonii* IPNI=*34030-2* 282
Iresine surinamensis IPNI=*34033-2* 282
Iresine vermicularis IPNI=*34033-2* 282
Iresine vermicularis var. *aggregata* IPNI=*34033-2* 282
Iresine vermicularis var. *longispicata* IPNI=*34033-2* 282
Iresine verticillata IPNI=*129900-2* 283
Iria aestivalis IPNI=*307664-1* 232
Iria capillaris IPNI=*1186511-2* 229
Iria caribea IPNI=*104972-2* 232
Iria castanea IPNI=*308274-1* 232
Iria indica IPNI=*104972-2* 232
Iria monandra IPNI=*307702-1* 232
Iria monostachya IPNI=*104972-2* 232
Iria muelleriana IPNI=*307702-1* 232
Iria spadicea IPNI=*308274-1* 232
Iria spadicea var. *nigra* IPNI=*308274-1* 232
Iria spadicea var. *normalis* IPNI=*308274-1* 232
Iria spadicea var. *pallida* IPNI=*308274-1* 232
Iria squarrosa IPNI=*308286-1* 232
Iria umbellata IPNI=*308274-1* 232
Iriartea andicola IPNI=*30119294-2* 207
Iriartea durissima IPNI=*1200702-2* 212
Iriartea exorrhiza IPNI=*1200702-2* 212
Iriartea exorrhiza var. *elegans* IPNI=*1200702-2* 212
Iriartea exorrhiza var. *orbignyana* IPNI=*1200702-2* 212
Iriartea klopstockia IPNI=*53436-2* 207
Iriartea lamarckiana IPNI=*666671-1* 208
Iriartea megalocarpa IPNI=*667594-1* 210
Iriartea nivea IPNI=*53436-2* 207
Iriartea orbignyana IPNI=*1200702-2* 212
Iriartea phaeocarpa IPNI=*667594-1* 210
Iriartea philonotia IPNI=*1200702-2* 212
Iriartea pruriens IPNI=*667622-1* 210
Iriartea setigera IPNI=*667622-1* 210
Iriartea spruceana IPNI=*667622-1* 210
Iriartea stenocarpa IPNI=*130009-2* 210

Iriartea ventricosa IPNI=*667594-1* 210
Iriartea weberbaueri IPNI=*667594-1* 210
Iriartea xanthorhiza IPNI=*667594-1* 210
Iriartella ferreyrae IPNI=*130009-2* 210
Iriartella setigera var. *pruriens* IPNI=*667622-1* 210
Irio ruderalis IPNI=*237940-2* 278
Iris × *alba* IPNI=*438637-1* 216
Iris × *amoena* IPNI=*438637-1* 216
Iris × *atroviolacea* IPNI=*438637-1* 216
Iris × *australis* IPNI=*438637-1* 216
Iris × *belouinii* IPNI=*438637-1* 216
Iris × *biliottii* IPNI=*438637-1* 216
Iris × *buiana* IPNI=*438637-1* 216
Iris × *croatica* IPNI=*438637-1* 216
Iris × *cypriana* IPNI=*438637-1* 216
Iris × *deflexa* IPNI=*438637-1* 216
Iris × *germanica* var. *gypsea* IPNI=*438637-1* 216
Iris × *glauca* IPNI=*438637-1* 216
Iris × *humei* IPNI=*438637-1* 216
Iris × *laciniata* IPNI=*438637-1* 216
Iris × *latifolia* IPNI=*438637-1* 216
Iris × *macrantha* IPNI=*438637-1* 216
Iris × *mesopotamica* IPNI=*438637-1* 216
Iris × *murorum* IPNI=*438637-1* 216
Iris × *neglecta* IPNI=*438637-1* 216
Iris × *nepalensis* IPNI=*438637-1* 216
Iris × *nostras* IPNI=*438637-1* 216
Iris × *nyaradyana* IPNI=*438637-1* 216
Iris × *pallida* IPNI=*438637-1* 216
Iris × *redouteana* IPNI=*438637-1* 216
Iris × *repanda* IPNI=*438637-1* 216
Iris × *rothschildii* IPNI=*438637-1* 216
Iris × *sambucina* IPNI=*438637-1* 216
Iris × *spectabilis* IPNI=*438637-1* 216
Iris × *squalens* IPNI=*438637-1* 216
Iris × *superba* IPNI=*438637-1* 216
Iris × *tardiflora* IPNI=*438637-1* 216
Iris × *trojana* IPNI=*438637-1* 216
Iris × *varbossania* IPNI=*438637-1* 216
Iris × *venusta* IPNI=*438637-1* 216
Iris × *violacea* IPNI=*438637-1* 216
Iris × *vulgaris* IPNI=*438637-1* 216
Iris chinensis IPNI=*438741-1* 216
Iris fimbriata IPNI=*438741-1* 216
Iris japonica f. *pallescens* IPNI=*438741-1* 216
Iris martinicensis IPNI=*441642-1* 217
Iris squalens IPNI=*438741-1* 216
Irlbachia alata IPNI=*118811-2* 345
Irlbachia bonplandiana IPNI=*130185-2* 345
Irlbachia cardonae IPNI=*130185-2* 345
Irlbachia elegans IPNI=*130185-2* 345
Irlbachia paruana IPNI=*130185-2* 345
Irlbachia purpurascens IPNI=*1002418-1* 344
Irlbachia recurva IPNI=*130185-2* 345
Irlbachia subcordata IPNI=*130185-2* 345
Iryanthera congestiflora IPNI=*585810-1* 191
Iryanthera densiflora IPNI=*130210-2* 191
Iryanthera elongata IPNI=*130222-2* 191
Iryanthera grandiflora IPNI=*130210-2* 191
Iryanthera krukovii IPNI=*586380-1* 191
Iryanthera leptoclada IPNI=*585810-1* 191
Iryanthera porcata IPNI=*130213-2* 191
Iryanthera sessilis IPNI=*130222-2* 191
Iryanthera trigona IPNI=*130210-2* 191
Iryanthera ulei IPNI=*585810-1* 191
Isachne mayocoensis IPNI=*324582-2* 242
Isandrina maxonii IPNI=*234410-2* 332
Ischaemum colladoa IPNI=*406281-1* 240
Ischaemum glabrum IPNI=*424990-1* 246
Ischaemum hispidum IPNI=*21052-2* 234
Ischaemum indicum IPNI=*963150-1* 244
Ischaemum pallidum IPNI=*21052-2* 234
Ischaemum royleanum IPNI=*406281-1* 240
Ischaemum rugosum var. *segetum* IPNI=*406281-1* 240
Ischaemum secundatum IPNI=*423051-1* 246
Ischaemum segetum IPNI=*406281-1* 240
Ischaemum tashiroi IPNI=*406281-1* 240
Ischurochloa floribunda IPNI=*1097548-2* 234
Ischyranthera laevigata IPNI=*1157663-2* 418
Isertia alba IPNI=*278892-2* 351
Isertia breviflora IPNI=*753635-1* 351
Isertia coccinea var. *hypoleuca* IPNI=*753635-1* 351
Isertia coccinea var. *pentamera* IPNI=*30281446-2* 351
Isertia flava IPNI=*30281446-2* 351
Isertia glabra IPNI=*753641-1* 352
Isertia hoehnei IPNI=*753635-1* 351
Isertia juruana IPNI=*278892-2* 351
Isertia panamensis IPNI=*278892-2* 351
Isertia parviflora var. *hirta* IPNI=*753641-1* 352
Isertia parvifolia IPNI=*278892-2* 351
Isertia purdiei IPNI=*278892-2* 351
Isertia scarlatina IPNI=*753635-1* 351
Isertia spraguei IPNI=*278892-2* 351
Isertia weberbaueri IPNI=*278892-2* 351
Ismelia frutescens IPNI=*178745-1* 257
Isnardia ascendens IPNI=*30031106-2* 429
Isnardia capitata IPNI=*144385-2* 429
Isnardia discolor IPNI=*144351-2* 429
Isnardia nitida IPNI=*30031106-2* 429
Isnardia palustris f. *submersa* IPNI=*30031106-2* 429
Isnardia palustris IPNI=*30031106-2* 429
Isnardia palustris var. *americana* IPNI=*30031106-2* 429
Isocarpha amellus IPNI=*241903-1* 269
Isocarpha angustata IPNI=*1008406-1* 264
Isocarpha pyrethraria IPNI=*914778-1* 255
Isocarpha tricephala IPNI=*1008406-1* 264
Isochilus prolifer IPNI=*228831-2* 223
Isochilus ramosus IPNI=*92214-2* 219
Isolepis acicularis IPNI=*1169169-2* 231
Isolepis anceps IPNI=*1111946-2* 232
Isolepis berlandieri IPNI=*38049-2* 229
Isolepis capillaris IPNI=*1186511-2* 229
Isolepis complanata IPNI=*1111946-2* 232
Isolepis congesta IPNI=*38049-2* 229
Isolepis dichotoma IPNI=*307853-1* 232
Isolepis echinulata IPNI=*324890-2* 231
Isolepis junciformis IPNI=*38049-2* 229
Isolepis lanata IPNI=*30060098-2* 229
Isolepis willdenowii IPNI=*1111946-2* 232
Isoloma bogotense var. *amabile* IPNI=*133755-2* 367
Isoloma breviflorum IPNI=*381628-1* 367
Isoloma ceciliae IPNI=*133755-2* 367
Isoloma cordifolium IPNI=*381638-1* 367
Isoloma erianthum IPNI=*381572-1* 367
Isoloma hirsutum IPNI=*381572-1* 367
Isoloma hondense var. *warszewiczii* IPNI=*381638-1* 367
Isoloma ignoratum IPNI=*381628-1* 367
Isoloma incanum IPNI=*381628-1* 367
Isoloma incurvum IPNI=*381638-1* 367
Isoloma kramerianum IPNI=*381628-1* 367
Isoloma linkianum IPNI=*381628-1* 367
Isoloma longifolium IPNI=*381628-1* 367
Isoloma molle IPNI=*381572-1* 367
Isoloma petiolare IPNI=*381628-1* 367
Isoloma pictum IPNI=*381638-1* 367
Isoloma pilosum IPNI=*381638-1* 367

Isoloma pycnosuzygium IPNI=*381572-1* 367
Isoloma rhynchocarpum IPNI=*381638-1* 367
Isoloma rubricaule IPNI=*381572-1* 367
Isoloma rupestre IPNI=*381628-1* 367
Isoloma schiedeanum IPNI=*381628-1* 367
Isoloma seemannii IPNI=*381628-1* 367
Isoloma spicatum IPNI=*381628-1* 367
Isoloma strictum IPNI=*381638-1* 367
Isoloma tetragonum IPNI=*381628-1* 367
Isoloma tubiflorum IPNI=*381638-1* 367
Isoloma tydaeum IPNI=*133755-2* 367
Isoloma venosum IPNI=*133755-2* 367
Isoloma vestitum IPNI=*381572-1* 367
Isomeria chinensis IPNI=*962413-1* 260
Isomeria cinerea IPNI=*962413-1* 260
Isopteryx longirostris IPNI=*105087-1* 292
Isotoma longiflora IPNI=*1133957-2* 273
Isotoma longiflora var. *runcinata* IPNI=*1133957-2* 273
Isotoma runcinata IPNI=*1133957-2* 273
Isotropis argentea IPNI=*322601-2* 316
Isotypus onoseroides IPNI=*174455-2* 267
Isotypus rosiflorus IPNI=*174455-2* 267
Itea caroliniana IPNI=*319176-2* 298
Itea cyrilla IPNI=*319176-2* 298
Itea floribunda IPNI=*319176-2* 298
Itea japonica IPNI=*835147-1* 431
Itoasia crassa IPNI=*103322-1* 448
Itoasia sphaerica IPNI=*103322-1* 448
Iva monophylla IPNI=*176058-1* 256
Ivira pruriens IPNI=*825412-1* 414
Ixia americana IPNI=*89940-2* 216
Ixophorus glaucus IPNI=*77105978-1* 235
Ixophorus italicus IPNI=*1207294-2* 245
Ixophorus verticillatus IPNI=*421857-1* 245
Ixora americana IPNI=*1187456-2* 350
Ixora armeniaca IPNI=*753844-1* 352
Ixora blanda IPNI=*753834-1* 352
Ixora crocata IPNI=*753834-1* 352
Ixora denticulata IPNI=*753956-1* 352
Ixora dubia IPNI=*753834-1* 352
Ixora findlayana IPNI=*753956-1* 352
Ixora flammea IPNI=*753834-1* 352
Ixora incarnata IPNI=*753834-1* 352
Ixora kroneana IPNI=*753834-1* 352
Ixora occidentalis IPNI=*1187456-2* 350
Ixora pallida IPNI=*753834-1* 352
Ixora pentandra IPNI=*752006-1* 351
Ixora rosea IPNI=*753834-1* 352
Ixora speciosa IPNI=*753834-1* 352
Ixora spinosa IPNI=*746662-1* 348
Ixora stricta IPNI=*753834-1* 352

J
Jacaranda atropurpurea IPNI=*130921-2* 364
Jacaranda caucana subsp. *calycina* IPNI=*130903-2* 364
Jacaranda caucana subsp. *glabrata* IPNI=*130903-2* 364
Jacaranda caucana subsp. *sandwithiana* IPNI=*130903-2* 364
Jacaranda cauliflora IPNI=*130921-2* 364
Jacaranda chelonia IPNI=*130936-2* 364
Jacaranda echinata IPNI=*110483-1* 365
Jacaranda gualanday IPNI=*130903-2* 364
Jacaranda intermedia IPNI=*130921-2* 364
Jacaranda ovalifolia IPNI=*130936-2* 364
Jacaranda procera IPNI=*109789-1* 364
Jacaranda rachidoptera IPNI=*130921-2* 364
Jacaranda trianae IPNI=*130903-2* 364
Jacaratia boliviana IPNI=*130970-2* 279
Jacaratia costaricensis IPNI=*30104966-2* 279
Jacaratia dodecaphylla f. *longiflora* IPNI=*30104966-2* 279
Jacaratia dodecaphylla var. *lucida* IPNI=*30104966-2* 279
Jacaratia spinosa var. *digitata* IPNI=*130970-2* 279
Jacea segetalis IPNI=*190310-1* 259
Jacea segetum IPNI=*190310-1* 259
Jacea tricolor IPNI=*869428-1* 406
Jacobaea canyza IPNI=*60478388-2* 267
Jacobaea vulgaris IPNI=*248251-1* 270
Jacobinia atramentaria IPNI=*133119-2* 360
Jacobinia carnea IPNI=*132826-2* 360
Jacobinia coccinea IPNI=*943504-1* 361
Jacobinia ghiesbreghtiana IPNI=*133119-2* 360
Jacobinia magnifica IPNI=*132826-2* 360
Jacobinia magnifica var. *carnea* IPNI=*132826-2* 360
Jacobinia mexicana IPNI=*1140399-2* 359
Jacobinia mohintli IPNI=*133119-2* 360
Jacobinia neglecta IPNI=*133119-2* 360
Jacobinia obtusior IPNI=*132826-2* 360
Jacobinia pohliana f. *velutina* IPNI=*132826-2* 360
Jacobinia pohliana IPNI=*132826-2* 360
Jacobinia scarlatina IPNI=*133119-2* 360
Jacobinia spicigera IPNI=*133119-2* 360
Jacobinia tinctoria IPNI=*907564-1* 360
Jacquinia aciculata IPNI=*60430434-2* 303
Jacquinia aristata IPNI=*60430434-2* 303
Jacquinia brasiliensis IPNI=*312066-2* 304
Jacquinia caracasana IPNI=*60430434-2* 303
Jacquinia frutescens IPNI=*60430434-2* 303
Jacquinia geniculata IPNI=*60430434-2* 303
Jacquinia gracilis IPNI=*60430434-2* 303
Jacquinia mucronulata IPNI=*312066-2* 304
Jacquinia ovalifolia IPNI=*60430434-2* 303
Jacquinia revoluta IPNI=*312066-2* 304
Jacquinia ruscifolia IPNI=*319112-2* 377
Jacquinia ruscifolia var. *frutescens* IPNI=*60430434-2* 303
Jacquinia venosa IPNI=*56650-2* 348
Jaegeria uliginosa IPNI=*174556-1* 256
Jaegeria urticifolia IPNI=*208557-1* 263
Jahnia meridensis IPNI=*785554-1* 292
Jalapa congesta IPNI=*162591-2* 287
Jalapa dichotoma IPNI=*162591-2* 287
Jalapa officinalis IPNI=*162591-2* 287
Jalapa officinarum IPNI=*162591-2* 287
Jalapa undulata IPNI=*162591-2* 287
Jambolifera chinensis IPNI=*601603-1* 428
Jambolifera coromandelica IPNI=*601603-1* 428
Jambolifera pedunculata IPNI=*601603-1* 428
Jambosa domestica IPNI=*601907-1* 428
Jambosa jambos IPNI=*601802-1* 428
Jambosa laevis IPNI=*601907-1* 428
Jambosa leptostachya IPNI=*601802-1* 428
Jambosa macrophylla IPNI=*601907-1* 428
Jambosa malaccensis IPNI=*601907-1* 428
Jambosa palembanica IPNI=*601802-1* 428
Jambosa purpurascens IPNI=*601907-1* 428
Jambosa purpurea IPNI=*601907-1* 428
Jambosa vulgaris IPNI=*601802-1* 428
Janipha aipi IPNI=*351790-1* 392
Janipha manihot IPNI=*351790-1* 392
Janipha urens IPNI=*61899-2* 388
Japanobotrychum virginianum IPNI=*1106719-2* 173
Japarandiba angustifolia IPNI=*114913-2* 301
Japarandiba antillana IPNI=*114915-2* 301
Japarandiba augusta IPNI=*114915-2* 301
Japarandiba brasiliana IPNI=*596743-1* 302
Japarandiba dubia IPNI=*596736-1* 302
Japarandiba fastuosa IPNI=*596743-1* 302
Japarandiba gracillima IPNI=*114943-2* 302

Japarandiba hexapetala IPNI=*596743-1* 302
Japarandiba longifolia IPNI=*114952-2* 302
Japarandiba marcgraaviana IPNI=*114915-2* 301
Japarandiba poeppigiana IPNI=*114966-2* 302
Japarandiba pterocarpa IPNI=*596743-1* 302
Japarandiba ruiziana IPNI=*114913-2* 301
Japarandiba speciosa IPNI=*596758-1* 302
Japarandiba spruceana IPNI=*114952-2* 302
Japarandiba superba IPNI=*323304-2* 302
Jasminonerium africanum IPNI=*77727-1* 340
Jasminonerium grandiflorum IPNI=*77727-1* 340
Jasminonerium macrocarpum IPNI=*77727-1* 340
Jasminum bicorollatum IPNI=*609755-1* 375
Jasminum blancoi IPNI=*609755-1* 375
Jasminum blinii IPNI=*609699-1* 375
Jasminum capense IPNI=*751073-1* 350
Jasminum delafieldii IPNI=*609699-1* 375
Jasminum excellens IPNI=*609699-1* 375
Jasminum fragrans IPNI=*609755-1* 375
Jasminum gracillimum IPNI=*609641-1* 375
Jasminum humile IPNI=*77144222-1* 374
Jasminum multiflorum var. *nicobaricum* IPNI=*609641-1* 375
Jasminum odoratum IPNI=*609755-1* 375
Jasminum officinale f. *grandiflorum* IPNI=*328130-2* 375
Jasminum officinale subsp. *grandiflorum* IPNI=*328130-2* 375
Jasminum officinale var. *piliferum* IPNI=*609672-1* 375
Jasminum officinale var. *tibeticum* IPNI=*609672-1* 375
Jasminum pubescens IPNI=*609641-1* 375
Jasminum pubescens IPNI=*609755-1* 375
Jasminum quadrifolium IPNI=*609755-1* 375
Jasminum quinqueflorum IPNI=*609755-1* 375
Jasminum sambac var. *heyneanum* IPNI=*609755-1* 375
Jasminum sanjurium IPNI=*609755-1* 375
Jasminum suaveolens IPNI=*609354-1* 374
Jasminum trifoliatum IPNI=*609354-1* 374
Jasminum undulatum IPNI=*609755-1* 375
Jasminum viminale IPNI=*609672-1* 375
Jasminum vulgatius IPNI=*609672-1* 375
Jasminum zambac IPNI=*609755-1* 375
Jasminum zeylanicum IPNI=*82097-1* 343
Jatropha acerifolia IPNI=*131462-2* 391
Jatropha aconitifolia IPNI=*61821-2* 388
Jatropha aconitifolia var. *genuina* IPNI=*61821-2* 388
Jatropha acuminata IPNI=*350290-1* 391
Jatropha adenophila IPNI=*61899-2* 388
Jatropha afrocurcas IPNI=*131462-2* 391
Jatropha aipi IPNI=*351790-1* 392
Jatropha carthagenensis IPNI=*1100922-2* 392
Jatropha coccinea IPNI=*350290-1* 391
Jatropha cordifolia IPNI=*61897-2* 388
Jatropha diffusa IPNI=*351790-1* 392
Jatropha digitiformis IPNI=*351790-1* 392
Jatropha diversifolia IPNI=*350290-1* 391
Jatropha diversifolia var. *pandurifolia* IPNI=*350290-1* 391
Jatropha dulcis IPNI=*351790-1* 392
Jatropha edulis IPNI=*131462-2* 391
Jatropha flabellifolia IPNI=*351790-1* 392
Jatropha glauca IPNI=*351790-1* 392
Jatropha gossypiifolia var. *typica* IPNI=*323930-2* 391
Jatropha hastata IPNI=*350290-1* 391
Jatropha herbacea IPNI=*61899-2* 388
Jatropha integerrima var. *coccinea* IPNI=*350290-1* 391
Jatropha integerrima var. *hastata* IPNI=*350290-1* 391
Jatropha integerrima var. *latifolia* IPNI=*350290-1* 391
Jatropha janipha IPNI=*314308-2* 391
Jatropha janipha IPNI=*351790-1* 392
Jatropha jurgensenii IPNI=*61897-2* 388
Jatropha longipes IPNI=*61855-2* 388
Jatropha loureiroi IPNI=*351790-1* 392
Jatropha manihot IPNI=*351790-1* 392
Jatropha mitis IPNI=*351790-1* 392
Jatropha moluccana IPNI=*338514-1* 387
Jatropha moluensis IPNI=*350290-1* 391
Jatropha osteocarpa IPNI=*61899-2* 388
Jatropha pandurifolia IPNI=*350290-1* 391
Jatropha pandurifolia var. *coccinea* IPNI=*350290-1* 391
Jatropha pandurifolia var. *latifolia* IPNI=*350290-1* 391
Jatropha paniculata IPNI=*351790-1* 392
Jatropha quinqueloba IPNI=*61897-2* 388
Jatropha silvestris IPNI=*351790-1* 392
Jatropha stipulata IPNI=*351790-1* 392
Jatropha tomentosa IPNI=*865678-1* 373
Jatropha tubulosa IPNI=*61897-2* 388
Jatropha tubulosa var. *septemloba* IPNI=*61897-2* 388
Jatropha tubulosa var. *triloba* IPNI=*61899-2* 388
Jatropha urens f. *osteocarpa* IPNI=*61899-2* 388
Jatropha urens IPNI=*61899-2* 388
Jatropha urens var. *brachyloba* IPNI=*61899-2* 388
Jatropha urens var. *genuina* IPNI=*61899-2* 388
Jatropha urens var. *herbacea* IPNI=*61899-2* 388
Jatropha urens var. *marcgravii* IPNI=*61899-2* 388
Jatropha urens var. *osteocarpa* IPNI=*61899-2* 388
Jatropha vitifolia var. *repanda* IPNI=*61899-2* 388
Jatropha yucatanensis IPNI=*131462-2* 391
Jatus grandis IPNI=*864923-1* 373
Jeilium tricolor IPNI=*10291-2* 281
Jessenia amazonum IPNI=*134734-3* 204
Jessenia bataua IPNI=*668510-1* 210
Jessenia glazioviana IPNI=*664083-1* 204
Joannesia insignis IPNI=*194359-1* 260
Johannia insignis IPNI=*194359-1* 260
Johannia lancifolia IPNI=*194359-1* 260
Johannia microphylla IPNI=*194359-1* 260
Joncquetia paniculata IPNI=*71559-1* 451
Josepha augusta IPNI=*604358-1* 286
Jossinia cotinifolia IPNI=*594126-1* 425
Jossinia pinnata IPNI=*203871-2* 428
Jububa mediterranea IPNI=*719213-1* 442
Jucunda bredemeyriana IPNI=*572641-1* 421
Jucunda rudolphiana IPNI=*572641-1* 421
Jucunda tomentosa IPNI=*1088633-2* 423
Jugastrum coriaceum IPNI=*947902-1* 301
Juglans asplenifolia IPNI=*442427-1* 339
Juglans catappa IPNI=*171034-1* 416
Juglans columbiensis IPNI=*131772-2* 338
Juglans dissecta IPNI=*442427-1* 339
Juglans duclouxiana IPNI=*442427-1* 339
Juglans equatoriensis IPNI=*131772-2* 338
Juglans fallax IPNI=*442427-1* 339
Juglans fertilis IPNI=*442427-1* 339
Juglans frutescens IPNI=*442427-1* 339
Juglans fruticosa IPNI=*442427-1* 339
Juglans granatensis IPNI=*131772-2* 338
Juglans heterophylla IPNI=*442427-1* 339
Juglans honorei IPNI=*131772-2* 338
Juglans kamaonia IPNI=*442427-1* 339
Juglans longirostris IPNI=*442427-1* 339
Juglans orientis IPNI=*442427-1* 339
Juglans quercifolia IPNI=*442427-1* 339
Juglans regia subsp. *sinensis* IPNI=*442427-1* 339
Juglans regia var. *laciniata* IPNI=*442427-1* 339
Juglans salicifolia IPNI=*442427-1* 339
Juglans sinensis IPNI=*442427-1* 339
Julocroton argenteus IPNI=*69894-2* 388
Julocroton argenteus var. *managuensis* IPNI=*69894-2* 388
Julocroton camporum IPNI=*69894-2* 388

Julocroton elaeagnoides IPNI=*69894-2* 388
Julocroton integer f. *parvifolia* IPNI=*69894-2* 388
Julocroton integer IPNI=*69894-2* 388
Julocroton linearifolius IPNI=*69894-2* 388
Julocroton montevidensis IPNI=*69894-2* 388
Julocroton montevidensis var. *elata* IPNI=*69894-2* 388
Julocroton montevidensis var. *genuinus* IPNI=*69894-2* 388
Julocroton montevidensis var. *glabra* IPNI=*69894-2* 388
Julocroton montevidensis var. *guatemalensis* IPNI=*69894-2* 388
Julocroton montevidensis var. *lanceolatus* IPNI=*69894-2* 388
Julocroton montevidensis var. *linearifolius* IPNI=*69894-2* 388
Julocroton montevidensis var. *pilosus* IPNI=*69894-2* 388
Julocroton montevidensis var. *virgatus* IPNI=*69894-2* 388
Julocroton pilosus IPNI=*69894-2* 388
Julocroton quinquenervius IPNI=*69894-2* 388
Juncellus laevigatus IPNI=*304938-1* 231
Juncus achalensis IPNI=*443313-1* 233
Juncus aletaiensis IPNI=*30079279-2* 233
Juncus andreanus IPNI=*132109-2* 233
Juncus arcticus subsp. *balticus* IPNI=*324783-2* 233
Juncus bicornis IPNI=*304453-2* 233
Juncus bicornis var. *williamsii* IPNI=*304453-2* 233
Juncus bilineatus IPNI=*30079279-2* 233
Juncus bufonius subsp. *juzepczukii* IPNI=*30079279-2* 233
Juncus bufonius var. *congestus* IPNI=*30079279-2* 233
Juncus bufonius var. *juzepczukii* IPNI=*30079279-2* 233
Juncus bufonius var. *pumilio* IPNI=*30079279-2* 233
Juncus bufonum IPNI=*30079279-2* 233
Juncus cespifolius IPNI=*30079279-2* 233
Juncus chloroticus IPNI=*304453-2* 233
Juncus coarctatus IPNI=*132109-2* 233
Juncus commixtus IPNI=*443313-1* 233
Juncus communis IPNI=*442917-1* 233
Juncus communis var. *effusus* IPNI=*442917-1* 233
Juncus creticus IPNI=*30079279-2* 233
Juncus divaricatus IPNI=*30079279-2* 233
Juncus dombeyanus IPNI=*443313-1* 233
Juncus dombeyanus var. *elatus* IPNI=*443313-1* 233
Juncus dombeyanus var. *pycnanthus* IPNI=*443313-1* 233
Juncus erythropodus IPNI=*30079279-2* 233
Juncus fasciatus IPNI=*30079279-2* 233
Juncus fernandezianus IPNI=*443313-1* 233
Juncus floribundus IPNI=*443213-1* 233
Juncus fuscocapitatus IPNI=*443313-1* 233
Juncus germanorum IPNI=*304453-2* 233
Juncus gesneri IPNI=*304453-2* 233
Juncus glaucus var. *littoralis* IPNI=*324783-2* 233
Juncus gracilis IPNI=*304453-2* 233
Juncus helodes IPNI=*324783-2* 233
Juncus involucratus IPNI=*304453-2* 233
Juncus involucratus IPNI=*443213-1* 233
Juncus istriacus IPNI=*30079279-2* 233
Juncus juzepczukii IPNI=*30079279-2* 233
Juncus laevis IPNI=*442917-1* 233
Juncus leptocladus IPNI=*30079279-2* 233
Juncus longifolius IPNI=*443313-1* 233
Juncus lucidus IPNI=*304453-2* 233
Juncus luzuloxiphium IPNI=*443213-1* 233
Juncus macer f. *williamsii* IPNI=*304453-2* 233
Juncus macer IPNI=*304453-2* 233
Juncus macer var. *williamsii* IPNI=*304453-2* 233
Juncus megakoleos IPNI=*443313-1* 233
Juncus microcephalus f. *conglobatus* IPNI=*443213-1* 233
Juncus microcephalus var. *floribundus* IPNI=*443213-1* 233
Juncus microcephalus var. *virens* IPNI=*443213-1* 233
Juncus pallescens var. *achalensis* IPNI=*443313-1* 233
Juncus platycaulos IPNI=*132109-2* 233
Juncus plebeius IPNI=*30079279-2* 233
Juncus prolifer IPNI=*30079279-2* 233
Juncus proximus IPNI=*443313-1* 233
Juncus pumilus IPNI=*30079279-2* 233
Juncus rudis IPNI=*443213-1* 233
Juncus sellowianus IPNI=*443213-1* 233
Juncus smithii IPNI=*304453-2* 233
Juncus tenuis f. *victorinii* IPNI=*304453-2* 233
Juncus tenuis f. *williamsii* IPNI=*304453-2* 233
Juncus tenuis var. *bicornis* IPNI=*304453-2* 233
Juncus tenuis var. *dudleyi* IPNI=*132109-2* 233
Juncus tenuis var. *multicornis* IPNI=*304453-2* 233
Juncus tenuis var. *platycaulos* IPNI=*132109-2* 233
Juncus tenuis var. *uniflorus* IPNI=*132109-2* 233
Juncus tenuis var. *williamsii* IPNI=*304453-2* 233
Juncus timotensis IPNI=*443213-1* 233
Juncus tristanianus IPNI=*304453-2* 233
Juncus vacillans IPNI=*304453-2* 233
Jungia splendens IPNI=*60466516-2* 373
Juniperus ericoides IPNI=*263956-1* 180
Juniperus ericoides IPNI=*677124-1* 180
Juniperus uhdeana IPNI=*60451554-2* 180
Juniperus whitleyana IPNI=*261974-1* 180
Jupunba jupunba IPNI=*469538-1* 310
Jurgensenia mexicana IPNI=*317630-2* 298
Jurgensia psylloides IPNI=*317930-2* 357
Jussiaea acuminata IPNI=*144351-2* 429
Jussiaea acuminata var. *latifolia* IPNI=*144351-2* 429
Jussiaea acuminata var. *longifolia* IPNI=*144351-2* 429
Jussiaea adscendens IPNI=*144324-2* 429
Jussiaea affinis IPNI=*144328-2* 429
Jussiaea affinis var. *genuina* IPNI=*144328-2* 429
Jussiaea affinis var. *multiflora* IPNI=*144328-2* 429
Jussiaea alata IPNI=*613333-1* 429
Jussiaea alternifolia IPNI=*144324-2* 429
Jussiaea altissima IPNI=*144351-2* 429
Jussiaea aluligera IPNI=*144366-2* 429
Jussiaea amasonica IPNI=*613361-1* 429
Jussiaea angustifolia IPNI=*144385-2* 429
Jussiaea berteroana IPNI=*613407-1* 429
Jussiaea bertonii IPNI=*613333-1* 429
Jussiaea biacuminata IPNI=*144366-2* 429
Jussiaea blumeana IPNI=*144385-2* 429
Jussiaea boydiana IPNI=*613407-1* 429
Jussiaea burmanni IPNI=*144385-2* 429
Jussiaea california IPNI=*613407-1* 429
Jussiaea calycina IPNI=*144385-2* 429
Jussiaea clavata IPNI=*144385-2* 429
Jussiaea commersonii IPNI=*144384-2* 429
Jussiaea declinata IPNI=*144351-2* 429
Jussiaea decumbens IPNI=*144385-2* 429
Jussiaea decurrens IPNI=*613333-1* 429
Jussiaea diffusa var. *californica* IPNI=*613407-1* 429
Jussiaea distans IPNI=*144328-2* 429
Jussiaea erecta IPNI=*144351-2* 429
Jussiaea erecta IPNI=*613333-1* 429
Jussiaea erecta var. *plumeriana* IPNI=*144351-2* 429
Jussiaea erecta var. *sebana* IPNI=*144351-2* 429
Jussiaea erythrocaulis IPNI=*144385-2* 429
Jussiaea exaltata IPNI=*144385-2* 429
Jussiaea ferruginea IPNI=*144328-2* 429
Jussiaea fissendocarpa IPNI=*613360-1* 429
Jussiaea floribunda IPNI=*144324-2* 429
Jussiaea fluitans IPNI=*144324-2* 429
Jussiaea fluitans IPNI=*613407-1* 429
Jussiaea fluviatilis IPNI=*144324-2* 429
Jussiaea frutescens IPNI=*144385-2* 429
Jussiaea fruticosa IPNI=*144385-2* 429
Jussiaea gomezii IPNI=*613407-1* 429

Jussiaea grandiflora IPNI=*144406-2* 429
Jussiaea heterophylla IPNI=*144366-2* 429
Jussiaea hexamera IPNI=*144328-2* 429
Jussiaea hirsuta IPNI=*144385-2* 429
Jussiaea hirsuta IPNI=*144406-2* 429
Jussiaea hirta IPNI=*144406-2* 429
Jussiaea hyssopifolia IPNI=*613360-1* 429
Jussiaea inclinata IPNI=*613361-1* 429
Jussiaea inclinata IPNI=*613407-1* 429
Jussiaea isnardia IPNI=*30031106-2* 429
Jussiaea junghuhniana IPNI=*144385-2* 429
Jussiaea lanceolata IPNI=*144384-2* 429
Jussiaea leptocarpa f. *biacuminata* IPNI=*144366-2* 429
Jussiaea leptocarpa IPNI=*144366-2* 429
Jussiaea leptocarpa var. *genuina* IPNI=*144366-2* 429
Jussiaea leptocarpa var. *meyeriana* IPNI=*144366-2* 429
Jussiaea ligustrifolia IPNI=*144385-2* 429
Jussiaea linearis IPNI=*144385-2* 429
Jussiaea linifolia IPNI=*613360-1* 429
Jussiaea longifolia IPNI=*144385-2* 429
Jussiaea longipes IPNI=*144385-2* 429
Jussiaea macrocarpa IPNI=*144406-2* 429
Jussiaea macropoda IPNI=*144385-2* 429
Jussiaea marcgravii IPNI=*144366-2* 429
Jussiaea marginata IPNI=*144384-2* 429
Jussiaea maypurensis IPNI=*144384-2* 429
Jussiaea mendocinensis IPNI=*613407-1* 429
Jussiaea micrantha IPNI=*613360-1* 429
Jussiaea micropetala IPNI=*144328-2* 429
Jussiaea miquelii IPNI=*144366-2* 429
Jussiaea mollis IPNI=*144406-2* 429
Jussiaea nervosa f. *foliosa* IPNI=*144384-2* 429
Jussiaea nervosa f. *microphylla* IPNI=*144384-2* 429
Jussiaea nervosa f. *salicina* IPNI=*144384-2* 429
Jussiaea nervosa IPNI=*144384-2* 429
Jussiaea nervosa stirps lanceolata IPNI=*144384-2* 429
Jussiaea nervosa stirps stricta IPNI=*144384-2* 429
Jussiaea nervosa var. *glaberrima* IPNI=*144384-2* 429
Jussiaea nervosa var. *marginata* IPNI=*144384-2* 429
Jussiaea nervosa var. *microphylla* IPNI=*144384-2* 429
Jussiaea nervosa var. *pubescens* IPNI=*144384-2* 429
Jussiaea nervosa var. *typica* IPNI=*144384-2* 429
Jussiaea nubica IPNI=*144351-2* 429
Jussiaea oblonga IPNI=*613407-1* 429
Jussiaea octofila IPNI=*144385-2* 429
Jussiaea octonervia IPNI=*144385-2* 429
Jussiaea octovalvis IPNI=*144385-2* 429
Jussiaea onagra IPNI=*144351-2* 429
Jussiaea ovalifolia IPNI=*144385-2* 429
Jussiaea palmitensis IPNI=*144384-2* 429
Jussiaea palustris IPNI=*613333-1* 429
Jussiaea paniculata IPNI=*613333-1* 429
Jussiaea parviflora IPNI=*144351-2* 429
Jussiaea parviflora IPNI=*144385-2* 429
Jussiaea patibilcensis IPNI=*613407-1* 429
Jussiaea pauciflora IPNI=*144384-2* 429
Jussiaea peploides IPNI=*613407-1* 429
Jussiaea persicariifolia f. *major* IPNI=*144385-2* 429
Jussiaea persicariifolia f. *minor* IPNI=*144385-2* 429
Jussiaea persicariifolia IPNI=*144385-2* 429
Jussiaea peruviana f. *hirsuta* IPNI=*144406-2* 429
Jussiaea peruviana f. *tomentosa* IPNI=*144406-2* 429
Jussiaea peruviana IPNI=*144406-2* 429
Jussiaea peruviana stirps macrophylla IPNI=*144406-2* 429
Jussiaea peruviana var. *australis* IPNI=*144406-2* 429
Jussiaea peruviana var. *glaberrima* IPNI=*144406-2* 429
Jussiaea peruviana var. *macrocarpa* IPNI=*144406-2* 429
Jussiaea peruviana var. *octofila* IPNI=*144385-2* 429
Jussiaea peruviana var. *typica* IPNI=*144406-2* 429
Jussiaea pilosa IPNI=*144366-2* 429
Jussiaea pilosa var. *glabra* IPNI=*144366-2* 429
Jussiaea pilosa var. *pterocarpa* IPNI=*144366-2* 429
Jussiaea pilosa var. *robustior* IPNI=*144366-2* 429
Jussiaea plumeriana IPNI=*144351-2* 429
Jussiaea polygonifolia IPNI=*613407-1* 429
Jussiaea polygonoides IPNI=*613407-1* 429
Jussiaea pseudoheterophylla IPNI=*144366-2* 429
Jussiaea pterophora IPNI=*613333-1* 429
Jussiaea pubescens IPNI=*144385-2* 429
Jussiaea pumila IPNI=*613407-1* 429
Jussiaea ramosa IPNI=*144351-2* 429
Jussiaea ramulosa IPNI=*613407-1* 429
Jussiaea repens f. *brevipes* IPNI=*144324-2* 429
Jussiaea repens f. *longipes* IPNI=*144324-2* 429
Jussiaea repens f. *typica* IPNI=*144324-2* 429
Jussiaea repens IPNI=*144324-2* 429
Jussiaea repens subf. orientalis IPNI=*144324-2* 429
Jussiaea repens subf. paraguayensis IPNI=*144324-2* 429
Jussiaea repens subsp. *glabrata* IPNI=*144324-2* 429
Jussiaea repens subsp. *hirsuta* IPNI=*144324-2* 429
Jussiaea repens var. *californica* IPNI=*613407-1* 429
Jussiaea repens var. *clarenciana* IPNI=*144324-2* 429
Jussiaea repens var. *grandiflora* IPNI=*144324-2* 429
Jussiaea repens var. *inflata* IPNI=*613361-1* 429
Jussiaea repens var. *minor* IPNI=*613407-1* 429
Jussiaea repens var. *peploides* IPNI=*613407-1* 429
Jussiaea repens var. *ramulosa* IPNI=*613407-1* 429
Jussiaea repens var. *typica* IPNI=*144324-2* 429
Jussiaea rigida IPNI=*144328-2* 429
Jussiaea sagrana IPNI=*144385-2* 429
Jussiaea salicifolia IPNI=*144385-2* 429
Jussiaea scabra IPNI=*144385-2* 429
Jussiaea schottii IPNI=*144366-2* 429
Jussiaea seminuda IPNI=*144366-2* 429
Jussiaea sessiliflora IPNI=*144351-2* 429
Jussiaea speciosa IPNI=*144406-2* 429
Jussiaea sprengeri IPNI=*144406-2* 429
Jussiaea suffruticosa f. *genuina* IPNI=*144385-2* 429
Jussiaea suffruticosa f. *grandiflora* IPNI=*144385-2* 429
Jussiaea suffruticosa f. *latifrons* IPNI=*144385-2* 429
Jussiaea suffruticosa f. *nerviseta* IPNI=*144385-2* 429
Jussiaea suffruticosa IPNI=*144385-2* 429
Jussiaea suffruticosa proles salicifolia IPNI=*144385-2* 429
Jussiaea suffruticosa subsp. *octonervia* IPNI=*144385-2* 429
Jussiaea suffruticosa var. *acuminata* IPNI=*144385-2* 429
Jussiaea suffruticosa var. *angustifolia* IPNI=*144385-2* 429
Jussiaea suffruticosa var. *brevifolia* IPNI=*144385-2* 429
Jussiaea suffruticosa var. *genuina* IPNI=*144385-2* 429
Jussiaea suffruticosa var. *grandiflora* IPNI=*144385-2* 429
Jussiaea suffruticosa var. *interstans* IPNI=*144385-2* 429
Jussiaea suffruticosa var. *ligustrifolia* IPNI=*144385-2* 429
Jussiaea suffruticosa var. *linearifolia* IPNI=*144385-2* 429
Jussiaea suffruticosa var. *macropoda* IPNI=*144385-2* 429
Jussiaea suffruticosa var. *octofila* IPNI=*144385-2* 429
Jussiaea suffruticosa var. *octonervia* IPNI=*144385-2* 429
Jussiaea suffruticosa var. *sintenisii* IPNI=*144385-2* 429
Jussiaea surinamensis IPNI=*144366-2* 429
Jussiaea tenuifolia IPNI=*613333-1* 429
Jussiaea tetragona IPNI=*144385-2* 429
Jussiaea variabilis IPNI=*144366-2* 429
Jussiaea variabilis var. *affinis* IPNI=*144328-2* 429
Jussiaea variabilis var. *meyeriana* IPNI=*144366-2* 429
Jussiaea variabilis var. *pilosa* IPNI=*144366-2* 429
Jussiaea velutina IPNI=*144366-2* 429
Jussiaea venosa IPNI=*144385-2* 429
Jussiaea ventiliginum IPNI=*144385-2* 429

Jussiaea verticillata IPNI=*613361-1* 429
Jussiaea weddelii IPNI=*613360-1* 429
Jussiaea yacumensis IPNI=*144384-2* 429
Jussieuia herbacea IPNI=*61899-2* 388
Justicia acuminata IPNI=*78920-2* 359
Justicia ajugoides IPNI=*52558-1* 360
Justicia angustibracteata IPNI=*1140399-2* 359
Justicia annua IPNI=*1114607-2* 359
Justicia antidota IPNI=*52612-1* 360
Justicia assurgens IPNI=*1114607-2* 359
Justicia atramentaria IPNI=*133119-2* 360
Justicia bengalensis IPNI=*52558-1* 360
Justicia betonica IPNI=*52612-1* 360
Justicia betonicifolia IPNI=*52612-1* 360
Justicia bicolor IPNI=*48879-1* 360
Justicia bracteolata IPNI=*172440-2* 360
Justicia candelariae IPNI=*1140399-2* 359
Justicia canescens IPNI=*52558-1* 360
Justicia caripensis IPNI=*47308-1* 359
Justicia carnea IPNI=*943504-1* 361
Justicia charthaginensis IPNI=*50468-1* 360
Justicia cheiranthifolia IPNI=*52612-1* 360
Justicia chiapensis IPNI=*1140399-2* 359
Justicia chlorostachya IPNI=*1140399-2* 359
Justicia coccinea IPNI=*943504-1* 361
Justicia colorata IPNI=*48879-1* 360
Justicia colorifera IPNI=*907564-1* 360
Justicia comata IPNI=*78920-2* 359
Justicia corynimorpha IPNI=*50468-1* 360
Justicia costata IPNI=*49672-1* 360
Justicia eustachiana IPNI=*1114607-2* 359
Justicia furcata IPNI=*54389-1* 361
Justicia gangetica IPNI=*45742-1* 359
Justicia gendarussa IPNI=*48879-1* 360
Justicia geniculata IPNI=*47308-1* 359
Justicia ghiesbreghtiana IPNI=*133119-2* 360
Justicia ghiesbreghtii IPNI=*133119-2* 360
Justicia humifusa IPNI=*78920-2* 359
Justicia humilis IPNI=*78920-2* 359
Justicia imbricata IPNI=*1167021-2* 360
Justicia ladanoides IPNI=*1114607-2* 359
Justicia lamiifolia IPNI=*52558-1* 360
Justicia leptospermifolia IPNI=*1114607-2* 359
Justicia liebmanii IPNI=*133119-2* 360
Justicia lindenii IPNI=*50468-1* 360
Justicia lithospermifolia IPNI=*49672-1* 360
Justicia longiflora IPNI=*132826-2* 360
Justicia lucida IPNI=*47308-1* 359
Justicia lupulina IPNI=*224052-2* 361
Justicia lupulina IPNI=*52612-1* 360
Justicia lutea IPNI=*46114-1* 359
Justicia macrophylla IPNI=*315544-2* 359
Justicia maculata IPNI=*77158889-1* 361
Justicia magnifica IPNI=*132826-2* 360
Justicia marrubiastrum IPNI=*49672-1* 360
Justicia martinicensis IPNI=*224052-2* 361
Justicia mexicana IPNI=*1140399-2* 359
Justicia mohintli IPNI=*133119-2* 360
Justicia moricandiana IPNI=*47308-1* 359
Justicia nemorosa IPNI=*56148-1* 361
Justicia neotinctoria IPNI=*907564-1* 360
Justicia nervosa IPNI=*48879-1* 360
Justicia nummulariifolia IPNI=*52558-1* 360
Justicia origanoides IPNI=*52558-1* 360
Justicia oxyphylla IPNI=*46114-1* 359
Justicia pallidior IPNI=*52612-1* 360
Justicia papilionacea IPNI=*56148-1* 361
Justicia pectoralis IPNI=*47290-1* 359
Justicia pectoralis var. *latifolia* IPNI=*47290-1* 359
Justicia pectoralis var. *macrophyllus* IPNI=*47290-1* 359
Justicia pectoralis var. *ovata* IPNI=*47290-1* 359
Justicia pectoralis var. *stenophylla* IPNI=*47290-1* 359
Justicia pentasticha IPNI=*52558-1* 360
Justicia planchonii IPNI=*50468-1* 360
Justicia plumbaginea IPNI=*45742-1* 359
Justicia procumbens IPNI=*51951-1* 360
Justicia pseudobetonica IPNI=*52612-1* 360
Justicia pusilla IPNI=*52558-1* 360
Justicia ramosissima IPNI=*52612-1* 360
Justicia resupinata IPNI=*941389-1* 359
Justicia retusa IPNI=*50468-1* 360
Justicia rostrata IPNI=*941389-1* 359
Justicia sanctae-martae IPNI=*50468-1* 360
Justicia scabra IPNI=*941389-1* 359
Justicia scandens IPNI=*1114607-2* 359
Justicia scarlatina IPNI=*133119-2* 360
Justicia secunda IPNI=*47308-1* 359
Justicia secunda var. *leucantha* IPNI=*47308-1* 359
Justicia sexangularis IPNI=*1114607-2* 359
Justicia spathulata IPNI=*50468-1* 360
Justicia spicata IPNI=*943504-1* 361
Justicia strobiloglossa IPNI=*1140399-2* 359
Justicia stuebelii IPNI=*47290-1* 359
Justicia suaveolens IPNI=*51367-1* 360
Justicia tinctoria IPNI=*907564-1* 360
Justicia tomentosa IPNI=*52558-1* 360
Justicia trinervia IPNI=*52612-1* 360
Justicia tubaeformis IPNI=*172486-2* 360
Justicia tuerckheimiana IPNI=*1140399-2* 359
Justicia uninervis IPNI=*52612-1* 360
Justicia variabilis IPNI=*54389-1* 361
Justicia variegata IPNI=*52612-1* 360
Justicia vestita IPNI=*52558-1* 360

K

Kadalia rotundifolia IPNI=*908858-1* 419
Kaempferia hedychium IPNI=*796836-1* 251
Kallias ovata IPNI=*213524-1* 264
Karatas carolinae IPNI=*169003-2* 229
Karatas guianensis IPNI=*36836-2* 228
Karatas karatas IPNI=*328361-2* 228
Karatas lagopus IPNI=*328361-2* 228
Karatas meyendorffii IPNI=*169003-2* 229
Karatas nidus-puellae IPNI=*122363-1* 228
Karatas olens IPNI=*169003-2* 229
Karatas pinguin IPNI=*319113-2* 228
Karatas plumieri IPNI=*328361-2* 228
Kardanoglyphos nana IPNI=*222164-2* 277
Karkandela × malabarica IPNI=*797595-1* 250
Karwinskia affinis IPNI=*717579-1* 442
Karwinskia biniflora IPNI=*717579-1* 442
Karwinskia parvifolia IPNI=*717579-1* 442
Karwinskia parvifolia var. *pubescens* IPNI=*717579-1* 442
Karwinskia pubescens IPNI=*717579-1* 442
Karwinskia sessilifolia IPNI=*717579-1* 442
Kefersteinia bicallosa IPNI=*154952-2* 220
Kegelia ruderalis IPNI=*202984-1* 261
Kellettia odorata IPNI=*321266-2* 404
Kennedia speciosa IPNI=*948198-1* 334
Kentia forsteriana IPNI=*667434-1* 209
Kentia macarthurii IPNI=*55245-3* 211
Kentia oleracea IPNI=*222805-2* 212
Kepa esculenta IPNI=*527795-1* 213
Kepa fistulosa IPNI=*528046-1* 213
Keraselma cyparissias IPNI=*346205-1* 390
Keraselma lathyris IPNI=*347072-1* 390

Keraselma peplus IPNI=*347724-1* 390
Kerneria bipinnata IPNI=*184769-1* 258
Kerneria coreopsoides IPNI=*32497-2* 258
Kerneria dubia IPNI=*32564-2* 258
Kerneria helianthoides IPNI=*32497-2* 258
Kerneria leucanthema IPNI=*184711-1* 258
Kerneria pilosa IPNI=*32564-2* 258
Kerneria pilosa var. *discoidea* IPNI=*32564-2* 258
Kerneria pilosa var. *radiata* IPNI=*184711-1* 258
Kerneria tetragona IPNI=*32564-2* 258
Ketmia arborea IPNI=*560890-1* 410
Ketmia glandulosa IPNI=*560068-1* 409
Ketmia mutabilis IPNI=*560549-1* 409
Ketmia syriaca IPNI=*560890-1* 410
Ketmia syrorum IPNI=*560890-1* 410
Kibera officinalis IPNI=*289589-1* 278
Kiesera sericea IPNI=*520444-1* 335
Kikuyuochloa clandestina IPNI=*77105989-1* 235
Kimnachia ramulosa f. *angustissima* IPNI=*963240-1* 284
Kimnachia ramulosa IPNI=*963240-1* 284
Kiosmina hispanica IPNI=*456385-1* 372
Kippistia cognata IPNI=*55844-2* 290
Kirganelia floribunda IPNI=*354635-1* 402
Kirganelia salviifolia IPNI=*354635-1* 402
Kleinia adenocalyx IPNI=*1011411-1* 260
Kleinia glandulosa IPNI=*208446-2* 268
Kleinia porophyllum IPNI=*208446-2* 268
Kleinia quadricolor IPNI=*77101543-1* 416
Kleinia radicans IPNI=*1011411-1* 260
Kleinia ruderalis IPNI=*208446-2* 268
Klopstockia cerifera IPNI=*53436-2* 207
Klopstockia coarctata IPNI=*665703-1* 207
Klopstockia interrupta IPNI=*53436-2* 207
Klopstockia parvifrons IPNI=*665695-1* 207
Klopstockia quindiuensis IPNI=*665697-1* 207
Klopstockia vogeliana IPNI=*665703-1* 207
Klugiodendron laetum IPNI=*989371-1* 310
Klugiodendron umbrianum IPNI=*989371-1* 310
Knafia helix IPNI=*301321-2* 404
Knafia purpurea IPNI=*301321-2* 404
Knautia sagittata IPNI=*232743-1* 266
Knavel annuum IPNI=*300494-2* 286
Kniphofia alooides IPNI=*537112-1* 216
Kniphofia bachmannii IPNI=*537112-1* 216
Kniphofia burchellii IPNI=*537112-1* 216
Kniphofia occidentalis IPNI=*537112-1* 216
Kniphofia odorata IPNI=*537112-1* 216
Kobresia cyperina IPNI=*74768-2* 230
Koeleria aristata IPNI=*258855-2* 246
Koeleria dactylis IPNI=*397309-1* 236
Koeleria spicata IPNI=*258855-2* 246
Koeleria subspicata IPNI=*258855-2* 246
Kohleria anonyma IPNI=*381638-1* 367
Kohleria brachycalyx IPNI=*381572-1* 367
Kohleria chiapensis IPNI=*381628-1* 367
Kohleria colorata IPNI=*381638-1* 367
Kohleria cordifolia IPNI=*381638-1* 367
Kohleria eriantha IPNI=*381572-1* 367
Kohleria guazumifolia IPNI=*381388-1* 367
Kohleria hookeri IPNI=*381638-1* 367
Kohleria ignorata IPNI=*381628-1* 367
Kohleria incana IPNI=*381628-1* 367
Kohleria krameriana IPNI=*381628-1* 367
Kohleria lanigera IPNI=*381572-1* 367
Kohleria leucomallos IPNI=*381572-1* 367
Kohleria linkiana IPNI=*381628-1* 367
Kohleria longiflora IPNI=*381572-1* 367
Kohleria longifolia IPNI=*381628-1* 367
Kohleria longifolia var. *petiolaris* IPNI=*381628-1* 367
Kohleria longipedunculata IPNI=*381572-1* 367
Kohleria mollis IPNI=*381572-1* 367
Kohleria moritziana IPNI=*381572-1* 367
Kohleria petiolaris IPNI=*381628-1* 367
Kohleria picta IPNI=*381638-1* 367
Kohleria pilosa IPNI=*381638-1* 367
Kohleria pycnosuzygium IPNI=*381572-1* 367
Kohleria rhodomallos IPNI=*381572-1* 367
Kohleria rubricaulis IPNI=*381572-1* 367
Kohleria rupestris IPNI=*381628-1* 367
Kohleria schiedeana IPNI=*381628-1* 367
Kohleria seemannii IPNI=*381628-1* 367
Kohleria spicata var. *hispida* IPNI=*381628-1* 367
Kohleria spicata var. *wageneri* IPNI=*381628-1* 367
Kohleria straussiana IPNI=*381572-1* 367
Kohleria stricta IPNI=*381638-1* 367
Kohleria tetragona IPNI=*381628-1* 367
Kohleria tomentosa IPNI=*381628-1* 367
Kohleria ventricosa IPNI=*381572-1* 367
Kohleria vestita IPNI=*381572-1* 367
Kohleria wageneri IPNI=*381628-1* 367
Kokera paniculata IPNI=*60077-1* 282
Kolobochilus leiorhachis IPNI=*60478486-2* 361
Komana patula IPNI=*927423-1* 394
Koniga halimifolia IPNI=*1085072-2* 277
Koniga maritima IPNI=*1085072-2* 277
Koniga maroccana IPNI=*1085072-2* 277
Koniga strigulosa IPNI=*1085072-2* 277
Kopsia cochinchinensis IPNI=*82097-1* 343
Kosteletzkya acuminata IPNI=*562981-1* 412
Kosteletzkya asterocarpa IPNI=*133833-2* 410
Kosteletzkya cordata IPNI=*133833-2* 410
Kosteletzkya hastata IPNI=*133833-2* 410
Kosteletzkya hastata IPNI=*30143354-2* 410
Kosteletzkya hispida IPNI=*133833-2* 410
Kosteletzkya pentasperma IPNI=*133833-2* 410
Kosteletzkya sagittata IPNI=*133833-2* 410
Kosteletzkya stellata IPNI=*133833-2* 410
Kosteletzkya tampicensis IPNI=*133833-2* 410
Kosteletzkya violacea IPNI=*133833-2* 410
Kowalewskia integerrima IPNI=*327610-1* 298
Krugella hartii IPNI=*788972-1* 307
Krugiodendron acuminatum IPNI=*133956-2* 442
Kua domestica IPNI=*796451-1* 250
Kuhlia mollis IPNI=*779396-1* 402
Kuhnia spicata IPNI=*230634-1* 265
Kunokale carneum IPNI=*694526-1* 289
Kunthia deppei IPNI=*54865-2* 207
Kunthia montana IPNI=*54901-2* 207
Kurkas populifolium IPNI=*70599-2* 389
Kutchubaea speciosa IPNI=*133858-2* 352
Kyllinga aphylla IPNI=*1009998-1* 231
Kyllinga blepharinota IPNI=*77144505-1* 230
Kyllinga brevifolia IPNI=*30313989-2* 230
Kyllinga bulbosa IPNI=*304256-1* 230
Kyllinga caespitosa IPNI=*77144505-1* 230
Kyllinga caespitosa subvar. *debilis* IPNI=*77144505-1* 230
Kyllinga caespitosa var. *elatior* IPNI=*77144505-1* 230
Kyllinga caespitosa var. *pumila* IPNI=*77144505-1* 230
Kyllinga capitata IPNI=*74934-2* 231
Kyllinga cephalotes IPNI=*77140538-1* 231
Kyllinga colorata IPNI=*311841-1* 232
Kyllinga curvispiculosa IPNI=*77140538-1* 231
Kyllinga cyperina IPNI=*304256-1* 230
Kyllinga erecta var. *africana* IPNI=*74934-2* 231
Kyllinga flexuosa IPNI=*77144505-1* 230
Kyllinga fraterna IPNI=*77144505-1* 230

Kyllinga globosa IPNI=*74934-2* 231
Kyllinga gracilis IPNI=*77140538-1* 231
Kyllinga hortensis IPNI=*77144505-1* 230
Kyllinga incompleta IPNI=*74768-2* 230
Kyllinga intermedia IPNI=*77140538-1* 231
Kyllinga mindorensis IPNI=*77140538-1* 231
Kyllinga monocephala IPNI=*311841-1* 232
Kyllinga monocephala IPNI=*77140538-1* 231
Kyllinga monocephala IPNI=*77144505-1* 230
Kyllinga naumanniana IPNI=*77144505-1* 230
Kyllinga nemoralis IPNI=*77140538-1* 231
Kyllinga nemoralis var. *subnemoralis* IPNI=*77140538-1* 231
Kyllinga obtusata IPNI=*74934-2* 231
Kyllinga obtusata var. *aphylla* IPNI=*74934-2* 231
Kyllinga obtusata var. *cylindrostachya* IPNI=*74934-2* 231
Kyllinga obtusata var. *subaphylla* IPNI=*74934-2* 231
Kyllinga odorata IPNI=*77144505-1* 230
Kyllinga peruviana IPNI=*74934-2* 231
Kyllinga peruviana var. *foliata* IPNI=*1009998-1* 231
Kyllinga pluristaminea IPNI=*30313989-2* 230
Kyllinga polycephala IPNI=*74851-2* 231
Kyllinga pumila IPNI=*77144505-1* 230
Kyllinga pumila var. *elatior* IPNI=*77144505-1* 230
Kyllinga pungens IPNI=*74934-2* 231
Kyllinga repens IPNI=*74934-2* 231
Kyllinga rigida IPNI=*304256-1* 230
Kyllinga rigida IPNI=*74934-2* 231
Kyllinga rigidula IPNI=*77144505-1* 230
Kyllinga scirpina IPNI=*74851-2* 231
Kyllinga sesquiflora IPNI=*75123-2* 231
Kyllinga stricta IPNI=*74934-2* 231
Kyllinga tenuis IPNI=*74934-2* 231
Kyllinga tibialis IPNI=*1009998-1* 231
Kyllinga uncinata IPNI=*77144505-1* 230
Kyllinga vaginata IPNI=*74934-2* 231
Kyllinga viridiflora IPNI=*77144505-1* 230
Kyrstenia aborginum IPNI=*6815-2* 256
Kyrstenia altissima IPNI=*6815-2* 256
Kyrstenia borealis IPNI=*6815-2* 256
Kyrstenia ciliata IPNI=*6988-2* 256
Kyrstenia guadalupensis IPNI=*105109-2* 263
Kyrtanthus longiflorus IPNI=*761025-1* 355

L
Labatia bilocularis IPNI=*209700-2* 306
Labatia caimito IPNI=*788845-1* 306
Labatia glomorata IPNI=*788965-1* 307
Labatia laevigata IPNI=*789009-1* 307
Labatia lasiocarpa IPNI=*788845-1* 306
Labatia macrocarpa IPNI=*789045-1* 307
Labatia pedunculata IPNI=*788972-1* 307
Labatia reticulata IPNI=*788845-1* 306
Labatia torta IPNI=*789233-1* 308
Labatia tovarensis IPNI=*788972-1* 307
Labradia pruriens IPNI=*509275-1* 328
Labradia urens IPNI=*1197939-2* 328
Lacepedea pinnata IPNI=*785554-1* 292
Lachemilla achilleifolia IPNI=*720867-1* 443
Lachemilla aphanoides IPNI=*720378-1* 442
Lachemilla aphanoides var. *tripartita* IPNI=*720378-1* 442
Lachemilla costaricensis IPNI=*8135-2* 443
Lachemilla fulvescens IPNI=*8079-2* 442
Lachemilla galioides IPNI=*720586-1* 442
Lachemilla glandulosa IPNI=*720378-1* 442
Lachemilla hirta IPNI=*8090-2* 442
Lachemilla hispidula IPNI=*8091-2* 442
Lachemilla holosericea IPNI=*8093-2* 442
Lachemilla hultenii IPNI=*720378-1* 442
Lachemilla imbricata IPNI=*720798-1* 442
Lachemilla killipii IPNI=*8097-2* 442
Lachemilla laxa IPNI=*8135-2* 443
Lachemilla moritziana IPNI=*8105-2* 442
Lachemilla mutisii IPNI=*8106-2* 442
Lachemilla nivalis IPNI=*720798-1* 442
Lachemilla orbiculata IPNI=*720816-1* 442
Lachemilla paludicola IPNI=*8117-2* 442
Lachemilla pectinata IPNI=*720846-1* 442
Lachemilla polylepis IPNI=*720877-1* 443
Lachemilla procumbens IPNI=*8135-2* 443
Lachemilla radicans IPNI=*720586-1* 442
Lachemilla subalpestris IPNI=*720378-1* 442
Lachemilla tripartita IPNI=*720378-1* 442
Lachenalia ramosa IPNI=*539926-1* 216
Lachnostoma maritimum IPNI=*98854-1* 341
Lacinaria elongata IPNI=*230634-1* 265
Lacinaria spicata f. *albiflora* IPNI=*230634-1* 265
Lacinaria spicata IPNI=*230634-1* 265
Lacinaria spicata var. *albiflora* IPNI=*230634-1* 265
Lacinaria spicata var. *foliacea* IPNI=*230634-1* 265
Lacinaria spicata var. *montana* IPNI=*230634-1* 265
Lacinaria spicata var. *pumila* IPNI=*230634-1* 265
Lacinaria vittata IPNI=*230634-1* 265
Lacistema aggregatum var. *elongatum* IPNI=*134457-2* 394
Lacistema alterum IPNI=*166220-2* 339
Lacistema angustum IPNI=*134457-2* 394
Lacistema berteroanum IPNI=*166220-2* 339
Lacistema bolivianum IPNI=*134457-2* 394
Lacistema coriaceum IPNI=*134457-2* 394
Lacistema curtum IPNI=*134457-2* 394
Lacistema elongatum IPNI=*134457-2* 394
Lacistema floribundum IPNI=*134457-2* 394
Lacistema guyanense IPNI=*134457-2* 394
Lacistema myricoides IPNI=*134457-2* 394
Lacistema myricoides var. *stipitatum* IPNI=*134457-2* 394
Lacistema oblongum IPNI=*134457-2* 394
Lacistema occidentale IPNI=*134457-2* 394
Lacistema orinocense IPNI=*134457-2* 394
Lacistema pacificum IPNI=*134457-2* 394
Lacistema poeppigii IPNI=*134457-2* 394
Lacistema purpureum IPNI=*134457-2* 394
Lacistema rosidiscum IPNI=*134457-2* 394
Lacistema weberbaueri IPNI=*134457-2* 394
Lacmellea aculeata var. *inermis* IPNI=*134481-2* 341
Lacmellea aculeata var. *surinamensis* IPNI=*134481-2* 341
Lacmellea arborescens var. *monosperma* IPNI=*79531-1* 341
Lacmellea arborescens var. *muelleri* IPNI=*79531-1* 341
Lacmellea armata IPNI=*79551-1* 341
Lacmellea grandiflora IPNI=*134498-2* 341
Lacmellea lactescens IPNI=*79551-1* 341
Lacmellea lucida IPNI=*945268-1* 342
Lacmellea ramosissima var. *hirtella* IPNI=*79547-1* 341
Lactuca arabica IPNI=*229022-1* 264
Lactuca capitata IPNI=*228239-1* 264
Lactuca crispa IPNI=*228239-1* 264
Lactuca esculenta IPNI=*228239-1* 264
Lactuca goraeensis IPNI=*229022-1* 264
Lactuca heyneana IPNI=*229022-1* 264
Lactuca intybacea IPNI=*229022-1* 264
Lactuca japonica IPNI=*1124129-2* 272
Lactuca laciniata IPNI=*228239-1* 264
Lactuca napifolia IPNI=*1124129-2* 272
Lactuca nubica IPNI=*229022-1* 264
Lactuca octophylla IPNI=*229022-1* 264
Lactuca palmata IPNI=*228239-1* 264
Lactuca pinnatifida IPNI=*229022-1* 264
Lactuca remotiflora IPNI=*229022-1* 264

Lactuca runcinata IPNI=*229022-1* 264
Lactuca sativa convar. *asparagina* IPNI=*228239-1* 264
Lactuca sativa subsp. *capitata* IPNI=*228239-1* 264
Lactuca sativa subsp. *crispa* IPNI=*228239-1* 264
Lactuca sativa subsp. *minii* IPNI=*228239-1* 264
Lactuca sativa subsp. *romana* IPNI=*228239-1* 264
Lactuca sativa var. *angustana* IPNI=*228239-1* 264
Lactuca sativa var. *asparagina* IPNI=*228239-1* 264
Lactuca sativa var. *capitata* IPNI=*228239-1* 264
Lactuca sativa var. *crispa* IPNI=*228239-1* 264
Lactuca sativa var. *yemensis* IPNI=*228239-1* 264
Lactuca scariola var. *sativa* IPNI=*228239-1* 264
Lactuca schimperi IPNI=*229022-1* 264
Lactuca taquetii IPNI=*1124129-2* 272
Lactuca taraxacum IPNI=*1124129-2* 272
Lactuca virgata IPNI=*229022-1* 264
Lacunaria acreana IPNI=*134578-2* 397
Lacunaria colonensis IPNI=*429118-1* 397
Lacunaria coriacea IPNI=*134578-2* 397
Lacunaria minor IPNI=*134578-2* 397
Lacunaria panamensis IPNI=*134573-2* 397
Lacunaria pauciflora IPNI=*134573-2* 397
Lacunaria silvatica IPNI=*134577-2* 397
Lacunaria spruceana IPNI=*429129-1* 397
Ladenbergia crassifolia IPNI=*1215741-2* 352
Ladenbergia gavanensis IPNI=*1106418-2* 352
Ladenbergia hookeriana IPNI=*134610-2* 352
Ladenbergia magnifolia IPNI=*1106418-2* 352
Ladenbergia magnifolia var. *rostrata* IPNI=*1106418-2* 352
Ladenbergia malacophylla IPNI=*754813-1* 352
Ladenbergia nitida IPNI=*1106418-2* 352
Ladenbergia pedunculata IPNI=*77065717-1* 348
Ladenbergia ulei IPNI=*134610-2* 352
Laennecia parvifolia IPNI=*228376-1* 264
Laennecia pinnatifida IPNI=*228375-1* 264
Laetia acuminata IPNI=*77197250-1* 403
Laetia americana IPNI=*77197250-1* 403
Laetia apetala IPNI=*77197250-1* 403
Laetia apetala var. *pubescens* IPNI=*77197250-1* 403
Laetia casearioides IPNI=*47957-2* 403
Laetia crenata IPNI=*48022-2* 403
Laetia glandulosa IPNI=*28885-2* 402
Laetia guidonia IPNI=*48022-2* 403
Laetia longifolia IPNI=*48022-2* 403
Laetia procera IPNI=*47957-2* 403
Laetia serrulata IPNI=*77197250-1* 403
Lafoensia mexicana IPNI=*553519-1* 417
Lafoensia speciosa IPNI=*1054383-2* 417
Lagascea campestris IPNI=*134795-2* 264
Lagascea kunthiana IPNI=*134795-2* 264
Lagascea parvifolia IPNI=*134795-2* 264
Lagenaria bicornuta IPNI=*134809-2* 294
Lagenaria cochinchinensis IPNI=*134809-2* 294
Lagenaria hispida IPNI=*134809-2* 294
Lagenaria idolatrica IPNI=*134809-2* 294
Lagenaria lagenaria IPNI=*134809-2* 294
Lagenaria leucantha IPNI=*134809-2* 294
Lagenaria microcarpa IPNI=*134809-2* 294
Lagenaria siceraria f. *depressa* IPNI=*134809-2* 294
Lagenaria siceraria f. *microcarpa* IPNI=*134809-2* 294
Lagenaria siceraria subsp. *asiatica* IPNI=*134809-2* 294
Lagenaria siceraria var. *laevisperma* IPNI=*134809-2* 294
Lagenaria vittata IPNI=*134809-2* 294
Lagenaria vulgaris IPNI=*134809-2* 294
Lagenaria vulgaris var. *clavata* IPNI=*134809-2* 294
Lagenaria vulgaris var. *gourda* IPNI=*134809-2* 294
Lagenaria vulgaris var. *viscosa* IPNI=*134809-2* 294
Lagerstroemia augusta IPNI=*553618-1* 417
Lagerstroemia chinensis IPNI=*553576-1* 417
Lagerstroemia elegans IPNI=*553576-1* 417
Lagerstroemia flos-reginae IPNI=*553618-1* 417
Lagerstroemia hirsuta IPNI=*553618-1* 417
Lagerstroemia indica var. *grafii* IPNI=*553576-1* 417
Lagerstroemia major IPNI=*553618-1* 417
Lagerstroemia minor IPNI=*553576-1* 417
Lagerstroemia munchausia IPNI=*553618-1* 417
Lagerstroemia munchhausia IPNI=*553618-1* 417
Lagerstroemia plicifolia IPNI=*553618-1* 417
Lagerstroemia pulchra IPNI=*553576-1* 417
Lagerstroemia punctata IPNI=*553618-1* 417
Lagerstroemia reginae IPNI=*553618-1* 417
Laggera aurita IPNI=*211317-2* 268
Laggera lyrata IPNI=*211317-2* 268
Laggera viscosa IPNI=*211317-2* 268
Lagopus lanceolatus IPNI=*321285-2* 376
Lagopus pratensis IPNI=*523575-1* 335
Lagopus timbali IPNI=*321285-2* 376
Laguncularia glabriflora IPNI=*170672-1* 416
Laguncularia martii IPNI=*170672-1* 416
Laguncularia obovata IPNI=*170672-1* 416
Laguncularia racemosa f. *longifolia* IPNI=*170672-1* 416
Laguncularia racemosa var. *glabriflora* IPNI=*170672-1* 416
Lagurus dalmaticus IPNI=*406799-1* 240
Lagurus dimorphus IPNI=*406799-1* 240
Lagurus freynii IPNI=*406799-1* 240
Lagurus humilis IPNI=*406799-1* 240
Lagurus longifolius IPNI=*406799-1* 240
Lagurus nitens IPNI=*406799-1* 240
Lagurus ovatus var. *subglaber* IPNI=*406799-1* 240
Lagurus paniculatus IPNI=*396972-1* 236
Lagurus siculus IPNI=*406799-1* 240
Lalda communis IPNI=*228732-1* 264
Lamarckia hookeriana IPNI=*406803-1* 240
Lamiastrum galeobdolon IPNI=*60452701-2* 370
Lamiella amplexicaulis IPNI=*448797-1* 370
Lamiopsis amplexicaulis IPNI=*448797-1* 370
Lamium moluccella IPNI=*451989-1* 371
Lamourouxia loxensis IPNI=*804205-1* 375
Lamourouxia serratifolia IPNI=*804205-1* 375
Lampranthus blandus IPNI=*362274-1* 280
Lampranthus incurvus IPNI=*362274-1* 280
Lampranthus macrocarpus IPNI=*362323-1* 280
Lampranthus macrosepalus IPNI=*362323-1* 280
Lampranthus mutans IPNI=*362301-1* 280
Lampranthus roseus IPNI=*362274-1* 280
Lamprophyllum gardnerianum IPNI=*972610-1* 385
Landersia pervaga IPNI=*293293-1* 294
Landolphia guianensis IPNI=*80734-1* 342
Landoltia punctata IPNI=*242868-2* 202
Lanessania turbinata IPNI=*317009-2* 441
Langsdorffia janeirensis IPNI=*315558-2* 448
Langsdorffia moritziana IPNI=*315558-2* 448
Langsdorffia rubiginosa IPNI=*315558-2* 448
Languas purpurata IPNI=*795383-1* 250
Languas schumanniana IPNI=*872083-1* 250
Languas speciosa IPNI=*872083-1* 250
Lansbergia caracasana IPNI=*441642-1* 217
Lansbergia martinicensis IPNI=*441642-1* 217
Lansbergia meridensis IPNI=*441642-1* 217
Lantana × mutabilis IPNI=*60444377-2* 378
Lantana achyranthifolia f. *grandifolia* IPNI=*324301-2* 378
Lantana achyranthifolia f. *lilacina* IPNI=*324301-2* 378
Lantana alba IPNI=*141020-2* 378
Lantana annua IPNI=*135333-2* 378
Lantana aristata var. *cabrerae* IPNI=*863236-1* 378
Lantana arubensis IPNI=*135206-2* 378

Lantana bernardinensis IPNI=*324301-2* 378
Lantana bipinnatifida IPNI=*135333-2* 378
Lantana cabrerae IPNI=*324301-2* 378
Lantana canescens f. *grandifolia* IPNI=*863236-1* 378
Lantana canescens f. *parvifolia* IPNI=*863236-1* 378
Lantana canescens f. *pluripedunculata* IPNI=*863236-1* 378
Lantana canescens var. *integrifolia* IPNI=*863236-1* 378
Lantana celtidifolia IPNI=*135333-2* 378
Lantana citrosa IPNI=*863236-1* 378
Lantana costaricensis IPNI=*30105602-2* 378
Lantana cubensis IPNI=*135333-2* 378
Lantana cujabensis var. *parvifolia* IPNI=*324302-2* 378
Lantana cujabensis var. *punctata* IPNI=*287900-2* 378
Lantana cujabensis var. *scabrifolia* IPNI=*324302-2* 378
Lantana cuneatifolia IPNI=*141020-2* 378
Lantana czermakii IPNI=*324303-2* 378
Lantana decumbens IPNI=*135260-2* 378
Lantana delicatissima IPNI=*135260-2* 378
Lantana dubia IPNI=*135333-2* 378
Lantana fiebrigii IPNI=*135333-2* 378
Lantana fiebrigii var. *puberulenta* IPNI=*135333-2* 378
Lantana fucata f. *albiflora* IPNI=*324303-2* 378
Lantana fucata var. *antillana* IPNI=*324303-2* 378
Lantana fucata var. *longipes* IPNI=*324303-2* 378
Lantana fuscata IPNI=*324303-2* 378
Lantana geminata IPNI=*141020-2* 378
Lantana geroldiana IPNI=*30105602-2* 378
Lantana gogchana IPNI=*135333-2* 378
Lantana hintonii IPNI=*30105602-2* 378
Lantana hirta f. *caerulea* IPNI=*30105602-2* 378
Lantana hirta var. *pubescens* IPNI=*30105602-2* 378
Lantana incana IPNI=*135206-2* 378
Lantana inconspicua IPNI=*324303-2* 378
Lantana involucrata f. *candida* IPNI=*135206-2* 378
Lantana involucrata f. *leucocarpa* IPNI=*135206-2* 378
Lantana involucrata f. *rubella* IPNI=*135206-2* 378
Lantana involucrata var. *floridana* IPNI=*135206-2* 378
Lantana involucrata var. *kuhnholtziana* IPNI=*135206-2* 378
Lantana involucrata var. *odorata* IPNI=*135206-2* 378
Lantana involucrata var. *socorrensis* IPNI=*135206-2* 378
Lantana involucrata var. *velutina* IPNI=*135206-2* 378
Lantana lamiifolia IPNI=*863236-1* 378
Lantana lanuginosa IPNI=*135206-2* 378
Lantana lilacina IPNI=*324303-2* 378
Lantana lilacina var. *media* IPNI=*324303-2* 378
Lantana lilacina var. *mollis* IPNI=*324303-2* 378
Lantana lilacina var. *parvifolia* IPNI=*324303-2* 378
Lantana lippioides IPNI=*141020-2* 378
Lantana lippioides IPNI=*863236-1* 378
Lantana macropoda f. *albiflora* IPNI=*324301-2* 378
Lantana macropoda f. *parvula* IPNI=*324301-2* 378
Lantana macropoda IPNI=*324301-2* 378
Lantana macropodioides IPNI=*324301-2* 378
Lantana malabarica IPNI=*141020-2* 378
Lantana maxima f. *alba* IPNI=*135333-2* 378
Lantana maxima IPNI=*135333-2* 378
Lantana mearnsii IPNI=*135333-2* 378
Lantana microcephala IPNI=*863236-1* 378
Lantana mollissima IPNI=*141020-2* 378
Lantana montevidensis f. *albiflora* IPNI=*135260-2* 378
Lantana odorata IPNI=*135206-2* 378
Lantana odorata IPNI=*141020-2* 378
Lantana odorata var. *berlandieri* IPNI=*135206-2* 378
Lantana origanoides IPNI=*863752-1* 379
Lantana origanoides var. *sampaionis* IPNI=*863752-1* 379
Lantana parvifolia IPNI=*135206-2* 378
Lantana pilosa IPNI=*135333-2* 378
Lantana pittieri IPNI=*135333-2* 378
Lantana purpurea IPNI=*324301-2* 378
Lantana recta IPNI=*324303-2* 378
Lantana reticulata IPNI=*135206-2* 378
Lantana riedeliana IPNI=*324302-2* 378
Lantana rugulosa f. *albiflora* IPNI=*863431-1* 378
Lantana rugulosa var. *parvipedunculata* IPNI=*863431-1* 378
Lantana salviifolia IPNI=*324303-2* 378
Lantana selloi IPNI=*135260-2* 378
Lantana sellowiana f. *planifolia* IPNI=*135260-2* 378
Lantana sellowiana f. *rugosa* IPNI=*135260-2* 378
Lantana sellowiana IPNI=*135260-2* 378
Lantana sellowiana var. *lanceolata* IPNI=*135260-2* 378
Lantana tenuifolia IPNI=*324302-2* 378
Lantana teucriifolia IPNI=*30105602-2* 378
Lantana trifolia f. *albiflora* IPNI=*135333-2* 378
Lantana trifolia f. *brevipes* IPNI=*135333-2* 378
Lantana trifolia f. *congolensis* IPNI=*135333-2* 378
Lantana trifolia f. *hirsuta* IPNI=*135333-2* 378
Lantana trifolia f. *pluripedunculata* IPNI=*135333-2* 378
Lantana trifolia f. *rubierensis* IPNI=*135333-2* 378
Lantana trifolia var. *geminata* IPNI=*135333-2* 378
Lantana trifolia var. *quadriverticillata* IPNI=*135333-2* 378
Lantana trifolia var. *rigidiuscula* IPNI=*135333-2* 378
Lantana trifolia var. *vulgata* IPNI=*135333-2* 378
Lantana undulata var. *saltensis* IPNI=*135333-2* 378
Lapathum acetosella IPNI=*224391-2* 289
Lapathum arvense IPNI=*224391-2* 289
Lapathum conglomeratum IPNI=*697003-1* 289
Lapathum crispum IPNI=*224413-2* 289
Lapathum glomeratum IPNI=*697003-1* 289
Lapathum hortense IPNI=*697388-1* 290
Lapathum obtusifolium IPNI=*310739-2* 290
Laphangium luteoalbum IPNI=*212994-1* 264
Laplacea cameliifolia IPNI=*112597-2* 309
Laplacea camenioides IPNI=*112597-2* 309
Laplacea fruticosa IPNI=*112597-2* 309
Laplacea fruticosa var. *chimantae* IPNI=*112597-2* 309
Laplacea fruticosa var. *pulcherrima* IPNI=*112597-2* 309
Laplacea fruticosa var. *sericea* IPNI=*112597-2* 309
Laplacea fruticosa var. *symplocoides* IPNI=*112597-2* 309
Laplacea inaequalilatera IPNI=*112597-2* 309
Laplacea intermedia IPNI=*112597-2* 309
Laplacea obovata IPNI=*112597-2* 309
Laplacea praemorsa IPNI=*112597-2* 309
Laplacea pubescens IPNI=*112597-2* 309
Laplacea pubescens var. *camelliifolia* IPNI=*112597-2* 309
Laplacea pubescens var. *minor* IPNI=*112597-2* 309
Laplacea pubescens var. *subcaudata* IPNI=*112597-2* 309
Laplacea pulcherrima IPNI=*112597-2* 309
Laplacea quinoderma IPNI=*112597-2* 309
Laplacea raimondiana IPNI=*112597-2* 309
Laplacea semiserrata IPNI=*112597-2* 309
Laplacea semiserrata var. *communis* IPNI=*112597-2* 309
Laplacea semiserrata var. *microphylla* IPNI=*112597-2* 309
Laplacea semiserrata var. *obovata* IPNI=*112597-2* 309
Laplacea semiserrata var. *sericea* IPNI=*112597-2* 309
Laplacea semiserrata var. *typica* IPNI=*112597-2* 309
Laplacea speciosa IPNI=*112597-2* 309
Laplacea speciosa var. *intermedia* IPNI=*112597-2* 309
Laplacea symplocoides IPNI=*112597-2* 309
Laplacea tomentosa IPNI=*112597-2* 309
Laplacea tomentosa var. *glabrata* IPNI=*112597-2* 309
Laplacea tomentosa var. *typica* IPNI=*112597-2* 309
Laportea bathiei IPNI=*135476-2* 446
Laportea glandulosa IPNI=*135476-2* 446
Lappago aliena IPNI=*211125-2* 244
Lappago berteroniana IPNI=*256481-2* 246
Lappago occidentalis IPNI=*256481-2* 246

Lappago oplismenoides IPNI=*211125-2* 244
Lappago phleoides IPNI=*256481-2* 246
Lappago racemosa var. *erecta* IPNI=*256481-2* 246
Lapsana capillaris IPNI=*199753-1* 260
Laricorchis fulgens IPNI=*155059-2* 220
Larnastyra claytonii IPNI=*457495-1* 373
Larnastyra verbenaca IPNI=*457495-1* 373
Laseguea bicolor IPNI=*152052-2* 341
Laserpitium ternatum IPNI=*170084-2* 252
Lasiacis acuminata IPNI=*135707-2* 240
Lasiacis guaranitica IPNI=*135707-2* 240
Lasiacis patentiflora IPNI=*135707-2* 240
Lasiacis sorghoidea IPNI=*135707-2* 240
Lasiacis sorghoidea var. *patentiflora* IPNI=*135707-2* 240
Lasiacis swartziana IPNI=*135707-2* 240
Lasiandra ledifolia IPNI=*566862-1* 418
Lasiandra macrantha IPNI=*77206410-1* 424
Lasiandra urvilleana IPNI=*77206410-1* 424
Lasiocephalus doryphyllus IPNI=*6086-2* 256
Lasiocephalus patens IPNI=*6096-2* 256
Lasiocorys poggeana IPNI=*447982-1* 370
Lasiolepis aquatica IPNI=*335722-1* 233
Lasionema cinchonioides IPNI=*755513-1* 352
Lasionema glabrescens IPNI=*755513-1* 352
Lasionema roseum IPNI=*755513-1* 352
Lasiostoma bredemeyeri IPNI=*547083-1* 346
Lasiostoma cirrhosa IPNI=*547220-1* 346
Lasiostoma curare IPNI=*547220-1* 346
Lasiostoma divaricata IPNI=*547220-1* 346
Lasiostoma glabrum IPNI=*547083-1* 346
Lasiostoma rouhamon IPNI=*547220-1* 346
Lasiostoma ruhamon IPNI=*547220-1* 346
Lass cordifolia IPNI=*562981-1* 412
Lass fruticosa IPNI=*185868-2* 412
Lass sepium IPNI=*186058-2* 412
Lass spinifex IPNI=*186083-2* 412
Lassonia heptapeta IPNI=*554678-1* 190
Lassonia quinquepeta IPNI=*554759-1* 190
Lastrea adiantiformis IPNI=*224543-2* 176
Lastrea aspidioides IPNI=*224543-2* 176
Lastrea caduca IPNI=*1148359-2* 175
Lastrea cosmopolita IPNI=*1019960-2* 174
Lastrea filix-femina IPNI=*30055316-2* 174
Lastrea filix-mas var. *paleacea* IPNI=*17392950-1* 175
Lastrea flaccida IPNI=*1019960-2* 174
Lastrea funesta IPNI=*286436-2* 177
Lastrea kunthii IPNI=*251400-2* 174
Lastrea malaccensis IPNI=*17460390-1* 174
Lastrea multijuga IPNI=*17460390-1* 174
Lastrea normalis IPNI=*251400-2* 174
Lastrea oligophlebia IPNI=*1019960-2* 174
Lastrea semicordata IPNI=*1148359-2* 175
Lastrea tenericaulis IPNI=*1019960-2* 174
Lastrea torresiana IPNI=*1019960-2* 174
Latania chinensis IPNI=*668021-1* 210
Laterifissum rivulare IPNI=*311155-2* 286
Lathyrus armitageanus IPNI=*323132-2* 325
Lathyrus asiaticus IPNI=*501990-1* 325
Lathyrus azureus IPNI=*501990-1* 325
Lathyrus cyprius IPNI=*501909-1* 325
Lathyrus lens IPNI=*524872-1* 336
Lathyrus maccaguenii IPNI=*501909-1* 325
Lathyrus magellanicus var. *normalis* IPNI=*501845-1* 325
Lathyrus odoratus-zeylanicus IPNI=*501909-1* 325
Lathyrus oleraceus subsp. *biflorus* IPNI=*501912-1* 325
Lathyrus oleraceus subsp. *oleraceus* IPNI=*501912-1* 325
Lathyrus oleraceus var. *arvense* IPNI=*501912-1* 325
Lathyrus sativus f. *azureus* IPNI=*501990-1* 325
Lathyrus sativus f. *biflorus* IPNI=*501990-1* 325
Lathyrus sativus f. *chlorospermus* IPNI=*501990-1* 325
Lathyrus sativus subsp. *albus* IPNI=*501990-1* 325
Lathyrus sativus var. *comitans* IPNI=*501990-1* 325
Lathyrus sativus var. *depressus* IPNI=*501990-1* 325
Lathyrus sativus var. *orbiculatus* IPNI=*501990-1* 325
Lathyrus sativus var. *parviflorus* IPNI=*501990-1* 325
Lathyrus sativus var. *pisiformis* IPNI=*501990-1* 325
Lathyrus sativus var. *platyspermus* IPNI=*501990-1* 325
Lathyrus sativus var. *pulchrus* IPNI=*501990-1* 325
Lathyrus sativus var. *variegatus* IPNI=*501990-1* 325
Lathyrus sativus var. *violascens* IPNI=*501990-1* 325
Lathyrus schaeferi IPNI=*501912-1* 325
Lathyrus spectabilis IPNI=*486606-1* 315
Lathyrus trigonus IPNI=*323132-2* 325
Latreillea dicarpa IPNI=*127605-2* 264
Latreillea glabrata IPNI=*127605-2* 264
Latreillea serrata IPNI=*127605-2* 264
Laugieria edechia IPNI=*752006-1* 351
Laugieria odorata IPNI=*752006-1* 351
Launaea goraeensis IPNI=*229022-1* 264
Launaea kuriensis IPNI=*229022-1* 264
Launaea remotiflora IPNI=*229022-1* 264
Laurentia erinoides IPNI=*143240-1* 273
Laurentia longiflora IPNI=*1133957-2* 273
Laurentia maximiliana IPNI=*79151-2* 273
Laurentia michoacana IPNI=*79151-2* 273
Laurentia michoacana var. *ovatifolia* IPNI=*79151-2* 273
Laurentia micrantha IPNI=*79151-2* 273
Laurentia micrantha var. *longibracteata* IPNI=*79151-2* 273
Laurentia micrantha var. *ovatifolia* IPNI=*79151-2* 273
Laurentia ovatifolia IPNI=*79151-2* 273
Laurentia pedunculata IPNI=*79151-2* 273
Lauro-cerasus opaca IPNI=*730059-1* 443
Laurus acutifolia IPNI=*167746-2* 183
Laurus aestivalis IPNI=*466894-1* 184
Laurus americana IPNI=*76687-2* 415
Laurus angusta IPNI=*465049-1* 183
Laurus armeniaca IPNI=*465248-1* 183
Laurus aurantiodora IPNI=*171526-2* 184
Laurus belloto IPNI=*191376-2* 185
Laurus bofo IPNI=*467161-1* 184
Laurus borbonia IPNI=*974281-1* 184
Laurus caerulea IPNI=*190613-2* 185
Laurus calycina IPNI=*463336-1* 183
Laurus camphora IPNI=*463336-1* 183
Laurus camphorifera IPNI=*463336-1* 183
Laurus canellila IPNI=*13816-2* 182
Laurus cassia IPNI=*463752-1* 183
Laurus cinnamifera IPNI=*463752-1* 183
Laurus cinnamomea IPNI=*463752-1* 183
Laurus cinnamomoides IPNI=*77198017-1* 183
Laurus cinnamomum IPNI=*463752-1* 183
Laurus coerulea IPNI=*190613-2* 185
Laurus coruscans IPNI=*77198011-1* 183
Laurus culitlaban IPNI=*463752-1* 183
Laurus discolor IPNI=*1203274-2* 184
Laurus elongata IPNI=*77173628-1* 182
Laurus exaltata IPNI=*1146025-2* 184
Laurus exaltata IPNI=*466885-1* 184
Laurus floribunda IPNI=*1146025-2* 184
Laurus globosa IPNI=*167822-2* 184
Laurus globosa IPNI=*974281-1* 184
Laurus gracilis IPNI=*463336-1* 183
Laurus hihua IPNI=*974281-1* 184
Laurus hundensis IPNI=*77173628-1* 182
Laurus indica IPNI=*325643-2* 185
Laurus laevigata IPNI=*190613-2* 185

Laurus lanceifolia IPNI=*466885-1* 184
Laurus leptobotra IPNI=*171727-2* 184
Laurus leucoxylon IPNI=*77173628-1* 182
Laurus limbosa IPNI=*465312-1* 183
Laurus lingui IPNI=*191376-2* 185
Laurus longifolia IPNI=*171526-2* 184
Laurus martinicensis IPNI=*171727-2* 184
Laurus membranacea IPNI=*466773-1* 184
Laurus montana IPNI=*463752-1* 183
Laurus montana IPNI=*77173628-1* 182
Laurus multiflora IPNI=*466626-1* 183
Laurus multiglandulosa IPNI=*1146025-2* 184
Laurus nutans IPNI=*171727-2* 184
Laurus nutans IPNI=*77173628-1* 182
Laurus ocotea IPNI=*467336-1* 184
Laurus opifera IPNI=*171526-2* 184
Laurus paniculata IPNI=*77173628-1* 182
Laurus parviflora IPNI=*465312-1* 183
Laurus parviflora IPNI=*77173628-1* 182
Laurus pendula IPNI=*1057986-2* 182
Laurus persea IPNI=*325643-2* 185
Laurus persea IPNI=*467161-1* 184
Laurus pichurim IPNI=*1171177-2* 184
Laurus pruinosa IPNI=*190613-2* 185
Laurus puberula IPNI=*1032261-2* 185
Laurus purpurea IPNI=*77158867-1* 183
Laurus pyrifolia IPNI=*190613-2* 185
Laurus quixos IPNI=*77198017-1* 183
Laurus retroflexa IPNI=*1146025-2* 184
Laurus rigida IPNI=*463752-1* 183
Laurus salicifolia IPNI=*1146025-2* 184
Laurus sericea IPNI=*464535-1* 183
Laurus splendens IPNI=*464535-1* 183
Laurus sulcata IPNI=*31062-2* 182
Laurus surca IPNI=*467336-1* 184
Laurus tenuifolia IPNI=*465049-1* 183
Laurus tigricida IPNI=*1203274-2* 184
Laurus triandra IPNI=*465312-1* 183
Laurus triplinervis IPNI=*77173628-1* 182
Laurus undulata IPNI=*465049-1* 183
Laurus venosa IPNI=*466894-1* 184
Laurus viburnoides IPNI=*190613-2* 185
Laurus villosa IPNI=*466894-1* 184
Lavandula angustifolia IPNI=*449051-1* 370
Lavandula cladophora IPNI=*449051-1* 370
Lavandula decipiens IPNI=*449051-1* 370
Lavandula erigens IPNI=*449051-1* 370
Lavandula guinardii IPNI=*449051-1* 370
Lavandula hybrida IPNI=*449051-1* 370
Lavandula inclinans IPNI=*449051-1* 370
Lavandula interrupta IPNI=*449051-1* 370
Lavandula major IPNI=*449051-1* 370
Lavandula ovata IPNI=*449051-1* 370
Lavandula spica IPNI=*449051-1* 370
Lavatera americana IPNI=*1055043-2* 406
Lavatera arborea IPNI=*561541-1* 410
Lavatera assurgentiflora IPNI=*1003822-1* 410
Lavatera assurgentiflora subsp. *dendroidea* IPNI=*1003822-1* 410
Lavatera assurgentiflora subsp. *glabra* IPNI=*1003822-1* 410
Lavatera eriocalyx IPNI=*561541-1* 410
Lavatera veneta IPNI=*561541-1* 410
Lavenia alba IPNI=*175030-1* 256
Lavenia aquatica IPNI=*175030-1* 256
Lavenia carnosa IPNI=*175030-1* 256
Lavenia divaricata IPNI=*175030-1* 256
Lavenia elata IPNI=*175030-1* 256
Lavenia fastigiata IPNI=*175030-1* 256
Lavenia glutinosa IPNI=*175030-1* 256
Lavenia latifolia IPNI=*175030-1* 256
Lavenia nudiflora IPNI=*175030-1* 256
Lavenia viscida IPNI=*175030-1* 256
Lawsonia alba IPNI=*553638-1* 417
Lawsonia alba var. *flavescens* IPNI=*553638-1* 417
Lawsonia alba var. *miniata* IPNI=*553638-1* 417
Lawsonia coccinea IPNI=*553638-1* 417
Lawsonia falcifolia IPNI=*553638-1* 417
Lawsonia inermis var. *spinosa* IPNI=*553638-1* 417
Lawsonia purpurea IPNI=*553638-1* 417
Lawsonia spinosa IPNI=*553638-1* 417
Leandra aristigera IPNI=*77207374-1* 419
Leandra caquetana IPNI=*77207884-1* 422
Leandra caquetensis IPNI=*77207690-1* 419
Leandra ciliata IPNI=*572590-1* 421
Leandra cinnamomea IPNI=*158912-2* 420
Leandra dichotoma IPNI=*77207668-1* 423
Leandra dichotoma var. *villosissima* IPNI=*77207668-1* 423
Leandra divaricata IPNI=*77207868-1* 422
Leandra fimbriata IPNI=*77164247-1* 420
Leandra granatensis IPNI=*77207722-1* 420
Leandra heterobasis IPNI=*77207474-1* 420
Leandra heterobasis var. *angustifolia* IPNI=*77207474-1* 420
Leandra mucida IPNI=*77207614-1* 423
Leandra rhamnifolia var. *macrodon* IPNI=*77207868-1* 422
Leandra secunda IPNI=*77207935-1* 421
Leandra strigosa IPNI=*77207722-1* 420
Leandra subseriata IPNI=*77207614-1* 423
Lecointea tango IPNI=*136896-2* 325
Lecostomon terniflorum IPNI=*324587-1* 431
Lecythis acuminatissima IPNI=*947902-1* 301
Lecythis amapaensis IPNI=*597315-1* 302
Lecythis amazonum IPNI=*597315-1* 302
Lecythis ampullaria IPNI=*597240-1* 302
Lecythis armilensis IPNI=*597240-1* 302
Lecythis attenuata IPNI=*596736-1* 302
Lecythis basizone IPNI=*276907-2* 301
Lecythis bipartita IPNI=*597290-1* 302
Lecythis biserrata IPNI=*597275-1* 302
Lecythis boyacensis IPNI=*597240-1* 302
Lecythis bracteata IPNI=*592385-1* 301
Lecythis calycocarpa IPNI=*136980-2* 302
Lecythis chaffanjonii IPNI=*592624-1* 301
Lecythis cordata IPNI=*77176926-1* 302
Lecythis coriacea IPNI=*947902-1* 301
Lecythis costaricensis IPNI=*597240-1* 302
Lecythis couratari IPNI=*592371-1* 301
Lecythis coxiana IPNI=*597275-1* 302
Lecythis crassinoda IPNI=*597352-1* 302
Lecythis cupularis IPNI=*597250-1* 302
Lecythis curranii IPNI=*597240-1* 302
Lecythis davisii IPNI=*597352-1* 302
Lecythis davisii var. *gracilipes* IPNI=*597352-1* 302
Lecythis densa IPNI=*597315-1* 302
Lecythis dubia IPNI=*596736-1* 302
Lecythis elata IPNI=*136980-2* 302
Lecythis elegans IPNI=*592624-1* 301
Lecythis elliptica IPNI=*597290-1* 302
Lecythis grandifolia IPNI=*947902-1* 301
Lecythis hians IPNI=*597352-1* 302
Lecythis hoppiana IPNI=*597315-1* 302
Lecythis idatimonoides IPNI=*276907-2* 301
Lecythis laevicula IPNI=*597250-1* 302
Lecythis langsdorffii IPNI=*597250-1* 302
Lecythis lecomtei IPNI=*597352-1* 302
Lecythis limbata IPNI=*597275-1* 302
Lecythis longipes f. *genuina* IPNI=*276907-2* 301
Lecythis longipes f. *platycarpa* IPNI=*276907-2* 301

Lecythis longipes IPNI=*276907-2* 301
Lecythis macrophylla IPNI=*276907-2* 301
Lecythis madagascariensis IPNI=*597275-1* 302
Lecythis magdalenica IPNI=*597290-1* 302
Lecythis marawynensis IPNI=*597250-1* 302
Lecythis melliana IPNI=*136980-2* 302
Lecythis miersii IPNI=*597275-1* 302
Lecythis minor IPNI=*597275-1* 302
Lecythis monosperma IPNI=*597250-1* 302
Lecythis odora IPNI=*947902-1* 301
Lecythis ollaria IPNI=*597275-1* 302
Lecythis ollaria IPNI=*597315-1* 302
Lecythis ollaria IPNI=*77176926-1* 302
Lecythis parvifolia IPNI=*592624-1* 301
Lecythis pedicellaris IPNI=*276907-2* 301
Lecythis pedicellata IPNI=*276907-2* 301
Lecythis peruviana IPNI=*947902-1* 301
Lecythis pilaris IPNI=*597315-1* 302
Lecythis pilosa IPNI=*276907-2* 301
Lecythis pisonis subsp. *usitata* IPNI=*597315-1* 302
Lecythis platycarpa IPNI=*276907-2* 301
Lecythis platyzone IPNI=*597275-1* 302
Lecythis pohlii IPNI=*597275-1* 302
Lecythis purdiei IPNI=*597290-1* 302
Lecythis retroflexa IPNI=*947902-1* 301
Lecythis rigida IPNI=*597250-1* 302
Lecythis rorida IPNI=*597250-1* 302
Lecythis setifera IPNI=*597315-1* 302
Lecythis sphaeroides IPNI=*597315-1* 302
Lecythis spruceana IPNI=*597250-1* 302
Lecythis tuberculata IPNI=*597275-1* 302
Lecythis tumefacta IPNI=*597352-1* 302
Lecythis urnigera IPNI=*597315-1* 302
Lecythis usitata IPNI=*597315-1* 302
Lecythis usitata var. *tenuifolia* IPNI=*597315-1* 302
Lecythis validissima IPNI=*597352-1* 302
Lecythis variolata IPNI=*597275-1* 302
Lecythis vasiformis IPNI=*597275-1* 302
Lecythis velloziana IPNI=*597315-1* 302
Lecythis wullschlaegeliana IPNI=*276907-2* 301
Lecythopsis guianensis IPNI=*592371-1* 301
Leersia abyssinica IPNI=*137026-2* 241
Leersia aegyptiaca IPNI=*137026-2* 241
Leersia australis IPNI=*137026-2* 241
Leersia brasiliensis IPNI=*137026-2* 241
Leersia capensis IPNI=*137026-2* 241
Leersia ciliaris IPNI=*137026-2* 241
Leersia ciliata IPNI=*137026-2* 241
Leersia compressa IPNI=*137026-2* 241
Leersia contracta IPNI=*137026-2* 241
Leersia dubia IPNI=*137026-2* 241
Leersia elongata IPNI=*137026-2* 241
Leersia ferox IPNI=*137026-2* 241
Leersia glaberrima IPNI=*137026-2* 241
Leersia gouinii IPNI=*137026-2* 241
Leersia gracilis IPNI=*137026-2* 241
Leersia griffithiana IPNI=*137026-2* 241
Leersia luzonensis IPNI=*137026-2* 241
Leersia mauritanica IPNI=*137026-2* 241
Leersia mexicana IPNI=*137026-2* 241
Leersia parviflora IPNI=*137026-2* 241
Leersia triniana IPNI=*137026-2* 241
Legnotis cassipaurea IPNI=*719636-1* 402
Legnotis macrophylla IPNI=*719636-1* 402
Legousia perfoliata IPNI=*258447-2* 273
Leighia speciosa IPNI=*254724-2* 271
Leimanisa pratensis IPNI=*108639-2* 344
Leiphaimos aphylla IPNI=*1167766-2* 345
Leiphaimos costaricensis IPNI=*1167766-2* 345
Leiphaimos eximia IPNI=*1167766-2* 345
Leiphaimos feildenii IPNI=*1167766-2* 345
Leiphaimos lutea IPNI=*1167766-2* 345
Leiphaimos pallidus IPNI=*1167766-2* 345
Leiphaimos uniflorus IPNI=*1167766-2* 345
Leleba amakusensis IPNI=*1097548-2* 234
Leleba breviflora IPNI=*392553-1* 235
Leleba dolichomerithalla IPNI=*1097548-2* 234
Leleba elegans IPNI=*1097548-2* 234
Leleba faurei IPNI=*392553-1* 235
Leleba floribunda IPNI=*1097548-2* 234
Leleba liukiuensis IPNI=*1097548-2* 234
Leleba multiplex IPNI=*1097548-2* 234
Leleba oldhamii IPNI=*392445-1* 234
Leleba shimadae IPNI=*1097548-2* 234
Leleba tuldoides IPNI=*392553-1* 235
Lemaireocereus deficiens IPNI=*244634-2* 285
Lemaireocereus griseus IPNI=*244634-2* 285
Lemna angolensis IPNI=*526159-1* 200
Lemna aoukikusa IPNI=*526159-1* 200
Lemna aoukikusa subsp. *hokurikuensis* IPNI=*526159-1* 200
Lemna bannatica IPNI=*526250-1* 202
Lemna blatteri IPNI=*526159-1* 200
Lemna conjugata IPNI=*526194-1* 200
Lemna cordata IPNI=*526178-1* 200
Lemna eleanorae IPNI=*526159-1* 200
Lemna globosa IPNI=*1135607-2* 202
Lemna major IPNI=*526250-1* 202
Lemna maxima IPNI=*526250-1* 202
Lemna minima IPNI=*526159-1* 200
Lemna minuta IPNI=*279392-2* 200
Lemna obcordata IPNI=*526194-1* 200
Lemna orbicularis IPNI=*526250-1* 202
Lemna orbiculata IPNI=*526250-1* 202
Lemna ovata IPNI=*526194-1* 200
Lemna palustris IPNI=*526194-1* 200
Lemna parodiana IPNI=*526178-1* 200
Lemna polyrhiza IPNI=*526250-1* 202
Lemna punctata IPNI=*242868-2* 202
Lemna rwandensis IPNI=*526194-1* 200
Lemna thermalis IPNI=*526250-1* 202
Lemna torreyi IPNI=*279392-2* 200
Lemna transsilvanica IPNI=*526250-1* 202
Lemna trichorrhiza IPNI=*526178-1* 200
Lemna trinervis IPNI=*526159-1* 200
Lemna umbonata IPNI=*526250-1* 202
Lemna valdiviana var. *pellucida* IPNI=*279392-2* 200
Lemna valdiviana var. *robusta* IPNI=*279392-2* 200
Lemna vulgaris IPNI=*526194-1* 200
Lemna yungensis IPNI=*279392-2* 200
Lendneria humilis IPNI=*244449-2* 377
Lendneria verticillata IPNI=*244449-2* 377
Lennoa caerulea IPNI=*526306-1* 275
Lennoa madreporoides f. *caerulea* IPNI=*526306-1* 275
Lennoa madreporoides subsp. *australis* IPNI=*526306-1* 275
Lennoa madreporoides subsp. *pringlei* IPNI=*526306-1* 275
Lennoa madreporoides subsp. *reichei* IPNI=*526306-1* 275
Lennoa madreporoides subsp. *schaffneri* IPNI=*526306-1* 275
Lennoa madreporoides var. *caerulea* IPNI=*526306-1* 275
Lens culinaris IPNI=*524872-1* 336
Lens esculenta IPNI=*524872-1* 336
Lens lens IPNI=*524872-1* 336
Lens soloniensis IPNI=*524865-1* 336
Lenticula cyclostasa IPNI=*279392-2* 200
Lenticula gibba IPNI=*526178-1* 200
Lenticula gibbosa IPNI=*526178-1* 200
Lenticula minima IPNI=*526194-1* 200

Lenticula minor IPNI=*526194-1* 200
Lenticula palustris IPNI=*526194-1* 200
Lenticula polyrrhiza IPNI=*526250-1* 202
Lenticula vulgaris IPNI=*526194-1* 200
Lentilla lens IPNI=*524872-1* 336
Leonia glycycarpa var. *racemosa* IPNI=*866960-1* 405
Leonia melinoniana IPNI=*866956-1* 405
Leonia sprucei IPNI=*866955-1* 405
Leonohebe odora IPNI=*812400-1* 377
Leontodon chillensis IPNI=*126428-2* 264
Leontodon ciliatus IPNI=*225575-1* 264
Leontodon glaber IPNI=*225504-1* 264
Leontophthalmum peruvianum IPNI=*41956-2* 259
Leontopodium gnaphalioides IPNI=*77197189-1* 259
Leontopodium monoicum IPNI=*77197189-1* 259
Leonurus africanus IPNI=*449133-1* 370
Leonurus altissimus IPNI=*449195-1* 370
Leonurus artemisia IPNI=*449195-1* 370
Leonurus artemisia var. *albiflorus* IPNI=*449195-1* 370
Leonurus cuneifolius IPNI=*449195-1* 370
Leonurus galeobdolon IPNI=*60452701-2* 370
Leonurus globosus IPNI=*127310-3* 370
Leonurus grandiflorus IPNI=*449133-1* 370
Leonurus heterophyllus IPNI=*449195-1* 370
Leonurus intermedius IPNI=*449195-1* 370
Leonurus japonicus f. *niveus* IPNI=*449195-1* 370
Leonurus mexicanus IPNI=*449195-1* 370
Leonurus nepetifolius IPNI=*127310-3* 370
Leonurus sibiricus f. *niveus* IPNI=*449195-1* 370
Leonurus superbus IPNI=*449133-1* 370
Leopoldinia insignis IPNI=*667794-1* 210
Leopoldinia major IPNI=*667794-1* 210
Lepanthes modesta IPNI=*1147281-2* 217
Lepechinia alpina IPNI=*449287-1* 370
Lepechinia hirsuta IPNI=*449287-1* 370
Lepechinia humilis IPNI=*449287-1* 370
Lepechinia procumbens IPNI=*449287-1* 370
Lepeocercis annulata IPNI=*398528-1* 237
Lepeocercis bladhii IPNI=*392701-1* 235
Lepeocercis digitata IPNI=*398529-1* 237
Lepeocercis mollissima IPNI=*398529-1* 237
Lepeocercis pertusa IPNI=*65406-3* 235
Lepeocercis serrata IPNI=*398532-1* 237
Lepia multiflora IPNI=*261363-1* 272
Lepia pauciflora IPNI=*261363-1* 272
Lepianthes granulatum IPNI=*680296-1* 195
Lepianthes maculosum IPNI=*679043-1* 194
Lepianthes peltata IPNI=*316334-2* 197
Lepianthes umbellata IPNI=*316339-2* 197
Lepicystis incana IPNI=*1039912-2* 176
Lepicystis lanceolata IPNI=*17448650-1* 176
Lepicystis percussa IPNI=*17546710-1* 175
Lepicystis vacciniifolia IPNI=*17406250-1* 176
Lepidagathis hyssopifolia IPNI=*51951-1* 360
Lepidagathis laciniata IPNI=*45927-1* 359
Lepidagathis laguroidea IPNI=*51951-1* 360
Lepidagathis lanceolata IPNI=*51951-1* 360
Lepidagathis laurentii IPNI=*51951-1* 360
Lepidaploa ruiz-teranii IPNI=*948527-1* 265
Lepidaploa scorpioides IPNI=*944993-1* 261
Lepidium americanum IPNI=*286148-1* 277
Lepidium anglicum IPNI=*286148-1* 277
Lepidium angustifolium IPNI=*1085072-2* 277
Lepidium arcuatum IPNI=*138164-2* 277
Lepidium auritum IPNI=*137955-2* 277
Lepidium bonariense IPNI=*286148-1* 277
Lepidium bursa-pastoris IPNI=*30092589-2* 276
Lepidium carrerasii IPNI=*286421-1* 277
Lepidium cornutum IPNI=*286421-1* 277
Lepidium coronopifolium IPNI=*137955-2* 277
Lepidium danielsii IPNI=*138164-2* 277
Lepidium diandrum IPNI=*138164-2* 277
Lepidium exiguiflorum IPNI=*138164-2* 277
Lepidium fragrans IPNI=*1085072-2* 277
Lepidium gerloffianum IPNI=*138164-2* 277
Lepidium gerloffianum var. *microcarpum* IPNI=*138164-2* 277
Lepidium horstii IPNI=*138164-2* 277
Lepidium humboldtii IPNI=*137955-2* 277
Lepidium iberis IPNI=*138164-2* 277
Lepidium kalenbornii IPNI=*137955-2* 277
Lepidium majus IPNI=*138164-2* 277
Lepidium micropterum IPNI=*138164-2* 277
Lepidium praecox IPNI=*138164-2* 277
Lepidium prostratum IPNI=*286148-1* 277
Lepidium sativum IPNI=*286421-1* 277
Lepidium sectifolium IPNI=*137955-2* 277
Lepidium thlaspi IPNI=*290477-1* 278
Lepidium virginicum f. *pusillum* IPNI=*138164-2* 277
Lepidium virginicum subsp. *euvirginicum* IPNI=*138164-2* 277
Lepidium virginicum var. *durangense* IPNI=*138164-2* 277
Lepidium virginicum var. *kashmirianum* IPNI=*138164-2* 277
Lepidium virginicum var. *kashmiricum* IPNI=*138164-2* 277
Lepidium virginicum var. *linearifolium* IPNI=*138164-2* 277
Lepidium virginicum var. *macropetalum* IPNI=*138164-2* 277
Lepidium virginicum var. *pavonianum* IPNI=*138164-2* 277
Lepidium virginicum var. *ramosum* IPNI=*138164-2* 277
Lepidium virginicum var. *tepicense* IPNI=*138164-2* 277
Lepidium virginicum var. *typicum* IPNI=*138164-2* 277
Lepidococca sieberi IPNI=*44564-2* 388
Lepidococcus aculeatus IPNI=*154896-2* 210
Lepidococcus armatus IPNI=*154897-2* 210
Lepidococcus duckei IPNI=*154897-2* 210
Lepidococcus huebneri IPNI=*154897-2* 210
Lepidococcus intermedius IPNI=*154897-2* 210
Lepidococcus macrocladus IPNI=*154903-2* 210
Lepidococcus martianus IPNI=*154897-2* 210
Lepidococcus peruvianus IPNI=*154897-2* 210
Lepidococcus pumilus IPNI=*154909-2* 210
Lepidococcus subinermis IPNI=*154909-2* 210
Lepidotis alopecuroides IPNI=*146929-2* 171
Lepidotis caroliniana IPNI=*146931-2* 171
Lepidotis cernua IPNI=*1054587-2* 171
Lepidotis ciliata IPNI=*304307-2* 171
Lepidotis clavata IPNI=*304307-2* 171
Lepidotis complanata IPNI=*147262-2* 171
Lepidotis inflexa IPNI=*304307-2* 171
Lepigonum arenarium IPNI=*159015-1* 286
Lepigonum caninum IPNI=*241336-2* 286
Lepigonum glandulosum IPNI=*159015-1* 286
Lepigonum largiflorum IPNI=*158983-1* 286
Lepigonum leiospermum IPNI=*241336-2* 286
Lepigonum liebmannianum IPNI=*159015-1* 286
Lepigonum marinum IPNI=*241336-2* 286
Lepigonum medium IPNI=*241336-2* 286
Lepigonum neglectum IPNI=*241336-2* 286
Lepigonum paradoxum IPNI=*30040285-2* 285
Lepigonum radicans IPNI=*158983-1* 286
Lepigonum rubrum IPNI=*158983-1* 286
Lepigonum rubrum var. *perennans* IPNI=*158983-1* 286
Lepigonum sperguloides IPNI=*158983-1* 286
Lepigonum tenue IPNI=*241336-2* 286
Lepigonum trachyspermum IPNI=*159015-1* 286
Lepigonum trachyspermum subsp. *murale* IPNI=*241336-2* 286
Lepigonum villosum IPNI=*159015-1* 286
Leptanthus peruvianus IPNI=*698080-1* 225
Leptanthus reniformis IPNI=*698080-1* 225

Leptanthus virginicus IPNI=*698080-1* 225
Leptilon bonariense IPNI=*93603-2* 261
Leptilon canadense IPNI=*203653-1* 261
Leptilon canadense var. *pusillum* IPNI=*203653-1* 261
Leptilon linifolium IPNI=*93603-2* 261
Leptilon pusillum IPNI=*203653-1* 261
Leptinella bogotensis IPNI=*238661-1* 268
Leptochloa digitatiformis IPNI=*60458636-2* 238
Leptochloa dubia IPNI=*60458636-2* 238
Leptochloa dubia var. *humboldtiana* IPNI=*60458636-2* 238
Leptochloa dubia var. *pringleana* IPNI=*60458636-2* 238
Leptochloa fascicularis IPNI=*399115-1* 238
Leptochloa filiformis var. *humilior* IPNI=*407138-1* 241
Leptochloa filiformis var. *remota* IPNI=*407138-1* 241
Leptochloa fusca IPNI=*60458632-2* 238
Leptochloa obtusiflora IPNI=*60458636-2* 238
Leptochloa patens IPNI=*60458636-2* 238
Leptochloa pectinata IPNI=*74675-3* 238
Leptochloa pringlei IPNI=*60458636-2* 238
Leptochloa tenerrima IPNI=*407138-1* 241
Leptocionium fucoides IPNI=*1009149-2* 173
Leptocoryphium penicilligerum IPNI=*1050107-2* 237
Leptodaphne pseudopsychotria IPNI=*171727-2* 184
Leptodaphne subalpina IPNI=*171727-2* 184
Leptoglottis leptocarpa IPNI=*1014032-1* 327
Leptolobium leiocarpum IPNI=*16766-2* 311
Leptolobium nitidulum IPNI=*502385-1* 325
Leptoloma dichotomiflora IPNI=*325362-2* 242
Leptorhoeo floribunda IPNI=*77184055-1* 224
Leptorkis nervosa IPNI=*641061-1* 219
Leptospermum bullatum IPNI=*597505-1* 426
Leptospermum floribundum IPNI=*597505-1* 426
Leptospermum humifusum IPNI=*597505-1* 426
Leptospermum linifolium IPNI=*597505-1* 426
Leptospermum multiflorum IPNI=*597505-1* 426
Leptospermum nichollsii IPNI=*597505-1* 426
Leptospermum obliquum IPNI=*597505-1* 426
Leptospermum oxycedrus IPNI=*597505-1* 426
Leptospermum pungens IPNI=*597505-1* 426
Leptospermum scoparium var. *bullatum* IPNI=*597505-1* 426
Leptospermum scoparium var. *confertifolium* IPNI=*597505-1* 426
Leptospermum scoparium var. *eximea* IPNI=*597505-1* 426
Leptospermum scoparium var. *forsteri* IPNI=*597505-1* 426
Leptospermum scoparium var. *linearifolium* IPNI=*597505-1* 426
Leptospermum scoparium var. *linifolium* IPNI=*597505-1* 426
Leptospermum scoparium var. *myrtifolium* IPNI=*597505-1* 426
Leptospermum scoparium var. *nichollsii* IPNI=*597505-1* 426
Leptospermum scoparium var. *sericeum* IPNI=*597505-1* 426
Leptospermum scoparium var. *vulgare* IPNI=*597505-1* 426
Leptostachya comata IPNI=*78920-2* 359
Leptostachya comata var. *humifusa* IPNI=*78920-2* 359
Leptostachya martiana var. *hispida* IPNI=*78920-2* 359
Leptostachya martiana var. *macrophylla* IPNI=*78920-2* 359
Leptostachya parviflora IPNI=*78920-2* 359
Leptostachya pectoralis IPNI=*47290-1* 359
Leptostachys dubia IPNI=*60458636-2* 238
Leptrina autumnalis IPNI=*311155-2* 286
Lepturus fasciculatus IPNI=*404684-1* 240
Leretia ampla IPNI=*963946-1* 359
Leretia glabrata IPNI=*607989-1* 359
Leretia nitida IPNI=*607989-1* 359
Leretia poeppigiana IPNI=*607989-1* 359
Leretia vellozoi IPNI=*607989-1* 359
Leria lyrata IPNI=*192653-1* 259
Leria nutans f. *lyrata* IPNI=*192653-1* 259
Leria nutans IPNI=*192653-1* 259
Lessingianthus intermedius IPNI=*962014-1* 265
Lestibudesia paniculata IPNI=*129900-2* 283
Lestibudesia paniculata IPNI=*60077-1* 282
Lestibudesia virgata IPNI=*60011-1* 282
Lettsomia dombeyana IPNI=*106032-2* 303
Lettsomia tomentosa IPNI=*106032-2* 303
Leucacantha cyanus IPNI=*190310-1* 259
Leucadendron adscendens IPNI=*704928-1* 433
Leucadendron cuspidatum IPNI=*704928-1* 433
Leucadendron eriocladum IPNI=*704928-1* 433
Leucadendron inflexum IPNI=*704928-1* 433
Leucadendron involucratum IPNI=*704928-1* 433
Leucadendron salignum IPNI=*704928-1* 433
Leucadendron virgatum IPNI=*704928-1* 433
Leucaena boliviana IPNI=*518289-1* 332
Leucanthemum chamaemelum IPNI=*154715-2* 265
Leucanthemum lawatum IPNI=*230047-1* 265
Leucanthemum odoratum IPNI=*252460-1* 271
Leucanthemum parthenium IPNI=*252460-1* 271
Leucanthemum segetum IPNI=*209007-1* 263
Leucanthemum vulgare subsp. *maximum* IPNI=*230047-1* 265
Leucas globulifera IPNI=*447982-1* 370
Leucas microscypha IPNI=*77120747-1* 368
Leucas poggeana IPNI=*447982-1* 370
Leuce alba IPNI=*776573-1* 404
Leucocasia esculenta IPNI=*1170772-2* 199
Leucochlamys callacea IPNI=*88955-1* 202
Leucodonium arvense IPNI=*329786-2* 285
Leucoium incanum IPNI=*286910-1* 277
Leucojum capitulatum IPNI=*64141-1* 216
Leuconymphaea amazonum IPNI=*605491-1* 181
Leuconymphaea ampla IPNI=*605494-1* 181
Leuconymphaea blanda IPNI=*281434-2* 181
Leuconymphaea elegans IPNI=*171178-2* 181
Leuconymphaea gibertii IPNI=*605588-1* 181
Leuconymphaea goudotiana IPNI=*605491-1* 181
Leuconymphaea jamesoniana IPNI=*605588-1* 181
Leuconymphaea lotus IPNI=*605604-1* 181
Leucophyllum frutescens f. *albiflorum* IPNI=*139187-2* 377
Leucophyllum frutescens f. *albineum* IPNI=*139187-2* 377
Leucophyllum texanum IPNI=*139187-2* 377
Leucosinapis alba IPNI=*288952-1* 278
Leucoxylon acuminata IPNI=*248689-2* 365
Leucoxylon riparia IPNI=*248689-2* 365
Leveillea chinensis IPNI=*962413-1* 260
Lewisia megarhiza IPNI=*698305-1* 286
Lewisia mexicana IPNI=*698305-1* 286
Liabum biattenuatum IPNI=*173688-2* 267
Liabum bonplandii IPNI=*230435-1* 265
Liabum hastifolium IPNI=*164998-2* 267
Liabum laticiferum IPNI=*164998-2* 267
Liabum lehmannii IPNI=*230435-1* 265
Liabum sessiliflorum IPNI=*173688-2* 267
Liabum tovarense IPNI=*173688-2* 267
Liabum valerii IPNI=*173688-2* 267
Liatris macrostachya IPNI=*230634-1* 265
Liatris magnifica IPNI=*230634-1* 265
Liatris oppositifolia IPNI=*57410-2* 259
Liatris picta IPNI=*230634-1* 265
Liatris pumila IPNI=*230634-1* 265
Liatris sessiliflora IPNI=*230634-1* 265
Liatris spicata f. *albiflora* IPNI=*230634-1* 265
Liatris spicata f. *montana* IPNI=*230634-1* 265
Liatris spicata var. *macrostachya* IPNI=*230634-1* 265
Liatris spicata var. *montana* IPNI=*230634-1* 265
Liatris spicata var. *typica* IPNI=*230634-1* 265
Liatris turbinata IPNI=*230634-1* 265
Liatris vittata IPNI=*230634-1* 265
Libadion variabile IPNI=*326619-2* 344
Libanothamnus occultus IPNI=*97736-2* 263

Libidibia corymbosa IPNI=*77130395-1* 325
Libidibia ebano IPNI=*139695-2* 325
Libidibia granadillo IPNI=*139695-2* 325
Licania apetala IPNI=*60472826-2* 382
Licania arachnoidea IPNI=*77156954-1* 382
Licania arborea IPNI=*77156125-1* 383
Licania aspera IPNI=*60472875-2* 382
Licania bracteosa IPNI=*77156118-1* 383
Licania bullatifolia IPNI=*77156125-1* 383
Licania capinensis IPNI=*77156118-1* 383
Licania egleri IPNI=*60472909-2* 383
Licania glabriflora IPNI=*77156959-1* 382
Licania heteromorpha IPNI=*77156960-1* 382
Licania heteromorpha var. *grandifolia* IPNI=*77156976-1* 382
Licania huberiana IPNI=*77156118-1* 383
Licania hypargyrea IPNI=*315580-2* 382
Licania intrapetiolaris IPNI=*77156976-1* 382
Licania intrapetiolaris var. *brevis* IPNI=*77156976-1* 382
Licania laurifolia IPNI=*726129-1* 382
Licania laxa IPNI=*726129-1* 382
Licania licaniiflora IPNI=*77156118-1* 383
Licania longistyla IPNI=*60472856-2* 382
Licania macrocarpa IPNI=*77200625-1* 383
Licania maritima IPNI=*60472931-2* 383
Licania micrantha var. *triandra* IPNI=*726172-1* 382
Licania myristicoides IPNI=*60472875-2* 382
Licania obovatifolia IPNI=*77156118-1* 383
Licania obtusifolia IPNI=*77156118-1* 383
Licania octandra IPNI=*60472870-2* 382
Licania parinarioides IPNI=*77156118-1* 383
Licania parinarioides var. *latifolia* IPNI=*77156118-1* 383
Licania parviflora IPNI=*77156118-1* 383
Licania parviflora var. *conduplicata* IPNI=*726118-1* 382
Licania parviflora var. *submembranacea* IPNI=*315580-2* 382
Licania platypus IPNI=*726774-1* 383
Licania poeppigii IPNI=*726129-1* 382
Licania pulchravenia IPNI=*726172-1* 382
Licania pyrifolia IPNI=*164227-2* 383
Licania retusa IPNI=*77156125-1* 383
Licania scabra IPNI=*60472875-2* 382
Licania sclerophylla IPNI=*60472875-2* 382
Licania sclerophylla var. *myristicoides* IPNI=*60472875-2* 382
Licania sclerophylla var. *scabra* IPNI=*60472875-2* 382
Licania seleriana IPNI=*77156125-1* 383
Licania tomentosa IPNI=*726782-1* 383
Licania turiuva IPNI=*60472870-2* 382
Licania wilson-brownei IPNI=*77156118-1* 383
Licaria appelii IPNI=*465275-1* 183
Licaria aritu IPNI=*465275-1* 183
Licaria cayennensis IPNI=*139921-2* 183
Licaria cymbarum IPNI=*77198011-1* 183
Licaria duartei IPNI=*465275-1* 183
Licaria fluminensis IPNI=*465275-1* 183
Licaria jamaicensis IPNI=*465312-1* 183
Licaria limbosa IPNI=*465312-1* 183
Licaria parviflora IPNI=*465248-1* 183
Licaria pittieri IPNI=*465312-1* 183
Licaria puchury-major IPNI=*13872-2* 182
Licaria quixos IPNI=*77198017-1* 183
Licaria reclinata IPNI=*465312-1* 183
Licaria reitzkleiniana IPNI=*465248-1* 183
Licaria rigida IPNI=*465260-1* 183
Licaria schultesii IPNI=*465280-1* 183
Licaria tikalana IPNI=*465312-1* 183
Licuala rotundifolia IPNI=*669832-1* 212
Ligeria caulescens IPNI=*382786-1* 368
Ligeria digitaliflora IPNI=*382786-1* 368
Ligeria discolor IPNI=*382786-1* 368
Ligeria maximiliana IPNI=*382786-1* 368
Ligeria menziesiana IPNI=*382786-1* 368
Ligeria speciosa IPNI=*382786-1* 368
Ligeria speciosa var. *caulescens* IPNI=*382786-1* 368
Ligeria speciosa var. *rubra* IPNI=*382786-1* 368
Lightia lemniscata IPNI=*824024-1* 409
Lignonia monodynama IPNI=*867031-1* 405
Ligusticum angelica IPNI=*837560-1* 251
Ligusticum cuminum IPNI=*840882-1* 252
Ligusticum foeniculum IPNI=*842680-1* 252
Ligusticum montanum IPNI=*170084-2* 252
Ligusticum peregrinum IPNI=*60442790-2* 252
Ligusticum ternatum IPNI=*170084-2* 252
Ligusticum tolucense IPNI=*838268-1* 251
Ligustridium japonicum IPNI=*610001-1* 375
Ligustrum amamianum IPNI=*610001-1* 375
Ligustrum amamianum var. *rotundifolium* IPNI=*610001-1* 375
Ligustrum angustifolium IPNI=*610130-1* 375
Ligustrum cotinifolium IPNI=*319141-2* 374
Ligustrum decipiens IPNI=*610130-1* 375
Ligustrum esquirolii IPNI=*610017-1* 375
Ligustrum glabrum IPNI=*610001-1* 375
Ligustrum hookeri IPNI=*610017-1* 375
Ligustrum insulare IPNI=*610130-1* 375
Ligustrum insulense IPNI=*610130-1* 375
Ligustrum italicum IPNI=*610130-1* 375
Ligustrum japonicum f. *pubescens* IPNI=*610001-1* 375
Ligustrum japonicum f. *rotundifolium* IPNI=*610001-1* 375
Ligustrum kellerianum IPNI=*610001-1* 375
Ligustrum latifolium IPNI=*610001-1* 375
Ligustrum lodense IPNI=*610130-1* 375
Ligustrum lucidum var. *xideense* IPNI=*610017-1* 375
Ligustrum macrophyllum IPNI=*610001-1* 375
Ligustrum magnoliifolium IPNI=*610017-1* 375
Ligustrum ovatum IPNI=*610001-1* 375
Ligustrum oviforme IPNI=*610130-1* 375
Ligustrum roxburghii IPNI=*610017-1* 375
Ligustrum sempervirens IPNI=*610130-1* 375
Ligustrum sieboldii IPNI=*610001-1* 375
Ligustrum syringiflorum IPNI=*610001-1* 375
Ligustrum syringifolium IPNI=*610001-1* 375
Ligustrum taquetii IPNI=*610001-1* 375
Ligustrum vicinum IPNI=*610130-1* 375
Ligustrum vulgare var. *coombei* IPNI=*610130-1* 375
Ligustrum wallichii IPNI=*610017-1* 375
Lilaeopsis andina f. *subulata* IPNI=*140089-2* 252
Lilaeopsis andina IPNI=*140089-2* 252
Lilaeopsis andina var. *inundata* IPNI=*140089-2* 252
Lilaeopsis andina var. *microcarpa* IPNI=*140089-2* 252
Lilaeopsis andina var. *multivittata* IPNI=*140089-2* 252
Lilaeopsis exigua IPNI=*140089-2* 252
Lilaeopsis hillii IPNI=*140089-2* 252
Lilaeopsis patagonica IPNI=*140089-2* 252
Lilaeopsis sinuata IPNI=*140089-2* 252
Lilium album IPNI=*537512-1* 226
Lilium peregrinum IPNI=*537512-1* 226
Lilium striatum IPNI=*537512-1* 226
Limnanthemum calycinum IPNI=*1182548-2* 273
Limnanthemum ecklonianum IPNI=*1182548-2* 273
Limnanthemum esquirolii IPNI=*1182548-2* 273
Limnanthemum forsteri IPNI=*1182548-2* 273
Limnanthemum fraserianum IPNI=*1182548-2* 273
Limnanthemum humboldtianum IPNI=*171243-2* 273
Limnanthemum indicum IPNI=*1182548-2* 273
Limnanthemum kleinianum IPNI=*1182548-2* 273
Limnanthemum microphyllum IPNI=*171243-2* 273
Limnanthemum wightianum IPNI=*1182548-2* 273
Limnobium boscii IPNI=*431804-1* 203

Limnobium spongia IPNI=*431804-1* 203
Limnocharis commersonii IPNI=*58399-1* 198
Limnocharis emarginata IPNI=*58404-1* 198
Limnocharis flava var. *indica* IPNI=*58404-1* 198
Limnocharis haenkei IPNI=*796953-1* 248
Limnocharis humboldtii IPNI=*58399-1* 198
Limnocharis nymphoides IPNI=*58399-1* 198
Limnocharis plumieri IPNI=*58404-1* 198
Limnochloa acicularis IPNI=*1169169-2* 231
Limnochloa acutangula IPNI=*157356-3* 231
Limnochloa articulata IPNI=*60444117-2* 232
Limnochloa constricta IPNI=*1029981-2* 232
Limnochloa crassiculmis IPNI=*306803-1* 231
Limnochloa densa IPNI=*1029981-2* 232
Limnochloa geniculata IPNI=*1029981-2* 232
Limnochloa obsoleta IPNI=*60444117-2* 232
Limnonesis commutata IPNI=*317684-2* 202
Limnonesis friedrichsthaliana IPNI=*317684-2* 202
Limodorum campanulatum IPNI=*617639-1* 217
Limodorum lankesteri IPNI=*617639-1* 217
Limodorum purpureum IPNI=*1218040-2* 217
Limodorum trifidum IPNI=*1218040-2* 217
Limodorum tuberosum IPNI=*1218040-2* 217
Limodorum verecundum IPNI=*1218040-2* 217
Limosella americana f. *natans* IPNI=*329497-2* 376
Limosella americana f. *terrestris* IPNI=*329497-2* 376
Limosella americana IPNI=*329497-2* 376
Limosella aquatica f. *natans* IPNI=*329497-2* 376
Limosella aquatica f. *terrestris* IPNI=*329497-2* 376
Limosella aquatica var. *americana* IPNI=*329497-2* 376
Limosella brachystema IPNI=*329497-2* 376
Limosella calycina IPNI=*1072674-2* 376
Limosella coerulea IPNI=*329497-2* 376
Limosella lineata f. *natans* IPNI=*329497-2* 376
Limosella lineata f. *submersa* IPNI=*329497-2* 376
Limosella lineata f. *terrestris* IPNI=*329497-2* 376
Limosella lineata IPNI=*329497-2* 376
Limosella lineata var. *brevistylis* IPNI=*329497-2* 376
Limosella lineata var. *spathulata* IPNI=*329497-2* 376
Limosella maritima IPNI=*329497-2* 376
Limosella minuta IPNI=*329497-2* 376
Limosella monticola IPNI=*329497-2* 376
Limosella subulata f. *maritima* IPNI=*329497-2* 376
Limosella subulata f. *submersa* IPNI=*329497-2* 376
Limosella subulata f. *terrestris* IPNI=*329497-2* 376
Limosella subulata IPNI=*329497-2* 376
Linaria acutangula IPNI=*325773-2* 376
Linaria canadensis f. *albina* IPNI=*281417-2* 376
Linaria canadensis f. *cleistogama* IPNI=*281417-2* 376
Linaria canadensis IPNI=*281417-2* 376
Linaria canadensis var. *typica* IPNI=*281417-2* 376
Linaria cymbalaria IPNI=*325773-2* 376
Linaria hederifolia IPNI=*325773-2* 376
Linaria hexandra IPNI=*1189750-2* 374
Linaria subandina IPNI=*281417-2* 376
Linczevskia sinuata IPNI=*686979-1* 288
Lindackeria maynensis IPNI=*140493-2* 380
Lindackeria nitida IPNI=*111811-1* 380
Lindackeria vageleri IPNI=*140493-2* 380
Lindackeria vernicosa IPNI=*111811-1* 380
Lindenbergia petelotii IPNI=*244449-2* 377
Lindenia acutiflora IPNI=*998196-1* 347
Lindenia rivalis f. *glabra* IPNI=*998196-1* 347
Lindenia rivalis IPNI=*998196-1* 347
Lindeniopiper williamsii IPNI=*681674-1* 196
Lindernia antipoda IPNI=*799766-1* 374
Lindernia antipoda subsp. *verbenifolia* IPNI=*799766-1* 374
Lindernia crustacea IPNI=*1189750-2* 374
Lindernia crustacea var. *clarkei* IPNI=*1189750-2* 374
Lindernia crustacea var. *verticillata* IPNI=*1189750-2* 374
Lindernia dianthera IPNI=*155661-2* 376
Lindernia diffusa IPNI=*262512-2* 374
Lindernia gracilis IPNI=*1189750-2* 374
Lindernia minuta IPNI=*1189750-2* 374
Lindernia obtusa IPNI=*1189750-2* 374
Lindernia pachypoda IPNI=*799766-1* 374
Lindernia racemosa IPNI=*1189750-2* 374
Lindernia thouarsii IPNI=*255156-2* 374
Lindernia verbenifolia IPNI=*799766-1* 374
Lindleya semiserrata IPNI=*112597-2* 309
Lingoum acuminatum IPNI=*214040-2* 331
Lingoum esculentum IPNI=*516568-1* 331
Lingoum officinale IPNI=*214060-2* 331
Lingoum rohrii IPNI=*516561-1* 331
Lingoum rufescens IPNI=*516561-1* 331
Lingoum villosum IPNI=*516561-1* 331
Lingoum violaceum IPNI=*516561-1* 331
Linnaeobreynia indica IPNI=*77217338-1* 278
Linnaeobreynia pulcherrima IPNI=*77183588-1* 278
Linnaeobreynia tonduzii IPNI=*77217338-1* 278
Linochilus costaricensis IPNI=*230916-1* 265
Linociera alba IPNI=*301087-2* 374
Linociera caribaea IPNI=*301087-2* 374
Linociera compacta IPNI=*301087-2* 374
Linociera cotinifolia IPNI=*319141-2* 374
Linociera pubescens IPNI=*300138-2* 374
Linociera santamartensis IPNI=*301087-2* 374
Linum arvense IPNI=*544772-1* 394
Linum crepitans IPNI=*544772-1* 394
Linum grandiflorum IPNI=*544772-1* 394
Linum humile IPNI=*544772-1* 394
Linum indehiscens IPNI=*544772-1* 394
Linum monadelphum IPNI=*544772-1* 394
Linum moroderorum IPNI=*544772-1* 394
Linum mucronatum IPNI=*544772-1* 394
Linum reuteri IPNI=*544772-1* 394
Linum sativum IPNI=*544772-1* 394
Linum trinervium IPNI=*544772-1* 394
Linum usitatissimum subsp. *humile* IPNI=*544772-1* 394
Linum usitatissimum subsp. *intermedium* IPNI=*544772-1* 394
Linum usitatissimum subsp. *latifolium* IPNI=*544772-1* 394
Linum utile IPNI=*544772-1* 394
Lippia affinis IPNI=*863752-1* 379
Lippia alba f. *intermedia* IPNI=*141020-2* 378
Lippia alba f. *macrophylla* IPNI=*141020-2* 378
Lippia alba f. *scabra* IPNI=*141020-2* 378
Lippia alba var. *carterae* IPNI=*141020-2* 378
Lippia alba var. *globiflora* IPNI=*141020-2* 378
Lippia americana f. *hyptoides* IPNI=*863509-1* 379
Lippia americana f. *pilosa* IPNI=*863509-1* 379
Lippia asperifolia IPNI=*141020-2* 378
Lippia asperifolia IPNI=*141103-2* 379
Lippia berlandieri IPNI=*863752-1* 379
Lippia berteroi IPNI=*863752-1* 379
Lippia briquetii IPNI=*863659-1* 379
Lippia candicans IPNI=*863752-1* 379
Lippia carterae IPNI=*141020-2* 378
Lippia citrata IPNI=*141020-2* 378
Lippia citrodora IPNI=*1015994-1* 377
Lippia crenata IPNI=*141020-2* 378
Lippia cuneifolia IPNI=*30135678-2* 379
Lippia cylindrica IPNI=*1126344-2* 379
Lippia elegans IPNI=*863752-1* 379
Lippia elegans var. *macrophylla* IPNI=*863752-1* 379
Lippia elegans var. *obtusifolia* IPNI=*863752-1* 379
Lippia fimbriata IPNI=*324301-2* 378

Lippia floribunda IPNI=*863509-1* 379
Lippia floribunda IPNI=*863659-1* 379
Lippia geminata IPNI=*141020-2* 378
Lippia geminata var. *microphylla* IPNI=*141020-2* 378
Lippia glandulosa IPNI=*863752-1* 379
Lippia globiflora f. *lilacina* IPNI=*141020-2* 378
Lippia globiflora IPNI=*141020-2* 378
Lippia globiflora var. *geminata* IPNI=*141020-2* 378
Lippia globiflora var. *normalis* IPNI=*141020-2* 378
Lippia graveolens f. *loeseneriana* IPNI=*863752-1* 379
Lippia graveolens f. *microphylla* IPNI=*863752-1* 379
Lippia graveolens IPNI=*863752-1* 379
Lippia guayaquilensis IPNI=*863509-1* 379
Lippia havanensis IPNI=*141020-2* 378
Lippia helleri IPNI=*30135678-2* 379
Lippia hemisphaerica IPNI=*863509-1* 379
Lippia hirsuta var. *moritzii* IPNI=*863659-1* 379
Lippia hirsuta var. *sphacelifolia* IPNI=*863659-1* 379
Lippia hyptoides IPNI=*863509-1* 379
Lippia imbricata IPNI=*324301-2* 378
Lippia lantanifolia IPNI=*141020-2* 378
Lippia lantanoides IPNI=*141020-2* 378
Lippia mattogrossensis IPNI=*863752-1* 379
Lippia micromera var. *helleri* IPNI=*30135678-2* 379
Lippia microphylla IPNI=*863752-1* 379
Lippia montevidensis IPNI=*135260-2* 378
Lippia moritzii IPNI=*863659-1* 379
Lippia nodiflora IPNI=*194567-2* 379
Lippia obovata IPNI=*141020-2* 378
Lippia obscura IPNI=*863752-1* 379
Lippia origanoides var. *sampaionis* IPNI=*863752-1* 379
Lippia pallescens IPNI=*863236-1* 378
Lippia palmeri f. *spicata* IPNI=*863752-1* 379
Lippia palmeri IPNI=*863752-1* 379
Lippia palmeri var. *spicata* IPNI=*863752-1* 379
Lippia panamensis IPNI=*141020-2* 378
Lippia pauciserrata IPNI=*863509-1* 379
Lippia pendula IPNI=*863752-1* 379
Lippia polycephala IPNI=*863752-1* 379
Lippia polycephala var. *aemilii* IPNI=*863752-1* 379
Lippia purpurea IPNI=*135333-2* 378
Lippia purpurea IPNI=*324301-2* 378
Lippia pyramidata IPNI=*863509-1* 379
Lippia rigida IPNI=*863752-1* 379
Lippia rivalis IPNI=*863752-1* 379
Lippia rondonensis IPNI=*141020-2* 378
Lippia rubiginosa IPNI=*863752-1* 379
Lippia rubiginosa var. *dives* IPNI=*863752-1* 379
Lippia rubiginosa var. *pauper* IPNI=*863752-1* 379
Lippia salviifolia IPNI=*863752-1* 379
Lippia schlechtendalii IPNI=*863659-1* 379
Lippia schliebenii IPNI=*135333-2* 378
Lippia schomburgkiana IPNI=*863752-1* 379
Lippia sidoides f. *flaccida* IPNI=*863752-1* 379
Lippia sidoides IPNI=*863752-1* 379
Lippia sphacelifolia IPNI=*863659-1* 379
Lippia triphylla IPNI=*1015994-1* 377
Lippia unica IPNI=*141020-2* 378
Lippia velutina IPNI=*863752-1* 379
Lipusa formosa IPNI=*513754-1* 330
Lipusa multiflora IPNI=*513754-1* 330
Liriosma inopiflora IPNI=*77183948-1* 449
Liriosma macrophylla IPNI=*608118-1* 449
Liriosma micrantha IPNI=*77183948-1* 449
Liriosma ovata IPNI=*77183948-1* 449
Lisianthius acutilobus IPNI=*130185-2* 345
Lisianthius alatus IPNI=*118811-2* 345
Lisianthius amoenus IPNI=*1002418-1* 344
Lisianthius campanulaceus IPNI=*118811-2* 345
Lisianthius cardonae IPNI=*130185-2* 345
Lisianthius exaltatus IPNI=*367440-1* 344
Lisianthius exaltatus var. *albiflorus* IPNI=*367440-1* 344
Lisianthius glaucifolius IPNI=*367440-1* 344
Lisianthius gracilis IPNI=*1002418-1* 344
Lisianthius grandiflorus IPNI=*1002418-1* 344
Lisianthius paniculatus IPNI=*130185-2* 345
Lisianthius purpurascens IPNI=*1002418-1* 344
Lisianthius recurvus IPNI=*130185-2* 345
Lisianthius spruceanus IPNI=*1002418-1* 344
Lisianthius subcordatus IPNI=*130185-2* 345
Lisianthius uliginosus IPNI=*1002418-1* 344
Lisianthius uliginosus var. *grandiflorus* IPNI=*1002418-1* 344
Lisianthius uliginosus var. *guianensis* IPNI=*1002418-1* 344
Lita aphylla IPNI=*1167766-2* 345
Lithagrostis lacryma-jobi IPNI=*30100521-2* 236
Lithocardium adnatum IPNI=*262677-2* 276
Lithocardium alliodorum IPNI=*113943-1* 273
Lithocardium angustifolium IPNI=*262640-2* 275
Lithocardium bicolor IPNI=*274895-2* 274
Lithocardium bogotense IPNI=*274898-2* 274
Lithocardium bracteatum IPNI=*65321-2* 274
Lithocardium brevispicatum IPNI=*262640-2* 275
Lithocardium canescens IPNI=*262640-2* 275
Lithocardium canum IPNI=*946177-1* 276
Lithocardium caracasanum IPNI=*262641-2* 276
Lithocardium collococca IPNI=*114058-1* 274
Lithocardium cordifolium IPNI=*275071-2* 274
Lithocardium corylifolium IPNI=*927551-1* 274
Lithocardium corymbosum f. *glabriusculum* IPNI=*262677-2* 276
Lithocardium corymbosum IPNI=*262677-2* 276
Lithocardium corymbosum var. *humile* IPNI=*262677-2* 276
Lithocardium corymbosum var. *lineatum* IPNI=*262677-2* 276
Lithocardium corymbosum var. *monogynum* IPNI=*262677-2* 276
Lithocardium corymbosum var. *patens* IPNI=*262677-2* 276
Lithocardium corymbosum var. *subedentatum* IPNI=*262677-2* 276
Lithocardium crenulatum IPNI=*946226-1* 276
Lithocardium cujabense IPNI=*113943-1* 273
Lithocardium cuneiforme IPNI=*262640-2* 275
Lithocardium curassavicum IPNI=*262640-2* 275
Lithocardium cylindrostachyum IPNI=*262641-2* 276
Lithocardium cylindrostachyum var. *cinerascens* IPNI=*262641-2* 276
Lithocardium cylindrostachyum var. *interruptum* IPNI=*262641-2* 276
Lithocardium cylindrostachyum var. *platyphyllum* IPNI=*262641-2* 276
Lithocardium dichotomum IPNI=*121424-1* 276
Lithocardium divaricatum IPNI=*262640-2* 275
Lithocardium diversifolium IPNI=*274931-2* 274
Lithocardium exaltatum IPNI=*114135-1* 274
Lithocardium ferrugineum IPNI=*946226-1* 276
Lithocardium ferrugineum var. *subcordatum* IPNI=*946226-1* 276
Lithocardium flavum IPNI=*274993-2* 274
Lithocardium gerascanthodes IPNI=*65321-2* 274
Lithocardium gerascanthus IPNI=*65321-2* 274
Lithocardium glabrum IPNI=*1088317-2* 273
Lithocardium grandiflorum IPNI=*121434-1* 276
Lithocardium guazumifolium IPNI=*262677-2* 276
Lithocardium hartwigsiana IPNI=*113943-1* 273
Lithocardium hebecarpum IPNI=*275076-2* 274
Lithocardium hermanniifolium IPNI=*77084349-1* 276
Lithocardium heterophyllum IPNI=*275076-2* 274
Lithocardium hirsutum IPNI=*114394-1* 274
Lithocardium hispidissimum IPNI=*114394-1* 274
Lithocardium hispidum IPNI=*262640-2* 275
Lithocardium laeve IPNI=*65551-2* 274
Lithocardium laxiflorum IPNI=*946226-1* 276
Lithocardium lineare IPNI=*262640-2* 275
Lithocardium macrostachyum IPNI=*262640-2* 275

Lithocardium mariquitense IPNI=*262677-2* 276
Lithocardium mirandum IPNI=*114394-1* 274
Lithocardium mucronatum IPNI=*114135-1* 274
Lithocardium nodosum IPNI=*114394-1* 274
Lithocardium oxyphyllum IPNI=*262640-2* 275
Lithocardium poeppigii IPNI=*946226-1* 276
Lithocardium pubescens IPNI=*275076-2* 274
Lithocardium riparium IPNI=*946226-1* 276
Lithocardium rotundifolium IPNI=*274993-2* 274
Lithocardium salicinum IPNI=*262640-2* 275
Lithocardium scabridum IPNI=*114135-1* 274
Lithocardium sericicalyx IPNI=*275053-2* 274
Lithocardium tenuifolium IPNI=*927551-1* 274
Lithocardium tetrandrum IPNI=*275071-2* 274
Lithocardium tetraphyllum IPNI=*77164595-1* 416
Lithocardium thibaudianum IPNI=*946226-1* 276
Lithocardium toqueve IPNI=*275076-2* 274
Lithocardium verbenaceum IPNI=*262640-2* 275
Lithophila vermicularis IPNI=*34033-2* 282
Lithospermum erectum IPNI=*118163-1* 275
Lithospermum majus IPNI=*118163-1* 275
Lithospermum ochroleucum IPNI=*118163-1* 275
Lithospermum officinale var. *stewartii* IPNI=*118163-1* 275
Lithraea chichita IPNI=*71069-1* 451
Litobrochia biaurita IPNI=*213840-2* 178
Litobrochia divaricata IPNI=*17217170-1* 178
Litobrochia grandifolia IPNI=*313710-2* 178
Litobrochia intermedia IPNI=*17217170-1* 178
Litobrochia marginata IPNI=*17217170-1* 178
Litobrochia microdictyon IPNI=*17217170-1* 178
Litobrochia milneana IPNI=*17217170-1* 178
Litobrochia pseudolonchitis IPNI=*17217170-1* 178
Litobrochia tripartita IPNI=*17217170-1* 178
Litocarpus cordifolius IPNI=*322566-2* 280
Litsea cervantesii IPNI=*465312-1* 183
Litsea glaucescens var. *schaffneri* IPNI=*465312-1* 183
Litsea matudae IPNI=*465312-1* 183
Litsea schaffneri IPNI=*465312-1* 183
Litsea villosa IPNI=*466894-1* 184
Livistona altissima IPNI=*669832-1* 212
Livistona filamentosa IPNI=*267788-2* 212
Livistona microcarpa IPNI=*669832-1* 212
Livistona mindorensis IPNI=*669832-1* 212
Livistona oliviformis IPNI=*668021-1* 210
Livistona robinsoniana IPNI=*669832-1* 212
Livistona rotundifolia IPNI=*669832-1* 212
Livistona sinensis IPNI=*668021-1* 210
Livistona subglobosa IPNI=*668021-1* 210
Loasa grandis IPNI=*77188059-1* 291
Loasa pinnata IPNI=*1014841-1* 292
Loasa triphylla IPNI=*1014841-1* 292
Loasa triphylla var. *genuina* IPNI=*1014841-1* 292
Loasa wallisii IPNI=*1014841-1* 292
Lobelia acutangula IPNI=*143240-1* 273
Lobelia alexia IPNI=*142908-1* 273
Lobelia algoensis IPNI=*143240-1* 273
Lobelia altimontis IPNI=*143240-1* 273
Lobelia andropogon IPNI=*51553-2* 272
Lobelia bellis IPNI=*143584-1* 273
Lobelia benguellensis IPNI=*143240-1* 273
Lobelia bicolor IPNI=*143240-1* 273
Lobelia bonplandiana IPNI=*51553-2* 272
Lobelia bracteolata IPNI=*143240-1* 273
Lobelia candida IPNI=*143240-1* 273
Lobelia cardinalis f. *alba* IPNI=*142115-2* 273
Lobelia cardinalis f. *cordigera* IPNI=*142115-2* 273
Lobelia cardinalis f. *hispidula* IPNI=*142115-2* 273
Lobelia cardinalis f. *rosea* IPNI=*142115-2* 273
Lobelia cardinalis subsp. *graminea* IPNI=*142115-2* 273
Lobelia cardinalis var. *alba* IPNI=*142115-2* 273
Lobelia cardinalis var. *angustifolia* IPNI=*142115-2* 273
Lobelia cardinalis var. *candida* IPNI=*142115-2* 273
Lobelia cardinalis var. *glandulosa* IPNI=*142115-2* 273
Lobelia cardinalis var. *graminea* IPNI=*142115-2* 273
Lobelia cardinalis var. *hispidula* IPNI=*142115-2* 273
Lobelia cardinalis var. *integerrima* IPNI=*142115-2* 273
Lobelia cardinalis var. *meridionalis* IPNI=*142115-2* 273
Lobelia cardinalis var. *multiflora* IPNI=*142115-2* 273
Lobelia cardinalis var. *phyllostachya* IPNI=*142115-2* 273
Lobelia cardinalis var. *propinqua* IPNI=*142115-2* 273
Lobelia cardinalis var. *pseudosplendens* IPNI=*142115-2* 273
Lobelia cardinalis var. *texensis* IPNI=*142115-2* 273
Lobelia chilawana IPNI=*143240-1* 273
Lobelia coccinea IPNI=*142115-2* 273
Lobelia columnae IPNI=*145124-1* 273
Lobelia cordigera IPNI=*142115-2* 273
Lobelia cordigera var. *fatalis* IPNI=*142115-2* 273
Lobelia cordigera var. *marryattiae* IPNI=*142115-2* 273
Lobelia cordigera var. *multiflora* IPNI=*142115-2* 273
Lobelia cornuta IPNI=*51553-2* 272
Lobelia cymbalaria IPNI=*143584-1* 273
Lobelia cymbalarioides IPNI=*143584-1* 273
Lobelia dortmannii IPNI=*143240-1* 273
Lobelia draba IPNI=*79151-2* 273
Lobelia elliptica IPNI=*145166-1* 273
Lobelia erinifolia IPNI=*143240-1* 273
Lobelia erinoides IPNI=*143240-1* 273
Lobelia erinus var. *gracilis* IPNI=*143240-1* 273
Lobelia fulgens f. *atrosanguinea* IPNI=*142115-2* 273
Lobelia fulgens IPNI=*142115-2* 273
Lobelia fulgens var. *multiflora* IPNI=*142115-2* 273
Lobelia fulgens var. *propinqua* IPNI=*142115-2* 273
Lobelia fulgens var. *pyramidalis* IPNI=*142115-2* 273
Lobelia gigantea IPNI=*145189-1* 273
Lobelia gracilis IPNI=*1020602-1* 273
Lobelia gracilis var. *major* IPNI=*1020602-1* 273
Lobelia graminea f. *kerneri* IPNI=*142115-2* 273
Lobelia graminea IPNI=*142115-2* 273
Lobelia graminea var. *intermedia* IPNI=*142115-2* 273
Lobelia graminea var. *phyllostachya* IPNI=*142115-2* 273
Lobelia ignea IPNI=*142115-2* 273
Lobelia jugosa IPNI=*143240-1* 273
Lobelia keilhackii IPNI=*143240-1* 273
Lobelia kerneri IPNI=*142115-2* 273
Lobelia kohautiana IPNI=*143240-1* 273
Lobelia laevigata IPNI=*51553-2* 272
Lobelia lavendulacea IPNI=*143240-1* 273
Lobelia longiflora IPNI=*1133957-2* 273
Lobelia longifolia IPNI=*142115-2* 273
Lobelia lydenburgensis IPNI=*143240-1* 273
Lobelia marryattiae IPNI=*142115-2* 273
Lobelia melsetteria IPNI=*143240-1* 273
Lobelia micrantha IPNI=*79151-2* 273
Lobelia minutiflora IPNI=*79151-2* 273
Lobelia montaguensis IPNI=*143240-1* 273
Lobelia mucronata IPNI=*142115-2* 273
Lobelia nana var. *cymbalarioides* IPNI=*143584-1* 273
Lobelia nana var. *flagelliformis* IPNI=*143584-1* 273
Lobelia natalensis IPNI=*143240-1* 273
Lobelia nivea IPNI=*145280-1* 273
Lobelia nuda IPNI=*143240-1* 273
Lobelia nuzana IPNI=*143240-1* 273
Lobelia obscura IPNI=*51553-2* 272
Lobelia oranjensis IPNI=*143240-1* 273
Lobelia parviflora IPNI=*79151-2* 273
Lobelia parvifolia IPNI=*143240-1* 273

Lobelia phyllostachya IPNI=*142115-2* 273
Lobelia polygalifolia IPNI=*143931-1* 273
Lobelia polyodon IPNI=*143240-1* 273
Lobelia porphyrantha IPNI=*142115-2* 273
Lobelia princeps IPNI=*142115-2* 273
Lobelia propinqua IPNI=*142115-2* 273
Lobelia punicea IPNI=*142115-2* 273
Lobelia punicea var. *kerneri* IPNI=*142115-2* 273
Lobelia purpurea IPNI=*51553-2* 272
Lobelia ramosa IPNI=*142115-2* 273
Lobelia raridentata IPNI=*143240-1* 273
Lobelia rosulata IPNI=*143240-1* 273
Lobelia ruderalis IPNI=*79151-2* 273
Lobelia rupestris IPNI=*143931-1* 273
Lobelia schiedeana IPNI=*142115-2* 273
Lobelia schrankii IPNI=*143240-1* 273
Lobelia senegalensis IPNI=*143240-1* 273
Lobelia spectabilis IPNI=*51553-2* 272
Lobelia splendens IPNI=*142115-2* 273
Lobelia splendens var. *atrosanguinea* IPNI=*142115-2* 273
Lobelia splendens var. *fulgens* IPNI=*142115-2* 273
Lobelia splendens var. *ignea* IPNI=*142115-2* 273
Lobelia subpubera var. *weberbaueri* IPNI=*143931-1* 273
Lobelia subtilis IPNI=*79151-2* 273
Lobelia surinamensis IPNI=*51553-2* 272
Lobelia tenera var. *belladonna* IPNI=*143931-1* 273
Lobelia tenera var. *tenuicaulis* IPNI=*143931-1* 273
Lobelia texensis IPNI=*142115-2* 273
Lobelia transvaalensis IPNI=*143240-1* 273
Lobelia trierarchii IPNI=*143240-1* 273
Lobelia turgida IPNI=*143240-1* 273
Lobelia veronicifolia IPNI=*143931-1* 273
Lobelia weberbaueri IPNI=*143931-1* 273
Lobelia wildii IPNI=*143240-1* 273
Lobularia halimifolia IPNI=*1085072-2* 277
Lobularia maritima f. *rubescens* IPNI=*1085072-2* 277
Lobularia maritima var. *rubescens* IPNI=*1085072-2* 277
Lobularia strigulosa IPNI=*1085072-2* 277
Lochnera rosea IPNI=*77880-1* 340
Lockhartia hologlossa IPNI=*142885-2* 219
Lockhartia lasseri IPNI=*641704-1* 219
Lockhartia pallida IPNI=*641704-1* 219
Lockhartia platyglossa IPNI=*142885-2* 219
Lockhartia unicornis IPNI=*142885-2* 219
Lodicularia capensis IPNI=*404684-1* 240
Lodicularia fasciculata IPNI=*404684-1* 240
Loeflingia hispanica IPNI=*300494-2* 286
Loeflingia ramosissima IPNI=*45444-2* 285
Loeflingia renifolia IPNI=*30172747-2* 285
Loeseneriella ovata IPNI=*161653-1* 290
Loghania pentacrina IPNI=*829493-1* 302
Loiseleria indica IPNI=*1137397-2* 300
Loliola pratensis IPNI=*407516-1* 241
Lolium aechicum IPNI=*407493-1* 241
Lolium aegyptiacum IPNI=*407516-1* 241
Lolium agreste IPNI=*407493-1* 241
Lolium album IPNI=*407516-1* 241
Lolium ambiguum IPNI=*407486-1* 241
Lolium annuum IPNI=*407493-1* 241
Lolium annuum IPNI=*407516-1* 241
Lolium arlstatum IPNI=*407486-1* 241
Lolium arvense IPNI=*407516-1* 241
Lolium asperum IPNI=*407516-1* 241
Lolium berteronianum IPNI=*407516-1* 241
Lolium brasilianum IPNI=*407493-1* 241
Lolium bromoides IPNI=*402472-1* 239
Lolium canadense IPNI=*407493-1* 241
Lolium cechicum IPNI=*407493-1* 241
Lolium compositum IPNI=*407486-1* 241
Lolium compressum IPNI=*407493-1* 241
Lolium cuneatum IPNI=*407516-1* 241
Lolium decipiens IPNI=*407516-1* 241
Lolium elongatum IPNI=*407486-1* 241
Lolium felix IPNI=*407493-1* 241
Lolium gaudinii IPNI=*407486-1* 241
Lolium giganteum IPNI=*407516-1* 241
Lolium glumosum IPNI=*407493-1* 241
Lolium gmelinii IPNI=*407493-1* 241
Lolium gracile IPNI=*407516-1* 241
Lolium gussonei IPNI=*407516-1* 241
Lolium halleri IPNI=*407493-1* 241
Lolium infelix IPNI=*407516-1* 241
Lolium jechelianum IPNI=*407493-1* 241
Lolium latum IPNI=*407493-1* 241
Lolium lesdainii IPNI=*407486-1* 241
Lolium longiglume IPNI=*407516-1* 241
Lolium lucidum IPNI=*407516-1* 241
Lolium marschallii IPNI=*407493-1* 241
Lolium maximum IPNI=*407516-1* 241
Lolium multiflorum f. *submuticum* IPNI=*407486-1* 241
Lolium multiflorum var. *laeviculme* IPNI=*407486-1* 241
Lolium multiflorum var. *latifolium* IPNI=*407486-1* 241
Lolium osiridis IPNI=*407486-1* 241
Lolium perenne subsp. *italicum* IPNI=*407486-1* 241
Lolium perenne subsp. *stoloniferum* IPNI=*407493-1* 241
Lolium perenne subsp. *trabutii* IPNI=*407493-1* 241
Lolium perenne var. *brasilianum* IPNI=*407493-1* 241
Lolium perenne var. *cristatum* IPNI=*407493-1* 241
Lolium perenne var. *monstrosum* IPNI=*407493-1* 241
Lolium perenne var. *ramosum* IPNI=*407493-1* 241
Lolium perenne var. *scabriculme* IPNI=*407493-1* 241
Lolium perenne var. *stickneiensis* IPNI=*407493-1* 241
Lolium perenne var. *stoloniferum* IPNI=*407493-1* 241
Lolium pseudoitalicum IPNI=*407493-1* 241
Lolium pseudolinicola IPNI=*407516-1* 241
Lolium remotum f. *asperum* IPNI=*407516-1* 241
Lolium repens IPNI=*407493-1* 241
Lolium robustum IPNI=*407516-1* 241
Lolium rosetlanum IPNI=*407493-1* 241
Lolium scabrum IPNI=*407486-1* 241
Lolium siculum IPNI=*407486-1* 241
Lolium speciosum IPNI=*407516-1* 241
Lolium temulentum f. *leve* IPNI=*407516-1* 241
Lolium temulentum IPNI=*407486-1* 241
Lolium temulentum subsp. *arvense* IPNI=*407516-1* 241
Lolium temulentum subsp. *cuneatum* IPNI=*407516-1* 241
Lolium temulentum var. *arvense* IPNI=*407516-1* 241
Lolium temulentum var. *leptochaeton* IPNI=*407516-1* 241
Lolium temulentum var. *macrochaeton* IPNI=*407516-1* 241
Lolium temulentum var. *multiflorum* IPNI=*407486-1* 241
Lolium tenue IPNI=*407493-1* 241
Lolium trabutii IPNI=*407493-1* 241
Lolium triticoides IPNI=*407516-1* 241
Lolium vulgare IPNI=*407493-1* 241
Lomake brachiata IPNI=*35430-2* 378
Lomaresis alba IPNI=*539581-1* 215
Lomaria arborescens IPNI=*33533-2* 174
Lomaria brasiliensis IPNI=*33533-2* 174
Lomaria campylotis IPNI=*33632-2* 174
Lomaria chilensis IPNI=*33533-2* 174
Lomaria cordata IPNI=*33533-2* 174
Lomaria gilliesii IPNI=*33533-2* 174
Lomaria loxensis IPNI=*77186436-1* 174
Lomaria ornifolia IPNI=*33533-2* 174
Lomaria regnelliana IPNI=*33533-2* 174
Lomaria serrulosa IPNI=*33533-2* 174

Lomaria spissa IPNI=*33533-2* 174
Lomaria squamulosa IPNI=*77186436-1* 174
Lomariopsis phlebodes IPNI=*17442740-1* 175
Lomatophyllum reflexum IPNI=*534338-1* 215
Lomenia borbonica IPNI=*944889-1* 217
Lonchitis repens IPNI=*1031307-2* 175
Lonchitis-aspera capensis var. *chilensis* IPNI=*33533-2* 174
Lonchocarpus albiflorus IPNI=*143380-2* 325
Lonchocarpus amerimnum IPNI=*75919-2* 317
Lonchocarpus benthamianus IPNI=*503074-1* 325
Lonchocarpus boliviensis IPNI=*60453072-2* 317
Lonchocarpus broadwayi IPNI=*60458868-2* 328
Lonchocarpus caribaeus IPNI=*503074-1* 325
Lonchocarpus densiflorus IPNI=*60453072-2* 317
Lonchocarpus densiflorus var. *moritzianus* IPNI=*60453072-2* 317
Lonchocarpus discolor IPNI=*502971-1* 325
Lonchocarpus domingensis IPNI=*143380-2* 325
Lonchocarpus fendleri IPNI=*77118389-1* 328
Lonchocarpus fendleri subsp. *pubescens* IPNI=*77118389-1* 328
Lonchocarpus fendleri var. *pubescens* IPNI=*77118389-1* 328
Lonchocarpus floribundus IPNI=*77143238-1* 317
Lonchocarpus formosianus IPNI=*143380-2* 325
Lonchocarpus fruticosus IPNI=*520444-1* 335
Lonchocarpus glabrescens IPNI=*60453072-2* 317
Lonchocarpus guaricensis IPNI=*316422-2* 330
Lonchocarpus guineensis IPNI=*143380-2* 325
Lonchocarpus latifolius IPNI=*502971-1* 325
Lonchocarpus lucidus IPNI=*503009-1* 325
Lonchocarpus monilis IPNI=*77118402-1* 328
Lonchocarpus negrensis IPNI=*77178-2* 317
Lonchocarpus nicou IPNI=*502971-1* 325
Lonchocarpus nicou var. *languidus* IPNI=*1005531-1* 317
Lonchocarpus nicou var. *typicus* IPNI=*502971-1* 325
Lonchocarpus nicou var. *utilis* IPNI=*1005531-1* 317
Lonchocarpus nitidulus IPNI=*77143238-1* 317
Lonchocarpus nitidulus var. *schomburgkii* IPNI=*77143238-1* 317
Lonchocarpus oxycarpus IPNI=*502971-1* 325
Lonchocarpus pallidus IPNI=*77118402-1* 328
Lonchocarpus pentaphyllus IPNI=*502971-1* 325
Lonchocarpus pictus IPNI=*77185872-1* 317
Lonchocarpus pterocarpus IPNI=*491205-1* 317
Lonchocarpus pyxidarius IPNI=*143380-2* 325
Lonchocarpus rariflorus IPNI=*77116882-1* 317
Lonchocarpus roseus IPNI=*496639-1* 320
Lonchocarpus scandens IPNI=*491205-1* 317
Lonchocarpus sepium IPNI=*496639-1* 320
Lonchocarpus sericeus var. *glabrescens* IPNI=*503009-1* 325
Lonchocarpus sericeus var. *jamaicensis* IPNI=*143380-2* 325
Lonchocarpus stramineus IPNI=*503128-1* 325
Lonchocarpus swartzii IPNI=*502971-1* 325
Lonchocarpus tomentosus IPNI=*143380-2* 325
Lonchocarpus turpinii IPNI=*143380-2* 325
Lonchocarpus utilis IPNI=*1005531-1* 317
Lonchocarpus violaceus IPNI=*503074-1* 325
Loncomelos caudatum IPNI=*77097737-1* 214
Lonicera alba IPNI=*56535-2* 348
Lonicera aureoreticulata IPNI=*148850-1* 296
Lonicera brachypoda IPNI=*148850-1* 296
Lonicera chinensis IPNI=*148850-1* 296
Lonicera cochinchinensis IPNI=*148850-1* 296
Lonicera confusa IPNI=*148850-1* 296
Lonicera diversifolia IPNI=*148850-1* 296
Lonicera fauriei IPNI=*148850-1* 296
Lonicera finlaysoniana IPNI=*148850-1* 296
Lonicera flexuosa IPNI=*148850-1* 296
Lonicera japonica f. *quercifolia* IPNI=*148850-1* 296
Lonicera miyagusukiana IPNI=*148850-1* 296
Lonicera nigra IPNI=*148850-1* 296
Lonicera repens IPNI=*148850-1* 296
Lonicera shintenensis IPNI=*148850-1* 296
Lophalix coccinea IPNI=*84424-2* 367
Lophanthus formosanus IPNI=*444504-1* 368
Lophanthus rugosus IPNI=*444504-1* 368
Lophiaris fragrans IPNI=*1021269-1* 223
Lophiaris lanceana IPNI=*1021269-1* 223
Lophidium elegans IPNI=*17220460-1* 179
Lophidium flabellum IPNI=*17220460-1* 179
Lophidium latifolium IPNI=*17220460-1* 179
Lophidium pacificans IPNI=*17220460-1* 179
Lophidium spectabile IPNI=*17220460-1* 179
Lophiocarpus guayanensis IPNI=*58497-1* 198
Lopholepis vacciniifolia IPNI=*17406250-1* 176
Lophosoria acaulis IPNI=*143624-2* 173
Lophosoria affinis IPNI=*143624-2* 173
Lophosoria brasiliensis IPNI=*143624-2* 173
Lophosoria caesia IPNI=*143624-2* 173
Lophosoria densa IPNI=*143624-2* 173
Lophosoria discolor IPNI=*143624-2* 173
Lophosoria frigida IPNI=*143624-2* 173
Lophosoria glauca IPNI=*143624-2* 173
Lophosoria glaucescens IPNI=*143624-2* 173
Lophosoria polypodioides IPNI=*143624-2* 173
Lophosoria prostrata IPNI=*143624-2* 173
Lophosoria pruinata IPNI=*143624-2* 173
Lophosoria schaffneriana IPNI=*143624-2* 173
Lophosoria warscewiczii IPNI=*143624-2* 173
Lophotocarpus guayanensis IPNI=*58497-1* 198
Lophotocarpus guayanensis var. *typicus* IPNI=*58497-1* 198
Lophoxera comosa IPNI=*316350-2* 282
Lophoxera paniculata IPNI=*129956-2* 283
Lophoxera racemosa IPNI=*316350-2* 282
Loranthos stelum IPNI=*77147531-1* 448
Loranthos taguum IPNI=*548708-1* 448
Loranthus acutifolius IPNI=*258636-2* 449
Loranthus aduncus IPNI=*77147531-1* 448
Loranthus alternatus IPNI=*258636-2* 449
Loranthus alveolatus IPNI=*176899-2* 448
Loranthus amplexicaulis IPNI=*176899-2* 448
Loranthus anceps IPNI=*77147531-1* 448
Loranthus arboreus IPNI=*548708-1* 448
Loranthus attenuatus IPNI=*258636-2* 449
Loranthus avicularis IPNI=*550656-1* 448
Loranthus avium IPNI=*551718-1* 449
Loranthus bracteatus IPNI=*548708-1* 448
Loranthus bracteatus IPNI=*551367-1* 449
Loranthus calyculatus IPNI=*551347-1* 448
Loranthus chordocladus IPNI=*551737-1* 449
Loranthus conduplicatus IPNI=*77147531-1* 448
Loranthus cucullaris IPNI=*551367-1* 449
Loranthus cupulifer IPNI=*551367-1* 449
Loranthus domingensis IPNI=*1075239-2* 450
Loranthus ellipticus IPNI=*548708-1* 448
Loranthus eugenioides IPNI=*258636-2* 449
Loranthus falcifrons IPNI=*551367-1* 449
Loranthus granularis IPNI=*194192-2* 450
Loranthus hoffmannseggianus IPNI=*185262-2* 448
Loranthus immersus IPNI=*550656-1* 448
Loranthus jacquinii IPNI=*551347-1* 448
Loranthus lanceolatus IPNI=*548708-1* 448
Loranthus lancifolius IPNI=*548708-1* 448
Loranthus laurifolius IPNI=*548708-1* 448
Loranthus laurinus IPNI=*548708-1* 448
Loranthus lepidobotrys IPNI=*185262-2* 448
Loranthus ligustrifolius IPNI=*258636-2* 449
Loranthus ligustrinus IPNI=*258636-2* 449
Loranthus lucarquensis IPNI=*551737-1* 449

Loranthus magdalenae IPNI=*77147531-1* 448
Loranthus marginatus IPNI=*551718-1* 449
Loranthus marginatus IPNI=*77147531-1* 448
Loranthus membranaceus IPNI=*551367-1* 449
Loranthus mexicanus IPNI=*551367-1* 449
Loranthus multiflorus IPNI=*258636-2* 449
Loranthus nitidus IPNI=*548708-1* 448
Loranthus nitidus IPNI=*551367-1* 449
Loranthus obvallatus IPNI=*548708-1* 448
Loranthus occidentalis IPNI=*550656-1* 448
Loranthus orbicularis IPNI=*551737-1* 449
Loranthus orinocensis IPNI=*77147531-1* 448
Loranthus paniculatus IPNI=*77147531-1* 448
Loranthus pedunculatus IPNI=*551347-1* 448
Loranthus pedunculatus IPNI=*77147531-1* 448
Loranthus perrottetii IPNI=*185262-2* 448
Loranthus piperoides IPNI=*194192-2* 450
Loranthus plumieri IPNI=*551347-1* 448
Loranthus punctatus IPNI=*548708-1* 448
Loranthus puracensis IPNI=*548708-1* 448
Loranthus pyrifolius IPNI=*185262-2* 448
Loranthus quadrangularis IPNI=*60443757-2* 450
Loranthus retroflexus IPNI=*77147531-1* 448
Loranthus rotundatus IPNI=*176899-2* 448
Loranthus rotundifolius IPNI=*176899-2* 448
Loranthus sessilis IPNI=*77147531-1* 448
Loranthus spicatus IPNI=*550656-1* 448
Loranthus stelis IPNI=*77147531-1* 448
Loranthus suaveolens IPNI=*258636-2* 449
Loranthus tagua IPNI=*548708-1* 448
Loranthus tetragonus IPNI=*77147531-1* 448
Loranthus torulosus IPNI=*194192-2* 450
Loranthus verticillatus IPNI=*185262-2* 448
Loranthus viscifolius IPNI=*60443757-2* 450
Loranthus vulgaris IPNI=*551718-1* 449
Lorentea polycephala IPNI=*236598-1* 267
Lorentea prostrata IPNI=*236681-1* 267
Loretoa peruviana IPNI=*77210192-1* 347
Loreya acutifolia IPNI=*566597-1* 417
Loreya arborescens IPNI=*566597-1* 417
Loreya brunnescens IPNI=*31112-2* 418
Loreya collatata IPNI=*31113-2* 418
Loreya grandiflora IPNI=*31113-2* 418
Loreya maguirei IPNI=*566597-1* 417
Loreya mespiloides IPNI=*31112-2* 418
Loreya spruceana IPNI=*31113-2* 418
Loricaria microphylla IPNI=*231180-1* 265
Loricaria reticulosa IPNI=*183309-1* 257
Loricaria stenophylla IPNI=*231180-1* 265
Loricaria stenophylla var. *microphylla* IPNI=*231180-1* 265
Loricaria stenophylla var. *vernicosa* IPNI=*231180-1* 265
Loricaria thuyoides f. *major* IPNI=*231180-1* 265
Loricaria thuyoides var. *laxifolia* IPNI=*231180-1* 265
Loricaria thuyoides var. *microphylla* IPNI=*231180-1* 265
Loroma amethystina IPNI=*134734-3* 204
Loroma cunninghamiana IPNI=*134734-3* 204
Lortetia biflora IPNI=*674824-1* 398
Lortetia vespertilio IPNI=*675584-1* 400
Lotodes divaricatum IPNI=*281771-2* 329
Lotodes mutisii IPNI=*281771-2* 329
Lotodes pinnatum IPNI=*516230-1* 331
Lotoxalis barrelieri IPNI=*177777-2* 432
Lotoxalis fasciculata IPNI=*374932-1* 432
Lotoxalis frutescens IPNI=*374932-1* 432
Lotoxalis glabrata IPNI=*374932-1* 432
Lotoxalis pentantha IPNI=*374932-1* 432
Lotoxalis psilotricha IPNI=*374932-1* 432
Lotoxalis tephrodes IPNI=*374932-1* 432
Lotus maritimus IPNI=*149066-2* 327
Loxogramme lineata IPNI=*17209080-1* 178
Loxopterygium gutierrezii IPNI=*275630-2* 450
Loxostachys uncinata IPNI=*211125-2* 244
Lucilia paramora IPNI=*77154313-1* 268
Lucuma acreana IPNI=*209808-2* 307
Lucuma anibifolia IPNI=*209864-2* 307
Lucuma arguacoensium IPNI=*209690-2* 306
Lucuma bifera IPNI=*209804-2* 307
Lucuma bonplandii IPNI=*1110421-2* 307
Lucuma buchaniifolia IPNI=*943696-1* 307
Lucuma caimito IPNI=*788845-1* 306
Lucuma campechiana IPNI=*209714-2* 306
Lucuma crotonoides IPNI=*160368-2* 306
Lucuma deliciosa IPNI=*209690-2* 306
Lucuma dolichophylla IPNI=*788972-1* 307
Lucuma durlandii IPNI=*788920-1* 306
Lucuma elongata IPNI=*209714-2* 306
Lucuma ephedrantha IPNI=*948081-1* 307
Lucuma excelsa IPNI=*943696-1* 307
Lucuma glabrifolia IPNI=*209714-2* 306
Lucuma glomerata IPNI=*788965-1* 307
Lucuma gomphiifolia IPNI=*788967-1* 307
Lucuma gomphiifolia var. *blepharantha* IPNI=*788967-1* 307
Lucuma goudotiana IPNI=*209808-2* 307
Lucuma gutta IPNI=*788972-1* 307
Lucuma hartii IPNI=*788972-1* 307
Lucuma heyderi IPNI=*209714-2* 306
Lucuma huallagae IPNI=*788972-1* 307
Lucuma inseparabilis IPNI=*209714-2* 306
Lucuma jenmanii IPNI=*788972-1* 307
Lucuma krukovii IPNI=*209797-2* 307
Lucuma laeteviridis IPNI=*209714-2* 306
Lucuma laevigata IPNI=*789009-1* 307
Lucuma lasiocarpa IPNI=*788845-1* 306
Lucuma laurifolia IPNI=*788845-1* 306
Lucuma laurifolia var. *reticulata* IPNI=*788845-1* 306
Lucuma leptocarpa IPNI=*209758-2* 306
Lucuma lucentifolia IPNI=*948088-1* 305
Lucuma macrocarpa IPNI=*209824-2* 307
Lucuma mammosa IPNI=*152641-2* 306
Lucuma martinicensis IPNI=*209824-2* 307
Lucuma melinonii IPNI=*948066-1* 306
Lucuma minutiflora IPNI=*209864-2* 307
Lucuma multiflora IPNI=*209824-2* 307
Lucuma multiflora var. *urbanii* IPNI=*209824-2* 307
Lucuma nervosa IPNI=*209714-2* 306
Lucuma oblonga IPNI=*943696-1* 307
Lucuma obovata IPNI=*209804-2* 307
Lucuma obovata var. *ruizii* IPNI=*209804-2* 307
Lucuma palmeri IPNI=*209714-2* 306
Lucuma paraensis IPNI=*789045-1* 307
Lucuma pentasperma IPNI=*948088-1* 305
Lucuma peruviana IPNI=*209804-2* 307
Lucuma procera IPNI=*943696-1* 307
Lucuma procera var. *cuspidata* IPNI=*943696-1* 307
Lucuma quadrifida IPNI=*209824-2* 307
Lucuma rivicoa IPNI=*209808-2* 307
Lucuma rivicoa var. *angustifolia* IPNI=*209714-2* 306
Lucuma rivicoa var. *glaucophylla* IPNI=*209808-2* 307
Lucuma salicifolia IPNI=*209714-2* 306
Lucuma sclerocarpa IPNI=*209883-2* 308
Lucuma sericea IPNI=*209907-2* 308
Lucuma speciosa IPNI=*209892-2* 308
Lucuma sphaerocarpa IPNI=*209714-2* 306
Lucuma stahliana IPNI=*209824-2* 307
Lucuma temare IPNI=*788845-1* 306
Lucuma tomentosa IPNI=*786784-1* 305

Lucuma torta IPNI=*789233-1* 308
Lucuma turbinata IPNI=*209804-2* 307
Lucuma urbanii IPNI=*209824-2* 307
Lucuma venosa IPNI=*209920-2* 308
Lucuma venulosa IPNI=*160417-2* 306
Ludolfia glaucescens IPNI=*1097548-2* 234
Ludovia crenifolia IPNI=*297687-1* 227
Ludovia funifera IPNI=*102548-2* 227
Ludovia palmata IPNI=*297578-1* 227
Ludwigia acuminata IPNI=*144351-2* 429
Ludwigia adscendens var. *peploides* IPNI=*613407-1* 429
Ludwigia angustifolia IPNI=*144385-2* 429
Ludwigia apetala IPNI=*30031106-2* 429
Ludwigia brachycarpa IPNI=*144385-2* 429
Ludwigia capitata IPNI=*144385-2* 429
Ludwigia capitata var. *pubens* IPNI=*144385-2* 429
Ludwigia clavellina var. *peploides* IPNI=*613407-1* 429
Ludwigia diffusa IPNI=*613407-1* 429
Ludwigia diffusa var. *californica* IPNI=*613407-1* 429
Ludwigia hirta IPNI=*144406-2* 429
Ludwigia hondurensis IPNI=*144384-2* 429
Ludwigia inclinata var. *verticillata* IPNI=*613361-1* 429
Ludwigia jussiaeoides IPNI=*613333-1* 429
Ludwigia leptocarpa var. *meyeriana* IPNI=*144366-2* 429
Ludwigia linifolia IPNI=*613360-1* 429
Ludwigia micrantha IPNI=*613360-1* 429
Ludwigia nitida IPNI=*30031106-2* 429
Ludwigia octovalvis subsp. *macropoda* IPNI=*144385-2* 429
Ludwigia octovalvis var. *ligustrifolia* IPNI=*144385-2* 429
Ludwigia octovalvis var. *macropoda* IPNI=*144385-2* 429
Ludwigia octovalvis var. *octofila* IPNI=*144385-2* 429
Ludwigia palustris f. *elongata* IPNI=*30031106-2* 429
Ludwigia palustris f. *submersa* IPNI=*30031106-2* 429
Ludwigia palustris IPNI=*144324-2* 429
Ludwigia palustris var. *americana* IPNI=*30031106-2* 429
Ludwigia palustris var. *inundata* IPNI=*30031106-2* 429
Ludwigia palustris var. *liebmannii* IPNI=*30031106-2* 429
Ludwigia palustris var. *nana* IPNI=*30031106-2* 429
Ludwigia palustris var. *pacifica* IPNI=*30031106-2* 429
Ludwigia palustris var. *typica* IPNI=*30031106-2* 429
Ludwigia perennis IPNI=*144385-2* 429
Ludwigia peruviana var. *glaberrima* IPNI=*144406-2* 429
Ludwigia pubescens IPNI=*144385-2* 429
Ludwigia pubescens var. *ligustrifolia* IPNI=*144385-2* 429
Ludwigia pubescens var. *linearifolia* IPNI=*144385-2* 429
Ludwigia pubescens var. *macropoda* IPNI=*144385-2* 429
Ludwigia ramulosa IPNI=*613407-1* 429
Ludwigia sagrana IPNI=*144385-2* 429
Ludwigia suffruticosa IPNI=*144385-2* 429
Ludwigia turbinata IPNI=*144384-2* 429
Ludwigia uniflora IPNI=*613333-1* 429
Ludwigia verticillata IPNI=*613361-1* 429
Lueddemannia wallisii IPNI=*168902-2* 222
Luehea alternifolia IPNI=*834759-1* 410
Luehea endopogon IPNI=*834727-1* 410
Luehea ferruginea IPNI=*834759-1* 410
Luehea nobilis IPNI=*834727-1* 410
Luehea platypetala IPNI=*834759-1* 410
Luehea rufescens IPNI=*834758-1* 410
Luehea rufescens IPNI=*834759-1* 410
Luehea scabrifolia IPNI=*834727-1* 410
Luehea seemannii var. *chocoensis* IPNI=*834758-1* 410
Luehea speciosa var. *laxiflora* IPNI=*834759-1* 410
Luehea tarapotina IPNI=*834759-1* 410
Luerssenia cyminum IPNI=*840882-1* 252
Luffa acutangula f. *amara* IPNI=*105280-3* 294
Luffa acutangula var. *amara* IPNI=*105280-3* 294
Luffa aegyptiaca f. *sylvestris* IPNI=*293060-1* 294
Luffa aegyptiaca var. *leiocarpa* IPNI=*292238-1* 293
Luffa aegyptiaca var. *peramara* IPNI=*293060-1* 294
Luffa amara IPNI=*105280-3* 294
Luffa cattu-picinna IPNI=*105280-3* 294
Luffa clavata IPNI=*292238-1* 293
Luffa cylindrica IPNI=*292238-1* 293
Luffa cylindrica var. *insularum* IPNI=*293060-1* 294
Luffa cylindrica var. *multiflora* IPNI=*292238-1* 293
Luffa drastica IPNI=*105280-3* 294
Luffa fluminensis IPNI=*105280-3* 294
Luffa foetida IPNI=*105280-3* 294
Luffa foetida IPNI=*292238-1* 293
Luffa forskalii IPNI=*105280-3* 294
Luffa fricatoria IPNI=*293060-1* 294
Luffa gosa IPNI=*105280-3* 294
Luffa hederacea IPNI=*292238-1* 293
Luffa hermaphrodita IPNI=*105280-3* 294
Luffa insularum IPNI=*293060-1* 294
Luffa jacquinii IPNI=*292238-1* 293
Luffa kleinii IPNI=*105280-3* 294
Luffa leiocarpa IPNI=*292238-1* 293
Luffa leucosperma IPNI=*293060-1* 294
Luffa luffa IPNI=*292238-1* 293
Luffa moghadd IPNI=*291972-1* 293
Luffa operculata var. *intermedia* IPNI=*293097-1* 294
Luffa parvala IPNI=*292238-1* 293
Luffa pentandra IPNI=*293060-1* 294
Luffa petola IPNI=*293060-1* 294
Luffa plukenetiana IPNI=*105280-3* 294
Luffa purgans IPNI=*1076251-2* 294
Luffa racemosa IPNI=*292238-1* 293
Luffa satpatia IPNI=*292238-1* 293
Luffa scabra IPNI=*292238-1* 293
Luffa striata IPNI=*292238-1* 293
Luffa subangulata IPNI=*293060-1* 294
Luffa sylvestris IPNI=*293060-1* 294
Luffa tenera IPNI=*105280-3* 294
Luffa umbellata IPNI=*105280-3* 294
Luffa veitchii IPNI=*292238-1* 293
Luffa vittata IPNI=*292238-1* 293
Luma arechavaletae IPNI=*596263-1* 425
Luma costata IPNI=*596263-1* 425
Luma dasyblasta IPNI=*596263-1* 425
Luma strigosa IPNI=*596263-1* 425
Lumnitzera tenuiflora IPNI=*453130-1* 371
Lunania cuspidata IPNI=*779887-1* 404
Lunania pittieri IPNI=*779887-1* 404
Lunaria annua f. *elliptica* IPNI=*286647-1* 277
Lunaria annua f. *orbiculata* IPNI=*286647-1* 277
Lunaria annua var. *oppositifolia* IPNI=*286647-1* 277
Lunaria biennis IPNI=*286647-1* 277
Lunaria biennis var. *elliptica* IPNI=*286647-1* 277
Lunaria biennis var. *orbiculata* IPNI=*286647-1* 277
Lunaria halimifolia IPNI=*1085072-2* 277
Lunaria inodora IPNI=*286647-1* 277
Lundia chica IPNI=*77142115-1* 363
Lundia schomburgkii IPNI=*42392-2* 362
Luntia sericea IPNI=*70513-2* 389
Lupinaster amabilis IPNI=*522840-1* 335
Lupinus ambaticus IPNI=*505001-1* 326
Lupinus angustifolius IPNI=*30026493-2* 337
Lupinus bogotensis var. *parvior* IPNI=*504211-1* 325
Lupinus cruckshanksii IPNI=*504808-1* 326
Lupinus gracilis IPNI=*505050-1* 326
Lupinus lindenianus IPNI=*504972-1* 326
Lupinus mutabilis var. *cruckshanksii* IPNI=*504808-1* 326
Lupinus mutabilis var. *roseus* IPNI=*504808-1* 326
Lupinus paniculatus var. *pubescens* IPNI=*504972-1* 326

Lupinus pubescens var. *suberectus* IPNI=*504972-1* 326
Lupinus ramosissimus var. *ambaticus* IPNI=*505001-1* 326
Lupinus ramosissimus var. *platyphyllus* IPNI=*505001-1* 326
Lupinus subvolutus IPNI=*505001-1* 326
Lupularia parviflora IPNI=*506268-1* 327
Lupulina aurata IPNI=*506268-1* 327
Lupulus amarus IPNI=*303502-2* 436
Lupulus communis IPNI=*303502-2* 436
Lupulus humulus IPNI=*303502-2* 436
Lupulus lupuloides IPNI=*112858-2* 441
Lupulus lupuloides var. *domingensis* IPNI=*112858-2* 441
Lupulus lupuloides var. *pubescens* IPNI=*112865-2* 442
Lupulus lupuloides var. *tomentosus* IPNI=*112865-2* 442
Lupulus scandens IPNI=*303502-2* 436
Lussacia buxifolia IPNI=*858406-1* 299
Lustrinia geniculata IPNI=*47308-1* 359
Luteola resedoides IPNI=*715535-1* 280
Luteola tinctoria IPNI=*715535-1* 280
Lutheria splendens var. *chlorostachya* IPNI=*77158988-1* 229
Lutheria splendens var. *formosa* IPNI=*77158988-1* 229
Lutheria splendens var. *oinochroma* IPNI=*77158988-1* 229
Lutheria splendens var. *striatifolia* IPNI=*77158988-1* 229
Luziola leiocarpa IPNI=*289770-2* 241
Luziola mexicana IPNI=*289770-2* 241
Lycaste aciantha IPNI=*643235-1* 220
Lycaste crocea IPNI=*77096738-1* 223
Lycaste farinosa IPNI=*643478-1* 220
Lycaste flavescens IPNI=*77096738-1* 223
Lycaste fulvescens IPNI=*77096738-1* 223
Lycaste gigantea IPNI=*77096741-1* 223
Lycaste gigantea var. *labelloviridis* IPNI=*77096741-1* 223
Lycaste grandiflora IPNI=*643478-1* 220
Lycaste longipetala IPNI=*155147-2* 221
Lychnis armoraria IPNI=*77093328-1* 285
Lychnis cerastoides IPNI=*332106-2* 286
Lychnis coriacea IPNI=*60461121-2* 286
Lychnis coronaria IPNI=*60461121-2* 286
Lychnis vulnerata IPNI=*332106-2* 286
Lychnophora vanisschoti IPNI=*194359-1* 260
Lyciodes angustifolium IPNI=*948032-1* 308
Lyciodes candolleanum IPNI=*948032-1* 308
Lyciodes ferox IPNI=*948032-1* 308
Lyciodes hayesii IPNI=*948032-1* 308
Lyciodes obtusifolium IPNI=*1008737-2* 308
Lyciodes spiniflorum IPNI=*948032-1* 308
Lycopodiella cernua var. *caboverdeana* IPNI=*1054587-2* 171
Lycopodiella cernua var. *capillacea* IPNI=*1054587-2* 171
Lycopodiella cernua var. *sikkimensis* IPNI=*1054587-2* 171
Lycopodiella eichleri IPNI=*279987-2* 171
Lycopodioides amazonica IPNI=*231969-2* 171
Lycopodioides ambigua IPNI=*232097-2* 172
Lycopodioides articulata IPNI=*77186609-1* 171
Lycopodioides asperulum IPNI=*77186584-1* 171
Lycopodioides convoluta IPNI=*77142531-1* 171
Lycopodioides cuspidata IPNI=*30016320-2* 172
Lycopodioides erythropus IPNI=*30243959-2* 171
Lycopodioides exaltata IPNI=*77186589-1* 171
Lycopodioides flabellata IPNI=*30226649-2* 172
Lycopodioides flagellata IPNI=*232097-2* 172
Lycopodioides galeottii IPNI=*77186626-1* 172
Lycopodioides geniculata IPNI=*77186554-1* 172
Lycopodioides leptostachya IPNI=*232097-2* 172
Lycopodioides lindigii IPNI=*232147-2* 172
Lycopodioides lingulata IPNI=*232147-2* 172
Lycopodioides macrocladum IPNI=*77186636-1* 172
Lycopodioides mnioides IPNI=*77186646-1* 171
Lycopodioides orbigniana IPNI=*77142531-1* 171
Lycopodioides ovalis IPNI=*77186584-1* 171
Lycopodioides parkeri IPNI=*77186636-1* 172
Lycopodioides polycephalum IPNI=*77186646-1* 171
Lycopodioides regularis IPNI=*232097-2* 172
Lycopodioides rhizophora IPNI=*232097-2* 172
Lycopodioides rionegrensis IPNI=*231969-2* 171
Lycopodioides subarborescens IPNI=*77186636-1* 172
Lycopodium alopecuroides IPNI=*146929-2* 171
Lycopodium alopecuroides var. *pseudoreflexum* IPNI=*146929-2* 171
Lycopodium aristatum IPNI=*304307-2* 171
Lycopodium aristatum var. *desvauxianum* IPNI=*304307-2* 171
Lycopodium aristatum var. *incurvum* IPNI=*304307-2* 171
Lycopodium aristatum var. *robustius* IPNI=*304307-2* 171
Lycopodium asperulum IPNI=*77186584-1* 171
Lycopodium attenuatum IPNI=*124842-2* 171
Lycopodium attenuatum var. *crassum* IPNI=*124842-2* 171
Lycopodium axillare IPNI=*60447820-2* 171
Lycopodium barbatum IPNI=*1090207-2* 171
Lycopodium bonae-voluntatis IPNI=*124844-2* 171
Lycopodium canceris IPNI=*147420-2* 171
Lycopodium capillaceum IPNI=*1054587-2* 171
Lycopodium carolinianum IPNI=*146931-2* 171
Lycopodium carolinianum var. *brevipedunculatum* IPNI=*146931-2* 171
Lycopodium carolinianum var. *longepedunculatum* IPNI=*146931-2* 171
Lycopodium carolinianum var. *springii* IPNI=*146931-2* 171
Lycopodium carolinianum var. *typicum* IPNI=*304307-2* 171
Lycopodium cernuum f. *durissimum* IPNI=*1054587-2* 171
Lycopodium cernuum f. *ingens* IPNI=*1054587-2* 171
Lycopodium cernuum f. *macrostachyum* IPNI=*1054587-2* 171
Lycopodium cernuum f. *sikkimense* IPNI=*1054587-2* 171
Lycopodium cernuum IPNI=*1054587-2* 171
Lycopodium cernuum var. *blancheanum* IPNI=*1054587-2* 171
Lycopodium cernuum var. *brevifolium* IPNI=*1054587-2* 171
Lycopodium cernuum var. *caboverdeanum* IPNI=*1054587-2* 171
Lycopodium cernuum var. *capillaceum* IPNI=*1054587-2* 171
Lycopodium cernuum var. *chaminadeanum* IPNI=*1054587-2* 171
Lycopodium cernuum var. *crassifolium* IPNI=*1054587-2* 171
Lycopodium cernuum var. *dussii* IPNI=*1054587-2* 171
Lycopodium cernuum var. *eichleri* IPNI=*279987-2* 171
Lycopodium cernuum var. *majus* IPNI=*1054587-2* 171
Lycopodium cernuum var. *medium* IPNI=*1054587-2* 171
Lycopodium cernuum var. *minus* IPNI=*1054587-2* 171
Lycopodium cernuum var. *panamense* IPNI=*1054587-2* 171
Lycopodium cernuum var. *watsonianum* IPNI=*1054587-2* 171
Lycopodium chamaepeuce IPNI=*1090207-2* 171
Lycopodium ciliatum IPNI=*304307-2* 171
Lycopodium clavatum f. *brevipedunculatum* IPNI=*304307-2* 171
Lycopodium clavatum f. *decapitatum* IPNI=*304307-2* 171
Lycopodium clavatum f. *frondescens* IPNI=*304307-2* 171
Lycopodium clavatum f. *frondosum* IPNI=*304307-2* 171
Lycopodium clavatum f. *incurvum* IPNI=*304307-2* 171
Lycopodium clavatum f. *robustinus* IPNI=*304307-2* 171
Lycopodium clavatum f. *sterilis* IPNI=*304307-2* 171
Lycopodium clavatum f. *trilobum* IPNI=*304307-2* 171
Lycopodium clavatum subvar. *trilobum* IPNI=*304307-2* 171
Lycopodium clavatum var. *aristatum* IPNI=*304307-2* 171
Lycopodium clavatum var. *asiaticum* IPNI=*304307-2* 171
Lycopodium clavatum var. *borbonicum* IPNI=*304307-2* 171
Lycopodium clavatum var. *desvauxianum* IPNI=*304307-2* 171
Lycopodium clavatum var. *distachyum* IPNI=*304307-2* 171
Lycopodium clavatum var. *equisetoides* IPNI=*304307-2* 171
Lycopodium clavatum var. *eriostachys* IPNI=*304307-2* 171
Lycopodium clavatum var. *inflexum* IPNI=*304307-2* 171
Lycopodium clavatum var. *jamaicense* IPNI=*304307-2* 171
Lycopodium clavatum var. *laurentianum* IPNI=*304307-2* 171
Lycopodium clavatum var. *minarum* IPNI=*304307-2* 171
Lycopodium clavatum var. *montanum* IPNI=*304307-2* 171
Lycopodium clavatum var. *nataliense* IPNI=*304307-2* 171
Lycopodium clavatum var. *paucidivisum* IPNI=*304307-2* 171

Lycopodium clavatum var. *piliferum* IPNI=*304307-2* 171
Lycopodium clavatum var. *preslianum* IPNI=*304307-2* 171
Lycopodium clavatum var. *raddianum* IPNI=*304307-2* 171
Lycopodium clavatum var. *robustius* IPNI=*304307-2* 171
Lycopodium clavatum var. *serpens* IPNI=*304307-2* 171
Lycopodium clavatum var. *sitchaense* IPNI=*304307-2* 171
Lycopodium clavatum var. *subremotum* IPNI=*304307-2* 171
Lycopodium clavatum var. *trichiatum* IPNI=*304307-2* 171
Lycopodium clavatum var. *trichophyllum* IPNI=*304307-2* 171
Lycopodium complanatum f. *brachypodum* IPNI=*147262-2* 171
Lycopodium complanatum f. *canadense* IPNI=*147262-2* 171
Lycopodium complanatum f. *simplex* IPNI=*147262-2* 171
Lycopodium complanatum f. *typicum* IPNI=*147262-2* 171
Lycopodium complanatum subsp. *montellii* IPNI=*147262-2* 171
Lycopodium complanatum subvar. *foliaceum* IPNI=*147262-2* 171
Lycopodium complanatum subvar. *proliferum* IPNI=*147262-2* 171
Lycopodium complanatum subvar. *recurvatum* IPNI=*147262-2* 171
Lycopodium complanatum var. *adpressifolium* IPNI=*147665-2* 171
Lycopodium complanatum var. *canadense* IPNI=*147262-2* 171
Lycopodium complanatum var. *thyoides* IPNI=*147665-2* 171
Lycopodium complanatum var. *validum* IPNI=*147665-2* 171
Lycopodium convolutum IPNI=*77142531-1* 171
Lycopodium crassum IPNI=*124844-2* 171
Lycopodium crassum var. *springii* IPNI=*124844-2* 171
Lycopodium cuspidatum IPNI=*30016320-2* 172
Lycopodium dichotomum IPNI=*1090207-2* 171
Lycopodium dichotomum var. *maximum* IPNI=*1090207-2* 171
Lycopodium dichotomum var. *pusillum* IPNI=*1090207-2* 171
Lycopodium dichotomus var. *spegazzinii* IPNI=*1090207-2* 171
Lycopodium eichleri IPNI=*279987-2* 171
Lycopodium elongatum IPNI=*60447820-2* 171
Lycopodium eriostachys IPNI=*304307-2* 171
Lycopodium erythraeum IPNI=*124844-2* 171
Lycopodium erythropus IPNI=*30243959-2* 171
Lycopodium flabellatum IPNI=*30226649-2* 172
Lycopodium geniculatum IPNI=*77186554-1* 172
Lycopodium gramineum IPNI=*1090207-2* 171
Lycopodium haenkei var. *ramosissimum* IPNI=*147420-2* 171
Lycopodium heeschii IPNI=*1054587-2* 171
Lycopodium heterophyllum IPNI=*147420-2* 171
Lycopodium holtonii IPNI=*147420-2* 171
Lycopodium hupeanum IPNI=*1054587-2* 171
Lycopodium hygrometicum IPNI=*77142531-1* 171
Lycopodium inflexum IPNI=*304307-2* 171
Lycopodium insulare IPNI=*60447820-2* 171
Lycopodium inundatum var. *alopecuroides* IPNI=*146929-2* 171
Lycopodium jussiaei var. *microphyllum* IPNI=*147420-2* 171
Lycopodium lindseaceum IPNI=*147420-2* 171
Lycopodium marianum IPNI=*1054587-2* 171
Lycopodium mnioides IPNI=*77186646-1* 171
Lycopodium nesselianum IPNI=*124844-2* 171
Lycopodium nudum IPNI=*212103-2* 179
Lycopodium pallescens IPNI=*30016320-2* 172
Lycopodium parkeri IPNI=*77186636-1* 172
Lycopodium pendulinum IPNI=*279987-2* 171
Lycopodium piliferum IPNI=*304307-2* 171
Lycopodium preslii IPNI=*304307-2* 171
Lycopodium sanctae-barbarae IPNI=*60447820-2* 171
Lycopodium saururus IPNI=*60447820-2* 171
Lycopodium scariosum var. *heterophyllum* IPNI=*147420-2* 171
Lycopodium scariosum var. *holtonii* IPNI=*147420-2* 171
Lycopodium scariosum var. *jussiaei* IPNI=*147420-2* 171
Lycopodium scariosum var. *microphyllum* IPNI=*147420-2* 171
Lycopodium scariosum var. *ramosissimum* IPNI=*147420-2* 171
Lycopodium secundum IPNI=*1054587-2* 171
Lycopodium serpens IPNI=*304307-2* 171
Lycopodium sikkimense IPNI=*1054587-2* 171
Lycopodium tobarii IPNI=*124842-2* 171
Lycopodium torridum IPNI=*304307-2* 171
Lycopodium trichiatum IPNI=*304307-2* 171
Lycopodium trichiatum var. *desvauxianum* IPNI=*304307-2* 171
Lycopodium trichophyllum IPNI=*304307-2* 171
Lycopodium wilceae IPNI=*147262-2* 171
Lycopodium willdenowii IPNI=*60471496-2* 172
Lycoseris latifolia IPNI=*231347-1* 265
Lycoseris oblongifolia IPNI=*231347-1* 265
Lycurus muticus IPNI=*90411-2* 238
Lygistum ignitum IPNI=*755652-1* 352
Lygistum ignitum var. *micans* IPNI=*755652-1* 352
Lygistum micans IPNI=*755652-1* 352
Lygistum pleiodon IPNI=*755756-1* 352
Lygistum reclinatum IPNI=*755756-1* 352
Lygistum uniflorum IPNI=*755756-1* 352
Lygodium acuminatum IPNI=*17143520-1* 179
Lygodium commutatum IPNI=*17143510-1* 179
Lygodium expansum IPNI=*17143520-1* 179
Lygodium hastatum IPNI=*17143510-1* 179
Lygodium heptaphyllum IPNI=*17143520-1* 179
Lygodium hirtum IPNI=*17143520-1* 179
Lygodium intermedium IPNI=*17143520-1* 179
Lygodium lucens IPNI=*17143520-1* 179
Lygodium mexicanum IPNI=*17143510-1* 179
Lygodium micans IPNI=*17143520-1* 179
Lygodium mucronulatum IPNI=*17143510-1* 179
Lygodium palmatilobum IPNI=*17143510-1* 179
Lygodium pedatum IPNI=*17143520-1* 179
Lygodium pedicellatum IPNI=*17143520-1* 179
Lygodium pohlianum IPNI=*17143510-1* 179
Lygodium polymorphum f. *glabra* IPNI=*17143510-1* 179
Lygodium polymorphum f. *hirsuta* IPNI=*17143510-1* 179
Lygodium polymorphum var. *hirsutum* IPNI=*17143510-1* 179
Lygodium polymorphum var. *palmatilobum* IPNI=*17143510-1* 179
Lygodium puberulum IPNI=*17143520-1* 179
Lygodium scandens IPNI=*17143520-1* 179
Lygodium schiedeanum IPNI=*17143510-1* 179
Lygodium surinamense IPNI=*17143520-1* 179
Lygodium varium IPNI=*17143520-1* 179
Lygodium venustum f. *glabra* IPNI=*17143510-1* 179
Lygodium venustum f. *hirsuta* IPNI=*17143510-1* 179
Lygodium venustum var. *granatense* IPNI=*17143510-1* 179
Lygodium volubile var. *acuminatum* IPNI=*17143520-1* 179
Lygodium volubile var. *hirsutum* IPNI=*17143510-1* 179
Lygodium volubile var. *hirtum* IPNI=*17143520-1* 179
Lygodium volubile var. *intermedium* IPNI=*17143520-1* 179
Lygodium volubile var. *lucens* IPNI=*17143520-1* 179
Lygodium volubile var. *wrightii* IPNI=*17143520-1* 179
Lygodium wrightii IPNI=*17143520-1* 179
Lyonnetia abrotanifolia IPNI=*192489-1* 259
Lyperodendron coccoloboides IPNI=*693560-1* 288
Lyprolepis denudata IPNI=*74934-2* 231
Lysiloma guachapele IPNI=*211561-2* 331
Lysimachia arvensis f. *latifolia* IPNI=*77100386-1* 304
Lysimachia arvensis subsp. *latifolia* IPNI=*77100386-1* 304
Lysimachia monnieri IPNI=*1072674-2* 376
Lythrum alatum IPNI=*553756-1* 417
Lythrum albicaule IPNI=*553756-1* 417
Lythrum campestre IPNI=*553756-1* 417
Lythrum carthagenense IPNI=*72326-2* 416
Lythrum ciliatum IPNI=*552959-1* 416
Lythrum cistifolium IPNI=*552931-1* 416
Lythrum dipetalum IPNI=*552967-1* 417
Lythrum domingense IPNI=*553109-1* 417
Lythrum graefferi IPNI=*553756-1* 417
Lythrum lycioides IPNI=*553756-1* 417
Lythrum racemosum IPNI=*290454-2* 417
Lytocaryum weddellianum IPNI=*246979-2* 212

Lytocaryum weddellianum var. *cinereum* IPNI=*246979-2* 212
Lytocaryum weddellianum var. *pynaertii* IPNI=*246979-2* 212

M

Maba inconstans IPNI=*322526-1* 298
Maba melinonii IPNI=*322202-1* 298
Maba myristicoides IPNI=*322748-1* 298
Maba myrmecocarpa IPNI=*322748-1* 298
Maba sericea IPNI=*323004-1* 298
Mabea acutissima IPNI=*280157-2* 392
Mabea argutissima IPNI=*148081-2* 392
Mabea belizensis IPNI=*280157-2* 392
Mabea brasiliensis IPNI=*350987-1* 392
Mabea brasiliensis var. *genuina* IPNI=*350987-1* 392
Mabea brasiliensis var. *intermedia* IPNI=*350987-1* 392
Mabea depauperata IPNI=*280156-2* 392
Mabea glauca IPNI=*280157-2* 392
Mabea lactescens IPNI=*350987-1* 392
Mabea maynensis IPNI=*350987-1* 392
Mabea microcarpa IPNI=*280157-2* 392
Mabea muricata IPNI=*280156-2* 392
Mabea nitida var. *purpurascens* IPNI=*280156-2* 392
Mabea occidentalis var. *concolor* IPNI=*350987-1* 392
Mabea occidentalis var. *genuina* IPNI=*280157-2* 392
Mabea pallida IPNI=*280156-2* 392
Mabea pallida IPNI=*280157-2* 392
Mabea parvifolia IPNI=*148215-2* 392
Mabea piriri var. *concolor* IPNI=*350987-1* 392
Mabea piriri var. *genuina* IPNI=*350987-1* 392
Mabea piririoides IPNI=*350987-1* 392
Mabea speciosa subsp. *concolor* IPNI=*350987-1* 392
Mabea surinamensis IPNI=*350987-1* 392
Mabola edulis IPNI=*322328-1* 298
Macadamia ternifolia var. *integrifolia* IPNI=*705192-1* 433
Macaglia alba IPNI=*77325-1* 339
Macaglia bicolor IPNI=*77347-1* 339
Macaglia decipiens IPNI=*77347-1* 339
Macaglia desmantha IPNI=*77353-1* 339
Macaglia excelsa IPNI=*77367-1* 339
Macaglia megalocarpa IPNI=*77405-1* 339
Macaglia melanocalyx IPNI=*22069-2* 339
Macaglia spruceana IPNI=*22069-2* 339
Macaglia vargasii IPNI=*315552-2* 339
Macaglia verruculosa IPNI=*22069-2* 339
Macfadyena bangii IPNI=*60474996-2* 362
Macfadyena bifolia IPNI=*108550-1* 362
Macfadyena coito IPNI=*60474996-2* 362
Macfadyena corymbosa IPNI=*60474996-2* 362
Macfadyena fenzliana IPNI=*77100511-1* 363
Macfadyena fockeana IPNI=*153710-2* 364
Macfadyena guatemalensis IPNI=*77100511-1* 363
Macfadyena hassleri IPNI=*77100511-1* 363
Macfadyena japurensis IPNI=*77142016-1* 363
Macfadyena laurifolia IPNI=*60474996-2* 362
Macfadyena obovata IPNI=*153710-2* 364
Macfadyena orinocensis IPNI=*60474996-2* 362
Macfadyena ovata IPNI=*60474996-2* 362
Macfadyena platypoda IPNI=*60474996-2* 362
Macfadyena riparia IPNI=*60474996-2* 362
Macfadyena uncata IPNI=*77100511-1* 363
Macfadyena uncinata IPNI=*77100511-1* 363
Macfadyena undulata IPNI=*77100466-1* 363
Macfadyena unguis-cati IPNI=*77100466-1* 363
Machaerium acuminatum IPNI=*505373-1* 326
Machaerium armatum IPNI=*148544-2* 326
Machaerium bangii IPNI=*148544-2* 326
Machaerium brachycarpum IPNI=*505373-1* 326
Machaerium campylothyrsum IPNI=*148542-2* 326
Machaerium campylothyrsum var. *crebrifoliolatum* IPNI=*148542-2* 326
Machaerium costaricanum IPNI=*148544-2* 326
Machaerium fruticetorum IPNI=*505373-1* 326
Machaerium longifolium IPNI=*148544-2* 326
Machaerium marginatum IPNI=*148544-2* 326
Machaerium microphyllum IPNI=*148544-2* 326
Machaerium pseudotipe IPNI=*516619-1* 331
Machaerium schunkei IPNI=*148544-2* 326
Machaerium setulosum IPNI=*148544-2* 326
Machaerium stenophyllum IPNI=*505373-1* 326
Machaerium tobagense IPNI=*148544-2* 326
Macleania alpicola IPNI=*148722-2* 299
Macleania attenuata IPNI=*148722-2* 299
Macleania costaricensis IPNI=*148722-2* 299
Macleania ecuadorensis IPNI=*148722-2* 299
Macleania elliptica IPNI=*148722-2* 299
Macleania euryphylla IPNI=*148722-2* 299
Macleania glabra IPNI=*148722-2* 299
Macleania humboldtiana IPNI=*252411-2* 300
Macleania irazuensis IPNI=*148722-2* 299
Macleania nervosa IPNI=*148722-2* 299
Macleania nitida IPNI=*50474-2* 298
Macleania pilgeriana IPNI=*148722-2* 299
Macleania punctata var. *puberula* IPNI=*148722-2* 299
Macleania racemosa IPNI=*148722-2* 299
Macleania recurva IPNI=*148722-2* 299
Macleania reducta IPNI=*148722-2* 299
Macleania sodiroi IPNI=*148722-2* 299
Macleania tovarensis IPNI=*858621-1* 300
Macleania trianae IPNI=*148722-2* 299
Macleania turrialbana IPNI=*148722-2* 299
Maclura affinis IPNI=*1151750-2* 440
Maclura chlorocarpa IPNI=*1151750-2* 440
Maclura mora IPNI=*1151750-2* 440
Maclura plumiera IPNI=*1151750-2* 440
Maclura polyneura IPNI=*1151750-2* 440
Maclura sempervirens IPNI=*1151750-2* 440
Maclura sieberi IPNI=*1151750-2* 440
Maclura subintegerrima IPNI=*1151750-2* 440
Maclura tinctoria subsp. *mora* IPNI=*1151750-2* 440
Maclura tinctoria subvar. *lobata* IPNI=*1151750-2* 440
Maclura tinctoria subvar. *quercina* IPNI=*1151750-2* 440
Maclura tinctoria subvar. *sinuata* IPNI=*1151750-2* 440
Maclura tinctoria var. *affinis* IPNI=*1151750-2* 440
Maclura tinctoria var. *chlorocarpa* IPNI=*1151750-2* 440
Maclura tinctoria var. *ovata* IPNI=*1151750-2* 440
Maclura tinctoria var. *polyneura* IPNI=*1151750-2* 440
Maclura tinctoria var. *subcuneata* IPNI=*1151750-2* 440
Maclura tinctoria var. *subintegerrima* IPNI=*1151750-2* 440
Maclura trilobata IPNI=*1151750-2* 440
Maclura velutina IPNI=*1151750-2* 440
Macnemaraea opulina IPNI=*791637-1* 291
Macoubea fasciculata IPNI=*182946-2* 342
Macoubea guianensis var. *pubiflora* IPNI=*79852-1* 341
Macoubea guianensis var. *reticulata* IPNI=*79852-1* 341
Macoubea paucifolia IPNI=*79852-1* 341
Macoubea reticulata IPNI=*79852-1* 341
Macoubea sinuosa IPNI=*79852-1* 341
Macoubea sprucei var. *paucifolia* IPNI=*79852-1* 341
Macoubea witotorum IPNI=*79852-1* 341
Macoucoua guianensis IPNI=*127855-2* 255
Macreightia inconstans IPNI=*322526-1* 298
Macreightia myristicoides IPNI=*322748-1* 298
Macroblepharus contractus IPNI=*1156566-2* 238
Macrobriza maxima IPNI=*393293-1* 235
Macrocatalpa longissima IPNI=*77176826-1* 362
Macrochlamys miquelii IPNI=*84424-2* 367
Macrochlamys patrisii IPNI=*84424-2* 367

Macrochordion bromeliifolium IPNI=*121574-1* 227
Macrocnemum candidissimum IPNI=*745241-1* 347
Macrocnemum cinchonioides IPNI=*755513-1* 352
Macrocnemum coccineum IPNI=*770293-1* 358
Macrocnemum corymbosum IPNI=*1010049-2* 348
Macrocnemum glabrescens IPNI=*755513-1* 352
Macrocnemum hirsutum IPNI=*755513-1* 352
Macrocnemum pastoense IPNI=*755513-1* 352
Macrocnemum pilosinervium IPNI=*755513-1* 352
Macrocnemum speciosum IPNI=*204802-2* 355
Macrocnemum sprucei IPNI=*755513-1* 352
Macrocroton cuneatus IPNI=*342381-1* 388
Macrocroton surinamensis IPNI=*342381-1* 388
Macrolobium acaciifolium var. *vestitum* IPNI=*505635-1* 326
Macrolobium elegans IPNI=*30147160-2* 326
Macrolobium hymenaeoides IPNI=*30147160-2* 326
Macrolobium stamineum IPNI=*30147160-2* 326
Macrolobium vuapa IPNI=*30147160-2* 326
Macrolomia bracteata IPNI=*314601-1* 232
Macrophora sanguinea IPNI=*30044451-2* 400
Macrophthalma elastica IPNI=*60458499-2* 438
Macropiper latum IPNI=*198988-2* 197
Macropsychanthus densiflorus IPNI=*77212324-1* 326
Macroptilium lathyroides var. *bustarretianum* IPNI=*149066-2* 327
Macroptilium lathyroides var. *semierectum* IPNI=*149066-2* 327
Macrosamanea pedicellaris IPNI=*989349-1* 311
Macrostegia ruiziana IPNI=*866038-1* 374
Macrothelypteris changshaensis IPNI=*1019960-2* 174
Macrothelypteris oligophlebia IPNI=*1019960-2* 174
Macrothelypteris oligophlebia var. *changshaensis* IPNI=*1019960-2* 174
Macrothelypteris oligophlebia var. *elegans* IPNI=*1019960-2* 174
Macrothelypteris torresiana monstr. lophaea IPNI=*1019960-2* 174
Macrozamia tridentata var. *pungens* IPNI=*668912-1* 211
Madorius giganteus IPNI=*95173-1* 339
Madorius procerus IPNI=*1004515-2* 339
Maerlensia hirsuta IPNI=*320795-2* 408
Magnolia × brozzonii IPNI=*60468302-2* 190
Magnolia × lenneana IPNI=*60468302-2* 190
Magnolia × lennei IPNI=*60468302-2* 190
Magnolia alexandrina IPNI=*554678-1* 190
Magnolia angustifolia IPNI=*554723-1* 190
Magnolia atropurpurea IPNI=*554759-1* 190
Magnolia citriodora IPNI=*554678-1* 190
Magnolia conspicua IPNI=*554678-1* 190
Magnolia cuneatifolia IPNI=*554678-1* 190
Magnolia cyathiformis IPNI=*554678-1* 190
Magnolia denudata var. *angustitepala* IPNI=*554678-1* 190
Magnolia denudata var. *glabrata* IPNI=*554678-1* 190
Magnolia denudata var. *pyriformis* IPNI=*554678-1* 190
Magnolia discolor IPNI=*554759-1* 190
Magnolia elliptica IPNI=*554723-1* 190
Magnolia exoniensis IPNI=*554723-1* 190
Magnolia ferruginea IPNI=*554723-1* 190
Magnolia foetida f. *margaretta* IPNI=*554723-1* 190
Magnolia foetida f. *parvifolia* IPNI=*554723-1* 190
Magnolia foetida IPNI=*554723-1* 190
Magnolia galissoniensis IPNI=*554723-1* 190
Magnolia glabra IPNI=*554723-1* 190
Magnolia gloriosa IPNI=*554723-1* 190
Magnolia gracilis IPNI=*554759-1* 190
Magnolia grandiflora var. *elliptica* IPNI=*554723-1* 190
Magnolia grandiflora var. *ferruginea* IPNI=*554723-1* 190
Magnolia grandiflora var. *lanceolata* IPNI=*554723-1* 190
Magnolia grandiflora var. *obovata* IPNI=*554723-1* 190
Magnolia heptapeta IPNI=*554678-1* 190
Magnolia lacunosa IPNI=*554723-1* 190
Magnolia lanceolata IPNI=*554723-1* 190
Magnolia liliiflora f. *nigra* IPNI=*554759-1* 190
Magnolia longifolia IPNI=*554723-1* 190
Magnolia maxima IPNI=*554723-1* 190
Magnolia obovata IPNI=*554723-1* 190
Magnolia plena IPNI=*554759-1* 190
Magnolia polytepala IPNI=*554759-1* 190
Magnolia praecox IPNI=*554723-1* 190
Magnolia pravertiana IPNI=*554723-1* 190
Magnolia precia IPNI=*554678-1* 190
Magnolia purpurea IPNI=*554759-1* 190
Magnolia quinquepeta IPNI=*554759-1* 190
Magnolia rotundifolia IPNI=*554723-1* 190
Magnolia virginiana var. *foetida* IPNI=*554723-1* 190
Magnolia virginiana var. *grisea* IPNI=*554723-1* 190
Magnolia yulan IPNI=*554678-1* 190
Magonia coriacea IPNI=*693983-1* 288
Magonia cumingii IPNI=*693983-1* 288
Magonia martii IPNI=*697785-1* 290
Magonia ramiflora IPNI=*693983-1* 288
Maieta argentea IPNI=*573207-1* 422
Maieta dentata IPNI=*77207920-1* 422
Maieta dependens IPNI=*77196165-1* 420
Maieta dispar IPNI=*77164249-1* 421
Maieta glandulifera IPNI=*77207761-1* 423
Maieta guianensis IPNI=*77164249-1* 421
Maieta guianensis var. *leticiana* IPNI=*77164249-1* 421
Maieta guianensis var. *peruviana* IPNI=*77164249-1* 421
Maieta heterophylla IPNI=*77207843-1* 420
Maieta hirta IPNI=*77164247-1* 420
Maieta hispida IPNI=*77164249-1* 421
Maieta hypophysca IPNI=*77164249-1* 421
Maieta pilosa IPNI=*77207921-1* 421
Maieta rubra IPNI=*77164250-1* 422
Maieta spicata IPNI=*77196165-1* 420
Maieta testiculata IPNI=*77207921-1* 421
Maieta tococoidea IPNI=*77207700-1* 423
Maieta tococoidea var. *watsonii* IPNI=*77207700-1* 423
Maieta valenzuelana IPNI=*77196163-1* 421
Maizilla stolonifera IPNI=*414634-1* 243
Majana acuminata IPNI=*446165-1* 369
Majana amboinica IPNI=*445902-1* 369
Majana blancoi IPNI=*446165-1* 369
Majana carnosa IPNI=*445902-1* 369
Majana grandifolia IPNI=*446165-1* 369
Majana multiflora IPNI=*446165-1* 369
Majana paniculata IPNI=*446113-1* 369
Majana pumila IPNI=*446165-1* 369
Majana scutellariodes IPNI=*446165-1* 369
Majana secundiflora IPNI=*446165-1* 369
Majana suganda IPNI=*445902-1* 369
Majorana × majorica IPNI=*453305-1* 372
Majorana dubia IPNI=*453303-1* 372
Majorana fragrans IPNI=*453303-1* 372
Majorana hortensis IPNI=*453303-1* 372
Majorana majorana IPNI=*453303-1* 372
Majorana mexicana IPNI=*453303-1* 372
Majorana ovalifolia IPNI=*453303-1* 372
Majorana ovatifolia IPNI=*453303-1* 372
Majorana suffruticosa IPNI=*453303-1* 372
Majorana tenuifolia IPNI=*453303-1* 372
Majorana uncinata IPNI=*453303-1* 372
Majorana vulgaris IPNI=*453303-1* 372
Malache cordifolia IPNI=*562981-1* 412
Malache flava IPNI=*186058-2* 412
Malache fruticosa IPNI=*185868-2* 412
Malache leptocarpa IPNI=*185868-2* 412
Malache rosea IPNI=*563076-1* 412
Malache sepium IPNI=*186058-2* 412

Malache spinifex IPNI=*186083-2* 412
Malache stellata IPNI=*185868-2* 412
Malache typhalaea f. *albiflora* IPNI=*185868-2* 412
Malache typhalaea f. *aurantiaca* IPNI=*185868-2* 412
Malache typhalaea f. *rosea* IPNI=*185868-2* 412
Malache typhalaea IPNI=*185868-2* 412
Malache typhalaeoides IPNI=*185868-2* 412
Malachodendron corchoroides IPNI=*235838-2* 413
Malachra alceifolia f. *hispida* IPNI=*561421-1* 410
Malachra alceifolia f. *hispidissima* IPNI=*561421-1* 410
Malachra alceifolia var. *conglomerata* IPNI=*561421-1* 410
Malachra alceifolia var. *fasciata* IPNI=*561430-1* 410
Malachra alceifolia var. *rotundifolia* IPNI=*561421-1* 410
Malachra alceifolia var. *typica* IPNI=*561421-1* 410
Malachra berteroi IPNI=*60439289-2* 415
Malachra bracteata IPNI=*561446-1* 410
Malachra capitata var. *albiflora* IPNI=*561421-1* 410
Malachra capitata var. *flavida* IPNI=*561424-1* 410
Malachra capitata var. *lasiocarpa* IPNI=*561424-1* 410
Malachra capitata var. *microcalyx* IPNI=*561424-1* 410
Malachra capitata var. *mollis* IPNI=*561424-1* 410
Malachra capitata var. *normalis* IPNI=*561424-1* 410
Malachra capitata var. *triloba* IPNI=*561424-1* 410
Malachra conglomerata IPNI=*561421-1* 410
Malachra gaudichaudiana IPNI=*561446-1* 410
Malachra heptaphylla IPNI=*561424-1* 410
Malachra hispida IPNI=*561421-1* 410
Malachra horrida IPNI=*561430-1* 410
Malachra humilis IPNI=*561430-1* 410
Malachra kegeliana IPNI=*561430-1* 410
Malachra lineariloba IPNI=*561430-1* 410
Malachra mexicana IPNI=*561424-1* 410
Malachra officinalis IPNI=*561421-1* 410
Malachra palmata IPNI=*561424-1* 410
Malachra rotundifolia IPNI=*561421-1* 410
Malachra urena IPNI=*564786-1* 415
Malachra velutina IPNI=*561424-1* 410
Malacochaete assimilis IPNI=*229810-2* 232
Malacochaete chilensis IPNI=*229810-2* 232
Malacochaete oligostachya IPNI=*229810-2* 232
Malacochaete riparia IPNI=*229810-2* 232
Malacochaete riparia var. *chamissoi* IPNI=*229810-2* 232
Malacochaete sanguinolenta IPNI=*229810-2* 232
Malacochaete tatora IPNI=*229810-2* 232
Malacothrix crepoides IPNI=*199753-1* 260
Malaxis nervosa IPNI=*641061-1* 219
Mallostoma caracasanum IPNI=*743608-1* 347
Mallostoma confertum IPNI=*17992-2* 347
Mallostoma hyssopifolium IPNI=*18012-2* 347
Mallostoma juniperifolium IPNI=*18012-2* 347
Mallostoma laricifolium IPNI=*18012-2* 347
Mallostoma nitidum IPNI=*743608-1* 347
Mallostoma thymifolium IPNI=*18012-2* 347
Mallotonia gnaphalodes IPNI=*120998-1* 275
Mallotus moluccanus IPNI=*338514-1* 387
Malmea anomala IPNI=*1003565-1* 189
Malocchia ensiformis IPNI=*1209500-2* 313
Malope lutea IPNI=*561999-1* 410
Malope malacoides IPNI=*561999-1* 410
Malouetia cubana IPNI=*79896-1* 341
Malouetia jasminoides IPNI=*81996-1* 343
Malouetia lactiflua IPNI=*182946-2* 342
Malouetia mexiae IPNI=*79896-1* 341
Malouetia retroflexa IPNI=*79896-1* 341
Malouetia riparia IPNI=*82139-1* 343
Malouetia schomburgkiana IPNI=*79896-1* 341
Malouetia tetrastachya IPNI=*20001045-1* 344
Malpighia altissima IPNI=*556192-1* 395
Malpighia argentea IPNI=*555883-1* 395
Malpighia armeniaca IPNI=*555884-1* 395
Malpighia berteroana IPNI=*323670-2* 396
Malpighia bicolor IPNI=*556209-1* 396
Malpighia biflora IPNI=*557048-1* 396
Malpighia cinerea IPNI=*556033-1* 395
Malpighia coccigera var. *coccigrya* IPNI=*323776-2* 396
Malpighia coccigera var. *microphylla* IPNI=*323776-2* 396
Malpighia coccigrya IPNI=*323776-2* 396
Malpighia coccolobifolia IPNI=*556020-1* 395
Malpighia coriacea IPNI=*556033-1* 395
Malpighia cotinifolia IPNI=*556033-1* 395
Malpighia crassifolia IPNI=*556033-1* 395
Malpighia crassifolia IPNI=*556209-1* 396
Malpighia dubia IPNI=*556563-1* 396
Malpighia elegans IPNI=*557422-1* 396
Malpighia fallax IPNI=*557048-1* 396
Malpighia glabra var. *acuminata* IPNI=*557048-1* 396
Malpighia glabra var. *antillana* IPNI=*557048-1* 396
Malpighia glabra var. *guatemalensis* IPNI=*557048-1* 396
Malpighia glabra var. *lancifolia* IPNI=*557048-1* 396
Malpighia glabra var. *typica* IPNI=*557048-1* 396
Malpighia glabra var. *undulata* IPNI=*557048-1* 396
Malpighia glandulifera IPNI=*555908-1* 395
Malpighia glandulosa IPNI=*555908-1* 395
Malpighia guadalupensis IPNI=*556192-1* 395
Malpighia heterantha IPNI=*323776-2* 396
Malpighia japurensis IPNI=*556093-1* 395
Malpighia lanceolata IPNI=*556033-1* 395
Malpighia lancifolia IPNI=*323670-2* 396
Malpighia laurifolia IPNI=*556033-1* 395
Malpighia lucida IPNI=*557048-1* 396
Malpighia macrophylla IPNI=*557048-1* 396
Malpighia montana IPNI=*556033-1* 395
Malpighia moureila IPNI=*556033-1* 395
Malpighia myrtifolia IPNI=*557048-1* 396
Malpighia myrtoides IPNI=*557048-1* 396
Malpighia neumanniana IPNI=*557048-1* 396
Malpighia nitida IPNI=*1122895-2* 395
Malpighia nitida IPNI=*555969-1* 395
Malpighia nitida IPNI=*557048-1* 396
Malpighia obovata IPNI=*556914-1* 396
Malpighia odorata IPNI=*555944-1* 395
Malpighia oxycocca var. *biflora* IPNI=*557048-1* 396
Malpighia peruviana IPNI=*557048-1* 396
Malpighia platyphylla IPNI=*555908-1* 395
Malpighia pruinosa IPNI=*556020-1* 395
Malpighia pulchra IPNI=*556033-1* 395
Malpighia punicifolia IPNI=*557048-1* 396
Malpighia punicifolia var. *lancifolia* IPNI=*323670-2* 396
Malpighia punicifolia var. *obovata* IPNI=*323670-2* 396
Malpighia punicifolia var. *vulgaris* IPNI=*323670-2* 396
Malpighia reticulata IPNI=*1079574-2* 396
Malpighia reticulata IPNI=*556192-1* 395
Malpighia retusa IPNI=*323670-2* 396
Malpighia rhopalifolia IPNI=*556033-1* 395
Malpighia rufa IPNI=*556033-1* 395
Malpighia semeruco IPNI=*557048-1* 396
Malpighia spicata IPNI=*1122895-2* 395
Malpighia spicata IPNI=*556192-1* 395
Malpighia umbellata IPNI=*323670-2* 396
Malpighia undulata IPNI=*557048-1* 396
Malpighia verbascifolia IPNI=*556209-1* 396
Malpighiodes benthamiana IPNI=*60443063-2* 396
Malus anisophylla IPNI=*726282-1* 443
Malus apetala IPNI=*726282-1* 443
Malus astracanica IPNI=*726282-1* 443
Malus batavica IPNI=*726282-1* 443

Malus calvilla IPNI=*726282-1* 443
Malus chitralensis IPNI=*726282-1* 443
Malus communis IPNI=*30065762-2* 444
Malus communis IPNI=*726282-1* 443
Malus cylindracea IPNI=*726282-1* 443
Malus dasyphylla IPNI=*726282-1* 443
Malus dioica IPNI=*726282-1* 443
Malus dolichomorpha IPNI=*726282-1* 443
Malus domestica subsp. *caucasica* IPNI=*726282-1* 443
Malus domestica subsp. *hybrida* IPNI=*726282-1* 443
Malus domestica subsp. *intermedia* IPNI=*726282-1* 443
Malus domestica subsp. *italotaurica* IPNI=*726282-1* 443
Malus domestica subsp. *macrocarpa* IPNI=*726282-1* 443
Malus domestica subsp. *medioasiatica* IPNI=*726282-1* 443
Malus domestica subsp. *occidentalieuropaea* IPNI=*726282-1* 443
Malus domestica subsp. *pumila* IPNI=*726282-1* 443
Malus domestica subsp. *rossica* IPNI=*726282-1* 443
Malus domestica var. *calvillea* IPNI=*726282-1* 443
Malus domestica var. *dasyphylla* IPNI=*726282-1* 443
Malus domestica var. *parmaenea* IPNI=*726282-1* 443
Malus domestica var. *renettea* IPNI=*726282-1* 443
Malus fontanesiana IPNI=*726282-1* 443
Malus frutescens IPNI=*726282-1* 443
Malus glabra IPNI=*726282-1* 443
Malus kirghisorum IPNI=*726282-1* 443
Malus kudrjaschevii IPNI=*726282-1* 443
Malus linczevskii IPNI=*726282-1* 443
Malus livida IPNI=*726282-1* 443
Malus malus IPNI=*726282-1* 443
Malus niedzwetzkyana IPNI=*726282-1* 443
Malus paradisiaca IPNI=*726282-1* 443
Malus passipoma IPNI=*726282-1* 443
Malus persicifolia IPNI=*726282-1* 443
Malus praecox IPNI=*726282-1* 443
Malus pumila IPNI=*726282-1* 443
Malus rambura IPNI=*726282-1* 443
Malus santonum IPNI=*726282-1* 443
Malus sieversii f. *cerea* IPNI=*726282-1* 443
Malus sieversii f. *flava* IPNI=*726282-1* 443
Malus sieversii f. *flavorosea* IPNI=*726282-1* 443
Malus sieversii f. *flavorubens* IPNI=*726282-1* 443
Malus sieversii f. *gilva* IPNI=*726282-1* 443
Malus sieversii f. *insipida* IPNI=*726282-1* 443
Malus sieversii f. *laetevirida* IPNI=*726282-1* 443
Malus sieversii f. *media* IPNI=*726282-1* 443
Malus sieversii f. *ochrorubra* IPNI=*726282-1* 443
Malus sieversii f. *pomacea* IPNI=*726282-1* 443
Malus sieversii f. *purpureofasciata* IPNI=*726282-1* 443
Malus sieversii f. *rubrifasciata* IPNI=*726282-1* 443
Malus sieversii f. *rubrolateralis* IPNI=*726282-1* 443
Malus sieversii f. *suffusa* IPNI=*726282-1* 443
Malus sieversii f. *viridicerea* IPNI=*726282-1* 443
Malus sieversii f. *viridifasciata* IPNI=*726282-1* 443
Malus sieversii f. *viridirubra* IPNI=*726282-1* 443
Malus sieversii f. *viridis* IPNI=*726282-1* 443
Malus sieversii f. *viridula* IPNI=*726282-1* 443
Malus sieversii f. *xanthella* IPNI=*726282-1* 443
Malus sieversii IPNI=*726282-1* 443
Malus sieversii subsp. *hissarica* IPNI=*726282-1* 443
Malus sieversii subsp. *kirghisorum* IPNI=*726282-1* 443
Malus sieversii var. *anisophylla* IPNI=*726282-1* 443
Malus sieversii var. *caspiriensis* IPNI=*726282-1* 443
Malus sieversii var. *chitralensis* IPNI=*726282-1* 443
Malus sieversii var. *hissarica* IPNI=*726282-1* 443
Malus sieversii var. *jarmolenkoi* IPNI=*726282-1* 443
Malus sieversii var. *juzepczukii* IPNI=*726282-1* 443
Malus sieversii var. *kirghisorum* IPNI=*726282-1* 443
Malus sieversii var. *kudrjaschevii* IPNI=*726282-1* 443
Malus sieversii var. *linczevskii* IPNI=*726282-1* 443
Malus sieversii var. *niedzwetzkyana* IPNI=*726282-1* 443
Malus sieversii var. *persicifolia* IPNI=*726282-1* 443
Malus sieversii var. *schischkinii* IPNI=*726282-1* 443
Malus sieversii var. *tianschanica* IPNI=*726282-1* 443
Malus spumosa IPNI=*726282-1* 443
Malus sylvestris subsp. *paradisiaca* IPNI=*726282-1* 443
Malus sylvestris var. *dasyphylla* IPNI=*726282-1* 443
Malus sylvestris var. *praecox* IPNI=*726282-1* 443
Malus tatarica IPNI=*726282-1* 443
Malus tianschanica IPNI=*726282-1* 443
Malus turbinoides IPNI=*726282-1* 443
Malus umbellata IPNI=*726282-1* 443
Malus upsalensis IPNI=*726282-1* 443
Malus violacea IPNI=*726282-1* 443
Malus viridis IPNI=*726282-1* 443
Malva acutiloba IPNI=*561932-1* 410
Malva albiflora IPNI=*561932-1* 410
Malva althaea IPNI=*558872-1* 407
Malva altissima IPNI=*561932-1* 410
Malva ambigua IPNI=*561932-1* 410
Malva americana IPNI=*561999-1* 410
Malva antillarum IPNI=*1063417-2* 410
Malva aragonensis IPNI=*561932-1* 410
Malva arvensis IPNI=*561798-1* 410
Malva astrolasia IPNI=*561999-1* 410
Malva betuloides IPNI=*1063417-2* 410
Malva bivoniana IPNI=*561830-1* 410
Malva blumeana IPNI=*561999-1* 410
Malva borbonica IPNI=*1063417-2* 410
Malva borealis IPNI=*561868-1* 410
Malva brachystachya IPNI=*561999-1* 410
Malva bracteata IPNI=*561868-1* 410
Malva caroliniana IPNI=*30017389-2* 412
Malva carpinifolia IPNI=*1063417-2* 410
Malva ciliata IPNI=*561932-1* 410
Malva cognata IPNI=*561798-1* 410
Malva coromandeliana IPNI=*1063417-2* 410
Malva coromandelica IPNI=*1063417-2* 410
Malva coronata IPNI=*561830-1* 410
Malva costata IPNI=*993881-1* 408
Malva crenata IPNI=*561868-1* 410
Malva cristata IPNI=*561830-1* 410
Malva curassavica IPNI=*561999-1* 410
Malva decumbens IPNI=*30017389-2* 412
Malva dendromorpha IPNI=*561541-1* 410
Malva domingensis IPNI=*1063417-2* 410
Malva elata IPNI=*561932-1* 410
Malva equina IPNI=*561932-1* 410
Malva erevaniana IPNI=*561932-1* 410
Malva eriocalyx IPNI=*561541-1* 410
Malva eriocarpa IPNI=*30017389-2* 412
Malva europaea IPNI=*1063417-2* 410
Malva exasperata IPNI=*993881-1* 408
Malva excelsa IPNI=*561798-1* 410
Malva fastuosa IPNI=*561541-1* 410
Malva flexuosa IPNI=*561830-1* 410
Malva florida IPNI=*558713-1* 407
Malva fluminensis IPNI=*561999-1* 410
Malva gangetica IPNI=*561999-1* 410
Malva glabra IPNI=*561932-1* 410
Malva grossheimii IPNI=*561932-1* 410
Malva gymnocarpa IPNI=*561932-1* 410
Malva haenkeana IPNI=*993885-1* 408
Malva havanensis IPNI=*1063417-2* 410
Malva hawaiensis IPNI=*30017389-2* 412
Malva henningii IPNI=*561868-1* 410
Malva heterocarpa IPNI=*561830-1* 410

Malva hirsuta IPNI=*561932-1* 410
Malva hirsuta IPNI=*563888-1* 413
Malva hispida IPNI=*1063417-2* 410
Malva horrida IPNI=*561430-1* 410
Malva hortensus IPNI=*558713-1* 407
Malva humifusa IPNI=*561868-1* 410
Malva juvenalis IPNI=*30017389-2* 412
Malva lactea IPNI=*133329-2* 410
Malva laevis IPNI=*993881-1* 408
Malva lignescens IPNI=*561868-1* 410
Malva limensis IPNI=*993881-1* 408
Malva lindheimeriana IPNI=*1063417-2* 410
Malva longilobata IPNI=*561932-1* 410
Malva lucida IPNI=*561932-1* 410
Malva luzonica IPNI=*1063417-2* 410
Malva macrostachya IPNI=*561999-1* 410
Malva major IPNI=*561932-1* 410
Malva mareotica IPNI=*561830-1* 410
Malva maritima IPNI=*558872-1* 407
Malva martrinii IPNI=*561932-1* 410
Malva mathewsii IPNI=*993885-1* 408
Malva mauritanica IPNI=*561932-1* 410
Malva mauritiana IPNI=*561932-1* 410
Malva microcalyx IPNI=*561830-1* 410
Malva microcarpa IPNI=*561830-1* 410
Malva minor IPNI=*561868-1* 410
Malva moschata IPNI=*376654-1* 358
Malva musiana IPNI=*561830-1* 410
Malva nicaeensis var. *liocarpa* IPNI=*561798-1* 410
Malva obtusa IPNI=*561868-1* 410
Malva obtusa IPNI=*561932-1* 410
Malva officinalis IPNI=*558872-1* 407
Malva orientalis IPNI=*561932-1* 410
Malva ovata IPNI=*561999-1* 410
Malva parviflora IPNI=*561868-1* 410
Malva peruviana IPNI=*993885-1* 408
Malva plebeia IPNI=*561932-1* 410
Malva polycarpa IPNI=*561830-1* 410
Malva polymorpha IPNI=*561932-1* 410
Malva polystachya IPNI=*561999-1* 410
Malva prostrata IPNI=*30017389-2* 412
Malva pseudoborealis IPNI=*561868-1* 410
Malva pumila IPNI=*561932-1* 410
Malva racemosa IPNI=*561932-1* 410
Malva radicans IPNI=*30017389-2* 412
Malva recta IPNI=*561932-1* 410
Malva repens IPNI=*561868-1* 410
Malva rhombifolia IPNI=*235798-2* 413
Malva rosea IPNI=*558713-1* 407
Malva rosea IPNI=*561868-1* 410
Malva rosea IPNI=*563228-1* 412
Malva rotundifolia IPNI=*561868-1* 410
Malva rotundifolia IPNI=*561932-1* 410
Malva ruderalis IPNI=*1063417-2* 410
Malva ruderalis IPNI=*561932-1* 410
Malva setosa IPNI=*561798-1* 410
Malva simpliciuscula IPNI=*561830-1* 410
Malva spicata IPNI=*561999-1* 410
Malva spinosa IPNI=*235838-2* 413
Malva subhastata IPNI=*1063417-2* 410
Malva sublobata IPNI=*561999-1* 410
Malva sylvestris f. *albiflora* IPNI=*561932-1* 410
Malva sylvestris f. *vestita* IPNI=*561932-1* 410
Malva sylvestris var. *mauritiana* IPNI=*561932-1* 410
Malva timoriensis IPNI=*561999-1* 410
Malva tomentella IPNI=*561932-1* 410
Malva trachelifolia IPNI=*561999-1* 410
Malva tricuspidata var. *subtriloba* IPNI=*1063417-2* 410
Malva trionoides IPNI=*561830-1* 410
Malva ulmifolia IPNI=*1063417-2* 410
Malva umbellata IPNI=*561541-1* 410
Malva urticifolia IPNI=*30017389-2* 412
Malva veneta IPNI=*561541-1* 410
Malva vitifolia IPNI=*133329-2* 410
Malva viviniana IPNI=*561932-1* 410
Malva vulgaris IPNI=*561932-1* 410
Malvastrum americanum var. *stellatum* IPNI=*561999-1* 410
Malvastrum cordifolium IPNI=*563600-1* 413
Malvastrum coromandelianum f. *breviaristata* IPNI=*1063417-2* 410
Malvastrum lacteum IPNI=*133329-2* 410
Malvastrum limense IPNI=*993881-1* 408
Malvastrum lindheimerianum IPNI=*1063417-2* 410
Malvastrum linearifolium IPNI=*313702-2* 413
Malvastrum macrostachyum IPNI=*561999-1* 410
Malvastrum peruvianum IPNI=*993885-1* 408
Malvastrum peruvianum var. *limense* IPNI=*993881-1* 408
Malvastrum roseum IPNI=*563228-1* 412
Malvastrum ruderale IPNI=*1063417-2* 410
Malvastrum spicatum IPNI=*561999-1* 410
Malvastrum trachelifolium IPNI=*561999-1* 410
Malvastrum vitifolium IPNI=*133329-2* 410
Malvaviscus acapulcensis IPNI=*30041958-2* 410
Malvaviscus arboreus subsp. *cubensis* IPNI=*30041958-2* 410
Malvaviscus arboreus subsp. *penduliflorus* IPNI=*150478-2* 411
Malvaviscus arboreus var. *brihondus* IPNI=*30041958-2* 410
Malvaviscus arboreus var. *cubensis* IPNI=*30041958-2* 410
Malvaviscus arboreus var. *grisebachii* IPNI=*30041958-2* 410
Malvaviscus arboreus var. *lobatus* IPNI=*30041958-2* 410
Malvaviscus arboreus var. *longifolius* IPNI=*150478-2* 411
Malvaviscus arboreus var. *mexicanus* IPNI=*30041958-2* 410
Malvaviscus arboreus var. *parviflorus* IPNI=*30041958-2* 410
Malvaviscus arboreus var. *penduliflorus* IPNI=*150478-2* 411
Malvaviscus arboreus var. *pilosus* IPNI=*30041958-2* 410
Malvaviscus arboreus var. *sepium* IPNI=*30041958-2* 410
Malvaviscus arboreus var. *sloanei* IPNI=*30041958-2* 410
Malvaviscus balbisii IPNI=*30041958-2* 410
Malvaviscus brevibracteatus IPNI=*30041958-2* 410
Malvaviscus brevipes IPNI=*30041958-2* 410
Malvaviscus ciliatus IPNI=*30041958-2* 410
Malvaviscus coccineus IPNI=*30041958-2* 410
Malvaviscus concinnus IPNI=*30041958-2* 410
Malvaviscus conzattii IPNI=*30041958-2* 410
Malvaviscus cordifolius IPNI=*30041958-2* 410
Malvaviscus grandiflorus IPNI=*30041958-2* 410
Malvaviscus jordan-mottii IPNI=*30041958-2* 410
Malvaviscus longifolius IPNI=*150478-2* 411
Malvaviscus malvaviscus IPNI=*30041958-2* 410
Malvaviscus mollis IPNI=*30041958-2* 410
Malvaviscus oligotrichus IPNI=*30041958-2* 410
Malvaviscus pilosus IPNI=*30041958-2* 410
Malvaviscus polakowskyi IPNI=*30041958-2* 410
Malvaviscus populneus IPNI=*564713-1* 415
Malvaviscus pulvinatus IPNI=*30041958-2* 410
Malvaviscus rivularis IPNI=*30041958-2* 410
Malvaviscus sagranus IPNI=*30041958-2* 410
Malvaviscus sepium IPNI=*30041958-2* 410
Malveopsis americana IPNI=*561999-1* 410
Malveopsis coromandeliana IPNI=*1063417-2* 410
Malveopsis coromandeliana var. *normalis* IPNI=*1063417-2* 410
Malveopsis coromandeliana var. *parvifolia* IPNI=*1063417-2* 410
Malveopsis macrostachya IPNI=*561999-1* 410
Malveopsis peruviana f. *albiflora* IPNI=*993885-1* 408
Malveopsis peruviana f. *coerulea* IPNI=*993885-1* 408
Malveopsis peruviana IPNI=*993885-1* 408
Malveopsis rosea IPNI=*563228-1* 412
Malveopsis spicata f. *aurantiaca* IPNI=*157043-2* 412

Malveopsis spicata f. *flava* IPNI=*157043-2* 412
Malveopsis spicata f. *rubra* IPNI=*561999-1* 410
Malveopsis spicata IPNI=*561999-1* 410
Malveopsis spicata var. *normalis* IPNI=*561999-1* 410
Malveopsis spicata var. *ovata* IPNI=*157043-2* 412
Malveopsis spicata var. *pauciflora* IPNI=*157043-2* 412
Malveopsis vitifolia IPNI=*133329-2* 410
Malvinda alnifolia IPNI=*235838-2* 413
Malvinda cordifolia IPNI=*563600-1* 413
Malvinda spinosa IPNI=*235838-2* 413
Mammillaria bogotensis IPNI=*314810-2* 284
Mammillaria caracassana IPNI=*134797-1* 284
Mammillaria columbiana subsp. *tamayonis* IPNI=*314810-2* 284
Mammillaria columbiana var. *albescens* IPNI=*314810-2* 284
Mammillaria columbiana var. *bogotensis* IPNI=*314810-2* 284
Mammillaria conica IPNI=*134797-1* 284
Mammillaria fuliginosa IPNI=*134797-1* 284
Mammillaria hennisii IPNI=*314810-2* 284
Mammillaria micracantha IPNI=*134797-1* 284
Mammillaria micrantha IPNI=*134797-1* 284
Mammillaria obvallata IPNI=*134797-1* 284
Mammillaria parvimamma IPNI=*134797-1* 284
Mammillaria pseudosimplex IPNI=*134797-1* 284
Mammillaria simplex IPNI=*134797-1* 284
Mammillaria simplex var. *albidispina* IPNI=*134797-1* 284
Mammillaria simplex var. *flavescens* IPNI=*134797-1* 284
Mammillaria simplex var. *parvimamma* IPNI=*134797-1* 284
Mammillaria soehlemannii IPNI=*314810-2* 284
Mammillaria tamayonis IPNI=*314810-2* 284
Manabea arborescens IPNI=*860891-1* 368
Mancinella venenata IPNI=*349995-1* 391
Mandevilla auriculata IPNI=*151979-2* 341
Mandevilla bella IPNI=*151935-2* 341
Mandevilla canescens IPNI=*152052-2* 341
Mandevilla cereola IPNI=*151935-2* 341
Mandevilla denticulata IPNI=*151979-2* 341
Mandevilla dielsiana IPNI=*80081-1* 342
Mandevilla fluminensis IPNI=*151979-2* 341
Mandevilla glabra IPNI=*80081-1* 342
Mandevilla hispida IPNI=*151979-2* 341
Mandevilla lasiocarpa IPNI=*151979-2* 341
Mandevilla macrophylla IPNI=*151979-2* 341
Mandevilla palustris IPNI=*151979-2* 341
Mandevilla permixta IPNI=*151935-2* 341
Mandevilla pittieri IPNI=*151935-2* 341
Mandevilla rusbyi IPNI=*151979-2* 341
Mandevilla rutila IPNI=*152052-2* 341
Mandevilla spruceana IPNI=*151987-2* 341
Mandevilla stephanotidifolia IPNI=*151987-2* 341
Mandevilla tomentosa IPNI=*151979-2* 341
Mandevilla tomentosa var. *hirsuta* IPNI=*151979-2* 341
Mandevilla tomentosa var. *hispida* IPNI=*151979-2* 341
Mandevilla tomentosa var. *vahleana* IPNI=*151979-2* 341
Mandioca aipi IPNI=*351790-1* 392
Mandioca dulcis IPNI=*351790-1* 392
Mandioca utilissima IPNI=*351790-1* 392
Manettia asperula IPNI=*224945-2* 357
Manettia attenuata IPNI=*755652-1* 352
Manettia boliviana IPNI=*755652-1* 352
Manettia capitata IPNI=*746945-1* 348
Manettia cordifolia f. *boliviana* IPNI=*755652-1* 352
Manettia cordifolia f. *incana* IPNI=*755652-1* 352
Manettia cordifolia var. *attenuata* IPNI=*755652-1* 352
Manettia cordifolia var. *filiformis* IPNI=*755652-1* 352
Manettia costaricensis IPNI=*755756-1* 352
Manettia cuspidata IPNI=*755756-1* 352
Manettia glabra IPNI=*755652-1* 352
Manettia grandiflora IPNI=*755652-1* 352
Manettia havanensis IPNI=*755756-1* 352
Manettia ignita IPNI=*755652-1* 352
Manettia ignita var. *angustifolia* IPNI=*755652-1* 352
Manettia ignita var. *cordifolia* IPNI=*755652-1* 352
Manettia ignita var. *glabra* IPNI=*755652-1* 352
Manettia ignita var. *incana* IPNI=*755652-1* 352
Manettia ignita var. *micans* IPNI=*755652-1* 352
Manettia lanceolata IPNI=*760275-1* 355
Manettia leianthiflora IPNI=*755652-1* 352
Manettia micans IPNI=*755652-1* 352
Manettia orbifera IPNI=*755756-1* 352
Manettia panamensis IPNI=*755756-1* 352
Manettia pleiodon IPNI=*755756-1* 352
Manettia seleriana IPNI=*755756-1* 352
Manettia stipulosa IPNI=*755652-1* 352
Manettia tarapotensis IPNI=*755756-1* 352
Manettia uniflora IPNI=*755756-1* 352
Manganaroa paniculata IPNI=*234346-2* 332
Mangifera amba IPNI=*69913-1* 450
Mangifera anisodora IPNI=*69913-1* 450
Mangifera austroyunnanensis IPNI=*69913-1* 450
Mangifera balba IPNI=*69913-1* 450
Mangifera cambodiana IPNI=*69913-1* 450
Mangifera domestica IPNI=*69913-1* 450
Mangifera equina IPNI=*69913-1* 450
Mangifera gladiata IPNI=*69913-1* 450
Mangifera kukulu IPNI=*69913-1* 450
Mangifera laxiflora IPNI=*69913-1* 450
Mangifera linnaei IPNI=*69913-1* 450
Mangifera maritima IPNI=*69913-1* 450
Mangifera mekongensis IPNI=*69913-1* 450
Mangifera montana IPNI=*69913-1* 450
Mangifera oryza IPNI=*69913-1* 450
Mangifera racemosa IPNI=*69913-1* 450
Mangifera rostrata IPNI=*69913-1* 450
Mangifera rubra IPNI=*69913-1* 450
Mangifera sativa IPNI=*69913-1* 450
Mangifera siamensis IPNI=*69913-1* 450
Mangifera viridis IPNI=*69913-1* 450
Manglilla dependens IPNI=*1119456-2* 304
Manglilla ferruginea IPNI=*1028287-2* 304
Manglilla latifolia IPNI=*1033305-2* 304
Mangostana garcinia IPNI=*428073-1* 385
Manicaria atricha IPNI=*668116-1* 210
Manicaria plukenetii IPNI=*668116-1* 210
Manicaria saccifera var. *mediterranea* IPNI=*668116-1* 210
Manicaria saccifera var. *plukenetii* IPNI=*668116-1* 210
Manihot aipi IPNI=*351790-1* 392
Manihot amazonica IPNI=*351747-1* 392
Manihot aypi IPNI=*351790-1* 392
Manihot cannabina IPNI=*351790-1* 392
Manihot cassava IPNI=*351790-1* 392
Manihot curcas IPNI=*131462-2* 391
Manihot diffusa IPNI=*351790-1* 392
Manihot digitiformis IPNI=*351790-1* 392
Manihot dulcis IPNI=*351790-1* 392
Manihot dulcis var. *aipi* IPNI=*351790-1* 392
Manihot dulcis var. *diffusa* IPNI=*351790-1* 392
Manihot dulcis var. *ferruginea* IPNI=*152449-2* 392
Manihot dulcis var. *flabellifolia* IPNI=*351790-1* 392
Manihot edule IPNI=*351790-1* 392
Manihot esculenta subsp. *flabellifolia* IPNI=*351790-1* 392
Manihot esculenta subsp. *grandifolia* IPNI=*351790-1* 392
Manihot esculenta var. *argentea* IPNI=*351790-1* 392
Manihot esculenta var. *coalescens* IPNI=*351790-1* 392
Manihot esculenta var. *communis* IPNI=*351790-1* 392
Manihot esculenta var. *debilis* IPNI=*351790-1* 392
Manihot esculenta var. *digitifolia* IPNI=*351790-1* 392

Manihot esculenta var. *domingensis* IPNI=*351790-1* 392
Manihot esculenta var. *fertilis* IPNI=*351790-1* 392
Manihot esculenta var. *flavicaulis* IPNI=*351790-1* 392
Manihot esculenta var. *fuscescens* IPNI=*351790-1* 392
Manihot esculenta var. *hispaniolensis* IPNI=*351790-1* 392
Manihot esculenta var. *jamaicensis* IPNI=*351790-1* 392
Manihot esculenta var. *luteola* IPNI=*351790-1* 392
Manihot esculenta var. *mutabilis* IPNI=*351790-1* 392
Manihot esculenta var. *nodosa* IPNI=*351790-1* 392
Manihot esculenta var. *pohlii* IPNI=*351790-1* 392
Manihot esculenta var. *ramosissima* IPNI=*351790-1* 392
Manihot esculenta var. *rufescens* IPNI=*351790-1* 392
Manihot esculenta var. *sprucei* IPNI=*351790-1* 392
Manihot esculenta var. *zimmermannii* IPNI=*351790-1* 392
Manihot flabellifolia IPNI=*351790-1* 392
Manihot flexuosa IPNI=*351790-1* 392
Manihot gossypiifolia IPNI=*323930-2* 391
Manihot guyanensis IPNI=*351790-1* 392
Manihot herbacea IPNI=*61899-2* 388
Manihot loureiroi IPNI=*351790-1* 392
Manihot manihot IPNI=*351790-1* 392
Manihot melanobasis IPNI=*351790-1* 392
Manihot moluccana IPNI=*338514-1* 387
Manihot multifida IPNI=*314308-2* 391
Manihot palmata var. *aipi* IPNI=*351790-1* 392
Manihot palmata var. *diffusa* IPNI=*351790-1* 392
Manihot palmata var. *digitiformis* IPNI=*351790-1* 392
Manihot palmata var. *ferruginea* IPNI=*152449-2* 392
Manihot palmata var. *flabellifolia* IPNI=*351790-1* 392
Manihot rusbyi IPNI=*351747-1* 392
Manihot sprucei IPNI=*351790-1* 392
Manihot urens IPNI=*61899-2* 388
Manihot utilissima IPNI=*351790-1* 392
Manilkara achras IPNI=*152641-2* 306
Manilkara breviloba IPNI=*152641-2* 306
Manilkara calcicola IPNI=*152591-2* 305
Manilkara calcicola var. *colombiana* IPNI=*152591-2* 305
Manilkara calderonii IPNI=*152641-2* 306
Manilkara conzattii IPNI=*152641-2* 306
Manilkara gaumeri IPNI=*152641-2* 306
Manilkara grisebachii IPNI=*152641-2* 306
Manilkara meridionalis IPNI=*152641-2* 306
Manilkara meridionalis var. *caribbensis* IPNI=*152641-2* 306
Manilkara rojasii IPNI=*152641-2* 306
Manilkara striata IPNI=*152641-2* 306
Manilkara tabogaensis IPNI=*152641-2* 306
Manilkara zapotilla IPNI=*152641-2* 306
Manilkariopsis lobulata IPNI=*152641-2* 306
Manilkariopsis meridionalis IPNI=*152641-2* 306
Manilkariopsis petenensis IPNI=*152641-2* 306
Manilkariopsis rojasii IPNI=*152641-2* 306
Manilkariopsis striata IPNI=*152641-2* 306
Manilkariopsis tabogaensis IPNI=*152641-2* 306
Manisuris altissima IPNI=*404684-1* 240
Manisuris exaltata IPNI=*909653-1* 244
Manisuris fasciculata IPNI=*404684-1* 240
Manisuris granularis IPNI=*116848-2* 240
Manisuris loricata IPNI=*325559-2* 244
Manisuris polystachya IPNI=*116848-2* 240
Mansana arborea IPNI=*719213-1* 442
Mapa graveolens IPNI=*323288-2* 287
Mapinguari desvauxianus IPNI=*643410-1* 220
Mapinguari neophyllus IPNI=*643647-1* 221
Mapinguari verrucifer IPNI=*643410-1* 220
Mapouria alba f. *intermedia* IPNI=*761655-1* 355
Mapouria alba IPNI=*212398-2* 355
Mapouria alba var. *tristis* IPNI=*761655-1* 355
Mapouria australis IPNI=*761655-1* 355
Mapouria catharinensis IPNI=*761655-1* 355
Mapouria chionantha IPNI=*60473461-2* 355
Mapouria crassa IPNI=*761655-1* 355
Mapouria ficigemma IPNI=*761655-1* 355
Mapouria fockeana IPNI=*761655-1* 355
Mapouria guianensis IPNI=*60473461-2* 355
Mapouria langsdorffiana IPNI=*77161077-1* 354
Mapouria luschnathiana IPNI=*60473461-2* 355
Mapouria luschnathii IPNI=*60473461-2* 355
Mapouria martiana IPNI=*761655-1* 355
Mapouria micrantha IPNI=*762666-1* 355
Mapouria microdon IPNI=*77154321-1* 350
Mapouria opaca IPNI=*60473461-2* 355
Mapouria pallescens IPNI=*761655-1* 355
Mapouria pohliana IPNI=*761655-1* 355
Mapouria rabeniana IPNI=*761655-1* 355
Mapouria remota IPNI=*763194-1* 356
Mapouria remota var. *angustifolia* IPNI=*763194-1* 356
Mapouria remota var. *genuina* IPNI=*763194-1* 356
Mapouria remota var. *longifolia* IPNI=*763194-1* 356
Mapouria remota var. *ovata* IPNI=*763194-1* 356
Mapouria riedeliana IPNI=*761655-1* 355
Mapouria rigida IPNI=*761655-1* 355
Mapouria schlechtendaliana IPNI=*761655-1* 355
Mapouria subsessilis IPNI=*764729-1* 356
Mapouria subsessilis var. *angustifolia* IPNI=*764729-1* 356
Mapouria subsessilis var. *genuina* IPNI=*764729-1* 356
Mapouria subsessilis var. *latifolia* IPNI=*764729-1* 356
Mapouria tomentosa IPNI=*762666-1* 355
Mapouria tristis IPNI=*761655-1* 355
Mapouria velhana IPNI=*761655-1* 355
Mappia ampla IPNI=*963946-1* 359
Mappia cordata IPNI=*607989-1* 359
Mappia glabrata IPNI=*607989-1* 359
Mappia nitida IPNI=*607989-1* 359
Mappia poeppigiana IPNI=*607989-1* 359
Maprounea guianensis var. *genuina* IPNI=*352054-1* 392
Maprounea guianensis var. *nervosa* IPNI=*352054-1* 392
Maprounea guianensis var. *undulata* IPNI=*352054-1* 392
Maquira granatensis IPNI=*855107-1* 440
Maquira grandifolia IPNI=*854491-1* 440
Marah rusbyi IPNI=*88306-2* 294
Maranta allouia IPNI=*60458762-2* 345
Maranta amabilis IPNI=*60461053-2* 249
Maranta arouma IPNI=*797103-1* 249
Maranta arundinacea f. *sylvestris* IPNI=*280408-2* 250
Maranta arundinacea IPNI=*798255-1* 250
Maranta arundinacea var. *indica* IPNI=*280408-2* 250
Maranta bachemiana IPNI=*60458927-2* 249
Maranta bicolor IPNI=*796810-1* 249
Maranta cachibou IPNI=*60458866-2* 249
Maranta casupo IPNI=*60458866-2* 249
Maranta coriifolia IPNI=*60461169-2* 249
Maranta cuyabensis IPNI=*797633-1* 250
Maranta depressa IPNI=*797361-1* 250
Maranta disticha IPNI=*60458866-2* 249
Maranta divaricata f. *major* IPNI=*280409-2* 250
Maranta fascinator IPNI=*797361-1* 250
Maranta flexuosa IPNI=*798255-1* 250
Maranta geniculata IPNI=*798255-1* 250
Maranta illustris IPNI=*60461319-2* 249
Maranta imperialis IPNI=*60461040-2* 249
Maranta indica IPNI=*280408-2* 250
Maranta jacquinii IPNI=*246147-2* 250
Maranta juncea IPNI=*246147-2* 250
Maranta juncea IPNI=*797103-1* 249
Maranta kegeliana IPNI=*60458927-2* 249
Maranta kerchoveana IPNI=*797361-1* 250

Maranta leuconeura var. *erythroneura* IPNI=*797361-1* 250
Maranta leuconeura var. *massangeana* IPNI=*797361-1* 250
Maranta lietzei IPNI=*77174968-1* 249
Maranta lutea IPNI=*246147-2* 250
Maranta lutea IPNI=*60458866-2* 249
Maranta majestica IPNI=*60461040-2* 249
Maranta massangeana IPNI=*797361-1* 250
Maranta micans IPNI=*60461053-2* 249
Maranta moritziana IPNI=*797633-1* 250
Maranta myrosma IPNI=*797633-1* 250
Maranta nana IPNI=*60461053-2* 249
Maranta niveiflora IPNI=*60458762-2* 345
Maranta ornata IPNI=*60461169-2* 249
Maranta princeps IPNI=*60461040-2* 249
Maranta pubescens IPNI=*280409-2* 250
Maranta pulchella IPNI=*796810-1* 249
Maranta ramosissima IPNI=*280408-2* 250
Maranta roseopicta IPNI=*60461319-2* 249
Maranta rotundifolia IPNI=*797124-1* 249
Maranta semperflorens IPNI=*60458762-2* 345
Maranta strobilifera IPNI=*218597-2* 251
Maranta sylvatica IPNI=*280408-2* 250
Maranta wagneri IPNI=*60461319-2* 249
Maranta zebrina IPNI=*796810-1* 249
Marantopsis lutea IPNI=*246147-2* 250
Marara aculeata IPNI=*7909-2* 204
Marara bicuspidata IPNI=*7909-2* 204
Marara caryotifolia IPNI=*7909-2* 204
Marathrum allenii IPNI=*688390-1* 402
Marathrum indifferens IPNI=*688390-1* 402
Marathrum leptophyllum IPNI=*688390-1* 402
Marathrum minutiflorum f. *allenii* IPNI=*688390-1* 402
Marathrum minutiflorum f. *diversifolium* IPNI=*688390-1* 402
Marathrum minutiflorum f. *indifferens* IPNI=*688390-1* 402
Marathrum minutiflorum f. *intermedium* IPNI=*688390-1* 402
Marathrum minutiflorum IPNI=*688390-1* 402
Marathrum modestum IPNI=*688390-1* 402
Marathrum oxycarpum IPNI=*688390-1* 402
Marathrum pusillum IPNI=*688390-1* 402
Marathrum schiedeanum var. *modestum* IPNI=*688390-1* 402
Marathrum schiedeanum var. *stenocarpum* IPNI=*688390-1* 402
Marathrum stenocarpum IPNI=*688390-1* 402
Marcelia aurea IPNI=*192489-1* 259
Marcgravia ampulligera IPNI=*152975-2* 302
Marcgravia dubia IPNI=*163921-2* 200
Marcgravia rectiflora var. *brownei* IPNI=*152975-2* 302
Marcorella colubrina IPNI=*63423-2* 441
Margaritaria adelioides IPNI=*352077-1* 401
Margaritaria alternifolia IPNI=*352077-1* 401
Margaritaria nobilis var. *antillana* IPNI=*352077-1* 401
Margaritaria nobilis var. *hypomalaca* IPNI=*352077-1* 401
Margaritopsis microdon IPNI=*77154321-1* 350
Margarocarpus procridioides IPNI=*856206-1* 448
Margarospermum officinale IPNI=*118163-1* 275
Marginaria angustifolia IPNI=*17063000-1* 175
Marginaria cardiophylla IPNI=*17406250-1* 176
Marginaria ceteracina IPNI=*1039912-2* 176
Marginaria dimorpha IPNI=*17063000-1* 175
Marginaria diversifolia IPNI=*160142-2* 175
Marginaria ensifolia IPNI=*17063000-1* 175
Marginaria gladiata IPNI=*77075127-1* 177
Marginaria incana IPNI=*1039912-2* 176
Marginaria lanceolata IPNI=*17448650-1* 176
Marginaria macrocarpa IPNI=*17448650-1* 176
Marginaria marginata IPNI=*17448650-1* 176
Marginaria neriifolia IPNI=*77075127-1* 177
Marginaria oleandrifolia IPNI=*77075087-1* 176
Marginaria polypodioides IPNI=*1039912-2* 176
Marginaria rhoifolia IPNI=*77075127-1* 177
Marginaria taeniosa IPNI=*17063000-1* 175
Marginaria vacciniifolia IPNI=*17406250-1* 176
Marginaria vacciniifolia monstr. furcata IPNI=*17406250-1* 176
Margyricarpus imberbis IPNI=*153124-2* 443
Margyricarpus laevis IPNI=*153124-2* 443
Margyricarpus patagonicus IPNI=*153124-2* 443
Margyricarpus pinnatus var. *hirsutus* IPNI=*153124-2* 443
Margyricarpus setosus IPNI=*153124-2* 443
Margyricarpus setosus var. *patagonicus* IPNI=*153124-2* 443
Margyricarpus setosus var. *paucifoliolatus* IPNI=*153124-2* 443
Mariacantha maculosa IPNI=*249211-1* 270
Marialva brasiliensis IPNI=*429349-1* 385
Marialva calyptrata IPNI=*972610-1* 385
Marialva fructipendula IPNI=*429349-1* 385
Mariana lactea IPNI=*249211-1* 270
Mariana mariana IPNI=*249211-1* 270
Marianthemum grandiflorum IPNI=*140670-1* 272
Marianthemum medium IPNI=*140670-1* 272
Marica bogotensis IPNI=*441310-1* 216
Marica bulbosa IPNI=*77182480-1* 217
Marica convoluta IPNI=*285111-2* 216
Marica humilis var. *lutea* IPNI=*77182480-1* 217
Marica iridifolia IPNI=*330260-2* 216
Marica longifolia IPNI=*77182480-1* 217
Marica martinicensis IPNI=*441642-1* 217
Marica micrantha IPNI=*330260-2* 216
Marica palmifolia IPNI=*237727-2* 216
Marica plicata IPNI=*89940-2* 216
Marica semiaperta IPNI=*441642-1* 217
Marica tigridia IPNI=*441553-1* 216
Marica tinctoria IPNI=*441310-1* 216
Mariscus acicularis IPNI=*1169169-2* 231
Mariscus acutus IPNI=*305060-1* 231
Mariscus aggregatus IPNI=*303704-1* 230
Mariscus alpinus IPNI=*74768-2* 230
Mariscus ambiguus IPNI=*74768-2* 230
Mariscus aphyllus IPNI=*1009998-1* 231
Mariscus aristatus IPNI=*324890-2* 231
Mariscus aximensis IPNI=*75178-2* 231
Mariscus bojeri IPNI=*301880-2* 230
Mariscus conflexus IPNI=*74768-2* 230
Mariscus coriaceus IPNI=*305019-1* 231
Mariscus cyperinus IPNI=*304256-1* 230
Mariscus cyperinus subsp. *laxatus* IPNI=*304256-1* 230
Mariscus cyperinus var. *laxatus* IPNI=*304256-1* 230
Mariscus cyperinus var. *pictus* IPNI=*304256-1* 230
Mariscus diciger IPNI=*75178-2* 231
Mariscus dissitiflorus IPNI=*75184-2* 231
Mariscus dissitiflorus var. *lucidior* IPNI=*75184-2* 231
Mariscus dissolutus IPNI=*305019-1* 231
Mariscus eurystachys IPNI=*75178-2* 231
Mariscus flabelliformis IPNI=*75178-2* 231
Mariscus flabelliformis var. *aximensis* IPNI=*75178-2* 231
Mariscus glandulosus IPNI=*305019-1* 231
Mariscus haematodes IPNI=*305060-1* 231
Mariscus hermaphroditus IPNI=*74768-2* 230
Mariscus intricatus IPNI=*324890-2* 231
Mariscus jacquinii IPNI=*74768-2* 230
Mariscus jacquinii subsp. *angustior* IPNI=*74768-2* 230
Mariscus jacquinii var. *angustior* IPNI=*74768-2* 230
Mariscus keniensis IPNI=*301880-2* 230
Mariscus leptochloides IPNI=*304793-1* 230
Mariscus ligularis IPNI=*305019-1* 231
Mariscus longibracteatus IPNI=*301880-2* 230
Mariscus longibracteatus var. *keniensis* IPNI=*301880-2* 230
Mariscus longiradiatus IPNI=*74768-2* 230
Mariscus lucidus IPNI=*305060-1* 231

Mariscus mariscus IPNI=*303489-1* 230
Mariscus pallens IPNI=*75184-2* 231
Mariscus parviflorus IPNI=*74851-2* 231
Mariscus patulus IPNI=*74768-2* 230
Mariscus pictus IPNI=*304256-1* 230
Mariscus polycephalus IPNI=*74851-2* 231
Mariscus rubrotinctus IPNI=*301880-2* 230
Mariscus rufus IPNI=*305019-1* 231
Mariscus sanguineofuscus IPNI=*305060-1* 231
Mariscus sertularinus IPNI=*74768-2* 230
Mariscus squarrosus IPNI=*324890-2* 231
Mariscus sundaicus IPNI=*304256-1* 230
Mariscus tenuifolius IPNI=*304256-1* 230
Mariscus tenuis IPNI=*75178-2* 231
Mariscus thyrsiflorus IPNI=*75184-2* 231
Mariscus tovarii IPNI=*74768-2* 230
Mariscus tribrachiatus IPNI=*75184-2* 231
Marlierea brachymischa IPNI=*599327-1* 427
Marlierea glomerata IPNI=*203871-2* 428
Marlierea insculpta IPNI=*77160867-1* 427
Marlierea macedoi IPNI=*166146-2* 427
Marlierea montana IPNI=*77160848-1* 426
Marlierea obtusa IPNI=*77160848-1* 426
Marlierea richardiana IPNI=*77160848-1* 426
Marlierea schomburgkiana IPNI=*77160852-1* 426
Marlierea spruceana IPNI=*77160821-1* 426
Marlierea spruceana var. *angustifolia* IPNI=*77160821-1* 426
Marlierea spruceana var. *latifolia* IPNI=*77160821-1* 426
Marlierea squarrosa IPNI=*77160867-1* 427
Marlierea suffruticosa IPNI=*77160848-1* 426
Marlierea uaupensis IPNI=*77160821-1* 426
Marlierea umbraticola IPNI=*77160867-1* 427
Marmaroxylon racemosum IPNI=*996977-1* 337
Marrubium album IPNI=*449990-1* 371
Marrubium apulum IPNI=*449990-1* 371
Marrubium aquaticum IPNI=*450192-1* 371
Marrubium ballotoides IPNI=*449990-1* 371
Marrubium germanicum IPNI=*449990-1* 371
Marrubium hamatum IPNI=*449990-1* 371
Marrubium hyperleucum IPNI=*449990-1* 371
Marrubium indicum IPNI=*158297-2* 371
Marrubium uncinatum IPNI=*449990-1* 371
Marrubium vaillantii IPNI=*449990-1* 371
Marsdenia altissima IPNI=*60478412-2* 343
Marsdenia altissima subsp. *faucinuda* IPNI=*60478412-2* 343
Marsdenia burchellii IPNI=*60478412-2* 343
Marsdenia caulantha IPNI=*60478412-2* 343
Marsdenia caurensis IPNI=*60478412-2* 343
Marsdenia ecorpuscula IPNI=*60478412-2* 343
Marsdenia gracilis IPNI=*94927-1* 339
Marsdenia imthurnii IPNI=*60478412-2* 343
Marsdenia macrophylla IPNI=*100911-1* 343
Marsdenia maculata IPNI=*100911-1* 343
Marsdenia mollissima IPNI=*60478412-2* 343
Marsdenia reichenbachii IPNI=*99126-1* 342
Marsdenia warmingii IPNI=*100911-1* 343
Marsea bonariensis IPNI=*93603-2* 261
Marsea bonariensis var. *leiotheca* IPNI=*60473415-2* 261
Marsea canadensis IPNI=*203653-1* 261
Marsea gnaphaliodes IPNI=*228376-1* 264
Marsea gnaphalioides IPNI=*228376-1* 264
Marsea lyrata IPNI=*211317-2* 268
Marsea sophiifolia IPNI=*947840-1* 264
Marsea viscosa IPNI=*211317-2* 268
Marsilea auriculata IPNI=*17219800-1* 179
Marsilea brasiliensis IPNI=*280439-2* 179
Marsilea natans IPNI=*17219800-1* 179
Marsupiaria angustata IPNI=*155338-2* 222
Marsupiaria iridifolia IPNI=*155338-2* 222
Marsupiaria valenzuelana IPNI=*155338-2* 222
Marsupiaria valenzuelana var. *angustata* IPNI=*155338-2* 222
Marsypianthes arenosa IPNI=*449997-1* 371
Marsypianthes hyptoides IPNI=*449997-1* 371
Marsypianthes hyptoides var. *arenosa* IPNI=*449997-1* 371
Marsypianthes hyptoides var. *bracteosa* IPNI=*449997-1* 371
Marsypianthes hyptoides var. *calycina* IPNI=*449997-1* 371
Marsypianthes hyptoides var. *elatior* IPNI=*449997-1* 371
Marsypianthes hyptoides var. *eriocephala* IPNI=*449997-1* 371
Marsypianthes hyptoides var. *umbrosa* IPNI=*449997-1* 371
Marsypianthes secundiflora IPNI=*449997-1* 371
Marsypianthes sessiliflora IPNI=*449997-1* 371
Marsypianthes viscosa IPNI=*449997-1* 371
Martinella gollmeri IPNI=*153710-2* 364
Martinella insculpta IPNI=*153710-2* 364
Martinezia aculeata IPNI=*7909-2* 204
Martinezia aiphanes IPNI=*7909-2* 204
Martinezia caryotifolia IPNI=*7909-2* 204
Martinezia elegans IPNI=*7909-2* 204
Martinezia ensiformis IPNI=*210063-2* 211
Martinezia ernestii IPNI=*7909-2* 204
Martinezia interrupta IPNI=*667119-1* 209
Martinezia lanceolata IPNI=*665826-1* 207
Martinezia lindeniana IPNI=*664018-1* 204
Martinezia linearis IPNI=*54901-2* 207
Martinezia ulei IPNI=*7909-2* 204
Martinezia ulei IPNI=*7926-2* 204
Martynia angulosa IPNI=*675840-1* 374
Martynia caulescens IPNI=*382786-1* 368
Martynia craniolaria IPNI=*675800-1* 374
Martynia diandra IPNI=*675840-1* 374
Martynia jussieui IPNI=*675840-1* 374
Martynia pallida IPNI=*675840-1* 374
Martynia perennis IPNI=*381282-1* 367
Martynia spathacea IPNI=*675800-1* 374
Martynia speciosa IPNI=*382786-1* 368
Marupa francoana IPNI=*71559-1* 451
Maruta cotula var. *lithuanica* IPNI=*1186403-2* 261
Masakia japonica IPNI=*160944-1* 290
Masakia yoshinagae IPNI=*160944-1* 290
Mascagnia anisopetala var. *macrodisca* IPNI=*60442801-2* 394
Mascagnia benthamiana IPNI=*60443063-2* 396
Mascagnia macrodisca f. *angustifolia* IPNI=*60442801-2* 394
Mascagnia macrodisca IPNI=*60442801-2* 394
Mascagnia nervosa IPNI=*30237926-2* 396
Mascagnia ovatifolia f. *cordata* IPNI=*30237926-2* 396
Mascagnia ovatifolia f. *oblongifolia* IPNI=*30237926-2* 396
Mascagnia ovatifolia f. *typica* IPNI=*30237926-2* 396
Mascagnia psilophylla f. *peruviana* IPNI=*273852-2* 396
Mascagnia psilophylla var. *antifebrilis* IPNI=*273852-2* 396
Mascagnia sericans IPNI=*60443063-2* 396
Mascagnia sericans subsp. *buricana* IPNI=*60443063-2* 396
Mascagnia sericans var. *boliviensis* IPNI=*60443063-2* 396
Mascagnia sericans var. *paraguariensis* IPNI=*60443063-2* 396
Mascagnia stannea IPNI=*60443063-2* 396
Mascarena verschaffeltii IPNI=*667456-1* 209
Masdevallia backhousiana IPNI=*83967-2* 218
Masdevallia bella IPNI=*83961-2* 218
Masdevallia bogotensis IPNI=*642756-1* 219
Masdevallia bruchmulleri IPNI=*642756-1* 219
Masdevallia caudata f. *xanthocorys* IPNI=*642726-1* 219
Masdevallia caudata var. *gudotii* IPNI=*642726-1* 219
Masdevallia caudata var. *shuttleworthii* IPNI=*642726-1* 219
Masdevallia caudata var. *xanthocorys* IPNI=*642726-1* 219
Masdevallia chimaera IPNI=*83967-2* 218
Masdevallia chimaera var. *backhousiana* IPNI=*83967-2* 218
Masdevallia chimaera var. *senilis* IPNI=*83967-2* 218

Masdevallia chimaera var. *stupenda* IPNI=*83967-2* 218
Masdevallia coccinea f. *alba* IPNI=*642746-1* 219
Masdevallia coccinea var. *conchiflora* IPNI=*642746-1* 219
Masdevallia coccinea var. *harryana* IPNI=*642746-1* 219
Masdevallia coccinea var. *lindenii* IPNI=*642746-1* 219
Masdevallia coccinea var. *militaris* IPNI=*642746-1* 219
Masdevallia colibri IPNI=*643178-1* 220
Masdevallia cucutillensis IPNI=*642726-1* 219
Masdevallia denisonii IPNI=*642746-1* 219
Masdevallia diversifolia IPNI=*154214-2* 220
Masdevallia echinata IPNI=*643089-1* 220
Masdevallia ephippium IPNI=*643178-1* 220
Masdevallia expansa IPNI=*642726-1* 219
Masdevallia harryana IPNI=*642746-1* 219
Masdevallia harryana var. *atrosanguinea* IPNI=*642746-1* 219
Masdevallia harryana var. *decora* IPNI=*642746-1* 219
Masdevallia harryana var. *miniata* IPNI=*642746-1* 219
Masdevallia lindenii IPNI=*642746-1* 219
Masdevallia lindenii var. *grandiflora* IPNI=*642746-1* 219
Masdevallia lindenii var. *harryana* IPNI=*642746-1* 219
Masdevallia militaris IPNI=*642746-1* 219
Masdevallia ochracea IPNI=*642756-1* 219
Masdevallia polyantha IPNI=*643106-1* 220
Masdevallia rodrigueziana IPNI=*154376-2* 220
Masdevallia rosea subsp. *echinata* IPNI=*643089-1* 220
Masdevallia schlimii var. *polyantha* IPNI=*643106-1* 220
Masdevallia senilis IPNI=*83967-2* 218
Masdevallia shuttleworthii IPNI=*642726-1* 219
Masdevallia shuttleworthii var. *xanthocorys* IPNI=*642726-1* 219
Masdevallia tricolor IPNI=*642726-1* 219
Masdevallia ulei IPNI=*154376-2* 220
Masdevallia wallisii var. *stupenda* IPNI=*83967-2* 218
Masdevallia yauaperyensis IPNI=*154376-2* 220
Massangea musaica IPNI=*115185-2* 229
Massowia cannifolia IPNI=*88955-1* 202
Massowia canniformis IPNI=*88955-1* 202
Massowia friedrichsthalii IPNI=*88964-1* 202
Massowia lanceifolia IPNI=*88983-1* 202
Massula latifolia IPNI=*836870-1* 247
Mastichodendron colombianum IPNI=*77176719-1* 308
Matamoria spicata IPNI=*239804-1* 268
Matelea denticulata IPNI=*77206440-1* 340
Matelea maritima IPNI=*98854-1* 341
Matelea maritima subsp. *andina* IPNI=*98854-1* 341
Matelea shortii IPNI=*154570-2* 342
Matelea viridiflora IPNI=*77206440-1* 340
Mathewsia boliviana IPNI=*222164-2* 277
Mathewsia diffusa IPNI=*222164-2* 277
Mathiolaria annua IPNI=*286910-1* 277
Mathiolaria glabrata IPNI=*286910-1* 277
Mathiolaria incana IPNI=*286910-1* 277
Matisia apaporiensis IPNI=*154702-2* 411
Matisia asymmetrica IPNI=*154700-2* 411
Matisia inaequalis IPNI=*154700-2* 411
Matisia inaequilatera IPNI=*154700-2* 411
Matisia muricata IPNI=*154702-2* 411
Matisia paraensis IPNI=*185557-2* 412
Matisia rhombifolia IPNI=*154672-2* 411
Matisia schultesii IPNI=*154675-2* 411
Matisia wurdackii IPNI=*154692-2* 411
Matonia cardamomum IPNI=*796556-1* 250
Matrella juncea IPNI=*1214917-2* 247
Matricaria × morifolia IPNI=*77111328-1* 259
Matricaria alba IPNI=*252460-1* 271
Matricaria bayeri IPNI=*154715-2* 265
Matricaria capitellata IPNI=*154715-2* 265
Matricaria ceratophylloides IPNI=*230047-1* 265
Matricaria chamaemilla IPNI=*154715-2* 265
Matricaria chamomilla f. *courrantiana* IPNI=*154715-2* 265
Matricaria chamomilla f. *kochiana* IPNI=*154715-2* 265
Matricaria chamomilla f. *suaveolens* IPNI=*154715-2* 265
Matricaria chamomilla subsp. *pusilla* IPNI=*154715-2* 265
Matricaria chamomilla var. *coronata* IPNI=*154715-2* 265
Matricaria chamomilla var. *courrantiana* IPNI=*154715-2* 265
Matricaria chamomilla var. *pappulosa* IPNI=*154715-2* 265
Matricaria chamomilla var. *pusilla* IPNI=*154715-2* 265
Matricaria chamomilla var. *recutita* IPNI=*154715-2* 265
Matricaria coronaria IPNI=*1013480-1* 263
Matricaria coronata IPNI=*154715-2* 265
Matricaria courrantiana IPNI=*154715-2* 265
Matricaria deflexa IPNI=*154715-2* 265
Matricaria exigua IPNI=*154715-2* 265
Matricaria eximia IPNI=*252460-1* 271
Matricaria florida IPNI=*252460-1* 271
Matricaria indica IPNI=*193510-1* 259
Matricaria kochiana IPNI=*154715-2* 265
Matricaria latifolia IPNI=*252460-1* 271
Matricaria littoralis IPNI=*154715-2* 265
Matricaria maxima IPNI=*230047-1* 265
Matricaria nobilis IPNI=*192489-1* 259
Matricaria obliqua IPNI=*154715-2* 265
Matricaria odorata IPNI=*252460-1* 271
Matricaria oleracea IPNI=*1013480-1* 263
Matricaria partheniifolia IPNI=*252460-1* 271
Matricaria parthenioides IPNI=*192489-1* 259
Matricaria parthenium IPNI=*252460-1* 271
Matricaria patens IPNI=*154715-2* 265
Matricaria praealta IPNI=*252460-1* 271
Matricaria procumbens IPNI=*193510-1* 259
Matricaria pulverulenta IPNI=*252460-1* 271
Matricaria pusilla IPNI=*154715-2* 265
Matricaria pyrethroides IPNI=*154715-2* 265
Matricaria recutita f. *kochiana* IPNI=*154715-2* 265
Matricaria recutita IPNI=*154715-2* 265
Matricaria recutita var. *coronata* IPNI=*154715-2* 265
Matricaria recutita var. *kochiana* IPNI=*154715-2* 265
Matricaria recutita var. *pappulosa* IPNI=*154715-2* 265
Matricaria salina IPNI=*154715-2* 265
Matricaria segetum IPNI=*209007-1* 263
Matricaria suaveolens f. *macrocephala* IPNI=*154715-2* 265
Matricaria suaveolens IPNI=*154715-2* 265
Matricaria tenuifolia f. *nikolicii* IPNI=*154715-2* 265
Matricaria tenuifolia IPNI=*154715-2* 265
Matricaria valentina IPNI=*208446-2* 268
Matricaria vulgaris IPNI=*252460-1* 271
Matthiola annua IPNI=*286910-1* 277
Matthiola arborescens IPNI=*286910-1* 277
Matthiola armata IPNI=*746662-1* 348
Matthiola crucigera IPNI=*286910-1* 277
Matthiola dichotoma IPNI=*114528-2* 351
Matthiola elliptica IPNI=*114528-2* 351
Matthiola fasciculata IPNI=*286910-1* 277
Matthiola fenestralis IPNI=*286910-1* 277
Matthiola glabra IPNI=*286910-1* 277
Matthiola glabrata IPNI=*286910-1* 277
Matthiola graeca IPNI=*286910-1* 277
Matthiola masguindali IPNI=*286910-1* 277
Matthiola odorata IPNI=*752006-1* 351
Matthiola paniculata IPNI=*57096-2* 348
Matthiola parviflora IPNI=*752006-1* 351
Matthiola parvifolia IPNI=*752006-1* 351
Matthiola patens IPNI=*286910-1* 277
Matthiola reticulata IPNI=*114528-2* 351
Matthiola scabra IPNI=*30058289-2* 351
Matthiola simplicicaulis IPNI=*286910-1* 277
Matthiola stephanos-mariae IPNI=*286910-1* 277

Matthiola vulgaris IPNI=*286910-1* 277
Maughania strobilifera f. *rubriflora* IPNI=*495307-1* 320
Maughania strobilifera IPNI=*495307-1* 320
Mauhlia africana IPNI=*527258-1* 213
Mauhlia linearis IPNI=*527258-1* 213
Mauhlia umbellata IPNI=*527258-1* 213
Maurandella hederifolia IPNI=*805502-1* 376
Maurandya antirrhiniflora IPNI=*805502-1* 376
Maurandya antirrhiniflora subsp. *hederifolia* IPNI=*805502-1* 376
Maurandya erubescens IPNI=*279831-2* 376
Maurandya erubescens var. *typica* IPNI=*279831-2* 376
Maurandya personata IPNI=*805502-1* 376
Maurandya scandens var. *erubescens* IPNI=*279831-2* 376
Mauria aurantiodora IPNI=*70012-1* 451
Mauria bijuga IPNI=*71559-1* 451
Mauria biringo IPNI=*70000-1* 451
Mauria biringo var. *granatensis* IPNI=*70000-1* 451
Mauria biringo var. *ruizii* IPNI=*70012-1* 451
Mauria biringo var. *weberbaueri* IPNI=*70000-1* 451
Mauria cyclocarpa IPNI=*71559-1* 451
Mauria decandra IPNI=*70000-1* 451
Mauria dugandii IPNI=*70000-1* 451
Mauria glauca IPNI=*70000-1* 451
Mauria heterophylla var. *contracta* IPNI=*70000-1* 451
Mauria heterophylla var. *humboldtii* IPNI=*70000-1* 451
Mauria heterophylla var. *puberula* IPNI=*70000-1* 451
Mauria juglandifolia IPNI=*71480-1* 451
Mauria multiflora IPNI=*71559-1* 451
Mauria obtusa IPNI=*970707-1* 451
Mauria ovalifolia IPNI=*70000-1* 451
Mauria puberula IPNI=*70000-1* 451
Mauria suaveolens IPNI=*70000-1* 451
Mauria subbijuga IPNI=*71559-1* 451
Mauritia aculeata IPNI=*154896-2* 210
Mauritia aculeata IPNI=*154897-2* 210
Mauritia armata IPNI=*154897-2* 210
Mauritia campylostachys IPNI=*154897-2* 210
Mauritia cataractarum IPNI=*154896-2* 210
Mauritia duckei IPNI=*154897-2* 210
Mauritia flexuosa var. *venezuelana* IPNI=*668158-1* 210
Mauritia gracilis IPNI=*154896-2* 210
Mauritia huebneri IPNI=*154897-2* 210
Mauritia intermedia IPNI=*154897-2* 210
Mauritia limnophylla IPNI=*154896-2* 210
Mauritia macroclada IPNI=*154903-2* 210
Mauritia macrospadix IPNI=*154897-2* 210
Mauritia martiana IPNI=*154897-2* 210
Mauritia minor IPNI=*668158-1* 210
Mauritia nannostachys IPNI=*154897-2* 210
Mauritia pacifica IPNI=*154903-2* 210
Mauritia peruviana IPNI=*154897-2* 210
Mauritia pumila IPNI=*154909-2* 210
Mauritia sagus IPNI=*668158-1* 210
Mauritia setigera IPNI=*668158-1* 210
Mauritia sphaerocarpa IPNI=*668158-1* 210
Mauritia subinermis IPNI=*154909-2* 210
Mauritia tenuis IPNI=*667804-1* 210
Mauritia vinifera IPNI=*668158-1* 210
Mauritiella campylostachys IPNI=*154897-2* 210
Mauritiella cataractarum IPNI=*154896-2* 210
Mauritiella duckei IPNI=*154897-2* 210
Mauritiella huebneri IPNI=*154897-2* 210
Mauritiella intermedia IPNI=*154897-2* 210
Mauritiella macrospadix IPNI=*154897-2* 210
Mauritiella martiana IPNI=*154897-2* 210
Mauritiella nannostachys IPNI=*154897-2* 210
Mauritiella pacifica IPNI=*154903-2* 210
Mauritiella peruviana IPNI=*154897-2* 210
Mauritiella subinermis IPNI=*154909-2* 210
Maurocenia heterophylla IPNI=*785554-1* 292
Maurocenia occidentalis IPNI=*785554-1* 292
Maurocenia occidentalis var. *leucosperma* IPNI=*785554-1* 292
Maurocenia pinnata IPNI=*785554-1* 292
Maxillaria abelei IPNI=*643781-1* 221
Maxillaria amparoana IPNI=*318680-2* 221
Maxillaria ampliflora IPNI=*643344-1* 220
Maxillaria angustifolia IPNI=*30044409-2* 222
Maxillaria articulata IPNI=*643781-1* 221
Maxillaria attenuata IPNI=*643901-1* 222
Maxillaria aurantiaca IPNI=*643731-1* 221
Maxillaria aurea var. *gigantea* IPNI=*984957-1* 220
Maxillaria brachyglossa IPNI=*77155180-1* 220
Maxillaria brenesii IPNI=*643731-1* 221
Maxillaria brevipedunculata IPNI=*643645-1* 221
Maxillaria brunnea IPNI=*643731-1* 221
Maxillaria caespitosa IPNI=*154952-2* 220
Maxillaria calendulina IPNI=*643731-1* 221
Maxillaria camaridii IPNI=*643592-1* 221
Maxillaria chiriquiensis IPNI=*30044409-2* 222
Maxillaria coniformis IPNI=*643585-1* 221
Maxillaria coriacea IPNI=*643410-1* 220
Maxillaria cyperifolia IPNI=*154952-2* 220
Maxillaria dalstromii IPNI=*643743-1* 221
Maxillaria danielae IPNI=*77149732-1* 221
Maxillaria fulvescens IPNI=*77096738-1* 223
Maxillaria gatunensis IPNI=*643387-1* 220
Maxillaria gigantea IPNI=*984957-1* 220
Maxillaria glauca IPNI=*643731-1* 221
Maxillaria glumacea IPNI=*643645-1* 221
Maxillaria guadalupensis IPNI=*643901-1* 222
Maxillaria hedyosma IPNI=*643250-1* 220
Maxillaria hirtilabia IPNI=*643684-1* 221
Maxillaria histiolugorum IPNI=*643821-1* 221
Maxillaria hoehneana IPNI=*643592-1* 221
Maxillaria huebneri IPNI=*643410-1* 220
Maxillaria inaudita var. *minor* IPNI=*643519-1* 220
Maxillaria iridifolia IPNI=*155338-2* 222
Maxillaria kalbreyeri IPNI=*643893-1* 222
Maxillaria lactea IPNI=*643731-1* 221
Maxillaria lactiflora IPNI=*318680-2* 221
Maxillaria langlassei IPNI=*643585-1* 221
Maxillaria lepidota var. *albida* IPNI=*643558-1* 221
Maxillaria leptosepala var. *subintegerrima* IPNI=*643559-1* 221
Maxillaria longifolia IPNI=*643845-1* 221
Maxillaria loretoensis IPNI=*643684-1* 221
Maxillaria lorifolia IPNI=*643684-1* 221
Maxillaria luteola IPNI=*654273-1* 222
Maxillaria lyonii IPNI=*30044409-2* 222
Maxillaria multiflora IPNI=*643684-1* 221
Maxillaria mutabilis IPNI=*643731-1* 221
Maxillaria nana IPNI=*155331-2* 222
Maxillaria oxysepala IPNI=*643645-1* 221
Maxillaria panamensis IPNI=*30044409-2* 222
Maxillaria pertusa IPNI=*643558-1* 221
Maxillaria pubilabia IPNI=*318680-2* 221
Maxillaria purpurea IPNI=*654273-1* 222
Maxillaria revoluta IPNI=*30044409-2* 222
Maxillaria rolfei IPNI=*155165-2* 221
Maxillaria rousseauae IPNI=*318680-2* 221
Maxillaria ruberrima var. *sophronitis* IPNI=*155290-2* 221
Maxillaria rubrofusca IPNI=*643650-1* 221
Maxillaria rubropunctata IPNI=*155165-2* 221
Maxillaria rugosa IPNI=*643781-1* 221
Maxillaria saxatilis IPNI=*643736-1* 221
Maxillaria sessilis IPNI=*643387-1* 220
Maxillaria setigera var. *carolii* IPNI=*643559-1* 221

Maxillaria shepheardii IPNI=*155084-2* 220
Maxillaria squamata IPNI=*155331-2* 222
Maxillaria stenostele IPNI=*155331-2* 222
Maxillaria striatella IPNI=*155331-2* 222
Maxillaria tarumaensis IPNI=*643845-1* 221
Maxillaria triloris IPNI=*155147-2* 221
Maxillaria trinitatis IPNI=*643731-1* 221
Maxillaria tuerckheimii IPNI=*318680-2* 221
Maxillaria valenzuelana subsp. *angustata* IPNI=*155338-2* 222
Maxillaria vanillodora IPNI=*643781-1* 221
Maxillaria variabilis var. *unipunctata* IPNI=*30044409-2* 222
Maxillaria verrucifera IPNI=*643410-1* 220
Maxillaria yzabalana IPNI=*318680-2* 221
Maxillariella alba IPNI=*643250-1* 220
Maxillariella chiriquiensis IPNI=*30044409-2* 222
Maxillariella guareimensis IPNI=*643483-1* 220
Maxillariella panamensis IPNI=*30044409-2* 222
Maxillariella procurrens IPNI=*643736-1* 221
Maxillariella variabilis IPNI=*30044409-2* 222
Maxillariella variabilis var. *unipunctata* IPNI=*30044409-2* 222
Maximiliana caribaea IPNI=*664452-1* 205
Maximiliana elegans IPNI=*664452-1* 205
Maximiliana inajai IPNI=*246946-2* 212
Maximiliana insignis IPNI=*26571-2* 205
Maximiliana longirostrata IPNI=*664452-1* 205
Maximiliana macrogyne IPNI=*664452-1* 205
Maximiliana macropetala IPNI=*664452-1* 205
Maximiliana maripa IPNI=*664452-1* 205
Maximiliana martiana IPNI=*664452-1* 205
Maximiliana orinocensis IPNI=*246957-2* 212
Maximiliana princeps IPNI=*26580-2* 205
Maximiliana regia IPNI=*664452-1* 205
Maximiliana stenocarpa IPNI=*664452-1* 205
Maximiliana tetrasticha IPNI=*664452-1* 205
Maximilianea codinae IPNI=*111539-1* 406
Maximilianea hibiscodes IPNI=*111539-1* 406
Maximilianea orinocensis IPNI=*111532-1* 406
Maximilianea parkeri IPNI=*111532-1* 406
Maximilianea triphylla IPNI=*111539-1* 406
Maximilianea vitifolia IPNI=*111539-1* 406
Mayepea caribaea IPNI=*301087-2* 374
Mayna amazonica IPNI=*365030-1* 380
Mayna apeibifolia IPNI=*111855-1* 380
Mayna denticulata IPNI=*111855-1* 380
Mayna echinata IPNI=*111855-1* 380
Mayna glomerata IPNI=*111855-1* 380
Mayna grandiflora IPNI=*111509-1* 380
Mayna integrifolia IPNI=*47696-2* 380
Mayna laurina IPNI=*111811-1* 380
Mayna laxiflora IPNI=*140493-2* 380
Mayna linguifolia IPNI=*111511-1* 380
Mayna longicuspis IPNI=*111855-1* 380
Mayna longifolia IPNI=*111511-1* 380
Mayna longifolia var. *heliocarpa* IPNI=*111511-1* 380
Mayna longifolia var. *phasmatocarpa* IPNI=*111511-1* 380
Mayna microphylla IPNI=*111855-1* 380
Mayna muricida IPNI=*111511-1* 380
Mayna pacifica IPNI=*47701-2* 380
Mayna pacifica var. *brachycarpa* IPNI=*47701-2* 380
Mayna pacifica var. *pusilla* IPNI=*47701-2* 380
Mayna paludosa IPNI=*140493-2* 380
Mayna ramosii IPNI=*47705-2* 380
Mayna sericea IPNI=*76256-1* 190
Mayna toxica IPNI=*365030-1* 380
Mayna vernicosa IPNI=*111855-1* 380
Mayna zuliana IPNI=*111509-1* 380
Mays americana IPNI=*426810-1* 247
Mays zea IPNI=*426810-1* 247
Maytenus amygdalina IPNI=*162852-1* 291
Maytenus chuchuhuasha IPNI=*155542-2* 291
Maytenus cuezzoi IPNI=*162118-1* 291
Maytenus guyanensis f. *crenulata* IPNI=*161926-1* 291
Maytenus jauaensis IPNI=*161953-1* 291
Maytenus krukovii IPNI=*155542-2* 291
Maytenus longipes IPNI=*155604-2* 291
Maytenus micrantha IPNI=*161926-1* 291
Maytenus multiflora IPNI=*155542-2* 291
Maytenus sapotiformis IPNI=*161953-1* 291
Maytenus tarapotensis IPNI=*155542-2* 291
Maytenus verticillata var. *latifolia* IPNI=*162118-1* 291
Maytenus verticillata var. *lehmannii* IPNI=*162118-1* 291
Mayzea cerealis IPNI=*426810-1* 247
Mecardonia dianthera IPNI=*155661-2* 376
Mecardonia divaricata IPNI=*155661-2* 376
Mecardonia montevidensis IPNI=*155661-2* 376
Mecardonia ovata IPNI=*155661-2* 376
Mecardonia peduncularis IPNI=*155661-2* 376
Mecardonia tenuis IPNI=*155661-2* 376
Mecardonia viridis IPNI=*155661-2* 376
Medemia nobilis IPNI=*664859-1* 207
Medica apiculata IPNI=*60465997-2* 327
Medica bimuricata IPNI=*60465997-2* 327
Medica denticulata IPNI=*60465997-2* 327
Medica lupulina IPNI=*506268-1* 327
Medica pentacycla IPNI=*60465997-2* 327
Medica polymorpha IPNI=*60465997-2* 327
Medica reticulata IPNI=*60465997-2* 327
Medicago aculeata IPNI=*60465997-2* 327
Medicago alba IPNI=*156656-2* 327
Medicago apennina IPNI=*506268-1* 327
Medicago apiculata IPNI=*60465997-2* 327
Medicago berteroana IPNI=*60465997-2* 327
Medicago bimuricata IPNI=*60465997-2* 327
Medicago breviflora IPNI=*506268-1* 327
Medicago canescens IPNI=*506268-1* 327
Medicago canescens IPNI=*60465997-2* 327
Medicago caspia IPNI=*60465997-2* 327
Medicago caspica IPNI=*60465997-2* 327
Medicago catalonica IPNI=*60465997-2* 327
Medicago ciliaris IPNI=*506268-1* 327
Medicago ciliaris IPNI=*60465997-2* 327
Medicago confinis IPNI=*60465997-2* 327
Medicago conica IPNI=*60465997-2* 327
Medicago corymbifera IPNI=*506268-1* 327
Medicago cupianiana IPNI=*506268-1* 327
Medicago cyrenaea IPNI=*60465997-2* 327
Medicago denticulata IPNI=*60465997-2* 327
Medicago distans IPNI=*60465997-2* 327
Medicago echinata IPNI=*60465997-2* 327
Medicago flexuosa IPNI=*60465997-2* 327
Medicago gracillima IPNI=*60465997-2* 327
Medicago hamata IPNI=*60465997-2* 327
Medicago heterocarpa IPNI=*60465997-2* 327
Medicago hirsuta IPNI=*60465997-2* 327
Medicago hispida f. *apiculata* IPNI=*60465997-2* 327
Medicago hispida f. *denticulata* IPNI=*60465997-2* 327
Medicago hispida f. *inermis* IPNI=*60465997-2* 327
Medicago hispida IPNI=*60465997-2* 327
Medicago hispida subsp. *denticulata* IPNI=*60465997-2* 327
Medicago hispida var. *confinis* IPNI=*60465997-2* 327
Medicago hystrix IPNI=*60465997-2* 327
Medicago intermedia IPNI=*60465997-2* 327
Medicago lappacea IPNI=*60465997-2* 327
Medicago loretii IPNI=*60465997-2* 327
Medicago lupulina subvar. *eriocarpa* IPNI=*506268-1* 327
Medicago microdon IPNI=*60465997-2* 327

Medicago mniocarpa IPNI=*506268-1* 327
Medicago neurocarpos IPNI=*60465997-2* 327
Medicago nigra IPNI=*60465997-2* 327
Medicago nigra subsp. *microcarpa* IPNI=*60465997-2* 327
Medicago nigra var. *apiculata* IPNI=*60465997-2* 327
Medicago nigra var. *denticulata* IPNI=*60465997-2* 327
Medicago nigra var. *pentacycla* IPNI=*60465997-2* 327
Medicago nigra var. *polygyra* IPNI=*60465997-2* 327
Medicago nigra var. *vulgaris* IPNI=*60465997-2* 327
Medicago obscura IPNI=*60465997-2* 327
Medicago oligocarpa IPNI=*60465997-2* 327
Medicago parviflora IPNI=*1207360-2* 327
Medicago pentacycla IPNI=*60465997-2* 327
Medicago persica IPNI=*60465997-2* 327
Medicago polyantha IPNI=*60465997-2* 327
Medicago polycarpa IPNI=*60465997-2* 327
Medicago polycarpa var. *apiculata* IPNI=*60465997-2* 327
Medicago polycarpa var. *denticulata* IPNI=*60465997-2* 327
Medicago polycarpa var. *tuberculata* IPNI=*60465997-2* 327
Medicago polymorpha subsp. *hispida* IPNI=*60465997-2* 327
Medicago polymorpha subsp. *microcarpa* IPNI=*60465997-2* 327
Medicago polymorpha subsp. *polycarpa* IPNI=*60465997-2* 327
Medicago polymorpha var. *brevispina* IPNI=*60465997-2* 327
Medicago polymorpha var. *denticulata* IPNI=*60465997-2* 327
Medicago polymorpha var. *nigra* IPNI=*60465997-2* 327
Medicago polymorpha var. *vulgaris* IPNI=*60465997-2* 327
Medicago reniformis IPNI=*506268-1* 327
Medicago reticulata IPNI=*60465997-2* 327
Medicago revoilii IPNI=*506268-1* 327
Medicago reynieri IPNI=*60465997-2* 327
Medicago rigidula var. *eriocarpa* IPNI=*506268-1* 327
Medicago sardoa IPNI=*60465997-2* 327
Medicago sativa subsp. *vulgaris* IPNI=*30234961-2* 327
Medicago stipularis IPNI=*506268-1* 327
Medicago stricta IPNI=*60465997-2* 327
Medicago terebellum IPNI=*60465997-2* 327
Medicago tuberculata IPNI=*60465997-2* 327
Medicago willdenowii IPNI=*506268-1* 327
Medicula lupulina IPNI=*506268-1* 327
Medicula lupulina subsp. *jalasii* IPNI=*506268-1* 327
Medicula lupulina subsp. *willdenowiana* IPNI=*506268-1* 327
Mediocactus megalanthus IPNI=*232386-2* 285
Medium grandiflorum IPNI=*140670-1* 272
Megadenus capitatus IPNI=*1029981-2* 232
Megapterium macrocarpum IPNI=*173157-2* 430
Megapterium missouriense IPNI=*173157-2* 430
Megapterium nuttallianum IPNI=*173157-2* 430
Megastachya amabilis IPNI=*401703-1* 239
Megastachya aturensis IPNI=*401189-1* 239
Megastachya boryana IPNI=*1156566-2* 238
Megastachya ciliaris IPNI=*1156566-2* 238
Megastachya eragrostis IPNI=*1104599-2* 238
Megastachya glomerata IPNI=*401189-1* 239
Megastachya leersioides IPNI=*1104599-2* 238
Megastachya maypurensis IPNI=*1147585-2* 239
Megastachya oblonga IPNI=*1104599-2* 238
Megastachya obtusa IPNI=*1104599-2* 238
Megastachya oxylepis var. *capitata* IPNI=*322572-2* 239
Megastachya panamensis IPNI=*1147585-2* 239
Megastachya prostrata IPNI=*82343-2* 238
Megastachya purpurascens IPNI=*1104599-2* 238
Megastachya swainsonii IPNI=*1147585-2* 239
Megastachya tenella IPNI=*1088685-2* 239
Megastachya thalassica IPNI=*82343-2* 238
Megastachya triticea IPNI=*1104599-2* 238
Megathyrsus bivonanus IPNI=*60426565-2* 241
Megathyrsus maximus var. *coloratus* IPNI=*60426565-2* 241
Megathyrsus maximus var. *pubiglumis* IPNI=*60426565-2* 241
Megema macrura IPNI=*642941-1* 219
Megopiza fornicata IPNI=*526777-1* 374
Megopiza integra IPNI=*526777-1* 374
Megopiza longirostris IPNI=*526777-1* 374
Meibomia adhaesiva IPNI=*134286-3* 318
Meibomia adscendens IPNI=*77187388-1* 320
Meibomia adscendens var. *incana* IPNI=*1015219-2* 318
Meibomia adscendens var. *obovata* IPNI=*77187388-1* 320
Meibomia affinis IPNI=*78342-2* 317
Meibomia albida IPNI=*491595-1* 317
Meibomia albiflora IPNI=*78342-2* 317
Meibomia andina IPNI=*491595-1* 317
Meibomia angustifolia IPNI=*78351-2* 317
Meibomia aparines IPNI=*134286-3* 318
Meibomia arenaria IPNI=*135026-3* 318
Meibomia aspera IPNI=*78415-2* 317
Meibomia axillaris IPNI=*491595-1* 317
Meibomia axillaris var. *acutifolia* IPNI=*491595-1* 317
Meibomia axillaris var. *obtusifoliola* IPNI=*491595-1* 317
Meibomia balaensis IPNI=*134286-3* 318
Meibomia barbata IPNI=*77187394-1* 320
Meibomia cajanifolia IPNI=*1108524-2* 317
Meibomia cana IPNI=*1015219-2* 318
Meibomia distorta IPNI=*78415-2* 317
Meibomia distorta var. *macrophylla* IPNI=*78415-2* 317
Meibomia distorta var. *pabularis* IPNI=*78415-2* 317
Meibomia distorta var. *perrottetii* IPNI=*78415-2* 317
Meibomia diversifolia IPNI=*1015219-2* 318
Meibomia elata IPNI=*78415-2* 317
Meibomia emarginata IPNI=*1130757-2* 317
Meibomia glabra IPNI=*1130757-2* 317
Meibomia graminea IPNI=*78351-2* 317
Meibomia heterocarpon IPNI=*77187416-1* 320
Meibomia hjalmarsonii IPNI=*1109494-2* 318
Meibomia incana IPNI=*1015219-2* 318
Meibomia incana var. *angustifolia* IPNI=*78351-2* 317
Meibomia intorta IPNI=*134286-3* 318
Meibomia limensis var. *pilosiuscula* IPNI=*1109494-2* 318
Meibomia linearifolia IPNI=*78351-2* 317
Meibomia martensii IPNI=*135026-3* 318
Meibomia matto-grossensis IPNI=*78415-2* 317
Meibomia mauritiana IPNI=*1015219-2* 318
Meibomia mollicula IPNI=*1124646-2* 318
Meibomia mollis IPNI=*1130757-2* 317
Meibomia multicaulis IPNI=*135026-3* 318
Meibomia orbicularis IPNI=*491958-1* 318
Meibomia ovalis IPNI=*1015219-2* 318
Meibomia pabularis IPNI=*78415-2* 317
Meibomia physocarpa IPNI=*1028008-2* 318
Meibomia pilosiuscula IPNI=*1109494-2* 318
Meibomia procumbens IPNI=*78541-2* 318
Meibomia prorepens IPNI=*491595-1* 317
Meibomia purpurea IPNI=*1028008-2* 318
Meibomia racemifera IPNI=*77187388-1* 320
Meibomia reptans IPNI=*491595-1* 317
Meibomia scorpiurus IPNI=*135026-3* 318
Meibomia sericophylla IPNI=*78576-2* 318
Meibomia sinclairii IPNI=*1109494-2* 318
Meibomia sintenisii IPNI=*491595-1* 317
Meibomia sonorae IPNI=*134286-3* 318
Meibomia spiralis IPNI=*78541-2* 318
Meibomia stipulacea IPNI=*1028008-2* 318
Meibomia subspicata IPNI=*78351-2* 317
Meibomia supina IPNI=*1015219-2* 318
Meibomia terminalis IPNI=*1130757-2* 317
Meibomia thwaitesii IPNI=*77187388-1* 320
Meibomia tortuosa IPNI=*1028008-2* 318
Meibomia triflora IPNI=*77187443-1* 320

Meibomia trifoliastra IPNI=*77187388-1* 320
Meibomia trigona IPNI=*134286-3* 318
Meibomia umbrosa IPNI=*491595-1* 317
Meibomia uncinata IPNI=*1109494-2* 318
Meibomia variifolia IPNI=*1015219-2* 318
Meibomia violacea IPNI=*78415-2* 317
Meioluma guianensis IPNI=*160417-2* 306
Meioperis angustifolia IPNI=*321973-2* 400
Meioperis hederacea IPNI=*321973-2* 400
Meioperis minima IPNI=*185104-2* 399
Meioperis pallida IPNI=*185104-2* 399
Meioperis peltata IPNI=*321973-2* 400
Meioperis suberosa IPNI=*321973-2* 400
Meiota campechana IPNI=*54954-2* 208
Meisteria anonyma IPNI=*608256-1* 359
Mekistus sinensis IPNI=*77101543-1* 416
Melaleuca affinis IPNI=*597912-1* 426
Melaleuca decussata var. *ovoidea* IPNI=*597912-1* 426
Melaleuca elegans IPNI=*597912-1* 426
Melaleuca gummifera IPNI=*593305-1* 425
Melaleuca leucadendra var. *albida* IPNI=*598073-1* 426
Melaleuca leucadendra var. *angustifolia* IPNI=*598073-1* 426
Melaleuca leucadendra var. *coriacea* IPNI=*598073-1* 426
Melaleuca maidenii IPNI=*598073-1* 426
Melaleuca parviflora IPNI=*597912-1* 426
Melaleuca pumila IPNI=*597912-1* 426
Melaleuca scoparia IPNI=*597505-1* 426
Melaleuca smithii IPNI=*598073-1* 426
Melaleuca tenuifolia IPNI=*597505-1* 426
Melaleuca tetragona IPNI=*597912-1* 426
Melaleuca viridiflora var. *angustifolia* IPNI=*598073-1* 426
Melaleuca viridiflora var. *rubriflora* IPNI=*598073-1* 426
Melampodium australe IPNI=*1692-2* 255
Melampodium baranguillae IPNI=*230784-2* 269
Melampodium baranquillae IPNI=*230784-2* 269
Melampodium berteroanum IPNI=*156179-2* 265
Melampodium bonairense IPNI=*28847-2* 258
Melampodium camphoratum IPNI=*257035-1* 272
Melampodium copiosum IPNI=*156179-2* 265
Melampodium digynum IPNI=*257035-1* 272
Melampodium flaccidum IPNI=*156179-2* 265
Melampodium geminatum IPNI=*28847-2* 258
Melampodium hirsutum IPNI=*257035-1* 272
Melampodium ovatifolium IPNI=*156179-2* 265
Melampodium paludosum IPNI=*156179-2* 265
Melampodium panamense IPNI=*156179-2* 265
Melampodium pumilum IPNI=*156179-2* 265
Melampodium rhombifolium IPNI=*156179-2* 265
Melampodium ruderale IPNI=*202984-1* 261
Melampodium tenellum var. *flaccidum* IPNI=*156179-2* 265
Melananthera deltoidea IPNI=*156327-2* 266
Melananthera hastata IPNI=*156327-2* 266
Melanchrysum rigens IPNI=*208720-1* 263
Melanchrysum spinulosum IPNI=*941423-1* 263
Melanium hirtum IPNI=*553109-1* 417
Melanosinapis communis IPNI=*279422-1* 276
Melanosinapis nigra IPNI=*279422-1* 276
Melanthera amellus IPNI=*241903-1* 269
Melanthera amellus var. *subhastata* IPNI=*156327-2* 266
Melanthera amethystina IPNI=*156327-2* 266
Melanthera angustifolia IPNI=*156327-2* 266
Melanthera angustifolia var. *subhastata* IPNI=*156327-2* 266
Melanthera aspera IPNI=*156327-2* 266
Melanthera aspera var. *canescens* IPNI=*156327-2* 266
Melanthera aspera var. *glabriuscula* IPNI=*156327-2* 266
Melanthera aspera var. *subhastata* IPNI=*156327-2* 266
Melanthera brevifolia IPNI=*156327-2* 266
Melanthera buchii IPNI=*156327-2* 266
Melanthera calcicola IPNI=*156327-2* 266
Melanthera canescens f. *bicolor* IPNI=*156327-2* 266
Melanthera canescens IPNI=*156327-2* 266
Melanthera carpenteri IPNI=*156327-2* 266
Melanthera confusa IPNI=*156327-2* 266
Melanthera corymbosa IPNI=*156327-2* 266
Melanthera crenata IPNI=*156327-2* 266
Melanthera deltoidea IPNI=*156327-2* 266
Melanthera discoidea IPNI=*156327-2* 266
Melanthera hastata IPNI=*156327-2* 266
Melanthera hastata subsp. *cubensis* IPNI=*156327-2* 266
Melanthera hastata subsp. *lobata* IPNI=*156327-2* 266
Melanthera hastata var. *cubensis* IPNI=*156327-2* 266
Melanthera hastata var. *lobata* IPNI=*156327-2* 266
Melanthera hastata var. *pandurata* IPNI=*156327-2* 266
Melanthera hastata var. *parvifolia* IPNI=*156327-2* 266
Melanthera hastifolia IPNI=*156327-2* 266
Melanthera lanceolata IPNI=*156327-2* 266
Melanthera ligulata IPNI=*156327-2* 266
Melanthera linearis IPNI=*156327-2* 266
Melanthera linnaei IPNI=*156327-2* 266
Melanthera lobata IPNI=*156327-2* 266
Melanthera longipes IPNI=*156327-2* 266
Melanthera microphylla IPNI=*156327-2* 266
Melanthera molliuscula IPNI=*156327-2* 266
Melanthera montana IPNI=*156327-2* 266
Melanthera oxycarpha IPNI=*156327-2* 266
Melanthera oxylepis IPNI=*156327-2* 266
Melanthera panduriformis IPNI=*156327-2* 266
Melanthera parviceps IPNI=*156327-2* 266
Melanthera purpurascens IPNI=*156327-2* 266
Melanthera trilobata IPNI=*156327-2* 266
Melanthera urticifolia IPNI=*156327-2* 266
Melanthesa neocaledonica IPNI=*340073-1* 401
Melanthium damascenum IPNI=*711645-1* 436
Melargyra rosea IPNI=*158983-1* 286
Melargyra rubra IPNI=*158983-1* 286
Melastoma acuminatum IPNI=*1069702-2* 422
Melastoma acuminatum IPNI=*572641-1* 421
Melastoma adscendens IPNI=*566051-1* 417
Melastoma albicans IPNI=*1125568-2* 419
Melastoma ambiguum IPNI=*572641-1* 421
Melastoma aplostachyum IPNI=*1039241-2* 419
Melastoma aquaticum IPNI=*573916-1* 424
Melastoma arborescens IPNI=*566597-1* 417
Melastoma arboreum IPNI=*326247-2* 419
Melastoma argenteum IPNI=*572109-1* 419
Melastoma aspersum IPNI=*572641-1* 421
Melastoma astrolasion IPNI=*1155116-2* 422
Melastoma auriculatum IPNI=*572590-1* 421
Melastoma balbisianum IPNI=*1000103-2* 422
Melastoma brachiatum IPNI=*572641-1* 421
Melastoma bracteatum IPNI=*1024004-2* 421
Melastoma bracteolatum IPNI=*572199-1* 419
Melastoma bubalinum IPNI=*572209-1* 419
Melastoma bullosum IPNI=*77207390-1* 419
Melastoma calvescens IPNI=*326247-2* 419
Melastoma calyptratum IPNI=*158886-2* 420
Melastoma capitellatum IPNI=*77196165-1* 420
Melastoma caudatum IPNI=*572256-1* 420
Melastoma ciliatum IPNI=*1108920-2* 420
Melastoma crenatum IPNI=*77164247-1* 420
Melastoma croceum IPNI=*1158780-2* 422
Melastoma croceum IPNI=*572353-1* 420
Melastoma cupreum IPNI=*1155116-2* 422
Melastoma cuspidatum IPNI=*572641-1* 421
Melastoma decussatum IPNI=*1108920-2* 420
Melastoma dentatum IPNI=*77207400-1* 420

Melastoma dependens IPNI=*77196165-1* 420
Melastoma detergibile IPNI=*1125568-2* 419
Melastoma dimorphum IPNI=*1110588-2* 423
Melastoma diversifolium IPNI=*77144083-1* 422
Melastoma dodecandrum IPNI=*572411-1* 420
Melastoma egense IPNI=*77112284-1* 418
Melastoma elatum IPNI=*1092872-2* 420
Melastoma elegans IPNI=*77164247-1* 420
Melastoma expansum IPNI=*572641-1* 421
Melastoma fastigiatum IPNI=*159763-2* 422
Melastoma ferruginatum IPNI=*572470-1* 420
Melastoma ferrugineum IPNI=*572470-1* 420
Melastoma glabrum IPNI=*1069702-2* 422
Melastoma glutinosum IPNI=*566936-1* 418
Melastoma grandiflorum IPNI=*575460-1* 424
Melastoma granulosum IPNI=*158784-2* 419
Melastoma grossularia IPNI=*1157663-2* 418
Melastoma grossularioides IPNI=*1157663-2* 418
Melastoma heterophyllum IPNI=*77207843-1* 420
Melastoma hirtum IPNI=*77164247-1* 420
Melastoma holosericeum IPNI=*1024004-2* 421
Melastoma holosericeum IPNI=*1125568-2* 419
Melastoma holosericeum IPNI=*159763-2* 422
Melastoma ibaguense IPNI=*572590-1* 421
Melastoma icosandrum IPNI=*159783-2* 423
Melastoma impetiginosum IPNI=*1018450-2* 422
Melastoma impetiolare IPNI=*572600-1* 421
Melastoma laevigatum IPNI=*1069702-2* 422
Melastoma laevigatum IPNI=*572641-1* 421
Melastoma lasiopetalum IPNI=*1000103-2* 422
Melastoma ledifolium IPNI=*566862-1* 418
Melastoma lepidotum IPNI=*572671-1* 421
Melastoma leucanthum IPNI=*1069702-2* 422
Melastoma ligustrinum IPNI=*572682-1* 421
Melastoma lineatum IPNI=*572590-1* 421
Melastoma longifolium IPNI=*1220088-2* 421
Melastoma macrophyllum IPNI=*1000103-2* 422
Melastoma macrophyllum IPNI=*572600-1* 421
Melastoma maieta IPNI=*77164249-1* 421
Melastoma majeta IPNI=*77164249-1* 421
Melastoma mandiocanum IPNI=*326247-2* 419
Melastoma mandioccanum IPNI=*326247-2* 419
Melastoma megalophyllum IPNI=*1088633-2* 423
Melastoma milleflorum IPNI=*1110588-2* 423
Melastoma minutiflorum IPNI=*1098452-2* 421
Melastoma montanum IPNI=*1069702-2* 422
Melastoma mucronatum IPNI=*1024004-2* 421
Melastoma nervosum IPNI=*572835-1* 421
Melastoma nicotianifolium IPNI=*572641-1* 421
Melastoma nitidum IPNI=*1125568-2* 419
Melastoma octonum IPNI=*77196163-1* 421
Melastoma parviflorum IPNI=*1069702-2* 422
Melastoma patens IPNI=*1000103-2* 422
Melastoma pauciflorum IPNI=*77164247-1* 420
Melastoma paulense IPNI–*77112284-1* 418
Melastoma pendulifolium IPNI=*1069702-2* 422
Melastoma pendulifolium IPNI=*572641-1* 421
Melastoma pendulum IPNI=*1098452-2* 421
Melastoma physiphorum IPNI=*77207761-1* 423
Melastoma pilosum IPNI=*572590-1* 421
Melastoma pllosum IPNI=*77207921-1* 421
Melastoma piluliferum IPNI=*572641-1* 421
Melastoma plantaginifolium IPNI=*1143367-2* 422
Melastoma plumosum IPNI=*908858-1* 419
Melastoma portoricense IPNI=*572641-1* 421
Melastoma prasinum IPNI=*1069702-2* 422
Melastoma pulchrum IPNI=*1069702-2* 422
Melastoma punctatum IPNI=*1069703-2* 422
Melastoma purpurascens IPNI=*1031600-2* 417
Melastoma purpureum IPNI=*1031600-2* 417
Melastoma quinquenervium IPNI=*1069702-2* 422
Melastoma quinquenervium IPNI=*77144083-1* 422
Melastoma racemosum IPNI=*1158780-2* 422
Melastoma ramiflorum IPNI=*569004-1* 418
Melastoma revolutum IPNI=*158784-2* 419
Melastoma rosmarinifolium IPNI=*573124-1* 422
Melastoma rostratum IPNI=*159943-2* 423
Melastoma rubiginosum IPNI=*1155116-2* 422
Melastoma rubrum IPNI=*77164250-1* 422
Melastoma rufescens IPNI=*1018450-2* 422
Melastoma salicifolium IPNI=*573124-1* 422
Melastoma secundum IPNI=*77207935-1* 421
Melastoma sepiarium IPNI=*1069702-2* 422
Melastoma serrulatum IPNI=*1000103-2* 422
Melastoma sessiliflorum IPNI=*1108920-2* 420
Melastoma sessiliflorum IPNI=*77164250-1* 422
Melastoma setinode IPNI=*572353-1* 420
Melastoma spicatum IPNI=*572835-1* 421
Melastoma spicatum IPNI=*77196165-1* 420
Melastoma splendens IPNI=*1143367-2* 422
Melastoma spondylanthum IPNI=*572835-1* 421
Melastoma squamulosum IPNI=*573207-1* 422
Melastoma stenostachyum IPNI=*159763-2* 422
Melastoma strangulatum IPNI=*1000103-2* 422
Melastoma striatum IPNI=*572641-1* 421
Melastoma strigillosum IPNI=*77196164-1* 422
Melastoma strigosum IPNI=*566891-1* 418
Melastoma subtriplinervium IPNI=*569110-1* 419
Melastoma succosum IPNI=*569017-1* 418
Melastoma swartzianum IPNI=*572411-1* 420
Melastoma tamonea IPNI=*572411-1* 420
Melastoma theaezans IPNI=*1110588-2* 423
Melastoma tococa IPNI=*77207761-1* 423
Melastoma tomentosum IPNI=*1088633-2* 423
Melastoma trinervium IPNI=*1069702-2* 422
Melastoma triste IPNI=*77196164-1* 422
Melastoma uliginosum IPNI=*1031600-2* 417
Melastoma umbilicatum IPNI=*159943-2* 423
Melastoma umbonatum IPNI=*77196164-1* 422
Melastoma velutinum IPNI=*1125568-2* 419
Melastoma venosum IPNI=*1158780-2* 422
Melastoma villosum IPNI=*1168615-2* 418
Melastoma volubile IPNI=*60465985-2* 417
Melastoma xalapense IPNI=*159943-2* 423
Melia azadirachta IPNI=*1213180-2* 454
Melia fraxinifolia IPNI=*1213180-2* 454
Melia hasskarlii IPNI=*1213180-2* 454
Melia indica IPNI=*1213180-2* 454
Melia japonica IPNI=*1213180-2* 454
Melia parviflora IPNI=*1213180-2* 454
Melia pinnata IPNI=*1213180-2* 454
Melica altissima IPNI=*395086-1* 235
Melica fluitans IPNI=*404100-1* 239
Melica triflora IPNI=*258855-2* 246
Melicocca geniculata IPNI=*128278-1* 452
Melidiscus gigantea IPNI=*147057-1* 279
Melilotus albus f. *prolifer* IPNI=*156656-2* 327
Melilotus albus IPNI=*156656-2* 327
Melilotus angulatus IPNI=*156656-2* 327
Melilotus argutus IPNI=*156656-2* 327
Melilotus berardii IPNI=*1207360-2* 327
Melilotus bonplandii IPNI=*1207360-2* 327
Melilotus boumettii IPNI=*1207360-2* 327
Melilotus brachystachyus IPNI=*1207360-2* 327
Melilotus exaltatus IPNI=*1207360-2* 327
Melilotus giganteus IPNI=*156656-2* 327

Melilotus kotschyi IPNI=*156656-2* 327
Melilotus leucanthus IPNI=*156656-2* 327
Melilotus lupulinus IPNI=*506268-1* 327
Melilotus medicaginoides IPNI=*506268-1* 327
Melilotus melanospermus IPNI=*156656-2* 327
Melilotus melilotus-indicus IPNI=*1207360-2* 327
Melilotus minimus IPNI=*1207360-2* 327
Melilotus occidentalis IPNI=*1207360-2* 327
Melilotus officinalis subsp. *albus* IPNI=*156656-2* 327
Melilotus parviflorus IPNI=*1207360-2* 327
Melilotus permixtus IPNI=*1207360-2* 327
Melilotus reticulatus IPNI=*1207360-2* 327
Melilotus robinii IPNI=*1207360-2* 327
Melilotus rugulosus IPNI=*1207360-2* 327
Melilotus rugulosus IPNI=*156656-2* 327
Melilotus strictus IPNI=*156656-2* 327
Melilotus tenellus IPNI=*156656-2* 327
Melilotus tommasinii IPNI=*1207360-2* 327
Melilotus urbanii IPNI=*156656-2* 327
Melilotus vulgaris IPNI=*156656-2* 327
Melinis maitlandii IPNI=*408232-1* 241
Melinis minutiflora var. *glutinosa* IPNI=*408232-1* 241
Melinis minutiflora var. *inodora* IPNI=*408232-1* 241
Melinis purpurea IPNI=*408232-1* 241
Melinis tenuinervis IPNI=*408232-1* 241
Meliosma arenosa IPNI=*156682-2* 433
Meliphlea vitifolia IPNI=*563228-1* 412
Melissa pulegium IPNI=*451022-1* 371
Melloa duseniana IPNI=*77100464-1* 363
Melloa fraxinifolia IPNI=*77100464-1* 363
Melloa populifolia IPNI=*77100464-1* 363
Melloa quadrivalvis IPNI=*77100464-1* 363
Melloca peruviana IPNI=*167633-1* 283
Melloca tuberosa IPNI=*167633-1* 283
Melo adana IPNI=*292238-1* 293
Melo adzhur IPNI=*292238-1* 293
Melo agrestis IPNI=*292238-1* 293
Melo ambiguus IPNI=*292238-1* 293
Melo ameri IPNI=*292238-1* 293
Melo cantalupa IPNI=*292238-1* 293
Melo cassaba IPNI=*292238-1* 293
Melo chandalak IPNI=*292238-1* 293
Melo chate IPNI=*292238-1* 293
Melo chinensis IPNI=*292238-1* 293
Melo conomon IPNI=*292238-1* 293
Melo figari IPNI=*292238-1* 293
Melo flexuosus IPNI=*292238-1* 293
Melo microcarpus IPNI=*292238-1* 293
Melo monoclinus IPNI=*292238-1* 293
Melo orientalis IPNI=*292238-1* 293
Melo persicus IPNI=*292238-1* 293
Melo sativus IPNI=*292238-1* 293
Melo vulgaris IPNI=*292238-1* 293
Melo zard IPNI=*292238-1* 293
Melocactus actinacanthus IPNI=*156850-2* 284
Melocactus brongnartii IPNI=*305646-2* 284
Melocactus crassicostatus IPNI=*305646-2* 284
Melocactus curvispinus subsp. *cucutensis* IPNI=*305646-2* 284
Melocactus curvispinus subsp. *loboguerreroi* IPNI=*305646-2* 284
Melocactus curvispinus subsp. *obtusipetalus* IPNI=*305646-2* 284
Melocactus delessertianus IPNI=*305646-2* 284
Melocactus guatemalensis IPNI=*305646-2* 284
Melocactus guitarti IPNI=*305646-2* 284
Melocactus holguinensis IPNI=*305646-2* 284
Melocactus humilis IPNI=*305646-2* 284
Melocactus jakusii IPNI=*305646-2* 284
Melocactus loboguerreroi IPNI=*305646-2* 284
Melocactus matanzanus subsp. *actinacanthus* IPNI=*156850-2* 284
Melocactus maxonii IPNI=*305646-2* 284
Melocactus maxonii var. *sanctae-rosae* IPNI=*305646-2* 284
Melocactus monvilleanus IPNI=*305646-2* 284
Melocactus oaxacensis IPNI=*305646-2* 284
Melocactus obtusipetalus IPNI=*305646-2* 284
Melocactus obtusipetalus var. *crassicostatus* IPNI=*305646-2* 284
Melocactus ruestii IPNI=*305646-2* 284
Melocactus ruestii subsp. *maxonii* IPNI=*305646-2* 284
Melocactus ruestii subsp. *oaxacensis* IPNI=*305646-2* 284
Melocactus ruestii subsp. *sanctae-rosae* IPNI=*305646-2* 284
Melocactus salvador IPNI=*305646-2* 284
Melocanna bambusoides IPNI=*408271-1* 241
Melocanna excelsa IPNI=*404020-1* 239
Melochia arida IPNI=*157023-2* 412
Melochia borbonica IPNI=*824514-1* 411
Melochia carpinifolia IPNI=*824514-1* 411
Melochia cephalodes IPNI=*157043-2* 412
Melochia clinopodium IPNI=*157043-2* 412
Melochia concinna IPNI=*824503-1* 411
Melochia conglobata IPNI=*824514-1* 411
Melochia corchorifolia IPNI=*320781-2* 415
Melochia crenata IPNI=*157023-2* 412
Melochia crinita IPNI=*824503-1* 411
Melochia densiflora IPNI=*157043-2* 412
Melochia depressa IPNI=*133833-2* 410
Melochia diffusa IPNI=*133833-2* 410
Melochia diffusa IPNI=*157023-2* 412
Melochia domingensis IPNI=*157004-2* 412
Melochia frutescens IPNI=*157023-2* 412
Melochia gracilis IPNI=*157002-2* 412
Melochia hirsuta f. *albiflora* IPNI=*157043-2* 412
Melochia hirsuta f. *ferruginea* IPNI=*157043-2* 412
Melochia hirsuta f. *minus-induta* IPNI=*157043-2* 412
Melochia hirsuta IPNI=*157043-2* 412
Melochia hirsuta var. *calophylla* IPNI=*157043-2* 412
Melochia hirsuta var. *glabrata* IPNI=*157043-2* 412
Melochia hirsuta var. *grandiflora* IPNI=*157043-2* 412
Melochia hirsuta var. *macrophylla* IPNI=*157043-2* 412
Melochia hirsuta var. *paraguayensis* IPNI=*157043-2* 412
Melochia hirsuta var. *rotundifolia* IPNI=*157043-2* 412
Melochia hypoleuca IPNI=*157023-2* 412
Melochia interrupta IPNI=*157002-2* 412
Melochia jamaicensis IPNI=*157002-2* 412
Melochia jurgensenii IPNI=*157043-2* 412
Melochia lilacina IPNI=*157043-2* 412
Melochia macrophylla IPNI=*824432-1* 411
Melochia nitidula IPNI=*157002-2* 412
Melochia nodiflora var. *longipetiolata* IPNI=*824514-1* 411
Melochia plicata IPNI=*157023-2* 412
Melochia polystachya IPNI=*157002-2* 412
Melochia portoricensis IPNI=*157023-2* 412
Melochia pyramidata var. *normalis* IPNI=*157004-2* 412
Melochia scutellarioides IPNI=*157043-2* 412
Melochia sericea IPNI=*157002-2* 412
Melochia serrata IPNI=*157043-2* 412
Melochia spicata IPNI=*561999-1* 410
Melochia subcordata IPNI=*157043-2* 412
Melochia tenella IPNI=*157043-2* 412
Melochia tomentosa var. *bahiensis* IPNI=*157023-2* 412
Melochia tomentosa var. *frutescens* IPNI=*157023-2* 412
Melochia tomentosa var. *turpiniana* IPNI=*157023-2* 412
Melochia tomentosa var. *typica* IPNI=*157023-2* 412
Melochia turpiniana IPNI=*157023-2* 412
Melochia ulmarioides IPNI=*157002-2* 412
Melochia venosa IPNI=*157002-2* 412
Melochia venosa var. *betonicifolia* IPNI=*157002-2* 412
Melochia venosa var. *flava* IPNI=*157002-2* 412
Melochia venosa var. *polystachya* IPNI=*157002-2* 412

Melochia venosa var. *purpurea* IPNI=*157002-2* 412
Melochia venosa var. *sericea* IPNI=*157002-2* 412
Melochia venosa var. *squarrosa* IPNI=*157002-2* 412
Melochia venosa var. *typica* IPNI=*157002-2* 412
Melochia venosa var. *widgrenii* IPNI=*157002-2* 412
Melochia vestita IPNI=*157043-2* 412
Melothria angustiloba IPNI=*293334-1* 294
Melothria candolleana IPNI=*293334-1* 294
Melothria chlorocarpa IPNI=*293293-1* 294
Melothria congestiflora IPNI=*592558-1* 294
Melothria costensis IPNI=*293320-1* 294
Melothria crassifolia IPNI=*293293-1* 294
Melothria donnell-smithii var. *hirtella* IPNI=*293320-1* 294
Melothria donnell-smithii var. *rotundifolia* IPNI=*293320-1* 294
Melothria edulis IPNI=*293293-1* 294
Melothria fluminensis IPNI=*293293-1* 294
Melothria fluminensis var. *hydrocotylifolia* IPNI=*293293-1* 294
Melothria fluminensis var. *ovata* IPNI=*293293-1* 294
Melothria fluminensis var. *triangularis* IPNI=*293293-1* 294
Melothria guadalupensis IPNI=*293293-1* 294
Melothria heteroloba IPNI=*293334-1* 294
Melothria microcarpa IPNI=*293293-1* 294
Melothria monoica IPNI=*293293-1* 294
Melothria nashii IPNI=*293293-1* 294
Melothria nigra IPNI=*293293-1* 294
Melothria pendula var. *aspera* IPNI=*293293-1* 294
Melothria pendula var. *chlorocarpa* IPNI=*293293-1* 294
Melothria pendula var. *crassifolia* IPNI=*293293-1* 294
Melothria pendula var. *microcarpa* IPNI=*293293-1* 294
Melothria pervaga IPNI=*293293-1* 294
Melothria punctata IPNI=*293293-1* 294
Melothria repanda IPNI=*293293-1* 294
Melothria touchanensis IPNI=*293060-1* 294
Melothria trifoliata IPNI=*592558-1* 294
Melothria trilobata var. *costaricensis* IPNI=*293334-1* 294
Melothria trilobata var. *sphaerocarpa* IPNI=*293334-1* 294
Memora bilabiata IPNI=*77100467-1* 365
Memora cladotricha IPNI=*77100436-1* 362
Memora consanguinea IPNI=*77100441-1* 362
Memora diffusa IPNI=*77100436-1* 362
Memora flaviflora IPNI=*77100438-1* 362
Memora fulgens var. *elongata* IPNI=*77100441-1* 362
Memora nobilis IPNI=*77100467-1* 365
Memora ovata IPNI=*77100441-1* 362
Memora pseudopatula IPNI=*77100440-1* 362
Memora schomburgkii IPNI=*77100441-1* 362
Menadena parkeri IPNI=*643684-1* 221
Menispermum abuta IPNI=*580080-1* 434
Menispermum domingense IPNI=*580887-1* 435
Menispermum tamoides IPNI=*581190-1* 435
Mentha × *balsamea* IPNI=*450969-1* 371
Mentha × *banatica* IPNI=*450969-1* 371
Mentha × *concinna* IPNI=*450969-1* 371
Mentha × *crispula* IPNI=*450969-1* 371
Mentha × *durandoana* IPNI=*450969-1* 371
Mentha × *exaltata* IPNI=*450969-1* 371
Mentha × *fraseri* IPNI=*450969-1* 371
Mentha × *glabrata* IPNI=*450969-1* 371
Mentha × *heuffelii* IPNI=*450969-1* 371
Mentha × *hircina* IPNI=*450969-1* 371
Mentha × *hortensis* IPNI=*450969-1* 371
Mentha × *hudsoniana* IPNI=*450969-1* 371
Mentha × *kahirina* IPNI=*450969-1* 371
Mentha × *langii* IPNI=*450969-1* 371
Mentha × *napolitana* IPNI=*450969-1* 371
Mentha × *nigricans* IPNI=*450969-1* 371
Mentha × *odora* IPNI=*450969-1* 371
Mentha × *officinalis* IPNI=*450969-1* 371
Mentha × *pimentum* IPNI=*450969-1* 371
Mentha × *piperita* var. *hispidula* IPNI=*450969-1* 371
Mentha × *piperita* var. *officinalis* IPNI=*450969-1* 371
Mentha × *piperita* var. *ouweneelii* IPNI=*450969-1* 371
Mentha × *piperita* var. *pennsylvanica* IPNI=*450969-1* 371
Mentha × *schultzii* IPNI=*450969-1* 371
Mentha × *tenuis* IPNI=*450969-1* 371
Mentha acuta IPNI=*450192-1* 371
Mentha acutata IPNI=*450192-1* 371
Mentha adspersa IPNI=*450192-1* 371
Mentha affinis IPNI=*450192-1* 371
Mentha agrestis IPNI=*30026217-2* 371
Mentha agrestris IPNI=*30026217-2* 371
Mentha albae-carolinae IPNI=*30026217-2* 371
Mentha albarracinensis IPNI=*451022-1* 371
Mentha alberti IPNI=*30026217-2* 371
Mentha allionii IPNI=*30026217-2* 371
Mentha angustifolia IPNI=*30026217-2* 371
Mentha anomala IPNI=*30026217-2* 371
Mentha approximata IPNI=*30026217-2* 371
Mentha aquatica f. *aquatica* IPNI=*450192-1* 371
Mentha aquatica var. *citrata* IPNI=*450192-1* 371
Mentha aquatica var. *subdentata* IPNI=*450192-1* 371
Mentha aquatica var. *subspicata* IPNI=*450192-1* 371
Mentha arenaria IPNI=*30026217-2* 371
Mentha arguta IPNI=*30026217-2* 371
Mentha argutissima IPNI=*30026217-2* 371
Mentha argutissima var. *subpilosa* IPNI=*30026217-2* 371
Mentha aromatica IPNI=*450192-1* 371
Mentha aromatica IPNI=*451022-1* 371
Mentha arvensis subsp. *parietariifolia* IPNI=*30026217-2* 371
Mentha arvensis var. *caespitosa* IPNI=*30026217-2* 371
Mentha arvensis var. *lanceolata* IPNI=*30026217-2* 371
Mentha arvensis var. *legitima* IPNI=*30026217-2* 371
Mentha arvensis var. *minor* IPNI=*30026217-2* 371
Mentha arvensis var. *parietariifolia* IPNI=*30026217-2* 371
Mentha arvensis var. *parviflora* IPNI=*30026217-2* 371
Mentha arvensis var. *procumbens* IPNI=*30026217-2* 371
Mentha arvensis var. *radicans* IPNI=*30026217-2* 371
Mentha atrovirens IPNI=*30026217-2* 371
Mentha aucheri IPNI=*451022-1* 371
Mentha augusta IPNI=*450192-1* 371
Mentha auneticensis IPNI=*450192-1* 371
Mentha aurita IPNI=*450192-1* 371
Mentha austriaca IPNI=*30026217-2* 371
Mentha avellinii IPNI=*450192-1* 371
Mentha badensis IPNI=*30026217-2* 371
Mentha baguetiana IPNI=*30026217-2* 371
Mentha barbata IPNI=*30026217-2* 371
Mentha brachiata IPNI=*450192-1* 371
Mentha bracteolata IPNI=*30026217-2* 371
Mentha bugulifolia IPNI=*450192-1* 371
Mentha calaminthifolia IPNI=*450192-1* 371
Mentha campestris IPNI=*30026217-2* 371
Mentha campicola IPNI=*30026217-2* 371
Mentha cetica IPNI=*450192-1* 371
Mentha chaixii IPNI=*450192-1* 371
Mentha citrata IPNI=*450192-1* 371
Mentha collina IPNI=*30026217-2* 371
Mentha cordata IPNI=*450192-1* 371
Mentha crenatodentata IPNI=*450192-1* 371
Mentha cuneifolia IPNI=*30026217-2* 371
Mentha daghestanica IPNI=*451022-1* 371
Mentha deflexa IPNI=*30026217-2* 371
Mentha densiflora IPNI=*30026217-2* 371
Mentha densifoliata IPNI=*30026217-2* 371
Mentha denticulata IPNI=*450192-1* 371
Mentha deseglisei IPNI=*450192-1* 371

Mentha diffusa IPNI=*30026217-2* 371
Mentha dissitiflora IPNI=*30026217-2* 371
Mentha divaricata IPNI=*30026217-2* 371
Mentha divergens IPNI=*30026217-2* 371
Mentha dubia IPNI=*30026217-2* 371
Mentha dubia IPNI=*450192-1* 371
Mentha duffourii IPNI=*30026217-2* 371
Mentha duftschmidii IPNI=*30026217-2* 371
Mentha dunensis IPNI=*450192-1* 371
Mentha duriuscula IPNI=*450192-1* 371
Mentha ehrhartiana IPNI=*30026217-2* 371
Mentha elongata IPNI=*450192-1* 371
Mentha eriantha IPNI=*450192-1* 371
Mentha exigua IPNI=*30026217-2* 371
Mentha exigua IPNI=*451022-1* 371
Mentha flagellifera IPNI=*30026217-2* 371
Mentha flexuosa IPNI=*30026217-2* 371
Mentha florida IPNI=*30026217-2* 371
Mentha fochii IPNI=*30026217-2* 371
Mentha fontana IPNI=*30026217-2* 371
Mentha fontana var. *brevibracteata* IPNI=*30026217-2* 371
Mentha fontana var. *conferta* IPNI=*30026217-2* 371
Mentha fontqueri IPNI=*30026217-2* 371
Mentha fossicola IPNI=*30026217-2* 371
Mentha gallica IPNI=*30026217-2* 371
Mentha gentiliformis IPNI=*30026217-2* 371
Mentha gentilis IPNI=*30026217-2* 371
Mentha gibraltarica IPNI=*451022-1* 371
Mentha glomerata IPNI=*450192-1* 371
Mentha gracilescens IPNI=*30026217-2* 371
Mentha grandidentata IPNI=*450192-1* 371
Mentha graveolens IPNI=*30026217-2* 371
Mentha hillebrandtii IPNI=*30026217-2* 371
Mentha hirsuta IPNI=*450192-1* 371
Mentha hirta IPNI=*450192-1* 371
Mentha hirtiflora IPNI=*451022-1* 371
Mentha hortensis var. *citrata* IPNI=*450969-1* 371
Mentha hostii IPNI=*30026217-2* 371
Mentha hostii var. *arvina* IPNI=*30026217-2* 371
Mentha hybrida IPNI=*450192-1* 371
Mentha hygrophila IPNI=*450192-1* 371
Mentha hyptiformis IPNI=*77120819-1* 369
Mentha hystrix IPNI=*450192-1* 371
Mentha incisoserrata IPNI=*450192-1* 371
Mentha intermedia IPNI=*30026217-2* 371
Mentha intermedia IPNI=*450192-1* 371
Mentha intricata IPNI=*450192-1* 371
Mentha joffrei IPNI=*30026217-2* 371
Mentha kitaibeliana IPNI=*30026217-2* 371
Mentha lamiifolia IPNI=*30026217-2* 371
Mentha lanceolata IPNI=*30026217-2* 371
Mentha lapponica IPNI=*30026217-2* 371
Mentha lata IPNI=*30026217-2* 371
Mentha lateovata IPNI=*450192-1* 371
Mentha latifolia IPNI=*30026217-2* 371
Mentha latifolia IPNI=*450192-1* 371
Mentha latissima IPNI=*30026217-2* 371
Mentha laxa IPNI=*30026217-2* 371
Mentha limicola IPNI=*450192-1* 371
Mentha limnetes IPNI=*450192-1* 371
Mentha limosa IPNI=*450192-1* 371
Mentha littoralis IPNI=*450192-1* 371
Mentha lloydii IPNI=*450192-1* 371
Mentha lloydii var. *opaca* IPNI=*450192-1* 371
Mentha lobeliana IPNI=*450192-1* 371
Mentha longibracteata IPNI=*30026217-2* 371
Mentha macrocephala IPNI=*450192-1* 371
Mentha maculata IPNI=*30026217-2* 371
Mentha melissifolia IPNI=*30026217-2* 371
Mentha microcephala IPNI=*450192-1* 371
Mentha minor IPNI=*30026217-2* 371
Mentha moenchii IPNI=*30026217-2* 371
Mentha mollis IPNI=*451957-1* 371
Mentha montana IPNI=*451022-1* 371
Mentha mosana IPNI=*30026217-2* 371
Mentha multiflora IPNI=*30026217-2* 371
Mentha multiflora var. *serpentina* IPNI=*30026217-2* 371
Mentha mutabilis IPNI=*30026217-2* 371
Mentha nederheimensis IPNI=*450192-1* 371
Mentha nemorosa IPNI=*30026217-2* 371
Mentha nemorum IPNI=*30026217-2* 371
Mentha nigrescens IPNI=*450192-1* 371
Mentha nobilis IPNI=*30026217-2* 371
Mentha numidica IPNI=*451022-1* 371
Mentha nummularia IPNI=*30026217-2* 371
Mentha obliqua IPNI=*450192-1* 371
Mentha obtusata IPNI=*30026217-2* 371
Mentha obtuseserrata IPNI=*450192-1* 371
Mentha obtusifolia IPNI=*450192-1* 371
Mentha obtusodentata IPNI=*30026217-2* 371
Mentha ocymoides IPNI=*30026217-2* 371
Mentha odorata IPNI=*30026217-2* 371
Mentha odorata IPNI=*450192-1* 371
Mentha origanifolia IPNI=*30026217-2* 371
Mentha origanoides IPNI=*450192-1* 371
Mentha ortmanniana IPNI=*450192-1* 371
Mentha ovata IPNI=*30026217-2* 371
Mentha palitzensis IPNI=*30026217-2* 371
Mentha paludosa IPNI=*30026217-2* 371
Mentha paludosa IPNI=*450192-1* 371
Mentha palustris IPNI=*30026217-2* 371
Mentha palustris IPNI=*450192-1* 371
Mentha parvifolia IPNI=*30026217-2* 371
Mentha parvula IPNI=*30026217-2* 371
Mentha pascuorum IPNI=*30026217-2* 371
Mentha pastoris IPNI=*30026217-2* 371
Mentha perilloides IPNI=*158242-2* 371
Mentha piersiana IPNI=*30026217-2* 371
Mentha pilosa IPNI=*30026217-2* 371
Mentha pilosella IPNI=*30026217-2* 371
Mentha pireana IPNI=*450192-1* 371
Mentha plagensis IPNI=*30026217-2* 371
Mentha plicata IPNI=*30026217-2* 371
Mentha polyanthetica IPNI=*450192-1* 371
Mentha polymorpha IPNI=*30026217-2* 371
Mentha praeclara IPNI=*30026217-2* 371
Mentha praecox IPNI=*30026217-2* 371
Mentha praticola IPNI=*30026217-2* 371
Mentha probabilis IPNI=*450192-1* 371
Mentha procumbens IPNI=*30026217-2* 371
Mentha prostrata IPNI=*30026217-2* 371
Mentha pulchella IPNI=*30026217-2* 371
Mentha pulegiformis IPNI=*30026217-2* 371
Mentha pulegioides IPNI=*451022-1* 371
Mentha pulegium f. *alba* IPNI=*451022-1* 371
Mentha pulegium f. *albiflora* IPNI=*451022-1* 371
Mentha pulegium var. *heterophylla* IPNI=*451022-1* 371
Mentha pulegium var. *hyperiantha* IPNI=*451022-1* 371
Mentha pulegium var. *pubescens* IPNI=*451022-1* 371
Mentha pumila IPNI=*30026217-2* 371
Mentha purpurea IPNI=*450192-1* 371
Mentha pyrifolia IPNI=*450192-1* 371
Mentha ramosissima IPNI=*450192-1* 371
Mentha ranina IPNI=*450192-1* 371
Mentha rauscheri IPNI=*450192-1* 371
Mentha rigida IPNI=*30026217-2* 371

Mentha riparia IPNI=*450192-1* 371
Mentha rothii IPNI=*30026217-2* 371
Mentha rotundata IPNI=*30026217-2* 371
Mentha rudaeana IPNI=*450192-1* 371
Mentha ruderalis IPNI=*30026217-2* 371
Mentha salebrosa IPNI=*30026217-2* 371
Mentha sativa IPNI=*30026217-2* 371
Mentha sativa IPNI=*450192-1* 371
Mentha schreberi IPNI=*30026217-2* 371
Mentha scrophulariifolia IPNI=*30026217-2* 371
Mentha segetalis IPNI=*30026217-2* 371
Mentha silvicola IPNI=*30026217-2* 371
Mentha simplex IPNI=*30026217-2* 371
Mentha slichoviensis IPNI=*30026217-2* 371
Mentha soleana IPNI=*450192-1* 371
Mentha sparsiflora IPNI=*30026217-2* 371
Mentha sparsiflora var. *pascuorum* IPNI=*30026217-2* 371
Mentha spicata var. *longifolia* IPNI=*450735-1* 371
Mentha stagnalis IPNI=*450192-1* 371
Mentha stolonifera IPNI=*450192-1* 371
Mentha subcollina IPNI=*30026217-2* 371
Mentha subcordata IPNI=*30026217-2* 371
Mentha subfontanea IPNI=*30026217-2* 371
Mentha subinodora IPNI=*30026217-2* 371
Mentha subspicata IPNI=*450192-1* 371
Mentha subthermalis IPNI=*450192-1* 371
Mentha subtomentella IPNI=*451022-1* 371
Mentha sylvatica IPNI=*30026217-2* 371
Mentha tenuicaulis IPNI=*30026217-2* 371
Mentha tenuifolia IPNI=*30026217-2* 371
Mentha thayana IPNI=*30026217-2* 371
Mentha tinantiana IPNI=*450192-1* 371
Mentha tomentella IPNI=*451022-1* 371
Mentha tomentosa IPNI=*451022-1* 371
Mentha trojana IPNI=*450192-1* 371
Mentha uliginosa IPNI=*30026217-2* 371
Mentha umbrosa IPNI=*450192-1* 371
Mentha urticifolia IPNI=*450192-1* 371
Mentha vanhaesendonckii IPNI=*30026217-2* 371
Mentha varians IPNI=*30026217-2* 371
Mentha verisimilis IPNI=*30026217-2* 371
Mentha viennensis IPNI=*450192-1* 371
Mentha villosa IPNI=*30026217-2* 371
Mentha viridula IPNI=*30026217-2* 371
Mentha weiheana IPNI=*450192-1* 371
Mentha weissenburgensis IPNI=*450192-1* 371
Mentzelia cordifolia IPNI=*545481-1* 291
Mentzelia fragilis IPNI=*545374-1* 291
Mentzelia fulva IPNI=*545481-1* 291
Mentzelia pedicellata IPNI=*545374-1* 291
Mentzelia propinqua IPNI=*545374-1* 291
Mentzelia soratensis IPNI=*545481-1* 291
Mentzelia stipitata IPNI=*545374-1* 291
Mentzelia triloba IPNI=*545374-1* 291
Menyanthes brasilica IPNI=*1182548-2* 273
Menyanthes indica IPNI=*1182548-2* 273
Menyanthes macrophylla IPNI=*1182548-2* 273
Menyanthes meridionalis IPNI=*1182548-2* 273
Menyanthes petioliflora IPNI=*1182548-2* 273
Meoschium arnottianum IPNI=*406281-1* 240
Meoschium griffithii IPNI=*406281-1* 240
Meoschium royleanum IPNI=*406281-1* 240
Meoschium rugosum IPNI=*406281-1* 240
Meoschium wightii IPNI=*406281-1* 240
Meridiana kraussii IPNI=*208700-1* 263
Meridiana linearis IPNI=*208700-1* 263
Meridiana longiscapa IPNI=*208700-1* 263
Meridiana pinnata IPNI=*941423-1* 263
Meridiana ringens IPNI=*208720-1* 263
Meridiana tesselata IPNI=*208720-1* 263
Meringium fucoides IPNI=*1009149-2* 173
Merione villosa IPNI=*318778-1* 226
Merizadenia amplifolia IPNI=*79852-1* 341
Merizadenia arcuata IPNI=*82084-1* 343
Merizadenia sananho IPNI=*82327-1* 344
Merostachys racemiflora IPNI=*219234-2* 244
Mertensia bifida IPNI=*245596-2* 173
Mertensia brasiliana IPNI=*110733-2* 173
Mertensia canescens IPNI=*110733-2* 173
Mertensia citrifolia IPNI=*1213898-2* 436
Mertensia dichotoma var. *lherminieri* IPNI=*110733-2* 173
Mertensia elata IPNI=*110733-2* 173
Mertensia emarginata IPNI=*110733-2* 173
Mertensia flexuosa IPNI=*79839-2* 173
Mertensia glaucescens IPNI=*110733-2* 173
Mertensia glaucescens var. *cubense* IPNI=*110733-2* 173
Mertensia glaucescens var. *mexicanum* IPNI=*110733-2* 173
Mertensia goudotii IPNI=*1213898-2* 436
Mertensia hookeri IPNI=*110733-2* 173
Mertensia iguanea IPNI=*1213898-2* 436
Mertensia immersa IPNI=*245596-2* 173
Mertensia laevigata IPNI=*1213898-2* 436
Mertensia nitida IPNI=*110733-2* 173
Mertensia nuda IPNI=*245632-2* 173
Mertensia pectinata IPNI=*110733-2* 173
Mertensia pubescens IPNI=*1213898-2* 436
Mertensia pubescens IPNI=*245596-2* 173
Mertensia pumila IPNI=*79839-2* 173
Mertensia rhamnoides IPNI=*1213898-2* 436
Mertensia rigida IPNI=*79839-2* 173
Mertensia scalpturata IPNI=*79839-2* 173
Mertensia zizyphoides IPNI=*1213898-2* 436
Mesechites acutisepalus IPNI=*30132318-2* 342
Mesechites bicorniculatus IPNI=*30132318-2* 342
Mesechites guayaquilensis IPNI=*152067-2* 342
Mesechites hastatus IPNI=*152067-2* 342
Mesechites hirtellus IPNI=*152067-2* 342
Mesechites japurensis IPNI=*30132318-2* 342
Mesechites jasminiflorus IPNI=*152067-2* 342
Mesechites sanctae-crucis IPNI=*30132318-2* 342
Mesechites siphiliticus IPNI=*20001045-1* 344
Mesechites subcarnosus IPNI=*80081-1* 342
Mesechites trifidus var. *tomentulosus* IPNI=*30132318-2* 342
Mesembryanthemum acinaciforme IPNI=*360773-1* 280
Mesembryanthemum aureum IPNI=*362165-1* 280
Mesembryanthemum blandum IPNI=*362274-1* 280
Mesembryanthemum confertum IPNI=*362274-1* 280
Mesembryanthemum cooperi IPNI=*361479-1* 280
Mesembryanthemum croceum IPNI=*1011059-2* 280
Mesembryanthemum edule IPNI=*1200427-2* 280
Mesembryanthemum glaucum IPNI=*1011059-2* 280
Mesembryanthemum incurvum IPNI=*362274-1* 280
Mesembryanthemum laevigatum IPNI=*360773-1* 280
Mesembryanthemum lepidum IPNI=*362301-1* 280
Mesembryanthemum macrocarpum IPNI=*362323-1* 280
Mesembryanthemum macrosepalum IPNI=*362323-1* 280
Mesembryanthemum multiradiatum IPNI=*362274-1* 280
Mesembryanthemum mutans IPNI=*362301-1* 280
Mesembryanthemum productum IPNI=*362301-1* 280
Mesembryanthemum roseum IPNI=*362274-1* 280
Mesembryanthemum rubrocinctum IPNI=*360773-1* 280
Mesembryanthemum spectabile IPNI=*362323-1* 280
Mesembryanthemum subalatum IPNI=*360773-1* 280
Mesembryanthemum tetrasepalum IPNI=*322566-2* 280
Mesoligus subulatus IPNI=*981909-1* 271
Mesopanax capitatus IPNI=*30050876-2* 254

Mesopanax liebmannii IPNI=*30050876-2* 254
Mesosphaerum affine IPNI=*448211-1* 370
Mesosphaerum arboreum IPNI=*278655-2* 369
Mesosphaerum arboreum var. *bracteosum* IPNI=*278655-2* 369
Mesosphaerum atrorubens IPNI=*447964-1* 369
Mesosphaerum barbatum IPNI=*77120760-1* 368
Mesosphaerum brachiatum IPNI=*126717-2* 369
Mesosphaerum brevipes IPNI=*447982-1* 370
Mesosphaerum canescens IPNI=*77120760-1* 368
Mesosphaerum capitatum IPNI=*447996-1* 370
Mesosphaerum confertum IPNI=*448022-1* 370
Mesosphaerum constrictum IPNI=*448022-1* 370
Mesosphaerum dilatatum IPNI=*448057-1* 370
Mesosphaerum excelsum IPNI=*448022-1* 370
Mesosphaerum ferrugineum IPNI=*448022-1* 370
Mesosphaerum gonocephalum IPNI=*77120747-1* 368
Mesosphaerum graveolens IPNI=*77121103-1* 371
Mesosphaerum hirsutum IPNI=*448128-1* 370
Mesosphaerum jurgensenii IPNI=*448022-1* 370
Mesosphaerum lanceolatum IPNI=*448173-1* 370
Mesosphaerum lantanifolium IPNI=*448177-1* 370
Mesosphaerum lophanthus IPNI=*77120747-1* 368
Mesosphaerum lundii IPNI=*448022-1* 370
Mesosphaerum lutescens IPNI=*448211-1* 370
Mesosphaerum marrubiastrum IPNI=*448405-1* 370
Mesosphaerum melanostictum IPNI=*447982-1* 370
Mesosphaerum menthodorum IPNI=*77120747-1* 368
Mesosphaerum mutabile f. *albiflorum* IPNI=*77120760-1* 368
Mesosphaerum mutabile f. *coeruleum* IPNI=*77120760-1* 368
Mesosphaerum mutabile IPNI=*77120760-1* 368
Mesosphaerum mutabile var. *polystachyum* IPNI=*77120760-1* 368
Mesosphaerum obtusiflorum IPNI=*127032-2* 370
Mesosphaerum pallidum IPNI=*127032-2* 370
Mesosphaerum pittieri IPNI=*447996-1* 370
Mesosphaerum polyanthum IPNI=*77121103-1* 371
Mesosphaerum recurvatum IPNI=*448358-1* 370
Mesosphaerum rhombodeum IPNI=*447996-1* 370
Mesosphaerum savannarum IPNI=*127129-2* 370
Mesosphaerum siderotrichum IPNI=*448128-1* 370
Mesosphaerum spicatum IPNI=*77120760-1* 368
Mesosphaerum spicigerum IPNI=*77120747-1* 368
Mesosphaerum subverticillatum IPNI=*77120747-1* 368
Mesosphaerum tenellum IPNI=*447964-1* 369
Mesosphaerum trachychroum IPNI=*448211-1* 370
Mesosphaerum verticillatum IPNI=*77120819-1* 369
Mesothema campylotis IPNI=*33632-2* 174
Mespilodaphne aurantiodora IPNI=*171526-2* 184
Mespilodaphne nitida IPNI=*171727-2* 184
Mespilodaphne novogranatensis IPNI=*466885-1* 184
Mespilodaphne oblonga IPNI=*171797-2* 184
Mespilodaphne opifera IPNI=*171526-2* 184
Mespilodaphne parviflora IPNI=*465248-1* 183
Mespilodaphne pretiosa IPNI=*77198017-1* 183
Mespilodaphne pretiosa var. *angustifolia* IPNI=*77198017-1* 183
Mespilus crenulata IPNI=*730613-1* 444
Mespilus ferruginea IPNI=*725473-1* 443
Mespilus glabrata IPNI=*725484-1* 443
Mespilus indica IPNI=*731389-1* 444
Mespilus japonica IPNI=*77209608-1* 444
Mespilus lalandei IPNI=*730611-1* 444
Mespilus lanuginosa IPNI=*725473-1* 443
Mespilus loureiroi IPNI=*730613-1* 444
Mespilus obtusifolia IPNI=*725484-1* 443
Mespilus pauciflora IPNI=*730611-1* 444
Mespilus persoonii IPNI=*725484-1* 443
Mespilus pyracantha IPNI=*730611-1* 444
Mespilus racemiflora IPNI=*77120298-1* 443
Mespilus rubra IPNI=*731389-1* 444
Mespilus sieboldii IPNI=*731389-1* 444
Mespilus sinensis IPNI=*731389-1* 444
Messerschmidia ferruginea IPNI=*119329-1* 275
Messerschmidia floribunda IPNI=*119329-1* 275
Messerschmidia gnaphalodes IPNI=*120998-1* 275
Messerschmidia hirsutissima IPNI=*77069564-1* 275
Messerschmidia laurifolia IPNI=*60449244-2* 275
Messerschmidia microphylla IPNI=*119329-1* 275
Messerschmidia milleri IPNI=*119329-1* 275
Messerschmidia minor IPNI=*119329-1* 275
Messerschmidia minuta IPNI=*119329-1* 275
Messerschmidia polystachya IPNI=*121103-1* 275
Messerschmidia psilostachya IPNI=*60430568-2* 275
Messerschmidia punctata IPNI=*60430568-2* 275
Messerschmidia scandens IPNI=*119329-1* 275
Messerschmidia sericea IPNI=*119329-1* 275
Messerschmidia syringifolia IPNI=*60449244-2* 275
Messerschmidia tomentosa IPNI=*77069564-1* 275
Messerschmidia velutina IPNI=*119329-1* 275
Messerschmidia volubilis IPNI=*119329-1* 275
Metagonia acuminata IPNI=*82286-2* 299
Metagonia alaternoides IPNI=*82288-2* 299
Metagonia crenulata IPNI=*858930-1* 300
Metagonia empetrifolia IPNI=*327994-1* 299
Metagonia marginata IPNI=*858930-1* 300
Metagonia meridionalis IPNI=*859020-1* 300
Metagonia penaeoides IPNI=*327994-1* 299
Metagonia prostrata IPNI=*261842-2* 300
Metagonia villosa IPNI=*858406-1* 299
Metastelma woodsonii IPNI=*74165-2* 340
Meterana castaneifolia IPNI=*44529-2* 388
Meterana palustris IPNI=*44564-2* 388
Methonica doniana IPNI=*535953-1* 226
Methonica gloriosa IPNI=*535953-1* 226
Methonica superba IPNI=*535953-1* 226
Metopium brownei var. *brachycarpum* IPNI=*158605-2* 451
Metopium linnaei IPNI=*158605-2* 451
Metopium metopium IPNI=*158605-2* 451
Metrosideros albida IPNI=*598073-1* 426
Metrosideros angustifolia IPNI=*77108602-1* 426
Metrosideros armillaris IPNI=*597845-1* 426
Metrosideros citrina IPNI=*77108602-1* 426
Metrosideros coriacea IPNI=*598073-1* 426
Metrosideros falcata IPNI=*77108602-1* 426
Metrosideros lanceolata IPNI=*77108602-1* 426
Metrosideros latifolia IPNI=*77108602-1* 426
Metrosideros longifolia IPNI=*77108602-1* 426
Metrosideros marginata IPNI=*77108602-1* 426
Metrosideros myrtifolia IPNI=*77108602-1* 426
Metrosideros quinquenervia IPNI=*598073-1* 426
Metrosideros semperflorens IPNI=*77108602-1* 426
Metrosideros tomentosa IPNI=*598207-1* 426
Metrosideros viminalis IPNI=*914221-1* 426
Metroxylon taedigerum IPNI=*669539-1* 211
Meum foeniculum IPNI=*842680-1* 252
Meum piperatum IPNI=*842680-1* 252
Meum piperinum IPNI=*842680-1* 252
Meum piperitum IPNI=*842680-1* 252
Meyenia erecta IPNI=*56259-1* 361
Meyenia longiflora IPNI=*56270-1* 362
Meyera capitata IPNI=*993359-1* 271
Meyera fluctuans IPNI=*203207-1* 261
Meyera helioncha IPNI=*203207-1* 261
Mezia anacardiodes IPNI=*158649-2* 183
Mezia ita-uba IPNI=*158649-2* 183
Mezilaurus anacardioides IPNI=*158649-2* 183
Mezoneuron benguetense IPNI=*481887-1* 312
Mezoneuron grande IPNI=*472105-1* 310

Micalia grandiflora IPNI=*97540-2* 375
Michelia yulan IPNI=*554678-1* 190
Miconia affinis IPNI=*572641-1* 421
Miconia ambigua IPNI=*572641-1* 421
Miconia amplexans IPNI=*1088633-2* 423
Miconia andina IPNI=*572660-1* 421
Miconia angustispica IPNI=*159880-2* 423
Miconia arborea IPNI=*326247-2* 419
Miconia astrolasia IPNI=*1155116-2* 422
Miconia attenuata IPNI=*1069702-2* 422
Miconia attenuata var. *subquintuplinervia* IPNI=*1069702-2* 422
Miconia aubletiana IPNI=*1069702-2* 422
Miconia auriculata IPNI=*572590-1* 421
Miconia auriculata var. *quintuplinervia* IPNI=*572590-1* 421
Miconia barbigera IPNI=*1108920-2* 420
Miconia belizensis IPNI=*572209-1* 419
Miconia bergii IPNI=*159943-2* 423
Miconia beurlingii IPNI=*158685-2* 419
Miconia bifrons IPNI=*572671-1* 421
Miconia boissieriana IPNI=*1000103-2* 422
Miconia borealis IPNI=*1098452-2* 421
Miconia bourgaeana IPNI=*572388-1* 420
Miconia brachypoda IPNI=*1158780-2* 422
Miconia buchtienii IPNI=*158869-2* 420
Miconia calophylla IPNI=*1108920-2* 420
Miconia calyptrata IPNI=*158886-2* 420
Miconia carioana IPNI=*572388-1* 420
Miconia cayumbensis IPNI=*158685-2* 419
Miconia ceanothina IPNI=*572641-1* 421
Miconia cecidophora IPNI=*158685-2* 419
Miconia centrandra IPNI=*158869-2* 420
Miconia ciliata IPNI=*1158780-2* 422
Miconia ciliata var. *attenuata* IPNI=*1108920-2* 420
Miconia ciliata var. *congestiflora* IPNI=*1108920-2* 420
Miconia ciliata var. *plurilocularis* IPNI=*1108920-2* 420
Miconia ciliata var. *serrata* IPNI=*1108920-2* 420
Miconia circumsecta IPNI=*1158780-2* 422
Miconia collina IPNI=*1069702-2* 422
Miconia congesta IPNI=*159532-2* 422
Miconia congestiflora IPNI=*1108920-2* 420
Miconia corazonica IPNI=*572118-1* 419
Miconia crispula IPNI=*1069702-2* 422
Miconia cristulata IPNI=*1069702-2* 422
Miconia crocea var. *setosa* IPNI=*572353-1* 420
Miconia cruegeri IPNI=*1108920-2* 420
Miconia darienensis IPNI=*159532-2* 422
Miconia decussata IPNI=*1108920-2* 420
Miconia densiflora IPNI=*572256-1* 420
Miconia detergibilis IPNI=*1125568-2* 419
Miconia dimorpha IPNI=*1110588-2* 423
Miconia discolor IPNI=*1069703-2* 422
Miconia dodecandra var. *gracilipes* IPNI=*572411-1* 420
Miconia dodecandra var. *longifolia* IPNI=*572411-1* 420
Miconia elliptica IPNI=*1143367-2* 422
Miconia epiphytica IPNI=*572660-1* 421
Miconia eriophora f. *minor* IPNI=*572590-1* 421
Miconia eriophora IPNI=*572590-1* 421
Miconia eurychaenioides IPNI=*1092872-2* 420
Miconia eurychinioides IPNI=*1092872-2* 420
Miconia fleischeriana IPNI=*1069702-2* 422
Miconia fockeana IPNI=*572835-1* 421
Miconia fothergilla var. *gracilipes* IPNI=*572411-1* 420
Miconia glossocentra IPNI=*1098452-2* 421
Miconia granulosa IPNI=*158784-2* 419
Miconia granulosa var. *biappendiculata* IPNI=*158784-2* 419
Miconia habrolepis IPNI=*1069703-2* 422
Miconia haplobotrya IPNI=*159880-2* 423
Miconia havanensis IPNI=*572641-1* 421
Miconia heterochroa IPNI=*1125568-2* 419
Miconia hexaptera IPNI=*159880-2* 423
Miconia holosericea var. *acuminata* IPNI=*1125568-2* 419
Miconia holosericea var. *bracteata* IPNI=*1024004-2* 421
Miconia holosericea var. *montana* IPNI=*1125568-2* 419
Miconia holosericea var. *mucronata* IPNI=*1024004-2* 421
Miconia holosericea var. *oblongata* IPNI=*1125568-2* 419
Miconia holosericea var. *obtusiuscula* IPNI=*1125568-2* 419
Miconia holosericea var. *subquintuplinervia* IPNI=*1024004-2* 421
Miconia hypargyrea IPNI=*159763-2* 422
Miconia ibaguensis var. *glabrata* IPNI=*572590-1* 421
Miconia impetiginosa IPNI=*1018450-2* 422
Miconia impetiolaris var. *spruceana* IPNI=*572600-1* 421
Miconia involucrata IPNI=*572080-1* 419
Miconia lambertiana IPNI=*1220088-2* 421
Miconia langlassei IPNI=*158886-2* 420
Miconia lepidota var. *grandifolia* IPNI=*1069703-2* 422
Miconia leucocephala IPNI=*1000103-2* 422
Miconia ligustrina var. *setulinodis* IPNI=*572682-1* 421
Miconia lindenii IPNI=*158684-2* 419
Miconia lineata IPNI=*572590-1* 421
Miconia longipedunculata IPNI=*159286-2* 421
Miconia longispicata var. *minor* IPNI=*159286-2* 421
Miconia longistyla IPNI=*572109-1* 419
Miconia macdanielii IPNI=*158746-2* 419
Miconia macrophylla IPNI=*1000103-2* 422
Miconia macrophylla IPNI=*1069702-2* 422
Miconia macrophylla IPNI=*572600-1* 421
Miconia macrophylla var. *hostmanni* IPNI=*1000103-2* 422
Miconia macrophylla var. *latifolia* IPNI=*1000103-2* 422
Miconia macrophylla var. *leucocephala* IPNI=*1000103-2* 422
Miconia macrophylla var. *serrulata* IPNI=*1000103-2* 422
Miconia macrostachya IPNI=*1039241-2* 419
Miconia magnifica IPNI=*326247-2* 419
Miconia magnifica var. *velutina* IPNI=*326247-2* 419
Miconia markgrafii IPNI=*572470-1* 420
Miconia megaphylla IPNI=*572080-1* 419
Miconia melaodendron IPNI=*1098452-2* 421
Miconia microcarpa IPNI=*158685-2* 419
Miconia microcarpa IPNI=*572641-1* 421
Miconia milleflora IPNI=*1110588-2* 423
Miconia minutiflora var. *latifolia* IPNI=*1098452-2* 421
Miconia montana IPNI=*1125568-2* 419
Miconia mucronata IPNI=*1024004-2* 421
Miconia mucronulata IPNI=*1069702-2* 422
Miconia multinervulosa IPNI=*1110588-2* 423
Miconia naudiniana IPNI=*572671-1* 421
Miconia nemoralis IPNI=*1069702-2* 422
Miconia nervosa var. *mediana* IPNI=*572835-1* 421
Miconia nicotianifolia IPNI=*572641-1* 421
Miconia obovalis IPNI=*1143367-2* 422
Miconia otodesma IPNI=*573207-1* 422
Miconia paludosa IPNI=*1110588-2* 423
Miconia palustris IPNI=*1069702-2* 422
Miconia panamensis IPNI=*158886-2* 420
Miconia parviflora IPNI=*572641-1* 421
Miconia pendulifolia IPNI=*572641-1* 421
Miconia pennellii IPNI=*572120-1* 419
Miconia planinervia IPNI=*158685-2* 419
Miconia platyhedra IPNI=*1000103-2* 422
Miconia prasina var. *angustifolia* IPNI=*1069702-2* 422
Miconia prasina var. *attenuata* IPNI=*1069702-2* 422
Miconia prasina var. *crispula* IPNI=*1069702-2* 422
Miconia procera IPNI=*572109-1* 419
Miconia pseudorubiginosa IPNI=*159311-2* 421
Miconia pteropoda IPNI=*1069702-2* 422
Miconia pulverulenta IPNI=*572575-1* 421
Miconia punctata var. *brevifolia* IPNI=*1069703-2* 422

Miconia punctata var. *latifolia* IPNI=*1069703-2* 422
Miconia purpurascens IPNI=*1031600-2* 417
Miconia pusilliflora IPNI=*158685-2* 419
Miconia racemosa var. *barbigera* IPNI=*1108920-2* 420
Miconia racemosa var. *brachypoda* IPNI=*1158780-2* 422
Miconia racemosa var. *lanceolata* IPNI=*1158780-2* 422
Miconia racemosa var. *plurilocularis* IPNI=*1108920-2* 420
Miconia racemosa var. *urbaniana* IPNI=*1108920-2* 420
Miconia renggeri IPNI=*1125568-2* 419
Miconia repandocrenata IPNI=*1069702-2* 422
Miconia revoluta IPNI=*1069702-2* 422
Miconia rhamnoidea IPNI=*572682-1* 421
Miconia rubiginosa var. *cinerea* IPNI=*1155116-2* 422
Miconia rubiginosa var. *glabrior* IPNI=*1155116-2* 422
Miconia rubiginosa var. *platyura* IPNI=*1155116-2* 422
Miconia rubiginosa var. *rufa* IPNI=*1155116-2* 422
Miconia rufescens IPNI=*1125568-2* 419
Miconia rufescens var. *grandifolia* IPNI=*1018450-2* 422
Miconia sancti-ramoni IPNI=*572590-1* 421
Miconia scandens IPNI=*566051-1* 417
Miconia schomburgkii IPNI=*572671-1* 421
Miconia sepiaria IPNI=*1069702-2* 422
Miconia septuplinervis IPNI=*572835-1* 421
Miconia serrulata IPNI=*1158780-2* 422
Miconia setinodis IPNI=*572353-1* 420
Miconia spicata IPNI=*159880-2* 423
Miconia spondylantha IPNI=*572835-1* 421
Miconia stricta IPNI=*573368-1* 423
Miconia subtriloba IPNI=*158869-2* 420
Miconia surinamensis IPNI=*159532-2* 422
Miconia sylvestris IPNI=*158869-2* 420
Miconia theizans subsp. *flavescens* IPNI=*1110588-2* 423
Miconia theizans subsp. *viridis* IPNI=*1110588-2* 423
Miconia theizans var. *cuneata* IPNI=*1110588-2* 423
Miconia theizans var. *genuina* IPNI=*1110588-2* 423
Miconia theizans var. *glaberrima* IPNI=*1110588-2* 423
Miconia theizans var. *glazioviana* IPNI=*1110588-2* 423
Miconia theizans var. *integrifolia* IPNI=*1110588-2* 423
Miconia theizans var. *lanceolata* IPNI=*1110588-2* 423
Miconia theizans var. *longifolia* IPNI=*1110588-2* 423
Miconia theizans var. *milleflora* IPNI=*1110588-2* 423
Miconia theizans var. *minutiflora* IPNI=*1110588-2* 423
Miconia theizans var. *paludosa* IPNI=*1110588-2* 423
Miconia theizans var. *parvifolia* IPNI=*1110588-2* 423
Miconia theizans var. *saldanhaei* IPNI=*1110588-2* 423
Miconia theizans var. *setulosa* IPNI=*1110588-2* 423
Miconia theizans var. *spinulosa* IPNI=*1110588-2* 423
Miconia theizans var. *subtriplinervia* IPNI=*1110588-2* 423
Miconia theizans var. *tetragona* IPNI=*1110588-2* 423
Miconia theizans var. *triplinervia* IPNI=*1110588-2* 423
Miconia theizans var. *vulgaris* IPNI=*1110588-2* 423
Miconia theresiae IPNI=*572353-1* 420
Miconia tomentosa var. *ovata* IPNI=*1088633-2* 423
Miconia trichotoma IPNI=*1069702-2* 422
Miconia trinervis IPNI=*159880-2* 423
Miconia trinitatis IPNI=*1098452-2* 421
Miconia triplinervis subsp. *exalluvia* IPNI=*159880-2* 423
Miconia umbilicata IPNI=*159943-2* 423
Miconia umbrifera IPNI=*1088633-2* 423
Miconia velutina IPNI=*326247-2* 419
Miconia verruculosa IPNI=*1158780-2* 422
Miconia verticillata IPNI=*1158780-2* 422
Miconia virgulata IPNI=*1108920-2* 420
Miconia wydleriana IPNI=*572600-1* 421
Micrampelis araneosa IPNI=*88306-2* 294
Micrampelis lanata IPNI=*88306-2* 294
Micrampelis macrocarpa IPNI=*88306-2* 294
Micrampelis polycarpa IPNI=*88306-2* 294
Micrampelis rusbyi IPNI=*88306-2* 294
Micrandra australis IPNI=*159993-2* 392
Micrandra cunuri IPNI=*159993-2* 392
Micrandra glabra IPNI=*159993-2* 392
Micrandra heterophylla IPNI=*352260-1* 392
Micrandra lopezii var. *microcarpa* IPNI=*352260-1* 392
Micrandra major IPNI=*352260-1* 392
Micrandra siphonioides var. *genuina* IPNI=*352260-1* 392
Micrandra siphonioides var. *major* IPNI=*352260-1* 392
Micrandra siphonioides var. *minor* IPNI=*352256-1* 392
Micrandra ternata IPNI=*121724-2* 391
Micrelium asteroides IPNI=*1186403-2* 261
Micrelium tolak IPNI=*1186403-2* 261
Microbignonia auristellae IPNI=*77100466-1* 363
Microcarpaea americana IPNI=*155661-2* 376
Microchaete corymbosa IPNI=*963576-1* 266
Microchaete vaccinioides IPNI=*963618-1* 267
Microchilus debilis IPNI=*77125687-1* 217
Microchilus querceticola IPNI=*77125705-1* 217
Microchloa indica var. *kunthii* IPNI=*408467-1* 241
Microchloa indica var. *tenuis* IPNI=*408465-1* 241
Microchloa kandalaensis IPNI=*408467-1* 241
Microchloa setacea IPNI=*408465-1* 241
Microcoelum martianum IPNI=*246979-2* 212
Microcoelum weddellianum IPNI=*246979-2* 212
Microelus roeperianus IPNI=*339921-1* 401
Microgramma lycopodioides f. *subdimorpha* IPNI=*160142-2* 175
Microlepia splendens IPNI=*17080730-1* 175
Micromeria bahamensis IPNI=*61179-2* 369
Micromeria brownei IPNI=*61179-2* 369
Micromeria brownei var. *pilosiuscula* IPNI=*61179-2* 369
Micromeria dalmatica IPNI=*451022-1* 371
Micromeria domingensis IPNI=*61179-2* 369
Micromeria fenzlii IPNI=*451022-1* 371
Micromeria maritima IPNI=*451022-1* 371
Micromeria montana IPNI=*457736-1* 373
Micromeria nubigena IPNI=*61198-2* 369
Micromeria nubigena subvar. *glabrescens* IPNI=*61198-2* 369
Micromeria nubigena var. *angustifolia* IPNI=*61198-2* 369
Micromeria nubigena var. *cordifolia* IPNI=*61198-2* 369
Micromeria nubigena var. *glabrescens* IPNI=*61198-2* 369
Micromeria obovata IPNI=*61205-2* 369
Micromeria pilosiuscula IPNI=*61179-2* 369
Micromeria stolonifera IPNI=*61179-2* 369
Micromeria viminea IPNI=*61205-2* 369
Micromeria xalapensis IPNI=*61179-2* 369
Micropetalon lanuginosum IPNI=*30040285-2* 285
Micropholis burchelliana IPNI=*787792-1* 306
Micropholis calophylloides IPNI=*160417-2* 306
Micropholis glazioviana IPNI=*787792-1* 306
Micropholis guatemalensis IPNI=*160393-2* 306
Micropholis martiana IPNI=*160374-2* 306
Micropholis mexicana IPNI=*160393-2* 306
Micropholis mucronata IPNI=*160417-2* 306
Micropholis paraensis IPNI=*160358-2* 306
Micropholis rigida IPNI=*160417-2* 306
Micropholis ulei IPNI=*160374-2* 306
Micropholis wurdackii IPNI=*160374-2* 306
Micropiper alatum IPNI=*677941-1* 193
Micropiper blandum IPNI=*678107-1* 193
Micropiper langsdorffii IPNI=*678107-1* 193
Micropiper macrostachyum IPNI=*30130843-2* 194
Micropiper melanostigma IPNI=*30001267-2* 193
Micropiper pellucidum IPNI=*679378-1* 194
Micropiper pusillum IPNI=*1068802-2* 194
Micropiper tenellum IPNI=*679378-1* 194
Micropteryx crista-galli IPNI=*494400-1* 319
Micropteryx fasciculata IPNI=*494400-1* 319

Micropteryx laurifolia IPNI=*494400-1* 319
Micropteryx poeppigiana IPNI=*96637-2* 319
Micropyxis exigua IPNI=*77100404-1* 304
Microrhynchus octophyllus IPNI=*229022-1* 264
Microrhynchus surinamensis IPNI=*229022-1* 264
Microsorum irioides IPNI=*17341250-1* 176
Microsorum irregulare IPNI=*17341250-1* 176
Microsorum neoguineense IPNI=*17341250-1* 176
Microsorum polycarpon IPNI=*17341250-1* 176
Microsorum sessile IPNI=*17341250-1* 176
Microsorum subirideum IPNI=*17341250-1* 176
Microsorum surinamense IPNI=*160142-2* 175
Microsorum validum IPNI=*17341250-1* 176
Microstigma incanum IPNI=*286910-1* 277
Microtea debilis var. *ovata* IPNI=*676291-1* 286
Microtea debilis var. *rhombifolia* IPNI=*676291-1* 286
Mieria virgata IPNI=*242907-1* 269
Mikania acrensis IPNI=*233001-1* 266
Mikania affinis IPNI=*232743-1* 266
Mikania almagroi IPNI=*232743-1* 266
Mikania amara IPNI=*232826-1* 266
Mikania amara IPNI=*233110-1* 266
Mikania amara var. *guaco* IPNI=*232913-1* 266
Mikania ambigens IPNI=*161327-2* 266
Mikania angustifolia IPNI=*6821-2* 256
Mikania antioquiensis IPNI=*232743-1* 266
Mikania antioquiensis var. *subcuneata* IPNI=*232743-1* 266
Mikania anzoatiguensis IPNI=*161327-2* 266
Mikania archidonensis IPNI=*232913-1* 266
Mikania argyrostigma IPNI=*232913-1* 266
Mikania aschersonii var. *induta* IPNI=*160759-2* 266
Mikania aschersonii var. *typica* IPNI=*160759-2* 266
Mikania aspera IPNI=*232913-1* 266
Mikania atriplicifolia IPNI=*232821-1* 266
Mikania attenuata IPNI=*232913-1* 266
Mikania baccharoidea IPNI=*160759-2* 266
Mikania badieri f. *broadwayana* IPNI=*232937-1* 266
Mikania badieri f. *kittsiana* IPNI=*232937-1* 266
Mikania badieri IPNI=*232937-1* 266
Mikania badieri var. *kittsiana* IPNI=*232937-1* 266
Mikania banisteriae f. *lanuginosa* IPNI=*232743-1* 266
Mikania banisteriae f. *lehmanniana* IPNI=*232743-1* 266
Mikania banisteriae f. *subglabra* IPNI=*232743-1* 266
Mikania banisteriae var. *gabrielii* IPNI=*232743-1* 266
Mikania banisteriae var. *umbrosa* IPNI=*232743-1* 266
Mikania bergantinensis IPNI=*232743-1* 266
Mikania berteroana IPNI=*105109-2* 263
Mikania biformis IPNI=*232826-1* 266
Mikania boliviensis IPNI=*161327-2* 266
Mikania brachiata IPNI=*232913-1* 266
Mikania brooksii IPNI=*233110-1* 266
Mikania canaguensis IPNI=*232743-1* 266
Mikania carnosula IPNI=*232826-1* 266
Mikania caudata IPNI=*232743-1* 266
Mikania caustolepis IPNI=*161327-2* 266
Mikania cissampelina IPNI=*232826-1* 266
Mikania cocaensis IPNI=*232826-1* 266
Mikania consanguinea IPNI=*233144-1* 266
Mikania convolvulacea IPNI=*232826-1* 266
Mikania convolvulacea var. *portoriccensis* IPNI=*232826-1* 266
Mikania cordifolia var. *carnosula* IPNI=*232826-1* 266
Mikania cordifolia var. *tomentosa* IPNI=*232826-1* 266
Mikania cordifolia var. *umbrosa* IPNI=*232826-1* 266
Mikania cornifolia IPNI=*233110-1* 266
Mikania cryptodonta IPNI=*161327-2* 266
Mikania cuneata IPNI=*232913-1* 266
Mikania deltoides IPNI=*161051-2* 266
Mikania denticulata IPNI=*161051-2* 266
Mikania divaricata IPNI=*233110-1* 266
Mikania eupatorioides IPNI=*160759-2* 266
Mikania fendleri IPNI=*232826-1* 266
Mikania ferruginea IPNI=*232743-1* 266
Mikania ferruginea var. *subglabra* IPNI=*232743-1* 266
Mikania fimbriata IPNI=*232826-1* 266
Mikania fockeana IPNI=*233144-1* 266
Mikania fockeana var. *pubescens* IPNI=*233144-1* 266
Mikania gabrielii IPNI=*232743-1* 266
Mikania glechomifolia IPNI=*161051-2* 266
Mikania gonoclada IPNI=*232826-1* 266
Mikania gonoclada var. *ambigua* IPNI=*232826-1* 266
Mikania gracilis IPNI=*232937-1* 266
Mikania grandifolia IPNI=*161327-2* 266
Mikania granulata var. *integra* IPNI=*160925-2* 266
Mikania granulata var. *typica* IPNI=*160925-2* 266
Mikania hirsutissima IPNI=*232743-1* 266
Mikania hirsutissima var. *margarethiae* IPNI=*232743-1* 266
Mikania hirsutissima var. *martiana* IPNI=*232743-1* 266
Mikania hookeriana var. *badieri* IPNI=*232937-1* 266
Mikania hookeriana var. *cardiophylla* IPNI=*232937-1* 266
Mikania hookeriana var. *crassicaulis* IPNI=*232937-1* 266
Mikania hookeriana var. *kittsiana* IPNI=*232937-1* 266
Mikania hookeriana var. *platyphylla* IPNI=*232937-1* 266
Mikania hookeriana var. *typica* IPNI=*232937-1* 266
Mikania hostmanii IPNI=*232826-1* 266
Mikania huaco IPNI=*232913-1* 266
Mikania imrayana IPNI=*232937-1* 266
Mikania jamesonii IPNI=*160925-2* 266
Mikania karsteniana IPNI=*232743-1* 266
Mikania karuaiensis IPNI=*233144-1* 266
Mikania kavanayensis IPNI=*232743-1* 266
Mikania lanuginosa IPNI=*232743-1* 266
Mikania lechleri IPNI=*161327-2* 266
Mikania lima IPNI=*233144-1* 266
Mikania loretensis IPNI=*233110-1* 266
Mikania loxensis IPNI=*232826-1* 266
Mikania macbridei IPNI=*232743-1* 266
Mikania martiana IPNI=*232743-1* 266
Mikania mazaruniensis IPNI=*232743-1* 266
Mikania meridana IPNI=*232743-1* 266
Mikania miconioides IPNI=*161279-2* 266
Mikania micrantha f. *congesta* IPNI=*232821-1* 266
Mikania micrantha f. *hirsuta* IPNI=*161051-2* 266
Mikania micrantha f. *typica* IPNI=*161051-2* 266
Mikania micrantha var. *cynanchifolia* IPNI=*161051-2* 266
Mikania molinensis IPNI=*232743-1* 266
Mikania mollis IPNI=*232826-1* 266
Mikania moyobambensis IPNI=*233144-1* 266
Mikania mutisiana IPNI=*160806-2* 266
Mikania napensis IPNI=*232913-1* 266
Mikania olivacea IPNI=*232913-1* 266
Mikania oniaensis IPNI=*232743-1* 266
Mikania opifera IPNI=*232826-1* 266
Mikania orinocensis IPNI=*161051-2* 266
Mikania parkeriana IPNI=*232821-1* 266
Mikania pastazae IPNI=*233001-1* 266
Mikania pilosa IPNI=*232743-1* 266
Mikania platyphylla IPNI=*232937-1* 266
Mikania poeppigii IPNI=*232826-1* 266
Mikania polybotrys IPNI=*233144-1* 266
Mikania polystachya IPNI=*233144-1* 266
Mikania psilostachya var. *albicans* IPNI=*233144-1* 266
Mikania psilostachya var. *racemulosa* IPNI=*233144-1* 266
Mikania psilostachya var. *scabra* IPNI=*233144-1* 266
Mikania ptaretepuiensis IPNI=*232743-1* 266
Mikania puberula IPNI=*233209-1* 266
Mikania pubescens IPNI=*232826-1* 266

Mikania pulverulenta var. *lasiophlebia* IPNI=*161279-2* 266
Mikania punctata IPNI=*161327-2* 266
Mikania racemulosa IPNI=*233144-1* 266
Mikania rotunda IPNI=*233144-1* 266
Mikania rubiginosa IPNI=*232826-1* 266
Mikania ruiziana IPNI=*232743-1* 266
Mikania ruiziana var. *lehmanniana* IPNI=*232743-1* 266
Mikania ruiziana var. *typica* IPNI=*232743-1* 266
Mikania sanjacintensis IPNI=*232937-1* 266
Mikania scabra IPNI=*233144-1* 266
Mikania scandens var. *congesta* IPNI=*232821-1* 266
Mikania scandens var. *hirsuta* IPNI=*161051-2* 266
Mikania scandens var. *pubescens* IPNI=*232826-1* 266
Mikania scandens var. *rhodotricha* IPNI=*232826-1* 266
Mikania scandens var. *subcymosa* IPNI=*161051-2* 266
Mikania scandens var. *umbellifera* IPNI=*161051-2* 266
Mikania scandens var. *villosa* IPNI=*161051-2* 266
Mikania sepiaria IPNI=*232743-1* 266
Mikania sieberiana IPNI=*232821-1* 266
Mikania sinuata IPNI=*161051-2* 266
Mikania skutchii IPNI=*232743-1* 266
Mikania stipitata IPNI=*233110-1* 266
Mikania suaveolens IPNI=*232826-1* 266
Mikania subcrenata IPNI=*161051-2* 266
Mikania subcymosa IPNI=*161051-2* 266
Mikania surinamensis IPNI=*232826-1* 266
Mikania tamoides IPNI=*161051-2* 266
Mikania thunbergiifolia IPNI=*232826-1* 266
Mikania umbellifera IPNI=*161051-2* 266
Mikania ursina IPNI=*232743-1* 266
Mikania variabilis IPNI=*161051-2* 266
Mikania variabilis IPNI=*232821-1* 266
Mikania vellosiana IPNI=*232826-1* 266
Mikania vitifolia f. *boliviensis* IPNI=*161327-2* 266
Mikania vitifolia f. *cryptodonta* IPNI=*161327-2* 266
Mikania vitifolia f. *typica* IPNI=*161327-2* 266
Mikania vitrea IPNI=*232937-1* 266
Mikania yapasensis IPNI=*232826-1* 266
Mikania zonensis IPNI=*232913-1* 266
Milium arundinaceum IPNI=*60426565-2* 241
Milium bicolor IPNI=*422090-1* 245
Milium ciliatum IPNI=*1038558-2* 237
Milium colonum IPNI=*399655-1* 238
Milium compactum IPNI=*422090-1* 245
Milium compressum IPNI=*1103484-2* 245
Milium compressum IPNI=*392087-1* 234
Milium crus-galli IPNI=*399660-1* 238
Milium dactylon IPNI=*397065-1* 236
Milium digitatum IPNI=*329018-2* 237
Milium distichum IPNI=*184275-2* 243
Milium effusum IPNI=*410542-1* 242
Milium esculentum IPNI=*412217-1* 242
Milium filiforme IPNI=*398933-1* 237
Milium filiforme IPNI=*80280-2* 237
Milium halepense IPNI=*422139-1* 245
Milium hirsutum IPNI=*398888-1* 237
Milium hispidum IPNI=*30197865-2* 233
Milium latifolium IPNI=*414634-1* 243
Milium maritimum IPNI=*386353-1* 233
Milium maximum IPNI=*422090-1* 245
Milium membranaceum IPNI=*399115-1* 238
Milium natans IPNI=*289770-2* 241
Milium nigricans IPNI=*422090-1* 245
Milium orixense IPNI=*402044-1* 239
Milium paniceum IPNI=*1048074-2* 244
Milium panicum IPNI=*412217-1* 242
Milium paspalodes IPNI=*184275-2* 243
Milium polystachyon IPNI=*77164258-1* 246
Milium punctatum IPNI=*1021229-2* 239
Milium radiatum IPNI=*398933-1* 237
Milium ramosum IPNI=*402044-1* 239
Milium sorghum IPNI=*422090-1* 245
Milium sorgo IPNI=*422090-1* 245
Milium stoloniferum IPNI=*386353-1* 233
Milium strictum IPNI=*424391-1* 246
Milium triracemosum IPNI=*398933-1* 237
Milium undulatifolium IPNI=*1004002-1* 242
Milium velutinum IPNI=*80280-2* 237
Milium villosum IPNI=*398888-1* 237
Milium zonatum IPNI=*402044-1* 239
Milla macrostemon var. *gracilis* IPNI=*130726-3* 214
Millefolium officinale IPNI=*2294-2* 255
Millefolium vulgare IPNI=*2294-2* 255
Milleria dichotoma IPNI=*315761-2* 266
Milleria glandulosa IPNI=*315761-2* 266
Milleria maculata IPNI=*315761-2* 266
Milleria perfoliata IPNI=*315761-2* 266
Milleria peruviana IPNI=*315761-2* 266
Milleria quinqueflora var. *maculata* IPNI=*315761-2* 266
Milleria trifolia IPNI=*315761-2* 266
Millettia luzonensis IPNI=*496639-1* 320
Millettia speciosa IPNI=*143380-2* 325
Miltonia vexillaria IPNI=*161453-2* 222
Miltonia vexillaria var. *leopoldii* IPNI=*161453-2* 222
Miltonia vexillaria var. *rubella* IPNI=*161453-2* 222
Miltonia vexillaria var. *stupenda* IPNI=*161453-2* 222
Mimosa acacioides IPNI=*12268-2* 311
Mimosa adenanthera IPNI=*1056199-2* 328
Mimosa adhaerens IPNI=*77085269-1* 332
Mimosa aeschynomenes IPNI=*507771-1* 327
Mimosa alba IPNI=*1129244-2* 321
Mimosa alba IPNI=*500286-1* 321
Mimosa angulata IPNI=*470138-1* 310
Mimosa angustisiliqua IPNI=*1125234-2* 317
Mimosa angustissima IPNI=*712-2* 310
Mimosa apodocarpa var. *hostilis* IPNI=*508775-1* 328
Mimosa aquatica IPNI=*509612-1* 328
Mimosa arborea IPNI=*62882-2* 315
Mimosa asperata var. *pigra* IPNI=*313003-2* 327
Mimosa atomaria IPNI=*60441223-2* 336
Mimosa bipinnata IPNI=*1009172-2* 319
Mimosa blancoana IPNI=*493866-1* 319
Mimosa bourgoni IPNI=*500286-1* 321
Mimosa bracaatinga f. *paucijuga* IPNI=*508647-1* 328
Mimosa bracaatinga IPNI=*508647-1* 328
Mimosa bracaatinga var. *aspericarpa* IPNI=*508647-1* 328
Mimosa brachystachys IPNI=*1183336-2* 330
Mimosa buceragenia IPNI=*77208976-1* 330
Mimosa cabrera IPNI=*508775-1* 328
Mimosa capitata IPNI=*128414-2* 321
Mimosa caracasana IPNI=*287164-2* 337
Mimosa carbonalis IPNI=*77208976-1* 330
Mimosa caudata IPNI=*1009172-2* 319
Mimosa chiliantha IPNI=*1009172-2* 319
Mimosa colubrina IPNI=*12264-2* 311
Mimosa coruscans IPNI=*1199545-2* 323
Mimosa coruscans IPNI=*128458-2* 322
Mimosa corymbosa IPNI=*989341-1* 320
Mimosa crocea IPNI=*520020-1* 335
Mimosa cyclocarpa IPNI=*1037928-2* 319
Mimosa cylindrica IPNI=*1090709-2* 322
Mimosa dealbata IPNI=*470130-1* 310
Mimosa decurrens IPNI=*470138-1* 310
Mimosa depressa IPNI=*1125234-2* 317
Mimosa dominiciana IPNI=*313177-2* 327
Mimosa dulcis IPNI=*1003508-2* 330

Mimosa dulcis IPNI=*128358-2* 321
Mimosa fagifolia IPNI=*1199545-2* 323
Mimosa falcata IPNI=*470782-1* 310
Mimosa farnesiana IPNI=*1099713-2* 336
Mimosa fasciculata IPNI=*316281-2* 313
Mimosa fastuosa IPNI=*1040744-2* 322
Mimosa flavescens IPNI=*507771-1* 327
Mimosa flaviseta IPNI=*507771-1* 327
Mimosa flexuosa IPNI=*77100888-1* 336
Mimosa flexuosa IPNI=*99109-3* 311
Mimosa fraxinea IPNI=*1129244-2* 321
Mimosa gigas IPNI=*1078643-2* 319
Mimosa glauca IPNI=*77077786-1* 310
Mimosa glockeri IPNI=*508520-1* 327
Mimosa glomerata IPNI=*962821-1* 337
Mimosa guadalupensis IPNI=*515137-1* 330
Mimosa hexaphylla IPNI=*508520-1* 327
Mimosa hostilis IPNI=*508775-1* 328
Mimosa hymenaeifolia IPNI=*1183336-2* 330
Mimosa inaequalis IPNI=*271431-2* 337
Mimosa inga IPNI=*1006087-2* 324
Mimosa ingoides IPNI=*1131764-2* 322
Mimosa juliflora IPNI=*1002432-2* 331
Mimosa jupunba IPNI=*469538-1* 310
Mimosa lacustris IPNI=*509612-1* 328
Mimosa lanceolata IPNI=*514978-1* 330
Mimosa latifolia IPNI=*271432-2* 337
Mimosa laurina IPNI=*1199545-2* 323
Mimosa laxa IPNI=*1161746-2* 313
Mimosa lebbeck IPNI=*99109-3* 311
Mimosa leucocephala IPNI=*138955-2* 325
Mimosa ligustrina IPNI=*271434-2* 337
Mimosa ligustrina IPNI=*514978-1* 330
Mimosa limana IPNI=*508775-1* 328
Mimosa longifolia IPNI=*271434-2* 337
Mimosa longifolia IPNI=*470782-1* 310
Mimosa lophantha IPNI=*912944-1* 329
Mimosa lucida IPNI=*128675-2* 323
Mimosa lutea IPNI=*60441223-2* 336
Mimosa lycopodioides IPNI=*1056199-2* 328
Mimosa macracantha IPNI=*60441223-2* 336
Mimosa macroloba IPNI=*188076-2* 330
Mimosa macrophylla IPNI=*128596-2* 323
Mimosa macrostachya IPNI=*470782-1* 310
Mimosa macrostachys IPNI=*1183336-2* 330
Mimosa mangensis IPNI=*56985-2* 315
Mimosa martiusiana IPNI=*77085269-1* 332
Mimosa microphylla IPNI=*762-2* 310
Mimosa mollissima IPNI=*470138-1* 310
Mimosa monilifera IPNI=*1003508-2* 330
Mimosa natans IPNI=*509612-1* 328
Mimosa nigra IPNI=*508775-1* 328
Mimosa nilotica IPNI=*77089275-1* 336
Mimosa niopo IPNI=*12268-2* 311
Mimosa nitida IPNI=*128675-2* 323
Mimosa obtusa IPNI=*60441223-2* 336
Mimosa pacay IPNI=*128414-2* 321
Mimosa paniculata IPNI=*234346-2* 332
Mimosa parae IPNI=*128535-2* 322
Mimosa parvifolia IPNI=*12268-2* 311
Mimosa pectinata IPNI=*512223-1* 329
Mimosa pellacantha IPNI=*60441223-2* 336
Mimosa peregrina IPNI=*12268-2* 311
Mimosa pernambucana IPNI=*1125234-2* 317
Mimosa pilosula IPNI=*128675-2* 323
Mimosa plena IPNI=*1056199-2* 328
Mimosa polyantha var. *levior* IPNI=*508515-1* 327
Mimosa polyphylla IPNI=*518289-1* 332
Mimosa polystachya IPNI=*1009172-2* 319
Mimosa portoricensis IPNI=*943976-1* 337
Mimosa prostrata IPNI=*509612-1* 328
Mimosa pulcherrima IPNI=*519643-1* 334
Mimosa pulchra IPNI=*470138-1* 310
Mimosa punctata IPNI=*1056199-2* 328
Mimosa pungens IPNI=*1003508-2* 330
Mimosa pusilla IPNI=*507771-1* 327
Mimosa quadrivalvis var. *leptocarpa* IPNI=*1014032-1* 327
Mimosa quassiifolia IPNI=*128675-2* 323
Mimosa retusa IPNI=*77208976-1* 330
Mimosa reversa IPNI=*148561-2* 326
Mimosa rosea IPNI=*996931-1* 330
Mimosa rubiginosa IPNI=*1068738-2* 323
Mimosa saligna IPNI=*471383-1* 310
Mimosa salinarum IPNI=*60441241-2* 336
Mimosa saman IPNI=*227126-2* 331
Mimosa sapindoides IPNI=*128732-2* 324
Mimosa scabrella var. *aspericarpa* IPNI=*508647-1* 328
Mimosa scandens IPNI=*493866-1* 319
Mimosa seeressa IPNI=*99109-3* 311
Mimosa semialata IPNI=*128599-2* 323
Mimosa sericea IPNI=*128693-2* 323
Mimosa setigera IPNI=*508515-1* 327
Mimosa sinemariensis IPNI=*1199545-2* 323
Mimosa sirissa IPNI=*99109-3* 311
Mimosa speciosa IPNI=*99109-3* 311
Mimosa spectabilis IPNI=*1122303-2* 324
Mimosa splendens IPNI=*128749-2* 324
Mimosa stipitata IPNI=*508515-1* 327
Mimosa stolonifera IPNI=*509612-1* 328
Mimosa subnervosa IPNI=*42066-2* 313
Mimosa tamarindifolia IPNI=*234370-2* 332
Mimosa tetraphylla IPNI=*128599-2* 323
Mimosa thyrsoidea IPNI=*1129244-2* 321
Mimosa tortuosa IPNI=*60441241-2* 336
Mimosa umbellata IPNI=*128358-2* 321
Mimosa umbellifera IPNI=*1035636-2* 324
Mimosa unguis-cati IPNI=*515137-1* 330
Mimosa velutina IPNI=*128810-2* 324
Mimosa verticillata IPNI=*471751-1* 310
Mimosa villosa IPNI=*762-2* 310
Mimosa virescens IPNI=*60441241-2* 336
Mimosa virgata IPNI=*1125234-2* 317
Mimosa wilsonii IPNI=*493928-1* 319
Mimosa ynga IPNI=*128478-2* 322
Mimulus andicola IPNI=*77120251-1* 375
Mimulus glabratus IPNI=*77120251-1* 375
Mimulus glabratus subsp. *typicus* IPNI=*77120251-1* 375
Mimulus glabratus var. *externus* IPNI=*77120251-1* 375
Mimulus glabratus var. *parviflorus* IPNI=*77120251-1* 375
Mimulus javanicus IPNI=*1189750-2* 374
Mimulus kingii IPNI=*77120251-1* 375
Mimulus parviflorus IPNI=*77120251-1* 375
Mimulus parviflorus var. *externus* IPNI=*77120251-1* 375
Mimulus parviflorus var. *nanus* IPNI=*77120251-1* 375
Mimulus pilosiusculus IPNI=*77120251-1* 375
Mimulus propinquus IPNI=*77120251-1* 375
Mimulus sylvaticus IPNI=*77120251-1* 375
Mimulus tener IPNI=*77120251-1* 375
Mimusops bidentata IPNI=*1032531-2* 305
Mimusops grisebachii IPNI=*152641-2* 306
Mimusops huberi IPNI=*152609-2* 305
Minquartia macrophylla IPNI=*110121-1* 449
Minquartia parvifolia IPNI=*110121-1* 449
Minquartia punctata IPNI=*110121-1* 449
Minthe pulegia IPNI=*451022-1* 371
Minyranthes heterophylla IPNI=*249084-1* 270

Mirabilis ambigua IPNI=*162591-2* 287
Mirabilis cordata IPNI=*162591-2* 287
Mirabilis dichotoma IPNI=*162591-2* 287
Mirabilis divaricata IPNI=*162591-2* 287
Mirabilis divaricata var. *uniflora* IPNI=*162591-2* 287
Mirabilis hybrida IPNI=*162591-2* 287
Mirabilis jalapa f. *eujalapa* IPNI=*162591-2* 287
Mirabilis jalapa f. *parviflora* IPNI=*162591-2* 287
Mirabilis jalapa subsp. *ciliata* IPNI=*162591-2* 287
Mirabilis jalapa subsp. *lindheimeri* IPNI=*162591-2* 287
Mirabilis jalapa subsp. *volcanica* IPNI=*162591-2* 287
Mirabilis jalapa var. *lindheimeri* IPNI=*162591-2* 287
Mirabilis jalapa var. *odorata* IPNI=*162591-2* 287
Mirabilis lindheimeri IPNI=*162591-2* 287
Mirabilis odorata IPNI=*162591-2* 287
Mirabilis parviflora IPNI=*162591-2* 287
Mirabilis pedunculata IPNI=*162591-2* 287
Mirabilis planiflora IPNI=*162591-2* 287
Mirabilis procera IPNI=*162591-2* 287
Mirabilis pubescens IPNI=*162591-2* 287
Mirabilis undulata IPNI=*162591-2* 287
Mirabilis uniflora IPNI=*162591-2* 287
Mirabilis xalapa IPNI=*162591-2* 287
Mirasolia diversifolia IPNI=*255747-1* 271
Misandra magellanica IPNI=*430163-1* 358
Misanteca anacardioides IPNI=*158649-2* 183
Misanteca appelii IPNI=*465275-1* 183
Misanteca aritu IPNI=*465275-1* 183
Misanteca armeniaca IPNI=*465248-1* 183
Misanteca cannella IPNI=*139921-2* 183
Misanteca cayennensis IPNI=*139921-2* 183
Misanteca chrysophylla IPNI=*465260-1* 183
Misanteca costaricensis IPNI=*465312-1* 183
Misanteca cubensis IPNI=*465312-1* 183
Misanteca cymbarum IPNI=*77198011-1* 183
Misanteca duartei IPNI=*465275-1* 183
Misanteca juergensenii IPNI=*465312-1* 183
Misanteca limbosa IPNI=*465312-1* 183
Misanteca macrophylla IPNI=*465280-1* 183
Misanteca pittieri IPNI=*465312-1* 183
Misanteca puchury-major IPNI=*13872-2* 182
Misanteca quixos IPNI=*77198017-1* 183
Misanteca reclinata IPNI=*465312-1* 183
Misanteca rigida IPNI=*465260-1* 183
Misanteca schultesii IPNI=*465280-1* 183
Misanteca tikalana IPNI=*465312-1* 183
Misanteca triandra IPNI=*465312-1* 183
Miscanthus polydactylos IPNI=*60469006-2* 245
Mitesia punctata IPNI=*190855-2* 289
Mitostax pallida IPNI=*1175026-2* 331
Mitozus tenellus IPNI=*1054185-2* 342
Mitozus tenuicaulis IPNI=*152052-2* 341
Mitozus versicolor IPNI=*152052-2* 341
Mitracarpus bakeri IPNI=*756228-1* 352
Mitracarpus diffusus IPNI=*756228-1* 352
Mitracarpus hirtus var. *remotiflorus* IPNI=*756228-1* 352
Mitracarpus hirtus var. *sessilis* IPNI=*756228-1* 352
Mitracarpus pallidus IPNI=*756228-1* 352
Mitracarpus puberulus IPNI=*766900-1* 357
Mitracarpus scaber IPNI=*756228-1* 352
Mitracarpus senegalensis IPNI=*756228-1* 352
Mitracarpus simplex IPNI=*756228-1* 352
Mitracarpus stylosus IPNI=*756228-1* 352
Mitracarpus torresianus IPNI=*756228-1* 352
Mitracarpus verticillatus IPNI=*756228-1* 352
Mitracarpus villosus IPNI=*756228-1* 352
Mitracarpus villosus var. *glabrior* IPNI=*756228-1* 352
Mitranthes eugenioides IPNI=*600952-1* 428
Mitranthes eugenioides var. *oblongifolia* IPNI=*600952-1* 428
Mitranthes eugenioides var. *ovata* IPNI=*600952-1* 428
Mitranthes gardneriana IPNI=*600952-1* 428
Mitranthes sartoriana IPNI=*600952-1* 428
Mitropsidium eugenioides IPNI=*600952-1* 428
Mitropsidium gardnerianum IPNI=*600952-1* 428
Mitropsidium oblanceolatum IPNI=*600952-1* 428
Mitropsidium oligospermum IPNI=*600952-1* 428
Mitropsidium pittieri IPNI=*600952-1* 428
Mitropsidium sartorianum IPNI=*600952-1* 428
Mnasium theophrasti IPNI=*584810-1* 250
Mnemion agreste IPNI=*291320-2* 406
Mnemion arvense IPNI=*291320-2* 406
Mnemion elegans IPNI=*869428-1* 406
Mnemion gracilescens IPNI=*291320-2* 406
Mnemion nemausense IPNI=*869428-1* 406
Mnemion obtusifolium IPNI=*291320-2* 406
Mnemion sagotii IPNI=*869428-1* 406
Mnemion segetale IPNI=*291320-2* 406
Mnemion tricolor IPNI=*869428-1* 406
Mnemion variatum IPNI=*869428-1* 406
Mnesithea granularis IPNI=*116848-2* 240
Modanthos caroliniana IPNI=*30017389-2* 412
Modanthos decumbens IPNI=*30017389-2* 412
Modanthos eriocarpa IPNI=*30017389-2* 412
Modanthos prostrata IPNI=*30017389-2* 412
Modanthos reptans IPNI=*30017389-2* 412
Modanthos urticifolia IPNI=*30017389-2* 412
Modiola caroliniana f. *parvula* IPNI=*30017389-2* 412
Modiola caroliniana var. *erecta* IPNI=*30017389-2* 412
Modiola decumbens IPNI=*30017389-2* 412
Modiola erecta IPNI=*30017389-2* 412
Modiola eriocarpa IPNI=*30017389-2* 412
Modiola fissistipula IPNI=*30017389-2* 412
Modiola prostrata IPNI=*30017389-2* 412
Modiola reptans IPNI=*30017389-2* 412
Modiola urticifolia IPNI=*30017389-2* 412
Modiolastrum jaeggianum IPNI=*30017389-2* 412
Moehringia muscosa IPNI=*302082-2* 285
Moehringia nemorosa IPNI=*30040285-2* 285
Moenchia obliqua IPNI=*528506-1* 213
Moeroris stipulata IPNI=*195027-2* 402
Mogiphanes brasiliensis IPNI=*10294-2* 281
Mogiphanes dunaliana IPNI=*191400-2* 283
Mogiphanes glauca IPNI=*191400-2* 283
Mogiphanes paniculata IPNI=*10447-2* 281
Mogiphanes philoxeroides IPNI=*1024156-2* 281
Mogiphanes soratensis IPNI=*10378-2* 281
Mogiphanes straminea IPNI=*10294-2* 281
Mogorium gimea IPNI=*609755-1* 375
Mogorium goaense IPNI=*609755-1* 375
Mogorium multiflorum IPNI=*609641-1* 375
Mogorium pubescens IPNI=*609641-1* 375
Mogorium sambac IPNI=*609755-1* 375
Mogorium undulatum IPNI=*609755-1* 375
Mohlana guineensis IPNI=*122647-2* 287
Mohlana latifolia IPNI=*122647-2* 287
Mohlana nemoralis IPNI=*122647-2* 287
Mohlana nemoralis var. *macrophylla* IPNI=*122647-2* 287
Moldenhawera mollis IPNI=*67201-2* 315
Molina incana IPNI=*231180-1* 265
Molina latifolia IPNI=*183399-1* 257
Molina nitida IPNI=*183511-1* 257
Molina reticulata IPNI=*183309-1* 257
Molina rhexioides IPNI=*183821-1* 258
Molina sagittalis IPNI=*30159244-2* 258
Molina salicifolia IPNI=*183704-1* 258
Molina venosa IPNI=*183309-1* 257

Molineria capitulata IPNI=*64141-1* 216
Molineria hortensis IPNI=*64141-1* 216
Molineria plicata IPNI=*64141-1* 216
Molineria recurvata IPNI=*64141-1* 216
Molineria sulcata IPNI=*64141-1* 216
Molineria sumatrana IPNI=*64141-1* 216
Molinia aquatica IPNI=*395086-1* 235
Molinia fluitans IPNI=*404100-1* 239
Molinia plicata IPNI=*404100-1* 239
Molinia retusa IPNI=*60458636-2* 238
Mollia sprucei IPNI=*834881-1* 412
Mollinedia casca IPNI=*582116-1* 185
Mollinedia cunninghamii IPNI=*582116-1* 185
Mollinedia cyathantha IPNI=*582116-1* 185
Mollinedia glabricaulis IPNI=*582116-1* 185
Mollinedia grandifolia IPNI=*582116-1* 185
Mollinedia grosseserrata IPNI=*582116-1* 185
Mollinedia hylophila IPNI=*582116-1* 185
Mollinedia krukovii IPNI=*582116-1* 185
Mollinedia latifolia IPNI=*582116-1* 185
Mollinedia laurina IPNI=*582116-1* 185
Mollinedia laurina var. *friburgensis* IPNI=*582116-1* 185
Mollinedia leiantha IPNI=*582116-1* 185
Mollinedia lepidota IPNI=*582354-1* 186
Mollinedia neblinensis IPNI=*582116-1* 185
Mollinedia oligotricha IPNI=*582116-1* 185
Mollinedia polycarpa IPNI=*582116-1* 185
Mollinedia ptariensis IPNI=*582116-1* 185
Mollinedia puberula IPNI=*582116-1* 185
Mollinedia rusbyana IPNI=*582116-1* 185
Mollinedia tessmannii IPNI=*582116-1* 185
Mollinedia wikstroemioides IPNI=*582116-1* 185
Mollugo araucana IPNI=*1008757-2* 286
Mollugo arenaria IPNI=*280992-2* 286
Mollugo axillaris IPNI=*280992-2* 286
Mollugo berteroana IPNI=*280992-2* 286
Mollugo cambessedesii IPNI=*1008757-2* 286
Mollugo chevalieri IPNI=*280992-2* 286
Mollugo costata IPNI=*280992-2* 286
Mollugo dichotoma IPNI=*280992-2* 286
Mollugo diffusa IPNI=*280992-2* 286
Mollugo glinoides IPNI=*1008757-2* 286
Mollugo hoffmannseggiana IPNI=*280992-2* 286
Mollugo juncea IPNI=*280992-2* 286
Mollugo radiata IPNI=*1008757-2* 286
Mollugo schrankii IPNI=*280992-2* 286
Mollugo spergulifolia IPNI=*280992-2* 286
Mollugo verticillata subsp. *subsessile* IPNI=*280992-2* 286
Mollugo verticillata var. *latifolia* IPNI=*280992-2* 286
Mollugo verticillata var. *linearis* IPNI=*280992-2* 286
Mollugo verticillata var. *longifolia* IPNI=*280992-2* 286
Mollugo verticillata var. *scrobiculata* IPNI=*280992-2* 286
Mollugo verticillata var. *vulgaris* IPNI=*280992-2* 286
Molongum cuneatum IPNI=*945268-1* 342
Molongum macrophyllum IPNI=*945265-1* 343
Molubda scandens IPNI=*687109-1* 288
Molucca laevis IPNI=*451989-1* 371
Moluchia frutescens IPNI=*157023-2* 412
Moluchia fruticosa IPNI=*157004-2* 412
Moluchia herbacea IPNI=*157004-2* 412
Moluchia pyramidata IPNI=*157004-2* 412
Moluchia tomentosa IPNI=*157023-2* 412
Momisia aculeata IPNI=*1213898-2* 436
Momisia alnifolia IPNI=*1213898-2* 436
Momisia anfractuosa IPNI=*1213898-2* 436
Momisia brevifolia IPNI=*1213898-2* 436
Momisia dichotoma IPNI=*1213898-2* 436
Momisia ehrenbergiana IPNI=*1213898-2* 436
Momisia iguanaea IPNI=*1213898-2* 436
Momisia laevigata IPNI=*1213898-2* 436
Momisia membranacea IPNI=*1213898-2* 436
Momisia platycaulis IPNI=*1213898-2* 436
Momisia pubescens IPNI=*1213898-2* 436
Momisia spinifera IPNI=*1213898-2* 436
Momisia spinosissima IPNI=*1213898-2* 436
Momisia tala IPNI=*1213898-2* 436
Momisia tarijensis IPNI=*1213898-2* 436
Momisia triflora IPNI=*1213898-2* 436
Momisia zizyphoides IPNI=*1213898-2* 436
Momordica anthelmintica IPNI=*293413-1* 295
Momordica bicolor IPNI=*291972-1* 293
Momordica brachybotrys IPNI=*292472-1* 294
Momordica carinata IPNI=*293060-1* 294
Momordica charantia f. *abbreviata* IPNI=*293413-1* 295
Momordica charantia f. *muricata* IPNI=*293413-1* 295
Momordica charantia subsp. *abbreviata* IPNI=*293413-1* 295
Momordica charantia var. *abbreviata* IPNI=*293413-1* 295
Momordica charantia var. *muricata* IPNI=*293413-1* 295
Momordica charantia var. *pseudobalsamina* IPNI=*293413-1* 295
Momordica charantia var. *zeylanica* IPNI=*293413-1* 295
Momordica covel IPNI=*291972-1* 293
Momordica cylindrica IPNI=*292238-1* 293
Momordica cylindrica IPNI=*293413-1* 295
Momordica dasycarpa IPNI=*292180-1* 293
Momordica elegans IPNI=*293413-1* 295
Momordica glauca IPNI=*292473-1* 294
Momordica humilis IPNI=*293413-1* 295
Momordica indica IPNI=*293413-1* 295
Momordica jagorana IPNI=*293413-1* 295
Momordica lambertiana IPNI=*292238-1* 293
Momordica lanata IPNI=*291938-1* 293
Momordica luffa IPNI=*293060-1* 294
Momordica monadelpha IPNI=*291972-1* 293
Momordica muricata IPNI=*293413-1* 295
Momordica muricata IPNI=*88306-2* 294
Momordica operculata IPNI=*293060-1* 294
Momordica operculata IPNI=*293097-1* 294
Momordica palmata IPNI=*292814-1* 294
Momordica papillosa IPNI=*293413-1* 295
Momordica pedata IPNI=*292512-1* 294
Momordica purgans IPNI=*1076251-2* 294
Momordica quinquefida IPNI=*293097-1* 294
Momordica racemosa IPNI=*88306-2* 294
Momordica reticulata IPNI=*293060-1* 294
Momordica roxburghiana IPNI=*293413-1* 295
Momordica sativa IPNI=*292238-1* 293
Momordica senegalensis IPNI=*293413-1* 295
Momordica thollonii IPNI=*293413-1* 295
Momordica tubiflora IPNI=*105280-3* 294
Momordica zeylanica IPNI=*293413-1* 295
Monachanthus bushnanii IPNI=*621811-1* 217
Monachanthus discolor IPNI=*621811-1* 217
Monachanthus discolor var. *bushnanii* IPNI=*621811-1* 217
Monachanthus discolor var. *viridiflorus* IPNI=*621811-1* 217
Monachanthus viridis IPNI=*1092236-2* 217
Monachne punctata IPNI=*1021229-2* 239
Monachne subglabra IPNI=*77164258-1* 246
Monachne unilateralis IPNI=*398888-1* 237
Monactineirma angustifolia IPNI=*321973-2* 400
Monactineirma coriacea IPNI=*674920-1* 398
Monactineirma hederacea IPNI=*321973-2* 400
Monactineirma minima IPNI=*185104-2* 399
Monactineirma peltata IPNI=*321973-2* 400
Monactineirma suberosa IPNI=*321973-2* 400
Monadelphanthus floridus IPNI=*77210192-1* 347
Moniera africana IPNI=*1072674-2* 376

Moniera axillaris IPNI=*28282-2* 375
Moniera brownei IPNI=*1072674-2* 376
Moniera caprarioides IPNI=*155661-2* 376
Moniera chamaedryoides var. *peduncularis* IPNI=*155661-2* 376
Moniera cuneifolia IPNI=*1072674-2* 376
Moniera dianthera IPNI=*155661-2* 376
Moniera diffusa IPNI=*244449-2* 377
Moniera monnieri IPNI=*1072674-2* 376
Moniera montevidensis IPNI=*155661-2* 376
Moniera pedunculosa IPNI=*1072674-2* 376
Moniera procumbens f. *albiflora* IPNI=*155661-2* 376
Moniera procumbens f. *bicolor* IPNI=*155661-2* 376
Moniera procumbens f. *flavida* IPNI=*155661-2* 376
Moniera procumbens IPNI=*155661-2* 376
Moniera procumbens var. *montevidensis* IPNI=*155661-2* 376
Moniera procumbens var. *normalis* IPNI=*155661-2* 376
Monnina aestuans var. *polystachya* IPNI=*163646-2* 338
Monnina angustata IPNI=*690538-1* 338
Monnina arborescens IPNI=*690538-1* 338
Monnina comata IPNI=*77143255-1* 338
Monnina crotalariifolia IPNI=*690654-1* 338
Monnina crotalarioides IPNI=*690654-1* 338
Monnina crotalarioides var. *glabrescens* IPNI=*690654-1* 338
Monnina crotalarioides var. *leptostachys* IPNI=*690654-1* 338
Monnina crotalarioides var. *macrophylla* IPNI=*690654-1* 338
Monnina cuspidata IPNI=*163646-2* 338
Monnina cuspidata var. *patula* IPNI=*163646-2* 338
Monnina elliptica IPNI=*77143255-1* 338
Monnina floribunda IPNI=*77143255-1* 338
Monnina hirta IPNI=*163646-2* 338
Monnina hirta subsp. *cuspidata* IPNI=*163646-2* 338
Monnina media IPNI=*163646-2* 338
Monnina multicomata IPNI=*163536-2* 338
Monnina obtusifolia IPNI=*690654-1* 338
Monnina obtusifolia var. *oblongifolia* IPNI=*690654-1* 338
Monnina paniculata IPNI=*163646-2* 338
Monnina patula IPNI=*163646-2* 338
Monnina phillyreoides IPNI=*690654-1* 338
Monnina phytolaccifolia IPNI=*77143255-1* 338
Monnina pilosa IPNI=*163646-2* 338
Monnina pilosa var. *glabrescens* IPNI=*163646-2* 338
Monnina pseudoaestuans IPNI=*690504-1* 338
Monnina revoluta IPNI=*690504-1* 338
Monnina trianae IPNI=*163646-2* 338
Monochaetum guatemalense IPNI=*121047-2* 419
Monochlaena sinuosa IPNI=*17084790-1* 175
Monolena sprucei IPNI=*323262-2* 423
Monoplegma sphaerospermum IPNI=*179469-2* 329
Monoplegma trinervium IPNI=*179469-2* 329
Monopsis conspicua IPNI=*143240-1* 273
Monosis elaeagnoides IPNI=*962413-1* 260
Monoteles paradoxa IPNI=*30129-2* 312
Monotropa morisoniana IPNI=*281016-2* 300
Monotropa morisonii IPNI=*281016-2* 300
Monotropa uniflora f. *rosea* IPNI=*281016-2* 300
Monotropa uniflora var. *typica* IPNI=*281016-2* 300
Monstera acreana IPNI=*163921-2* 200
Monstera amomifolia IPNI=*89035-1* 202
Monstera borsigiana IPNI=*87478-1* 200
Monstera caudata IPNI=*87046-1* 200
Monstera deliciosa var. *borsigiana* IPNI=*87478-1* 200
Monstera deliciosa var. *sierrana* IPNI=*87478-1* 200
Monstera dilacerata IPNI=*87046-1* 200
Monstera henry-pittieri IPNI=*87509-1* 200
Monstera irritans IPNI=*163921-2* 200
Monstera maxima IPNI=*87509-1* 200
Monstera moritziana IPNI=*88631-1* 202
Monstera pertusa IPNI=*87464-1* 200
Monstera pertusa IPNI=*88567-1* 202
Monstera pinnata IPNI=*87046-1* 200
Monstera tacanaensis IPNI=*87478-1* 200
Montanoa excelsa IPNI=*233605-1* 266
Montanoa lehmannii IPNI=*233605-1* 266
Montanoa moritziana IPNI=*233605-1* 266
Montanoa tamayonis IPNI=*233605-1* 266
Montbretia × *crocosmiiflora* IPNI=*436427-1* 216
Montbretia odorata IPNI=*437030-1* 216
Montbretia pottsii IPNI=*436437-1* 216
Montbretia refracta IPNI=*437030-1* 216
Monteverdia amazonica IPNI=*161805-1* 291
Monteverdia guyanensis IPNI=*161926-1* 291
Monteverdia krukovii IPNI=*155542-2* 291
Monteverdia laevis IPNI=*161953-1* 291
Monteverdia macrocarpa IPNI=*155542-2* 291
Monteverdia sapotiformis IPNI=*161953-1* 291
Monteverdia sieberiana IPNI=*155604-2* 291
Montia alsine-facie IPNI=*311155-2* 286
Montia andina IPNI=*311155-2* 286
Montia chabertii IPNI=*311155-2* 286
Montia decumbens IPNI=*311155-2* 286
Montia dipetala IPNI=*311155-2* 286
Montia erecta IPNI=*311155-2* 286
Montia fontana subsp. *lamprosperma* IPNI=*311155-2* 286
Montia fontana var. *lamprosperma* IPNI=*311155-2* 286
Montia fontana var. *minor* IPNI=*311155-2* 286
Montia fontana var. *tenerrima* IPNI=*311155-2* 286
Montia funstonii IPNI=*311155-2* 286
Montia gibba IPNI=*311155-2* 286
Montia hallii IPNI=*311155-2* 286
Montia humilis IPNI=*311155-2* 286
Montia lamprosperma IPNI=*311155-2* 286
Montia linearifolia IPNI=*311155-2* 286
Montia major IPNI=*311155-2* 286
Montia minima IPNI=*311155-2* 286
Montia minor var. *lamprosperma* IPNI=*311155-2* 286
Montia pentandra IPNI=*311155-2* 286
Montia repens IPNI=*311155-2* 286
Montia rivularis var. *lamprosperma* IPNI=*311155-2* 286
Montia stenophylla IPNI=*311155-2* 286
Montia tenella IPNI=*311155-2* 286
Montia verrucosa IPNI=*311155-2* 286
Montjolya angustifolia IPNI=*262640-2* 275
Montjolya bullata IPNI=*262640-2* 275
Montjolya cylindrostachya IPNI=*262641-2* 276
Montjolya ferruginea IPNI=*946226-1* 276
Montjolya floribunda IPNI=*262641-2* 276
Montjolya guianensis IPNI=*262640-2* 275
Montjolya macrostachya IPNI=*262641-2* 276
Montjolya mollis IPNI=*262641-2* 276
Montrichardia aculeatum IPNI=*87563-1* 200
Montrichardia arborea IPNI=*87563-1* 200
Montrichardia arborescens var. *aculeata* IPNI=*87563-1* 200
Montrichardia fendleri IPNI=*87563-1* 200
Montrichardia splitgerberi IPNI=*87563-1* 200
Monvillea alticostata IPNI=*1001185-1* 284
Monvillea campinensis IPNI=*1001185-1* 284
Monvillea euchlora IPNI=*1001185-1* 284
Monvillea lauterbachii IPNI=*1001185-1* 284
Monvillea leucantha IPNI=*1001185-1* 284
Monvillea piedadensis IPNI=*1001185-1* 284
Moorea argentea IPNI=*66317-2* 236
Moorea irrorata IPNI=*168902-2* 222
Mopania calcicola IPNI=*152591-2* 305
Mopania chicle IPNI=*152591-2* 305
Mopex sinensis IPNI=*320787-2* 415
Moquilea bracteosa IPNI=*722743-1* 381

Moquilea chrysocalyx IPNI=*722748-1* 381
Moquilea couepia IPNI=*67154-2* 381
Moquilea licaniiflora IPNI=*77156118-1* 383
Moquilea longistyla IPNI=*60472856-2* 382
Moquilea macrocarpa IPNI=*164227-2* 383
Moquilea paraensis IPNI=*722782-1* 381
Moquilea parviflora IPNI=*77156118-1* 383
Moquilea rufa IPNI=*722743-1* 381
Moquilea sclerophylla IPNI=*60472875-2* 382
Moquilea sclerophylla var. *myristicoides* IPNI=*60472875-2* 382
Moquilea sclerophylla var. *pohliana* IPNI=*60472875-2* 382
Moquilea sclerophylla var. *scabra* IPNI=*60472875-2* 382
Moquilea tomentosa var. *angustifolia* IPNI=*726782-1* 383
Moquilea tomentosa var. *latifolia* IPNI=*726782-1* 383
Moquilea turiuva IPNI=*60472870-2* 382
Mora megistosperma IPNI=*164265-2* 328
Moraea chimboracensis IPNI=*440470-1* 216
Moraea fimbriata IPNI=*438741-1* 216
Moraea gladioloides IPNI=*440470-1* 216
Moraea grandiflora IPNI=*441553-1* 216
Moraea lutea IPNI=*77182480-1* 217
Moraea palmifolia IPNI=*237727-2* 216
Moraea pavonia IPNI=*441553-1* 216
Moraea plicata IPNI=*89940-2* 216
Moraea tigridia IPNI=*441553-1* 216
Morella cerifera IPNI=*166220-2* 339
Morella domingana IPNI=*166220-2* 339
Morella parvifolia IPNI=*326824-2* 339
Morella pubescens IPNI=*326827-2* 339
Morella pumila IPNI=*166220-2* 339
Morelosia beurera IPNI=*1088317-2* 273
Morelosia domingensis IPNI=*1088317-2* 273
Morelosia ovata IPNI=*1088317-2* 273
Morelosia recurva IPNI=*1088317-2* 273
Morelosia revoluta IPNI=*1088317-2* 273
Morenia caudata IPNI=*54901-2* 207
Morenia corallina IPNI=*54901-2* 207
Morenia ernesti-augustii IPNI=*54867-2* 207
Morenia fragrans IPNI=*54901-2* 207
Morenia integrifolia IPNI=*54928-2* 207
Morenia integrifolia var. *nigricans* IPNI=*54928-2* 207
Morenia latisecta IPNI=*54901-2* 207
Morenia lindeniana IPNI=*54901-2* 207
Morenia linearis IPNI=*54901-2* 207
Morenia macrocarpa IPNI=*54901-2* 207
Morenia microspadix IPNI=*54901-2* 207
Morenia montana IPNI=*54901-2* 207
Morenia pauciflora IPNI=*54928-2* 207
Morenia poeppigiana IPNI=*54901-2* 207
Morenia robusta IPNI=*54901-2* 207
Morgania aspera IPNI=*1189750-2* 374
Morgania lucida IPNI=*1189750-2* 374
Morgania rotundifolia IPNI=*1061345-2* 374
Morinda angustifolia IPNI=*756359-1* 352
Morinda aspera IPNI=*756359-1* 352
Morinda asperula IPNI=*756446-1* 352
Morinda chachuca IPNI=*756359-1* 352
Morinda citrifolia f. *potteri* IPNI=*756359-1* 352
Morinda coreia var. *stenophylla* IPNI=*756359-1* 352
Morinda ferruginea IPNI=*756470-1* 352
Morinda ligulata IPNI=*756359-1* 352
Morinda littoralis IPNI=*756359-1* 352
Morinda longiloba IPNI=*1061341-2* 347
Morinda macrophylla IPNI=*756359-1* 352
Morinda mudia IPNI=*756359-1* 352
Morinda multiflora IPNI=*756359-1* 352
Morinda nodosa IPNI=*756359-1* 352
Morinda pubescens var. *aspera* IPNI=*756359-1* 352
Morinda quadrangularis IPNI=*756359-1* 352
Morinda royoc var. *ferruginea* IPNI=*756470-1* 352
Morinda royoc var. *latifolia* IPNI=*756470-1* 352
Morinda seibertii IPNI=*16747-2* 347
Morinda sparsiflora IPNI=*1061341-2* 347
Morinda stenophylla IPNI=*756359-1* 352
Morinda tenuiflora IPNI=*1061341-2* 347
Morinda tenuiflora var. *leiophylla* IPNI=*1061341-2* 347
Morinda teysmanniana IPNI=*756359-1* 352
Morinda tinctoria IPNI=*756359-1* 352
Morinda tomentosa IPNI=*756359-1* 352
Morinda yucatanensis IPNI=*756470-1* 352
Morinda zollingeriana IPNI=*756359-1* 352
Moringa amara IPNI=*584736-1* 280
Moringa domestica IPNI=*584736-1* 280
Moringa edulis IPNI=*584736-1* 280
Moringa erecta IPNI=*584736-1* 280
Moringa moringa IPNI=*584736-1* 280
Moringa nux-eben IPNI=*584736-1* 280
Moringa octogona IPNI=*584736-1* 280
Moringa parvifolia IPNI=*584736-1* 280
Moringa polygona IPNI=*584736-1* 280
Moringa pterygosperma IPNI=*584736-1* 280
Moringa robusta IPNI=*584736-1* 280
Moringa sylvestris IPNI=*584736-1* 280
Moringa zeylanica IPNI=*584736-1* 280
Morisonia americana var. *subpeltata* IPNI=*316366-2* 278
Morisonia breynia IPNI=*77217338-1* 278
Morisonia elegans IPNI=*77183570-1* 278
Morisonia elliptica IPNI=*316366-2* 278
Morisonia guayaquilensis IPNI=*316361-2* 278
Morisonia hagenii IPNI=*77183738-1* 278
Morisonia imrayi IPNI=*316366-2* 278
Morisonia johnstonii IPNI=*316366-2* 278
Mormodes amazonica IPNI=*644946-1* 222
Mormodes brachystachya IPNI=*644946-1* 222
Mormodes buccinator var. *major* IPNI=*644946-1* 222
Mormodes flavida IPNI=*644946-1* 222
Mormodes lentiginosa IPNI=*644946-1* 222
Mormodes leucochila IPNI=*644946-1* 222
Mormodes marmorea IPNI=*644946-1* 222
Mormodes stenoglossa IPNI=*644946-1* 222
Mormodes vitellina IPNI=*644946-1* 222
Mormodes wageneriana IPNI=*644946-1* 222
Mormolyca gracilipes IPNI=*77155011-1* 221
Mormolyca hedwigiae IPNI=*902533-1* 220
Mormolyca rufescens IPNI=*643781-1* 221
Mormolyca tenuibulba IPNI=*320703-2* 221
Moronobea coccinea var. *attenuata* IPNI=*428883-1* 385
Moronobea esculenta IPNI=*326389-2* 385
Moronobea exserens IPNI=*429281-1* 385
Moronobea globulifera IPNI=*429281-1* 385
Moronobea grandiflora IPNI=*326389-2* 385
Moronobea montana IPNI=*326389-2* 385
Morus alba f. *skeletoniana* IPNI=*30051955-2* 440
Morus alba pendula IPNI=*30051955-2* 440
Morus alba var. *emarginata* IPNI=*30051955-2* 440
Morus alba var. *laciniata* IPNI=*30051955-2* 440
Morus alba var. *skeletoniana* IPNI=*30051955-2* 440
Morus albida IPNI=*854634-1* 440
Morus alpina IPNI=*30051955-2* 440
Morus arabica IPNI=*30051955-2* 440
Morus arbuscula IPNI=*854634-1* 440
Morus atrata IPNI=*854688-1* 440
Morus atropurpurea IPNI=*30051955-2* 440
Morus betulifolia IPNI=*854634-1* 440
Morus brasiliensis IPNI=*1151750-2* 440
Morus bullata IPNI=*30051955-2* 440

Morus byzantina IPNI=*30051955-2* 440
Morus canina IPNI=*854634-1* 440
Morus colombassa IPNI=*30051955-2* 440
Morus constantinopolitana IPNI=*30051955-2* 440
Morus corylifolia IPNI=*854634-1* 440
Morus crataegifolia IPNI=*854634-1* 440
Morus cretica IPNI=*854688-1* 440
Morus cucullata IPNI=*30051955-2* 440
Morus dioica IPNI=*1151750-2* 440
Morus dulcis IPNI=*30051955-2* 440
Morus exsucca IPNI=*316991-2* 440
Morus fastigiata IPNI=*30051955-2* 440
Morus furcata IPNI=*30051955-2* 440
Morus goldmanii IPNI=*854634-1* 440
Morus guzziola IPNI=*30051955-2* 440
Morus heterophylla IPNI=*30051955-2* 440
Morus hispanica IPNI=*30051955-2* 440
Morus intermedia IPNI=*30051955-2* 440
Morus kaki IPNI=*30051955-2* 440
Morus laciniata IPNI=*854688-1* 440
Morus levasseurei IPNI=*30051955-2* 440
Morus lhou IPNI=*30051955-2* 440
Morus lucida IPNI=*30051955-2* 440
Morus mariettii IPNI=*30051955-2* 440
Morus marmolii IPNI=*316991-2* 440
Morus membranacea IPNI=*30051955-2* 440
Morus mexicana IPNI=*854634-1* 440
Morus microphilyra IPNI=*854634-1* 440
Morus mollis IPNI=*854634-1* 440
Morus morettii IPNI=*30051955-2* 440
Morus multicaulis IPNI=*30051955-2* 440
Morus nana IPNI=*30051955-2* 440
Morus nigriformis IPNI=*30051955-2* 440
Morus pandurata IPNI=*854634-1* 440
Morus patavia IPNI=*30051955-2* 440
Morus patavina IPNI=*30051955-2* 440
Morus peruviana IPNI=*316991-2* 440
Morus petiolaris IPNI=*854688-1* 440
Morus plumiera IPNI=*1151750-2* 440
Morus pumila IPNI=*30051955-2* 440
Morus romana IPNI=*30051955-2* 440
Morus serotina IPNI=*30051955-2* 440
Morus siciliana IPNI=*854688-1* 440
Morus spinosa IPNI=*1151750-2* 440
Morus subalba IPNI=*30051955-2* 440
Morus tataiba IPNI=*1151750-2* 440
Morus tatarica IPNI=*30051955-2* 440
Morus tatayiba IPNI=*1151750-2* 440
Morus tinctoria IPNI=*1151750-2* 440
Morus tokwa IPNI=*30051955-2* 440
Morus tortuosa IPNI=*30051955-2* 440
Morus trianae IPNI=*316991-2* 440
Morus venassainii IPNI=*30051955-2* 440
Morus venosa IPNI=*30051955-2* 440
Morus vernonii IPNI=*854634-1* 440
Morus vitifolia IPNI=*854634-1* 440
Morus zanthoxylon IPNI=*1151750-2* 440
Moschosma tenuiflorum IPNI=*453130-1* 371
Mosiera guineensis IPNI=*211886-2* 428
Mosquitoxylum jamaicense var. *panamense* IPNI=*164514-2* 451
Mougeotia caracasana IPNI=*824432-1* 411
Mougeotia hirsuta IPNI=*157043-2* 412
Mougeotia nodiflora IPNI=*824514-1* 411
Mougeotia polystachya IPNI=*157002-2* 412
Mouriri acutiflora var. *oligantha* IPNI=*164540-2* 423
Mouriri anomala IPNI=*573797-1* 423
Mouriri ciliata IPNI=*573777-1* 423
Mouriri graveolens IPNI=*573863-1* 424
Mouriri guianensis IPNI=*573777-1* 423
Mouriri pendulifolia IPNI=*573777-1* 423
Mouriri plasschaerti IPNI=*164583-2* 424
Mouriri polyantha IPNI=*306748-2* 423
Mouriri rivularis IPNI=*573790-1* 423
Mouriri ulei IPNI=*306748-2* 423
Mouriri weddellii IPNI=*306748-2* 423
Moutabea acostae IPNI=*690705-1* 338
Moutabea dibotrya IPNI=*690705-1* 338
Moutabea longifolia IPNI=*690705-1* 338
Moutouchi crispata IPNI=*214060-2* 331
Moutouchi draco IPNI=*214060-2* 331
Moutouchi suberosa IPNI=*214060-2* 331
Mucuna altissima IPNI=*1197939-2* 328
Mucuna altissima var. *pilosula* IPNI=*1197939-2* 328
Mucuna andreana IPNI=*164648-2* 328
Mucuna comosa IPNI=*77212307-1* 326
Mucuna prurita IPNI=*509275-1* 328
Mucuna rutilans IPNI=*30187778-2* 313
Mucuna spuria IPNI=*486481-1* 315
Mucuna umbellata IPNI=*1197939-2* 328
Muehlenbeckia benthamii IPNI=*694785-1* 289
Muehlenbeckia leptobotrys IPNI=*694785-1* 289
Muehlenbeckia peruviana IPNI=*694785-1* 289
Muehlenbeckia rupestris var. *volcanica* IPNI=*1022243-2* 289
Muehlenbeckia tamnifolia f. *tenuifolia* IPNI=*694785-1* 289
Muehlenbeckia tamnifolia var. *hartwegii* IPNI=*694785-1* 289
Muehlenbeckia tamnifolia var. *humboldtii* IPNI=*694785-1* 289
Muehlenbeckia tamnifolia var. *laxiflora* IPNI=*694785-1* 289
Muehlenbeckia tamnifolia var. *oligobotrys* IPNI=*694785-1* 289
Muehlenbeckia vulcanica IPNI=*1022243-2* 289
Muellera frutescens IPNI=*77118402-1* 328
Muellera moniliformis IPNI=*77118402-1* 328
Muhlenbergia affinis IPNI=*60445990-2* 241
Muhlenbergia berlandieri IPNI=*60445990-2* 241
Muhlenbergia brasiliensis IPNI=*408232-1* 241
Muhlenbergia elegans IPNI=*60445990-2* 241
Muhlenbergia elegans var. *atroviolacea* IPNI=*60445990-2* 241
Muhlenbergia elegans var. *subviridis* IPNI=*60445990-2* 241
Muhlenbergia fasciculata IPNI=*409081-1* 241
Muhlenbergia glabrata IPNI=*60445990-2* 241
Muhlenbergia metcalfei IPNI=*60445990-2* 241
Muhlenbergia phragmitoides IPNI=*60445990-2* 241
Muhlenbergia purpurea IPNI=*409081-1* 241
Muhlenbergia ramosissima IPNI=*409081-1* 241
Muhlenbergia setosa IPNI=*409081-1* 241
Muhlenbergia spicata IPNI=*115792-2* 240
Muhlenbergia tenuiramea IPNI=*60445990-2* 241
Muldera multinervis IPNI=*682369-1* 196
Muldera wightiana IPNI=*682369-1* 196
Munchausia ovata IPNI=*553618-1* 417
Muntingia calabura var. *trinitensis* IPNI=*320779-2* 415
Muntingia glabra IPNI=*320779-2* 415
Muntingia rosea IPNI=*320779-2* 415
Murdannia malabarica IPNI=*172873-1* 224
Murtughas hirsuta IPNI=*553618-1* 417
Murtughas indica IPNI=*553576-1* 417
Murtughas punctata IPNI=*553618-1* 417
Murtughas speciosa IPNI=*553618-1* 417
Murucuia mollissima IPNI=*185076-2* 399
Murucuia speciosa IPNI=*675224-1* 399
Musa × *acutibracteata* IPNI=*797595-1* 250
Musa × *alphurica* IPNI=*797595-1* 250
Musa × *aphurica* IPNI=*797595-1* 250
Musa × *bacoba* IPNI=*797595-1* 250
Musa × *berteroi* IPNI=*797595-1* 250
Musa × *bidigitalis* IPNI=*797595-1* 250
Musa × *carolinae* IPNI=*797595-1* 250

Musa × *chapara* IPNI=*797595-1* 250
Musa × *chiliocarpa* IPNI=*797595-1* 250
Musa × *consociata* IPNI=*797595-1* 250
Musa × *corbieri* IPNI=*797595-1* 250
Musa × *corniculata* IPNI=*797595-1* 250
Musa × *dacca* IPNI=*797595-1* 250
Musa × *discolor* IPNI=*797595-1* 250
Musa × *dulcissima* IPNI=*797595-1* 250
Musa × *emasculata* IPNI=*797595-1* 250
Musa × *humilis* IPNI=*797595-1* 250
Musa × *ingrata* IPNI=*797595-1* 250
Musa × *jaheri* IPNI=*797595-1* 250
Musa × *maculata* IPNI=*797595-1* 250
Musa × *megalocarpa* IPNI=*797595-1* 250
Musa × *mensaria* IPNI=*797595-1* 250
Musa × *mirabilis* IPNI=*797595-1* 250
Musa × *nigra* IPNI=*797595-1* 250
Musa × *odorata* IPNI=*797595-1* 250
Musa × *oleracea* IPNI=*797595-1* 250
Musa × *pallida* IPNI=*797595-1* 250
Musa × *polycarpa* IPNI=*797595-1* 250
Musa × *prematura* IPNI=*797595-1* 250
Musa × *protractorachis* IPNI=*797595-1* 250
Musa × *purpureotomentosa* IPNI=*797595-1* 250
Musa × *sapidisiaca* IPNI=*797595-1* 250
Musa × *sapientum* f. *dubia* IPNI=*797595-1* 250
Musa × *sapientum* IPNI=*797595-1* 250
Musa × *sapientum* subsp. *paradisiaca* IPNI=*797595-1* 250
Musa × *sapientum* var. *champa* IPNI=*797595-1* 250
Musa × *sapientum* var. *dacca* IPNI=*797595-1* 250
Musa × *sapientum* var. *martabarica* IPNI=*797595-1* 250
Musa × *sapientum* var. *mensaria* IPNI=*797595-1* 250
Musa × *sapientum* var. *odorata* IPNI=*797595-1* 250
Musa × *sapientum* var. *oleracea* IPNI=*797595-1* 250
Musa × *sapientum* var. *regia* IPNI=*797595-1* 250
Musa × *sapientum* var. *violacea* IPNI=*797595-1* 250
Musa × *sapientum* var. *vittata* IPNI=*797595-1* 250
Musa × *trichocarpa* IPNI=*797595-1* 250
Musa × *vittata* IPNI=*797595-1* 250
Musa africana IPNI=*584810-1* 250
Musa arnoldiana IPNI=*584810-1* 250
Musa bagshawei IPNI=*584810-1* 250
Musa balbisiana var. *vittata* IPNI=*797595-1* 250
Musa bihai IPNI=*796925-1* 248
Musa buchananii IPNI=*584810-1* 250
Musa dasycarpa IPNI=*797625-1* 250
Musa davyae IPNI=*584810-1* 250
Musa ensete IPNI=*584810-1* 250
Musa fecunda IPNI=*584810-1* 250
Musa holstii IPNI=*584810-1* 250
Musa humilis IPNI=*796972-1* 248
Musa kaguna IPNI=*584810-1* 250
Musa laurentii IPNI=*584810-1* 250
Musa martretiana IPNI=*584810-1* 250
Musa mexicana IPNI=*584973-1* 250
Musa proboscidea IPNI=*584810-1* 250
Musa ruandensis IPNI=*584810-1* 250
Musa rubronervata IPNI=*584810-1* 250
Musa schweinfurthii IPNI=*584810-1* 250
Musa speciosa IPNI=*584973-1* 250
Musa ulugurensis IPNI=*584810-1* 250
Musa velutina var. *variegata* IPNI=*797625-1* 250
Musa ventricosa IPNI=*584810-1* 250
Musidendron amazonicum IPNI=*585013-1* 250
Mussaenda aegyptiaca IPNI=*760275-1* 355
Mussaenda candida IPNI=*745241-1* 347
Mussaenda candidissima IPNI=*745241-1* 347
Mussaenda chinensis IPNI=*751073-1* 350
Mussaenda coccinea IPNI=*770293-1* 358
Mussaenda formosa IPNI=*222508-2* 356
Mussaenda fulgens IPNI=*756618-1* 352
Mussaenda glomerulata IPNI=*77124830-1* 347
Mussaenda humboldtiana IPNI=*764273-1* 356
Mussaenda lanceolata IPNI=*760275-1* 355
Mussaenda luteola IPNI=*760275-1* 355
Mussaenda nitida IPNI=*764273-1* 356
Mussaenda pubescens IPNI=*764273-1* 356
Mussaenda speciosa IPNI=*204802-2* 355
Mussaenda splendida IPNI=*756618-1* 352
Mussatia hyacinthina IPNI=*77100458-1* 362
Mussinia incisa IPNI=*941423-1* 263
Mussinia linearis IPNI=*208700-1* 263
Mussinia pinnata IPNI=*941423-1* 263
Mussinia speciosa IPNI=*941423-1* 263
Musteron pulchellum IPNI=*947840-1* 264
Mutarda nigra IPNI=*279422-1* 276
Mutelia officinalis IPNI=*450084-1* 371
Mutisia caldasiana IPNI=*233864-1* 267
Mutisia caldasiana var. *coccinea* IPNI=*233864-1* 267
Mutisia clematis var. *caldasiana* IPNI=*233864-1* 267
Mutisia santanderana IPNI=*233864-1* 267
Mutisia viridis IPNI=*233864-1* 267
Mutisiopersea cuneata IPNI=*190632-2* 185
Myagrum clavatum IPNI=*1061021-2* 277
Myagrum hispanicum IPNI=*1061021-2* 277
Myagrum monospermum IPNI=*1061021-2* 277
Myagrum orientale IPNI=*1061021-2* 277
Myagrum palustre IPNI=*288669-1* 277
Myagrum procumbens IPNI=*1061021-2* 277
Myagrum pumilum IPNI=*288669-1* 277
Myagrum rugosum IPNI=*1061021-2* 277
Myagrum stylosum IPNI=*1061021-2* 277
Myagrum venosum IPNI=*1061021-2* 277
Myanthus barbatus IPNI=*1092236-2* 217
Myanthus barbatus var. *immaculatus* IPNI=*1092236-2* 217
Myanthus spinosus IPNI=*1092236-2* 217
Mygalurus bromoides IPNI=*402472-1* 239
Mygalurus caudatus IPNI=*403134-1* 239
Mygalurus uniglumis IPNI=*402472-1* 239
Myginda integrifolia IPNI=*133956-2* 442
Myginda pallens IPNI=*60458018-2* 290
Myginda rhacoma IPNI=*60458018-2* 290
Myginda rotundata IPNI=*60458018-2* 290
Myogalum coarctatum IPNI=*539581-1* 215
Myogalum thyrsoides IPNI=*539581-1* 215
Myosotis acutata IPNI=*118892-1* 275
Myosotis adpressa IPNI=*118892-1* 275
Myosotis alpina IPNI=*118892-1* 275
Myosotis imitata subsp. *pseudoimitata* IPNI=*118892-1* 275
Myosotis linearis IPNI=*329786-2* 285
Myosotis odorata IPNI=*118892-1* 275
Myosotis pantocsekii IPNI=*118892-1* 275
Myosotis prostrata IPNI=*118892-1* 275
Myosotis repens IPNI=*329786-2* 285
Myosotis ronnigeri IPNI=*118892-1* 275
Myosotis rupicola IPNI=*118892-1* 275
Myosotis superalpina IPNI=*118892-1* 275
Myosotis vulgaris IPNI=*152274-1* 285
Myrcia acuminata IPNI=*599170-1* 427
Myrcia acuminata var. *bullata* IPNI=*599170-1* 427
Myrcia acuminata var. *meridensis* IPNI=*599170-1* 427
Myrcia acuminata var. *peruviana* IPNI=*599170-1* 427
Myrcia acuminata var. *tovarensis* IPNI=*599170-1* 427
Myrcia acutata IPNI=*599170-1* 427
Myrcia acutiloba IPNI=*599170-1* 427
Myrcia adpressepilosa IPNI=*598851-1* 426

Myrcia aguitensis IPNI=*599170-1* 427
Myrcia alagoensis IPNI=*599170-1* 427
Myrcia alagoensis var. *intermedia* IPNI=*599170-1* 427
Myrcia alagoensis var. *oblongata* IPNI=*599170-1* 427
Myrcia alagoensis var. *ovata* IPNI=*599170-1* 427
Myrcia alloiota IPNI=*599215-1* 427
Myrcia alloiota var. *obovata* IPNI=*599215-1* 427
Myrcia alloiota var. *subcordata* IPNI=*599215-1* 427
Myrcia alternifolia IPNI=*598851-1* 426
Myrcia ambigua IPNI=*599199-1* 427
Myrcia ambigua var. *dives* IPNI=*599199-1* 427
Myrcia ambigua var. *latifolia* IPNI=*599199-1* 427
Myrcia ambigua var. *multiflora* IPNI=*599199-1* 427
Myrcia ambigua var. *pauciflora* IPNI=*599199-1* 427
Myrcia ambigua var. *rostrata* IPNI=*599199-1* 427
Myrcia ambigua var. *sylvatica* IPNI=*599199-1* 427
Myrcia androsaemoides IPNI=*598851-1* 426
Myrcia androsaemoides var. *parvifolia* IPNI=*598851-1* 426
Myrcia angustifolia IPNI=*598851-1* 426
Myrcia arimensis IPNI=*598851-1* 426
Myrcia augustana IPNI=*599170-1* 427
Myrcia ayabambensis IPNI=*598986-1* 426
Myrcia ayresiana IPNI=*599170-1* 427
Myrcia balbisiana IPNI=*166091-2* 427
Myrcia barrensis IPNI=*599170-1* 427
Myrcia belizensis IPNI=*599170-1* 427
Myrcia berberis IPNI=*599170-1* 427
Myrcia berberis var. *angustifolia* IPNI=*599170-1* 427
Myrcia berberis var. *latifolia* IPNI=*599170-1* 427
Myrcia botrys IPNI=*598851-1* 426
Myrcia brachylopadia IPNI=*599170-1* 427
Myrcia bracteolaris IPNI=*599170-1* 427
Myrcia brandamii IPNI=*599170-1* 427
Myrcia camapuana IPNI=*598851-1* 426
Myrcia capivarhyensis IPNI=*599215-1* 427
Myrcia caracasana IPNI=*599215-1* 427
Myrcia cassinioides IPNI=*598851-1* 426
Myrcia catharinae IPNI=*599170-1* 427
Myrcia chilensis IPNI=*599170-1* 427
Myrcia ciarensis IPNI=*599170-1* 427
Myrcia coccolobifolia IPNI=*166115-2* 427
Myrcia collina IPNI=*598851-1* 426
Myrcia communis IPNI=*599170-1* 427
Myrcia communis var. *glabrata* IPNI=*599170-1* 427
Myrcia communis var. *latifolia* IPNI=*599170-1* 427
Myrcia complicata IPNI=*599170-1* 427
Myrcia compressa IPNI=*599170-1* 427
Myrcia corcovadensis IPNI=*599170-1* 427
Myrcia coroicensis IPNI=*599170-1* 427
Myrcia costaricensis IPNI=*599170-1* 427
Myrcia costata IPNI=*599170-1* 427
Myrcia costata var. *bahiensis* IPNI=*599170-1* 427
Myrcia costata var. *minensis* IPNI=*599170-1* 427
Myrcia coumetoides IPNI=*599170-1* 427
Myrcia crassicaulis IPNI=*598851-1* 426
Myrcia crassinervia IPNI=*598743-1* 426
Myrcia cucullata IPNI=*599170-1* 427
Myrcia cuneata IPNI=*598851-1* 426
Myrcia curatellifolia IPNI=*599215-1* 427
Myrcia curatellifolia var. *grandifolia* IPNI=*599215-1* 427
Myrcia curatellifolia var. *parvifolia* IPNI=*599215-1* 427
Myrcia cymosopaniculata IPNI=*598851-1* 426
Myrcia daphnoides IPNI=*598851-1* 426
Myrcia daphnoides var. *nervosa* IPNI=*598851-1* 426
Myrcia decrescens IPNI=*598851-1* 426
Myrcia deflexa var. *dussii* IPNI=*598743-1* 426
Myrcia desertorum IPNI=*598851-1* 426
Myrcia diaphanosticta IPNI=*598851-1* 426
Myrcia dictyoneura IPNI=*599170-1* 427
Myrcia didrichseniana IPNI=*598851-1* 426
Myrcia discolor IPNI=*599170-1* 427
Myrcia divaricata IPNI=*599170-1* 427
Myrcia divergens IPNI=*165926-2* 427
Myrcia doloresensis IPNI=*594452-1* 425
Myrcia duchassaingiana IPNI=*598743-1* 426
Myrcia elaeodendra IPNI=*598851-1* 426
Myrcia elegans IPNI=*598851-1* 426
Myrcia elongata IPNI=*599170-1* 427
Myrcia elongata var. *brunnea* IPNI=*599170-1* 427
Myrcia elongata var. *grandifolia* IPNI=*599170-1* 427
Myrcia elongata var. *ochracea* IPNI=*599170-1* 427
Myrcia emarginata IPNI=*598851-1* 426
Myrcia erythroxylon IPNI=*599170-1* 427
Myrcia erythroxylon var. *caerulescens* IPNI=*599170-1* 427
Myrcia erythroxylon var. *virescens* IPNI=*593768-1* 425
Myrcia exsucca IPNI=*598851-1* 426
Myrcia fallax IPNI=*599170-1* 427
Myrcia fastigiata IPNI=*598851-1* 426
Myrcia fastigiata var. *coriacea* IPNI=*598851-1* 426
Myrcia ferruginea IPNI=*598743-1* 426
Myrcia ferruginea var. *domingensis* IPNI=*598743-1* 426
Myrcia ferruginea var. *martinicensis* IPNI=*598743-1* 426
Myrcia floribunda IPNI=*599215-1* 427
Myrcia formosiana IPNI=*599170-1* 427
Myrcia friburgensis IPNI=*599170-1* 427
Myrcia frontinensis IPNI=*599026-1* 427
Myrcia frontinensis var. *gamaeana* IPNI=*599026-1* 427
Myrcia gardneriana IPNI=*599170-1* 427
Myrcia gatunensis IPNI=*599170-1* 427
Myrcia glandulosa IPNI=*598851-1* 426
Myrcia glandulosa var. *obovata* IPNI=*598851-1* 426
Myrcia gracilis IPNI=*599170-1* 427
Myrcia gracilis var. *opaca* IPNI=*599170-1* 427
Myrcia gracilis var. *prasina* IPNI=*599170-1* 427
Myrcia gracilis var. *sessiliflora* IPNI=*599170-1* 427
Myrcia guajavifolia f. *grandifolia* IPNI=*599170-1* 427
Myrcia guajavifolia IPNI=*599170-1* 427
Myrcia guajavifolia var. *bullata* IPNI=*599170-1* 427
Myrcia guajavifolia var. *impunctata* IPNI=*599170-1* 427
Myrcia guajavifolia var. *perforata* IPNI=*599170-1* 427
Myrcia guianensis var. *cuneata* IPNI=*598851-1* 426
Myrcia hayneana IPNI=*599170-1* 427
Myrcia hayneana var. *paraensis* IPNI=*599170-1* 427
Myrcia hepatica IPNI=*598851-1* 426
Myrcia hiemalis IPNI=*598851-1* 426
Myrcia hirtellifolia IPNI=*598646-1* 426
Myrcia huanocensis IPNI=*598986-1* 426
Myrcia huanocensis var. *glazioviana* IPNI=*598986-1* 426
Myrcia humboldtiana IPNI=*599170-1* 427
Myrcia humboldtiana var. *caribaea* IPNI=*599170-1* 427
Myrcia humboldtiana var. *orinocensis* IPNI=*599170-1* 427
Myrcia imperatoris-maximiliani IPNI=*599170-1* 427
Myrcia impressa IPNI=*599170-1* 427
Myrcia incisa IPNI=*598851-1* 426
Myrcia jaboticaba IPNI=*203847-2* 428
Myrcia kegeliana IPNI=*599170-1* 427
Myrcia kegeliana var. *angustifolia* IPNI=*599170-1* 427
Myrcia kegeliana var. *latifolia* IPNI=*599170-1* 427
Myrcia kegeliana var. *longifolia* IPNI=*599170-1* 427
Myrcia kegeliana var. *pendula* IPNI=*599170-1* 427
Myrcia kegeliana var. *vulgaris* IPNI=*599170-1* 427
Myrcia klotzschiana IPNI=*599170-1* 427
Myrcia klotzschiana var. *impellucida* IPNI=*599170-1* 427
Myrcia kunthiana IPNI=*599170-1* 427
Myrcia laevigata IPNI=*599170-1* 427
Myrcia laevigata var. *brunnea* IPNI=*599170-1* 427

Myrcia laevigata var. *canescens* IPNI=*599170-1* 427
Myrcia laevis IPNI=*598883-1* 426
Myrcia lamprosericea IPNI=*599170-1* 427
Myrcia lancea IPNI=*599215-1* 427
Myrcia lanceolata IPNI=*598646-1* 426
Myrcia lanceolata var. *angustifolia* IPNI=*598646-1* 426
Myrcia lanceolata var. *avenia* IPNI=*598646-1* 426
Myrcia lanceolata var. *grandifolia* IPNI=*598646-1* 426
Myrcia lanceolata var. *latifolia* IPNI=*598646-1* 426
Myrcia lanceolata var. *racemosa* IPNI=*598646-1* 426
Myrcia langsdorffii IPNI=*599170-1* 427
Myrcia lanuginosa IPNI=*599215-1* 427
Myrcia latifolia IPNI=*599170-1* 427
Myrcia lauriflora IPNI=*598851-1* 426
Myrcia lehmannii IPNI=*598851-1* 426
Myrcia leucadendron IPNI=*598851-1* 426
Myrcia leucantha IPNI=*599215-1* 427
Myrcia lindeniana IPNI=*328930-2* 426
Myrcia lingua IPNI=*598851-1* 426
Myrcia lingua var. *rufa* IPNI=*598851-1* 426
Myrcia linkiana IPNI=*599170-1* 427
Myrcia longicaudata IPNI=*599170-1* 427
Myrcia longipes f. *obovata* IPNI=*599215-1* 427
Myrcia longipes IPNI=*599215-1* 427
Myrcia lucida IPNI=*598883-1* 426
Myrcia lucida var. *attenuata* IPNI=*598883-1* 426
Myrcia luetzelburgii IPNI=*599170-1* 427
Myrcia macrophylla IPNI=*599170-1* 427
Myrcia magnoliifolia IPNI=*599170-1* 427
Myrcia magnoliifolia var. *angustifolia* IPNI=*599170-1* 427
Myrcia magnoliifolia var. *latifolia* IPNI=*599170-1* 427
Myrcia magnoliifolia var. *parvifolia* IPNI=*599170-1* 427
Myrcia mansoni IPNI=*598851-1* 426
Myrcia marahanensis IPNI=*599215-1* 427
Myrcia martiana IPNI=*599170-1* 427
Myrcia martinicensis IPNI=*599170-1* 427
Myrcia melanoclada IPNI=*599170-1* 427
Myrcia melanosticta IPNI=*599170-1* 427
Myrcia membranacea IPNI=*599215-1* 427
Myrcia micrantha IPNI=*599170-1* 427
Myrcia microcarpa IPNI=*598851-1* 426
Myrcia mikaniana IPNI=*599170-1* 427
Myrcia mikaniana var. *angustifolia* IPNI=*599170-1* 427
Myrcia mikaniana var. *latifolia* IPNI=*599170-1* 427
Myrcia mini IPNI=*593768-1* 425
Myrcia myoporina IPNI=*598851-1* 426
Myrcia negrensis IPNI=*599170-1* 427
Myrcia neomacrophylla IPNI=*77191461-1* 426
Myrcia nigrescens IPNI=*77160867-1* 427
Myrcia nigropunctata IPNI=*598851-1* 426
Myrcia nitens IPNI=*599170-1* 427
Myrcia obscura IPNI=*598851-1* 426
Myrcia obscura var. *longipes* IPNI=*598851-1* 426
Myrcia obtecta IPNI=*598851-1* 426
Myrcia obtusa IPNI=*598851-1* 426
Myrcia oerstediana IPNI=*599170-1* 427
Myrcia oocarpa IPNI–*599170-1* 427
Myrcia opaca IPNI=*599170-1* 427
Myrcia opaca var. *latifolia* IPNI=*599170-1* 427
Myrcia oxyoentophylla IPNI=*599170-1* 427
Myrcia paivae var. *gracilis* IPNI=*599026-1* 427
Myrcia pallens IPNI=*598851-1* 426
Myrcia pallens var. *ovalis* IPNI=*598851-1* 426
Myrcia pallens var. *subcordata* IPNI=*598851-1* 426
Myrcia paracatuensis IPNI=*598851-1* 426
Myrcia pellucida IPNI=*599170-1* 427
Myrcia phaeoclada IPNI=*599170-1* 427
Myrcia phaeoclada var. *alagoensis* IPNI=*599170-1* 427
Myrcia phaeoclada var. *guyanensis* IPNI=*599170-1* 427
Myrcia piauhiensis IPNI=*599215-1* 427
Myrcia pilosa IPNI=*599215-1* 427
Myrcia plicatocostata IPNI=*599170-1* 427
Myrcia plumbea IPNI=*598851-1* 426
Myrcia poeppigiana IPNI=*598851-1* 426
Myrcia pohliana IPNI=*599170-1* 427
Myrcia polyantha IPNI=*599170-1* 427
Myrcia prunifolia IPNI=*599215-1* 427
Myrcia prunifolia var. *angustior* IPNI=*599215-1* 427
Myrcia prunifolia var. *obovata* IPNI=*599215-1* 427
Myrcia prunifolia var. *ovata* IPNI=*599215-1* 427
Myrcia pseudomini IPNI=*599170-1* 427
Myrcia puberula IPNI=*599215-1* 427
Myrcia punctata IPNI=*166091-2* 427
Myrcia pusilla IPNI=*598851-1* 426
Myrcia pycnantha IPNI=*77198138-1* 427
Myrcia queimadensis IPNI=*598851-1* 426
Myrcia renatoana IPNI=*598851-1* 426
Myrcia reticulata IPNI=*599170-1* 427
Myrcia rhabdoides IPNI=*598851-1* 426
Myrcia rhodeosepala IPNI=*599215-1* 427
Myrcia riedeliana IPNI=*599170-1* 427
Myrcia riparia IPNI=*599170-1* 427
Myrcia roraimae IPNI=*598851-1* 426
Myrcia roraimensis IPNI=*598851-1* 426
Myrcia rostrata f. *flexuosa* IPNI=*599170-1* 427
Myrcia rostrata IPNI=*599170-1* 427
Myrcia rostrata var. *brunea* IPNI=*599170-1* 427
Myrcia rosulans IPNI=*599215-1* 427
Myrcia rotundifolia IPNI=*598851-1* 426
Myrcia rufidula IPNI=*599170-1* 427
Myrcia rufula IPNI=*599170-1* 427
Myrcia rufula var. *martiana* IPNI=*599170-1* 427
Myrcia sartoriana IPNI=*599170-1* 427
Myrcia saxicola IPNI=*599170-1* 427
Myrcia saxicola var. *grandifolia* IPNI=*599170-1* 427
Myrcia schippii IPNI=*599170-1* 427
Myrcia schomburgkiana IPNI=*593768-1* 425
Myrcia schrankiana IPNI=*598851-1* 426
Myrcia schuechiana IPNI=*599170-1* 427
Myrcia scrobiculata IPNI=*598851-1* 426
Myrcia seleriana IPNI=*166091-2* 427
Myrcia sellowiana IPNI=*599170-1* 427
Myrcia sellowiana var. *bullata* IPNI=*599170-1* 427
Myrcia sellowiana var. *costata* IPNI=*599170-1* 427
Myrcia sepiaria IPNI=*599170-1* 427
Myrcia sericea IPNI=*599170-1* 427
Myrcia sericiflora IPNI=*599170-1* 427
Myrcia sintenisii IPNI=*328930-2* 426
Myrcia sororia IPNI=*599170-1* 427
Myrcia spathulata IPNI=*599215-1* 427
Myrcia spixiana IPNI=*598851-1* 426
Myrcia splendens var. *chrysocoma* IPNI=*599170-1* 427
Myrcia splendens var. *guantanamana* IPNI=*599170-1* 427
Myrcia splendens var. *micropora* IPNI=*599170-1* 427
Myrcia splendens var. *mini* IPNI=*593768-1* 425
Myrcia splendens var. *obscura* IPNI=*599170-1* 427
Myrcia splendens var. *robustior* IPNI=*599170-1* 427
Myrcia spruceana IPNI=*599170-1* 427
Myrcia stemmeriana IPNI=*598851-1* 426
Myrcia suaveolens IPNI=*598851-1* 426
Myrcia subsessilis var. *ovalis* IPNI=*599191-1* 427
Myrcia subsessilis var. *subcordata* IPNI=*599191-1* 427
Myrcia sulcata IPNI=*598743-1* 426
Myrcia superba IPNI=*599170-1* 427
Myrcia surinamensis IPNI=*598851-1* 426
Myrcia taubatensis var. *ovalis* IPNI=*598851-1* 426

Myrcia tenuiflora IPNI=*77199448-1* 426
Myrcia terebinthacea IPNI=*598851-1* 426
Myrcia tingens IPNI=*599170-1* 427
Myrcia umbellulifera IPNI=*596253-1* 425
Myrcia vacciniifolia IPNI=*598851-1* 426
Myrcia velhensis IPNI=*598851-1* 426
Myrcia velutina IPNI=*599170-1* 427
Myrcia velutina var. *canescens* IPNI=*599170-1* 427
Myrcia velutina var. *ochracea* IPNI=*599170-1* 427
Myrcia venezuelensis IPNI=*599170-1* 427
Myrcia verruculata IPNI=*599215-1* 427
Myrcia ypanemensis IPNI=*599170-1* 427
Myrcia yungasensis IPNI=*598851-1* 426
Myrcianthes compressa IPNI=*166091-2* 427
Myrcianthes dicrana IPNI=*166091-2* 427
Myrcianthes dugandii IPNI=*166091-2* 427
Myrcianthes elegans IPNI=*598851-1* 426
Myrcianthes foliosa IPNI=*166103-2* 427
Myrcianthes fragrans subsp. *simpsonii* IPNI=*166091-2* 427
Myrcianthes fragrans var. *hispidula* IPNI=*166091-2* 427
Myrcianthes fragrans var. *simpsonii* IPNI=*166091-2* 427
Myrcianthes gracilipes IPNI=*166084-2* 427
Myrcianthes krugii IPNI=*596253-1* 425
Myrcianthes limbata IPNI=*166091-2* 427
Myrcianthes montana IPNI=*77160848-1* 426
Myrcianthes simpsonii IPNI=*166091-2* 427
Myrcianthes spixiana IPNI=*598851-1* 426
Myrcianthes terminalis IPNI=*598851-1* 426
Myrcianthes ternifolia IPNI=*974565-1* 427
Myrcianthes umbellulifera IPNI=*596253-1* 425
Myrciaria amazonica IPNI=*599327-1* 427
Myrciaria arborea IPNI=*599327-1* 427
Myrciaria arborea var. *rostrata* IPNI=*599327-1* 427
Myrciaria axillaris IPNI=*599327-1* 427
Myrciaria cauliflora IPNI=*203847-2* 428
Myrciaria caurensis IPNI=*166146-2* 427
Myrciaria ciliolata IPNI=*599327-1* 427
Myrciaria ciliolata var. *warmingiana* IPNI=*599327-1* 427
Myrciaria divaricata IPNI=*166146-2* 427
Myrciaria guapurium IPNI=*203847-2* 428
Myrciaria jaboticaba IPNI=*203847-2* 428
Myrciaria lanceolata IPNI=*166146-2* 427
Myrciaria lanceolata var. *angustifolia* IPNI=*166146-2* 427
Myrciaria lanceolata var. *glomerata* IPNI=*166146-2* 427
Myrciaria lanceolata var. *laxa* IPNI=*166146-2* 427
Myrciaria leucadendron IPNI=*599327-1* 427
Myrciaria leucophloea IPNI=*599327-1* 427
Myrciaria leucophloea var. *conferta* IPNI=*599327-1* 427
Myrciaria leucophloea var. *laxa* IPNI=*599327-1* 427
Myrciaria leucophloea var. *warmingiana* IPNI=*599327-1* 427
Myrciaria longipes IPNI=*599327-1* 427
Myrciaria longipes var. *opaca* IPNI=*599327-1* 427
Myrciaria longipes var. *pellucida* IPNI=*599327-1* 427
Myrciaria maragnanensis IPNI=*599327-1* 427
Myrciaria maranhensis IPNI=*599327-1* 427
Myrciaria maximiliana IPNI=*599327-1* 427
Myrciaria mexicana IPNI=*599327-1* 427
Myrciaria nitida IPNI=*598883-1* 426
Myrciaria nitida var. *chartacea* IPNI=*598883-1* 426
Myrciaria nitida var. *coriacea* IPNI=*598883-1* 426
Myrciaria nitida var. *dives* IPNI=*598883-1* 426
Myrciaria obscura IPNI=*166146-2* 427
Myrciaria oneillii IPNI=*599327-1* 427
Myrciaria paraensis IPNI=*166146-2* 427
Myrciaria peruviana IPNI=*203847-2* 428
Myrciaria phillyraeoides IPNI=*166146-2* 427
Myrciaria polyantha IPNI=*598883-1* 426
Myrciaria prasina IPNI=*599327-1* 427
Myrciaria protracta IPNI=*599327-1* 427
Myrciaria riedeliana IPNI=*166146-2* 427
Myrciaria salzmannii IPNI=*599327-1* 427
Myrciaria schuechiana IPNI=*599327-1* 427
Myrciaria schuechiana var. *deflexa* IPNI=*599327-1* 427
Myrciaria schuechiana var. *latifolia* IPNI=*599327-1* 427
Myrciaria sellowiana IPNI=*599327-1* 427
Myrciaria splendens IPNI=*599327-1* 427
Myrciaria spruceana IPNI=*166146-2* 427
Myrciaria tenuiramis IPNI=*599327-1* 427
Myrciaria tolypantha IPNI=*599327-1* 427
Myrciaria tolypantha var. *angustifolia* IPNI=*599327-1* 427
Myrciaria tolypantha var. *latifolia* IPNI=*599327-1* 427
Myrciaria trinitatis IPNI=*203871-2* 428
Myrciaria uliginosa IPNI=*599327-1* 427
Myrciaria verticillata IPNI=*599327-1* 427
Myriaspora decipiens IPNI=*77112284-1* 418
Myriaspora egensis IPNI=*77112284-1* 418
Myriaspora integrifolia IPNI=*77112284-1* 418
Myriaspora paulensis IPNI=*77112284-1* 418
Myriaspora surinamensis IPNI=*77112284-1* 418
Myrica altera IPNI=*166220-2* 339
Myrica apiculata IPNI=*166220-2* 339
Myrica arguta IPNI=*326827-2* 339
Myrica arguta var. *macrocarpa* IPNI=*326827-2* 339
Myrica arguta var. *peruviana* IPNI=*326827-2* 339
Myrica arguta var. *tinctoria* IPNI=*326827-2* 339
Myrica caracasana IPNI=*326827-2* 339
Myrica cerifera var. *arborescens* IPNI=*166220-2* 339
Myrica cerifera var. *dubia* IPNI=*166220-2* 339
Myrica cerifera var. *pumila* IPNI=*166220-2* 339
Myrica costata IPNI=*326827-2* 339
Myrica domingana IPNI=*166220-2* 339
Myrica jamaicensis IPNI=*166220-2* 339
Myrica macrocarpa IPNI=*326827-2* 339
Myrica mexicana IPNI=*166220-2* 339
Myrica mexicana var. *fastuosa* IPNI=*166220-2* 339
Myrica mexicana var. *subglabra* IPNI=*166220-2* 339
Myrica microcarpa IPNI=*166220-2* 339
Myrica microcarpa var. *angustifolia* IPNI=*166220-2* 339
Myrica microstachya IPNI=*166220-2* 339
Myrica parvifolia var. *confusa* IPNI=*326824-2* 339
Myrica parvifolia var. *longipedunculata* IPNI=*326824-2* 339
Myrica parvifolia var. *lucens* IPNI=*326824-2* 339
Myrica parvifolia var. *major* IPNI=*326824-2* 339
Myrica parvifolia var. *obtusa* IPNI=*326824-2* 339
Myrica polycarpa IPNI=*166220-2* 339
Myrica polycarpa IPNI=*326827-2* 339
Myrica pubescens var. *caracasana* IPNI=*326827-2* 339
Myrica pubescens var. *tomentosa* IPNI=*326827-2* 339
Myrica pumila IPNI=*166220-2* 339
Myrica pusilla IPNI=*166220-2* 339
Myrica reticulata IPNI=*166220-2* 339
Myrica verrucosa IPNI=*166220-2* 339
Myrica xalapensis IPNI=*166220-2* 339
Myriocarpa densiflora IPNI=*854765-1* 446
Myriocarpa densiflora var. *dombeyana* IPNI=*854765-1* 446
Myriocarpa dombeyana IPNI=*854765-1* 446
Myriocarpa humilis IPNI=*854765-1* 446
Myriocarpa inaequilatera IPNI=*854756-1* 446
Myriocarpa longipes var. *yzabalensis* IPNI=*854756-1* 446
Myriocarpa magnifica IPNI=*854756-1* 446
Myriocarpa polystachya IPNI=*854765-1* 446
Myriocarpa stipitata var. *ambigua* IPNI=*854765-1* 446
Myriocarpa yzabalensis IPNI=*854756-1* 446
Myriopteris elegans IPNI=*77183191-1* 178
Myriopteris intermedia IPNI=*77183191-1* 178
Myriopteris microphylla IPNI=*77183177-1* 178

Myriopteris myriophylla IPNI=77183191-1 178
Myriopteris paleacea IPNI=77183191-1 178
Myristica amboinensis IPNI=586076-1 191
Myristica americana IPNI=267202-2 192
Myristica aromatica IPNI=586076-1 191
Myristica calophylla IPNI=267140-2 191
Myristica capitellata IPNI=64004-2 190
Myristica carinata IPNI=267142-2 191
Myristica cordifolia IPNI=267199-2 192
Myristica cumaru IPNI=177341-2 191
Myristica elongata IPNI=267153-2 192
Myristica fatua IPNI=267202-2 192
Myristica gracilis IPNI=267202-2 192
Myristica hostmannii IPNI=585810-1 191
Myristica laurella IPNI=586076-1 191
Myristica laurifolia IPNI=64010-2 190
Myristica macrophylla IPNI=585818-1 191
Myristica mocoa IPNI=267199-2 192
Myristica moschata IPNI=586076-1 191
Myristica officinalis IPNI=586076-1 191
Myristica otoba IPNI=177341-2 191
Myristica otoba var. glaucescens IPNI=586389-1 191
Myristica panamensis IPNI=267199-2 192
Myristica pavonis IPNI=267190-2 192
Myristica peruviana IPNI=267191-2 192
Myristica philippinensis IPNI=586076-1 191
Myristica platysperma IPNI=586380-1 191
Myristica rufula IPNI=267203-2 192
Myristica sebifera IPNI=267199-2 192
Myristica sebifera var. cordifolia IPNI=267199-2 192
Myristica sebifera var. curvinervia IPNI=267199-2 192
Myristica sprucei IPNI=64010-2 190
Myristica surinamensis IPNI=267202-2 192
Myristica theiodora IPNI=267203-2 192
Myristica virola IPNI=267199-2 192
Myrmecodendron collinsii IPNI=60441206-2 336
Myrmecodendron costaricense IPNI=60441206-2 336
Myrmidone peruviana IPNI=77164249-1 421
Myrobalanus buceras IPNI=171012-1 416
Myrobalanus catappa IPNI=171034-1 416
Myrobalanus commersonii IPNI=171034-1 416
Myrobalanus excelsa IPNI=250537-2 416
Myrobalanus lucida IPNI=171207-1 416
Myrobalanus oblonga IPNI=171280-1 416
Myrobalanus obovata IPNI=250537-2 416
Myrobalanus rubrigemmis IPNI=171034-1 416
Myrobroma fragrans IPNI=262578-2 223
Myrodendrum amplexicaule IPNI=431431-1 393
Myrodendrum balsamiferum IPNI=431431-1 393
Myrodendrum subvaginale IPNI=431431-1 393
Myrosma boliviana IPNI=797633-1 250
Myrosma boliviana var. acreana IPNI=797633-1 250
Myrosma canniformis IPNI=797633-1 250
Myrosma cuyabensis IPNI=797633-1 250
Myrosma lutea IPNI=246147-2 250
Myrospermum arboreum IPNI=509454-1 328
Myrospermum balsamiferum IPNI=166483-2 328
Myrospermum emarginatum IPNI=509434-1 328
Myrospermum erythroxylum IPNI=509454-1 328
Myrospermum frutescens var. emarginatum IPNI=509434-1 328
Myrospermum pedicellatum IPNI=509454-1 328
Myrospermum pereirae IPNI=166483-2 328
Myrospermum peruiferum IPNI=509454-1 328
Myrospermum pubescens IPNI=509454-1 328
Myrospermum punctatum IPNI=166483-2 328
Myrospermum robiniifolium IPNI=509454-1 328
Myrospermum secundum IPNI=509434-1 328
Myrospermum toluiferum IPNI=166483-2 328
Myroxylon balsamum var. genuinum IPNI=166483-2 328
Myroxylon balsamum var. pereirae IPNI=166483-2 328
Myroxylon balsamum var. punctatum IPNI=166483-2 328
Myroxylon calophyllum IPNI=112146-1 405
Myroxylon celastrinum IPNI=112138-1 405
Myroxylon dicline IPNI=365348-1 404
Myroxylon flexuosum IPNI=112138-1 405
Myroxylon frutescens IPNI=509434-1 328
Myroxylon hanburyanum IPNI=166483-2 328
Myroxylon intermedium IPNI=112146-1 405
Myroxylon paliurus IPNI=112138-1 405
Myroxylon pereirae IPNI=166483-2 328
Myroxylon pringlei IPNI=112138-1 405
Myroxylon prunifolium IPNI=112146-1 405
Myroxylon pubescens IPNI=509454-1 328
Myroxylon punctatum IPNI=166483-2 328
Myroxylon robiniifolium IPNI=509454-1 328
Myroxylon seemannii IPNI=112138-1 405
Myrrhina inodora IPNI=372219-1 358
Myrrhis andicola IPNI=77085100-1 252
Myrsine ardisioides IPNI=247441-2 304
Myrsine baccata IPNI=166565-2 304
Myrsine berteroi IPNI=1028287-2 304
Myrsine brasiliensis IPNI=166565-2 304
Myrsine caballeria IPNI=1033305-2 304
Myrsine capororoca IPNI=1028287-2 304
Myrsine ciliata IPNI=1119456-2 304
Myrsine ferruginea f. glabra IPNI=1028287-2 304
Myrsine ferruginea IPNI=1028287-2 304
Myrsine flocculosa IPNI=1028287-2 304
Myrsine floribunda IPNI=166565-2 304
Myrsine floridana IPNI=589091-1 304
Myrsine guatemalensis IPNI=1028287-2 304
Myrsine icacorea IPNI=18168-2 303
Myrsine jelskii IPNI=1028287-2 304
Myrsine laeta IPNI=1028287-2 304
Myrsine maritima IPNI=166565-2 304
Myrsine microcalyx IPNI=1028287-2 304
Myrsine myricoides IPNI=1028287-2 304
Myrsine myrtifolia IPNI=1119456-2 304
Myrsine myrtoides IPNI=1119456-2 304
Myrsine paulensis IPNI=1028287-2 304
Myrsine peruviana IPNI=1119456-2 304
Myrsine pittieri IPNI=1119456-2 304
Myrsine popayanensis IPNI=1028287-2 304
Myrsine punctata IPNI=589091-1 304
Myrsine rapanea IPNI=166565-2 304
Myrsine rapanea var. major IPNI=166565-2 304
Myrsine rapanea var. minor IPNI=166565-2 304
Myrsine rapanea var. subacuta IPNI=166565-2 304
Myrsine rufa IPNI=1028287-2 304
Myrsine rufescens IPNI=1028287-2 304
Myrsine rufescens var. angustifolia IPNI=1028287-2 304
Myrsine salicifolia IPNI=1028287-2 304
Myrsine saligna IPNI=1028287-2 304
Myrsine schomburgkiana IPNI=89010-2 305
Myrsine spicata IPNI=73568-2 304
Myrsine tetrandra IPNI=1119456-2 304
Myrsine tomentosa IPNI=1028287-2 304
Myrsine umbrosa IPNI=166565-2 304
Myrsine verae-crucis IPNI=589091-1 304
Myrsine vestita IPNI=1028287-2 304
Myrsine viridis IPNI=1028287-2 304
Myrstiphyllum brachiatum IPNI=77115595-1 353
Myrstiphyllum horizontalis IPNI=762253-1 355
Myrstiphyllum marginatum IPNI=213065-2 355
Myrstiphyllum undatum IPNI=329821-2 355
Myrteola barneoudii IPNI=599434-1 427

Myrteola barneoudii var. *humifusa* IPNI=*599434-1* 427
Myrteola bullata IPNI=*599434-1* 427
Myrteola bullata var. *pentamera* IPNI=*599434-1* 427
Myrteola bullata var. *tetramera* IPNI=*599434-1* 427
Myrteola humifusa IPNI=*599434-1* 427
Myrteola leucomyrtillus IPNI=*599434-1* 427
Myrteola myrsinoides IPNI=*974565-1* 427
Myrteola nannophylla IPNI=*599434-1* 427
Myrteola nummularia var. *barneoudii* IPNI=*599434-1* 427
Myrteola nummularia var. *repens* IPNI=*599434-1* 427
Myrteola oxycoccoides IPNI=*599434-1* 427
Myrteola repens IPNI=*599434-1* 427
Myrteola vaccinioides IPNI=*599434-1* 427
Myrteola vaccinioides var. *carabaya* IPNI=*599434-1* 427
Myrtoleucodendron armillare IPNI=*597845-1* 426
Myrtoleucodendron decussatum IPNI=*597912-1* 426
Myrtus acuminata IPNI=*599170-1* 427
Myrtus ambrosiaca IPNI=*43033-2* 424
Myrtus arbutifolia IPNI=*1046564-2* 425
Myrtus aromatica IPNI=*196799-2* 428
Myrtus aubletii IPNI=*599215-1* 427
Myrtus axillaris IPNI=*593690-1* 425
Myrtus balsamica IPNI=*166091-2* 427
Myrtus barneoudii IPNI=*599434-1* 427
Myrtus baruensis IPNI=*593690-1* 425
Myrtus bergiana IPNI=*600884-1* 428
Myrtus berteroana IPNI=*593768-1* 425
Myrtus bicolor IPNI=*166084-2* 427
Myrtus biflora f. *subsericea* IPNI=*593768-1* 425
Myrtus biflora IPNI=*593768-1* 425
Myrtus biflora var. *salicifolia* IPNI=*593768-1* 425
Myrtus biflora var. *yapacani* IPNI=*593768-1* 425
Myrtus brachystemon IPNI=*595618-1* 425
Myrtus bracteolaris IPNI=*599170-1* 427
Myrtus brasiliana IPNI=*596263-1* 425
Myrtus brasiliana var. *diversifolia* IPNI=*596263-1* 425
Myrtus brasiliana var. *lanceolata* IPNI=*596263-1* 425
Myrtus brasiliana var. *lucida* IPNI=*596263-1* 425
Myrtus brasiliana var. *normalis* IPNI=*596263-1* 425
Myrtus brasiliana var. *parkeriana* IPNI=*595618-1* 425
Myrtus calophylla IPNI=*1061290-2* 424
Myrtus cauliflora IPNI=*203847-2* 428
Myrtus chytraculia IPNI=*77191371-1* 426
Myrtus citrifolia IPNI=*1061290-2* 424
Myrtus claraensis IPNI=*600952-1* 428
Myrtus clusiifolia IPNI=*1005985-2* 426
Myrtus coccolobifolia IPNI=*166115-2* 427
Myrtus complicata IPNI=*599170-1* 427
Myrtus compressa IPNI=*166091-2* 427
Myrtus corticosa IPNI=*601603-1* 428
Myrtus cotinifolia IPNI=*594126-1* 425
Myrtus cumini IPNI=*601603-1* 428
Myrtus daphnoides IPNI=*595618-1* 425
Myrtus deflexa IPNI=*599170-1* 427
Myrtus dichotoma IPNI=*166091-2* 427
Myrtus dioica IPNI=*196799-2* 428
Myrtus discolor IPNI=*166084-2* 427
Myrtus divaricata IPNI=*593690-1* 425
Myrtus elegans IPNI=*598851-1* 426
Myrtus erythroxyloides IPNI=*596253-1* 425
Myrtus floribunda IPNI=*599327-1* 427
Myrtus foliosa IPNI=*166103-2* 427
Myrtus formosa IPNI=*600884-1* 428
Myrtus fragrans IPNI=*166091-2* 427
Myrtus friedrichsthalii IPNI=*602600-1* 428
Myrtus friedrichsthalii var. *brevipes* IPNI=*602600-1* 428
Myrtus goetheana IPNI=*591482-1* 424
Myrtus guajava IPNI=*600841-1* 428
Myrtus guajava var. *pyrifera* IPNI=*600841-1* 428
Myrtus guayabillo IPNI=*166091-2* 427
Myrtus guayabo IPNI=*166091-2* 427
Myrtus guianensis IPNI=*598851-1* 426
Myrtus guineensis IPNI=*211886-2* 428
Myrtus hauthalii IPNI=*211886-2* 428
Myrtus heteroclita IPNI=*166091-2* 427
Myrtus jaboticaba IPNI=*203847-2* 428
Myrtus jambos IPNI=*601802-1* 428
Myrtus lancea IPNI=*593768-1* 425
Myrtus leucomyrtillus IPNI=*599434-1* 427
Myrtus leucoxyla IPNI=*166103-2* 427
Myrtus limbata IPNI=*166091-2* 427
Myrtus lucida IPNI=*599199-1* 427
Myrtus macrophylla IPNI=*601907-1* 428
Myrtus malaccensis IPNI=*601907-1* 428
Myrtus maritima IPNI=*593490-1* 425
Myrtus matudae IPNI=*602600-1* 428
Myrtus micrantha IPNI=*595185-1* 425
Myrtus micrantha IPNI=*599327-1* 427
Myrtus mini IPNI=*593768-1* 425
Myrtus mollis IPNI=*598986-1* 426
Myrtus montana IPNI=*602600-1* 428
Myrtus monticola IPNI=*595185-1* 425
Myrtus myricoides IPNI=*602600-1* 428
Myrtus myricoides var. *roraimensis* IPNI=*602600-1* 428
Myrtus myricoides var. *stenophylla* IPNI=*602600-1* 428
Myrtus myricoides var. *turumiquirensis* IPNI=*602600-1* 428
Myrtus myrsinoides IPNI=*974565-1* 427
Myrtus nummularia IPNI=*599434-1* 427
Myrtus nummularia var. *major* IPNI=*599434-1* 427
Myrtus obovata IPNI=*601603-1* 428
Myrtus oerstedii IPNI=*602600-1* 428
Myrtus oleifolia IPNI=*1046564-2* 425
Myrtus oxycoccoides IPNI=*599434-1* 427
Myrtus pallens IPNI=*593768-1* 425
Myrtus patrisii IPNI=*595457-1* 425
Myrtus phillyraeoides IPNI=*166146-2* 427
Myrtus pimenta IPNI=*196799-2* 428
Myrtus pimenta var. *longifolia* IPNI=*196799-2* 428
Myrtus piperita IPNI=*196799-2* 428
Myrtus polyantha IPNI=*599170-1* 427
Myrtus procera IPNI=*595618-1* 425
Myrtus punctata IPNI=*166091-2* 427
Myrtus punicifolia IPNI=*1046564-2* 425
Myrtus pyrifolia IPNI=*598851-1* 426
Myrtus repens IPNI=*599434-1* 427
Myrtus rhopaloides IPNI=*166115-2* 427
Myrtus roraimensis IPNI=*602600-1* 428
Myrtus salutaris IPNI=*1115374-2* 428
Myrtus splendens IPNI=*599170-1* 427
Myrtus stoupii IPNI=*599170-1* 427
Myrtus sylvatica IPNI=*599199-1* 427
Myrtus umbellulifera IPNI=*596253-1* 425
Myrtus umbraticola IPNI=*77160867-1* 427
Myrtus vaccinioides IPNI=*599434-1* 427
Myrtus verticillata IPNI=*599327-1* 427
Myrtus virgultosa IPNI=*593768-1* 425
Myrtus willdenowii IPNI=*596263-1* 425

N
Nacibea reclinata IPNI=*755756-1* 352
Nacibea uniflora IPNI=*755756-1* 352
Nageia angustifolia IPNI=*204598-2* 180
Nageia macrostachya IPNI=*263553-1* 180
Nageia oleifolia IPNI=*263553-1* 180
Nageia rospigliosii IPNI=*946471-1* 180
Naletonia violacea IPNI=*759372-1* 354

Napaea crispa IPNI=*120630-2* 409
Napaea incurva IPNI=*238-2* 406
Napaea latifolia IPNI=*561424-1* 410
Napaea rhombifolia IPNI=*235798-2* 413
Napaea scabra IPNI=*561430-1* 410
Napeanthus bullatus IPNI=*167091-2* 367
Napellus vulgaris IPNI=*707615-1* 436
Napimoga guianensis IPNI=*365491-1* 404
Napus campestris IPNI=*279485-1* 276
Napus leucosinapsis IPNI=*288952-1* 278
Napus oleifera IPNI=*279419-1* 276
Napus oleracea IPNI=*279435-1* 276
Napus rapa IPNI=*279485-1* 276
Nardus dactyloides IPNI=*398888-1* 237
Nardus indica IPNI=*408465-1* 241
Narega coduva IPNI=*745904-1* 348
Narukila cordata IPNI=*283610-2* 225
Narukila cordata var. *lancifolia* IPNI=*283610-2* 225
Nassella ligularis IPNI=*948410-1* 241
Nasturtiastrum virginicum IPNI=*138164-2* 277
Nasturtiolum castratum IPNI=*286148-1* 277
Nasturtiolum pinnatum IPNI=*286148-1* 277
Nasturtium alexandrinum IPNI=*287534-1* 277
Nasturtium americanum IPNI=*286148-1* 277
Nasturtium apetalum IPNI=*222144-2* 277
Nasturtium aquaticum IPNI=*287534-1* 277
Nasturtium arvense IPNI=*288669-1* 277
Nasturtium atrovirens IPNI=*222144-2* 277
Nasturtium bipinnatifidum IPNI=*137955-2* 277
Nasturtium bursa-pastoris IPNI=*30092589-2* 276
Nasturtium carrerasii IPNI=*286421-1* 277
Nasturtium cernuum IPNI=*288669-1* 277
Nasturtium clavatum IPNI=*288669-1* 277
Nasturtium densiflorum IPNI=*288669-1* 277
Nasturtium diandrum IPNI=*138164-2* 277
Nasturtium dictyotum IPNI=*288669-1* 277
Nasturtium diffusum IPNI=*222144-2* 277
Nasturtium divaricatum IPNI=*222144-2* 277
Nasturtium diversifolium IPNI=*288669-1* 277
Nasturtium dubium IPNI=*137955-2* 277
Nasturtium eggersii IPNI=*288736-1* 277
Nasturtium flexuosum IPNI=*287534-1* 277
Nasturtium fontanum IPNI=*287534-1* 277
Nasturtium fontanum var. *longisiliquum* IPNI=*287534-1* 277
Nasturtium gangeticum IPNI=*222144-2* 277
Nasturtium heterophyllum IPNI=*222144-2* 277
Nasturtium heterophyllum IPNI=*288735-1* 277
Nasturtium homalospermum IPNI=*288669-1* 277
Nasturtium iberis IPNI=*138164-2* 277
Nasturtium indicum IPNI=*222144-2* 277
Nasturtium indicum IPNI=*310974-2* 280
Nasturtium insipidum IPNI=*287534-1* 277
Nasturtium japonicum IPNI=*222144-2* 277
Nasturtium majus IPNI=*138164-2* 277
Nasturtium mexicanum IPNI=*222160-2* 277
Nasturtium montanum IPNI=*222144-2* 277
Nasturtium nanum IPNI=*222164-2* 277
Nasturtium nasturtium IPNI=*289589-1* 278
Nasturtium nasturtium-aquaticum IPNI=*287534-1* 277
Nasturtium obtusulum IPNI=*222144-2* 277
Nasturtium officinale subsp. *rotundifolium* IPNI=*287534-1* 277
Nasturtium palustre IPNI=*288669-1* 277
Nasturtium palustre var. *dasycarpum* IPNI=*288669-1* 277
Nasturtium parvifolium IPNI=*287534-1* 277
Nasturtium plebeium f. *latifolium* IPNI=*222160-2* 277
Nasturtium plebeium IPNI=*222160-2* 277
Nasturtium radicans IPNI=*280189-1* 276
Nasturtium semipinnatifidum IPNI=*288669-1* 277
Nasturtium siifolium IPNI=*287534-1* 277
Nasturtium sinapis IPNI=*222144-2* 277
Nasturtium spinosum IPNI=*286421-1* 277
Nasturtium sublyratum IPNI=*288735-1* 277
Nasturtium terrestre var. *occidentale* IPNI=*288669-1* 277
Nasturtium terrestre var. *semipinnatifidum* IPNI=*288669-1* 277
Nasturtium turfosum IPNI=*280189-1* 276
Nasturtium turfosum var. *radicans* IPNI=*280189-1* 276
Nasturtium virginicum IPNI=*138164-2* 277
Nasturtium williamsii IPNI=*288669-1* 277
Nastus baccifera IPNI=*408271-1* 241
Nastus carolinianus IPNI=*395325-1* 236
Nastus chusque IPNI=*1135523-2* 236
Nastus guadua IPNI=*1074014-2* 239
Nastus thouarsii IPNI=*392574-1* 235
Nastus verticillatus IPNI=*404020-1* 239
Nastus viviparus IPNI=*392574-1* 235
Nauchea bracteata IPNI=*486606-1* 315
Nauchea brasiliana IPNI=*485555-1* 314
Nauchea plumieri IPNI=*485609-1* 314
Nauchea pudica IPNI=*485609-1* 314
Nauchea ternatea IPNI=*486606-1* 315
Nauchea virginiana IPNI=*51897-2* 314
Nauclea aculeata IPNI=*768250-1* 358
Nauclea aculeata IPNI=*768322-1* 358
Nauclea guianensis IPNI=*768250-1* 358
Nauclea polycephala IPNI=*768322-1* 358
Nauclea surinamensis IPNI=*768322-1* 358
Naucleopsis mello-barretoi IPNI=*1182949-2* 440
Naucleopsis pauciflora IPNI=*584418-1* 440
Naucorephes pubescens IPNI=*693617-1* 288
Naucorephes punctata IPNI=*693617-1* 288
Nautilocalyx minutiflorus IPNI=*284180-2* 368
Nazia aliena IPNI=*211125-2* 244
Nazia occidentalis IPNI=*256481-2* 246
Nazia racemosa var. *aliena* IPNI=*211125-2* 244
Neanotis monosperma var. *tirunelvelica* IPNI=*766800-1* 357
Neanthe bella IPNI=*54865-2* 207
Neanthe elegans IPNI=*54865-2* 207
Neanthe neesiana IPNI=*54865-2* 207
Nebra rudifolia IPNI=*604762-1* 287
Nectandra acutangula IPNI=*171503-2* 184
Nectandra acutissima IPNI=*466894-1* 184
Nectandra albiflora IPNI=*974281-1* 184
Nectandra amazonum var. *oerstedii* IPNI=*974282-1* 184
Nectandra ambigua IPNI=*466626-1* 183
Nectandra antillana IPNI=*974281-1* 184
Nectandra aurantiodora IPNI=*171526-2* 184
Nectandra barcellensis IPNI=*77198011-1* 183
Nectandra berchemiifolia IPNI=*167746-2* 183
Nectandra berchemiifolia var. *caucana* IPNI=*974282-1* 184
Nectandra berchemiifolia var. *pavonii* IPNI=*167746-2* 183
Nectandra bombycina IPNI=*466626-1* 183
Nectandra bondarii IPNI=*466773-1* 184
Nectandra bredemeyeriana IPNI=*974281-1* 184
Nectandra canescens IPNI=*466626-1* 183
Nectandra caparrapii IPNI=*77198011-1* 183
Nectandra caucana IPNI=*974282-1* 184
Nectandra cinnamomoides IPNI=*77198017-1* 183
Nectandra comasensis IPNI=*1203274-2* 184
Nectandra concinna IPNI=*466885-1* 184
Nectandra coto IPNI=*13822-2* 182
Nectandra cuspidata var. *dubia* IPNI=*281229-2* 184
Nectandra cuspidata var. *macrocarpa* IPNI=*466773-1* 184
Nectandra cuspidata var. *strumosa* IPNI=*466773-1* 184
Nectandra cymbarum IPNI=*77198011-1* 183
Nectandra discolor var. *subvenosa* IPNI=*466894-1* 184
Nectandra effusa IPNI=*466885-1* 184

Nectandra elaiophora IPNI=*77198011-1* 183
Nectandra floribunda IPNI=*1146025-2* 184
Nectandra fuscobarbata IPNI=*974282-1* 184
Nectandra gentlei IPNI=*281229-2* 184
Nectandra glabrescens IPNI=*974281-1* 184
Nectandra glabrescens var. *fuscobarbata* IPNI=*974282-1* 184
Nectandra glandulifolia IPNI=*77158859-1* 183
Nectandra globosa var. *barbeyana* IPNI=*167822-2* 184
Nectandra grandis IPNI=*284250-2* 185
Nectandra grandis IPNI=*466894-1* 184
Nectandra guanaiensis IPNI=*974281-1* 184
Nectandra guaporensis IPNI=*466626-1* 183
Nectandra laevis IPNI=*466773-1* 184
Nectandra larensis IPNI=*466885-1* 184
Nectandra latifolia IPNI=*77158867-1* 183
Nectandra laurel var. *glabrescens* IPNI=*281234-2* 184
Nectandra laurel var. *triquetra* IPNI=*466894-1* 184
Nectandra leucantha IPNI=*974281-1* 184
Nectandra leucantha var. *attenuata* IPNI=*974281-1* 184
Nectandra leucantha var. *guianensis* IPNI=*974281-1* 184
Nectandra leucantha var. *peruviana* IPNI=*974281-1* 184
Nectandra leucothyrsus IPNI=*466773-1* 184
Nectandra leucoxyla IPNI=*466894-1* 184
Nectandra limbata IPNI=*465312-1* 183
Nectandra lucida IPNI=*90974-2* 183
Nectandra macrophylla IPNI=*326508-2* 184
Nectandra magnoliifolia IPNI=*974281-1* 184
Nectandra maranonensis IPNI=*974281-1* 184
Nectandra martinicensis IPNI=*77158859-1* 183
Nectandra membranacea subsp. *cuspidata* IPNI=*281229-2* 184
Nectandra mollis IPNI=*466894-1* 184
Nectandra mollis subsp. *laurel* IPNI=*281234-2* 184
Nectandra mollis var. *attenuata* IPNI=*466894-1* 184
Nectandra mollis var. *intermedia* IPNI=*466894-1* 184
Nectandra mollis var. *villosa* IPNI=*466894-1* 184
Nectandra mopanensis IPNI=*466885-1* 184
Nectandra nervosa IPNI=*466885-1* 184
Nectandra nigra IPNI=*466894-1* 184
Nectandra oleifera IPNI=*77198011-1* 183
Nectandra oligoneura IPNI=*466885-1* 184
Nectandra olivacea IPNI=*281229-2* 184
Nectandra pallida IPNI=*167822-2* 184
Nectandra pallida IPNI=*466626-1* 183
Nectandra panamensis IPNI=*466885-1* 184
Nectandra pendula IPNI=*171727-2* 184
Nectandra petenensis IPNI=*974282-1* 184
Nectandra pichurim var. *cuprea* IPNI=*281229-2* 184
Nectandra pisi IPNI=*167822-2* 184
Nectandra pittieri IPNI=*466894-1* 184
Nectandra polita IPNI=*77158867-1* 183
Nectandra polita var. *oerstedii* IPNI=*77158867-1* 183
Nectandra puchury-major IPNI=*13872-2* 182
Nectandra puchury-minor IPNI=*13872-2* 182
Nectandra purpurea IPNI=*77158867-1* 183
Nectandra raimondii IPNI=*466773-1* 184
Nectandra rectinervia IPNI=*466885-1* 184
Nectandra regnellii IPNI=*171499-2* 184
Nectandra reticulata IPNI=*466894-1* 184
Nectandra rhynchophylla IPNI=*171499-2* 184
Nectandra rubra IPNI=*999503-1* 185
Nectandra sawadae IPNI=*466885-1* 184
Nectandra schomburgkii IPNI=*974281-1* 184
Nectandra sintenisii IPNI=*466885-1* 184
Nectandra skutchii IPNI=*466773-1* 184
Nectandra standleyi IPNI=*466773-1* 184
Nectandra tabascensis IPNI=*77158859-1* 183
Nectandra tessmannii IPNI=*974281-1* 184
Nectandra tovarensis IPNI=*281234-2* 184
Nectandra trinitatis IPNI=*466773-1* 184
Nectandra turbacensis var. *mexicana* IPNI=*466885-1* 184
Nectandra turbacensis var. *schottii* IPNI=*466885-1* 184
Nectandra urophylla IPNI=*466626-1* 183
Nectandra vaga IPNI=*167822-2* 184
Nectandra vaga var. *major* IPNI=*167822-2* 184
Nectandra vaga var. *sprucei* IPNI=*167822-2* 184
Nectandra villosa var. *venosa* IPNI=*466894-1* 184
Nectandra williamsii IPNI=*466773-1* 184
Nectandra woodsoniana IPNI=*77158859-1* 183
Nectris caroliniana IPNI=*273581-2* 181
Neea belizensis IPNI=*168038-2* 287
Neea cauliflora IPNI=*604762-1* 287
Neea choriophylla IPNI=*168038-2* 287
Neea choriophylla var. *hirtella* IPNI=*168038-2* 287
Neea gentlei IPNI=*168038-2* 287
Neea hirtella IPNI=*168038-2* 287
Neea laetevirens IPNI=*168038-2* 287
Neea paraensis IPNI=*604789-1* 287
Neea parvifolia IPNI=*168038-2* 287
Neea popenoei IPNI=*168038-2* 287
Neea pycnantha IPNI=*168038-2* 287
Neea sphaerantha IPNI=*168038-2* 287
Neea xanthina IPNI=*168038-2* 287
Needhamia littoralis IPNI=*1071691-2* 335
Negretia mollis IPNI=*509248-1* 328
Negretia mutisiana IPNI=*509253-1* 328
Negretia pruriens IPNI=*509275-1* 328
Negretia sericea IPNI=*1197939-2* 328
Negretia urens IPNI=*1197939-2* 328
Neleixa prostrata IPNI=*1187456-2* 350
Nelsonia albicans IPNI=*52558-1* 360
Nelsonia campestris IPNI=*52558-1* 360
Nelsonia canescens var. *vestita* IPNI=*52558-1* 360
Nelsonia lamiifolia IPNI=*52558-1* 360
Nelsonia nummulariifolia IPNI=*52558-1* 360
Nelsonia origanoides IPNI=*52558-1* 360
Nelsonia pohlii IPNI=*52558-1* 360
Nelsonia rotundifolia IPNI=*52558-1* 360
Nelsonia senegalensis IPNI=*52558-1* 360
Nelsonia tomentosa IPNI=*52558-1* 360
Nelsonia vestita IPNI=*52558-1* 360
Nelsonia villosa IPNI=*52558-1* 360
Neltuma arenosa IPNI=*507659-1* 327
Neltuma juliflora IPNI=*1002432-2* 331
Nelumbium codophyllum IPNI=*605419-1* 433
Nelumbium jamaicense IPNI=*605419-1* 433
Nelumbium luteum IPNI=*605419-1* 433
Nelumbium pentapetalum IPNI=*605419-1* 433
Nelumbium reniforme IPNI=*605419-1* 433
Nelumbo nucifera subsp. *lutea* IPNI=*605419-1* 433
Nelumbo nucifera var. *lutea* IPNI=*605419-1* 433
Nelumbo pentapetala IPNI=*605419-1* 433
Nemaluma engleri IPNI=*291747-2* 307
Nematanthus heterophyllus IPNI=*63508-2* 366
Nematospermum laevigatum IPNI=*134457-2* 394
Nemepiodon caribeum IPNI=*65212-1* 213
Nemolepis canescens IPNI=*213524-1* 264
Neoblechnum brasiliense IPNI=*17058320-1* 174
Neobotrydium ambrosioides IPNI=*1198060-2* 282
Neocapparis pachaca IPNI=*77184114-1* 278
Neoceis carduifolia IPNI=*1152557-2* 261
Neoceis hieraciifolia IPNI=*1152557-2* 261
Neoceis rigidula IPNI=*1152557-2* 261
Neocleome spinosa IPNI=*60589-2* 279
Neocouma duckei IPNI=*281069-2* 342
Neocupressus arizonica IPNI=*60451546-2* 180
Neocupressus lusitanica IPNI=*60451554-2* 180

Neocupressus lusitanica var. lindleyi IPNI=60451554-2 180
Neocupressus macrocarpa IPNI=60451558-2 180
Neodypsis decaryi IPNI=985350-1 208
Neogossypium barbadense IPNI=559677-1 408
Neogriseocereus griseus IPNI=244634-2 285
Neolabatia glomerata IPNI=788965-1 307
Neolabatia macrocarpa IPNI=789045-1 307
Neolemaireocereus griseus IPNI=244634-2 285
Neomammillaria columbiana IPNI=314810-2 284
Neomammillaria mammillaris IPNI=134797-1 284
Neomammillaria parvimamma IPNI=134797-1 284
Neomarica bulbosa IPNI=77182480-1 217
Neomarica longifolia IPNI=77182480-1 217
Neomarica lutea IPNI=77182480-1 217
Neomolina bogotensis IPNI=300387-2 257
Neomoorea irrorata IPNI=168902-2 222
Neonelsonia ovata IPNI=168905-2 252
Neoraimondia peruviana IPNI=130627-1 283
Neoregelia carolinae f. tricolor IPNI=169003-2 229
Neoregelia carolinae var. tricolor IPNI=169003-2 229
Neosciadium glochidiatum IPNI=1197718-2 252
Neosinocalamus beecheyanus IPNI=392260-1 234
Neosinocalamus beecheyanus var. pubescens IPNI=392260-1 234
Neosinocalamus revolutus IPNI=392445-1 234
Neottia elata IPNI=73688-2 218
Neottia minor IPNI=73688-2 218
Neottia speciosa IPNI=659300-1 223
Neo-urbania cordyline IPNI=984957-1 220
Neo-urbania gigantea IPNI=984957-1 220
Neowashingtonia filamentosa IPNI=267788-2 212
Neowashingtonia filifera IPNI=267788-2 212
Neowashingtonia robusta IPNI=670278-1 212
Neowashingtonia sonorae IPNI=670278-1 212
Neoxythece cladantha IPNI=209727-2 306
Neoxythece dura IPNI=209738-2 306
Neoxythece elegans IPNI=209758-2 306
Neoxythece robusta IPNI=209738-2 306
Neoxythece schulzii IPNI=209738-2 306
Neoxythece wurdackii IPNI=209738-2 306
Nepeta americana IPNI=452333-1 371
Nepeta americana IPNI=77120747-1 368
Nepeta aristata IPNI=158242-2 371
Nepeta bodinieri IPNI=452333-1 371
Nepeta ceretana IPNI=452333-1 371
Nepeta citriodora IPNI=452333-1 371
Nepeta laurentii IPNI=452333-1 371
Nepeta macrura IPNI=452333-1 371
Nepeta minor IPNI=452333-1 371
Nepeta mollis IPNI=452333-1 371
Nepeta mutabilis IPNI=77120760-1 368
Nepeta pauciflora IPNI=77121103-1 371
Nepeta pectinata IPNI=158242-2 371
Nepeta persica IPNI=158242-2 371
Nepeta ruderalis IPNI=452333-1 371
Nepeta sidifolia IPNI=77121103-1 371
Nepeta tomentosa IPNI=452333-1 371
Nepeta vulgaris IPNI=452333-1 371
Nephelea arborea IPNI=17071820-1 172
Nephelea aureonitens IPNI=9971-2 172
Nephelea cuspidata IPNI=9954-2 172
Nephelea erinacea IPNI=9971-2 172
Nephelea erinacea var. purpurascens IPNI=9971-2 172
Nephelea mexicana IPNI=9976-2 172
Nephelea patellaris IPNI=9976-2 172
Nephelea purpurascens IPNI=9971-2 172
Nephelea tenerifrons IPNI=9976-2 172
Nephrodium acuminatum IPNI=17160710-1 176
Nephrodium acutum IPNI=17160710-1 176
Nephrodium andinum IPNI=17226640-1 177
Nephrodium asplenioides IPNI=17054310-1 174
Nephrodium auriculatum IPNI=17160730-1 176
Nephrodium biserratum IPNI=17160710-1 176
Nephrodium brownii IPNI=77069554-1 176
Nephrodium conionevron IPNI=17460390-1 174
Nephrodium consanguineum IPNI=17458930-1 174
Nephrodium contiguum IPNI=17458930-1 174
Nephrodium continuum IPNI=17458930-1 174
Nephrodium cordifolium IPNI=17160730-1 176
Nephrodium dentatum IPNI=251308-2 174
Nephrodium dicarpum IPNI=251308-2 174
Nephrodium dissectum IPNI=17458930-1 174
Nephrodium duriusculum IPNI=224543-2 176
Nephrodium ecklonianum IPNI=17458930-1 174
Nephrodium ecklonii IPNI=17458930-1 174
Nephrodium edule IPNI=17160730-1 176
Nephrodium exaltatum IPNI=17160830-1 176
Nephrodium expansum IPNI=17226640-1 177
Nephrodium extensum IPNI=17460390-1 174
Nephrodium filix-femina IPNI=30055316-2 174
Nephrodium funestum IPNI=286436-2 177
Nephrodium gibbosum IPNI=17160710-1 176
Nephrodium gongylodes IPNI=251353-2 174
Nephrodium gongylodes var. ecklonii IPNI=17458930-1 174
Nephrodium gongyloides IPNI=17458930-1 174
Nephrodium guianense IPNI=17160710-1 176
Nephrodium hoffmanseggii IPNI=17160710-1 176
Nephrodium imbricatum IPNI=17161310-1 176
Nephrodium impressum IPNI=17460390-1 174
Nephrodium inaequilaterum IPNI=17458930-1 174
Nephrodium kunthii IPNI=251400-2 174
Nephrodium lucens IPNI=17458930-1 174
Nephrodium lucidulum IPNI=17080730-1 175
Nephrodium macrophyllum IPNI=17226640-1 177
Nephrodium macrophyllum var. hirsutum IPNI=17226640-1 177
Nephrodium malabariense IPNI=251308-2 174
Nephrodium mauritianum IPNI=17160710-1 176
Nephrodium microcarpum IPNI=17458930-1 174
Nephrodium molle IPNI=251308-2 174
Nephrodium molle var. mollissimum IPNI=251308-2 174
Nephrodium molle var. pohlianum IPNI=251308-2 174
Nephrodium nymphale IPNI=251308-2 174
Nephrodium obtusatum IPNI=17458930-1 174
Nephrodium oligophlebium IPNI=1019960-2 174
Nephrodium opulentum IPNI=17460390-1 174
Nephrodium palmoides IPNI=17160710-1 176
Nephrodium paludosum IPNI=17458930-1 174
Nephrodium pectinatum IPNI=17161100-1 176
Nephrodium plantianum IPNI=17458930-1 174
Nephrodium pohlianum IPNI=251353-2 174
Nephrodium prionophyllum IPNI=17460390-1 174
Nephrodium propinquum IPNI=17458930-1 174
Nephrodium pteroides IPNI=17458930-1 174
Nephrodium punctatum IPNI=17460390-1 174
Nephrodium punctulatum IPNI=17160710-1 176
Nephrodium regulare IPNI=77069554-1 176
Nephrodium resiniferum IPNI=17458930-1 174
Nephrodium rufescens IPNI=17160710-1 176
Nephrodium schkuhrii IPNI=17161100-1 176
Nephrodium semicordatum IPNI=1148359-2 175
Nephrodium serratum IPNI=17458930-1 174
Nephrodium tenericaule IPNI=1019960-2 174
Nephrodium terminans IPNI=17460390-1 174
Nephrodium tetragonum var. marginale IPNI=252198-2 174
Nephrodium timoriense IPNI=17160710-1 176
Nephrodium tuberosum IPNI=17160730-1 176
Nephrodium unitum var. glabrum IPNI=251353-2 174

Nephrodium unitum var. *propinquum* IPNI=*17458930-1* 174
Nephrodium venulosum IPNI=*17458930-1* 174
Nephrodium violascens IPNI=*251308-2* 174
Nephrodium wakefieldii IPNI=*17460390-1* 174
Nephrolepis acuminata IPNI=*17160710-1* 176
Nephrolepis acuta IPNI=*17160710-1* 176
Nephrolepis acuta var. *laurifolia* IPNI=*17160710-1* 176
Nephrolepis acutangula IPNI=*77069554-1* 176
Nephrolepis auriculata IPNI=*17160730-1* 176
Nephrolepis biaurita IPNI=*17160710-1* 176
Nephrolepis biserrata subsp. *punctulatum* IPNI=*17160710-1* 176
Nephrolepis biserrata var. *glabrata* IPNI=*17160710-1* 176
Nephrolepis biserrata var. *normalis* IPNI=*17160710-1* 176
Nephrolepis biserrata var. *rufescens* IPNI=*17160710-1* 176
Nephrolepis biserrata var. *subintegrifolia* IPNI=*17160710-1* 176
Nephrolepis caudata IPNI=*17160710-1* 176
Nephrolepis cisneiriana IPNI=*17161100-1* 176
Nephrolepis cisneiriana var. *obtusata* IPNI=*17161100-1* 176
Nephrolepis cisneiriana var. *occidentalis* IPNI=*17161100-1* 176
Nephrolepis cisneiriana var. *pectinata* IPNI=*17161100-1* 176
Nephrolepis cordifolia var. *occidentalis* IPNI=*17161310-1* 176
Nephrolepis cordifolia var. *pectinata* IPNI=*17161100-1* 176
Nephrolepis cordifolia var. *tambourinensis* IPNI=*17160730-1* 176
Nephrolepis cordifolia var. *tuberosa* IPNI=*17160730-1* 176
Nephrolepis dayakorum IPNI=*17160710-1* 176
Nephrolepis depauperata IPNI=*17160710-1* 176
Nephrolepis ensifolia IPNI=*17160710-1* 176
Nephrolepis exaltata var. *biserrata* IPNI=*17160710-1* 176
Nephrolepis exaltata var. *pectinata* IPNI=*17161100-1* 176
Nephrolepis exaltata var. *tuberosa* IPNI=*17160730-1* 176
Nephrolepis filipes IPNI=*17161310-1* 176
Nephrolepis floccigera IPNI=*77069554-1* 176
Nephrolepis gibbosa IPNI=*17160710-1* 176
Nephrolepis glabra IPNI=*17161310-1* 176
Nephrolepis hirsutula var. *acuta* IPNI=*17160710-1* 176
Nephrolepis hirsutula var. *biserrata* IPNI=*17160710-1* 176
Nephrolepis hirsutula var. *ensifolia* IPNI=*17160710-1* 176
Nephrolepis hirsutula var. *rufescens* IPNI=*17160710-1* 176
Nephrolepis imbricata IPNI=*17161310-1* 176
Nephrolepis laurifolia IPNI=*17160710-1* 176
Nephrolepis lunulata IPNI=*17084790-1* 175
Nephrolepis macrophylla IPNI=*17160710-1* 176
Nephrolepis mauritiana IPNI=*17160710-1* 176
Nephrolepis mollis IPNI=*17160710-1* 176
Nephrolepis multiflora IPNI=*77069554-1* 176
Nephrolepis occidentalis IPNI=*17161310-1* 176
Nephrolepis paleacea IPNI=*17160730-1* 176
Nephrolepis paraensis IPNI=*17160710-1* 176
Nephrolepis pectinata var. *occidentalis* IPNI=*17161310-1* 176
Nephrolepis persicifolia IPNI=*17160710-1* 176
Nephrolepis pilosula IPNI=*17160710-1* 176
Nephrolepis pluma IPNI=*17161310-1* 176
Nephrolepis pubescens IPNI=*77069554-1* 176
Nephrolepis punctulata f. *genuina* IPNI=*17160710-1* 176
Nephrolepis punctulata IPNI=*17160710-1* 176
Nephrolepis punctulata var. *rufescens* IPNI=*17160710-1* 176
Nephrolepis radicans var. *cavernicola* IPNI=*17160730-1* 176
Nephrolepis rhizodes IPNI=*17160730-1* 176
Nephrolepis rufescens IPNI=*17160710-1* 176
Nephrolepis schkuhrii IPNI=*17161100-1* 176
Nephrolepis splendens IPNI=*17160710-1* 176
Nephrolepis subcordata IPNI=*17160710-1* 176
Nephrolepis tomentosa IPNI=*77069554-1* 176
Nephrolepis tuberosa IPNI=*17160730-1* 176
Nephrolepis zollingeriana IPNI=*17160710-1* 176
Nephromeria axillaris f. *robustior* IPNI=*491595-1* 317
Nephromeria axillaris IPNI=*491595-1* 317
Nephromeria axillaris var. *acutifolia* IPNI=*491595-1* 317
Nephromeria axillaris var. *obtusifoliola* IPNI=*491595-1* 317
Nephromeria axillaris var. *sintenisii* IPNI=*491595-1* 317
Nephromeria sintenisii IPNI=*491595-1* 317
Nephromeria umbrosa IPNI=*491595-1* 317
Nephrosis aculeata IPNI=*148561-2* 326
Nephthytis triphylla IPNI=*89127-1* 202
Neptunia natans IPNI=*509612-1* 328
Neptunia oleracea IPNI=*509612-1* 328
Neptunia plena f. *lumbricoides* IPNI=*1056199-2* 328
Neptunia polyphylla IPNI=*1056199-2* 328
Neptunia stolonifera IPNI=*509612-1* 328
Neptunia surinamensis IPNI=*1056199-2* 328
Nerion oleandrum IPNI=*80460-1* 342
Nerium carneum IPNI=*80460-1* 342
Nerium coronarium IPNI=*82097-1* 343
Nerium divaricatum IPNI=*82097-1* 343
Nerium flavescens IPNI=*80460-1* 342
Nerium floridum IPNI=*80460-1* 342
Nerium grandiflorum IPNI=*80460-1* 342
Nerium grandiflorum IPNI=*95962-1* 340
Nerium indicum IPNI=*80460-1* 342
Nerium indicum subsp. *kotschyi* IPNI=*80460-1* 342
Nerium japonicum IPNI=*80460-1* 342
Nerium kotschyi IPNI=*80460-1* 342
Nerium latifolium IPNI=*80460-1* 342
Nerium lauriforme IPNI=*80460-1* 342
Nerium luteum IPNI=*80460-1* 342
Nerium mascatense IPNI=*80460-1* 342
Nerium odoratissimum IPNI=*80460-1* 342
Nerium odoratum IPNI=*80460-1* 342
Nerium odorum IPNI=*80460-1* 342
Nerium oleander subsp. *kurdicum* IPNI=*80460-1* 342
Nerium splendens IPNI=*80460-1* 342
Nerium thyrsiflorum IPNI=*80460-1* 342
Nerium verecundum IPNI=*80460-1* 342
Nesaea dinteri subsp. *elata* IPNI=*30030870-2* 416
Nesaea transvaalica IPNI=*30030870-2* 416
Neurocarpaea lanceolata IPNI=*760275-1* 355
Neurocarpum javitense IPNI=*486540-1* 315
Neurogramma calomelanos IPNI=*17345640-1* 178
Neurogramma peruviana IPNI=*200935-2* 178
Neurogramma tartarea IPNI=*200935-2* 178
Neurogramma tartareum IPNI=*200935-2* 178
Neurolaena fulva IPNI=*234386-1* 267
Neurolaena integrifolia IPNI=*234386-1* 267
Neurolaena lobata var. *indivisa* IPNI=*234386-1* 267
Neurolaena suriani IPNI=*234386-1* 267
Neuromanes hedwigii IPNI=*320804-2* 173
Neuromanes immersum IPNI=*320804-2* 173
Neuromanes kaulfussii IPNI=*320804-2* 173
Neuromanes schomburgkianum IPNI=*320804-2* 173
Neurophyllum hedwigianum IPNI=*320804-2* 173
Neurophyllum hedwigii IPNI=*320804-2* 173
Neurophyllum pennatum IPNI=*320804-2* 173
Neurophyllum pinnatum IPNI=*320804-2* 173
Neurophyllum rhizophyllum IPNI=*320804-2* 173
Neurophyllum thecaphyllum IPNI=*320804-2* 173
Neustanthus sericans IPNI=*509661-1* 328
Neves-armondia cordifolia IPNI=*77100444-1* 362
Nevosmila arborea IPNI=*316363-2* 278
Nevrolis fuscata IPNI=*60011-1* 282
Newtonia psilostachya IPNI=*962800-1* 331
Nhandiroba cordifolia IPNI=*104422-2* 294
Nhandiroba karstenii IPNI=*104422-2* 294
Nhandiroba moorei IPNI=*104422-2* 294
Nhandiroba scandens IPNI=*104422-2* 294
Nicandra amara IPNI=*546882-1* 345
Nicodemia rondeletiiflora IPNI=*545779-1* 377

Nicolaia elatior IPNI=*942355-1* 251
Nicolaia imperialis IPNI=*942355-1* 251
Nicolaia intermedia IPNI=*942355-1* 251
Nicolaia magnifica IPNI=*942355-1* 251
Nicolaia speciosa IPNI=*942355-1* 251
Nicolsonia barbata IPNI=*77187394-1* 320
Nicolsonia reptans IPNI=*77187443-1* 320
Nicolsonia triflora IPNI=*77187443-1* 320
Nicoteba trinervia IPNI=*52612-1* 360
Nicotiana urens IPNI=*114242-3* 276
Nidularium carolinae IPNI=*169003-2* 229
Nidularium karatas IPNI=*328361-2* 228
Nidularium meyendorffii IPNI=*169003-2* 229
Nierembergia prunellifolia IPNI=*55411-1* 361
Nigella bithynica IPNI=*711645-1* 436
Nigella bourgaei IPNI=*711645-1* 436
Nigella coarctata IPNI=*711645-1* 436
Nigella coerulea IPNI=*711645-1* 436
Nigella elegans IPNI=*711645-1* 436
Nigella involucrata IPNI=*711645-1* 436
Nigella multifida IPNI=*711645-1* 436
Nigella nana IPNI=*711645-1* 436
Nigella pygmaea IPNI=*711645-1* 436
Nigella romana IPNI=*711645-1* 436
Nigella taurica IPNI=*711645-1* 436
Nintooa japonica IPNI=*148850-1* 296
Niopa peregrina IPNI=*12268-2* 311
Niphidium albopunctatissimum IPNI=*77183020-1* 175
Niphidium crassifolium IPNI=*77183236-1* 175
Niphobolus lanceolatus IPNI=*17448650-1* 176
Niphobolus lycopodioides IPNI=*160142-2* 175
Niphobolus percussus IPNI=*17546710-1* 175
Niphobolus polycarpus IPNI=*17341250-1* 176
Niphobolus vacciniifolius IPNI=*17406250-1* 176
Niphogeton ternata var. *orientalis* IPNI=*170084-2* 252
Nipponocalamus dimorphophyllus IPNI=*416271-1* 244
Nipponocalamus fortunei IPNI=*416271-1* 244
Nipponocalamus pygmaeus IPNI=*416271-1* 244
Nipponocalamus ramosissimus IPNI=*416271-1* 244
Nipponocalamus shibuyanus IPNI=*416271-1* 244
Nipponocalamus tosaensis IPNI=*416271-1* 244
Nipponocalamus tsukubensis IPNI=*416271-1* 244
Nipponocalamus uyenoensis IPNI=*416271-1* 244
Nipponocalamus yasuianus IPNI=*416271-1* 244
Nipponocalamus yoshidake IPNI=*416271-1* 244
Niruris annua IPNI=*194900-2* 402
Nisoralis jamaicensis IPNI=*823472-1* 408
Nispero achras IPNI=*152641-2* 306
Nissolia aculeata IPNI=*148544-2* 326
Nissolia acuminata IPNI=*505373-1* 326
Nissolia arborea IPNI=*505373-1* 326
Nissolia glabrata IPNI=*505373-1* 326
Nissolia reticulata IPNI=*516561-1* 331
Nissoloides cylindrica IPNI=*135026-3* 318
Nitidobulbon nasutum IPNI=*643645-1* 221
Nitidobulbon oxysepalum IPNI=*643645-1* 221
Nocca mollis IPNI=*134795-2* 264
Nochotta oleracea IPNI=*486336-1* 315
Noisettia frangulifolia IPNI=*12522-2* 405
Noisettia orinocensis IPNI=*77144891-1* 405
Nonatelia cataractarum IPNI=*759121-1* 354
Nonatelia grandiflora IPNI=*181012-2* 354
Nonatelia guianensis IPNI=*759121-1* 354
Nonatelia officinalis IPNI=*77161098-1* 354
Nonatelia panamensis IPNI=*77115627-1* 354
Nonatelia pubescens IPNI=*77115593-1* 353
Nonatelia racemosa IPNI=*77115627-1* 354
Nonatelia violacea IPNI=*759372-1* 354
Nopalea brittonii IPNI=*1008816-2* 284
Nopalea brittonii var. *variegata* IPNI=*1008816-2* 284
Nopalea cochenillifera IPNI=*1008816-2* 284
Nopalea nuda IPNI=*1008816-2* 284
Norantea curva IPNI=*997443-1* 302
Norantea mixta IPNI=*997443-1* 302
Norantea uleana IPNI=*997443-1* 302
Normanbya merrillii IPNI=*663990-1* 204
Norta altissima IPNI=*289144-1* 278
Norta irio IPNI=*237940-2* 278
Nortenia thouarsii IPNI=*255156-2* 374
Norysca calycina IPNI=*433287-1* 393
Norysca patula IPNI=*927423-1* 394
Norysca venosa IPNI=*433287-1* 393
Nostelis viminea IPNI=*61205-2* 369
Notanthera conduplicatus IPNI=*77147531-1* 448
Notanthera ellipticus IPNI=*548708-1* 448
Notanthera lanceolatus IPNI=*548708-1* 448
Notanthera ligustrifolius IPNI=*258636-2* 449
Notanthera orbicularis IPNI=*551737-1* 449
Notanthera suaveolens IPNI=*258636-2* 449
Nothoholcus lanatus IPNI=*405034-1* 240
Notholaena albidula IPNI=*77182941-1* 178
Notholaena balansae IPNI=*77182303-1* 178
Notholaena cheilanthoides IPNI=*77183177-1* 178
Notholaena chrysophylla IPNI=*77182941-1* 178
Notholaena flavens IPNI=*77182941-1* 178
Notholaena herzogii IPNI=*77182303-1* 178
Notholaena microphylla IPNI=*77183177-1* 178
Notholaena myriophylla IPNI=*77183191-1* 178
Notholaena nivea IPNI=*77182941-1* 178
Notholaena nivea var. *flava* IPNI=*77182941-1* 178
Notholaena nivea var. *tenera* IPNI=*77182941-1* 178
Notholaena obducta IPNI=*77182303-1* 178
Notholaena tenera IPNI=*77182941-1* 178
Notholaena tenera var. *major* IPNI=*77182941-1* 178
Notholaena ternifolia IPNI=*77182326-1* 178
Notholcus lanatus IPNI=*405034-1* 240
Notholcus mollis IPNI=*123530-2* 240
Nothopanax cochleatus IPNI=*91817-1* 254
Nothopanax crispatus IPNI=*91749-1* 254
Nothopanax fruticosus IPNI=*847351-1* 254
Nothopanax fruticosus var. *plumatus* IPNI=*847351-1* 254
Nothopanax fruticosus var. *victoriae* IPNI=*91773-1* 254
Nothopanax guilfoylei IPNI=*91773-1* 254
Nothopanax pinnatus IPNI=*91749-1* 254
Nothopanax scutellarius IPNI=*91817-1* 254
Nothopanax tricochleatus IPNI=*91817-1* 254
Notoceras cardaminifolium IPNI=*286421-1* 277
Notylia planifolia IPNI=*262578-2* 223
Notylia pompona IPNI=*661259-1* 223
Notylia sativa IPNI=*262578-2* 223
Notylia sylvestris IPNI=*262578-2* 223
Notylia vanilla IPNI=*661225-1* 223
Noyera mollis IPNI=*1008109-1* 440
Nunnezharia amazonica IPNI=*54928-2* 207
Nunnezharia bartlingiana IPNI=*665826-1* 207
Nunnezharia bifurcata IPNI=*665826-1* 207
Nunnezharia bracteata IPNI=*665826-1* 207
Nunnezharia brevifrons IPNI=*665826-1* 207
Nunnezharia casperiana IPNI=*54964-2* 208
Nunnezharia concolor IPNI=*665826-1* 207
Nunnezharia conocarpa IPNI=*665826-1* 207
Nunnezharia corallina IPNI=*54901-2* 207
Nunnezharia elegans IPNI=*54865-2* 207
Nunnezharia ernesti-augustii IPNI=*54867-2* 207
Nunnezharia flavovirens IPNI=*665826-1* 207
Nunnezharia formosa IPNI=*54901-2* 207

Nunnezharia geonomoides IPNI=*665826-1* 207
Nunnezharia integrifolia IPNI=*54928-2* 207
Nunnezharia lanceolata IPNI=*665826-1* 207
Nunnezharia lindeniana IPNI=*54901-2* 207
Nunnezharia linearis IPNI=*54901-2* 207
Nunnezharia membranacea IPNI=*665826-1* 207
Nunnezharia montana IPNI=*54901-2* 207
Nunnezharia morenia IPNI=*54901-2* 207
Nunnezharia pacaya IPNI=*665826-1* 207
Nunnezharia pauciflora IPNI=*54928-2* 207
Nunnezharia pinnatifrons IPNI=*665826-1* 207
Nunnezharia poeppigiana IPNI=*54901-2* 207
Nunnezharia pulchella IPNI=*54865-2* 207
Nunnezharia simplicifrons IPNI=*54867-2* 207
Nunnezharia tepejilote IPNI=*54964-2* 208
Nunnezharia wendlandiana IPNI=*54964-2* 208
Nuttallanthus canadensis f. *albinus* IPNI=*281417-2* 376
Nuttallanthus subandinus IPNI=*281417-2* 376
Nuxia elata IPNI=*860847-1* 368
Nyctago jalapae IPNI=*162591-2* 287
Nyctago mirabilis IPNI=*162591-2* 287
Nyctago versicolor IPNI=*162591-2* 287
Nyctanthes acuminata IPNI=*82097-1* 343
Nyctanthes goa IPNI=*609755-1* 375
Nyctanthes multiflora IPNI=*609641-1* 375
Nyctanthes pubescens IPNI=*609641-1* 375
Nyctanthes sambac IPNI=*609755-1* 375
Nyctanthes undulata IPNI=*609755-1* 375
Nyctosma nocturna IPNI=*30010607-2* 219
Nymanina leichtlinii IPNI=*437022-1* 216
Nymanina refracta IPNI=*437030-1* 216
Nymphaea acutidens IPNI=*605604-1* 181
Nymphaea aegyptiaca IPNI=*605604-1* 181
Nymphaea alboviridis IPNI=*605491-1* 181
Nymphaea amazonum f. *goudotiana* IPNI=*605491-1* 181
Nymphaea amazonum var. *forma-submersa* IPNI=*605491-1* 181
Nymphaea amazonum var. *goudotiana* IPNI=*605491-1* 181
Nymphaea ampla var. *hookeri* IPNI=*605675-1* 181
Nymphaea ampla var. *parviflora* IPNI=*605494-1* 181
Nymphaea ampla var. *plumieri* IPNI=*605494-1* 181
Nymphaea ampla var. *pulchella* IPNI=*605675-1* 181
Nymphaea ampla var. *rudgeana* IPNI=*281442-2* 181
Nymphaea ampla var. *salzmannii* IPNI=*605675-1* 181
Nymphaea ampla var. *speciosa* IPNI=*605675-1* 181
Nymphaea blanda f. *amazonum* IPNI=*605491-1* 181
Nymphaea blanda f. *genuina* IPNI=*281434-2* 181
Nymphaea blanda IPNI=*281434-2* 181
Nymphaea blanda IPNI=*281442-2* 181
Nymphaea blanda IPNI=*605491-1* 181
Nymphaea blanda var. *fenzliana* IPNI=*281434-2* 181
Nymphaea boucheana IPNI=*605604-1* 181
Nymphaea candolleana IPNI=*605494-1* 181
Nymphaea coteka IPNI=*605604-1* 181
Nymphaea dentata IPNI=*605604-1* 181
Nymphaea edulis IPNI=*605604-1* 181
Nymphaea esculenta IPNI=*605604-1* 181
Nymphaea fenzliana IPNI=*281434-2* 181
Nymphaea foetida IPNI=*605491-1* 181
Nymphaea fragrantissima IPNI=*605604-1* 181
Nymphaea gibertii IPNI=*605588-1* 181
Nymphaea goudotiana IPNI=*605491-1* 181
Nymphaea hypotricha IPNI=*605604-1* 181
Nymphaea integrifolia IPNI=*605491-1* 181
Nymphaea leiboldiana IPNI=*605675-1* 181
Nymphaea leucantha IPNI=*605604-1* 181
Nymphaea liberiensis IPNI=*605604-1* 181
Nymphaea lineata IPNI=*605675-1* 181
Nymphaea mexicana IPNI=*171178-2* 181
Nymphaea nervosa IPNI=*605675-1* 181
Nymphaea nocturna IPNI=*605491-1* 181
Nymphaea ortgiesiana IPNI=*605604-1* 181
Nymphaea pentapetala IPNI=*605419-1* 433
Nymphaea regina IPNI=*605776-1* 181
Nymphaea reichardiana IPNI=*605604-1* 181
Nymphaea rudgeana var. *amazonum* IPNI=*605491-1* 181
Nymphaea sagittariifolia IPNI=*605588-1* 181
Nymphaea sagittata IPNI=*605604-1* 181
Nymphaea sinuata IPNI=*605494-1* 181
Nymphaea speciosa IPNI=*605675-1* 181
Nymphaea trisepala IPNI=*605675-1* 181
Nymphaea tropaeolifolia IPNI=*605675-1* 181
Nymphaea victoria IPNI=*605776-1* 181
Nymphaea zenkeri IPNI=*605604-1* 181
Nymphanthus niruri IPNI=*194900-2* 402
Nymphoides ecklonianа IPNI=*1182548-2* 273
Nymphoides microphylla IPNI=*171243-2* 273

O

Ochlopoa annua IPNI=*320035-2* 244
Ochlopoa annua subsp. *notabilis* IPNI=*320035-2* 244
Ochlopoa annua subsp. *pilantha* IPNI=*320035-2* 244
Ochlopoa annua subsp. *raniglumis* IPNI=*320035-2* 244
Ochlopoa notabilis IPNI=*320035-2* 244
Ochlopoa pilantha IPNI=*320035-2* 244
Ochlopoa raniglumis IPNI=*320035-2* 244
Ochna guatemalensis IPNI=*607063-1* 397
Ochna hildebrandtii IPNI=*606804-1* 397
Ochna lucens IPNI=*607063-1* 397
Ochna nitida IPNI=*607105-1* 397
Ochroma bicolor IPNI=*171458-2* 412
Ochroma boliviana IPNI=*171458-2* 412
Ochroma concolor IPNI=*171458-2* 412
Ochroma grandiflora IPNI=*171458-2* 412
Ochroma lagopodum IPNI=*171458-2* 412
Ochroma lagopus IPNI=*171458-2* 412
Ochroma limonensis IPNI=*171458-2* 412
Ochroma obtusa IPNI=*171458-2* 412
Ochroma peruviana IPNI=*171458-2* 412
Ochroma pyramidale var. *bicolor* IPNI=*171458-2* 412
Ochroma pyramidale var. *concolor* IPNI=*171458-2* 412
Ochroma tomentosa IPNI=*171458-2* 412
Ochroma velutina IPNI=*171458-2* 412
Ocimum album IPNI=*316218-2* 371
Ocimum brachiatum IPNI=*316218-2* 371
Ocimum canum IPNI=*316218-2* 371
Ocimum dichotomum IPNI=*316218-2* 371
Ocimum dinteri IPNI=*316218-2* 371
Ocimum flexuosum IPNI=*452886-1* 371
Ocimum flexuosum IPNI=*453130-1* 371
Ocimum fluminense IPNI=*316218-2* 371
Ocimum fruticulosum IPNI=*316218-2* 371
Ocimum guatemalense IPNI=*452886-1* 371
Ocimum hirsutum IPNI=*453130-1* 371
Ocimum hispidulum IPNI=*316218-2* 371
Ocimum incanescens IPNI=*316218-2* 371
Ocimum inodorum IPNI=*453130-1* 371
Ocimum micranthum IPNI=*452886-1* 371
Ocimum monachorum IPNI=*453130-1* 371
Ocimum montanum IPNI=*452886-1* 371
Ocimum odorum IPNI=*452874-1* 371
Ocimum peltatum IPNI=*446165-1* 369
Ocimum pubescens IPNI=*452886-1* 371
Ocimum racemosum IPNI=*454770-1* 372
Ocimum sanctum IPNI=*453130-1* 371
Ocimum sanctum var. *cubensis* IPNI=*453130-1* 371
Ocimum sanctum var. *hirsutum* IPNI=*453130-1* 371

Ocimum scutellarioides IPNI=*446165-1* 369
Ocimum scutellarioides IPNI=*453130-1* 371
Ocimum stamineum IPNI=*316218-2* 371
Ocimum subserratum IPNI=*453130-1* 371
Ocimum tenuiflorum f. *villicaulis* IPNI=*453130-1* 371
Ocimum thymoides IPNI=*316218-2* 371
Ocimum tomentosum IPNI=*453130-1* 371
Ocimum vaalae IPNI=*445902-1* 369
Ocimum verticillatum IPNI=*454770-1* 372
Ocimum villosum IPNI=*453130-1* 371
Ocotea aciphylla var. *chimantaensis* IPNI=*171499-2* 184
Ocotea aciphylla var. *robusta* IPNI=*171499-2* 184
Ocotea alloiophylla IPNI=*3010840-1* 182
Ocotea amara IPNI=*77198011-1* 183
Ocotea amazonum IPNI=*466626-1* 183
Ocotea architectorum IPNI=*1146025-2* 184
Ocotea arechavaletae IPNI=*1032261-2* 185
Ocotea arenaensis IPNI=*1146025-2* 184
Ocotea bakeri IPNI=*77198019-1* 183
Ocotea barcellensis IPNI=*77198011-1* 183
Ocotea baturitensis IPNI=*1032261-2* 185
Ocotea botryophylla IPNI=*1146025-2* 184
Ocotea calophylla IPNI=*77167373-1* 184
Ocotea caniflora IPNI=*1146025-2* 184
Ocotea caparrapii IPNI=*77198011-1* 183
Ocotea castanea IPNI=*171883-2* 185
Ocotea caudata IPNI=*171727-2* 184
Ocotea cernua IPNI=*171727-2* 184
Ocotea cinnamomoides IPNI=*77198017-1* 183
Ocotea citrifolia IPNI=*171904-2* 185
Ocotea compactiflora IPNI=*203826-2* 185
Ocotea costulata IPNI=*171499-2* 184
Ocotea cuspidata IPNI=*281229-2* 184
Ocotea cymbarum IPNI=*77198011-1* 183
Ocotea discolor IPNI=*1203274-2* 184
Ocotea dissimilis IPNI=*171904-2* 185
Ocotea elongata IPNI=*77173628-1* 182
Ocotea endlicheriopsis IPNI=*307011-2* 183
Ocotea escuintlensis IPNI=*77198019-1* 183
Ocotea finium IPNI=*171499-2* 184
Ocotea flavescens IPNI=*77173628-1* 182
Ocotea flexuosa IPNI=*77158867-1* 183
Ocotea floccifera IPNI=*171697-2* 184
Ocotea fulvifolia IPNI=*171499-2* 184
Ocotea globifera IPNI=*171904-2* 185
Ocotea globosa IPNI=*167822-2* 184
Ocotea grandifolia IPNI=*171526-2* 184
Ocotea guianensis var. *aurea* IPNI=*467336-1* 184
Ocotea guianensis var. *subsericea* IPNI=*467336-1* 184
Ocotea hartiana IPNI=*171797-2* 184
Ocotea helicterifolia IPNI=*326508-2* 184
Ocotea hookeriana IPNI=*158649-2* 183
Ocotea ira IPNI=*171697-2* 184
Ocotea latifolia IPNI=*77158867-1* 183
Ocotea lineata IPNI=*974282-1* 184
Ocotea longifolia IPNI=*171526-2* 184
Ocotea lucida IPNI=*90974-2* 183
Ocotea maguireana IPNI=*171499-2* 184
Ocotea marmellensis var. *acrensis* IPNI=*171727-2* 184
Ocotea marowynensis IPNI=*171727-2* 184
Ocotea martiniana IPNI=*1032261-2* 185
Ocotea matensis IPNI=*171727-2* 184
Ocotea mayana IPNI=*171797-2* 184
Ocotea maynensis IPNI=*77173628-1* 182
Ocotea membranacea IPNI=*466773-1* 184
Ocotea mexicana IPNI=*326508-2* 184
Ocotea mexicana var. *longipes* IPNI=*326508-2* 184
Ocotea mexicana var. *subsessilis* IPNI=*326508-2* 184
Ocotea microcarpa IPNI=*171526-2* 184
Ocotea mollis IPNI=*466894-1* 184
Ocotea multiglandulosa IPNI=*1146025-2* 184
Ocotea oligoneura IPNI=*466885-1* 184
Ocotea opifera IPNI=*171526-2* 184
Ocotea otuzcensis IPNI=*171684-2* 184
Ocotea paradoxa IPNI=*77198019-1* 183
Ocotea paraensis IPNI=*1032261-2* 185
Ocotea paranapiacabensis IPNI=*1032261-2* 185
Ocotea parviflora IPNI=*203826-2* 185
Ocotea pichurim IPNI=*1171177-2* 184
Ocotea poeppigiana IPNI=*1146025-2* 184
Ocotea pseudocoto IPNI=*31062-2* 182
Ocotea psychotrioides IPNI=*466626-1* 183
Ocotea puberula var. *arechavaletae* IPNI=*1032261-2* 185
Ocotea puberula var. *truncata* IPNI=*1032261-2* 185
Ocotea puchury-major IPNI=*13872-2* 182
Ocotea puchury-minor IPNI=*13872-2* 182
Ocotea pyramidata IPNI=*1032261-2* 185
Ocotea quixos IPNI=*77198017-1* 183
Ocotea ramentacea IPNI=*465312-1* 183
Ocotea rhynchophylla IPNI=*171499-2* 184
Ocotea riparia IPNI=*281229-2* 184
Ocotea roraimae IPNI=*171499-2* 184
Ocotea rubra IPNI=*999503-1* 185
Ocotea rusbyana IPNI=*171526-2* 184
Ocotea salicifolia IPNI=*974281-1* 184
Ocotea sericiflora IPNI=*171499-2* 184
Ocotea sieberi IPNI=*171727-2* 184
Ocotea simulans IPNI=*464481-1* 183
Ocotea sintenisii IPNI=*466885-1* 184
Ocotea subalpina IPNI=*171727-2* 184
Ocotea subalveolata IPNI=*171883-2* 185
Ocotea subglabra IPNI=*1032261-2* 185
Ocotea tabascensis IPNI=*77158859-1* 183
Ocotea tenejapensis IPNI=*326508-2* 184
Ocotea trianae IPNI=*171526-2* 184
Ocotea turbacensis IPNI=*466885-1* 184
Ocotea turbacensis IPNI=*90974-2* 183
Ocotea umbrosa var. *bullata* IPNI=*326508-2* 184
Ocotea urophylla IPNI=*171727-2* 184
Ocotea venenosa IPNI=*274529-2* 183
Ocotea veraguensis IPNI=*77198019-1* 183
Ocotea wachenheimii IPNI=*1146025-2* 184
Octadenia maritima IPNI=*1085072-2* 277
Octanema crotonoides IPNI=*77183834-1* 278
Octelisia aurea IPNI=*518364-1* 333
Octomeris bonplandii IPNI=*77196163-1* 421
Octonum humboldtii IPNI=*77196163-1* 421
Octopleura diversifolia IPNI=*77144083-1* 422
Octopleura macrophylla IPNI=*77144075-1* 421
Octopleura quinquenervia IPNI=*77144083-1* 422
Octopleura robusta IPNI=*77144090-1* 423
Ocymastrum rubrum IPNI=*860141-1* 296
Odina francoana IPNI=*71559-1* 451
Odontadenia angustifolia IPNI=*1054185-2* 342
Odontadenia boliviana IPNI=*291380-2* 342
Odontadenia caudigera IPNI=*77191232-1* 342
Odontadenia cordata IPNI=*1054185-2* 342
Odontadenia cururu IPNI=*291380-2* 342
Odontadenia dusendschoenii IPNI=*1054185-2* 342
Odontadenia formosa IPNI=*77191232-1* 342
Odontadenia grandiflora IPNI=*77191232-1* 342
Odontadenia harrisii IPNI=*77191232-1* 342
Odontadenia lucida IPNI=*1054185-2* 342
Odontadenia macrantha IPNI=*77191232-1* 342
Odontadenia macrocalyx IPNI=*285812-2* 343
Odontadenia nitida var. *acuminata* IPNI=*1054185-2* 342

Odontadenia nitida var. *angustifolia* IPNI=*1054185-2* 342
Odontadenia nitida var. *oblongifolia* IPNI=*1054185-2* 342
Odontadenia schippii IPNI=*291380-2* 342
Odontadenia speciosa IPNI=*77191232-1* 342
Odontadenia spoliata IPNI=*291380-2* 342
Odontadenia sprucei IPNI=*291380-2* 342
Odontadenia sylvestris IPNI=*77191232-1* 342
Odontites santolinifolius IPNI=*77160774-1* 375
Odontocarya hederifolia IPNI=*581190-1* 435
Odontocarya nitida IPNI=*230118-2* 435
Odontocarya sagittata IPNI=*581190-1* 435
Odontoglossum alexandrae IPNI=*77101415-1* 222
Odontoglossum alexandrae var. *cooksonii* IPNI=*77101415-1* 222
Odontoglossum alexandrae var. *flaveolum* IPNI=*77101415-1* 222
Odontoglossum alexandrae var. *outramii* IPNI=*77101415-1* 222
Odontoglossum alexandrae var. *stevensii* IPNI=*77101415-1* 222
Odontoglossum alexandrae var. *trianae* IPNI=*77101415-1* 222
Odontoglossum alexandrae var. *wilsonii* IPNI=*77101415-1* 222
Odontoglossum angustatum IPNI=*1021207-1* 218
Odontoglossum bluntii IPNI=*77101415-1* 222
Odontoglossum bogotense IPNI=*1021208-1* 218
Odontoglossum crispum f. *hyperxanthum* IPNI=*77101415-1* 222
Odontoglossum crispum IPNI=*77101415-1* 222
Odontoglossum crispum var. *bluntii* IPNI=*77101415-1* 222
Odontoglossum crispum var. *chestertonii* IPNI=*77101415-1* 222
Odontoglossum crispum var. *flaveolum* IPNI=*77101415-1* 222
Odontoglossum crispum var. *hyperxanthum* IPNI=*77101415-1* 222
Odontoglossum crispum var. *latranum* IPNI=*77101415-1* 222
Odontoglossum crispum var. *lehmannii* IPNI=*77101415-1* 222
Odontoglossum crispum var. *lyntanthum* IPNI=*77101415-1* 222
Odontoglossum crispum var. *plumatum* IPNI=*77101415-1* 222
Odontoglossum crispum var. *reginae* IPNI=*77101415-1* 222
Odontoglossum crispum var. *trianae* IPNI=*77101415-1* 222
Odontoglossum crispum var. *wolstenholmiae* IPNI=*77101415-1* 222
Odontoglossum crispum var. *wrigleyanum* IPNI=*77101415-1* 222
Odontoglossum edithiae IPNI=*77101415-1* 222
Odontoglossum jamaicense IPNI=*1021208-1* 218
Odontoglossum latimaculatum IPNI=*77101415-1* 222
Odontoglossum lindenii IPNI=*1021208-1* 218
Odontoglossum platyodon IPNI=*1021208-1* 218
Odontoglossum platyotum IPNI=*1021208-1* 218
Odontoglossum ramosissimum f. *albomaculatum* IPNI=*1021207-1* 218
Odontoglossum ramosissimum f. *xanthinum* IPNI=*1021207-1* 218
Odontoglossum ramosissimum IPNI=*1021207-1* 218
Odontoglossum ramosissimum var. *albomaculatum* IPNI=*1021207-1* 218
Odontoglossum ramosissimum var. *coeleste* IPNI=*1021207-1* 218
Odontoglossum ramosissimum var. *liliiflorum* IPNI=*1021207-1* 218
Odontoglossum ramosissimum var. *xanthinum* IPNI=*1021207-1* 218
Odontoglossum reichenbachianum IPNI=*77101415-1* 222
Odontoglossum revolutum IPNI=*1021208-1* 218
Odontoglossum vexillarium IPNI=*161453-2* 222
Odontoglossum vexillarium var. *lehmannii* IPNI=*161453-2* 222
Odontoglossum vexillarium var. *purpureum* IPNI=*161453-2* 222
Odontoglossum vexillarium var. *superbum* IPNI=*161453-2* 222
Odontoglossum warocqueanum IPNI=*77101415-1* 222
Odontoloma acuminata IPNI=*988439-1* 268
Odontonema adenostachyum IPNI=*913240-1* 361
Odontonema amicorum IPNI=*172486-2* 360
Odontonema bracteolatum var. *parviflorum* IPNI=*172440-2* 360
Odontonema flagellum IPNI=*172486-2* 360
Odontonema longifolium IPNI=*172486-2* 360
Odontonema lucidum IPNI=*172440-2* 360
Odontonema pantasmense IPNI=*172486-2* 360
Odontonema strictum IPNI=*172486-2* 360
Odontonema venezuelense IPNI=*172440-2* 360
Odontostephana obliqua IPNI=*154570-2* 342
Odontostephana shortii IPNI=*154570-2* 342
Odontostigma galeottianum IPNI=*285810-2* 343
Oeceoclades mackenii IPNI=*1073117-2* 222
Oeceoclades maculata var. *pterocarpa* IPNI=*1073117-2* 222
Oeceoclades monophylla IPNI=*1073117-2* 222
Oedematopus duidae IPNI=*1190784-2* 384
Oedematopus obovatus IPNI=*1082351-2* 384
Oedematopus octandrus IPNI=*1174011-2* 384
Oedematopus quadratus IPNI=*1082351-2* 384
Oedipachne punctata IPNI=*1021229-2* 239
Oenocarpus bacaba var. *grandis* IPNI=*668509-1* 210
Oenocarpus bacaba var. *parvus* IPNI=*281514-2* 210
Oenocarpus dealbatus IPNI=*154897-2* 210
Oenocarpus dryanderae IPNI=*668532-1* 211
Oenocarpus grandis IPNI=*668509-1* 210
Oenocarpus hoppii IPNI=*668509-1* 210
Oenocarpus huebneri IPNI=*668534-1* 211
Oenocarpus intermedius IPNI=*668534-1* 211
Oenocarpus macrocalyx IPNI=*668532-1* 211
Oenocarpus mapora subsp. *dryanderae* IPNI=*668532-1* 211
Oenocarpus microspadix IPNI=*668534-1* 211
Oenocarpus minor subsp. *intermedius* IPNI=*668534-1* 211
Oenocarpus multicaulis IPNI=*668532-1* 211
Oenocarpus panamanus IPNI=*668532-1* 211
Oenocarpus regius IPNI=*222809-2* 212
Oenocarpus sancona IPNI=*670066-1* 212
Oenothera aberrans IPNI=*317809-2* 429
Oenothera alata IPNI=*173157-2* 430
Oenothera alata IPNI=*613333-1* 429
Oenothera albida IPNI=*317809-2* 429
Oenothera aurantia IPNI=*173202-2* 430
Oenothera barbeyana IPNI=*173202-2* 430
Oenothera beckeri IPNI=*172755-2* 429
Oenothera biennis f. *leptomeres* IPNI=*172755-2* 429
Oenothera biennis f. *muricata* IPNI=*172755-2* 429
Oenothera biennis subf. *latifolia* IPNI=*172755-2* 429
Oenothera biennis subsp. *biennis* IPNI=*172755-2* 429
Oenothera biennis subsp. *caeciarum* IPNI=*172755-2* 429
Oenothera biennis subsp. *centralis* IPNI=*172755-2* 429
Oenothera biennis subsp. *chicaginensis* IPNI=*172755-2* 429
Oenothera biennis subsp. *grandiflora* IPNI=*172755-2* 429
Oenothera biennis subsp. *rubricaulis* IPNI=*172755-2* 429
Oenothera biennis subsp. *suaveolens* IPNI=*172755-2* 429
Oenothera biennis var. *angustifolia* IPNI=*172755-2* 429
Oenothera biennis var. *cantabrigiana* IPNI=*172755-2* 429
Oenothera biennis var. *leptomeres* IPNI=*172755-2* 429
Oenothera biennis var. *muricata* IPNI=*172755-2* 429
Oenothera biennis var. *pycnocarpa* IPNI=*172755-2* 429
Oenothera biennis var. *rubricaulis* IPNI=*172755-2* 429
Oenothera biennis var. *sulphurea* IPNI=*172755-2* 429
Oenothera biennis var. *vulgaris* IPNI=*172755-2* 429
Oenothera bipartita IPNI=*317809-2* 429
Oenothera blandina IPNI=*317809-2* 429
Oenothera brasiliensis IPNI=*144384-2* 429
Oenothera brevicapsula IPNI=*172755-2* 429
Oenothera brevispicata IPNI=*172755-2* 429
Oenothera brevistylis IPNI=*317809-2* 429
Oenothera cambrica IPNI=*172755-2* 429
Oenothera cambrica var. *impunctata* IPNI=*172755-2* 429
Oenothera cana IPNI=*317809-2* 429
Oenothera candida IPNI=*614141-1* 430
Oenothera capensis IPNI=*614141-1* 430
Oenothera carinthiaca IPNI=*172755-2* 429
Oenothera casimiri IPNI=*172755-2* 429
Oenothera chicaginensis IPNI=*172755-2* 429
Oenothera chicaginensis var. *bartlettii* IPNI=*172755-2* 429
Oenothera chicaginensis var. *minutiflora* IPNI=*172755-2* 429
Oenothera chicaginensis var. *parviflora* IPNI=*172755-2* 429
Oenothera communis f. *canescens* IPNI=*172755-2* 429

Oenothera compacta IPNI=*172755-2* 429
Oenothera coronifera IPNI=*317809-2* 429
Oenothera crassifolia var. *leptocarpa* IPNI=*317809-2* 429
Oenothera dubia f. *latifolia* IPNI=*614141-1* 430
Oenothera dubia IPNI=*614141-1* 430
Oenothera editicaulis IPNI=*172755-2* 429
Oenothera elliptica IPNI=*317809-2* 429
Oenothera ersteinensis IPNI=*172755-2* 429
Oenothera erythrosepala IPNI=*317809-2* 429
Oenothera erythrosepala var. *azorica* IPNI=*317809-2* 429
Oenothera fatua IPNI=*317809-2* 429
Oenothera flaemingina IPNI=*172755-2* 429
Oenothera furca IPNI=*172755-2* 429
Oenothera fusiformis IPNI=*317809-2* 429
Oenothera gauroides IPNI=*172755-2* 429
Oenothera gauroides var. *brevicapsula* IPNI=*172755-2* 429
Oenothera gigantea IPNI=*317809-2* 429
Oenothera glabra IPNI=*172755-2* 429
Oenothera glazioviana f. *rubricalyx* IPNI=*317809-2* 429
Oenothera grandiflora subsp. *coronifera* IPNI=*317809-2* 429
Oenothera grandiflora var. *tracyi* IPNI=*172755-2* 429
Oenothera grandifolia IPNI=*172755-2* 429
Oenothera graveolens IPNI=*172755-2* 429
Oenothera hirta IPNI=*144406-2* 429
Oenothera hirtella IPNI=*172755-2* 429
Oenothera inconspecta IPNI=*172755-2* 429
Oenothera indivisa IPNI=*172755-2* 429
Oenothera jueterbogensis IPNI=*172755-2* 429
Oenothera jueterbogensis var. *macrosperma* IPNI=*172755-2* 429
Oenothera laciniata var. *mexicana* IPNI=*613817-1* 430
Oenothera laciniata var. *typica* IPNI=*613817-1* 430
Oenothera laevifolia IPNI=*317809-2* 429
Oenothera lamarckiana IPNI=*317809-2* 429
Oenothera lata IPNI=*317809-2* 429
Oenothera latiflora IPNI=*614141-1* 430
Oenothera latipetala IPNI=*172755-2* 429
Oenothera leptocarpa IPNI=*317809-2* 429
Oenothera liquida IPNI=*317809-2* 429
Oenothera lyrata IPNI=*173348-2* 430
Oenothera macrocarpa var. *missouriensis* IPNI=*173157-2* 430
Oenothera macrosperma IPNI=*172755-2* 429
Oenothera marinellae IPNI=*172755-2* 429
Oenothera media IPNI=*172755-2* 429
Oenothera mediomarchica IPNI=*172755-2* 429
Oenothera mexicana IPNI=*613817-1* 430
Oenothera micrantha IPNI=*613817-1* 430
Oenothera minima IPNI=*613817-1* 430
Oenothera missouriensis f. *elongata* IPNI=*173157-2* 430
Oenothera missouriensis IPNI=*173157-2* 430
Oenothera missouriensis var. *latifolia* IPNI=*173157-2* 430
Oenothera multicaulis var. *petiolaris* IPNI=*173202-2* 430
Oenothera multicaulis var. *typica* IPNI=*173202-2* 430
Oenothera multiflora IPNI=*317809-2* 429
Oenothera multiflora var. *elliptica* IPNI=*317809-2* 429
Oenothera muricata IPNI=*172755-2* 429
Ocnothera muricata var. *latifolia* IPNI=*172755-2* 429
Oenothera muricata var. *rhodoneura* IPNI=*172755-2* 429
Oenothera muricata var. *rubricaulis* IPNI=*172755-2* 429
Oenothera mutabilis IPNI=*614141-1* 430
Oenothera nanella IPNI=*317809-2* 429
Oenothera nissensis IPNI=*172755-2* 429
Oenothera nissensis var. *fiedleri* IPNI=*172755-2* 429
Oenothera novae-scotiae var. *serratifolia* IPNI=*172755-2* 429
Oenothera nuda IPNI=*172755-2* 429
Oenothera numismatica IPNI=*172755-2* 429
Oenothera oblonga IPNI=*317809-2* 429
Oenothera obscurifolia IPNI=*172755-2* 429
Oenothera octolineata IPNI=*172755-2* 429
Oenothera octovalvis IPNI=*144385-2* 429
Oenothera oxypetala IPNI=*317809-2* 429
Oenothera pallescens IPNI=*317809-2* 429
Oenothera paludosa IPNI=*613333-1* 429
Oenothera paradoxa IPNI=*172755-2* 429
Oenothera paralamarckiana IPNI=*172755-2* 429
Oenothera parviflora var. *muricata* IPNI=*172755-2* 429
Oenothera pedemontana IPNI=*172755-2* 429
Oenothera pellegrinii IPNI=*172755-2* 429
Oenothera plicatula IPNI=*317809-2* 429
Oenothera pratincola f. *typica* IPNI=*172755-2* 429
Oenothera pratincola IPNI=*172755-2* 429
Oenothera pratincola var. *numismatica* IPNI=*172755-2* 429
Oenothera psychrophila IPNI=*173348-2* 430
Oenothera pubescens IPNI=*172755-2* 429
Oenothera punctulata IPNI=*172755-2* 429
Oenothera purpurata IPNI=*172755-2* 429
Oenothera purpurea IPNI=*173348-2* 430
Oenothera pycnocarpa IPNI=*172755-2* 429
Oenothera pycnocarpa var. *cleistogama* IPNI=*172755-2* 429
Oenothera pycnocarpa var. *parviflora* IPNI=*172755-2* 429
Oenothera pyramidiflora IPNI=*172755-2* 429
Oenothera renneri IPNI=*172755-2* 429
Oenothera repanda IPNI=*613817-1* 430
Oenothera reynoldsii f. *semialta* IPNI=*172755-2* 429
Oenothera reynoldsii f. *typica* IPNI=*172755-2* 429
Oenothera reynoldsii IPNI=*172755-2* 429
Oenothera rosea f. *epilobiifolia* IPNI=*173348-2* 430
Oenothera rosea var. *helleri* IPNI=*173348-2* 430
Oenothera rosea var. *parvifolia* IPNI=*173348-2* 430
Oenothera rostanskii IPNI=*172755-2* 429
Oenothera royfraseri IPNI=*172755-2* 429
Oenothera rubiennis IPNI=*172755-2* 429
Oenothera rubra IPNI=*173348-2* 430
Oenothera rubricalyx IPNI=*317809-2* 429
Oenothera rubricaulis IPNI=*172755-2* 429
Oenothera rubricaulis var. *dentifolia* IPNI=*172755-2* 429
Oenothera rubricaulis var. *longistylis* IPNI=*172755-2* 429
Oenothera rubricauloides IPNI=*172755-2* 429
Oenothera rubrinervis IPNI=*317809-2* 429
Oenothera rubrinervoides IPNI=*317809-2* 429
Oenothera rubritincta IPNI=*317809-2* 429
Oenothera ruderalis IPNI=*172755-2* 429
Oenothera sabulosa IPNI=*172755-2* 429
Oenothera sackvillensis IPNI=*172755-2* 429
Oenothera sackvillensis var. *albiviridis* IPNI=*172755-2* 429
Oenothera sackvillensis var. *royfraseri* IPNI=*172755-2* 429
Oenothera salicastrum IPNI=*172755-2* 429
Oenothera scindens IPNI=*317809-2* 429
Oenothera scintillans IPNI=*317809-2* 429
Oenothera semilata IPNI=*317809-2* 429
Oenothera sesitensis IPNI=*172755-2* 429
Oenothera shulliana IPNI=*172755-2* 429
Oenothera simplex IPNI=*317809-2* 429
Oenothera sinuata IPNI=*613817-1* 430
Oenothera sinuata subvar. *helleriana* IPNI=*613817-1* 430
Oenothera sinuata var. *bicolor* IPNI=*613817-1* 430
Oenothera sinuata var. *hirsuta* IPNI=*613817-1* 430
Oenothera spathulata IPNI=*317809-2* 429
Oenothera stenomeres f. *typica* IPNI=*172755-2* 429
Oenothera stenomeres IPNI=*172755-2* 429
Oenothera stucchii IPNI=*172755-2* 429
Oenothera suaveolens IPNI=*172755-2* 429
Oenothera suaveolens var. *latipetala* IPNI=*172755-2* 429
Oenothera sublinearis IPNI=*317809-2* 429
Oenothera subovata IPNI=*317809-2* 429
Oenothera superflua IPNI=*317809-2* 429
Oenothera tacikii IPNI=*172755-2* 429

Oenothera tarda IPNI=*317809-2* 429
Oenothera tardiflora IPNI=*317809-2* 429
Oenothera tetraptera var. *chenopodifolia* IPNI=*614141-1* 430
Oenothera tetraptera var. *depauperata* IPNI=*614141-1* 430
Oenothera tetraptera var. *immutabilis* IPNI=*614141-1* 430
Oenothera tetraptera var. *roseoidea* IPNI=*614141-1* 430
Oenothera tracyi IPNI=*172755-2* 429
Oenothera turoviensis IPNI=*172755-2* 429
Oenothera victorinii f. *rostanskii* IPNI=*172755-2* 429
Oenothera victorinii IPNI=*172755-2* 429
Oenothera victorinii var. *intermedia* IPNI=*172755-2* 429
Oenothera victorinii var. *parviflora* IPNI=*172755-2* 429
Oenothera victorinii var. *undulata* IPNI=*172755-2* 429
Oenothera virgata IPNI=*173348-2* 430
Oenothera viscosa IPNI=*613817-1* 430
Oenothera vrieseana IPNI=*317809-2* 429
Oenothera walpersii IPNI=*613817-1* 430
Oenothera wratislaviensis IPNI=*172755-2* 429
Oerstedella pseudowallisii IPNI=*633025-1* 219
Oerstedella wallisii IPNI=*633025-1* 219
Oetosis filiformis IPNI=*17234040-1* 178
Oetosis lineata IPNI=*17234040-1* 178
Oetosis lineata var. *filifolia* IPNI=*266216-2* 178
Oetosis lineata var. *filiformis* IPNI=*17234040-1* 178
Ogcodeia acreana IPNI=*167395-2* 440
Ogcodeia glabra IPNI=*167395-2* 440
Ogcodeia imitans IPNI=*167397-2* 440
Ogcodeia naga IPNI=*167403-2* 440
Ogcodeia oblongifolia IPNI=*1182949-2* 440
Ogcodeia pallescens IPNI=*167395-2* 440
Ogcodeia pseudonaga IPNI=*167405-2* 440
Ogcodeia sandwithiana IPNI=*167395-2* 440
Ogcodeia tamamuri IPNI=*167395-2* 440
Ogcodeia ternstroemiiflora IPNI=*584418-1* 440
Ogcodeia tessmannii IPNI=*167395-2* 440
Ogcodeia ulei IPNI=*167409-2* 440
Ogiera eleuthranthera IPNI=*202984-1* 261
Ogiera leiocarpa IPNI=*202984-1* 261
Ogiera ruderalis IPNI=*202984-1* 261
Ogiera triplinervis IPNI=*202984-1* 261
Ogiera triplinervis var. *leiocarpa* IPNI=*202984-1* 261
Oldenlandia cordata IPNI=*746945-1* 348
Olea chinensis IPNI=*610017-1* 375
Olea clavata IPNI=*610017-1* 375
Olea humilis IPNI=*610130-1* 375
Olea pallida IPNI=*610675-1* 375
Olea sativa IPNI=*610675-1* 375
Oleander indica IPNI=*80460-1* 342
Oleander vulgaris IPNI=*80460-1* 342
Oleandra articulata f. *eglandulosa* IPNI=*1138771-2* 176
Oleandra micans IPNI=*173662-2* 176
Oleandra neriiformis var. *pilosa* IPNI=*173662-2* 176
Oleandra nodosa IPNI=*1138771-2* 176
Oleandra nodosa var. *magalhaesii* IPNI=*1138771-2* 176
Oleandra trinitensis IPNI=*173662-2* 176
Oleandra trujillensis IPNI=*173662-2* 176
Oleiocarpon panamense IPNI=*492770-1* 318
Olfersia dissimilis IPNI=*89270-2* 175
Olfersia japurensis IPNI=*17442740-1* 175
Olfersia neriifolia IPNI=*89270-2* 175
Olfersia viscosa IPNI=*89270-2* 175
Oligactis apodocephala IPNI=*173688-2* 267
Oligactis biattenuata IPNI=*173688-2* 267
Oligactis valerii IPNI=*173688-2* 267
Oliganthes acuminata IPNI=*988439-1* 268
Oliganthes condensatus IPNI=*988439-1* 268
Oliganthes corei IPNI=*988437-1* 268
Oliganthes discolor IPNI=*988437-1* 268
Oliganthes ferruginea IPNI=*988437-1* 268
Oliganthes karstenii IPNI=*988437-1* 268
Oligosporus condimentarius IPNI=*307848-2* 257
Oligosporus crithmifolius IPNI=*307848-2* 257
Oligosporus dracunculiformis IPNI=*307848-2* 257
Oligosporus dracunculinus IPNI=*307848-2* 257
Oligosporus dracunculoides IPNI=*307848-2* 257
Oligosporus dracunculus IPNI=*307848-2* 257
Oligosporus nuttallianus IPNI=*307848-2* 257
Olmedia affinis IPNI=*584303-1* 440
Olmedia angustifolia IPNI=*190063-2* 440
Olmedia armata IPNI=*1059330-2* 441
Olmedia aspera IPNI=*945890-1* 441
Olmedia asperula IPNI=*584303-1* 440
Olmedia boyacana IPNI=*119542-2* 440
Olmedia calophylla IPNI=*152821-2* 440
Olmedia caucana IPNI=*945890-1* 441
Olmedia caurensis IPNI=*211411-2* 441
Olmedia erythrorhiza IPNI=*851229-1* 437
Olmedia falcifolia IPNI=*945890-1* 441
Olmedia ferruginea IPNI=*211411-2* 441
Olmedia grandifolia IPNI=*854491-1* 440
Olmedia guianensis IPNI=*854491-1* 440
Olmedia habas IPNI=*855103-1* 440
Olmedia laevis IPNI=*211411-2* 441
Olmedia laurina IPNI=*152822-2* 440
Olmedia maquira IPNI=*854491-1* 440
Olmedia maxima IPNI=*152822-2* 440
Olmedia mollis IPNI=*1008109-1* 440
Olmedia obliqua IPNI=*152822-2* 440
Olmedia poeppigiana IPNI=*584303-1* 440
Olmedia poeppigiana IPNI=*945890-1* 441
Olmedia polycephala IPNI=*584303-1* 440
Olmedia rigida IPNI=*211419-2* 441
Olmedia tomentosa IPNI=*584303-1* 440
Olmedia tovarensis IPNI=*119542-2* 440
Olmedia virgata IPNI=*240971-2* 441
Olmedioperebea calophylla IPNI=*152821-2* 440
Olmediophaena coriacea IPNI=*152822-2* 440
Olmediophaena maxima IPNI=*152822-2* 440
Olmediophaena obliqua IPNI=*152822-2* 440
Olmediopsis lanceolata IPNI=*211411-2* 441
Olmediopsis obliqua IPNI=*211411-2* 441
Oluntos laevigata IPNI=*316340-2* 438
Oluntos trigona IPNI=*853876-1* 439
Olyra arundinacea IPNI=*316519-2* 241
Olyra brasiliensis IPNI=*316519-2* 241
Olyra brevifolia IPNI=*316519-2* 241
Olyra cordifolia IPNI=*316519-2* 241
Olyra cordifolia var. *scabriuscula* IPNI=*316519-2* 241
Olyra guineensis IPNI=*316519-2* 241
Olyra latifolia var. *arundinacea* IPNI=*316519-2* 241
Olyra latifolia var. *glabriuscula* IPNI=*316519-2* 241
Olyra latifolia var. *pubescens* IPNI=*316519-2* 241
Olyra latifolia var. *vestita* IPNI=*316519-2* 241
Olyra media IPNI=*316519-2* 241
Olyra paniculata IPNI=*316519-2* 241
Olyra pubescens IPNI=*316519-2* 241
Olyra scabra IPNI=*316519-2* 241
Olyra surinamensis IPNI=*316519-2* 241
Omentaria alliacea IPNI=*542757-1* 214
Omentaria violacea IPNI=*542801-1* 214
Omonoia californica IPNI=*30043364-2* 435
Omoscleria flagellum IPNI=*230730-2* 233
Omphalandria diandra IPNI=*352565-1* 392
Omphalea cordata IPNI=*352565-1* 392
Omphalea diandra var. *genuina* IPNI=*352565-1* 392
Omphalea diandra var. *panamensis* IPNI=*352565-1* 392

Omphalea diandra var. *paraensis* IPNI=*352565-1* 392
Omphalea glandulata IPNI=*227503-2* 392
Omphalea guyanensis IPNI=*352565-1* 392
Omphalea lactescens IPNI=*350987-1* 392
Omphalea megacarpa IPNI=*352565-1* 392
Omphalea panamensis IPNI=*352565-1* 392
Omphalobium lambertii IPNI=*64181-2* 431
Omphalobium rubrum IPNI=*264496-1* 431
Onagra biennis IPNI=*172755-2* 429
Onagra chrysantha var. *latifolia* IPNI=*172755-2* 429
Onagra erythrosepala IPNI=*317809-2* 429
Onagra europaea IPNI=*172755-2* 429
Onagra gigas IPNI=*317809-2* 429
Onagra laevifolia IPNI=*317809-2* 429
Onagra lamarckiana IPNI=*317809-2* 429
Onagra media IPNI=*172755-2* 429
Onagra muricata IPNI=*172755-2* 429
Onagra nanella IPNI=*317809-2* 429
Onagra rubrinervis IPNI=*317809-2* 429
Onagra sinuata IPNI=*613817-1* 430
Onagra vulgaris IPNI=*172755-2* 429
Oncerum gallicum IPNI=*332106-2* 286
Oncidium alexandrae f. *hyperxanthum* IPNI=*77101415-1* 222
Oncidium allemanii IPNI=*1021256-1* 219
Oncidium halteratum IPNI=*75479-2* 218
Oncidium halteratum var. *fuscum* IPNI=*75479-2* 218
Oncidium halteratum var. *luteum* IPNI=*75479-2* 218
Oncidium iridifolium IPNI=*1021256-1* 219
Oncidium kramerianum IPNI=*212378-2* 222
Oncidium lanceanum f. *aureum* IPNI=*1021269-1* 223
Oncidium lanceanum IPNI=*1021269-1* 223
Oncidium nodosum IPNI=*212378-2* 222
Oncidium papilio f. *latourae* IPNI=*212379-2* 222
Oncidium papilio IPNI=*212379-2* 222
Oncidium papilio var. *albiflorum* IPNI=*212379-2* 222
Oncidium papilio var. *latourae* IPNI=*212379-2* 222
Oncidium papilio var. *limbatum* IPNI=*212379-2* 222
Oncidium papilioniforme IPNI=*212378-2* 222
Oncidium pusillum IPNI=*1021256-1* 219
Oncidium ramosissimum IPNI=*1021207-1* 218
Oncidium revolutum IPNI=*1021208-1* 218
Oncidium schrautianum IPNI=*323062-2* 222
Oncidium sphacelatum f. *xanthinum* IPNI=*323062-2* 222
Oncidium sphacelatum var. *majus* IPNI=*323062-2* 222
Oncidium sphacelatum var. *minus* IPNI=*323062-2* 222
Oncidium stenostalix IPNI=*323062-2* 222
Oncidium superbiens IPNI=*75479-2* 218
Oncoba laurina IPNI=*111811-1* 380
Oncoba maynensis IPNI=*140493-2* 380
Oncoba maynensis var. *laxiflora* IPNI=*140493-2* 380
Oncoba paludosa IPNI=*140493-2* 380
Oncorhiza esculentus IPNI=*317946-1* 225
Oncostylis junciformis f. *laxa* IPNI=*38049-2* 229
Oncostylis junciformis IPNI=*38049-2* 229
Oncostylis junciformis var. *ambigua* IPNI=*38049-2* 229
Oncostylis junciformis var. *congesta* IPNI=*38049-2* 229
Oncostylis junciformis var. *humboldtiana* IPNI=*38049-2* 229
Oncostylis junciformis var. *speciosior* IPNI=*38049-2* 229
Oncostylis junciformis var. *stricta* IPNI=*38049-2* 229
Oncus esculentus IPNI=*317946-1* 225
Onobrychis heterophylla IPNI=*77187443-1* 320
Onohualcoa fissa IPNI=*152688-2* 364
Onohualcoa helicocalyx IPNI=*152688-2* 364
Onohualcoa seleri IPNI=*152688-2* 364
Onohualcoa verrucifera IPNI=*152688-2* 364
Ononis anil IPNI=*500063-1* 321
Ononis cerrifolia IPNI=*247544-2* 334
Ononis conjugata IPNI=*54479-2* 314
Ononis crotalarioides IPNI=*486336-1* 315
Onoseris conspicua IPNI=*174455-2* 267
Onoseris isotypus IPNI=*174455-2* 267
Onoseris latifolia IPNI=*231347-1* 265
Onoseris mexicana IPNI=*231341-1* 265
Onoseris paniculata IPNI=*174455-2* 267
Onoseris purpurata IPNI=*77181848-1* 267
Onoseris rupestris IPNI=*174455-2* 267
Onoseris silvatica var. *colombiana* IPNI=*174463-2* 267
Onoseris triplinervia IPNI=*231347-1* 265
Onoseris turbacensis IPNI=*231341-1* 265
Onosuris acuminata IPNI=*172755-2* 429
Onychacanthus arboreus IPNI=*36223-2* 359
Onychacanthus cumingii IPNI=*36223-2* 359
Onychacanthus speciosus IPNI=*36223-2* 359
Onychium capense IPNI=*17210060-1* 175
Onychium flavescens IPNI=*654273-1* 222
Ooclinium clavatum IPNI=*57410-2* 259
Opa japonica IPNI=*731389-1* 444
Opa metrosideros IPNI=*731389-1* 444
Opetiola myosuroides IPNI=*304256-1* 230
Ophioglossum austroasiaticum IPNI=*174612-2* 173
Ophioglossum cognatum IPNI=*174612-2* 173
Ophioglossum cordifolium IPNI=*174612-2* 173
Ophioglossum cumingianum IPNI=*174612-2* 173
Ophioglossum holm-nielsenii IPNI=*174612-2* 173
Ophioglossum moluccanum f. *dilatatum* IPNI=*174612-2* 173
Ophioglossum obovatum IPNI=*174612-2* 173
Ophioglossum ovatum IPNI=*174612-2* 173
Ophioglossum pedunculatum IPNI=*174612-2* 173
Ophioglossum peruvianum IPNI=*174612-2* 173
Ophioglossum raciborskii IPNI=*174612-2* 173
Ophioglossum ramosii IPNI=*174612-2* 173
Ophioglossum reticulatum f. *dilatatum* IPNI=*174612-2* 173
Ophioglossum reticulatum var. *acutius* IPNI=*174612-2* 173
Ophioglossum reticulatum var. *polyangium* IPNI=*174612-2* 173
Ophioglossum scandens IPNI=*17143520-1* 179
Ophioglossum timorense IPNI=*174612-2* 173
Ophioglossum usterianum IPNI=*174612-2* 173
Ophioglossum vulgatum var. *minutum* IPNI=*174612-2* 173
Ophioglossum vulgatum var. *reticulatum* IPNI=*174612-2* 173
Ophiorrhiza lanceolata IPNI=*760275-1* 355
Ophiuros appendiculatus IPNI=*909653-1* 244
Ophiuros loricatus IPNI=*325559-2* 244
Ophiuros plicatus IPNI=*325559-2* 244
Ophryoscleria asperrima IPNI=*314849-1* 232
Ophryoscleria paludosa IPNI=*314849-1* 232
Ophrys nervosa IPNI=*641061-1* 219
Opizia bursoides IPNI=*30092589-2* 276
Oplismenus abortivus IPNI=*396213-1* 236
Oplismenus acuminatus IPNI=*1004002-1* 242
Oplismenus aemulus IPNI=*1004002-1* 242
Oplismenus aemulus var. *densiflorus* IPNI=*1004002-1* 242
Oplismenus aemulus var. *flaccidus* IPNI=*1004002-1* 242
Oplismenus aemulus var. *lasiorhachis* IPNI=*1004002-1* 242
Oplismenus affinis IPNI=*30453766-2* 242
Oplismenus affinis var. *humboldtianus* IPNI=*30453766-2* 242
Oplismenus africanus IPNI=*1004002-1* 242
Oplismenus albus IPNI=*30453766-2* 242
Oplismenus angustifolius IPNI=*30149942-2* 238
Oplismenus aristulatus IPNI=*1004002-1* 242
Oplismenus baronii IPNI=*30453766-2* 242
Oplismenus borhidii IPNI=*1004002-1* 242
Oplismenus brasiliensis IPNI=*1004002-1* 242
Oplismenus brevisetus IPNI=*1004002-1* 242
Oplismenus bromoides IPNI=*30453766-2* 242
Oplismenus burmanni var. *multisetus* IPNI=*30453766-2* 242
Oplismenus burmanni var. *nudicaulis* IPNI=*30453766-2* 242

Oplismenus chondrosioides IPNI=*1004002-1* 242
Oplismenus colonum IPNI=*399655-1* 238
Oplismenus compositus f. *vittatus* IPNI=*409823-1* 242
Oplismenus compositus subsp. *patens* IPNI=*409823-1* 242
Oplismenus compositus var. *angustifolius* IPNI=*409823-1* 242
Oplismenus compositus var. *formosanus* IPNI=*409823-1* 242
Oplismenus compositus var. *loliaceus* IPNI=*1004002-1* 242
Oplismenus compositus var. *rariflorus* IPNI=*409823-1* 242
Oplismenus compositus var. *submuticus* IPNI=*409823-1* 242
Oplismenus compositus var. *sylvaticus* IPNI=*409823-1* 242
Oplismenus compositus var. *vittatus* IPNI=*409823-1* 242
Oplismenus cristatus IPNI=*30453766-2* 242
Oplismenus crus-galli IPNI=*399660-1* 238
Oplismenus crus-pavonis IPNI=*30149942-2* 238
Oplismenus cubensis IPNI=*1004002-1* 242
Oplismenus cuspidatus IPNI=*399655-1* 238
Oplismenus decompositus IPNI=*409823-1* 242
Oplismenus depauperatus IPNI=*1004002-1* 242
Oplismenus elatior IPNI=*409823-1* 242
Oplismenus flaccidus IPNI=*1004002-1* 242
Oplismenus foliaceus IPNI=*1004002-1* 242
Oplismenus formosanus IPNI=*409823-1* 242
Oplismenus gracilis IPNI=*30453766-2* 242
Oplismenus gracillimus IPNI=*30453766-2* 242
Oplismenus helvolus IPNI=*1207294-2* 245
Oplismenus hirtellus f. *imbecillis* IPNI=*1004002-1* 242
Oplismenus hirtellus f. *lanceolatus* IPNI=*1004002-1* 242
Oplismenus hirtellus subsp. *acuminatus* IPNI=*1004002-1* 242
Oplismenus hirtellus subsp. *capensis* IPNI=*1004002-1* 242
Oplismenus hirtellus subsp. *fasciculatus* IPNI=*1004002-1* 242
Oplismenus hirtellus subsp. *japonicus* IPNI=*30453766-2* 242
Oplismenus hirtellus subsp. *microphyllus* IPNI=*1004002-1* 242
Oplismenus hirtellus subsp. *psilostachys* IPNI=*1004002-1* 242
Oplismenus hirtellus subsp. *tsushimensis* IPNI=*1004002-1* 242
Oplismenus hirtellus var. *imbecillis* IPNI=*1004002-1* 242
Oplismenus hirtellus var. *microphyllus* IPNI=*1004002-1* 242
Oplismenus hookeri IPNI=*30453766-2* 242
Oplismenus humbertianus IPNI=*30453766-2* 242
Oplismenus humboldtianus IPNI=*30453766-2* 242
Oplismenus humboldtianus var. *muticus* IPNI=*30453766-2* 242
Oplismenus humboldtianus var. *nudicaulis* IPNI=*30453766-2* 242
Oplismenus imbecillis IPNI=*1004002-1* 242
Oplismenus indicus IPNI=*409823-1* 242
Oplismenus intermedius IPNI=*1207294-2* 245
Oplismenus jamaicensis IPNI=*30149942-2* 238
Oplismenus japonicus IPNI=*30453766-2* 242
Oplismenus junghuhnii IPNI=*409823-1* 242
Oplismenus lanceolatus IPNI=*409823-1* 242
Oplismenus latifolius IPNI=*1004002-1* 242
Oplismenus liebmannii IPNI=*409823-1* 242
Oplismenus loliaceus IPNI=*1004002-1* 242
Oplismenus margaritaceus IPNI=*399655-1* 238
Oplismenus mollissimus IPNI=*30453766-2* 242
Oplismenus multisetus IPNI=*30453766-2* 242
Oplismenus muticus IPNI=*399655-1* 238
Oplismenus oahuaensis IPNI=*1004002-1* 242
Oplismenus oplismenoides IPNI=*211125-2* 244
Oplismenus owatarii IPNI=*409823-1* 242
Oplismenus parvifolius IPNI=*1004002-1* 242
Oplismenus patens IPNI=*409823-1* 242
Oplismenus patens var. *angustifolius* IPNI=*409823-1* 242
Oplismenus patens var. *yunnanensis* IPNI=*409823-1* 242
Oplismenus polliniifolius IPNI=*409823-1* 242
Oplismenus polystachyus IPNI=*88097-2* 238
Oplismenus pratensis IPNI=*409823-1* 242
Oplismenus preslii IPNI=*30453766-2* 242
Oplismenus pseudocolonus IPNI=*399655-1* 238
Oplismenus pseudoundulatifolius IPNI=*77105897-1* 241
Oplismenus psilostachys IPNI=*1004002-1* 242
Oplismenus rariflorus IPNI=*409823-1* 242
Oplismenus repens IPNI=*399655-1* 238
Oplismenus sabulicola IPNI=*30149942-2* 238
Oplismenus setarius IPNI=*409823-1* 242
Oplismenus spectabilis IPNI=*88097-2* 238
Oplismenus sylvaticus IPNI=*409823-1* 242
Oplismenus thiebautii IPNI=*409823-1* 242
Oplismenus tsushimensis IPNI=*1004002-1* 242
Oplismenus undulatifolius var. *imbecillis* IPNI=*1004002-1* 242
Oplismenus velutinus IPNI=*1004002-1* 242
Opsantha amarella IPNI=*108639-2* 344
Opuntia airampo IPNI=*137078-1* 284
Opuntia apurimacensis IPNI=*1151735-2* 284
Opuntia arborea IPNI=*129102-1* 283
Opuntia arcei IPNI=*1151735-2* 284
Opuntia argentina IPNI=*129102-1* 283
Opuntia bahamana IPNI=*137078-1* 284
Opuntia bahiensis IPNI=*129102-1* 283
Opuntia balearica IPNI=*137078-1* 284
Opuntia bartramii IPNI=*137078-1* 284
Opuntia bella IPNI=*175432-2* 284
Opuntia bentonii IPNI=*137078-1* 284
Opuntia bergeriana IPNI=*136561-1* 284
Opuntia bisetosa IPNI=*136428-1* 284
Opuntia boldinghii IPNI=*136561-1* 284
Opuntia bradleyi IPNI=*128942-1* 283
Opuntia brasiliensis IPNI=*129102-1* 283
Opuntia brasiliensis subsp. *bahiensis* IPNI=*129102-1* 283
Opuntia brasiliensis subsp. *subacarpa* IPNI=*129102-1* 283
Opuntia caribaea IPNI=*73852-2* 283
Opuntia castillae IPNI=*1151735-2* 284
Opuntia chinensis IPNI=*1151735-2* 284
Opuntia coccinea IPNI=*136561-1* 284
Opuntia coccinellifera IPNI=*1008816-2* 284
Opuntia cordobensis IPNI=*1151735-2* 284
Opuntia cylindrica IPNI=*128942-1* 283
Opuntia dillenii var. *orbiculata* IPNI=*136534-1* 284
Opuntia dillenii var. *reitzii* IPNI=*137078-1* 284
Opuntia ellemetiana IPNI=*27133-2* 283
Opuntia exuviata IPNI=*73905-2* 283
Opuntia exuviata var. *angustior* IPNI=*73905-2* 283
Opuntia exuviata var. *spinosior* IPNI=*73905-2* 283
Opuntia exuviata var. *viridior* IPNI=*73905-2* 283
Opuntia ficus-indica f. *reticulata* IPNI=*1151735-2* 284
Opuntia ficus-indica var. *elongata* IPNI=*1151735-2* 284
Opuntia ficus-indica var. *saboten* IPNI=*1151735-2* 284
Opuntia flexibilis IPNI=*136561-1* 284
Opuntia furiosa IPNI=*73905-2* 283
Opuntia fusicaulis IPNI=*1151735-2* 284
Opuntia hieronymi IPNI=*129102-1* 283
Opuntia horrida IPNI=*136561-1* 284
Opuntia humilis IPNI=*136561-1* 284
Opuntia incarnadilla IPNI=*1151735-2* 284
Opuntia inermis IPNI=*137078-1* 284
Opuntia jamaicensis IPNI=*137125-1* 284
Opuntia joconostle IPNI=*1151735-2* 284
Opuntia keyensis IPNI=*137078-1* 284
Opuntia ledienii IPNI=*136561-1* 284
Opuntia leiascheinvariana IPNI=*1151735-2* 284
Opuntia leucostata IPNI=*1151735-2* 284
Opuntia longiclada IPNI=*137078-1* 284
Opuntia macbridei IPNI=*175475-2* 284
Opuntia macbridei var. *orbicularis* IPNI=*175475-2* 284
Opuntia macrocalyx IPNI=*136806-1* 284
Opuntia magnifica IPNI=*137078-1* 284
Opuntia magnifolia IPNI=*1008816-2* 284
Opuntia maidenii IPNI=*136561-1* 284

Opuntia maritima IPNI=*136534-1* 284
Opuntia megacantha IPNI=*1151735-2* 284
Opuntia megacantha var. *tenuispina* IPNI=*1151735-2* 284
Opuntia megalantha IPNI=*136561-1* 284
Opuntia melanosperma IPNI=*136534-1* 284
Opuntia metuenda IPNI=*73852-2* 283
Opuntia mexicana IPNI=*1151735-2* 284
Opuntia microdasys f. *minor* IPNI=*136806-1* 284
Opuntia microdasys subvar. *albiflora* IPNI=*136806-1* 284
Opuntia microdasys var. *albispina* IPNI=*136806-1* 284
Opuntia microdasys var. *laevior* IPNI=*136806-1* 284
Opuntia microdasys var. *minor* IPNI=*136806-1* 284
Opuntia multiflora IPNI=*136561-1* 284
Opuntia nejapensis IPNI=*137078-1* 284
Opuntia neoargentina IPNI=*129102-1* 283
Opuntia nigricans IPNI=*136561-1* 284
Opuntia nuda IPNI=*1008816-2* 284
Opuntia obovata IPNI=*1151735-2* 284
Opuntia paraguayensis IPNI=*1151735-2* 284
Opuntia parva IPNI=*137078-1* 284
Opuntia pennellii IPNI=*136428-1* 284
Opuntia phyllanthus IPNI=*132939-1* 283
Opuntia polyantha IPNI=*136561-1* 284
Opuntia pseudotuna IPNI=*136561-1* 284
Opuntia puelchana IPNI=*73905-2* 283
Opuntia pulvinata IPNI=*136806-1* 284
Opuntia robusta var. *megacantha* IPNI=*1151735-2* 284
Opuntia salvadorensis IPNI=*136428-1* 284
Opuntia schulzii IPNI=*129102-1* 283
Opuntia segethii IPNI=*27133-2* 283
Opuntia spinaurea IPNI=*136561-1* 284
Opuntia stapeliae IPNI=*73905-2* 283
Opuntia stricta subsp. *esparzae* IPNI=*137078-1* 284
Opuntia stricta var. *dillenii* IPNI=*136534-1* 284
Opuntia stricta var. *reitzii* IPNI=*137078-1* 284
Opuntia subinermis IPNI=*1008816-2* 284
Opuntia subulata IPNI=*27133-2* 283
Opuntia tenuiflora IPNI=*137078-1* 284
Opuntia tribuloides IPNI=*1151735-2* 284
Opuntia tuna var. *cubaguensis* IPNI=*136561-1* 284
Opuntia tuna var. *manse* IPNI=*136561-1* 284
Opuntia tuna-blanca IPNI=*1151735-2* 284
Opuntia tunicata IPNI=*73905-2* 283
Opuntia tunicata var. *chilensis* IPNI=*73905-2* 283
Opuntia tunicata var. *laevior* IPNI=*73905-2* 283
Opuntia undosa IPNI=*1151735-2* 284
Opuntia undulata IPNI=*1151735-2* 284
Opuntia undulata IPNI=*73905-2* 283
Opuntia vulgaris var. *balaerica* IPNI=*137078-1* 284
Opuntia wentiana IPNI=*136428-1* 284
Opuntia zebrina IPNI=*136534-1* 284
Orbignya agrestis IPNI=*319686-2* 205
Orbignya cohune IPNI=*26558-2* 205
Orbignya cuatrecasasiana IPNI=*77177001-1* 205
Orbignya dammeriana IPNI=*26558-2* 205
Orbignya luetzelburgii IPNI=*961936-1* 205
Orbignya microcarpa IPNI=*319686-2* 205
Orbignya polysticha IPNI=*319686-2* 205
Orbignya racemosa IPNI=*664463-1* 205
Orbignya sabulosa IPNI=*319686-2* 205
Orbignya sagotii IPNI=*319686-2* 205
Orchiodes quercilobum IPNI=*77125705-1* 217
Orcya adhaerens IPNI=*1692-2* 255
Oreinotinus ferrugineus IPNI=*264840-2* 296
Oreinotinus glabratus IPNI=*149702-1* 296
Oreinotinus lasiophyllus IPNI=*149770-1* 296
Oreinotinus laurifolius IPNI=*149985-1* 297
Oreinotinus pichinchensis IPNI=*30356705-2* 296
Oreinotinus tinoides IPNI=*149971-1* 297
Oreinotinus triphyllus IPNI=*149985-1* 297
Orelia grandiflora IPNI=*319065-2* 339
Orellana americana IPNI=*33335-2* 406
Orellana americana var. *leiocarpa* IPNI=*33335-2* 406
Orellana americana var. *normalis* IPNI=*33335-2* 406
Orellana americana var. *urucurana* IPNI=*111504-1* 406
Orellana orellana IPNI=*33335-2* 406
Oreobroma acaule IPNI=*698305-1* 286
Oreobroma megarhizum IPNI=*698305-1* 286
Oreobroma mexicanum IPNI=*698305-1* 286
Oreodaphne aciphylla IPNI=*171499-2* 184
Oreodaphne acutangula IPNI=*171503-2* 184
Oreodaphne acutifolia var. *latifolia* IPNI=*1032261-2* 185
Oreodaphne alba IPNI=*77173628-1* 182
Oreodaphne bofo IPNI=*467161-1* 184
Oreodaphne caracasana IPNI=*171568-2* 184
Oreodaphne caudata IPNI=*171727-2* 184
Oreodaphne cernua IPNI=*171727-2* 184
Oreodaphne coriacea IPNI=*281229-2* 184
Oreodaphne costulata IPNI=*171499-2* 184
Oreodaphne domingensis IPNI=*1146025-2* 184
Oreodaphne evonymodaphne IPNI=*465248-1* 183
Oreodaphne floribunda IPNI=*1146025-2* 184
Oreodaphne glomerata IPNI=*171672-2* 184
Oreodaphne grandifolia IPNI=*171526-2* 184
Oreodaphne guianensis IPNI=*467336-1* 184
Oreodaphne guianensis var. *argentea* IPNI=*467336-1* 184
Oreodaphne guianensis var. *aurea* IPNI=*467336-1* 184
Oreodaphne helicterifolia IPNI=*326508-2* 184
Oreodaphne hookeriana IPNI=*158649-2* 183
Oreodaphne hostmanniana IPNI=*1032261-2* 185
Oreodaphne leptobotra IPNI=*171727-2* 184
Oreodaphne leptobotra var. *attenuata* IPNI=*171727-2* 184
Oreodaphne leptobotra var. *petiolaris* IPNI=*171727-2* 184
Oreodaphne longifolia IPNI=*171526-2* 184
Oreodaphne marowynensis IPNI=*171727-2* 184
Oreodaphne martiniana IPNI=*1032261-2* 185
Oreodaphne martiniana var. *dubia* IPNI=*1032261-2* 185
Oreodaphne martiniana var. *latifolia* IPNI=*1032261-2* 185
Oreodaphne martiniana var. *opaca* IPNI=*1032261-2* 185
Oreodaphne mexicana IPNI=*326508-2* 184
Oreodaphne mexicana var. *longipes* IPNI=*326508-2* 184
Oreodaphne mexicana var. *subsessilis* IPNI=*326508-2* 184
Oreodaphne moritziana IPNI=*171672-2* 184
Oreodaphne neesiana IPNI=*171499-2* 184
Oreodaphne oleifera IPNI=*77198011-1* 183
Oreodaphne opifera IPNI=*171526-2* 184
Oreodaphne perseoides IPNI=*1032261-2* 185
Oreodaphne poeppigiana IPNI=*1146025-2* 184
Oreodaphne retroflexa IPNI=*1146025-2* 184
Oreodaphne schomburgkiana IPNI=*171883-2* 185
Oreodaphne schomburgkiana var. *sparsiflora* IPNI=*171883-2* 185
Oreodaphne sieberi IPNI=*171727-2* 184
Oreodaphne splendens IPNI=*171904-2* 185
Oreodaphne strumosa IPNI=*466773-1* 184
Oreodaphne subalpina IPNI=*171727-2* 184
Oreodaphne terminalis IPNI=*171727-2* 184
Oreodaphne terminalis IPNI=*465248-1* 183
Oreodaphne umbrosa var. *bullata* IPNI=*326508-2* 184
Oreodaphne urophylla IPNI=*171727-2* 184
Oreodaphne warmingii IPNI=*1032261-2* 185
Oreodaphne willdenoviana IPNI=*1146025-2* 184
Oreodoxa acuminata IPNI=*210057-2* 211
Oreodoxa caribaea IPNI=*222805-2* 212
Oreodoxa oleracea IPNI=*222805-2* 212
Oreodoxa regia IPNI=*222809-2* 212
Oreodoxa regia var. *jenmanii* IPNI=*222805-2* 212

Oreodoxa sancona IPNI=*670066-1* 212
Oreomyrrhis andicola IPNI=*77085100-1* 252
Oreomyrrhis andicola var. *tomentosa* IPNI=*77085100-1* 252
Oreomyrrhis andina IPNI=*77085100-1* 252
Oreomyrrhis brachycarpa IPNI=*77085100-1* 252
Oreomyrrhis lasiopetala IPNI=*77085100-1* 252
Oreopanax capitatus var. *catalpifolius* IPNI=*91160-1* 254
Oreopanax capitatus var. *minor* IPNI=*30050876-2* 254
Oreopanax capitatus var. *multiflorus* IPNI=*30050876-2* 254
Oreopanax destructor IPNI=*30050876-2* 254
Oreopanax floribundus IPNI=*91222-1* 254
Oreopanax liebmannii IPNI=*30050876-2* 254
Oreopanax meiocephalus IPNI=*30050876-2* 254
Oreopanax microcephalus IPNI=*30050876-2* 254
Oreopanax morototoni IPNI=*90355-1* 253
Oreopanax septemnervius IPNI=*91160-1* 254
Oreophila sessiliflora IPNI=*317594-2* 264
Oreophila taraxacifolia IPNI=*296211-2* 264
Oreosciadium andinum IPNI=*840899-1* 252
Oreosciadium glaucescens IPNI=*170074-2* 252
Oreosciadium montanum IPNI=*170084-2* 252
Oreosciadium montanum subvar. *glaucescens* IPNI=*170074-2* 252
Oreosciadium montanum var. *ranunculifolium* IPNI=*170084-2* 252
Orestion vaccinioides IPNI=*599434-1* 427
Oribasia officinalis IPNI=*77161098-1* 354
Oribasia racemosa IPNI=*77115627-1* 354
Oribasia violacea IPNI=*759372-1* 354
Origanum × *acinifolium* IPNI=*453305-1* 372
Origanum × *applii nothosubsp. majoricum* IPNI=*453305-1* 372
Origanum × *balearicum* IPNI=*453305-1* 372
Origanum × *lusitanicum* IPNI=*453305-1* 372
Origanum × *paui* IPNI=*453305-1* 372
Origanum confertum IPNI=*453303-1* 372
Origanum dubium IPNI=*453303-1* 372
Origanum floridum IPNI=*453395-1* 372
Origanum majoranoides IPNI=*453303-1* 372
Origanum odorum IPNI=*453303-1* 372
Origanum salvifolium IPNI=*453303-1* 372
Origanum thymus IPNI=*461765-1* 373
Ormenis aurea IPNI=*192489-1* 259
Ormenis nobilis IPNI=*192489-1* 259
Ormenis nobilis subsp. *aurea* IPNI=*192489-1* 259
Ormenis nobilis var. *discoidea* IPNI=*192489-1* 259
Ormosia amazonica var. *venenifera* IPNI=*176344-2* 329
Ormosia chlorocalyx IPNI=*176377-2* 329
Ormosia cuneata IPNI=*510681-1* 329
Ormosia euneura IPNI=*176344-2* 329
Ormosia glazioviana IPNI=*510703-1* 329
Ormosia heterophylla IPNI=*176389-2* 329
Ormosia pacimonensis IPNI=*524446-1* 336
Ormosia toledoana IPNI=*176377-2* 329
Ornithidium album IPNI=*643592-1* 221
Ornithidium bicolor IPNI=*155165-2* 221
Ornithidium cordyline IPNI=*984957-1* 220
Ornithidium cyperifolium IPNI=*154952-2* 220
Ornithidium dolichophyllum IPNI=*154952-2* 220
Ornithidium fulgens IPNI=*155059-2* 220
Ornithidium giganteum IPNI=*984957-1* 220
Ornithidium miniatum IPNI=*155165-2* 221
Ornithidium nanum IPNI=*155331-2* 222
Ornithidium ruberrimum IPNI=*155262-2* 221
Ornithidium sophronitis IPNI=*155290-2* 221
Ornithidium squamatum IPNI=*155331-2* 222
Ornithidium vagans IPNI=*155262-2* 221
Ornithogalum bicolor IPNI=*539581-1* 215
Ornithogalum bracteatum IPNI=*77097737-1* 214
Ornithogalum caudatum IPNI=*77097737-1* 214
Ornithogalum ceresianum IPNI=*539581-1* 215
Ornithogalum coarctatum IPNI=*539581-1* 215
Ornithogalum conicum IPNI=*539581-1* 215
Ornithogalum gilgianum IPNI=*539581-1* 215
Ornithogalum grimaldiae IPNI=*539581-1* 215
Ornithogalum hermannii IPNI=*539581-1* 215
Ornithogalum longibracteatum IPNI=*77097737-1* 214
Ornithogalum massonii IPNI=*77097737-1* 214
Ornithogalum revolutum IPNI=*539581-1* 215
Ornithogalum rubrum IPNI=*304977-2* 224
Ornithogalum scilloides IPNI=*77097737-1* 214
Ornithophora quadricolor IPNI=*77101300-1* 219
Ornithophora radicans f. *alba* IPNI=*77101300-1* 219
Ornithophora radicans IPNI=*77101300-1* 219
Ornithopteris adiantifolia IPNI=*1031990-2* 179
Ornithopteris aquilina IPNI=*17210060-1* 175
Ornithopteris caudata IPNI=*213799-2* 175
Ornus europaea IPNI=*609128-1* 374
Orobanche abyssinica IPNI=*53607-3* 375
Orobanche ambigua IPNI=*53607-3* 375
Orobanche angelicifixa IPNI=*53607-3* 375
Orobanche apiculata IPNI=*53607-3* 375
Orobanche arvensis IPNI=*53607-3* 375
Orobanche barbata IPNI=*53607-3* 375
Orobanche boissieri IPNI=*53607-3* 375
Orobanche bovei IPNI=*53607-3* 375
Orobanche cernua IPNI=*382786-1* 368
Orobanche columbariae IPNI=*53607-3* 375
Orobanche columbiana IPNI=*53607-3* 375
Orobanche concolor IPNI=*53607-3* 375
Orobanche crithmi IPNI=*53607-3* 375
Orobanche crithmi-maritimi IPNI=*53607-3* 375
Orobanche densiflora IPNI=*53607-3* 375
Orobanche hians IPNI=*53607-3* 375
Orobanche hyalina IPNI=*53607-3* 375
Orobanche hydrocotylei IPNI=*53607-3* 375
Orobanche langei IPNI=*53607-3* 375
Orobanche leonuri IPNI=*53607-3* 375
Orobanche leucantha IPNI=*53607-3* 375
Orobanche major f. *hypochoeridis* IPNI=*53607-3* 375
Orobanche maritima IPNI=*53607-3* 375
Orobanche minor var. *hypochoeridis* IPNI=*53607-3* 375
Orobanche nudiflora IPNI=*53607-3* 375
Orobanche paralias IPNI=*53607-3* 375
Orobanche polymorpha IPNI=*53607-3* 375
Orobanche pyrrha IPNI=*53607-3* 375
Orobanche racemosa IPNI=*53607-3* 375
Orobanche reuteri IPNI=*53607-3* 375
Orobanche salisii IPNI=*53607-3* 375
Orobanche scabiosae-columbariae IPNI=*53607-3* 375
Orobanche speciosa IPNI=*53607-3* 375
Orobanche trifolifixa IPNI=*53607-3* 375
Orobanche trifolii-pratensis IPNI=*53607-3* 375
Orobanche unicolor IPNI=*53607-3* 375
Orobus bimarginatus IPNI=*501990-1* 325
Orobus faba IPNI=*524737-1* 336
Orobus lens IPNI=*524872-1* 336
Orobus magellanicus IPNI=*501845-1* 325
Orobus procumbens IPNI=*1071691-2* 335
Orobus sericeus IPNI=*250423-2* 335
Orobus trifoliatus IPNI=*491595-1* 317
Orobus uninervius IPNI=*323132-2* 325
Orontium majus IPNI=*799180-1* 375
Orophoma carana IPNI=*319735-2* 210
Orophoma subinermis IPNI=*154909-2* 210
Orsopea fetida IPNI=*245150-2* 414
Ortgiesia tillandsioides IPNI=*121847-1* 228
Ortgiesia tillandsioides var. *nidulans* IPNI=*121847-1* 228
Orthaea lehmannii IPNI=*1064712-2* 298

Orthaea panurensis IPNI=*77163666-1* 300
Orthanthe fyfiana IPNI=*382786-1* 368
Orthogramma gilliesii IPNI=*33533-2* 174
Ortholoma acuminatum IPNI=*63508-2* 366
Ortholoma anisophyllum IPNI=*63508-2* 366
Ortholoma calotrichum IPNI=*63523-2* 366
Ortholoma dictyophyllum IPNI=*63557-2* 366
Ortholoma parviflorum IPNI=*63636-2* 366
Ortholoma vestitum IPNI=*63508-2* 366
Ortholoma warscewiczianum IPNI=*63508-2* 366
Ortholoma warszewiczii IPNI=*63508-2* 366
Orthopogon aemulus IPNI=*1004002-1* 242
Orthopogon africanus IPNI=*1004002-1* 242
Orthopogon albus IPNI=*30453766-2* 242
Orthopogon burmanni IPNI=*30453766-2* 242
Orthopogon compositus IPNI=*409823-1* 242
Orthopogon crus-galli IPNI=*399660-1* 238
Orthopogon cubensis IPNI=*1004002-1* 242
Orthopogon dichotomus IPNI=*399655-1* 238
Orthopogon flaccidus IPNI=*1004002-1* 242
Orthopogon gonyrrhizus IPNI=*409823-1* 242
Orthopogon hirsutus IPNI=*88097-2* 238
Orthopogon hirtellus IPNI=*1004002-1* 242
Orthopogon imbecillis IPNI=*1004002-1* 242
Orthopogon junghuhnii IPNI=*409823-1* 242
Orthopogon loliaceus IPNI=*1004002-1* 242
Orthopogon longeracemosus IPNI=*409823-1* 242
Orthopogon parvifolius IPNI=*1004002-1* 242
Orthopogon pratensis IPNI=*409823-1* 242
Orthopogon remotus IPNI=*409823-1* 242
Orthopogon setarius IPNI=*409823-1* 242
Orthopogon subverticillatus IPNI=*399655-1* 238
Orthopogon sylvaticus IPNI=*409823-1* 242
Orthopogon velutinus IPNI=*1004002-1* 242
Orthosiphon spicatus IPNI=*1150341-2* 368
Orthosporum ambrosioides IPNI=*1198060-2* 282
Orthosporum anthelminticum IPNI=*1090820-2* 282
Orthosporum suffruticosum IPNI=*1198060-2* 282
Orthostemon erectus IPNI=*366552-1* 344
Orthostemon huegelii IPNI=*366552-1* 344
Orthostemon kirkii IPNI=*366552-1* 344
Orthostemon obovatus IPNI=*596467-1* 426
Orthostemon sellowianus IPNI=*596467-1* 426
Orthotactus ciliatus IPNI=*132826-2* 360
Orthotactus felisbertianus IPNI=*50468-1* 360
Orthotactus pohlianus IPNI=*132826-2* 360
Orthotactus pohlianus var. *angustifolius* IPNI=*132826-2* 360
Orthrosanthus chimboracensis var. *centroamericanus* IPNI=*176894-2* 216
Orthrosanthus chimboracensis var. *gladioloides* IPNI=*440470-1* 216
Orthrosanthus chimboracensis var. *intermedius* IPNI=*176894-2* 216
Orthrosanthus chimboracensis var. *typicus* IPNI=*440470-1* 216
Orthrosanthus moritzianus IPNI=*440470-1* 216
Orucaria lunata IPNI=*148561-2* 326
Oryctanthes ligustrina IPNI=*176899-2* 448
Oryctanthus alveolatus var. *amplexicaulis* IPNI=*176899-2* 448
Oryctanthus alveolatus var. *kuijtii* IPNI=*176899-2* 448
Oryctanthus amazonicus IPNI=*176899-2* 448
Oryctanthus amplexicaulis IPNI=*176899-2* 448
Oryctanthus botryostachys IPNI=*176899-2* 448
Oryctanthus botryostachyus IPNI=*176899-2* 448
Oryctanthus laceratus IPNI=*176899-2* 448
Oryctanthus ligustrinus IPNI=*176899-2* 448
Oryctanthus lucarquensis IPNI=*551737-1* 449
Oryza alta IPNI=*316809-2* 242
Oryza aristata IPNI=*316812-2* 242
Oryza communissima IPNI=*316812-2* 242
Oryza cubensis IPNI=*410115-1* 242
Oryza denudata IPNI=*316812-2* 242
Oryza elongata IPNI=*316812-2* 242
Oryza fatua IPNI=*410115-1* 242
Oryza formosana IPNI=*316812-2* 242
Oryza glutinosa IPNI=*316812-2* 242
Oryza hexandra IPNI=*137026-2* 241
Oryza jeyporensis IPNI=*410115-1* 242
Oryza latifolia var. *grandispiculis* IPNI=*316809-2* 242
Oryza marginata IPNI=*316812-2* 242
Oryza meridionalis IPNI=*410115-1* 242
Oryza mexicana IPNI=*137026-2* 241
Oryza montana IPNI=*316812-2* 242
Oryza mutica IPNI=*316812-2* 242
Oryza nivara IPNI=*410115-1* 242
Oryza palustris IPNI=*316812-2* 242
Oryza paraguayensis IPNI=*410115-1* 242
Oryza parviflora IPNI=*316812-2* 242
Oryza perennis IPNI=*316812-2* 242
Oryza platyphylla IPNI=*316809-2* 242
Oryza plena IPNI=*316812-2* 242
Oryza praecox IPNI=*316812-2* 242
Oryza pubescens IPNI=*316812-2* 242
Oryza pumila IPNI=*316812-2* 242
Oryza repens IPNI=*316812-2* 242
Oryza rubribarbis IPNI=*316812-2* 242
Oryza sativa subsp. *fatua* IPNI=*410115-1* 242
Oryza sativa subsp. *indica* IPNI=*316812-2* 242
Oryza sativa subsp. *japonica* IPNI=*316812-2* 242
Oryza sativa subsp. *rufipogon* IPNI=*410115-1* 242
Oryza sativa var. *fatua* IPNI=*410115-1* 242
Oryza sativa var. *grandiglumis* IPNI=*176926-2* 242
Oryza sativa var. *latifolia* IPNI=*316809-2* 242
Oryza sativa var. *paraguayensis* IPNI=*410115-1* 242
Oryza sativa var. *sundensis* IPNI=*410115-1* 242
Oryza segetalis IPNI=*316812-2* 242
Osbeckia aubletiana IPNI=*575460-1* 424
Osbeckia glutinosa IPNI=*566936-1* 418
Osbeckia microphylla IPNI=*567329-1* 418
Osbeckia pumila IPNI=*1131052-2* 424
Osbeckia rotundifolia IPNI=*908858-1* 419
Oscaria chinensis IPNI=*702446-1* 304
Oskampia hirsuta IPNI=*77069564-1* 275
Oskampia scandens IPNI=*119329-1* 275
Osmhydrophora nocturna IPNI=*77100460-1* 362
Osmia alternans IPNI=*57414-2* 259
Osmia atriplicifolia IPNI=*57443-2* 259
Osmia caleoides IPNI=*57489-2* 259
Osmia clematitis IPNI=*57443-2* 259
Osmia conyzoides IPNI=*57443-2* 259
Osmia crenata IPNI=*57489-2* 259
Osmia divergens IPNI=*57443-2* 259
Osmia floribunda IPNI=*57443-2* 259
Osmia furcata IPNI=*57410-2* 259
Osmia graciliflora IPNI=*57443-2* 259
Osmia gracillima IPNI=*57410-2* 259
Osmia ivifolia IPNI=*57410-2* 259
Osmia laevigata IPNI=*57414-2* 259
Osmia martiusii IPNI=*57489-2* 259
Osmia obscura IPNI=*57410-2* 259
Osmia odorata IPNI=*57443-2* 259
Osmia propinqua IPNI=*57414-2* 259
Osmia ramosissima IPNI=*57489-2* 259
Osmia scabra IPNI=*57486-2* 259
Osmia squalida IPNI=*57489-2* 259
Osmia stenophylla IPNI=*57410-2* 259
Osmia subvelutina IPNI=*57489-2* 259
Osmia tomentosa IPNI=*57489-2* 259
Osmia trifoliata IPNI=*57489-2* 259

Osmunda adiantifolia IPNI=*1031990-2* 179
Osmunda aspleniifolia IPNI=*1031990-2* 179
Osmunda brasiliensis IPNI=*1202412-2* 179
Osmunda brayeri IPNI=*305721-2* 173
Osmunda cinnamomea IPNI=*60447721-2* 173
Osmunda cinnamomea var. *typica* IPNI=*60447721-2* 173
Osmunda filiculifolia IPNI=*30017613-2* 176
Osmunda leschenaultii IPNI=*305721-2* 173
Osmunda longifolia IPNI=*305721-2* 173
Osmunda phyllitidis IPNI=*1202412-2* 179
Osmunda plumieri IPNI=*305721-2* 173
Osmunda regalis var. *longifolia* IPNI=*305721-2* 173
Osmunda scandens IPNI=*17143520-1* 179
Osmunda spicata IPNI=*305721-2* 173
Osmunda virginiana IPNI=*1106719-2* 173
Osmunda virginica IPNI=*1106719-2* 173
Osmundopteris virginiana IPNI=*1106719-2* 173
Ossaea diversifolia IPNI=*77144083-1* 422
Ossaea laxivenula IPNI=*77144073-1* 421
Ossaea laxivenula var. *grandifolia* IPNI=*77144073-1* 421
Ossaea laxivenula var. *ovalifolia* IPNI=*77144073-1* 421
Ossaea macrophylla IPNI=*77144075-1* 421
Ossaea quadrisulca IPNI=*77207567-1* 422
Ossaea quinquenervia IPNI=*77144083-1* 422
Ossaea robusta f. *glabrata* IPNI=*77144073-1* 421
Ossaea robusta IPNI=*77144090-1* 423
Ossaea trichocalyx IPNI=*77196168-1* 422
Osteomeles cordata IPNI=*725473-1* 443
Osteomeles escalloniifolia IPNI=*725484-1* 443
Osteomeles ferruginea IPNI=*725473-1* 443
Osteomeles glabrata IPNI=*725484-1* 443
Osteomeles goudotiana IPNI=*120960-2* 443
Osteomeles intermedia IPNI=*725484-1* 443
Osteomeles latifolia IPNI=*725473-1* 443
Osteomeles obovata IPNI=*725484-1* 443
Osteomeles obtusifolia IPNI=*725484-1* 443
Osteomeles pernettyoides IPNI=*725484-1* 443
Osteomeles pyracantha IPNI=*730613-1* 444
Osteospermum ecklonis IPNI=*201468-1* 261
Osteospermum jucundum IPNI=*201475-1* 261
Osterdamia matrella IPNI=*1214917-2* 247
Osterdamia zoysia IPNI=*1214917-2* 247
Oswalda baillierioides IPNI=*319147-2* 260
Othonna maritima IPNI=*77075194-1* 264
Othonna rigens IPNI=*208720-1* 263
Otoba incolor IPNI=*267140-2* 191
Otoba otoba IPNI=*177341-2* 191
Ottonia burchellii IPNI=*682138-1* 196
Ottonia carpunya IPNI=*680810-1* 195
Ottonia glaucescens IPNI=*681128-1* 196
Ottonia plantaginea IPNI=*680359-1* 195
Ottonia warakabacoura IPNI=*680574-1* 195
Ottonia warakabacoura var. *latifolia* IPNI=*680574-1* 195
Ouratea cubensis IPNI=*607105-1* 397
Ouratea guatemalensis IPNI=*607063-1* 397
Ouratea isthmica IPNI=*607063-1* 397
Ouratea ligans IPNI=*607063-1* 397
Ouratea lucens var. *podogyna* IPNI=*607063-1* 397
Ouratea panamica IPNI=*607105-1* 397
Ouratea peckii IPNI=*607063-1* 397
Ouratea podocarpa IPNI=*607063-1* 397
Ouratea podogyna IPNI=*607063-1* 397
Ouratea stenobotrys IPNI=*607063-1* 397
Ouratea wrightii IPNI=*607063-1* 397
Ourouparia guianensis IPNI=*768250-1* 358
Ourouparia polycephala IPNI=*768322-1* 358
Ourouparia tomentosa IPNI=*768322-1* 358
Outea acaciifolia IPNI=*505635-1* 326
Outea multijuga IPNI=*505740-1* 326
Ovidia gracilis IPNI=*302798-2* 224
Ovieda aculeata IPNI=*866088-1* 374
Ovieda fragrans IPNI=*946989-1* 369
Oxalis adscendens IPNI=*178220-2* 432
Oxalis alba IPNI=*30012601-2* 432
Oxalis albicans var. *sericea* IPNI=*177893-2* 432
Oxalis amazonica IPNI=*177777-2* 432
Oxalis amplifolia IPNI=*374872-1* 432
Oxalis aracatcha IPNI=*178566-2* 432
Oxalis arechavaletae IPNI=*374673-1* 432
Oxalis arracacha IPNI=*178566-2* 432
Oxalis articulata f. *crassipes* IPNI=*374673-1* 432
Oxalis articulata f. *guttata* IPNI=*374673-1* 432
Oxalis articulata f. *halophila* IPNI=*374673-1* 432
Oxalis articulata subsp. *floribunda* IPNI=*374673-1* 432
Oxalis articulata subsp. *nodulosa* IPNI=*374673-1* 432
Oxalis articulata subsp. *rubra* IPNI=*374673-1* 432
Oxalis articulata var. *hirsuta* IPNI=*374673-1* 432
Oxalis articulata var. *sericea* IPNI=*374673-1* 432
Oxalis atroglandulosa IPNI=*375051-1* 432
Oxalis avilensis IPNI=*374932-1* 432
Oxalis bahiensis IPNI=*374932-1* 432
Oxalis barrelieri subsp. *blumenavensis* IPNI=*177777-2* 432
Oxalis barrelieri subsp. *ianeirensis* IPNI=*177777-2* 432
Oxalis barrelieri var. *glabrescens* IPNI=*177777-2* 432
Oxalis bifida IPNI=*178480-2* 432
Oxalis binervis IPNI=*375051-1* 432
Oxalis bipunctata IPNI=*374833-1* 432
Oxalis boreaui IPNI=*177893-2* 432
Oxalis boridiensis IPNI=*177893-2* 432
Oxalis bradei IPNI=*178000-2* 432
Oxalis brevipes IPNI=*177883-2* 432
Oxalis buchtienii IPNI=*375051-1* 432
Oxalis bulbifera IPNI=*374833-1* 432
Oxalis caespitosa IPNI=*177883-2* 432
Oxalis cajanifolia IPNI=*177777-2* 432
Oxalis calcaria IPNI=*375051-1* 432
Oxalis camporum IPNI=*374932-1* 432
Oxalis canterae IPNI=*374673-1* 432
Oxalis caripensis IPNI=*374833-1* 432
Oxalis catharinensis IPNI=*375478-1* 432
Oxalis ceratilis IPNI=*177893-2* 432
Oxalis chacoensis IPNI=*374932-1* 432
Oxalis charaguensis IPNI=*374932-1* 432
Oxalis chicligastensis IPNI=*178566-2* 432
Oxalis chilensis IPNI=*374673-1* 432
Oxalis chiriquensis IPNI=*375051-1* 432
Oxalis chrysantha IPNI=*177883-2* 432
Oxalis chrysantha var. *pusilla* IPNI=*177883-2* 432
Oxalis ciliifera IPNI=*177893-2* 432
Oxalis cineracea IPNI=*177883-2* 432
Oxalis coarctata IPNI=*374932-1* 432
Oxalis coccinea IPNI=*374932-1* 432
Oxalis colombiensis IPNI=*177777-2* 432
Oxalis commersonii IPNI=*177883-2* 432
Oxalis cordobensis IPNI=*177883-2* 432
Oxalis cordobensis var. *humilior* IPNI=*177883-2* 432
Oxalis cordobensis var. *typica* IPNI=*177883-2* 432
Oxalis corniculata rubra IPNI=*177893-2* 432
Oxalis corniculata var. *atropurpurea* IPNI=*177893-2* 432
Oxalis corniculata var. *corcovadensis* IPNI=*177893-2* 432
Oxalis corniculata var. *glabrocapsula* IPNI=*177893-2* 432
Oxalis corniculata var. *langloisii* IPNI=*177893-2* 432
Oxalis corniculata var. *macrophylla* IPNI=*177893-2* 432
Oxalis corniculata var. *pilosiuscula* IPNI=*177893-2* 432
Oxalis corniculata var. *pubescens* IPNI=*177893-2* 432
Oxalis corniculata var. *radicosa* IPNI=*177893-2* 432

Oxalis corniculata var. *repens* IPNI=*177893-2* 432
Oxalis corniculata var. *serpens* IPNI=*177883-2* 432
Oxalis corniculata var. *villosa* IPNI=*177893-2* 432
Oxalis corniculata var. *viscidula* IPNI=*177893-2* 432
Oxalis corumbaensis IPNI=*375478-1* 432
Oxalis corymbosa IPNI=*374833-1* 432
Oxalis crassicaulis IPNI=*178566-2* 432
Oxalis crassipes IPNI=*374673-1* 432
Oxalis crenata IPNI=*178566-2* 432
Oxalis cruzeirensis IPNI=*374932-1* 432
Oxalis cuernavacana IPNI=*375466-1* 432
Oxalis debilis subsp. *corymbosa* IPNI=*374833-1* 432
Oxalis debilis var. *corymbosa* IPNI=*374833-1* 432
Oxalis delta IPNI=*375478-1* 432
Oxalis dendroides IPNI=*371953-1* 432
Oxalis deppei IPNI=*375466-1* 432
Oxalis descolei IPNI=*374932-1* 432
Oxalis diffusa IPNI=*177893-2* 432
Oxalis diffusa IPNI=*374932-1* 432
Oxalis distans IPNI=*374932-1* 432
Oxalis divergens var. *amplifolia* IPNI=*374872-1* 432
Oxalis dumicola IPNI=*374673-1* 432
Oxalis eggersii IPNI=*375051-1* 432
Oxalis fasciculata IPNI=*374932-1* 432
Oxalis ferae IPNI=*178000-2* 432
Oxalis floribunda IPNI=*374673-1* 432
Oxalis foliosa IPNI=*177893-2* 432
Oxalis fragrans IPNI=*30012601-2* 432
Oxalis frutescens IPNI=*374932-1* 432
Oxalis frutescens subsp. *pentantha* IPNI=*374932-1* 432
Oxalis frutescens var. *cicatricosa* IPNI=*374932-1* 432
Oxalis frutescens var. *margaritensis* IPNI=*374932-1* 432
Oxalis fruticetorum IPNI=*178186-2* 432
Oxalis furcata IPNI=*177893-2* 432
Oxalis gemella IPNI=*374833-1* 432
Oxalis glaberrima IPNI=*375478-1* 432
Oxalis glabrata IPNI=*374932-1* 432
Oxalis glandulosa IPNI=*374673-1* 432
Oxalis glauca IPNI=*177893-2* 432
Oxalis glaucifolia IPNI=*374932-1* 432
Oxalis grandifolia IPNI=*374833-1* 432
Oxalis grenadensis IPNI=*177893-2* 432
Oxalis guatemalensis IPNI=*374932-1* 432
Oxalis guttata IPNI=*374673-1* 432
Oxalis haenkeana IPNI=*178225-2* 432
Oxalis halophila IPNI=*374673-1* 432
Oxalis harmsiana IPNI=*374932-1* 432
Oxalis hassleriana IPNI=*177883-2* 432
Oxalis hayi IPNI=*375466-1* 432
Oxalis herpestica IPNI=*177893-2* 432
Oxalis herzogii IPNI=*374932-1* 432
Oxalis hirtella IPNI=*178000-2* 432
Oxalis hispida IPNI=*374932-1* 432
Oxalis huilensis IPNI=*374833-1* 432
Oxalis humistrata IPNI=*177893-2* 432
Oxalis incisa IPNI=*374872-1* 432
Oxalis insignis IPNI=*374990-1* 432
Oxalis jahnii IPNI=*374932-1* 432
Oxalis jamaicensis IPNI=*177893-2* 432
Oxalis javanica IPNI=*177893-2* 432
Oxalis knuthii IPNI=*178480-2* 432
Oxalis langloisii IPNI=*177893-2* 432
Oxalis lasiandra IPNI=*374673-1* 432
Oxalis lasiopetala var. *pubescens* IPNI=*374673-1* 432
Oxalis lasiophylla IPNI=*374673-1* 432
Oxalis latipes IPNI=*177883-2* 432
Oxalis leonis IPNI=*374872-1* 432
Oxalis leptophylla IPNI=*374932-1* 432
Oxalis lilacina IPNI=*375051-1* 432
Oxalis lilloana IPNI=*177777-2* 432
Oxalis linnaeiformis IPNI=*177883-2* 432
Oxalis littoralis IPNI=*374932-1* 432
Oxalis longiscapa IPNI=*178322-2* 432
Oxalis lujula IPNI=*177893-2* 432
Oxalis lupulina IPNI=*177893-2* 432
Oxalis lutea IPNI=*177893-2* 432
Oxalis mairaryensis IPNI=*374932-1* 432
Oxalis martiana IPNI=*374833-1* 432
Oxalis mauritiana IPNI=*375051-1* 432
Oxalis megapotamica IPNI=*177883-2* 432
Oxalis melilotoides var. *argentina* IPNI=*178566-2* 432
Oxalis meridensis IPNI=*177893-2* 432
Oxalis micrantha IPNI=*177893-2* 432
Oxalis mimosoides IPNI=*371953-1* 432
Oxalis minima IPNI=*177893-2* 432
Oxalis monadelpha IPNI=*177893-2* 432
Oxalis monticola var. *sericea* IPNI=*374673-1* 432
Oxalis morelosensis IPNI=*375051-1* 432
Oxalis mucronata IPNI=*375466-1* 432
Oxalis multibulbosa IPNI=*374833-1* 432
Oxalis multipes IPNI=*375051-1* 432
Oxalis neaei var. *glabrata* IPNI=*374932-1* 432
Oxalis nematodes IPNI=*178000-2* 432
Oxalis nemoralis IPNI=*30012601-2* 432
Oxalis nodulosa IPNI=*374932-1* 432
Oxalis norlindiana IPNI=*177893-2* 432
Oxalis oblongiformis IPNI=*178524-2* 432
Oxalis oxyptera IPNI=*375478-1* 432
Oxalis palustris IPNI=*375478-1* 432
Oxalis palustris var. *major* IPNI=*375478-1* 432
Oxalis papilionacea IPNI=*375478-1* 432
Oxalis parviflora IPNI=*30012601-2* 432
Oxalis parvifolia IPNI=*178000-2* 432
Oxalis parvifolia subsp. *aureoflava* IPNI=*177771-2* 432
Oxalis pentantha IPNI=*374932-1* 432
Oxalis pentantha var. *humboldtii* IPNI=*374932-1* 432
Oxalis phellandrioides IPNI=*178245-2* 432
Oxalis pichinchensis IPNI=*178186-2* 432
Oxalis pilosissima IPNI=*374932-1* 432
Oxalis pilosiuscula IPNI=*177893-2* 432
Oxalis platensis IPNI=*374673-1* 432
Oxalis plumieri IPNI=*374932-1* 432
Oxalis praecox IPNI=*374673-1* 432
Oxalis pseudostipulata IPNI=*374673-1* 432
Oxalis pseudotetraphylla IPNI=*375466-1* 432
Oxalis psilotricha IPNI=*374932-1* 432
Oxalis pubescens IPNI=*177893-2* 432
Oxalis pubescens IPNI=*178480-2* 432
Oxalis pubescens var. *glandulosa* IPNI=*178480-2* 432
Oxalis pusilla IPNI=*177893-2* 432
Oxalis pynaertii IPNI=*374833-1* 432
Oxalis racemosa IPNI=*374673-1* 432
Oxalis radiata IPNI=*177883-2* 432
Oxalis radicosa IPNI=*177893-2* 432
Oxalis recisa IPNI=*177893-2* 432
Oxalis regnellii IPNI=*375478-1* 432
Oxalis regnellii var. *catharinensis* IPNI=*375478-1* 432
Oxalis repens IPNI=*177893-2* 432
Oxalis reptans IPNI=*177893-2* 432
Oxalis rivalis IPNI=*374673-1* 432
Oxalis rubra IPNI=*374673-1* 432
Oxalis saulierei IPNI=*178480-2* 432
Oxalis schomburgkiana IPNI=*374932-1* 432
Oxalis schomburgkiana var. *lasiocarpa* IPNI=*374932-1* 432
Oxalis schomburgkiana var. *leiocarpa* IPNI=*374932-1* 432
Oxalis scopulorum IPNI=*375466-1* 432

Oxalis sepium var. *picta* IPNI=*177777-2* 432
Oxalis sericea IPNI=*374673-1* 432
Oxalis sexenata IPNI=*177883-2* 432
Oxalis simulans IPNI=*177893-2* 432
Oxalis spiralis var. *glandulosa* IPNI=*178480-2* 432
Oxalis sternbergii IPNI=*177883-2* 432
Oxalis steudeliana IPNI=*177893-2* 432
Oxalis stipulata IPNI=*375051-1* 432
Oxalis sublignosa IPNI=*374932-1* 432
Oxalis taiwanensis IPNI=*177893-2* 432
Oxalis taquetii IPNI=*30012601-2* 432
Oxalis tenuicaulis IPNI=*177893-2* 432
Oxalis tenuiloba IPNI=*375051-1* 432
Oxalis tenuiscaposa IPNI=*375478-1* 432
Oxalis tephrodes IPNI=*374932-1* 432
Oxalis thunbergiana IPNI=*177893-2* 432
Oxalis tlalpamensis IPNI=*375466-1* 432
Oxalis triangularis f. *glabrifolia* IPNI=*375478-1* 432
Oxalis triangularis subsp. *papilionacea* IPNI=*375478-1* 432
Oxalis trinidadensis IPNI=*177893-2* 432
Oxalis truncata IPNI=*375478-1* 432
Oxalis tubistipula IPNI=*374673-1* 432
Oxalis uittienii IPNI=*177893-2* 432
Oxalis umbellata IPNI=*178322-2* 432
Oxalis umbrosa IPNI=*374833-1* 432
Oxalis urbica IPNI=*374833-1* 432
Oxalis urvillei IPNI=*177893-2* 432
Oxalis vallicola IPNI=*375051-1* 432
Oxalis venturiana IPNI=*375478-1* 432
Oxalis vernalis IPNI=*375478-1* 432
Oxalis versicolor IPNI=*30012601-2* 432
Oxalis villosa IPNI=*177893-2* 432
Oxalis viva IPNI=*371953-1* 432
Oxalis vulgaris IPNI=*30012601-2* 432
Oxalis yapacaniensis IPNI=*375478-1* 432
Oxandra aromatica IPNI=*1013110-1* 189
Oxandra major IPNI=*178642-2* 189
Oxandra nitida IPNI=*1013110-1* 189
Oxandra opaca IPNI=*1013110-1* 189
Oxandra ovata IPNI=*1013110-1* 189
Oxisma arborescens IPNI=*566597-1* 417
Oxodium callosum IPNI=*680748-1* 195
Oxyacantha amygdalifolia IPNI=*730611-1* 444
Oxydectes albida IPNI=*342537-1* 388
Oxydectes althaeifolia IPNI=*342537-1* 388
Oxydectes argyrophylloides IPNI=*343663-1* 389
Oxydectes astroites IPNI=*342537-1* 388
Oxydectes benthamiana IPNI=*343150-1* 389
Oxydectes bixoides IPNI=*70536-2* 389
Oxydectes caryophyllus IPNI=*70513-2* 389
Oxydectes chamaedrifolia IPNI=*71013-2* 389
Oxydectes citrifolia IPNI=*343083-1* 389
Oxydectes cuneata IPNI=*342381-1* 388
Oxydectes flavens IPNI=*342537-1* 388
Oxydectes flocculosa IPNI=*342537-1* 388
Oxydectes fragilis IPNI=*342550-1* 388
Oxydectes fragrans IPNI=*342551-1* 388
Oxydectes funckiana IPNI=*342669-1* 388
Oxydectes glabella IPNI=*1029203-2* 388
Oxydectes glandulosa IPNI=*70233-2* 388
Oxydectes glandulosa var. *genuina* IPNI=*70233-2* 388
Oxydectes glandulosa var. *hirta* IPNI=*70337-2* 389
Oxydectes gossypifolia var. *heterophylla* IPNI=*70274-2* 388
Oxydectes gossypiifolia IPNI=*70274-2* 388
Oxydectes isertii IPNI=*325164-2* 388
Oxydectes lechleri IPNI=*342842-1* 389
Oxydectes leptostachya IPNI=*342854-1* 389
Oxydectes lobata IPNI=*339515-1* 388
Oxydectes magdalenensis IPNI=*343058-1* 389
Oxydectes malambo IPNI=*342933-1* 389
Oxydectes martii IPNI=*342381-1* 388
Oxydectes matourensis IPNI=*70513-2* 389
Oxydectes menthodora IPNI=*342979-1* 389
Oxydectes micans IPNI=*70536-2* 389
Oxydectes morifolia IPNI=*70562-2* 389
Oxydectes mutisiana IPNI=*343058-1* 389
Oxydectes nivea IPNI=*70599-2* 389
Oxydectes ovalifolia IPNI=*325185-2* 389
Oxydectes palanostigma IPNI=*343150-1* 389
Oxydectes polycarpa IPNI=*343058-1* 389
Oxydectes populifolia IPNI=*70599-2* 389
Oxydectes rufoargentea IPNI=*343663-1* 389
Oxydectes schiedeana IPNI=*70846-2* 389
Oxydectes schlechtendaliana IPNI=*342550-1* 388
Oxydectes strigosa IPNI=*325185-2* 389
Oxydectes surinamensis IPNI=*342381-1* 388
Oxydectes tricolor IPNI=*343663-1* 389
Oxydectes variegata IPNI=*85073-3* 388
Oxygraphis cymbalaria IPNI=*711250-1* 436
Oxygraphis nubigena IPNI=*713340-1* 436
Oxygraphis tridentata IPNI=*711250-1* 436
Oxymeris aristigera IPNI=*77207374-1* 419
Oxymeris calyptrata IPNI=*158886-2* 420
Oxymeris ciliata IPNI=*158869-2* 420
Oxymeris cinnamomea IPNI=*158912-2* 420
Oxymeris dichotoma IPNI=*77207668-1* 423
Oxymeris furfuracea IPNI=*159532-2* 422
Oxymeris heterobasis IPNI=*77207474-1* 420
Oxymeris macrophylla IPNI=*77144075-1* 421
Oxymeris secunda IPNI=*77207935-1* 421
Oxymeris subseriata IPNI=*77207614-1* 423
Oxymyrsine pungens IPNI=*540443-1* 215
Oxys acetosella IPNI=*30012601-2* 432
Oxys alba IPNI=*30012601-2* 432
Oxys corniculata IPNI=*177893-2* 432
Oxys lutea IPNI=*177893-2* 432
Oxys pliniana IPNI=*30012601-2* 432
Oxys vulgaris IPNI=*30012601-2* 432
Oxytenanthera auriculata IPNI=*392574-1* 235
Oxythece caracasana IPNI=*948100-1* 308
Oxythece ferreirii IPNI=*89008-2* 305
Oxythece glabrescens IPNI=*89008-2* 305
Oxythece glaucescens IPNI=*209758-2* 306
Oxythece leptocarpa IPNI=*209758-2* 306
Oxythece pseudosideroxylon IPNI=*209758-2* 306
Oxythece robusta IPNI=*209738-2* 306
Oxythece robusta var. *brevipetiolata* IPNI=*209738-2* 306
Oxythece schomburgkiana IPNI=*89010-2* 305
Oxythece steyermarkiana IPNI=*89010-2* 305
Ozodia foeniculacea IPNI=*842680-1* 252

P

Pachira affinis IPNI=*562668-1* 412
Pachira alba IPNI=*211230-2* 412
Pachira angusta IPNI=*281912-2* 412
Pachira aquatica var. *manausensis* IPNI=*179762-2* 412
Pachira aquatica var. *occidentalis* IPNI=*179762-2* 412
Pachira aquatica var. *surinamensis* IPNI=*179762-2* 412
Pachira barrigon IPNI=*211241-2* 413
Pachira commersonii IPNI=*211230-2* 412
Pachira cyathophora IPNI=*211230-2* 412
Pachira decaphylla IPNI=*211230-2* 412
Pachira grandiflora IPNI=*179762-2* 412
Pachira loddigesii IPNI=*562668-1* 412
Pachira longifolia IPNI=*179762-2* 412
Pachira macrocarpa IPNI=*179762-2* 412

Pachira oleagina IPNI=*562664-1* 412
Pachira pustulifera IPNI=*179762-2* 412
Pachira quinata IPNI=*559120-1* 407
Pachira spruceana IPNI=*562668-1* 412
Pachira villosula IPNI=*179762-2* 412
Pachyanthus corymbifer IPNI=*567365-1* 418
Pachyanthus corymbosus IPNI=*567365-1* 418
Pachygone domingensis IPNI=*580887-1* 435
Pachygone graciliflora IPNI=*580887-1* 435
Pachylecythis egleri IPNI=*597315-1* 302
Pachylobus peruvianus IPNI=*75732-2* 452
Pachypodium pannonicum IPNI=*289144-1* 278
Pachyptera alliacea IPNI=*152672-2* 364
Pachyptera hymenaea IPNI=*152677-2* 364
Pachyptera puberula IPNI=*77100511-1* 363
Pachyptera standleyi IPNI=*152684-2* 364
Pachyrhizus angulatus var. *integrifolius* IPNI=*945254-1* 329
Pachyrhizus mollis IPNI=*509661-1* 328
Pachyrhizus montanus IPNI=*509661-1* 328
Pachyrhizus teres IPNI=*509661-1* 328
Pachyrhizus vernalis IPNI=*945254-1* 329
Pachyrhizus vernalis var. *angustilobatus* IPNI=*945254-1* 329
Pachyrhizus vernalis var. *integrifolius* IPNI=*945254-1* 329
Pachyrhizus vernalis var. *linearifoliolus* IPNI=*945254-1* 329
Pachyrhizus vernalis var. *typicus* IPNI=*945254-1* 329
Pachystachys albiflora IPNI=*52814-1* 360
Pachystachys asperula IPNI=*943504-1* 361
Pachystachys riedeliana IPNI=*943504-1* 361
Pacourina cirsiifolia IPNI=*236351-1* 267
Pacourina edulis var. *spinosissima* IPNI=*236351-1* 267
Pacourinopsis dentata IPNI=*236351-1* 267
Pacourinopsis integrifolia IPNI=*236351-1* 267
Padus cartilaginea IPNI=*730262-1* 444
Padus eximia IPNI=*730262-1* 444
Padus hirsuta IPNI=*730262-1* 444
Padus retusa IPNI=*730262-1* 444
Padus serotina IPNI=*730262-1* 444
Padus serotina var. *asplenifolia* IPNI=*730262-1* 444
Padus serotina var. *cartilaginea* IPNI=*730262-1* 444
Padus serotina var. *pendula* IPNI=*730262-1* 444
Padus serotina var. *pyramidalis* IPNI=*730262-1* 444
Paepalanthus barkleyi IPNI=*336162-1* 233
Paepalanthus columbiensis IPNI=*336004-1* 233
Paepalanthus karstenii f. *corei* IPNI=*336162-1* 233
Paepalanthus karstenii var. *corei* IPNI=*336162-1* 233
Paesia aquilina IPNI=*17210060-1* 175
Pagaea nemorosa IPNI=*130185-2* 345
Pagaea recurva IPNI=*130185-2* 345
Pagaea subcordata IPNI=*130185-2* 345
Pagamea coriacea var. *acuta* IPNI=*546862-1* 353
Pagamea coriacea var. *pubescens* IPNI=*546862-1* 353
Pagesia dianthera IPNI=*155661-2* 376
Pagesia peduncularis IPNI=*155661-2* 376
Pagesia procumbens IPNI=*155661-2* 376
Paiva verticillata IPNI=*765972-1* 357
Palala calophylla IPNI=*267140-2* 191
Palala capitellata IPNI=*64004-2* 190
Palala carinata IPNI=*267142-2* 191
Palala elongata IPNI=*267153-2* 192
Palala fragrans IPNI=*586076-1* 191
Palala gracilis IPNI=*267202-2* 192
Palala hostmannii IPNI=*585810-1* 191
Palala mocoa IPNI=*267199-2* 192
Palala otoba IPNI=*177341-2* 191
Palala panamensis IPNI=*267199-2* 192
Palala pavonis IPNI=*267190-2* 192
Palala peruviana IPNI=*267191-2* 192
Palala platysperma IPNI=*586380-1* 191
Palala sebifera IPNI=*267199-2* 192
Palala sprucei IPNI=*64010-2* 190
Palala surinamensis IPNI=*267202-2* 192
Palala theiodora IPNI=*267203-2* 192
Palanostigma crotonoides IPNI=*343150-1* 389
Palanostigma hibiscifolium IPNI=*342669-1* 388
Palanostigma martiana IPNI=*343150-1* 389
Palaua scabra IPNI=*1183910-2* 297
Palaua tomentosa IPNI=*829436-1* 297
Paleista brachypoda IPNI=*1186403-2* 261
Paleista procumbens IPNI=*1186403-2* 261
Palhinhaea brevibracteata IPNI=*1054587-2* 171
Palhinhaea capillacea IPNI=*1054587-2* 171
Palhinhaea cernua f. *ciliatomarginata* IPNI=*1054587-2* 171
Palhinhaea cernua f. *laxa* IPNI=*1054587-2* 171
Palhinhaea cernua f. *sikkimensis* IPNI=*1054587-2* 171
Palhinhaea cernua IPNI=*1054587-2* 171
Palhinhaea cernua var. *lanceolata* IPNI=*1054587-2* 171
Palhinhaea cernua var. *pendula* IPNI=*1054587-2* 171
Palhinhaea cernua var. *sikkimensis* IPNI=*1054587-2* 171
Palhinhaea eichleri IPNI=*279987-2* 171
Palhinhaea lufengensis IPNI=*1054587-2* 171
Palhinhaea mariana IPNI=*1054587-2* 171
Palhinhaea pendulina IPNI=*279987-2* 171
Palhinhaea polycephala IPNI=*1054587-2* 171
Palicourea acanthaceoides IPNI=*759203-1* 354
Palicourea angustifolia var. *querceticola* IPNI=*319913-2* 353
Palicourea aurata IPNI=*759300-1* 354
Palicourea barbinervia IPNI=*759121-1* 354
Palicourea capitata IPNI=*759372-1* 354
Palicourea chionantha IPNI=*60473461-2* 355
Palicourea coccinea IPNI=*759061-1* 353
Palicourea contracta IPNI=*77161077-1* 354
Palicourea crocea f. *citriflora* IPNI=*759061-1* 353
Palicourea cyclotis IPNI=*759242-1* 354
Palicourea discolor IPNI=*759203-1* 354
Palicourea diuretica IPNI=*759300-1* 354
Palicourea elongata IPNI=*759353-1* 354
Palicourea ensiformis IPNI=*319913-2* 353
Palicourea furcata IPNI=*77115613-1* 354
Palicourea guianensis f. *glabra* IPNI=*759121-1* 354
Palicourea guianensis subsp. *occidentalis* IPNI=*759121-1* 354
Palicourea guianensis var. *glabrescens* IPNI=*759121-1* 354
Palicourea guianensis var. *tetramera* IPNI=*759121-1* 354
Palicourea guianensis var. *trimera* IPNI=*759121-1* 354
Palicourea hassleriana IPNI=*77161098-1* 354
Palicourea hyacinthiflora IPNI=*759242-1* 354
Palicourea hypomalaca IPNI=*759121-1* 354
Palicourea intermedia IPNI=*319913-2* 353
Palicourea izabalensis IPNI=*77115593-1* 353
Palicourea klugii IPNI=*759234-1* 354
Palicourea lanceolata IPNI=*319913-2* 353
Palicourea lasioneura IPNI=*759203-1* 354
Palicourea longibracteata IPNI=*759353-1* 354
Palicourea longipes IPNI=*759353-1* 354
Palicourea lupulina IPNI=*77161077-1* 354
Palicourea molliramis IPNI=*77115609-1* 354
Palicourea mollis IPNI=*759353-1* 354
Palicourea moritzii IPNI=*319913-2* 353
Palicourea nicotianifolia IPNI=*759203-1* 354
Palicourea obscurata IPNI=*181012-2* 354
Palicourea ornata IPNI=*759203-1* 354
Palicourea pacapacensis IPNI=*77115593-1* 353
Palicourea papyracea IPNI=*181012-2* 354
Palicourea parviflora IPNI=*759353-1* 354
Palicourea perotensis IPNI=*281967-2* 353
Palicourea platypodina IPNI=*759061-1* 353
Palicourea querceticola IPNI=*319913-2* 353

Palicourea quindiensis IPNI=*319913-2* 353
Palicourea rigida f. *angustior* IPNI=*759300-1* 354
Palicourea rigida f. *brevistyla* IPNI=*759300-1* 354
Palicourea rigida f. *longistyla* IPNI=*759300-1* 354
Palicourea rigida f. *microphylla* IPNI=*759300-1* 354
Palicourea rigida subsp. *hirtella* IPNI=*759300-1* 354
Palicourea rigida var. *amazonica* IPNI=*759300-1* 354
Palicourea rigida var. *aurata* IPNI=*759300-1* 354
Palicourea rigida var. *hirtibacca* IPNI=*759300-1* 354
Palicourea rigida var. *paraguariensis* IPNI=*759300-1* 354
Palicourea rigida var. *stenophylla* IPNI=*759300-1* 354
Palicourea rosea IPNI=*77161054-1* 353
Palicourea rosea var. *trichophora* IPNI=*77161054-1* 353
Palicourea sonans IPNI=*759300-1* 354
Palicourea stenostachys IPNI=*759353-1* 354
Palicourea stevensonii IPNI=*759372-1* 354
Palicourea strepens IPNI=*759300-1* 354
Palicourea subrubra IPNI=*281971-2* 354
Palicourea subulata IPNI=*759061-1* 353
Palicourea swartziana IPNI=*77161098-1* 354
Palicourea ternata IPNI=*759353-1* 354
Palicourea tricholoba IPNI=*77161076-1* 354
Palicourea viridis IPNI=*763751-1* 356
Paliurus colubrina IPNI=*63423-2* 441
Paliurus inermis IPNI=*63435-2* 441
Paliurus lucidus IPNI=*719213-1* 442
Paliurus mairei IPNI=*719349-1* 442
Pallasia serratifolia IPNI=*30147294-2* 272
Pallavia aculeata IPNI=*326897-2* 287
Palma argentata IPNI=*62341-2* 208
Palma cocos IPNI=*666160-1* 208
Palma dactylifera IPNI=*668912-1* 211
Palma elata IPNI=*222809-2* 212
Palma major IPNI=*668912-1* 211
Palma maripa IPNI=*664452-1* 205
Palma mocaia IPNI=*663914-1* 203
Palma oleosa IPNI=*666802-1* 208
Palma sancona IPNI=*670066-1* 212
Palma spinosa IPNI=*663914-1* 203
Palmifolium chigua IPNI=*297256-1* 179
Palmifolium latifolium IPNI=*270552-2* 179
Palmifolium muricatum IPNI=*270552-2* 179
Palmifolium poeppigianum IPNI=*270556-2* 179
Panax aureus IPNI=*847351-1* 254
Panax bandanensis IPNI=*91749-1* 254
Panax chrysophyllus IPNI=*90355-1* 253
Panax cochleatus IPNI=*91817-1* 254
Panax conchifolius IPNI=*91817-1* 254
Panax crispatus IPNI=*91749-1* 254
Panax cumingianus IPNI=*91749-1* 254
Panax dissectus IPNI=*847351-1* 254
Panax dumosus IPNI=*847351-1* 254
Panax fissus IPNI=*847351-1* 254
Panax fruticosus IPNI=*847351-1* 254
Panax guilfoylei IPNI=*91773-1* 254
Panax heyneanus IPNI=*91817-1* 254
Panax morototoni IPNI=*90355-1* 253
Panax papyrifer IPNI=*86417-3* 254
Panax pinnatus IPNI=*91749-1* 254
Panax plumatus IPNI=*847351-1* 254
Panax rotundatus IPNI=*91817-1* 254
Panax rumphii IPNI=*91817-1* 254
Panax scutellarioides IPNI=*91817-1* 254
Panax secundus IPNI=*91749-1* 254
Panax speciosus IPNI=*90355-1* 253
Panax spinosus IPNI=*90355-1* 253
Panax splendens IPNI=*90355-1* 253
Panax undulatus IPNI=*90355-1* 253
Panax verus IPNI=*91472-1* 254
Panax victoriae IPNI=*91773-1* 254
Pancratium americanum IPNI=*30048918-2* 213
Pancratium amoenum IPNI=*65212-1* 213
Pancratium angustum IPNI=*65212-1* 213
Pancratium caribaeum IPNI=*65212-1* 213
Pancratium declinatum IPNI=*65212-1* 213
Pancratium distichum IPNI=*30048918-2* 213
Pancratium dryanderi IPNI=*30048918-2* 213
Pancratium excisum IPNI=*65212-1* 213
Pancratium fragrans IPNI=*65212-1* 213
Pancratium littorale IPNI=*30048918-2* 213
Pancratium littorale var. *dryanderi* IPNI=*30048918-2* 213
Pancratium obtusatum IPNI=*65212-1* 213
Pancratium patens IPNI=*65212-1* 213
Pancratium pedale IPNI=*30048918-2* 213
Pancratium recurvum IPNI=*65212-1* 213
Pancratium staplesii IPNI=*30048918-2* 213
Pancratium tenuiflorum IPNI=*30048918-2* 213
Pandanus absonus IPNI=*895770-1* 227
Pandanus adscendens IPNI=*895770-1* 227
Pandanus aequor IPNI=*895770-1* 227
Pandanus aitutakiensis IPNI=*895770-1* 227
Pandanus akiakiensis IPNI=*895770-1* 227
Pandanus ala-kai IPNI=*895770-1* 227
Pandanus amplexus IPNI=*895770-1* 227
Pandanus angulosus IPNI=*895770-1* 227
Pandanus anisos IPNI=*895770-1* 227
Pandanus aoraiensis IPNI=*895770-1* 227
Pandanus apionops IPNI=*895770-1* 227
Pandanus asauensis IPNI=*895770-1* 227
Pandanus ater IPNI=*895770-1* 227
Pandanus baptistii IPNI=*895770-1* 227
Pandanus bassus IPNI=*895770-1* 227
Pandanus bathys IPNI=*895770-1* 227
Pandanus benignus IPNI=*895770-1* 227
Pandanus bergmanii IPNI=*895770-1* 227
Pandanus bicurvatus IPNI=*895770-1* 227
Pandanus blakei IPNI=*895770-1* 227
Pandanus boraboraensis IPNI=*895770-1* 227
Pandanus bothreus IPNI=*895770-1* 227
Pandanus bowenensis IPNI=*895770-1* 227
Pandanus brachypodus IPNI=*895770-1* 227
Pandanus brownii IPNI=*895770-1* 227
Pandanus cacuminatus IPNI=*895770-1* 227
Pandanus chamissonis IPNI=*895770-1* 227
Pandanus charancanus IPNI=*895770-1* 227
Pandanus christophersenii IPNI=*895770-1* 227
Pandanus collatus IPNI=*895770-1* 227
Pandanus complanatus IPNI=*895770-1* 227
Pandanus cooperi IPNI=*895770-1* 227
Pandanus coronatus IPNI=*895770-1* 227
Pandanus crassiaculeatus IPNI=*895770-1* 227
Pandanus crassus IPNI=*895770-1* 227
Pandanus crustatus IPNI=*895770-1* 227
Pandanus cylindricus IPNI=*895770-1* 227
Pandanus cymatilis IPNI=*895770-1* 227
Pandanus decorus IPNI=*895770-1* 227
Pandanus dicheres IPNI=*895770-1* 227
Pandanus dilatatus IPNI=*895770-1* 227
Pandanus distinctus IPNI=*895770-1* 227
Pandanus divaricatus IPNI=*895770-1* 227
Pandanus divergens IPNI=*895770-1* 227
Pandanus dotyi IPNI=*895770-1* 227
Pandanus douglasii IPNI=*895770-1* 227
Pandanus drakei IPNI=*895770-1* 227
Pandanus drolletianus IPNI=*895770-1* 227
Pandanus duriocarpoides IPNI=*895770-1* 227

Pandanus duriocarpus IPNI=*895770-1* 227
Pandanus edwinii IPNI=*895770-1* 227
Pandanus elevatus IPNI=*895770-1* 227
Pandanus enchabiensis IPNI=*895770-1* 227
Pandanus erythrophloeus IPNI=*895770-1* 227
Pandanus evexus IPNI=*895770-1* 227
Pandanus extralittoralis IPNI=*895770-1* 227
Pandanus eyesyes IPNI=*895770-1* 227
Pandanus fahina IPNI=*895770-1* 227
Pandanus faramaa IPNI=*895770-1* 227
Pandanus fatuhivaensis IPNI=*895770-1* 227
Pandanus fatyanion IPNI=*895770-1* 227
Pandanus ferulifer IPNI=*895770-1* 227
Pandanus filiciatilis IPNI=*895770-1* 227
Pandanus fischerianus IPNI=*895770-1* 227
Pandanus flintinsularis IPNI=*895770-1* 227
Pandanus fragrans IPNI=*895770-1* 227
Pandanus futunaensis IPNI=*895770-1* 227
Pandanus gambierensis IPNI=*895770-1* 227
Pandanus glomerosus IPNI=*895770-1* 227
Pandanus grantii IPNI=*895770-1* 227
Pandanus guamensis IPNI=*895770-1* 227
Pandanus haapaiensis IPNI=*895770-1* 227
Pandanus hendersonensis IPNI=*895770-1* 227
Pandanus heronensis IPNI=*895770-1* 227
Pandanus hivaoaensis IPNI=*895770-1* 227
Pandanus horneinsularum IPNI=*895770-1* 227
Pandanus hosinoi IPNI=*895770-1* 227
Pandanus hosokawae IPNI=*895770-1* 227
Pandanus houmaensis IPNI=*895770-1* 227
Pandanus hubbardii IPNI=*895770-1* 227
Pandanus humericus IPNI=*895770-1* 227
Pandanus inarmatus IPNI=*895770-1* 227
Pandanus inermis IPNI=*534103-1* 215
Pandanus inermis IPNI=*895770-1* 227
Pandanus inflexus IPNI=*895770-1* 227
Pandanus infundibuliformis IPNI=*895770-1* 227
Pandanus insularis IPNI=*895770-1* 227
Pandanus interior IPNI=*895770-1* 227
Pandanus intralaevis IPNI=*895770-1* 227
Pandanus jaluitensis IPNI=*895770-1* 227
Pandanus jonesii IPNI=*895770-1* 227
Pandanus kafu IPNI=*895770-1* 227
Pandanus kamptos IPNI=*895770-1* 227
Pandanus katensis IPNI=*895770-1* 227
Pandanus koidzumii IPNI=*895770-1* 227
Pandanus korrensis IPNI=*895770-1* 227
Pandanus kraussii IPNI=*895770-1* 227
Pandanus kusaiensis IPNI=*895770-1* 227
Pandanus laculatus IPNI=*895770-1* 227
Pandanus laevis IPNI=*895770-1* 227
Pandanus lakatwa IPNI=*895770-1* 227
Pandanus lambasaensis IPNI=*895770-1* 227
Pandanus laticanaliculatus IPNI=*895770-1* 227
Pandanus latior IPNI=*895770-1* 227
Pandanus lauensis IPNI=*895770-1* 227
Pandanus licinus IPNI=*895770-1* 227
Pandanus limitaris IPNI=*895770-1* 227
Pandanus littoralis IPNI=*895770-1* 227
Pandanus longifolius IPNI=*895770-1* 227
Pandanus macfarlanei IPNI=*895770-1* 227
Pandanus macrocephalus IPNI=*895770-1* 227
Pandanus makateaensis IPNI=*895770-1* 227
Pandanus malatensis IPNI=*895770-1* 227
Pandanus mangarevaensis IPNI=*895770-1* 227
Pandanus mariaensis IPNI=*895770-1* 227
Pandanus marquesasensis IPNI=*895770-1* 227
Pandanus matukuensis IPNI=*895770-1* 227
Pandanus mbalawa IPNI=*895770-1* 227
Pandanus meetiaensis IPNI=*895770-1* 227
Pandanus mei IPNI=*895770-1* 227
Pandanus mendanensis IPNI=*895770-1* 227
Pandanus menne IPNI=*895770-1* 227
Pandanus menziesii IPNI=*895770-1* 227
Pandanus metius IPNI=*895770-1* 227
Pandanus minysocephalus IPNI=*895770-1* 227
Pandanus montaguei IPNI=*895770-1* 227
Pandanus mooreaensis IPNI=*895770-1* 227
Pandanus moschatus IPNI=*895770-1* 227
Pandanus motuensis IPNI=*895770-1* 227
Pandanus nandiensis IPNI=*895770-1* 227
Pandanus niueensis IPNI=*895770-1* 227
Pandanus notialis IPNI=*895770-1* 227
Pandanus oblatiapicalis IPNI=*895770-1* 227
Pandanus oblaticonvexus IPNI=*895770-1* 227
Pandanus obliquus IPNI=*895770-1* 227
Pandanus odontoides IPNI=*895770-1* 227
Pandanus okamotoi IPNI=*895770-1* 227
Pandanus onoilauensis IPNI=*895770-1* 227
Pandanus orarius IPNI=*895770-1* 227
Pandanus otemanuensis IPNI=*895770-1* 227
Pandanus ovalauensis IPNI=*895770-1* 227
Pandanus pachys IPNI=*895770-1* 227
Pandanus pakari IPNI=*895770-1* 227
Pandanus palkilensis IPNI=*895770-1* 227
Pandanus palmyraensis IPNI=*895770-1* 227
Pandanus pansus IPNI=*895770-1* 227
Pandanus paogo IPNI=*895770-1* 227
Pandanus papeariensis IPNI=*895770-1* 227
Pandanus papenooensis IPNI=*895770-1* 227
Pandanus parhamii IPNI=*895770-1* 227
Pandanus parksii IPNI=*895770-1* 227
Pandanus patulior IPNI=*895770-1* 227
Pandanus pedunculatus IPNI=*895770-1* 227
Pandanus pedunculatus var. *insularis* IPNI=*895770-1* 227
Pandanus pedunculatus var. *malagunensis* IPNI=*895770-1* 227
Pandanus pedunculatus var. *rendovensis* IPNI=*895770-1* 227
Pandanus planus IPNI=*895770-1* 227
Pandanus politus IPNI=*895770-1* 227
Pandanus ponapensis IPNI=*895770-1* 227
Pandanus prismaticus IPNI=*895770-1* 227
Pandanus prolixus IPNI=*895770-1* 227
Pandanus pseudomenne IPNI=*895770-1* 227
Pandanus pulposus IPNI=*895770-1* 227
Pandanus pusillus IPNI=*895770-1* 227
Pandanus pyriformis IPNI=*895770-1* 227
Pandanus radiatus IPNI=*895770-1* 227
Pandanus raiateaensis IPNI=*895770-1* 227
Pandanus raivavaensis IPNI=*895770-1* 227
Pandanus raroiaensis IPNI=*895770-1* 227
Pandanus rectangulatus IPNI=*895770-1* 227
Pandanus repens IPNI=*895770-1* 227
Pandanus rhizophorensis IPNI=*895770-1* 227
Pandanus rhombocarpus IPNI=*895770-1* 227
Pandanus rikiteaensis IPNI=*895770-1* 227
Pandanus rimataraensis IPNI=*895770-1* 227
Pandanus rockii IPNI=*895770-1* 227
Pandanus rotensis IPNI=*895770-1* 227
Pandanus rotundatus IPNI=*895770-1* 227
Pandanus rurutuensis IPNI=*895770-1* 227
Pandanus sabotan IPNI=*895770-1* 227
Pandanus saipanensis IPNI=*895770-1* 227
Pandanus saltuarius IPNI=*895770-1* 227
Pandanus samak IPNI=*895770-1* 227
Pandanus samoanus IPNI=*895770-1* 227
Pandanus sanderi IPNI=*895770-1* 227

Pandanus savaiensis IPNI=*895770-1* 227
Pandanus schizocarpus IPNI=*895770-1* 227
Pandanus seruaensis IPNI=*895770-1* 227
Pandanus sinuosus IPNI=*895770-1* 227
Pandanus sinuvadosus IPNI=*895770-1* 227
Pandanus smithii IPNI=*895770-1* 227
Pandanus spurius IPNI=*895770-1* 227
Pandanus stradbrookeensis IPNI=*895770-1* 227
Pandanus subaequalis IPNI=*895770-1* 227
Pandanus subhumerosus IPNI=*895770-1* 227
Pandanus subradiatus IPNI=*895770-1* 227
Pandanus suvaensis IPNI=*895770-1* 227
Pandanus sykesii IPNI=*895770-1* 227
Pandanus taepa IPNI=*895770-1* 227
Pandanus tahaaensis IPNI=*895770-1* 227
Pandanus tahitensis IPNI=*895770-1* 227
Pandanus takaroaensis IPNI=*895770-1* 227
Pandanus tamaruensis IPNI=*895770-1* 227
Pandanus tapeinos IPNI=*895770-1* 227
Pandanus taravaiensis IPNI=*895770-1* 227
Pandanus tauensis IPNI=*895770-1* 227
Pandanus tectorius f. *convexus* IPNI=*895770-1* 227
Pandanus tectorius var. *australianus* IPNI=*895770-1* 227
Pandanus tectorius var. *cocosensis* IPNI=*895770-1* 227
Pandanus tectorius var. *drolletianus* IPNI=*895770-1* 227
Pandanus tectorius var. *exiguus* IPNI=*895770-1* 227
Pandanus tectorius var. *heronensis* IPNI=*895770-1* 227
Pandanus tectorius var. *incrassatus* IPNI=*895770-1* 227
Pandanus tectorius var. *laevigatus* IPNI=*895770-1* 227
Pandanus tectorius var. *oahuensis* IPNI=*895770-1* 227
Pandanus tectorius var. *pedunculatus* IPNI=*895770-1* 227
Pandanus tectorius var. *stradbrookensis* IPNI=*895770-1* 227
Pandanus tectorius var. *tubuaiensis* IPNI=*895770-1* 227
Pandanus tectorius var. *yorkensis* IPNI=*895770-1* 227
Pandanus temehaniensis IPNI=*895770-1* 227
Pandanus terrireginae IPNI=*895770-1* 227
Pandanus tessellatus IPNI=*895770-1* 227
Pandanus tikeiensis IPNI=*895770-1* 227
Pandanus tima IPNI=*895770-1* 227
Pandanus timoeensis IPNI=*895770-1* 227
Pandanus tolotomensis IPNI=*895770-1* 227
Pandanus tomilensis IPNI=*895770-1* 227
Pandanus tongaensis IPNI=*895770-1* 227
Pandanus trapaneus IPNI=*895770-1* 227
Pandanus tritosphaericus IPNI=*895770-1* 227
Pandanus trukensis IPNI=*895770-1* 227
Pandanus tuamotensis IPNI=*895770-1* 227
Pandanus tubuaiensis IPNI=*895770-1* 227
Pandanus tupaiensis IPNI=*895770-1* 227
Pandanus uea IPNI=*895770-1* 227
Pandanus utiyamae IPNI=*895770-1* 227
Pandanus vahitahiensis IPNI=*895770-1* 227
Pandanus vandra IPNI=*895770-1* 227
Pandanus variegatus IPNI=*895770-1* 227
Pandanus virginalis IPNI=*895770-1* 227
Pandanus viri IPNI=*895770-1* 227
Pandanus viridinsularis IPNI=*895770-1* 227
Pandanus volkensii IPNI=*895770-1* 227
Pandanus wilderi IPNI=*895770-1* 227
Pandanus yorkensis IPNI=*895770-1* 227
Pandanus yunckeri IPNI=*895770-1* 227
Pandorea ricasoliana IPNI=*110440-1* 364
Paneguia palmifolia IPNI=*237727-2* 216
Paneion pratense IPNI=*417792-1* 244
Panicastrella muricata IPNI=*316518-2* 235
Panicularia brachyphylla IPNI=*404100-1* 239
Panicularia fluitans IPNI=*404100-1* 239
Panicum abyssinicum IPNI=*1209084-2* 237
Panicum acuminatissimum IPNI=*1004002-1* 242
Panicum acuminatum IPNI=*405429-1* 240
Panicum acuminatum IPNI=*79328-2* 237
Panicum acuminatum var. *thurowii* IPNI=*79328-2* 237
Panicum acutiglumum IPNI=*405429-1* 240
Panicum adhaerens IPNI=*421428-1* 245
Panicum adscendens IPNI=*1038558-2* 237
Panicum aegyptiacum IPNI=*1207294-2* 245
Panicum aegyptiacum IPNI=*399115-1* 238
Panicum aegyptium IPNI=*399115-1* 238
Panicum aemulum IPNI=*1004002-1* 242
Panicum affine IPNI=*979871-1* 245
Panicum africanum IPNI=*1004002-1* 242
Panicum agglutinans IPNI=*135707-2* 240
Panicum agrostidiforme IPNI=*1156911-2* 246
Panicum agrostis IPNI=*1156911-2* 246
Panicum agrostoides IPNI=*1156911-2* 246
Panicum airoides IPNI=*60426565-2* 241
Panicum akoense IPNI=*410542-1* 242
Panicum albovellereum IPNI=*992521-1* 246
Panicum album IPNI=*30453766-2* 242
Panicum alectorocnemum IPNI=*399660-1* 238
Panicum alectromerum IPNI=*399660-1* 238
Panicum alopecuroides IPNI=*77105978-1* 235
Panicum alopecuroideum IPNI=*943811-1* 245
Panicum alopecuros IPNI=*235234-2* 245
Panicum alternatum IPNI=*325362-2* 242
Panicum altissimum IPNI=*77076421-1* 247
Panicum ambiguum IPNI=*397065-1* 236
Panicum ambiguum IPNI=*399115-1* 238
Panicum ambiguum IPNI=*421857-1* 245
Panicum americanum IPNI=*77105978-1* 235
Panicum ammophilum IPNI=*325362-2* 242
Panicum amphibium IPNI=*426266-1* 246
Panicum amplectens IPNI=*325362-2* 242
Panicum amplexicaule IPNI=*405429-1* 240
Panicum amplexicaule var. *deflexum* IPNI=*405429-1* 240
Panicum amplexicaule var. *erectum* IPNI=*405429-1* 240
Panicum amplifolium IPNI=*1182687-2* 245
Panicum amplissimum IPNI=*421711-1* 245
Panicum anceps var. *minus* IPNI=*1156911-2* 246
Panicum andrewsii IPNI=*324582-2* 242
Panicum annulatum IPNI=*402044-1* 239
Panicum aparine IPNI=*421428-1* 245
Panicum apiculatum IPNI=*77144198-1* 244
Panicum appressum IPNI=*1218147-2* 237
Panicum appressum IPNI=*426266-1* 246
Panicum appressum IPNI=*979871-1* 245
Panicum aquaticum var. *cartagoense* IPNI=*325362-2* 242
Panicum aquaticum var. *chloroticum* IPNI=*325362-2* 242
Panicum arborescens IPNI=*135707-2* 240
Panicum aristatum IPNI=*1048074-2* 244
Panicum aristatum IPNI=*30149942-2* 238
Panicum aristatum IPNI=*409823-1* 242
Panicum ascendens IPNI=*943811-1* 245
Panicum asiaticum IPNI=*1207294-2* 245
Panicum asperrimum IPNI=*412217-1* 242
Panicum asperum IPNI=*421857-1* 245
Panicum auburne IPNI=*79328-2* 237
Panicum aurelianum IPNI=*426280-1* 246
Panicum aureum IPNI=*321761-2* 234
Panicum auriculatum IPNI=*125410-2* 240
Panicum auritum IPNI=*405429-1* 240
Panicum avenaceum IPNI=*77076421-1* 247
Panicum axipilium IPNI=*411699-1* 242
Panicum balbisianum IPNI=*3538-2* 233
Panicum balfourii IPNI=*1004002-1* 242
Panicum bambusoides IPNI=*413385-1* 242

Panicum barbatum IPNI=*1172068-2* 245
Panicum barbatum IPNI=*1218147-2* 237
Panicum barbifultum IPNI=*1004002-1* 242
Panicum barbinode IPNI=*426266-1* 246
Panicum barbipedum IPNI=*426280-1* 246
Panicum basisetum IPNI=*1172068-2* 245
Panicum beccabunga IPNI=*943811-1* 245
Panicum beckmanniiforme IPNI=*979871-1* 245
Panicum berteronianum IPNI=*943811-1* 245
Panicum beyrichii IPNI=*412226-1* 242
Panicum bicorne IPNI=*1218147-2* 237
Panicum bidentatum IPNI=*409823-1* 242
Panicum biforme IPNI=*1218147-2* 237
Panicum bivonianum IPNI=*60426565-2* 241
Panicum bongaense IPNI=*1172068-2* 245
Panicum bonplandianum IPNI=*88097-2* 238
Panicum brachiariiforme IPNI=*399655-1* 238
Panicum brachiatum IPNI=*325362-2* 242
Panicum brachyclados IPNI=*77105897-1* 241
Panicum brachyphyllum IPNI=*1038558-2* 237
Panicum brachythyrsum IPNI=*426280-1* 246
Panicum brachytrichum IPNI=*943811-1* 245
Panicum brasiliense IPNI=*943811-1* 245
Panicum brevifolium IPNI=*1038558-2* 237
Panicum brevifolium IPNI=*410542-1* 242
Panicum brevisetum IPNI=*1004002-1* 242
Panicum briziforme IPNI=*979871-1* 245
Panicum brizoides IPNI=*399655-1* 238
Panicum brizoides IPNI=*979871-1* 245
Panicum bromoides IPNI=*30453766-2* 242
Panicum bulbosum IPNI=*77076421-1* 247
Panicum bulbosum subsp. *sciaphilum* IPNI=*77076421-1* 247
Panicum bulbosum subvar. *violaceum* IPNI=*77076421-1* 247
Panicum bulbosum var. *avenaceum* IPNI=*77076421-1* 247
Panicum burmanni IPNI=*30453766-2* 242
Panicum caatingense IPNI=*182558-2* 242
Panicum caesium IPNI=*399655-1* 238
Panicum caespitosum IPNI=*426280-1* 246
Panicum caffrorum IPNI=*422090-1* 245
Panicum calaccanzense IPNI=*426280-1* 246
Panicum californicum IPNI=*80220-2* 237
Panicum capillaceum IPNI=*324582-2* 242
Panicum capillaceum var. *strictius* IPNI=*324582-2* 242
Panicum capillare var. *hirticaule* IPNI=*411695-1* 242
Panicum capillare var. *miliaceum* IPNI=*411695-1* 242
Panicum capillare var. *stramineum* IPNI=*182558-2* 242
Panicum carnosum IPNI=*979871-1* 245
Panicum caroniense IPNI=*1156911-2* 246
Panicum carthaginense IPNI=*1156554-2* 246
Panicum caudatum IPNI=*405429-1* 240
Panicum cayennense var. *divaricatum* IPNI=*412983-1* 242
Panicum cenchroides IPNI=*51261-2* 235
Panicum certificandum IPNI=*409823-1* 242
Panicum chartaginense IPNI=*1156554-2* 246
Panicum chinense IPNI=*1207294-2* 245
Panicum chloroticum IPNI=*325362-2* 242
Panicum chloroticum var. *luxurians* IPNI=*325362-2* 242
Panicum chloroticum var. *pingue* IPNI=*325362-2* 242
Panicum chloroticum var. *sylvestre* IPNI=*325362-2* 242
Panicum chrysites IPNI=*321761-2* 234
Panicum chrysodactylon IPNI=*321761-2* 234
Panicum chrysostachyum IPNI=*321761-2* 234
Panicum ciliare IPNI=*1038558-2* 237
Panicum ciliatifolium IPNI=*79395-2* 237
Panicum ciliatum IPNI=*79395-2* 237
Panicum ciliosum IPNI=*79328-2* 237
Panicum coenosum IPNI=*77144198-1* 244
Panicum coeruleum IPNI=*77105978-1* 235
Panicum colonum f. *maculatum* IPNI=*399655-1* 238
Panicum colonum IPNI=*399655-1* 238
Panicum colonum var. *zonale* IPNI=*399655-1* 238
Panicum columbiense IPNI=*27415-2* 234
Panicum commutatum IPNI=*1218147-2* 237
Panicum comophyllum IPNI=*79328-2* 237
Panicum compactum IPNI=*1207294-2* 245
Panicum compositum IPNI=*1182687-2* 245
Panicum compositum IPNI=*409823-1* 242
Panicum compressum IPNI=*60426565-2* 241
Panicum compressum IPNI=*77105978-1* 235
Panicum confine IPNI=*60426565-2* 241
Panicum confusum IPNI=*77076421-1* 247
Panicum congestum IPNI=*943811-1* 245
Panicum conjugatum IPNI=*30011315-2* 242
Panicum cordatum IPNI=*125410-2* 240
Panicum costatum IPNI=*1172068-2* 245
Panicum cristatum IPNI=*30453766-2* 242
Panicum crus-ardeae IPNI=*421828-1* 245
Panicum crus-ardeae var. *brevisetum* IPNI=*421828-1* 245
Panicum crus-ardeae var. *longisetum* IPNI=*421828-1* 245
Panicum crus-galli IPNI=*399660-1* 238
Panicum crus-galli subsp. *colonum* IPNI=*399655-1* 238
Panicum crus-pavonis IPNI=*30149942-2* 238
Panicum ctenodes IPNI=*77105897-1* 241
Panicum ctenodes var. *majus* IPNI=*77105897-1* 241
Panicum cubense IPNI=*1004002-1* 242
Panicum cumingianum IPNI=*399655-1* 238
Panicum curvinerve IPNI=*80233-2* 237
Panicum cuspidatum IPNI=*399655-1* 238
Panicum dactylon IPNI=*397065-1* 236
Panicum daltonii IPNI=*399655-1* 238
Panicum dasytrichum IPNI=*412983-1* 242
Panicum dasyurum IPNI=*943811-1* 245
Panicum decipiens IPNI=*421857-1* 245
Panicum densepilosum IPNI=*412217-1* 242
Panicum densiflorum IPNI=*77144198-1* 244
Panicum diamesum IPNI=*329018-2* 237
Panicum diandrum IPNI=*1156911-2* 246
Panicum dichotomiflorum subsp. *puritanorum* IPNI=*325362-2* 242
Panicum dichotomiflorum var. *chloroticum* IPNI=*325362-2* 242
Panicum dichotomiflorum var. *geniculatum* IPNI=*325362-2* 242
Panicum dichotomiflorum var. *puritanorum* IPNI=*325362-2* 242
Panicum dichotomum var. *glabrescens* IPNI=*79395-2* 237
Panicum dictyoneurum IPNI=*992521-1* 246
Panicum didistichum IPNI=*972086-1* 246
Panicum digitaria IPNI=*184275-2* 243
Panicum dioecum IPNI=*3538-2* 233
Panicum disciferum IPNI=*286530-2* 246
Panicum disjunctum IPNI=*211125-2* 244
Panicum dispersum IPNI=*412226-1* 242
Panicum dissitiflorum IPNI=*319037-2* 245
Panicum distachyon IPNI=*426250-1* 246
Panicum distans IPNI=*286530-2* 246
Panicum distans IPNI=*77144198-1* 244
Panicum distichum IPNI=*77144198-1* 244
Panicum distichum var. *luxurians* IPNI=*77144198-1* 244
Panicum divaricatum var. *lanatum* IPNI=*135707-2* 240
Panicum divaricatum var. *latifolium* IPNI=*135707-2* 240
Panicum donacifolium IPNI=*125410-2* 240
Panicum duchaissingii IPNI=*398888-1* 237
Panicum eburneum IPNI=*60426565-2* 241
Panicum echinatum IPNI=*211125-2* 244
Panicum echinochloa IPNI=*399655-1* 238
Panicum echinolaena IPNI=*88264-2* 238
Panicum echinolaena var. *ciliatum* IPNI=*88264-2* 238
Panicum echinolaena var. *glabrum* IPNI=*88264-2* 238
Panicum echinolaena var. *tomentellum* IPNI=*88264-2* 238

Panicum echinolaena var. *velutinum* IPNI=*88264-2* 238
Panicum echinolaena var. *villosum* IPNI=*88264-2* 238
Panicum elatius IPNI=*409823-1* 242
Panicum elephantipes IPNI=*77147487-1* 241
Panicum elliottii IPNI=*410542-1* 242
Panicum elongatum IPNI=*1207294-2* 245
Panicum emergens IPNI=*979871-1* 245
Panicum eminii IPNI=*974586-1* 246
Panicum equinum IPNI=*426266-1* 246
Panicum equitans IPNI=*399655-1* 238
Panicum eriogonum IPNI=*399115-1* 238
Panicum erythrospermum IPNI=*1207294-2* 245
Panicum expansum IPNI=*412226-1* 242
Panicum extensum IPNI=*426280-1* 246
Panicum fallax IPNI=*399115-1* 238
Panicum falsum IPNI=*398888-1* 237
Panicum fasciculatum IPNI=*1156554-2* 246
Panicum fasciculatum var. *carthaginense* IPNI=*1156554-2* 246
Panicum fasciculatum var. *flavescens* IPNI=*1156554-2* 246
Panicum fasciculatum var. *fuscum* IPNI=*1156554-2* 246
Panicum fasciculatum var. *genuinum* IPNI=*1156554-2* 246
Panicum fasciculatum var. *reticulatum* IPNI=*1156554-2* 246
Panicum fasciculatum var. *strictum* IPNI=*1156554-2* 246
Panicum fastigiatum IPNI=*1156554-2* 246
Panicum filamentosum IPNI=*324582-2* 242
Panicum filiculme IPNI=*79328-2* 237
Panicum fimbriatum IPNI=*1038558-2* 237
Panicum fimbriatum var. *setigerum* IPNI=*1203032-2* 238
Panicum fistulosum IPNI=*77147487-1* 241
Panicum flabellatum IPNI=*411695-1* 242
Panicum flaccidum IPNI=*1004002-1* 242
Panicum flaccidum IPNI=*399655-1* 238
Panicum flavescens IPNI=*1156554-2* 246
Panicum flavescens IPNI=*421764-1* 245
Panicum flavescens IPNI=*77164258-1* 246
Panicum flavomarginatum IPNI=*413385-1* 242
Panicum flavum IPNI=*943811-1* 245
Panicum floribundum IPNI=*421857-1* 245
Panicum fluitans IPNI=*979871-1* 245
Panicum foliaceum IPNI=*1004002-1* 242
Panicum francavillanum IPNI=*325362-2* 242
Panicum francoi IPNI=*30453766-2* 242
Panicum frondescens IPNI=*77105897-1* 241
Panicum frumentaceum IPNI=*422090-1* 245
Panicum funckianum IPNI=*412456-1* 242
Panicum fuscatum IPNI=*1156554-2* 246
Panicum fuscescens IPNI=*943811-1* 245
Panicum fuscorubens IPNI=*1156554-2* 246
Panicum fuscum IPNI=*1156554-2* 246
Panicum fuscum var. *reticulatum* IPNI=*1156554-2* 246
Panicum gavanianum IPNI=*398888-1* 237
Panicum geminatum IPNI=*979871-1* 245
Panicum geniculatum IPNI=*399655-1* 238
Panicum geniculatum IPNI=*943811-1* 245
Panicum germanicum IPNI=*1207294-2* 245
Panicum giganteum IPNI=*27415-2* 234
Panicum giganteum IPNI=*60426565-2* 241
Panicum glaucescens IPNI=*1218147-2* 237
Panicum glaucum IPNI=*77105978-1* 235
Panicum glaucum IPNI=*943811-1* 245
Panicum glaucum var. *elongatum* IPNI=*421764-1* 245
Panicum glaucum var. *laevigatum* IPNI=*421764-1* 245
Panicum glaucum var. *purpurascens* IPNI=*943811-1* 245
Panicum globulare IPNI=*1207294-2* 245
Panicum glomeratum IPNI=*1207294-2* 245
Panicum glomeratum IPNI=*979871-1* 245
Panicum glutinoscabrum IPNI=*79328-2* 237
Panicum glutinosum IPNI=*135707-2* 240
Panicum glutinosum IPNI=*278239-2* 240
Panicum gobariense IPNI=*943811-1* 245
Panicum golae IPNI=*992521-1* 246
Panicum gongylodes IPNI=*77076421-1* 247
Panicum gonyrrhizum IPNI=*409823-1* 242
Panicum gracile IPNI=*399115-1* 238
Panicum grande IPNI=*77200715-1* 240
Panicum grandiflorum IPNI=*3538-2* 233
Panicum grisebachianum IPNI=*405429-1* 240
Panicum grossarium IPNI=*426280-1* 246
Panicum grossum IPNI=*399660-1* 238
Panicum guaraniticum IPNI=*135707-2* 240
Panicum guayaquilense IPNI=*413376-1* 242
Panicum guluense IPNI=*3538-2* 233
Panicum gussonei IPNI=*399115-1* 238
Panicum hackelii IPNI=*1209084-2* 237
Panicum haematodes IPNI=*399655-1* 238
Panicum hamiltonii IPNI=*329014-2* 237
Panicum helvolum IPNI=*1207294-2* 245
Panicum henryi IPNI=*1038558-2* 237
Panicum heteranthum IPNI=*211125-2* 244
Panicum heterochlamys IPNI=*211125-2* 244
Panicum heterophyllum IPNI=*77144198-1* 244
Panicum heterophyllum IPNI=*79391-2* 237
Panicum heterostachyum IPNI=*411699-1* 242
Panicum heynei IPNI=*60426565-2* 241
Panicum hirsutissimum IPNI=*60426565-2* 241
Panicum hirsutum IPNI=*412226-1* 242
Panicum hirsutum IPNI=*412983-1* 242
Panicum hirtellum IPNI=*1004002-1* 242
Panicum hirticaule subsp. *sonorum* IPNI=*411695-1* 242
Panicum hirticaule subsp. *stramineum* IPNI=*182558-2* 242
Panicum hirticaule var. *miliaceum* IPNI=*411695-1* 242
Panicum hirticaule var. *stramineum* IPNI=*182558-2* 242
Panicum hirticaule var. *verrucosum* IPNI=*411695-1* 242
Panicum holcoides IPNI=*77105978-1* 235
Panicum holochrysum IPNI=*321761-2* 234
Panicum hondurense IPNI=*1156911-2* 246
Panicum hookeri IPNI=*399655-1* 238
Panicum horizontale IPNI=*329014-2* 237
Panicum horridum IPNI=*30149942-2* 238
Panicum humidicola IPNI=*992521-1* 246
Panicum hygrophilum IPNI=*325362-2* 242
Panicum illinoniense IPNI=*1156554-2* 246
Panicum imbecille IPNI=*1004002-1* 242
Panicum imberbe IPNI=*943811-1* 245
Panicum imberbe var. *dasyurum* IPNI=*943811-1* 245
Panicum imberbe var. *glaucocaesium* IPNI=*943811-1* 245
Panicum imberbe var. *latifolium* IPNI=*943811-1* 245
Panicum imberbe var. *pumilum* IPNI=*421764-1* 245
Panicum imberbe var. *purpurascens* IPNI=*943811-1* 245
Panicum inaequale IPNI=*1038558-2* 237
Panicum incanum IPNI=*1004002-1* 242
Panicum incertum IPNI=*399655-1* 238
Panicum indicum IPNI=*77105978-1* 235
Panicum insulare IPNI=*398888-1* 237
Panicum insulare IPNI=*426280-1* 246
Panicum insulare var. *lachnanthum* IPNI=*80220-2* 237
Panicum insulare var. *leucophaeum* IPNI=*398888-1* 237
Panicum insulare var. *penicilligerum* IPNI=*1050107-2* 237
Panicum insulare var. *sacchariflorum* IPNI=*1050107-2* 237
Panicum insulare var. *typicum* IPNI=*398888-1* 237
Panicum intermedium IPNI=*1207294-2* 245
Panicum involucratum IPNI=*77105978-1* 235
Panicum iowense IPNI=*79328-2* 237
Panicum ischnocaulon IPNI=*30453766-2* 242
Panicum italicum convar. *moharia* IPNI=*1207294-2* 245
Panicum italicum IPNI=*1207294-2* 245

Panicum italicum IPNI=*421857-1* 245
Panicum italicum var. *brevisetum* IPNI=*1207294-2* 245
Panicum italicum var. *longisetum* IPNI=*1207294-2* 245
Panicum italicum var. *mitis* IPNI=*1207294-2* 245
Panicum italicum var. *praecox* IPNI=*1207294-2* 245
Panicum itieri IPNI=*1207294-2* 245
Panicum jacobinae IPNI=*88264-2* 238
Panicum jamaicense IPNI=*30149942-2* 238
Panicum japonicum IPNI=*30453766-2* 242
Panicum jumentorum IPNI=*60426565-2* 241
Panicum junceum IPNI=*413385-1* 242
Panicum junceum var. *subnutans* IPNI=*413385-1* 242
Panicum kafuroense IPNI=*1209084-2* 237
Panicum kalmii IPNI=*79391-2* 237
Panicum kegelii IPNI=*77105897-1* 241
Panicum kleinianum IPNI=*421711-1* 245
Panicum lachnanthum IPNI=*80220-2* 237
Panicum laeve IPNI=*60426565-2* 241
Panicum laevigatum IPNI=*943811-1* 245
Panicum lagascae var. *schreberianum* IPNI=*414431-1* 243
Panicum lagascae var. *virgatum* IPNI=*414854-1* 243
Panicum lanatum IPNI=*398888-1* 237
Panicum lanceolatum IPNI=*409823-1* 242
Panicum lappaceum IPNI=*30453766-2* 242
Panicum lasianthum IPNI=*412226-1* 242
Panicum latifolium IPNI=*135721-2* 241
Panicum laxiflorum var. *pubescens* IPNI=*79395-2* 237
Panicum laxum IPNI=*1156911-2* 246
Panicum laxum var. *pubescens* IPNI=*1156911-2* 246
Panicum laxum var. *vestitum* IPNI=*1156911-2* 246
Panicum leandrii IPNI=*286530-2* 246
Panicum leiogonum IPNI=*426266-1* 246
Panicum lene IPNI=*421711-1* 245
Panicum leprieurii IPNI=*77105897-1* 241
Panicum leptanthum IPNI=*401189-1* 239
Panicum leptomerum IPNI=*1156911-2* 246
Panicum leucites IPNI=*80280-2* 237
Panicum leucoblepharis IPNI=*79395-2* 237
Panicum leucoblepharis var. *pubescens* IPNI=*79395-2* 237
Panicum leucophaeum var. *sacchariflorum* IPNI=*1050107-2* 237
Panicum limnaeum IPNI=*426266-1* 246
Panicum lineare IPNI=*397065-1* 236
Panicum lineatum IPNI=*1172068-2* 245
Panicum linkianum IPNI=*1038558-2* 237
Panicum linkianum var. *minus* IPNI=*1038558-2* 237
Panicum loliaceum IPNI=*1004002-1* 242
Panicum longeracemosum IPNI=*409823-1* 242
Panicum longiflorum IPNI=*398933-1* 237
Panicum longipedunculatum IPNI=*79395-2* 237
Panicum lutetense IPNI=*3538-2* 233
Panicum luteum IPNI=*421764-1* 245
Panicum luticola IPNI=*1156911-2* 246
Panicum macrochaetum IPNI=*1207294-2* 245
Panicum macrospermum IPNI=*412456-1* 242
Panicum macrostachyum IPNI=*1218147-2* 237
Panicum macrostachyum IPNI=*235234-2* 245
Panicum macrostachyum var. *patens* IPNI=*235234-2* 245
Panicum macrourum IPNI=*1182687-2* 245
Panicum maculatum IPNI=*135707-2* 240
Panicum maculatum var. *pilosum* IPNI=*135707-2* 240
Panicum mahafalense IPNI=*60426565-2* 241
Panicum mananarense IPNI=*60426565-2* 241
Panicum mangaloricum IPNI=*225088-2* 244
Panicum margaritaceum IPNI=*399655-1* 238
Panicum marginale IPNI=*88264-2* 238
Panicum maritimum IPNI=*1207294-2* 245
Panicum martinicense IPNI=*135707-2* 240
Panicum mauritianum IPNI=*1172068-2* 245
Panicum maximum IPNI=*60426565-2* 241
Panicum maximum subsp. *pubescens* IPNI=*60426565-2* 241
Panicum maximum var. *coloratum* IPNI=*60426565-2* 241
Panicum maximum var. *gongylodes* IPNI=*77076421-1* 247
Panicum maximum var. *pubiglume* IPNI=*60426565-2* 241
Panicum maximum var. *trichoglume* IPNI=*60426565-2* 241
Panicum melfrugum IPNI=*1207294-2* 245
Panicum melicoides IPNI=*3538-2* 233
Panicum melinis IPNI=*408232-1* 241
Panicum melinis var. *inerme* IPNI=*408232-1* 241
Panicum meneri IPNI=*410542-1* 242
Panicum mexicanum IPNI=*421711-1* 245
Panicum microphyllum IPNI=*79328-2* 237
Panicum miliaceum IPNI=*1207294-2* 245
Panicum miliaceum IPNI=*325362-2* 242
Panicum miliaceum subsp. *agricola* IPNI=*412217-1* 242
Panicum miliaceum subsp. *ruderale* IPNI=*412217-1* 242
Panicum miliaceum var. *aerugineum* IPNI=*412217-1* 242
Panicum miliaceum var. *anthracinum* IPNI=*412217-1* 242
Panicum miliaceum var. *aquilum* IPNI=*412217-1* 242
Panicum miliaceum var. *atrobrunneum* IPNI=*412217-1* 242
Panicum miliaceum var. *coffeatum* IPNI=*412217-1* 242
Panicum miliaceum var. *corsinum* IPNI=*412217-1* 242
Panicum miliaceum var. *densobrunneum* IPNI=*412217-1* 242
Panicum miliaceum var. *fuscum* IPNI=*412217-1* 242
Panicum miliaceum var. *glaucum* IPNI=*412217-1* 242
Panicum miliaceum var. *nicotianum* IPNI=*412217-1* 242
Panicum miliaceum var. *ruderale* IPNI=*412217-1* 242
Panicum miliaceum var. *subvitellinotephrum* IPNI=*412217-1* 242
Panicum miliaceum var. *virescens* IPNI=*412217-1* 242
Panicum miliare IPNI=*410542-1* 242
Panicum miliiforme IPNI=*943744-1* 246
Panicum milium IPNI=*412217-1* 242
Panicum milleflorum IPNI=*77144198-1* 244
Panicum minutiflorum IPNI=*408232-1* 241
Panicum mississippiense IPNI=*79391-2* 237
Panicum moharicum IPNI=*1207294-2* 245
Panicum molle IPNI=*972086-1* 246
Panicum mollissimum IPNI=*80280-2* 237
Panicum monobotrys IPNI=*77144198-1* 244
Panicum monostachyum IPNI=*77144198-1* 244
Panicum monostachyum var. *robustius* IPNI=*77144198-1* 244
Panicum moritzii IPNI=*972086-1* 246
Panicum multiflorum IPNI=*325362-2* 242
Panicum multifolium IPNI=*972082-1* 246
Panicum multinodosum IPNI=*412226-1* 242
Panicum musciparum IPNI=*410542-1* 242
Panicum musei IPNI=*399655-1* 238
Panicum muticum IPNI=*426266-1* 246
Panicum myosurus IPNI=*225088-2* 244
Panicum myrianthum IPNI=*77200715-1* 240
Panicum myuros IPNI=*225088-2* 244
Panicum myurum IPNI=*405429-1* 240
Panicum nakaianum IPNI=*1172068-2* 245
Panicum neesii IPNI=*1218147-2* 237
Panicum nemorosum IPNI=*77105897-1* 241
Panicum nepalense IPNI=*421711-1* 245
Panicum nervosum IPNI=*399115-1* 238
Panicum nervosum IPNI=*421711-1* 245
Panicum neurodes IPNI=*421711-1* 245
Panicum nigricans IPNI=*1156554-2* 246
Panicum nilagiricum IPNI=*426280-1* 246
Panicum nitidum var. *pubescens* IPNI=*79328-2* 237
Panicum nodosum IPNI=*77076421-1* 247
Panicum numidianum IPNI=*399655-1* 238
Panicum numidianum IPNI=*426266-1* 246
Panicum nuttallianum IPNI=*1004002-1* 242
Panicum oahuaense IPNI=*1004002-1* 242

Panicum occidentale IPNI=*943811-1* 245
Panicum ogowense IPNI=*3538-2* 233
Panicum olivaceum IPNI=*79328-2* 237
Panicum olyrifolium IPNI=*77105897-1* 241
Panicum olyroides var. *denudatum* IPNI=*412456-1* 242
Panicum olyroides var. *fimbriatum* IPNI=*412456-1* 242
Panicum olyroides var. *hirsutum* IPNI=*412456-1* 242
Panicum orinocense IPNI=*135707-2* 240
Panicum ornatum IPNI=*79328-2* 237
Panicum ornithopus IPNI=*1038558-2* 237
Panicum oryzoides IPNI=*3538-2* 233
Panicum pallens IPNI=*405749-1* 240
Panicum pallens var. *swartzii* IPNI=*405749-1* 240
Panicum pallide-fuscum IPNI=*421764-1* 245
Panicum palmatum IPNI=*421711-1* 245
Panicum palmifolium IPNI=*421711-1* 245
Panicum paludicola IPNI=*125410-2* 240
Panicum paludivagum IPNI=*979871-1* 245
Panicum palustre IPNI=*125410-2* 240
Panicum pamplemoussense IPNI=*60426565-2* 241
Panicum paniculatum IPNI=*184526-2* 243
Panicum paniculiferum IPNI=*421711-1* 245
Panicum panis IPNI=*1207294-2* 245
Panicum parciflorum IPNI=*409823-1* 242
Panicum parvulum IPNI=*398933-1* 237
Panicum parvum IPNI=*426280-1* 246
Panicum paspaliforme IPNI=*184275-2* 243
Panicum paspaloides IPNI=*426250-1* 246
Panicum paspaloides IPNI=*979871-1* 245
Panicum patentissimum IPNI=*412226-1* 242
Panicum patulum IPNI=*426280-1* 246
Panicum paucifolium IPNI=*77076421-1* 247
Panicum penicillatum IPNI=*943811-1* 245
Panicum penicilligerum IPNI=*1050107-2* 237
Panicum peninsularum IPNI=*409823-1* 242
Panicum pennisetum IPNI=*77144198-1* 244
Panicum perdensum IPNI=*405429-1* 240
Panicum pertenue IPNI=*399226-1* 238
Panicum petiveri IPNI=*399655-1* 238
Panicum petrosum IPNI=*60440769-2* 243
Panicum phleiforme IPNI=*225088-2* 244
Panicum phyllanthum IPNI=*88097-2* 238
Panicum pictigluma IPNI=*426266-1* 246
Panicum pilgeri IPNI=*413385-1* 242
Panicum pilisparsum IPNI=*77144198-1* 244
Panicum pilosissimum IPNI=*412983-1* 242
Panicum pilosum IPNI=*412226-1* 242
Panicum pilosum IPNI=*77144198-1* 244
Panicum pilosum var. *distachyum* IPNI=*77144198-1* 244
Panicum pilosum var. *epilosum* IPNI=*1156911-2* 246
Panicum pilosum var. *lancifolium* IPNI=*77144198-1* 244
Panicum plantagineum IPNI=*286530-2* 246
Panicum platense IPNI=*414070-1* 243
Panicum platicaulon IPNI=*392087-1* 234
Panicum plicatulum IPNI=*184571-2* 243
Panicum plicatum IPNI=*421711-1* 245
Panicum poiforme IPNI=*3538-2* 233
Panicum poiforme IPNI=*60426565-2* 241
Panicum polycaulon IPNI=*79395-2* 237
Panicum polygamum IPNI=*60426565-2* 241
Panicum polygamum var. *gongylodes* IPNI=*77076421-1* 247
Panicum polygamum var. *hirticaule* IPNI=*411695-1* 242
Panicum polyrrhizum IPNI=*184275-2* 243
Panicum polystachion IPNI=*943811-1* 245
Panicum polystachyum IPNI=*125410-2* 240
Panicum polystachyum IPNI=*88097-2* 238
Panicum polytrichum IPNI=*972086-1* 246
Panicum porranthum IPNI=*329014-2* 237
Panicum praegnans IPNI=*135707-2* 240
Panicum praelongum IPNI=*60426565-2* 241
Panicum pratense IPNI=*409823-1* 242
Panicum praticola IPNI=*60426565-2* 241
Panicum proboscideum IPNI=*412456-1* 242
Panicum procerrimum IPNI=*135714-2* 241
Panicum procumbens IPNI=*426280-1* 246
Panicum proliferum IPNI=*410542-1* 242
Panicum proliferum var. *chloroticum* IPNI=*325362-2* 242
Panicum proliferum var. *geniculatum* IPNI=*325362-2* 242
Panicum proliferum var. *pilosum* IPNI=*325362-2* 242
Panicum proliferum var. *richardii* IPNI=*325362-2* 242
Panicum proliferum var. *strictum* IPNI=*325362-2* 242
Panicum proliferum var. *xanthochlorum* IPNI=*325362-2* 242
Panicum propinquum IPNI=*398933-1* 237
Panicum prorepens IPNI=*399655-1* 238
Panicum prostratum IPNI=*426280-1* 246
Panicum prostratum var. *pilosum* IPNI=*426280-1* 246
Panicum provinciale IPNI=*412217-1* 242
Panicum pseudocolonum IPNI=*399655-1* 238
Panicum pseudodistachyum IPNI=*426250-1* 246
Panicum pseudodurva IPNI=*399226-1* 238
Panicum pseudoholcus IPNI=*943811-1* 245
Panicum pseudoischaemum IPNI=*80254-2* 237
Panicum pseudoryzoides IPNI=*3538-2* 233
Panicum pseudoundulatifolium IPNI=*77105897-1* 241
Panicum psilanthum IPNI=*1156911-2* 246
Panicum pubescens IPNI=*278239-2* 240
Panicum pulchrum IPNI=*321761-2* 234
Panicum pumilum IPNI=*1207294-2* 245
Panicum pumilum IPNI=*421764-1* 245
Panicum punctulatum IPNI=*426266-1* 246
Panicum purpurascens IPNI=*426266-1* 246
Panicum racemiferum IPNI=*421828-1* 245
Panicum raddianum IPNI=*1004002-1* 242
Panicum ramuliflorum IPNI=*1156911-2* 246
Panicum rariflorum IPNI=*319037-2* 245
Panicum raripilum IPNI=*943811-1* 245
Panicum rarisetum IPNI=*1172068-2* 245
Panicum rautanenii IPNI=*992521-1* 246
Panicum reptans IPNI=*426280-1* 246
Panicum respiciens IPNI=*421428-1* 245
Panicum reticulatum IPNI=*1156554-2* 246
Panicum retrofractum IPNI=*325362-2* 242
Panicum rhachitrichum IPNI=*1172068-2* 245
Panicum rhigiophyllum IPNI=*412983-1* 242
Panicum rigens IPNI=*412983-1* 242
Panicum rottleri IPNI=*399115-1* 238
Panicum rottleri IPNI=*421428-1* 245
Panicum rubiginosum IPNI=*421764-1* 245
Panicum rude IPNI=*421857-1* 245
Panicum ruderale IPNI=*412217-1* 242
Panicum rudgei var. *brasiliense* IPNI=*412983-1* 242
Panicum rugulosum f. *effusum* IPNI=*412226-1* 242
Panicum rugulosum IPNI=*412226-1* 242
Panicum rugulosum var. *glabrescens* IPNI=*412226-1* 242
Panicum rugulosum var. *lasianthum* IPNI=*412226-1* 242
Panicum rugulosum var. *pubescens* IPNI=*412226-1* 242
Panicum rugulosum var. *subvelutinum* IPNI=*412226-1* 242
Panicum ruscifolium IPNI=*406853-1* 241
Panicum ruscifolium subvar. *glabra* IPNI=*406853-1* 241
Panicum ruscifolium var. *amblyoides* IPNI=*406853-1* 241
Panicum sabiense IPNI=*324582-2* 242
Panicum sabulicola IPNI=*30149942-2* 238
Panicum saccharatum IPNI=*80220-2* 237
Panicum sacchariferum IPNI=*422090-1* 245
Panicum sacchariflorum IPNI=*1050107-2* 237
Panicum saccharoides IPNI=*398888-1* 237

Panicum sanctae-marthae IPNI=*30453766-2* 242
Panicum sanguinale IPNI=*399115-1* 238
Panicum sanguinale subsp. *horizontale* IPNI=*329014-2* 237
Panicum sanguinale subsp. *marginatum* IPNI=*1038558-2* 237
Panicum sanguinale subvar. *anisotrichum* IPNI=*399115-1* 238
Panicum sanguinale subvar. *leptostachyon* IPNI=*399115-1* 238
Panicum sanguinale subvar. *marginatum* IPNI=*1038558-2* 237
Panicum sanguinale subvar. *platystachyon* IPNI=*399115-1* 238
Panicum sanguinale var. *barcaldinense* IPNI=*1218147-2* 237
Panicum sanguinale var. *breviglume* IPNI=*399115-1* 238
Panicum sanguinale var. *digitatum* IPNI=*329018-2* 237
Panicum sanguinale var. *horizontale* IPNI=*329014-2* 237
Panicum sanguinale var. *longiglume* IPNI=*1038558-2* 237
Panicum sanguineum IPNI=*399115-1* 238
Panicum sanguinolentum IPNI=*399115-1* 238
Panicum scaberrimum IPNI=*60426565-2* 241
Panicum scalarum IPNI=*1209084-2* 237
Panicum scariosum IPNI=*135714-2* 241
Panicum schultesii IPNI=*30453766-2* 242
Panicum sciaphilum IPNI=*77076421-1* 247
Panicum scoparium IPNI=*412983-1* 242
Panicum sellowii var. *subvelutinum* IPNI=*412226-1* 242
Panicum sempervirens IPNI=*413385-1* 242
Panicum serotinum IPNI=*1207294-2* 245
Panicum serrulatum IPNI=*225088-2* 244
Panicum setaceum IPNI=*1207294-2* 245
Panicum setarium IPNI=*409823-1* 242
Panicum setosum IPNI=*1207294-2* 245
Panicum shastense IPNI=*79328-2* 237
Panicum sibiricum IPNI=*1207294-2* 245
Panicum sieberi IPNI=*426280-1* 246
Panicum sloanei IPNI=*135721-2* 241
Panicum sonorum IPNI=*411695-1* 242
Panicum sorghoideum IPNI=*135707-2* 240
Panicum sparsum IPNI=*60426565-2* 241
Panicum speciosum IPNI=*1103484-2* 245
Panicum spectabile IPNI=*88097-2* 238
Panicum sphaerocarpon IPNI=*79391-2* 237
Panicum spicatum IPNI=*77105978-1* 235
Panicum spithameum IPNI=*1156554-2* 246
Panicum spithamineum IPNI=*1156554-2* 246
Panicum sprengelianum IPNI=*77144198-1* 244
Panicum steudelianum IPNI=*399226-1* 238
Panicum stipatum IPNI=*329014-2* 237
Panicum stoloniferum IPNI=*77105897-1* 241
Panicum stoloniferum var. *frondosum* IPNI=*77105897-1* 241
Panicum stoloniferum var. *majus* IPNI=*77105897-1* 241
Panicum strigosum IPNI=*79395-2* 237
Panicum strigosum var. *glabrescens* IPNI=*79395-2* 237
Panicum strigosum var. *leucoblepharis* IPNI=*79395-2* 237
Panicum strigosum var. *puberulum* IPNI=*79395-2* 237
Panicum subcordatum IPNI=*426280-1* 246
Panicum subglobosum IPNI=*412226-1* 242
Panicum subquadriparum IPNI=*943744-1* 246
Panicum subsphaerocarpum IPNI=*1182687-2* 245
Panicum subspicatum IPNI=*3538-2* 233
Panicum sucosum IPNI=*77147487-1* 241
Panicum sulcatum IPNI=*421711-1* 245
Panicum sulcatum IPNI=*421828-1* 245
Panicum swartzianum IPNI=*135707-2* 240
Panicum sylvaticum IPNI=*409823-1* 242
Panicum taitense IPNI=*426280-1* 246
Panicum tectum IPNI=*79328-2* 237
Panicum teff IPNI=*60426565-2* 241
Panicum tenuiculmum IPNI=*1156911-2* 246
Panicum tenuiflorum IPNI=*243044-2* 245
Panicum tenuiflorum IPNI=*398933-1* 237
Panicum tenuissimum IPNI=*1048074-2* 244
Panicum tenuissimum IPNI=*243044-2* 245
Panicum tephrosanthum IPNI=*60426565-2* 241
Panicum tetrastichum IPNI=*399655-1* 238
Panicum thollonii IPNI=*1172068-2* 245
Panicum thouarsianum IPNI=*1156554-2* 246
Panicum thurowii IPNI=*79328-2* 237
Panicum trichanthum var. *modestum* IPNI=*413376-1* 242
Panicum trichocladum IPNI=*77105897-1* 241
Panicum trichocondylum IPNI=*60426565-2* 241
Panicum trichoglume IPNI=*60426565-2* 241
Panicum tricholaenoides var. *flavomarginatum* IPNI=*413385-1* 242
Panicum trichophorum IPNI=*77144198-1* 244
Panicum trichostachyum IPNI=*399115-1* 238
Panicum trinii IPNI=*211125-2* 244
Panicum triphellon IPNI=*412217-1* 242
Panicum truncatum IPNI=*979871-1* 245
Panicum umbraticola IPNI=*329014-2* 237
Panicum umbrosum IPNI=*426280-1* 246
Panicum umbrosum IPNI=*77105897-1* 241
Panicum uncinatum IPNI=*211125-2* 244
Panicum undatum IPNI=*409823-1* 242
Panicum unguinosum IPNI=*409823-1* 242
Panicum vaginatum IPNI=*184742-2* 243
Panicum vagum IPNI=*1048074-2* 244
Panicum velutinum IPNI=*1004002-1* 242
Panicum velutinum IPNI=*79328-2* 237
Panicum velutinum IPNI=*972086-1* 246
Panicum ventenatii IPNI=*943811-1* 245
Panicum versicolor IPNI=*943811-1* 245
Panicum verticillatum IPNI=*386353-1* 233
Panicum verticillatum IPNI=*421857-1* 245
Panicum verticillatum var. *ambiguum* IPNI=*421857-1* 245
Panicum verticillatum var. *parviflorum* IPNI=*421428-1* 245
Panicum vexillare IPNI=*992521-1* 246
Panicum viaticum IPNI=*1172068-2* 245
Panicum vicarium IPNI=*79391-2* 237
Panicum villiferum IPNI=*1038558-2* 237
Panicum violascens IPNI=*399226-1* 238
Panicum violascens var. *glabrescens* IPNI=*399226-1* 238
Panicum violascens var. *remotiusculum* IPNI=*399226-1* 238
Panicum virescens IPNI=*426280-1* 246
Panicum virescens IPNI=*943811-1* 245
Panicum viride IPNI=*421857-1* 245
Panicum vulgare IPNI=*1207294-2* 245
Panicum vulpinum IPNI=*943811-1* 245
Panicum vulpisetum IPNI=*1182687-2* 245
Panicum zizanioides IPNI=*3538-2* 233
Panicum zizanioides var. *microphyllum* IPNI=*3538-2* 233
Panicum zonale IPNI=*399655-1* 238
Panke tinctoria IPNI=*430194-1* 359
Panmorphia brevipes IPNI=*1147281-2* 217
Panopsis acuminata IPNI=*182723-2* 433
Panopsis cuaensis IPNI=*182723-2* 433
Panopsis hameliifolia IPNI=*182728-2* 433
Panopsis polystachya IPNI=*1076423-2* 433
Panopsis rubescens var. *sprucei* IPNI=*182723-2* 433
Panopsis sprucei IPNI=*182723-2* 433
Panoxis salicifolia IPNI=*812616-1* 377
Panterpa mollissima IPNI=*77142038-1* 363
Panzera falcata IPNI=*493934-1* 319
Panzhuyuia omeiensis IPNI=*84151-1* 198
Papaver aegadicum IPNI=*306058-2* 435
Papaver agrivagum IPNI=*306058-2* 435
Papaver album IPNI=*673724-1* 435
Papaver album-nigrum IPNI=*673724-1* 435
Papaver ameristophyllum IPNI=*306058-2* 435
Papaver amoenum IPNI=*673724-1* 435
Papaver amplexicaule IPNI=*673724-1* 435

Papaver anisotrichum IPNI=*306058-2* 435
Papaver arvatecum IPNI=*306058-2* 435
Papaver arvense IPNI=*306058-2* 435
Papaver atropurpureum IPNI=*306058-2* 435
Papaver balanocarpum IPNI=*306058-2* 435
Papaver caespitosum IPNI=*306058-2* 435
Papaver caudatifolium IPNI=*306058-2* 435
Papaver cereale IPNI=*306058-2* 435
Papaver chanceliae IPNI=*306058-2* 435
Papaver chelidonioides IPNI=*306058-2* 435
Papaver cinerascens IPNI=*306058-2* 435
Papaver commixtum IPNI=*306058-2* 435
Papaver commutatum subsp. *euxinum* IPNI=*306058-2* 435
Papaver cruciatum IPNI=*306058-2* 435
Papaver dodonei IPNI=*306058-2* 435
Papaver dubium var. *stipitatum* IPNI=*306058-2* 435
Papaver dzeghamicum IPNI=*673641-1* 435
Papaver erraticum IPNI=*306058-2* 435
Papaver erucifolium IPNI=*306058-2* 435
Papaver feddeanum IPNI=*306058-2* 435
Papaver fuchsii IPNI=*306058-2* 435
Papaver gabrielianae IPNI=*306058-2* 435
Papaver glabrum IPNI=*673724-1* 435
Papaver graecum IPNI=*306058-2* 435
Papaver grandiflorum IPNI=*673641-1* 435
Papaver guerlekense IPNI=*306058-2* 435
Papaver hookeri IPNI=*306058-2* 435
Papaver hortense IPNI=*673724-1* 435
Papaver humifusum IPNI=*306058-2* 435
Papaver indehiscens IPNI=*673724-1* 435
Papaver insignitum IPNI=*306058-2* 435
Papaver integrifolium IPNI=*306058-2* 435
Papaver interjectum IPNI=*306058-2* 435
Papaver intermedium IPNI=*306058-2* 435
Papaver lateritium subsp. *monanthum* IPNI=*673641-1* 435
Papaver mexicanum IPNI=*18816-2* 435
Papaver monanthum IPNI=*673641-1* 435
Papaver montenegrinum IPNI=*306058-2* 435
Papaver nigrum IPNI=*673724-1* 435
Papaver officinale IPNI=*673724-1* 435
Papaver omphalodeum IPNI=*306058-2* 435
Papaver opiiferum IPNI=*673724-1* 435
Papaver osswaldii IPNI=*306058-2* 435
Papaver paeoniiflorum IPNI=*673724-1* 435
Papaver paucifoliatum IPNI=*673641-1* 435
Papaver paucisetum IPNI=*306058-2* 435
Papaver propinquum IPNI=*306058-2* 435
Papaver pseudohaussknechtii IPNI=*306058-2* 435
Papaver rapiferum IPNI=*306058-2* 435
Papaver rhoeas f. *cruciatum* IPNI=*306058-2* 435
Papaver rhoeas f. *erythrotrichum* IPNI=*306058-2* 435
Papaver rhoeas f. *hoffmanianum* IPNI=*306058-2* 435
Papaver rhoeas var. *erythrotrichum* IPNI=*306058-2* 435
Papaver rhoeas var. *hoffmanianum* IPNI=*306058-2* 435
Papaver rhoeas var. *submamillatum* IPNI=*306058-2* 435
Papaver rhopalothece IPNI=*306058-2* 435
Papaver robertianella IPNI=*306058-2* 435
Papaver rumelicum IPNI=*306058-2* 435
Papaver rusticum IPNI=*306058-2* 435
Papaver segetale IPNI=*306058-2* 435
Papaver somniferum convar. *drogist* IPNI=*673724-1* 435
Papaver somniferum convar. *indicum* IPNI=*673724-1* 435
Papaver somniferum convar. *opiiferum* IPNI=*673724-1* 435
Papaver somniferum convar. *tarbagataicum* IPNI=*673724-1* 435
Papaver somniferum var. *album* IPNI=*673724-1* 435
Papaver somniferum var. *austroviolaceum* IPNI=*673724-1* 435
Papaver somniferum var. *genuinovilaceum* IPNI=*673724-1* 435
Papaver somniferum var. *leucum* IPNI=*673724-1* 435
Papaver somniferum var. *nigrum* IPNI=*673724-1* 435
Papaver somniferum var. *pullatum* IPNI=*673724-1* 435
Papaver somniferum var. *rubellum* IPNI=*673724-1* 435
Papaver somniferum var. *violiflorum* IPNI=*673724-1* 435
Papaver spectabile IPNI=*673641-1* 435
Papaver spurium IPNI=*306058-2* 435
Papaver stipitatum IPNI=*306058-2* 435
Papaver subumbilicatum IPNI=*306058-2* 435
Papaver sylvestre IPNI=*673724-1* 435
Papaver tenuissimum IPNI=*306058-2* 435
Papaver thaumasiosepalum IPNI=*306058-2* 435
Papaver tumidulum IPNI=*306058-2* 435
Papaver uniflorum IPNI=*306058-2* 435
Papaya bourgeaui IPNI=*675769-1* 279
Papaya carica IPNI=*30011248-2* 279
Papaya cauliflora IPNI=*675769-1* 279
Papaya cestriflora IPNI=*675782-1* 279
Papaya citriformis IPNI=*30011248-2* 279
Papaya communis IPNI=*30011248-2* 279
Papaya cubensis IPNI=*30011248-2* 279
Papaya cucumerina IPNI=*30011248-2* 279
Papaya cundinamarcensis IPNI=*675782-1* 279
Papaya digitata IPNI=*130970-2* 279
Papaya dodecaphylla IPNI=*30104966-2* 279
Papaya edulis IPNI=*30011248-2* 279
Papaya erythrocarpa IPNI=*675779-1* 279
Papaya goudotiana IPNI=*675773-1* 279
Papaya gracilis IPNI=*675773-1* 279
Papaya manihot IPNI=*675778-1* 279
Papaya microcarpa IPNI=*675778-1* 279
Papaya monoica IPNI=*675779-1* 279
Papaya papaya IPNI=*30011248-2* 279
Papaya peltata IPNI=*30011248-2* 279
Papaya posopora IPNI=*30011248-2* 279
Papaya pubescens IPNI=*675782-1* 279
Papaya pyriformis IPNI=*30011248-2* 279
Papaya rochefortii IPNI=*30011248-2* 279
Papaya sativa IPNI=*30011248-2* 279
Papaya spinosa IPNI=*30104966-2* 279
Papaya vulgaris IPNI=*30011248-2* 279
Paphiopedilum caudatum IPNI=*194355-2* 222
Paphiopedilum cothurnum IPNI=*325522-2* 217
Paphiopedilum lindenii IPNI=*1119571-2* 222
Papiliopsis nodosus IPNI=*212378-2* 222
Pappophorum alopecuroideum IPNI=*182865-2* 242
Pappophorum alopecuroideum var. *glabrum* IPNI=*182865-2* 242
Pappophorum alopecuroideum var. *pilosiusculum* IPNI=*182865-2* 242
Pappophorum arabicum IPNI=*400626-1* 238
Pappophorum boreale IPNI=*400626-1* 238
Pappophorum brachystachyum IPNI=*400626-1* 238
Pappophorum bulbosum IPNI=*400626-1* 238
Pappophorum elongatum IPNI=*182865-2* 242
Pappophorum fasciculatum IPNI=*400626-1* 238
Pappophorum figarianum IPNI=*400626-1* 238
Pappophorum jaminianum IPNI=*400626-1* 238
Pappophorum laguroideum IPNI=*182865-2* 242
Pappophorum macrostachyum IPNI=*182865-2* 242
Pappophorum mexicanum IPNI=*400626-1* 238
Pappophorum nanum IPNI=*400626-1* 238
Pappophorum pappiferum var. *alopecuroides* IPNI=*182865-2* 242
Pappophorum pappiferum var. *laguroideum* IPNI=*182865-2* 242
Pappophorum phleoides IPNI=*400626-1* 238
Pappophorum polystachyum IPNI=*182865-2* 242
Pappophorum pusillum IPNI=*400626-1* 238
Pappophorum saccharoides IPNI=*182865-2* 242
Pappophorum vincentianum IPNI=*400626-1* 238
Pappophorum wrightii IPNI=*400626-1* 238
Papyrus elegans IPNI=*304595-1* 230

Papyrus giganteus IPNI=*304595-1* 230
Papyrus odorata IPNI=*74939-2* 231
Papyrus odoratus IPNI=*304595-1* 230
Papyrus pangorei IPNI=*303822-1* 230
Papyrus spectabilis IPNI=*304595-1* 230
Parabesleria costaricensis IPNI=*378279-1* 366
Parablechnum chilense IPNI=*33533-2* 174
Parablechnum cordatum IPNI=*33533-2* 174
Parablechnum loxense IPNI=*77186436-1* 174
Parahancornia amapa IPNI=*182946-2* 342
Parahancornia tabernaemontana IPNI=*79852-1* 341
Parahemionitis arifolia IPNI=*319061-2* 177
Paralabatia acutangula IPNI=*160358-2* 306
Paralabatia casiquiarensis IPNI=*948065-1* 307
Paralabatia durlandii IPNI=*788920-1* 306
Paramicropholis acutangula IPNI=*160358-2* 306
Paramiflos glandulosus IPNI=*97673-2* 262
Paramyrciaria ciliolata IPNI=*599327-1* 427
Paraprotium amazonicum IPNI=*1063489-2* 452
Parascheelea anchistropetala IPNI=*961936-1* 205
Parascheelea luetzelburgii IPNI=*961936-1* 205
Parathesis viridis IPNI=*247472-2* 304
Paratropia cumingiana IPNI=*91749-1* 254
Paratropia latifolia IPNI=*91817-1* 254
Parestia caudata IPNI=*17080730-1* 175
Parietaria appendiculata IPNI=*854908-1* 446
Parietaria assurgens IPNI=*854958-1* 446
Parietaria australis IPNI=*854908-1* 446
Parietaria australis var. *glandulifera* IPNI=*854908-1* 446
Parietaria canescens IPNI=*854958-1* 446
Parietaria carnosula IPNI=*854908-1* 446
Parietaria coreana IPNI=*854908-1* 446
Parietaria cretica IPNI=*856545-1* 448
Parietaria debilis var. *australis* IPNI=*854908-1* 446
Parietaria debilis var. *gracilis* IPNI=*854908-1* 446
Parietaria debilis var. *micrantha* IPNI=*854908-1* 446
Parietaria debilis var. *squalida* IPNI=*854908-1* 446
Parietaria diffusa IPNI=*854958-1* 446
Parietaria erecta IPNI=*854958-1* 446
Parietaria fernandeziana IPNI=*854908-1* 446
Parietaria gracilis IPNI=*854908-1* 446
Parietaria humifusa IPNI=*854908-1* 446
Parietaria laxiflora IPNI=*854908-1* 446
Parietaria micrantha IPNI=*854908-1* 446
Parietaria microphylla IPNI=*1162156-2* 447
Parietaria muralis IPNI=*854958-1* 446
Parietaria nitens IPNI=*854958-1* 446
Parietaria nitida IPNI=*854958-1* 446
Parietaria populifolia IPNI=*854958-1* 446
Parietaria repens IPNI=*856545-1* 448
Parietaria rubicunda IPNI=*854958-1* 446
Parietaria ruwenzoriensis IPNI=*854908-1* 446
Parietaria ruwenzoriensis subsp. *keniensis* IPNI=*854908-1* 446
Parietaria scandens IPNI=*854908-1* 446
Parietaria soleirolii IPNI=*856545-1* 448
Parietaria squalida IPNI=*854908-1* 446
Parinari amazonica IPNI=*727101-1* 383
Parinari brachystachya IPNI=*727101-1* 383
Parinari elliottii IPNI=*727101-1* 383
Parinari glazioviana IPNI=*727101-1* 383
Parinari holstii IPNI=*727101-1* 383
Parinari laxiflora IPNI=*727101-1* 383
Parinari laxiflora var. *lata* IPNI=*727101-1* 383
Parinari liberica IPNI=*727101-1* 383
Parinari mildbraedii IPNI=*727101-1* 383
Parinari nalaensis IPNI=*727101-1* 383
Parinari pajura IPNI=*727140-1* 383
Parinari riparia IPNI=*727101-1* 383
Parinari salicifolia IPNI=*727101-1* 383
Parinari silvestris IPNI=*727101-1* 383
Parinari tenuifolia IPNI=*727101-1* 383
Parinari verdickii IPNI=*727101-1* 383
Parinari whytei IPNI=*727101-1* 383
Parita populneus IPNI=*564713-1* 415
Parita tiliaceus IPNI=*30143354-2* 410
Pariti boninense IPNI=*30143354-2* 410
Pariti circinnatum IPNI=*30143354-2* 410
Pariti grande IPNI=*30143354-2* 410
Pariti guineense IPNI=*30143354-2* 410
Pariti hastatum IPNI=*30143354-2* 410
Pariti purpurascens IPNI=*30143354-2* 410
Pariti tiliaceum IPNI=*30143354-2* 410
Pariti tiliifolium IPNI=*30143354-2* 410
Pariti tortuosum IPNI=*30143354-2* 410
Pariti tricuspe IPNI=*30143354-2* 410
Parkeria lockhartii IPNI=*17242590-1* 178
Parkeria pteridoides IPNI=*17242590-1* 178
Parkia alliodora IPNI=*512215-1* 329
Parkia arborea IPNI=*512215-1* 329
Parkia auriculata IPNI=*512191-1* 329
Parkia igneiflora var. *aurea* IPNI=*183268-2* 329
Parkia igneiflora var. *purpurea* IPNI=*183268-2* 329
Parkia ingens IPNI=*512215-1* 329
Parkia inundabilis IPNI=*512215-1* 329
Parkia oppositifolia IPNI=*512215-1* 329
Parkia panurensis IPNI=*512223-1* 329
Parkia paryphosphaera IPNI=*512215-1* 329
Parkia sylvatica IPNI=*512215-1* 329
Parkinsonia inermis IPNI=*512242-1* 329
Parkinsonia spinosa IPNI=*512242-1* 329
Parkinsonia thornberi IPNI=*512242-1* 329
Parmentiera edulis IPNI=*110241-1* 364
Parmentiera foliolosa IPNI=*110241-1* 364
Parmentiera lanceolata IPNI=*110241-1* 364
Paronychia achyrantha IPNI=*327342-2* 281
Paronychia confertissima IPNI=*45444-2* 285
Paronychia dubia IPNI=*1069792-2* 281
Paronychia ficoidea IPNI=*1078972-2* 281
Paronychia frutescens IPNI=*10362-2* 281
Paronychia paranensis IPNI=*435017-1* 285
Paronychia pauciflora IPNI=*45444-2* 285
Paronychia polycnemoides IPNI=*45444-2* 285
Paronychia ramosissima IPNI=*45444-2* 285
Paronychia rosetta IPNI=*45444-2* 285
Paronychia sessilis IPNI=*1069792-2* 281
Paronychia tenella IPNI=*1078972-2* 281
Paronychia tetragona IPNI=*1069792-2* 281
Parosela acutifolia IPNI=*76085-2* 317
Parosela cliffortiana IPNI=*76085-2* 317
Parosela coerulea IPNI=*76086-2* 317
Parosela dalea IPNI=*76085-2* 317
Parosela dalea var. *robusta* IPNI=*76085-2* 317
Parosela glandulosa IPNI=*76085-2* 317
Parosela nigra IPNI=*76085-2* 317
Parosela pumila IPNI=*76085-2* 317
Parosela robusta IPNI=*76085-2* 317
Parsonsia balsamona IPNI=*72326-2* 416
Parsonsia calophylla IPNI=*552922-1* 416
Parsonsia ciliata IPNI=*552959-1* 416
Parsonsia hyssopifolia IPNI=*553043-1* 417
Parsonsia ignea IPNI=*553048-1* 417
Parsonsia liebmannii IPNI=*553048-1* 417
Parsonsia micropetala IPNI=*553111-1* 417
Parsonsia pinto IPNI=*72326-2* 416
Parsonsia racemosa IPNI=*290454-2* 417
Parsonsia watsonii IPNI=*553048-1* 417

Parsonsia wrightii IPNI=*553315-1* 417
Parthenium lobatum IPNI=*236510-1* 267
Parthenium matricaria IPNI=*252460-1* 271
Parthenium pinnatifidum IPNI=*236510-1* 267
Paryphosphaera arborea IPNI=*512215-1* 329
Pascalia baccata IPNI=*993359-1* 271
Paspalanthium stoloniferum IPNI=*414634-1* 243
Paspalidium geminatum IPNI=*979871-1* 245
Paspalidium geminatum subsp. *paludivagum* IPNI=*979871-1* 245
Paspalidium geminatum var. *paludivagum* IPNI=*979871-1* 245
Paspalidium paludivagum IPNI=*979871-1* 245
Paspalidium pilgeri IPNI=*979871-1* 245
Paspalum abrahamii IPNI=*414411-1* 243
Paspalum adelogaeum IPNI=*414702-1* 243
Paspalum aegyptiacum IPNI=*399115-1* 238
Paspalum affine IPNI=*184526-2* 243
Paspalum africanum IPNI=*30011315-2* 242
Paspalum akoense IPNI=*414702-1* 243
Paspalum amazonicum IPNI=*414702-1* 243
Paspalum ambiguum IPNI=*397065-1* 236
Paspalum ancylocarpum IPNI=*414002-1* 243
Paspalum annulatum IPNI=*402044-1* 239
Paspalum antillense IPNI=*184571-2* 243
Paspalum appressum IPNI=*979871-1* 245
Paspalum argenteum IPNI=*398933-1* 237
Paspalum atrocarpum IPNI=*184571-2* 243
Paspalum aureum IPNI=*321761-2* 234
Paspalum auriculatum IPNI=*414702-1* 243
Paspalum berteroanum IPNI=*184275-2* 243
Paspalum bicorne IPNI=*1218147-2* 237
Paspalum bicrurum IPNI=*30011315-2* 242
Paspalum bifarium IPNI=*398933-1* 237
Paspalum biglume IPNI=*414634-1* 243
Paspalum bistipulatum IPNI=*30475806-2* 243
Paspalum borbonicum IPNI=*414702-1* 243
Paspalum boryanum IPNI=*184742-2* 243
Paspalum boscianum IPNI=*414702-1* 243
Paspalum brachiatum IPNI=*184742-2* 243
Paspalum bracteatum IPNI=*184275-2* 243
Paspalum brevifolium IPNI=*398933-1* 237
Paspalum brevifolium var. *propinquum* IPNI=*398933-1* 237
Paspalum caespitosum IPNI=*184571-2* 243
Paspalum campestre IPNI=*184571-2* 243
Paspalum caperatum IPNI=*184571-2* 243
Paspalum carinatovaginatum IPNI=*321761-2* 234
Paspalum cartilagineum IPNI=*414702-1* 243
Paspalum cartilagineum var. *biglumaceum* IPNI=*414702-1* 243
Paspalum chepica IPNI=*184275-2* 243
Paspalum chrysites IPNI=*321761-2* 234
Paspalum chrysocomum IPNI=*321761-2* 234
Paspalum chrysodactylon IPNI=*321761-2* 234
Paspalum chrysodactylon var. *glabratum* IPNI=*321761-2* 234
Paspalum chrysodactylon var. *psilachne* IPNI=*321761-2* 234
Paspalum chrysodactylon var. *villosum* IPNI=*321761-2* 234
Paspalum chrysostachyum IPNI=*321761-2* 234
Paspalum ciliare IPNI=*1038558-2* 237
Paspalum ciliatum IPNI=*30011315-2* 242
Paspalum coloratum IPNI=*414702-1* 243
Paspalum columbiense IPNI=*27415-2* 234
Paspalum commersonii IPNI=*414702-1* 243
Paspalum commutatum IPNI=*414702-1* 243
Paspalum compressicaule IPNI=*184526-2* 243
Paspalum compressum var. *arenarium* IPNI=*27338-2* 234
Paspalum confertum IPNI=*414702-1* 243
Paspalum conjugatum f. *tristachyum* IPNI=*30011315-2* 242
Paspalum conjugatum var. *parviflorum* IPNI=*30011315-2* 242
Paspalum conjugatum var. *pubescens* IPNI=*30011315-2* 242
Paspalum cordovense IPNI=*184526-2* 243
Paspalum coromandelinum IPNI=*414702-1* 243
Paspalum dactylon IPNI=*397065-1* 236
Paspalum decumbens IPNI=*184571-2* 243
Paspalum deightonii IPNI=*414702-1* 243
Paspalum densum var. *elliptico-oblongum* IPNI=*184394-2* 243
Paspalum depressum IPNI=*392087-1* 234
Paspalum didactylum IPNI=*184742-2* 243
Paspalum digitaria IPNI=*184275-2* 243
Paspalum digitatum IPNI=*329018-2* 237
Paspalum dilatatum f. *paucispica* IPNI=*414070-1* 243
Paspalum dilatatum subsp. *flavescens* IPNI=*414070-1* 243
Paspalum dilatatum var. *decumbens* IPNI=*414070-1* 243
Paspalum dilatatum var. *parviflorum* IPNI=*414828-1* 243
Paspalum dilatatum var. *sacchariferum* IPNI=*414070-1* 243
Paspalum dissectum IPNI=*74675-3* 238
Paspalum distachyon IPNI=*414477-1* 243
Paspalum distichum subsp. *paucispicatum* IPNI=*184275-2* 243
Paspalum distichum subsp. *vaginatum* IPNI=*184742-2* 243
Paspalum distichum var. *digitaria* IPNI=*184275-2* 243
Paspalum distichum var. *indutum* IPNI=*184275-2* 243
Paspalum distichum var. *littorale* IPNI=*184742-2* 243
Paspalum distichum var. *longirepens* IPNI=*184275-2* 243
Paspalum distichum var. *microstachyum* IPNI=*184275-2* 243
Paspalum distichum var. *nanum* IPNI=*184742-2* 243
Paspalum distichum var. *paspalodes* IPNI=*184275-2* 243
Paspalum distichum var. *vaginatum* IPNI=*184742-2* 243
Paspalum dolichopus IPNI=*30011315-2* 242
Paspalum elliottii IPNI=*184275-2* 243
Paspalum eriophorum IPNI=*414070-1* 243
Paspalum exasperatum IPNI=*321761-2* 234
Paspalum fernandezianum IPNI=*184275-2* 243
Paspalum filiculme IPNI=*398933-1* 237
Paspalum filifolium IPNI=*27338-2* 234
Paspalum filostachyum IPNI=*392087-1* 234
Paspalum firmum IPNI=*414702-1* 243
Paspalum fissifolium IPNI=*184742-2* 243
Paspalum fissifolium IPNI=*27338-2* 234
Paspalum fluctuans IPNI=*30475806-2* 243
Paspalum fluitans IPNI=*30475806-2* 243
Paspalum foliosum IPNI=*184742-2* 243
Paspalum formosum IPNI=*184571-2* 243
Paspalum fragile IPNI=*80253-2* 237
Paspalum frankii IPNI=*30475806-2* 243
Paspalum furcatum var. *fissum* IPNI=*184742-2* 243
Paspalum furcatum var. *parviflorum* IPNI=*392087-1* 234
Paspalum fuscescens IPNI=*80254-2* 237
Paspalum fuscum IPNI=*399226-1* 238
Paspalum galmarra IPNI=*184526-2* 243
Paspalum gardnerianum var. *oligostachyum* IPNI=*414188-1* 243
Paspalum gardnerianum var. *vestitum* IPNI=*414188-1* 243
Paspalum gayanum IPNI=*184742-2* 243
Paspalum germanicum IPNI=*1207294-2* 245
Paspalum glaucifolium IPNI=*184275-2* 243
Paspalum gnaphalioideum IPNI=*321761-2* 234
Paspalum gracile IPNI=*30475806-2* 243
Paspalum guadaloupense IPNI=*392087-1* 234
Paspalum guineense IPNI=*184526-2* 243
Paspalum hemicryptum IPNI=*414002-1* 243
Paspalum hemisphericum IPNI=*184526-2* 243
Paspalum hieronymi IPNI=*414056-1* 243
Paspalum hirtiglumum IPNI=*414411-1* 243
Paspalum horneri IPNI=*414702-1* 243
Paspalum inaequale IPNI=*1038558-2* 237
Paspalum inflatum IPNI=*184742-2* 243
Paspalum inops IPNI=*414002-1* 243
Paspalum iridifolium IPNI=*27415-2* 234
Paspalum jaguaense IPNI=*184742-2* 243
Paspalum jardinii IPNI=*414702-1* 243

Paspalum jujuyense IPNI=*414056-1* 243
Paspalum karwinskyi IPNI=*414431-1* 243
Paspalum kisantuense IPNI=*392087-1* 234
Paspalum kleinianum IPNI=*184742-2* 243
Paspalum kora IPNI=*184742-2* 243
Paspalum kora IPNI=*414702-1* 243
Paspalum lachneum IPNI=*184571-2* 243
Paspalum lanatum IPNI=*414070-1* 243
Paspalum lanceifolium IPNI=*1207790-2* 245
Paspalum larranagae IPNI=*414828-1* 243
Paspalum laticulmum IPNI=*392087-1* 234
Paspalum latifolium IPNI=*414854-1* 243
Paspalum ledermannii IPNI=*414702-1* 243
Paspalum lenticulare IPNI=*184571-2* 243
Paspalum leucocheilum IPNI=*414854-1* 243
Paspalum littorale IPNI=*184742-2* 243
Paspalum lividum IPNI=*414056-1* 243
Paspalum loliaceum IPNI=*408465-1* 241
Paspalum longiflorum IPNI=*398933-1* 237
Paspalum longissimum IPNI=*30011315-2* 242
Paspalum macropodium IPNI=*392087-1* 234
Paspalum manabiense IPNI=*414634-1* 243
Paspalum marginatum IPNI=*184571-2* 243
Paspalum mathewsii IPNI=*27415-2* 234
Paspalum mauritanicum IPNI=*414702-1* 243
Paspalum metabolon IPNI=*414702-1* 243
Paspalum metzii IPNI=*414702-1* 243
Paspalum meyerianum IPNI=*1021229-2* 239
Paspalum michauxianum IPNI=*184275-2* 243
Paspalum micranthum IPNI=*80254-2* 237
Paspalum miliare IPNI=*410542-1* 242
Paspalum minutiflorum IPNI=*399226-1* 238
Paspalum moluccanum IPNI=*414070-1* 243
Paspalum montevidense IPNI=*184571-2* 243
Paspalum moratii IPNI=*414702-1* 243
Paspalum mucronatum IPNI=*30475806-2* 243
Paspalum multiflorum IPNI=*184571-2* 243
Paspalum multispica IPNI=*184526-2* 243
Paspalum natans IPNI=*30475806-2* 243
Paspalum nematodes IPNI=*398933-1* 237
Paspalum notatum var. *latiflorum* IPNI=*414477-1* 243
Paspalum notatum var. *typicum* IPNI=*414477-1* 243
Paspalum oajacense IPNI=*184275-2* 243
Paspalum oaxacense IPNI=*184275-2* 243
Paspalum oriziforme IPNI=*27415-2* 234
Paspalum orthos IPNI=*184571-2* 243
Paspalum ovatum IPNI=*414070-1* 243
Paspalum ovatum var. *grandiflorum* IPNI=*414070-1* 243
Paspalum ovatum var. *parviflorum* IPNI=*414828-1* 243
Paspalum oxyanthum IPNI=*329014-2* 237
Paspalum paludosum IPNI=*184571-2* 243
Paspalum palustre IPNI=*414702-1* 243
Paspalum paniculatum IPNI=*30475806-2* 243
Paspalum paniculatum var. *rigidum* IPNI=*184526-2* 243
Paspalum parviflorum IPNI=*1207790-2* 245
Paspalum paspalodes IPNI=*184275-2* 243
Paspalum paucisplcatum IPNI=*184275-2* 243
Paspalum pauperculum IPNI=*184571-2* 243
Paspalum pauperculum var. *altius* IPNI=*184571-2* 243
Paspalum pedunculare IPNI=*414070-1* 243
Paspalum pertenue IPNI=*399226-1* 238
Paspalum platense IPNI=*414070-1* 243
Paspalum platicaulon IPNI=*392087-1* 234
Paspalum platicaulon var. *gracilius* IPNI=*27338-2* 234
Paspalum platicaulon var. *pygmaeum* IPNI=*27338-2* 234
Paspalum platyculmum IPNI=*392087-1* 234
Paspalum platyphyllum IPNI=*414854-1* 243
Paspalum plicatulum var. *cinereum* IPNI=*184571-2* 243
Paspalum plicatulum var. *ellipticum* IPNI=*184571-2* 243
Paspalum plicatulum var. *genuinum* IPNI=*184571-2* 243
Paspalum plicatulum var. *glabrum* IPNI=*184571-2* 243
Paspalum plicatulum var. *intumescens* IPNI=*184571-2* 243
Paspalum plicatulum var. *longipilum* IPNI=*184571-2* 243
Paspalum plicatulum var. *oblongum* IPNI=*184571-2* 243
Paspalum plicatulum var. *rigidum* IPNI=*184571-2* 243
Paspalum plicatulum var. *subrotundum* IPNI=*414002-1* 243
Paspalum plicatulum var. *villosissimum* IPNI=*184571-2* 243
Paspalum plicatum IPNI=*184571-2* 243
Paspalum polo IPNI=*414702-1* 243
Paspalum polystachyum IPNI=*184526-2* 243
Paspalum polystachyum IPNI=*414702-1* 243
Paspalum polystachyum IPNI=*77164258-1* 246
Paspalum preslii IPNI=*398933-1* 237
Paspalum proliferum IPNI=*414056-1* 243
Paspalum pseudodurva IPNI=*399226-1* 238
Paspalum puberulum IPNI=*414702-1* 243
Paspalum pubescens IPNI=*398933-1* 237
Paspalum pubescens IPNI=*414702-1* 243
Paspalum pubifolium IPNI=*184571-2* 243
Paspalum pulchrum IPNI=*321761-2* 234
Paspalum pulchrum var. *angustifolium* IPNI=*321761-2* 234
Paspalum pulchrum var. *planifolium* IPNI=*321761-2* 234
Paspalum pulchrum var. *rigidiusculum* IPNI=*321761-2* 234
Paspalum punctatum IPNI=*1021229-2* 239
Paspalum purpurascens IPNI=*414702-1* 243
Paspalum purpureum IPNI=*414634-1* 243
Paspalum pyramidale IPNI=*30475806-2* 243
Paspalum radiatum IPNI=*321761-2* 234
Paspalum ramosissimum IPNI=*321761-2* 234
Paspalum ramosum IPNI=*184571-2* 243
Paspalum raunkiaeri IPNI=*392087-1* 234
Paspalum reimarioides IPNI=*184742-2* 243
Paspalum renggeri IPNI=*30011315-2* 242
Paspalum repens var. *fluitans* IPNI=*30475806-2* 243
Paspalum reptans IPNI=*184742-2* 243
Paspalum saltense IPNI=*414477-1* 243
Paspalum sanguinale IPNI=*399115-1* 238
Paspalum saxatile IPNI=*184571-2* 243
Paspalum schaffneri IPNI=*184275-2* 243
Paspalum schreberianum IPNI=*414431-1* 243
Paspalum scoparium IPNI=*27415-2* 234
Paspalum scrobiculatum var. *gracillimum* IPNI=*414702-1* 243
Paspalum scrobiculatum var. *horneri* IPNI=*414702-1* 243
Paspalum selloi IPNI=*414070-1* 243
Paspalum serpens IPNI=*414702-1* 243
Paspalum setaceum IPNI=*408465-1* 241
Paspalum sieberianum IPNI=*30011315-2* 242
Paspalum squamatum IPNI=*184742-2* 243
Paspalum stoloniferum IPNI=*414634-1* 243
Paspalum strictum IPNI=*184526-2* 243
Paspalum supinum IPNI=*184526-2* 243
Paspalum taphrophyllum IPNI=*414477-1* 243
Paspalum tenue IPNI=*184571-2* 243
Paspalum tenue IPNI=*30011315-2* 242
Paspalum tenue IPNI=*392087-1* 234
Paspalum tenuissimum IPNI=*408467-1* 241
Paspalum texanum IPNI=*184571-2* 243
Paspalum triglume IPNI=*414634-1* 243
Paspalum tripinnatum IPNI=*27415-2* 234
Paspalum tristachyum IPNI=*184742-2* 243
Paspalum tristachyum IPNI=*392087-1* 234
Paspalum underwoodii IPNI=*414431-1* 243
Paspalum undulatum IPNI=*184571-2* 243
Paspalum uniflorum IPNI=*392087-1* 234
Paspalum uruguayense IPNI=*414477-1* 243
Paspalum vaginatum subsp. *nanum* IPNI=*184742-2* 243

Paspalum vaginatum var. *nanum* IPNI=*184742-2* 243
Paspalum vaginatum var. *pubescens* IPNI=*184275-2* 243
Paspalum vaseyanum IPNI=*414828-1* 243
Paspalum velutinum IPNI=*80280-2* 237
Paspalum villifolium IPNI=*414002-1* 243
Paspalum virgatum var. *ciliatum* IPNI=*414854-1* 243
Paspalum virgatum var. *jacquinianum* IPNI=*414854-1* 243
Paspalum virgatum var. *linnaeanum* IPNI=*414854-1* 243
Paspalum virgatum var. *parviflorum* IPNI=*414828-1* 243
Paspalum virgatum var. *pubiflorum* IPNI=*414828-1* 243
Paspalum virgatum var. *purpurascens* IPNI=*414702-1* 243
Paspalum virgatum var. *schreberianum* IPNI=*414431-1* 243
Paspalum virgatum var. *willdenowianum* IPNI=*414854-1* 243
Paspalum vulnerans IPNI=*414431-1* 243
Paspalum wettsteinii IPNI=*414854-1* 243
Paspalum wullschlaegelii IPNI=*414411-1* 243
Paspalum xizangense IPNI=*27338-2* 234
Paspalum zollingeri IPNI=*414702-1* 243
Passaveria lanceolata IPNI=*86808-2* 305
Passaveria obovata IPNI=*786784-1* 305
Passaveria obovata var. *tomentosa* IPNI=*786784-1* 305
Passiflora acerifolia IPNI=*674754-1* 397
Passiflora adenophylla IPNI=*675494-1* 400
Passiflora alata var. *brasiliana* IPNI=*321312-2* 397
Passiflora alata var. *latifolia* IPNI=*321312-2* 397
Passiflora alba IPNI=*675494-1* 400
Passiflora alnifolia IPNI=*674832-1* 398
Passiflora angustifolia IPNI=*321973-2* 400
Passiflora appendiculata IPNI=*674805-1* 398
Passiflora aspera IPNI=*674754-1* 397
Passiflora asperata IPNI=*674754-1* 397
Passiflora atomaria IPNI=*675494-1* 400
Passiflora bauhiniifolia IPNI=*674768-1* 397
Passiflora bicolorata IPNI=*321312-2* 397
Passiflora biformis IPNI=*674920-1* 398
Passiflora bifurca IPNI=*674933-1* 398
Passiflora bilobata IPNI=*321950-2* 398
Passiflora brachychlamys IPNI=*675224-1* 399
Passiflora brasiliana IPNI=*321312-2* 397
Passiflora brighamii IPNI=*674824-1* 398
Passiflora bryonifolia IPNI=*674847-1* 398
Passiflora buchananii IPNI=*30044451-2* 400
Passiflora caerulea var. *angustifolia* IPNI=*321949-2* 398
Passiflora caerulea var. *glauca* IPNI=*321949-2* 398
Passiflora caerulea var. *imbricata* IPNI=*321949-2* 398
Passiflora caerulea var. *regnellii* IPNI=*321949-2* 398
Passiflora calcarata IPNI=*675494-1* 400
Passiflora calliaquatica IPNI=*321973-2* 400
Passiflora caloneura IPNI=*319092-2* 193
Passiflora capsularis var. *acutiloba* IPNI=*321950-2* 398
Passiflora caracasana IPNI=*675238-1* 399
Passiflora caucaense IPNI=*674975-1* 398
Passiflora caudata IPNI=*324148-2* 399
Passiflora cayaponioides IPNI=*674805-1* 398
Passiflora cearaensis IPNI=*675456-1* 400
Passiflora ceratosepala IPNI=*674754-1* 397
Passiflora chilensis IPNI=*675348-1* 399
Passiflora chinensis IPNI=*321949-2* 398
Passiflora ciliaris IPNI=*184857-2* 398
Passiflora ciliata var. *polyadena* IPNI=*184857-2* 398
Passiflora ciliata var. *quinqueloba* IPNI=*184857-2* 398
Passiflora ciliata var. *riparia* IPNI=*184857-2* 398
Passiflora cincinnata var. *imbricata* IPNI=*321963-2* 398
Passiflora cincinnata var. *minor* IPNI=*321963-2* 398
Passiflora cinerea IPNI=*674805-1* 398
Passiflora citrifolia IPNI=*321312-2* 397
Passiflora clypeata IPNI=*674920-1* 398
Passiflora coccinea var. *minor* IPNI=*674909-1* 398
Passiflora coccinea var. *velutina* IPNI=*674909-1* 398
Passiflora crenata IPNI=*675414-1* 400
Passiflora cryptopetala IPNI=*674805-1* 398
Passiflora cuneifolia IPNI=*321964-2* 398
Passiflora cyathophora IPNI=*674805-1* 398
Passiflora diaden IPNI=*321964-2* 398
Passiflora difformis IPNI=*674920-1* 398
Passiflora digitata IPNI=*675456-1* 400
Passiflora discolor IPNI=*321967-2* 399
Passiflora divaricata IPNI=*674821-1* 398
Passiflora dumetosa IPNI=*321969-2* 399
Passiflora ecuadorica IPNI=*184886-2* 398
Passiflora edulis var. *pomifera* IPNI=*321964-2* 398
Passiflora edulis var. *rubricaulis* IPNI=*321964-2* 398
Passiflora edulis var. *verrucifera* IPNI=*321964-2* 398
Passiflora emiliae IPNI=*675414-1* 400
Passiflora erosa IPNI=*321969-2* 399
Passiflora erubescens IPNI=*675423-1* 400
Passiflora europhylla IPNI=*675584-1* 400
Passiflora fernandezii IPNI=*675414-1* 400
Passiflora flexicaulis IPNI=*674933-1* 398
Passiflora foetida f. *latifolia* IPNI=*321966-2* 398
Passiflora foetida f. *suberecta* IPNI=*321966-2* 398
Passiflora foetida var. *ciliata* IPNI=*184857-2* 398
Passiflora foetida var. *eliasii* IPNI=*184857-2* 398
Passiflora foetida var. *ellisonii* IPNI=*321966-2* 398
Passiflora foetida var. *fluminensis* IPNI=*321966-2* 398
Passiflora foetida var. *gardneri* IPNI=*321966-2* 398
Passiflora foetida var. *glaziovii* IPNI=*321966-2* 398
Passiflora foetida var. *gossypiifolia* IPNI=*321966-2* 398
Passiflora foetida var. *hastata* IPNI=*184857-2* 398
Passiflora foetida var. *hirsutissima* IPNI=*184857-2* 398
Passiflora foetida var. *hispida* IPNI=*675582-1* 400
Passiflora foetida var. *isthmia* IPNI=*675582-1* 400
Passiflora foetida var. *lanuginosa* IPNI=*184857-2* 398
Passiflora foetida var. *longipedunculata* IPNI=*321966-2* 398
Passiflora foetida var. *maxonii* IPNI=*184857-2* 398
Passiflora foetida var. *mayarum* IPNI=*184857-2* 398
Passiflora foetida var. *moritziana* IPNI=*321966-2* 398
Passiflora foetida var. *muralis* IPNI=*321966-2* 398
Passiflora foetida var. *nicaraguensis* IPNI=*184857-2* 398
Passiflora foetida var. *parvifolia* IPNI=*184857-2* 398
Passiflora foetida var. *pectinata* IPNI=*321966-2* 398
Passiflora foetida var. *polyadenia* IPNI=*184857-2* 398
Passiflora foetida var. *quinqueloba* IPNI=*184857-2* 398
Passiflora foetida var. *riparia* IPNI=*184857-2* 398
Passiflora foetida var. *salvadorensis* IPNI=*184857-2* 398
Passiflora foetida var. *sanctae-martae* IPNI=*184857-2* 398
Passiflora foetida var. *sericea* IPNI=*321966-2* 398
Passiflora foetida var. *strigosa* IPNI=*321966-2* 398
Passiflora foetida var. *subintegra* IPNI=*184857-2* 398
Passiflora foetida var. *subpalmata* IPNI=*184857-2* 398
Passiflora foetida var. *tainaniana* IPNI=*321966-2* 398
Passiflora fulgens IPNI=*674909-1* 398
Passiflora furcata IPNI=*674933-1* 398
Passiflora geminiflora IPNI=*675584-1* 400
Passiflora gigantifolia IPNI=*675177-1* 399
Passiflora glaberrima var. *cumbalensis* IPNI=*184886-2* 398
Passiflora glabra IPNI=*185104-2* 399
Passiflora glabrata IPNI=*674824-1* 398
Passiflora glandulosa var. *canaliculata* IPNI=*675022-1* 398
Passiflora glandulosa var. *stoupyana* IPNI=*675022-1* 398
Passiflora glaucophylla IPNI=*675010-1* 398
Passiflora gossypiifolia IPNI=*321966-2* 398
Passiflora granadilla IPNI=*321949-2* 398
Passiflora grandiflora IPNI=*675390-1* 400
Passiflora guianensis IPNI=*675392-1* 400
Passiflora hartwiesiana IPNI=*321949-2* 398

Passiflora hassleriana IPNI=*321950-2* 398
Passiflora hassleriana var. *grandifolia* IPNI=*321950-2* 398
Passiflora hassleriana var. *paraguariensis* IPNI=*321950-2* 398
Passiflora hastata IPNI=*184857-2* 398
Passiflora hastata var. *nicaraguensis* IPNI=*184857-2* 398
Passiflora hederacea IPNI=*321973-2* 400
Passiflora hederifolia IPNI=*321973-2* 400
Passiflora helleborifolia IPNI=*674426-1* 397
Passiflora hemicycla IPNI=*675584-1* 400
Passiflora hermannii IPNI=*321966-2* 398
Passiflora heterophylla IPNI=*185104-2* 399
Passiflora hexangularis IPNI=*675390-1* 400
Passiflora heydei IPNI=*321969-2* 399
Passiflora hirsuta IPNI=*184857-2* 398
Passiflora hirsuta IPNI=*185104-2* 399
Passiflora hispida IPNI=*675582-1* 400
Passiflora hispidula IPNI=*321973-2* 400
Passiflora holosericea IPNI=*675494-1* 400
Passiflora hullettii IPNI=*675390-1* 400
Passiflora imthurnii IPNI=*675022-1* 398
Passiflora inamoena IPNI=*674847-1* 398
Passiflora incana IPNI=*675446-1* 400
Passiflora incarnata IPNI=*321964-2* 398
Passiflora iodocarpa IPNI=*321964-2* 398
Passiflora karsteniana IPNI=*674847-1* 398
Passiflora kegeliana IPNI=*674805-1* 398
Passiflora kohautiana IPNI=*321973-2* 400
Passiflora laticaulis IPNI=*321967-2* 399
Passiflora latifolia IPNI=*321312-2* 397
Passiflora laurifolia var. *tinifolia* IPNI=*675132-1* 399
Passiflora liebmannii IPNI=*184857-2* 398
Passiflora ligularis var. *geminiflora* IPNI=*675142-1* 399
Passiflora lockhartii IPNI=*675392-1* 400
Passiflora longiflora IPNI=*675224-1* 399
Passiflora longifolia IPNI=*321973-2* 400
Passiflora longipes var. *retusa* IPNI=*321967-2* 399
Passiflora lorifera IPNI=*675177-1* 399
Passiflora loureiroi IPNI=*321949-2* 398
Passiflora lowei IPNI=*675142-1* 399
Passiflora luciensis IPNI=*674933-1* 398
Passiflora lunata IPNI=*321950-2* 398
Passiflora lunata IPNI=*674824-1* 398
Passiflora lunata IPNI=*77063117-1* 398
Passiflora lunulata IPNI=*674824-1* 398
Passiflora lutea IPNI=*675494-1* 400
Passiflora macrocarpa IPNI=*675390-1* 400
Passiflora maculata IPNI=*321973-2* 400
Passiflora maculifolia IPNI=*321967-2* 399
Passiflora maliformis IPNI=*321312-2* 397
Passiflora marigouja IPNI=*675582-1* 400
Passiflora mascarensis IPNI=*321312-2* 397
Passiflora mauritiana IPNI=*321312-2* 397
Passiflora maximiliana IPNI=*321967-2* 399
Passiflora maximiliana var. *acutiloba* IPNI=*321967-2* 399
Passiflora maximiliana var. *expansa* IPNI=*321967-2* 399
Passiflora maximiliana var. *retusa* IPNI=*321967-2* 399
Passiflora mayana IPNI=*321949-2* 398
Passiflora menispermifolia var. *cuellensis* IPNI=*675211-1* 399
Passiflora meridensis IPNI=*675190-1* 399
Passiflora microcarpa IPNI=*321967-2* 399
Passiflora middletoniana IPNI=*321964-2* 398
Passiflora minima IPNI=*185104-2* 399
Passiflora minima IPNI=*675457-1* 400
Passiflora mixta var. *eriantha* IPNI=*675224-1* 399
Passiflora mixta var. *subquinqueloba* IPNI=*675224-1* 399
Passiflora monnieri IPNI=*674909-1* 398
Passiflora moritziana IPNI=*321966-2* 398
Passiflora muchronata IPNI=*674821-1* 398
Passiflora muralis IPNI=*321966-2* 398
Passiflora neillii IPNI=*675494-1* 400
Passiflora nigra IPNI=*185104-2* 399
Passiflora nitens IPNI=*675457-1* 400
Passiflora nymphaeoides IPNI=*675263-1* 399
Passiflora oblongifolia IPNI=*675132-1* 399
Passiflora obscura IPNI=*321966-2* 398
Passiflora oliviformis IPNI=*321973-2* 400
Passiflora ornata IPNI=*324148-2* 399
Passiflora oviformis IPNI=*321312-2* 397
Passiflora pallidiflora IPNI=*321964-2* 398
Passiflora palmata IPNI=*675456-1* 400
Passiflora paraguayensis IPNI=*321950-2* 398
Passiflora parviflora IPNI=*185104-2* 399
Passiflora pediculata IPNI=*60461856-2* 399
Passiflora pedunculata IPNI=*321312-2* 397
Passiflora peltata IPNI=*321973-2* 400
Passiflora pennipes IPNI=*675348-1* 399
Passiflora pergrandis IPNI=*675414-1* 400
Passiflora perlobata IPNI=*321963-2* 398
Passiflora phoenicea IPNI=*321312-2* 397
Passiflora physocalymma IPNI=*675238-1* 399
Passiflora picroderma IPNI=*321964-2* 398
Passiflora piligera IPNI=*321950-2* 398
Passiflora pinnatifolia IPNI=*675348-1* 399
Passiflora pisonis IPNI=*321963-2* 398
Passiflora polyadena IPNI=*321966-2* 398
Passiflora pomifera IPNI=*321964-2* 398
Passiflora pruinosa IPNI=*675010-1* 398
Passiflora pseudociliata IPNI=*184857-2* 398
Passiflora psilantha IPNI=*675547-1* 400
Passiflora pubescens IPNI=*321950-2* 398
Passiflora pulchella IPNI=*674821-1* 398
Passiflora pulchella var. *bifidata* IPNI=*674821-1* 398
Passiflora pyriformis IPNI=*321312-2* 397
Passiflora quadrangularis var. *sulcata* IPNI=*675390-1* 400
Passiflora quadriglandulosa var. *involucrata* IPNI=*185013-2* 399
Passiflora retusa IPNI=*321967-2* 399
Passiflora rhodantha IPNI=*675190-1* 399
Passiflora rohrii IPNI=*674805-1* 398
Passiflora rotundifolia IPNI=*674821-1* 398
Passiflora rotundifolia var. *jacquinii* IPNI=*674821-1* 398
Passiflora rubricaulis IPNI=*321964-2* 398
Passiflora salmonea IPNI=*185110-2* 399
Passiflora sanguinea IPNI=*30044451-2* 400
Passiflora sarcosepala IPNI=*321312-2* 397
Passiflora scabra IPNI=*674754-1* 397
Passiflora selloi IPNI=*321949-2* 398
Passiflora serrata IPNI=*675456-1* 400
Passiflora serrata var. *digitata* IPNI=*675456-1* 400
Passiflora serratistipula IPNI=*675142-1* 399
Passiflora servitensis IPNI=*30044451-2* 400
Passiflora servitensis var. *bracteosa* IPNI=*30044451-2* 400
Passiflora silvicola IPNI=*675022-1* 398
Passiflora spathulata IPNI=*674824-1* 398
Passiflora stenophylla IPNI=*321949-2* 398
Passiflora stipulata IPNI=*675494-1* 400
Passiflora stipulata var. *atomaria* IPNI=*675494-1* 400
Passiflora suberosa var. *angustifolia* IPNI=*321973-2* 400
Passiflora suberosa var. *hederacea* IPNI=*321973-2* 400
Passiflora suberosa var. *hirsuta* IPNI=*185104-2* 399
Passiflora suberosa var. *minima* IPNI=*185104-2* 399
Passiflora suberosa var. *pallida* IPNI=*185104-2* 399
Passiflora sulcata IPNI=*675390-1* 400
Passiflora surinamensis IPNI=*675584-1* 400
Passiflora tacso IPNI=*675224-1* 399
Passiflora tetradena IPNI=*321312-2* 397
Passiflora tetragona IPNI=*675390-1* 400

Passiflora theobromifolia IPNI=*675041-1* 399
Passiflora tiliifolia IPNI=*675142-1* 399
Passiflora tinifolia IPNI=*675132-1* 399
Passiflora tomentosa IPNI=*675224-1* 399
Passiflora tomentosa var. *mollissima* IPNI=*185076-2* 399
Passiflora torta IPNI=*674805-1* 398
Passiflora toxicaria IPNI=*674909-1* 398
Passiflora translinearis IPNI=*321967-2* 399
Passiflora tripartita var. *mollissima* IPNI=*185076-2* 399
Passiflora urceolata IPNI=*675224-1* 399
Passiflora vanvolxemii IPNI=*674787-1* 398
Passiflora variegata IPNI=*321966-2* 398
Passiflora velata IPNI=*321973-2* 400
Passiflora velutina IPNI=*674909-1* 398
Passiflora verecunda IPNI=*675423-1* 400
Passiflora vernicosa IPNI=*321964-2* 398
Passiflora verrucifera IPNI=*321964-2* 398
Passiflora viridiflora IPNI=*321973-2* 400
Passiflora vitifolia var. *bracteosa* IPNI=*30044451-2* 400
Passiflora vitifolia var. *involucrata* IPNI=*185013-2* 399
Passiflora vitifolia var. *minor* IPNI=*675392-1* 400
Passiflora warei IPNI=*185104-2* 399
Passiflora warmingii IPNI=*321969-2* 399
Passiflora warmingii subsp. *chacoensis* IPNI=*321969-2* 399
Passiflora yacumensis IPNI=*675392-1* 400
Passovia erythrocarpa IPNI=*77147531-1* 448
Passovia jamaicensis IPNI=*77147531-1* 448
Passovia lepidobotrys IPNI=*185262-2* 448
Passovia magdalenae IPNI=*77147531-1* 448
Passovia odorata IPNI=*77147531-1* 448
Passovia orinocensis IPNI=*77147531-1* 448
Passovia polystachya IPNI=*77147531-1* 448
Passovia stelis IPNI=*77147531-1* 448
Passovia suaveolens IPNI=*77147531-1* 448
Passovia theobromae IPNI=*77147531-1* 448
Passovia virgata IPNI=*77147531-1* 448
Pastinaca anethum IPNI=*837530-1* 251
Pastinaca graveolens IPNI=*837530-1* 251
Patabea alba IPNI=*77161077-1* 354
Patabea capitellata IPNI=*77115613-1* 354
Patabea tenerior IPNI=*77161098-1* 354
Patania erosa IPNI=*17081410-1* 175
Patrisa acuminata IPNI=*224736-2* 404
Patrisa pyrifera IPNI=*365862-1* 404
Patrisa sagotiana IPNI=*365862-1* 404
Pauladolfia acetosella IPNI=*224391-2* 289
Pauletia bicuspidata IPNI=*30050541-2* 311
Pauletia breviloba IPNI=*481315-1* 311
Pauletia forficata IPNI=*481315-1* 311
Pauletia forficata var. *latifolia* IPNI=*481315-1* 311
Pauletia forficata var. *longiflora* IPNI=*481315-1* 311
Pauletia geminata IPNI=*481456-1* 312
Pauletia longicuspis IPNI=*30050541-2* 311
Pauletia longiflora IPNI=*481315-1* 311
Pauletia longifolia IPNI=*481456-1* 312
Pauletia obtusata IPNI=*481456-1* 312
Pauletia picta IPNI=*481565-1* 312
Pauletia tarapotensis IPNI=*481718-1* 312
Pauletia ungulata IPNI=*316241-2* 312
Paulownia coreana IPNI=*185784-2* 401
Paulownia glabrata IPNI=*185784-2* 401
Paulownia grandifolia IPNI=*185784-2* 401
Paulownia imperialis IPNI=*185784-2* 401
Paulownia intermedia IPNI=*185784-2* 401
Paulownia lilacina IPNI=*185784-2* 401
Paulownia recurva IPNI=*185784-2* 401
Paulownia tomentosa f. *virginea* IPNI=*185784-2* 401
Paulownia tomentosa var. *glabrata* IPNI=*185784-2* 401
Paurotis androsana IPNI=*663912-1* 203
Paurotis arborescens IPNI=*663912-1* 203
Paurotis psilocalyx IPNI=*663912-1* 203
Paurotis schippii IPNI=*663912-1* 203
Paurotis wrightii IPNI=*663912-1* 203
Pavetta arborea IPNI=*753834-1* 352
Pavetta chinensis IPNI=*753834-1* 352
Pavetta coccinea IPNI=*753844-1* 352
Pavetta kroneana IPNI=*753834-1* 352
Pavetta macrothyrsa IPNI=*60459342-2* 352
Pavetta stricta IPNI=*753834-1* 352
Pavonia acuminata IPNI=*562981-1* 412
Pavonia albiflora IPNI=*562981-1* 412
Pavonia aristata IPNI=*186083-2* 412
Pavonia begoniifolia IPNI=*563076-1* 412
Pavonia berteroi IPNI=*133833-2* 410
Pavonia brachypoda IPNI=*185868-2* 412
Pavonia brachysepala IPNI=*563076-1* 412
Pavonia cordifolia IPNI=*562981-1* 412
Pavonia flava IPNI=*186058-2* 412
Pavonia flava var. *grandifolia* IPNI=*186058-2* 412
Pavonia fruticosa var. *angustiloba* IPNI=*563076-1* 412
Pavonia fruticosa var. *typhalaea* IPNI=*185868-2* 412
Pavonia gardneriana IPNI=*563076-1* 412
Pavonia glomerata IPNI=*185868-2* 412
Pavonia grandifolia IPNI=*186058-2* 412
Pavonia guianensis IPNI=*185868-2* 412
Pavonia hastata IPNI=*133833-2* 410
Pavonia hispida IPNI=*561446-1* 410
Pavonia kunthii IPNI=*562981-1* 412
Pavonia kunthii var. *cordifolia* IPNI=*562981-1* 412
Pavonia lappacea IPNI=*563076-1* 412
Pavonia leptocarpa IPNI=*185868-2* 412
Pavonia malvacea IPNI=*186058-2* 412
Pavonia monatherica IPNI=*185868-2* 412
Pavonia paraibica IPNI=*185868-2* 412
Pavonia perinii IPNI=*560723-1* 409
Pavonia regnelliana IPNI=*186058-2* 412
Pavonia rosea IPNI=*563076-1* 412
Pavonia rosea var. *rigida* IPNI=*563076-1* 412
Pavonia sepium subsp. *macrocarpa* IPNI=*186058-2* 412
Pavonia sepium var. *silvatica* IPNI=*186058-2* 412
Pavonia silvatica IPNI=*563076-1* 412
Pavonia spinifex subsp. *genuina* IPNI=*186083-2* 412
Pavonia spinifex var. *grandiflora* IPNI=*186083-2* 412
Pavonia spinifex var. *ovalifolia* IPNI=*186083-2* 412
Pavonia spinigera IPNI=*186083-2* 412
Pavonia spiralis IPNI=*30041958-2* 410
Pavonia stellata IPNI=*185868-2* 412
Pavonia surinamensis IPNI=*185868-2* 412
Pavonia typhalaea IPNI=*185868-2* 412
Pavonia typhalaea var. *cavanillesii* IPNI=*185868-2* 412
Pavonia typhalaeoides IPNI=*185868-2* 412
Pavonia umbrosa IPNI=*563076-1* 412
Payrola guianensis IPNI=*867031-1* 405
Peckia guyanensis IPNI=*1166848-2* 304
Pectidium punctatum IPNI=*236639-1* 267
Pectis bipinnata IPNI=*242907-1* 269
Pectis capillaris IPNI=*236637-1* 267
Pectis costata IPNI=*236681-1* 267
Pectis densa IPNI=*236637-1* 267
Pectis elongata var. *divaricata* IPNI=*236598-1* 267
Pectis elongata var. *floribunda* IPNI=*236598-1* 267
Pectis elongata var. *oerstediana* IPNI=*236598-1* 267
Pectis elongata var. *schottii* IPNI=*236637-1* 267
Pectis fasciculata IPNI=*236681-1* 267
Pectis febrifuga IPNI=*236637-1* 267
Pectis floribunda IPNI=*236598-1* 267

Pectis graveolens IPNI=*236637-1* 267
Pectis linifolia var. *hirtella* IPNI=*236639-1* 267
Pectis linifolia var. *marginalis* IPNI=*236639-1* 267
Pectis multifida IPNI=*242907-1* 269
Pectis multisetosa IPNI=*236681-1* 267
Pectis oerstediana IPNI=*236598-1* 267
Pectis pinnata IPNI=*242907-1* 269
Pectis plumieri IPNI=*236598-1* 267
Pectis polycephala IPNI=*236598-1* 267
Pectis portoricensis IPNI=*236681-1* 267
Pectis prostrata var. *urceolata* IPNI=*236681-1* 267
Pectis punctata IPNI=*236639-1* 267
Pectis ramosissima IPNI=*202984-1* 261
Pectis rosea IPNI=*236637-1* 267
Pectis schottii IPNI=*236637-1* 267
Pectis stricta IPNI=*236598-1* 267
Pectis taliscana IPNI=*236598-1* 267
Pectis urceolata IPNI=*236681-1* 267
Pedersenia canescens IPNI=*129900-2* 283
Pedicellaria gynandra IPNI=*60518-2* 279
Pedicellaria pentaphylla IPNI=*60518-2* 279
Pedicularis sessilis IPNI=*49672-1* 360
Pedilanthus tithymaloides IPNI=*102036-2* 391
Peiranisia falconiensis IPNI=*911412-1* 333
Peiranisia multijuga IPNI=*234544-2* 333
Pekea couroupita IPNI=*592385-1* 301
Pekea ternata IPNI=*828599-1* 381
Pekea tuberculosa IPNI=*828602-1* 381
Pelargonium × coccineum IPNI=*376335-1* 358
Pelargonium × hortorum IPNI=*376335-1* 358
Pelargonium × hybridum var. *roseum* IPNI=*376335-1* 358
Pelargonium × roseum IPNI=*376335-1* 358
Pelargonium amoenum IPNI=*377229-1* 358
Pelargonium angulatum IPNI=*164626-3* 358
Pelargonium anosma IPNI=*164626-3* 358
Pelargonium asperum IPNI=*376259-1* 358
Pelargonium bachmannii IPNI=*376728-1* 358
Pelargonium citriodorum IPNI=*164626-3* 358
Pelargonium citrosum IPNI=*164626-3* 358
Pelargonium clarum IPNI=*377229-1* 358
Pelargonium clypeatum IPNI=*376728-1* 358
Pelargonium cochleatum IPNI=*164626-3* 358
Pelargonium concavifolium IPNI=*164626-3* 358
Pelargonium crenatum IPNI=*377229-1* 358
Pelargonium densiflorum IPNI=*377229-1* 358
Pelargonium hederifolium IPNI=*376728-1* 358
Pelargonium intermedium IPNI=*376259-1* 358
Pelargonium leptopetalum IPNI=*377229-1* 358
Pelargonium leucanthum IPNI=*377229-1* 358
Pelargonium marginatum IPNI=*377229-1* 358
Pelargonium novum IPNI=*377229-1* 358
Pelargonium odoratum IPNI=*376654-1* 358
Pelargonium odorum IPNI=*376654-1* 358
Pelargonium placidum IPNI=*377229-1* 358
Pelargonium plicatum IPNI=*164626-3* 358
Pelargonium rosula IPNI=*377229-1* 358
Pelargonium scutatum IPNI=*376728-1* 358
Pelargonium siccifolium IPNI=*377229-1* 358
Pelargonium stenopetalum IPNI=*377229-1* 358
Pelargonium terebinthinaceum IPNI=*376259-1* 358
Pelargonium tricuspidatum IPNI=*164626-3* 358
Pelargonium variegatum IPNI=*377229-1* 358
Pelargonium zonale var. *lacunosum* IPNI=*377229-1* 358
Pelargonium zonale var. *marginatum* IPNI=*377229-1* 358
Pellaea angustifolia var. *cuneata* IPNI=*77182463-1* 178
Pellaea brandegeei IPNI=*77182326-1* 178
Pellaea concolor IPNI=*77182373-1* 178
Pellaea cordata var. *flexuosa* IPNI=*77183106-1* 178
Pellaea cuneata IPNI=*77182463-1* 178
Pellaea flavens IPNI=*77182941-1* 178
Pellaea flexuosa IPNI=*77183106-1* 178
Pellaea lanuginosa IPNI=*77182326-1* 178
Pellaea marginata IPNI=*77182463-1* 178
Pellaea nivea f. *flavens* IPNI=*77182941-1* 178
Pellaea nivea f. *tenera* IPNI=*77182941-1* 178
Pellaea nivea IPNI=*77182941-1* 178
Pellaea ovata IPNI=*77183106-1* 178
Pellaea palmescens IPNI=*77182373-1* 178
Pellaea peruviana IPNI=*77182941-1* 178
Pellaea pulcherrima IPNI=*77182808-1* 178
Pellaea tenera IPNI=*77182941-1* 178
Pellaea ternifolia IPNI=*77182326-1* 178
Pellaea ternifolia subsp. *brandegeei* IPNI=*77182326-1* 178
Pellaea ternifolia subsp. *weddelliana* IPNI=*77182326-1* 178
Pellaea ternifolia var. *petiolulata* IPNI=*77182326-1* 178
Pellaea weddelliana IPNI=*77182326-1* 178
Pellegrinia colombiana IPNI=*204021-2* 300
Pellegrinia guascensis IPNI=*204021-2* 300
Pellegrinia paramensis IPNI=*204021-2* 300
Pellegrinia parvifolia IPNI=*252657-2* 300
Pelta palustris IPNI=*603593-1* 203
Peltobryon callosum IPNI=*680748-1* 195
Peltobryon exserens IPNI=*681041-1* 195
Peltobryon guilleminianum IPNI=*681041-1* 195
Peltobryon hookeri IPNI=*682305-1* 196
Peltobryon poeppigii IPNI=*680748-1* 195
Peltogyne confertifolia IPNI=*125420-2* 320
Peltophorum acutifolium IPNI=*67201-2* 315
Peltophorum ferrugineum IPNI=*513021-1* 330
Peltophorum inerme IPNI=*513021-1* 330
Peltophorum roxburghii IPNI=*513021-1* 330
Peltophorum suringari IPNI=*67201-2* 315
Peltospermum latisiliquum IPNI=*77325-1* 339
Peltospermum patrisii IPNI=*77325-1* 339
Penicillaria arabica IPNI=*77105978-1* 235
Penicillaria deflexa IPNI=*77105978-1* 235
Penicillaria elongata IPNI=*77105978-1* 235
Penicillaria involucrata IPNI=*77105978-1* 235
Penicillaria italica IPNI=*1207294-2* 245
Penicillaria nigritarum IPNI=*77105978-1* 235
Penicillaria plukenetii IPNI=*77105978-1* 235
Penicillaria roxburghii IPNI=*77105978-1* 235
Penicillaria solitaria IPNI=*77105978-1* 235
Penicillaria spicata IPNI=*77105978-1* 235
Pennisetum albicauda IPNI=*77105978-1* 235
Pennisetum alopecuros IPNI=*235234-2* 245
Pennisetum americanum convar. *spicatum* IPNI=*77105978-1* 235
Pennisetum americanum convar. *typhoides* IPNI=*77105978-1* 235
Pennisetum ancylochaete IPNI=*77105978-1* 235
Pennisetum angustifolium IPNI=*51255-2* 235
Pennisetum aureum IPNI=*77105978-1* 235
Pennisetum benthamii IPNI=*77106033-1* 235
Pennisetum blepharideum IPNI=*77106033-1* 235
Pennisetum cenchroides IPNI=*395214-1* 235
Pennisetum cereale IPNI=*77105978-1* 235
Pennisetum ciliare IPNI=*395214-1* 235
Pennisetum cinereum IPNI=*77105978-1* 235
Pennisetum clandestinum IPNI=*77105989-1* 235
Pennisetum crus-galli IPNI=*399660-1* 238
Pennisetum distylum IPNI=*395214-1* 235
Pennisetum echinurus IPNI=*77105978-1* 235
Pennisetum erythraeum IPNI=*77106046-1* 236
Pennisetum flavicomum IPNI=*77106033-1* 235
Pennisetum flexispica IPNI=*77106033-1* 235
Pennisetum gambiense IPNI=*77105978-1* 235
Pennisetum geniculatum IPNI=*943811-1* 245

Pennisetum germanicum IPNI=*1207294-2* 245
Pennisetum gibbosum IPNI=*77105978-1* 235
Pennisetum giganteum IPNI=*77105978-1* 235
Pennisetum giganteum IPNI=*77106033-1* 235
Pennisetum glaucum IPNI=*77105978-1* 235
Pennisetum gossweileri IPNI=*77106033-1* 235
Pennisetum helvolum IPNI=*1207294-2* 245
Pennisetum inclusum IPNI=*77105989-1* 235
Pennisetum incomptum IPNI=*395214-1* 235
Pennisetum indicum subvar. *parviflorum* IPNI=*943811-1* 245
Pennisetum italicum IPNI=*1207294-2* 245
Pennisetum lachnorrhachis IPNI=*77106033-1* 235
Pennisetum laevigatum IPNI=*943811-1* 245
Pennisetum leonis IPNI=*77105978-1* 235
Pennisetum longifolium IPNI=*395214-1* 235
Pennisetum macrochaetum IPNI=*1207294-2* 245
Pennisetum macrostachyum IPNI=*77106033-1* 235
Pennisetum maiwa IPNI=*77105978-1* 235
Pennisetum malacochaete IPNI=*77105978-1* 235
Pennisetum megastachyum IPNI=*77105978-1* 235
Pennisetum myosuroides IPNI=*51261-2* 235
Pennisetum nigritarum IPNI=*77105978-1* 235
Pennisetum nigritarum var. *deflexum* IPNI=*77105978-1* 235
Pennisetum nigritarum var. *macrostachyum* IPNI=*77105978-1* 235
Pennisetum nitens IPNI=*77106033-1* 235
Pennisetum numidicum IPNI=*77106046-1* 236
Pennisetum oxyphyllum IPNI=*395214-1* 235
Pennisetum pallescens IPNI=*77106033-1* 235
Pennisetum panormitanum IPNI=*395214-1* 235
Pennisetum parisii IPNI=*77106046-1* 236
Pennisetum parviflorum IPNI=*943811-1* 245
Pennisetum petraeum IPNI=*395214-1* 235
Pennisetum phalaroides IPNI=*77106046-1* 236
Pennisetum plukenetii IPNI=*77105978-1* 235
Pennisetum polycladum IPNI=*395214-1* 235
Pennisetum polystachion IPNI=*943811-1* 245
Pennisetum pruinosum IPNI=*77106033-1* 235
Pennisetum pungens IPNI=*51261-2* 235
Pennisetum purpureum IPNI=*77106033-1* 235
Pennisetum pycnostachyum IPNI=*77105978-1* 235
Pennisetum rangei IPNI=*395214-1* 235
Pennisetum respiciens IPNI=*421428-1* 245
Pennisetum rufescens IPNI=*395214-1* 235
Pennisetum ruppellii IPNI=*77106046-1* 236
Pennisetum scoparium IPNI=*77106046-1* 236
Pennisetum setaceum IPNI=*77106046-1* 236
Pennisetum spectabile IPNI=*77106046-1* 236
Pennisetum spicatum IPNI=*77105978-1* 235
Pennisetum spicatum var. *typhoideum* IPNI=*77105978-1* 235
Pennisetum teneriffae IPNI=*395214-1* 235
Pennisetum tiberiadis IPNI=*77106046-1* 236
Pennisetum verticillatum IPNI=*421857-1* 245
Pennisetum villosum IPNI=*51255-2* 235
Penstemon gentianoides IPNI=*808129-1* 376
Penstemon giganteus IPNI=*808129-1* 376
Penstemon puniceus IPNI=*808129-1* 376
Pentacaena polycnemoides IPNI=*45444-2* 285
Pentacaena ramosissima IPNI=*45444-2* 285
Pentacaena rosetta IPNI=*45444-2* 285
Pentacaena tenuior IPNI=*45444-2* 285
Pentacalia abietina IPNI=*963566-1* 266
Pentacalia corymbosa IPNI=*963576-1* 266
Pentacalia guicanensis IPNI=*963584-1* 267
Pentacalia ledifolia IPNI=*963586-1* 267
Pentacalia leioclada IPNI=*963587-1* 267
Pentacalia nitida IPNI=*963597-1* 267
Pentacalia pulchella IPNI=*963604-1* 267
Pentacalia vaccinioides IPNI=*963618-1* 267
Pentaclethra brevifila IPNI=*188076-2* 330
Pentaclethra filamentosa IPNI=*188076-2* 330
Pentadenia aurantiaca IPNI=*379021-1* 367
Pentadenia strigosa IPNI=*379021-1* 367
Pentagonia alba IPNI=*760198-1* 355
Pentagonia alfaroana IPNI=*760198-1* 355
Pentagonia brachyotis IPNI=*760190-1* 355
Pentagonia cuatrecasasii IPNI=*760184-1* 355
Pentagonia gigantophylla IPNI=*188107-2* 355
Pentagonia grantii IPNI=*188124-2* 355
Pentagonia orthoneura IPNI=*760184-1* 355
Pentagonia perfoliata IPNI=*258447-2* 273
Pentagonia peruviana IPNI=*188124-2* 355
Pentagonia pittieri IPNI=*760190-1* 355
Pentagonia pubescens IPNI=*760184-1* 355
Pentagonia sprucei IPNI=*760184-1* 355
Pentagonia velutina IPNI=*188124-2* 355
Pentagonia veraguensis IPNI=*760184-1* 355
Pentagonia woronovii IPNI=*188124-2* 355
Pentagonocarpus cordatus IPNI=*133833-2* 410
Pentagonocarpus hastatus IPNI=*133833-2* 410
Pentagonocarpus hispidus IPNI=*133833-2* 410
Pentagonocarpus pentaspermus IPNI=*133833-2* 410
Pentagonocarpus sagittatus IPNI=*133833-2* 410
Pentanema conyzae IPNI=*60478388-2* 267
Pentapanax granatensis IPNI=*20012157-1* 253
Pentapetes angustifolia IPNI=*824676-1* 412
Pentapetes cebuana IPNI=*824676-1* 412
Pentapetes coccinea IPNI=*824676-1* 412
Peperidia afzeliana IPNI=*316339-2* 197
Peperidia peltata IPNI=*316334-2* 197
Peperidia sidifolia IPNI=*316339-2* 197
Peperidia subpeltata IPNI=*316339-2* 197
Peperidia umbellata IPNI=*316339-2* 197
Peperidium tubulatum IPNI=*218583-2* 251
Peperomia acuminata f. *rubra* IPNI=*677918-1* 193
Peperomia acuminatifolia IPNI=*677918-1* 193
Peperomia aemula IPNI=*1068802-2* 194
Peperomia affinis IPNI=*1068802-2* 194
Peperomia agapatensis IPNI=*678659-1* 193
Peperomia alata var. *angustifolia* IPNI=*677941-1* 193
Peperomia alata var. *pterocaulis* IPNI=*677941-1* 193
Peperomia allenii IPNI=*679591-1* 194
Peperomia alpina IPNI=*679597-1* 194
Peperomia amantlanensis IPNI=*679080-1* 194
Peperomia americana IPNI=*1068802-2* 194
Peperomia amphoterophylla IPNI=*678659-1* 193
Peperomia amphoterophylla var. *glutineofructa* IPNI=*678659-1* 193
Peperomia andinacea IPNI=*678531-1* 193
Peperomia angularis var. *bethaniana* IPNI=*282355-2* 193
Peperomia angulata IPNI=*679591-1* 194
Peperomia angulata var. *orbicans* IPNI=*679591-1* 194
Peperomia angulata var. *parvifolia* IPNI=*679591-1* 194
Peperomia anisophylla IPNI=*678659-1* 193
Peperomia antonii IPNI=*679047-1* 194
Peperomia antonii var. *fertilior* IPNI=*679047-1* 194
Peperomia antonii var. *lutea* IPNI=*679047-1* 194
Peperomia antonii var. *reducta* IPNI=*679269-1* 194
Peperomia apoda IPNI=*678659-1* 193
Peperomia apodostachya IPNI=*30130843-2* 194
Peperomia argyraea IPNI=*678021-1* 193
Peperomia arifolia var. *argyreia* IPNI=*678021-1* 193
Peperomia artatiflora IPNI=*678659-1* 193
Peperomia atabapoensis IPNI=*679591-1* 194
Peperomia atocongona IPNI=*678848-1* 194
Peperomia atropunctata IPNI=*30001267-2* 193
Peperomia balbisii IPNI=*1090908-2* 194
Peperomia balineorum IPNI=*30001267-2* 193

Peperomia bartlettii IPNI=*189588-2* 194
Peperomia basellifolia IPNI=*677918-1* 193
Peperomia baturiteana IPNI=*1068802-2* 194
Peperomia berroi IPNI=*1068802-2* 194
Peperomia blanda f. *blandiformis* IPNI=*678107-1* 193
Peperomia blanda f. *dissimilis* IPNI=*678107-1* 193
Peperomia blanda var. *dissimilis* IPNI=*678107-1* 193
Peperomia blanda var. *magnifolia* IPNI=*678107-1* 193
Peperomia blanda var. *parvifolia* IPNI=*678107-1* 193
Peperomia blanda var. *reticulata* IPNI=*678107-1* 193
Peperomia blanda var. *sericea* IPNI=*678107-1* 193
Peperomia brachyiula IPNI=*678659-1* 193
Peperomia brevipes IPNI=*189588-2* 194
Peperomia buchii IPNI=*30001267-2* 193
Peperomia buchii var. *macrostachya* IPNI=*30001267-2* 193
Peperomia cabaiana IPNI=*679597-1* 194
Peperomia cacuminicola IPNI=*677918-1* 193
Peperomia cangrejalana IPNI=*679269-1* 194
Peperomia cardiophylla IPNI=*679080-1* 194
Peperomia carlo-wrightii IPNI=*679591-1* 194
Peperomia cartagoana IPNI=*1068802-2* 194
Peperomia caudulilimba var. *cylindribacca* IPNI=*30130843-2* 194
Peperomia caulibarbis IPNI=*30001267-2* 193
Peperomia caulibarbis var. *jimenesana* IPNI=*30001267-2* 193
Peperomia ceapanana IPNI=*678659-1* 193
Peperomia chagalana IPNI=*30001267-2* 193
Peperomia chillonensis IPNI=*678659-1* 193
Peperomia chrysotricha IPNI=*678848-1* 194
Peperomia ciliata IPNI=*678107-1* 193
Peperomia ciliatifolia var. *iquitosana* IPNI=*30130843-2* 194
Peperomia ciliatifolia var. *santarosana* IPNI=*30130843-2* 194
Peperomia circulifolia IPNI=*679597-1* 194
Peperomia circulifolia var. *eciliata* IPNI=*679597-1* 194
Peperomia circulifolia var. *flava* IPNI=*679597-1* 194
Peperomia circumscissa IPNI=*30130843-2* 194
Peperomia coliblancoana IPNI=*189825-2* 194
Peperomia collicola IPNI=*679597-1* 194
Peperomia commutata IPNI=*679047-1* 194
Peperomia concinna IPNI=*679378-1* 194
Peperomia condormiens IPNI=*189825-2* 194
Peperomia conjungens IPNI=*30001267-2* 193
Peperomia conocarpa IPNI=*679597-1* 194
Peperomia constanzana IPNI=*678107-1* 193
Peperomia contraria IPNI=*679047-1* 194
Peperomia controversa IPNI=*30130843-2* 194
Peperomia crassiuscula IPNI=*677986-1* 193
Peperomia cruciata IPNI=*679269-1* 194
Peperomia cuneata IPNI=*679269-1* 194
Peperomia cuneifolia IPNI=*679269-1* 194
Peperomia cuspidigera IPNI=*30130843-2* 194
Peperomia cylindribacca IPNI=*30130843-2* 194
Peperomia daiquiriana IPNI=*679269-1* 194
Peperomia decipiens IPNI=*678107-1* 193
Peperomia defluens IPNI=*30130843-2* 194
Peperomia defrenata IPNI=*679080-1* 194
Peperomia delicatissima IPNI=*678531-1* 193
Peperomia delicatissima var. *venusta* IPNI=*189588-2* 194
Peperomia dendroides IPNI=*678659-1* 193
Peperomia dendromorphis IPNI=*678659-1* 193
Peperomia dendrophila IPNI=*677918-1* 193
Peperomia dichroophylla IPNI=*677918-1* 193
Peperomia diehliana IPNI=*1068802-2* 194
Peperomia dissimilis IPNI=*678107-1* 193
Peperomia distractiflora IPNI=*678659-1* 193
Peperomia dodecatheontophylla IPNI=*679269-1* 194
Peperomia earlei IPNI=*679269-1* 194
Peperomia ekmanii IPNI=*677918-1* 193
Peperomia ellipticifolia IPNI=*678107-1* 193
Peperomia elongata IPNI=*30130843-2* 194
Peperomia elongata var. *guianensis* IPNI=*30130843-2* 194
Peperomia elongata var. *piliramea* IPNI=*30130843-2* 194
Peperomia emarginella var. *exilis* IPNI=*678531-1* 193
Peperomia emarginella var. *glabrior* IPNI=*678531-1* 193
Peperomia emarginella var. *ovalis* IPNI=*678531-1* 193
Peperomia ephemera IPNI=*679378-1* 194
Peperomia erecta IPNI=*680084-1* 194
Peperomia erythrophlebia IPNI=*679080-1* 194
Peperomia euosma IPNI=*679047-1* 194
Peperomia euryanthelia IPNI=*679909-1* 194
Peperomia exigua IPNI=*678531-1* 193
Peperomia exilis IPNI=*678531-1* 193
Peperomia fieldiana IPNI=*679269-1* 194
Peperomia fimbriata IPNI=*1090908-2* 194
Peperomia fimbriata var. *gracilior* IPNI=*1090908-2* 194
Peperomia floridana IPNI=*679269-1* 194
Peperomia fontinalis IPNI=*1068802-2* 194
Peperomia fortipes IPNI=*678532-1* 193
Peperomia fraseri var. *peltata* IPNI=*678641-1* 193
Peperomia friabilis IPNI=*677986-1* 193
Peperomia frigidula IPNI=*679909-1* 194
Peperomia fuscociliata IPNI=*30001267-2* 193
Peperomia galeottii IPNI=*679269-1* 194
Peperomia galiifolia IPNI=*678659-1* 193
Peperomia galioides var. *aromatica* IPNI=*678659-1* 193
Peperomia galioides var. *longifolia* IPNI=*678659-1* 193
Peperomia galioides var. *menkeana* IPNI=*678659-1* 193
Peperomia galioides var. *saxicola* IPNI=*678659-1* 193
Peperomia gallitoensis IPNI=*678659-1* 193
Peperomia garrapatilla IPNI=*678659-1* 193
Peperomia glabella var. *brachyphylla* IPNI=*30001267-2* 193
Peperomia glabella var. *caulibarbis* IPNI=*30001267-2* 193
Peperomia glabella var. *eustatiana* IPNI=*30001267-2* 193
Peperomia glabella var. *jimenesana* IPNI=*30001267-2* 193
Peperomia glabella var. *melanostigma* IPNI=*30001267-2* 193
Peperomia glabella var. *nervulosa* IPNI=*30001267-2* 193
Peperomia glabella var. *nigropunctata* IPNI=*30001267-2* 193
Peperomia glabricaulis IPNI=*30130843-2* 194
Peperomia glanduligera IPNI=*678107-1* 193
Peperomia glanduligera var. *villosissima* IPNI=*678107-1* 193
Peperomia glutinosa IPNI=*677986-1* 193
Peperomia gollii IPNI=*679047-1* 194
Peperomia granata IPNI=*678659-1* 193
Peperomia grandifolia IPNI=*316339-2* 197
Peperomia guadaloupensis var. *emarginulata* IPNI=*679080-1* 194
Peperomia guayabillosana IPNI=*678659-1* 193
Peperomia guayaquilensis IPNI=*189588-2* 194
Peperomia guildingiana IPNI=*679841-1* 194
Peperomia helminthostachya IPNI=*679909-1* 194
Peperomia hemionitidifolia IPNI=*679269-1* 194
Peperomia hirsuta IPNI=*678107-1* 193
Peperomia houelmonte IPNI=*30001267-2* 193
Peperomia huacachiana IPNI=*677918-1* 193
Peperomia huitzensis IPNI=*679909-1* 194
Peperomia inaequalifolia var. *emarginata* IPNI=*678842-1* 194
Peperomia inaequalifolia var. *galioides* IPNI=*678659-1* 193
Peperomia inaudax IPNI=*189825-2* 194
Peperomia ioeides IPNI=*679705-1* 194
Peperomia ionophylla IPNI=*679841-1* 194
Peperomia jamesoniana var. *microphylla* IPNI=*679597-1* 194
Peperomia jamesonii IPNI=*678659-1* 193
Peperomia japurensis IPNI=*30130843-2* 194
Peperomia jimenesana IPNI=*30001267-2* 193
Peperomia knoblecheriana IPNI=*679378-1* 194
Peperomia koepperi IPNI=*189588-2* 194
Peperomia laevis IPNI=*677941-1* 193
Peperomia lancetillana IPNI=*679269-1* 194

Peperomia lancetillana var. *spathifolia* IPNI=*679047-1* 194
Peperomia langsdorffii IPNI=*678107-1* 193
Peperomia langsdorffii var. *dissimilis* IPNI=*678107-1* 193
Peperomia langsdorffii var. *resacana* IPNI=*678107-1* 193
Peperomia lanjouwii IPNI=*189588-2* 194
Peperomia larecajana IPNI=*677918-1* 193
Peperomia larecajana var. *angustifolia* IPNI=*677918-1* 193
Peperomia leridana IPNI=*679043-1* 194
Peperomia leucandra IPNI=*30001267-2* 193
Peperomia limaensis IPNI=*678848-1* 194
Peperomia linatifolia IPNI=*30001267-2* 193
Peperomia longeacuminata IPNI=*30001267-2* 193
Peperomia longespicata IPNI=*677986-1* 193
Peperomia longispica IPNI=*678659-1* 193
Peperomia lunana IPNI=*679269-1* 194
Peperomia lundellii IPNI=*677986-1* 193
Peperomia luxii IPNI=*679597-1* 194
Peperomia macorisiana IPNI=*30001267-2* 193
Peperomia macropoda IPNI=*679269-1* 194
Peperomia macrostachyos var. *nematostachya* IPNI=*30130843-2* 194
Peperomia macrostachyos var. *rupestris* IPNI=*30130843-2* 194
Peperomia magnoliifolia var. *brevirostrata* IPNI=*679047-1* 194
Peperomia magnoliifolia var. *emarginulata* IPNI=*679269-1* 194
Peperomia magnoliifolia var. *microphylla* IPNI=*679047-1* 194
Peperomia magnoliifolia var. *parvifolia* IPNI=*679047-1* 194
Peperomia magnoliifolia var. *sintenisiana* IPNI=*679047-1* 194
Peperomia maleuvreana IPNI=*30001267-2* 193
Peperomia mararyna IPNI=*189588-2* 194
Peperomia maribiana IPNI=*30130843-2* 194
Peperomia martiana var. *nervosa* IPNI=*679080-1* 194
Peperomia martinicensis IPNI=*30001267-2* 193
Peperomia medianiana IPNI=*678659-1* 193
Peperomia melanostigma IPNI=*30001267-2* 193
Peperomia melanostigma var. *angustifolia* IPNI=*30001267-2* 193
Peperomia melanostigma var. *glabrior* IPNI=*30001267-2* 193
Peperomia melanostigma var. *nervulosa* IPNI=*30001267-2* 193
Peperomia menkeana IPNI=*678659-1* 193
Peperomia mentiens IPNI=*679269-1* 194
Peperomia mentiens var. *lata* IPNI=*679047-1* 194
Peperomia metapalcoensis IPNI=*189588-2* 194
Peperomia microreticulata IPNI=*677941-1* 193
Peperomia minima IPNI=*678531-1* 193
Peperomia monsterifolia IPNI=*679043-1* 194
Peperomia muscicola IPNI=*678659-1* 193
Peperomia muscisedens IPNI=*678659-1* 193
Peperomia muscosa IPNI=*679591-1* 194
Peperomia myosuroides IPNI=*30130843-2* 194
Peperomia myosurus IPNI=*679841-1* 194
Peperomia myriocarpa IPNI=*30130843-2* 194
Peperomia myriocarpa var. *tenuifolia* IPNI=*30130843-2* 194
Peperomia nana IPNI=*679378-1* 194
Peperomia nematostachya IPNI=*30130843-2* 194
Peperomia nematostachya var. *genuina* IPNI=*30130843-2* 194
Peperomia nemorosa IPNI=*677918-1* 193
Peperomia nigrescens IPNI=*30001267-2* 193
Peperomia nigropunctata IPNI=*30001267-2* 193
Peperomia nilssonii IPNI=*677941-1* 193
Peperomia novae-zelandiae IPNI=*1068802-2* 194
Peperomia novella IPNI=*678659-1* 193
Peperomia nummulariifolia IPNI=*189588-2* 194
Peperomia nummulariifolia var. *pubescens* IPNI=*189588-2* 194
Peperomia oblongibacca IPNI=*30130843-2* 194
Peperomia oblongibacca var. *subvillosa* IPNI=*30130843-2* 194
Peperomia oblongifolia IPNI=*678659-1* 193
Peperomia obovata IPNI=*1090908-2* 194
Peperomia obtusifolia f. *pusilla* IPNI=*678532-1* 193
Peperomia obtusifolia IPNI=*678532-1* 193
Peperomia obtusifolia var. *emarginulata* IPNI=*679269-1* 194
Peperomia obtusifolia var. *floridana* IPNI=*679269-1* 194
Peperomia obtusifolia var. *genuina* IPNI=*679269-1* 194
Peperomia obtusifolia var. *grandifolia* IPNI=*679047-1* 194
Peperomia obtusifolia var. *longibracteata* IPNI=*679047-1* 194
Peperomia obtusifolia var. *macropoda* IPNI=*679269-1* 194
Peperomia obtusifolia var. *oblongifolia* IPNI=*679269-1* 194
Peperomia obtusifolia var. *papyracea* IPNI=*679269-1* 194
Peperomia obtusifolia var. *parvifolia* IPNI=*679269-1* 194
Peperomia obtusifolia var. *petiolata* IPNI=*679269-1* 194
Peperomia obversa IPNI=*1090908-2* 194
Peperomia okarana IPNI=*678659-1* 193
Peperomia olens IPNI=*679705-1* 194
Peperomia oleracea IPNI=*679378-1* 194
Peperomia omnicola IPNI=*679909-1* 194
Peperomia omnicola var. *oblanceolata* IPNI=*679909-1* 194
Peperomia opaca IPNI=*1068802-2* 194
Peperomia opaca var. *ciliata* IPNI=*1068802-2* 194
Peperomia ornata IPNI=*679095-1* 194
Peperomia ovalifolia IPNI=*1090908-2* 194
Peperomia parkeriana IPNI=*30130843-2* 194
Peperomia parmata IPNI=*679043-1* 194
Peperomia parvula IPNI=*1068802-2* 194
Peperomia pellucida var. *minor* IPNI=*679378-1* 194
Peperomia peltata IPNI=*316334-2* 197
Peperomia pendula IPNI=*30130843-2* 194
Peperomia peninsularis IPNI=*679269-1* 194
Peperomia percuneata IPNI=*30001267-2* 193
Peperomia petenensis IPNI=*679047-1* 194
Peperomia petenensis var. *hondurensis* IPNI=*679047-1* 194
Peperomia piperea IPNI=*30130843-2* 194
Peperomia pirrisana IPNI=*679591-1* 194
Peperomia pololensis IPNI=*30001267-2* 193
Peperomia polystachya var. *godetiana* IPNI=*678107-1* 193
Peperomia portulacifolia IPNI=*679597-1* 194
Peperomia pothifolia IPNI=*679909-1* 194
Peperomia praetenuis IPNI=*679378-1* 194
Peperomia pruinosa IPNI=*316334-2* 197
Peperomia pseudoamplexicaulis IPNI=*679047-1* 194
Peperomia pseudogalapagensis IPNI=*678848-1* 194
Peperomia pseudotetraphylla IPNI=*679597-1* 194
Peperomia pseudotetraphylla var. *juvenalis* IPNI=*679597-1* 194
Peperomia pterocaulis f. *stipiticarpa* IPNI=*677941-1* 193
Peperomia pterocaulis IPNI=*677941-1* 193
Peperomia pterocaulis var. *filispica* IPNI=*677941-1* 193
Peperomia pterocaulis var. *palmamochana* IPNI=*677941-1* 193
Peperomia pulicaris IPNI=*679841-1* 194
Peperomia pustulatibacca IPNI=*679047-1* 194
Peperomia pyrolifolia IPNI=*679269-1* 194
Peperomia quadrifolia f. *angusta* IPNI=*679597-1* 194
Peperomia quadrivii IPNI=*30001267-2* 193
Peperomia quaternata IPNI=*1068802-2* 194
Peperomia quatrometralis IPNI=*30130843-2* 194
Peperomia quicheensis IPNI=*677986-1* 193
Peperomia quindioensis IPNI=*679705-1* 194
Peperomia quiriguana IPNI=*30001267-2* 193
Peperomia quitensis IPNI=*678107-1* 193
Peperomia reflexa f. *brevipes* IPNI=*1068802-2* 194
Peperomia reflexa f. *forsteriana* IPNI=*1068802-2* 194
Peperomia reflexa f. *longipes* IPNI=*1068802-2* 194
Peperomia reflexa f. *variipila* IPNI=*1068802-2* 194
Peperomia reflexa IPNI=*1068802-2* 194
Peperomia reflexa var. *americana* IPNI=*1068802-2* 194
Peperomia reflexa var. *angustifolia* IPNI=*1068802-2* 194
Peperomia reflexa var. *argentina* IPNI=*1068802-2* 194
Peperomia reflexa var. *coriacea* IPNI=*1068802-2* 194
Peperomia reflexa var. *enervulosa* IPNI=*1068802-2* 194
Peperomia reflexa var. *opaca* IPNI=*1068802-2* 194
Peperomia reflexa var. *pallida* IPNI=*1068802-2* 194

Peperomia reflexa var. *robustior* IPNI=*1068802-2* 194
Peperomia reflexa var. *rotundilimba* IPNI=*1068802-2* 194
Peperomia reflexa var. *subemarginulata* IPNI=*1068802-2* 194
Peperomia reflexa var. *valantoides* IPNI=*1068802-2* 194
Peperomia reflexifolia IPNI=*1068802-2* 194
Peperomia reniformis IPNI=*316339-2* 197
Peperomia reniformis IPNI=*679841-1* 194
Peperomia repens IPNI=*679841-1* 194
Peperomia reptabunda IPNI=*679080-1* 194
Peperomia resediflora IPNI=*678641-1* 193
Peperomia rhodophlebia IPNI=*677986-1* 193
Peperomia romaensis IPNI=*679047-1* 194
Peperomia roqueana IPNI=*679080-1* 194
Peperomia rotundifolia f. *obovata* IPNI=*189588-2* 194
Peperomia rotundifolia f. *pubescens* IPNI=*189588-2* 194
Peperomia rotundifolia IPNI=*189588-2* 194
Peperomia rotundifolia var. *glabrilimba* IPNI=*189588-2* 194
Peperomia rotundifolia var. *nummulariifolia* IPNI=*189588-2* 194
Peperomia rotundifolia var. *obcordata* IPNI=*189588-2* 194
Peperomia rotundifolia var. *ovata* IPNI=*189588-2* 194
Peperomia rotundifolia var. *pilosior* IPNI=*189588-2* 194
Peperomia rotundifolia var. *subglabrilimba* IPNI=*189588-2* 194
Peperomia ruiziana IPNI=*677918-1* 193
Peperomia rupestris f. *cordifolia* IPNI=*30130843-2* 194
Peperomia rupestris IPNI=*30130843-2* 194
Peperomia rupestris var. *cordifolia* IPNI=*30130843-2* 194
Peperomia salvaje IPNI=*678107-1* 193
Peperomia san-buenaventurana IPNI=*678659-1* 193
Peperomia sandersii IPNI=*678021-1* 193
Peperomia santa-rosana IPNI=*679597-1* 194
Peperomia sarcocarpa IPNI=*677986-1* 193
Peperomia saxatilis IPNI=*1068802-2* 194
Peperomia scandens IPNI=*679841-1* 194
Peperomia scandens var. *fertilior* IPNI=*679841-1* 194
Peperomia scandens var. *longispica* IPNI=*679841-1* 194
Peperomia scheryi IPNI=*679043-1* 194
Peperomia schomburgkii IPNI=*679597-1* 194
Peperomia scutata IPNI=*316334-2* 197
Peperomia semisupina IPNI=*30130843-2* 194
Peperomia septuplinervia IPNI=*679043-1* 194
Peperomia serpens IPNI=*290058-2* 193
Peperomia sidifolia IPNI=*316339-2* 197
Peperomia sintenisii IPNI=*30001267-2* 193
Peperomia speciosa IPNI=*316334-2* 197
Peperomia spectabilis IPNI=*679909-1* 194
Peperomia standleyi IPNI=*679597-1* 194
Peperomia stenocaulis IPNI=*679597-1* 194
Peperomia suavis IPNI=*679043-1* 194
Peperomia suavis var. *hondurensis* IPNI=*679043-1* 194
Peperomia subamplexicaulis IPNI=*679047-1* 194
Peperomia subbasellifolia IPNI=*30001267-2* 193
Peperomia subcorymbosa IPNI=*678659-1* 193
Peperomia subemarginulata IPNI=*1068802-2* 194
Peperomia subgeminispica IPNI=*679269-1* 194
Peperomia subpeltata IPNI=*316339-2* 197
Peperomia subpeltata IPNI=*679597-1* 194
Peperomia subquadrifolia IPNI=*679597-1* 194
Peperomia subrenifolia IPNI=*679597-1* 194
Peperomia subrenifolia var. *microphylla* IPNI=*679597-1* 194
Peperomia substriata IPNI=*679909-1* 194
Peperomia suizana IPNI=*678363-1* 193
Peperomia sumichrastii IPNI=*679597-1* 194
Peperomia surinamensis IPNI=*30130843-2* 194
Peperomia tacuariana IPNI=*1068802-2* 194
Peperomia taylorii IPNI=*30001267-2* 193
Peperomia tenebraegaudens IPNI=*679043-1* 194
Peperomia tenella var. *tyleri* IPNI=*189825-2* 194
Peperomia tenuicaulis IPNI=*679597-1* 194
Peperomia tetraphylla f. *protractifolia* IPNI=*1068802-2* 194
Peperomia tetraphylla var. *americana* IPNI=*1068802-2* 194
Peperomia tetraphylla var. *sinensis* IPNI=*1068802-2* 194
Peperomia tetraphylla var. *valantoides* IPNI=*1068802-2* 194
Peperomia thionvilleana IPNI=*30001267-2* 193
Peperomia tithymaloides IPNI=*679269-1* 194
Peperomia tortugana IPNI=*678107-1* 193
Peperomia treleasei IPNI=*678641-1* 193
Peperomia triadophylla IPNI=*679378-1* 194
Peperomia trifolia f. *pilosior* IPNI=*1090908-2* 194
Peperomia trinervula var. *suboppositifolia* IPNI=*679705-1* 194
Peperomia trullifolia IPNI=*678659-1* 193
Peperomia umbellata IPNI=*316339-2* 197
Peperomia valantoides IPNI=*1068802-2* 194
Peperomia valerioi IPNI=*679269-1* 194
Peperomia variegata IPNI=*679043-1* 194
Peperomia velloziana var. *polysticta* IPNI=*30001267-2* 193
Peperomia velutina var. *lanceolata* IPNI=*680160-1* 195
Peperomia vernouana IPNI=*189588-2* 194
Peperomia victoriana IPNI=*677986-1* 193
Peperomia victoriana var. *margaritana* IPNI=*677986-1* 193
Peperomia viridispica IPNI=*677986-1* 193
Peperomia viridispica var. *perejil* IPNI=*677986-1* 193
Peperomia vogelii IPNI=*679378-1* 194
Peperomia wagneri IPNI=*677986-1* 193
Peperomia wilsonii IPNI=*30001267-2* 193
Peperomia yapensis IPNI=*679378-1* 194
Peperomia yaquena IPNI=*679597-1* 194
Peperomia yousei IPNI=*679047-1* 194
Peplis portula IPNI=*30031106-2* 429
Pepo aurantius IPNI=*292416-1* 294
Pepo citrullus IPNI=*292416-1* 294
Pepo clypeiformis IPNI=*292416-1* 294
Pepo eximius IPNI=*320034-2* 293
Pepo ficifolia IPNI=*292369-1* 293
Pepo indicus IPNI=*320034-2* 293
Pepo lagenarius IPNI=*134809-2* 294
Pepo macrocarpus IPNI=*320034-2* 293
Pepo malabaricus IPNI=*292369-1* 293
Pepo maximus IPNI=*30006538-2* 293
Pepo melopepo IPNI=*292238-1* 293
Pepo moschatus IPNI=*320034-2* 293
Pepo oblongus IPNI=*292416-1* 294
Pepo potiron IPNI=*292416-1* 294
Pepo verrucosus IPNI=*292238-1* 293
Pepo vulgaris IPNI=*292416-1* 294
Pera arborea IPNI=*353360-1* 401
Pera bailloniana IPNI=*353360-1* 401
Pera bumeliifolia IPNI=*353360-1* 401
Pera ferruginea IPNI=*353360-1* 401
Pera glabrata var. *parvifolia* IPNI=*353360-1* 401
Pera klotzschiana IPNI=*353360-1* 401
Pera obovata IPNI=*353360-1* 401
Pera obtusifolia IPNI=*353360-1* 401
Pera parvifolia IPNI=*353360-1* 401
Perdicium corymbosum IPNI=*256729-1* 272
Perdicium divaricatum IPNI=*1207471-2* 272
Perdicium havanense IPNI=*256729-1* 272
Perdicium laevigatum IPNI=*256729-1* 272
Perdicium radiale IPNI=*256729-1* 272
Perebea australis IPNI=*855103-1* 440
Perebea calophylla IPNI=*152821-2* 440
Perebea chimiqua IPNI=*855107-1* 440
Perebea concinna IPNI=*167394-2* 440
Perebea elegans IPNI=*119532-2* 439
Perebea glabrata IPNI=*855107-1* 440
Perebea integrifolia IPNI=*855107-1* 440
Perebea laevigata IPNI=*855107-1* 440

Perebea laurifolia IPNI=*854491-1* 440
Perebea lecointei IPNI=*1008109-1* 440
Perebea markhamiana IPNI=*855107-1* 440
Perebea molliflora IPNI=*855107-1* 440
Perebea paraensis IPNI=*1008109-1* 440
Perebea standleyi IPNI=*85710-2* 188
Perenideboles ciliatum IPNI=*155854-2* 360
Pereskia acardia IPNI=*137452-1* 284
Pereskia aculeata f. *rubescens* IPNI=*137452-1* 284
Pereskia aculeata var. *brasiliensis* IPNI=*137452-1* 284
Pereskia aculeata var. *longispina* IPNI=*137452-1* 284
Pereskia aculeata var. *rubescens* IPNI=*137452-1* 284
Pereskia affinis IPNI=*27133-2* 283
Pereskia bleo IPNI=*60461844-2* 284
Pereskia brasiliensis IPNI=*137452-1* 284
Pereskia colombiana IPNI=*60461845-2* 284
Pereskia corrugata IPNI=*60461844-2* 284
Pereskia cruenta IPNI=*60461844-2* 284
Pereskia grandiflora IPNI=*219420-2* 284
Pereskia grandifolia IPNI=*219420-2* 284
Pereskia guamacho IPNI=*60461845-2* 284
Pereskia lanceolata IPNI=*137452-1* 284
Pereskia longispina IPNI=*137452-1* 284
Pereskia ochnacarpa IPNI=*219420-2* 284
Pereskia panamensis IPNI=*60461844-2* 284
Pereskia paniculata IPNI=*161653-1* 290
Pereskia peireskia IPNI=*137452-1* 284
Pereskia plantaginea IPNI=*137452-1* 284
Pereskia rosea IPNI=*219420-2* 284
Pereskia rubescens IPNI=*137452-1* 284
Pereskia subulata IPNI=*27133-2* 283
Pereskia tampicana IPNI=*219420-2* 284
Pereskia undulata IPNI=*137452-1* 284
Pereskiopsis subulata IPNI=*27133-2* 283
Perezia multiflora var. *typica* IPNI=*237026-1* 268
Perfoliata rotundifolia IPNI=*38361-2* 252
Perfolisa obtusifolia IPNI=*38361-2* 252
Perfonon ferrugineum IPNI=*63423-2* 441
Periarrabidaea duckei IPNI=*153710-2* 364
Pericalia cordifolia IPNI=*232826-1* 266
Pericampylus membranaceus IPNI=*59219-2* 434
Periclistia latifolia IPNI=*867031-1* 405
Periclistia longifolia IPNI=*867031-1* 405
Peridium ferrugineum IPNI=*353360-1* 401
Peridium glabratum IPNI=*353360-1* 401
Peridium oblongifolium IPNI=*353360-1* 401
Peridium obovatum IPNI=*353360-1* 401
Peridium obtusifolium IPNI=*353360-1* 401
Peridium ovale IPNI=*353360-1* 401
Peridium parvifolium IPNI=*353360-1* 401
Perilla nankinensis IPNI=*446165-1* 369
Perima odorata IPNI=*493866-1* 319
Periphragmos dependens IPNI=*688721-1* 303
Periploca chinensis IPNI=*95173-1* 339
Periploca cochinchinensis IPNI=*95173-1* 339
Periploca fruticosa IPNI=*98854-1* 341
Peristeria humboldtii IPNI=*614722-1* 217
Peristeria humboldtii var. *fulva* IPNI=*614722-1* 217
Peritassa granulata IPNI=*190399-2* 291
Peritassa rorida IPNI=*190399-2* 291
Peritassa rubicunda IPNI=*190399-2* 291
Peritassa smaragdina IPNI=*190399-2* 291
Perlebia purpurea IPNI=*481605-1* 312
Perlebia purpurea var. *corneri* IPNI=*481605-1* 312
Perlebia purpurea var. *violacea* IPNI=*481605-1* 312
Perlebia variegata IPNI=*481756-1* 312
Pernettya albiflora IPNI=*330618-1* 299
Pernettya angustata IPNI=*330618-1* 299
Pernettya buxifolia IPNI=*330618-1* 299
Pernettya caracasana IPNI=*330618-1* 299
Pernettya cavanillesiana IPNI=*330618-1* 299
Pernettya ciliaris IPNI=*330618-1* 299
Pernettya ciliaris var. *alpina* IPNI=*330618-1* 299
Pernettya ciliata IPNI=*330618-1* 299
Pernettya congesta IPNI=*330618-1* 299
Pernettya coriacea IPNI=*330618-1* 299
Pernettya densa IPNI=*330618-1* 299
Pernettya elliptica IPNI=*330618-1* 299
Pernettya halliana IPNI=*330618-1* 299
Pernettya hirsuta IPNI=*330618-1* 299
Pernettya humilis IPNI=*330618-1* 299
Pernettya mexicana IPNI=*330618-1* 299
Pernettya myrsinoides IPNI=*330618-1* 299
Pernettya nitida IPNI=*330618-1* 299
Pernettya obovata IPNI=*330618-1* 299
Pernettya oerstediana IPNI=*330618-1* 299
Pernettya parvifolia IPNI=*330618-1* 299
Pernettya pentlandii IPNI=*330618-1* 299
Pernettya pentlandii subvar. *humilis* IPNI=*330618-1* 299
Pernettya pentlandii subvar. *myrtilloides* IPNI=*330618-1* 299
Pernettya pentlandii var. *parvifolia* IPNI=*330618-1* 299
Pernettya pilosa IPNI=*330618-1* 299
Pernettya prostrata IPNI=*330618-1* 299
Pernettya prostrata subsp. *pentlandii* IPNI=*330618-1* 299
Pernettya prostrata var. *elliptica* IPNI=*330618-1* 299
Pernettya prostrata var. *euprostrata* IPNI=*330618-1* 299
Pernettya prostrata var. *myrsinoides* IPNI=*330618-1* 299
Pernettya prostrata var. *pentlandii* IPNI=*330618-1* 299
Pernettya prostrata var. *purpurea* IPNI=*330618-1* 299
Pernettya purpurea IPNI=*330618-1* 299
Pernettya purpurea var. *angustata* IPNI=*330618-1* 299
Pernettya purpurea var. *parvifolia* IPNI=*330618-1* 299
Pernettya purpurea var. *pentlandii* IPNI=*330618-1* 299
Pernettya purpurea var. *subsetosa* IPNI=*330618-1* 299
Pernettya robusta IPNI=*330618-1* 299
Pernettya saxicola IPNI=*330618-1* 299
Pernettya schizostigma IPNI=*330618-1* 299
Pernettya setigera IPNI=*330618-1* 299
Pernettya thomasii IPNI=*330618-1* 299
Perrottetia barbata IPNI=*77187394-1* 320
Persea americana var. *angustifolia* IPNI=*325643-2* 185
Persea americana var. *drimyfolia* IPNI=*325643-2* 185
Persea americana var. *floccosa* IPNI=*325643-2* 185
Persea americana var. *guatemalensis* IPNI=*325643-2* 185
Persea americana var. *leiogyna* IPNI=*325643-2* 185
Persea americana var. *nubigena* IPNI=*325643-2* 185
Persea americana var. *steyermarkii* IPNI=*325643-2* 185
Persea americana var. *tolimanensis* IPNI=*325643-2* 185
Persea americana var. *zentmyeri* IPNI=*325643-2* 185
Persea argentea IPNI=*467336-1* 184
Persea bofo IPNI=*467161-1* 184
Persea caustica IPNI=*70012-1* 451
Persea cinnamomifolia IPNI=*77173628-1* 182
Persea cinnamomum IPNI=*463752-1* 183
Persea drimyfolia IPNI=*325643-2* 185
Persea edulis IPNI=*325643-2* 185
Persea floccosa IPNI=*325643-2* 185
Persea floribunda IPNI=*1146025-2* 184
Persea gigantea IPNI=*325643-2* 185
Persea globosa IPNI=*167822-2* 184
Persea grandis IPNI=*466894-1* 184
Persea gratissima IPNI=*325643-2* 185
Persea gratissima var. *drimyfolia* IPNI=*325643-2* 185
Persea gratissima var. *macrophylla* IPNI=*325643-2* 185
Persea gratissima var. *melanocarpa* IPNI=*325643-2* 185
Persea gratissima var. *oblonga* IPNI=*325643-2* 185

Persea gratissima var. *schiedeana* IPNI=*190745-2* 185
Persea humboldtii IPNI=*468017-1* 185
Persea incana IPNI=*466894-1* 184
Persea laevigata IPNI=*190613-2* 185
Persea laevigata var. *caerulea* IPNI=*190613-2* 185
Persea lancifolia IPNI=*466885-1* 184
Persea latifolia IPNI=*77158867-1* 183
Persea leiogyna IPNI=*325643-2* 185
Persea lignitepala IPNI=*190613-2* 185
Persea lineata IPNI=*974282-1* 184
Persea longifolia IPNI=*171526-2* 184
Persea macrocarpa IPNI=*325643-2* 185
Persea macrophylla IPNI=*326508-2* 184
Persea marginata IPNI=*1032261-2* 185
Persea membranacea IPNI=*466773-1* 184
Persea mexicana IPNI=*77173628-1* 182
Persea mollis IPNI=*466894-1* 184
Persea montana IPNI=*77173628-1* 182
Persea nubigena IPNI=*325643-2* 185
Persea nubigena var. *guatemalensis* IPNI=*325643-2* 185
Persea nutans IPNI=*77173628-1* 182
Persea paucitriplinervia IPNI=*325643-2* 185
Persea persea IPNI=*325643-2* 185
Persea petiolaris IPNI=*190613-2* 185
Persea pichurim IPNI=*1171177-2* 184
Persea pittieri IPNI=*190745-2* 185
Persea praecox IPNI=*325643-2* 185
Persea retroflexa IPNI=*1146025-2* 184
Persea richardiana IPNI=*1032261-2* 185
Persea skutchii IPNI=*190613-2* 185
Persea steyermarkii IPNI=*325643-2* 185
Persea sulcata IPNI=*31062-2* 182
Persea sylvestris IPNI=*325643-2* 185
Persea tigricida IPNI=*1203274-2* 184
Persea triplinervis var. *valenzuelana* IPNI=*77173628-1* 182
Persica chinensis IPNI=*1212858-2* 444
Persica domestica IPNI=*1212858-2* 444
Persica laevis IPNI=*1212858-2* 444
Persica mammillata IPNI=*1212858-2* 444
Persica nucipersica IPNI=*1212858-2* 444
Persica pendula IPNI=*1212858-2* 444
Persica platycarpa IPNI=*1212858-2* 444
Persica vulgaris convar. *nucipersica* IPNI=*1212858-2* 444
Persica vulgaris IPNI=*1212858-2* 444
Persica vulgaris subsp. *atropurpurea* IPNI=*1212858-2* 444
Persica vulgaris subsp. *erythrocarpa* IPNI=*1212858-2* 444
Persica vulgaris var. *alba* IPNI=*1212858-2* 444
Persica vulgaris var. *albaplena* IPNI=*1212858-2* 444
Persica vulgaris var. *atropurpurea* IPNI=*1212858-2* 444
Persica vulgaris var. *camelliiflora* IPNI=*1212858-2* 444
Persica vulgaris var. *densa* IPNI=*1212858-2* 444
Persica vulgaris var. *dianthiflora* IPNI=*1212858-2* 444
Persica vulgaris var. *duplex* IPNI=*1212858-2* 444
Persica vulgaris var. *magnifica* IPNI=*1212858-2* 444
Persica vulgaris var. *nectarina* IPNI=*1212858-2* 444
Persica vulgaris var. *pendula* IPNI=*1212858-2* 444
Persica vulgaris var. *pyramidalis* IPNI=*1212858-2* 444
Persicaria acris IPNI=*190855-2* 289
Persicaria acris IPNI=*77069479-1* 289
Persicaria alata IPNI=*60433514-2* 289
Persicaria alata var. *violascens* IPNI=*60433514-2* 289
Persicaria bioritsuensis IPNI=*77069480-1* 289
Persicaria brachypoda IPNI=*1050850-2* 289
Persicaria brittingeri IPNI=*77069480-1* 289
Persicaria densiflora IPNI=*1005755-2* 289
Persicaria dolichopoda IPNI=*695052-1* 289
Persicaria fallax IPNI=*695052-1* 289
Persicaria fastigiatoramosa IPNI=*77069479-1* 289
Persicaria fusiformis IPNI=*695052-1* 289
Persicaria giranensis IPNI=*695113-1* 289
Persicaria glabra var. *scabrinervis* IPNI=*1005755-2* 289
Persicaria glandulosa IPNI=*77069480-1* 289
Persicaria granulata IPNI=*695052-1* 289
Persicaria hydropiper subsp. *hispida* IPNI=*77069479-1* 289
Persicaria hydropiper subsp. *laetivirens* IPNI=*77069479-1* 289
Persicaria hydropiper subsp. *maximowiczii* IPNI=*77069479-1* 289
Persicaria hydropiper var. *angustifolia* IPNI=*77069479-1* 289
Persicaria hydropiper var. *densiflora* IPNI=*77069479-1* 289
Persicaria hydropiper var. *glandulosissima* IPNI=*77069479-1* 289
Persicaria hydropiper var. *longistachya* IPNI=*77069479-1* 289
Persicaria hydropiper var. *projecta* IPNI=*77069479-1* 289
Persicaria hydropiperoides var. *breviciliata* IPNI=*190811-2* 289
Persicaria hydropiperoides var. *euronotorum* IPNI=*190811-2* 289
Persicaria hydropiperoides var. *strigosa* IPNI=*190811-2* 289
Persicaria hypanica IPNI=*77069480-1* 289
Persicaria incana IPNI=*695052-1* 289
Persicaria incarnata IPNI=*77069480-1* 289
Persicaria interrupta IPNI=*695052-1* 289
Persicaria kawakamii IPNI=*695113-1* 289
Persicaria komarovii IPNI=*77069480-1* 289
Persicaria koreana IPNI=*77069479-1* 289
Persicaria koreensis f. *viridiflora* IPNI=*77069479-1* 289
Persicaria krugii IPNI=*77069480-1* 289
Persicaria lapathifolia f. *alba* IPNI=*77069480-1* 289
Persicaria lapathifolia f. *baumgarteniana* IPNI=*77069480-1* 289
Persicaria lapathifolia subsp. *andrzejowskiana* IPNI=*77069480-1* 289
Persicaria lapathifolia subsp. *brittingeri* IPNI=*77069480-1* 289
Persicaria lapathifolia subsp. *glandulosa* IPNI=*77069480-1* 289
Persicaria lapathifolia subsp. *hypanica* IPNI=*77069480-1* 289
Persicaria lapathifolia subsp. *klokovii* IPNI=*77069480-1* 289
Persicaria lapathifolia subsp. *leptoclada* IPNI=*77069480-1* 289
Persicaria lapathifolia subsp. *mesomorpha* IPNI=*77069480-1* 289
Persicaria lapathifolia subsp. *nodosa* IPNI=*77069480-1* 289
Persicaria lapathifolia subsp. *oligoclada* IPNI=*77069480-1* 289
Persicaria lapathifolia subsp. *saporoviensis* IPNI=*77069480-1* 289
Persicaria lapathifolia subsp. *syringifolia* IPNI=*77069480-1* 289
Persicaria lapathifolia subsp. *turgida* IPNI=*77069480-1* 289
Persicaria lapathifolia var. *nodosa* IPNI=*77069480-1* 289
Persicaria lapathifolia var. *rubra* IPNI=*77069480-1* 289
Persicaria lapathifolia var. *salicifolia* IPNI=*77069480-1* 289
Persicaria laxa IPNI=*77069480-1* 289
Persicaria lorinseri IPNI=*695052-1* 289
Persicaria ludoviciana IPNI=*190869-2* 289
Persicaria lyrata IPNI=*60433514-2* 289
Persicaria maculata IPNI=*77069480-1* 289
Persicaria maculosa var. *biformis* IPNI=*695052-1* 289
Persicaria maculosa var. *ruderalis* IPNI=*695052-1* 289
Persicaria maximowiczii IPNI=*77069479-1* 289
Persicaria mesomorpha IPNI=*77069480-1* 289
Persicaria mexicana IPNI=*190869-2* 289
Persicaria mitis IPNI=*695052-1* 289
Persicaria nodosa IPNI=*77069480-1* 289
Persicaria oneillii IPNI=*77069480-1* 289
Persicaria opaca IPNI=*695052-1* 289
Persicaria opelousana IPNI=*190811-2* 289
Persicaria paludicola IPNI=*190811-2* 289
Persicaria perforata IPNI=*60433514-2* 289
Persicaria persicaria IPNI=*695052-1* 289
Persicaria persicarioides IPNI=*190811-2* 289
Persicaria portoricensis IPNI=*1005755-2* 289
Persicaria posumbu f. *albiflora* IPNI=*695113-1* 289
Persicaria posumbu var. *albiflora* IPNI=*695113-1* 289
Persicaria posumbu var. *stenophylla* IPNI=*695113-1* 289
Persicaria pseudojaponica IPNI=*77069480-1* 289
Persicaria pseudonodosa IPNI=*77069480-1* 289
Persicaria punctata var. *confertiflora* IPNI=*190855-2* 289

Persicaria punctata var. *eciliata* IPNI=*190855-2* 289
Persicaria punctata var. *leptostachya* IPNI=*190855-2* 289
Persicaria punctata var. *tacubayana* IPNI=*190855-2* 289
Persicaria pusilla IPNI=*695052-1* 289
Persicaria rivularis IPNI=*695052-1* 289
Persicaria ruderalis f. *angustifolia* IPNI=*695052-1* 289
Persicaria ruderalis IPNI=*695052-1* 289
Persicaria ruderalis var. *vulgaris* IPNI=*695052-1* 289
Persicaria salicifolia IPNI=*695052-1* 289
Persicaria saporoviensis IPNI=*77069480-1* 289
Persicaria tenuiflora IPNI=*77069480-1* 289
Persicaria tomentosa var. *glabrior* IPNI=*77069480-1* 289
Persicaria turgida IPNI=*77069480-1* 289
Persicaria urens IPNI=*77069479-1* 289
Persicaria vaniotiana IPNI=*77069480-1* 289
Persicaria vernalis IPNI=*77069479-1* 289
Persicaria vulgaris IPNI=*695052-1* 289
Persicaria yokusaiana IPNI=*695113-1* 289
Perula benghalensis IPNI=*852482-1* 438
Pervinca heterophyla IPNI=*82701-1* 344
Pervinca major IPNI=*82697-1* 344
Pervinca minor IPNI=*82701-1* 344
Pervinca procumbens IPNI=*82701-1* 344
Pervinca repens IPNI=*82701-1* 344
Pervinca rosea IPNI=*77880-1* 340
Peschiera arborea IPNI=*248839-2* 343
Peschiera arcuata IPNI=*82084-1* 343
Peschiera blanda IPNI=*285821-2* 344
Peschiera buchtienii IPNI=*82084-1* 343
Peschiera concinna IPNI=*82084-1* 343
Peschiera cuspidata IPNI=*82146-1* 343
Peschiera cymosa IPNI=*82084-1* 343
Peschiera diversifolia IPNI=*82146-1* 343
Peschiera echinata IPNI=*20001045-1* 344
Peschiera heterophylla IPNI=*82146-1* 343
Peschiera laevifructa IPNI=*82146-1* 343
Peschiera lingulata IPNI=*285821-2* 344
Peschiera litoralis IPNI=*285810-2* 343
Peschiera multiflora IPNI=*285812-2* 343
Peschiera psychotriifolia IPNI=*82084-1* 343
Peschiera puberiflora IPNI=*82084-1* 343
Peschiera schippii IPNI=*248839-2* 343
Peschiera stenoloba IPNI=*82146-1* 343
Peschiera surinamensis IPNI=*82371-1* 344
Peschiera tenuiflora IPNI=*82146-1* 343
Peschiera umbrosa IPNI=*82084-1* 343
Peschiera vanheurckii IPNI=*285821-2* 344
Pessopteris albopunctatissima IPNI=*77183020-1* 175
Pessopteris crassifolia IPNI=*77183236-1* 175
Pessopteris crassifolia var. *albopunctatissima* IPNI=*77183020-1* 175
Pessopteris crassifolia var. *angusta* IPNI=*77183236-1* 175
Petagomoa atricapilla IPNI=*77161076-1* 354
Petagomoa horridula IPNI=*77161076-1* 354
Petagomoa iodotricha IPNI=*77161076-1* 354
Petagomoa tricholoba IPNI=*77161076-1* 354
Petaloma cauliflorum IPNI=*573790-1* 423
Petaloma guianense IPNI=*306748-2* 423
Petaloma mouriri IPNI=*306748-2* 423
Petastoma langlasseanum IPNI=*152677-2* 364
Petastoma phaseoloides IPNI=*77100444-1* 362
Petastoma pubescens IPNI=*77142059-1* 363
Petastoma tonduzianum IPNI=*152677-2* 364
Peteniodendron durlandii IPNI=*788920-1* 306
Petesia repens IPNI=*746963-1* 348
Petitia domingensis var. *ekmanii* IPNI=*864000-1* 372
Petitia domingensis var. *poeppigii* IPNI=*864000-1* 372
Petitia poeppigii IPNI=*864000-1* 372
Petiveria alliacea var. *grandifolia* IPNI=*323288-2* 287
Petiveria alliacea var. *octandra* IPNI=*323288-2* 287
Petiveria corrientina IPNI=*323288-2* 287
Petiveria foetida IPNI=*323288-2* 287
Petiveria hexandria IPNI=*323288-2* 287
Petiveria ochroleuca IPNI=*323288-2* 287
Petiveria octandra IPNI=*323288-2* 287
Petiveria paraguayensis IPNI=*323288-2* 287
Petrea algentryi IPNI=*300267-2* 379
Petrea amazonica IPNI=*330591-2* 379
Petrea arborea f. *albiflora* IPNI=*330591-2* 379
Petrea arborea f. *broadwayi* IPNI=*330591-2* 379
Petrea arborea IPNI=*330591-2* 379
Petrea arborea var. *broadwayi* IPNI=*330591-2* 379
Petrea arborescens IPNI=*330591-2* 379
Petrea aspera f. *albiflora* IPNI=*330591-2* 379
Petrea aspera IPNI=*330591-2* 379
Petrea atrocoerulea IPNI=*330591-2* 379
Petrea colombiana IPNI=*330591-2* 379
Petrea erecta IPNI=*330591-2* 379
Petrea fragrantissima IPNI=*330591-2* 379
Petrea kohautiana f. *alba* IPNI=*330591-2* 379
Petrea kohautiana IPNI=*330591-2* 379
Petrea kohautiana var. *anomala* IPNI=*330591-2* 379
Petrea kohautiana var. *pilosula* IPNI=*330591-2* 379
Petrea mexicana IPNI=*330591-2* 379
Petrea morii IPNI=*300267-2* 379
Petrea nitidula IPNI=*330591-2* 379
Petrea obtusifolia IPNI=*300390-2* 379
Petrea ovata IPNI=*330591-2* 379
Petrea peruviana IPNI=*300267-2* 379
Petrea peruviana var. *acuminata* IPNI=*300267-2* 379
Petrea racemosa IPNI=*330591-2* 379
Petrea racemosa var. *alba* IPNI=*330591-2* 379
Petrea retusa IPNI=*330591-2* 379
Petrea riparia IPNI=*330591-2* 379
Petrea rivularis IPNI=*330591-2* 379
Petrea rugosa f. *casta* IPNI=*300390-2* 379
Petrea serrata IPNI=*330591-2* 379
Petrea stapeliae IPNI=*330591-2* 379
Petrea subserrata IPNI=*330591-2* 379
Petrea swallenii IPNI=*330591-2* 379
Petrea vincentina IPNI=*330591-2* 379
Petrea volubilis f. *pubescens* IPNI=*330591-2* 379
Petrea volubilis var. *alba* IPNI=*330591-2* 379
Petrea volubilis var. *albiflora* IPNI=*330591-2* 379
Petrea volubilis var. *pubescens* IPNI=*330591-2* 379
Petrocarya campestris IPNI=*727083-1* 383
Petrocarya excelsa IPNI=*727101-1* 383
Petrocarya montana IPNI=*727140-1* 383
Petrocosmea ionantha IPNI=*77117375-1* 368
Petroselinum anatolicum IPNI=*60442790-2* 252
Petroselinum crispum f. *angustifolium* IPNI=*60442790-2* 252
Petroselinum crispum f. *breve* IPNI=*60442790-2* 252
Petroselinum crispum f. *hispanicum* IPNI=*60442790-2* 252
Petroselinum crispum f. *tenuisectum* IPNI=*60442790-2* 252
Petroselinum crispum f. *variegatum* IPNI=*60442790-2* 252
Petroselinum crispum subsp. *tuberosum* IPNI=*60442790-2* 252
Petroselinum crispum var. *angustifolium* IPNI=*60442790-2* 252
Petroselinum crispum var. *erfurtense* IPNI=*60442790-2* 252
Petroselinum crispum var. *neapolitanum* IPNI=*60442790-2* 252
Petroselinum crispum var. *petroselinum* IPNI=*60442790-2* 252
Petroselinum crispum var. *radicosum* IPNI=*60442790-2* 252
Petroselinum crispum var. *vulgare* IPNI=*60442790-2* 252
Petroselinum fractophyllum IPNI=*60442790-2* 252
Petroselinum hortense f. *tenuisectum* IPNI=*60442790-2* 252
Petroselinum hortense IPNI=*60442790-2* 252
Petroselinum macedonicum IPNI=*60442790-2* 252
Petroselinum peregrinum IPNI=*60442790-2* 252

Petroselinum romanum IPNI=*60442790-2* 252
Petroselinum sativum convar. *radicosum* IPNI=*60442790-2* 252
Petroselinum sativum IPNI=*60442790-2* 252
Petroselinum sativum var. *breve* IPNI=*60442790-2* 252
Petroselinum sativum var. *hispanicum* IPNI=*60442790-2* 252
Petroselinum sativum var. *longum* IPNI=*60442790-2* 252
Petroselinum sativum var. *silvestre* IPNI=*60442790-2* 252
Petroselinum sativum var. *variegatum* IPNI=*60442790-2* 252
Petroselinum sativum var. *vulgare* IPNI=*60442790-2* 252
Petroselinum selinoides IPNI=*60442790-2* 252
Petroselinum thermoeri IPNI=*60442790-2* 252
Petroselinum vulgare IPNI=*60442790-2* 252
Peucedanum anethum IPNI=*837530-1* 251
Peucedanum graveolens IPNI=*837530-1* 251
Peucedanum intermedium IPNI=*60442790-2* 252
Peucedanum petroselinum IPNI=*60442790-2* 252
Peucedanum sowa IPNI=*837530-1* 251
Peumus fragrans IPNI=*191376-2* 185
Peutalis hydropiper IPNI=*77069479-1* 289
Peutalis incana IPNI=*77069480-1* 289
Peutalis nodosa IPNI=*77069480-1* 289
Peutalis persicaria IPNI=*695052-1* 289
Peutalis scabra IPNI=*77069480-1* 289
Pfaffia aurata IPNI=*1002675-1* 283
Pfaffia divergens IPNI=*191400-2* 283
Pfaffia dunaliana IPNI=*191400-2* 283
Pfaffia fiebrigii IPNI=*191400-2* 283
Pfaffia glabrescens IPNI=*191400-2* 283
Pfaffia glauca IPNI=*191400-2* 283
Pfaffia glomerata var. *squarrosa* IPNI=*191400-2* 283
Pfaffia iresinoides var. *angustifolia* IPNI=*191400-2* 283
Pfaffia iresinoides var. *luzuliflora* IPNI=*191400-2* 283
Pfaffia iresinoides var. *parvifolia* IPNI=*61055-1* 283
Pfaffia iresinoides var. *villosa* IPNI=*61055-1* 283
Pfaffia luzuliflora f. *gracilis* IPNI=*191400-2* 283
Pfaffia luzuliflora f. *virgata* IPNI=*191400-2* 283
Pfaffia luzuliflora IPNI=*191400-2* 283
Pfaffia luzuliflora subsp. *squarrosa* IPNI=*191400-2* 283
Pfaffia luzuliflora var. *colombiana* IPNI=*191400-2* 283
Pfaffia luzuliflora var. *elliptica* IPNI=*191400-2* 283
Pfaffia luzuliflora var. *microcephala* IPNI=*191400-2* 283
Pfaffia luzuliflora var. *paniculata* IPNI=*191400-2* 283
Pfaffia meridensis IPNI=*10378-2* 281
Pfaffia sericea f. *eusericea* IPNI=*1219488-2* 283
Pfaffia sericea f. *glabrescens* IPNI=*1219488-2* 283
Pfaffia sericea f. *vestita* IPNI=*1219488-2* 283
Pfaffia sericea f. *villosissima* IPNI=*1219488-2* 283
Pfaffia sericea var. *foliosa* IPNI=*1219488-2* 283
Pfaffia sericea var. *glabrescens* IPNI=*1219488-2* 283
Pfaffia sericea var. *paraguariensis* IPNI=*1219488-2* 283
Pfaffia sericea var. *vestita* IPNI=*1219488-2* 283
Pfaffia stenophylla IPNI=*191400-2* 283
Pfaffia stenophylla var. *albiramea* IPNI=*191400-2* 283
Pfaffia stenophylla var. *basilignosa* IPNI=*191400-2* 283
Pfaffia stenophylla var. *foliosa* IPNI=*191400-2* 283
Pfaffia tuberosa f. *rubella* IPNI=*1219488-2* 283
Pfaffia tuberosa subsp. *goiana* IPNI=*1219488-2* 283
Pfaffia vana IPNI=*191400-2* 283
Phacosperma peruviana IPNI=*698357-1* 286
Phaeneilema diversifolium IPNI=*172873-1* 224
Phaeneilema malabaricum IPNI=*172873-1* 224
Phaeneilema nudiflorum IPNI=*172873-1* 224
Phaenixopus intybaceus IPNI=*229022-1* 264
Phaeomeria magnifica IPNI=*942355-1* 251
Phaeomeria speciosa IPNI=*942355-1* 251
Phaeosphaerion mathewsii IPNI=*63900-2* 224
Phaeosphaerion persicariifolium var. *scabratum* IPNI=*63900-2* 224
Phalacraea coelestina IPNI=*7086-2* 256
Phalacrodiscus ceratophylloides IPNI=*230047-1* 265
Phalacrodiscus maximus IPNI=*230047-1* 265
Phalacromesus riparia IPNI=*30283369-2* 271
Phalangium comosum IPNI=*532810-1* 215
Phalangium viviparum IPNI=*532810-1* 215
Phalaridantha arundinacea IPNI=*415571-1* 243
Phalaris ambigua IPNI=*415632-1* 243
Phalaris americana var. *picta* IPNI=*415571-1* 243
Phalaris aquatica IPNI=*415588-1* 243
Phalaris aquatica var. *hirtiglumis* IPNI=*415560-1* 243
Phalaris aristata IPNI=*1048074-2* 244
Phalaris arundinacea subsp. *hispanica* IPNI=*415571-1* 243
Phalaris arundinacea subsp. *oehleri* IPNI=*415571-1* 243
Phalaris arundinacea subsp. *rotgesii* IPNI=*415571-1* 243
Phalaris arundinacea var. *picta* IPNI=*415571-1* 243
Phalaris arundinacea var. *variegata* IPNI=*415571-1* 243
Phalaris avicularis IPNI=*415588-1* 243
Phalaris brevis IPNI=*415632-1* 243
Phalaris bulbosa IPNI=*415560-1* 243
Phalaris caesia IPNI=*415571-1* 243
Phalaris caespitosa IPNI=*415560-1* 243
Phalaris capensis IPNI=*415632-1* 243
Phalaris chilensis IPNI=*311946-2* 243
Phalaris ciliata IPNI=*389034-1* 234
Phalaris colorata IPNI=*415571-1* 243
Phalaris cristata IPNI=*1048074-2* 244
Phalaris decumbens IPNI=*415632-1* 243
Phalaris elongata IPNI=*415560-1* 243
Phalaris glomerata IPNI=*397309-1* 236
Phalaris gracilis IPNI=*415632-1* 243
Phalaris hirtiglumis IPNI=*415560-1* 243
Phalaris hispanica IPNI=*415571-1* 243
Phalaris intermedia var. *angusta* IPNI=*311946-2* 243
Phalaris japonica IPNI=*415571-1* 243
Phalaris laxa IPNI=*311946-2* 243
Phalaris ludoviciana IPNI=*311946-2* 243
Phalaris monspeliensis IPNI=*415571-1* 243
Phalaris nepalensis IPNI=*415632-1* 243
Phalaris nodosa IPNI=*415560-1* 243
Phalaris ovata IPNI=*415588-1* 243
Phalaris picta IPNI=*415571-1* 243
Phalaris segetalis IPNI=*311946-2* 243
Phalaris semineutra IPNI=*320346-2* 244
Phalaris setacea IPNI=*77106046-1* 236
Phalaris stenoptera IPNI=*415560-1* 243
Phalaris trabutiana IPNI=*415560-1* 243
Phalaris tuberosa IPNI=*415560-1* 243
Phalaris tuberosa var. *hirtiglumis* IPNI=*415560-1* 243
Phalaris zizanioides IPNI=*396213-1* 236
Phalaroides arundinacea convar. *picta* IPNI=*415571-1* 243
Phalaroides arundinacea IPNI=*415571-1* 243
Phalaroides arundinacea subsp. *caesia* IPNI=*415571-1* 243
Phalaroides arundinacea subsp. *japonica* IPNI=*415571-1* 243
Phalaroides arundinacea subsp. *oehleri* IPNI=*415571-1* 243
Phalaroides arundinacea subsp. *rotgesii* IPNI=*415571-1* 243
Phalaroides arundinacea var. *kuzenevae* IPNI=*415571-1* 243
Phalaroides arundinacea var. *picta* IPNI=*415571-1* 243
Phalaroides caesia IPNI=*415571-1* 243
Phalaroides hispanica IPNI=*415571-1* 243
Phalaroides japonica IPNI=*415571-1* 243
Phalaroides picta IPNI=*415571-1* 243
Phalaroides rotgesii IPNI=*415571-1* 243
Phanera glabra IPNI=*229758-2* 332
Phanera guianensis IPNI=*77111925-1* 332
Phanera kunthiana IPNI=*77111927-1* 332
Phanera kurzii IPNI=*481605-1* 312
Phanera microstachya IPNI=*229762-2* 332
Phanera punctata IPNI=*229758-2* 332

Phanera purpurea IPNI=*481605-1* 312
Phanera rosea IPNI=*481605-1* 312
Phanera splendens IPNI=*518025-1* 332
Phanera variegata IPNI=*481756-1* 312
Phanerophlebia semicordata IPNI=*1148359-2* 175
Pharmacosycea anthelmintica IPNI=*853024-1* 438
Pharmacosycea brittonii IPNI=*853024-1* 438
Pharmacosycea glaucescens IPNI=*853200-1* 439
Pharmacosycea grandaeva IPNI=*853200-1* 439
Pharmacosycea guyanensis IPNI=*853200-1* 439
Pharmacosycea hernandezii IPNI=*853200-1* 439
Pharmacosycea laurifolia IPNI=*853405-1* 439
Pharmacosycea mexicana IPNI=*853212-1* 439
Pharmacosycea peruviana IPNI=*853405-1* 439
Pharmacosycea pseudoradula IPNI=*853200-1* 439
Pharmacosycea radula IPNI=*853200-1* 439
Pharmacosycea rigida IPNI=*853200-1* 439
Pharnaceum arenarium IPNI=*280992-2* 286
Pharnaceum berteroanum IPNI=*280992-2* 286
Pharnaceum galioides IPNI=*280992-2* 286
Pharnaceum hoffmannseggianum IPNI=*280992-2* 286
Pharnaceum lineare IPNI=*280992-2* 286
Pharnaceum verticillatum IPNI=*280992-2* 286
Pharus angustifolius IPNI=*415708-1* 243
Pharus brasiliensis IPNI=*415708-1* 243
Pharus brasiliensis var. *angustifolius* IPNI=*415708-1* 243
Pharus brasiliensis var. *latifolius* IPNI=*415708-1* 243
Pharus ciliatus IPNI=*137026-2* 241
Pharus glaber IPNI=*415708-1* 243
Pharus glaber var. *pubescens* IPNI=*415708-1* 243
Pharus glochidiatus IPNI=*316520-2* 243
Pharus lancifolius IPNI=*415708-1* 243
Pharus latifolius var. *angustifolius* IPNI=*415708-1* 243
Pharus latifolius var. *elegantissimus* IPNI=*316520-2* 243
Pharus latifolius var. *parvifolius* IPNI=*415708-1* 243
Pharus micranthus IPNI=*415708-1* 243
Pharus micranthus var. *concolor* IPNI=*415708-1* 243
Pharus micranthus var. *discolor* IPNI=*415708-1* 243
Pharus ovalifolius IPNI=*316520-2* 243
Pharus pubescens IPNI=*415708-1* 243
Pharus scaber IPNI=*316520-2* 243
Pharus scaber var. *pictus* IPNI=*316520-2* 243
Phasellus lathyroides IPNI=*149066-2* 327
Phasellus violaceus IPNI=*525419-1* 336
Phaseolus × *flavescens* IPNI=*513799-1* 330
Phaseolus × *harmsianus* IPNI=*513799-1* 330
Phaseolus × *leucanthus* IPNI=*513799-1* 330
Phaseolus × *polyanthus* IPNI=*513799-1* 330
Phaseolus aborigineus IPNI=*514191-1* 330
Phaseolus abyssinicus IPNI=*525419-1* 336
Phaseolus adenanthus f. *bahiensis* IPNI=*60457245-2* 325
Phaseolus adenanthus f. *guaraniticus* IPNI=*60457245-2* 325
Phaseolus adenanthus f. *janeirensis* IPNI=*60457245-2* 325
Phaseolus adenanthus f. *subtortus* IPNI=*60457245-2* 325
Phaseolus adenanthus f. *velutinus* IPNI=*60457245-2* 325
Phaseolus adenanthus IPNI=*60457245-2* 325
Phaseolus adenanthus var. *caeduorum* IPNI=*60457245-2* 325
Phaseolus adenanthus var. *genuinus* IPNI=*60457245-2* 325
Phaseolus adenanthus var. *latifolius* IPNI=*60457245-2* 325
Phaseolus adenanthus var. *radicans* IPNI=*60457245-2* 325
Phaseolus adenanthus var. *truxillensis* IPNI=*60457245-2* 325
Phaseolus affinis IPNI=*57174-3* 327
Phaseolus alatus IPNI=*60457245-2* 325
Phaseolus albus IPNI=*514191-1* 330
Phaseolus amarus IPNI=*60457245-2* 325
Phaseolus amazonicus IPNI=*513918-1* 330
Phaseolus amoenus IPNI=*60457245-2* 325
Phaseolus angularis IPNI=*525291-1* 336
Phaseolus angulosus IPNI=*514191-1* 330
Phaseolus angustifolius IPNI=*525554-1* 337
Phaseolus asparagioides IPNI=*514191-1* 330
Phaseolus asper IPNI=*60457240-2* 320
Phaseolus atropurpureus f. *intermedius* IPNI=*57174-3* 327
Phaseolus atropurpureus IPNI=*57174-3* 327
Phaseolus atropurpureus var. *canescens* IPNI=*57174-3* 327
Phaseolus atropurpureus var. *genuinus* IPNI=*57174-3* 327
Phaseolus atropurpureus var. *pseuderythroloma* IPNI=*57174-3* 327
Phaseolus atropurpureus var. *sericeus* IPNI=*57174-3* 327
Phaseolus atropurpureus var. *vestitus* IPNI=*57174-3* 327
Phaseolus atsuki IPNI=*525291-1* 336
Phaseolus barbulatus IPNI=*60457245-2* 325
Phaseolus bertonii f. *albus* IPNI=*60471746-2* 315
Phaseolus bertonii IPNI=*60471746-2* 315
Phaseolus bicolor IPNI=*513754-1* 330
Phaseolus bipunctatus IPNI=*513918-1* 330
Phaseolus brevipes IPNI=*60457245-2* 325
Phaseolus bulae IPNI=*489042-1* 316
Phaseolus caeduorum IPNI=*60457245-2* 325
Phaseolus calcaratus IPNI=*525554-1* 337
Phaseolus candidus f. *maritimus* IPNI=*149066-2* 327
Phaseolus canescens IPNI=*57174-3* 327
Phaseolus capensis IPNI=*514191-1* 330
Phaseolus caracalla IPNI=*60471746-2* 315
Phaseolus caracalla var. *brachystachys* IPNI=*60471746-2* 315
Phaseolus caracallensis IPNI=*60471746-2* 315
Phaseolus chrysanthos IPNI=*525554-1* 337
Phaseolus cinnabarinus IPNI=*149066-2* 327
Phaseolus cirrosus IPNI=*60457245-2* 325
Phaseolus clitorius IPNI=*486606-1* 315
Phaseolus coccineus IPNI=*513754-1* 330
Phaseolus coccineus subsp. *formosus* IPNI=*513754-1* 330
Phaseolus coccineus subsp. *obvallatus* IPNI=*513754-1* 330
Phaseolus coccineus subsp. *polyanthus* IPNI=*513799-1* 330
Phaseolus coccineus var. *argenteus* IPNI=*513754-1* 330
Phaseolus coccineus var. *condensatus* IPNI=*513754-1* 330
Phaseolus coccineus var. *griseus* IPNI=*513754-1* 330
Phaseolus coccineus var. *guatemalensis* IPNI=*513754-1* 330
Phaseolus coccineus var. *lineatibracteolatus* IPNI=*513754-1* 330
Phaseolus coccineus var. *minuticicatricatus* IPNI=*513754-1* 330
Phaseolus coccineus var. *parvibracteolatus* IPNI=*513754-1* 330
Phaseolus coccineus var. *pringlei* IPNI=*513754-1* 330
Phaseolus coccineus var. *pubescens* IPNI=*513754-1* 330
Phaseolus coccineus var. *purpurascens* IPNI=*513754-1* 330
Phaseolus coccineus var. *rigidicaulis* IPNI=*513754-1* 330
Phaseolus coccineus var. *semperbracteolatus* IPNI=*513754-1* 330
Phaseolus coccineus var. *splendens* IPNI=*513754-1* 330
Phaseolus coccineus var. *strigillosus* IPNI=*513754-1* 330
Phaseolus coccineus var. *timilpanensis* IPNI=*513754-1* 330
Phaseolus coccineus var. *tridentatus* IPNI=*513754-1* 330
Phaseolus coccineus var. *zongolicensis* IPNI=*513754-1* 330
Phaseolus cochleatus IPNI=*60457245-2* 325
Phaseolus cochleatus IPNI=*60471746-2* 315
Phaseolus cochleatus var. *pallidus* IPNI=*60457245-2* 325
Phaseolus cochleatus var. *violaceus* IPNI=*60457245-2* 325
Phaseolus communis IPNI=*514191-1* 330
Phaseolus compressus IPNI=*513918-1* 330
Phaseolus compressus IPNI=*514191-1* 330
Phaseolus coriaceus IPNI=*60457240-2* 320
Phaseolus crotalarioides IPNI=*149066-2* 327
Phaseolus cruentus IPNI=*514191-1* 330
Phaseolus cuernavacanus IPNI=*60457245-2* 325
Phaseolus cummingii IPNI=*60457245-2* 325
Phaseolus cytisoides IPNI=*60471746-2* 315
Phaseolus decurrens IPNI=*509661-1* 328
Phaseolus deliciosus IPNI=*514191-1* 330
Phaseolus derasus IPNI=*513918-1* 330

Phaseolus dimidiatus IPNI=*514191-1* 330
Phaseolus domingensis IPNI=*514191-1* 330
Phaseolus dysophyllus IPNI=*57174-3* 327
Phaseolus ellipticus IPNI=*514191-1* 330
Phaseolus ellipticus var. *atrofuscus* IPNI=*514191-1* 330
Phaseolus ellipticus var. *niger* IPNI=*514191-1* 330
Phaseolus elongatus IPNI=*60457240-2* 320
Phaseolus erythroloma IPNI=*149061-2* 327
Phaseolus esculentus IPNI=*514191-1* 330
Phaseolus falcatus IPNI=*513918-1* 330
Phaseolus ferrugineus IPNI=*514191-1* 330
Phaseolus foecundus IPNI=*513918-1* 330
Phaseolus formosus IPNI=*513754-1* 330
Phaseolus gonospermos IPNI=*514191-1* 330
Phaseolus gracilis IPNI=*149063-2* 327
Phaseolus grandiflorus IPNI=*60457240-2* 320
Phaseolus griseus IPNI=*513754-1* 330
Phaseolus haematocarpus IPNI=*514191-1* 330
Phaseolus hastifolius IPNI=*149066-2* 327
Phaseolus hispidulus IPNI=*525554-1* 337
Phaseolus humilis IPNI=*525554-1* 337
Phaseolus ilocanus IPNI=*513918-1* 330
Phaseolus inamoenus IPNI=*513918-1* 330
Phaseolus ionocarpus IPNI=*514191-1* 330
Phaseolus lathyroides f. *chacoensis* IPNI=*149066-2* 327
Phaseolus lathyroides f. *hirsutus* IPNI=*149066-2* 327
Phaseolus lathyroides f. *repandus* IPNI=*149066-2* 327
Phaseolus lathyroides f. *typicus* IPNI=*149066-2* 327
Phaseolus lathyroides IPNI=*149066-2* 327
Phaseolus lathyroides subf. *nanus* IPNI=*149066-2* 327
Phaseolus lathyroides var. *genuinus* IPNI=*149066-2* 327
Phaseolus lathyroides var. *hastifolius* IPNI=*149066-2* 327
Phaseolus lathyroides var. *semierectus* IPNI=*149066-2* 327
Phaseolus latifolius IPNI=*60457245-2* 325
Phaseolus latisiliquus IPNI=*513918-1* 330
Phaseolus leiosepalus IPNI=*513754-1* 330
Phaseolus leptospermus IPNI=*525554-1* 337
Phaseolus lilac IPNI=*514191-1* 330
Phaseolus limensis IPNI=*513918-1* 330
Phaseolus linearis f. *euviolaceus* IPNI=*60457240-2* 320
Phaseolus linearis f. *intermedius* IPNI=*60457240-2* 320
Phaseolus linearis f. *subrhombeus* IPNI=*60457240-2* 320
Phaseolus linearis f. *transiens* IPNI=*60457240-2* 320
Phaseolus linearis f. *typicus* IPNI=*60457240-2* 320
Phaseolus linearis IPNI=*60457240-2* 320
Phaseolus linearis var. *coriaceus* IPNI=*60457240-2* 320
Phaseolus linearis var. *genuinus* IPNI=*60457240-2* 320
Phaseolus linearis var. *latifolius* IPNI=*60457240-2* 320
Phaseolus linearis var. *violaceus* IPNI=*60457240-2* 320
Phaseolus longepedunculatus var. *linearifoliolatus* IPNI=*149063-2* 327
Phaseolus longirostratus IPNI=*60471746-2* 315
Phaseolus lunatus f. *limenanus* IPNI=*513918-1* 330
Phaseolus lunatus f. *lunonanus* IPNI=*513918-1* 330
Phaseolus lunatus f. *macrocarpus* IPNI=*513918-1* 330
Phaseolus lunatus f. *salicis* IPNI=*513918-1* 330
Phaseolus lunatus f. *solanoides* IPNI=*513918-1* 330
Phaseolus lunatus f. *vulgaris* IPNI=*513918-1* 330
Phaseolus lunatus var. *macrocarpus* IPNI=*513918-1* 330
Phaseolus lunatus var. *silvester* IPNI=*513918-1* 330
Phaseolus lupinoides IPNI=*514191-1* 330
Phaseolus luteolus IPNI=*525419-1* 336
Phaseolus macfadyenii IPNI=*60457245-2* 325
Phaseolus macrocarpus IPNI=*513918-1* 330
Phaseolus marinus IPNI=*525419-1* 336
Phaseolus marinus IPNI=*525428-1* 336
Phaseolus maritimus IPNI=*149066-2* 327
Phaseolus max IPNI=*60450240-2* 320
Phaseolus maximus IPNI=*513918-1* 330
Phaseolus melanospermus IPNI=*514191-1* 330
Phaseolus mesoleucus IPNI=*514191-1* 330
Phaseolus mexicanus IPNI=*514191-1* 330
Phaseolus multiflorus IPNI=*513754-1* 330
Phaseolus multiflorus IPNI=*514191-1* 330
Phaseolus multiflorus var. *albiflorus* IPNI=*513754-1* 330
Phaseolus multiflorus var. *coccineus* IPNI=*513754-1* 330
Phaseolus nanus IPNI=*514191-1* 330
Phaseolus nigerrimus IPNI=*514191-1* 330
Phaseolus nigricans IPNI=*514191-1* 330
Phaseolus nipponensis IPNI=*525291-1* 336
Phaseolus oblongus IPNI=*514191-1* 330
Phaseolus obovatus IPNI=*525428-1* 336
Phaseolus obvallatus IPNI=*513754-1* 330
Phaseolus occidentalis IPNI=*60457245-2* 325
Phaseolus ovalispermus IPNI=*514191-1* 330
Phaseolus parviflorus IPNI=*513918-1* 330
Phaseolus pascuorum IPNI=*60457232-2* 311
Phaseolus peduncularis f. *intermedius* IPNI=*60457232-2* 311
Phaseolus peduncularis f. *subhastatus* IPNI=*60457232-2* 311
Phaseolus peduncularis IPNI=*60457232-2* 311
Phaseolus peduncularis var. *genuinus* IPNI=*60457232-2* 311
Phaseolus pedunculatus IPNI=*60457232-2* 311
Phaseolus phanerophlebius IPNI=*60457240-2* 320
Phaseolus pictus IPNI=*514191-1* 330
Phaseolus pisiformis IPNI=*514191-1* 330
Phaseolus platyspermus IPNI=*513918-1* 330
Phaseolus portoricensis IPNI=*513918-1* 330
Phaseolus praecox IPNI=*514191-1* 330
Phaseolus principissae IPNI=*514191-1* 330
Phaseolus prorifirus IPNI=*513754-1* 330
Phaseolus pruriens IPNI=*1197939-2* 328
Phaseolus psoraleoides IPNI=*149066-2* 327
Phaseolus puberulus IPNI=*513918-1* 330
Phaseolus pubescens IPNI=*525554-1* 337
Phaseolus radicans IPNI=*60457245-2* 325
Phaseolus ricciardianus IPNI=*525554-1* 337
Phaseolus rigidus IPNI=*60457240-2* 320
Phaseolus romanus IPNI=*514191-1* 330
Phaseolus rostratus IPNI=*60457245-2* 325
Phaseolus rubidus IPNI=*149061-2* 327
Phaseolus rufus IPNI=*149061-2* 327
Phaseolus rufus IPNI=*513918-1* 330
Phaseolus saccharatus IPNI=*513918-1* 330
Phaseolus saccharatus IPNI=*514191-1* 330
Phaseolus sanctorum IPNI=*513814-1* 330
Phaseolus saponaceus IPNI=*514191-1* 330
Phaseolus savannarum IPNI=*149063-2* 327
Phaseolus schiedeanus IPNI=*57174-3* 327
Phaseolus semierectus IPNI=*149066-2* 327
Phaseolus semierectus var. *angustifolius* IPNI=*149066-2* 327
Phaseolus semierectus var. *atropurpureus* IPNI=*57174-3* 327
Phaseolus semierectus var. *gracilis* IPNI=*149066-2* 327
Phaseolus semierectus var. *nanus* IPNI=*149066-2* 327
Phaseolus semierectus var. *subhastatus* IPNI=*149066-2* 327
Phaseolus senegalensis IPNI=*60457245-2* 325
Phaseolus sinensis IPNI=*514191-1* 330
Phaseolus sordidus IPNI=*60450240-2* 320
Phaseolus sphaericus IPNI=*514191-1* 330
Phaseolus spiralis IPNI=*60457245-2* 325
Phaseolus spixianus IPNI=*60457232-2* 311
Phaseolus spixianus var. *schottianus* IPNI=*60457232-2* 311
Phaseolus stenolobus IPNI=*60457240-2* 320
Phaseolus stipularis IPNI=*514191-1* 330
Phaseolus striatus IPNI=*513754-1* 330
Phaseolus strictus IPNI=*149066-2* 327
Phaseolus strigillosus IPNI=*513754-1* 330
Phaseolus subglobosus IPNI=*514191-1* 330

Phaseolus subtortus IPNI=*60457245-2* 325
Phaseolus superbus IPNI=*513754-1* 330
Phaseolus surinamensis IPNI=*60457245-2* 325
Phaseolus sylvestris IPNI=*513754-1* 330
Phaseolus tenuiflorus IPNI=*513918-1* 330
Phaseolus torosus IPNI=*525554-1* 337
Phaseolus triangularis IPNI=*514191-1* 330
Phaseolus triangulus IPNI=*514191-1* 330
Phaseolus truxillensis f. *velutina* IPNI=*60457245-2* 325
Phaseolus truxillensis var. *grandiflorus* IPNI=*60457245-2* 325
Phaseolus truxillensis var. *minor* IPNI=*60457245-2* 325
Phaseolus tumidus IPNI=*514191-1* 330
Phaseolus tunkinensis IPNI=*513918-1* 330
Phaseolus umbilicalis IPNI=*514191-1* 330
Phaseolus unguiculatus IPNI=*1127257-2* 337
Phaseolus unilobatus IPNI=*149063-2* 327
Phaseolus vestitus IPNI=*57174-3* 327
Phaseolus vexillatus IPNI=*525561-1* 337
Phaseolus violaceus IPNI=*525419-1* 336
Phaseolus violaceus IPNI=*60457240-2* 320
Phaseolus virosus IPNI=*1209500-2* 313
Phaseolus vulgaris var. *aborigineus* IPNI=*514191-1* 330
Phaseolus wrightii IPNI=*513814-1* 330
Phaseolus xuarezii IPNI=*513918-1* 330
Phaseolus zebra IPNI=*514191-1* 330
Phaseolus zembra IPNI=*514191-1* 330
Phegopteris alsophiloides IPNI=*1019960-2* 174
Phegopteris montevidensis IPNI=*208026-2* 176
Phegopteris nicaraguensis IPNI=*252198-2* 174
Phegopteris oligophlebia IPNI=*1019960-2* 174
Phegopteris trichodes IPNI=*1019960-2* 174
Phegopyrum emarginatum IPNI=*694526-1* 289
Phegopyrum esculentum IPNI=*694526-1* 289
Phellocarpus acutus IPNI=*214040-2* 331
Phellocarpus amazonum IPNI=*214040-2* 331
Phellocarpus floridus IPNI=*516561-1* 331
Phellocarpus laxiflorus IPNI=*516561-1* 331
Phenakospermum amazonicum IPNI=*585013-1* 250
Phenax ballotifolius IPNI=*855180-1* 446
Phenax rugosus f. *uberrima* IPNI=*855180-1* 446
Phenax rugosus var. *serratus* IPNI=*855180-1* 446
Phenax tomentosus IPNI=*855189-1* 447
Phenax uliginosus var. *rotundatus* IPNI=*855189-1* 447
Phenax uliginosus var. *tomentosus* IPNI=*855189-1* 447
Phenax uliginosus var. *velutinus* IPNI=*855189-1* 447
Philenoptera schimperi IPNI=*143380-2* 325
Philibertella clausa IPNI=*106366-2* 340
Philibertella crassifolia IPNI=*106366-2* 340
Philibertella cumanensis IPNI=*106366-2* 340
Philibertella lasiantha IPNI=*106366-2* 340
Philibertella ovalifolia IPNI=*106382-2* 340
Philibertella pallida IPNI=*106366-2* 340
Philibertella palmeri IPNI=*106366-2* 340
Philibertella pedunculata IPNI=*106366-2* 340
Philibertella riparia IPNI=*106366-2* 340
Philibertia anomala IPNI=*94927-1* 339
Philibertia bonariensis IPNI=*106366-2* 340
Philibertia clausa IPNI=*106366-2* 340
Philibertia crassifolia IPNI=*106366-2* 340
Philibertia cumanensis IPNI=*106366-2* 340
Philibertia cuspidata IPNI=*106366-2* 340
Philibertia gardneri IPNI=*106366-2* 340
Philibertia palmeri IPNI=*106366-2* 340
Philibertia riparia IPNI=*106366-2* 340
Philibertia viminalis IPNI=*106366-2* 340
Phillyrea paniculata IPNI=*610017-1* 375
Philodendron accrescens IPNI=*87769-1* 201
Philodendron acutatum IPNI=*87967-1* 201
Philodendron affine IPNI=*88045-1* 201
Philodendron amazonicum IPNI=*87928-1* 201
Philodendron anatomicum IPNI=*87478-1* 200
Philodendron andreanum IPNI=*87882-1* 201
Philodendron arborescens IPNI=*87563-1* 200
Philodendron asperatum IPNI=*87918-1* 201
Philodendron belizense IPNI=*87815-1* 201
Philodendron bipinnatifidum IPNI=*77178497-1* 202
Philodendron bissectum IPNI=*252718-2* 227
Philodendron brevilaminatum IPNI=*87769-1* 201
Philodendron calophyllum IPNI=*87819-1* 201
Philodendron cannifolium IPNI=*88955-1* 202
Philodendron clementis IPNI=*87769-1* 201
Philodendron coerulescens IPNI=*87815-1* 201
Philodendron cyclophyllum IPNI=*87565-1* 200
Philodendron cyclops IPNI=*87967-1* 201
Philodendron daemonum IPNI=*87989-1* 201
Philodendron dagilla IPNI=*192963-2* 200
Philodendron demerarae IPNI=*87769-1* 201
Philodendron dolosum IPNI=*87918-1* 201
Philodendron duisbergii IPNI=*87928-1* 201
Philodendron fenzlii IPNI=*88045-1* 201
Philodendron fenzlii var. *anisotomum* IPNI=*192963-2* 200
Philodendron fontanesii IPNI=*269304-2* 202
Philodendron ghiesbrechtii IPNI=*87989-1* 201
Philodendron goeldii IPNI=*89184-1* 202
Philodendron guaraense IPNI=*87967-1* 201
Philodendron guatemalense IPNI=*87815-1* 201
Philodendron haematinum IPNI=*87819-1* 201
Philodendron hastatum IPNI=*289907-2* 201
Philodendron holtonianum IPNI=*88045-1* 201
Philodendron imperiale IPNI=*87918-1* 201
Philodendron inaequilaterum subsp. *zulianum* IPNI=*87815-1* 201
Philodendron laciniatum IPNI=*87928-1* 201
Philodendron laciniatum var. *palmatisectum* IPNI=*87928-1* 201
Philodendron laciniatum var. *weddellianum* IPNI=*87928-1* 201
Philodendron laciniosum IPNI=*87928-1* 201
Philodendron lancigerum IPNI=*87989-1* 201
Philodendron latipes IPNI=*87769-1* 201
Philodendron muschlerianum IPNI=*87918-1* 201
Philodendron nechodomae IPNI=*87046-1* 200
Philodendron nigrum IPNI=*89373-1* 203
Philodendron niveochermesinum IPNI=*87819-1* 201
Philodendron peregrinum IPNI=*60444565-2* 198
Philodendron pertusum IPNI=*87478-1* 200
Philodendron poeppigii IPNI=*87769-1* 201
Philodendron polypodioides IPNI=*87928-1* 201
Philodendron prieureanum IPNI=*87819-1* 201
Philodendron punctatum IPNI=*60444565-2* 198
Philodendron pygmaeum IPNI=*77178497-1* 202
Philodendron quercifolium IPNI=*87928-1* 201
Philodendron rubens IPNI=*87918-1* 201
Philodendron sanguineum IPNI=*87989-1* 201
Philodendron silvaticum IPNI=*87815-1* 201
Philodendron solimoesense IPNI=*77178490-1* 202
Philodendron spruceanum IPNI=*89184-1* 202
Philodendron tanyphyllum IPNI=*87989-1* 201
Philodendron tobagense IPNI=*87918-1* 201
Philodendron trilobatum IPNI=*88045-1* 201
Philodendron tripartitum var. *holtonianum* IPNI=*88045-1* 201
Philodendron tripartitum var. *tricuspidatum* IPNI=*88045-1* 201
Philodendron trisectum IPNI=*192963-2* 200
Philodendron tuxtlanum IPNI=*87989-1* 201
Philodendron weddellianum IPNI=*87928-1* 201
Philodendron zulianum IPNI=*87815-1* 201
Philotria densa IPNI=*431710-1* 203
Philoxerus aggregatus IPNI=*34033-2* 282
Philoxerus brasiliana IPNI=*10294-2* 281

Philoxerus brasiliensis IPNI=*10294-2* 281
Philoxerus carnosus IPNI=*34030-2* 282
Philoxerus crassifolius IPNI=*34033-2* 282
Philoxerus litoralis IPNI=*34033-2* 282
Philoxerus portulacoides IPNI=*34030-2* 282
Philoxerus portulacoides var. *commersonii* IPNI=*34030-2* 282
Philoxerus surinamensis IPNI=*34033-2* 282
Philoxerus vermicularis IPNI=*34030-2* 282
Philoxerus vermicularis IPNI=*34033-2* 282
Philoxerus vermicularis var. *microcephalus* IPNI=*34033-2* 282
Phlebodium araneosum IPNI=*17175380-1* 176
Phlebodium aureum var. *araneosum* IPNI=*17175380-1* 176
Phlebodium aureum var. *areolatum* IPNI=*17175380-1* 176
Phlebodium aureum var. *pulvinatum* IPNI=*17175390-1* 176
Phlebodium aureum var. *trilobatum* IPNI=*17175380-1* 176
Phlebodium dictyocallis IPNI=*17175400-1* 176
Phlebodium fulvum IPNI=*17175380-1* 176
Phlebodium grande IPNI=*17175390-1* 176
Phlebodium lanceolatum IPNI=*17448650-1* 176
Phlebodium lanosum IPNI=*17175380-1* 176
Phlebodium lycopodioides IPNI=*160142-2* 175
Phlebodium macrocarpum IPNI=*17448650-1* 176
Phlebodium microdictyum IPNI=*17175390-1* 176
Phlebodium multiseriale IPNI=*17175400-1* 176
Phlebodium nitidum IPNI=*17444940-1* 175
Phlebodium percussum IPNI=*17546710-1* 175
Phlebodium pseudoaureum IPNI=*17175380-1* 176
Phlebodium pulvinatum IPNI=*17175390-1* 176
Phlebodium salicifolium IPNI=*160142-2* 175
Phlebodium sporadocarpum IPNI=*17175380-1* 176
Phlebodium venosum IPNI=*160142-2* 175
Phlegmariurus attenuatus IPNI=*124842-2* 171
Phlegmariurus crassus IPNI=*124844-2* 171
Phlegmariurus dichotomus IPNI=*1090207-2* 171
Phlegmariurus saururus IPNI=*60447820-2* 171
Phleum africanum IPNI=*77105978-1* 235
Phleum crinitum IPNI=*1048074-2* 244
Phleum dactylon IPNI=*397065-1* 236
Phleum indicum IPNI=*963150-1* 244
Phleum monspeliense IPNI=*1048074-2* 244
Phlomis leonurus IPNI=*449133-1* 370
Phlomis nepetifolia IPNI=*127310-3* 370
Phlomis speciosa IPNI=*449133-1* 370
Phlox drummondii f. *albiflora* IPNI=*193463-2* 303
Phlox drummondii subsp. *eudrummondii* IPNI=*193463-2* 303
Phlox drummondii var. *goldsmithii* IPNI=*193463-2* 303
Phlox drummondii var. *peregrina* IPNI=*193463-2* 303
Phlox drummondii var. *typica* IPNI=*193463-2* 303
Phlox goldsmithii IPNI=*193463-2* 303
Phlox heynholdii IPNI=*193463-2* 303
Phoebe antillana IPNI=*77173628-1* 182
Phoebe antillana var. *cubensis* IPNI=*77173628-1* 182
Phoebe antillana var. *portoricensis* IPNI=*77173628-1* 182
Phoebe areolatocostae IPNI=*282532-2* 185
Phoebe brasiliensis IPNI=*77173628-1* 182
Phoebe cinnamomifolia IPNI=*77173628-1* 182
Phoebe corzoana IPNI=*326508-2* 184
Phoebe cubensis IPNI=*77173628-1* 182
Phoebe elongata IPNI=*77173628-1* 182
Phoebe elongata var. *elliptica* IPNI=*77173628-1* 182
Phoebe elongata var. *lanceolata* IPNI=*77173628-1* 182
Phoebe filamentosa IPNI=*77173628-1* 182
Phoebe fruticosa IPNI=*77173628-1* 182
Phoebe granatensis IPNI=*77173628-1* 182
Phoebe granatensis var. *oerstedii* IPNI=*77173628-1* 182
Phoebe helicterifolia IPNI=*326508-2* 184
Phoebe heterotepala IPNI=*77173628-1* 182
Phoebe insularis IPNI=*171697-2* 184
Phoebe johnstonii IPNI=*77173628-1* 182
Phoebe mayana IPNI=*171797-2* 184
Phoebe maynensis IPNI=*77173628-1* 182
Phoebe membranacea IPNI=*466773-1* 184
Phoebe mexicana IPNI=*77173628-1* 182
Phoebe mexicana var. *bourgeauana* IPNI=*77173628-1* 182
Phoebe montana IPNI=*77173628-1* 182
Phoebe montana var. *rigida* IPNI=*77173628-1* 182
Phoebe nectandroides IPNI=*326508-2* 184
Phoebe obtusata IPNI=*326508-2* 184
Phoebe paraguariensis IPNI=*77173628-1* 182
Phoebe peruviana IPNI=*77173628-1* 182
Phoebe peruviana var. *glabriflora* IPNI=*77173628-1* 182
Phoebe pichisensis IPNI=*77173628-1* 182
Phoebe pickellii IPNI=*77173628-1* 182
Phoebe poeppigii IPNI=*77173628-1* 182
Phoebe poiretiana IPNI=*77173628-1* 182
Phoebe rivularis IPNI=*171499-2* 184
Phoebe triplinervis IPNI=*77173628-1* 182
Phoebe triplinervis var. *cubensis* IPNI=*77173628-1* 182
Phoebe valenzuelana IPNI=*77173628-1* 182
Phoenix abyssinica IPNI=*668943-1* 211
Phoenix baoulensis IPNI=*668943-1* 211
Phoenix chevalieri IPNI=*668912-1* 211
Phoenix comorensis IPNI=*668943-1* 211
Phoenix cycadifolia IPNI=*668909-1* 211
Phoenix djalonensis IPNI=*668943-1* 211
Phoenix dybowskii IPNI=*668943-1* 211
Phoenix equinoxialis IPNI=*668943-1* 211
Phoenix erecta IPNI=*668909-1* 211
Phoenix excelsior IPNI=*668912-1* 211
Phoenix humilis IPNI=*665917-1* 208
Phoenix iberica IPNI=*668912-1* 211
Phoenix jubae IPNI=*668909-1* 211
Phoenix leonensis IPNI=*668943-1* 211
Phoenix macrocarpa IPNI=*668909-1* 211
Phoenix spinosa IPNI=*668943-1* 211
Phoenix vigieri IPNI=*668909-1* 211
Phoradendron antillarum f. *longum* IPNI=*60443757-2* 450
Phoradendron antillarum IPNI=*60443757-2* 450
Phoradendron antillarum var. *orientale* IPNI=*60443757-2* 450
Phoradendron apertum IPNI=*1075239-2* 450
Phoradendron appuni IPNI=*1075239-2* 450
Phoradendron ayliffii IPNI=*1006438-2* 450
Phoradendron baileyae IPNI=*1075239-2* 450
Phoradendron baileyae var. *exile* IPNI=*1075239-2* 450
Phoradendron belizense IPNI=*60443757-2* 450
Phoradendron biolleyi IPNI=*194192-2* 450
Phoradendron brachypodum IPNI=*550970-1* 450
Phoradendron caerulescens IPNI=*194156-2* 450
Phoradendron caracasanum IPNI=*194156-2* 450
Phoradendron cayanum IPNI=*60443757-2* 450
Phoradendron cearense IPNI=*194156-2* 450
Phoradendron ceibanum IPNI=*60443757-2* 450
Phoradendron chaguaramasanum IPNI=*194156-2* 450
Phoradendron chrysocarpum var. *stehlei* IPNI=*1006438-2* 450
Phoradendron chrysostachyum IPNI=*77735-2* 449
Phoradendron clavatum IPNI=*77736-2* 449
Phoradendron cocquericotanum IPNI=*60443757-2* 450
Phoradendron commutatum IPNI=*60443757-2* 450
Phoradendron crassifolium var. *multiflorum* IPNI=*550798-1* 449
Phoradendron crassifolium var. *parvifolium* IPNI=*550798-1* 449
Phoradendron crassifolium var. *pittieri* IPNI=*550798-1* 449
Phoradendron crassuloides IPNI=*77736-2* 449
Phoradendron degenianum IPNI=*194156-2* 450
Phoradendron dichotomum IPNI=*194192-2* 450
Phoradendron domingense IPNI=*1075239-2* 450
Phoradendron emarginatum IPNI=*194156-2* 450

Phoradendron emarginatum var. *obovatifolium* IPNI=*194156-2* 450
Phoradendron exile IPNI=*1075239-2* 450
Phoradendron ficulneum IPNI=*194192-2* 450
Phoradendron flavens var. *australe* IPNI=*329775-2* 449
Phoradendron flavens var. *longispicum* IPNI=*329775-2* 449
Phoradendron flavum IPNI=*329775-2* 449
Phoradendron franciscanum IPNI=*60443757-2* 450
Phoradendron fuscoerubescens IPNI=*1075239-2* 450
Phoradendron gaumeri IPNI=*60443757-2* 450
Phoradendron gentlei IPNI=*185262-2* 448
Phoradendron glaucolutescens IPNI=*194192-2* 450
Phoradendron globuliflorum IPNI=*77735-2* 449
Phoradendron glyptoneurum IPNI=*329775-2* 449
Phoradendron gracile IPNI=*60443757-2* 450
Phoradendron gracile var. *ballii* IPNI=*60443757-2* 450
Phoradendron gracilispicum IPNI=*1006438-2* 450
Phoradendron granaticola IPNI=*550970-1* 450
Phoradendron granaticola var. *taeniicaule* IPNI=*550970-1* 450
Phoradendron guazumae IPNI=*60443757-2* 450
Phoradendron hamatifolium IPNI=*1006438-2* 450
Phoradendron herminieri IPNI=*1006438-2* 450
Phoradendron herrerense IPNI=*60443757-2* 450
Phoradendron huehuetecum IPNI=*550970-1* 450
Phoradendron huricola IPNI=*60443757-2* 450
Phoradendron knoopii IPNI=*329775-2* 449
Phoradendron latifolium IPNI=*194192-2* 450
Phoradendron laurifolium IPNI=*194192-2* 450
Phoradendron libertadanum IPNI=*60443757-2* 450
Phoradendron ligustrinum IPNI=*176899-2* 448
Phoradendron lyonii IPNI=*194156-2* 450
Phoradendron macrophyllum IPNI=*1006438-2* 450
Phoradendron manatense IPNI=*60443757-2* 450
Phoradendron mandonii IPNI=*550970-1* 450
Phoradendron martianum IPNI=*60443757-2* 450
Phoradendron martinicense IPNI=*550798-1* 449
Phoradendron membranaceum IPNI=*329775-2* 449
Phoradendron millspaughii IPNI=*60443757-2* 450
Phoradendron morazanense IPNI=*194192-2* 450
Phoradendron moringicola IPNI=*60443757-2* 450
Phoradendron mucronatum var. *emarginatum* IPNI=*194156-2* 450
Phoradendron mucronatum var. *truncatoretusum* IPNI=*194156-2* 450
Phoradendron myrtilloides IPNI=*1075239-2* 450
Phoradendron oblongifolium IPNI=*1075239-2* 450
Phoradendron obovatifolium IPNI=*194156-2* 450
Phoradendron ottonis IPNI=*194156-2* 450
Phoradendron ottonis var. *oblongifolium* IPNI=*194156-2* 450
Phoradendron pachanum IPNI=*550970-1* 450
Phoradendron pepericarpum IPNI=*550798-1* 449
Phoradendron petenense IPNI=*60443757-2* 450
Phoradendron piauhyanum IPNI=*1075239-2* 450
Phoradendron pifoense IPNI=*550970-1* 450
Phoradendron piperoides f. *compositum* IPNI=*194192-2* 450
Phoradendron quadrangulare var. *gracile* IPNI=*60443757-2* 450
Phoradendron quinquenervium IPNI=*329775-2* 449
Phoradendron randiae f. *longum* IPNI=*60443757-2* 450
Phoradendron randiae IPNI=*60443757-2* 450
Phoradendron randiae var. *orientale* IPNI=*60443757-2* 450
Phoradendron rehderianum IPNI=*60443757-2* 450
Phoradendron rensonii IPNI=*60443757-2* 450
Phoradendron reticulatum IPNI=*329775-2* 449
Phoradendron roseanum IPNI=*550970-1* 450
Phoradendron rubrum var. *brevispicum* IPNI=*1075239-2* 450
Phoradendron rubrum var. *gracile* IPNI=*60443757-2* 450
Phoradendron rubrum var. *latifolium* IPNI=*1075239-2* 450
Phoradendron rubrum var. *longipedunculatum* IPNI=*1075239-2* 450
Phoradendron rubrum var. *longispicum* IPNI=*1075239-2* 450
Phoradendron rusbyanum IPNI=*60443757-2* 450
Phoradendron salicifolium IPNI=*60443757-2* 450
Phoradendron sanctae-martae IPNI=*1075239-2* 450
Phoradendron sauruoides IPNI=*194192-2* 450
Phoradendron saxicola IPNI=*1075239-2* 450
Phoradendron semiteres IPNI=*550970-1* 450
Phoradendron sonanum IPNI=*60443757-2* 450
Phoradendron supravenulosum IPNI=*329775-2* 449
Phoradendron taeniicaule IPNI=*550970-1* 450
Phoradendron tafallaeoides IPNI=*77147531-1* 448
Phoradendron tamaulipense IPNI=*60443757-2* 450
Phoradendron tereticaule IPNI=*194192-2* 450
Phoradendron tetragonum IPNI=*194156-2* 450
Phoradendron theloneuron IPNI=*1075239-2* 450
Phoradendron torulosum IPNI=*194192-2* 450
Phoradendron townsendii IPNI=*60443757-2* 450
Phoradendron treleasei IPNI=*1075239-2* 450
Phoradendron trianae IPNI=*550970-1* 450
Phoradendron trinervium var. *domingense* IPNI=*1075239-2* 450
Phoradendron trisulcatum IPNI=*329775-2* 449
Phoradendron tubulosum IPNI=*550970-1* 450
Phoradendron turbinispicum IPNI=*550970-1* 450
Phoradendron uleanum IPNI=*1006438-2* 450
Phoradendron urbanianum IPNI=*329775-2* 449
Phoradendron venezuelense IPNI=*60443757-2* 450
Phoradendron venezuelense var. *filispicum* IPNI=*60443757-2* 450
Phoradendron verleysenii IPNI=*550970-1* 450
Phoradendron verticillatum var. *domingense* IPNI=*1075239-2* 450
Phoradendron viscifolium IPNI=*60443757-2* 450
Phoradendron yucatanum IPNI=*194156-2* 450
Phoradendron zacapanum IPNI=*60443757-2* 450
Phoradendron zuloagae IPNI=*1075239-2* 450
Phormium ramosum IPNI=*539926-1* 216
Photinia crenatoserrata IPNI=*730613-1* 444
Photinia fortuneana IPNI=*730613-1* 444
Photinia sieboldii IPNI=*731389-1* 444
Phragmocassia skinneri IPNI=*911412-1* 333
Phryganocydia coito IPNI=*60474996-2* 362
Phryganocydia coito var. *platypoda* IPNI=*60474996-2* 362
Phryganocydia corymbosa IPNI=*60474996-2* 362
Phryganocydia orinocensis IPNI=*60474996-2* 362
Phrygilanthus acutifolius IPNI=*258636-2* 449
Phrygilanthus acutifolius subsp. *coriaceus* IPNI=*258636-2* 449
Phrygilanthus acutifolius subsp. *eugenioides* IPNI=*258636-2* 449
Phrygilanthus bolivianus IPNI=*258636-2* 449
Phrygilanthus cordifolius IPNI=*258636-2* 449
Phrygilanthus ellipticus IPNI=*548708-1* 448
Phrygilanthus eugenioides IPNI=*258636-2* 449
Phrygilanthus eugenioides var. *robustus* IPNI=*258636-2* 449
Phrygilanthus falcatus IPNI=*551367-1* 449
Phrygilanthus falcatus var. *macrocalyx* IPNI=*551367-1* 449
Phrygilanthus lanceolatus IPNI=*548708-1* 448
Phrygilanthus mapirensis IPNI=*258636-2* 449
Phrygilanthus punctatus IPNI=*548708-1* 448
Phrygilanthus puracensis IPNI=*548708-1* 448
Phrygilanthus suaveolens IPNI=*258636-2* 449
Phrygilanthus tagua IPNI=*548708-1* 448
Phryne laxata IPNI=*237940-2* 278
Phryne vulgaris IPNI=*289589-1* 278
Phrynium allouia IPNI=*60458762-2* 345
Phrynium bicolor IPNI=*796810-1* 249
Phrynium casupo IPNI=*60458866-2* 249
Phrynium luteum IPNI=*60458866-2* 249
Phrynium micans IPNI=*60461053-2* 249
Phrynium myrosma IPNI=*797633-1* 250
Phrynium ornatum IPNI=*60461169-2* 249
Phrynium propinquum IPNI=*60461303-2* 249
Phrynium pulchellum IPNI=*796810-1* 249
Phrynium pumilum IPNI=*60461053-2* 249
Phrynium pusillum IPNI=*60461053-2* 249

Phrynium reniforme IPNI=*698080-1* 225
Phrynium reniforme var. *acutum* IPNI=*698080-1* 225
Phrynium rotundifolium IPNI=*797124-1* 249
Phrynium strobiliferum IPNI=*218597-2* 251
Phrynium trifasciatum IPNI=*60461303-2* 249
Phrynium variegatum IPNI=*280408-2* 250
Phrynium zebrinum IPNI=*796810-1* 249
Phthirusa abdita IPNI=*77147531-1* 448
Phthirusa adenostemon IPNI=*77147531-1* 448
Phthirusa adenostemon var. *huberi* IPNI=*77147531-1* 448
Phthirusa adunca f. *magnifolia* IPNI=*77147531-1* 448
Phthirusa adunca IPNI=*77147531-1* 448
Phthirusa adunca var. *magdalenae* IPNI=*77147531-1* 448
Phthirusa adunca var. *orinocensis* IPNI=*77147531-1* 448
Phthirusa adunca var. *rigidifolia* IPNI=*77147531-1* 448
Phthirusa amazonensis IPNI=*77147531-1* 448
Phthirusa anastyla IPNI=*77147531-1* 448
Phthirusa bauhiniae IPNI=*77147531-1* 448
Phthirusa brasiliensis IPNI=*185262-2* 448
Phthirusa caucana IPNI=*77147531-1* 448
Phthirusa cochliostyla IPNI=*77147531-1* 448
Phthirusa cothurnata IPNI=*77147531-1* 448
Phthirusa domingensis IPNI=*1075239-2* 450
Phthirusa elongata IPNI=*77147531-1* 448
Phthirusa erythrocarpus IPNI=*77147531-1* 448
Phthirusa geniculifera IPNI=*77147531-1* 448
Phthirusa gonioclada IPNI=*77147531-1* 448
Phthirusa heterophylla IPNI=*185262-2* 448
Phthirusa jamaicensis IPNI=*77147531-1* 448
Phthirusa krukovii IPNI=*77147531-1* 448
Phthirusa lepidobotrys IPNI=*185262-2* 448
Phthirusa lucarquensis IPNI=*551737-1* 449
Phthirusa magdalenae IPNI=*77147531-1* 448
Phthirusa maritima IPNI=*77147531-1* 448
Phthirusa orbicularis IPNI=*551737-1* 449
Phthirusa orinocensis IPNI=*77147531-1* 448
Phthirusa paniculata IPNI=*77147531-1* 448
Phthirusa papillosa IPNI=*77147531-1* 448
Phthirusa perdivergens IPNI=*77147531-1* 448
Phthirusa platyclada IPNI=*185262-2* 448
Phthirusa polystachya IPNI=*77147531-1* 448
Phthirusa punctata IPNI=*77147531-1* 448
Phthirusa pyramidalis IPNI=*77147531-1* 448
Phthirusa pyrifolia IPNI=*185262-2* 448
Phthirusa pyrifolia var. *grandifolia* IPNI=*185262-2* 448
Phthirusa pyrifolia var. *intermedia* IPNI=*185262-2* 448
Phthirusa pyrifolia var. *parvifolia* IPNI=*185262-2* 448
Phthirusa pyrifolia var. *terminalis* IPNI=*185262-2* 448
Phthirusa retroflexa IPNI=*77147531-1* 448
Phthirusa rubromicans IPNI=*77147531-1* 448
Phthirusa seitzii IPNI=*77147531-1* 448
Phthirusa stelis IPNI=*77147531-1* 448
Phthirusa theobromae IPNI=*77147531-1* 448
Phthirusa tortuosa IPNI=*77147531-1* 448
Phthirusa virgata IPNI=*77147531-1* 448
Phyla dulcis IPNI=*141103-2* 379
Phyla scaberrima IPNI=*141103-2* 379
Phyllactis mutisiana IPNI=*860264-1* 296
Phyllactis rigida IPNI=*286650-2* 296
Phyllanthus acidissimus IPNI=*194610-2* 401
Phyllanthus acuminatus var. *paraguariensis* IPNI=*194611-2* 402
Phyllanthus amarus subsp. *sanyaensis* IPNI=*326408-2* 402
Phyllanthus antillanus IPNI=*352077-1* 401
Phyllanthus antillanus var. *concolor* IPNI=*352077-1* 401
Phyllanthus antillanus var. *hypomalacus* IPNI=*352077-1* 401
Phyllanthus antillanus var. *pedicellaris* IPNI=*352077-1* 401
Phyllanthus aquaticus IPNI=*195027-2* 402
Phyllanthus averrhoifolius IPNI=*194611-2* 402
Phyllanthus brasiliensis var. *genuinus* IPNI=*353613-1* 402
Phyllanthus cicca IPNI=*194610-2* 401
Phyllanthus cochinchinensis IPNI=*194610-2* 401
Phyllanthus conami IPNI=*353613-1* 402
Phyllanthus cumanensis IPNI=*194611-2* 402
Phyllanthus diffusus IPNI=*195027-2* 402
Phyllanthus diffusus var. *genuinus* IPNI=*195027-2* 402
Phyllanthus distichus IPNI=*194610-2* 401
Phyllanthus floribundus IPNI=*354635-1* 402
Phyllanthus foetidus IPNI=*194611-2* 402
Phyllanthus gymnanthus IPNI=*339921-1* 401
Phyllanthus heteromorphus IPNI=*352077-1* 401
Phyllanthus hoffmannseggii IPNI=*195027-2* 402
Phyllanthus hoffmeisteri IPNI=*195027-2* 402
Phyllanthus ibonensis IPNI=*352077-1* 401
Phyllanthus kunthiana IPNI=*354635-1* 402
Phyllanthus laurifolius IPNI=*356125-1* 402
Phyllanthus longifolius IPNI=*194610-2* 401
Phyllanthus lycioides IPNI=*194611-2* 402
Phyllanthus microphyllus IPNI=*195027-2* 402
Phyllanthus mucronatus IPNI=*194611-2* 402
Phyllanthus nanus IPNI=*326408-2* 402
Phyllanthus neogranatensis IPNI=*156150-2* 401
Phyllanthus niruri var. *genuinus* IPNI=*194900-2* 402
Phyllanthus niruri var. *scabrellus* IPNI=*326408-2* 402
Phyllanthus nivosus IPNI=*340073-1* 401
Phyllanthus nobilis IPNI=*352077-1* 401
Phyllanthus nobilis var. *antillanus* IPNI=*352077-1* 401
Phyllanthus nobilis var. *genuinus* IPNI=*352077-1* 401
Phyllanthus nobilis var. *guyanensis* IPNI=*352077-1* 401
Phyllanthus nobilis var. *hypomalacus* IPNI=*352077-1* 401
Phyllanthus nobilis var. *martii* IPNI=*352077-1* 401
Phyllanthus nobilis var. *panamensis* IPNI=*352077-1* 401
Phyllanthus nobilis var. *pavonianus* IPNI=*352077-1* 401
Phyllanthus nobilis var. *peruvianus* IPNI=*352077-1* 401
Phyllanthus nobilis var. *riedelianus* IPNI=*352077-1* 401
Phyllanthus orbiculatus var. *acutifolius* IPNI=*194933-2* 402
Phyllanthus orbiculatus var. *genuinus* IPNI=*194933-2* 402
Phyllanthus orbiculatus var. *intermedius* IPNI=*194933-2* 402
Phyllanthus orbiculatus var. *lignescens* IPNI=*194933-2* 402
Phyllanthus orbiculatus var. *rupestris* IPNI=*194933-2* 402
Phyllanthus poiretianus IPNI=*194933-2* 402
Phyllanthus poiretianus var. *acutifolius* IPNI=*194933-2* 402
Phyllanthus pseudonobilis IPNI=*352077-1* 401
Phyllanthus pubiger IPNI=*356125-1* 402
Phyllanthus ruscoides IPNI=*194611-2* 402
Phyllanthus salviifolius var. *floribundus* IPNI=*354635-1* 402
Phyllanthus salviifolius var. *genuinus* IPNI=*354635-1* 402
Phyllanthus salviifolius var. *glabrescens* IPNI=*354635-1* 402
Phyllanthus scabrellus IPNI=*326408-2* 402
Phyllanthus sessei IPNI=*194611-2* 402
Phyllanthus sinicus IPNI=*352077-1* 401
Phyllanthus swartzii IPNI=*326408-2* 402
Phyllaurea variegata IPNI=*85073-3* 388
Phyllimena paludosa IPNI=*203207-1* 261
Phyllocactus gaillardae IPNI=*132939-1* 283
Phyllocactus hookeri IPNI=*132926-1* 283
Phyllocactus marginatus IPNI=*132926-1* 283
Phyllocactus phyllanthus IPNI=*132939-1* 283
Phyllocactus phyllanthus var. *boliviensis* IPNI=*132939-1* 283
Phyllocactus phyllanthus var. *paraguayensis* IPNI=*132939-1* 283
Phyllocactus ruestii IPNI=*132926-1* 283
Phyllocactus speciosus IPNI=*132939-1* 283
Phyllocactus stenopetalus IPNI=*132926-1* 283
Phyllocactus strictus IPNI=*132926-1* 283
Phyllocarpus pterocarpus IPNI=*491205-1* 317
Phyllodes albicans IPNI=*60461053-2* 249
Phyllodes allouia IPNI=*60458762-2* 345

Phyllodes arrecta IPNI=*60461169-2* 249
Phyllodes bachemiana IPNI=*60458927-2* 249
Phyllodes insignis IPNI=*273648-2* 249
Phyllodes lietzei IPNI=*77174968-1* 249
Phyllodes lutea IPNI=*60458866-2* 249
Phyllodes micans IPNI=*60461053-2* 249
Phyllodes myrosma IPNI=*797633-1* 250
Phyllodes ornata IPNI=*60461169-2* 249
Phyllodes platyphylla IPNI=*60459373-2* 249
Phyllodes princeps IPNI=*60461040-2* 249
Phyllodes propinqua IPNI=*60461303-2* 249
Phyllodes pulchella IPNI=*796810-1* 249
Phyllodes roseopicta IPNI=*60461319-2* 249
Phyllodes strobilifera IPNI=*218597-2* 251
Phyllodes trifasciata IPNI=*60461303-2* 249
Phyllodes zebrina IPNI=*796810-1* 249
Phyllodoce longifolia IPNI=*470782-1* 310
Phyllodoce verticillata IPNI=*471751-1* 310
Phyllodolon fistulosum IPNI=*528046-1* 213
Phyllostachys breviligula IPNI=*415983-1* 244
Phyllostachys formosana IPNI=*415983-1* 244
Phyllostachys meyeri f. *sphaeroidea* IPNI=*416023-1* 244
Phyllostachys mitis IPNI=*392574-1* 235
Phyllostylon brasiliense IPNI=*855199-1* 445
Phyllostylon orthopterum IPNI=*855199-1* 445
Phyllotaenium lindenii IPNI=*40313-2* 199
Phymaspermum carnosulum IPNI=*178745-1* 257
Phymatodes acrosora IPNI=*77183236-1* 175
Phymatodes crassifolia IPNI=*77183236-1* 175
Phymatodes irioides IPNI=*17341250-1* 176
Phymatodes lingulata IPNI=*17341250-1* 176
Phymatodes lycopodioides IPNI=*160142-2* 175
Phymatodes palmeri IPNI=*17444940-1* 175
Phymatodes polycarpa IPNI=*17341250-1* 176
Phymatodes porrecta IPNI=*77183236-1* 175
Phymatodes prominula IPNI=*160142-2* 175
Phymatodes secunda IPNI=*17458930-1* 174
Phymatodes sessilis IPNI=*17341250-1* 176
Phyodina gracilis IPNI=*42443-2* 223
Physidium gardneri IPNI=*798952-1* 375
Physurus commelinifolius IPNI=*77125705-1* 217
Physurus debilis IPNI=*77125687-1* 217
Physurus humidicola IPNI=*77125705-1* 217
Physurus jamaicensis IPNI=*77125705-1* 217
Physurus macer IPNI=*77125687-1* 217
Physurus maculatus IPNI=*77125705-1* 217
Physurus mayorianus IPNI=*77125705-1* 217
Physurus parviflorus IPNI=*77125705-1* 217
Physurus querceticola IPNI=*77125705-1* 217
Physurus sagranus IPNI=*77125705-1* 217
Physurus trilobulatus IPNI=*77125705-1* 217
Physurus vaginatus IPNI=*77125705-1* 217
Phytelephas brachelus IPNI=*668996-1* 211
Phytelephas brachinus IPNI=*668996-1* 211
Phytelephas brevipes IPNI=*668996-1* 211
Phytelephas cornutus IPNI=*668996-1* 211
Phytelephas dasyneura IPNI=*11322-2* 204
Phytelephas decasperma IPNI=*11322-2* 204
Phytelephas karstenii IPNI=*668996-1* 211
Phytelephas longiflora IPNI=*668996-1* 211
Phytelephas macrocarpa subsp. *schottii* IPNI=*668996-1* 211
Phytelephas macrocarpa subsp. *tenuicaulis* IPNI=*981519-1* 211
Phytelephas microcarpa IPNI=*668996-1* 211
Phytelephas pittieri IPNI=*668996-1* 211
Phytelephas schottii IPNI=*668996-1* 211
Phytelephas seemannii IPNI=*668996-1* 211
Phytelephas seemannii subsp. *brevipes* IPNI=*668996-1* 211
Phytolacca acuminata IPNI=*323292-2* 287
Phytolacca altamiranii IPNI=*323292-2* 287
Phytolacca americana var. *lancifolia* IPNI=*323290-2* 287
Phytolacca arborea IPNI=*676348-1* 287
Phytolacca australis IPNI=*323291-2* 287
Phytolacca costaricensis IPNI=*676398-1* 288
Phytolacca decandra IPNI=*323290-2* 287
Phytolacca dioica var. *ovatifolia* IPNI=*676348-1* 287
Phytolacca erubescens IPNI=*323292-2* 287
Phytolacca icosandra var. *angustitepala* IPNI=*323292-2* 287
Phytolacca icosandra var. *fraseri* IPNI=*676397-1* 287
Phytolacca icosandra var. *sessiliflora* IPNI=*323292-2* 287
Phytolacca javanica IPNI=*171034-1* 416
Phytolacca longespica IPNI=*323292-2* 287
Phytolacca macrostachya IPNI=*323292-2* 287
Phytolacca malabarica IPNI=*323292-2* 287
Phytolacca mexicana IPNI=*323292-2* 287
Phytolacca micrantha IPNI=*323291-2* 287
Phytolacca nova-hispania IPNI=*323292-2* 287
Phytolacca octandra var. *angustifolia* IPNI=*323292-2* 287
Phytolacca octandra var. *grandifolia* IPNI=*323293-2* 287
Phytolacca parviflora IPNI=*323291-2* 287
Phytolacca polystigma IPNI=*323292-2* 287
Phytolacca polystyla IPNI=*676397-1* 287
Phytolacca populifolia IPNI=*676348-1* 287
Phytolacca purpurascens IPNI=*323292-2* 287
Phytolacca sessiliflora IPNI=*323292-2* 287
Phytolacca thyrsiflora var. *reducta* IPNI=*30307474-2* 288
Phytolacca triquetra IPNI=*323292-2* 287
Phytolacca venezuelensis IPNI=*323292-2* 287
Phytolacca vulgaris IPNI=*323290-2* 287
Phytolacca weberbaueri IPNI=*676348-1* 287
Piaropus azureus IPNI=*698138-1* 225
Piaropus crassipes IPNI=*310928-2* 225
Piaropus diversifolius IPNI=*77188085-1* 225
Piaropus mesomelas IPNI=*310928-2* 225
Piaropus paniculatus IPNI=*698175-1* 225
Piaropus tricolor IPNI=*698138-1* 225
Piaropus undulatus IPNI=*698138-1* 225
Picramnia antidesma var. *carpinterae* IPNI=*813938-1* 433
Picramnia antidesma var. *microphylla* IPNI=*813869-1* 433
Picramnia antidesma var. *normalis* IPNI=*813869-1* 433
Picramnia antidesmoides IPNI=*813923-1* 433
Picramnia carpinterae IPNI=*813938-1* 433
Picramnia cooperi IPNI=*813902-1* 433
Picramnia corallodendron IPNI=*813935-1* 433
Picramnia corallodendron var. *peruviana* IPNI=*813935-1* 433
Picramnia cubensis IPNI=*813923-1* 433
Picramnia dwyeri IPNI=*813896-1* 433
Picramnia eosina IPNI=*813902-1* 433
Picramnia kunthii IPNI=*813935-1* 433
Picramnia longifolia IPNI=*813938-1* 433
Picramnia longissima IPNI=*813902-1* 433
Picramnia macrostachys IPNI=*813902-1* 433
Picramnia micrantha IPNI=*813923-1* 433
Picramnia monninifolia IPNI=*813896-1* 433
Picramnia oblongifolia IPNI=*813923-1* 433
Picramnia pendula IPNI=*301399-2* 433
Picramnia sellowii f. *glabrescens* IPNI=*301399-2* 433
Picramnia sellowii f. *hirsuta* IPNI=*301399-2* 433
Picramnia sellowii f. *intermedia* IPNI=*301399-2* 433
Picramnia sellowii var. *latifolia* IPNI=*301399-2* 433
Picramnia triandra IPNI=*813869-1* 433
Picramnia umbrosa IPNI=*813902-1* 433
Picrium spicatum IPNI=*367036-1* 344
Piercea acuminata IPNI=*221479-2* 287
Piercea glabra IPNI=*221479-2* 287
Piercea obliquata IPNI=*221479-2* 287
Piercea tomentosa IPNI=*221479-2* 287

Pilderia hirsuta IPNI=*30121655-2* 292
Pilea aripoensis IPNI=*1162156-2* 447
Pilea callitrichoides IPNI=*1162156-2* 447
Pilea chloroclada IPNI=*30064315-2* 447
Pilea chrysosplenioides IPNI=*77091436-1* 447
Pilea ciliaris IPNI=*30064315-2* 447
Pilea ciliaris var. *tumida* IPNI=*30064315-2* 447
Pilea ciliaris var. *wydleri* IPNI=*30064315-2* 447
Pilea dendrophila IPNI=*321319-2* 447
Pilea dendrophila var. *canescens* IPNI=*321319-2* 447
Pilea dendrophila var. *subintegra* IPNI=*321319-2* 447
Pilea diffusa IPNI=*855698-1* 447
Pilea elongata IPNI=*855573-1* 447
Pilea fuscata IPNI=*855698-1* 447
Pilea gaudichaudiana IPNI=*855698-1* 447
Pilea grossecrenata IPNI=*855698-1* 447
Pilea guyanensis IPNI=*855698-1* 447
Pilea herniarifolia IPNI=*1162156-2* 447
Pilea hyalina var. *longipes* IPNI=*855481-1* 447
Pilea imparifolia var. *canescens* IPNI=*321319-2* 447
Pilea integrifolia IPNI=*30064315-2* 447
Pilea involucrata IPNI=*855698-1* 447
Pilea litoralis IPNI=*77091436-1* 447
Pilea lundii IPNI=*855481-1* 447
Pilea microphylla var. *longifolia* IPNI=*1162156-2* 447
Pilea microphylla var. *portulacoides* IPNI=*1162156-2* 447
Pilea muscosa IPNI=*1162156-2* 447
Pilea muscosa var. *microphylla* IPNI=*1162156-2* 447
Pilea muscosa var. *portulacoides* IPNI=*1162156-2* 447
Pilea oreophila IPNI=*855573-1* 447
Pilea ovatinucula IPNI=*855573-1* 447
Pilea parietaria var. *hispaniolensis* IPNI=*30064315-2* 447
Pilea parietaria var. *wydleri* IPNI=*30064315-2* 447
Pilea pellucida IPNI=*855573-1* 447
Pilea peperomiifolia IPNI=*1162156-2* 447
Pilea portula IPNI=*1162156-2* 447
Pilea portulacina IPNI=*1162156-2* 447
Pilea pubescens var. *fuscata* IPNI=*855698-1* 447
Pilea pubescens var. *guyanensis* IPNI=*855698-1* 447
Pilea pubescens var. *involucrata* IPNI=*77091436-1* 447
Pilea pubescens var. *rupicola* IPNI=*855698-1* 447
Pilea puncticaulis IPNI=*30064315-2* 447
Pilea rubiifolia IPNI=*30064315-2* 447
Pilea rupicola IPNI=*855698-1* 447
Pilea scrobiculata IPNI=*855481-1* 447
Pilea subcrenata IPNI=*1162156-2* 447
Pilea succulenta IPNI=*1162156-2* 447
Pilea trianthmoides var. *microphylla* IPNI=*1162156-2* 447
Pilea trinervia IPNI=*855573-1* 447
Pilea wydleri IPNI=*30064315-2* 447
Pilea xalapensis IPNI=*855698-1* 447
Pileocalyx elegans IPNI=*30006538-2* 293
Pilocereus atroviridis IPNI=*130627-1* 283
Pilocereus backebergii IPNI=*196706-2* 284
Pilocereus campinensis IPNI=*1001185-1* 284
Pilocereus colombianus IPNI=*196706-2* 284
Pilocereus consolei IPNI=*196706-2* 284
Pilocereus deficiens IPNI=*244634-2* 285
Pilocereus flavispinus IPNI=*137078-1* 284
Pilocereus fricii IPNI=*130627-1* 283
Pilocereus lanuginosus IPNI=*196706-2* 284
Pilocereus lanuginosus var. *virens* IPNI=*196706-2* 284
Pilocereus nobilis IPNI=*137078-1* 284
Pilocereus perlucens IPNI=*130283-1* 283
Pilocereus remolinensis IPNI=*130627-1* 283
Pilocereus repandus IPNI=*130627-1* 283
Pilocereus russelianus IPNI=*52982-2* 283
Pilocereus russelianus subsp. *margaritensis* IPNI=*130627-1* 283
Pilocereus swartzii IPNI=*196706-2* 284
Pilophora saccifera IPNI=*668116-1* 210
Pilophora testicularis IPNI=*668116-1* 210
Pilosocereus panchesiorum IPNI=*196706-2* 284
Pilosocereus perlucens IPNI=*130283-1* 283
Pilumna fragrans IPNI=*660593-1* 223
Pilumna nobilis IPNI=*660593-1* 223
Pimecaria odorata IPNI=*166220-2* 339
Pimenta aromatica IPNI=*196799-2* 428
Pimenta communis IPNI=*196799-2* 428
Pimenta officinalis IPNI=*196799-2* 428
Pimenta officinalis var. *longifolia* IPNI=*196799-2* 428
Pimenta officinalis var. *ovalifolia* IPNI=*196799-2* 428
Pimenta pimenta IPNI=*196799-2* 428
Pimenta vulgaris IPNI=*196799-2* 428
Pimentus aromatica IPNI=*196799-2* 428
Pimentus geminata IPNI=*196799-2* 428
Pimentus vera IPNI=*196799-2* 428
Pimpinele anisa IPNI=*846658-1* 252
Pimpinella capillacea IPNI=*837486-1* 251
Pimpinella dichotoma IPNI=*837486-1* 251
Pimpinella dioica IPNI=*837486-1* 251
Pimpinella domingensis IPNI=*840899-1* 252
Pimpinella lateriflora IPNI=*837486-1* 251
Pimpinella leptophylla IPNI=*840899-1* 252
Pimpinella minor IPNI=*741394-1* 445
Pimpinella officinalis IPNI=*741402-1* 445
Pimpinella palmetorum IPNI=*846658-1* 252
Pimpinella ramosissima IPNI=*837486-1* 251
Pimpinella sanguisorba IPNI=*741394-1* 445
Pinanga purpurea IPNI=*666411-1* 208
Pinanga rubricaulis IPNI=*666411-1* 208
Pinardia coronaria IPNI=*1013480-1* 263
Pinardia frutescens IPNI=*178745-1* 257
Pinardia segetum IPNI=*209007-1* 263
Pinckneya ionantha IPNI=*204802-2* 355
Pineda lindenii IPNI=*202149-2* 404
Pingraea latifolia IPNI=*183399-1* 257
Pingraea rhexioides IPNI=*183821-1* 258
Pingraea sagittalis IPNI=*30159244-2* 258
Pingraea salicifolia IPNI=*183704-1* 258
Pinus araucana IPNI=*261681-1* 180
Pinus ayacahuite subsp. *neorecurvata* IPNI=*314973-2* 180
Pinus ayacahuite subsp. *oaxacana* IPNI=*314973-2* 180
Pinus ayacahuite var. *oaxacana* IPNI=*314973-2* 180
Pinus borealis IPNI=*263353-1* 180
Pinus colorado IPNI=*314973-2* 180
Pinus columbaria IPNI=*676669-1* 180
Pinus digenea IPNI=*677083-1* 180
Pinus dioica IPNI=*17518-2* 180
Pinus don-pedrii IPNI=*314973-2* 180
Pinus durangensis IPNI=*314973-2* 180
Pinus echinata IPNI=*677083-1* 180
Pinus glomerata IPNI=*263216-1* 180
Pinus hamata IPNI=*314973-2* 180
Pinus magellensis IPNI=*677083-1* 180
Pinus montana IPNI=*677083-1* 180
Pinus mughus IPNI=*677083-1* 180
Pinus mugo f. *applanata* IPNI=*677083-1* 180
Pinus pumilio IPNI=*677083-1* 180
Pinus rostrata IPNI=*677083-1* 180
Pinus rubriflora IPNI=*677083-1* 180
Pinus sylvestris var. *montana* IPNI=*677083-1* 180
Pinus wettsteinii IPNI=*677083-1* 180
Pinzona calineoides IPNI=*317113-1* 295
Piofontia anactinota IPNI=*77203458-1* 265
Piofontia costaricensis IPNI=*230916-1* 265
Piofontia cyparissias IPNI=*77203471-1* 265

Piofontia nevadensis IPNI=*77203490-1* 265
Piofontia revoluta IPNI=*77203500-1* 265
Piofontia rosmarinifolia IPNI=*230916-1* 265
Piofontia weddellii IPNI=*77203515-1* 265
Piper abditum IPNI=*681027-1* 195
Piper acuminatissimum IPNI=*681128-1* 196
Piper acuminatum IPNI=*30001267-2* 193
Piper acuminatum IPNI=*677918-1* 193
Piper acutifolium IPNI=*683806-1* 197
Piper acutifolium var. *membranaceum* IPNI=*680296-1* 195
Piper adenophorum IPNI=*681251-1* 196
Piper aduncifolium IPNI=*680296-1* 195
Piper aduncum IPNI=*681674-1* 196
Piper aduncum subsp. *ossanum* IPNI=*680296-1* 195
Piper aduncum var. *brachyarthrum* IPNI=*680296-1* 195
Piper aduncum var. *laevifolium* IPNI=*680296-1* 195
Piper aduncum var. *laevilimbum* IPNI=*680296-1* 195
Piper aduncum var. *sigualense* IPNI=*680296-1* 195
Piper aegrum IPNI=*681674-1* 196
Piper aemulum IPNI=*1068802-2* 194
Piper aequale f. *acutispicum* IPNI=*680300-1* 195
Piper aequale var. *dussii* IPNI=*680300-1* 195
Piper aequale var. *fimbribracteatum* IPNI=*680300-1* 195
Piper aequale var. *hirtinervulum* IPNI=*680300-1* 195
Piper aequale var. *lasiocarpum* IPNI=*680300-1* 195
Piper aequale var. *latum* IPNI=*680300-1* 195
Piper aequale var. *ovalifolium* IPNI=*680300-1* 195
Piper aequale var. *santa-martanum* IPNI=*680300-1* 195
Piper aequale var. *typicum* IPNI=*680300-1* 195
Piper aequale var. *variifolium* IPNI=*680300-1* 195
Piper aequilaterum IPNI=*682429-1* 196
Piper affine IPNI=*679597-1* 194
Piper afzelianum IPNI=*316339-2* 197
Piper aggregatum IPNI=*134457-2* 394
Piper alare IPNI=*682138-1* 196
Piper alatum IPNI=*677941-1* 193
Piper alluvicola IPNI=*681674-1* 196
Piper alpinum IPNI=*680559-1* 195
Piper alstonii IPNI=*680520-1* 195
Piper alveolatifolium IPNI=*680680-1* 195
Piper amalago var. *medium* IPNI=*680359-1* 195
Piper amazonicum IPNI=*681027-1* 195
Piper ambiguum IPNI=*680359-1* 195
Piper amnigaudens IPNI=*681674-1* 196
Piper andicola IPNI=*680559-1* 195
Piper angremondii IPNI=*681674-1* 196
Piper angulatum IPNI=*679591-1* 194
Piper angustatum IPNI=*677986-1* 193
Piper angustifolium IPNI=*680296-1* 195
Piper angustifolium var. *ossanum* IPNI=*680296-1* 195
Piper angustilimbum IPNI=*681193-1* 196
Piper anisatum IPNI=*682138-1* 196
Piper annulatum IPNI=*681041-1* 195
Piper anonifolium f. *angustifolium* IPNI=*1003227-1* 195
Piper anonifolium f. *parvifolium* IPNI=*1003227-1* 195
Piper anonifolium var. *latifolium* IPNI=*1003227-1* 195
Piper apiculatum IPNI=*681224-1* 196
Piper aragonense IPNI=*682301-1* 196
Piper arbelaezii var. *flavibracteum* IPNI=*682429-1* 196
Piper arboreum f. *geniculatum* IPNI=*680435-1* 195
Piper arboreum subsp. *stamineum* IPNI=*680435-1* 195
Piper arboreum subsp. *tuberculatum* IPNI=*683806-1* 197
Piper arboreum var. *giganteum* IPNI=*680435-1* 195
Piper arboreum var. *paucinervium* IPNI=*680435-1* 195
Piper arboreum var. *stamineum* IPNI=*680435-1* 195
Piper archidonense IPNI=*198173-2* 196
Piper arieianum f. *barbulaense* IPNI=*683782-1* 197
Piper arieianum f. *falconense* IPNI=*683782-1* 197
Piper arieianum f. *puberulum* IPNI=*683782-1* 197
Piper arieianum var. *chimantanum* IPNI=*683782-1* 197
Piper arieianum var. *duidaense* IPNI=*683782-1* 197
Piper aripoense IPNI=*681157-1* 196
Piper aristeguietae IPNI=*682429-1* 196
Piper aristolochiifolium IPNI=*682301-1* 196
Piper aromaticum IPNI=*678848-1* 194
Piper aromaticum IPNI=*682369-1* 196
Piper articulosum IPNI=*681674-1* 196
Piper asperifolium IPNI=*681674-1* 196
Piper asperrimicaule IPNI=*681674-1* 196
Piper asperum IPNI=*681674-1* 196
Piper atlantidanum var. *yoroense* IPNI=*682899-1* 197
Piper atrorameum IPNI=*680559-1* 195
Piper attenuatum IPNI=*681876-1* 196
Piper augustum var. *magnifolium* IPNI=*680507-1* 195
Piper augustum var. *pubinerve* IPNI=*680507-1* 195
Piper auristellanum IPNI=*681251-1* 196
Piper auritilaminum IPNI=*680520-1* 195
Piper auritilimbum IPNI=*680520-1* 195
Piper auritum var. *amplifolium* IPNI=*680520-1* 195
Piper auritum var. *schiedeanum* IPNI=*680520-1* 195
Piper auritum var. *seemannianum* IPNI=*680520-1* 195
Piper baccatum IPNI=*682369-1* 196
Piper bahiense IPNI=*680296-1* 195
Piper balbisianum IPNI=*681193-1* 196
Piper barbatum var. *andicola* IPNI=*680559-1* 195
Piper bartlingianum var. *parimanum* IPNI=*680574-1* 195
Piper baryanum IPNI=*681027-1* 195
Piper basellifolium IPNI=*677918-1* 193
Piper bayamonanum IPNI=*681674-1* 196
Piper belemense IPNI=*681224-1* 196
Piper benianum IPNI=*680748-1* 195
Piper benthamianum IPNI=*683316-1* 197
Piper berlandieri f. *nova* IPNI=*680359-1* 195
Piper bermudezense IPNI=*680507-1* 195
Piper bigelovii IPNI=*682899-1* 197
Piper blandum IPNI=*678107-1* 193
Piper bogotense IPNI=*680559-1* 195
Piper bogotense var. *foetans* IPNI=*680559-1* 195
Piper bogotense var. *ovalilimbum* IPNI=*680559-1* 195
Piper borucanum IPNI=*681559-1* 196
Piper brachypodon f. *hirtellum* IPNI=*680665-1* 195
Piper brachypodon var. *hirsuticaule* IPNI=*680665-1* 195
Piper brachystachyum IPNI=*1003227-1* 195
Piper bracteatum IPNI=*679841-1* 194
Piper bredemeyeri f. *escabridum* IPNI=*680680-1* 195
Piper bredemeyeri var. *angustifolium* IPNI=*680680-1* 195
Piper bredemeyeri var. *flexuosum* IPNI=*680680-1* 195
Piper brittonorum IPNI=*680435-1* 195
Piper buchtienii IPNI=*681651-1* 196
Piper buchtienii var. *charopampanum* IPNI=*681651-1* 196
Piper calceolatum IPNI=*680737-1* 195
Piper caledonianum IPNI=*680300-1* 195
Piper callosum var. *franciscoanum* IPNI=*680748-1* 195
Piper canelosense IPNI=*198173-2* 196
Piper canescenticaule IPNI=*681674-1* 196
Piper canescenticaule var. *expansum* IPNI=*681674-1* 196
Piper canyazasense IPNI=*682899-1* 197
Piper caracasanum IPNI=*682429-1* 196
Piper caracasanum var. *firmius* IPNI=*682429-1* 196
Piper caripense IPNI=*680300-1* 195
Piper carpupija IPNI=*680810-1* 195
Piper catalpifolium IPNI=*682138-1* 196
Piper catucheanum IPNI=*681674-1* 196
Piper caudatum IPNI=*682138-1* 196
Piper caudefactum IPNI=*682429-1* 196
Piper cauranum IPNI=*681157-1* 196

Piper celtidifolium IPNI=*680296-1* 195
Piper celtidifolium IPNI=*680359-1* 195
Piper charopampanum IPNI=*681651-1* 196
Piper chimantanum IPNI=*683782-1* 197
Piper ciliatum IPNI=*678107-1* 193
Piper cinereoramulum IPNI=*681674-1* 196
Piper citrifolium IPNI=*681224-1* 196
Piper claudicans IPNI=*680435-1* 195
Piper cocherense IPNI=*680810-1* 195
Piper cojimaranum IPNI=*680359-1* 195
Piper colombianum IPNI=*680810-1* 195
Piper colonum IPNI=*682369-1* 196
Piper colubrinum var. *adenophyllum* IPNI=*681224-1* 196
Piper concinnum IPNI=*679378-1* 194
Piper concinnum var. *trinitense* IPNI=*680300-1* 195
Piper coruscans var. *aureum* IPNI=*681027-1* 195
Piper coruscans var. *membranaceum* IPNI=*681027-1* 195
Piper crassamentum IPNI=*681041-1* 195
Piper crassifolium IPNI=*681224-1* 196
Piper crassinervium var. *guilleminianum* IPNI=*681041-1* 195
Piper crassinervium var. *tocotanum* IPNI=*681041-1* 195
Piper crassispicatum IPNI=*681559-1* 196
Piper crassum IPNI=*681224-1* 196
Piper cryptocarpum IPNI=*681193-1* 196
Piper cuernavacanum IPNI=*316339-2* 197
Piper cujabanum IPNI=*683806-1* 197
Piper cuneatum IPNI=*679269-1* 194
Piper cuneifolium IPNI=*679269-1* 194
Piper darienense var. *tricuspe* IPNI=*681128-1* 196
Piper decorticans IPNI=*682429-1* 196
Piper decumanum IPNI=*682138-1* 196
Piper demeraranum var. *rubescens* IPNI=*681157-1* 196
Piper densespicatum IPNI=*681224-1* 196
Piper denudatum IPNI=*682369-1* 196
Piper diazanum var. *viae-kobeanae* IPNI=*682899-1* 197
Piper diguetii IPNI=*680359-1* 195
Piper dilatatifolium IPNI=*681193-1* 196
Piper dilatatum f. *calcicolens* IPNI=*681193-1* 196
Piper dilatatum f. *diamantense* IPNI=*681193-1* 196
Piper dilatatum f. *dilatatifolium* IPNI=*681193-1* 196
Piper dilatatum f. *diversifolium* IPNI=*681193-1* 196
Piper dilatatum f. *guayranum* IPNI=*681193-1* 196
Piper dilatatum f. *longipilum* IPNI=*681193-1* 196
Piper dilatatum f. *magnifolium* IPNI=*681193-1* 196
Piper dilatatum f. *medium* IPNI=*681193-1* 196
Piper dilatatum f. *naris-fractae* IPNI=*681193-1* 196
Piper dilatatum f. *regresivanum* IPNI=*681193-1* 196
Piper dilatatum f. *vauclinii* IPNI=*681193-1* 196
Piper dilatatum var. *vincentianum* IPNI=*681193-1* 196
Piper dimidiatum IPNI=*680435-1* 195
Piper discophorum IPNI=*198988-2* 197
Piper dispansum IPNI=*681651-1* 196
Piper disparispicum IPNI=*680296-1* 195
Piper dissimile IPNI=*678107-1* 193
Piper dombeyanum IPNI=*316339-2* 197
Piper dominicanum IPNI=*680300-1* 195
Piper donnell-smithii var. *longipetiolatum* IPNI=*680300-1* 195
Piper dubium IPNI=*680296-1* 195
Piper duchassaingii IPNI=*198988-2* 197
Piper duidaense IPNI=*683782-1* 197
Piper duidaense var. *bolivianum* IPNI=*683782-1* 197
Piper duidaense var. *eciliatum* IPNI=*683782-1* 197
Piper dumetorum IPNI=*681041-1* 195
Piper durirameum IPNI=*680559-1* 195
Piper eggersii IPNI=*681193-1* 196
Piper elliptico-oblongifolium IPNI=*681651-1* 196
Piper ellipticum IPNI=*683806-1* 197
Piper elmetanum IPNI=*682429-1* 196
Piper elongatifolium IPNI=*680296-1* 195
Piper elongatum f. *subglabrum* IPNI=*680296-1* 195
Piper elongatum IPNI=*30130843-2* 194
Piper elongatum IPNI=*680296-1* 195
Piper elongatum var. *brachyarthrum* IPNI=*680296-1* 195
Piper elongatum var. *laevifolium* IPNI=*680296-1* 195
Piper elongatum var. *mocco-mocco* IPNI=*680296-1* 195
Piper elongatum var. *ossanum* IPNI=*680296-1* 195
Piper elongatum var. *pampayacusum* IPNI=*680296-1* 195
Piper elongatum var. *salviifolium* IPNI=*680296-1* 195
Piper emarginellum IPNI=*678531-1* 193
Piper enckea IPNI=*681876-1* 196
Piper escasuense IPNI=*681041-1* 195
Piper exsectum IPNI=*680435-1* 195
Piper exserens IPNI=*681041-1* 195
Piper exserens var. *miraflorensanum* IPNI=*681041-1* 195
Piper exserens var. *tocotanum* IPNI=*681041-1* 195
Piper extensum IPNI=*30130843-2* 194
Piper fagopyricarpum IPNI=*681128-1* 196
Piper fasciculare IPNI=*134457-2* 394
Piper fasciculatum IPNI=*316334-2* 197
Piper fatoanum IPNI=*680296-1* 195
Piper fistulosum IPNI=*680559-1* 195
Piper flagellare IPNI=*682429-1* 196
Piper flexuosum IPNI=*680680-1* 195
Piper florencianum IPNI=*198599-2* 197
Piper fontinale IPNI=*1068802-2* 194
Piper freirifolium IPNI=*679378-1* 194
Piper fruticosum IPNI=*681674-1* 196
Piper fuscescens IPNI=*680559-1* 195
Piper galioides IPNI=*678659-1* 193
Piper gemellum IPNI=*316334-2* 197
Piper geniculatum f. *puberulum* IPNI=*683806-1* 197
Piper geniculatum var. *stamineum* IPNI=*680435-1* 195
Piper geniculatum var. *verrucosum* IPNI=*680435-1* 195
Piper glabellum IPNI=*30001267-2* 193
Piper glabellum IPNI=*681224-1* 196
Piper glaberrimicaule IPNI=*681674-1* 196
Piper glabrilimbum IPNI=*681224-1* 196
Piper glabrirameum IPNI=*680810-1* 195
Piper glabrispica IPNI=*682369-1* 196
Piper gleasonii var. *wonotoboense* IPNI=*681692-1* 196
Piper glyphicum IPNI=*682369-1* 196
Piper gorgonae-insulae IPNI=*683782-1* 197
Piper gracilispicum IPNI=*681193-1* 196
Piper gracillimum var. *dactoense* IPNI=*680359-1* 195
Piper grahamii IPNI=*681193-1* 196
Piper granatense IPNI=*681559-1* 196
Piper grande var. *puberulentum* IPNI=*681559-1* 196
Piper grandifolium IPNI=*316339-2* 197
Piper grandifolium IPNI=*681559-1* 196
Piper granulosum IPNI=*680296-1* 195
Piper griseovenosum IPNI=*198599-2* 197
Piper guavanum IPNI=*680435-1* 195
Piper guayranum IPNI=*681193-1* 196
Piper guayranum var. *humilius* IPNI=*681193-1* 196
Piper guildingianum IPNI=*679841-1* 194
Piper hebecarpum IPNI=*680296-1* 195
Piper hemionitidifolium IPNI=*679269-1* 194
Piper henshenii IPNI=*680359-1* 195
Piper henshenii var. *barretoi* IPNI=*680359-1* 195
Piper heraldii IPNI=*680520-1* 195
Piper heraldii var. *amplius* IPNI=*680520-1* 195
Piper heraldii var. *cocleanum* IPNI=*680520-1* 195
Piper herbaceum IPNI=*679841-1* 194
Piper hirsutum IPNI=*681674-1* 196
Piper hirsutum var. *fuscopunctulatum* IPNI=*681674-1* 196
Piper hirsutum var. *jamaicense* IPNI=*681674-1* 196

Piper hirsutum var. *magnifolium* IPNI=*681674-1* 196
Piper hirsutum var. *olfersianum* IPNI=*681674-1* 196
Piper hirtulum IPNI=*681193-1* 196
Piper hispaniolae IPNI=*680359-1* 195
Piper hispidinervum IPNI=*681674-1* 196
Piper hispidum f. *lanceolatum* IPNI=*681674-1* 196
Piper hispidum f. *patulipilum* IPNI=*681674-1* 196
Piper hispidum f. *surinamense* IPNI=*681674-1* 196
Piper hispidum IPNI=*681674-1* 196
Piper hispidum var. *albescens* IPNI=*681674-1* 196
Piper hispidum var. *lanceolatum* IPNI=*681674-1* 196
Piper hispidum var. *magnifolium* IPNI=*681674-1* 196
Piper hispidum var. *obliquum* IPNI=*681674-1* 196
Piper hispidum var. *plurinerve* IPNI=*681674-1* 196
Piper hohenackeri IPNI=*1003227-1* 195
Piper holtonii var. *parvispicum* IPNI=*681685-1* 196
Piper hostmannianum var. *glabrirameum* IPNI=*681692-1* 196
Piper huberi IPNI=*681041-1* 195
Piper humile IPNI=*679269-1* 194
Piper ibaguense IPNI=*680300-1* 195
Piper ignacioanum IPNI=*681193-1* 196
Piper impunctatum IPNI=*681224-1* 196
Piper inaequalifolium IPNI=*678848-1* 194
Piper incisum IPNI=*679597-1* 194
Piper infidele IPNI=*682429-1* 196
Piper insigne IPNI=*682429-1* 196
Piper intersitum IPNI=*680296-1* 195
Piper isernii IPNI=*198173-2* 196
Piper jamesonii IPNI=*680559-1* 195
Piper johnstonii IPNI=*680300-1* 195
Piper lacunosum var. *angustifolium* IPNI=*681864-1* 196
Piper lacunosum var. *glabricaule* IPNI=*681864-1* 196
Piper lambeauense IPNI=*681193-1* 196
Piper lanceatum IPNI=*680296-1* 195
Piper lancifolium IPNI=*680296-1* 195
Piper latifolium IPNI=*316339-2* 197
Piper latum IPNI=*198988-2* 197
Piper laxispicum IPNI=*681128-1* 196
Piper laxispicum var. *latifolium* IPNI=*681128-1* 196
Piper laxum IPNI=*682369-1* 196
Piper leiophyllum IPNI=*681224-1* 196
Piper lenormandianum IPNI=*681157-1* 196
Piper leprieurii IPNI=*680988-1* 195
Piper leucofustum IPNI=*681674-1* 196
Piper ligatispicum IPNI=*682429-1* 196
Piper lineatum var. *hirtipetiolatum* IPNI=*680296-1* 195
Piper linguifolium IPNI=*680296-1* 195
Piper linguifolium var. *brevifolium* IPNI=*680435-1* 195
Piper linkii IPNI=*683806-1* 197
Piper longegerminatum IPNI=*681674-1* 196
Piper lucaeanum var. *magnifolium* IPNI=*683782-1* 197
Piper macintoshii IPNI=*680359-1* 195
Piper macrophyllum IPNI=*1003227-1* 195
Piper maculosum IPNI=*679043-1* 194
Piper maestranum IPNI=*681674-1* 196
Piper magnoliifolium IPNI=*679047-1* 194
Piper magnoliifolium IPNI=*679269-1* 194
Piper malabarense IPNI=*682369-1* 196
Piper mandinganum IPNI=*682429-1* 196
Piper maravalanum IPNI=*681041-1* 195
Piper marcescentispicum IPNI=*681251-1* 196
Piper margaritanum IPNI=*681193-1* 196
Piper marginatum IPNI=*679269-1* 194
Piper marginatum var. *anisatum* IPNI=*682138-1* 196
Piper marginatum var. *catalpifolium* IPNI=*682138-1* 196
Piper marginatum var. *clausum* IPNI=*682138-1* 196
Piper marginatum var. *domatiiferum* IPNI=*682138-1* 196
Piper marginatum var. *firmum* IPNI=*682138-1* 196
Piper marmoreum IPNI=*681559-1* 196
Piper marsupiatum IPNI=*198403-2* 196
Piper martinicense IPNI=*680296-1* 195
Piper martinicense var. *genuinum* IPNI=*680296-1* 195
Piper martinicense var. *montis-pilati* IPNI=*680296-1* 195
Piper masiseanum IPNI=*681674-1* 196
Piper maypurense IPNI=*679095-1* 194
Piper medium IPNI=*680359-1* 195
Piper medium var. *laxiflorum* IPNI=*680359-1* 195
Piper medium var. *leonardii* IPNI=*680359-1* 195
Piper megalophyllum IPNI=*683316-1* 197
Piper megalophyllum var. *connivens* IPNI=*683316-1* 197
Piper meraense IPNI=*198173-2* 196
Piper miersinum var. *magnifolium* IPNI=*682399-1* 196
Piper milleri IPNI=*679269-1* 194
Piper miquelii IPNI=*1003227-1* 195
Piper molanoi IPNI=*682429-1* 196
Piper mollipilum IPNI=*683758-1* 197
Piper monostachyos IPNI=*677918-1* 193
Piper moritzianum IPNI=*682429-1* 196
Piper multinervium IPNI=*680296-1* 195
Piper multinervium var. *amplum* IPNI=*680296-1* 195
Piper multinervium var. *cayoense* IPNI=*680296-1* 195
Piper multinervium var. *flavicans* IPNI=*680296-1* 195
Piper multinervium var. *hirsuticaule* IPNI=*680296-1* 195
Piper multinervium var. *kantelolense* IPNI=*680296-1* 195
Piper multinervium var. *paralense* IPNI=*680296-1* 195
Piper multinervium var. *peracutum* IPNI=*680296-1* 195
Piper multinervium var. *productipes* IPNI=*680296-1* 195
Piper multinervium var. *protractifolium* IPNI=*680296-1* 195
Piper multinervium var. *puberulipedunculum* IPNI=*680296-1* 195
Piper multinervium var. *pubescenticaule* IPNI=*680296-1* 195
Piper multinervium var. *skutchii* IPNI=*680296-1* 195
Piper multinervium var. *tamashense* IPNI=*680296-1* 195
Piper multinervium var. *telanum* IPNI=*680296-1* 195
Piper munchanum var. *latum* IPNI=*682305-1* 196
Piper murinum IPNI=*677918-1* 193
Piper muscosum IPNI=*679591-1* 194
Piper myosuroides IPNI=*30130843-2* 194
Piper myosuros IPNI=*679841-1* 194
Piper nanegalense IPNI=*682429-1* 196
Piper nematostachyum IPNI=*30130843-2* 194
Piper nemorale IPNI=*681692-1* 196
Piper nemorosum IPNI=*677918-1* 193
Piper nhandi IPNI=*682138-1* 196
Piper niceforoi var. *glabricaule* IPNI=*682138-1* 196
Piper nieblyanum IPNI=*680810-1* 195
Piper nigricans IPNI=*682369-1* 196
Piper nigropunctatum var. *franciscoanum* IPNI=*681251-1* 196
Piper nigrum var. *hirtellosum* IPNI=*682369-1* 196
Piper nitidum IPNI=*681224-1* 196
Piper nonconformans IPNI=*680296-1* 195
Piper novae-helvetiae IPNI=*681041-1* 195
Piper nudibracteatum IPNI=*682429-1* 196
Piper nudibracteatum var. *micropuberulum* IPNI=*682429-1* 196
Piper nudibracteatum var. *pubescens* IPNI=*682429-1* 196
Piper nummulariifolium IPNI=*189588-2* 194
Piper nutans IPNI=*683806-1* 197
Piper obesum IPNI=*681674-1* 196
Piper oblanceolatum var. *fragilicaule* IPNI=*680296-1* 195
Piper obliquum IPNI=*677918-1* 193
Piper obliquum IPNI=*683806-1* 197
Piper obliquum var. *apterum* IPNI=*682429-1* 196
Piper obliquum var. *puberulidrupum* IPNI=*682429-1* 196
Piper obliquum var. *sprucei* IPNI=*682429-1* 196
Piper obliquum var. *subeximium* IPNI=*682429-1* 196
Piper oblitum IPNI=*680559-1* 195
Piper oblongilimbum IPNI=*681224-1* 196

Piper obovatum IPNI=*1090908-2* 194
Piper obovatum var. *bolivianum* IPNI=*681651-1* 196
Piper obtusifolium IPNI=*679269-1* 194
Piper obumbratum var. *parvifolium* IPNI=*680435-1* 195
Piper obversum IPNI=*1090908-2* 194
Piper olens IPNI=*680435-1* 195
Piper oparumaense IPNI=*682429-1* 196
Piper opizii IPNI=*681674-1* 196
Piper orbiculatum IPNI=*1090908-2* 194
Piper orenocanum IPNI=*681027-1* 195
Piper orenocanum var. *koriaboense* IPNI=*681027-1* 195
Piper ossanum IPNI=*680296-1* 195
Piper ottonis IPNI=*316334-2* 197
Piper ovale IPNI=*681041-1* 195
Piper ovalifolium IPNI=*1090908-2* 194
Piper pachystachyon IPNI=*681864-1* 196
Piper pallidirameum IPNI=*680810-1* 195
Piper pangoense IPNI=*198988-2* 197
Piper papyraceum IPNI=*681041-1* 195
Piper paracaimanum IPNI=*681692-1* 196
Piper paranapuranum IPNI=*681674-1* 196
Piper parvibracteatum IPNI=*680296-1* 195
Piper patulipilum IPNI=*681674-1* 196
Piper pellucidum IPNI=*679378-1* 194
Piper peltatum var. *colombianum* IPNI=*316334-2* 197
Piper peltatum var. *hirtellum* IPNI=*316334-2* 197
Piper pennellii IPNI=*682429-1* 196
Piper pereskiifolium IPNI=*679404-1* 194
Piper perlongipes IPNI=*680520-1* 195
Piper permari IPNI=*681128-1* 196
Piper perobumbratum IPNI=*680435-1* 195
Piper perpuberulum IPNI=*682301-1* 196
Piper perspicuibracteum IPNI=*681674-1* 196
Piper pichisense IPNI=*198173-2* 196
Piper pilipedunculum IPNI=*681674-1* 196
Piper planifolium IPNI=*677918-1* 193
Piper plantagineum IPNI=*680359-1* 195
Piper platannanum IPNI=*681674-1* 196
Piper poeppigii IPNI=*680748-1* 195
Piper poiretianum IPNI=*680748-1* 195
Piper portulacifolium IPNI=*679597-1* 194
Piper pothomorphe IPNI=*683316-1* 197
Piper praeclarum IPNI=*682429-1* 196
Piper praegnans IPNI=*681692-1* 196
Piper praemorsum IPNI=*681224-1* 196
Piper propinquum IPNI=*681041-1* 195
Piper propinquum var. *brevistylum* IPNI=*681041-1* 195
Piper propinquum var. *grande* IPNI=*681041-1* 195
Piper pruinosum IPNI=*316334-2* 197
Piper pseudoamalago IPNI=*680359-1* 195
Piper pseudochurumayu IPNI=*681027-1* 195
Piper pseudochurumayu var. *membranaceum* IPNI=*681027-1* 195
Piper pseudomarginatum IPNI=*682138-1* 196
Piper pseudonobile IPNI=*682429-1* 196
Piper pseudopropinquum IPNI=*681041-1* 195
Piper pseudopsis IPNI=*680680-1* 195
Piper puberulibaccum IPNI=*681692-1* 196
Piper puberulibracteum IPNI=*682429-1* 196
Piper puberulibracteum var. *monticola* IPNI=*682429-1* 196
Piper pulicare IPNI=*679841-1* 194
Piper pullispicum IPNI=*680435-1* 195
Piper punctatum IPNI=*681651-1* 196
Piper punctulatifolium IPNI=*681251-1* 196
Piper purpurascens IPNI=*680296-1* 195
Piper pusillum IPNI=*1068802-2* 194
Piper quadrangulare IPNI=*679591-1* 194
Piper quadrangulum IPNI=*679591-1* 194
Piper quadrifolium IPNI=*1068802-2* 194
Piper quadrifolium IPNI=*679597-1* 194
Piper quiriguanum IPNI=*682138-1* 196
Piper racemosum IPNI=*680359-1* 195
Piper radula IPNI=*680680-1* 195
Piper ratisportanum IPNI=*680507-1* 195
Piper readii IPNI=*681193-1* 196
Piper reciprocum IPNI=*680296-1* 195
Piper reflexum IPNI=*1068802-2* 194
Piper regresivanum IPNI=*681193-1* 196
Piper reniforme IPNI=*316339-2* 197
Piper repens IPNI=*679841-1* 194
Piper resatum IPNI=*681674-1* 196
Piper reticulatum IPNI=*680359-1* 195
Piper reticulatum var. *santa-rosanum* IPNI=*198988-2* 197
Piper retropuberulum IPNI=*198173-2* 196
Piper richardianum IPNI=*680359-1* 195
Piper richardianum var. *glabrifolium* IPNI=*680359-1* 195
Piper richardianum var. *latilimbum* IPNI=*680359-1* 195
Piper rigidum IPNI=*680435-1* 195
Piper rigidum var. *verdeanum* IPNI=*680435-1* 195
Piper riparium IPNI=*681559-1* 196
Piper rivialbi IPNI=*681674-1* 196
Piper rivulare IPNI=*680300-1* 195
Piper rivulare var. *hygrophilum* IPNI=*680300-1* 195
Piper rohrii IPNI=*1003227-1* 195
Piper rohrii var. *crassum* IPNI=*1003227-1* 195
Piper romboutsii IPNI=*681224-1* 196
Piper rotundatum IPNI=*679705-1* 194
Piper rotundifolium IPNI=*189588-2* 194
Piper rotundistipulum IPNI=*198388-2* 196
Piper rotundum IPNI=*682369-1* 196
Piper ruizianum IPNI=*680359-1* 195
Piper ruizianum IPNI=*682429-1* 196
Piper rupestre IPNI=*30130843-2* 194
Piper rurrenbaqueanum IPNI=*681651-1* 196
Piper sabanillanum IPNI=*681674-1* 196
Piper salinasanum IPNI=*682899-1* 197
Piper salinasanum var. *parvifolium* IPNI=*682899-1* 197
Piper saltuum IPNI=*683782-1* 197
Piper salutatrix IPNI=*682899-1* 197
Piper salviifolium IPNI=*680296-1* 195
Piper sanctum IPNI=*680520-1* 195
Piper sanguineispicum IPNI=*682429-1* 196
Piper san-joseanum IPNI=*682138-1* 196
Piper san-joseanum var. *aristolochiifolium* IPNI=*682301-1* 196
Piper san-joseanum var. *chiriquinum* IPNI=*682138-1* 196
Piper san-joseanum var. *kobense* IPNI=*682138-1* 196
Piper san-joseanum var. *panamanum* IPNI=*682138-1* 196
Piper san-joseanum var. *remediosense* IPNI=*682138-1* 196
Piper san-joseanum var. *tabogense* IPNI=*682138-1* 196
Piper san-luisense IPNI=*681041-1* 195
Piper santa-martanum IPNI=*680300-1* 195
Piper santiaganum IPNI=*681027-1* 195
Piper santo-domingense IPNI=*683806-1* 197
Piper saramaccanum IPNI=*682429-1* 196
Piper sasaimanum IPNI=*682429-1* 196
Piper saururoides IPNI=*680359-1* 195
Piper saxatile IPNI=*1068802-2* 194
Piper scabricaule IPNI=*681674-1* 196
Piper scabriseptum var. *reductum* IPNI=*681674-1* 196
Piper scabrum IPNI=*680296-1* 195
Piper scabrum IPNI=*681674-1* 196
Piper scabrum var. *kalacroixense* IPNI=*681674-1* 196
Piper scandens IPNI=*30001267-2* 193
Piper scandens IPNI=*679378-1* 194
Piper scandens IPNI=*679841-1* 194
Piper schultesii IPNI=*198599-2* 197
Piper scutatum IPNI=*316334-2* 197

Piper scutiphyllum IPNI=*316334-2* 197
Piper seitzii IPNI=*681193-1* 196
Piper serpens IPNI=*679841-1* 194
Piper sidifolium IPNI=*316339-2* 197
Piper sieberi IPNI=*680359-1* 195
Piper silvioberrans IPNI=*681674-1* 196
Piper silvivagum f. *peruvianum* IPNI=*681674-1* 196
Piper sinuosirameum IPNI=*197302-2* 195
Piper smilacifolium IPNI=*198988-2* 197
Piper smilacifolium IPNI=*681876-1* 196
Piper speciosum IPNI=*316334-2* 197
Piper spoliatum IPNI=*682429-1* 196
Piper stamineum IPNI=*680435-1* 195
Piper stamineum var. *longifolium* IPNI=*680435-1* 195
Piper staphyle IPNI=*680359-1* 195
Piper statum IPNI=*682429-1* 196
Piper stehleorum IPNI=*680296-1* 195
Piper striatinervosum IPNI=*682429-1* 196
Piper striatum IPNI=*679909-1* 194
Piper subaequale IPNI=*681224-1* 196
Piper subbrunneispicum IPNI=*198173-2* 196
Piper subcrassifolium IPNI=*681692-1* 196
Piper submelanostictum IPNI=*682429-1* 196
Piper submelanostictum var. *amelanostictum* IPNI=*682429-1* 196
Piper submultiplinerve IPNI=*681041-1* 195
Piper subpeltatum IPNI=*316339-2* 197
Piper subpeltatum var. *sidifolium* IPNI=*316339-2* 197
Piper subpuberulum IPNI=*680435-1* 195
Piper subpurpureum IPNI=*680871-1* 195
Piper subrectinerve IPNI=*680296-1* 195
Piper subscabridum IPNI=*681193-1* 196
Piper subsessile IPNI=*681251-1* 196
Piper subtropicum IPNI=*682429-1* 196
Piper subtropicum var. *tomentellum* IPNI=*682429-1* 196
Piper sulcatum IPNI=*1068802-2* 194
Piper sumideranum IPNI=*681674-1* 196
Piper surinamense IPNI=*680988-1* 195
Piper suspectum IPNI=*681651-1* 196
Piper syringifolium IPNI=*680359-1* 195
Piper taboganum var. *rugosum* IPNI=*682899-1* 197
Piper tarapotianum IPNI=*198988-2* 197
Piper tardum IPNI=*681674-1* 196
Piper tejucense IPNI=*681224-1* 196
Piper tenuiamentum IPNI=*681193-1* 196
Piper tetraphyllum IPNI=*1068802-2* 194
Piper thebaudianum IPNI=*681674-1* 196
Piper tiliifolium IPNI=*683316-1* 197
Piper tingens IPNI=*681027-1* 195
Piper tithymaloides IPNI=*679269-1* 194
Piper tobagoanum IPNI=*681193-1* 196
Piper tobagoanum var. *humilius* IPNI=*681193-1* 196
Piper trifolium IPNI=*1090908-2* 194
Piper trinitatis IPNI=*681193-1* 196
Piper trioicum IPNI=*682369-1* 196
Piper tristachyum IPNI=*680084-1* 194
Piper tuberculatum var. *allenii* IPNI=*680435-1* 195
Piper tuberculatum var. *cangrejalense* IPNI=*683806-1* 197
Piper tuberculatum var. *majus* IPNI=*683806-1* 197
Piper tuberculatum var. *minus* IPNI=*683806-1* 197
Piper tuberculatum var. *obtusifolium* IPNI=*683806-1* 197
Piper tuberculatum var. *rigidomembranaceum* IPNI=*683806-1* 197
Piper tuberculatum var. *scandens* IPNI=*683806-1* 197
Piper tumidicondyli IPNI=*681224-1* 196
Piper ulmifolium IPNI=*681193-1* 196
Piper umbellatum var. *geminatum* IPNI=*316339-2* 197
Piper umbellatum var. *glabrum* IPNI=*316339-2* 197
Piper umbellatum var. *majus* IPNI=*316339-2* 197
Piper umbellatum var. *tomentellum* IPNI=*316339-2* 197
Piper umbellatum var. *vestitum* IPNI=*316339-2* 197
Piper uncatum IPNI=*682138-1* 196
Piper uncatum var. *levyanum* IPNI=*682138-1* 196
Piper undeninervium IPNI=*682138-1* 196
Piper variegatum IPNI=*679043-1* 194
Piper verrucosum IPNI=*680435-1* 195
Piper verrucosum IPNI=*681193-1* 196
Piper villibracteum IPNI=*680680-1* 195
Piper villosulum IPNI=*680359-1* 195
Piper virgulticola IPNI=*681674-1* 196
Piper willdenowii IPNI=*681224-1* 196
Piper wulfschlaegelii IPNI=*680359-1* 195
Piper wurdackii IPNI=*681027-1* 195
Piper yaranum IPNI=*680435-1* 195
Piper yurimaguasanum IPNI=*683806-1* 197
Piper zacuapanum IPNI=*683316-1* 197
Piptadenia biuncifera IPNI=*77178320-1* 325
Piptadenia cateniformis IPNI=*51040-2* 314
Piptadenia cobi IPNI=*519643-1* 334
Piptadenia colubrina IPNI=*12264-2* 311
Piptadenia falcata IPNI=*12268-2* 311
Piptadenia flava IPNI=*77208976-1* 330
Piptadenia foliolosa IPNI=*519643-1* 334
Piptadenia leptocarpa IPNI=*77208976-1* 330
Piptadenia opacifolia IPNI=*514434-1* 330
Piptadenia peregrina IPNI=*12268-2* 311
Piptadenia peregrina var. *falcata* IPNI=*12268-2* 311
Piptadenia psilostachya IPNI=*962800-1* 331
Piptadenia pteroclada var. *klugii* IPNI=*514434-1* 330
Piptadenia recurva IPNI=*962800-1* 331
Piptadenia rubescens IPNI=*77178320-1* 325
Piptadenia speciosa IPNI=*77178320-1* 325
Piptadenia suaveolens IPNI=*77208976-1* 330
Piptadenia subtilifolia IPNI=*77178320-1* 325
Piptadenia viridiflora IPNI=*77178320-1* 325
Piptanthocereus peruvianus IPNI=*130627-1* 283
Piptatherum annulatum IPNI=*77164258-1* 246
Piptatherum confine IPNI=*21052-2* 234
Piptatherum flavescens IPNI=*1021229-2* 239
Piptatherum punctatum IPNI=*1021229-2* 239
Piratinera acutifolia IPNI=*37254-2* 437
Piratinera alicastrum IPNI=*850824-1* 437
Piratinera angustifolia IPNI=*37310-2* 437
Piratinera discolor IPNI=*582693-1* 437
Piratinera guianensis IPNI=*582693-1* 437
Piratinera lancifera IPNI=*37310-2* 437
Piratinera lecointei IPNI=*582693-1* 437
Piratinera lemeei IPNI=*582693-1* 437
Piratinera mollis IPNI=*582693-1* 437
Piratinera panamensis IPNI=*582693-1* 437
Piratinera paraensis IPNI=*37310-2* 437
Piratinera parinarioides IPNI=*37302-2* 437
Piratinera potabilis IPNI=*37305-2* 437
Piratinera rubescens IPNI=*37310-2* 437
Piratinera scabridula IPNI=*582693-1* 437
Piratinera terrabana IPNI=*850824-1* 437
Piratinera utilis IPNI=*582724-1* 437
Piratinera velutina IPNI=*582693-1* 437
Pircunia dioica IPNI=*676348-1* 287
Piresodendron ucuqui IPNI=*209914-2* 308
Pirigara hexapetala IPNI=*596743-1* 302
Pirigara insignis IPNI=*323304-2* 302
Pirigara speciosa IPNI=*596758-1* 302
Pirigara superba IPNI=*323304-2* 302
Pirigara tetrapetala IPNI=*114915-2* 301
Piriqueta caroliniana var. *exasperata* IPNI=*30068867-2* 400
Piriqueta cistoides var. *bracteolata* IPNI=*30068867-2* 400
Piriqueta cistoides var. *foliosa* IPNI=*30068867-2* 400

Piriqueta cistoides var. *genuina* IPNI=*30068867-2* 400
Piriqueta cistoides var. *glabrescens* IPNI=*30068867-2* 400
Piriqueta cistoides var. *latifolia* IPNI=*30068867-2* 400
Piriqueta foliosa IPNI=*30068867-2* 400
Piriqueta ionidioides IPNI=*30068867-2* 400
Piriqueta lanceolata IPNI=*30068867-2* 400
Piriqueta longifolia IPNI=*30068867-2* 400
Piriqueta ovata IPNI=*836478-1* 401
Piriqueta seticarpa IPNI=*30068867-2* 400
Piriqueta stenophylla IPNI=*30068867-2* 400
Piriqueta tomentosa IPNI=*30068867-2* 400
Piriqueta villosa IPNI=*30068867-2* 400
Piscidia acuminata IPNI=*316422-2* 330
Piscidia americana IPNI=*316422-2* 330
Piscidia corallodendrum IPNI=*316421-2* 319
Piscidia guaricensis IPNI=*316422-2* 330
Pisonia aculeata var. *guaranitica* IPNI=*326897-2* 287
Pisonia aculeata var. *hirsutissima* IPNI=*326897-2* 287
Pisonia aculeata var. *typica* IPNI=*326897-2* 287
Pisonia acuminata IPNI=*326897-2* 287
Pisonia anisophylla IPNI=*326897-2* 287
Pisonia comosa IPNI=*326897-2* 287
Pisonia cordifolia IPNI=*326897-2* 287
Pisonia coriifolia IPNI=*113934-2* 287
Pisonia crenulata IPNI=*326897-2* 287
Pisonia cuspidata IPNI=*113926-2* 286
Pisonia dussii IPNI=*113934-2* 287
Pisonia elliptica IPNI=*326897-2* 287
Pisonia eucalyptifolia IPNI=*113934-2* 287
Pisonia florida var. *pabstiana* IPNI=*326897-2* 287
Pisonia fragrans IPNI=*113934-2* 287
Pisonia fragrans var. *oblanceolata* IPNI=*113934-2* 287
Pisonia georgina IPNI=*326897-2* 287
Pisonia grandifolia IPNI=*326897-2* 287
Pisonia harrisiana IPNI=*113934-2* 287
Pisonia helleri IPNI=*326897-2* 287
Pisonia inermis IPNI=*113934-2* 287
Pisonia limonella IPNI=*326897-2* 287
Pisonia loranthoides IPNI=*326897-2* 287
Pisonia minor IPNI=*326897-2* 287
Pisonia mitis IPNI=*326897-2* 287
Pisonia monotaxadenia IPNI=*326897-2* 287
Pisonia nitida IPNI=*113934-2* 287
Pisonia pallavia IPNI=*326897-2* 287
Pisonia schomburgkiana IPNI=*113926-2* 286
Pisonia tellulata IPNI=*326897-2* 287
Pisonia uniseriata IPNI=*326897-2* 287
Pisonia villosa IPNI=*326897-2* 287
Pisonia yagua-pinda IPNI=*326897-2* 287
Pistacia simaruba IPNI=*127217-1* 451
Pistia aegyptiaca IPNI=*317684-2* 202
Pistia aethiopica IPNI=*317684-2* 202
Pistia africana IPNI=*317684-2* 202
Pistia amazonica IPNI=*317684-2* 202
Pistia brasiliensis IPNI=*317684-2* 202
Pistia commutata IPNI=*317684-2* 202
Pistia crispata IPNI=*317684-2* 202
Pistia cumingii IPNI=*317684-2* 202
Pistia gardneri IPNI=*317684-2* 202
Pistia horkeliana IPNI=*317684-2* 202
Pistia leprieuri IPNI=*317684-2* 202
Pistia linguiformis IPNI=*317684-2* 202
Pistia minor IPNI=*317684-2* 202
Pistia natalensis IPNI=*317684-2* 202
Pistia obcordata IPNI=*317684-2* 202
Pistia occidentalis IPNI=*317684-2* 202
Pistia schleideniana IPNI=*317684-2* 202
Pistia spathulata IPNI=*317684-2* 202
Pistia stratiotes var. *obcordata* IPNI=*317684-2* 202
Pistia stratiotes var. *spathulata* IPNI=*317684-2* 202
Pistia texensis IPNI=*317684-2* 202
Pistia turpinii IPNI=*317684-2* 202
Pistia weigeltiana IPNI=*317684-2* 202
Pisum abyssinicum IPNI=*501912-1* 325
Pisum album IPNI=*501912-1* 325
Pisum arvense IPNI=*501912-1* 325
Pisum baclium IPNI=*501912-1* 325
Pisum biflorum IPNI=*501912-1* 325
Pisum borussicum IPNI=*501912-1* 325
Pisum coccineum IPNI=*501912-1* 325
Pisum coerulescens IPNI=*501912-1* 325
Pisum commune IPNI=*501912-1* 325
Pisum elatius IPNI=*501912-1* 325
Pisum elatum IPNI=*501912-1* 325
Pisum excorticatum IPNI=*501912-1* 325
Pisum fertile IPNI=*501912-1* 325
Pisum granulatum IPNI=*501912-1* 325
Pisum hortense IPNI=*501912-1* 325
Pisum humile IPNI=*501912-1* 325
Pisum jomardii IPNI=*501912-1* 325
Pisum lathyrus IPNI=*501990-1* 325
Pisum leptolobum IPNI=*501912-1* 325
Pisum macrocarpum IPNI=*501912-1* 325
Pisum macrospermum IPNI=*501912-1* 325
Pisum odoratum IPNI=*501909-1* 325
Pisum praecox IPNI=*501912-1* 325
Pisum prolificum IPNI=*501912-1* 325
Pisum pumilio IPNI=*501912-1* 325
Pisum quadratum IPNI=*501912-1* 325
Pisum ramulare IPNI=*501912-1* 325
Pisum roseum IPNI=*501912-1* 325
Pisum rugosum IPNI=*501912-1* 325
Pisum saccharatum IPNI=*501912-1* 325
Pisum sativum IPNI=*501912-1* 325
Pisum sativum subsp. *arvense* IPNI=*501912-1* 325
Pisum sativum subsp. *biflorum* IPNI=*501912-1* 325
Pisum sativum subsp. *brevipedunculatum* IPNI=*501912-1* 325
Pisum sativum subsp. *elatius* IPNI=*501912-1* 325
Pisum sativum subsp. *hortense* IPNI=*501912-1* 325
Pisum sativum subsp. *humile* IPNI=*501912-1* 325
Pisum sativum subsp. *pumilio* IPNI=*501912-1* 325
Pisum sativum var. *elatius* IPNI=*501912-1* 325
Pisum sativum var. *hortense* IPNI=*501912-1* 325
Pisum sativum var. *quadratum* IPNI=*501912-1* 325
Pisum sativum var. *umbellatum* IPNI=*501912-1* 325
Pisum sibiricum IPNI=*501912-1* 325
Pisum smyrnense IPNI=*501912-1* 325
Pisum syriacum IPNI=*501912-1* 325
Pisum tetragonum IPNI=*501912-1* 325
Pisum thebaicum IPNI=*501912-1* 325
Pisum transcaucasicum IPNI=*501912-1* 325
Pisum tuffetii IPNI=*501912-1* 325
Pisum umbellatum IPNI=*501912-1* 325
Pisum uniflorum IPNI=*501912-1* 325
Pisum variegatum IPNI=*501912-1* 325
Pisum viride IPNI=*501912-1* 325
Pisum vulgare IPNI=*501912-1* 325
Pisum zeylanicum IPNI=*501912-1* 325
Pitcairnia cernua IPNI=*123989-1* 229
Pitcairnia exscapa IPNI=*123989-1* 229
Pitcairnia heterophylla f. *albiflora* IPNI=*123989-1* 229
Pitcairnia heterophylla var. *exscapa* IPNI=*123989-1* 229
Pitcairnia lepidota IPNI=*123813-1* 229
Pitcairnia liebmannii IPNI=*123989-1* 229
Pitcairnia longifolia IPNI=*123989-1* 229
Pitcairnia morrenii IPNI=*123989-1* 229

Pithecellobium albicaule IPNI=*514978-1* 330
Pithecellobium andaquiense IPNI=*962818-1* 337
Pithecellobium arboreum IPNI=*62882-2* 315
Pithecellobium auriculatum IPNI=*989358-1* 309
Pithecellobium bertolonii IPNI=*1003508-2* 330
Pithecellobium bicolor IPNI=*271431-2* 337
Pithecellobium calostachys IPNI=*514978-1* 330
Pithecellobium campechense IPNI=*514978-1* 330
Pithecellobium carbonaria IPNI=*8016-2* 311
Pithecellobium cataractae IPNI=*962821-1* 337
Pithecellobium coccineum IPNI=*962823-1* 337
Pithecellobium corymbosum IPNI=*989341-1* 320
Pithecellobium cyclocarpum IPNI=*1037928-2* 319
Pithecellobium divaricatum IPNI=*962821-1* 337
Pithecellobium flavovirens IPNI=*515137-1* 330
Pithecellobium foreroi IPNI=*271431-2* 337
Pithecellobium forfex IPNI=*996931-1* 330
Pithecellobium glaucescens IPNI=*996931-1* 330
Pithecellobium glomeratum IPNI=*962821-1* 337
Pithecellobium glomeratum var. *spicatum* IPNI=*271434-2* 337
Pithecellobium guachapele IPNI=*211561-2* 331
Pithecellobium guadalupense IPNI=*515137-1* 330
Pithecellobium guaricense IPNI=*515137-1* 330
Pithecellobium hondurense IPNI=*1183336-2* 330
Pithecellobium inaequale IPNI=*271431-2* 337
Pithecellobium insigne IPNI=*514978-1* 330
Pithecellobium jupunba IPNI=*469538-1* 310
Pithecellobium juruanum IPNI=*962830-1* 337
Pithecellobium laetum IPNI=*989371-1* 310
Pithecellobium latifolium IPNI=*271432-2* 337
Pithecellobium latifolium subsp. *tayronense* IPNI=*271432-2* 337
Pithecellobium leucophyllum IPNI=*989368-1* 310
Pithecellobium ligustrinum IPNI=*271434-2* 337
Pithecellobium littorale IPNI=*1003508-2* 330
Pithecellobium longepedatum IPNI=*211561-2* 331
Pithecellobium longifolium IPNI=*271434-2* 337
Pithecellobium longiramosum IPNI=*271431-2* 337
Pithecellobium macrosiphon IPNI=*1183336-2* 330
Pithecellobium macrostachys IPNI=*1183336-2* 330
Pithecellobium malacotrichum IPNI=*989358-1* 309
Pithecellobium mangense IPNI=*56985-2* 315
Pithecellobium marginatum IPNI=*989342-1* 320
Pithecellobium microchlamys IPNI=*515137-1* 330
Pithecellobium multiflorum IPNI=*989448-1* 311
Pithecellobium niopoides IPNI=*8032-2* 311
Pithecellobium pachypus IPNI=*514978-1* 330
Pithecellobium panamense IPNI=*1183336-2* 330
Pithecellobium paniculatum IPNI=*515137-1* 330
Pithecellobium parviflorum IPNI=*996931-1* 330
Pithecellobium pedicellare IPNI=*989349-1* 311
Pithecellobium pilosulum IPNI=*271431-2* 337
Pithecellobium polycarpum IPNI=*989371-1* 310
Pithecellobium pubescens IPNI=*996931-1* 330
Pithecellobium pulchellum IPNI=*515137-1* 330
Pithecellobium racemiflorum IPNI=*996977-1* 337
Pithecellobium racemosum IPNI=*996977-1* 337
Pithecellobium roseum var. *glaucescens* IPNI=*996931-1* 330
Pithecellobium rufescens IPNI=*62892-2* 315
Pithecellobium salutare IPNI=*227126-2* 331
Pithecellobium saman IPNI=*227126-2* 331
Pithecellobium samaningua IPNI=*211561-2* 331
Pithecellobium saxosum IPNI=*515137-1* 330
Pithecellobium schomburgkii IPNI=*493928-1* 319
Pithecellobium seleri IPNI=*515137-1* 330
Pithecellobium spinulosum IPNI=*514978-1* 330
Pithecellobium splitgerberianum IPNI=*99109-3* 311
Pithecellobium subcorymbosum IPNI=*989341-1* 320
Pithecellobium unguis-cati var. *forfex* IPNI=*996931-1* 330
Pithecellobium vahlianum IPNI=*271434-2* 337
Pithecellobium zuliaense IPNI=*996931-1* 330
Pithecoctenium aubletii IPNI=*77100444-1* 362
Pithecoctenium botryoides IPNI=*77100444-1* 362
Pithecoctenium catharinae IPNI=*77100444-1* 362
Pithecoctenium cordifolium IPNI=*77100444-1* 362
Pithecoctenium crucigerum IPNI=*77100444-1* 362
Pithecoctenium echinatum f. *grandiflorum* IPNI=*77100444-1* 362
Pithecoctenium echinatum IPNI=*77100444-1* 362
Pithecoctenium echinatum var. *intermedium* IPNI=*77100444-1* 362
Pithecoctenium echinatum var. *tomentosum* IPNI=*77100444-1* 362
Pithecoctenium echinatum var. *typicum* IPNI=*77100444-1* 362
Pithecoctenium glaucum IPNI=*77100444-1* 362
Pithecoctenium hexagonum IPNI=*77100444-1* 362
Pithecoctenium hexagonum var. *brasilianum* IPNI=*77100444-1* 362
Pithecoctenium lundii IPNI=*77100444-1* 362
Pithecoctenium muricatum IPNI=*77100444-1* 362
Pithecoctenium phaseoloides IPNI=*77100444-1* 362
Pithecoctenium squalus IPNI=*77100444-1* 362
Pithecoctenium tribrachiatum IPNI=*77100444-1* 362
Pithecoctenium vitalba f. *glabratum* IPNI=*77100444-1* 362
Pithecoctenium vitalba f. *intermedium* IPNI=*77100444-1* 362
Pithecoctenium vitalba IPNI=*77100444-1* 362
Pithecoctenium vitalba var. *rostratum* IPNI=*77100444-1* 362
Pithecoctenium xanthophyllum IPNI=*77142096-1* 365
Pittonia hirsuta IPNI=*77069564-1* 275
Pittosporum chinense IPNI=*684716-1* 254
Pittosporum japonicum IPNI=*835147-1* 431
Pittosporum tobira var. *chinense* IPNI=*684716-1* 254
Pitumba edulis IPNI=*297979-2* 403
Pitumba guianensis IPNI=*297979-2* 403
Pityranthus crassifolius IPNI=*327342-2* 281
Pityrocarpa flava IPNI=*77208976-1* 330
Pityrocarpa pteroclada IPNI=*514434-1* 330
Pityrocarpa viridiflora IPNI=*77178320-1* 325
Pityrogramma × *dualis* IPNI=*200935-2* 178
Pityrogramma albicans IPNI=*17345640-1* 178
Pityrogramma calomelanos var. *distans* IPNI=*17345640-1* 178
Pityrogramma calomelanos var. *integrifoliolata* IPNI=*17345640-1* 178
Pityrogramma calomelanos var. *peruviana* IPNI=*200935-2* 178
Pityrogramma chamaesorbus IPNI=*17345640-1* 178
Pityrogramma consimilis IPNI=*17345640-1* 178
Pityrogramma hybrida IPNI=*17345640-1* 178
Pityrogramma insularis IPNI=*17345640-1* 178
Pityrogramma intermedia IPNI=*17345640-1* 178
Pityrogramma mira IPNI=*17345640-1* 178
Pityrogramma peruviana IPNI=*200935-2* 178
Pityrogramma tartarea IPNI=*200935-2* 178
Pityrogramma tartarea var. *arcuata* IPNI=*200935-2* 178
Pityrogramma tartarea var. *delicatula* IPNI=*200935-2* 178
Pityrogramma tartarea var. *fallax* IPNI=*200935-2* 178
Pityrogramma tartarea var. *kalbreyeri* IPNI=*200935-2* 178
Pityrogramma tartarea var. *mathewsii* IPNI=*200935-2* 178
Pityrogramma tartarea var. *pteridioides* IPNI=*200935-2* 178
Pityrogramma tenerrima IPNI=*17345640-1* 178
Pityrogramma tripinnata IPNI=*200935-2* 178
Placus chinensis IPNI=*962413-1* 260
Placus floresiana IPNI=*60473415-2* 261
Placus foetidus IPNI=*238826-1* 268
Placus odoratus IPNI=*1099254-2* 268
Placus purpurascens IPNI=*1099254-2* 268
Placus purpurascens var. *glabratum* IPNI=*1099254-2* 268
Placus suaveolens IPNI=*203916-2* 268
Plagioceltis dichotoma IPNI=*1213898-2* 436
Plagiocheilus prostratus IPNI=*238661-1* 268
Plagiostigma pumila IPNI=*853513-1* 439
Plagiostigma stipulata IPNI=*853513-1* 439
Planchonella elegans IPNI=*209758-2* 306

Planchonella guianensis IPNI=*948086-1* 305
Plantago accrescens IPNI=*684911-1* 376
Plantago adriatica IPNI=*321286-2* 376
Plantago altissima IPNI=*321286-2* 376
Plantago ambigua IPNI=*321285-2* 376
Plantago amblyacme IPNI=*685346-1* 376
Plantago angustata IPNI=*321286-2* 376
Plantago argentea IPNI=*321285-2* 376
Plantago argyrophylla IPNI=*685641-1* 376
Plantago asplundii IPNI=*684911-1* 376
Plantago attenuata IPNI=*321285-2* 376
Plantago australis subsp. *ecuadorensis* IPNI=*684911-1* 376
Plantago australis subsp. *macrostachys* IPNI=*684911-1* 376
Plantago australis subsp. *orbignyana* IPNI=*201466-2* 376
Plantago azorica IPNI=*321285-2* 376
Plantago barneoudii IPNI=*685346-1* 376
Plantago borysthenica IPNI=*321286-2* 376
Plantago bracteata IPNI=*321286-2* 376
Plantago byzantina IPNI=*321285-2* 376
Plantago candollei IPNI=*684911-1* 376
Plantago capensis IPNI=*321285-2* 376
Plantago capitata IPNI=*321285-2* 376
Plantago chilensis IPNI=*684911-1* 376
Plantago compressiscapa IPNI=*321286-2* 376
Plantago congesta IPNI=*685641-1* 376
Plantago contorta IPNI=*321285-2* 376
Plantago crenata IPNI=*321286-2* 376
Plantago dalmatica IPNI=*321285-2* 376
Plantago dentata IPNI=*321286-2* 376
Plantago denudata IPNI=*684911-1* 376
Plantago dilatata IPNI=*321286-2* 376
Plantago dombeyi IPNI=*685641-1* 376
Plantago dostalii IPNI=*321286-2* 376
Plantago dregeana IPNI=*321286-2* 376
Plantago dubia IPNI=*321285-2* 376
Plantago durvillei subsp. *mollior* IPNI=*684911-1* 376
Plantago durvillei var. *angustata* IPNI=*684911-1* 376
Plantago ecuadorensis IPNI=*684911-1* 376
Plantago elata IPNI=*321285-2* 376
Plantago eriophora IPNI=*321285-2* 376
Plantago eriorrhiza IPNI=*685346-1* 376
Plantago exaltata IPNI=*321286-2* 376
Plantago flexuosa IPNI=*321285-2* 376
Plantago fonticola IPNI=*321286-2* 376
Plantago fontis-curvae IPNI=*321285-2* 376
Plantago gigantea IPNI=*684911-1* 376
Plantago gigas IPNI=*321286-2* 376
Plantago glabriflora IPNI=*321285-2* 376
Plantago glareosa IPNI=*321285-2* 376
Plantago glauca IPNI=*321285-2* 376
Plantago gouanii IPNI=*321286-2* 376
Plantago gracilis IPNI=*321286-2* 376
Plantago hillii IPNI=*685641-1* 376
Plantago hirsuta IPNI=*321285-2* 376
Plantago hirtella var. *orbignyana* IPNI=*201466-2* 376
Plantago hirtella var. *platensis* IPNI=*684911-1* 376
Plantago humifusa IPNI=*321286-2* 376
Plantago hungarica IPNI=*321285-2* 376
Plantago hypolasia IPNI=*684911-1* 376
Plantago intermedia IPNI=*321285-2* 376
Plantago irrigua IPNI=*321285-2* 376
Plantago japonica f. *polystachya* IPNI=*321286-2* 376
Plantago japonica IPNI=*321286-2* 376
Plantago kurdica IPNI=*321285-2* 376
Plantago kurtzii IPNI=*684911-1* 376
Plantago laciniosa IPNI=*321286-2* 376
Plantago lanata IPNI=*321285-2* 376
Plantago lanceifolia IPNI=*321285-2* 376
Plantago lanceolata f. *composita* IPNI=*321285-2* 376
Plantago lanceolata subsp. *eriophora* IPNI=*321285-2* 376
Plantago lanceolata var. *angustifolia* IPNI=*321285-2* 376
Plantago lanceolata var. *communis* IPNI=*321285-2* 376
Plantago lanceolata var. *dubia* IPNI=*321285-2* 376
Plantago lanceolata var. *erecta* IPNI=*321285-2* 376
Plantago lanceolata var. *eriophylla* IPNI=*321285-2* 376
Plantago lanceolata var. *lanuginosa* IPNI=*321285-2* 376
Plantago lanceolata var. *minor* IPNI=*321285-2* 376
Plantago lanceolata var. *sphaerostachya* IPNI=*321285-2* 376
Plantago lanuginosa IPNI=*321285-2* 376
Plantago latifolia IPNI=*321286-2* 376
Plantago leiopetala IPNI=*321285-2* 376
Plantago limosa IPNI=*321286-2* 376
Plantago lindeniana IPNI=*685641-1* 376
Plantago linearis f. *barneoudii* IPNI=*685346-1* 376
Plantago linearis f. *supina* IPNI=*685346-1* 376
Plantago linearis var. *capitata* IPNI=*685346-1* 376
Plantago linearis var. *eulinearis* IPNI=*685346-1* 376
Plantago linearis var. *lasiophylla* IPNI=*685346-1* 376
Plantago linearis var. *stenophylla* IPNI=*685346-1* 376
Plantago linearis var. *villosa* IPNI=*685346-1* 376
Plantago linkii IPNI=*321285-2* 376
Plantago littoralis IPNI=*321286-2* 376
Plantago longiscapa IPNI=*321285-2* 376
Plantago longiscapa IPNI=*321286-2* 376
Plantago longistipes IPNI=*321285-2* 376
Plantago loureiroi IPNI=*321286-2* 376
Plantago luzuloidea IPNI=*685346-1* 376
Plantago macbridei IPNI=*685641-1* 376
Plantago macronipponica IPNI=*321286-2* 376
Plantago macropus IPNI=*684911-1* 376
Plantago macrostachys IPNI=*684911-1* 376
Plantago macrostachys subvar. *catamarcensis* IPNI=*684911-1* 376
Plantago macrostachys var. *accrescens* IPNI=*684911-1* 376
Plantago macrostachys var. *denudata* IPNI=*684911-1* 376
Plantago macrostachys var. *gigantea* IPNI=*684911-1* 376
Plantago macrostachys var. *platensis* IPNI=*684911-1* 376
Plantago macrostachys var. *stuckertii* IPNI=*684911-1* 376
Plantago macrostachys var. *typica* IPNI=*684911-1* 376
Plantago major f. *brachystachya* IPNI=*321286-2* 376
Plantago major f. *crispa* IPNI=*321286-2* 376
Plantago major f. *dura* IPNI=*321286-2* 376
Plantago major f. *lancifolia* IPNI=*321286-2* 376
Plantago major f. *latissima* IPNI=*321286-2* 376
Plantago major f. *leptostachya* IPNI=*321286-2* 376
Plantago major f. *maxima* IPNI=*321286-2* 376
Plantago major f. *megastachya* IPNI=*321286-2* 376
Plantago major f. *minima* IPNI=*321286-2* 376
Plantago major f. *minor* IPNI=*321286-2* 376
Plantago major f. *nervata* IPNI=*321286-2* 376
Plantago major f. *pleiosperma* IPNI=*321286-2* 376
Plantago major f. *polystachya* IPNI=*321286-2* 376
Plantago major f. *rotundata* IPNI=*321286-2* 376
Plantago major f. *scopulorum* IPNI=*321286-2* 376
Plantago major f. *sinuata* IPNI=*321286-2* 376
Plantago major subf. *dura* IPNI=*321286-2* 376
Plantago major subf. *subintegra* IPNI=*321286-2* 376
Plantago major subsp. *pachystachya* IPNI=*321286-2* 376
Plantago major subsp. *rigidifolia* IPNI=*321286-2* 376
Plantago major subsp. *sinuata* IPNI=*321286-2* 376
Plantago major subvar. *incisa* IPNI=*321286-2* 376
Plantago major var. *angustata* IPNI=*321286-2* 376
Plantago major var. *brachystachya* IPNI=*321286-2* 376
Plantago major var. *brasiliensis* IPNI=*321286-2* 376
Plantago major var. *compressiscapa* IPNI=*321286-2* 376
Plantago major var. *crispa* IPNI=*321286-2* 376
Plantago major var. *dumanii* IPNI=*321286-2* 376

Plantago major var. *flavovirens* IPNI=*321286-2* 376
Plantago major var. *japonica* IPNI=*321286-2* 376
Plantago major var. *jehohlensis* IPNI=*321286-2* 376
Plantago major var. *leptostachya* IPNI=*321286-2* 376
Plantago major var. *luxuriosa* IPNI=*321286-2* 376
Plantago major var. *maxima* IPNI=*321286-2* 376
Plantago major var. *megastachya* IPNI=*321286-2* 376
Plantago major var. *microstachya* IPNI=*321286-2* 376
Plantago major var. *minima* IPNI=*321286-2* 376
Plantago major var. *pachyphylla* IPNI=*321286-2* 376
Plantago major var. *pachystachya* IPNI=*321286-2* 376
Plantago major var. *pilgeri* IPNI=*321286-2* 376
Plantago major var. *pilosiuscula* IPNI=*321286-2* 376
Plantago major var. *polystachya* IPNI=*321286-2* 376
Plantago major var. *pubescens* IPNI=*321286-2* 376
Plantago major var. *rigidifolia* IPNI=*321286-2* 376
Plantago major var. *runcinata* IPNI=*321286-2* 376
Plantago major var. *scopulorum* IPNI=*321286-2* 376
Plantago major var. *sinuata* IPNI=*321286-2* 376
Plantago major var. *tropica* IPNI=*321286-2* 376
Plantago major var. *ungavensis* IPNI=*321286-2* 376
Plantago maxima IPNI=*321286-2* 376
Plantago microcephala IPNI=*321285-2* 376
Plantago minima IPNI=*321286-2* 376
Plantago minor IPNI=*321285-2* 376
Plantago minor IPNI=*321286-2* 376
Plantago monticola subsp. *angusta* IPNI=*685641-1* 376
Plantago monticola subsp. *eumonticola* IPNI=*685440-1* 376
Plantago monticola subsp. *sericans* IPNI=*685440-1* 376
Plantago monticola var. *humillima* IPNI=*685641-1* 376
Plantago multistachya IPNI=*321286-2* 376
Plantago myosuros var. *latifolia* IPNI=*684911-1* 376
Plantago namikawae IPNI=*321286-2* 376
Plantago nana IPNI=*321286-2* 376
Plantago nigricans IPNI=*321285-2* 376
Plantago nubigena var. *candida* IPNI=*685641-1* 376
Plantago nubigena var. *macrocephala* IPNI=*685641-1* 376
Plantago nutans IPNI=*321285-2* 376
Plantago officinarum IPNI=*321286-2* 376
Plantago orbignyana var. *lasiantha* IPNI=*201466-2* 376
Plantago orientalis IPNI=*321285-2* 376
Plantago pachyneura subsp. *mollior* IPNI=*684911-1* 376
Plantago pauciflora IPNI=*321286-2* 376
Plantago pflanzii var. *mollior* IPNI=*684911-1* 376
Plantago polystachia IPNI=*321286-2* 376
Plantago polystachya IPNI=*321286-2* 376
Plantago preslii IPNI=*321285-2* 376
Plantago pseudopatagonica IPNI=*321285-2* 376
Plantago quinquenervis IPNI=*321286-2* 376
Plantago refracta IPNI=*684911-1* 376
Plantago rocae IPNI=*321286-2* 376
Plantago rosea IPNI=*321286-2* 376
Plantago sawadae IPNI=*321286-2* 376
Plantago schottii IPNI=*321285-2* 376
Plantago scopulorum IPNI=*321286-2* 376
Plantago sericea subsp. *argyrophylla* IPNI=*685641-1* 376
Plantago sericea subsp. *sericans* IPNI=*685440-1* 376
Plantago sericea var. *congesta* IPNI=*685641-1* 376
Plantago sericea var. *linearis* IPNI=*685346-1* 376
Plantago sericea var. *luzuloidea* IPNI=*685346-1* 376
Plantago sericea var. *monticola* IPNI=*685440-1* 376
Plantago sinuata IPNI=*321286-2* 376
Plantago sorokini IPNI=*321286-2* 376
Plantago sphaerostachya IPNI=*321285-2* 376
Plantago stuckertii IPNI=*684911-1* 376
Plantago stuckertii subsp. *catamarcensis* IPNI=*684911-1* 376
Plantago stylosa IPNI=*321286-2* 376
Plantago subsinuata IPNI=*321286-2* 376
Plantago succisa IPNI=*321285-2* 376
Plantago sylvatica IPNI=*321285-2* 376
Plantago tabernaemontani IPNI=*321286-2* 376
Plantago togashii IPNI=*321286-2* 376
Plantago tomentosa IPNI=*321285-2* 376
Plantago tomentosa subsp. *hypolasia* IPNI=*684911-1* 376
Plantago trinervis IPNI=*321285-2* 376
Plantago urvillei IPNI=*684911-1* 376
Plantago variabilis IPNI=*321285-2* 376
Plantago vestita IPNI=*685641-1* 376
Plantago villifera IPNI=*321286-2* 376
Plantago vulgaris IPNI=*321286-2* 376
Plantago xorulensis IPNI=*685346-1* 376
Plantago yezomaritima IPNI=*321285-2* 376
Platanocephalus orientalis IPNI=*756359-1* 352
Platanthera garberi IPNI=*637263-1* 219
Platantheroides floribunda IPNI=*637263-1* 219
Platantheroides herzogii IPNI=*637263-1* 219
Platantheroides odontopetala IPNI=*637263-1* 219
Platanus densicoma IPNI=*685871-1* 377
Platanus excelsa IPNI=*685871-1* 377
Platanus integrifolia IPNI=*685871-1* 377
Platanus lobata IPNI=*685871-1* 377
Platanus occidentalis f. *attenuata* IPNI=*685871-1* 377
Platanus orientalis var. *occidentalis* IPNI=*685871-1* 377
Platanus pyramidalis IPNI=*685871-1* 377
Platanus vulgaris var. *angulosa* IPNI=*685871-1* 377
Platenia chiragua IPNI=*670066-1* 212
Platonia esculenta IPNI=*326389-2* 385
Platonia grandiflora IPNI=*326389-2* 385
Platonia nodiflora IPNI=*194567-2* 379
Platorheedia arborea IPNI=*972610-1* 385
Platunum rubrum IPNI=*863073-1* 369
Platyaechmea chantinii IPNI=*5385-2* 228
Platyaechmea tillandsioides IPNI=*121847-1* 228
Platyaechmea zebrina IPNI=*271782-2* 228
Platycarpum schultesii var. *zarucchii* IPNI=*201723-2* 355
Platycentrum rex IPNI=*105573-1* 292
Platycerium angolense IPNI=*17177470-1* 176
Platycerium angustatum IPNI=*17177460-1* 176
Platycerium bifurcatum var. *normale* IPNI=*17177460-1* 176
Platycerium velutinum IPNI=*17177470-1* 176
Platycladus chengii IPNI=*677124-1* 180
Platycladus orientalis f. *pendula* IPNI=*677124-1* 180
Platycladus orientalis subsp. *chengii* IPNI=*677124-1* 180
Platycladus stricta IPNI=*677124-1* 180
Platyloma flexuosum IPNI=*77183106-1* 178
Platyloma ternifolium IPNI=*77182326-1* 178
Platyluma calophylloides IPNI=*160417-2* 306
Platymiscium ellipticum IPNI=*16766-2* 311
Platyopuntia apurimacensis IPNI=*1151735-2* 284
Platyopuntia cordobensis IPNI=*1151735-2* 284
Platyopuntia ficus-indica IPNI=*1151735-2* 284
Platyopuntia quitensis IPNI=*175475-2* 284
Platyruscus hypoglossum IPNI=*540452-1* 215
Platyspermum alatum IPNI=*841063-1* 252
Platystachys complanata IPNI=*125000-1* 229
Platystachys ehrenbergiana IPNI=*254406-2* 229
Platystachys ehrenbergii IPNI=*254406-2* 229
Platystachys gracilis IPNI=*121847-1* 228
Platystachys incarnata IPNI=*253931-2* 229
Platystachys utriculata IPNI=*254406-2* 229
Platystephium graveolens IPNI=*1037395-2* 261
Platythelys debilis IPNI=*77125687-1* 217
Platythelys maculata IPNI=*77125705-1* 217
Platythelys mayoriana IPNI=*77125705-1* 217
Platythelys querceticola IPNI=*77125705-1* 217
Platythelys sagrana IPNI=*77125705-1* 217

Platythelys vaginata IPNI=*77125705-1* 217
Plectranthus amboinicus IPNI=*445902-1* 369
Plectranthus aromaticus IPNI=*446165-1* 369
Plectranthus bernardii IPNI=*446113-1* 369
Plectranthus blumei IPNI=*446165-1* 369
Plectranthus coleoides IPNI=*446113-1* 369
Plectranthus forsteri IPNI=*77201075-1* 369
Plectranthus glabratus IPNI=*446113-1* 369
Plectranthus ingratus IPNI=*446165-1* 369
Plectranthus laciniatus IPNI=*446165-1* 369
Plectranthus merrillii IPNI=*446165-1* 369
Plectranthus monachorum IPNI=*453130-1* 371
Plectranthus nudiflorus IPNI=*446165-1* 369
Plectranthus nummularius IPNI=*454770-1* 372
Plectranthus scutellarioides IPNI=*446165-1* 369
Plectranthus thunbergii IPNI=*454770-1* 372
Pleiarina humboldtiana IPNI=*777818-1* 404
Pleioblastus argenteostriatus f. *glaber* IPNI=*416271-1* 244
Pleioblastus dimorphophyllus IPNI=*416271-1* 244
Pleioblastus pubescens IPNI=*416271-1* 244
Pleioblastus pygmaeus f. *ramosissimus* IPNI=*416271-1* 244
Pleioblastus pygmaeus IPNI=*416271-1* 244
Pleioblastus ramosissimus IPNI=*416271-1* 244
Pleioblastus shibuyanus f. *pubescens* IPNI=*416271-1* 244
Pleioblastus shibuyanus IPNI=*416271-1* 244
Pleioblastus shibuyanus var. *basihirsutus* IPNI=*416271-1* 244
Pleioblastus tectus IPNI=*416271-1* 244
Pleioblastus tosaensis IPNI=*416271-1* 244
Pleioblastus tsukubensis IPNI=*416271-1* 244
Pleioblastus uyenoensis IPNI=*416271-1* 244
Pleioblastus variegatus IPNI=*416271-1* 244
Pleioblastus yasuianus IPNI=*416271-1* 244
Pleioblastus yoshidake IPNI=*416271-1* 244
Pleomele aloifolia IPNI=*77164234-1* 215
Pleomele angustifolia IPNI=*534103-1* 215
Pleomele angustifolia var. *angustior* IPNI=*534103-1* 215
Pleomele australasica IPNI=*534103-1* 215
Pleomele braunii IPNI=*534129-1* 215
Pleomele deremensis IPNI=*534207-1* 215
Pleomele flexuosa IPNI=*534103-1* 215
Pleomele fragrans IPNI=*534207-1* 215
Pleomele fruticosa IPNI=*534103-1* 215
Pleomele reflexa IPNI=*534338-1* 215
Pleomele reflexa var. *angustifolia* IPNI=*534103-1* 215
Pleomele smithii IPNI=*534207-1* 215
Pleomele ugandensis IPNI=*534207-1* 215
Pleonotoma brittonii IPNI=*202082-2* 364
Pleonotoma decomposita IPNI=*77100463-1* 363
Pleonotoma inaequalis IPNI=*77100463-1* 363
Pleonotoma tetragonocaulis IPNI=*110423-1* 364
Pleonotoma triternata IPNI=*77100463-1* 363
Pleopeltis antrophyoides IPNI=*17341250-1* 176
Pleopeltis araneosa IPNI=*17175380-1* 176
Pleopeltis areolata IPNI=*17175380-1* 176
Pleopeltis aurea IPNI=*17175390-1* 176
Pleopeltis corcovadensis IPNI=*17448650-1* 176
Pleopeltis crassifolia IPNI=*77183236-1* 175
Pleopeltis cuspidata IPNI=*17546710-1* 175
Pleopeltis decumana IPNI=*17175400-1* 176
Pleopeltis ensifolia IPNI=*17448650-1* 176
Pleopeltis glaucina IPNI=*17175380-1* 176
Pleopeltis helenae IPNI=*17448650-1* 176
Pleopeltis incana IPNI=*1039912-2* 176
Pleopeltis irioides IPNI=*17341250-1* 176
Pleopeltis kaulfussiana IPNI=*17448650-1* 176
Pleopeltis lanceolata IPNI=*17448650-1* 176
Pleopeltis lepidota IPNI=*17448650-1* 176
Pleopeltis leprieurii IPNI=*160142-2* 175
Pleopeltis linearis IPNI=*17448650-1* 176
Pleopeltis longifolia IPNI=*17341250-1* 176
Pleopeltis lycopodioides IPNI=*160142-2* 175
Pleopeltis lycopodioides var. *salicifolia* IPNI=*160142-2* 175
Pleopeltis marginalis IPNI=*17448650-1* 176
Pleopeltis megalosoroides IPNI=*17341250-1* 176
Pleopeltis millisora IPNI=*17341250-1* 176
Pleopeltis neoguineensis IPNI=*17341250-1* 176
Pleopeltis obtusata IPNI=*17341250-1* 176
Pleopeltis percussa IPNI=*17546710-1* 175
Pleopeltis polycarpa IPNI=*17341250-1* 176
Pleopeltis pulverulenta IPNI=*17341250-1* 176
Pleopeltis punctata IPNI=*17341250-1* 176
Pleopeltis salicifolia IPNI=*160142-2* 175
Pleopeltis sessilis IPNI=*17341250-1* 176
Pleopeltis stigmatica IPNI=*160142-2* 175
Pleopeltis surinamensis IPNI=*160142-2* 175
Pleopeltis valida IPNI=*17341250-1* 176
Pleopeltis viridis IPNI=*17341250-1* 176
Pleroma holosericeum IPNI=*1024004-2* 421
Pleroma ledifolium IPNI=*566862-1* 418
Pleroma lepidotum IPNI=*77205900-1* 417
Pleroma strigosum IPNI=*566891-1* 418
Plesisa bipartita IPNI=*526777-1* 374
Plethyrsis glauca IPNI=*764679-1* 356
Pleuradena coccinea IPNI=*30176564-2* 390
Pleuranthodendron mexicanum IPNI=*202149-2* 404
Pleuremidis grandiflora IPNI=*117460-3* 362
Pleuridium albopunctatissimum IPNI=*77183020-1* 175
Pleuridium angustum IPNI=*77183236-1* 175
Pleuridium crassifolium IPNI=*77183236-1* 175
Pleurolobus heterocarpon IPNI=*77187416-1* 320
Pleurolobus spicatus IPNI=*1015219-2* 318
Pleurolobus triflorus IPNI=*77187443-1* 320
Pleuropteropyrum ochreatum IPNI=*77069480-1* 289
Pleurospa reticulata IPNI=*87563-1* 200
Pleurospermum toluccense IPNI=*838268-1* 251
Pleurothallis brevipes IPNI=*1147281-2* 217
Pleurothallis coccinea IPNI=*655797-1* 223
Pleurothallis modesta IPNI=*1147281-2* 217
Pleurothallis ospinae IPNI=*655513-1* 222
Pleurothallis valenzuelana IPNI=*155338-2* 222
Pleurothyrium densiflorum IPNI=*203826-2* 185
Pleurothyrium velutinum IPNI=*77167373-1* 184
Pleuteron breynia IPNI=*77217338-1* 278
Pleuteron frondosa IPNI=*77183570-1* 278
Pleuteron linearis IPNI=*77184255-1* 278
Plinia acutissima IPNI=*599327-1* 427
Plinia crocea IPNI=*203871-2* 428
Plinia jaboticaba IPNI=*203847-2* 428
Plinia jambos IPNI=*601802-1* 428
Plinia pedunculata IPNI=*596263-1* 425
Plinia pentapetala IPNI=*203871-2* 428
Plinia peruviana IPNI=*203847-2* 428
Plinia petiolata IPNI=*596263-1* 425
Plinia rubra IPNI=*596263-1* 425
Plinia tetrapetala IPNI=*596263-1* 425
Pluchea alopecuroides IPNI=*240147-1* 269
Pluchea amorifera IPNI=*1099254-2* 268
Pluchea bifrons IPNI=*238826-1* 268
Pluchea camphorata var. *angustifolia* IPNI=*1099254-2* 268
Pluchea chapmanii IPNI=*238826-1* 268
Pluchea cortesii IPNI=*238803-1* 268
Pluchea decussata IPNI=*27895-2* 257
Pluchea foetida IPNI=*238826-1* 268
Pluchea foetida var. *imbricata* IPNI=*238826-1* 268
Pluchea glabra IPNI=*183399-1* 257
Pluchea glabrata IPNI=*1099254-2* 268

Pluchea imbricata IPNI=*238826-1* 268
Pluchea kotschyi IPNI=*211317-2* 268
Pluchea marilandica IPNI=*1099254-2* 268
Pluchea montana IPNI=*183580-1* 257
Pluchea odorata var. *brevifolia* IPNI=*238803-1* 268
Pluchea odorata var. *normalis* IPNI=*1099254-2* 268
Pluchea odorata var. *succulenta* IPNI=*1099254-2* 268
Pluchea petiolata IPNI=*1099254-2* 268
Pluchea purpurascens f. *obovata* IPNI=*1099254-2* 268
Pluchea purpurascens IPNI=*1099254-2* 268
Pluchea purpurascens var. *glabrata* IPNI=*1099254-2* 268
Pluchea purpurascens var. *succulenta* IPNI=*1099254-2* 268
Pluchea quitoc IPNI=*203916-2* 268
Pluchea sagittalis IPNI=*30159244-2* 258
Pluchea senegalensis IPNI=*1099254-2* 268
Pluchea spicata IPNI=*30022662-2* 269
Pluchea suaveolens IPNI=*203916-2* 268
Pluchea symphytifolia IPNI=*234386-1* 267
Pluchea tenuifolia IPNI=*238826-1* 268
Pluchea tweediei IPNI=*203916-2* 268
Pluchea virgata IPNI=*30022662-2* 269
Plukenetia abutifolia IPNI=*203935-2* 392
Plukenetia macrostyla IPNI=*354970-1* 392
Plukenetia peruviana IPNI=*354970-1* 392
Plumaria vulgaris IPNI=*153717-1* 285
Plumbagidium auriculatum IPNI=*687056-1* 288
Plumbagidium scandens IPNI=*687109-1* 288
Plumbago alba IPNI=*687056-1* 288
Plumbago americana IPNI=*687109-1* 288
Plumbago auriculata f. *alba* IPNI=*687056-1* 288
Plumbago capensis IPNI=*687056-1* 288
Plumbago flaccida IPNI=*687109-1* 288
Plumbago floridana IPNI=*687109-1* 288
Plumbago glandulosa IPNI=*687057-1* 288
Plumbago grandiflora IPNI=*687056-1* 288
Plumbago humboldtiana IPNI=*687057-1* 288
Plumbago juncea IPNI=*687109-1* 288
Plumbago lactea IPNI=*687109-1* 288
Plumbago littoralis IPNI=*687109-1* 288
Plumbago maximowiczii IPNI=*687109-1* 288
Plumbago mexicana IPNI=*687109-1* 288
Plumbago occidentalis IPNI=*687109-1* 288
Plumbago rhomboidea IPNI=*687057-1* 288
Plumbago sarmentosa IPNI=*687109-1* 288
Plumbago scandens IPNI=*687109-1* 288
Plumbago scandens var. *densiflora* IPNI=*687109-1* 288
Plumbago scandens var. *erecta* IPNI=*687109-1* 288
Plumbago scandens var. *normalis* IPNI=*687109-1* 288
Plumbago toxicaria IPNI=*687109-1* 288
Plumbago viscosa IPNI=*687109-1* 288
Plumeria acuminata IPNI=*81275-1* 342
Plumeria acutifolia IPNI=*81275-1* 342
Plumeria alba var. *jacquiniana* IPNI=*81171-1* 342
Plumeria ambigua IPNI=*122690-2* 340
Plumeria angustifolia IPNI=*81275-1* 342
Plumeria arborea IPNI=*81275-1* 342
Plumeria arborescens IPNI=*81275-1* 342
Plumeria articulata IPNI=*122688-2* 340
Plumeria aurantia IPNI=*81275-1* 342
Plumeria aurantiaca IPNI=*81275-1* 342
Plumeria bicolor IPNI=*81275-1* 342
Plumeria bracteata IPNI=*122690-2* 340
Plumeria caracasana IPNI=*81268-1* 342
Plumeria carinata IPNI=*81275-1* 342
Plumeria cochleata IPNI=*81268-1* 342
Plumeria conspicua IPNI=*81275-1* 342
Plumeria floribunda IPNI=*122688-2* 340
Plumeria floribunda var. *acutifolia* IPNI=*122690-2* 340
Plumeria floribunda var. *calycina* IPNI=*122690-2* 340
Plumeria floribunda var. *crassipes* IPNI=*122690-2* 340
Plumeria gouanii IPNI=*81275-1* 342
Plumeria incarnata IPNI=*81275-1* 342
Plumeria incarnata var. *milleri* IPNI=*81275-1* 342
Plumeria jamesonii IPNI=*81275-1* 342
Plumeria kerrii IPNI=*81275-1* 342
Plumeria kunthiana IPNI=*81275-1* 342
Plumeria lambertiana IPNI=*81275-1* 342
Plumeria lancifolia IPNI=*122690-2* 340
Plumeria lancifolia var. *major* IPNI=*122690-2* 340
Plumeria lancifolia var. *microphylla* IPNI=*122690-2* 340
Plumeria loranthifolia IPNI=*81275-1* 342
Plumeria lutea IPNI=*81275-1* 342
Plumeria mariaelenae IPNI=*81275-1* 342
Plumeria martii IPNI=*122690-2* 340
Plumeria megaphylla IPNI=*81275-1* 342
Plumeria mexicana IPNI=*81275-1* 342
Plumeria microcalyx IPNI=*122688-2* 340
Plumeria milleri IPNI=*81275-1* 342
Plumeria mollis IPNI=*81275-1* 342
Plumeria northiana IPNI=*81275-1* 342
Plumeria phagedaenica IPNI=*122695-2* 340
Plumeria purpurea IPNI=*81275-1* 342
Plumeria revolutifolia IPNI=*81171-1* 342
Plumeria rubra f. *acutifolia* IPNI=*81275-1* 342
Plumeria rubra f. *lutea* IPNI=*81275-1* 342
Plumeria rubra f. *tricolor* IPNI=*81275-1* 342
Plumeria rubra f. *typica* IPNI=*81275-1* 342
Plumeria rubra var. *acuminata* IPNI=*81275-1* 342
Plumeria speciosa IPNI=*122690-2* 340
Plumeria sucuuba IPNI=*122688-2* 340
Plumeria tarapotensis IPNI=*962262-1* 340
Plumeria tenuifolia IPNI=*81275-1* 342
Plumeria tricolor IPNI=*81275-1* 342
Plumeriopsis ahouai IPNI=*82440-1* 344
Pneumonanthe exaltata IPNI=*367440-1* 344
Pneumonanthe remota IPNI=*108639-2* 344
Poa aestivalis IPNI=*320035-2* 244
Poa airoides IPNI=*395086-1* 235
Poa alata IPNI=*1147585-2* 239
Poa alba IPNI=*1141212-2* 239
Poa algida IPNI=*320035-2* 244
Poa amabilis IPNI=*401703-1* 239
Poa amboinensis IPNI=*1156566-2* 238
Poa amboinica IPNI=*1156566-2* 238
Poa amoena IPNI=*1147585-2* 239
Poa angustifolia IPNI=*320346-2* 244
Poa annua f. *purpurea* IPNI=*320035-2* 244
Poa annua f. *reptans* IPNI=*320035-2* 244
Poa annua subsp. *notabilis* IPNI=*320035-2* 244
Poa annua var. *eriolepis* IPNI=*320035-2* 244
Poa annua var. *pilantha* IPNI=*320035-2* 244
Poa annua var. *raniglumis* IPNI=*320035-2* 244
Poa annua var. *rigidiuscula* IPNI=*320035-2* 244
Poa ariguensis IPNI=*320346-2* 244
Poa atrovirens IPNI=*400813-1* 238
Poa attica IPNI=*320346-2* 244
Poa aturensis IPNI=*401189-1* 239
Poa aurea IPNI=*401189-1* 239
Poa bahiensis IPNI=*93057-2* 238
Poa barrelieri IPNI=*404100-1* 239
Poa biflora IPNI=*320346-2* 244
Poa biflora IPNI=*401189-1* 239
Poa biformis IPNI=*400813-1* 238
Poa bipollicaris IPNI=*320035-2* 244
Poa borealis IPNI=*82343-2* 238
Poa boryana IPNI=*1156566-2* 238

Poa brasiliensis IPNI=*401189-1* 239
Poa cachectica IPNI=*1104599-2* 238
Poa callida IPNI=*320346-2* 244
Poa cambessediana IPNI=*401103-1* 239
Poa capensis IPNI=*1169955-2* 239
Poa caroliniana IPNI=*1015280-2* 239
Poa chariis IPNI=*400813-1* 238
Poa chilensis IPNI=*1180620-2* 239
Poa cilianensis IPNI=*1104599-2* 238
Poa ciliaris IPNI=*1156566-2* 238
Poa compacta IPNI=*322572-2* 239
Poa compta IPNI=*1156566-2* 238
Poa conferta IPNI=*401189-1* 239
Poa curvula IPNI=*1169955-2* 239
Poa dactyloides IPNI=*82343-2* 238
Poa delicatior IPNI=*1180620-2* 239
Poa despiciens IPNI=*1088685-2* 239
Poa diandra IPNI=*1015280-2* 239
Poa diandra IPNI=*401189-1* 239
Poa diarrhena IPNI=*401189-1* 239
Poa domingensis IPNI=*1075146-2* 239
Poa dubia IPNI=*320346-2* 244
Poa dulcis IPNI=*395086-1* 235
Poa elegans IPNI=*1156566-2* 238
Poa elegans IPNI=*400813-1* 238
Poa elegans IPNI=*401189-1* 239
Poa elegantula IPNI=*400813-1* 238
Poa elliptica IPNI=*1156566-2* 238
Poa eragrostis IPNI=*1104599-2* 238
Poa eragrostis IPNI=*1180620-2* 239
Poa expansa IPNI=*93057-2* 238
Poa fascicularis IPNI=*1075146-2* 239
Poa feratiana IPNI=*320346-2* 244
Poa filiformis IPNI=*1169955-2* 239
Poa flava IPNI=*1104599-2* 238
Poa flexuosa IPNI=*1104599-2* 238
Poa floribunda IPNI=*401103-1* 239
Poa fluitans IPNI=*404100-1* 239
Poa fusca IPNI=*60458632-2* 238
Poa gangetica IPNI=*401103-1* 239
Poa glomerata IPNI=*401189-1* 239
Poa grandiflora IPNI=*395086-1* 235
Poa haenkei IPNI=*1075146-2* 239
Poa hansiana IPNI=*320035-2* 244
Poa hohenackeri IPNI=*320035-2* 244
Poa humilis IPNI=*320035-2* 244
Poa indica IPNI=*1088685-2* 239
Poa indica IPNI=*1180620-2* 239
Poa interrupta IPNI=*322572-2* 239
Poa interrupta IPNI=*401189-1* 239
Poa japonica IPNI=*401189-1* 239
Poa kitaibelii IPNI=*320346-2* 244
Poa koenigii IPNI=*401189-1* 239
Poa leersioides IPNI=*1104599-2* 238
Poa linearis IPNI=*1075146-2* 239
Poa linkii IPNI=*1180620-2* 239
Poa lobata IPNI=*1156566-2* 238
Poa maxima IPNI=*393293-1* 235
Poa maypurensis IPNI=*1147585-2* 239
Poa megastachya IPNI=*1104599-2* 238
Poa meratiana IPNI=*1147585-2* 239
Poa mexicana IPNI=*1141212-2* 239
Poa mexicana IPNI=*1180620-2* 239
Poa meyenii IPNI=*320035-2* 244
Poa michauxii IPNI=*82343-2* 238
Poa microstachya IPNI=*93057-2* 238
Poa modesta IPNI=*320346-2* 244
Poa monandra IPNI=*77116853-1* 236
Poa nuttallii IPNI=*1015280-2* 239
Poa nuttallii IPNI=*1104599-2* 238
Poa oblonga IPNI=*1104599-2* 238
Poa obtusa IPNI=*1104599-2* 238
Poa ovalis IPNI=*320035-2* 244
Poa ovina IPNI=*401103-1* 239
Poa oxylepis IPNI=*322572-2* 239
Poa palustris IPNI=*320346-2* 244
Poa panicea IPNI=*320346-2* 244
Poa panicea IPNI=*407138-1* 241
Poa pectinacea IPNI=*1015280-2* 239
Poa pennsylvanica IPNI=*1104599-2* 238
Poa perouinii IPNI=*320346-2* 244
Poa philadelphica IPNI=*1104599-2* 238
Poa pichinchensis IPNI=*320346-2* 244
Poa pilosa IPNI=*1180620-2* 239
Poa plumosa IPNI=*1088685-2* 239
Poa plumosa IPNI=*1180620-2* 239
Poa poiretii IPNI=*1180620-2* 239
Poa pollinensis IPNI=*320346-2* 244
Poa polymorpha IPNI=*1104599-2* 238
Poa portalii IPNI=*320346-2* 244
Poa prolifera IPNI=*1075146-2* 239
Poa prostrata IPNI=*82343-2* 238
Poa psammodes IPNI=*93057-2* 238
Poa pseudairoides IPNI=*395086-1* 235
Poa pseudamabilis IPNI=*1088685-2* 239
Poa pseudohybrida IPNI=*320346-2* 244
Poa puberula IPNI=*320035-2* 244
Poa punctata IPNI=*1180620-2* 239
Poa racemosa IPNI=*1147585-2* 239
Poa repens IPNI=*82343-2* 238
Poa rivularis IPNI=*395086-1* 235
Poa rothii IPNI=*401189-1* 239
Poa roxburghiana IPNI=*1104599-2* 238
Poa sabauda IPNI=*395086-1* 235
Poa scabra IPNI=*320346-2* 244
Poa secundiflora IPNI=*322572-2* 239
Poa semineutra IPNI=*320346-2* 244
Poa senegalensis IPNI=*1180620-2* 239
Poa setacea IPNI=*320346-2* 244
Poa speciosa IPNI=*1088685-2* 239
Poa spicata IPNI=*74675-3* 238
Poa sporoboloides IPNI=*401189-1* 239
Poa squarrosa IPNI=*1147585-2* 239
Poa stenoclada IPNI=*401103-1* 239
Poa stolonifera IPNI=*320346-2* 244
Poa stricta IPNI=*401189-1* 239
Poa subulata IPNI=*1075146-2* 239
Poa subulata IPNI=*82343-2* 238
Poa sylvicola IPNI=*320346-2* 244
Poa tenella IPNI=*1015280-2* 239
Poa tenella IPNI=*1088685-2* 239
Poa tenella IPNI=*1180620-2* 239
Poa tenella IPNI=*401189-1* 239
Poa tenuifolia IPNI=*1064189 2* 239
Poa tenuifolia var. *oregona* IPNI=*30010036-2* 239
Poa tenuissima IPNI=*1088685-2* 239
Poa thalassica IPNI=*82343-2* 238
Poa tortuosa IPNI=*1104599-2* 238
Poa trachyphylla IPNI=*320346-2* 244
Poa tristriata IPNI=*320346-2* 244
Poa triticea IPNI=*1104599-2* 238
Poa trivialis f. *humilis* IPNI=*320346-2* 244
Poa trivialis f. *prorepens* IPNI=*320346-2* 244
Poa trivialis subsp. *feratiana* IPNI=*320346-2* 244
Poa trivialis subsp. *latifolia* IPNI=*320346-2* 244
Poa trivialis subsp. *semineutra* IPNI=*320346-2* 244

Poa trivialis var. *holsatia* IPNI=*320346-2* 244
Poa trivialis var. *latifolia* IPNI=*320346-2* 244
Poa trivialis var. *obtusata* IPNI=*320346-2* 244
Poa uda IPNI=*320346-2* 244
Poa unioloides IPNI=*1088685-2* 239
Poa vahlii IPNI=*1147585-2* 239
Poa variegata IPNI=*320035-2* 244
Poa verticillata IPNI=*1180620-2* 239
Poa viscosa IPNI=*401703-1* 239
Poa walteri IPNI=*401189-1* 239
Poa woronowii IPNI=*320346-2* 244
Poarchon fluminensis IPNI=*441642-1* 217
Poarium veronicoides IPNI=*244449-2* 377
Poarium verticillatum IPNI=*244449-2* 377
Pochota glabra IPNI=*562664-1* 412
Pochota quinata IPNI=*559120-1* 407
Pochota sessilis IPNI=*281912-2* 412
Pochota speciosa IPNI=*281913-2* 412
Pocilla persica IPNI=*30198355-2* 377
Podachaenium andinum IPNI=*238923-1* 268
Podachaenium paniculatum IPNI=*238923-1* 268
Podaechmea mexicana IPNI=*271722-2* 228
Podocarpus allenii IPNI=*204582-2* 180
Podocarpus angustifolius IPNI=*204598-2* 180
Podocarpus ballivianensis IPNI=*263553-1* 180
Podocarpus guatemalensis subsp. *allenii* IPNI=*204582-2* 180
Podocarpus guatemalensis subsp. *pinetorum* IPNI=*204582-2* 180
Podocarpus guatemalensis var. *allenii* IPNI=*204582-2* 180
Podocarpus harmsianus IPNI=*77203787-1* 180
Podocarpus ingensis IPNI=*263553-1* 180
Podocarpus macrostachys IPNI=*263553-1* 180
Podocarpus oleifolius subsp. *equadorensis* IPNI=*263553-1* 180
Podocarpus oleifolius subsp. *trujillensis* IPNI=*263553-1* 180
Podocarpus oleifolius var. *equadorensis* IPNI=*263553-1* 180
Podocarpus oleifolius var. *macrostachys* IPNI=*263553-1* 180
Podocarpus oleifolius var. *trujillensis* IPNI=*263553-1* 180
Podocarpus pinetorum IPNI=*204582-2* 180
Podocarpus rospigliosii IPNI=*946471-1* 180
Podocarpus utilior IPNI=*77203787-1* 180
Podogyne pentaphylla IPNI=*60518-2* 279
Podosemum affine IPNI=*60445990-2* 241
Podosemum debile IPNI=*409081-1* 241
Podosemum elegans IPNI=*60445990-2* 241
Podosemum glabratum IPNI=*60445990-2* 241
Podosemum rigidum IPNI=*60445990-2* 241
Podosemum setosum IPNI=*409081-1* 241
Podosemum virginicum IPNI=*423010-1* 245
Podospadix reticulata IPNI=*84707-1* 198
Podospermum taraxacifolium IPNI=*225504-1* 264
Podostachys hirta IPNI=*70337-2* 389
Poecilodermis populnea IPNI=*822268-1* 407
Pogoblephis missurica IPNI=*108639-2* 344
Pogomesia erecta IPNI=*173146-1* 224
Pogomesia undata IPNI=*173146-1* 224
Pogonatherum amaurum IPNI=*963150-1* 244
Pogonema carinata IPNI=*270933-2* 214
Pogonophora cunuri IPNI=*159993-2* 392
Pogonophora schomburgkiana f. *elliptica* IPNI=*354987-1* 401
Pogonophora schomburgkiana var. *longifolia* IPNI=*354987-1* 401
Pogonopus caracasensis IPNI=*204802-2* 355
Pogonopus ottonis IPNI=*204802-2* 355
Pogonopus speciosus subsp. *sandwithianus* IPNI=*204802-2* 355
Pogonorhynchus amplexans IPNI=*1088633-2* 423
Pogonorhynchus sessilifolius IPNI=*1088633-2* 423
Pogonostylis squarrosus IPNI=*308286-1* 232
Pogospermum berteronianum IPNI=*50112-2* 228
Pogostemon perilloides IPNI=*158242-2* 371
Poidium monandrum IPNI=*77116853-1* 236
Poinciana alata IPNI=*1164658-2* 312
Poinciana bijuga IPNI=*1164658-2* 312
Poinciana cassioides IPNI=*482664-1* 312
Poinciana coriaria IPNI=*1010207-2* 325
Poinciana hispida IPNI=*1164658-2* 312
Poinciana insignis IPNI=*482664-1* 312
Poinciana mucronata IPNI=*249443-2* 335
Poinciana pulcherrima f. *flava* IPNI=*1164658-2* 312
Poinciana pulcherrima IPNI=*1164658-2* 312
Poinciana pulcherrima var. *flava* IPNI=*1164658-2* 312
Poinciana pulchra IPNI=*1164658-2* 312
Poinciana punctata IPNI=*139695-2* 325
Poinciana regia IPNI=*491231-1* 317
Poinciana roxburghii IPNI=*513021-1* 330
Poinciana spinosa IPNI=*249443-2* 335
Poinciana tara IPNI=*249443-2* 335
Poincianella pluviosa IPNI=*77158081-1* 314
Poinsettia heterophylla IPNI=*30062970-2* 390
Poinsettia lancifolia IPNI=*347056-1* 390
Poinsettia pulcherrima IPNI=*30176564-2* 390
Poinsettia pulcherrima var. *albida* IPNI=*30176564-2* 390
Poinsettia pulcherrima var. *roseocarminata* IPNI=*30176564-2* 390
Poivrea alternifolia IPNI=*170030-1* 416
Poivrea phaneropetala IPNI=*63788-2* 416
Poivrea roxburghii IPNI=*170030-1* 416
Polanisia icosandra IPNI=*320133-2* 280
Polanisia microphylla IPNI=*320133-2* 280
Polanisia viscosa IPNI=*320133-2* 280
Polanisia viscosa var. *genuina* IPNI=*320133-2* 280
Polanisia viscosa var. *speciosa* IPNI=*320133-2* 280
Polemonium drummondii IPNI=*193463-2* 303
Polianthes gracilis IPNI=*77153983-1* 214
Polianthes tuberosa f. *plena* IPNI=*77153983-1* 214
Polianthes tuberosa IPNI=*77153983-1* 214
Pollalesta acuminata IPNI=*988439-1* 268
Pollalesta argentea IPNI=*988437-1* 268
Pollalesta brasiliana IPNI=*988437-1* 268
Pollalesta colombiana IPNI=*988437-1* 268
Pollalesta condensata IPNI=*988439-1* 268
Pollalesta corei IPNI=*988437-1* 268
Pollalesta discolor IPNI=*988437-1* 268
Pollalesta ecuatoriana IPNI=*988437-1* 268
Pollalesta ferruginea IPNI=*988437-1* 268
Pollalesta karstenii IPNI=*988437-1* 268
Pollalesta klugii IPNI=*988437-1* 268
Pollalesta lilacea IPNI=*988437-1* 268
Pollalesta peruviana IPNI=*988437-1* 268
Pollichia amplexicaulis IPNI=*448797-1* 370
Pollichia galeobdolon IPNI=*60452701-2* 370
Pollinia brevifolia IPNI=*1126606-2* 244
Pollinia fastigiata IPNI=*398608-1* 237
Pollinia fulta IPNI=*420951-1* 244
Pollinia jacquinii IPNI=*409823-1* 242
Pollinia microstachya IPNI=*420918-1* 244
Pollinia praemorsa IPNI=*963150-1* 244
Pollinia undata IPNI=*409823-1* 242
Pollinia vaginata IPNI=*1126606-2* 244
Polyboea crassifolia IPNI=*60458979-2* 298
Polyboea quereme IPNI=*1064712-2* 298
Polyboea velutina IPNI=*60458979-2* 298
Polybotrya aristeguietae IPNI=*30017613-2* 176
Polybotrya crassa IPNI=*17181340-1* 176
Polybotrya filiculifolia IPNI=*30017613-2* 176
Polybotrya hartii IPNI=*17181340-1* 176
Polybotrya trapezoides IPNI=*30017613-2* 176
Polybotrya vareschii IPNI=*30017613-2* 176
Polycarpon uniflorum IPNI=*235145-2* 280
Polycnemum bonariense IPNI=*45444-2* 285

Polycnemum scleranthoides IPNI=*45444-2* 285
Polycodium arcuatum IPNI=*261823-2* 300
Polycodium sericeum f. *chrysocarpon* IPNI=*261823-2* 300
Polycodium sericeum IPNI=*261823-2* 300
Polycodium sericeum var. *eburneum* IPNI=*261823-2* 300
Polycodium sericeum var. *elongatum* IPNI=*261823-2* 300
Polygala acuminata IPNI=*77121649-1* 337
Polygala aestuans IPNI=*690457-1* 338
Polygala alopecurus IPNI=*691063-1* 338
Polygala amaniensis IPNI=*330527-2* 338
Polygala americana var. *angustifolia* IPNI=*77121670-1* 337
Polygala anatina IPNI=*77119197-1* 337
Polygala angustifolia f. *ramosa* IPNI=*77121670-1* 337
Polygala angustifolia IPNI=*691242-1* 338
Polygala angustifolia IPNI=*77121670-1* 337
Polygala angustifolia var. *linearifolia* IPNI=*77121670-1* 337
Polygala bahiensis IPNI=*77121670-1* 337
Polygala brasiliensis IPNI=*330527-2* 338
Polygala brizoides var. *latifolia* IPNI=*77121670-1* 337
Polygala bukobensis IPNI=*1003233-1* 338
Polygala camporum IPNI=*77121670-1* 337
Polygala caripensis IPNI=*77121670-1* 337
Polygala carlotina IPNI=*330527-2* 338
Polygala chinensis var. *normalis* IPNI=*691242-1* 338
Polygala chinensis var. *pygmaea* IPNI=*691242-1* 338
Polygala chinensis var. *villosa* IPNI=*691242-1* 338
Polygala chuiti IPNI=*691063-1* 338
Polygala cineraria IPNI=*77121670-1* 337
Polygala cinerea IPNI=*77121670-1* 337
Polygala densiflora IPNI=*691242-1* 338
Polygala depauperata IPNI=*691242-1* 338
Polygala diversifolia f. *elliptica* IPNI=*231560-2* 338
Polygala diversifolia IPNI=*231560-2* 338
Polygala exigua IPNI=*691242-1* 338
Polygala fimbriata IPNI=*77121649-1* 337
Polygala funkii IPNI=*691063-1* 338
Polygala galioides var. *asperuloides* IPNI=*205398-2* 338
Polygala glomerata IPNI=*691242-1* 338
Polygala glomerata var. *pygmaea* IPNI=*691242-1* 338
Polygala glomerata var. *villosa* IPNI=*691242-1* 338
Polygala gracilis IPNI=*691063-1* 338
Polygala grandiflora IPNI=*691242-1* 338
Polygala hostmannii IPNI=*692483-1* 338
Polygala huberiana IPNI=*77121670-1* 337
Polygala humilis IPNI=*330527-2* 338
Polygala lagoana IPNI=*691063-1* 338
Polygala leptocaulis IPNI=*691063-1* 338
Polygala leptocaulis var. *glochidiata* IPNI=*691063-1* 338
Polygala linariifolia IPNI=*691242-1* 338
Polygala linearis IPNI=*691242-1* 338
Polygala macerrima IPNI=*691063-1* 338
Polygala martii IPNI=*77121670-1* 337
Polygala monticola var. *brizoides* IPNI=*77121670-1* 337
Polygala mucronata IPNI=*690504-1* 338
Polygala nelsonii IPNI=*692483-1* 338
Polygala orobus IPNI=*77121670-1* 337
Polygala paludosa var. *angustocarpa* IPNI=*691063-1* 338
Polygala paludosa var. *exappendiculata* IPNI=*691063-1* 338
Polygala paludosa var. *glochidiata* IPNI=*691063-1* 338
Polygala paludosa var. *longe-spicata* IPNI=*691063-1* 338
Polygala paludosa var. *subsessiliflora* IPNI=*691063-1* 338
Polygala paniculata f. *humilis* IPNI=*330527-2* 338
Polygala paniculata f. *leucoptera* IPNI=*330527-2* 338
Polygala paniculata f. *violacea* IPNI=*330527-2* 338
Polygala paniculata var. *ambigua* IPNI=*330527-2* 338
Polygala paniculata var. *brasiliana* IPNI=*330527-2* 338
Polygala paniculata var. *normalis* IPNI=*330527-2* 338
Polygala paniculata var. *verticillata* IPNI=*330527-2* 338
Polygala pringlei IPNI=*691063-1* 338
Polygala procumbens IPNI=*691242-1* 338
Polygala psammophila IPNI=*692483-1* 338
Polygala puberula IPNI=*77121670-1* 337
Polygala pubescens IPNI=*77121670-1* 337
Polygala ramosa IPNI=*691242-1* 338
Polygala ramosissima IPNI=*330527-2* 338
Polygala remansoensis IPNI=*692483-1* 338
Polygala rothiana IPNI=*691242-1* 338
Polygala rottleriana IPNI=*77121670-1* 337
Polygala senduaris IPNI=*691242-1* 338
Polygala sherardioides IPNI=*205398-2* 338
Polygala sinensis IPNI=*691242-1* 338
Polygala smithii IPNI=*77121649-1* 337
Polygala socotrana IPNI=*691242-1* 338
Polygala sororia IPNI=*692483-1* 338
Polygala spectabilis IPNI=*77119197-1* 337
Polygala spectabilis var. *minor* IPNI=*77119197-1* 337
Polygala stenocaulon IPNI=*691063-1* 338
Polygala sulcata IPNI=*692483-1* 338
Polygala telephioides IPNI=*691242-1* 338
Polygala tenella IPNI=*330527-2* 338
Polygala thwaitesii IPNI=*691242-1* 338
Polygala toxoptera IPNI=*691242-1* 338
Polygala tranquebarica IPNI=*691242-1* 338
Polygala unguiculata IPNI=*77121670-1* 337
Polygala variabilis f. *leucanthema* IPNI=*692483-1* 338
Polygala variabilis f. *typica* IPNI=*692483-1* 338
Polygala variabilis IPNI=*692483-1* 338
Polygala varians IPNI=*691242-1* 338
Polygala vauthieri IPNI=*77121649-1* 337
Polygala violacea IPNI=*77121670-1* 337
Polygala zindae IPNI=*692483-1* 338
Polygonon persicarium IPNI=*695052-1* 289
Polygonum acmophyllum IPNI=*695422-1* 289
Polygonum acre f. *albiflorum* IPNI=*190855-2* 289
Polygonum acre IPNI=*190855-2* 289
Polygonum acre IPNI=*77069479-1* 289
Polygonum acre var. *aquatile* IPNI=*190855-2* 289
Polygonum acre var. *brachystachyum* IPNI=*190855-2* 289
Polygonum acre var. *confertiflorum* IPNI=*190855-2* 289
Polygonum acre var. *leptostachyum* IPNI=*190855-2* 289
Polygonum acre var. *riparium* IPNI=*190855-2* 289
Polygonum acuminatum IPNI=*1005755-2* 289
Polygonum acuminatum IPNI=*190869-2* 289
Polygonum acuminatum IPNI=*694896-1* 289
Polygonum acuminatum var. *glabrescens* IPNI=*694896-1* 289
Polygonum acuminatum var. *humboldtii* IPNI=*694896-1* 289
Polygonum acuminatum var. *microstemon* IPNI=*694896-1* 289
Polygonum acuminatum var. *stenophylla* IPNI=*694896-1* 289
Polygonum acutifolium IPNI=*695422-1* 289
Polygonum adenophorum IPNI=*77069480-1* 289
Polygonum aequale subsp. *oedocarpum* IPNI=*695422-1* 289
Polygonum agrestinum IPNI=*695422-1* 289
Polygonum alatum IPNI=*60433514-2* 289
Polygonum alatum var. *arenarium* IPNI=*60433514-2* 289
Polygonum alatum var. *metzianum* IPNI=*60433514-2* 289
Polygonum alatum var. *nepalense* IPNI=*60433514-2* 289
Polygonum alatum var. *parviflorum* IPNI=*60433514-2* 289
Polygonum alatum var. *rigidulum* IPNI=*60433514-2* 289
Polygonum alatum var. *tenuicaule* IPNI=*60433514-2* 289
Polygonum albescens IPNI=*695052-1* 289
Polygonum alfredi IPNI=*694896-1* 289
Polygonum andrzejowskianum IPNI=*77069480-1* 289
Polygonum antihaemorrhoidale f. *riparium* IPNI=*190855-2* 289
Polygonum antihaemorrhoidale IPNI=*190855-2* 289
Polygonum antihaemorrhoidale var. *aquatile* IPNI=*190855-2* 289
Polygonum araucanum IPNI=*77069480-1* 289

Polygonum arenarium IPNI=*695422-1* 289
Polygonum arnassense IPNI=*695052-1* 289
Polygonum australe IPNI=*190811-2* 289
Polygonum aviculare f. *erectum* IPNI=*695422-1* 289
Polygonum aviculare subsp. *heterophyllum* IPNI=*695422-1* 289
Polygonum aviculare subsp. *monspeliense* IPNI=*695422-1* 289
Polygonum aviculare subsp. *nanum* IPNI=*695422-1* 289
Polygonum aviculare subsp. *neglectum* IPNI=*695422-1* 289
Polygonum aviculare subsp. *rectum* IPNI=*695422-1* 289
Polygonum aviculare subvar. *denudatum* IPNI=*695422-1* 289
Polygonum aviculare subvar. *humifusum* IPNI=*695422-1* 289
Polygonum aviculare var. *agrestinum* IPNI=*695422-1* 289
Polygonum aviculare var. *angustifolium* IPNI=*695422-1* 289
Polygonum aviculare var. *denudatum* IPNI=*695422-1* 289
Polygonum aviculare var. *diffusum* IPNI=*695422-1* 289
Polygonum aviculare var. *gracilis* IPNI=*695422-1* 289
Polygonum aviculare var. *heterophyllum* IPNI=*695422-1* 289
Polygonum aviculare var. *latifolium* IPNI=*695422-1* 289
Polygonum aviculare var. *laxum* IPNI=*695422-1* 289
Polygonum aviculare var. *littorale* IPNI=*695422-1* 289
Polygonum aviculare var. *longifolium* IPNI=*695422-1* 289
Polygonum aviculare var. *microphyllum* IPNI=*695422-1* 289
Polygonum aviculare var. *monspeliense* IPNI=*695422-1* 289
Polygonum aviculare var. *nanum* IPNI=*695422-1* 289
Polygonum aviculare var. *neglectum* IPNI=*695422-1* 289
Polygonum aviculare var. *oedocarpum* IPNI=*695422-1* 289
Polygonum aviculare var. *polycnemiforme* IPNI=*695422-1* 289
Polygonum aviculare var. *scabrellum* IPNI=*695422-1* 289
Polygonum aviculare var. *vegetum* IPNI=*695422-1* 289
Polygonum aviculare var. *vulgare* IPNI=*695422-1* 289
Polygonum baltistanicum IPNI=*695422-1* 289
Polygonum barbatum IPNI=*694896-1* 289
Polygonum barbatum IPNI=*77069480-1* 289
Polygonum baumgartenianum IPNI=*77069480-1* 289
Polygonum bellardii IPNI=*695052-1* 289
Polygonum bellardii var. *patulum* IPNI=*696222-1* 289
Polygonum bettfreundianum IPNI=*694896-1* 289
Polygonum beyrichianum IPNI=*1050850-2* 289
Polygonum biforme IPNI=*695052-1* 289
Polygonum bioritauense IPNI=*77069480-1* 289
Polygonum bordzilowskii IPNI=*696222-1* 289
Polygonum brachypodum IPNI=*1050850-2* 289
Polygonum bracteosum IPNI=*77069480-1* 289
Polygonum brittingeri IPNI=*77069480-1* 289
Polygonum buxifolium IPNI=*695422-1* 289
Polygonum caballeroi IPNI=*695422-1* 289
Polygonum caducifolium IPNI=*695422-1* 289
Polygonum camptocladum IPNI=*695052-1* 289
Polygonum caniusculum IPNI=*695052-1* 289
Polygonum capitatum IPNI=*60433514-2* 289
Polygonum capitatum IPNI=*60433515-2* 289
Polygonum caucanum IPNI=*1005755-2* 289
Polygonum centinodium IPNI=*695422-1* 289
Polygonum cereale IPNI=*694526-1* 289
Polygonum chamaechyton IPNI=*695422-1* 289
Polygonum chamissoanum IPNI=*1050850-2* 289
Polygonum chamissonianum IPNI=*1050850-2* 289
Polygonum chloroleucon IPNI=*77069479-1* 289
Polygonum chymophyllum IPNI=*695422-1* 289
Polygonum ciliare IPNI=*77069479-1* 289
Polygonum cirrhosum IPNI=*15860-2* 288
Polygonum cretaceum IPNI=*696222-1* 289
Polygonum crispatum IPNI=*695422-1* 289
Polygonum cumingianum IPNI=*77069480-1* 289
Polygonum cuspidatum IPNI=*694896-1* 289
Polygonum danubiale IPNI=*77069480-1* 289
Polygonum debilius IPNI=*695052-1* 289
Polygonum decumbens IPNI=*695113-1* 289
Polygonum decumbens IPNI=*77069480-1* 289
Polygonum deloynei IPNI=*695422-1* 289
Polygonum densiflorum f. *ciliolatum* IPNI=*1005755-2* 289
Polygonum densiflorum f. *imberbe* IPNI=*1005755-2* 289
Polygonum densiflorum IPNI=*1005755-2* 289
Polygonum densiflorum var. *ciliolatum* IPNI=*1005755-2* 289
Polygonum densiflorum var. *imberbe* IPNI=*1005755-2* 289
Polygonum denudatum IPNI=*695422-1* 289
Polygonum dioicum IPNI=*694526-1* 289
Polygonum dissitiflorum IPNI=*695422-1* 289
Polygonum distachyum IPNI=*190855-2* 289
Polygonum dombeyanum IPNI=*694785-1* 289
Polygonum donii IPNI=*695113-1* 289
Polygonum dregeanum IPNI=*695422-1* 289
Polygonum eciliatum IPNI=*190855-2* 289
Polygonum elatiusculum IPNI=*695052-1* 289
Polygonum elegans IPNI=*694526-1* 289
Polygonum emarginatum IPNI=*694526-1* 289
Polygonum epilobioides IPNI=*190855-2* 289
Polygonum erectum IPNI=*694896-1* 289
Polygonum erythranthum IPNI=*695422-1* 289
Polygonum erythrocladum IPNI=*695052-1* 289
Polygonum eurybelonum IPNI=*77069480-1* 289
Polygonum euryphyllum IPNI=*77069480-1* 289
Polygonum fagopyrum IPNI=*694526-1* 289
Polygonum fallacinum IPNI=*77069479-1* 289
Polygonum fallax IPNI=*695052-1* 289
Polygonum fallax IPNI=*695422-1* 289
Polygonum fastigiatoramosum IPNI=*77069479-1* 289
Polygonum flexuosum IPNI=*694785-1* 289
Polygonum flexuosum IPNI=*77069479-1* 289
Polygonum floribundum IPNI=*694896-1* 289
Polygonum floridum IPNI=*695422-1* 289
Polygonum franktonii IPNI=*695422-1* 289
Polygonum fusiforme IPNI=*695052-1* 289
Polygonum ganderbalense IPNI=*695422-1* 289
Polygonum geniculatum IPNI=*1050850-2* 289
Polygonum geniculatum IPNI=*695422-1* 289
Polygonum giranense IPNI=*695113-1* 289
Polygonum glabrum IPNI=*1005755-2* 289
Polygonum glabrum IPNI=*77069480-1* 289
Polygonum glabrum var. *caribaeum* IPNI=*1005755-2* 289
Polygonum glandulosum IPNI=*77069479-1* 289
Polygonum glandulosum IPNI=*77069480-1* 289
Polygonum glaucochloros IPNI=*695422-1* 289
Polygonum glutinosum IPNI=*77069480-1* 289
Polygonum gracile IPNI=*694526-1* 289
Polygonum gracile IPNI=*695113-1* 289
Polygonum gracile IPNI=*77069479-1* 289
Polygonum gracilescens IPNI=*190855-2* 289
Polygonum gracilius IPNI=*696222-1* 289
Polygonum gramineum IPNI=*77069479-1* 289
Polygonum guatemalense IPNI=*694896-1* 289
Polygonum guttuliferum IPNI=*60433514-2* 289
Polygonum hassanabdalicum IPNI=*77069480-1* 289
Polygonum hastatum IPNI=*1050850-2* 289
Polygonum hecasanthum IPNI=*77069479-1* 289
Polygonum heterophyllum f. *rubescens* IPNI=*695422-1* 289
Polygonum heterophyllum IPNI=*695422-1* 289
Polygonum heterophyllum var. *caespitosum* IPNI=*695422-1* 289
Polygonum heterophyllum var. *eximium* IPNI=*695422-1* 289
Polygonum heterophyllum var. *rubescens* IPNI=*695422-1* 289
Polygonum hippopotami IPNI=*1005755-2* 289
Polygonum hirtovaginum IPNI=*695052-1* 289
Polygonum hispidum IPNI=*694999-1* 289
Polygonum hispidum IPNI=*77069479-1* 289
Polygonum hispidum var. *glabrescens* IPNI=*694999-1* 289
Polygonum humifixum IPNI=*695052-1* 289

Polygonum humifusum IPNI=*695422-1* 289
Polygonum hydropiper f. *ciliare* IPNI=*77069479-1* 289
Polygonum hydropiper IPNI=*77069479-1* 289
Polygonum hydropiper var. *densiflorum* IPNI=*77069479-1* 289
Polygonum hydropiper var. *projectum* IPNI=*77069479-1* 289
Polygonum hydropiper var. *sarbhanganicum* IPNI=*77069480-1* 289
Polygonum hydropiperoides f. *leucochranthum* IPNI=*190811-2* 289
Polygonum hydropiperoides f. *rubrum* IPNI=*190811-2* 289
Polygonum hydropiperoides f. *strigosum* IPNI=*190811-2* 289
Polygonum hydropiperoides IPNI=*190811-2* 289
Polygonum hydropiperoides subsp. *opelousanum* IPNI=*190811-2* 289
Polygonum hydropiperoides var. *adenocalyx* IPNI=*190811-2* 289
Polygonum hydropiperoides var. *asperifolium* IPNI=*190811-2* 289
Polygonum hydropiperoides var. *breviciliatum* IPNI=*190811-2* 289
Polygonum hydropiperoides var. *bushianum* IPNI=*190811-2* 289
Polygonum hydropiperoides var. *digitatum* IPNI=*190811-2* 289
Polygonum hydropiperoides var. *euronotorum* IPNI=*190811-2* 289
Polygonum hydropiperoides var. *macerum* IPNI=*190811-2* 289
Polygonum hydropiperoides var. *macounii* IPNI=*190811-2* 289
Polygonum hydropiperoides var. *persicarioides* IPNI=*190811-2* 289
Polygonum hydropiperoides var. *psilostachyum* IPNI=*190811-2* 289
Polygonum hydropiperoides var. *sanibelense* IPNI=*190811-2* 289
Polygonum hydropiperoides var. *strigosum* IPNI=*190811-2* 289
Polygonum hygrogenes IPNI=*695422-1* 289
Polygonum hypanicum IPNI=*77069480-1* 289
Polygonum ilophilum IPNI=*695052-1* 289
Polygonum imberbe IPNI=*1005755-2* 289
Polygonum incarnatum IPNI=*77069480-1* 289
Polygonum intermixtum IPNI=*695052-1* 289
Polygonum interruptellum IPNI=*695052-1* 289
Polygonum kawakamii IPNI=*695113-1* 289
Polygonum komarovii IPNI=*77069480-1* 289
Polygonum koreense IPNI=*77069479-1* 289
Polygonum kotovii IPNI=*696222-1* 289
Polygonum kumaonum IPNI=*60433514-2* 289
Polygonum lacunosum IPNI=*695052-1* 289
Polygonum lamellosum IPNI=*695052-1* 289
Polygonum lanceifolium IPNI=*77069480-1* 289
Polygonum lanigerum var. *glutinosum* IPNI=*77069480-1* 289
Polygonum lapathifolium f. *pallidum* IPNI=*77069480-1* 289
Polygonum lapathifolium IPNI=*695052-1* 289
Polygonum lapathifolium IPNI=*77069480-1* 289
Polygonum lapathifolium proles incarnatum IPNI=*77069480-1* 289
Polygonum lapathifolium proles vernicosum IPNI=*77069480-1* 289
Polygonum lapathifolium subsp. *maculatum* IPNI=*77069480-1* 289
Polygonum lapathifolium subsp. *mesomorphum* IPNI=*77069480-1* 289
Polygonum lapathifolium var. *euryphyllum* IPNI=*77069480-1* 289
Polygonum lapathifolium var. *genystachium* IPNI=*77069480-1* 289
Polygonum lapathifolium var. *incarnatum* IPNI=*77069480-1* 289
Polygonum lapathifolium var. *laxum* IPNI=*77069480-1* 289
Polygonum lapathifolium var. *maculatum* IPNI=*77069480-1* 289
Polygonum lapathifolium var. *salicifolium* IPNI=*77069480-1* 289
Polygonum latifolium IPNI=*695422-1* 289
Polygonum laxum IPNI=*77069480-1* 289
Polygonum leptobotrys IPNI=*694785-1* 289
Polygonum leptocladum IPNI=*695422-1* 289
Polygonum ligerinum IPNI=*695422-1* 289
Polygonum littorale IPNI=*695422-1* 289
Polygonum longipilum IPNI=*695052-1* 289
Polygonum lorinseri IPNI=*695052-1* 289
Polygonum ludovicianum IPNI=*190869-2* 289
Polygonum lugdunense IPNI=*695052-1* 289
Polygonum lyratum IPNI=*60433514-2* 289
Polygonum macropterum IPNI=*694526-1* 289
Polygonum maculatum IPNI=*77069480-1* 289
Polygonum maritimum IPNI=*190855-2* 289
Polygonum maximowiczii IPNI=*77069479-1* 289
Polygonum meisnerianum f. *intermedia* IPNI=*1050850-2* 289
Polygonum meisnerianum f. *petiolatum* IPNI=*1050850-2* 289
Polygonum meisnerianum IPNI=*1050850-2* 289
Polygonum meisnerianum var. *beyrichianum* IPNI=*1050850-2* 289
Polygonum meisnerianum var. *jalapense* IPNI=*1050850-2* 289
Polygonum meisnerianum var. *setosum* IPNI=*1050850-2* 289
Polygonum meisnerianum var. *usterianum* IPNI=*1050850-2* 289
Polygonum metzianum IPNI=*60433514-2* 289
Polygonum mexicanum IPNI=*190869-2* 289
Polygonum meyenii IPNI=*695422-1* 289
Polygonum micranthum IPNI=*694896-1* 289
Polygonum microcephalum IPNI=*60433514-2* 289
Polygonum millepunctatum IPNI=*695052-1* 289
Polygonum mite IPNI=*190811-2* 289
Polygonum mite IPNI=*77069479-1* 289
Polygonum monspeliense IPNI=*695422-1* 289
Polygonum multangulare IPNI=*1050850-2* 289
Polygonum myriocladum IPNI=*695422-1* 289
Polygonum nanum IPNI=*695422-1* 289
Polygonum neglectum IPNI=*695422-1* 289
Polygonum neglectum subsp. *rectum* IPNI=*695422-1* 289
Polygonum neolittorale IPNI=*695422-1* 289
Polygonum nepalense IPNI=*60433514-2* 289
Polygonum nepalense IPNI=*694526-1* 289
Polygonum nervosum IPNI=*695422-1* 289
Polygonum niloticum IPNI=*695052-1* 289
Polygonum nisus IPNI=*695052-1* 289
Polygonum nodosum IPNI=*77069480-1* 289
Polygonum nodosum var. *incarnatum* IPNI=*77069480-1* 289
Polygonum novoascanicum subsp. *cretaceum* IPNI=*696222-1* 289
Polygonum nutans IPNI=*77069480-1* 289
Polygonum obliquum IPNI=*77069480-1* 289
Polygonum obtusatum IPNI=*77069480-1* 289
Polygonum obtusifolium IPNI=*77069479-1* 289
Polygonum ochreatum IPNI=*77069480-1* 289
Polygonum oedocarpum IPNI=*695422-1* 289
Polygonum oenophyton IPNI=*695422-1* 289
Polygonum opacum IPNI=*695052-1* 289
Polygonum opelousanum IPNI=*190811-2* 289
Polygonum opelousanum var. *adenocalyx* IPNI=*190811-2* 289
Polygonum orogenes IPNI=*695422-1* 289
Polygonum orthocladum IPNI=*695052-1* 289
Polygonum ovatolanceolatum IPNI=*695052-1* 289
Polygonum pallens IPNI=*695422-1* 289
Polygonum pallens IPNI=*77069480-1* 289
Polygonum pallidiflorum IPNI=*695052-1* 289
Polygonum pallidium IPNI=*77069479-1* 289
Polygonum paniculatum IPNI=*77069480-1* 289
Polygonum parviflorum IPNI=*695422-1* 289
Polygonum patulum f. *budense* IPNI=*696222-1* 289
Polygonum patulum f. *nyirsegense* IPNI=*696222-1* 289
Polygonum patulum f. *transsilvanicum* IPNI=*696222-1* 289
Polygonum patulum subsp. *kotovii* IPNI=*696222-1* 289
Polygonum perforatum IPNI=*60433514-2* 289
Polygonum persicaria f. *albiflorum* IPNI=*695052-1* 289
Polygonum persicaria f. *angustifolium* IPNI=*695052-1* 289
Polygonum persicaria f. *prostratum* IPNI=*695052-1* 289
Polygonum persicaria IPNI=*695052-1* 289
Polygonum persicaria var. *lapathifolium* IPNI=*77069480-1* 289
Polygonum persicaria var. *minus* IPNI=*695052-1* 289
Polygonum persicaria var. *opacum* IPNI=*695052-1* 289
Polygonum persicaria var. *ruderale* IPNI=*695052-1* 289
Polygonum persicaria var. *vulgare* IPNI=*695052-1* 289
Polygonum persicarioides IPNI=*190811-2* 289
Polygonum persoonii IPNI=*190811-2* 289
Polygonum petecticale IPNI=*77069480-1* 289
Polygonum phyllophorum IPNI=*695422-1* 289
Polygonum pictaviense IPNI=*695422-1* 289

Polygonum platycladum IPNI=*694774-1* 289
Polygonum pleianthum IPNI=*695422-1* 289
Polygonum podophyllum IPNI=*77069479-1* 289
Polygonum polycnemiforme IPNI=*695422-1* 289
Polygonum portoricense IPNI=*1005755-2* 289
Polygonum posumbu f. *albiflorum* IPNI=*695113-1* 289
Polygonum posumbu IPNI=*695113-1* 289
Polygonum posumbu var. *stenophyllum* IPNI=*695113-1* 289
Polygonum praelongum IPNI=*695052-1* 289
Polygonum procumbens IPNI=*695113-1* 289
Polygonum procumbens IPNI=*695422-1* 289
Polygonum provinciale IPNI=*695422-1* 289
Polygonum pseudojaponicum IPNI=*77069480-1* 289
Polygonum pseudonodosum IPNI=*77069480-1* 289
Polygonum punctatum f. *longicollum* IPNI=*190855-2* 289
Polygonum punctatum f. *stipitatum* IPNI=*190855-2* 289
Polygonum punctatum IPNI=*190811-2* 289
Polygonum punctatum IPNI=*190855-2* 289
Polygonum punctatum IPNI=*60433514-2* 289
Polygonum punctatum IPNI=*695052-1* 289
Polygonum punctatum IPNI=*77069479-1* 289
Polygonum punctatum IPNI=*77069480-1* 289
Polygonum punctatum var. *alatum* IPNI=*60433514-2* 289
Polygonum punctatum var. *aquatile* IPNI=*190855-2* 289
Polygonum punctatum var. *confertiflorum* IPNI=*190855-2* 289
Polygonum punctatum var. *eciliatum* IPNI=*190855-2* 289
Polygonum punctatum var. *ellipticum* IPNI=*190855-2* 289
Polygonum punctatum var. *leptostachyum* IPNI=*190855-2* 289
Polygonum punctatum var. *littorale* IPNI=*190855-2* 289
Polygonum punctatum var. *mexicanum* IPNI=*190855-2* 289
Polygonum punctatum var. *parviflorum* IPNI=*190855-2* 289
Polygonum punctatum var. *parvum* IPNI=*190855-2* 289
Polygonum punctatum var. *tacubayanum* IPNI=*190855-2* 289
Polygonum punctatum var. *typicum* IPNI=*190855-2* 289
Polygonum purpuratum IPNI=*77069479-1* 289
Polygonum pyramidale IPNI=*77069480-1* 289
Polygonum pyramidatum IPNI=*694526-1* 289
Polygonum quadrangulatum IPNI=*694785-1* 289
Polygonum quadrifidum IPNI=*60433514-2* 289
Polygonum rectum IPNI=*695422-1* 289
Polygonum refractum IPNI=*1050850-2* 289
Polygonum repens IPNI=*60433515-2* 289
Polygonum rhizocaulon IPNI=*695113-1* 289
Polygonum rhodanense IPNI=*695422-1* 289
Polygonum rhombaeum IPNI=*695052-1* 289
Polygonum riparium IPNI=*190855-2* 289
Polygonum rivulare IPNI=*190811-2* 289
Polygonum robynsii IPNI=*60433515-2* 289
Polygonum rubelliflorum IPNI=*695052-1* 289
Polygonum rubescens IPNI=*695422-1* 289
Polygonum ruderale IPNI=*695052-1* 289
Polygonum rufescens IPNI=*695052-1* 289
Polygonum sagittatum IPNI=*1050850-2* 289
Polygonum salinum IPNI=*696222-1* 289
Polygonum saporoviense IPNI=*77069480-1* 289
Polygonum sarbhanganicum IPNI=*77069480-1* 289
Polygonum scabridum IPNI=*190811-2* 289
Polygonum scabrinervium IPNI=*1005755-2* 289
Polygonum schimperi IPNI=*1005755-2* 289
Polygonum schindleri IPNI=*77069479-1* 289
Polygonum schinzii IPNI=*77069479-1* 289
Polygonum segetum IPNI=*190869-2* 289
Polygonum segetum var. *lindenii* IPNI=*190869-2* 289
Polygonum segetum var. *stamineum* IPNI=*190869-2* 289
Polygonum segetum var. *verrucosum* IPNI=*190869-2* 289
Polygonum serotinum IPNI=*695052-1* 289
Polygonum serrulatum IPNI=*1005755-2* 289
Polygonum setigerum IPNI=*694896-1* 289
Polygonum simlense IPNI=*77069480-1* 289
Polygonum somphocarpum IPNI=*77069480-1* 289
Polygonum spectabile IPNI=*696222-1* 289
Polygonum spectabile var. *patagonica* IPNI=*77069480-1* 289
Polygonum striatum IPNI=*695422-1* 289
Polygonum stricticaule IPNI=*695422-1* 289
Polygonum stuckertii IPNI=*694896-1* 289
Polygonum subcanum IPNI=*695052-1* 289
Polygonum subsimplex IPNI=*695052-1* 289
Polygonum tamnifolium IPNI=*694785-1* 289
Polygonum tamnifolium var. *xerocarpum* IPNI=*694785-1* 289
Polygonum tarmense IPNI=*190811-2* 289
Polygonum telonense IPNI=*695422-1* 289
Polygonum tenuiflorum IPNI=*77069480-1* 289
Polygonum triandrum IPNI=*258621-2* 290
Polygonum truncatum IPNI=*1005755-2* 289
Polygonum turgidum IPNI=*77069480-1* 289
Polygonum uruguense IPNI=*695422-1* 289
Polygonum utriculatum IPNI=*77069480-1* 289
Polygonum uviferum IPNI=*693918-1* 288
Polygonum vanictianum IPNI=*77069480-1* 289
Polygonum vellosoi IPNI=*694896-1* 289
Polygonum vernale IPNI=*77069479-1* 289
Polygonum vernum IPNI=*695052-1* 289
Polygonum virgatum var. *latifolium* IPNI=*190811-2* 289
Polygonum viridiflorum IPNI=*306087-2* 436
Polygonum viscosum IPNI=*77069480-1* 289
Polygonum volcanicum IPNI=*1022243-2* 289
Polygonum yokusaianum IPNI=*695113-1* 289
Polygonum yokusaianum var. *stenophyllum* IPNI=*695113-1* 289
Polygyne inconspicua IPNI=*1186403-2* 261
Polylepis boyacensis IPNI=*206511-2* 443
Polylepis cocuyensis IPNI=*206511-2* 443
Polymnia arborea IPNI=*237495-2* 270
Polymnia cocuyensis IPNI=*77137314-1* 270
Polymnia crenata IPNI=*1093589-2* 270
Polymnia edulis IPNI=*237510-2* 270
Polymnia eurylepis IPNI=*237506-2* 270
Polymnia fruticosa IPNI=*237495-2* 270
Polymnia odoratissima IPNI=*249077-1* 270
Polymnia pyramidalis IPNI=*237506-2* 270
Polymnia riparia IPNI=*237508-2* 270
Polymnia sonchifolia IPNI=*237510-2* 270
Polymnia wedelia IPNI=*60468294-2* 272
Polynome alata IPNI=*317625-1* 225
Polynome bulbifera IPNI=*323908-2* 225
Polyphema champeden IPNI=*582622-1* 436
Polyphema jaca IPNI=*850389-1* 436
Polypodium acuminatum IPNI=*77075127-1* 177
Polypodium acutum IPNI=*305721-2* 173
Polypodium adianthoides IPNI=*224543-2* 176
Polypodium adiantiforme IPNI=*224543-2* 176
Polypodium adspersum IPNI=*17448650-1* 176
Polypodium albidum IPNI=*1039912-2* 176
Polypodium albopunctatum IPNI=*77075127-1* 177
Polypodium altum IPNI=*17341250-1* 176
Polypodium ambiguum IPNI=*17341250-1* 176
Polypodium ampliatum IPNI=*77075127-1* 177
Polypodium angustifolium f. *densifolium* IPNI=*274004-2* 175
Polypodium angustifolium IPNI=*17063000-1* 175
Polypodium angustifolium var. *gramineum* IPNI=*17063000-1* 175
Polypodium angustifolium var. *latum* IPNI=*17063000-1* 175
Polypodium angustifolium var. *squamulosum* IPNI=*17063000-1* 175
Polypodium angustifolium var. *taeniosum* IPNI=*17063000-1* 175
Polypodium anthriscifolium IPNI=*30013429-2* 174
Polypodium antrophyoides IPNI=*17341250-1* 176
Polypodium araneosum IPNI=*17175380-1* 176
Polypodium arboreum IPNI=*17071820-1* 172

Polypodium areolatum IPNI=*17175380-1* 176
Polypodium areolatum var. *loreum* IPNI=*17175380-1* 176
Polypodium argentatum IPNI=*224543-2* 176
Polypodium articulatum IPNI=*1138771-2* 176
Polypodium articulatum var. *angustifolium* IPNI=*1138771-2* 176
Polypodium attenuatum IPNI=*17175390-1* 176
Polypodium attenuatum IPNI=*77075087-1* 176
Polypodium auratum IPNI=*17175390-1* 176
Polypodium aureum IPNI=*17175390-1* 176
Polypodium aureum subsp. *pseudoaureum* IPNI=*17175380-1* 176
Polypodium aureum var. *araneosum* IPNI=*17175380-1* 176
Polypodium aureum var. *areolatum* IPNI=*17175380-1* 176
Polypodium aureum var. *leucatomos* IPNI=*17175390-1* 176
Polypodium aureum var. *pulvinatum* IPNI=*17175390-1* 176
Polypodium aureum var. *reductum* IPNI=*17175380-1* 176
Polypodium auriculatum IPNI=*17160730-1* 176
Polypodium austriacum IPNI=*17210060-1* 175
Polypodium avenium IPNI=*17546710-1* 175
Polypodium balansae IPNI=*1019960-2* 174
Polypodium balaonense IPNI=*77141144-1* 176
Polypodium brasiliense f. *giganteum* IPNI=*77183236-1* 175
Polypodium brasiliense IPNI=*77075127-1* 177
Polypodium brasiliense var. *attenuatum* IPNI=*77075087-1* 176
Polypodium brasiliense var. *gladiatum* IPNI=*77075127-1* 177
Polypodium brasiliense var. *neriifolium* IPNI=*77075127-1* 177
Polypodium brasiliense var. *pleiosorum* IPNI=*77075127-1* 177
Polypodium brasiliense var. *subcordatum* IPNI=*77075127-1* 177
Polypodium caducum IPNI=*1148359-2* 175
Polypodium caesium IPNI=*143624-2* 173
Polypodium calaguala IPNI=*274004-2* 175
Polypodium campylopodum IPNI=*77075127-1* 177
Polypodium cardiophyllum IPNI=*17406250-1* 176
Polypodium ceteraccinum IPNI=*1039912-2* 176
Polypodium chrysoconion IPNI=*1039912-2* 176
Polypodium cinereum IPNI=*143624-2* 173
Polypodium comosum IPNI=*17063380-1* 175
Polypodium cordifolium IPNI=*17160730-1* 176
Polypodium cordifolium IPNI=*249899-2* 177
Polypodium coriaceum IPNI=*224543-2* 176
Polypodium coriaceum IPNI=*77183236-1* 175
Polypodium crassifolium IPNI=*77183236-1* 175
Polypodium crassifolium var. *longipes* IPNI=*77183020-1* 175
Polypodium crassinerve IPNI=*17341250-1* 176
Polypodium cuspidatum IPNI=*17546710-1* 175
Polypodium decumanum IPNI=*17175400-1* 176
Polypodium dentatum IPNI=*251308-2* 174
Polypodium dictyocallis IPNI=*17175400-1* 176
Polypodium difforme IPNI=*17063000-1* 175
Polypodium dimorphum IPNI=*17063000-1* 175
Polypodium dissimile IPNI=*77075087-1* 176
Polypodium drynarioides IPNI=*17175380-1* 176
Polypodium exaltatum IPNI=*17160830-1* 176
Polypodium expansum IPNI=*17226640-1* 177
Polypodium filix-femina IPNI=*30055316-2* 174
Polypodium fragile IPNI=*30013429-2* 174
Polypodium fulvum IPNI=*17175380-1* 176
Polypodium funiculosum IPNI=*160142-2* 175
Polypodium gauthieri IPNI=*77075127-1* 177
Polypodium giganteum IPNI=*77183236-1* 175
Polypodium glabrum IPNI=*17160710-1* 176
Polypodium glabrum IPNI=*17341250-1* 176
Polypodium gladiatum IPNI=*17063380-1* 175
Polypodium gladiatum IPNI=*77075127-1* 177
Polypodium glaucinum IPNI=*17175380-1* 176
Polypodium glaucophyllum IPNI=*77075105-1* 177
Polypodium glaucum IPNI=*143624-2* 173
Polypodium gramineum IPNI=*17234040-1* 178
Polypodium griseum IPNI=*143624-2* 173
Polypodium haenkeanum IPNI=*160142-2* 175
Polypodium haenkeanum IPNI=*17406250-1* 176
Polypodium haenkei IPNI=*77075127-1* 177
Polypodium hastatum IPNI=*17226640-1* 177
Polypodium henchmannii IPNI=*17189260-1* 176
Polypodium hendersonii IPNI=*17063000-1* 175
Polypodium incanoides IPNI=*1039912-2* 176
Polypodium incanum IPNI=*1039912-2* 176
Polypodium incanum var. *oblongum* IPNI=*1039912-2* 176
Polypodium integerrimum IPNI=*17160710-1* 176
Polypodium irioides IPNI=*17341250-1* 176
Polypodium kuhnii IPNI=*77075087-1* 176
Polypodium laciniatum IPNI=*87046-1* 200
Polypodium laetum IPNI=*30055316-2* 174
Polypodium laevigatum var. *rigidum* IPNI=*17063380-1* 175
Polypodium lanceolatum IPNI=*17448650-1* 176
Polypodium lanceolatum var. *araucanum* IPNI=*17448650-1* 176
Polypodium lepidopteris IPNI=*77141144-1* 176
Polypodium lepidotum IPNI=*17448650-1* 176
Polypodium lepidotum var. *latifolium* IPNI=*17448650-1* 176
Polypodium lepidotum var. *macrocarpum* IPNI=*17448650-1* 176
Polypodium leucatomos IPNI=*17175390-1* 176
Polypodium leuconeuron var. *angustifolium* IPNI=*17063000-1* 175
Polypodium levigatum IPNI=*77075105-1* 177
Polypodium lingulatum IPNI=*17341250-1* 176
Polypodium longifolium IPNI=*77075127-1* 177
Polypodium loreum IPNI=*17063000-1* 175
Polypodium lucens IPNI=*77075127-1* 177
Polypodium lusitanicum IPNI=*17078300-1* 175
Polypodium lycopodioides f. *obtusum* IPNI=*160142-2* 175
Polypodium lycopodioides IPNI=*160142-2* 175
Polypodium lycopodioides var. *longipes* IPNI=*160142-2* 175
Polypodium lycopodioides var. *salicifolium* IPNI=*160142-2* 175
Polypodium lycopodioides var. *stipitatum* IPNI=*160142-2* 175
Polypodium lycopodioides var. *subdimorphum* IPNI=*160142-2* 175
Polypodium lycopodioides var. *surinamense* IPNI=*160142-2* 175
Polypodium lycopodioides var. *venulosum* IPNI=*160142-2* 175
Polypodium macrocarpum IPNI=*17448650-1* 176
Polypodium maculatum IPNI=*143624-2* 173
Polypodium marginale IPNI=*17448650-1* 176
Polypodium maritimum IPNI=*77075108-1* 177
Polypodium megalosoroides IPNI=*17341250-1* 176
Polypodium menisciifolium var. *albopunctatum* IPNI=*77075127-1* 177
Polypodium mesetae IPNI=*1039912-2* 176
Polypodium microdontum IPNI=*73288-2* 172
Polypodium microlepidum IPNI=*17546710-1* 175
Polypodium microsorum IPNI=*17341250-1* 176
Polypodium millisorum IPNI=*17341250-1* 176
Polypodium molle IPNI=*251308-2* 174
Polypodium montevidense IPNI=*208026-2* 176
Polypodium nemorale IPNI=*1019960-2* 174
Polypodium neoguineense IPNI=*17341250-1* 176
Polypodium nephrolepioides IPNI=*17160710-1* 176
Polypodium nerifolium var. *acuminatissimum* IPNI=*77075127-1* 177
Polypodium nerifolium var. *heterophyllum* IPNI=*77075127-1* 177
Polypodium neriifolium IPNI=*77075127-1* 177
Polypodium nitidulum IPNI=*17063000-1* 175
Polypodium nymphale IPNI=*251308-2* 174
Polypodium officinale IPNI=*300720-2* 176
Polypodium palmatum IPNI=*77075127-1* 177
Polypodium palmeri IPNI=*17444940-1* 175
Polypodium palmoides IPNI=*17160710-1* 176
Polypodium parallelinerve IPNI=*17063380-1* 175
Polypodium pellitum IPNI=*160142-2* 175
Polypodium pennigerum IPNI=*17458930-1* 174
Polypodium percussum IPNI=*17546710-1* 175
Polypodium percussum var. *squamosum* IPNI=*17546710-1* 175
Polypodium petiolatum IPNI=*17189260-1* 176

Polypodium phyllitidis IPNI=*17063380-1* 175
Polypodium phyllitidis var. *elongatum* IPNI=*17063380-1* 175
Polypodium phyllitidis var. *linnaeanum* IPNI=*17063380-1* 175
Polypodium phyllitidis var. *swartzianum* IPNI=*17063380-1* 175
Polypodium politum IPNI=*224543-2* 176
Polypodium polycarpon IPNI=*17341250-1* 176
Polypodium polycarpon var. *obtusatum* IPNI=*17341250-1* 176
Polypodium polycephalum IPNI=*17341250-1* 176
Polypodium polypodioides IPNI=*1039912-2* 176
Polypodium porrectum IPNI=*77183236-1* 175
Polypodium preslianum IPNI=*77075127-1* 177
Polypodium preslianum var. *immersum* IPNI=*77075127-1* 177
Polypodium preslii IPNI=*17546710-1* 175
Polypodium prominulum IPNI=*160142-2* 175
Polypodium pruinatum IPNI=*143624-2* 173
Polypodium pseudoaureum IPNI=*17175380-1* 176
Polypodium pulvinatum IPNI=*17175390-1* 176
Polypodium punctatum IPNI=*17341250-1* 176
Polypodium punctatum subsp. *subirideum* IPNI=*17341250-1* 176
Polypodium punctulatum IPNI=*17160710-1* 176
Polypodium pycnosorum IPNI=*77075127-1* 177
Polypodium quadripinnatum IPNI=*143624-2* 173
Polypodium repandum IPNI=*17226640-1* 177
Polypodium restingae IPNI=*17406250-1* 176
Polypodium revertens IPNI=*17189260-1* 176
Polypodium rostratum IPNI=*17546710-1* 175
Polypodium russelianum IPNI=*1019960-2* 174
Polypodium salicifolium IPNI=*160142-2* 175
Polypodium salicinum IPNI=*160142-2* 175
Polypodium secundum IPNI=*17458930-1* 174
Polypodium semicordatum IPNI=*1148359-2* 175
Polypodium serratum IPNI=*22222-2* 174
Polypodium simplicifolium IPNI=*17063380-1* 175
Polypodium sporadocarpum IPNI=*17175380-1* 176
Polypodium stigmaticum IPNI=*160142-2* 175
Polypodium subalatum IPNI=*77075127-1* 177
Polypodium subirideum IPNI=*17341250-1* 176
Polypodium surinamense IPNI=*160142-2* 175
Polypodium taeniosum IPNI=*17063000-1* 175
Polypodium tenericaule IPNI=*1019960-2* 174
Polypodium thomassetii IPNI=*224543-2* 176
Polypodium tottum IPNI=*17458930-1* 174
Polypodium trichodes IPNI=*1019960-2* 174
Polypodium triseriale IPNI=*77075127-1* 177
Polypodium triseriale var. *gladiatum* IPNI=*77075127-1* 177
Polypodium vacciniifolium f. *incisofimbriata* IPNI=*17406250-1* 176
Polypodium vacciniifolium IPNI=*17406250-1* 176
Polypodium vacciniifolium var. *cordifolium* IPNI=*17406250-1* 176
Polypodium validum IPNI=*17341250-1* 176
Polypodium variolatum IPNI=*77075127-1* 177
Polypodium velatum IPNI=*1039912-2* 176
Polypodium venosum IPNI=*160142-2* 175
Polypodium venulosum IPNI=*160142-2* 175
Polypodium vulgare f. *attenuatum* IPNI=*300720-2* 176
Polypodium vulgare f. *pygmaeum* IPNI=*300720-2* 176
Polypodium xiphophoron IPNI=*77075127-1* 177
Polypodium zeyheri IPNI=*17448650-1* 176
Polypogon alopecurus IPNI=*1048074-2* 244
Polypogon crinitus IPNI=*1048074-2* 244
Polypogon flavescens IPNI=*1048074-2* 244
Polypogon melillensis IPNI=*1048074-2* 244
Polypogon monspeliensis f. *argentinus* IPNI=*1048074-2* 244
Polypogon monspeliensis f. *nanus* IPNI=*1048074-2* 244
Polypogon monspeliensis var. *indicus* IPNI=*1048074-2* 244
Polypogon paniceus IPNI=*1048074-2* 244
Polypogon polysetus IPNI=*1048074-2* 244
Polypogon spicatus IPNI=*115792-2* 240
Polypogon zeylanicus IPNI=*1048074-2* 244
Polyscias crispata IPNI=*91749-1* 254
Polyscias cumingii IPNI=*91749-1* 254
Polyscias guilfoylei var. *monstrosa* IPNI=*91773-1* 254
Polyscias guilfoylei var. *victoriae* IPNI=*91773-1* 254
Polyscias rumphiana IPNI=*91749-1* 254
Polyscias scutellarioides IPNI=*91817-1* 254
Polyscias tricochleata IPNI=*91817-1* 254
Polystachya buchananii IPNI=*654273-1* 222
Polystachya caquetana IPNI=*654273-1* 222
Polystachya colombiana IPNI=*654273-1* 222
Polystachya cubensis IPNI=*654273-1* 222
Polystachya dagremondiana IPNI=*654273-1* 222
Polystachya extinctoria IPNI=*654273-1* 222
Polystachya flavescens IPNI=*654273-1* 222
Polystachya gracilis IPNI=*654273-1* 222
Polystachya hypocrita IPNI=*654273-1* 222
Polystachya jussieuana IPNI=*654273-1* 222
Polystachya kraenzliniana IPNI=*654273-1* 222
Polystachya latifolia IPNI=*654273-1* 222
Polystachya lehmbachiana IPNI=*654273-1* 222
Polystachya lepidantha IPNI=*654273-1* 222
Polystachya lettowiana IPNI=*654273-1* 222
Polystachya luteola IPNI=*654273-1* 222
Polystachya mauritiana IPNI=*654273-1* 222
Polystachya minuta IPNI=*654273-1* 222
Polystachya penangensis IPNI=*654273-1* 222
Polystachya pleistantha IPNI=*654273-1* 222
Polystachya praealta IPNI=*654273-1* 222
Polystachya purpurea IPNI=*654273-1* 222
Polystachya reichenbachiana IPNI=*654273-1* 222
Polystachya rufinula IPNI=*654273-1* 222
Polystachya siamensis IPNI=*654273-1* 222
Polystachya singapurensis IPNI=*654273-1* 222
Polystachya tessellata IPNI=*654273-1* 222
Polystachya tricruris IPNI=*654273-1* 222
Polystachya tricuspidata IPNI=*654273-1* 222
Polystachya zanguebarica IPNI=*654273-1* 222
Polystachya zeylanica IPNI=*654273-1* 222
Polystachya zollingeri IPNI=*654273-1* 222
Polystichum aculeatum var. *microsorium* IPNI=*208026-2* 176
Polystichum aculeatum var. *montevidense* IPNI=*208026-2* 176
Polystichum acutum IPNI=*17160710-1* 176
Polystichum adiantiforme IPNI=*224543-2* 176
Polystichum auriculatum IPNI=*17160730-1* 176
Polystichum coriaceum IPNI=*224543-2* 176
Polystichum coriaceum var. *acuminatum* IPNI=*224543-2* 176
Polystichum coriaceum var. *callochlamys* IPNI=*224543-2* 176
Polystichum coriaceum var. *platychlamys* IPNI=*224543-2* 176
Polystichum cumingianum IPNI=*224543-2* 176
Polystichum dissimulans IPNI=*319061-2* 177
Polystichum duriusculum IPNI=*224543-2* 176
Polystichum emarginatum IPNI=*319061-2* 177
Polystichum gongylodes IPNI=*251353-2* 174
Polystichum macroporum IPNI=*224543-2* 176
Polystichum microsorium IPNI=*208026-2* 176
Polystichum molle IPNI=*251308-2* 174
Polystichum montevidense f. *genuina* IPNI=*208026-2* 176
Polystichum montevidense f. *imbricata* IPNI=*208026-2* 176
Polystichum nymphale IPNI=*251308-2* 174
Polystichum propinquum IPNI=*17458930-1* 174
Polystichum quadrangulare IPNI=*224543-2* 176
Polystichum remotum IPNI=*224543-2* 176
Polystichum sellowianum IPNI=*208026-2* 176
Polystichum semicordatum IPNI=*1148359-2* 175
Polystichum serratum IPNI=*17458930-1* 174
Polystichum torresianum IPNI=*1019960-2* 174
Polytaenium lanceolatum IPNI=*17209080-1* 178
Polythecandra schomburgkiana IPNI=*61594-2* 385

Polythecandra spruceana IPNI=*61594-2* 385
Polythecium hildebrandtii IPNI=*606804-1* 397
Polythecium thomasianum IPNI=*606804-1* 397
Polytrias amaura IPNI=*963150-1* 244
Polytrias amaura var. *nana* IPNI=*963150-1* 244
Polytrias diversiflora IPNI=*963150-1* 244
Polytrias indica var. *nana* IPNI=*963150-1* 244
Polytrias praemorsa IPNI=*963150-1* 244
Pombalia hieronymi IPNI=*77143566-1* 405
Pombalia ipecacuanha IPNI=*77143549-1* 405
Pombalia itoubou IPNI=*77143549-1* 405
Ponera mapiriensis IPNI=*228831-2* 223
Pontederia angustifolia IPNI=*283610-2* 225
Pontederia aquatica IPNI=*698138-1* 225
Pontederia azurea IPNI=*698080-1* 225
Pontederia cordata f. *albiflora* IPNI=*283610-2* 225
Pontederia cordata f. *angustifolia* IPNI=*283610-2* 225
Pontederia cordata f. *bernardii* IPNI=*283610-2* 225
Pontederia cordata f. *brasiliensis* IPNI=*283610-2* 225
Pontederia cordata f. *latifolia* IPNI=*283610-2* 225
Pontederia cordata f. *taenia* IPNI=*283610-2* 225
Pontederia cordata var. *albiflora* IPNI=*283610-2* 225
Pontederia cordata var. *angustifolia* IPNI=*283610-2* 225
Pontederia cordata var. *lanceolata* IPNI=*283610-2* 225
Pontederia crassicaulis IPNI=*310928-2* 225
Pontederia elongata IPNI=*310928-2* 225
Pontederia lanceolata f. *brasiliensis* IPNI=*283610-2* 225
Pontederia lanceolata f. *trullifolia* IPNI=*283610-2* 225
Pontederia lanceolata IPNI=*283610-2* 225
Pontederia lanceolata var. *vichadensis* IPNI=*283610-2* 225
Pontederia maculata IPNI=*283610-2* 225
Pontederia martiana IPNI=*698175-1* 225
Pontederia martiusiana IPNI=*698175-1* 225
Pontederia mucronata IPNI=*283610-2* 225
Pontederia nymphaeifolia IPNI=*283610-2* 225
Pontederia obtusifolia IPNI=*283610-2* 225
Pontederia rotundifolia var. *nymphaeifolia* IPNI=*283610-2* 225
Pontederia tricolor IPNI=*698175-1* 225
Pontederia tumida IPNI=*698138-1* 225
Pontia grandis IPNI=*230047-1* 265
Pontia matricaria IPNI=*252460-1* 271
Pontia vulgaris IPNI=*30296198-2* 265
Poponax cowellii IPNI=*60441223-2* 336
Poponax farnesiana IPNI=*1099713-2* 336
Poponax lutea IPNI=*60441223-2* 336
Poponax macracantha IPNI=*60441223-2* 336
Poponax macracanthoides IPNI=*60441223-2* 336
Poponax pennatula IPNI=*60441228-2* 336
Poponax tortuosa IPNI=*60441241-2* 336
Poponax venosa IPNI=*60441241-2* 336
Poppya fabiana IPNI=*293060-1* 294
Poppya operculata IPNI=*293097-1* 294
Populus × *canadensis* var. *virginiana* IPNI=*30102841-2* 404
Populus acerifolia IPNI=*776573-1* 404
Populus aegyptiaca IPNI=*776573-1* 404
Populus afghanica IPNI=*776790-1* 404
Populus alba f. *pendula* IPNI=*776573-1* 404
Populus alba var. *caspica* IPNI=*776573-1* 404
Populus alba var. *croatica* IPNI=*776790-1* 404
Populus alba var. *nutans* IPNI=*776573-1* 404
Populus angulata IPNI=*30102841-2* 404
Populus angulata var. *missouriensis* IPNI=*30102841-2* 404
Populus angulosa IPNI=*30102841-2* 404
Populus arbeel IPNI=*776573-1* 404
Populus arembergiana IPNI=*776573-1* 404
Populus arembergica IPNI=*776573-1* 404
Populus arizonica var. *jonesii* IPNI=*30102841-2* 404
Populus bachofenii IPNI=*776573-1* 404
Populus balsamifera var. *missouriensis* IPNI=*30102841-2* 404
Populus balsamifera var. *virginiana* IPNI=*30102841-2* 404
Populus belgica IPNI=*776573-1* 404
Populus berkarensis IPNI=*776573-1* 404
Populus bethmontiana IPNI=*776790-1* 404
Populus carolinensis IPNI=*30102841-2* 404
Populus caspica IPNI=*776573-1* 404
Populus charkowiensis IPNI=*776790-1* 404
Populus croatica IPNI=*776790-1* 404
Populus deltoides f. *pilosa* IPNI=*30102841-2* 404
Populus deltoides var. *angulata* IPNI=*30102841-2* 404
Populus deltoides var. *carolinensis* IPNI=*30102841-2* 404
Populus deltoides var. *missouriensis* IPNI=*30102841-2* 404
Populus deltoides var. *pilosa* IPNI=*30102841-2* 404
Populus deltoides var. *virginiana* IPNI=*30102841-2* 404
Populus elaverensis IPNI=*776790-1* 404
Populus excelsa IPNI=*776573-1* 404
Populus fastigiata IPNI=*776790-1* 404
Populus flexibilis IPNI=*776790-1* 404
Populus flexilis IPNI=*776790-1* 404
Populus grisea IPNI=*776573-1* 404
Populus helvetica IPNI=*776790-1* 404
Populus hickeliana IPNI=*776573-1* 404
Populus hudsoniana IPNI=*776790-1* 404
Populus hudsonica IPNI=*776790-1* 404
Populus hybrida IPNI=*776573-1* 404
Populus intermedia IPNI=*776573-1* 404
Populus kanjilaliana IPNI=*776790-1* 404
Populus laevigata IPNI=*30102841-2* 404
Populus lombardica IPNI=*776790-1* 404
Populus major IPNI=*776573-1* 404
Populus morisetiana IPNI=*776573-1* 404
Populus nigra f. *parvifolia* IPNI=*776790-1* 404
Populus nigra f. *pseudocaudina* IPNI=*776790-1* 404
Populus nigra f. *torulosa* IPNI=*776790-1* 404
Populus nigra IPNI=*776573-1* 404
Populus nigra var. *charkowiensis* IPNI=*776790-1* 404
Populus nigra var. *helvetica* IPNI=*776790-1* 404
Populus nigra var. *hudsonica* IPNI=*776790-1* 404
Populus nigra var. *narentana* IPNI=*776790-1* 404
Populus nigra var. *pendula* IPNI=*776790-1* 404
Populus nigra var. *sosnowskyi* IPNI=*776790-1* 404
Populus nigra var. *virginiana* IPNI=*30102841-2* 404
Populus nigra var. *vistulensis* IPNI=*776790-1* 404
Populus nivea IPNI=*776573-1* 404
Populus nolestii IPNI=*776790-1* 404
Populus paletskyana IPNI=*776573-1* 404
Populus palmata IPNI=*776573-1* 404
Populus palmeri IPNI=*30102841-2* 404
Populus pannonica IPNI=*776790-1* 404
Populus pendula IPNI=*776573-1* 404
Populus polonica IPNI=*776790-1* 404
Populus pubescens var. *pendula* IPNI=*776790-1* 404
Populus pyramidata IPNI=*776790-1* 404
Populus quercifolia IPNI=*776573-1* 404
Populus regenerata IPNI=*30102841-2* 404
Populus rubra IPNI=*776790-1* 404
Populus sosnowskyi IPNI=*776790-1* 404
Populus thevestina IPNI=*776790-1* 404
Populus triloba IPNI=*776573-1* 404
Populus versicolor IPNI=*776790-1* 404
Populus viminalis IPNI=*776573-1* 404
Populus viminea IPNI=*776790-1* 404
Populus virginiana IPNI=*30102841-2* 404
Populus virginiana var. *pilosa* IPNI=*30102841-2* 404
Populus viridis IPNI=*776790-1* 404
Populus vistulensis IPNI=*776790-1* 404
Poraqueiba acuminata IPNI=*608257-1* 359

Poraqueiba surinamensis IPNI=*608256-1* 359
Poraresia anomala IPNI=*354987-1* 401
Porcellites brasiliensis IPNI=*126428-2* 264
Porcellites brasiliensis var. *hirta* IPNI=*126428-2* 264
Porcellites brasiliensis var. *integrifolia* IPNI=*126428-2* 264
Porcellites brasiliensis var. *pinnatifida* IPNI=*126428-2* 264
Porcellites radicata IPNI=*225575-1* 264
Porophyllum ellipticum IPNI=*208446-2* 268
Porophyllum ellipticum var. *genuinum* IPNI=*208446-2* 268
Porophyllum ellipticum var. *intermedium* IPNI=*208446-2* 268
Porophyllum ellipticum var. *ruderale* IPNI=*208446-2* 268
Porophyllum macrolepidium IPNI=*208446-2* 268
Porophyllum oblanceolatum IPNI=*208446-2* 268
Porophyllum porophyllum IPNI=*208446-2* 268
Porophyllum ruderale f. *suffruticosum* IPNI=*208446-2* 268
Porophyllum ruderale var. *angustifolium* IPNI=*208446-2* 268
Porophyllum ruderale var. *ellipticum* IPNI=*208446-2* 268
Porophyllum ruderale var. *glandulosum* IPNI=*208446-2* 268
Porophyllum ruderale var. *macrolepidium* IPNI=*208446-2* 268
Porostema guianensis IPNI=*467336-1* 184
Porphyrocoma pohliana IPNI=*132826-2* 360
Porrum amethystinum IPNI=*332079-2* 213
Porrum ampeloprasum IPNI=*332079-2* 213
Porrum cepa IPNI=*527795-1* 213
Porrum commune IPNI=*332079-2* 213
Porrum fistulosum IPNI=*528046-1* 213
Porrum ophioscorodon IPNI=*528796-1* 213
Porrum sativum IPNI=*332079-2* 213
Porrum sativum IPNI=*528796-1* 213
Porrum schoenoprasum IPNI=*528823-1* 213
Porrum sibiricum IPNI=*528823-1* 213
Portea tillandsioides IPNI=*121847-1* 228
Porteria oblongifolia IPNI=*860264-1* 296
Porteria pubescens IPNI=*859894-1* 296
Portlandia acuminata IPNI=*67491-2* 349
Portlandia hexandra IPNI=*67491-2* 349
Portlandia speciosa IPNI=*67491-2* 349
Portulaca advena IPNI=*316354-2* 290
Portulaca aurea IPNI=*323270-2* 290
Portulaca axilliflora IPNI=*322565-2* 280
Portulaca brevioperculata IPNI=*323271-2* 290
Portulaca cincta IPNI=*316354-2* 290
Portulaca consanguinea IPNI=*323270-2* 290
Portulaca crassicaulis IPNI=*1182764-2* 290
Portulaca crassifolia IPNI=*1182764-2* 290
Portulaca cryptopetala var. *poellnitziana* IPNI=*323270-2* 290
Portulaca cyanosperma IPNI=*316354-2* 290
Portulaca cyathostyla IPNI=*316354-2* 290
Portulaca denudata IPNI=*323272-2* 290
Portulaca dichotoma IPNI=*323270-2* 290
Portulaca dubia IPNI=*316354-2* 290
Portulaca ehrenbergii IPNI=*316354-2* 290
Portulaca foliosa IPNI=*316354-2* 290
Portulaca fosbergii IPNI=*323270-2* 290
Portulaca fruticosa IPNI=*1182764-2* 290
Portulaca gagatosperma IPNI=*316354-2* 290
Portulaca helianthemoides IPNI=*316354-2* 290
Portulaca hortensis IPNI=*323270-2* 290
Portulaca involucrata IPNI=*316354-2* 290
Portulaca javanensis IPNI=*316354-2* 290
Portulaca karwinskii IPNI=*316354-2* 290
Portulaca laevis IPNI=*323270-2* 290
Portulaca lakhonensis IPNI=*316354-2* 290
Portulaca lanata IPNI=*316354-2* 290
Portulaca lanceolata var. *minor* IPNI=*323272-2* 290
Portulaca lanceolata var. *versicolor* IPNI=*323272-2* 290
Portulaca lanuginosa IPNI=*316354-2* 290
Portulaca latifolia IPNI=*323270-2* 290
Portulaca marginata IPNI=*323270-2* 290
Portulaca maritima IPNI=*323270-2* 290
Portulaca microsperma IPNI=*316354-2* 290
Portulaca mundula IPNI=*316354-2* 290
Portulaca napiformis IPNI=*316354-2* 290
Portulaca neglecta IPNI=*323270-2* 290
Portulaca officinarum IPNI=*323270-2* 290
Portulaca oleracea subsp. *stellata* IPNI=*323270-2* 290
Portulaca oleracea var. *gracilescens* IPNI=*323270-2* 290
Portulaca oleracea var. *grandiflora* IPNI=*323270-2* 290
Portulaca oleracea var. *linearifolia* IPNI=*323270-2* 290
Portulaca oleracea var. *sativa* IPNI=*323270-2* 290
Portulaca olitoria IPNI=*323270-2* 290
Portulaca opposita IPNI=*323270-2* 290
Portulaca pachyrrhiza IPNI=*316354-2* 290
Portulaca paniculata IPNI=*1007146-2* 290
Portulaca paniculata IPNI=*1182764-2* 290
Portulaca papillosa IPNI=*323270-2* 290
Portulaca parviflora IPNI=*323270-2* 290
Portulaca parvifolia IPNI=*323270-2* 290
Portulaca patens IPNI=*1007146-2* 290
Portulaca phaeosperma IPNI=*323271-2* 290
Portulaca pilosa f. *aurantiaca* IPNI=*316354-2* 290
Portulaca pilosa f. *mexicana* IPNI=*316354-2* 290
Portulaca pilosa f. *procumbens* IPNI=*316354-2* 290
Portulaca pilosa var. *guyannensis* IPNI=*316354-2* 290
Portulaca pilosa var. *marginata* IPNI=*323270-2* 290
Portulaca pilosa var. *mundula* IPNI=*316354-2* 290
Portulaca pilosa var. *normalis* IPNI=*316354-2* 290
Portulaca pilosa var. *paraguayensis* IPNI=*316354-2* 290
Portulaca pilosissima IPNI=*316354-2* 290
Portulaca plano-operculata IPNI=*323272-2* 290
Portulaca poellnitziana IPNI=*323270-2* 290
Portulaca portulacastrum IPNI=*235145-2* 280
Portulaca procumbens IPNI=*316354-2* 290
Portulaca racemosa IPNI=*1182764-2* 290
Portulaca reflexa IPNI=*1007146-2* 290
Portulaca remota IPNI=*316354-2* 290
Portulaca repens IPNI=*323270-2* 290
Portulaca retusa IPNI=*323270-2* 290
Portulaca rostellata IPNI=*323270-2* 290
Portulaca sativa IPNI=*323270-2* 290
Portulaca sedoides IPNI=*316354-2* 290
Portulaca stellata IPNI=*323270-2* 290
Portulaca suffruticosa IPNI=*323270-2* 290
Portulaca sylvestris IPNI=*323270-2* 290
Portulaca talmyana IPNI=*316354-2* 290
Portulaca terraereginae IPNI=*323270-2* 290
Portulaca toston IPNI=*322565-2* 280
Portulaca triangularis IPNI=*1182764-2* 290
Portulaca vilmoriana IPNI=*316354-2* 290
Portulaca viridis IPNI=*323270-2* 290
Portulacastrum monogynum IPNI=*322565-2* 280
Posadaea sphaerocarpa IPNI=*77110571-1* 294
Posoqueria acuminata IPNI=*761025-1* 355
Posoqueria dumetorum IPNI=*745904-1* 348
Posoqueria floribunda IPNI=*745904-1* 348
Posoqueria obliquinervia IPNI=*254985-2* 358
Posoqueria pittieri IPNI=*254985-2* 358
Possira arborescens IPNI=*519836-1* 334
Possira dodecandra IPNI=*519836-1* 334
Possira grandiflora IPNI=*520020-1* 335
Possira pinnata IPNI=*519989-1* 334
Possira simplex IPNI=*520020-1* 335
Possira triphylla IPNI=*519836-1* 334
Possura guianensis IPNI=*519836-1* 334
Potalia amara var. *resinifera* IPNI=*546884-1* 345
Potamocharis mamei IPNI=*428743-1* 381

Potamogeton acifolius IPNI=*77099639-1* 203
Potamogeton angustifolius f. *longipedunculatus* IPNI=*77099639-1* 203
Potamogeton angustifolius var. *connecticutensis* IPNI=*208837-2* 203
Potamogeton angustissimus IPNI=*77099639-1* 203
Potamogeton austriacus IPNI=*323116-2* 203
Potamogeton balatonicus IPNI=*77099639-1* 203
Potamogeton borealis IPNI=*77099639-1* 203
Potamogeton bracteatus IPNI=*77099639-1* 203
Potamogeton brasiliensis IPNI=*208837-2* 203
Potamogeton chakassiensis IPNI=*77099639-1* 203
Potamogeton chakassiensis var. *tenuior* IPNI=*77099639-1* 203
Potamogeton columbianus IPNI=*77099639-1* 203
Potamogeton concinnitus IPNI=*323116-2* 203
Potamogeton crenulatus IPNI=*323116-2* 203
Potamogeton crispatus IPNI=*323116-2* 203
Potamogeton curvatus IPNI=*208837-2* 203
Potamogeton diffusus IPNI=*77099639-1* 203
Potamogeton drupaceus IPNI=*77099639-1* 203
Potamogeton dunicola IPNI=*208837-2* 203
Potamogeton elipticus IPNI=*208837-2* 203
Potamogeton erhaiensis IPNI=*77099639-1* 203
Potamogeton filicaulis IPNI=*77099639-1* 203
Potamogeton filiformis IPNI=*77099639-1* 203
Potamogeton filiformis var. *borealis* IPNI=*77099639-1* 203
Potamogeton flabellatus IPNI=*77099639-1* 203
Potamogeton fragillimus IPNI=*208837-2* 203
Potamogeton helveticus IPNI=*77099639-1* 203
Potamogeton hohenackeri IPNI=*323116-2* 203
Potamogeton hungaricus IPNI=*323116-2* 203
Potamogeton illinoensis f. *homophyllus* IPNI=*208837-2* 203
Potamogeton illinoensis f. *rosulatus* IPNI=*208837-2* 203
Potamogeton interruptus IPNI=*77099639-1* 203
Potamogeton lanceolatus IPNI=*77099639-1* 203
Potamogeton leptanthus IPNI=*77099639-1* 203
Potamogeton leptophyllus IPNI=*323116-2* 203
Potamogeton linearis IPNI=*208837-2* 203
Potamogeton livingstonei IPNI=*77099639-1* 203
Potamogeton lucens f. *longipedunculatus* IPNI=*77099639-1* 203
Potamogeton lucens subsp. *brasiliensis* IPNI=*208837-2* 203
Potamogeton lucens var. *connecticutensis* IPNI=*208837-2* 203
Potamogeton lucens var. *floridanus* IPNI=*208837-2* 203
Potamogeton macrophylloides IPNI=*208837-2* 203
Potamogeton macrorrhynchus IPNI=*323116-2* 203
Potamogeton marinus IPNI=*77099639-1* 203
Potamogeton miniatus IPNI=*77099639-1* 203
Potamogeton nanus IPNI=*77099639-1* 203
Potamogeton notarisii IPNI=*323116-2* 203
Potamogeton pallidior IPNI=*323116-2* 203
Potamogeton pectinatus f. *brachypus* IPNI=*77099639-1* 203
Potamogeton pectinatus f. *geminatus* IPNI=*77099639-1* 203
Potamogeton pectinatus f. *pseudomarinus* IPNI=*77099639-1* 203
Potamogeton pectinatus IPNI=*77099639-1* 203
Potamogeton pectinatus subsp. *chakassiensis* IPNI=*77099639-1* 203
Potamogeton pectinatus subsp. *mongolicus* IPNI=*77099639-1* 203
Potamogeton pectinatus var. *diffusus* IPNI=*77099639-1* 203
Potamogeton pectinatus var. *mongolicus* IPNI=*77099639-1* 203
Potamogeton pectinatus var. *ungulatus* IPNI=*77099639-1* 203
Potamogeton pedersenii IPNI=*208837-2* 203
Potamogeton pulchelliformis IPNI=*208837-2* 203
Potamogeton rubricans IPNI=*323116-2* 203
Potamogeton rubrinaevus IPNI=*323116-2* 203
Potamogeton serrulatus IPNI=*323116-2* 203
Potamogeton siculus IPNI=*77099639-1* 203
Potamogeton tenuifolius IPNI=*77099639-1* 203
Potamogeton tuberosus IPNI=*323116-2* 203
Potamogeton vaillantii IPNI=*77099639-1* 203
Potamogeton zizii var. *connecticutensis* IPNI=*208837-2* 203
Potamogeton zizii var. *gracilis* IPNI=*208837-2* 203
Potamogeton zizii var. *porrectifolius* IPNI=*208837-2* 203
Potamogeton ziziiformis IPNI=*208837-2* 203
Potamogeton zosteraceus IPNI=*77099639-1* 203
Potentilla chiloensis IPNI=*724988-1* 443
Potentilla denticulosa IPNI=*728289-1* 443
Potentilla durandii IPNI=*728289-1* 443
Potentilla indica f. *albicaput* IPNI=*728289-1* 443
Potentilla indica var. *microphylla* IPNI=*728289-1* 443
Potentilla khasiana IPNI=*728289-1* 443
Potentilla trifida IPNI=*728289-1* 443
Potentilla vesca IPNI=*30074127-2* 443
Potentilla virginiana IPNI=*318974-2* 443
Poterion sanguisorbens IPNI=*741394-1* 445
Poterium anceps IPNI=*741394-1* 445
Poterium boreale IPNI=*741402-1* 445
Poterium collinum IPNI=*741394-1* 445
Poterium dictyocarpum IPNI=*741394-1* 445
Poterium duriaei IPNI=*741394-1* 445
Poterium glaucescens IPNI=*741394-1* 445
Poterium guestphalicum IPNI=*741394-1* 445
Poterium maroccanum IPNI=*741394-1* 445
Poterium minus IPNI=*741394-1* 445
Poterium officinale IPNI=*741402-1* 445
Poterium sanguisorba IPNI=*741394-1* 445
Poterium sanguisorba var. *garganicum* IPNI=*741394-1* 445
Pothomorphe allenii IPNI=*316339-2* 197
Pothomorphe almirantensis IPNI=*316334-2* 197
Pothomorphe baileyorum IPNI=*316334-2* 197
Pothomorphe baileyorum var. *paucispica* IPNI=*316334-2* 197
Pothomorphe dombeyana IPNI=*316339-2* 197
Pothomorphe esmeraldana IPNI=*683758-1* 197
Pothomorphe hirsuta IPNI=*316334-2* 197
Pothomorphe lessertiana IPNI=*316334-2* 197
Pothomorphe ottonis IPNI=*316334-2* 197
Pothomorphe peltata IPNI=*316334-2* 197
Pothomorphe peltata var. *colombiana* IPNI=*316334-2* 197
Pothomorphe pruinosa IPNI=*316334-2* 197
Pothomorphe scutata IPNI=*316334-2* 197
Pothomorphe sidifolia IPNI=*316339-2* 197
Pothomorphe speciosa IPNI=*316334-2* 197
Pothomorphe subpeltata IPNI=*316339-2* 197
Pothomorphe tecumensis IPNI=*316334-2* 197
Pothomorphe tecumensis var. *grandis* IPNI=*316334-2* 197
Pothomorphe umbellata f. *glabra* IPNI=*316339-2* 197
Pothomorphe umbellata IPNI=*316339-2* 197
Pothomorphe umbellata var. *cuernavacana* IPNI=*316339-2* 197
Pothomorphe umbellata var. *glabra* IPNI=*316339-2* 197
Pothomorphe umbellata var. *glabrior* IPNI=*316339-2* 197
Pothomorphe umbellata var. *vestita* IPNI=*316339-2* 197
Pothos breviscapus IPNI=*84611-1* 198
Pothos canniformis IPNI=*88955-1* 202
Pothos caudatus IPNI=*87046-1* 200
Pothos crassinervius IPNI=*84707-1* 198
Pothos flexuosus IPNI=*121108-2* 200
Pothos gracilis IPNI=*84849-1* 198
Pothos hederaceus IPNI=*87797-1* 201
Pothos leucophaeus IPNI=*88955-1* 202
Pothos odoratus IPNI=*88955-1* 202
Pothos palmatus IPNI=*85170-1* 199
Pothos pentaphyllus IPNI=*85201-1* 199
Pothos pertusus IPNI=*88567-1* 202
Pothos pinnatifidus IPNI=*87046-1* 200
Pothos pinnatus IPNI=*87046-1* 200
Pothos rubrinervius IPNI=*85321-1* 199
Pothos sagittatus IPNI=*85321-1* 199
Pothos sagittifolius IPNI=*89296-1* 202
Pothos scolopendrinus IPNI=*84849-1* 198
Pothos scolopendrioides IPNI=*84849-1* 198

Pothos suaveolens IPNI=*85321-1* 199
Pothuava nudicaulis IPNI=*121752-1* 228
Poupartia amazonica IPNI=*16050-2* 450
Poupartia dulcis IPNI=*71455-1* 451
Pourouma acutifolia IPNI=*856084-1* 447
Pourouma albistipulata IPNI=*283703-2* 447
Pourouma apaporiensis f. *macrophylla* IPNI=*209652-2* 447
Pourouma apaporiensis IPNI=*209652-2* 447
Pourouma aspera IPNI=*856077-1* 447
Pourouma aurea IPNI=*283697-2* 447
Pourouma camaratana IPNI=*856077-1* 447
Pourouma cinerascens IPNI=*856084-1* 447
Pourouma crassivenosa IPNI=*856077-1* 447
Pourouma cuatrecasasii IPNI=*283697-2* 447
Pourouma folleata IPNI=*283697-2* 447
Pourouma fuliginea IPNI=*856084-1* 447
Pourouma garciana IPNI=*209636-2* 447
Pourouma heterophylla IPNI=*856084-1* 447
Pourouma isophlebia IPNI=*283697-2* 447
Pourouma lawrancei IPNI=*856077-1* 447
Pourouma longipendula IPNI=*856089-1* 447
Pourouma multifida IPNI=*2856079-1* 447
Pourouma palmata IPNI=*856084-1* 447
Pourouma populifolia IPNI=*856074-1* 447
Pourouma radula IPNI=*856084-1* 447
Pourouma sapida IPNI=*2856079-1* 447
Pourouma scabra IPNI=*856084-1* 447
Pourouma schultesii IPNI=*856077-1* 447
Pourouma subplicata IPNI=*283697-2* 447
Pourouma substrigosa IPNI=*856084-1* 447
Pourouma subtriloba IPNI=*856084-1* 447
Pourouma tergoscabra IPNI=*283688-2* 447
Pourouma umbellata IPNI=*283697-2* 447
Pourouma umbellifera IPNI=*283697-2* 447
Pourouma uvifera IPNI=*2856079-1* 447
Pourretia platanifolia IPNI=*559273-1* 407
Pourretia sympaganthera IPNI=*115178-2* 229
Pourretia tuberculata IPNI=*559273-1* 407
Pouteria acutangula IPNI=*160358-2* 306
Pouteria albiflora IPNI=*209738-2* 306
Pouteria anibifolia IPNI=*209864-2* 307
Pouteria arbuscula IPNI=*209758-2* 306
Pouteria arenicola IPNI=*787792-1* 306
Pouteria benoistii IPNI=*788972-1* 307
Pouteria bouffardiana IPNI=*789009-1* 307
Pouteria bulliformis IPNI=*283713-2* 306
Pouteria burchelliana IPNI=*787792-1* 306
Pouteria caimito var. *strigosa* IPNI=*788972-1* 307
Pouteria caimito var. *typica* IPNI=*788845-1* 306
Pouteria calophylloides IPNI=*160393-2* 306
Pouteria campechiana var. *nervosa* IPNI=*209714-2* 306
Pouteria campechiana var. *palmeri* IPNI=*209714-2* 306
Pouteria campechiana var. *salicifolia* IPNI=*209714-2* 306
Pouteria campechiana var. *typica* IPNI=*209714-2* 306
Pouteria carabobensis IPNI=*209824-2* 307
Pouteria cataractae IPNI=*160417-2* 306
Pouteria cearensis IPNI=*209864-2* 307
Pouteria crassifolia IPNI=*209758-2* 306
Pouteria crotonoides IPNI=*160368-2* 306
Pouteria cuatrecasasii IPNI=*788972-1* 307
Pouteria cuspidata subsp. *dura* IPNI=*209738-2* 306
Pouteria cuspidata subsp. *robusta* IPNI=*209738-2* 306
Pouteria demerarae IPNI=*788972-1* 307
Pouteria dibrachiata IPNI=*948088-1* 305
Pouteria dura IPNI=*209738-2* 306
Pouteria dussiana IPNI=*209824-2* 307
Pouteria egensis IPNI=*160374-2* 306
Pouteria elegans var. *micrantha* IPNI=*209738-2* 306
Pouteria elegans var. *typica* IPNI=*209758-2* 306
Pouteria elongata IPNI=*209714-2* 306
Pouteria excelsa IPNI=*943696-1* 307
Pouteria flava IPNI=*160417-2* 306
Pouteria glabrescens IPNI=*89008-2* 305
Pouteria glabrifolia IPNI=*209714-2* 306
Pouteria glazioviana IPNI=*788972-1* 307
Pouteria glomerata var. *typica* IPNI=*788965-1* 307
Pouteria gomphiifolia var. *typica* IPNI=*788967-1* 307
Pouteria gutta IPNI=*788972-1* 307
Pouteria hartii IPNI=*788972-1* 307
Pouteria hispida IPNI=*788972-1* 307
Pouteria hondurensis IPNI=*209864-2* 307
Pouteria huberiana IPNI=*209824-2* 307
Pouteria insignis IPNI=*209804-2* 307
Pouteria jenmanii IPNI=*788972-1* 307
Pouteria laeteviridis IPNI=*209714-2* 306
Pouteria lasiocarpa IPNI=*788845-1* 306
Pouteria laurifolia IPNI=*788845-1* 306
Pouteria leucophaea IPNI=*788845-1* 306
Pouteria loretensis IPNI=*209727-2* 306
Pouteria lucentifolia IPNI=*948088-1* 305
Pouteria macao IPNI=*209907-2* 308
Pouteria macrocarpa IPNI=*209824-2* 307
Pouteria mammosa IPNI=*152641-2* 306
Pouteria mante IPNI=*209714-2* 306
Pouteria martinicensis IPNI=*209824-2* 307
Pouteria mayana IPNI=*948088-1* 305
Pouteria melinoniana IPNI=*160393-2* 306
Pouteria meyeri IPNI=*209864-2* 307
Pouteria micans IPNI=*948066-1* 306
Pouteria minutiflora IPNI=*209732-2* 306
Pouteria multiflora var. *quadrifida* IPNI=*209824-2* 307
Pouteria multiflora var. *typica* IPNI=*209824-2* 307
Pouteria multiflora var. *urbanii* IPNI=*209824-2* 307
Pouteria obidensis IPNI=*788972-1* 307
Pouteria obtusifolia IPNI=*787792-1* 306
Pouteria obtusifolia var. *typica* IPNI=*787792-1* 306
Pouteria officinalis IPNI=*209824-2* 307
Pouteria paraensis IPNI=*789045-1* 307
Pouteria pentasperma IPNI=*948088-1* 305
Pouteria petenensis IPNI=*948088-1* 305
Pouteria polyneura IPNI=*160417-2* 306
Pouteria pomifera IPNI=*948086-1* 305
Pouteria reticulata subsp. *surinamensis* IPNI=*209864-2* 307
Pouteria rivicoa IPNI=*209808-2* 307
Pouteria rivularis IPNI=*160417-2* 306
Pouteria robusta IPNI=*209738-2* 306
Pouteria robusta var. *longifolia* IPNI=*209738-2* 306
Pouteria robusta var. *typica* IPNI=*209738-2* 306
Pouteria rubriflora IPNI=*307657-2* 305
Pouteria scytalophora IPNI=*948066-1* 306
Pouteria setosa IPNI=*160374-2* 306
Pouteria solimoesensis IPNI=*788972-1* 307
Pouteria steyermarkiana IPNI=*89010-2* 305
Pouteria temare IPNI=*788845-1* 306
Pouteria torta subsp. *glabra* IPNI=*788972-1* 307
Pouteria tovarensis IPNI=*788972-1* 307
Pouteria trichopoda IPNI=*788972-1* 307
Pouteria ulei IPNI=*160374-2* 306
Pouteria unilocularis IPNI=*209864-2* 307
Pouteria venulosa IPNI=*160417-2* 306
Pouzolzia phenacoides IPNI=*856206-1* 448
Pouzolzia procridioides IPNI=*856206-1* 448
Praealstonia arechea IPNI=*827005-1* 309
Praealstonia mathewsii IPNI=*827005-1* 309
Praealstonia rufescens IPNI=*827299-1* 309
Praealstonia serrulata IPNI=*827299-1* 309

Praealstonia theiformis IPNI=*827837-1* 309
Praealstonia tomentosa IPNI=*827299-1* 309
Praecereus campinensis IPNI=*1001185-1* 284
Pragmotessara japonica IPNI=*160944-1* 290
Prasium marrubium IPNI=*449990-1* 371
Pratia acuminata IPNI=*142908-1* 273
Pratia boliviensis IPNI=*143584-1* 273
Premna arborea IPNI=*77159191-1* 369
Premna reticulata IPNI=*864000-1* 372
Prenanthes fastigiata IPNI=*1124129-2* 272
Prenanthes fruticosa IPNI=*256729-1* 272
Prenanthes japonica IPNI=*1124129-2* 272
Prenanthes multiflora IPNI=*1124129-2* 272
Prenanthes napifolia IPNI=*1124129-2* 272
Prenanthes poosia IPNI=*1124129-2* 272
Prenanthes sarmentosa IPNI=*203080-1* 261
Prenanthes sonchifolia IPNI=*229022-1* 264
Prenanthes striata IPNI=*1124129-2* 272
Preslianthus detonsus IPNI=*77183701-1* 278
Preslianthus pittieri IPNI=*77184198-1* 278
Prestoea asplundii IPNI=*210069-2* 211
Prestoea darienensis IPNI=*210063-2* 211
Prestoea integrifolia IPNI=*210063-2* 211
Prestoea sejuncta IPNI=*210063-2* 211
Prestonia acutifolia IPNI=*81405-1* 342
Prestonia acutifolia var. *latissima* IPNI=*81405-1* 342
Prestonia ecuadorensis IPNI=*81387-1* 342
Prestonia evansii IPNI=*210147-2* 342
Prestonia glabrata IPNI=*81387-1* 342
Prestonia marginata IPNI=*81405-1* 342
Prestonia nutans IPNI=*81405-1* 342
Prestonia obovata IPNI=*210147-2* 342
Prestonia pachyphylla IPNI=*81405-1* 342
Prestonia pickelii IPNI=*81405-1* 342
Prestonia simulans IPNI=*81405-1* 342
Prestonia venosa IPNI=*81405-1* 342
Prestonia weberbaueri IPNI=*81387-1* 342
Prieurella colombiana IPNI=*948084-1* 305
Prieurella lanceolata IPNI=*941286-1* 305
Prieurella maguirei IPNI=*948078-1* 307
Prieurella manaosensis IPNI=*948085-1* 305
Prieurella prieurii IPNI=*57816-2* 305
Prieurella wurdackii IPNI=*941286-1* 305
Primula bonatii IPNI=*702340-1* 304
Primula brandisii IPNI=*702751-1* 304
Primula cordifolia IPNI=*702751-1* 304
Primula coronaria IPNI=*702751-1* 304
Primula discolor IPNI=*702751-1* 304
Primula inflata IPNI=*702751-1* 304
Primula lasiopetala IPNI=*702751-1* 304
Primula mandarina IPNI=*702446-1* 304
Primula mutabilis IPNI=*791637-1* 291
Primula officinalis IPNI=*702751-1* 304
Primula pannonica IPNI=*702751-1* 304
Primula petitmenginii IPNI=*702340-1* 304
Primula pistillaris IPNI=*702751-1* 304
Primula poculiformis IPNI=*702340-1* 304
Primula praticola IPNI=*702751-1* 304
Primula pyrenaica IPNI=*702751-1* 304
Primula semperflorens IPNI=*702446-1* 304
Primula sertulosa IPNI=*702446-1* 304
Primula sinensis IPNI=*702446-1* 304
Primula velenovskyi IPNI=*702751-1* 304
Primulidium sinense IPNI=*702446-1* 304
Prinos myricoides IPNI=*83545-1* 255
Prionostemma malpighiifolia IPNI=*162403-1* 291
Prionostemma scabridula IPNI=*162403-1* 291
Prismatocarpus perfoliatus IPNI=*258447-2* 273
Pristimera andina IPNI=*210308-2* 291
Pristimera apiculata IPNI=*72123-2* 290
Pristimera caribaea IPNI=*210308-2* 291
Pristimera granulosa IPNI=*72123-2* 290
Pristimera oblongata IPNI=*72123-2* 290
Pristimera tabascensis IPNI=*210308-2* 291
Pristimera tenella IPNI=*210308-2* 291
Pritchardia robusta IPNI=*670278-1* 212
Pritzelia coccinea IPNI=*315561-2* 292
Priva echinata IPNI=*1194372-2* 379
Priva lamiifolia IPNI=*1194372-2* 379
Priva lappulacea f. *albiflora* IPNI=*1194372-2* 379
Proboscidea jussieui IPNI=*675840-1* 374
Prockia acuta IPNI=*321266-2* 404
Prockia bahiensis IPNI=*321266-2* 404
Prockia completa IPNI=*321266-2* 404
Prockia crucifolia IPNI=*321266-2* 404
Prockia crucis var. *cordata* IPNI=*321266-2* 404
Prockia crucis var. *septemnervia* IPNI=*321266-2* 404
Prockia glabra IPNI=*321266-2* 404
Prockia grandiflora IPNI=*321266-2* 404
Prockia hassleri IPNI=*321266-2* 404
Prockia lobata IPNI=*321266-2* 404
Prockia lutea IPNI=*321266-2* 404
Prockia macrostachya IPNI=*321266-2* 404
Prockia mexicana IPNI=*321266-2* 404
Prockia morifolia IPNI=*321266-2* 404
Prockia obovata IPNI=*112138-1* 405
Prockia odorata IPNI=*779396-1* 402
Prockia septemnervia IPNI=*321266-2* 404
Prockia subcordata IPNI=*321266-2* 404
Prockia tomentosa IPNI=*321266-2* 404
Prockia villosa IPNI=*321266-2* 404
Proclesia acuminata IPNI=*60458979-2* 298
Proclesia benthamiana IPNI=*60458979-2* 298
Proclesia bracteata IPNI=*60458979-2* 298
Proclesia cordata IPNI=*60458979-2* 298
Proclesia cordifolia IPNI=*60458979-2* 298
Proclesia hartwegiana IPNI=*60458979-2* 298
Proclesia pseudopubescens IPNI=*60458979-2* 298
Proclesia pubescens IPNI=*858236-1* 298
Proclesia scabriuscula IPNI=*60458979-2* 298
Proclesia splendens IPNI=*60458979-2* 298
Proclesia strobolifera IPNI=*60458979-2* 298
Proclesia veraguensis IPNI=*60458979-2* 298
Proclesia warscewiczii IPNI=*60458979-2* 298
Procris caudata IPNI=*850541-1* 445
Procris nivea IPNI=*55676-3* 445
Procris ramiflora IPNI=*34156-2* 445
Procris rugosa IPNI=*855180-1* 446
Procris sieberi IPNI=*1151750-2* 440
Pronacron ramosissimum IPNI=*257035-1* 272
Prosanerpis panamensis IPNI=*77207810-1* 421
Prosanerpis trichocalyx IPNI=*77207810-1* 421
Prosopis bonplandiana IPNI=*1009172-2* 319
Prosopis dubia IPNI=*1037928-2* 319
Prosopis microphylla IPNI=*60441241-2* 336
Prosopis scandens IPNI=*493866-1* 319
Prosthechea cochleata var. *grandiflora* IPNI=*1001247-1* 222
Prosthechea fragrans f. *alba* IPNI=*1001253-1* 222
Protasparagus aethiopicus IPNI=*530994-1* 214
Protasparagus densiflorus IPNI=*531072-1* 214
Protasparagus plumosus IPNI=*531312-1* 214
Protasparagus setaceus IPNI=*531312-1* 214
Protasparagus virgatus IPNI=*531381-1* 214
Protea diversifolia IPNI=*704928-1* 433
Protea inflexa IPNI=*704928-1* 433
Protea involucrata IPNI=*704928-1* 433

Protea pallens IPNI=*704928-1* 433
Protea saligna IPNI=*704928-1* 433
Protea sericea IPNI=*704928-1* 433
Protea stricta IPNI=*704928-1* 433
Protium almacega IPNI=*128420-1* 453
Protium altissimum IPNI=*77183601-1* 452
Protium ambrosiacum IPNI=*128337-1* 452
Protium angustifolium IPNI=*128338-1* 453
Protium aracouchini var. *angustifolium* IPNI=*128278-1* 452
Protium aromaticum IPNI=*128338-1* 453
Protium asperum IPNI=*256487-2* 453
Protium belemense IPNI=*210488-2* 452
Protium benthamii IPNI=*128430-1* 453
Protium decandrum var. *breviacuminatum* IPNI=*128316-1* 452
Protium divaricatum var. *intermedium* IPNI=*128278-1* 452
Protium divaricatum var. *trijugum* IPNI=*128318-1* 452
Protium fimbriatum IPNI=*1063489-2* 452
Protium firmum IPNI=*210491-2* 452
Protium heptaphyllum var. *angustifolium* IPNI=*128338-1* 453
Protium heptaphyllum var. *aromaticum* IPNI=*128338-1* 453
Protium heptaphyllum var. *floribundum* IPNI=*128338-1* 453
Protium heptaphyllum var. *grandiflorum* IPNI=*128338-1* 453
Protium heptaphyllum var. *multiflorum* IPNI=*128338-1* 453
Protium heptaphyllum var. *puberulum* IPNI=*128337-1* 452
Protium heptaphyllum var. *surinamense* IPNI=*128338-1* 453
Protium heptaphyllum var. *unifoliolatum* IPNI=*128338-1* 453
Protium hostmannii IPNI=*128337-1* 452
Protium hostmannii var. *brasiliense* IPNI=*128338-1* 453
Protium insigne IPNI=*128413-1* 453
Protium joannis IPNI=*128429-1* 453
Protium juruense IPNI=*210625-2* 453
Protium llewelynii IPNI=*128335-1* 452
Protium lucidum IPNI=*70000-1* 451
Protium marthae IPNI=*210542-2* 452
Protium martianum IPNI=*128429-1* 453
Protium medianum IPNI=*128316-1* 452
Protium multiflorum IPNI=*128338-1* 453
Protium neglectum IPNI=*128424-1* 453
Protium neglectum var. *robustum* IPNI=*210629-2* 453
Protium octandrum IPNI=*128338-1* 453
Protium orinocense IPNI=*128316-1* 452
Protium paraense IPNI=*210488-2* 452
Protium pauciflorum IPNI=*128429-1* 453
Protium picramnioides IPNI=*77202519-1* 453
Protium poeppigianum IPNI=*210595-2* 453
Protium polybotryum var. *robustum* IPNI=*128400-1* 453
Protium puberulentum IPNI=*210488-2* 452
Protium pubescens IPNI=*128430-1* 453
Protium pubescens var. *glabrescens* IPNI=*128430-1* 453
Protium schomburgkianum IPNI=*128316-1* 452
Protium tacamahaca IPNI=*128338-1* 453
Protium titubans IPNI=*128429-1* 453
Protium unifoliolatum var. *inconforme* IPNI=*128430-1* 453
Protium unifoliolatum var. *macrophyllum* IPNI=*128430-1* 453
Protium unifoliolatum var. *puberulum* IPNI=*128430-1* 453
Protium unifoliolatum var. *subserratum* IPNI=*128430-1* 453
Prumnopitys harmsiana IPNI=*77203787-1* 180
Prumnopitys utilior IPNI=*77203787-1* 180
Prunus acinaria IPNI=*30221374-2* 443
Prunus amygdalina IPNI=*30221374-2* 443
Prunus augustana IPNI=*30221374-2* 443
Prunus ballota IPNI=*30221374-2* 443
Prunus brignola IPNI=*30221374-2* 443
Prunus carolinae IPNI=*210915-2* 444
Prunus cartilaginea IPNI=*730262-1* 444
Prunus cerasiflora IPNI=*30221374-2* 443
Prunus cereola IPNI=*30221374-2* 443
Prunus communis IPNI=*30221374-2* 443
Prunus daemonifuga IPNI=*1212858-2* 444
Prunus domestica subsp. *cyclopetala* IPNI=*30221374-2* 443
Prunus domestica var. *claudiana* IPNI=*30221374-2* 443
Prunus domestica var. *oeconomica* IPNI=*30221374-2* 443
Prunus enucleata IPNI=*30221374-2* 443
Prunus erubescens IPNI=*30221374-2* 443
Prunus eximia IPNI=*730262-1* 444
Prunus ferganensis IPNI=*1212858-2* 444
Prunus galatensis IPNI=*30221374-2* 443
Prunus hirsuta IPNI=*730262-1* 444
Prunus latifolia IPNI=*30221374-2* 443
Prunus maliformis IPNI=*30221374-2* 443
Prunus oeconomica IPNI=*30221374-2* 443
Prunus oxycarpa IPNI=*30221374-2* 443
Prunus perdigona IPNI=*30221374-2* 443
Prunus persica subsp. *ferganensis* IPNI=*1212858-2* 444
Prunus persica subsp. *floriplena* IPNI=*1212858-2* 444
Prunus persica subsp. *nucipersica* IPNI=*1212858-2* 444
Prunus persica subsp. *vulgaris* IPNI=*1212858-2* 444
Prunus persica var. *aposarca* IPNI=*1212858-2* 444
Prunus persica var. *nucipersica* IPNI=*1212858-2* 444
Prunus rosiflora IPNI=*30221374-2* 443
Prunus rubescens IPNI=*30221374-2* 443
Prunus sana IPNI=*210756-2* 443
Prunus sanctae-catharinae IPNI=*30221374-2* 443
Prunus sativa IPNI=*30221374-2* 443
Prunus saxicola IPNI=*30221374-2* 443
Prunus sepivaga IPNI=*30221374-2* 443
Prunus serotina subsp. *eximia* IPNI=*730262-1* 444
Prunus serotina subsp. *hirsuta* IPNI=*730262-1* 444
Prunus serotina var. *cartilaginea* IPNI=*730262-1* 444
Prunus serotina var. *eximia* IPNI=*730262-1* 444
Prunus serotina var. *montana* IPNI=*730262-1* 444
Prunus serotina var. *neomontana* IPNI=*730262-1* 444
Prunus vana IPNI=*210756-2* 443
Prunus vapincensis IPNI=*30221374-2* 443
Prunus varactensis IPNI=*30221374-2* 443
Prunus virescens IPNI=*30221374-2* 443
Prunus williamsii IPNI=*210756-2* 443
Psacadocalymma comatum IPNI=*78920-2* 359
Psacadocalymma pectorale IPNI=*47290-1* 359
Psacalium glabratum IPNI=*963618-1* 267
Psacalium vaccinioides IPNI=*963618-1* 267
Psammanthe marina IPNI=*235145-2* 280
Psammisia alpicola IPNI=*148722-2* 299
Psammisia costaricensis IPNI=*148722-2* 299
Psammisia cyathifera IPNI=*858617-1* 300
Psammisia glabra IPNI=*148722-2* 299
Psammisia jessicae IPNI=*858621-1* 300
Psammisia lehmannii IPNI=*211039-2* 300
Psammisia longifolia IPNI=*858728-1* 300
Psammisia nitida IPNI=*50474-2* 298
Psammisia pichinchensis IPNI=*858621-1* 300
Psammisia rupestris IPNI=*148722-2* 299
Psammisia sarcantha IPNI=*858621-1* 300
Psammisia sclerophylla IPNI=*858621-1* 300
Psammisia tovarensis IPNI=*858621-1* 300
Psathura obovata IPNI=*315444-2* 350
Pseudabutilon spicatum IPNI=*36556-2* 407
Pseudanamomis umbellulifera IPNI=*596253-1* 425
Pseudechinolaena inflexa IPNI=*88264-2* 238
Pseudechinolaena spectabilis IPNI=*88097-2* 238
Pseudelephantopus crispus IPNI=*176829-2* 267
Pseudelephantopus funckii IPNI=*211400-2* 268
Pseuderanthemum asperum IPNI=*48879-1* 360
Pseuderanthemum atropurpureum IPNI=*77158889-1* 361
Pseuderanthemum carruthersii IPNI=*77158889-1* 361
Pseuderanthemum carruthersii var. *reticulatum* IPNI=*77158889-1* 361

Pseuderanthemum eldorado IPNI=*77158889-1* 361
Pseuderanthemum jaluitense IPNI=*77158889-1* 361
Pseuderanthemum kewense IPNI=*77158889-1* 361
Pseuderanthemum majus IPNI=*77158889-1* 361
Pseuderanthemum moorei IPNI=*77158889-1* 361
Pseuderanthemum poecilanthum IPNI=*913240-1* 361
Pseuderanthemum pulchellum IPNI=*48879-1* 360
Pseuderanthemum reticulatum IPNI=*77158889-1* 361
Pseuderanthemum tricolor IPNI=*77158889-1* 361
Pseuderanthemum versicolor IPNI=*77158889-1* 361
Pseudobaccharis cotinifolia IPNI=*28074-2* 257
Pseudobaccharis rhexioides IPNI=*183821-1* 258
Pseudobaccharis trinervis IPNI=*183821-1* 258
Pseudobastardia crispa IPNI=*120630-2* 409
Pseudobombax amapaense IPNI=*211238-2* 413
Pseudocalymma alliaceum IPNI=*152672-2* 364
Pseudocalymma alliaceum var. *macrocalyx* IPNI=*152672-2* 364
Pseudocalymma alliaceum var. *microcalyx* IPNI=*152677-2* 364
Pseudocalymma hymenaeum IPNI=*152677-2* 364
Pseudocalymma kuhlmannii IPNI=*77100467-1* 365
Pseudocalymma laevigatum IPNI=*152677-2* 364
Pseudocalymma langlasseanum IPNI=*152677-2* 364
Pseudocalymma macrocarpum IPNI=*152677-2* 364
Pseudocalymma pachypus IPNI=*152672-2* 364
Pseudocalymma pohlianum IPNI=*152677-2* 364
Pseudocalymma sagotii IPNI=*152672-2* 364
Pseudocalymma sagotii var. *macrocalyx* IPNI=*152672-2* 364
Pseudocalymma standleyi IPNI=*152684-2* 364
Pseudocaryophyllus coccolobifolius IPNI=*166115-2* 427
Pseudocaryophyllus heteroclitus IPNI=*166091-2* 427
Pseudocassia spectabilis IPNI=*234663-2* 333
Pseudochimarrhis diformis IPNI=*308173-2* 348
Pseudochimarrhis turbinata IPNI=*308173-2* 348
Pseudocladia colombiana IPNI=*948066-1* 306
Pseudocladia melinonii IPNI=*948066-1* 306
Pseudocladia minutiflora IPNI=*209732-2* 306
Pseudocladia neblinaensis IPNI=*209700-2* 306
Pseudocladia orinocoensis IPNI=*948065-1* 307
Pseudocladia scytalophora IPNI=*948066-1* 306
Pseudoconnarus fecundus IPNI=*211312-2* 431
Pseudoconyza lyrata IPNI=*211317-2* 268
Pseudoconyza viscosa var. *lyrata* IPNI=*211317-2* 268
Pseudocroton tinctorius IPNI=*77217338-1* 278
Pseudognaphalium antennarioides IPNI=*77197189-1* 259
Pseudognaphalium austrobrasilicum IPNI=*283850-2* 268
Pseudognaphalium crenatum IPNI=*283895-2* 269
Pseudognaphalium graveolens IPNI=*77142354-1* 268
Pseudognaphalium luteoalbum IPNI=*212994-1* 264
Pseudognaphalium luteoalbum subsp. *turcicum* IPNI=*212994-1* 264
Pseudognaphalium luteoalbum var. *pallidum* IPNI=*212994-1* 264
Pseudognaphalium mendocinum IPNI=*283850-2* 268
Pseudognaphalium obtusifolium subsp. *praecox* IPNI=*909626-1* 268
Pseudognaphalium obtusifolium subsp. *saxicola* IPNI=*909626-1* 268
Pseudognaphalium resedifolium IPNI=*283846-2* 268
Pseudognaphalium saxicola IPNI=*909626-1* 268
Pseudogynoxys berlandieri IPNI=*211361-2* 269
Pseudogynoxys chenopodioides var. *cummingii* IPNI=*211365-2* 269
Pseudogynoxys cordifolia IPNI=*211361-2* 269
Pseudogynoxys hoffmannii IPNI=*211365-2* 269
Pseudogynoxys volubilis IPNI=*211361-2* 269
Pseudohomalomena pastoensis IPNI=*89403-1* 203
Pseudolabatia filipes IPNI=*209766-2* 307
Pseudolmedia alnifolia IPNI=*211411-2* 441
Pseudolmedia brosimifolia IPNI=*856393-1* 441
Pseudolmedia bucidifolia IPNI=*77164595-1* 416
Pseudolmedia coriacea IPNI=*152822-2* 440
Pseudolmedia ferruginea IPNI=*211411-2* 441
Pseudolmedia guaranitica IPNI=*856393-1* 441
Pseudolmedia hirsuta IPNI=*211411-2* 441
Pseudolmedia hirtellifolia IPNI=*211411-2* 441
Pseudolmedia huberi IPNI=*152822-2* 440
Pseudolmedia karstenii IPNI=*260203-2* 441
Pseudolmedia krukovii IPNI=*167401-2* 440
Pseudolmedia mildbraedii IPNI=*856393-1* 441
Pseudolmedia multinervis IPNI=*211411-2* 441
Pseudolmedia murure IPNI=*856394-1* 441
Pseudolmedia obliqua IPNI=*152822-2* 440
Pseudolmedia sagotii IPNI=*211411-2* 441
Pseudolmedia scabra IPNI=*119540-2* 439
Pseudolycopodiella caroliniana IPNI=*146931-2* 171
Pseudomalachra ciliaris IPNI=*313702-2* 413
Pseudomalachra plumosa IPNI=*313702-2* 413
Pseudomalachra tridentata IPNI=*313702-2* 413
Pseudomarsdenia cundurango IPNI=*99126-1* 342
Pseudomussaenda lanceolata IPNI=*760275-1* 355
Pseudomyrcianthes kochiana IPNI=*1046564-2* 425
Pseudo-oenothera virginiana IPNI=*172755-2* 429
Pseudorhipsalis amazonica subsp. *panamensis* IPNI=*60426912-2* 284
Pseudorhipsalis ramulosa f. *angustissima* IPNI=*963240-1* 284
Pseudoryza ciliata IPNI=*137026-2* 241
Pseudosasa variegata IPNI=*416271-1* 244
Pseudosorocea poeppigii IPNI=*59683-2* 437
Pseudosorocea sprucei IPNI=*240971-2* 441
Pseudrachicallis floribunda IPNI=*743608-1* 347
Pseusmagennetus equatoriensis IPNI=*99126-1* 342
Psidiastrum dubium IPNI=*593690-1* 425
Psidiopsis moritzianum IPNI=*43033-2* 424
Psidium acidum IPNI=*600701-1* 428
Psidium acutangulum var. *crassirame* IPNI=*600701-1* 428
Psidium acutangulum var. *oblongatum* IPNI=*600701-1* 428
Psidium acutangulum var. *tenuirame* IPNI=*600701-1* 428
Psidium albidum IPNI=*211886-2* 428
Psidium angustifolium IPNI=*600841-1* 428
Psidium araca IPNI=*211886-2* 428
Psidium arboreum IPNI=*600760-1* 428
Psidium aromaticum IPNI=*600841-1* 428
Psidium atiraense IPNI=*211886-2* 428
Psidium basanthum IPNI=*600884-1* 428
Psidium benthamianum IPNI=*211886-2* 428
Psidium bergianum IPNI=*600884-1* 428
Psidium bergianum var. *verticillata* IPNI=*600884-1* 428
Psidium buxifolium IPNI=*600760-1* 428
Psidium campomanesioides IPNI=*1061290-2* 424
Psidium capibaryense IPNI=*600884-1* 428
Psidium cattleyanum f. *lucidum* IPNI=*600760-1* 428
Psidium cattleyanum var. *littorale* IPNI=*600760-1* 428
Psidium cattleyanum var. *purpureum* IPNI=*600760-1* 428
Psidium caudatum IPNI=*43033-2* 424
Psidium chodatianum IPNI=*600841-1* 428
Psidium claraense IPNI=*600952-1* 428
Psidium cordillerense IPNI=*211886-2* 428
Psidium coriaceum IPNI=*600857-1* 428
Psidium coriaceum var. *grandifolium* IPNI=*600760-1* 428
Psidium coriaceum var. *longipes* IPNI=*600760-1* 428
Psidium coriaceum var. *obovatum* IPNI=*600760-1* 428
Psidium costaricense IPNI=*211886-2* 428
Psidium crispum IPNI=*600841-1* 428
Psidium cujavillus IPNI=*600841-1* 428
Psidium cujavus IPNI=*600841-1* 428
Psidium dichotomum IPNI=*211886-2* 428
Psidium dubium IPNI=*166146-2* 427
Psidium ellipticum IPNI=*600841-1* 428
Psidium ferrugineum IPNI=*600760-1* 428
Psidium fluviatile IPNI=*600845-1* 428
Psidium formosum IPNI=*600884-1* 428

Psidium fragrans IPNI=*600841-1* 428
Psidium galapagaeum IPNI=*600952-1* 428
Psidium glandulosum IPNI=*211886-2* 428
Psidium glaucescens IPNI=*600884-1* 428
Psidium globosum IPNI=*600841-1* 428
Psidium guajava var. *cujavillum* IPNI=*600841-1* 428
Psidium guajava var. *minor* IPNI=*600841-1* 428
Psidium hians IPNI=*211886-2* 428
Psidium hians var. *cuneatum* IPNI=*211886-2* 428
Psidium hians var. *truncatum* IPNI=*211886-2* 428
Psidium igatemyense IPNI=*600841-1* 428
Psidium indicum IPNI=*600760-1* 428
Psidium intermedium IPNI=*600841-1* 428
Psidium jollyanum IPNI=*211886-2* 428
Psidium laurifolium IPNI=*211886-2* 428
Psidium lehmannii IPNI=*211886-2* 428
Psidium lineatifolium IPNI=*592044-1* 424
Psidium littorale IPNI=*600760-1* 428
Psidium littorale var. *longipes* IPNI=*600760-1* 428
Psidium longifolium IPNI=*600841-1* 428
Psidium macrophyllum IPNI=*211886-2* 428
Psidium microphyllum IPNI=*600952-1* 428
Psidium minus IPNI=*211886-2* 428
Psidium minutiflorum IPNI=*600952-1* 428
Psidium molinae IPNI=*600952-1* 428
Psidium molle IPNI=*211886-2* 428
Psidium molle var. *gracile* IPNI=*211886-2* 428
Psidium molle var. *robustum* IPNI=*211886-2* 428
Psidium monticola IPNI=*211886-2* 428
Psidium monticola var. *gracile* IPNI=*211886-2* 428
Psidium monticola var. *robustum* IPNI=*211886-2* 428
Psidium multiflorum IPNI=*211886-2* 428
Psidium obovatum IPNI=*600760-1* 428
Psidium ooideum IPNI=*211886-2* 428
Psidium ooideum var. *grandifolium* IPNI=*211886-2* 428
Psidium ooideum var. *intermedium* IPNI=*211886-2* 428
Psidium ooideum var. *longipedunculatum* IPNI=*211886-2* 428
Psidium ooideum var. *parvifolium* IPNI=*211886-2* 428
Psidium ovatifolium IPNI=*600795-1* 428
Psidium ovatifolium var. *glabrum* IPNI=*600795-1* 428
Psidium persoonii IPNI=*600701-1* 428
Psidium polycarpon IPNI=*211886-2* 428
Psidium pomiferum IPNI=*600841-1* 428
Psidium pomiferum var. *sapidissimum* IPNI=*600841-1* 428
Psidium prostratum IPNI=*600841-1* 428
Psidium protractum IPNI=*600952-1* 428
Psidium pumilum IPNI=*600841-1* 428
Psidium pumilum var. *guadalupense* IPNI=*600841-1* 428
Psidium pyriferum IPNI=*600841-1* 428
Psidium pyriferum var. *glabrum* IPNI=*600841-1* 428
Psidium quinquedentatum IPNI=*600884-1* 428
Psidium richardianum IPNI=*600845-1* 428
Psidium rivulare IPNI=*592044-1* 424
Psidium rotundifolium IPNI=*211886-2* 428
Psidium rufinervum IPNI=*211886-2* 428
Psidium salutare var. *strictum* IPNI=*1115374-2* 428
Psidium sapidissimum IPNI=*600841-1* 428
Psidium sartorianum IPNI=*600952-1* 428
Psidium sartorianum var. *yucatanense* IPNI=*600952-1* 428
Psidium savannarum IPNI=*600884-1* 428
Psidium schiedeanum IPNI=*211886-2* 428
Psidium schippii IPNI=*211886-2* 428
Psidium sellowianum IPNI=*600760-1* 428
Psidium sericiflorum IPNI=*211886-2* 428
Psidium socorrense IPNI=*600952-1* 428
Psidium solisii IPNI=*600952-1* 428
Psidium sprucei IPNI=*600845-1* 428
Psidium subcrenatum IPNI=*600841-1* 428
Psidium tomasense IPNI=*211886-2* 428
Psidium ubatubense IPNI=*600760-1* 428
Psidium umbrosum IPNI=*600845-1* 428
Psidium variabile IPNI=*600760-1* 428
Psidium vulgare IPNI=*600841-1* 428
Psidium warmingianum IPNI=*600884-1* 428
Psidium warmingianum var. *verticillatum* IPNI=*600884-1* 428
Psidium ypanemense IPNI=*211886-2* 428
Psidium yucatanense IPNI=*600952-1* 428
Psiguria bignoniacea IPNI=*114819-2* 294
Psiguria diversifolia IPNI=*211990-2* 295
Psiguria dunlapii IPNI=*211990-2* 295
Psiguria longipedunculata IPNI=*211990-2* 295
Psila cinnamomifolia IPNI=*28074-2* 257
Psila trinervis IPNI=*183821-1* 258
Psilogramme flexuosa IPNI=*77110633-1* 178
Psilogramme haematodes IPNI=*77110633-1* 178
Psilogramme villosula IPNI=*77110633-1* 178
Psilorhegma planisiliqua IPNI=*1117728-2* 333
Psilorhegma suffruticosa IPNI=*911423-1* 333
Psilotrichum malaccense IPNI=*10294-2* 281
Psilotum complanatum var. *flaccidum* IPNI=*114220-3* 179
Psilotum complanatum var. *latissimum* IPNI=*114220-3* 179
Psilotum complanatum var. *mexicanum* IPNI=*114220-3* 179
Psilotum domingense IPNI=*212103-2* 179
Psilotum flabellatum IPNI=*212103-2* 179
Psilotum flaccidum IPNI=*114220-3* 179
Psilotum floridanum IPNI=*212103-2* 179
Psilotum heterocarpum IPNI=*212103-2* 179
Psilotum novae-zelandiae IPNI=*212103-2* 179
Psilotum nudum var. *fallicinum* IPNI=*212103-2* 179
Psilotum nudum var. *molesworthiae* IPNI=*212103-2* 179
Psilotum nudum var. *normale* IPNI=*212103-2* 179
Psilotum triquetrum f. *insulare* IPNI=*212103-2* 179
Psilotum triquetrum IPNI=*212103-2* 179
Psilotum triquetrum var. *fallacinum* IPNI=*212103-2* 179
Psilotum triquetrum var. *gracile* IPNI=*212103-2* 179
Psilotum triquetrum var. *latum* IPNI=*212103-2* 179
Psittacanthus acutifolius IPNI=*258636-2* 449
Psittacanthus cupulifer IPNI=*551367-1* 449
Psittacanthus falcifrons IPNI=*551367-1* 449
Psittacanthus hoffmannseggianus IPNI=*185262-2* 448
Psittacanthus jacquinii IPNI=*551347-1* 448
Psittacanthus mexicanus IPNI=*551367-1* 449
Psittacanthus plumieri IPNI=*551347-1* 448
Psittacanthus smithii IPNI=*551367-1* 449
Psophocarpus cienkowskii IPNI=*1209500-2* 313
Psoralea altissima IPNI=*516230-1* 331
Psoralea annua IPNI=*76085-2* 317
Psoralea conferta IPNI=*516230-1* 331
Psoralea dalea IPNI=*76085-2* 317
Psoralea decidua IPNI=*516230-1* 331
Psoralea divaricata IPNI=*281771-2* 329
Psoralea foliolosa IPNI=*76146-2* 317
Psoralea maleolens IPNI=*281771-2* 329
Psoralea mexicana IPNI=*281771-2* 329
Psoralea mexicana var. *trianae* IPNI=*281771-2* 329
Psoralea mutisii IPNI=*281771-2* 329
Psoralea tenuifolia IPNI=*516230-1* 331
Psoralea trianae IPNI=*281771-2* 329
Psychopsis latourae IPNI=*212379-2* 222
Psychopsis papilio f. *latourae* IPNI=*212379-2* 222
Psychopsis papilio var. *albiflora* IPNI=*212379-2* 222
Psychopsis papilio var. *latourae* IPNI=*212379-2* 222
Psychopsis picta IPNI=*212379-2* 222
Psychotria acuminata IPNI=*77115588-1* 353
Psychotria alba f. *glabrata* IPNI=*761655-1* 355
Psychotria alba var. *tonsa* IPNI=*761655-1* 355

Psychotria albonervia IPNI=759372-1 354
Psychotria alboviridula IPNI=763194-1 356
Psychotria amplifolia IPNI=759372-1 354
Psychotria arcuata IPNI=759372-1 354
Psychotria ardisiifolia IPNI=212398-2 355
Psychotria aurantiaca IPNI=759061-1 353
Psychotria axillaris IPNI=764729-1 356
Psychotria bahiensis IPNI=77115613-1 354
Psychotria barbiflora IPNI=77115613-1 354
Psychotria barbiflora var. amazonica IPNI=77161098-1 354
Psychotria barbiflora var. genuina IPNI=77115613-1 354
Psychotria barcellana IPNI=77115635-1 354
Psychotria berteroana IPNI=77115593-1 353
Psychotria berteroana subsp. venezuelica IPNI=77115593-1 353
Psychotria bertiera IPNI=744444-1 347
Psychotria biattenuata IPNI=77115613-1 354
Psychotria bicolor IPNI=759061-1 353
Psychotria bimea IPNI=762253-1 355
Psychotria blepharophora IPNI=77161948-1 353
Psychotria boqueronensis IPNI=77202779-1 353
Psychotria brachiata IPNI=77115595-1 353
Psychotria brachybotrya IPNI=77161069-1 353
Psychotria bracteocardia IPNI=60473604-2 353
Psychotria bracteolata IPNI=77115609-1 354
Psychotria burchelliana IPNI=60473604-2 353
Psychotria byrsophylla IPNI=759300-1 354
Psychotria caerulea IPNI=759372-1 354
Psychotria calviflora IPNI=77161054-1 353
Psychotria capitata f. tomentosa IPNI=759372-1 354
Psychotria capitata f. trichophora IPNI=759372-1 354
Psychotria capitata IPNI=759372-1 354
Psychotria capitata subsp. amplifolia IPNI=759372-1 354
Psychotria capitata subsp. fissistipularis IPNI=759372-1 354
Psychotria capitata subsp. inundata IPNI=759372-1 354
Psychotria capitata subsp. rubra IPNI=759372-1 354
Psychotria capitata var. roraimensis IPNI=759372-1 354
Psychotria capitata var. septentrionalis IPNI=759372-1 354
Psychotria capitellata IPNI=77115595-1 353
Psychotria capitellata IPNI=77115613-1 354
Psychotria cataractarum IPNI=759121-1 354
Psychotria chimarrhoides IPNI=329821-2 355
Psychotria chionantha IPNI=60473461-2 355
Psychotria chlorotica IPNI=759372-1 354
Psychotria chlorotica var. bahiensis IPNI=759372-1 354
Psychotria chlorotica var. lanceolata IPNI=759372-1 354
Psychotria chlorotica var. obovata IPNI=759372-1 354
Psychotria cincta IPNI=77161078-1 354
Psychotria coccinea IPNI=759061-1 353
Psychotria colorata IPNI=77161054-1 353
Psychotria colorata subsp. megapontica IPNI=77161054-1 353
Psychotria contracta IPNI=77161077-1 354
Psychotria cordifolia IPNI=1080914-2 351
Psychotria crebrinervia IPNI=77115593-1 353
Psychotria crocea IPNI=759061-1 353
Psychotria ctenophora IPNI=60473604-2 353
Psychotria cubitalis IPNI=759372-1 354
Psychotria cujabensis IPNI=759061-1 353
Psychotria cuspidata IPNI=77148083-1 353
Psychotria cuspidata var. genuina IPNI=77148083-1 353
Psychotria cuspidata var. vulgatior IPNI=77115613-1 354
Psychotria cyanea IPNI=749186-1 350
Psychotria cyrilli-nelsonii IPNI=77154321-1 350
Psychotria decidua IPNI=761655-1 355
Psychotria deflexa IPNI=77115599-1 353
Psychotria densiflora IPNI=212398-2 355
Psychotria divaricata IPNI=762253-1 355
Psychotria egensis IPNI=77161866-1 353
Psychotria elata IPNI=77115600-1 353
Psychotria elliptica IPNI=761655-1 355
Psychotria emetica IPNI=764727-1 356
Psychotria erecta f. fluctuans IPNI=764729-1 356
Psychotria erecta IPNI=764729-1 356
Psychotria erythrophylla IPNI=77161098-1 354
Psychotria fadyenii IPNI=329821-2 355
Psychotria ficigemma IPNI=761655-1 355
Psychotria flavicans IPNI=77115613-1 354
Psychotria flavovirens IPNI=77161077-1 354
Psychotria fluctuans IPNI=764729-1 356
Psychotria fluctuans var. angustifolia IPNI=764729-1 356
Psychotria fluctuans var. latifolia IPNI=764729-1 356
Psychotria fockeana IPNI=761655-1 355
Psychotria foveolata IPNI=761655-1 355
Psychotria furcata IPNI=77115613-1 354
Psychotria galeottiana IPNI=281967-2 353
Psychotria glabrescens IPNI=77161054-1 353
Psychotria glaucescens IPNI=762253-1 355
Psychotria glomerata IPNI=763751-1 356
Psychotria glomerulata IPNI=77115607-1 353
Psychotria gracilenta IPNI=77161069-1 353
Psychotria guianensis IPNI=759121-1 354
Psychotria hebeclada IPNI=77115609-1 354
Psychotria hebecladoides IPNI=77115593-1 353
Psychotria hederacea IPNI=1080914-2 351
Psychotria heterocarpa IPNI=759372-1 354
Psychotria heterocephala IPNI=77115613-1 354
Psychotria hindsii IPNI=762666-1 355
Psychotria hirta IPNI=329821-2 355
Psychotria hoffmannseggiana f. pubescens IPNI=77115613-1 354
Psychotria hoffmannseggiana IPNI=77115613-1 354
Psychotria hoffmannseggiana subsp. tribracteata IPNI=77115613-1 354
Psychotria hoffmannseggiana var. celsa IPNI=77115613-1 354
Psychotria hoffmannseggiana var. erythrophylla IPNI=77161098-1 354
Psychotria hoffmannseggiana var. tribracteata IPNI=77115613-1 354
Psychotria horizontalis subsp. basicordata IPNI=762253-1 355
Psychotria horizontalis subsp. glaucescens IPNI=762253-1 355
Psychotria horizontalis subsp. glaucina IPNI=762253-1 355
Psychotria horizontalis var. glaucescens IPNI=762253-1 355
Psychotria horizontalis var. psilophylla IPNI=762253-1 355
Psychotria horridula IPNI=77161076-1 354
Psychotria humboldtiana IPNI=77161073-1 354
Psychotria humboldtiana var. caudata IPNI=77161073-1 354
Psychotria humboldtiana var. genuina IPNI=77161073-1 354
Psychotria humboldtiana var. ornata IPNI=77161073-1 354
Psychotria hundensis IPNI=212398-2 355
Psychotria hyacinthiflora IPNI=759242-1 354
Psychotria hyacinthoides IPNI=759242-1 354
Psychotria ilheosana IPNI=761655-1 355
Psychotria inundata IPNI=759372-1 354
Psychotria involucrata IPNI=77161098-1 354
Psychotria iodotricha IPNI=77161076-1 354
Psychotria iodotricha subsp. atricapilla IPNI=77161076-1 354
Psychotria iodotricha subsp. multiflora IPNI=77161076-1 354
Psychotria ipecacuanha IPNI=1214950-2 347
Psychotria iquitosensis IPNI=77161069-1 353
Psychotria izabalensis IPNI=77115593-1 353
Psychotria japurensis IPNI=77161076-1 354
Psychotria justicioides IPNI=77115609-1 354
Psychotria laevis IPNI=170751-2 353
Psychotria lanceolata IPNI=329821-2 355
Psychotria langsdorffiana IPNI=77161077-1 354
Psychotria leucophaea IPNI=77161077-1 354
Psychotria longicuspis IPNI=77161078-1 354
Psychotria longirostris IPNI=77161818-1 354
Psychotria longistipula IPNI=77115627-1 354
Psychotria loretensis IPNI=77161078-1 354

Psychotria lupulina f. *pubens* IPNI=*77161077-1* 354
Psychotria lupulina IPNI=*77161077-1* 354
Psychotria lupulina subsp. *rhodoleuca* IPNI=*77161077-1* 354
Psychotria lupulina var. *genuina* IPNI=*77161077-1* 354
Psychotria lupulina var. *maypurensis* IPNI=*77161077-1* 354
Psychotria lupulina var. *stipulacea* IPNI=*77161077-1* 354
Psychotria luschnathiana IPNI=*60473461-2* 355
Psychotria luschnathii IPNI=*60473461-2* 355
Psychotria macrobotrys IPNI=*759203-1* 354
Psychotria macropoda IPNI=*1080914-2* 351
Psychotria magnoliifolia IPNI=*212751-2* 355
Psychotria mapouria IPNI=*60473461-2* 355
Psychotria mapourioides IPNI=*60473461-2* 355
Psychotria mapourioides var. *opaca* IPNI=*60473461-2* 355
Psychotria mapourioides var. *tobagensis* IPNI=*60473461-2* 355
Psychotria mapuria IPNI=*60473461-2* 355
Psychotria mathewsii IPNI=*212751-2* 355
Psychotria maypurensis IPNI=*77161077-1* 354
Psychotria megapontica IPNI=*77161054-1* 353
Psychotria mesomorpha IPNI=*77161098-1* 354
Psychotria mexicana IPNI=*281971-2* 354
Psychotria microcephala IPNI=*77161098-1* 354
Psychotria microdesmia IPNI=*763751-1* 356
Psychotria microdon IPNI=*77154321-1* 350
Psychotria microdon var. *meridionalis* IPNI=*77154321-1* 350
Psychotria molliramis IPNI=*77115609-1* 354
Psychotria morindoides IPNI=*764729-1* 356
Psychotria muelleriana IPNI=*77161076-1* 354
Psychotria nervosa IPNI=*77161077-1* 354
Psychotria nervosa subsp. *rufescens* IPNI=*329821-2* 355
Psychotria nervosa var. *lanceolata* IPNI=*329821-2* 355
Psychotria nervosa var. *rufescens* IPNI=*329821-2* 355
Psychotria neurothricha IPNI=*77115595-1* 353
Psychotria nitida IPNI=*60473461-2* 355
Psychotria obovata IPNI=*759242-1* 354
Psychotria obscurata IPNI=*181012-2* 354
Psychotria officinalis subsp. *wilhelminensis* IPNI=*77161098-1* 354
Psychotria oligotricha IPNI=*329821-2* 355
Psychotria olyphylla IPNI=*759353-1* 354
Psychotria padifolia IPNI=*281971-2* 354
Psychotria palicourea IPNI=*759121-1* 354
Psychotria pallescens IPNI=*761655-1* 355
Psychotria pallida IPNI=*759061-1* 353
Psychotria pardina IPNI=*77115635-1* 354
Psychotria perotensis IPNI=*281967-2* 353
Psychotria persimilis IPNI=*77161077-1* 354
Psychotria phoeniciana IPNI=*281967-2* 353
Psychotria phytolacca IPNI=*170751-2* 353
Psychotria pinularis IPNI=*77154321-1* 350
Psychotria platyphylla IPNI=*77115593-1* 353
Psychotria platyphylla var. *angustior* IPNI=*77115593-1* 353
Psychotria platypodina IPNI=*759061-1* 353
Psychotria poeppigiana IPNI=*77115635-1* 354
Psychotria poeppigiana subsp. *barcellana* IPNI=*77115635-1* 354
Psychotria polyantha IPNI=*762156-1* 355
Psychotria portoricensis IPNI=*329821-2* 355
Psychotria proxima IPNI=*761655-1* 355
Psychotria pseudojapurensis IPNI=*77161076-1* 354
Psychotria pubescens IPNI=*77115609-1* 354
Psychotria pulverulenta IPNI=*213510-2* 356
Psychotria punicea IPNI=*759286-1* 354
Psychotria pyramidalis IPNI=*77115593-1* 353
Psychotria quiniifolia IPNI=*329821-2* 355
Psychotria quinquecuspis IPNI=*77115627-1* 354
Psychotria racemosa IPNI=*77115627-1* 354
Psychotria repens IPNI=*109024-2* 351
Psychotria rhodoleuca IPNI=*77161077-1* 354
Psychotria rhodophylla IPNI=*77161077-1* 354
Psychotria rigida IPNI=*759300-1* 354
Psychotria rigida var. *aurata* IPNI=*759300-1* 354
Psychotria rigida var. *genuina* IPNI=*759300-1* 354
Psychotria rigida var. *hirtella* IPNI=*759300-1* 354
Psychotria rigida var. *intermedia* IPNI=*759300-1* 354
Psychotria rigida var. *oligoneura* IPNI=*759300-1* 354
Psychotria rigida var. *strepens* IPNI=*759300-1* 354
Psychotria rosea f. *calvescens* IPNI=*77161054-1* 353
Psychotria rosea IPNI=*77161054-1* 353
Psychotria rosea var. *trichophora* IPNI=*77161054-1* 353
Psychotria rubra IPNI=*77115613-1* 354
Psychotria rubra var. *gracilis* IPNI=*77115613-1* 354
Psychotria rubra var. *lanceolata* IPNI=*77115613-1* 354
Psychotria rubra var. *virens* IPNI=*77115613-1* 354
Psychotria rudgei IPNI=*77161077-1* 354
Psychotria rufescens IPNI=*329821-2* 355
Psychotria rufescens IPNI=*762666-1* 355
Psychotria rufescens var. *ferruginea* IPNI=*329821-2* 355
Psychotria rufescens var. *haenkeana* IPNI=*329821-2* 355
Psychotria rufescens var. *hirta* IPNI=*329821-2* 355
Psychotria sagrana IPNI=*761655-1* 355
Psychotria sessilifolia IPNI=*213510-2* 356
Psychotria setifera IPNI=*759372-1* 354
Psychotria simulans IPNI=*60473604-2* 353
Psychotria sinuensis IPNI=*213293-2* 356
Psychotria stenophylla IPNI=*319913-2* 353
Psychotria stenostachya IPNI=*77160493-1* 354
Psychotria stipulacea IPNI=*329821-2* 355
Psychotria subcrocea var. *confusa* IPNI=*759061-1* 353
Psychotria subcrocea var. *intermedia* IPNI=*759061-1* 353
Psychotria sulzneri IPNI=*213510-2* 356
Psychotria tabacifolia IPNI=*759203-1* 354
Psychotria tarapotensis IPNI=*77161077-1* 354
Psychotria tenerior IPNI=*77161098-1* 354
Psychotria tenuiramea IPNI=*77115613-1* 354
Psychotria tobagensis IPNI=*60473461-2* 355
Psychotria tomentosa IPNI=*762666-1* 355
Psychotria tomentosa IPNI=*77115635-1* 354
Psychotria tribracteata IPNI=*77115613-1* 354
Psychotria tricholoba IPNI=*77161076-1* 354
Psychotria trifolia IPNI=*81670-1* 343
Psychotria trinitensis IPNI=*60473461-2* 355
Psychotria triphylla IPNI=*759353-1* 354
Psychotria trispicata IPNI=*763751-1* 356
Psychotria tristicula IPNI=*761655-1* 355
Psychotria uliginosa IPNI=*170751-2* 353
Psychotria ulviformis IPNI=*60473597-2* 353
Psychotria undata IPNI=*329821-2* 355
Psychotria undata var. *chimarrhoides* IPNI=*329821-2* 355
Psychotria undata var. *poitaei* IPNI=*329821-2* 355
Psychotria undulata IPNI=*329821-2* 355
Psychotria urophylla IPNI=*77115588-1* 353
Psychotria vauthieri IPNI=*77115599-1* 353
Psychotria villosa IPNI=*77161098-1* 354
Psychotria violacea IPNI=*109024-2* 351
Psychotria zevallosii IPNI=*77153569-1* 354
Psygmorchis allemanii IPNI=*1021256-1* 219
Psygmorchis pusilla IPNI=*1021256-1* 219
Ptarmica borealis IPNI=*2294-2* 255
Pteridium aquilinum f. *arachnoideum* IPNI=*213798-2* 175
Pteridium aquilinum f. *glabrata* IPNI=*213799-2* 175
Pteridium aquilinum f. *pubescens* IPNI=*17210060-1* 175
Pteridium aquilinum subsp. *atlanticum* IPNI=*17210060-1* 175
Pteridium aquilinum subsp. *brevipes* IPNI=*17210060-1* 175
Pteridium aquilinum subsp. *caudatum* IPNI=*213799-2* 175
Pteridium aquilinum subsp. *fulvum* IPNI=*17210060-1* 175
Pteridium aquilinum subsp. *heredia* IPNI=*17210060-1* 175
Pteridium aquilinum subsp. *osmundaceum* IPNI=*17210060-1* 175

Pteridium aquilinum subsp. *tauricum* IPNI=*17210060-1* 175
Pteridium aquilinum subsp. *typicum* IPNI=*17210060-1* 175
Pteridium aquilinum var. *arachnoideum* IPNI=*213798-2* 175
Pteridium aquilinum var. *caudatum* IPNI=*213799-2* 175
Pteridium aquilinum var. *osmundaceum* IPNI=*17210060-1* 175
Pteridium aquilinum var. *umbrosum* IPNI=*17210060-1* 175
Pteridium caudatum subsp. *arachnoideum* IPNI=*213798-2* 175
Pteridium cehegínense IPNI=*17210060-1* 175
Pteridium esculentum subsp. *arachnoideum* IPNI=*213798-2* 175
Pteridium heredia IPNI=*17210060-1* 175
Pteridium pinetorum subsp. *osmundaceum* IPNI=*17210060-1* 175
Pteridium pinetorum subsp. *sajanense* IPNI=*17210060-1* 175
Pteridium pinetorum subsp. *sibiricum* IPNI=*17210060-1* 175
Pteridium psittacinum IPNI=*213798-2* 175
Pteridium tauricum IPNI=*17210060-1* 175
Pteris aquilina IPNI=*17210060-1* 175
Pteris aquilina var. *arachnoidea* IPNI=*213798-2* 175
Pteris aquilina var. *caudata* IPNI=*213799-2* 175
Pteris aquilina var. *glabra* IPNI=*17210060-1* 175
Pteris aquilina var. *psittacina* IPNI=*213798-2* 175
Pteris aquilina var. *transcaucasica* IPNI=*17210060-1* 175
Pteris arachnoidea IPNI=*213798-2* 175
Pteris aspera IPNI=*17011270-1* 177
Pteris attenuata IPNI=*17217170-1* 178
Pteris auriculata IPNI=*17210060-1* 175
Pteris biauriculata IPNI=*213840-2* 178
Pteris biaurita var. *quadriaurita* IPNI=*17215900-1* 178
Pteris borealis IPNI=*17210060-1* 175
Pteris brevipes IPNI=*17210060-1* 175
Pteris brownii IPNI=*77182373-1* 178
Pteris caudata IPNI=*213799-2* 175
Pteris concolor IPNI=*77182373-1* 178
Pteris connexa IPNI=*17217170-1* 178
Pteris crenata IPNI=*17458930-1* 174
Pteris farinosa IPNI=*77182808-1* 178
Pteris feliciennae IPNI=*17217170-1* 178
Pteris firma IPNI=*17210060-1* 175
Pteris flexuosa IPNI=*77183106-1* 178
Pteris foemina IPNI=*17210060-1* 175
Pteris gardneri IPNI=*17210060-1* 175
Pteris grandifolia var. *campanae* IPNI=*313710-2* 178
Pteris grandifolia var. *socorrensis* IPNI=*313710-2* 178
Pteris guilleminei IPNI=*17217170-1* 178
Pteris heredia IPNI=*17210060-1* 175
Pteris intermedia IPNI=*17217170-1* 178
Pteris interrupta IPNI=*17458930-1* 174
Pteris laciniata IPNI=*77182373-1* 178
Pteris linearis IPNI=*17217170-1* 178
Pteris lineata IPNI=*17234040-1* 178
Pteris longipes IPNI=*17217170-1* 178
Pteris lucida IPNI=*17011270-1* 177
Pteris marginata IPNI=*17217170-1* 178
Pteris microdictyon IPNI=*17217170-1* 178
Pteris milneana IPNI=*17217170-1* 178
Pteris mysurensis IPNI=*77182373-1* 178
Pteris nivea IPNI=*77182941-1* 178
Pteris nudicaulis IPNI=*17210060-1* 175
Pteris ovata IPNI=*77183106-1* 178
Pteris pectinata IPNI=*77186436-1* 174
Pteris pedatifida IPNI=*17217170-1* 178
Pteris peruviana IPNI=*77182326-1* 178
Pteris piloselloides IPNI=*17160710-1* 176
Pteris pohliana IPNI=*77182373-1* 178
Pteris psittacina IPNI=*213798-2* 175
Pteris quadrialata IPNI=*17215900-1* 178
Pteris quadriaurita var. *affluentior* IPNI=*17215900-1* 178
Pteris quadriaurita var. *biaurita* IPNI=*213840-2* 178
Pteris quadriaurita var. *christii* IPNI=*17215900-1* 178
Pteris quadriaurita var. *fluminensis* IPNI=*17215900-1* 178
Pteris recurvata IPNI=*17210060-1* 175
Pteris revolvens IPNI=*17217170-1* 178
Pteris semiovata IPNI=*17217170-1* 178
Pteris socorrensis IPNI=*313710-2* 178
Pteris subpedata IPNI=*17217170-1* 178
Pteris subverticillata IPNI=*77182326-1* 178
Pteris ternifolia IPNI=*77182326-1* 178
Pteris thalictroides IPNI=*17064190-1* 178
Pteris tripartita var. *feliciennae* IPNI=*17217170-1* 178
Pteris tripartita var. *milneana* IPNI=*17217170-1* 178
Pteris triphylla IPNI=*77182326-1* 178
Pteris uniseriata IPNI=*17217170-1* 178
Pteris vittata IPNI=*313710-2* 178
Pterium elegans IPNI=*406803-1* 240
Pterocarpus amazonicus IPNI=*516568-1* 331
Pterocarpus amerimnon IPNI=*75919-2* 317
Pterocarpus ancylocalyx IPNI=*516568-1* 331
Pterocarpus ancylocalyx var. *angustifolius* IPNI=*214040-2* 331
Pterocarpus anterus IPNI=*148561-2* 326
Pterocarpus apterus IPNI=*148561-2* 326
Pterocarpus berteroi IPNI=*77117114-1* 317
Pterocarpus crispatus IPNI=*214060-2* 331
Pterocarpus cultratus IPNI=*516561-1* 331
Pterocarpus draco IPNI=*214060-2* 331
Pterocarpus ecastaphyllum IPNI=*77117114-1* 317
Pterocarpus esculentus IPNI=*516568-1* 331
Pterocarpus floribundus IPNI=*516561-1* 331
Pterocarpus frutescens IPNI=*75939-2* 317
Pterocarpus grandis IPNI=*516568-1* 331
Pterocarpus guianensis IPNI=*491205-1* 317
Pterocarpus hayesii IPNI=*516561-1* 331
Pterocarpus hemipterus IPNI=*214060-2* 331
Pterocarpus heterophyllus IPNI=*214038-2* 331
Pterocarpus javanicus IPNI=*486606-1* 315
Pterocarpus latifolius IPNI=*502971-1* 325
Pterocarpus lunatus IPNI=*148561-2* 326
Pterocarpus magnicarpus IPNI=*516561-1* 331
Pterocarpus mexicanus IPNI=*75919-2* 317
Pterocarpus michelii IPNI=*516568-1* 331
Pterocarpus minimus IPNI=*214038-2* 331
Pterocarpus moutouchi IPNI=*214060-2* 331
Pterocarpus ovalis IPNI=*77117114-1* 317
Pterocarpus paraguayensis IPNI=*516561-1* 331
Pterocarpus plumieri IPNI=*77117114-1* 317
Pterocarpus podocarpus IPNI=*214038-2* 331
Pterocarpus quercinus IPNI=*77117114-1* 317
Pterocarpus reticulatus IPNI=*516561-1* 331
Pterocarpus richardii IPNI=*77117114-1* 317
Pterocarpus rohrii var. *rubiginosus* IPNI=*516561-1* 331
Pterocarpus rufescens IPNI=*516561-1* 331
Pterocarpus rupestris IPNI=*516561-1* 331
Pterocarpus sanguis-draconis IPNI=*214060-2* 331
Pterocarpus suberosus IPNI=*214060-2* 331
Pterocarpus ternatus IPNI=*77117114-1* 317
Pterocarpus ulei IPNI=*214040-2* 331
Pterocarpus vernalis IPNI=*214038-2* 331
Pterocarpus violaceus IPNI=*516561-1* 331
Pterocaulon alopecuroides var. *glabrescens* IPNI=*240147-1* 269
Pterocaulon alopecuroides var. *mollis* IPNI=*240147-1* 269
Pterocaulon alopecuroides var. *polystachyum* IPNI=*240147-1* 269
Pterocaulon alopecuroides var. *salicifolium* IPNI=*240147-1* 269
Pterocaulon interruptum IPNI=*240147-1* 269
Pterocaulon interruptum var. *monostachyum* IPNI=*240147-1* 269
Pterocaulon interruptum var. *polystachyum* IPNI=*240147-1* 269
Pterocaulon latifolium IPNI=*240147-1* 269
Pterocaulon pilcomayense IPNI=*30022662-2* 269
Pterocaulon pompilianum IPNI=*30022662-2* 269

Pterocaulon spicatum IPNI=*30022662-2* 269
Pterocaulon spicatum var. *cylindricum* IPNI=*30022662-2* 269
Pterocaulon subspicatum IPNI=*30022662-2* 269
Pterocaulon subvirgatum IPNI=*30022662-2* 269
Pterocaulon virgatum f. *alopecuroides* IPNI=*240147-1* 269
Pterocaulon virgatum f. *subcorymbosa* IPNI=*240147-1* 269
Pterocaulon virgatum f. *subvirgatum* IPNI=*30022662-2* 269
Pterocaulon virgatum var. *capitatum* IPNI=*30022662-2* 269
Pterocladis genistelloides IPNI=*183309-1* 257
Pterocyperus esculentus IPNI=*300668-2* 230
Pterogyne nitens f. *parvifolia* IPNI=*516619-1* 331
Pterolepis egleri IPNI=*1131052-2* 424
Pterolepis exigua IPNI=*1131052-2* 424
Pterolepis goudotiana IPNI=*1131052-2* 424
Pterolepis hispida IPNI=*1131052-2* 424
Pterolepis ladanoides IPNI=*1131052-2* 424
Pterolepis lanceolata IPNI=*1131052-2* 424
Pterolepis pohliana IPNI=*1131052-2* 424
Pterolepis pumila IPNI=*1131052-2* 424
Pterolepis pumila var. *procera* IPNI=*1131052-2* 424
Pterolepis pumila var. *ramosa* IPNI=*1131052-2* 424
Pterolepis pumila var. *robusta* IPNI=*1131052-2* 424
Pterolepis pusilla IPNI=*1131052-2* 424
Pteronia tomentosa IPNI=*962413-1* 260
Pterophyllus salisburiensis aurea IPNI=*262125-1* 179
Pteropodium glabrum IPNI=*130921-2* 364
Pterozonium reniforme var. *ulei* IPNI=*17217830-1* 178
Ptileris hieraciifolius IPNI=*1152557-2* 261
Ptilophyllum crispum IPNI=*257306-2* 173
Ptilophyllum elegans IPNI=*17228970-1* 173
Ptilophyllum pinnatum IPNI=*320804-2* 173
Ptilophyllum schomburgkianum IPNI=*320804-2* 173
Ptilotrichum strigulosum IPNI=*1085072-2* 277
Ptychosperma alexandrae IPNI=*664083-1* 204
Ptychosperma beatriceae IPNI=*664083-1* 204
Ptychosperma bleeseri IPNI=*55245-3* 211
Ptychosperma capitis-yorki IPNI=*669378-1* 211
Ptychosperma coccinea IPNI=*666411-1* 208
Ptychosperma cunninghamianum IPNI=*134734-3* 204
Ptychosperma drudei IPNI=*664083-1* 204
Ptychosperma elegans var. *sphaerocarpum* IPNI=*669378-1* 211
Ptychosperma hospitum IPNI=*55245-3* 211
Ptychosperma jardinei IPNI=*669378-1* 211
Ptychosperma julianettii IPNI=*55245-3* 211
Ptychosperma seaforthii IPNI=*669378-1* 211
Ptychosperma wendlandianum IPNI=*669378-1* 211
Ptychosperma wendlandianum var. *sphaerocarpum* IPNI=*669378-1* 211
Ptychotis leptophylla IPNI=*840899-1* 252
Ptychotis vargasiana IPNI=*846658-1* 252
Pueraria lobata var. *montana* IPNI=*516708-1* 331
Pueraria phaseoloides IPNI=*509661-1* 328
Pulegium aromaticum IPNI=*451022-1* 371
Pulegium daghestanicum IPNI=*451022-1* 371
Pulegium erectum IPNI=*451022-1* 371
Pulegium heterophyllum IPNI=*451022-1* 371
Pulegium micranthum IPNI=*451022-1* 371
Pulegium pubescens IPNI=*451022-1* 371
Pulegium pulegium IPNI=*451022-1* 371
Pulegium tomentellum IPNI=*451022-1* 371
Pulegium vulgare IPNI=*451022-1* 371
Pulicaria rufescens IPNI=*93603-2* 261
Pulsatilla coronaria IPNI=*708314-1* 436
Punica florida IPNI=*554129-1* 417
Punica grandiflora IPNI=*554129-1* 417
Punica nana IPNI=*554129-1* 417
Punica spinosa IPNI=*554129-1* 417
Punjuba killipii IPNI=*1096490-2* 310
Pupalia densiflora IPNI=*60139-1* 282
Pupalia prostrata IPNI=*60185-1* 282
Pupalia sordida IPNI=*327342-2* 281
Purkinjia nodosa IPNI=*65321-2* 274
Pusaetha acaciifolia IPNI=*1009172-2* 319
Pusaetha entada IPNI=*1009172-2* 319
Pusaetha polyphylla IPNI=*91067-2* 319
Pusaetha scandens IPNI=*493866-1* 319
Pusaetha scandens var. *acutifolia* IPNI=*514978-1* 330
Puya heterophylla IPNI=*123989-1* 229
Puya longifolia IPNI=*123989-1* 229
Puya trianae var. *amplior* IPNI=*214698-2* 229
Pycnanthemum decurrens IPNI=*447996-1* 370
Pycnanthemum elongatum IPNI=*77120747-1* 368
Pycnanthemum subulatum IPNI=*447982-1* 370
Pycnodoria cretica IPNI=*17211870-1* 178
Pycnodoria vittata IPNI=*17217500-1* 178
Pycreus bathiei IPNI=*74698-2* 230
Pycreus cimicinus IPNI=*74915-2* 231
Pycreus debilissimus IPNI=*74698-2* 230
Pycreus esculentus IPNI=*300668-2* 230
Pycreus fallaciosus IPNI=*74698-2* 230
Pycreus flavescens f. *caucasicus* IPNI=*74698-2* 230
Pycreus flavescens f. *gracillimus* IPNI=*74698-2* 230
Pycreus flavescens f. *minor* IPNI=*74698-2* 230
Pycreus flavescens f. *monostachyus* IPNI=*74698-2* 230
Pycreus flavescens f. *setaceus* IPNI=*74698-2* 230
Pycreus flavescens IPNI=*74698-2* 230
Pycreus flavescens subsp. *fallaciosus* IPNI=*74698-2* 230
Pycreus flavescens subsp. *intermedius* IPNI=*74698-2* 230
Pycreus flavescens subsp. *microglumis* IPNI=*74698-2* 230
Pycreus flavescens subsp. *tanaensis* IPNI=*74698-2* 230
Pycreus flavescens var. *castaneus* IPNI=*74698-2* 230
Pycreus flavescens var. *rehmannianus* IPNI=*74698-2* 230
Pycreus intermedius IPNI=*74698-2* 230
Pycreus laevigatus IPNI=*304938-1* 231
Pycreus melanostachyus IPNI=*74915-2* 231
Pycreus melanostachyus var. *variegatus* IPNI=*74915-2* 231
Pycreus niger IPNI=*74915-2* 231
Pycreus paraensis IPNI=*74698-2* 230
Pycreus piceus IPNI=*74698-2* 230
Pycreus polystachyos IPNI=*75004-2* 231
Pycreus polystachyos var. *typicus* IPNI=*75004-2* 231
Pycreus potosiensis IPNI=*74698-2* 230
Pycreus rehmannianus IPNI=*74698-2* 230
Pycreus rotundus IPNI=*305797-1* 231
Pycreus squarrosus IPNI=*324890-2* 231
Pycreus tristachyus IPNI=*74698-2* 230
Pycreus vicinus IPNI=*74698-2* 230
Pyracantha atalantioides IPNI=*730613-1* 444
Pyracantha chinensis IPNI=*730613-1* 444
Pyracantha crenatoserrata IPNI=*730613-1* 444
Pyracantha crenulata var. *kansuensis* IPNI=*730613-1* 444
Pyracantha crenulata var. *rogersiana* IPNI=*730613-1* 444
Pyracantha crenulata var. *yunnanensis* IPNI=*730613-1* 444
Pyracantha discolor IPNI=*730613-1* 444
Pyracantha fortuneana IPNI=*730613-1* 444
Pyracantha gibbsii IPNI=*730613-1* 444
Pyracantha gibbsii var. *yunnanensis* IPNI=*730613-1* 444
Pyracantha loureiroi IPNI=*730613-1* 444
Pyracantha pauciflora IPNI=*730611-1* 444
Pyracantha pyracantha IPNI=*730611-1* 444
Pyracantha rogersiana IPNI=*730613-1* 444
Pyracantha yunnanensis IPNI=*730613-1* 444
Pyramidostylium grandifolium IPNI=*225455-2* 291
Pyramidostylium parviflorum IPNI=*162581-1* 291
Pyrenia malus IPNI=*726282-1* 443
Pyrenia pyrus IPNI=*30065762-2* 444

Pyrenoglyphis aristata IPNI=*664596-1* 206
Pyrenoglyphis bifida IPNI=*664520-1* 206
Pyrenoglyphis brongniartii IPNI=*664527-1* 206
Pyrenoglyphis concinna IPNI=*664555-1* 206
Pyrenoglyphis concinna var. *depauperata* IPNI=*664555-1* 206
Pyrenoglyphis concinna var. *inundata* IPNI=*664555-1* 206
Pyrenoglyphis major IPNI=*664668-1* 206
Pyrenoglyphis maraja IPNI=*28503-2* 206
Pyrenoglyphis microcarpa IPNI=*664527-1* 206
Pyrenoglyphis pallidispina IPNI=*664527-1* 206
Pyrenoglyphis piscatorum IPNI=*664527-1* 206
Pyrenoglyphis rivularis IPNI=*664527-1* 206
Pyrenoglyphis tenera IPNI=*664527-1* 206
Pyrethraria dichotoma IPNI=*914778-1* 255
Pyrethrum × *sinense* IPNI=*77111328-1* 259
Pyrethrum × *sinense* var. *plenum* IPNI=*77111328-1* 259
Pyrethrum argenteum IPNI=*252460-1* 271
Pyrethrum buschianum IPNI=*252460-1* 271
Pyrethrum cinerariifolium IPNI=*252275-1* 271
Pyrethrum demetrii IPNI=*252460-1* 271
Pyrethrum divaricatum IPNI=*252460-1* 271
Pyrethrum elongatum IPNI=*252275-1* 271
Pyrethrum eximium IPNI=*252460-1* 271
Pyrethrum frutescens IPNI=*178745-1* 257
Pyrethrum fruticosum IPNI=*252460-1* 271
Pyrethrum glanduliferum IPNI=*252460-1* 271
Pyrethrum grossheimii IPNI=*252460-1* 271
Pyrethrum hispanicum IPNI=*154715-2* 265
Pyrethrum indicum IPNI=*1013480-1* 263
Pyrethrum indicum IPNI=*193510-1* 259
Pyrethrum kubense IPNI=*252460-1* 271
Pyrethrum latilobum IPNI=*252460-1* 271
Pyrethrum matricarioides IPNI=*193510-1* 259
Pyrethrum niveum var. *divaricatum* IPNI=*252460-1* 271
Pyrethrum partheniifolium f. *divaricatum* IPNI=*252460-1* 271
Pyrethrum partheniifolium IPNI=*252460-1* 271
Pyrethrum partheniifolium var. *latilobum* IPNI=*252460-1* 271
Pyrethrum parthenium IPNI=*252460-1* 271
Pyrethrum persicum IPNI=*252460-1* 271
Pyrethrum praealtum IPNI=*252460-1* 271
Pyrethrum procumbens IPNI=*193510-1* 259
Pyrethrum pulverulentum IPNI=*252460-1* 271
Pyrethrum segetum IPNI=*209007-1* 263
Pyrethrum sevanense IPNI=*252460-1* 271
Pyrethrum spilanthus IPNI=*914778-1* 255
Pyrethrum turreyanum IPNI=*252275-1* 271
Pyrethrum umbrosum IPNI=*209007-1* 263
Pyrethrum vulgare IPNI=*252568-1* 271
Pyrethrum willemotii IPNI=*252275-1* 271
Pyrostegia amabilis IPNI=*110483-1* 365
Pyrostegia ignea IPNI=*110483-1* 365
Pyrostegia ornata IPNI=*110483-1* 365
Pyrostegia pallida IPNI=*110483-1* 365
Pyrostegia parvifolia IPNI=*110483-1* 365
Pyrostegia puberula IPNI=*110483-1* 365
Pyrostegia reticulata IPNI=*110483-1* 365
Pyrostegia tecomiflora IPNI=*110483-1* 365
Pyrostegia tubulosa IPNI=*110483-1* 365
Pyrostegia venusta var. *typica* IPNI=*110483-1* 365
Pyrostegia venusta var. *villosa* IPNI=*110483-1* 365
Pyrostoma ternatum IPNI=*866038-1* 374
Pyrrosia macrocarpa IPNI=*17448650-1* 176
Pyrus achras IPNI=*30065762-2* 444
Pyrus ambrosiaca IPNI=*30065762-2* 444
Pyrus amphigenea IPNI=*30065762-2* 444
Pyrus andina IPNI=*725473-1* 443
Pyrus anglica IPNI=*30065762-2* 444
Pyrus apiosa IPNI=*726282-1* 443
Pyrus armeniacifolia IPNI=*726282-1* 443
Pyrus astracanica IPNI=*726282-1* 443
Pyrus balansae IPNI=*30065762-2* 444
Pyrus bibas IPNI=*77209608-1* 444
Pyrus bombycilla IPNI=*722623-1* 443
Pyrus borealichinensis IPNI=*722592-1* 443
Pyrus caucasica var. *schuntukensis* IPNI=*30065762-2* 444
Pyrus cavillea IPNI=*726282-1* 443
Pyrus cestiana IPNI=*726282-1* 443
Pyrus chitralensis IPNI=*726282-1* 443
Pyrus communis subsp. *australeuropaea* IPNI=*30065762-2* 444
Pyrus communis subsp. *medioasiatica* IPNI=*30065762-2* 444
Pyrus communis subsp. *orientaleuropaea* IPNI=*30065762-2* 444
Pyrus communis subsp. *transcaucasica* IPNI=*30065762-2* 444
Pyrus communis var. *pompejana* IPNI=*30065762-2* 444
Pyrus communis var. *pyraster* IPNI=*30065762-2* 444
Pyrus communis var. *sativa* IPNI=*30065762-2* 444
Pyrus communis var. *sylvestris* IPNI=*30065762-2* 444
Pyrus communis var. *volema* IPNI=*30065762-2* 444
Pyrus coriacea IPNI=*722459-1* 443
Pyrus cydonia IPNI=*724549-1* 443
Pyrus dioica IPNI=*726282-1* 443
Pyrus divisa IPNI=*726282-1* 443
Pyrus domestica IPNI=*30065762-2* 444
Pyrus epirotica IPNI=*726282-1* 443
Pyrus eryostyla IPNI=*726282-1* 443
Pyrus escalloniifolia IPNI=*725484-1* 443
Pyrus fontanesiana IPNI=*726282-1* 443
Pyrus frigida IPNI=*722495-1* 443
Pyrus gamblei IPNI=*722495-1* 443
Pyrus karpatiana IPNI=*30065762-2* 444
Pyrus kirghisorum IPNI=*726282-1* 443
Pyrus lactea IPNI=*722459-1* 443
Pyrus lidjiangensis IPNI=*722623-1* 443
Pyrus lithomelo IPNI=*725484-1* 443
Pyrus lusitanica IPNI=*724549-1* 443
Pyrus magnifica IPNI=*722592-1* 443
Pyrus magyarica IPNI=*30065762-2* 444
Pyrus maliformis IPNI=*724549-1* 443
Pyrus malus IPNI=*726282-1* 443
Pyrus malus var. *cestiana* IPNI=*726282-1* 443
Pyrus malus var. *paradisiaca* IPNI=*726282-1* 443
Pyrus malus var. *prasomila* IPNI=*726282-1* 443
Pyrus malus var. *rubelliana* IPNI=*726282-1* 443
Pyrus mekongensis IPNI=*731389-1* 444
Pyrus moschata IPNI=*30065762-2* 444
Pyrus multiflora IPNI=*722592-1* 443
Pyrus niedzwetzkyana IPNI=*726282-1* 443
Pyrus oblonga IPNI=*724549-1* 443
Pyrus obtusifolia IPNI=*725484-1* 443
Pyrus pannona IPNI=*722623-1* 443
Pyrus papulata IPNI=*30065762-2* 444
Pyrus paris IPNI=*726282-1* 443
Pyrus pollveria IPNI=*726282-1* 443
Pyrus praecox IPNI=*726282-1* 443
Pyrus prasomila IPNI=*726282-1* 443
Pyrus pulverulenta IPNI=*726282-1* 443
Pyrus pumila IPNI=*726282-1* 443
Pyrus pyraster IPNI=*30065762-2* 444
Pyrus pyraster subsp. *achras* IPNI=*30065762-2* 444
Pyrus pyraster var. *achras* IPNI=*30065762-2* 444
Pyrus pyraster var. *relicta* IPNI=*30065762-2* 444
Pyrus pyraster var. *rossica* IPNI=*30065762-2* 444
Pyrus pyraster var. *tomentosa* IPNI=*30065762-2* 444
Pyrus racemiflora IPNI=*77120298-1* 443
Pyrus reflexa IPNI=*722592-1* 443
Pyrus rhaphiolepis IPNI=*731389-1* 444
Pyrus rossica IPNI=*30065762-2* 444

Pyrus rubelliana IPNI=*726282-1* 443
Pyrus salviati IPNI=*30065762-2* 444
Pyrus saxatilis IPNI=*726282-1* 443
Pyrus smithii IPNI=*722459-1* 443
Pyrus songarica IPNI=*726282-1* 443
Pyrus sumboshia IPNI=*724549-1* 443
Pyrus vernae IPNI=*722623-1* 443
Pyrus wilsonii IPNI=*722592-1* 443
Pyxidaria crustacea IPNI=*1189750-2* 374
Pyxidaria diffusa IPNI=*262512-2* 374
Pyxidaria minuta IPNI=*1189750-2* 374
Pyxidaria obtusa IPNI=*1189750-2* 374
Pyxidaria pachypoda IPNI=*799766-1* 374
Pyxidium gangeticum IPNI=*327386-2* 282
Pyxidium graecizans IPNI=*10619-2* 281
Pyxidium melancholicum IPNI=*327386-2* 282
Pyxidium viride IPNI=*316349-2* 282
Pyxipoma polyandrum IPNI=*235145-2* 280

Q
Quadrella breynia IPNI=*77217338-1* 278
Quadrella crotonoides IPNI=*77183834-1* 278
Quadrella furfuracea IPNI=*77217338-1* 278
Quadrella indica IPNI=*77217338-1* 278
Quadrella intermedia IPNI=*77184072-1* 278
Quadrella odoratissima IPNI=*77184072-1* 278
Quadrella sidifolia IPNI=*77183834-1* 278
Quadricasaea caquetensis IPNI=*285812-2* 343
Quadricasaea inaequilateralis IPNI=*285812-2* 343
Quadricosta palustris IPNI=*30031106-2* 429
Qualea calcarata IPNI=*870864-1* 430
Qualea speciosa IPNI=*318965-2* 430
Quapoya peruviana IPNI=*1195925-2* 384
Quapoya peruviana var. *guayanensis* IPNI=*1195925-2* 384
Quapoya peruviana var. *occidentalis* IPNI=*1195925-2* 384
Quapoya robusta IPNI=*61594-2* 385
Quararibea alata IPNI=*154666-2* 411
Quararibea apaporiensis IPNI=*154702-2* 411
Quararibea asymmetrica IPNI=*154700-2* 411
Quararibea bicolor IPNI=*154672-2* 411
Quararibea bracteolosa IPNI=*154675-2* 411
Quararibea castano IPNI=*112441-1* 411
Quararibea chodatii IPNI=*215362-2* 413
Quararibea cordata IPNI=*562392-1* 411
Quararibea dowdingii IPNI=*154683-2* 411
Quararibea idroboi IPNI=*154689-2* 411
Quararibea inaequalis IPNI=*154700-2* 411
Quararibea inaequilatera IPNI=*154700-2* 411
Quararibea lasiocalyx IPNI=*154692-2* 411
Quararibea lomensis IPNI=*154695-2* 411
Quararibea longiflora IPNI=*154696-2* 411
Quararibea longipes IPNI=*280582-2* 411
Quararibea malacocalyx IPNI=*946934-1* 411
Quararibea muricata IPNI=*154702-2* 411
Quararibea obliquifolia IPNI=*154700-2* 411
Quararibea ochrocalyx IPNI=*154702-2* 411
Quararibea paraensis IPNI=*185557-2* 412
Quararibea rhombifolia IPNI=*154672-2* 411
Quararibea schultesii IPNI=*154675-2* 411
Quararibea sclerophylla IPNI=*154709-2* 411
Quararibea soegengii IPNI=*154710-2* 411
Quararibea wurdackii IPNI=*154692-2* 411
Quarena spinescens IPNI=*946226-1* 276
Quelusia regia IPNI=*106266-2* 429
Quercus almaguerensis IPNI=*296268-1* 338
Quercus boyacensis IPNI=*296268-1* 338
Quercus colombiana IPNI=*296268-1* 338
Quercus humboldtii IPNI=*296268-1* 338
Quercus humboldtii var. *lehmanniana* IPNI=*296268-1* 338
Quercus lindenii IPNI=*296268-1* 338
Quercus tolimensis IPNI=*296268-1* 338
Quesnelia lamarckii IPNI=*797554-1* 250
Quiina albiflora IPNI=*429133-1* 397
Quiina cajambrensis IPNI=*429118-1* 397
Quiina colonensis IPNI=*429118-1* 397
Quiina crenata IPNI=*134573-2* 397
Quiina guaporensis IPNI=*429118-1* 397
Quiina macrostachya IPNI=*134578-2* 397
Quiina micrantha IPNI=*429136-1* 397
Quiina multiflora IPNI=*429121-1* 397
Quiina nitens IPNI=*429121-1* 397
Quiina oblanceolata IPNI=*429133-1* 397
Quiina panamensis IPNI=*134573-2* 397
Quiina peruviana IPNI=*429118-1* 397
Quiina poeppigiana IPNI=*429121-1* 397
Quiina schippii IPNI=*429129-1* 397
Quiina silvatica IPNI=*134577-2* 397
Quiina spruceana IPNI=*429129-1* 397
Quinquina calisaya IPNI=*746718-1* 348
Quinquina carabayensis IPNI=*746718-1* 348
Quinquina carabayensis var. *villosa* IPNI=*746718-1* 348
Quinquina lancifolia IPNI=*746806-1* 348
Quinquina ledgeriana IPNI=*746718-1* 348
Quinquina obovata IPNI=*746872-1* 348
Quinquina officinalis IPNI=*58723-2* 348
Quinquina ovata IPNI=*746872-1* 348
Quinquina palton IPNI=*58723-2* 348
Quinquina pubescens IPNI=*746872-1* 348
Quinquina succirubra IPNI=*746872-1* 348
Quirina microphyla IPNI=*552931-1* 416
Quirosia anceps IPNI=*489042-1* 316
Quisqualis ebracteata IPNI=*77101543-1* 416
Quisqualis glabra IPNI=*77101543-1* 416
Quisqualis grandiflora IPNI=*77101543-1* 416
Quisqualis indica IPNI=*77101543-1* 416
Quisqualis longiflora IPNI=*77101543-1* 416
Quisqualis loureiroi IPNI=*77101543-1* 416
Quisqualis madagascariensis IPNI=*77101543-1* 416
Quisqualis obovata IPNI=*77101543-1* 416
Quisqualis pubescens IPNI=*77101543-1* 416
Quisqualis sinensis IPNI=*77101543-1* 416
Quisqualis spinosa IPNI=*77101543-1* 416
Quisqualis villosa IPNI=*77101543-1* 416

R
Rabdochloa mucronata IPNI=*397387-1* 236
Rabdochloa virgata IPNI=*396063-1* 236
Rachicallis caracasana IPNI=*743608-1* 347
Rachicallis nitida IPNI=*743608-1* 347
Racosperma baileyanum IPNI=*469798-1* 310
Racosperma dealbatum IPNI=*470130-1* 310
Racosperma decurrens IPNI=*470138-1* 310
Racosperma longifolium IPNI=*470782-1* 310
Racosperma mangium IPNI=*470839-1* 310
Racosperma melanoxylon IPNI=*470873-1* 310
Racosperma podalyriifolium IPNI=*76522-3* 310
Racosperma retinodes IPNI=*471313-1* 310
Racosperma salignum IPNI=*471383-1* 310
Racosperma verticillatum IPNI=*471751-1* 310
Racoubea guianensis IPNI=*365491-1* 404
Radackia amicorum IPNI=*1164658-2* 312
Raddia impressifolia IPNI=*225455-2* 291
Radermachia incisa IPNI=*582598-1* 436
Radermachia integra IPNI=*582622-1* 436
Radicula clavata IPNI=*288669-1* 277
Radicula dictyota IPNI=*288669-1* 277

Radicula indica IPNI=*222144-2* 277
Radicula mexicana IPNI=*222160-2* 277
Radicula nana IPNI=*222164-2* 277
Radicula nasturtium IPNI=*287534-1* 277
Radicula nasturtium-aquaticum IPNI=*287534-1* 277
Radicula officinalis IPNI=*287534-1* 277
Radicula pacifica IPNI=*288669-1* 277
Radicula palustris IPNI=*288669-1* 277
Radicula terrestris IPNI=*288669-1* 277
Radicula williamsii IPNI=*288669-1* 277
Radlkoferella arguacoensium IPNI=*209690-2* 306
Radlkoferella dussiana IPNI=*209824-2* 307
Radlkoferella glabrifolia IPNI=*209714-2* 306
Radlkoferella guadelupensis IPNI=*209824-2* 307
Radlkoferella inseparabilis IPNI=*209714-2* 306
Radlkoferella latifolia IPNI=*209824-2* 307
Radlkoferella macrocarpa IPNI=*209824-2* 307
Radlkoferella martinicensis IPNI=*209824-2* 307
Radlkoferella multiflora IPNI=*209824-2* 307
Radlkoferella quadrifida IPNI=*209824-2* 307
Radlkoferella sphaerocarpa IPNI=*209714-2* 306
Radlkoferella urbanii IPNI=*209824-2* 307
Radlkoferella venosa IPNI=*209920-2* 308
Ragala sanguinolenta IPNI=*57826-2* 305
Ragatelus eriophorus IPNI=*257306-2* 173
Raimannia laciniata IPNI=*613817-1* 430
Raimannia mexicana IPNI=*613817-1* 430
Raimondia cherimolioides IPNI=*300526-2* 186
Raimondia monoica IPNI=*300526-2* 186
Raimondia quinduensis IPNI=*72292-1* 187
Raimondia tenuiflora IPNI=*72324-1* 187
Ramium ampullaceum IPNI=*850541-1* 445
Ramium caudatum IPNI=*850541-1* 445
Ramium compactum IPNI=*55676-3* 445
Ramium mollicoma IPNI=*55676-3* 445
Ramium niveum IPNI=*55676-3* 445
Ramium ramiflorum IPNI=*34156-2* 445
Ramium rhynchophyllum IPNI=*34156-2* 445
Randia armata var. *pubescens* IPNI=*764273-1* 356
Randia brandisii IPNI=*745904-1* 348
Randia caracasana IPNI=*764082-1* 356
Randia densiflora IPNI=*222507-2* 356
Randia dumetorum IPNI=*745904-1* 348
Randia floribunda IPNI=*745904-1* 348
Randia formosa IPNI=*222508-2* 356
Randia formosa var. *densiflora* IPNI=*222507-2* 356
Randia formosa var. *longiflora* IPNI=*222509-2* 356
Randia gaumeri IPNI=*217048-2* 356
Randia humboldtiana IPNI=*764273-1* 356
Randia lachnosiphonium IPNI=*745904-1* 348
Randia mussaenda IPNI=*222508-2* 356
Randia oblanceolata IPNI=*764273-1* 356
Randia oppositiflora IPNI=*222507-2* 356
Randia oxypetala IPNI=*745904-1* 348
Randia rhagocarpa IPNI=*217048-2* 356
Randia rottleri IPNI=*745904-1* 348
Randia ruiziana IPNI=*222509-2* 356
Randia speciosa IPNI=*222509-2* 356
Randia spinosa IPNI=*745904-1* 348
Randia stipulosa IPNI=*745904-1* 348
Randia texensis IPNI=*217048-2* 356
Randia tomentosa IPNI=*745904-1* 348
Randia uniflora IPNI=*745904-1* 348
Randia williamsii IPNI=*222509-2* 356
Ranunculus amarillo IPNI=*713577-1* 436
Ranunculus argemonifolius IPNI=*713577-1* 436
Ranunculus aschenbornianus IPNI=*713491-1* 436
Ranunculus bonplandianus IPNI=*713577-1* 436
Ranunculus brevipes IPNI=*713577-1* 436
Ranunculus chiclensis IPNI=*713577-1* 436
Ranunculus cymbalaria f. *hebecaulis* IPNI=*711250-1* 436
Ranunculus cymbalaria f. *microphyllus* IPNI=*711250-1* 436
Ranunculus cymbalaria IPNI=*711250-1* 436
Ranunculus cymbalaria subsp. *saximontanus* IPNI=*711250-1* 436
Ranunculus cymbalaria var. *alpinus* IPNI=*711250-1* 436
Ranunculus cymbalaria var. *americanus* IPNI=*711250-1* 436
Ranunculus cymbalaria var. *saximontanus* IPNI=*711250-1* 436
Ranunculus cymbalaria var. *typicus* IPNI=*711250-1* 436
Ranunculus galeottii IPNI=*713491-1* 436
Ranunculus geoides var. *standleyi* IPNI=*713491-1* 436
Ranunculus geoides var. *typicus* IPNI=*713491-1* 436
Ranunculus halophilus IPNI=*711250-1* 436
Ranunculus hemignostus IPNI=*711250-1* 436
Ranunculus heterophyllus IPNI=*713340-1* 436
Ranunculus hookeri IPNI=*713491-1* 436
Ranunculus kunthii IPNI=*713340-1* 436
Ranunculus limoselloides f. *minor* IPNI=*713095-1* 436
Ranunculus llaveanus IPNI=*713491-1* 436
Ranunculus magellanicus IPNI=*713491-1* 436
Ranunculus mandonianus f. *minor* IPNI=*713095-1* 436
Ranunculus maritimus IPNI=*711250-1* 436
Ranunculus mexicanus IPNI=*713491-1* 436
Ranunculus microcarpus IPNI=*711250-1* 436
Ranunculus minutus IPNI=*711250-1* 436
Ranunculus minutus var. *microphylla* IPNI=*711250-1* 436
Ranunculus nanus IPNI=*711250-1* 436
Ranunculus petiolaris var. *hookeri* IPNI=*713491-1* 436
Ranunculus petiolaris var. *standleyi* IPNI=*713491-1* 436
Ranunculus petiolaris var. *typicus* IPNI=*713491-1* 436
Ranunculus pilosus IPNI=*713577-1* 436
Ranunculus pozoifolius IPNI=*711250-1* 436
Ranunculus praemorsus subvar. *pusillus* IPNI=*713577-1* 436
Ranunculus praemorsus var. *bonplandianus* IPNI=*713577-1* 436
Ranunculus pseudophilonotis IPNI=*713577-1* 436
Ranunculus setosopilosus IPNI=*713577-1* 436
Ranunculus sprucei IPNI=*713577-1* 436
Ranunculus tridentatus IPNI=*711250-1* 436
Ranunculus tridentatus var. *major* IPNI=*711250-1* 436
Ranunculus tridentatus var. *minor* IPNI=*711250-1* 436
Rapa napus IPNI=*279419-1* 276
Rapa oblonga IPNI=*279485-1* 276
Rapa rotunda IPNI=*279435-1* 276
Rapanea ambigua IPNI=*1028287-2* 304
Rapanea chartacea IPNI=*247483-2* 305
Rapanea ciliata IPNI=*1119456-2* 304
Rapanea ciliata var. *pentandra* IPNI=*1119456-2* 304
Rapanea coriacea IPNI=*1028287-2* 304
Rapanea dependens IPNI=*1119456-2* 304
Rapanea dependens var. *myrtoides* IPNI=*1119456-2* 304
Rapanea dependens var. *pentandra* IPNI=*1119456-2* 304
Rapanea dependens var. *saxatilis* IPNI=*1119456-2* 304
Rapanea ferruginea IPNI=*1028287-2* 304
Rapanea ferruginea var. *spathulata* IPNI=*1028287-2* 304
Rapanea glaucorubens IPNI=*166565-2* 304
Rapanea guianensis IPNI=*166565-2* 304
Rapanea guianensis var. *andicola* IPNI=*166565-2* 304
Rapanea guianensis var. *bogotensis* IPNI=*166565-2* 304
Rapanea guianensis var. *calensis* IPNI=*166565-2* 304
Rapanea jelskii IPNI=*1028287-2* 304
Rapanea latifolia IPNI=*1033305-2* 304
Rapanea mandonii IPNI=*1028287-2* 304
Rapanea microcalyx IPNI=*1028287-2* 304
Rapanea myricoides IPNI=*1028287-2* 304
Rapanea myrtoides IPNI=*1119456-2* 304
Rapanea oblonga IPNI=*166565-2* 304
Rapanea pallens IPNI=*247441-2* 304

Rapanea paulensis IPNI=*1028287-2* 304
Rapanea peruviana IPNI=*1119456-2* 304
Rapanea pittieri IPNI=*1119456-2* 304
Rapanea pittieri var. *chirripoensis* IPNI=*1119456-2* 304
Rapanea punctata IPNI=*589091-1* 304
Rapanea rufa IPNI=*1028287-2* 304
Rapanea savannarum IPNI=*166565-2* 304
Rapanea schwackeana IPNI=*1028287-2* 304
Rapanea sprucei IPNI=*1028287-2* 304
Rapanea umbrosa IPNI=*166565-2* 304
Rapanea verae-crucis IPNI=*589091-1* 304
Rapanea vestita IPNI=*1028287-2* 304
Raphanistrum fugax IPNI=*217826-2* 277
Raphanistrum innocuum IPNI=*217826-2* 277
Raphanistrum lampsana IPNI=*217826-2* 277
Raphanistrum microcarpum IPNI=*217826-2* 277
Raphanistrum raphanistrum IPNI=*217826-2* 277
Raphanistrum segetale IPNI=*217826-2* 277
Raphanistrum segetum IPNI=*217826-2* 277
Raphanistrum silvestre IPNI=*217826-2* 277
Raphanistrum vulgare IPNI=*217826-2* 277
Raphanus albus IPNI=*288952-1* 278
Raphanus amplexicaulis IPNI=*279485-1* 276
Raphanus articulatus IPNI=*217826-2* 277
Raphanus brassica IPNI=*279435-1* 276
Raphanus brassica-officinalis IPNI=*279435-1* 276
Raphanus campestris IPNI=*279485-1* 276
Raphanus candidus IPNI=*217826-2* 277
Raphanus chinensis IPNI=*279485-1* 276
Raphanus cornutus IPNI=*217826-2* 277
Raphanus eruca IPNI=*1137625-2* 277
Raphanus erucoides IPNI=*279435-1* 276
Raphanus fugax IPNI=*217826-2* 277
Raphanus infestus IPNI=*217826-2* 277
Raphanus junceus IPNI=*60442520-2* 276
Raphanus lanceolatus IPNI=*118574-3* 276
Raphanus landroides IPNI=*217826-2* 277
Raphanus longistylus IPNI=*217826-2* 277
Raphanus microcarpus IPNI=*217826-2* 277
Raphanus napus IPNI=*279419-1* 276
Raphanus rapa IPNI=*279485-1* 276
Raphanus raphanellus IPNI=*217826-2* 277
Raphanus raphanistrum var. *hispidus* IPNI=*217826-2* 277
Raphanus raphanistrum var. *intermedius* IPNI=*217826-2* 277
Raphanus segetum IPNI=*217826-2* 277
Raphanus sinapis-officinalis IPNI=*279422-1* 276
Raphanus stylosus IPNI=*217826-2* 277
Raphanus sylvestris IPNI=*217826-2* 277
Raphanus vesicarius IPNI=*1137625-2* 277
Raphia aulacolepis IPNI=*669539-1* 211
Raphia nicaraguensis IPNI=*669539-1* 211
Raphia vinifera var. *taedigera* IPNI=*669539-1* 211
Rapistrum arvense IPNI=*217826-2* 277
Rapistrum clavatum IPNI=*1061021-2* 277
Rapistrum confusum IPNI=*1061021-2* 277
Rapistrum conoideum IPNI=*1061021-2* 277
Rapistrum glabrum IPNI=*1061021-2* 277
Rapistrum hirsutum IPNI=*1061021-2* 277
Rapistrum hirtum IPNI=*1061021-2* 277
Rapistrum hispanicum IPNI=*1061021-2* 277
Rapistrum hispidum IPNI=*1061021-2* 277
Rapistrum linnaeanum IPNI=*1061021-2* 277
Rapistrum longiracemosum IPNI=*1061021-2* 277
Rapistrum microcarpum IPNI=*1061021-2* 277
Rapistrum nemausense IPNI=*1061021-2* 277
Rapistrum orientale IPNI=*1061021-2* 277
Rapistrum raphanistrum IPNI=*217826-2* 277
Rapistrum rugosum subsp. *linnaeanum* IPNI=*1061021-2* 277
Rapistrum rugosum subsp. *orientale* IPNI=*1061021-2* 277
Rapistrum rugosum var. *stylosum* IPNI=*1061021-2* 277
Rapistrum rugosum var. *venosum* IPNI=*1061021-2* 277
Rapistrum strictissimum IPNI=*1061021-2* 277
Rapistrum venosum IPNI=*1061021-2* 277
Rapuntia media IPNI=*140670-1* 272
Rapuntium acuminatum IPNI=*142908-1* 273
Rapuntium acutangulum IPNI=*143240-1* 273
Rapuntium bellidifolium IPNI=*143240-1* 273
Rapuntium bicolor IPNI=*143240-1* 273
Rapuntium cardinale IPNI=*142115-2* 273
Rapuntium coccineum IPNI=*142115-2* 273
Rapuntium cordigerum IPNI=*142115-2* 273
Rapuntium erinoides IPNI=*143240-1* 273
Rapuntium erinus IPNI=*143240-1* 273
Rapuntium fulgens IPNI=*142115-2* 273
Rapuntium gracile IPNI=*1020602-1* 273
Rapuntium gramineum IPNI=*142115-2* 273
Rapuntium kohautianum IPNI=*143240-1* 273
Rapuntium krebsianum IPNI=*143240-1* 273
Rapuntium laxiflorum IPNI=*142260-2* 273
Rapuntium longiflorum IPNI=*1133957-2* 273
Rapuntium longifolium IPNI=*142115-2* 273
Rapuntium micranthum IPNI=*79151-2* 273
Rapuntium nanum IPNI=*143584-1* 273
Rapuntium procumbens IPNI=*143240-1* 273
Rapuntium rupestre IPNI=*143931-1* 273
Rapuntium splendens IPNI=*142115-2* 273
Rapuntium subtile IPNI=*79151-2* 273
Rapuntium tenerum IPNI=*143931-1* 273
Rapuntium veronicifolium IPNI=*143931-1* 273
Rathbunia deficiens IPNI=*244634-2* 285
Rathbunia grisea f. *cristata* IPNI=*244634-2* 285
Rathbunia grisea IPNI=*244634-2* 285
Rathbunia grisea var. *monstrosa* IPNI=*244634-2* 285
Rauvolfia alphonsiana IPNI=*81565-1* 343
Rauvolfia blanchetii IPNI=*81565-1* 343
Rauvolfia canescens IPNI=*81663-1* 343
Rauvolfia canescens var. *glabra* IPNI=*81663-1* 343
Rauvolfia canescens var. *intermedia* IPNI=*81663-1* 343
Rauvolfia canescens var. *tomentosa* IPNI=*81663-1* 343
Rauvolfia canescens var. *typica* IPNI=*81663-1* 343
Rauvolfia divergens IPNI=*81565-1* 343
Rauvolfia glabra IPNI=*82668-1* 344
Rauvolfia heterophylla IPNI=*81663-1* 343
Rauvolfia hirsuta IPNI=*81663-1* 343
Rauvolfia hirsuta var. *glabra* IPNI=*81663-1* 343
Rauvolfia indecora IPNI=*81565-1* 343
Rauvolfia laevigata IPNI=*81996-1* 343
Rauvolfia lamarckii IPNI=*81670-1* 343
Rauvolfia latifolia IPNI=*81670-1* 343
Rauvolfia latifolia var. *minor* IPNI=*81663-1* 343
Rauvolfia leptophylla var. *orientalis* IPNI=*217918-2* 343
Rauvolfia macrocarpa IPNI=*217921-2* 343
Rauvolfia mollis IPNI=*81565-1* 343
Rauvolfia mollissima IPNI=*81670-1* 343
Rauvolfia multiflora IPNI=*217921-2* 343
Rauvolfia nitida IPNI=*81663-1* 343
Rauvolfia odontophora IPNI=*81663-1* 343
Rauvolfia parvifolia IPNI=*81565-1* 343
Rauvolfia parvifolia var. *cubana* IPNI=*81565-1* 343
Rauvolfia psychotrioides IPNI=*81670-1* 343
Rauvolfia subpubescens IPNI=*81663-1* 343
Rauvolfia ternifolia IPNI=*81565-1* 343
Rauvolfia tomentosa IPNI=*81663-1* 343
Ravenala guyannensis IPNI=*585013-1* 250
Razisea spicata IPNI=*60478486-2* 361
Razisea spicata var. *pulchra* IPNI=*60478486-2* 361

Rechsteineria aurantiaca IPNI=*237093-2* 368
Rechsteineria caracasana IPNI=*237093-2* 368
Rechsteineria chelonoides IPNI=*237093-2* 368
Rechsteineria elliptica IPNI=*237093-2* 368
Rechsteineria erubescens IPNI=*237093-2* 368
Rechsteineria faucidens IPNI=*237093-2* 368
Rechsteineria faucidens var. *parvifolia* IPNI=*237093-2* 368
Rechsteineria flavescens IPNI=*237093-2* 368
Rechsteineria gollmeriana IPNI=*237093-2* 368
Rechsteineria gracilis IPNI=*237093-2* 368
Rechsteineria guazumifolia IPNI=*381388-1* 367
Rechsteineria incarnata IPNI=*237093-2* 368
Rechsteineria rutila IPNI=*237093-2* 368
Rechsteineria vargasii IPNI=*237093-2* 368
Rechsteineria warczewiczii IPNI=*237093-2* 368
Rectanthera fragrans IPNI=*42441-2* 223
Regelia meyendorffii IPNI=*169003-2* 229
Regina gallorum IPNI=*319112-2* 377
Regnellia purpurea IPNI=*617644-1* 217
Reichardia decapetala IPNI=*481887-1* 312
Reichardia grandiflora IPNI=*82097-1* 343
Reichardia jasminoides IPNI=*82097-1* 343
Reichertia linarifolia IPNI=*230042-2* 345
Reichertia rosea IPNI=*230042-2* 345
Reigera maritima IPNI=*298094-1* 229
Reimaria purpurascens IPNI=*399226-1* 238
Reineckea triandra IPNI=*248082-2* 212
Relbunium alpicola IPNI=*749770-1* 350
Relbunium chloranthum IPNI=*749770-1* 350
Relbunium ciliatum IPNI=*749770-1* 350
Relbunium compactum IPNI=*749770-1* 350
Relbunium corymbosum IPNI=*749770-1* 350
Relbunium croceum subsp. *involucratum* IPNI=*749770-1* 350
Relbunium hirsutum IPNI=*749770-1* 350
Relbunium hypocarpium IPNI=*750043-1* 350
Relbunium involucratum IPNI=*749770-1* 350
Relbunium tarmense IPNI=*749770-1* 350
Relbunium tuberculosum IPNI=*749770-1* 350
Remaclea funebris IPNI=*441642-1* 217
Remijia bracteata IPNI=*77065718-1* 348
Remijia involucrata IPNI=*77065718-1* 348
Remijia megistocaula IPNI=*969466-1* 347
Remijia pedunculata IPNI=*77065717-1* 348
Remijia peruviana IPNI=*77065717-1* 348
Remijia prismatostylis IPNI=*1215741-2* 352
Remijia purdieana IPNI=*77065718-1* 348
Renealmia arundinaria IPNI=*798069-1* 251
Renealmia bracteosa IPNI=*218583-2* 251
Renealmia coelobractea IPNI=*218583-2* 251
Renealmia domingensis IPNI=*798014-1* 251
Renealmia erecta IPNI=*798255-1* 250
Renealmia exaltata IPNI=*218583-2* 251
Renealmia exaltata var. *gracilis* IPNI=*218583-2* 251
Renealmia foliosa IPNI=*218583-2* 251
Renealmia geniculata IPNI=*798255-1* 250
Renealmia goyazensis IPNI=*218583-2* 251
Renealmia holdenii IPNI=*798014-1* 251
Renealmia humilis IPNI=*798069-1* 251
Renealmia hygrophila IPNI=*218597-2* 251
Renealmia jalapensis IPNI=*218639-2* 251
Renealmia lativagina IPNI=*218583-2* 251
Renealmia lehmannii IPNI=*798014-1* 251
Renealmia lutea IPNI=*798014-1* 251
Renealmia macrantha IPNI=*218583-2* 251
Renealmia monostachia IPNI=*115178-2* 229
Renealmia nutans IPNI=*872083-1* 250
Renealmia occidentalis IPNI=*798014-1* 251
Renealmia occidentalis var. *cylindrica* IPNI=*798014-1* 251
Renealmia occidentalis var. *longipes* IPNI=*218616-2* 251
Renealmia occidentalis var. *macrocarpa* IPNI=*798014-1* 251
Renealmia pedicellaris IPNI=*218616-2* 251
Renealmia pendula IPNI=*50112-2* 228
Renealmia pilosa IPNI=*798014-1* 251
Renealmia raja IPNI=*218583-2* 251
Renealmia recurvata IPNI=*1194855-2* 229
Renealmia rubroflava IPNI=*218583-2* 251
Renealmia spectabilis IPNI=*872083-1* 250
Renealmia spicata IPNI=*798020-1* 251
Renealmia strobilifera IPNI=*218597-2* 251
Renealmia usneoides IPNI=*1169665-2* 229
Renggeria acuminata IPNI=*427431-1* 383
Renggeria montana IPNI=*427431-1* 383
Rengifa peruviana IPNI=*1195925-2* 384
Reseda crispata IPNI=*715535-1* 280
Reseda dimerocarpa IPNI=*715535-1* 280
Reseda glauca IPNI=*715535-1* 280
Reseda gussonei IPNI=*715535-1* 280
Reseda luteola subsp. *dimerocarpa* IPNI=*715535-1* 280
Reseda luteola subsp. *gussonei* IPNI=*715535-1* 280
Reseda neilgherrensis IPNI=*715569-1* 280
Reseda odorata f. *pilosa* IPNI=*715569-1* 280
Reseda odorata var. *neilgherrensis* IPNI=*715569-1* 280
Reseda salicifolia IPNI=*715535-1* 280
Reseda tinctoria IPNI=*715535-1* 280
Reseda undulata IPNI=*715535-1* 280
Reseda virescens IPNI=*715535-1* 280
Restrepia antennifera subsp. *erythroxantha* IPNI=*655533-1* 222
Restrepia antennifera subsp. *hemsleyana* IPNI=*655513-1* 222
Restrepia antennifera subsp. *klabochorum* IPNI=*655513-1* 222
Restrepia erythroxantha IPNI=*655533-1* 222
Restrepia hemsleyana IPNI=*655513-1* 222
Resupinaria grandiflora IPNI=*518476-1* 334
Retiniphyllum pallidum IPNI=*764635-1* 356
Retiniphyllum schomburgkii subsp. *occidentale* IPNI=*764631-1* 356
Retiniphyllum schomburgkii var. *angustifolium* IPNI=*764631-1* 356
Retiniphyllum schomburgkii var. *hirticalyx* IPNI=*764631-1* 356
Retiniphyllum truncatum var. *angustifolium* IPNI=*764635-1* 356
Retinispora decurvata IPNI=*677124-1* 180
Retinispora decussata IPNI=*677124-1* 180
Retinispora devriesiana IPNI=*263956-1* 180
Retinispora dubia IPNI=*263956-1* 180
Retinispora ellwangeriana IPNI=*263956-1* 180
Retinispora ericoides IPNI=*677124-1* 180
Retinispora flavescens IPNI=*677124-1* 180
Retinispora glauca IPNI=*677124-1* 180
Retinispora glaucescens IPNI=*263956-1* 180
Retinispora juniperoides IPNI=*677124-1* 180
Retinispora keteleeri IPNI=*263956-1* 180
Retinispora lawsoniana IPNI=*261839-1* 180
Retinispora meldensis IPNI=*263956-1* 180
Retinispora nobleana IPNI=*263956-1* 180
Retinispora pygmaea IPNI=*263956-1* 180
Retinispora recurvata IPNI=*677124-1* 180
Retinispora rigida IPNI=*677124-1* 180
Reynoldsia americana IPNI=*20012157-1* 253
Reynosia laevigata IPNI=*133956-2* 442
Rhabarbarum palmatum IPNI=*696772-1* 289
Rhabarbarum rhaponticum IPNI=*696786-1* 289
Rhabdadenia campestris IPNI=*151979-2* 341
Rhabdadenia cordata IPNI=*81684-1* 343
Rhabdadenia ehrenbergii IPNI=*81684-1* 343
Rhabdadenia lucida IPNI=*1054185-2* 342
Rhabdadenia macrantha IPNI=*81684-1* 343
Rhabdadenia nervosa IPNI=*81684-1* 343
Rhabdadenia paludosa IPNI=*81684-1* 343
Rhacodiscus lucidus IPNI=*47308-1* 359

Rhacodiscus moricandianus IPNI=*47308-1* 359
Rhacodiscus secundus IPNI=*47308-1* 359
Rhacoma crossopetalum f. *longifolia* IPNI=*60458018-2* 290
Rhacoma crossopetalum f. *rotundata* IPNI=*60458018-2* 290
Rhacoma crossopetalum f. *typica* IPNI=*60458018-2* 290
Rhacoma crossopetalum IPNI=*60458018-2* 290
Rhacoma pallens IPNI=*60458018-2* 290
Rhadinocarpus multiflorus IPNI=*77191263-1* 329
Rhagodia baccata var. *congesta* IPNI=*77121007-1* 282
Rhagodia billardierei var. *congesta* IPNI=*77121007-1* 282
Rhagodia congesta IPNI=*77121007-1* 282
Rhamnidium ferreum IPNI=*133956-2* 442
Rhamnidium revolutum IPNI=*133956-2* 442
Rhamnus alnifolia IPNI=*30426867-2* 441
Rhamnus atroviridis IPNI=*105605-2* 441
Rhamnus biniflora IPNI=*717579-1* 442
Rhamnus brandegeeana IPNI=*133956-2* 442
Rhamnus chimantensis IPNI=*105605-2* 441
Rhamnus circumcissa IPNI=*719213-1* 442
Rhamnus colmeiroi IPNI=*718405-1* 442
Rhamnus colubrina IPNI=*63423-2* 441
Rhamnus domingensis IPNI=*112858-2* 441
Rhamnus elegans IPNI=*1013652-1* 442
Rhamnus elliptica IPNI=*63435-2* 441
Rhamnus ferrea IPNI=*133956-2* 442
Rhamnus ferruginea IPNI=*63423-2* 441
Rhamnus goudotiana IPNI=*105605-2* 441
Rhamnus grangenos IPNI=*1213898-2* 436
Rhamnus granulosa IPNI=*105606-2* 441
Rhamnus humboldtiana IPNI=*717579-1* 442
Rhamnus humifusa IPNI=*718405-1* 442
Rhamnus iguanaea IPNI=*1213898-2* 436
Rhamnus jujuba IPNI=*719349-1* 442
Rhamnus lucida IPNI=*719213-1* 442
Rhamnus lycium IPNI=*718405-1* 442
Rhamnus maculata IPNI=*717579-1* 442
Rhamnus mauritiana IPNI=*719213-1* 442
Rhamnus micrantha IPNI=*1018059-2* 436
Rhamnus minor IPNI=*718405-1* 442
Rhamnus oleoides IPNI=*718405-1* 442
Rhamnus polygama IPNI=*112865-2* 442
Rhamnus pubescens IPNI=*718405-1* 442
Rhamnus purpusii IPNI=*133956-2* 442
Rhamnus purshiana IPNI=*30426867-2* 441
Rhamnus purshiana var. *hirtella* IPNI=*30426867-2* 441
Rhamnus saxatilis var. *nana* IPNI=*718405-1* 442
Rhamnus soporifera IPNI=*719213-1* 442
Rhamnus spicata IPNI=*1013652-1* 442
Rhamnus tinctoria IPNI=*718405-1* 442
Rhamnus venosa IPNI=*63435-2* 441
Rhamnus vulgaris IPNI=*719213-1* 442
Rhamnus zizyphus IPNI=*719213-1* 442
Rhamphospermum album IPNI=*288952-1* 278
Rhamphospermum volgense IPNI=*60442520-2* 276
Rhaphidophora caudata IPNI=*87046-1* 200
Rhaphidophora crassifolia IPNI=*87046-1* 200
Rhaphidophora cunninghamii IPNI=*87046-1* 200
Rhaphidophora dilacerata IPNI=*87046-1* 200
Rhaphidophora formosana IPNI=*87046-1* 200
Rhaphidophora lacera IPNI=*88567-1* 202
Rhaphidophora laciniata IPNI=*87046-1* 200
Rhaphidophora lovellae IPNI=*87046-1* 200
Rhaphidophora merrillii IPNI=*87046-1* 200
Rhaphidophora neocaledonica IPNI=*87046-1* 200
Rhaphidophora pinnata IPNI=*87046-1* 200
Rhaphidophora pinnatifida IPNI=*87046-1* 200
Rhaphidophora rosenburghii IPNI=*87046-1* 200
Rhaphidophora vitiensis IPNI=*87046-1* 200
Rhaphidophora wallichii IPNI=*87046-1* 200
Rhaphidospora vestita IPNI=*52558-1* 360
Rhaphiolepis crataegoides IPNI=*731389-1* 444
Rhaphiolepis fragrans IPNI=*731389-1* 444
Rhaphiolepis gracilis IPNI=*731389-1* 444
Rhaphiolepis indica var. *insularis* IPNI=*731389-1* 444
Rhaphiolepis indica var. *mekongensis* IPNI=*731389-1* 444
Rhaphiolepis japonica IPNI=*731389-1* 444
Rhaphiolepis kerrii IPNI=*731389-1* 444
Rhaphiolepis loquata IPNI=*77209608-1* 444
Rhaphiolepis loureiroi IPNI=*731389-1* 444
Rhaphiolepis mekongensis IPNI=*731389-1* 444
Rhaphiolepis parvibracteolata IPNI=*731389-1* 444
Rhaphiolepis pheostemon IPNI=*731389-1* 444
Rhaphiolepis pheostemonia IPNI=*731389-1* 444
Rhaphiolepis rubra IPNI=*731389-1* 444
Rhaphiolepis rugosa IPNI=*731389-1* 444
Rhaphiolepis sieboldii IPNI=*731389-1* 444
Rhaphiolepis sinensis IPNI=*731389-1* 444
Rhaphis aciculata IPNI=*396110-1* 236
Rhaphis arundinacea IPNI=*422078-1* 245
Rhaphis halepensis IPNI=*422139-1* 245
Rhaphis sorghum IPNI=*422090-1* 245
Rhaphis stricta IPNI=*392701-1* 235
Rhaphis trivalvis IPNI=*396110-1* 236
Rhaphis trivialis IPNI=*396110-1* 236
Rhaphis zizanioides IPNI=*396213-1* 236
Rhapis aspera IPNI=*669577-1* 211
Rhapis cordata IPNI=*669577-1* 211
Rhapis divaricata IPNI=*669577-1* 211
Rhapis javanica IPNI=*669577-1* 211
Rhapis major IPNI=*669577-1* 211
Rhaptostylum acuminatum IPNI=*70029599-1* 449
Rhaptostylum cyanocarpum IPNI=*70029599-1* 449
Rhaptostylum kappleri IPNI=*55672-2* 449
Rhaptostylum pallidum IPNI=*70029599-1* 449
Rheedia acuminata IPNI=*277321-2* 385
Rheedia acuminata var. *floribunda* IPNI=*277321-2* 385
Rheedia arubayensis IPNI=*972610-1* 385
Rheedia benthamiana IPNI=*77100471-1* 385
Rheedia brasiliensis IPNI=*427844-1* 385
Rheedia brasiliensis var. *salicifolia* IPNI=*427844-1* 385
Rheedia calyptrata IPNI=*972610-1* 385
Rheedia chocoensis IPNI=*277318-2* 385
Rheedia edulis IPNI=*277320-2* 385
Rheedia floribunda IPNI=*277321-2* 385
Rheedia gardneriana IPNI=*972610-1* 385
Rheedia gardneriana var. *glaziovii* IPNI=*972610-1* 385
Rheedia gardneriana var. *parvifolia* IPNI=*972610-1* 385
Rheedia intermedia IPNI=*277320-2* 385
Rheedia kappleri IPNI=*277321-2* 385
Rheedia macrantha IPNI=*428065-1* 385
Rheedia macrophylla IPNI=*428065-1* 385
Rheedia macrophylla var. *benthamiana* IPNI=*428065-1* 385
Rheedia madruno IPNI=*277321-2* 385
Rheedia madruno subsp. *bituberculata* IPNI=*277321-2* 385
Rheedia madruno subsp. *ovata* IPNI=*277321-2* 385
Rheedia magnifolia IPNI=*277318-2* 385
Rheedia rostrata IPNI=*277321-2* 385
Rheedia sagotiana IPNI=*428065-1* 385
Rheedia spruceana IPNI=*972610-1* 385
Rheedia spruceana var. *cuneata* IPNI=*972610-1* 385
Rheedia tenuifolia IPNI=*972610-1* 385
Rheedia tonduziana IPNI=*277320-2* 385
Rhetinantha aciantha IPNI=*643235-1* 220
Rhetinophloeum viride IPNI=*1006943-1* 330
Rheum baillonii IPNI=*696768-1* 289
Rheum crispum IPNI=*224413-2* 289

Rheum esculentum IPNI=*696786-1* 289
Rheum potaninii IPNI=*696772-1* 289
Rheum qinlingense IPNI=*696772-1* 289
Rheum rotundatum IPNI=*696786-1* 289
Rheum rotundifolium IPNI=*696786-1* 289
Rheum undulatum IPNI=*696786-1* 289
Rhexia aquatica IPNI=*573916-1* 424
Rhexia bogotensis IPNI=*566936-1* 418
Rhexia canescens IPNI=*566862-1* 418
Rhexia diversifolia IPNI=*60465985-2* 417
Rhexia elegans IPNI=*121047-2* 419
Rhexia fragilis IPNI=*569110-1* 419
Rhexia glutinosa IPNI=*566936-1* 418
Rhexia hispida IPNI=*1131052-2* 424
Rhexia hypericoides IPNI=*1168615-2* 418
Rhexia inaequalifolia IPNI=*60465985-2* 417
Rhexia jussiaeoides IPNI=*144384-2* 429
Rhexia ladanoides IPNI=*1131052-2* 424
Rhexia lepidota IPNI=*77205900-1* 417
Rhexia lychnitoides IPNI=*1168615-2* 418
Rhexia microphylla IPNI=*567329-1* 418
Rhexia myrtoidea IPNI=*573702-1* 423
Rhexia pumila IPNI=*1131052-2* 424
Rhexia sileniflora IPNI=*1031600-2* 417
Rhexia stachyoides IPNI=*1168615-2* 418
Rhexia stricta IPNI=*566891-1* 418
Rhexia strigosa IPNI=*1131052-2* 424
Rhexia trichotoma IPNI=*1131052-2* 424
Rhexia villosissima IPNI=*1168615-2* 418
Rhigospira paucifolia IPNI=*79852-1* 341
Rhigospira reticulata IPNI=*79852-1* 341
Rhigospira sinuosa IPNI=*79852-1* 341
Rhigospira sprucei IPNI=*148754-2* 341
Rhigospira ternstroemiacea IPNI=*168521-2* 342
Rhigospira venulosa IPNI=*945265-1* 343
Rhinanthus virginicus IPNI=*804205-1* 375
Rhinocarpus excelsus IPNI=*12224-2* 450
Rhipsalis aethiopica IPNI=*219250-2* 284
Rhipsalis angustissima IPNI=*963240-1* 284
Rhipsalis baccifera f. *zanzibarica* IPNI=*219250-2* 284
Rhipsalis baccifera subsp. *fasciculata* IPNI=*219250-2* 284
Rhipsalis baccifera subsp. *rhodocarpa* IPNI=*219250-2* 284
Rhipsalis bartlettii IPNI=*219250-2* 284
Rhipsalis caripensis IPNI=*219250-2* 284
Rhipsalis cassytha var. *dichotoma* IPNI=*219250-2* 284
Rhipsalis cassytha var. *hookeriana* IPNI=*219250-2* 284
Rhipsalis cassytha var. *mocinoana* IPNI=*219250-2* 284
Rhipsalis cassytha var. *rhodocarpa* IPNI=*219250-2* 284
Rhipsalis cassythoides IPNI=*219250-2* 284
Rhipsalis cassythopsis IPNI=*219250-2* 284
Rhipsalis comorensis IPNI=*219250-2* 284
Rhipsalis coriacea IPNI=*963240-1* 284
Rhipsalis dichotoma IPNI=*219250-2* 284
Rhipsalis fasciculata IPNI=*219250-2* 284
Rhipsalis guineensis IPNI=*219250-2* 284
Rhipsalis heptagona IPNI=*219250-2* 284
Rhipsalis hookeriana IPNI=*219250-2* 284
Rhipsalis kirbergii IPNI=*138328-1* 284
Rhipsalis leiophloea IPNI=*963240-1* 284
Rhipsalis macrocarpa IPNI=*132939-1* 283
Rhipsalis madagascariensis IPNI=*219250-2* 284
Rhipsalis micrantha f. *kirbergii* IPNI=*138328-1* 284
Rhipsalis micrantha f. *rauhiorum* IPNI=*138328-1* 284
Rhipsalis minutiflora IPNI=*219250-2* 284
Rhipsalis neocassutha IPNI=*219250-2* 284
Rhipsalis parasitica IPNI=*219250-2* 284
Rhipsalis pendula IPNI=*219250-2* 284
Rhipsalis phyllanthus IPNI=*132939-1* 283
Rhipsalis purpusii IPNI=*963240-1* 284
Rhipsalis ramulosa IPNI=*963240-1* 284
Rhipsalis roseana IPNI=*138328-1* 284
Rhipsalis sansibarica IPNI=*219250-2* 284
Rhipsalis suareziana IPNI=*219250-2* 284
Rhipsalis tetragona IPNI=*219250-2* 284
Rhipsalis tonduzii IPNI=*138328-1* 284
Rhipsalis turpinii IPNI=*138328-1* 284
Rhipsalis undulata IPNI=*219250-2* 284
Rhipsalis wercklei IPNI=*138328-1* 284
Rhipsalis zanzibarica IPNI=*219250-2* 284
Rhizaeris alba IPNI=*170672-1* 416
Rhizakenia ovata IPNI=*431804-1* 203
Rhizobolus amygdalifer IPNI=*47856-2* 381
Rhizobolus glaber IPNI=*828599-1* 381
Rhizobolus pekea IPNI=*828602-1* 381
Rhizobolus saouvari IPNI=*828599-1* 381
Rhizobolus souari IPNI=*828599-1* 381
Rhizobolus tomentosus IPNI=*828602-1* 381
Rhizobolus tuberculosum IPNI=*828602-1* 381
Rhizophora × brevistyla IPNI=*219395-2* 402
Rhizophora americana IPNI=*30061148-2* 402
Rhizophora mangle var. *racemosa* IPNI=*719838-1* 402
Rhodanthe atrosanguinea IPNI=*241451-1* 269
Rhodanthe immaculata IPNI=*241451-1* 269
Rhodanthe maculata IPNI=*241451-1* 269
Rhodanthe manglesii var. *alba* IPNI=*241451-1* 269
Rhodanthe manglesii var. *maculata* IPNI=*241451-1* 269
Rhodanthe manglesii var. *sanguinea* IPNI=*241451-1* 269
Rhodanthe manglesii var. *typica* IPNI=*241451-1* 269
Rhodocactus bleo IPNI=*60461844-2* 284
Rhodocactus colombianus IPNI=*60461845-2* 284
Rhodocactus corrugatus IPNI=*60461844-2* 284
Rhodocactus guamacho IPNI=*60461845-2* 284
Rhodocactus tampicanus IPNI=*219420-2* 284
Rhodocalyx crassipes IPNI=*78821-1* 340
Rhododendron annamense IPNI=*333359-1* 300
Rhododendron austrokiusianum IPNI=*1137397-2* 300
Rhododendron balsaminiflorum IPNI=*1137397-2* 300
Rhododendron bellum IPNI=*333359-1* 300
Rhododendron bicolor IPNI=*333359-1* 300
Rhododendron brachycarpum IPNI=*1137397-2* 300
Rhododendron breynii IPNI=*333359-1* 300
Rhododendron buergeri IPNI=*1137397-2* 300
Rhododendron calleryi IPNI=*333359-1* 300
Rhododendron crispiflorum IPNI=*1137397-2* 300
Rhododendron danielsianum IPNI=*1137397-2* 300
Rhododendron decumbens IPNI=*1137397-2* 300
Rhododendron hannoense IPNI=*333359-1* 300
Rhododendron hispidum IPNI=*1137397-2* 300
Rhododendron kiusianum var. *sataense* IPNI=*1137397-2* 300
Rhododendron lateritium IPNI=*1137397-2* 300
Rhododendron macranthum IPNI=*1137397-2* 300
Rhododendron obtusum IPNI=*1137397-2* 300
Rhododendron petilum IPNI=*333359-1* 300
Rhododendron ramentaceum IPNI=*1137397-2* 300
Rhododendron sataense IPNI=*1137397-2* 300
Rhododendron simsii var. *albiflorum* IPNI=*333359-1* 300
Rhododendron simsii var. *bellum* IPNI=*333359-1* 300
Rhododendron simsii var. *strigosistylum* IPNI=*333359-1* 300
Rhododendron tamurae IPNI=*333359-1* 300
Rhododendron thunbergii IPNI=*1137397-2* 300
Rhododendron viburnifolium IPNI=*333359-1* 300
Rhododendron vittatum IPNI=*333359-1* 300
Rhodoseris conspicua IPNI=*174455-2* 267
Rhodospatha amomifolia IPNI=*89035-1* 202
Rhodospatha blanda IPNI=*88634-1* 202
Rhodospatha heliconiifolia IPNI=*88624-1* 202

Rivina bengalensis IPNI=*221479-2* 287
Rivina brasiliensis IPNI=*221479-2* 287
Rivina brasiliensis var. *viridis* IPNI=*221479-2* 287
Rivina canescens IPNI=*221479-2* 287
Rivina dodecandra IPNI=*257665-2* 287
Rivina ehrenbergiana IPNI=*257665-2* 287
Rivina glabrata IPNI=*221479-2* 287
Rivina gracilis IPNI=*221479-2* 287
Rivina herbacea IPNI=*221479-2* 287
Rivina humilis var. *bracteata* IPNI=*221479-2* 287
Rivina humilis var. *canescens* IPNI=*221479-2* 287
Rivina humilis var. *glabra* IPNI=*221479-2* 287
Rivina humilis var. *laevis* IPNI=*221479-2* 287
Rivina humilis var. *plumbaginifolia* IPNI=*221479-2* 287
Rivina humilis var. *puberula* IPNI=*221479-2* 287
Rivina humilis var. *scandens* IPNI=*257665-2* 287
Rivina inaequalis IPNI=*122647-2* 287
Rivina laevis f. *brasiliensis* IPNI=*221479-2* 287
Rivina laevis f. *humilis* IPNI=*221479-2* 287
Rivina laevis f. *viridiflora* IPNI=*221479-2* 287
Rivina laevis IPNI=*221479-2* 287
Rivina laevis var. *pubescens* IPNI=*221479-2* 287
Rivina lanceolata IPNI=*221479-2* 287
Rivina latifolia IPNI=*122647-2* 287
Rivina mexicana IPNI=*221479-2* 287
Rivina moritziana IPNI=*257665-2* 287
Rivina obliquata IPNI=*221479-2* 287
Rivina octandra IPNI=*257665-2* 287
Rivina octandra var. *obtusifolia* IPNI=*257665-2* 287
Rivina orientalis IPNI=*221479-2* 287
Rivina pallida IPNI=*221479-2* 287
Rivina paraguayensis IPNI=*221479-2* 287
Rivina peruviana IPNI=*257667-2* 287
Rivina plumbaginifolia IPNI=*221479-2* 287
Rivina portulaccoides IPNI=*221479-2* 287
Rivina procumbens IPNI=*221479-2* 287
Rivina puberula IPNI=*221479-2* 287
Rivina purpurascens IPNI=*221479-2* 287
Rivina purpurascens var. *mollis* IPNI=*221479-2* 287
Rivina scandens IPNI=*257665-2* 287
Rivina tetrandra IPNI=*221479-2* 287
Rivina tinctaria IPNI=*221479-2* 287
Rivina tinctoria IPNI=*257667-2* 287
Rivina vernalis IPNI=*221479-2* 287
Rivina viridiflora IPNI=*221479-2* 287
Rivina viridis IPNI=*221479-2* 287
Robbia nitida IPNI=*79913-1* 341
Robergia glabra IPNI=*264767-1* 431
Robinia argentiflora IPNI=*143380-2* 325
Robinia candida IPNI=*520444-1* 335
Robinia coccinea IPNI=*510682-1* 329
Robinia ferruginea IPNI=*944647-1* 315
Robinia grandiflora IPNI=*518476-1* 334
Robinia javanica IPNI=*520444-1* 335
Robinia leucantha IPNI=*520444-1* 335
Robinia mollis IPNI=*143380-2* 325
Robinia nicou IPNI=*502971-1* 325
Robinia rosea IPNI=*496639-1* 320
Robinia scandens IPNI=*502971-1* 325
Robinia sepium IPNI=*496639-1* 320
Robinia sepium IPNI=*502971-1* 325
Robinia sericea IPNI=*143380-2* 325
Robinia sericea IPNI=*520444-1* 335
Robinia striata IPNI=*496501-1* 320
Robinia variegata IPNI=*496639-1* 320
Robinia violacea IPNI=*143380-2* 325
Robinia violacea IPNI=*503131-1* 325
Roccardia manglesii IPNI=*241451-1* 269
Roccardia manglesii subsp. *alba* IPNI=*241451-1* 269
Rodrigoa diversifolia IPNI=*154214-2* 220
Rodrigoa parvula IPNI=*154214-2* 220
Rodriguezia macrantha IPNI=*655792-1* 222
Rodriguezia secunda IPNI=*655797-1* 223
Rodriguezia secunda var. *panamensis* IPNI=*655797-1* 223
Rodschiedia bursa-pastoris IPNI=*30092589-2* 276
Roettlera capsularis IPNI=*799766-1* 374
Rolandra argentea IPNI=*221674-2* 269
Rolandra diacantha IPNI=*221674-2* 269
Rolandra monacantha IPNI=*221674-2* 269
Rolandra terminalis IPNI=*127605-2* 264
Rollinia biflora IPNI=*72259-1* 187
Rollinia brevipes IPNI=*72201-1* 186
Rollinia broadwayi IPNI=*72201-1* 186
Rollinia broadwayi var. *cuneata* IPNI=*72201-1* 186
Rollinia cardiantha IPNI=*77080775-1* 186
Rollinia cherimolioides IPNI=*77080795-1* 187
Rollinia cordifolia IPNI=*72182-1* 186
Rollinia curvipetala IPNI=*72259-1* 187
Rollinia cuspidata IPNI=*77080775-1* 186
Rollinia deliciosa IPNI=*72259-1* 187
Rollinia edulis IPNI=*77080778-1* 186
Rollinia edulis var. *acuta* IPNI=*77080778-1* 186
Rollinia edulis var. *macropus* IPNI=*77080778-1* 186
Rollinia exsucca IPNI=*72201-1* 186
Rollinia exsucca subsp. *elongata* IPNI=*72201-1* 186
Rollinia exsucca subsp. *resinosa* IPNI=*72201-1* 186
Rollinia gardneri IPNI=*72201-1* 186
Rollinia glaucescens IPNI=*72201-1* 186
Rollinia incurva IPNI=*72201-1* 186
Rollinia insignis IPNI=*77080806-1* 187
Rollinia insignis var. *pallida* IPNI=*77080806-1* 187
Rollinia jimenezii IPNI=*72259-1* 187
Rollinia jimenezii var. *nelsonii* IPNI=*72259-1* 187
Rollinia membranacea IPNI=*77080795-1* 187
Rollinia mexicana IPNI=*77080795-1* 187
Rollinia micrantha IPNI=*72259-1* 187
Rollinia microsepala IPNI=*77080793-1* 187
Rollinia mucosa IPNI=*72259-1* 187
Rollinia mucosa subsp. *aequatorialis* IPNI=*72259-1* 187
Rollinia mucosa subsp. *portoricensis* IPNI=*72259-1* 187
Rollinia mucosa var. *macropoda* IPNI=*72259-1* 187
Rollinia mucosa var. *neglecta* IPNI=*72259-1* 187
Rollinia multiflora IPNI=*72201-1* 186
Rollinia neglecta IPNI=*72259-1* 187
Rollinia orthopetala IPNI=*72259-1* 187
Rollinia pachyptera IPNI=*77080775-1* 186
Rollinia papilionella IPNI=*77080793-1* 187
Rollinia permensis IPNI=*72259-1* 187
Rollinia pittieri IPNI=*77080793-1* 187
Rollinia procera IPNI=*72201-1* 186
Rollinia pterocarpa IPNI=*72259-1* 187
Rollinia puberula IPNI=*72201-1* 186
Rollinia pulchrinervia IPNI=*72259-1* 187
Rollinia rensoniana IPNI=*77080795-1* 187
Rollinia rensoniana subsp. *granatensis* IPNI=*77080795-1* 187
Rollinia resinosa IPNI=*72201-1* 186
Rollinia rufinervis IPNI=*77080796-1* 187
Rollinia sieberi IPNI=*72259-1* 187
Rollinia sphaerantha IPNI=*77080775-1* 186
Rollinia standleyi IPNI=*77080795-1* 187
Rollinia surinamensis IPNI=*72201-1* 186
Rollinia tinifolia IPNI=*72201-1* 186
Rollinia uniflora IPNI=*77080775-1* 186
Romualdea apiculata IPNI=*72123-2* 290
Romualdea lancifolia IPNI=*72123-2* 290
Romualdea malpighifolia IPNI=*162403-1* 291

Romualdea wrightiana IPNI=*162403-1* 291
Ronabea erecta IPNI=*764729-1* 356
Ronabea latifolia var. *hispidula* IPNI=*764729-1* 356
Ronabea morindoides IPNI=*764729-1* 356
Ronabea viscoides IPNI=*764729-1* 356
Ronconia triflora IPNI=*30030870-2* 416
Rondeletia capitata IPNI=*746936-1* 348
Rondeletia caracasana IPNI=*743570-1* 347
Rondeletia glomerulata IPNI=*77124830-1* 347
Rondeletia inconstans IPNI=*743570-1* 347
Rondeletia microdon IPNI=*77154321-1* 350
Rondeletia microdonta IPNI=*77154321-1* 350
Rondeletia orthoneura IPNI=*743570-1* 347
Rondeletia orthoneura var. *angustior* IPNI=*743570-1* 347
Rondeletia reflexa IPNI=*743570-1* 347
Rondeletia repens IPNI=*109024-2* 351
Rondeletia spinescens IPNI=*764273-1* 356
Rondeletia trichotoma IPNI=*743570-1* 347
Rorippa arvensis IPNI=*288669-1* 277
Rorippa atrovirens IPNI=*222144-2* 277
Rorippa coloradensis IPNI=*288952-1* 278
Rorippa dictyota IPNI=*288669-1* 277
Rorippa dubia f. *elliptifolia* IPNI=*288735-1* 277
Rorippa heterophylla IPNI=*288735-1* 277
Rorippa hispida var. *glabrata* IPNI=*288669-1* 277
Rorippa islandica subsp. *fernaldiana* IPNI=*288669-1* 277
Rorippa islandica var. *fernaldiana* IPNI=*288669-1* 277
Rorippa islandica var. *glabrata* IPNI=*288669-1* 277
Rorippa islandica var. *occidentalis* IPNI=*288669-1* 277
Rorippa montana IPNI=*222144-2* 277
Rorippa nasturtioides IPNI=*288669-1* 277
Rorippa nasturtium IPNI=*287534-1* 277
Rorippa nasturtium-aquaticum IPNI=*287534-1* 277
Rorippa nasturtium-aquaticum var. *longisiliqua* IPNI=*287534-1* 277
Rorippa officinalis IPNI=*287534-1* 277
Rorippa palustris f. *aquatica* IPNI=*288669-1* 277
Rorippa palustris f. *inundata* IPNI=*288669-1* 277
Rorippa palustris subsp. *fernaldiana* IPNI=*288669-1* 277
Rorippa palustris subsp. *occidentalis* IPNI=*288669-1* 277
Rorippa palustris var. *cernua* IPNI=*288669-1* 277
Rorippa palustris var. *clavata* IPNI=*288669-1* 277
Rorippa palustris var. *dictyota* IPNI=*288669-1* 277
Rorippa palustris var. *elongata* IPNI=*288669-1* 277
Rorippa palustris var. *fernaldiana* IPNI=*288669-1* 277
Rorippa palustris var. *glabrata* IPNI=*288669-1* 277
Rorippa palustris var. *occidentalis* IPNI=*288669-1* 277
Rorippa palustris var. *pacifica* IPNI=*288669-1* 277
Rorippa palustris var. *williamsii* IPNI=*288669-1* 277
Rorippa sinapis IPNI=*222144-2* 277
Rorippa sublyrata IPNI=*288735-1* 277
Rorippa terrestris var. *globosa* IPNI=*288669-1* 277
Rorippa williamsii IPNI=*288669-1* 277
Rosa × *gechouitangensis* IPNI=*733521-1* 444
Rosa × *oulengensis* IPNI=*733521-1* 444
Rosa × *thea* IPNI=*733521-1* 444
Rosa × *tongchouanensis* IPNI=*733521-1* 444
Rosa × *tongtchouanensis* IPNI=*733521-1* 444
Rosa bengalensis chinensis IPNI=*732029-1* 444
Rosa bengalensis IPNI=*732029-1* 444
Rosa blinii IPNI=*927374-1* 444
Rosa chinensis var. *spontanea* IPNI=*732029-1* 444
Rosa diffusa IPNI=*927374-1* 444
Rosa diversifolia IPNI=*732029-1* 444
Rosa florida IPNI=*927374-1* 444
Rosa formosana IPNI=*927374-1* 444
Rosa franchetii IPNI=*927374-1* 444
Rosa grevillei IPNI=*927374-1* 444
Rosa intermedia IPNI=*927374-1* 444
Rosa lawranceana IPNI=*732029-1* 444
Rosa lebrunei IPNI=*927374-1* 444
Rosa longifolia IPNI=*732029-1* 444
Rosa monstrosa IPNI=*732029-1* 444
Rosa multiflora f. *inermis* IPNI=*927374-1* 444
Rosa multiflora f. *rosipetala* IPNI=*927374-1* 444
Rosa multiflora var. *alboplena* IPNI=*927374-1* 444
Rosa multiflora var. *nanningensis* IPNI=*927374-1* 444
Rosa multiflora var. *taoyuanensis* IPNI=*927374-1* 444
Rosa mutabilis IPNI=*732029-1* 444
Rosa nankinensis IPNI=*732029-1* 444
Rosa nanula IPNI=*732029-1* 444
Rosa polyantha f. *roseipetala* IPNI=*927374-1* 444
Rosa polyantha IPNI=*927374-1* 444
Rosa polyanthos IPNI=*927374-1* 444
Rosa prolifera IPNI=*927374-1* 444
Rosa quelpaertensis IPNI=*927374-1* 444
Rosa rouletii IPNI=*732029-1* 444
Rosa semperflorens IPNI=*732029-1* 444
Rosa semperflorens var. *molinerii* IPNI=*732029-1* 444
Rosa sinica IPNI=*732029-1* 444
Rosa thoryi IPNI=*927374-1* 444
Rosa thunbergii IPNI=*927374-1* 444
Rosa thyrsiflora IPNI=*927374-1* 444
Rosa trichogyna IPNI=*927374-1* 444
Rosa watsoniana IPNI=*927374-1* 444
Rosa wichurae IPNI=*927374-1* 444
Roscoea villosa IPNI=*862610-1* 369
Rosea poeppigiana IPNI=*129900-2* 283
Rosea virgata IPNI=*129900-2* 283
Rosenbergia scandens IPNI=*61922-2* 303
Roseodendron millsii IPNI=*222518-2* 365
Roshevitzia diarrhena IPNI=*401189-1* 239
Roshevitzia diplachnoides IPNI=*401189-1* 239
Roshevitzia glomerata IPNI=*401189-1* 239
Roshevitzia japonica IPNI=*401189-1* 239
Rosmarinus × *lavandulaceus* var. *trogloditarum* IPNI=*457138-1* 372
Rosmarinus angustifolius IPNI=*457138-1* 372
Rosmarinus communis IPNI=*457138-1* 372
Rosmarinus flexuosus IPNI=*457138-1* 372
Rosmarinus latifolius IPNI=*457138-1* 372
Rosmarinus laxiflorus IPNI=*457138-1* 372
Rosmarinus ligusticus IPNI=*457138-1* 372
Rosmarinus officinalis IPNI=*457138-1* 372
Rosmarinus officinalis subsp. *palaui* IPNI=*457138-1* 372
Rosmarinus officinalis subsp. *valentinus* IPNI=*457138-1* 372
Rosmarinus officinalis subvar. *macrocalyx* IPNI=*457138-1* 372
Rosmarinus officinalis var. *palaui* IPNI=*457138-1* 372
Rosmarinus officinalis var. *prostratus* IPNI=*457138-1* 372
Rosmarinus officinalis var. *serotinus* IPNI=*457138-1* 372
Rosmarinus palaui IPNI=*457138-1* 372
Rosmarinus rigidus IPNI=*457138-1* 372
Rosmarinus serotinus IPNI=*457138-1* 372
Rosmarinus tenuifolius IPNI=*457138-1* 372
Rossmannia repens IPNI=*105608-1* 293
Rostellularia benthainensis IPNI=*52558-1* 360
Rotantha combretoides IPNI=*553638-1* 417
Rothia pinnata IPNI=*242907-1* 269
Rothia pinnata var. *pallida* IPNI=*242907-1* 269
Rothia pinnata var. *purpurascens* IPNI=*242907-1* 269
Rottboellia altissima IPNI=*404684-1* 240
Rottboellia articulata IPNI=*404684-1* 240
Rottboellia arundinacea IPNI=*909653-1* 244
Rottboellia caespitosa IPNI=*325559-2* 244
Rottboellia denudata IPNI=*909653-1* 244
Rottboellia exaltata var. *robusta* IPNI=*909653-1* 244
Rottboellia fasciculata IPNI=*404684-1* 240
Rottboellia filifolia IPNI=*325559-2* 244

Rottboellia glabrata IPNI=*406281-1* 240
Rottboellia granularis IPNI=*116848-2* 240
Rottboellia heterochroa IPNI=*404684-1* 240
Rottboellia hispida IPNI=*909653-1* 244
Rottboellia loricata IPNI=*325559-2* 244
Rottboellia paleacea IPNI=*423051-1* 246
Rottboellia rhytachne IPNI=*325559-2* 244
Rottboellia sanguinea IPNI=*420951-1* 244
Rottboellia setacea IPNI=*408465-1* 241
Rottboellia setifolia IPNI=*325559-2* 244
Rottboellia setosa IPNI=*909653-1* 244
Rottboellia spathacea IPNI=*404684-1* 240
Rottboellia stipoides IPNI=*12758-2* 233
Rottboellia stolonifera IPNI=*423051-1* 246
Rottboellia tripsacoides IPNI=*423051-1* 246
Rottlera moluccana IPNI=*338514-1* 387
Roubieva anthelmintica IPNI=*1090820-2* 282
Roubieva querciformis IPNI=*1198060-2* 282
Roucheria parviflora IPNI=*544878-1* 394
Rouhamon bredemeyeri IPNI=*547083-1* 346
Rouhamon brevifolium IPNI=*547493-1* 346
Rouhamon curare IPNI=*547220-1* 346
Rouhamon divaricatum IPNI=*547220-1* 346
Rouhamon glabrum IPNI=*547207-1* 346
Rouhamon guianense var. *ecirrhosum* IPNI=*547220-1* 346
Rouhamon pedunculatum IPNI=*547083-1* 346
Roulinia acuta IPNI=*74165-2* 340
Roulinia convolvulacea IPNI=*74165-2* 340
Roulinia cordata IPNI=*74165-2* 340
Roulinia fluminensis IPNI=*74165-2* 340
Roulinia jamaicensis IPNI=*74165-2* 340
Roulinia mannii IPNI=*74165-2* 340
Roulinia modesta IPNI=*74165-2* 340
Roulinia montevidensis IPNI=*74165-2* 340
Roulinia riedelii IPNI=*74165-2* 340
Roulinia selloana IPNI=*74165-2* 340
Roulinia tamifolia IPNI=*74165-2* 340
Roulinia versicolor IPNI=*74165-2* 340
Roulinia voglii IPNI=*74165-2* 340
Rouliniella jamaicensis IPNI=*74165-2* 340
Roupala adiantifolia IPNI=*706342-1* 433
Roupala adiantifolia var. *grandidentata* IPNI=*706342-1* 433
Roupala arvensis IPNI=*706342-1* 433
Roupala boissieriana IPNI=*706342-1* 433
Roupala borealis IPNI=*706342-1* 433
Roupala complicata IPNI=*706342-1* 433
Roupala darienensis IPNI=*706342-1* 433
Roupala dentata IPNI=*706342-1* 433
Roupala dentigera IPNI=*706342-1* 433
Roupala discolor IPNI=*706342-1* 433
Roupala dissimilis IPNI=*706342-1* 433
Roupala frondosa IPNI=*706342-1* 433
Roupala gardneri IPNI=*706342-1* 433
Roupala gardneri var. *angustata* IPNI=*706342-1* 433
Roupala gardneri var. *dentata* IPNI=*706342-1* 433
Roupala gardneri var. *integrifolia* IPNI=*706342-1* 433
Roupala glabrata IPNI=*706342-1* 433
Roupala hameliifolia IPNI=*182728-2* 433
Roupala heterophylla IPNI=*706342-1* 433
Roupala heterophylla var. *pinnata* IPNI=*706342-1* 433
Roupala longipetiolata IPNI=*706342-1* 433
Roupala macropoda IPNI=*706342-1* 433
Roupala martii IPNI=*706342-1* 433
Roupala martii var. *pinnata* IPNI=*706342-1* 433
Roupala martii var. *simplicifolia* IPNI=*706342-1* 433
Roupala media IPNI=*706342-1* 433
Roupala mollis IPNI=*222667-2* 433
Roupala montana var. *dentata* IPNI=*706342-1* 433
Roupala montana var. *heterophylla* IPNI=*706342-1* 433
Roupala mucronulata IPNI=*706342-1* 433
Roupala nervosa IPNI=*706342-1* 433
Roupala obovata IPNI=*222667-2* 433
Roupala organensis IPNI=*706342-1* 433
Roupala ovalis IPNI=*706342-1* 433
Roupala pamplonensis IPNI=*706342-1* 433
Roupala panamensis IPNI=*706342-1* 433
Roupala peruviana IPNI=*222667-2* 433
Roupala polystachya IPNI=*1076423-2* 433
Roupala pyrifolia IPNI=*706342-1* 433
Roupala raimondii IPNI=*706342-1* 433
Roupala repanda IPNI=*706342-1* 433
Roupala rhombifolia IPNI=*706342-1* 433
Roupala saxicola IPNI=*102211-2* 433
Roupala sessilifolia IPNI=*182728-2* 433
Roupala tomentosa var. *integrifolia* IPNI=*706342-1* 433
Roupala tomentosa var. *sellovii* IPNI=*706342-1* 433
Roupala veraguensis IPNI=*706342-1* 433
Roupala yauaperyensis IPNI=*182723-2* 433
Rourea cubensis IPNI=*264767-1* 431
Rourea glabra var. *floribunda* IPNI=*264767-1* 431
Rourea granatensis IPNI=*264767-1* 431
Rourea macrophylla IPNI=*211312-2* 431
Rourea oblongifolia IPNI=*264767-1* 431
Rourea oblongifolia var. *floribunda* IPNI=*264767-1* 431
Rourea paucifoliata IPNI=*264767-1* 431
Rourea sympetala IPNI=*264767-1* 431
Roystonea caribaea IPNI=*222805-2* 212
Roystonea elata IPNI=*222809-2* 212
Roystonea floridana IPNI=*222809-2* 212
Roystonea jenmanii IPNI=*222809-2* 212
Roystonea oleracea var. *excelsior* IPNI=*222805-2* 212
Roystonea oleracea var. *jenmanii* IPNI=*222805-2* 212
Roystonea regia var. *hondurensis* IPNI=*222809-2* 212
Roystonea venezuelana IPNI=*222805-2* 212
Roystonea ventricosa IPNI=*222809-2* 212
Rubia aparine IPNI=*30007294-2* 350
Rubia ciliata IPNI=*749770-1* 350
Rubia corymbosa IPNI=*749770-1* 350
Rubia cruciata IPNI=*747959-1* 349
Rubia hirsuta IPNI=*749770-1* 350
Rubia hirta IPNI=*749770-1* 350
Rubia hypocarpia IPNI=*750043-1* 350
Rubia monantha IPNI=*749770-1* 350
Rubia montana IPNI=*749770-1* 350
Rubia mucronata IPNI=*749770-1* 350
Rubia tuberculosa IPNI=*749770-1* 350
Rubus abundus IPNI=*736846-1* 444
Rubus adenotrichos subsp. *leptaleos* IPNI=*734894-1* 444
Rubus apoensis IPNI=*739800-1* 445
Rubus aurantiacus IPNI=*298089-2* 444
Rubus borealis IPNI=*298089-2* 444
Rubus buschii IPNI=*298089-2* 444
Rubus chagalensis IPNI=*735458-1* 444
Rubus chinensis IPNI=*739800-1* 445
Rubus comintanus IPNI=*739800-1* 445
Rubus commersonii IPNI=*739800-1* 445
Rubus coronarius IPNI=*739800-1* 445
Rubus diamantinus IPNI=*298089-2* 444
Rubus dosedlae IPNI=*739800-1* 445
Rubus eglanteria IPNI=*739800-1* 445
Rubus euroasiaticus IPNI=*298089-2* 444
Rubus floribundus var. *seleri* IPNI=*736846-1* 444
Rubus fragrans IPNI=*298089-2* 444
Rubus frambaesianus IPNI=*298089-2* 444
Rubus gigantiflorus IPNI=*741240-1* 445
Rubus glaber IPNI=*298089-2* 444

Rubus glandulosopunctatus IPNI=*739800-1* 445
Rubus greeneanus IPNI=*298089-2* 444
Rubus hassleri IPNI=*741023-1* 445
Rubus helioscopus IPNI=*736846-1* 444
Rubus hirsutus var. *glabellus* IPNI=*739800-1* 445
Rubus hookeri IPNI=*741240-1* 445
Rubus hopingensis IPNI=*739800-1* 445
Rubus idaeus f. *erythrochlamydeus* IPNI=*298089-2* 444
Rubus idaeus f. *marmoratus* IPNI=*298089-2* 444
Rubus idaeus f. *niveus* IPNI=*298089-2* 444
Rubus idaeus f. *succineus* IPNI=*298089-2* 444
Rubus idaeus f. *warei* IPNI=*298089-2* 444
Rubus idaeus subsp. *maritimus* IPNI=*298089-2* 444
Rubus idaeus subsp. *melanotrachys* IPNI=*298089-2* 444
Rubus idaeus subsp. *nipponicus* IPNI=*298089-2* 444
Rubus idaeus subsp. *vulgatus* IPNI=*298089-2* 444
Rubus idaeus var. *borealisinensis* IPNI=*298089-2* 444
Rubus idaeus var. *glabratus* IPNI=*298089-2* 444
Rubus idaeus var. *nesophilus* IPNI=*298089-2* 444
Rubus iloensis IPNI=*739815-1* 445
Rubus jamaicensis IPNI=*739800-1* 445
Rubus javanicus IPNI=*739800-1* 445
Rubus jelskii IPNI=*734809-1* 444
Rubus jizogatakensis IPNI=*298089-2* 444
Rubus kanayamensis IPNI=*298089-2* 444
Rubus karafutoanus IPNI=*298089-2* 444
Rubus leesii IPNI=*298089-2* 444
Rubus macrocarpus IPNI=*738683-1* 445
Rubus macrocarpus IPNI=*741240-1* 445
Rubus marmoratus IPNI=*298089-2* 444
Rubus matsumuranus IPNI=*298089-2* 444
Rubus mingendensis IPNI=*739800-1* 445
Rubus nipponicus IPNI=*298089-2* 444
Rubus norikurensis IPNI=*298089-2* 444
Rubus nubigenus var. *subinermis* IPNI=*738683-1* 445
Rubus obtusifolius IPNI=*298089-2* 444
Rubus occidentalis subsp. *eriocarpus* IPNI=*60464737-2* 444
Rubus parvirosifolius IPNI=*739800-1* 445
Rubus polyphyllarius IPNI=*739800-1* 445
Rubus praecipuus IPNI=*734894-1* 444
Rubus robustus IPNI=*736846-1* 444
Rubus roseorum IPNI=*734894-1* 444
Rubus rosifolius var. *commersonii* IPNI=*739800-1* 445
Rubus rosifolius var. *coronarius* IPNI=*739800-1* 445
Rubus rosifolius var. *rubrocarpus* IPNI=*739800-1* 445
Rubus rosifolius var. *wuyishanensis* IPNI=*739800-1* 445
Rubus ruizii IPNI=*223690-2* 445
Rubus schomburgkii IPNI=*737191-1* 444
Rubus sinensis IPNI=*739800-1* 445
Rubus stipularis IPNI=*738683-1* 445
Rubus stuebellii IPNI=*736037-1* 444
Rubus tagallus IPNI=*739800-1* 445
Rubus taiwanianus IPNI=*739800-1* 445
Rubus trichomallus IPNI=*741023-1* 445
Rubus urticifolius var. *rosiflorus* IPNI=*741023-1* 445
Rubus yabei IPNI=*298089-2* 444
Rubus yatsugatakensis IPNI=*298089-2* 444
Rudbeckia dentata IPNI=*259782-1* 272
Rudbeckia digitata IPNI=*241716-1* 269
Rudbeckia laciniata var. *digitata* IPNI=*241716-1* 269
Rudbeckia laciniata var. *gaspereauensis* IPNI=*241716-1* 269
Rudbeckia laciniata var. *hortensia* IPNI=*241716-1* 269
Rudbeckia oppositifolia IPNI=*213534-1* 264
Rudbeckia pauciflora IPNI=*261363-1* 272
Rudbeckia quinata IPNI=*241716-1* 269
Rudgea alvarezii IPNI=*77154321-1* 350
Rudgea calophylla IPNI=*765569-1* 357
Rudgea japurensis IPNI=*765569-1* 357
Rudgea longirostris IPNI=*77161818-1* 354
Rudgea racemosa IPNI=*1114920-2* 357
Rudgea spicata IPNI=*77202779-1* 353
Rudgea traillii IPNI=*765569-1* 357
Rudolphia dubia IPNI=*485618-1* 314
Ruehssia estebanensis IPNI=*100911-1* 343
Ruehssia glauca IPNI=*100911-1* 343
Ruehssia pubescens IPNI=*100911-1* 343
Ruehssia purpurea IPNI=*100911-1* 343
Ruellia acaulis IPNI=*55411-1* 361
Ruellia albicans IPNI=*54193-1* 361
Ruellia albicaulis IPNI=*54193-1* 361
Ruellia alopecuroidea IPNI=*51951-1* 360
Ruellia alternata IPNI=*77147688-1* 361
Ruellia antipoda IPNI=*799766-1* 374
Ruellia arborescens IPNI=*36223-2* 359
Ruellia articulata IPNI=*1072674-2* 376
Ruellia barbillana IPNI=*1014800-1* 361
Ruellia brasiliensis IPNI=*49672-1* 360
Ruellia brittoniana IPNI=*54585-1* 361
Ruellia chenopodiifolia IPNI=*51951-1* 360
Ruellia clandestina IPNI=*54671-1* 361
Ruellia coerulea IPNI=*54585-1* 361
Ruellia colorata IPNI=*284442-2* 361
Ruellia colorata IPNI=*77147688-1* 361
Ruellia comosa IPNI=*315544-2* 359
Ruellia consocialis IPNI=*1014800-1* 361
Ruellia cooperi IPNI=*53850-1* 361
Ruellia coromandeliana IPNI=*45742-1* 359
Ruellia coromandelina IPNI=*45742-1* 359
Ruellia dichotoma IPNI=*54389-1* 361
Ruellia diffusa IPNI=*52558-1* 360
Ruellia dulcis IPNI=*55411-1* 361
Ruellia eriostachya IPNI=*52558-1* 360
Ruellia filicifolia IPNI=*49672-1* 360
Ruellia filiformis IPNI=*45742-1* 359
Ruellia gairae IPNI=*224052-2* 361
Ruellia gangetica IPNI=*45742-1* 359
Ruellia gigantea IPNI=*56505-1* 362
Ruellia graecizans IPNI=*947317-1* 361
Ruellia ignorantiae IPNI=*54585-1* 361
Ruellia justiciiflora IPNI=*49672-1* 360
Ruellia lacustris IPNI=*49672-1* 360
Ruellia lagopodes IPNI=*51951-1* 360
Ruellia longifolia IPNI=*947317-1* 361
Ruellia longipes IPNI=*54585-1* 361
Ruellia longissima IPNI=*53850-1* 361
Ruellia longissima var. *glabra* IPNI=*53850-1* 361
Ruellia lorentziana IPNI=*53920-1* 361
Ruellia lysimachia IPNI=*54671-1* 361
Ruellia malacosperma IPNI=*54585-1* 361
Ruellia metallica IPNI=*1014800-1* 361
Ruellia obliqua IPNI=*45742-1* 359
Ruellia obtusa f. *albiflora* IPNI=*54363-1* 361
Ruellia parviflora IPNI=*224052-2* 361
Ruellia pennellii IPNI=*1014800-1* 361
Ruellia picta IPNI=*54671-1* 361
Ruellia quadrangularis IPNI=*1053173-2* 359
Ruellia rhamnifolia IPNI=*54193-1* 361
Ruellia riopalenquensis IPNI=*1014800-1* 361
Ruellia rivularis IPNI=*49672-1* 360
Ruellia rohrii IPNI=*224052-2* 361
Ruellia secunda IPNI=*45742-1* 359
Ruellia serrata IPNI=*799766-1* 374
Ruellia serratitheca IPNI=*947317-1* 361
Ruellia spectabilis IPNI=*54585-1* 361
Ruellia thyrsacanthoides IPNI=*54157-1* 361
Ruellia tuberosa var. *humilis* IPNI=*54671-1* 361

Ruellia turbacensis IPNI=*54389-1* 361
Ruellia turbascensis IPNI=*54389-1* 361
Ruellia tweedyana IPNI=*54585-1* 361
Ruellia uliginosa IPNI=*224052-2* 361
Ruellia umbrosa IPNI=*45742-1* 359
Ruellia variabilis IPNI=*45742-1* 359
Ruellia varians IPNI=*48879-1* 360
Ruellia ventricosa IPNI=*947317-1* 361
Ruellia verticillata IPNI=*49672-1* 360
Ruellia verticilliflora IPNI=*49672-1* 360
Ruellia viscosa IPNI=*54389-1* 361
Ruellia zeylanica IPNI=*45742-1* 359
Ruizia fragrans IPNI=*191376-2* 185
Rumea hebecarpa IPNI=*111561-1* 404
Rumex acetosella var. *tenuifolius* IPNI=*224391-2* 289
Rumex acutus IPNI=*697003-1* 289
Rumex agrestis IPNI=*310739-2* 290
Rumex arvensis IPNI=*224391-2* 289
Rumex callosus IPNI=*697388-1* 290
Rumex campestris IPNI=*697003-1* 289
Rumex coreanus IPNI=*224413-2* 289
Rumex crispatulus IPNI=*310739-2* 290
Rumex crispus subsp. *fauriei* IPNI=*224413-2* 289
Rumex crispus var. *fauriei* IPNI=*224413-2* 289
Rumex divaricatus IPNI=*310739-2* 290
Rumex divaricatus IPNI=*697003-1* 289
Rumex dubius IPNI=*697003-1* 289
Rumex ecuadoriensis IPNI=*310739-2* 290
Rumex elongatus IPNI=*224413-2* 289
Rumex exsanguis IPNI=*697003-1* 289
Rumex falcarius IPNI=*224391-2* 289
Rumex fascilobus IPNI=*224391-2* 289
Rumex fauriei IPNI=*224413-2* 289
Rumex ferrugineus IPNI=*697003-1* 289
Rumex foliosus IPNI=*697003-1* 289
Rumex glomeratus IPNI=*697003-1* 289
Rumex infestus IPNI=*224391-2* 289
Rumex kunthianus IPNI=*224413-2* 289
Rumex kunthii IPNI=*224413-2* 289
Rumex lachanus IPNI=*697003-1* 289
Rumex laevigatus IPNI=*310739-2* 290
Rumex lingulatus IPNI=*224413-2* 289
Rumex litoralis IPNI=*697003-1* 289
Rumex longifolius IPNI=*224413-2* 289
Rumex luederi IPNI=*224413-2* 289
Rumex macropterus IPNI=*310739-2* 290
Rumex megalophyllus IPNI=*310739-2* 290
Rumex nemolapathum IPNI=*697003-1* 289
Rumex nemorosus IPNI=*697003-1* 289
Rumex nevadensis IPNI=*697003-1* 289
Rumex oxylapathum IPNI=*310739-2* 290
Rumex paludosus IPNI=*697003-1* 289
Rumex parvifolius IPNI=*697003-1* 289
Rumex patens IPNI=*310739-2* 290
Rumex patientia var. *crispus* IPNI=*224413-2* 289
Rumex purpureus IPNI=*310739-2* 290
Rumex pusillus IPNI=*697003-1* 289
Rumex pycnocarpus IPNI=*697003-1* 289
Rumex sepium IPNI=*310739-2* 290
Rumex subulatus IPNI=*310739-2* 290
Rumex sylvaticus IPNI=*697003-1* 289
Rumex tauricus IPNI=*697003-1* 289
Rumex tenuifolius IPNI=*224391-2* 289
Rumex turcicus IPNI=*224413-2* 289
Rumex undulatus IPNI=*697003-1* 289
Rumex virgatus IPNI=*697003-1* 289
Rumex wallrothii IPNI=*310739-2* 290
Rumex winterlii IPNI=*697003-1* 289
Rumohra aspidioides IPNI=*224543-2* 176
Rumohra quadrangularis IPNI=*224543-2* 176
Rumputris fasciculata IPNI=*463202-1* 182
Rupestrina pubescens IPNI=*258855-2* 246
Ruppia andina IPNI=*224565-2* 203
Ruppia brachypus IPNI=*224565-2* 203
Ruppia brevipes IPNI=*224565-2* 203
Ruppia cirrhosa subsp. *longipes* IPNI=*224565-2* 203
Ruppia curvicarpa IPNI=*224565-2* 203
Ruppia intermedia IPNI=*224565-2* 203
Ruppia maritima f. *aculeata* IPNI=*224565-2* 203
Ruppia maritima f. *curvirostris* IPNI=*224565-2* 203
Ruppia maritima f. *pectinata* IPNI=*224565-2* 203
Ruppia maritima proles curvicarpa IPNI=*224565-2* 203
Ruppia maritima subsp. *brevirostris* IPNI=*224565-2* 203
Ruppia maritima subsp. *intermedia* IPNI=*224565-2* 203
Ruppia maritima subsp. *obliqua* IPNI=*224565-2* 203
Ruppia maritima subsp. *rostellata* IPNI=*224565-2* 203
Ruppia maritima subsp. *rostrata* IPNI=*224565-2* 203
Ruppia maritima var. *brachypus* IPNI=*224565-2* 203
Ruppia maritima var. *curvicarpa* IPNI=*224565-2* 203
Ruppia maritima var. *exigua* IPNI=*224565-2* 203
Ruppia maritima var. *intermedia* IPNI=*224565-2* 203
Ruppia maritima var. *japonica* IPNI=*224565-2* 203
Ruppia maritima var. *longipes* IPNI=*224565-2* 203
Ruppia maritima var. *onondagensis* IPNI=*224565-2* 203
Ruppia maritima var. *rostrata* IPNI=*224565-2* 203
Ruppia maritima var. *subcapitata* IPNI=*224565-2* 203
Ruppia obliqua IPNI=*224565-2* 203
Ruppia pectinata IPNI=*224565-2* 203
Ruppia rostellata IPNI=*224565-2* 203
Ruppia salina IPNI=*224565-2* 203
Ruppia spiralis subsp. *longipes* IPNI=*224565-2* 203
Ruppia subsessilis IPNI=*224565-2* 203
Ruppia taquetii IPNI=*224565-2* 203
Ruppia transsilvanica IPNI=*224565-2* 203
Ruppia trichodes IPNI=*224565-2* 203
Ruppia zosteroides IPNI=*224565-2* 203
Ruprechtia colorata IPNI=*693983-1* 288
Ruprechtia coriacea IPNI=*693983-1* 288
Ruprechtia cumingii IPNI=*693983-1* 288
Ruprechtia hamanii IPNI=*693983-1* 288
Ruprechtia martii IPNI=*697785-1* 290
Ruprechtia oxyphylla IPNI=*693983-1* 288
Ruprechtia ramiflora IPNI=*693983-1* 288
Ruscus alexandrinus IPNI=*540452-1* 215
Ruscus dumosus IPNI=*540443-1* 215
Ruscus flexuosus IPNI=*540443-1* 215
Ruscus humilis IPNI=*540452-1* 215
Ruscus laxus IPNI=*540443-1* 215
Ruscus parasiticus IPNI=*540443-1* 215
Ruscus ponticus IPNI=*540443-1* 215
Ruscus troadensis IPNI=*540452-1* 215
Russelia americana IPNI=*808935-1* 376
Russelia colombiana IPNI=*808935-1* 376
Russelia flammea IPNI=*808935-1* 376
Russelia flavoviridis IPNI=*808935-1* 376
Russelia juncea IPNI=*808903-1* 376
Russelia oxyphylla IPNI=*808935-1* 376
Russelia sarmentosa f. *eglandulata* IPNI=*808935-1* 376
Russelia sarmentosa f. *furfuracea* IPNI=*808935-1* 376
Russelia sarmentosa f. *pubescens* IPNI=*808935-1* 376
Russelia sarmentosa f. *velutina* IPNI=*808935-1* 376
Russelia sarmentosa var. *nicaraguensis* IPNI=*808935-1* 376
Russelia sarmentosa var. *oaxacensis* IPNI=*808935-1* 376
Russelia sarmentosa var. *parviflora* IPNI=*808935-1* 376
Russelia tabascensis IPNI=*808935-1* 376
Rustia isernii IPNI=*1095647-2* 346

Rustia rosea IPNI=*134610-2* 352
Ruteria africana IPNI=*516230-1* 331
Ruteria pinnata IPNI=*516230-1* 331
Ruyschia bahiensis IPNI=*829493-1* 302
Ruyschia corallina IPNI=*829489-1* 302
Ruyschia lepidota IPNI=*829493-1* 302
Ruyschia souroubea IPNI=*829493-1* 302
Ruyschia surubea IPNI=*829493-1* 302
Ryania acuminata IPNI=*224736-2* 404
Ryania kunthii IPNI=*112045-1* 404
Ryania patrisii IPNI=*112052-1* 404
Ryania pyrifera var. *tomentosa* IPNI=*365862-1* 404
Ryania sagotiana IPNI=*365862-1* 404
Rytidostylis amazonica IPNI=*77110356-1* 294
Rytidostylis carthagenensis IPNI=*77110356-1* 294
Rytidostylis ciliata IPNI=*77110356-1* 294
Rytidostylis cordata IPNI=*77110356-1* 294
Rytidostylis filiformis IPNI=*77110356-1* 294
Rytidostylis gracilis IPNI=*77110356-1* 294
Rytidostylis macrophylla IPNI=*77110356-1* 294
Rytidostylis quadrifida IPNI=*77110356-1* 294
Rytidostylis quinqueloba IPNI=*77110356-1* 294
Rytidostylis trilobata IPNI=*77110356-1* 294
Rytilix granularis IPNI=*116848-2* 240

S
Sabal allenii IPNI=*669714-1* 212
Sabal glaucescens IPNI=*669714-1* 212
Sabal morrisiana IPNI=*669714-1* 212
Sabal nematoclada IPNI=*669714-1* 212
Sabal serrulata IPNI=*941782-1* 212
Sabazia anomala IPNI=*1186403-2* 261
Sabazia leiachaenia IPNI=*1186403-2* 261
Sabazia microglossa IPNI=*208554-1* 263
Sabazia microglossa var. *puberula* IPNI=*208554-1* 263
Sabazia urticifolia IPNI=*208557-1* 263
Sabbata polyphylla IPNI=*192653-1* 259
Sabdariffa digitata IPNI=*326388-2* 409
Sabdariffa rubra IPNI=*326388-2* 409
Sabicea ambigua IPNI=*11534-2* 347
Sabicea aspera var. *scandens* IPNI=*765972-1* 357
Sabicea aurea IPNI=*746936-1* 348
Sabicea colombiana IPNI=*224977-2* 357
Sabicea costaricensis IPNI=*224977-2* 357
Sabicea domingensis IPNI=*765972-1* 357
Sabicea edulis IPNI=*743001-1* 347
Sabicea eriantha IPNI=*765830-1* 357
Sabicea hirsuta IPNI=*765972-1* 357
Sabicea hirsuta var. *adpressa* IPNI=*765972-1* 357
Sabicea hirsuta var. *sellowii* IPNI=*765972-1* 357
Sabicea paraensis IPNI=*224977-2* 357
Sabicea reflexa IPNI=*224977-2* 357
Sabicea umbellata var. *paraensis* IPNI=*224977-2* 357
Sabicea villosa var. *adpressa* IPNI=*765972-1* 357
Sabicea villosa var. *sellowii* IPNI=*765972-1* 357
Saccharifera officinalis IPNI=*419977-1* 244
Saccharum alopecuroides subvar. *angustifolium* IPNI=*419853-1* 244
Saccharum alopecuroides subvar. *trinii* IPNI=*420046-1* 244
Saccharum atrorubens IPNI=*419977-1* 244
Saccharum balansae IPNI=*420046-1* 244
Saccharum bicorne IPNI=*12758-2* 233
Saccharum caudatum IPNI=*405817-1* 240
Saccharum contractum IPNI=*405817-1* 240
Saccharum cylindricum var. *contractum* IPNI=*405817-1* 240
Saccharum cylindricum var. *longifolium* IPNI=*405817-1* 240
Saccharum dubium IPNI=*405817-1* 240
Saccharum fasciculatum IPNI=*12920-2* 234
Saccharum fragile IPNI=*419977-1* 244
Saccharum hybridum IPNI=*419977-1* 244
Saccharum infirmum IPNI=*419977-1* 244
Saccharum monandrum IPNI=*405817-1* 240
Saccharum occidentale IPNI=*419977-1* 244
Saccharum officinale IPNI=*419977-1* 244
Saccharum officinarum var. *brevipedicellatum* IPNI=*419977-1* 244
Saccharum pappiferum IPNI=*182865-2* 242
Saccharum polydactylon IPNI=*60469006-2* 245
Saccharum polystachyum IPNI=*398888-1* 237
Saccharum repens IPNI=*2945817-1* 241
Saccharum rubicundum IPNI=*419977-1* 244
Saccharum sagittatum IPNI=*404486-1* 240
Saccharum sape IPNI=*405817-1* 240
Saccharum trinii IPNI=*420046-1* 244
Saccharum violaceum IPNI=*419977-1* 244
Sacciolepis donacifolia IPNI=*125410-2* 240
Saccus champeden IPNI=*582622-1* 436
Saccus communis IPNI=*582598-1* 436
Saccus heterophyllus IPNI=*850389-1* 436
Saccus integer IPNI=*582622-1* 436
Saccus laevis IPNI=*582598-1* 436
Saccus mariannensis IPNI=*582598-1* 436
Sacoglottis amazonica IPNI=*431492-1* 393
Sacoglottis diguensis IPNI=*124798-2* 393
Sacoglottis excelsa var. *colombiana* IPNI=*124794-2* 393
Sacoglottis guianensis f. *dolichocarpa* IPNI=*431492-1* 393
Sacoglottis guianensis f. *glabra* IPNI=*431492-1* 393
Sacoglottis guianensis f. *sphaerocarpa* IPNI=*225114-2* 393
Sacoglottis mattogrossensis f. *glabra* IPNI=*225114-2* 393
Sacoglottis mattogrossensis f. *puberula* IPNI=*225114-2* 393
Sacoglottis procera IPNI=*124807-2* 393
Sacoglottis retusa IPNI=*229304-2* 393
Sagenia ampla IPNI=*17226640-1* 177
Sagenia angustior IPNI=*249914-2* 177
Sagenia latifolia IPNI=*249914-2* 177
Sagenia macrophylla IPNI=*17226640-1* 177
Sagenia martinicensis IPNI=*17226640-1* 177
Sagenia mexicana IPNI=*249914-2* 177
Sageretia salamensis IPNI=*1013652-1* 442
Sagina anomala IPNI=*302082-2* 285
Sagina boydii IPNI=*302082-2* 285
Sagina bryoides IPNI=*302082-2* 285
Sagina corsica IPNI=*302082-2* 285
Sagina fasciculata IPNI=*302082-2* 285
Sagina muscoides IPNI=*302082-2* 285
Sagina pentamera IPNI=*302082-2* 285
Sagina procumbens f. *daviesii* IPNI=*302082-2* 285
Sagina procumbens lusus daviesii IPNI=*302082-2* 285
Sagina procumbens subsp. *bryoides* IPNI=*302082-2* 285
Sagina procumbens subsp. *littoralis* IPNI=*302082-2* 285
Sagina procumbens var. *compacta* IPNI=*302082-2* 285
Sagina procumbens var. *luxurians* IPNI=*302082-2* 285
Sagina truncata IPNI=*302082-2* 285
Sagina viscida IPNI=*302082-2* 285
Sagitta latifolia IPNI=*225280-2* 198
Sagittaria affinis IPNI=*58564-1* 198
Sagittaria amazonica IPNI=*58564-1* 198
Sagittaria brasiliensis IPNI=*77183476-1* 197
Sagittaria brevipedicellata IPNI=*58564-1* 198
Sagittaria cerulea IPNI=*283610-2* 225
Sagittaria cordifolia IPNI=*58564-1* 198
Sagittaria cordifolia IPNI=*77182733-1* 197
Sagittaria engelmanniana subsp. *longirostra* IPNI=*225280-2* 198
Sagittaria esculenta IPNI=*225280-2* 198
Sagittaria gigantea IPNI=*225280-2* 198
Sagittaria gracilis IPNI=*225280-2* 198
Sagittaria hastata IPNI=*225280-2* 198
Sagittaria hastata var. *gracilis* IPNI=*225280-2* 198

Sagittaria lagoensis IPNI=*58564-1* 198
Sagittaria latifolia f. *diversifolia* IPNI=*225280-2* 198
Sagittaria latifolia f. *gracilis* IPNI=*225280-2* 198
Sagittaria latifolia f. *hastata* IPNI=*225280-2* 198
Sagittaria latifolia var. *glabra* IPNI=*225280-2* 198
Sagittaria latifolia var. *major* IPNI=*225280-2* 198
Sagittaria latifolia var. *obtusa* IPNI=*225280-2* 198
Sagittaria latifolia var. *pubescens* IPNI=*225280-2* 198
Sagittaria longirostra IPNI=*225280-2* 198
Sagittaria obtusa IPNI=*225280-2* 198
Sagittaria ornithorhyncha IPNI=*225280-2* 198
Sagittaria planipes IPNI=*225280-2* 198
Sagittaria pseudohermaphroditica IPNI=*58564-1* 198
Sagittaria pugioniformis f. *macrophylla* IPNI=*58564-1* 198
Sagittaria pugioniformis var. *affinis* IPNI=*58564-1* 198
Sagittaria pugioniformis var. *platyphylla* IPNI=*58564-1* 198
Sagittaria pugioniformis var. *rhombifolia* IPNI=*58564-1* 198
Sagittaria radicans IPNI=*77182733-1* 197
Sagittaria rhomboidalis IPNI=*58564-1* 198
Sagittaria sagittifolia f. *angustifolia* IPNI=*225280-2* 198
Sagittaria sagittifolia f. *gracilis* IPNI=*225280-2* 198
Sagittaria sagittifolia f. *hastata* IPNI=*225280-2* 198
Sagittaria sagittifolia f. *latifolia* IPNI=*225280-2* 198
Sagittaria sagittifolia f. *obtusa* IPNI=*225280-2* 198
Sagittaria sagittifolia var. *angustifolia* IPNI=*225280-2* 198
Sagittaria sagittifolia var. *gracilis* IPNI=*225280-2* 198
Sagittaria sagittifolia var. *hastata* IPNI=*225280-2* 198
Sagittaria sagittifolia var. *longirostra* IPNI=*225280-2* 198
Sagittaria sagittifolia var. *macrophylla* IPNI=*225280-2* 198
Sagittaria sagittifolia var. *pubescens* IPNI=*225280-2* 198
Sagittaria sagittifolia var. *variabilis* IPNI=*225280-2* 198
Sagittaria sagittifolia var. *vulgaris* IPNI=*225280-2* 198
Sagittaria simplex IPNI=*225280-2* 198
Sagittaria tenella IPNI=*118649-2* 197
Sagittaria variabilis f. *angustifolia* IPNI=*225280-2* 198
Sagittaria variabilis f. *hastata* IPNI=*225280-2* 198
Sagittaria variabilis f. *obtusa* IPNI=*225280-2* 198
Sagittaria variabilis IPNI=*225280-2* 198
Sagittaria variabilis var. *angustifolia* IPNI=*225280-2* 198
Sagittaria variabilis var. *diversifolia* IPNI=*225280-2* 198
Sagittaria variabilis var. *gracilis* IPNI=*225280-2* 198
Sagittaria variabilis var. *latifolia* IPNI=*225280-2* 198
Sagittaria variabilis var. *obtusa* IPNI=*225280-2* 198
Sagittaria variabilis var. *pubescens* IPNI=*225280-2* 198
Sagittaria variabilis var. *sagittifolia* IPNI=*225280-2* 198
Sagittaria viscosa IPNI=*225280-2* 198
Saglorithys polygonoides IPNI=*51367-1* 360
Sagotanthus kappleri IPNI=*55672-2* 449
Sagotia racemosa var. *ligularis* IPNI=*355722-1* 392
Sagotia racemosa var. *macrocarpa* IPNI=*355722-1* 392
Sagotia racemosa var. *microsepala* IPNI=*355722-1* 392
Sagotia tafelbergii IPNI=*355722-1* 392
Sagotia triflora IPNI=*77187443-1* 320
Sagraea cognata IPNI=*77164250-1* 422
Sagraea columneifolia IPNI=*77207578-1* 422
Sagraea epiphytica IPNI=*77207810-1* 421
Sagraea platyphylla IPNI=*77164250-1* 422
Sagraea rubra IPNI=*77164250-1* 422
Sagraea sessiliflora IPNI=*77164250-1* 422
Saguaster capitis-yorki IPNI=*669378-1* 211
Saguaster drudei IPNI=*664083-1* 204
Saguaster elegans IPNI=*669378-1* 211
Saguaster macarthurii IPNI=*55245-3* 211
Saguerus americanus IPNI=*668158-1* 210
Sagus taedigera IPNI=*669539-1* 211
Sahagunia colombiana IPNI=*851228-1* 437
Sahagunia paraensis IPNI=*59683-2* 437
Sahagunia racemifera IPNI=*59683-2* 437
Sahagunia strepitans IPNI=*59683-2* 437
Sahagunia urophylla IPNI=*260203-2* 441
Saintpaulia ionantha IPNI=*77117375-1* 368
Saintpaulia kewensis IPNI=*77117375-1* 368
Saintpaulia tongwensis IPNI=*77117375-1* 368
Sajorium volubile IPNI=*354970-1* 392
Salacia acreana IPNI=*964785-1* 291
Salacia affinis IPNI=*162581-1* 291
Salacia amazonica IPNI=*190399-2* 291
Salacia amygdalina IPNI=*162581-1* 291
Salacia anomala IPNI=*430947-1* 290
Salacia articulata IPNI=*55842-2* 290
Salacia attenuata IPNI=*162852-1* 291
Salacia blainii IPNI=*161653-1* 290
Salacia brasiliensis IPNI=*162581-1* 291
Salacia capilliflora IPNI=*161663-1* 291
Salacia catalinensis IPNI=*162403-1* 291
Salacia cognata IPNI=*55844-2* 290
Salacia cognata var. *egensis* IPNI=*55844-2* 290
Salacia cognata var. *genuina* IPNI=*55844-2* 290
Salacia divaricata IPNI=*55849-2* 290
Salacia duckei IPNI=*162581-1* 291
Salacia elliptica var. *apiculata* IPNI=*162581-1* 291
Salacia elliptica var. *genuina* IPNI=*162581-1* 291
Salacia erythroxyloides IPNI=*162581-1* 291
Salacia gleasoniana IPNI=*55849-2* 290
Salacia glomerata IPNI=*162581-1* 291
Salacia grandiflora IPNI=*225455-2* 291
Salacia granulata IPNI=*190399-2* 291
Salacia guianensis IPNI=*162581-1* 291
Salacia hippocrateoides IPNI=*55849-2* 290
Salacia induta IPNI=*162581-1* 291
Salacia krukovii IPNI=*55849-2* 290
Salacia lacunosa IPNI=*162581-1* 291
Salacia laevigata IPNI=*190399-2* 291
Salacia laevigata subsp. *acuminata* IPNI=*190399-2* 291
Salacia laevigata subsp. *obtusa* IPNI=*162329-1* 291
Salacia laevigata subsp. *ovata* IPNI=*190399-2* 291
Salacia lineolata IPNI=*55844-2* 290
Salacia minutiflora IPNI=*55849-2* 290
Salacia multiflora IPNI=*161653-1* 290
Salacia obtusifolia IPNI=*162581-1* 291
Salacia polyanthomaniaca IPNI=*225455-2* 291
Salacia praecelsa IPNI=*125377-2* 291
Salacia pruinosa IPNI=*190402-2* 291
Salacia radula IPNI=*162403-1* 291
Salacia siputa IPNI=*190399-2* 291
Salacia sphaerocarpa IPNI=*55844-2* 290
Salacia vahliana IPNI=*162403-1* 291
Salicornia indica IPNI=*165886-1* 283
Salisburia adiantifolia IPNI=*262125-1* 179
Salisburia biloba IPNI=*262125-1* 179
Salisburia ginkgo IPNI=*262125-1* 179
Salisburia macrophylla IPNI=*262125-1* 179
Salisia gloxiniiflora IPNI=*381282-1* 367
Salisia maculata IPNI=*381282-1* 367
Salisia pallidiflora IPNI=*381282-1* 367
Salisia suaveolens IPNI=*381282-1* 367
Salix angustior IPNI=*301321-2* 404
Salix arborescens IPNI=*301321-2* 404
Salix auritoides IPNI=*301321-2* 404
Salix babylonica f. *pekinensis* IPNI=*777133-1* 404
Salix babylonica f. *tortuosa* IPNI=*777133-1* 404
Salix babylonica f. *villosa* IPNI=*777133-1* 404
Salix babylonica var. *glandulipilosa* IPNI=*777133-1* 404
Salix babylonica var. *matsudana* IPNI=*777133-1* 404
Salix baumgarteniana IPNI=*301321-2* 404
Salix bifurcata IPNI=*301321-2* 404

Salix caesifolia IPNI=*301321-2* 404
Salix cantoniensis IPNI=*777133-1* 404
Salix capitata IPNI=*777133-1* 404
Salix capraeformis IPNI=*779152-1* 404
Salix carniolica IPNI=*301321-2* 404
Salix chilensis var. *crispa* IPNI=*777818-1* 404
Salix chinensis IPNI=*777133-1* 404
Salix dahurica IPNI=*301321-2* 404
Salix dependens IPNI=*777133-1* 404
Salix elaeagnifolia IPNI=*301321-2* 404
Salix falcata IPNI=*777818-1* 404
Salix heliciflora IPNI=*301321-2* 404
Salix helix IPNI=*301321-2* 404
Salix hippophaeifolia IPNI=*301321-2* 404
Salix hoffmanniana IPNI=*301321-2* 404
Salix humboldtiana var. *fastigiata* IPNI=*777818-1* 404
Salix humboldtiana var. *fruticosa* IPNI=*777818-1* 404
Salix humboldtiana var. *pyramidalis* IPNI=*777818-1* 404
Salix humboldtiana var. *stipulacea* IPNI=*777818-1* 404
Salix jeholensis IPNI=*777133-1* 404
Salix jishiensis IPNI=*777133-1* 404
Salix kochiana IPNI=*301321-2* 404
Salix lasiogyne IPNI=*777133-1* 404
Salix lenta IPNI=*777133-1* 404
Salix longifolia IPNI=*779152-1* 404
Salix magellanica IPNI=*777818-1* 404
Salix matsudana f. *rubriflora* IPNI=*777133-1* 404
Salix matsudana IPNI=*777133-1* 404
Salix matsudana var. *anshanensis* IPNI=*777133-1* 404
Salix matsudana var. *pseudomatsudana* IPNI=*777133-1* 404
Salix mauternensis IPNI=*301321-2* 404
Salix mollissima IPNI=*301321-2* 404
Salix monadelpha IPNI=*301321-2* 404
Salix monandra IPNI=*301321-2* 404
Salix monandra var. *sericea* IPNI=*301321-2* 404
Salix monandra var. *subverticillata* IPNI=*301321-2* 404
Salix multiformis IPNI=*301321-2* 404
Salix mutabilis IPNI=*301321-2* 404
Salix napoleona IPNI=*777133-1* 404
Salix napoleonis IPNI=*777133-1* 404
Salix neilreichii IPNI=*301321-2* 404
Salix neolasiogyne IPNI=*777133-1* 404
Salix nitens IPNI=*779152-1* 404
Salix ohsidare IPNI=*777133-1* 404
Salix oleifolia IPNI=*301321-2* 404
Salix olivacea IPNI=*301321-2* 404
Salix oppositifolia IPNI=*301321-2* 404
Salix oxyphylla IPNI=*777818-1* 404
Salix pendula IPNI=*777133-1* 404
Salix perpendens IPNI=*777133-1* 404
Salix pingliensis IPNI=*777133-1* 404
Salix polia IPNI=*779152-1* 404
Salix pontederana IPNI=*301321-2* 404
Salix pratensis IPNI=*301321-2* 404
Salix propendens IPNI=*777133-1* 404
Salix pseudogilgiana IPNI=*777133-1* 404
Salix pseudolasiogyne IPNI=*777133-1* 404
Salix pseudolasiogyne var. *bilofolia* IPNI=*777133-1* 404
Salix pseudolasiogyne var. *erythrantha* IPNI=*777133-1* 404
Salix pseudomatsudana IPNI=*777133-1* 404
Salix purpurea f. *eriantha* IPNI=*301321-2* 404
Salix purpurea f. *furcata* IPNI=*301321-2* 404
Salix purpurea f. *sericea* IPNI=*301321-2* 404
Salix purpurea f. *styligera* IPNI=*301321-2* 404
Salix purpurea helix IPNI=*301321-2* 404
Salix purpurea subsp. *angustior* IPNI=*301321-2* 404
Salix purpurea uralensis IPNI=*301321-2* 404
Salix ritschelii IPNI=*301321-2* 404
Salix rosea IPNI=*301321-2* 404
Salix rossica IPNI=*779152-1* 404
Salix rossmaessleri IPNI=*301321-2* 404
Salix sansoniana IPNI=*301321-2* 404
Salix semihelix IPNI=*301321-2* 404
Salix serotina IPNI=*779152-1* 404
Salix siegertii IPNI=*301321-2* 404
Salix sphaerocephala IPNI=*301321-2* 404
Salix splendens IPNI=*779152-1* 404
Salix stipulacea IPNI=*777818-1* 404
Salix strobilacea IPNI=*779152-1* 404
Salix subfragilis IPNI=*777133-1* 404
Salix tenuis IPNI=*301321-2* 404
Salix traunsteineri IPNI=*301321-2* 404
Salix uralensis IPNI=*301321-2* 404
Salix viadrina IPNI=*779152-1* 404
Salix viminalis aurea IPNI=*779152-1* 404
Salix viminalis cannabina IPNI=*779152-1* 404
Salix viminalis regalis IPNI=*779152-1* 404
Salix viminalis var. *genuina* IPNI=*779152-1* 404
Salix viminalis var. *latifolia* IPNI=*779152-1* 404
Salix viminalis var. *longistyla* IPNI=*779152-1* 404
Salix viminalis var. *pallida* IPNI=*779152-1* 404
Salix viminalis var. *pyramidalis* IPNI=*779152-1* 404
Salix viminalis var. *rossica* IPNI=*779152-1* 404
Salix viminalis var. *strobilacea* IPNI=*779152-1* 404
Salix virescens IPNI=*779152-1* 404
Salix yuhkii IPNI=*777133-1* 404
Salmalia malabarica IPNI=*559120-1* 407
Salmasia multiculmis IPNI=*386633-1* 233
Salmasia vulgaris IPNI=*386633-1* 233
Salmea eupatoria IPNI=*241903-1* 269
Salmea eupatoria var. *intermedia* IPNI=*241903-1* 269
Salmea grandiceps IPNI=*241903-1* 269
Salmea nitida IPNI=*241903-1* 269
Salmea oppositiceps IPNI=*241903-1* 269
Salmea parviceps IPNI=*241903-1* 269
Salmea pubescens IPNI=*241903-1* 269
Salmea salicifolia IPNI=*241903-1* 269
Salmea scandens subsp. *paraguariensis* IPNI=*241903-1* 269
Salmea scandens var. *amellus* IPNI=*241903-1* 269
Salmea scandens var. *genuina* IPNI=*241903-1* 269
Salmea scandens var. *obtusata* IPNI=*241903-1* 269
Salmea scandens var. *pubescens* IPNI=*241903-1* 269
Salmea sessilifolia IPNI=*241903-1* 269
Salmea tomentosa IPNI=*241903-1* 269
Salmeopsis claussenii IPNI=*241903-1* 269
Salmia funifera IPNI=*102548-2* 227
Salmia guineensis IPNI=*77164234-1* 215
Salmia jamaicensis IPNI=*297578-1* 227
Salmia palmata IPNI=*297578-1* 227
Salmia spicata IPNI=*77164234-1* 215
Salpichlaena brasiliensis IPNI=*17058320-1* 174
Salpichlaena serrulata IPNI=*77142412-1* 174
Salsola stricta IPNI=*55993-2* 282
Salvia acutata IPNI=*457495-1* 373
Salvia agrestis IPNI=*457495-1* 373
Salvia ambigua IPNI=*457495-1* 373
Salvia anglica IPNI=*457495-1* 373
Salvia anselmii IPNI=*457495-1* 373
Salvia barcinonensis IPNI=*457495-1* 373
Salvia basilii IPNI=*457495-1* 373
Salvia betonicifolia IPNI=*457495-1* 373
Salvia brasiliensis IPNI=*60466516-2* 373
Salvia byzantina IPNI=*457495-1* 373
Salvia candollei IPNI=*457495-1* 373
Salvia chia IPNI=*456385-1* 372
Salvia ciliata IPNI=*455978-1* 372

Salvia clandestina IPNI=*457495-1* 373
Salvia clandestinoides IPNI=*457495-1* 373
Salvia claytonii IPNI=*457495-1* 373
Salvia cleistogama IPNI=*457495-1* 373
Salvia coccinea f. *pseudococcinea* IPNI=*455978-1* 372
Salvia coccinea IPNI=*455978-1* 372
Salvia coccinea var. *pseudococcinea* IPNI=*455978-1* 372
Salvia collina IPNI=*457495-1* 373
Salvia colorans IPNI=*60466516-2* 373
Salvia controversa IPNI=*457495-1* 373
Salvia debilis IPNI=*455889-1* 372
Salvia discolor IPNI=*457495-1* 373
Salvia disermas IPNI=*457495-1* 373
Salvia domenechii IPNI=*457495-1* 373
Salvia dubia IPNI=*457495-1* 373
Salvia electa IPNI=*457495-1* 373
Salvia elsholtzioides IPNI=*456526-1* 372
Salvia erosa IPNI=*457495-1* 373
Salvia fasciculata IPNI=*457138-1* 372
Salvia filamentosa IPNI=*455978-1* 372
Salvia fimbriata IPNI=*457397-1* 373
Salvia fontii IPNI=*457495-1* 373
Salvia fracta IPNI=*456526-1* 372
Salvia galeottii IPNI=*455978-1* 372
Salvia galinsogifolia IPNI=*456526-1* 372
Salvia glaucescens IPNI=*455978-1* 372
Salvia gracilis IPNI=*455889-1* 372
Salvia gracilis IPNI=*457495-1* 373
Salvia hiemalis IPNI=*457495-1* 373
Salvia horminioides IPNI=*457495-1* 373
Salvia hyptoides IPNI=*456526-1* 372
Salvia illyrica IPNI=*457495-1* 373
Salvia intricata IPNI=*457495-1* 373
Salvia iodochroa IPNI=*455889-1* 372
Salvia irazuensis IPNI=*455889-1* 372
Salvia killipiana IPNI=*455889-1* 372
Salvia laciniata IPNI=*457495-1* 373
Salvia lateriflora IPNI=*456822-1* 372
Salvia linnaei IPNI=*457495-1* 373
Salvia lowei IPNI=*457495-1* 373
Salvia macrophylla var. *malacophylla* IPNI=*456630-1* 372
Salvia martensii IPNI=*455889-1* 372
Salvia martinicensis IPNI=*456822-1* 372
Salvia mediterranea IPNI=*457495-1* 373
Salvia membranacea IPNI=*455889-1* 372
Salvia mollissima IPNI=*455978-1* 372
Salvia moschata IPNI=*457410-1* 373
Salvia multifida IPNI=*457495-1* 373
Salvia multispicata IPNI=*456526-1* 372
Salvia myriantha IPNI=*455889-1* 372
Salvia natalis IPNI=*455889-1* 372
Salvia neglecta IPNI=*457495-1* 373
Salvia oblongata IPNI=*457495-1* 373
Salvia obtusata IPNI=*457495-1* 373
Salvia obvallata IPNI=*457397-1* 373
Salvia ochroleuca IPNI=*457495-1* 373
Salvia pallidiflora IPNI=*457495-1* 373
Salvia parviflora IPNI=*457495-1* 373
Salvia petiolaris IPNI=*457216-1* 372
Salvia polymorpha IPNI=*457495-1* 373
Salvia praecox IPNI=*457495-1* 373
Salvia privoides var. *garberi* IPNI=*456822-1* 372
Salvia procumbens IPNI=*456822-1* 372
Salvia prysmatica IPNI=*456385-1* 372
Salvia pseudococcinea IPNI=*455978-1* 372
Salvia pseudogracilis IPNI=*455889-1* 372
Salvia pseudoprivoides IPNI=*456822-1* 372
Salvia psilophylla IPNI=*457397-1* 373
Salvia purpurascens IPNI=*455889-1* 372
Salvia pyrenaica IPNI=*457495-1* 373
Salvia radicans IPNI=*456822-1* 372
Salvia rhodantha IPNI=*457495-1* 373
Salvia rosea IPNI=*455978-1* 372
Salvia rosmarinus subsp. *palaui* IPNI=*457138-1* 372
Salvia rosmarinus subsp. *valentina* IPNI=*457138-1* 372
Salvia sabulicola IPNI=*457495-1* 373
Salvia sapinea IPNI=*455889-1* 372
Salvia schiedeana IPNI=*456385-1* 372
Salvia sennenii IPNI=*457495-1* 373
Salvia setosa IPNI=*456822-1* 372
Salvia sibthorpii IPNI=*457495-1* 373
Salvia sidifolia IPNI=*455889-1* 372
Salvia simulans IPNI=*455889-1* 372
Salvia spielmanniana IPNI=*457495-1* 373
Salvia spielmannii IPNI=*457495-1* 373
Salvia subscaposa IPNI=*457495-1* 373
Salvia superba IPNI=*455978-1* 372
Salvia tetragona IPNI=*456385-1* 372
Salvia theodori IPNI=*457495-1* 373
Salvia variabilis IPNI=*457495-1* 373
Salvia verbenaca var. *anglica* IPNI=*457495-1* 373
Salvia verbenaca var. *delicatula* IPNI=*457495-1* 373
Salvia verbenacoides IPNI=*457495-1* 373
Salvia verbenifolia IPNI=*457495-1* 373
Salvia vivianii IPNI=*457495-1* 373
Salvia weihaiensis IPNI=*457495-1* 373
Salvinia affinis IPNI=*17219800-1* 179
Salvinia auriculata var. *olfersiana* IPNI=*17219800-1* 179
Salvinia azolla IPNI=*17056910-1* 179
Salvinia hispida IPNI=*17219800-1* 179
Salvinia olfersiana IPNI=*17219800-1* 179
Salvinia radula IPNI=*17219800-1* 179
Salvinia rotundifolia IPNI=*17219800-1* 179
Samama citrifolia IPNI=*756359-1* 352
Samanea arborea IPNI=*62882-2* 315
Samanea corymbosa IPNI=*989341-1* 320
Samanea marginata IPNI=*989342-1* 320
Samanea samaningua IPNI=*211561-2* 331
Samara coriacea IPNI=*1028287-2* 304
Samara floribunda IPNI=*166565-2* 304
Samara guianensis IPNI=*166565-2* 304
Samara laeta IPNI=*1028287-2* 304
Samara pentandra IPNI=*166565-2* 304
Samaroceltis rhamnoides IPNI=*855199-1* 445
Sambucus alba IPNI=*30122169-2* 296
Sambucus arborescens IPNI=*30122169-2* 296
Sambucus aurea IPNI=*30122169-2* 296
Sambucus bipinnata IPNI=*227173-2* 296
Sambucus bipinnata IPNI=*321978-2* 296
Sambucus canadensis f. *atroflavula* IPNI=*321978-2* 296
Sambucus canadensis f. *aurea* IPNI=*321978-2* 296
Sambucus canadensis f. *chlorocarpa* IPNI=*321978-2* 296
Sambucus canadensis f. *delicatissima* IPNI=*321978-2* 296
Sambucus canadensis f. *engelmannii* IPNI=*321978-2* 296
Sambucus canadensis f. *repens* IPNI=*321978-2* 296
Sambucus canadensis f. *rosicalyx* IPNI=*321978-2* 296
Sambucus canadensis f. *rubra* IPNI=*321978-2* 296
Sambucus canadensis subsp. *laciniata* IPNI=*321978-2* 296
Sambucus canadensis var. *laciniata* IPNI=*321978-2* 296
Sambucus canadensis var. *mexicana* IPNI=*227173-2* 296
Sambucus canadensis var. *oreopola* IPNI=*321979-2* 296
Sambucus canadensis var. *rubella* IPNI=*321978-2* 296
Sambucus canadensis var. *submollis* IPNI=*321978-2* 296
Sambucus canadensis var. *tarda* IPNI=*321978-2* 296
Sambucus cerulea var. *arizonica* IPNI=*321978-2* 296
Sambucus cerulea var. *mexicana* IPNI=*227173-2* 296

Sambucus columnaris IPNI=*30122169-2* 296
Sambucus dissecta IPNI=*30122169-2* 296
Sambucus eberhardtii IPNI=*321978-2* 296
Sambucus elegans IPNI=*30122169-2* 296
Sambucus florida IPNI=*30122169-2* 296
Sambucus graveolens IPNI=*321979-2* 296
Sambucus hirta IPNI=*30122169-2* 296
Sambucus humilis IPNI=*321978-2* 296
Sambucus intermedia var. *insularis* IPNI=*321978-2* 296
Sambucus laciniata IPNI=*30122169-2* 296
Sambucus leucocarpa IPNI=*30122169-2* 296
Sambucus lucida IPNI=*30122169-2* 296
Sambucus medullosa IPNI=*30122169-2* 296
Sambucus mexicana f. *caudata* IPNI=*227173-2* 296
Sambucus mexicana var. *bipinnata* IPNI=*227173-2* 296
Sambucus mexicana var. *plantierensis* IPNI=*227173-2* 296
Sambucus mexicana var. *typica* IPNI=*227173-2* 296
Sambucus monstrosa IPNI=*30122169-2* 296
Sambucus nigra f. *albomarmorata* IPNI=*30122169-2* 296
Sambucus nigra f. *chlorocarpa* IPNI=*30122169-2* 296
Sambucus nigra f. *luteomarmorata* IPNI=*30122169-2* 296
Sambucus nigra f. *porphyrophylla* IPNI=*30122169-2* 296
Sambucus nigra subsp. *canadensis* IPNI=*321978-2* 296
Sambucus nigra subsp. *peruviana* IPNI=*321979-2* 296
Sambucus nigra var. *canadensis* IPNI=*321978-2* 296
Sambucus nigra var. *laciniata* IPNI=*30122169-2* 296
Sambucus nigra var. *rotundifolia* IPNI=*30122169-2* 296
Sambucus orbiculata IPNI=*321978-2* 296
Sambucus orbiculata var. *glabra* IPNI=*321978-2* 296
Sambucus orbiculata var. *puberula* IPNI=*321978-2* 296
Sambucus oreopola IPNI=*321979-2* 296
Sambucus planteriensis IPNI=*321978-2* 296
Sambucus plantierensis IPNI=*321978-2* 296
Sambucus pulverulenta IPNI=*30122169-2* 296
Sambucus pyramidata IPNI=*30122169-2* 296
Sambucus rehderiana IPNI=*321978-2* 296
Sambucus repens IPNI=*321978-2* 296
Sambucus rotundifolia IPNI=*30122169-2* 296
Sambucus simpsonii IPNI=*321978-2* 296
Sambucus suaveolens IPNI=*321979-2* 296
Sambucus virescens IPNI=*30122169-2* 296
Sambucus vulgaris IPNI=*30122169-2* 296
Samyda affinis IPNI=*47939-2* 403
Samyda arborea IPNI=*47945-2* 403
Samyda decandra IPNI=*779541-1* 403
Samyda enneandria IPNI=*47939-2* 403
Samyda icosandra IPNI=*48022-2* 403
Samyda lancifolia IPNI=*779541-1* 403
Samyda multiflora IPNI=*47939-2* 403
Samyda niviana IPNI=*47945-2* 403
Samyda parviflora IPNI=*274165-2* 403
Samyda parviflora IPNI=*779541-1* 403
Samyda parvifolia IPNI=*779541-1* 403
Samyda pitumba IPNI=*297979-2* 403
Samyda procera IPNI=*47957-2* 403
Samyda spinosa IPNI=*47939-2* 403
Samyda stipularis IPNI=*47945-2* 403
Samyda sylvestris IPNI=*274165-2* 403
Samyda tomentosa IPNI=*47939-2* 403
Samyda virgata IPNI=*47945-2* 403
Sanchezia habra IPNI=*1014801-1* 361
Sanchezia helophila IPNI=*54890-1* 361
Sanchezia hirsuta IPNI=*54890-1* 361
Sanchezia macbridei IPNI=*54890-1* 361
Sanchezia nobilis IPNI=*54890-1* 361
Sanchezia nobilis var. *glaucophylla* IPNI=*54890-1* 361
Sanchezia pennellii IPNI=*1014801-1* 361
Sanchezia peruviana IPNI=*54890-1* 361
Sanchezia sprucei var. *salvadorensis* IPNI=*227313-2* 361
Sandwithiodoxa egregia IPNI=*209757-2* 306
Sanguinaria ciliaris IPNI=*1038558-2* 237
Sanguinaria nevenarae IPNI=*399115-1* 238
Sanguinaria vaginata IPNI=*184742-2* 243
Sanguisorba altissima IPNI=*741402-1* 445
Sanguisorba andersonii IPNI=*741402-1* 445
Sanguisorba auriculata IPNI=*741402-1* 445
Sanguisorba baicalensis IPNI=*741402-1* 445
Sanguisorba bracteosa IPNI=*741402-1* 445
Sanguisorba carnea IPNI=*741402-1* 445
Sanguisorba cernua IPNI=*741402-1* 445
Sanguisorba cordifolia IPNI=*741402-1* 445
Sanguisorba cylindrica IPNI=*741402-1* 445
Sanguisorba dictyocarpa IPNI=*741394-1* 445
Sanguisorba glandulosa IPNI=*741402-1* 445
Sanguisorba guestphalica IPNI=*741394-1* 445
Sanguisorba hispanica IPNI=*741402-1* 445
Sanguisorba komaroviana IPNI=*741402-1* 445
Sanguisorba longa IPNI=*741402-1* 445
Sanguisorba major IPNI=*741402-1* 445
Sanguisorba maroccana IPNI=*741394-1* 445
Sanguisorba media IPNI=*741402-1* 445
Sanguisorba menziesii IPNI=*741402-1* 445
Sanguisorba microcephala IPNI=*741402-1* 445
Sanguisorba minor var. *psiloritica* IPNI=*741394-1* 445
Sanguisorba montana IPNI=*741402-1* 445
Sanguisorba neglecta IPNI=*741402-1* 445
Sanguisorba officinalis f. *pallescens* IPNI=*741402-1* 445
Sanguisorba officinalis f. *pilosella* IPNI=*741402-1* 445
Sanguisorba officinalis subsp. *glandulosa* IPNI=*741402-1* 445
Sanguisorba officinalis subsp. *microcephala* IPNI=*741402-1* 445
Sanguisorba officinalis var. *dilutiflora* IPNI=*741402-1* 445
Sanguisorba officinalis var. *globularis* IPNI=*741402-1* 445
Sanguisorba officinalis var. *latifoliata* IPNI=*741402-1* 445
Sanguisorba pimpinella IPNI=*741394-1* 445
Sanguisorba polygama IPNI=*741402-1* 445
Sanguisorba praecox IPNI=*741402-1* 445
Sanguisorba rubra IPNI=*741402-1* 445
Sanguisorba sabauda IPNI=*741402-1* 445
Sanguisorba sanguisorba IPNI=*741394-1* 445
Sanguisorba serotina IPNI=*741402-1* 445
Sanguisorba taurica IPNI=*741402-1* 445
Sanguisorba unsanensis IPNI=*741402-1* 445
Sanicula mexicana IPNI=*847852-1* 253
Sanseverinia thyrsiflora IPNI=*77164234-1* 215
Sansevieria angustiflora IPNI=*77164234-1* 215
Sansevieria angustifolia IPNI=*77164234-1* 215
Sansevieria aureovariegata IPNI=*77164235-1* 215
Sansevieria ceylonica IPNI=*77164234-1* 215
Sansevieria flexuosa IPNI=*534103-1* 215
Sansevieria fragrans IPNI=*534207-1* 215
Sansevieria fruticosa IPNI=*534103-1* 215
Sansevieria fulvocincta IPNI=*77164234-1* 215
Sansevieria grandis IPNI=*77164234-1* 215
Sansevieria guineensis IPNI=*77164234-1* 215
Sansevieria hyacinthoides IPNI=*77164234-1* 215
Sansevieria jacquinii IPNI=*77164235-1* 215
Sansevieria laetevirens IPNI=*77164234-1* 215
Sansevieria latifolia IPNI=*77164234-1* 215
Sansevieria laurentii IPNI=*77164235-1* 215
Sansevieria polyphylla IPNI=*77164234-1* 215
Sansevieria rufocincta IPNI=*77164234-1* 215
Sansevieria spicata IPNI=*77164234-1* 215
Sansevieria stenophylla IPNI=*77164234-1* 215
Sansevieria thyrsiflora IPNI=*77164234-1* 215
Sansevieria trifasciata IPNI=*77164235-1* 215
Santalodes glabrum IPNI=*264767-1* 431

Santalodes macrophyllum IPNI=*211312-2* 431
Santalodes puberulum IPNI=*264853-1* 431
Santia monspeliensis IPNI=*1048074-2* 244
Santia plumosa IPNI=*1048074-2* 244
Santolina amellus IPNI=*241903-1* 269
Santolina brevidentata IPNI=*241927-1* 269
Santolina cupressiformis IPNI=*241927-1* 269
Santolina dentata IPNI=*241927-1* 269
Santolina jamaicensis IPNI=*1157019-2* 259
Santolina linearifolia IPNI=*241927-1* 269
Santolina lobata IPNI=*241927-1* 269
Santolina marchii IPNI=*241927-1* 269
Santolina millefolium IPNI=*2294-2* 255
Santolina oppositifolia IPNI=*1008406-1* 264
Santolina pallida IPNI=*241927-1* 269
Santolina pecten IPNI=*241927-1* 269
Santolina provincialis IPNI=*241927-1* 269
Santolina ruscinonensis IPNI=*241927-1* 269
Santolina sericea IPNI=*241927-1* 269
Santolina tomentosa IPNI=*241927-1* 269
Sanvitalia longepedunculata IPNI=*271548-2* 256
Saouari glabra IPNI=*828599-1* 381
Sapiopsis cremostachys IPNI=*227503-2* 392
Sapium aereum IPNI=*227503-2* 392
Sapium alainianum IPNI=*227503-2* 392
Sapium albomarginatum IPNI=*227503-2* 392
Sapium aubletianum IPNI=*227503-2* 392
Sapium aucuparium IPNI=*227503-2* 392
Sapium aucuparium var. *hippomane* IPNI=*227503-2* 392
Sapium aucuparium var. *moritzianum* IPNI=*227503-2* 392
Sapium aucuparium var. *petiolare* IPNI=*227503-2* 392
Sapium aureum IPNI=*227503-2* 392
Sapium biglandulosum f. *longifolium* IPNI=*227503-2* 392
Sapium biglandulosum f. *minus* IPNI=*227503-2* 392
Sapium biglandulosum f. *oblongatum* IPNI=*227503-2* 392
Sapium biglandulosum f. *obovatum* IPNI=*227503-2* 392
Sapium biglandulosum f. *pavonianum* IPNI=*227503-2* 392
Sapium biglandulosum var. *aubletianum* IPNI=*227503-2* 392
Sapium biglandulosum var. *aucuparium* IPNI=*227503-2* 392
Sapium biglandulosum var. *bogotense* IPNI=*227503-2* 392
Sapium biglandulosum var. *cuneatum* IPNI=*227503-2* 392
Sapium biglandulosum var. *hamatum* IPNI=*227503-2* 392
Sapium biglandulosum var. *klotzschianum* IPNI=*227503-2* 392
Sapium biglandulosum var. *lanceolatum* IPNI=*227503-2* 392
Sapium biglandulosum var. *meyerianum* IPNI=*227503-2* 392
Sapium biglandulosum var. *moritzianum* IPNI=*227503-2* 392
Sapium biglandulosum var. *oligoneurum* IPNI=*227503-2* 392
Sapium biglandulosum var. *salicifolium* IPNI=*227503-2* 392
Sapium biglandulosum var. *serratum* IPNI=*227503-2* 392
Sapium biglandulosum var. *sulciferum* IPNI=*227503-2* 392
Sapium bogotense IPNI=*227503-2* 392
Sapium caribaeum IPNI=*227503-2* 392
Sapium claussenianum IPNI=*227503-2* 392
Sapium cremostachyum IPNI=*227503-2* 392
Sapium cuatrecasasii IPNI=*355950-1* 393
Sapium endlicherianum IPNI=*227503-2* 392
Sapium fendleri IPNI=*227503-2* 392
Sapium fragile IPNI=*227503-2* 392
Sapium giganteum IPNI=*227503-2* 392
Sapium glandulatum IPNI=*227503-2* 392
Sapium guaricense IPNI=*227503-2* 392
Sapium haematospermum var. *saltense* IPNI=*227503-2* 392
Sapium hamatum IPNI=*227503-2* 392
Sapium hemsleyanum IPNI=*227503-2* 392
Sapium hippomane IPNI=*227503-2* 392
Sapium integrifolium IPNI=*227503-2* 392
Sapium intercedens IPNI=*227503-2* 392
Sapium itzanum IPNI=*227503-2* 392
Sapium ixiamasense IPNI=*227503-2* 392
Sapium izabalense IPNI=*227503-2* 392
Sapium jamaicense IPNI=*227503-2* 392
Sapium klotzschianum IPNI=*227503-2* 392
Sapium klotzschianum var. *glaziovii* IPNI=*227503-2* 392
Sapium lanceolatum IPNI=*227503-2* 392
Sapium leitera IPNI=*355874-1* 393
Sapium leptadenium IPNI=*227503-2* 392
Sapium marginatum f. *majus* IPNI=*227503-2* 392
Sapium marginatum IPNI=*227503-2* 392
Sapium marginatum var. *conjungens* IPNI=*227503-2* 392
Sapium marginatum var. *grandifolium* IPNI=*227503-2* 392
Sapium marginatum var. *lanceolatum* IPNI=*227503-2* 392
Sapium marginatum var. *longifolium* IPNI=*227503-2* 392
Sapium marginatum var. *spathulatum* IPNI=*227503-2* 392
Sapium moaense IPNI=*227503-2* 392
Sapium montevidense IPNI=*227503-2* 392
Sapium moritzianum IPNI=*227503-2* 392
Sapium muelleri IPNI=*227503-2* 392
Sapium myrmecophilum IPNI=*355950-1* 393
Sapium naiguatense IPNI=*227503-2* 392
Sapium obtusatum IPNI=*227503-2* 392
Sapium obtusilobum IPNI=*227503-2* 392
Sapium occidentale IPNI=*227503-2* 392
Sapium oligoneurum IPNI=*227503-2* 392
Sapium paranaense IPNI=*227503-2* 392
Sapium paucistamineum IPNI=*227503-2* 392
Sapium pavonianum IPNI=*227503-2* 392
Sapium peloto IPNI=*355874-1* 393
Sapium petiolare IPNI=*227503-2* 392
Sapium pittieri IPNI=*227503-2* 392
Sapium poeppigii IPNI=*227503-2* 392
Sapium pohlianum IPNI=*227503-2* 392
Sapium prunifolium IPNI=*227503-2* 392
Sapium punctatum IPNI=*227503-2* 392
Sapium putumayense IPNI=*355950-1* 393
Sapium pycnostachys IPNI=*227503-2* 392
Sapium salicifolium IPNI=*227503-2* 392
Sapium saltense IPNI=*227503-2* 392
Sapium schippii IPNI=*227503-2* 392
Sapium serratum IPNI=*227503-2* 392
Sapium solisii IPNI=*355950-1* 393
Sapium suberosum IPNI=*227503-2* 392
Sapium submarginatum IPNI=*227503-2* 392
Sapium subserratum IPNI=*227503-2* 392
Sapium sulciferum IPNI=*227503-2* 392
Sapium taburu IPNI=*227503-2* 392
Sapium thomsonii IPNI=*355950-1* 393
Sapium tolimense IPNI=*355950-1* 393
Sapium verum IPNI=*355950-1* 393
Sapota achras IPNI=*152641-2* 306
Sapota achras var. *lobata* IPNI=*152641-2* 306
Sapota achras var. *sphaerica* IPNI=*152641-2* 306
Sapota elongata IPNI=*209714-2* 306
Sapota latifolia IPNI=*209824-2* 307
Sapota mammosa IPNI=*1110421-2* 307
Sapota nigra IPNI=*322764-1* 298
Sapota venosa IPNI=*209920-2* 308
Saranthe cuyabensis IPNI=*797633-1* 250
Saranthe marcgravii IPNI=*797633-1* 250
Saranthe moritziana IPNI=*797633-1* 250
Sarcinanthus utilis IPNI=*22967-2* 227
Sarcoca dioica IPNI=*676348-1* 287
Sarcoca icosandra IPNI=*323292-2* 287
Sarcocephalus leichhardtii IPNI=*756359-1* 352
Sarcogonum platycladum IPNI=*694774-1* 289
Sarcogonum tamnifolium IPNI=*694785-1* 289
Sarcogonum volcanicum IPNI=*1022243-2* 289

Sarcomphalus mauritianus IPNI=*719349-1* 442
Sarcoperis bibracteata IPNI=*1191409-2* 224
Sarcostemma apiculatum IPNI=*106366-2* 340
Sarcostemma barbatum IPNI=*106366-2* 340
Sarcostemma bifidum IPNI=*106366-2* 340
Sarcostemma bonariense IPNI=*106366-2* 340
Sarcostemma brownii IPNI=*106366-2* 340
Sarcostemma clausum IPNI=*106366-2* 340
Sarcostemma crassifolium IPNI=*106366-2* 340
Sarcostemma cumanense IPNI=*106366-2* 340
Sarcostemma cuspidatum IPNI=*106366-2* 340
Sarcostemma dombeyanum IPNI=*106366-2* 340
Sarcostemma gardneri IPNI=*106366-2* 340
Sarcostemma glaucum IPNI=*106382-2* 340
Sarcostemma glaziovii IPNI=*106366-2* 340
Sarcostemma jacquinii IPNI=*106366-2* 340
Sarcostemma lineare IPNI=*106366-2* 340
Sarcostemma pallidum IPNI=*106366-2* 340
Sarcostemma pedunculatum IPNI=*106366-2* 340
Sarcostemma pubescens IPNI=*106366-2* 340
Sarcostemma riparium IPNI=*106366-2* 340
Sarcostemma schottii IPNI=*106366-2* 340
Sarcostemma swartzianum IPNI=*106366-2* 340
Sarcotheca bahiensis IPNI=*71069-1* 451
Sarcotheca philippica IPNI=*371870-1* 432
Sarcoyucca baccata IPNI=*270284-2* 215
Sarcoyucca elephantipes IPNI=*543700-1* 215
Saribus chinensis IPNI=*668021-1* 210
Saribus oliviformis IPNI=*668021-1* 210
Saribus subglobosus IPNI=*668021-1* 210
Saritaea magnifica IPNI=*108858-1* 362
Sarmentaria decora IPNI=*566051-1* 417
Sarotheca rostrata IPNI=*941389-1* 359
Sarothra juniperina IPNI=*433524-1* 393
Sarothra mutila IPNI=*126289-2* 394
Sarsaparilla acuminata IPNI=*315317-2* 226
Sasa fortunei IPNI=*416271-1* 244
Sasa variegata IPNI=*416271-1* 244
Sassafridium macrophyllum IPNI=*974281-1* 184
Sassafridium veraguense IPNI=*77198019-1* 183
Sassia debilis IPNI=*374833-1* 432
Sassia latifolia IPNI=*375051-1* 432
Sassia tetraphylla IPNI=*375466-1* 432
Satureja andrei IPNI=*1009966-1* 369
Satureja brownei IPNI=*61179-2* 369
Satureja brownei subsp. *eubrownei* IPNI=*61179-2* 369
Satureja brownei var. *pilosiuscula* IPNI=*61179-2* 369
Satureja caerulescens IPNI=*1001364-1* 371
Satureja fenzlii IPNI=*451022-1* 371
Satureja glabrata IPNI=*1009966-1* 369
Satureja lindeniana IPNI=*1001364-1* 371
Satureja lineata IPNI=*1009966-1* 369
Satureja nubigena IPNI=*61198-2* 369
Satureja stolonifera IPNI=*61179-2* 369
Satureja taxifolia IPNI=*1009966-1* 369
Satureja viminea IPNI=*61205-2* 369
Satureja xalapensis IPNI=*61179-2* 369
Saturiastrum montanum IPNI=*457736-1* 373
Satyria clonantha IPNI=*858655-1* 300
Satyria elongata IPNI=*77163666-1* 300
Satyria ovata IPNI=*858655-1* 300
Satyria triloba IPNI=*77163666-1* 300
Satyrium elatum IPNI=*73688-2* 218
Saurauia alvaroi IPNI=*228008-2* 297
Saurauia belizensis IPNI=*228173-2* 297
Saurauia brachybotrys f. *veranii* IPNI=*829301-1* 297
Saurauia brachybotrys var. *macrantha* IPNI=*829301-1* 297
Saurauia brachybotrys var. *scabra* IPNI=*829301-1* 297
Saurauia consimilis IPNI=*228097-2* 297
Saurauia echinosepala IPNI=*228008-2* 297
Saurauia excelsa var. *moritziana* IPNI=*829326-1* 297
Saurauia excelsa var. *xanthotricha* IPNI=*829326-1* 297
Saurauia floribunda IPNI=*829326-1* 297
Saurauia garcia-barrigae IPNI=*228016-2* 297
Saurauia goudotiana IPNI=*829301-1* 297
Saurauia herbert-smithii IPNI=*228054-2* 297
Saurauia humboldtiana IPNI=*1183910-2* 297
Saurauia hypomalla IPNI=*228016-2* 297
Saurauia mojandensis IPNI=*228016-2* 297
Saurauia moritziana IPNI=*829326-1* 297
Saurauia narcissifragrans IPNI=*1183910-2* 297
Saurauia peduncularis var. *veraniana* IPNI=*829301-1* 297
Saurauia prainiana var. *humboldtiana* IPNI=*1183910-2* 297
Saurauia pruinosa IPNI=*829436-1* 297
Saurauia pseudopittieri IPNI=*228173-2* 297
Saurauia pycnotricha IPNI=*829326-1* 297
Saurauia roseotincta IPNI=*228016-2* 297
Saurauia ruiziana var. *tomentosa* IPNI=*829436-1* 297
Saurauia scabra f. *moritziana* IPNI=*1183910-2* 297
Saurauia scabra f. *schlimiana* IPNI=*1183910-2* 297
Saurauia scabra f. *veranii* IPNI=*1183910-2* 297
Saurauia schlimii IPNI=*829326-1* 297
Saurauia smithiana IPNI=*228054-2* 297
Saurauia sydowii IPNI=*228097-2* 297
Saurauia ursina f. *strigosa* IPNI=*228162-2* 297
Saurauia ursina f. *veranii* IPNI=*228162-2* 297
Saurauia xanthotricha IPNI=*829326-1* 297
Saurauia yasicae f. *veranii* IPNI=*228173-2* 297
Saurauia yasicae var. *laevigata* IPNI=*228173-2* 297
Saurauia zetekiana IPNI=*228173-2* 297
Saurobroma iguanense IPNI=*1213898-2* 436
Sauvagesia adima IPNI=*867452-1* 397
Sauvagesia erecta var. *parvifolia* IPNI=*867452-1* 397
Sauvagesia erecta var. *sincorensis* IPNI=*867452-1* 397
Sauvagesia floribunda IPNI=*867452-1* 397
Sauvagesia geminiflora IPNI=*867452-1* 397
Sauvagesia grandifolia IPNI=*867452-1* 397
Sauvagesia laxa IPNI=*867452-1* 397
Sauvagesia nutans IPNI=*867452-1* 397
Sauvagesia peruviana IPNI=*867452-1* 397
Sauvagesia rubiginosa var. *luxurians* IPNI=*867469-1* 397
Sauvagesia salicifolia IPNI=*867452-1* 397
Saviniona assurgentiflora IPNI=*1003822-1* 410
Saviniona clementinae IPNI=*1003822-1* 410
Saviniona dendroidea IPNI=*1003822-1* 410
Saviniona reticulata IPNI=*1003822-1* 410
Saviniona suspensa IPNI=*1003822-1* 410
Scabiosa caucasea IPNI=*915212-1* 296
Scabiosa caucasica IPNI=*915212-1* 296
Scabiosa connata IPNI=*915212-1* 296
Scabiosa elegans IPNI=*915212-1* 296
Scabiosa fruticulosa IPNI=*319717-1* 296
Scabiosa grandiflora IPNI=*915212-1* 296
Scabiosa major IPNI=*319717-1* 296
Scadianus multiflorus IPNI=*319172-2* 213
Scaphyglottis cuneata IPNI=*228831-2* 223
Scaphyglottis gracilis IPNI=*228831-2* 223
Scaphyglottis ruberrima IPNI=*155262-2* 221
Scaphyglottis wercklei IPNI=*228831-2* 223
Sceptridium pusillum IPNI=*35345-2* 173
Sceptridium schaffneri IPNI=*35345-2* 173
Sceptridium ternatum var. *underwoodianum* IPNI=*35373-2* 173
Sceptridium underwoodianum IPNI=*35373-2* 173
Schachtia dioica IPNI=*86135-2* 349
Schaefferia berteroi IPNI=*162760-1* 291
Schaefferia buxifolia IPNI=*162760-1* 291

Schaefferia completa IPNI=*162760-1* 291
Schaefferia elliptica IPNI=*162760-1* 291
Schaefferia frutescens var. *buxifolia* IPNI=*162760-1* 291
Schaefferia lanceifolia IPNI=*162760-1* 291
Schaefferia viridescens IPNI=*162760-1* 291
Schaetzellia deckeri IPNI=*174455-2* 267
Schaueria graveolens IPNI=*158297-2* 371
Schedonorus arundinaceus IPNI=*973267-1* 241
Schedonorus dumetorum IPNI=*30010036-2* 239
Schedonorus unioloides IPNI=*37001-2* 235
Scheelea attaleoides IPNI=*26571-2* 205
Scheelea blepharopus IPNI=*1172851-2* 205
Scheelea butyracea IPNI=*1172851-2* 205
Scheelea corumbaensis IPNI=*26580-2* 205
Scheelea dryanderae IPNI=*1172851-2* 205
Scheelea excelsa IPNI=*1172851-2* 205
Scheelea fairchildensis IPNI=*20008131-1* 205
Scheelea goeldiana IPNI=*26571-2* 205
Scheelea gomphococca IPNI=*1172851-2* 205
Scheelea humboldtiana IPNI=*1172851-2* 205
Scheelea insignis IPNI=*26571-2* 205
Scheelea macrocarpa IPNI=*1172851-2* 205
Scheelea magdalenica IPNI=*20008141-1* 205
Scheelea maripa IPNI=*664452-1* 205
Scheelea martiana IPNI=*26580-2* 205
Scheelea microspadix IPNI=*26580-2* 205
Scheelea passargei IPNI=*1172851-2* 205
Scheelea phalerata IPNI=*26580-2* 205
Scheelea plowmanii IPNI=*20008146-1* 205
Scheelea princeps var. *corumbaensis* IPNI=*26580-2* 205
Scheelea regia IPNI=*1172851-2* 205
Scheelea tetrasticha IPNI=*664452-1* 205
Scheelea wallisii IPNI=*1172851-2* 205
Schefflera actinophylla IPNI=*77213393-1* 253
Schefflera arborea IPNI=*1133937-2* 253
Schefflera arboricola IPNI=*90845-1* 253
Schefflera bogotensis IPNI=*77210463-1* 254
Schefflera elegantissima IPNI=*77131172-1* 254
Schefflera faguetii IPNI=*77131172-1* 254
Schefflera lehmannii IPNI=*92546-1* 254
Schefflera minutiflora IPNI=*92546-1* 254
Schefflera morototoni IPNI=*90355-1* 253
Schefflera morototoni var. *sessiliflorus* IPNI=*90355-1* 253
Schefflera multiflora IPNI=*92546-1* 254
Schefflera sanquininensis IPNI=*92546-1* 254
Schefflera sphaerocoma IPNI=*92546-1* 254
Schefflera splendens IPNI=*90355-1* 253
Schefflera trianae IPNI=*230167-2* 254
Schefflera uribei IPNI=*230167-2* 254
Schiekia congesta IPNI=*429893-1* 224
Schiekia flavescens IPNI=*429893-1* 224
Schiekia orinocensis subsp. *savannarum* IPNI=*429893-1* 224
Schilleria aequalis IPNI=*680300-1* 195
Schilleria aurita IPNI=*680520-1* 195
Schilleria barbata IPNI=*680559-1* 195
Schilleria callosa IPNI=*680748-1* 195
Schilleria carpunya IPNI=*680810-1* 195
Schilleria catalpifolia IPNI=*682138-1* 196
Schilleria caudata IPNI=*682138-1* 196
Schilleria coruscans IPNI=*681027-1* 195
Schilleria corylifolia IPNI=*681193-1* 196
Schilleria grandifolia IPNI=*681559-1* 196
Schilleria heterophylla IPNI=*681651-1* 196
Schilleria inaequalifolia IPNI=*678848-1* 194
Schilleria marginata IPNI=*682138-1* 196
Schilleria obovata IPNI=*1090908-2* 194
Schilleria riparia IPNI=*681559-1* 196
Schilleria ulmifolia IPNI=*681193-1* 196
Schilleria variegata IPNI=*679043-1* 194
Schinus aroeira IPNI=*71069-1* 451
Schinus aurantiodora IPNI=*70012-1* 451
Schinus bituminosa IPNI=*71044-1* 451
Schinus chichita IPNI=*71069-1* 451
Schinus maurioides IPNI=*70000-1* 451
Schinus mellisii IPNI=*71069-1* 451
Schinus molle var. *areira* IPNI=*71006-1* 451
Schinus molle var. *argentifolia* IPNI=*71044-1* 451
Schinus mucronulata IPNI=*71069-1* 451
Schinus occidentalis IPNI=*71044-1* 451
Schinus rhoifolia IPNI=*71069-1* 451
Schinus terebinthifolia var. *damaziana* IPNI=*71069-1* 451
Schinus terebinthifolia var. *glazioviana* IPNI=*71069-1* 451
Schinus terebinthifolia var. *pohliana* IPNI=*71069-1* 451
Schinus terebinthifolia var. *raddiana* IPNI=*71069-1* 451
Schinus terebinthifolia var. *rhoifolia* IPNI=*71069-1* 451
Schinus terebinthifolia var. *selloana* IPNI=*71069-1* 451
Schismus patens IPNI=*60458636-2* 238
Schizachyrium ambalavense IPNI=*420951-1* 244
Schizachyrium bimucronatum IPNI=*420877-1* 244
Schizachyrium brevifolium var. *maclaudii* IPNI=*229330-2* 244
Schizachyrium condensatum var. *paniculatum* IPNI=*420918-1* 244
Schizachyrium domingense IPNI=*420951-1* 244
Schizachyrium fasciculatum IPNI=*1126606-2* 244
Schizachyrium feense IPNI=*420951-1* 244
Schizachyrium griseum IPNI=*420951-1* 244
Schizachyrium hirtiflorum IPNI=*420951-1* 244
Schizachyrium inspersum IPNI=*420951-1* 244
Schizachyrium lactiflorum IPNI=*420877-1* 244
Schizachyrium lindiense IPNI=*420951-1* 244
Schizachyrium minutum IPNI=*1126606-2* 244
Schizachyrium myosurus IPNI=*420951-1* 244
Schizachyrium neoscoparium IPNI=*420918-1* 244
Schizachyrium oligostachyum IPNI=*420951-1* 244
Schizachyrium paniculatum IPNI=*420918-1* 244
Schizachyrium paradoxum IPNI=*1126606-2* 244
Schizachyrium paucispiculatum IPNI=*420951-1* 244
Schizachyrium plumigerum IPNI=*420918-1* 244
Schizachyrium riedelii IPNI=*420951-1* 244
Schizachyrium riedelii var. *multirameum* IPNI=*420951-1* 244
Schizachyrium sanguineum subvar. *domingense* IPNI=*420951-1* 244
Schizachyrium sanguineum subvar. *feense* IPNI=*420951-1* 244
Schizachyrium sanguineum subvar. *hirtiflorum* IPNI=*420951-1* 244
Schizachyrium sanguineum subvar. *multirameum* IPNI=*420951-1* 244
Schizachyrium sanguineum subvar. *riedelii* IPNI=*420951-1* 244
Schizachyrium sanguineum subvar. *semiberbe* IPNI=*420951-1* 244
Schizachyrium sanguineum var. *brevipedicellatum* IPNI=*420951-1* 244
Schizachyrium sanguineum var. *hirtiflorum* IPNI=*420951-1* 244
Schizachyrium sanguineum var. *oligostachyum* IPNI=*420951-1* 244
Schizachyrium scoparium subvar. *lolioides* IPNI=*420874-1* 244
Schizachyrium semiberbe IPNI=*420951-1* 244
Schizachyrium semiglabrum IPNI=*420951-1* 244
Schizachyrium tenerum var. *cirratum* IPNI=*420874-1* 244
Schizachyrium tenuispicatum IPNI=*420951-1* 244
Schizachyrium weberbaueri IPNI=*420951-1* 244
Schizaea attenuata IPNI=*17220460-1* 179
Schizaea elegans var. *attenuata* IPNI=*17220460-1* 179
Schizaea elegans var. *flabellum* IPNI=*17220460-1* 179
Schizaea elegans var. *latifolia* IPNI=*17220460-1* 179
Schizaea flabellum IPNI=*17220460-1* 179
Schizaea pacificans IPNI=*17220460-1* 179
Schizaea spectabilis IPNI=*17220460-1* 179
Schizocalyx hirsuta IPNI=*766156-1* 357
Schizocentron elegans IPNI=*121047-2* 419
Schizoglossum filiforme IPNI=*106366-2* 340
Schizolobium amazonicum IPNI=*517988-1* 332
Schizolobium excelsum IPNI=*517988-1* 332

Schizolobium glutinosum IPNI=*517988-1* 332
Schizolobium parahyba var. *amazonicum* IPNI=*517988-1* 332
Schizopogon decumbens IPNI=*1126606-2* 244
Schizopsis polyantha IPNI=*259274-2* 365
Schizopsis polyantha var. *boliviana* IPNI=*259274-2* 365
Schizostachyum bitung IPNI=*397862-1* 237
Schizostachyum loriforme IPNI=*397862-1* 237
Schkuhria abrotanoides IPNI=*242907-1* 269
Schkuhria abrotanoides var. *pomasquiensis* IPNI=*242907-1* 269
Schkuhria advena IPNI=*242907-1* 269
Schkuhria anthemoidea f. *flava* IPNI=*242907-1* 269
Schkuhria anthemoidea IPNI=*242907-1* 269
Schkuhria anthemoidea var. *guatemalensis* IPNI=*242907-1* 269
Schkuhria anthemoidea var. *wislizeni* IPNI=*242907-1* 269
Schkuhria anthemoidea var. *wrightii* IPNI=*242907-1* 269
Schkuhria bonariensis IPNI=*242907-1* 269
Schkuhria coquimbana IPNI=*242907-1* 269
Schkuhria glabrescens IPNI=*242907-1* 269
Schkuhria guatemalensis IPNI=*242907-1* 269
Schkuhria hopkirkia IPNI=*242907-1* 269
Schkuhria isopappa IPNI=*242907-1* 269
Schkuhria octoaristata IPNI=*242907-1* 269
Schkuhria pinnata f. *pringlei* IPNI=*242907-1* 269
Schkuhria pinnata var. *abrotanoides* IPNI=*242907-1* 269
Schkuhria pinnata var. *guatemalensis* IPNI=*242907-1* 269
Schkuhria pinnata var. *octoaristata* IPNI=*242907-1* 269
Schkuhria pinnata var. *virgata* IPNI=*242907-1* 269
Schkuhria pinnata var. *wislizeni* IPNI=*242907-1* 269
Schkuhria pringlei IPNI=*242907-1* 269
Schkuhria virgata IPNI=*242907-1* 269
Schkuhria wislizeni f. *flava* IPNI=*242907-1* 269
Schkuhria wislizeni IPNI=*242907-1* 269
Schkuhria wislizeni var. *frustrata* IPNI=*242907-1* 269
Schkuhria wislizeni var. *guatemalensis* IPNI=*242907-1* 269
Schkuhria wislizeni var. *wrightii* IPNI=*242907-1* 269
Schkuhria wrightii IPNI=*242907-1* 269
Schlegelia gentlei IPNI=*229565-2* 377
Schlegelia organensis IPNI=*229565-2* 377
Schlegelia parviflora var. *trichandra* IPNI=*229565-2* 377
Schlegelia ramizii IPNI=*229565-2* 377
Schlegelia ramizii var. *costaricensis* IPNI=*229565-2* 377
Schlegelia ramizii var. *macrandra* IPNI=*229565-2* 377
Schlegelia ramizii var. *trichandra* IPNI=*229565-2* 377
Schleidenia elliptica IPNI=*60430582-2* 274
Schleidenia inundata IPNI=*60430582-2* 274
Schleidenia leptostachya IPNI=*60430582-2* 274
Schleidenia longepetiolata IPNI=*60430582-2* 274
Schlumbergera truncata var. *altensteinii* IPNI=*229597-2* 284
Schlumbergera truncata var. *delicata* IPNI=*229597-2* 284
Schnella bicornata IPNI=*77111925-1* 332
Schnella brachystachya IPNI=*229758-2* 332
Schnella columbiensis IPNI=*229758-2* 332
Schnella guianensis var. *splendens* IPNI=*518025-1* 332
Schnella heterophylla IPNI=*229758-2* 332
Schnella longipetala IPNI=*229758-2* 332
Schnella macrophylla IPNI=*77111925-1* 332
Schnella rosea IPNI=*77111927-1* 332
Schnella storkii IPNI=*229758-2* 332
Schnella trinitensis IPNI=*229758-2* 332
Schobera angiosperma IPNI=*116662-1* 274
Schobera hirsuta IPNI=*116662-1* 274
Schoenia chlorocephala IPNI=*969106-1* 269
Schoenissa rosea IPNI=*528823-1* 213
Schoenissa schoenoprasa IPNI=*528823-1* 213
Schoenleinia thyrsoidea IPNI=*752194-1* 351
Schoenoplectus californicus var. *spoliatus* IPNI=*229810-2* 232
Schoenoplectus cernuus IPNI=*309238-1* 232
Schoenoplectus chamissoi IPNI=*229810-2* 232
Schoenoplectus lacustris f. *glaucus* IPNI=*1019164-1* 232
Schoenoplectus lacustris subsp. *creber* IPNI=*1019164-1* 232
Schoenoplectus lacustris subsp. *glaucus* IPNI=*1019164-1* 232
Schoenoplectus lacustris subsp. *tabernaemontani* IPNI=*1019164-1* 232
Schoenoplectus lacustris subsp. *validus* IPNI=*1019164-1* 232
Schoenoplectus maritimus IPNI=*298094-1* 229
Schoenoplectus pseudotriqueter IPNI=*229810-2* 232
Schoenoplectus riparius IPNI=*229810-2* 232
Schoenoplectus tabernaemontani f. *paludosus* IPNI=*1019164-1* 232
Schoenoplectus tabernaemontani f. *sub-duvalii* IPNI=*1019164-1* 232
Schoenoplectus tatora IPNI=*229810-2* 232
Schoenoplectus validus IPNI=*1019164-1* 232
Schoenoplectus validus subsp. *creber* IPNI=*1019164-1* 232
Schoenoplectus validus subsp. *luxurians* IPNI=*1019164-1* 232
Schoenoprasum vulgare IPNI=*528823-1* 213
Schoenus angularis IPNI=*74934-2* 231
Schoenus articulatus IPNI=*220271-2* 232
Schoenus cephalotes IPNI=*312067-1* 232
Schoenus ciliatus IPNI=*312402-1* 232
Schoenus coloratus IPNI=*311841-1* 232
Schoenus corymbosus IPNI=*220271-2* 232
Schoenus fabri IPNI=*398888-1* 237
Schoenus ferrugineus IPNI=*30079279-2* 233
Schoenus floridus IPNI=*220271-2* 232
Schoenus globosus IPNI=*312402-1* 232
Schoenus lanatus IPNI=*30060098-2* 229
Schoenus mariscus IPNI=*303489-1* 230
Schoenus secans IPNI=*230730-2* 233
Schoenus spadiceus IPNI=*308274-1* 232
Schoenus stellatus IPNI=*311841-1* 232
Schoenus surinamensis IPNI=*220271-2* 232
Schoenus tenuifolius IPNI=*312330-1* 232
Schoenus tuberosus IPNI=*305797-1* 231
Schoepfia americana IPNI=*608391-1* 450
Schoepfia angustata IPNI=*608391-1* 450
Schoepfia arborescens IPNI=*608391-1* 450
Schoepfia chrysophylloides IPNI=*608391-1* 450
Schoepfia marchii IPNI=*608391-1* 450
Schoepfia mexicana IPNI=*608391-1* 450
Schollera debilis IPNI=*676291-1* 286
Schollera reniformis IPNI=*698080-1* 225
Schollia carnosa IPNI=*107308-3* 341
Schollia chinensis IPNI=*107308-3* 341
Schollia crassifolia IPNI=*107308-3* 341
Schollia viminalis IPNI=*106366-2* 340
Schomburgkia crispa var. *alba* IPNI=*134675-2* 219
Schomburgkia crispa var. *wallisii* IPNI=*134675-2* 219
Schomburgkia fimbriata IPNI=*134675-2* 219
Schomburgkia gloriosa f. *alba* IPNI=*134675-2* 219
Schomburgkia gloriosa IPNI=*134675-2* 219
Schomburgkia wallisii IPNI=*77090555-1* 219
Schousboea coccinea IPNI=*63781-2* 415
Schousboea commutata IPNI=*170672-1* 416
Schranckia quinquefaria IPNI=*161264-1* 393
Schrankia leptocarpa IPNI=*1014032-1* 327
Schrankia rugosa IPNI=*1061021-2* 277
Schrankia sulcata IPNI=*1061021-2* 277
Schrankia trijuga IPNI=*1014032-1* 327
Schubertia japonica IPNI=*261870-1* 180
Schubertia sempervirens IPNI=*263741-1* 180
Schufia arborescens IPNI=*106136-2* 429
Schultesia apiculata IPNI=*370581-1* 345
Schultesia aubletii IPNI=*230042-2* 345
Schultesia chiapensis IPNI=*230042-2* 345
Schultesia hassleriana IPNI=*370581-1* 345
Schultesia hayesii IPNI=*370581-1* 345
Schultesia heterophylla IPNI=*370581-1* 345

Schultesia lisianthoides IPNI=*371484-1* 345
Schultesia mexicana IPNI=*230042-2* 345
Schultesia neuroptera IPNI=*370581-1* 345
Schultesia pallens IPNI=*230042-2* 345
Schultesia peckiana IPNI=*371484-1* 345
Schultesia stenophylla IPNI=*230042-2* 345
Schwaegrichenia flavida IPNI=*429575-1* 224
Schwenkfelda aspera IPNI=*315609-2* 348
Schwenkfelda aurea IPNI=*746936-1* 348
Schwenkfelda cinerea IPNI=*765830-1* 357
Schwenkfelda eriantha IPNI=*765830-1* 357
Schwenkfelda hirsuta IPNI=*765972-1* 357
Sciadodendron excelsum IPNI=*20012157-1* 253
Sciadotenia ramiflora var. *colombiana* IPNI=*230118-2* 435
Sciadotenia similis IPNI=*230126-2* 435
Sciaphila elata IPNI=*836332-1* 227
Sciaphila richardii IPNI=*836332-1* 227
Scindapsus bipinnatifidus IPNI=*87046-1* 200
Scindapsus caudatus IPNI=*87046-1* 200
Scindapsus decursivus IPNI=*87046-1* 200
Scindapsus dilaceratus IPNI=*87046-1* 200
Scindapsus forsteri IPNI=*87046-1* 200
Scindapsus peepla IPNI=*88567-1* 202
Scindapsus pertusus IPNI=*88567-1* 202
Scindapsus pinnatifidus IPNI=*87046-1* 200
Scindapsus pinnatus IPNI=*87046-1* 200
Sciodaphyllum arboreum IPNI=*1133937-2* 253
Sciodaphyllum capitatum IPNI=*30050876-2* 254
Sciodaphyllum goudotii IPNI=*77210463-1* 254
Sciodaphyllum jacquinii IPNI=*1133937-2* 253
Sciodaphyllum lehmannii IPNI=*92546-1* 254
Sciodaphyllum ochroleucum IPNI=*230167-2* 254
Sciodaphyllum paniculatum IPNI=*90355-1* 253
Sciodaphyllum samydifolium IPNI=*1133937-2* 253
Scirpidium aciculare IPNI=*1169169-2* 231
Scirpus acicularis IPNI=*1169169-2* 231
Scirpus acutangulus IPNI=*157356-3* 231
Scirpus aestivalis IPNI=*307664-1* 232
Scirpus amentaceus IPNI=*1111946-2* 232
Scirpus argenteus IPNI=*307702-1* 232
Scirpus atrofactus IPNI=*1029981-2* 232
Scirpus atrosanguineus IPNI=*229810-2* 232
Scirpus autumnalis IPNI=*304701-1* 230
Scirpus bangalorensis IPNI=*220271-2* 232
Scirpus berlandieri IPNI=*38049-2* 229
Scirpus brownii IPNI=*1029981-2* 232
Scirpus californicus IPNI=*229810-2* 232
Scirpus californicus subsp. *tatora* IPNI=*229810-2* 232
Scirpus californicus var. *chamissoi* IPNI=*229810-2* 232
Scirpus californicus var. *spoliatus* IPNI=*229810-2* 232
Scirpus californicus var. *tatora* IPNI=*229810-2* 232
Scirpus capillaris IPNI=*1186511-2* 229
Scirpus capitatus IPNI=*1029981-2* 232
Scirpus caribaeus IPNI=*1029981-2* 232
Scirpus castaneus IPNI=*308274-1* 232
Scirpus cephalotes IPNI=*312067-1* 232
Scirpus cephalotes IPNI=*77140538-1* 231
Scirpus cernuus IPNI=*309238-1* 232
Scirpus chamissoi IPNI=*229810-2* 232
Scirpus complanatus IPNI=*1111946-2* 232
Scirpus congestus IPNI=*38049-2* 229
Scirpus constrictus IPNI=*1029981-2* 232
Scirpus corymbosus IPNI=*220271-2* 232
Scirpus cylindraceus IPNI=*308274-1* 232
Scirpus decipiens IPNI=*229810-2* 232
Scirpus depressus IPNI=*1029981-2* 232
Scirpus dichotomus IPNI=*307853-1* 232
Scirpus dichotomus IPNI=*308274-1* 232
Scirpus domingensis IPNI=*308274-1* 232
Scirpus elegans IPNI=*306803-1* 231
Scirpus elegans IPNI=*74851-2* 231
Scirpus fuirena IPNI=*308552-1* 232
Scirpus geniculatus IPNI=*1029981-2* 232
Scirpus globifer IPNI=*1019164-1* 232
Scirpus hispidulus IPNI=*912172-1* 229
Scirpus humboldtii IPNI=*38049-2* 229
Scirpus interstinctus IPNI=*1029981-2* 232
Scirpus interstinctus IPNI=*60444117-2* 232
Scirpus intricatus IPNI=*324890-2* 231
Scirpus junciformis f. *robustus* IPNI=*38049-2* 229
Scirpus junciformis IPNI=*38049-2* 229
Scirpus juncoides IPNI=*38049-2* 229
Scirpus kielmeyeri IPNI=*1111946-2* 232
Scirpus lacustris f. *pictus* IPNI=*1019164-1* 232
Scirpus lacustris subsp. *creber* IPNI=*1019164-1* 232
Scirpus lacustris var. *validus* IPNI=*1019164-1* 232
Scirpus lanatus IPNI=*30060098-2* 229
Scirpus lappaceus IPNI=*324890-2* 231
Scirpus lusulinus IPNI=*38049-2* 229
Scirpus luzulae IPNI=*74851-2* 231
Scirpus maritimus IPNI=*298094-1* 229
Scirpus monander IPNI=*307702-1* 232
Scirpus monostachyus IPNI=*104972-2* 232
Scirpus nanus IPNI=*307702-1* 232
Scirpus paschalis IPNI=*229810-2* 232
Scirpus plantagineus IPNI=*60444117-2* 232
Scirpus pseudotriqueter IPNI=*229810-2* 232
Scirpus rigidus IPNI=*229810-2* 232
Scirpus riparius IPNI=*229810-2* 232
Scirpus riparius var. *tereticulmis* IPNI=*229810-2* 232
Scirpus rivularis IPNI=*308552-1* 232
Scirpus schoenoides IPNI=*104972-2* 232
Scirpus siculus IPNI=*1019164-1* 232
Scirpus spadiceus IPNI=*308274-1* 232
Scirpus squarrosus IPNI=*308286-1* 232
Scirpus sterilis IPNI=*308274-1* 232
Scirpus tabernaemontani f. *luxurians* IPNI=*1019164-1* 232
Scirpus tabernaemontani IPNI=*1019164-1* 232
Scirpus tatora IPNI=*229810-2* 232
Scirpus tereticulmis IPNI=*229810-2* 232
Scirpus uliginosus IPNI=*1019164-1* 232
Scirpus umbellatus IPNI=*220271-2* 232
Scirpus umbellatus IPNI=*308552-1* 232
Scirpus validus f. *creber* IPNI=*1019164-1* 232
Scirpus validus IPNI=*1019164-1* 232
Scirpus validus IPNI=*229810-2* 232
Scirpus validus var. *creber* IPNI=*1019164-1* 232
Scirpus validus var. *laeviglumis* IPNI=*1019164-1* 232
Scirpus viridis IPNI=*306803-1* 231
Scirpus welwitschii IPNI=*1019164-1* 232
Sclarea decidua IPNI=*457495-1* 373
Sclarea rhodantha IPNI=*457495-1* 373
Sclarea verbenaca IPNI=*457495-1* 373
Sclarea viscosissima IPNI=*457495-1* 373
Scleranthus abyssinicus IPNI=*300494-2* 286
Scleranthus adsurgens IPNI=*300494-2* 286
Scleranthus aequidens IPNI=*300494-2* 286
Scleranthus anisodus IPNI=*300494-2* 286
Scleranthus argillophilus IPNI=*300494-2* 286
Scleranthus ascriviensis IPNI=*300494-2* 286
Scleranthus augustae IPNI=*300494-2* 286
Scleranthus banaticus IPNI=*300494-2* 286
Scleranthus biennis IPNI=*300494-2* 286
Scleranthus borborygius IPNI=*300494-2* 286
Scleranthus brachycarpus IPNI=*300494-2* 286
Scleranthus brachyodon IPNI=*300494-2* 286

Scleranthus brumalis IPNI=*300494-2* 286
Scleranthus byzantinus IPNI=*300494-2* 286
Scleranthus campestris IPNI=*300494-2* 286
Scleranthus canescens IPNI=*300494-2* 286
Scleranthus carthusianorum IPNI=*300494-2* 286
Scleranthus cattaroanus IPNI=*300494-2* 286
Scleranthus caudiger IPNI=*300494-2* 286
Scleranthus chaitophyllus IPNI=*300494-2* 286
Scleranthus chersonensis IPNI=*300494-2* 286
Scleranthus cinereus IPNI=*300494-2* 286
Scleranthus clevemontanus IPNI=*300494-2* 286
Scleranthus comatus IPNI=*300494-2* 286
Scleranthus comosus IPNI=*300494-2* 286
Scleranthus compactus IPNI=*300494-2* 286
Scleranthus concinnus IPNI=*300494-2* 286
Scleranthus crinitus IPNI=*300494-2* 286
Scleranthus crispatulus IPNI=*300494-2* 286
Scleranthus debilis IPNI=*300494-2* 286
Scleranthus decipiens IPNI=*300494-2* 286
Scleranthus densiramaeus IPNI=*300494-2* 286
Scleranthus diffusus IPNI=*300494-2* 286
Scleranthus dissitiflorus IPNI=*300494-2* 286
Scleranthus divaricatus IPNI=*300494-2* 286
Scleranthus durandoi IPNI=*300494-2* 286
Scleranthus echinophorus IPNI=*300494-2* 286
Scleranthus echinulifer IPNI=*300494-2* 286
Scleranthus engadineus IPNI=*300494-2* 286
Scleranthus expansus IPNI=*300494-2* 286
Scleranthus ferrugineus IPNI=*300494-2* 286
Scleranthus filiferricaulis IPNI=*300494-2* 286
Scleranthus floribundus IPNI=*300494-2* 286
Scleranthus folliotii IPNI=*300494-2* 286
Scleranthus fritzscheanus IPNI=*300494-2* 286
Scleranthus fuscus IPNI=*300494-2* 286
Scleranthus gemmulosus IPNI=*300494-2* 286
Scleranthus glomerifer IPNI=*300494-2* 286
Scleranthus graniti IPNI=*300494-2* 286
Scleranthus haepei IPNI=*300494-2* 286
Scleranthus hamatus IPNI=*300494-2* 286
Scleranthus havranus IPNI=*300494-2* 286
Scleranthus hoemeanus IPNI=*300494-2* 286
Scleranthus hohenackeri IPNI=*300494-2* 286
Scleranthus holanthus IPNI=*300494-2* 286
Scleranthus holubyi IPNI=*300494-2* 286
Scleranthus horridus IPNI=*300494-2* 286
Scleranthus implicatus IPNI=*300494-2* 286
Scleranthus inflexus IPNI=*300494-2* 286
Scleranthus intricatus IPNI=*300494-2* 286
Scleranthus johannis IPNI=*300494-2* 286
Scleranthus kessneri IPNI=*300494-2* 286
Scleranthus kittelii IPNI=*300494-2* 286
Scleranthus knazicanus IPNI=*300494-2* 286
Scleranthus kohlianus IPNI=*300494-2* 286
Scleranthus kotschyanus IPNI=*300494-2* 286
Scleranthus laxus IPNI=*300494-2* 286
Scleranthus leptodus IPNI=*300494-2* 286
Scleranthus leucoperos IPNI=*300494-2* 286
Scleranthus longidens IPNI=*300494-2* 286
Scleranthus longirameus IPNI=*300494-2* 286
Scleranthus longiramosus IPNI=*300494-2* 286
Scleranthus longisetus IPNI=*300494-2* 286
Scleranthus macilentus IPNI=*300494-2* 286
Scleranthus macronemus IPNI=*300494-2* 286
Scleranthus marginellus IPNI=*300494-2* 286
Scleranthus medusellus IPNI=*300494-2* 286
Scleranthus micranthus IPNI=*300494-2* 286
Scleranthus microdus IPNI=*300494-2* 286
Scleranthus myosoteidos IPNI=*300494-2* 286
Scleranthus myriocarpus IPNI=*300494-2* 286
Scleranthus obscurus IPNI=*300494-2* 286
Scleranthus obsoletus IPNI=*300494-2* 286
Scleranthus oligoconus IPNI=*300494-2* 286
Scleranthus orientalis IPNI=*300494-2* 286
Scleranthus orthocarpus IPNI=*300494-2* 286
Scleranthus orthoxys IPNI=*300494-2* 286
Scleranthus orzinii IPNI=*300494-2* 286
Scleranthus oxystephanus IPNI=*300494-2* 286
Scleranthus paganus IPNI=*300494-2* 286
Scleranthus paradoxus IPNI=*300494-2* 286
Scleranthus petronellae IPNI=*300494-2* 286
Scleranthus picturatus IPNI=*300494-2* 286
Scleranthus pityophilus IPNI=*300494-2* 286
Scleranthus polybotrys IPNI=*300494-2* 286
Scleranthus polyclados IPNI=*300494-2* 286
Scleranthus poscharskyi IPNI=*300494-2* 286
Scleranthus profusus IPNI=*300494-2* 286
Scleranthus ptyocarpus IPNI=*300494-2* 286
Scleranthus pycnocarpus IPNI=*300494-2* 286
Scleranthus reichenbachii IPNI=*300494-2* 286
Scleranthus remotiflorus IPNI=*300494-2* 286
Scleranthus riparius IPNI=*300494-2* 286
Scleranthus rusticus IPNI=*300494-2* 286
Scleranthus sanctus IPNI=*300494-2* 286
Scleranthus schimperi IPNI=*300494-2* 286
Scleranthus schroederi IPNI=*300494-2* 286
Scleranthus scoparius IPNI=*300494-2* 286
Scleranthus sekerae IPNI=*300494-2* 286
Scleranthus semiglaber IPNI=*300494-2* 286
Scleranthus serotinus IPNI=*300494-2* 286
Scleranthus serpyllaceus IPNI=*300494-2* 286
Scleranthus setibracteatus IPNI=*300494-2* 286
Scleranthus setifer IPNI=*300494-2* 286
Scleranthus sordidus IPNI=*300494-2* 286
Scleranthus soroanus IPNI=*300494-2* 286
Scleranthus sparsiflorus IPNI=*300494-2* 286
Scleranthus spectabilis IPNI=*300494-2* 286
Scleranthus stenoleucus IPNI=*300494-2* 286
Scleranthus stipatus IPNI=*300494-2* 286
Scleranthus striatipes IPNI=*300494-2* 286
Scleranthus subspathaceus IPNI=*300494-2* 286
Scleranthus sulcifer IPNI=*300494-2* 286
Scleranthus tauricus IPNI=*300494-2* 286
Scleranthus tauscheri IPNI=*300494-2* 286
Scleranthus tenellus IPNI=*300494-2* 286
Scleranthus tenuis IPNI=*300494-2* 286
Scleranthus theodori IPNI=*300494-2* 286
Scleranthus umagellus IPNI=*300494-2* 286
Scleranthus vagans IPNI=*300494-2* 286
Scleranthus vallis-demonis IPNI=*300494-2* 286
Scleranthus variflexus IPNI=*300494-2* 286
Scleranthus velebiticus IPNI=*300494-2* 286
Scleranthus verecundus IPNI=*300494-2* 286
Scleranthus visianii IPNI=*300494-2* 286
Scleria bracteata f. *simplicior* IPNI=*314601-1* 232
Scleria bracteata f. *supra-gynaecea* IPNI=*314601-1* 232
Scleria bracteata var. *floribunda* IPNI=*314601-1* 232
Scleria centralis IPNI=*314862-1* 233
Scleria enchroides IPNI=*329993-2* 232
Scleria fasciculata IPNI=*312067-1* 232
Scleria flagellum IPNI=*230730-2* 233
Scleria floribunda IPNI=*314601-1* 232
Scleria grandis IPNI=*320885-2* 233
Scleria hirtella var. *glabra* IPNI=*329993-2* 232
Scleria hirtella var. *pubescens* IPNI=*329993-2* 232
Scleria hirtella var. *strigosa* IPNI=*329993-2* 232
Scleria interrupta IPNI=*329993-2* 232

Senecio tabacon IPNI=*245177-1* 269
Senecio vaccinioides IPNI=*963618-1* 267
Senecio vaccinioides var. *pruinosus* IPNI=*963618-1* 267
Senecio valerianifolius IPNI=*203365-1* 261
Senecio verrucosus IPNI=*970931-1* 256
Senecio volubilis IPNI=*211361-2* 269
Senecio vukotinovicii IPNI=*1152557-2* 261
Senecio vulgarihumilis IPNI=*248251-1* 270
Senecio zollingerianus IPNI=*211219-1* 263
Seneciodes cinereum IPNI=*962413-1* 260
Senefeldera inclinata var. *macrophylla* IPNI=*1008822-1* 392
Senefeldera macrophylla IPNI=*1008822-1* 392
Senegalia angustissima IPNI=*712-2* 310
Senegalia cordobana IPNI=*234346-2* 332
Senegalia glomerosa IPNI=*518289-1* 332
Senegalia grenadensis IPNI=*234370-2* 332
Senegalia langlassei IPNI=*518289-1* 332
Senegalia lobana IPNI=*518289-1* 332
Senegalia multiflora IPNI=*989448-1* 311
Senegalia multipinnata IPNI=*234346-2* 332
Senegalia podadenia IPNI=*77085269-1* 332
Senegalia scandens IPNI=*234370-2* 332
Senegalia tomentella IPNI=*518289-1* 332
Senftenbergia moritziana IPNI=*315558-2* 448
Senkenbergia erecta IPNI=*319111-2* 286
Senna × floribunda var. *pubescens* IPNI=*234652-2* 333
Senna absus IPNI=*54310-2* 314
Senna aurata IPNI=*234652-2* 333
Senna esculenta IPNI=*518364-1* 333
Senna lanceolata IPNI=*518364-1* 333
Senna obovata IPNI=*518341-1* 332
Senna obtusa IPNI=*518341-1* 332
Senna occidentalis var. *sophera* IPNI=*518364-1* 333
Senna officinalis IPNI=*518364-1* 333
Senna orientalis IPNI=*1117728-2* 333
Senna purpurea IPNI=*518364-1* 333
Senna septemtrionalis var. *pubescens* IPNI=*234652-2* 333
Senna socotrana IPNI=*518364-1* 333
Senna sophera var. *purpurea* IPNI=*518364-1* 333
Senna sumatrana IPNI=*911410-1* 333
Senna tora var. *obtusifolia* IPNI=*234560-2* 333
Senna toroides IPNI=*234560-2* 333
Septas repens IPNI=*1072674-2* 376
Septilia repens IPNI=*1072674-2* 376
Sequoia pyramidata IPNI=*263741-1* 180
Sequoia religiosa IPNI=*263741-1* 180
Sequoia taxifolia IPNI=*263741-1* 180
Serapias diphylla IPNI=*617639-1* 217
Serapias purpurea IPNI=*1218040-2* 217
Serapias speciosa IPNI=*659300-1* 223
Serenoa arborescens IPNI=*663912-1* 203
Serenoa repens f. *glauca* IPNI=*941782-1* 212
Serenoa serrulata IPNI=*941782-1* 212
Sericandra lophantha IPNI=*912944-1* 329
Sericographis caripensis IPNI=*47308-1* 359
Sericographis ghiesbreghtiana IPNI=*133119-2* 360
Sericographis mohintli IPNI=*133119-2* 360
Sericographis neglecta IPNI=*133119-2* 360
Sericographis tinctoria IPNI=*907564-1* 360
Sericola brachiata IPNI=*1024004-2* 421
Seriola brasiliensis IPNI=*126428-2* 264
Seriola caespitosa IPNI=*225575-1* 264
Seriola taraxacifolia IPNI=*225575-1* 264
Seriola tweediei IPNI=*126428-2* 264
Seris conspicua IPNI=*174455-2* 267
Seris onoseroides IPNI=*174455-2* 267
Seris purpurea IPNI=*77181848-1* 267
Seris rupestris IPNI=*174455-2* 267
Serpicula verticillata IPNI=*114224-3* 203
Serpocaulon giganteum IPNI=*77183236-1* 175
Serrafalcus divaricatus IPNI=*394095-1* 235
Serrafalcus gmelinii IPNI=*394095-1* 235
Serrafalcus secalinus IPNI=*394095-1* 235
Serrafalcus unioloides IPNI=*37001-2* 235
Serratula cinerea IPNI=*962413-1* 260
Serratula spicata IPNI=*230634-1* 265
Sertuernera glauca IPNI=*191400-2* 283
Sertuernera iresinoides IPNI=*61055-1* 283
Sertuernera luzuliflora IPNI=*191400-2* 283
Sertuernera meridensis IPNI=*10378-2* 281
Sertula alba IPNI=*156656-2* 327
Sertula indica IPNI=*1207360-2* 327
Seruneum acapulcense IPNI=*30439090-2* 272
Seruneum frutescens IPNI=*60468294-2* 272
Seruneum paludosum IPNI=*1093589-2* 270
Seruneum trilobatum IPNI=*1093589-2* 270
Sesamum africanum IPNI=*675971-1* 375
Sesamum auriculatum IPNI=*675971-1* 375
Sesamum brasiliense IPNI=*675971-1* 375
Sesamum edule IPNI=*675971-1* 375
Sesamum foetidum IPNI=*675971-1* 375
Sesamum hopkinsii IPNI=*675971-1* 375
Sesamum indicum subsp. *malabaricum* IPNI=*675971-1* 375
Sesamum luteum IPNI=*675971-1* 375
Sesamum malabaricum IPNI=*675971-1* 375
Sesamum mulayanum IPNI=*675971-1* 375
Sesamum occidentalis IPNI=*675971-1* 375
Sesamum oleiferum IPNI=*675971-1* 375
Sesamum orientale IPNI=*675971-1* 375
Sesamum somalense IPNI=*675971-1* 375
Sesamum tavakarii IPNI=*675971-1* 375
Sesamum trifoliatum IPNI=*675971-1* 375
Sesbania cassioides IPNI=*284993-2* 334
Sesbania emerus IPNI=*284993-2* 334
Sesbania exaltata IPNI=*284993-2* 334
Sesbania laevigata IPNI=*1015399-2* 334
Sesbania macrocarpa var. *picta* IPNI=*284993-2* 334
Sesbania mannii IPNI=*518476-1* 334
Sesbania occidentalis IPNI=*284993-2* 334
Sesbania occidentalis var. *multijuga* IPNI=*284993-2* 334
Sesbania occidentalis var. *pubescens* IPNI=*284993-2* 334
Sesbania pubescens IPNI=*1015399-2* 334
Seseli ammi IPNI=*840899-1* 252
Seseli ammoides IPNI=*837486-1* 251
Seseli dulce IPNI=*842680-1* 252
Seseli foeniculum IPNI=*842680-1* 252
Seseli graveolens IPNI=*838067-1* 251
Seseli piperitum IPNI=*842680-1* 252
Sesuvium acutifolium IPNI=*235145-2* 280
Sesuvium brevifolium IPNI=*235145-2* 280
Sesuvium distylum IPNI=*235145-2* 280
Sesuvium edule IPNI=*235145-2* 280
Sesuvium longifolium IPNI=*235145-2* 280
Sesuvium microphylium IPNI=*235145-2* 280
Sesuvium ortegae IPNI=*235145-2* 280
Sesuvium pedunculatum IPNI=*235145-2* 280
Sesuvium portulaca IPNI=*235145-2* 280
Sesuvium portulacastrum f. *albiflorum* IPNI=*235145-2* 280
Sesuvium portulacastrum var. *pedunculatum* IPNI=*235145-2* 280
Sesuvium portulacastrum var. *subsessile* IPNI=*235145-2* 280
Sesuvium repens IPNI=*235145-2* 280
Sesuvium revolutifolium IPNI=*235145-2* 280
Sesuvium sessiliflorum IPNI=*235145-2* 280
Sesuvium spathulatum IPNI=*235145-2* 280
Sesuvium uvifolium IPNI=*235145-2* 280
Setachna cyanus IPNI=*190310-1* 259

Setaria adhaerens subsp. *verticillata* IPNI=*421857-1* 245
Setaria adhaerens var. *ambigua* IPNI=*421857-1* 245
Setaria adhaerens var. *antrorsa* IPNI=*421857-1* 245
Setaria affinis IPNI=*943811-1* 245
Setaria alopecurus IPNI=*1182687-2* 245
Setaria ambigua IPNI=*421857-1* 245
Setaria ambigua IPNI=*943811-1* 245
Setaria aparine IPNI=*421428-1* 245
Setaria asiatica IPNI=*1207294-2* 245
Setaria barretoi IPNI=*943811-1* 245
Setaria basiseta IPNI=*1172068-2* 245
Setaria berteroniana IPNI=*943811-1* 245
Setaria bongaensis IPNI=*1172068-2* 245
Setaria boninensis IPNI=*421764-1* 245
Setaria brachiariiformis IPNI=*399655-1* 238
Setaria breviglumis IPNI=*421764-1* 245
Setaria brevispica IPNI=*421428-1* 245
Setaria californica IPNI=*1207294-2* 245
Setaria carnei IPNI=*421857-1* 245
Setaria caudata var. *pauciflora* IPNI=*235234-2* 245
Setaria compacta IPNI=*1207294-2* 245
Setaria composita IPNI=*1182687-2* 245
Setaria conglomerata IPNI=*421857-1* 245
Setaria crus-ardeae IPNI=*421828-1* 245
Setaria dasyura IPNI=*421764-1* 245
Setaria decipiens f. *latifolia* IPNI=*421857-1* 245
Setaria decipiens f. *major* IPNI=*421857-1* 245
Setaria decipiens f. *schultheissii* IPNI=*421857-1* 245
Setaria decipiens subf. *ramiflora* IPNI=*421857-1* 245
Setaria decipiens var. *schulteissii* IPNI=*421857-1* 245
Setaria depauperata IPNI=*421428-1* 245
Setaria discolor IPNI=*943811-1* 245
Setaria disticha IPNI=*77144198-1* 244
Setaria effusa IPNI=*421711-1* 245
Setaria elliottiana IPNI=*51261-2* 235
Setaria erythraeae IPNI=*421764-1* 245
Setaria erythrosperma IPNI=*1207294-2* 245
Setaria fieldingii IPNI=*943811-1* 245
Setaria flaccifolia IPNI=*1172068-2* 245
Setaria flava IPNI=*943811-1* 245
Setaria flava var. *pumila* IPNI=*943811-1* 245
Setaria flavida IPNI=*1207294-2* 245
Setaria floriana IPNI=*943811-1* 245
Setaria floribunda IPNI=*421857-1* 245
Setaria geminata var. *paludivaga* IPNI=*979871-1* 245
Setaria geniculata IPNI=*943811-1* 245
Setaria geniculata var. *latifolia* IPNI=*943811-1* 245
Setaria geniculata var. *purpurascens* IPNI=*943811-1* 245
Setaria germanica IPNI=*1207294-2* 245
Setaria gibbosa IPNI=*235234-2* 245
Setaria glauca IPNI=*77105978-1* 235
Setaria glauca subsp. *subtesselata* IPNI=*421764-1* 245
Setaria glauca var. *elongata* IPNI=*421764-1* 245
Setaria glauca var. *geniculata* IPNI=*943811-1* 245
Setaria glauca var. *imberbis* IPNI=*943811-1* 245
Setaria glauca var. *laevigata* IPNI=*943811-1* 245
Setaria glauca var. *minutissima* IPNI=*421764-1* 245
Setaria glauca var. *penicillata* IPNI=*943811-1* 245
Setaria glauca var. *purpurascens* IPNI=*943811-1* 245
Setaria globularis IPNI=*1207294-2* 245
Setaria gracilipes IPNI=*1172068-2* 245
Setaria gracilis f. *breviglumis* IPNI=*943811-1* 245
Setaria gracilis f. *longiseta* IPNI=*943811-1* 245
Setaria gracilis IPNI=*943811-1* 245
Setaria gracilis var. *glaucocaesia* IPNI=*943811-1* 245
Setaria gracilis var. *pilosissima* IPNI=*943811-1* 245
Setaria gracilis var. *purpurascens* IPNI=*943811-1* 245
Setaria gussonei IPNI=*421857-1* 245
Setaria helvola IPNI=*1207294-2* 245
Setaria hirtella IPNI=*1004002-1* 242
Setaria humifusa IPNI=*421764-1* 245
Setaria imberbis f. *brevispica* IPNI=*943811-1* 245
Setaria imberbis f. *flaviseta* IPNI=*943811-1* 245
Setaria imberbis f. *longiseta* IPNI=*943811-1* 245
Setaria imberbis f. *radicans* IPNI=*943811-1* 245
Setaria imberbis f. *setislongioribus* IPNI=*943811-1* 245
Setaria imberbis IPNI=*943811-1* 245
Setaria imberbis var. *gracilis* IPNI=*943811-1* 245
Setaria imberbis var. *purpurascens* IPNI=*943811-1* 245
Setaria inopinata IPNI=*235234-2* 245
Setaria italica subsp. *colchica* IPNI=*1207294-2* 245
Setaria italica subsp. *moharia* IPNI=*1207294-2* 245
Setaria italica subsp. *stramineofructa* IPNI=*1207294-2* 245
Setaria italica subvar. *densior* IPNI=*1207294-2* 245
Setaria italica subvar. *germanica* IPNI=*1207294-2* 245
Setaria itieri IPNI=*1207294-2* 245
Setaria japonica IPNI=*1207294-2* 245
Setaria kwamouthensis IPNI=*1172068-2* 245
Setaria laeta IPNI=*421764-1* 245
Setaria laevigata IPNI=*943811-1* 245
Setaria laevis IPNI=*319034-2* 245
Setaria lancifolia IPNI=*1182687-2* 245
Setaria leiantha f. *subhirsuta* IPNI=*421857-1* 245
Setaria lenis IPNI=*421711-1* 245
Setaria liebmannii f. *trichorhachis* IPNI=*1182687-2* 245
Setaria lutescens f. *longiseta* IPNI=*943811-1* 245
Setaria lutescens f. *pallens* IPNI=*943811-1* 245
Setaria lutescens f. *pumila* IPNI=*421764-1* 245
Setaria macrochaeta IPNI=*1207294-2* 245
Setaria macrostachya IPNI=*1182687-2* 245
Setaria maritima IPNI=*1207294-2* 245
Setaria mauritiana IPNI=*1172068-2* 245
Setaria maxima IPNI=*1182687-2* 245
Setaria melinis IPNI=*1207294-2* 245
Setaria membranifolia IPNI=*319034-2* 245
Setaria meyeri IPNI=*77144198-1* 244
Setaria multiseta IPNI=*1207294-2* 245
Setaria nakaiana IPNI=*1172068-2* 245
Setaria nubica IPNI=*421857-1* 245
Setaria ocreata IPNI=*1004002-1* 242
Setaria pallide-fusca IPNI=*421764-1* 245
Setaria panicea IPNI=*1048074-2* 244
Setaria paniculifera IPNI=*421711-1* 245
Setaria panis IPNI=*1207294-2* 245
Setaria parviflora var. *brachytricha* IPNI=*943811-1* 245
Setaria parviflora var. *pilosissima* IPNI=*943811-1* 245
Setaria penicillata IPNI=*943811-1* 245
Setaria pennicillata IPNI=*421764-1* 245
Setaria pennisetum IPNI=*77144198-1* 244
Setaria perennis IPNI=*943811-1* 245
Setaria persica IPNI=*1207294-2* 245
Setaria pilosa IPNI=*77144198-1* 244
Setaria polystachya IPNI=*235234-2* 245
Setaria pratensis IPNI=*421857-1* 245
Setaria pseudoverticillata IPNI=*319034-2* 245
Setaria pseudoverticillata IPNI=*421857-1* 245
Setaria pumila subsp. *pallide-fusca* IPNI=*421764-1* 245
Setaria pumila subsp. *subtesselata* IPNI=*421764-1* 245
Setaria purpurascens IPNI=*943811-1* 245
Setaria rariflora IPNI=*319037-2* 245
Setaria respiciens IPNI=*421428-1* 245
Setaria reversipila IPNI=*1182687-2* 245
Setaria rhachitricha IPNI=*1172068-2* 245
Setaria rottleri IPNI=*421428-1* 245
Setaria rubiginosa IPNI=*421764-1* 245
Setaria rufa IPNI=*77105978-1* 235

Setaria schraderi IPNI=*77144198-1* 244
Setaria sciuroidea IPNI=*421764-1* 245
Setaria sericea IPNI=*77105978-1* 235
Setaria streptobotrys IPNI=*943811-1* 245
Setaria subcordata IPNI=*426280-1* 246
Setaria sulcata IPNI=*421711-1* 245
Setaria tejucensis IPNI=*943811-1* 245
Setaria tenella IPNI=*943811-1* 245
Setaria teysmannii IPNI=*421857-1* 245
Setaria thollonii IPNI=*1172068-2* 245
Setaria trichorhachis IPNI=*1182687-2* 245
Setaria umbrosa IPNI=*426280-1* 246
Setaria ustilata IPNI=*421764-1* 245
Setaria verticillata f. *ambigua* IPNI=*421857-1* 245
Setaria verticillata f. *arenosa* IPNI=*421857-1* 245
Setaria verticillata subsp. *adhaerens* IPNI=*421428-1* 245
Setaria verticillata var. *ambigua* IPNI=*421857-1* 245
Setaria verticillata var. *font-queri* IPNI=*421428-1* 245
Setaria verticilliformis IPNI=*421857-1* 245
Setaria violacea IPNI=*1207294-2* 245
Setaria viridis var. *ambigua* IPNI=*421857-1* 245
Setaria viscidula IPNI=*1172068-2* 245
Setaria vulpina IPNI=*943811-1* 245
Setaria vulpiseta var. *reversipila* IPNI=*1182687-2* 245
Setaria yucatana IPNI=*319034-2* 245
Setariopsis glauca IPNI=*77105978-1* 235
Setariopsis italica IPNI=*1207294-2* 245
Setariopsis verticillata IPNI=*421857-1* 245
Setcreasea jaumavensis IPNI=*256037-2* 224
Setcreasea lanceolata IPNI=*256037-2* 224
Setcreasea pallida IPNI=*256037-2* 224
Setcreasea purpurea IPNI=*256037-2* 224
Seticereus roezlii IPNI=*35149-2* 283
Setilobus vicentinus IPNI=*152688-2* 364
Shallonium serrulatum IPNI=*317388-2* 299
Sickingia japurensis IPNI=*236591-2* 357
Sickingia japurensis var. *glaberrima* IPNI=*236591-2* 357
Sickingia maxonii IPNI=*236581-2* 357
Sickingia tinctoria IPNI=*236591-2* 357
Sickingia xanthostema IPNI=*236591-2* 357
Sicyomorpha pruinosa IPNI=*190402-2* 291
Sicyos depauperatus IPNI=*294002-1* 295
Sicyos fauriei IPNI=*293413-1* 295
Sicyos laciniatus IPNI=*293946-1* 295
Sicyos quinquelobatus IPNI=*294002-1* 295
Sida abutiloides IPNI=*1055043-2* 406
Sida acuta var. *stipulata* IPNI=*329574-2* 414
Sida adjusta IPNI=*235798-2* 413
Sida alba IPNI=*235798-2* 413
Sida albida IPNI=*1101632-2* 406
Sida alpestris IPNI=*235979-2* 414
Sida americana IPNI=*1055043-2* 406
Sida amplexicaulis IPNI=*120630-2* 409
Sida amplexicaulis IPNI=*36556-2* 407
Sida amplissima IPNI=*268896-2* 415
Sida andicola IPNI=*235798-2* 413
Sida angustifolia IPNI=*235838-2* 413
Sida angustifolia var. *major* IPNI=*235838-2* 413
Sida angustissima IPNI=*563888-1* 413
Sida arabica IPNI=*563387-1* 413
Sida arbuscula IPNI=*235798-2* 413
Sida arguta IPNI=*329579-2* 413
Sida aristata IPNI=*563600-1* 413
Sida arnottiana IPNI=*1026852-2* 406
Sida arnottiana IPNI=*1101632-2* 406
Sida arrudiana IPNI=*235518-2* 413
Sida asiatica IPNI=*1101632-2* 406
Sida atrosanguinea IPNI=*235979-2* 414
Sida aurea IPNI=*1101632-2* 406
Sida avicularioides IPNI=*313702-2* 413
Sida balbisiana IPNI=*329574-2* 414
Sida bastardia IPNI=*77146518-1* 407
Sida bellidifolia IPNI=*313702-2* 413
Sida beloere IPNI=*238-2* 406
Sida berlandieri IPNI=*235518-2* 413
Sida berteroana IPNI=*329545-2* 413
Sida berteroana IPNI=*329574-2* 414
Sida betonicifolia IPNI=*235838-2* 413
Sida betonicifolia IPNI=*563387-1* 413
Sida betulifolia IPNI=*1063417-2* 410
Sida betulifolia IPNI=*561999-1* 410
Sida betulina IPNI=*235518-2* 413
Sida bicolor IPNI=*235838-2* 413
Sida bicuspidata IPNI=*235838-2* 413
Sida biflora IPNI=*323515-2* 406
Sida bodinieri IPNI=*235518-2* 413
Sida boivinii IPNI=*564343-1* 414
Sida borbonica IPNI=*563600-1* 413
Sida boriara IPNI=*235838-2* 413
Sida brachypetala IPNI=*329574-2* 414
Sida bracteolata IPNI=*329545-2* 413
Sida bradei IPNI=*235518-2* 413
Sida brasila IPNI=*235518-2* 413
Sida breviflora IPNI=*564343-1* 414
Sida brevipes IPNI=*77146518-1* 407
Sida brevipubens IPNI=*563741-1* 413
Sida buettneriacea IPNI=*235977-2* 414
Sida byssina IPNI=*563600-1* 413
Sida caespitosa IPNI=*563387-1* 413
Sida callifera IPNI=*563820-1* 413
Sida campii IPNI=*563888-1* 413
Sida canariensis IPNI=*235798-2* 413
Sida canescens IPNI=*235798-2* 413
Sida capensis IPNI=*235518-2* 413
Sida capensis IPNI=*235798-2* 413
Sida capillaris IPNI=*235979-2* 414
Sida capitata IPNI=*561424-1* 410
Sida carnea IPNI=*133833-2* 410
Sida carpinifolia f. *antilliana* IPNI=*235518-2* 413
Sida carpinifolia f. *balbisiana* IPNI=*329574-2* 414
Sida carpinifolia f. *spiraeifolia* IPNI=*235518-2* 413
Sida carpinifolia IPNI=*329574-2* 414
Sida carpinifolia var. *acuta* IPNI=*235518-2* 413
Sida carpinifolia var. *brevicuspidata* IPNI=*329574-2* 414
Sida carpinoides IPNI=*1063417-2* 410
Sida cearensis IPNI=*329579-2* 413
Sida centrota IPNI=*1218005-2* 407
Sida chanetii IPNI=*235518-2* 413
Sida ciliaris var. *fasciculata* IPNI=*313702-2* 413
Sida ciliaris var. *fulva* IPNI=*313702-2* 413
Sida ciliaris var. *mexicana* IPNI=*313702-2* 413
Sida ciliosa IPNI=*563600-1* 413
Sida circinata IPNI=*77146518-1* 407
Sida collina IPNI=*563820-1* 413
Sida commixta IPNI=*235518-2* 413
Sida compressa IPNI=*235798-2* 413
Sida conferta IPNI=*563600-1* 413
Sida conferta IPNI=*564343-1* 414
Sida congensis IPNI=*564343-1* 414
Sida contracta IPNI=*268909-2* 415
Sida cordifolia IPNI=*1101632-2* 406
Sida cordifolia var. *conferta* IPNI=*563600-1* 413
Sida coromandelina IPNI=*1063417-2* 410
Sida coronata IPNI=*1101632-2* 406
Sida corymbosa IPNI=*563820-1* 413
Sida costata IPNI=*563820-1* 413

Sida crassa IPNI=*235518-2* 413
Sida crassifolia IPNI=*1055043-2* 406
Sida crispa IPNI=*120630-2* 409
Sida cristata IPNI=*1218005-2* 407
Sida cubensis IPNI=*564343-1* 414
Sida dasytricha IPNI=*564343-1* 414
Sida debilis IPNI=*564343-1* 414
Sida decagyna IPNI=*563600-1* 413
Sida deltoidea IPNI=*1218005-2* 407
Sida densiflora IPNI=*564343-1* 414
Sida diffusa IPNI=*563387-1* 413
Sida diffusa var. *setosa* IPNI=*563387-1* 413
Sida dilleniana IPNI=*1218005-2* 407
Sida disticha IPNI=*235518-2* 413
Sida domingensis IPNI=*564343-1* 414
Sida dumosa IPNI=*945985-1* 407
Sida dumosa var. *fendleri* IPNI=*945985-1* 407
Sida editorum IPNI=*563387-1* 413
Sida emarginata IPNI=*235838-2* 413
Sida endlicheriana IPNI=*329579-2* 413
Sida erosa IPNI=*313702-2* 413
Sida eteromischos IPNI=*1101632-2* 406
Sida fasciculata IPNI=*313702-2* 413
Sida fasciculata IPNI=*329579-2* 413
Sida fasciculiflora IPNI=*329579-2* 413
Sida fauriei IPNI=*1063417-2* 410
Sida fiebrigii IPNI=*563888-1* 413
Sida filicaulis IPNI=*563387-1* 413
Sida filicaulis var. *setosa* IPNI=*563387-1* 413
Sida filiformis IPNI=*120630-2* 409
Sida filiformis IPNI=*563387-1* 413
Sida floribunda IPNI=*235979-2* 414
Sida foetida IPNI=*77146518-1* 407
Sida foliosa IPNI=*235518-2* 413
Sida forsteri IPNI=*235798-2* 413
Sida frutescens IPNI=*235518-2* 413
Sida frutescens IPNI=*235798-2* 413
Sida fruticosa IPNI=*185868-2* 412
Sida fryxellii IPNI=*235798-2* 413
Sida fulva IPNI=*313702-2* 413
Sida garckeana IPNI=*235518-2* 413
Sida geminiflora IPNI=*323515-2* 406
Sida glabra IPNI=*235518-2* 413
Sida glandulosa IPNI=*235838-2* 413
Sida glechomifolia IPNI=*563387-1* 413
Sida glutinosa var. *antiguensis* IPNI=*563741-1* 413
Sida glutinosa var. *cinerea* IPNI=*563741-1* 413
Sida gorgonea IPNI=*235798-2* 413
Sida gracilis IPNI=*120630-2* 409
Sida graminifolia IPNI=*563888-1* 413
Sida grandifolia IPNI=*1026852-2* 406
Sida grandior IPNI=*235518-2* 413
Sida grata IPNI=*235798-2* 413
Sida graveolens IPNI=*238-2* 406
Sida guilleminiana IPNI=*1101632-2* 406
Sida hamulosa IPNI=*563600-1* 413
Sida hastata IPNI=*1218005-2* 407
Sida herbacea IPNI=*563600-1* 413
Sida heterosperma IPNI=*268896-2* 415
Sida hirta IPNI=*238-2* 406
Sida hispida IPNI=*563387-1* 413
Sida holosericea IPNI=*563600-1* 413
Sida hondensis IPNI=*235798-2* 413
Sida hongkongensis IPNI=*563600-1* 413
Sida hookeri IPNI=*1101632-2* 406
Sida humboldtii IPNI=*235979-2* 414
Sida ibarrensis IPNI=*323515-2* 406
Sida imberbis IPNI=*120630-2* 409
Sida incerta IPNI=*235798-2* 413
Sida indica IPNI=*1101632-2* 406
Sida indica IPNI=*120630-2* 409
Sida insperata IPNI=*329579-2* 413
Sida insularis IPNI=*235798-2* 413
Sida involucrata IPNI=*313702-2* 413
Sida jacquinii IPNI=*1055043-2* 406
Sida jaliscensis IPNI=*313702-2* 413
Sida jamaicensis IPNI=*1063417-2* 410
Sida jamaicensis IPNI=*235518-2* 413
Sida kohautiana IPNI=*235798-2* 413
Sida kohautiana var. *corchorifolia* IPNI=*564227-1* 413
Sida lancea IPNI=*235518-2* 413
Sida lanceolata IPNI=*235518-2* 413
Sida lanceolata IPNI=*329574-2* 414
Sida lasiostega IPNI=*120630-2* 409
Sida laxiflora IPNI=*268896-2* 415
Sida leiophylla IPNI=*945985-1* 407
Sida leschenaultiana IPNI=*268909-2* 415
Sida leucanthema IPNI=*268909-2* 415
Sida lignosa IPNI=*1055043-2* 406
Sida linearifolia IPNI=*563888-1* 413
Sida longifolia IPNI=*563888-1* 413
Sida longistipula IPNI=*313702-2* 413
Sida luciana IPNI=*268909-2* 415
Sida magdalenae IPNI=*77146518-1* 407
Sida malifolia IPNI=*235518-2* 413
Sida malvacea IPNI=*186058-2* 412
Sida martinicensis IPNI=*235518-2* 413
Sida martiusiana IPNI=*235518-2* 413
Sida meridionalis IPNI=*1101632-2* 406
Sida mexicana IPNI=*1218005-2* 407
Sida micans IPNI=*563600-1* 413
Sida micrantha IPNI=*235977-2* 414
Sida microtricha IPNI=*313702-2* 413
Sida milleri IPNI=*235838-2* 413
Sida minor IPNI=*235838-2* 413
Sida minutifolia IPNI=*313702-2* 413
Sida mocinoana IPNI=*133833-2* 410
Sida mollis IPNI=*1026852-2* 406
Sida mollis IPNI=*563600-1* 413
Sida montana IPNI=*120630-2* 409
Sida mucronulata IPNI=*1063417-2* 410
Sida multiflora IPNI=*563600-1* 413
Sida muricata IPNI=*313702-2* 413
Sida nervosa IPNI=*563741-1* 413
Sida nudata IPNI=*235798-2* 413
Sida oahuensis IPNI=*1063417-2* 410
Sida obtusa IPNI=*329574-2* 414
Sida olens IPNI=*329579-2* 413
Sida ononidifolia IPNI=*313702-2* 413
Sida orientalis IPNI=*235518-2* 413
Sida ovata IPNI=*235518-2* 413
Sida ovata IPNI=*563387-1* 413
Sida paniculata IPNI=*235979-2* 414
Sida paniculata IPNI=*268896-2* 415
Sida paniculata var. *floribunda* IPNI=*235979-2* 414
Sida paniculata var. *normalis* IPNI=*235979-2* 414
Sida paniculata var. *rufescens* IPNI=*235979-2* 414
Sida pannosa IPNI=*77146518-1* 407
Sida parens IPNI=*235518-2* 413
Sida parviflora IPNI=*563741-1* 413
Sida patula IPNI=*329545-2* 413
Sida pauciflora IPNI=*268896-2* 415
Sida pellita IPNI=*235979-2* 414
Sida periplocifolia IPNI=*1213890-2* 415
Sida periplocifolia var. *peruviana* IPNI=*1213890-2* 415
Sida periplocifolia var. *zeylanica* IPNI=*1213890-2* 415

Sida philippica IPNI=*235798-2* 413
Sida phlebococca IPNI=*235977-2* 414
Sida picta IPNI=*77130215-1* 407
Sida pilosa IPNI=*157002-2* 412
Sida pilosa IPNI=*238-2* 406
Sida pilosa IPNI=*563387-1* 413
Sida pimpinellifolia IPNI=*235838-2* 413
Sida planicaulis IPNI=*235518-2* 413
Sida plumosa IPNI=*313702-2* 413
Sida pohliana IPNI=*564227-1* 413
Sida polyantha IPNI=*268896-2* 415
Sida polycarpa IPNI=*1101632-2* 406
Sida populifolia IPNI=*1101632-2* 406
Sida portoricensis IPNI=*563600-1* 413
Sida praelonga IPNI=*235798-2* 413
Sida pringlei IPNI=*235798-2* 413
Sida procumbens IPNI=*563387-1* 413
Sida prostrata IPNI=*235518-2* 413
Sida pubescens IPNI=*1101632-2* 406
Sida pungens IPNI=*563600-1* 413
Sida pyramidata IPNI=*945985-1* 407
Sida quinqueangulata IPNI=*1218005-2* 407
Sida quinqueloba IPNI=*1218005-2* 407
Sida racemosa IPNI=*268896-2* 415
Sida racemosa IPNI=*563741-1* 413
Sida radiata IPNI=*561446-1* 410
Sida rajoides IPNI=*157023-2* 412
Sida ramosa IPNI=*563387-1* 413
Sida recisa IPNI=*235798-2* 413
Sida repanda IPNI=*235518-2* 413
Sida retrofracta IPNI=*120630-2* 409
Sida retusa IPNI=*235838-2* 413
Sida retusifolia IPNI=*235798-2* 413
Sida rhombifolia subsp. *insularis* IPNI=*235798-2* 413
Sida rhombifolia var. *insularis* IPNI=*235798-2* 413
Sida rhombifolia var. *minor* IPNI=*235798-2* 413
Sida rhombifolia var. *poeppigiana* IPNI=*969172-1* 413
Sida rhombifolia var. *pohliana* IPNI=*564227-1* 413
Sida rhombifolia var. *surinamensis* IPNI=*564227-1* 413
Sida rhomboidea IPNI=*235798-2* 413
Sida rotundifolia IPNI=*563600-1* 413
Sida rubra IPNI=*235977-2* 414
Sida ruderata IPNI=*235798-2* 413
Sida rugosa IPNI=*235518-2* 413
Sida sabeana IPNI=*157004-2* 412
Sida scabra IPNI=*235838-2* 413
Sida schmitzii IPNI=*133329-2* 410
Sida schrankii IPNI=*235518-2* 413
Sida schweinfurthii IPNI=*235979-2* 414
Sida scoparia IPNI=*235518-2* 413
Sida scoparia IPNI=*235798-2* 413
Sida semicrenata IPNI=*235798-2* 413
Sida sericea IPNI=*1026852-2* 406
Sida sessiliflora IPNI=*564343-1* 414
Sida sessilis IPNI=*120630-2* 409
Sida sinensis IPNI=*235798-2* 413
Sida sordida IPNI=*77146518-1* 407
Sida spicata IPNI=*36556-2* 407
Sida spiciflora IPNI=*36556-2* 407
Sida spinosa f. *albiflora* IPNI=*235838-2* 413
Sida spinosa var. *angustifolia* IPNI=*235838-2* 413
Sida spinosa var. *kazmii* IPNI=*235838-2* 413
Sida spiraeifolia IPNI=*235518-2* 413
Sida stauntaniana IPNI=*235518-2* 413
Sida stipulata IPNI=*235798-2* 413
Sida stipulata IPNI=*329574-2* 414
Sida striata IPNI=*77130224-1* 407
Sida subdistans IPNI=*235838-2* 413
Sida suberosa IPNI=*1218005-2* 407
Sida subspicata IPNI=*563600-1* 413
Sida sundensis IPNI=*268896-2* 415
Sida supina IPNI=*563387-1* 413
Sida surinamensis IPNI=*564227-1* 413
Sida swartzii IPNI=*329579-2* 413
Sida tenax IPNI=*329579-2* 413
Sida tenuicaulis IPNI=*235838-2* 413
Sida tetracocca IPNI=*268896-2* 415
Sida tomentosa IPNI=*1055043-2* 406
Sida tomentosa IPNI=*238-2* 406
Sida tomentosa IPNI=*563600-1* 413
Sida triangularis IPNI=*1218005-2* 407
Sida trichoda IPNI=*120630-2* 409
Sida tricuspidata IPNI=*1055043-2* 406
Sida tridentata IPNI=*313702-2* 413
Sida trinervia IPNI=*235798-2* 413
Sida trivialis IPNI=*329574-2* 414
Sida troyana IPNI=*235798-2* 413
Sida truncata IPNI=*235838-2* 413
Sida truncata IPNI=*563600-1* 413
Sida truncatula IPNI=*235798-2* 413
Sida ulmifolia IPNI=*1063417-2* 410
Sida ulmifolia IPNI=*235838-2* 413
Sida ulmifolia IPNI=*329579-2* 413
Sida velloziana IPNI=*563600-1* 413
Sida velutina IPNI=*563600-1* 413
Sida velutinifolia IPNI=*563741-1* 413
Sida verruculata IPNI=*329579-2* 413
Sida versatilis IPNI=*235518-2* 413
Sida verticillata IPNI=*564343-1* 414
Sida vesicaria IPNI=*1101632-2* 406
Sida vestita IPNI=*563600-1* 413
Sida villosa IPNI=*157043-2* 412
Sida viminea IPNI=*563888-1* 413
Sida viridis IPNI=*329545-2* 413
Sida viscidula IPNI=*329579-2* 413
Sida viscosa IPNI=*77146518-1* 407
Sida vogelii IPNI=*235518-2* 413
Sida waltheriifolia IPNI=*563600-1* 413
Sida willdenowii IPNI=*329579-2* 413
Sida zelotes IPNI=*235518-2* 413
Sideritis bullata IPNI=*137860-2* 370
Sideritis salviifolia IPNI=*137894-2* 370
Sideroxylon acutangulum IPNI=*160358-2* 306
Sideroxylon amygdalicarpum IPNI=*948062-1* 306
Sideroxylon bangii IPNI=*948066-1* 306
Sideroxylon calophylloides IPNI=*160393-2* 306
Sideroxylon calophylloides IPNI=*160417-2* 306
Sideroxylon campestre IPNI=*209714-2* 306
Sideroxylon colombianum IPNI=*77176719-1* 308
Sideroxylon cuspidatum IPNI=*209738-2* 306
Sideroxylon cuspidatum var. *crassifolium* IPNI=*209738-2* 306
Sideroxylon cuspidatum var. *ellipticum* IPNI=*209738-2* 306
Sideroxylon durum IPNI=*209738-2* 306
Sideroxylon egense IPNI=*160374-2* 306
Sideroxylon elegans IPNI=*209758-2* 306
Sideroxylon elegans var. *micranthum* IPNI=*209758-2* 306
Sideroxylon ellipticum IPNI=*209738-2* 306
Sideroxylon gardnerianum IPNI=*787792-1* 306
Sideroxylon gardnerianum var. *glabratum* IPNI=*787792-1* 306
Sideroxylon guianense IPNI=*160417-2* 306
Sideroxylon guyanense IPNI=*160386-2* 306
Sideroxylon hondurense IPNI=*209864-2* 307
Sideroxylon meyeri IPNI=*209864-2* 307
Sideroxylon micranthum IPNI=*209758-2* 306
Sideroxylon robustum IPNI=*209738-2* 306
Sideroxylon rugosum var. *egense* IPNI=*160374-2* 306

Sideroxylon sapota IPNI=*1110421-2* 307
Sideroxylon ulei IPNI=*160374-2* 306
Sideroxylon uniloculare IPNI=*209864-2* 307
Sigesbeckia bogotensis IPNI=*249077-1* 270
Sigesbeckia brachiata IPNI=*249084-1* 270
Sigesbeckia caspia IPNI=*249084-1* 270
Sigesbeckia cordifolia IPNI=*249077-1* 270
Sigesbeckia droseroides IPNI=*249084-1* 270
Sigesbeckia esquirolii IPNI=*249084-1* 270
Sigesbeckia glutinosa IPNI=*249084-1* 270
Sigesbeckia gracilis IPNI=*249084-1* 270
Sigesbeckia humilis IPNI=*249084-1* 270
Sigesbeckia iberica IPNI=*249084-1* 270
Sigesbeckia melanolepsis IPNI=*249077-1* 270
Sigesbeckia microcephala IPNI=*249084-1* 270
Sigesbeckia orientalis f. *angustifolia* IPNI=*249084-1* 270
Sigesbeckia orientalis var. *tenggerensis* IPNI=*249084-1* 270
Sigesbeckia portoriccensis IPNI=*202984-1* 261
Sigesbeckia triangularis IPNI=*249084-1* 270
Sigmatostalix radicans IPNI=*77101300-1* 219
Sigmoidotropis grandiflora IPNI=*60457240-2* 320
Sigmoidotropis spixiana IPNI=*60457232-2* 311
Silene agrestina IPNI=*332106-2* 286
Silene agricola IPNI=*332106-2* 286
Silene anglica IPNI=*332106-2* 286
Silene anglica var. *quinquevulnera* IPNI=*332106-2* 286
Silene armeria f. *albiflora* IPNI=*77093328-1* 285
Silene armeria IPNI=*77093328-1* 285
Silene armeria var. *sparsiflora* IPNI=*77093328-1* 285
Silene arvensis IPNI=*332106-2* 286
Silene barbata IPNI=*306674-2* 285
Silene berdaui IPNI=*77093328-1* 285
Silene bohemica IPNI=*332106-2* 286
Silene caesia IPNI=*153408-1* 285
Silene candollei IPNI=*332106-2* 286
Silene canescens IPNI=*332106-2* 286
Silene cantabrica IPNI=*332106-2* 286
Silene caryophylla IPNI=*153205-1* 285
Silene cerastoides IPNI=*332106-2* 286
Silene chia IPNI=*332106-2* 286
Silene chilensis IPNI=*332106-2* 286
Silene clusii IPNI=*332106-2* 286
Silene colorata IPNI=*332106-2* 286
Silene congesta IPNI=*332106-2* 286
Silene conica IPNI=*332106-2* 286
Silene cruentata IPNI=*332106-2* 286
Silene cuneifolia IPNI=*332106-2* 286
Silene distycha IPNI=*332106-2* 286
Silene elegans IPNI=*154453-1* 285
Silene gallica subsp. *sylvestris* IPNI=*332106-2* 286
Silene gallica var. *quinquevulnera* IPNI=*332106-2* 286
Silene glandulosa IPNI=*332106-2* 286
Silene glauca IPNI=*77093328-1* 285
Silene glomerata IPNI=*332106-2* 286
Silene haenkeana IPNI=*332106-2* 286
Silene hirsuta IPNI=*332106-2* 286
Silene hispida IPNI=*332106-2* 286
Silene jucunda IPNI=*332106-2* 286
Silene latifolia IPNI=*77093328-1* 285
Silene linifolia IPNI=*332106-2* 286
Silene linophila IPNI=*332106-2* 286
Silene littoralis IPNI=*332106-2* 286
Silene littorea IPNI=*332106-2* 286
Silene lituanica IPNI=*77093328-1* 285
Silene lusitanica IPNI=*332106-2* 286
Silene marginata IPNI=*332106-2* 286
Silene micropetala IPNI=*332106-2* 286
Silene minutiflora IPNI=*332106-2* 286
Silene mixta IPNI=*77093328-1* 285
Silene modesta IPNI=*332106-2* 286
Silene myloptera IPNI=*332106-2* 286
Silene occitanica IPNI=*332106-2* 286
Silene oliveriana IPNI=*332106-2* 286
Silene pachyphylla IPNI=*332106-2* 286
Silene parvula IPNI=*332106-2* 286
Silene plumaria IPNI=*153717-1* 285
Silene pygmaea IPNI=*332106-2* 286
Silene quinquevulnera IPNI=*332106-2* 286
Silene reflexa IPNI=*332106-2* 286
Silene rigidula IPNI=*332106-2* 286
Silene rosella IPNI=*332106-2* 286
Silene rostrata IPNI=*332106-2* 286
Silene sabuletorum IPNI=*332106-2* 286
Silene sardoa IPNI=*332106-2* 286
Silene sciotica IPNI=*332106-2* 286
Silene semiglabra IPNI=*332106-2* 286
Silene spatulata IPNI=*332106-2* 286
Silene subleopoliensis IPNI=*77093328-1* 285
Silene suboccultata IPNI=*332106-2* 286
Silene sylvestris IPNI=*332106-2* 286
Silene tergestina IPNI=*332106-2* 286
Silene transtagana IPNI=*332106-2* 286
Silene tricuspidata IPNI=*332106-2* 286
Silene tridentata IPNI=*332106-2* 286
Silene umbellata IPNI=*77093328-1* 285
Silene undulatifolia IPNI=*332106-2* 286
Silene vaga IPNI=*310755-2* 285
Silene variegata IPNI=*77093328-1* 285
Silene vernalis IPNI=*332106-2* 286
Siler japonicum IPNI=*60442790-2* 252
Silphium edule IPNI=*237510-2* 270
Silphium helianthoides IPNI=*213534-1* 264
Silphium solidaginoides IPNI=*213534-1* 264
Silphium trilobatum IPNI=*1093589-2* 270
Silvia anacardioides IPNI=*158649-2* 183
Silvia curialis IPNI=*97540-2* 375
Silvia ita-uba IPNI=*158649-2* 183
Silybum intermedium IPNI=*249211-1* 270
Silybum leucanthum IPNI=*249211-1* 270
Silybum maculatum IPNI=*249211-1* 270
Silybum mariae IPNI=*249211-1* 270
Silybum marianum subsp. *anatolicum* IPNI=*249211-1* 270
Silybum marianum var. *albiflorum* IPNI=*249211-1* 270
Silybum pygmaeum IPNI=*249211-1* 270
Simira erythroxylon var. *saxicola* IPNI=*236591-2* 357
Simira nitida IPNI=*60473461-2* 355
Simira palicourea IPNI=*759121-1* 354
Sinabraca juncea IPNI=*60442520-2* 276
Sinapis arvensis IPNI=*217826-2* 277
Sinapis auriculata IPNI=*279485-1* 276
Sinapis benghalensis IPNI=*222144-2* 277
Sinapis bracteolata IPNI=*279422-1* 276
Sinapis brassica IPNI=*60442520-2* 276
Sinapis campestris IPNI=*279485-1* 276
Sinapis caspica IPNI=*999093-1* 276
Sinapis cernua IPNI=*60442520-2* 276
Sinapis chinensis IPNI=*60442520-2* 276
Sinapis communis IPNI=*279485-1* 276
Sinapis cuneifolia IPNI=*60442520-2* 276
Sinapis dichotoma IPNI=*279485-1* 276
Sinapis divaricata IPNI=*222144-2* 277
Sinapis eruca IPNI=*1137625-2* 277
Sinapis erysimoides IPNI=*279422-1* 276
Sinapis exotica IPNI=*1137625-2* 277
Sinapis foliosa IPNI=*288952-1* 278
Sinapis glauca IPNI=*279419-1* 276

Sinapis gorraea IPNI=*279422-1* 276
Sinapis integrifolia IPNI=*60442520-2* 276
Sinapis juncea IPNI=*60442520-2* 276
Sinapis juncea var. *napiformis* IPNI=*60442520-2* 276
Sinapis lanceolata IPNI=*60442520-2* 276
Sinapis napus IPNI=*279419-1* 276
Sinapis nigra IPNI=*279422-1* 276
Sinapis oleracea IPNI=*60442520-2* 276
Sinapis oliveriana IPNI=*289144-1* 278
Sinapis orgyalis IPNI=*279422-1* 276
Sinapis patens IPNI=*222144-2* 277
Sinapis patens IPNI=*60442520-2* 276
Sinapis pekinensis IPNI=*279485-1* 276
Sinapis pekinensis var. *cai-ben* IPNI=*279485-1* 276
Sinapis pekinensis var. *pe-tsai* IPNI=*279485-1* 276
Sinapis persoonii IPNI=*279422-1* 276
Sinapis pusilla IPNI=*222144-2* 277
Sinapis ramosa IPNI=*60442520-2* 276
Sinapis rapa IPNI=*279485-1* 276
Sinapis raphanistrum IPNI=*217826-2* 277
Sinapis rugosa IPNI=*60442520-2* 276
Sinapis sinensis IPNI=*60442520-2* 276
Sinapis tenella IPNI=*60442520-2* 276
Sinapis timoriana IPNI=*60442520-2* 276
Sinapis torulosa IPNI=*279422-1* 276
Sinapis trilocularis IPNI=*279485-1* 276
Sinapis tuberosa IPNI=*279485-1* 276
Sinapis turgida IPNI=*279422-1* 276
Sinapistrum giganteum IPNI=*147057-1* 279
Sinapistrum pentaphyllum IPNI=*60518-2* 279
Sinapistrum viscosum IPNI=*320133-2* 280
Sinningia discolor IPNI=*382786-1* 368
Sinningia floribunda IPNI=*382786-1* 368
Sinningia menziesiana IPNI=*382786-1* 368
Sinningia regina IPNI=*382786-1* 368
Sinningia warszewiczii IPNI=*237093-2* 368
Sinocalamus beecheyanus IPNI=*392260-1* 234
Sinocalamus flagellifer IPNI=*397862-1* 237
Sinocalamus oldhamii f. *revolutus* IPNI=*392445-1* 234
Sinocalamus oldhamii IPNI=*392445-1* 234
Siparuna amazonica IPNI=*582354-1* 186
Siparuna amplifolia IPNI=*582223-1* 185
Siparuna andina IPNI=*582392-1* 186
Siparuna apicifera IPNI=*582257-1* 185
Siparuna apicifera var. *acutissima* IPNI=*582257-1* 185
Siparuna archeri IPNI=*582285-1* 185
Siparuna argyrochrysea IPNI=*582248-1* 185
Siparuna arianeae IPNI=*582285-1* 185
Siparuna budlejifolia IPNI=*582257-1* 185
Siparuna camporum IPNI=*582285-1* 185
Siparuna cauliflora IPNI=*582342-1* 185
Siparuna cavalcantei IPNI=*582285-1* 185
Siparuna chiridota IPNI=*582392-1* 186
Siparuna colimensis IPNI=*582392-1* 186
Siparuna crassiflora IPNI=*582248-1* 185
Siparuna dasyantha IPNI=*582392-1* 186
Siparuna davillifolia IPNI=*582392-1* 186
Siparuna diandra IPNI=*582392-1* 186
Siparuna discolor IPNI=*582285-1* 185
Siparuna domatiata IPNI=*582392-1* 186
Siparuna duckeana IPNI=*582285-1* 185
Siparuna ecuadorica IPNI=*582223-1* 185
Siparuna elliptica IPNI=*582223-1* 185
Siparuna emarginata IPNI=*237245-2* 185
Siparuna eriocalyx IPNI=*582223-1* 185
Siparuna espinhacensis IPNI=*582354-1* 186
Siparuna exsculpta IPNI=*582223-1* 185
Siparuna foetida IPNI=*582285-1* 185
Siparuna geniculata IPNI=*582392-1* 186
Siparuna gigantophylla IPNI=*582223-1* 185
Siparuna gilgiana IPNI=*582392-1* 186
Siparuna grisea IPNI=*582392-1* 186
Siparuna guianensis var. *glabrescens* IPNI=*582285-1* 185
Siparuna guianensis var. *longifolia* IPNI=*582285-1* 185
Siparuna guianensis var. *nitens* IPNI=*582285-1* 185
Siparuna harongifolia IPNI=*582223-1* 185
Siparuna heteropoda IPNI=*582392-1* 186
Siparuna holtonii IPNI=*582257-1* 185
Siparuna insculpta IPNI=*582223-1* 185
Siparuna itacaiunensis IPNI=*582285-1* 185
Siparuna killipii IPNI=*582257-1* 185
Siparuna lagopus IPNI=*582392-1* 186
Siparuna lepidantha IPNI=*582253-1* 185
Siparuna lepidiflora IPNI=*582248-1* 185
Siparuna limoniodora IPNI=*582223-1* 185
Siparuna loretensis IPNI=*582392-1* 186
Siparuna lyrata IPNI=*582223-1* 185
Siparuna macra IPNI=*582392-1* 186
Siparuna macrophylla IPNI=*582223-1* 185
Siparuna magnifica IPNI=*582373-1* 186
Siparuna manarae IPNI=*582392-1* 186
Siparuna metensis IPNI=*582392-1* 186
Siparuna microphylla IPNI=*582392-1* 186
Siparuna mollicoma IPNI=*582373-1* 186
Siparuna mollis IPNI=*582373-1* 186
Siparuna nicaraguensis IPNI=*582392-1* 186
Siparuna nigra IPNI=*582392-1* 186
Siparuna oblongifolia IPNI=*582354-1* 186
Siparuna olivaceovelutina IPNI=*582257-1* 185
Siparuna panamensis IPNI=*582285-1* 185
Siparuna paniculata IPNI=*90974-2* 183
Siparuna parviflora IPNI=*582392-1* 186
Siparuna phaneroneura IPNI=*582257-1* 185
Siparuna pittieri IPNI=*582373-1* 186
Siparuna podocarpa IPNI=*582223-1* 185
Siparuna pubancura IPNI=*582392-1* 186
Siparuna pyricarpa IPNI=*582223-1* 185
Siparuna quadrangularis IPNI=*582223-1* 185
Siparuna radiata IPNI=*582373-1* 186
Siparuna rimbachii IPNI=*582223-1* 185
Siparuna riparia IPNI=*582392-1* 186
Siparuna riparia var. *macrophylla* IPNI=*582392-1* 186
Siparuna riparia var. *sumichrastii* IPNI=*582392-1* 186
Siparuna savanicola IPNI=*582285-1* 185
Siparuna spectabilis IPNI=*582223-1* 185
Siparuna sprucei IPNI=*582354-1* 186
Siparuna steleandra IPNI=*582373-1* 186
Siparuna stellulata var. *manarae* IPNI=*582392-1* 186
Siparuna sumichrastii IPNI=*582392-1* 186
Siparuna tabacifolia IPNI=*582223-1* 185
Siparuna tapatana IPNI=*582392-1* 186
Siparuna ternata IPNI=*582392-1* 186
Siparuna tetradenia IPNI=*582392-1* 186
Siparuna tulasnei IPNI=*582223-1* 185
Siparuna ucayaliensis IPNI=*582248-1* 185
Siparuna valenzuelae IPNI=*582257-1* 185
Siparuna venezuelensis IPNI=*582392-1* 186
Siparuna verticillata IPNI=*582392-1* 186
Siphocampylus acuminatus IPNI=*142908-1* 273
Siphocampylus acuminatus IPNI=*236827-2* 273
Siphocampylus andropogon IPNI=*51553-2* 272
Siphocampylus asper IPNI=*145335-1* 273
Siphocampylus coccineus IPNI=*77083518-1* 272
Siphocampylus cordifolius IPNI=*145082-1* 273
Siphocampylus ellipticus var. *funckeanus* IPNI=*145166-1* 273
Siphocampylus ellipticus var. *peritornus* IPNI=*145166-1* 273

Siphocampylus erianthus IPNI=*141468-1* 272
Siphocampylus eximius IPNI=*145335-1* 273
Siphocampylus funckeanus IPNI=*145166-1* 273
Siphocampylus funckeanus var. *subcanus* IPNI=*145166-1* 273
Siphocampylus giganteus f. *brachyodon* IPNI=*145189-1* 273
Siphocampylus giganteus var. *angustifolius* IPNI=*145189-1* 273
Siphocampylus lanatus IPNI=*145280-1* 273
Siphocampylus longipes IPNI=*141634-1* 273
Siphocampylus macranthus IPNI=*51553-2* 272
Siphocampylus mollis IPNI=*145312-1* 273
Siphocampylus pennellii IPNI=*145280-1* 273
Siphocampylus peritornus IPNI=*145166-1* 273
Siphocampylus planchonis f. *flaviflorus* IPNI=*145312-1* 273
Siphocampylus planchonis IPNI=*145312-1* 273
Siphocampylus planchonis var. *daweanus* IPNI=*145312-1* 273
Siphocampylus planchonis var. *polyphyllus* IPNI=*145312-1* 273
Siphocampylus planchonis var. *truxillensis* IPNI=*145312-1* 273
Siphocampylus radicans IPNI=*77083518-1* 272
Siphocampylus reflexifolius IPNI=*145335-1* 273
Siphocampylus retrorsus var. *asper* IPNI=*145335-1* 273
Siphocampylus retrorsus var. *eximius* IPNI=*145335-1* 273
Siphocampylus retrorsus var. *semiasper* IPNI=*145335-1* 273
Siphocampylus roseus IPNI=*77083518-1* 272
Siphocampylus rugosus IPNI=*77083518-1* 272
Siphocampylus spectabilis IPNI=*51553-2* 272
Siphocampylus surinamensis IPNI=*51553-2* 272
Siphonanthus elasticus IPNI=*121724-2* 391
Siphoneranthemum atropurpureum IPNI=*77158889-1* 361
Siphoneranthemum eldorado IPNI=*77158889-1* 361
Siphoneranthemum moorei IPNI=*77158889-1* 361
Siphoneranthemum nervosum IPNI=*48879-1* 360
Siphoneranthemum reticulatum IPNI=*77158889-1* 361
Siphoneranthemum tricolor IPNI=*77158889-1* 361
Siphoneranthemum versicolor IPNI=*77158889-1* 361
Siphoneugena cantareirae IPNI=*599327-1* 427
Siphoneugena micrantha IPNI=*599327-1* 427
Siphonia brasiliensis IPNI=*349913-1* 391
Siphonia cahuchu IPNI=*121724-2* 391
Siphonia discolor IPNI=*349947-1* 391
Siphonia elastica IPNI=*121724-2* 391
Siphonia guianensis IPNI=*121724-2* 391
Siphonia janeirensis IPNI=*349913-1* 391
Siphonia pauciflora IPNI=*349941-1* 391
Siphonia ridleyana IPNI=*349913-1* 391
Siphonia rigidifolia IPNI=*349944-1* 391
Siphonia rivalis IPNI=*998196-1* 347
Siphonia spruceana IPNI=*349947-1* 391
Siphotria squamosa IPNI=*218583-2* 251
Sismondaea dioscoreoides IPNI=*323920-2* 226
Sison anisum IPNI=*846658-1* 252
Sison fasciculatum IPNI=*837486-1* 251
Sison haenkei IPNI=*837486-1* 251
Sison lateriflorum IPNI=*837486-1* 251
Sison majus IPNI=*837486-1* 251
Sison peregrinum IPNI=*60442790-2* 252
Sison pumilum IPNI=*837486-1* 251
Sison ranunculifolium IPNI=*840899-1* 252
Sison ruta IPNI=*838067-1* 251
Sison trifidum IPNI=*838067-1* 251
Sisymbrianthus palustris IPNI=*288669-1* 277
Sisymbrianthus pusillus IPNI=*288669-1* 277
Sisymbrium amarum IPNI=*287534-1* 277
Sisymbrium amphibium IPNI=*288669-1* 277
Sisymbrium apetalum IPNI=*222144-2* 277
Sisymbrium atrovirens IPNI=*222144-2* 277
Sisymbrium barbareifolium IPNI=*288669-1* 277
Sisymbrium brachypetalum IPNI=*289144-1* 278
Sisymbrium capillaceum IPNI=*288669-1* 277
Sisymbrium cardaminefolium IPNI=*287534-1* 277
Sisymbrium dissectum IPNI=*288669-1* 277
Sisymbrium dubium IPNI=*288735-1* 277
Sisymbrium erucifolium IPNI=*237940-2* 278
Sisymbrium erysimastrum IPNI=*237940-2* 278
Sisymbrium filiforme IPNI=*237940-2* 278
Sisymbrium fluviatile IPNI=*287534-1* 277
Sisymbrium gangeticum IPNI=*222144-2* 277
Sisymbrium glabrum IPNI=*237940-2* 278
Sisymbrium hispidum IPNI=*288669-1* 277
Sisymbrium hungaricum IPNI=*289144-1* 278
Sisymbrium hybridum IPNI=*288669-1* 277
Sisymbrium indicum IPNI=*222144-2* 277
Sisymbrium irioides IPNI=*237940-2* 278
Sisymbrium latifolium IPNI=*237940-2* 278
Sisymbrium mandonii IPNI=*284416-2* 277
Sisymbrium mexicanum IPNI=*288669-1* 277
Sisymbrium nasturtium IPNI=*287534-1* 277
Sisymbrium nasturtium-aquaticum IPNI=*287534-1* 277
Sisymbrium niagarense IPNI=*289589-1* 278
Sisymbrium nigrum IPNI=*279422-1* 276
Sisymbrium obtusangulum IPNI=*289144-1* 278
Sisymbrium officinale subsp. *leiocarpum* IPNI=*289589-1* 278
Sisymbrium officinale var. *leiocarpum* IPNI=*289589-1* 278
Sisymbrium officinarum IPNI=*289589-1* 278
Sisymbrium palustre IPNI=*288669-1* 277
Sisymbrium pannonicum IPNI=*289144-1* 278
Sisymbrium pinnatifidum IPNI=*237940-2* 278
Sisymbrium pusillum IPNI=*288669-1* 277
Sisymbrium ramulosum IPNI=*237940-2* 278
Sisymbrium rigidulum IPNI=*289144-1* 278
Sisymbrium setosum IPNI=*999093-1* 276
Sisymbrium sinapifolium IPNI=*222144-2* 277
Sisymbrium sinapios IPNI=*289144-1* 278
Sisymbrium sinapis IPNI=*222144-2* 277
Sisymbrium sinapistrum IPNI=*289144-1* 278
Sisymbrium subulatum IPNI=*237940-2* 278
Sisymbrium sylvestre IPNI=*288669-1* 277
Sisyrinchium berteroi IPNI=*330256-2* 216
Sisyrinchium bogotense IPNI=*441310-1* 216
Sisyrinchium bulbosum IPNI=*89940-2* 216
Sisyrinchium capitatum IPNI=*89940-2* 216
Sisyrinchium congestum IPNI=*89940-2* 216
Sisyrinchium dichroum IPNI=*330260-2* 216
Sisyrinchium elatum IPNI=*89940-2* 216
Sisyrinchium fimbriatum IPNI=*330260-2* 216
Sisyrinchium fluminense IPNI=*441642-1* 217
Sisyrinchium galaxioides IPNI=*441642-1* 217
Sisyrinchium graminifolium IPNI=*330256-2* 216
Sisyrinchium grandiflorum IPNI=*441553-1* 216
Sisyrinchium guatemalense IPNI=*285111-2* 216
Sisyrinchium herrerae IPNI=*330256-2* 216
Sisyrinchium iridifolium IPNI=*330260-2* 216
Sisyrinchium iridifolium subsp. *valdivianum* IPNI=*330256-2* 216
Sisyrinchium latifolium IPNI=*89940-2* 216
Sisyrinchium luteum IPNI=*285111-2* 216
Sisyrinchium micranthemum IPNI=*330260-2* 216
Sisyrinchium micranthum f. *eburneochraceum* IPNI=*330260-2* 216
Sisyrinchium micranthum f. *flavum* IPNI=*330260-2* 216
Sisyrinchium micranthum f. *luteum* IPNI=*330260-2* 216
Sisyrinchium micranthum f. *purpureum* IPNI=*330260-2* 216
Sisyrinchium micranthum subsp. *scudiculare* IPNI=*330260-2* 216
Sisyrinchium micranthum subsp. *valdivianum* IPNI=*330256-2* 216
Sisyrinchium moritzianum IPNI=*440470-1* 216
Sisyrinchium palmifolium IPNI=*89940-2* 216
Sisyrinchium palmifolium var. *congestum* IPNI=*89940-2* 216
Sisyrinchium plicatum IPNI=*89940-2* 216
Sisyrinchium quinquevulnerum IPNI=*330256-2* 216

Sisyrinchium racemosum IPNI=*89940-2* 216
Sisyrinchium rigidum IPNI=*441310-1* 216
Sisyrinchium uniflorum IPNI=*330256-2* 216
Sisyrinchium valdivianum IPNI=*330256-2* 216
Sisyrinchium vesiculatum IPNI=*285111-2* 216
Sisyrinchium vulgare IPNI=*330256-2* 216
Sitobolium cicutarium IPNI=*17081410-1* 175
Sitobolium rubiginosum IPNI=*17081410-1* 175
Sitodium altile IPNI=*582598-1* 436
Sitodium cauliflorum IPNI=*850389-1* 436
Sitodium macrocarpon IPNI=*582622-1* 436
Sium apium IPNI=*838067-1* 251
Sium conium IPNI=*840668-1* 252
Sium coriandrum IPNI=*840760-1* 252
Sium graveolens IPNI=*838067-1* 251
Sium oppositifolium IPNI=*60442790-2* 252
Sium petroselinum IPNI=*60442790-2* 252
Sium visnaga IPNI=*849956-1* 253
Sixalix atropurpurea IPNI=*319717-1* 296
Sixalix atropurpurea subsp. *amansii* IPNI=*319717-1* 296
Skiophila pulchella IPNI=*378695-1* 366
Sloanea alnifolia IPNI=*835145-1* 431
Sloanea alnifolia var. *lancea* IPNI=*835145-1* 431
Sloanea alnifolia var. *ovalis* IPNI=*835145-1* 431
Sloanea caudata IPNI=*835157-1* 431
Sloanea chrysocarpa IPNI=*140493-2* 380
Sloanea cuneifolia IPNI=*835145-1* 431
Sloanea fasciculata IPNI=*835147-1* 431
Sloanea filiformis IPNI=*835186-1* 431
Sloanea gladysiae IPNI=*835157-1* 431
Sloanea guianensis var. *microcarpa* IPNI=*835145-1* 431
Sloanea lasiocoma IPNI=*835147-1* 431
Sloanea longicuspis IPNI=*111855-1* 380
Sloanea macrantha IPNI=*835194-1* 431
Sloanea macropoda IPNI=*237492-2* 431
Sloanea megacarpa IPNI=*835194-1* 431
Sloanea microcarpa IPNI=*835145-1* 431
Sloanea microcephala IPNI=*237492-2* 431
Sloanea monosperma IPNI=*835147-1* 431
Sloanea monosperma var. *coriacea* IPNI=*835147-1* 431
Sloanea monosperma var. *hirsutissima* IPNI=*835147-1* 431
Sloanea monosperma var. *ovalis* IPNI=*835147-1* 431
Sloanea monosperma var. *pulverulenta* IPNI=*835147-1* 431
Sloanea monosperma var. *virgata* IPNI=*835147-1* 431
Sloanea obtusa IPNI=*237456-2* 431
Sloanea ochrocarpa IPNI=*835147-1* 431
Sloanea porphyrocarpa IPNI=*324587-1* 431
Sloanea ptariana IPNI=*835145-1* 431
Sloanea pubescens IPNI=*835145-1* 431
Sloanea pulverulenta IPNI=*835147-1* 431
Sloanea quadrivalvis IPNI=*324587-1* 431
Sloanea regelii IPNI=*835145-1* 431
Sloanea riparia IPNI=*835147-1* 431
Sloanea sinemariensis var. *melinonis* IPNI=*237404-2* 431
Smilax acuminata IPNI=*315317-2* 226
Smilax aequatorialis IPNI=*541854-1* 226
Smilax balbisiana IPNI=*541406-1* 226
Smilax barbillana IPNI=*541685-1* 226
Smilax bernhardii IPNI=*541685-1* 226
Smilax berteroi IPNI=*541406-1* 226
Smilax canaliculata IPNI=*541406-1* 226
Smilax caudata IPNI=*541406-1* 226
Smilax chiapensis IPNI=*541406-1* 226
Smilax chiriquensis IPNI=*541685-1* 226
Smilax colubrina IPNI=*541406-1* 226
Smilax costaricae IPNI=*541862-1* 227
Smilax cumanensis IPNI=*541862-1* 227
Smilax domingensis IPNI=*541406-1* 226
Smilax domingensis var. *microscola* IPNI=*541406-1* 226
Smilax domingensis var. *sagrana* IPNI=*541406-1* 226
Smilax duidae IPNI=*541854-1* 226
Smilax engleriana IPNI=*541406-1* 226
Smilax eucalyptifolia IPNI=*541406-1* 226
Smilax febrifuga IPNI=*541778-1* 226
Smilax febrifuga var. *aequatoris* IPNI=*541778-1* 226
Smilax flavescens IPNI=*315317-2* 226
Smilax floribunda IPNI=*541388-1* 226
Smilax floribunda IPNI=*541406-1* 226
Smilax gaumeri IPNI=*541862-1* 227
Smilax gilgiana IPNI=*541685-1* 226
Smilax gilva IPNI=*541406-1* 226
Smilax graciliflora IPNI=*541778-1* 226
Smilax grandifolia IPNI=*541778-1* 226
Smilax immersa IPNI=*541837-1* 226
Smilax insignis IPNI=*541778-1* 226
Smilax jauaensis IPNI=*541406-1* 226
Smilax krukovii IPNI=*541406-1* 226
Smilax kunthii IPNI=*541406-1* 226
Smilax lasseriana IPNI=*541406-1* 226
Smilax latipes IPNI=*541837-1* 226
Smilax luculenta IPNI=*541862-1* 227
Smilax lundellii IPNI=*541862-1* 227
Smilax macrophylla IPNI=*541778-1* 226
Smilax mazatlanensis IPNI=*541862-1* 227
Smilax mexicana IPNI=*541862-1* 227
Smilax mexicana var. *costaricae* IPNI=*541862-1* 227
Smilax microscola IPNI=*541406-1* 226
Smilax multiflora IPNI=*541406-1* 226
Smilax munda IPNI=*541862-1* 227
Smilax obtusa IPNI=*541862-1* 227
Smilax officinalis IPNI=*315317-2* 226
Smilax panamensis IPNI=*541778-1* 226
Smilax papyracea IPNI=*315317-2* 226
Smilax parvifolia IPNI=*541406-1* 226
Smilax poeppigii IPNI=*541778-1* 226
Smilax pseudosyphilitica IPNI=*541837-1* 226
Smilax pseudosyphilitica var. *foliosa* IPNI=*541837-1* 226
Smilax ramonensis IPNI=*541778-1* 226
Smilax reticulata IPNI=*541406-1* 226
Smilax robusta IPNI=*541406-1* 226
Smilax ruiziana IPNI=*541778-1* 226
Smilax scabriuscula IPNI=*541862-1* 227
Smilax scabriuscula var. *fendleri* IPNI=*541862-1* 227
Smilax schlechtendalii IPNI=*541406-1* 226
Smilax schlechtendalii var. *lindenii* IPNI=*541406-1* 226
Smilax schomburgkiana var. *foliosa* IPNI=*541837-1* 226
Smilax schomburgkiana var. *gracilis* IPNI=*541837-1* 226
Smilax siphilitica var. *aequatorialis* IPNI=*541854-1* 226
Smilax staminea f. *obtusata* IPNI=*541406-1* 226
Smilax staminea IPNI=*541406-1* 226
Smilax standleyi IPNI=*541685-1* 226
Smilax syphilitica IPNI=*315317-2* 226
Smilax syphilitica IPNI=*541837-1* 226
Smilax tonduzii IPNI=*541685-1* 226
Smilax vaga IPNI=*541862-1* 227
Smilax vanilliodora IPNI=*541685-1* 226
Smilax wagneriana IPNI=*541862-1* 227
Smilax williamsii IPNI=*541862-1* 227
Smithia bigeminata IPNI=*30026493-2* 337
Smyrnium laterale IPNI=*838067-1* 251
Soaresia ilicifolia IPNI=*59683-2* 437
Soaresia nitida IPNI=*851229-1* 437
Sobralia aurantiaca IPNI=*657915-1* 223
Sobralia chlorantha IPNI=*657915-1* 223
Sobralia eublepharis IPNI=*657888-1* 223
Sobralia neglecta IPNI=*657877-1* 223

Sobralia panamensis IPNI=*657877-1* 223
Sobralia panamensis var. *albiflos* IPNI=*657877-1* 223
Sobralia rolfeana IPNI=*657915-1* 223
Socratea albolineata IPNI=*1200702-2* 212
Socratea durissima IPNI=*1200702-2* 212
Socratea elegans IPNI=*1200702-2* 212
Socratea gracilis IPNI=*1200702-2* 212
Socratea hoppii IPNI=*1200702-2* 212
Socratea macrochlamys IPNI=*1200702-2* 212
Socratea montana IPNI=*238168-2* 212
Socratea orbignyana IPNI=*1200702-2* 212
Socratea philonotia IPNI=*1200702-2* 212
Soja max IPNI=*60450240-2* 320
Solanoides laevis IPNI=*221479-2* 287
Solanoides pubescens IPNI=*221479-2* 287
Solanoides undulata IPNI=*221479-2* 287
Solanum mucronatum IPNI=*113943-1* 273
Solea anomala IPNI=*77143573-1* 405
Solea calceolaria IPNI=*77143549-1* 405
Solea circaeoides IPNI=*77143566-1* 405
Solea ipecacuanha IPNI=*77143549-1* 405
Solea itoubou IPNI=*77143549-1* 405
Solea microphylla IPNI=*77143568-1* 405
Solea oppositifolia IPNI=*77143566-1* 405
Solea parietarifolia IPNI=*77143566-1* 405
Solea parviflora IPNI=*77143568-1* 405
Solea parvifolia IPNI=*77143568-1* 405
Solea urticifolia IPNI=*77143566-1* 405
Solea viscidula IPNI=*77143566-1* 405
Soleirolia corsica IPNI=*856545-1* 448
Soleirolia repens IPNI=*856545-1* 448
Solena armata IPNI=*1115268-2* 356
Solena calycina IPNI=*764017-1* 356
Solena dumetorum IPNI=*745904-1* 348
Solena floribunda IPNI=*745904-1* 348
Solena humboldtiana IPNI=*764273-1* 356
Solena latifolia IPNI=*1079607-2* 355
Solena longiflora IPNI=*761025-1* 355
Solena longispina IPNI=*745904-1* 348
Solena nitida IPNI=*764273-1* 356
Solena nutans IPNI=*745904-1* 348
Solena randia IPNI=*763958-1* 356
Solena ruiziana IPNI=*222509-2* 356
Solena speciosa IPNI=*222509-2* 356
Solenochasma assurgens IPNI=*1114607-2* 359
Solenomelus chimboracensis IPNI=*440470-1* 216
Solenomelus gladioloides IPNI=*440470-1* 216
Solenopsis longiflora IPNI=*1133957-2* 273
Solenostemon blumei IPNI=*446165-1* 369
Solenostemon paniculatus IPNI=*446113-1* 369
Solenostemon scutellarioides IPNI=*446165-1* 369
Solenotheca tenella IPNI=*252097-1* 271
Solidago bonariensis IPNI=*249473-1* 270
Solidago cleliae IPNI=*249575-1* 270
Solidago coquimbana IPNI=*249473-1* 270
Solidago deflexa IPNI=*249575-1* 270
Solidago dumetorum IPNI=*249575-1* 270
Solidago fragrans IPNI=*249575-1* 270
Solidago fruticosa IPNI=*256729-1* 272
Solidago gigantea subsp. *serotina* IPNI=*249575-1* 270
Solidago gigantea var. *leiophylla* IPNI=*249575-1* 270
Solidago gigantea var. *pitcheri* IPNI=*249575-1* 270
Solidago gigantea var. *serotina* IPNI=*249575-1* 270
Solidago gigantea var. *shinnersii* IPNI=*249575-1* 270
Solidago glabra IPNI=*249575-1* 270
Solidago linearifolia IPNI=*249473-1* 270
Solidago linearifolia var. *brachypoda* IPNI=*249473-1* 270
Solidago linearifolia var. *poeppigii* IPNI=*249473-1* 270
Solidago marginella IPNI=*249473-1* 270
Solidago marginella var. *sublanceolata* IPNI=*249473-1* 270
Solidago microglossa var. *linearifolia* IPNI=*249473-1* 270
Solidago microglossa var. *ramosa* IPNI=*249473-1* 270
Solidago nitidula IPNI=*249473-1* 270
Solidago odora IPNI=*249473-1* 270
Solidago pitcheri IPNI=*249575-1* 270
Solidago repens IPNI=*249473-1* 270
Solidago sera IPNI=*249575-1* 270
Solidago serotina f. *huntingdonensis* IPNI=*249575-1* 270
Solidago serotina IPNI=*249575-1* 270
Solidago serotina var. *dumertorum* IPNI=*249575-1* 270
Solidago serotina var. *gigantea* IPNI=*249575-1* 270
Solidago serotina var. *minor* IPNI=*249575-1* 270
Solidago serotinoides IPNI=*249575-1* 270
Solidago shinnersii IPNI=*249575-1* 270
Solidago somesii IPNI=*249575-1* 270
Solidago vulneraria IPNI=*249473-1* 270
Soliva mutisii IPNI=*249968-1* 270
Sommerfeldtia obovata IPNI=*148561-2* 326
Sonchus agrestis IPNI=*1152557-2* 261
Sonchus angustissimus IPNI=*250268-1* 270
Sonchus asper f. *adenotrichus* IPNI=*1003017-2* 270
Sonchus asper f. *glandulosus* IPNI=*1003017-2* 270
Sonchus asper f. *laevis* IPNI=*1003017-2* 270
Sonchus asper f. *lignosus* IPNI=*1003017-2* 270
Sonchus asper var. *glanduliferus* IPNI=*1003017-2* 270
Sonchus asper var. *sabulosus* IPNI=*1003017-2* 270
Sonchus australis IPNI=*1003017-2* 270
Sonchus borderi IPNI=*1003017-2* 270
Sonchus brasiliensis IPNI=*1152557-2* 261
Sonchus carolinianus IPNI=*1003017-2* 270
Sonchus ciliatus IPNI=*229022-1* 264
Sonchus ciliatus IPNI=*250268-1* 270
Sonchus ciliatus var. *subbipinnatifidus* IPNI=*250268-1* 270
Sonchus crocifolius IPNI=*1003017-2* 270
Sonchus cuspidatus IPNI=*1003017-2* 270
Sonchus decipiens IPNI=*1003017-2* 270
Sonchus eryngioides IPNI=*1003017-2* 270
Sonchus erythropappus IPNI=*203365-1* 261
Sonchus fabrae IPNI=*250268-1* 270
Sonchus fallax IPNI=*1003017-2* 270
Sonchus fallax var. *asper* IPNI=*1003017-2* 270
Sonchus fallax var. *laevis* IPNI=*1003017-2* 270
Sonchus ferox IPNI=*1003017-2* 270
Sonchus glaber IPNI=*1003017-2* 270
Sonchus glaber IPNI=*250268-1* 270
Sonchus goraeensis IPNI=*229022-1* 264
Sonchus gracilis IPNI=*250268-1* 270
Sonchus infestus IPNI=*1003017-2* 270
Sonchus lacerus IPNI=*250268-1* 270
Sonchus laevis IPNI=*250268-1* 270
Sonchus longifolius IPNI=*250268-1* 270
Sonchus macrotus IPNI=*250268-1* 270
Sonchus mairei IPNI=*250268-1* 270
Sonchus maritimus IPNI=*250268-1* 270
Sonchus occidentalis IPNI=*1152557-2* 261
Sonchus oleraceus f. *runcinatus* IPNI=*250268-1* 270
Sonchus oleraceus subsp. *angustissimus* IPNI=*250268-1* 270
Sonchus oleraceus var. *asper* IPNI=*1003017-2* 270
Sonchus oleraceus var. *integrifolius* IPNI=*250268-1* 270
Sonchus oleraceus var. *lacerus* IPNI=*250268-1* 270
Sonchus oleraceus var. *litoralis* IPNI=*250268-1* 270
Sonchus oleraceus var. *triangularis* IPNI=*250268-1* 270
Sonchus pallescens IPNI=*250268-1* 270
Sonchus paniculatus IPNI=*229022-1* 264
Sonchus parviflorus IPNI=*250268-1* 270
Sonchus plumieri IPNI=*250268-1* 270

Sonchus reversus IPNI=*250268-1* 270
Sonchus rivularis IPNI=*250268-1* 270
Sonchus roseus IPNI=*250268-1* 270
Sonchus royleanus IPNI=*250268-1* 270
Sonchus runcinatus IPNI=*250268-1* 270
Sonchus schimperi IPNI=*250268-1* 270
Sonchus schmidianus IPNI=*250268-1* 270
Sonchus spachii IPNI=*229022-1* 264
Sonchus spinosus IPNI=*1003017-2* 270
Sonchus spinulifolius IPNI=*250268-1* 270
Sonchus spinulosus IPNI=*1003017-2* 270
Sonchus subbipinnatifidus IPNI=*250268-1* 270
Sonchus sulphureus IPNI=*1003017-2* 270
Sonchus sundaicus IPNI=*250268-1* 270
Sonchus taraxacifolius IPNI=*229022-1* 264
Sonchus tenerrimus IPNI=*250268-1* 270
Sonchus tibesticus IPNI=*1003017-2* 270
Sonchus umbellatus IPNI=*1003017-2* 270
Sonchus umbellifer IPNI=*250268-1* 270
Sonchus viridis IPNI=*1003017-2* 270
Sonchus zacinthoides IPNI=*250268-1* 270
Sophoclesia cordifolia IPNI=*858679-1* 300
Sophoclesia cordifolia var. *normalis* IPNI=*858679-1* 300
Sophoclesia cordifolia var. *oophylla* IPNI=*858679-1* 300
Sophoclesia domingensis IPNI=*858679-1* 300
Sophoclesia flaccida IPNI=*858679-1* 300
Sophoclesia klotzschiana IPNI=*858679-1* 300
Sophoclesia major IPNI=*858679-1* 300
Sophoclesia mirandensis IPNI=*858679-1* 300
Sophoclesia nummulariifolia IPNI=*858679-1* 300
Sophoclesia ovata IPNI=*858679-1* 300
Sophoclesia robusta IPNI=*858679-1* 300
Sophoclesia subscandens IPNI=*858679-1* 300
Sophoclesia trinitensis IPNI=*858679-1* 300
Sophoclesia weberbaueri IPNI=*858679-1* 300
Sorbus cydonia IPNI=*724549-1* 443
Sorbus malus IPNI=*726282-1* 443
Sorbus pyrus IPNI=*30065762-2* 444
Sorghastrum bipennatum IPNI=*240818-2* 245
Sorghastrum canescens IPNI=*240818-2* 245
Sorghastrum incompletum var. *bipennatum* IPNI=*240818-2* 245
Sorghastrum liebmannianum IPNI=*240818-2* 245
Sorghastrum trichopus IPNI=*240834-2* 245
Sorghastrum trollii IPNI=*240834-2* 245
Sorghum abyssinicum IPNI=*422078-1* 245
Sorghum abyssinicum IPNI=*422090-1* 245
Sorghum aciculatum IPNI=*396110-1* 236
Sorghum aethiopicum IPNI=*422078-1* 245
Sorghum album IPNI=*422090-1* 245
Sorghum ankolib IPNI=*422090-1* 245
Sorghum annulatum IPNI=*398528-1* 237
Sorghum anomalum IPNI=*422090-1* 245
Sorghum arduinoi IPNI=*422090-1* 245
Sorghum basiplicatum IPNI=*422090-1* 245
Sorghum basutorum IPNI=*422090-1* 245
Sorghum bicolor subsp. *verticilliflorum* IPNI=*422078-1* 245
Sorghum bicolor var. *arduinii* IPNI=*422090-1* 245
Sorghum bicolor var. *arundinaceum* IPNI=*422078-1* 245
Sorghum bicolor var. *cafer* IPNI=*422090-1* 245
Sorghum bicolor var. *caffrorum* IPNI=*422090-1* 245
Sorghum bicolor var. *cernuum* IPNI=*422090-1* 245
Sorghum bicolor var. *exaristatum* IPNI=*422090-1* 245
Sorghum bicolor var. *miliiforme* IPNI=*422090-1* 245
Sorghum bicolor var. *obovatum* IPNI=*422090-1* 245
Sorghum bicolor var. *rotundulum* IPNI=*422090-1* 245
Sorghum bicolor var. *saccharatum* IPNI=*422090-1* 245
Sorghum bicolor var. *sikkimense* IPNI=*422090-1* 245
Sorghum bicolor var. *subglabrescens* IPNI=*422090-1* 245
Sorghum bicorne IPNI=*12758-2* 233
Sorghum bipennatum IPNI=*240818-2* 245
Sorghum brevicarinatum IPNI=*422078-1* 245
Sorghum brevifolium IPNI=*1126606-2* 244
Sorghum caffrorum IPNI=*422090-1* 245
Sorghum caffrorum var. *bicarinatum* IPNI=*422090-1* 245
Sorghum campanum IPNI=*422090-1* 245
Sorghum canescens IPNI=*240818-2* 245
Sorghum caricosum IPNI=*398532-1* 237
Sorghum castaneum IPNI=*422078-1* 245
Sorghum caucasicum IPNI=*392701-1* 235
Sorghum caudatum IPNI=*422090-1* 245
Sorghum caudatum var. *aristatum* IPNI=*422090-1* 245
Sorghum caudatum var. *coffeatum* IPNI=*422090-1* 245
Sorghum caudatum var. *dicarpum* IPNI=*422090-1* 245
Sorghum caudatum var. *purpureum* IPNI=*422090-1* 245
Sorghum centroplicatum IPNI=*422090-1* 245
Sorghum cirratum IPNI=*420874-1* 244
Sorghum commune IPNI=*422090-1* 245
Sorghum compactum IPNI=*422090-1* 245
Sorghum condensatum IPNI=*420877-1* 244
Sorghum condensatum var. *lactiflorum* IPNI=*420877-1* 244
Sorghum conspicuum IPNI=*422090-1* 245
Sorghum contortum IPNI=*404772-1* 240
Sorghum coriaceum IPNI=*422090-1* 245
Sorghum crupina IPNI=*422139-1* 245
Sorghum decolor IPNI=*422139-1* 245
Sorghum decolorans IPNI=*422139-1* 245
Sorghum dochna IPNI=*422090-1* 245
Sorghum dochna var. *corymbosum* IPNI=*422090-1* 245
Sorghum dochna var. *technicum* IPNI=*422090-1* 245
Sorghum dubium IPNI=*422139-1* 245
Sorghum dulcicaule IPNI=*422090-1* 245
Sorghum dulcicaule var. *griseolilacinum* IPNI=*422090-1* 245
Sorghum durra IPNI=*422090-1* 245
Sorghum elegans IPNI=*422090-1* 245
Sorghum eplicatum IPNI=*422090-1* 245
Sorghum exsertum IPNI=*422090-1* 245
Sorghum fastigiatum IPNI=*398608-1* 237
Sorghum flavescens IPNI=*420951-1* 244
Sorghum gambicum IPNI=*422090-1* 245
Sorghum gayanum IPNI=*387915-1* 233
Sorghum giganteum IPNI=*422090-1* 245
Sorghum glomeratum IPNI=*12850-2* 233
Sorghum glycychylum IPNI=*422090-1* 245
Sorghum guineense IPNI=*422090-1* 245
Sorghum halepense f. *muticum* IPNI=*422139-1* 245
Sorghum halepense var. *genuinum* IPNI=*422139-1* 245
Sorghum halepense var. *latifolium* IPNI=*422139-1* 245
Sorghum hirtiflorum IPNI=*420951-1* 244
Sorghum incompletum IPNI=*240818-2* 245
Sorghum intermedium IPNI=*392701-1* 235
Sorghum ischaemum IPNI=*1191417-2* 235
Sorghum japonicum IPNI=*422090-1* 245
Sorghum lanceolatum IPNI=*422078-1* 245
Sorghum leucostachyum IPNI=*12920-2* 234
Sorghum macrochaetum IPNI=*422078-1* 245
Sorghum macrolepis IPNI=*126405-2* 240
Sorghum margaritiferum IPNI=*422090-1* 245
Sorghum medioplicatum IPNI=*422090-1* 245
Sorghum melaleucum IPNI=*422090-1* 245
Sorghum melanocarpum IPNI=*422090-1* 245
Sorghum mellitum IPNI=*422090-1* 245
Sorghum membranaceum IPNI=*422090-1* 245
Sorghum miliaceum IPNI=*422139-1* 245
Sorghum miliiforme IPNI=*422090-1* 245
Sorghum montanum IPNI=*392701-1* 235
Sorghum myosurus IPNI=*420951-1* 244

Sorghum nankinense IPNI=*422090-1* 245
Sorghum nardus IPNI=*396972-1* 236
Sorghum negrosense IPNI=*422090-1* 245
Sorghum nervosum IPNI=*422090-1* 245
Sorghum nigericum IPNI=*422090-1* 245
Sorghum nigricans IPNI=*422090-1* 245
Sorghum nigrum IPNI=*422090-1* 245
Sorghum notabile IPNI=*422090-1* 245
Sorghum nutans subvar. *canescens* IPNI=*240818-2* 245
Sorghum nutans var. *burchellii* IPNI=*240834-2* 245
Sorghum nutans var. *stipoides* IPNI=*240834-2* 245
Sorghum pallidum IPNI=*422090-1* 245
Sorghum panicoides IPNI=*422078-1* 245
Sorghum papyrascens IPNI=*422090-1* 245
Sorghum pertusum IPNI=*65406-3* 235
Sorghum piptatherum IPNI=*97916-2* 239
Sorghum plumosum IPNI=*424391-1* 246
Sorghum pugionifolium IPNI=*422078-1* 245
Sorghum pyramidale IPNI=*422090-1* 245
Sorghum riedelii IPNI=*420951-1* 244
Sorghum rigidifolium IPNI=*240834-2* 245
Sorghum rigidum IPNI=*422090-1* 245
Sorghum rollii IPNI=*422090-1* 245
Sorghum roxburghii IPNI=*422090-1* 245
Sorghum rubens IPNI=*422090-1* 245
Sorghum rufum IPNI=*405583-1* 240
Sorghum ruprechtii IPNI=*126405-2* 240
Sorghum saccharatum IPNI=*422090-1* 245
Sorghum saccharatum var. *atrum* IPNI=*422090-1* 245
Sorghum saccharatum var. *bicolor* IPNI=*422090-1* 245
Sorghum saccharatum var. *burmanicum* IPNI=*422090-1* 245
Sorghum saccharatum var. *corymbosum* IPNI=*422090-1* 245
Sorghum saccharatum var. *formosum* IPNI=*422090-1* 245
Sorghum saccharatum var. *giganteum* IPNI=*422090-1* 245
Sorghum saccharatum var. *halepense* IPNI=*422139-1* 245
Sorghum saccharatum var. *irungu* IPNI=*422090-1* 245
Sorghum saccharatum var. *melliferum* IPNI=*422090-1* 245
Sorghum saccharatum var. *obovatum* IPNI=*422090-1* 245
Sorghum saccharatum var. *papyraceum* IPNI=*422090-1* 245
Sorghum saccharatum var. *pulchrum* IPNI=*422090-1* 245
Sorghum saccharatum var. *technicum* IPNI=*422090-1* 245
Sorghum saccharatum var. *vulgare* IPNI=*422090-1* 245
Sorghum saccharatum var. *wightii* IPNI=*422090-1* 245
Sorghum saccharoides IPNI=*35205-2* 235
Sorghum saccharoides var. *leucopogon* IPNI=*35184-2* 235
Sorghum sanguineum IPNI=*420951-1* 244
Sorghum sativum IPNI=*422090-1* 245
Sorghum schimperi IPNI=*422090-1* 245
Sorghum schlumbergeri IPNI=*35184-2* 235
Sorghum schreberi IPNI=*422139-1* 245
Sorghum semiberbe IPNI=*420951-1* 244
Sorghum simulans IPNI=*422090-1* 245
Sorghum somaliense IPNI=*422078-1* 245
Sorghum spathiflorum IPNI=*13112-2* 234
Sorghum splendidum IPNI=*422090-1* 245
Sorghum stapfii IPNI=*422078-1* 245
Sorghum subglabrescens IPNI=*422090-1* 245
Sorghum tataricum IPNI=*422090-1* 245
Sorghum technicum IPNI=*422090-1* 245
Sorghum thonizzii IPNI=*422090-1* 245
Sorghum trichopus IPNI=*240834-2* 245
Sorghum truchmenorum IPNI=*422090-1* 245
Sorghum usambarense IPNI=*422078-1* 245
Sorghum usorum IPNI=*422090-1* 245
Sorghum verticilliflorum IPNI=*422078-1* 245
Sorghum vestitum IPNI=*424391-1* 246
Sorghum vogelianum IPNI=*422078-1* 245
Sorghum vulgare IPNI=*422090-1* 245
Sorghum vulgare var. *bicolor* IPNI=*422090-1* 245
Sorghum vulgare var. *rubens* IPNI=*422090-1* 245
Sorghum vulgare var. *saccharatum* IPNI=*422090-1* 245
Sorindeia aurantiodora IPNI=*70012-1* 451
Sorindeia biringo IPNI=*70000-1* 451
Sorindeia glaberrima IPNI=*70000-1* 451
Sorindeia heterophylla IPNI=*70000-1* 451
Sorindeia ovalifolia IPNI=*70000-1* 451
Sorindeia puberula IPNI=*70000-1* 451
Sorindeia simplicifolia IPNI=*70012-1* 451
Sorindeia suaveolens IPNI=*70000-1* 451
Sorindeia venulosa IPNI=*70000-1* 451
Sorocea amazonica IPNI=*316993-2* 441
Sorocea arnoldoi IPNI=*240971-2* 441
Sorocea castaneifolia IPNI=*856566-1* 441
Sorocea colombiana IPNI=*260203-2* 441
Sorocea cufodontii IPNI=*856575-1* 441
Sorocea dentata IPNI=*316993-2* 441
Sorocea faustiniana IPNI=*856575-1* 441
Sorocea grandifolia IPNI=*856566-1* 441
Sorocea hilariana IPNI=*856566-1* 441
Sorocea houllettiana IPNI=*856566-1* 441
Sorocea ilicifolia var. *grandifolia* IPNI=*856566-1* 441
Sorocea macrogyna IPNI=*856566-1* 441
Sorocea martineziana IPNI=*240975-2* 441
Sorocea nitida IPNI=*851229-1* 437
Sorocea rhodorachis IPNI=*240975-2* 441
Sorocea stenophylla IPNI=*59683-2* 437
Sorocea trophoides subsp. *rhodorachis* IPNI=*240975-2* 441
Soromanes serratifolia IPNI=*17181340-1* 176
Sotularia malabarica IPNI=*553618-1* 417
Souroubea aubletii IPNI=*829493-1* 302
Souroubea auriculata IPNI=*829493-1* 302
Souroubea bahiensis IPNI=*829493-1* 302
Souroubea guianensis var. *bahiensis* IPNI=*829493-1* 302
Souroubea guianensis var. *corallina* IPNI=*829489-1* 302
Souroubea guianensis var. *lepidota* IPNI=*829493-1* 302
Souroubea lepidota IPNI=*829493-1* 302
Souroubea meyeri IPNI=*829493-1* 302
Spachea elegans var. *obovata* IPNI=*557422-1* 396
Spachea littoralis IPNI=*557422-1* 396
Spachea ossana IPNI=*557422-1* 396
Spachea parviflora IPNI=*557422-1* 396
Sparattanthelium amazonum subsp. *guatemalense* IPNI=*170915-1* 182
Sparattanthelium guatemalense IPNI=*170915-1* 182
Sparattanthelium septentrionale IPNI=*170915-1* 182
Sparattosperma neurocalyx IPNI=*248789-2* 365
Sparattosperma roseum IPNI=*111027-1* 365
Sparaxis thoubertii IPNI=*437022-1* 216
Sparganium americanum var. *rigidum* IPNI=*836744-1* 247
Sparganium americanum var. *typicum* IPNI=*836744-1* 247
Sparganium ligulare IPNI=*836744-1* 247
Sparganophorus africanus IPNI=*1059151-2* 270
Sparganophorus ethulia IPNI=*1059151-2* 270
Sparganophorus fasciatus IPNI=*1059151-2* 270
Sparganophorus fasciculatus IPNI=*1059151-2* 270
Sparganophorus fuscatus IPNI=*1059151-2* 270
Sparganophorus obtusifolius IPNI=*7086-2* 256
Sparganophorus sparganophora IPNI=*1059151-2* 270
Sparganophorus struchium IPNI=*1059151-2* 270
Sparganophorus vaillantii IPNI=*1059151-2* 270
Sparganophorus vaillantii var. *longifolius* IPNI=*1059151-2* 270
Sparrea schippii IPNI=*51219-2* 436
Sparrmannia acerifolia IPNI=*835200-1* 414
Spartianthus americanus IPNI=*519057-1* 334
Spartianthus junceus IPNI=*519057-1* 334
Spartina pubera IPNI=*1038558-2* 237

Spartium acutifolium IPNI=*519057-1* 334
Spartium albicans IPNI=*495939-1* 320
Spartium americanum IPNI=*519057-1* 334
Spartium japonicum IPNI=*519057-1* 334
Spartium odoratissimum IPNI=*519057-1* 334
Spartium odoratum IPNI=*519057-1* 334
Spartium terminale IPNI=*495939-1* 320
Spathicalyx xanthophylla IPNI=*77142096-1* 365
Spathiger strobilifer IPNI=*92214-2* 219
Spathiphyllum amazonicum IPNI=*89296-1* 202
Spathiphyllum bonplandii IPNI=*88955-1* 202
Spathiphyllum candicans IPNI=*88955-1* 202
Spathiphyllum cannifolium var. *nanum* IPNI=*88955-1* 202
Spathiphyllum canniforme IPNI=*88955-1* 202
Spathiphyllum caudatum IPNI=*89280-1* 202
Spathiphyllum dechardii IPNI=*88955-1* 202
Spathiphyllum fendleri IPNI=*88964-1* 202
Spathiphyllum friedrichsthalii var. *angustifolium* IPNI=*88964-1* 202
Spathiphyllum friedrichsthalii var. *brevifolium* IPNI=*88964-1* 202
Spathiphyllum friedrichsthalii var. *latifolium* IPNI=*88964-1* 202
Spathiphyllum kochii IPNI=*88983-1* 202
Spathiphyllum sagittifolium IPNI=*89296-1* 202
Spathodea coito IPNI=*60474996-2* 362
Spathodea corymbosa IPNI=*60474996-2* 362
Spathodea danckelmaniana IPNI=*110661-1* 365
Spathodea fenzliana IPNI=*77100511-1* 363
Spathodea flaviflora IPNI=*77100438-1* 362
Spathodea fraxinifolia IPNI=*77100464-1* 363
Spathodea kohautiana IPNI=*77100466-1* 363
Spathodea laurifolia IPNI=*60474996-2* 362
Spathodea mansoana IPNI=*42392-2* 362
Spathodea nilotica f. *bryanii* IPNI=*110661-1* 365
Spathodea obovata IPNI=*153710-2* 364
Spathodea orinocensis IPNI=*60474996-2* 362
Spathodea pisoniana IPNI=*77100464-1* 363
Spathodea platypoda IPNI=*60474996-2* 362
Spathodea schomburgkii IPNI=*77100441-1* 362
Spathodea tulipifera IPNI=*110661-1* 365
Spathodea uncata IPNI=*77100511-1* 363
Spathodea uncinata IPNI=*77100511-1* 363
Specklinia bradei IPNI=*1147281-2* 217
Specklinia modesta IPNI=*1147281-2* 217
Specularia perfoliata IPNI=*258447-2* 273
Spennera acuminifolia IPNI=*1025557-2* 417
Spennera alata IPNI=*1101289-2* 417
Spennera anomala IPNI=*1025557-2* 417
Spennera aquatica IPNI=*573916-1* 424
Spennera asphalti IPNI=*573916-1* 424
Spennera caulialata IPNI=*1101289-2* 417
Spennera dichotoma IPNI=*1025557-2* 417
Spennera dichotoma var. *lanceolata* IPNI=*1025557-2* 417
Spennera fragilis IPNI=*1031600-2* 417
Spennera grandifolia IPNI=*1031600-2* 417
Spennera hydrophila IPNI=*573916-1* 424
Spennera lancifolia IPNI=*1031600-2* 417
Spennera martinicensis IPNI=*1031600-2* 417
Spennera panicularis IPNI=*1101289-2* 417
Spennera pellucida IPNI=*1031600-2* 417
Spennera pellucida var. *subcordata* IPNI=*1031600-2* 417
Spennera quintuplinervia IPNI=*77207474-1* 420
Spennera rubricaulis IPNI=*1101289-2* 417
Spennera sieberi IPNI=*573916-1* 424
Spennera sileniflora IPNI=*1031600-2* 417
Spennera sphaeranthera IPNI=*1025557-2* 417
Spennera uliginosa IPNI=*1031600-2* 417
Spergula arvensis f. *rubra* IPNI=*158725-1* 286
Spergula arvensis subsp. *nana* IPNI=*158725-1* 286
Spergula arvensis var. *sativa* IPNI=*158725-1* 286
Spergula camarae IPNI=*158725-1* 286
Spergula campestris IPNI=*158983-1* 286
Spergula coreana IPNI=*241336-2* 286
Spergula decandra IPNI=*158725-1* 286
Spergula geniculata IPNI=*158725-1* 286
Spergula heterosperma IPNI=*241336-2* 286
Spergula linicola IPNI=*158725-1* 286
Spergula marina IPNI=*241336-2* 286
Spergula maxima IPNI=*158725-1* 286
Spergula maxima IPNI=*158983-1* 286
Spergula maxima subsp. *linicola* IPNI=*158725-1* 286
Spergula nordica IPNI=*158725-1* 286
Spergula praevisa IPNI=*158725-1* 286
Spergula pseudomorisonii IPNI=*158725-1* 286
Spergula radicans IPNI=*158983-1* 286
Spergula ramosissima IPNI=*158725-1* 286
Spergula refracta IPNI=*158725-1* 286
Spergula rubra IPNI=*158983-1* 286
Spergula salina IPNI=*241336-2* 286
Spergula sativa IPNI=*158725-1* 286
Spergula seminulifera IPNI=*241336-2* 286
Spergula villosa IPNI=*159015-1* 286
Spergula vulgaris IPNI=*158725-1* 286
Spergularia alata IPNI=*241336-2* 286
Spergularia alpina IPNI=*158983-1* 286
Spergularia aprica IPNI=*159015-1* 286
Spergularia arenosa IPNI=*158983-1* 286
Spergularia arvensis IPNI=*158725-1* 286
Spergularia bofillii IPNI=*241336-2* 286
Spergularia borealis IPNI=*158983-1* 286
Spergularia bungeana IPNI=*241336-2* 286
Spergularia caballeroi IPNI=*241336-2* 286
Spergularia campestris f. *pubescens* IPNI=*158983-1* 286
Spergularia campestris IPNI=*158983-1* 286
Spergularia dillenii IPNI=*241336-2* 286
Spergularia fauriei IPNI=*241336-2* 286
Spergularia halophila IPNI=*241336-2* 286
Spergularia heterosperma IPNI=*241336-2* 286
Spergularia leiosperma IPNI=*241336-2* 286
Spergularia liebmanniana IPNI=*159015-1* 286
Spergularia longicaulis IPNI=*241336-2* 286
Spergularia longipes IPNI=*158983-1* 286
Spergularia lycia IPNI=*158983-1* 286
Spergularia marina var. *cleistogama* IPNI=*241336-2* 286
Spergularia marina var. *densiflora* IPNI=*241336-2* 286
Spergularia marina var. *simonii* IPNI=*241336-2* 286
Spergularia marina var. *tenuis* IPNI=*241336-2* 286
Spergularia maxima IPNI=*158725-1* 286
Spergularia miquelonensis IPNI=*241336-2* 286
Spergularia neglecta IPNI=*241336-2* 286
Spergularia pinguis IPNI=*241336-2* 286
Spergularia polyphylla IPNI=*158983-1* 286
Spergularia radicans IPNI=*158983-1* 286
Spergularia remotiflora IPNI=*159015-1* 286
Spergularia rodrigoi IPNI=*241336-2* 286
Spergularia rubra subsp. *alpina* IPNI=*158983-1* 286
Spergularia rubra var. *campestris* IPNI=*158983-1* 286
Spergularia rubra var. *elongata* IPNI=*158983-1* 286
Spergularia rubra var. *marina* IPNI=*241336-2* 286
Spergularia rubra var. *perennans* IPNI=*158983-1* 286
Spergularia rupestris IPNI=*158983-1* 286
Spergularia rupestris IPNI=*159015-1* 286
Spergularia rupestris var. *elongata* IPNI=*158983-1* 286
Spergularia salina IPNI=*241336-2* 286
Spergularia salina var. *involucrata* IPNI=*241336-2* 286
Spergularia salina var. *minor* IPNI=*241336-2* 286
Spergularia salina var. *sordida* IPNI=*241336-2* 286
Spergularia salina var. *tenuis* IPNI=*241336-2* 286

Spergularia seminulifera IPNI=*241336-2* 286
Spergularia sparsiflora IPNI=*241336-2* 286
Spergularia sperguloides IPNI=*158983-1* 286
Spergularia sperguloides IPNI=*241336-2* 286
Spergularia tenuis IPNI=*241336-2* 286
Spergularia tenuis var. *involucrata* IPNI=*241336-2* 286
Spergularia uliginosa IPNI=*241336-2* 286
Spergularia urbica IPNI=*241336-2* 286
Spergularia villosa var. *berteroana* IPNI=*159015-1* 286
Spergularia villosa var. *rupestris* IPNI=*159015-1* 286
Spergulastrum lanuginosum IPNI=*30040285-2* 285
Spermacoce aculeata IPNI=*317930-2* 357
Spermacoce alternans IPNI=*766862-1* 357
Spermacoce anthospermodes IPNI=*999397-1* 350
Spermacoce arenicola IPNI=*945476-1* 357
Spermacoce aspera IPNI=*766800-1* 357
Spermacoce aspera IPNI=*766863-1* 357
Spermacoce aspera var. *latifolia* IPNI=*766800-1* 357
Spermacoce asperula IPNI=*766862-1* 357
Spermacoce assurgens IPNI=*319724-2* 357
Spermacoce aturensis IPNI=*317930-2* 357
Spermacoce bartlingiana IPNI=*766800-1* 357
Spermacoce bogotensis IPNI=*999397-1* 350
Spermacoce caerulescens IPNI=*766800-1* 357
Spermacoce capitata var. *ferruginea* IPNI=*317930-2* 357
Spermacoce capitata var. *normalis* IPNI=*317930-2* 357
Spermacoce capitellata IPNI=*241474-2* 357
Spermacoce cephalophora IPNI=*766900-1* 357
Spermacoce cephalotes IPNI=*756228-1* 352
Spermacoce chapmanii IPNI=*319724-2* 357
Spermacoce coerulea IPNI=*766862-1* 357
Spermacoce commutata IPNI=*77154345-1* 351
Spermacoce cornifolia IPNI=*317930-2* 357
Spermacoce crassifolia IPNI=*756228-1* 352
Spermacoce declinata IPNI=*756228-1* 352
Spermacoce dichotoma IPNI=*319724-2* 357
Spermacoce diffusa IPNI=*756228-1* 352
Spermacoce distans IPNI=*317930-2* 357
Spermacoce domingensis IPNI=*766900-1* 357
Spermacoce echioides IPNI=*319724-2* 357
Spermacoce elliptica IPNI=*766862-1* 357
Spermacoce ferruginea IPNI=*317930-2* 357
Spermacoce fruticosa IPNI=*767031-1* 357
Spermacoce garro IPNI=*77154345-1* 351
Spermacoce globosa IPNI=*767031-1* 357
Spermacoce gracilis IPNI=*756228-1* 352
Spermacoce gracilis IPNI=*766900-1* 357
Spermacoce graminifolia IPNI=*767031-1* 357
Spermacoce guianensis IPNI=*241474-2* 357
Spermacoce hebecarpa IPNI=*319724-2* 357
Spermacoce heterophylla IPNI=*766900-1* 357
Spermacoce hirsuta IPNI=*764679-1* 356
Spermacoce hirta IPNI=*756228-1* 352
Spermacoce hirta IPNI=*766863-1* 357
Spermacoce hondurensis IPNI=*319724-2* 357
Spermacoce hyssopifolia IPNI=*767031-1* 357
Spermacoce involucrata IPNI=*764679-1* 356
Spermacoce laeta IPNI=*767031-1* 357
Spermacoce laevigata IPNI=*767031-1* 357
Spermacoce laevis IPNI=*241474-2* 357
Spermacoce laevis IPNI=*766863-1* 357
Spermacoce lanceolata IPNI=*319724-2* 357
Spermacoce linifolia IPNI=*241474-2* 357
Spermacoce longiflora IPNI=*766862-1* 357
Spermacoce longifolia f. *glabrata* IPNI=*766862-1* 357
Spermacoce maritima IPNI=*77154345-1* 351
Spermacoce minima IPNI=*767031-1* 357
Spermacoce molleri IPNI=*767031-1* 357
Spermacoce mucronata IPNI=*767031-1* 357
Spermacoce neovirgata IPNI=*766862-1* 357
Spermacoce nervata IPNI=*766862-1* 357
Spermacoce oaxacana IPNI=*767031-1* 357
Spermacoce oligodonta IPNI=*767031-1* 357
Spermacoce parviflora IPNI=*319724-2* 357
Spermacoce patula IPNI=*766900-1* 357
Spermacoce pilosa IPNI=*77154343-1* 351
Spermacoce podocephala IPNI=*767031-1* 357
Spermacoce polycephala IPNI=*767031-1* 357
Spermacoce portoricensis IPNI=*766862-1* 357
Spermacoce procera IPNI=*317930-2* 357
Spermacoce pygmaea IPNI=*767031-1* 357
Spermacoce radicans IPNI=*77154345-1* 351
Spermacoce radula IPNI=*77154343-1* 351
Spermacoce ramosa IPNI=*766863-1* 357
Spermacoce reclinata IPNI=*767031-1* 357
Spermacoce rotata IPNI=*766900-1* 357
Spermacoce scabrida IPNI=*766800-1* 357
Spermacoce serrulata IPNI=*77154345-1* 351
Spermacoce sexangularis IPNI=*319724-2* 357
Spermacoce spinosa IPNI=*77154343-1* 351
Spermacoce stellata IPNI=*767031-1* 357
Spermacoce stricta IPNI=*766863-1* 357
Spermacoce stylosa IPNI=*756228-1* 352
Spermacoce suffrutescens IPNI=*319724-2* 357
Spermacoce suffruticosa IPNI=*319724-2* 357
Spermacoce suffruticosa IPNI=*77154343-1* 351
Spermacoce tenera IPNI=*766863-1* 357
Spermacoce tenuior var. *angustifolia* IPNI=*241474-2* 357
Spermacoce tenuior var. *commersonii* IPNI=*319724-2* 357
Spermacoce terminalis IPNI=*317930-2* 357
Spermacoce tetragona IPNI=*319724-2* 357
Spermacoce tomentosa IPNI=*317930-2* 357
Spermacoce trichantha IPNI=*766863-1* 357
Spermacoce vaginata IPNI=*319724-2* 357
Spermacoce villosa IPNI=*756228-1* 352
Spermacoce virgata IPNI=*766862-1* 357
Spermacoce wydleriana IPNI=*319724-2* 357
Spermacoceodes tenuior IPNI=*241474-2* 357
Sphacele alpina IPNI=*449287-1* 370
Sphacele bullata IPNI=*137860-2* 370
Sphacele conferta IPNI=*137869-2* 370
Sphacele hirsuta IPNI=*449287-1* 370
Sphacele intermedia IPNI=*137859-2* 370
Sphacele jamesonii IPNI=*137859-2* 370
Sphacele lindeniana IPNI=*137860-2* 370
Sphacele parviflora IPNI=*137860-2* 370
Sphacele salviifolia IPNI=*137894-2* 370
Sphacele sprucei IPNI=*137859-2* 370
Sphaeralcea americana IPNI=*561999-1* 410
Sphaeralcea filicaulis IPNI=*563387-1* 413
Sphaeralcea lactea IPNI=*133329-2* 410
Sphaeralcea nutans IPNI=*563228-1* 412
Sphaeralcea rosea IPNI=*563228-1* 412
Sphaeralcea schenckii IPNI=*563228-1* 412
Sphaeralcea umbellata var. *rosea* IPNI=*563228-1* 412
Sphaeralcea vitifolia IPNI=*563228-1* 412
Sphaerium lacryma IPNI=*30100521-2* 236
Sphaerocionium fragile IPNI=*125766-2* 173
Sphaerocionium raddianum IPNI=*125766-2* 173
Sphaerogyne cinnamomea IPNI=*158746-2* 419
Sphaeroma filicaule IPNI=*563387-1* 413
Sphaeroma nutans IPNI=*563228-1* 412
Sphaeroma rosea IPNI=*563228-1* 412
Sphaeroma vitifolium IPNI=*563228-1* 412
Sphaeropteris aterrima IPNI=*73117-2* 172
Sphaerostigma minimum IPNI=*144406-2* 429

Sphaerotheca scoparioides IPNI=*801672-1* 376
Sphagneticola ulei IPNI=*1093589-2* 270
Sphincterostigma bipinnatifidum IPNI=*77178497-1* 202
Sphinctolobium glaucescens IPNI=*77118402-1* 328
Sphyrospermum cordifolium IPNI=*858679-1* 300
Sphyrospermum ellipticum IPNI=*858679-1* 300
Sphyrospermum klotzschianum IPNI=*858679-1* 300
Sphyrospermum longifolium IPNI=*858679-1* 300
Sphyrospermum majus IPNI=*858679-1* 300
Sphyrospermum roraimae IPNI=*858679-1* 300
Sphyrospermum standleyi IPNI=*858679-1* 300
Sphyrospermum tuberculatum IPNI=*858679-1* 300
Sphyrospermum weberbaueri IPNI=*858679-1* 300
Spicanta capensis var. *gilliesii* IPNI=*33533-2* 174
Spicanta occidentalis IPNI=*33632-2* 174
Spicanta serrulata IPNI=*77142412-1* 174
Spigelia anthelmia var. *nervosa* IPNI=*318920-2* 345
Spigelia anthelmia var. *obliquinervia* IPNI=*318920-2* 345
Spigelia anthelmia var. *peruviana* IPNI=*318920-2* 345
Spigelia australis IPNI=*546947-1* 345
Spigelia breviflora IPNI=*546947-1* 345
Spigelia chamaedryoides IPNI=*546947-1* 345
Spigelia domingensis IPNI=*318920-2* 345
Spigelia filipes IPNI=*546941-1* 345
Spigelia fruticulosa IPNI=*318920-2* 345
Spigelia gilgii IPNI=*546941-1* 345
Spigelia guaranitica IPNI=*546947-1* 345
Spigelia guaranitica var. *interrupta* IPNI=*546947-1* 345
Spigelia hassleriana f. *roseiflora* IPNI=*546947-1* 345
Spigelia hassleriana IPNI=*546947-1* 345
Spigelia hassleriana var. *major* IPNI=*546947-1* 345
Spigelia humboldtiana var. *breviflora* IPNI=*546947-1* 345
Spigelia humboldtiana var. *obtusifolia* IPNI=*546947-1* 345
Spigelia humboldtiana var. *pubescens* IPNI=*546947-1* 345
Spigelia intermedia IPNI=*546947-1* 345
Spigelia killipii IPNI=*546941-1* 345
Spigelia laevigata IPNI=*546941-1* 345
Spigelia multispica IPNI=*546941-1* 345
Spigelia multispica var. *concolor* IPNI=*546941-1* 345
Spigelia multispica var. *discolor* IPNI=*546941-1* 345
Spigelia nervosa IPNI=*318920-2* 345
Spigelia nervosa IPNI=*546941-1* 345
Spigelia oppositifolia IPNI=*318920-2* 345
Spigelia peduncularis IPNI=*546982-1* 345
Spigelia pedunculata IPNI=*546982-1* 345
Spigelia persicarioides var. *insularis* IPNI=*546941-1* 345
Spigelia quadrifolia IPNI=*318920-2* 345
Spigelia quaternata IPNI=*546947-1* 345
Spigelia rubelliana IPNI=*546947-1* 345
Spigelia scabra IPNI=*546947-1* 345
Spigelia scabra var. *angustata* IPNI=*546947-1* 345
Spigelia schomburgkiana IPNI=*546947-1* 345
Spigelia stipularis IPNI=*318920-2* 345
Spigelia uruguaya IPNI=*546947-1* 345
Spilanthes acmella var. *oleracea* IPNI=*914778-1* 255
Spilanthes acmella var. *uliginosa* IPNI=*174556-1* 256
Spilanthes alba IPNI=*271528-2* 255
Spilanthes americana f. *ciliatifolia* IPNI=*174552-1* 256
Spilanthes americana f. *lanitecta* IPNI=*174552-1* 256
Spilanthes americana f. *longiinternodiata* IPNI=*174552-1* 256
Spilanthes americana f. *parvifolia* IPNI=*271533-2* 255
Spilanthes americana IPNI=*213524-1* 264
Spilanthes americana var. *parvula* IPNI=*174552-1* 256
Spilanthes americana var. *repens* IPNI=*174552-1* 256
Spilanthes arrayana IPNI=*174530-1* 255
Spilanthes beccabunga IPNI=*174552-1* 256
Spilanthes beccabunga var. *parvula* IPNI=*174552-1* 256
Spilanthes bicolor IPNI=*250890-1* 270
Spilanthes botterii IPNI=*271548-2* 256
Spilanthes caespitosa IPNI=*174530-1* 255
Spilanthes calva var. *oleracea* IPNI=*914778-1* 255
Spilanthes charitopis IPNI=*174556-1* 256
Spilanthes ciliata IPNI=*174534-1* 255
Spilanthes ciliata var. *diffusa* IPNI=*174552-1* 256
Spilanthes cocuyensis IPNI=*174552-1* 256
Spilanthes commutata IPNI=*250890-1* 270
Spilanthes debilis IPNI=*271548-2* 256
Spilanthes diffusa IPNI=*174552-1* 256
Spilanthes disciformis IPNI=*213524-1* 264
Spilanthes disciformis var. *phaneractis* IPNI=*174552-1* 256
Spilanthes eggersii IPNI=*174530-1* 255
Spilanthes exasperata IPNI=*271548-2* 256
Spilanthes exasperata var. *cayennensis* IPNI=*271548-2* 256
Spilanthes filipes IPNI=*271533-2* 255
Spilanthes fimbriata IPNI=*174534-1* 255
Spilanthes fusca IPNI=*914778-1* 255
Spilanthes grandifolia IPNI=*914780-1* 256
Spilanthes guatemalensis IPNI=*156179-2* 265
Spilanthes iabadicensis IPNI=*174556-1* 256
Spilanthes iolepis IPNI=*271528-2* 255
Spilanthes lateraliflora IPNI=*174552-1* 256
Spilanthes lehmanniana IPNI=*174552-1* 256
Spilanthes leucophaea IPNI=*271548-2* 256
Spilanthes limonica IPNI=*174530-1* 255
Spilanthes lobata IPNI=*914780-1* 256
Spilanthes lundii IPNI=*174556-1* 256
Spilanthes mandonii IPNI=*271548-2* 256
Spilanthes melampodioides IPNI=*174534-1* 255
Spilanthes muticus IPNI=*202984-1* 261
Spilanthes mutisii IPNI=*213524-1* 264
Spilanthes nitida IPNI=*241903-1* 269
Spilanthes nuttallii IPNI=*174552-1* 256
Spilanthes ocymifolia f. *radiifera* IPNI=*174530-1* 255
Spilanthes ocymifolia IPNI=*271528-2* 255
Spilanthes ocymifolia var. *acutiserrata* IPNI=*271548-2* 256
Spilanthes oleracea IPNI=*914778-1* 255
Spilanthes oleracea var. *fusca* IPNI=*914778-1* 255
Spilanthes oppositifolia IPNI=*213524-1* 264
Spilanthes orizabaensis IPNI=*174552-1* 256
Spilanthes palustris IPNI=*250890-1* 270
Spilanthes pammicrophylla IPNI=*271533-2* 255
Spilanthes paniculata IPNI=*914780-1* 256
Spilanthes parvifolia IPNI=*271533-2* 255
Spilanthes phaneractis IPNI=*174552-1* 256
Spilanthes poeppigii IPNI=*174534-1* 255
Spilanthes popayanensis IPNI=*174534-1* 255
Spilanthes pseudoacmella IPNI=*1186403-2* 261
Spilanthes radicans IPNI=*271548-2* 256
Spilanthes radicans IPNI=*914778-1* 255
Spilanthes repens IPNI=*174552-1* 256
Spilanthes repens IPNI=*250890-1* 270
Spilanthes salivaria IPNI=*271528-2* 255
Spilanthes salzmannii IPNI=*174556-1* 256
Spilanthes subhirsuta IPNI=*174552-1* 256
Spilanthes tenella IPNI=*271548-2* 256
Spilanthes tinctoria IPNI=*175030-1* 256
Spilanthes uliginosa IPNI=*174556-1* 256
Spilanthes uliginosa var. *discoidea* IPNI=*174556-1* 256
Spilanthes urens var. *megalophylla* IPNI=*250890-1* 270
Spilanthes urens var. *typica* IPNI=*250890-1* 270
Spilanthes volubilis IPNI=*241903-1* 269
Spilanthes wedelioides IPNI=*1186403-2* 261
Spinacia domestica IPNI=*167384-1* 283
Spinacia glabra IPNI=*167384-1* 283
Spinacia inermis IPNI=*167384-1* 283
Spinacia oleracea subsp. *orientalis* IPNI=*167384-1* 283

Spinacia oleracea var. *bullata* IPNI=*167384-1* 283
Spinacia oleracea var. *subspontanea* IPNI=*167384-1* 283
Spinacia sessiliflora IPNI=*167384-1* 283
Spinacia spinosa IPNI=*167384-1* 283
Spiraea caerulescens IPNI=*742160-1* 445
Spiraea corymbosa IPNI=*742160-1* 445
Spiraea humanni IPNI=*742160-1* 445
Spiraea japonica IPNI=*742160-1* 445
Spiraea lanceolata IPNI=*742160-1* 445
Spiraea pubescens IPNI=*742160-1* 445
Spiraea reevesiana IPNI=*742160-1* 445
Spiraea roxburghiana IPNI=*742160-1* 445
Spiraea sinensis IPNI=*742160-1* 445
Spiranthes coccinea IPNI=*61935-2* 218
Spiranthes colorans IPNI=*659300-1* 223
Spiranthes colorata IPNI=*659300-1* 223
Spiranthes corymbosa IPNI=*61935-2* 218
Spiranthes elata IPNI=*73688-2* 218
Spiranthes elata var. *foliosa* IPNI=*73688-2* 218
Spiranthes elata var. *humilis* IPNI=*73688-2* 218
Spiranthes elata var. *longipetiolata* IPNI=*73688-2* 218
Spiranthes elata var. *minutiflora* IPNI=*73688-2* 218
Spiranthes elata var. *ovata* IPNI=*73688-2* 218
Spiranthes elata var. *parvifolia* IPNI=*73688-2* 218
Spiranthes nutans IPNI=*659300-1* 223
Spiranthes variegata IPNI=*73688-2* 218
Spirillus interruptus IPNI=*77099639-1* 203
Spirillus pectiniformis IPNI=*77099639-1* 203
Spirochaeta funckii IPNI=*211400-2* 268
Spirodela biperforata IPNI=*242868-2* 202
Spirodela intermedia IPNI=*242868-2* 202
Spirodela maxima IPNI=*526250-1* 202
Spiroloba fastuosa IPNI=*1040744-2* 322
Spiroloba unguis IPNI=*515137-1* 330
Spironema aphylla IPNI=*463202-1* 182
Spironema fragrans IPNI=*42441-2* 223
Spironema orthandrum IPNI=*42441-2* 223
Spirostachys ritteriana IPNI=*165886-1* 283
Spirotheca allenii IPNI=*60446802-2* 414
Spirotheca rimbachii IPNI=*60446802-2* 414
Spixia violacea IPNI=*51409-2* 259
Spondias acida IPNI=*71455-1* 451
Spondias aurantiaca IPNI=*71480-1* 451
Spondias brasiliensis IPNI=*71480-1* 451
Spondias cirouella IPNI=*71480-1* 451
Spondias crispula IPNI=*71494-1* 451
Spondias cytherea IPNI=*71455-1* 451
Spondias cytherea IPNI=*71480-1* 451
Spondias dubia IPNI=*71480-1* 451
Spondias dulcis var. *mucroniserrata* IPNI=*71455-1* 451
Spondias edmonstonei IPNI=*127138-1* 451
Spondias fragrans IPNI=*71455-1* 451
Spondias glabra IPNI=*71480-1* 451
Spondias graveolens IPNI=*71480-1* 451
Spondias guianensis IPNI=*71559-1* 451
Spondias jocote-amarillo IPNI=*71494-1* 451
Spondias longifolia IPNI=*71455-1* 451
Spondias lucida IPNI=*71480-1* 451
Spondias lutea IPNI=*71480-1* 451
Spondias lutea var. *glabra* IPNI=*71480-1* 451
Spondias lutea var. *maxima* IPNI=*71480-1* 451
Spondias macrophylla IPNI=*71559-1* 451
Spondias maxima IPNI=*71480-1* 451
Spondias mexicana IPNI=*71494-1* 451
Spondias negrosensis IPNI=*71494-1* 451
Spondias nigrescens IPNI=*242911-2* 451
Spondias parviflora IPNI=*71559-1* 451
Spondias pseudomyrobalanus IPNI=*71480-1* 451
Spondias purpurea f. *lutea* IPNI=*71494-1* 451
Spondias purpurea var. *munita* IPNI=*71494-1* 451
Spondias rugosa IPNI=*71559-1* 451
Spondias surinamensis IPNI=*71559-1* 451
Spondias therebintoides IPNI=*71455-1* 451
Spongostemma atropurpureum IPNI=*319717-1* 296
Sponia canescens IPNI=*1018059-2* 436
Sponia chichilea IPNI=*1018059-2* 436
Sponia crassifolia IPNI=*1018059-2* 436
Sponia grisea IPNI=*1018059-2* 436
Sponia lima IPNI=*1018059-2* 436
Sponia macrophylla IPNI=*1018059-2* 436
Sponia micrantha IPNI=*1018059-2* 436
Sponia mollis IPNI=*1018059-2* 436
Sponia riparia IPNI=*1018059-2* 436
Sponia rufescens IPNI=*1018059-2* 436
Sponia rugosa IPNI=*1018059-2* 436
Sponia schiedeana IPNI=*1018059-2* 436
Sponia stipellata IPNI=*1018059-2* 436
Sporobolus affinis IPNI=*243026-2* 245
Sporobolus angustus IPNI=*1207790-2* 245
Sporobolus argutus f. *purpurascens* IPNI=*243026-2* 245
Sporobolus argutus IPNI=*243026-2* 245
Sporobolus argutus var. *tuberculatus* IPNI=*243026-2* 245
Sporobolus benthamii var. *robustus* IPNI=*423010-1* 245
Sporobolus berteroanus IPNI=*1207790-2* 245
Sporobolus bibansiensis IPNI=*243044-2* 245
Sporobolus confertiflorus IPNI=*401189-1* 239
Sporobolus confertus IPNI=*423010-1* 245
Sporobolus cryptandrus var. *giganteus* IPNI=*242986-2* 245
Sporobolus cryptandrus var. *robustus* IPNI=*242986-2* 245
Sporobolus effusus IPNI=*243044-2* 245
Sporobolus exilis IPNI=*1207790-2* 245
Sporobolus gaditanus IPNI=*386353-1* 233
Sporobolus geniculatus IPNI=*423010-1* 245
Sporobolus humifusus IPNI=*243026-2* 245
Sporobolus humilis subsp. *minor* IPNI=*423010-1* 245
Sporobolus hypseloteros IPNI=*422898-1* 245
Sporobolus indicus f. *africanoides* IPNI=*1207790-2* 245
Sporobolus indicus f. *microspiculus* IPNI=*1207790-2* 245
Sporobolus indicus var. *andinus* IPNI=*1207790-2* 245
Sporobolus indicus var. *exilis* IPNI=*1207790-2* 245
Sporobolus indicus var. *pyramidalis* IPNI=*422898-1* 245
Sporobolus indicus var. *saxicola* IPNI=*422898-1* 245
Sporobolus jacquemontii IPNI=*422898-1* 245
Sporobolus littoralis IPNI=*423010-1* 245
Sporobolus mangaloricus IPNI=*243044-2* 245
Sporobolus minutiflorus IPNI=*243044-2* 245
Sporobolus minutus subsp. *confertus* IPNI=*423010-1* 245
Sporobolus muhlenbergii IPNI=*423010-1* 245
Sporobolus muralis IPNI=*243044-2* 245
Sporobolus orientalis IPNI=*1207790-2* 245
Sporobolus patens IPNI=*243026-2* 245
Sporobolus poiretii IPNI=*1103484-2* 245
Sporobolus pulvinatus IPNI=*243026-2* 245
Sporobolus pyramidalis var. *jacquemontii* IPNI=*422898-1* 245
Sporobolus rueppellianus IPNI=*422898-1* 245
Sporobolus sundaicus IPNI=*423010-1* 245
Sporobolus tenacissimus IPNI=*1207790-2* 245
Sporobolus tremulus IPNI=*423010-1* 245
Sporobolus tuberculatus IPNI=*243026-2* 245
Sporobolus verticillatus IPNI=*401189-1* 239
Sporobolus villosus IPNI=*405429-1* 240
Sporobolus virginicus var. *gaditanus* IPNI=*386353-1* 233
Sporobolus virginicus var. *littoralis* IPNI=*423010-1* 245
Sporobolus virginicus var. *minor* IPNI=*423010-1* 245
Sporobolus virginicus var. *pallidus* IPNI=*423010-1* 245
Sporobolus wombaliensis IPNI=*422898-1* 245

Sportella atalantioides IPNI=*730613-1* 444
Sprucea rubescens IPNI=*236591-2* 357
Stachyanthemum schomburgkii IPNI=*319176-2* 298
Stachyarpagophora aspera IPNI=*2468-2* 280
Stachycarpus harmsianus IPNI=*77203787-1* 180
Stachycarpus utilior IPNI=*77203787-1* 180
Stachys artemisia IPNI=*449195-1* 370
Stachys arvensis var. *bermudiana* IPNI=*459319-1* 373
Stachys arvensis var. *lamiiformis* IPNI=*459319-1* 373
Stachys arvensis var. *typica* IPNI=*459319-1* 373
Stachys boliviana IPNI=*459558-1* 373
Stachys brasiliensis IPNI=*459319-1* 373
Stachys elliptica var. *aristata* IPNI=*30008354-2* 373
Stachys elliptica var. *grandiflora* IPNI=*30008354-2* 373
Stachys galanderi IPNI=*459558-1* 373
Stachys karasmontana IPNI=*459988-1* 373
Stachys lanata IPNI=*459393-1* 373
Stachys lorentzii IPNI=*459558-1* 373
Stachys mayorii IPNI=*459558-1* 373
Stachys micheliana IPNI=*459558-1* 373
Stachys micrantha IPNI=*459558-1* 373
Stachys minima IPNI=*459988-1* 373
Stachys olympica IPNI=*459393-1* 373
Stachys pachycalymna IPNI=*459988-1* 373
Stachys patens IPNI=*77120819-1* 369
Stachys petiolosa IPNI=*459558-1* 373
Stachys schiedeana IPNI=*449287-1* 370
Stachys taurica IPNI=*459393-1* 373
Stachys tucumanensis IPNI=*459558-1* 373
Stachytarpheta angustifolia f. *jenmanii* IPNI=*864749-1* 379
Stachytarpheta angustifolia f. *rionegrensis* IPNI=*864749-1* 379
Stachytarpheta angustifolia IPNI=*864749-1* 379
Stachytarpheta angustissima IPNI=*864749-1* 379
Stachytarpheta aristata IPNI=*864784-1* 379
Stachytarpheta australis f. *albiflora* IPNI=*1126344-2* 379
Stachytarpheta australis IPNI=*1126344-2* 379
Stachytarpheta bogoriensis IPNI=*1092351-2* 379
Stachytarpheta bracteosa IPNI=*864784-1* 379
Stachytarpheta cayennensis f. *albiflora* IPNI=*1126344-2* 379
Stachytarpheta cayennensis f. *purpurea* IPNI=*1126344-2* 379
Stachytarpheta cayennensis var. *candicans* IPNI=*1126344-2* 379
Stachytarpheta cayennensis var. *schiedeana* IPNI=*243385-2* 379
Stachytarpheta cayennensis var. *virescens* IPNI=*1126344-2* 379
Stachytarpheta dichotoma f. *albiflora* IPNI=*1126344-2* 379
Stachytarpheta dichotoma IPNI=*1126344-2* 379
Stachytarpheta dichotoma var. *neocaledonica* IPNI=*1126344-2* 379
Stachytarpheta elatior IPNI=*864749-1* 379
Stachytarpheta elatior var. *jenmanii* IPNI=*864749-1* 379
Stachytarpheta fallax IPNI=*1126344-2* 379
Stachytarpheta gibberosa IPNI=*1126344-2* 379
Stachytarpheta guatemalensis f. *albiflora* IPNI=*1126344-2* 379
Stachytarpheta guatemalensis IPNI=*1126344-2* 379
Stachytarpheta hirta IPNI=*1126344-2* 379
Stachytarpheta jamaicensis f. *albiflora* IPNI=*1092351-2* 379
Stachytarpheta jamaicensis f. *atrocoerulea* IPNI=*1092351-2* 379
Stachytarpheta jamaicensis f. *parviflora* IPNI=*864749-1* 379
Stachytarpheta jamaicensis var. *longifolia* IPNI=*1092351-2* 379
Stachytarpheta laxiflora IPNI=*35430-2* 378
Stachytarpheta maximiliani var. *ciliaris* IPNI=*1126344-2* 379
Stachytarpheta mutabilis var. *violacea* IPNI=*243385-2* 379
Stachytarpheta patens IPNI=*1126344-2* 379
Stachytarpheta pilosiuscula IPNI=*1092351-2* 379
Stachytarpheta prismatica IPNI=*35430-2* 378
Stachytarpheta purpurea IPNI=*243385-2* 379
Stachytarpheta roraimensis IPNI=*243419-2* 379
Stachytarpheta roraimensis var. *pubescens* IPNI=*243419-2* 379
Stachytarpheta subulata IPNI=*1126344-2* 379
Stachytarpheta tabascana IPNI=*1126344-2* 379
Stachytarpheta theezans IPNI=*1126344-2* 379
Stachytarpheta umbrosa IPNI=*1126344-2* 379
Stachytarpheta veronicifolia IPNI=*1126344-2* 379
Stachytarpheta zuccagnii IPNI=*243385-2* 379
Staehelina corymbosa IPNI=*913807-1* 260
Staehelina solidaginoides IPNI=*944993-1* 261
Stanhopea bucephalus var. *jenischiana* IPNI=*243490-2* 223
Stanhopea bucephalus var. *roezlii* IPNI=*243490-2* 223
Stanhopea grandiflora IPNI=*243490-2* 223
Stanhopea grandiflora var. *jenischiana* IPNI=*243490-2* 223
Stanhopea hoppii IPNI=*243490-2* 223
Stapelia chinensis IPNI=*107308-3* 341
Staphidiastrum cognatum IPNI=*77164250-1* 422
Staphidiastrum impetiolare IPNI=*77207921-1* 421
Staphidiastrum platyphyllum IPNI=*77164250-1* 422
Staphidiastrum rubrum IPNI=*77164250-1* 422
Staphidiastrum rubrum var. *biacutum* IPNI=*77164250-1* 422
Staphidiastrum rubrum var. *microphyllum* IPNI=*77164250-1* 422
Staphidium anceps IPNI=*77164247-1* 420
Staphidium benthamianum IPNI=*77164247-1* 420
Staphidium brachystephanum IPNI=*77207400-1* 420
Staphidium bracteosum IPNI=*77196165-1* 420
Staphidium chrysanthum IPNI=*77164247-1* 420
Staphidium confertiflorum IPNI=*77196165-1* 420
Staphidium dispar IPNI=*77207368-1* 420
Staphidium diversifolium IPNI=*77144083-1* 422
Staphidium elegans IPNI=*77164247-1* 420
Staphidium glabratum IPNI=*572590-1* 421
Staphidium hostmannii IPNI=*77164247-1* 420
Staphidium naevulum IPNI=*77207474-1* 420
Staphidium octonum IPNI=*77196163-1* 421
Staphidium pauciflorum IPNI=*77164247-1* 420
Staphidium pauciflorum var. *alboroseum* IPNI=*77207400-1* 420
Staphidium pauciflorum var. *calcaratum* IPNI=*77164247-1* 420
Staphidium pauciflorum var. *stellulatum* IPNI=*77164247-1* 420
Staphidium pliostemon IPNI=*77207390-1* 419
Staphidium spicatum IPNI=*77196164-1* 422
Staphidium tiliifolium IPNI=*77164247-1* 420
Staphylea corymbosa IPNI=*785554-1* 292
Staphylea heterophylla IPNI=*785554-1* 292
Staphylea occidentalis IPNI=*785554-1* 292
Statice armeria IPNI=*19793-2* 288
Statice armeria subsp. *maritima* IPNI=*19793-2* 288
Statice capitata IPNI=*19793-2* 288
Statice floribunda IPNI=*686979-1* 288
Statice hirsuta IPNI=*686979-1* 288
Statice linearifolia IPNI=*19793-2* 288
Statice major IPNI=*19793-2* 288
Statice maritima IPNI=*19793-2* 288
Statice planifolia IPNI=*19793-2* 288
Statice pubescens IPNI=*19793-2* 288
Statice pulchella IPNI=*19793-2* 288
Statice sinuata IPNI=*686979-1* 288
Statice vulgaris IPNI=*19793-2* 288
Staurospermum verticillatum IPNI=*756228-1* 352
Steffensia adunca IPNI=*680296-1* 195
Steffensia angustifolia IPNI=*680296-1* 195
Steffensia anonifolia IPNI=*1003227-1* 195
Steffensia asperifolia IPNI=*681674-1* 196
Steffensia caracasana IPNI=*682429-1* 196
Steffensia carpunya IPNI=*680810-1* 195
Steffensia celtidifolia IPNI=*680296-1* 195
Steffensia consanguinea IPNI=*680988-1* 195
Steffensia crassinervia IPNI=*681041-1* 195
Steffensia flexuosa IPNI=*680680-1* 195
Steffensia hirsuta IPNI=*681674-1* 196
Steffensia hydrangeifolia IPNI=*681041-1* 195
Steffensia insignis IPNI=*682429-1* 196

Steffensia lacunosa IPNI=*681864-1* 196
Steffensia luschnathiana f. *glabrata* IPNI=*680435-1* 195
Steffensia luschnathiana IPNI=*680435-1* 195
Steffensia nubigena IPNI=*682399-1* 196
Steffensia obliqua IPNI=*682429-1* 196
Steffensia olfersiana IPNI=*681674-1* 196
Steffensia opizii IPNI=*681674-1* 196
Steffensia pseudochurumayu IPNI=*681027-1* 195
Steffensia radula IPNI=*680680-1* 195
Steffensia scabra IPNI=*681674-1* 196
Steffensia sellowiana IPNI=*681041-1* 195
Steffensia tiliifolia IPNI=*683316-1* 197
Steffensia tuberculata IPNI=*683806-1* 197
Steffensia tuberculata var. *glabrata* IPNI=*683806-1* 197
Steffensia verrucosa IPNI=*680435-1* 195
Steffensia xylopioides IPNI=*680435-1* 195
Steganotropis conjugata IPNI=*51897-2* 314
Stegosia cochinchinensis IPNI=*909653-1* 244
Stegosia cochinchinensis var. *robusta* IPNI=*909653-1* 244
Stegosia exaltata IPNI=*909653-1* 244
Steinhauera sempervirens IPNI=*263741-1* 180
Steiremis ficoidea IPNI=*1078972-2* 281
Steiremis repens IPNI=*327342-2* 281
Steiremis sessilifolia IPNI=*1069792-2* 281
Stellaria adenophora IPNI=*30172747-2* 285
Stellaria adnophora IPNI=*30172747-2* 285
Stellaria alpicola IPNI=*159330-1* 286
Stellaria alsinoides IPNI=*159330-1* 286
Stellaria arvensis IPNI=*158725-1* 286
Stellaria arvensis IPNI=*329786-2* 285
Stellaria baldwinii IPNI=*244185-2* 286
Stellaria bertolae IPNI=*159330-1* 286
Stellaria caroliniana IPNI=*159330-1* 286
Stellaria cerastium IPNI=*159330-1* 286
Stellaria chlorotica IPNI=*159330-1* 286
Stellaria ciliata IPNI=*159379-1* 286
Stellaria ciliata var. *cuspidata* IPNI=*244185-2* 286
Stellaria cordata IPNI=*30172747-2* 285
Stellaria cucubaloides IPNI=*159330-1* 286
Stellaria decandra IPNI=*159330-1* 286
Stellaria dichotoma IPNI=*159330-1* 286
Stellaria duthiei IPNI=*159330-1* 286
Stellaria elisabethiae IPNI=*159330-1* 286
Stellaria elongata IPNI=*30040285-2* 285
Stellaria glabella IPNI=*159330-1* 286
Stellaria glabra IPNI=*159330-1* 286
Stellaria glomerata IPNI=*152274-1* 285
Stellaria grandiflora IPNI=*159330-1* 286
Stellaria hiemalis IPNI=*159330-1* 286
Stellaria lagunensis IPNI=*30040285-2* 285
Stellaria lanuginosa IPNI=*30040285-2* 285
Stellaria latifolia IPNI=*159330-1* 286
Stellaria latisepala IPNI=*159330-1* 286
Stellaria laxa IPNI=*30040285-2* 285
Stellaria media subsp. *latisepala* IPNI=*159330-1* 286
Stellaria media subsp. *typica* IPNI=*159330-1* 286
Stellaria media var. *apetala* IPNI=*159330-1* 286
Stellaria media var. *gymnocalyx* IPNI=*159330-1* 286
Stellaria media var. *nana* IPNI=*159330-1* 286
Stellaria media var. *normalis* IPNI=*159330-1* 286
Stellaria media var. *oligandra* IPNI=*159330-1* 286
Stellaria media var. *subgymnocalyx* IPNI=*159330-1* 286
Stellaria media var. *succulenta* IPNI=*159330-1* 286
Stellaria media var. *transiens* IPNI=*159330-1* 286
Stellaria media var. *trichocalyx* IPNI=*159330-1* 286
Stellaria micrantha IPNI=*244185-2* 286
Stellaria microphylla IPNI=*159330-1* 286
Stellaria minor IPNI=*159330-1* 286
Stellaria modesta IPNI=*159330-1* 286
Stellaria monogyna IPNI=*159330-1* 286
Stellaria murmuria IPNI=*159330-1* 286
Stellaria neogaea IPNI=*244185-2* 286
Stellaria oligandra IPNI=*159330-1* 286
Stellaria ovata f. *paradoxa* IPNI=*159379-1* 286
Stellaria ovata var. *palmeri* IPNI=*159379-1* 286
Stellaria pilosa IPNI=*159330-1* 286
Stellaria pubescens IPNI=*30040285-2* 285
Stellaria trivialis IPNI=*152274-1* 285
Stellaria umbrosa IPNI=*159330-1* 286
Stellaria vernalis IPNI=*159330-1* 286
Stellaria viscosa IPNI=*152312-1* 285
Stellaria vulgaris IPNI=*159330-1* 286
Stellaria vulgata IPNI=*152274-1* 285
Stellaria xanthanthera IPNI=*159330-1* 286
Stellarioides longibracteata IPNI=*77097737-1* 214
Stellularia media IPNI=*159330-1* 286
Stellularia media var. *bononiensis* IPNI=*159330-1* 286
Stellularia media var. *calabrica* IPNI=*159330-1* 286
Stellularia media var. *chinensis* IPNI=*159330-1* 286
Stellularia media var. *holosteoides* IPNI=*159330-1* 286
Stellularia media var. *pygmaea* IPNI=*159330-1* 286
Stellularia media var. *vulgaris* IPNI=*159330-1* 286
Stemmadenia bella IPNI=*285810-2* 343
Stemmadenia bignoniiflora IPNI=*285810-2* 343
Stemmadenia galeottiana IPNI=*285810-2* 343
Stemmadenia grandiflora IPNI=*82139-1* 343
Stemmadenia greenmanii IPNI=*285810-2* 343
Stemmadenia insignis IPNI=*285810-2* 343
Stemmadenia lagunae IPNI=*77105798-1* 344
Stemmadenia litoralis IPNI=*285810-2* 343
Stemmadenia macrantha IPNI=*77105798-1* 344
Stemmadenia macrophylla IPNI=*285810-2* 343
Stemmadenia minima IPNI=*77105798-1* 344
Stemmadenia nervosa IPNI=*82371-1* 344
Stemmadenia pauciflora IPNI=*82139-1* 343
Stemmadenia pennellii IPNI=*82139-1* 343
Stemmadenia robinsonii IPNI=*77105798-1* 344
Stemmatella lehmannii IPNI=*208557-1* 263
Stemmatella sodiroi IPNI=*208554-1* 263
Stemmatella urticifolia IPNI=*208557-1* 263
Stemmatella urticifolia var. *eglandulosa* IPNI=*208557-1* 263
Stemmodontia carnosa IPNI=*1093589-2* 270
Stemmodontia elongata IPNI=*213524-1* 264
Stemmodontia trilobata IPNI=*1093589-2* 270
Stemodia arenaria IPNI=*244449-2* 377
Stemodia arizonica IPNI=*328499-2* 377
Stemodia berteroana IPNI=*328499-2* 377
Stemodia bissei IPNI=*328499-2* 377
Stemodia durantifolia var. *angustifolia* IPNI=*328499-2* 377
Stemodia ehrenbergiana IPNI=*328499-2* 377
Stemodia erecta f. *bracteosa* IPNI=*328499-2* 377
Stemodia erecta IPNI=*328499-2* 377
Stemodia glabra IPNI=*803515-1* 376
Stemodia humilis IPNI=*244449-2* 377
Stemodia macrotricha IPNI=*244449-2* 377
Stemodia maritima var. *rigida* IPNI=*308488-2* 377
Stemodia neglecta IPNI=*244449-2* 377
Stemodia parviflora IPNI=*244449-2* 377
Stemodia subhastata IPNI=*328499-2* 377
Stemodia verticillaris IPNI=*328499-2* 377
Stemodiacra berteroana IPNI=*328499-2* 377
Stemodiacra durantifolia IPNI=*328499-2* 377
Stemodiacra ehrenbergiana IPNI=*328499-2* 377
Stemodiacra maritima IPNI=*308488-2* 377
Stemodiacra subhastata IPNI=*328499-2* 377
Stemodiacra verticillata IPNI=*244449-2* 377

Stemonacanthus humboldtianus IPNI=*54157-1* 361
Stemonacanthus thyrsacanthoides IPNI=*54157-1* 361
Stenandrium fasciculare IPNI=*55411-1* 361
Stenandrium guatemalense IPNI=*55411-1* 361
Stenandrium mexicanum IPNI=*55411-1* 361
Stenandrium parodii IPNI=*55411-1* 361
Stenandrium trinerve IPNI=*55411-1* 361
Stenandrium trinerve var. *exscapum* IPNI=*55411-1* 361
Stenandrium trinerve var. *subexscapum* IPNI=*55411-1* 361
Stenocalyx affinis IPNI=*596263-1* 425
Stenocalyx brunneus IPNI=*596263-1* 425
Stenocalyx costatus IPNI=*596263-1* 425
Stenocalyx dasyblastus IPNI=*596263-1* 425
Stenocalyx glaber IPNI=*596263-1* 425
Stenocalyx impunctatus IPNI=*596263-1* 425
Stenocalyx lucidus IPNI=*596263-1* 425
Stenocalyx michelii IPNI=*596263-1* 425
Stenocalyx michelii var. *membranacea* IPNI=*596263-1* 425
Stenocalyx michelii var. *rigida* IPNI=*596263-1* 425
Stenocalyx oblongifolius IPNI=*596263-1* 425
Stenocalyx patrisii IPNI=*595457-1* 425
Stenocalyx patrisii var. *grandifolius* IPNI=*595457-1* 425
Stenocalyx patrisii var. *parvifolius* IPNI=*595457-1* 425
Stenocalyx plumieri IPNI=*203871-2* 428
Stenocalyx rhampiri IPNI=*596263-1* 425
Stenocalyx ruber IPNI=*596263-1* 425
Stenocalyx strigosus IPNI=*596263-1* 425
Stenocalyx uniflorus IPNI=*596263-1* 425
Stenocereus deficiens IPNI=*244634-2* 285
Stenocereus griseus f. *cristatus* IPNI=*244634-2* 285
Stenocereus griseus var. *monstrosus* IPNI=*244634-2* 285
Stenocereus peruvianus IPNI=*130627-1* 283
Stenochlaena japurensis IPNI=*17442740-1* 175
Stenocline saturejoides IPNI=*174407-1* 255
Stenolobium brachycarpum IPNI=*483437-1* 313
Stenolobium brachycarpum var. *brachystachyum* IPNI=*483437-1* 313
Stenolobium caeruleum IPNI=*60471356-2* 313
Stenolobium caeruleum var. *glabrescens* IPNI=*60471356-2* 313
Stenolobium caeruleum var. *villosius* IPNI=*60471356-2* 313
Stenolobium stans IPNI=*111284-1* 365
Stenolobus kunzeanus IPNI=*17080730-1* 175
Stenolobus ornatus IPNI=*17080730-1* 175
Stenolobus solidus IPNI=*17080730-1* 175
Stenophyllus capillaris IPNI=*1186511-2* 229
Stenophyllus junciformis IPNI=*38049-2* 229
Stenophyllus lanatus IPNI=*30060098-2* 229
Stenorrhynchos cernuum IPNI=*61935-2* 218
Stenorrhynchos millei IPNI=*659300-1* 223
Stenorrhynchos nutans IPNI=*659300-1* 223
Stenorrhynchos speciosum var. *nutans* IPNI=*659300-1* 223
Stenosiphanthus duckei IPNI=*153710-2* 364
Stenosolen eggersii f. *pubescens* IPNI=*82146-1* 343
Stenosolen eggersii IPNI=*82146-1* 343
Stenosolen grandifolius IPNI=*82146-1* 343
Stenosolen heterophyllus IPNI=*82146-1* 343
Stenosolen macrosiphon IPNI=*285821-2* 344
Stenosolen stenolobus IPNI=*82146-1* 343
Stenosolen vanheurckii IPNI=*285821-2* 344
Stenospermation ammiticum subsp. *neblinae* IPNI=*244978-2* 202
Stenostomum paniculatum IPNI=*57096-2* 348
Stenotaphrum americanum IPNI=*423051-1* 246
Stenotaphrum dimidiatum var. *americanum* IPNI=*423051-1* 246
Stenotaphrum dimidiatum var. *secundatum* IPNI=*423051-1* 246
Stenotaphrum glabrum var. *americanum* IPNI=*423051-1* 246
Stenotaphrum sarmentosum IPNI=*423051-1* 246
Stenouratea wrightii IPNI=*607063-1* 397
Stephania convolvulacea IPNI=*59219-2* 434
Stephanoluma rugosa IPNI=*160393-2* 306
Stephanophysum brevifolium IPNI=*947317-1* 361
Stephanophysum longifolium IPNI=*947317-1* 361
Stephanophysum macrandrum IPNI=*947317-1* 361
Stephanophysum ventricosum IPNI=*947317-1* 361
Stephanostachys casperiana IPNI=*54964-2* 208
Stephanostachys tepejilote IPNI=*54964-2* 208
Stephanostachys wendlandiana IPNI=*54964-2* 208
Stephanotrichum octonum IPNI=*77196163-1* 421
Sterculia carthaginensis IPNI=*245150-2* 414
Sterculia chicha IPNI=*245150-2* 414
Sterculia convoluta IPNI=*245150-2* 414
Sterculia crinita IPNI=*825412-1* 414
Sterculia diguensis IPNI=*245153-2* 414
Sterculia diversifolia IPNI=*822268-1* 407
Sterculia diversifolia var. *longipes* IPNI=*822268-1* 407
Sterculia helicteres IPNI=*245150-2* 414
Sterculia ivira IPNI=*825412-1* 414
Sterculia pruriens var. *grandiflora* IPNI=*825412-1* 414
Sterculia punctata IPNI=*245150-2* 414
Sterculia solitudinis IPNI=*245153-2* 414
Stereoxylon corymbosum IPNI=*791067-1* 309
Stereoxylon discolor IPNI=*791015-1* 309
Stereoxylon paniculatum IPNI=*791071-1* 309
Stereoxylon pendulum IPNI=*791075-1* 309
Stereoxylon rubrum IPNI=*791101-1* 309
Stethoma comata IPNI=*78920-2* 359
Stethoma pectoralis IPNI=*47290-1* 359
Steudelia cymosa IPNI=*866955-1* 405
Steudelia racemosa IPNI=*866956-1* 405
Stevia benthamiana IPNI=*30284853-2* 270
Stevia benthamiana var. *pesarensis* IPNI=*30284853-2* 270
Stevia canescens IPNI=*251571-1* 270
Stevia collina var. *hyptifolia* IPNI=*251300-1* 270
Stevia coriacea IPNI=*251437-1* 270
Stevia ehrenbergiana IPNI=*30284853-2* 270
Stevia elliptica IPNI=*251284-1* 270
Stevia elongata var. *caracasana* IPNI=*251284-1* 270
Stevia elongata var. *pallida* IPNI=*293446-2* 270
Stevia erecta IPNI=*30284853-2* 270
Stevia fascicularis IPNI=*30284853-2* 270
Stevia glutinosa IPNI=*251437-1* 270
Stevia glutinosa var. *oaxacana* IPNI=*251437-1* 270
Stevia hirtiflora IPNI=*251284-1* 270
Stevia hymenopappa IPNI=*251437-1* 270
Stevia hyptifolia IPNI=*251300-1* 270
Stevia hyssopifolia IPNI=*30284853-2* 270
Stevia ivifolia IPNI=*251571-1* 270
Stevia ivifolia var. *bogotensis* IPNI=*251571-1* 270
Stevia jorullensis var. *ehrenbergiana* IPNI=*30284853-2* 270
Stevia lanceolata IPNI=*251571-1* 270
Stevia leucantha IPNI=*251582-1* 270
Stevia linoides IPNI=*251571-1* 270
Stevia lucida var. *angustifolia* IPNI=*251437-1* 270
Stevia lucida var. *latifolia* IPNI=*251437-1* 270
Stevia lucida var. *oaxacana* IPNI=*251437-1* 270
Stevia lucida var. *typica* IPNI=*251437-1* 270
Stevia microcephala IPNI=*251571-1* 270
Stevia mollis IPNI=*251582-1* 270
Stevia nepetifolia f. *leucantha* IPNI=*251582-1* 270
Stevia nepetifolia f. *mollis* IPNI=*251582-1* 270
Stevia nepetifolia IPNI=*251582-1* 270
Stevia nervosa IPNI=*30284853-2* 270
Stevia oaxacana IPNI=*251437-1* 270
Stevia ovata var. *reglensis* IPNI=*30284853-2* 270
Stevia pallida IPNI=*251300-1* 270
Stevia paniculata IPNI=*30284853-2* 270
Stevia pubescens IPNI=*251571-1* 270
Stevia pubescens IPNI=*251582-1* 270

Stevia punctata IPNI=*251571-1* 270
Stevia quitensis IPNI=*30284853-2* 270
Stevia rebaudiana IPNI=*245468-2* 270
Stevia reglensis IPNI=*30284853-2* 270
Stevia rhombifolia f. *colorata* IPNI=*30284853-2* 270
Stevia rhombifolia f. *glandulifera* IPNI=*30284853-2* 270
Stevia rhombifolia IPNI=*30284853-2* 270
Stevia rhombifolia var. *typica* IPNI=*30284853-2* 270
Stevia rhombifolia var. *uniaristata* IPNI=*30284853-2* 270
Stevia serrata var. *arguta* IPNI=*251571-1* 270
Stevia serrata var. *haplopappa* IPNI=*251571-1* 270
Stevia serrata var. *ivifolia* IPNI=*251571-1* 270
Stevia serrata var. *linoides* IPNI=*251571-1* 270
Stevia serrata var. *typica* IPNI=*251571-1* 270
Stevia ternifolia IPNI=*30284853-2* 270
Stevia tomentosa IPNI=*251582-1* 270
Stevia uniaristata IPNI=*30284853-2* 270
Stevia virgata IPNI=*251571-1* 270
Stewartia corchoroides IPNI=*235838-2* 413
Sticherus pubescens IPNI=*245596-2* 173
Stickmannia gracilis IPNI=*79497-2* 224
Stickmannia guyanensis IPNI=*172873-1* 224
Stickmannia hexandra IPNI=*79497-2* 224
Stickmannia inaequalis IPNI=*79497-2* 224
Stickmannia longicollis IPNI=*172873-1* 224
Stickmannia mexicana IPNI=*79497-2* 224
Stickmannia mosaica IPNI=*172628-1* 224
Stickmannia ovalifolia IPNI=*79497-2* 224
Stickmannia tenuior IPNI=*79497-2* 224
Stigmaphyllon brachiatum IPNI=*557592-1* 396
Stigmaphyllon ellipticum var. *intermedium* IPNI=*30241255-2* 396
Stigmaphyllon ellipticum var. *nicaraguense* IPNI=*30241255-2* 396
Stigmaphyllon felixii IPNI=*30241255-2* 396
Stigmaphyllon fulgens IPNI=*557592-1* 396
Stigmaphyllon humboldtianum IPNI=*557494-1* 396
Stigmaphyllon hypoleucum IPNI=*557592-1* 396
Stigmaphyllon monancistrum IPNI=*557592-1* 396
Stigmaphyllon mucronatum IPNI=*30241255-2* 396
Stigmaphyllon mucronatum var. *intermedium* IPNI=*30241255-2* 396
Stigmaphyllon mucronatum var. *nicaraguense* IPNI=*30241255-2* 396
Stigmaphyllon mucronatum var. *ternatum* IPNI=*30241255-2* 396
Stigmaphyllon purpureum IPNI=*557592-1* 396
Stigmaphyllon sagittatum var. *sinuatum* IPNI=*557592-1* 396
Stigmaphyllon splendens IPNI=*557592-1* 396
Stigmaphyllon ternatum IPNI=*30241255-2* 396
Stigmaphyllon tiliifolium IPNI=*557494-1* 396
Stigmaphyllon tiliifolium var. *typicum* IPNI=*557494-1* 396
Stigmarota africana IPNI=*365348-1* 404
Stigmarota edulis IPNI=*365348-1* 404
Stigmatococca solanacea IPNI=*18168-2* 303
Stilaginella benthamii IPNI=*350078-1* 401
Stilaginella blanchetiana IPNI=*350078-1* 401
Stilaginella oblonga IPNI=*350078-1* 401
Stillingia aucuparia IPNI=*227503-2* 392
Stillingia biglandulosa IPNI=*227503-2* 392
Stillingia brasiliensis IPNI=*356183-1* 393
Stillingia concolor IPNI=*1187218-2* 387
Stillingia dracunculoides IPNI=*227503-2* 392
Stillingia guianensis IPNI=*352054-1* 392
Stillingia haematantha IPNI=*227503-2* 392
Stillingia hilariana IPNI=*352054-1* 392
Stillingia marginata IPNI=*227503-2* 392
Stilpnophyllum elasticum IPNI=*60458499-2* 438
Stipa cirrosa IPNI=*281184-2* 241
Stipa contracta IPNI=*60445990-2* 241
Stipa contracta IPNI=*948410-1* 241
Stipa eminens IPNI=*948410-1* 241
Stipa eminens var. *micrantha* IPNI=*948410-1* 241
Stipa fernandeziana IPNI=*948410-1* 241
Stipa geniculata IPNI=*281184-2* 241
Stipa hispida IPNI=*948410-1* 241
Stipa intricata IPNI=*948410-1* 241
Stipa latifolia IPNI=*316519-2* 241
Stipa ligularis IPNI=*948410-1* 241
Stipa longiflora IPNI=*948410-1* 241
Stipa melanocarpa IPNI=*404802-1* 240
Stipa microsperma IPNI=*240834-2* 245
Stipa neesiana f. *contorta* IPNI=*948410-1* 241
Stipa neesiana f. *depauperata* IPNI=*948410-1* 241
Stipa neesiana IPNI=*948410-1* 241
Stipa neesiana var. *fernandeziana* IPNI=*948410-1* 241
Stipa neesiana var. *formicarioides* IPNI=*948410-1* 241
Stipa neesiana var. *glabrata* IPNI=*948410-1* 241
Stipa neesiana var. *gracilior* IPNI=*948410-1* 241
Stipa neesiana var. *hirsuta* IPNI=*948410-1* 241
Stipa neesiana var. *ligularis* IPNI=*948410-1* 241
Stipa neesiana var. *longiaristata* IPNI=*948410-1* 241
Stipa oreophila IPNI=*281184-2* 241
Stipa setigera f. *pallida* IPNI=*948410-1* 241
Stipa setigera f. *purpurascens* IPNI=*948410-1* 241
Stipa setigera f. *versicolor* IPNI=*948410-1* 241
Stipa setigera var. *hispidula* IPNI=*948410-1* 241
Stipa skottsbergii IPNI=*948410-1* 241
Stipa spicata IPNI=*424391-1* 246
Stipa sublaevis IPNI=*948410-1* 241
Stipa subulata IPNI=*281184-2* 241
Stipa tenuissima f. *colorata* IPNI=*281184-2* 241
Stipa tenuissima IPNI=*281184-2* 241
Stipa tenuissima var. *oreophila* IPNI=*281184-2* 241
Stipa tenuissima var. *planicola* IPNI=*281184-2* 241
Stipa trachysperma IPNI=*948410-1* 241
Stipa virletii IPNI=*30065892-2* 235
Stipularia marina IPNI=*241336-2* 286
Stipularia rubra IPNI=*158983-1* 286
Stizolobium altissimum IPNI=*1197939-2* 328
Stizolobium holtonii IPNI=*164648-2* 328
Stizolobium meyeri IPNI=*486481-1* 315
Stizolobium molle IPNI=*509248-1* 328
Stizolobium montanum IPNI=*509661-1* 328
Stizolobium mutisianum IPNI=*509253-1* 328
Stizolobium pruriens IPNI=*509275-1* 328
Stizolobium pruritum IPNI=*509275-1* 328
Stizolobium spurium IPNI=*486481-1* 315
Stizolobium tuberosum IPNI=*512027-1* 329
Stizolobium urens IPNI=*1197939-2* 328
Stizophyllum flos-ardeae IPNI=*246385-2* 365
Stizophyllum punctifolium IPNI=*246385-2* 365
Stratiotes nymphoides IPNI=*58399-1* 198
Strelitzia quensonii IPNI=*798185-1* 250
Strelitzia regalis IPNI=*798194-1* 250
Strepsia usneoides IPNI=*1169665-2* 229
Strepsilobus scandens IPNI=*493866-1* 319
Streptachne domingensis IPNI=*420951-1* 244
Streptocalyx hoppii IPNI=*5475-2* 228
Streptocaulon cochinchinense IPNI=*95173-1* 339
Striga esquirolii IPNI=*366552-1* 344
Strigilia argentea IPNI=*327046-2* 308
Strigilia cordata IPNI=*826890-1* 308
Strigilia guyanensis IPNI=*327042-2* 308
Strigilia ovata IPNI=*826890-1* 308
Strigilia pavonii IPNI=*826890-1* 308
Strigilia pohlii IPNI=*826900-1* 308
Strigilia punctata IPNI=*826900-1* 308
Strigilia tomentosa IPNI=*826958-1* 309
Striolaria amazonica IPNI=*77065394-1* 354
Strobilanthes gentiliana IPNI=*675971-1* 375

Stromanthe lutea IPNI=*246147-2* 250
Strongylodon megaphyllus IPNI=*519595-1* 334
Strongylodon warburgii IPNI=*519595-1* 334
Strophacanthus bonthainensis IPNI=*52558-1* 360
Struchium africanum IPNI=*1059151-2* 270
Struchium americanum IPNI=*1059151-2* 270
Struthanthus aduncus IPNI=*77147531-1* 448
Struthanthus affinis IPNI=*185262-2* 448
Struthanthus alveolatus IPNI=*176899-2* 448
Struthanthus amplexicaulis IPNI=*176899-2* 448
Struthanthus anceps IPNI=*77147531-1* 448
Struthanthus avicularius IPNI=*77147531-1* 448
Struthanthus belizensis IPNI=*551737-1* 449
Struthanthus chordocladus IPNI=*551737-1* 449
Struthanthus conduplicatus IPNI=*77147531-1* 448
Struthanthus domingensis IPNI=*1075239-2* 450
Struthanthus eichleri IPNI=*77147531-1* 448
Struthanthus ellipticus IPNI=*548708-1* 448
Struthanthus erythrocarpus IPNI=*77147531-1* 448
Struthanthus escuintlensis IPNI=*551737-1* 449
Struthanthus lancifolius IPNI=*548708-1* 448
Struthanthus laurifolius IPNI=*548708-1* 448
Struthanthus ligustrifolius IPNI=*258636-2* 449
Struthanthus lucarquensis IPNI=*551737-1* 449
Struthanthus magdalenae IPNI=*77147531-1* 448
Struthanthus mapirensis IPNI=*258636-2* 449
Struthanthus marginatus var. *friburgensis* IPNI=*551718-1* 449
Struthanthus minutiflorus IPNI=*185262-2* 448
Struthanthus nitidus IPNI=*548708-1* 448
Struthanthus occidentalis IPNI=*550656-1* 448
Struthanthus orinocensis IPNI=*77147531-1* 448
Struthanthus paniculatus IPNI=*77147531-1* 448
Struthanthus pedunculatus IPNI=*77147531-1* 448
Struthanthus perrottetii IPNI=*185262-2* 448
Struthanthus platycladus IPNI=*185262-2* 448
Struthanthus punctatus IPNI=*548708-1* 448
Struthanthus pyrifolius IPNI=*185262-2* 448
Struthanthus pyrifolius var. *terminalis* IPNI=*185262-2* 448
Struthanthus retroflexus IPNI=*77147531-1* 448
Struthanthus sessilis IPNI=*77147531-1* 448
Struthanthus stelis IPNI=*77147531-1* 448
Struthanthus suaveolens IPNI=*258636-2* 449
Struthanthus subcampestris IPNI=*185262-2* 448
Struthanthus theobromae IPNI=*77147531-1* 448
Struthanthus virgatus IPNI=*77147531-1* 448
Struthanthus vulgaris IPNI=*551718-1* 449
Struthiopteris appendiculata IPNI=*33632-2* 174
Struthiopteris brasiliensis IPNI=*33533-2* 174
Struthiopteris caudata IPNI=*33632-2* 174
Struthiopteris cinnamomea IPNI=*60447721-2* 173
Struthiopteris gilliesii IPNI=*33533-2* 174
Struthiopteris lineata IPNI=*1148359-2* 175
Struthiopteris loxensis IPNI=*77186436-1* 174
Struthiopteris regalis IPNI=*305721-2* 173
Strychnodaphne floribunda IPNI=*1146025-2* 184
Strychnodaphne puberula IPNI=*1032261-2* 185
Strychnodaphne puberula var. *angustata* IPNI=*1032261-2* 185
Strychnodaphne puberula var. *truncata* IPNI=*1032261-2* 185
Strychnos blackii IPNI=*547077-1* 346
Strychnos bovetiana IPNI=*547183-1* 346
Strychnos brevifolia IPNI=*547493-1* 346
Strychnos calantha IPNI=*547220-1* 346
Strychnos concinna IPNI=*547398-1* 346
Strychnos cordifolia IPNI=*547493-1* 346
Strychnos crevauxiana IPNI=*547207-1* 346
Strychnos crevauxii IPNI=*547207-1* 346
Strychnos curare IPNI=*547220-1* 346
Strychnos depauperata IPNI=*547493-1* 346
Strychnos geoffraeana IPNI=*547220-1* 346
Strychnos guianensis var. *trichina* IPNI=*547220-1* 346
Strychnos hachensis IPNI=*547390-1* 346
Strychnos hirtiflora IPNI=*547390-1* 346
Strychnos lanceolata IPNI=*547220-1* 346
Strychnos longissima IPNI=*547390-1* 346
Strychnos marginata IPNI=*547398-1* 346
Strychnos oblonga IPNI=*547220-1* 346
Strychnos panamensis var. *hirtiflora* IPNI=*547390-1* 346
Strychnos panurensis IPNI=*547217-1* 346
Strychnos pedunculata IPNI=*547083-1* 346
Strychnos petrophila IPNI=*547493-1* 346
Strychnos placida IPNI=*547217-1* 346
Strychnos rouhamon IPNI=*547220-1* 346
Strychnos ruizii IPNI=*547077-1* 346
Strychnos schomburgkiana IPNI=*547083-1* 346
Strychnos smilacina IPNI=*547341-1* 346
Strychnos smithiana IPNI=*547453-1* 346
Strychnos syntoxica IPNI=*547525-1* 346
Strychnos tepicensis IPNI=*547390-1* 346
Strychnos trichostyla IPNI=*547247-1* 346
Strychnos trinitensis IPNI=*547083-1* 346
Strychnos ulei IPNI=*547083-1* 346
Strychnos yapurensis IPNI=*547247-1* 346
Stryphnodendron angustum IPNI=*519643-1* 334
Stryphnodendron floribundum IPNI=*519643-1* 334
Stryphnodendron guianense f. *floribundum* IPNI=*519643-1* 334
Stryphnodendron melinonis IPNI=*519643-1* 334
Stuckenia borealis IPNI=*77099639-1* 203
Stuckenia chakassiensis IPNI=*77099639-1* 203
Stuckenia filiformis subsp. *borealis* IPNI=*77099639-1* 203
Stuckenia helvetica IPNI=*77099639-1* 203
Stuckenia intramongolica IPNI=*77099639-1* 203
Stuckenia marina IPNI=*77099639-1* 203
Stuckenia mongolica IPNI=*77099639-1* 203
Stuckenia pectinata var. *interrupta* IPNI=*77099639-1* 203
Stuckenia zosteracea IPNI=*77099639-1* 203
Stuebelia nemorosa IPNI=*77183738-1* 278
Stuebelia nitida IPNI=*77183738-1* 278
Sturmia nervosa IPNI=*641061-1* 219
Styloceras dombeyanum IPNI=*356634-1* 280
Stylodiscus trifoliatus IPNI=*339921-1* 401
Stylogyne amazonica IPNI=*247474-2* 304
Stylogyne amplifolia IPNI=*247441-2* 304
Stylogyne balaensis IPNI=*247490-2* 305
Stylogyne brasiliensis IPNI=*247474-2* 304
Stylogyne brunnescens IPNI=*247472-2* 304
Stylogyne chartacea IPNI=*247483-2* 305
Stylogyne gentryi IPNI=*247441-2* 304
Stylogyne guatemalensis IPNI=*247490-2* 305
Stylogyne kappleri IPNI=*247474-2* 304
Stylogyne latifolia IPNI=*247469-2* 304
Stylogyne micans IPNI=*247474-2* 304
Stylogyne oaxacana IPNI=*247490-2* 305
Stylogyne perpunctata IPNI=*247490-2* 305
Stylogyne poeppigii IPNI=*247474-2* 304
Stylogyne schomburgkiana IPNI=*247474-2* 304
Stylogyne surinamensis IPNI=*247474-2* 304
Stylogyne turumiquirensis IPNI=*247490-2* 305
Stylogyne venezuelana IPNI=*247469-2* 304
Stylogyne viridis IPNI=*247472-2* 304
Styloma pacifica IPNI=*669288-1* 211
Stylosanthes biflora var. *guianensis* IPNI=*1020451-2* 334
Stylosanthes debilis IPNI=*1176701-2* 334
Stylosanthes diarthra IPNI=*304693-2* 334
Stylosanthes eriocarpa IPNI=*247544-2* 334
Stylosanthes figueroae IPNI=*519681-1* 334
Stylosanthes gloiodes IPNI=*304693-2* 334

Stylosanthes glutinosa IPNI=*1176701-2* 334
Stylosanthes humilis IPNI=*247544-2* 334
Stylosanthes pilosa IPNI=*1176701-2* 334
Stylosanthes plicata IPNI=*304693-2* 334
Stylosanthes procumbens IPNI=*247544-2* 334
Stylosanthes prostrata IPNI=*1176701-2* 334
Stylosanthes scoparia IPNI=*304693-2* 334
Stylosanthes subsericea IPNI=*304693-2* 334
Stylosanthes sundaica IPNI=*519681-1* 334
Stylosanthes viscosa var. *minor* IPNI=*1176701-2* 334
Stylurus robustus IPNI=*50798-3* 433
Styrax ambiguus IPNI=*826900-1* 308
Styrax ambiguus var. *apiculatus* IPNI=*826900-1* 308
Styrax andinus IPNI=*826890-1* 308
Styrax anomalus IPNI=*826900-1* 308
Styrax argyrophyllus IPNI=*826890-1* 308
Styrax bogotensis IPNI=*826900-1* 308
Styrax buchtienii IPNI=*327039-2* 308
Styrax cordatus IPNI=*826890-1* 308
Styrax discolor IPNI=*826900-1* 308
Styrax ferax IPNI=*327039-2* 308
Styrax ferrugineus IPNI=*826900-1* 308
Styrax heterotrichus IPNI=*826890-1* 308
Styrax japurensis IPNI=*327042-2* 308
Styrax leptactinosus IPNI=*327039-2* 308
Styrax limoncillo IPNI=*827186-1* 309
Styrax longifolius IPNI=*327042-2* 308
Styrax loxensis IPNI=*826958-1* 309
Styrax macrotrichus IPNI=*247727-2* 309
Styrax mathewsii IPNI=*826890-1* 308
Styrax myristicifolius IPNI=*327046-2* 308
Styrax ovatus IPNI=*826890-1* 308
Styrax pachyphyllus IPNI=*826900-1* 308
Styrax pohlii f. *calvescens* IPNI=*826900-1* 308
Styrax poissonianus IPNI=*826958-1* 309
Styrax pseudargyrophyllus IPNI=*826890-1* 308
Styrax punctatus IPNI=*826900-1* 308
Styrax rigidifolius f. *yapobodensis* IPNI=*247710-2* 309
Styrax socialis IPNI=*327039-2* 308
Styrax subheterotrichus IPNI=*327039-2* 308
Styrax tessmannii IPNI=*327042-2* 308
Styrax vidalianus IPNI=*826890-1* 308
Styrax weberbaueri IPNI=*826890-1* 308
Styrax yapobodensis IPNI=*247710-2* 309
Suardia picta IPNI=*408232-1* 241
Sublimia areca IPNI=*664107-1* 204
Subpilocereus atroviridis IPNI=*130627-1* 283
Subpilocereus fricii IPNI=*52982-2* 283
Subpilocereus grenadensis IPNI=*130627-1* 283
Subpilocereus margaritensis IPNI=*130627-1* 283
Subpilocereus margaritensis var. *micranthus* IPNI=*52982-2* 283
Subpilocereus remolinensis IPNI=*130627-1* 283
Subpilocereus repandus IPNI=*130627-1* 283
Subpilocereus repandus subsp. *micracanthus* IPNI=*52982-2* 283
Subpilocereus repandus var. *weberi* IPNI=*130627-1* 283
Subpilocereus russelianus IPNI=*52982-2* 283
Subpilocereus russelianus var. *margaritensis* IPNI=*130627-1* 283
Subpilocereus russelianus var. *micracanthus* IPNI=*52982-2* 283
Sudamerlycaste longipetala IPNI=*155147-2* 221
Sulla amoena IPNI=*519725-1* 334
Suprago spicata IPNI=*230634-1* 265
Swainsona atrococcinea IPNI=*948198-1* 334
Swallenochloa tessellata IPNI=*274648-2* 236
Swartzia bifida IPNI=*519836-1* 334
Swartzia crocea IPNI=*520020-1* 335
Swartzia darienensis IPNI=*520020-1* 335
Swartzia discolor IPNI=*519927-1* 334
Swartzia dodecandra IPNI=*519836-1* 334
Swartzia erythrocarpa IPNI=*520018-1* 335
Swartzia fugax IPNI=*519927-1* 334
Swartzia macrosperma IPNI=*520020-1* 335
Swartzia melanoxylon IPNI=*519927-1* 334
Swartzia ochnacea IPNI=*520020-1* 335
Swartzia opacifolia IPNI=*519994-1* 334
Swartzia parviflora IPNI=*519836-1* 334
Swartzia platygyne IPNI=*519994-1* 334
Swartzia rariflora IPNI=*519836-1* 334
Swartzia rotundata IPNI=*519927-1* 334
Swartzia simplex var. *continentalis* IPNI=*520020-1* 335
Swartzia simplex var. *darienensis* IPNI=*520020-1* 335
Swartzia simplex var. *genuina* IPNI=*520020-1* 335
Swartzia simplex var. *grandiflora* IPNI=*520020-1* 335
Swartzia simplex var. *jamaicensis* IPNI=*520020-1* 335
Swartzia simplex var. *ochnacea* IPNI=*520020-1* 335
Swartzia simplicifolia IPNI=*520020-1* 335
Swartzia trifolia IPNI=*520020-1* 335
Swartzia triphylla IPNI=*519836-1* 334
Swartzia triphylla var. *grandiflora* IPNI=*520020-1* 335
Swartzia urubuensis IPNI=*519994-1* 334
Sweetia nitens IPNI=*502385-1* 325
Sweetia panamensis IPNI=*77091925-1* 325
Sweetia tomentella IPNI=*502385-1* 325
Swertia asclepiadea IPNI=*369443-1* 345
Swertia quadricornis IPNI=*369443-1* 345
Swietenia alternifolia IPNI=*834759-1* 410
Swietenia krukovii IPNI=*519937-1* 334
Swietenia macrophylla var. *marabaensis* IPNI=*519937-1* 334
Syagrus allenii IPNI=*246957-2* 212
Syagrus argentea IPNI=*670066-1* 212
Syagrus chavesiana IPNI=*246946-2* 212
Syagrus chiragua IPNI=*670066-1* 212
Syagrus chloroleuca IPNI=*664513-1* 206
Syagrus ecuadorensis IPNI=*670066-1* 212
Syagrus tessmannii IPNI=*670066-1* 212
Syagrus weddelliana var. *cinereus* IPNI=*246979-2* 212
Syagrus weddelliana var. *pynaertii* IPNI=*246979-2* 212
Sykesia hongkongensis IPNI=*753834-1* 352
Sykoraea hortensis IPNI=*140670-1* 272
Symphonia coccinea IPNI=*428883-1* 385
Symphonia gabonensis IPNI=*429281-1* 385
Symphonia globulifera var. *angustifolia* IPNI=*429281-1* 385
Symphonia globulifera var. *macoubea* IPNI=*429281-1* 385
Symphonia globulifera var. *major* IPNI=*429281-1* 385
Symphonia grandiflora IPNI=*326389-2* 385
Symphonia moronobea IPNI=*326389-2* 385
Symphysodaphne cubensis IPNI=*465312-1* 183
Symphytum album IPNI=*120815-1* 275
Symphytum ambiguum IPNI=*120815-1* 275
Symphytum besseri IPNI=*120815-1* 275
Symphytum bohemicum IPNI=*120815-1* 275
Symphytum coccineum IPNI=*120815-1* 275
Symphytum commune IPNI=*120815-1* 275
Symphytum consolida IPNI=*120815-1* 275
Symphytum elatum IPNI=*120815-1* 275
Symphytum majus IPNI=*120815-1* 275
Symphytum microcalyx IPNI=*120815-1* 275
Symphytum molle IPNI=*120815-1* 275
Symphytum patens IPNI=*120815-1* 275
Symphytum peregrinum IPNI=*120815-1* 275
Symphytum rakosiense IPNI=*120815-1* 275
Symphytum stenophyllum IPNI=*120815-1* 275
Symplocos alstonia IPNI=*827837-1* 309
Symplocos apiculata IPNI=*827186-1* 309
Symplocos apolis IPNI=*827005-1* 309
Symplocos bradei IPNI=*827299-1* 309
Symplocos castanea IPNI=*827005-1* 309

Symplocos gibraltarica IPNI=*827299-1* 309
Symplocos irazuensis IPNI=*827299-1* 309
Symplocos martinicensis var. *apiculata* IPNI=*827186-1* 309
Symplocos mathewsii IPNI=*827005-1* 309
Symplocos rufescens IPNI=*827299-1* 309
Symplocos tomentosa IPNI=*827299-1* 309
Synaphlebium tetraphyllum IPNI=*315261-2* 177
Synechanthus angustifolius IPNI=*248082-2* 212
Synechanthus ecuadorensis IPNI=*248082-2* 212
Synechanthus panamensis IPNI=*248082-2* 212
Syngonium albolineatum IPNI=*89127-1* 202
Syngonium oerstedianum IPNI=*89127-1* 202
Syngonium podophyllum var. *albolineatum* IPNI=*89127-1* 202
Syngonium podophyllum var. *oerstedianum* IPNI=*89127-1* 202
Syngonium podophyllum var. *typicum* IPNI=*89152-1* 202
Synnema verticillatum IPNI=*49672-1* 360
Synoptera laevigata IPNI=*572641-1* 421
Syntherisma aegyptiaca IPNI=*397387-1* 236
Syntherisma argillacea IPNI=*398643-1* 237
Syntherisma barbata IPNI=*1218147-2* 237
Syntherisma digitata IPNI=*329018-2* 237
Syntherisma fimbriata IPNI=*1038558-2* 237
Syntherisma fusca IPNI=*399226-1* 238
Syntherisma fuscescens IPNI=*80254-2* 237
Syntherisma helleri IPNI=*399226-1* 238
Syntherisma longiflora IPNI=*398933-1* 237
Syntherisma marginata IPNI=*1038558-2* 237
Syntherisma praecox IPNI=*399115-1* 238
Syntherisma pubescens IPNI=*398933-1* 237
Syntherisma sasakii IPNI=*1038558-2* 237
Syntherisma sericea IPNI=*1038558-2* 237
Syntherisma setosa IPNI=*329014-2* 237
Syntherisma velutina IPNI=*80280-2* 237
Syntherisma velutina subsp. *glabella* IPNI=*80280-2* 237
Syntherisma violascens IPNI=*399226-1* 238
Synzistachium peruvianum IPNI=*116670-1* 275
Synzyganthera purpurea IPNI=*134457-2* 394
Syzygium brachiatum IPNI=*601603-1* 428
Syzygium caryophyllifolium IPNI=*601603-1* 428
Syzygium cumini var. *caryophyllifolium* IPNI=*601603-1* 428
Syzygium cumini var. *obtusifolium* IPNI=*601603-1* 428
Syzygium cumini var. *tsoi* IPNI=*601603-1* 428
Syzygium ellipticum IPNI=*600841-1* 428
Syzygium fruticosum IPNI=*601603-1* 428
Syzygium jambolanum IPNI=*601603-1* 428
Syzygium jambolanum var. *axillare* IPNI=*601603-1* 428
Syzygium jambos var. *linearilimbum* IPNI=*601802-1* 428
Syzygium laeve IPNI=*601907-1* 428
Syzygium leptostachyum IPNI=*601802-1* 428
Syzygium merrillii IPNI=*601802-1* 428
Syzygium michelii IPNI=*596263-1* 425
Syzygium monanthum IPNI=*601802-1* 428
Syzygium obovatum IPNI=*601603-1* 428
Syzygium obtusifolium IPNI=*601603-1* 428
Syzygium pseudojambolana IPNI=*601603-1* 428
Syzygium tenue IPNI=*601603-1* 428

T

Tabebuia aesculifolia IPNI=*111572-2* 363
Tabebuia araliacea IPNI=*77085096-1* 364
Tabebuia arenicola IPNI=*248689-2* 365
Tabebuia avellanedae IPNI=*117330-2* 363
Tabebuia barbata IPNI=*109687-1* 363
Tabebuia beyeri IPNI=*248689-2* 365
Tabebuia billbergii IPNI=*77085061-1* 363
Tabebuia brigandina IPNI=*248689-2* 365
Tabebuia camagueyensis IPNI=*248689-2* 365
Tabebuia capitata IPNI=*109688-1* 363
Tabebuia capotei IPNI=*248689-2* 365
Tabebuia chrysantha IPNI=*77085066-1* 363
Tabebuia chrysea IPNI=*222517-2* 365
Tabebuia chrysotricha IPNI=*109691-1* 363
Tabebuia chrysotricha var. *obtusata* IPNI=*109691-1* 363
Tabebuia coralibe IPNI=*77085069-1* 363
Tabebuia cordata IPNI=*153710-2* 364
Tabebuia curtissii IPNI=*248689-2* 365
Tabebuia dictyophylla IPNI=*248689-2* 365
Tabebuia donnell-smithii IPNI=*222518-2* 365
Tabebuia dugandii IPNI=*117330-2* 363
Tabebuia dura IPNI=*248703-2* 365
Tabebuia fuscata IPNI=*111572-2* 363
Tabebuia geronensis IPNI=*248689-2* 365
Tabebuia globiflora IPNI=*111572-2* 363
Tabebuia glomerata IPNI=*109688-1* 363
Tabebuia gonavensis IPNI=*248689-2* 365
Tabebuia guayacan IPNI=*77085089-1* 363
Tabebuia heterophylla subsp. *genuina* IPNI=*248689-2* 365
Tabebuia hypolepra IPNI=*109688-1* 363
Tabebuia impetiginosa IPNI=*117330-2* 363
Tabebuia incana IPNI=*77085090-1* 363
Tabebuia insignis var. *monophylla* IPNI=*248703-2* 365
Tabebuia insignis var. *pacimonensis* IPNI=*248703-2* 365
Tabebuia ipe var. *integra* IPNI=*117330-2* 363
Tabebuia japurensis IPNI=*77142016-1* 363
Tabebuia latifolia IPNI=*42392-2* 362
Tabebuia leptopoda IPNI=*248689-2* 365
Tabebuia lindahlii IPNI=*248689-2* 365
Tabebuia longipes IPNI=*248703-2* 365
Tabebuia lucida IPNI=*248689-2* 365
Tabebuia mansoana IPNI=*42392-2* 362
Tabebuia mexicana IPNI=*111027-1* 365
Tabebuia micrantha IPNI=*229565-2* 377
Tabebuia millsii IPNI=*222518-2* 365
Tabebuia nicaraguensis IPNI=*117330-2* 363
Tabebuia obscura IPNI=*109700-1* 364
Tabebuia obscura var. *schultesiana* IPNI=*109700-1* 364
Tabebuia ochracea IPNI=*117333-2* 364
Tabebuia odontodiscus IPNI=*248789-2* 365
Tabebuia odontodiscus var. *violascens* IPNI=*248789-2* 365
Tabebuia pallida subsp. *heterophylla* IPNI=*248689-2* 365
Tabebuia palmeri IPNI=*117330-2* 363
Tabebuia papyrophloios IPNI=*248789-2* 365
Tabebuia pentaphylla IPNI=*248689-2* 365
Tabebuia pentaphylla var. *leucoxylon* IPNI=*111027-1* 365
Tabebuia pentaphylla var. *normalis* IPNI=*111027-1* 365
Tabebuia pisoniana IPNI=*77100464-1* 363
Tabebuia piutinga IPNI=*248789-2* 365
Tabebuia punctatissima IPNI=*111027-1* 365
Tabebuia riparia IPNI=*248689-2* 365
Tabebuia roraimae IPNI=*248703-2* 365
Tabebuia rufinervis IPNI=*42392-2* 362
Tabebuia schunkevigoi IPNI=*117330-2* 363
Tabebuia serratifolia IPNI=*77085096-1* 364
Tabebuia speciosa IPNI=*42392-2* 362
Tabebuia subtilis var. *schultesiana* IPNI=*109700-1* 364
Tabebuia triphylla IPNI=*248689-2* 365
Tabebuia xanthophylla IPNI=*77142096-1* 365
Taberna cymosa IPNI=*82084-1* 343
Taberna discolor IPNI=*82097-1* 343
Taberna poeppigii IPNI=*82327-1* 344
Tabernaemontana acapulcensis IPNI=*81996-1* 343
Tabernaemontana albescens IPNI=*82371-1* 344
Tabernaemontana amblyblasta IPNI=*81985-1* 343
Tabernaemontana amblyocarpa IPNI=*81985-1* 343
Tabernaemontana amygdalifolia var. *glaucophylla* IPNI=*81996-1* 343
Tabernaemontana amygdalifolia var. *obtusiloba* IPNI=*81996-1* 343

Tabernaemontana arcuata IPNI=*82084-1* 343
Tabernaemontana aubletii IPNI=*79852-1* 341
Tabernaemontana berteroi var. *parviflora* IPNI=*81985-1* 343
Tabernaemontana buchtienii IPNI=*82084-1* 343
Tabernaemontana chrysocarpa IPNI=*81985-1* 343
Tabernaemontana chrysocarpa var. *costaricensis* IPNI=*248876-2* 343
Tabernaemontana citrifolia IPNI=*82097-1* 343
Tabernaemontana concinna IPNI=*82084-1* 343
Tabernaemontana coronaria IPNI=*82097-1* 343
Tabernaemontana costaricensis IPNI=*248876-2* 343
Tabernaemontana cuyabensis IPNI=*20001045-1* 344
Tabernaemontana deamii IPNI=*81996-1* 343
Tabernaemontana discolor IPNI=*82097-1* 343
Tabernaemontana duckei IPNI=*20001045-1* 344
Tabernaemontana echinata IPNI=*20001045-1* 344
Tabernaemontana echites IPNI=*78821-1* 340
Tabernaemontana fasciculata IPNI=*182946-2* 342
Tabernaemontana flabelliformis IPNI=*82097-1* 343
Tabernaemontana flavescens IPNI=*79896-1* 341
Tabernaemontana gentryana IPNI=*248876-2* 343
Tabernaemontana gratissima IPNI=*82097-1* 343
Tabernaemontana guianensis IPNI=*20001045-1* 344
Tabernaemontana guyanensis IPNI=*20001045-1* 344
Tabernaemontana hirtula IPNI=*20001045-1* 344
Tabernaemontana hirtula var. *maynensis* IPNI=*20001045-1* 344
Tabernaemontana holothuria IPNI=*82146-1* 343
Tabernaemontana inaequilateralis IPNI=*285812-2* 343
Tabernaemontana jasminoides IPNI=*81996-1* 343
Tabernaemontana juruana IPNI=*20001045-1* 344
Tabernaemontana killipii IPNI=*20001045-1* 344
Tabernaemontana lactiflua IPNI=*182946-2* 342
Tabernaemontana laurifolia IPNI=*285810-2* 343
Tabernaemontana longifolia IPNI=*20001045-1* 344
Tabernaemontana lucida IPNI=*945268-1* 342
Tabernaemontana lurida IPNI=*82097-1* 343
Tabernaemontana macrophylla IPNI=*79852-1* 341
Tabernaemontana macrosiphon IPNI=*285821-2* 344
Tabernaemontana mapirensis IPNI=*82084-1* 343
Tabernaemontana martensii IPNI=*81985-1* 343
Tabernaemontana meyeri IPNI=*82371-1* 344
Tabernaemontana nereifolia IPNI=*81996-1* 343
Tabernaemontana obliqua IPNI=*82371-1* 344
Tabernaemontana occidentalis IPNI=*81996-1* 343
Tabernaemontana paisavelensis IPNI=*81985-1* 343
Tabernaemontana paucifolia IPNI=*79852-1* 341
Tabernaemontana pendula IPNI=*248876-2* 343
Tabernaemontana perrottetii IPNI=*82371-1* 344
Tabernaemontana poeppigii IPNI=*82327-1* 344
Tabernaemontana psychotriifolia IPNI=*82084-1* 343
Tabernaemontana psychotrioides IPNI=*82084-1* 343
Tabernaemontana recurva IPNI=*82097-1* 343
Tabernaemontana reticulata IPNI=*79852-1* 341
Tabernaemontana riparia IPNI=*82139-1* 343
Tabernaemontana schippii IPNI=*248839-2* 343
Tabernaemontana siamensis IPNI=*82097-1* 343
Tabernaemontana speciosa IPNI=*20001045-1* 344
Tabernaemontana sprucei IPNI=*148754-2* 341
Tabernaemontana stenoloba IPNI=*82146-1* 343
Tabernaemontana tenuiflora IPNI=*82146-1* 343
Tabernaemontana tenuis IPNI=*248891-2* 344
Tabernaemontana ternstroemiacea IPNI=*168521-2* 342
Tabernaemontana tessmannii IPNI=*20001045-1* 344
Tabernaemontana tetrastachya IPNI=*20001045-1* 344
Tabernaemontana tuxtlensis IPNI=*81985-1* 343
Tabernaemontana umbellata IPNI=*81985-1* 343
Tabernaemontana umbrosa IPNI=*82084-1* 343
Tabernaemontana urguiculata IPNI=*82146-1* 343
Tabernaemontana veracrucensis IPNI=*81985-1* 343
Tachia paniculata IPNI=*520138-1* 335
Tachia trigona IPNI=*520138-1* 335
Tachigali angustifolia IPNI=*520138-1* 335
Tachigali carinata IPNI=*520138-1* 335
Tachigali eriocalyx IPNI=*520138-1* 335
Tachigali grandiflora IPNI=*520138-1* 335
Tachigali paniculata var. *angustifolia* IPNI=*520138-1* 335
Tachigali paniculata var. *cavipes* IPNI=*248955-2* 335
Tachigali paniculata var. *comosa* IPNI=*520138-1* 335
Tachigali pulchra IPNI=*520138-1* 335
Tachigali purpurea IPNI=*492750-1* 318
Tachigali reticulosa IPNI=*974515-1* 335
Tachigali rusbyi IPNI=*520138-1* 335
Tachigali sericea IPNI=*520138-1* 335
Tachigali subbullata IPNI=*974508-1* 335
Tachigali trigona IPNI=*520138-1* 335
Tachigali uleana IPNI=*974508-1* 335
Tachigali ulei IPNI=*520138-1* 335
Tacsonia bicoronata IPNI=*675224-1* 399
Tacsonia buchananii IPNI=*30044451-2* 400
Tacsonia canaliculata IPNI=*675022-1* 398
Tacsonia coccinea IPNI=*674909-1* 398
Tacsonia cumbalensis IPNI=*184886-2* 398
Tacsonia cyanea IPNI=*184886-2* 398
Tacsonia cyanea var. *insignis* IPNI=*184886-2* 398
Tacsonia eriantha IPNI=*675224-1* 399
Tacsonia flexipes IPNI=*674996-1* 398
Tacsonia fockeana IPNI=*675022-1* 398
Tacsonia fulgens IPNI=*674909-1* 398
Tacsonia glaberrima var. *cumbalensis* IPNI=*184886-2* 398
Tacsonia glandulosa IPNI=*675022-1* 398
Tacsonia glandulosa var. *canaliculata* IPNI=*675022-1* 398
Tacsonia grandis IPNI=*674787-1* 398
Tacsonia hederacea IPNI=*184886-2* 398
Tacsonia ignea IPNI=*675190-1* 399
Tacsonia longifolia IPNI=*321973-2* 400
Tacsonia manicata IPNI=*675190-1* 399
Tacsonia micradena IPNI=*675348-1* 399
Tacsonia mixta IPNI=*675224-1* 399
Tacsonia mixta subsp. *quitensis* IPNI=*675224-1* 399
Tacsonia mixta subsp. *tomentosa* IPNI=*675224-1* 399
Tacsonia mixta var. *eriantha* IPNI=*675224-1* 399
Tacsonia mixta var. *longiflora* IPNI=*675224-1* 399
Tacsonia mixta var. *speciosa* IPNI=*675224-1* 399
Tacsonia mollissima IPNI=*185076-2* 399
Tacsonia mollissima IPNI=*675224-1* 399
Tacsonia mollissima var. *glabrescens* IPNI=*185076-2* 399
Tacsonia parritae IPNI=*185110-2* 399
Tacsonia pennipes IPNI=*675348-1* 399
Tacsonia pinnatistipula IPNI=*675348-1* 399
Tacsonia pinnatistipula var. *pennipes* IPNI=*675348-1* 399
Tacsonia psilantha IPNI=*675547-1* 400
Tacsonia pubescens IPNI=*675392-1* 400
Tacsonia purupuru IPNI=*675348-1* 399
Tacsonia quadridentata IPNI=*675392-1* 400
Tacsonia quadriglandulosa IPNI=*675392-1* 400
Tacsonia quitensis IPNI=*675224-1* 399
Tacsonia quitensis var. *eriantha* IPNI=*675224-1* 399
Tacsonia rohriana IPNI=*675022-1* 398
Tacsonia sanguinea IPNI=*30044451-2* 400
Tacsonia sanguinea IPNI=*675392-1* 400
Tacsonia serrata IPNI=*675224-1* 399
Tacsonia speciosa IPNI=*675224-1* 399
Tacsonia splendens IPNI=*675190-1* 399
Tacsonia stoupyana IPNI=*675022-1* 398
Tacsonia subcoriacea IPNI=*675022-1* 398
Tacsonia tacso IPNI=*675224-1* 399
Tacsonia tomentosa IPNI=*675224-1* 399

Tacsonia tomentosa var. *speciosa* IPNI=*675224-1* 399
Tacsonia tripartita IPNI=*675547-1* 400
Tacsonia tungurahuae IPNI=*184886-2* 398
Tacsonia urceolata IPNI=*675224-1* 399
Tacsonia vanvolxemii IPNI=*674787-1* 398
Tacsonia volxemii IPNI=*674787-1* 398
Taenais ampla IPNI=*63934-1* 213
Taenais zeylanica IPNI=*64125-1* 213
Taenianthera acaulis IPNI=*667158-1* 209
Taenianthera camana IPNI=*667031-1* 209
Taenianthera gracilis IPNI=*667158-1* 209
Taenianthera lagesiana IPNI=*667031-1* 209
Taenianthera macrostachys IPNI=*667158-1* 209
Taenianthera minor IPNI=*667158-1* 209
Taenianthera multisecta IPNI=*108938-2* 209
Taenianthera oligosticha IPNI=*667158-1* 209
Taenianthera tamandua IPNI=*667158-1* 209
Taenianthera tapajotensis IPNI=*667158-1* 209
Taeniopsis graminifolia IPNI=*266216-2* 178
Taeniopsis lineata IPNI=*17234040-1* 178
Taenitis linearis IPNI=*266216-2* 178
Taetsia ferrea IPNI=*533580-1* 215
Taetsia fruticosa IPNI=*533580-1* 215
Taetsia terminalis IPNI=*533580-1* 215
Tafalla angustifolia IPNI=*167832-1* 181
Tafalla artocarpus IPNI=*167876-1* 181
Tafalla bonplandiana IPNI=*277969-2* 181
Tafalla callososerrata IPNI=*277969-2* 181
Tafalla cumbalensis IPNI=*167850-1* 181
Tafalla glabrata IPNI=*167884-1* 181
Tafalla glauca IPNI=*167884-1* 181
Tafalla glauca IPNI=*277969-2* 181
Tafalla goudotiana IPNI=*277980-2* 181
Tafalla integra IPNI=*167884-1* 181
Tafalla laciniata IPNI=*167832-1* 181
Tafalla mandonii IPNI=*167887-1* 182
Tafalla maxima IPNI=*118340-2* 181
Tafalla mexicana IPNI=*167876-1* 181
Tafalla racemosa IPNI=*167884-1* 181
Tafalla scabra IPNI=*167887-1* 182
Tafalla sprucei IPNI=*277998-2* 182
Tafalla thuyoides IPNI=*231180-1* 265
Tagera filiformis IPNI=*518369-1* 333
Tagetes anethina IPNI=*252121-1* 271
Tagetes anisata IPNI=*252097-1* 271
Tagetes bonariensis IPNI=*252132-1* 271
Tagetes cabrerae IPNI=*60461729-2* 271
Tagetes caracasana IPNI=*252181-1* 271
Tagetes congesta IPNI=*252097-1* 271
Tagetes corymbosa IPNI=*252092-1* 271
Tagetes crassiceps IPNI=*252181-1* 271
Tagetes dichotoma IPNI=*252097-1* 271
Tagetes elongata IPNI=*252092-1* 271
Tagetes erecta f. *pleniflora* IPNI=*252092-1* 271
Tagetes ernstii IPNI=*252177-1* 271
Tagetes florida IPNI=*252121-1* 271
Tagetes foeniculacea IPNI=*252097-1* 271
Tagetes fragrantissima IPNI=*252097-1* 271
Tagetes gigantea IPNI=*60461729-2* 271
Tagetes gilletii IPNI=*252121-1* 271
Tagetes glandulifera IPNI=*252132-1* 271
Tagetes glandulosa IPNI=*252132-1* 271
Tagetes graveolens IPNI=*60461729-2* 271
Tagetes heterocarpha IPNI=*252092-1* 271
Tagetes integrifolia IPNI=*208446-2* 268
Tagetes jaliscensis var. *minor* IPNI=*252177-1* 271
Tagetes lucida f. *florida* IPNI=*252121-1* 271
Tagetes macroglossa IPNI=*252177-1* 271
Tagetes major IPNI=*252092-1* 271
Tagetes microglossa IPNI=*252177-1* 271
Tagetes montana IPNI=*252132-1* 271
Tagetes multifida IPNI=*252097-1* 271
Tagetes oligocephala IPNI=*252177-1* 271
Tagetes patula IPNI=*252092-1* 271
Tagetes pectinata IPNI=*252183-1* 271
Tagetes peduncularis IPNI=*252177-1* 271
Tagetes perretii IPNI=*252097-1* 271
Tagetes pineda IPNI=*252121-1* 271
Tagetes porophyllum IPNI=*252132-1* 271
Tagetes pseudomicrantha IPNI=*252097-1* 271
Tagetes punctata IPNI=*252121-1* 271
Tagetes pusilla IPNI=*252097-1* 271
Tagetes remotiflora IPNI=*252092-1* 271
Tagetes rotundifolia IPNI=*254724-2* 271
Tagetes scabra IPNI=*252097-1* 271
Tagetes schiedeana IPNI=*252121-1* 271
Tagetes seleri IPNI=*252121-1* 271
Tagetes silenoides IPNI=*252097-1* 271
Tagetes tenuifolia IPNI=*252092-1* 271
Tagetes tinctoria IPNI=*252132-1* 271
Taguaria elliptica IPNI=*548708-1* 448
Taguaria laurifolia IPNI=*548708-1* 448
Taguaria nitida IPNI=*548708-1* 448
Taguaria punctata IPNI=*548708-1* 448
Taguaria puracensis IPNI=*548708-1* 448
Taguaria vera IPNI=*548708-1* 448
Talanelis medium IPNI=*140670-1* 272
Talauma cespedesii IPNI=*991410-1* 190
Talauma chocoensis IPNI=*991411-1* 190
Talauma espinalii IPNI=*991415-1* 190
Talauma gilbertoi IPNI=*991417-1* 190
Talauma hernandezii IPNI=*991421-1* 190
Talauma sambuensis IPNI=*991437-1* 190
Talauma sieboldii IPNI=*554759-1* 190
Talauma silvioi IPNI=*991440-1* 190
Talinum andrewsii IPNI=*1182764-2* 290
Talinum attenuatum IPNI=*1182764-2* 290
Talinum caulescens IPNI=*698357-1* 286
Talinum chrysanthum IPNI=*1007146-2* 290
Talinum ciliatum IPNI=*698357-1* 286
Talinum confusum IPNI=*1182764-2* 290
Talinum crassifolium IPNI=*1182764-2* 290
Talinum crassifolium var. *albiflorum* IPNI=*1182764-2* 290
Talinum dichotomum IPNI=*1007146-2* 290
Talinum diffusum IPNI=*1182764-2* 290
Talinum fruticosum IPNI=*1182764-2* 290
Talinum grandiflorum IPNI=*1182764-2* 290
Talinum menziesii IPNI=*698357-1* 286
Talinum mucronatum IPNI=*1182764-2* 290
Talinum paniculatum f. *variegatum* IPNI=*1007146-2* 290
Talinum paniculatum var. *sarmentosum* IPNI=*1007146-2* 290
Talinum patens IPNI=*1007146-2* 290
Talinum patens IPNI=*1182764-2* 290
Talinum purpureum IPNI=*1007146-2* 290
Talinum racemosum IPNI=*1182764-2* 290
Talinum reflexum f. *sarmentosum* IPNI=*1007146-2* 290
Talinum reflexum IPNI=*1007146-2* 290
Talinum revolutum IPNI=*1182764-2* 290
Talinum roseum IPNI=*1007146-2* 290
Talinum sarmentosum IPNI=*1007146-2* 290
Talinum spathulatum IPNI=*1007146-2* 290
Talinum triangulare IPNI=*1182764-2* 290
Talinum triangulare var. *purpureum* IPNI=*1182764-2* 290
Talipariti hastatum IPNI=*30143354-2* 410
Talipariti tiliaceum IPNI=*30143354-2* 410
Talipariti tortuosum IPNI=*30143354-2* 410

Talisia peruviana IPNI=*77115312-1* 433
Tamarindus erythraeus IPNI=*520167-1* 335
Tamarindus occidentalis IPNI=*520167-1* 335
Tamarindus officinalis IPNI=*520167-1* 335
Tamarindus somalensis IPNI=*520167-1* 335
Tamarindus umbrosa IPNI=*520167-1* 335
Tamonea albicans IPNI=*1125568-2* 419
Tamonea andina IPNI=*572660-1* 421
Tamonea argyrophylla IPNI=*158734-2* 419
Tamonea ciliata IPNI=*1108920-2* 420
Tamonea epiphytica IPNI=*572660-1* 421
Tamonea holosericea IPNI=*1024004-2* 421
Tamonea ibaguensis IPNI=*572590-1* 421
Tamonea impetiolaris IPNI=*572600-1* 421
Tamonea laevigata IPNI=*572641-1* 421
Tamonea lappulacea IPNI=*1194372-2* 379
Tamonea lasiopetala IPNI=*1000103-2* 422
Tamonea longifolia IPNI=*1220088-2* 421
Tamonea macrophylla IPNI=*1000103-2* 422
Tamonea minutiflora IPNI=*1098452-2* 421
Tamonea mutica IPNI=*864913-1* 379
Tamonea nervosa IPNI=*572835-1* 421
Tamonea prasina IPNI=*1069702-2* 422
Tamonea racemosa IPNI=*1158780-2* 422
Tamonea rubiginosa IPNI=*1155116-2* 422
Tamonea stenostachya IPNI=*159763-2* 422
Tamonea theaezans IPNI=*1110588-2* 423
Tamonea tomentosa IPNI=*1088633-2* 423
Tamonea tomentosa var. *auriculata* IPNI=*1000103-2* 422
Tanacetum × morifolium IPNI=*77111328-1* 259
Tanacetum × sinense IPNI=*77111328-1* 259
Tanacetum demetrii IPNI=*252460-1* 271
Tanacetum glanduliferum IPNI=*252460-1* 271
Tanacetum grossheimii IPNI=*252460-1* 271
Tanacetum indicum IPNI=*193510-1* 259
Tanacetum kubense IPNI=*252460-1* 271
Tanacetum maximum IPNI=*230047-1* 265
Tanacetum partheniifolium IPNI=*252460-1* 271
Tanacetum parthenium var. *praealtum* IPNI=*252460-1* 271
Tanacetum persicum IPNI=*252460-1* 271
Tanacetum sevanense IPNI=*252460-1* 271
Tanaecium albiflorum IPNI=*111083-1* 365
Tanaecium exsertum IPNI=*111083-1* 365
Tanaecium nocturnum IPNI=*77100460-1* 362
Tanaecium ovatum IPNI=*77100441-1* 362
Tanaecium praelongum IPNI=*111083-1* 365
Tapeinochilos australis IPNI=*873025-1* 248
Tapeinochilos pungens IPNI=*873025-1* 248
Tapeinochilos pungens var. *queenslandiae* IPNI=*873025-1* 248
Tapeinochilos queenslandiae IPNI=*873025-1* 248
Tapeinochilos teysmannianus IPNI=*873025-1* 248
Tapina splendens IPNI=*380772-1* 367
Tapina variegata IPNI=*380772-1* 367
Tapirira bijuga IPNI=*71559-1* 451
Tapirira fanshawei IPNI=*71559-1* 451
Tapirira guianensis var. *cuneata* IPNI=*71559-1* 451
Tapirira guianensis var. *elliptica* IPNI=*71559-1* 451
Tapirira marchandii IPNI=*970707-1* 451
Tapirira myriantha IPNI=*71559-1* 451
Tapirira pao-pombo IPNI=*71559-1* 451
Tapirira pao-pombo var. *major* IPNI=*970707-1* 451
Tapirira pearcei IPNI=*71559-1* 451
Tapirira peckoltiana IPNI=*970707-1* 451
Tapirira velutinifolia IPNI=*275630-2* 450
Tapogomea alba IPNI=*60473597-2* 353
Tapogomea elata IPNI=*77115600-1* 353
Tapogomea punicea IPNI=*77115600-1* 353
Tapogomea tomentosa IPNI=*77115635-1* 354
Tapura amazonica var. *ciliata* IPNI=*163302-1* 385
Tapura amazonica var. *cuspidata* IPNI=*163302-1* 385
Tapura amazonica var. *dasyphylla* IPNI=*163302-1* 385
Tapura amazonica var. *sublanceolata* IPNI=*163302-1* 385
Tapura amazonica var. *typica* IPNI=*163302-1* 385
Tapura ciliata IPNI=*163302-1* 385
Tara tinctoria IPNI=*249443-2* 335
Tara tinctoria var. *horrida* IPNI=*249443-2* 335
Tarachia cuneata IPNI=*328699-2* 174
Taraktogenos macrocarpa IPNI=*111756-1* 380
Taralea oppositifolia var. *parviflora* IPNI=*520179-1* 335
Tardavel bartlingiana IPNI=*766800-1* 357
Tardavel laevis IPNI=*241474-2* 357
Tardavel latifolia IPNI=*766800-1* 357
Tardavel ocymoides IPNI=*766863-1* 357
Tardavel podocephala IPNI=*767031-1* 357
Tardavel spinosa IPNI=*77154343-1* 351
Tardavel stricta IPNI=*766863-1* 357
Tardavel verticillata IPNI=*767031-1* 357
Tarenaya aculeata IPNI=*316362-2* 279
Tarenaya hassleriana IPNI=*147105-1* 279
Tarenaya hassleriana var. *parviflora* IPNI=*147105-1* 279
Tarenaya parviflora IPNI=*147226-1* 279
Tarenaya spinosa IPNI=*60589-2* 279
Tariri antidesma IPNI=*813869-1* 433
Tariri pentandra IPNI=*813923-1* 433
Tartagalia roseorum IPNI=*95985-2* 408
Tartagalia rubra IPNI=*95985-2* 408
Tattia ignota IPNI=*365491-1* 404
Taxanthema armeria IPNI=*19793-2* 288
Taxanthema sinuatum IPNI=*686979-1* 288
Taxodium japonicum IPNI=*261870-1* 180
Taxodium nutkaense IPNI=*263741-1* 180
Taxodium sempervirens IPNI=*263741-1* 180
Tecoma adenophylla IPNI=*117330-2* 363
Tecoma aesculifolia IPNI=*111572-2* 363
Tecoma africana IPNI=*109874-1* 364
Tecoma albiflora IPNI=*248703-2* 365
Tecoma araliacea IPNI=*77085096-1* 364
Tecoma atractocarpa IPNI=*77085096-1* 364
Tecoma avellanedae IPNI=*117330-2* 363
Tecoma avellanedae var. *alba* IPNI=*117330-2* 363
Tecoma barbata IPNI=*109687-1* 363
Tecoma bernoullii IPNI=*222518-2* 365
Tecoma billbergii IPNI=*77085061-1* 363
Tecoma capensis IPNI=*111333-1* 365
Tecoma capitata IPNI=*109688-1* 363
Tecoma chrysantha IPNI=*77085066-1* 363
Tecoma chrysea IPNI=*222517-2* 365
Tecoma chrysotricha IPNI=*109691-1* 363
Tecoma chrysotricha var. *obtusata* IPNI=*109691-1* 363
Tecoma conspicua IPNI=*77085096-1* 364
Tecoma digitata IPNI=*111572-2* 363
Tecoma dura IPNI=*248703-2* 365
Tecoma eggersii IPNI=*248689-2* 365
Tecoma flavescens IPNI=*109691-1* 363
Tecoma fuscata IPNI=*111572-2* 363
Tecoma grandis IPNI=*109691-1* 363
Tecoma guayacan IPNI=*77085089-1* 363
Tecoma impetiginosa IPNI=*117330-2* 363
Tecoma insignis IPNI=*248703-2* 365
Tecoma integra IPNI=*117330-2* 363
Tecoma ipe f. *leucotricha* IPNI=*117330-2* 363
Tecoma ipe var. *integra* IPNI=*117330-2* 363
Tecoma leucoxylon var. *miquelii* IPNI=*109688-1* 363
Tecoma leucoxylon var. *pentaphylla* IPNI=*109688-1* 363
Tecoma leucoxylon var. *salpingophora* IPNI=*248703-2* 365
Tecoma loxensis IPNI=*109429-1* 363

Tecoma mattogrossensis IPNI=*248789-2* 365
Tecoma mexicana IPNI=*111027-1* 365
Tecoma nigricans IPNI=*77085096-1* 364
Tecoma obscura IPNI=*109700-1* 364
Tecoma obtusata IPNI=*109691-1* 363
Tecoma ochracea IPNI=*117333-2* 364
Tecoma ochracea var. *denudata* IPNI=*109691-1* 363
Tecoma odontodiscus f. *leucotricha* IPNI=*248789-2* 365
Tecoma odontodiscus IPNI=*248789-2* 365
Tecoma odontodiscus var. *paraguariensis* IPNI=*248789-2* 365
Tecoma papyrophloios IPNI=*248789-2* 365
Tecoma parviflora IPNI=*109921-1* 364
Tecoma patrisiana IPNI=*77085096-1* 364
Tecoma pentaphylla IPNI=*248689-2* 365
Tecoma petersii IPNI=*111333-1* 365
Tecoma pisoniana IPNI=*77100464-1* 363
Tecoma piutinga IPNI=*248789-2* 365
Tecoma punctatissima IPNI=*111027-1* 365
Tecoma ricasoliana IPNI=*110440-1* 364
Tecoma rosea IPNI=*111027-1* 365
Tecoma schumanni IPNI=*248789-2* 365
Tecoma serratifolia IPNI=*77085096-1* 364
Tecoma toxophora IPNI=*109687-1* 363
Tecoma triphylla IPNI=*248689-2* 365
Tecoma venusta IPNI=*110483-1* 365
Tecoma villosa IPNI=*108550-1* 362
Tecomaria capensis var. *flava* IPNI=*111333-1* 365
Tecomaria krebsii IPNI=*111333-1* 365
Tecomaria petersii IPNI=*111333-1* 365
Tecomaria ricasoliana IPNI=*110440-1* 364
Tectaria acutiloba IPNI=*249914-2* 177
Tectaria andina IPNI=*17226640-1* 177
Tectaria angustior IPNI=*249914-2* 177
Tectaria angustius IPNI=*249914-2* 177
Tectaria calahuala IPNI=*224543-2* 176
Tectaria coriacea IPNI=*224543-2* 176
Tectaria ferruginea IPNI=*224543-2* 176
Tectaria filix-femina IPNI=*30055316-2* 174
Tectaria fragilis IPNI=*30013429-2* 174
Tectaria fraxinea IPNI=*17160710-1* 176
Tectaria marginalis IPNI=*17458930-1* 174
Tectaria martinicensis IPNI=*17226640-1* 177
Tectaria martinicensis var. *pittieri* IPNI=*17226640-1* 177
Tectaria martinicensis var. *salicifolia* IPNI=*17226640-1* 177
Tectaria mexicana var. *pilosula* IPNI=*249914-2* 177
Tectaria trifoliata var. *heracleifolia* IPNI=*249899-2* 177
Tectona grandis f. *canescens* IPNI=*864923-1* 373
Tectona grandis f. *pilosula* IPNI=*864923-1* 373
Tectona grandis f. *punctata* IPNI=*864923-1* 373
Tectona grandis f. *tomentella* IPNI=*864923-1* 373
Tegularia adiantifolia IPNI=*17084790-1* 175
Telanthera amoena IPNI=*10291-2* 281
Telanthera arenaria IPNI=*327343-2* 281
Telanthera aristata IPNI=*10378-2* 281
Telanthera bangii IPNI=*10447-2* 281
Telanthera bettzickiana IPNI=*10291-2* 281
Telanthera brasiliana IPNI=*10294-2* 281
Telanthera brasiliensis IPNI=*10294-2* 281
Telanthera caracasana IPNI=*327344-2* 281
Telanthera cinerascens IPNI=*10378-2* 281
Telanthera crucis IPNI=*10362-2* 281
Telanthera cyclophylla IPNI=*10291-2* 281
Telanthera densiflora IPNI=*10362-2* 281
Telanthera dentata IPNI=*10294-2* 281
Telanthera dolichocephala IPNI=*10362-2* 281
Telanthera elongata var. *nigriceps* IPNI=*10447-2* 281
Telanthera eupatorioides IPNI=*10450-2* 281
Telanthera ficoidea IPNI=*1078972-2* 281
Telanthera flavogrisea IPNI=*10362-2* 281
Telanthera frutescens IPNI=*10362-2* 281
Telanthera frutescens var. *acutifolia* IPNI=*10362-2* 281
Telanthera frutescens var. *manillensis* IPNI=*10362-2* 281
Telanthera glomerata IPNI=*10450-2* 281
Telanthera gomphrenoides IPNI=*10447-2* 281
Telanthera gomphrenoides var. *ovata* IPNI=*10447-2* 281
Telanthera halimifolia IPNI=*10362-2* 281
Telanthera junciflora IPNI=*10447-2* 281
Telanthera lanceolata IPNI=*10378-2* 281
Telanthera luzuloides IPNI=*10447-2* 281
Telanthera manillensis IPNI=*10362-2* 281
Telanthera martinicensis IPNI=*10362-2* 281
Telanthera mexicana IPNI=*10378-2* 281
Telanthera meyeriana IPNI=*10378-2* 281
Telanthera microcephala IPNI=*10378-2* 281
Telanthera microcephala var. *hispida* IPNI=*10378-2* 281
Telanthera paniculata IPNI=*10447-2* 281
Telanthera philoxeroides IPNI=*1024156-2* 281
Telanthera philoxeroides var. *acutifolia* IPNI=*1024156-2* 281
Telanthera philoxeroides var. *carnosa* IPNI=*34030-2* 282
Telanthera philoxeroides var. *denticulata* IPNI=*1024156-2* 281
Telanthera philoxeroides var. *linearifolia* IPNI=*1024156-2* 281
Telanthera philoxeroides var. *obtusifolia* IPNI=*1024156-2* 281
Telanthera philoxeroides var. *phyllantha* IPNI=*1024156-2* 281
Telanthera picta IPNI=*10291-2* 281
Telanthera polygonoides IPNI=*1069792-2* 281
Telanthera polygonoides var. *brachiata* IPNI=*1078972-2* 281
Telanthera polygonoides var. *compacta* IPNI=*1069792-2* 281
Telanthera polygonoides var. *diffusa* IPNI=*327343-2* 281
Telanthera polygonoides var. *major* IPNI=*1069792-2* 281
Telanthera polygonoides var. *radicans* IPNI=*1078972-2* 281
Telanthera porrigens IPNI=*10447-2* 281
Telanthera pubiflora IPNI=*10450-2* 281
Telanthera pubiflora var. *glomerata* IPNI=*10450-2* 281
Telanthera pubiflora var. *monocephala* IPNI=*10450-2* 281
Telanthera pungens IPNI=*327342-2* 281
Telanthera rivetii IPNI=*10447-2* 281
Telanthera sintenisii IPNI=*10362-2* 281
Telestria purpurea IPNI=*481605-1* 312
Telinaria canariensis IPNI=*495939-1* 320
Telinaria candicans IPNI=*496189-1* 320
Telinaria ramosissima IPNI=*495939-1* 320
Teline canariensis IPNI=*495939-1* 320
Teline canariensis var. *lanuginosa* IPNI=*495939-1* 320
Teline canariensis var. *ramosissima* IPNI=*495939-1* 320
Teline candicans IPNI=*496189-1* 320
Teline eriocarpa IPNI=*496189-1* 320
Teline hillebrandii IPNI=*495939-1* 320
Teline medicagioides IPNI=*496189-1* 320
Teline monspessulana IPNI=*496189-1* 320
Teline ramosissima IPNI=*495939-1* 320
Teliostachya alopecuroidea IPNI=*51951-1* 360
Teliostachya hyssopifolia IPNI=*51951-1* 360
Teliostachya laguroidea IPNI=*51951-1* 360
Teliostachya lanceolata IPNI=*51951-1* 360
Teliostachya lanceolata var. *crispa* IPNI=*51951-1* 360
Teliostachya petraea IPNI=*51951-1* 360
Telipogon angustifolius IPNI=*659882-1* 223
Telipogon steyermarkii IPNI=*659882-1* 223
Telmatophace arrhiza IPNI=*526178-1* 200
Telmatophace cylindracea IPNI=*1135607-2* 202
Telmatophace generalis IPNI=*526178-1* 200
Telmatophace gibba IPNI=*526178-1* 200
Telmatophace orbicularis IPNI=*526250-1* 202
Telmatophace polyrhiza IPNI=*526250-1* 202
Telopea perspicua IPNI=*338514-1* 387
Teloxys ambrosioides IPNI=*1198060-2* 282

Temenia regia IPNI=*664452-1* 205
Temnadenia lasiocarpa IPNI=*151979-2* 341
Temnadenia leptoloba IPNI=*81405-1* 342
Temnadenia lobbiana IPNI=*151979-2* 341
Temnadenia palustris IPNI=*151979-2* 341
Temnadenia quinquangularis IPNI=*81405-1* 342
Temnadenia secundiflora IPNI=*152067-2* 342
Temnadenia semidigyna IPNI=*77191232-1* 342
Temnadenia tomentosa IPNI=*151979-2* 341
Tenageia bufonia IPNI=*30079279-2* 233
Tenorea berteroi IPNI=*256729-1* 272
Tenorea calyculata IPNI=*256729-1* 272
Tenorea heterophylla IPNI=*853513-1* 439
Tenoria romana IPNI=*842680-1* 252
Tenoria rotundifolia IPNI=*38361-2* 252
Tephrosia adunca f. *glabrior* IPNI=*520366-1* 335
Tephrosia adunca f. *grandifolia* IPNI=*520366-1* 335
Tephrosia adunca f. *paraguayensis* IPNI=*520366-1* 335
Tephrosia adunca f. *pseudomarginata* IPNI=*520366-1* 335
Tephrosia adunca var. *genuina* IPNI=*520366-1* 335
Tephrosia adunca var. *guaranitica* IPNI=*520366-1* 335
Tephrosia adunca var. *intermedia* IPNI=*520366-1* 335
Tephrosia adunca var. *rufescens* IPNI=*520366-1* 335
Tephrosia adunca var. *subglabrata* IPNI=*520366-1* 335
Tephrosia caribaea IPNI=*275194-2* 315
Tephrosia cinerea var. *typica* IPNI=*1071691-2* 335
Tephrosia coerulea IPNI=*76086-2* 317
Tephrosia emarginata IPNI=*250423-2* 335
Tephrosia gynothrix IPNI=*1071691-2* 335
Tephrosia inebrians IPNI=*521002-1* 335
Tephrosia littoralis IPNI=*1071691-2* 335
Tephrosia megalantha IPNI=*521002-1* 335
Tephrosia penicillata IPNI=*520366-1* 335
Tephrosia periculosa IPNI=*521002-1* 335
Tephrosia procumbens IPNI=*1071691-2* 335
Tephrosia rufescens IPNI=*520366-1* 335
Tephrosia rufescens var. *paraguayensis* IPNI=*520366-1* 335
Tephrosia schiedeana IPNI=*250423-2* 335
Tephrosia scopulorum IPNI=*1071691-2* 335
Tephrosia sericea IPNI=*520444-1* 335
Tephrosia toxicaria IPNI=*250423-2* 335
Tephrosia venustula IPNI=*1071691-2* 335
Terania frutescens IPNI=*139187-2* 377
Terebinthus arborea IPNI=*127217-1* 451
Terebinthus brownei IPNI=*158605-2* 451
Terebinthus graveolens IPNI=*127138-1* 451
Terebinthus pubescens IPNI=*127138-1* 451
Terebinthus tomentosa IPNI=*127234-1* 451
Terera triplinervia IPNI=*159880-2* 423
Terminalia burmanica IPNI=*171034-1* 416
Terminalia cacoucia IPNI=*63781-2* 415
Terminalia capitata IPNI=*77164595-1* 416
Terminalia chiriquensis IPNI=*171280-1* 416
Terminalia erecta IPNI=*319151-2* 416
Terminalia erecta var. *procumbens* IPNI=*319151-2* 416
Terminalia excelsa IPNI=*250537-2* 416
Terminalia firma IPNI=*171207-1* 416
Terminalia hayesii IPNI=*171207-1* 416
Terminalia hilariana IPNI=*77164595-1* 416
Terminalia intermedia IPNI=*171034-1* 416
Terminalia kydiana IPNI=*171034-1* 416
Terminalia latifolia IPNI=*171034-1* 416
Terminalia mauritiana IPNI=*171034-1* 416
Terminalia moluccana IPNI=*171034-1* 416
Terminalia nyssifolia IPNI=*171207-1* 416
Terminalia obovata IPNI=*250537-2* 416
Terminalia obovata IPNI=*77164595-1* 416
Terminalia odontoptera IPNI=*250537-2* 416
Terminalia ovatifolia IPNI=*171034-1* 416
Terminalia paraensis IPNI=*171034-1* 416
Terminalia rubrigemmis IPNI=*171034-1* 416
Terminalia subcordata IPNI=*171034-1* 416
Terminalia tarapotensis IPNI=*171280-1* 416
Terminalis angustifolia IPNI=*534103-1* 215
Terminalis australis IPNI=*533544-1* 215
Terminalis fruticosa IPNI=*533580-1* 215
Terminalis rubra IPNI=*97540-3* 215
Termontis racemosa IPNI=*799180-1* 375
Ternatea arborescens IPNI=*486481-1* 315
Ternatea grandiflora IPNI=*485611-1* 314
Ternatea indica IPNI=*486606-1* 315
Ternatea javitensis IPNI=*486540-1* 315
Ternatea racemosa IPNI=*525491-1* 336
Ternatea schiedeana IPNI=*485611-1* 314
Ternatea ternatea IPNI=*486606-1* 315
Ternatea virginiana IPNI=*51897-2* 314
Ternatea vulgaris IPNI=*486606-1* 315
Teruncius arvensis IPNI=*290477-1* 278
Tessaria absinthoides IPNI=*30283369-2* 271
Tessaria ambigua var. *uniflora* IPNI=*30283369-2* 271
Tessaria dentata IPNI=*30283369-2* 271
Tessaria integrifolia var. *obtusifolia* IPNI=*30283369-2* 271
Tessaria legitima IPNI=*30283369-2* 271
Tessaria mucronata IPNI=*30283369-2* 271
Tessenia canadensis IPNI=*203653-1* 261
Tessiera pubescens IPNI=*319724-2* 357
Tessmanniophoenix longibracteata IPNI=*55986-2* 208
Testudinaria cocolmeca IPNI=*318262-1* 225
Testudinaria macrostachya IPNI=*318262-1* 225
Testudipes recurva IPNI=*82097-1* 343
Tetracanthus linearifolius IPNI=*236639-1* 267
Tetracarpum anthemoideum IPNI=*242907-1* 269
Tetracarpum flavum IPNI=*242907-1* 269
Tetracarpum guatemalense IPNI=*242907-1* 269
Tetracarpum pinnatum IPNI=*242907-1* 269
Tetracarpum pringlei IPNI=*242907-1* 269
Tetracarpum virgatum IPNI=*242907-1* 269
Tetracarpum wislizeni IPNI=*242907-1* 269
Tetracarpum wrightii IPNI=*242907-1* 269
Tetracellion ellipsoideum IPNI=*288669-1* 277
Tetracera acuminata IPNI=*332108-2* 295
Tetracera alata IPNI=*332108-2* 295
Tetracera calophylla IPNI=*316577-1* 295
Tetracera castaneifolia IPNI=*332108-2* 295
Tetracera cuspidata IPNI=*82916-2* 295
Tetracera doliocarpus IPNI=*82916-2* 295
Tetracera erecta IPNI=*332108-2* 295
Tetracera fagifolia IPNI=*317513-1* 295
Tetracera jamaicensis IPNI=*332108-2* 295
Tetracera japurensis IPNI=*332108-2* 295
Tetracera lima IPNI=*328431-2* 295
Tetracera mexicana IPNI=*332108-2* 295
Tetracera multiflora IPNI=*77025-2* 295
Tetracera nitida IPNI=*77025-2* 295
Tetracera plicata IPNI=*316577-1* 295
Tetracera poeppigiana IPNI=*332108-2* 295
Tetracera rhamnifolia IPNI=*317513-1* 295
Tetracera riedeliana IPNI=*317513-1* 295
Tetracera rotundifolia var. *angustata* IPNI=*317513-1* 295
Tetracera rugosa IPNI=*316577-1* 295
Tetracera salicifolia IPNI=*332108-2* 295
Tetracera sessiliflora IPNI=*317470-1* 295
Tetracera stricta IPNI=*30327947-2* 295
Tetracera surinamensis var. *reticulata* IPNI=*317513-1* 295
Tetracera tomentosa IPNI=*82916-2* 295
Tetracera volubilis IPNI=*317470-1* 295

Tetracera williamsii IPNI=*332108-2* 295
Tetradapa javanorum IPNI=*494614-1* 320
Tetragamestus gracilis IPNI=*228831-2* 223
Tetragastris altissima IPNI=*77183601-1* 452
Tetragastris mucronata IPNI=*210584-2* 453
Tetragastris panamensis IPNI=*77202519-1* 453
Tetragastris panamensis var. *grandifolia* IPNI=*77202519-1* 453
Tetragastris panamensis var. *hirtella* IPNI=*77202519-1* 453
Tetragastris paraensis IPNI=*77202519-1* 453
Tetragastris phanerosepala IPNI=*77183601-1* 452
Tetragastris stevensonii IPNI=*77202519-1* 453
Tetragastris tomentosa IPNI=*210615-2* 453
Tetragastris unifoliolata IPNI=*128430-1* 453
Tetragonanthus elatus IPNI=*369463-1* 345
Tetragonanthus hypericoides IPNI=*117020-2* 345
Tetragonia chisimajensis IPNI=*322565-2* 280
Tetragonia peruviana IPNI=*698357-1* 286
Tetragonocalamus angulatus IPNI=*392553-1* 235
Tetramerium angustius IPNI=*56148-1* 361
Tetramerium calderonii IPNI=*56148-1* 361
Tetramerium geniculatum IPNI=*1140399-2* 359
Tetramerium hispidum IPNI=*56148-1* 361
Tetramerium hispidum var. *greenmanii* IPNI=*56148-1* 361
Tetramerium latifolium IPNI=*1039243-2* 350
Tetramerium nervosum var. *angustifolium* IPNI=*56148-1* 361
Tetramerium nervosum var. *hispidum* IPNI=*56148-1* 361
Tetramerium occidentale IPNI=*1187456-2* 350
Tetramerium odoratissimum IPNI=*1187456-2* 350
Tetramerium ovalifolium IPNI=*56148-1* 361
Tetramerium ovatum IPNI=*56148-1* 361
Tetramerium polystachyum IPNI=*56148-1* 361
Tetramerium scabrum IPNI=*56148-1* 361
Tetranthera cervantesii IPNI=*465312-1* 183
Tetranthera pichurim IPNI=*1171177-2* 184
Tetranthus ruderalis IPNI=*202984-1* 261
Tetraotis longifolia IPNI=*203207-1* 261
Tetraotis paludosa IPNI=*203207-1* 261
Tetrapterys alternifolia IPNI=*557660-1* 396
Tetrapterys benthamiana IPNI=*60443063-2* 396
Tetrapterys boliviensis IPNI=*557785-1* 396
Tetrapterys boliviensis var. *granatensis* IPNI=*557785-1* 396
Tetrapterys boliviensis var. *typica* IPNI=*557785-1* 396
Tetrapterys crispa subsp. *pseudotriopterys* IPNI=*557660-1* 396
Tetrapterys crispa subsp. *typica* IPNI=*557660-1* 396
Tetrapterys crispa var. *ovata* IPNI=*557660-1* 396
Tetrapterys crispa var. *punicans* IPNI=*557660-1* 396
Tetrapterys crispa var. *subcordata* IPNI=*557660-1* 396
Tetrapterys cuspidata IPNI=*557670-1* 396
Tetrapterys discolor var. *andina* IPNI=*557670-1* 396
Tetrapterys discolor var. *brownsbergensis* IPNI=*557670-1* 396
Tetrapterys discolor var. *lanuginosa* IPNI=*557670-1* 396
Tetrapterys discolor var. *surinamensis* IPNI=*557670-1* 396
Tetrapterys discolor var. *trinitensis* IPNI=*557670-1* 396
Tetrapterys discolor var. *typica* IPNI=*557670-1* 396
Tetrapterys elliptica IPNI=*557660-1* 396
Tetrapterys macrophylla IPNI=*557670-1* 396
Tetrapterys methystica IPNI=*557785-1* 396
Tetrapterys ovalifolia IPNI=*557670-1* 396
Tetrapterys peruviana IPNI=*557660-1* 396
Tetrapterys puberula IPNI=*557670-1* 396
Tetrapterys reticulata IPNI=*557785-1* 396
Tetrapterys schiedeana f. *grandifolia* IPNI=*557670-1* 396
Tetrapterys squarrosa f. *lanceolata* IPNI=*557785-1* 396
Tetrapterys squarrosa f. *ovata* IPNI=*557785-1* 396
Tetrapterys squarrosa IPNI=*557785-1* 396
Tetrapterys surinamensis IPNI=*557670-1* 396
Tetrapterys tenuistachys IPNI=*557785-1* 396
Tetrapterys trinitensis IPNI=*557670-1* 396
Tetrapterys triopteris IPNI=*557660-1* 396
Tetrastichella inaequalis IPNI=*77100463-1* 363
Tetrathylacium costaricense IPNI=*867537-1* 404
Tetrathylacium macrophyllum var. *pacificum* IPNI=*867537-1* 404
Tetrathylacium nutans IPNI=*867537-1* 404
Tetrathylacium pacificum IPNI=*867537-1* 404
Tetratome latifolia IPNI=*582116-1* 185
Tetratome lepidota IPNI=*582354-1* 186
Tetrorchidium boyacanum IPNI=*356873-1* 393
Tetrorchidium gorgonae IPNI=*356853-1* 393
Tetrorchidium rubrivenium var. *fendleri* IPNI=*356873-1* 393
Tetrorchidium rubrivenium var. *genuinum* IPNI=*356873-1* 393
Tetrorchidium rubrivenium var. *integrifolium* IPNI=*356873-1* 393
Tetrorchidium trigynum IPNI=*356873-1* 393
Teucrium angustifolium IPNI=*460416-1* 373
Teucrium fruticans subsp. *latifolium* IPNI=*460416-1* 373
Teucrium genevense IPNI=*444576-1* 368
Teucrium glomeratum IPNI=*460416-1* 373
Teucrium hirtum IPNI=*460785-1* 373
Teucrium inflatum IPNI=*460785-1* 373
Teucrium latifolium IPNI=*460416-1* 373
Teucrium mollifolium IPNI=*460785-1* 373
Teucrium palustre IPNI=*460785-1* 373
Teucrium picardae IPNI=*460785-1* 373
Teucrium pyramidale IPNI=*444657-1* 368
Teucrium reptans IPNI=*444664-1* 368
Teucrium rhombifolium IPNI=*77120760-1* 368
Teucrium tenuipes IPNI=*460785-1* 373
Teucrium tomentosum IPNI=*460416-1* 373
Teucrium vesicarium var. *palustre* IPNI=*460785-1* 373
Teucrium villosum IPNI=*460785-1* 373
Thalestris graminiformis IPNI=*78920-2* 359
Thalia altissima IPNI=*798255-1* 250
Thalia angustifolia IPNI=*798255-1* 250
Thalia caerulea IPNI=*798255-1* 250
Thalia coarctata IPNI=*797633-1* 250
Thalia dipetala IPNI=*798255-1* 250
Thalia divaricata IPNI=*798255-1* 250
Thalia erecta IPNI=*798255-1* 250
Thalia geniculata f. *rheumoides* IPNI=*798255-1* 250
Thalia latifolia IPNI=*60459373-2* 249
Thalia lutea IPNI=*246147-2* 250
Thalia nemorosa IPNI=*796953-1* 248
Thalia schumanniana IPNI=*798255-1* 250
Thalia trichocalyx IPNI=*798255-1* 250
Thalia welwitschii IPNI=*798255-1* 250
Thalianthus macropus IPNI=*797633-1* 250
Thalysia mays IPNI=*426810-1* 247
Thea bohea IPNI=*828548-1* 309
Thea bohea var. *stricta* IPNI=*828548-1* 309
Thea camellia IPNI=*828524-1* 309
Thea cantoniensis IPNI=*828548-1* 309
Thea chinensis IPNI=*828548-1* 309
Thea formosensis IPNI=*828548-1* 309
Thea grandifolia IPNI=*828548-1* 309
Thea hozanensis IPNI=*828524-1* 309
Thea japonica IPNI=*828524-1* 309
Thea latifolia IPNI=*828548-1* 309
Thea longifolia IPNI=*828548-1* 309
Thea macrophylla IPNI=*828548-1* 309
Thea nakaii IPNI=*828524-1* 309
Thea olearia IPNI=*828548-1* 309
Thea oleosa IPNI=*828548-1* 309
Thea parvifolia IPNI=*828548-1* 309
Thea sinensis IPNI=*828548-1* 309
Thea stricta IPNI=*828548-1* 309
Thea viridis IPNI=*828548-1* 309
Theaphylla anamensis IPNI=*828548-1* 309

Theaphylla cantoniensis IPNI=*828548-1* 309
Theaphylla laxa IPNI=*828548-1* 309
Theaphylla viridis IPNI=*828548-1* 309
Theka grandis IPNI=*864923-1* 373
Thela alba IPNI=*687109-1* 288
Thelechitonia trilobata IPNI=*1093589-2* 270
Thelepogon sanguineus IPNI=*420951-1* 244
Thelypteris dentata var. *violascens* IPNI=*251308-2* 174
Thelypteris dicarpa IPNI=*251308-2* 174
Thelypteris extensa IPNI=*17460390-1* 174
Thelypteris funesta IPNI=*286436-2* 177
Thelypteris impressa IPNI=*17460390-1* 174
Thelypteris incerta IPNI=*17460390-1* 174
Thelypteris macrorhizoma IPNI=*251400-2* 174
Thelypteris malabariensis IPNI=*251308-2* 174
Thelypteris mindanaensis IPNI=*251308-2* 174
Thelypteris normalis IPNI=*251400-2* 174
Thelypteris nymphalis IPNI=*251308-2* 174
Thelypteris oligophlebia IPNI=*1019960-2* 174
Thelypteris orbicularis IPNI=*17460390-1* 174
Thelypteris reducta IPNI=*251308-2* 174
Thelypteris saxatilis IPNI=*251400-2* 174
Thelypteris sulfurea IPNI=*17460390-1* 174
Thelypteris taprobanica IPNI=*251308-2* 174
Thelypteris terminans IPNI=*17460390-1* 174
Thelypteris torresiana IPNI=*1019960-2* 174
Thelypteris totta IPNI=*17458930-1* 174
Thelypteris totta var. *hirsuta* IPNI=*17458930-1* 174
Thelypteris totta var. *longipinna* IPNI=*251353-2* 174
Thelypteris uliginosa IPNI=*1019960-2* 174
Thelypteris unca IPNI=*251400-2* 174
Themistoclesia humboldtiana IPNI=*252411-2* 300
Themistoclesia lehmannii IPNI=*252411-2* 300
Themistoclesia pendula IPNI=*252411-2* 300
Themistoclesia valerioi IPNI=*858679-1* 300
Theobroma albiflorum IPNI=*824010-1* 409
Theobroma balaense IPNI=*824014-1* 409
Theobroma cacao f. *lacandonense* IPNI=*320783-2* 414
Theobroma cacao f. *leiocarpum* IPNI=*320783-2* 414
Theobroma cacao f. *pentagonum* IPNI=*320783-2* 414
Theobroma cacao subsp. *leiocarpum* IPNI=*320783-2* 414
Theobroma cacao subsp. *pentagonum* IPNI=*320783-2* 414
Theobroma cacao subsp. *sphaerocarpum* IPNI=*320783-2* 414
Theobroma cacao var. *leiocarpum* IPNI=*320783-2* 414
Theobroma calodesme IPNI=*825579-1* 414
Theobroma camargoanum IPNI=*120754-2* 409
Theobroma caribaeum IPNI=*320783-2* 414
Theobroma celtifolium IPNI=*823393-1* 408
Theobroma chocoense IPNI=*825610-1* 414
Theobroma chocoense var. *bullatum* IPNI=*825610-1* 414
Theobroma ferrugineum IPNI=*825619-1* 414
Theobroma guazuma IPNI=*823393-1* 408
Theobroma guyanense IPNI=*825619-1* 414
Theobroma integerrimum IPNI=*320783-2* 414
Theobroma kalaguum IPNI=*320783-2* 414
Theobroma laciniifolium IPNI=*824022-1* 409
Theobroma leiocarpum IPNI=*320783-2* 414
Theobroma macranthum IPNI=*825580-1* 414
Theobroma martianum IPNI=*825619-1* 414
Theobroma ovatifolium IPNI=*825565-1* 414
Theobroma pentagonum IPNI=*320783-2* 414
Theobroma pulcherrimum IPNI=*824028-1* 409
Theobroma purpureum IPNI=*120768-2* 409
Theobroma sagittatum IPNI=*824026-1* 409
Theobroma salzmannianum IPNI=*320783-2* 414
Theobroma sapidum IPNI=*320783-2* 414
Theobroma sativum IPNI=*320783-2* 414
Theobroma speciosum IPNI=*825580-1* 414
Theobroma sphaerocarpum IPNI=*320783-2* 414
Theobroma stipulatum IPNI=*825610-1* 414
Theobroma tomentosum IPNI=*823393-1* 408
Theophrasta glycycarpa IPNI=*866956-1* 405
Theophrasta latifolia IPNI=*587931-1* 303
Theophrasta longifolia IPNI=*587939-1* 303
Theophrasta macrocarpa IPNI=*311911-2* 303
Theophrasta pungens IPNI=*587939-1* 303
Theophrasta sparsifolia IPNI=*866956-1* 405
Theophrasta spathulata IPNI=*311911-2* 303
Thermophila cordata IPNI=*964785-1* 291
Thespesia altissima IPNI=*564713-1* 415
Thespesia macrophylla IPNI=*564713-1* 415
Thespesia populnea var. *bynoeana* IPNI=*564713-1* 415
Thespesia populnea var. *typica* IPNI=*564713-1* 415
Thevetia ahuai IPNI=*82440-1* 344
Thevetia calophylla IPNI=*82440-1* 344
Thevetia linearis IPNI=*47927-2* 340
Thevetia neriifolia IPNI=*47927-2* 340
Thevetia neriifolia var. *hirsuta* IPNI=*47927-2* 340
Thevetia neriifolia var. *leucantha* IPNI=*47927-2* 340
Thevetia neriifolia var. *pubescens* IPNI=*47927-2* 340
Thevetia nitida IPNI=*82440-1* 344
Thevetia peruviana f. *aurantiaca* IPNI=*47927-2* 340
Thevetia peruviana IPNI=*47927-2* 340
Thibaudia acuminata IPNI=*60458979-2* 298
Thibaudia bracteata IPNI=*60458979-2* 298
Thibaudia cinnamomifolia IPNI=*858236-1* 298
Thibaudia cordifolia IPNI=*60458979-2* 298
Thibaudia crassifolia IPNI=*60458979-2* 298
Thibaudia cyathifera IPNI=*858617-1* 300
Thibaudia dependens IPNI=*252411-2* 300
Thibaudia falcata IPNI=*858617-1* 300
Thibaudia glandulosa IPNI=*858406-1* 299
Thibaudia hendersonii IPNI=*60458979-2* 298
Thibaudia hirtiflora IPNI=*148696-2* 299
Thibaudia hookeri IPNI=*60458979-2* 298
Thibaudia jessicae IPNI=*858621-1* 300
Thibaudia lehmannii IPNI=*858728-1* 300
Thibaudia longifolia IPNI=*858728-1* 300
Thibaudia macrophylla IPNI=*858626-1* 300
Thibaudia melastomoides IPNI=*60458979-2* 298
Thibaudia mexicana IPNI=*60458979-2* 298
Thibaudia microphylla IPNI=*82286-2* 299
Thibaudia nitida IPNI=*50474-2* 298
Thibaudia oblongifolia IPNI=*858236-1* 298
Thibaudia ocanensis IPNI=*60458979-2* 298
Thibaudia panurensis IPNI=*77163666-1* 300
Thibaudia pichinchensis IPNI=*858728-1* 300
Thibaudia pichinchensis var. *glabra* IPNI=*858621-1* 300
Thibaudia pubescens IPNI=*60458979-2* 298
Thibaudia pubescens IPNI=*858236-1* 298
Thibaudia quereme IPNI=*1064712-2* 298
Thibaudia rupestris IPNI=*148722-2* 299
Thibaudia sarcantha IPNI=*858621-1* 300
Thibaudia scabriuscula IPNI=*60458979-2* 298
Thibaudia septemnervia IPNI=*858236-1* 298
Thibaudia strobilifera IPNI=*60458979-2* 298
Thibaudia tarapotana IPNI=*858240-1* 298
Thibaudia tetragona IPNI=*60458979-2* 298
Thibaudia wardii IPNI=*858211-1* 298
Thieleodoxa sorbilis IPNI=*8417-2* 347
Thieleodoxa verticillata IPNI=*943695-1* 347
Thilcum tinctorium IPNI=*612572-1* 429
Thiodia laetioides IPNI=*48022-2* 403
Thlaspi amarum IPNI=*324660-2* 277
Thlaspi baicalense IPNI=*290477-1* 278
Thlaspi bispinosum IPNI=*286421-1* 277

Thlaspi bursa-pastoris IPNI=*30092589-2* 276
Thlaspi bursetta IPNI=*30092589-2* 276
Thlaspi collinum IPNI=*290477-1* 278
Thlaspi cuneatum IPNI=*30092589-2* 276
Thlaspi infestum IPNI=*30092589-2* 276
Thlaspi latifolium IPNI=*290477-1* 278
Thlaspi lutescens IPNI=*290477-1* 278
Thlaspi polymorphum IPNI=*30092589-2* 276
Thlaspi schrankii IPNI=*30092589-2* 276
Thlaspi spinosum IPNI=*286421-1* 277
Thlaspi strictum IPNI=*290477-1* 278
Thlaspi umbellatum IPNI=*324671-2* 277
Thlaspi virginianum IPNI=*138164-2* 277
Thlaspi virginicum IPNI=*138164-2* 277
Thlaspidea arvensis IPNI=*290477-1* 278
Thlaspidium arvense IPNI=*290477-1* 278
Thoa leyboldii IPNI=*111368-2* 179
Thoa nigra IPNI=*111373-2* 179
Thoa nodiflora IPNI=*111373-2* 179
Thoa paniculata IPNI=*111375-2* 179
Thoa schwackeana IPNI=*111377-2* 180
Thoa urens IPNI=*383578-1* 180
Thonningia janeirensis IPNI=*315558-2* 448
Thonningia mexicana IPNI=*315558-2* 448
Thornbera dalea IPNI=*76085-2* 317
Thornbera pumila IPNI=*76085-2* 317
Thornbera robusta IPNI=*76085-2* 317
Thrasya petrosa IPNI=*60440769-2* 243
Thrasya petrosa var. *ciliata* IPNI=*60440769-2* 243
Thryallis gracilis IPNI=*556349-1* 396
Thryallis humboldtiana IPNI=*556362-1* 396
Thryallis latifolia IPNI=*556362-1* 396
Thryallis paniculata IPNI=*556362-1* 396
Thryocephalon nemorale IPNI=*77140538-1* 231
Thryocephalon pumilum IPNI=*77144505-1* 230
Thuessinkia speciosa IPNI=*665616-1* 207
Thuja acuta IPNI=*677124-1* 180
Thuja antarctica IPNI=*677124-1* 180
Thuja argentea IPNI=*677124-1* 180
Thuja aurea IPNI=*677124-1* 180
Thuja australis IPNI=*677124-1* 180
Thuja bodmeri IPNI=*263956-1* 180
Thuja canadensis IPNI=*263956-1* 180
Thuja caucasica IPNI=*263956-1* 180
Thuja chengii IPNI=*677124-1* 180
Thuja compacta IPNI=*263956-1* 180
Thuja decora IPNI=*677124-1* 180
Thuja devriesiana IPNI=*263956-1* 180
Thuja dumosa IPNI=*677124-1* 180
Thuja elegantissima IPNI=*677124-1* 180
Thuja ellwangeriana IPNI=*263956-1* 180
Thuja ericoides IPNI=*263956-1* 180
Thuja ericoides IPNI=*677124-1* 180
Thuja expansa IPNI=*677124-1* 180
Thuja falcata IPNI=*677124-1* 180
Thuja filiformis IPNI=*677124-1* 180
Thuja fortunei IPNI=*677124-1* 180
Thuja funiculata IPNI=*677124-1* 180
Thuja glauca IPNI=*677124-1* 180
Thuja globosa IPNI=*263956-1* 180
Thuja gracilifolia IPNI=*677124-1* 180
Thuja hoveyi IPNI=*263956-1* 180
Thuja hybrida IPNI=*677124-1* 180
Thuja intermedia IPNI=*677124-1* 180
Thuja meldensis IPNI=*677124-1* 180
Thuja minor IPNI=*263956-1* 180
Thuja minor IPNI=*677124-1* 180
Thuja monstrosa IPNI=*677124-1* 180
Thuja nana IPNI=*263956-1* 180
Thuja obtusa IPNI=*263956-1* 180
Thuja occidentalis f. *gaspensis* IPNI=*263956-1* 180
Thuja occidentalis f. *prostrata* IPNI=*263956-1* 180
Thuja occidentalis var. *douglasii* IPNI=*263956-1* 180
Thuja odorata IPNI=*263956-1* 180
Thuja orientalis IPNI=*677124-1* 180
Thuja orientalis var. *mexicana* IPNI=*677124-1* 180
Thuja procera IPNI=*263956-1* 180
Thuja pyramidalis IPNI=*677124-1* 180
Thuja semperaurescens IPNI=*677124-1* 180
Thuja sibirica IPNI=*263956-1* 180
Thuja stricta IPNI=*677124-1* 180
Thuja tatarica IPNI=*263956-1* 180
Thuja tatarica IPNI=*677124-1* 180
Thuja theophrasti IPNI=*263956-1* 180
Thuja variegata IPNI=*263956-1* 180
Thuja vervaeneana IPNI=*263956-1* 180
Thunbergia adenophora IPNI=*117460-3* 362
Thunbergia albiflora IPNI=*56177-1* 361
Thunbergia angustifolia IPNI=*56270-1* 362
Thunbergia aurantiaca IPNI=*56177-1* 361
Thunbergia bikimaensis IPNI=*56177-1* 361
Thunbergia bodinieri IPNI=*56270-1* 362
Thunbergia chinensis IPNI=*117460-3* 362
Thunbergia cordata IPNI=*56270-1* 362
Thunbergia cordifolia IPNI=*117460-3* 362
Thunbergia delamerei IPNI=*56177-1* 361
Thunbergia doddsii IPNI=*56177-1* 361
Thunbergia fragrans f. *javanica* IPNI=*56270-1* 362
Thunbergia fragrans subsp. *hainanensis* IPNI=*56270-1* 362
Thunbergia fuscata IPNI=*56177-1* 361
Thunbergia grandiflora f. *alba* IPNI=*117460-3* 362
Thunbergia grandiflora f. *citrina* IPNI=*117460-3* 362
Thunbergia hainanensis IPNI=*56270-1* 362
Thunbergia heterophylla IPNI=*56270-1* 362
Thunbergia ikbaliana IPNI=*56259-1* 361
Thunbergia java IPNI=*56270-1* 362
Thunbergia kamatembica IPNI=*56177-1* 361
Thunbergia lacei IPNI=*117460-3* 362
Thunbergia linearifolia IPNI=*56270-1* 362
Thunbergia loheri IPNI=*56270-1* 362
Thunbergia longiflora IPNI=*56270-1* 362
Thunbergia malvifolia IPNI=*117460-3* 362
Thunbergia manganjensis IPNI=*56177-1* 361
Thunbergia mestdaghi IPNI=*56259-1* 361
Thunbergia mindanaensis IPNI=*56270-1* 362
Thunbergia nymphaeifolia IPNI=*56177-1* 361
Thunbergia oculata IPNI=*56177-1* 361
Thunbergia roxburghia IPNI=*56270-1* 362
Thunbergia scandens IPNI=*56270-1* 362
Thunbergia stolonifera IPNI=*1072674-2* 376
Thunbergia subsagittata IPNI=*56270-1* 362
Thunbergia talbotiae IPNI=*117460-3* 362
Thunbergia volubilis IPNI=*56270-1* 362
Thurnheyssera tragopyrum IPNI=*697710-1* 290
Thylacantha campestris IPNI=*798952-1* 375
Thymus bidentatus IPNI=*451022-1* 371
Thymus biserratus IPNI=*447982-1* 370
Thymus brownei IPNI=*61179-2* 369
Thymus collinus IPNI=*461765-1* 373
Thymus gracilis IPNI=*61179-2* 369
Thymus humifusus IPNI=*61198-2* 369
Thymus majorana IPNI=*453303-1* 372
Thymus melissa IPNI=*450084-1* 371
Thymus montanus IPNI=*457736-1* 373
Thymus nubigenus IPNI=*61198-2* 369
Thymus origanum IPNI=*453395-1* 372

Thymus taxifolius IPNI=*1009966-1* 369
Thymus virginicus IPNI=*447996-1* 370
Thymus xalapensis IPNI=*61179-2* 369
Thyroma bicolor IPNI=*77347-1* 339
Thyroma decipiens IPNI=*77347-1* 339
Thyroma lhotzkiana IPNI=*77347-1* 339
Thyroma nitida IPNI=*77367-1* 339
Thyroma parvifolia IPNI=*315552-2* 339
Thyroma polyneura IPNI=*77433-1* 339
Thyroma sessiliflora IPNI=*77347-1* 339
Thyrsacanthus bracteolatus IPNI=*172440-2* 360
Thyrsacanthus flagellum IPNI=*172486-2* 360
Thyrsacanthus longifolius IPNI=*172486-2* 360
Thyrsacanthus pantasmensis IPNI=*172486-2* 360
Thyrsacanthus strictus f. *evolutior* IPNI=*172486-2* 360
Thyrsacanthus strictus IPNI=*172486-2* 360
Thyrsacanthus tubaeformis IPNI=*172486-2* 360
Thyrsanthema ebracteata IPNI=*192653-1* 259
Thyrsanthema nutans IPNI=*192653-1* 259
Thyrsanthus fasciculatus IPNI=*182946-2* 342
Thyrsodium dasytrichum IPNI=*71652-1* 451
Thyrsodium giganteum IPNI=*71652-1* 451
Thyrsodium paraense IPNI=*71652-1* 451
Thyrsodium salzmannianum IPNI=*71652-1* 451
Thyrsodium schomburgkianum IPNI=*71652-1* 451
Thyrsodium schomburgkianum var. *salzmannianum* IPNI=*71652-1* 451
Thysanachne scoparia IPNI=*21052-2* 234
Thysanolaena procera IPNI=*402044-1* 239
Tiaridium anisophyllum IPNI=*116938-1* 275
Tiaridium indicum IPNI=*116938-1* 275
Tiaridium indicum var. *mexicanum* IPNI=*116938-1* 275
Tiaridium velutinum IPNI=*116938-1* 275
Tibouchina aquatica IPNI=*573916-1* 424
Tibouchina fragilis IPNI=*60465985-2* 417
Tibouchina heteronoma IPNI=*60465985-2* 417
Tibouchina lepidota IPNI=*77205900-1* 417
Tibouchina macrostachya IPNI=*569110-1* 419
Tibouchina paulistana IPNI=*77206410-1* 424
Tibouchina urceolaris var. *papillosa* IPNI=*77206410-1* 424
Tibouchina urvilleana IPNI=*77206410-1* 424
Tibouchina urvilleana var. *glandulifera* IPNI=*77206410-1* 424
Tieghemopanax fruticosus IPNI=*847351-1* 254
Tigarea dentata IPNI=*82916-2* 295
Tigridia conchiflora IPNI=*441553-1* 216
Tigridia grandiflora IPNI=*441553-1* 216
Tigridia lutea IPNI=*441553-1* 216
Tigridia oxypetala IPNI=*441553-1* 216
Tigridia pavonia f. *conchiflora* IPNI=*441553-1* 216
Tigridia pavonia var. *conchiflora* IPNI=*441553-1* 216
Tigridia pringlei IPNI=*441553-1* 216
Tigridia speciosa IPNI=*441553-1* 216
Tilesia capitata IPNI=*993359-1* 271
Tillandsia appuniana IPNI=*77158988-1* 229
Tillandsia axillaris IPNI=*125000-1* 229
Tillandsia bartramii IPNI=*254406-2* 229
Tillandsia berteroniana IPNI=*50112-2* 228
Tillandsia brevibracteata IPNI=*254406-2* 229
Tillandsia brevifolia IPNI=*253931-2* 229
Tillandsia bromeliifolia IPNI=*121574-1* 227
Tillandsia clavata IPNI=*115161-2* 228
Tillandsia clavata IPNI=*115178-2* 229
Tillandsia comosa IPNI=*115178-2* 229
Tillandsia complanata subsp. *latifolia* IPNI=*125000-1* 229
Tillandsia crinita IPNI=*1169665-2* 229
Tillandsia ehrenbergiana IPNI=*254406-2* 229
Tillandsia filiformis IPNI=*1169665-2* 229
Tillandsia flexuosa var. *pallida* IPNI=*254406-2* 229
Tillandsia gymnophylla IPNI=*115178-2* 229
Tillandsia incarnata var. *margaritacea* IPNI=*253931-2* 229
Tillandsia lingulata IPNI=*115161-2* 228
Tillandsia lingulata IPNI=*254406-2* 229
Tillandsia longibracteata IPNI=*77158988-1* 229
Tillandsia monostachia IPNI=*115178-2* 229
Tillandsia monostachya IPNI=*1194855-2* 229
Tillandsia musaica IPNI=*115185-2* 229
Tillandsia nuttalliana IPNI=*254406-2* 229
Tillandsia pachycarpa IPNI=*115178-2* 229
Tillandsia pauciflora IPNI=*1194855-2* 229
Tillandsia pendula IPNI=*50112-2* 228
Tillandsia picta IPNI=*77158988-1* 229
Tillandsia ramosa IPNI=*254406-2* 229
Tillandsia ramosa var. *cingstuniensis* IPNI=*254406-2* 229
Tillandsia recurvata f. *argentea* IPNI=*1194855-2* 229
Tillandsia recurvata f. *brevifolia* IPNI=*1194855-2* 229
Tillandsia recurvata f. *caespitosa* IPNI=*1194855-2* 229
Tillandsia recurvata f. *contorta* IPNI=*1194855-2* 229
Tillandsia recurvata f. *elongata* IPNI=*1194855-2* 229
Tillandsia recurvata f. *major* IPNI=*1194855-2* 229
Tillandsia recurvata f. *minor* IPNI=*1194855-2* 229
Tillandsia recurvata f. *minuta* IPNI=*1194855-2* 229
Tillandsia recurvata var. *ciliata* IPNI=*1194855-2* 229
Tillandsia recurvata var. *genuina* IPNI=*1194855-2* 229
Tillandsia recurvata var. *minuta* IPNI=*1194855-2* 229
Tillandsia sintenisii IPNI=*254406-2* 229
Tillandsia speciosa IPNI=*77158988-1* 229
Tillandsia spinosa IPNI=*121570-1* 227
Tillandsia splendens IPNI=*77158988-1* 229
Tillandsia striata IPNI=*253931-2* 229
Tillandsia tetrantha IPNI=*970621-1* 229
Tillandsia tetrantha var. *typica* IPNI=*970621-1* 229
Tillandsia trichoides IPNI=*1169665-2* 229
Tillandsia uniflora IPNI=*1194855-2* 229
Tillandsia usneoides f. *cretacea* IPNI=*1169665-2* 229
Tillandsia usneoides f. *crispa* IPNI=*1169665-2* 229
Tillandsia usneoides f. *ferruginea* IPNI=*1169665-2* 229
Tillandsia usneoides f. *filiformis* IPNI=*1169665-2* 229
Tillandsia usneoides f. *longissima* IPNI=*1169665-2* 229
Tillandsia usneoides f. *major* IPNI=*1169665-2* 229
Tillandsia utriculata f. *variegata* IPNI=*254406-2* 229
Tilmia caryotifolia IPNI=*7909-2* 204
Timbalia pyracantha IPNI=*730611-1* 444
Tinaea elegans IPNI=*406803-1* 240
Tinantia erecta f. *puberula* IPNI=*173146-1* 224
Tinantia fugax IPNI=*173146-1* 224
Tinantia latifolia IPNI=*173146-1* 224
Tinantia modesta IPNI=*173146-1* 224
Tinantia undata IPNI=*173146-1* 224
Tinea triplinervia IPNI=*321266-2* 404
Tingulonga almacega IPNI=*128420-1* 453
Tingulonga altissima IPNI=*77183601-1* 452
Tingulonga aracouchini IPNI=*128278-1* 452
Tingulonga carana IPNI=*128295-1* 452
Tingulonga caudata IPNI=*128413-1* 453
Tingulonga divaricata IPNI=*128318-1* 452
Tingulonga elegans IPNI=*128322-1* 452
Tingulonga enneandra IPNI=*128316-1* 452
Tingulonga grandifolia IPNI=*128335-1* 452
Tingulonga guianensis IPNI=*128337-1* 452
Tingulonga heptaphylla IPNI=*128338-1* 453
Tingulonga hostmannii IPNI=*128337-1* 452
Tingulonga insignis IPNI=*128413-1* 453
Tingulonga macrophylla IPNI=*128356-1* 453
Tingulonga martiana IPNI=*128429-1* 453
Tingulonga multiflora IPNI=*128338-1* 453
Tingulonga polybotrya IPNI=*128400-1* 453
Tingulonga simplicifolia IPNI=*128430-1* 453

Tingulonga spruceana IPNI=*128420-1* 453
Tingulonga subserrata IPNI=*128422-1* 453
Tingulonga tenuifolia IPNI=*128424-1* 453
Tinus angustifolia IPNI=*18168-2* 303
Tinus ardisioides IPNI=*247441-2* 304
Tinus brasiliensis IPNI=*18168-2* 303
Tinus capollina IPNI=*586863-1* 303
Tinus compressa IPNI=*586863-1* 303
Tinus coriacea IPNI=*1028287-2* 304
Tinus cuspidata IPNI=*586863-1* 303
Tinus guadelupensis IPNI=*1028287-2* 304
Tinus guianensis IPNI=*18168-2* 303
Tinus hostmannii IPNI=*247474-2* 304
Tinus irazuensis IPNI=*586863-1* 303
Tinus longifolia IPNI=*247465-2* 304
Tinus micrantha IPNI=*247469-2* 304
Tinus nicaraguensis IPNI=*586863-1* 303
Tinus nigricans IPNI=*247472-2* 304
Tinus orinocensis IPNI=*247474-2* 304
Tinus schomburgkiana IPNI=*247474-2* 304
Tinus semicrenata IPNI=*18168-2* 303
Tinus surinamensis IPNI=*247474-2* 304
Tinus turbacensis IPNI=*247490-2* 305
Tipuana fusca IPNI=*262720-2* 336
Tiricta daucoides IPNI=*841063-1* 252
Tirucalia enterophora IPNI=*348758-1* 391
Tirucalia indica IPNI=*348517-1* 391
Tirucalia mauritanica IPNI=*347292-1* 390
Tirucalia paxiana IPNI=*347292-1* 390
Tirucalia sarcostemmatoides IPNI=*347292-1* 390
Tirucalia tirucalli IPNI=*348517-1* 391
Tirucalia xylophylloides IPNI=*348758-1* 391
Tissa aprica IPNI=*159015-1* 286
Tissa coreana IPNI=*241336-2* 286
Tissa fauriei IPNI=*241336-2* 286
Tissa leiosperma IPNI=*241336-2* 286
Tissa marina IPNI=*241336-2* 286
Tissa marina var. *minor* IPNI=*241336-2* 286
Tissa media var. *berteroana* IPNI=*159015-1* 286
Tissa paradoxa IPNI=*30040285-2* 285
Tissa rubra IPNI=*158983-1* 286
Tissa rubra var. *perennans* IPNI=*158983-1* 286
Tissa salina IPNI=*241336-2* 286
Tissa salina var. *involucrata* IPNI=*241336-2* 286
Tissa salina var. *sanfordii* IPNI=*241336-2* 286
Tissa salina var. *sordida* IPNI=*241336-2* 286
Tissa salina var. *tenuis* IPNI=*241336-2* 286
Tissa sparsiflora IPNI=*241336-2* 286
Tissa tenuis IPNI=*241336-2* 286
Tissa villosa IPNI=*159015-1* 286
Tithonia aristata IPNI=*254724-2* 271
Tithonia diversifolia subsp. *glabriuscula* IPNI=*255747-1* 271
Tithonia heterophylla IPNI=*254724-2* 271
Tithonia humilis IPNI=*221479-2* 287
Tithonia laciniata IPNI=*241716-1* 269
Tithonia macrophylla IPNI=*254724-2* 271
Tithonia speciosa IPNI=*254724-2* 271
Tithonia tagetiflora IPNI=*254724-2* 271
Tithonia uniflora IPNI=*254724-2* 271
Tithonia vilmoriniana IPNI=*254724-2* 271
Tithymalus acicularis IPNI=*346205-1* 390
Tithymalus angustifolius IPNI=*346205-1* 390
Tithymalus brachypus IPNI=*347292-1* 390
Tithymalus cataputia IPNI=*347072-1* 390
Tithymalus cinerea IPNI=*346205-1* 390
Tithymalus cotinifolius IPNI=*346143-1* 389
Tithymalus cyparissias IPNI=*346205-1* 390
Tithymalus edulis IPNI=*347489-1* 390
Tithymalus fischerianus IPNI=*346510-1* 390
Tithymalus flaccidus IPNI=*347292-1* 390
Tithymalus heterophyllus IPNI=*30062970-2* 390
Tithymalus lathyris IPNI=*347072-1* 390
Tithymalus maculatus IPNI=*347236-1* 390
Tithymalus mauritanicus IPNI=*347292-1* 390
Tithymalus nutans IPNI=*101777-2* 390
Tithymalus pallasii IPNI=*346510-1* 390
Tithymalus prostratus IPNI=*101913-2* 390
Tithymalus seguierianus IPNI=*348221-1* 391
Tithymalus tithymaloides IPNI=*102036-2* 391
Tithymalus zeyheri IPNI=*347292-1* 390
Tittelbachia fuchsioides IPNI=*104715-1* 292
Tittelbachia miniata IPNI=*104715-1* 292
Tittmannia colemannii IPNI=*799766-1* 374
Tittmannia ovata IPNI=*1189750-2* 374
Tittmannia scabra IPNI=*1189750-2* 374
Tococa acuminata IPNI=*77207761-1* 423
Tococa aubletii IPNI=*77207761-1* 423
Tococa barbata IPNI=*77207761-1* 423
Tococa cachimbensis IPNI=*77207761-1* 423
Tococa cardiophylla IPNI=*77207761-1* 423
Tococa chuivensis IPNI=*77207761-1* 423
Tococa chuivensis var. *ecuadorensis* IPNI=*77207761-1* 423
Tococa coriacea IPNI=*77207761-1* 423
Tococa coronata IPNI=*77207926-1* 423
Tococa didymophysca IPNI=*77207761-1* 423
Tococa discolor IPNI=*77207761-1* 423
Tococa disolenia IPNI=*77207926-1* 423
Tococa egensis IPNI=*77207926-1* 423
Tococa ferruginea IPNI=*77207761-1* 423
Tococa formicaria IPNI=*77207761-1* 423
Tococa formicaria var. *didymophysca* IPNI=*77207761-1* 423
Tococa formicaria var. *gardneri* IPNI=*77207761-1* 423
Tococa glandulosa IPNI=*77207926-1* 423
Tococa grandifolia IPNI=*159460-2* 421
Tococa guianensis IPNI=*77207761-1* 423
Tococa guianensis var. *orinocensis* IPNI=*77207761-1* 423
Tococa guyanensis var. *macrophysca* IPNI=*77207761-1* 423
Tococa heterophylla IPNI=*77207843-1* 420
Tococa hirsuta IPNI=*77207761-1* 423
Tococa juruensis IPNI=*77207761-1* 423
Tococa kuhlmannii IPNI=*77207761-1* 423
Tococa latifolia IPNI=*77207761-1* 423
Tococa longisepala IPNI=*77207926-1* 423
Tococa loretensis IPNI=*77207761-1* 423
Tococa mayeta IPNI=*77164249-1* 421
Tococa mexiae IPNI=*77207761-1* 423
Tococa neocinnamomea IPNI=*158746-2* 419
Tococa occidentalis IPNI=*77207761-1* 423
Tococa orcheophora IPNI=*77207761-1* 423
Tococa peckiana IPNI=*77207761-1* 423
Tococa platyphysca IPNI=*77207761-1* 423
Tococa pubescens IPNI=*77207926-1* 423
Tococa roraimae IPNI=*77207761-1* 423
Tococa subglabrata IPNI=*77207761-1* 423
Tococa trailii IPNI=*77207926-1* 423
Tococa truncata IPNI=*77207926-1* 423
Tococa vesiculosa IPNI=*77207761-1* 423
Tocoyena obliquinervia IPNI=*254985-2* 358
Tocoyena undulatifolia IPNI=*761025-1* 355
Toddavaddia dendrodes IPNI=*371953-1* 432
Toddavaddia mimosoides IPNI=*371953-1* 432
Toluifera balsamum IPNI=*166483-2* 328
Toluifera balsamum var. *peruifera* IPNI=*509454-1* 328
Toluifera erythroxyla IPNI=*509454-1* 328
Toluifera frutescens IPNI=*509434-1* 328
Toluifera pedicellata IPNI=*509454-1* 328

Toluifera pereirae IPNI=*166483-2* 328
Toluifera peruifera IPNI=*509454-1* 328
Toluifera pubescens IPNI=*509454-1* 328
Toluifera punctata IPNI=*166483-2* 328
Tolumnia pusilla IPNI=*1021256-1* 219
Tomoxis coarctata IPNI=*539581-1* 215
Tonsella brasiliensis IPNI=*162581-1* 291
Tonsella elliptica IPNI=*162581-1* 291
Tonsella glomerata IPNI=*162581-1* 291
Tonsella grandiflora IPNI=*225455-2* 291
Tonsella laevigata IPNI=*190399-2* 291
Tonsella malpighifolia IPNI=*162403-1* 291
Tonsella radula IPNI=*162403-1* 291
Tontanea canescens IPNI=*315609-2* 348
Tontanea glabra IPNI=*61955-2* 348
Tontanea guianensis IPNI=*61955-2* 348
Tontanea hirsuta IPNI=*746963-1* 348
Tontelea brasiliensis IPNI=*162581-1* 291
Tontelea elliptica IPNI=*162581-1* 291
Tontelea erythroxyloides IPNI=*162581-1* 291
Tontelea glomerata IPNI=*162581-1* 291
Tontelea hookeriana IPNI=*210308-2* 291
Tontelea levigata IPNI=*161653-1* 290
Tontelea longifolia IPNI=*162852-1* 291
Tontelea malpighiifolia IPNI=*162403-1* 291
Tontelea oblongifolia IPNI=*162581-1* 291
Tontelea obtusifolia IPNI=*162581-1* 291
Tontelea radula IPNI=*162403-1* 291
Topobea alternifolia IPNI=*33356-2* 418
Topobea reducta IPNI=*33356-2* 418
Toquera tomentosa IPNI=*275076-2* 274
Torenia alata IPNI=*810361-1* 374
Torenia chamaedrys IPNI=*255156-2* 374
Torenia flaccida IPNI=*1189750-2* 374
Torenia gavottiana IPNI=*255156-2* 374
Torenia glabra IPNI=*810361-1* 374
Torenia gracilis IPNI=*799766-1* 374
Torenia hexandra IPNI=*1189750-2* 374
Torenia kiusiana IPNI=*810361-1* 374
Torenia lucida IPNI=*1189750-2* 374
Torenia minuta IPNI=*1189750-2* 374
Torenia nana IPNI=*255156-2* 374
Torenia nortenia IPNI=*255156-2* 374
Torenia obtusa IPNI=*1189750-2* 374
Torenia obtusifolia IPNI=*810361-1* 374
Torenia ovata IPNI=*1189750-2* 374
Torenia peduncularifomis IPNI=*255156-2* 374
Torenia pedunculata IPNI=*255156-2* 374
Torenia ramosissima IPNI=*255156-2* 374
Torenia rotundifolia IPNI=*1061345-2* 374
Torenia rubella IPNI=*1189750-2* 374
Torenia scabra IPNI=*1189750-2* 374
Torenia thouarsii var. *nivea* IPNI=*255156-2* 374
Torenia varians IPNI=*1189750-2* 374
Torenia veronicifolia IPNI=*255156-2* 374
Torenia viguieri IPNI=*255156-2* 374
Torilis peruviana IPNI=*841167-1* 252
Tornelia dilacerata IPNI=*87046-1* 200
Tornelia fragrans IPNI=*87478-1* 200
Torrentia quinquenervis IPNI=*127605-2* 264
Torresia redolens IPNI=*389043-1* 234
Torreya bogotensis IPNI=*946471-1* 180
Torrubia coriifolia IPNI=*113934-2* 287
Torrubia cuspidata IPNI=*113926-2* 286
Torrubia dussii IPNI=*113934-2* 287
Torrubia fragrans IPNI=*113934-2* 287
Torrubia harrisiana IPNI=*113934-2* 287
Torrubia inermis IPNI=*113934-2* 287
Torrubia schomburgkiana IPNI=*113926-2* 286
Torulinium odoratum IPNI=*74939-2* 231
Torulinium samoense IPNI=*304256-1* 230
Tounatea arborescens IPNI=*519836-1* 334
Tounatea benthamiana IPNI=*519841-1* 334
Tounatea crocea IPNI=*520020-1* 335
Tounatea cuneata IPNI=*520020-1* 335
Tounatea darienensis IPNI=*520020-1* 335
Tounatea discolor IPNI=*519927-1* 334
Tounatea dodecandra IPNI=*519836-1* 334
Tounatea fugax IPNI=*519927-1* 334
Tounatea gatunensis IPNI=*520020-1* 335
Tounatea grandiflora IPNI=*520020-1* 335
Tounatea hayesii IPNI=*520020-1* 335
Tounatea jamaicensis IPNI=*520020-1* 335
Tounatea leptopetala IPNI=*519927-1* 334
Tounatea macrosperma IPNI=*520020-1* 335
Tounatea ochnacea IPNI=*520020-1* 335
Tounatea penomenensis IPNI=*520020-1* 335
Tounatea pinnata IPNI=*519989-1* 334
Tounatea pittieri IPNI=*520020-1* 335
Tounatea polyphylla IPNI=*519994-1* 334
Tounatea racemosa IPNI=*520000-1* 334
Tounatea schomburgkii IPNI=*520015-1* 334
Tounatea sericea IPNI=*520018-1* 335
Tounatea simplex IPNI=*520020-1* 335
Tounatea subcoriacea IPNI=*520020-1* 335
Tounatea trifolia IPNI=*520020-1* 335
Tounatea williamsii IPNI=*520020-1* 335
Tournefortia alba IPNI=*77069564-1* 275
Tournefortia angustiflora IPNI=*77192212-1* 274
Tournefortia asperrima IPNI=*77069564-1* 275
Tournefortia barbadensis IPNI=*119329-1* 275
Tournefortia bicolor IPNI=*77069564-1* 275
Tournefortia bicolor var. *calycosa* IPNI=*77069564-1* 275
Tournefortia billbergiana IPNI=*77069564-1* 275
Tournefortia calycosa IPNI=*77069564-1* 275
Tournefortia carnosa IPNI=*77069564-1* 275
Tournefortia cirrhosa IPNI=*60430568-2* 275
Tournefortia cuspidata IPNI=*77203945-1* 275
Tournefortia difformis IPNI=*60430568-2* 275
Tournefortia elliptica IPNI=*77069564-1* 275
Tournefortia ferruginea IPNI=*119329-1* 275
Tournefortia floribunda IPNI=*119329-1* 275
Tournefortia glabra IPNI=*77210153-1* 275
Tournefortia guadalupensis IPNI=*60449244-2* 275
Tournefortia hirsutissima IPNI=*77069564-1* 275
Tournefortia jamaicensis IPNI=*60449244-2* 275
Tournefortia laevigata IPNI=*119329-1* 275
Tournefortia latifolia IPNI=*60449244-2* 275
Tournefortia laurifolia IPNI=*60449244-2* 275
Tournefortia laxiflora IPNI=*119329-1* 275
Tournefortia leonis IPNI=*119329-1* 275
Tournefortia loxensis IPNI=*121103-1* 275
Tournefortia lucida IPNI=*119329-1* 275
Tournefortia lucida IPNI=*60449244-2* 275
Tournefortia macrophylla IPNI=*77210153-1* 275
Tournefortia maculata IPNI=*60449244-2* 275
Tournefortia minuta IPNI=*119329-1* 275
Tournefortia miquelii IPNI=*60449244-2* 275
Tournefortia oblongata IPNI=*119329-1* 275
Tournefortia obscura IPNI=*77203945-1* 275
Tournefortia octostachya IPNI=*121103-1* 275
Tournefortia odorata IPNI=*77069564-1* 275
Tournefortia peruviana IPNI=*60449244-2* 275
Tournefortia picardae IPNI=*119329-1* 275
Tournefortia polystachya IPNI=*946226-1* 276
Tournefortia potosina IPNI=*119329-1* 275

Tournefortia psilostachya IPNI=*60430568-2* 275
Tournefortia psilostachya var. *caribaea* IPNI=*60430568-2* 275
Tournefortia psilostachya var. *guayaquilensis* IPNI=*60430568-2* 275
Tournefortia psilostachya var. *rubicunda* IPNI=*60430568-2* 275
Tournefortia punctata IPNI=*119329-1* 275
Tournefortia sagrana IPNI=*60449244-2* 275
Tournefortia scandens IPNI=*119329-1* 275
Tournefortia scandens IPNI=*60449244-2* 275
Tournefortia schomburgkii IPNI=*77069564-1* 275
Tournefortia sericea IPNI=*119329-1* 275
Tournefortia setifera IPNI=*77203945-1* 275
Tournefortia strigosa IPNI=*60430568-2* 275
Tournefortia suaveolens IPNI=*121103-1* 275
Tournefortia surinamensis IPNI=*60449244-2* 275
Tournefortia synzystachya IPNI=*116662-1* 274
Tournefortia syringifolia IPNI=*60449244-2* 275
Tournefortia tepicana IPNI=*77069564-1* 275
Tournefortia tomentosa IPNI=*77069564-1* 275
Tournefortia trinitatis IPNI=*119329-1* 275
Tournefortia tristachya IPNI=*60430568-2* 275
Tournefortia velutina IPNI=*119329-1* 275
Tournefortia volubilis IPNI=*119329-1* 275
Tournefortia volubilis IPNI=*60449244-2* 275
Tournefortia volubilis var. *glabrata* IPNI=*119329-1* 275
Tournefortia volubilis var. *hirsuta* IPNI=*119329-1* 275
Tournefortia volubilis var. *oblongifolia* IPNI=*119329-1* 275
Tournesol castaneifolia IPNI=*44529-2* 388
Touroulia jenmanii IPNI=*134577-2* 397
Touroulia solitaria IPNI=*429315-1* 397
Tovomita alatopetiolata IPNI=*427694-1* 385
Tovomita brasiliensis var. *secunda* IPNI=*429349-1* 385
Tovomita calyptrata IPNI=*972610-1* 385
Tovomita hameliifolia IPNI=*429403-1* 385
Tovomita jenmanii IPNI=*255606-2* 385
Tovomita ligustrina IPNI=*429349-1* 385
Tovomita membranifolia IPNI=*429411-1* 385
Tovomita schomburgkiana IPNI=*429403-1* 385
Tovomita stenantha IPNI=*429411-1* 385
Tovomita triflora IPNI=*255606-2* 385
Tovomitopsis colombiana IPNI=*57606-2* 383
Tovomitopsis membranifolia IPNI=*429411-1* 385
Toxicaria americana IPNI=*547220-1* 346
Tozzettia fulva IPNI=*387005-1* 233
Tozzettia pratensis IPNI=*387176-1* 233
Tozzettia vulgaris IPNI=*387176-1* 233
Tracaulon beyrichianum IPNI=*1050850-2* 289
Tracaulon meisnerianum IPNI=*1050850-2* 289
Tracaulon multangulare IPNI=*1050850-2* 289
Trachodes paniculata IPNI=*229022-1* 264
Trachycarpus caespitosus IPNI=*670182-1* 212
Trachycarpus excelsus IPNI=*669577-1* 211
Trachycarpus wagnerianus IPNI=*670182-1* 212
Trachylobium martianum IPNI=*498789-1* 321
Trachypoa vulgaris IPNI=*397309-1* 236
Trachypogon angustifolius IPNI=*424391-1* 246
Trachypogon avenaceus IPNI=*422139-1* 245
Trachypogon canescens IPNI=*424391-1* 246
Trachypogon capensis IPNI=*424391-1* 246
Trachypogon capensis subvar. *mollis* IPNI=*424391-1* 246
Trachypogon dactyloides IPNI=*424391-1* 246
Trachypogon densus IPNI=*424391-1* 246
Trachypogon dissolutus IPNI=*424391-1* 246
Trachypogon durus IPNI=*424391-1* 246
Trachypogon glaucescens IPNI=*424391-1* 246
Trachypogon gouinii IPNI=*424391-1* 246
Trachypogon gracilis IPNI=*424391-1* 246
Trachypogon involutus IPNI=*424391-1* 246
Trachypogon karwinskyi IPNI=*424391-1* 246
Trachypogon ligularis IPNI=*424391-1* 246
Trachypogon mayaensis IPNI=*424391-1* 246
Trachypogon micans IPNI=*424391-1* 246
Trachypogon mollis IPNI=*424391-1* 246
Trachypogon montufarii IPNI=*424391-1* 246
Trachypogon montufarii var. *bolivianus* IPNI=*424391-1* 246
Trachypogon montufarii var. *mollis* IPNI=*424391-1* 246
Trachypogon muelleri IPNI=*424391-1* 246
Trachypogon palmeri IPNI=*424391-1* 246
Trachypogon parviflorus IPNI=*424391-1* 246
Trachypogon planifolius IPNI=*424391-1* 246
Trachypogon plumosus IPNI=*424391-1* 246
Trachypogon plumosus subvar. *secundus* IPNI=*424391-1* 246
Trachypogon plumosus var. *secundus* IPNI=*424391-1* 246
Trachypogon polymorphus IPNI=*424391-1* 246
Trachypogon polymorphus subvar. *dactyloides* IPNI=*424391-1* 246
Trachypogon polymorphus subvar. *gracilis* IPNI=*424391-1* 246
Trachypogon polymorphus subvar. *mollis* IPNI=*424391-1* 246
Trachypogon polymorphus subvar. *secundus* IPNI=*424391-1* 246
Trachypogon polymorphus var. *bolivianus* IPNI=*424391-1* 246
Trachypogon polymorphus var. *canescens* IPNI=*424391-1* 246
Trachypogon polymorphus var. *dissolutus* IPNI=*424391-1* 246
Trachypogon polymorphus var. *gouinii* IPNI=*424391-1* 246
Trachypogon polymorphus var. *karwinskyi* IPNI=*424391-1* 246
Trachypogon polymorphus var. *ligularis* IPNI=*424391-1* 246
Trachypogon polymorphus var. *montufarii* IPNI=*424391-1* 246
Trachypogon polymorphus var. *plumosus* IPNI=*424391-1* 246
Trachypogon preslii IPNI=*424391-1* 246
Trachypogon preslii var. *secundus* IPNI=*424391-1* 246
Trachypogon ramosus IPNI=*424391-1* 246
Trachypogon rigidifolius IPNI=*424391-1* 246
Trachypogon rufus IPNI=*405583-1* 240
Trachypogon scrobiculatus IPNI=*404802-1* 240
Trachypogon secundus IPNI=*424391-1* 246
Trachypogon stenophyllus IPNI=*424391-1* 246
Trachypogon stipoides IPNI=*240834-2* 245
Trachypogon thollonii IPNI=*424391-1* 246
Trachypogon truncatus IPNI=*424391-1* 246
Trachypogon violaceus IPNI=*424391-1* 246
Trachysperma humboldtiana IPNI=*171243-2* 273
Tradescantia albiflora IPNI=*30090792-2* 224
Tradescantia aubletii IPNI=*79497-2* 224
Tradescantia balbisii IPNI=*77184055-1* 224
Tradescantia bicolor IPNI=*42443-2* 223
Tradescantia bifida IPNI=*173146-1* 224
Tradescantia callisia IPNI=*171826-1* 224
Tradescantia capitata IPNI=*1191409-2* 224
Tradescantia commelina IPNI=*173146-1* 224
Tradescantia congesta IPNI=*77184055-1* 224
Tradescantia cumanensis IPNI=*77184055-1* 224
Tradescantia cumanensis var. *glabrior* IPNI=*77184055-1* 224
Tradescantia debilis IPNI=*42443-2* 223
Tradescantia diaphana IPNI=*42443-2* 223
Tradescantia dilecta IPNI=*172628-1* 224
Tradescantia discolor IPNI=*173471-1* 224
Tradescantia discolor var. *concolor* IPNI=*173471-1* 224
Tradescantia discolor var. *variegata* IPNI=*173471-1* 224
Tradescantia divaricata IPNI=*79497-2* 224
Tradescantia elongata IPNI=*77184055-1* 224
Tradescantia elongata var. *schlechtendalii* IPNI=*77184055-1* 224
Tradescantia elongata var. *triandra* IPNI=*77184055-1* 224
Tradescantia erecta IPNI=*173146-1* 224
Tradescantia geniculata var. *botterii* IPNI=*109899-2* 224
Tradescantia geniculata var. *schiedeana* IPNI=*109899-2* 224
Tradescantia gentianifolia IPNI=*1191409-2* 224
Tradescantia gonatandra IPNI=*1191409-2* 224
Tradescantia gracilis IPNI=*42443-2* 223
Tradescantia gracilis var. *bicolor* IPNI=*42443-2* 223

Tradescantia guianensis IPNI=*77184055-1* 224
Tradescantia laekenensis IPNI=*30090792-2* 224
Tradescantia latifolia IPNI=*173146-1* 224
Tradescantia lundellii IPNI=*109899-2* 224
Tradescantia malabarica IPNI=*172873-1* 224
Tradescantia malabarica IPNI=*173146-1* 224
Tradescantia mexicana IPNI=*77184055-1* 224
Tradescantia multiflora IPNI=*77205602-1* 224
Tradescantia multiflora var. *linnaei* IPNI=*77205602-1* 224
Tradescantia multiflora var. *parviflora* IPNI=*77205602-1* 224
Tradescantia multiflora var. *tobagensis* IPNI=*77205602-1* 224
Tradescantia nervosa IPNI=*659882-1* 223
Tradescantia parviflora IPNI=*77205602-1* 224
Tradescantia pellucida IPNI=*109899-2* 224
Tradescantia procumbens IPNI=*77205602-1* 224
Tradescantia recta IPNI=*173146-1* 224
Tradescantia reginae IPNI=*172628-1* 224
Tradescantia schiedeana IPNI=*109899-2* 224
Tradescantia schlechtendalii IPNI=*77184055-1* 224
Tradescantia schomburgkiana IPNI=*77184055-1* 224
Tradescantia striata IPNI=*256037-2* 224
Tradescantia triandra IPNI=*77184055-1* 224
Tradescantia undata IPNI=*173146-1* 224
Tradescantia undulata IPNI=*173146-1* 224
Tradescantia velutina IPNI=*256037-2* 224
Tradescantia versicolor IPNI=*173471-1* 224
Tragia amoena IPNI=*315307-2* 393
Tragia diffusa IPNI=*315307-2* 393
Tragia gayana IPNI=*315307-2* 393
Tragia haguensis IPNI=*315307-2* 393
Tragia ibaguensis IPNI=*315307-2* 393
Tragia infesta IPNI=*315307-2* 393
Tragia monandra IPNI=*315307-2* 393
Tragia ovata IPNI=*315307-2* 393
Tragia pedicillaris IPNI=*315307-2* 393
Tragia pedunculata IPNI=*315307-2* 393
Tragia plumosa IPNI=*315307-2* 393
Tragia serra IPNI=*315307-2* 393
Tragia triangularis IPNI=*315307-2* 393
Tragia virgata IPNI=*315307-2* 393
Tragia volubilis var. *genuina* IPNI=*315307-2* 393
Tragia volubilis var. *grandifolia* IPNI=*315307-2* 393
Tragia volubilis var. *guaranitica* IPNI=*315307-2* 393
Tragia volubilis var. *lanceolata* IPNI=*315307-2* 393
Tragia volubilis var. *longifolia* IPNI=*315307-2* 393
Tragia volubilis var. *pedicellaris* IPNI=*315307-2* 393
Tragia volubilis var. *serra* IPNI=*315307-2* 393
Tragia volubilis var. *tenuifolia* IPNI=*315307-2* 393
Tragia volubilis var. *triangularis* IPNI=*315307-2* 393
Tragium anisum IPNI=*846658-1* 252
Tragium palmetorum IPNI=*846658-1* 252
Tragopogon clavulatus IPNI=*256109-1* 271
Tragularia horrida IPNI=*326897-2* 287
Tragus alienus IPNI=*211125-2* 244
Tragus occidentalis IPNI=*256481-2* 246
Tragus racemosus var. *berteronianus* IPNI=*256481-2* 246
Tragus racemosus var. *brevispiculus* IPNI=*256481-2* 246
Tragus tcheliensis IPNI=*256481-2* 246
Tragus unioloides IPNI=*37001-2* 235
Trattinnickia barbourii IPNI=*256487-2* 453
Trattinnickia burserifolia var. *lindemanii* IPNI=*128598-1* 453
Trattinnickia burserifolia var. *obtusa* IPNI=*128598-1* 453
Trattinnickia burserifolia var. *quinquejuga* IPNI=*128598-1* 453
Trattinnickia guianensis IPNI=*128598-1* 453
Trattinnickia humilis IPNI=*128598-1* 453
Trattinnickia multiflora IPNI=*128598-1* 453
Trattinnickia rhoifolia var. *pubescens* IPNI=*256503-2* 453
Trattinnickia rhoifolia var. *sprucei* IPNI=*128602-1* 454
Trattinnickia rhoifolia var. *willdenowii* IPNI=*128602-1* 454
Trattinnickia ryanii IPNI=*128602-1* 454
Trattinnickia schomburgkiana IPNI=*128598-1* 453
Trattinnickia schomburgkii IPNI=*128598-1* 453
Trattinnickia subchoripetala IPNI=*128598-1* 453
Trema canescens IPNI=*1018059-2* 436
Trema chichileum IPNI=*1018059-2* 436
Trema lima IPNI=*1018059-2* 436
Trema macrophyllum IPNI=*1018059-2* 436
Trema melinonum IPNI=*1018059-2* 436
Trema micranthum var. *floridanum* IPNI=*1018059-2* 436
Trema micranthum var. *obtusatum* IPNI=*1018059-2* 436
Trema micranthum var. *strigillosum* IPNI=*1018059-2* 436
Trema molle IPNI=*1018059-2* 436
Trema orientale f. *micranthum* IPNI=*1018059-2* 436
Trema orientale var. *rugosum* IPNI=*1018059-2* 436
Trema riparium IPNI=*1018059-2* 436
Trema rufescens IPNI=*1018059-2* 436
Trema rugosum IPNI=*1018059-2* 436
Trema schiedeanum IPNI=*1018059-2* 436
Trema strigillosum IPNI=*1018059-2* 436
Tremanthus cordata IPNI=*826890-1* 308
Tremanthus ovata IPNI=*826890-1* 308
Trevirana coccinea IPNI=*2352-2* 365
Trevirana longiflora IPNI=*377615-1* 365
Trevirana pulchella IPNI=*2352-2* 365
Trevirana tenella IPNI=*2352-2* 365
Triacis microphylla IPNI=*836578-1* 401
Trianaeopiper confertinodum IPNI=*77061871-1* 195
Trianaeopiper contraverrugosa IPNI=*77146701-1* 195
Trianaeopiper killipii IPNI=*683758-1* 197
Trianaeopiper mexiae IPNI=*683758-1* 197
Trianaeopiper trianae IPNI=*683758-1* 197
Trianosperma glandulosum IPNI=*291801-1* 293
Trianosperma heterophyllum IPNI=*291801-1* 293
Trianosperma racemosum IPNI=*291845-1* 293
Trianthema flexuosum IPNI=*322565-2* 280
Trianthema griseum IPNI=*235145-2* 280
Trianthema littorale IPNI=*322565-2* 280
Trianthema monogynum IPNI=*322565-2* 280
Trianthema obcordatum IPNI=*322565-2* 280
Trianthema polyandrum IPNI=*235145-2* 280
Trianthema procumbens IPNI=*322565-2* 280
Triathera aristidoides IPNI=*1049662-2* 235
Tribroma bicolor IPNI=*825565-1* 414
Tricarium cochinchinense IPNI=*194610-2* 401
Triceros occidentalis IPNI=*785554-1* 292
Triceros paniculata IPNI=*785554-1* 292
Trichachne affinis IPNI=*80323-2* 238
Trichachne californica IPNI=*80220-2* 237
Trichachne californica var. *villosissima* IPNI=*80220-2* 237
Trichachne insularis IPNI=*398888-1* 237
Trichachne penicilligera IPNI=*1050107-2* 237
Trichachne saccharata IPNI=*80220-2* 237
Trichachne sacchariflora IPNI=*1050107-2* 237
Trichachne sellowii IPNI=*80323-2* 238
Trichantha acuminata IPNI=*63508-2* 366
Trichantha anisophylla IPNI=*63508-2* 366
Trichantha brenneri IPNI=*378787-1* 366
Trichantha calotricha IPNI=*63523-2* 366
Trichantha dictyophylla IPNI=*63557-2* 366
Trichantha gracilis IPNI=*969025-1* 367
Trichantha parviflora IPNI=*63636-2* 366
Trichantha tenella IPNI=*969025-1* 367
Trichanthera gigantea var. *guianensis* IPNI=*56505-1* 362
Trichelostylis capillaris IPNI=*1186511-2* 229
Trichelostylis complanata IPNI=*1111946-2* 232
Trichelostylis kielmeyeri IPNI=*1111946-2* 232

Trophaeum pinnatum IPNI=*310974-2* 280
Trophaeum scheunerianum IPNI=*310974-2* 280
Trophaeum tuberosum IPNI=*310353-2* 280
Trophaeum zanderi IPNI=*310974-2* 280
Trophis americana IPNI=*260203-2* 441
Trophis americana var. *meridionalis* IPNI=*260203-2* 441
Trophis americana var. *ramon* IPNI=*260203-2* 441
Trophis aurantiaca IPNI=*945890-1* 441
Trophis hilariana IPNI=*856566-1* 441
Trophis laurifolia IPNI=*356634-1* 280
Trophis macrostachya IPNI=*856575-1* 441
Trophis racemosa subsp. *meridionalis* IPNI=*260203-2* 441
Trophis racemosa subsp. *ramon* IPNI=*260203-2* 441
Trophis racemosa var. *meridionalis* IPNI=*260203-2* 441
Trophis ramon IPNI=*260203-2* 441
Troschelia orinocensis IPNI=*429893-1* 224
Troxirum blandum IPNI=*678107-1* 193
Troxirum pereskia IPNI=*679404-1* 194
Troxirum reflexum IPNI=*1068802-2* 194
Troxirum trifolium IPNI=*1090908-2* 194
Troxirum villosum IPNI=*678107-1* 193
Troxistemon distichus IPNI=*30048918-2* 213
Troxistemon dryanderi IPNI=*30048918-2* 213
Troxistemon fragrans IPNI=*65212-1* 213
Troxistemon littoralis IPNI=*30048918-2* 213
Truellum capitatum IPNI=*60433515-2* 289
Truellum meisnerianum IPNI=*1050850-2* 289
Truellum nepalense IPNI=*60433514-2* 289
Trymatococcus guanabarinus IPNI=*584303-1* 440
Trymatococcus paraensis IPNI=*317009-2* 441
Trymatococcus turbinatus IPNI=*317009-2* 441
Tschudya ibaguensis IPNI=*572590-1* 421
Tschudya spondylantha IPNI=*572835-1* 421
Tsiangia hongkongensis IPNI=*753834-1* 352
Tuberosa amica IPNI=*77153983-1* 214
Tubiflora pachystachya IPNI=*1167021-2* 360
Tubiflora squamosa IPNI=*1167021-2* 360
Tulbaghia affinis IPNI=*542757-1* 214
Tulbaghia africana IPNI=*527258-1* 213
Tulbaghia brachystemma IPNI=*542757-1* 214
Tulbaghia daviesii IPNI=*542797-1* 214
Tulbaghia heisteri IPNI=*527258-1* 213
Tulbaghia inodora IPNI=*542757-1* 214
Tulbaghia minor IPNI=*527258-1* 213
Tulbaghia narcissifolia IPNI=*542757-1* 214
Tulbaghia praecox IPNI=*527283-1* 213
Tunica barbata IPNI=*306674-2* 285
Tunica caryophyllus IPNI=*153205-1* 285
Tunica morrisii IPNI=*153205-1* 285
Tunica plumaria IPNI=*153717-1* 285
Tupa acuminata IPNI=*142908-1* 273
Tupa ignescens IPNI=*142115-2* 273
Tupa laxiflora IPNI=*142260-2* 273
Tupistra esquirolii IPNI=*64141-1* 216
Turia cordata IPNI=*293060-1* 294
Turia cylindrica IPNI=*293060-1* 294
Turia sativa IPNI=*293060-1* 294
Turnera alba IPNI=*836702-1* 401
Turnera aphrodisiaca IPNI=*836578-1* 401
Turnera aspera IPNI=*30068867-2* 400
Turnera aturensis IPNI=*30068867-2* 400
Turnera betonicifolia IPNI=*30068867-2* 400
Turnera carpinifolia IPNI=*259040-2* 401
Turnera chamaedrys IPNI=*836702-1* 401
Turnera cistoides IPNI=*30068867-2* 400
Turnera cistoides IPNI=*836702-1* 401
Turnera corchoroides IPNI=*836702-1* 401
Turnera desvauxii IPNI=*30068867-2* 400
Turnera diffusa var. *aphrodisiaca* IPNI=*836578-1* 401
Turnera elegans IPNI=*836689-1* 401
Turnera frutescens IPNI=*836702-1* 401
Turnera helianthemoides IPNI=*30068867-2* 400
Turnera hirsuta IPNI=*30068867-2* 400
Turnera hirsutissima IPNI=*836658-1* 401
Turnera hirta IPNI=*30068867-2* 400
Turnera hispidissima IPNI=*836478-1* 401
Turnera humifusa IPNI=*836578-1* 401
Turnera microphylla IPNI=*836578-1* 401
Turnera mollis IPNI=*836689-1* 401
Turnera obtusifolia IPNI=*836702-1* 401
Turnera opifera IPNI=*836702-1* 401
Turnera ovata IPNI=*836478-1* 401
Turnera pilosula IPNI=*259040-2* 401
Turnera pringlei IPNI=*836578-1* 401
Turnera racemosa IPNI=*836478-1* 401
Turnera rugosa IPNI=*30068867-2* 400
Turnera sericea IPNI=*836689-1* 401
Turnera surinamensis IPNI=*836702-1* 401
Turnera trioniflora IPNI=*836689-1* 401
Turnera ulmifolia var. *acuta* IPNI=*259040-2* 401
Turnera ulmifolia var. *alba* IPNI=*836702-1* 401
Turnera ulmifolia var. *elegans* IPNI=*836689-1* 401
Turnera ulmifolia var. *intermedia* IPNI=*259112-2* 401
Turnera villosa IPNI=*30068867-2* 400
Turnera virgata IPNI=*836702-1* 401
Turpinia corymbosa IPNI=*785554-1* 292
Turpinia glandulosa IPNI=*785554-1* 292
Turpinia heterophylla IPNI=*785554-1* 292
Turpinia jimenezii IPNI=*785554-1* 292
Turpinia multijuga IPNI=*785554-1* 292
Turpinia paniculata IPNI=*785554-1* 292
Turpinia pinnata IPNI=*785554-1* 292
Turpinia sambucina IPNI=*785554-1* 292
Turpinia schiedeana IPNI=*785554-1* 292
Tursitis cymbalaria IPNI=*325773-2* 376
Tussacia friedrichsthaliana IPNI=*58213-2* 366
Tussacia sanguinea IPNI=*378985-1* 367
Tussacia villosa IPNI=*378695-1* 366
Tussacia woodsonii IPNI=*378695-1* 366
Tussilago nutans IPNI=*192653-1* 259
Tussilago rubra IPNI=*237108-1* 268
Tussilago vaccina IPNI=*192653-1* 259
Tydaea amabilis IPNI=*133755-2* 367
Tydaea ceciliae IPNI=*133755-2* 367
Tydaea venosa IPNI=*133755-2* 367
Tyloderma praecelsa IPNI=*125377-2* 291
Tylopetalum abutiforme IPNI=*581368-1* 435
Tylothrasya petrosa IPNI=*60440769-2* 243
Tynanthus caryophylleus IPNI=*259274-2* 365
Tynanthus hyacinthinus IPNI=*77100458-1* 362
Tynanthus igneus IPNI=*110483-1* 365
Tynanthus lindmannii IPNI=*259274-2* 365
Tynanthus myrianthus IPNI=*259274-2* 365
Tynanthus villosus IPNI=*259274-2* 365
Tynus ellipticus IPNI=*149971-1* 297
Typha abyssinica IPNI=*836837-1* 247
Typha aequalis IPNI=*836837-1* 247
Typha aethiopica IPNI=*836837-1* 247
Typha ambigua IPNI=*836870-1* 247
Typha americana IPNI=*836837-1* 247
Typha angustata IPNI=*836837-1* 247
Typha angustifolia var. *brownii* IPNI=*836837-1* 247
Typha angustifolia var. *virginica* IPNI=*836837-1* 247
Typha australis IPNI=*836837-1* 247
Typha basedowii IPNI=*836837-1* 247
Typha bracteata IPNI=*836837-1* 247

Typha brownii IPNI=*836837-1* 247
Typha crassa IPNI=*836870-1* 247
Typha damiattica IPNI=*836837-1* 247
Typha domingensis f. *strimonii* IPNI=*836837-1* 247
Typha domingensis subsp. *australis* IPNI=*836837-1* 247
Typha domingensis var. *eudomingensis* IPNI=*836837-1* 247
Typha domingensis var. *sachetiae* IPNI=*836837-1* 247
Typha ehrenbergii IPNI=*836837-1* 247
Typha elatior IPNI=*836870-1* 247
Typha elongata IPNI=*836870-1* 247
Typha engelmannii IPNI=*836870-1* 247
Typha essequeboensis IPNI=*836837-1* 247
Typha gigantea IPNI=*836837-1* 247
Typha gracilis IPNI=*836837-1* 247
Typha intermedia IPNI=*836870-1* 247
Typha javanica IPNI=*836837-1* 247
Typha latifolia f. *divisa* IPNI=*836870-1* 247
Typha latifolia subsp. *eulatifolia* IPNI=*836870-1* 247
Typha latifolia var. *elongata* IPNI=*836870-1* 247
Typha latifolia var. *obconica* IPNI=*836870-1* 247
Typha latifolia var. *typica* IPNI=*836870-1* 247
Typha macranthelia IPNI=*836837-1* 247
Typha major IPNI=*836870-1* 247
Typha maxima IPNI=*836837-1* 247
Typha media IPNI=*836837-1* 247
Typha media IPNI=*836870-1* 247
Typha palustris IPNI=*836870-1* 247
Typha pendula IPNI=*836870-1* 247
Typha remotiuscula IPNI=*836870-1* 247
Typha salgirica IPNI=*836837-1* 247
Typha spathulifolia IPNI=*836870-1* 247
Typha spiralis IPNI=*836837-1* 247
Typha tenuifolia IPNI=*836837-1* 247
Typha truxillensis IPNI=*836837-1* 247
Typhalaea fruticosa IPNI=*185868-2* 412
Typhalea fruticosa IPNI=*185868-2* 412
Typhalea malvacea IPNI=*186058-2* 412
Typhalea monatherica IPNI=*185868-2* 412
Typhalea spinifex IPNI=*186083-2* 412
Typhalea umbrosa IPNI=*563076-1* 412
Typhoides arundinacea f. *colorata* IPNI=*415571-1* 243
Typhoides arundinacea f. *schwarzii* IPNI=*415571-1* 243
Typhoides arundinacea f. *umbrosa* IPNI=*415571-1* 243
Typhoides arundinacea IPNI=*415571-1* 243
Typhoides arundinacea subsp. *rotgesii* IPNI=*415571-1* 243

U

Ubochea dichotoma IPNI=*1126344-2* 379
Ucacou nodiflorum IPNI=*252024-1* 271
Ucnopsolen cordatum IPNI=*1189750-2* 374
Udora densa IPNI=*431710-1* 203
Udora lithuanica IPNI=*114224-3* 203
Udora occidentalis IPNI=*114224-3* 203
Udora pomeranica IPNI=*114224-3* 203
Udora verticillata IPNI=*114224-3* 203
Ugni angustifolia IPNI=*602600-1* 428
Ugni disterigmoides IPNI=*602600-1* 428
Ugni friedrichsthalii IPNI=*602600-1* 428
Ugni friedrichsthalii var. *brevipes* IPNI=*602600-1* 428
Ugni friedrichsthalii var. *longipes* IPNI=*602600-1* 428
Ugni montana IPNI=*602600-1* 428
Ugni myricoides f. *bifaria* IPNI=*602600-1* 428
Ugni myricoides f. *grossa* IPNI=*602600-1* 428
Ugni myricoides f. *oligandra* IPNI=*602600-1* 428
Ugni myricoides f. *stenophylla* IPNI=*602600-1* 428
Ugni myricoides var. *longipes* IPNI=*602600-1* 428
Ugni myricoides var. *oerstedii* IPNI=*602600-1* 428
Ugni myricoides var. *roraimensis* IPNI=*602600-1* 428
Ugni oerstedii IPNI=*602600-1* 428
Ugni roraimensis IPNI=*602600-1* 428
Ugni stenophylla IPNI=*602600-1* 428
Ugni vaccinioides IPNI=*602600-1* 428
Ugni warscewiczii IPNI=*602600-1* 428
Uleodendron ulei IPNI=*167409-2* 440
Ullucus aborigineus IPNI=*167633-1* 283
Ullucus kunthii IPNI=*167633-1* 283
Ullucus tuberosus f. *albiflorus* IPNI=*167633-1* 283
Ullucus tuberosus f. *rubriflorus* IPNI=*167633-1* 283
Ullucus tuberosus subsp. *aborigineus* IPNI=*167633-1* 283
Ulmarronia corymbosa IPNI=*262677-2* 276
Ulmarronia dichotoma IPNI=*121424-1* 276
Ulmarronia mariquitensis IPNI=*262677-2* 276
Ulmarronia patens IPNI=*262677-2* 276
Umsema mucronata IPNI=*283610-2* 225
Umsema obtusifolia IPNI=*283610-2* 225
Uncaria aculeata IPNI=*768250-1* 358
Uncaria spinosa IPNI=*768250-1* 358
Uncaria surinamensis IPNI=*768322-1* 358
Uncaria tomentosa var. *dioica* IPNI=*768322-1* 358
Uncasia pallescens IPNI=*27149-2* 257
Uncinia galeottii IPNI=*46311-2* 230
Uncinia hamata f. *angustifolia* IPNI=*46311-2* 230
Uncinia hamata IPNI=*46311-2* 230
Uncinia hamata var. *mexicana* IPNI=*46311-2* 230
Uncinia jamaicensis IPNI=*46311-2* 230
Uncinia mexicana IPNI=*46311-2* 230
Uncinia multifolia IPNI=*46311-2* 230
Uncinia phleoides IPNI=*46311-2* 230
Uniola distichophylla IPNI=*82343-2* 238
Uniola flexuosa IPNI=*82343-2* 238
Uniola multiflora IPNI=*82343-2* 238
Uniola prostrata IPNI=*82343-2* 238
Uniola pungens IPNI=*82343-2* 238
Uniola spicata IPNI=*82343-2* 238
Uniola stricta IPNI=*82343-2* 238
Uniola thalassica IPNI=*82343-2* 238
Unisema acutifolia IPNI=*283610-2* 225
Unisema cordata f. *angustifolia* IPNI=*283610-2* 225
Unisema cordata f. *latifolia* IPNI=*283610-2* 225
Unisema cordata IPNI=*283610-2* 225
Unisema deltifolia IPNI=*283610-2* 225
Unisema heterophylla IPNI=*283610-2* 225
Unisema lancifolia f. *trullifolia* IPNI=*283610-2* 225
Unisema lancifolia IPNI=*283610-2* 225
Unisema latifolia IPNI=*283610-2* 225
Unisema media IPNI=*283610-2* 225
Unisema mucronata IPNI=*283610-2* 225
Unisema obliquata IPNI=*283610-2* 225
Unisema obtusifolia IPNI=*283610-2* 225
Unisema peduncularis IPNI=*283610-2* 225
Unisema purshiana IPNI=*283610-2* 225
Unisema rotundifolia IPNI=*283610-2* 225
Unisema sagittata IPNI=*283610-2* 225
Unona aromatica IPNI=*76067-1* 189
Unona cananga IPNI=*72580-1* 187
Unona carminativa IPNI=*76256-1* 190
Unona concolor IPNI=*76202-1* 190
Unona cubensis IPNI=*76067-1* 189
Unona discreta IPNI=*76119-1* 189
Unona fitzgeraldii IPNI=*72580-1* 187
Unona fluminensis IPNI=*76256-1* 190
Unona leptopetala IPNI=*72580-1* 187
Unona lucida IPNI=*76067-1* 189
Unona odorata IPNI=*72580-1* 187
Unona ossea IPNI=*72580-1* 187
Unona panamensis IPNI=*78729-2* 187

Unona parviflora IPNI=*76067-1* 189
Unona uniflora IPNI=*72920-1* 188
Unona xylopioides IPNI=*76067-1* 189
Unonopsis angustifolia IPNI=*259508-2* 189
Unonopsis antillana IPNI=*259508-2* 189
Unonopsis boliviensis IPNI=*259508-2* 189
Unonopsis buchtienii IPNI=*259508-2* 189
Unonopsis gracilis IPNI=*259508-2* 189
Unonopsis grandis IPNI=*259508-2* 189
Unonopsis guaraya IPNI=*259508-2* 189
Unonopsis guatterioides f. *elongata* IPNI=*259508-2* 189
Unonopsis lindmanii IPNI=*259508-2* 189
Unonopsis matthewsii IPNI=*259508-2* 189
Unonopsis obovata IPNI=*259508-2* 189
Unonopsis velutina IPNI=*259528-2* 189
Unonopsis williamsii IPNI=*259508-2* 189
Unxia digyna IPNI=*257035-1* 272
Unxia hirsuta IPNI=*257035-1* 272
Upudalia pulchella IPNI=*48879-1* 360
Urachne longiflora IPNI=*948410-1* 241
Uragoga acuminata IPNI=*1214950-2* 347
Uragoga acuminata IPNI=*77115588-1* 353
Uragoga alba IPNI=*212398-2* 355
Uragoga alba IPNI=*60473597-2* 353
Uragoga arcuata IPNI=*759372-1* 354
Uragoga australis IPNI=*761655-1* 355
Uragoga barbiflora IPNI=*77115613-1* 354
Uragoga barbinervia IPNI=*759121-1* 354
Uragoga barcellana IPNI=*77115635-1* 354
Uragoga berteroana IPNI=*77115593-1* 353
Uragoga biattenuata IPNI=*77115613-1* 354
Uragoga brachiata IPNI=*77115595-1* 353
Uragoga brachybotrya IPNI=*77161069-1* 353
Uragoga bracteocardia IPNI=*60473604-2* 353
Uragoga bracteolata IPNI=*77115609-1* 354
Uragoga britoguianensis IPNI=*77161077-1* 354
Uragoga candelariae IPNI=*319913-2* 353
Uragoga capitata IPNI=*759372-1* 354
Uragoga capitellata IPNI=*77115613-1* 354
Uragoga carapichea IPNI=*77115613-1* 354
Uragoga carpinterensis IPNI=*319913-2* 353
Uragoga carthagenensis IPNI=*761655-1* 355
Uragoga catharinensis IPNI=*761655-1* 355
Uragoga chinantlensis IPNI=*77115635-1* 354
Uragoga chionantha IPNI=*60473461-2* 355
Uragoga chlorotica IPNI=*759372-1* 354
Uragoga colorata IPNI=*77161054-1* 353
Uragoga compaginata IPNI=*761655-1* 355
Uragoga contracta IPNI=*77161077-1* 354
Uragoga cordillerae IPNI=*281971-2* 354
Uragoga costaricensis IPNI=*281971-2* 354
Uragoga crocea IPNI=*759061-1* 353
Uragoga cujabensis IPNI=*759061-1* 353
Uragoga cuspidata IPNI=*77148083-1* 353
Uragoga deflexa IPNI=*77115599-1* 353
Uragoga egensis IPNI=*77161866-1* 353
Uragoga elata IPNI=*77115600-1* 353
Uragoga elliptica IPNI=*761655-1* 355
Uragoga elongata IPNI=*329821-2* 355
Uragoga emetica IPNI=*764727-1* 356
Uragoga erythrantha IPNI=*759061-1* 353
Uragoga erythrophylla IPNI=*77161098-1* 354
Uragoga ficigemma IPNI=*761655-1* 355
Uragoga flavicans IPNI=*77115613-1* 354
Uragoga fockeana IPNI=*761655-1* 355
Uragoga furcata IPNI=*77115613-1* 354
Uragoga galeottiana IPNI=*281967-2* 353
Uragoga glabrescens IPNI=*77161054-1* 353
Uragoga glaucescens IPNI=*762253-1* 355
Uragoga glomerata IPNI=*763751-1* 356
Uragoga gracilenta IPNI=*77161069-1* 353
Uragoga granadensis IPNI=*329821-2* 355
Uragoga granatensis IPNI=*1214950-2* 347
Uragoga grandis IPNI=*762156-1* 355
Uragoga hebeclada IPNI=*77115609-1* 354
Uragoga hebecladoides IPNI=*77115593-1* 353
Uragoga hemsleyana IPNI=*762666-1* 355
Uragoga heterocephala IPNI=*77115613-1* 354
Uragoga hoffmannseggiana IPNI=*77115613-1* 354
Uragoga horizontalis IPNI=*762253-1* 355
Uragoga horridula IPNI=*77161076-1* 354
Uragoga humboldtiana IPNI=*77161073-1* 354
Uragoga hyacinthiflora IPNI=*759242-1* 354
Uragoga ipecacuanha IPNI=*1214950-2* 347
Uragoga jacobaschii IPNI=*761655-1* 355
Uragoga japurensis IPNI=*765569-1* 357
Uragoga justicioides IPNI=*77115609-1* 354
Uragoga laevis IPNI=*170751-2* 353
Uragoga langsdorffiana IPNI=*77161077-1* 354
Uragoga latifolia IPNI=*764729-1* 356
Uragoga longicollis IPNI=*762253-1* 355
Uragoga longicuspis IPNI=*77161078-1* 354
Uragoga lupulina IPNI=*77161077-1* 354
Uragoga macrobotrys IPNI=*759203-1* 354
Uragoga macropoda IPNI=*1080914-2* 351
Uragoga magnoliifolia IPNI=*212751-2* 355
Uragoga mapouria IPNI=*60473461-2* 355
Uragoga marginata IPNI=*213065-2* 355
Uragoga megapontica IPNI=*77161054-1* 353
Uragoga mesomorpha IPNI=*77161098-1* 354
Uragoga mexicana IPNI=*281971-2* 354
Uragoga micrantha IPNI=*762666-1* 355
Uragoga microdesmia IPNI=*763751-1* 356
Uragoga neurothricha IPNI=*77115595-1* 353
Uragoga nicaraguensis IPNI=*213065-2* 355
Uragoga nicotianifolia IPNI=*759203-1* 354
Uragoga obovata IPNI=*759242-1* 354
Uragoga obscurata IPNI=*181012-2* 354
Uragoga officinalis IPNI=*77161098-1* 354
Uragoga oligotricha IPNI=*329821-2* 355
Uragoga padifolia IPNI=*281971-2* 354
Uragoga palicourea IPNI=*759121-1* 354
Uragoga panurensis IPNI=*765569-1* 357
Uragoga pardina IPNI=*77115635-1* 354
Uragoga parviflora IPNI=*759353-1* 354
Uragoga persimilis IPNI=*77161077-1* 354
Uragoga platyphylla IPNI=*77115593-1* 353
Uragoga poeppigiana IPNI=*77115635-1* 354
Uragoga pohliana IPNI=*761655-1* 355
Uragoga polyantha IPNI=*762156-1* 355
Uragoga pseudojapurensis IPNI=*77161076-1* 354
Uragoga psychotriifolia IPNI=*213293-2* 356
Uragoga pubescens IPNI=*60473604-2* 353
Uragoga punicea IPNI=*77115600-1* 353
Uragoga pyramidalis IPNI=*77115593-1* 353
Uragoga quinquecuspis IPNI=*77115627-1* 354
Uragoga rabeniana IPNI=*761655-1* 355
Uragoga racemosa IPNI=*77115627-1* 354
Uragoga reevesii IPNI=*761655-1* 355
Uragoga remota IPNI=*763194-1* 356
Uragoga reniformis IPNI=*109024-2* 351
Uragoga rhodoleuca IPNI=*77161077-1* 354
Uragoga rigida IPNI=*759300-1* 354
Uragoga rosea IPNI=*77161054-1* 353
Uragoga rufescens IPNI=*329821-2* 355
Uragoga sessilifolia IPNI=*213510-2* 356

Uragoga setifera IPNI=*759372-1* 354
Uragoga sieberi IPNI=*759372-1* 354
Uragoga simulans IPNI=*60473604-2* 353
Uragoga solmsiana IPNI=*77161076-1* 354
Uragoga stenophylla IPNI=*319913-2* 353
Uragoga subrubra IPNI=*281971-2* 354
Uragoga tapa-camino IPNI=*759061-1* 353
Uragoga tenerior IPNI=*77161098-1* 354
Uragoga tenuifolia IPNI=*213510-2* 356
Uragoga tomentosa IPNI=*77115635-1* 354
Uragoga tonsa IPNI=*761655-1* 355
Uragoga tribracteata IPNI=*77115613-1* 354
Uragoga tricholoba IPNI=*77161076-1* 354
Uragoga trichophylla IPNI=*759353-1* 354
Uragoga triphylla IPNI=*759353-1* 354
Uragoga triplocephala IPNI=*170751-2* 353
Uragoga trispicata IPNI=*763751-1* 356
Uragoga tristis IPNI=*761655-1* 355
Uragoga uliginosa IPNI=*170751-2* 353
Uragoga undata IPNI=*329821-2* 355
Uragoga undulata IPNI=*329821-2* 355
Uragoga urophylla IPNI=*77115588-1* 353
Uragoga vauthieri IPNI=*77115599-1* 353
Uragoga velhana IPNI=*761655-1* 355
Uragoga viridis IPNI=*763751-1* 356
Uragoga watsoniana IPNI=*761655-1* 355
Uragoga wittmackiana IPNI=*77161077-1* 354
Urananthus chironioides IPNI=*367440-1* 344
Urananthus glaucifolius IPNI=*367440-1* 344
Urania amazonica IPNI=*585013-1* 250
Urania guyannensis IPNI=*585013-1* 250
Urania madagascariensis IPNI=*60466138-2* 250
Urania ravenala IPNI=*60466138-2* 250
Urania speciosa IPNI=*60466138-2* 250
Uraria barbata IPNI=*77187394-1* 320
Urbanella buchananiifolia IPNI=*943696-1* 307
Urbanella cuspidata IPNI=*943696-1* 307
Urbanella excelsa IPNI=*943696-1* 307
Urbanella oblonga IPNI=*209738-2* 306
Urbanella procera IPNI=*943696-1* 307
Urbanisol aristatus IPNI=*254724-2* 271
Urbanisol heterophyllus IPNI=*254724-2* 271
Urbanisol tagetiflorus IPNI=*254724-2* 271
Urbanisol tagetiflorus var. *diversifolius* IPNI=*255747-1* 271
Urbanisol tagetiflorus var. *normalis* IPNI=*254724-2* 271
Urbanisol tagetiflorus var. *speciosus* IPNI=*254724-2* 271
Urceolina × lowii IPNI=*261579-2* 214
Urceolina × mastersii IPNI=*261579-2* 214
Urceolina fosteri IPNI=*67017-1* 214
Urceolina sanderi subsp. *multiflora* IPNI=*261595-2* 214
Urena aculeata IPNI=*564786-1* 415
Urena americana IPNI=*564786-1* 415
Urena blumei IPNI=*564786-1* 415
Urena callifera IPNI=*564786-1* 415
Urena cana IPNI=*564786-1* 415
Urena capitata IPNI=*561424-1* 410
Urena capitata var. *alceifolia* IPNI=*561421-1* 410
Urena capitata var. *rotundifolia* IPNI=*561424-1* 410
Urena diversifolia IPNI=*564786-1* 415
Urena grandiflora IPNI=*564786-1* 415
Urena haenkeana IPNI=*564786-1* 415
Urena heteromorpha IPNI=*564786-1* 415
Urena heterophylla IPNI=*564786-1* 415
Urena leptocarpa IPNI=*185868-2* 412
Urena lobata f. *swartzii* IPNI=*564786-1* 415
Urena lobata subsp. *alba* IPNI=*564786-1* 415
Urena lobata var. *americana* IPNI=*564786-1* 415
Urena lobata var. *nummularia* IPNI=*564786-1* 415
Urena lobata var. *reticulata* IPNI=*564786-1* 415
Urena lobata var. *trilobata* IPNI=*564786-1* 415
Urena lobata var. *viminea* IPNI=*564786-1* 415
Urena loureiroi IPNI=*564786-1* 415
Urena mauritiana IPNI=*564786-1* 415
Urena microcarpa IPNI=*564786-1* 415
Urena moenchii IPNI=*561424-1* 410
Urena monopetala IPNI=*564786-1* 415
Urena obtusata IPNI=*564786-1* 415
Urena phyllomorpha IPNI=*564786-1* 415
Urena polyflora IPNI=*320787-2* 415
Urena radiata IPNI=*561446-1* 410
Urena repanda IPNI=*564786-1* 415
Urena reticulata IPNI=*564786-1* 415
Urena ribesia IPNI=*564786-1* 415
Urena sieberi IPNI=*564786-1* 415
Urena sinuata IPNI=*564786-1* 415
Urena stellata IPNI=*185868-2* 412
Urena swartzii IPNI=*564786-1* 415
Urena tomentosa IPNI=*564786-1* 415
Urena trichocarpa IPNI=*564786-1* 415
Urena trilobata IPNI=*564786-1* 415
Urena typhalaea IPNI=*185868-2* 412
Urena viminea IPNI=*564786-1* 415
Urena virgata IPNI=*564786-1* 415
Urera acuminata IPNI=*856982-1* 448
Urera alceifolia IPNI=*856982-1* 448
Urera armigera IPNI=*1044269-2* 448
Urera aurantiaca var. *scandens* IPNI=*856972-1* 448
Urera baccifera var. *horrida* IPNI=*1044269-2* 448
Urera capitata var. *pavonii* IPNI=*856982-1* 448
Urera caracasana var. *miquelii* IPNI=*856982-1* 448
Urera caracasana var. *mitis* IPNI=*856982-1* 448
Urera caracasana var. *subpeltata* IPNI=*856982-1* 448
Urera caracasana var. *tomentosa* IPNI=*856982-1* 448
Urera corallina IPNI=*856982-1* 448
Urera densiflora IPNI=*856982-1* 448
Urera gaudichaudiana IPNI=*135476-2* 446
Urera girardinioides IPNI=*857027-1* 448
Urera horrida IPNI=*1044269-2* 448
Urera jacquinii IPNI=*856982-1* 448
Urera jacquinii var. *corallina* IPNI=*856982-1* 448
Urera jacquinii var. *miquelii* IPNI=*856982-1* 448
Urera jacquinii var. *subpeltata* IPNI=*856982-1* 448
Urera jacquinii var. *tomentosa* IPNI=*856982-1* 448
Urera jacquinii var. *ulmifolia* IPNI=*856982-1* 448
Urera lamiifolia IPNI=*55676-3* 445
Urera magna IPNI=*856982-1* 448
Urera mitis IPNI=*856982-1* 448
Urera rugosa IPNI=*1044269-2* 448
Urera scandens IPNI=*856972-1* 448
Urera subpeltata IPNI=*856982-1* 448
Urera subpeltata var. *morifolia* IPNI=*856982-1* 448
Urera tuberculata IPNI=*856998-1* 448
Urera viridisetosa IPNI=*1044269-2* 448
Urginea mouretii IPNI=*77097737-1* 214
Urocarpidium limense IPNI=*993881-1* 408
Urocarpidium mathewsii IPNI=*993885-1* 408
Urocarpidium peruvianum IPNI=*993885-1* 408
Urochloa brizantha var. *ciliata* IPNI=*943736-1* 246
Urochloa decumbens IPNI=*974586-1* 246
Urochloa discifera IPNI=*286530-2* 246
Urochloa fasciculata IPNI=*1156554-2* 246
Urochloa fasciculata var. *reticulata* IPNI=*1156554-2* 246
Urochloa fusca var. *reticulata* IPNI=*1156554-2* 246
Urochloa humidicola IPNI=*992521-1* 246
Urochloa mauritiana IPNI=*1172068-2* 245
Urochloa maxima IPNI=*60426565-2* 241

Urochloa reptans var. *glabra* IPNI=*426280-1* 246
Urochloa ruziziensis IPNI=*974586-1* 246
Uropedium lindenii IPNI=*1119571-2* 222
Urophyllon caudatum IPNI=*77097737-1* 214
Urospatha affinis IPNI=*89280-1* 202
Urospatha angusta IPNI=*89296-1* 202
Urospatha decipiens IPNI=*89296-1* 202
Urospatha dubia IPNI=*89296-1* 202
Urospatha friedrichsthalii IPNI=*225280-2* 198
Urospatha hostmannii IPNI=*89296-1* 202
Urospatha langsdorffiana IPNI=*89296-1* 202
Urospatha maculata IPNI=*89296-1* 202
Urospatha marmorea IPNI=*89296-1* 202
Urospatha poeppigiana IPNI=*89296-1* 202
Urospatha quinquenervis IPNI=*89296-1* 202
Urospatha savannarum IPNI=*89296-1* 202
Urospatha spruceana IPNI=*89296-1* 202
Urostachys attenuatus IPNI=*124842-2* 171
Urostachys barbatus IPNI=*1090207-2* 171
Urostachys bonae-voluntatis IPNI=*124844-2* 171
Urostachys bonae-voluntatis var. *minor* IPNI=*124844-2* 171
Urostachys chamaepeuce IPNI=*1090207-2* 171
Urostachys chamaepeuce var. *maior* IPNI=*1090207-2* 171
Urostachys chamaepeuce var. *maximus* IPNI=*1090207-2* 171
Urostachys chamaepeuce var. *urbanianus* IPNI=*1090207-2* 171
Urostachys crassus IPNI=*124844-2* 171
Urostachys crassus var. *herteri* IPNI=*124844-2* 171
Urostachys dichotomus IPNI=*1090207-2* 171
Urostachys dichotomus var. *pusillus* IPNI=*1090207-2* 171
Urostachys elongatus IPNI=*60447820-2* 171
Urostachys erythraeus IPNI=*124844-2* 171
Urostachys gramineus IPNI=*1090207-2* 171
Urostachys kupperi IPNI=*124844-2* 171
Urostachys lindeneri IPNI=*1090207-2* 171
Urostachys mortonii IPNI=*1090207-2* 171
Urostachys orionis IPNI=*124844-2* 171
Urostachys pilgerianus IPNI=*124844-2* 171
Urostachys plutonis IPNI=*304307-2* 171
Urostachys saururus IPNI=*60447820-2* 171
Urostachys saururus var. *elongatus* IPNI=*60447820-2* 171
Urostachys schlechtendalii IPNI=*1090207-2* 171
Urostachys springii IPNI=*124844-2* 171
Urostachys tobarii IPNI=*124842-2* 171
Urostigma affine IPNI=*853563-1* 439
Urostigma altissimum IPNI=*852369-1* 437
Urostigma amazonicum IPNI=*852372-1* 437
Urostigma amblyphyllum IPNI=*853217-1* 439
Urostigma angustifolium IPNI=*852372-1* 437
Urostigma arbutifolium IPNI=*853405-1* 439
Urostigma baccatum IPNI=*853405-1* 439
Urostigma bengalense IPNI=*852482-1* 438
Urostigma benjaminum IPNI=*852486-1* 438
Urostigma benjaminum var. *nudum* IPNI=*852486-1* 438
Urostigma bonplandianum IPNI=*853317-1* 439
Urostigma botryapioides IPNI=*316340-2* 438
Urostigma casapiense IPNI=*852560-1* 438
Urostigma chiriquianum IPNI=*316336-2* 437
Urostigma circumscissum IPNI=*60458499-2* 438
Urostigma complicatum IPNI=*853405-1* 439
Urostigma consanguineum IPNI=*316336-2* 437
Urostigma costaricanum IPNI=*852672-1* 438
Urostigma costatum IPNI=*853876-1* 439
Urostigma crocatum IPNI=*852689-1* 438
Urostigma cystopodum IPNI=*856394-1* 441
Urostigma elasticum IPNI=*60458499-2* 438
Urostigma erythrostictum IPNI=*853405-1* 439
Urostigma eugeniifolium IPNI=*316336-2* 437
Urostigma euomphalum IPNI=*853876-1* 439
Urostigma eximium IPNI=*852825-1* 438
Urostigma fagifolium IPNI=*853876-1* 439
Urostigma fagineum IPNI=*850824-1* 437
Urostigma gardnerianum IPNI=*853317-1* 439
Urostigma geminum IPNI=*853405-1* 439
Urostigma giganteum IPNI=*316340-2* 438
Urostigma glabrum IPNI=*852825-1* 438
Urostigma glycicarpum IPNI=*853933-1* 439
Urostigma haematocarpum IPNI=*852486-1* 438
Urostigma hartwegii IPNI=*852958-1* 438
Urostigma involutum IPNI=*853317-1* 439
Urostigma karet IPNI=*60458499-2* 438
Urostigma lacciferum IPNI=*852369-1* 437
Urostigma laevigatum IPNI=*316340-2* 438
Urostigma lancifolium IPNI=*853405-1* 439
Urostigma laurifolium IPNI=*316340-2* 438
Urostigma lentiginosum IPNI=*316340-2* 438
Urostigma liebmannianum IPNI=*316336-2* 437
Urostigma littorale IPNI=*853217-1* 439
Urostigma mathewsii IPNI=*853198-1* 438
Urostigma maximilianum IPNI=*852689-1* 438
Urostigma microcarpum IPNI=*853217-1* 439
Urostigma neglectum IPNI=*852486-1* 438
Urostigma nitidum IPNI=*852486-1* 438
Urostigma nudum IPNI=*852486-1* 438
Urostigma nymphifolium IPNI=*853301-1* 439
Urostigma odoratum IPNI=*60458499-2* 438
Urostigma oerstediana IPNI=*316336-2* 437
Urostigma padifolium IPNI=*853405-1* 439
Urostigma pallidum IPNI=*853355-1* 439
Urostigma paraense IPNI=*1004295-2* 439
Urostigma pedunculatum IPNI=*316340-2* 438
Urostigma pertusum IPNI=*853405-1* 439
Urostigma pisiferum IPNI=*853217-1* 439
Urostigma planicostatum IPNI=*316340-2* 438
Urostigma populneum IPNI=*316340-2* 438
Urostigma prinoides IPNI=*853355-1* 439
Urostigma protensum IPNI=*853200-1* 439
Urostigma religiosum IPNI=*853563-1* 439
Urostigma rolandri IPNI=*853405-1* 439
Urostigma sapidum IPNI=*853405-1* 439
Urostigma scandens IPNI=*853513-1* 439
Urostigma schiedeanum IPNI=*853405-1* 439
Urostigma subpanduriforme IPNI=*852970-1* 438
Urostigma subtriplinervium f. *ellipticum* IPNI=*853405-1* 439
Urostigma subtriplinervium IPNI=*853405-1* 439
Urostigma sulcipes IPNI=*853405-1* 439
Urostigma syringifolium IPNI=*316340-2* 438
Urostigma tomentellum IPNI=*852689-1* 438
Urostigma trigonatum IPNI=*316338-2* 439
Urostigma trigonum IPNI=*853876-1* 439
Urostigma turbinatum IPNI=*853405-1* 439
Urostigma velutinum IPNI=*853933-1* 439
Urostigma williamsii IPNI=*1004295-2* 439
Urtica aestuans IPNI=*135476-2* 446
Urtica alceifolia IPNI=*856982-1* 448
Urtica alnifolia IPNI=*1018059-2* 436
Urtica andicola IPNI=*857552-1* 448
Urtica armigera IPNI=*1044269-2* 448
Urtica arvensis IPNI=*855481-1* 447
Urtica atlantica IPNI=*857987-1* 448
Urtica baccifera IPNI=*1044269-2* 448
Urtica ballotifolia var. *macrostachya* IPNI=*857434-1* 448
Urtica candicans IPNI=*55676-3* 445
Urtica caracasana IPNI=*856982-1* 448
Urtica caravellana IPNI=*135476-2* 446
Urtica cataractae IPNI=*855573-1* 447
Urtica chichicaztli IPNI=*856982-1* 448

Urtica ciliaris IPNI=*30064315-2* 447
Urtica copeyana IPNI=*857681-1* 448
Urtica corallina IPNI=*856982-1* 448
Urtica corylifolia IPNI=*135476-2* 446
Urtica cubensis IPNI=*857987-1* 448
Urtica cymosa IPNI=*135476-2* 446
Urtica debilis IPNI=*854908-1* 446
Urtica dioecia IPNI=*260630-2* 448
Urtica dioica subsp. *eudioica* IPNI=*260630-2* 448
Urtica divaricata IPNI=*135476-2* 446
Urtica divergens IPNI=*135476-2* 446
Urtica echinata var. *punctulata* IPNI=*857552-1* 448
Urtica eckloniana IPNI=*260630-2* 448
Urtica elata IPNI=*856998-1* 448
Urtica elongata IPNI=*855573-1* 447
Urtica fernandeziana IPNI=*854908-1* 446
Urtica freireiformis IPNI=*854908-1* 446
Urtica gibbosa IPNI=*855573-1* 447
Urtica grandidentata IPNI=*1044269-2* 448
Urtica grandifolia IPNI=*855698-1* 447
Urtica grossecrenata IPNI=*855698-1* 447
Urtica haussknechtii IPNI=*260630-2* 448
Urtica herniarifolia IPNI=*1162156-2* 447
Urtica hirsuta IPNI=*135476-2* 446
Urtica horrida IPNI=*1044269-2* 448
Urtica involucrata IPNI=*77091436-1* 447
Urtica laciniata IPNI=*857027-1* 448
Urtica lamiifolia IPNI=*55676-3* 445
Urtica latifolia IPNI=*135476-2* 446
Urtica longipes IPNI=*855481-1* 447
Urtica longispica IPNI=*857434-1* 448
Urtica major IPNI=*260630-2* 448
Urtica melastomoides IPNI=*855573-1* 447
Urtica microphylla IPNI=*1162156-2* 447
Urtica microphylla IPNI=*260630-2* 448
Urtica millettii IPNI=*55676-3* 445
Urtica minor IPNI=*857987-1* 448
Urtica mitis IPNI=*856982-1* 448
Urtica monoica IPNI=*857987-1* 448
Urtica muralis IPNI=*856206-1* 448
Urtica nemorosa IPNI=*135476-2* 446
Urtica nicaraguensis IPNI=*857681-1* 448
Urtica nivea IPNI=*55676-3* 445
Urtica oreophila IPNI=*855573-1* 447
Urtica ovalifolia IPNI=*857987-1* 448
Urtica parasitica IPNI=*856206-1* 448
Urtica parietaria IPNI=*30064315-2* 447
Urtica parietariifolia IPNI=*854908-1* 446
Urtica parvula IPNI=*857987-1* 448
Urtica peduncularis IPNI=*855573-1* 447
Urtica petiolata IPNI=*135476-2* 446
Urtica portulacina IPNI=*1162156-2* 447
Urtica portulacoides IPNI=*1162156-2* 447
Urtica procridioides IPNI=*856206-1* 448
Urtica pumila IPNI=*855481-1* 447
Urtica quadristipulata IPNI=*857987-1* 448
Urtica racemosa IPNI=*135476-2* 446
Urtica radicans IPNI=*260630-2* 448
Urtica radicans IPNI=*855698-1* 447
Urtica schimperiana IPNI=*135476-2* 446
Urtica sicula IPNI=*260630-2* 448
Urtica simsii IPNI=*77091436-1* 447
Urtica spicata IPNI=*850541-1* 445
Urtica spiralis IPNI=*854765-1* 446
Urtica submitis IPNI=*260630-2* 448
Urtica succulenta IPNI=*855481-1* 447
Urtica tenacissima IPNI=*55676-3* 445
Urtica tibetica IPNI=*260630-2* 448
Urtica tiliifolia IPNI=*856982-1* 448
Urtica trianae IPNI=*857987-1* 448
Urtica trinervia IPNI=*855573-1* 447
Urtica tuberculata IPNI=*135476-2* 446
Urtica ulmifolia IPNI=*856982-1* 448
Urtica utilis IPNI=*55676-3* 445
Urtica verrucosa IPNI=*60439275-2* 448
Uruparia versicolor IPNI=*768250-1* 358
Usteria barclayana IPNI=*280591-2* 376
Uterveria breynia IPNI=*77217338-1* 278
Uterveria eustachiana IPNI=*316361-2* 278
Uterveria frondosa IPNI=*77183570-1* 278
Uterveria linearis IPNI=*77184255-1* 278
Uterveria verrucosa IPNI=*77183644-1* 278
Utricularia alba IPNI=*526777-1* 374
Utricularia ambigua IPNI=*526777-1* 374
Utricularia anomala IPNI=*526777-1* 374
Utricularia aphylla IPNI=*526777-1* 374
Utricularia bifidocalcar IPNI=*526777-1* 374
Utricularia biflora IPNI=*526777-1* 374
Utricularia bipartita IPNI=*526777-1* 374
Utricularia conferta IPNI=*526777-1* 374
Utricularia crenata IPNI=*526777-1* 374
Utricularia diantha IPNI=*526777-1* 374
Utricularia diflora IPNI=*526777-1* 374
Utricularia elegans IPNI=*526777-1* 374
Utricularia emarginata IPNI=*526777-1* 374
Utricularia exoleta IPNI=*526777-1* 374
Utricularia fibrosa IPNI=*526777-1* 374
Utricularia fornicata IPNI=*526777-1* 374
Utricularia furcata IPNI=*526777-1* 374
Utricularia gayana IPNI=*526777-1* 374
Utricularia gibba f. *natans* IPNI=*526777-1* 374
Utricularia gibba subsp. *exoleta* IPNI=*526777-1* 374
Utricularia gibbosa IPNI=*526777-1* 374
Utricularia gracilis IPNI=*526777-1* 374
Utricularia integra IPNI=*526777-1* 374
Utricularia kalmaloensis IPNI=*526777-1* 374
Utricularia khasiana IPNI=*526777-1* 374
Utricularia longirostris IPNI=*526777-1* 374
Utricularia macrorhyncha IPNI=*526777-1* 374
Utricularia nagurae IPNI=*526777-1* 374
Utricularia natans IPNI=*526777-1* 374
Utricularia obtusa IPNI=*526777-1* 374
Utricularia pallens IPNI=*526777-1* 374
Utricularia pallens var. *natans* IPNI=*526777-1* 374
Utricularia parkeriana IPNI=*526777-1* 374
Utricularia pauciflora IPNI=*526777-1* 374
Utricularia pterosperma IPNI=*526777-1* 374
Utricularia pumila IPNI=*526777-1* 374
Utricularia riccioides IPNI=*526777-1* 374
Utricularia roxburghii IPNI=*526777-1* 374
Utricularia saharunporensis IPNI=*526777-1* 374
Utricularia secunda IPNI=*526777-1* 374
Utricularia spirandra IPNI=*526777-1* 374
Utricularia sumatrana IPNI=*526777-1* 374
Utricularia tenuifolia IPNI=*526777-1* 374
Utricularia tenuis IPNI=*526777-1* 374
Utricularia tenuis var. *poeppigii* IPNI=*526777-1* 374
Utricularia tricrenata IPNI=*526777-1* 374
Uva brasiliensis IPNI=*72706-1* 187
Uvaria aberemoa IPNI=*114255-2* 188
Uvaria aromatica IPNI=*76067-1* 189
Uvaria axillaris IPNI=*72580-1* 187
Uvaria brasiliensis IPNI=*72706-1* 187
Uvaria farcta IPNI=*72580-1* 187
Uvaria febrifuga IPNI=*269786-2* 189
Uvaria fluminensis IPNI=*76256-1* 190

Uvaria guatterioides IPNI=259508-2 189
Uvaria hortensis IPNI=72580-1 187
Uvaria javanica IPNI=72580-1 187
Uvaria odorata IPNI=72580-1 187
Uvaria ossea IPNI=72580-1 187
Uvaria spectabilis IPNI=106440-2 188
Uvaria trifoliata IPNI=72580-1 187
Uvaria undulata IPNI=72580-1 187
Uvaria unona IPNI=76119-1 189
Uvaria zeylanica IPNI=76202-1 190
Uvifera acuminata IPNI=693560-1 288
Uvifera alagoensis IPNI=693788-1 288
Uvifera ascendens IPNI=693572-1 288
Uvifera barbeyana IPNI=693638-1 288
Uvifera billbergii IPNI=693788-1 288
Uvifera bracteolosa IPNI=693654-1 288
Uvifera caracasana IPNI=693597-1 288
Uvifera densifrons IPNI=693638-1 288
Uvifera dioica IPNI=693617-1 288
Uvifera excelsa IPNI=693654-1 288
Uvifera fagifolia IPNI=693617-1 288
Uvifera guianensis IPNI=693777-1 288
Uvifera latifolia IPNI=693712-1 288
Uvifera marginata IPNI=693777-1 288
Uvifera microneura IPNI=693777-1 288
Uvifera nitida IPNI=693777-1 288
Uvifera novogranatensis IPNI=693617-1 288
Uvifera obovata IPNI=693787-1 288
Uvifera obtusifolia IPNI=693788-1 288
Uvifera ovata IPNI=693795-1 288
Uvifera polystachya IPNI=693754-1 288
Uvifera sparsifolia IPNI=693572-1 288
Uvifera strobilulifera IPNI=693560-1 288
Uvifera trinitatis IPNI=693777-1 288

V
Vacciniopsis tetramera IPNI=82288-2 299
Vaccinium acuminatum IPNI=82286-2 299
Vaccinium affine IPNI=261842-2 300
Vaccinium alaternoides IPNI=82288-2 299
Vaccinium alaternoides var. parvifolium IPNI=82288-2 299
Vaccinium albiflorum IPNI=261823-2 300
Vaccinium amoenum IPNI=261823-2 300
Vaccinium arcuatum IPNI=261823-2 300
Vaccinium arkansanum IPNI=261823-2 300
Vaccinium ashei IPNI=261823-2 300
Vaccinium atrococcum f. leucococcum IPNI=261823-2 300
Vaccinium atrococcum IPNI=261823-2 300
Vaccinium atrococcum var. longianum IPNI=261823-2 300
Vaccinium attenuatum IPNI=261842-2 300
Vaccinium australe IPNI=261823-2 300
Vaccinium callosum IPNI=858930-1 300
Vaccinium caracasanum f. acuminatum IPNI=859020-1 300
Vaccinium caracasanum f. lanceolatum IPNI=859020-1 300
Vaccinium caracasanum f. subellipticum IPNI=859020-1 300
Vaccinium caracasanum IPNI=859020-1 300
Vaccinium constablaei IPNI=261823-2 300
Vaccinium corymbosum f. albiflorum IPNI=261823-2 300
Vaccinium corymbosum f. glabrum IPNI=261823-2 300
Vaccinium corymbosum f. laeve IPNI=261823-2 300
Vaccinium corymbosum f. leucococcum IPNI=261823-2 300
Vaccinium corymbosum f. typicum IPNI=261823-2 300
Vaccinium corymbosum f. viride IPNI=261823-2 300
Vaccinium corymbosum var. albiflorum IPNI=261823-2 300
Vaccinium corymbosum var. amoenum IPNI=261823-2 300
Vaccinium corymbosum var. atrocarpum IPNI=261823-2 300
Vaccinium corymbosum var. atrococcum IPNI=261823-2 300
Vaccinium corymbosum var. disocarpum IPNI=261823-2 300
Vaccinium corymbosum var. fuscatum IPNI=261823-2 300
Vaccinium corymbosum var. glabrum IPNI=261823-2 300
Vaccinium corymbosum var. virgatum IPNI=261823-2 300
Vaccinium crenulatum IPNI=858930-1 300
Vaccinium cuthbertii IPNI=261823-2 300
Vaccinium dasygynum IPNI=858930-1 300
Vaccinium dendrophilum IPNI=82286-2 299
Vaccinium disocarpum IPNI=261823-2 300
Vaccinium disomorphum IPNI=261823-2 300
Vaccinium elliottii IPNI=261823-2 300
Vaccinium elongatum IPNI=261823-2 300
Vaccinium empetrifolium IPNI=327994-1 299
Vaccinium epacridifolium IPNI=327994-1 299
Vaccinium escallonioides IPNI=261842-2 300
Vaccinium floribundum var. marginatum IPNI=858930-1 300
Vaccinium floribundum var. polystachyum IPNI=858930-1 300
Vaccinium floribundum var. ramosissimum IPNI=858930-1 300
Vaccinium floribundum var. splendens IPNI=858930-1 300
Vaccinium floribundum var. tatei IPNI=858930-1 300
Vaccinium formosum IPNI=261823-2 300
Vaccinium fuscatum IPNI=261823-2 300
Vaccinium fuscatum var. angustifolium IPNI=261823-2 300
Vaccinium fuscatum var. pullum IPNI=261823-2 300
Vaccinium galezans IPNI=261823-2 300
Vaccinium galiformis IPNI=261823-2 300
Vaccinium grandiflorum IPNI=261823-2 300
Vaccinium holophyllum IPNI=261823-2 300
Vaccinium humboldtii IPNI=82301-2 299
Vaccinium latifolium IPNI=261823-2 300
Vaccinium leiandrum IPNI=858930-1 300
Vaccinium marginatum IPNI=858930-1 300
Vaccinium marylandicum IPNI=261823-2 300
Vaccinium melanocarpum var. sericeum IPNI=261823-2 300
Vaccinium moritzianum IPNI=858930-1 300
Vaccinium mortinia IPNI=858930-1 300
Vaccinium myrtifolium IPNI=82301-2 299
Vaccinium myrtilloides IPNI=261823-2 300
Vaccinium ottonis IPNI=261842-2 300
Vaccinium pachycardium IPNI=858679-1 300
Vaccinium pachyphyllum IPNI=82301-2 299
Vaccinium penaeoides f. empetrifolia IPNI=327994-1 299
Vaccinium penaeoides IPNI=327994-1 299
Vaccinium penaeoides var. epacridifolium IPNI=327994-1 299
Vaccinium polystachyum IPNI=858930-1 300
Vaccinium ramosissimum IPNI=858930-1 300
Vaccinium reclinatum IPNI=261842-2 300
Vaccinium reflexum IPNI=261842-2 300
Vaccinium retifolium IPNI=261842-2 300
Vaccinium sericeum f. chrysocarpon IPNI=261823-2 300
Vaccinium sericeum IPNI=261823-2 300
Vaccinium sericeum var. eburneum IPNI=261823-2 300
Vaccinium sericeum var. elongatum IPNI=261823-2 300
Vaccinium simulatum IPNI=261823-2 300
Vaccinium stamineum var. sericeum IPNI=261823-2 300
Vaccinium tatei IPNI=858930-1 300
Vaccinium thymifolium IPNI=858930-1 300
Vaccinium trinerve IPNI=60458979-2 298
Vaccinium valerioi IPNI=858679-1 300
Vaccinium vicinum IPNI=261823-2 300
Vaccinium villosum IPNI=858406-1 299
Vaccinium virgatum IPNI=261823-2 300
Vaccinium virgatum var. angustifolium IPNI=261823-2 300
Vaccinium virgatum var. latifolium IPNI=261823-2 300
Vaccinium virgatum var. ozarkense IPNI=261823-2 300
Vaccinium virgatum var. parvifolium IPNI=261823-2 300
Vaccinium virgatum var. speciosum IPNI=261823-2 300
Vachellia farnesiana f. typica IPNI=1099713-2 336
Vachellia lutea IPNI=60441223-2 336

Vailia anomala IPNI=*94927-1* 339
Valantia crucialis IPNI=*747959-1* 349
Valantia cruciata IPNI=*747959-1* 349
Valantia hirsuta IPNI=*747959-1* 349
Valantia hypocarpia IPNI=*750043-1* 350
Valarum officinale IPNI=*289589-1* 278
Valdesia rosea IPNI=*323201-2* 418
Valdesia subconnata IPNI=*566768-1* 418
Valentiana volubilis IPNI=*56177-1* 361
Valeriana alba IPNI=*860141-1* 296
Valeriana alophis IPNI=*286639-2* 296
Valeriana altissima IPNI=*860012-1* 296
Valeriana amphilophis IPNI=*286639-2* 296
Valeriana arsenei IPNI=*286659-2* 296
Valeriana baltica IPNI=*860012-1* 296
Valeriana bonariensis IPNI=*30039915-2* 296
Valeriana bonplandiana IPNI=*286639-2* 296
Valeriana chaerophylla IPNI=*859680-1* 296
Valeriana clematoides IPNI=*859734-1* 296
Valeriana coreana subsp. *pseudoumbrosa* IPNI=*860012-1* 296
Valeriana decussata var. *acuminatissima* IPNI=*859734-1* 296
Valeriana decussata var. *normalis* IPNI=*859734-1* 296
Valeriana erysimoides IPNI=*286659-2* 296
Valeriana exaltata IPNI=*860012-1* 296
Valeriana florida IPNI=*860141-1* 296
Valeriana germainii IPNI=*286641-2* 296
Valeriana ghiesbrechtii IPNI=*286620-2* 296
Valeriana gonatolophis IPNI=*286639-2* 296
Valeriana gracilis IPNI=*286655-2* 296
Valeriana hieronymi IPNI=*286639-2* 296
Valeriana hispida IPNI=*286620-2* 296
Valeriana hortensis IPNI=*860141-1* 296
Valeriana laciniata IPNI=*859680-1* 296
Valeriana latifolia IPNI=*319111-2* 286
Valeriana laurifolia IPNI=*286620-2* 296
Valeriana longifolia IPNI=*286641-2* 296
Valeriana longifolia var. *pilosa* IPNI=*286641-2* 296
Valeriana lucida IPNI=*860012-1* 296
Valeriana major IPNI=*860012-1* 296
Valeriana melanocarpa IPNI=*859734-1* 296
Valeriana multiceps IPNI=*860012-1* 296
Valeriana mutisiana IPNI=*860264-1* 296
Valeriana oblongifolia IPNI=*860264-1* 296
Valeriana officinalis var. *sarkanyi* IPNI=*860012-1* 296
Valeriana origanifolia IPNI=*286639-2* 296
Valeriana palustris IPNI=*860012-1* 296
Valeriana pavonii IPNI=*286620-2* 296
Valeriana pavonii var. *yungasensis* IPNI=*286620-2* 296
Valeriana phaseoli IPNI=*30039915-2* 296
Valeriana pilosa IPNI=*286641-2* 296
Valeriana pinnata IPNI=*860012-1* 296
Valeriana plantaginea IPNI=*286641-2* 296
Valeriana plantaginea var. *minor* IPNI=*286641-2* 296
Valeriana pseudoumbrosa IPNI=*860012-1* 296
Valeriana revoluta IPNI=*286639-2* 296
Valeriana rhomboidea IPNI=*286659-2* 296
Valeriana sallei IPNI=*286659-2* 296
Valeriana scandens IPNI=*63937-2* 286
Valeriana scandens var. *angustiloba* IPNI=*30039915-2* 296
Valeriana scandens var. *genuina* IPNI=*30039915-2* 296
Valeriana scorpioides IPNI=*286659-2* 296
Valeriana sinensis IPNI=*860012-1* 296
Valeriana subincisa IPNI=*286620-2* 296
Valeriana sylvestris IPNI=*860012-1* 296
Valeriana tenella IPNI=*286655-2* 296
Valeriana toluccana IPNI=*286655-2* 296
Valeriana trichiata IPNI=*286620-2* 296
Valeriana urticifolia var. *scorpioides* IPNI=*286659-2* 296
Valeriana veronicifolia IPNI=*286659-2* 296
Valeriana volubilis IPNI=*30039915-2* 296
Valeriana vulgaris IPNI=*860012-1* 296
Valerianoides cayennensis IPNI=*1126344-2* 379
Valerianoides dichotoma IPNI=*1126344-2* 379
Valerianoides indica IPNI=*864749-1* 379
Valerianoides jamaicensis f. *glabra* IPNI=*1092351-2* 379
Valerianoides jamaicensis f. *strigosa* IPNI=*1092351-2* 379
Valerianoides jamaicensis IPNI=*1092351-2* 379
Valerianoides jamaicensis var. *angustifolia* IPNI=*1092351-2* 379
Valerianoides jamaicensis var. *linearifolia* IPNI=*864749-1* 379
Valerianoides jamaicensis var. *spathulata* IPNI=*1092351-2* 379
Valerianoides mutabilis IPNI=*243385-2* 379
Valerianoides orubica IPNI=*864784-1* 379
Vallea cernua IPNI=*836044-1* 431
Vallea cordifolia IPNI=*836044-1* 431
Vallea glabra IPNI=*836044-1* 431
Vallea macrophylla IPNI=*836044-1* 431
Vallea ovata IPNI=*836044-1* 431
Vallea pubescens IPNI=*836044-1* 431
Vallea pyrifolia IPNI=*836044-1* 431
Vallea schlimii IPNI=*836044-1* 431
Vallea stipularis var. *parvifolia* IPNI=*836044-1* 431
Vallea stipularis var. *pyrifolia* IPNI=*836044-1* 431
Vallesia chiococcoides IPNI=*82668-1* 344
Vallesia cymbifolia IPNI=*82668-1* 344
Vallesia dichotoma IPNI=*82668-1* 344
Vallesia inedita IPNI=*82668-1* 344
Vallesia punctata IPNI=*82668-1* 344
Vallisneria neotropicalis IPNI=*431972-1* 203
Vallisneria verticillata IPNI=*114224-3* 203
Vallota miniata IPNI=*994513-1* 213
Valota insularis IPNI=*398888-1* 237
Valota pennicilligera IPNI=*1050107-2* 237
Valota saccharata IPNI=*80220-2* 237
Vandellia alba IPNI=*1189750-2* 374
Vandellia antipoda IPNI=*799766-1* 374
Vandellia bodinieri IPNI=*1189750-2* 374
Vandellia brownii IPNI=*1189750-2* 374
Vandellia crustacea IPNI=*1189750-2* 374
Vandellia diffusa IPNI=*255156-2* 374
Vandellia gracilis IPNI=*1189750-2* 374
Vandellia minuta IPNI=*1189750-2* 374
Vandellia obtusa IPNI=*1189750-2* 374
Vandellia pachypoda IPNI=*799766-1* 374
Vandellia racemosa IPNI=*1189750-2* 374
Vandellia repens IPNI=*262512-2* 374
Vandellia serrata IPNI=*799766-1* 374
Vandellia varians IPNI=*1189750-2* 374
Vandellia verbenifolia IPNI=*799766-1* 374
Vandesia edulis IPNI=*63283-1* 226
Vanessa cordifolia IPNI=*755652-1* 352
Vanilla anaromatica IPNI=*661225-1* 223
Vanilla angustifolia IPNI=*661225-1* 223
Vanilla aromatica IPNI=*262578-2* 223
Vanilla axillaris IPNI=*661225-1* 223
Vanilla bahiana IPNI=*661245-1* 223
Vanilla bampsiana IPNI=*262578-2* 223
Vanilla calyculata IPNI=*262547-2* 223
Vanilla carinata IPNI=*661245-1* 223
Vanilla claviculata IPNI=*661161-1* 223
Vanilla cochlearilabia IPNI=*661259-1* 223
Vanilla denticulata IPNI=*661231-1* 223
Vanilla duckei IPNI=*262578-2* 223
Vanilla ensifolia IPNI=*661231-1* 223
Vanilla epidendrum IPNI=*661225-1* 223
Vanilla esquipulensis IPNI=*262547-2* 223
Vanilla fragrans IPNI=*262578-2* 223

Vanilla gardneri IPNI=*661245-1* 223
Vanilla grandiflora IPNI=*661259-1* 223
Vanilla gratiosa IPNI=*661161-1* 223
Vanilla guatemalensis IPNI=*661225-1* 223
Vanilla hamata IPNI=*661259-1* 223
Vanilla latisegmenta IPNI=*661200-1* 223
Vanilla lindmaniana IPNI=*661237-1* 223
Vanilla lutea IPNI=*661161-1* 223
Vanilla lutescens IPNI=*661259-1* 223
Vanilla martinezii IPNI=*661225-1* 223
Vanilla ovata IPNI=*661225-1* 223
Vanilla palmarum var. *grandifolia* IPNI=*661237-1* 223
Vanilla paludosa IPNI=*661225-1* 223
Vanilla pittieri IPNI=*661259-1* 223
Vanilla planifolia var. *gigantea* IPNI=*661259-1* 223
Vanilla planifolia var. *macrantha* IPNI=*661245-1* 223
Vanilla pleei IPNI=*661259-1* 223
Vanilla pompona subsp. *grandiflora* IPNI=*661259-1* 223
Vanilla pompona subsp. *pittieri* IPNI=*661259-1* 223
Vanilla pompona subsp. *pompona* IPNI=*661259-1* 223
Vanilla purusara IPNI=*262538-2* 223
Vanilla rubra IPNI=*262578-2* 223
Vanilla sativa IPNI=*262578-2* 223
Vanilla schwackeana IPNI=*262578-2* 223
Vanilla sotoarenasii IPNI=*262578-2* 223
Vanilla surinamensis IPNI=*661200-1* 223
Vanilla sylvestris IPNI=*262578-2* 223
Vanilla vanilla IPNI=*661225-1* 223
Vanilla viridiflora IPNI=*262578-2* 223
Vanilla wrightii IPNI=*661161-1* 223
Vanilla yanesha IPNI=*661225-1* 223
Vantanea cupularis IPNI=*431534-1* 393
Vantanea guianensis IPNI=*431534-1* 393
Varasia podocarpa IPNI=*317620-2* 344
Varasia prostrata IPNI=*108504-2* 344
Varasia sedifolia IPNI=*317620-2* 344
Varasia sessilis IPNI=*108504-2* 344
Vargasia caracasana IPNI=*208557-1* 263
Varinga repens IPNI=*853513-1* 439
Varronia alba IPNI=*1018059-2* 436
Varronia angustifolia IPNI=*262640-2* 275
Varronia asperrima IPNI=*318022-2* 275
Varronia bifurcata IPNI=*121424-1* 276
Varronia boliviana IPNI=*262677-2* 276
Varronia brevispicata IPNI=*262640-2* 275
Varronia calyptrata IPNI=*927551-1* 274
Varronia calyptrata var. *hartwegii* IPNI=*927551-1* 274
Varronia cana IPNI=*946177-1* 276
Varronia caracasana IPNI=*262641-2* 276
Varronia chacoensis IPNI=*262640-2* 275
Varronia coloradiphila IPNI=*946226-1* 276
Varronia corymbosa IPNI=*262677-2* 276
Varronia costaricensis IPNI=*946226-1* 276
Varronia crenulata IPNI=*946226-1* 276
Varronia cuneiformis IPNI=*262640-2* 275
Varronia curassavica IPNI=*262641-2* 276
Varronia divaricata IPNI=*262640-2* 275
Varronia ferruginea IPNI=*946226-1* 276
Varronia flava IPNI=*274993-2* 274
Varronia floribunda IPNI=*262641-2* 276
Varronia globosa IPNI=*262677-2* 276
Varronia grandiflora var. *glabrata* IPNI=*121434-1* 276
Varronia grandiflora var. *sprucei* IPNI=*121434-1* 276
Varronia graveolens IPNI=*262640-2* 275
Varronia guazumifolia IPNI=*262677-2* 276
Varronia guianensis IPNI=*262640-2* 275
Varronia hermanniifolia IPNI=*262677-2* 276
Varronia hispida IPNI=*262640-2* 275
Varronia humboldtii IPNI=*262641-2* 276
Varronia interrupta IPNI=*262640-2* 275
Varronia intonsa IPNI=*262640-2* 275
Varronia lanata IPNI=*262641-2* 276
Varronia lanceolata IPNI=*262677-2* 276
Varronia lantanoides IPNI=*121434-1* 276
Varronia laxiflora IPNI=*946226-1* 276
Varronia linearis IPNI=*262640-2* 275
Varronia lineata IPNI=*77084349-1* 276
Varronia macrostachya IPNI=*262640-2* 275
Varronia macrostachya IPNI=*262641-2* 276
Varronia martinicensis IPNI=*262640-2* 275
Varronia martinicensis IPNI=*946226-1* 276
Varronia monosperma IPNI=*262677-2* 276
Varronia oxyphylla IPNI=*262640-2* 275
Varronia paniculata IPNI=*262677-2* 276
Varronia paraguariensis IPNI=*262677-2* 276
Varronia parviflora IPNI=*121424-1* 276
Varronia patens IPNI=*262677-2* 276
Varronia pauciflora IPNI=*946226-1* 276
Varronia poeppigii IPNI=*946226-1* 276
Varronia pyrifolia IPNI=*946226-1* 276
Varronia riparia IPNI=*946226-1* 276
Varronia rotundifolia IPNI=*274993-2* 274
Varronia rugosa IPNI=*262640-2* 275
Varronia rusbyi IPNI=*113943-1* 273
Varronia salicina IPNI=*262640-2* 275
Varronia salviifolia IPNI=*262641-2* 276
Varronia spicata IPNI=*262641-2* 276
Varronia tuberosa IPNI=*113943-1* 273
Varronia ulmifolia IPNI=*262677-2* 276
Varronia urticifolia IPNI=*262677-2* 276
Varronia verbenacea IPNI=*262640-2* 275
Vasargia caracasana IPNI=*208557-1* 263
Vasconcellea × heilbornii IPNI=*77164259-1* 279
Vasconcellea × heilbornii var. *chrysopetala* IPNI=*77164259-1* 279
Vasconcellea × heilbornii var. *fructifragrans* IPNI=*77164259-1* 279
Vasconcellea boissieri IPNI=*675769-1* 279
Vasconcellea cestriflora IPNI=*675782-1* 279
Vasconcellea cundinamarcensis IPNI=*675782-1* 279
Vasconcellea hookeri IPNI=*675779-1* 279
Vasconcellea manihot IPNI=*675778-1* 279
Vasconcellea peltata IPNI=*30011248-2* 279
Vasconcellia acutifolia IPNI=*77142115-1* 363
Vatairea surinamensis IPNI=*524446-1* 336
Vatkea diandra IPNI=*675840-1* 374
Veitchia merrillii IPNI=*663990-1* 204
Velaea toluccensis IPNI=*838268-1* 251
Velaga globosa IPNI=*553576-1* 417
Velarum leiocarpum IPNI=*289589-1* 278
Velarum officinale IPNI=*289589-1* 278
Velarum officinale subsp. *leiocarpum* IPNI=*289589-1* 278
Velarum tzvelevii IPNI=*289589-1* 278
Vellasquezia melaenodendron IPNI=*258609-2* 290
Velleruca longistyla IPNI=*1137625-2* 277
Velleruca vesicaria IPNI=*1137625-2* 277
Veltheimia speciosa IPNI=*537112-1* 216
Veltheimia uvaria IPNI=*537112-1* 216
Veratrum mairei IPNI=*64141-1* 216
Verbascum annuum IPNI=*770211-1* 377
Verbascum blattarioides IPNI=*770211-1* 377
Verbascum celsiae IPNI=*770211-1* 377
Verbascum glabrum IPNI=*770211-1* 377
Verbascum glandulosum IPNI=*770211-1* 377
Verbascum grandiflorum IPNI=*770211-1* 377
Verbascum laevigatum IPNI=*770211-1* 377
Verbascum lanceolatum IPNI=*770211-1* 377
Verbascum viscidulum IPNI=*770211-1* 377

Verbena affinis IPNI=*30007120-2* 379
Verbena americana IPNI=*1092351-2* 379
Verbena angustifolia IPNI=*864749-1* 379
Verbena aristata IPNI=*864784-1* 379
Verbena bangiana IPNI=*865215-1* 379
Verbena brasiliensis var. *subglabrata* IPNI=*30007120-2* 379
Verbena cajamarcensis IPNI=*30007120-2* 379
Verbena caracasana IPNI=*30007120-2* 379
Verbena caudata IPNI=*864749-1* 379
Verbena cayennensis IPNI=*1126344-2* 379
Verbena citrodora IPNI=*1015994-1* 377
Verbena dichotoma IPNI=*1126344-2* 379
Verbena elegans var. *asperata* IPNI=*263032-2* 379
Verbena fragrans IPNI=*1015994-1* 377
Verbena gentryi IPNI=*30007120-2* 379
Verbena glabrata var. *tenuispicata* IPNI=*30007120-2* 379
Verbena glandulosa IPNI=*865215-1* 379
Verbena globiflora IPNI=*141020-2* 378
Verbena globulifera IPNI=*141020-2* 378
Verbena hispida var. *obovata* IPNI=*865215-1* 379
Verbena indica IPNI=*864749-1* 379
Verbena integrifolia IPNI=*30007120-2* 379
Verbena jamaicensis IPNI=*1092351-2* 379
Verbena lanceolata IPNI=*30007120-2* 379
Verbena lancifolia IPNI=*864749-1* 379
Verbena lantanoides IPNI=*141020-2* 378
Verbena lappulacea IPNI=*1194372-2* 379
Verbena litoralis f. *albiflora* IPNI=*30007120-2* 379
Verbena litoralis var. *albiflora* IPNI=*30007120-2* 379
Verbena litoralis var. *caracasana* IPNI=*30007120-2* 379
Verbena litoralis var. *leptostachya* IPNI=*30007120-2* 379
Verbena litoralis var. *melanopotamica* IPNI=*30007120-2* 379
Verbena litoralis var. *portoricensis* IPNI=*30007120-2* 379
Verbena litoralis var. *pycnostachya* IPNI=*30007120-2* 379
Verbena litoralis var. *subglabrata* IPNI=*30007120-2* 379
Verbena longifolia f. *albiflora* IPNI=*30007120-2* 379
Verbena longifolia IPNI=*30007120-2* 379
Verbena longifolia var. *pubescens* IPNI=*30007120-2* 379
Verbena minutiflora var. *peruviana* IPNI=*30007120-2* 379
Verbena moranensis IPNI=*263032-2* 379
Verbena mutabilis IPNI=*243385-2* 379
Verbena nodiflora IPNI=*194567-2* 379
Verbena nudiflora IPNI=*30007120-2* 379
Verbena orubica IPNI=*864784-1* 379
Verbena parvula f. *albiflora* IPNI=*865182-1* 379
Verbena parvula IPNI=*865182-1* 379
Verbena parvula var. *gigas* IPNI=*865182-1* 379
Verbena parvula var. *obovata* IPNI=*865215-1* 379
Verbena paucifolia IPNI=*30007120-2* 379
Verbena pilosiuscula IPNI=*1092351-2* 379
Verbena prismatica IPNI=*35430-2* 378
Verbena sedula IPNI=*30007120-2* 379
Verbena sedula var. *fournieri* IPNI=*30007120-2* 379
Verbena triphylla IPNI=*1015994-1* 377
Verbesina alba IPNI=*1186403-2* 261
Verbesina australis IPNI=*30147294-2* 272
Verbesina carnosa IPNI=*1093589-2* 270
Verbesina carnosa var. *triloba* IPNI=*1093589-2* 270
Verbesina conyzoides IPNI=*1186403-2* 261
Verbesina debilis IPNI=*1186403-2* 261
Verbesina dichotoma IPNI=*252024-1* 271
Verbesina encelioides subsp. *exauriculata* IPNI=*30147294-2* 272
Verbesina encelioides var. *cana* IPNI=*30147294-2* 272
Verbesina encelioides var. *exauriculata* IPNI=*30147294-2* 272
Verbesina exalata IPNI=*257696-1* 272
Verbesina exauriculata IPNI=*30147294-2* 272
Verbesina foliacea IPNI=*202984-1* 261
Verbesina lavenia IPNI=*175003-1* 256
Verbesina linifolia IPNI=*236639-1* 267
Verbesina membranifolia IPNI=*993359-1* 271
Verbesina microcephala IPNI=*257696-1* 272
Verbesina microptera IPNI=*30147294-2* 272
Verbesina mutica IPNI=*993359-1* 271
Verbesina nicaraguensis IPNI=*257696-1* 272
Verbesina nodiflora IPNI=*252024-1* 271
Verbesina oppositiflora IPNI=*993359-1* 271
Verbesina ovata IPNI=*213524-1* 264
Verbesina prostrata IPNI=*1186403-2* 261
Verbesina pseudoacmella IPNI=*1186403-2* 261
Verbesina pusilla IPNI=*1186403-2* 261
Verbesina sartori IPNI=*257696-1* 272
Verbesina scabra IPNI=*30147294-2* 272
Verbesina scandens IPNI=*241903-1* 269
Verbesina spicata IPNI=*1186403-2* 261
Verbesina sublobata IPNI=*257696-1* 272
Verbesina szyszylowiczii IPNI=*254724-2* 271
Verbesina tridentata IPNI=*1093589-2* 270
Verbesina verbascifolia IPNI=*257696-1* 272
Verdcourtia lignosa IPNI=*492756-1* 318
Verhuellia knoblocheriana IPNI=*679378-1* 194
Verhuellia serpens IPNI=*679841-1* 194
Vermicularia decurrens IPNI=*1092351-2* 379
Vermicularia lancifolia IPNI=*864749-1* 379
Vernonanthura yurimaguasensis IPNI=*969218-1* 272
Vernonia abbreviata IPNI=*962413-1* 260
Vernonia acuminata IPNI=*948489-1* 265
Vernonia albifila IPNI=*969218-1* 272
Vernonia apocynifolia IPNI=*962413-1* 260
Vernonia arborescens var. *corrientensis* IPNI=*944993-1* 261
Vernonia arborescens var. *cuneifolia* IPNI=*948489-1* 265
Vernonia aschenborniana IPNI=*969218-1* 272
Vernonia assans IPNI=*969180-1* 272
Vernonia baccharoides IPNI=*969218-1* 272
Vernonia bangii IPNI=*969218-1* 272
Vernonia betonicifolia IPNI=*962413-1* 260
Vernonia bogotana IPNI=*200446-1* 260
Vernonia brachiata IPNI=*944995-1* 261
Vernonia brachylepis IPNI=*51409-2* 259
Vernonia brasiliana IPNI=*969180-1* 272
Vernonia brasiliensis IPNI=*1038787-2* 272
Vernonia bullata IPNI=*948489-1* 265
Vernonia calerana IPNI=*200446-1* 260
Vernonia canescens IPNI=*948489-1* 265
Vernonia canescens var. *pilata* IPNI=*948489-1* 265
Vernonia cardonae IPNI=*993359-1* 271
Vernonia centriflora IPNI=*944993-1* 261
Vernonia chromolepis IPNI=*962014-1* 265
Vernonia cinerea IPNI=*962413-1* 260
Vernonia cinerea var. *erigeroides* IPNI=*962413-1* 260
Vernonia cinerea var. *lentii* IPNI=*962413-1* 260
Vernonia cirsiifolia IPNI=*236351-1* 267
Vernonia corcovadensis IPNI=*1038787-2* 272
Vernonia cuneifolia IPNI=*948489-1* 265
Vernonia cyanonioides IPNI=*962413-1* 260
Vernonia dendigulensis IPNI=*962413-1* 260
Vernonia dichotoma IPNI=*944995-1* 261
Vernonia diffusa IPNI=*962413-1* 260
Vernonia edulis IPNI=*236351-1* 267
Vernonia erigeroides IPNI=*962413-1* 260
Vernonia exilis IPNI=*962413-1* 260
Vernonia fasciculata IPNI=*962413-1* 260
Vernonia flavescens IPNI=*944993-1* 261
Vernonia floribunda IPNI=*969218-1* 272
Vernonia geminata IPNI=*948489-1* 265
Vernonia geminata var. *aprica* IPNI=*948489-1* 265
Vernonia geminata var. *glabrescens* IPNI=*948489-1* 265

Vernonia glandulata IPNI=*200452-1* 260
Vernonia haenkeana IPNI=*969218-1* 272
Vernonia hirsutivena IPNI=*948489-1* 265
Vernonia intermedia IPNI=*962014-1* 265
Vernonia intermedia var. *ramosior* IPNI=*962014-1* 265
Vernonia karstenii IPNI=*948527-1* 265
Vernonia kroneana IPNI=*962413-1* 260
Vernonia lanceolaris IPNI=*969218-1* 272
Vernonia larensis IPNI=*948528-1* 265
Vernonia lastii IPNI=*962413-1* 260
Vernonia laxiflora IPNI=*962413-1* 260
Vernonia lehmannii IPNI=*948528-1* 265
Vernonia lentii IPNI=*962413-1* 260
Vernonia leptophylla IPNI=*962413-1* 260
Vernonia linearifolia IPNI=*962014-1* 265
Vernonia linearis IPNI=*962014-1* 265
Vernonia mattogrossensis IPNI=*969218-1* 272
Vernonia medialis IPNI=*948489-1* 265
Vernonia micradenia IPNI=*969218-1* 272
Vernonia micrantha IPNI=*948489-1* 265
Vernonia mollis IPNI=*948489-1* 265
Vernonia monsonensis IPNI=*969218-1* 272
Vernonia odoratissima IPNI=*969180-1* 272
Vernonia odoratissima var. *caracasana* IPNI=*969180-1* 272
Vernonia odoratissima var. *guianensis* IPNI=*969180-1* 272
Vernonia otavalensis IPNI=*183399-1* 257
Vernonia pacchensis IPNI=*969218-1* 272
Vernonia pacchensis var. *tambillensis* IPNI=*969218-1* 272
Vernonia patens IPNI=*1038787-2* 272
Vernonia patens IPNI=*969218-1* 272
Vernonia patuliflora IPNI=*948489-1* 265
Vernonia physalifolia IPNI=*962413-1* 260
Vernonia polyantha IPNI=*1038787-2* 272
Vernonia pseudomollis IPNI=*948489-1* 265
Vernonia psittacorum IPNI=*1038787-2* 272
Vernonia purpusii IPNI=*948489-1* 265
Vernonia reticulata IPNI=*948489-1* 265
Vernonia rhomboidea IPNI=*962413-1* 260
Vernonia rubricaulis f. *genuina* IPNI=*962014-1* 265
Vernonia rubricaulis IPNI=*962014-1* 265
Vernonia rubricaulis var. *australis* IPNI=*962014-1* 265
Vernonia rubricaulis var. *dentata* IPNI=*962014-1* 265
Vernonia rubricaulis var. *denudata* IPNI=*962014-1* 265
Vernonia rubricaulis var. *floribunda* IPNI=*962014-1* 265
Vernonia rubricaulis var. *glomerata* IPNI=*962014-1* 265
Vernonia ruiz-teranii IPNI=*948527-1* 265
Vernonia rusbyi IPNI=*948489-1* 265
Vernonia salamana IPNI=*969218-1* 272
Vernonia salicifolia IPNI=*962014-1* 265
Vernonia salviifolia IPNI=*969180-1* 272
Vernonia scabra IPNI=*969180-1* 272
Vernonia scabra var. *acuminata* IPNI=*969180-1* 272
Vernonia scorpioides IPNI=*944993-1* 261
Vernonia scorpioides var. *centriflora* IPNI=*944993-1* 261
Vernonia scorpioides var. *longeracemosa* IPNI=*944993-1* 261
Vernonia scorpioides var. *longifolia* IPNI=*944993-1* 261
Vernonia scorpioides var. *subrepanda* IPNI=*944993-1* 261
Vernonia scorpioides var. *subtomentosa* IPNI=*944993-1* 261
Vernonia sinulata IPNI=*962413-1* 260
Vernonia sodiroi IPNI=*948489-1* 265
Vernonia spiritus-sancti IPNI=*948489-1* 265
Vernonia stuebelii IPNI=*969218-1* 272
Vernonia subrepanda IPNI=*944993-1* 261
Vernonia tournefortioides IPNI=*944993-1* 261
Vernonia unillensis IPNI=*948489-1* 265
Vernonia vargasii IPNI=*969218-1* 272
Vernonia viarum IPNI=*969218-1* 272
Vernonia villosa IPNI=*962413-1* 260
Vernonia volubilis IPNI=*948489-1* 265
Vernonia weberbaueri IPNI=*969218-1* 272
Vernonia yurimaguasensis IPNI=*969218-1* 272
Vernonia zollingeriana IPNI=*962413-1* 260
Veronica acinifolia IPNI=*811649-1* 377
Veronica afrochamaedrys IPNI=*812151-1* 377
Veronica alpestris IPNI=*812650-1* 377
Veronica alpiphila IPNI=*30198355-2* 377
Veronica anomala IPNI=*812400-1* 377
Veronica areolata IPNI=*30198355-2* 377
Veronica arvensis subsp. *pennellii* IPNI=*811649-1* 377
Veronica arvensis var. *reniformis* IPNI=*811649-1* 377
Veronica bolivari IPNI=*812650-1* 377
Veronica borealis IPNI=*812650-1* 377
Veronica brevipedunculata IPNI=*811649-1* 377
Veronica bungabecca IPNI=*812650-1* 377
Veronica buxbaumii IPNI=*30198355-2* 377
Veronica buxifolia IPNI=*812400-1* 377
Veronica byzantina IPNI=*30198355-2* 377
Veronica capensis IPNI=*812650-1* 377
Veronica carnosula IPNI=*812460-1* 377
Veronica caroliniana IPNI=*812460-1* 377
Veronica chamaedryoides IPNI=*812151-1* 377
Veronica chillensis IPNI=*812460-1* 377
Veronica cymbalariifolia IPNI=*30198355-2* 377
Veronica decussata IPNI=*811951-1* 377
Veronica demissa IPNI=*811649-1* 377
Veronica depressa IPNI=*811649-1* 377
Veronica diffusa IPNI=*30198355-2* 377
Veronica filiformis IPNI=*30198355-2* 377
Veronica fonkii IPNI=*812616-1* 377
Veronica forsteri IPNI=*812616-1* 377
Veronica funesta IPNI=*812650-1* 377
Veronica haustrata IPNI=*812400-1* 377
Veronica hawaiensis IPNI=*811649-1* 377
Veronica hirsuta IPNI=*811649-1* 377
Veronica hospita IPNI=*30198355-2* 377
Veronica integerrima IPNI=*812650-1* 377
Veronica kermesina IPNI=*812675-1* 377
Veronica koeleri IPNI=*811649-1* 377
Veronica laevis IPNI=*812460-1* 377
Veronica lindleyana IPNI=*812616-1* 377
Veronica longiracemosa IPNI=*811649-1* 377
Veronica longistyla IPNI=*812650-1* 377
Veronica macrocalyx IPNI=*812650-1* 377
Veronica maddenii IPNI=*812151-1* 377
Veronica magellanica IPNI=*811951-1* 377
Veronica marginata IPNI=*811951-1* 377
Veronica marilandica IPNI=*812460-1* 377
Veronica maximowicziana IPNI=*812460-1* 377
Veronica menziesii IPNI=*811951-1* 377
Veronica meskhetica IPNI=*30198355-2* 377
Veronica microphylla IPNI=*812650-1* 377
Veronica murorum IPNI=*812151-1* 377
Veronica neglecta IPNI=*812650-1* 377
Veronica nova IPNI=*811649-1* 377
Veronica nummulariifolia IPNI=*812650-1* 377
Veronica pallescens IPNI=*812460-1* 377
Veronica peregrina subsp. *asiatica* IPNI=*812460-1* 377
Veronica peregrina subsp. *typica* IPNI=*812460-1* 377
Veronica peregrina subsp. *xalapensis* IPNI=*812460-1* 377
Veronica peregrina var. *laurentiana* IPNI=*812460-1* 377
Veronica peregrina var. *xalapensis* IPNI=*812460-1* 377
Veronica persica var. *aschersoniana* IPNI=*30198355-2* 377
Veronica persica var. *corrensiana* IPNI=*30198355-2* 377
Veronica peruviana IPNI=*812460-1* 377
Veronica polyanthos IPNI=*811649-1* 377
Veronica polygonoides IPNI=*811649-1* 377

Veronica precox IPNI=*30198355-2* 377
Veronica pseudoarvensis IPNI=*811649-1* 377
Veronica racemifoliata IPNI=*811649-1* 377
Veronica reniformis IPNI=*811649-1* 377
Veronica riederiana IPNI=*812650-1* 377
Veronica romana IPNI=*812460-1* 377
Veronica rotundifolia IPNI=*812650-1* 377
Veronica ruderalis IPNI=*812650-1* 377
Veronica rugulosella IPNI=*812650-1* 377
Veronica serpyllifolia subsp. *trichocaulis* IPNI=*812650-1* 377
Veronica serpyllifolia var. *alpestris* IPNI=*812650-1* 377
Veronica serpyllifolia var. *decipiens* IPNI=*812650-1* 377
Veronica serpyllifolia var. *pubescens* IPNI=*812650-1* 377
Veronica sherwoodii IPNI=*812460-1* 377
Veronica simpsonii IPNI=*811951-1* 377
Veronica spicatoracemosa IPNI=*812650-1* 377
Veronica tournefortii IPNI=*30198355-2* 377
Veronica tournefortii subsp. *aschersoniana* IPNI=*30198355-2* 377
Veronica tournefortii subsp. *corrensiana* IPNI=*30198355-2* 377
Veronica tournefortii subsp. *restrictior* IPNI=*30198355-2* 377
Veronica tournefortii var. *aschersoniana* IPNI=*30198355-2* 377
Veronica tournefortii var. *corrensiana* IPNI=*30198355-2* 377
Veronica villosiuscula IPNI=*812650-1* 377
Veronica vogerensis IPNI=*812151-1* 377
Veronica wogerensis IPNI=*812151-1* 377
Veronica wolffiana IPNI=*812650-1* 377
Veronica xalapensis IPNI=*812460-1* 377
Veronicastrum serpyllifolium IPNI=*812650-1* 377
Verrucaria volubilis IPNI=*119329-1* 275
Verticillaria acuminata IPNI=*277321-2* 385
Verticillaria madruno IPNI=*277321-2* 385
Verticillaria peruviana IPNI=*277321-2* 385
Verticillaria rostrata IPNI=*277321-2* 385
Vesiculina gibba IPNI=*526777-1* 374
Vespuccia humboldtii IPNI=*58399-1* 198
Vetiveria arundinacea IPNI=*396213-1* 236
Vetiveria muricata IPNI=*396213-1* 236
Vetiveria odorata IPNI=*396213-1* 236
Vetiveria odoratissima IPNI=*396213-1* 236
Vetiveria zizanioides IPNI=*396213-1* 236
Vetrix helix IPNI=*301321-2* 404
Vetrix purpurea IPNI=*301321-2* 404
Vetrix sicula IPNI=*301321-2* 404
Vexillaria brasiliana IPNI=*485555-1* 314
Vexillaria virginiana IPNI=*51897-2* 314
Viburnum cornifolium IPNI=*264984-2* 297
Viburnum ferrugineum IPNI=*264840-2* 296
Viburnum glabrum IPNI=*149702-1* 296
Viburnum leptophyllum IPNI=*149971-1* 297
Viburnum macrophyllum IPNI=*791637-1* 291
Viburnum parviflorum IPNI=*149971-1* 297
Vicia andicola var. *galbina* IPNI=*524519-1* 336
Vicia andicola var. *mathewsii* IPNI=*524519-1* 336
Vicia angustifolia IPNI=*524865-1* 336
Vicia arenivaga IPNI=*524865-1* 336
Vicia arietina IPNI=*486336-1* 315
Vicia bogotensis IPNI=*524519-1* 336
Vicia equina IPNI=*524737-1* 336
Vicia esculenta IPNI=*524737-1* 336
Vicia faba var. *minor* IPNI=*524737-1* 336
Vicia godronii IPNI=*525259-1* 336
Vicia littoralis IPNI=*1071691-2* 335
Vicia mathewsii IPNI=*524519-1* 336
Vicia minima IPNI=*524865-1* 336
Vicia multicaulis IPNI=*524865-1* 336
Vicia pauciflora IPNI=*524865-1* 336
Vicia pliniana IPNI=*524737-1* 336
Vicia porphyrea IPNI=*524737-1* 336
Vicia selloi IPNI=*524789-1* 336
Vicia vulgaris IPNI=*524737-1* 336
Victoria regalis IPNI=*605776-1* 181
Victoria regia IPNI=*605776-1* 181
Victoria regina IPNI=*605776-1* 181
Victoria reginae IPNI=*605776-1* 181
Vieusseuxia martinicensis IPNI=*441642-1* 217
Vieusseuxia pavonia IPNI=*441553-1* 216
Vigna acuminata IPNI=*525419-1* 336
Vigna adenantha IPNI=*60457245-2* 325
Vigna anomala IPNI=*525428-1* 336
Vigna brachycalyx IPNI=*525554-1* 337
Vigna brachystachys IPNI=*525419-1* 336
Vigna brasiliensis IPNI=*525419-1* 336
Vigna bukobensis IPNI=*525419-1* 336
Vigna calcarata IPNI=*525554-1* 337
Vigna caracalla IPNI=*60471746-2* 315
Vigna donii IPNI=*525491-1* 336
Vigna gangetica IPNI=*525419-1* 336
Vigna glabra IPNI=*525419-1* 336
Vigna glabra var. *angustifolia* IPNI=*525419-1* 336
Vigna helicopus IPNI=*525419-1* 336
Vigna hirsuta IPNI=*525419-1* 336
Vigna holstii IPNI=*525419-1* 336
Vigna linearis IPNI=*60457240-2* 320
Vigna linearis var. *latifolia* IPNI=*60457240-2* 320
Vigna longepedunculata IPNI=*525419-1* 336
Vigna lutea IPNI=*525428-1* 336
Vigna luteola var. *angustifolia* IPNI=*525419-1* 336
Vigna marina subsp. *oblonga* IPNI=*525428-1* 336
Vigna nigerica IPNI=*525419-1* 336
Vigna nilotica IPNI=*525419-1* 336
Vigna oblonga IPNI=*525428-1* 336
Vigna papuana IPNI=*525554-1* 337
Vigna peduncularis IPNI=*60457232-2* 311
Vigna repens IPNI=*525419-1* 336
Vigna repens var. *angustifolia* IPNI=*525419-1* 336
Vigna repens var. *glabra* IPNI=*525419-1* 336
Vigna repens var. *lutea* IPNI=*525428-1* 336
Vigna repens var. *luteola* IPNI=*525419-1* 336
Vigna repens var. *villosa* IPNI=*525419-1* 336
Vigna retusa IPNI=*525428-1* 336
Vigna ricciardiana IPNI=*525554-1* 337
Vigna ricciardiana var. *macrocarpa* IPNI=*525554-1* 337
Vigna rotundifolia IPNI=*525428-1* 336
Vigna strophiolata IPNI=*525491-1* 336
Vigna unguiculata subvar. *typica* IPNI=*1127257-2* 337
Vigna vexillata f. *typica* IPNI=*525561-1* 337
Vigna villosa IPNI=*525419-1* 336
Vignaldia luteola IPNI=*760275-1* 355
Vigolina acmella IPNI=*208554-1* 263
Vigolina armella IPNI=*208554-1* 263
Viguiera brevipes IPNI=*259782-1* 272
Viguiera canescens IPNI=*259782-1* 272
Viguiera dentata var. *brevipes* IPNI=*259782-1* 272
Viguiera dentata var. *canescens* IPNI=*259782-1* 272
Viguiera dentata var. *helianthoides* IPNI=*259782-1* 272
Viguiera helianthoides IPNI=*259782-1* 272
Viguiera laxa IPNI=*259782-1* 272
Viguiera laxa var. *brevipes* IPNI=*259782-1* 272
Viguiera longiligula IPNI=*259782-1* 272
Viguiera microcline IPNI=*259782-1* 272
Viguiera nelsonii IPNI=*259782-1* 272
Viguiera oppositipes IPNI=*259782-1* 272
Viguiera pedunculata IPNI=*259782-1* 272
Viguiera sagrana IPNI=*259782-1* 272
Viguiera texana IPNI=*259782-1* 272
Viguiera triquetra IPNI=*259782-1* 272

Vilfa affinis IPNI=*243026-2* 245
Vilfa agrostidea IPNI=*243026-2* 245
Vilfa alopecuroides IPNI=*1048074-2* 244
Vilfa arguta IPNI=*243026-2* 245
Vilfa arkansana IPNI=*243026-2* 245
Vilfa barbata IPNI=*423010-1* 245
Vilfa berteroana IPNI=*1207790-2* 245
Vilfa conferta IPNI=*423010-1* 245
Vilfa confertiflora IPNI=*401189-1* 239
Vilfa decumbens IPNI=*386353-1* 233
Vilfa divaricata IPNI=*30197865-2* 233
Vilfa dulcis IPNI=*386353-1* 233
Vilfa dura IPNI=*423010-1* 245
Vilfa elatior IPNI=*243026-2* 245
Vilfa elegans IPNI=*1128639-2* 233
Vilfa elongata IPNI=*1207790-2* 245
Vilfa exilis IPNI=*1207790-2* 245
Vilfa familiaris IPNI=*1103484-2* 245
Vilfa fasciculata IPNI=*1128639-2* 233
Vilfa fulvescens IPNI=*1103484-2* 245
Vilfa gaditana IPNI=*386353-1* 233
Vilfa gaudichaudii IPNI=*386353-1* 233
Vilfa gavana IPNI=*1156911-2* 246
Vilfa geniculata IPNI=*423010-1* 245
Vilfa gigantea IPNI=*385783-1* 233
Vilfa glaucescens IPNI=*386353-1* 233
Vilfa hispida IPNI=*30197865-2* 233
Vilfa humifusa IPNI=*243026-2* 245
Vilfa indica IPNI=*1207790-2* 245
Vilfa intermedia IPNI=*423010-1* 245
Vilfa jacquemontii IPNI=*422898-1* 245
Vilfa juncea IPNI=*1103484-2* 245
Vilfa lenta IPNI=*398933-1* 237
Vilfa linearis IPNI=*397065-1* 236
Vilfa littoralis IPNI=*423010-1* 245
Vilfa luxurians IPNI=*423010-1* 245
Vilfa mangalorica IPNI=*243044-2* 245
Vilfa maritima IPNI=*386353-1* 233
Vilfa matrella IPNI=*423010-1* 245
Vilfa minutiflora IPNI=*243044-2* 245
Vilfa muhlenbergii IPNI=*423010-1* 245
Vilfa murina IPNI=*423010-1* 245
Vilfa orientalis IPNI=*1207790-2* 245
Vilfa patula IPNI=*386353-1* 233
Vilfa procera IPNI=*402044-1* 239
Vilfa pumila IPNI=*30197865-2* 233
Vilfa pyramidalis IPNI=*422898-1* 245
Vilfa richardii IPNI=*243026-2* 245
Vilfa rueppelliana IPNI=*422898-1* 245
Vilfa sabeana IPNI=*243026-2* 245
Vilfa schiedeana IPNI=*1103484-2* 245
Vilfa stellata IPNI=*397065-1* 236
Vilfa stolonifera IPNI=*386353-1* 233
Vilfa stolonifera var. *marina* IPNI=*386353-1* 233
Vilfa subpyramidata IPNI=*243026-2* 245
Vilfa subsetacea IPNI=*1103484-2* 245
Vilfa sylvatica IPNI=*30197865-2* 233
Vilfa tenacissima IPNI=*1207790-2* 245
Vilfa tenacissima var. *exilis* IPNI=*1207790-2* 245
Vilfa tremula IPNI=*423010-1* 245
Vilfa verticillata IPNI=*401189-1* 239
Vilfa vinzentii IPNI=*1103484-2* 245
Vilfa virginica IPNI=*423010-1* 245
Vilfa vulgaris IPNI=*30197865-2* 233
Villamilla octandra IPNI=*257665-2* 287
Villamilla peruviana IPNI=*257667-2* 287
Villamilla tinctoria IPNI=*257667-2* 287
Villarsia communis IPNI=*1182548-2* 273
Villarsia eglandulosa IPNI=*1182548-2* 273
Villarsia glandulosa IPNI=*1182548-2* 273
Villarsia humboldtiana IPNI=*171243-2* 273
Villarsia indica IPNI=*1182548-2* 273
Villarsia macrophylla IPNI=*1182548-2* 273
Villarsia microphylla IPNI=*171243-2* 273
Villarsia nymphaeifolia IPNI=*1182548-2* 273
Villarsia platiphylla IPNI=*1182548-2* 273
Villarsia rheedei IPNI=*1182548-2* 273
Villarsia simsii IPNI=*1182548-2* 273
Villarsia swartzii IPNI=*1182548-2* 273
Villarsia trachysperma IPNI=*1182548-2* 273
Vinca acutiflora IPNI=*82701-1* 344
Vinca alba IPNI=*82097-1* 343
Vinca ellipticifolia IPNI=*82701-1* 344
Vinca grandiflora IPNI=*82697-1* 344
Vinca humilis IPNI=*82701-1* 344
Vinca intermedia IPNI=*82701-1* 344
Vinca minor var. *sabinka* IPNI=*82701-1* 344
Vinca rosea IPNI=*77880-1* 340
Vinca speciosa IPNI=*77880-1* 340
Vincentia caffra IPNI=*834119-1* 408
Vincetoxicum montevidense IPNI=*74165-2* 340
Vincetoxicum obliquum IPNI=*154570-2* 342
Vincetoxicum shortii IPNI=*154570-2* 342
Vincetoxicum viridiflorum IPNI=*77206440-1* 340
Vinticena caffra IPNI=*834119-1* 408
Viola agrestis IPNI=*291320-2* 406
Viola anglica IPNI=*291320-2* 406
Viola arborea IPNI=*77144891-1* 405
Viola arborescens IPNI=*77144891-1* 405
Viola arguta subsp. *meridionalis* IPNI=*867656-1* 405
Viola arguta subsp. *typica* IPNI=*867656-1* 405
Viola arguta var. *glaberrima* IPNI=*867656-1* 405
Viola arguta var. *meridionalis* IPNI=*867656-1* 405
Viola arvatica IPNI=*291320-2* 406
Viola arvensis subsp. *andina* IPNI=*291320-2* 406
Viola arvensis var. *anglica* IPNI=*291320-2* 406
Viola arvensis var. *arvatica* IPNI=*291320-2* 406
Viola arvensis var. *deseglisei* IPNI=*291320-2* 406
Viola arvensis var. *latifolia* IPNI=*291320-2* 406
Viola arvensis var. *segetalis* IPNI=*291320-2* 406
Viola atlantica IPNI=*291320-2* 406
Viola attenuata IPNI=*77143566-1* 405
Viola aubletii IPNI=*77144891-1* 405
Viola aurasiaca IPNI=*291320-2* 406
Viola austriaca IPNI=*868828-1* 406
Viola banatica IPNI=*869428-1* 406
Viola begoniifolia IPNI=*266985-2* 406
Viola bella IPNI=*869428-1* 406
Viola bicolor IPNI=*869428-1* 406
Viola calcarea IPNI=*868309-1* 406
Viola calceolaria IPNI=*77143549-1* 405
Viola callosa IPNI=*266985-2* 406
Viola cantiana IPNI=*869428-1* 406
Viola carpatica IPNI=*869428-1* 406
Viola ciliaris IPNI=*266985-2* 406
Viola circaeoides IPNI=*77143566-1* 405
Viola colombiana IPNI=*869203-1* 406
Viola confinis IPNI=*869428-1* 406
Viola consimilis IPNI=*868828-1* 406
Viola corchorifolia IPNI=*867656-1* 405
Viola cyclophylla IPNI=*868828-1* 406
Viola deminuta IPNI=*869428-1* 406
Viola derelicta IPNI=*869428-1* 406
Viola deseglisei IPNI=*291320-2* 406
Viola dichotama IPNI=*869203-1* 406
Viola drymeia IPNI=*868309-1* 406

Viola dumetorum IPNI=*868828-1* 406
Viola favratii IPNI=*868828-1* 406
Viola flavida IPNI=*869428-1* 406
Viola floribunda IPNI=*868828-1* 406
Viola foucaudii IPNI=*869428-1* 406
Viola foudrasii IPNI=*868309-1* 406
Viola frutescens IPNI=*77143566-1* 405
Viola garganica IPNI=*869428-1* 406
Viola glutinosa IPNI=*77143568-1* 405
Viola gonzalii IPNI=*868828-1* 406
Viola gracilescens IPNI=*291320-2* 406
Viola grandiflora IPNI=*869428-1* 406
Viola hebecarpa IPNI=*868309-1* 406
Viola hirsuta IPNI=*868309-1* 406
Viola hirta var. *gymnocarpa* IPNI=*868309-1* 406
Viola hirtoalba IPNI=*868309-1* 406
Viola hortensis IPNI=*868828-1* 406
Viola humboldtii IPNI=*868059-1* 406
Viola humboldtii var. *cuneata* IPNI=*868059-1* 406
Viola humboldtii var. *renifolia* IPNI=*868059-1* 406
Viola ignobilis IPNI=*868828-1* 406
Viola incompta IPNI=*868828-1* 406
Viola ipecacuanha IPNI=*77143549-1* 405
Viola itoubou IPNI=*77143549-1* 405
Viola jucunda IPNI=*868828-1* 406
Viola kalbreyeri IPNI=*868344-1* 406
Viola kupffieri IPNI=*869428-1* 406
Viola latifolia IPNI=*291320-2* 406
Viola laurifolia IPNI=*77144891-1* 405
Viola lehmannii subsp. *cordifolia* IPNI=*869203-1* 406
Viola lejeunei IPNI=*869428-1* 406
Viola lepida IPNI=*869428-1* 406
Viola leptoceras IPNI=*869428-1* 406
Viola leucantha IPNI=*868828-1* 406
Viola lilascens IPNI=*869428-1* 406
Viola lindeniana IPNI=*869203-1* 406
Viola lloydii IPNI=*869428-1* 406
Viola loefflingiana IPNI=*77143549-1* 405
Viola longifolia IPNI=*291320-2* 406
Viola lunaria IPNI=*286647-1* 277
Viola luteola IPNI=*869428-1* 406
Viola maderensis IPNI=*868828-1* 406
Viola maillardii IPNI=*868309-1* 406
Viola martia IPNI=*868828-1* 406
Viola media IPNI=*869428-1* 406
Viola meduanensis IPNI=*869428-1* 406
Viola melanosperma IPNI=*77143566-1* 405
Viola mentita IPNI=*869428-1* 406
Viola micrantha IPNI=*869428-1* 406
Viola mucuzii IPNI=*869428-1* 406
Viola mutabilis IPNI=*869428-1* 406
Viola nelsonii IPNI=*869203-1* 406
Viola nemausensis IPNI=*869428-1* 406
Viola obliquifolia IPNI=*266985-2* 406
Viola odora IPNI=*868828-1* 406
Viola odorata f. *albiflora* IPNI=*868828-1* 406
Viola odorata subsp. *maderensis* IPNI=*868828-1* 406
Viola odorata var. *albiflora* IPNI=*868828-1* 406
Viola odorata var. *barraui* IPNI=*868828-1* 406
Viola odorata var. *incompta* IPNI=*868828-1* 406
Viola officinalis IPNI=*868828-1* 406
Viola olonnensis IPNI=*869428-1* 406
Viola oppositifolia IPNI=*77143566-1* 405
Viola orcadensis IPNI=*869428-1* 406
Viola pallescens IPNI=*869428-1* 406
Viola pallida IPNI=*868309-1* 406
Viola parviflora IPNI=*77143568-1* 405
Viola parviflora IPNI=*869428-1* 406
Viola parvifolia IPNI=*77143568-1* 405
Viola peregrina IPNI=*869428-1* 406
Viola perennis IPNI=*868828-1* 406
Viola perrobusta IPNI=*869428-1* 406
Viola persicariifolia IPNI=*266985-2* 406
Viola pichinchensis IPNI=*869203-1* 406
Viola pirottae IPNI=*868828-1* 406
Viola plantaginea IPNI=*868828-1* 406
Viola propinqua IPNI=*868828-1* 406
Viola provostii IPNI=*869428-1* 406
Viola prunellifolia IPNI=*868344-1* 406
Viola prunifolia IPNI=*77143573-1* 405
Viola pseudolutea IPNI=*869428-1* 406
Viola pseudosuavis IPNI=*868828-1* 406
Viola punicea IPNI=*867656-1* 405
Viola ramosissima IPNI=*868059-1* 406
Viola reptans IPNI=*868828-1* 406
Viola reticulata IPNI=*868344-1* 406
Viola rubra IPNI=*868309-1* 406
Viola ruralis IPNI=*869428-1* 406
Viola sagotii IPNI=*869428-1* 406
Viola sarmentosa IPNI=*868828-1* 406
Viola scandens var. *hirtella* IPNI=*869203-1* 406
Viola scandens var. *integristipula* IPNI=*869203-1* 406
Viola segetalis IPNI=*291320-2* 406
Viola semperflorens IPNI=*868828-1* 406
Viola sicula IPNI=*869428-1* 406
Viola soongorica IPNI=*869428-1* 406
Viola spectabilis IPNI=*868828-1* 406
Viola stevenii IPNI=*868828-1* 406
Viola suaveolens IPNI=*868828-1* 406
Viola suavissima IPNI=*868828-1* 406
Viola subcarnea IPNI=*868828-1* 406
Viola subtilis IPNI=*869428-1* 406
Viola sulfurea IPNI=*868828-1* 406
Viola tenerrima IPNI=*868309-1* 406
Viola timbalii IPNI=*291320-2* 406
Viola tricolor f. *alba* IPNI=*869428-1* 406
Viola tricolor f. *tenella* IPNI=*869428-1* 406
Viola tricolor var. *cantiana* IPNI=*869428-1* 406
Viola tricolor var. *carpatica* IPNI=*869428-1* 406
Viola tricolor var. *lejeunei* IPNI=*869428-1* 406
Viola tricolor var. *orcadensis* IPNI=*869428-1* 406
Viola tricolor var. *raiblensis* IPNI=*869428-1* 406
Viola variata IPNI=*869428-1* 406
Viola variegata IPNI=*869428-1* 406
Viola vectensis IPNI=*869428-1* 406
Viola venezuelensis IPNI=*77143568-1* 405
Viola versicolor IPNI=*869428-1* 406
Viola vilmoriniana IPNI=*868828-1* 406
Viola vinealis IPNI=*868828-1* 406
Viola virgata IPNI=*869428-1* 406
Viola vivariensis IPNI=*869428-1* 406
Viola wiedemannii IPNI=*868828-1* 406
Virchowia cubensis IPNI=*1061345-2* 374
Virecta pratensis IPNI=*766491-1* 357
Vireya sanguinolenta IPNI=*378985-1* 367
Virgilia helioides IPNI=*329185-2* 263
Virgularia lanceolata IPNI=*6185-2* 375
Virgularia revoluta IPNI=*6185-2* 375
Virola atopa IPNI=*64003-2* 190
Virola boliviensis IPNI=*267199-2* 192
Virola carinata var. *gracilis* IPNI=*267202-2* 192
Virola elliptica IPNI=*267190-2* 192
Virola glaziovii IPNI=*267202-2* 192
Virola glaziovii var. *latifolia* IPNI=*267202-2* 192
Virola incolor IPNI=*267140-2* 191
Virola kukachkana IPNI=*267133-2* 191

Virola lepidota IPNI=*267140-2* 191
Virola merendonis IPNI=*267165-2* 192
Virola mocoa IPNI=*267199-2* 192
Virola mycetis IPNI=*267199-2* 192
Virola panamensis IPNI=*267199-2* 192
Virola papillosa IPNI=*130202-2* 190
Virola peruviana var. *tomentosa* IPNI=*267199-2* 192
Virola rufula IPNI=*267203-2* 192
Virola venezuelensis IPNI=*267199-2* 192
Virola venosa var. *martii* IPNI=*267142-2* 191
Virola venosa var. *pavonis* IPNI=*267190-2* 192
Virola villosa IPNI=*267168-2* 192
Virola warburgii IPNI=*267199-2* 192
Viscum angustifolium IPNI=*60443757-2* 450
Viscum buxifolium IPNI=*194192-2* 450
Viscum chrysostachyum IPNI=*77735-2* 449
Viscum clavatum IPNI=*77736-2* 449
Viscum crassifolium IPNI=*550798-1* 449
Viscum crassuloides IPNI=*77736-2* 449
Viscum dichotomum IPNI=*194192-2* 450
Viscum fockeanum IPNI=*194192-2* 450
Viscum globuliflorum IPNI=*77735-2* 449
Viscum jamaicense IPNI=*1075239-2* 450
Viscum kunthianum IPNI=*60443757-2* 450
Viscum latifolium IPNI=*194192-2* 450
Viscum laurifolium IPNI=*194192-2* 450
Viscum ligustrinum IPNI=*176899-2* 448
Viscum macrophyllum IPNI=*329775-2* 449
Viscum martinicense IPNI=*550798-1* 449
Viscum mucronatum IPNI=*194156-2* 450
Viscum myrtilloides IPNI=*1075239-2* 450
Viscum oblongifolium IPNI=*1075239-2* 450
Viscum piperoides IPNI=*194192-2* 450
Viscum portulacoides IPNI=*77736-2* 449
Viscum quadrangulare IPNI=*60443757-2* 450
Viscum randiae IPNI=*60443757-2* 450
Viscum salicifolium IPNI=*60443757-2* 450
Viscum saururoides IPNI=*194192-2* 450
Viscum schottii IPNI=*194192-2* 450
Viscum tereticaule IPNI=*194192-2* 450
Viscum tereticaule var. *cubense* IPNI=*194192-2* 450
Viscum tetragonum IPNI=*1075239-2* 450
Viscum toruloides IPNI=*194192-2* 450
Viscum torulosum IPNI=*194192-2* 450
Viscum trigonum IPNI=*60443757-2* 450
Viscum trinervium IPNI=*1075239-2* 450
Viscum undulatum IPNI=*1006438-2* 450
Viscum verticillatum IPNI=*1075239-2* 450
Visenia caracasana IPNI=*824432-1* 411
Visenia crenata IPNI=*157023-2* 412
Visenia depressa IPNI=*133833-2* 410
Visenia hirsuta IPNI=*157043-2* 412
Visenia nodiflora IPNI=*824514-1* 411
Visenia polystachya IPNI=*157002-2* 412
Visenia sericea IPNI=*157002-2* 412
Visenia serrata IPNI=*157043-2* 412
Visenia spicata IPNI=*561999-1* 410
Visenia ulmariae IPNI=*157002-2* 412
Visenia venosa IPNI=*157002-2* 412
Visiania elastica IPNI=*60458499-2* 438
Visiania paniculata IPNI=*610017-1* 375
Vismia acuminata IPNI=*30051774-2* 394
Vismia acuminata var. *caparosa* IPNI=*30051774-2* 394
Vismia angusta IPNI=*434144-1* 394
Vismia baccifera subsp. *ferruginea* IPNI=*434114-1* 394
Vismia baccifera subsp. *subcuneata* IPNI=*434138-1* 394
Vismia baccifera var. *angustifolia* IPNI=*30051774-2* 394
Vismia calvescens IPNI=*434138-1* 394
Vismia caparosa IPNI=*30051774-2* 394
Vismia cearensis IPNI=*30051774-2* 394
Vismia guaramirangae IPNI=*30051774-2* 394
Vismia guianensis var. *acuminata* IPNI=*30051774-2* 394
Vismia guianensis var. *glabrata* IPNI=*30051774-2* 394
Vismia guttifera IPNI=*434100-1* 394
Vismia latifolia IPNI=*30051774-2* 394
Vismia latifolia var. *acuminata* IPNI=*30051774-2* 394
Vismia latifolia var. *reticulata* IPNI=*434144-1* 394
Vismia laurifolia IPNI=*434138-1* 394
Vismia macrophylla var. *glabrescens* IPNI=*434144-1* 394
Vismia reichardtiana IPNI=*30051774-2* 394
Vismia reichardtiana var. *polymera* IPNI=*30051774-2* 394
Vismia reticulata IPNI=*434144-1* 394
Vismia rufescens IPNI=*434162-1* 394
Vismia rufescens var. *sessilifolia* IPNI=*434162-1* 394
Vismia schultesii IPNI=*434114-1* 394
Vismia sieberiana IPNI=*30051774-2* 394
Vismia subcuneata IPNI=*434138-1* 394
Visnaga major IPNI=*837486-1* 251
Visnaga vera IPNI=*849956-1* 253
Visnaga vulgaris IPNI=*837486-1* 251
Vitellaria campechiana IPNI=*209714-2* 306
Vitellaria glabrescens IPNI=*89008-2* 305
Vitellaria glaucophylla IPNI=*209808-2* 307
Vitellaria mammosa IPNI=*152641-2* 306
Vitellaria multiflora IPNI=*209824-2* 307
Vitellaria nervosa IPNI=*209714-2* 306
Vitellaria procera IPNI=*943696-1* 307
Vitellaria pulverulenta IPNI=*209723-2* 306
Vitellaria rivicoa IPNI=*209808-2* 307
Vitellaria salicifolia IPNI=*209714-2* 306
Vitellaria sphaerocarpa IPNI=*209714-2* 306
Vitellaria tenuifolia IPNI=*209714-2* 306
Vitellaria venosa IPNI=*209920-2* 308
Vitex agnus IPNI=*865568-1* 373
Vitex agnus-castus f. *caerulea* IPNI=*865568-1* 373
Vitex agnus-castus f. *laciniosa* IPNI=*865568-1* 373
Vitex berteroana IPNI=*865663-1* 373
Vitex bignonioides IPNI=*865638-1* 373
Vitex brasiliensis IPNI=*865638-1* 373
Vitex brittoniana IPNI=*865663-1* 373
Vitex capitata f. *albiflora* IPNI=*865638-1* 373
Vitex compressa f. *angustifolia* IPNI=*865663-1* 373
Vitex cymosa f. *albiflora* IPNI=*865678-1* 373
Vitex discolor IPNI=*865678-1* 373
Vitex divaricata var. *cubensis* IPNI=*865688-1* 373
Vitex divaricata var. *haitiensis* IPNI=*865688-1* 373
Vitex geniculata IPNI=*865916-1* 374
Vitex gigantea var. *congestiflora* IPNI=*865730-1* 374
Vitex glaberrima IPNI=*865916-1* 374
Vitex glandulosa IPNI=*865916-1* 374
Vitex grandiflora IPNI=*77141989-1* 363
Vitex hybrida IPNI=*865568-1* 373
Vitex integra IPNI=*865568-1* 373
Vitex latifolia IPNI=*865568-1* 373
Vitex leucoxylon IPNI=*865916-1* 374
Vitex littoralis IPNI=*865916-1* 374
Vitex longiracemosa IPNI=*265966-2* 374
Vitex lupinifolia IPNI=*865568-1* 373
Vitex moronensis IPNI=*77085096-1* 364
Vitex odorata IPNI=*865718-1* 373
Vitex petitia IPNI=*864000-1* 372
Vitex pseudonegundo IPNI=*865568-1* 373
Vitex robusta IPNI=*865568-1* 373
Vitex rusbyi IPNI=*865718-1* 373
Vitex timoriensis IPNI=*865916-1* 374
Vitex triflora f. *quinquefoliolata* IPNI=*866038-1* 374

Vitex triflora var. *angustiloba* IPNI=*866038-1* 374
Vitex triflora var. *coriacea* IPNI=*866038-1* 374
Vitex triflora var. *floribunda* IPNI=*866038-1* 374
Vitex triflora var. *hirsuta* IPNI=*866038-1* 374
Vitex triflora var. *kraatzii* IPNI=*866038-1* 374
Vitex triflora var. *quinquefoliolata* IPNI=*866038-1* 374
Vitex triflora var. *tenuifolia* IPNI=*866038-1* 374
Vitex verticillata IPNI=*865568-1* 373
Vitex wittrockiana IPNI=*865638-1* 373
Vittaria angustifrons IPNI=*17234040-1* 178
Vittaria costaricensis IPNI=*17234040-1* 178
Vittaria curvidentata IPNI=*17234040-1* 178
Vittaria deppeana IPNI=*17234040-1* 178
Vittaria dimorpha IPNI=*266216-2* 178
Vittaria filifolia IPNI=*266216-2* 178
Vittaria filiformis IPNI=*17234040-1* 178
Vittaria lanceolata IPNI=*17209080-1* 178
Vittaria lineata var. *curvidentata* IPNI=*17234040-1* 178
Vittaria lineata var. *graminifolia* IPNI=*266216-2* 178
Vittaria pachydictyon IPNI=*17234040-1* 178
Vittaria schkuhrii IPNI=*17234040-1* 178
Vittaria setacea IPNI=*17234040-1* 178
Vittaria spartimorpha IPNI=*266216-2* 178
Vochysia aurea IPNI=*266313-2* 430
Vochysia hondurensis IPNI=*266313-2* 430
Vochysia hondurensis var. *parvifolia* IPNI=*266313-2* 430
Vochysia macrophylla IPNI=*266313-2* 430
Vochysia polyantha IPNI=*266333-2* 430
Vochysia tintin IPNI=*266331-2* 430
Vochysia urubuensis IPNI=*871092-1* 430
Volkameria angulata IPNI=*862301-1* 369
Volkameria angustifolia IPNI=*866088-1* 374
Volkameria cordifolia IPNI=*862461-1* 369
Volkameria diversifolia IPNI=*862301-1* 369
Volkameria fragrans IPNI=*946989-1* 369
Volkameria linifolia IPNI=*866088-1* 374
Volkameria orientalis IPNI=*675971-1* 375
Volkameria sesamoides IPNI=*675971-1* 375
Volkmannia japonica IPNI=*946989-1* 369
Volutella aphylla IPNI=*463202-1* 182
Vouacapoua amazonum IPNI=*524446-1* 336
Vouacapoua inermis IPNI=*1075415-2* 311
Vouacapoua retusa IPNI=*12607-2* 311
Vouacapoua surinamensis IPNI=*12607-2* 311
Vouapa acaciifolia IPNI=*505635-1* 326
Vouapa bifolia IPNI=*30147160-2* 326
Vouapa elegans IPNI=*30147160-2* 326
Vouapa multijuga IPNI=*505740-1* 326
Vouapa staminea IPNI=*30147160-2* 326
Voyria araguensis IPNI=*1167766-2* 345
Voyria pallida IPNI=*1167766-2* 345
Voyria uniflora var. *pallida* IPNI=*1167766-2* 345
Vriesea longibracteata IPNI=*77158988-1* 229
Vriesea speciosa IPNI=*77158988-1* 229
Vriesea splendens IPNI=*77158988-1* 229
Vriesea splendens var. *chlorostachya* IPNI=*77158988-1* 229
Vriesea splendens var. *formosa* IPNI=*77158988-1* 229
Vriesea splendens var. *longibracteata* IPNI=*77158988-1* 229
Vriesea splendens var. *oinochroma* IPNI=*77158988-1* 229
Vriesea splendens var. *striatifolia* IPNI=*77158988-1* 229
Vriesea utriculata IPNI=*254406-2* 229
Vulpia alpina IPNI=*403134-1* 239
Vulpia bromoides f. *gracilis* IPNI=*402472-1* 239
Vulpia bromoides IPNI=*402472-1* 239
Vulpia ciliata IPNI=*403134-1* 239
Vulpia crinita IPNI=*403134-1* 239
Vulpia dertonensis IPNI=*402472-1* 239
Vulpia exserta IPNI=*402472-1* 239
Vulpia granulata IPNI=*402472-1* 239
Vulpia longivaginata IPNI=*403134-1* 239
Vulpia major IPNI=*403134-1* 239
Vulpia megalura IPNI=*403134-1* 239
Vulpia murorum IPNI=*403134-1* 239
Vulpia myuros f. *exserens* IPNI=*403134-1* 239
Vulpia myuros f. *megalura* IPNI=*403134-1* 239
Vulpia myuros f. *racemifera* IPNI=*403134-1* 239
Vulpia myuros IPNI=*403134-1* 239
Vulpia myuros subsp. *megalura* IPNI=*403134-1* 239
Vulpia myuros var. *megalura* IPNI=*403134-1* 239
Vulpia panormitana IPNI=*402472-1* 239
Vulpia pilosa IPNI=*403134-1* 239
Vulpia reclinata IPNI=*403134-1* 239
Vulpia sciuroides IPNI=*402472-1* 239
Vulpia uniglumis IPNI=*402472-1* 239
Vulpia vaginata IPNI=*403134-1* 239
Vulvaria albescens IPNI=*55993-2* 282
Vulvaria ambrosioides IPNI=*1198060-2* 282
Vulvaria trachisperma IPNI=*77121007-1* 282

W

Wachendorfia orinocensis IPNI=*429893-1* 224
Wageneria deflexa IPNI=*104759-1* 292
Wageneria glabra IPNI=*104759-1* 292
Wageneria lucida IPNI=*104759-1* 292
Wageneria montana IPNI=*104759-1* 292
Wahlenbergia globularis IPNI=*203207-1* 261
Waitzia odorata IPNI=*437030-1* 216
Waitzia refracta IPNI=*437030-1* 216
Waldensia lupulina IPNI=*303502-2* 436
Wallenia angustifolia IPNI=*18168-2* 303
Wallenia laxiflora IPNI=*73568-2* 304
Waltheria africana IPNI=*320781-2* 415
Waltheria americana IPNI=*320781-2* 415
Waltheria americana var. *elliptica* IPNI=*320781-2* 415
Waltheria americana var. *glandulosa* IPNI=*320781-2* 415
Waltheria angustifolia IPNI=*320781-2* 415
Waltheria arborescens IPNI=*320781-2* 415
Waltheria corchorifolia IPNI=*320781-2* 415
Waltheria debilis IPNI=*320781-2* 415
Waltheria elliptica IPNI=*320781-2* 415
Waltheria erioclada IPNI=*320781-2* 415
Waltheria guineensis IPNI=*320781-2* 415
Waltheria hirsuta IPNI=*825848-1* 415
Waltheria indica var. *prostrata* IPNI=*320781-2* 415
Waltheria laxa IPNI=*320781-2* 415
Waltheria machrisiana IPNI=*825848-1* 415
Waltheria makinoi IPNI=*320781-2* 415
Waltheria martii IPNI=*320781-2* 415
Waltheria obtusa IPNI=*320781-2* 415
Waltheria paniculata IPNI=*320781-2* 415
Waltheria pauciflora IPNI=*320781-2* 415
Waltheria pedunculata IPNI=*320781-2* 415
Waltheria prostrata IPNI=*320781-2* 415
Waltheria subcordata IPNI=*60439289-2* 415
Waltheria terminalis IPNI=*836702-1* 401
Waltheria tubiflora IPNI=*825848-1* 415
Waltheria wildii IPNI=*320781-2* 415
Warczewiczella velata IPNI=*661384-1* 223
Waria zeylanica IPNI=*76202-1* 190
Warmingia pauciflora IPNI=*71494-1* 451
Warneria augusta IPNI=*751073-1* 350
Warrea marginata IPNI=*661384-1* 223
Warrea quadrata IPNI=*661384-1* 223
Warszewiczia macrophylla IPNI=*770293-1* 358
Warszewiczia maynensis IPNI=*770293-1* 358
Warszewiczia obovata IPNI=*43052-2* 347

Warszewiczia poeppigiana IPNI=*770293-1* 358
Warszewiczia pulcherrima IPNI=*770293-1* 358
Warszewiczia schomburgkiana IPNI=*770293-1* 358
Warszewiczia splendens IPNI=*770293-1* 358
Washingtonia filamentosa IPNI=*267788-2* 212
Washingtonia filifera var. *gracilis* IPNI=*670278-1* 212
Washingtonia filifera var. *microsperma* IPNI=*267788-2* 212
Washingtonia filifera var. *robusta* IPNI=*670278-1* 212
Washingtonia filifera var. *sonorae* IPNI=*670278-1* 212
Washingtonia filifera var. *typica* IPNI=*267788-2* 212
Washingtonia gracilis IPNI=*670278-1* 212
Washingtonia mexicana IPNI=*845560-1* 252
Washingtonia pacifica IPNI=*669288-1* 211
Washingtonia robusta var. *gracilis* IPNI=*670278-1* 212
Washingtonia sonorae IPNI=*670278-1* 212
Watsonamra alfaroana IPNI=*760198-1* 355
Watsonamra brachyotis IPNI=*760190-1* 355
Watsonamra macrophylla IPNI=*760184-1* 355
Watsonamra pinnatifida IPNI=*760190-1* 355
Watsonamra pittieri IPNI=*760190-1* 355
Watsonamra pubescens IPNI=*760184-1* 355
Watsonamra spathicalyx IPNI=*188124-2* 355
Watsonamra tinajita IPNI=*760198-1* 355
Webera quinquenervia IPNI=*1157663-2* 418
Wedelia acapulcensis var. *cintalapana* IPNI=*30439090-2* 272
Wedelia acapulcensis var. *parviceps* IPNI=*30439090-2* 272
Wedelia acapulcensis var. *ramosissima* IPNI=*30439090-2* 272
Wedelia adhaerens IPNI=*30439090-2* 272
Wedelia annua IPNI=*213524-1* 264
Wedelia brasiliensis IPNI=*1093589-2* 270
Wedelia carnosa IPNI=*1093589-2* 270
Wedelia carnosa triloba IPNI=*1093589-2* 270
Wedelia crenata IPNI=*1093589-2* 270
Wedelia cristata IPNI=*8606-2* 286
Wedelia cryptocephala IPNI=*252024-1* 271
Wedelia discoidea IPNI=*202984-1* 261
Wedelia frutescens IPNI=*60468294-2* 272
Wedelia hispida var. *ramosissima* IPNI=*30439090-2* 272
Wedelia incarnata IPNI=*8606-2* 286
Wedelia incarnata subsp. *anodonta* IPNI=*8606-2* 286
Wedelia malachroides IPNI=*8606-2* 286
Wedelia minor IPNI=*156179-2* 265
Wedelia ovatifolia IPNI=*156179-2* 265
Wedelia paludicola IPNI=*1093589-2* 270
Wedelia paludosa IPNI=*1093589-2* 270
Wedelia paludosa var. *latifolia* IPNI=*1093589-2* 270
Wedelia paludosa var. *vialis* IPNI=*1093589-2* 270
Wedelia parviceps IPNI=*30439090-2* 272
Wedelia pedunculata IPNI=*1093589-2* 270
Wedelia pinetorum IPNI=*30439090-2* 272
Wedelia psammophila IPNI=*1186403-2* 261
Wedelia pulchella IPNI=*60468294-2* 272
Wedelia tannensis IPNI=*1093589-2* 270
Wedelia trilobata IPNI=*1093589-2* 270
Wedelia trilobata var. *hirtella* IPNI=*1093589-2* 270
Wedelia trilobata var. *pilosissima* IPNI=*1093589-2* 270
Wedeliella cristata IPNI=*8606-2* 286
Wedeliella incarnata IPNI=*8606-2* 286
Wedeliella incarnata var. *anodonta* IPNI=*8606-2* 286
Weigeltia albiflora IPNI=*73545-2* 304
Weigeltia kayapii IPNI=*1006171-1* 304
Weigeltia panamensis IPNI=*73569-2* 304
Weigeltia purpurea IPNI=*73569-2* 304
Weigeltia schlimii IPNI=*73559-2* 304
Weigeltia schlimii var. *intermedia* IPNI=*73559-2* 304
Weigeltia spectabilis IPNI=*73559-2* 304
Weigeltia sprucei IPNI=*73569-2* 304
Weigeltia triandra IPNI=*73559-2* 304
Weilbachia pustulata IPNI=*105525-1* 292
Weinmannia alpestris IPNI=*268273-2* 431
Weinmannia antioquensis IPNI=*268273-2* 431
Weinmannia auriculata var. *glabra* IPNI=*268187-2* 431
Weinmannia baccariniana var. *caracasana* IPNI=*268190-2* 431
Weinmannia baezana IPNI=*268284-2* 431
Weinmannia balbisiana var. *bangiana* IPNI=*319003-2* 431
Weinmannia balbisiana var. *lamprophylla* IPNI=*319003-2* 431
Weinmannia balbisiana var. *moritzii* IPNI=*319003-2* 431
Weinmannia balbisiana var. *ptariana* IPNI=*268187-2* 431
Weinmannia balbisiana var. *roraimensis* IPNI=*268187-2* 431
Weinmannia balbisiana var. *subovata* IPNI=*319003-2* 431
Weinmannia cardonae IPNI=*268187-2* 431
Weinmannia caripensis IPNI=*268273-2* 431
Weinmannia caucana IPNI=*794947-1* 431
Weinmannia engleriana IPNI=*268241-2* 431
Weinmannia fagaroides var. *bogotensis* IPNI=*268190-2* 431
Weinmannia glabra var. *antioquensis* IPNI=*268273-2* 431
Weinmannia glabra var. *caripensis* IPNI=*268273-2* 431
Weinmannia glabra var. *mexicana* IPNI=*268273-2* 431
Weinmannia guanacasana IPNI=*268241-2* 431
Weinmannia hirta f. *nitida* IPNI=*268273-2* 431
Weinmannia hirta IPNI=*268273-2* 431
Weinmannia hirta var. *antillana* IPNI=*268273-2* 431
Weinmannia hirta var. *brasiliensis* IPNI=*268273-2* 431
Weinmannia hirta var. *glabra* IPNI=*268195-2* 431
Weinmannia hirtella IPNI=*268273-2* 431
Weinmannia intermedia var. *pittieri* IPNI=*268273-2* 431
Weinmannia lamprophylla IPNI=*319003-2* 431
Weinmannia laxiflora f. *minor* IPNI=*268190-2* 431
Weinmannia laxiflora IPNI=*268273-2* 431
Weinmannia liebmannii IPNI=*268273-2* 431
Weinmannia microphylla var. *caracasana* IPNI=*268190-2* 431
Weinmannia microphylla var. *fagaroides* IPNI=*268190-2* 431
Weinmannia moritzii IPNI=*319003-2* 431
Weinmannia neblinensis IPNI=*268284-2* 431
Weinmannia nebularum IPNI=*794947-1* 431
Weinmannia nervosa IPNI=*268273-2* 431
Weinmannia nitida IPNI=*268273-2* 431
Weinmannia ovalis var. *elliptica* IPNI=*268187-2* 431
Weinmannia ovalis var. *penicillata* IPNI=*268187-2* 431
Weinmannia ovalis var. *petiolata* IPNI=*268187-2* 431
Weinmannia ovalis var. *roraimensis* IPNI=*268187-2* 431
Weinmannia penicillata IPNI=*268187-2* 431
Weinmannia pinnata f. *nitida* IPNI=*268273-2* 431
Weinmannia pinnata var. *caripensis* IPNI=*268273-2* 431
Weinmannia roraimensis IPNI=*268190-2* 431
Weinmannia schomburgkii IPNI=*268187-2* 431
Weinmannia silvatica IPNI=*794947-1* 431
Weinmannia stuebelii IPNI=*268273-2* 431
Weinmannia tamana IPNI=*268190-2* 431
Weinmannia weberbaueri IPNI=*268273-2* 431
Welfia microcarpa IPNI=*670285-1* 212
Werauhia splendens IPNI=*77158988-1* 229
Werneria brachypappa IPNI=*260433-1* 272
Werneria cherlerioides IPNI=*260433-1* 272
Werneria cornea IPNI=*260433-1* 272
Werneria minima IPNI=*260433-1* 272
Werneria minima var. *pygmaea* IPNI=*260433-1* 272
Werneria pygmaea var. *caespitosa* IPNI=*260433-1* 272
Werneria pygmaea var. *cylindrica* IPNI=*260433-1* 272
Werneria pygmaea var. *iodopappus* IPNI=*260433-1* 272
Werneria pygmaea var. *praemorsa* IPNI=*260433-1* 272
Werneria pygmaea var. *remyi* IPNI=*260433-1* 272
Werneria rhizoma IPNI=*260433-1* 272
Wettinella maynensis IPNI=*670297-1* 213
Wettinella quinaria IPNI=*268491-2* 213
Wettinia cladospadix IPNI=*268487-2* 212

Wettinia illaqueans IPNI=*670297-1* 213
Wettinia poeppigii IPNI=*670291-1* 212
Wettinia utilis IPNI=*268491-2* 213
Wettinia weberbaueri IPNI=*670291-1* 212
Wettiniicarpus cladospadix IPNI=*268487-2* 212
Wettiniicarpus fascicularis IPNI=*268487-2* 212
Wibelia guianensis IPNI=*867031-1* 405
Wiborgia acmelea IPNI=*208554-1* 263
Wiborgia brachystephana IPNI=*208557-1* 263
Wiborgia oblongifolia IPNI=*1186403-2* 261
Wiborgia parviflora IPNI=*208554-1* 263
Wiborgia urticifolia IPNI=*208557-1* 263
Widdringtonia glauca IPNI=*677124-1* 180
Wigandia caracasana f. *viscosa* IPNI=*114242-3* 276
Wigandia caracasana IPNI=*114242-3* 276
Wigandia caracasana var. *calycina* IPNI=*114242-3* 276
Wigandia caracasana var. *eucaracasana* IPNI=*114242-3* 276
Wigandia caracasana var. *macrophylla* IPNI=*114242-3* 276
Wigandia caracasana var. *viscosa* IPNI=*114242-3* 276
Wigandia darii IPNI=*114242-3* 276
Wigandia kunthii var. *eukunthii* IPNI=*114242-3* 276
Wigandia kunthii var. *intermedia* IPNI=*114242-3* 276
Wigandia kunthii var. *macrophylla* IPNI=*114242-3* 276
Wigandia kunthii var. *viscosa* IPNI=*114242-3* 276
Wigandia macrophylla IPNI=*114242-3* 276
Wigandia peruviana IPNI=*114242-3* 276
Wigandia scorpioides IPNI=*114242-3* 276
Wigandia urens f. *africana* IPNI=*114242-3* 276
Wigandia urens IPNI=*114242-3* 276
Wigandia urens var. *caracasana* IPNI=*114242-3* 276
Wigandia vigieri IPNI=*114242-3* 276
Wiggersia lathyroides IPNI=*524865-1* 336
Wiggersia minima IPNI=*524865-1* 336
Wikstroemia camelliifolia IPNI=*112597-2* 309
Wikstroemia fruticosa IPNI=*112597-2* 309
Wikstroemia fruticosa var. *acutifolia* IPNI=*112597-2* 309
Wikstroemia fruticosa var. *communis* IPNI=*112597-2* 309
Wikstroemia fruticosa var. *microphylla* IPNI=*112597-2* 309
Wikstroemia fruticosa var. *obovata* IPNI=*112597-2* 309
Wikstroemia fruticosa var. *sericea* IPNI=*112597-2* 309
Wikstroemia fruticosa var. *typica* IPNI=*112597-2* 309
Wikstroemia intermedia IPNI=*112597-2* 309
Wikstroemia parviflora IPNI=*112597-2* 309
Wikstroemia pubescens IPNI=*112597-2* 309
Wikstroemia quinoderma IPNI=*112597-2* 309
Wikstroemia speciosa IPNI=*112597-2* 309
Wikstroemia symplocoides IPNI=*112597-2* 309
Willdampia formosa IPNI=*948198-1* 334
Willdenovia schreberi IPNI=*761025-1* 355
Willoughbya argyrostigma IPNI=*232913-1* 266
Willoughbya aspera IPNI=*232913-1* 266
Willoughbya attenuata IPNI=*232913-1* 266
Willoughbya badieri IPNI=*232937-1* 266
Willoughbya banisteriae IPNI=*232743-1* 266
Willoughbya brachiata IPNI=*232913-1* 266
Willoughbya caudata IPNI=*232743-1* 266
Willoughbya cissampelina IPNI=*232826-1* 266
Willoughbya cordifolia IPNI=*232826-1* 266
Willoughbya divaricata IPNI=*233110-1* 266
Willoughbya ferruginea IPNI=*232743-1* 266
Willoughbya gabrieli IPNI=*232743-1* 266
Willoughbya gracilis IPNI=*232937-1* 266
Willoughbya guaco IPNI=*232913-1* 266
Willoughbya halei IPNI=*232826-1* 266
Willoughbya heterophylla IPNI=*161051-2* 266
Willoughbya hirsutissima IPNI=*232743-1* 266
Willoughbya imrayana IPNI=*232937-1* 266
Willoughbya lanuginosa IPNI=*232743-1* 266
Willoughbya lehmannii IPNI=*161007-2* 266
Willoughbya lindleyana IPNI=*233001-1* 266
Willoughbya loxensis IPNI=*232826-1* 266
Willoughbya micrantha IPNI=*161051-2* 266
Willoughbya parviflora IPNI=*233110-1* 266
Willoughbya parviflora var. *guaco* IPNI=*232913-1* 266
Willoughbya pilosa IPNI=*232743-1* 266
Willoughbya platyphylla IPNI=*232937-1* 266
Willoughbya polystachya IPNI=*233144-1* 266
Willoughbya rotunda IPNI=*233144-1* 266
Willoughbya ruiziana IPNI=*232743-1* 266
Willoughbya scabra IPNI=*233144-1* 266
Willoughbya scandens var. *congesta* IPNI=*232821-1* 266
Willoughbya scandens var. *orinocensis* IPNI=*161051-2* 266
Willoughbya scandens var. *pubescens* IPNI=*232826-1* 266
Willoughbya selloi IPNI=*233209-1* 266
Willoughbya stipitata IPNI=*233110-1* 266
Willoughbya variabilis IPNI=*161051-2* 266
Willoughbya vitifolia IPNI=*161327-2* 266
Willughbeia guianensis IPNI=*80734-1* 342
Willughbeia scandens IPNI=*80734-1* 342
Wilmattea venezuelensis IPNI=*77163180-1* 285
Windmannia balbisiana IPNI=*319003-2* 431
Windmannia elliptica IPNI=*268187-2* 431
Windmannia engleriana IPNI=*268241-2* 431
Windmannia fagaroides IPNI=*268190-2* 431
Windmannia glabra IPNI=*268195-2* 431
Windmannia guanacasana IPNI=*268241-2* 431
Windmannia hirtella IPNI=*268273-2* 431
Windmannia nitida IPNI=*268273-2* 431
Windmannia pinnata IPNI=*268273-2* 431
Windmannia pubescens IPNI=*268284-2* 431
Windmannia tomentosa IPNI=*268326-2* 431
Wintera aromatica IPNI=*330748-2* 181
Wintera granadensis IPNI=*84100-2* 181
Wissadula amplissima f. *lobulata* IPNI=*268896-2* 415
Wissadula amplissima var. *typica* IPNI=*268896-2* 415
Wissadula elongata IPNI=*36556-2* 407
Wissadula gymnostachya IPNI=*36556-2* 407
Wissadula heterosperma IPNI=*268896-2* 415
Wissadula hirsuta IPNI=*268896-2* 415
Wissadula jamesonii IPNI=*36556-2* 407
Wissadula leschenaultiana IPNI=*268909-2* 415
Wissadula periplocifolia var. *antillarum* IPNI=*1213890-2* 415
Wissadula periplocifolia var. *fadyenii* IPNI=*1213890-2* 415
Wissadula periplocifolia var. *genuina* IPNI=*1213890-2* 415
Wissadula periplocifolia var. *gracillima* IPNI=*1213890-2* 415
Wissadula periplocifolia var. *heterosperma* IPNI=*1213890-2* 415
Wissadula periplocifolia var. *luciana* IPNI=*268909-2* 415
Wissadula spicata IPNI=*36556-2* 407
Wissadula spiciflora IPNI=*36556-2* 407
Wissadula zeylanica IPNI=*1213890-2* 415
Wisteria dubia IPNI=*486606-1* 315
Witheringia salicifolia IPNI=*800802-1* 377
Wittelsbachia orinocensis IPNI=*111532-1* 406
Wittelsbachia vitifolia IPNI=*111539-1* 406
Wittia amazonica IPNI=*60426912-2* 284
Wittia panamensis IPNI=*60426912-2* 284
Wittiocactus amazonicus IPNI=*60426912-2* 284
Wittiocactus panamensis IPNI=*60426912-2* 284
Wolffia schleidenii IPNI=*1135607-2* 202
Wulffia angustifolia IPNI=*156327-2* 266
Wulffia baccata IPNI=*993359-1* 271
Wulffia baccata var. *vincentina* IPNI=*993359-1* 271
Wulffia blanchetii IPNI=*993359-1* 271
Wulffia capitata IPNI=*993359-1* 271
Wulffia deltoidea IPNI=*156327-2* 266
Wulffia elongata IPNI=*993359-1* 271

Wulffia hastata IPNI=*156327-2* 266
Wulffia havanensis IPNI=*993359-1* 271
Wulffia longifolia IPNI=*993359-1* 271
Wulffia maculata IPNI=*993359-1* 271
Wulffia maculata var. *oblongifolia* IPNI=*993359-1* 271
Wulffia oblongifolia IPNI=*993359-1* 271
Wulffia platyglossa IPNI=*993359-1* 271
Wulffia quitensis IPNI=*993359-1* 271
Wulffia salzmannii IPNI=*993359-1* 271
Wulffia scandens IPNI=*993359-1* 271
Wulffia sodiroi IPNI=*60769-2* 260
Wulffia stenoglossa IPNI=*993359-1* 271
Wulffia strigosa IPNI=*993359-1* 271
Wulffia suffruticosa IPNI=*993359-1* 271
Wydleria portoricensis IPNI=*60442790-2* 252

X
Xamesike depressa IPNI=*347236-1* 390
Xamesike littoralis IPNI=*347236-1* 390
Xamesike maculata IPNI=*347236-1* 390
Xamesike supina IPNI=*347236-1* 390
Xanthidium artemisioides IPNI=*176055-1* 256
Xanthium fruticosum IPNI=*176055-1* 256
Xanthium spinosum f. *typicum* IPNI=*260892-1* 272
Xanthocromyon herbertii IPNI=*441642-1* 217
Xanthonanthos odoratus IPNI=*389034-1* 234
Xanthosoma angustisectum IPNI=*89333-1* 203
Xanthosoma appendiculatum IPNI=*89373-1* 203
Xanthosoma atrovirens IPNI=*89373-1* 203
Xanthosoma atrovirens var. *appendiculatum* IPNI=*89373-1* 203
Xanthosoma atrovirens var. *hybridum* IPNI=*89373-1* 203
Xanthosoma atrovirens var. *kochii* IPNI=*89373-1* 203
Xanthosoma atrovirens var. *moritzii* IPNI=*89373-1* 203
Xanthosoma atrovirens var. *panduriforme* IPNI=*89373-1* 203
Xanthosoma atrovirens var. *versicolor* IPNI=*89373-1* 203
Xanthosoma aureum IPNI=*89373-1* 203
Xanthosoma belophyllum var. *caracassanum* IPNI=*89310-1* 202
Xanthosoma belophyllum var. *kunthii* IPNI=*89310-1* 202
Xanthosoma blandum IPNI=*89373-1* 203
Xanthosoma caracasanum IPNI=*89310-1* 202
Xanthosoma croatanum IPNI=*89347-1* 203
Xanthosoma edule IPNI=*89373-1* 203
Xanthosoma hastatum IPNI=*269304-2* 202
Xanthosoma hastifolium IPNI=*269304-2* 202
Xanthosoma helleborifolium f. *variegatum* IPNI=*89333-1* 203
Xanthosoma helleborifolium f. *viride* IPNI=*89333-1* 203
Xanthosoma helleborifolium var. *angustisectum* IPNI=*89333-1* 203
Xanthosoma helleborifolium var. *weberbaueri* IPNI=*89333-1* 203
Xanthosoma holtonianum IPNI=*89347-1* 203
Xanthosoma ianthinum IPNI=*89373-1* 203
Xanthosoma jacquinii IPNI=*89373-1* 203
Xanthosoma kerberi IPNI=*89347-1* 203
Xanthosoma lindenii IPNI=*40313-2* 199
Xanthosoma maculatum IPNI=*89373-1* 203
Xanthosoma mafaffa IPNI=*89373-1* 203
Xanthosoma mafaffa var. *blandum* IPNI=*89373-1* 203
Xanthosoma mafaffa var. *typicum* IPNI=*89373-1* 203
Xanthosoma monstruosum IPNI=*89373-1* 203
Xanthosoma nigrum IPNI=*89373-1* 203
Xanthosoma obtusilobum IPNI=*89373-1* 203
Xanthosoma panduriforme IPNI=*89373-1* 203
Xanthosoma peregrinum IPNI=*89373-1* 203
Xanthosoma pilosum IPNI=*89347-1* 203
Xanthosoma plumieri IPNI=*89333-1* 203
Xanthosoma poeppigii var. *mafaffa* IPNI=*89373-1* 203
Xanthosoma purpuratum IPNI=*269320-2* 203
Xanthosoma roseum IPNI=*89373-1* 203
Xanthosoma sagittifolium IPNI=*89373-1* 203
Xanthosoma sagittifolium IPNI=*89384-1* 203
Xanthosoma sylvestre IPNI=*30010088-2* 199
Xanthosoma utile IPNI=*89373-1* 203
Xanthosoma variegatum IPNI=*89333-1* 203
Xanthosoma violaceum IPNI=*89373-1* 203
Xanthosoma xantharrhizon IPNI=*89373-1* 203
Xanthoxalis aureoflava IPNI=*177771-2* 432
Xanthoxalis chrysantha IPNI=*177883-2* 432
Xanthoxalis cordobensis IPNI=*177883-2* 432
Xanthoxalis corniculata IPNI=*177893-2* 432
Xanthoxalis corniculata subsp. *repens* IPNI=*177893-2* 432
Xanthoxalis corniculata var. *aureoflava* IPNI=*177771-2* 432
Xanthoxalis crassicaulis IPNI=*178566-2* 432
Xanthoxalis filiformis IPNI=*178000-2* 432
Xanthoxalis grenadensis IPNI=*177893-2* 432
Xanthoxalis langloisii IPNI=*177893-2* 432
Xanthoxalis lotoides IPNI=*178186-2* 432
Xanthoxalis medicaginea IPNI=*178220-2* 432
Xanthoxalis parvifolia IPNI=*178000-2* 432
Xanthoxalis repens IPNI=*177893-2* 432
Xanthoxalis spiralis IPNI=*178480-2* 432
Xanthoxalis sternbergii IPNI=*177883-2* 432
Xanthoxalis trinidadensis IPNI=*177893-2* 432
Xanthoxalis tuberosa IPNI=*178566-2* 432
Xanthoxerampellia hedwigiae IPNI=*902533-1* 220
Xanthoxerampellia rufescens IPNI=*643781-1* 221
Xanthoxerampellia tenuibulba IPNI=*320703-2* 221
Xantolis palmeri IPNI=*209714-2* 306
Xantolis venulosa IPNI=*160417-2* 306
Xantophtalmum coronarium IPNI=*1013480-1* 263
Xantophtalmum coronarium var. *discolor* IPNI=*1013480-1* 263
Xantophtalmum segetum IPNI=*209007-1* 263
Xenopoma obovatum IPNI=*61205-2* 369
Xeracina aquatica IPNI=*573916-1* 424
Xeraea albiflora IPNI=*34033-2* 282
Xeraea decumbens IPNI=*112097-2* 282
Xeraea globosa IPNI=*327333-2* 282
Xeraea lancifolia IPNI=*112097-2* 282
Xeraea serrata IPNI=*112097-2* 282
Xerandra celosioides IPNI=*129956-2* 283
Xeranthemum bracteatum IPNI=*962307-1* 272
Xeranthemum lucidum IPNI=*962307-1* 272
Xerogona biloba IPNI=*321950-2* 398
Xeromphis retzii IPNI=*745904-1* 348
Xeromphis spinosa IPNI=*745904-1* 348
Ximenesia australis IPNI=*30147294-2* 272
Ximenesia encelioides IPNI=*30147294-2* 272
Ximenesia encelioides var. *exauriculata* IPNI=*30147294-2* 272
Ximenesia encelioides var. *pachyptera* IPNI=*30147294-2* 272
Ximenesia exauriculata IPNI=*30147294-2* 272
Ximenesia microptera IPNI=*30147294-2* 272
Ximenia aculeata IPNI=*316341-2* 449
Ximenia americana f. *inermis* IPNI=*316341-2* 449
Ximenia americana var. *argentinensis* IPNI=*316341-2* 449
Ximenia americana var. *oblonga* IPNI=*316341-2* 449
Ximenia americana var. *ovata* IPNI=*316341-2* 449
Ximenia arborescens IPNI=*316341-2* 449
Ximenia elliptica IPNI=*316341-2* 449
Ximenia exarmata IPNI=*316341-2* 449
Ximenia fluminensis IPNI=*316341-2* 449
Ximenia gabonensis IPNI=*316341-2* 449
Ximenia inermis IPNI=*316341-2* 449
Ximenia laurina IPNI=*316341-2* 449
Ximenia loranthifolia IPNI=*316341-2* 449
Ximenia montana IPNI=*316341-2* 449
Ximenia multiflora IPNI=*316341-2* 449
Ximenia oblonga IPNI=*316341-2* 449
Ximenia rogersii IPNI=*316341-2* 449

Ximenia russeliana IPNI=*316341-2* 449
Ximenia spinosa IPNI=*316341-2* 449
Ximenia subscandens IPNI=*316341-2* 449
Ximenia verrucosa IPNI=*316341-2* 449
Xiphidium albidum IPNI=*304977-2* 224
Xiphidium album IPNI=*304977-2* 224
Xiphidium angustifolium IPNI=*429893-1* 224
Xiphidium floribundum IPNI=*304977-2* 224
Xiphidium giganteum IPNI=*304977-2* 224
Xiphidium loeflingii IPNI=*304977-2* 224
Xiphidium rubrum IPNI=*304977-2* 224
Xiphion fimbriatum IPNI=*438741-1* 216
Xiphocarpus candidus IPNI=*520444-1* 335
Xiphocarpus martinicensis IPNI=*520444-1* 335
Xolemia palustris IPNI=*326619-2* 344
Xuarezia biflora IPNI=*800773-1* 377
Xuarezia hirsuta IPNI=*800773-1* 377
Xuarezia semiserrata IPNI=*800773-1* 377
Xyladenius glandulosus IPNI=*779396-1* 402
Xylon burchellii IPNI=*51083-2* 408
Xylon hirsutum IPNI=*30042843-2* 408
Xylon indicum IPNI=*559664-1* 408
Xylon leoninum IPNI=*559730-1* 408
Xylon pentandrum IPNI=*1166232-2* 407
Xylon phaeosanthum IPNI=*51083-2* 408
Xylon religiosum IPNI=*30042843-2* 408
Xylon roseum IPNI=*60446802-2* 414
Xylon samauma IPNI=*51083-2* 408
Xylon strictum IPNI=*559730-1* 408
Xylophragma xanthophyllum IPNI=*77142096-1* 365
Xylopia acuminata IPNI=*72023-1* 186
Xylopia aromatica IPNI=*76202-1* 190
Xylopia carminativa IPNI=*76256-1* 190
Xylopia cinerea IPNI=*76202-1* 190
Xylopia cubensis IPNI=*76067-1* 189
Xylopia dunaliana IPNI=*76067-1* 189
Xylopia emarginata var. *duckei* IPNI=*76126-1* 189
Xylopia frutescens IPNI=*72319-1* 187
Xylopia frutescens var. *ferruginea* IPNI=*76133-1* 189
Xylopia frutescens var. *glabra* IPNI=*76133-1* 189
Xylopia glabra IPNI=*72319-1* 187
Xylopia grandiflora IPNI=*76067-1* 189
Xylopia intermedia IPNI=*76256-1* 190
Xylopia longifolia IPNI=*76067-1* 189
Xylopia lucida IPNI=*76067-1* 189
Xylopia meridensis IPNI=*76133-1* 189
Xylopia muricata IPNI=*76133-1* 189
Xylopia neglecta IPNI=*76217-1* 190
Xylopia nitida var. *nervosa* IPNI=*76202-1* 190
Xylopia poeppigii IPNI=*269786-2* 189
Xylopia polyantha var. *nicaraguensis* IPNI=*76133-1* 189
Xylopia salicifolia IPNI=*76119-1* 189
Xylopia setosa IPNI=*76133-1* 189
Xylopia ulei IPNI=*269770-2* 189
Xylopia xylopioides IPNI=*76067-1* 189
Xylopicron glabrum IPNI=*72319-1* 187
Xylopicrum acuminatum IPNI=*72023-1* 186
Xylopicrum aromaticum IPNI=*76067-1* 189
Xylopicrum carminativum IPNI=*76256-1* 190
Xylopicrum emarginatum IPNI=*76126-1* 189
Xylopicrum frutescens IPNI=*76133-1* 189
Xylopicrum grandiflorum IPNI=*76067-1* 189
Xylopicrum longifolium IPNI=*76067-1* 189
Xylopicrum neglectum IPNI=*76217-1* 190
Xylopicrum nitidum IPNI=*76202-1* 190
Xylopleurum multicaule IPNI=*173202-2* 430
Xylopleurum roseum IPNI=*173348-2* 430
Xylopleurum tetrapterum IPNI=*614141-1* 430
Xylosma albida IPNI=*112146-1* 405
Xylosma arnoldii IPNI=*112138-1* 405
Xylosma blepharodes IPNI=*112138-1* 405
Xylosma calophylla IPNI=*112146-1* 405
Xylosma celastrina IPNI=*112138-1* 405
Xylosma elliptica IPNI=*365348-1* 404
Xylosma excelsa IPNI=*112146-1* 405
Xylosma heterophylla IPNI=*112178-1* 405
Xylosma paliurus IPNI=*112138-1* 405
Xylosma palmeri IPNI=*112138-1* 405
Xylosma pringlei IPNI=*112138-1* 405
Xylosma prunifolia IPNI=*112146-1* 405
Xylosma seemannii IPNI=*112138-1* 405
Xylosma turrialbana IPNI=*47939-2* 403
Xylosteon flexuosum IPNI=*148850-1* 296
Xylosterculia rugosa IPNI=*927898-1* 414
Xyphostylis angustifolia IPNI=*796016-1* 247
Xyphostylis lutea IPNI=*319130-2* 247
Xyris acuminata IPNI=*315094-2* 247
Xyris americana IPNI=*319776-2* 247
Xyris anceps IPNI=*315094-2* 247
Xyris arenicola IPNI=*315094-2* 247
Xyris brasiliensis IPNI=*104972-2* 232
Xyris communis IPNI=*315094-2* 247
Xyris glabrata IPNI=*319776-2* 247
Xyris guianensis IPNI=*315094-2* 247
Xyris gymnoptera IPNI=*315094-2* 247
Xyris jupicae IPNI=*315094-2* 247
Xyris jupicai var. *brachylepis* IPNI=*315094-2* 247
Xyris jupicai var. *humilis* IPNI=*315094-2* 247
Xyris laxifolia var. *minor* IPNI=*315094-2* 247
Xyris laxifolia var. *sellowiana* IPNI=*315094-2* 247
Xyris macrocephala f. *minor* IPNI=*315094-2* 247
Xyris macrocephala IPNI=*315094-2* 247
Xyris macrocephala var. *minor* IPNI=*315094-2* 247
Xyris partita IPNI=*315094-2* 247
Xyris pumila IPNI=*319776-2* 247
Xyris savanensis f. *primaria* IPNI=*319776-2* 247
Xyris savanensis f. *procera* IPNI=*319776-2* 247
Xyris savanensis var. *glabrata* IPNI=*319776-2* 247
Xyris sellowiana f. *humilis* IPNI=*315094-2* 247
Xyris sellowiana IPNI=*315094-2* 247
Xyris surinamensis IPNI=*315094-2* 247

Y

Yarina microcarpa IPNI=*668996-1* 211
Ygramela maritima IPNI=*329497-2* 376
Ynesa colenda IPNI=*272952-2* 205
Youngia ambigua IPNI=*1124129-2* 272
Youngia fastigiata IPNI=*1124129-2* 272
Youngia formosana IPNI=*1124129-2* 272
Youngia gracilis IPNI=*1124129-2* 272
Youngia integrifolia IPNI=*1124129-2* 272
Youngia japonica var. *formosana* IPNI=*1124129-2* 272
Youngia mauritiana IPNI=*1124129-2* 272
Youngia multiflora IPNI=*1124129-2* 272
Youngia napifolia IPNI=*1124129-2* 272
Youngia poosia IPNI=*1124129-2* 272
Youngia runcinata IPNI=*1124129-2* 272
Youngia striata IPNI=*1124129-2* 272
Yucca baccata f. *genuina* IPNI=*270284-2* 215
Yucca eleana IPNI=*543700-1* 215
Yucca elephantipes IPNI=*543700-1* 215
Yucca ghiesbreghtii IPNI=*543700-1* 215
Yucca guatemalensis IPNI=*543700-1* 215
Yucca lenneana IPNI=*543700-1* 215
Yucca mazelii IPNI=*543700-1* 215
Yucca mooreana IPNI=*543700-1* 215

Yucca roezlii IPNI=*543700-1* 215
Yulania × *lenneana* IPNI=*60468302-2* 190
Yulania × *soulangeana* IPNI=*60468302-2* 190
Yulania conspicua IPNI=*554678-1* 190
Yulania cuneatifolia IPNI=*554678-1* 190
Yulania denudata IPNI=*554678-1* 190
Yulania japonica IPNI=*554759-1* 190
Yulania liliiflora IPNI=*554759-1* 190
Yulania pyriformis IPNI=*554678-1* 190
Yutajea liesneri IPNI=*753641-1* 352
Yuyba dakamana IPNI=*664750-1* 207
Yuyba essequiboensis IPNI=*664750-1* 207
Yuyba gleasonii IPNI=*664750-1* 207
Yuyba maguirei IPNI=*664750-1* 207
Yuyba paula IPNI=*664697-1* 207
Yuyba schultesii IPNI=*28556-2* 207
Yuyba simplicifrons IPNI=*664750-1* 207
Yuyba simplicifrons var. *acanthocnemis* IPNI=*664750-1* 207
Yuyba simplicifrons var. *subpinnata* IPNI=*664750-1* 207
Yuyba stahelii IPNI=*664750-1* 207
Yuyba trinitensis IPNI=*664750-1* 207

Z
Zahlbrucknera conferta IPNI=*49672-1* 360
Zahlbrucknera longifolia IPNI=*49672-1* 360
Zahlbrucknera micrantha IPNI=*49672-1* 360
Zahlbrucknera oblongifolia IPNI=*49672-1* 360
Zahlbrucknera repens IPNI=*52558-1* 360
Zala asiatica IPNI=*317684-2* 202
Zaluzania eminens IPNI=*238923-1* 268
Zaluzianskia polycarpa IPNI=*280439-2* 179
Zamia corsoniana IPNI=*326820-2* 179
Zamia gutierrezii IPNI=*270552-2* 179
Zamia lindleyana IPNI=*297256-1* 179
Zamia media var. *gutierrezii* IPNI=*270552-2* 179
Zamia muricata var. *angustifolia* IPNI=*270552-2* 179
Zamia muricata var. *obtusifolia* IPNI=*270552-2* 179
Zamia pallida IPNI=*668912-1* 211
Zamia parasitica IPNI=*270556-2* 179
Zamia wielandii IPNI=*270556-2* 179
Zannichellia pedunculata IPNI=*224565-2* 203
Zanonia bibracteata IPNI=*1191409-2* 224
Zantedeschia asperata IPNI=*87918-1* 201
Zappania angustifolia IPNI=*864749-1* 379
Zappania aristata IPNI=*864784-1* 379
Zappania cayennensis IPNI=*1126344-2* 379
Zappania citrodora IPNI=*1015994-1* 377
Zappania dichotoma IPNI=*1126344-2* 379
Zappania geminata IPNI=*141020-2* 378
Zappania globiflora IPNI=*141020-2* 378
Zappania indica IPNI=*864749-1* 379
Zappania jamaicensis IPNI=*1092351-2* 379
Zappania lantanoides IPNI=*141020-2* 378
Zappania lappulacea IPNI=*1194372-2* 379
Zappania mutabilis IPNI=*243385-2* 379
Zappania nodiflora IPNI=*194567-2* 379
Zappania odorata IPNI=*141020-2* 378
Zappania orubica IPNI=*864784-1* 379
Zappania prismatica IPNI=*35430-2* 378
Zappania scaberrima IPNI=*141103-2* 379
Zarabellia rhomboidea IPNI=*156179-2* 265
Zea mays subsp. *luxurians* IPNI=*270861-2* 247
Zea mexicana subsp. *luxurians* IPNI=*270861-2* 247
Zea perennis subsp. *diploperennis* IPNI=*306962-2* 246
Zea segetalis IPNI=*426810-1* 247
Zenkeria dalbergioides IPNI=*16766-2* 311
Zenkeria lundii IPNI=*16766-2* 311
Zephyranthes andersonii IPNI=*67314-1* 214
Zephyranthes andersonii var. *rosea* IPNI=*67314-1* 214
Zephyranthes carnea IPNI=*67293-1* 214
Zephyranthes commersoniana IPNI=*67314-1* 214
Zephyranthes taubertiana IPNI=*270986-2* 214
Zephyranthes taubertii IPNI=*270986-2* 214
Zephyranthes texana IPNI=*67314-1* 214
Zephyranthes tsouii IPNI=*270933-2* 214
Zerna unioloides IPNI=*37001-2* 235
Zerumbet speciosum IPNI=*872083-1* 250
Zexmenia epapposa IPNI=*30439090-2* 272
Zexmenia hispida var. *ramosissima* IPNI=*30439090-2* 272
Zexmenia longipes IPNI=*30439090-2* 272
Zexmenia pinetorum IPNI=*30439090-2* 272
Zeyheria barbata IPNI=*109687-1* 363
Ziegera laevigata IPNI=*572641-1* 421
Zingiber aromaticum IPNI=*798372-1* 251
Zingiber cholmondeleyi IPNI=*798372-1* 251
Zingiber minus IPNI=*796556-1* 250
Zingiber missionis IPNI=*798372-1* 251
Zingiber officinale f. *macrorhizonum* IPNI=*798372-1* 251
Zingiber officinale f. *rubens* IPNI=*798372-1* 251
Zingiber officinale var. *cholmondeleyi* IPNI=*798372-1* 251
Zingiber officinale var. *rubrum* IPNI=*798372-1* 251
Zingiber sichuanense IPNI=*798372-1* 251
Zinnia australis IPNI=*261331-1* 272
Zinnia elegans var. *alba* IPNI=*261331-1* 272
Zinnia elegans var. *coccinea* IPNI=*261331-1* 272
Zinnia elegans var. *purpurascens* IPNI=*261331-1* 272
Zinnia elegans var. *violacea* IPNI=*261331-1* 272
Zinnia florida IPNI=*261363-1* 272
Zinnia floridana IPNI=*261363-1* 272
Zinnia fulva IPNI=*261363-1* 272
Zinnia grandiflora IPNI=*261363-1* 272
Zinnia hybrida IPNI=*261363-1* 272
Zinnia intermedia IPNI=*261363-1* 272
Zinnia leptopoda IPNI=*261363-1* 272
Zinnia linearifolia IPNI=*261363-1* 272
Zinnia linearis IPNI=*261331-1* 272
Zinnia linearis var. *latifolia* IPNI=*261331-1* 272
Zinnia lutea IPNI=*261363-1* 272
Zinnia mendocina IPNI=*261363-1* 272
Zinnia multiflora IPNI=*261363-1* 272
Zinnia multiflora var. *flava* IPNI=*261363-1* 272
Zinnia multiflora var. *rubra* IPNI=*261363-1* 272
Zinnia pauciflora IPNI=*261363-1* 272
Zinnia revoluta IPNI=*261363-1* 272
Zinnia tenuiflora IPNI=*261363-1* 272
Zinnia verticillata IPNI=*261363-1* 272
Zinnia violacea IPNI=*261331-1* 272
Zinnia violacea var. *coccinea* IPNI=*261331-1* 272
Zizania ciliata IPNI=*137026-2* 241
Ziziphus acidojujuba f. *granulata* IPNI=*719213-1* 442
Ziziphus acidojujuba f. *infecunda* IPNI=*719213-1* 442
Ziziphus acidojujuba f. *trachysperma* IPNI=*719213-1* 442
Ziziphus acidojujuba IPNI=*719213-1* 442
Ziziphus angolito IPNI=*77154251-1* 442
Ziziphus chinensis IPNI=*719213-1* 442
Ziziphus cinnamomum IPNI=*77154236-1* 442
Ziziphus commutata IPNI=*1213898-2* 436
Ziziphus cyclocardia IPNI=*77154237-1* 442
Ziziphus domingensis IPNI=*63435-2* 441
Ziziphus emarginata IPNI=*133956-2* 442
Ziziphus flexuosa IPNI=*719213-1* 442
Ziziphus hysundrica IPNI=*719349-1* 442
Ziziphus iguanea IPNI=*1213898-2* 436
Ziziphus insularis IPNI=*719349-1* 442
Ziziphus itacaiunensis IPNI=*77154236-1* 442
Ziziphus jujuba f. *allochroa* IPNI=*719213-1* 442

Ziziphus jujuba f. *apyrena* IPNI=*719213-1* 442
Ziziphus jujuba f. *carnosicalycis* IPNI=*719213-1* 442
Ziziphus jujuba f. *heteroformis* IPNI=*719213-1* 442
Ziziphus jujuba f. *tortuosa* IPNI=*719213-1* 442
Ziziphus littorea IPNI=*316341-2* 449
Ziziphus mairei IPNI=*719349-1* 442
Ziziphus mauritiana var. *hysundrica* IPNI=*719349-1* 442
Ziziphus melanogona IPNI=*719213-1* 442
Ziziphus muratiana IPNI=*719349-1* 442
Ziziphus natsme IPNI=*719213-1* 442
Ziziphus nitida IPNI=*719213-1* 442
Ziziphus officinarum IPNI=*719213-1* 442
Ziziphus orthacantha IPNI=*719349-1* 442
Ziziphus poiretii IPNI=*719349-1* 442
Ziziphus rotundata IPNI=*719349-1* 442
Ziziphus saeri IPNI=*77154251-1* 442
Ziziphus sativa IPNI=*719213-1* 442
Ziziphus sinensis IPNI=*719213-1* 442
Ziziphus soporifera IPNI=*719213-1* 442
Ziziphus sororia IPNI=*719213-1* 442
Ziziphus spinosa IPNI=*719213-1* 442
Ziziphus tomentosa IPNI=*719349-1* 442
Ziziphus trinervia IPNI=*719213-1* 442
Ziziphus vulgaris IPNI=*719213-1* 442
Ziziphus vulgaris var. *spinosa* IPNI=*719213-1* 442
Ziziphus zizyphus IPNI=*719213-1* 442
Zizyphon jujubum IPNI=*719213-1* 442
Zollernia tango IPNI=*136896-2* 325
Zornia amaralaensis IPNI=*271272-2* 337
Zornia barbata IPNI=*271325-2* 337
Zornia ceylanica IPNI=*30026493-2* 337
Zornia conjugata IPNI=*30026493-2* 337
Zornia cuyabensis IPNI=*271325-2* 337
Zornia diphylla f. *ciliata* IPNI=*271325-2* 337
Zornia diphylla f. *diversifolia* IPNI=*271325-2* 337
Zornia diphylla f. *intermedia* IPNI=*271325-2* 337
Zornia diphylla subsp. *cuyabensis* IPNI=*271325-2* 337
Zornia diphylla subsp. *gracilis* IPNI=*30156577-2* 337
Zornia diphylla subsp. *reticulata* IPNI=*271325-2* 337
Zornia diphylla var. *bernardinensis* IPNI=*30156577-2* 337
Zornia diphylla var. *conjugata* IPNI=*30026493-2* 337
Zornia diphylla var. *elatior* IPNI=*271325-2* 337
Zornia diphylla var. *gracilis* IPNI=*30156577-2* 337
Zornia diphylla var. *leptophylla* IPNI=*526041-1* 337
Zornia diphylla var. *paraguariensis* IPNI=*271325-2* 337
Zornia diphylla var. *pubescens* IPNI=*271325-2* 337
Zornia diphylla var. *reticulata* IPNI=*271325-2* 337
Zornia diphylla var. *rupestris* IPNI=*271325-2* 337
Zornia diphylla var. *stenophylla* IPNI=*271325-2* 337
Zornia diphylla var. *stricta* IPNI=*271325-2* 337
Zornia diphylla var. *thymifolia* IPNI=*526091-1* 337
Zornia echinata IPNI=*271325-2* 337
Zornia gavilanesii IPNI=*271272-2* 337
Zornia gemella IPNI=*30156577-2* 337
Zornia glabella IPNI=*30026493-2* 337
Zornia gracilis IPNI=*30156577-2* 337
Zornia havanensis IPNI=*271325-2* 337
Zornia inermis IPNI=*271325-2* 337
Zornia latifolia var. *bernardinensis* IPNI=*30156577-2* 337
Zornia maranhamensis IPNI=*30156577-2* 337
Zornia nuda IPNI=*30026493-2* 337
Zornia ovata IPNI=*271325-2* 337
Zornia pubescens IPNI=*271325-2* 337
Zornia reticulata var. *puberula* IPNI=*271325-2* 337
Zornia reticulata var. *punctata* IPNI=*271325-2* 337
Zornia strobilifera IPNI=*495307-1* 320
Zornia subperforata IPNI=*271325-2* 337
Zornia surinamensis IPNI=*30156577-2* 337
Zornia venusta IPNI=*30026493-2* 337
Zornia zeylonensis IPNI=*30026493-2* 337
Zosterophyllanthos baudoensis IPNI=*316910-2* 222
Zschokkea aculeata IPNI=*134481-2* 341
Zschokkea arborescens IPNI=*79531-1* 341
Zschokkea armata IPNI=*79551-1* 341
Zschokkea floribunda IPNI=*79552-1* 341
Zschokkea foxii IPNI=*134491-2* 341
Zschokkea gracilis IPNI=*79536-1* 341
Zschokkea lactescens IPNI=*79551-1* 341
Zschokkea monosperma IPNI=*79531-1* 341
Zschokkea panamensis IPNI=*134499-2* 341
Zschokkea ramosissima IPNI=*79547-1* 341
Zuelania belizensis IPNI=*48022-2* 403
Zuelania crenata IPNI=*48022-2* 403
Zuelania guidonia IPNI=*48022-2* 403
Zuelania icosandra IPNI=*48022-2* 403
Zuelania laetioides IPNI=*48022-2* 403
Zuelania roussoviae IPNI=*48022-2* 403
Zygalchemilla pinnata IPNI=*720867-1* 443
Zygia divaricata IPNI=*962821-1* 337
Zygia dulcis IPNI=*1003508-2* 330
Zygia foreroi IPNI=*271431-2* 337
Zygia glomerata IPNI=*962821-1* 337
Zygia guadalupensis IPNI=*515137-1* 330
Zygia longiramosa IPNI=*271431-2* 337
Zygia saman IPNI=*227126-2* 331
Zygia unguis-cati IPNI=*515137-1* 330
Zygis aromatica IPNI=*61205-2* 369
Zygocactus altensteinii IPNI=*229597-2* 284
Zygocactus delicatus IPNI=*229597-2* 284
Zygocactus truncatus IPNI=*229597-2* 284
Zygopetalum bicallosum IPNI=*154952-2* 220
Zygopetalum fragrans IPNI=*661384-1* 223
Zygopetalum lalindei IPNI=*60440679-2* 222
Zygopetalum marginatum IPNI=*661384-1* 223
Zygopetalum meleagris IPNI=*639074-1* 219
Zygopetalum patinii IPNI=*60440679-2* 222
Zygopetalum quadratum IPNI=*661384-1* 223
Zygopetalum velatum IPNI=*661384-1* 223

Spondias mombin figurines. Six figures of various types of natives in Colombia. Made from corky outgrowths of the bark of 'Hobor' tree. Note the details of the carving on the faces.
Laura Green -Digital Collection – RBG Kew

Index of Vernacular/Common Names

A
á 384
aace 191
aayae 210
ababol 435
abaca 371
abacachirana 182
abacaraá 414
abacú 205
abalazo 200
abalazos 200
abanico 211, 281, 282
abarco 301
abataque 457
abay 388
abayao 292
abebe 251
abebe grande 251
abejín 332
abejón 332, 363, 370, 450
abejorro 354
abekará 414
abeto 317, 333
abey 364
abina 328
abiorama 306
abiu 306
abiurama 306
aborrecedora 360
abortaganado 339
abrazadera 434
abrazapalo 198, 200, 201, 438, 439
abrecaminos 171, 360
abrejillo 286
abrojillo 285, 286
abrojita 285, 286
abrojito 285, 286
abrojo 235, 255, 258, 280, 281, 282, 318, 412, 440, 442, 472, 473
abrojo blanco 280
abrojo rojo 281
abrotano 442
abubuchique 368
abuelo de cacao 411, 413
abuelo de higuerón 439
abuelo del cacao 413
abugua 282
abuichillo 406
abujo 406
abujua 258
abujumujariyocù 406
abundancia 370
abuta 434
abutilón 406, 409, 410
abutilón rojo 406
acacia 310, 312, 313, 314, 315, 317, 320, 325, 329, 330, 333, 364
acacia alemana 310
acacia amarilla 310, 333
acacia blanca 325
acacia bracatinga 329
acacia brasileña 314
acacia ceniza 310
acacia de Girardot 317
acacia de jardín 312
acacia de Siam 333
acacia de tierra fría 310
acacia forrajera 325
acacia gris 310
acacia japonesa 310
acacia lluvia de oro 313
acacia negra 310
acacia roja 317
acacia rosada 314
acacia tunuda 330
acacio 310, 312, 317, 333
acacio amarillo 333
acacio extranjero 317
acacio ordinario 317
acacio rojo 317
acacio santandereano 333
acaena 442
acahapo rosado 314
acaí 209
acaneboto 230
acaño de aceite 300
acanto 359
acapón 266
acapro 364
acapro blanco 364
acapu 374
acapú 381, 449
acapú macho 381
acapú negro 449
acaracay 453
acarana 452
acaricuara 449
acaruco 415
acebuche 375
acedera 289, 290, 432
acederilla 289, 432
acederita 289, 290, 432
aceite 183, 200, 208, 280, 300, 315, 321, 331, 380, 385
aceite canime 315, 331
aceite de Ben 280
aceite de caparrapí 183
aceite de maría 380
aceite de María 380
aceite maría 380
aceitillo 385, 402
aceituna 373, 375
aceituno 183, 184, 255, 318, 330, 347, 364, 373, 374, 375, 385, 393, 428, 449, 460
aceituno dulce 428
aceituno montés 330
aceituno rojo 393
acerola 396
achapo 314
achapo blanco 314
achapo colorado 314
achapo rojo 314
achapo rosado 314
achera 247
achi 301
achí 301, 337
achicora 264
achicoria 251, 260, 264, 270, 271
achikaira 196
achil 247, 337
achila 247
achín 199
achiote 184, 380, 406, 431
achiote amarillo 406
achiote cimarrón 406
achiote de bija 406
achiote de cholo 406
achiote de monte 406
achiote de venado 380
achiote pepa roja 406
achiote rojo 406
achiote verde 406
achiotillo 380, 381, 394, 406, 421
achiotillo mani chalmugra 381
achira 247, 250, 251, 412
achira blanca 247
achira bugueña 247
achira colorada 247
achira criolla 247
achira de monte 251
achira morada 247
achira natagueña 247
achira oscura 247
achira roja 247
achira verde 247
achirilla 247
achita 281, 282
achocha 294, 295
achocha cimarrona 294
achocho 294
achón 189, 190
achón sembe 190
achoncha 294
achoote 406
achote 393, 406, 408
achote amarillo 406
achote casero 406
achote cimarrón 406, 408
achote colorado 406
achote de bija 406
achote de monte 393
achote montañero 406

achotillo 335, 380, 381, 394, 457
achotillo crespo 380
achuapo 314
achulunga 344
aclaraboyos 290
acquai 448
acrorutú 325
acuapa 391
acuapar 391
acuasia 460
acui huipo 200
acupú 449
acurruco 415
acuruco 415
acurucu 415
acurucú 415
acurutú 325
adelfa 342
adelfo 342
adelicia 343
admirable 246
adonis 378
adormidera 310, 314, 327, 328, 435
adormidera dulce 310
adormidera rosada 328
adorote 227, 287, 362, 363, 434, 442
adura 247
afelandra 361
africano 250, 365
afuera 199
afuga 368
agapaca 372
agave 214
agey+ 184
aglomena 198
ago 364
agras 300
agrás 300
agraz 288, 295, 300, 442, 472
agráz 300
agraz coronel 300
agrumo 446
agua bendita 345
agua miel 341
aguacacho 394
aguacarillo 185
aguacate 182, 184, 185
aguacate búho 185
aguacate cimarrón 182, 185
aguacate corneto 185
aguacate cura 185
aguacate de anís 185
aguacate de danta 185
aguacate de fantasma 185
aguacate de monte 185
aguacate de perico 185
aguacate Lorena 185
aguacate macho 184, 185
aguacate mariquiteño 185
aguacate morado 185
aguacate negro 185
aguacate papelillo 185
aguacatillo 182, 184, 185, 306, 394
aguacatillo colorado 185
aguacato blanco 185
aguacatón 182, 253
aguacerito 310, 319
aguacolda 223
aguadeño 330
aguadija 218
aguadulce 255, 300, 353, 369
aguaje 210
agualibia 269
agualivia 269
aguamiel 191
aguanchinoso 446
aguanillo 191
aguanosa 446
aguanoso 181, 191, 297, 369, 421, 445, 446
aguaqu 304
aguaquín 182
aguardientillo 196
aguarrás 183, 184, 295
aguarrás amarillo 184
aguarrús 469
aguasio 460
aguayusa 255
aguileña 436
aguire 381
agüire 381
aguja de arria 405
agujas 358
aheko 210
ahi 210
ahitera 332, 333
ahogacachaco 250
ahumado 330, 449
ahumado caqueteño 449
ahumado de hoja pequeña 449
ahumado pajarito 449
ahumado pedro 449
ahumao 449
ahumapescao 369
ahuyama 293
ahuyama burrera 293
ahuyama candelaria 293
ahuyama chica 293
ahuyama cintura 293
ahuyama peruana 293
ahuyamo 297, 465
ahuyano 292
aichwua 391
air plant 461
airambo 287
aire 192, 461
airo kajo 202
airon 197
aitera 333
aití 383
ajedrea 369
ajengible 251
ajengibre 251
ajenje 467
ajenjo 257
ajenjo rucio 257
ajenjol 257
ají 180, 181, 184, 188, 255, 311, 312, 334, 335, 350, 354, 355, 357, 386, 393, 397, 415, 431, 440, 460, 465, 466, 467
ají - pique 465
ají ajuja 465
ají amarillo 465
ají amarillo de culebra 465
ají bejuco 465
ají blanco 465
ají botón 465
ají bravo 465
ají camarón 465
ají carcarón de tiro 465
ají caribe 465
ají cascarón 465
ají catuto 465
ají cerbatano 465
aji charapilla 465
aji cheroso 465
ají chiche de perro 465
ají chirel 465
ají chivato 465
ají cimarrón 465
ají de babilla 465
ají de barbasco 312, 334
ají de guacamayo 465
ají de guara 188
ají de la capitanía 465
ají de monte 184, 350, 355, 397
ají de pájaro 386
ají de páramo 181
ají de pescado 460
ají del diablo 354
ají dulce 465
ají huevo de aragua ají 465
ají huevo de araguana 465
ají mora 440, 465
ají morado 465
ají paja 465
ají pajarito 465
ají pimienta 465
ají pimiento 465
ají pique 465
aji piquiucho 465
ají rocote 465
ají rocoto 465
ají silvestre 466
ají vocato 465
ajicillo 181, 184, 185, 287, 297, 369, 386, 415, 441
ajícillo 185
ajicito 303, 351, 374, 451
ajicito del diablo 303
ajínae 407
ajingible 251
ajito con pelo 387
ajizuelo 181
ajo 199, 213, 214, 273, 362, 364, 383, 402
ajo de loma 383
ajo de montaña 273
ajo de monte 199, 364, 402
ajo de perdiz 214
ajo macho 213
ajo sacha 364
ajo-ajo 273
ajonjolí 375
ajunado 311
akanepona 232
al pia 331
ala de ángel 292
ala de loro 282
ala de murciélago 398
ala de panguana 251
alacoque 385
alacrán 195, 274, 275, 282, 334, 340, 343, 438
alacrana 275
álamo 387, 390, 404
álamo de monte 387
álamo negro 404
alas de ángel 292
alazano 347
alazor 259
albahaca 369, 371, 378, 379, 410
albahaca blanca 371
albahaca blanca del Santísimo 371
albahaca brasilera 379
albahaca canela 371
albahaca chiracán 371
albahaca chirarán 371
albahaca cimarrona 369, 371
albahaca común 371
albahaca de anís 371
albahaca de Castilla 371

albahaca de gallinaza 371
albahaca de monte 371
albahaca de páramo 378
albahaca del Santísimo 371
albahaca morada 371
albahaca toronjil 371, 378
albahaca vaca 371
albahaquilla 371
albajaca 371
albarica 206
albarico 204, 206
albarico montañero 204
albaricoque 381
albarina 259
albarracín 435
albarrana 370
albarrana verde 370
albarrecín 435
albérchigo 444
albicia 416
alcachofa 260
alcafeto negro 402
alcanfor 183, 257, 319, 371, 452
alcanfor de monte 371
alcanzaviejo 291
alcaparrillo 333
alcaparrito 333
alcaparro 332, 333, 334
alcaparro chico 333
alcaparro de Bogotá 334
alcaparro enano 333
alcaparro florón 333
alcaparro grande 334
alcaparro macho 333
alcaparro pálido 332
alchocha 294
alchonchas 295
alchucha 294, 295
alcoparro 333
alcornoco 312, 395
alcornoque 312, 328, 381, 460
alcornoque sabanero 312
alegría 292, 297, 338, 344, 373, 377, 395
alegría de páramo 344
alegría del duende 338
alejandra 204
alejandría 342
alejandro 397
alelí 285, 342, 454
alemán 238
alemana 238, 471
alfalfa 327, 334, 394
alfalfa argenti herb amarilla 394
alfalfa argentina 394
alfalfa del Brasil 334
alfalfa peruana 327
alfandoquillo 408
alfandoquitos 408
alfiler 358
alfileres 327, 358
alfilerillo 327, 358
alfiletero 296
alfombra de agua 179
alfonsolópez 390
algalia 406
algarroba 320, 321
algarrobeño 330
algarrobilla 315, 321
algarrobillo 315, 318, 320, 321, 331, 334
algarrobito 284, 330
algarrobo 312, 316, 320, 321, 330, 331, 332, 381, 385
algayubo 388
algodón 255, 339, 381, 387, 388, 406, 408, 409, 410, 412, 413, 415
algodón andariego 339
algodón chino 339
algodón de árbol 408, 413
algodón de Castilla 339
algodón de monte 410, 415
algodón de seda 339
algodón de tierra firme 408
algodón extranjero 340
algodón fino 340
algodón higuerillo 339
algodón lechero 340
algodón macho 408
algodón morado 408
algodón pajarito 408
algodón silvestre 408, 415
algodoncillo 257, 258, 339, 353, 387, 388, 407, 410, 412, 415, 441, 451
algodonero 408
algodonero pajarito 408
algofar 232
alhelí 277, 340, 454
alhelí amarillo 277
alhelí blanco 277
alhelí morado 277
alhucema 257
ali a 451
alia 250
aliento de niño 176
aligustrc 375
alín 203
aliso 184, 268, 271, 338, 436, 440, 460
aliso calentano 271
aliso de tierra caliente 271
aliso negro 268
aliso playero 271
alitas de ángel 292
aliviadolor 378
allita 363
alma negra 190
almácigo 451
almanegra 190, 302, 311, 329, 330, 363, 451
almanegra de urrao 190
almanegra del calima 190
almanga 257
almendra 193, 381, 416, 440
almendrillo 311, 381, 416
almendro 292, 295, 302, 318, 381, 388, 414, 415, 416, 443
almendro de Indias 416
almendro de la costa 416
almendro de la India 416
almendro de las Indias 416
almendro de playa 416
almendrón 205, 381, 416
almendrón de mariquita 381
almidón 304, 306, 359, 430
almirajo 412
almirajó 411, 412
almirajó macho 411
almizclado 375
almizcle 181, 358, 412
almizclera 358
almizclillo 267, 375, 388, 406
almondo 416
almoraduz 257, 260, 272
almoraús 258
almorrana 281, 418
alnus 338
aloe 215
alondra 465
alorva 316
aloukemasu 388, 389
alpaquina 282
alpaquinoa 282
alpaquinua 282
alparada 239
alpargate 362
alpargatera 281
alpichala 252
alpiste 243, 276, 277, 278
alpiste de canarios 243, 276
alpiste de forraje 243
alpistillo 234, 276
alquenque 468
alsaciana 409
altamisa 256, 257, 267, 271, 371
altamisa blanca 257
altamisa chiquita 257
altamisa grande 256
altamisa negra 257
altamisa pastusa 257
altasara 287
alternatera 281
altramuz 325
altusa 287
altusara morado 287
alubeno 371
alumbra a las once 282
alúnka 205
alvarrana 370
alvarrana verde 370
alverja 318, 325
alverja de monte 318, 325
alverjilla 314, 336, 337
alverjillo 325, 335
alverjita 337
alverjo 330
alverjón 283
alverjones 324
amabaibillo 447
amacá 440
amacay 340
amacey 183
amacise 319
amainae 420
amaine 420
amajagua 410
amaji 437, 440
amamor 456
amanague 313
amancai 340
amancayo 342, 394
amancayo colorado 394
amansa mujer 331
amansaguapo 359
amansaguapos 359, 360
amansajusticia 359
amansamacho 359
amansamujer 315, 331
amansapeon 329
amansapeón 329
amapa 185
amapá 342
amapola 258, 267, 269, 271, 272, 401, 407, 435
amapola amarilla 271
amapola común 435
amapola de jardín 271
amapola de monte 267
amapola del camino 269
amapola silvestre 258
amapolo 312
amaranta 250
amaranto 281, 282, 283, 289, 342, 369, 387
amaranto blanco 281
amaranto de arbolito 387

amaranto espinoso 281
amarga 212, 240, 256, 257, 259, 260, 348, 376, 377, 386, 413, 460, 472
amargo 184, 212, 242, 244, 246, 256, 257, 259, 294, 300, 311, 339, 349, 354, 355, 376, 404, 433, 451, 454, 455, 460, 470
amargo pajarito 355
amargoandré 345
amargoandrés 345
amargón 259, 268, 271
amargosa 257
amargosita 258, 259
amargosito 388, 389
amargoso 216, 256, 294, 339, 389, 434, 435, 460
amarguero 256, 259, 260
amarguero amarillo 260
amarguero negro 256
amargura 351
amarilla 188, 215, 217, 235, 244, 247, 251, 252, 257, 258, 259, 260, 261, 264, 270, 271, 278, 283, 310, 333, 353
amarillo 180, 182, 183, 184, 185, 191, 192, 193, 203, 206, 215, 219, 247, 254, 260, 264, 273, 274, 276, 277, 280, 305, 306, 307, 310, 314, 315, 318, 330, 331, 332, 333, 335, 336, 339, 341, 342, 348, 350, 352, 354, 355, 358, 359, 360, 363, 364, 365, 367, 374, 380, 381, 406, 409, 412, 413, 416, 417, 420, 423, 429, 430, 432, 434, 435, 437, 440, 441, 443, 450, 455, 460, 465
amarillo baboso 183, 184
amarillo comino 183
amarillo corazón negro 182
amarillo de mata 330
amarillo del blanco 184
amarillo espino 441
amarillo jigua 185
amarillo laurel 182
amarillo oloroso 182
amarillo peludo 441
amarillo real 184
amarillo tara 182
amarillo toloncher 183
amarillón 416
amarillón serrano 416
amarón 465, 466
amarón borrachero 466
amarradiablo 362
amasisa 319, 325
amasise 319
amayuyo 271
ambaiba uva 447
ambar 363, 365
ámbar 279, 320
ambarina 406
ambarinas 259
ambe mora 295
ambirá 467
ambra 406
ambrosía 282
ambrosia silvestre 257
ambuca 336
ambuque 310
amé 337
américa 330
americana 185, 203, 214, 218, 263, 278, 287, 290, 295, 310, 318, 351, 368, 377, 378, 379, 381, 382, 403, 415, 437, 449, 454, 457, 462, 465, 468, 471
amígdala 257
amigo del diablo 349
amistad del día 409
amolada 203
amolado 203
amolao 203
amole 287
amor al día 416
amor al uso 409
amor antioqueño 344
amor ardiente 297, 352
amor de campo 320
amor de madre 214, 352, 361, 362, 379, 391
amor de mico 278
amor de zumba 416
amor do campo 318
amor encarnado 297
amor escondido 361
amor fino 285
amor seco 318, 320
amoraús 417
amorseca 320
amorseco 258, 268, 280, 316, 317, 318, 320, 442, 445
amorseco del valle 318
amorseco pequeño 320
ampiapúne 210
ampó 453
anacao 418
anacardo 450
anaco 319
anaco rojo 319
añadidor 367
anaku tsatsavó 212
añame 451
anamos 326
anamú 252, 287, 296
ananios 326
anapa 199
anavaco 351
ancabesux 434
anchamba 205
anchambe 205
ancharma 205
anchipia guaira 331
anchique 307
ancla 227, 388
ancú 359
andalucía 378
andaquí 193, 250
andaquí de loma 193
andelfo 342
andino trepador 330
andiroba 454
anea 247, 357
aneiba 216
aneldo 251
anelvia 303
anémona 436
anémone 436
angarilla 315
angarillo 315
ángela 280
angélica 251, 338
angelina 311, 351
angelino 311, 330, 404
angelita 277, 312, 338
angelito 312, 351, 360, 418, 423
angelito clavellino 312
angelón 375
angelonia 406
angle 237
ángleto 237
ángleton 237
ánglica 237
ánglico 237
ánglito 237
ángliton 237
angola 238, 356
angolito 316, 318
angostura 455
angucha 298
anguche 298
angucho 298
anguito 343
angujo 298
angulo 330
añí 376
anica 200
añil 321
añil indigo 321
añil silvestre 321
anilina 321
añilina 321
añilón 321
anime 270, 452, 453, 454
anime alcanfor 452
anime blanco 452, 453
anime macho 453
anime rojo 452
anime ruda 453
anime toroó 452
aninga 200
anís 181, 196, 252, 271, 371, 374
anís anís 271
anís cimarrón 271
anís común 252
anís del monte 271
anís estrella 181
anís estrellado 181
anís silvestre 271
anís verde 252, 271
anisillo 181, 195, 196, 197, 252, 264, 267, 271, 286, 312
anizo 298
anjengibre 251
anjera 319
anjero 319
anjigible 251
año nuevo 377
anocuchuelo 252
anón 186, 187, 188, 189, 190, 404, 440
anón borroñoso 187
anón caucano 187
anón cienaguero 186
anón cimarrón 187
anón de caimán 186
anón de cerdo 186
anón de charapa 186
anón de corcho 187
anón de espino 187
anón de guara 188
anón de monte 186, 187, 188, 189, 404, 440
anón de pescado 186
anón de riñón 186
anón de verruga 187
anón enredadera 186
anón lechoso 440
anón liso 186
anón montañero 440
anón morado 187
anón negro 188

anón pelado 186
anón pelón 187
anón rosado 187
anón silvestre 187
anón verrugoso 187
anona 187
anoncillo 187, 188, 274
anoncito 187
anonilla 186
anselmo 418
ansermo 424
antá 204, 211
antañawi 430
anteojo de poeta 361
antioqueña 280, 398
antioqueño 297, 330, 404
antioquia 330
antoniodíaz 261
antra blanca 292
antra blanco 292
antra roja 367
anturio 198, 199, 201, 202
anturio de tres dedos 201
anturio garcita 202
anturio garza 202
anturio negro 199
anturio rojo 198
anturio tresdedos 201
antusara 287
anu wan u 290
ao 414
apacharama 382, 383
apacharana 382
apacín 287
apamate 363, 365
apamu 431
apararrayo 215
aparoje 224
apazote 282
ape-takii 380
apichu 463
apilja 432
apilla 432
apiña 432
apio 251, 252, 260, 430, 432
apio blanco 432
apio criollo 251
apio de monte 252, 260
apio de páramo 251, 252
apio grande 251
apio pequeño 251
aporrejao 188
aquí 457
ar 210, 364
ara 189, 213, 218, 222, 248, 265, 274, 282, 288, 335, 339, 347, 350, 351, 387, 404, 420, 422, 423
ara de la altura 420
arabia 280
arábigo 348
arabio 348
aracá-boi 425
arachigue 422
araco 212
araguá 240
araguaney 363, 365
araguaney bobo 365
araguaney sabanero 363
arajuimuci 426
aralea 254, 335
aralén 256, 263, 434
aralén de monte 434
aramina 415
araña 219, 234, 280, 301, 305, 383, 387, 436
araña de perro 280
arañagato 279, 469, 471
arañaplato 469
arañas 436
arancel 281
araño 413
arapara 326
araparí 326
araque 212
arara 278
ararà 335
araratucupé 329
araratucupí 329
araribis 314
arato 274
araucaria 180, 257
araucaria crespa 180
araucaria de páramo 257
araza 425
arazá 425
arbaquita 371
árbol baboso 407
árbol de aceite 315
árbol de algodón 413
árbol de alivio 188
árbol de barbasco falso 318
árbol de borugo 191, 451, 452
árbol de carambolas 432
árbol de caraña 328
árbol de caucho 437
árbol de caucho negro 437
árbol de cera 350, 427
árbol de chucha 185
árbol de chupaflor 423
árbol de churuco 307
árbol de coca 403
árbol de coral 433
árbol de Cúcuta 383
árbol de cuentas 433
árbol de fruta venenosa 318
árbol de hormiga 420
árbol de indio 458
árbol de jabón 459
arbol de jobo 450
árbol de la copaiba 315
árbol de la cruz 312
árbol de la lluvia 331
árbol de la verrugosa 197
árbol de la vida 454
árbol de las salchichas 364
árbol de lima 456
árbol de lluvia 331
árbol de los lamentos 393
árbol de nuche 388, 389
árbol de oreja 313
árbol de orejas 313
árbol de pan 436, 437
árbol de pava 349
árbol de pilón 381
árbol de seda 339, 433
árbol de Tolú 328
árbol de tomar 315
árbol de velas 364
árbol de zancudo 430
árbol del caucho 437
árbol del dinero 310
árbol del pan 436, 437
árbol del pan de danta 437
árbol del pan de tigre 437
árbol del pan de zambico 437
árbol del pan propio 437
árbol del paraíso 454
árbol del viajero 250
árbol dulce 392
árbol guaca 364
árbol guaruro 329
árbol guaruto 329
árbol orquídea 312
árbol salchicha 364
árbol vaca 341, 437
árbol vaco 437
árbol vela 280, 364
arbolito 298, 315, 389, 390, 396, 417
arbolito de navidad 389, 390, 396, 417
arboloco 266, 270
archucha 294, 295
archurchis 295
arciniegas 369
arco 188, 282, 331, 334, 363, 364, 367, 374
ardeverde 347
ardilla de montaña 410
areca 208
are-de-yé 191
arenillas 332
arenillero 391
arenillo 184, 255, 310, 331, 391, 393, 401, 430, 460
arenillo blanco 430, 460
arenillo de monte 310
arenillo de paloma 430
arenoso 287, 311, 314, 387
arepero 439
arepito 325, 326
arepo 284, 288, 330
arete 292
arevalo 457
arévalo 457
argalia 406
argentina 236, 286, 330, 394, 471
argentino 236, 246, 331
argüello 330
aria 249
arirai 393
arisa 312
arisco 305, 451
aristoloquia 192
arito 410
ariza 312, 439
arizá 312, 317
arizá amapolo 312
arizal 312
arnica 258
árnica 181, 254, 261, 267, 269, 271, 272, 366, 388
árnica blanca 269
árnica ceniza 269
árnica colombiana 269
árnica de árbol 181
árnica de bogotá 269
árnica de Bogotá 269
árnica de flor morada 269
árnica de la tierra 269
árnica de montaña 261
árnica de monte 269, 272
árnica de párampo 269
árnica morada 366
árnica santamaría 267
aro 362
aroeira 451
aroma 358
aromática 234, 376
aromático 257, 424
aromo 258, 331, 336, 339, 358
arquina 276
arrá 210
arra sĩi 212
arracacha 251, 252, 406, 437
arracacha amarilla 251
arracacha de monte 252
arracacha de tarro 251
arracachilla 252

arracachillo 391
arracacho 200, 332, 411, 430, 437
arracachuela 268, 289, 436
arracachuelo 259, 289
arraiján 424
arrasa con todo 282
arrasquillo 434
arrayan 427
arrayán 298, 304, 307, 326, 392, 396, 416, 417, 421, 424, 425, 426, 427, 428, 433, 449, 455
arrayán blanco 326, 392, 426
arrayán chizo 425
arrayán colorado 396
arrayán de hoja pequeña 427
arrayán de río 304
arrayán escobo 417
arrayán grande 427
arrayán guayabo 424, 427
arrayán negro 427
arrayana 388, 424
arrayancillo 428
arrayancito 396, 427
arrebatamacho 409
arrebianca 347
arrecate 251
arrecheche 382, 423
arrematamacho 409
arrocillo 232, 238, 240, 241, 242, 282, 336, 378
arrocito 461
arrojasangre 367
arroz 242, 246, 261, 277, 282, 311, 318, 321, 324, 331, 336, 390, 447
arroz blanco 242
arroz chino 242
arroz cimarrón 242
arroz colorado 242
arroz con coco 311, 318, 321, 324, 331
arroz del perú 282
arroz fortuna 242
arroz garrapata 242
arroz llanero 242, 261
arroz lucinda 242
arroz macho 242
arroz mira 242
arroz miramono 242
arroz monito 242
arroz monolaya 242
arroz rojo 242
arroz rongoy 242
arroz sanjerónimo 242
arroz silvestre 246
arruruz 250
artamisa 256, 257
artemisa 256, 257
artemisia 256, 257
artillero 250
arú 210
aruñagato 287
arupo 374
aruwalo 371
arveja 320, 324, 325, 330, 336
arveja de monte 336
arveja del diablo 336
arveja parda 325
arveja real 324
arveja soya 320
arvejilla 336
arvejón blanco 325
asaí 208, 209
asaí de Pará 209
asaí de sabana 208
asaí falso 208
asaí rojo 208
asaí sabanero 209
asaí verdadero 209
asaó 458
asaquiro 351
asaquirú 351
asawín 258
asentao 250
aserrín 310, 329, 330
asesina 339
asna cedro 454
asnayuyo 271
aso-ya-ee 414
asparcuarguat 279
aspave 450
aspavé 450
aspingo 364
assu 450
astillo 454
astroloja 193
astromelia 226, 409, 417
astromelia púrpura 417
astromelia roja 417
astromelio 409, 410, 417
astromelio rosado 417
ãtá 211
ataa 320
ataco 281, 282
ataco morado 281
atacorral 227
atadijo 436
ataiboto 210
ataja sangre 366
atajadanta 208
atajasangre 365, 366, 367
atakco 281
atarraya 290, 472
atiyayo 371
atonsora 287
atrapabruja 257
atu rui 390
atupa 315
atusara 287
aujilla 358
auranazeiro 458
auraucho 257
aurora 303, 409
aurorita 303
aurorita crespa 303
auroro 410
auso 278
auyama 293, 294
auyama blanca 293
auyama candelaria 293
auyama caturra 293
auyama común 293
auyama corazón de res 293
auyama lagartija 293
auyama pirul 293
auyama poteca 293
auyamo 298
ave 248, 250, 287
ave del paraíso 248, 250
avellana 391
avellano 391
avellano del Nare 391
avellano purgante 391
avena 234, 240
avena choclo 240
avena cimarrona 234
avena silvestre 234
avenilla 241
aviata 294
aviatá 294
avichucha 294
avichure 340, 437
avichuri 437
avichurí 437
avinge 311
avinje 440
avispa 409
avispita 273
avispitas 273, 368
avocado 185
ayahuasca 394, 395, 467
ayahuasca de los brujos 395
ayahuasca negra 395
ayahuasca rosada 395
ayainae 469
ayak pilli 264
ayako 210
ayota 295
ayubara 256
ayuela 418
ayuelo 303
ayuirai 381
ayullo 457
ayuno 457
ayurai 381
azabache 330
azafrán 250, 259, 287, 298, 375
azafrán bastardo 259
azafrán cimarrón 250
azafrán criollo 375
azafrán de la tierra 375
azafrán de montaña 375
azafrán de Perú 375
azafrán de tierra 375
azafrancillo 259
azafrás 182
azahar 352, 375, 456
azahar de indio 375
azahar de la india 456
azahar de la India 456
azahar de novia 375
azahar macho 352
azalea 298, 300
azalea de monte 298
azalea del monte 298
azapaca 372
azarito 448
azaritos 214
azola 179
azotacaballo 189, 190
azúcar 244, 321, 344
azúcar hayo 321
azúcar huayo 321
azucarerito 458
azucena 213, 214, 215, 226, 273, 342
azucena amarilla 215
azucena chirosa 213
azucena de la costa 214
azucena de La Habana 342
azucena de Obando 226
azucena de Quito 226
azucena roja 213
azucenillo 392
azuceno 342, 343, 348, 349, 350, 351, 352, 355, 388, 405
azuceno amarillo 342
azuceno blanco 342
azuceno cascarillo 352
azuceno colorado 342, 348
azuceno de La Habana 342
azuceno de monte 342, 355
azuceno rojo 342
azufre 385, 433
azul 184, 214, 236, 239, 240, 244, 249, 256, 288, 310, 330, 360, 379

azulejo 315, 330, 349, 353, 401, 422, 430
azulejos 259, 436
azulín 330
azulina 288, 379
azulita 238, 377, 471
azulito 247, 330, 379, 437
azurinas 259

B
baba 355, 408
baba india 408
babaco 279
babacú 332
babaimio 349
babasaino 274
babasaíno 414
babayhu igai 210
babillo 433
babo 408
babosa 412, 413, 415
babosa amarilla 412, 415
babosa hojimenuda 415
babosilla 413
babosita 410
baboso 183, 184, 253, 274, 407, 409, 413, 433, 445
baby blue 424
bacaba 210, 211
bacaíto 411
bacalao 280, 281
bacao 414
bacao de monte 414
bacaria 234, 246
bacaria lisa 246
bachaco 340
bachizomo 430
bacoe 447
bacopa 376
bacú 205
bacuma 227
bacuri 385
bacurí 385
badea 400
badizomo 430
bafacona 305
bagala 251
bagatá 311, 319, 403
bagazo 410
bagre pajo 175
bagtzue 437
bagua 332
baha 209
bahía 243
baho blanco 408
bahtápohkoño 210
bailador 454
bailador trompillo 454
bailarina 339, 409, 428, 429
baiuc 414
bajagua 332, 333
bajaguo 333
bajé-guayaba 426
bajman 296
ba-ki-rá 335
bala 199, 301, 330
bala de cañón 301
bala de zaino 199
bala rojo 330
balata 225, 305, 306
balata blanca 306
balata roja 305
balato 305, 392
balauruma 416
balaúste 314
balaústre 314
balaústre colorado 314
balay 249, 319
balayo 249
balazo 200, 201, 428
balazos 200
balguerajalé 265
balín 330
balita 330
balitas 330
ballico 241
balsa 389, 412
balsa macho 389
balsamín 295
balsamina 295, 297
balsamino 295
balsamira 295
balsamito 328
bálsamo 273, 315, 318, 328, 336, 343, 380, 451, 458
bálsamo blanco 328
bálsamo de canime 315
bálsamo de copaiba 315
bálsamo de María 380
bálsamo de tolú 328
bálsamo de Tolú 328
bálsamo del Perú 328
bálsamo hediondo 315
bálsamo negro 318
bálsamo rubio 328
bálsamo sanalotodo 451
bálsamo tolutiano 328
balsero 184
balsilla 257, 259, 272, 402, 412
balsillo 412
balso 266, 305, 388, 407, 408, 412, 436
balso algodón 412
balso baboso 409
balso blanco 409, 412
balso cadillo 409
balso colorado 412
balso menudito 409
balso negro 408, 412
balso real 412
balso tambor 412
balsudo 274, 451
balú 319
baluy 319
bamba 334, 438
bamba de alacrán 334
bambadu 463
bambaro 330
bámbaro 250
bambaro negro 330
bambil 209
bambou-bananier 280
bambú 235, 239, 244
bambú de la india 235
bambú de tallos amarillos 235
bambú guadua 239
bambú nativo 239
bambudo 331
bamneé 441
banano 250
banano amazónico 250
banano cerenano 250
banano enano 250
banano indio 250
banano indio rojo 250
banano maduraverde 250
banano manchoso 250
banano manzano 250
banco 182
bandera 219, 247, 390, 471
bandera amarilla 247
bandera colorada 219
bandera de cola 219
bandera de guerra 390
bandera de Pizarro 390
bandera de Uribe 247
banderilla 372
banderitas 219
bangaña 294
bangaño 294
bangaño de bejuco 294
baqueta 213
barajo 332
baranó 325
baranoa 311
barapaná 326
baratsui naekono 259
barayá 209
barba de ángel 330
barba de burro 337
barba de chivo 256, 279, 378
barba de gallo 358
barba de gallo colorado 358
barba de indio 232, 238
barba de mono 290
barba de palo 229
barba de sapo 279
barba de tigre 206
barba de viejo 229
barba roja 238
barbacoa 234
barbacobo 317
barbacona 305
barbaegallo 358
barbagallo 358
barbarito 441
barbas 180, 229, 283, 358, 387
barbas de chivo 180
barbas de gallo 358
barbas de guasco 387
barbas de palo 229
barbas de viejo 229, 283
barbascajo 402
barbasco 208, 260, 264, 269, 287, 289, 293, 303, 304, 312, 315, 317, 318, 321, 323, 325, 326, 330, 332, 334, 335, 340, 341, 343, 344, 372, 381, 388, 389, 391, 393, 395, 401, 402, 415, 431, 434, 449, 454, 456, 458, 459, 460, 462
barbásco 188
barbasco ahumado 449
barbasco amarillo 260, 330
barbasco de ají 431
barbasco de árbol 335, 402
barbasco de bejuco 325
barbasco de Castilla 335
barbasco de costilla 325
barbasco de dormilón 325
barbasco de escoba 208
barbasco de hoja 260, 335, 402
barbasco de lagartija 260
barbasco de loma 335
barbasco de monte 381
barbasco de pepa grande 381
barbasco de playa 335
barbasco de púa 303
barbasco de raíz 335
barbasco del río 381
barbasco del sol 325
barbasco fantasma 325
barbasco huaranga 335
barbasco huitoto 260
barbasco lechoso 341
barbasco mediano 381
barbasco negro 449
barbasco patagón 402

barbasco propio 325
barbasco rosado 289
barbasco sembrado 325
barbasco tinta 402
barbascoerraíz 335
barbascos 343
barbasquillo 289, 310, 402, 403, 415
barbasquito 402
barbato 454
barbatuscas 319
barbatusco 319
barbatuzcas 319
barbechivo 229
barbegallo 282, 358, 390, 409
barbero 287, 462
barboquillo 311
barbuches 330
barcino 246, 273, 330, 380
barcino del amarillo 273
barco 224
bardon 437
bardón 437
barelita 272
bareta 436
bareu puori 210
barie 205
bario 380
barniz 345, 350
barniz blanco 350
barniz de pasto 350
barniz de Pasto 350
barqui 362, 363
barqui blanco 362
barqui de bejuco 362
barqui negro 362
barquito morado 224
barquito morao 224
barquitos 365
barrabacillo 289
barrabás 289, 290, 303
barrabasillo 289
barranquillo 250
barredera 271
barrenillo 236
barrero 330
barrigón 407, 413
barrigona 208, 210, 413
barrigona blanca 208
barrigona negra 210
barrigonilla 210
barriguda 210
barril 279, 407
barroso 430
barsalito 257
barzalita 257, 258
barzalito 257, 258
barzalito de monte 258
basee 437
basen 371
basgüín 266, 267
basijo 321
basil 371
basilik 371
basquín 267
bastardo 259, 454
bastimento 405
bastón de obispo 251
bastón del diablo 274
bastón del emperador 251
bastón del rey 251
bastón imperial 251
basura prieta 275
basuriña 376
batabana 406
batata 203, 225, 226, 266, 337, 462, 463, 464, 471
batata amarilla 463
batata blanca 463
batata de monte 463
batata de sapo venenoso 266
batata morada 463
batata roja 463
batatilla 396, 434, 462, 463, 464
batatilla amarilla 462
batatilla blanca 462
batatilla de playa 464
batatilla de puerco 463
batatilla lila 463
batatilla morada 462, 463
batatilla roja 463
batatilla rosada 463
batatillo 193, 325, 396, 462, 463, 464
batato 434
batato de monte 434
bateí 210
bateíto 384
bateo 393
batú 210
batundo 444
bauk 414
baurá 325, 330
bay run 428
bay run bwaden 428
bayawond blan 331
bayeto 311, 313
bayo 330, 430
bayo blanco 430
bayrum 428
bebasojo 385
becerra 375
bedecapîcù 347
begonia 292, 293, 369
begonia chiquita 292
begonia cruz de hierro 292
begonia de hoja 292
begonia de pantano 292
begonia de salón 292
begonia regia 292
begonia sencilla 292, 369
behtáñó 204
beijoropenaci 382
bejuco 175, 179, 182, 192, 193, 196, 200, 201, 202, 208, 223, 226, 227, 256, 266, 267, 276, 278, 281, 282, 287, 288, 289, 290, 291, 293, 294, 295, 310, 312, 313, 314, 315, 317, 319, 322, 323, 325, 326, 328, 330, 331, 332, 333, 334, 337, 339, 340, 341, 342, 343, 346, 348, 358, 359, 361, 362, 363, 364, 365, 367, 368, 391, 394, 396, 398, 400, 407, 414, 429, 431, 434, 435, 436, 441, 442, 457, 458, 459, 462, 463, 464, 465, 472
bejuco agraz 295
bejuco ajo 364
bejuco alambre 179, 290, 340
bejuco alcalde 208
bejuco almendro 295
bejuco amarillo 341
bejuco arauato 200
bejuco badeo 400
bejuco barbasco 325
bejuco barbero 462
bejuco batatero 462
bejuco batatilla 463
bejuco batatillo 464
bejuco blanco 256, 363, 365, 436
bejuco bravo 394, 434, 435
bejuco buendía 459
bejuco burro 200, 202
bejuco cade bohobejuco 332
bejuco cadena 332
bejuco cadeno 312, 319, 332
bejuco calabacilla 363, 365
bejuco campana 463
bejuco campanita 429, 464
bejuco canasta 193
bejuco canastero 365
bejuco canastilla 362, 398
bejuco canastillo 398
bejuco canasto 226
bejuco candela 365
bejuco cangrejo 200
bejuco carare 192, 193
bejuco carey 192
bejuco caro 472
bejuco castro 472
bejuco Castro 472
bejuco cebollo 364
bejuco cestillo 227
bejuco chaparro 295
bejuco chimbilá 398
bejuco chinche 201
bejuco chirriador 472
bejuco chivo 289
bejuco choco 276
bejuco clavellino 267
bejuco coloradito 289
bejuco colorado 289, 295, 364
bejuco colorao 364
bejuco comemano 472
bejuco conchita 315
bejuco cortapicos 226
bejuco costillo 458
bejuco cruceta 458
bejuco culatero 363
bejuco de adorote 363
bejuco de agua 295, 358, 463, 472
bejuco de ajo 362, 364
bejuco de almidón 359
bejuco de anzuelo 358
bejuco de bicho 333
bejuco de bruja 462
bejuco de burro 287
bejuco de cadena 332
bejuco de cadena marrón 332
bejuco de canastillo 193
bejuco de cáncer 196, 343
bejuco de candela 295
bejuco de cangrejo 363
bejuco de carnero 313
bejuco de caro 472
bejuco de cerca 434
bejuco de chapolas 362
bejuco de chivo 266, 337
bejuco de chucha 362
bejuco de chupachupa 363
bejuco de cojé 295
bejuco de costilla de chimbe 200
bejuco de costillas de chimbe 200
bejuco de cruz 362
bejuco de culata 363
bejuco de culebra 266, 295, 435
bejuco de currucu 330
bejuco de estribo 319
bejuco de garrote 431
bejuco de garza 319

bejuco de globo 457
bejuco de granadillo 400
bejuco de guacharaca 282
bejuco de Guayaquil 193
bejuco de jabón 441
bejuco de jirigüelo 330
bejuco de lagartija 200, 227
bejuco de las peonias 331
bejuco de leche 340
bejuco de mamure 200
bejuco de maracure 346
bejuco de monte 295
bejuco de oro 362, 394
bejuco de orote 362
bejuco de oroto 362
bejuco de ortiga 313
bejuco de penda blanca 317
bejuco de peonía 310
bejuco de perdiz 361
bejuco de pescado 339, 340
bejuco de puño 364
bejuco de rana 362
bejuco de rejo 398
bejuco de rema 182, 442
bejuco de reuma 441
bejuco de riñoncito 367
bejuco de salud 434
bejuco de San Juan 396
bejuco de San Pedro 458
bejuco de sapo 293, 339, 340, 342
bejuco de sarna 202
bejuco de sonajera 314
bejuco de tierra 363
bejuco de tinto 462
bejuco de uñegato 227
bejuco de Urabá 442
bejuco de venado 431
bejuco de zaragoza 193
bejuco del curare 346
bejuco del diablo 340
bejuco del sol 192, 193
bejuco dulce 394
bejuco escalera 332
bejuco escalero 312
bejuco espinoso 407
bejuco espuelo 226
bejuco espuelón 227
bejuco esquinero 362, 459
bejuco esquinero macho 459
bejuco estrellita 463
bejuco falso 368
bejuco garcero 319
bejuco gonorrea 459
bejuco grande 346
bejuco guaco 192, 266
bejuco guasagil 193
bejuco guayacán 434
bejuco guayaquil 193
bejuco güendía 459
bejuco haparro 295
bejuco huavo 434
bejuco inglés 278
bejuco jabón 319
bejuco lechoso 340
bejuco mamure 200
bejuco mataganado 362
bejuco mechocuán 342
bejuco mercade 358
bejuco mollantín 289
bejuco mora dulce 295
bejuco mulato 458
bejuco murcia 435
bejuco nacundo 276
bejuco negro 226, 276, 287, 330, 332, 348
bejuco panzeburro 202
bejuco para nube de ojo 391
bejuco parta 319
bejuco pecoso 226
bejuco pegahueso 459
bejuco peinemico 362
bejuco penda blanca 317
bejuco pende 317
bejuco picapica 328
bejuco pintabollo 363
bejuco pintapollo 363
bejuco pitilla 434
bejuco pito 314
bejuco potré 227
bejuco prieto 458
bejuco pupú 463
bejuco quemador 295
bejuco rascarrasca 435
bejuco ratón 434
bejuco rayado 295
bejuco reuma 281
bejuco rosa 295
bejuco san juan 396
bejuco San Juan 396
bejuco sangre 326
bejuco sapo 266, 339, 340
bejuco sardinato 362
bejuco sierra 459
bejuco siete llagas 398
bejuco silvestre 202
bejuco soldador 367
bejuco tapabotijo 463
bejuco tomé 295
bejuco tripa de perro 201
bejuco tripa de pollo 200
bejuco tripadanta 458
bejuco tripaeperro 201
bejuco tripedanta 458
bejuco ubí 472
bejuco uñita 363
bejuco vainilla 223
bejuco vainillo 223
bejuco yaré 200
bejuco yute 348
bejuco zaragozo 434
bejuco zorro 319
bejuquillo 223, 335, 362, 364, 462, 463
bejuquillo de la hembra 364
bejuquito 223, 294
bejuquito maraquero 294
belanae 458
beleño 333
belero 342
bella 218, 287, 288, 303, 372, 401, 431, 464
bella de día 464
bella de noche 287
bella emilia 288
bella hasta las once 401
bella maria 431
bella salvia 372
bella sombra 287
bellaalasnueve 401
bellaalasonce 290
bellaco 250
belladona 436, 467, 468
bellaemilia 288
bellahelena 297
bellaonce 290
bellísima 277, 286, 288, 312, 366
bellísimo 288
bellita 288
bellota 294, 295
belmania 219
ben 224, 280, 336, 438
ben olifere 280
bencenuco 339, 351
bengible 251
benjamín 438, 439
benjuí 309
beno 253
beñoó 409
benzolive 280
berada 240
berba de chivo 256
berbería 342
berebere 251
berengüela 291
berenjena 294, 470
berenjena de monte 470
beringuera 359
berlinda 215
bermuda 236
bermudera 312
berraquillo 404, 426, 436
berriadores 469, 471
berrito caminante 194
berro 225, 252, 276, 277, 376
berro de Castilla 276
berro de huerta 277
berro de laguna 225
berro macho 252
berrón 428
berros 253, 264, 276, 277, 286
besa 228
besito 297
besito de jardín 297
besito de monte 297
besitos 297, 303
beso 290, 297, 303, 354, 391
beso de antioqueño 297
beso de Gaitán 290
beso de madre 391
beso de negra 354
beso de novia 297
besos 297
betaca 471
betacacù 471
betaño 204
betañő 204
beto 204
betoñi 204
betónica 370
biao 249
bibimiku háheku 210
bibisinae 458
bibosi 437, 438
bibrán 438
bicco 414
bichachá 299
biche 294, 332
bichichinejo 186
bichinche 424
bicho 276, 316, 317, 332, 333, 337
bicho abejón 332
bicho bombito 332, 333
bicho de cafe 333
bicho de café 332, 333
bicho macho 333
bicho negro 333
bichuela 294, 330
bien te veo 294
bienmesabe 457
bienteveo 294
biernesabe 457
biflora 300, 343, 377, 398, 425, 437
biflora reina de Hungría 300
bigarro 341
bignonia 365
bihao 248, 249, 250, 345

bihao blanco 249
bihao negro 249
bihaos 248
bihua junu 362
bija 363, 406, 451
bijá 451
bija amarilla 451
bijagüillo 248, 250, 251
bijao 247, 248, 249, 250
bijao cuero 249
bijao de agua 247, 249, 250
bijao de fardo 250
bijao de palito 250
bijao de puerco 248
bijao mantagorda 249
bijao rosario 247
bijaos 248
bijo 192, 385, 451
bijuacá 289, 290
bilanda 455
bilasonce 290
bilibil 454, 455
bilibile 454
bilibili 454
billalasonce 290
billete 214
biota 295
birajo 447
birch 451
biribá 187
biringo 451
birringo 451
bismarckia 207
bispera 418
bita 191
bitamorrial 391
bitibiti 454
bitora 294
bitoria 294
bitoriera 294
biumía 327
biyuyo 274
bizcochito 430
bizcocho 315, 391, 392
blanco 178, 182, 184, 185, 186, 188, 191, 192, 193, 195, 196, 203, 204, 209, 218, 226, 229, 232, 234, 235, 236, 241, 242, 247, 249, 250, 254, 255, 256, 257, 261, 264, 265, 266, 267, 268, 270, 271, 272, 273, 274, 276, 277, 278, 280, 281, 282, 292, 294, 295, 296, 297, 301, 302, 303, 304, 305, 306, 308, 309, 310, 311, 313, 314, 315, 319, 320, 322, 323, 324, 325, 326, 327, 328, 329, 330, 331, 334, 335, 336, 339, 340, 341, 342, 348, 350, 352, 358, 359, 362, 363, 364, 365, 367, 369, 370, 371, 374, 377, 378, 379, 380, 383, 384, 386, 391, 392, 393, 394, 395, 397, 401, 403, 405, 407, 408, 412, 414, 415, 416, 420, 421, 422, 423, 424, 426, 427, 430, 432, 436, 437, 439, 441, 442, 445, 446, 448, 450, 451, 452, 453, 454, 455, 456, 457, 460, 463, 465, 466, 472, 473
blanquecino 447
blanquillo 254
blanquísimo 447
blanquito 469
blanquizco 269
blas del mar 390
bledo 281, 282, 289
bledo blanco 281, 282
bledo chico 281, 282
bledo colorado 281
bledo colorao 281
bledo de espina 281
bledo de puerco 281, 282
bledo espinoso 281
bledo hembra 281
bledo liso 281
bledo macho 281
bledo maleza 281
bledo manso 282
bledo rojizo 281
bledochupa 284
bleo 281, 282, 284, 331, 468
bleo blanco 282
bleo de chupa 284
bleo de gallinazo 468
bleo de golero 468
bleo de puerco 281
bleo negro 282
bleochupa 284
boba 173, 364
bobena 414
bobi 206
bobil 205
bobila 205
bobo 200, 330, 365, 404, 445
bobolo 471
bobotomobu 447
boca caliente 354
boca de diablo 353
boca de dragón 360, 375
boca de monte 314
boca de negro 354
boca de sapo 354
bocachica 250
bocachico 325
bocadillo 181, 250
bocado de culebra 295
bocaoeculebra 295
bocarique 416
bocas de dragón 375
bocavieja 355
bóchoa tsáha 447
boco 384
boco-á 384
bodó 251
bodoquera 210, 338
bodoquero 296
bogotana 260, 289, 471
bogotano 250, 338, 454
bogüe 413
bohobejuco 332
bohsomuhĩ́ 210
bohsomũĩ 210
boighe 181
boique 181
bois mabi 441
bois pilori 292
boj de la China 456
boj de tierra caliente 456
bokohañu 210
bola 213, 278, 294, 308, 344, 379, 472, 473
bola de gato 344
bola de perro 278, 344
bola de toro 344
bolaegato 469
bolaina 408
bolas de gato 274, 371
boldo 185
bolegato 469
bolenillo 191, 209
bolero 302
bolinillo 413
bolito 294
boliviana 255, 365, 429
bollo blanco 331
bollo de puerco 287, 387
bollo de puerco rojo 387
bollo limpio 325
bollo mohoso 458
bollolimpio 458
bolo 293, 294
bolombolo 311
boloquera 210
bolos 294
bolsa de amor 468
bolsa de lobo 340
bolsa de mono 340
bolsa de pastor 340
bombas 340
bombillo 409
bombito 332, 333, 359
bombo 210, 454
bombolo 471
bómbolo 471
bombona 208, 210
bomboná 210
bombona falsa 208
bombonaje 210
bon am 191
bonbon 212
bonche 409
boncheaí 409
bonga 407, 413
bonga bruja 407
bonga colorada 413
bongo 185, 311, 407
bongo de agua 311
bonguito 332, 412
bonifacio 411
bonitel colorado 274
boque 413
boquelión 375
boquiabierto 375
boquiabiertos 375
boquiblanco 321
bora 225
borbón 348
borbón amarillo 348
bordón 437
bordoncillo 250
bore 199, 200, 203
bore de raya 200
bore morado 199
boreyabeñu 210
boriabea 210
boricá 199
boriche 383
borkai 451
borla 261, 292
borla de San Pedro 292
borlas 281, 282, 387
borlita 261
borlitas 261
borojó 346, 347, 349, 351, 355, 380
borojó de monte 380
borojó del pacífico 346, 347
borojó del Pacífico 346
borojó montano 347
borojocillo 355
borombolo 294, 318
borrachera 281, 299, 369, 444, 465, 466
borrachera de la morada 369
borrachera del yagé 281

borrachero 281, 283, 293, 299, 323, 359, 394, 404, 465, 466, 467, 471
borrachero amarillo 465
borrachero blanco 465, 466
borrachero colorado 465
borrachero montero 359
borrachero rojo 465
borrachero rosado 465
borracho 181, 330
borrachuelo 241
borragón 276
borraja 273, 274
borraja blanca 273
borrajo 275, 369
borrajón 275, 276
borreguitas 325
bórúá 210
bosa de pastor 276
bosopa 448
botagajo 318, 430
botaguajo 318
botecito 362
botellón 351
botijita 286
botijitas 286
botón 255, 256, 259, 261, 264, 265, 266, 269, 270, 271, 272, 282, 298, 330, 370, 371, 392, 413, 416, 465
botón blanco 261, 266
botón de oro 255, 256, 259, 264, 265, 269, 270, 271, 272
botón de plata 272
botón de saco 330
botón de soltero 259
botón dorado 272
botón morado 371
botón negro 370
botón-botón 298
botoncillo 230, 255, 256, 258, 259, 261, 264, 270, 271, 281, 356, 357, 369, 370, 416
botoncillo amarillo 264
botoncillo blanco 264
botoncillo de Castilla 255, 264
botoncillo de monte 270
botoncillo hembra 255
botoncillo macho 264
botoncito 280, 350
botoncito blanco 280
botondioro 271
botones 396
bototo 406
botumbo 443
botundo 444
botuto 406
boul di mas 224
bouquet de novia 352
bovista 470
boya 203, 337, 341
bracaria 246
bragueta de viejo 252
brailejón 391
braquiara amargo 246
braquiaria 246
braquiaria amargo 246
braquiaria común 246
braquiaria dulce 246
braquipará 246
brasil 320, 321, 339, 349, 357
brasileña 283, 314, 425
brasilera 209, 362
brasilete 320, 330
brasileto 320, 330
brasilito 320
brazanegra 430
brazil 301, 320, 349, 352
brazilete 320
brazilito 320
brazo de negro 276
brazonegro 276
brea 330, 385
brecharia 246
breo 306, 385, 452, 453, 454, 455
breo amarillo 306
breo blanco 452, 453
breo de cucaracha 453
breo para calafatear 385
bretaño 274
bretónica 368, 370, 371, 372, 410, 411, 412
bretónica morada 412
breva 438
breva de monte 438
brevo 438
brigua 247
brijolito 323
brillaalasonce 290
brillalasonce 290, 401
brisa 214, 286
brisa de Rin 214
brisa del río 214
brisas del mar 277
brisas del Rin 214
brocas 413
bruca 333
brucha macho 332
bruja 224, 267, 271, 300, 350, 461
brujita 350
brujita negra 350
brujitos 310
brujo 308, 350, 449
bruma 214
bruma del mar 214
brusca 272, 332, 333
brusca macho 333
brusca negra 333
brusca sabanera 272
brusco 215, 333
bruscón 332
brusquillo 316
buandn-doch 275
buatensmesh 247
buba 267
buburi 306
bucago 319
bucare 319
búcare 319
bucare peonía 319
bucaro 319
búcaro 319
búcaro de pantano 319
buche 193, 225, 311, 330
buche blanco 330
buche colorado 330
buche de gallina 225
buche de pavo 193
buchegallina 288
buchegallino 288
buchesapo 393
buchón 198, 225
buchón de agua 225
buchona 202, 468
buchuba 468
buchuva 468
buenamigo 358, 451
buenas tardes 287
buenastardes 265, 272, 287, 290, 297, 340, 344, 430
buenasuerte 456
buéntsamo 257
buganvil 286, 369
buganvil anaranjado 286
buganvil rojo 369
buganvilia 286
bugueño 324
buha tuuko 205
búho bombito 332
buii 392
buitsaha 211
bujeque 311
bujurqui 439
bulletigre 405
bulto 463
bumbuhe 210
bunquín 379
bünü taju boto 320
buñuelo 392
buqué 352, 469
buqué costeño 352
buqué de novia 352, 469
buquecito 343
buria 247
burikui 201
burilico 190
buriligo 190
burriadá 445, 446
burriará 446
burrillo 415
burucuña 399
burücunae 332, 333, 408
buru-moj-quere 206
busaca 333
busú 210
busukobenaeji 171
butsiboto 210
buuyutoci 434
buyecacu 391
buzunuco 351
bwa dom 408
bwa kanpéch 320
bwa wa tau 378

C
caá 392
caabaza 363
caanuci 188, 189
caapi 394, 396
caaporopita 344
caballito 363, 364, 365
caballo 173, 197, 231, 233, 240, 255, 261, 268, 284, 315, 316, 333, 335, 376
caballo caspi 335
cabalonga 340
cabalongo 340, 365
cabalongo extranjero 365
cabaret 362
cabari 315
cabazo 294
cabecenegro 205
cabecera 414
cabecerá 414
cabecetigre 412
cabecilla 204
cabecinegro 205, 210, 284
cabecita 204
cabecitanegra 337
cabellito de ángel 279
cabello de ángel 279, 280, 380, 469
cabello de indio 376
cabellos de ángel 279, 280, 312
cabeza de churuco 302, 414

cabeza de indio 284
cabeza de negrito 408
cabeza de negro 211, 225, 229, 284
cabeza de picure 393
cabeza de tigre 412
cabezona 187, 237, 243, 258, 259
cabezoncillo 231
cabiche blanco 331
cabima 315
cabo de hacha 190, 191, 192, 255, 296, 297, 301, 339, 351, 441, 449
cabo de indio 191
cabo de judío 191
cabo de oro 311
cabo hacha 191
cabrito 448
cabro 255, 256
cabuya 214, 215
cabuya azul 214
cabuya común 215
cabuya con espinas 215
cabuya de México 214
cabuya hembra 215
cabuya macho 215
cabuya montua 214
cabuya sin espinas 215
cabuyo 301
cacabel 316
cacachilla 442
cacaco 311
cacagúillo 414
cacagüillo 412
cacagüito 298, 299
cacahua caspi 414
cacahuate 311
cacahuete 311
cacahuillo 380, 409
cacahuito 414
cacaíto 353, 380, 409, 414
cacaíto de monte 409
cacambal 415
cacao 299, 334, 348, 364, 380, 383, 409, 411, 412, 414, 454, 465, 466
cacao amarillo 409
cacao blanco 380
cacao cahuí 409
cacao caturro 414
cacao cimarrón 348, 409, 414
cacao cojón 414
cacao criollo 414
cacao cuadrado 409
cacao de agua 412
cacao de altura 409
cacao de boruga 409
cacao de iguana 414
cacao de indio 414
cacao de mico 414
cacao de monte 380, 409, 412, 414, 454
cacao de montero 454
cacao de muerciélago 409
cacao de murciélago 409
cacao de perro 299
cacao de rama 414
cacao de semilla 414
cacao del monte 409
cacao dulce 414
cacao hartón 414
cacao higuillo 409
cacao indio 414
cacao maduraverde 414
cacao manavén 414
cacao maní 409
cacao maracos 414
cacao marraco 414
cacao montañero 409, 454
cacao monte 409
cacao morado 414
cacao sabanero 465, 466
cacao silvestre 409, 414, 454
cacaona 332
cacaotero 414
cacau 414
cacauillo 409
cacay 388, 409
cacha 330
cacha amarilla 330
cacha azul 330
cacha blanco 330
cacha calentano 330
cachaco 250, 280, 355
cachaco de muladar 280
cacharro 255
cachas 330
cachetón 330
cachetona 201
cachicamo 380
cachicarnero 458
cachimbo 319
cachimolivo 340
cachimonda 278
cachingo 319
cachinguba 438
cachipá 206
cachipae 206
cachipay 206
cachipay montañero 206
cachito 255, 315, 336, 398
cachito de cabro 255
cachito de venado 398
cachitos 280, 345
cacho 171, 176, 247, 258, 274, 280, 305, 311, 318, 329, 330, 336, 356, 358, 363, 378, 405, 420, 430, 457
cacho cabra 311
cacho de cabra 258, 336
cacho de chivo 363
cacho de vaca 363
cacho de venado 171, 176, 274, 280, 356, 358, 363, 378, 405, 420, 430, 457
cacho de venado sabanero 363
cacho de venao 171
cachoecabro 255
cachoevenado 176
cachorro 235
cachovenado 430
cachuda 210, 212
cachuda barriguda 210
cachuda zancona 212
cachudita 432
cachudo 210
cachungo 319
cacique 250, 268, 318, 455
caco 364
caco castañete 364
cacto 283, 284, 285
cactus 283, 284, 391
cactus de flor roja 283
cadena de amor 288
cadeno 312, 319, 332
cadilla 415
cadillo 224, 235, 257, 258, 259, 260, 263, 268, 271, 272, 280, 281, 282, 291, 303, 310, 317, 318, 320, 327, 331, 372, 379, 387, 407, 409, 412, 415, 442, 445
cadillo arroz 235
cadillo azul 407
cadillo blanco 415, 442, 445
cadillo bola 379
cadillo carretón 235
cadillo carretón morado 235
cadillo chisacá 271
cadillo cimarrón 415
cadillo común 415
cadillo correntino 235
cadillo de bejuco 282
cadillo de bestia 263, 412, 415
cadillo de bolsa 282, 379
cadillo de chipaca 271
cadillo de huerta 258
cadillo de indio 258
cadillo de mazorca 280, 387
cadillo de mula 415
cadillo de paja 235
cadillo de perro 327, 415
cadillo de tierra 327
cadillo grande 407
cadillo juancho 318
cadillo lanudo 415
cadillo mulato 415
cadillo pegajoso 379
cadillo pipa 320
cadillo puya 318
cadillo sacalotodo 282
cadillo tigre 235
cadmia 187
cadmio 187
cadulo 442
café 332, 333, 334, 348, 349, 353, 355, 394, 395, 403, 425, 454
café arábico 348
café borbón 348
café borbón amarillo 348
café chilinchil 333
café churrusco 333
café Colombia 348
café común 348
café de brusca 333
café de brusco 333
café de Colombia 348
café de montaña 353
café de monte 349, 353, 394, 395
café de páramo 353
café de pobre 332, 333
café de pobres 333
café del diablo 403
café enano 348
café excelso 348
cafe furrusco 333
café furrusco 332, 333
café libérico 348
café macho 425
café moucha 333
café San Lorenzo 348
café San Ramón 348
cafecillo 333, 347, 366, 394, 449
cafecito 333, 386
cafelillo 333
cafetana 388
cafeterana 388
cafetero 362
cafetillo 350, 354, 455
cafeto 198, 253, 324, 348, 353, 354, 355, 362
cafeto arábico 348
cafeto de monte 353, 354, 355

cafuche 306
cagá de pájaro 448
cagada de pajarito 450
cagalera 354
cagalero 436
cagaloduro 248
cagalungasé 442
cagao de lombriz 305
cagua 295
caguaneje 388
caguanejo 388
cagüi 381
cagüí 381, 382
cahaco de muladar 280
cahay 388
cahimolivo 340
cahu 450
cahuaia 210
cahuayo 210
cahuí 381
caica 348
caico 455
caigua 294
caima 447
caimancillo 449
caimaron 447
caimarón 447
caimarón de mico 447
caimarón de mono 447
caimarón de montaña 447
caimarón de monte 447
caimaron silvestre 447
caimarón silvestre 447
caimarona 447
caimillo 291, 305, 306, 307, 308, 349
caimillo fino 307
caimillo ordinario 349
caimitico 308
caimitillo 305, 306, 307, 308, 347, 353
caimitillo balso 305
caimito 305, 306, 307, 308, 339, 340, 341, 355, 383, 394, 424, 427, 441, 449
caimito amarillo 306
caimito arisco 305
caimito blanco 305
caimito canelo 339, 427
caimito colorado 307
caimito de montaña 305
caimito de monte 305, 306, 449
caimito de rebalse 308
caimito demonte 449
caimito fundipaila 305
caimito jiowoo-mesese-e 305
caimito leche 305, 441
caimito leche de perra 441
caimito lirio 340
caimito mora 305, 306
caimito morado 305
caimito pelao 308, 355
caimito plátano 340
caimito popa 305, 340
caimito Pumarejo 341
caimito rucio 306
caimito sabanero 308
caimito silbador 306
caimo 291, 305, 306, 307, 308, 340, 342, 343, 349, 381, 388, 392, 405, 440, 441
caimo amarillo 305, 306, 307
caimo balata 305
caimo blanco 305
caimo buñuelo 381
caimo cascabel 306
caimo cerezo 305
caimo chicle 440
caimo churuco 305
caimo colorado 305, 306
caimo colorao 305
caimo danto 308
caimo de borugo 306
caimo de charapa 306
caimo de churruco 305
caimo de churuco 305
caimo de lombriz 305, 308
caimo de monte 305, 306, 307, 308, 349
caimo de pescado 349
caimo de plátano 340
caimo de sol 306
caimo fierro 441
caimo hembra 305
caimo lechoso 441
caimo liso de monte 305
caimo maduroverde 306
caimo morado 305, 307, 342, 343
caimo negro 305, 307
caimo peludo 308
caimo pequeñito 306
caimo plátano 340
caimo sapo 306
caimo silvestre 405
caimo vela 306
caimo verde 305
caimo zanca de venado 308
caimón 307
cainica 348
caipe 312
cairasco 227
cairo 312
cairuro 329
cairurú 329
caiuia 420
cajao-amo 290
cajaró 250
cajete 253
cajeto 198, 250, 253, 287, 291, 323, 324, 359, 362, 363, 378, 409, 437, 441
cajeto arco 362
cajeto crecedor 363
cajeto-arco 362
caji 394
caji-idirecaii 394
cajón 291, 331, 344, 347, 356
cajón del amo 344
cajú 450
cajuaro 274
cajuche 436
cajuil 450
calaba 380
calabacera 294
calabacilla 293, 363, 365
calabacillo 286, 294, 355, 362, 365
calabacillo prieto 365
calabacín 294
calabacita 294
calabacito 294, 309
calabacito amargo 294
calabash tree 363
calabatá 300
calabaza 293, 294, 363, 365
calabaza biche 294
calabaza bonetera 294
calabaza chura 294
calabaza churita 294
calabaza de guasca 294
calabaza dulce 294
calabaza en bichuela 294
calabaza guagua 294
calabaza vinatera 294
calabazo 293, 294, 355, 363, 433
calabazo amargoso 294
calabazo bangaño 294
calabazo blanco 294
calabazo candongo 363
calabazo cimarrón 363
calabazo cinturo 294
calabazo de bangaño 294
calabazo de palo 363
calabazo de pico 363
calabazuelo 278
calabomba 305
calaguala 175, 176, 177, 178, 283, 284
calahuala 176
calalú 281
calamanta 287
calamayó 317
calambayó 317
calambolo 432
calambombo 283, 359
calanguala 283
calarqueño 330
calastoliendra 406
calatea 249
calavera 215, 217, 362, 453
calaveras 217
calavero 374
calbón 209
calchilla 275
caléndula 259, 272
calentana 276, 413, 451
calentano 271, 363
calentuno 434
calicá 212
calistemo blanco 426
cáliz 276
calla 199
callpachina yuyu 344
caló 209
calopo 313
calostro 441
calú 319
calzón 276, 373
calzón del diablo 373
calzoncillo 227, 410
calzoncillo del diablo 411
calzoncitos 276
calzones del diablo 373
camacho 354
camahua 208
camajón 414
camajonduro 414
camajorú 414
camajurú 414
camalete 274
camalitenae 451
camándula 236
camándulo 402
camaní 250
camaparu 318
camaperro 243
camarero 299
camargo 266, 270
camarón 191, 282, 291, 312, 359, 360, 447, 465
camarón amarillo 360
camarón silvestre 447
camarona 298, 299, 300, 360
camaroncillo 387
camasa 363
camasey 420
camatai 282

camay 187
camazo 363
cambombulo 333
cambuayo 278
cambul negro 250
cámbulo 319
cámbulo rosado 319
cambur 250
cambur guajibo 250
cambur sopero 250
cambure negro 250
camburú 250
cambusil 253
camelia 309
camerona 298
camina con pie 242
caminacompié 234, 242, 244
caminadera 171, 240
caminadora 235, 240, 244
caminera 235
camita 297
camivor 331
cammon 279
camomilla 265
camoruco 414
camote 201, 463
camote blanco 463
camote de sargotín 201
camote morado 463
campana 272, 303, 316, 339, 362, 406, 463
campana amarilla 316
campana de oro 339
campana morada 362
campanilla 255, 315, 349, 407, 435, 463, 464, 465
campanilla de oro 255
campanilla encarnada 465
campanilla morada 407
campanillas 272
campanillo 310, 319
campanita 315, 406, 462, 463, 464
campanita blanca 462, 463
campanita morada 464
campanitas 272, 361
campanito 431
campano 315, 331, 414, 431, 470
campaño 331
campano bleo 331
campano común 414
campano hembra 331
campanola 272
campánula 272
campesina 465
camucamu 427
camuruco 414
can 193, 341, 409
cana 254, 283
caña 199, 206, 207, 215, 234, 236, 239, 240, 244, 247, 248, 417
caña agria 248, 417
caña agria chiquita 417
caña amarga 240
caña amarilla 244
caña canapoy 244
caña cimarrona 199
caña criolla 244
caña de adorno 215
caña de azúcar 244
caña de azúcar altacumbre 244
caña de azúcar amarilla 244
caña de azúcar amarilla rayuna 244
caña de azúcar americana 244
caña de azúcar azúcar 244
caña de azúcar azucarera 244
caña de azúcar bella 244
caña de Castilla 234, 240
caña de flecha 240
caña de la pasión 247
caña de lata 206
caña de molinillo 207
caña de puente 199
caña de queco 234
caña de San Pablo 207
caña de tierra fría 234
caña de víbora 207
cana de vieja 254, 283
caña hueca fina 239
caña india 240
caña limonaria 236
caña muda 199
caña negra 206
caña selvática 248
cañablanca 240
cañaboba 240
cañabrava 190, 206, 234, 239, 240, 241, 247, 248
cañabrava de Castilla 234
cañabrava playera 240
cañabravo 190, 311
cañadonga 314
cañaelaindia 215
cañafina 240
cañafisto 314, 315
cañafístol 314, 333
cañafístola 313, 314, 333
cañafístola delgada 314
cañafístola gruesa 314
cañafístola macho 333
cañafístole 314
cañafístolo 314, 333
cañafistula 314
cañafístula 313, 314, 332, 333
cañafístula burrero 314
cañafístula de caballo 333
cañafístula delgada 314
cañafístula gruesa 314
cañafístula macho 333
cañafístula pequeña 314
cañafistulillo 333
cañafístulo 313, 314, 333
cañafístulo amarillo 314
cañafístulo de altura 333
cañafistulo macho 333
cañafístulo macho 314, 333
cañafístulo pequeño 314
cañaflecha 240
cañaflete 240
cañaflota 314
cañaflote 314
cañafuerte 248
cañagri 417
cañagria 224, 247, 248, 251, 383, 417
cañagria cimarrona 251
cañagria menuda 247
cañagua 224
cañaguadua 239
cañaguate 247, 248, 363, 364, 365
cañaguatos 247, 248
cañaguauda 239
cañahua 282
cañahuate 365
canajá 242, 244
cañalanuda 248
canalete 273, 274, 364
canalete común 364
canalete de humo 273
canalete de río 274
canalete prieto 273
cañamansa 240
cañamazo 234
canambo 205
cañamenuda 240
cáñamo 215, 436
cáñamo de África 215
cáñamo de la India 436
cáñamo índico 436
cananá 209
cananco 454
cañandonga 313
cañandronga 330
cañandunga 313
cananeo 330, 454
cananga 187
canangucha 210, 392
canangucha de sabana 210
canangucha falsa 210
canangucha macho 210
cananguche 210
cananguche falso 210
cananguchilla 210
cananguchillo 210
cananguchillo de sabana 210
canangucho 210
canangucho bravo 210
canangucho de sabana 210
cañapángala 240, 248
cañapopa 234
canapoy 244
cañarote 246
canasperro 372
canasta 193, 409
canastilla 193, 362, 398, 409
canastillo 193, 395, 409
canavalia 313
canavalia Saldaña 313
canayuyo 270
can-cay 341
cáncer 196, 255, 267, 281, 283, 285, 320, 343, 370, 373, 380, 387, 464
cancer blanco 281
cáncer blanco 281
cáncer de cocha 464
cáncer de laguna 281
cáncer de loma 255, 320
cáncer espinoso 281
cancer morado 281
cáncer morado 281, 373
cáncer morado silvestre 373
cáncer rojo 373
cáncer sol 283
cancerín 371
canchalagua 338, 390
canchalagüe 390
cancharama 451
cancharana 451
cancho 386
canción de la India 215
candango 187
candela 295, 365
candelabro 180, 248, 251, 254, 360
candelabro de agua 251
candelero 254, 274, 388, 389, 395
candelero real 388
candelillo 333, 401
candelo 309, 401, 441
candelo blanco 401
candelo rojo 401
candia 406
candil 455
cane 339

cané 339
cañegua 347
cañeja 247, 248
cañejo 248
canela 181, 182, 183, 371, 389
canela de páramo 181
canela de Quijos 182
canela falsa 389
canela muena 182
caneletillo 416
canelo 181, 182, 183, 184, 194, 339, 401, 418, 427
canelo cenizo 184
canelo de Andaquíes 182, 183
canelo de los Andaquíes 183
canelo de monte 181
canelo de páramo 181
canelo real caparrapí 183
canelo real hoja chica 183
canelón 181, 182, 183, 193, 194, 195, 285, 350
cañena 347
cañeno 409
caney 339
cangalla 375
cangil 247
cangreja 359
cangrejo 191, 200, 305, 306, 363
cangrio 325
canguil 247
canigua 282
canihua 282
canilla de golero 280
canilla de grulla 198
canilla de mula 260
canilla de tente 397
canilla de vieja blanco 306
canillo 442
canime 315, 318, 328, 331
canjara 274
canjí 415
canllarana 451
canoa 224
canoita 202
canola 276
canopio 251
cañote 234
canotillo 195, 415
canot-macaque 420
cansacerro 443
cansamozo 357
cansamuelas 323
cansamulato 276
cansaperro 372
cansin 209
cantagallo 319, 334
cantagallo blanco 334
cantagallo colorado 319
cantagallos 319
cántaro 423
cantimplora 375
cañuela 239, 241, 242, 251, 417
cañuela trepadora 241
canutillo 173, 224, 238, 240, 243, 245, 297, 302
canutillo morado 224
canuto 351
canyarana 451
canyí 415
caoba 454
caoba americana 454
caoba de interior 454
caobo 454
capá 372
capachí 190
capacho 332, 416
capacira 351
capaelarreina 215, 369
capafraile 398
caparo apedomu 338
caparo opedoamo 291
caparrosa 394
cape 190, 349, 383, 384, 385
capé 384
cape grande 190, 383, 385
capecape 349, 384
capecito 349
capenegro 349
capey 383
capi 247, 394
capí 247
capi o caapi 394
capilera 177
capín venezuela 234
capinurí 382, 440, 441
capiria 471
capirón 347, 393
capirona 347, 358
capirona de altura 347
capitán 310, 311
capitana 192, 193, 259, 266, 267, 283, 339, 434
capitana blanca 434
capitana de corazón 193
capitana macho 283
capitancillo 330
capitiú 185
capón 330
caporuto 319
capote 312, 314, 326
capotillo colorado 198
capuchina 280
capuí 414
capulí 299, 304, 444, 457, 468
capulí silvestre 299
capulín 415, 436
capullo 339
capure 306
capure blanco 306
capure negro 306
capuri balachina 306
caquezá 388
caqui 381
cara de mico 408
carabajo 465
carabiri 433
caraca 215
caracaro 319
caracasuna 276
caracha caspi 335
caracol 250, 279, 292, 367, 450
caracol de jardín 279
caracola 365, 366, 367
caracola roja 367
caracolí 191, 450
caracolí amarillo 450
caracolí blanco 450
caracolito 275, 377
carácter 416
carácter de hombre 416
carácter de los hombres 416
carácter del hombre 416
caracucha 297, 313
caracucho 280, 297
caracucho blanco 297
caracucho grande 280
caracuchos 297
carafiambre 433
caragay 378
caraguala 283, 284
câraiñu 210
carajirú 363
caramacate 404
carambola 432
carambolas 432
carambolí 274
carambolo 432
carambombo 332
caramelo 388
caramo 175
caraná 210, 227
caraña 328, 383, 451, 452, 453
caraná bravo 210
caranaí 210
caranganito 332
carano 328
caraño 451, 452, 453, 454
caranuca 335
carao 314
carapa 320, 454
carapacha 441
carapacunga 258
carapate 392
carapo 454, 457
caraqueña 340, 379
caraqueño 320
carare 192, 193
carate 263, 318, 351, 394, 451
carate blanco 394
carate de mariposa 263
carate negro 394
carate rojo 394
carateja 199, 369
caratejo 330, 455
caratero 451, 460
caratillo 394
carató 390
caratoso 451
carauta 330
carauta blanca 330
carauta silvestre 330
carbón 298, 309, 313, 326, 337, 441
carbón mosco 298
carbonal 328
carboncillo 382
carbonera 383
carbonerito 298, 313
carbonero 274, 297, 298, 310, 311, 313, 315, 321, 325, 331, 337, 380, 382, 383, 397, 405, 431, 457
carbonero amarillo 310
carbonero bayeto 313
carbonero blanco 311, 325, 397
carbonero cáscara 405
carbonero de frío 298
carbonero de monte 298
carbonero de sombrío 311
carbonero de tierra fría 298
carbonero dormilón 311, 315
carbonero gigante 311
carbonero hojicenizo 383
carbonero morado 313
carbonero negro 315
carbonero quiebrahacha 325
carbonero rojo 313
carbonero zorro 315
carbunquera 296
carcanapire 388
carcha 330
carcha rojo 330
carchas 330

carchilla 185
carchita 330
carcho blanco 330
carcho rojo 330
carcoma 358
carcomo 387
carda 295, 466
cardamomo 250
cardenal 390
cardencha 295
cardencho 295
cardenillo 255
cardillo 442, 472
cardita 233
cardo 228, 229, 252, 260, 270, 272, 283, 295
cardo blanco 270
cardo cuatrofilos 283
cardo de María 270
cardo de páramo 229
cardo mariano 270
cardo pitajayo 283
cardo pitaya 283
cardol 295
cardón 228, 229, 252, 283, 284, 285, 295
cardón de arenales 285
cardón guajiro 285
cardón higo 283
cardón marica 285
cardón peludo 283
cardón real 285
cardoncillo 233, 252
cardoncillo espinoso 233
cardos 229
cardosanto 270, 368, 435
careto 351
carevaca 330
carey 192, 215, 456
carey morado 215
carey verde 215
cargadera 188
cargadero 188, 189, 190, 301, 436
cargadero blanco 301
cargadero negro 188, 190
cargadero pita 188
cargadita 316, 320, 334, 336, 337
cargadita de frío 337
cargagua 297
cargamanta 287
cargamanto 287, 401
cargamento 401
cargamuchacho 288
cargarrocío 310, 314, 442
cargero 187
carguero 186, 187, 188, 189, 190, 191, 274, 291, 301, 302, 320, 412, 413, 454
carguero balsudo de rebalse 274
carguero blanco 301
carguero colorado 301
carguero de araña 301
carguero de gavilán 301
carguero de guacamaya 301
carguero de guacamayo 301
carguero de hormiga 189, 190
carguero de lombriz 301
carguero de mojarra 301
carguero de pava 186
carguero de pescado 189
carguero de puerco 187
carguero de ratón 186
carguero de rebalse 188, 189
carguero guacamaya 301
carguero lombriz 301
carguero negro 186, 188, 189, 301
carguero popay 301
carguero rosado 291
carguero yolombillo 413
cari?ca 306
cariaco 247, 274
cariaco blanco 274
carianabo 447
cariaño 453
cariaquita 378
cariaquito 266, 275, 378
cariaquito blanco 266
cariaquito morado 378
cariaquito negro 275
cariaqyuito 275
carica 317, 438
caricapuñoruaci 405
caricarito 451
caricia 345
caricillo 287
carimagua 233
cariñito 359
cariño de suegra 327
carinta 351
caripacunga 258, 263, 442, 443
cariquiay 451
cariseco 457
carita 311, 319, 382, 405
carita de negro 382
carita de santo 405
carito 319
carleya amarilla 217
carleya blanca 217
carmaná 207, 209, 211
carmín 287, 333
carnaval 362, 390
carne de bagre 192
carne de fiambre 295, 433
carne de vaca 191, 192, 341, 351, 452
carneasada 288, 459
carneasado 406
carneasao 433
carneazada 362
carnefiambre 433
carnefresca 330, 352, 369, 450
carnegallina 387
carnero 328
carnestolenda 406
carnestolendas 406
carnestolendo 406
carnevaca 191, 192, 333
caro 319, 364
caroba 364
caroba manaca 364
carocaro 319
caro-caro 319
carolia 362
carolina 362
carpeta 319
carpintero 172, 255, 359, 367, 417
carpintero trenza 367
carrá 410
carranconcha 177
carrapato 325
carrapo 325, 404, 473
carraquilla 378
carraquillo 266, 378
carrás 410
carraspique 277
carrasposa 256, 258, 259, 266, 370, 378, 394
carrasposa de tierra caliente 259
carrasquillo 258, 259, 378
carrecillo 407
carrestoliendo 406
carretillo 309, 339, 391, 407, 438
carreto 339, 351
carreto de vela 339
carreto mameyón 339
carretón 235, 327, 335, 336, 415, 472
carretón amarillo 335
carretón blanco 336
carretón cadillo 327
carretón centella 335
carretón colorado 335
carretón de caballo 327
carretón de huerta 327
carretón dulce blanco 327
carretón híbrido 335
carretón morado 235
carretón rojo 335
carretoncito 327
carretosendro 406
carricillo 241
carriel 330
carrielito 330, 365
carrielito amarillo 365
carrillo 261
carrión 250
carripaco 311
carrizo 216, 234, 236, 240, 241, 343
carrizo agrio 234
carrizo blanco 241
cartage chuchunchullo 405
cartagena 260, 272, 281, 282, 340, 348, 369, 370, 371, 381, 405, 413, 432
cartagenero 176, 438
cartageno 413
cartagueño 454
cartagüeño 454
cártamo 259
cartucho 203
cartucho amarillo 203
cartucho blanco 203
cartur 314
cãrucù 186
carupo 351
carurú 287
caruta 351
carutillo 351
caruto 349, 351
cas 428
casaco 308, 397
casamarucha 272
casanareña 392
cascabel 186, 253, 265, 273, 285, 286, 306, 316, 340, 355, 377
cascabel de tierra caliente 265
cascabel del diablo 377
cascabeles 316
cascabelillo 316, 377
cascabelillos 313
cascabelito 316, 377
cascabelitos 316
cascajero 253, 352, 353, 355, 422
cascajo flor 224
cáscara 182, 305, 385, 387, 389, 405, 441, 460
cáscara amarga 460
cáscara de cangrejo 305
cáscara de lombrices 389
cáscara de yuca 387

cáscara preciosa 182
cáscara sagrada 441
cascaraeyuca 192
cascarilla 181, 348, 352, 388, 455
cascarilla amarga 348
cascarilla amarilla 348
cascarilla colorada 348
cascarilla fina 348
cascarilla gallinazo 348
cascarilla gris 348
cascarillero 388
cascarillo 304, 348, 349, 352, 362, 378, 427, 436, 454
cascarillo amarillo 352
cascarillo blanco 352
cascarillo hojigrande 352
casco de buey 312
casco de vaca 311, 312, 313, 332
cascoebuey 330
cascoevaca 312
cascotillo 342
cascujalé 265, 296, 348
casepache 388
casero 403, 458
casha 434
cashco 181
cashew 450
casia amarilla 333
casia de Java 314
casia rosada 314
casima 186
casima casima 186
casimana 186
casisé 463
caspe 451
caspi 188, 191, 368, 451
caspicaracha 451
caspicaracho 451
caspicolorado 451
casposo 422
casquillo 430
castaña 190, 301, 302, 381, 383, 413, 436
castaña de macaco 301
castaña de mico 381
castaña de paca 413
castaña de picure 383
castaña espinosa 381
castañetc 414
castañeto 340, 391
castañito 381
castaño 190, 301, 302, 381, 388, 411, 412, 413, 414, 434, 439, 449
castaño de Brasil 301
castaño de Indias 414
castaño de paca 301, 413
castaño del Marañón 301
castaño del rojo 413
castaño espinoso 449
castaño hojicolorado 190
castaño querré 190
castilla 206, 241, 252, 271, 296, 336, 369, 376, 378, 399, 413, 437, 444, 451
castillejo 330
castillera 245
castillo 214, 350
castillo de monte 350
castor 392
casú 450
casuarina 338
catagua 391, 455
catahua 391
catalina 260, 263
catalpa 362
catano 379
catape 340
catapes 340
catapis 340
catay 289
catechu 204
cativo 331
cativo cucharo 331
catleya 217
catleya blanca 217
catleya de Dabeiba 217
catleya de oro 217
catleya dorada 217
catreche 186
catuche 187
catura 332
caturra 293, 348
caucana 218, 364, 441, 465
caucano 187
caucapito 424
caucha 439
cauchillo 340, 392, 393, 437, 440, 441
cauchillo colorao 437
cauchillo de loma 437, 440
cauchillo de monte 440, 441
cauchillo de rastrojo 437
cauchillo serrano 393
cauchito 438
caucho 182, 273, 305, 340, 380, 384, 385, 389, 390, 391, 392, 393, 437, 438, 439, 440, 441, 446
caucho amazónico 391
caucho balato 392
caucho benjamina 438
caucho blanco 392, 393, 437, 439
caucho cartagenero 438
caucho común 439
caucho de la India 438, 439
caucho de pará 391
caucho de Pará 391, 437
caucho gaque 384
caucho hembra 437
caucho higuerón 438
caucho lechoso 389
caucho lechudo 340
caucho lirata 438
caucho macho 390
caucho matapalo 439
caucho menudito 438, 439
caucho menudo 439
caucho negro 392, 437
caucho piñique 392
caucho rey 305
caucho rojo 389
caucho rosado 438
cauchorana 393
cauchos 439
cauco 470
caue 382
caujaro 274
cauje 306
caují 450
caujil 450
caujir 450
caulote 408
caumano 353
caunce 397
caupí 337
cautivo 331
cavadonga hembra 340
caviche blanco 331
cawapat 392
cay 341
caya negra 266
cayandonga 314
cayena 365, 409
cayeno 390, 409
cayú 450
cayuasú 450
cayuvana 450
cazuelo 192, 301
cebada 234, 240
cebada anchuela 240
cebada santafereña 234
cebadilla 235, 237, 238
cebadilla criolla 235
cebo 235, 393
cebo de Cuba 235
cebolla 213, 214, 217, 239, 454
cebolla blanca 213
cebolla cabezona 213
cebolla común 213
cebolla de a ciento 213
cebolla de huevo 213
cebolla de perdiz 214
cebolla de rabo 213
cebolla de rama 213
cebolla junca 213
cebolla junco 213
cebolla larga 213
cebolla morada 213
cebolla peruana 213
cebolleta 213, 214, 217, 224, 225, 231, 232
cebolleta blanca 213, 214, 224
cebolleta colorada 213
cebolleta de pantano 231, 232
cebolleta de pegar 217
cebolleta sabanera 224
cebollín 213, 224, 454
cebollina 213, 214
cebollita 214
cecilia 209, 211, 472
cecilio 209
cedera 432
cedrillo 402, 430, 431, 451, 454, 455, 460
cedrito 273, 402, 457
cedrito de fucha 402
cedro 314, 331, 338, 401, 402, 407, 433, 451, 454, 455, 460
cedro achapo 314
cedro amargo 454
cedro amarillo 331
cedro bastardo 454
cedro bateo 454
Cedro bayo 454
cedro blanco 454
cedro caoba 454
cedro caobo 454
cedro cebolla 454
cedro cebollín 454
cedro cebollino 454
cedro cebollo 454
cedro clavel 454
cedro colorado 454
Cedro coya 454
Cedro de altura 454
cedro de montaña 454
cedro de tierra fría 454
Cedro del cerro 454
cedro dulce 454
cedro espino 407
cedro espinoso 407
cedro grande 338
cedro guamo 454
cedro güino 454
cedro hembra 433, 454

cedro jaspeado 454
cedro macho 407, 451, 454, 455
cedro monde 401, 454
cedro mondé 454
cedro mondey 401
cedro mure 314
cedro negro 338
cedro nogal 338
cedro oloroso 454
cedro pataegallo 454
Cedro peludo 454
cedro playero 338
cedro rojo 454, 460
cedro rosado 454
Cedro salteno 454
cedro tigre 454
cedro tolúa 407
cedro trompillo 454, 455
Cedro vírgen 454
cedro zambo 454
cedromacho 451
cedrón 377, 391, 455, 460
cedrón rojo 460
cedrorana 314
cedrotigre 454
ceiba 279, 391, 392, 406, 407, 410, 412, 413, 414
ceiba acuática 413
ceiba amarilla 391
ceiba arracacha 406
ceiba barrigón 413
ceiba barrigona 413
ceiba blanca 391, 407
ceiba boba 279
ceiba bomba 413
ceiba bonga 407
ceiba bongo 407
ceiba brava 391
ceiba bruja 391, 407
ceiba Cartagena 407
ceiba colorá 407
ceiba colorada 407
ceiba de agua 407
ceiba de lana 407
ceiba de leche 391
ceiba de majagua 413
ceiba espinosa 407
ceiba lechera 391
ceiba lechosa 391
ceiba lechuda 391
ceiba majagua 413
ceiba menché 413
ceiba pacua 391
ceiba roja 407
ceiba sama 410
ceiba samá 410
ceiba tarro 391
ceiba tolú 407
ceiba tolúa 407
ceiba tronadora 391
ceiba vaquera 413
ceiba verde 413
ceiba yuca 406
ceibaebruja 407
ceibito 304
ceibo 319, 391, 406, 407, 412, 413
ceibo barrigón 413
ceibo de agua 412
ceibo macho 407
ceibo yuco 406
celedonia 194, 285, 390, 402
celestina azul 256
celidonia 285
celosia 282
cemento 209, 291, 382, 383
cemeruco 396
cenascuras 430
cenascuro 430
cenescudo 430
cenicero 331, 362
cenicero suiban 362
cenizo 184, 255, 263, 268, 271, 282, 381, 392, 419, 420, 441
cenizo blanco 282
cenizo playero 271
cenizoso 192
cenón 325
cenoscuro 430
centaura 257, 344, 461
centaura menor 344
centavito 448
centella 252, 264, 279, 280, 286, 287, 288, 335, 436
centella asiatica 252
centello 430, 436
centeno 244
centrosema 314
cepa 213, 298, 350
cepa de caballo 350
cepo 298
cera 207, 250, 339, 340, 341, 350, 427
cera de rayo 350
cerbatana 261, 271, 272
cerebril 359
cerecilla 396
cerecillo 343, 396
cerecito 396
cereipo 328
cerera 292
cereza 396, 401, 425, 444
cereza cuadrada 425
cereza de Cayena 425
cereza de morrocoy 425
cereza de Surinam 425
cereza montañera 396
cerezo 303, 305, 338, 380, 395, 396, 403, 404, 425, 440, 443, 444
cerezo blanco 303
cerezo criollo 444
cerezo de monte 303
cerezo de morrocoy 425
cerezo del gobernador 404
cerezo real 338
cerezo rojo 395
cerezo silvestre 396
cerilla 341
cerillo 385
ceriman 200
cerizo 420
cernilla 266
cero 350
cero mopa 350
cerote 443
ceroto 439
cerraja 261, 264, 270
cerraja áspera 270
cerraja espinosa 270
cerraja lechosa 270
cerrajilla 270
cerrate puta 327
cerrilla 406
cervate 344
cervera 441
cerveza 367
cesticas 193
cestillo 227
cestillo hembra 227
cetico 446
chacarita 294
chacarrá 206, 207
chacarra de agua 206
chacarrá macho 207
chacarrás 206, 207
chacarrita 294
chachaajo 364
chachafruta 319
chachafruto 319
chachajo 182, 184
chachique 451
chacla 240
chaco 463
chacra 349
chadwoin beni 252
chaforuto 319
chagará 455
chagraquigua 263
chagro panga 396
chagropanga 356
chagrupanga 396
chaguaitera 213
chagualito 304
chagualo 280, 304, 305, 380, 383, 384, 385
chagualón 383, 384
chaguanoso 445
chaguar 214
chaguarpanga 396
chaguato 304
chala 389
chalá 211
chalar 209, 211
chalare 240
chalde 431
chaldé 201
chalmagra 381
chalmugra 381
chalupo 400
chalviande 192
chamana 457
chamano 457
chamarro 457
chamba 424
chambimba 459
chambimbe 459
chambimbo 459
chambira 204
chambolán de lo frío 182
chambul 319
chamburo 279
chamico 376, 465, 466, 467, 468, 472
chamico morado 466
chamiza 256
chamizo 404
champa 424
champe 424
champito 424
champo 424
chanana 401
chanca piedra 402
chanchama 438, 439
chanchán 330
chanchi 293
chanchí 293
chanchinguba 438
chandeperro 396
chandul 231
chandur 230, 231
changa dulce 376
changropanga 394
changuelito 304
chankira 196
chanó 393
chantre 303
chanu 383
chanú 393
chanud 303

chanul 207, 393
chapaja 205
chapaja de bajo 205
chapantye 359
chaparral 258, 416
chaparral de bestia 258
chaparrillo 354, 379
chaparrito 349, 379
chaparro 295, 379, 395, 401, 416, 460, 472
chaparro amargoso 460
chaparro bobo 295
chaparro de agua 295
chaparro de monte 401
chaparro de vega 401
chaparro manteco 395
chaparro montañero 401
chaparro sabanero 295
chaparromanteco 395
chaparrón 295
chapí 207
chapil 210, 211
chapilde 211
chapililla 211
chapín 209, 211
chapingo 211
chaporuto 319
chapul 460
chaque 431
chaquén 431
chaquero 338
chaquilulo 298, 299, 300
chaquiro 180, 184, 338, 393, 455
chaquiro amargo 455
chaquiro dulce 180
chaquiro marfil 455
charambirá 190
charapa huasca 175
charapilla 191, 318, 465
charapillo 335
charara 257
charechuelo 385
charichuelo 385
charmolán 303
charmuelán 303
charne 418, 423
charne saltón 418
charo 437, 440, 441
charrado 380
charrul 468
chascaraiza 204
chascaral 204
chascaray 204, 206
chascaraza 204
chascarrá 206, 207
chascarrás 206
chascarray 206
chascarrial 207
chasmiande 268
chauacha 471
chauchilla 294, 295
chava 471
chaviaco 182, 395
chavica 195
chavito 371
chayota 295
chchillamatu 264
cheflera 253
chemen contra 282
chengue 250, 319
chengué 319
chepesi 331
cherche 468
cherte 422
chiah 447
chiata 268
chibará 351
chibato 332
chibechi 438
chibechí 438
chibichibe 228
chibillo 301
chibo 289
chibú 451
chibugá 301
chica 183, 207, 293, 304, 363
chicaipe 359
chicalá 363, 364, 365
chicalá polvillo 365
chícalo 199
chicapoy 199
chícaro 363
chicarrón 408
chicató 415
chichacún 185
chichagua 206
chichaja 299
chicharrón 177, 257, 258, 259, 274, 318, 330, 378, 379, 380, 388, 441, 457
chicható 415
chiche 280
chiche de borugo 280
chichegato 280
chichera 277
chichí 414
chichiborugo 282
chichiburra 348
chichiburrú 211
chíchicù 393
chichicuvo 447
chichipince 351
chichique 393
chichire 205
chichirimico 437
chicho 302, 325
chicho borugo 302
chichoborugo 282
chichu 469
chichuá 291
chici ordinario 251
chickenweed 390
chiclayo 337
chicle 272, 273, 305, 306, 340, 341, 440
chicle con espinas 341
chicle espinoso 341
chicle fino 272
chiclo colorado 309
chico 281, 282
chícol 199
chícolo 199
chicón 208
chicopia chiguave 196
chicoria 224, 259, 261, 264, 268, 270, 271, 286, 361
chicoria amarilla 271
chicoria blanca 264
chicoria ñinguicue 286
chicorrón 228
chicul 199
chicuria 264, 272
chicuria cucui 264
chiegua 338
chifla 353
chiflachifla 181
chiflador 181
chiflichifli 181
chiga 424
chigo 313
chigua 179
chigua macho 179
chiguamacho 177
chihichuine 228
chii 346
chiki-chiki 332
chìkichìki 210
chikku chikku 264
chilacuan 279
chilacuán 279
chilangua 252
chilangué 212
chilapo 302
chilca 256, 257, 258, 259, 260, 270, 309
chilca amarga 256, 257
chilca blanca 257, 258, 270
chilca de teñir 258
chilca dulce 257
chilca morada 257
chilca negra 257
chilca rucia 257
chilca verde 257
chilchi 332
chilco 256, 257, 258, 259, 309, 393, 405, 419, 420, 421, 423
chilco amargo 256
chilco blanco 257, 405, 423
chilco colorado 309, 421
chilco común 257
chilco de páramo 258, 309
chilco de teñir 257
chilco dorado 309
chilco grande 257
chilco negro 257, 420
chile 207, 432
chileno 250
chilguacán 279
chilhuacán 279
chilicaspi 434
chilicuan 279
chilifruta 332
chilinchele 332
chilinchil 333
chilinchile 314, 332, 333
chilinchín 316
chilipique 335
chilka 258
chillalde 415
chillangua 252
chilma 225
chilopo 258
chilquilla 258, 267
chilquilla menuda 267
chilquita 257, 258
chiluanga 252
chima 206
chimachimalí 418
chimane 440
chimar 183
chimar-karoa 183
chimbalo 469
chimbilla 323
chimbillo 315, 323, 324
chimbillo bravo largo 315
chimbo poruto 319
chimboeborugo 280
chimbolo 469
chimbombó 406
chimbusa 184
chime 199
chimico 437
chimigua 439
chimilo 250
chiminango 315, 330
chimpamplena 226
chimú 287
chin 234, 240, 434
chin caspi 434

chin de Castilla 234, 240
china 210, 226, 227, 258, 362, 365, 372, 444, 456
china-thula 258
chincheno 283
chinchile 332
chinchimali 258
chinchimalí 258, 271, 418
chinchimalía 418
chinchimaní 338, 417, 418
chinchinali 394
chinchira 277
chinchoncho 313
chinese root 227
chingale 205
chingalé 205, 364
chingo 230
chingongo 303, 437
chinia 434
chinicua 441
chino 250, 339, 428, 451
chío 289
chipa 258
chipaca 258, 261, 263, 268, 271, 361
chipaco 258
chipaque 289
chipará 351
chiparo 331, 337
chíparo 337
chipati 211
chipatoria 270
chipeci 453
chipero 337
chípero 337
chipó 416
chipro 393
chipuelo 457
chipuelo aterciopelado 457
chipuelo hediondo 457
chiquichique 332, 333
chiquichiqui 210
chiquinquireño 330
chiquiragua 260
chiquito 204, 257, 266, 280, 281, 338, 369, 422
chiraco 402, 451
chiracó 330
chiracru 376
chiracu 376
chiragua 212
chirai 311
chirarán 371
chirarán morado 371
chirateté 402
chirca 204
chircagüero 457
chirco 258, 370, 420, 422
chirco de montaña 420
chirco de sabana 422
chirco sabanero 422
chiri 471
chiriaco 402
chiricaspi 465
chiriguaco 297
chiriguey 465
chirigüey 465
chirilla 402
chirimorriñón 186
chirimoya 186, 187
chirimoya de monte 186, 187
chirimoya montañera 187
chirimoya roja 187
chirimoyitos parchas 398, 399
chirimoyo 186, 187, 188
chirimoyo de monte 186, 187
chirineto 250
chiripique 317
chirivaña 321
chirlobirlo 365, 402
chirlobirlos 365, 375
chirobiro 365
chirosa 213
chirquí 206
chirquía 204
chirriadera 299
chirriador 362, 402, 415, 431, 472
chirridera 299
chirrinchado 333
chirrinchao 310, 333, 402, 454
chirrinche 402
chirrinchinchao 402
chirrincho 333, 402
chirrinchoga 402
chirte 422
chisa 427
chisacá 255, 256, 258, 261, 264, 265, 271
chisca 226
chisgo 309
chisgua 247, 251
chiso 427
chiso sabanero 427
chispare 191
chispeador 418
chíspero 418
chitacá 266, 267
chitato 415
chitató 415
chite 347, 393, 394
chitón 267
chitú 410
chitua 413
chiva 256, 378, 437, 441
chivaco 300
chivasa 231, 232
chivata 268
chivatera 256
chivato 317, 332, 339
chivaza 256
chivecha 407
chivichivi 228
chiviná 254
chivita 256
chivo 173, 180, 255, 256, 260, 266, 279, 289, 337, 359, 363, 378, 440, 441
chivo macho 256
choapo 210, 212
choba 428
chobo 428
choca 378
chocanari 433
chocha 319, 325
chochas 319
chochilla 294
chochitas 310, 354
chochito 310, 316
chochito de indio 310
chochitos 310, 316, 319, 325
chochitos de indio 310
chocho 310, 319, 325, 326, 327, 328, 329, 459
chocho azul 310
chocho de árbol 319
chocho de jardín 325
chocho de sabana 327
chocho pequeño 310
chocho trepador 310
chochos 310, 311, 319, 325, 326, 331
chochos de bejuco 310
chochos de flor 325
chochos de hoja abanicada 325
chochos de pinta negra 310
chochos de tree 319
choclo 247
choco 276, 325, 428
chocó 181, 304, 367, 454
chocó blanco 367
chocolata 340
chocolate 356, 392, 409, 414, 428
chocolate corona 392
chocolate de monte 409, 414
chocolatí 249
chocolatil 302
chocolatillo 249, 348, 403, 409, 414
chocolatín de monte 318
chocorrón 228
choibá 318
cholo 250, 406
cholupa 399
cholupo 399
chompa 418
choncho 250, 313
choncho plátano 250
chonque 199
chonta 203, 204, 206, 207, 210, 212
chonta de castilla 206
chonta de mico 212
chonta negra 210
chonta pambil 212
chontadura 206
chontadurillo 206, 207
chontadurillo liso 207
chontaduro 206, 207, 426
chontaduro amarillo 206
chontaduro blanco 426
chontaduro colorado 206
chontaduro de búho 206
chontaduro de grillo 206
chontaduro de lagartija 207
chontaduro de los peces 206
chontaduro de monte 206
chontaduro de pescado 206, 207
chontaduro de rana 206, 207
chontaduro de rana de rastrojo 206, 207
chontaduro del diablo 207
chontaduro falso 206
chontaduro mantecoso 206
chontaduro rojo 206
chontaduro sin pepa 206
chonte 199
chontica 210
chontilla 204, 206, 207, 209
chontilla coco 206
chontilla sitanó 206
chontula 208
chopín 241
chopín gordura 241
chopo 250, 404
chopuré 227
chópure 227
chorontas 300
chorota 300
chorotas 300
chorote 422
choroticos 286, 377, 420
chorr 213
chorrillo 282
chorro de oro 313
chota 295
choublak 409
choyota 295

chuagua 283
chuapo 210, 212
chubilla 263
chucha 185, 190, 191, 200, 268, 272, 329, 349, 362, 372
chucha panga 185
chuchana 204, 205
chuchapanga 185, 186
chuchapanga blanca 186
chuchapanga negra 186
chucharrón 266
chuchi 385
chucho 185, 186, 187, 188, 256, 316, 319, 385, 466, 468, 470, 471
chucho fruteador 468
chucho hediondo 471
chucho lanudo 471
chuchoquemao 264
chuchos 316
chuchuaza 291
chuchugasi 291
chuchuguaca 401
chuchuguacha 291
chuchuguache 291
chuchuguasa 181, 291, 346
chuchuguasa de rebalse 291
chuchuguasa macho 346
chuchuguasi 291
chuchuhasa de tierra baja 385
chuchuhaso 385
chuchuhuasa 291
chuchuhuasha 291
chuchul 385
chuchunchullo 405
chuchuuaso 291
chuco 264
chucú 414
chucua 297
chucunda 326
chueca 289
chufa 230
chugchuguao 291
chugua 283
chuguacá 296, 359, 401
chuguacá hojiancho 401
chuguas 283
chuhuacá 401
chula 200, 403
chulco 293, 417, 423, 432
chulo 183, 200, 268, 316, 359, 412
chulpa 417
chulpe 247
chulquillo 432
chulquín 227, 240
chulquito macho 432
chulupa 398, 399, 400
chulupa de loma 398
chulupa de mico 398, 400
chulupe 399
chulupo 398, 400
chumbaguás 362
chumbalo 468
chumbera 284
chumbimbe 459
chumbimbo 459
chumbimbomichú 459
chumbo 284, 363
chumillo 322
chunche 451
chunchu vainilla 332
chundul 231
chundún 231
chundur 231
chundur de gente 231
chunduy 186
chunga 203, 205
chunque 199
chúntaro 207
chupa 284, 299, 302, 363, 383, 450
chupa chupa 363
chupa melón 284
chupa quinde 299
chupachupa 284, 297, 348, 363, 411, 416, 432
chupadera 200, 260, 270
chupador del grande 360
chupador del pequeño 359
chupaflor 367, 423
chupahuevo 280, 460, 461
chupahuevos 460, 461
chupahuevos de tejados 460
chupaila 228
chupalla 200, 225, 229
chupamelón 284
chupana 253, 263, 357
chupapanga 199
chuparán 253
chupayo 200
chupequinde 299
chupeta 409
chupico 279
chupín 241
chupo 301, 302
chupo de montaña 302
chupo de monte 302
chuque 297
chuquiraga cuasco 260
chuquiragua 260
chura 294, 450
churá 450
churapanga 443
churca 248, 369
churco 176, 293, 423
churí 293
churijuique 421
churima 322, 324
churimba 324
churimbo 321, 331
churimito 322, 323
churimo 297, 313, 321, 322, 323, 324, 337
churimo de monte 323
churimo de orilla 323
churimo rosario 322
churquillo 293
churri 243
churrichurri 243
churro 240, 242
churrubay 212
churrubay coco 212
churruhay 212
churrusco 173, 333, 387
churrusquillo 387
churrusquillos 387
churumbelo 355
churumuyuí 187
chusco 234
chusque 236
chusquejón 236
chusquito 241
cibulé 213
cica 179
cícara 179
cicilio 209
ciclamen 304
cicuta 252
cidra 295, 456
cidrapapa 295
cidrayota 295
cidrero 456
cidro 456
cidrón 377
cielito 462, 465
cieloazul 274
ciempatas 198
cierra tus puertas 327
ciérrate 327
cierrateciérrate 327
cierratepula 327
cigao háheku 210
cigarrilla 231
cigarrillito 429
cigarrillitos 417
cigarrón 332
cilanda 457
cilantrillo 177, 252, 401
cilantro 252, 360
cilantro ancho 252
cilantro castilla 252
cilantro chicorio 252
cilantro cimarrón 252
cilantro común 252
cilantro de Castilla 252
cilantro de indio 252
cilantro de la tierra 252
cilantro de monte 252, 360
cilantro de paisa 252
cilantro de sabana 252
cilantrón 456, 457
cimarrón 182, 185, 186, 187, 242, 250, 252, 271, 293, 299, 315, 348, 358, 363, 379, 388, 395, 406, 408, 409, 414, 415, 418, 424, 454, 458, 465, 466
cimarrona 186, 187, 199, 234, 251, 264, 265, 280, 282, 283, 294, 310, 339, 371
cimarrú 460
cimbidio 286
cimbrapotro 337
cinaro 424, 427
cinco llagas 398
cinco negritos 378
cincocentavos 326
cincodedos 253, 254
cincollagas 398
cincuentacentavos 319, 326, 328
cineraria 264, 372
cinta 215, 224, 243, 289, 388
cintilla 215
cintura de mono 430
cipó 226, 363, 434
cipó cruz 363
cipó pequeño 226
ciprés 180
cira 234
cira junca 234
cirgüelo 451
cirio 283
ciro 257, 258
cirpe 447
cirpe macho 447
cirpo 446, 447
cirpo macho 447
cirquí 204
cirro 257
ciruela 381, 395, 396, 401, 404, 443, 451
ciruela calentana 451
ciruela chilena 443
ciruela Claudia 443
ciruela de algodón 381
ciruela de fraile 395
ciruela de perro 396
ciruela del gobernador 404
ciruela europea 443

ciruela silvestre 395
ciruelillo 395
ciruelo 311, 395, 396, 444, 449, 451
ciruelo blanco 451
ciruelo calentano 451
ciruelo cimarrón 395
ciruelo de castilla 451
ciruelo de Castilla 451
ciruelo de fraile 395
ciruelo de loma 395
ciruelo de perro 395, 396
ciruelo de playa 311
ciruelo dulce 451
ciruelo hobo 451
ciruelo rojo 451
cisne 192, 218
citron vert 455
citronela 236
ciuca 330
cizaña 241, 255
cizaño 241
claro de lu muérdago 448
claro de luna 448
clavel 267, 280, 285, 290, 298, 312, 313, 317, 342, 401, 409, 454, 466
clavel chino 280, 285, 290
clavel de día 280
clavel de oro 401
clavel habano 342
clavel japonés 409
clavel reventón 285
clavel rojo 285
clavelillo 261, 298, 312
clavelina 271
clavelino 301, 312, 317
clavelito 261, 298
clavellina 267, 272, 285, 312
clavellina de bejuco 267
clavellina doble 285
clavellina pintada 285
clavellino 267, 285, 298, 312, 313, 317, 466
clavellino de bejuco 267
clavellino verde 466
clavelón 272
clavillo 368, 429, 465
clavillo rojo 429
clavito 285, 371, 429
clavito de agua 429
clavito de pantano 429
clavito kirukise 285
clavitos 391
clavo 194, 261, 278, 330, 355, 387, 390, 391, 392, 428, 429
clavo amarillo 429
clavo de agua 278, 429
clavo de cristo 390, 391
clavo de huerta 194
clavo de huerto 391
clavo de laguna 429
clavo de pozo 261
clavo pasado 387
clavos 390, 391, 429
clavos de Cristo 390, 391
clemón 415
climacuna 237
coacha 401
cobalonga 340, 344, 388
cobalonga extranjera 388
cobalongo 319, 340
cobán 467
cobre 311
cobuia 471
coca 208, 227, 321, 386, 387, 392, 403, 405, 437, 457
coca cimarrona 386
coca de águila 405
coca de antiguo 386
coca de danta 386
coca de mico 321
coca de monte 386
coca de pescado 386
coca de trujillo 386
coca del creador 405
coca del sur 386
coca macho 392
coca mono 387
coca montana 386
coca peruana 386
coca propia 386
coca rayo 386
coca silvestre 457
coca suave 386
coca tinga 386
coca zambico 386
coca-á 386
cocacá 280
cocaemico 288
cocaito 380
cocamono 387
cochayuyo 362
cochilla 294
cochinito 309
cocla 312
coclí 472
coco 204, 205, 206, 208, 211, 212, 230, 231, 301, 302, 311, 318, 321, 324, 331, 410, 449
coco blanco 301
coco caballero 208
coco cabuyo 301
coco cazuelo 301
coco chambira 204
coco colorado 301
coco cristal 301, 302
coco cumbe 211
coco cuna 302
coco de mono 301, 302
coco de monte 204, 311
coco de puerco 204
coco de tigre 204
coco espárrago 301
coco mabaco 205
coco manteco 301
coco mojagua 301
coco mono 301, 302
coco peludo 204
coco picho 301
coco plumoso 212
coco redondo 301
coco volador 410
cocoa plum 381
cocoabarco 301
cocoba 438
cocochunduro 230
coco-chunduro 230
cococristal 302
cococu joanicu 437
cocohui 402
cocombro 293
cocomono 302
cocomuerto 302
cocona 471
coconilla 471
coconuco 377
cocopi mataefinae 189
cocopicho 301
cocorilla 398, 400
cocorilla de monte 400
cocorillo 397, 398
cocoro 301, 397, 405
cocoromo 211
cocorote 206
cocosapo 307
cocota 451
cocote 451
cocotera 208
cocotero 208
cocoto 451
cocu 311
cocua 266, 441
cocuá 441
cocúa 280
cocubo 469, 470
cocuelo 302
cocuiza 215
cocuy 214, 341, 470
cocuyina 235
codillo 224, 445
coemeni 291
cofrey 275
cofrón 343
cogote 309
cogotillo 442
cohitre 224
cohitre morado 224
cohoba 311
cohombro 293
cóijibo 228
coijol 332, 333
coime 281
coirdeimiento 380, 387
cõja 210
cojaca 469
cojé 295
cojoba 311, 315
cojoco 467
cojón 328, 340, 343, 344
cojón de amo 344
cojón de burro 343
cojón de cabrito 340, 343
cojón de fraile 328, 343
cojón de gato 343
cojón de perro 340
cojón de puerco 343
cojón de verraco 343
cojoncito 343
cojueiro de campo 450
col 198, 199, 200, 276
col de monte 198
col del diablo 200
col rizada 276
cola de armadillo 379
cola de babilla 224
cola de burro 238, 268
cola de caballo 173, 231, 233, 240, 284, 376
cola de caimán 280, 283, 284
cola de gallo 204, 209
cola de gato 215, 387
cola de guacamaya 358
cola de gurre 347
cola de morada 209
cola de oveja 412
cola de paloma 224
cola de pato 206, 227
cola de pescado 207, 215
cola de reina 255
cola de vaca 233, 234
cola de venado 233
cola de zorro 213, 233, 237, 245
colación 277
coladera 293
colaecaballo 376
colaepuerco 261
colagoge 338
colchón de niño 447

colchon de pobre 171
colchón de pobre 171, 181, 255
coleburro 261
colegallo 207, 209
colegial 358
colegiala 272, 340
colegiales 272
colegianta 272
colegiante 272
colemula 211
coleo 369
coleos 369
colepato 227
colibrí 219, 220
colicero 250
colinabo 276
colizorro 233
colla 261, 266, 270
colla negra 266
colluco 283
colmillo de danta 350
colmillo de león 271
colmillo de perro 227
colocación 282
colombia 348
colombian coca 386
colombian waxweed 416
colombiana 218, 219, 269, 303, 308, 329, 361, 368, 383, 414, 461, 462
colombina 436
colombo 294, 433
colomboro 204
colopa 399
color 249, 375, 376
color azul 249
colorada 191, 211, 213, 219, 236, 247, 263, 301, 305, 378
coloradito 289, 383
colorado 181, 182, 185, 198, 204, 206, 211, 234, 238, 242, 274, 277, 280, 281, 289, 295, 298, 299, 301, 304, 305, 306, 307, 309, 314, 319, 323, 330, 335, 342, 343, 347, 348, 350, 351, 358, 364, 365, 377, 381, 384, 391, 394, 396, 397, 401, 402, 404, 406, 407, 408, 410, 412, 416, 419, 421, 428, 437, 441, 443, 445, 451, 454, 455, 457, 458, 465
colorado chilacó 416
colorado clavel 401
coloraíto 387
colosoana 235
colosuana 235
colpachi 388
coluda 234
colupa 399
colusuana 235
combia 445
combito 274, 331
comblilla 445
combo 339, 387
come 210
comedera 359
comedero 309, 359, 416
comedero de iguana 359
cómenecù 322
cometa 362
comida de danta 350, 353, 357, 441
comida de morrocoy 197
comida de murciélago 333
comida paloma 415
cominillo 263, 376
cominillo rosado 263
comino 182, 183, 184, 252, 267, 386
comino baboso 183
comino crespo 182
comino real 182, 184
comino silvestre 267
comiture 423
como 280
compadre 406, 451
compadre bototo 406
compadre chiraco 451
comulá 339
conami 260
concha 259, 278
concha prieta 278
conchas 193
conchita 259, 266, 315, 378, 460
conchita blanca 266
conchitas 292, 315, 461
conchito 449
concombro 293
cóndor 288
condurango 342
coneja 237, 244, 261, 374, 422
conejera 224
conejillo 389, 393
conejitos 367
conejo 173, 237, 238, 244, 258, 293, 315, 360, 378, 428
confite 302
confrey 275
conga 249, 251, 366, 385, 399
congas 193
congo 197, 249, 250, 251, 268, 311, 415
congo de agua 311
congola 328
congolala 261
congolo 319, 328
cóngolo 319, 328, 399
congona 194
congona arisca 194
congona perfumada 194
congonilla 447
congora 287
congüe 332
conguerán 287
conguito 311, 465
coninae 363
conopio 251
conservadora 467
conservo 438
consuelda 224, 261, 263, 268, 269, 448, 450
consuelda menor 263
contentica 373
contra 193, 267
contraarará 278
contracapitana 192, 193, 266
contracapitana de mompós 192
contracapitana de Mompós 192
contraconga 227
contracruceta 460
contracruceto 460
contrafuego 368, 372
contragavilana 259, 267
contrahierba 259, 437
contraincendio 241
contrapito 227
contraprieta 193, 278
contrarreuma 332
contraveneno 339
contraverrugosa 195, 197
contrayerba 437
convólvulo rojo 463
coomeji 210
copa de David 377
copa de oro 339
copachí 190, 389, 393
copaia 364
copaiba 315
copaíba 315
copaibí 315
copaiva 315
copal 320, 321, 350, 404, 451, 452, 453
copal americano 320
copalchi 389
copaseca 397
copay 301, 382
copayero 315
cope 438, 439
copé 192, 383, 384, 385, 438, 439
cope higuerón 438
cope negro 439
copé negro 439
copei 385, 438
copelia 374
copetica 361
copetona 256
copey 383, 384, 385, 438, 439
copey cucharo 384
copey de tierra fría 439
copialila 361
copialina 361
copijuique 278
copita de oro 339
copo de nieve 256
copo morado 379
copoazú 414
copoazú de altura 414
copoazú de monte 414
copoazú silvestre 414
copuazú 414
copudo 176
coquera 208
coquero 208
coqueta 280, 465
coquillo 204, 206, 212, 231, 302, 391
coquindillo 453
coquindo cimarrón 454, 458
coquito 204, 206, 212, 230, 231, 232, 302
coquito blanco 232
coquito espinoso 206
coquito espinoso de lago 206
coral 287, 298, 300, 350, 351, 352, 365, 376, 391, 392, 393, 394, 433
coral de agua 352
coral de la India 352
coral doble 352
coral extranjero 352
coral gaitán 352
coral rojo 352
coral sencillo 352
corales 392, 433
coralibe 363
coralillo 281, 287, 319, 351, 352
coralitas 351
coralito 350, 351, 352, 376, 394, 416, 433, 460, 464

coralito de agua 352
coralitos 350, 352
coralplant 391
coraltree 319
coramina 360, 373
corango sempreviva 283
corazón 182, 187, 192, 193, 194, 196, 198, 199, 201, 289, 292, 293, 318, 331, 340, 369, 441, 467
corazón de envuelto 441
corazón de hombre 194
corazón de jesús 199
corazón de Jesús 192, 199, 340
corazón de la virgen 196
corazón de la Virgen 196
corazón de pollo 467
corazón herido 199, 289, 369
corazón negro 182, 318, 331
corazoncillo 267, 394
corazoncito 235
corazoncitos 235
corazones partidos 292
corazonfino 330
corazóno herido 199
corbata de Gaitán 390
corbata de rico 367
corbata de Uribe 390
corbatas 367
corbatín 215
corbón 441
corchito 351
corcho 187, 331, 407, 426
corcho blanco 407
corderillo 381
cordillero 182, 368
cordon de fraile 357
cordón de fraile 370
cordón de novia 461
cordón de obispo 372, 387
cordón umbilical 394
cordoncillo 193, 194, 195, 196, 197, 224, 275, 280, 353, 387, 388
cordoncillo blanco 195, 196
cordoncillo chinibo 197
cordoncillo de bejuco 196
cordoncillo de hoja ancha 195
cordoncillo de hoja pequeña 195
cordoncillo grande 195
cordoncillo hojancha 196
cordoncillo morado 224
cordoncillo negro 196
cordoncillo paramero 353
cordoncillo rayo 197
corianobo 447
corme 447
corneta 465
corneto 185, 210, 465
cornique 364
cornoco 312
coroba 210, 251
corobá 191
corobore 320
corobore. 320
corocilla 177
corocillo 230, 231, 423
corocita 179
corocito 179, 204, 206, 230, 291
corocito de agua 204
corocito de monte 204
coroiba 210
corombolo 204
coromo 336
corona 209, 284, 388, 390, 405
corona de cristo 390
corona de Cristo 390
corona de espinas 284, 390, 405
corona de espino 405
corona de oliva 390
corona de príncipe 388
coroncorito 387
coronilla 200, 300, 418
coronillo 289, 404, 418
coronito 298
corono 258, 405, 442
corono espina 405
corono espino 405
corono espinoso 405
coropo 319, 349, 395, 457
corotú 319
coroza 205, 208
corozo 177, 203, 204, 205, 206, 208
corozo amolado 203
corozo chiquito 204
corozo colorado 204
corozo de agüita 204
corozo de amolao 203
corozo de ciénaga 177
corozo de gallinazo 206
corozo de lata 206
corozo de manteca 208
corozo de marrano 205
corozo de monte 204
corozo de puerco 205
corozo de vaca 205
corozo de vino 203
corozo grande 203
corozo hembra 206
corozo pajarito 205
corozo redondo 203
corozo rojo 204
corpus 354
corralero 288
corregidor 303
correo 260
corrimiento 340
corro 259
cortacorta 231
cortadera 204, 209, 216, 230, 231, 232, 233, 236
cortadero 236
cortalonso 364
cortapico 226
cortapicos 226
corte 340
cortecorte 361
cortejo 316, 340, 344, 358
cortés 340, 445
cortesita 340
cortezo 407, 445
cortina africana 362
cortina china 362
cortizo 407
corúa 205
corunta 213
corvejo 400
cosedera negra 445
cosia 236
costeña 197, 442
costeño 176, 197, 308, 352
costilla de Adán 200
costillera 245
costillero de sangre 341
costillo 334, 335, 339, 341, 347, 380, 385, 393, 437, 445
costillo acanalado 339
costillo blanco 334, 339
costillo de mal olor 334
costillo propio 341
costillo redondo 339
costillo verdadero 339
costillocaspi 339
coto 182, 355
cotoperi 458
cotoplí 458
cotoplís 458
cotoprí 458
cotoprís 458
cotorera 371
cotorrera 388
cotorrero 278
covalonga 340, 344
covalongo 340
coya 386, 387, 455
coya blanco 455
coya colorada 455
coyo de campo 450
cozoiba 385
crecedor 363
creceto morado 460
crenitaria 286
creolino 328
crepuscularia 430
crespa 212, 213
crespín 451
crespino 451
crespo 176, 182, 252, 294, 443
crespón 411
cresta de gallo 230, 281, 282, 319, 351, 358, 387, 416, 437
cresta de guacamayo 441
crestagallo 319, 355
crestas 282
crestegallo 282, 358, 390
cri 404
cri-jé-ric 404
criolla 234, 235, 244, 247, 274, 325, 443
criollo 180, 215, 236, 251
crispeta 355
crispetas 362
crispín 451, 452
cristal blanco 441
crítica 236
cromos 369
crotalaria 315, 316
croto 224, 388
croto palmita 224
croton 388
crotos 388
crucecita 348
crucero 330, 383
cruceta 236, 308, 356, 458, 460
cruceta negra 356
crucete 460
crucetillo 356
cruceto 278, 312, 343, 351, 356, 378, 460
cruceto blanco 460
cruceto morado 460
cruceto real 356
crucito 353
cruz de Malta 286
cu?va 210
cuacho amazónico 391
cuadrada 417
cuadrado 409
cuajaro 274
cuajo 191, 192
cuajó 191
cuala 333

cualecuina 196
cuángare 190, 191, 192
cuángare blanco 191
cuángare cholo 191
cuángare chucha 191
cuángare de loma 191, 192
cuángare de mangual 191
cuángare otobo 191
cuángare punta de lanza 192
cuangata 227
cuarenta días 272
cuarentana 272
cuarentena 272
cuarque 229
cuartillita 327
cuartillito 194, 327
cuartillo 194, 374, 462
cuas 348
cuasa 309
cuasbil 212
cuasco 258
cuasia 460
cuasia amarga 460
cuasio 451, 460
cuatrofilos 250, 324
cubano 359
cubaro 206
cubarra 204, 206
cubarrillo 204
cubarro 204, 206
cubarro pintón 206
cubarru de espinas 206
cube 317, 335
cube ordinario 335
cubeo 227
cubia 280
cubia cimarrona 280
cubias 280
cubio 280, 432
cubios 280
cubiyú 471
cubo 363
cuca 211
cucama 382
cucaná 378
cucanao 378
cúcane 378
cúcano 378
cucaracha 224, 282, 453
cucarachero 200
cucaracho 198, 200, 326, 340, 365, 403, 404
cucarrón 223, 414
cucas 378
cuchanguilla 360
cuchara 304, 341, 453
cuchara caspi 341
cucharacaspi 341
cucharero 304
cucharilla 389
cucharillo 304
cucharillo blanco 304
cucharitas 193
cucharito 363
cucharo 280, 298, 304, 305, 331, 350, 363, 383, 384, 385, 402, 430, 445
cucharo blanco 304
cucharo colorado 280
cucharo pequeño 304
cucharro 304
cuchayuyo 362
cuchilla 248, 370
cuchilla de perdomo 370
cuchilla de Perdomo 370
cuchillito 332
cuchillo de los Perdomo 370
cuchimbo 459
cuchipén 234
cuchiyuyo 362
cucho 391
cuchuyuyo 362
cuco 185
cucu 470
cucuá 441
cucua kjun 282
cucubo 274, 469, 470
cucubo espinoso 470
cucuma 471
cucuná 378
cucunanay 471
cucuo 468, 469
cucupa 447
cucura 447
cucurita 205
cucurito 205
cucutún 358
cucuva 447
cucuy 341
cucuy de monte 341
cucuye 341
cucuyina 235
cudio 363
cuen 380
cuenta de milagro 236
cuentas de milagro 236
cuentecita 433
cuentecito 433
cuerda de violín 462
cuerdas de violín 462
cuerdilla 462
cuerdillas 462
cueresapo 459
cuerinegro 189
cuerito rubio 276
cuerno 176
cuero 249, 321, 355, 372, 394, 404, 436, 437, 451, 457
cuero de gallina 394
cuero de marrano 321, 404, 436
cuero de sapo 372, 437, 457
cuero de vieja 451
cueroesapo 290, 372
cueroevieja 442
cueronegro 188
cuerosapo 393
cueruda 418
cuesco 203, 205
cuetillo 351
cui 392, 407
cuiba 432
cuibangui jalé 326
cuica 330
cuiche rojo 390
cuico 330
cuicuillá 431
cuipo 407
cuit 346
cuitote 408
cujaca 330, 471
cujaco 469, 471
cují 331, 336
cují negro 331
cujicabro 189
cujicillo 310
cujita 206
cujupa 192
culantrillo 177, 252, 267, 376
culantrillo negro 177
culantro 252, 361
culantro de burro 252
culantro doméstico 252
culape 369
culatero 363
culebra 215, 348, 359, 439, 465, 472
culebra borrachera 465
culebra borrachero 465
culebra sabanera 465
culebrero 215
culebrilla 215, 472
culebrillera 472
culebrina 215
culebro 215
culeco 258
culefierro 306, 381
culegato 357
culen 329
cullera 341
culnicre 311
culo de hierro 318
culupa 399
culupo 399
cumaca 331, 407
cumacanae 438
cumala 190, 191, 192, 380
cumalá blanca 191, 192
cumalá blanco 192
cumala colorada 191
cumala roja 191
cumaná 270, 388, 401
cumara 463
cumarana 343
cumare 204
cumaré 204
cumare de guara 204
cumaricama 318
cumarillo 204
cumaruco 414
cumataci 391
cumé de rebalse 416
cumulá 339
cumula blanca 192
cun 199
cuna 202, 217, 302, 308, 366
cuña 334
cuna de Venus 217
cundamor 463
cunde 442
cundeamor 295, 463, 464
cundidora 233
cundurango 342
cunite 364
cunula 274
cuomulá 339
cupana 458
cupani 458
cupania 458
cupay 385
cupea 417
cupey 385
cupiagua 436
cupis 181
cupón 409
cupuassú 414
cupuazú 414
cura 185, 401
curaarador 435
curacarate 389
curachimici 197
curadientes 196
curador 435
curahígado 257
curalina 345
curalotodo 378
curanina 272
curapia 457
curarador 435
curare 338, 346, 357, 362, 434, 435

curare de pescado 346
curare ordinario 357
curarín 192
curarina 192, 193, 215, 345, 359
curarina de selva 192
curarina vegetal 192
curarire 364
curata 295
curatodo 359
curaúbo 332
curaverrugas 378
curayá 378
curazao 286
curazao rojo 286
curcuma 250
cúrcuma 250
curcuncha 194
curí 321, 363, 365
curi guasca 365
curia 359
curía 359
curiaci 191
curíbano 232
curibano amararillo 416
curichí 416
curichichi 449
curí-guasca 365
curí-huasca 363
curillo 197
curito 454
curiza 381
curnique 364
curnite 364
curomacho 185
currucay 423, 453
curruco 422
currusay 453
curse 420
curtidera 293
curúa 205
curuba 398, 399, 400, 432
curuba antioqueña 398
curuba de indio 399
curuba de monte 398
curuba quiteña 398
curuba silvestre 398, 399
curubito 399
curubito de indio 399
curubo 398, 399, 401, 403
curubo antioqueño 398
curubo de árbol 398
curubo de castilla 399
curubo de indio 399
curubo de monte 399
curubo quiteño 398
curucujurro 228
curujujul 228
curujul 228
curumacho 185
curumuta 205
curunta 213
curura 447
curutú 319
cuscuta 462
cusoiritu 447
cuspa 339
cuspare 455
cusú 328
cusuca 228
cutuma 260
cutupén 245
cutuplí 458
cúvañù 210
cuya 363
cuyanguilla 193, 360
cuyanguillo 193
cuyanquillo 223
cuyara 209
cuyarocrí 351
cuychunchullo 373
cuyo seuyodopirai 301
cuyubí 383, 449
cuyubí de rebalse 383
cuzú 274, 328

D

daápi 200
dadeató 360
dakali 418
dakalinae 418
dale dale 345
daledale 345
dalia 261
dalia silvestre 261
dama de noche 462, 466
damagua 438, 441
damiana 401
danda kidai 195
dandeleón 332, 333
danta chondur 231
danto 308, 388, 421, 423
dapa 394
daruja 439
dátil 211
datilera 211
datira 466
dauri [red fruit] 210
de flor naranja 247
de la potra 464
decalinae 418
dedalera 376
dedimicù 412
deene háhe 210
degonflé 387
deixa 358
dekuyukai 191
delfinio 436
delfo 342
demaguo 441
dentón 398
desbaratabaile 280
desbaratadora 367
desbaratadora blanca 367
desbaratadora colorada 367
desbaratadora morada 367
descancé 281, 283, 360
descancé común 283
descancé de Castilla 281
descancé gigante 283
descancel 280, 281
descancer común 283
descansé 281, 360
descansé menudo 281
descansel 281
desdicha 280
desempreñadora 446
desgarretadera 398
deshinchadera 317
deshinchadora 195, 196
deshollinador 245
desinflamador 197
desinflamadora 370
desjarretadera 399
desnudo florecido 320
destrancadera 360, 371
desvanecedera 196, 197
desvanecedora 195, 196, 197, 420
diablo 177, 183, 200, 207, 274, 280, 288, 289, 293, 303, 328, 335, 336, 340, 344, 349, 353, 354, 355, 359, 361, 365, 367, 373, 377, 390, 403, 411, 440
diablorrojo 395
diamante 450
diapechimuci 449
dicadá 383
dicharachero 355
dictamo 277, 345, 391
díctamo 234, 264, 345, 390, 391
dictamo real 345, 391
díctamo real 264, 345, 390
dídao 453
diego 287
diente de culebra 208, 402
diente de león 259, 261, 264, 270, 271
diente de león rojo 261
diente de perro 226, 287
diez del día 290
digital 376
digitalis 376
diku háheku 210
dikuháhe imoháhe 210
dinde 274, 326, 330, 437, 440
dinde mora 437
dinde negro 437
dioimujiriñame 435
diomante 450
diomate 395, 450
diomato 450
diosma 417
diosme 456
diplodemia 341
discancer 280, 281
discancer chiquito 280, 281
discancer grande 280, 281
disciplina 284
disciplinas 284, 390
ditamo real 391
dítamo real 391
divá 209
dividive 311
dividivi 312, 325, 335
dividivi calentano 325
dividivi de tierra caliente 325
dividivi de tierra fría 335
divina verbena 379
diwita 210
dm 325
dm-dm 325
do 201, 318
doblador 296
dobómaña 210
dobonare 306
doce apóstoles 213
dócoro 227
doctorina 359
dodake 228
dogonpapoe 191
dogonpapohue 191
dokobiijerojero 193
dólar 198
dolor de muelas 197
domingo feliz 403
dominico 372
dompedrito 211
don 263, 268, 282, 287, 345
don alonso 263, 268
don diego de noche 287
don diego morado 282
doña juana 266
doñaelena 436
doñajuana 256, 266
donalonso 263, 264, 268
doncel 433, 456, 457
doncella 194, 357, 367, 384
doncella negra 357

doncenón 315, 325
dondequiera 403
dondiego 287
dondiego de noche 287
donduada 270
donguaudú 358
donjuán de noche 287
dontagua 448
doopoje 393
dopa 311
dopamagerná 311
dopanae 311
dopane 311
doradilla 172, 174, 417, 418, 432
doradillo 418
doradita 418
dorado 272, 309, 403
dorancé 332, 333
dorancé flor amarillo 333
dorancel 332, 333
dorchoro 236
dorida 212
dormidera 265, 270, 310, 326, 327, 328, 329
dormidera acuática 328
dormidera cimarrona 310
dormidera espinosa 327
dormideras 310
dormidero 313, 319, 326, 329
dormidero blanco 329
dormilón 311, 313, 315, 319, 325, 326, 329, 330, 430
dormilona 327, 344
dorokolle 228
dorotiango 387
dosdeoros 272
dovida 191
dowatjiitajutokuerenae 354
drago 331, 388, 389
drago sangregado 389
dragón 375
drosella 265
duaná 456
duarte 258
ducu 301
duebokohañu 210
duende caspi 408
dueño de pajarito 422
duermeboca 269
duibokane 210
duirá 326
dulomoco 280, 297
dulomoco blanco 297
dulonsoga 469
dulumoco 297
dulunsoga 469
dumestri 204
dunpahue 191
duraznillo 402, 443, 444
durazno 303, 444
durazno camueso 444
durazno común 444
durazno de carne amarilla 444
durazno de carne blanca 444
durazno de Israel 444
durazno de monte 303
durazno melocotón 444
duricke 191
duro 457
düü 210

E

ea 246, 247, 255, 465
ebáñi maihóbariño 212
ébano 325, 342
echaduy 440
economía 216
eekegua 321
ẽeñá 212
eguielin 338
eiñaphí 210
ejedje 250
ekuku 188
ekuray 188
elamoroso 463
elástica 390
elemento 403
elemí 328, 452, 453
eletroespermo 426
emali 306
emalinae 306
embagatao 319
embiandé 412
embisca 328
embrujo de la India 417
emege 182
emehe 182
emético 391
emimoe 209
êmiyapochîabo 205
empanadilla 318
empanaditas 276, 316, 318
empella 256
emperatriz 362
emumeke 205
enano 333, 348
encaje 175, 176, 285, 447
encarrugada 337
encenillo 395, 431
encenillo crespo 431
encenillo negro 431
encenillo peludo 431
enchama 205
enchamba 205
enchambe 205
encillo 298
encino 431
endiquero 403
enea 247
enebro 340
eneldo 251, 252
engajado 433
enredacotorra 462
enredadera 186, 283, 341, 350, 369, 459
enredadera cera 341
enredadera pava 369
enredalocas 236
ensueño 214
enuraci 402
envidia 289
envira 187
enviroleira 191
eomu 325
epadu 386
epasota 282
epazota 282
epidendro 218
equis 193, 351
equisetum 173
erchu chuá 362
erequí 346
ereri 210
eriza café 380
erizo 258, 271, 306, 362, 380, 407
erizo café 380
ermaño caspi 363
escabiosa 296
escalera de mico 332
escalera de mono 332
escalero 312
escancé gigante 281
escancel 281
escancerillo 417
escances morado 281
escandalosa 409
escansé espinoso 281
escansé morado 281
escansel 281
escarcha 280
escarola 270
escoba 208, 268, 319, 376, 377, 407, 410, 411, 412, 413, 415, 417
escoba amarilla 413
escoba babosa 411, 412, 413
escoba blanca 412, 413
escoba cimarrona 413
escoba de babosa 413
escoba de castilla 376
escoba de Dios 376
escoba de macaco 319
escoba de monte 377
escoba de paraguay 376
escoba de puerco 415
escoba de San Antonio 268
escoba dulce 376
escoba hedionda 407
escoba negra 413
escoba peluda 377, 413
escoba pestañosa 413
escoba real 410, 415
escobababosa 413
escobabosa 413
escobadura 413
escobadura calentana 413
escobilla 189, 268, 361, 370, 376, 407, 408, 410, 412, 413, 415, 417
escobilla amarga 376
escobilla amarga granito de aníz 376
escobilla amarilla 410, 413
escobilla arisca 370
escobilla babosa 413
escobilla blanca 268, 407, 412, 413
escobilla de ratón 361
escobilla menuda 376
escobilla menudita 376
escobilla morada 412
escobilla negra 408, 413
escobilla real 268
escobillita 407
escobillo 189, 190, 334, 364, 410, 425
escobito 258, 333
escobo 184, 216, 276, 347, 379, 387, 393, 403, 407, 410, 412, 413, 414, 417, 457
escobo amarillo 413
escobo de jardín 412
escobo macho 379, 413
escobo rosado 412
escopetica 361
escorzonera 268, 270, 374, 392
escubilla 189, 410, 413, 417
escubilla babosa 413
escubilla macha 413
escubilla pimentona 189
escubillo 189, 190
escubillo negro 189
escudilla 283, 417
escupidijo 341
esmeralda 310, 419, 422
esmeraldo 260, 275, 298, 420, 421, 422, 423
espada 215
espada de Bolívar 215

espada de Judas 215
espadaña 247
espadero 184, 280, 304, 403
espadero blanco 304
espadero colorado 304
espadero de hoja chica 304
espadero real 184
espadero rubio 304
espadilla 216, 230, 231, 263, 408
espadilla blanca 263
espadillón 231
espaletaria 446
espaletario 446
espalitaria 447
espárrago 176, 214
espárrago de jardín 214
espárrago fino 176
espartillo 232, 238, 242, 245
esparto 233, 242, 413
espatifilo 202
espavé 450
esperanza 364
esperma 364, 391
espina 204, 206, 207, 225, 258, 281, 308, 356, 379, 390, 405, 471
espina de Cristo 308, 390
espina de raya 379
espinaca 283, 290
espinaca de enredadera 283
espinaca de filipinas 290
espinaca de java 290
espinaca de surinam 290
espinas de Cristo 356
espinazo de picure 427
espingo 183
espinillo 271, 329
espinito blanco 315
espino 187, 215, 225, 258, 308, 329, 330, 336, 337, 341, 356, 378, 405, 407, 433, 434, 440, 441, 449, 456
espino amarillo 336, 434
espino blanco 336, 378
espino cabro 405
espino cruz 378
espino de brujo 308, 449
espino de cabra 405
espino de cabras 405
espino de garbanzo 378
espino de jardín 336
espino de mono 337
espino de oro 336, 433, 434
espino mora 440
espino negro 378
espino tabio 378
espino totumito 356
espinosa 270, 358
espinoso mono 330
espintana 189, 190
espintaza 189
espíritu santo 222, 461
espiritusanto 222, 461
esponja 293, 294
esponjilla 293, 294, 340, 398, 407
esponjillas 294
esponjillas indígenas 294
espuela 255, 280, 434, 436
espuela de caballero 255, 436
espuela de caballo 255
espuela de galán 280, 436
espuelo 226, 433, 434
espuma de mar 214
estable 272
estancadera 256, 261, 410, 472
estatis 288
esterilla 216, 337
esterillo 216
estevia 270
estiladera 201
estopa 293
estoraque 272, 297, 308, 309, 328, 342, 389
estoraque negro 297
estramonio 466
estrancadera 360
estrella 181, 195, 214, 215, 231, 232, 236, 252, 273, 367, 390, 392, 401, 407
estrella africana 236
estrella caracha morada 367
estrella caracha roja 367
estrella de Belén 214, 215
estrella de David 215
estrella del Caquetá 214
estrella polar 273
estrella tote 232
estrellita 216, 232, 263, 273, 463
estrellita de Japón 232
estrellitas 352, 461
estromelia 417
estropajo 293
eta 212, 282
etá 210, 212
etá irakü 212
etamo real 234
eto-a 471
etopaa 471
e-to-pa-a 471
eú 210
eucalipto 424, 425
eucalipto aromático 424
eucalipto azul 424
eucalipto común 424
eucalipto plateado 424
eucalipto pomarroso 424
eucaliptus 424
eugenia 428
eugenio 428
evening primrose 429
éwani kíhiñi 212
explotaexplota 361
extraña de páramo 269
extranjero 313, 317, 331, 338, 340, 352, 365, 388, 472

F
faca-naca o aría 451
falsa árnica 269
falsa balata 305
falsa poa 240
falsa quina 352
falsa sangretoro 191, 192
falsa verbena 268
falso biribá 188
falso borojó 355
falso buchú 456
falso caraño 454
falso caucho 438
falso cocoro 397
falso guascayuca 274
falso helecho 293, 447
falso helecho de arroz 447
falso juansoco 342
falso laurel 438
falso loto 198
falso pimiento 451
falso pino 338
falso puy 209
falso samán 331
falso sangretoro 192
falso tabaquillo 354
familia de kajao-jufaifimuo 191
fara 302
faragua 240
fariña seca 392, 403
fariñero 315
fariñoso 315
faroles 272
farolillo 272
farolito 272, 457
farolitos 272, 457
favoranto 332
fegona 210
feijoa 426
feliz blanco 416
feliz torcido 416
felpa 281, 348, 442
festuca azul 239
fever grass 236
fey choche 461
fey doulé 352
fey sezi 340
fiafianae 390
fiambre 295, 433
fibra 210, 216
fibra de María 216
fichacha tasa 441
fico 438
ficus 284, 438
fideos 462
fierrito 310, 393
fierro amarillo 191, 192
fierro lanzo 394
figueroa 454
fiífia 209
filú 317
fimercarra 287
fique 214, 215, 318
fique americano 215
fique canalito 215
fique común 215
fique criollo 215
fique de Castilla 215
fique espada 215
fique espino 215
fique espinoso 215
fique fique 215
fique hembra 215
fique hembrita 215
fique hembro 215
fique liso 215
fique macho 215
fique pelón 215
fique perolero 215
fique rayado 215
fitofito 216
fito-fito 216
flaco maluco 464
flamboyán 317
flauta 234
flautilla 466
flecha 205, 240
flechero 406
floating seedbox 429
flooded gum 424
flor amarilla 255, 364
flor amarillo 364
flor candelabro 248
flor de alcatraz 192
flor de alcatraz de mompós 192
flor de amor 187
flor de ángel 312, 317
flor de azar 348
flor de baile 283
flor de cagalera 354

flor de calentura 339
flor de casa 215
flor de cera 340, 341
flor de cinco 430
flor de coral 391
flor de difunto 215
flor de fuego 317
flor de grama 232
flor de Jamaica 409
flor de Jericó 372, 460
flor de la cagalera 354
flor de la cruz 312
flor de la culebra 359
flor de La Habana 342
flor de la maravilla 216
flor de la pasión 398, 400
flor de la pasión roja 400
flor de la playa 342
flor de la reina 417
flor de las once 290
flor de madera 462
flor de mar 263, 432
flor de mariposa 263
flor de mayo 217, 342, 351, 354, 397
flor de Mompós 192
flor de muerto 259, 268, 271, 280, 302, 312, 339, 353, 355, 418, 430
flor de muerto o amapola 271
flor de navidad 390
flor de nube 465
flor de octubre 256
flor de pajarito 280
flor de palomita 192
flor de Panamá 287
flor de paraguas 270
flor de pascua 362
flor de pato 193
flor de pavo 312, 317
flor de pila 213
flor de playa 342, 464
flor de pluma 255
flor de quinde 373
flor de quinde pequeño 373
flor de reina 417
flor de sabana 232
flor de salmón 410
flor de San Juan 217
flor de sangre 378
flor de Santana 465
flor de santos 411
flor de seda 339
flor de tarde 287
flor de tierra 345
flor de totumo 202
flor de un día 216
flor de verano 286
flor del cardo 295
flor del día 409
flor del diablo 280
flor escondida 402
flor morado 365
flor pajarito 248
flor roja 283, 354
flor rosada 312
flor rosao 365
flor variante 410
flora amarilla 271
floral 342
floramarilla 265, 271, 365
floramarillo 313, 332, 333, 354, 363, 365, 397, 406
floramarillo montero 333
floranta 264
florazul 256, 379, 472
florblanca 365
florblanco 365
florecilla 263
florella 379
florida hopbush 457
floripón 465
floripondio 465, 466
floripondio blanco 465
floripondio colorado 465
floripondio encarna 465
floripondio encarnado 465
floripondio rojo 465
florisanto 312
florito 312, 313, 317
flormarina 315
flormorado 364, 365, 374, 430
flormorao 365, 430
floro 365
floroazul 288
florón 333, 342
florota 256
flotilla de gaitán 224
fodder burnet 445
foei 202
foeí 202
foiye 181
folla furada 200
fonbazen 371
fono 301, 302, 405, 455
fono blanco 301, 302
fono carguero 301
fono cocoro 405
fono colorado 301
fono hediondo 301
fono medicines 301
fono negro 301, 455
fono podrido 301
fontana 286
forastera 286
fosforito 230, 232
fósforo 230, 373
fostón 258
foua 191
fragua 298
fraijoa 426
fraile 417
frailecillo 343, 388, 391
frailecillo cimarrón 388
frailejón 262, 263, 377, 391, 417
frailejón calentano 391
frailejón negro 262
frailejón perrito 262
frailejón plateado 262
frailejoncito 262
frambuesa 443, 444, 445
francés 290, 333
francesina 465
francesino 465
frangipán 342
frángula 442
fredolina 289
frega plato 469
frega platos 469
fregaplato 469
fregaplatos 469, 471
freijoa 426
frente de toro 326
frente de toro pico de loro 326
frenteloro 261, 467
frentín 333
fresa 443, 445
fresa de Chile 443
fresa de monte 443, 445
fresa silvestre 443
fresal de Chile 443
fresco 303, 440
fresia 216
fresnillo 365
fresno 338, 363, 365, 374, 387, 433, 451, 460
fresno amargo 433, 460
fresno urapán 374
fresón 443
friega plato 469
friegaplato 469, 471
friegaplatos 275, 470, 471
fríjol 313, 314, 319, 320, 324, 328, 329, 330, 336, 337, 389
fríjol arrocillo 336
fríjol arroz 336
fríjol arveja 324
fríjol arveja real 324
fríjol azul 330
fríjol azulejo 330
fríjol batata 337
fríjol bichuela 330
fríjol bichuela rojo 330
frijol blanco 313
fríjol blanco 313
fríjol botón de saco 330
fríjol bugueño 324
fríjol burro 313
fríjol cabecenegra 337
fríjol cabecitanegra 337
fríjol cabezanegra 337
fríjol cacao 313
fríjol cacha 330
fríjol cacha amarilla 330
fríjol cachas 330
fríjol cachetón 330
fríjol cacho 330
fríjol calentano 324
fríjol canavalia 313
fríjol caupí 337
fríjol de año 330
fríjol de cabeza negra 337
fríjol de enredadera 330
fríjol de hormiga 314
fríjol de monte 319
fríjol de playa 313
fríjol de puerco 313
fríjol de todo el año 330
fríjol de tubérculo 313, 329
fríjol extranjero 313
fríjol gualuy 319
fríjol guamo 313
fríjol guandó 313
fríjol guandú 313
fríjol guandul 313
fríjol guandús 313
fríjol guanduz 313
fríjol guisante 324
fríjol jacinto 324
frijol linaza 337
fríjol machete 313
fríjol mataarriera 313
fríjol morado 389
fríjol morro 313
fríjol nopas 319
fríjol nopás 319
fríjol nopaz 319
fríjol ojinegro 337
fríjol peruano 324
fríjol pispo 336
fríjol quinchancho 313
fríjol quinchoncho 313
fríjol risueño 324
fríjol sobaco 324
fríjol soya 320, 324
frijol terciopelo 328
fríjol terciopelo 328
fríjol veleño 337
frijolato 333

frijolero 332
frijolillo 289, 311, 314, 315, 317, 320, 325, 327, 331, 332, 333, 337
frijolillo dos pepas 331
frijolillo nativo 311
frijolillo rastrero 337
frijolillo silvestre 314, 315
frijolito 313, 314, 327, 330, 332, 336, 337
frijolito azul 313
frijolito pijindí 337
frijolón 313
frisol 313, 319, 320, 324, 330, 336
frisol cachetón 330
frisol calu 319
frisol calú 319
frisol carriel 330
frisol de monte 330
frisol de palo 313
frisol de vaca 336
frisol de vida 330
frisol guandús 313
frisol sapo 330
frisol sojo 320
frisol tropical 324
frisol tropical o mulato 324
frisolato 332, 333
frisoles 324
frisolillo 317, 334
frisolillo de llano 317
frisolillos 332
frisolito 332
frisolo 189
frito 311, 318, 361
fruncida 417
fruta bomba 279
fruta de burro 189, 278
fruta de capuchino 189, 190
fruta de culebra 294, 348, 357
fruta de indio 447, 457
fruta de loro 453
fruta de molinillo 190
fruta de mono 355, 385
fruta de murciélago 333
fruta de pan 436
fruta de pava 287, 305, 309, 343, 348, 437
fruta de pavo 348
fruta de tórtolas 378
fruta de zorro 278
fruta del capuchino 189
fruta del pan 436
fruta sábalo 331
frutaeburro 189
frutaepava 348
frutapán 436
fruteburro 189
fruteloro 278
frutemato 193
frutenevado 449
frutepan 436
frutevenado 449
frutilla 443
frutillo 275, 469, 470, 471
frutillo hembra 470
frutillo macho 469
frutipan de monte 437
fruto 189, 278, 295, 309, 355, 384, 425, 436, 470
fruto de burra 189
fruto de burro 189
fruto de la culebra 295
fruto de lora 384
fruto de morrocón 425
fruto de pava 309, 355
fruto de zorro 278
fruto del pan 436
frutogallino 467
fucsia 428, 429
fucsia arbustiva 429
fuente 472
fumaria 252, 435
fumo 289, 451
fumo bravo 289
funa 348
funche 231
fune 362
fuñe 181
fuñe boighe 181
funeraria 303
funita 278
furrusca 333
fustete 440
fut-avá-a 465

G

gabeche 332
gabilla 294
gabui 191
gache 431
gachico 319, 325
gachingo 319
gaco 431
gahuar-chuchuncca 430
gaitana 364
galán de la noche 466
galán de noche 307, 466
galapo 311
galba 380
galbecillo 333
galicosa 264
galleta 407
gallina 187, 234, 272, 293, 317, 345, 394, 404, 467
gallina de pobre 272
gallina gorda 187
gallinaza 190, 206, 268, 271, 371, 422
gallinaza de azotea 268
gallinazo 181, 185, 190, 206, 207, 268, 271, 293, 314, 315, 333, 336, 348, 359, 364, 369, 371, 379, 380, 468, 469, 470, 471
gallinazo blanco 268, 369
gallinazo común 271
gallinazo negro 268, 369
gallinazo ordinario 336
gallinazo pago 471
gallinazo pajo 470
gallinazo panga 185, 469
gallinero 173, 330
gallinita 283
gallino 317, 467
gallita 222, 223
gallito 189, 192, 193, 272, 312, 314, 319, 334, 340, 391, 429, 434
gallito colorado 391
gallito de laguna 429
gallito de pantano 319
gallito rosado 272
gallitos 193
gallo 192, 193, 272, 358
gallogiro 229
galve 332, 333
galve amarillo 332
galvecillo 333
galves 332, 333
galves arisco 333
galvillo 333
galvis 333
gamboa 260, 264, 270
gamuzón 424
gancho 442
gapuy 344
gaque 383, 384, 385, 431
gaque de monte 384
gaque rosado 384
garabato 358
garabato amarillo 358
garagay 378
garata 451
garatero 451
garbancillo 378
garbanzo 315
garcero 383
garcita 202, 280
gargantilla 387
gargantillo 387
garlipo 339
garlucho 402
garopa 287
garotá 451
garra de canguro 224
garrapatilla 320
garrapato 188, 189, 312, 325, 331, 382, 383, 427
garrapato lanudo 188
garrapato negro 188
garrapo 382
garretón 235
garrocho 297, 298, 304, 339
garrotillo 362
garrucho 190, 304
garve 333
garza 202, 234, 279, 319, 390, 460
garzovia 217
gasa 285, 286
gasilla 286
gaspadillo 423
gasparillo 423
gatara 329
gateadera 171
gateadero 299
gateado 450
gateadora 234, 242, 244, 367
gateadora blanca 242, 244
gateadora morada 234
gatiao 450
gatico 285, 417
gaticos 285
gato 171, 215, 226, 227, 233, 234, 245, 258, 274, 280, 287, 291, 297, 310, 326, 335, 343, 358, 363, 371, 380, 387, 406, 417, 433, 434, 436, 439, 442, 445, 456, 457, 469, 470, 471
gaudín 237
gavilán 301, 368, 377
gavilana 192, 257, 258, 267, 378
gavilana negra 378
gavilla 294
gediondito 470
gemena 437
genciana 344
gender 212
genene 381
genené 381
gengibrillo 232
genio 416
genízaro 331
geranio 358
geranio blanco 358
geranio de olor 358
geranio de páramo 358

geranio de rosa 358
geranio de yedra 358
geranio rojo 358
geranio rosado 358
gérbera 263
gerillo 260
german rampion 429
gibiño 210
gicombo 406
gigante 311, 450
gigantón morado 466
ginesta 334
ginkgo biloba 179
ginseng 254
girasol 263, 271
gitana 369
glaves 332
gleo 281
globito 340, 457
globitos 340, 457
globos 340
gloria 289
gobernador 404
gõháño 212
golden shower 313
gólgota 417
golocera 194
golondrina 242, 253, 277, 285, 357, 366, 372, 377, 390, 397, 436, 448, 450
golondrino 188, 189, 448, 449
golupo 399
goma 274, 280, 430
gombo 406
gomelero 438
gomero 424, 438
gomero azul 424
gomo 274, 276, 414, 430, 449
gomo amarillo 274
gomo blanco 274, 414, 430
gomoso 274
gondu 313
gordura 241
gorretas 354, 355
gorrito rojo 367
gorro de obispo 284
gorro del diablo 355
gorro liberal 367
gõsa 210
gota de aceite 200
gota de Cristo 369
gota de sangre 366, 369, 390
gota de tinta 361
gotas de amor 367
gotelia 374
gourd 185, 363
gourda 185
gracia 198, 376
grado 389
grago 389
graines benné 280
grama 233, 234, 236, 237, 238, 239, 241, 242, 243, 244, 245, 246
grama alta 237
grama andadora 242
grama azul 244
grama blanca 233, 234, 242, 243, 245
grama china 245
grama común 242
grama criolla 234
grama de gallina 236
grama de horqueta 238, 246
grama de matojo 238
grama de moño 238
grama de trenza 243
grama dulce 234, 236
grama dura 245
grama egipcia 236
grama extranjera de hoja ancha 234
grama horquetilla 234
grama ilusión 239
grama macho 233
grama natural 234
grama olorosa 234
grama salada 238
grama seda 241
grama trencilla 243
grama trenza 243
gramablanca 238
gramalote 234, 238, 240, 241, 243, 246
gramalote blanco 234
gramalote colorado 234
gramalote grueso 241
gramilla 234, 236, 238, 243
gramilla blanca 243
gramilla brava 236
gramilla colorada 236
gramilla rastrera 234
gramilla zacate 238
gramínea de sabana 234
gramote 245
granada 417, 431
granada de monte 431
granadilla 246, 330, 397, 398, 399, 400
granadilla babosa 399
granadilla chica negra 398
granadilla de árbol 398, 399
granadilla de monte 398, 399, 400
granadilla de murciélago 400
granadilla de piedra 399
granadilla de quijos 400
granadilla real 400
granadilla roja 398
granadilla silvestre 397, 398, 399, 400
granadillo 318, 325, 326, 330, 336, 398, 399, 400, 401, 402, 403, 416, 437
granadillo arbóreo 398
granadillo de árbol 398
granadillo de conga 399
granadillo de monte 400
granadillo grande 400
granadillo rojo 437
granadillo rosado 437
granadillo silvestre 400
granado 351, 417
granavi 378
granecillo 181
granicillo 181, 347, 353, 449
granito de arroz 390
granito de oro 388
granizo 177, 181, 182, 247, 355, 393, 401, 447
grano de oro 246, 388, 470
grao 389
grape point 380
grasoso 380
grau blanco 331
greifruta 456
grosella 401, 418, 425
grosella roja 425
grosellero 401
grosello 396, 401
gua 192, 193, 418, 441, 449
guaba 287, 288, 305, 322, 323, 324
guaba churimba 324
guaba de bejuco 322
guaba del Putumayo 322
guaba manza 322
guabillo 318, 322, 364
guabito 460
guabo 288, 322, 323, 324, 332
guabo macheto 324
guabo mico 322
guabo tirisio 332
guabra 239
guaca 255, 256, 258, 260, 264, 364, 368
guaca cimarrona 264
guaca de marrano 264
guaca negra 255, 258
guacamacho 284
guacamacho falso 284
guacamari 321, 330
guacamaya 287, 301, 312, 358
guacamaya roja 358
guacamayo 191, 228, 290, 301, 310, 311, 318, 319, 321, 325, 352, 358, 368, 388, 389, 441, 453, 457, 465
guacamayo blanco 310
guacamayo caspi 191, 368
guacamure 292
guacapurana 449
guacapurano 313
guacarí 449
guachamaca 341, 344
guachamacá 341, 342
guachamaca de tree 344
guachamacán 341
guachamacay 341
guacharaca 191, 282, 368, 436, 455
guacharaco 182, 184, 253, 288, 329, 334, 339, 343, 344, 351, 378, 436, 453, 457, 458
guacharaco colorado 457
guacharaco hediondo 453
guacharaco pavero 351
guache 254, 316
guachecito 316
guachecitos 316
guácima 408
guacimo 408
guácimo 274, 275, 276, 408, 410, 437, 451
guácimo colorado 408, 410
guácimo colorao 410
guácimo coloro 410
guácimo peludo 274
guaco 192, 193, 226, 266, 270, 301, 380, 381, 434, 466
guaco blanco 193, 226, 266
guaco morado 266
guaco negro 193
guacuna 323
guacuray 454
guacure 359, 381, 426
guacure de chimbe 381
guacure de danta 359
guacure de monte 359
guacure morado 359
guacure negro 359
guacure verde 359
guacuri 359, 382
guacurí 382, 383
guacuri amarillo 359
guacuri anaranjado 359

guacuri danta 382
guacuri de carne gruesa 359
guacurí de danta 382, 383
guacuri grasoso 359
guacuri negro 359
guada 239
guadín 238
guadua 235, 239
guadúa 239
guadua amarilla 235
guadua cebolla 239
guadua lisa 239
guadua macana 239
guadua macha 239
guadua rayada 239
guaduilla 240, 244
guaduín 237
guafa 239, 387
guafilla 239
guagaripo 447
guagay 447
guagua 287
guagualín 209
guaguarín 209, 212
guagüera 209
guagüilla 258
guaguín huashüin 266
guahibo 311, 420
guaimarillo 441
guaimaro 439
guáimaro 395, 437, 439, 440, 441
guaimarón 437
guaipán 436
guaira bombón 213
guaira caspi 454
guaira sacha 241
guaisulidia 356
guaitil 351
guájaro 339
guaje 199
guají 377
guajirilla 255
guajo 205
guajuro 339
guakaria 210
gualanday 364
gualba 248
gualicón 298
gualola 289
gualpante 182
gualpite 402
gualte 212, 213
gualte bola 213
gualte macho 212
gualte patudo 212
gualte upián 213
guama 321, 322, 323, 324, 329, 334, 337
guama coperuna 324
guama de altura 321
guama de antiguo 323
guama de barbasco 323
guama de boruga 324
guama de borugo 324
guama de chucha 329
guama de churuco 321, 323, 324
guama de hoja grande 321
guama de mico 324
guama de montaña 324
guama macheta 324
guama machete 323
guama negra 329
guama peluda 323
guama seco 337
guamache 284
guamachito 284
guamacho 284
guamacho extranjero 284
guamacho serrano 284
guambi 345
guambia 345
guambía 345
guamilla 313, 323, 337
guamillo 319, 321, 322, 323, 326, 329, 335, 337
guamillo blanco 335
guamillo de rastrojo 322
guamita 323, 329
guamito 321, 322, 323, 324, 327
guamito de danta 323
guamito de muinanes 323
guamito de orilla 322
guamito silvestre 324
guamo 315, 321, 322, 323, 324, 326, 329, 330, 331, 333, 334, 335, 337, 347, 382, 424, 452, 459
guamo arroyero 337
guamo bejuco 322, 323
guamo blanco 322, 323, 324
guamo borrachero 323
guamo borugo 329
guamo cafetero 322
guamo cafeto 324
guamo cajeto 323, 324
guamo caliente 321
guamo calzamuelas 323
guamo cansamuelas 323, 324
guamo cerindo 323
guamo chairus 322
guamo cheneve 324
guamo chigo 424
guamo chinibo 322
guamo chirimo 321
guamo churimbo 324
guamo churimo 321, 322, 323, 324
guamo churimo hojiamarilllo 323
guamo churito 323, 324
guamo copero 324
guamo cuatrofilos 324
guamo cupero 324
guamo de arqueado 323
guamo de barbasco 323
guamo de bejuco 334
guamo de borugo 324
guamo de chupar 324
guamo de guacamayo 321
guamo de güio 324
guamo de loma 324
guamo de mico 321, 322, 333
guamo de montana 324
guamo de montaña 323, 324
guamo de monte 322
guamo de osito hormiguero 322
guamo de pepa negra 329
guamo de Popayán 322
guamo de río 337
guamo de sapo 452
guamo de semilla 322
guamo de venado 334
guamo hembra 323
guamo hojiancho 323, 335
guamo lanoso 322
guamo largo 322
guamo loro 322, 324, 330
guamo machete 321, 322, 323, 324
guamo machetero 321
guamo macheto 322, 324
guamo macho 337
guamo mene 323
guamo mestizo 382
guamo mico 323
guamo negrito 323
guamo negro 329, 347
guamo pepaloro 322
guamo plancheto 321
guamo playero 337
guamo querré 322
guamo rabo de mico 322
guamo rabo de mono 322, 324
guamo racero 324
guamo rosado 322, 326
guamo rosario 323
guamo rosarito 323
guamo santafereño 323, 324
guamo serindo 323
guamo venenoso 337
guamos 321, 323
guampana 399
guamsian borrachero 466
guamuca 465
guamuca borrachera 465
guamuche 311, 335
guamuco 345, 465
guan 363
guaná 433
guanábana 186, 187, 190
guanábana arisca 186
guanábana cabezona 187
guanabana cimarrona 187
guanábana cimarrona 186, 187
guanábana de cabeza negra 187
guanábana de monte 186, 187, 190
guanabana de pozo 187
guanábana de pozo 186, 187
guanábana matimbá 187
guanábana ñeca 187
guanabanadillo 190
guanabanilla 187
guanabanillo 187, 190, 192, 305
guanabanito 186, 187, 278
guanabanito de playa 186
guanábano 186, 187, 190
guanábano brasilero 186
guanábano cimarrona 186
guanábano de monte 186, 190
guanábano de pozo 186
guanábano pun 187
guanabarjo 187
guanacaste 319
guanacón 187
guanacona 187
guanano 319
guanapay 351
guandú 313
guandul 313
guandumbo 313
guandús 313
guanduz 313
guaney 319
guango 311, 331
guango muche 311
guanto 465
guañusa 255
guapa 192, 319
guapante 378
guapanto 378
guapichuna 452, 453
guapinol 320
guapo 250

guaque 384
guaquilla 264
guaquillo 294
guaquito 266, 338, 347, 367
guara 183, 188, 204, 210, 457
guará 430
guara blanca 457
guarabilla 413
guarabillo 413
guaraí 383
guaral 201, 203
guaraná 458
guaranazeiro 458
guarango 326, 329, 333, 334, 335
guarango barbasco 335
guarango blanco 329
guarango negro 333
guarango rayao 329
guaranguilla 402
guarano blanco 319
guarapero 288
guarapita 300
guarapo 363
guararo 328
guarataro 374
guarauno 414
guardaparque 277, 281
guardaparque blanco 277
guardarocío 314
guardarrocío 237, 238, 314, 393, 394
guardarrocío colchado 237
guardasereno 238
guardasireno 242
guarderiana 249
guaré 383
guaretero 374
guargo 248
guarguerón 376
guargüerón 375, 376
guarguerón o digital 376
guargüerones 376
guaricamo 404
guaricha irayol 351
guaritoto 388
guariuba 437
guarno arroyero 324
guarnul 213
guaro 381
gua-roo'-ta-ta 192
guarrás 321
guaruchí 189
guaruma 249
guarumá 249
guarumero 253
guarumo 249, 445, 446, 447
guarumo blanco 446
guarumo negro 445
guarumo uva 447
guarumón 253
guarupa 364
guarupayo 451
guarutata 192
guasábara 283
guasaí 209
guasaí de sabana 209
guasaí pequeño 209
guasaina 212
guasay 209
guasca 188, 192, 193, 208, 256, 263, 264, 275, 276, 292, 294, 301, 365, 409
guasca amarilla 188
guasca colorada 301
guasca dulce 409
guasca matamba 208
guasca nato 301
guasca nato pedo 301
guasca negra 275, 276, 301
guasca peo 301
guascas 263
guascayuca 274, 412
guasco 256, 258, 275, 276, 301, 387, 416
guasco blanco 416
guasco colorado 301
guascua 193
guasgüín 266, 267
guasgüís blanco 267
guasicaspi blanco 430
guásima 408
guasimillo 408
guásimo 273, 274, 275, 407, 408, 410
guásimo baba 407
guásimo blanco 407, 408
guásimo colorado 407, 408, 410
guásimo colorao 410
guásimo macho 408
guásimo negro 274, 408
guásimo rojo 408
guásimo serrano 410
guásimo torcido 408
guásimo vegetal 273
guasquila 295
guasquilla 295, 402
guasquín 266, 267
guata roja 471
guatamare 328
guatamo 225
guataro 374
guate 250, 351
guatemala 246, 260, 471, 472
guatemalan rhubard 392
guatemalteca 471, 472
guatila 295
guatilla 295
guatinajo 205
guatinajo hembra 205
guauda 239
guaúda 239
guaudua 239
guaudúa 239
guavito amargo 460
guawmoamú 364
guayaba 342, 347, 356, 396, 412, 418, 423, 424, 425, 426, 427, 428
guayaba agria 428
guayaba brasileña 425
guayaba champa 424
guayaba choba 428
guayaba común 428
guayaba de choco 428
guayaba de coronilla 418
guayaba de danta 342, 418
guayaba de leche 424
guayaba de monte 418, 423
guayaba de pava 418
guayaba de perro 396
guayaba dulce 428
guayaba feijoa 426
guayaba manzana 428
guayaba pera 428
guayaba perulera 356
guayaba perurera 347
guayaba sabanera 428
guayabero 437
guayabete 347
guayabeto de altura 347
guayabilla 299, 404, 412, 418, 420, 426, 427, 428, 438
guayabilla agria 412
guayabilla negra 299
guayabillo 295, 303, 333, 347, 416, 418, 425, 426, 428, 429
guayabillo negro 416
guayabita de conejo 428
guayabita peruana 418
guayabita sabanera 428
guayabito 347, 378, 388, 389, 416, 424, 427, 428, 429
guayabito agrio 428
guayabito de agua 429
guayabito de cerro 428
guayabito liso 427
guayabito montañero 347
guayabo 290, 297, 304, 343, 347, 350, 351, 355, 356, 380, 388, 396, 416, 417, 418, 422, 423, 424, 425, 426, 427, 428, 438, 441, 443, 455
guayabo agrio 426, 428
guayabo amarillo 380
guayabo anselmo 424
guayabo arrayán 427
guayabo blanco 427
guayabo brasilero 424
guayabo cascarillo 427
guayabo champo 424
guayabo chobo 428
guayabo cimarrón 418, 424
guayabo colorado 347, 428
guayabo colorao 297, 416
guayabo coronillo 418
guayabo de Anselmo 424
guayabo de Castilla 427, 428
guayabo de danta 343
guayabo de flor 355
guayabo de hierro 416
guayabo de leche 424
guayabo de león 388, 416
guayabo de mico 355, 418
guayabo de mono 418, 422, 424
guayabo de monte 418, 423, 425, 426, 427
guayabo de páramo 443
guayabo de pava 418
guayabo de perro 396
guayabo de tejado 351
guayabo del brazil 428
guayabo dulce 428
guayabo hojiancho 426
guayabo hojimenudo 426
guayabo león 416
guayabo liso 424, 427
guayabo macanillo 347
guayabo macho 347
guayabo manizaleño 417
guayabo mico 418
guayabo pera 428
guayabo perulero 356
guayabo rebalsero 428
guayabo rojo 425
guayabo silvestre 418
guayabo ulanda 455
guayabo zancón 290
guayabón 427
guayacán 245, 268, 308, 311, 314, 319, 325, 330, 331, 362, 363, 364, 365, 374, 383, 416, 417, 423, 424, 434, 437, 441, 472, 473
guayacán amarillo 363, 416, 417

guayacán azul 472
guayacán blanco 472, 473
guayacán carrapo 472, 473
guayacán cavío 319
guayacán chaparro 472
guayacán chicalá 365
guayacán de bola 308, 472, 473
guayacán de Manizales 417
guayacán de piedra 472
guayacán de playa 472
guayacán del Cauca 417
guayacán extranjero 472
guayacán flor azul 472
guayacán hobo 314
guayacán lila 365
guayacán llanero 423
guayacán mochilero 416
guayacán morado 363, 365
guayacán negro 472
guayacan polvillo 364
guayacán polvillo 363, 364, 383, 472
guayacán rosado 365
guayacán sabanero 331
guayacán trébol 330
guayacán verdadero 472
guayacana 226, 227, 236, 240
guayacancito 374
guayaco 472
guayaquil 193, 314
guayoba silvestre 458
guayusa 181, 253, 255
guayusa andina 181
guayusa verde 253
guayuyo 436
guayuyu 415, 436
guazumilla 412
guchero 468
guchuba 468
guchubo 468
guchunchullo 373
guchuva de perro 467
guchuvo 468
güeba 190
güéguerre 205
güeraca 451
guergüerón 376
guerra 336, 388
guerra santa 388
güerre 205
güérregue 205
güeva de arriera 186
güeva de burro 278
güeva de gato 344
gueva de indio 406
güeva de tigre 291
güeva de toro 341
güeva peluda 343
guevas de chucha 349
güevas de chucha 349
güevas de perro 340
guevo de burro 278
güevo de indio 406
güevo de varraco 343
güevoetigre 206
guevotigre 206
güibán 362
guiche 229
güichire 205
güichirre 205
guicombo 406
güicungo 204
güiesquín 267
guilache 433
güimo 319
güimó 319
guina 454
güina 454
güinche 242
guincho 229
guinchos 229
guinda 298, 401, 442, 444, 471
guindaguinda 442
guinea 241, 287, 333, 461
guinea hen 287
guinea pajarito 241
guinedón 241
guino 454
güino 288, 454
güino cedro 454
güino rojo 454
guinul 205
güinul 205
guira 247, 453
güira 247, 363
güiridima 204
guisador 250
guisante 325
guisante de olor 325
guiso 369
gulupa 398, 399, 400
gulupo 398, 400
gumán 424
gumbía 345
gungo 313
guntsiangre 258
gupediaque 453
gurango rayao 329
gurapo 185
gure 440
gurrubo 257, 467
gurrubu 347
gurumo 467
gurupacha 378
gusanera 450
gusanero 258, 295, 368, 450
gusanero morado 450
gusanillo 245, 317, 368, 387
gusanito 245, 367
gusano 301, 346, 387, 441, 451
guypani 209

H

haadió kuume 210
haba 313, 320, 336
haba soya 320
habana 342
habano 342, 344, 361, 467
habano morado 361
habilla 294
habillo 318
habo 319
hacha cabo 192
hachón 189
hacillo 394
há-ha 414
háheku 210
hala 320
halconcito 192, 193
hamamelis 461
hambi 388
hañõn 210
harecha 191
harino 347, 457
hartón 414, 429
hawapi para mujer 349
haya 228
hayaku 210
hayo 386, 457
hayuelo 180, 303, 386, 457
hebukanu 210
hedionda 246, 407
hediondilla 333
hediondo 185, 301, 302, 315, 319, 453, 457, 466, 471
heé-a 414
helecha 175
helechillo 177, 178
helechillo negro 178
helecho 171, 172, 173, 174, 175, 176, 177, 178, 179, 214, 224, 267
helecho acuático 178
helecho arborescente 172
helecho blanco 178
helecho chiguamacho 177
helecho churco 176
helecho cilantro 177
helecho cola de pez 176
helecho costeño 176
helecho crespo 176
helecho cuerno 176
helecho cuero 176
helecho de árbol 173
helecho de bejuco 175
helecho de Castilla 214
helecho de cuero 176
helecho de diablo 177
helecho de espiga 173
helecho de la suerte 175
helecho de madremonte 179
helecho de marrano 175
helecho de perfume 177
helecho de pescado 173
helecho encaje 175
helecho estacoso 172
helecho gallinero 173
helecho gateadero 171
helecho granizo 177
helecho liso 175
helecho macho 172, 174, 175
helecho mano de tigre 177
helecho marranero 175
helecho negro 178
helecho pata de gallina 173
helecho pategallina 173
helecho peine 176
heliconia 248, 249
heliconia enana 249
heliotoropo de olor a vainilla 275
heliotropo 251, 275
henequén 214
hepe 202
hepe-kuu-tata 202
heralda 264
herb amarga 259
herb amarilla 394
herb cebolla 217
herb de arroz 230
herb de bruja 350
herb de caballo 268
herb de camarón 359
herb de canotillo 195
herb de clavo 261
herb de conejo 173
herb de eripsepela 351
herb de estrella 232
herb de golpe 268
herb de la cagalera 354
herb de la gracia 376
herb de la pastora 401
herb de la pastorcita 401
herb de la trinidad 360
herb de la verrugosa 195
herb de lagartija 175
herb de limón 236
herb de marrano 268
herb de oveja 369

herb de pollo 390
herb de puerco 200
herb de San Antonio 359
herb de san juan 271
herb de San Juan 271
herb de santamaría 267
herb de sapo 289, 370
herb de venado 316
herb del cáncer 380
herb del dinero 376
herb del soldado 195
herb del venado 401
herb dulce 238, 379
herb hedionda 287, 466
herb mora 470
herbbuena 468
herbbuenilla 416
herbmora 470
hermosito 421
hernandia 182
herrerún 439
hiabetoñi 204
hiaigin 210
hiaukiña 210
hibia 432
hicaco 381
hidotropo 251
hidrópica 202
híduiki 210
hiebra de gallinazo 314
hiedra 253, 261, 376, 448
hiedra amarilla 261
hiedra de tiesto 376
hiedra de tiestos 376
hierba alacrán 274
hierba alegre 373
hierba bicho de platanito 333
hierba bicho guinea 333
hierba blanca 266
hierba borriquera 360
hierba caimán 289
hierba cana 267
hierba capitana 266
hierba cebolla 217
hierba centella 286
hierba cuna 366
hierba damiana 401
hierba de arco 282
hierba de arroz 230, 241
hierba de bruja 350
hierba de buey 472
hierba de caballo 261
hierba de cabro 256
hierba de camarón 359
hierba de chancho 264
hierba de chivo 256
hierba de chucha 372
hierba de chucho 256
hierba de chulo 268
hierba de clavo 261, 429
hierba de conejo 238
hierba de cotorra 388
hierba de culebra 287
hierba de danta 350
hierba de estrella 232
hierba de gallinazo 268
hierba de golpe 268, 282
hierba de indio 376
hierba de la luna 316
hierba de la potra 464
hierba de lagartija 462
hierba de leche 339, 390
hierba de limón 236
hierba de los Uribe 265
hierba de marrano 268
hierba de oveja 369
hierba de perro 231, 235
hierba de pisingo 238
hierba de playa 332
hierba de pollo 390, 472, 473
hierba de puerco 261, 263, 282
hierba de san juan 394
hierba de San Juan 394
hierba de Santa Lucía 256
hierba de Santa María 267
hierba de sapo 289, 290, 370, 376
hierba de sarna 367
hierba de tortero 357
hierba de umarí 441
hierba de venado 316, 345
hierba de vidrio 283, 290
hierba de zorra 237
hierba del cáncer 387, 464
hierba del carpintero 255
hierba del caspe 402
hierba del cuartillo 462
hierba del dedo 358
hierba del melado 241
hierba del vidrio 290
hierba dulce 238
hierba golondrina 390
hierba golpe 268
hierba india 241
hierba lechera 391
hierba limón 236
hierba luisa 236
hierba maldita 197
hierba sagrada 282
hierba santa 267
hierbabuena 468
hierbabuenilla 417
hierbadulce 349
hierbaechulo 268
hierbamora 423, 468, 470, 472
hierbamora blanca 468
hierbasanta 461
hierro 292, 308, 394
hierrolanzo 394
hígado de diablo 328
hígado del diablo 328
higantón 466
higo 279, 283, 284, 305, 331, 338, 391, 438, 439, 466
higo amarillo 331
higo chumbo 284
higo de infierno 391
higo de monte 279
higo hembra 439
higuá 205
higuatá 205
higuera 392, 438
higuereta 392
higuereto 392
higuerilla 391, 392
higuerillo 254, 339, 380, 392
higuerillo blanco 392
higuerillo rojo 392
higuerito 392
higüero 363
higuero de lata 206
higueron 439
higuerón 254, 305, 392, 437, 438, 439, 441
higuerón grande 439
higuerón guayabo 438
higuerón pequeño 439
higuerón raspador 439
higuerón tunoso 441
higuerona 439
higueroncillo 437, 438, 439
higuillo 279, 409, 413
higuito 438, 439
hiidíko paña 210
hiimasa 204
hija de sapo 463
hikatatano 201
hiliotropo 251
hilo de oro 462
hilotropo 251
himuntẽgn 212
hincahinca 303
hinkató 209
hinojo 252
hinojo dulce 252
hipérico 393, 394
hira 212
hiriwa 212
hirwa-kowa 258
hisopo de agua costera 376
hobito 274
hobo 306, 314, 451
hobo arisco 451
hobo blanco 451
hobo colorado 451
hobo manso 451
hog doctor 451
hog plum 451
hoja ancha 192, 195, 234, 267, 359
hoja cucua 282
hoja de aguacero 211
hoja de algodón 408
hoja de arroz 290
hoja de boa 326, 370
hoja de buitre 198
hoja de canangucha 392
hoja de cáncer 285
hoja de capa 209
hoja de carpintero 351
hoja de chucha 199
hoja de cuero 355
hoja de difunto 271
hoja de garza 202
hoja de higo 338
hoja de la estrella 195
hoja de la fortuna 461
hoja de lapa 227
hoja de luna 468, 471
hoja de mano 197
hoja de mariposa 263
hoja de mono 466
hoja de murciélago 398
hoja de palma 215
hoja de pantano 358, 359
hoja de piedra 194, 198, 200
hoja de pulir 274
hoja de raya 196, 197, 273
hoja de rayo 197
hoja de saíno 199
hoja de sal 248, 249, 345, 402
hoja de sangre 366
hoja de Santa María 195, 267
hoja de santamaría 267
hoja de sebo 407
hoja de solimán 417
hoja de teñir 363
hoja de tortuga 209
hoja de tresdedos 197
hoja de tunda 194
hoja de vaso 195
hoja de venado 332
hoja del aire 461
hoja del soldado 461
hoja dulce 417
hoja inmortal 461
hoja limón 423
hoja mano 197
hoja morada 402
hoja para la viruela 197

hoja redonda 197
hoja santa 461
hojaancha 288, 348
hojablanca 249
hojadillo 200
hojamano 197
hojancho 387
hojarancín 461
hojaras de gilberto 190
hojarasco 190, 288, 393, 441, 451
hojarrota 200, 201
hojas de corazón 198
hojas de granizo 181
hojas de pantano 358
hojas de sen 332
hojasanta 461
hojenegro 439
hojiancha 234
hojiancho negro 401
hojicenizoso 191
hojita de teñir 363
hojitas de sapo 348
hókiki 204
holguín 433
holliday 396
holly brillante 396
holly piracanta 396
hombre grande 460
hombregrande 460
hopshrup 457
horcaperros 255
hormigo 404, 405
hormigón 404
hormiguero 290, 322, 423
horqueta 227
horquetilla 209, 227, 231, 232, 238, 242
hortensia 291
hortiga 373, 448
hortiga blanca 373
hosmeca 364
hota mohe 210
houiri 454
hoyo de zorro 315
huaca 260
huacapú 380
huacapurana 313
huacimo 408
hualanday 364
hualicón llucho 299
huambia 345
huambía 345
huambía o yerba de huambía 345
huangana coca 386
huapa 192
huarme cuyanguilla 360
huarumá 249
huaruro 329
huasaí 209
huasca 301, 363
huasguin 267
huasgüín 267
huashuín 266
huasipanga 209
huayro 329
huayusa 255
huba 212
huéguerre 205
huella de tigre 250
hueporo 441
huerre 205
huérregue 205
huesillo 302, 380, 403
huesito 254, 265, 347, 350, 378, 380, 395, 403, 404, 405, 426, 440, 458
hueso 288, 318, 319, 347, 362, 404, 449, 458
hueso de pescado 319
hueso negro 288
hueva de verraco 343
huevas de Adán 344
huevas de tigre 349
huevita de gato 344
huevito de gato 344
huevo abajo 402
huevo burro 278
huevo de berraco 344
huevo de burro 278, 308
huevo de danta 381
huevo de gallo 272, 434
huevo de gato 291, 343, 380, 469, 470, 471
huevo de hicotea 303
huevo de mono 344
huevo de morroco 303, 380, 449
huevo de morrocoy 303, 380, 449
huevo de morrocoyo 380, 449
huevo de perro 343, 471
huevo de sapo 468
huevo de tigre 344
huevo de toro 471
huevo de varraco 343
huevo de verraco 343
huevo morroco 303
huevo morrocoy 303
huevo vegetal 457
huevo verraco 198
huevoburro 278
huevoegato 469
huevoemorrocoy 303
huevos de danta 381
huevos de perro 343
huevos de verraco 343
hugacal 453
hugucal 453
huichira 205
huichu caspi 380
huiconga 204
huicongo 204
huicungo 204
huilla 467
huimba 412
huina 454
huinar 413
huino 454
huira guayo cacar-blanco 380
huira-huira 255
huiririma 204
huisquila 295
huisquilla 295
huitillo 251
huito 351
huitoto 226, 260, 436
hule 391, 392
humari 359
humarí 359
humaria 252
huo-ca 371
hurón 181
huru 212
hurudã 212
husquilete 281
husquilla 281
huyama 293
hypericón 394

I

iabeai 184
iachapo 314
ibacaba 210
ibapichuna 451, 452
ibapichuna de pepa pequeña 451
ibapichuna grande 452
ibapichuna pequeño 451
ibapurú 428
ibaracú 179
ibe 210
ibia 432
ibias 432
ibilán 338
icaco 352, 381
icainé 383
ichanga 446, 447, 448
ichí 325
ichil 325
ictericio 332
idéwi 210
igalán 199
iguá 318, 331
iguá amarillo 331
iguamarillo 331
iguanero 359, 416
iguano 404, 457
iguanul 207
iguilan 338
iguilan phytolaca 338
iiamaiki 210
iidikuumee 210
ijebioici 454
ikutiñu 205
ilán ilán 187
ilang ilang 187
ilanga ilang 187
ilanilán 187
ilusión 235, 242, 283
ilusión de hojita ancha 242
ilusión golondrina 242
imán 461
imbiana 412
imbiande 412
imbira 186, 187, 188, 189, 413
imbira blanco 186, 188
imbira negra 189
impa 384
impamo 383
impar 309
impar colorado 309
imparmo 384
imperial 234
impérico 393
imuño 364
inaciabi 267
inayá 205
incana 262, 277, 316, 377
inche 388
inche hembra 388
inchi 388
incibe 184
incienso 384, 452, 453
incienso blanco 452
incienso hembra 453
india 241, 408, 456
indiecito 433
indígena 433
índigo 325
indio 191, 197, 232, 238, 241, 250, 252, 258, 267, 272, 284, 300, 302, 310, 347, 375, 376, 399, 406, 414, 447, 451, 457, 458, 469
indio desnudo 451
indio en cuero 451
indio pelo 302
indio viejo 197, 267, 272
indiodesnudo 451

indioencuero 451
indioencueros 451
indioviejo 257, 261, 272
indioviejo blanco 257
indira 188
iñéhe 210
ínȅkáhá 210
ingerta de shrub 338
inguanul 207
inguanul grande 207
inguanul pequeño 207
inhoboto 210
inibue 209
injerta 338, 418, 430
injerta de tierra 418
injerto 430, 448, 449, 450, 455
injerto de tierra 430
inmortal 269, 272, 282, 461
inmortal blanca 282
inmortales 272, 282
inóho 210
inohoboto 210
inopropio 251
inosadilla 281
inse 385
insignia 272
insince 452
insinia 272
insira 440
insulina 283, 359
intetano 200
ipacina 287
ipadá 386
iparo 437
ipatú 386
ipeacuana 338
ipeca 347, 356, 391
ipeca de agua 356
ipeca falsa 356
ipecacua raicilla 347
ipecacuana 340, 347, 355, 356, 391
ipecacuana blanca o amilácea 356
ipecacuana de cartage ipeca 348
ipecacuana de cartagena 348
ira 233, 234, 268, 438, 462
iraca 210, 227
iracana 388
irapay 210
irimaʔucoro 449
irirochí 346
irisipela rosada 366
iro 346
iro-chii 346
iron odd 440
iruwa 373
iscancel chiquito 281
iscancer 281
iscansé 281
iscanser 280, 281
isfuque 385
isimalinae 261
isinde 436
isira 248
isiraachon 316
isirra 248
islugu 305
isoi 210
isora 352
ispaguilla 216
istoraque 298
istri 251
itamo real 277, 391
itamorreal 391
itamorrial 391
itató 415
itauba 183
ite 319
itebi 210
itewi 210
itewída 210
itsonae 395
ivira 413
iwana alu 340
iwitsulinae 350
ixbut 390
iyá 271
iyõ 210

J

jabara 203
jabilla 294
jabillo 407
jabo 459
jabón 224, 319, 352, 441, 459
jaboncilla 442
jaboncillo 287, 288, 293, 319, 351, 352, 354, 441, 459
jaboncillo de campo 293
jabonsillo 351
jaborandi 456
jabotí 188
jaboticaba 428
jacintica 412
jacinto 214, 225, 280, 324
jacinto de agua 225
jade 334
jade verdeazul 334
jadebano 387
jadzami jinguaque 221
jaen sananho 344
jagalana 218
jago 351
jagua 351
jagua maluco 351
jaguarito 351
jagüillo 404
jagüito 351
jaguo 351
jaico 363
jâjodedimicù 408
jaladanta 208
jalamayó 317
jalapa 287, 339
jalapatrás 208, 226, 358
jalapo 311, 436
jalechepil 224
jalelo 332
jam plum 451
jamachipeke 250
jambolán 428
janami unüboto 332
jandiroba 454
janeiro 246
jansoco 340
jantukuru 278
jaolu 200
japahue 445
jaqui 224
jaquillo 359, 391
jaramago 279
jarapijinae 266
jardín 214, 265, 271, 279, 297, 360, 454, 465
jardín de mayo 465
jardín de seda 360
jarekita 366
jarijana 311
jarilla 256, 257, 259, 260, 270
jarilla blanca 270
jarilla jumarunna 270
jarillo 270
jarito 387
jarra nasitsinae 333
jarrilla 270
jaspe 200
jaspeado 199, 454
jatum kimsa nkuchu 257
jaulo 387
javiamena 332
javilla 294
javo 386
jawaapia 412
jayito 386
jayo 386, 455, 457
jayo colorado 455
jayuelo 386
jazmillo 287
jazmín 254, 273, 280, 287, 288, 303, 339, 343, 349, 350, 352, 355, 365, 375, 379, 417, 454, 456, 465, 466
jazmín amarillo 339
jazmín australiano 254
jazmín azul 288, 379
jazmin chino 273
jazmín chino 273
jazmín de Arabia 375
jazmín de árbol 355
jazmín de caballo 454
jazmín de Calamar 343
jazmín de embarcadero 355
jazmín de España 454
jazmín de fuego 365
jazmín de la India 456
jazmín de luto 454
jazmín de Malabar 343, 465
jazmín de mariangola 349
jazmín de México 287
jazmín de montaña 343
jazmín de monte 343, 355, 466
jazmín de noche 466
jazmín de novio 352
jazmín de tarde 287
jazmín de tree 355
jazmín del río 280
jazmín estrellado 375
jazmín lavanda 465
jazmín malabar 343
jazmín nocturno 466
jazmín rojo 352
jazmin silvestre de estrella 273
jé 404
jebe 330
jenené 381
jengibre 251
jengibre de monte 251
jengibrillo 242
jenine 381
jepe 201
jepe-kutata 201
jequirite 310
jerecú 190
jero 311
jeta de perro 457
jetiabierto 375
jeyutse 274, 275
jeyutse pal 274
jícama 266, 270, 329
jícara 210
jiclilla 210
jicra 210
jidadiapique 440
jidonuma 437
jieratapi 392
jifeo 385
jificona 306

jigra 210
jigracaspi 433
jigua 182, 183, 184, 185, 273, 274, 306, 308, 320, 331
jigüa 190
jigua amarillo 183, 184
jigua baboso 184
jigua blanco 184, 308
jigua canelo 183
jigua laurel 182, 184, 273, 274
jigua negro 182, 183, 184
jigua palialte 183
jigua pava 184
jigua piedra 306
jigua sombrillo 184
jilotropos 251
jilrawal 278
jimaboto 202
jimo 454
jincha 212
jingapá 206, 207
jingurú 249
jinisico 383
jinnuttureo 336
jinollu 200
jipijapa 227
jipió 319
jipioca 319
jíquera 210
jíquima 270
jiquimilla 266, 270
jiquimillo 270
jíquina 266
jiraca 227
jiromon 293
jiroumou 293
jite 319
jitomagaro 359
jiura 282
jiwa jiwa 241
jiwaboto 241
jo 236, 293, 328, 338, 342, 343
joba 451
jobito 274, 451
jobo 451, 455
jobo arisco 451
jobo blanco 451
jobo colorado 451
jobo de Castilla 451
jobo macho 455
jobo verde 455
joetor+mu 339
joli 396
jolomocox 271
jome 266
jomokwei 316
jon 232
joquetaramo 417
jordanillo 436
jorojoro 332
joropenaque 382
joropo 227, 458
jororo unubot 332
jororo unüboto 332
jororounüboto 332
josebé 190
josefina 222
josefino 317
josefita 222
jotduna 393
joumou 293
joural 374
joveroso 308
joyapa 298, 299
joyo 241
jpayandé 331
juagüito 363
juajuá 249, 250
juambautista 347
juamblanco 255, 267
juamparao 261
juampintao 451
juan blanco 255
juan eliberto 261
juan garrote 288
juan soco 340
juan solo 340
juan viejo 185
juana mestiza 331
juanajuana 387
juandediós 298
juanita 361, 387
juanjuán 387
juanlanas 436
juanramos 465
juansoco 340, 341, 342, 343, 437, 438
juansoco con espinas 341
juansoco corriente 340
juansoco de alacrán 340, 343, 438
juansoco de danta 343
juansoco de la sabana 340, 342, 437
juansoco de monte 340
juansoco de rebalse 340
juansoco de sabana 437
juansoco propio 340
juansoquillo 339, 340, 341, 342, 343
juansoquillo de alacrán 340
juansoquillo de alacrán pendare 340
juaquín 365
juari corobeabo 216
jubuicacurameaci 458
juco 234, 297
judío errante 224
jufaifim +o 191
jufigü yeneckke 393
jufigüna 393
juíbaro 287
juino 348
juji? 448
jujucapeci 423
jujuli 428
jujulinae 305, 428
jujure 369
julio 340, 378, 391
juluwa 460
juma?dacoci 397
jumangue 424
jumaria 252
jumbá 440
jumeca 364
jumeni tataboa 362
jumo 191, 410
juncia 232
junco 230, 231, 232, 233, 247
junco de coquito 231
junco de estera 247
junco de la pasión 247
junco de seda 232
junco papuche 231
junco redondo 232
junco sudadero 247
junjún 372
juntalli 173
jupe 249
júpiter 417
juque 260
jurmmejeguio 453
jurva 460
justarrazón 437, 440, 456, 457
justicia 360
justo sin razón 332
juufaifim 191

K

ká 201, 383
kaai 349
kaarú 210
ka-heé 386
ka-hee-wa-ka 383
kãhíka tsiambo 204
kaĩngát 210
kaĩntokëgn 210
kajae 308
kajáe 308
kaka 436
kakarí 388
kalaipata 339
kalaitpana 339
kalawasu 293
kalbas 363
kama kama 399
kaméla 204
kamomy 269
kamururu 385
kamuwe 233
kanangucho 210
kandjí 415
kanongocho 210
kaphensayo 347
kapi 394
kapuchiru 278
karapa 454
karatón 390
karkimchu 295
karoa 183
kartusan 295
karugiri 210
kash kol 343
kasho 319, 325
kashushira 407
kasirusa outka 389
kasisu 463
kasyalata 332
katalawanae 448
kateesua 278
katsubarunae 364
kawi 432
kawinanae 363
k-do-fru 308
kea-kea 255
kéd 210
keiyá 424
kencia 209
kentia 209
kerantë˝gn 209
kidi 357
kikuyo 235
killi-killi 268
kimula flojo 191
kinde borrachero 466
kinena 210
kino 402, 425
kino gum 425
kinobeay 430
kiriko 191
kita jampaway 399
kitha anís 271
kiwicha 281
kiwincha 281
k-na-pla 308
kóchbot 205
koko 250
kokopi mata efinae 189
kò-ma-ma 318

kö-ma-na 335
komtë?gn 204
konéna 210
konta 210
koorí 363
kopajhiri 295
koray 351
koshot 274
kosters curse 420
kotoloinsu 413
kotu 210
koujourouk 287
koulit 288
koumboto 295
krapa 454
kuamna 280
kuaso 325
kudzú 313, 328
kudzú americano 328
kudzú tropical 328
kugna 308
kuipirai 392
kukuma 470
kulupaneekono 341
kuluuala 205
kumá 465
kumaki 204
kumali 204
kumaliboto 204
kumaliñu 204
kumaliphé 204
kumari 204
kumaria 204
kumula blanca 191
kunda 307
kurua 205
kutata 201
kute ena 451
kuu 202
kuyawisi unuboto 434

L

la brea 385
la diabla 179
la mora 445
labicá 207
labio de negra 354
labio de niña 430
labiorin 431
labios de negra 421
labios de novia 354
labranced 333
labrea 385
lacre 192, 350, 381, 394
lacre hojiancho 394
lagaña de gallina 345
lagaña de perro 286
lagaña de vieja 283
lagartija 175, 200, 207, 227, 260, 293, 363, 462
lagartijo 355
lagarto 227, 259, 344, 380, 406
lagarto micunam 344
lagarto micunan 344
lágrima de madre 391
lágrimas de bebé 448
lágrimas de Job 236
lágrimas de maría 199
lágrimas de María 199, 348
lágrimas de niño 447
lágrimas de Obando 369
lágrimas de San Pedro 236, 275, 433
lagunero 331, 403
laguno 430
lairén 345
lakayote 293
lamaki 432
lambedera 232, 318
lambedora 241
lambrán 338
lameardita 404
lamparilla 289
lamparina 382
lampazo 370
lana 229, 407, 412
lance 394
lanceta 205
lancetilla 287
lancillo 394
lang chat 259
lanillo 338
lano 407, 412
lanzeta 205
lanzo 394, 420
largo 305, 315, 322, 430
lasil me meju 191
lata 206, 207, 213, 234
lata arroyera 206
lata blanca 207
lata de arroyo 206
lata de Castilla 234
lata de corozo 206
lata de gallinazo 206
lata de playa 206
lata hembra 206
lata macho 206, 207
lata montan?era 206
lata montañera 206, 207
lata peluda 207
lata playera 206
lata prieta 206
lata sabanera 206
latape 314, 454
latape altura 454
latape de altura 314
látigo 189, 326, 393, 436
laurel 182, 183, 184, 185, 253, 254, 273, 274, 288, 330, 339, 342, 394, 403, 404, 412, 433, 438, 439, 440, 451, 452, 453, 460
laurel abrazapalo 439
laurel ají 184
laurel aliso 184
laurel amargo 184
laurel amarillo 183, 184
laurel baboso 184, 253
laurel baboso blanco 184
laurel blanco 184, 273
laurel bongo 185
laurel bracilargo 412
laurel canelo 183, 184
laurel chaquiro 184
laurel colorado 182
laurel comino 182, 184
laurel comino de caquetá 184
laurel de búho 453
laurel de cera 339
laurel de Dafne 342
laurel de Judea 342
laurel de la India 438
laurel de mierda de perico 452
laurel de monte 451
laurel de pajarito 453
laurel de pepa pequeña 451
laurel de perico 451, 452
laurel de puná 273
laurel guacharaco 184
laurel hojipequeño 339
laurel huesito 254
laurel jigua 183
laurel llanero 184
laurel mierda 184
laurel molinillo 184
laurel mora 440
laurel negro 184, 273, 339
laurel oloroso 182, 184
laurel ordinario 453
laurel pajita 184
laurel perico 451
laurel propio 452
laurel real 183
laurel rosa 342, 394
laurel rosado 394
laurel silvestre 452
laurel tabaidá 184
laurel tuno 433
laurel vava 184
laurelillo 184
laurelito 273
lavallaga 313
lavaplato 293
lavaplatos 260, 369, 408, 414
lazo de amor 215
leaf of live 461
leche 238, 305, 307, 390, 392, 393, 437, 438, 439, 440, 441
leche chiva 441
leche chivo 440
leche de cabra 307, 439
leche de casingua 438
leche de cerca 390
leche de chiva 437, 441
leche de chivo 441
leche de miel 238
leche de perra 439, 440, 441
leche eterna 390
leche perra 439, 441
lechecaspi 340
lechechiva 440, 441
lechemiel 341
lecheperra 437, 440, 441
lechera 339, 391, 437
lecherito 390
lechero 272, 306, 339, 340, 341, 343, 381, 389, 390, 391, 392, 393, 437, 438, 439, 440, 441
lechero amarillo 381, 437
lechero blanco 437
lechero colorado 437, 441
lechero colorao 389
lechero de cerca 390
lechero de cercas 390
lechero de hoja ancha 439
lechero de palito 391
lechero de plomo 306
lechero del blanco 393
lechero espinoso 390
lechero hojas de lanza 390
lechero pequeño 390
lechero plomo 437, 441
lechero rojo 389
lechero verde 389
lecherote 340
lechesapo 441
lechimullo 344
lechiterna 391
lechito 344, 392
lecho 172, 173, 343
lecho de mar 343
lechón 438
lechosa 270, 279, 390, 391
lechoso 279, 339, 340, 341, 343, 344, 389, 391, 437, 438, 440, 441
lechudo 341, 389, 393, 439, 440
lechudo colorao 389

lechuga 202, 264, 267, 270, 290
lechuga amarilla 270
lechuga arrepollada 264
lechuga crespa 264
lechuga de agua 202
lechuga de platanal 290
lechuga de río 202
lechuga platanera 290
lechugilla 202, 259, 263
lechuguilla 202, 225, 255, 259, 261, 263, 269, 270, 287
lechuguita 259
leguminosa 242
lembo 253
lemon verbena 377
lengua de buey 448
lengua de gallo 358
lengua de gato 258
lengua de lombriz 430
lengua de perro 306
lengua de pichón 197
lengua de picón 195
lengua de suegra 215
lengua de vaca 248, 249, 259, 260, 261, 289, 290, 307, 389, 393, 397
lengua de vaca cebo 393
lengua de venado 184, 278, 339, 356
lengua del diablo 390
lenguadevaca 397
lenguaediablo 354
lenguaesuegra 215, 461
lengüebuey 417
lengüegato 258
lenguevaca 332
lengüevaca 215, 250, 289, 290, 307, 332, 397
lengüevaco 272, 397
lengüevenado 392, 421
lengüilla 289
lenteja 336
leoncico 351
leoncito 351
leptospermon 426
lerén 345
lerene 345
letoespermo 426
leucaena 325
levaduro 427
levantapalos 227
liana del napo 394
liana iguana 363
liberal 367, 373, 375, 389, 390
liberalito 390
libidivi 325
liccha 282
licuala 210
liendra de puerco 238, 282
liendre de puerco 238
liga 392
ligustro 375
lija 295, 447
lija vegetal 295
lila 277, 361, 365, 463
lilas 277
lima 247, 455, 456
lima ácida 456
lima agria 455
lima dulce 456
limanajú 346
lime 455
limón 185, 186, 236, 267, 278, 423, 455, 456
limón cimarrón 186
limón criollo 455
limón de monte 185, 186
limón dulce 456
limón hediondo 185
limón suingla 456
limón suingle 456
limón swinglia 456
limón toco 278
limonacho 280
limonaria 236
limoncillo 185, 186, 188, 236, 280, 291, 293, 298, 403, 437, 448, 449, 456
limoncillo de cachos 280
limoncillo de monte 186
limoncillo de playa 449
limoncito 280, 449
limoncito de cachos 280
limonera 236
limonero 455
limoracho 449
limpiaculo 260
limpiadiente 298
limpiadientes 441, 442
limpiafrascos 245, 373
limpiaplato 471
limpión 293
linacilla 286
linaza 394
linda 207, 215
linda palmera 215
lindachina 226
lindona 209, 211
lino 216, 394
lino enano 216
lino morado 216
liquidambar 460
lirata 438
lirgua 281
lirio 197, 213, 214, 215, 216, 217, 222, 226, 251, 340, 341, 342, 343, 352, 358
lirio amarillo 215
lirio colorado 358
lirio de agua 197
lirio de los pantanos 215
lirio de mayo 217
lirio de monte 216
lirio de páramo 216
lirio macho 343
lirio rojo 213
lirio rosado 213
lisas 283
liscano 309
listón 189
litamo 277
lítamo 277
litamo real 277
lítamo real 277
litamo rial 277
lítamo rial 277
little tree mano de sapo 171
lizcano 309
llaga de Cristo 390
llagas de Cristo 280
llama 216
llamarada 365
llamas 216
llambu 255
llamitas 216
llanchamilla 394
llano 245, 317
llantén 376
llantén de manga 376
llantén de monte 376
llantén de páramo 376
llanten del cuichi 376
llantén del cuichi 376
llanto de niño 176
llausa 412
llorón 181, 239, 351, 404, 469, 471
llorones 469
llovizna de enero 277
llovizno 310, 331
lluvia de coral 365
lluvia de estrellas 390
lluvia de fuego 376
lluvia de oro 313, 365
lluvia de perlas 284
lluvia de verano 313
lluviadeoro 388, 462
lobo 255, 287, 455
loiro 182
lola 469
lolium 241
lombricera 339, 345, 447
lombricero 339, 387
lombrirera 355
lombriz 301, 305, 308, 364
lomo de babilla 184
lomo de caimán 311, 331
lomo de machete 404
longapijo 362
lonicera 296
loqueta 392
loqueto 309
lorencillo 389
loro 171, 173, 282, 305, 322, 324, 326, 330, 331, 366, 367, 452, 453, 457
loto 181, 198, 273
loto silvestre 273
lou gawou 461
lozanelia 448
lúa 411
lucaica 227
lucaina 227
lucateba 227
lucero 283
lucmo 307
lucua 227
lúcuma 307
lucumo 381
lúcumo 307
lucutema 332
lucutena 332
luisa 236, 377
luisa de Chile 377
lujuji 437
lulillo 471
lulo 468, 469, 470, 471
lulo de Castilla 471
lulo de charapa 471
lulo de monte 469, 471
lulo de perro 469, 470, 471
lulo dulce 471
lulo grande 471
lulo morado 471
lulo pequeño 470, 471
lulo silvestre 469, 471
luloeperro 470
lulumoco 297
lumbaga 332
lumbi 472
lumbí 416, 472
luna 460
lunaria 394
lupino 325
lúpulo 436
lupuna 407
lusara 287
lusesteban 372
lustrosa grande 290

lyann mol 472
lyann panye 282
lycopodio 171

M
ma 212, 231, 232, 251, 298, 370, 384, 407, 415, 471
maache 395
maawui 408
mabaco 205
mabái 205
mabi 441
mabizou 378
macael 234
macambo 414
macambu 414
macán borrachero 283
macana 204, 206, 207, 209, 212, 213, 239, 243, 434
macana abanico 212
macanilla 204, 206
macanillo 204, 347, 378, 420, 422
macanillo negro 420
macano 231, 393, 416
macano amarillo 416
macano negro 416
macao 414, 458
macetilla 285
maceto 199
machaco 364, 460
macha-macha 299
machare 385
machare de loma 385
machare de mangual 385
machare hojiancho 385
machare hojimenudo 385
machare teu 385
machari 385
macharre 385
macheco 378
machimbi 439
machimbí 439
machucha 258
machucho 258
maciega 243
maciga 243
maco 307, 454, 458
maco serrano 307
maconda 407
macondo 407
macoya guafa 387
macucú 329
macuiro 315, 339
maculculu champu 472
macuma 227
macuná 334
macuripí 385
macurutú 315, 325, 328
macuy 470
madera blanca 389
madera de hierro 308
madiagola 349
madibia 460
madre agua 362
madre de agua 191, 362
madre de la yuca 283
madre de yuca 359
madre del cacao 320
madre moderna 215
madreagua 362
madrecacao 320
madredeagua 297
madrejuano 387
madrevieja 177
madrona 385
madroña 385, 392
madroñito 380
madrono 385
madroño 304, 384, 385, 440
madroño de cáscara lisa 385
madroño de monte 385
madroño liso 385
madroño rugoso 385
madroño silvestre 385
madruño 385
madura 305, 364
madura plátano 364
madura verde 305
madurachontaduro 364
maduraplátano 364, 402
maduraverde 191, 250, 305, 409, 414, 451
maduro 402
maduroverde 305
mafa 432
mafafa 199, 203
mafafa blanca 203
mafafa de charapa 199
mafafa de murciélago 199
mafafa mija 203
mafafa morada 203
mafafa pata de gallineta 203
magaleto 403
magameai 387
magnolia 190
magnolia blanca 190
magnolio 190
magué 214
maguey 214, 215
magueyito 224
maicillo 232, 245, 247
maicito 179
maiki 247
maiku jat 175
maispepe 212
maíz 245, 247, 267, 275, 277, 287, 299, 304, 378
maíz azulito 247
maíz berrendo 247
maíz blanco 247
maíz brisa 247
maíz cacho de buey 247
maíz capio 247
maíz cariaco 247
maíz cariaco amarillo 247
maíz cariaco rayado 247
maíz de gallo 275
maíz de Guinea 245
maíz de lobo 287
maíz de perro 287, 299
maíz millo 245
maíz tostado 267, 277, 304
maíz tostao 378
maíz Yucatán 247
maizcocho 288
maizorro 378
maizpepe 209
maiztostado 275, 276, 277, 304, 356, 378, 408, 436
maiztostao 288, 304, 356, 387, 436, 455
majagua 181, 186, 315, 317, 409, 410, 412, 413, 415, 436, 441
majagua colorada 436
majagua de gallina 317
majagua de playa 410
majagua melada 409
majagüillo 187, 189, 415, 436
majagüito 406, 410, 415, 436
majagüito de playa 406, 410
majaguo 408, 409, 410, 411
majaguo de playa 408
majaguo melao 409
majao 275, 409, 412, 414
majao negro 275
majaquillo 415
majate 181
majua 280, 431
majuo 409
makeru 211
maki 257
makibosoto 285
makukula 186
makuri 363
mal de ojo 448
malaatapui 466
malabar 343
malabarí 343
malagano 407, 410
malagre 185
malagueño macho 190
malagueta 428
malagueto 190
malaguito 190
malaire 195, 196, 197
malaire kirú 195, 197
malakala 209
malama 210
malamadre 217, 461
malambito de perro 339
malambo 389, 407, 412
malamujer 360
malanga 199, 203
malangay 203
malaqueta 428
malauya 410
malavecina 289
malcasada 339
maleticas 276
maleza de perro 235
malgano 407
malibu 205
malibú 205, 290, 350
malla 433
mallorquí 276
mallorquín 272, 275, 276, 415
maloine 437
maluua 451
malva 280, 358, 401, 406, 407, 408, 409, 410, 412, 413, 415
malva áspera 410
malva blanca 408, 410, 415
malva brusca 410
malva cimarrona 410
malva común 410
malva de castilla 413
malva de Castilla 413
malva de olor 358
malva grande 408
malva morada 410
malva olorosa 358
malva peluda 410
malva pequeña 410
malva pintada 410
malva silvestre 410
malva tendida 410
malva vagabunda 280
malva vagamunda 280
malvaloca 407
malvarrosa 407, 409, 410
malvavisco 407, 408, 409, 410, 411
malvavisco de flor 407
malvilla 358, 410
malvisco 406, 408, 409, 410
malvón 410
mamajuana 256, 289
mamajuanita 289
mamarro?n 205

mamarrón 205
mamaubo 279
mamei 381
mameicillo 339
ma-mê-tê-tê' 384
mamey 301, 305, 306, 307, 339, 381, 385, 395
mamey blanco de tierra fría 395
mamey colorado 306, 307, 381
mamey de cartagena 381
mamey de monte 385
mamey de tierra fria 395
mamey de tierra fría 395
mamey hediondo 301
mamey rojo 307
mamey sapote 307
mamey zapote 307
mamilla 191
mamira 227
mamirí 200
mamita 191, 192
mamita blanco 191
mamita de boa 192
mamita de cananguchal 192
mamita de cangrejo 191
mamita grande 191
mamita mediana 191
mamita pequeña 192
mamita roja 191
mamito 190, 191, 192, 264
mamito blanco 191
mamito de hoja ancha 192
mamito del pequeño 191
mamito pequeño 191
mamo 188, 191
mamo de tigre 188
mamón 278, 306, 308, 341, 441, 458, 469
mamón común 458
mamón cotopli 458
mamón cotoprí 458
mamón cotoprix 458
mamón cutuplí 458
mamón de María 458
mamón de mico 306, 458
mamón de monte 469
mamón de puerco 278, 458
mamón de tigre 308
mamón leche 441
mamona 279
mamoncillo 440, 458, 469
mamoncillo de monte 458
mamoncillo macho 458
mamoncillo silvestre 440, 458
mamure 200
mana 209
manaca 209, 364
manaca brasilera 209
manaco 209
manáka 209
manakai 209
manakái 209
manakaiboto 209
manana 315
manangola 356
manaqué 209
manaquita 209
manca mulo 303
mancacaballo 331
mancadera 468
mancamano 468
mancamulo 303, 460
mancatigre 471, 472, 473
mancayo 342
mancha 287, 366, 367, 394
mancha de sangre 366
mancha mancha 394
mancha mano 287
manchada 369
manchado 394
manchador 394
manchados 369
manchamancha 394
mancharà 394
mancharí 394
mancharra 394
mancharropa 394
mancidibo 402
manciribo 460
mandaguasca 258
mandarina 456
mandarina arrayana 456
mandarina cleopatra 456
mandarina común 456
mandarina israelita 456
mandarina oneco 456
mandarina reina 456
mandarina satsuma 456
mandarina tangolea 456
mandarina volkameriana 456
mandarino 456
mandevila 341
mandibá 302
mandigüiche 258
mandingo 312
mandioca 392
mandrágora 236
mandul 303
mandur 394
mandurillo blanco 384
manduro 384
mandús 394
mané 337
manea blanca 357
manelito 408
manero 395
manga 450
manga albania 450
manga de Mariquita 450
manga mariquiteña 450
manga tomi 450
mangapaca 372
mangel 309
manglar 348
mangle 290, 309, 316, 320, 328, 343, 359, 385, 387, 402, 416, 455, 458
mangle amarillo 416
mangle arroyero 458
mangle blanco 359, 416, 455
mangle bobo 290, 416
mangle botón 416
mangle botoncillo 416
mangle cativo 402
mangle colorado 343, 402, 455
mangle comedero 309
mangle de Buenaventura 309
mangle de montaña 385
mangle duro 316, 320
mangle garbancillo 416
mangle jelí 416
mangle nato 328
mangle negro 359, 416
mangle picudo 309
mangle piñón 402
mangle piñuelo 309
mangle prieto 359
mangle rojo 290, 402, 455
mangle rosado 402
mangle salado 359, 402, 416
mangle salao 359
mangle zaragoza 416
manglecita 455
manglecito 455
manglillo 303, 304
mango 302, 383, 450
mango canime 450
mango chancleto 450
mango de azúcar 450
mango de chupa 450
mango de masa 450
mango de monte 383
mango de rosa 450
mango farchil 450
mango filipino 450
mango tommy 450
mangosán 455
mangostán 385
mangostín 385
mangostín de Mariquita 385
mangostino 385
mangué 205
manguengue 211
manguepaque 372
manguipaca 372
manguipaque 372
manguito 388, 451
manguito extranjero 388
maní 311, 350, 381, 388, 392, 409, 449
maní blanco 311
maní blanco de picón 311
maní de huerta 311
maní de perico 350, 449
maní de picón 311
maní estrella 392
maní forrajero 311
maní rojo 311
manicuera 392
manie 385
manil 385
manirita 186
manirito 186, 415
manitas 282
manito de dios 224
manito de Dios 224, 272, 281
manito de oso 264
manjó 200
manna 473
mano de ángel 224
mano de Dios 224, 271, 272
mano de lagarto 259
mano de lagrato 267
mano de león 253, 264, 414
mano de oso 253, 254, 274
mano de pilón 328
mano de tigre 177, 188, 249, 250, 253, 259, 261, 267, 318, 437, 440
manodioso 253
manoedió́s 224
manoeleón 264
manoelión 264
manoetigre 176
manolión 264
manolo 388
mañolo 391
manquenque 211
manquetillo macho 404
manrubio 256, 260, 371
manrubio blanco 371
manta 367
mantagorda 249
mantavieja 287
manteco 190, 280, 295, 301, 304, 393, 395, 396, 403, 407, 430, 451
manteco blanco 393
manteco orejón 396

manteco peralejo 395
manteco yuco 295
mantecona 294, 392
mantecoso 206, 388, 392
mantequera 395
mantequero 395
mantequilla 307, 393, 451, 455
mantequillo 292, 368, 393, 451
manto de cristo 282
manto de la Virgen 214, 469, 471
manto de María 469
manto de novia 469
manto del padre 263
mantoevieja 372
manubo 433
manuelión 253, 264
manyok 392
manzana 315, 352, 380, 401, 428, 443, 470
manzana apí 443
manzana camuera 443
manzana común 443
manzana criolla 443
manzana de agua 428
manzana de amor 470
manzana de cerdo 352
manzana de estrella 401
manzana de puerco 352
manzana de Saint-Valery 443
manzana de venado 380
manzana reineta 443
manzanilla 255, 259, 265, 271, 357, 391, 401, 418
manzanilla chiquita 265
manzanilla cimarrona 265
manzanilla común 265
manzanilla de Castilla 265
manzanilla de los campos 401
manzanilla de los montes 255
manzanilla dulce 265
manzanilla grande 271
manzanilla loca 265
manzanilla matricaria 259, 265, 271
manzanilla romana 259, 265
manzanillo 258, 309, 317, 331, 389, 391, 392, 411, 451, 457
manzanillo caucano 389
manzanillo de cerca 389
manzanillo de playa 391
manzanillón 271
manzanita 428, 442
manzanita costeña 442
manzanita de rosa 428
manzanito 442
manzano 250, 296, 297, 298, 303, 305, 306, 395, 397, 428, 431, 443, 457, 466
manzano chino 428
manzano de monte 395, 457
manzano espinoso 466
manzano montañero 297
maogüa 346
maoguatsegesesa 349
maoo-hee-rée 414
mapán 352
mapaná 200, 215
mapanae 438
mapanare 215
mapanaré 205
mapaso 407
mapola 265, 269, 272
mapolilla 261
mapora 208, 211, 212
maprara 260
mapua 330
mapuja 329
mapurite 287
mapurito 287, 433, 456, 457, 460
mapurito prieto 456
mapuro 468
mapuso 330
maquenco 209, 211
maquenque 206, 209, 211, 212, 213
maqui 311
maquionso 275
mara 231, 232, 320
ma-ra 471
maraca 251, 409, 411, 414
maraca de armadillo 414
maraca de bejuco 414
maraca de boruga 409, 414
maraca de búho 414
maraca de cucarrón 414
maraca de danta 414
maraca de iguana 414
maraca de lorito 414
maraca de monte 414
maraca de ratón 409
maraca silvestre 411, 414
maracachafa 436
maraco 301, 316
maracuyá 397, 398, 399, 400
maracuyá de monte 397, 398, 399, 400
maracuyá dulce 397
maracuyá silvestre 398
maradague 407
maragojito 348
maragua 441
marama 210
maramá 210
maráma 210
maramapé [palm] 210
maramathe [fruit] 210
maramo 315
maranae 192
marañón 303, 428, 450
marañón amarillo 450
marañón de breo 450
marañón de monte 450
marañón gigante 450
marañón grande 450
marañón monte 450
marañón rojo 450
maranta 250
marapiranga 452
maraquita 316
maraquitas 316
mararave 204
mararay 204, 206
maravilla 216, 224, 259, 260, 263, 287, 295, 389
maraya 209
marayón 209
marcela 255
marcela hembra 255
marcelo 255, 403, 404
marciega 243
marco 256
mare 306, 392, 437, 440, 441
mare blanco 392
mare casaca 440, 441
mare dinde 437
marecasaca 440
marengo 246
marey 450
marfil 350, 359, 393, 404, 422, 455, 460
marfil otara 460
marfilito amarillo 350
margarita 257, 263, 264, 265, 340
margarita africana 263
margarita amarilla 257, 264
margarita blanca 257
margarita de jardín 265
margaritón 260, 264, 265, 271
margaritón de monte 260
mari 359
marí 359
maría 195, 197, 199, 260, 290, 328, 356, 380
maría angola 356
marialuisa 236
mariangola 348, 349, 356
mariangola de espina 356
mariangola macho 349
marianita 294
marigold 271
mariguana 436
marihuana 436
marija 205
marilope 401
marimá 437
marimari 321, 336, 437
marimba 436
marimoña 259
marina 286, 336, 437
maringol 271
mario 380
marío 380
maripa 205, 208
mariposa 207, 222, 263, 316
marira 425
mariramba 381
marirana 381
marju 257
markju 257
marmelo 395
mármol 200
marmolejo 397
marranera 282
marroquina 387
marrubio 256, 371, 379
marsamila 295
martiguaja 345
martiguaje 345
martín doble 385
martín galve 465
martinica 378
martirios 390
maru 457
marubio 257
marucha 258
maruchonga 265
marupá 191, 392, 460
marupa blanco 191
marupá blanco 191
marure 436
masa de pan 436
masaba 393
masábalo 393, 410, 454
masábolo 454
masaguaro 331
masano 315
masarandi 305
masarandua 305
masato 383, 387
mascapalo 442
mascarey 401
maschactie 471
mascuriaga 318
masequia 258
masequía 258

mashua 280
mashula 398
masipapa 201
masiquía 258
masiquiara 258
masketi 392
masoko 226
masquiaunsabe 434
mastardo 370
mastoko 453
mastranco 368, 371
mastrantillo 369, 371
mastranto 260, 368, 369, 370, 371, 372, 373, 378
mastranto blanco 370
mastranto de perro 371
mastranto de sabana 371
mastranto grande 372
mastranto morado 371
mastruerzo 277
mastuerzo 277
mastuerzo de agua 277
mastuerzo de huerta 277
mata andrea 251
mata cuy 266
mata de borugo 309
mata de calabazo 363
mata de coco 208
mata de conga 366
mata de jumo 410
mata de mora 415
mata de quincha 379
mata de San Pedro 466
mata mata 301
mata negro 431
mata puerco 343
matacaimán 319
matacangrejo 363
matacolibrí 251
mataconga 380
matacucaracha 404
matacucarachas 404, 470
mataculebra 207, 209
mataculebra blanco 209
mataculebra verde 207
mataganado 339, 365, 380
mataganao 362, 365
matagente 342
mataguacuco 311
matagusanos 260
matahijo 457
matahijón 457
matáigahiba 204
matalimón 462
matamaleza 349
matamatá 301, 302, 394
matamatá blanco 301, 302
matamatá colorado 301
matamatá de altura 301
matamatá pequeño 301
matamatá ripero 394
matamba 207, 208
matamonte 349
matamorro 439
matamosca 298
matandrea 251
matandrea blanca 251
matanegro 431
matano waü ünuboto 472
mataopez 325
matapaja 272
matapalaa 392
matapalo 347, 349, 383, 384, 385, 418, 437, 438, 439, 446, 448, 449, 450
matapalo de agua 439
matapalos 439, 446, 448, 449, 450
matapasto 332
matapasto bravo 332
mataperros 346
matapescado 260
matapez 325, 330
matapuerco 199, 201, 343, 460
matapuercos 430
matarratón 313, 315, 317, 320, 331
matarratón cimarrón 315
matarratón criollo 320
matarratón de Castilla 320
matarratón extranjero 313, 331
matasano 455
matasardina 402
matatigre 177, 184
matatodo 349
matatoro 397
matatotumo 222, 223
matavieja 287
mataviri 342
matay 407
mate 320, 363
matecillo 349, 362
matecito 362
matico 195, 257, 469
maticuaja 345
matiguaja 345
matiguaje 345
matijón 456, 457
mato 193
matojiro 319
matojo dulce 238
matricaria 259, 271
matricaria doble 271
matrimonio 292, 401
matú 339
matupa 453
maturia 339
maturranga 268
maucuari 341
mauricio 412
maurú 248
mavé 210
maveci 312
mavilla 295
max soeo 434
maya 228
mayadena 379
maycha 298
mayi 247
mayita 228
mayita chivichivi 228
mayo 217, 222, 342, 351, 354, 397, 417, 448, 465
mayorquín 275
mazábalo 454
mazamorra roja 352
mazamorrera 413
mazamorro 348, 352, 387
mazarandua 306
mazorca 190, 213, 280, 358, 359, 361, 387
mazorca de agua 359
mazorca de la parra 358, 359
mazorco playero 466
mazorquilla 361
mazorquitas 448
me 191
mê 384
meaparado 325
meaparao 325
mebaganoó 458
mecanito 232
mechá 299
mechino 439
mechudo 242
media lu mal de ojo 448
media luna 448
mediacara 307, 438
mediacaro 307
mediagola 349
medialuna 278
medio amarillo 184
medio comino 182, 184
mediocomino 182, 184
medsiyen 391
mee 453
mégeguio 453
megua 316
mejeguío 453
mejicano 215, 293
mejorada 392
mejorana 359, 372, 378
mejorana blanca 372
mejorana silvestre 372
mekwáibak 210
melá 407
melao 409, 415
melena 219, 229
melina 369
mella 227
melliza 227
melloco 283
melocotón 293, 295, 444
melón 279, 293, 295
melón de agua 293
melón de conejo 293
melón de culebra 293
melón de gallinazo 293
melón de golero 293
melón de olor 295
meloncillo 293
melosa 375, 410
melosita 263, 412
meloso 263
melote 332
melusca 336
membrillero 443
membrillo 301, 302, 303, 445
membrillo hembra 302
membrillo macho 302
meme 213
memé 213
memecucú 437
memeo 202
menea blanca 357
menejùmecacù 335
menta 338, 371, 378, 379
menta americana 378
mentol 338
mentolín 267, 338
mercader 358
mercy 450
merecure 382, 383
mereicillo 416
meretiro 455
merey 450
merigol 271
mermelada 317, 472
mesquite 331
mestiza 349
mestizo 184, 382, 404, 437, 457, 459
meta] 250
metisinajalé 357
metró-reemooeé 414
metsajakobejapanae 353
mevoycontigo 280
mezcla 382
mi 342, 343, 396
mĩã phã pú 205

miao de arco pequeño 447
miaparao 325
micay 234
micay blanco 234
michichijo 346
michu 459
michú 459
mico 175, 212, 245, 248, 278, 279, 306, 387, 407, 458
micongo 288
miel de angelita 277
mielenrama 255
mielito 417
mielquemada 408, 464, 465
mierdegato 333
miflora 300
miguel colorado 419
migüito 419
miirui 191
mija 203, 450
mijagua 450
mijao 450
mijí 258
mijo 245
mikura 287
mil de rama 255
milagro 236
milbalazos 200
milderama 255
milefolio 255
milenrama 255
milflora 300
milflores 472
milhoja 461
milile 175
mililí 175
milkweed 339
mille fleurs 378
millo 242
millón 254
millonaria 254, 372
millonario 254
millonario blanco 254
milo 415
milpesilla 211
milpesillo 210, 211
milpesillo de rebalse 211
milpesillo de sabana 211
milpesillos 210, 211
milpesos 393
milpesos de cimbe 393
milpo 430
mimbre 201, 429
mimito 191
mi-mi-wa-jo-ka-ki 342, 343
mimosa 315, 327, 328, 401
min 454
minás 365
mindaca 236, 246
mindacá 236
mínigo 467
mínigua 467
míniguo 467
minipapiro 231
mint 428
minuake 427
miø juji? 448
mión 418
miona 261, 286, 365, 418
mioncita 418
miosotis 274
mirabira colorada 263
mirable 246
miralsol 210
mirapiranga 437, 441
mirasol 210, 263, 271
miratava 182
miroco 191
mirsimarika 434
mirto 343, 353, 355, 417, 425, 427, 445, 456, 470
mirto blanco 456
mirto de la India 456
mirto de Malabar 456
mishi-polé 432
mi-sigó 396
misiraca 227, 251
mismia 279, 280
mismiche 227
míspero 395, 418, 444
míyacù 188
miyaibo 199
mo 239, 253, 273, 312, 453
mobil 205
mocaiyeibu 447
mocantia 289
mochila 257, 294
mochila vieja 257, 294
mochilero 330
mochilita 294
mochilitas 294
moco 238, 281, 282, 297
moco de pavo 238, 281, 282
mocoepavo 282
mocuillibua 247
mocuya 186
moena 183, 184
mogote 238
moho 273, 409
moisés 236
mojarra 301, 403
mokochira 284
molde 363
molenillo 191, 207
molenillo colorao 191
molinilla 207
molinillo 184, 190, 191, 207, 208, 209, 411, 412, 413, 458
molinillo del río cauca 190
mollantín 289
molle 314, 451
molleja 283, 426
molleja de paujil 426
mollentín 289
moló 457
moló blanco 457
molojillo criollo 236
molonga 341
molongo 341
mona 212, 242, 302, 436
monablanca 392
monachino 342
monacillo 411
monben bata 455
móncoro 274
moncro 273
monde 454
monde bogotano 454
mondey 401
moneda 194, 277, 367
moneda del Papa 277
monita 392
monito 242, 406
monkey apple 352
mono 185, 186, 238, 322, 324, 330, 332, 337, 340, 344, 355, 357, 385, 387, 407, 418, 422, 424, 430, 441, 447, 466
moño 238, 282, 390, 461
mono alemán 238
moño de seda 282
montañero 180, 204, 206, 296, 297
monte del soldado 195
montefrío 387
montenegro 272
monterillo 312
moo-há-hê-tê 424
mopa 350
mopamopa 350
mopo 388
moque 384
moquillo 297
moquillo de páramo 297
mora 295, 306, 395, 417, 420, 421, 422, 423, 437, 440, 444, 445, 470
mora andi mora blanca 444
mora andina 444
mora blanca 420, 444, 445
mora borrachera 444
mora común 444
mora de castilla 444
mora de Castilla 444, 445
mora de cerro 444
mora de la virgen 444
mora de los andes 444
mora de monte 445
mora de páramo 444
mora silvestre 444, 445
moracotura 423
morada 259, 362, 367, 369, 371, 391, 402, 406, 407, 410, 412, 448, 462, 463, 464, 465
moradilla 281, 373, 432
moradillo 330, 422
moradita 260, 283, 286, 327, 340, 370, 416, 417, 432
moradito 382, 418
morado 185, 187, 193, 199, 215, 216, 224, 225, 229, 235, 266, 277, 281, 282, 287, 305, 307, 313, 325, 342, 343, 359, 361, 363, 365, 371, 377, 378, 379, 389, 392, 408, 414, 418, 422, 432, 433, 450, 460, 463, 465, 466, 470, 471
morajil 440
moral 440, 444, 445
moral de Castilla 444
moral de tierra fría 444
moral fustete 440
moramo 315
morao 224, 353
mordaza 366
morera 440
moriche 210
moriche macho 210
morichito 210
moridera 299
morilla 417
morisca 443
morita 276, 351, 417, 420, 440, 444, 445
morita de alambre 417
morita de cercas 445
morita de palo 420
morita espumosa 420
morito 406, 440
morochillo 304, 394, 418, 419, 420, 421, 422, 423
morochillo colorado 394
morochillo guayaba 422
morón 440, 445
morrocota 444
morrocoy 197, 199, 303, 341, 380, 425, 440

mortecino 301, 302, 392, 414
mortillo 443
mortin 288
mortín 288
mortiño 276, 293, 299, 300, 304, 305, 348, 354, 419, 420, 421, 422, 423, 427, 443
mortiño amargo 300
mortiño borrachero 293, 299
mortiño cimarrón 299
mortiño de alambre 422
mortiño de árbol 300
mortiño de culebra 348
mortiño de monte 423
mortiño falso 299, 300
mortiño lanudo 423
mortiño negro 420
mortiño rojo 422
mortiño venenoso 299, 300
mortiño verdadero 443
mortiño zumaque 293
morupacha 378
mosca 258
mosco 298, 416
mosquerillo 389
mosquerito 389
mosquero 388, 389, 408
mosquita 258, 357
mosquito 298, 299, 363
mostacilla 276, 278
mostacilla silvestre 276
mostaza 267, 276, 277, 278, 376, 467
mostaza amarilla 278
mostaza blanca 278
mostaza de canarios 276
mostaza extranjera 467
mostaza negra 276
mote 303, 443, 455, 457
mote mote 443
motemote 443
motilón 181, 401
motilón colorado 401
motilón dulce 401
motilón rosado 401
motilón silvestre 401
motua 214, 215
moutaraton 320
moyejo 185
mozgal 298
mu 224, 229, 271, 273, 339, 409
mũã pũrĩ 210
muche 311, 440
mucheblanco 311
muchila 294
muchilero 301, 330, 416
muchilita 294
muco 301, 458
mucuna 323, 334
mucuná 334
mucunana 331
mucura 287
múcura 420
mucuracá 287
mucurita 420
mucuritos 420
mucurutu 301
muela 423, 457
muela de vaca 423
muelavaca 423
muelle 451
muena 182
muérdago 395, 448
muguel colorado 397
muhí 210
muí 437
muichiãbo 210
muiĩa 210
mũĩũká 204
mulamuerta 302
mulata 268
mulato 246, 255, 268, 274, 275, 276, 324, 369, 387, 401, 415, 451, 458
mulatón 401
mulchis 425
mulí 416
mull kketo-kketo 268
mullaca 467
mulli 451
mun 293
munchira 245, 465, 466
munchira roja 245
munchirero 423
mundani 310
munde 454
muñeco 274, 275, 394
muñeco blanco 274
muñeco prieto 274
muñequito 401
mun-teka 293
munuaniitabukuerunae 264
muñuño 298
muopicoroboado 303
muqu-muqu 257
murapo 266
murcia 435
murciélago 199, 274, 279, 333, 348, 355, 363, 398, 400, 409, 449
mure 314
murillo 393
murraco 227
murrai 336
murrapa 227
murrapho 209, 211
murrapo 209, 213, 227, 248, 273
murta 288
murtá 309
murupacha 289
murupacha de frío 289
murure 202
musaenda 352
musaendo 352
musana 182
musandra 352
muscuai borrachera 466
mu-se-ge-ke 409
mu-se-na 409
musgo 229, 260
musgo blanco 229
musgo florido 260
mute 402
mutsibulibulinae 421
mutsualiunubo 459
mutundó 457
muu 248
muuru 191
muy 210
muyaña 298
muyo pava 354
muyo-pava 354

N

naaĩ 204
nabalá 263
nabio 280
nabios 280
nabo 276, 277, 278, 280
nabo amarillo 276
nabo de canarios 276
nabo silvestre 278
nabos 280
nabueno 186, 187
nabugui 177
nacancio 276
nacansio 276
nacedera 362, 390
nacedero 359, 362, 363, 392
naceró 362
nacha 294
ñachag 258
nacuma 227
naecono 472
naguala 227
naguare 410
nagüí 188, 335
nagüí chirimoyo 188
nãhõemimueñi 209
naĩ 204
nãi 204
naichinupa 336
naidi 209
naidí 209
naikú 251
naipe 200
naiquí 191
najoemimoeñù 209
ñajú 284
nalga de araña pollera 305
namagua 441
nambi 388
ñambo 428
ñame 199, 225, 226, 436
ñame blanco 226
ñame común 226
ñame congo 225
ñame de cotudo 226
ñame de espina 225
ñame de milpesillo 226
ñame de monte 226
ñame espina 225
ñame espino 225
ñame espinoso 225
ñame huitoto 226
ñame morado 225
ñame negro 226
ñame papa 225
ñame redondito 226
ñame redondo 226
ñame silvestre 226
ñame volador 225
ñame yota 199
namemo 208
ñampí 225
namú 287
ñamu 226
nanache 395
nance 395
nandiroba 454
nanio 200
ñanta juju 202
napa 227
napaima 349
napán blanco 377
napanga 185
naquinso 276
naquinzo 276
naquiunso 276
naquiunzo 276
naranja 247, 335, 456
naranja agria 456
naranja agrio 456
naranja china 456
naranja común 456
naranja swinglia 456
naranjilla 471
naranjillo 278, 280, 385, 470, 471

naranjillo de pájaro 471
naranjito 188, 278, 388, 391, 403, 455
naranjito quigua 455
naranjo 342, 385, 441
naranjo de monte 385
naranjo podrido 342
naranjuelo 278, 385, 402, 441, 443, 455, 470
naranjuelo blanco 278
narciso 214, 349
narcizo 213
nardo 214, 216, 226
narizlargo 272
nasacatá 434
nashau 305
ñasmande 272
nat sin ñemé 191
natema 394
natiunsa 276
natiunza 276
nato 328
nato rojo 328
naumo 331
naune 311
nauno 331
navajuelo 363
navegadora 173
navidad 378, 387, 417
nazarena 340
nazareno 320, 321, 330, 351, 369, 424
neacurí 363
neblina 334
necha 294
neche 294
ñee-au 414
néejogua 453
neem 454
negrillo 326, 350
negrito 275, 304, 323, 335, 383, 408, 455
negro 177, 178, 180, 182, 183, 184, 185, 186, 188, 189, 190, 193, 195, 196, 199, 200, 211, 224, 225, 226, 229, 234, 243, 245, 249, 250, 254, 256, 257, 262, 265, 268, 273, 274, 275, 276, 277, 278, 282, 284, 287, 288, 297, 298, 301, 305, 306, 307, 330, 332, 335, 339, 348, 369, 408, 416, 420, 437, 455
negro caspi 335
negro en cuero 437
neguilla 463
néhe 204
nejilla 207
nenakirú 196
neoduricu 191
neptunia 328
neralla 194
nero 395
nescafé 328
nevado 359
newajü 392
newajüsinae 392
newaüibü 392
newütjümujüyorotonae 315
ngumü chi 211
ngupe 210
niaara 440
niba johuimeo 191
nibagaigaraoau 392
nicha 294
nido de paloma 454
nieve 200, 256, 277
nieve tropical 200
nigua 415, 421, 435, 443
nigua de puerco 435
nigüita 423, 443
nigüitas 443
niguito 422
nigüito 403, 415, 418, 419, 420, 421, 422, 423
nigüito chamicero 423
nigüitos 423
niguo 275, 403
ñijuto 436
nim 454
nimé kogore 426
nimekogore 426
niñabonita 340
niniche 193
niño de cuna 202
niñomuerto 339
niopo 311
nipachire 203
nishé 295
nisperillo 305, 306
nisperito 444
níspero 278, 298, 302, 305, 306, 308, 418, 444
níspero caimito 305
níspero de anís 306
níspero de monte 306
níspero de saíno 278
níspero del japón 444
níspero del Japón 444
níspero japonés 444
níspero macho 298
niubajalé 461
nivaoue 191
niya háheku 210
noairay 188
noay 188
nocua 467
nocuito 374
nocuito amarillo 374
nocuito blanco 374
ñodku 202
nogal 274, 338, 339, 388, 410, 460
nogal bogotano 338
nogal sabanero 338
noiguita 366
noikai 191
nolí 208, 392
ñolí 208
nolocrea 179, 216
nolocrea macho 179
nomana 302
nomeolvides 225, 274
nongué 466
ñongué 466
noni 352
ñoño 401
noorijujuique 421
nopal 284
nopo 311
ñoque 466
noro 395, 443
noro lanoso 395
noro rubio 395
norte 436
noucda 453
novio 358, 392
novios 358
noyocaricaci 307
nuámano 191, 192
nuánamo 190, 191, 192
nuánamo colorao 190
nuánamo puntelanza 192
nuánamo sangregallo 190, 191, 192
nucalé 345
nuchu-kuchu 257
nudillo 195, 224, 233, 234, 243
nudillo acuático 243
nudo-nudo 257
nuez 191, 301, 388
nucz de barinas 388
nuez de barquisimeto 388
nuez de Pará 301
nuez del Brasil 301
nuez del brazil 301
nuez llanera 388
nuez moscada 191
nuicuna 334
nujü wukarto 200
nujübosonae 200
nujubujupiali 177
nukcalá 375
nulyi 205
numa cucú 437
ñume 204
ñume blanco 204
ñumeroca 428
ñuñucurá 401
nupa 208
nupe 329
nutmeg 191

O

o 201, 266, 287, 302, 324, 345, 346, 347, 355, 356, 364, 376, 380, 394, 412, 449, 451, 453, 472
oakta 325
obelisco de la sierra 411
oca 432
oca de chile 432
ocal 424, 425
ocalipto 424
ocalita 424
ocalito 424
ocas 432
ochaiteiba 200
ochor 236
ochoro 236
ochorróneta 368
ochuva 468
ocoba 191
ocobo 365
ocoó 396
ocoó-yagé 396
ocra 406, 432
ocro 406
ocumo 203
odoká 201
o-do-ká 201
odorate 387
ofaebü wonotakirinae 448
ofoma terikai 191
oití 383
oivo 278
ojalillo 200
ojera 224
ojerón 361
ojito de nena 330
ojito de nene 458
ojito de Santa Lucía 224, 310
ojitos 361
ojo de buey 291, 319, 327, 328
ojo de culebra 439
ojo de Dios 236
ojo de gato 297, 406
ojo de nene 458, 459
ojo de perdiz 458

ojo de pichón 292
ojo de poeta 361
ojo de Santa Lucía 224
ojo de sapo 463
ojo de sardina 389
ojo de tucano 304
ojo de tucano picón 304
ojo de vaca 328
ojo de venado 307, 328, 331, 338, 449
ojo de vendao 338
ojo de zamuro 328
ojovenao 327
okka 432
olivo 278, 339, 375, 416, 460
olivo de cera 339
olivo europeo 375
olivo hembra 278
olivo macho 278
olivo negro 278, 416
olivo santo 278
olivo silvestre 460
olivón 272
olla de mono 294, 302
olla mono 302
olleta de mono 302
olleto 301, 302
ollita 301, 302
ollita de mona 302
ollita de mono 302
olloco 283
olmeca 364
olor 197, 234, 275, 279, 295, 328
olorosa 235, 358
oloroso 182, 184, 234, 267, 393, 454
olot 278
omachi 200
omonoes 312
o-mo-ta' 453
onagra 429
ondequera 403
ondequero 403
ong 346
ong-uué 346
onotillo 394
onoto 406
oobeoba 458
oofa 394
opachiro 182
opio 457
oquendo 437, 440, 441
oquilla 432
orchoro 236
oreganillo 378
oreganito 378
oreganito de La Sierra 378
orégano 350, 369, 371, 372, 373, 378, 379
orégano ancho 379
orégano calentano 378
orégano cimarrón 379
orégano de cerro 378, 379
orégano de España 372
orégano de monte 379
orégano de tierra fría 373
orégano griego 372
orégano orejón 369
oreja 192, 199, 202, 224, 225, 253, 261, 263, 268, 302, 331, 335, 348, 361, 364, 367, 369, 370, 372, 374, 395, 401, 430, 434, 436, 442, 450
oreja de burro 202, 395, 434, 450
oreja de chimbe 430
oreja de coyote 401
oreja de elefante 192
oreja de guapore 331
oreja de león 370
oreja de mula 199, 202, 225, 268, 302, 348, 361
oreja de mulo 202, 225, 361
oreja de negro 224
oreja de pellejo 335
oreja de perro 367, 372
oreja de puerco 261
oreja de ratón 253, 263, 374, 434, 436, 442
oreja de tigre 192
oreja micu albahaca cimarrona 369
oreja micuna 369
orejas de burro 262
orejas del diablo 289
orejemula 184, 225, 261
orejera 319
orejero 319
orejero piñón 319
orejita de ratón 311
orejitas 253
orejo 319
orejón 361, 364
orejones 342
orejuela 442
orejuela de ratón 442
orejuelita 442
orgullo 277, 417
orgullo de la India 417
origa de buho 448
orijuela 442
oro azul 379
orochoa 236
oropel 444
oropo 455
orosú 379
orosul 379
orosús 342
orozú 379
orozul 379
orozús 342, 379
orozuz 379
orquídea 218, 312
ortegón 401
ortiga 281, 291, 292, 313, 388, 445, 446, 448
ortiga blanca 388, 446, 448
ortiga de búho 448
ortiga de flor 448
ortiga lisa 448
ortiga macho 448
ortiga mansa 445
ortiga mayor 448
ortiga menor 448
ortiga morada 281, 448
ortiga negra 448
ortiga pequeña 448
ortiga roja 448
ortigo 446, 447, 448
ortigo blanco 448
ortigo macho 446
ortigo negro 448
ortiguilla 393
ortiguillo 387, 409, 445, 448
ortiguillo rojo 448
orumo 446
osa de muerto 271
osapanga 209
oso 253, 254, 264, 274, 403, 431, 455
oso blanco 403
osopanga 209
oti 321
otiba 191
oticón 439
otinae 321
otó 203
otoba 191
otobo 190, 191, 192
otobo pajarito 191
otonasaré 409
otonosare 409
ovegue 318, 335
ovegue de rebalse 335
ovejitos 254

P

pa 205, 226, 260, 298, 306, 308, 311, 312, 330, 332, 384, 386, 405, 460, 471
pa pa 205
pa?coro 308
pachaca 278
pachaco 329
pacharama 382
pachín 299
pachipa 339
pachita amarilla 398
pachuba 210, 212
pachubita 210
pachulí 236
paciencia 289, 290
pacífico 250, 346, 347
paco 302
pacó 301, 302, 355, 380, 397
pacó amarillo 355
pacó casaco 397
pacó de monte 302, 397
pacó guitarro 397
pacocito 274, 412
pacocito de monte 412
pacora 397
pacoyuyo 263
pactá 208
pacua 441, 451
pacualito 391
pacunga 258, 263, 270
pacunga pequeña 263
pacurú 440, 441
pacurú-niaara 440, 441
pada 385
padilla 224
paduana 471
pagoda 309
pagüis 309
paico 282
paimbo 233, 234
paio 427
paja 210, 216, 227, 230, 231, 232, 233, 234, 235, 236, 238, 239, 241, 242, 243, 244, 245, 246, 250
paja aromática 234
paja blanca 236, 242
paja brava 243, 244
paja brillante 241
paja churcada 242
paja ciempiés 234
paja cola de mula 244
paja colchón 238
paja cortadera 231
paja de arroz 238
paja de burro 238, 245
paja de camino 243
paja de empacar 242
paja de empajar 242
paja de grama 245
paja de la Virgen 236
paja de laguna 246
paja de limón 236

paja de Páez 246
paja de panela 242
paja de pantano 230
paja de páramo 241, 242
paja de pato 238
paja de pay 246
paja de perro 235
paja de pisco 230, 231
paja de ratón 242
paja de tigre 242
paja de zorro 233
paja del camino 243
paja dulce 245
paja garnona 216
paja hoz 243
paja ilusión 239
paja judía 232
paja lanosa 242
paja limón 236
paja lisa 246
paja mona 242
paja montosa 233
paja páez 246
paja pará 246
paja peluda 243
paja puyita 245
paja rosada 241
paja salada 238
paja tetera 250
paja tigre 242
paja tresdedos 236
paja zancona 243
pajacana 373
pajaepaje 246
pajaepay 246
pajaeperro 235
pajaezorro 233
pajarera 286
pajarillo 448, 449
pajarita 386
pajarito 187, 191, 193, 205, 218, 219, 241, 248, 280, 289, 292, 315, 349, 355, 391, 404, 408, 412, 422, 436, 448, 449, 450, 453, 465, 472
pajarito abejucado 448
pajarito amarillo 219, 280, 412
pajarito blanco 218
pajaritos 193, 218, 219, 263
pajaritos verdes 193
pájaro árbol 387
pájaro bobo 404
pajas 242
paje 226
pajilla blanca 245
pajillita rosada 241
pajita del Niño Dios 236
pajita del pesebre 236
pajón 243
pakuash 258
palacón ajicito 386
palama ñoli 208
palancabueno 188
palanco 437, 439
palchas 399
palchiguaca 399
paleta 318, 329, 334
paleta de perico 318, 334
paletaria 446, 447
palialte 184
paliaría 290
palindera 287
palitaria 194, 446
palito equis 351
palm bruja 224
palm de coco 208
palm de dátil 211
palm de jícara 210
palm kristi 392
palm mazorca 213
palm real de tumaco 205
palma 172, 179, 203, 204, 205, 206, 207, 208, 209, 210, 211, 212, 213, 214, 215, 217, 227, 250, 251
palma abanico 210, 211, 212
palma abanico de la China 210
palma africana 208
palma africana de aceite 208
palma Alejandra 204
palma almendrón 205
palma amarga 212
palma amargo 212
palma areca 208
palma bambú 208
palma barbasco 208
palma bella 207
palma Bismarck 207
palma bismarckia 207
palma blanca 209
palma boba 172
palma bobila 205
palma bolillo 213
palma bombona 208
palma botella 204, 212
palma cabeza de negro 211
palma carey 215
palma chará 208
palma chingalé 205
palma chiquichiqui 210
palma chonta 203, 206, 212
palma chontilla 209
palma churrubay 212
palma cica 179
palma coco 204, 205, 208
palma cocotera 208
palma cola de pescado 207
palma colepato 209
palma colorada 211
palma corozo 204, 205
palma corúa 205
palma crespa 205
palma cubarro 206
palma cucurita 205
palma cuesco 205
palma curunta 213
palma datilera 211
palma de abanico 212
palma de aceite 208
palma de cabeza 204
palma de Canarias 211
palma de carey 215
palma de cemento 209
palma de cera 207
palma de cerbatana 210
palma de cerrillo 204
palma de coco 208
palma de coquito 212
palma de corona 209
palma de corozo 203, 206, 208
palma de Cristo 215
palma de cubarro 206
palma de cuesco 205
palma de culebra 207
palma de dátil 211
palma de dátil de las Canarias 211
palma de estera 205
palma de fibra 210
palma de grulla 207
palma de jícara 210
palma de jíquera 210
palma de lata 206
palma de Manila 204
palma de mararay 204
palma de nolí 208
palma de ñolí 208
palma de plata 208
palma de ramo 205, 207
palma de sagú 179
palma de sal 205
palma de San José 215
palma de San Pablo 209
palma de sarare 212
palma de tunda 206
palma de verano 209
palma de vi 205, 212
palma de viento 212
palma de virote 205
palma del viajero 250
palma dompedrito 211
palma enchama 205
palma escoba 208
palma estera 205
palma fénix 211
palma funeral 179
palma güinul 205
palma huichire 205
palma inayá 205
palma iraca 227
palma jícara 210
palma lanceta 205
palma lata 206
palma macana 204
palma macuma 209
palma malibú 205
palma manila 204
palma mapora 212
palma maquenque 211
palma mararay 204
palma mariposa 207
palma mariposa chica 207
palma mariposa grande 207
palma mazorca 213
palma memé 213
palma molino de viento chino 212
palma mona 212
palma montañera 214
palma murrapo 209
palma naidí 209
palma negra 207
palma nolí 208
palma ñolí 208
palma ñonce 209
palma panameña 208
palma pángana 211
palma payanesa 204
palma pintalabios 208
palma Popayán 204
palma puy 209
palma rafis 211
palma ramo 207
palma real 205, 207, 212
palma real de Cuba 212
palma redonda 212
palma reina 204, 212
palma reina Alejandra 204
palma robeleni 211
palma robelini 211
palma roja 208
palma sagú 179
palma sanjuán 212
palma sará 208
palma sarare 212
palma saraza 208
palma solita 211
palma solitaria 211
palma surca 251

palma tamaco 203
palma táparo 205
palma tetera 250
palma triangular 208
palma triste 209
palma washingtonia 212
palma wasintonia 212
palma yagua 205
palma zancona 212
palmacristi 392
palmaeñolí 208
palmalito 210
palmarito 210
palmchristis 392
palmera 245
palmicha 208, 209, 211, 227
palmiche 207, 208, 209, 210, 211, 212, 227
palmiche colorado 211
palmicho 209, 211, 212, 227
palmilla 207, 209, 212
palmillo 209, 310
palmita 215, 224
palmita roja 215
palmito 207, 209, 210, 211, 212
palmito real 207
palo 181, 182, 183, 184, 188, 189, 190, 191, 200, 208, 229, 254, 255, 258, 261, 268, 272, 274, 275, 276, 278, 288, 290, 291, 292, 296, 301, 302, 304, 305, 306, 308, 309, 311, 312, 313, 315, 317, 318, 319, 320, 321, 322, 325, 329, 330, 331, 334, 335, 336, 337, 339, 340, 341, 343, 346, 347, 348, 349, 350, 352, 356, 357, 358, 359, 362, 363, 364, 365, 374, 378, 380, 381, 382, 383, 385, 386, 387, 388, 389, 392, 393, 394, 397, 402, 405, 407, 409, 411, 412, 413, 414, 416, 421, 423, 424, 425, 427, 429, 430, 431, 433, 435, 437, 438, 440, 441, 445, 446, 449, 450, 451, 452, 453, 455, 456, 458, 460, 472
palo aceite 315
palo achiote 412
palo am 318, 339, 423, 430, 435, 437, 440, 441, 451, 455, 460
palo amargo 339, 451
palo amarillo 318, 423, 430, 435, 437, 440, 441, 455, 460
palo anís 184
palo astilloso 189
palo bálsamo 343
palo balsudo 451, 455
palo barbasco 381
palo blanco 182, 278, 309, 383, 386, 424, 427, 453, 460
palo bobo 445
palo boroqueño 296
palo boya 337
palo brasil 320, 339, 437
palo brazil 437
palo breo 385
palo calavera 453
palo carey 456
palo cebo 413
palo chino 451
palo cincuentacentavos 319
palo colorado 365, 451
palo cruz 312
palo de aceite 183, 315, 380
palo de achiote 184, 431
palo de achote 412
palo de agua 181, 261, 272, 274, 359, 362, 429
palo de ají 181, 255, 335
palo de alacrán 334
palo de algodón 340, 413
palo de almidón 306
palo de almizcle 412
palo de arco 188, 331, 363, 364, 374
palo de ardilla 330
palo de barbasco 330
palo de batidor 392
palo de bejuco 288, 331
palo de boa 325, 341
palo de boba 364
palo de bostezo 433
palo de boya blanco 341
palo de Brasil 320, 416
palo de brazil 320, 416
palo de breo 452, 453
palo de bufeo 313, 334
palo de buho 188
palo de búho 188
palo de burro 278
palo de caballo 335
palo de candela 312
palo de cangrejo 305, 306
palo de caparrapí 183
palo de carapa 321
palo de carguero 320
palo de casabe de trueno 305
palo de cemento 291, 382, 383
palo de chicharra 394
palo de chicle 305
palo de chimbe 409, 430
palo de churuco 305
palo de coco 208
palo de cruz 312, 317, 356
palo de cuchara 341, 453
palo de cucharo 350
palo de diablo 365
palo de drago 389
palo de erizo 306
palo de gallinazo 336
palo de garrapata 388
palo de gavilán 301
palo de grulla 318, 336, 382
palo de guara 183
palo de guayabilla 425
palo de gusano 301, 346, 451
palo de hormiga 349
palo de hueso 362
palo de imbira 188
palo de incienso 451, 453
palo de indio 347
palo de indira 188
palo de inflamación 306
palo de josé 325
palo de José 325
palo de la cruz 312
palo de la equis 200
Palo de la vida 172
palo de laurel 452
palo de leche 392, 393, 438
palo de lombriz 441
palo de loro 331
palo de majiña 350, 421
palo de manteca 183, 382, 383
palo de María 290, 380
palo de mico 329
palo de monte 394
palo de mora 440
palo de murciélago 409
palo de nuez 388
palo de oso caballuno 411
palo de pelejo 318
palo de pelusa 331
palo de perezoso 394
palo de pescado 318, 335
palo de picón 430
palo de piedra 188, 322, 382
palo de piedra de rebalse 188
palo de pintura 321
palo de pollo 331
palo de puerco 446
palo de ratón 301
palo de raya 380
palo de rayo 329
palo de rosa 182, 312
palo de sábalo 321
palo de sal 302, 359, 402
palo de sangre 311, 437
palo de Santa María 290
palo de santamaría 272
palo de sapo 405
palo de sardina 357
palo de sarno 451
palo de sol 358
palo de tabaco 191, 393
palo de tigre 349
palo de trueno venenoso 381
palo de tunda 393
palo de vaca 437
palo de vela 364
palo de venado 397
palo de veneno 381
palo de yuca 392, 412
palo diablo 288, 440
palo espino 341
palo fara 301
palo gateado 450
palo gavilán 301
palo guitarra 378
palo guitarro 347
palo hediondo 301
palo jobo 407
palo largo 305, 430
palo loro 305
palo mamito 191
palo maní 388
palo maría 290, 380
palo mazamorra 352
palo mono 430
palo monte 275
palo mora 440
palo moro 440
palo morrocoy 341
palo mosquito 363
palo mulato 255, 451
palo negrito 335
palo negro 254, 268, 275
palo papa 190
palo pegante 319
palo pelusa 331
palo perico 430
palo picure 393
palo piedra 382
palo pilón 311, 339
palo prieto 288
palo quemado 275, 276
palo quinde 268
palo rosa 182, 312, 387
palo roso 431
palo sal 449
palo sanguíneo campeche 320

palo sano 472
palo santillo 430
palo santo 290, 292, 430, 451, 455, 472
palo santo de montaña 292
palo sapo 306
palo sarna 347
palo seco 430
palo sin rama 304
palo tábano 388
palo tigre 451, 458
palo tintín 383
palo tumba 393
palo vela 364
palo veneno 315, 346
palo venenoso 315
palo verde 329, 330
palo verde mexicano 329
palo verraco 343
palo yanaconas 309
palo yuca 308
paloamargo 339
palobobo 409
palocruz 312
palocruz hembra 312
palocruz macho 312
palodulce 451
paloepiedra 181
paloerrayo 458
paloesal 362
paloetigre 403
paloma 209, 222, 224, 305, 342, 389, 404, 415, 430, 454
palomaní 466
palomatías 389
palometa huaiyú 287
palometa huelto 287
palomilla 342, 435
palomita 263, 273
palomitas 273, 339, 343, 390, 435
palomitas de seda 339
palomuerto 302
palonegro 275, 276
palopescado 335
palopintura 321
paloroso 431
palorrejo 436
palosangre 437
palosanto 258, 290, 312, 414, 472
palosinrama 304
palosol 347
palotal 272
palotalo 359
paloverde 329, 330
pama 440
pamá 440
pamitano 200
pampa anís 271
pampadur 369
pampanillo 440
pampera 281
pamplemusa 456
pamplonita 260
pan 234, 263, 276, 318, 414, 436, 437, 457
pan con queso 263, 276
pan de árbol 436, 437
pan de ñame 436
pan de pobre 436
pan de queso 263
pan de todo el año 436
pan de tree 437
pan del árbol 436
pan del norte 436
pan y cacao 414
pan y quesito 276, 457
pana 367, 369
pana churca 369
panal 260, 266
panamá 388, 414
panameña 224
panameño 388
panas 369
pancayate 278
panceburro 234
panchecito 401
pánciga 360
panda 204
pandaré 423
pandeaño 412
panderitas 277
pandiaño 436
pandurata 438
panecito 406, 409
panelo 325
panga 204, 209, 211, 227, 363, 396
panga negra 209
pángala 227, 248
pangalleta 278
pángama 211
pángana 211
pangola 237
panguita 209
panocho 420
pañoleta 199
panque 266, 267
panque romero 266
pánsiga 360
pantanillo 401
pantano 230, 231, 232, 292, 401
pañueleta 199
pañuelito 277
panza de burro 202
panzaeburro 202
paotati-mosha 358
papa 190, 199, 200, 213, 225, 294, 295, 343, 432, 463, 468, 469, 470, 471
papa abondaria 471
papa alemana 471
papa americana 471
papa arbolona 471
papa argentina 471
papa azul 471
papa azulita 471
papa bambera 471
papa bandera 471
papa batata 471
papa berlín 295
papa blanca 471
papa bogotana 471
papa boliviana 471
papa brucha 471
papa cebolla 213
papa cun 199
papa de aire 225
papa de cruz 294
papa de culebra 200, 343
papa de duende 469
papa de indio 469
papa de la vieja 469
papa de monte 469
papa de pobre 295
papa de vieja 469
papa dulce 463
papa silvestre 468, 469, 470
papabü emalinae 305
papabü natsaketsiwaü 320
papachina 199
papacú 279
papaculebra 200
papaepobre 295
papagayo 260
papaguasca 226
papaíbia 432
papaindia 200
papama 394
papamo 394
papas lisas 283
papasacha 226
papaya 279, 380
papaya calentana 279
papaya de gallinazo 380
papaya de monte 279
papaya de olor 279
papaya de tierra fría 279
papaya del monte 279
papaya hawaiana 279
papaya hembra 279
papaya macho 279
papaya maradol 279
papaya melona 279
papaya silvestre 279
papayito 279, 391
papayito de los andes 279
papayo 279
papayo de montaña 279
papayote 226, 279, 388, 406
papayuela 279
papayuela maligna 279
papayuelo 200, 279, 388, 406
papelilla 451
papelillo 185, 267, 300, 301, 364, 368, 460
papepobre 295
papiro 231
papiro enano 231
papito 315
pap-ká 383
papún 355
papunga 258
pará 209, 246, 364, 391
para todo 282
paraa 472
paracelsa 402
parada 239
paraejo 295
paraguas 253
paraguas de sapo 253
paraguatán 357
paraguay 374
paraguaycito 425
paragüitas 171, 253
paraíso 248, 250, 454
parajuansoco 341
paramuerto 421
parapalo 360
parapara 459
para-pará 364
parasiempre 416
parásita 217, 222, 284, 344
parature 183, 393
parature negro 183
parcha 399
parchita amarilla 399
pardillo 274
pardillo blanco 274
pareira 434
pargua 402
pari 392
parica 295
parietaria 445, 446, 447
parinari 326, 383
parindera 372
paripiri 230
parra 359, 472
parra de monte 472

parra de pantano 359
parta 319, 387, 388
pasaganado 242
pasai 437
pasca 282
pascua 258, 362
pascuala 263, 406
pascualito 258, 391
pascuitas 258
pashaco 319
pasiflora 398
pasionaria 398, 436
pasita 347
pasito 308, 382, 415
pasmo 466
pasota 282
pasote 282
pasta 294
pastel 261
pastelillo 288
pastilla 309
pastillas de chocolate 319
pasto 203, 230, 231, 232, 233, 234, 235, 236, 237, 238, 239, 240, 241, 242, 243, 244, 245, 246, 317, 350
pasto aguja 244
pasto alambre 246
pasto alemán 238
pasto alfombra 234
pasto alpiste 243
pasto amargo 242, 244, 246
pasto amor 239
pasto ángleton 237
pasto argentina 236
pasto argentino 236
pasto azul 236, 239
pasto bahía 243
pasto bermuda 236
pasto blanco 235, 236
pasto braquiaria 246
pasto brasilero 243
pasto brillante 241
pasto bufel 235
pasto cañitas 242
pasto castilla 241
pasto chato 238
pasto chigüiro 234, 241
pasto chino 239
pasto chopín 241
pasto cinta 243
pasto colorado 238
pasto conejo 237
pasto cordobés 235
pasto de agua 203, 241
pasto de cuaresma 238
pasto de estero 238
pasto de gallina 233, 236
pasto de laguna 246
pasto de olor 234
pasto de palma 245
pasto de pisingo 238
pasto de salitre 238
pasto de trenza 234
pasto de yulo 238
pasto dominícola 238
pasto dulce 234, 238, 246
pasto elefante 233
pasto espiga 244
pasto estrada 238
pasto estrella 231, 236
pasto fino 244
pasto gallina 233, 244
pasto gordura 241
pasto guatemala 246
pasto guinea 241
pasto horqueta 242
pasto imperial 234
pasto indio 241
pasto janeiro 238, 246
pasto johnson 245
pasto Johnson 245
pasto lamedora 241
pasto lanoso 240
pasto lanudo 240
pasto liendre de puerco 238
pasto llanero 246
pasto llorón 239
pasto macana 243
pasto maíz 246
pasto micael 234
pasto micay 234
pasto muczuro 240
pasto natal 241
pasto negro 234, 243, 245
pasto nudillo 233
pasto nulludo 233
pasto ochor 236
pasto ochoro 236
pasto orchoro 236
pasto orillo 236
pasto pangola 237
pasto pará 246
pasto peludo 246
pasto plateado 237
pasto poa 240
pasto puntero 233, 240
pasto rabo de zorro 233
pasto rescate 235
pasto romano 243
pasto rosado 241
pasto ruso 245
pasto sabana 234, 242
pasto simple 244
pasto trejos 244
pasto trencilla 243
pasto unca 317
pasto víbora 230
pasto yaraguá 240, 241
pastoazul 236
pastorcita 401
patá 269
pata de babilla 270
pata de buey 312, 332
pata de chulo 412
pata de cuervo 175
pata de danta 201, 414
pata de gallina 173, 234, 238, 244, 253, 254, 263, 275, 299, 412
pata de gallinazo 412
pata de gallo 238, 299
pata de lancha 270
pata de lancho 292
pata de loro 457
pata de morrocoy 440
pata de paloma 389
pata de paujil 457
pata de perdiz 236
pata de pichón 292
pata de picón 353
pata de torcaza 386
pata de totoco 351
pata de vaca 311, 312, 332
pata de venado 209
patabá 210
patabá de guara 210
patacón 401, 402
pataegallina 238, 404
pataepajuil 457
pataepichón 292
patagallina 253
patebuey 312
patechula 412
pategallina 238, 253, 254
patelancha 267
patelanche 292
patemorroco 440
patemorrocoy 440
patepalomo 417
patepaujil 458
patepava 283
pateperro 395
patepichona 292
patetórtola 392
patevaca 298, 311, 312, 330, 332
patevenado 209, 405
patevenao 405
patía 293
patica de paloma 305
patico 193, 315
patico azul 315
paticos 193
patihinchado 392
patilla 293, 294
patilla de golero 293
patilla de monte 293, 294
patilla de ratón 293
patillita 294, 398
patillita de monte 294
patinegra 260
patinegro 260
pátiu 211
pato 193, 197, 225, 362
patodiagua 273
pa-too 386
patriota 461
patsua 319
patú 386
patudo 212, 442, 446
pau darco 364
pauche 266, 270
pauche blanco 266
pauji 450
paují 450
pausa 270
páutta 208
pava 184, 186, 209, 210, 211, 287, 305, 309, 343, 348, 351, 355, 437
pavito 364, 382, 460
pavo 193, 238, 281, 282, 364, 460
pavonia 216
pavonilla 367
paya 315
payaca 470
payaco 194
payaco-vinan 194
payandé 315, 330, 337
payandé bobo 330
payandé clavo 330
payanesa 204
payo 425, 427
pazota 282
pazote 282
pebetera 261, 272
peca 226
pecado mortal 255, 473
pecao mortal 255
pecas 226, 360
pechiche 290
pechuga 404, 451
pechuga de gallina 404
pechugo 451, 457
pecoje 393
pecosa 175, 226
pecosita 226
pecueca 320
pecuequilla 321

pecunga 258
pedo de padre 364
pedorrera 290
pedralejo 295
pedro noche 466
pedro tomín 397
pedrofernández 451
pedrohernández 451
pedrojucho 451
pedronoche 470
pedrote 451
pega 235, 257, 282, 286, 298, 310, 318, 320
pega pega 282, 286, 298, 310, 318, 320
pegadera 318
pegahueso 282, 459
pegajosa 257, 270, 274, 286, 341, 371, 391
pegajosita 286
pegajoso 379, 391
pegamosca 298
pegamosco 298, 299
pegamosco de arbolito 298
pegamoscos 298
pegamoso 298
pegante 319
pegaparuma 320, 415
pegapega 235, 257, 258, 270, 280, 281, 291, 298, 310, 317, 318, 320, 341, 363, 415
pega-pega 235
pega-pega 318
pegapega de bejuco 291
pegasaya 318
pegosa 288
peine 176, 362, 381, 407
peine de abuela 381
peine de mico 362, 407
peine de mono 407
peine mico 407
peinecillo 317, 416
peinemico 362, 407
peinemono 380, 407, 410, 440
peineto 416
peinicillo 416
pejí 394
pekuoino 226
pelá 336
pelá de castilla 336
pelacar 437
pelacara 437
pelador 190
pelamán 415
pelamano 415
pelamanos 415
peledillo 285
pelícano 218
pelillo 242
pelincú 417
pelitre 271
Pelma 172
pelo de ángel 436
pelo de saíno 207
pelo de vieja 236, 245
pelo de zaino 206
peloepuerco 468
pelón 186
pelones 471
pelotillo 296, 301, 380, 471
peloto 311
peludo 204, 238, 240, 246, 274, 283, 308, 407, 431, 441, 454
pelusa 339
pematatsabanaewa 422
penacho 236, 264
peñaemico 436
penca 214, 215, 284, 390
penca de cabuya 214
penca de Castilla 284
penca de cercas 390
penca de sábila 215
penca sábila 215
pencón 174
penda 275, 317
penda blanca 317
pendare 308, 340, 341, 342
pendaré 340
pendare grande 340
pendarillo 342
pendarito 340, 341
pendaya 248
pendejón 208
penemono 407
penga pienqa 344
penga-penga 344
penicilina 281
peñico 392
peñique 392
penitente 199, 367
penoncahue 344
pensamiento 297, 314, 340, 374, 406, 465
pensamiento de Bolívar 374
pensamiento de pobre 465
pentagrama 200
peona 375
peonia 310, 331
peonía 310, 319, 329
peonio 329
peonío 329
pepa de cruz 294
pepa de culebra 294, 343, 466
pepa de loro 452
pepa de monte 349
pepa de nigua 430
pepa de pava 309, 437
pepa de sábalo 278
pepa de sierra 324
pepaemato 193
pepaepán 436
pepapán 436
pepeguará 301
pepeloro 452
pepemono 453
pepero 404, 437
pepi 191, 409
pepiadero 191
pepilla 295
pepinillo 293, 294, 295
pepinillo de monte 294
pepinillo silvestre 294
pepinito de monte 293
pepinito redondo 470
pepino 186, 293, 294, 295, 468, 469, 470, 471
pepino cimarrón 293
pepino cocombro 293
pepino cohombro 293
pepino crespo 294
pepino de árbol 468
pepino de monte 293, 294
pepino de olor 295
pepino de picar 294
pepino de rellenar 294
pepino de relleno 294
pepino de tree 468
pepino del diablo 293
pepino diablito 294
pepino dulce 470
pepino dulce morado 470
pepino hueco 294
pepino llorón 469, 471
pepino mataculebra 186
pepino morado 470
pepino redondo 470
pepino redondo morado 470
pepino silvestre 470
pepita 351
pepito 280, 405, 469, 470, 471
pepo 294, 307, 433, 459, 470, 471
pepo cujaco 471
pera 347, 428, 444
pera de agua 428
pera de Malasia 428
pera de Sogamoso 444
pera montañera 347
pera o perita 347
peraco 330, 455
peral 444
peralejo 184, 295, 395, 403
peralejo blanco 295
peralejo de montaña 395
peralejo macho 295
peralejo montañero 395, 403
peralonso 263, 269
perdomo 370
peregüétano 383
peregüetero 383
perehuétano 383
pereirana 461
perejil 194, 252
perejil crespo 252
perejil liso 252
perenne 241
perfume de tigre 188
pergüétano 383
peri 453
perico 185, 280, 318, 334, 350, 355, 403, 430, 449, 451, 452, 455, 460
pericoaguao 407
pericón 261, 271, 377
perilla 347
perillo 340, 341, 437
perillo blanco 340
perillo negro 340
perillo rosado 340
periquera 308
periquito 415
perita 347, 428, 458
perita de monte 458
perito 428
perla 252
perlito 403
pernilla del monte 335
pero 347, 428
pero de agua 428
perón 326
peronia 319
peronilla 310
peronio 319
perpetua 272, 282
perpetuas 282
perrito 262, 375, 472, 473
perritos 375
peruana 213, 293, 386, 472
pervinca 344
pesá 393
peseta 200
pestaña 254, 409, 454
pestaña de mula 409
pestaña de vieja 254
pestañaevieja 254
petacas 226
petaquín 184
petaquita 226

petronita 272
petunia 379, 467
petunia de monte 379
peuma huasca 192
pezuño 311
pialde 454
piande 393
piangüita 344
pianida 310
piasaba 210
piaste 308
piaste caimito 308
piaunde 393
pibal 401
pica pica 328, 338, 441
picaci 305
picadora 448
picantico 390
picantilla 289
picantillo 289
picantina 390
picapica 292, 328, 338, 389, 441
picatón 200
pichanga 413, 418
pichanquilla 413
pichaque 298
pichi 425
pichincha 286
pichindé 337
pichindecillo 313, 417
pichirina 394
picho 301
pichón 292
pichón de ángel 292
pico de loro 330, 366, 367
pico de loro amarillo 367
pico de pato 362
pico de paujil 346, 367, 434
picón 179, 195, 203, 304, 311, 435
picuyo 258
pidoquera 328
pie de cabra 312
pie de pajarito 449
pie de paloma 404
piedeyerro 339
piedrito 350
piedro 255, 416
piedro amarillo 416
piel de pescado 344
piel de tigre 199
piernaepisco 437
piga-mishi 432
piganá 207
pigllo 390
pijao 320
pildé 394
piletre 271
piligan 171
pilón 311, 339, 356
piloncillo 301, 403
pilpe 375, 376, 417
piman 465
pimenteira 451
pimentera 451
pimentero 196
pimento 267
pimentón 465
pimienta 196, 261, 428, 465
pimienta blanca 196
pimienta de Castilla 196
pimienta de olor 428
pimienta dulce 428
pimienta negra 196
pimienta picante 196
pimiento 194, 428, 451, 465
pimpí 210
pimpín 287
pimpinela 390, 445
pimpinela blanca 390
pimpinella 445, 465
pimpinella mayor 445
pimpinilla 465
pimpinillo 441
pimpollo 281
piña 200, 228, 357, 430, 460
piña anona 200
piña banqueña 228
piña brava 228
piña de abeja 357
piña de agua 228
piña de borugo 228
piña de charapa 228
piña de chucha 228
piña de cumare 228
piña de cuzumbo 228
piña de danta 228
piña de guacamayo 228
piña de iguana 228
piña de mono 357
piña de monte 228
piña de morroco 430
piña de omima 228
piña de panguana 228
piña de paujil 228
piña silvestre 228
piñanona 200
pincel 249, 261
pincel de amor 261
pincelito 261
pinde 394
pinga de gato 280
pinganá 207
pinguasí 364
pinguin 228
piñi 392
piñico 392
pinillo 390
piñique 343, 392
piñique piñique 343
piñiquepiñique 343, 392
piñiqui 343
piñita 352
pinllo 390
pino 180, 214, 266, 269, 280, 315, 331, 338, 403, 405, 465
piño 188, 471
pino amarillo 180
pino australiano 338
pino azul 331
pino blanco 315, 465
pino brasilero 269
pino candelabro 180
pino cansabrazo 266
pino chaquiro 180
pino colombiano 180
pino común 180
pino criollo 180
pino de Australia 338
pino de montaña 180
pino de Pacho 180
pino de peña 180
pino espátula 180
pino extranjero 338
pino falso australiano 338
pino hayuelo 180
pino libro 180
pino manteco 280
pino marítimo 180
pino montañero 180
pino negral 180
pino negro 180
pino ocarpo 180
pino pátula 180
pino real 180
pino romerillo 180
pino romerillo caleño 180
pino romerón 180
piñón 319, 320, 364, 391, 409, 414
piñón blanco 391
piñón botija 391
piñón de cerca 319
piñón de indias 391
piñón de la India 391
piñón de oreja 364
piñón de puerco 414
piñón de purga 391
piñón florido 320
piñón negro 391
piñoncito de sabana 402
piñones 414
pinta bollo 282, 363
pinta bollo real 363
pinta mono 330
pintabollo 282, 363
pintabollo colatero 363
pintabollo culatero 363
pintabollo macho 363
pintabollo real 363
pintacanillo 330
pintadilla 260
pintado 455
pintamono 330
pintauñas 386
pintor 360
pintura de danta 354
pintura de zorra 343, 349, 353
pintura negra 196, 251
piñuela 228, 352
piñuela de tierra 228
piñuela silvestre 228
piñuelo 228, 309
piñuelo picudo 309
pío 437
piojillo 244
piojito 272, 347, 370, 387, 445
piojo 272, 347, 387, 412, 442
piojo de marrano 412
pionia 331
pionía 320, 329
pionio 329
pionío 319, 329
pipa 199, 300, 320, 454
pipa de indio 300
pipí 287, 365
pipiboto 248
pipica 471
pipichuelo 284
pipilanga 270
pipilongo 194, 195
pipita 259
pipo 459
pipón desgranador 242
pipote 195
pique 465
piquigua 200
piquimullo 443
piquisique 299
pira 281
piragua 200, 202
piranga 363
pirawa 202
piretro 271
pirigallo 314, 334
pirigua 200
piripiri 230, 231

piru 451
pisajachu 287
písamo 319
pisca 330
piscande 412
piscandé 412
pischicachilla 339
pisco 232
piso 231
pispo 336
pispura 317
pispure 317
pispusa 317
pisquín 311, 331
pistacho 442
pisuilse 425
pita 214, 215, 228, 296, 297
pitá 296
pita floja 228
pitahaya 215, 283, 284, 285
pitahaya roja 285
pitajaya 283
pitanga 425, 463
pitaya 283, 285
pitaya amarilla 283
pitaya roja 283, 285
pitayo 283
pitigua 200
pitillo 241
pito 181, 314, 319
pito jigante 319
pitongo 463
pitopito 280
pituya 390
piula 315
pivijay 438, 439
piyandé 208
piyu 437, 439
piyuque 437
pizarro 249
placo 224
plan tana 343
planchido 387
planchira 387
planchito 314
planta de arroyo 213
planta de cominos 252
planta de guineíto 321
planta de maní 311
planta de sanguijuela 175
planta de Santa María 267
planta de santamaría 267
planta de soldado 181
planta de soledad 181
planta del pajaro azul 184
planta del soldado 181
planta mano de danta 353
platanilla 248, 251, 430
platanillo 199, 223, 248, 249, 250, 251, 276, 304, 325, 340, 352, 364, 430, 432
platanillo de altura 251
platanillo de ciénaga 250
platanillo de monte 250
platanillo fosforito 248
platanillo macho 250
platanillo rojo 248
platanito 276, 314, 317, 321, 331, 332, 333, 429, 432, 461
platanito filú 317
platanito salado 276
platanitos 432, 461
plátano 179, 250, 340, 364, 402
plátano de Abisinia 250
plátano de machin 364
platano de monte 250
plátano de monte 250
plátano guineo 250
plátano ornamental 250
platanote 340
plateada 224
plateado 237, 262, 389
plateao 389
platero 253, 380, 389, 448
platiado 389
platiao 389
platillo 194, 197, 199
platino 199, 389
platino rosado 199
playero 271, 337
plaza 197
plegadera 253, 442
plomo 306, 403
plonío 319
pluma 234, 236, 379
pluma de garza 234
pluma de reina 379
plumaria 264, 283, 435
plumaria chiquita 264
plumaria cimarrona 283
plumilla 282, 283, 387
plumito 379
plumón 236
plumosa 282
poa 240, 244
poa azul 240
poajá 348
pobo 439
podai ijie juici 458
po-ham-bó-raek 191
polca 360
polea 369
poleo 285, 369, 370, 371
poleo chiquito 369
poleo de castilla 369
poleo de monte 370
poleo grande 371
poleo macho 285
poleo morado 371
poleo orejón 369
poleo pequeño 369
polipodio 176
pollitos 435
polo 472
polo o guayacán de playa 472
poloó 472
polvillo 363, 365
poma 270, 418, 428
pomarrosa 418, 428
pomarrosa común 428
pomarrosa del Brasil 428
pomarroso 428, 436
pomarroso brasileño 428
pomarroso brasilero 428
pomelo 456
pomo 269, 428, 441
pomorroso 428
pomorroso brasilero 428
pomorroso de Castilla 428
pompadur 286, 369
pómpulo 471
pona 210, 301
poná 191
ponedora 470
ponetukuerenae 232
ponilla 210, 335
ponni inna 305
ponowaiinboto 201
ponqué 469
poogu 451
popa 305, 340, 341
popay 301, 306, 307, 339
popay carguero 301
popay de muchilero 301
popay de rebalse 301
popay de sapo 306, 307, 339
popo 234, 392
popo de chumbe 234
popó de mula 405
popocho 250
poprillo 300
popuyo 420
poraquero 393
porcelana 292, 461
porita 401
poro 319, 426
poroporo 406
porotillo 319, 335
poroto 319, 320
porrejao 394
porson 415
portadora 337
portal 272
portulaca 290
portulaca de playa 290
pos 202
potacaci 440
potamé 210
potra 333
potré 227
potrillo 363
ppenccacuc 344
prendeamor 320
pretino 407
prieto 273, 274, 288, 359
primavera 286, 297, 304, 340
primaveras 304
prímula 304
princesa 297, 340, 344
princesita 340
pringamaría 386
pringamosa 388, 448
pringamosa de flores 448
pringamosa morada 448
pringamoso 448
pringamoza 276, 388, 391, 446, 448
pringamoza arbórea 448
pringamoza blanca 448
pringamoza de caballo 448
pringamoza dientona 448
pringamoza extranjera 388
pringamoza macho 391
pringamoza negra 448
promento 428
promontorio 347
pronto alivio 378
prontoalivio 259, 371, 378, 379
proporó 460
protea 433
púa 303, 417
puca 350, 363
puca panga 363
pucarón 341
pucasacha 417
puchato 298
puchicama 363
pucre 437
pucsato 298
pucuna caspi 191
puercoespín 380
pui espinoso 210
puike 184
puila 409
puilá 409
puirasado 305
pujamo 204
pukarón 341
pulen 329

púlen 329
pulga 364
pulguito 317
pulijpanaa 448
pulisa 257
puliz 257
pulmón 310
pulmonaria 282, 283, 360, 417, 420
pulsario caspi 354
pulvide 388
pumarejo 341
pun 305
puná 191
púnciga 321
pun-nashau 305
puño huasca 227
punta de lanza 192, 394
punta lanza 394
puntaelanza 394
puntalanza 388
punte 441
punteamarillo 305
punteblanco 422
puntegranado 305
puntelanza 192, 394, 421
puntelanzo 394, 420
puntero 233, 240, 244, 358
puntero del llano 244
puntero tenedor 358
punula 412
punuyoro 320
punuyoronaekono 259
pupunga 258
pupuyo 423
purga 294, 319, 389, 391
purga de fraile 391
purga de Juan Páez 294
purgación 326
purgante de perro 232
purgante de picón 179
purgo 305
purichingo 186
purranga 268
purula 235
pururug colorado 181
pusana 194
pusanga 250
pusilde 335
pussachpan 386
pusuy 211
putarsaniko 284
puy 209, 210, 211, 320, 363, 364
puy de paloma 209
puy de pava 210, 211
puy de rajar 210
puy falso 209
puya 229, 318, 407
puyón 405

Q
quebracho 450, 451
quebradura 261, 391
quebrahacha 315
quebrajacho 313
quebrajancho 313
quebrantadera 260
quebraolla 418
quebraplatos 316
quebrollo 418
quebroollo 418
queco 234
quemacarate 201
quemadera 256, 264, 298
quemadero 298
quemador 295, 328
quematatabro 439, 440, 441
quende 298
quenepa 458
querabare 363
queredora 194
quereme 194, 222, 223, 251, 298, 300, 418
quereme contra 194
quereme de las quebradas 251
quereme menudito 418
quereme rosa 298
querida 387
querido 387
querrico 382
quesero 274
quesillo 412
quesito 276, 411, 422, 457
queso 182, 440
queso fresco 440
quibru negro 200
quica 330
quiche 229
quiche morado 229
quicugau 308
quiebra olla 351
quiebrabarrigo 359, 442
quiebracha 315, 449
quiebracho 313
quiebrahacha 450
quiebrajacho 313
quiebramachete 304, 457
quiebraolla 351
quiebraollo 418
quiebrapatas 387
quiebrapiedra 402
quiebraplatos 316, 340, 471
quiebrapuro 316
quielete 470
quii 392
quijadas 220
quijón 181
quilache 434
quillo 420
quillotocto 365
quimbombó 406
quimsa-kuchu 257
quimulá 309, 339, 378
quin 348, 442
quin felpa 348, 442
quina 305, 348, 349, 350, 352, 356, 388, 395, 433, 455
quina amarilla 348, 352, 356
quina anaranjada 348
quina anaranjada y cascarilla 348
quina blanca 348, 352, 388, 455
quina boliviana 348
quina canchosa 348
quina cuprea 348
quina indígena 433
quina orinoco 455
quina punta de lanza 348
quina roja 348, 352
quina roja de pitayó 348
quina rosada 348
quina tuna 348, 352
quina túnica 348
quina tunita 348
quinaquina 190
quinceañera 362
quincedías 451
quinde 268, 299, 373, 465
quinde borrachera 465
quingombó 293
quinilla 305
quinilla colorada 305
quino 288, 348, 351, 352, 356
quino blanco 348
quino rojo 352
quinoa 282
quinocera 350
quinocero 350
quinón 181
quiñón 181
quinoy 298
quinua 282
quinua cimarrona 282
quinua verde 282
quipo 407
quiquiijej 430
quirache 212
quiripa 179
quiripiti 385
quirón 181
quitadolor 378
quitamanteca 413
quitasol 210, 274, 359
quiteria 345
quituá 204
quivá 392
quiveche 438
quivechi 438

R
rabanillo 277
rabanit 277
rabanito colorado 277
rábano 200, 276, 277
rábano blanco 277
rábano de canarios 276
rábano maleza 277
rábano morado 277
rábano negro 277
rábano silvestre 277
raberruncho 347
rabichucha 424
rabiguano 283
rabihorcao 204, 209, 211
rabito 440
rabito de mora 440
rabo de alacran 274
rabo de alacrán 274, 275, 379
rabo de ardilla 387
rabo de ardita 176
rabo de burro 245
rabo de caimán 275, 280
rabo de chucha 173
rabo de gallo 209, 211, 281, 361
rabo de gato 233, 234, 245, 387, 445
rabo de gurre 350
rabo de iguana 280, 283, 287, 313, 337, 362, 458, 459
rabo de mico 175, 245, 279, 322, 387
rabo de mono 441
rabo de mula 173, 245
rabo de pavo 433
rabo de puerco 261, 268, 408
rabo de ratón 331
rabo de runcho 216
rabo de tucán 366
rabo de vaca 233, 234
rabo de zorra 245, 379
rabo de zorro 204, 209, 214, 233, 234, 236, 237, 242, 261, 275, 378, 379
rabo pelado 301
raboeburro 245
raboemarrano 261

raboemico 245
raboevaca 233
raborratón 320
racacha 251
racancha 177
racimo de San José 461
raco 431
racu 226
radical 369
radio 369
raiberana 439
raicilla 348, 350, 355, 356, 375
raicilla de agua 356
raicilla gateadora 348
raicilla preta 356
raicillo 196
raigrás 236, 241
raigrás anual 241
raigrás inglés 241
raigrás italiano 241
raiján 424
raíz 216, 226, 227, 232, 287, 289, 321, 335, 348, 375, 437
raíz china 227
raíz colorada 321
raíz de cepillos 216
raíz de china 226
raíz de color 375
raíz de ipeca 348
raíz de mara 232
raíz de montaña 348
raíz de muerciélago 348
raíz de murciélago 348
raíz de oro 348
raíz de pipí 287
raíz de resfrío 232, 437
raíz de ruibarbo 289
rajan 342
rajarraja 227
rajateta 279, 469
rallador 212
rallo de mico 362
ramillete 377
ramio 445
ramo 207
ramón de conejo 315
ramoncillo 315
rampacho colorado 384
rampayuco 315
ranconcha 177
rangaya 358
rapabarba 383
rapabarbo 304, 383
rapancho 384
rapis 211
raposapanga 185
raque 431
rasca 192
rascadera 199, 200, 203, 224, 358
rascadera blanca 199
rascadera morada 203
rascadera negra 203
rascadera rosada 199, 203
rascaderilla 199
rascador 203, 339, 451
rascaor 200
rascarrasca 435
rase ñemee 192
rasete 435
rasgarropa 275, 276
rasgateta 469
rasguñagato 326
raso 435
raspador 436, 439
raspayuco 315
rastrera 234, 327
ratania 411, 472
ratania de nueva granada 472
ratania de sabanilla 472
ratón 186, 242, 253, 263, 293, 301, 311, 331, 374, 403, 434, 436, 442
ray grass inglés 241
raya 196, 197, 200, 273, 355, 379, 380, 407, 459
rayado 190, 215, 247, 295, 313
rayo 197, 199, 310, 329, 350, 386, 415, 457
rayo caspi 415
realidad 235
rebancá 276, 277
rebelén 211
rebelín 211
recuerdo 288, 376
red mahogany 425
redículo 365
redonditas de agua 253
ree -ka-ne'-to-ma 383
regalgar 471
regencia 344
regencias 344
reigrás 241
reina 204, 212, 214, 255, 300, 303, 315, 340, 367, 379, 417, 443, 456
reina Alejandra 204
reina Claudia 443
reina de las flores 417
reina del Caquetá 214
rejalgar 392, 470, 471
rejilla 462
rejillo 462
rejito 398
rejoncito 436
rejudo 455
relajar 392
relojito 358
rema 182, 418, 436, 442, 462
remedio de raya 355
remiendoepobre 388
remocaspi 339
remolacha 282, 368
remolina 243
renaco 439
repollito de agua 202
requesón 387
resbalamono 319, 347
resbalamonos 347
reseda 277, 417
resedá 417
reseda blanca 277
resedad 417
resedo 417
resino 350
resucitado 410, 411
resucitado de monte 411
resucito 411
resume 350
resune 350
retaima 329
retama 320, 329, 334, 336, 460
retama amarilla 336
retama blanca 320
retama de jardín 320
retama de olor 334
retama espinosa 336
retama junca 334
retamo 320, 329, 334, 336
retamo blanco 320
retamo calentano 329
retamo de jardín 320
retamo de olor 334
retamo espinoso 329, 336
retamo junca 334
retamo liso 320
retamo pequeño 320
reventadera 293, 299, 300, 471
reventador 260, 352
reventadora 299
reventilla 179, 392
reventillo 191, 306, 382, 387, 388, 392
revienta puerco 460
revientacaballos 273
revientapuerco 460
revientatetas 319
revolcado 209
rey 251, 305, 309
ribarbo 289
ric 404
rica-netoma 383
ricaré-tumisi-nama 383
ricin 392
ricino 392
rin 214
rinchao 379
riñón 186, 187, 364, 377, 430, 469, 470
riñón de oreja 364
riñón morado 377
riñonada 367, 468, 469
riñoncito 367
riñoncitos 319
riobarbo 289
riogrande 392
risacá 255, 256, 264
riso 427
risueña 258
robalgodón 318
robelini 211
robertico 391
roble 321, 328, 330, 331, 338, 363, 365, 416, 417, 433
roble africano 365
roble amarillo 331, 363, 365, 417
roble australiano 433
roble blanco 330
roble de río 365
roble de tierra fría 338
roble maría 328
roble morado 365
roble playero 365
roble sabanero 365
rocío 200, 277, 280
rocote 465
rocoto 465
rodamonte 309
rodilla de pollo 181, 196, 286, 361
rodillo de pollo 286
rodillón 292, 354
rododendro 300
roja 187, 191, 208, 213, 215, 238, 245, 247, 283, 285, 288, 305
rojitos 350
rojo 198, 204, 206, 208, 213, 242, 248, 250, 261, 281, 285, 286, 290, 307, 311, 387, 402, 409, 431, 455
rollera 195
romacilla 289
romacillo 289

romadizo 185, 186
romaza 289, 290
romazo 289
romerillo 180, 258, 265, 266, 267, 369
romerillo blanco 267
romerillo rucio 267
romerito 267
romerito de hoja ancha 267
romero 265, 266, 267, 331, 347, 372
romero blanco 265, 267
romero chiquito 266
romero común 372
romero de monte 265, 266, 267
romero de páramo 265, 267, 347
romero negro 265
romero ninyucué 265
rompe camisa macho 401
rompe zaragüey 259
rompehacha 288
rompeplato 226
rompeplatos 226
rompepollera 280
romperropa 276
rosa 182, 258, 259, 260, 271, 288, 295, 298, 312, 342, 358, 365, 387, 409, 428, 444, 450, 460
rosa amarilla 258, 259, 260, 271
rosa amarilla común 258, 259, 271
rosa bebé 444
rosa china 365, 444
rosa de cruz 312
rosa de engaño 409
rosa de Jericó 460
rosa de la cruz 312
rosa de La Habana 342
rosa de montaña 312
rosa de monte 312
rosa de venezuela 312
rosa del cielo 288
rosa japonesa 444
rosa té 444
rosadita 417
rosadito 298, 431
rosal de enredadera 444
rose gum 424
rosella 409
roseta 235
roseto 258
rosicenizo 185
roso 309, 312, 317, 365, 423, 431
roso de monte 312, 317
rosquil 295
roya 471
rozavieja 259, 265
ruba 283, 432
rubacá 334
rubas 283
rubia 350
rubiarbo 448
ruchica 329
rúchica 329
rúchico 329
rúchigo 329
rucia 292
ruda 257, 268, 271, 453, 455, 456, 460
ruda común 456
ruda de arado 271
ruda de arao 271, 456
ruda de castilla 271
ruda de Castilla 271, 456
ruda de gallina 268, 271
ruda de gallinaza 271
ruda de gallinazo 268
ruda de guala 456
ruda de huerta 456
ruda de Quito 271
ruda dulce 271
ruda gallinaza 268, 271
ruda silvestre 271, 456
rudón 272
rudu huequi aho 310
ruhua 283
ruibarbo 289
ruibarbo de huerta 289
ruibarbo del tíbet 289
rumalda 359
rumaza 290
rumbo 334
rumia 446
rungumá 284
rupiña caspi 425
rurruco 269
rusco 215
rústica 338
rústico 338
ruutka 303

S

sabaleto 389
sabana 242, 245, 327, 337, 340, 342, 371, 391, 392, 402, 422, 437
sabana de Castilla 245
sabanagrande 195
sabanerito 423
sabanero 209, 295, 308, 420, 422
sabanilla 472
sabañón 351, 417
sabañón doble 351
sábila 215
sabonero 445
saboya 241
saboyá 240
sabroso 331
sabuguera 296
saca ojo 354, 377
sacabuche 467
sacaestaca 471
sacalaestaca 471
sacaojo 354, 379
sacaojo blanco 379
sacaojos 370, 402
sacaojos colorado 402
sacatrapos 408
sacatropis 408
sacatsun 296
sacha 194, 228, 241, 292, 344, 379, 392
sacha inchi 392
sacha piña 228
sacha tos de guasca 292
sacha verbena 379
sacha yuyu 194
sachacebolla 214
sachapapa 469
sacha-sananjo 344
sachaulluco 283
sachauva 253
saeta 246
saeta lisa 246
sagalita 298
sago 250
sagu 211
sagú 250
saibo 319
saino 393
saíno 199, 200, 207, 278, 393, 403
saíno blanco 403
saino negro 393
saíno negro 393
saivo 319
sajalagua 227
sajino 393, 403
sajino blanco 403
sajino negro 393
sajo 450
sajo conchecaimán 450
sal de lagartija 227
sal de raspar 185
saladillo 381, 430
saladillo blanco 430
saladillo colorado 381
saladillo colorao 381
saladillo rojo 381
salado 276, 387, 430
salamán borrachero 466
salao 381
salapa 298
salbuguera 296
salchicho 433
salchucha 294
salcillo 383
salero 302, 412
saligna gum 424
salivón 342
salsa 327
saltafuera 269
saltón 418, 423
salvaje 229, 471
salvajina 229
salvarrecín 435
salvia 256, 257, 259, 260, 264, 265, 266, 267, 275, 276, 370, 371, 372, 373, 377, 379, 413, 415
salvia amarga 256, 257, 259, 260
salvia blanca 257, 259, 265, 377, 379, 415
salvia china 372
salvia chiquita 372
salvia de Castilla 257
salvia de hora 257
salvia de monte 275, 276
salvia de montesurundé 276
salvia gavilana 267
salvia macho 257
salvia morada 259
salvia negra 259, 275, 276, 370
salvia real 371, 372
salvia real silvestre 371
salvia santa 379
salvialugo 370
salviarrial 372
salviarroja 373
salviecita 373
salviecita de bogotá 373
salvielugo 370
salvio 257, 259, 260, 275, 276, 369, 370, 440
salvio amargo 257, 259
salvio blanco 276
salvio negro 275, 276
salvión 265, 272
sama 412
samaguare 331
samán 331
samán iguá 331
sámano 331
sambo 467

samo 412
sampablito 209
sampablo 209, 211
sampaulo 209
sampedrito 209, 211, 236
sampedro blanco 348
samuda negra 371
samut apal 287
San Antonio 268, 416
San Diego 282
San Francisco 394
San Joaquín 274
san josé 213, 400
San José 213, 214, 215, 217, 222, 400
San Juanito 251
sanagua 210
sanaguaré 331
sanaguaro 331
sanaguas 210
sanagustín 274, 282
sanahuesos 449
sanalotodo 258, 259, 265, 267, 347, 350, 353, 357, 379, 416, 417, 451, 465
sanalotototal 282
sanangillo 344
sanango 303, 343, 344
sanango de altura 344
sananho 344
sanantonio 416
sanapechos 251
sancia 293
sancie 293
sande 437, 440
sandé 437
sande mora 437
sandia 293
sandía 293
sandiego 472
sandillita 294
sanfrancisco 268, 272
sangre 190, 191, 192, 215, 272, 289, 290, 311, 315, 331, 339, 366, 367, 384, 387, 388, 389, 394, 437
sangre de carey 215
sangre de cristo 366, 388
sangre de Cristo 366, 367
sangre de doncella 384
sangre de drago 331, 388, 389
sangre de dragón 388
sangre de gallo 192, 331
sangre de grado 389
sangre de pescado 191, 192
sangre de toro 191, 192, 289, 290, 315, 339, 387
sangre de toro blanco 191
sangre drago 331
sangre toro 191
sangrecorteza 331
sangrecristo 390
sangregadillo 388
sangregado 331, 388, 389
sangregallina 191, 331
sangregallo 190, 191, 192, 331, 394
sangregao 326, 331, 387, 388, 389
sangregao blanco 331
sangregrado 388, 389
sangregrago 388
sangrerillo 389
sangrero 389
sangretoro 190, 191, 192, 289, 290, 311, 329, 393
sangretoro bajito 192
sangretoro blanco 191
sangretoro copé 192
sangretoro de bajo 192
sangretoro de montaña 191
sangrevaca 191
sangrito 322, 359
sangro 331, 388, 389
sangrón 388
sanguijuela 201
sanguinaria 281, 366, 367, 378
sanguisorba 445
sanjoaquín 274, 410
sanjosé 217, 343
sanjuán 212, 375
sanjuanico 250
sanjuanito 251, 417, 431
sanramón 348
sansá 293
sansí 293
sansú 293
Santa Juana 232, 265
santa maría 195, 197
Santa María 191, 196, 197, 260
santa maría boba 197
Santa María de anís 196
santa maría macho 197
Santa María siete venas 191
santacrú 450
santacruz 450
santainés 211
santalucía 224, 401
santamaría 195, 196, 197, 260, 261, 265, 267, 272, 368, 377
santamaría boba 197
santamaría de anís 195
santamaría negra 196
santamarianí 195
santamarianís 195
santaura 269
santocristo 406
santodomingo 365
sapa 213
sapá 446
sapan 315
sapán 315
sape 362, 384
sape mandur 362
sapepá 434
sapito 198
sapo 171, 253, 266, 270, 279, 289, 290, 293, 306, 307, 316, 330, 339, 340, 342, 343, 348, 352, 354, 370, 372, 376, 390, 405, 437, 452, 457, 463, 468
sapote 306, 307, 308, 411, 414
sapote colorado 308
sapote costeño 308
sapote cuna 308
sapote de carne 308
sapote de monte 308, 411
sapote mamey 306
sapotillo 306, 308, 315, 405, 408, 411
sapotolongo 412
sapotón 412
sapotorongo 412
saquisaqui 407
sará 208
saramán 331
saramuyo 187
sarandago gigante 251
sarandango 251
sarare 212
saray 212
sardina 389
sardinaprieta 392
sardinato 339, 459
sare 411
saril 409
sarmento 358
sarna de Castilla 370
sarnícalo 303
sarniclo 303
sarnisclo 303
sarno 435, 451
sarpoleta 338
sarracacha 251
sarrapia 318, 319
sarrapia real 318
sarrapio 318, 319, 328, 335, 441
sarya 430
sasafrán hembra 451
sasafrás 182, 183, 451, 453
sauce 268, 271, 329, 335, 404
sauce amargo 404
sauce blanco 271
sauce bobo 271
sauce colorado 404
sauce de playa 271
sauce espino 329
sauce espinoso 329
sauce guajiro 329
sauce llorón 404
sauce macho 404
sauce playero 271
saucito 268
sauco 296, 297, 403, 466, 469, 470
saúco 274, 287, 296
sauco amargo 470
sauco blanco 296
sauco cimarrón 466
sauco comun 470
sauco común 470
sauco de castilla 296
sauco de Castilla 296
sauco de montaña 296
sauco de monte 296, 297
sauco hediondo 466
sauco macho 469
sauco montañero 296, 297
sauco negro 466
sausa 404
sauye 240
savonette 325
saw palmeto 212
saya 430
saya saya 430
saya-saya 430
sayota 295
schasha 281
se in 332
sebestén 274
sebillo 191, 192
sebo 192, 407
sebujá 216
seca 319, 320
secá 319
secúa 294
séhéñeede 191
seique 314
seje 210, 211
seje pequeño 210, 211
selam 370
semana santa 186
semen contra 282
semilla lisa 308

semilla peluda 440
sen 332, 333, 335, 337, 367
sen de la tierra 367
sen de sabana 337
sen de tierra 367
sen españa 332
sena alejandrina 332
senescudo 430
señora maría anís 195
señora maría boba 197
señorita 465
sensitiva 311, 327, 328
sensitiva acuática 328
sensitiva de agua 328
serenilla 443
serensere 387
serindo colorado 323
serpentina 305
serpiente 380
serrano 284, 307, 454
serrapia 318
serrapia chibá 318
sésamo 375
seso vegetal 457
sevilla 471
sewi 345
shaashal 286
shiatushi 289
shiiyo u muusa 357
shipiwai 282
shoponoy 288
shot-shot 276
shullu amarillo 430
shuru wina 457
shushucu 194
sibari huipo 202
sichaca 319
sidra 295
siempreviva 193, 194, 223, 224, 259, 269, 272, 277, 281, 282, 370
siempreviva chocoana 224
siemprevivas 269
siete cueros 298, 394
siete de bastos 365
siete sangrías 416
sietecapas 443
sietecáscaras 326
sietecuero 394
sietecueros 298, 314, 326, 340, 394, 417, 418, 419, 420, 423, 424, 443, 465
sietecueros brasilero 424
sietecueros nazareno 424
sietecueros real 417
sietehermanas 444
sietelenguas 384
sietellagas 398
sieteluchas 417
sietenudos 278
sigó 396
sígueme 371, 447
sígueme caleño 447
sihuaruma montañera 414
sijipe 346
sikisikiboto 210
silbadera 320
silbadero 320
silbador 181
silbasilba 181, 182
silbato 181
silbo silbo 181
silbosilbo 181
silva silva 181
silvador 181
simen contra 282
simetría 390
siminduñe 459
siminuñe 459
simón 350
simpática 303, 369
sinaro 424
sincogollo 278
singamochila 359
singonio 202
singuinaria cabé 256
singuinaria chabama 375
sinigual 365
sinnoches 218
sinvergüenza 369, 372
siquita pacay 344
siratro 327
sirena 202, 213
sirigual 319
siriguay 319
sirigüé 465
sirimbo 186
sirina de sabana 392
sirindango 251
siringa 305, 307, 341, 391, 392
siringa blanca 392
siringa cauchosa 391
siringa de hoja larga 391
siringa de panguana 391, 392
siringa de pescado 307, 392
siringa de sabana 391, 392
siringa de sol 305
siringa hoja de ratón 391
siringa pegajosa 341, 391
siringa propia 391
siringalo 391
siringarana 391, 393
siringay 319
siringo 391
sirpo 447
sisipiyanae 354
sitron 455
sitwonel 236
sixe 236
snissn wood 456
so ulua 302
soapbush 420
sobralia 223
sódanie 340
soja 320
sol de mediodía 290
solana 237
soldado parado 344, 360
solimán 349
solimón 201
solita 343
solito 343
sombo 467
sombra 214, 287, 361, 464
sombra de mar 214
sombrerito 253, 309, 367
sombrerito de agua 253
sombrerito de sapo 253
sombrerito del diablo 367
sombrero 292, 315, 369, 384, 385
sombrero de la China 369
sombrero de obispo 292
sombrilla 369, 465
sombrilla japonesa 369
sombrilla morada 465
sombrillita 179
sonajera 314
sonoscuro 430
sopoca 308
sopo-ka 308
soragá 430
sorba 340
sorbetana 234, 338, 371
sordé 318
sordera 227
sorgo 236, 245
sorgo forrajero 245
sorguillo 245
soro sacha 369
soroga 430
sororokinha 248
sorrel 409
sorrocloco 278
sorva 340
sota 192
soto 190, 191, 192
soto sangre 190, 191
soya 320
sserrín 330
suaso 420
suaye 240
suba 282
sucupira 312, 318
sudán 245
suela 331
suelda 224, 261, 268, 269, 446, 448, 449, 450
suelda chicoria 261
suelda con suelda 224
suelda consuelda 224, 261, 269
suelda de marrano 261
suelda de roca 446
suelda de sabana 261, 268
suelda de tierra 268
suelda espigada 261
suelda marranera 261, 268
suelda redonda 448
sueldaconsuelda 224, 261, 268, 269, 359, 448, 449, 450
suelda-consuelda 268
sueldaconsuelda mayor 261
sueldaquesuelda 224
sueldasuelda 224
sueldilla 363
sueldo 437, 442
sueldoconsueldo 197
sueño 176
sueño de niño 176
suerpo 439
sugiscala 460
suingla 456
suingle 456
sukumusha 236
sul maduno 364
sulfatillo 329
sulsul 347
sulú 250
sunikila 287
sunila 258
sunshulá 429
supi 284
supire 284
surapanga 185
surca 251
surco 251
suribio 337
surundé 436
susacá 268
susana 361
susana amarilla 361
susanita 361
susanita amarilla 361
susca 184
suso 255, 268
suspiro 282, 284, 387, 467
suspiros 470
susque 184
susuca 295

susumuco 258
suulishaya pa a 332
swinglia 456
symploco 309

T
ta 192, 202, 289, 300, 440, 453, 464
taaudau 211
tabaca 331
tabaco 191, 195, 273, 275, 315, 368, 377, 380, 393, 466, 467, 469
tabaco calzao 467
tabaco cuba 467
tabaco de alacrán 195
tabaco de monte 377, 469
tabaco de puntas 467
tabaco de serpiente 380
tabaco frenteloro 467
tabaco habano 467
tabaco macho 273
tabaco negrolindo 467
tabaco pescador 275
tabaco puntalargo 467
tabaco sabanero 467
tabaco yuá 467
tabacón 269
tabacote 269
tabak 467
tabalgue 467
tábano 403, 448
tabaqueros 262
tabaquillo 261, 263, 265, 269, 287, 289, 315, 351, 354, 368, 389, 466, 467, 469
tabaquillos 262
tabardillo 436, 437
tabarí 300, 301
tabarique 300
tabena 225, 226
tablemina 213
tablero 389
tablón 187, 189, 278, 331, 380
tabloncillo 278
tacaloa 288, 403
tacamaca 451
tacamaco 451
tacamahaco 453
tacamahaco colombiano 453
tacana 248, 250
tacarigua 412
tacay 388
tache 263, 328, 432
tacheulo 341
tacho 456
tachuelo 326, 356, 405, 433, 434, 456, 457, 469, 470
tachuelo blanco 457
tachuelo de tierra fría 433
tacis 172
tacomocho 469
tacotaco 370
tacso 399
taebü 238
taecillo 314
taforita 390
tagua 204, 208, 211, 448
tagua blanca 204
tagua de monte 204
tagua de tumaco 211
taguapiche 448
taguo 448
tail 363
tainde 267
taka taka 332
talhui 334
talla 202
tamaca 203
tamaco 203
tamadero 299
tamagaco 451
tamajecua 184
tamamuri 437
tamara 392
tamarillo 468
tamarinda 432
tamarindo 335, 336, 432, 472
tamarindo de mico 336
tamarindo de montaña 336
tamarindo llanero 432
tamaris 472
tambe 427
tambolero 332
tambor 182, 332, 387, 412
tamborero 332
tamboro 332
tamishi 200
tampaco 384
tampeque 385
tanamé 325
tanané 330
tananeo 330
tangarana 290, 335
tangarana negro 335
tangare 454
tangare pipa 454
tango 360, 365
tangue 267
tanimboca 416
tanimbuco 416
tanta 294, 436
tanta jabilla 294
tapa hueco 195
tapabotija 415, 463
tapabotijo 415, 463
tapacú 279
tapaculo 279, 301, 355, 451
tapafrío 209
tapaliso 383
tapallito 398
tapamuerto 227, 248
taparín 205
taparito maraquero 294
táparo 205, 380, 431
táparo baudoseño 380
táparo calimeño 205
taparón 301
tapas 384
tapás 438, 439
tapecue 255
tapete 292, 367
tapiz 367
tapoca 197
tapon 356
tapón 225, 280, 356
tapunga 258
taque 388, 434
taquecurauqe 434
taque-puraque 434
tara 182, 319, 335, 460
tarábwa mĩhĩ́ 209
taracana 393
taracua 194
taraira 414
tarapấnẽ 209
taray 330
taro 199
tarqui 181
tarrali 294
tarralí 294
tarriago 250
tarta 329
tártago 391, 392
tártago emético 391
tártaro 390
taruquillo 349
taruya 225, 273
taruya raya 273
tarwi 326
tasajo 330, 331
taspín 451
tata 202
tataba 209
tatamaco 451
tatibuco 190
tauli 326
tauquin 457
tauri 326
taurutoto 354
tausa 398
tausilla 397
tauso 398
tautano 455
tawaku 467
taya 210, 335
tayanqa 258
té 236, 281, 299, 309, 329, 377, 408, 444
tê 384
té criollo 377
té de bogotá 309
té de Bogotá 309
té de china 309
té de popayan 329
té de popayán 329
té de Popayán 329
té del canadá 299
té limón 236
té verde 281
teack teca 350
teatina 405
teca 373
teco 373
tedbai 200
tee-shai-pee-ka 266
tefé 351
tefé-capacira 351
tejedora 236
teka 293
telembí 234
tembladera 173
tembladera pequeña 173
temutorabueaci 457
teñidera 293
teñidero 402
teñidor 402, 421, 442
teñir de negro 195, 196
tento 329
tento malungo 329
teología 390
teosinte 247
tepá 315
tepaci 315
terciopelo 184, 198, 199, 266, 282, 328, 372
terebinto 363, 451
teresita 273, 344, 465
teresita azul 465
teresitas 344
terriago en San Martín 250
teta churuco 305
teta de bruja 267, 300
tete 301, 335
tetera 250
teticas 470
tetilla 470
tetón 226
tetona 226
tewpede 206
tiba 309

tibabana 202
tíbar 309
tíber 309
tibigaro 450
tibligaro 450
tica 400, 403
tienderropa 332, 333
tierra amarillo 191
tiga 311
tigre 188, 192, 193, 199, 204, 206, 235, 242, 249, 250, 253, 259, 261, 267, 291, 308, 318, 437, 440
tigre guasca 192, 193
tigre guasca amarillo 193
tigre guasca morado 193
tigre guasca verde 192
tigrillo 450, 470
tigrito 184, 311
tiirai 458
tijereta 227
tijereto 280
tila 359
tilia 248
tilo 359
tilo criollo 359
tilo cubano 359
timbo 317, 335, 432
timbo nocou 317
tinajero 254
tinga 386
tinguitingui 393
tinracha 199
tinta 249, 293, 361, 402, 466
tinta barbasco 402
tinta pango 466
tinta verde 249
tintamorada 287
tintero 356
tintillo 338
tintin 399
tinto 183, 278, 338, 353, 354, 359, 432, 462, 466
tinto blanco 466
tinto chiquito 338
tinto macho 338
tinto negro 466
tintorera 280
típico 208
tiraagua 418
tiraco 330
tiramiaos 418
tirapatrás 208
tirateté 329
tiriaodes 275
tiribuche 330
tiribuchi 330
tiricio 416
tiririca 216
tirisia 333, 334
tirita 249
tirraya 392
tisaca 260
tisis 293
tissly 435
tisuapo 340
tiza 363
tjalatjala 200
to hua ri 364
tobaco 331
tobo 309, 382, 467
toborachi 407
tocá 459
tocino 405
toco 278
toco negro 278
tocua 467
tocumo 412
todikai 393
todoelaño 391
tofetorofe 467
toje 368
tola 257
tolonchero 183
tolú 328, 407, 437
tolúa 407
tomasuco 339
tomate 344, 468, 469, 470, 471
tomate común 470
tomate de árbol 344, 468, 471
tomate de árbol silvestre 471
tomate de montaña 468
tomate de monte 344, 468, 469, 471
tomate de riñón 470
tomate de tree 468, 469, 471
tomate de tree silvestre 471
tomate del diablo 344
tomate grande 470
tomate guiteño 468
tomate quiteño 468
tomate salvaje 471
tomate silvestre 468, 469, 471
tomatico de monte 344
tomaticos 468
tomatillo 344, 352, 467
tomatón 344
tomatoquín 467
tomatoquina 467
tomé 295
tometo 385
tomillo 267, 271, 286, 373
tomillo oloroso 267
tominejo 220
tonca 318, 422, 465
tonchuré 294
toñeca 383
toñeca propia 383
tonga 318, 465
tonmía 371
tonto grande 471
topa 389, 407, 412
topa roja 389
topaz 438
topo 457, 468
topo toropo 457
topotopo 457, 467
topo-topo 468
topotoropo 457, 467
toquín 467
torbellino 387
torcacero 441
torcazo 253
torito 219
toritos 289
tornillo 227, 314
torombolo 432
toroncoyo 347
toronja 456, 470, 471
toronjil 236, 358, 371, 378
toronjil americano 378
toronjil cimarrón 358
toronjil de paja 236
toronjil morado 371
toronjil silvestre 371
toronjilla 470
toronjo 456, 470
toropate 436
torrijao 188
torta 181, 308
tortolito 253, 423
tórtolo 253
tortuga 188, 209
tortuga caspi 188
toscana 361
tosilla 367
tostado 253, 267, 277, 304, 455
tostao 378
tot 465
tote 232, 316, 361, 467
tot-ea-kumá 465
totecito 232, 350, 357, 362
totes 232
totesito 232
totiadera 299
totiadera blanca 299
totiadora 299
totocá 280, 378
totocal 280, 378
totocul 378
totojando 278
to-to'-ko 324
totora 227, 232, 247
totorilla 231
totumillo 362
totumito 278, 280, 356, 362
totumo 202, 261, 268
toullí 406
towanae 452
trabuclo 405
trampolín 370
trancadera 224
trapiche 402
trapichero 305, 308
traquitraqui 361, 408
trébol 179, 260, 327, 330, 335, 336, 432
trébol amarillo 432
trébol aromatizador 260
trébol blanco 336
trébol cadillo 327
trébol de agua 179
trébol de carretilla 327
trébol de cuatro hojas 432
trébol de olor 327
trébol dorado 335
trébol dulce 327
trébol híbrido 335
trébol morado 432
trébol negro 330
trébol rojo 335
tree camarón o cabo de hacha 191
tree de aceite 315
tree de algodón 413
tree de borugo 452
tree de chucha 185
tree de churuco 307
tree de copaiba 315
tree de la verrugosa 197
tree de la vida 451
tree de los lamentos 393
tree de pava 349
tree de vaca 342
tree del pan propio 437
tree del toro 192
tree dulce 392
tree lengua de juagar 271
tree santo 391
tree vaca 341
treeoco 266, 270
treeoco de bogotá 266
tregua 318
trementina 304
trementino 304, 315, 331, 453
trementinos 262
trencilla 243, 337, 387
trencita 243, 337, 461

trenza 243, 367
trenza acuática 243
tres dedo 197, 201, 224
tres dedos 197, 201, 224
tres espigas 257
tres puntas 267
tresbolas 370
trescáscaras 191
tresdedos 197, 201, 225, 236
tresfilos 231
treshojas 457
trespelotas 370
tresrayas 419, 434
trestablas 458
triangularis 208, 432
trichi o tricho 172
trífer 214
trigo 236, 246, 289
trigo almequemé 246
trigo amazónico 236
trigo argentino 246
trigo arroceño 246
trigo barcino 246
trigo bolapicota 246
trigo bolarraspa 246
trigo bolas 246
trigo bonza 246
trigo brasilero 236
trigo cañablanca 246
trigo cartagua 246
trigo centeno 246
trigo colorao 246
trigo cuanca 246
trigo diacó 246
trigo florel marengo 246
trigo peruano 236
trigo sarraceno 289
triguillo 235
trinitaria 286
tripa de babilla 328
tripa de buey 411
tripa de perro 201
tripa de pollo 200, 280, 281, 328, 357, 390, 429
tripaeperro 201
tripaepollo 390
tripapollo 281
tripepollo 447
tripita de pollo 390
tripona 399
triqui traque copia-lila 361
triquitraque 361
triqui-traque 361
tristeza 280, 447
trompa de cusumbo 409
trompeta 339
trompetero 434
trompetica roja 463, 464
trompetilla 264, 270, 287
trompetillo 375
trompeto 435
trompeto curador 435
trompillo 454, 455
trompos 399
tronco 199, 215
tronco de zaino 199
truco 443
trueno 305, 361, 375
trujillo coca 386
trúntago 374
trúntago blanco 374
trupa 212
trupí 331
trupillo 331
tsakeke 327
tsakenae 327
tsakewaül 327
tsakibeiwa 392
tsalikali 258
tsemane 448
tsematsemanea 368
tsica tabütsujunae 453
tsitsinae 425
tsukuiiniibo 458
tsukunae 350
tsun 293
tsun-jo 293
tú y yo 390
túa tua 391
túa túa 391
tuabe 182, 183
tuabe canelo 182, 183
tua-tua 391
túa-túa 391
túa-túa morada 391
tuave 182
tubaci 452
tuchica 399
tucumá 204
tucumo 412
tucuno 412
tucutaco 370
tucutuco 272
tuercemadre 408
tuétano 455
tuhui yasi 256
tuiyó 390
tukulipainae 446
tukupe 228
tũkür 210
tula 191
tulá 427
tula pueta 191
tulapueta 191
tulipá 222
tulipán 213, 216, 217, 365
tulipán africano 365
tulipán de Gabón 365
tulipán rojo 365
tulipancillo 411
tulla 298
tullá 298
tullidora 442
tumapá 209
tumasuco 339
tumatumale 185
tumbabobos 236
tumbaguay 260
tumbamaco 253
tumbatoro 274
tumbatoro peludo 274
tumbaviejos 236
tumbis 399
tumbo 399
tumbué 370
tumbues 193
tumoho-screbano 453
tumpu 399
tun 200
tuna 284, 348, 352
tuna de la sabana 284
tuna guajira 284
tunas 284
tunchín 420
tunda 194, 206, 234, 393
tundushé 201
tuni esmeraldo 275
tuniesmeraldo 276
tunilla 271
tunillo 271
tunisco 364
tunita 348
tunito 278, 284, 419, 469
tunito espinoso 469
tuno 182, 284, 356, 418, 419, 420, 421, 422, 423, 433, 435, 447
tuno blanco 420, 421, 422
tuno coludo 420
tuno esmeralda 422
tuno esmeraldo 421, 422
tuno morado 422
tuno negro 421, 423, 447
tuno o tunas 284
tuno rosado 422, 423
tuno roso 423
tuno sabanero 419
tuno vejlgoso 423
tunoso 315
tupirijo 341
tupiro 469, 471
tupiru 471
tupirú 471
tupu 302
turbemono 274
turco 329
turi 301
turí 301
turma 343, 349
turma de mico 349
turma de perro 343
turmaemico 349
turmamico 347, 349
turmas de gato 469
turmas de toro 341
turmemico 349
turmemono 274, 311, 349
turmeperro 344
turmequé 349
turmetoro 274
turmo 449
turquete 388
turriago 250
turu 224
turuca yoco 459
turumuto 207
tururi 437
tururí 437
tusa 263, 269, 282, 367
tusílago 369
tusilago chabama 371
tusilla 202, 367, 368
tusillo 245
tute 234
tutuma 268
tutumilla 294
tuya 180
tuya del Canadá 180
tuyo 390

U

u 210, 290, 380, 468
uamao 185
ubasoshi 457
ubejuici 353
ubí 209, 210, 472
ubre de vaca 186, 274
ubrevaco 274
uca 392, 432
ucaena 325
ucal 424
ucapón 266
uchu sananho 344
uchuba 468
uchubo 468
uchuva 467, 468
uchuva blanca 467
uchuva de perro 467
uchuva de perros 467
uchuva del duende 467
uchuva negra 467
uchuvilla 467

uchuvo 468
ucucal 191, 192
ucuquí 308
ucuuba 191
ucuúba 191
ucuye 341
uflé 182
uflé-aguacatillo 182
uguray 452
ùjùmedù 172
ulanda 455
ulando 457
ulloco 283
ulluco 283
ullucu 283
ulluma 283
ullus 326
umarí 359, 381, 382
umarí de chimbe 382
umarí de monte 359
umbe 200
uña de águila 358, 436, 442
uña de caballo 197
uña de cabra 336
uña de gato 171, 226, 227, 287, 310, 326, 335, 358, 363, 433, 434, 436, 439, 442, 456, 457, 469, 471
uña de gato de altura 358
uña de gavilán 317, 456
uña de perro 287
uñaegato 456
uñas de tigre 330
unca 317
uncucha 358
undequiera 403
uñedanta 200
uñegato 227, 436, 456
unganangi 358
uñita 439
upa 266
upacón 266
upo 266
upujicabue 447
upvonia 380
urapán 374
urapo 365
urero 331
urgat 287
uribe 240
urinaria 337, 402
urrapia 312
úrtugo 302
uruapo 378
urucú 406
urucurú 406
urugá 191
urunta 213
usiyab-ai 392
usmillo 427
usmillo de corteza 427
usper 414
uué 346
uva 274, 288, 298, 299, 300, 302, 370, 428, 446, 447, 467, 472
uva blanca 298
uva caleta 288
uva camarera 299
uva camarona 298, 299, 300
uva cimarrona 472
uva de anis 298
uva de anís 298, 299
uva de culebra 472
uva de gorro 472
uva de la playa 288
uva de mar 288
uva de monte 298, 300, 446, 447, 472
uva de orilla 447
uva de pajarito 472
uva de perro 299, 370
uva de playa 288, 472
uva montañera 302
uva monte 298
uva playa 288
uva silvestre 447, 472
uvero 274, 288
uvero de playa 288
uvero macho 288
uvero uvita 288
uvilla 288, 293, 447, 451, 466, 467, 468, 471, 472
uvilla buchona 468
uvilla lija 447
uvillas 467
uvillo 274, 288, 447, 466, 468
uvita 206, 274, 288, 298, 352, 467
uvita de agua 352
uvita de gallinaza 206
uvita de monte 298, 467
uvita de playa 288
uvita de tigre 206
uvita gallinaza 206
uvita mocosa 274
uvita pegajosa 274
uvita playera 288
uvita roja 288
uvito 274, 288, 298, 299, 300, 415, 428, 467
uvito de barrancos 298
uvito de gallina 467
uvito de Guasca 300
uvito de monte 298, 299, 300
uvito de murciélago 274
uvito de playa 288
uvito mocoso 274
uvito noble 298
uvo 288, 298, 299, 300, 304, 350, 423, 447
uvo caimarón 447
uvo de mar 288
uvo de monte 298, 300
uvo falso 299
uvo silvestre 447
uyama 293
uyamera 293
uyamo 387, 433

V

vacaricuara 416
vacaricuara amarillo 416
vacavieja 331
vaco 437
vacùcù 381
vaga local 302
vagamundo 340
vaguerá 450
vaina 411, 412
vainilla 223, 275, 332
vainillo 223, 315, 364, 430
vainillón 223
vaino 315
valbuena 379
valdivia 460
valerian 443
valeriana 267, 296, 373, 443
vangolalá 261
vanne van 371
vara blanca 408, 454
vara de alcalde 218, 387
vara de humo 274
vara de león 416
vara de mico 248
vara de pescar 189
vara de piedra 288, 339, 347, 395, 403, 404, 455
vara de piedra de humo 403
vara de San José 213, 222
vara de sarna 347
vara negra 387, 422
vara santa 265, 352
varablanca 290, 352, 388, 402, 403, 457
varaepiedra 403
varajón 265
varasana 388
varasanta 265, 290, 335, 352, 403, 440
varazón 403
varecasa 403
varejón 259, 261, 265, 267, 272
varejón blanco 267, 272
varejón de caballo 261
varejón de Castilla 272
varejón recio 265
variedad 409, 416
variedades 416
varilla de vidrio 276
varita de lazo 379
varita de san josé 213
varita de San José 213, 214, 217
varito 265
varo blanco 445
varoblanco 403, 445
varote 272
varraco 276, 343
varsovia 217
vayavacù 428
veinte de julio 340, 391
vejiga 255, 457
vejiga de lobo 255
vejigo 339, 340
vejigon 468
vejigón 339, 340, 415, 467, 468
vejigón negro 415
vela 260, 280, 306, 364
vela de páramo 260
velacho 448
velero 332, 342, 372
velillo 347, 394, 422
velita 260, 379
velita de páramo 260
velitas 360
velito 347
velo de novia 469
vena 204, 234
venadillo 255, 261, 262, 264, 267, 268, 269, 344, 347, 375, 418
venadillo grande 269
vendeaguja 235, 240
vendeaguja de montaña 235
vendiaguja 240
venenillo 471
venenito 343, 466
veneno 299, 313, 346, 381, 405, 434, 435
veneno de churuco 434
venenosa 359
vengueverraco 343
venita 204
venjués 343
venturino 348
venturosa 258, 265, 271, 378
venturosa morada 265
veranera 286, 369
veranera morada 286

verano 209, 286, 313
verbena 259, 268, 274, 275, 368, 377, 378, 379, 380
verbena arisca 259
verbena blanca 259, 274, 275, 379
verbena chiquita 379
verbena criolla 274
verbena de castilla 378
verbena de Castilla 378
verbena del campo 379
verbena macha 379
verbena macho 379
verbena morada 275, 379
verbena negra 368, 378, 379, 380
verbena rabo de alacrán 275
verbena verde 379
verbenilla 274, 357, 379, 417
verde vainilla 223
verde y negro 274
verdenegro 275, 276
verdinegro 442
verdolaga 261, 268, 280, 281, 286, 290, 472
verdolaga blanca 281
verdolaga de césped 286
verdolaga de playa 290
verdolaga francés 290
verdolaga montañera 290
verdolago 290
verduguillo 209
vergonzosa 327
vergüenza 367
vero 288, 364
verónica 377, 465
verraco 198, 343, 433, 436, 445
verraquillo 445
verruga 187, 437
verruguero 390
vetamo 439
veterina 281
vetiao 292
vetiver 236
víbora 207, 224, 240
vibuajeiria 434
victoria 181, 293, 304
victoria regia 181
vida del hombre 214, 217
vidita mía 266, 389, 401
vidriosa 281
vieja 267, 451, 469
viejo barbudo 222
viernes santo 390, 402
vinagrera 400, 432
vinan 194
vinca 340
vino 203, 288, 409
vinotinto 281
violeta 225, 297, 304, 368, 406
violeta africana 368
violeta de agua 225
violeta de los Alpes 304
violeta de páramo 406
violeta morada 406
violeta silvestre 406
violetilla 361, 377
vira 251, 255, 263, 268, 269, 462
vira vira 269
viravilona 255
viravira 255, 258, 263, 264, 265, 268, 269, 275, 282
vira-vira 255
vira-vira 255
vira-vira 255
vira-vira 263
vira-vira 263
vira-vira 268
vira-vira 268
vira-vira 268
vira-vira 268
vira-vira 462
viravira blanca 275
viravirón 255
virginia 379
virola 191, 192
viroviro blanco 265
visco 448
vitoria 293
vitoriera 293
viuda 296, 303, 340, 436
viudalegre 373
viudaloca 373
viudita 340
viuditas 344
vivaseca 315, 318
volado 407
volador 182, 225, 288, 410
volandero 407
volantona 272
volao 407
volaó 407
vua 390

W

wabama 439
waiarima 305
waja 200
wajonae 330
wajuñe 206
wakarabiijiiterenae 287
wakonae 330
waliruwain 308
wanalienae 420
wanalijorobarii 201
wara atsu 321
waraara 192
wareara 267
wasái do pará 209
washa washa 372
wáshington 200
washingtonia 212
wasintonia 212
wayusa 255
wéger 205
wejerinae 453
wepe taxusinae 320
werre 205
wetsuenae 407
wewena 251
white rhubard 392
wigirbü 205
wild mango 302
wild pop thatch 208
win-cheék 414
wira kupa 472
wormweed 282

Y

yaba 329
yabaquillo 261
yabo 329, 330
yaca 314, 436
yacaeruba 380
yacana 311
yacayaca 314
yachitara 208
yachoro 206
yaco 289, 394, 436
yaco borrachero 394
yacón 270
yaegaseii 191
yagará 455
yagé 281, 394, 395, 396, 459
yagé de cacería 394
yagé de monte 394
yagé-yoco 459
yagrume 253
yagrumo 446
yagua 205
yaguache 262
yaguará 240, 241
yaguará uribe 240
yaguaro 315
ya-hi-ae-o 346
yahuar-shonka 430
yaigue 412
yairén 345
yajé 395
yakee 191
yalar 211
yamba 441
yambó 437
yambolán 428
yamichuí 178
yamitá 435
yanchama 415, 439, 441
yanchama blanca 439
yanchama bujurqui 439
yanchama colorada 441
yanchama de pelazón 415
yansaquitup 192
yape 318
yapima 463
yapurá 430
yará 210
yaraguá 240, 241, 246
yaraguá falso 241
yaraguá legítimo 240
yaraguá peluda 241
yaraguá peludo 240
yaraguá Uribe 240, 241
yaré 200, 227
yarina 204, 211
yarina blanca 204
yaripa 210, 239
yarrow 255
yaruma 249
yarumo 445, 446
yarumo blanco 445, 446
yarumo colorado 445
yarumo hembra 445
yarumo macho 446
yarumo negro 445, 446
yarumo plateado 446
yarumo rojo 446
yashitara 208
yasitara 208
yasmande 272
yasmiande 272
yasu 285
yataro 452, 453
yató 191
yatomay 312
yatonay 312
yatornay 312
yatorona 312
yaucasi 200
yaurero 283
yavarí 204
yavímajuna 266
yawa 388
yawati 402
yawatinae 434
yaya 187, 188, 189, 190, 192, 347, 449, 455
yaya blanca 190
yaya de cajón 347
yaya de sangre 449

yaya negra 188, 190
yaya sangre de pescado 192
yayajalé 265
yayo 329, 455
yayo blanco 455
yayo colorado 455
yayu-ae-o 346
yedr 472
yedra 253, 358, 376, 472
yedra geranio 358
yedrita 376
yedro 439
yellow milkbush 390
yema 182, 332, 333, 405, 406, 439, 455, 460
yema de huevo 182, 332, 333, 405, 406, 439, 455
yemaehuevo 405
yemahuevo 185, 298
yemehuevo 298, 435
yeneckke 393
yerba agria 242
yerba amarga 377, 413, 472
yerba anís 374
yerba blanca 265, 266, 267
yerba clara 373
yerba culebrillera 472
yerba cundidora 233
yerba damia damia herb de la pastora 401
yerba damiana 401
yerba de adán 345
yerba de almizcle 358
yerba de arco 367
yerba de barbasco 335
yerba de bruja 271, 350
yerba de caballo 261, 268, 315, 316
yerba de cáncer 267, 380, 387
yerba de carpintero 255
yerba de chivo 173, 255, 256, 260, 359
yerba de chivo blanca 256
yerba de chulo 316
yerba de conejo 244, 360, 378
yerba de culebra 195, 287
yerba de Egipto 236
yerba de gallina 244, 271, 333
yerba de gallinaza 271
yerba de gallinazo 271, 333
yerba de Guinea 241
yerba de hierro 357
yerba de hormiga mediana 366
yerba de huambía 345
yerba de jabotí 194
yerba de la leche 338
yerba de la potra 464
yerba de la verrugosa 195, 197
yerba de la virgen 178
yerba de la Virgen 178, 338
yerba de laguna 261
yerba de las almorranas 270
yerba de las muelas 371
yerba de las tres flores 430
yerba de limón 267
yerba de loro 171
yerba de los Uribe 265
yerba de los uribes 265
yerba de milagro 236
yerba de murciélago 279, 333
yerba de nigua 443
yerba de ojo 176, 281, 447
yerba de pájaro 286
yerba de Pará 246
yerba de pasmo 472
yerba de perro 235
yerba de pollito 374
yerba de pollo 374, 390
yerba de puerco 263
yerba de rejo 429
yerba de reuma 371
yerba de San Agustín 246
yerba de sangre 367
yerba de Santa María 265
yerba de sapo 270, 289, 316, 352, 372, 376, 390
yerba de venado 345
yerba de vidrio 283
yerba de zaíno 247
yerba del cáncer 255, 464
yerba del carpintero 172, 255
yerba del comején 227
yerba del cuartillo 194
yerba del dedo 358
yerba del pecado mortal 255
yerba del sol 442, 445
yerba del susto 360
yerba del toro 357
yerba dulce 417
yerba envidia 289
yerba fina 232
yerba garza 279
yerba gorda 252
yerba india 241
yerba janeiro 246
yerba lechera 390, 391
yerba luisa 377
yerba mara 320
yerba micay 234
yerba mora 347, 468, 470, 472
yerba pajarilla 263
yerba pollo 374
yerba santa 282
yerba socialista 261
yerbabue menta 371
yerbabuena 371, 417
yerbabuena lisa 371
yerbabuenilla 369, 416, 417
yerbadiarco 367
yerbaeleche 390
yerbaepuerco 263
yerbaesiote 417
yerbaevidrio 283
yerbalancha 240
yerbaluisa 377
yerbamora 347, 358, 465, 467, 468, 470, 472
yerbandrés 345
yerbarrucia 229
yerbasanta 282
yerbavirgen 282
yerbepollo 290
yerda 358
yévaro 319
yeyomo 457
yinye 250
yínyer 250
yoco 458, 459
yocó 459
yoco colorado 458
yoco cusumbe 458
yocoo 458
yocoó 458
yodo 394
yolanda 342
yolí 208
yolombo 413, 433
yolombó 433
yomate 450
yompa 311
yopa 311, 465
yopo 311
yopokolo 399
yota 199, 203, 295
yota de picón 203
yota de sapito 203
yoto 311
yotojoro 285
yotójoro 285
yoyo 227, 241
ysi 453
yuca 179, 283, 308, 361, 374, 391, 392, 471
yuca brava 392
yuca de espíritu 471
yuca de la India 391
yuca de monte 179, 392
yuca de puerco 361
yuca escorzonera 374, 392
yuca guajira 392
yuca nativa 392
yuca silvestre 392
yucatán 214, 342
yuco 254, 295, 363, 379, 392, 402, 406, 422, 430
yuco riñón 430
yuco serrano 402
yucu 308
yúcucù 306
yucutare 392
yugo 305, 306, 308
yugo chicle 305
yugo de charapa 305
yugo de pescado 306
yumal 204
yumbá 437, 441
yumbá colorado 441
yumbé 182, 404
yumbé aguacatillo y zumbí 182
yume 204
yumpa 311
yundul 232
yunka oqa 250
yupa 311
yupatí 210, 211
yuque 407
yuquilla 250, 364
yurajicù 182
yurape 200
yurune 347
yute 406, 408, 409, 415
yute congo 415
yute de China 406
yute de Java 409
yuya 264
yuyito 255, 270
yuyo 255, 256, 263, 264, 265, 266, 271, 281, 417, 432, 448
yuyo de flor amarilla 255
yuyo de sábalo 266
yuyo hembra 281
yuyo quemado 255, 256
yuyo quemao 255

Z

zábila 215, 387
zábila de monte 387
zácata 293
zafiro 419
zafrán 250, 375
zaina 241
zaino 199, 206, 404
zambo 433
zamia 179
zanahoria 251, 252
zanahoria amarilla 252
zanahoria blanca 251
zanango 465
zanca de araña 383, 387

zanca de guacharaca 436
zanca de mula 298, 387, 445
zanca de mula blanca 387
zanca de mulo 420
zanca de pava 209
zancaemula 387
zancamula 387
zancaraño 213
zancazanca 181
zanco de araña 383
zanco de mula 312, 387, 420
zancoaraño 359
zancón 290
zancona 212
zanconcita 210
zancudo 222, 430
zanquemula 387
zapallo 293
zapallo marranero 293
zapata 470
zapatero 319, 403
zapatico 315, 362, 470
zapatico de la reina 315
zapatico de la Virgen 315
zapatico de reina 315
zapatillo 319, 334, 413, 455
zapato 222, 303
zapato de obispo 303
zapato de Venus 222
zapayo 293
zapote 279, 298, 305, 306, 307, 308, 381, 411, 412, 414, 455
zapote blanco 455
zapote chupachupa 411
zapote colorado 308
zapote costeño 308
zapote cuna 308
zapote de agua 412
zapote de carne 308
zapote de monte 305, 308, 411, 414
zapote negro 298
zapoterana 414
zapotillo 308, 342, 411, 413, 414, 445
zapotolongo 412
zara 402
zaragocilla 331
zaragoza 192, 193, 311, 314, 360, 392, 416
zaragoza almendra 193
zaral 417
zarallejo 377
zarcilla 226
zarcillejo 267, 272, 273, 336, 377, 417, 418, 428
zarcillejo morado 418
zarcillero 429
zarcillo 418, 468
zarcillo de bruja 468
zarza 310, 314, 317, 318, 326, 327, 328, 330, 331, 407, 436, 444
zarza blanca 444
zarza colorada 330
zarza de mora 444
zarza de vega 327, 328
zarza hueca 407
zarza menor 314
zarza prieta 327
zarzamora 444, 445
zarzamora azul 444
zarzaparrilla 192, 225, 226, 227, 442, 445
zarzaparrilla de páramo 227
zarzilla 226
zarzón 328
zazafrás 183
zeb ame 470
zebra 224
zeu 293
ziinijava 188
ziipaye 350
zoquete 191
zorrillo 287, 377, 378
zorrito 244
zorro 204, 209, 213, 214, 233, 234, 236, 237, 242, 245, 256, 258, 261, 275, 278, 310, 315, 319, 332, 378, 379
zorrocloco 278
zucarito 357
zulia 273, 465
zulmatuno 364
zumaque 293
zumbito 467
zuná 315
zurrumbo 436

Tropical rain forest near Mocoa, Putumayo, Colombia.
Mauricio Diazgranados